BERGEY'S MANUAL® OF
Systematic
Bacteriology
Second Edition

Volume Two
The *Proteobacteria*

Part B
The *Gammaproteobacteria*

BERGEY'S MANUAL® OF
Systematic Bacteriology
Second Edition

Volume Two
The *Proteobacteria*

Part B
The *Gammaproteobacteria*

Don J. Brenner
Noel R. Krieg
James T. Staley
EDITORS, VOLUME TWO

George M. Garrity
EDITOR-IN-CHIEF

Springer

George M. Garrity, Sc.D.
Bergey's Manual Trust
Department of Microbiology and Molecular Genetics
Michigan State University
East Lansing, MI 48824-4320
USA

ISBN-10: 0-387-24144-2   Printed on acid-free paper.
ISBN-13: 978-0387-24144-9

Printed in the United States of America.   (IMP/MVY)

9 8 7 6 5 4 3 2   SPIN 10993729

springer.com

Preface to Volume Two of the Second Edition of *Bergey's Manual® of Systematic Bacteriology*

There is a long-standing tradition for the Editors of *Bergey's Manual* to open their respective editions with the observation that the new edition is a departure from the earlier ones. As this volume goes to press, however, we recognize a need to deviate from this practice, by offering a separate preface to each volume within this edition. In part, this departure is necessary because the size and complexity of this edition far exceeded our expectations, as has the amount of time that has elapsed between publication of the first volume of this edition and this volume.

Earlier, we noted that systematic procaryotic biology is a dynamic field, driven by constant theoretical and methodological advances that will ultimately lead to a more perfect and useful classification scheme. Clearly, the pace has been accelerating as evidenced in the super-linear rate at which new taxa are being described. Much of the increase can be attributed to rapid advances in sequencing technology, which has brought about a major shift in how we view the relationships among *Bacteria* and *Archaea*. While the possibility of a universally applicable natural classification was evident as the First Edition was in preparation, it is only recently that the sequence databases became large enough, and the taxonomic coverage broad enough to make such an arrangement feasible. We have relied heavily upon these data in organizing the contents of this edition of *Bergey's Manual of Systematic Bacteriology,* which will follow a phylogenetic framework based on analysis of the nucleotide sequence of the small ribosomal subunit RNA, rather than a phenotypic structure. This departs from the First Edition, as well as the Eighth and Ninth Editions of the *Determinative Manual.* While the rationale for presenting the content of this edition in such a manner should be evident to most readers, they should bear in mind that this edition, as in all preceding ones represents a progress report, rather than a final classification of procaryotes.

The Editors remind the readers that the *Systematics Manual* is a peer-reviewed collection of chapters, contributed by authors who were invited by the Trust to share their knowledge and expertise of specific taxa. Citation should refer to the author, the chapter title, and inclusive pages rather than to the Editors. The Trust is indebted to all of the contributors and reviewers, without whom this work would not be possible. The Editors are grateful for the time and effort that each expended on behalf of the entire scientific community. We also thank the authors for their good grace in accepting comments, criticisms, and editing of their manuscripts. We would also like to thank Drs. Hans Trüper, Brian Tindall, and Jean Euzéby for their assistance on matters of nomenclature and etymology.

We would like to express our thanks to the Department of Microbiology and Molecular Genetics at Michigan State University for housing our headquarters and editorial office and for providing a congenial and supportive environment for microbial systematics. We would also like to thank Connie Williams not only for her expert secretarial assistance, but also for unflagging dedication to the mission of Bergey's Manual Trust and Drs. Julia Bell and Denise Searles for their expert editorial assistance and diligence in verifying countless pieces of critical information and to Dr. Timothy G. Lilburn for constructing many of the phylogenetic trees used in this volume. We also extend our thanks to Alissa Wesche, Matt Chval and Kristen Johnson for their assistance in compilation of the bibliography.

A project such as the *Systematics Manual* also requires the strong and continued support of a dedicated publisher, and we have been most fortunate in this regard. We would also like to express our gratitude to Springer-Verlag for supporting our efforts and for the development of the Bergey's Document Type Definition (DTD). We would especially like to thank our Executive Editor, Dr. William Curtis for his courage, patience, understanding, and support; Catherine Lyons for her expertise in designing and developing our DTD, and Jeri Lambert and Leslie Grossberg of Impressions Book and Journal Services for their efforts during the pre-production and production phases. We would also like to acknowledge the support of ArborText, Inc., for providing us with state-of-the-art SGML development and editing tools at reduced cost. Lastly, I would like to express my personal thanks to my fellow trustees for providing me with the opportunity to participate in this effort, to Drs. Don Brenner, Noel Krieg, and James Staley for their enormous efforts as volume editors and to my wife, Nancy, and daughter, Jane, for their continued patience, tolerance and support.

Comments on this edition are welcomed and should be directed to Bergey's Manual Trust, Department of Microbiology and Molecular Genetics, 6162 Biomedical and Physical Sciences Building, Michigan State University, East Lansing, MI, USA 48824-4320. Email: garrity@msu.edu

George M. Garrity

Preface to the First Edition of *Bergey's Manual® of Systematic Bacteriology*

Many microbiologists advised the Trust that a new edition of the *Manual* was urgently needed. Of great concern to us was the steadily increasing time interval between editions; this interval reached a maximum of 17 years between the seventh and eighth editions. To be useful the *Manual* must reflect relatively recent information; a new edition is soon dated or obsolete in parts because of the nearly exponential rate at which new information accumulates. A new approach to publication was needed, and from this conviction came our plan to publish the *Manual* as a sequence of four subvolumes concerned with systematic bacteriology as it applies to taxonomy. The four subvolumes are divided roughly as follows: (a) the Gram-negatives of general, medical or industrial importance; (b) the Gram-positives other than actinomycetes; (c) the archaeobacteria, cyanobacteria and remaining Gram-negatives; and (d) the actinomycetes. The Trust believed that more attention and care could be given to preparation of the various descriptions within each subvolume, and also that each subvolume could be prepared, published, and revised as the area demanded, more rapidly than could be the case if the *Manual* were to remain as a single, comprehensive volume as in the past. Moreover, microbiologists would have the option of purchasing only that particular subvolume containing the organisms in which they were interested.

The Trust also believed that the scope of the *Manual* needed to be expanded to include more information of importance for systematic bacteriology and bring together information dealing with ecology, enrichment and isolation, descriptions of species and their determinative characters, maintenance and preservation, all focused on the illumination of bacterial taxonomy. To reflect this change in scope, the title of the *Manual* was changed and the primary publication becomes *Bergey's Manual of Systematic Bacteriology*. This contains not only determinative material such as diagnostic keys and tables useful for identification, but also all of the detailed descriptive information and taxonomic comments. Upon completion of each subvolume, the purely determinative information will be assembled for eventual incorporation into a much smaller publication which will continue the original name of the *Manual, Bergey's Manual of Determinative Bacteriology*, which will be a similar but improved version of the present *Shorter Bergey's Manual*. So, in the end there will be two publications, one systematic and one determinative in character.

An important task of the Trust was to decide which genera should be covered in the first and subsequent subvolumes. We were assisted in this decision by the recommendations of our Advisory Committees, composed of prominent taxonomic authorities to whom we are most grateful. Authors were chosen on the basis of constant surveillance of the literature of bacterial systematics and by recommendations from our Advisory Committees.

The activation of the 1976 Code had introduced some novel problems. We decided to include not only those genera that had been published in the Approved Lists of Bacterial Names in January 1980 or that had been subsequently validly published, but also certain genera whose names had no current standing in nomenclature. We also decided to include descriptions of certain organisms which had no formal taxonomic nomenclature, such as the endosymbionts of insects. Our goal was to omit no important group of cultivated bacteria and also to stimulate taxonomic research on "neglected" groups and on some groups of undoubted bacteria that have not yet been cultivated and subjected to conventional studies.

The invited authors were provided with instructions and exemplary chapters in June 1980 and, although the intended deadline for receipt of manuscripts was March 1981, all contributions were assembled in January 1982 for the final preparations. The *Manual* was forwarded to the publisher in June 1982.

Some readers will note the consistent use of the stem -var instead of -type in words such as biovar, serovar and pathovar. This is in keeping with the recommendations of the Bacteriological Code and was done against the wishes of some of the authors.

We have deleted much of the synonymy of scientific names which was contained in past editions. The adoption of the new starting date of January 1, 1980 and publication of the Approved Lists of Bacterial Names has made mention of past synonymy obsolete. We have included synonyms of a name only if they have been published since the new starting date, or if they were also on the Approved Lists and, in rare cases with certain pathogens, if the mention of an old name would help readers associate the organism with a clinical problem. If the reader is interested in tracing the history of a name we suggest he or she consult past editions of the *Manual* or the *Index Bergeyana* and its *Supplement*. In citations of names we have used the abbreviation AL to denote the inclusion of the name on the Approved Lists of Bacterial Names and VP to show the name has been validly published.

In the matter of citation of the *Manual* in the scientific literature we again stress the fact that the *Manual* is a collection of authored chapters and the citation should refer to the author, the chapter title and its inclusive pages, not the Editor.

To all contributors, the sincere thanks of the Trust is due; the Editor is especially grateful for the good grace with which the authors accepted comments, criticisms and editing of their manuscripts. It is only because of the voluntary and dedicated efforts of these authors that the *Manual* can continue to serve the science of bacteriology on an international basis.

A number of institutions and individuals deserve special acknowledgment from the Trust for their help in bringing about the publication of this volume. We are grateful to the Department of Biology of the Virginia Polytechnic Institute and State University for providing space, facilities and, above all, tolerance for the diverted time taken by the Editor during the preparation of the book. The Department of Microbiology at Iowa State University of Science and Technology continues to provide a welcome home for the main editorial offices and archives of the Trust and we acknowledge their continued support. A grant (LM-03707) from the National Library of Medicine, National Institutes of Health to assist in the preparation of this and the next volume of the *Manual* is gratefully acknowledged.

A number of individuals deserve special mention and thanks for their help. Professor Thomas O. McAdoo of the Department of Foreign Languages and Literatures at the Virginia Polytechnic Institute and State University has given invaluable advice on the etymology and correctness of scientific names. Those assisting the Editor in the Blacksburg office were R. Martin Roop II, Don D. Lee, Eileen C. Falk and Michael W. Friedman and their help is sincerely appreciated. In the Ames office we were ably assisted by Gretchen Colletti and Diane Triggs during the early period of preparation and by Cynthia Pease during the major portion of the editing process. Mrs. Pease has been responsible for the construction of the List of References and her willingness to handle the cumbersome details of text editing on a big computer is gratefully acknowledged.

John G. Holt

Preface to the First Edition of *Bergey's Manual®* of *Determinative Bacteriology*

The elaborate system of classification of the bacteria into families, tribes and genera by a Committee on Characterization and Classification of the Society of American Bacteriologists (1911, 1920) has made it very desirable to be able to place in the hands of students a more detailed key for the identification of species than any that is available at present. The valuable book on "Determinative Bacteriology" by Professor F. D. Chester, published in 1901, is now of very little assistance to the student, and all previous classifications are of still less value, especially as earlier systems of classification were based entirely on morphologic characters.

It is hoped that this manual will serve to stimulate efforts to perfect the classification of bacteria, especially by emphasizing the valuable features as well as the weaker points in the new system which the Committee of the Society of American Bacteriologists has promulgated. The Committee does not regard the classification of species offered here as in any sense final, but merely a progress report leading to more satisfactory classification in the future.

The Committee desires to express its appreciation and thanks to those members of the society who gave valuable aid in the compilation of material and the classification of certain species. . . .

The assistance of all bacteriologists is earnestly solicited in the correction of possible errors in the text; in the collection of descriptions of all bacteria that may have been omitted from the text; in supplying more detailed descriptions of such organisms as are described incompletely; and in furnishing complete descriptions of new organisms that may be discovered, or in directing the attention of the Committee to publications of such newly described bacteria.

David H. Bergey, *Chairman*
Francis C. Harrison
Robert S. Breed
Bernard W. Hammer
Frank M. Huntoon
Committee on Manual.
August, 1923.

Contents

Contributors

Sharon L. Abbott
Microbial Diseases Laboratory, Communicable Disease Control, California Department of Health Services, Berkeley, CA 94704-1011, USA

Wolf-Rainer Abraham
Chemical Microbiology Group, GBF-National Research Centre for Biotechnology, Mascheroder Weg 1, D-38124 Braunschweig, Germany

Paula Aguiar
Portland State University, Portland, OR 97207-0751, USA

Azeem Ahmad
Dept. Organismic and Evolutionary Biology, Harvard University Biolabs., Cambridge, MA 02144, USA

Raymond J. Akhurst
Division of Entomology, Commonwealth Scientific and Industrial Research Organization (CSIRO), Canberra Australian Cap. Terr. 2601, Australia

Serap Aksoy
Department of Epidemiology and Public Health, Yale University School of Medicine, New Haven, CT 06520-8034, USA

Milton J. Allison
Department of Microbiology, Iowa State University, Ames, IA 50011-3211, USA

Rudolf Amann
Nachwuchsgruppe Molekulare Ökologie, Max Planck-Institute für Marine Mikrobiologie, Celsiusstrasse 1, D-28359 Bremen, Germany

Øystein Angen
Danish Veterinary Laboratory, Bülowsvej 27, 1790 Copenhagen V, Denmark

Jacint Arnau
Meat Technology Centre, Institut for Food & Agricultural Res. & Tech., Granja Camps i Armet s/n, 17121 Monells, Spain

Georg Auling
Institute für Mikrobiologie, Universität Hannover, Schneiderberg 50, D-30167 Hannover, Germany

Dawn A. Austin
Department of Biological Sciences, School of Life Sciences, Heriot-Watt University, Riccarton, Edinburgh EH14 4AS, United Kingdom

Hans-Dietrich Babenzien
Department of Limnology of Stratified Lakes, Inst. of Freshwater Ecology & Inland Fisheries, Alte Fischerhütte 2, D-16775 Neuglobsow, Germany

Marcie L. Baer
Biology Department, Shippensburg University, Shippensburg, PA 17257, USA

Simon C. Baker
Birkbeck College, Malet Street, Bloomsbury, London WC1E 7HX, United Kingdom

José Ivo Baldani
Centro Nacional de Pesquisa de Agrobiologia, Empresa Brasileira de Pesquisa Agropecuária, Room 247-23851-970 Seropédica, Caixa Postal 74.505, Rio de Janeiro 465, Brazil

Vera Lúcia Divan Baldani
Centro Nacional de Pesquisa de Agrobiologia, Empresa Brasileira de Pesquisa Agropecuária, Room 247-23851-970 Seropédica, Caixa Postal 74.505, Rio de Janeiro 465, Brazil

David L. Balkwill
Department of Biological Science, Florida State University, Tallahassee, FL 32306-4470, USA

Menachem Banai
Ministry of Agriculture, Veterinary Services & Animal Health, Kimron Veterinary Institute, P.O. Box 12, Bet Dagan 50 250, Israel

Claudio Bandi
Dipartimento di Patologia Animale, Igiene e Sanità Pubblica Veterinaria, Sezione di Patologia Generale e Parassitologia, Università degli Studi di Milano, Via Celoria 10 20133 Milano, Italy

Ellen Jo Baron
Clinical Microbiology/Virology Laboratory, Stanford University Medical Center, Stanford, CA 94305-5250, USA

Linda Baumann
Department of Microbiology #0875, College of Letters and Science, University of California, Davis, CA 95616-8665, USA

Paul Baumann
Department of Microbiology #0875, College of Letters and Science, University of California, Davis, CA 95616-8665, USA

Janiche Beeder
Section for Biotechnology, Novsk Hydro ASA Research Centre, P. O. Box 2560, N-3901 Porsgruun, Norway

Julia A. Bell
Dept. of Microbiology and Molecular Genetics, Michigan State University, East Lansing, MI 48824-4320, USA

Hervé Bercovier
Hadassah Medical School, The Hebrew University, Jerusalem, Israel

Karen M. Birkhead
Foodborne & Diarrheal Diseases Lab. Section, Division of Bacterial and Mycotic Diseases, Centers for Disease Control and Prevention, Atlanta, GA 30333, USA

Magne Bisgaard
Department of Veterinary Microbiology, The Royal Veterinary & Agricultural University, Bulowsvej 13, DK-1870 Frederiksberg C, Denmark

Judith A. Bland
Merck and Company, Inc., WS2F-45, Whitehouse Station, NJ 08889-0100, USA

Nancy M.C. Bleumink-Pluym
Dept. of Bacteriology, Inst. of Infectious Diseases & Immunology, Vet. Medicine, Universität Utrecht, Yalelaan 1, 3584 CL Utrecht, The Netherlands

Eberhard Bock
Inst. für Allgemeine Botanik und Botanischer Garten, Universität Hamburg, Ohnhorststrasse 18, D-22609 Hamburg, Germany

Noël E. Boemare
Laboratoire de Pathologie comparée, C.P. 101, Laboratoire associé, Université Montpellier II, NRA-CNRS-UM II, 34095 Montpellier Cedex 05, France

David R. Boone
Department of Environmental Biology, Portland State University, Portland, OR 97207-0751, USA

Edward J. Bottone
Department of Infectious Diseases, The Mount Sinai Hospital, New York, NY 10029-6574, USA

Kjell Bøvre (Deceased)
Kaptein W. Wilhelmsen og Frues Mikrobiologis-ke Institutt, University of Oslo, Rikshospitalet, N-0027 Oslo, Norway

John P. Bowman
School of Agricultural Science, University of Tasmania, Antartic CRC, Private Bag 54, Hobart 7001, Tasmania, Australia

John F. Bradbury
CABI Bioscience, Bakeham Lane, Egham, Surrey TW20 9TY, United Kingdom

Kristian K. Brandt
Section of Genetics and Microbiology, Department of Ecology, Royal Veterinary and Agricultural University, DK-1871 Frederiksberg, Denmark

Don J. Brenner
Meningitis & Special Pathogens Branch Laboratory Section, Centers for Disease Control & Prevention, Atlanta, GA 30333, USA

Frances W. Brenner
Foodborne & Diarrheal Diseases Lab. Section, Division of Bacterial and Mycotic Diseases, Centers for Disease Control and Prevention, Atlanta, GA 30333, USA

Thorsten Brinkhoff
Inst. für Chemie und Biol. des Meeres (ICBM), Carl von Ossietzky Universität Oldenburg, D-26111 Oldenburg, Germany

Thomas D. Brock
Department of Bacteriology, University of Wisconsin, Madison, WI 53706, USA

George H. Brownell
Department of Biochemistry & Molecular Biology, Medical College of Georgia, Augusta, GA 30912-2100, USA

Marvin P. Bryant (Deceased)
Department of Animal Science, University of Illinois, Urbana, IL 61801-3838, USA

Hans-Jürgen Busse
Institut für Bakteriologie, Mykologie und Hygiene, Veterinärmedizinische Universität Wien, Veterinärplatz 1, A-1210 Wien, Austria

Douglas E. Caldwell
Dept. of Applied Microbiology and Food Science, University of Saskatchewan, Saskatoon, 51 Campus Drive, Saskatchewan S7N 5A8 SK, Canada

Daniel N. Cameron
Foodborne & Diarrheal Diseases Lab. Section, Division of Bacterial and Mycotic Diseases, Centers for Disease Control and Prevention, Atlanta, GA 30333, USA

Pierre Caumette
Department Debiologie L.E.M, Universite de Pau, Av de L'Universite BP1155, Pau F-64013, France

Wen Xin Chen
Department of Microbiology, Biology College, Beijing Agricultural University, Beijing, P.R. China

Henrik Christensen
Department of Veterinary Microbiology, Stigbøjlen 4, Frederiksberg C 1870, Denmark

Penelope Christensen
National Institute for Genealogical Studies, Faculty of Information Studies, University of Toronto, Toronto, Ontario, Canada

John D. Coates
Plant and Microbial Biology, University of California, Berkeley, Berkeley, CA 94720-3102. USA

Matthew D. Collins
Department of Food Science and Technology, University of Reading, Earley Gate-White-knights Rd., Reading RG6 6AP, United Kingdom

Michael J. Corbel
National Institute for Biol. Standards & Control, Blanche Lane, South Mimms, Potters Bar, Hertfordshire EN6 3QG, United Kingdom

Heribert Cypionka
Inst. für Biol. und Chemie des Meeres (ICBM), Universitat Oldenburg, Oldenburg, PFS 2503, D-26111, Germany

Milton S. da Costa
Departamento de Zoologia, Centro de Neurociências, Universidade de Coimbra, Apartado 3126, P-3004-517 Coimbra, Portugal

Colin Dale
Botany and Microbiology, Auburn University, Auburn, AL 36849-5407, USA

Subrata K. Das
Institute of Life Sciences, Nalco square, Bhubaneswar 751 023, India

Gregory A. Dasch
Division of Viral and Rickettsial Diseases, Viral and Rickettsial Zoonoses Branch, National Center for Infectious Diseases, Centers for Disease Control and Prevention, Atlanta, GA 30333, USA

Catherine Dauga
Génopole de l'Institut Pasteur, Plateau Technique 4, Bât Le Pasteur, Institut Pasteur, 28 rue du Docteur Roux, 75724 Paris Cedex 15, France

Frank B. Dazzo
Department of Microbiology and Molecular Genetics, Michigan State University. East Lansing, MI 48824-4320, USA

Jody W. Deming
School of Oceanography, University of Washington, Seattle, WA 98195-0001, USA

Ewald B.M. Denner
Abteilung Mikrobiologie und Biotechnologie, Institut für Mikrobiologie und Genetik, Dr. Bohr-Gasse 9, A-1030 Wein, Austria

Richard Devereux
NHEERL, Gulf Ecology Division, U.S.E.P.A., Gulf Breeze, FL 32561, USA

Paul De Vos
Dept. Biochem., Physiology & Micro. (WE 10V), University of Gent, K.L. Ledeganckstraat 35, B-9000 Gent, Belgium

Kim A. DeWeerd
Department of Chemistry, State University of New York, University at Albany, Albany, NY 12222, USA

Floyd E. Dewhirst
Department of Molecular Genetics, The Forsyth Institute, 140 The Fenway, Boston, MA 02115-3799, USA

Johanna Döbereiner (Deceased)
Centro Nacional de Pesquisa de Agrobiologia, Empresa Brasiliera de Pesquisa Agropecuária, Room 247, 23851-970 Seropédica, Caixa Postal 74.505, Rio de Janeiro 465, Brazil

Nina V. Doronina
Inst. of Biochemistry & Physiology of Micro-organisms RAS, Laboratory of Methylotrophy, Russian Academy of Sciences, Pushchino-on-the-Oka, Moscow Region 142290, Russia

Michel Drancourt
Faculté de Médecine, Unité des Rickettsies, 27 Boulevard Jean Moulin, 13385 Marseille Cedex 05, France

Galina A. Dubinina
Institute of Microbiology, Russia Academy of Sciences, Prospect 6—let. Oktyabrya 7/2, Moscow, Russia

J. Stephen Dumler
Division of Microbiology, Department of Pathology, The Johns Hopkins Hospital, Univ. School of Medicine, Baltimore, MD 21287-7093, USA

Jürgen Eberspächer
Institut für Mikrobiologie (250), Universität Hohenheim, Garbenstrasse 30, D-70599 Stuttgart, Germany

Thomas W. Egli
Department of Microbiology, EAWAG, Überlandstrasse 133, CH 8600 Düebendorf, Switzerland

Matthias A. Ehrmann
Lehrstuhl für Mikrobiologie, Technische Universität München, Weihenstephan, Freising 85350, Germany

Stefanie J.W.H. Oude Elferink
ID TNO Animal Nutrition, P.O. Box 65, 8200 AB Lelystad, The Netherlands

Takayuki T. Ezaki
Department of Microbiology and Bioinformatics, Regeneration and Advanced Medical Science, Gifu University School of Medicine, 40 Tsukasa-machi, Gifu 500 8705, Japan

J.J. Farmer III
Foodborne & Diarrheal Diseases Lab. Section, Division of Bacterial and Mycotic Diseases, Centers for Disease Control and Prevention, Atlanta, GA 30333, USA

Mark Fegan
Coop. Research Centre for Tropical Plant Protection, Dept. of Micro. & Parasitology, The University of Queensland, St. Lucia, Brisbane, Queensland 4072, Australia

Andreas Fesefeldt
Geibelallee 12a, 24116 Kiel, Germany

Kai W. Finster
Department of Microbial Ecology, Institute of Biological Sciences, University of Aarhus, Building 540, Ny, Munkegade, DK-8000 Åarhus C, Denmark

Carmen Fischer-Romero
Institut für Medizinische Mikrobiologie, Universität Zürich, Gloriastrasse 30/32, CH-8028 Zürich, Switzerland

Geoffrey Foster
Veterinary Division, Scottish Agricultural College, Drummondhill, Stratherrick Road, Inverness IV2 4JZ, United Kingdom

Pierre-Edouard Fournier
Faculté de Médecine, Unité des Rickettsies, 27, Boulevard Jean Moulin, 13385 Marseille Cedex 05, France

James G. Fox
Department of Comparative Medicine, Massachusetts Institute of Technology, Cambridge, MA 02139, USA

Wilhelm Frederiksen
Dept. of Diagnostic Bacteriology and Antibiotics, Statens Seruminstitut, DK-2300 Copenhagen S, Denmark

Michael Friedrich
Abteilung Biogeochemie, Max Planck-Institut für Terrestrische Mikrobiologie, Karl-von-Frisch-Strasse, D-35043 Marburg, Germany

John L. Fryer
Dept. of Microbiology Ctr./Salmon Disease Research, Oregon State University, Corvallis, OR 97331-3804, USA

Georg Fuchs
Mikrobiologie, Institut für Biologie II, Albert-Ludwigs-Universität Freiburg, D-79104 Freiburg, Germany

John A. Fuerst
Center for Bacterial Diversity and Identification, Department of Microbiology, University of Queensland, Brisbane, Queensland 4072, Australia

Tateo Fujii
Department of Food Science and Technology, Tokyo University of Fisheries, 4-5-7 Konan, Minato-ku, Tokyo 108-8477, Japan

Jean-Louis Garcia
Laboratoire de Microbiologie, ORSTOM-ESIL-Case 925, Université de Provence, 163, Avenue de Luminy, 13288 Marseille, Cédex 9, France

Monique Garnier (Deceased)
Institut National de la Recherche Agronomique et Université Victor Ségalen, Laboratoire de Biologie Cellulaire et Moléculaire, Bordeaux 2, 33883, BP 81, Villenave d'Ormon Cedex, France

Margarita Garriga
Centro de Tecnologia de la Carne, Inst. de Recerca i Tecnologia Agroalimentàries, Granja Camps i Armet s/n, 17121 Monells (Girona) España, Spain

George M. Garrity
Dept. of Microbiology and Molecular Genetics, Michigan State University, East Lansing, MI 48824, USA

Rainer Gebers
Depenweg 12, D-24217 Schönberg/Holstein, Germany

Connie J. Gebhart
Division of Comparative Medicine, University of Minnesota Health Center, Minneapolis, MN 55455, USA

Allison D. Geiselbrecht
Floyd Snider McCarthy, Inc, Seattle, WA 98104-2851, USA

Barbara R. Sharak Genthner
Center for Environmental Diagnostics and Bioremediation, University of West Florida, Pensacola, FL 32514, USA

Peter Gerner-Smidt
Department of Gastrointestinal Infections, Statens Serum Institut, Artillerivej 5, DK-2300 Copenhagen S, Denmark

Monique Gillis
Laboratorium voor Microbiologie Vakgroep WE 10V, Universiteit Gent, K.L. Ledeganckstraat 35, B-9000 Gent, Belgium

Christian Gliesche
Institut für Ökologie, Ernst-Moritz-Arndt-Universität, Greifswald Schwedenhagen 6, D-18565 Kloster/Hiddensee, Germany

Frank Oliver Glöckner
Max Planck-Institute for Marine Microbiology, Celsuisstrasse 1, Bremen D-28359, Germany

Peter N. Golyshin
Division of Microbiology, GBF-Natl. Research Centre for Biotechnology, Mascheroder Weg 1, 38124 Braunschweig, Germany

José M. González
Departamento de Microbiologia y Biologia Celular, Facultad de Farmacia, Universidad de La Laguna, 38071 La Laguna. Tenerife, Spain

Yvonne E. Goodman
Department of Medical Bacteriology, University of Alberta, Medical Services Building, Edmonton, Alberta, Canada

Vladimir M. Gorlenko
Institute of Microbiology, Russian Academy of Sciences, Prospect 60-letiya, Oktyabrya 7, korpus 2, Moscow 117312, Russia

Hans-Dieter Görtz
Department of Zoology, Biologisches Institut, Universität Stuttgart, Pfaffenwaldring 57, D-70550 Stuttgart, Germany

John J. Gosink
Amgen, Inc., Seattle, WA 98101, USA

Jennifer Gossling
8401 University Drive, St. Louis, MO 63105-3641, USA

Masao Goto
Plant Pathology Laboratory, Faculty of Agriculture, Shizuoka University, 836 Ohya, Shizuoka 422-8017, Japan

Peter N. Green
National Collection of Industrial & Marine Bacteria, 23 St. Machar Drive, Aberdeen AB24 3RY, United Kingdom

Francine Grimont
Unité des Entérobactéries, Inst. Natl. de la Santé et de la Recherce Médicale, Institut Pasteur, 28 rue du Docteur Roux, Unité 389 75724 Paris Cedex 15, France

Patrick A.D. Grimont
Unité des Entérobactéries, Inst. Natl. de la Santé et de la Recherce Medicale, Institut Pasteur, 28 rue du Docteur Roux, F-75724 Paris Cedex 15, France

Rémi Guyoneaud
Institut d'Ecologica Aquatica-Microbiologia, Campus de Montilivi, E-17071 Girona, Spain

Lotta E-L. Hallbeck
Department of Cell and Molecular Biology, Göteborg University, Medicinaregatan 9 C, Box 462, S-405 30 Göteborg, Sweden

Theo A. Hansen
Department of Microbial Physiology, Groningen Biomolecular Sci. & Biotech. Inst., University of Groningen, P. O. Box 14, 9750 AA Haren, The Netherlands

Shigeaki Harayama
Marine Biotechnology Institute, 3-75-1 Heita, Kamaishi, Ivate 026-001, Japan

Anton Hartmann
Institute of Soil Ecology, Rhizosphere Biology Division, GSF Research Center, PO Box 1129, D-85764 Neuherberg, München, Germany

Fawzy M. Hashem
Sustainable Agriculture Laboratory, Animal and Natural Resources Institute, Beltsville Agricultural Research Institute,USDA-ARS, Beltsville, MD 20705, USA

Lysiane Hauben
Applied Maths BVBA, Keistraat 120, B-9830 Sint Martens-Latem, Belgium

Ian M. Head
Fossil Fuels & Environ. Geochem.Postgraduate Inst. (NRG), University of Newcastle-upon-Tyne, Newcastle-upon-Tyne NE1 7RU, United Kingdom

Brian P. Hedlund
Department of Biological Sciences, University of Nevada, Las Vegas, Las Vegas, NV 89154-4004, USA

Johann Heider
Mikrobiologie, Institut für Biologie II, Universität Freiburg, Schänzlestrasse 1, D-79104 Freiburg, Germany

Robert B. Hespell (Deceased)
Natl. Center of Agricultural Utilization Research, Agricultural Research Service, United States Department of Agriculture, Peoria, IL 61604-3902, USA

Karl-Heinz Hinz
Klinik für Geflügel der Tierärztlichen Hochschule, Bünteweg 17, D-30559 Hannover, Germany

Akira Hiraishi
Department of Ecological Engineering, Toyohashi University of Technology, Tempaku-cho, Toyohashi 441-8580, Japan

Peter Hirsch
Institut für Allgemeine Mikrobiologie der Biozentrum, Universität Kiel, Am Botanischen Garten 1-9, D-24118 Kiel, Germany

Becky Hollen
Department of Biological Sciences, Louisiana State University, Baton Rouge, LA 70803, US

Barry Holmes
Public Health Laboratory Service, Central Public Health Laboratory, National Collection of Type Cultures, 61 Colindale Avenue, London NW9 5HT, United Kingdom

John Holt
Department of Microbiology and Molecular Genetics, Michigan State University, East Lansing, MI 48824-1101, USA

Marta Hugas
Meat Technology Centre, Inst. for Food & Agricultural Research & Tech., Granja Camps i Armet s/n, 17121 Monells, Spain

Philip Hugenholtz
Ecosystem Sciences Division, Department of Environmental Science, Policy, and Management, University of California, Berkeley, Berkeley, CA 94720-3110, USA

Thomas Hurek
Arbeitsgruppe Symbioseforschung, Planck-Institut für Terrestrische Mikrobiologie, Karl-von-Frisch-Strasse, D-35043 Marburg, Germany

Johannes F. Imhoff
Institut für Meereskunde, Abt. Marine Mikrobiologie, Universität Kiel, Düsternbrooker Weg 20, D-24105 Kiel, Germany

Kjeld Ingvorsen
Department of Microbial Ecology, Institute of Biological Sciences, University of Aarhus, Building 540, Ny Munkegade, DK-8000 Aarhus C, Denmark

Francis L. Jackson
Medical Microbiology and Immunology, University of Alberta, 1-41-Medical Sciences Building, Edmonton, Alberta AB T6G 2H7, Canada

J. Michael Janda
Microbial Diseases Laboratory, Communicable Disease Control, California Department of Health Services, Richmond, CA 94804, USA

Holger W. Jannasch (Deceased)
Department of Biology, Woods Hole Oceanographic Institution, Woods Hole, MA 02543, USA

Cheryl Jenkins
Department of Microbiology, University of Washington, Seattle, WA 98195-0001, USA

Bo Barker Jorgensen
Max Planck-Institute, Celsuisstrasse 1, Bremen 28359, Germany

Samuel W. Joseph
Microbiology Department, University of Maryland, College Park, MD 20742, USA

Karen Junge
School of Oceanography, University of Washington, Seattle, WA 98195-0001, USA

Elliot Juni
Department of Microbiology and Immunology, University of Michigan Medical School, Ann Arbor, MI 48109-0620, USA

Sibylle Kalmbach
Studienstiftung des Deutschen Volkes, Mirbachstrasse 7, D-53173 Bonn, Germany

Peter Kämpfer
Institut für Angewandte Mikrobiologie, Justus-Liebig-Universität Giessen, Heinrich-Buff-Ring 26-32, IFZ, D-35392 Giessen, Germany

Yoshiaki Kawamura
Department of Microbiology, Gifu University School of Medicine, 40 Tsukasa-machi, Gifu 500 8705, Japan

Donovan P. Kelly
Department of Biological Sciences, University of Warwick, Coventry CV4 7AL, United Kingdom

Suzanne V. Kelly
Professor of Biology, Scottsdale Community College, Scottsdale, AZ 85250, USA

Christina Kennedy
Department of Plant Pathology, College of Agriculture, The University of Arizona, Tucson, AZ 85721-0036, USA

Allen Kerr
Waite Agricultural Research Institute, The University of Adelaide, Glen Osmond 5064, South Australia

Karel Kersters
Lab. voor Microbiologie, Vakgroep Biochemie, Fysiologie en Microbiologie, Rijksuniversiteit Gent, K.L. Ledeganckstraat 35, B-9000 Gent, Belgium

Mogens Kilian
Dept. of Medical Microbiology & Immunology, The University of Aarhus, DK-8000 Aarhus C, Denmark

Bon Kimura
Department of Food Science and Technology, Tokyo University of Fisheries, 4-5-7 Konan, Minato-ku, Tokyo 108-8477, Japan

Hans-Peter Klenk
VP Genomics, Epidauros Biotechnology Inc., Am Neuland 1, D-82347 Bernried, Germany

Oliver Klimmek
Biozentrum Niederursel, Institut für Mikrobiologie der Johann Wolfgang Goethe-Universität, Marie-Curie-Strasse 9, D-60439 Frankfurt am Main, Germany

Allan E. Konopka
Department of Biological Science, Purdue University, West Lafayette, IN 47907-2054, USA

Hans-Peter Koops
Abteilung Mikrobiologie, Inst. für Allgemeine Botanik und Botanischer Garten, Universität Hamburg, Ohnhorststrasse 18, D-22609 Hamburg, Germany

Yoshimasa Kosako
The Institute of Physical and Chemical Research, Japan Collection of Microorganisms, RIKEN, Wako-shi, Saitama 351-0198, Japan

Julius P. Kreier
Department of Microbiology, The Ohio State University, Columbus, OH 43201, USA

Noel R. Krieg
Department of Biology, Virginia Polytechnic Institute & State University, Blacksburg, VA 24061-0406, USA

Achim Kröger (Deceased)
Biozentrum Niederursel, Institut für Mikrobiologie der Johann Wolfgang Goethe-Universität, Marie-Curie-Strasse 9, D-60439 Frankfurt am Main, Germany

J. Gijs Kuenen
Faculty of Chemical Tech. & Materials Science, Kluyver Laboratory for Biotechnology, Delft University of Technology, 2628 BC Delft, The Netherlands

Jan Kuever
Department of Microbiology, Institute for Material Testing, Foundation Institute for Materials Science, D-28199 Bremen, Germany

Hiroshi Kuraishi
1-29-10 Kamiikeburkuro, Toshima-ku, Tokyo 170-0012, Japan

L. David Kuykendall
Molecular Plant Pathology Laboratory, Plant Sciences Institute, United States Department of Agriculture, Beltsville, MD 20705-2350, USA

David P. Labeda
Natl. Ctr. For Agricultural Utilization Research, Microbial Properties Research, U.S. Department of Agriculture, Peoria, IL 61604-3999, USA

Matthias Labrenz
Institut für Allgemeine Mikrobiologie, Biologiezentrum, University of Kiel, Am Botanischen Garten 1-9, 24118 Kiel, Germany

Catherine N. Lannan
Department of Microbiology, Ctr./Salmon Disease Research, Oregon State University, Corvallis, OR 97331-3804, USA

Bernard La Scola
CNRS UMR6020, Unité des Rickettsies, 27 Boulevard Jean Moulin, 13385 Marseille Cedex 05, France

Adrian Lee
School of Microbiology and Immunology, University of New South Wales, Kensington, Sydney, Australia

Léon E. Le Minor
Entérobactéries, Institut Pasteur, 28 Rue du Docteur Roux, 75724 Paris Cedex 15, France

Werner Liesack
Max Planck-Institut für Terrestrische Mikrobiologie, Karl-von-Frisch-Strasse, D-35043 Marburg, Germany

Timothy Lilburn
ATCC Bioinformatics, Manassas, VA 20110-2209, USA

John A. Lindquist
Department of Bacteriology, University of Wisconsin, Madison, WI 53706, USA

André Lipski
Abteilung Mikrobiologie, Fachbereich Biologie/Chemie, Universität Osnabrück, 49069 Osnabrück, Germany

Niall A. Logan
School of Biological and Biomedical Sciences, Glasgow Caledonian University, Cowcaddens Road, Glasgow G4 0BA, United Kingdom

Derek R. Lovley
Department of Microbiology, University of Massachusetts, Physiology & Ecology of Anaerobic Micro., Amherst, MA 01003, USA

Wolfgang Ludwig
Lehrstuhl für Mikrobiologie, Technische Universität München, Am Hochanger 4, D-85350 Freising, Germany

Melanie L. MacDonald
Guilford College, Greensboro, NC 27410, USA

Barbara J. MacGregor
Max Planck-Institute for Marine Microbiology, Celsiusstrasse 1, D-28359 Bremen, Germany

Michael T. Madigan
Department of Microbiology, Life Science II, Southern Illinois University, Carbondale, IL 62901-6508, USA

Åsa Malmqvist
ANOX AB, Klosterangsvagen 11A, S-226 47 Lund, Sweden

Henry Malnick
Laboratory of Hospital Infection, Central Public Health Laboratory, London NW9 5HT, United Kingdom

Werner Manz
Section G3, Ecotoxicology and Biochemistry, German Federal Institute of Hydrology, Kaiserin-Augusta-Anlagen 15-17, P. O. Box 20 02 53, D-56002 Koblenz, Germany

Amy Martin-Carnahan
Dept. of Epidemiology and Preventive Medicine, University of Maryland School of Medicine, Baltimore, MD 21201, USA

Esperanza Martínez-Romero
Centro de Investigación sobre Fijación de Nitrógeno, UNAM, Ap Postal 565–A, Cuernavaca, Morelos, México

Abdul M. Maszenan
Environmental Engineering Research Centre, School of Civil and Structural Engineering, Nanyang Technological University, Block N1, #1a-29, 50 Nanyang Avenue, Singapore 639798

Ian Maudlin
Sir Alexander Robinson Ctr. for Trop. Vet. Med., Royal Dick School of Vet. Stud., University of Edinburgh, Easter Bush, Roslin, Midlothian EH25 9RG, United Kingdom

Anthony T. Maurelli
Department of Microbiology and Immunology, Uniformed Services Univ. of the Health Sciences, F. Edward Hébert School. of Medicine, Bethesda, MD 20814-4799, USA

Michael J. McInerney
Department of Botany and Microbiology, The University of Oklahoma, Norman, OK 73019-6131, USA

Thomas A. McMeekin
Inst. for Antartic and Southern Ocean Studies, University of Tasmania, Antartic CRC, GPO Box 252-80, Hobart, Tasmania 7001, Australia

Steven McOrist
Department of Biomedical Sciences, Tufts University College of Veterinary Medicine, North Grafton, MA 01536, USA

Thoyd T. Melton (Deceased)
North Carolina A&T State University, Greensboro, NC 27411, USA

Roy D. Meredith (Deceased)

Joris Mergaert
Laboratorium voor Microbiologie Vakgroep Biochemie, Fysiologie en Microbiol., Uni-versiteit Gent, K.L. Ledeganckstraat 35, B-9000 Gent, Belgium

Ortwin D. Meyer
Lehrstuhl für Mikrobiologie, Universität Bayreuth, Universitätsstrasse 30, D-95440 Bayreuth, Germany

Henri H. Mollaret
Institut Pasteur, 28 Rue du Docteur Roux, 75724 Paris Cedex 15, France

Kristian Møller
Department of Microbiology, Danish Veterinary Laboratory, Bulousvej 27, DK-1790 Copenhagen V, Denmark

Edward R.B. Moore
Programme of Soil Quality and Protection, The Macaulay Research Institute, Macaulay Dr., Craigiebuckler, AB15 8QH Aberdeen, United Kingdom

Nancy A. Moran
Dept. of Ecology and Evolutionary Biology, University of Arizona, Tucson, AZ 85721-0088, USA

Maurice O. Moss
Department of Microbiology, School of Biological Sciences, University of Surrey, Guildford, Surrey GU2 5XH, United Kingdom

R.G.E. Murray
Department of Microbiology and Immunology, The University of Western Ontario, London, Ontario N6A 5C1, Canada

Reinier Mutters
Institut für Medizinische Mikrobiologie und Krankenhaushygiene, Klinikum der Philipps-Universität Marburg, D-35037 Marburg, Germany

Gerard Muyzer
Kluyver Laboratory for Biotechnology, Department of Microbiology, Delft University of Technology, 2628 BC Delft, The Netherlands

Yasuyoshi Nakagawa
Biological Resource Center (NBRC), Department of Biotechnology, National Institute of Technology and Evaluation, 2-5-8, Kazusakamatari, Kisarazu, Chiba 292-0818, Japan

Hirofumi Nishihara
School of Agriculture, Ibaraki University, 3-21-1 Chu-ou, Ami-machi, Inashiki-gun, Ibaraki 300-0393, Japan

M. Fernanda Nobre
Departmento de Zoologia, Universidade de Coimbra, Apartado 3126, P-3000 Coimbra, Portugal

Caroline M. O'Hara
Diagnostic Microbiology Section, Division of Healthcare Quality Promotion, Centers for Disease Control and Prevention, Atlanta, GA 30333, USA

Tomoyuki Okamoto
Research and Development Center, Kirin Brewery Company, Ltd., 100-1 Hagiwara-machi, Takasaki-shi, Gunma 370-0013, Japan

Frans Ollevier
Laboratorium voor Aquatische Ecologie, Zoological Institute, Ch. de Bériotstraat 32, Leuven B-3000, Belgium

Bernard Ollivier
Laboratoirede Microbiologie—LMI, ORSTOM, Case 925, Université de Provence, ESIL, 163 Avenue de Luminy, Marseille 13288 Cedex 09, France

Ingar Olsen
Det Odontologiske Fakultet, Institutt for oral biologi, Moltke Moesvei 30/32, Universitetet I Oslo, Postboks 1052 Blindern, N-0316 Oslo, Norway

Stephen L.W. On
Danish Veterinary Institute, Bülowsvej 27, DK-1790, Copenhagen V, Denmark

Ronald S. Oremland
Water Research Division, U.S. Geological Survey, Menlo Park, CA 94025-3591, USA

Aharon Oren
Division of Microbial and Molecular Ecology, The Institute of Life Science, and the Moshe Shilo Minerva Center for Marine Biogeochemistry, The Hebrew University of Jerusalem, Givat Ram, Jerusalem 91904, Israel

Jani L. O'Rourke
School of Microbiology and Immunology, University of New South Wales, Kensington, Sydney, Australia

Ro Osawa
Division of Bioscience, Grad. Sch. of Science & Tech., Kobe University, Rokko-dai 1-1, Nada-ku, Kobe City 657-8501, Japan

Dr. Jörg Overmann
Inst. für Chemie & Biologie des Meeres (ICBM), Universität Oldenburg, Postfach 25 03, D026111, Oldenburg, Germany

Norberto J. Palleroni
Rutgers, North Caldwell, NJ 07006-4146, USA

Bruce J. Paster
Department of Molecular Genetics, The Forsyth Institute, 140 The Fenway, Boston, MA 02115-3799, USA

Bharat K.C. Patel
Microbial Discovery Research Unit, School of Biomolecular Sciences, Griffith University, Nathan Campus, Kessels Road, Brisbane, Queensland 4111, Australia

Dominique Patureau
Laboratoire de Biotechnologie de l'Environnement, INRA Narbonne, avenue des étangs, 11 100 Narbonne, France

Karsten Pedersen
Department of Cell and Molecular Biology, Göteborg University, Medicinaregatan 9 C, Box 462, S-405 30 Göteborg, Sweden

John L. Penner
Dept. of Medical Genetics & Microbiology Grad. Dept./Mol. & Med. Genet., University of Toronto, Toronto, Ontario M5S 3E2, Canada

Jeanne S. Poindexter
Department of Biological Sciences, Barnard College, Columbia University, New York, NY 10027-6598, USA

Andreas Pommerening-Röser
Abteilung Mikrobiologiem, Inst. für Allgemeine Botanik und Botanischer Garten, Universität Hamburg, Ohnhorststrasse 18, D-22609 Hamburg, Germany

Michel Y. Popoff
Unite de Génétique des Bactéries Intracellulaires, Institut Pasteur, 28 rue du Docteur Roux, F-75724 Paris Cedex 15, France

Bruno Pot
Science Department, Yakult Belgium, Joseph Wybranlaan 40, B-1070 Brussels, Belgium

Fred A. Rainey
Department of Biological Sciences, Louisiana State University, Baton Rouge, LA 70803, USA

Didier Raoult
Faculté de Médecine, CNRS, Unité des Rickettsies, 27 Boulevard Jean Moulin, 13385 Marseille Cedex 05, France

Christopher Rathgeber
Department of Microbiology, The University of Manitoba, Winnipeg, Manitoba R3T 2N2, Canada

Gavin N. Rees
Murray-Darling Freshwater Research Centre, CRC Freshwater Ecology, Ellis Street, Thurgoona, PO Box 921, Albury NSW 2640, Australia

Hans Reichenbach
Arbeitsgruppe Mikrobielle Sekundärstoffe, Gesellschaft für Biotechnologische Forschung mbH, Mascheroder Weg 1, D-38124 Braunschweig, Germany

Barbara Reinhold-Hurek
Universität Bremen, Fachbereich 2, Allgemeine Mikrobiologie, P. O. Box 330440, D-28334 Bremen, Germany

Anna-Louise Reysenbach
Department of Environmental Biology, Portland State University, Portland, OR 97207, USA

Yasuko Rikihisa
Department of Veterinary Biosciences, The Ohio State University, 1925 Coffey Road, Columbus, OH 43210-1093, USA

Lesley A. Robertson
Kluyver Laboratory for Biotechnology, Delft University of Technology, Julianalaan 67, P. O. Box 5057, 2628BC Delft, The Netherlands

Julian I. Rood
Monash University, Bacterial Pathogenesis Research Group, Department of Microbiology, Clayton 3168, Australia

Ramon A. Rosselló-Mora
Inst. Mediterrani d'Estudis Avançats (CSIC-UIB), C/Miquel Marque's 21, E-07290 Esporles, Mallorca, Spain

Paul Rudnick
Maryland Technology Development Center, SAIC, Rockville, MD 20850, USA

Gerard S. Saddler
Scottish Agricultural Science Agency, 82 Craigs Road, East Craigs, Edinburgh EH12 8NJ, United Kingdom

Takeshi Sakane
Institute for Fermentation, Osaka, Yodogawa-ku, Osaka 532-8686, Japan

Riichi Sakazaki (Deceased)
Nippon Institute of Biological Sciences, 9-2221-1 Sinmachi, Oume, Tokyo 198-0024, Japan

Abigail A. Salyers
Department of Microbiology, University of Illinois-Urbana, Champaign, Urbana, IL 61801-3704, USA

Antonio Sanchez-Amat
Faculty of Biology, Department of Genetics and Microbiology, University of Murcia, Murcia 30100, Spain

Gary N. Sanden
Epidemic Investigations Laboratory, Meningitis and Special Pathogens Branch, Division of Bacterial and Mycotic Diseases, Centers for Disease Control and Prevention, Atlanta, GA 30333, USA

Masataka Satomi
National Research Institute of Fisheries Science, 2-12-4 Fukuura, Knazawa-ku, Yokohama, Kanagawa 236-8648, Japan

Hiroyuki Sawada
National Institute of Agro-Environmental Sciences, 3-1-1 Kannondai, Tsukuba, Ibaraki 305-8604, Japan

Flemming Scheutz
WHO, The Int. Escherichia & Klebsiella Centre, Statens Seruminstitut, Artillerivej 5, DK-2300 Copenhagen S, Denmark

Jiri Schindler, Sr.
Clinical Microbiology Group and Natl. Ctr. of Surveillance of Antibiotic Resistance, National Institute of Public Health, Prague 10 10042, Czech Republic

Bernhard H. Schink
Fakultät für Biologie, Lehrstuhl für Mikrobielle Ökologie, Universität Konstanz, Postfach 55 60, D-78457 Konstanz, Germany

Karl-Heinz Schleifer
Lehrstuhl für Mikrobiologie, Technische Universität München, Am Hochanger 4, D-85350 Freising, Germany

Heinz Schlesner
Institut für Allgemeine Mikrobiologie, Universität Kiel, Am Botanischen Garten 1-9, Biologiezentrum, D-24118 Kiel, Germany

Helmut J. Schmidt
Biological Faculty, University of Kaiserslautern, Building 14, Pf 3049, D-67653 Kaiserslautern, Germany

Jean M. Schmidt
Department of Microbiology, Arizona State University, Tempe, AZ 85287-2701, USA

Dirk Schüler
Max Planck-Institute for Marine Microbiology, Celsiusstrasse 1, D-28359 Bremen, Germany

Heide N. Schulz
Section of Microbiology, University of California, Davis, Davis, CA 95616, USA

Paul Segers
Lab. voor Microbiologie Vakgroep WE 10V, Universiteit Gent, K.L. Ledeganckstraat 35, B-9000 Gent, Belgium

Robert J. Seviour
Biotechnology Research Centre, La Trobe University, P.O. Box 199, Bendigo VIC 3550, Australia

Richard Sharp
School of Applied Sciences, South Bank University, 103 Borough Road, London SE1 0AA, United Kingdom

Tsuneo Shiba
Shimonoseki University of Fisheries, Dept. of Food Science and Technology, Yoshimi-Nagatahoncho Shimonose, Yamaguchi 759-65, Japan

Martin Sievers
University of Applied Sciences, Department of Biotechnology, Molecular Biology, CH 8820 Wädenswil, Switzerland

Anders B. Sjöstedt
Department of Microbiology, National Defense Research Establishment, Cementvagen 20, S-901 82 Umeå, Sweden

Lindsay I. Sly
Centre for Bacterial Diversity and Identification, Department of Microbiology and Parasitology, University of Queensland, St. Lucia, Brisbane, Queensland 4072, Australia

Peter H.A. Sneath
Department of Microbiology and Immunology, School of Medicine, University of Leicester, P.O. Box 138, Leicester LE1 9HN, United Kingdom

Martin Sobieraj
Department of Environmental Biology, Portland State University, P. O. Box 751, Portland, OR 97207-0751, USA

Francisco Solano
Dept. of Biochemistry and Molecular Biology B, School of Medicine, University of Murcia, Murcia 30100, Spain

Dimitry Y. Sorokin
Dept. of Biochemistry and Molecular Biology B, School of Medicine, University of Murcia, Murcia 30100, Spain

Eva Spieck
Inst. für Allgemeine Botanik und Botanischer Garten, Universität Hamburg, Ohnhorststrasse 18, D-22609 Hamburg, Germany

Georg A. Sprenger
Forschungszentrum Jülich GmbH, Institut für Biotechnologie 1, P. O. Box 1913, D-52425 Jülich, Germany

Stefan Spring
DSM-Deutsche Sammlung von Mikroorganismen und Zellkulturen, GmbH, D-38124 Braunschweig, Germany

Erko S. Stackebrandt
Deutsche Sammlung von Mikroorganismen und Zellkulturen, GmbH, and GBF, Forschung GmbH2, Mascheroder Weg 1b, D-38124 Braunschweig, Germany

David A. Stahl
Civil and Environmental Engineering, University of Washington, Seattle, WA 98195-2700, USA

James T. Staley
Department of Microbiology, University of Washington, Seattle, WA 98195-0001, USA

Alfons J.M. Stams
Department of Microbiology, Wageningen Agricultural University, Hesselink Van Suchtelenweg 4, NL-6703 CT Wageningen, The Netherlands

Patricia M. Stanley
Minntech Corporation, North, Minneapolis, MN 55447-4822, USA

David J. Stewart
CSIRO, Australian Animal Health Laboratory, Private Bag 24, 5 Portarlington Road, Geelong Victoria 3220, Australia

John F. Stolz
Department of Biological Sciences, Duquesne University, Pittsburgh, PA 15282-2504, USA

Adriaan H. Stouthamer
Dept. of Molecular Cell Physiology/Molecular Microbial Ecology, Vrije Universiteit, De Boelelaan 1087, NL-1081 HV Amsterdam, The Netherlands

Nancy A. Strockbine
Foodborne and Diarrheal Diseases Branch, Division of Bacterial and Mycotic Diseases, Centers for Disease Control and Prevention, Atlanta, GA 30333, USA

William R. Strohl
Merck & Company, Rahway, NJ 07065-0900, USA

Joseph M. Suflita
Environmental and General Applied Microbiology, Department of Botany & Micro., The University of Oklahoma, Norman, OK 73019-0245, USA

Jörg Süling
Institut für Meereskunde, Abt. Marine Mikrobiologie, Universität Kiel, Düsternbrooker Weg 20, D-24105 Kiel, Germany

Jean Swings
Laboratorium voor Microbiologie Vakgroep WE10V, Fysiologie en Microbiologie, Universiteit of Gent, K.L. Ledeganckstraat 35, B-9000 Gent, Belgium

Ulrich Szewzyk
Department of Microbial Ecology, Technical University Berlin, Franklinstrasse 29, Secr. OE 5, D-10587 Berlin, Germany

Zhiyuan Tan
Department of Microbiology and Molecular Genetics, College of Agronomy, South China Agricultural University, 510642, China

Ralph S. Tanner
Department of Botany and Microbiology, University of Oklahoma, Norman, OK 73019-6131, USA

Anders Ternström
ANOX AB, Klosterangsvagen 11A, S-226 47 Lund, Sweden

Andreas Teske
Department of Biology, Woods Hole Oceanographic Institution, Woods Hole, MA 02543, USA

An Thyssen
GCPCP, Johnson & Johnson Pharm. Res. & Develop., Turnhoutseweg 30, B-4320 Beerse, Belgium

Kenneth N. Timmis
National Research Centre for Biotechnology, Division of Microbiology, Gesellschaft/Biotech- nologische Forschung mbH, Mascheroder Weg 1b, D-38124 Braunschweig, Germany

Brian J. Tindall
Deutsche Sammlung von Mikroorganismen und Zellkulturen, GmbH, Mascheroder Weg 1b, D-38124 Braunschweig, Germany

Tone Tønjum
Institute of Microbiology, Section of Molecular Microbiology A3, Rikshospitalet (National Hospital), Pilestredet 32, N-0027 Olso, Norway

G. Todd Townsend
University of Oklahoma, Norman, OK 73072, USA

Yuri A. Trotsenko
Institute of Biochemistry and Physiology of Microorganisms RAS, Laboratory of Methylotrophy, Prospekt Nauki, 5, Moscow Region 142290, Russia

Hans G. Trüper
Institut für Mikrobiologie und Biotechnologie, Universität Bonn, Mechenheimer Allee 168, W-53115 Bonn, Germany

John J. Tudor
Department of Biology, St. Joseph's University Philadelphia, PA 19131-1308, USA

Richard F. Unz
Department of Civil Engineering, The Pennsylvania State University, University Park, PA 16802-1408, USA

Teizi Urakami
Biochemicals Development Div., Mitsubishi Building, Mitsubishi Gas Chemical Company, 5-2, Marunouchi 2-chome, Chiyoda-ku, Tokyo 100-8324, Japan

Marc Vancanneyt
Laboratorium voor Microbiologie, Universiteit Gent, K.L. Ledeganckstraat 35, B-9000 Gent, Belgium

Peter Vandamme
Lab. voor Microbiologie en Microbiele Genetica, Univeristeit of Gent, Faculteit Wetenschappen, K.L. Ledeganckstraat 35, B-9000 Gent, Belgium

Bernard A.M. van der Zeijst
National Institute of Public Health and Environ., Antonie van Leeuwenhoeklaan 9, P. O. Box 1, P. O. Box 80.165, 3720 BA Bilthoven, The Netherlands

Frederique Van Gijsegem
Laboratorium Moleculaire Genetica, Universiteit Gent, K.L. Ledeganckstraat 35, B-9000 Gent, Belgium

Rob J.M. van Spanning
Department of Molecular Cell Physiology/Mole-cular Microbial Ecology, Vrije Universiteit, De Boelelaan 1087, NL-1081 HV Amsterdam, The Netherlands

Henk W. van Verseveld
Dept. of Molecular Cell Physiology, Molecular and Microbial Ecology, Vrije Universiteit, De Boelelaan 1087, NL-1081 HV Amsterdam, The Netherlands

Leana V. Vasilyeva
Institute of Microbiology RAN, 117811, Russian Academy of Sciences, 60-let. Oktyabrya 7 build. 2, Moscow, Russia

Jill A. Vaughan
CSIRO, Australian Animal Health Laboratory, Private Bag 24, 5 Portarlington Road, Geelong Victoria 3220, Australia

Antonio Ventosa
Departamento de Microbiologia y Parasitologia, Facultad de Farmacia, Universidad de Sevilla, Apdo. 874, 41080 Sevilla, Spain

Rudi F. Vogel
Lehrstuhl für Mikrobiologie, Technische Universität München, Freising-Wihen 85350, Germany

Russell H. Vreeland
Department of Biology, West Chester University, West Chester, PA 19383, USA

David H. Walker
Department of Pathology, University of Texas Medical Branch, 301 University Boulevard, Galveston, TX 77555-0609, USA

En Tao Wang
Departamento de Microbiologia, Escuela Nacional de Ciencias Biológicas, Instituto Politécnico Nacional, Carpio y Plan de Ayala S/N, México D.F. 11340, México

Naomi L. Ward
The Institute for Genomic Research, Rockville, MD 20850, USA

Richard I. Webb
Department of Microbiology, University of Queensland, Brisbane, Queensland 4072, Australia

Ronald M. Weiner
Cell Biology Cluster, Division of Molecular and Cellular Biosciences, National Science Foundation, Arlington, VA 22230, USA

Susan C. Welburn
Sir Alexander Robinson Ctr. for Trop. Vet. Med., Royal Dick School of Vet. Stud., University of Edinburgh, Easter Bush, Roslin, Midlothian EH25 9RG, United Kingdom

David F. Welch
Laboratory Corporation of America, Dallas, Texas 75230, USA

Aimin Wen
Food Science and Technology Program, Pacific Agri-Food Research Centre, Summerland BC V0H 1Z0, Canada

John H. Werren
Department of Biology, University of Rochester, Rochester, NY 14627-0211, USA

Hannah M. Wexler
Department of Veterans Affairs, West Los Angeles Medical Ctr., UCLA School of Medicine, 11301 Wilshire Boulevard, Los Angeles, CA 90073, USA

Robbin S. Weyant
Meningitis & Special Pathogens Branch, Centers for Disease Control and Prevention, Atlanta, GA 30333, USA

Anne M. Whitney
Meningitis & Special Pathogens Branch Lab. Section, MS D-11, Centers for Disease Control & Prevention, Atlanta, GA 30303, USA

Friedrich W. Widdel
Abteilung Mikrobiologie, Max Planck-Institut für Marine Mikrobiologie, Celsiusstrasse 1, D-28359 Bremen, Germany

Jürgen K.W. Wiegel
Department of Microbiology, University of Georgia, Athens, GA 30602-2605, USA

Anne Willems
Laboratorium voor Microbiologie, Universiteit Gent, K.L. Ledeganckstraat 35, B-9000 Gent, Belgium

Henry N. Williams
Department of OCBS, Dental School, University of Maryland at Baltimore, Baltimore, MD 21201-1510, USA

Washington C. Winn, Jr.
Microbiology Laboratory, Medical Center Hospital of Vermont DVE, Fletcher Allen Health Care, UHC Campus, Burlington, VT 05401-3456, USA

Ann P. Wood
Microbiology Research Group, King's College, London Div. of Life Sciences, Franklin-Wilkins Building, 150 Stamford Street, London SE1 8WA, United Kingdom

Eiko Yabuuchi
Aichi Medical University, Omiya 4-19-18, Asahi-ku, Osaka 535-0002, Japan

Michail M. Yakimov
Istituto Sperimentale Talassografico-CNR, Spianata S. Raineri, 86, 98122 Messina, Italy

Kazuhide Yamasato
Department of Fermentation Science, Faculty of Applied Bioscience, Tokyo University of Agriculture, Sakuragaoka, Setagaya-ku, Tokyo 158-0852, Japan

Akira Yokota
Institute of Molecular and Cellular Biosciences, The University of Tokyo, Yayoi 1-1-1, Bunkyo-ku, Tokyo 113-0032, Japan

John M. Young
Mt. Albert Research Centre, Landcare Research New Zealand Ltd., Private Bage 92 170, Auckland, New Zealand

Xue-jie Yu
Department of Pathology, University of Texas Medical Branch, 301 University Boulevard, Galveston, TX 77555-0609

Vladimir V. Yurkov
Department of Microbiology, The University of Manitoba, Winnipeg, Manitoba R3T 2N2, Canada

George A. Zavarzin
Institute of Microbiology, Russian Academy of Sciences, Building 2, Prospect 60-letja Oktyabrya 7a, Moscow 117312, Russia

Tatjana N. Zhilina
Institute of Microbiology, Russian Academy of Sciences, Prospect 60-letja Oktyabrya 7a, Moscow 117312, Russia

Stephen H. Zinder
Department of Microbiology, Cornell University, Ithaca, NY 14853-0001, USA

Phylum XIV. Proteobacteria *phyl. nov.*

GEORGE M. GARRITY, JULIA A. BELL AND TIMOTHY LILBURN

Pro.te.o.bac.te' ri.a. Gr. n. *Proteus* ocean god able to change shape; Gr. n. *bakterion* a small rod; M.L. fem. pl. n. *Proteobacteria* the phylum of bacteria having 16S rRNA gene sequences related to those of the members of the order *Pseudomonadales*.

The phylum *Proteobacteria* was circumscribed for this volume on the basis of phylogenetic analysis of 16S rRNA gene sequences. The phylum contains Gram-negative bacteria in the classes *Alpha*-*proteobacteria*, *Betaproteobacteria*, *Gammaproteobacteria*, *Deltaproteobacteria*, and *Epsilonproteobacteria*.

Type order: **Pseudomonadales** Orla-Jensen 1921, 270.

Class III. **Gammaproteobacteria** *class. nov.*

GEORGE M. GARRITY, JULIA A. BELL AND TIMOTHY LILBURN

Gam.ma.pro.te.o.bac.te' ri.a. Gr. n. *gamma* name of third letter of Greek alphabet; Gr. n. *Proteus* ocean god able to change shape; Gr. n. *bakterion* a small rod; M.L. fem. pl. n. *Gammaproteobacteria* class of bacteria having 16S rRNA gene sequences related to those of the members of the order *Pseudomonadales*.

The class *Gammaproteobacteria* was circumscribed for this volume on the basis of phylogenetic analysis of 16S rRNA sequences; the class contains the orders *Acidithiobacillales*, *Aeromonadales*, *Alteromonadales*, *Cardiobacteriales*, *Chromatiales*, *Enterobacteriales*, *Legionellales*, *Methylococcales*, *Oceanospirillales*, *Pasteurellales*, *Pseudomonadales*, *Thiotrichales*, *Vibrionales*, and *Xanthomonadales*.

Type order: **Pseudomonadales** Orla-Jensen 1921, 270.

Order I. **Chromatiales** *ord. nov.*

JOHANNES F. IMHOFF

Chro.ma.ti.a' les. M.L. neut. n. *Chromatium* type genus of the order; *-ales* ending to denote an order; M.L. fem. pl. n. *Chromatiaceae* the *Chromatium* order.

The order *Chromatiales* consists of unicellular Gram-negative bacteria belonging to the *Gammaproteobacteria*. Cells are spherical, vibrioid, spiral, or rod shaped. Multiply by binary fission, motile by flagella or nonmotile; some contain gas vesicles.

The order *Chromatiales* contains anoxygenic phototrophic bacteria, known as phototrophic purple sulfur bacteria, able to perform photosynthesis under anoxic conditions without oxygen production. In addition, the order contains nonphototrophic, purely chemotrophic relatives. Color of cell suspensions of the phototrophic species ranges from purple-violet to purple-red, red, orange-brown, yellowish brown, brownish red, brown, and green. Common to all phototrophic species is the **presence of bacteriochlorophyll *a* or *b*** (see Table BXII.γ.1), and various carotenoids of the spirilloxanthin, okenone, or rhodopinal groups (Table BXII.γ.2). **Photosynthetic pigments are located in the cy**toplasmic membrane, and in internal membrane systems of different fine structure which originate from, and are continuous with, the cytoplasmic membrane.

Phototrophic species are able to grow by a photolithoautotrophic or photoorganoheterotrophic metabolism under anoxic conditions. The photosynthetic metabolism differs from that of the cyanobacteria, algae, and green plants in that water cannot serve as electron donor substrate. Photosynthetic CO_2 assimilation depends on the utilization of external electron donors such

TABLE BXII.γ.1. Bacteriochlorophylls of phototrophic purple bacteria

Pigment	Absorption maxima (nm) in living cells
Bacteriochlorophyll *a*	375–378, 590, 796–805, 820–898
Bacteriochlorophyll *b*	400, 605, 835–850, 980–1030

TABLE BXII.γ.2. Carotenoid groups of phototrophic purple bacteria[a]

Carotenoid group	Major components
Normal spirilloxanthin group	spirilloxanthin, lycopene, rhodopin
Alternative spirilloxanthin group	spheroidene, spheroidenone, chloroxanthin, (spirilloxanthin)
Okenone group	okenone
Rhodopinal group	rhodopinal, rhodopin, lycopenal, lycopene, lycopenol, spirilloxanthin

[a]Data from Schmidt (1978).

as reduced sulfur compounds, reduced iron, molecular hydrogen, or organic compounds. In the presence of sulfide and light, cells of phototrophic members of the *Chromatiales* (purple sulfur bacteria) form highly refractile globules of elemental sulfur either inside or outside their cells. CO_2 is photoassimilated through the reductive pentose phosphate cycle. Fixation of N_2 has been demonstrated in many species. All species contain cytochromes, quinones, and non-heme iron proteins of the ferredoxin type. In addition to their phototrophic capabilities, many species of this order can grow chemotrophically under microoxic to oxic conditions in the dark, but some species, such as *Arhodomonas aquaeolei* (see family *Ectothiorhodospiraceae*), may be purely chemotrophic and unable to grow phototrophically.

Type genus: **Chromatium** Perty 1852, 174 emend. Imhoff, Süling and Petri 1998b, 1138.

FURTHER DESCRIPTIVE INFORMATION

The available 16S rDNA sequence information for *Ectothiorhodospiraceae* (Imhoff and Süling, 1996) and *Chromatiaceae* species (DeWeerd et al., 1990; Wahlund et al., 1991; Caumette et al., 1997; Guyoneaud et al., 1997, 1998b; Imhoff et al., 1998b) supports the classification of these bacteria in two different families (Imhoff, 1984b). At the same time, it indicates the existence of different genera within the family *Ectothiorhodospiraceae* (Imhoff and Süling, 1996), and the inconsistency of the previous taxonomic classification of the family *Chromatiaceae* (see Pfennig and Trüper, 1989) with its phylogeny (Imhoff et al., 1998b). In addition, the phylogenetic relatedness of both families and their distance to other bacteria of the *Gammaproteobacteria* support their treatment within one order, for which the name *Chromatiales* is proposed. The order *Chromatiales* includes the phototrophic purple sulfur bacteria of the *Ectothiorhodospiraceae* and *Chromatiaceae*

(Imhoff, 1984b, 1989; Pfennig and Trüper, 1989) and their purely chemotrophic relatives (see Figs. BXII.γ.3 and BXII.γ.4 of the chapter describing the family *Chromatiaceae*.). The purely chemotrophic bacteria known so far are specifically related to *Ectothiorhodospiraceae*. The 16S rDNA sequence of *Arhodomonas aquaeolei* (Adkins et al., 1993; Imhoff and Süling, 1996) is similar to those of *Halorhodospira* species, but clearly distinct at the genus level. In addition, sequences of *Nitrococcus mobilis* (Teske et al., 1994) and *Nitrosococcus oceani* (Head et al., 1993) place these bacteria in phylogenetic proximity to the *Ectothiorhodospiraceae*, though on a more distant level than the genus *Arhodomonas*.

Differentiation of **Chromatiaceae** *and* **Ectothiorhodospiraceae**
Traditionally, the most important and easily recognized distinguishing property within the phototrophic members of the order *Chromatiales* is the deposition of S^0 during growth on sulfide. Sulfur globules appear exclusively inside the cells (*Chromatiaceae*) or outside the cells (*Ectothiorhodospiraceae*). A clear distinction between the two families is also possible based on differences in quinone, lipid, and fatty acid composition (Imhoff and Bias-Imhoff, 1995). Several glucolipids are present in *Chromatiaceae* species, but absent from members of *Ectothiorhodospiraceae* (Imhoff et al., 1982a). In addition, the lipopolysaccharides show significant differences between members of the two families (Weckesser et al., 1979, 1995). The lipid A of all investigated *Chromatiaceae* species (*Allochromatium vinosum, Thermochromatium tepidum, Thiocystis violacea, Thiocapsa roseopersicina, Thiococcus pfennigii*) is characterized by a phosphate-free backbone with D-glucosamine as the only amino sugar, having terminally attached D-mannose and amide-bound $C_{14:0\ 3OH}$. In the lipid A of all tested *Ectothiorhodospiraceae* species (*Ectothiorhodospira vacuolata, E. shaposhnikovii, E. haloalkaliphila,* and *Halorhodospira halophila*), phosphate is present, 2,3-diamino-2,3-dideoxy-D-glucose is the major amino sugar (D-glucosamine is also present), D-mannose is lacking (D-galacturonic acid, and D-glucuronic acid are present instead) and, quite remarkably, $C_{10:0\ 3OH}$ is present as an amide-bound fatty acid (Zahr et al., 1992; Weckesser et al., 1995; Imhoff and Süling, 1996). These distinctive properties of the lipid A appear to be characteristic features of the two families. In addition, sequences of 16S rDNA are significantly different in representatives of the two families, as demonstrated by a number of characteristic signatures (Table BXII.γ.3) and by their overall dissimilarity (Figs. BXII.γ.3 and BXII.γ.4 of the chapter describing the family *Chromatiaceae*).

Key to the phototrophic members of the families of the Chromatiales

I. *Gammaproteobacteria* that contain bacteriochlorophyll *a* or *b* and different carotenoid pigments, have internal photosynthetic membrane structures and are able to grow phototrophically under anoxic conditions in the light.

    A. In the presence of both sulfide and light, globules of S^0 are formed inside the cells and further oxidized to sulfate.

Chromatiaceae

    B. In the presence of both sulfide and light, globules of S^0 are formed outside the cells (only exceptionally also in the peripheral part of the cell) and may be further oxidized to sulfate. Growth is dependent on saline and alkaline growth conditions.

Ectothiorhodospiraceae

TABLE BXII.γ.3. Differential characteristics of the phototrophic members of the families *Chromatiaceae* and *Ectothiorhodospiraceae*

Characteristic	*Chromatiaceae*	*Ectothiorhodospiraceae*
Sulfur deposition	inside cells	outside cells
Characteristics of polar lipids[a]	several glycolipids	glycolipids absent
Major fatty quinones[a]	Q8/MK8	Q7 or Q8/MK7 or MK8
Major fatty acids[a]:		
$C_{16:0}$	20–35%	11–25%
$C_{16:1}$	25–37%	<10%
$C_{18:1}$	38–45%	50–75%
Properties of lipopolysaccharides [b]:		
Major aminosugar	glucosamine	diamino-dideoxy-glucose
Phosphate	present	absent
D-Mannose	+	−
D-Galacturonic acid	−	+
D-Glucuronic acid	−	+
Amide-bound fatty acid	$C_{14:0\ 3OH}$	$C_{12:0\ 3OH}$
Signature 16S rDNA sequence [c]:		
Position 217	T	G/A[d]
Position 234–235	CA	TG/CG[d]
Position 821–822	TC	AG
Position 876–877	GA	CT
Position 985	A	T/G[d]

[a]Data from Imhoff and Bias-Imhoff (1995).

[b]Data from Weckesser et al. (1995).

[c]Numbering based on *E. coli* 16S rDNA sequence.

[d]Bases above the line are found in the genera *Ectothiorhodospira* and *Thiorhodospira*; bases below the line are found in the genera *Arhodomonas* and *Halorhodospira*.

Family I. **Chromatiaceae** Bavendamm 1924, 125[AL] emend. Imhoff 1984b, 339

JOHANNES F. IMHOFF

Chro.ma.ti.a' ce.ae. M.L. neut. n. *Chromatium* type genus of the family; *-aceae* ending to denote a family; M.L. fem. pl. n. *Chromatiaceae* the *Chromatium* family.

Cells are spherical, ovoid, vibrioid, spiral, or rod shaped, multiply by binary fission, may contain gas vesicles, may be motile by flagella or nonmotile. Motile forms have either monotrichous or multitrichous polar flagella. **Gram negative, belong to the *Gammaproteobacteria*, and contain internal photosynthetic membranes,** which are continuous with the cytoplasmic membrane and of vesicular type in most species (Fig. BXII.γ.1), though some species have tubular internal membranes (Fig. BXII.γ.2) or lamellae. **As photosynthetic pigments, bacteriochlorophyll *a* or *b* and carotenoids of the spirilloxanthin, okenone, or rhodopinal groups are present** in most species, but other carotenoids such as tetrahydrospirilloxanthin, may occur. In general, the color of cultures of strains with carotenoids of the spirilloxanthin group is orange-brown to brownish red or pink, of those with okenone, purple-red, and of those with carotenoids of the rhodopinal group, purple-violet.

Under anoxic conditions in the light, all species are capable of photolithoautotrophic growth with sulfide or S[0] as electron donor. S[0] globules accumulate as an intermediate oxidation product inside the cells. Many species are able to use H_2 as an electron donor under reducing culture conditions, and some use reduced iron ions. All species are potentially mixotrophic and photoassimilate a number of simple organic compounds, of which acetate and pyruvate are the most widely used. Some species are able to photoassimilate organic substances in the absence of sulfide or sulfur, or grow photoorganoheterotrophically. Many species are strictly anaerobic and obligately phototrophic, while others are capable of chemolithoautotrophic or chemoorganohet-

erotrophic growth under microoxic to oxic conditions in the dark.

The pathway of autotrophic CO_2 assimilation was shown to be the reductive pentose phosphate cycle in all species tested so far. The fixation of dinitrogen has been demonstrated in several species. Storage materials are polysaccharides, poly-β-hydroxybutyrate, elemental sulfur, and polyphosphate. Vitamin B_{12} is required by several species.

In nature, members of the *Chromatiaceae* occur in the illuminated, anoxic, and sulfide-containing parts of a large variety of aquatic environments and sediments from moist and muddy soils, ditches, ponds, lakes, rivers, sulfur springs, estuaries, marine habitats, and even salt and soda lakes.

The mol% G + C of the DNA is: 45.5–70.4.

Type genus: **Chromatium** Perty 1852, 174 emend. Imhoff, Süling and Petri 1998b, 1138.

TAXONOMIC COMMENTS

In the eighth edition of the *Determinative Manual*, the family *Chromatiaceae* was described to include the genus *Ectothiorhodospira*, purple sulfur bacteria that form sulfur globules outside the cells (Pfennig and Trüper, 1974). With the establishment of the new family *Ectothiorhodospiraceae* and an emended description of the family *Chromatiaceae* Bavendamm 1924 (Imhoff, 1984b), the family *Chromatiaceae* exclusively comprises phototrophic sulfur bacteria capable of depositing globules of S[0] inside the cells. This description agrees with Molisch's (1907) definition of the *"Thiorhodaceae"* and is in line with their phylogenetic separation from

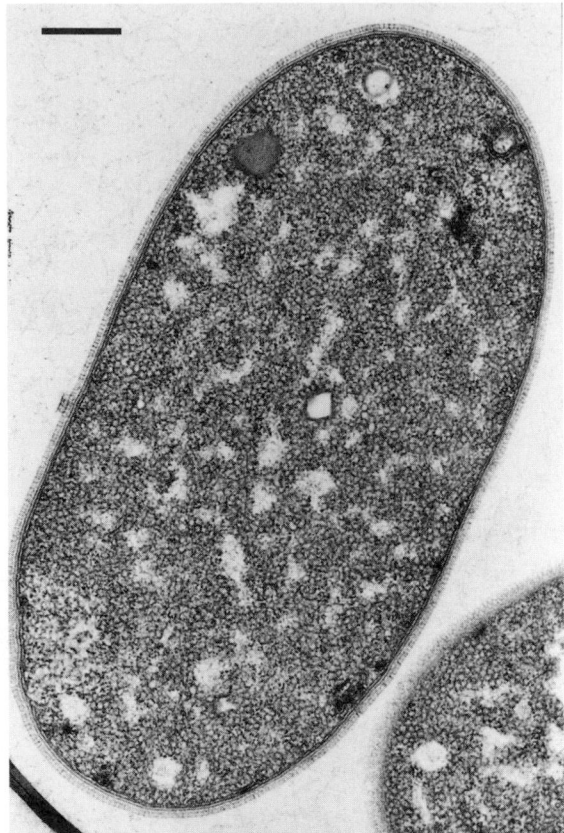

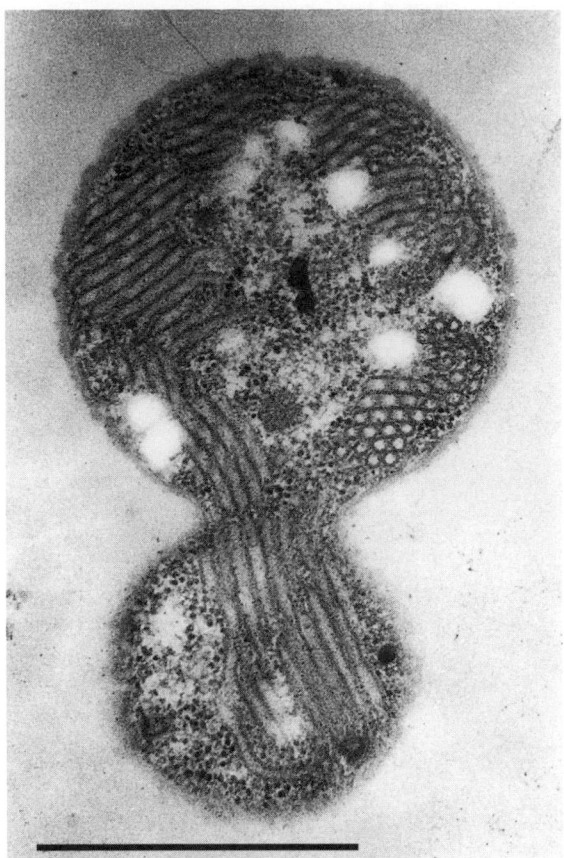

FIGURE BXII.γ.1. Ultrathin section of a cell of *Thiocystis gelatinosa* DSM 215 showing internal photosynthetic membranes as numerous small vesicles, which are evenly distributed throughout the cell. Bar = 1 μm (Courtesy of H.G. Trüper and J.B. Waterbury.)

FIGURE BXII.γ.2. Ultrathin section of a cell of *Thiococcus pfennigii* strain 9111 showing tubular internal photosynthetic membranes in longitudinal and cross section. Bar = 0.5 μm (Courtesy of H.G. Trüper and J.B. Waterbury.)

other phototrophic purple sulfur bacteria (Fowler et al., 1984; Imhoff and Süling, 1996; Imhoff et al., 1998b).

In the past, the taxonomy of the phototrophic purple bacteria, including the family *Chromatiaceae*, was entirely based on simple phenotypic characteristics (see the previous editions of this *Manual*, Pfennig and Trüper, 1974, 1989). Until recently, little molecular and sequence-based information, such as rRNA oligonucleotide data for selected species (Fowler et al., 1984) and a few 16S rDNA sequences (DeWeerd et al., 1990; Wahlund et al., 1991), was available for *Chromatiaceae* species. During the past few years, complete 16S rDNA sequences for most of the *Chromatiaceae* species have been determined (see Caumette et al., 1997; Guyoneaud et al., 1998b; Imhoff et al., 1998b). At the time of writing, 16S rDNA sequences of quite a few type strains of purple sulfur bacteria have not yet been determined or are not available: *Chromatium weissei*, *Thiospirillum jenense*, *Thiodictyon elegans*, *Thiodictyon bacillosum*, *Lamprobacter modestohalophilus*, and *Thiopedia rosea*.

The results of 16S rDNA sequence analysis, shown in Figs. BXII.γ.3 and BXII.γ.4, revealed that separate major phylogenetic branches of the family *Chromatiaceae* contain i) truly marine and halophilic species; ii) species that are motile by polar flagella, do not contain gas vesicles, and are primarily freshwater species; and iii) species with ovoid to spherical cells that are primarily freshwater species (some also occur in brackish and marine habitats), the majority of which are nonmotile (except *Lamprocystis roseopersicina*), and most of which contain gas vesicles. Because the phylogenetic relationships were not in congruence with tra-

ditional classification, a reclassification based on 16S rDNA sequence similarity, supported by selected phenotypic properties, was proposed (Guyoneaud et al., 1998b; Imhoff et al., 1998b). Positive correlation between 16S rDNA sequence similarity values and particular phenotypic properties of the species were taken as an indication of the importance of these properties in achieving a phylogenetically oriented taxonomy that includes both genetic and phenotypic information (Imhoff et al., 1998b).

Ecological source and adaptation to specific environmental conditions are of importance for phylogeny and taxonomy. Evolution occurs in particular natural environments, and separate phylogenetic lineages are expected to result from bacterial evolution in distinct habitats that, due to their particular properties (e.g., extremes of temperature, salinity, and pH), allow only the development of specifically adapted bacteria. In principle, major phylogenetic branches of the family *Chromatiaceae* can be distinguished on the basis of their salt requirement, which implies a separate evolutionary development in the marine and in the freshwater environment. This suggests that the salt response is an important taxonomic criterion in a phylogenetically oriented taxonomic system of these bacteria. The salt requirement has been considered in the taxonomy of purple nonsulfur bacteria and *Ectothiorhodospiraceae*. It proved to be of relevance for distinguishing the genera *Halorhodospira* and *Ectothiorhodospira* (Imhoff and Süling, 1996), *Rhodovulum*, *Rhodobacter* (Hiraishi and Ueda, 1994), and genera of the spiral, purple nonsulfur *Alphaproteobacteria*, the former *Rhodospirillum* species (Imhoff et al., 1998b).

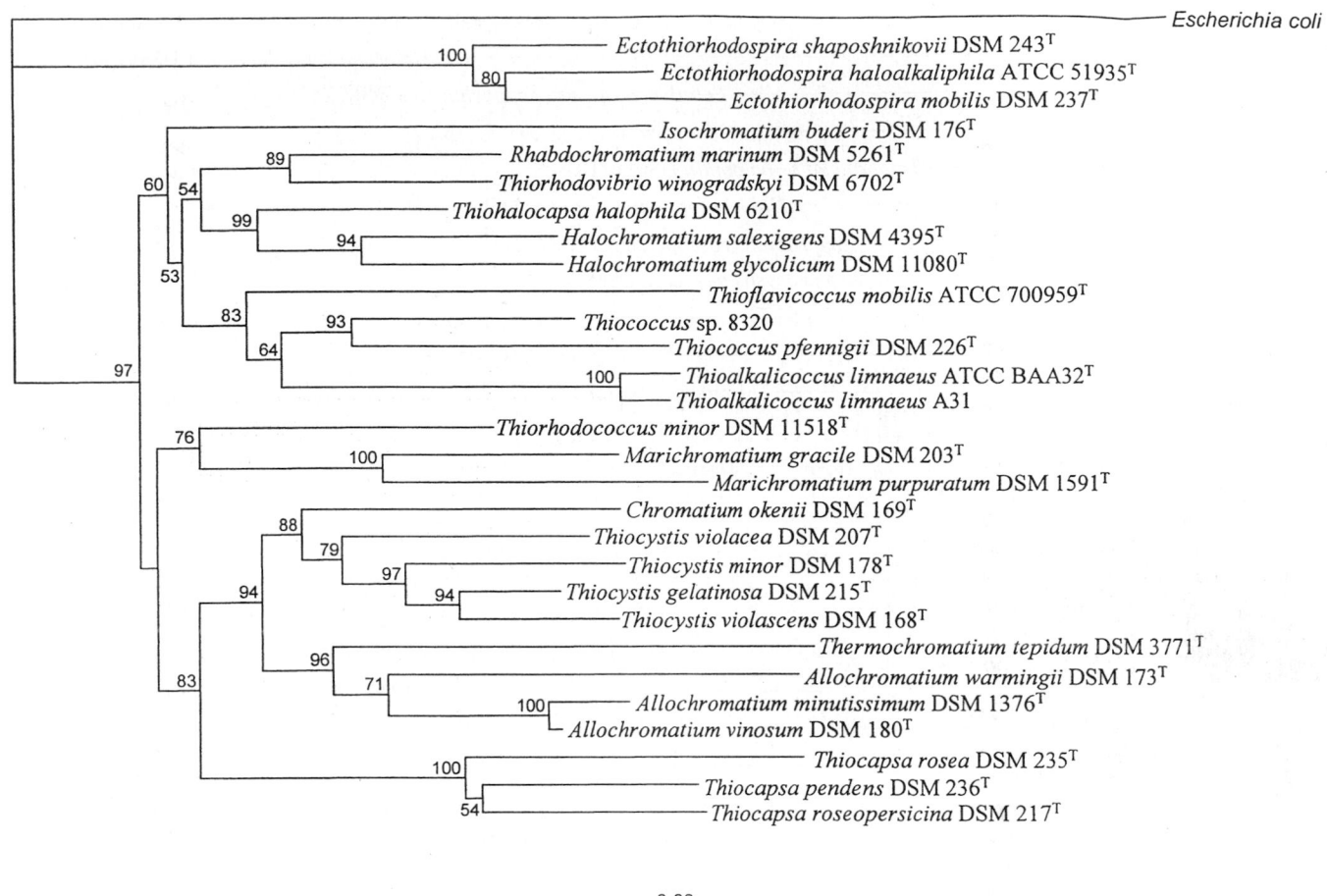

0.02

FIGURE BXII.γ.3. 16S rDNA-derived phylogenetic tree of the family *Chromatiaceae* including species of *Chromatium*, *Thiocystis*, *Allochromatium*, *Thermochromatium*, and the halophilic species *Isochromatium*, *Rhabdochromatium*, *Thiorhodovibrio*, *Thiohalocapsa*, *Thiococcus*, *Halochromatium*, *Thiorhodococcus* and *Marichromatium*. Representatives of *Thiocapsa* and *Thiolamprovum* and *Ectothiorhodospira* were included to complete the view on purple sulfur bacteria. The phylogeny of these genera is shown in more detail in Fig. BXII.γ.4 and in Fig. BXII.γ.19 in the chapter on *Ectothiorhodospiraceae*. Sequence data used were published previously (Imhoff et al., 1998b). Sequence alignments, calculation of distance data and tree construction were performed as described earlier (Petri and Imhoff 2000).

Within the family *Chromatiaceae*, both the phylogenetic relationship and the salt responses enable us to distinguish the halophilic *Halochromatium* species from the marine *Marichromatium* and the freshwater *Allochromatium* species, all of which were previously known as species of the genus *Chromatium* (Imhoff et al., 1998b). A differentiation of groups of species is also possible based on compatible solute production. Different compatible solutes are accumulated as osmotica by *Chromatiaceae* species compared to *Ectothiorhodospiraceae* species (Severin et al., 1992; Imhoff, 1993; Welsh and Herbert, 1993). Ectoine, trehalose, and glycine betaine are common among *Ectothiorhodospira* and *Halorhodospira* species, but not found in *Chromatiaceae*. Instead of trehalose, the less compatible sugar sucrose was found as the major component in freshwater species including *Allochromatium vinosum*, *Thiocystis violacea*, *Thiocystis minor*, *Thiocapsa roseopersicina*, and *Thiocapsa rosea* (Welsh and Herbert, 1993). Among the halophilic and marine species (*Halochromatium salexigens*, *Thiohalocapsa halophila*, and *Marichromatium purpuratum*), glycine betaine appears to occur as the major component, along with sucrose and *N*-acetyl-glutaminylglutamine amide (Severin et al., 1992; Imhoff, 1993).

Although morphological properties of the cells appear to be of lower phylogenetic significance, it is apparent that one group of freshwater *Chromatiaceae* species contains motile single cells without gas vesicles (e.g., *Chromatium* and *Allochromatium* species, see Fig. BXII.γ.3), and the possession of gas vesicles is restricted to a second group with predominantly spherical to ovoid cells that are nonmotile (e.g., *Thiocapsa* and *Thiolamprovum* species, with the exception of *Lamprocystis roseopersicina*, see Fig. BXII.γ.4).

The DNA base ratio within the family *Chromatiaceae*, expressed as the mol% G + C content, has been found to span a large scale. Thus, in the previous edition of the *Manual* the genus *Chromatium* (Pfennig, 1989b) had values ranging from 48.0 to 70.4. Because the mol% G + C content is crudely indicative of the genetic relatedness of bacteria, these values already suggested excessive diversity within the genus *Chromatium*. A phylogenetic system would be expected to group bacteria that have much less variation in their mol% G + C content together. Indeed, after rearrangement of the species of the family *Chromatiaceae* according to their genetic relatedness on the basis of 16S rDNA sequence analysis (Guyoneaud et al., 1998b; Imhoff et al., 1998b), a quite narrow range of mol% G + C content is found within most of the newly established genera. The ranges are ~63–66 in the cluster including species of *Thiocapsa*, *Lamprocystis*, and *Thiolamprovum*, 69–70 in *Marichromatium* species, 64–66 in *Halochromatium* species, and 48–50 in the true *Chromatium* species. This correlation appears of such significance, that any case of high

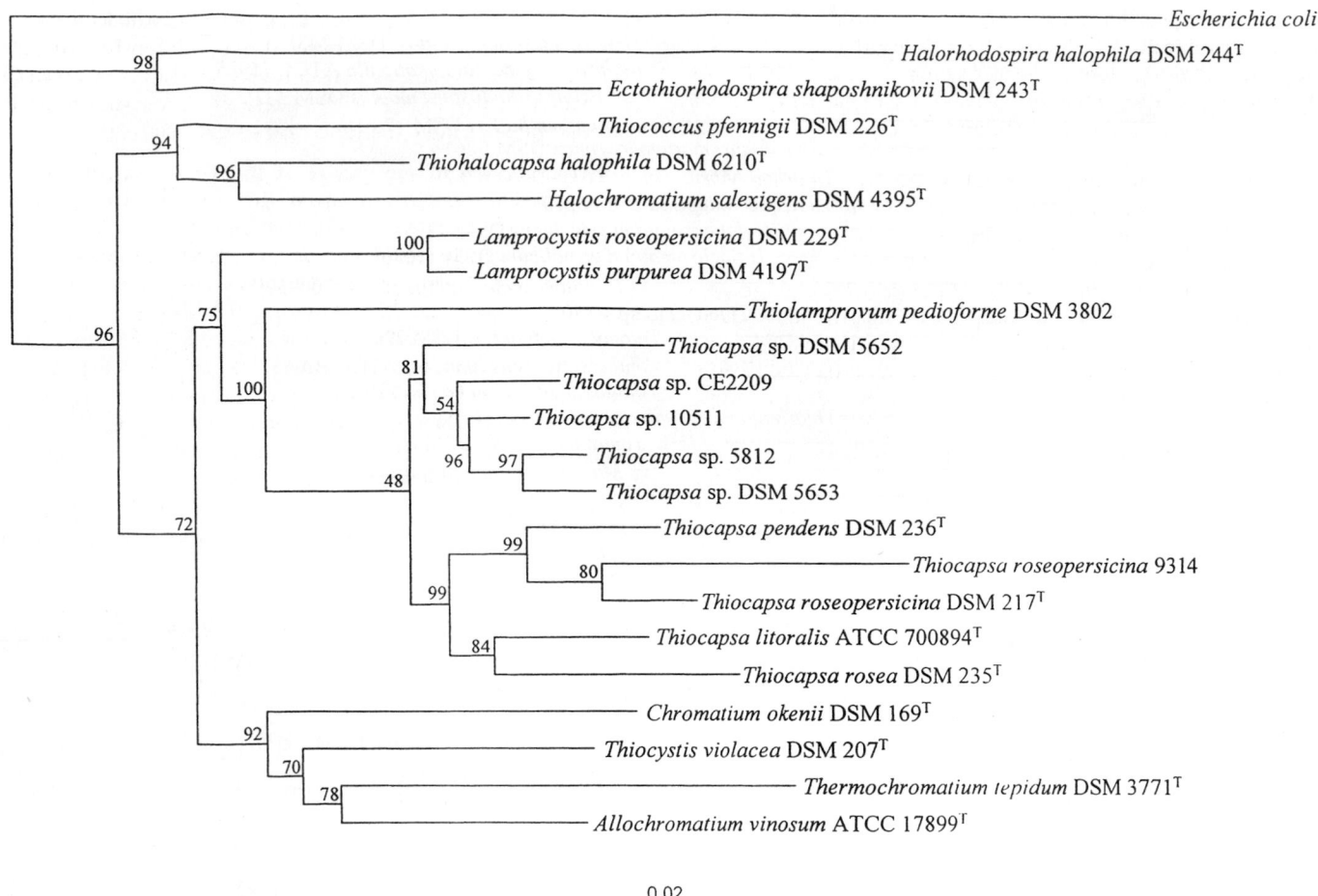

Escherichia coli
Halorhodospira halophila DSM 244[T]
Ectothiorhodospira shaposhnikovii DSM 243[T]
Thiococcus pfennigii DSM 226[T]
Thiohalocapsa halophila DSM 6210[T]
Halochromatium salexigens DSM 4395[T]
Lamprocystis roseopersicina DSM 229[T]
Lamprocystis purpurea DSM 4197[T]
Thiolamprovum pedioforme DSM 3802
Thiocapsa sp. DSM 5652
Thiocapsa sp. CE2209
Thiocapsa sp. 10511
Thiocapsa sp. 5812
Thiocapsa sp. DSM 5653
Thiocapsa pendens DSM 236[T]
Thiocapsa roseopersicina 9314
Thiocapsa roseopersicina DSM 217[T]
Thiocapsa litoralis ATCC 700894[T]
Thiocapsa rosea DSM 235[T]
Chromatium okenii DSM 169[T]
Thiocystis violacea DSM 207[T]
Thermochromatium tepidum DSM 3771[T]
Allochromatium vinosum ATCC 17899[T]

0.02

FIGURE BXIIγ.4. 16S rDNA-derived phylogenetic tree of the family *Chromatiaceae* including species of *Thiocapsa*, *Lamprocystis* and *Thiolamprovum*. Representatives of *Ectothiorhodospiraceae* and other *Chromatiaceae* genera were included to complete the view on purple sulfur bacteria. The phylogeny of these genera is shown in more detail in Fig. BXII.γ.3 and in Figure BXII.γ.19 in the chapter on *Ectothiorhodospiraceae*. Sequence data used were published previously (Guyoneaud et al., 1998b; Imhoff et al., 1998b). Sequence alignments, calculation of distance data and tree construction were performed as described earlier (Petri and Imhoff 2000).

intrageneric and intraspecies variation of the mol% G + C content necessitates experimental examination. For example, the high intraspecies variation that has been noted for *Allochromatium vinosum* (61.3–66.3), *Allochromatium warmingii* (55.1–60.2), and *Thiocystis violacea* (62.8–67.9) suggests misclassification of some strains. Further studies are needed to clarify these problems.

Although fatty acid and quinone composition is of great significance for classification of purple nonsulfur bacteria of the *Alphaproteobacteria* and *Betaproteobacteria*, and these compounds are well-accepted tools for taxonomic differentiation of phototrophic bacteria (Imhoff, 1991, 1984a; Imhoff and Bias-Imhoff, 1995; Thiemann and Imhoff, 1996), both show only minor variation and therefore are of minor significance for differentiation within the family *Chromatiaceae* (Imhoff and Bias-Imhoff, 1995). Polar lipids have been analyzed so far in only a limited number of *Chromatiaceae* species. First results, obtained with several freshwater species, showed significant correlations with the new classification, i.e., identical polar lipid composition was found i) in *Allochromatium vinosum* and *Allochromatium warmingii*, ii) in all four *Thiocystis* species, and iii) in the two *Chromatium* species (Imhoff, unpublished results). Thus, polar lipid composition may well turn out to be a relevant property for distinguishing between genera or groups of genera of the family *Chromatiaceae*. This implies that the polar lipid composition reflects phylogenetic relationships and is a rather stable property in evolutionary

terms. More analytical data on members of the family *Chromatiaceae* are required to support this concept.

Further Comments Physiological properties have always been important characteristics in taxonomy, although their taxonomic significance is ambiguous. On the one hand, members of the family *Chromatiaceae* are quite conservative in their use of sulfur sources and their restricted use of organic carbon sources: all species are able to grow photolithoautotrophically under anoxic conditions in the light using sulfide or S^0 as an electron donor. On the other hand, we can distinguish two major physiological groups, versatile and specialized species. (i) The specialized species depend on strictly anoxic conditions and are obligately phototrophic. Sulfide is required; thiosulfate and hydrogen are not used as electron donors. Only acetate and pyruvate (or propionate) are photoassimilated in the presence of sulfide and CO_2. These bacteria do not grow with organic electron donors, chemotrophic growth is not possible, and sulfate is not assimilated as a sulfur source. Among these species are *Chromatium okenii*, *Chromatium weissei*, *Allochromatium warmingii*, *Isochromatium buderi*, *Thiospirillum jenense*, and *Thiococcus pfennigii*. (ii) The versatile species photoassimilate a larger variety of organic substrates. Most of them are able to grow in the absence of reduced sulfur sources with organic substrates as electron donors for photosynthesis, and to assimilate sulfate as the sole sulfur source. Some species even grow chemoautotrophically or chemoheterotrophically (Gorlenko, 1974; Kondrat'eva et al., 1976; Kämpf and Pfennig, 1980, 1986). Among these species are *Allochromatium vinosum*, *Thiocystis violacea*, *Thiocapsa roseopersicina*, *Thiocapsa rosea*, *Thiocapsa pendens*, and *Lamprobacter modestohalophilus*.

Depending on the environmental conditions in nature or on the culture conditions, all species are able to develop either in the form of single cells or in immobile cell aggregates of variable size and shape embedded in slime. The different modes of growth described by Winogradsky (1888) may be obtained experimentally by varying sulfide concentration, light intensity, pH, salinity, temperature, and oxygen tension. At high sulfide concentration (2–4 mM) and high light intensity (2000–4000 lux), all flagellated species become immobile as the result of slime formation that embeds groups of cells. For a given light intensity, the various species differ from each other with respect to the sulfide concentration at which the cells develop functional flagella and become motile. For a given sulfide concentration, the species differ with respect to the light intensity and mode of

illumination (diurnal light and dark periods) which allow the cells to become motile. The lower the light intensity (100–200 lux), the higher the sulfide concentration at which a given strain can persist in the motile stage. These characteristics must be considered for reliable identification of pure cultures.

Differentiation of the genera of the *Chromatiaceae* Three major genetically distinct groups of the family *Chromatiaceae* are known: i) a group of truly halophilic and salt-requiring species of the genera, including *Thiorhodovibrio*, *Marichromatium*, *Isochromatium*, *Rhabdochromatium*, *Thiohalocapsa*, *Halochromatium*, *Thiorhodococcus*, *Thiococcus*, *Thioflavicoccus*, and *Thioalkalicoccus*; ii) a group that is dominated by spherical or rod-shaped nonmotile species, many of which produce gas vesicles and that are primarily freshwater bacteria (including species of the genera *Thiocapsa*, *Lamprocystis*, and *Thiolamprovum*); and iii) a group of motile species without gas vesicles that are primarily freshwater bacteria and include the genera *Chromatium*, *Allochromatium*, *Thermochromatium*, and *Thiocystis*. In both of the freshwater clusters, strains or species that may behave indifferently to low salt concentra-

TABLE BXII.γ.4. Differential characteristics of the motile *Chromatiaceae* without gas vesicles typically found in freshwater habitats[a]

Characteristic	*Chromatium okenii*	*Chromatium weissei*	*Allochromatium vinosum*	*Allochromatium minutissimum*	*Allochromatium warmingii*	*Thermochromatium tepidum*	*Thiocystis violacea*	*Thiocystis gelatinosa*	*Thiocystis minor*	*Thiocystis violascens*	*Thiospirillum jenense*
Cell shape	rod	rod	rod	rod	rod	rod	sphere	sphere	rod	rod	spiral
Cell size (μm)	4.5–6.0 × 8.0–16.0	3.5–4.5 × 7.0–14.0	2.0 × 2.5–6.0	1.0–1.2 × 2	3.5–4.0 × 5.0–11.0	1.0–2.0 × 2.8–3.2	2.5–3.0	3.0	2.0 × 2.5–6.0	2.0 × 2.5–6.0	2.5–4.5 × 30–40
Motile by flagella	+	+	+	+	+	+	+	+	+	+	+
Color of cell suspensions[b]	Pur	Pur	Br	Br	Puv	R	Puv	Pur	Pur	Puv	Ob
Carotenoid group	ok	ok	sp	sp	ra	sp	ra	ok	ok	ra	ra
Mol% G + C of DNA	48.0	48.0	61.3–66.3	63.7	55.1–60.2	61.5	62.8–67.9	61.3	62.2	61.8–64.3	45.5
B$_{12}$ requirement	+	+	−	−	+	−	−	−	−	−	+
Sulfate assimilation	−	−	+	+	−	nd	±	−	−	+	nd
Chemotrophic growth	−	−	+	+	−	−	+	+	+	+	−
pH optimum	7.0	7.0	7.0–7.3	7.0–7.3	7.0	7.0	7.0–7.3	7.0–7.3	7.0–7.3	7.0–7.3	7.0
Temperature optimum, °C	20–35	20–35	25–35	25–35	25–30	48–50	25–35	30	30	25–35	20–25
Salt requirement	none	none	none[c]	none	none	none	0–2%	0–1%	none	none	none
Substrates used:											
Hydrogen	−	−	+	+	−	−	+	nd	nd	+	nd
Sulfide	+	+	+	+	+	+	+	+	+	+	nd
Thiosulfate	−	−	+	+	−	−	+	−	+	+	nd
Sulfur	+	+	+	+	+	+	+	+	+	+	nd
Sulfite	nd	nd	+	+	nd	−	±	−	nd	+	nd
Formate	−	−	+	+	−	−	nd	nd	−	+	nd
Acetate	+	+	+	+	+	+	+	+	+	+	+
Pyruvate	+	+	+	+	+	+	+	+	+	+	nd
Propionate	−	−	+	+	nd	−	±	−	−	+	nd
Butyrate	−	−	±	±	−	−	nd	nd	−	±	nd
Lactate	−	−	nd	nd	−	nd	−	−	nd	nd	nd
Fumarate	−	−	+	+	−	−	+	−	−	−	nd
Succinate	−	−	+	+	−	−	±	−	−	−	nd
Malate	−	−	+	+	−	−	nd	nd	−	−	nd
Fructose	−	−	−	−	−	−	±	−	−	−	nd
Glucose	−	−	−	−	−	−	±	−	+	−	nd
Ethanol	−	−	−	−	−	nd	−	−	−	−	nd
Propanol	−	−	−	−	−	nd	±	−	−	−	nd
Glycerol	−	−	−	−	−	nd	±	−	−	−	nd

[a]Symbols: +, positive in most strains; −, negative in most strains; ±, variable; nd, not determined; ok, okenone; sp, spirilloxanthin; ra, rhodopinal.

[b]Br, brownish red; Ob, orange-brown; Pur, purple red; Puv, purple violet; R, red.

[c]Marine strains may tolerate low concentrations of NaCl.

tions, or require only small amounts of salts occur. Major differentiating characteristics of species and genera of the family *Chromatiaceae* are presented in Tables BXII.γ.4, BXII.γ.5, and BXII.γ.6. Their genetic relationship is shown in Figs. BXII.γ.3 and BXII.γ.4.

ACKNOWLEDGMENTS

Sequence alignments and the construction of phylogenetic trees by Dr. J. Süling (IFM Kiel, Germany) are gratefully acknowledged.

FURTHER READING

Drews, G. and Imhoff, J.F. 1991. Phototrophic purple bacteria. *In* Shively, and Barton, (Editors), Variations in Autotrophic Life, Academic Press, London. p. 51–97.

Fowler, V.J., Pfennig, N., Schubert, W. and Stackbrandt, E. 1984. Towards a phylogeny of phototrophic purple sulfur bacteria – 16S rRNA oligonucleotide cataloguing of 11 species of *Chromatiaceae*. Arch. Microbiol. *139*: 382–387.

Guyoneaud, R., Süling, J., Petri, R., Matheron, R., Caumette, P., Pfennig, N. and Imhoff, J.F. 1998. Taxonomic rearrangements of the genera *Thiocapsa* and *Amoebobacter* on the basis of 16S rDNA sequence analyses and description of *Thiolamprovum* gen. nov. Int. J. Syst. Bacteriol. *48*: 957–964.

Imhoff, J.F. 1984. Reassignment of the genus *Ectothiorhodospira* Pelsh 1936 to a new family, *Ectothiorhodospiraceae* fam. nov., and emended description of the *Chromatiaceae* Bavendamm 1924. Int. J. Syst. Bacteriol. *134*: 338–339.

Imhoff, J.F. 1988. Anoxygenic phototrophic bacteria. *In* Austin, (Editor),

TABLE BXII.γ.5. Differential characteristics of the truly marine and halophilic *Chromatiaceae* [a]

Characteristic	Halochromatium salexigens	Halochromatium glycolicum	Isochromatium buderi	Marichromatium gracile	Marichromatium purpuratum	Rhabdochromatium marinum	Thioalkalicoccus limnaeus	Thiococcus pfennigii	Thioflavicoccus mobilis	Thiohalocapsa halophila	Thiorhodococcus minor	Thiorhodovibrio winogradskyi
Cell shape	rod	rod	rod	rod	rod	rod	sphere	sphere	sphere	sphere	sphere	vibrio
Cell size (µm)	2.0–2.5 × 4.0–7.5	0.8–1.0 × 2.0–4.0	3.5–4.5 × 4.5–9.0	1.0–1.3 × 2.0–6.0	1.2–1.7 × 3.0–4.0	1.5–1.7 × 16–32	1.3–1.8	1.2–1.5	0.8–1.0	1.5–2.5	1.0–2.0	1.2–1.6 × 2.6–4.0
Motility	+	+	+	+	+	+		−	−	+	−	+
Color of cell suspensions	P, Rr	P, Pr	P, Puv	Br	Pur	B, Ob	Yb, Ob	Ob	Yb, Ob	Pur	Bo	P, Pr
Carotenoid group	sp	sp	ra	sp	ok	lycopene	ths	ths	ths	ok	sp	sp
Mol% G + C of DNA	64.6	66.1–66.5	62.2–62.8	68.9–70.4	68.4–68.9	60.4	63.6–64.8	69.4–69.9	66.5	65.9–66.6	66.9	61.0
B_{12} requirement	+	−	+	−	−	−	−	−	−	+	−	−
Sulfate assimilation	−	−	−	+	nd	nd	nd	−	ndo	−	−	nd
Chemolithotrophic growth	+	+	−	+	−	−	−	−	−	+	+	+
pH optimum	7.4–7.6	7.2–7.4	7.2–7.4	7.2–7.4	7.2–7.6	7.2–7.3	8.8–9.5	7.2–7.4	7.2–7.4	7.0	7.0–7.2	7.0–7.4
Temperature optimum (°C)	20–30	25–35	25–35	20–35	25–30	30	20–25	20–35	25–30	20–30	30–35	33
Salinity range (%)	4–20	2–20	1–5	0.5–8	2–7	1–6.5	0.5–7	0–3	1–3	3–20	0.5–9	up to 7.2
NaCl optimum (%)	8–11	4–6	2–3	2–3	5	1.5–5	1–6	0.5–2	0.5–2	4–8	2	2.2–3.2
Substrates used:												
Hydrogen	+	+	−	+	nd	+	nd	nd	nd	+	+	nd
Sulfide	+	+	+	+	+	+	+	+	+	+	+	+
Thiosulfate	+	+	−	+	+	+	−	−	−	+	+	−
Sulfur	+	+	+	+	+	+	+	+	+	+	+	+
Sulfite	+	+	−	+	nd	nd	nd	nd	nd	+	−	nd
Formate	−	(+)	−	+	−	−	−	−	−	−	−	−
Acetate	+	(+)	+	+	+	+	+	+	+	+	+	+
Pyruvate	+	(+)	+	+	+	+	+	+	+	+	+	+
Propionate	−	−	−	+	+	+	+	+	−	−	+	+
Butyrate	−	−	−	+	+	−	−	−	−	−	−	−
Lactate	−	−	−	+	+	−	−	+	−	+	+	−
Fumarate	−	+	−	+	+	−	(+)	+	−	−	+	−
Succinate	−	+	−	+	+	−	+	+	−	−	+	−
Malate	−	−	−	+	+	−	+	+	−	−	(+)	−
Fructose	−	nd	−	−	(+)	−	−	+	−	+	+	−
Glucose	−	−	−	−	−	−	−	−	−	(+)	−	−
Ethanol	−	−	−	−	−	−	−	−	−	−	+	−
Propanol	−	nd	−	nd	−	−	nd	−	−	−	+	−
Glycerol	−	+	−	−	−	−	−	−	−	(+)	−	−
Glycolate	−	+	nd	nd	nd	−	−	−	−	−	+	−
Crotonate	−	−	−	+	nd	−	nd	−	−	−	−	−
Valerate	−	nd	−	nd	+	−	−	−	−	−	−	−
Casamino acids	−	(+)	−	+	(+)	−	−	(+)	−	−	−	−

[a]Symbols: +, positive in most strains; −, negative in most strains; ±, variable; (+), weak growth; nd, not determined; P, pink; Pr, pinkish red; Rr, rose red; Puv, purple-violet; Br, brownish red; Pur, purple-red; B, beige; Ob, orange-brown; Yb, yellowish brown; Bo, brown-orange; ok, okenone; sp, spirilloxanthin; ra, rhodopinal; ths, tetrahydrospirilloxanthin.

TABLE BXII.γ.6. Differential characteristics of the *Chromatiaceae* group characterised by nonmotile cells with gas vesicles typically found in freshwater habitats, including the motile *Lamprocystis roseopersicina* and *Lamprobacter modestohalophilus* [a]

Characteristic	*Lamprobacter modestohalophilus*	*Lamprocystis roseopersicina*	*Lamprocystis purpurea*	*Thiocapsa roseopersicina*	*Thiocapsa rosea*	*Thiocapsa pendens*	*Thiodictyon elegans*	*Thiodictyon bacillosum*	*Thiolamprovum pedioforme*	*Thiopedia rosea*
Cell shape	ovoid to rod	sphere	sphere	sphere	sphere	sphere	rod	rod	sphere	sphere
Motility	+	+	−	−	−	−	−	−	−	−
Cell diameter (μm)	2.0–2.5	2.0–3.5	1.9–2.3	1.2–3.0	2.0–3.0	1.5–2.0	1.5–2.0	1.5–2.0	2.0	2.0–2.5
Gas vesicles	+	+	+	−	+	+	+	+	+	+
Aggregate formation:										
Clumps of cells			+							
Irregular aggregates		+		+	+	+				
Irregular clumps								+		
Irregular net-ishlike aggregates							+			
Platelets									+	+
Tetrads				+						
Color of cell suspensions[b]	Pur	Pv	Pur	P, Rr	P, Rr	P, Rr	Puv	Puv	P, Rr	Pur
Carotenoid group	ok	ra	ok	sp	sp	sp	ra	ra	sp	ok
Mol% G + C of DNA	60–64	63.8	63.5–63.6	63.3–66.3	64.3	65.3	65.3–66.3	66.3	65.5	62.5–63.5
B_{12} requirement	+	nd	nd	−	+	+	nd	nd	−	−
Sulfate assimilation	−	nd	−	+	−	−	nd	nd	−	−
Chemolithotrophic growth	+	nd	+	+	+	−	nd	nd	+	−
pH optimum	7.4–7.6	7.0–7.3	7.0–7.3	7.3	6.7–7.5	6.7–7.5	6.7–7.3	6.7–7.3	7.4–7.6	7.3–7.5
Temperature optimum (°C)	23–27	20–30	23–25	20–35	20–35	20–35	20–25	20–30	37	20
NaCl optimum (%)	1–4	none	none	none[c]	none	none	none	none	none	none
Substrates used:										
Hydrogen	+	nd	−	+	−	−	nd	nd	−	−
Sulfide	+	+	+	+	+	+	+	+	+	+
Thiosulfate	+	nd	+	+	+	+	nd	nd	+	−
Sulfur	+	+	+	+	+	+	+	+	+	+
Formate	−	nd	+	−	−	−	nd	nd	−	nd
Acetate	+	+	+	+	+	+	+	+	+	+
Pyruvate	+	+	+	+	+	+	+	+	+	+
Propionate	−	nd	+	±	+	+	nd	nd	−	−
Butyrate	nd	nd	−	−	−	−	nd	nd	−	+
Lactate	+	nd	+	±	+	+	nd	nd	+	−
Fumarate	−	nd	−	+	−	−	nd	nd	−	±
Succinate	−	nd	−	+	−	−	nd	nd	−	±
Malate	−	nd	−	+	±	+	nd	nd	−	±
Fructose	−	nd	−	+	+	−	nd	nd	−	±
Glucose	nd	nd	+	−	−	+	nd	nd	−	+
Ethanol	+	nd	−	−	−	nd	nd	nd	−	−
Propanol	+	nd	−	−	−	nd	nd	nd	−	−
Glycerol	+	nd	−	+	−	−	nd	nd	−	−
Glycolate	nd	nd	−	−	nd	nd	nd	nd	−	−
Crotonate	nd	nd	−	nd	nd	nd	nd	nd	−	−
Valerate	nd	nd	−	nd	−	−	nd	nd	−	+
Casamino acids	nd	nd	−	nd	nd	nd	nd	nd	−	−

[a]Symbols: +, positive in most strains; −, negative in most strains; ±, variable; nd, not determined; ok, okenone; sp, spirilloxanthin; ra, rhodopinal.

[b]P, pinkish; Pur, purple red; Puv, purple violet; Pv, pinkish violet; Rr, rose-red.

[c]Marine strains may tolerate low concentrations of NaCl.

Methods in Aquatic Microbiology, Chapter 9, John Wiley & Sons Ltd., p. 207–240.

Imhoff, J.F. 1992. Taxonomy, phylogeny and general ecology of anoxygenic phototrophic bacteria. *In* Carr, and Mann (Editors), Biotechnology Handbook for Photosynthetic Prokaryotes, Plenum Press, London, New York. p. 83–92.

Imhoff, J.F., Süling, J. and Petri, R. 1998. Phylogenetic relationships among the *Chromatiaceae*, their taxonomic reclassification and description of the new genera *Allochromatium*, *Halochromatium*, *Isochromatium*, *Marichromatium*, *Thiococcus*, *Thiohalocapsa*, and *Thermochromatium*. Int. J. Syst. Bacteriol. *48*: 1129–1143.

Molisch, H. 1907. Die Purpurbakterien nach neuen Untersuchungen, G. Fischer, Jena. p. 1–95.

Pfennig, N. 1967. Photosynthetic bacteria. Annu. Rev. Microbiol. *21*: 285–324.

Pfennig, N. 1977. Phototrophic green and purple bacteria: a comparative, systematic survey. Annu. Rev. Microbiol. *31*: 275–290.

Winogradsky, S. 1888. Beiträge zur Morphologie und Physiologie der Bakterien. Heft 1. Zur Morphologie und Physiologie der Schwefelbakterien, Arthur Felix, Leipzig. p. 1–120.

Genus I. *Chromatium* Perty 1852, 174[AL] emend. Imhoff, Süling and Petri 1998b, 1138

JOHANNES F. IMHOFF

Chro.ma' ti.um. Gr. n. *chroma* color; M.L. neut. n. *Chromatium* one which is colored.

Cells are **straight or slightly curved rods**, more than 3 μm wide, single or in pairs, multiply by binary fission. **Motile** by a polar tuft of multitrichous flagella that can be seen in the light microscope, Gram negative, belong to the *Gammaproteobacteria*, and contain **internal photosynthetic membranes of vesicular type** in which the photosynthetic pigments bacteriochlorophyll *a* and carotenoids are located.

Strictly anaerobic, obligately phototrophic, and require sulfide-reduced media. **Photolithoautotrophic growth occurs under anoxic conditions in the light with sulfide and S^0 as electron donors.** Thiosulfate and hydrogen are not used. During oxidation of sulfide, S^0 is transiently stored inside the cells in the form of highly refractile globules. Sulfate is the final oxidation product. In the presence of sulfide or sulfur and bicarbonate, a few simple organic substrates are photoassimilated. Assimilatory sulfate reduction is lacking. Ammonia is used as nitrogen source, N_2 may be used, nitrate is not reduced. Storage materials: polysaccharides, poly-β-hydroxybutyrate, S^0, and polyphosphates. Vitamin B_{12} required.

Mesophilic freshwater bacteria with optimum growth temperatures from 20 to 35°C and without a distinct requirement for salt.

Habitat: water and sediment surface layers of stagnant freshwater habitats such as ditches, ponds, and lakes, containing hydrogen sulfide and exposed to light.

The mol% G + C of the DNA is: 48.0–50.0.

Type species: **Chromatium okenii** (Ehrenberg 1838) Perty 1852,174 (*Monas okenii* Ehrenberg 1838, 15.)

FURTHER DESCRIPTIVE INFORMATION

Under unfavorable culture conditions, cells of *Chromatium* species may not show active motility when viewed under the microscope. Diurnal light and dark phases, low concentrations of sulfide, and low light intensities may be required to sustain active flagellar motility. Optimal growth conditions are clearly recognized by the bioconvection patterns that arise because of the swarming activity of the cells and that can be seen with the naked eye (Pfennig, 1962).

ENRICHMENT AND ISOLATION PROCEDURES

Chromatium okenii and *C. weissei* have so far been found only in sulfide-containing freshwater habitats, which is of primary importance for the design of enrichment experiments. Because of the selectivity of any enrichment condition, even species and strains that were not detected in the original sample from the natural habitat may become dominant in liquid enrichment cultures. Therefore, enrichment cultures are only indicated when a certain species should be enriched and isolated from a natural sample, and selective conditions for this bacterium are known. When information on the diversity of culturable species in a natural sample is desired, enrichment cultures should be avoided and dilution series in agar deeps should be prepared directly from the natural sample. Agar cultures should be incubated under conditions similar to those used for liquid cultures of the desired bacteria. The defined medium[1] proved to be relatively nonspecific and may be used for the cultivation of not only *Chromatium* species, but also most of the freshwater and, with the addition of appropriate concentrations of NaCl, marine *Chro-matiaceae* species. A somewhat simpler, but less widely applicable, culture medium was described by Pfennig and Trüper (1981). In agar shake dilution cultures, growth may be enhanced by the addition of 3 mM acetate to the defined medium.

Chromatium species have a selective advantage when cultures with relatively low sulfide concentrations (1–2 mM) are incubated at about 20°C, at low light intensities of 50–300 lux with diurnal light and dark phases (e.g., 16 h light, 8 h dark). They keep swarming throughout the whole bottle and can be further enriched by using cell suspensions from the upper part of the enrichment culture as inoculum for subsequent enrichments (Pfennig, 1962).

The sulfide concentration initially provided in fresh culture medium does not support much growth. To achieve reasonably high population densities in enrichments or pure cultures, repeated additions of neutralized sodium sulfide solution are necessary. Additions are made when the previously added sulfide is consumed and the transiently stored S^0 is nearly oxidized. The sulfide solution for feeding of cultures can be prepared by neutralizing a stirred sodium sulfide solution (60 mM) with a certain amount of sterile 2 M sulfuric acid to a pH of ~7.5. This neutralized solution has to be applied immediately. Preferably, a special device for preparation of sterile sulfide solution, neutralized and stored under CO_2-pressure, would be used (Siefert and Pfennig, 1984).

Pure cultures are obtained by repeated application of the agar shake dilution method. Water agar is prepared with 1.8% (w/v) of agar. Before the solution is prepared, the agar should be washed several times in distilled water. Depending on the salinity of the sample from nature or the enrichment culture medium, appropriate amounts of NaCl and MgCl$_2$·6H$_2$O are added. The agar solution is dispensed in 3-ml amounts into test tubes, which are stoppered with cotton plugs and autoclaved. The agar tubes are kept molten in a water bath at 50°C. Ready-prepared, defined

---

1. Defined medium for *Chromatium* and other freshwater and marine *Chromatiaceae* species after Pfennig (Pfennig, 1965, 1989b; Pfennig and Trüper, 1981, 1992): The medium is prepared in an Erlenmeyer flask with an outlet near the bottom at one side. Connected to the outlet is a silicon rubber tube with a pinchcock, and a bell for aseptic distribution of the medium into bottles or tubes. The flask is closed by a silicon rubber stopper with an inlet and an outlet for gas, and a screw-capped glass tube through which additions can be made or samplings taken.

The defined basal medium has the following composition (per liter of distilled water): KH$_2$PO$_4$, 0.34 g; NH$_4$Cl, 0.34 g; KCl, 0.34 g; CaCl$_2$·2H$_2$O, 0.25 g; NaCl (only for seawater medium), 20.0 g; MgSO$_4$·7H$_2$O (3.0 g for seawater medium), 0.5 g; and trace element solution, 1 ml. After the medium has been autoclaved and cooled under an atmosphere of N_2, the following components are added aseptically per liter of medium from sterile stock solutions, while access of air is prevented by continuous flushing with N_2: 15 ml of a 10% (w/v) solution of NaHCO$_3$ (saturated with CO_2 and autoclaved under a CO_2 atmosphere), 4 ml of a 10% (w/v) solution of Na$_2$S·9H$_2$O (autoclaved under an N_2 atmosphere), and 1 ml of a vitamin B_{12} solution containing 2 mg vitamin B_{12} in 100 ml of distilled water. The pH of the medium is adjusted to pH 7.2 by stirring under an atmosphere of CO_2 (0.5 bar pressure) for approx. 40 min. The medium is then dispensed aseptically under pressure of N_2 into sterile 100-ml bottles with metal screw caps containing autoclavable rubber seals. A small pea-sized air bubble is left in each bottle to meet possible pressure changes. Trace element solution (SL12 of Pfennig and Trüper, 1992) contains (per liter of distilled water): ethylenediaminetetraacetate-Na$_2$, 3.0 g; FeSO$_4$·7H$_2$O, 1.1 g; CoCl$_2$·6H$_2$O, 190 mg; MnCl$_2$·4H$_2$O, 40 mg; ZnCl$_2$, 42 mg; Na$_2$MoO$_4$·2H$_2$O, 18 mg; NiCl$_2$·6H$_2$O, 24 mg; H$_3$BO$_3$, 300 mg; and CuCl$_2$·2H$_2$O, 2 mg. The EDTA-Na$_2$ is dissolved first, followed by the other components. More information on preparation of media and on isolation procedures is described in Pfennig and Trüper (1992).

culture medium is prewarmed to 50°C, and 6-ml amounts are added to the tubes of liquefied agar. Exposure to air is minimized by dipping the tip of the pipette into the agar medium. Starting with a few drops of sample from nature or an enrichment culture as inoculum and using 6–8 tubes, serial dilutions are made. All tubes are then hardened in cold water and immediately sealed with a sterile overlay consisting of 1 part paraffin wax and 3 parts paraffin oil. The overlay should be 2 cm thick. Alternatively, the tubes are finally flushed with N_2/CO_2 (90:10) and sealed with butyl rubber stoppers. The tubes are kept in the dark for 6–12 h and subsequently incubated at the desired light intensity and temperature. During the first 2 days of incubation, the paraffin overlay is gently reheated to achieve a complete sealing effect. Well-separated pink, purple-red, purple-violet, or orange-brown colonies that develop in the higher dilutions can be removed for microscopic examination and further purification with sterile Pasteur pipettes, without breaking the tube. The cells are suspended in 0.5–1.0 ml of anoxic medium and used as the inoculum for subsequent agar shake cultures. The process is repeated until a pure culture is achieved.

When pure agar cultures are obtained, individual colonies are isolated and inoculated into liquid medium. It is advisable to start with small-sized bottles or screw-capped tubes (10 or 25 ml), and then to scale up to the regularly used sizes. Purity is checked both microscopically and by use of AC medium (Difco), adjusted to the appropriate salinity.

Maintenance Procedures

Pure cultures of *Chromatium* species can be maintained in the defined mineral medium, preferably in 100-ml screw-capped bottles. The freshly grown cultures are supplemented with neutralized sulfide solution to a final concentration of 1.5 mM, and kept in the light for about 2 h until the cells have formed intracellular sulfur globules. At this stage, the stock cultures can be stored at 4°C for 2–3 months. The cultures keep well if they are put back under dim light at room temperature and supplemented again with neutralized sulfide solution. After formation of sulfur globules, the cultures are put back into the refrigerator.

Long-term preservation of *Chromatium* species is successfully performed by storage in liquid nitrogen. For this purpose, dense cell suspensions of liquid cultures are supplemented with DMSO as a protective agent to a final concentration of 5%, using an autoclaved 50% DMSO stock solution. Such suspensions are dispensed into 2-ml plastic ampules, which are then sealed and freeze-stored. *Chromatium* species do not survive lyophilization and cannot be preserved in this way.

Differentiation of the Genus *Chromatium* from Other Genera

According to 16S rDNA sequence analysis (Fig. BXII.γ.3 of the chapter describing the family *Chromatiaceae*), the genus *Chromatium* belongs to the branch of the family *Chromatiaceae* that characteristically contains freshwater species with flagellar motility. Species of this genus have very low mol% G + C values (48–50), large-sized cells (more than 3 µm wide), require vitamin B_{12}, and lack the capability for chemotrophic growth. Characteristics for differentiation of *Chromatium* species from other phototrophic members of the *Chromatiaceae* are shown in Table BXII.γ.4 of the chapter describing the family *Chromatiaceae*. The genetic relationship between *Chromatium* species and other members of the *Chromatiaceae*, based on 16S rDNA sequence comparison, is shown in Fig. BXII.γ.3 of the chapter describing the family *Chromatiaceae*.

Taxonomic Comments

In the past, all rod-shaped, motile, purple sulfur bacteria without gas vesicles that store S^0 inside their cells were classified as *Chromatium* species (Pfennig and Trüper, 1989). These species were not only phenotypically quite diverse, but also distantly related on a genetic basis, as indicated by the large variation in DNA base ratio within these bacteria (48.0–70.4 mol% G + C). 16S rDNA sequence analysis of recognized species of the family *Chromatiaceae* gave support for a new classification, in which the type species of the genus, *Chromatium okenii*, appeared distant and separate from most of the other species previously assigned to this genus. The only other species showing sufficient phenotypic similarity to *C. okenii* to allow assignment to this genus is *C. weissei* (Imhoff et al., 1998b), although 16S rDNA sequence data for this species are currently not available.

Further Reading

Fowler, V.J., Pfennig, N., Schubert, W. and Stackbrandt, E. 1984. Towards a phylogeny of phototrophic purple sulfur bacteria – 16S rRNA oligonucleotide cataloguing of 11 species of *Chromatiaceae*. Arch. Microbiol. *139*: 382–387.

Imhoff, J.F. 1992. Taxonomy, phylogeny and general ecology of anoxygenic phototrophic bacteria. *In* Carr, and Mann (Editors), Biotechnology Handbook for Photosynthetic Prokaryotes, Plenum Press, London, New York. p. 83–92.

Imhoff, J.F., Süling, J. and Petri, R. 1998. Phylogenetic relationships among the *Chromatiaceae*, their taxonomic reclassification and description of the new genera *Allochromatium*, *Halochromatium*, *Isochromatium*, *Marichromatium*, *Thiococcus*, *Thiohalocapsa*, and *Thermochromatium*. Int. J. Syst. Bacteriol. *48*: 1129–1143.

Imhoff, J.F. 2001a. True marine and halophilic anoxygenic phototrophic bacteria. Arch. Microbiol. *176*: 243–254.

Differentiation of the Species of the Genus *Chromatium*

The most significant known difference between *C. okenii* and *C. weissei* is the cell width (Table BXII.γ.4 of the chapter describing the family *Chromatiaceae*).

List of species of the genus Chromatium

1. **Chromatium okenii** (Ehrenberg 1838) Perty 1852, 174[AL] (*Monas okenii* Ehrenberg 1838, 15.)

   o.ken'i.i. M.L. gen. n. *okenii* of Oken, named for L. Oken, a German naturalist.

   Cells straight or slightly curved rods, 4.5–6.0 × 8–16 µm, occasionally longer, motile by a polar tuft of flagella which is 1.5–2 times the cell length and visible by brightfield (Fig. BXII.γ.5) or phase-contrast microscopy. In the presence of sulfide and light, globules of elemental sulfur appear evenly distributed within the cell. Color of individual cells and of cell suspensions is purple-red. Photosynthetic pigments are bacteriochlorophyll *a* and the carotenoid okenone.

   Strictly anaerobic and obligately phototrophic. Sulfide-reduced media required. Sulfide and S^0 used as photosynthetic electron donors. In the presence of sulfide and bi-

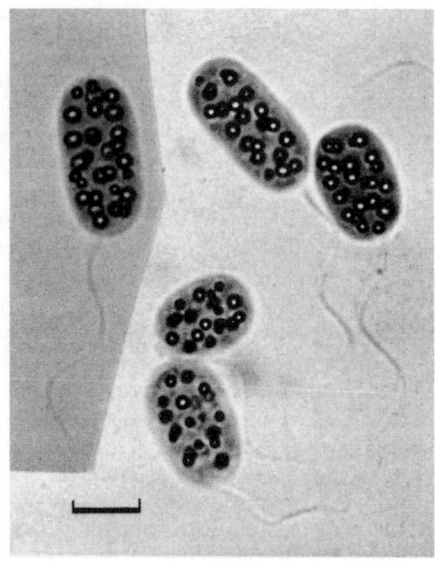

FIGURE BXII.γ.5. *Chromatium okenii* DSM 169 cultured photoautotrophically with sulfide. Note the evenly distributed intracellular globules of S⁰ and the spiral tufts of polar flagella. Brightfield micrograph. Bar = 5 μm.

carbonate, acetate and pyruvate are photoassimilated. Not utilized: thiosulfate, molecular hydrogen, sugars, sugar alcohols, alcohols, higher fatty acids, amino acids, benzoate, formate, and most intermediates of the tricarboxylic acid cycle. Assimilatory sulfate reduction is absent. Nitrogen sources: ammonium salts, and urea. Vitamin B_{12} required.

Mesophilic freshwater bacterium with optimum growth at 20–35°C and pH 7.0 (range 6.5–7.3).

Habitat: water and sediment surface layers of stagnant freshwater habitats such as ditches, ponds, and lakes, containing hydrogen sulfide and exposed to light.

The mol% G + C of the DNA is: 48.0 (Bd).

Type strain: 1111, BN 6010, DSM 169.

GenBank accession number (16S rRNA): AJ223234, Y12376.

2. **Chromatium weissei** Perty 1852, 174[AL]

weis′ se.i. M.L. gen. n. *weissei* of Weisse, named for J.F. Weisse, a German zoologist.

Cells are rod shaped, 3.5–4.5 × 7–14 μm. Nitrogenase activity is present. Other characteristics are as for *C. okenii*.

The mol% G + C of the DNA is: 48.0 (Bd).

Type strain: DSM 171.

Genus II. *Allochromatium* Imhoff, Süling and Petri 1998b, 1140[VP]

JOHANNES F. IMHOFF

Al′lo.chro.ma′ ti.um. Gr. adj. *allos* the other; *Chromatium* a genus name; M.L. neut. n. *Allochromatium* the other *Chromatium*.

Cells straight to slightly curved rods, single or in pairs, multiply by binary fission, **motile by polar flagella**, Gram negative, belong to the *Gammaproteobacteria*, and contain internal **photosynthetic membranes of vesicular type** in which the photosynthetic pigments bacteriochlorophyll *a* and carotenoids are located.

Photolithoautotrophic growth occurs under anoxic conditions in the light with sulfide and S⁰ as electron donors. During oxidation of sulfide, S⁰ is transiently stored inside the cells in the form of highly refractile globules. Sulfate is the final oxidation product. Molecular hydrogen and thiosulfate may be used as electron donors. In the presence of sulfide and bicarbonate, organic substrates can be photoassimilated, potentially mixotrophic. **Some species may grow chemoautotrophically or chemoorganotrophically under microoxic to oxic conditions in the dark**. Storage materials: polysaccharides, poly-β-hydroxybutyrate, S⁰, and polyphosphates. Vitamin B_{12} is required.

Mesophilic bacteria with growth at 25–35°C and pH 6.5–7.6. No distinct requirement for salt, except for strains that originate from the marine and brackish water environment and may tolerate or require low concentrations of salt.

Habitat: ditches, ponds, and lakes with stagnant freshwater containing hydrogen sulfide and exposed to the light; also sewage lagoons, estuaries, and salt marshes.

The mol% G + C of the DNA is: 55.1–66.3.

Type species: **Allochromatium vinosum** (Ehrenberg 1838) Imhoff, Süling and Petri 1998b, 1141 (*Chromatium vinosum* (Ehrenberg 1838) Winogradsky 1888, 99; *Monas vinosa* Ehrenberg 1838, 11.)

ENRICHMENT AND ISOLATION PROCEDURES

Media and culture conditions are the same as described for *Chromatium* species. *Allochromatium vinosum* is one of the most com-

mon purple sulfur bacteria in nature. Since it is a nonfastidious purple sulfur bacterium, it can be readily isolated. *Allochromatium vinosum* is enriched at high sulfide concentrations (3–4 mM), at high light intensities of more than 1000 lux and at incubation temperatures of 20–30°C. This species is among the dominant members of the *Chromatiaceae* in enrichment cultures under strictly autotrophic conditions, but also in the presence of organic substrates and complex nutrients, if present in the sample. *Allochromatium* species can also be isolated directly by agar shake dilution cultures from samples collected in nature, as described for *Chromatium* species.

MAINTENANCE PROCEDURES

Maintenance of strains in liquid culture medium and long-term preservation in liquid nitrogen are carried out as described for the genus *Chromatium*.

DIFFERENTIATION OF THE GENUS *ALLOCHROMATIUM* FROM OTHER GENERA

According to 16S rDNA sequence analysis, the genus *Allochromatium* belongs to the branch of the family *Chromatiaceae* containing motile freshwater species and characterized by rod-shaped bacteria. Characteristics for differentiation of *Allochromatium* species from other phototrophic members of the *Chromatiaceae* are shown in Table BXII.γ.4 of the chapter describing the family *Chromatiaceae*. The phylogenetic relationships of *Allochromatium* species to other members of the family *Chromatiaceae*, based on 16S rDNA sequence comparison, are shown in Fig. BXII.γ.3 of the chapter describing the family *Chromatiaceae*.

TAXONOMIC COMMENTS

Although salt requirements are obviously important properties for the taxonomic characterization of *Chromatiaceae* species, the congruence between phylogenetic position and salt response requires further attention in a few species. Species such as *Allochromatium vinosum* have been reported to occur in freshwater, brackish water, and marine habitats. Large differences in the mol% G + C content of the DNA of strains assigned to this species (61.3–66.3) may indicate genetic heterogeneity. So far, clear proof of identity of various isolates from marine habitats tentatively assigned to *Allochromatium vinosum* is lacking, because of improper physiological and genetic (DNA–DNA hybridization, 16S rDNA sequences) characterization. Further studies are needed, to show whether differences in base composition are related to the marine or freshwater nature of these bacteria.

The strain HPC, which was tentatively identified as a strain of "*Chromatium vinosum*" (Bauld et al., 1987), was certainly misidentified and does not belong to this species. This strain has an optimum salt concentration for growth of 2.5–4.5% NaCl, which may indicate membership to the genus *Halochromatium*. Further studies should allow its taxonomic position to be determined.

FURTHER READING

Imhoff, J.F., Süling, J. and Petri, R. 1998. Phylogenetic relationships among the *Chromatiaceae*, their taxonomic reclassification and description of the new genera *Allochromatium*, *Halochromatium*, *Isochromatium*, *Marichromatium*, *Thiococcus*, *Thiohalocapsa*, and *Thermochromatium*. Int. J. Syst. Bacteriol. *48*: 1129–1143.

DIFFERENTIATION OF THE SPECIES OF THE GENUS *ALLOCHROMATIUM*

The genus *Allochromatium* contains two easily distinguished groups of species. *A. vinosum* and *A. minutissimum* are physiologically versatile species that do not require growth factors, grow chemoautotrophically, have rather high tolerance towards sulfide, and use thiosulfate, hydrogen, and organic compounds as photosynthetic electron donors. *A. warmingii*, on the other hand, is a specialized species that requires vitamin B_{12} as a growth factor,

tolerates only rather low concentrations of sulfide, and is a strictly anaerobic and phototrophic bacterium. Differential characteristics of *Allochromatium* species are shown in Table BXII.γ.4 of the chapter describing the family *Chromatiaceae*. Presently known *Allochromatium* species can easily be distinguished on the basis of their cell diameter: Cells 2.0 μm wide, *A. vinosum*; cells 1–1.2 μm wide, *A. minutissimum*; cells 3.5–4.0 μm wide, *A. warmingii*.

List of species of the genus Allochromatium

1. **Allochromatium vinosum** (Ehrenberg 1838) Imhoff, Süling and Petri 1998b, 1141^{VP} (*Chromatium vinosum* (Ehrenberg 1838) Winogradsky 1888, 99; *Monas vinosa* Ehrenberg 1838,11.)

   vi.no' sum. L. neut. adj. *vinosum* full of wine.

   Cells are rod shaped, 2 × 2.5–6 μm, occasionally longer. Globules of S^0 evenly distributed within the cell (Fig. BXII.γ.6). Single cells are colorless, color of growing cultures first yellowish to orange-brown, later brownish red. Photosynthetic pigments are bacteriochlorophyll *a* and carotenoids of the normal spirilloxanthin group.

   Grow phototrophically under anoxic conditions in the light or chemotrophically under microoxic to semioxic conditions in the dark (Kämpf and Pfennig, 1980). Both chemotrophic and phototrophic growth is possible under autotrophic and heterotrophic conditions. Photosynthetic electron donors: sulfide, sulfur, thiosulfate, sulfite, molecular

   hydrogen, formate, acetate, propionate, pyruvate, fumarate, malate, and succinate. Some strains utilize butyrate. Not utilized: sugars, sugar alcohols, alcohols, benzoate, citrate, and amino acids. Assimilatory sulfate reduction occurs (Thiele, 1968). Nitrogen sources: ammonium salts and N_2.

   Mesophilic freshwater bacterium with optimum growth at 25–35°C and pH 7.0–7.3 (range 6.5–7.6). Marine isolates may tolerate or require low concentrations of NaCl.

   Habitat: ponds and lakes with stagnant freshwater, sewage lagoons, brackish waters, estuaries, salt marshes, and marine habitats containing hydrogen sulfide and exposed to light. Among the most widely occurring species of the family *Chromatiaceae*.

   The mol% G + C of the DNA is: 61.3–66.3 (Bd); type strain 64.3 (Bd).

   Type strain: D, BN 5110, ATCC 17899, DSM 180.

   GenBank accession number (16S rRNA): M26629.

2. **Allochromatium minutissimum** (Winogradsky 1888) Imhoff, Süling and Petri 1998b, 1141^{VP} (*Chromatium minutissimum* Winogradsky 1888, 100.)

   mi.nu.tis' si.mum. L. neut. sup. adj. *minutissimum* very small, smallest.

   Cells are rod shaped, 1–1.2 × 2 μm. Other characteristics are the same as for *A. vinosum*.

   The mol% G + C of the DNA is: 63.7 (Bd).

   Type strain: MSV, BN 5310, DSM 1376.

   GenBank accession number (16S rRNA): Y12369.

3. **Allochromatium warmingii** (Cohn 1875) Imhoff, Süling and Petri 1998b, 1141^{VP} (*Chromatium warmingii* (Cohn 1875) Migula 1900, 1048; *Monas warmingii* Cohn 1875, 167.)

   war.min' gi.i. M.L. gen. n. *warmingii* of Warming; named for E. Warming, a Danish botanist.

   Cells ovoid to rod shaped, 3.5–4.0 × 5–11 μm, sometimes longer, motile by a flagellar tuft which is usually 1.5–2 times the cell length and visible by brightfield or phase-contrast microscopy. In the presence of sulfide and light, globules of S^0 are predominantly located at the two poles

FIGURE BXII.γ.6. *Allochromatium vinosum* ATCC 17899 cultured photoautotrophically with sulfide. The cells contain sulfur globules. Brightfield micrograph. Bar = 5 μm.

of the cell. Dividing cells form additional sulfur globules near the central division plane (Fig. BXII.γ.7). Color of individual cells is grayish to slightly pink, color of cell suspensions pinkish to purple-violet. Photosynthetic pigments are bacteriochlorophyll *a* and carotenoids of the rhodopinal group.

Strictly anaerobic and obligately phototrophic. Sulfide-reduced media required. Sulfide and S^0 used as photosynthetic electron donors. In the presence of sulfide and bicarbonate, acetate and pyruvate are photoassimilated. Not utilized: thiosulfate, sugars, alcohols, higher fatty acids, amino acids, benzoate, formate, and most intermediates of the tricarboxylic acid cycle. Assimilatory sulfate reduction is absent. Nitrogen sources: ammonium salts, urea, and N_2. Vitamin B_{12} required.

Freshwater bacterium with optimum growth at 25–30°C and pH 7.0 (pH range 6.5–7.3).

Habitat: ditches, ponds, and lakes with stagnant freshwater containing hydrogen sulfide and exposed to light.

The mol% G + C of the DNA is: 55.1–60.2 (Bd); type strain 55.1 (Bd).

Type strain: 6512, ATCC 14959, BN 5810, DSM 173.

GenBank accession number (16S rRNA): Y12365.

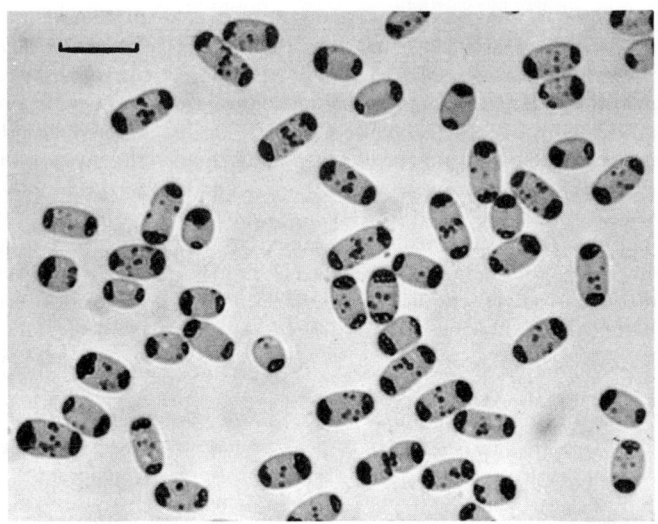

FIGURE BXII.γ.7. *Allochromatium warmingii* DSM 173 cultured photoautotrophically with sulfide. Note the polarly located globules of S^0. Dividing cells also deposited sulfur globules at the newly formed cross-walls. Brightfield micrograph. Bar = 5 μm.

Genus III. **Halochromatium** *Imhoff, Süling and Petri 1998b, 1139*[VP]

JOHANNES F. IMHOFF AND PIERRE CAUMETTE

Ha'lo.chro.ma' ti.um. Gr. gen. n. *halos* of the salt; *Chromatium* a genus name; M.L. neut. n. *Halochromatium* the *Chromatium* of the salt.

Cells straight to slightly curved rods, single or in pairs, multiply by binary fission. **Motile by polar flagella,** Gram negative, belong to the *Gammaproteobacteria*, and contain internal **photosynthetic membranes of vesicular type** with bacteriochlorophyll *a* and carotenoids as photosynthetic pigments.

Photolithoautotrophic growth under anoxic conditions in the light with sulfide and S^0 as electron donor. During oxidation of sulfide, S^0 is formed as an intermediate and stored in the form of highly refractile globules inside the cells. Sulfate is the final oxidation product. Hydrogen may be used as electron donor. In the presence of sulfide and bicarbonate, organic substrates can be photoassimilated. **Chemolithotrophic and chemoorganotrophic growth under microoxic conditions in the dark is possible.**

Mesophilic bacteria with optimum growth at 20–35°C, and pH 7.2–7.6. **Elevated salt concentrations are required for growth.**

Habitat: Reduced sediments in salinas, microbial mats in hypersaline environments, and coastal lagoons with elevated salt concentrations, which are exposed to the light.

The mol% G + C of the DNA is: 64.6–66.3.

Type species: **Halochromatium salexigens** (Caumette, Baulaigue and Matheron 1988) Imhoff, Süling and Petri 1998b, 1140 (*Chromatium salexigens* Caumette, Baulaigue and Matheron 1988, 291.)

ENRICHMENT AND ISOLATION PROCEDURES

Methods for enrichment and isolation are the same as those described for the genus *Chromatium*. *Halochromatium* species are generally found in microbial mats at the sediment surface of hypersaline environments, under salt (gypsum) deposits of salterns, in shallow salt lakes, and in evaporative lagoons with salinities of approx. 10–20%. From samples of purple layers in mats, these bacteria can be enriched or isolated by direct inoculation of deep agar dilution series in synthetic media containing 7–10% NaCl, as described by Caumette et al. (1988).

MAINTENANCE PROCEDURES

Maintenance of strains in liquid culture medium and long-term preservation in liquid nitrogen are carried out as described for the genus *Chromatium*.

DIFFERENTIATION OF THE GENUS *HALOCHROMATIUM* FROM OTHER GENERA

Halochromatium species are characterized by their requirement for elevated salt concentrations and by having motile rod-shaped cells. Characteristics for differentiation of the genus *Halochromatium* from other members of the *Chromatiaceae* are shown in Table BXII.γ.5 of the chapter describing the family *Chromatiaceae*. The phylogenetic relationships of *Halochromatium* species to other members of the family *Chromatiaceae*, based on 16S rDNA sequence comparison, are shown in Fig. BXII.γ.3of the chapter describing the family *Chromatiaceae*.

TAXONOMIC COMMENTS

The two species presently comprising the genus *Halochromatium* were originally described as species of the genus *Chromatium*, because they have rod-shaped, motile cells without gas vesicles, features that previously defined the species of the genus *Chromatium* (Pfennig and Trüper, 1974, 1989). As a consequence of 16S rDNA sequence analyses, which showed their phylogenetic distance from true *Chromatium* species and other members of the family *Chromatiaceae*, they were transferred to the new genus *Halochromatium* (Imhoff et al., 1998b). *Halochromatium* species are among the most halophilic species of the family *Chromatiaceae* and belong to the branch of truly marine and halophilic *Chromatiaceae* species.

FURTHER READING

Caumette, P. , Baulaigue, R. and Matheron, R. 1988. Characterization of *Chromatium salexignes* sp. nov., a halophilic *Chromatiaceae* isolated from Mediterranean salinas. System. Appl. Microbiol. *10*: 284–292.

Caumette, P., Imhoff, J.F., Süling, J. and Matheron, R. 1997. *Chromatium glycolicum* sp. nov., a moderately halophilic purple sulfter bacterium that uses glycolate as a substrate. Arch. Microbiol. *167*: 1129–1143.

Imhoff, J.F., Süling, J. and Petri, R. 1998. Phylogenetic relationships among the *Chromatiaceae*, their taxonomic reclassification and description of the new genera *Allochromatium, Halochromatium, Isochromatium, Marichromatium, Thiococcus, Thiohalocapsa,* and *Thermochromatium*. Int. J. Syst. Bacteriol. *48*: 1129–1143.

DIFFERENTIATION OF THE SPECIES OF THE GENUS *HALOCHROMATIUM*

Characteristic properties for distinguishing *Halochromatium* species are shown in Table BXII.γ.5 of the chapter describing the family *Chromatiaceae*.

List of species of the genus Halochromatium

1. **Halochromatium salexigens** (Caumette, Baulaigue and Matheron 1988) Imhoff, Süling and Petri 1998b, 1140[VP] (*Chromatium salexigens* Caumette, Baulaigue and Matheron 1988, 291.)

sal.ex' i.gens. L. n. *sal* salt; L. v. *exigo* demanding, to demand; M.L. part. adj. *salexigens* salt-demanding.

Cells straight or slightly curved rods, 2.0–2.5 × 4.0–7.5 μm, motile by means of a polar or subpolar tuft of flagella. Color of cell suspension pinkish red to slightly violet. Photosynthetic pigments are bacteriochlorophyll *a* and the carotenoids spirilloxanthin, anhydrorhodovibrin, rhodopin, lycopene, and rhodovibrin.

Grow photolithoautotrophically under anoxic conditions in the light with sulfide, S⁰, sulfite, thiosulfate, and H₂ as electron donors. In the presence of sulfide and carbonate, acetate and pyruvate are photoassimilated. Chemolithotrophic growth under microoxic conditions in the dark is possible with sulfide and thiosulfate, chemoorganotrophic growth with acetate and pyruvate. Assimilatory sulfate reduction is absent. Vitamin B$_{12}$ required.

Mesophilic and halophilic bacterium with optimum growth at 20–30°C and pH 7.4–7.6 (pH range 7.0–8.0). Sodium chloride required for growth, optimum salinity for growth at 8–11% NaCl, salinity range for growth from 4 to 20% NaCl.

Habitat: Reduced sediments in coastal salinas with salt deposits exposed to light.

The mol% G + C of the DNA is: 64.6 (Bd).

Type strain: SG3201, BN 6310, DSM 4395.

GenBank accession number (16S rRNA): X98597.

2. **Halochromatium glycolicum** (Caumette, Imhoff, Süling and Matheron 1997) Imhoff, Süling and Petri 1998b, 1140[VP] (*Chromatium glycolicum* Caumette, Imhoff, Süling and Matheron 1997, 17.)

gly.co' li.cum. M.L. neut. adj. *glycolicum* related to the ability to use glycolate as substrate.

Cells straight to slightly curved rods, motile by polar flagella. Autotrophically grown cells 0.8–1.0 × 2–4 μm; in the presence of organic substrates 1.0–3.0 × 2–11 μm. Color of cell suspension pinkish red. Photosynthetic pigments are bacteriochlorophyll *a* and the major carotenoids spirilloxanthin and anhydrorhodovibrin, together with small amounts of rhodopin, lycopene, and rhodovibrin.

Grow photolithoautotrophically under anoxic conditions in the light with hydrogen, sulfide, S⁰, sulfite, and thiosulfate as electron donors. In the presence of sulfide and hydrogen carbonate, glycolate, glycerol, fumarate, and succinate are photoassimilated. Formate, acetate, pyruvate, and casamino acids slightly increase the growth yield. Chemolithotrophic growth under microoxic conditions in the dark is possible with sulfide and thiosulfate, chemoorganotrophic growth with glycolate and glycerol. Assimilatory sulfate reduction is absent.

Mesophilic and halophilic bacterium with optimum growth at 25–35°C (no growth above 40°C) and pH 7.2–7.4 (pH range 6.2–9.0). Sodium chloride required for growth, optimum salinity at 4–6% NaCl, salinity range from 2–20% NaCl.

Habitat: Anoxic, sulfidic, and light-exposed habitats in hypersaline environments such as coastal lagoons subject to evaporative water losses, salt lakes, and salinas at elevated salt concentrations.

The mol% G + C of the DNA is: 66.3 (HPLC).

Type strain: SL3201, ATCC 700202, DSM 11080.

GenBank accession number (16S rRNA): X93472.

Genus IV. **Isochromatium** *Imhoff, Süling and Petri 1998b, 1141[VP]*

JOHANNES F. IMHOFF

I'so.chro.ma' ti.um. Gr. adj. *isos* equal, similar; *Chromatium* a genus name; M.L. neut. n. *Isochromatium* the similar *Chromatium*, the bacterium similar to *Chromatium*.

Cells straight to slightly curved rods, multiply by binary fission, **motile by a polar tuft of flagella**. Gram negative, belong to the *Gammaproteobacteria*, and contain internal **photosynthetic membranes of vesicular type** with bacteriochlorophyll *a* and carotenoids as photosynthetic pigments.

Obligately phototrophic and strictly anaerobic. Photolithoautotrophic growth under anoxic conditions in the light with sulfide and S⁰ as electron donors. During oxidation of sulfide, S⁰ stored inside the cells in the form of highly refractile globules. Final oxidation product is sulfate. In the presence of sulfide and bicarbonate, simple organic substrates are photoassimilated.

Mesophilic bacteria with optimum growth at 25–35°C and a pH range from 6.5 to 7.6. **Salt is required for growth** at concentrations typical for marine bacteria.

Habitat: estuarine salt flats and salt marshes.

The mol% G + C of the DNA is: 62.2–62.8.

Type species: **Isochromatium buderi** (Trüper and Jannasch 1968) Imhoff, Süling and Petri 1998b, 1141 (*Chromatium buderi* Trüper and Jannasch 1968, 364.)

ENRICHMENT AND ISOLATION PROCEDURES

The media described for *Chromatium* species can also be used for *Isochromatium* and other marine members of the family *Chromatiaceae*, if supplemented with 2–3% NaCl as required. The culture conditions and isolation procedures are the same as described for the genus *Chromatium*.

MAINTENANCE PROCEDURES

Maintenance of strains in liquid culture medium and long-term preservation in liquid nitrogen are carried out as described for the genus *Chromatium*.

DIFFERENTIATION OF THE GENUS *ISOCHROMATIUM* FROM OTHER GENERA

Characteristics for differentiation of the genus *Isochromatium* from other phototrophic members of the family *Chromatiaceae* are shown in Table BXII.γ.5 of the chapter describing the family *Chromatiaceae*. The genetic relationship between *Isochromatium buderi* and other *Chromatiaceae* species, based on 16S rDNA sequence comparison, is shown in Fig. BXII.γ.3 of the chapter describing the family *Chromatiaceae*.

TAXONOMIC COMMENTS

Isochromatium buderi was previously known as *Chromatium buderi* (Trüper and Jannasch, 1968). It was transferred to the new genus *Isochromatium* based on significant genetic and physiological differences from *Chromatium* species and other members of the *Chromatiaceae* (Imhoff et al., 1998b). *Isochromatium buderi* belongs to the branch of truly marine and halophilic *Chromatiaceae* species.

List of species of the genus Isochromatium

1. **Isochromatium buderi** (Trüper and Jannasch 1968) Imhoff, Süling and Petri 1998b, 1141[VP] (*Chromatium buderi* Trüper and Jannasch 1968, 364.)

   bu′der.i. M.L. gen. n. *buderi* of Buder; named for J. Buder, a German plant physiologist.

   Cells ovoid to rod shaped, 3.5–4.5 × 4.5–9 μm during exponential growth, about 3–4 μm in the stationary phase, motile by a polar tuft of flagella which is usually 1.5–2 times the cell length and may be visible in the light microscope. Globules of S⁰ appear evenly distributed within the cell. Color of individual cells grayish, color of cell suspensions pinkish violet to purple-violet. Photosynthetic pigments are bacteriochlorophyll *a* and carotenoids of the rhodopinal group.

   Obligately phototrophic and strictly anaerobic. Sulfide and S⁰ used as electron donors for photolithoautotrophic growth. In the presence of sulfide and bicarbonate, acetate and pyruvate are photoassimilated. Not utilized: thiosulfate, molecular hydrogen, sugars, alcohols, higher fatty acids, amino acids, benzoate, formate, and most intermediates of the tricarboxylic acid cycle. Assimilatory sulfate reduction is absent. Nitrogen sources: ammonium salts. Vitamin B_{12} required.

   Mesophilic marine bacterium requiring NaCl for growth. Optimum growth at 1–3% (w/v) NaCl, otherwise extremely pleomorphic, pH optimum for growth 7.2–7.4 (range 6.5–7.6), and temperature growth optimum 25–35°C.

   Habitat: estuarine salt flats and salt marshes.

   The mol% G + C of the DNA is: 62.2–62.8 (Bd), type strain: 62.2 (Bd).

   Type strain: 8111, ATCC 25588, DSM 176.

   GenBank accession number (16S rRNA): AJ224430.

Genus V. **Lamprobacter** *Gorlenko, Krasil'nikova, Kikina and Tatarinova 1988, 220[VP]*
(Effective publication: Gorlenko, Krasil'nikova, Kikina and Tatarinova 1979, 765)

VLADIMIR M. GORLENKO AND JOHANNES F. IMHOFF

Lam′pro.bac′ter. Gr. adj. *lampros* bright, brilliant; M.L. masc. n. *bacter* rod; M.L. masc. n. *Lamprobacter* brilliant rod.

Cells are rod shaped or ovoid, 2.0–2.5 × 4–5 μm, multiply by binary fission, do not form aggregates. **Gas vesicles formed** in cell periphery or in the entire cytoplasm. **Motile by flagella** during certain stages of development. Motile cells normally devoid of gas vesicles. Gram negative, belong to the *Gammaproteobacteria*, and contain internal photosynthetic membranes of vesicular type with bacteriochlorophyll *a* and carotenoids as photosynthetic pigments.

Photolithoautotrophic growth under anoxic conditions in the light with reduced sulfur compounds as electron donors. During oxidation of sulfide or thiosulfate, S⁰ transiently stored inside the cells in the form of highly refractile globules. In the presence of sulfide and bicarbonate, simple organic substrates are photoassimilated. Chemotrophic growth under microoxic conditions in the dark.

The mol% G + C of the DNA is: 60–64.

Type species: **Lamprobacter modestohalophilus** Gorlenko, Krasil'nikova, Kikina and Tatarinova 1988, 220 (Effective publication: Gorlenko, Krasil'nikova, Kikina and Tatarinova 1979, 765.)

FURTHER DESCRIPTIVE INFORMATION

L. modestohalophilus exhibits considerable polymorphism under various growth conditions (Gorlenko et al., 1979). On the mineral medium of Pfennig (1965) with sulfide or thiosulfate, rod-shaped or ovoid cells of different strains may be 0.5–1.5 × 2.5–5 μm. These cells display dimorphism during normal development of cultures on bicarbonate-containing media with acetate, glycerol, or other organic compounds. In freshly transferred cultures, cells are rod shaped, motile by means of flagella, generally devoid of gas vesicles, and contain globules of S⁰. Late in the exponential growth phase, cells lose their motility, gas vesicles appear, and slime capsules are formed. In stationary phase, slime capsules consolidate and cells are transformed into cysts. Cells with gas vesicles tend to concentrate in the upper part of the bottle at room temperature. Gas vesicles, short point-ended cylinders of 60×80 nm to 120×140 nm, are grouped near the cell periphery, while sulfur is deposited in the central vesicle-free part of the cells. Multiple passages on mineral salts medium result in a temporary loss of the ability to produce motile cells, and bacteria that are surrounded by loose or dense slime capsules and contain gas vesicles during all stages of growth. Cells are large and irregularly shaped at high NaCl concentrations and an acid pH of 5.9–6.6, but are thin, elongated, and slightly bent at a pH of 8.0 or higher.

The cell wall is of the Gram-negative type and, characteristically, the outer membrane is overlaid by an external layer (S-layer) consisting of hexagonal subunits which are 20 nm long. Vegetative cells possess a loose fibrous slime capsule. Slime capsules of the cyst-like cells are dense, with a clear margin. The photosynthetic membrane system occupies the major part of the cytoplasm in bacteria growing anaerobically in the light. Storage materials are abundantly produced during growth with glycerol, both under chemotrophic (aerobically in the dark) as well as under phototrophic (anaerobically in the light) growth conditions.

Suspensions of bacteria grown anaerobically in the light appear purple-pink. *In vivo* absorption maxima at 370, 655, 804, and 827 nm and a shoulder at 880 nm indicate the presence of bacteriochlorophyll *a*, while a major peak at 518 nm and shoulders at 486 and 549 nm reveal the presence of carotenoids of the okenone group. By chemical analysis, the major components were identified as okenone (64.2%), the compound thiothece-OH-484 (18.5%), also present in *Thiocystis gelatinosa* (Schmidt, 1978), and spirilloxanthin (10.5%) (Sidorova et al., 1998).

Among the organic compounds that are photoassimilated, glycerol, acetate, pyruvate, lactate, maltose, and lactose strongly increase the yield. Glycerol metabolism involves the functioning of glycerol kinase and α-glycerophosphate dehydrogenase independent of pyridine nucleotides (Krasil'nikova et al., 1979). Organic acids are assimilated through the incomplete tricarboxylic acid cycle. The glyoxylate cycle is absent (Krasil'nikova, 1985; Krasil'nikova and Kondrat'eva, 1979). Sugars are utilized through the Embden–Meyerhof pathway.

Optimum conditions for chemotrophic growth in the dark are ~10% O_2 in the gas phase. Phototrophic growth in the light is suppressed at O_2 concentrations above 2%. In a stab culture in 0.8% agar medium in the dark, growth occurs 1 cm below the surface. Thiosulfate or S^0 are required as electron donor and sulfur source for assimilatory purposes (Krasil'nikova, 1981). Optimum concentration of thiosulfate is 0.2%. At a concentration of 0.5% $Na_2S_2O_3 \cdot 5H_2O$, growth of aerobic cultures is strongly inhibited. In the dark, thiosulfate is oxidized to S^0 and sulfate. S^0 is deposited inside the cells as an intermediate product, accumulation of S^0 and generation of sulfate occur in parallel. Vitamin B_{12} is required for chemotrophic aerobic growth (Kondrat'eva et al., 1981). Best growth in the dark occurs with glycerol as carbon source, while lactate, pyruvate, and propionate also stimulate growth.

ENRICHMENT AND ISOLATION PROCEDURES

L. modestohalophilus is widely distributed in microbial mats of shallow, warm, and saline water bodies (0.5–12% NaCl) with high sulfide content of the southern Crimea, the Donetsk region (Ukraine), and Turkmenia (former U.S.S.R.). *Lamprobacter* was not found in the White Sea littoral zone.

When water and mud samples are inoculated into a medium containing 0.5–4% NaCl, 0.05% $Na_2S \cdot 9H_2O$, 0.1% $Na_2S_2O_3 \cdot 5H_2O$, and 0.1% sodium acetate, bacteria containing gas vesicles are concentrated in the upper part of the culture vessel. This fraction is removed by using a sterile capillary and is transferred into agar dilution series, as described for the genus *Lamprocystis*.

MAINTENANCE PROCEDURES

Maintenance of strains in liquid culture medium and long-term preservation in liquid nitrogen are carried out as described for the genus *Chromatium*. For intervals of 1–3 months, cultures can be maintained in liquid medium stored in a refrigerator at low light intensity.

DIFFERENTIATION OF THE GENUS *LAMPROBACTER* FROM OTHER GENERA

The genus *Lamprobacter* is characterized and distinguished from other members of the family *Chromatiaceae* by rod-shaped motile cells that develop gas vesicles under certain growth conditions. Characteristics used for differentiation of the genus *Lamprobacter* from other members of the *Chromatiaceae* are shown in Table BXII.γ.6 of the chapter describing the family *Chromatiaceae*.

List of species of the genus Lamprobacter

1. **Lamprobacter modestohalophilus** Gorlenko, Krasil'nikova, Kikina and Tatarinova 1988, 220[VP] (Effective publication: Gorlenko, Krasil'nikova, Kikina and Tatarinova 1979, 765.) *mo.des'to.ha.lo'phi.lus.* L. n. *modestus* moderate; Gr. n. *hals* salt; Gr. adj. *philos* loving; M.L. masc. adj. *modestohalophilus* moderate salt-loving.

Cells are rod shaped or ovoid, 2.0–2.5 $\times$ 4–5 µm, polymorphic under changing conditions of incubation (Fig. BXII.γ.8). Life cycle with two alternating morphological forms: nonmotile cells with gas vesicles and slime capsules, and cells motile by means of flagella and devoid of gas vesicles. Gas vesicles frequently located in the cell periphery, globules of S^0 in center of cells. Photosynthetic membranes of the vesicular type. Vesicles in cells grown in mineral medium measure about 70 nm in diameter. Photosynthetic pigments are bacteriochlorophyll *a* and carotenoids of the okenone group.

Growth under anoxic conditions in the light and under microoxic conditions in the dark. Under anoxic conditions in the light, cell suspensions appear purple to pink. Under aerobic conditions in the dark, synthesis of pigments is reduced and cultures become colorless. Photosynthetic electron donors used are sulfide, thiosulfate, S^0, and hydrogen, but not sulfite. Electron donors in the dark are thiosulfate and sulfur. Sulfide and thiosulfate are oxidized to S^0 and sulfate. Assimilatory sulfate reduction is absent and reduced sulfur compounds are required.

CO_2 assimilated via the Calvin cycle. In the presence of sulfide (or thiosulfate) and bicarbonate, alcohols (glycerol, *n*-propanol, *n*-butanol, ethanol, isobutanol, and isopropanol), and organic acids (lactate, pyruvate, and acetate) are photoassimilated. Methanol, isoamyl alcohol, sorbitol, dulcitol, mannitol, succinate, fumarate, malate, lactose, maltose, xylose, galactose, and fructose are not utilized. Growth inhibited by *n*-amyl alcohol, allyl alcohol, formate, propionate, benzoate, arabinose, rhamnose, and raffinose. Ni-

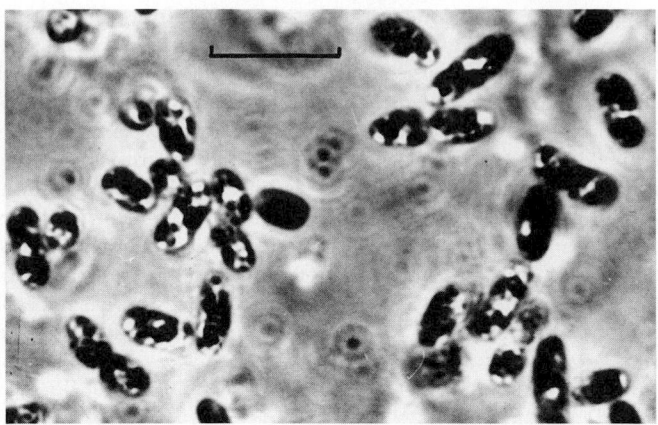

FIGURE BXII.γ.8.   *Lamprobacter modestohalophilus* strain RO-1. Phase-contrast micrograph. Bar = 10 μm. (Reproduced with permission from V.M. Gorlenko et al., Izvestiya Akademii Nauk S.S.S.R. Seriya Biologicheskaya (Moskva) *5:* 765, 1979© MAIK Nauka/Interperiodic Publishing, Moscow.)

trogen sources used: ammonium salts, urea, casein hydrolysate, glutamic acid, and N_2. Hydrogenase and catalase activity occur. Storage materials: S^0, poly-β-hydroxybutyrate, polyphosphate, and polysaccharide. Vitamin B_{12} required.

Optimum growth is at 23–27°C, pH 7.4–7.6, 1–4% NaCl (up to 9% NaCl). Sodium chloride is required for growth. Favorable light intensity is 3000 lux.

Habitat: hydrogen sulfide-containing mud and water of saline water bodies with 1–11% salinity.

The mol% G + C of the DNA is: 60–64.

Type strain: Type strain INMI RO-1 is lost; no neotype strain designated.

Additional Remarks: Available candidate for neotype: "Syvash" (Institute of Microbiology, Russia Academy of Sciences, Moscow, Russia), isolated from a lagoon of salt lake Syvash (Crimea) with 6.7% salinity, pH range for growth 6.9–7.6, temperature range for growth from 20 to 30°C.

Genus VI. **Lamprocystis** Schroeter 1886, 151[AL]

JOHANNES F. IMHOFF

Lam'pro.cys'tis. Gr. adj. *lampros* bright, brilliant; Gr. n. *cystis* the bladder, a bag; M.L. fem. n. *Lamprocystis* brilliant bag.

Cells spherical to ovoid, 1.9–3.8 μm in diameter, diplococcus-shaped before cell division and may occur in irregular aggregates, multiply by binary fission, **nonmotile or motile by means of a single flagellum, form gas vesicles** in central part. Gram negative and belong to the *Gammaproteobacteria*. At high sulfide concentration (4–6 mM) and light intensity (1000–2000 lux), cells may grow as long, branching cell aggregates embedded in slime. Under favorable growth conditions, cell aggregates may break up into smaller clusters and spheroidal microcolonies, which become motile by flagella; finally, individual motile cells are liberated. Internal **photosynthetic membranes are of vesicular type** and contain the photosynthetic pigments bacteriochlorophyll *a* and carotenoids.

Photolithoautotrophic growth under anoxic conditions in the light with reduced sulfur compounds such as sulfide and S^0 as electron donors. During oxidation of sulfide, S^0 is transiently stored in the peripheral part of the cells that is free of gas vesicles. Final oxidation product is sulfate. In the presence of sulfide and bicarbonate, simple organic substrates are photoassimilated. Obligately phototrophic or facultatively chemotrophic under microoxic to oxic conditions in the dark.

The mol% G + C of the DNA is: 63.4–64.1.

Type species: **Lamprocystis roseopersicina** (Kützing 1849) Schroeter 1886, 151 (*Protococcus roseopersicinus* Kützing 1849, 196.)

ENRICHMENT AND ISOLATION PROCEDURES

Media and culture conditions are the same as described for *Chromatium* species. *Lamprocystis* species, like other purple sulfur bacteria containing gas vesicles, may be selectively enriched by making use of the buoyancy of their cells at lower temperatures of 4–10°C (Pfennig et al., 1968). For enrichment cultures, an inoculum from a natural habitat, in which the presence of *Lamprocystis* cells and cell aggregates was established microscopically, should be used. The culture medium and incubation conditions are the same as described for the genus *Chromatium*. Low sulfide

concentrations (1–2 mM) and low light intensity (100–300 lux) with diurnal light and dark periods (e.g., 18 h light, 6 h dark) are employed at a room temperature of about 20°C. The bottles are incubated in a lying position to avoid accumulation of cells in the nonilluminated area under the screw cap. After two or three supplementations of the enrichment culture with neutralized sulfide solution, the well-developed culture is stored in a refrigerator in an upright position. This is not done until the various types of purple sulfur bacteria have almost completely oxidized the sulfur globules inside the cells. After 1–2 weeks of storage at a temperature between 4 and 10°C, purple sulfur bacteria with gas vesicles, including *Lamprocystis* species, accumulate at the surface of the liquid medium under the screw cap. A small amount of the floating material is removed with an inoculation loop and inspected microscopically under brightfield illumination. For further enrichment, the floating cell mass is carefully pipetted from the surface and transferred to fresh medium. Alternatively, or in addition, the enriched material is used to inoculate one or two series of agar shake cultures as described for the isolation of *Chromatium* species.

MAINTENANCE PROCEDURES

Maintenance of strains in liquid culture medium and long-term preservation in liquid nitrogen are carried out as described for the genus *Chromatium*.

DIFFERENTIATION OF THE GENUS *LAMPROCYSTIS* FROM OTHER GENERA

Characteristics used for differentiation of the genus *Lamprocystis* from other phototrophic members of the family *Chromatiaceae* are shown in Table BXII.γ.6 of the chapter describing the family *Chromatiaceae*. The phylogenetic relationship of *Lamprocystis* species to other members of the *Chromatiaceae*, based on 16S rDNA sequence comparison, is shown in Fig. BXII.γ.4 of the chapter describing the family *Chromatiaceae*.

TAXONOMIC COMMENTS

Only a few strains of *Lamprocystis roseopersicina* have thus far been studied in pure culture. From observations in natural samples, it is apparent that cells showing the characteristics of *Lamprocystis* are quite frequent and that cells of different diameter occur. Further studies using a greater variety of pure cultures of these bacteria are needed, to show whether clusters of strains of different cell diameter represent different species. According to 16S rRNA oligonucleotide pattern analysis, *Lamprocystis roseopersicina* is distantly related to *Thiocapsa* and *Thiodictyon* species (Fowler et al., 1984). Sequence analyses of 16S rDNA place this species into the group of spherical to ovoid, nonmotile freshwater *Chromatiaceae* species, many of which contain gas vesicles (Fig. BXII.γ.4 of the chapter describing the family *Chromatiaceae*), and in close proximity to the species originally described as *Amoebobacter purpureus* (Eichler and Pfennig, 1988). In accordance with the rules of the International Code of Nomenclature of Bacteria (Lapage et al., 1992), the transfer of this species to the genus *Lamprocystis* as the new combination *Lamprocystis purpurea* comb. nov. was proposed (Imhoff, 2001b).

The genus *Amoebobacter* comprised spherical, nonmotile species containing gas vesicles (Winogradsky, 1888; Pfennig, 1989a). Two species, including the type species, were transferred to the genus *Thiocapsa* and a third species to the new genus *Thiolamprovum* (Guyoneaud et al., 1998b; see genera *Thiocapsa* and *Thiolamprovum*). Because the type species had been removed from the genus *Amoebobacter*, the names of other species of this genus had to change, according to rule 37a, and it was illegitimate to retain the name *Amoebobacter purpureus* (Guyoneaud et al., 1998b). The transfer of this species to the new genus *Pfennigia* has been proposed (Tindall, 1999). However, the high degree of similarity between 16S rDNA sequences from this bacterium and *Lamprocystis roseopersicina* demonstrated the close relationship of the two bacteria, and precluded their separation into different genera (see Fig. BXII.γ.4 of the chapter describing the family *Chromatiaceae*). Because of the priority of the genus and species name of *Lamprocystis roseopersicina*, *Pfennigia purpurea* Tindall 1999 (*Amoebobacter purpureus* Eichler and Pfennig, 1988) was transferred to *Lamprocystis purpurea* comb. nov. (Imhoff, 2001b).

Lamprocystis purpurea (*Amoebobacter purpureus*) has been described based on six isolates, which were classified into two groups according to cell size and substrate utilization (Eichler and Pfennig, 1988). The described differences between the groups may be sufficient to allow the recognition of two different species. The description of *Lamprocystis purpurea* given below includes properties of the type strain and an additional strain (ThSchm4). Strains of the second group differ from *Pfennigia purpurea* in the following characteristics: cells are 3.3–3.8 × 3.5–4.5 µm, do not utilize formate, and have a mol% G + C of 63.4–64.1 (Eichler and Pfennig, 1988).

FURTHER READING

Pfennig, N., Markham, M.C. and Liaaen-Jensen, S. 1968. Carotenoids of *Thiorhodaceae*. 8. Isolation and characterization of a *Thiothece*, *Lamprocystis* and *Thiodictyon* strain and their carotenoid pigments. Arch. Mikrobiol. *62*: 178–191.

Eichler, B. and Pfennig, N. 1988. A new purple sulfur bacterium from stratified freshwater lakes, *Amoebobacter purpureus* sp. nov. Arch. Microbiol. *149*: 395–400.

Guyoneaud, R., Süling, J., Petri, R., Matheron, R., Caumette, P., Pfennig, N. and Imhoff, J.F. 1998. Taxonomic rearrangments of the genera *Thiocapsa* and *Amoebobacter* on the basis of 16S rRNA sequence analysis and description of *Thiolamprovum* gen. nov. Int. J. Syst. Bacteriol. *48*: 957–964.

Tindall, B.J. 1999. Transfer of *Amoebobacter purpureus* to the genus *Pfennigia* gen. nov., as *Pfennigia purpurea* comb. nov., on the basis of the illegitimate proposal to make *Amoebobacter purpureus* the type species of the genus *Amoebobacter*. Int. J. Syst. Bacteriol. *49*: 1307–1308.

Imhoff, J.F. 2001. Transfer of *Pfennigia purpurea* Tindall 1999 (*Amoebobacter purpureus* Eichler and Pfennig 1988) to the genus *Lamprocystis* as *Lamprocystis purpurea* comb. nov. Int. J. Syst. Evol. Microbiol. *51*: 1699–1701.

List of species of the genus Lamprocystis

1. **Lamprocystis roseopersicina** (Kützing 1849) Schroeter 1886,151[AL] (*Protococcus roseopersicinus* Kützing 1849, 196.) ro′ se.o.per.si.ci′ na. L. adj. *roseus* rosy; Gr. n. *persicus* the peach; M.L. fem. adj. *roseopersicina* rosy peach-colored.

Cells spherical to ovoid, 2.0–3.5 µm in diameter (Fig. BXII.γ.9), motile by flagella, contain gas vesicles. Color of cell suspensions pinkish violet to purple-violet. Photosynthetic pigments are bacteriochlorophyll *a* and carotenoids of the rhodopinal group, with lycopenal as main component (Pfennig et al., 1968).

Obligately phototrophic and strictly anaerobic. Photolithoautotrophic growth with sulfide and S⁰ as electron donors. In the presence of sulfide and bicarbonate, acetate and pyruvate are photoassimilated. Nitrogen sources: ammonium salts.

Mesophilic freshwater bacterium with optimum growth at 20–30°C and pH 7.0-7.3.

Habitat: mud and stagnant water of ponds and lakes containing hydrogen sulfide; common planktonic purple sulfur bacterium in the sulfide-containing hypolimnion of freshwater lakes.

The mol% G + C of the DNA is: 63.8 (Bd).

Type strain: 3012, BN 4510, DSM 229.

GenBank accession number (16S rRNA): AJ006063.

2. **Lamprocystis purpurea** *comb. nov.* (Eichler and Pfennig 1988) Imhoff 2001b, 1700[VP] (*Pfennigia purpurea* (Eichler and Pfennig 1988) Tindall 1999, 1308; *Amoebobacter purpureus* Eichler and Pfennig 1988, 399.)

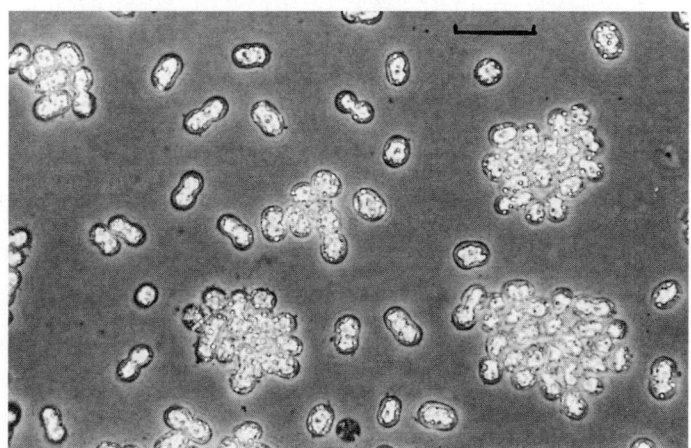

FIGURE BXII.γ.9. *Lamprocystis roseopersicina* DSM 229 cultured photoautotrophically with sulfide. Single cells and cell colonies are motile in liquid medium. The *irregular whitish areas* within the cells are the gas vacuoles. The *small spherical bodies* in the cells are the sulfur globules. Phase-contrast micrograph. Bar = 10 µm.

pur.pur' e.a. L. fem. adj. *purpurea* purple or purple-red.

Cells nearly spherical to oval, 1.9–2.3 $\times$ 2.0–3.2 µm, nonmotile, and contain gas vesicles in central parts of cells. Cells single or in irregular aggregates of up to 40 cells. Color of dense cell suspensions purple-red. Photosynthetic pigments are bacteriochlorophyll *a* and the carotenoid okenone is a major component.

Photolithoautotrophic growth under anoxic conditions in the light with sulfide, thiosulfate, and S^0 as electron donors. In the presence of sulfide and bicarbonate, acetate, propionate, pyruvate, lactate, glucose, and formate are photoassimilated. Chemolithoautotrophic growth possible with sulfide and thiosulfate as electron donors under microoxic conditions in the dark. Assimilatory sulfate reduction is absent.

Mesophilic freshwater bacterium with optimum growth at 23–25°C and pH 7.0–7.3.

Habitat: Anoxic sulfide-containing water of stratified freshwater lakes, mud, and stagnant water of ponds and lakes, may occur as dominant bloom-forming bacterium, together with *Thiocapsa roseopersicina*, in wastewater lagoons containing degradable organic substances.

The mol% G + C of the DNA is: 63.5–63.6 (Bd).

Type strain: ThSchl2, DSM 4197.

GenBank accession number (16S rRNA): Y12366, AJ223235.

Genus VII. Marichromatium *Imhoff, Süling and Petri 1998b, 1140*[VP]

JOHANNES F. IMHOFF

Ma' ri.chro.ma' ti.um. L. gen. n. *maris* of the sea; *Chromatium* a genus name; M.L. neut. n. *Marichromatium* the *Chromatium* of the sea, the truly marine *Chromatium*.

Cells straight to slightly curved rods, motile by polar flagella and multiply by binary fission. Cells single or in pairs, and may stick together and form clumps. Gram negative, belong to the *Gammaproteobacteria*, and contain internal **photosynthetic membranes of vesicular type** with bacteriochlorophyll *a* and carotenoids as photosynthetic pigments.

Grow photolithoautotrophically under anoxic conditions in the light with sulfide and S^0 as electron donors. During oxidation of sulfide, S^0 is formed as an intermediate and stored in the form of highly refractile globules inside the cells. Final oxidation product is sulfate. Molecular hydrogen may be used as electron donor. In the presence of sulfide and bicarbonate, organic substrates can be photoassimilated. **Chemolithotrophic and chemoorganotrophic growth may be possible under microoxic conditions in the dark**.

Mesophilic bacteria with optimum growth at 25–35°C, pH 6.5–7.6, and at salt concentrations typical of marine bacteria.

Habitat: anoxic marine sediments and stagnant waters containing hydrogen sulfide and exposed to the light, marine sponges and other marine invertebrates.

The mol% G + C of the DNA is: 68.9–70.4.

Type species: **Marichromatium gracile** (Strzeszewski 1913) Imhoff, Süling and Petri 1998b, 1140 (*Chromatium gracile* Strzeszewski 1913, 321.)

ENRICHMENT AND ISOLATION PROCEDURES

The media described for *Chromatium* species can also be used for *Marichromatium* species and other marine members of the family *Chromatiaceae*, if supplemented with appropriate amounts of salt as required. The culture conditions and isolation procedures are the same as described for the genus *Chromatium*.

MAINTENANCE PROCEDURES

Maintenance of strains in liquid culture medium and long-term preservation in liquid nitrogen are carried out as described for the genus *Chromatium*.

DIFFERENTIATION OF THE GENUS *MARICHROMATIUM* FROM OTHER GENERA

Marichromatium species are distinct from other members of the family *Chromatiaceae* by their specific salt requirement and the very high mol% G + C content of their DNA. Characteristics for differentiation of the genus *Marichromatium* from other phototrophic members of the family *Chromatiaceae* are shown in Table BXII.γ.5 of the chapter describing the family *Chromatiaceae*.

TAXONOMIC COMMENTS

Species of this genus have previously been classified as belonging to the genus *Chromatium* (Pfennig and Trüper, 1974; Pfennig, 1989b). However, because of significant phenotypic and genetic differences to true *Chromatium* species and other members of the family *Chromatiaceae*, they have been transferred to the new genus *Marichromatium* (Imhoff et al., 1998b).

FURTHER READING

Imhoff, J.F., Süling, J. and Petri, R. 1998. Phylogenetic relationships among the *Chromatiaceae*, their taxonomic reclassification and description of the new genera *Allochromatium, Halochromatium, Isochromatium, Marichromatium, Thiococcus, Thiohalocapsa*, and *Thermochromatium*. Int. J. Syst. Bacteriol. 48: 1129–1143.

DIFFERENTIATION OF THE SPECIES OF THE GENUS *MARICHROMATIUM*

Marichromatium species can be distinguished on the basis of cell suspension color and pigment content, differences in substrate utilization, as well as slight differences in cell size and salt response. Characteristic properties for distinguishing *Marichromatium* species are shown in Table BXII.γ.5 of the chapter describing the family *Chromatiaceae*. The phylogenetic relationships of *Marichromatium* species to other members of the family *Chromatiaceae* is shown in Fig. BXII.γ.3 of the chapter describing the family *Chromatiaceae*.

List of species of the genus Marichromatium

1. **Marichromatium gracile** (Strzeszewski 1913) Imhoff, Süling and Petri 1998b, 1140[VP] (*Chromatium gracile* Strzeszewski 1913, 321.)

gra' ci.le. L. neut. adj. *gracile* thin, slender.

Cells are rod shaped, 1.0–1.3 $\times$ 2–6 µm. Globules of S^0 evenly distributed within cells. Single cells colorless. Color of growing cultures orange-brown to brownish red. Photosynthetic pigments are bacteriochlorophyll *a* and carotenoids of the normal spirilloxanthin group.

Photolithoautotrophic growth under anoxic conditions in the light with sulfide, sulfur, thiosulfate, sulfite, and molecular hydrogen as electron donors. Organic substrates formate, acetate, propionate, pyruvate, fumarate, malate, succinate, lactate, propionate, butyrate, crotonate, and Casamino acids are used. Not utilized: sugars, sugar alcohols, alcohols, benzoate, citrate, and amino acids. Chemoautotrophic and mixotrophic growth under microoxic to semioxic conditions possible. Assimilatory sulfate reduction occurs (Thiele, 1968). Nitrogen sources: ammonium salts, N_2. Vitamins not required.

Habitat: brackish waters, estuaries, salt marshes, and marine habitats containing hydrogen sulfide and exposed to the light.

Mesophilic, marine bacterium with optimum growth at 2–3% NaCl, pH 7.2–7.4 (range 6.8–7.6), and a temperature range of 20–35°C.

The mol% G + C of the DNA is: 68.9–70.4 (Bd); type strain 69.9 (Bd).

Type strain: 8611, BN 5210, DSM 203.

GenBank accession number (16S rRNA): X93473.

2. **Marichromatium purpuratum** (Imhoff and Trüper 1980) Imhoff, Süling and Petri 1998b, 1140[VP] (*Chromatium purpuratum* Imhoff and Trüper 1980, 69.)

pur.pur.a' tum. L. neut. adj. *purpuratum* dressed in purple.

Cells are rod shaped, 1.2–1.7 × 3–4 μm, when grown under autotrophic conditions, motile by a single polar flagellum. Color of cell suspensions purple-red. Photosynthetic pigments are bacteriochlorophyll *a* and carotenoids of the okenone group.

Obligately phototrophic and strictly anaerobic. Photolithoautotrophic growth with sulfide, sulfur, and thiosulfate as electron donors; photoorganoheterotrophic growth with acetate, propionate, butyrate, valerate, lactate, pyruvate, fumarate, malate, and succinate in the presence of bicarbonate. Fructose and casamino acids are poor substrates. Not utilized: sugars, sugar-alcohols, alcohols, higher fatty acids, benzoate, and amino acids. Nitrogen sources: ammonium salts. Vitamins not required.

Mesophilic marine bacterium that requires NaCl for growth, optimum salinity 5% NaCl, salinity range 2–7% NaCl, pH range 7.2–7.6, temperature range 25–30°C.

Habitat: anoxic and sulfidic marine sediments and waters exposed to the light, marine sponges, and other marine invertebrates, such as didemnids and copepods.

The mol% G + C of the DNA is: 68.9 (T_m).

Type strain: BN 5500, DSM 1591.

GenBank accession number (16S rRNA): AF001580, AJ224439.

Genus VIII. **Nitrosococcus** *Winogradsky1892, 127*[AL] *(Nom. Cons. Opin. 23 Jud. Comm. 1958b, 169)*

HANS-PETER KOOPS AND ANDREAS POMMERENING-RÖSER

Ni.tro.so.coc' cus. M.L. adj. *nitrosus* nitrous; *coccus* a grain, berry; M.L. masc. n. *Nitrosococcus* nitrous sphere.

Cells are spherical to ellipsoidal (Fig. BXII.γ.10A). Typical Gram-negative cell wall, but **some strains have additional layers composed of repeating subunits arranged in a molecular array** (Watson and Remsen, 1969, 1970). **Extensive intracytoplasmic membrane systems arranged centrally in the protoplasm** (Fig. BXII.γ.10B). Carboxysomes not observed. **Obligate salt requirement. Urea** can be used as ammonia source by one of the two described species. Motile cells possess **a tuft of flagella or a single flagellum.** Species are **distributed in oceans and in salt lakes.**

Type species: **Nitrosococcus nitrosus** (Migula 1900) Buchanan 1925, 402 (Nom. Cons. Opin. 23 Jud. Comm. 1958b, 169) (*Micrococcus nitrosus* Migula 1900, 194.)

TAXONOMIC COMMENTS

Isolates of coccoid ammonia-oxidizers were first described by Winogradsky (1892, 1904). One of these strains has been named *Nitrosococcus nitrosus* by Migula (1900). However, this type species of the genus *Nitrosococcus* was not preserved in culture. Only data

on the cell shape and size are available from the original description (Winogradsky, 1892). Since this species cannot be identified with certainty, it is a candidate for placement on the list of rejected names.

Another coccoid ammonia-oxidizer, originally placed in *Nitrosococcus* as *N. mobilis* (Koops et al., 1976), has been proved via phylogenetic analyses to be a member of the genus *Nitrosomonas* (Woese et al., 1984b; Head et al., 1993; Teske et al., 1994).

The remaining two described species of the genus *Nitrosococcus*, *N. oceani* and "*N. halophilus*", are both members of the *Gammaproteobacteria*. They are closely related to each other, as demonstrated by DNA–DNA hybridizations. Using the S_1 nuclease technique, a DNA similarity of 9–10% was estimated (Pommerening-Röser, 1993).

In the future, the taxonomically revised genus *Nitrosococcus* should be reserved for representatives of ammonia-oxidizing bacteria belonging to the *Gammaproteobacteria* and being closely related to the grouping of *N. oceani* and "*N. halophilus*".

List of species of the genus Nitrosococcus

1. **Nitrosococcus nitrosus** (Migula 1900) Buchanan 1925, 402[AL] (Nom. Cons. Opin. 23 Jud. Comm. 1958b, 169) (*Micrococcus nitrosus* Migula 1900, 194.)
ni.tro' sus. M.L. adj. *nitrosus* nitrous.

Cells are spheres 1.5–1.7 μm in diameter. No further phenotypic characteristics are available.

The mol% G + C of the DNA is: unavailable.

Type strain: no longer in culture.

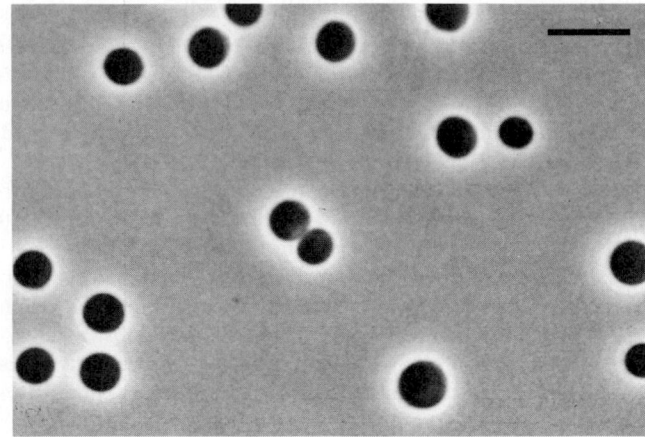

A

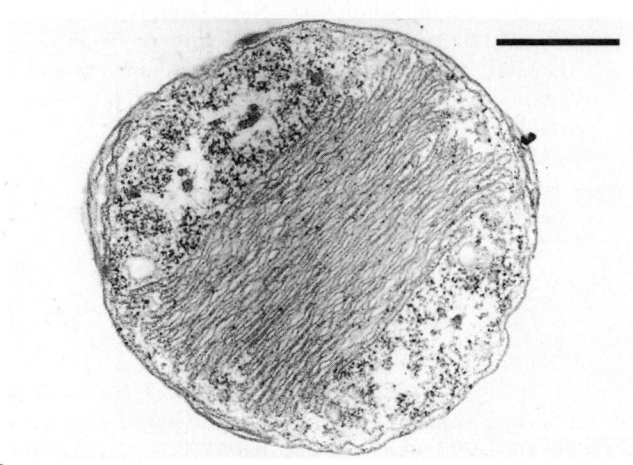

B

FIGURE BXII.γ.10. (*A*) Phase contrast micrograph of a thin section of cells of *Nitrosococcus oceani.* Bar = 5 μm. (*B*) Electron micrograph of a thin section of cells of *N. oceani.* Bar = 0.5 μm.

2. **Nitrosococcus oceani** (Watson 1965) Watson 1971, 267[AL] (*"Nitrosocystis oceanus"* Watson 1965, R279.)
   o.ce.a'nus. M.L. gen. n. *oceani* of the ocean.

   The cells are spherical to ellipsoidal, 1.8–2.2 μm in diameter. Cells generally occur singly or in pairs, but sometimes aggregates are formed via production of exopolymeric substances. Motile cells possess 1–20 flagella. Cells have a typical Gram-negative envelope, with two additional cell wall layers composed of subunits arranged in rectilinear and hexagonal arrays. Cells possess an extensive intracytoplasmic membrane system, arranged as a centrally located stack of parallel, flattened vesicles. Cells have an obligate salt requirement, with optimum NaCl concentrations for growth of 400–500 mM. The maximum concentration of ammonium salts is around 1000 mM at pH 7.8. Urease positive. All strains have been isolated from marine environments.

   The mol% G + C of the DNA is: 50.5 (T_m, Bd).
   Type strain: C-107, ATCC 19707.
   GenBank accession number (16S rRNA): M96395.

3. **"Nitrosococcus halophilus"** Koops, Böttcher, Möller, Pommerening-Röser and Stehr 1990, 247.
   ha.lo'phi.lus. Gr. n. *halos* salt; Gr. adj. *philos* loving; M.L. adj. *halophilus* salt loving.

   Cells are spherical to ellipsoidal, 1.8–2.5 μm in diameter. Gram-negative cell wall. Additional outer layers, as demonstrated for *N. oceani*, may exist, but have not been reported. Motile cells have a tuft of flagella. Intracytoplasmic membranes are arranged as a stack of centrally located flattened vesicles. Cells are obligately halophilic, optimum NaCl concentrations for growth are 600–800 mM, up to 1600 mM is tolerated. Ammonium compounds are tolerated up to concentrations around 600 mM. Urease negative. Isolates originate from salt lagoons and salt lakes.

   The mol% G + C of the DNA is: 50.5 (T_m).
   Deposited strain: Nc 4.

Genus IX. **Pfennigia** Tindall 1999, 1308[VP]*

BRIAN J. TINDALL

Pfen.nig'ia. M.L. *ia* ending; M.L. fem. n. *Pfennigia* named after Norbert Pfennig, in recognition of his contribution to the biology and taxonomy of anoxygenic phototrophic bacteria.

Oval to spherical cells, 1.9–3.8 × 2.0–4.5 μm. May form large aggregates (up to 40 cells). Gas vacuoles occur in the central part of the cytoplasm. The intracytoplasmic membrane system is of the vesicular type. **Nonmotile.** Gram negative. **Photolithotrophic under anaerobic conditions in the light**; chemoautotrophic under microaerobic conditions in the dark. **Hydrogen sulfide, thiosulfate, and S⁰ serve as electron donors. Globules of S⁰ appear in cells during oxidation of sulfide or thiosulfate to sulfate. Major photosynthetic pigments are bacteriochlorophyll *a* and the carotenoid okenone.** Isolated from the chemocline of freshwater lakes.

   The mol% G + C of the DNA is: 63.4–64.1.

   Type species: **Pfennigia purpurea** (Eichler and Pfennig 1988)

Tindall 1999, 1308 (*Amoebobacter purpureus* Eichler and Pfennig 1988, 399.)

TAXONOMIC COMMENTS

The trend towards the use of 16S rDNA sequences in the taxonomy of anoxygenic phototrophic bacteria has indicated that many genera, such as *Chromatium*, *Thiocapsa*, and *Amoebobacter*, which are phenotypically relatively heterogeneous are also phylogenetically heterogeneous. However, this has also led to a tendency to rely heavily on the 16S rDNA sequences and to reduce many phenotypic parameters to being "unreliable". 16S rDNA sequence analysis of the type strain of *Pfennigia purpurea* (DSM 4197) indicates that this strain has a high degree of sequence similarity to *Lamprocystis roseopersicina* DSM 229. However, two 16S rDNA sequences have been deposited as AJ223235 and Y12366, both being from the type strain of *Pfennigia purpurea* DSM 4197, although they are sufficiently different to infer that they come from different species. Thus, while *Pfennigia purpurea* (DSM 4197;

Editorial Note: The type species of *Pfennigia*, *P. purpurea*, has been transferred to the genus *Lamprocystis* as *L. purpurea* (Eichler and Pfennig 1988) Imhoff 2001b comb. nov.

GenBank accession no. AJ223235) shows less than 1% 16S rDNA sequence divergence from *Lamprocystis roseopersicina* (DSM 229; GenBank accession no. AJ006063), the degree of sequence divergence between the latter and *Pfennigia purpurea* (DSM 4197; GenBank accession no. Y12366) may be more than 3%. In the first case, it would be appropriate to examine the possibility that *Pfennigia purpurea* (DSM 4197; Y12366) and *Lamprocystis roseopersicina* (DSM 229; AJ006063) are members of the same species, whereas in the latter case their membership in the same genus is questionable. These two possibilities cannot be decided based on currently available sequence data. It should be pointed out that it is implicit in the evaluations of Guyoneaud et al. (1998b), Imhoff et al. (1998b), or Puchkova et al. (2000) that species sharing 5% or less 16S rDNA sequence divergence belong in the same genus. However, this is usually without adequate supporting data or in the light of other phenotypic data being "unreliable". In other genera, sequence similarities may be higher than 95% and rely on supporting phenotypic data (see, for example, Lincoln et al., 1999).

List of species of the genus Pfennigia

1. **Pfennigia purpurea** (Eichler and Pfennig 1988) Tindall 1999, 1308[VP] (*Amoebobacter purpureus* Eichler and Pfennig 1988, 399.)

    pur.pur′ea. L. fem. adj. *purpurea* purple or purple-red.

    The characteristics are as given for the genus.
    The mol% G + C of the DNA is: 63.4–64.1
    Type strain: ThSch12, DSM 4197.
    GenBank accession number (16S rRNA): AJ223235, Y12366.

Genus X. **Rhabdochromatium** Dilling, Liesack and Pfennig 1996, 362[VP] (Effective publication: Dilling, Liesack and Pfennig 1995, 130)

JOHANNES F. IMHOFF

Rhab′do.chro.ma′tium. Gr. n. *rhabdos* a rod; Gr. n. *chromatium* color, paint; M.L. neut. n. *Rhabdochromatium* colored rod.

Cells straight or slightly bent rods, motile by a polar tuft of flagella, multiply by binary fission. Gram negative, belong to the *Gammaproteobacteria*, and contain internal **photosynthetic membranes of vesicular type** in which the photosynthetic pigments bacteriochlorophyll *a* and carotenoids are located.

Metabolism obligately phototrophic and strictly anaerobic. Photolithoautotrophic growth with hydrogen, sulfide, S^0, and thiosulfate as electron donors. During oxidation of sulfide, globules of sulfur are stored inside the cells as intermediary product. Final oxidation product is sulfate. In the presence of sulfide and bicarbonate, simple organic substrates are photoassimilated.

Mesophilic marine bacteria that require NaCl for growth, optimum growth at neutral pH and 20–35°C.

The mol% G + C of the DNA is: 60.4.

Type species: **Rhabdochromatium marinum** Dilling, Liesack and Pfennig 1996, 362 (Effective publication: Dilling, Liesack and Pfennig 1995, 130.)

ENRICHMENT AND ISOLATION PROCEDURES

The media described for *Chromatium* species can also be used for *Rhabdochromatium* and other marine members of the family *Chromatiaceae*, if supplemented with 2–3% NaCl as required. The culture conditions and isolation procedures are the same as described for the genus *Chromatium*.

MAINTENANCE PROCEDURES

Maintenance of strains in liquid culture medium and long-term preservation in liquid nitrogen are as described for the genus *Chromatium*.

DIFFERENTIATION OF THE GENUS *RHABDOCHROMATIUM* FROM OTHER GENERA

Characteristics for differentiation of the genus *Rhabdochromatium* from other phototrophic members of the family *Chromatiaceae* are shown in Table BXII.γ.5 of the chapter describing the family *Chromatiaceae*. The genetic relationship of *Rhabdochromatium marinum* to other members of the family *Chromatiaceae*, based on 16S rDNA sequence comparison, is shown in Fig. BXII.γ.3 of the chapter describing the family *Chromatiaceae*.

TAXONOMIC COMMENTS

The genus *Rhabdochromatium* was established by Winogradsky (1888) for long, rod- or spindle-shaped, red sulfur bacteria observed in samples from natural habitats. From microscopic observations of reddish cells containing sulfur globules, he described three species that were distinguished by cell size: *R. roseum*, *R. minus*, and *R. fusiforme*. Two more species were described later (see Bavendamm, 1924). Since none of the described species has so far been isolated in pure culture or observed again in samples from nature and enrichment cultures, the genus was not included in the Approved Lists of Bacterial Names. Indeed, this genus was included in the genus *Chromatium* (Pfennig and Trüper, 1974), because long and irregularly swollen cells have been observed in pure cultures of *Chromatium okenii*, *Chromatium weissei*, *Isochromatium buderi*, and *Allochromatium warmingii*, when the mineral salts concentrations of the culture media were inappropriate. The name *Rhabdochromatium* was revived, according to Rule 33c of the International Code of Nomenclature of Bacteria (Lapage et al., 1992), for a long, rod-shaped, red sulfur bacterium, the morphology of which has proven to be stable in pure culture for several years. Only one species is currently known. *Rhabdochromatium marinum* belongs to the cluster of truly marine species of the family *Chromatiaceae*.

FURTHER READING

Dilling, W., Liesack, W. and Pfennig, N. 1995. *Rhabdochromatium marinum* gen. nom. rev., sp. nov., a purple sulfur bacterium from a salt marsh microbial mat. Arch. Microbiol. *164*: 125–131.

List of species of the genus Rhabdochromatium

1. **Rhabdochromatium marinum** Dilling, Liesack and Pfennig 1996, 362^{VP} (Effective publication: Dilling, Liesack and Pfennig 1995, 130.)

ma.ri' num. L. neut. adj. *marinum* marine.

Cells straight rods with slightly conical ends, sometimes slightly bent, 1.5–1.7 $\times$ 16–32 µm, motile by a polar tuft of flagella of about 10–20 bipolar fibrils, may be flagellated at both ends. Color of cell suspension beige to orange-brown. Photosynthetic pigments are bacteriochlorophyll *a* and carotenoids with lycopene as major component.

Obligately phototrophic and strictly anaerobic. Hydrogen, sulfide, S^0, and thiosulfate used as electron donors for photolithoautotrophic growth. In the presence of sulfide and bicarbonate, acetate, propionate, and pyruvate are photoassimilated. Growth factors not required.

Mesophilic marine bacterium requiring NaCl for growth: optimum growth at 1.5–5% NaCl, no growth below 1% and above 6.5%, optimum pH 7.2–7.3, optimum temperature 30°C, no growth below 8°C and above 35°C.

Habitat: laminated microbial mat at Great Sippewissett Salt Marsh, Cape Cod, USA.

The mol% G + C of the DNA is: 60.4 (Bd).

Type strain: 8315, DSM 5261.

GenBank accession number (16S rRNA): X84316.

Genus XI. **Thermochromatium** *Imhoff, Süling and Petri 1998b, 1140^{VP}*

JOHANNES F. IMHOFF AND MICHAEL T. MADIGAN

Ther' mo.chro.ma' ti.um. Gr. adj. *thermos* hot; *Chromatium* a genus name; M.L. neut. n. *Thermochromatium* the hot *Chromatium*.

Cells straight to slightly curved rods, multiply by binary fission, **motile**, most likely by polar flagella. Gram negative, belong to the *Gammaproteobacteria*, and contain internal **photosynthetic membranes of vesicular type** in which the photosynthetic pigments bacteriochlorophyll *a* and carotenoids are located.

Strictly anaerobic, obligately phototrophic, and require sulfide-reduced media. **Photolithoautotrophic growth under anoxic conditions in the light with sulfide and S^0 as electron donors.** During oxidation of sulfide, S^0 is transiently stored inside the cells in the form of highly refractile globules. Sulfate is the final oxidation product. In the presence of sulfide and bicarbonate, organic substrates can be photoassimilated.

Thermophilic freshwater bacteria with optimum growth at temperatures higher than 40°C, at neutral pH 7.0, and without a requirement for salt.

Habitat: sulfide-containing hot springs of neutral to alkaline pH having temperatures between 35 and 60°C.

The mol% G + C of the DNA is: 60–63.

Type species: **Thermochromatium tepidum** (Madigan 1986) Imhoff, Süling and Petri 1998b, 1140 (*Chromatium tepidum* Madigan 1986, 226.)

ENRICHMENT AND ISOLATION PROCEDURES

Media and culture conditions are the same as described for *Chromatium* species. Enrichment cultures with conditions suitable for *Chromatiaceae* species in general, in the presence of sulfide and acetate, and incubated at 45–55°C and neutral pH, give rise to the development of *Thermochromatium tepidum*, if present in the inoculum.

MAINTENANCE PROCEDURES

Maintenance of strains in liquid culture medium and long-term preservation in liquid nitrogen are carried out as described for the genus *Chromatium*.

DIFFERENTIATION OF THE GENUS *THERMOCHROMATIUM* FROM OTHER GENERA

According to the results of 16S rDNA sequence analyses, the genus *Thermochromatium* belongs to the branch of the family *Chro-matiaceae* containing motile freshwater bacteria, characterized by high temperature optima, above 40°C (*T. tepidum* at 48–50°C) and lack of capability for chemotrophic growth. The type strain of *T. tepidum* has an unusual and quite characteristic absorption spectrum. Intact cells and photosynthetic membrane preparations show a near-infrared absorption peak at 920 nm (Garcia et al., 1986). This absorption is due to a novel light-harvesting complex, and is the longest wavelength-absorbing band observed for a photosynthetic light-harvesting complex containing bacteriochlorophyll *a*. Whether this unusual complex is related to the thermophilic lifestyle of *T. tepidum* is unknown.

Morphological and physiological characteristics for the differentiation of *Thermochromatium* species from other phototrophic members of the family *Chromatiaceae* are shown in Table BXII.γ.4 of the chapter describing the family *Chromatiaceae*. The phylogenetic relationship of *Thermochromatium tepidum* to other members of the family *Chromatiaceae*, based on 16S rDNA sequence comparison, is shown in Fig. BXII.γ.3 of the chapter describing the family *Chromatiaceae*.

TAXONOMIC COMMENTS

The temperature range for growth of *Thermochromatium tepidum* restricts development of this bacterium to hot springs and other thermal environments with sulfide and light (Madigan, 1984; Castenholz and Pierson, 1995). The successful adaptation to growth at high temperatures was regarded as a taxonomically relevant property of *Thermochromatium tepidum*, which is also phylogenetically distinct from mesophilic freshwater *Chromatiaceae* species. Therefore, *Chromatium tepidum* was transferred to the new genus *Thermochromatium* (Imhoff et al., 1998b).

FURTHER READING

Imhoff, J.F., Süling, J. and Petri, R. 1998. Phylogenetic relationships among the *Chromatiaceae*, their taxonomic reclassification and description of the new genera *Allochromatium*, *Halochromatium*, *Isochromatium*, *Marichromatium*, *Thiococcus*, *Thiohalocapsa*, and *Thermochromatium*. Int. J. Syst. Bacteriol. 48: 1129–1143.

Madigan, M.T. 1986. *Chromatium tepidum* sp. nov., a thermophilic photosynthetic bacterium of the family *Chromatiaceae*. Int. J. Syst. Bacteriol. 36: 222–227.

List of species of the genus Thermochromatium

1. **Thermochromatium tepidum** (Madigan 1986) Imhoff, Süling and Petri 1998b, 1140[VP] (*Chromatium tepidum* Madigan 1986, 226.)

te' pi.dum. L. neut. adj. *tepidum* lukewarm.

Cells are rod shaped, 1–2 × 2.8–3.2 μm, occasionally motile by flagella. Photosynthetic pigments are bacteriochlorophyll *a*, esterified with phytol, and carotenoids of the normal spirilloxanthin group, with rhodovibrin and spirilloxanthin as predominant components.

Strictly anaerobic and obligately phototrophic. Growth occurs photolithoautotrophically in mineral media supplemented with sulfide as electron donor. Hydrogen and thiosulfate not utilized as electron donors. Acetate and pyruvate photoassimilated in the presence of sulfide. Citric acid cycle intermediates, other than acetate, fatty acids, and sugars not utilized. Ammonia, urea, and glutamine serve as nitrogen sources. No growth factors required. Poly-β-hydroxybutyrate is a storage material.

Thermophilic freshwater bacterium with optimum growth at 48–50°C, no growth below 34°C or above 57°C. Optimum pH is 7.0. NaCl is not required for growth and is inhibitory at concentrations above 1% (w/v).

Habitat: Sulfide-containing hot springs of neutral to alkaline pH at temperatures from 35 to 60°C.

The mol% G + C of the DNA is: 61.5 (T_m).

Type strain: MC, ATCC 43061, DSM 3771.

GenBank accession number (16S rRNA): M59150.

Genus XII. **Thioalkalicoccus** Bryantseva, Gorlenko, Kompantseva and Imhoff 2000b, 2161[VP]

JOHANNES F. IMHOFF

Thi.o.al' ka.li.coc' cus. Gr. n. *thios* sulfur; Arab n. *al kali* potash, soda; L. masc. n. *coccus* sphere; M.L.masc. n. *Thioalkalicoccus* sulfur sphere from soda.

Cells are spherical or oval, typically form diplococcus-shaped cells during cell division, multiply by binary fission, and are Gram negative. Internal membranes are of the tubular type. **Photosynthetic pigments are bacteriochlorophyll *b* and carotenoids with absorption properties similar to tetrahydrospirilloxanthin.** The metabolism is strictly anaerobic and obligately phototrophic. During photolithoautotrophic growth with sulfide as electron donor, globules of S⁰ are accumulated inside the cytoplasm. The final oxidation product is sulfate. In the presence of sulfide and bicarbonate, organic substrates are photoassimilated. **Mesophilic, obligate alkaliphilic bacterium** with optimum growth at 20–25°C. Optimal development is dependent on sodium salts in low concentrations and on alkaline conditions.

Habitat: Surface of sediments rich in organic matter, and microbial mats of soda lakes containing hydrogen sulfide and exposed to the light.

The mol% G + C of the DNA is: 63.6–64.8.

Type species: **Thioalkalicoccus limnaeus** Bryantseva, Gorlenko, Kompantseva and Imhoff 2000b, 2162.

DIFFERENTIATION OF THE GENUS *THIOALKALICOCCUS* FROM OTHER GENERA

According to the results of 16S rDNA sequence analyses (Fig. BXII.γ.3) of the chapter describing the family *Chromatiaceae*, the genus *Thioalkalicoccus* belongs to the branch of the family *Chromatiaceae* that characteristically contains marine and halophilic species. Characteristics for differentiation from other phototrophic *Chromatiaceae* species are shown in Table BXII.γ.5 of the chapter describing the family *Chromatiaceae*. *Thioalkalicoccus limnaeus* is one of the very few species (in addition to *Thiococcus pfennigii* and *Thioflavicoccus mobilis*) that contain tubular internal photosynthetic membranes. The phylogenetic relationships of *Thioalkalicoccus* species to other members of the family *Chromatiaceae*, based on 16S rDNA sequence comparison, are shown in Fig. BXII.γ.3 of the chapter describing the family *Chromatiaceae*.

List of species of the genus Thioalkalicoccus

1. **Thioalkalicoccus limnaeus** Bryantseva, Gorlenko, Kompantseva and Imhoff 2000b, 2162[VP]

lim.nae' us. Gr. fem. n. *limne* lake, pond, swamp; Gr. adj. *limnaios* pertaining to, living in lakes, swamps; M.L. masc. adj. *limnaeus* living in lakes and swamps.

Cells are spherical or oval in shape, multiply by binary fission, and are Gram negative. Internal membranes are of the tubular type. Cells are cocci of 1.3–1.8 μm diameter, usually nonmotile, and surrounded by a thin capsule. Occasionally cells with a single flagellum are observed. Internal photosynthetic membranes of tubular type are formed by invagination of the cell membrane and fill most of the cytoplasm. Color of cell suspensions is yellowish to orange-brown. The absorption spectrum of intact cells exhibits maxima at 410, 462, 492, 530, and 1030 nm, with shoulders at 602 and 835 nm. Photosynthetic pigments are bacteriochlorophyll *b* and carotenoids with absorption properties similar to tetrahydrospirilloxanthin.

The metabolism is strictly anaerobic. Photolithoautotrophic growth occurs in the light with hydrogen sulfide and S⁰ as electron donors. Thiosulfate was not used for phototrophic growth. During growth with sulfide as electron donor, globules of S⁰ are accumulated inside the cells. The final oxidation product is sulfate. In the presence of sulfide and sodium bicarbonate, acetate, yeast extract, malate, propionate, pyruvate, succinate, and fumarate were used as organic substrates for phototrophic growth. Growth factors were not required.

Mesophilic, obligate alkaliphilic bacterium with optimum growth at pH 8.8–9.5 (range from 8–10) and 20–25°C. Development is dependent on sodium salts in low concentrations, and good growth occurs over a broad range of salt concentrations without exhibiting a strong optimum, up to 6% NaCl (in the presence of 0.5% sodium carbonates) and up to 8.5% sodium carbonates (in the presence of 0.05% NaCl).

Habitat: Surface of sediments rich in organic matter, and microbial mats of soda lakes containing hydrogen sulfide and exposed to the light.

The mol% G + C of the DNA is: 64.0–64.5 (T_m).

Type strain: A26, ATCC BAA32.

GenBank accession number (16S rRNA): AJ277023.

Genus XIII. **Thiocapsa** Winogradsky 1888, 84[AL] emend Guyoneaud, Süling, Petri, Matheron, Caumette, Pfennig and Imhoff 1998b, 962

JOHANNES F. IMHOFF AND PIERRE CAUMETTE

Thi.o.cap′sa. Gr. n. *thios* sulfur; L. n. *capsa* box; M.L. fem. n. *Thiocapsa* sulfur box.

Cells spherical, 1.0–3.0 μm in diameter, diplococci before multiplication by binary fission, **nonmotile**. Tetrads may be formed after consecutive division in two perpendicular planes. Individual cells surrounded by a strong slime capsule and may contain gas vesicles. Gram negative, belong to the *Gammaproteobacteria*, and have internal **photosynthetic membranes of vesicular type** containing the photosynthetic pigments bacteriochlorophyll *a* and carotenoids.

Photolithotrophic growth under anoxic conditions in the light possible with sulfide, thiosulfate, and sulfur as electron donors. Hydrogen may also be used. During oxidation of sulfide, S^0 is transiently stored inside the cells in the form of highly refractile globules. Sulfate is the final oxidation product. In the presence of sulfide and bicarbonate, simple organic substrates can be photoassimilated. **Some species may grow chemoautotrophically or chemoheterotrophically under microoxic to oxic conditions in the dark**. Storage materials: polysaccharides, poly-β-hydroxybutyrate, S^0, polyphosphates. Vitamin B_{12} may be required.

The mol% G + C of the DNA is: 63.3–66.3.

Type species: **Thiocapsa roseopersicina** Winogradsky 1888, 84.

FURTHER DESCRIPTIVE INFORMATION

Under unfavorable growth conditions, *Thiocapsa* species may form aggregates of cells embedded in slime, similar to species of *Thiocystis, Allochromatium,* and *Marichromatium*. Depending on culture conditions, the cell diameter of *Thiocapsa* species may vary considerably. Two of the presently known species, *Thiocapsa rosea* and *Thiocapsa pendens*, contain gas vesicles; *Thiocapsa roseopersicina* does not.

Thiocapsa roseopersicina has been found in freshwater, brackish water, and marine habitats, and a characteristic of this species may be the rather indifferent response to low salt concentrations, allowing development and successful competition in freshwater, brackish water, and marine environments. However, comparative sequence analyses indicate that a number of strains isolated from the marine environment and tentatively assigned to this species, could possibly be distinguished at the species level (Guyoneaud et al., 1998b). These bacteria behave indifferently to low salt concentrations and may be regarded as halotolerant, but not halophilic.

ENRICHMENT AND ISOLATION PROCEDURES

The methods are the same as those described for the genus *Chromatium*. *Thiocapsa roseopersicina* is one of the most common purple sulfur bacteria in nature. It is particularly abundant in ponds, pools, or lagoons receiving sewage or wastewater rich in organic matter. Since this species is the least fastidious purple sulfur bacterium known, its isolation presents no problems. The other species of *Thiocapsa, Thiocapsa pendens* and *Thiocapsa rosea*, can be isolated directly from samples collected in nature by agar shake dilution cultures, as described for the *Chromatium* species.

The color of colonies in agar is characteristically pinkish. Because of their gas vesicle content, colonies of *Thiocapsa pendens* and *Thiocapsa rosea* maintain a chalky appearance, even if the cells are free of S^0.

MAINTENANCE PROCEDURES

Maintenance of strains in liquid culture medium and long-term preservation in liquid nitrogen are carried out as described for the genus *Chromatium*.

DIFFERENTIATION OF THE GENUS *THIOCAPSA* FROM OTHER GENERA

Characteristics for differentiation of the genus *Thiocapsa* from other phototrophic members of the family *Chromatiaceae* are shown in Table BXII.γ.6 of the chapter describing the family *Chromatiaceae*. *Thiocapsa* cells may form tetrads, and therefore can be mistaken for *Thiopedia rosea* or *Thiolamprovum pedioforme*. The cells of these latter species are, however, readily distinguishable from *Thiocapsa* by the formation of platelets, which usually consist of 8, 16, or more regularly arranged cells. Other differential phenotypic traits of these genera are listed in Table BXII.γ.6 of the chapter describing the family *Chromatiaceae*. The phylogenetic relationship of *Thiocapsa* species to other members of the family *Chromatiaceae*, based on 16S rDNA sequence comparison, is shown in Fig. BXII.γ.4 of the chapter describing the family *Chromatiaceae*.

TAXONOMIC COMMENTS

Previously, nonmotile, spherical, purple sulfur bacteria without gas vesicles have been united within the genus *Thiocapsa*. However, the results of genetic analyses indicate that two of the three species placed in this genus (the former *Thiocapsa halophila* and *Thiocapsa pfennigii*) were phylogenetically quite distant from the type species, *Thiocapsa roseopersicina*. Both species belong to the cluster containing truly marine *Chromatiaceae* species Figs. BXII.γ.3 and BXII.γ.4 of the chapter describing the family *Chromatiaceae*). Consequently, they were removed from the genus *Thiocapsa* and transferred to *Thiococcus* and *Thiohalocapsa* as *Thiococcus pfennigii* and *Thiohalocapsa halophila*, respectively (Imhoff et al., 1998b).

Furthermore, *Thiocapsa rosea* and *Thiocapsa pendens*, previously known as *Amoebobacter roseus* and *Amoebobacter pendens* (Pfennig and Trüper, 1971b; Pfennig, 1989a), are closely related genetically to *Thiocapsa roseopersicina* and were included in the genus *Thiocapsa* (Guyoneaud et al., 1998b).

The type species of the genera *Amoebobacter* and *Thiocapsa* were united and therefore a decision concerning the genus name had to be made. Because both genera and their type strains were included in the Approved List and had been legitimately published at the same date (Winogradsky, 1888), none of the genera and species had priority according to the rules of Nomenclature (rules 23a and 24b). *Thiocapsa* is the better-known genus and the more appropriate name, and *Thiocapsa roseopersicina* one of the best-studied species of purple sulfur bacteria. Therefore, according to rule 42 and recommendation 42 of the International Code of Nomenclature of Bacteria (Lapage et al., 1992), the two bacteria were united within the genus *Thiocapsa*, and *Thiocapsa roseopersicina* retained as the type species (Guyoneaud et al., 1998b). Because the nomenclatural type had been removed from the genus *Amoebobacter*, the names of other *Amoebobacter* species were changed, according to rule 37a (see also genera *Thiolamprovum* and *Lamprocystis*).

FURTHER READING

Guyoneaud, R., Süling, J., Petri, R., Matheron, R., Caumette, P., Pfennig, N. and Imhoff, J.F. 1998. Taxonomic rearrangements of the genera *Thiocapsa* and *Amoebobacter* on the basis of 16S rDNA sequence analyses and description of *Thiolamprovum* gen. nov. Int. J. Syst. Bacteriol. *48*: 957–964.

Pfennig, N. and Trüper, H.G. 1971. New nomenclatural combinations in the phototrophic sulfur bacteria. Int. J. Syst. Bacteriol. *21*: 11–14.

Takács, B.J. and Holt, S.C. 1971. *Thiocapsa floridana*; a cytological physiological and chemical characterization. I. Cytology of whole cells and isolated chromatophore membranes. Biochim. Biophys. Acta *233*: 258–277.

DIFFERENTIATION OF THE SPECIES OF THE GENUS *THIOCAPSA*

I. Cells without gas vesicles, capable of assimilatory sulfate reduction, able to use hydrogen as electron donor.
Thiocapsa roseopersicina

II. Cells with gas vesicles, not capable of assimilatory sulfate reduction, not able to use hydrogen as electron donor.

A. Able to grow under microoxic conditions in the dark, not able to use glucose.
Thiocapsa rosea

B. Not able to grow under microoxic conditions, strict anaerobe, able to use glucose.
Thiocapsa pendens

List of species of the genus Thiocapsa

1. **Thiocapsa roseopersicina** Winogradsky 1888, 84[AL]

ro.se.o.per.si.ci′ na. L. adj. *roseus* rosy; Gr. n. *persicus* the peach; M.L. fem. adj. *roseopersicina* rosy peach-colored.

Cells spherical, 1.2–3.0 μm, usually 1.5 μm in diameter (Fig. BXII.γ.11), and surrounded by a strong slime capsule. Aggregates of two, four, or more cells common; irregular clumps of cells usually surrounded by slime. Gas vesicles absent. Individual cells colorless, color of cell suspensions pink to rose-red. Photosynthetic pigments are bacteriochlorophyll *a* and carotenoids of the normal spirilloxanthin group (Schmidt et al., 1965).

Photolithoautotrophic growth under anoxic conditions in the light with hydrogen, sulfide, thiosulfate, and sulfur as electron donors. Acetate, pyruvate, fumarate, malate, succinate, glycerol, and fructose are photoassimilated. Not utilized: lactate, propionate, butyrate, tartrate, α-oxoglutarate, citrate, benzoate, and alcohols. Chemoautotrophic or chemoheterotrophic growth under microoxic to oxic conditions in the dark is possible. Most strains capable of assimilatory sulfate reduction. Nitrogen sources: ammonium salts. Nitrogenase activity present. Growth factors not required.

Mesophilic bacterium with optimum growth at 20–35°C and pH 7.3 (pH range 6.5–7.5).

Habitat: stagnant water and mud of ponds, pools, or wastewater lagoons containing degradable organic substances and hydrogen sulfide. Also common in estuaries and salt marshes.

The mol% G + C of the DNA is: 63.3–66.3 (Bd); type strain: 65.3 (Bd).

Type strain: 1711, BN 4210, DSM 217.

GenBank accession number (16S rRNA): Y12364.

2. **Thiocapsa litoralis** Puchkova, Imhoff and Gorlenko 2000, 1446[VP]

li.to.ra′ lis. L. adj. *litoralis* from the shore.

Cells are spherical and nonmotile, 1.5–2.5 μm in diameter and organized in regular platelets of 4, 8, 16, or more cells. Photosynthetic membranes are of vesicular type. Color of cell suspensions is pink to rose-red. Bacteriochlorophyll *a* and, most probably, pigments of the spirilloxanthin series are present as photosynthetic pigments.

Phototrophic growth occurs under anoxic conditions in the light and chemoautotrophic growth is possible under microoxic conditions in the dark. Electron donors used for photoautotrophic growth are hydrogen sulfide, elemental sulfur, thiosulfate, and sulfite. Globules of elemental sulfur are stored inside the cells as an intermediary product. In the presence of sulfide and bicarbonate, acetate, propionate, butyrate, valerate, pyruvate, lactate, malate, succinate, fumarate, fructose, glucose, Casamino acids, and yeast extract are photoassimilated. Caprylate, tartrate, formate, ascorbate, asparaginate, benzoate, malonate, citrate, maltose, galactose, mannitol, sorbitol, glycerol, methanol, and ethanol not utilized. Assimilatory sulfate reduction occurs. Chemolithoautotrophic growth occurs with hydrogen sulfide and thiosulfate as electron donors. Vitamin B_{12} is required as growth factor. Storage materials are poly-β-hydroxybutyrate and polyphosphates. Mesophilic bacterium with optimum growth at pH 6.5, 30°C, and 1% NaCl.

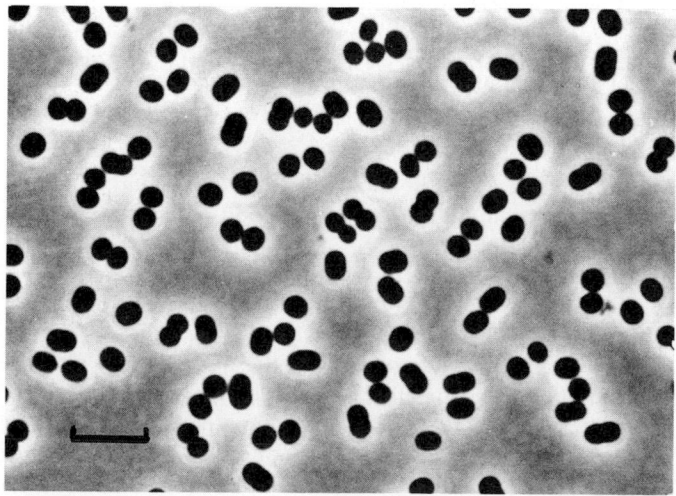

FIGURE BXII.γ.11. *Thiocapsa roseopersicina* DSM 219 cultured photoautotrophically with sulfide. Phase-contrast micrograph. Bar = 5 μm.

Habitat: Microbial mat communities of the White Sea littoral shore.

The mol% G + C of the DNA is: 64.0 (T_m).

Type strain: BM5, ATCC 700894.

GenBank accession number (16S rRNA): AJ24772.

3. **Thiocapsa pendens** (Molisch 1906) Guyoneaud, Süling, Petri, Matheron, Caumette, Pfennig and Imhoff 1998b, 962[VP] (*Amoebobacter pendens* (Molisch 1906) Pfennig and Trüper 1971b, 13; *Rhodothece pendens* Molisch 1906, 230 .)

pen' dens. L. part. adj. *pendens* hanging.

Cells spherical, 1.5–2.0 μm in diameter, sometimes up to 2.5 μm in diameter. Individual cells surrounded by slime capsules, which give rise to irregular cell aggregates of different size. Because of copious slime formation, the culture medium becomes viscous. Gas vesicles present. Individual cells colorless, color of cell suspensions pink to rose-red. Photosynthetic pigments are bacteriochlorophyll *a* and carotenoids of the spirilloxanthin group, with spirilloxanthin as the dominant component.

Strictly anaerobic and obligately phototrophic. No growth under microoxic to oxic conditions in the dark. Photolithoautotrophic growth under anoxic conditions in the light occurs with sulfide, thiosulfate, sulfite, and sulfur as electron donors. Acetate, pyruvate, propionate, lactate, malate, and glucose are the only organic substrates photoassimilated. Assimilatory sulfate reduction is absent. Nitrogen sources: ammonium salts. Vitamin B_{12} required for growth.

Mesophilic freshwater bacterium with optimum growth at 20–35°C, pH range 6.7–7.5.

Habitat: stagnant freshwater and mud containing sulfide and exposed to the light.

The mol% G + C of the DNA is: 65.3 (Bd).

Type strain: DSM 236 (neotype strain).

GenBank accession number (16S rRNA): AJ002797.

4. **Thiocapsa rosea** (Winogradsky 1888) Guyoneaud, Süling, Petri, Matheron, Caumette, Pfennig and Imhoff 1998b, 962[VP] (*Amoebobacter roseus* Winogradsky 1888, 77.)

ro' se.a. L. fem. adj. *rosea* rosy, rose-colored, pink.

Cells spherical, 2.0–3.0 μm in diameter. Individual cells and irregular cell aggregates surrounded by slime. Gas vesicles present. Individual cells colorless, color of cell suspensions pink to rose-red. Photosynthetic pigments are bacteriochlorophyll *a* and carotenoids of the spirilloxanthin group, with spirilloxanthin as the major component.

Photolithoautotrophic growth under anoxic conditions in the light occurs with sulfide, thiosulfate, and sulfur as electron donors. Acetate, pyruvate, propionate, lactate, malate, fructose, and Casamino acids are photoassimilated. Not utilized: other sugars, alcohols, other fatty acids, succinate, tartrate, benzoate, or citrate. Chemoautotrophic or mixotrophic growth under microoxic to oxic conditions in the dark is possible. Assimilatory sulfate reduction does not occur. Nitrogen source, ammonia. Vitamin B_{12} required.

Mesophilic freshwater bacterium with optimum growth at 20–35°C, pH range 6.7–7.5.

The mol% G + C of the DNA is: 64.3 (Bd).

Type strain: DSM 235.

GenBank accession number (16S rRNA): AJ002798, AJ006062.

Genus XIV. **Thiococcus** *Imhoff, Süling and Petri 1998b, 1139*[VP]

JOHANNES F. IMHOFF

Thi.o.coc' cus. Gr. n. *thios* sulfur; L. masc. n. *coccus* sphere; M.L. masc. n. *Thiococcus* sphere with sulfur.

Cells spherical, with diplococcus-shaped division stages, **nonmotile**, multiply by binary fission. Gram negative, belong to the *Gammaproteobacteria*, and contain internal **photosynthetic membranes consisting of bundles of ribbon-like branched tubes** continuous with the cytoplasmic membrane. Individual cells colorless, color of cell suspensions is yellowish to orange-brown. Photosynthetic pigments are bacteriochlorophyll *b* and carotenoids.

Obligately phototrophic and strictly anaerobic. Photolithoautotrophic growth under anoxic conditions in the light with sulfide and S^0 as electron donors. During oxidation of sulfide, S^0 transiently stored inside the cells in the form of highly refractile globules. Final oxidation product is sulfate. In the presence of sulfide and bicarbonate, simple organic substrates photoassimilated.

Mesophilic bacteria with optimum growth at 20–35°C and neutral pH (pH range 6.5–7.5). May require salt for optimum growth.

The mol% G + C of the DNA is: 69.4–69.9.

Type species: **Thiococcus pfennigii** (Eimhjellen 1970) Imhoff, Süling and Petri 1998b, 1139 (*Thiocapsa pfennigii* Eimhjellen 1970, 193.)

ENRICHMENT AND ISOLATION PROCEDURES

Thiococcus pfennigii is the only species of the family *Chromatiaceae* that contains bacteriochlorophyll *b* described so far. Owing to the presence of this bacteriochlorophyll, the bacterium has an

in vivo absorption maximum in the infrared region of the spectrum, between 1020 nm and 1030 nm. By use of selective filters, this specific absorption has been successfully applied for the selective enrichment culture of *Thiococcus* species (Eimhjellen et al., 1967; Eimhjellen, 1970). Otherwise, the culture conditions and isolation procedures are the same as described for the genus *Chromatium*. The media described for *Chromatium* species can also be used for *Thiococcus* species, if supplemented with 1–3% NaCl as required.

MAINTENANCE PROCEDURES

Maintenance of strains in liquid culture medium and long-term preservation in liquid nitrogen are carried out as described for the genus *Chromatium*.

DIFFERENTIATION OF THE GENUS *THIOCOCCUS* FROM OTHER GENERA

Characteristics for differentiation of the genus *Thiococcus* from other phototrophic members of the family *Chromatiaceae* are shown in Table BXII.γ.5 of the chapter describing the family *Chromatiaceae*. The phylogenetic relatedness of *Thiococcus pfennigii* to other members of the family *Chromatiaceae*, based on 16S rDNA analysis, is shown in Fig. BXII.γ.3 of the chapter describing the family *Chromatiaceae*.

TAXONOMIC COMMENTS

The genus name *Thiococcus* was proposed for this bacterium by Eimhjellen et al. (1967). However, this name was never validly published and later the bacterium was validly named *Thiocapsa pfennigii* (Eimhjellen, 1970; Approved List of Bacterial Names, Skerman et al., 1980). Because of the striking genetic and phenotypic difference between *Thiococcus pfennigii* and members of the genus *Thiocapsa*, it was transferred to a new genus and the name *Thiococcus* was revived (Imhoff et al., 1998b). A possible candidate for the neotype strain of *Thiococcus pfennigii* (strain DSM 226, Trüper 8013, Imhoff 4250) belongs to the cluster of truly marine species of the family *Chromatiaceae* (Fig. BXII.γ.3 of the chapter describing the family *Chromatiaceae*) and requires at least 1% NaCl for growth. The salt dependence and natural distribution of this species are poorly understood. The original type strain was isolated from a freshwater source (river water), but unfortunately was lost. Additional strains were isolated later from freshwater and marine sources. However, chemotaxonomic and genetic analyses have not been performed on these strains. In addition, salt dependence has not yet been determined for all of the isolates.

FURTHER READING

Eimhjellen, K.E. 1970. *Thiocapsa pfennigii* sp. nov., a new species of phototrophic sulfur bacteria. Arch. Microbiol. *73*: 193–194.

Eimhjellen, K.E., Steensland, H. and Traetteberg, H. 1967. A *Thiococcus* sp. nov. gen., its pigments and internal membrane system. Arch. Microbiol. *59*: 82–92.

Imhoff, J.F., Süling, J. and Petri, R. 1998. Phylogenetic relationships among the *Chromatiaceae*, their taxonomic reclassification and description of the new genera *Allochromatium*, *Halochromatium*, *Isochromatium*, *Marichromatium*, *Thiococcus*, *Thiohalocapsa*, and *Thermochromatium*. Int. J. Syst. Bacteriol. *48*: 1129–1143.

List of species of the genus Thiococcus

1. **Thiococcus pfennigii** (Eimhjellen 1970) Imhoff, Süling and Petri 1998b, 1139[VP] (*Thiocapsa pfennigii* Eimhjellen 1970, 193.)

pfen.nig' i.i. M.L. gen. n. *pfennigii* of Pfennig, named after N. Pfennig, a German microbiologist.

Cells spherical, 1.2–1.5 µm in diameter; diplococcus-shaped division stages about 2.5 µm long; stationary phase cells 0.8–1.0 µm in diameter. Internal photosynthetic membranes are bundles of ribbon-like branched tubes continuous with the cytoplasmic membrane. Individual cells colorless, color of cell suspensions yellowish to orange-brown. Photosynthetic pigments are bacteriochlorophyll *b* and the carotenoid 3,4,3′,4′-tetrahydrospirilloxanthin (Eimhjellen et al., 1967).

Obligately phototrophic and strictly anaerobic. Photolithoautotrophic growth with sulfide and S⁰ as electron donors. In the presence of sulfide and bicarbonate, acetate, pyruvate, propionate, lactate, fumarate, succinate, malate, and fructose are photoassimilated. Not utilized: thiosulfate and butyrate. Nitrogen sources: ammonium salts and N_2.

Mesophilic bacterium with optimum growth at 0.5–2% NaCl, 20-35°C, and pH 7.0 (range 6.5–7.5).

Habitat: freshwater, brackish, and marine waters and sediments with hydrogen sulfide and exposed to the light.

The mol% G + C of the DNA is: 69.4–69.9 (Bd).

Type strain: DSM 1375 (RG3, Nidelven) is lost; no neotype strain is designated.

GenBank accession number (16S rRNA): Y12373 (strain DSM 226).

Additional Remarks: The GenBank accession number for strain BN 4254 (Pfennig 8320) is AJ010125.

Genus XV. **Thiocystis** *Winogradsky 1888, 60*[AL] *emend. Imhoff, Süling and Petri 1998b, 1138*

JOHANNES F. IMHOFF

Thi.o.cys' tis. Gr. n. *thios* sulfur; Gr. n. *cystis* the bladder, a bag; M.L. fem. n. *Thiocystis* sulfur bag.

Cells are rod shaped or ovoid to spherical, before cell division often diplococcus-shaped, **motile by flagella**, occurring singly or in pairs; may grow in irregular aggregates surrounded by slime, multiply by binary fission. Gram negative, belong to the *Gammaproteobacteria*, and contain internal **photosynthetic membranes of vesicular type** in which the photosynthetic pigments bacteriochlorophyll *a* and carotenoids are located.

Photolithotrophic growth occurs under anoxic conditions in the light with sulfide and S⁰ as electron donors. Thiosulfate and hydrogen may also be used. During oxidation of sulfide, S⁰ is transiently stored inside the cells in the form of highly refractile globules. Sulfate is the final oxidation product. In the presence of sulfide and bicarbonate, simple organic substrates can be photoassimilated. **Some species grow chemoautotrophically or chemoorganotrophically under microoxic to oxic conditions in the dark**. Storage materials: polysaccharides, poly-β-hydroxybutyrate, S⁰, and polyphosphates.

Ubiquitous, mesophilic bacteria with optimum growth at 25–35°C and pH 6.5–7.6, that can be isolated from freshwater, brackish water, and marine environments.

Habitat: anoxic sulfide-containing water and mud of freshwater ponds and lakes, as well as brackish water and marine environments exposed to light.

The mol% G + C of the DNA is: 61.3–67.9.

Type species: **Thiocystis violacea** Winogradsky 1888, 65.

FURTHER DESCRIPTIVE INFORMATION

Species of the genus *Thiocystis* are frequently found in the same natural habitats as *Allochromatium* species. Bacteria of both genera share many physiological properties. This is true not only for the phototrophic metabolism of reduced sulfur compounds and organic substrates, but also for facultatively chemoautotrophic or chemoorganotrophic metabolism under microoxic to oxic conditions in the dark (Kämpf and Pfennig, 1980), which occurs in some species of this genus.

ENRICHMENT AND ISOLATION PROCEDURES

Media and culture conditions are the same as described for *Chromatium* species.

MAINTENANCE PROCEDURES

Maintenance of strains in liquid culture medium and long-term preservation in liquid nitrogen are carried out as described for the genus *Chromatium*.

DIFFERENTIATION OF THE GENUS *THIOCYSTIS* FROM OTHER GENERA

According to the results of 16S rDNA sequence analyses, the genus *Thiocystis* belongs to the branch of the family *Chromatiaceae* characterized by motile freshwater species. Characteristics for differentiation of *Thiocystis* species from other phototrophic members of the family *Chromatiaceae* are shown in Table BXII.γ.4 of the chapter describing the family *Chromatiaceae*. The phylogenetic relationship of *Thiocystis* species to other *Chromatiaceae* species, based on 16S rDNA sequence comparison, is shown in Fig. BXII.γ.3 of the chapter describing the family *Chromatiaceae*.

TAXONOMIC COMMENTS

Previously, the genus *Thiocystis* was distinguished from other genera of the family *Chromatiaceae* based on its spherical morphology, flagellar motility, and lack of gas vesicles (Pfennig and Trüper, 1989). Winogradsky (1888) distinguished *Thiothece* from *Thiocystis*

as a second genus of spherical, motile, purple sulfur bacteria, based on the location of the intracellular sulfur globules. In the monotypic genus *Thiothece*, the sulfur globules were located at the inner periphery of the cells, whereas in *Thiocystis* the globules were distributed at random in the cell (Winogradsky, 1888). Pfennig and Trüper (1971b) included *Thiothece gelatinosa* Winogradsky 1888 in the genus *Thiocystis* because they considered the maintenance of *Thiothece*, in addition to *Thiocystis*, unjustified.

Similarity studies on the 16S rRNA oligonucleotide catalogs of *Chromatiaceae* species (Fowler et al., 1984) and complete 16S rDNA sequence analyses (Imhoff et al., 1998b) not only substantiated the close relationship of *Thiocystis gelatinosa* and *Thiocystis violacea* (see Fig. BXII.γ.3 of the chapter describing the family *Chromatiaceae*), but also revealed a high degree of genetic relatedness between these species and *Thiocystis minor* (formerly *Chromatium minus*) and *Thiocystis violascens* (formerly *Chromatium violascens*) (Imhoff et al., 1998b).

DIFFERENTIATION OF THE SPECIES OF THE GENUS *THIOCYSTIS*

Characteristics for diagnosis of *Thiocystis* species are shown in Table BXII.γ.4 of the chapter describing the family *Chromatiaceae*.

I. Cells spherical.
  A. Sulfur globules appear randomly distributed within the cell, cells contain carotenoids of the rhodopinal group, and cultures are purple-violet.
           Thiocystis violacea
  B. Sulfur globules occur only at the inner periphery of the

cell, cells contain carotenoids of the okenone group, and cultures are purple-red.
           Thiocystis gelatinosa
II. Cells rod shaped.
  A. Carotenoids of the okenone group are present; only acetate, pyruvate, and glucose are utilized.
           Thiocystis minor
  B. Carotenoids of the rhodopinal group are present; hydrogen, but no sugars are utilized.
           Thiocystis violascens

List of species of the genus Thiocystis

1. **Thiocystis violacea** Winogradsky 1888, 65[AL]

vi.o.la′ce.a. L. fem. adj. *violacea* violet-colored.

Cells spherical and about 2.5–3 μm in diameter. Depending on the culture conditions, larger individual cells may occur. Irregular cell aggregates surrounded by slime are formed under unfavorable conditions. Sulfur globules appear randomly distributed within the cell (Fig. BXII.γ.12). Color of individual cells grayish, color of cell suspensions purple-violet. Photosynthetic pigments are bacteriochlorophyll *a* and carotenoids of the rhodopinal group (Schmidt et al., 1965).

Photoautotrophic growth occurs under anoxic conditions in the light with sulfide, sulfur, thiosulfate, sulfite (used by some strains), and molecular hydrogen as electron donors. In the presence of sulfide and bicarbonate, acetate, pyruvate, and fumarate are photoassimilated. In addition, some strains use propionate, succinate, oxoglutarate, glucose, fructose, propanol, or glycerol. Chemoautotrophic or mixotrophic growth possible under microoxic to oxic conditions in the dark. Some strains capable of assimilatory sulfate reduction. Nitrogen sources: ammonium salts, urea, and N_2.

Mesophilic bacterium with optimum growth at 25–35°C and pH 7.0–7.3 (range 6.5–7.6). Marine isolates may tolerate or require low concentrations of NaCl.

Habitat: ponds and lakes with stagnant freshwater, brackish water, or seawater containing hydrogen sulfide; sewage lagoons, estuaries, salt marshes, and sulfur springs.

The mol% G + C of the DNA is: 62.8–67.9 (Bd); type strain 63.1 (Bd).

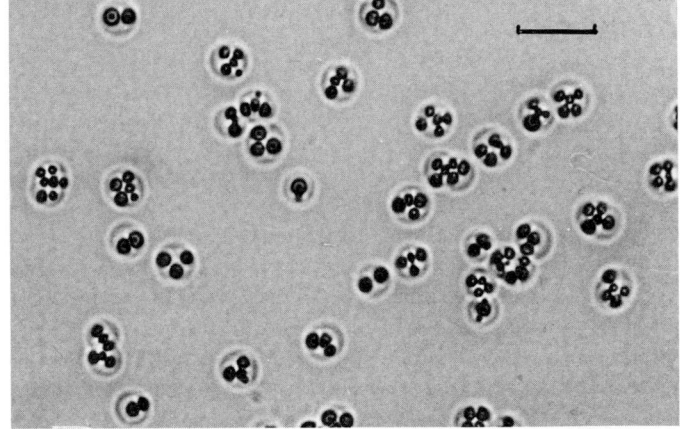

FIGURE BXII.γ.12. *Thiocystis violacea* DSM 207 cultured photoautotrophically with sulfide. The cells contain sulfur globules. Brightfield micrograph. Bar = 5 μm.

Type strain: 2711, BN 4340, DSM 207.
GenBank accession number (16S rRNA): Y11315.

2. **Thiocystis gelatinosa** (Winogradsky 1888) Pfennig and Trüper 1971b, 11[AL] (*Thiothece gelatinosa* Winogradsky 1888, 82.)

ge.la.ti.no′sa. L. part. adj. *gelatus* frozen, stiffened; M.L. n. *gelatinum* that which stiffens; M.L. fem. adj. *gelatinosa* gelatinous.

Cells spherical and about 3 μm in diameter. Under unfavorable conditions, elongated ovoid cells may occur. At high sulfide concentration (4–6 mM) and light intensity (>1000 lux), cells become immobile, growing in irregular aggregates surrounded by slime. Under optimum growth conditions, globules of S⁰ occur only in the peripheral part of the cytoplasm (Fig. BXII.γ.13). Color of individual cells slightly pink, color of cell suspensions purple-red. Photosynthetic pigments are bacteriochlorophyll *a* and carotenoids of the okenone group (Pfennig et al., 1968).

Photolithoautotrophic growth under anoxic conditions in the light with sulfide and sulfur as electron donors. In the presence of sulfide and bicarbonate, acetate and pyruvate are photoassimilated. Chemoautotrophic or mixotrophic growth is possible under microoxic to semioxic conditions. Nitrogen sources: ammonium salts.

Mesophilic bacterium with optimum growth at 30°C and pH 7.0–7.3 (range 6.5–7.6). The type strain requires 1% NaCl.

Habitat: stagnant water containing hydrogen sulfide, hypolimnion of meromictic lakes.

The mol% G + C of the DNA is: 61.3 (Bd).
Type strain: 2611, BN 4310, DSM 215.
GenBank accession number (16S rRNA): Y11317.

3. **Thiocystis minor** (Winogradsky 1888) Imhoff, Süling and Petri 1998b, 1139^VP (*Chromatium minus* Winogradsky 1888, 99.)

mi' nor. L. fem. comp. adj. *minor* less, smaller.

Cells are rod shaped, ~2 × 2.5–6 μm, with polar flagella.

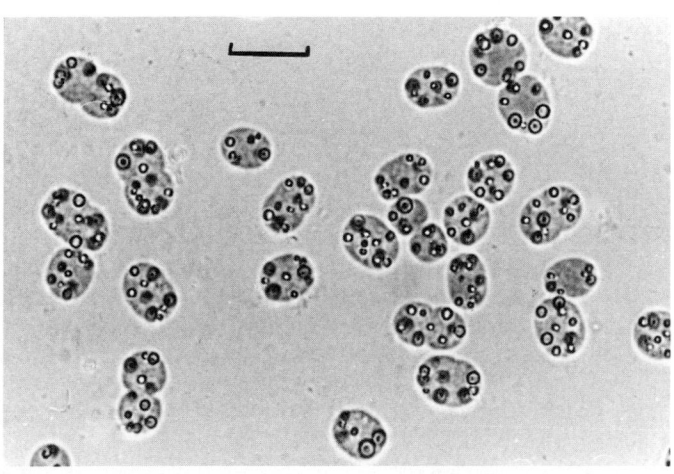

FIGURE BXII.γ.13. *Thiocystis gelatinosa* DSM 215 cultured photoautotrophically with sulfide. The peripheral deposition of the sulfur globules is clearly visible in some of the cells. Brightfield micrograph. Bar = 5 μm.

Globules of S⁰ appear evenly distributed within the cell. Single cells colorless to pinkish, color of cell suspensions purple-red. Photosynthetic pigments are bacteriochlorophyll *a* and the carotenoid okenone.

Photolithoautotrophic growth under anoxic conditions in the light with sulfide, thiosulfate, and sulfur as electron donors. In the presence of sulfide and bicarbonate, acetate, pyruvate, and glucose are photoassimilated. Not utilized: sugar-alcohols, alcohols, higher fatty acids, amino acids, benzoate, formate, and most intermediates of the tricarboxylic acid cycle (Thiele, 1968). Chemoautotrophic or mixotrophic growth possible under microoxic to semioxic conditions (Kämpf and Pfennig, 1980). Assimilatory sulfate reduction is absent. Nitrogen sources: ammonium salts and N₂.

Mesophilic freshwater bacterium with optimum growth at 30°C and pH 7.0–7.3 (pH range 6.5–7.6).

Habitat: freshwater ponds and lakes containing hydrogen sulfide and exposed to light.

The mol% G + C of the DNA is: 62.2 (Bd).
Type strain: 1711, BN 5710, DSM 178.
GenBank accession number (16S rRNA): Y12372.

4. **Thiocystis violascens** (Perty 1852) Imhoff, Süling and Petri 1998b, 1139^VP (*Chromatium violascens* Perty 1852, 174.)
vi.o.la' scens. L. part. *violascens* becoming violet.

Cells are rod shaped, ~2 × 2.5–6.0 μm, occasionally longer. Color of cell suspensions purple-violet. Photosynthetic pigments are bacteriochlorophyll *a* and carotenoids of the rhodopinal group.

Phototrophic growth under anoxic conditions in the light with sulfide, sulfur, thiosulfate, sulfite, H₂, formate, acetate, propionate, and pyruvate as electron donors. Some strains utilize butyrate. Not utilized: sugars, sugar alcohols, alcohols, benzoate, citrate, and amino acids. Chemoautotrophic or mixotrophic growth is possible under microoxic to semioxic conditions (Kämpf and Pfennig, 1980). Assimilatory sulfate reduction present (Thiele, 1968). Nitrogen sources: ammonium salts and N₂.

Mesophilic bacterium with optimum growth at 25–35°C and pH 7.0–7.3 (range 6.5–7.6). Marine isolates may require up to 2% NaCl.

Habitat: ponds and lakes with stagnant freshwater, sewage lagoons, brackish waters, estuaries, salt marshes, and marine habitats containing hydrogen sulfide and exposed to the light.

The mol% G + C of the DNA is: 61.8–64.3 (Bd); type strain: 62.2 (Bd).
Type strain: 6111, BN 5410, ATCC 17096, DSM 198.
GenBank accession number (16S rRNA): AJ224438.

Genus XVI. **Thiodictyon** *Winogradsky 1888, 80^AL*

JOHANNES F. IMHOFF

Thi' o.dic' ty.on. Gr. n. *thios* sulfur; Gr. n. *dictyon* a net; M.L. neut. n. *Thiodictyon* sulfur net.

Cells are rod shaped with rounded ends, sometimes appearing spindle-shaped, multiply by binary fission, **nonmotile under all conditions**, may form aggregates in which the cells are arranged end-to-end in an irregular netlike structure, the shape of which is variable. May also form clumps that are more compact or break up into individual cells. Contain large, **irregularly shaped gas** vesicles in the central part of the cell. Gram negative, belong to the *Gammaproteobacteria*, and contain internal **photosynthetic membranes of vesicular type** with bacteriochlorophyll *a* and carotenoids as photosynthetic pigments.

Obligately phototrophic and strictly anaerobic. Photolithoautotrophic growth under anoxic conditions in the light with sulfide and S⁰ as electron donors. During oxidation of sulfide, S^0 is transiently stored in the peripheral part of the cell that is free of gas vesicles. Final oxidation product is sulfate. In the presence of sulfide and bicarbonate, simple organic substrates are photoassimilated.

Habitat: mud and stagnant water of ponds and lakes containing hydrogen sulfide; common as planktonic bacterium in the sulfide-containing hypolimnion of freshwater lakes.

The mol% G + C of the DNA is: 65.3–66.3.

Type species: **Thiodictyon elegans** Winogradsky 1888, 82.

ENRICHMENT AND ISOLATION PROCEDURES

Media and culture conditions are the same as described for *Chromatium* species. Prerequisite for the successful enrichment and isolation of *Thiodictyon* from anoxic water or mud samples is the microscopic detection of the bacterium in the particular sample. Since in its natural habitats *Thiodictyon elegans* usually grows in the form of readily recognizable netlike cell aggregates, recognition of this bacterium is relatively easy.

For the isolation of pure cultures, it is advisable to prepare both liquid enrichment cultures and agar shake dilution cultures that are directly inoculated from the sample of the natural habitat. In liquid enrichment cultures, *Thiodictyon* may be outgrown by other purple sulfur bacteria containing gas vesicles. The culture medium and incubation conditions are the same as those described for *Lamprocystis* species.

MAINTENANCE PROCEDURES

Maintenance of strains in liquid culture medium and long-term preservation in liquid nitrogen are carried out as described for the genus *Chromatium*.

DIFFERENTIATION OF THE GENUS *THIODICTYON* FROM OTHER GENERA

Characteristics used for differentiation of the genus *Thiodictyon* from other phototrophic members of the family *Chromatiaceae* are shown in Table BXII.γ.6 of the chapter describing the family *Chromatiaceae*.

TAXONOMIC COMMENTS

The genus *Thiodictyon* presently comprises two species: *T. elegans*, the type species of the genus, and *T. bacillosum*. The latter species was originally described by Winogradsky (1888) as *Amoebobacter bacillosus*. Pfennig and Trüper (1971b) transferred *Amoebobacter bacillosus* to the genus *Thiodictyon*, because the rod-shaped cells are very similar to those of *Thiodictyon elegans*. In addition, according to the original description of the type species of the genus *Amoebobacter* (*Amoebobacter roseus* Winogradsky 1888), this genus should comprise only spherical, nonmotile species with gas vesicles. According to 16S rRNA oligonucleotide pattern analysis, *Thiodictyon elegans* is related to *Thiocapsa* species, including *Thiocapsa pendens* and *Thiocapsa rosea* (Fowler et al., 1984). Studies on 16S rDNA sequence similarities of *Thiodictyon* species are expected to confirm this relationship. Because of the lack of genetic sequence information, *Thiodictyon* species have not been included in a recent reclassification of *Chromatiaceae* species (Guyoneaud et al., 1998b; Imhoff et al., 1998b).

DIFFERENTIATION OF THE SPECIES OF THE GENUS *THIODICTYON*

I. Cells able to grow in form of typical netlike aggregates under certain culture conditions.

Thiodictyon elegans

II. Cells that do not form netlike aggregates, usually growing in the form of free single cells.

Thiodictyon bacillosum

List of species of the genus Thiodictyon

1. **Thiodictyon elegans** Winogradsky 1888, 82[AL]

   e'le.gans. L. adj. *elegans* choice, elegant.

   Cells are rod shaped, 1.5–2.0 × 3–8 μm, grow in the form of irregular netlike aggregates, dependent on the sulfide concentration and light intensity (Fig. BXII.γ.14). In cultures of higher population densities, cell chains break apart and the cultures contain predominantly single cells (Fig. BXII.γ.15). Color of single cells grayish, color of cell suspensions light violet to purple-violet. Photosynthetic pigments are bacteriochlorophyll *a* and carotenoids of the rhodopinal group, with rhodopinal and rhodopin as major components (Pfennig et al., 1968).

   Obligately phototrophic and strictly anaerobic. Photolithoautotrophic growth under anoxic conditions in the light with sulfide and S⁰ as electron donors. During oxidation of sulfide, S⁰ is transiently stored in the peripheral part of the cell that is free of gas vesicles. In the presence of sulfide and bicarbonate, acetate and pyruvate are photoassimilated. Nitrogen sources: ammonium salts.

   Mesophilic freshwater bacterium with optimum growth at 20–25°C, pH range 6.7–7.3.

   The mol% G + C of the DNA is: 65.3–66.3 (Bd); type strain 65.4 (Bd).

   Type strain: 3011, DSM 232.

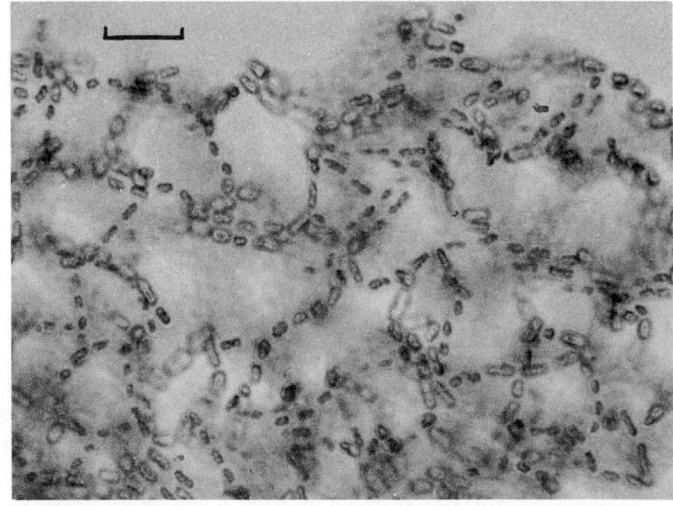

FIGURE BXII.γ.14. *Thiodictyon elegans* DSM 232 showing the typical, somewhat irregular netlike arrangement of the cells. Brightfield micrograph. Bar = 12 μm.

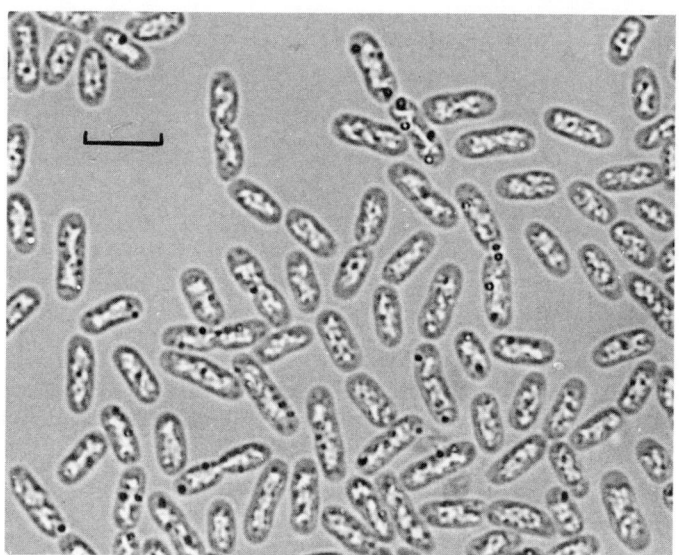

FIGURE BXII.γ.15. *Thiodictyon elegans* DSM 232 cultured photoautotrophically with sulfide. The *irregular whitish areas* within the cells are the gas vacuoles. The cells also contain small, blackish-appearing globules of S⁰. Brightfield micrograph. Bar = 5 μm.

2. **Thiodictyon bacillosum** (Winogradsky 1888) Pfennig and Trüper 1971b, 12[AL] (*Amoebobacter bacillosus* Winogradsky 1888, 78.)

ba.cil.lo' sum. L. dim. n. *bacillus* a small rod; M.L. neut. adj. *bacillosum* full of or made up of small rods.

Cells are rod shaped, 1.5–2.0 × 3–6 μm, may grow in irregular clumps surrounded by slime. Netlike aggregates not formed. Color of cell suspensions light violet to purple-violet. Photosynthetic pigments are bacteriochlorophyll *a* and carotenoids of the rhodopinal group, with rhodopinal as major component (Schmidt et al., 1965).

In the presence of sulfide and bicarbonate, acetate and pyruvate are photoassimilated. Nitrogen sources: ammonium salts.

Mesophilic freshwater bacterium with optimum growth at 20–30°C, pH range 6.7–7.3.

The mol% G + C of the DNA is: 66.3 (Bd).

Type strain: 1814, DSM 234.

Genus XVII. **Thioflavicoccus** *Imhoff and Pfennig 2001, 109[VP]*

JOHANNES F. IMHOFF

Th.io.fla' vi.coc' cus. Gr. n. *thios* sulfur; L. masc. adj. *flavus* golden-yellow, beige; L. masc. n. *coccus* sphere; M.L. masc. n. *Thioflavicoccus* beige-yellow coccus with sulfur.

Cells are spherical with diplococcus-shaped division stages, motile by flagella, multiply by binary fission. Gram negative; belong to the *Gammaproteobacteria.* Contain internal photosynthetic membranes of tubular shape. Photosynthetic pigments are bacteriochlorophyll *b* and carotenoids.

Obligately phototrophic and strictly anaerobic. Photolithoautotrophic growth under anoxic conditions in the light with sulfide and elemental sulfur as electron donors. During oxidation of sulfide, elemental sulfur transiently stored inside the cells in the form of highly refractile globules. Final oxidation product is sulfate. In the presence of sulfide and bicarbonate, simple organic substrates photoassimilated.

Mesophilic bacteria growing well at 20–30°C and neutral pH (pH range 6.5–7.5) and **requiring sodium chloride for optimum growth.**

The mol% G + C of the DNA is: 66.5.

Type species: **Thioflavicoccus mobilis** Imhoff and Pfennig 2001, 109.

DIFFERENTIATION OF THE GENUS *THIOFLAVICOCCUS* FROM OTHER GENERA

According to the results of 16S rDNA sequence analyses (Fig. BXII.γ.3 of the chapter describing the family *Chromatiaceae*), the genus *Thioflavicoccus* belongs to the branch of the family *Chromatiaceae* that characteristically contains marine and halophilic species. Characteristics for differentiation from other phototrophic members of the family *Chromatiaceae* are shown in Table BXII.γ.5 of the chapter describing the family *Chromatiaceae*. *Thioflavicoccus mobilis* is one of the very few species (in addition to *Thiococcus pfennigii* and *Thioalkalicoccus limnaeus*) that contain tubular internal photosynthetic membranes. The phylogenetic relationship of the *Thioflavicoccus* species to other *Chromatiaceae* species, based on 16S rDNA sequence comparison, is shown in Fig. BXII.γ.3 of the chapter describing the family *Chromatiaceae*.

List of species of the genus Thioflavicoccus

1. **Thioflavicoccus mobilis** Imhoff and Pfennig 2001, 109[VP]

mo' bi.lis. L. adj. *mobilis* mobile.

Cells are coccoid, short, rod-shaped to diplococcus-shaped before cell division. Cocci 0.8–1.0 μm in diameter. Motile by monopolar monotrichous flagella. Internal photosynthetic membrane system of tubular type. Color of cell suspensions is yellowish beige to orange-brown. The absorption spectrum of living cell suspensions exhibits maxima at 410, 462, 492, 530, and 1025 nm, with shoulders at 602 and 835 nm. Photosynthetic pigments are bacterio-

chlorophyll *b* and 3,4,3′,4′-tetrahydro-spirilloxanthin as the main carotenoid. The metabolism is obligately phototrophic and strictly anaerobic. Photolithoautotrophic growth occurs in the light with hydrogen sulfide as electron donor. Globules of elemental sulfur are accumulated inside the cells. The final oxidation product is sulfate. Thiosulfate is not used. In the presence of sulfide and bicarbonate, acetate, pyruvate, and ascorbate are used as organic substrates. Growth factors are not required. Sodium chloride is required for growth. Good growth occurs at 25–30°C, pH

7.2–7.4, with 1–3% NaCl (optimum is at 2%), and at 500 lux light intensity from a tungsten lamp. Habitat: laminated microbial mats of salt marshes.

The mol% G + C of the DNA is: 66.5 (Bd).
Type strain: 8321, ATCC 700959.
GenBank accession number (16S rRNA): AJ010126.

Genus XVIII. **Thiohalocapsa** Imhoff, Süling and Petri 1998b, 1139[VP]

JOHANNES F. IMHOFF AND PIERRE CAUMETTE

Thi'o.ha.lo.cap'sa. Gr. n. *thios* sulfur; Gr. gen. n. *halos* of the salt; L. fem. n. *capsa* capsule; M.L. fem. n. *Thiohalocapsa* the sulfur capsule of the salt.

Cells spherical, nonmotile, multiply by binary fission. Individual cells surrounded by a strong slime capsule and cell aggregates formed. Gram negative, belong to the *Gammaproteobacteria*, and contain internal **photosynthetic membranes of vesicular type** with bacteriochlorophyll *a* and carotenoids as photosynthetic pigments.

Photolithoautotrophic growth under anoxic conditions in the light with sulfide, S^0, and possibly other electron donors. During oxidation of sulfide, S^0 transiently stored inside the cells in the form of highly refractile globules. Final oxidation product is sulfate. In the presence of sulfide and bicarbonate, simple organic substrates are photoassimilated. **Chemolithotrophic or chemoorganotrophic growth is possible under microoxic to oxic conditions in the dark**.

Mesophilic bacteria with optimum growth at 20–30°C and pH 6.5–7.5. Elevated concentrations of sodium chloride required for growth.

Habitat: Sulfide-containing waters and sediments of saltern and other coastal habitats with elevated salt concentrations and exposed to the light.

The mol% G + C of the DNA is: 65.9–66.6.

Type species: **Thiohalocapsa halophila** (Caumette, Baulaigue and Matheron 1991) Imhoff, Süling and Petri 1998b, 1139 (*Thiocapsa halophila* Caumette, Baulaigue and Matheron 1991, 175.)

ENRICHMENT AND ISOLATION PROCEDURES

Methods for enrichment and isolation are the same as those described for the genus *Chromatium*. *Thiohalocapsa* species often form dense purple layers in microbial mats of hypersaline environments with shallow waters at salinities from 10 to 20%, such as salterns, shallow salt lakes, and evaporitic lagoons. They are also found at the surface of sediments below top layers of salt deposits. *Thiohalocapsa* species can be enriched from samples of the natural environment using media composed of natural brines, as described by Caumette et al. (1991). They can be purified by repeated deep-agar dilution series. After achievement of pure cultures, a single colony is transferred into synthetic media containing 7–10% NaCl (Caumette et al., 1991).

MAINTENANCE PROCEDURES

Maintenance of strains in liquid culture medium and long-term preservation in liquid nitrogen are carried out as described for the genus *Chromatium*. Maintenance is easier in media composed of natural brine than in synthetic media.

DIFFERENTIATION OF THE GENUS *THIOHALOCAPSA* FROM OTHER GENERA

Characteristics for differentiation of the genus *Thiohalocapsa* from other members of the family *Chromatiaceae* are shown in Table BXII.γ.5 of the chapter describing the family *Chromatiaceae*. The genetic relationship of *Thiohalocapsa halophila* to other *Chromatiaceae* species, based on 16S rDNA sequence comparison, is shown in Fig. BXII.γ.3 of the chapter describing the family *Chromatiaceae*.

TAXONOMIC COMMENTS

Thiohalocapsa halophila was originally described as a *Thiocapsa* species (Caumette et al., 1991), because its phenotypic properties were in agreement with this genus classification, according to Pfennig and Trüper (1989). Cells are spherical and nonmotile purple sulfur bacteria. Because phylogenetic relationships and salt response clearly distinguish *Thiohalocapsa halophila* from species of the genus *Thiocapsa* (Guyoneaud et al., 1998b; Imhoff et al., 1998b), this bacterium was transferred to the new genus *Thiohalocapsa* (Imhoff et al., 1998b).

FURTHER READING

Guyoneaud, R., Süling, J., Petri, R., Matheron, R., Caumette, P., Pfennig, N. and Imhoff, J.F. 1998. Taxonomic rearrangements of the genera *Thiocapsa* and *Amoebobacter* on the basis of 16S rDNA sequence analyses and description of *Thiolamprovum* gen. nov. Int. J. Syst. Bacteriol. 48: 957–964.

Caumette, P., Baulaigue, R. and Matheron, R. 1991. *Thiocapsa halophila*, sp. nov., a new halophilic phototrophic purple sulfur bacterium. Arch. Microbiol. *155:* 170–176.

Imhoff, J.F., Süling, J. and Petri, R. 1998. Phylogenetic relationships among the *Chromatiaceae*, their taxonomic reclassification and description of the new genera *Allochromatium, Halochromatium, Isochromatium, Marichromatium, Thiococcus, Thiohalocapsa* and *Thermochromatium*. Int. J. Syst. Bacteriol. 48: 1129–1143.

List of species of the genus Thiohalocapsa

1. **Thiohalocapsa halophila** (Caumette, Baulaigue and Matheron 1991) Imhoff, Süling and Petri 1998b, 1139[VP] (*Thiocapsa halophila* Caumette, Baulaigue and Matheron 1991, 175.)

   ha.lo'phi.la. Gr. n. *hals* salt; Gr. adj. *philos* loving; M.L. fem. adj. *halophila* salt-loving.

   Cells spherical, 1.5–2.5 μm in diameter, form aggregates of 2, 4, or more cells, nonmotile. Color of cell suspensions purple-red. Photosynthetic pigments are bacteriochlorophyll *a* and okenone.

   Photolithoautotrophic growth occurs with hydrogen, sulfide, S^0, sulfite, and thiosulfate as electron donors. In presence of sulfide and bicarbonate, acetate, pyruvate, lactate, and fructose are photoassimilated. Glycerol and glucose poorly used. Chemolithotrophic or chemoorganotrophic growth is possible under microoxic conditions in the dark. Assimilatory sulfate reduction is absent. Vitamin B_{12} required.

   Mesophilic and halophilic bacterium with optimum

growth at 20–30°C, pH 7.0 (pH range 6.0–8.0), and at 4–8% NaCl (salinity range 3–20% NaCl).

Habitat: Reduced sediments in coastal saltern with salt deposits exposed to light.

The mol% G + C of the DNA is: 65.9–66.6 (T_m).
Type strain: SG3202, ATCC 49740, DSM 6210.
GenBank accession number (16S rRNA): AJ002796.

Genus XIX. **Thiolamprovum** Guyoneaud, Süling, Petri, Matheron, Caumette, Pfennig and Imhoff 1998b, 963[VP]

PIERRE CAUMETTE, RÉMI GUYONEAUD AND JOHANNES F. IMHOFF

Thi'o.lam.pro' vum. Gr. n. *thios* sulfur; Gr. n. *lampros* bright, brilliant; L. n. *ovum* egg; M.L. neut. n. *Thiolamprovum* bright egg with sulfur.

Cells nearly spherical to oval, 2×2–3 μm, may occur in regular platelets of 4–16 cells, **nonmotile**, and multiply by binary fission. Gram negative, belong to the *Gammaproteobacteria*, and have internal photosynthetic membranes of vesicular type containing the photosynthetic pigments bacteriochlorophyll *a* and carotenoids.

Photolithoautotrophic growth under anoxic conditions in the light occurs with sulfide, thiosulfate, and sulfur as electron donors. During oxidation of sulfide, S^0 transiently stored inside the cells in the form of highly refractile globules. Sulfate is the final oxidation product. In the presence of sulfide and bicarbonate, simple organic substrates can be photoassimilated. **Chemoautotrophic or mixotrophic growth under microoxic to oxic conditions in the dark is possible.**

The mol% G + C of the DNA is: 64.5–66.5 (Bd).

Type species: **Thiolamprovum pedioforme** (Eichler and Pfennig 1986) Guyoneaud, Süling, Petri, Matheron, Caumette, Pfennig and Imhoff 1998b, 963 (*Amoebobacter pedioformis* Eichler and Pfennig 1986, 300.)

ENRICHMENT AND ISOLATION PROCEDURES

Media and culture conditions are the same as described for *Chromatium* species. No specific enrichment or isolation procedures are applicable for *Thiolamprovum* species. Inoculation of agar shake dilution cultures directly from samples collected in nature is recommended as described for the *Chromatium* species.

MAINTENANCE PROCEDURES

Maintenance of strains in liquid culture medium and long-term preservation in liquid nitrogen are carried out as described for the genus *Chromatium*.

DIFFERENTIATION OF THE GENUS *THIOLAMPROVUM* FROM OTHER GENERA

Characteristics for differentiation of *Thiolamprovum* species from other phototrophic members of the family *Chromatiaceae* are shown in Table BXII.γ.6 of the chapter describing the family *Chromatiaceae*. *Thiolamprovum* is characterized by the formation of platelets of 8–16 or more regularly arranged cells, similar to *Thiopedia rosea*. These bacteria can be distinguished by their different carotenoid content and color of cultures and colonies. *Thiolamprovum* cells contain carotenoids of the spirilloxanthin group with a pinkish color of the colonies, whereas *Thiopedia rosea* cells contain okenone with a purple color of the colonies. The phylogenetic relationship of *Thiolamprovum pedioforme* to other *Chromatiaceae* species, based on 16S rDNA sequence comparison, is shown in Fig. BXII.γ.4 of the chapter describing the family *Chromatiaceae*.

TAXONOMIC COMMENTS

At present, *Thiolamprovum pedioforme* is the only representative of this genus. It was previously known as *Amoebobacter pedioformis* (Eichler and Pfennig, 1986). Sequence analyses of 16S rRNA genes have shown that *Thiolamprovum pedioforme* is genetically distant from *Lamprocystis* and *Thiocapsa* species (Guyoneaud et al., 1998b). Therefore, it has been transferred to *Thiolamprovum pedioforme* (Guyoneaud et al., 1998b).

FURTHER READING

Eichler, B. and Pfennig, N. 1986. Characterization of a new platelet-forming purple sulfur bacterium, *Amoebobacter pedioformis* sp. nov. Arch. Microbiol. *146*: 295–300.

Guyoneaud, R., Süling, J., Petri, R., Matheron, R., Caumette, P., Pfennig, N. and Imhoff, J.F. 1998. Taxonomic rearrangements of the genera *Thiocapsa* and *Amoebobacter* on the basis of 16S rDNA sequence analyses and description of *Thiolamprovum* gen. nov. Int. J. Syst. Bacteriol. *48*: 957–964.

List of species of the genus Thiolamprovum

1. **Thiolamprovum pedioforme** (Eichler and Pfennig 1986) Guyoneaud, Süling, Petri, Matheron, Caumette, Pfennig and Imhoff 1998b, 963[VP] (*Amoebobacter pedioformis* Eichler and Pfennig 1986, 300.)

   pe' di.o.for' me. Gr. n. *pedion* a plain, a flat area; L. n. *forma* shape; M.L. neut. adj. *pedioforme* flat shaped.

   Spherical to oval cells occur singly or in regular platelets of 4–16 cells. Individual cells 2×2–3 μm, nonmotile, and contain gas vesicles in peripheral parts of the cell. Color of cell suspensions pinkish red. Photosynthetic pigments are bacteriochlorophyll *a* and the carotenoids spirilloxanthin, anhydrorhodovibrin, rhodopin, and OH-spirilloxanthin.

   Photolithoautotrophic growth with sulfide, thiosulfate, and S^0 as electron donors. In the presence of sulfide and bicarbonate, acetate, lactate, and pyruvate photoassimilated. Chemolithoautotrophic growth under microoxic conditions in the dark is possible with sulfide and thiosulfate as electron donors. Assimilatory sulfate reduction does not occur.

   Mesophilic freshwater bacterium with optimum growth at 37°C and pH 7.4–7.6.

   Habitat. freshwater and waste water ponds, type strain originates from a wastewater pond of a sugar factory.

   The mol% G + C of the DNA is: 65.5 (Bd).

   Type strain: Strain CML2, DSM 3802.

   GenBank accession number (16S rRNA): Y12297.

Genus XX. **Thiopedia** *Winogradsky 1888, 85*[AL]

JOHANNES F. IMHOFF

Thi.o.pe' di.a. G. n. *thios* sulfur; Gr. n. *pedium* a plain, a flat area; M.L. fem. n. *Thiopedia* sulfur plain.

Cells spherical to ovoid, nonmotile, and multiply by binary fission. Rectangular platelets formed with 4, 8, 16, 32, or more regularly arranged cells (up to 128 or 256 cells may stick together). Gram negative, **contain irregularly shaped gas vesicles in the central part of the cell and internal photosynthetic membranes of vesicular type** with bacteriochlorophyll *a* and carotenoids as photosynthetic pigments.

Photolithoautotrophic growth under anoxic conditions in the light with sulfide and S⁰ as electron donors. During oxidation of sulfide, S^0 transiently stored in the peripheral part of the cell in the form of highly refractile globules. Final oxidation product is sulfate. In the presence of sulfide and bicarbonate, simple organic substrates are photoassimilated. Reduced sulfur compounds required as sulfur source.

Habitat: mud and stagnant water of ponds and lakes containing hydrogen sulfide, common planktonic bacterium in the sulfide-containing hypolimnion of freshwater lakes.

The mol% G + C of the DNA is: 62.5–63.5.

Type species: **Thiopedia rosea** Winogradsky 1888, 85.

ENRICHMENT AND ISOLATION PROCEDURES

A prerequisite for the successful enrichment and isolation of *Thiopedia* species is the use of an inoculum from anoxic water or mud samples in which the characteristic *Thiopedia* platelets were microscopically detected. For the isolation of pure cultures, it is advisable to prepare enrichment cultures in liquid media, as well as agar shake dilution cultures that are directly inoculated from the sample of the natural habitat. In liquid media, *Thiopedia* may be outgrown by other purple sulfur bacteria containing gas vesicles.

The culture medium and incubation conditions are similar to those described for *Chromatium* species. Not more than 1.5 mM sulfide and 3 mM acetate should be used for growth of *Thiopedia*, and the trace element solution must not contain a chelating agent (e.g., EDTA). Growth is enhanced by the addition of 100 µM dithionite to the medium. Incubation is carried out at a temperature of 20–25°C and a light intensity of 100–200 lux. *Thiopedia* species grow in the form of individual cells, or platelets with 4–128 cells. Colonies and cell suspensions of *Thiopedia rosea* exhibit a unique bright purple-red color and can, therefore, be differentiated readily from other purple sulfur bacteria.

MAINTENANCE PROCEDURES

Stock cultures may be maintained in liquid medium in tightly closed 100-ml screw-capped bottles at 4–8°C. Such cultures should be transferred every 2 months. For long-term preservation, vials of cell suspensions in anoxic culture medium containing 5% (v/v) DMSO are stored in liquid nitrogen.

DIFFERENTIATION OF THE GENUS *THIOPEDIA* FROM OTHER GENERA

Characteristics that differentiate *Thiopedia* species from other phototrophic members of the family *Chromatiaceae* are: i) the arrangement of the cells in rectangular platelets, ii) possession of gas vesicles, and iii) the purple-red color of cultures and colonies, which is due to the carotenoid okenone. Platelets similar to those of *Thiopedia* species are formed by *Thiolamprovum pedi-*

oforme. Characteristics for differentiation of *Thiopedia* species from other spherical and nonmotile purple sulfur bacteria and from other phototrophic *Chromatiaceae* species are shown in Table BXII.γ.6 of the chapter describing the family *Chromatiaceae*.

TAXONOMIC COMMENTS

Purple sulfur bacteria having the morphology first described by Winogradsky for *Thiopedia rosea* are widely distributed in sulfide-containing ponds and lakes (Utermöhl, 1925; Eichler and Pfennig, 1991). Fine-structure studies of *Thiopedia*-like cell material collected from a purple bacteria bloom in a small forest pond (Michigan, USA) revealed the presence of at least two cell types with different internal membrane systems (Hirsch, 1973). One type contained a vesicular membrane system, while the other type contained a membrane system that consisted of stacks of disk-shaped lamellae. Thin section micrographs of the type strain of *Thiopedia rosea*, showing a vesicular membrane system, were published by Remsen (1978).

The description of *T. rosea* given here is based on information from five strains described by Eichler and Pfennig (1991). Little information on this species was previously available. It should be noted, however, that the properties of the five isolates differ in some aspects from earlier descriptions. Winogradsky (1888) indicated a cell diameter of 1–2 µm and Pfennig (1989c) reported cells being 1.2–1.6 × 1.5–2.0 µm and requiring vitamin B_{12}. The latter property was not substantiated by Eichler and Pfennig (1991) and cells were reported to be 2.0–2.5 × 2.5–3.0 µm by these authors.

Unfortunately, the ability of *Thiopedia* to multiply in rectangular platelets, as characteristically observed in many natural habitats, is gradually lost over periods of months or years in pure culture. Therefore, it is not a taxonomically reliable character. Because the formation of rectangular platelets was considered a genus-specific property, distinguishing *Thiopedia* from other *Chromatiaceae* species (Pfennig and Trüper, 1989), recognition of the taxonomic unreliability of this property raises questions about the existence of the genus *Thiopedia*. Further studies are needed to show whether *Thiopedia rosea* is a close relative of *Thiolamprovum pedioforme*, which also forms rectangular platelets (Eichler and Pfennig, 1986), or *Thiocapsa* species, that do not form large platelets, but may form tetrads of cells.

FURTHER READING

Eichler, B. and Pfennig, N. 1986. Characterization of a new platelet-forming purple sulfur bacterium, *Amoebobacter pedioformis* sp. nov. Arch. Microbiol. **146**: 295–300.

Eichler, B. and Pfennig, N. 1991. Isolation and characteristics of *Thiopedia rosea* (neotype). Arch. Microbiol. **155**: 210–216.

Hirsch, P. 1973. Fine structure of *Thiopedia* spp. *In* Drews (Editor), Abstracts of the Symposium on Prokaryotic Photosynthetic Organisms, Freiburg. p. 184–185.

Pfennig, N. 1973. Culture and ecology of *Thiopedia rosea*. *In* Drews (Editor), Abstracts of the Symposium on Prokaryotic Photosynthetic Organisms, Freiburg. p. 75–76.

Remsen, C.C. 1978. Comparative subcellular architecture of photosynthetic bacteria. *In* Clayton and Sistrom (Editors), The Photosynthetic Bacteria, Plenum Press, New York. p. 31–60.

Utermöhl, H. 1925. Limnologische Phytoplanktonstudien. Arch. Hydrobiol. Suppl. **5**: 251–277.

List of species of the genus Thiopedia

1. **Thiopedia rosea** Winogradsky 1888, 85[AL]

ro′se.a. L. adj. *rosea* rose-colored, pink.

Cells spherical to ovoid, 2.0–2.5 × 2.5–3.0 μm, with regular arrangements in rectangular platelets of up to 64 or more cells (Fig. BXII.γ.16). Nonmotile and contain gas vesicles in the central part of the cells. Individual cells colorless, color of cell suspensions bright purple-red. Internal photosynthetic membranes of vesicular type. Photosynthetic pigments are bacteriochlorophyll *a* and carotenoids with okenone as major component.

Obligately phototrophic and strictly anaerobic. Photolithoautotrophic growth with sulfide and S⁰ as electron donors. Sulfide concentrations above 0.6 mM are growth inhibitory. In the presence of sulfide and bicarbonate, the following substrates are photoassimilated: acetate, pyruvate, butyrate, valerate (not more than 0.03% (w/v) of each to be added), and glucose. Some strains also utilize succinate, fumarate, malate, and fructose (0.05%). Ammonium salts serve as nitrogen source. Reduced sulfur compounds required as sulfur source. Assimilatory sulfate reduction does not occur. Growth enhanced by the addition of 100 μM dithionite to growth media.

Mesophilic planktonic freshwater bacterium with optimum growth at 23°C and pH 7.3–7.5.

Habitat: anoxic and sulfidic water of stratified freshwater lakes.

The mol% G + C of the DNA is: 62.5–63.5 (Bd).

Type strain: 4711, DSM 1236 (neotype strain; no type strain available).

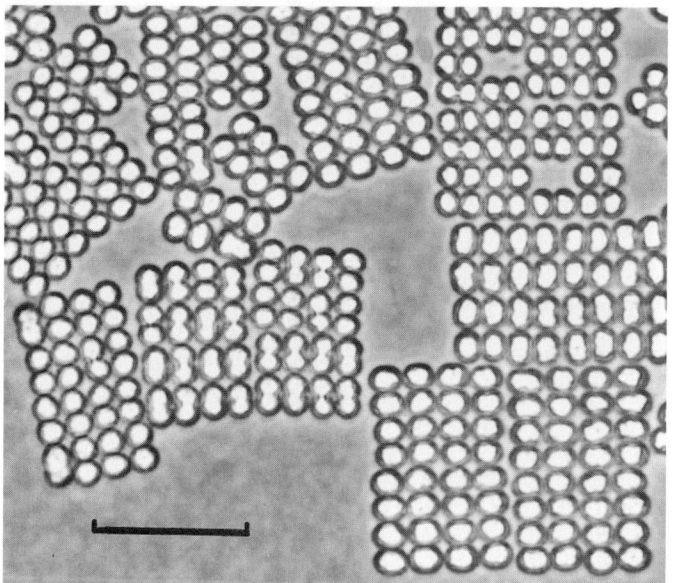

FIGURE BXII.γ.16. *Thiopedia rosea* strain 4211 cultured photoautotrophically with sulfide and acetate. The *light areas* inside the cells are the gas vacuoles. In some of the platelets, synchronously dividing cells can be recognized. Phase-contrast micrograph. Bar = 10 μm.

Genus XXI. **Thiorhodococcus** *Guyoneaud, Matheron, Liesack, Imhoff and Caumette 1998a, 328^VP (Effective publication: Guyoneaud, Matheron, Liesack, Imhoff and Caumette 1997, 22)*

RÉMI GUYONEAUD, PIERRE CAUMETTE AND JOHANNES F. IMHOFF

Thi′o.rho.do.co′ccus. Gr. n. *thios* sulfur; Gr. n. *rhodon* the rose; L. n. *coccus* sphere; M.L. masc. n. *Thiorhodococcus* rose sphere with sulfur.

Cells spherical to slightly ovoid, diplococcus-shaped before cell division, motile by flagella. Cells single or in pairs but may form irregular aggregates, multiply by binary fission. Gram negative, belong to the *Gammaproteobacteria*, **and contain internal photosynthetic membranes of vesicular type** with bacteriochlorophyll *a* and carotenoids as photosynthetic pigments.

Photolithoautotrophic growth under anoxic conditions in the light with sulfide and S⁰ as electron donors. During oxidation of sulfide, S⁰ transiently stored inside the cells in the form of highly refractile globules. Final oxidation product is sulfate. In the presence of sulfide and bicarbonate, simple organic substrates are photoassimilated. Chemotrophic growth is possible under microoxic conditions in the dark.

Habitat: water and sediment of marine and brackish environments containing hydrogen sulfide.

The mol% G + C of the DNA is: 66.8–67.0.

Type species: **Thiorhodococcus minor** Guyoneaud, Matheron, Liesack, Imhoff and Caumette 1998a, 328 (Effective publication: Guyoneaud, Matheron, Liesack, Imhoff and Caumette 1997, 22.)

ENRICHMENT AND ISOLATION PROCEDURES

The media described for *Chromatium* species can also be used for *Thiorhodococcus* and other marine *Chromatiaceae* species, if supplemented with 2–3% NaCl as required. The culture conditions and isolation procedures are the same as described for the genus *Chromatium*.

MAINTENANCE PROCEDURES

Maintenance of strains in liquid culture medium and long-term preservation in liquid nitrogen are carried out as described for the genus *Chromatium*. Long term preservation by freezing at −80°C in 40% glycerol is also possible.

DIFFERENTIATION OF THE GENUS *THIORHODOCOCCUS* FROM OTHER GENERA

The genus *Thiorhodococcus* is distinguished from other members of the family *Chromatiaceae* by phenotypic and genetic traits. The spherical, motile, purple sulfur bacteria of the genus *Thiorhodo-*

coccus are morphologically similar to *Thiocystis* species, but differ from them by their salt requirement and 16S rDNA sequence composition. *Thiorhodococcus minor* forms a separate lineage not specifically affiliated with other *Chromatiaceae* species (Imhoff et al., 1998b; see Fig. BXII.γ.3 of the chapter describing the family *Chromatiaceae*). Characteristics useful for the differentiation of the genus *Thiorhodococcus* from other members of the family *Chromatiaceae* are shown in Table BXII.γ.5 of the chapter describing the family *Chromatiaceae*.

FURTHER READING

Guyoneaud, R., Matheron, R., Liesack, W., Imhoff, J.F. and Caumette, P. 1997. *Thiorhodococcus minus*, gen. nov., sp. nov., a new purple sulfur bacterium isolated from coastal lagoon sediments. Arch. Microbiol. *168*: 16–23.

Guyoneaud, R., Matheron, R., Baulaigue, R., Podeur, K., Hirschler, A. and Caumette, P. 1996. Anoxygenic phototrophic bacteria in eutrophic coastal lagoons of the French Mediterranean and Atlantic Coast (Prévost Lagoon, Arcachon Bay, Certes Fishponds). Hydrobiologia *329*: 33–43.

List of species of the genus Thiorhodococcus

1. **Thiorhodococcus minor** Guyoneaud, Matheron, Liesack, Imhoff and Caumette 1998a, 328[VP] (Effective publication: Guyoneaud, Matheron, Liesack, Imhoff and Caumette 1997, 22.)

mi′ nor. L. masc. adj. *minor* small.

Cells spherical to slightly ovoid, 1.0–2.0 μm in diameter, highly motile by flagella, divide by binary fission, and form diplococcus-shaped division stages (Fig. BXII.γ.17). Color of individual cells grayish, color of cell suspensions brown-orange. Photosynthetic pigments are bacteriochlorophyll *a* and carotenoids of the normal spirilloxanthin group, with rhodopin as major pigment.

Photolithoautotrophic growth under anoxic conditions in the light with hydrogen, sulfide, thiosulfate, and S⁰ as electron donors. In the presence of sulfide and hydrogen carbonate, acetate, propionate, lactate, glycolate, pyruvate, malate, fumarate, succinate, fructose, sucrose, ethanol, and propanol are photoassimilated. Chemolithotrophic growth with sulfide and thiosulfate, and chemoorganotrophic growth with acetate and fructose are possible under microoxic conditions in the dark. Some strains (but not the type strain) assimilate sulfate. Nitrogen sources: ammonium salts and N₂.

Mesophilic marine bacterium requiring NaCl for growth. Optimal growth at 2% NaCl (salinity range 0.5–9% NaCl), pH 7.0–7.2 (pH range 6.0–8.0), and 30–35°C.

Habitat: anoxic sediments from coastal lagoons and marine environments rich in sulfide.

The mol% G + C of the DNA is: 66.9 (HPLC).
Type strain: CE2203, ATCC 700259, DSM 11518.
GenBank accession number (16S rRNA): Y11316.

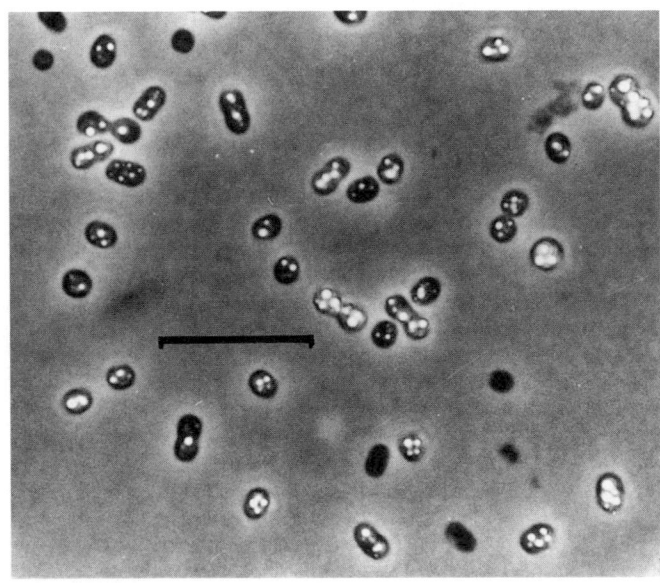

FIGURE BXII.γ.17. *Thiorhodococcus minor* DSM 11518[T] grown photolithoautotrophically with sulfide. Phase-contrast microphotograph. Bar = 10 μm.

Genus XXII. **Thiorhodovibrio** *Overmann, Fischer and Pfennig 1993, 188[VP] (Effective publication: Overmann, Fischer and Pfennig 1992, 334)*

JOHANNES F. IMHOFF AND JÖRG OVERMANN

Thi′ o.rho.do.vi′ brio. Gr. n. *thios* sulfur; Gr. n. *rhodon* the rose; L. v. *vibrio* vibrate; M.L. masc. n. *vibrio* that which vibrates; *Thiorhodovibrio* rose vibrio with sulfur.

Individual **cells vibrioid to spirilloid, motile by a single polar flagellum**, multiply by binary fission. Gram negative, belong to the *Gammaproteobacteria*, and contain **internal membranes of vesicular type** with bacteriochlorophyll *a* and carotenoids as photosynthetic pigments.

Photolithoautotrophic growth under anoxic conditions in the light with sulfide and S⁰ as electron donors. During oxidation of sulfide, S⁰ stored inside the cells in the form of highly refractile globules. Final oxidation product is sulfate. In the presence of sulfide and bicarbonate, simple organic substrates are photoassimilated. Chemotrophic growth at reduced oxygen partial pressure.

Habitat: surface layers of marine and salt lake sediments and microbial mats containing hydrogen sulfide.

The mol% G + C of the DNA is: 61–63.

Type species: **Thiorhodovibrio winogradskyi** Overmann, Fischer and Pfennig 1993, 188 (Effective publication: Overmann, Fischer and Pfennig 1992, 334.)

ENRICHMENT AND ISOLATION PROCEDURES

The media described for *Chromatium* species can also be used for *Thiorhodovibrio* and other marine members of the family *Chromatiaceae*, if supplemented with 2–3% NaCl as required. The culture conditions and isolation procedures are the same as described for the genus *Chromatium*.

MAINTENANCE PROCEDURES

Maintenance of strains in liquid culture medium and long-term preservation in liquid nitrogen are carried out as described for the genus *Chromatium*.

DIFFERENTIATION OF THE GENUS *THIORHODOVIBRIO* FROM OTHER GENERA

Characteristics for differentiation of the genus *Thiorhodovibrio* from other phototrophic *Chromatiaceae* species are shown in Table BXII.γ.5 of the chapter describing the family *Chromatiaceae*. The genetic relationship of *Thiorhodovibrio winogradskyi* to other members of the family *Chromatiaceae*, based on 16S rDNA sequence comparison, is shown in Fig. BXII.γ.3 of the chapter describing the family *Chromatiaceae*.

TAXONOMIC COMMENTS

Thiorhodovibrio winogradskyi was described based on two isolates that show significant phenotypic differences (Overmann et al., 1992). Both strains may be considered different species, and the description of *Thiorhodovibrio winogradskyi* given here is that of the type strain (DSM 6702, SSP1). The following properties distinguish strain 06511 from the type strain of *Thiorhodovibrio winogradskyi*: Cells are spirilloid, 0.7–0.9 × 2.5–3.9 µm. Thiosulfate is used as photosynthetic electron donor and the mol% G + C content of the DNA is 62.4. Propionate does not support growth of strain 06511. However, in addition to the substrates used by the type strain, succinate, fumarate, malate, tartrate, malonate, glycerol, and peptone stimulate growth slightly (Overmann et al., 1992) for strain O6511.

FURTHER READING

Overmann, J., Fischer, U. and Pfennig, N. 1992. A new purple sulfur bacterium from saline littoral sediments, *Thiorhodovibrio winogradsky* gen. nov. and sp. nov. Arch. Microbiol. *157*: 329–335.

List of species of the genus Thiorhodovibrio

1. **Thiorhodovibrio winogradskyi** Overmann, Fischer and Pfennig 1993, 188[VP] (Effective publication: Overmann, Fischer and Pfennig 1992, 334.)

   wi.no.grad' sky.i. M.L. gen. n. *winogradskyi* of Winogradsky, named for S.N. Winogradsky, a Russian microbiologist, who did the first comprehensive studies on the purple sulfur bacteria.

   Cells vibrioid, 1.2–1.6 × 2.6–4.0 µm, motile by a single polar flagellum. Gram negative and belong to the *Gammaproteobacteria*. Anaerobically grown cultures pink to light pinkish purple. Living cell suspensions have absorption maxima at 370–372, 483–484, 510, 590–591, 794–795, and 867 nm. Photosynthetic pigments are bacteriochlorophyll *a* and carotenoids of the spirilloxanthin group, with rhodopin as the major component, and spirilloxanthin, lycopene, anhydrorhodovibrin, and rhodovibrin.

   Photolithoautotrophic growth with sulfide and S[0] as photosynthetic electron donors. When growing on sulfide, up to ten small sulfur globules are formed inside the cell, often in a row along its long axis. High growth rates are obtained with S[0] as electron donor. In the presence of sulfide and hydrogen carbonate, acetate, pyruvate, and propionate are photoassimilated. Chemotrophic growth possible under mixotrophic conditions at reduced oxygen partial pressure ($\sim$1% O_2). No vitamins required.

   Mesophilic marine bacterium requiring NaCl, with optimum growth at 2.2–3.2% NaCl (range 0.2–7.2% NaCl), pH 7.0–7.4 and 33°C (temperature range 14–37°C).

   Habitat: sulfide-containing surface layers of salt lake sediments and marine microbial mats.

   The mol% G + C of the DNA is: 61.0 (HPLC).

   Type strain: SSP1, DSM 6702.

   GenBank accession number (16S rRNA): AB016986, AJ006214, Y12368.

Genus XXIII. **Thiospirillum** Winogradsky 1888, 104[AL]

JOHANNES F. IMHOFF

Thi' o.spi.ril' lum. Gr. n. *thios* sulfur; M.L. dim. neut. n. *Spirillum* a bacterial genus; M.L. neut. n. *Thiospirillum* sulfur *Spirillum*.

Cells **curved rods or spirals** (sigmoid), dividing by binary fission, **motile by (multitrichous) polar flagella**. Cells may be immobile and grow in irregular aggregates surrounded by slime. Gram negative, belong to the *Gammaproteobacteria*, and contain **internal photosynthetic membranes of vesicular type** in which the photosynthetic pigments bacteriochlorophyll *a* and carotenoids are located.

Obligately phototrophic and strictly anaerobic. Photolithoautotrophic growth under anoxic conditions in the light with sulfide and S[0] as electron donors. During oxidation of sulfide, S[0] transiently stored inside the cells in the form of highly refractile globules. Sulfate is the final oxidation product. Storage materials are polysaccharides, poly-β-hydroxybutyrate, S[0], and polyphosphate.

The mol% G + C of the DNA is: 45.5.

Type species: **Thiospirillum jenense** (Ehrenberg 1838) Migula 1900, 1050 (*Ophidomonas jenensis* Ehrenberg 1838, 44.)

FURTHER DESCRIPTIVE INFORMATION

Thiospirillum jenense is the only species of this genus studied in pure culture so far. *T. jenense* is a typical freshwater bacterium. Information obtained from rRNA oligonucleotide patterns (Fowler et al., 1984) indicates that *T. jenense* belongs to the family *Chromatiaceae* and is distantly related to *Chromatium* and *Allochromatium* species. Complete sequences of the 16S rDNA of this bacterium are not yet available.

ENRICHMENT AND ISOLATION PROCEDURES

Conditions for enrichment and cultivation of *Thiospirillum jenense* are essentially the same as those described for *Chromatium* species. The large cells of *T. jenense* (generally 30–40 μm in length) grow only in agar shake cultures, when the final concentration of thoroughly washed agar is below 0.6–0.8% (w/v). In such agar cultures, colonies of *Thiospirillum* appear irregular and wisp-shaped. Individual cells can be recognized by using a low power, binocular microscope. Cultures containing not more than 1–2 mM sulfide should be incubated at low light intensity (200–500 lux). Colonies from agar shake cultures grow well in liquid medium, provided culture vessels of small volume are used first, e.g., screw-capped test tubes of 10–15-ml volume.

MAINTENANCE PROCEDURES

Maintenance of strains in liquid culture medium and long-term preservation in liquid nitrogen are carried out as described for the genus *Chromatium*. *Thiospirillum* species cannot be preserved by lyophilization.

DIFFERENTIATION OF THE GENUS *THIOSPIRILLUM* FROM OTHER GENERA

Characteristics used for the differentiation of *Thiospirillum* species from other phototrophic members of the family *Chromatiaceae* are shown in Table BXII.γ.4 of the chapter describing the family *Chromatiaceae*.

TAXONOMIC COMMENTS

In addition to *Thiospirillum jenense*, two other species of *Thiospirillum*, "*Thiospirillum sanguineum*" and "*Thiospirillum rosenbergii*", have been described from observations of water and mud samples collected in nature. The former species was described as having the same size and morphology as *Thiospirillum jenense*, with which it could have been confused. More recent reports of the existence of "*Thiospirillum rosenbergii*" are lacking. Winogradsky (1888) originally designated *Thiospirillum sanguineum* as the type species of this genus. However, because this species, in contrast to *Thiospirillum jenense*, has never been obtained in pure culture, *Thiospirillum jenense* was later recognized as the type species (Pfennig, 1989d). Further studies of strains obtained in pure culture and 16S rDNA sequence analyses are required to determine the relatedness of *Thiospirillum* species with each other and with other members of the family *Chromatiaceae*. According to 16S rRNA oligonucleotide catalogs, *Thiospirillum jenense* is not similar to *Chromatium* and *Allochromatium* species (Fowler et al., 1984), though it may be distantly related to these bacteria and belong to the corresponding group of motile freshwater *Chromatiaceae* species.

List of species of the genus Thiospirillum

1. **Thiospirillum jenense** (Ehrenberg 1838) Migula 1900, 1050[AL] (*Ophidomonas jenensis* Ehrenberg 1838, 44.)
   je.nen' se. M.L. neut. adj. *jenense* jenense pertaining to Jena, Germany, the city where Ehrenberg discovered this organism.

   Cells curved rods, sigmoid or spiral-shaped, 2.5–4.0 μm wide. Sigmoid cells usually 30–40 μm long, spiral cells up to 100 μm long. Complete turns may measure 15–40 μm and have a coil depth of 3–7 μm. Motile by a polar tuft of flagella, which is usually 10–12 μm long and visible by brightfield or phase-contrast microscopy (Fig. BXII.γ.18). Cells rarely tufted at both ends. Color of individual cells is pale yellow, color of cell suspensions yellowish to orange-brown. Photosynthetic pigments are bacteriochlorophyll *a* and the carotenoids lycopene and rhodopin (Schmidt et al., 1965).

   Obligately phototrophic and strictly anaerobic. Photolithoautotrophic growth with sulfide and S^0 as electron donors. Thiosulfate not utilized. In the presence of sulfide and bicarbonate, acetate is photoassimilated. Nitrogen sources: ammonium salts. Vitamin B_{12} required for growth.

   Mesophilic freshwater bacterium with optimum growth at 20–25°C and pH 7.0 (range 6.5–7.5).

   Habitat: mud and stagnant water of ditches and freshwater ponds containing hydrogen sulfide.

   The mol% G + C of the DNA is: 45.5 (Bd).

   Type strain: 1112, DSM 216.

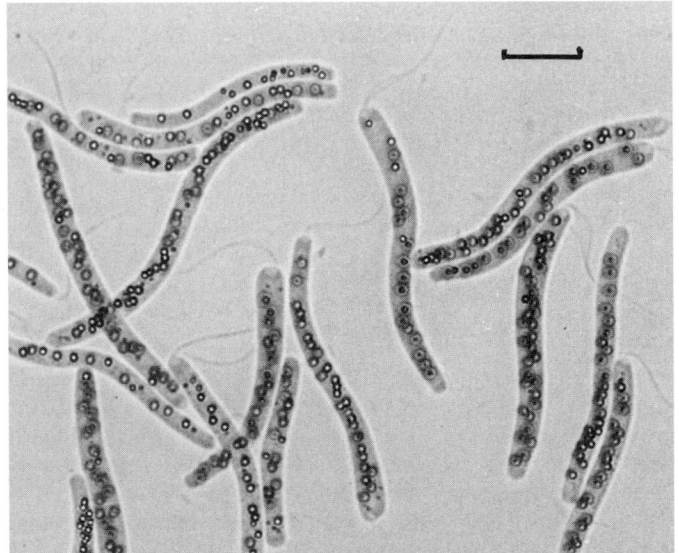

FIGURE BXII.γ.18. *Thiospirillum jenense* DSM 216 cultured photoautotrophically with sulfide. Note the intracellular sulfur globules and the short, slightly sigmoidal tufts of polar flagella. Brightfield micrograph. Bar = 10 μm.

Family II. **Ectothiorhodospiraceae** Imhoff 1984b, 339[VP]

JOHANNES F. IMHOFF

Ec' to.thi' o.rho' do.spi.ra' ce.ae. M.L. fem. n. *Ectothiorhodospira* type genus of the family; *-aceae* ending to denote a family; M.L. fem. pl. n. *Ectothiorhodospiraceae* the *Ectothiorhodospira* family.

This family contains **phototrophic purple sulfur bacteria** that perform **anoxygenic photosynthesis with bacteriochlorophylls and carotenoids as photosynthetic pigments and strictly chemotrophic bacteria.** Cells spiral, vibrioid, or rod-shaped, motile by means of polar flagella, divide by binary fission, with or without gas vesicles. **Gram negative** and **belong to the *Gammaproteobacteria*.** Phototrophic representatives preferably grow anaerobically in the light using reduced sulfur compounds as electron donors, **oxidize sulfide to S^0, which is deposited outside the cells, and eventually in the peripheral periplasmic space of the cell body,** forming sulfate as the final oxidation product. **Phototrophic *Ectothiorhodospiraceae* are found in marine to extremely saline environments containing sulfide and having an alkaline to extremely alkaline pH.** Species of this family are the most halophilic eubacteria.

The mol% G + C of the DNA is: 50.5–69.7.

Type genus: **Ectothiorhodospira** Pelsh 1936, 120.

TAXONOMIC COMMENTS

The taxonomic position of bacteria now classified as *Ectothiorhodospiraceae* has been disputed since their discovery by Pelsh (1936). He distinguished these bacteria, which he called "*Ectothiorhodaceae*", from those phototrophic purple bacteria with elemental sulfur inside their cells, which he called "*Endothiorhodaceae*" (Pelsh, 1937). Pelsh's isolates were poorly characterized and were lost soon after their isolation. Trüper (1968) reisolated *Ectothiorhodospira mobilis*, and Pfennig and Trüper (1971a) included *Ectothiorhodospira* as an exceptional genus into the *Chromatiaceae* because of its ability to perform an oxidative dissimilatory sulfur metabolism similar to that of other phototrophic purple sulfur bacteria. Because bacteria known at that time as *Ectothiorhodospira* species were significantly different from all other genera of the *Chromatiaceae*, and because representatives of the purple nonsulfur bacteria had been found to oxidize sulfide to extracellular elemental sulfur, Pfennig (1977) suggested the removal of *Ectothiorhodospira* from the *Chromatiaceae* and its placement in the purple nonsulfur bacteria. However, on the basis of significant differences in their polar lipid composition, three major groups of phototrophic purple bacteria were distinguished by Imhoff et al. (1982a): *Chromatiaceae*, the bacteria known at that time as *Ectothiorhodospira*, and the purple nonsulfur bacteria (see also Imhoff and Bias-Imhoff, 1995). A clear differentiation of *Ectothiorhodospira* species from *Chromatiaceae* was later demonstrated by the oligonucleotide patterns of 16S rRNA molecules (Stackebrandt et al., 1984). With the support of this information, it was proposed that the *Ectothiorhodospiraceae* be separated as a family distinct from the *Chromatiaceae* (Imhoff, 1984b). The separation of two groups of species within the family had

been recognized not only based on physiological properties and lipid composition (Tindall, 1980; Imhoff et al., 1982a; Imhoff, 1984a), but also by similarities of rRNA oligonucleotide patterns (Stackebrandt et al., 1984). However, a formal separation into genera was not proposed until later. The complete sequence analysis of the 16S rDNA from a large number of strains, including all available type strains of *Ectothiorhodospiraceae*, not only supported their separation from the *Chromatiaceae*, but also the divergence of the two groups within this family (Imhoff and Süling, 1996). Because of the large phylogenetic distance between these two groups, which is also reflected in a number of distinctive phenotypic properties, the extremely halophilic species were transferred to the new genus *Halorhodospira* (Imhoff and Süling, 1996). With this taxonomic reevaluation of the *Ectothiorhodospiraceae*, their taxonomy now follows their phylogenetic relationships, based on 16S rDNA sequences.

Two new isolates have expanded our knowledge of the *Ectothiorhodospiraceae*. A new phototrophic sulfur bacterium isolated from a Siberian soda lake, *Thiorhodospira sibirica*, has been shown to be genetically related to the genus *Ectothiorhodospira*, and many of its properties resemble those of *Ectothiorhodospira* species (Bryantseva et al., 1999b). This impressively large bacterium is unique in that it forms S^0 globules outside the cells as well as inside. Sulfur globules are located in the cells only in the very peripheral part, which is represented by an apparently large periplasmic space; they are not located in the cell interior. Therefore, the mechanism of S^0 formation and the location of the involved enzymes appear to be the same as in other *Ectothiorhodospiraceae*. However, particular care is needed to distinguish between internal and external deposition of S^0.

Another new halophilic and strictly chemotrophic bacterium has been found to be genetically related to the *Ectothiorhodospiraceae*, in particular to *Halorhodospira* (Adkins et al., 1993; Imhoff and Süling, 1996). This bacterium, *Arhodomonas aquaeolei*, is a close, purely chemotrophic relative to the phototrophic *Ectothiorhodospiraceae*. It uses a restricted spectrum of carbon sources for growth. It is considered as a member of the *Ectothiorhodospiraceae* primarily because of its phylogenetic relationship by 16S rDNA sequence.

Differentiation of the genera of *Ectothiorhodospiraceae* Characteristics for differentiation of the genera *Ectothiorhodospira*, *Halorhodospira*, *Thiorhodospira*, and *Arhodomonas* are summarized in Table BXII.γ.7. Characteristic fatty acid compositions of *Halorhodospira* and *Ectothiorhodospira* species are shown in Table BXII.γ.8. The genetic relationships of *Ectothiorhodospiraceae* based on 16S rDNA sequences are presented in Fig. BXII.γ.19.

Key to the genera of the Ectothiorhodospiraceae

I. Halophilic *Gammaproteobacteria* that require >10% total salts for optimum growth.
    A. Cells contain photosynthetic pigments typical of phototrophic purple bacteria, perform a strictly phototrophic metabolism, and require alkaline pH for optimum growth.
        Genus *Halorhodospira*
    B. Cells without photosynthetic pigments and with a strictly chemotrophic metabolism.
        Genus *Arhodomonas*

II. Halophilic phototrophic purple sulfur bacteria that require alkaline pH and salt concentrations below 10% total salts for optimum growth.

A. S⁰ globules deposited exclusively outside the cells.

Genus *Ectothiorhodospira*

B. S⁰ globules deposited both outside and inside the cells.

Genus *Thiorhodospira*

III. Halophilic chemolithotrophic bacteria that require a pH range of 6.8–8.0 and carry out aerobic oxidation of nitrate to nitrite.

Genus *Nitrococcus*

ACKNOWLEDGMENTS

Sequence alignments and the construction of the phylogenetic tree by Dr. J. Süling (IFM Kiel, Germany) are gratefully acknowledged.

FURTHER READING

Adkins, J.P., M.T. Madigan, L. Mandelco, C.R. Woese and R.S. Tanner. 1993. *Arhodomonas aquaeolei* gen. nov., sp. nov., an aerobic, halophilic bacterium isolated from a subterranean brine. Int. J. Syst. Bacteriol. *43*: 514–520.

Bryantseva, I., V.M. Gorlenko, E.I. Kompantseva, J.F. Imhoff, J. Süling and L. Mityushina. 1999. *Thiorhodospira sibirica* gen. nov., sp. nov., a new alkaliphilic purple sulfur bacterium from a Siberian soda lake. Int. J. Syst. Bacteriol. *49*: 697–703.

Imhoff, J.F. 1984. Reassignment of the genus *Ectothiorhodospira* Pelsh 1936 to a new family, *Ectothiorhodospiraceae* fam. nov., and emended description of the *Chromatiaceae* Bavendamm 1924. Int. J. Syst. Bacteriol. *34*: 338–339.

Imhoff, J.F. 1993. Osmotic adaptation in halophilic and halotolerant microorganisms. *In* Vreeland and Hochstein (Editors), The Biology of Halophilic Bacteria, CRC Press, Boca Raton. pp. 211–253.

Imhoff, J.F. and J. Süling. 1996. The phylogenetic relationship among *Ectothiorhodospiraceae*: a reevaluation of their taxonomy on the basis of 16S rDNA analyses. Arch. Microbiol. *165*: 106–113.

Imhoff, J.F. 2001. True marine and halophilic anoxygenic phototrophic bacteria. Arch. Microbiol. *176*: 243–254.

TABLE BXII.γ.7. Differential characteristics of the genera of the *Ectothiorhodospiraceae*

Characteristic	*Ectothiorhodospira*	*Arhodomonas*[a]	*Halorhodospira*	*Thiorhodospira*[a]
Gram stain	negative	negative	negative	negative
Phylogenetic group	*Gammaproteobacteria*	*Gammaproteobacteria*	*Gammaproteobacteria*	*Gammaproteobacteria*
Flagellation	polar tuft of flagella	polar flagellum	bipolar flagella	polar tuft
Internal membranes	lamellar stacks	nd[b]	lamellar stacks	lamellar stacks
Cell form	vibrioid to spirilloid	straight rods	vibrioid to spirilloid	vibrioid to spirilloid
Cell width (µm)	0.7–1.5	0.8–1.0	0.5–1.2	3–4
Growth medium[c]	EM 5% or mMPM	complex medium	EM 15–30%	see genus *Thiorhodospira*
Optimum salinity	1–7%	15%	15–25%	0.5–1.0%
Compatable solutes	glycine betaine, sucrose	nd	glycine betaine, ectoine, trehalose	nd
Fatty acid cluster[d]	V–VII	nd	I–IV	nd.
Major quinone components[e]	MK7 and Q7 or Q8	–	MK8, Q8, and MK4/5	–
Mol% G + C of DNA	61.4–68.4	67	50.5–69.7	56.0–57.4
16S rDNA signature sequences:				
E. coli position 92–96	CTGAC	CTGGC	GCGGC	CTGAC
E. coli position 135–137	TGG	CTT	CTT	TGG
E. coli position 149–151	CAA	TAG	TAG	TAA
E. coli position 221–225	TATCA	CGAAG	CGAAG	TATCA
E. coli position 667–669	AGA	GAG	GTA	AGG
E. coli position 761	T	T	C	T
E. coli position 1010–1011	CA	CA	TG	CA

[a]Data on *Arhodomonas* and *Thiorhodospira* are from Adkins et al. (1993) and Bryantseva et al. (1999b), respectively.

[b]nd, not determined.

[c]EM: standard medium for haloalkaliphilic *Ectothiorhodospiraceae* species according to Imhoff (1988a) with varying salt concentrations; mMPM: modified marine Pfennig's medium according to Trüper (1970).

[d]Fatty acid clusters according to Thiemann and Imhoff (1996).

[e]Major quinone components according to Imhoff (1984a) and Ventura et al. (1993).

TABLE BXII.γ.8. Fatty acid clusters of *Ectothiorhodospira* and *Halorhodospira* species[a,b]

Characteristic	*E. mobilis*	*E. haloalkaliphila*	*E. marina*	*E. marismortui*	*E. shaposhnikovii*	*E. vacuolata*	*H. halophila*	*H. halophila*	*H. abdelmalekii*	*H. halochloris*
Fatty acid cluster	V	VII	VII	V	VI	VI	I	II	III	IV
$C_{12:0}$	0.7–0.8	0.9–1.2	0.9	1.1	1.3–1.7	nd	2.4–3.4	2.8–5.0	2.4	1.1–2.9
$C_{16:0}$	35.1–36.4	29.0	24.8	33.2	24.4–26.4	21.6	9.2–11.1	10.8–14.5	21.4	21.0–24.0
$C_{16:1\,\omega7c}$	2.7–3.5	4.8–5.1	5.9	3.8	4.1–4.6	3.6	0.4–0.6	0.6–0.8	1.4	1.2–1.6
$C_{18:0}$	5.7–7.6	6.8–7.1	9.2	5.0	5.6–6.8	5.5	14.2–17.7	11.3–13.2	6.0	6.3–11.6
$C_{18:1\,\omega7c}$	12.1–16.2	38.2–38.6	29.2	18.5	57.1–59.0	65.3	62.8–65.5	58.6–62.8	62.8	57.7–57.9
$C_{19:0cyc}$	33.5–37.7	15.9–16.3	25.8	34.7	nd	nd	2.5–3.4	3.0–5.8	0.4	1.3–2.0

[a]nd, not determined.

[b]Data from Thiemann and Imhoff (1996).

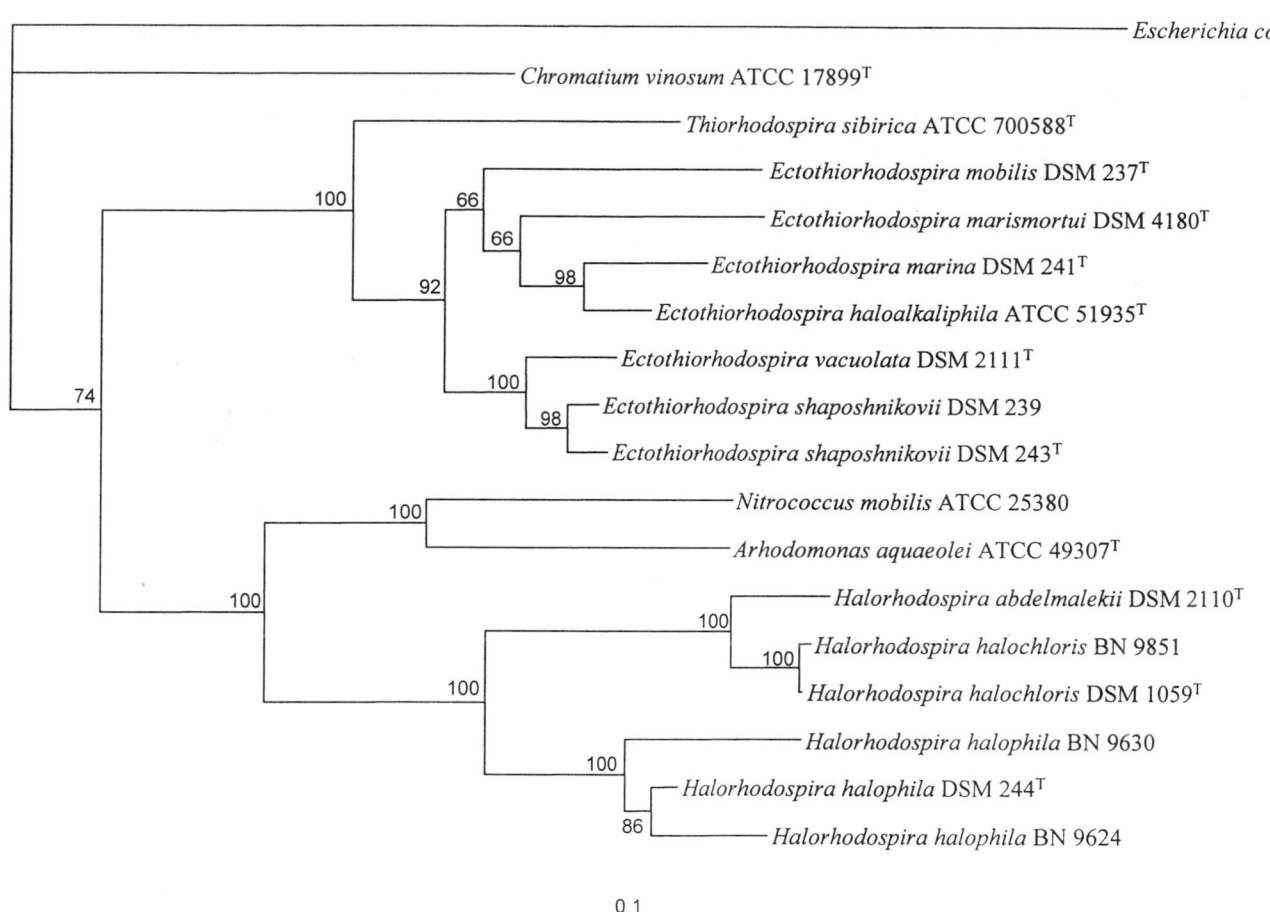

FIGURE BXII.γ.19. 16S rDNA-derived phylogenetic tree of the family *Ectothiorhodospiraceae* including species of *Ectothiorhodospira*, *Halorhodospira*, *Thiorhodospira* and *Arhodomonas*. Sequence data used were published previously (Imhoff and Süling, 1996). In addition, 16S rDNA sequences of *Arhodomonas aquaeolei* (M26631) and *Thiorhodospira sibirica* (AJ006530) were used. Sequence alignments, calculation of distance data and tree construction were performed as described earlier (Petri and Imhoff, 2000).

Stackebrandt, E., V.J. Fowler, W. Schubert and J.F. Imhoff. 1984. Towards a phylogeny of phototrophic purple sulfur bacteria: the genus *Ectothiorhodospira*. Arch. Microbiol. *137*: 366–370.

Thiemann, B. and J.F. Imhoff. 1996. Differentiation of *Ectothiorhodospiraceae* based on their fatty acid composition. Syst. Appl. Microbiol. *19*: 223–230.

Genus I. **Ectothiorhodospira** *Pelsh 1936, 120*[AL]

JOHANNES F. IMHOFF

Ec′ to.thi′ o.rho′ do.spi′ ra. Gr. prep. *ectos* outside; Gr. n. *thios* sulfur; Gr. n. *rhodon* the rose; Gr. n. *spira* the spiral; M.L. fem. n. *Ectothiorhodospira* spiral rose with sulfur outside.

Cells rod-shaped or vibrioid, also appearing as true spirals, 0.7–1.5 µm in diameter, **motile by a polar tuft of flagella**, multiply by binary fission, and may contain gas vesicles. **Gram-negative and belong to the** *Gammaproteobacteria*. **Internal photosynthetic membranes are lamellar stacks** that are continuous with the cytoplasmic membrane. Photosynthetic pigments are bacteriochlorophyll *a* and carotenoids.

Grow photoautotrophically under anoxic conditions with reduced sulfur compounds or hydrogen as electron donors, or photoheterotrophically with a limited number of organic compounds. **Sulfide oxidized to sulfate, with S⁰, which is deposited outside the cells, as an intermediary product.** Some species grow microaerobically to aerobically in the dark. **Sodium chloride is** **required for growth**, which is dependent on saline and alkaline conditions. Compatible solutes include glycine betaine, sucrose, and *N*-carbamoyl-L-glutamine amide. Growth factors are not required, but vitamin B_{12} enhances growth of some strains. Storage products are polysaccharides, poly-β-hydroxybutyrate, and polyphosphate.

Most species live in marine and saline environments that contain sulfide, have slightly to extremely alkaline pH, and are exposed to light, such as estuaries, salt flats, salt lakes, and soda lakes. Occasionally found in soil.

The mol% G + C of the DNA is: 61.4–68.4.

Type species: **Ectothiorhodospira mobilis** Pelsh 1936, 120.

FURTHER DESCRIPTIVE INFORMATION

Cell morphology of the *Ectothiorhodospira* species varies with the growth conditions, in particular with pH and salinity. Cells of *Ectothiorhodospira mobilis* are spirals under optimum growth conditions, but tend to form vibrioid cells or bent rods. *Ectothiorhodospira shaposhnikovii* and *Ectothiorhodospira vacuolata* form straight or slightly bent rods under suitable growth conditions. Carotenoids are of the normal spirilloxanthin series with spirilloxanthin as the predominant component (Schmidt and Trüper, 1971). The *in vivo* absorption spectra of *Ectothiorhodospira* species are characterized by long wavelength absorption bands at 798, 822–827, 851–855, and 892–895 nm. The band at ~850 nm is the absolute maximum; the band at ~890 nm is mostly recognized as a shoulder, sometimes as a maximum of similar strength to that at 850 nm; and the one at ~820–830 nm can be seen only in second derivative spectra (Imhoff, unpublished results).

All species grow well under anoxic conditions in the light with reduced sulfur compounds as photosynthetic electron donors and in the presence of organic carbon sources and inorganic carbonate. *Ectothiorhodospira mobilis*, *Ectothiorhodospira shaposhnikovii*, and *Ectothiorhodospira haloalkaliphila* also grow microaerobically in the dark if sulfide is present. During phototrophic growth with sulfide as electron donor, the oxidation of sulfide and S^0 strictly follow each other, as has been shown for *E. mobilis* (see Trüper, 1978). Under the alkaline growth conditions, which are optimal for *Ectothiorhodospira* species, polysulfides are stable intermediates in sulfide oxidation. As a result, polysulfides are the first measurable oxidation products, and the medium becomes translucent yellow at this stage. After sulfide depletion, S^0 droplets are formed rather rapidly, and the medium becomes opaque and whitish. Upon further growth, cultures become pinkish and finally red, if S^0 disappears. The knowledge of enzymes involved in the oxidation of reduced sulfur compounds by *Ectothiorhodospira* species has been summarized by Trüper and Fischer (1982).

The fixation of carbon dioxide via the Calvin cycle is apparently the major route of carbon assimilation in *Ectothiorhodospira* species under autotrophic growth conditions. High activities of ribulosebisphosphate carboxylase have been found in *E. shaposhnikovii* (Firsov et al., 1974) and *E. mobilis* (Sahl and Trüper, 1977). However, the assimilation of several organic carbon sources, such as acetate and propionate, depends on the presence of carbon dioxide and proceeds via several carboxylation reactions (Firsov and Ivanovskii, 1974, 1975). Under these conditions, a considerable proportion of the cellular carbon is therefore derived from carbon dioxide, which is not assimilated via the ribulose-bisphosphate pathway. Phosphoenolpyruvate carboxylase, ferredoxin-dependent pyruvate synthase and oxoglutarate synthase have been found in *E. shaposhnikovii* (Firsov et al., 1974); phosphoenolpyruvate carboxylase, phosphoenolpyruvate carboxykinase, and pyruvate carboxylase have been found in *E. mobilis* (Sahl and Trüper, 1977). All enzymes of the glycolytic pathway and the tricarboxylic acid cycle, with the exception of oxoglutarate dehydrogenase, are present in *E. shaposhnikovii* (Krasil'nikova, 1975). Cells grown on acetate demonstrate increased activities of isocitrate lyase, which indicates activity of the glyoxylic acid pathway. The major reserve material formed from acetate and butyrate in the absence of carbon dioxide is poly-β-hydroxybutyric acid. In the presence of carbon dioxide and from other organic carbon sources, carbohydrates are formed instead (Novikova, 1971).

Ammonia and glutamine are suitable nitrogen sources for all species. Dinitrogen fixation has been demonstrated for some species and is probably a property inherent to the genus as a whole. Glutamate dehydrogenase (NADH-dependent) and glutamine synthetase/glutamate synthase (NADH-dependent) have been found in *E. mobilis* (Bast, 1977). Nitrate can be used as a nitrogen source by *E. shaposhnikovii*, and its reduction is catalyzed by a ferredoxin-dependent nitrate reductase, which is associated with or bound to the membranes and induced during growth with nitrate (Malofeeva et al., 1975).

In regard to their salt responses and the salt concentrations required for optimum growth, *Ectothiorhodospira* species are slightly halophilic or marine. Based on the salt concentration at which optimum growth occurs, species are grouped as nonhalophilic bacteria (<0.1 M NaCl), brackish water bacteria (~0.1–0.35 M NaCl), marine or slightly halophilic bacteria (~0.35–1.2 M NaCl), moderately halophilic bacteria (~1.2–2.5 M NaCl), and extremely halophilic bacteria (>2.5 M NaCl) (see Imhoff, 1993, 2001d). Bacteria of each of these groups may show up with different degrees of salt tolerance. Even nonhalophilic bacteria may exhibit extreme salt tolerances and grow at very high concentration, though at a suboptimal level. In this respect, *Ectothiorhodospira* species have salt optima at the upper border of marine bacteria, and some, in particular *E. haloalkaliphila*, are moderately halotolerant. Because of its extended salt tolerance, this species may successfully compete with *Halorhodospira* species in moderately saline soda lakes. To cope with the salt and osmotic stress, *Ectothiorhodospira* species (*E. haloalkaliphila*, *E. marismortui*) accumulate glycine betaine, sucrose, and a third component (α-N-carbamoyl-L-glutamine amide) as osmotica (Oren et al., 1991; Severin et al., 1992; Imhoff, 1993; Imhoff and Riedel, unpublished results), but not ectoine and trehalose, which along with glycine betaine have been found as major components in *Halorhodospira* species (Galinski et al., 1985; Severin et al., 1992).

ENRICHMENT AND ISOLATION PROCEDURES

Ectothiorhodospira species can be selectively enriched under photoautotrophic conditions with sulfide as the electron donor in saline and alkaline media (3% NaCl and pH 7.6–8.5), even when high proportions of *Chromatiaceae* species occur in the natural sample. Using Pfennig's medium for purple sulfur bacteria with the addition of appropriate concentrations of NaCl and adjusted to an alkaline pH (Trüper, 1970) and alternatively with a medium based on the mineral composition of the soda lakes of the Wadi Natrun (Jannasch, 1957; Imhoff and Trüper, 1977; Imhoff et al., 1979) or modifications thereof, *Ectothiorhodospira* strains have been isolated from marine sources, hypersaline ponds, salt swamps, and salt lakes from many parts of the world. From such sources, isolation can be achieved in agar dilution series with natural samples without prior enrichment. A recipe of the medium specifically designed for species from alkaline lakes containing 5% (and 20%) salts is given below (see Imhoff, 1988a).[1]

---

1. The medium is prepared as a basal solution with (final content per liter medium): KH_2PO_4, 0.8 g; trace element solution SLA (see below), 1 ml; 1 M sodium carbonate buffer pH 9.0, 200 ml; and NaCl, 30 g. This basal solution is sterilized in an autoclave. Solutions of 20% NH_4Cl (4 ml/l), 1% $CaCl_2 \cdot 2H_2O$ (5ml/l), 2% $MgCl_2 \cdot 6H_2O$ (5 ml/l), and 5% $Na_2S \cdot 9H_2O$ (5–10 ml/l) are sterilized separately and added after cooling to room temperature. Sodium thiosulfate (1.0 g/l), if used as electron donor, and sodium acetate (1–2 g/l) may be added to the basal solution before sterilization. The final salinity is adjusted by changing the concentrations of sodium chloride, keeping under consideration the constant level of carbonate buffer, which is equivalent to ~2% total salts. The amounts yield a medium with 5% total salts. Part of the sodium chloride (10% of the total final salt content by weight) may be replaced by sodium sulfate. In order to adjust the salinity to a total

The recipe of the standard medium according to Pfennig is given in the description of the genus *Chromatium*.

MAINTENANCE PROCEDURES

Cultures can be maintained at the lab bench in dim light for several months. They can be stored in agar deeps kept in the dark at 4–10°C for up to a year. For long-term preservation, storage in liquid nitrogen is recommended. In this case, cell suspensions of well-grown cultures supplemented with DMSO to a final concentration of 5% are dispensed in plastic ampules, sealed, and frozen at the temperature of liquid nitrogen.

DIFFERENTIATION OF THE GENUS *ECTOTHIORHODOSPIRA* FROM OTHER GENERA

Species of the genus *Ectothiorhodospira* are differentiated from other purple sulfur bacteria of the *Chromatiaceae* family by the deposition of elemental sulfur outside their cells, when grown with sulfide as the photosynthetic electron donor. Differentiation of *Ectothiorhodospira* and *Halorhodospira* species is possible based on both molecular and physiological properties (Table BXII.γ.7). *Ectothiorhodospira* species contain MK-7 and either Q-7 or Q-8 as major components, whereas *Halorhodospira* species lack significant proportions of homologs with 7 isoprenoid units, but instead have MK-8, Q-8, and a short chain MK component as major components (Imhoff, 1984a; Ventura et al., 1993). Both genera form separate groups according to their fatty acid compositions (Thiemann and Imhoff, 1996; Table BXII.γ.8). They differ significantly according to their 16S rDNA sequences, as can be seen by a number of signature sequences and sequence similarities of 87.2–89.9% (Table BXII.γ.7). In contrast to *Ectothiorhodospira* species that form sulfur globules outside their cells, *Thiorhodospira* also forms sulfur globules enclosed by the outer membrane (Bryantseva et al., 1999b).

Metabolic properties such as utilization of carbon, sulfur, and nitrogen sources are very similar among the species of both genera. Whereas several species of the genus *Ectothiorhodospira* are able to grow under chemotrophic conditions in the dark, *Halorhodospira* species are obligately phototrophic. *Ectothiorhodospira* species have growth optima under conditions of 1–8% total salts, whereas *Halorhodospira* species do not grow below 10% total salts.

TAXONOMIC COMMENTS

After reevaluation of the taxonomy of *Ectothiorhodospiraceae* on the basis of complete 16S rDNA sequences, the extremely halophilic species were removed from *Ectothiorhodospira* and reassigned to the new genus *Halorhodospira* (Imhoff and Süling, 1996). Analyses of 16S rDNA sequences and fatty acid compositions supported the identification and species assignments of some strains, but demonstrated the misclassification of others (Imhoff and Süling, 1996; Thiemann and Imhoff, 1996). A number of strains have been tentatively assigned to *E. mobilis*. Strains 8112 (DSMZ 237), 8113 (DSMZ 238), and 8815 (DSMZ 240, Trüper, 1970; Mandel et al., 1971) were confirmed as strains of *E. mobilis*. Strain 8115 (DSMZ 239, Trüper, 1970; Mandel et al., 1971) was recognized as belonging to *E. shaposhnikovii*. Strains 51/7 (BN 9903, ATCC 51935) and C (BN 9902, Imhoff et al., 1978) were recognized as belonging to a new species, *E. haloalkaliphila*. Strain BA 1010 (DSMZ 241, Matheron and Baulaigue, 1972) was found to represent the new species *E. marina*, and strain BA 1011 (Matheron and Baulaigue, 1972) was found to be most closely related to strain BA 1010. The recognition of strain DSMZ 239 as belonging to *E. shaposhnikovii*, the close relationship between *E. shaposhnikovii* and *E. vacuolata*, and that between *E. mobilis* and *E. marismortui* are supported by the DNA–DNA hybridization studies of Ivanova et al. (1985) and DNA restriction pattern analysis by Ventura et al. (1993). These results were substantiated by DNA–DNA reassociation studies by Ventura et al. (2000); these authors proposed the transfer of *E. vacuolata* to *E. shaposhnikovii* and *E. marismortui* to *E. mobilis*. Given the present state of knowledge, we do not follow this proposal here because of clearly distinctive phenotypic properties and ribotyping patterns. In addition, strain BN 9903 (ATCC 51935) has been intensively studied and, though it is known from the literature as *E. mobilis* (Stackebrandt et al., 1984; Imhoff et al., 1982a; Imhoff, 1984a; Imhoff and Riedel, 1989; Imhoff and Thiemann, 1991; Imhoff et al., 1991; Zahr et al., 1992; Ventura et al., 1993), recent studies have led to the definition of this strain as the type strain of *E. haloalkaliphila* (Imhoff and Süling, 1996).

DIFFERENTIATION OF THE SPECIES OF THE GENUS *ECTOTHIORHODOSPIRA*

Characteristic differential properties of *Ectothiorhodospira* species are summarized in Table BXII.γ.9. The genetic relationship of *Ectothiorhodospira* species based on 16S rDNA sequences is shown in Fig. BXII.γ.19 of the chapter describing the family *Ectothiorhodospiraceae*.

List of species of the genus Ectothiorhodospira

1. **Ectothiorhodospira mobilis** Pelsh 1936, 120[AL] emend. Ventura, Viti, Pastorelli and Giovannetti 2000, 589.

    mo'bi.lis. L. adj. *mobilis* mobile.

    Cells vibrioid, curved in a short spiral or slightly bent rods, 0.7–1.0 × 2.0–2.6 μm (Fig. BXII.γ.20). Motile by

means of a polar tuft of flagella. Color of cell suspensions free of polysulfides and S⁰ is red. Absorption spectra of living cells show maxima at 378, 488, 516, 550, 590, 798, and 854 nm, with a shoulder at 892 nm. Photosynthetic pigments are bacteriochlorophyll *a* (esterified with phytol) and carotenoids of the spirilloxanthin series with spirilloxanthin as the major component.

Obligately phototrophic and strictly anaerobic. Grow under anoxic conditions in the light with reduced sulfur compounds or organic carbon sources as electron donors. Photoautotrophic growth occurs with sulfide, thiosulfate, S⁰, sulfite and hydrogen. Acetate, pyruvate, malate, succinate, and fumarate used as organic carbon sources and electron donors. Some strains also use fructose, glucose, lactate, butyrate, and propionate. Nitrogen sources are ammonia, glu-

---

of 20% salts, either 180 g NaCl or 20 g Na₂SO₄ and 160 g NaCl are added to a final volume of 1 liter medium. The medium is dispensed into tightly closed screw-capped bottles that are completely filled and stored until use. Prior to inoculation, it is necessary to preincubate the bottles, after removal of an appropriate small volume, at the incubation temperature in order to achieve volume expansion.

The trace element solution SLA has the following composition: FeCl₂·4H₂O, 1.8 g; CoCl₂·6H₂O, 250 mg; NiCl₂·6H₂O, 10 mg; CuCl₂·5H₂O, 10 mg; MnCl₂·4H₂O, 70 mg; ZnCl₂, 100 mg; H₃BO₃, 500 mg; Na₂MoO₄·2H₂O, 30 mg; and Na₂SeO₃·5H₂O, 10 mg. These components are dissolved in 1 liter of doubly distilled water. The pH of the solution is adjusted with HCl to 2–3.

TABLE BXII.γ.9. Differential characteristics of *Ectothiorhodospira* species[a]

Characteristics	*Ectothiorhodospira mobilis*	*Ectothiorhodospira haloalkaliphila*	*Ectothiorhodospira marina*	*Ectothiorhodospira marismortui*	*Ectothiorhodospira shaposhnikovii*	*Ectothiorhodospira vacuolata*
Cell diameter (μm)	0.7–1.0	0.7–1.2	0.8–1.2	0.9–1.3	0.8–0.9	1.5
Motility	+	+	+	+	+	+
Color of cell suspensions	red	red	red	red	red	red
NaCl optimum (%)	2–3	5	2–6	3-8	3	1–6
Salinity range (%)	1–5	2.5–15	0.5–10.0	1–20	0-7	0.5–10.0
pH optimum	7.6–8.0	8.5–10.0	7.5–8.5	7.0–8.0	8.0–8.5	7.5–9.5
Growth medium	mPF	mPF/EM5%	mPF	mPF	mPF/EM5%	mPF/EM5%
Chemolithotrophic growth	−	+	U	−	+	U
Sulfate assimilation	+	+	(+)	−	+	−
Fatty acid cluster	V	VII	VII	V	VI	VI
Major quinones	MK7/Q8	MK7/Q8	MK7/Q8	MK7/Q8	MK7/Q7	MK7/Q7
Mol% G + C of DNA of the species	67.3–68.4	62.2–63.5	62.8	65.0	62.0–64.0	61.4–63.6
Mol% G + C of DNA of the type strain	67.3 (Bd)	63.5 (T_m)	62.8 (Bd)	65.0 (T_m)	62.0 (T_m)	63.6 (T_m)
Substrates used:[b]						
Hydrogen	+	U	U	+	+	+
Sulfide	+	+	+	+	+	+
Thiosulfate	+	+	+	−	+	+
Sulfur	+	+	+	+	+	+
Sulfite	+	U	+	U	+	U
Acetate	+	+	+	+	+	+
Pyruvate	+	+	+	+	+	+
Propionate	±	U	+	(+)	+	+
Butyrate	±	U	−	−	+	−
Lactate	±	U	+	(+)	+	−
Fumarate	+	+	+	+	+	+
Succinate	+	+	+	+	+	+
Malate	+	+	+	+	+	+
Fructose	±	−	−	−	+	(+)
Glucose	±	−	−	−	U	−
Ethanol	−	−	−	−	−	−
Propanol	−	−	−	−	−	−

[a]Symbols: +, characteristic positive or substrate utilized by most strains; ±, characteristic positive or substrate utilized by some strains; −, characteristic negative or substrate not utilized by most strains; U, unknown—no data available; (+) weak activity or growth.

[b]None of the species uses formate, methanol, ethanol, propanol, citrate, or benzoate.

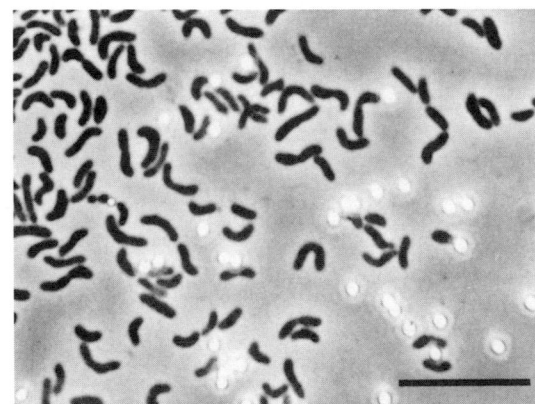

FIGURE BXII.γ.20. *Ectothiorhodospira mobilis* DSMZ 237. Note extracellular sulfur globules. Phase-contrast micrograph. Bar = 10 μm. (Reproduced with permission from H.G. Trüper and J.F. Imhoff. 1977. *In* Starr, Stolp, Trüper, Balows and Schlegel (Editors), The Prokaryotes. A Handbook on Habitats, Isolation, and Identification of Bacteria. Springer-Verlag, Berlin, p. 277.)

tamine, and some other amino acids. Sulfate is assimilated as sole sulfur source. Growth of some strains is enhanced by vitamin B_{12}.

Sodium chloride required for growth. Optimal devel-opment is at salt concentrations of 2–3% (salinity range 1–7%), pH 7.6–8.0, and 25–30°C.

Habitat: Salt marshes, marine coastal sediments containing sulfide and exposed to the light.

Major quinone components are MK-7 and Q-8.

The mol% G + C of the DNA is: 67.3–68.4 (Bd); type strain 67.3 (Bd).

Type strain: DSMZ 237.

GenBank accession number (16S rRNA): BN 9911, X93481.

2. **Ectothiorhodospira haloalkaliphila** Imhoff and Süling 1997,915[VP] (Effective publication: Imhoff and Süling 1996, 112.)

ha'lo.al.ka.li'phi.la. Gr. n. *hals* salt; Arabic n. *al kali* potash; Gr. adj. *philos* loving; M.L. fem. adj. *haloalkaliphila* loving salt and alkaline conditions.

Cells are vibrios or curved in a short spiral, 0.7–1.2 × 2.0–3.0 μm. Motile by means of a polar tuft of flagella. Color of cell suspensions free of polysulfides and S^0 is red. Absorption spectra of living cells show maxima at 379, 488, 513, 552, 593, 798, and 851 nm, with a shoulder at 892 nm. Photosynthetic pigments are bacteriochlorophyll *a* (esterified with phytol) and carotenoids of the spirilloxanthin series with spirilloxanthin as the major component.

Grow preferably anaerobically in the light with reduced sulfur compounds or organic carbon sources as electron donors. Photoautotrophic growth occurs with sulfide, thiosulfate, and S^0. Chemoautotrophic and chemohetero-trophic growth is possible under microoxic conditions in the presence of sulfide. Organic carbon sources and electron donors used are acetate, pyruvate, malate, succinate, and fumarate. Nitrogen sources are ammonia and some amino acids. Sulfate is assimilated as the sole sulfur source.

Sodium chloride required for growth. Optimal development is at salt concentrations of 5% (range 2.5–15%), pH 8.5–10, and 26–40°C. Habitat: alkaline salt and soda lakes containing sulfide and exposed to the light.

Major quinone components are MK-7 and Q-8.

The mol% G + C of the DNA is: 63.5 (T_m).

Type strain: ATCC 51935, BN 9903.

GenBank accession number (16S rRNA): X93479.

3. **Ectothiorhodospira marina** Imhoff and Süling 1997, 915[VP] (Effective publication: Imhoff and Süling 1996, 112.)

ma.ri′ na. L. fem. adj. *marina* marine.

Cells curved, sometimes slightly bent rods, 0.8–1.2 × 1.5–4.0 µm. Motile by means of a polar tuft of flagella. Color of cell suspensions that are free of polysulfides and S^0 is red. Absorption spectra of living cells show characteristic absorption bands at 379, 483, 513, 552, 593, 798, and 853 nm, with a shoulder at 893 nm. Photosynthetic pigments are bacteriochlorophyll *a* (esterified with phytol) and carotenoids of the spirilloxanthin series with spirilloxanthin as the major component.

Grow preferably anaerobically in the light under photoautotrophic and photoheterotrophic conditions. Reduced sulfur compounds or organic carbon sources serve as electron donors. Photoautotrophic growth is possible with sulfide, thiosulfate, S^0, and sulfite. Acetate, pyruvate, lactate, propionate, malate, succinate, and fumarate used as organic carbon sources and electron donors. Oxoglutarate and peptone also support growth. Nitrogen sources are ammonia, dinitrogen, and glutamine. Sulfate is assimilated as sole sulfur source.

Sodium chloride required for growth. Optimal development is at salt concentrations of 2–6% (salinity range 0.5–10%), pH 7.5–8.5, and 30–40°C. Habitat: Coastal marine sediments containing sulfide and exposed to the light.

Major quinone components are MK-7 and Q-8.

The mol% G + C of the DNA is: 62.8 (Bd).

Type strain: BA 1010, BN 9914, DSMZ 241.

GenBank accession number (16S rRNA): X93476.

4. **Ectothiorhodospira marismortui*** Oren, Kessel and Stackebrandt 1990, 105[VP] (Effective publication: Oren, Kessel and Stackebrandt 1989, 529.)

ma′ ris.mor′ tu.i. L. gen. n. *maris* of the sea; L. adj. *mortuus* dead; M.L. gen. n. *marismortui* of the Dead Sea.

Pleomorphic irregular short rods, often curved, 0.9–1.3 × 1.5–3.3 µm. Motile by means of a polar tuft of flagella. Color of liquid cultures is red to purple. *In vivo* absorption spectra show distinct maxima at 486, 515, 550, 591, 796, and 856 nm, with a shoulder at 885 nm. Photosynthetic pigments are bacteriochlorophyll *a* and carotenoids of the spirilloxanthin series with spirilloxanthin as the major component.

Strictly phototrophic and obligately anaerobic. Grow photoautotrophically with sulfide and S^0 as electron donors, or photoheterotrophically with simple organic substrates. Carbon sources used are acetate, succinate, fumarate, malate, and pyruvate; poor substrates are lactate, glycerol, propionate, and yeast extract. Not utilized: methanol, ethanol, *n*-propanol, isopropanol, *n*-butanol, hydrogen, formate, butyrate, isobutyrate, citrate, benzoate, glycine, L-alanine, L-glutamate, glucose, fructose, mannitol, and sorbitol. Ammonia and dinitrogen can serve as nitrogen sources. Assimilatory sulfate reduction absent. No growth factors required, but small concentrations of yeast extract stimulate growth.

Sodium chloride required for growth. Optimum development is at salt concentrations of 3–8% (salinity range 1–20%), pH 7–8, and 35–45°C.

Habitat: Microbial mats of hypersaline sulfur springs.

Major quinone components are MK-7 and Q-8.

The mol% G + C of the DNA is: 65.0 (T_m).

Type strain: EG-1, BN 9410, DSMZ 4180.

GenBank accession number (16S rRNA): X93482.

5. **Ectothiorhodospira shaposhnikovii** Cherni, Solovieva, Fedorova and Kondratieva 1969, 483[AL] emend. Ventura, Viti, Pastorelli and Giovannetti 2000, 589.

sha′ posh′ ni.kov′ i.i. M.L. gen. n. *shaposhnikovii* of Shaposhnikov, named for D. I. Shaposhnikov, a Russian microbiologist.

Cells rod-shaped, usually slightly bent vibrios or short spirilla. 0.8–0.9 × 1.5–2.5 µm when grown on propionate as carbon source. Motile by means of a polar tuft of flagella. Color of cell suspensions in the absence of polysulfides and S^0 is red. Absorption spectra of living cells show maxima at 378, 481, 513, 556, 593, 798, and 854 nm, with a shoulder at 892 nm. Photosynthetic pigments are bacteriochlorophyll *a* (esterified with phytol) and carotenoids of the spirilloxanthin series with spirilloxanthin as the major component.

Cells preferably grow under anoxic conditions in the light. Photoautotrophic growth occurs with reduced sulfur compounds or molecular hydrogen as the electron donor. Chemoautotrophic and chemoheterotrophic growth possible under microoxic conditions in the dark. Photosynthetic electron donors are sulfide, S^0, thiosulfate, sulfite, molecular hydrogen, acetate, propionate, butyrate, lactate, pyruvate, malate, succinate, fumarate, and fructose. Not used: formate, methanol, ethanol, propanol, glycerol, citrate, and benzoate. Ammonia, dinitrogen, nitrate, and some amino acids used as nitrogen sources. Sulfate is assimilated as the sole sulfur source under photoheterotrophic conditions. Growth factors are not required.

Optimal development is at salt concentrations of 3% (salinity range 0–7%), pH 8.0-9.0, and 30–35°C.

Habitat: Marine sediments containing sulfide and exposed to the light.

Major quinone components are MK-7 and Q-7.

The mol% G + C of the DNA is: 62.3 (Bd), 64.0 (chem. anal.), and 62.0 (T_m).

Type strain: BN 9711, DSMZ 243, KMMGU N1.

GenBank accession number (16S rRNA): M59151.

6. **Ectothiorhodospira vacuolata*** Imhoff, Tindall, Grant and Trüper 1982b, 266[VP] (Effective publication: Imhoff, Tindall, Grant and Trüper 1981, 240.)

va.cu.o.la′ ta. L. fem. adj. *vacuolata* containing vacuoles.

Editorial Note: Ventura et al. (2000) consider *E. marismortui* a junior heterotypic synonym of *E. mobilis*.

Cells rod-shaped, sometimes slightly bent, $1.5 \times 2–4\ \mu m$. Motile by means of a polar tuft of flagella. Motile forms often develop gas vesicles and later become immotile and float to the top because of the gas vesicles. At low sulfide concentrations and low light intensities, motile cells without gas vesicles predominate. Stationary-phase cells generally become immotile and form gas vesicles. Color of cell suspensions is pink to red. Absorption spectra of living cells show maxima at 378, 486, 513, 555, 593, 798, and 855 nm, with a shoulder at 893 nm. Photosynthetic pigments are bacteriochlorophyll *a* (esterified with phytol) and carotenoids of the spirilloxanthin series with spirilloxanthin as the major component.

Grow under anoxic conditions in the light. Photoautotrophic growth occurs with sulfide, thiosulfate, or S^0. Organic carbon sources and electron donors used are acetate, propionate, pyruvate, malate, succinate, and fumarate. Nitrogen sources are ammonia and dinitrogen. Cysteine, but not sulfate or methionine, serves as the sulfur source under photoheterotrophic growth conditions. Growth factors are not required.

Sodium chloride required for growth. Optimal development is at salt concentrations of 1–6% (salinity range 0.5–10%), pH 7.5–9.5, and 30–40°C.

Habitat: Salt flats and salt lakes containing sulfide and exposed to the light.

Major quinone components are MK-7 and Q-7.

The mol% G + C of the DNA is: 61.4–63.6 (T_m), type strain 63.6 (T_m).

Type strain: ATCC 43036, BN 9512, DSMZ 2111.

GenBank accession number (16S rRNA): X93478.

Genus II. *Arhodomonas* Adkins, Madigan, Mandelco, Woese and Tanner 1993, 518[VP]

RALPH S. TANNER AND JOHANNES F. IMHOFF

A.rho.do.mo'nas. Gr. pref. *a* not; Gr. n. *rhodon* the rose, also the color or odor of the rose; Gr. n. *monas* lonely, alone; M.L. n. *Arhodomonas* the *monas* that is not rose colored.

Cells are **rods that occur singly or in pairs, are motile by means of flagella**, and multiply by binary fission. Gram negative. Members of the *Ectothiorhodospiraceae* of the *Gammaproteobacteria*. **Gas vesicles and endospores are not formed.**

Growth occurs under chemoorganotrophic conditions with oxygen or nitrate as the electron acceptor. Phototrophic growth is not possible and photosynthetic pigments are not produced. Catalase and oxidase positive.

Mesophilic and halophilic bacteria that require sodium for growth.

The mol% G + C of the DNA is: 67 (T_m).

Type species: **Arhodomonas aquaeolei** Adkins, Madigan, Mandelco, Woese and Tanner 1993, 518.

FURTHER DESCRIPTIVE INFORMATION

Arhodomonas aquaeolei is one of the few purely chemotrophic bacteria known that is phylogenetically related to the phototrophic purple sulfur bacteria of the *Gammaproteobacteria* and that belongs to the phylogenetic rather coherent group of the *Ectothiorhodospiraceae*. According to its 16S rDNA sequence, it is genetically most similar to *Halorhodospira* species but is sufficiently different so as to be recognized as a separate genus (Fig. BXII.γ.19of the chapter describing the family *Ectothiorhodospiraceae* Imhoff and Süling, 1996). The characteristic loop at position 420 (UGCG) and the addition of C in the loop covering position 1361 are also present in the sequence of *Arhodomonas aquaeolei* (Adkins et al., 1993). *Arhodomonas* shares both the requirement for high salt concentrations and the restricted spectrum of carbon sources used with *Halorhodospira* and *Ectothiorhodospira* species (Imhoff, 1989). Most carbohydrates do not support growth, but other simple organic compounds are used instead. *Arhodomonas* does not have the ability to perform a phototrophic type of metabolism.

ENRICHMENT AND ISOLATION PROCEDURES

Arhodomonas aquaeolei was found in a subterranean brine produced from a petroleum reservoir, a habitat that has been without sunlight for very long time periods. It was isolated from this brine by direct plating on plate broth count agar (Difco) containing 15% NaCl and subsequent incubation under aerobic conditions at 23°C. Current information is based on a single isolate.

MAINTENANCE PROCEDURES

Cultures can be preserved by conventional freeze-drying.

DIFFERENTIATION OF THE GENUS *ARHODOMONAS* FROM OTHER GENERA

The 16S rDNA sequence similarity shows that *A. aquaeolei* is related to *Ectothiorhodospiraceae* and is most similar to *Halorhodospira* (Fig. BXII.γ.19 of the chapter describing the family *Ectothiorhodospiraceae*; Imhoff and Süling, 1996). It is clearly distinguished from these phototrophic bacteria by its strictly chemotrophic metabolism, its inability to grow phototrophically, and its lack of photosynthetic pigments. It shares the requirement for high salt concentrations with *Halorhodospira* species and in addition has a restricted spectrum of carbon sources, similar to those of its phototrophic relatives (Imhoff, 1989, 1992). Some properties of *A. aquaeolei* that differentiate it from other genera of the *Ectothiorhodospiraceae* are shown in Table BXII.γ.7 of the chapter describing the family *Ectothiorhodospiraceae*.

FURTHER READING

Adkins, J.P., M.T. Madigan, L. Mandelco, C.R. Woese and R.S. Tanner. 1993. *Arhodomonas aquaeolei* gen. nov., sp. nov., an aerobic, halophilic bacterium isolated from a subterranean brine. Int. J. Syst. Bacteriol. *43*: 514–520.

Imhoff, J.F. and J. Süling. 1996. The phylogenetic relationship among *Ectothiorhodospiraceae*: a reevaluation of their taxonomy on the basis of 16S rDNA analyses. Arch. Microbiol. *165*: 106–113.

Editorial Note: Ventura et al. (2000) consider *E. vacuolata* a junior heterotypic synonym of *E. shaposhnikovii*.

1. **Arhodomonas aquaeolei** Adkins, Madigan, Mandelco, Woese and Tanner 1993, 518^VP

a.quae.o' le.i. L. gen. n. *aquae* of the water; L. gen. n. *olei* of the oil; M.L. gen. n. *aquaeolei* from water of oil.

Cells are straight, short rods, 0.8–1.0 × 2.0–2.5 μm and motile by means of a single polar flagellum. Metabolism is strictly chemotrophic. Simple organic compounds such as acetate, propionate, lactate, butyrate, valerate, isovalerate, ethanol, glycerol, glutamic acid, glutamine, and xylose are used as substrates for aerobic chemotrophic growth. Glucose, fructose, sucrose, mannose, maltose, lactose, citrate, alanine, aspartic acid, asparagine, histidine, arginine, ser-ine, and tryptophan are not used. Nitrate, but not nitrite, is reduced. Oxidase, catalase, and urease are present, but arginine, lysine, and ornithine decarboxylases are absent. Hydrolysis of Tween 80 is positive, while casein, DNA, esculin, gelatin, and starch are not hydrolysed. Cells are sensitive to chloramphenicol, penicillin, and tetracycline.

Mesophilic and extremely halophilic bacterium with optimum growth at 15% NaCl (range: 6–20%).

Habitat: oil field brine.

The mol% G + C of the DNA is: 67 (T_m).

Type strain: HA-1, ATCC 49307.

GenBank accession number (16S rRNA): M26631.

Genus III. **Halorhodospira** Imhoff and Süling 1997, 915^VP (Effective publication: Imhoff and Süling 1996, 112.)

JOHANNES F. IMHOFF

Ha' lo.rho' do.spi' ra. Gr. gen. n. *halos* of the salt; Gr. n. *rhodon* the rose; Gr. n. *spira* the spiral; M.L. fem. n. *Halorhodospira* the spiral rose from salt lakes.

Cells spirals or rods, 0.5–1.2 μm in diameter, motile by means of bipolar flagella, multiply by binary fission. **Gram negative and belong to the** *Gammaproteobacteria.* **Internal photosynthetic membranes appear as lamellar stacks** that are continuous with the cytoplasmic membrane. **Photosynthetic pigments are bacteriochlorophyll *a* or *b* and carotenoids.**

Grow **photoautotrophically** under anoxic conditions with reduced sulfur compounds as electron donors, or photoheterotrophically with a limited number of simple organic compounds. **Sulfide oxidized to S⁰, which is deposited outside the cells** and may be further oxidized to sulfate.

Growth dependent on highly saline and alkaline conditions. Greater than 10% (w/v) total salt concentration required by all known species, some of which grow in saturated salt solutions. **Glycine betaine, ectoine, and trehalose accumulated as compatible solutes in response to salt and osmotic stress.** Growth factors not required. Storage products are polysaccharides, poly-β-hydroxybutyrate, and polyphosphate.

Halorhodospira **species are found in hypersaline and extremely saline environments** with slightly to extremely alkaline pH (up to pH 11–12) that contain sulfide and are exposed to light, such as salt flats, salt lakes, and soda lakes.

The mol% G + C of the DNA is: 50.5–69.7.

Type species: **Halorhodospira halophila** (Raymond and Sistrom 1969) Imhoff and Süling 1997, 915 (Effective publication: Imhoff and Süling 1996, 110) (*Ectothiorhodospira halophila* Raymond and Sistrom 1969, 125.)

FURTHER DESCRIPTIVE INFORMATION

Cells of *Halorhodospira* species are spirals, but under severe osmotic stress, i.e., at low salt concentrations, they tend to become vibrioid to rod-shaped and increase in volume. The photosynthetic membranes of *Halorhodospira halochloris* have been intensively studied. In negatively stained preparations under the electron microscope, a regularly granulated structure is observed (Imhoff and Trüper, 1977). More detailed work, including image analysis, has demonstrated that the photosynthetic complexes are arranged in a regular hexagonal array in *Halorhodospira halochloris* and *Halorhodospira abdelmalekii* (as well as in other bacteria containing bacteriochlorophyll *b*) (Engelhardt et al., 1983). Bacteriochlorophyll *b* of *H. halochloris* and *H. abdelmalekii* is esterified with Δ-2,10-phytadienol, not with phytol as is bacteriochlorophyll *a* of *H. halophila* (Steiner et al., 1981; R. Steiner, personal communication). Carotenoids of the normal spirilloxanthin series are present in *H. halophila* with spirilloxanthin as predominant component and negligible amounts of rhodopin (Schmidt and Trüper, 1971). The content of carotenoids in *H. halochloris* and *H. abdelmalekii* is low. Major components were methoxy-hydroxylycopene glucoside (methoxy-rhodopin glucoside), hydroxylycopene glucoside (rhodopin glucoside), and dihydroxylycopene diglucoside (Takaichi et al., 2001). Substantial amounts of dihydroxylycopene diglucoside diester were also found.

Halorhodospira species are among the most halophilic eubacteria known (Imhoff, 1988a, b). With regard to their salt responses and the salt concentrations required for optimum growth, *Halorhodospira* species are extremely halophilic (see Genus *Ectothiorhodospira* in this volume and Imhoff, 1993). In alkaline and highly hypersaline salt and soda lakes, they commonly develop massive blooms that cause red or green coloration of the water or sediment horizons (Jannasch, 1957; Imhoff et al., 1979). To adapt to high concentrations of salts and to cope with the high external osmotic pressure, bacteria and other unicellular microorganisms have to accumulate osmotically active molecules that are compatible with the molecular cell structures and metabolic processes; these are called compatible solutes or osmotica (see Imhoff, 1986, 1993). In contrast to halophilic algae and archaebacteria that selectively accumulate glycerol and potassium ions, respectively, *Halorhodospira* species accumulate glycine betaine, ectoine, and trehalose. In *H. halochloris*, glycine betaine was initially reported as the main osmotic active cytoplasmic component (Galinski and Trüper, 1982). More recently, a novel component that is active as an osmoprotectant and accumulates in high concentrations in cells of *H. halochloris* and other *Halorhodospira* species has been identified as 1,4,5,6-tetrahydro-2-methyl-4-pyrimidinecarboxylic acid (Galinski et al., 1985). The trivial name ectoine has been given to this component because of its first discovery in an *Ectothiorhodospiraceae* (at that time *Ectothiorhodospira*) species. This compound was later found to be widely distributed among the halophilic eubacteria (Severin et al., 1992). Unlike the *Halorhodospira* species, ectoine and trehalose have not been found as osmotica in *Ectothiorhodospira* species (*E. haloalkaliphila* and *E. marismortui*) that characteristically accumulate glycine betaine, sucrose, and an unidentified component (Oren et al., 1991; Imhoff and Riedel, unpublished. results).

ENRICHMENT AND ISOLATION PROCEDURES

Halorhodospira species have been isolated from various hypersaline environments, such as salt and soda lakes, from many parts of the world. Alkaline soda lakes show a natural abundance of *Ectothiorhodospira* species, which can be taken as proof of their successful adaptation to these environments. From such sources, isolation can be achieved by agar dilution series using natural samples without prior enrichment. The tolerance of and dependence on high salinity and alkalinity are strong selective conditions for the enrichment of *Halorhodospira* species.

A medium based on the mineral composition of the soda lakes of the Wadi Natrun (Jannasch, 1957; Imhoff and Trüper, 1977; Imhoff et al., 1979) is appropriate for the cultivation of all currently known *Halorhodospira* species. A recipe for this medium containing 20% total salts is given with the description of the genus *Ectothiorhodospira* (see also Imhoff, 1988a).

MAINTENANCE PROCEDURES

Most cultures can be maintained at the lab bench in dim light for several months; they can be stored for up to a year in agar deeps kept in the dark. Species that have bacteriochlorophyll *b* are more sensitive to prolonged storage, particularly with respect to oxygen and light. These should be transferred more frequently than *H. halophila*. For long-term preservation, storage in liquid nitrogen is recommended as the standard maintenance procedure.

DIFFERENTIATION OF THE GENUS *HALORHODOSPIRA* FROM OTHER GENERA

Halorhodospira species represent the most halophilic species known in the phototrophic purple sulfur bacteria, and they are well adapted to form massive blooms in hypersaline environments and in alkaline soda lakes. *Halorhodospira* species are extremely halophilic bacteria and do not grow at total salt concentrations below 10%, whereas *Ectothiorhodospira* species have growth optima at salt concentrations below 10%.

Species of the genus *Halorhodospira* are easily differentiated from other purple sulfur bacteria of the *Chromatiaceae* by the deposition of S^0 outside their cells, when grown with sulfide as the photosynthetic electron donor. Differentiation of *Ectothiorhodospira* and *Halorhodospira* species is possible based on both molecular and physiological properties (Table BXII.γ.7 of the chapter describing the family *Ectothiorhodospiraceae*). Absorption spectra of *H. halophila* have a clear maximum at approximately 890 nm, but do not show absorbance at 830 nm. The long wavelength absorption maximum of the species having bacteriochlorophyll *b* is between 1010 and 1020 nm. Whereas several species of the genus *Ectothiorhodospira* are able to grow under chemotrophic conditions in the dark, all *Halorhodospira* species are obligately phototrophic. In contrast to *Ectothiorhodospira* species, *Halorhodospira* species do not contain significant proportions of quinone homologs with 7 isoprenoid units, but instead have MK-8, Q-8, and a short chain MK component as major components (Imhoff, 1984a; Ventura et al., 1993). Both genera also form separate groups according to their fatty acid compositions (Thiemann and Imhoff, 1996; Table BXII.γ.8 of the chapter describing the family *Ectothiorhodospiraceae*). Their 16S rDNA sequences are significantly different; this is evident from a number of signature sequences as well as sequence similarities of 87.2–89.9% (Table BXII.γ.7 of the chapter describing the family *Ectothiorhodospiraceae*).

TAXONOMIC COMMENTS

After reevaluation of the taxonomy of *Ectothiorhodospiraceae* based on complete 16S rDNA sequence comparison, the extremely halophilic species were removed from *Ectothiorhodospira* and reassigned to the new genus *Halorhodospira* (Imhoff and Süling, 1996).

Two different but closely related groups of *H. halophila* strains (12 strains included in the study) have been recognized based on fatty acid composition (Thiemann and Imhoff, 1996), 16S rDNA sequences (Imhoff and Süling, 1996), and DNA restriction pattern analysis (Ventura et al., 1993). This diversity is considered to be intraspecies variation, most likely related to different salt responses. The mol% G + C content of the DNA of *H. halochloris* is surprisingly low when compared to that of the two other species of this genus. Such a difference of nearly 20% within the same genus is unusual, particularly when given the rather similar sequences of the 16S rRNA gene. However, the mol% G + C values in Table BXII.γ.9 of the chapter describing the family *Ectothiorhodospiraceae* have been confirmed by two independent analyses of a total of three strains (personal communication of H. Hippe, Braunschweig).

DIFFERENTIATION OF THE SPECIES OF THE GENUS *HALORHODOSPIRA*

Characteristic properties for differentiation of *Halorhodospira* species are summarized in Table BXII.γ.10 of the chapter describing the family *Ectothiorhodospiraceae*. The genetic relationship of *Ha-* *lorhodospira* species based on 16S rDNA sequences is shown in Fig. BXII.γ.19 of the chapter describing the family *Ectothiorhodospiraceae*.

List of species of the genus Halorhodospira

1. **Halorhodospira halophila** (Raymond and Sistrom 1969) Imhoff and Süling 1997, 915[VP] (Effective publication: Imhoff and Süling 1996, 110) (*Ectothiorhodospira halophila* Raymond and Sistrom 1969, 125.)

   ha.lo'phi.la. Gr. n. *hals* salt; Gr. adj. *philos* loving; M.L. fem. adj. *halophila* salt-loving.

   Cells curved in a spiral, 0.6–0.9 × 2–8 μm (Fig.

BXII.γ.21). Color of cell suspensions in the absence of polysulfides and S^0 is red. Absorption spectra of living cells show maxima at 378, 488, 520, 554, 590, 799, 831-835, and 886–889 nm. Photosynthetic pigments are bacteriochlorophyll *a* (esterified with phytol) and carotenoids of the spirilloxanthin series with spirilloxanthin as the predominant component.

TABLE BXII.γ.10. Differential characteristics of *Halorhodospira* species[a]

Characteristic	*Halorhodospira halophila*	*Halorhodospira abdelmalekii*	*Halorhodospira halochloris*
Cell diameter (μm)	0.6–0.9	0.9–1.2	0.5–0.6
Motility	+	+	+
Color of cell suspensions	red	green	green
Bchl	*a*	*b*	*b*
NaCl optimum (%)	11–32	12–18	14–27
pH optimum	8.5–9.0	8.0–9.2	8.1–9.1
Sulfate assimilation	−	−	+
Fatty acid cluster	I + II	III	IV
Major quinones	MK8/Q8[b]	MK8/Q8[b]	MK8/Q8[b]
Mol% G + C of DNA of the species	66.5–69.7	63.3–63.8	50.5–52.9
Mol% G + C of DNA of the type strain	68.4 (Bd)	63.8 (T_m)	52.9 (T_m)
Substrates used:[c]			
Hydrogen	U	U	U
Sulfide	+	+	+
Thiosulfate	+	−	−
Acetate	+	+	+
Pyruvate	+	+	+
Propionate	+	+	+
Lactate	±	−	−
Fumarate	+	+	+
Succinate	+	+	+
Malate	+	+	+

[a]Symbols: +, characteristic positive or substrate utilized by some strains; ±, characteristic positive or substrate utilized by some strains; −, characteristic negative or substrate not utilized by most strains; U, unknown—no data available; (+) weak activity or growth.

[b]In addition a short chain isoprenoid quinone is present in large amounts.

[c]No species uses formate, butyrate, glucose, fructose, glycerol, methanol, ethanol, or propanol.

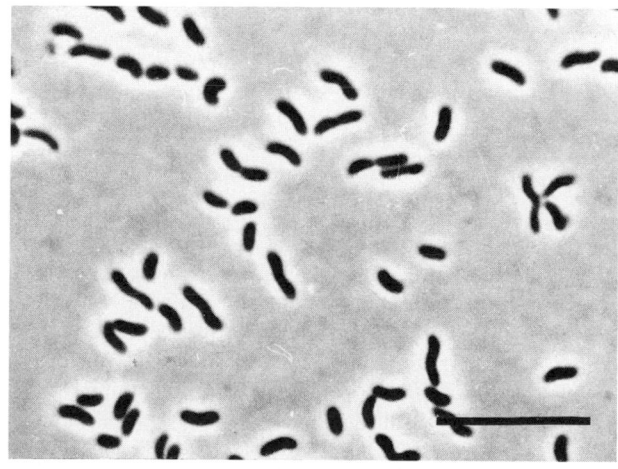

FIGURE BXII.γ.21. *Halorhodospira halophila* BN 9621. Phase-contrast micrograph. Bar = 10 μm. (Reproduced with permission from H.G. Trüper and J.F. Imhoff. 1977. *In* Starr, Stolp, Trüper, Balows and Schlegel (Editors), The Prokaryotes. A Handbook on Habitats, Isolation, and Identification of Bacteria. Springer-Verlag, Berlin, p. 277.)

Obligately phototrophic and strictly anaerobic. Photoautotrophic growth occurs with sulfide and thiosulfate as electron donors. When inorganic carbonate is present, acetate, propionate, pyruvate, malate, succinate, and fumarate are used as organic carbon sources. Ammonia, glutamate, and glutamine serve as nitrogen sources. Sulfate assimilation absent. Cysteine used as a sulfur source under photoheterotrophic growth conditions. Growth factors not required, but vitamin B_{12} enhances growth of some strains.

Extremely high concentrations of sodium chloride are required for growth. Optimal development is at 30–47°C, pH 8.5–9.0, and 11–32% salt concentration. Cells lyse at low salt concentrations (below approximately 0.5 M NaCl).

Habitat: mud and waters of highly saline and alkaline environments containing sulfide and exposed to the light, such as salt and soda lakes.

The mol% G + C of the DNA is: 66.5-69.7 (T_m), type strain 68.4 (Bd).

Type strain: SL1, BN 9610, DSMZ 244.

GenBank accession number (16S rRNA): M26630.

2. **Halorhodospira abdelmalekii** (Imhoff and Trüper 1982) Imhoff and Süling 1997, 915[VP] (Effective publication: Imhoff and Süling 1996, 111) (*Ectothiorhodospira abdelmalekii* Imhoff and Trüper 1982, 266.)

abd'el.ma.lek' i.i. M.L. gen. n. *abdelmalekii* of Abd-El-Malek, named for Yousef Abd-El-Malek, an Egyptian microbiologist.

Cells spirals or rods, 0.9–1.2 × 4–6 μm, or longer. Motile by means of bipolar, sheathed flagella. Color of cell suspensions is pale green, sometimes with a brownish tinge. Photosynthetic pigments are bacteriochlorophyll *b* (esterified with phytadienol) and small amounts of carotenoids.

Grow under strictly anoxic conditions in the presence of sulfide, inorganic carbonate, and an organic carbon source. Acetate, propionate, pyruvate, succinate, fumarate, and malate are used as carbon sources and electron donors. Growth factors are not required. Ammonia is used as nitrogen source.

Extremely high concentrations of sodium chloride are required for growth. Optimal development is at 30–44°C, pH 8.0-9.2, and 12–18% salts.

Habitat: mud and waters of highly saline and alkaline environments containing sulfide and exposed to the light, such as salt and soda lakes.

The mol% G + C of the DNA is: 63.3–63.8 (T_m), type strain 63.8 (T_m).

Type strain: BN 9840, ATCC 35917, DSMZ 2110.

GenBank accession number (16S rRNA): X93477.

3. **Halorhodospira halochloris** (Imhoff and Trüper 1979) Imhoff and Süling 1997, 915[VP] (Effective publication: Imhoff and Süling 1996, 111) (*Ectothiorhodospira halochloris* Imhoff and Trüper 1979, 79.)

ha.lo.chlo′ris. Gr. gen. n. *halos* of the salt; Gr. adj. *chloros* green; M.L. adj. *halochloris* green-colored and from salt lakes.

Cells spiral-shaped, 0.5–0.6 μm wide and, depending on the culture conditions, 2.5–8.0 μm long (Fig. BXII.γ.22). Color of cell suspensions is pale green to gooseberry-green, in dense populations with a brownish tinge. Living cells have absorption maxima at 374, 389, 598, 796, 884, and 1018 nm. Photosynthetic pigments are bacteriochlorophyll *b* (esterified with phytadienol) and small amounts of carotenoids (mainly rhodopin glucoside and its methoxy derivative).

Obligately phototrophic and strictly anaerobic. In the presence of sulfide as the photosynthetic electron donor and inorganic carbonate, acetate, propionate, pyruvate, succinate, fumarate, and malate are used. Nitrogen sources are ammonia and glutamine. Growth factors are not required.

Extremely high concentrations of sodium chloride required for growth. Optimal development is at 30–47°C, pH 8.1–9.1, and 14–27% salt concentration, no growth below approximately 10% total salts.

Habitat: mud and waters of highly saline and alkaline environments containing sulfide and exposed to the light, such as salt and soda lakes.

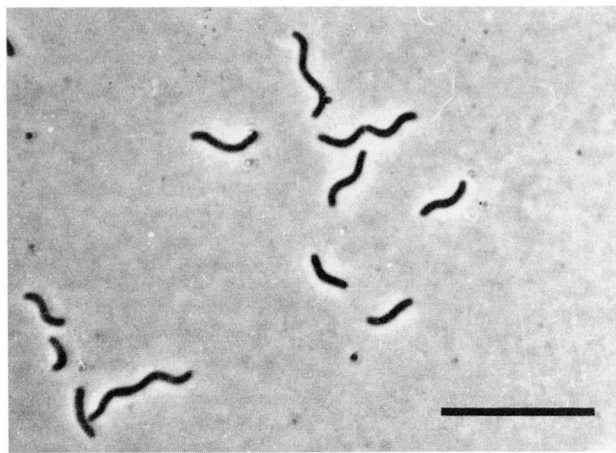

FIGURE BXII.γ.22. *Halorhodospira halochloris* DSMZ 1059. Phase-contrast micrograph. Bar = 10 μm. (Reproduced with permission from H.G. Trüper and J.F. Imhoff. 1977. *In* Starr, Stolp, Trüper, Balows and Schlegel (Editors), The Prokaryotes. A Handbook on Habitats, Isolation, and Identification of Bacteria. Springer-Verlag, Berlin, p. 277.)

The mol% G + C of the DNA is: 50.5–52.9 (T_m), type strain 52.9 (T_m).

Type strain: BN 9850, ATCC 35916, DSMZ 1059.

GenBank accession number (16S rRNA): M59152.

Genus IV. **Nitrococcus** Watson and Waterbury 1971, 224[AL]

EVA SPIECK AND EBERHARD BOCK

Ni.tro.coc′cus. L. n. *nitrum* nitrate; Gr. n. *coccus* a grain, berry; M.L. masc. n. *Nitrococcus* nitrate sphere.

Spherical cells 1.5 μm or more in diameter, occurring singly or in pairs. Cells reproduce by binary fission. **Intracytoplasmic membranes occur as tubes**, randomly arranged throughout the cytoplasm. Gram negative. **Motile** by 1 or 2 flagella. **Aerobic.** Energy derived from the **oxidation of nitrite to nitrate. Obligate lithoautotrophs.** Optimal growth in 70–100% seawater enriched with nitrite and other inorganic salts. No organic growth factors required. Isolated from marine environments.

The mol% G + C of the DNA is: 61.2.

Type species: **Nitrococcus mobilis** Watson and Waterbury 1971, 224[AL].

FURTHER DESCRIPTIVE INFORMATION

Nitrococcus mobilis belongs to the family *Ectothiorhodospiraceae* in the class *Gammaproteobacteria* (Woese et al., 1985). It is the only example of a nitrite-oxidizing bacterium that is phylogenetically related to ammonia-oxidizing bacteria (*Nitrosococcus oceani* and "*Nitrosococcus halophilus*") in the same class. All these organisms are from marine habitats. No growth occurs in freshwater media even if NaCl is included. The temperature range for growth is 15–30°C and the pH range is 6.8–8.0. No terrestrial strains are known so far. *Ectothiorhodospira, Arhodomonas, Halorhodospira*, and *Thiorhodospira* are the closest relatives of *Nitrococcus*.

Additional details and a comparison of the biochemical properties of *Nitrococcus* to those of other nitrite-oxidizing genera can be found in the introductory chapter "Lithoautotrophic Nitrite-Oxidizing Bacteria." Detailed treatments of the ecology of nitrite-oxidizing bacteria and of the phylogeny of these organisms can be found in the introductory chapter "Nitrifying Bacteria."

Because of the presence of extensive intracytoplasmic membrane systems in both genera, Teske et al. (1994) hypothesized that *Nitrobacter* and *Nitrococcus* derived from an immediate photosynthetic ancestry. Additional features that the two genera have in common include *a*-type cytochromes, flagella, carboxysomes, and storage materials. The tubular membranes of *Nitrococcus* are quite similar to the internal photosynthetic membrane system of *Thiocapsa pfennigii* (*Thiococcus pfennigii*).

A spherical cell of *Nitrococcus* with two flagella is shown in Fig. BXII.γ.23. The tubular intracytoplasmic membranes are visible in Fig. BXII.γ.24. Cells are elongated before division (1.8 × 3.5 μm). They may occur as small clumps of cells embedded in a slime matrix. The cell wall is similar to that seen in most other Gram-negative bacteria. The tubular membrane system arises by invaginations of the cytoplasmic membrane and has a branching nature. As in *Nitrobacter*, the cytoplasmic and intracytoplasmic membranes are asymmetric, with an electron-dense layer on the inner side; the structure of the membranes becomes apparent when cells are ruptured and negatively stained. The surface of the membrane is covered with particles, which are arranged in rows (Fig. BXII.γ.24 and BXII.γ.25). Watson and Waterbury (1971) described doughnut-shaped particles 6–8 nm in diameter, which occasionally form rectilinear arrays. The particulate intracytoplasmic membranes of *Nitrococcus* were labeled with gold particles using monoclonal antibodies (MAbs) that recognize the β-

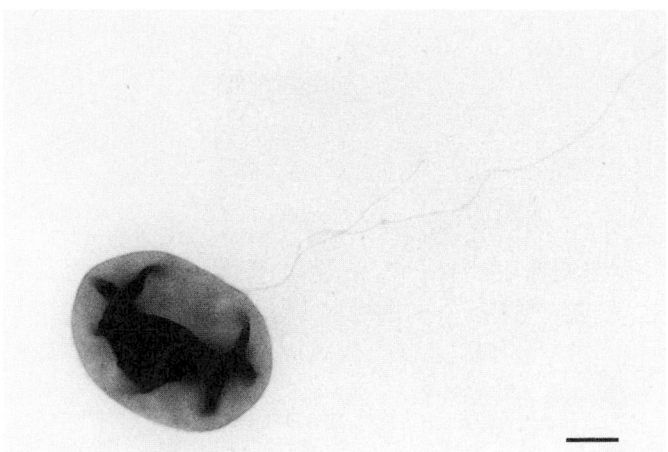

FIGURE BXII.γ.23. Negatively stained cell of *Nitrococcus mobilis* with flagella. Bar = 500 nm.

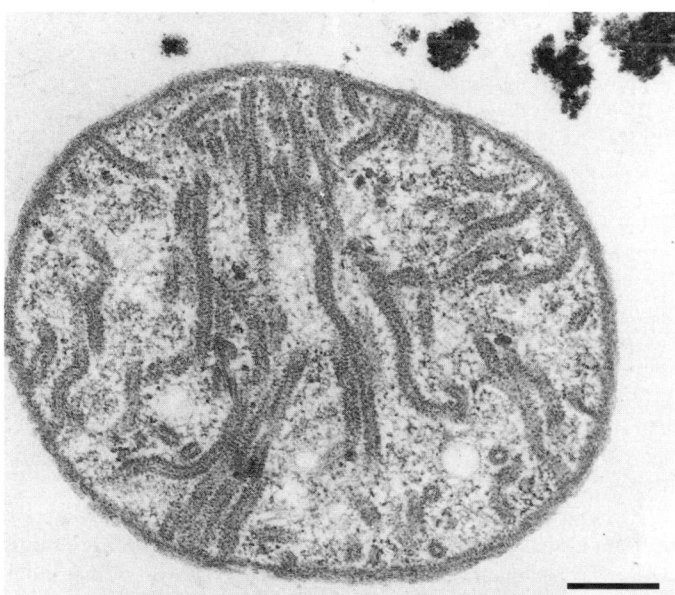

FIGURE BXII.γ.24. Ultrathin section of *Nitrococcus mobilis* with tubular intracytoplasmic membranes which surface is covered with particles. Bar = 250 nm.

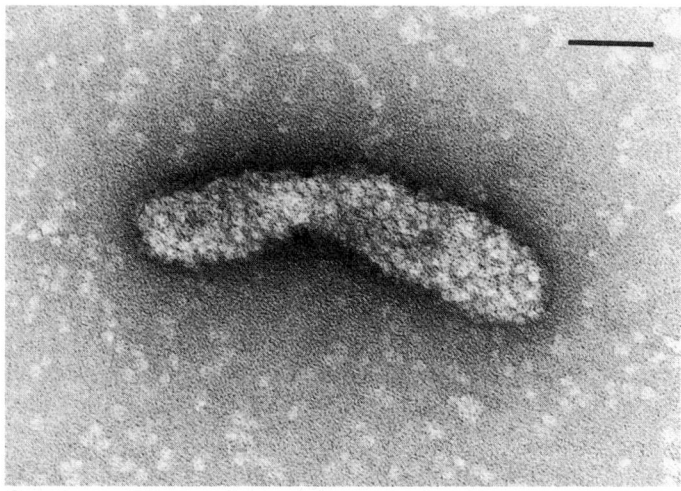

A

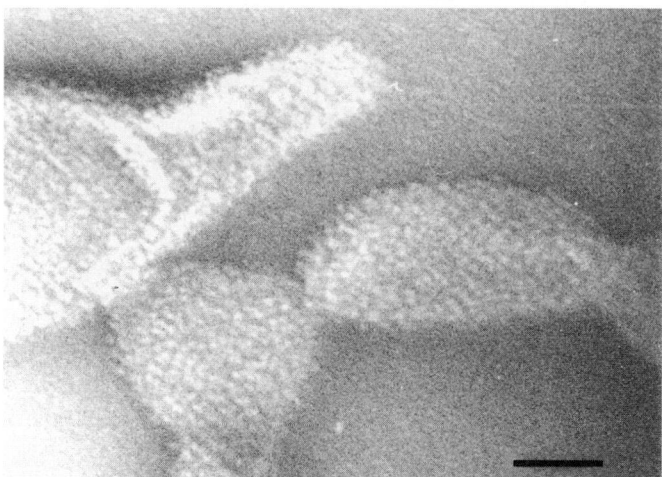

B

FIGURE BXII.γ.25. Isolated intracytoplasmic membranes of *Nitrococcus mobilis* in negative contrast. The membrane surface is densely packed with particles of about 6 nm in diameter (*A*) which can be regularly arranged in rows (*B*). Bars = 50 nm.

subunit of the nitrite oxidoreductase (β-NOR) of *Nitrobacter* (Fig. BXII.γ.26). The molecular mass of the β-subunit of the nitrite-oxidizing system (β-NOS) of *Nitrococcus* is in the same range as the β-NOR of *Nitrobacter* (65 KDa) (Bartosch et al., 1999). The size of this protein corresponds to that of one of the main proteins (135 and 61 KDa) found in crude extracts of *Nitrococcus* (Krause-Kupsch, personal communication). Cytochromes of the *a*- and *c*-types as well as several kinds of inclusion bodies such as carboxysomes and poly-β-hydroxybutyrate granules are present in *Nitrococcus*. Electron-dense bodies believed to be composed of glycogen are observed.

ENRICHMENT AND ISOLATION PROCEDURES

Nitrite oxidizers can be isolated using a mineral medium containing nitrite; the compositions of media for lithotrophic, mixotrophic, and heterotrophic growth are given in Table BXII.γ.11.

Serial dilutions of enrichment cultures must be incubated for one to several months in the dark. Since nitrite oxidizers are sensitive to high partial pressures of oxygen, cell growth on agar surfaces is limited. Nitrite oxidizers can be separated from heterotrophic contaminants by Percoll gradient centrifugation and subsequent serial dilution (Ehrich et al., 1995).

MAINTENANCE PROCEDURES

Nitrifying organisms can survive starvation for more than one year when kept at 17°C in liquid medium. Nevertheless, cells should be transferred to fresh media every four months. Table BXII.γ.11 lists three different growth media for nitrite oxidizers. Freezing in liquid nitrogen is a suitable technique for maintenance of stock cultures suspended in a cryoprotective buffer containing sucrose and histidine.

DIFFERENTIATION OF THE GENUS *NITROCOCCUS* FROM OTHER GENERA

The nitrite oxidizers are a diverse group of long or short rods, cocci, and spirillae. The genera *Nitrobacter* and *Nitrococcus* possess

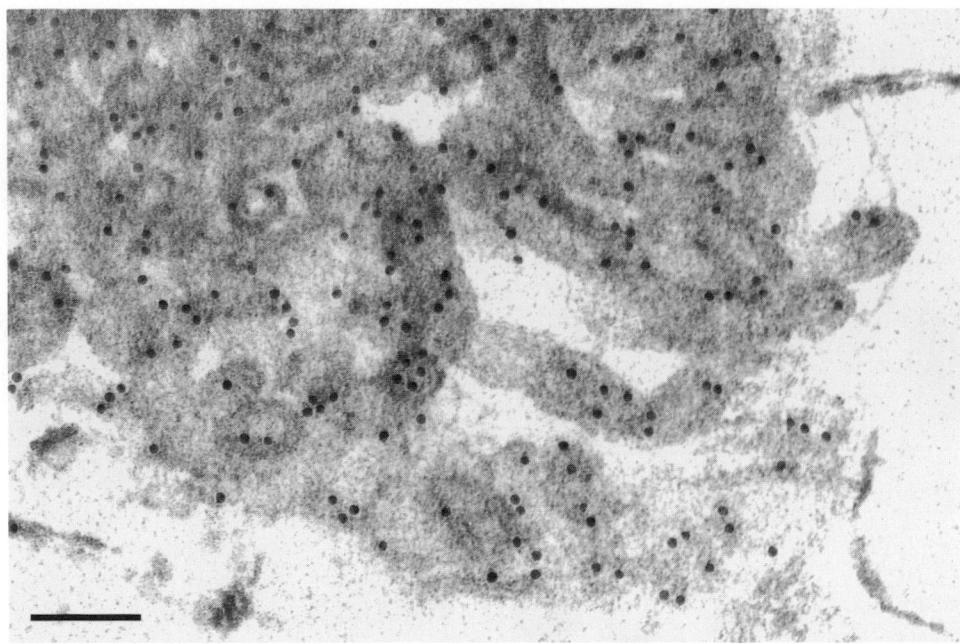

FIGURE BXII.γ.26. Location of the β-NOS in *Nitrococcus mobilis* on the cytoplasmic side of the tubular intracytoplasmic membranes shown by immunogoldlabeling (MAbs 153-1). The cell was partly destroyed by the embedding procedure. Bar = 100 nm.

TABLE BXII.γ.11. Three different media for lithoautotrophic (medium A for terrestrial strains; medium B for marine strains), mixotrophic (medium C), and heterotrophic (medium C without NaNO₂) growth of nitrite oxidizers

	Culture medium		
Ingredient	A[a,]	B[b]	C[c, d]
Distilled water (ml)	1000	300	1000
Seawater (ml)		700	
NaNO₂ (mg)	200–2000	69	200–2000
MgSO₄·7H₂O (mg)	50	100	50
CaCl₂·2H₂O (mg)		6	
CaCO₃ (mg)	3		3
KH₂PO₄ (mg)	150	1.7	150
FeSO₄·7H₂O (mg)	0.15		0.15
Chelated iron (13%, Geigy) (mg)		1	
Na₂MoO₄·2H₂O (µg)		30	
(NH₄)₂Mo₇O₂₄·4H₂O (µg)	50		50
MnCl₂·6H₂O (µg)		66	
CoCl₂·6H₂O (µg)		0.6	
CuSO₄·5H₂O (µg)		6	
ZnSO₄·7H₂O (µg)		30	
NaCl (mg)	500		500
Sodium pyruvate (mg)			550
Yeast extract (Difco) (mg)			1,500
Peptone (Difco) (mg)			1,500
pH adjusted to[e]	8.6	6	7.4

[a]For terrestrial strains from Bock et al. (1983).

[b]For marine strains modified from Watson and Waterbury (1971).

[c]For terrestrial strains from Bock et al. (1983).

[d]For heterotrophic growth medium C without NaNO₂ is used.

[e]After sterilization pH should be 7.4–7.8.

a complex arrangement of intracytoplasmic membranes in form of flattened vesicles or tubes. *Nitrococcus* is motile by 1–2 flagella and possesses a tubular membrane system as well as carboxysomes. The taxonomic categorization is based on the work of Sergei and Helene Winogradsky (Winogradsky, 1892). Traditionally, the classification of genera is performed primarily on cell shape and arrangement of intracytoplasmic membranes. So far, four morphologically distinct genera (*Nitrobacter, Nitrococcus, Nitrospina,* and *Nitrospira*) have been described; the four genera contain a total of eight species (Watson et al., 1989; Bock and Koops, 1992; Ehrich et al., 1995; Sorokin et al., 1998). Differential traits of the four genera of nitrite-oxidizing bacteria are given in Table BXII.γ.12, further properties in Table BXII.γ.13, and fatty acid profiles in Table BXII.γ.14. Motility and carboxysomes have been observed in *Nitrobacter* and *Nitrococcus* (Tables BXII.γ.12 and BXII.γ.13).

Nitrococcus can be separated from the other genera by means of monoclonal antibodies that recognize the key enzyme (Table BXII.γ.12). Aamand et al. (1996) developed three monoclonal antibodies (MAbs) that recognize the α- and the β-subunit of the NOR of *Nitrobacter*. MAb Hyb 153-2 was specific for the α-NOR of *Nitrobacter*, whereas MAb Hyb 153-1 recognized the β-NOSs of both *Nitrobacter* and *Nitrococcus*. The MAb Hyb 153-3 recognized the β-NOSs of *Nitrobacter, Nitrococcus, Nitrospina* and *Nitrospira* (Bartosch et al., 1999). Thus, the suite of MAbs permits detection of all known nitrite oxidizers and discrimination of *Nitrobacter* and *Nitrococcus* from *Nitrospina* and *Nitrospira*.

TAXONOMIC COMMENTS

Classification of *Nitrococcus* as a distinct genus of nitrite-oxidizing bacteria is based primarily on its unique morphology and ultrastructure. Sequence analysis of the 16S rRNA confirmed the separation.

FURTHER READING

Watson, S.W. and J.B. Waterbury. 1971. Characteristics of two marine nitrite oxidizing bacteria, *Nitrospina gracilis* nov. gen. nov. sp. and *Nitrococcus mobilis* nov. gen. nov. sp. Arch. Microbiol. *77*: 203–230.

Watson, S.W., E. Bock, H. Harms, H.P. Koops and A.B. Hooper. 1989. Nitrifying bacteria. *In* Staley, Bryant, Pfennig and Holt (Editors), Bergey's Manual of Systematic Bacteriology, 1st Ed., Vol. 3, The Williams & Wilkins Co., Baltimore. pp. 1808–1833.

TABLE BXII.γ.12. Differentiation of the four genera of nitrite-oxidizing bacteria

Characteristic	*Nitrobacter*	*Nitrococcus*	*Nitrospina*	*Nitrospira*
Phylogenetic position	*Alphaproteobacteria*	*Gammaproteobacteria*	*Deltaproteobacteria* (preliminary)	Phylum *Nitrospirae*
Morphology	Pleomorphic short rods	Coccoid cells	Straight rods	Curved rods to spirals
Intracytoplasmic membranes	Polar cap	Tubular	Lacking	Lacking
Size (μm)	0.5–0.9 × 1.0–2.0	1.5–1.8	0.3–0.5 × 1.7–6.6	0.2–0.4 × 0.9–2.2
Motility	+	+	−	−
Reproduction:	Budding or binary fission	Binary fission	Binary fission	Binary fission
Main cytochrome types[a]	*a, c*	*a, c*	*c*	*b, c*
Location of the nitrite oxidizing system on membranes	Cytoplasmic	Cytoplasmic	Periplasmic	Periplasmic
MAb-labeled subunits (KDa)[b]	130 and 65	65	48	46
Crystalline structure of membrane-bound particles	Rows of particle dimers	Particles in rows	Hexagonal pattern	Hexagonal pattern

[a]Lithoautotrophic growth.

[b]MAbs, monoclonal antibodies.

TABLE BXII.γ.13. Properties of the nitrite-oxidizing bacteria

Characteristic	*Nitrobacter winogradskyi*	*Nitrobacter alkalicus*	*Nitrobacter hamburgensis*	*Nitrobacter vulgaris*	*Nitrococcus mobilis*	*Nitrospina gracilis*	*Nitrospira marina*	*Nitrospira moscoviensis*
Mol% G + C of the DNA	61.7	62	61.6	59.4	61.2	57.7	50	56.9
Carboxysomes	+	−	+	+	+	−	−	−
Habitat:								
Fresh water	+			+				
Waste water	+			+				
Brackish water				+				
Oceans	+				+	+	+	
Soda lakes		+						
Soil	+		+	+				
Soda soil		+						
Stones	+			+				
Heating system								+

TABLE BXII.γ.14. Primary fatty acids of the described species of nitrite-oxidizing bacteria[a,b]

Fatty acid	*Nitrobacter winogradskyi* Engel	*Nitrobacter alkalicus* AN4	*Nitrobacter hamburgensis* X14	*Nitrobacter vulgaris* Z	*Nitrococcus mobilis* 231	*Nitrospina gracilis* 3	*Nitrospira marina* 295	*Nitrospira moscoviensis* M1
$C_{14:1\ \omega 5c}$						+		
$C_{14:0}$	+		+		+	+ + +	+	+
$C_{16:1\ \omega 9c}$							+ + +	+ +
$C_{16:1\ \omega 7c}$	+	+	+	+	+ + +	+ + +		
$C_{16:1\ \omega 5c}$							+ + +	+ + +
$C_{16:0\ 3\text{-OH}}$						+		+
$C_{16:0}$	+ +	+ +	+ +	+ +	+ + +	+ +	+ + +	+ + +
$C_{16:0\ 11\text{-CH}_3}$							+	+ + +
$C_{18:1\ \omega 9c}$	+		+			+	+	
$C_{18:1\ \omega 7}$	+ + + +	+ + + +	+ + + +	+ + + +	+ + +	+	+	+
$C_{18:0}$	+	+	+		+	+	+ +	+
$C_{19:0\ \omega cyclo7\text{-}8}$	+	+	+	+	+			

[a]Symbols: +, <5%; + +, 6–15%; + + +, 16–60%; + + + +, >60%.

[b]Stirred cultures were grown autotrophically at 28°C (*Nitrospira moscoviensis* at 37°C) and collected at the end of exponential growth. Modified from Lipski et al., (2001).

List of species of the genus Nitrococcus

1. **Nitrococcus mobilis** Watson and Waterbury 1971, 224[AL] *mo' bi.lis.* L. adj. *mobilis* movable, motile.

   The morphological, cultural and biochemical characteristics are as described for the genus. The minimum generation time is 10 h. Optimum temperature range, 25–30°C. Optimum pH range, 7.5–8.0. The type strain was isolated from equatorial Pacific seawater.

   The mol% G + C of the DNA is: 61.3 (Bd).

   Type strain: ATCC 25380.

Genus V. **Thioalkalivibrio** Sorokin, Lysenko, Mityushina, Tourova, Jones, Rainey, Robertson and Kuenen 2001a, 578[VP]

THE EDITORIAL BOARD

Thi.o.al.kal.i.vi′bri.o. Gr. n. *thios* sulfur; N.L. n. *alkali* Arabic *al* the; *qaliy* soda ash; L. v. *vibrio* vibrate; M.L. n. *Thioalkalivibrio* sulfur alkaline vibrio.

Motile Gram-negative curved rods or spirilla (0.4–0.6 × 0.8–3.0 μm). **One polar flagellum. Cell wall rippled.** Possess carboxysomes. **Obligate chemolithoautotrophs. Oxidize sulfide, thiosulfate, sulfur, and tetrathionate.** Carbon assimilation via Calvin cycle. **Halotolerant** up to 1.2–1.5 M Na^+; require at least 0.3 M Na^+.

The mol% G + C of the DNA is: 61–65.6.

Type species: **Thioalkalivibrio versutus** Sorokin, Lysenko, Mityushina, Tourova, Jones, Rainey, Robertson and Kuenen 2001a, 579 (*Thioalkalivibrio versutus* (sic) Sorokin, Lysenko, Mityushina, Tourova, Jones, Rainey, Robertson and Kuenen 2001a, 579.)

FURTHER DESCRIPTIVE INFORMATION

These organisms were found in the water and sediments of alkaline lakes in which sulfide is produced in the underlying sediments. Growth in the laboratory was studied in carbonate/bicarbonate buffered medium containing thiosulfate, sulfide, polysulfide, sulfur, and polythionates. All strains produce elemental sulfur from thiosulfate at the beginning of the exponential phase in batch culture; this sulfur vanishes later in growth. Tetrathionate and elemental sulfur were used by most strains. The pH optimum was 9–10. Ammonia, nitrite, and nitrate were used as nitrogen sources. Some strains reduced nitrate to nitrite under reduced O_2 levels. The type strain of *T. denitrificans*, ALJD, is a facultative anaerobe able to grow with thiosulfate and nitrous oxide as electron donor and acceptor, respectively (Sorokin et al., 2001a).

Analysis of the 16S rDNA sequence of the type strains of *T. versutus*, *T. denitrificans*, and *T. nitratis* showed that the genus *Thioalkalivibrio* belongs to the *Gammaproteobacteria*; two species of the genus *Ectothiorhodospira* (anaerobic purple sulfur bacteria) were their closest relatives (Sorokin et al., 2001a).

ENRICHMENT AND ISOLATION PROCEDURES

These bacteria were obtained by aerobic enrichment and isolation in a carbonate/bicarbonate buffered liquid salts medium containing thiosulfate. For some strains the amount of NaCl in the base medium had to be increased. Strains that reduce nitrate to nitrite can be obtained by adding 20 mM nitrate to the base medium. The single denitrifying strain was obtained using thiosulfate and N_2O in an anaerobic enrichment culture (Sorokin et al., 2001a).

MAINTENANCE PROCEDURES

Thioalkalivibrio spp. can be stored at 4°C for 2 months between transfers in liquid salts medium (pH 10) containing 5 mM $MgCl_2$ to stabilize the cell walls. They can also be stored in 10% glycerol (v/v) at –80°C (Sorokin et al., 2001a).

List of species of the genus Thioalkalivibrio

1. **Thioalkalivibrio versutus** Sorokin, Lysenko, Mityushina, Tourova, Jones, Rainey, Robertson and Kuenen 2001a, 579 (*Thioalkalivibrio versutus* (sic) Sorokin, Lysenko, Mityushina, Tourova, Jones, Rainey, Robertson and Kuenen 2001a, 579.) *ver.su′ tus*. L. adj. *versutus* versatile.

   Description as for the genus with the following additional characteristic. Best growth occurs under atmospheric levels of O_2.

   The mol% G + C of the DNA is: 63–65.6 (T_m).
   Type strain: AL 2, DSM 13738.
   GenBank accession number (16S rRNA): AF126546.

2. **Thioalkalivibrio denitrificans** Sorokin, Lysenko, Mityushina, Tourova, Jones, Rainey, Robertson and Kuenen 2001a, 579[VP] (*Thioalkalivibrio denitrificans* (sic) Sorokin, Lysenko, Mityushina, Tourova, Jones, Rainey, Robertson and Kuenen 2001a, 579.) *de.ni.tri′ fi.cans* M.L. v. *denitrifico* denitrify; M.L. part. adj. *denitrificans* denitrifying.

   Description as for the genus with the following additional characteristics. Facultatively anaerobic denitrifier. Best growth is obtained with thiosulfate and nitrous oxide under anaerobic conditions.

   The mol% G + C of the DNA is: 62.3–65. 0 (T_m).
   Type strain: ALJD, DSM 13742.
   GenBank accession number (16S rRNA): AF126545.

3. **Thioalkalivibrio nitratis** Sorokin, Lysenko, Mityushina, Tourova, Jones, Rainey, Robertson and Kuenen 2001a, 579[VP] (*Thioalkalivibrio nitratis* (sic) Sorokin, Lysenko, Mityushina, Tourova, Jones, Rainey, Robertson and Kuenen 2001a, 579.) *ni.tra′ tis*. M.L. adj. *nitratis* nitrate.

   Description as for the genus with the following additional characteristic. Nitrate reduced to nitrite under microaerophilic conditions.

   The mol% G + C of the DNA is: 61.3–62.1 (T_m).
   Type strain: ALJ 12, DSM 13741.
   GenBank accession number (16S rRNA): AF126547.

Other Organisms

Sorokin et al. (2001c) described two further alkaliphilic chemoautotrophs whose 16S rDNA sequences were related to *Thioalkalivibrio* spp.: one type utilized thiocyanate as a nitrogen source while using thiosulfate as an electron donor; the other utilized thiocyanate as both a source of nitrogen and of electrons.

Genus VI. **Thiorhodospira** Bryantseva, Gorlenko, Kompantseva, Imhoff, Süling and Mityushina 1999b, 700[VP]

JOHANNES F. IMHOFF AND VLADIMIR M. GORLENKO

Thi′o.rho′do.spi′ra. Gr. n. *thios* sulfur; Gr. n. *rhodon* the rose; Gr. n. *spira* spiral; M.L. fem. n. *Thiorhodospira* the spiral rose with sulfur.

Cells vibrioid or spiral-shaped, motile by means of a monopolar flagellar tuft, multiplies by binary fission. Gram-negative and belong to the *Gammaproteobacteria*. Internal photosynthetic membranes are parallel lamellae piercing the cytoplasm lengthwise or underlying the cytoplasmic membrane. **Photosynthetic pigments are bacteriochlorophyll *a* and carotenoids.**

Obligately phototrophic and strictly anaerobic. **During photolithoautotrophic growth with sulfide as the electron donor, globules of S^0 are formed outside the cytoplasm**, in the medium, attached to cells, or in the periplasm. The final oxidation product is sulfate. In the presence of sulfide, organic substances may be photoassimilated. Development is **dependent on sodium salts** in low concentrations and on alkaline conditions.

Habitat: surface of sediments rich in organic matter and microbial mats from brackish soda lakes that contain hydrogen sulfide.

The mol% G + C of the DNA is: 56.0–57.4 (T_m).

Type species: **Thiorhodospira sibirica** Bryantseva, Gorlenko, Kompantseva, Imhoff, Süling and Mityushina 1999b, 700.

ENRICHMENT AND ISOLATION PROCEDURES

Thiorhodospira species can be selectively enriched under photoautotrophic conditions with sulfide as the electron donor and in saline and alkaline media similar to those for *Ectothiorhodospira* and *Halorhodospira* species. From suitable environmental sources, isolation can be achieved using a basal medium[1] in agar dilution series inoculated with natural samples without prior enrichment. Cultures are grown phototrophically in 50-ml screw-capped bottles at 25–30°C and a light intensity of 2000 lux.

MAINTENANCE PROCEDURES

For long-term preservation, storage in liquid nitrogen is recommended as the standard maintenance procedure. For this purpose, cell suspensions of well-grown cultures supplemented with DMSO to a final concentration of 5% are dispensed in plastic ampules, sealed, and frozen at the temperature of liquid nitrogen.

DIFFERENTIATION OF THE GENUS *THIORHODOSPIRA* FROM OTHER GENERA

Thiorhodospira species may be differentiated from other purple sulfur bacteria by the deposition of S^0 both outside and inside the cells, as seen under the light microscope. Differentiation from *Ectothiorhodospira* and *Halorhodospira* species is possible on the basis of sulfur deposition both outside and inside the cells, the irregular arrangement of the membrane stacks of photosynthetic membranes, the low mol% G + C content, and some characteristic differences in the 16S rDNA sequence (Table BXII.γ.7 of the chapter describing the family *Ectothiorhodospiraceae*). The 16S rDNA sequence of *Thiorhodospira* is ~88–92% similar to those of *Ectothiorhodospira* species. Metabolic properties, such as utilization of carbon, sulfur, and nitrogen sources, are very similar to those of other *Ectothiorhodospiraceae*. *Thiorhodospira* species grow at salt concentrations between 0.5–8%, as do *Ectothiorhodospira* species.

List of species of the genus Thiorhodospira

1. **Thiorhodospira sibirica** Bryantseva, Gorlenko, Kompantseva, Imhoff, Süling and Mityushina 1999b, 700[VP]

   si.bi′ri.ca. M.L. fem. n. *sibirica* related or belonging to Siberia.

   Cells vibrioid or spirals, 3–4 × 7–20 μm, multiply by binary fission with formation of septa. Motile by means of a monopolar tuft of flagella, which is approximately the same length as the cells. Internal photosynthetic membranes are parallel lamellae piercing the cytoplasm lengthwise or underlying the cytoplasmic membrane. Color of cell suspensions is brownish-red to red. Suspensions of intact cells show absorption maxima at 377, 485, 509, 537–550, 591, 711, 799, 830, 858, and 901 nm. Photosynthetic pigments are bacteriochlorophyll *a* and carotenoids of the spirilloxanthin series, with anhydrorhodovibrin, rhodopin, and spirilloxanthin as major components.

   Strictly anaerobic and obligately phototrophic. Photolithoautotrophic growth occurs with sulfide and sulfur as electron donors. Thiosulfate is not oxidized. When sulfide and carbonates are present, acetate, pyruvate, malate, succinate, propionate, and fumarate are photoassimilated.

   Mesophilic and alkaliphilic brackish-water bacterium with optimum growth at pH 9.0–9.5 (pH-range 7.5–10.5) and 25–30°C. Requires sodium carbonate and/or sodium chloride for growth. Salinity range is 0.5–8% sodium bicarbonate or 0–6% NaCl (in the presence of 0.5% sodium carbonate).

   Habitat: Sulfide-containing surface layers of the Lake Malyi Kasytui sediment.

   The mol% G + C of the DNA is: 56.0–57.4 (T_m).

   Type strain: A12, ATCC 700588, BN 9312.

   GenBank accession number (16S rRNA): AJ006530.

---

1. The basal medium used for isolation and cultivation of *Thiorhodospira* contains (g/l distilled water): KH_2PO_4, 0.5 g; NH_4Cl, 0.5 g; NaCl, 0.5 g; $MgCl_2 \cdot 6H_2O$, 0.2 g; $CaCl_2 \cdot 2H_2O$, 0.05 g; $NaHCO_3$, 2.5 g; Na_2CO_3, 2.5 g; Na-acetate, 0.5 g; Na-malate, 0.5 g; yeast extract, 0.1 g; $Na_2S \cdot 9H_2O$, 0.7 g; and trace element solution SL8, 1 ml. The pH is adjusted to 9.0–9.5 (Bryantseva et al., 1999b).

Family III. **Halothiobacillaceae** *fam. nov.* Kelly and Wood 2003

DONOVAN P. KELLY AND ANN P. WOOD

Hal.o.thi.o.ba.cil' la' ce.ae. M.L. neut. n. *Halothiobacillus* type genus of the family; *-aceae* ending to denote a family; M.L. fem. pl. n. *Halothiobacillaceae* the *Halothiobacillus* family.

Cells are rods, motile by polar flagella, contain ubiquinone Q-8, and are obligately aerobic and chemolithotrophic. **Gram negative, belong to the *Gammaproteobacteria*, halotolerant and halophilic, obligately chemolithoautotrophic.** Aerobic chemolithoautotrophic growth is dependent on the oxidation of inorganic sulfur compounds or elemental sulfur and the fixation of carbon dioxide by the Calvin-Bassham-Benson reductive pentose phosphate cycle. In nature, members of the *Halothiobacillaceae* occur in seawater and marine mud, fresh water, salt lakes, and hydrothermal vents.

The mol% G + C of the DNA is: 56–58.

Type genus: **Halothiobacillus** Kelly and Wood 2000, 515 emend. Sievert, Heidorn and Kuever 2000, 1235.

TAXONOMIC COMMENTS

The family *Halothiobacillaceae* has been newly created to include the single genus *Halothiobacillus*, and the characteristics of the family are currently those given in the Genus *Halothiobacillus* chapter.

The four currently recognized and validly published species are physiologically and morphologically similar, and exhibit a relatively wide spread of mol% G + C content of their DNA. Their distinctive physiological properties and biochemical requirements distinguish them all other species. Analysis of nearly complete 16S rDNA sequences from all four species showed them to form a single phylogenetic cluster (Kelly et al., 1998; Kelly and Wood, 2000; Sievert et al., 2000).

Pairwise comparisons of the relative identities of their 16S rDNA sequences using BLASTN and BLAST2 (http://www.ncbi.nlm.gov) showed the type species, *H. neapolitanus*, to share 91.0–92.9% identity with the other three; *H. kellyi* showed 93.4% identity to both *H. halophilus* and *H. hydrothermalis*; and *H. halophilus* and *H. hydrothermalis* shared 98.6% identity. For comparison, 16S rDNA from *Allochromatium vinosum* ATCC 17899 (M26629) showed only 86.7% and 88.0% identity to *H. neapolitanus* and *H. halophilus*. Several other strains of *Halothiobacillus* have been identified by 16S rDNA sequencing (strains MS02, RA13, W5 and WJ18), and additional species are likely to be defined and validly published in due course.

Genus I. **Halothiobacillus** *Kelly and Wood 2000, 515^VP emend. Sievert, Heidorn and Kuever 2000, 1235*

DONOVAN P. KELLY AND ANN P. WOOD

Hal.o.thi.o.ba.cil' lus. Gr. n. *hals* sea, salt; Gr. n. *thios* sulfur; L. n. *bacillus* a small rod; L. masc. n. *Thiobacillus* salt-loving sulfur rodlet.

Obligately chemolithoautotrophic Gram-negative rods, 0.3–0.6 × 1.0–2.5 μm, occurring singly or in pairs. Energy for growth is obtained from the oxidation of reduced inorganic sulfur compounds (thiosulfate, tetrathionate, trithionate, sulfur, and sulfide, but not thiocyanate); unable to oxidize ferrous iron. Tolerant of high concentrations of solutes (e.g., 4 M NaCl; 0.25 M sodium thiosulfate), with some strains showing moderate halophily with a requirement for NaCl (optima 0.4–1.0 M NaCl). Optimum temperature for growth 30–42°C and optimum pH 6.5–8.0. During growth on reduced sulfur compounds, pH is changed from neutrality to pH 2.5–3.0. Contain ubiquinone Q-8. Member of the *Gammaproteobacteria*.

The mol% G + C of the DNA is: 56–67.

Type species: **Halothiobacillus neapolitanus** (Parker 1957) Kelly and Wood 2000, 515 (*Thiobacillus neapolitanus* Parker 1957, 86.)

FURTHER DESCRIPTIVE INFORMATION

This genus was created to accommodate the neutrophilic, obligately chemolithoautotrophic, and halophilic or halotolerant species previously described as *Thiobacillus*. Characteristics of the four species of *Halothiobacillus* currently recognized are shown in Table BXII.γ.15. The properties, taxonomy, and differentiation of the genera of sulfur-oxidizing, chemolithoautotrophic, Gram-negative, rod-shaped bacteria are summarized in the tables and the figure in the chapter on the genus *Thiobacillus* which appear in Volume 2 Part C in this *Manual*.

DIFFERENTIATION OF THE SPECIES OF THE GENUS *HALOTHIOBACILLUS*

The properties, similarity, and distinctive features of the species of *Halothiobacillus* are fully discussed elsewhere (Kelly et al., 1998; Kelly and Wood, 2000; Sievert et al., 2000).

List of species of the genus Halothiobacillus

1. **Halothiobacillus neapolitanus** (Parker 1957) Kelly and Wood 2000, 515^VP (*Thiobacillus neapolitanus* Parker 1957, 86.)

ne.a.po.li.ta' nus. L. adj. *neapolitanus* Neapolitan; pertaining to the seawater at Naples from which this species was probably first isolated by Nathansohn in 1902.

Small rods, 0.3–0.5 × 1.0–1.5 μm. Motile by means of a polar flagellum. Colonies grown on thiosulfate agar are small (1–2 mm), circular, convex, glistening, and whitish yellow due to precipitated sulfur. The center of old colonies becomes pink. In static culture in liquid thiosulfate medium, sulfur and polythionates may accumulate, and the medium becomes uniformly turbid with a sulfur pellicle. Aerated cultures may show transitory accumulation of tetrathionate and trithionate. pH drops to 2.8–3.3. Chemostat cultures do not accumulate intermediates and convert thiosulfate quantitatively to sulfate. This organism oxidizes sulfur, sulfide, thiosulfate, tetrathionate, and trithionate but

TABLE BXII.γ.15. Basic characteristics of species of the genus *Halothiobacillus*

Characteristics	1. *H. neapolitanus*	2. *H. halophilus*	3. *H. hydrothermalis*	4. *H. kellyi*
Mol% G + C	56	64	67–68	62
Cell size (μm)	0.3–0.5 × 1.0–1.5	0.5 × 1.0	0.5 × 1.0	0.4–0.6 × 1.2–1.5
Motility	+	+	+	+
Carboxysomes	+	nd	nd	nd
Obligately chemolithoautotrophic	+	+	+	+
Halotolerant	+	+	+	+
Optimum NaCl (M)	nd	0.8–1.0	0.43	0.4–0.5
Maximum NaCl (M)	0.86	4	2	2.5
Optimum pH	6.5–6.9	7.0–7.3	7.5–8.0	6.5
pH limits	4.5–8.5	6.0–8.0	6.0–9.0	3.5–8.5
Optimum temperature, °C	28–32	30–32	35–40	37–42
Nitrate reduction to N$_2$	−	−	−	−
Growth on/oxidation of:				
Sulfur	+	+	+	+
Thiosulfate	+	+	+	+
Tetrathionate	+	+	+	+
Trithionate	+	+	nd	nd

not thiocyanate, and uses ammonium salts or nitrates as nitrogen source. Obligately chemolithotrophic and autotrophic. Strict aerobe. Optimum temperature for growth: 28–32°C; growth range: 8–39°C. Optimum pH for growth: 6.5–6.9; growth range pH: 4.5–8.5. Isolated from seawater, canal water, and corroding concrete. Frequently, this organism is found in marine mud and in seawater. Presumably widely distributed in freshwater, soil, and marine environments. Member of the *Gammaproteobacteria*.

The mol% G + C of the DNA is: 56 (Bd).

Type strain: ATCC 23641, DSM 15147, NCIB 8539.

GenBank accession number (16S rRNA): AF173169.

2. **Halothiobacillus halophilus** (Wood and Kelly 1991) Kelly and Wood 2000, 515^{VP} (*Thiobacillus halophilus* Wood and Kelly 1991, 280.)

hal.o.phil.us. L. n. *halophil* salt-loving; M.L. adj. *halophilus* loving salt.

Cells are Gram-negative small rods. Obligately halophilic, optimum growth with 1 M NaCl and tolerating up to 4 M NaCl. Obligately chemolithoautotrophic; grows aerobically with thiosulfate, tetrathionate or sulfur; lowers the pH to 5.5–6. Also oxidizes sulfide, trithionate, and hexathionate. On thiosulfate agar, colonies are small (1–3 mm), circular, opaque, smooth, and convex, and become yellow/white with deposited sulfur. Optimum temperature for growth 30–32°C and optimum pH for growth 7.0–7.3. Nitrate reduced to nitrite but anaerobic growth with thiosulfate does not occur. Contains ubiquinone Q-8. Isolated from a salt lake in the wheat belt of Western Australia. Member of the *Gammaproteobacteria*.

The mol% G + C of the DNA is: 64.2 (HPLC).

Type strain: DSM 6132, ATCC 49870.

GenBank accession number (16S rRNA): U58020.

3. **Halothiobacillus hydrothermalis** (Durand, Reysenbach, Prieur and Pace 1993) Kelly and Wood 2000, 515^{VP} (*Thiobacillus hydrothermalis* Durand, Reysenbach, Prieur and Pace 1993, 43.)

hy.dro.ther.mal'is. M.L. *hydrothermalis* pertaining to a hydrothermal vent in the North Fiji basin.

Cells are Gram negative, motile, nonsporeforming, short rods, 0.5 × 1.0 μm. Obligately chemolithotrophic, using oxidation of thiosulfate, tetrathionate, sulfide, or sulfur for growth. Lowers the pH of thiosulfate medium to 4.8, without formation of tetrathionate. Growth occurs in the ranges

pH 6–9, 11–45°C, and 0–2 M NaCl; no growth at pH 5 or 9.5, or at 8°C or 50°C. Optimum pH 7.5–8 and temperature 35–40°C; optimum salinity 0.43 M NaCl, tolerates 2 M NaCl. Contains ubiquinone Q-8. Isolated from samples taken from a depth of 2 km in an active hydrothermal vent in a rift system of the North Fiji Basin. Member of the *Gammaproteobacteria*.

The mol% G + C of the DNA is: 67.1–67.4 (T_m, Bd, HPLC).

Type strain: ATCC 51453, DSM 7121.

GenBank accession number (16S rRNA): M90662.

4. **Halothiobacillus kellyi** Sievert, Heidorn and Kuever 2000, 1235^{VP}

kel'ly.i. M.L. gen. n. *kellyi* of Kelly; named after Donovan P. Kelly, a British microbiologist who has made important contributions to research on sulfur-oxidizing bacteria and their physiology.

Cells are Gram negative, motile, and rod shaped (0.4–0.6 × 1.2–2.5 μm). The organism is strictly aerobic and grows autotrophically on thiosulfate, tetrathionate, sulfur, and sulfide, but not thiocyanate. The organism does not grow heterotrophically. When thiosulfate is used as the primary energy source, a transient formation of sulfur occurs. During growth on reduced sulfur compounds, the pH decreases from neutrality to around pH 2.8. Thiosulfate is completely oxidized to sulfate. Autotrophic growth on thiosulfate occurs between pH 3.5 and 8.5 and at a temperature of 35–49°C; optimum growth occurs at pH 6.5 and at 37–42°C. The optimal NaCl concentration for growth is 0.4–0.5 M; growth is possible between NaCl concentrations of 0–2 M. Nitrate is not used as terminal electron acceptor. The average maximum specific growth rate on thiosulfate is 0.45 h^{-1} for the type strain, Milos-BIII1. On thiosulfate agar, cells produce white to yellowish, smooth, entire colonies (diameter on 1.2% w/v agar is 1–4 mm), in which sulfur is deposited and acid is produced. Ubiquinone Q-8 is present in the respiratory chain. As determined by 16S rRNA gene sequence analysis, *Halothiobacillus kellyi* belongs to the *Gammaproteobacteria* and is most closely related to other members of the *Halothiobacillus* genus. The type strain Milos-BIII1 was isolated from a marine, shallow-water, hydrothermal vent system and is deposited at the DSM.

The mol% G + C of the DNA is: 62 (HPLC).

Type strain: Milos-BIII, DSM 13162.

GenBank accession number (16S rRNA): AF170419.

Order II. **Acidithiobacillales** *ord. nov.*

GEORGE M. GARRITY, JULIA A. BELL AND TIMOTHY LILBURN

A.ci.di.thi.o.ba.cil.la' les. M.L. masc. n. *Acidothiobacillus* type genus of the order; *-ales* ending to denote order; M.L. fem. n. *Acidithiobacillales* the *Acidithiobacillus* order.

The order *Acidithiobacillales* was circumscribed for this volume on the basis of phylogenetic analysis of 16S rDNA sequences; the order contains the families *Acidithiobacillaceae* and *Thermithiobacillaceae*.

Autotrophs utilizing reduced sulfur compounds; mesophilic or moderately thermophilic.

Type genus: **Acidithiobacillus** Kelly and Wood 2000, 513[VP].

Family I. **Acidithiobacillaceae** *fam. nov.*

GEORGE M. GARRITY, JULIA A. BELL AND TIMOTHY LILBURN

A.ci.di.thi.o.ba.cil.la' ce.ae. M.L. masc. n. *Acidothiobacillus* type genus of the family; *-aceae* ending to denote family; M.L. fem. pl. n. *Acidithiobacillaceae* the *Acidithiobacillus* family.

The family *Acidithiobacillaceae* was circumscribed for this volume on the basis of phylogenetic analysis of 16S rDNA sequences; the family contains the genus *Acidithiobacillus* (type genus).

Description is the same as for the genus *Acidithiobacillus*.

Type genus: **Acidithiobacillus** Kelly and Wood 2000, 513[VP].

Genus I. **Acidithiobacillus** *Kelly and Wood 2000, 513*[VP]

DONOVAN P. KELLY AND ANN P. WOOD

A.ci.di.thi.o.ba.cil' lus. L. adj. *acidus* sour; Gr. n. *thios* sulfur; L. n. *bacillus* a small rod; M.L. masc. n. *Thiobacillus* small acid-loving sulfur rod.

Obligately acidophilic, aerobic, Gram-negative rods, 0.4×2.0 μm, motile by one or more flagella. Optimum growth below pH 4.0, using reduced sulfur compounds to support autotrophic growth. Some species oxidize ferrous iron or use natural and synthetic metal sulfides for energy generation; some species oxidize hydrogen. Optimum temperature 30–35°C for mesophilic species and 45°C for moderately thermophilic species. Contain ubiquinone Q-8. Member of the *Gammaproteobacteria*.

The mol% G + C of the DNA is: 52–64.

Type species: **Acidithiobacillus thiooxidans** (Waksman and Joffe 1922) Kelly and Wood 2000, 513 (*Thiobacillus thiooxidans* Waksman and Joffe 1922, 239.)

FURTHER DESCRIPTIVE INFORMATION

The characteristics of the four current recognized species of *Acidithiobacillus* are shown in Table BXII.γ.16. The properties, taxonomy, and differentiation of the genera of sulfur-oxidizing, chemolithoautotrophic, Gram-negative, rod-shaped bacteria are summarized in the chapter on the genus *Thiobacillus* which appears in Volume 2 Part C in this *Manual*. Some species of *Acidithiobacillus* grow chemolithoautotrophically on sulfide minerals, and *A. ferrooxidans* also uses the oxidation of ferrous iron [and probably uranium (IV) and copper (I) oxidation] as the source of metabolic energy. One strain of *A. ferrooxidans* (ATCC 21834) has been shown to have a circular chromosome (2.9 Mb) and one small plasmid (8.6 kb) (Irazabal et al., 1997). Other bacteria able to oxidize sulfide minerals or iron (II) that appear superficially similar to *Acidithiobacillus*, but that are taxonomically and biochemically very different, have been described. These include the Gram-negative genus *Leptospirillum* and the Gram-positive genera *Acidimicrobium* and *Sulfobacillus* (Balashova et al., 1974; Golovacheva and Karavaiko, 1979; Harrison and Norris, 1985; Clark and Norris, 1996; Norris et al., 1996).

TAXONOMIC COMMENTS

This genus was created to accommodate the obligately acidophilic *Thiobacillus* species that were shown by 16S rRNA gene sequence analysis to fall into the *Gammaproteobacteria* (McDonald et al., 1997; Kelly and Wood, 2000).

The assignment of *T. albertis* to the new genus as *Acidithiobacillus albertensis* is very tentative. The 16S rRNA gene sequence, required for placement in a family in the *Proteobacteria*, is not yet available. Its relatively high mol% G + C value (61.5) and the possession of a tuft of flagella and a glycocalyx could indicate significant differences from the other three species included here. Unfortunately, the original isolate may have been lost from culture (B.M. Goebel, personal communication).

Since the first description of *Thiobacillus ferrooxidans* (Temple and Colmer, 1951), numerous organisms from mineral leaching environments have been described as strains of *T. ferrooxidans*, because they were extremely acidophilic, obligately chemolithoautotrophic with ferrous iron as an energy substrate, and could degrade pyrite and various other sulfide minerals. It became clear that selective culture for organisms exhibiting this growth phenotype actually resulted in the isolation of morphologically and genomically diverse strains (Harrison, 1982; Kelly and Harrison, 1989; Kondrat'eva and Karavaiko, 1997; Goebel et al., 2000). Eight different DNA homology groups of isolates described as *T. ferrooxidans* were identified by Harrison (Harrison, 1982; Kelly and Harrison, 1989). One of these groups was subsequently identified as *Leptospirillum*, and DNA–DNA hybridization values between the group containing the type strain (ATCC 23270) and the other six groups ranged from 0–19% to 58–73%. Using PCR-based techniques to assess this genomic variability, similarity coefficients between various isolates ranged from nearly 0 to >98% (Novo et al., 1996), and while several culture collection isolates were very closely related to the type strain (ATCC 23270[T]), one

TABLE BXII.γ.16. Basic characteristics of species of the genus *Acidithiobacillus*

Characteristic	1. *A. thiooxidans*	2. *A. albertensis*	3. *A. caldus*	4. *A. ferrooxidans*
Mol% G + C	52	61–62	63–64	58–59
Cell size (μm)	0.5 × 1.0–2.0	0.45 × 1.2–1.3	0.7 × 1.2–1.8	0.5 × 1.0
Motility	+	+	+	+
Single polar flagellum	+	−	+	+
Tuft of polar flagella	−	+	−	−
Glycocalyx	−	+	−	−
Carboxysomes	+	nd	nd	+[a]
Obligately chemolithoautotrophic	+	+	+	+
Optimum pH	2.0–3.0	3.5–4.0	2.0–2.5	2.5
pH limits	0.5–5.5	2.0–4.5	1.0–3.5	1.3–4.5
Optimum temperature, °C	28–30	28–30	45	30–35
Nitrate reduction	−	−	−	−
Growth on:				
Sulfur	+	+	+	+
Thiosulfate	+	+	+	+
Metal sulfides	+	−	−	+
Ferrous iron	−	−	−	+
Methylated sulfides	−	nd	−	−
Complex media	−	−	−[b]	−

[a]Under carbon dioxide limitation.

[b]Grows mixotrophically with tetrathionate plus glucose or yeast extract.

strain (ATCC 33020) was not taxonomically related to the others (Selenska-Pobell et al., 1998). It is, therefore, inevitable that some strains assigned to *Acidithiobacillus ferrooxidans* will in due course be reassigned to new species or genera.

Thiobacillus concretivorus (Parker, 1945) was once listed as a distinct species in *Bergey's Manual* (Parker, 1957) and appeared in the Approved Lists of Bacterial Names (Skerman et al., 1980). It is, however, now recognized as a synonym of *T. thiooxidans* (now *Acidithiobacillus thiooxidans*) (Vishniac, 1974; Kelly and Harrison, 1989).

List of species of the genus Acidithiobacillus

1. **Acidithiobacillus thiooxidans** (Waksman and Joffe 1922) Kelly and Wood 2000, 513[VP] (*Thiobacillus thiooxidans* Waksman and Joffe 1922, 239.)

thio.ox′ i.dans. Gr. n. *thios* sulfur; M.L. v. *oxido* make acid, oxidize; M.L. part adj. *thiooxidans* oxidizing sulfur.

Short rods, single, paired or in short chains, 0.5 × 1.0–2.0 μm. Motile by means of a polar flagellum (Doetsch et al., 1967). Minute colonies (0.5–1.0 mm) grown on thiosulfate agar appear transparent or whitish yellow and clear on prolonged incubation; edges appear complete. This organism grows in liquid medium on S⁰, thiosulfate, or tetrathionate. Tetrathionate and sulfur may be produced transiently during growth on thiosulfate. This organism cannot oxidize iron or pyrite but has been shown to grow on sulfur from pyrite in co-culture with *"Leptospirillum ferrooxidans"*, an iron-oxidizing, non-sulfur-oxidizing vibrio. Reduces pH of sulfur media to values of 0.5–0.8.

Obligate chemolithotroph and autotroph; remarkable ability to oxidize S⁰ rapidly. Strictly aerobic. Ammonium sulfate used as nitrogen source. Optimum temperature: 28–30°C; temperature range: 10–37°C. Optimum pH: 2.0–3.0; growth range pH: 0.5–5.5. Isolated from soil, sulfur springs, acid mine drainage waters, and corroding concrete and steel environments. Likely to be widely distributed in acidic habitats with elemental sulfur or soluble reduced sulfur compounds. Member of the *Gammaproteobacteria*.

The mol% G + C of the DNA is: 52 (T_m, Bd).

Type strain: ATCC 19377, NCIMB 8343.

GenBank accession number (16S rRNA): Y11596.

2. **Acidithiobacillus albertensis** (Bryant, McGroarty, Costerton and Laishley 1988) Kelly and Wood 2000, 514[VP] (*Thiobacillus*

albertis Bryant, McGroarty, Costerton and Laishley 1988, 221.)

al.ber′ ten.sis. M.L. adj. *albertensis* named for Alberta, Canadian province of Alberta.

Rods, 0.45 × 1.2–1.5 μm. Motile by means of a tuft of polar flagella. Condensed glycocalyx present, extending outwards from outer membrane, apparently involved in cell adhesion to surfaces such as sulfur. This organism grows on sulfur, thiosulfate, or tetrathionate; tetrathionate is accumulated transiently during growth on thiosulfate but not on sulfur. Intracellular sulfur granules have been observed in stationary phase cells. Obligate chemolithotroph and autotroph. Aerobic. Ammonium sulfate is used as nitrogen source. Optimum temperature: 28–30°C. Optimum pH: 3.5–4.0; growth range pH: 2.0–4.5. Isolated from extremely acidic soil adjacent to a sulfur stockpile. Probably occurs in other similar environments.

The mol% G + C of the DNA is: 61.5 (UV spectrum).

Type strain: ATCC 35403.

3. **Acidithiobacillus caldus** (Hallberg and Lindström 1995) Kelly and Wood 2000, 514[VP] (*Thiobacillus caldus* Hallberg and Lindström 1995, 619.)

cal′ dus. M.L. adj. *caldus* warm, liking warmth.

Cells are short, motile, Gram-negative rods, 0.7–0.8 × 1.2–1.8 μm. Capable of chemolithoautotrophic growth with thiosulfate, tetrathionate, sulfide, sulfur, and molecular hydrogen; no growth with ferrous iron or sulfidic ores. Mixotrophic growth with tetrathionate and glucose or yeast extract. Colonies on tetrathionate agar are small, circular, convex, smooth, and transparent, with precipitated sulfur in their center. Growth occurs between 32 and 52°C and pH 1–3.5, with optimum growth at 45°C and pH 2–2.5. Contains ubiquinone Q-8 and a menaquinone. No significant DNA hybridization (2–20%) with *T. ferrooxidans*, *T. thiooxidans*, *T. cuprinus*, or *T. thioparus*. Member of the *Gammaproteobacteria*.

The mol% G + C of the DNA is: 63.1–63.9 (T_m).
Type strain: KU, ATCC 51756, DSM 8584.
GenBank accession number (16S rRNA): Z29975.

4. **Acidithiobacillus ferrooxidans** (Temple and Colmer 1951) Kelly and Wood 2000, 513[VP] (*Thiobacillus ferrooxidans* Temple and Colmer 1951, 605.)

ferro.ox′ i.dans. L. n. *ferrum* iron; M.L. v. *oxido* oxidize, make acid; M.L. part. adj. *ferrooxidans* iron-oxidizing.

Rods, usually single or in pairs, 0.5 × 1.0 µm. Motile; the type strain is motile by a single polar flagellum. Small colonies (0.5–1 mm) grown on thiosulfate or tetrathionate agar are round, sometimes with irregular margins, and white with sulfur. On solid media (agar or agarose) with ferrous sulfate, microscopic colonies with low iron concentrations are formed, resulting in the appearance of an amber zone in the medium around them. With higher iron concentrations, round colonies (1 mm) that become red, brown, and hard with deposited ferric salts are produced. Liquid medium with ferrous sulfate at pH 1.6 changes from clear pale green to amber to red-brown with ferric sulfate. At pH 1.9 and above, considerable precipitation and encrustation with basic ferric sulfates (jarosite) take place.

Obligate chemolithotroph and autotroph. Oxidizes and grows on ferrous iron, pyrite, numerous sulfide minerals, sulfur, thiosulfate, or tetrathionate. Strictly aerobic. Ammonium salts and probably nitrate can be used as nitrogen source; reported to be able to fix dinitrogen. Optimum temperature: 30–35°C; growth range temperature: 10–37°C (no growth at 42°C). Optimum pH: about 2.5; growth range pH: 1.3–4.5. Isolated from many locations worldwide where oxidizable iron, sulfide mineral, and sulfur are exposed; particularly prevalent in acid drainage waters of mines for sulfide minerals, mineral leach dumps, and drainage waters from coal mines and coal or spoil heaps. Member of the *Gammaproteobacteria*.

The mol% G + C of the DNA is: 58–59 (T_m).
Type strain: ATCC 23270, NCIB 8455.

Family II. **Thermithiobacillaceae** *fam. nov.*

GEORGE M. GARRITY, JULIA A. BELL AND TIMOTHY LILBURN

Therm.i.thi.o.ba.cil.la′ ce.ae. M.L. masc. n. *Thermithiobacillus* type genus of the family; *-aceae* ending to denote family; M.L. fem. pl. n. *Thermithiobacillaceae* the *Thermithiobacillus* family.

The family *Thermithiobacillaceae* was circumscribed for this volume on the basis of phylogenetic analysis of 16S rDNA sequences; the family contains the genus *Thermithiobacillus* (type genus).

Description is the same as for the genus *Thermithiobacillus*.
Type genus: **Thermithiobacillus** Kelly and Wood 2000, 515[VP].

Genus I. **Thermithiobacillus** *Kelly and Wood 2000, 515[VP]*

ANN P. WOOD AND DONOVAN P. KELLY

Therm.i.thi.o.ba.cil′ lus. L. n. *thermae* warm baths; Gr. n. *thios* sulfur; L. n. *bacillus* a small rod; L. masc. n. *Thermithiobacillus* warmth-loving sulfur rodlet.

Gram-negative, nonsporeforming rods, 0.5 × 1–2 µm. Obligate aerobes, moderately thermophilic and obligately chemolithoautotrophic on reduced inorganic sulfur compounds (thiosulfate, polythionates, sulfur, and sulfide, but not thiocyanate); unable to oxidize ferrous iron. Contain ubiquinone Q-8. Other general characteristics are those of *Thiobacillus*. Member of the *Gammaproteobacteria*.

The mol% G + C of the DNA is: 66–67.
Type species: **Thermithiobacillus tepidarius** (Wood and Kelly 1985) Kelly and Wood 2000, 515 (*Thiobacillus tepidarius* Wood and Kelly 1985, 436.)

FURTHER DESCRIPTIVE INFORMATION

This genus was created to accommodate the single species of moderate thermophile, *Thiobacillus tepidarius*. While this species showed many physiological similarities to *Thiobacillus neapolitanus* (now *Halothiobacillus neapolitanus*) (Wood and Kelly, 1985, 1986), it was shown by 16S rRNA gene sequence analysis to be unrelated to that genus (Kelly and Wood, 2000). The properties, taxonomy, and differentiation of the genera of sulfur-oxidizing chemolithoautotrophic Gram-negative rod-shaped bacteria are summarized in Tables 1 and 2 and Fig. 1 in the chapter on the genus *Thiobacillus* which appear in Volume 2 Part C in this *Manual*.

List of species of the genus Thermithiobacillus

1. **Thermithiobacillus tepidarius** (Wood and Kelly 1985) Kelly and Wood 2000, 515[VP] (*Thiobacillus tepidarius* Wood and Kelly 1985, 436.)

tep.i.dar′ ius. L. n. *tepidarium* a warm bath fed by natural thermal water; M.L. adj. *tepidarius* warm-bathing (*Thiobacillus*).

Small rods, 0.5 × 1.0–2.0 µm. Motile by a single polar flagellum (5–10 µm in length). Gram negative, nonsporeforming. Colonies grown on thiosulfate agar at 43°C are small (1–2 mm), circular, convex and smooth and become white or yellow with precipitated sulfur. In liquid batch culture on thiosulfate, quantitative production of tetrathionate takes place before growth continues to effect complete oxidation to sulfate. Sulfur may be precipitated. pH drops to 4.5–5.0. Chemostat cultures do not accumulate intermediates during growth on sulfide, thiosulfate, tetrathionate, or trithionate. This organism oxidizes sulfide, thiosulfate, trithionate, tetrathionate, hexathionate, heptathionate, and sulfite, but not thiocyanate. Obligately chemolithotrophic and autotrophic. Uses ammonium salts as nitrogen source.

Aerobic. Optimum temperature: 43–45°C; range: 20–52°C (no growth at 15 or 55°C). Optimum pH: 6.8–7.5; range: 5.2–8.0. Isolated from the Great Roman Bath at the Temple of Sulis Minerva, Bath, Avon, England; distribution unknown. Member of the *Gammaproteobacteria*.

The mol% G + C of the DNA is: 66.6 (UV ratios; Bd).

Type strain: DSM 3134, ATCC 43215.

Order III. **Xanthomonadales** *ord. nov.*

GERARD S. SADDLER AND JOHN F. BRADBURY

Xan.tho.mo.na.da' les. Xanthomonas type genus of the order; *-ales* ending to denote order; M.L. fem. pl. n. *Xanthomonadales* the *Xanthomonas* order.

Gram-negative straight rods lacking prosthecae; hyaline stalks found in one genus (*Nevskia*). Endospores are not produced, no accumulation of poly-β-hydroxybutyrate inclusions. Motile with polar flagellation, flexing/gliding (*Lysobacter*), or nonmotile. Obligate aerobes, having a strictly respiratory metabolism with oxygen as the terminal electron acceptor. Catalase positive. Nitrate not reduced, with the exception of *Stenotrophomonas*. Oxidase production varies. Fatty acid profiles are complex; branched chain and/or hydroxy fatty acids are found. Ubiquinone Q8 is generally found. Optimum temperature 20–35°C; except *Thermomonas*, 37–50°C.

Type genus: **Xanthomonas** Dowson 1939, 187, emend. Vauterin, Hoste, Kersters and Swings 1995, 483.

Family I. **Xanthomonadaceae** *fam. nov.*

GERARD S. SADDLER AND JOHN F. BRADBURY

Xan.tho.mo.na.da' ce.ae. M.L. fem. n. *Xanthomonas* type genus of the family; *-aceae* ending to denote a family; M.L. fem. pl. n. *Xanthomonadaceae* the *Xanthomonas* family.

The charactcristics arc as given for the order. There are twelve genera: *Xanthomonas, Frateuria, Fulvimonas**, *Luteimonas, Lysobacter, Nevskia, Pseudoxanthomonas, Rhodanobacter, Schineria, Stenotrophomonas, Thermomonas,* and *Xylella*. Differential characters are given in Table BXII.γ.17 of the chapter describing the genus *Xanthomonas*.

Type genus: **Xanthomonas** Dowson 1939, 187, emend. Vauterin, Hoste, Kersters and Swings 1995, 483.

Genus I. **Xanthomonas** *Dowson 1939, 187[AL] emend. Vauterin, Hoste, Kersters and Swings 1995, 483*

GERARD S. SADDLER AND JOHN F. BRADBURY

Xan.tho' mo.nas or Xan.tho.mo' nas.* Gr. adj. *xanthus* yellow; Gr. fem. n. *monas* unit, monad; M.L. fem. n. *Xanthomonas* yellow monad.

Straight rods, 0.4–0.6 × 0.8–2.0 μm, mostly single or in pairs, occasionally short chains, filaments rarely seen. Gram negative. Do not produce poly-β-hydroxybutyrate inclusions, nor have sheaths, prosthecae, or resting stages. **Motile by a single polar flagellum. Obligately aerobic, having a strictly respiratory type of metabolism** with oxygen as the terminal electron acceptor. **No denitrification or nitrate reduction occurs. Colonies are usually yellow,** smooth and butyrous, mucoid or viscid. The pigments are highly characteristic brominated aryl polyenes or "xanthomonadins". A characteristic extracellular acidic heteropolysaccharide called xanthan is produced by most strains giving the viscous consistency. Growth is inhibited by 6% NaCl, 30% glucose, 0.01% lead acetate, methyl green, or thionin, and by 0.1% (and usually by 0.02%) triphenyl tetrazolium chloride. **Catalase positive; oxidase negative or weak;** urease not produced. H_2S is usually produced, but not indole or acetoin. Acid is not produced in litmus milk or purple milk. Chemoorganotrophic; able to use various carbohydrates and salts of organic acids as sole carbon sources. Small amounts of acid are produced from many carbohydrates, but not from L-rhamnosc, adonitol, sorbose, D-sorbitol, *meso*-inositol, or *meso*-erythritol. Metabolic activity is shown in Biolog GN microplate tests with D-fructose, D-glucose, D-mannose and methylpyruvate, but not with α-cyclodextrin, adonitol, D-arabitol, *meso*-erythritol, *meso*-inositol, xylitol, D-glucosaminate, γ-hydroxybutyrate, itaconate, sebacate, L-ornithine, L-pyroglutamate, D-serine, D,L-carnitine, γ-aminobutyrate, phenylethylamine, putrescine, 2-aminoethanol, or 2,3-butanediol. L-asparagine, L-glutamine, and glycine cannot be used as sole sources of both

---

**Editorial Note:* The genus *Fulvimonas* was described after the cut-off date for inclusion of new taxa in this volume of the *Manual;* hence, there is no separate chapter devoted to it. Characteristics of the genus *Fulvimonas* are given in Table BXII.γ.17 of the chapter describing the Genus *Xanthomonas*.

**Editorial Note:* The first pronunciation is according to strict Latin usage while the second is more commonly used.

TABLE BXII.γ.17. Differential characteristics of the genera in the family *Xanthomonadaceae*[a]

Characteristic	Xanthomonas	Frateuria	Fulvimonas	Luteimonas	Lysobacter	Nevskia	Pseudoxanthomonas	Rhodanobacter	Schineria	Stenotrophomonas	Thermomonas	Xylella
Motility	+ (exceptions)	+	+	−	+ (flexing, gliding)	+	−	−	−	+	V	−
Flagellation	Single, polar	Single, polar	Single, polar	None		Single, polar		None	None	Multiple, polar	Single, polar	None
Cell shape	Rods	Rods	Rods		Rods	Rods	Rods	Rods	Rods	Rods	Rods	Rods
Cell size (mm)	0.4–0.6 × 1.0–2.9	0.5–0.7 × 0.7–3.5	0.5 × 2		0.2–0.5 × 1.0–15 (sometimes 70)				2–3 × 0.8–0.9	0.5 × 1.5	1.0–12.5 × 0.5–0.75	0.25–0.35 × 0.9–3.5
Temperature optimum	28					20–25	28	30	28–37		37–50	26–28
Oxidase	−	−	+		+		−	+	V	V+	+	−
Catalase	+	+	+		+			+	+	+	+	+
Nitrate reduction	−	−			−		−	−	−	+	−	−
Aesculin hydrolysis				−			−				−	−
Susceptibility to:												
Ampicillin		+		−			−			−	+	+
Penicillin G		−		−	V−		−				+	−
Erythromycin	+	+		+			−			V−	+	+
Kanamycin				V−			+				+	+
Neomycin		−		−	V−					−	+	−
Streptomycin												
Quinones	Q8	Q8		Q8			Q8	Q8	Q8	Q8	Q8	
Polyamines	Spermine									Spermidine and cadaverine	Spermadine	
Mol% G + C of the DNA	63.3–69.7	62–64	71.5–71.9		65.4–70.1	67–69		63	42	66.1–67.8	67.1–68.7	51–52.4
Ecology	Plant pathogens	Plant associated	Soil; plastic degrader	Biofilter	Soil, freshwater; lysing microbes	Freshwater surfaces	Biofilter	Soil; lindane degrader	Fly larvae	Clinical materials; widespread	Kaolin slurry	Plant xylem

Major fatty acids:
C$_{14:0}$
C$_{15:0}$ iso
C$_{15:0}$ anteiso
16:0
16:1
17:0 iso
17:1 iso
18:0
18:1
Major hydroxy fatty acids:
C$_{11:0}$ iso 2OH
C$_{11:0}$ iso 3OH
C$_{12:0}$ 3OH
C$_{12:0}$ iso 3OH
C$_{13:0}$ 2OH
C$_{13:0}$ iso 3OH
C$_{14:0}$ 3OH
Differential fatty acids:
C$_{11:0}$ iso
C$_{11:0}$ iso 2OH
C$_{11:0}$ iso 3OH
C$_{12:0}$ 2OH
C$_{13:0}$ iso 3OH
C$_{17:0}$ cyclo
C$_{18:0}$ cyclo
Major polar lipids:
Diphosphatidyl-glycerol
Phosphatidyl-ethanolamine
Phosphatidyl-glycerol
Phosphatidyl-monoethanolamine
Phosphatidylserine
Unknown polar lipid

[a]For symbols see standard definitions.

carbon and nitrogen. Among the nine fatty acids that predominate in whole cell preparations are 9-methyl decanoic acid ($C_{11:0\ iso}$), 3-hydroxy-9-methyl decanoic acid ($C_{11:0\ iso\ 3OH}$), and 3-hydroxy-11-methyl dodecanoic acid ($C_{13:0\ iso\ 3OH}$), which are highly characteristic of this genus. The ubiquinone that is present has eight isoprene units. Spermidine is the main polyamine; spermine is usually detectable, but not 2-hydroxyputrescine or 1,3-diaminopropane. Species so far described are plant pathogens or are plant associated.

The mol% G + C of the DNA is: 63.3–69.7 (T_m).

Type species: **Xanthomonas campestris** (Pammel 1895) Dowson 1939, 190, emend. Vauterin Hoste, Kersters and Swings 1995, 484 (*"Bacillus campestris"* Pammel 1895, 130.)

FURTHER DESCRIPTIVE INFORMATION

Cell morphology Cells examined directly from plant tissues, or from old cultures, may show irregularities in size and shape, but those from actively growing cultures are reasonably uniform rods with rounded ends. The vast majority of cells are within the range 0.4–0.6 × 0.8–2.0 µm and predominantly motile by a single polar flagellum. Capsules are visible in many isolates using simple staining techniques though results are variable and depend on handling and conditions prior to staining. The capsular polysaccharides are loosely associated with the cells and the production of large amounts of extracellular polysaccharide results in the slimy or mucoid colonies seen on carbohydrate-rich media.

Fine structure The structure of the cell envelope in *Xanthomonas* appears to be similar to that of other Gram-negative cells, as reviewed by Costerton et al. (1974). Schnaitman (1970) found that the cell wall morphology of a *X. campestris* isolate, as viewed with the electron microscope, was similar to that described for *Escherichia coli* by De Petris (1967). The cell wall is relatively electron transparent, with nuclear material, ribosomes, and granular-like inclusions visible within the cell (Horino, 1973). In cells actively growing in broth cultures, ribosomes are observed throughout the cytoplasm with DNA strands occupying the central part of the cell. Polyribosomes and osmophilic vacuoles were found to be present only in actively growing cells (Al-Mousawi and Richardson, 1990). Studies on *X. hyacinthi*, which causes yellow disease in *Hyacinthus*, have shown the presence of polar fimbriae of (5 nm × 6 µm), thought to be involved in the early stages of infection as they were found to attach selectively to stomatal cells (van Doorn et al., 1994). The single polar flagellum may be demonstrated by flagella staining or by electron microscopy of cells from the syneresis water of young cultures. The wavelength of normal flagella averages 1.79 µm, but occasionally variants with wavelengths of 2–3 µm occur. Very rarely cells with two polar flagella occur (Leifson, 1960).

Cell wall composition As with all Gram-negative bacteria, the cell wall of xanthomonads contains three distinct layers: cytoplasmic or inner membrane, peptidoglycan layer, and outer membrane. The inner and outer membranes are made up of proteins and phospholipids, in addition, lipopolysaccharides are found in the outer membrane. Lipopolysaccharides consist of a heteropolysaccharide portion to which "lipid A" is covalently bonded. The backbone of lipid A in strains from various genera, including *Xanthomonas*, has been shown to contain β-1',6-linked glucosamine disaccharides carrying two phosphate groups, one glycosidically linked and one with ester linkage (Hase and Rietschel, 1976). Hydroxy fatty acids are joined to the LPS backbone through ester and amide linkages (Rietschel et al., 1975). Early

studies of the polysaccharide portion of the LPS of *Xanthomonas* were conducted by Volk (1966, 1968a, b), who obtained D-galacturonic acid-1-phosphate, glucose, mannose, and 2-keto-3-deoxyoctonate. The latter, predictably, predominates in the outer membrane of the cell wall (dos Santos and Dianese, 1985). Volk (1968b) found different ratios of uronic acid:mannose, roughly 1.2:1 and 0.66:1. Most also contained xylose or fucose, but not both. Low amounts of rhamnose were present in two isolates. In addition, 4,7-anhydro- and 4,8-anhydro-3-deoxyoctulosonic acids (Volk et al., 1972) and 3-acetamido-3,6-dideoxy-D-galactose (Hickman and Ashwell, 1966) have been isolated. The high rhamnose content, low content of 2-keto-3-deoxyoctonate, presence of D-galacturonic acid, and absence of heptoses are also found in *Stenotrophomonas maltophilia* (Wilkinson, 1968b; Neal and Wilkinson, 1979).

In early studies on cell envelope proteins, Schnaitman (1970) found that the electrophoretic pattern from *Xanthomonas* was broadly similar to that from *E. coli*, but the major protein constituent had a slightly lower mobility. More detailed analysis demonstrated that strains of *X. campestris* pathovar *campestris* could be differentiated from other xanthomonads and a range of Gram-negative bacteria on the basis of membrane protein patterns (Dianese and Schaad, 1982; Minsavage and Schaad, 1983), and that the antigenic determinants of pathovar-specific peptides were resistant to a range of treatments, with the exception of protease Type 1 digestion (Thaveechai and Schaad, 1986). Analysis of the cell envelopes of cassava pathogens, *X. axonopodis* pathovar *manihotis* and *X. cassavae*, revealed two major components relating to the inner and outer membranes of the cell wall (dos Santos and Dianese, 1985). The lighter fraction showed high nicotinamide adenine dinucleotide oxidase and succinate dehydrogenase activities, and larger numbers of peptides than are usually found in the inner membrane of other bacteria. The heavier component showed low enzyme activity and was characteristic of an outer membrane pattern. Heat-modifiable proteins were found in the cell envelopes of both pathovars. Membrane-bound proteins associated with the pigments (xanthomonadins) in *Xanthomonas* have been found in several other genera including a number of Gram-positive bacteria (Lawani et al., 1990).

Studies of the fatty acid component of xanthomonad LPS gave preliminary data on the uniqueness of the fatty acid profile (Rietschel et al., 1975). The taxonomic significance of fatty acid profiles within the genus *Xanthomonas* was further established in a preliminary study of 14 *X. campestris* pathovars pathogenic for hosts in the family *Gramineae* (Stead, 1989). Profiles were found to be complex, containing up to 40 acids, most of which were *iso*- and *anteiso*-branched acids. Grouping of profile types generally correlated with pathovar, although pathovar *translucens*, pathovar *hordei*, pathovar *cerealis*, pathovar *secalis*, and pathovar *undulosa* could not be differentiated. In a comprehensive study of 975 *Xanthomonas* strains 65 different fatty acids were detected and nine were predominantly found in all members of the genus (Yang et al., 1993d): 9-methyl decanoic acid ($C_{11:0\ iso}$), 3-hydroxy-9-methyl decanoic acid ($C_{11:0\ iso\ 3OH}$), 3-hydroxydodecanoic acid ($C_{12:0\ 3OH}$), 3-hydroxy-11-methyl dodecanoic acid ($C_{13:0\ iso\ 3OH}$), 13-methyl tetradecanoic acid ($C_{15:0\ iso}$), *cis*-9-hexadecenoic acid ($C_{16:1\ \omega 7c}$), hexadecanoic acid ($C_{16:0}$), 15-methyl hexadecenoic acid isomer F ($C_{17:1\ iso\ F}$), and 15-methyl hexadecanoic acid ($C_{17:0\ iso}$), and three of which, $C_{11:0\ iso}$, $C_{11:0\ iso\ 3OH}$, and $C_{13:0\ iso\ 3OH}$, are useful in differentiating xanthomonads from other bacteria (Stead, 1992; Yang et al., 1993d). Fatty acids are a valuable tool in systematic studies and serve as a distinguishing characteristic in assigning new species to the genus (Hu et al., 1997), describing hitherto unrealized diversity within the species *X. fragariae* (Roberts et al., 1998), *X. albilineans* (Yang et al., 1993c), and *X. vesicatoria* (Stall et al., 1994), and differentiating between pathogenic and nonpathogenic strains isolated from the same host (Sahin and Miller, 1996; Vauterin et al., 1996a).

Pigmentation Yellow pigments are present in all species of *Xanthomonas*, but nonpigmented strains sometimes occur. *X. axonopodis* pathovar *manihotis* and some strains of pathovar *ricini* occur naturally as nonpigmented organisms. Occasionally nonpigmented mutants arise in culture (Bryan, 1932; Durgapal, 1977) and studies have shown that the absence of xanthomonadins, and other characters such as EPS production have a limiting effect on epiphytic survival and host infection (Jenkins and Starr, 1982a; Poplawsky and Chun, 1998). Although the presence of yellow pigment is an important characteristic for identification, its absence does not exclude an organism from the genus if other characteristics are in agreement. Starr and Stephens (1964), examining a number of isolates belonging to *X. campestris*, found that all contained one or more pigments that showed absorption maxima at 418 (a shoulder), 437, and 463 nm when dissolved in petroleum ether. The pigments were released from disrupted cells at the same rate as a component of the cytoplasmic membrane, suggesting their attachment to the cell envelope (Stephens and Starr, 1963). Later they were shown to be brominated aryl polyenes and were named xanthomonadins. From two to five different xanthomonadins are present in the various species examined (Andrewes et al., 1973; Starr et al., 1977). Xanthomonadin I, the main component of the pigment of *"Xanthomonas juglandis"*, now *X. arboricola* pathovar *juglandis*, was found to be 17 (4 bromo-3-methoxyphenyl) 17-bromo-heptadeca-2,4,6,8,10,12,14,16 octaenoic-acid (Andrewes et al., 1976). The detection of xanthomonadins is highly significant and their presence (Irey and Stall, 1982; Jenkins and Starr, 1982b; Gitaitis et al., 1987; Angeles-Ramos et al., 1991), or that of the gene cluster coding for their synthesis (Poplawsky et al., 1993), can be used to assign unknown strains to the genus *Xanthomonas*.

Quinones Isoprenoid quinones occur in the cytoplasmic membranes of most procaryotes, where they have an important role as electron carriers in electron transport. Members of the *Alphaproteobacteria*, *Betaproteobacteria*, and *Gammaproteobacteria* contain only ubiquinones, with the latter possessing Q7, Q8, Q9, and Q12 (Suzuki et al., 1993a). The only ubiquinone detected in members of the genus *Xanthomonas*, thus far, is ubiquinone Q8 (Ikemoto et al., 1980).

Colonial or cultural characteristics and life cycles Colonies of all species of *Xanthomonas* are normally smooth, round, entire, and butyrous, at least when young, but may show surface markings such as striations and become lobed when older. The production of mucoid, butyrous colonies is an essential feature for discriminating certain pathogenic xanthomonads from saprophytes (Samson et al., 1989). Mutant colonies of different appearance occasionally occur. These may be less mucoid and tend toward rough (Corey and Starr, 1957a) or may be crenated (Whitfield et al., 1981).

When they are first visible, colonies are transparent and buff to very pale yellow, but the color soon deepens and then varies according to the species and the medium used. Pathovars of *X. campestris* and *X. axonopodis* darken towards a buttercup yellow or light cadmium, especially on nutrient agar supplemented with 5% glucose or sucrose. *X. fragariae* remains small and pale on nutrient agar, but gives abundant, spreading, orange-yellow growth on addition of 1% glucose. *X. albilineans* grows on an agar medium containing 2% sucrose and 1% peptone, as buff-yellow or honey-colored colonies.

Growth rates vary widely in the genus. The faster-growing strains produce visible colonies from single cells in 24–36 h at 25°C. Slower-growing strains may take 2–3 days for single colonies to become visible. The time taken to produce colonies about 1 mm in diameter at this temperature varies from 2 to 3 days, to a week or 10 days. The slowest-growing species take about a week to appear, and often longer on isolation plates.

Differences in colonial morphology have been used for identification. CKTM medium[1] (Sijam et al., 1991) can be used to differentiate pepper and tomato strains of *X. vesicatoria* (Sijam et al., 1992). After 2–3 d growth on this medium, all strains produce circular, raised colonies that are yellow and are surrounded by a clear ring. Minute crystals can be seen within the clear ring within 4 d. Strains isolated from tomato then develop opaque haloes around the colonies during a further 2–6 d, unlike pepper strains. Of 69 strains studied, 60 were correctly identified to host by colony characteristics on this medium. Similarly, it is possible to differentiate colonies of *X. arboricola* pathovar *pruni* from other yellow-pigmented organisms by viewing colony morphology on nutrient agar through a lined template, subilluminated by a fluorescent light box (Gitaitis et al., 1988). The clarity and refractive quality of colonies of the pathogen contrasted with distorted and opaque patterns obtained from other organisms.

There is some evidence that colony morphology may be linked to pathogenicity and virulence characteristics. Early studies on *X. phaseoli* found that the size and number of lesions on the host was reduced when using rough colonial variants instead of the mucoid wild type (Corey and Starr, 1957a). Similarly, *X. oryzae* pathovar *oryzae* (*X. oryzae*) exhibited a range of colonial variants. Nwigwe (1973) described two such variants: yellow, with wax-yellow, entire, glistening colonies of 2.5–3.5 mm diameter, and white, with spherical, entire, butyrous colonies of 3.5–5.5 mm diameter. The nonpigmented larger colonies grew faster and were more apparent in cultures known to be less virulent. Subsequent reports have given details of further colonial variants, again linking these variations to differences in pathogenicity and virulence (Choi et al., 1981; Dath and Devadath, 1982; Tsuchiya et al., 1982). Similar findings linking colony morphology and virulence have been presented for *X. oryzae* pathovar *oryzicola* (Rao and Devadath, 1977), *X. axonopodis* pathovar *vasculorum* (Peros, 1988), and *X. axonopodis* pathovar *phaseoli* (Jindal and Patel, 1984). Plasmid loss has been implicated in this phenomenon (Chakrabarty et al., 1995; Ulaganathan and Mahadevan, 1991), as have the actions of transposable elements (Dai et al., 1991).

Nutrition and growth conditions *Xanthomonas* species are chemoorganotrophic. In purely synthetic media containing minerals, ammonium nitrogen, and a suitable carbon source such

---

1. CKTM medium consists of (g/900 ml, unless otherwise stated): Soy peptone, 2.0; Bacto tryptone, 2.0; dextrose, 1.0; L-glutamine 6.0; L-histidine, 1.0; $(NH_4)_2HPO_4$, 0.8; KH_2PO_4, 1.0; $MgSO_4 \cdot 7H_2O$, 0.4; anhydrous $CaCl$, 0.25; Bacto Agar (Difco), 12.0; Tween 80, 10 ml; Pourite (Analytical Products, Belmont CA), 1 drop. Subsequent to autoclaving, the medium is cooled to 45°C and the following are added: cycloheximide, 0.5 ml (1.0 g in 5 ml methanol); bacitracin, 2.0 ml (1.0 g in 20 ml distilled H_2O); neomycin sulfate, 0.5 ml (0.4 g in 20 ml distilled H_2O); cephalexin (65 mg); 5-fluorouracil (12 mg); and tobramycin (0.4 mg); all dissolved in 100 ml distilled H_2O.

as glucose, amino acid supplements are required by most strains of all species. Usually glutamate or methionine, but occasionally both, are needed. There is evidence that methionine may play a role in pathogenesis, as studies on methylthiopropionic acid (MTPA), a phytotoxin produced by *X. axonopodis* pathovar *manihotis*, have demonstrated that methionine is a precursor of MTPA (Ewbank and Maraite, 1990). *X. axonopodis* pathovar *axonopodis* and *X. fragariae* will grow when vitamin-free casein hydrolysate is added, but their exact amino acid requirements are unknown (Kennedy and King, 1962), and *X. arboricola* pathovar *pruni* also requires nicotinic acid (Starr, 1946). Experiments have shown that in this pathovar nicotinic acid can be replaced by tryptophan, which is then converted to nicotinic acid (Wilson and Henderson, 1963).

Ammonium salts, e.g., ammonium phosphate, can serve as nitrogen sources, but nitrates are often less acceptable. Patel and Kulkarni (1949) reported that *X. axonopodis* pathovar *malvacearum* uses potassium nitrate, but not sodium nitrate. Kotasthane et al. (1965) examined 16 pathovars of *X. campestris* and found that all would accept DL-alanine, L-glutamate, and L-proline as sole carbon sources. Glycine, L-leucine, DL-methionine, DL-serine, and DL-norleucine were not used. As sole nitrogen source, DL-alanine, L-glutamate, L-proline, DL-methionine, DL-threonine, DL-aspartate, L-asparagine, L-hydroxyproline, and L-histidine were accepted by all; DL-serine, DL-norleucine, and L-tyrosine by none; while results with glycine, DL-valine, L-tryptophan, L-leucine, DL-isoleucine, L-arginine, DL-lysine, and L-cystine varied with the pathovar. Asparagine can serve as a nitrogen source if glucose is supplied, but it cannot serve as a simultaneous source of both carbon and nitrogen (Starr and Weiss, 1943; Starr, 1946). This is used as a diagnostic test for *Xanthomonas* (Dye, 1962), since yellow *Enterobacteriaceae* and many *Pseudomonas* species will grow with asparagine as sole source of both elements (da Silva Romeiro and Moura, 1998; Schaad and Stall, 1988; Pernezny et al., 1995). Glutamate or alanine can serve in this way for *Xanthomonas* (Lewis, 1930).

Studies from the optimization of xanthan gum production have shown that production is influenced by the type and initial concentrations of the carbon and nitrogen sources (de Vuyst and Vermeire, 1994). The optimal sole carbon source was found to be either 4% glucose, 4–5% sucrose (optionally as 10% molasses) or 2.8% Sirodex A (a glucose syrup, of which the glucose content was 95–96%); other components were 2.0% corn steep liquor as a combined nitrogen-phosphate source. Similar studies found that a medium containing 8% sucrose and 4% peptone was optimal, and it was noted that the bacterium preferentially consumed glucose and fructose formed during sterilization of the sucrose-containing medium (Bruggeman et al., 1998).

The optimum growth temperature is usually in the range 25–30°C, but lower for *X. populi* at about 20°C. The minimum is above 4°C and the maximum varies from 27.5 to 39°C.

Respiratory chain Xanthomonads are obligately aerobic and use oxygen as the terminal electron acceptor. They are oxidase negative by the Kovacs test, indicating that cytochrome c is not present (Jones and Poole, 1985). Hochster and Nozzolillo (1960) postulated that the respiratory chain in *X. axonopodis* pathovar *phaseoli* is comprised of cytochrome a_1, a_2, b_1, and flavoprotein. The spectra found at liquid air temperature showed a slight shoulder at 549 nm, suggesting a very small cytochrome c content, and with cell-free extracts, they showed some reduced nicotinamide adenine dinucleotide (NADH)-cytochrome c reductase activity. Similar results were found by Rye et al. (1988), who

reported cytochromes b, c, aa_3, o type, and possibly a_1, present, with cytochrome c only present in reduced amounts. The difference spectra of Sands et al. (1967) for *Pseudomonas syringae*, and of Stanier et al. (1966) for *S. maltophilia*, are very similar, but do not show any noticeable peak or shoulder at or near 549 nm. Both of these latter species are oxidase negative and thought to contain no cytochrome c.

Carbohydrate metabolism Studies of eight of the currently accepted species indicate that glucose catabolism in *Xanthomonas* is similar to that in *Pseudomonas*. Glucose is oxidized readily, but not gluconate. No gluconate or 2-keto-gluconate is produced from glucose (Lockwood et al., 1941). The Entner–Doudoroff pathway is present (Katznelson, 1955, 1958; Whitfield et al., 1982) and is the predominant pathway of glucose catabolism, with only 8–16% of the glucose following the pentose phosphate pathway. The hexose cycle is not significantly used (Zagallo and Wang, 1967). The tricarboxylic acid and glyoxylate cycles are present in *X. axonopodis* pathovar *phaseoli* (Madsen and Hochster, 1959), and the presence of the TCA cycle confirmed in *X. campestris* (Whitfield et al., 1982). Two enzymes of the tricarboxylic acid cycle that show interesting variations of taxonomic significance are citrate synthase and succinate thiokinase (Weitzman, 1980). *X. hyacinthi* has a citrate synthase that is typical of Gram-negative aerobes: it is a large molecule (molecular weight is ~250,000), is inhibited by NADH, and is reactivated by adenosine monophosphate. The succinate thiokinase is also typical of Gram-negative bacteria, being large in size (molecular weight is 140,000–150,000).

With respect to the breakdown of oligosaccharides, Hayward (1977) assessed the glycosidase activity of various bacteria, mostly plant pathogens; of 39 strains of *Xanthomonas* from nine of the currently accepted species, all showed β-glucosidase activity, as did *Stenotrophomonas maltophilia*. α-Glucosidase activity was weak in *X. albilineans* and absent in both of the two strains of *X. hortorum* pathovar *vitians* and *X. vasicola* pathovar *holcicola*. It was also absent from the four strains of *X. arboricola* pathovar *pruni*, but present in *X. arboricola* pathovar *juglandis*. Dekker and Candy (1979) showed that β-mannanases and carboxymethylcellulases produced by *X. campestris* were present both as membrane-bound constituents and extracellularly. Fructose is phosphorylated by a specific phosphoenolpyruvate-dependent phosphotransferase system in *X. campestris* pathovar *campestris* (de Crecy-Lagard et al., 1991a, b). Transposon mutagenesis studies revealed that mutants incapable of metabolizing fructose also failed to utilize mannose, sucrose, and mannitol. Pectolytic activity has been shown for xanthomonads, but varies among strains. Small variations in methodology can also cause variation in results. Dye (1960) examined 142 isolates of various pathovars of *X. campestris* and 3 isolates of *X. albilineans*. Some of the *X. campestris* strains had as much activity as *Erwinia carotovora*, but many showed little or no activity. Two of the *X. albilineans* strains showed only weak polygalacturonase activity, while one also showed some pectin methylesterase activity. Using pectate gels at various pH levels, Hildebrand (1971) found that the pitting produced by two pathovars of *X. campestris* was strong and unaffected by pH, whereas with *X. arboricola* pathovar *juglandis* and *X. axonopodis* pathovar *phaseoli* and pathovar *vitians* there was greatest activity at pH 5. Two isolates of *X. fragariae* and one of *X. axonopodis* pathovar *dieffenbachiae* showed slight activity. Lange and Knösel (1970) examined eight strains belonging to three species and detected strong pectin methylesterase, pectic transeliminase, and some polygalacturonase activity in *X. vesicatoria* (one isolate); *X. axonopodis* pathovar *begoniae* and *X. hortorum* pathovar *pelargonii* showed slight or no activity. These authors also demonstrated significant C$_x$cellulase and proteolytic activity in all strains except one of pathovar *begoniae*.

Metabolism of aromatic compounds Degradation of protocatechuate, phenylalanine, and synephrine by six pathovars from four species of *Xanthomonas* and four pathovars of uncertain affiliation was found to occur by meta cleavage (William and Mahadevan, 1980). These compounds could serve as carbon sources for growth, but benzoate and *o*-hydroxybenzoate could not be used. More recently, studies on the degradation of xenobiotic compounds such as 1-chloro-2,4-dinitrobenzene and the herbicide alachlor, have demonstrated the presence of glutathione-S-transferase, a key enzyme in the detoxification of xenobiotics, in *X. campestris* (Zablotowicz et al., 1995).

In the synthesis of aromatic amino acids by *Xanthomonas*, the first step is the combination of the multifunctional metabolites D-erythrose-4-phosphate and phosphoenolpyruvate to form 3-deoxy-D-arabinoheptulosonate-7-phosphate (DAHP), the first metabolite that is specific to the pathway. The step is catalyzed by the enzyme DAHP synthetase. The subsequent pathway is multibranched and the activity of DAHP synthetase may be governed by various regulatory systems. Thirty-two genera were examined by Jensen et al. (1967) and found to have one of six different control patterns. Control in *X. campestris* and *X. hyacinthi* was unusual. It was identified as sequential feedback inhibition by chorismate. Arogenate also produced some inhibition, but prephenate, phenylalanine, and tyrosine had little or no effect. There was also some inhibition by tryptophan in *X. hyacinthi* but little or none in *X. campestris*. Whitaker et al. (1981a) examined, among other species, five isolates of *X. campestris* (pathovars not specified), one of *X. albilineans*, one of *X. axonopodis*, two of *S. maltophilia*, one of "*Pseudomonas gardneri*", and the type strain of *Pseudomonas geniculata*, and found that all had DAHP synthetases that showed sequential feedback inhibition by chorismate and lack of stimulation by divalent cations. These properties have not been found in any other groups of bacteria. It is interesting to note that Dye (1966b) found an isolate of "*P. gardneri*" to be a synonym of *X. vesicatoria*, and De Ley (1978) considered it *Xanthomonas* based on DNA–rRNA hybridization. Whitaker et al. (1981a) point out that the isolate of *P. geniculata* was originally deposited in the ATCC as a strain of *P. maltophilia*.

The enzymes involved in the biosynthesis of tyrosine and phenylalanine from prephenate also have taxonomic importance. The dehydrogenases of the two pathways to tyrosine were studied by Byng et al. (1980). They found that although both dehydrogenase activities (i.e., prephenate dehydrogenase and arogenate dehydrogenase) were shown by nearly all species of *Pseudomonas* and *Xanthomonas* studied, the enzymes varied in the cofactor required and in their inhibition by tyrosine (feedback control). Isolates of *X. axonopodis* (pathovar *axonopodis*, pathovar *malvacearum* and pathovar *phaseoli*) *X. hyacinthi*, *X. hortorum* pathovar *pelargonii*, *X. albilineans*, and *S. maltophilia* all showed an NAD-dependent prephenate dehydrogenase that was inhibited by tyrosine, and all showed an NAD-dependent arogenate dehydrogenase, except *X. axonopodis* pathovar *axonopodis*, which showed no activity.

The dehydratases of the two pathways to phenylalanine (i.e., prephenate dehydratase and arogenate dehydratase) were studied by Whitaker et al. (1981b). They found that not only was one or the other of these enzymes absent in some groups, but that the degree of activation of prephenate dehydratase by tyrosine

varied considerably between groups. The combination of these characteristics gave a fine taxonomic tool for distinguishing not only the five rRNA–DNA hybridization groups, but also two subgroups of group II and three subgroups of group III, along the same lines as found with DNA hybridization. Group V organisms contained arogenate dehydratase and prephenate dehydratase, which was not activated by tyrosine. They could not be distinguished from group Ib using these enzymes. Assigned to group V were *S. maltophilia*, *X. campestris* (three isolates), *X. albilineans*, *X. axonopodis*, "*P. gardneri*", and *P. geniculata*.

A further metabolic pathway involving aromatic amino acids is relatively well known, at least in one pathovar of *X. arboricola*, pathovar *pruni*, which requires nicotinic acid for growth, but can accept tryptophan or 3-hydroxyanthranilic acid as a substitute. This it does by synthesis of nicotinic acid from these substances by a pathway that leads to synthesis of nicotinamide adenine dinucleotide (NAD) from tryptophan (Wilson and Henderson, 1963). This pathway has been found in animals, the fungus *Neurospora crassa*, yeasts, and some *Streptomyces* species, but is absent from various bacteria that have been examined (Yanofsky, 1954; Lingens et al., 1966). The first three enzymes involved, tryptophan pyrrolase, kynurenine formamidase, and kynureninase, are coordinately induced by L-tryptophan (Brown and Wagner, 1970). Wagner and Brown (1970) presented evidence that the pathway is regulated by the first step from tryptophan. The activity of the allosteric enzyme involved, tryptophan pyrrolase, is under feedback control (inhibition) by the end products of the pathway, NADH and NADH phosphate.

Van Eys (1960) has shown that nicotinic acid may also be replaced by NAD for growth of *X. arboricola* pathovar *pruni*. The final step in degradation of NAD to nicotinic acid is catalyzed by a pyridine ribosidase that shows unusual preference for the pyridine-riboside linkage over the purine-riboside linkage.

Catabolism of nucleoside triphosphates Huang et al. (1975) found that ATPase is bound to the cell membrane of *X. oryzae* pathovar *oryzae* by a mechanism that involves magnesium ions, which also stabilize this enzyme. Nucleoside tri- and diphosphates are hydrolyzed, but not monophosphates. The pathway for degradation of ATP has been elucidated by Hochster and Madsen (1959). Working with *X. oryzae* pathovar *oryzae*, Yang et al. (1975) obtained evidence for the extracellular breakdown of deoxycytidine triphosphate to the diphosphate, monophosphate, and finally to deoxycytidine. The material was taken up by the cell and incorporated into DNA, probably by stepping up again through mono-, di-, and triphosphates.

Genetics Early studies demonstrated genetic transformation in *X. axonopodis* pathovar *phaseoli* using DNA from strains differing in colony type or in streptomycin resistance (Corey and Starr, 1957b, c). Subsequently, exchange of genetic material has been achieved by conjugal plasmids (Daniels et al., 1984) transformation by electroporation (Wang and Tseng, 1992; Ferreira et al., 1995) and transducing phage (Weiss et al., 1994). Studies using a transposon (Tn5-SC), constructed to quantify genetic deletions or amplifications, demonstrated that the genomic stability of *X. campestris* pathovar *campestris* was as great that of other Gram-negative bacteria (Martinez-Salazar et al., 1993). Similarly, homologous recombination between plasmid sequences was found to occur at a similar level to that reported for other Gram-negative bacteria.

Much study has been directed towards pathogenicity, xanthomonadin, and xanthan gum production (see Daniels and Leach,

1993). Transposon-induced mutagenesis has been employed widely to characterize components within the pathogenicity function. Hu et al. (1992) found that nonpathogenic mutants of *X. campestris* pathovar *campestris* accumulated extracellular polygalacturonate lyase, α-amylase, and endoglucanase in the periplasm, but were deficient in a signal peptidase involved in protein secretion. Similarly, Tn5-induced mutants of *X. campestris* pathovar *campestris* deficient in protease and extracellular polysaccharide production were found to be nonpathogenic to turnip seedlings (Rosato et al., 1994). In contrast, no differences in the production of protease, polygalacturonase, extracellular polysaccharides or growth rate were found between an avirulent, transposon-induced, mutant of *X. vesicatoria* and the wild type (Yun et al., 1995). A 5.4-kb DNA fragment from the type strain of *X. campestris* pathovar *campestris* was found to alter the phenotype of *X. campestris* pathovar *armoraciae* 417 when inoculated onto cabbage (Chen et al., 1994). This transfer conferred the ability to cause blight on the recipient, which generally causes a nonsystemic leaf spot of crucifers. It was suggested that the genes conferring the ability to induce blight symptoms may be unique to *X. campestris* pathovar *campestris* and can affect the apparent pathovar status of the recipient strain.

The *hrp* gene cluster determines functions necessary for pathogenicity on host plants.

Cross-complementation showed that *X. axonopodis* pathovar *vitians* and *X. campestris* pathovar *campestris hrp* sequences are functionally interchangeable, and there is evidence to suggest that there is some functional homology between the *hrp* gene cluster of *Ralstonia solanacearum* and a range of xanthomonads (Arlat et al., 1991). Other complementation studies have revealed that the *hrp* region is organized into at least six different groups (Bonas et al., 1991) and in some pathovars homology was high enough to permit marker gene exchange or complementation of nonpathogenic mutants. Further, *X. campestris* pathovar *campestris* and *X. oryzae* pathovar *oryzae* were found to contain a 1428-base pair *hrp*X gene, involved in the regulation of *hrp* genes (Oku et al., 1998). Sixteen distinct xanthomonad pathovars and 12 strains of *X. oryzae* pathovar *oryzae* were shown to contain homologs of *hrp*X, which were not apparent in heterologous bacteria such as *Agrobacterium*, *Erwinia*, and *Pseudomonas* spp.

Expression of *hrp* genes was found to be repressed when the organisms were grown on rich media, and in minimal medium the level of expression depended on the carbon source supplied to the cells (Arlat et al., 1991). Similar results were obtained by Schulte and Bonas (1992b) who found that expression of *hrp* genes and the hypersensitive response were suppressed in complex media, but induced in plants. Using a minimal medium, both sucrose and methionine were required for induction of *hrp* genes at expression levels similar to those found by induction in plants. Further studies showed that plant filtrates were able to induce *hrp* genes (Schulte and Bonas, 1992a); the inducing molecule(s) were heat stable and hydrophilic.

Seven transcriptional units (*pig*A through *pig*G) involved in xanthomonadin production have been identified (Poplawsky and Chun, 1997). Inactivation of the *pig*B unit resulted in reduced xanthomonadin and extracellular polysaccharide slime production. This phenomenon was reversible when *pig*B mutant strains were grown with other xanthomonadin transcriptional mutants, the wild type, or other xanthomonads. This finding suggests that *pig*B is involved in a regulatory function, which can be restored by a nontransforming diffusible factor (DF). The DF is thought

to be novel and different from other signal molecules, such as homoserine lactone derivatives found in *Vibrio, Agrobacterium, Erwinia, Pseudomonas* and *Burkholderia* spp. (Chun et al., 1997). This finding was similar to that obtained from a recent survey of 106 isolates from seven genera of Gram-negative plant-associated bacteria (Chung et al., 1998a), in which relatively few xanthomonads produced any detectable homoserine lactones.

Nonmucoid mutants deficient in xanthan gum production have been restored by conjugal mating (Thorne et al., 1987). Groups of clones that contained overlapping homologous DNA were found to complement specific mutations, suggesting clustering of the genetic loci involved in xanthan synthesis. Further work has shown that the xanthan gum gene cluster is made up of 12 genes, mainly expressed as an operon under the control of a promoter situated upstream from the first gene, *gum*B (Katzen et al., 1996).

The chromosome map of *X. campestris* pathovar *campestris* has recently been elucidated (Tseng et al., 1999) and the positions of eight loci involved in EPS (xanthan gum) production, nine auxotrophic markers, the gene cluster for xanthomonadin production, and two 16S rRNA operons have been determined.

Molecular biology   A range of fingerprinting methods have been applied to xanthomonads, usually at the species/pathovar/strain level. Hartung and Civerolo (1987) digested genomic DNA from a collection of *X. axonopodis* pathovar *citri* isolates with the restriction endonuclease *Eco*R1, the fragments were separated by PAGE, and the resulting genomic DNA fingerprints compared. Fingerprints varied among separate disease outbreaks and within single outbreaks. Subsequent studies using a conventional RFLP approach, probing *Eco*RI and *Pvu*II polymorphisms with seven cosmid clones selected from a library of *X. axonopodis* pathovar *citri* XC62, demonstrated that *X. axonopodis* pathovar *citri* has a clonal population structure, consistent with previous findings (Hartung and Civerolo, 1989). A relatively heterogeneous group of related strains was isolated only from Florida citrus nurseries, and these were circumscribed into two groups in a subsequent study (Hartung and Civerolo, 1991). RFLP analysis has formed the basis of characterization studies for "*X. campestris* pathovar *citrumelo*" (Gottwald et al., 1991), *X. axonopodis* pathovar *vignicola* (Verdier et al., 1998a), "*X. campestris* pathovar *zeae*" (Qhobela et al., 1990), *X. axonopodis* pathovar *manihotis* (Verdier et al., 1998b), *X. oryzae* pathovar *oryzae* (Leach et al., 1992), *X. campestris* pathovar *mangiferaeindicae* (Gagnevin et al., 1997), and *X. vesicatoria* (Chung et al., 1996). Variations on this basic method have been used to detect polymorphisms in *hrp* genes (Leite et al., 1994a, b) and insertion sequences (Berthier et al., 1994; Adhikari et al., 1995; Chen et al., 1999a), for strain characterization or in conjunction with pulsed-field gel electrophoresis (Cooksey and Graham, 1989; Davis et al., 1997; Roberts et al., 1998).

DNA sequences corresponding to conserved motifs of repetitive bacterial elements (REP, ERIC, and BOX) are widely distributed in phytopathogenic *Xanthomonas* and *Pseudomonas* strains (Louws et al., 1994) and have been used to characterize *X. hortorum* pathovar *pelargonii* (Sulzinski et al., 1995), *X. vesicatoria* (Louws et al., 1995), and *X. fragariae* (Opgenorth et al., 1996). The AFLP technique (Vos et al., 1995), based on the selective PCR amplification of restriction fragments from a total digest of genomic DNA, has been used to study *X. axonopodis* pathovars and *X. vasicola* (Janssen, et al., 1996; Restrepo et al., 1999). From the former analysis, pathovars of *X. axonopodis*, which mainly cause disease in citrus and grasses, were recovered in two groups separate from *X. vasicola*, which correlated well with studies using other techniques.

Plasmids Early studies by Lai et al. (1977a) demonstrated transmission of plasmids RP4 and RK2 from *Escherichia coli* to *X. vesicatoria*, conferring resistance to various antibiotics and production of penicillinase. The transconjugants were able to transmit the plasmids to other strains of the same pathovar, as well as to other pathovars, and also to other bacterial phytopathogens belonging to a number of Gram-negative genera. Stability of the plasmids in the transconjugants varied. A high proportion were retained in bacteria in leaf lesions on plants, in detached leaves kept moist at 5°C, and in dried leaves, but were rapidly lost from cells grown on agar (Lai et al., 1977b). More recently, copper resistance was found to be plasmid borne and self-transmissible in strains of *X. vesicatoria* (Stall et al., 1986). Copper resistance appears to be linked to avirulence, as avirulent, copper-resistant transconjugants were detected subsequent to mating copper-resistant strains with virulent, copper-sensitive strains. Further studies into the emergence of copper resistance have confirmed the wide transmissibility of plasmids coding for this attribute (Bender et al., 1990; Park and Cho, 1996). Copper resistance was transferred to sensitive strains of the same pathovar at a rate of 4.3 $\times$ 10^{-3} to 1.0 $\times$ 10^{-5}, and could be transferred to other pathovars of *X. campestris*. Within *X. vesicatoria*, there is also evidence that race determinants are coded on plasmids, as race shifts have been observed (Swanson et al., 1988; Kousik and Ritchie, 1996). Race shifting was thought to result from the loss of a plasmid carrying avirulence genes or the inactivation of the avirulence gene by an insertion element. In an extensive study of 522 strains of *X. vesicatoria*, the number and size of plasmids was found to vary widely, the latter ranging from 3 to 300 kb (Canteros et al., 1995). Seventy-one different profiles were observed, with a maximum of five plasmid classes present in any one profile. No plasmid or plasmid profile was found to be characteristic for the species (pathovar) and exchange of plasmids in natural populations was thought to be widespread.

Although plasmids are not found universally in xanthomonads (Lazo and Gabriel, 1987), they have been found in *X. axonopodis* pathovar *cyamopsidis*, pathovar *dieffenbachiae*, and pathovar *glycines*, and *X. hortorum* pathovar *hederae* (Lazo and Gabriel, 1987), *X. arboricola* pathovar *pruni* and *X. axonopodis* pathovar *vitians* (Kado and Liu, 1981; Lazo and Gabriel, 1987), *X. campestris* pathovar *sesami* (Sheela et al., 1994), *X. axonopodis* pathovar *manihotis* (Lin et al., 1979), *X. axonopodis* pathovar *malvacearum* (Lazo and Gabriel, 1987; de Feyter and Gabriel, 1991), *X. oryzae* pathovar *oryzae* (Choi et al., 1989b; Amuthan and Mahadevan, 1994), *X. axonopodis* pathovar *vignicola* (Lazo and Gabriel, 1987; Ulaganathan and Mahadevan, 1988), *X. axonopodis* pathovar *vasculorum* (Ezavin et al., 1992), and *X. axonopodis* pathovar *citri* (Civerolo, 1985; Lazo and Gabriel, 1987). In the case of the latter, the number of plasmids varies from 1 to 5, sizes ranging from 7 to 100 kb (Pruvost et al., 1992). In general, strains associated with the same form of citrus bacterial canker disease shared the same plasmid types and these exhibited similar restriction patterns. This uniformity, also present in other pathovars within the genus, has enabled the exploitation of plasmids as high-resolution markers in population studies of *X. vesicatoria* (Hwang et al., 1995), *X. axonopodis* pathovar *manihotis* (Restrepo and Verdier, 1997), and *X. axonopodis* pathovar *citri* itself (Hartung et al., 1996). In the case of the latter, a nested PCR approach relied on the use of labeled primers to allow detection of the amplification product in a microtiter plate. All strains of *X. axonopodis*

pathovar *citri* tested were detected with this system, as were some strains of *X. axonopodis* pathovar *aurantifolii* but not other xanthomonads such as *X. axonopodis* citrumelo, which causes citrus bacterial spot disease.

Antigenic structure Polyclonal antibodies as taxonomic tools have been limited in their impact in the study of *Xanthomonas*. Work has mostly resulted in multiple groupings of pathovars and many cross-reactions. Thus, Elrod and Braun (1947a), using whole cells as antigens for agglutination, found that 36 nomenspecies fell into five groups. Cross-reactions were reduced if the exopolysaccharides were removed by thorough washing, or if the cells were grown on medium low in carbohydrate. Further work on one group (Elrod and Braun, 1947b), using cross-absorbed antisera, separated two "serologically good species," *"Xanthomonas juglandis"* (four isolates) and *"Xanthomonas carotae"* (three isolates). With *X. vesicatoria*, results have varied. Lovrekovich and Klement (1965) found two serovars among 34 strains and reported correlation with phagovars and host of origin. Charudattan et al. (1973) also found two serovars in 72 strains and Schaad (1976), using ribosome preparations from 25 strains, found two serovars and a single isolated strain. Both used double gel-diffusion and neither found any correlation with host origin or pathogenicity. Yano et al. (1979) found nine serovars in 31 isolates belonging to 14 pathovars of *X. campestris*. They used whole cells washed in formalized saline for injection, preparations of extracellular polysaccharide as antigens, and indirect hemagglutination. Cross-reactions were evident, as serovars did not correspond to pathovars. Similar results were obtained by Anderson and Nameth (1990) using polyclonal antisera raised against *X. hortorum* pathovar *pelargonii*. The antisera, when used to develop an immunogold silver stain dot-immunobinding assay and ELISA, were found to be highly reactive for *X. hortorum* pathovar *pelargonii*, but five of seven other pathovars tested also gave moderate reactions. When used in conjunction with other methods, such as immunofluorescence microscopy (Schaad, 1978; Franken, 1992) or for immunocapture prior to use of a PCR assay (Lopes and Damann, 1996; Hartung et al., 1997), polyclonals clearly have value.

The use of monoclonal antibodies (MAbs) has greatly increased specificity. Thus Alvarez et al. (1985) were able to distinguish *Xanthomonas* spp. from a large number of other bacteria using one carefully selected MAb. Using three others they could separate *X. campestris* pathovar *campestris* from various other pathovars, and using a further five MAbs they separated pathovar *campestris* into six groups, finding interesting epidemiological correlations. Extension of these methods allowed identification of *X. albilineans* (Alvarez, et al., 1996), *X. axonopodis* pathovar *citri* (Benedict et al., 1985; Alvarez et al., 1991), *X. axonopodis* pathovar *citrumelo* (Alvarez et al., 1991; Gottwald et al., 1991), *X. axonopodis* pathovar *dieffenbachiae* (Lipp et al., 1992), *X. vesicatoria* (Bouzar et al., 1994), *X. oryzae* pathovar *oryzae* and pathovar *oryzicola* (Benedict et al., 1989), and *X. hortorum* pathovar *pelargonii* and *X. axonopodis* begoniae (Benedict et al., 1990). Franken et al. (1992), using polyclonal and monoclonal antibodies made with whole cells and with flagella preparations of *X. campestris* pathovar *campestris*, found that four MAbs reacted with lipopolysaccharides, whereas two reacted with proteins of 39 and 29 kDa. These proteins were apparently present only in *Xanthomonas* and may be the same proteins found in the outer membrane extracts (Minsavage and Schaad, 1983; Laakso et al., 1990). There was an interesting correlation with ability to hydrolyze starch, but again no correlation with pathogenicity. Out of 16 other *Xanthomonas* isolates included in the tests, only two isolates of *X. campestris* pathovar *campestris* cross-reacted with *X. vesicatoria.*

Antibiotic and drug sensitivity Isolates of *X. oryzae* pathovar *oryzae* were found to be sensitive to streptomycin, ampicillin, novobiocin, erythromycin, oleandomycin, vancomycin, rifampicin, and oxymycin (all at 5µg/ml). The MIC of penicillin G was 40 µg/ml and pimaricin did not inhibit growth up to 100 µg/ml (Choi et al., 1988). In a study of *X. campestris* pathovar *mangiferaeindicae*, strains were also found to be sensitive to streptomycin, in addition to kanamycin, gentamycin, oxytetracycline, and erythromycin (Pruvost et al., 1998). Other antibiotics active against some xanthomonads include ascamycin (Sudo et al., 1996), xanthobacidin (Huang and Chang, 1975), and zhongshenmycin (Zhang et al., 1996).

Pathogenicity and ecology Species and pathovars of *Xanthomonas* have been reported to cause disease in at least 124 monocotyledons and 268 dicotyledons, but no gymnosperms, ferns, or lower plants (Leyns et al., 1984). The host ranges and geographical distributions of these pathovars vary widely (Bradbury, 1986). The symptoms produced also vary widely and are helpful, sometimes essential, for identification of the pathogen involved. Most species and pathovars produce leaf spots, at least initially. These may spread very little, or be limited by the leaf veins, causing angular leaf spots in dicotyledons. They may spread to include the veins and then may develop systemically. Systemic infections produce wilts, death of shoots, or cankers on twigs and branches, or combinations of these symptoms, and if severe the whole plant may be killed. The extent of symptom development and type of symptoms expressed may also depend on the environmental conditions and the cultivar of the host plant (see Rudolph, 1993).

The mechanisms involved in symptom production are incompletely known. Metabolic products such as toxins and enzymes have been shown to be involved. In the case of the latter, protease-deficient mutants showed considerable loss of virulence in pathogenicity tests (Dow et al., 1990) and endoglucanase is thought to play a role, albeit minor, in the early stages of disease development (Gough et al., 1988). Certain enzymes involved in pectin and polygalacturonate degradation appear to have a limited role in pathogenesis (Dow et al., 1989; Beaulieu et al., 1991), though Boher et al. (1995) found that during the infection process the middle lamellae, primary, and secondary plant cell walls were degraded by cellulolytic and pectolytic activity. In addition, their results suggest that bacterial exocellular polysaccharides were involved in plant cell surface degradation (Boher et al., 1997). Similar findings have been made in other studies (Kingsley et al., 1993; Dow et al., 1995) and the involvement of extracellular polysaccharides has long been considered (Morris et al., 1977). El Banoby and Rudolph (1979) found that specific fractions of these polysaccharides from several xanthomonads produced persistent water-soaked spots when introduced into leaves of the host plants from which the bacteria came. Similar results were found with EPS prepared from culture filtrates of *X. vesicatoria*, in which watersoaking and, in some cases, foliar discoloration or necrosis, could be induced by EPS alone (Walkes and O'Garro, 1996). Ramirez et al. (1988) found a correlation between virulence and parameters such as the final viscosity of the culture, the viscosifying capacity of the polymer, and the amount of acetyl substituents in the gum. Infrared spectral analysis revealed that intramolecular interaction of gum constituents could play a significant role in virulence. Dow et al. (1995) found that a pathogenicity locus within *X. campestris* pathovar *campestris* comprises two genes governing the biosynthesis of lipopolysaccharide within the bacterial cell. They suggested that the LPS protects the bacteria against antibacterial substances produced by the plant early in the infection process. Subsequent studies on the early phases of infection detected the LPS produced by *X. axonopodis* pathovar *manihotis* on the outer surface of the bacterial envelope and in areas of the plant middle lamellae in the vicinity of the pathogen (Boher et al., 1997). Lipopolysaccharides are also thought to be involved in the induction of the hypersensitive response and resistance in susceptible hosts (Newman et al., 1995, 1997; Romeiro and Kimura, 1997); they may also have value in epidemiological studies, as differences in the LPS of *X. albilineans* isolated in different locations has been reported (Pillay et al., 1995).

Host specificity is high in pathovars of *Xanthomonas*. In the lists of pathovars below, only about half a dozen have been found to have hosts in more than one plant family. On the other hand, Schnathorst (1966) found that *X. axonopodis* pathovar *malvacearum*, when inoculated onto *Phaseolus* sp., produced early stages of infection very similar to those of pathovar *phaseoli*. The two pathovars were also serologically related, according to earlier experiments. His experiments helped to disprove earlier ideas that host specificity could be changed by artificial passage through nonhost plants.

A critical point in the life cycle of xanthomonads, as with other plant pathogens, is transmission to a new host, particularly if a period of survival in the absence of the host is necessary (see Stall et al., 1993). Such survival may be achieved in many ways, such as with seed, plant residues, perennial hosts, epiphytically, saprophytically in soil, and in insects (Schuster and Coyne, 1974). Many xanthomonads solve this problem by transmission with the seed of their host. These organisms are listed by Richardson (1990). The bacteria may be carried in detritus with the seed, on or in the seed coat, or deeper inside the tissues of the seed itself. Epiphytic survival is becoming well known, e.g., *X. axonopodis* pathovar *manihotis* and *X. arboricola* pathovar *pruni* are both known to spend the interseasonal time epiphytically (Persley, 1978; Young, 1978). In agriculture, many pathogens are most effectively carried by the activities of man, either in infected planting material, which may be symptomless (Hayward, 1974), or on tools, wheeled vehicles, and even grazing animals, as for *X. axonopodis* (Castaño et al., 1964). Survival in soil saprophytically is unusual, but has been suggested by the work of Goto et al. (1978) for *X. axonopodis* pathovar *citri*. Xanthomonads have occasionally been detected in run-off water and ditches around fields of infected plants. Results suggest that survival in this situation would be short (Steadman et al., 1975).

All organisms currently named in the genus *Xanthomonas* are plant pathogens, but isolates that are apparently non-pathogenic are also found (Vauterin et al. 1996a). Some have been identified as known pathogens growing epiphytically, e.g., *X. axonopodis* pathovar *phaseoli* isolated from symptomless weeds (Angeles-Ramos et al., 1991). Most of those examined by SDS-PAGE of proteins and fatty acid analysis can be identified with the currently accepted *Xanthomonas* spp. But identification to the pathovar level in the absence of a pathogenicity test is more difficult (Vauterin et al., 1996a). In any case, the naming of a nonpathogenic organism, with no known connection with disease, as a pathovar is unacceptable. The vast majority of *Xanthomonas* spp. found so far have been associated with plants, but recently organisms belonging to this genus or closely related to it have been shown to form a significant proportion of the bacteria present in microbial communities of hydrothermal vents. These organisms were also found to be related to several *Thiobacillus* spp. (Moyer et al., 1995).

ENRICHMENT AND ISOLATION PROCEDURES

With plant-pathogenic bacteria, the plant itself is probably the ideal "enrichment medium" and isolation is best made from plant material using a dilution procedure, unless contraindicated, onto a nonselective medium. Nutrient agar is usually satisfactory for isolation because it is not sufficiently rich to encourage growth of saprophytes. Supplementing Difco nutrient agar with yeast extract (5 g/l) has been recommended for general isolation work (Dye, 1980; Schaad and Stall, 1988). YSP[2] medium has been recommended for the isolation of *X. albilineans* (Dye, 1962) or SP[3] medium supplemented with cycloheximide and penicillin G (Hayward, 1979). Dye's GYCA medium[4] is useful for pigment production, and it is good for *X. cassavae*, which grows poorly and is short lived on media containing peptone. GYCA is also good for keeping stock cultures, as is YDC[5] (Schaad and Stall, 1988).

Semi-selective media for particular xanthomonads have also been described: CKTM for *X. vesicatoria* (Sijam et al., 1991). The selectivity depends on cycloheximide, bacitracin, neomycin, cephalexin, 5-fluorouracil, tobramycin, and Tween 80. Strains of *X. vesicatoria* can be distinguished from strains of other xanthomonads by the formation of a clear ring around their colonies. *X. arboricola* pathovar *pruni* has been recovered from soil and plant material using XPSM[6] medium (Civerolo et al., 1982), which also suppresses the growth of saprophytic bacteria and fungi generally found in soil samples. The use of nutrient agar supplemented with sodium deoxycholate (200 mg/l), has been used for the selective isolation of *X. vesicatoria* (Bashan and Assouline, 1983), as has a medium containing carboxymethyl cellulose and gelatin (Gitaitis et al., 1991). For a more detailed account, see Kado and Heskett (1970) and Schaad and Stall (1988).

A number of plating assays for detecting *X. campestris* pathovar *campestris* in crucifer seeds have been evaluated (Franken, et al., 1991). Centrifugation prior to plating had no appreciable effect, but the addition of Tween 20, benomyl, and chlorothalonil to the saline (0.85% NaCl) centrifugation solution generally resulted in lower recovery of the target organism. It was found that shaking at room temperature for 2.5 h and soaking at 4–6°C for

1.5 h increased the number of colony forming units recovered, but did not result in more infected seed lots being detected. No significant differences were found between the plating media tested: FS (Yuen et al., 1987), NSCA, and NSCAA (Randhawa and Schaad, 1984). In a similar study, CS20ABN medium was developed by incorporation of bacitracin, neomycin, and cycloheximide into CS20A (Chang et al., 1991). Comparison with CS20A, NSCA, NSCAA, and FS showed that all media recovered high numbers of saprophytic bacteria, that were too numerous to count and overgrew *X. campestris* pathovar *campestris* colonies, with the exception of CS20ABN, where 59–100% of recovered colonies were identified as *X. campestris* campestris.

MAINTENANCE PROCEDURES

Stock cultures may be maintained on GYCA for periods up to six months, with regular subculturing at approximately monthly intervals. Hokawat and Rudolph (1991) found that there was no appreciable decrease in virulence in cultures stored for two years on YDC agar slants. Less frequent subculturing is possible with most isolates at lower-than-room-temperature, e.g., in a refrigerator at ~7–10°C, but care should be taken to ensure that the isolate can tolerate the temperature. *X. axonopodis* pathovar *citri*, cultured on agar plates inverted and stored at 4°C in darkness, could be maintained for at least 26 weeks on a semi-enriched minimal medium (Wu et al., 1990), with no loss of virulence or alteration in colonial morphology. Cultures on richer media did not survive storage.

For longer periods of storage, most isolates can be freeze dried and will keep for years (Lelliott, 1965; Hokawat and Rudolph, 1991). Freeze-dried cultures of *X. oryzae* suspended in 10% sucrose and 1% gelatin retained viability and virulence in a suspending medium containing 2% dextrin, 0.5% ascorbic acid, 0.5% ammonium chloride, 0.5% thiourea, and 0.85% NaCl (Hwang and Cho, 1986). Liquid cultures, applied and then dried onto Whatman 3M paper, have also been used effectively to store xanthomonads (Kidby et al., 1977). *X. albilineans*, which does not keep well in the lyophilized condition, can be kept on an agar slope under sterile mineral oil for periods up to approximately two years.

PROCEDURES FOR TESTING SPECIAL CHARACTERS

The significance of polyamines as a tool in differentiating xanthomonads from other members of the *Gammaproteobacteria* has been established (Auling et al., 1991; Hamana and Matsuzaki, 1993), and in an extensive study of polyamine profiles of 140 *Xanthomonas* strains, spermidine was found to predominate in all members of the genus (Yang et al., 1993b). In contrast, *Stenotrophomonas maltophilia*—at that time encompassed within the genus *Xanthomonas*—contained spermidine and large amounts of cadaverine.

Exopolysaccharides are generally heteropolysaccharides that have repeat structures with carbohydrate and noncarbohydrate substituents (Leigh and Coplin, 1992). The EPS produced by *Xanthomonas* is similar in structure to class 1 capsular polysaccharides and is known commercially as xanthan gum. Xanthans are high-molecular-weight, acidic, anionic heteropolysaccharides composed mainly of D-glucose, D-mannose, and D-glucuronic acid in a ratio of 2:2:1, with small amounts of pyruvic and acetic acids that vary with strains and conditions (Cadmus et al., 1978; Whitfield et al., 1981; Sutherland, 1993). Because of their physical properties and lack of toxicity to man and animals, xanthans

---

2. YSP medium consists of (g/l): yeast extract, 5.0; sucrose, 20.0; peptone, 10.0; and agar 15.0.

3. SP medium consists of (g/l): peptone, 5.0; K_2HPO_4, 0.5; $MgSO_4 \cdot 7H_2O$, 0.25; sucrose, 10.0 or 20.0; agar, 15.0. To control growth of saprophytes, cycloheximide (100 mg/l) and penicillin G (200 IU/ml) can be added.

4. GYCA medium consists of (g/l): yeast extract, 5.0; glucose, 5.0, $CaCO_3$, 40.0, and agar, 15.0. The carbonate is mixed well into the medium just before solidification.

5. YDC medium consists of (g/l): yeast extract, 5.0; dextrose, 20.0, $CaCO_3$, 20.0, and agar, 15.0. The carbonate is mixed well into the medium just before solidification.

6. XPSM medium consists of (g/l except where indicated): alginic acid, 2.0; 8-azaguanine, 0.2; nicotinic acid, 2 mg; cysteine, 3 mg; KH_2PO_4, 0.8; K_2HPO_4, 0.8; $MgSO_4$, 0.1; Bacto agar, 15.0. Subsequent to autoclaving, the medium is cooled to 45°C and chlorothalonil (Diamond Shamrock, Cleveland, OH) and kasugamicin are added aseptically to a final concentration of 80 and 16 mg/ml, respectively.

have a wide range of uses in industry. They are gelling agents, emulsifiers, stabilizing agents, and plasticizers, particularly in the food industry (Jeannes, 1974; Sutherland and Ellwood, 1979). For the genetic control of xanthan synthesis, see Köplin et al. (1992).

DIFFERENTIATION OF THE GENUS *XANTHOMONAS* FROM OTHER GENERA

Characteristics useful for differentiating *Xanthomonas* from other genera are listed in Table BXII.γ.17.

TAXONOMIC COMMENTS

Xanthomonas is a well-defined, homogeneous genus. This is particularly true now that the closely related yet distinctive species, *Xanthomonas maltophilia*, as described by Swings et al. (1983), is accommodated in its own genus, *Stenotrophomonas* (Palleroni and Bradbury, 1993). In addition, *Xanthomonas ampelina* (Panagopoulos, 1969), as described in the previous edition, has been relocated to *Xylophilus* (Willems et al., 1987) in the family *Pseudomonadaceae*, from which *Xanthomonas* was removed. The exact position of *Xylella* may need further investigation, as its very different G + C ratio suggests a genus rather distantly related to other members of the family.

Phylogenetic treatment Our current understanding of the phylogenetic position of the genus *Xanthomonas* owes much to early nucleic acid hybridization studies (Palleroni et al., 1973; De Ley, 1978; De Vos and De Ley, 1983). In this work, the first indications of the multigeneric nature of the genus *Pseudomonas* were presented, as well as strong evidence that the genus *Xanthomonas*, traditionally associated with pseudomonads, was well defined and distinct (De Ley, 1992). The pioneering work of Palleroni and co-workers (1973) assigned the genus *Xanthomonas* to rRNA hybridization group V, distinct from the majority of fluorescent and nonfluorescent taxa of the genus *Pseudomonas*. Congruent results were presented in subsequent studies, and in these broader analyses encompassing a wider range of Gram-negative genera, both the genus *Xanthomonas* and *Pseudomonas sensu stricto* were placed in rRNA super-family II (De Ley, 1978). In addition, results from these studies demonstrated that *Stenotrophomonas maltophilia* (at that time considered a member of the genus *Pseudomonas*) was related to the genus *Xanthomonas* and was subsequently renamed as *Xanthomonas maltophilia* (Swings et al., 1983; see also Van Zyl and Steyn, 1992; Palleroni and Bradbury, 1993).

Subsequently, cataloging of 16S rRNA sequences placed the genus *Xanthomonas* in the *Gammaproteobacteria* (purple bacteria; Stackebrandt et al., 1988), along with *Pseudomonas sensu stricto* (Woese et al., 1985), the subdivision being broadly synonymous with rRNA super families I and II (De Ley, 1978). In a more detailed overview of the phylogeny of the genus *Pseudomonas* and traditionally associated genera and species, the genera *Xanthomonas*, *Stenotrophomonas*, and *Xylella* were recovered together based on 16S rDNA sequence information (Kersters et al., 1996). This grouping broadly equated with rRNA hybridization group V (Palleroni et al., 1973). In this analysis, however, some doubt was expressed as to the true phylogenetic position of the group, as it was not possible to unequivocally place it within either the *Gammaproteobacteria* or *Betaproteobacteria*. Subsequently, the position of all three genera was confirmed within the *Gammaproteobacteria* and the monophyletic lineage of the genus *Xanthomonas*

demonstrated (Moore et al., 1997). In this analysis, a core set of species was found to be closely related; *X. axonopodis* pathovar *axonopodis*, *X. campestris* pathovar *campestris*, *X. fragariae*, *X. populi* pathovar *populi*, and *X. oryzae* pathovar *oryzae* all clustered together, with *X. albilineans* more distant. A high degree of homogeneity was evident within the genus, and indeed *S. maltophilia* was found to share 95–97% sequence similarity with the xanthomonads studied here. In a subsequent, more detailed study by Hauben et al. (1997), similar findings were presented. The mean similarity value derived from the comparison of 16S rDNA sequence data from 46 xanthomonads, including 20 type strains, was found to be 98.2%. The genus was recovered in three clusters, one of which contained *X. albilineans*, *X. hyacinthi*, *X. theicola*, and *X. translucens*, while the second contained *X. sacchari*, and the remaining cluster the "core" species of the genus. In addition, *S. maltophilia*, although closely related to *Xanthomonas*, was clearly distinct. This finding, and similar results reported by Moore et al. (1997), was in keeping with an earlier study of restriction maps of the 16S rRNA gene, which demonstrated that the genera *Stenotrophomonas* and *Xanthomonas* were closely related, sharing 69 restriction sites, though a number of sites specific to *Stenotrophomonas* gave indications of the distinctiveness of the genera (Nesme et al., 1995).

Pathovars Within the genus *Xanthomonas*, it has long been known that *X. campestris*, with its many pathovars, was heterogeneous. Some pathovars are very similar; others show clear distinctions in phenotype and genotype, especially when sensitive methods are used. DNA–DNA hybridization studies of 54 pathovars then considered to belong to *X. campestris* indicated six groups and four ungrouped pathovars (Hildebrand et al., 1990; Vauterin et al., 1990; Palleroni et al., 1993). Extending these studies with more strains confirmed the six groups and added 14 more, to make 20 genomic species in the genus as a whole (Vauterin et al., 1995). Four of these equate to the existing species *X. albilineans*, *X. fragariae*, *X. oryzae*, and *X. populi*. The remaining 16 are derived from the heterogeneous *X. campestris*. Rather surprisingly, Vauterin et al. (1995) found that the phenotypically distinct species *X. axonopodis*, as studied by classical methods, belongs to the largest DNA hybridization group, which therefore takes its name. In the case of the latter, the emended species was found to encompass 37 pathovars, together with a further 11 included based on a previous study by Palleroni et al. (1993). In a similar way, *X. campestris* pathovar *plantaginis* has been added to the six pathovars in the emended description of *X. campestris*, and pathovar *carotae* and pathovar *taraxaci* added to the three already in *X. hortorum*. The inclusion of these pathovars on the basis of previously collected results is open to debate, as Vauterin et al. employed spectrophotometric determination of renaturation rates to measure DNA hybridization values (De Ley et al., 1970), whilst Palleroni et al. used an S1 nuclease radiolabeling procedure (Hildebrand et al., 1990). In addition, in the study of Palleroni et al., some of the individual reassociation values reported were very low. The reassociation values reported for *X. axonopodis* pathovar *betlicola* did not exceed 43% to any other member of the species *X. axonopodis*, as defined by Vauterin et al. (1995), and average reassociation values for each of the other pathovars studied were seldom in excess of 50% to other members of the species. These findings, coupled with ambiguous results derived from fatty acid analysis (Yang et al., 1993d) and SDS-PAGE analysis of whole cell proteins (Vauterin et al., 1991), further call into question the assumptions made about these previously studied pathovars in the current treatment of the genus. Clearly, further study is required to confirm the species affiliations of the pathovars studied by Palleroni et al. (1993); for the purposes of this review, we have highlighted those concerned.

Subsequent to the recent treatment of the genus by Vauterin et al. (1995), the number of recognized pathovars within *X. axonopodis* was reduced, as some of them were considered invalid in terms of the Standards for Naming Pathovars (Dye et al., 1980) *viz*: *X. axonopodis* pathovar *aurantifolii* and *X. axonopodis* pathovar *citrumelo* (See Young et al., 1991b), and *X. axonopodis* pathovar *cassavae*, *X. axonopodis* pathovar *phaseoli* biovar fuscans and *X. axonopodis* pathovar *vesicatoria* (Young et al., 1996b). Similarly, *X. arboricola* pathovar *poinsettiicola*, *X. hortorum* pathovar *vitians*, *X. translucens* pathovar *hordei*, and *X. vasicola* pathovar *vasculorum*, all described by Vauterin et al. (1995), were also considered defective in terms of the Standards for Naming Pathovars and not included in the list of Young et al. (1996b). In the case of the latter, by invalidating *X. vasicola* pathovar *vasculorum*, the need to maintain *X. vasicola* pathovar *holcicola* is removed and the pathovar is reduced to a synonym of *X. vasicola*. The remaining pathovars are listed as *Pathovars Incertae Sedis*, together with all other pathovars not studied by Vauterin et al. (1995) under their names as pathovars of *X. campestris sensu lato*.

In conclusion, the complexity of the genus *Xanthomonas* is recognized and few would argue with Vauterin and Swings (1997) that the genus represents a continuum of genotypic and phenotypic diversity, albeit encompassing a number of distinctive taxa. The heterogeneity within *X. campestris sensu lato* was self-evident and much valuable information has been produced over the years to support this. This said, it is our opinion that the classification produced by Vauterin et al. (1995) and described below is unconvincing, especially for some of the more heterogenous taxa, such as *X. axonopodis*. The power and value of DNA hybridization data for bacterial taxonomy is well established, but when the groupings derived from this method are not conclusively backed up by other techniques such as 16S rRNA gene sequencing (Hauben et al., 1997), SDS-PAGE analysis of whole-cell proteins (Vauterin et al., 1991), fatty acids (Yang et al., 1993d; Vauterin et al., 1996b), and biochemical tests (Vauterin et al., 1995), which is particularly true for *X. axonopodis*, a classification driven by these data alone is weakened, as the method cannot be used in any practical sense to perform identifications. Diagnosticians with poor resources will have problems operating within the current system, even when the polyphasic methods concur, as classical phenotypic data are of marginal relevance. Polyphasic methods will only work for the four existing species of *Xanthomonas*, *viz*: *X. albilineans*, *X. fragariae*, *X. oryzae*, and *X. populi* (see Table BXII.γ.18), as the 16 recently proposed species were derived from *X. campestris sensu lato* and classical phenotypic tests were not included in their description (Vauterin et al., 1995). It should be recognized that many diagnosticians are currently identifying putative xanthomonads to the genus level, noting the diseased host and identifying straight to the pathovar level, paying little heed to the species affiliation. Clearly, this situation is unacceptable in the long term, and much work remains to be done to develop an alternative strategy for characterization within the genus *Xanthomonas*.

TABLE BXII.γ.18. Characteristics of some *Xanthomonas* species that can be determined by classical methods[a,b]

Characteristic	X. albilineans	X. fragariae	X. oryzae	X. populi
Mucoid growth on nutrient agar + 5% glucose	−	+	+	+
Hydrolysis of gelatin	D	+	−[c]	−
H₂S from peptone	−	−	+	−
Maximum growth temperature, °C	37	33	32	27.5
Maximum salt tolerance (%, w/v)	0.5	0.5–1.0	0.5–2.0	0.4–0.6
Acid production from: [d]				
Arabinose	−	−	d[e]	−
Glucose, sucrose	+	+	+	+
Mannose	+	+	+	+
Galactose	d	−	+	+
Trehalose	−	−	+	+
Cellobiose	−	−	+	−
Fructose	−	+	+	+
Lactose	−	−	−	−
Maltose	−	−	−	−
Xylose	+	−	−	
Utilization of:				
Acetate, citrate, malate			+	
Propionate		−	+	
Succinate		+	+	
DL-Tartrate	−	−		
Benzoate	−	−		
Growth on Oxoid NA:				
None	+			
Good			+	
Poor to very poor		+		+

[a]for symbols see standard definitions.

[b]Data from Dye (1962, 1966b); Ridé and Ridé (1978, 1992; Swings et al. (1990); Vera Cruz et al. (1984).

[c]Reaction is positive for pathovar *oryzicola*.

[d]In the medium of Hayward (1964).

[e]In the medium of Starr (1946).

FURTHER READING

Bradbury, J.F. 1986. Guide to Plant Pathogenic Bacteria, CAB International Mycological Institute, Kew.

Swings, J.G. and E.L. Civerolo (Editors). 1993. *Xanthomonas*, Chapman & Hall, London.

Vauterin, L. and J. Swings. 1997. Are classification and phytopathological diversity compatible in *Xanthomonas*? J. Ind. Microbiol. Biotechnol. *19*: 77–82.

List of species of the genus Xanthomonas

1. **Xanthomonas campestris** (Pammel 1895) Dowson 1939, 190[AL] emend. Vauterin, Hoste, Kersters and Swings 1995, 484 (*"Bacillus campestris"* Pammel 1895, 130.)

cam.pes′ tris. L. gen. n. *campestris* of a level field; also the specific epithet of *Brassica campestris*, a host plant.

The characteristics are as described for the genus and as listed in Tables BXII.γ.19 and BXII.γ.20. The original isolates of this species caused a vascular disease of *Brassica* spp. Currently, the species includes a number of pathovars that cause diseases predominantly within the family Cruciferae. In general, pathovars are not distinguishable by phenotypic characterization and identification is reliant on knowledge of their hosts.

Seventeen strains isolated from host plants in the family Cruciferae fell into DNA hybridization group 15 with the type strain of *X. campestris* (Vauterin et al., 1995). The average DNA binding value within the group was 87% ± 7%. Analysis of strains by fatty acid methyl esters recovered the majority in Cluster 2 (Yang et al., 1993d), which also contains strains from *X. arboricola*, *X. axonopodis*, and *X. vesicatoria* (Table BXII.γ.19). Subsequent analysis of fatty acid data demonstrated a high degree of internal homogeneity within the species (Vauterin et al., 1996b). SDS-PAGE analysis of total proteins placed strains in cluster 1 (Vauterin et al., 1991), though only strains from pathovar *campestris* were examined in this study. Metabolic activity on carbon compounds, as shown by Biolog GN microplates, clearly distinguishes 14 of the other 19 species, but not the remaining five *viz X. hortorum*, *X. arboricola*, *X. cassavae*, *X. axonopodis*, and *X. vasicola* (see Table BXII.γ.20). Laboratory identification of the species without complex apparatus and processes would seem to be uncertain.

In the following list of the seven currently acceptable pathovars of *X. campestris* and their pathotype strains (Vauterin et al., 1995; Young et al., 1996b), names of host plants and their respective families are given. It should be noted that *X. campestris* pathovar *plantaginis* was not studied by Vauterin et al. (1995) and its inclusion in the species was based on the work of Palleroni et al. (1993). The validity of this action could be questioned, as DNA from this pathovar was not hybridized against the full range of pathovars covered in the study of Vauterin et al. and operational differences between the analyses may have occurred.

The mol% G + C of the DNA is: 65.8–66.6; 68.0–68.3 (T_m) for *Xanthomonas campestris* pathovar *incanae*.

Type strain: ATCC 33913, CFPB 2350, DSM 3586, ICMP 13, LMG 568, NCPPB 528.

GenBank accession number (16S rRNA): X95917.

 a. **Xanthomonas campestris** pathovar **campestris** (Pammel 1895) Dowson 1939.

  Hosts: *Brassica* spp., *Capsella bursa-pastoris, Lepidium sativum, Matthiola* spp., *Raphanus sativus, Amorica rusticana*, (fam. Cruciferae). The one report on *Boerhaavia*

erecta (fam. Nyctaginaceae) seems doubtful and has not been confirmed.

  Type strain: ATCC 33913, CFBP 2350; ICMP 13; LMG 568; NCPPB 528.

 b. **Xanthomonas campestris** pathovar **aberrans** (Knösel 1961) Dye 1978.

  Host: *Brassica oleracea* var. *botrytis*, (fam. Cruciferae).

  Type strain: ICMP 4805, LMG 9037, NCPPB 2986.

 c. **Xanthomonas campestris** pathovar **armoraciae** (McCulloch 1929) Dye 1978.

  Hosts: *Armoracia rusticana*. By inoculation *Brassica oleracea* var. *botrytis, B. oleracea* var. *capitata* (fam. Cruciferae), *Phaseolus vulgaris* (fam. Leguminosae).

  Type strain: ICMP 7, LMG 535, NCPPB 347.

 d. **Xanthomonas campestris** pathovar **barbareae** (Burkholder 1941) Dye 1978.

  Host: *Barbarea vulgaris* (fam. Cruciferae).

  Type strain: ATCC 13460, ICMP 438, LMG 547, NCPPB 983.

 e. **Xanthomonas campestris** pathovar **incanae** (Kendrick and Baker 1942) Dye 1978.

  Host: *Matthiola incanae* (fam. Cruciferae).

  Type strain: ATCC 13462, CFBP 2527, ICMP 574, LMG 7490, NCPPB 937.

 f. **Xanthomonas campestris** pathovar **plantaginis** (Thornberry and Anderson 1937) Dye 1978.

  Hosts: *Plantago* spp. (fam. Plantaginaceae).

  Type strain: ATCC 23382, ICMP 1028, LMG 848, NCPPB 1061.

 g. **Xanthomonas campestris** pathovar **raphani** (White 1930) Dye 1978.

  Hosts: *Brassica* spp., *Raphanus sativus*, (fam. Cruciferae), *Capsicurn annuum, Lycopersicon esculentum, Nicotiana tabacum* (fam. Solanaceae).

  Type strain: ATCC 49079, ICMP 1404, LMG 860, NCPPB 1946.

2. **Xanthomonas albilineans** (Ashby 1929) Dowson 1943, 11[AL] emend. van den Mooter and Swings 1990, 367 (*"Bacterium albilineans"* Ashby 1929, 135.)

al.bi.lin′ e.ans. L. adj. *albus* white; L. part. adj. *lineans* striping; M.L. adj. *albilineans* white striping.

The characteristics are as described for the genus and as listed in Tables BXII.γ.18, BXII.γ.19, and BXII.γ.20. Causes leaf scald disease of sugar cane (*Saccharum officinarum*); other natural hosts include *Brachiaria piligera, Imperata cylindrica* var. *major, Paspalum conjugatum* (fam. Gramineae). Colonies are yellowish buff on Dye′s YSP medium or similar sucrose-peptone media. Growth usually fails or is very poor on nutrient agar. Glutamate and methionine are required for growth. Esculin is hydrolyzed, but not starch.

TABLE BXII.γ.19. Percentage concentration of fatty acid methyl esters useful for species differentiation within the genus *Xanthomonas*[a,b]

Fatty Acid	1. X. campestris	2. X. albilineans	3. X. arboricola	4. X. axonopodis	5. X. bromi	6. X. cassavae	7. X. codiaei	8. X. cucurbitae	9. X. fragariae	10. X. hortorum	11. X. hyacinthi	12. X. melonis	13. X. oryzae	14. X. pisi	15. X. populi	16. X. sacchari	17. X. theicola	18. X. translucens	19. X. vasicola	20. X. vesicatoria
Saturated acids:																				
C$_{16:0}$	3.6	13.4	4	4.1	4.9	1.4	6.3	6.7	3.7	2	3.8	4.3	4.3	3.4	7.4	4.2	8.8	4.3	15.8	3.3
Unsaturated acids:																				
C$_{15:1}$ ω8c	0	0	0	0	0	0	0	0	0	0	0	0	**18.5**	0	**2.9**	0	0	0	0	0
C$_{16:1}$ ω9c	0.9	0	0.9	0.7	0.7	0.5	1.5	0.1	0	0.1	0	1.6	2.3	1	**16.9**	1.1	0.2	1	0.7	0.8
C$_{16:1}$ ω7c	12.7	13.2	15	16.2	8.9	10.3	15.1	16	11.7	18.6	12.1	10.4	24.9	13.8	**0**	15	12.9	21.1	25	15.6
Branched saturated acids:																				
C$_{14:0}$ iso	0.7	0	0.8	0.5	0	0	0.6	1.2	0.2	0.7	3.7	0.2	0	0.7	0	**14**	9.8	2.6	0	0.9
C$_{15:0}$ iso	26.5	**2.4**	31.1	28.2	21.3	31.8	26	31.5	36.8	33.2	11.1	28.5	6.3	31.8	23.4	17.3	12.9	29.4	15.5	24.6
C$_{15:0}$ anteiso	13.9	**0.1**	13.7	9.3	2.6	9.5	7.4	10.2	10.5	15.7	26.8	6.6	0.7	9.4	1.6	5.6	15.9	5	3	15.5
C$_{16:0}$ iso	3.2	4.7	2	2.6	1.7	1.1	3.6	3.2	0.9	1.7	6.6	1.9	0.2	1	0	**18.8**	9.9	3.6	1.3	4.1
C$_{17:0}$ iso	6.8	**19.3**	4.5	6.7	10.9	9.5	7.4	3.9	3.7	2.7	2.1	10.2	14.6	5.3	2.5	2.7	2.4	4.8	7.3	6.8
C$_{17:0}$ anteiso	0.8	**2.9**	0.3	0.4	0.1	1	0.5	0.2	0.2	0.1	1.1	0.2	0.3	0.5	0	0.2	1.2	0	1.1	1.1
C$_{19:0}$ iso	0	**1.6**	0	0	0	0	0	0	0	0	0	0	0	0	0	0	0	0	0	0
Branched unsaturated acids:																				
C$_{15:1}$ iso G	0	0	0	0	0	0	0	0	0	0	0	0	0	0	**1.6**	0	0	0	0	0
C$_{17:1}$ iso ω9c	8.5	23.3	3.8	6.5	25.7	17.5	7.5	6.1	3.5	3.9	1.7	10.4	9.1	6.2	0	2.2	2.4	7.4	11.2	8.3
Others:																				
C$_{16:0}$ alc	0	0.3	0	0	0	0	0	0	0	0	**1.6**	0	0	0	0	0	0	0	0	0
C$_{17:0}$ cyclo	0	0	0	0	0	0	0	0	0	0	1	0	0	0	0	0	0	0	0	0

[a]The most distinctive components are highlighted in bold.

[b]Adapted from Vauterin et al. (1996b).

TABLE BXII.γ.20. Some diagnostic tests for species of *Xanthomonas* with Biolog GN® plates[a,b]

Carbon source	1. X. campestris (19)	2. X. albilineans (5)	3. X. arboricola (12)	4. X. axonopodis (96)	5. X. bromi (2)	6. X. cassavae (5)	7. X. codiaei (2)	8. X. cucurbitae (5)	9. X. fragariae (2)	10. X. hortorum (29)	11. X. hyacinthi (5)	12. X. melonis (2)	13. X. oryzae (10)	14. X. pisi (2)	15. X. populi (2)	16. X. sacchari (2)	17. X. theicola (3)	18. X. translucens (32)	19. X. vasicola (9)	20. X. vesicatoria (7)
Dextrin	+	−	+	+	+	d	+	+	+	d	+	+	+	+	+	+	+	+	d[c]	+
N-Acetyl-D-galactosamine	−	−	−	d	−	d	−	−	−	d	−	−	−	+	+	−	−	+	−	d
Cellobiose	+	d	+	+	+	d	+	+	−	+	+	+	+	+	+	+	+	d	d[c]	d
L-Fucose	d[c]	d	+	d	d	+	+	d	d	+	+	+	d	+	−	+	+	d	d[c]	+
D-Galactose	+	−	+	d	+	d	+	d	d	d	+	+	d	+	d	+	+	d	d[c]	−
Gentiobiose	+	−	+	d	+	+	+	−	+	+	+	+	d	+	d	+	+	d	d[c]	+
α-D-Lactose	+	−	d	d	+	+	+	+	+	+	+	+	+	+	+	+	+	+	d[c]	+
Maltose	+	−	d	d	+	d	+	+	−	+	−	d	+	+	+	+	+	+	d[c]	+
Melibiose	+	−	+	−	d	+	+	+	−	−	D	d	−	+	d	+	−	+	−	d
β-Methyl-D-glucoside	−	−	−	−	−	−	−	−	−	−	−	−	−	−	−	−	−	−	−	−
D-Psicose	+	−	+	+	+	d	+	+	+	+	+	+	+	+	+	+	+	+	+	+
L-Rhamnose	−	−	−	d	d	−	−	−	−	d	−	−	+	+	d	+	d	−	d	d
Acetic acid	D	−	+	d	d	d	−	−	d	d	−	+	d	+	d	+	d	−	d	−
cis-Aconitic acid	D	−	d	d	−	d	+	+	+	d	+	+	d	+	d	−	d	−	d	+
D-Glacturonic acid	−	−	−	−	−	−	−	−	−	−	−	−	−	−	d	−	−	−	−	−
D-Glucuronic acid	−	−	−	+	d	−	−	d	−	d	−	d	−	+	−	d	−	−	d	−
β-Hydroxybutyrate	−	−	d	d	d	−	d	d	−	d	+	−	d	+	d	+	+	−	d	d
p-Hydroxyphenyl-acetate	−	−	−	−	−	−	−	−	−	−	−	−	−	−	−	−	−	−	−	−
α-Ketobutyric acid	D	−	d	d	d	d	+	−	−	d	−	+	d	+	d	+	−	−	d	d
α-Ketovaleric acid	−	−	−	−	+	−	−	−	−	−	−	d	−	+	d	+	+	−	−	−
DL-Lactic acid	D	+	+	d	d	d	−	+	−	+	−	d	−	+	d	−	+	d	d	+
Malonic acid	d[c]	−	d	d	−	−	−	d	d	d	−	d	d	−	d	d	−	d	d	d
D-Saccharic acid	d[c]	−	−	d	−	d	d	d	−	d	−	+	d	d	d	+	−	d	d[c]	+
Bromosuccinic acid	+	−	+	+	+	+	+	+	d	+	+	+	d	+	d	+	−	d	d[c]	+
Alaninamide	D	D	d	d[c]	+	+	+	d	−	d	−	+	d	d	+	+	+	+	d[c]	+
D-Alanine	D	−	d	+	+	+	+	−	d	d	−	+	d	+	d	+	+	d	d[c]	d
L-Alanine	D	−	d	+	d	+	+	d	d	+	−	d	d	+	d	+	+	d	d[c]	d
L-Alanylglycine	D	D	d	+	+	+	+	d	−	+	−	d	d	+	d	+	d	+	d[c]	d
L-Glutamic acid	+	D	+	+	+	+	+	+	−	+	D	+	−	+	−	+	+	+	+	d
Hydroxy-L-proline	D	−	d	d	d	d	−	−	−	d	D	+	−	d	d	+	+	−	d[c]	d
L-Leucine	−	−	d	d	−	d	d	−	−	+	D	+	d	d	d	+	−	−	d	d
L-Serine	D	−	d	d	d	+	+	d	+	+	D	d	−	d	d	+	d	−	d	d
Glycerol	D	−	+	d	+	+	+	d	−	d	+	−	−	+	d	−	d	−	d	d
Glucose-1-phosphate	D	−	d	d	d	d	+	−	−	d	+	−	−	+	d	+	d	−	−	d
Glucose-6-phosphate	D	−	−	d	d	d	−	−	−	d	−	−	−	+	−	−	d	d	−	−

[a]For symbols see standard definitions. Numbers in parentheses indicate the number of strains tested.

[b]Adapted from Vauterin et al. (1995).

[c]89% positive.

H₂S is not produced from peptone. The maximum sodium chloride tolerance is 0.5%. Maximum growth temperature is 37°C. Acid is produced within 21 days in Dye's medium C from glucose, sucrose, mannose, and xylose, but not from arabinose, fructose, trehalose, cellobiose, rhamnose, adonitol, mannitol, sorbitol, dulcitol, *meso*-inositol, salicin, inulin, or α-methylglucoside. Two strains isolated from the host plant *Saccharum officinarum* showed an average DNA binding value of 97% (Vauterin et al., 1995). The fatty acid methyl ester profile (Table BXII.γ.19) recovers the majority of strains studied in Cluster 31 of Yang et al. (1993d). Subsequent analysis of fatty acid data demonstrated a high degree of internal homogeneity within the species (Vauterin et al., 1996b). SDS-PAGE analysis of total proteins was unable to recover the strains in a single cluster, though the three strains included in this study showed greater similarity to each other than to any other xanthomonad examined (Vauterin et al. 1991). Metabolic activity on carbon compounds as shown by Biolog GN microplates is shown in Table BXII.γ.20.

Certain strains isolated from *Saccharum officinarum*, exclusively from Guadeloupe, were originally included in this species but are now encompassed within *X. sacchari*, as they are clearly distinct on the basis of DNA hybridization studies, fatty acids, and protein profiling.

The mol% G + C of the DNA is: 63.1–64.5 (Bd, T_m).

Type strain: ATCC 33915, CFBP 2523, ICMP 196, LMG 494, NCPPB 2969.

GenBank accession number (16S rRNA): X95918.

3. **Xanthomonas arboricola** Vauterin, Hoste, Kersters and Swings 1995, 484[VP]

ar.bo.ri' co.la. L. n. *arbor* a tree; L. suff. *-cola* dweller; M.L. adj. n. *arboricola* tree dweller.

The characteristics are as described for the genus and as listed in Tables BXII.γ.19 and BXII.γ.20. The isolates cause disease mainly in temperate trees, including hazelnut, walnut, and *Prunus* spp. In general, pathovars are not distinguishable by phenotypic characterization and identification is reliant on knowledge of their hosts. The species was recovered as DNA hybridization group 4 by Vauterin et al. (1995). The average DNA binding value within the group was 79% ± 15%. Three of the five pathovars were closely related, with hybridization values of 89%, and were distinguished from other strains by their quinate metabolism (Lee et al., 1992). *X. arboricola* pathovar *populi* and pathovar *poinsettiicola* were related to other group 4 pathovars at average DNA hybridization values of 57% and 68%, respectively, and ranged from 48 to 76%, indicating some lack of homogeneity in the species. Fatty acid methyl ester composition (Table BXII.γ.19) placed the strains in Cluster 2 of Yang et al. (1993d), a position shared by three pathovars of *X. axonopodis*, several pathovars of *X. campestris*, and *X. vesicatoria*. Subsequent analysis of fatty acid data demonstrated a high degree of homogeneity within the species (Vauterin et al., 1996b). Most strains were recovered in Cluster 7c by protein SDS-PAGE analysis (Vauterin et al., 1991), though some strains of *X. arboricola* pathovar *juglandis* were recovered in Cluster 2 and *X. arboricola* pathovar *poinsettiicola* LMG 5403 was unclustered in this study. Metabolic activity on various carbon sources, determined using GN microplates, distinguishes this species from 15 of the 20 in the genus (see Table BXII.γ.20), but distinction from *X. hor-*

torum, *X. cassavae*, *X. axonopodis*, *X. vasicola*, and *X. campestris* is not clear. Laboratory identification of the species without complex apparatus and processes would seem to be uncertain.

In the following list of currently acceptable pathovars of *X. arboricola* and their pathotype strains (Vauterin et al., 1995; Young et al., 1996b), the synonyms, names of host plants, and their respective families are given. It should be noted that *X. campestris* pathovar *celebensis* was not studied by Vauterin et al. (1995) and its inclusion in this species was based on the work of Palleroni et al. (1993). The validity of this action could be questioned, as DNA from this pathovar was not hybridized against the full range of pathovars covered in the study of Vauterin et al. and operational differences between the analyses may have occurred.

The mol% G + C of the DNA is: 66.0–67.0 (T_m).

Type strain: ATCC 49083, ICMP 35, LMG 747, NCPPB 411.

GenBank accession number (16S rRNA): Y10757.

Additional Remarks: This is also the pathotype strain of *X. arboricola* pathovar *juglandis* (there being no *X. arboricola* pathovar *arboricola*).

a. **Xanthomonas arboricola** *pathovar* **celebensis** (Gäumann 1923) Vauterin, Hoste, Kersters and Swings 1995 (*Xanthomonas campestris* pathovar *celebensis* (Gäumann 1923) Dye 1978.)

Hosts: *Musa* spp. (fam. Musaceae).

Type strain: ATCC 19045, ICMP 1488, LMG 677, NCPPB 1832.

b. **Xanthomonas arboricola** *pathovar* **corylina** (Miller, Bollen, Simmons, Gross and Barss 1940) Vauterin, Hoste, Kersters and Swings 1995 (*Xanthomonas campestris* pathovar *corylina* (Miller, Bollen, Simmons, Gross and Barss 1940) Dye 1978.)

Hosts: *Corylus* spp. (fam. Betulaceae).

Type strain: ATCC 19313, CFBP 1159, ICMP 5726, LMG 689, NCPPB 935.

c. **Xanthomonas arboricola** *pathovar* **juglandis** (Pierce 1901) Vauterin, Hoste, Kersters and Swings 1995 (*Xanthomonas campestris* pathovar *juglandis* (Pierce 1901) Dye 1978.)

Hosts: *Juglans* spp. (fam. Juglandaceae).

Type strain: ATCC 49083, CFBP 2528, ICMP 35, LMG 747, NCPPB 411.

d. **Xanthomonas arboricola** *pathovar* **populi** (ex de Kam 1984) Vauterin, Hoste, Kersters and Swings 1995 (*Xanthomonas campestris* pathovar *populi* (ex de Kam 1984) Young, Bradbury, Davis, Dickey, Ercolani, Hayward and Vidaver 1991a.)

Hosts: *Populus* spp. (fam. Salicaceae).

Type strain: ICMP 8923, LMG 12141.

e. **Xanthomonas arboricola** *pathovar* **pruni** (Smith 1903) Vauterin, Hoste, Kersters and Swings 1995 (*Xanthomonas campestris* pathovar *pruni* (Smith 1903) Dye 1978.)

Hosts: *Prunus* spp., *Sorbus japonica* (fam. Rosaceae).

Type strain: ATCC 19316, CFBP 2535, ICMP 51, LMG 852, NCPPB 416.

4. **Xanthomonas axonopodis** Starr and Garcés 1950, 81[AL] emend. Vauterin, Hoste, Kersters and Swings 1995, 484 (*Xanthomonas axonoperis* (sic) Starr and Garcés 1950, 81.)

a.xo.no' po.dis. M.L. n. *Axonopus* generic name of a grass; M.L. gen. n. *axonopodis* of *Axonopus*.

The characteristics are as described for the genus and as listed in Tables BXII.γ.19 and BXII.γ.20. As originally constituted, this species was slower growing and less metabolically active than *X. campestris*, from which it was clearly differentiated phenotypically. The emended description now includes isolates from a wide variety of host plants and *X. axonopodis* is the most heterogeneous species currently recognized in the genus. In general, pathovars are not distinguishable by phenotypic characterization and identification is reliant on knowledge of their hosts. The species was recovered as DNA hybridization group 9 by Vauterin et al. (1995). The average DNA binding value within the group was 77% ± 15%. The type strain of the species shows a mean DNA binding within the group of only 66%. *X. axonopodis* and *X. campestris* pathovar *vasculorum* type A were closely related and were assigned to Group 9 with an overall level of DNA hybridization of 58% with other strains of the group. The individual values ranged from 47 to 71%, indicating a lack of homogeneity in the group. Fatty acid methyl ester composition (Table BXII.γ.19) recovered the bulk of isolates in Cluster 1, but isolates were also found in Clusters 2, 4, 5, 15, 18, 21, 28, and 29, together with several pathovars belonging to different species, as currently described (Yang et al., 1993d). Subsequent analysis of fatty acid data demonstrated a low homogeneity within the species (Vauterin et al., 1996b). Most isolates were recovered in Cluster 3 by protein SDS-PAGE analysis (Vauterin et al., 1991), alongside isolates of *X. cucurbitae*. Other isolates were recovered in clusters 5, 6, 8, 9, and a sizeable number were unclustered by this method. Metabolic activity on various carbon sources, determined using Biolog GN microplates, distinguishes this species from 10 others, but not from the remaining nine, *viz X. hortorum*, *X. arboricola*, *X. cassavae*, *X. oryzae*, *X. vasicola*, *X. melonis*, *X. vesicatoria*, *X. campestris*, and *X. translucens* (Table BXII.γ.20).

In the following list of currently acceptable pathovars of *X. axonopodis* and their pathotype strains (Vauterin et al., 1995; Young et al., 1996b), the synonyms, names of host plants, and their respective families are given. It should be noted that *X. campestris* pathovar *betlicola*, *X. campestris* pathovar *biophyti*, *X. campestris* pathovar *fasicularis*, *X. campestris* pathovar *khayae*, *X. campestris* pathovar *maculifoliigardeniae*, *X. campestris* pathovar *martyniicola*, *X. campestris* pathovar *mehusii*, *X. campestris* pathovar *nakataecorchori*, *X. campestris* pathovar *pedalii*, *X. campestris* pathovar *physalidicola*, and *X. campestris* pathovar *punicae* were not studied by Vauterin et al., (1995) and their inclusion in this species was based on the work of Palleroni et al. (1993). The validity of this action could be questioned, as DNA from these pathovars were not hybridized against the full range of pathovars covered in the study of Vauterin et al. and operational differences between the analyses may have occurred.

The mol% G + C of the DNA is: 62.6–65 (Bd, T_m).

Type strain: ATCC 19312, ICMP 50, LMG 538, NCPPB 457.

GenBank accession number (16S rRNA): X95919.

a. **Xanthomonas axonopodis** *pathovar* **axonopodis** Starr and Garcés 1950.

Hosts: *Axonopus scoparius*, *A. micay*, *A. compressus*, and *A. affinis*. By inoculation, *Digitaria decumbens*, *Hypharrhenia rufa*, *Panicum* sp., and *Saccharum officinarum* (fam. Gramineae).

Type strain: ATCC 19312, ICMP 50, LMG 538, NCPPB 457.

b. **Xanthomonas axonopodis** *pathovar* **alfalfae** (Riker, Jones and Davis 1935) Vauterin, Hoste, Kersters and Swings 1995 (*Xanthomonas campestris* pathovar *alfalfae* (Riker, Jones and Davis 1935) Dye 1978.)

Hosts: *Medicago sativa*, *Melilotus indica*, *Pisum sativum*, *Phaseolus vulgaris*, *Trigonella foenum-graecum* (fam. Leguminosae).

Type strain: ICMP 5718, LMG 497, NCPPB 2062.

c. **Xanthomonas axonopodis** *pathovar* **bauhiniae** (Padhya, Patel and Kotasthane 1965a) Vauterin, Hoste, Kersters and Swings 1995 (*Xanthomonas campestris* pathovar *bauhiniae* (Padhya, Patel and Kotasthane 1965a) Dye 1978.)

Host: *Bauhinia racemosa* (fam. Leguminosae).

Type strain: ICMP 5720, LMG 548, NCPPB 1335.

d. **Xanthomonas axonopodis** *pathovar* **begoniae** (Takimoto 1934) Vauterin, Hoste, Kersters and Swings 1995 (*Xanthomonas campestris* pathovar *begoniae* (Takimoto 1934) Dye 1978.)

Host: *Begonia* spp. (fam. Begoniaceae).

Type strain: ATCC 49082, CFBP 2524, ICMP 194, LMG 7303, NCPPB 1926.

e. **Xanthomonas axonopodis** *pathovar* **betlicola** (Patel, Kulkarni and Dhande 1951b) Vauterin, Hoste, Kersters and Swings 1995 (*Xanthomonas campestris* pathovar *betlicola* (Patel, Kulkarni and Dhande 1951b) Dye 1978.)

Hosts: *Piper betle*, *P. hookeri*, *P. longum* (fam. Piperaceae).

Type strain: ATCC 11677, ICMP 312, LMG 555, NCPPB 2972.

f. **Xanthomonas axonopodis** *pathovar* **biophyti** (Patel, Chauhan, Kotasthane and Desai 1969) Vauterin, Hoste, Kersters and Swings 1995 (*Xanthomonas campestris* pathovar *biophyti* (Patel, Chauhan, Kotasthane and Desai 1969) Dye 1978.)

Host: *Biophytum sensitivum* (fam. Oxalidaceae).

Type strain: ICMP 2780, LMG 556, NCPPB 2228.

g. **Xanthomonas axonopodis** *pathovar* **cajani** (Kulkarni, Patel and Abhyankar 1950) Vauterin, Hoste, Kersters and Swings 1995 (*Xanthomonas campestris* pathovar *cajani* (Kulkarni, Patel and Abhyankar 1950) Dye 1978.)

Host: *Cajanus cajan* (fam. Leguminosae).

Type strain: ATCC 11639, ICMP 444, LMG 558, NCPPB 573.

h. **Xanthomonas axonopodis** *pathovar* **cassiae** (Kulkarni, Patel and Dhande 1951) Vauterin, Hoste, Kersters and Swings 1995 (*Xanthomonas campestris* pathovar *cassiae* (Kulkarni, Patel and Dhande 1951) Dye 1978.)

Hosts: *Cassia tora*, *C. occidentalis*, *Cicer arietinum*, *Pisum sativum* (fam. Leguminosae).

Type strain: ATCC 11638, ICMP 358, LMG 675, NCPPB 2973.

i. **Xanthomonas axonopodis** *pathovar* **citri** (Hasse 1915) Vauterin, Hoste, Kersters and Swings 1995 (*Xanthomonas citri* (ex Hasse 1915) Gabriel, Kingsley, Hunter and Gottwald 1989; *Xanthomonas campestris* pathovar *citri* (Hasse 1915) Dye 1978.)

Hosts: *Aegle marmelos*, *Atalantia* spp. *Balsamocitrus paniculata*, *Casimiroa edulis*, *Chaetospermum glutinosa*, *Citropsis schweinfurthii*, *Citrus* spp. and hybrids, *Clausena lansium*, *Eremocitrus glauca*, *Evodia* spp. *Feronia* spp., *Feroniella* spp., *Fortunelia* spp., *Hesperethusa crenulata*, *Limonia* spp., *Melicope triphylla*, *Microcitrus* spp., *Murraya erotica*, *Paramigyna longipedunculata*, *Poncirus trifoliata* and hybrids, *Severina buxifolia*, *Toddalia asiatica*, *Zanthoxylum* spp. (fam. Rutaceae).

Type strain: CFBP 2525, ICMP 24, LMG 682, NCPPB 409.

j. **Xanthomonas axonopodis** *pathovar* **clitoriae** (Pandit and Kulkarni 1979) Vauterin, Hoste, Kersters and Swings 1995 (*Xanthomonas campestris* pathovar *clitoriae* (Pandit and Kulkarni 1979) Dye, Bradbury, Goto, Hayward, Lelliott and Schroth 1980.)

Host: *Clitoria biflora* (fam. Leguminosae).
Type strain: ICMP 6574, LMG 9045, NCPPB 3092.

k. **Xanthomonas axonopodis** *pathovar* **coracanae** (Desai, Thirumalachar and Patel 1965) Vauterin, Hoste, Kersters and Swings 1995 (*Xanthomonas campestris* pathovar *coracanae* (Desai, Thirumalachar and Patel 1965) Dye 1978.)

Host: *Eleusine coracana* (fam. Gramineae).
Type strain: ICMP 5724, LMG 686; NCPPB 1786.

l. **Xanthomonas axonopodis** *pathovar* **cyamopsidis** (Patel, Dhande and Kulkarni 1953) Vauterin, Hoste, Kersters and Swings 1995 (*Xanthomonas campestris* pathovar *cyamopsidis* (Patel, Dhande and Kulkarni 1953) Dye 1978.)

Host: *Cyamopsis tetragonoloba* (fam. Leguminosae).
Type strain: ICMP 616, LMG 691, NCPPB 637.

m. **Xanthomonas axonopodis** *pathovar* **desmodii** (Patel 1949) Vauterin, Hoste, Kersters and Swings 1995 (*Xanthomonas campestris* pathovar *desmodii* (Patel 1949) Dye 1978.)

Host: *Desmodium diffusum* (fam. Leguminosae).
Type strain: ATCC 11640, ICMP 315, LMG 692, NCPPB 481.

n. **Xanthomonas axonopodis** *pathovar* **desmodiigangetici** (Patel and Moniz 1948) Vauterin, Hoste, Kersters and Swings 1995 (*Xanthomonas campestris* pathovar *desmodiigangetici* (Patel and Moniz 1948) Dye 1978.)

Host: *Desmodium gangeticum* (fam. Leguminosae).
Type strain: ATCC 11671, ICMP 577, LMG 693, NCPPB 577.

o. **Xanthomonas axonopodis** *pathovar* **desmodiilaxiflori** (Pant and Kulkarni 1976a) Vauterin, Hoste, Kersters and Swings 1995 (*Xanthomonas campestris* pathovar *desmodiilaxiflori* (Pant and Kulkarni 1976a.)

Hosts: *Desmodium laxiflorum*, *Tamarindus indica* (fam. Leguminosae).
Type strain: ICMP 6502, LMG 9046, NCPPB 3086.

p. **Xanthomonas axonopodis** *pathovar* **desmodiirotundifolii** (Desai and Shah 1960) Vauterin, Hoste, Kersters and Swings 1995 (*Xanthomonas campestris* pathovar *desmodiirotundifolii* (Desai and Shah 1960) Dye 1978.)

Host: *Desmodium rotundifolium* (fam. Leguminosae).
Type strain: ICMP 168, LMG 694, NCPPB 885).

q. **Xanthomonas axonopodis** *pathovar* **dieffenbachiae** (McCulloch and Pirone 1939) Vauterin, Hoste, Kersters and Swings 1995 (*Xanthomonas campestris* pathovar *dieffenbachiae* (McCulloch and Pirone 1939) Dye 1978.)

Hosts: *Aglaonema robellinii*, *Anthurium andraeanum*, *Dieffenbachia* spp. (fam. Araceae), *Dracaena fragrans* (fam. Liliaceae).
Type strain: ICMP 5727, LMG 695, NCPPB 1833.

r. **Xanthomonas axonopodis** *pathovar* **erythrinae** (Patel, Kulkarni and Dhande 1952a) Vauterin, Hoste, Kersters and Swings 1995 (*Xanthomonas campestris* pathovar *erythrinae* (Patel, Kulkarni and Dhande 1952a) Dye 1978.)

Host: *Erythrina indica* (fam. Leguminosae).
Type strain: ATCC 11679, ICMP 446, LMG 698, NCPPB 578.

s. **Xanthomonas axonopodis** *pathovar* **fascicularis** (Patel and Kotasthane 1969b) Vauterin, Hoste, Kersters and Swings 1995 (*Xanthomonas campestris* pathovar *fascicularis* (Patel and Kotasthane 1969b) Dye 1978.)

Host: *Corchorus fascicularis* (fam. Tiliaceae).
Type strain: ICMP 5731, LMG 9047, NCPPB 2230.

t. **Xanthomonas axonopodis** *pathovar* **glycines** (Nakano 1919) Vauterin, Hoste, Kersters and Swings 1995 (*Xanthomonas campestris* pathovar *glycines* (Nakano 1919) Dye 1978.)

Hosts: *Brunnichia cirrhosa* (fam. Polygonaceae), *Dolichos uniflorus*, *Glycine* spp. *Phaseolus lunatus*, *P. vulgaris* (fam. Leguminosae).
Type strain: ATCC 43911, CFBP 2526, ICMP 5732, LMG 712, NCPPB 554.

u. **Xanthomonas axonopodis** *pathovar* **khayae** (Sabet 1959) Vauterin, Hoste, Kersters and Swings 1995 (*Xanthomonas campestris* pathovar *khayae* (Sabet 1959) Dye 1978.)

Hosts: *Khaya senegalensis*, *K. grandifoliola* (fam. Meliaceae).
Type strain: ICMP 671, LMG 753, NCPPB 536.

v. **Xanthomonas axonopodis** *pathovar* **lespedezae** (Ayers, Lefebvre and Johnson 1939) Vauterin, Hoste, Kersters and Swings 1995 (*Xanthomonas campestris* pathovar *lespedezae* (Ayers, Lefebvre and Johnson 1939) Dye 1978.)

Host: *Lespedeza* spp. (fam. Leguminosae).
Type strain: ATCC 13463, ICMP 439, LMG 757, NCPPB 993.

w. **Xanthomonas axonopodis** *pathovar* **maculifoliigardeniae** (Ark and Barrett 1946) Vauterin, Hoste, Kersters and Swings 1995 (*Xanthomonas campestris* pathovar *maculifoliigardeniae* (Ark and Barrett 1946) Dye 1978.)

Hosts: *Gardenia* spp., *Ixora coccinea* (fam. Rubiaceae).
Type strain: CFBP 1155, ICMP 318, LMG 758, NCPPB 971.

x. **Xanthomonas axonopodis** *pathovar* **malvacearum** (Smith 1901) Vauterin, Hoste, Kersters and Swings 1995 (*Xanthomonas campestris* pathovar *malvacearum* (Smith 1901) Dye 1978.)

Hosts: *Gossypium* spp., *Thespesia lambas* (fam. Malvaceae), *Ceiba pentandra* (fam. Bombacaceae).
Type strain: CFBP 2530, ICMP 5739, LMG 761, NCPPB 633.

y. **Xanthomonas axonopodis** *pathovar* **manihotis** (Bondar 1915) Vauterin, Hoste, Kersters and Swings 1995 (*Xanthomonas campestris* pathovar *manihotis* (Bondar 1915) Dye 1978.)

Host: *Manihot* spp. (fam. Euphorbiaceae).
Type strain: ATCC 49073, CFBP 2603, ICMP 5741, LMG 773, NCPPB 1834.

z. **Xanthomonas axonopodis** *pathovar* **martyniicola** (Moniz and Patel 1958) Vauterin, Hoste, Kersters and Swings 1995 (*Xanthomonas campestris* pathovar *martyniicola* (Moniz and Patel 1958) Dye 1978.)

    Host: *Martynia diandra* (fam. Martyniaceae).

    Type strain: ICMP 82, LMG 9049, NCPPB 1148.

aa. **Xanthomonas axonopodis** *pathovar* **melhusii** (Patel, Kulkarni and Dhande 1952a) Vauterin, Hoste, Kersters and Swings 1995 (*Xanthomonas campestris* pathovar *melhusii* (Patel, Kulkarni and Dhande 1952a) Dye 1978.)

    Host: *Tectona grandis* (fam. Verbenaceae).

    Type strain: ATCC 11644, ICMP 619, LMG 9050, NCPPB 994.

ab. **Xanthomonas axonopodis** *pathovar* **nakataecorchori** (Padhya and Patel 1963b) Vauterin, Hoste, Kersters and Swings 1995 (*Xanthomonas campestris* pathovar *nakataecorchori* (Padhya and Patel 1963b) Dye 1978.)

    Host: *Corchorus acutangulus* (fam. Tiliaceae).

    Type strain: ICMP 5742, LMG 786, NCPPB 1337.

ac. **Xanthomonas axonopodis** *pathovar* **patelii** (Desai and Shah 1959) Vauterin, Hoste, Kersters and Swings 1995 (*Xanthomonas campestris* pathovar *patelii* (Desai and Shah 1959) Dye 1978.)

    Host: *Crotalaria juncea* (fam. Leguminosae).

    Type strain: ICMP 167, LMG 811, NCPPB 840.

ad. **Xanthomonas axonopodis** *pathovar* **pedalii** (Patel and Jindal 1972) Vauterin, Hoste, Kersters and Swings 1995 (*Xanthomonas campestris* pathovar *pedalii* (Patel and Jindal 1972) Dye 1978.)

    Host: *Pedalium murex* (fam. Pedaliaceae).

    Type strain: ICMP 3030, LMG 812, NCPPB 2368.

ae. **Xanthomonas axonopodis** *pathovar* **phaseoli** (Smith 1897) Vauterin, Hoste, Kersters and Swings 1995 (*Xanthomonas phaseoli* (ex Smith 1897) Gabriel, Kingsley, Hunter and Gottwald 1989; *Xanthomonas campestris* pathovar *phaseoli* (Smith 1897) Dye 1978.)

    Hosts: *Lablab purpureus* (syn. *Dolichos lablab*), *Lupinus polyphyllus, Phaseolus lunatus, Phaseolus vulgaris* (fam. Leguminosae).

    Type strain: ATCC 9563, CFBP 2534, ICMP 5834, LMG 7455, NCPPB 3035.

af. **Xanthomonas axonopodis** *pathovar* **phyllanthi** (Sabet, Ishag and Khalil 1969) Vauterin, Hoste, Kersters and Swings 1995 (*Xanthomonas campestris* pathovar *phyllanthi* (Sabet, Ishag and Khalil 1969) Dye 1978.)

    Host: *Phyllanthus niruri* (fam. Euphorbiaceae).

    Type strain: ICMP 5745, LMG 844, NCPPB 2066.

ag. **Xanthomonas axonopodis** *pathovar* **physalidicola** (Goto and Okabe 1958) Vauterin, Hoste, Kersters and Swings 1995 (*Xanthomonas campestris* pathovar *physalidicola* (Goto and Okabe 1958) Dye 1978.)

    Host: *Physalis alkekengi* var. *francheti* (fam. Solanaceae).

    Type strain: ATCC 49077, ICMP 586, LMG 845, NCPPB 761.

ah. **Xanthomonas axonopodis** *pathovar* **poinsettiicola** (Patel, Bhatt and Kulkarni 1951a) Vauterin, Hoste, Kersters and Swings 1995 (*Xanthomonas campestris* pathovar *poinsettiicola* (Patel, Bhatt and Kulkarni 1951a) Dye 1978.)

    Hosts: *Euphorbia pulcherrima, E. milii, Manihot escutenta* (fam. Euphorbiaceae).

    Type strain: ATCC 11643, ICMP 5779, LMG 849, NCPPB 581.

ai. **Xanthomonas axonopodis** *pathovar* **punicae** (Hingorani and Singh 1959) Vauterin, Hoste, Kersters and Swings 1995 (*Xanthomonas campestris* pathovar *punicae* (Hingorani and Singh 1959) Dye 1978.)

    Host: *Punica granatum* (fam. Punicaceae).

    Type strain: ICMP 360, LMG 859, NCPPB 466.

aj. **Xanthomonas axonopodis** *pathovar* **rhynchosiae** (Sabet, Ishag and Khalil 1969) Vauterin, Hoste, Kersters and Swings 1995 (*Xanthomonas campestris* pathovar *rhynchosiae* (Sabet, Ishag and Khalil 1969) Dye 1978.)

    Hosts: *Lupinus termis, Mucuna pruriens* (syn. *Stizolobium alterrimum*), *Rhynchosia memnonia* (fam. Leguminosae).

    Type strain: ICMP 5748, LMG 8021, NCPPB 1827.

ak. **Xanthomonas axonopodis** *pathovar* **ricini** (Yoshii and Takimoto 1928) Vauterin, Hoste, Kersters and Swings 1995 (*Xanthomonas campestris* pathovar *ricini* (Yoshii and Takimoto 1928) Dye 1978.)

    Host: *Ricinus communis* (fam. Euphorbiaceae).

    Type strain: ATCC 19317, ICMP 5747, LMG 861, NCPPB 1063.

al. **Xanthomonas axonopodis** *pathovar* **sesbaniae** (Patel, Kulkarni and Dhande 1952b) Vauterin, Hoste, Kersters and Swings 1995 (*Xanthomonas campestris* pathovar *sesbaniae* (Patel, Kulkarni and Dhande 1952b) Dye 1978.)

    Host: *Sesbania aegyptiaca* (fam. Leguminosae).

    Type strain: ATCC 11675, (ICMP 367, LMG 867, NCPPB 582.

am. **Xanthomonas axonopodis** *pathovar* **tamarindi** (Patel, Bhatt and Kulkarni 1951a) Vauterin, Hoste, Kersters and Swings 1995 (*Xanthomonas campestris* pathovar *tamarindi* (Patel, Bhatt and Kulkarni 1951a) Dye 1978.)

    Hosts: *Caesalpina sepiaria, Tamarindus indica* (fam. Leguminosae).

    Type strain: ATCC 11673, ICMP 572, LMG 955, NCPPB 584.

an. **Xanthomonas axonopodis** *pathovar* **vasculorum** (Cobb 1893) Vauterin, Hoste, Kersters and Swings 1995 (*Xanthomonas campestris* pathovar *vasculorum* (Cobb 1893) Dye 1978.)

    Hosts: *Bambusa vulgaris, Brachiaria mutica, Coix lacryma-jobi, Panicum maximum, Pennisetum purpureum, Saccharum officinarum, Sorghum* spp., *Thysanolaena maxima, Zea mays* (fam. Gramineae), *Cocos nucifera, Dictyosperma alba* (fam. Palmae).

    Type strain: ATCC 35938, CFBP 2602, ICMP 5757, LMG 901, NCPPB 796.

ao. **Xanthomonas axonopodis** *pathovar* **vignaeradiatae** (Sabet, Ishag and Khalil 1969) Vauterin, Hoste, Kersters and Swings 1995 (*Xanthomonas campestris* pathovar *vignaeradiatae* (Sabet, Ishag and Khalil 1969) Dye 1978.)

    Hosts: *Lablab purpureus* (syn. *Dolichos lablab*), *Vigna radiata* (fam. Leguminosae).

    Type strain: ICMP 5759, LMG 936, NCPPB 2058.

ap. **Xanthomonas axonopodis** *pathovar* **vignicola** (Burkholder 1944) Vauterin, Hoste, Kersters and Swings 1995 (*Xanthomonas campestris* pathovar *vignicola* (Burkholder 1944) Dye 1978.)

Hosts: *Phaseolus vulgaris, Vigna pubigera, V. unguiculata* (fam. Leguminosae).

Type strain: ATCC 11648, ICMP 333, LMG 8752, NCPPB 1838.

aq. **Xanthomonas axonopodis** *pathovar* **vitians** (Brown 1918) Vauterin, Hoste, Kersters and Swings 1995 (*Xanthomonas campestris* pathovar *vitians* (Brown 1918) Dye 1978.)

Host: *Lactuca* spp. (fam. Compositae).

Type strain: ATCC 19320, CFBP 2538, ICMP 336, LMG 937, NCPPB 976.

Additional Remarks: This strain may originally have been inappropriately chosen as the pathotype strain (Vauterin et al., 1995).

5. **Xanthomonas bromi** Vauterin, Hoste, Kersters and Swings 1995, 485[VP]

bro′mi. L. masc. gen. M.L. n. *Bromus* generic name for the bromegrass; M.L. gen. n. *bromi* of *Bromus*.

The characteristics are as described for the genus and as listed in Tables BXII.γ.19 and BXII.γ.20. The isolates were all recovered from *Bromus* spp. (fam. Gramineae) and cause a wilting disease. Three strains studied were recovered in DNA hybridization group 7 (Vauterin et al., 1995). The average DNA binding value within the group was 88% ± 6%. The fatty acid methyl ester profile (Table BXII.γ.19) belongs to cluster 25 of Yang et al. (1993d), which only contains isolates from *Bromus* spp. Subsequent analysis of fatty acid data demonstrated a very high degree of internal homogeneity within the species (Vauterin et al., 1996b). All *Bromus* isolates were recovered in Cluster 7 by protein SDS-PAGE analysis (Vauterin et al., 1992), distinct from other xanthomonads isolated from cereals and grasses. Metabolic activity on carbon compounds, determined using Biolog GN microplates, clearly distinguishes the two strains studied from all other species within the genus (Table BXII.γ.20).

The mol% G + C of the DNA is: 65.6 (T_m).

Type strain: CFBP 1976, ICMP 12545, LMG 947.

GenBank accession number (16S rRNA): Y10764.

6. **Xanthomonas cassavae** (ex Wiehe and Dowson 1953) Vauterin, Hoste, Kersters and Swings 1995, 485[VP]

cas.sa′vae. M.L. fem. gen. n. *Cassava* the English form of the flour made from *Manihot esculenta*.

The characteristics are as described for the genus and as listed in Tables BXII.γ.19 and BXII.γ.20. The isolates cause leaf spotting and necrosis of *Manihot* spp. (fam. Euphorbiaceae). Of the five strains received as this pathogen and studied in DNA hybridization experiments (Vauterin et al., 1995), four, including the pathotype strain, showed an average DNA binding value of 94% ± 2%. The remaining strain showed binding at 87% with group 9 and was included in *X. axonopodis* as pathovar *cassavae*. The fatty acid methyl ester profile (Table BXII.γ.19) of the majority of strains studied belongs to Cluster 5 of Yang et al. (1993d). Subsequent analysis of fatty acid data demonstrated a high degree of internal homogeneity within the species (Vauterin et al., 1996b). SDS-PAGE analysis of total proteins placed strains in cluster 5 (Vauterin et al., 1991). Metabolic activity on carbon compounds, determined using Biolog GN microplates, clearly distinguishes this species from 14 of the other 19 species, but not the remaining five *viz X. hortorum, X. arboricola, X. axonopodis, X. vesicatoria,* or *X. campestris* (Table BXII.γ.20).

The mol% G + C of the DNA is: 64.2–66.1 (T_m).

Type strain: ICMP 204, LMG 673, NCPPB 101.

GenBank accession number (16S rRNA): Y10762.

7. **Xanthomonas codiaei** Vauterin, Hoste, Kersters and Swings 1995, 485[VP]

co.di.ae′i. M.L. neut. n. *codiaeum* generic name of croton; M.L. gen. n. *codiaei* of *Codiaeum.*

The characteristics are as described for the genus and as listed in Tables BXII.γ.19 and BXII.γ.20. The isolates cause a wilt of *Codiaeum variegatum* (fam. Euphorbiaceae). Two of four strains of *X. campestris* pathovar *poinsettiicola* studied were recovered in DNA hybridization group 6 in the study of Vauterin et al. (1995), with an average DNA binding value of 99%. They were considered sufficiently distinct from the pathotype and from all other groups to be separated as a species. These strains were recovered in Cluster 2 by fatty acid methyl ester analysis (Table BXII.γ.19; Yang et al. 1993d), alongside strains of *X. arboricola, X. axonopodis, X. campestris,* and *X. vesicatoria.* Subsequent analysis of fatty acid data demonstrated a very high degree of internal homogeneity within the species, though it should be noted that only two strains were studied (Vauterin et al., 1996b). SDS-PAGE analysis of total proteins placed strains in Cluster 3c (Vauterin et al. 1991), alongside *X. axonopodis* pathovar *alfalfae* and *X. axonopodis* pathovar *ricini.* Metabolic activity on carbon compounds, determined using Biolog GN microplates, clearly distinguishes this species from all others (Table BXII.γ.20).

The mol% G + C of the DNA is: 66.3 (T_m).

Type strain: ICMP 9513, LMG 8678.

GenBank accession number (16S rRNA): Y10765.

8. **Xanthomonas cucurbitae** (ex Bryan 1926) Vauterin, Hoste, Kersters and Swings 1995, 485[VP]

cu.cur′ bit.ae. M.L. fem. n. *Cucurbita* generic name of squash; M.L. gen. n. *cucurbitae* of *Cucurbita.*

The characteristics are as described for the genus and as listed in Tables BXII.γ.19 and BXII.γ.20. The isolates cause necrotic spots on leaves, and occasionally stems, petioles, and fruits of *Citrullus vulgaris, Cucumis sativus,* and *Cucurbita* spp. (fam. Cucurbitaceae). Two strains of *X. campestris* pathovar *cucurbitae* were recovered in DNA hybridization group 8 in the study of Vauterin et al. (1995), with an average DNA binding value of 88%. They were considered sufficiently distinct from all other groups to be separated as a species. These strains were recovered in Cluster 1 by fatty acid methyl ester analysis (Table BXII.γ.19; Yang et al., 1993d), alongside many strains of *X. axonopodis.* Subsequent analysis of fatty acid data demonstrated a very high degree of internal homogeneity within the species, though it should be noted that only two strains were studied (Vauterin et al., 1996b). SDS-PAGE analysis of total proteins placed strains in Cluster 10 (Vauterin et al., 1991). Metabolic activity on carbon compounds, determined using Biolog GN microplates, clearly distinguishes this species from all others, except *X. translucens* from which there are no clear differences (Table BXII.γ.20). The negative reaction with gentiobiose is particularly useful for diagnosis.

The mol% G + C of the DNA is: 66.1–66.8 (T_m).

Type strain: ICMP 2299, LMG 690, NCPPB 2597.

GenBank accession number (16S rRNA): Y10760.

9. **Xanthomonas fragariae** Kennedy and King 1962, 875[AL] emend. van den Mooter and Swings 1990, 367.

fra.gar′i.ae. M.L. n. *Fragaria* generic name of strawberry; M.L. gen. n. *fragariae* of *Fragaria*.

The characteristics are as described for the genus and as listed in Tables BXII.γ.18, BXII.γ.19, and BXII.γ.20. The isolates cause leaf spot disease of *Fragaria* spp. (fam. Rosaceae). Two strains of *X. fragariae* studied were recovered in DNA hybridization group 1 in the study of Vauterin et al. (1995), with an average DNA binding value of 91%. These strains were recovered in Cluster 12 by fatty acid methyl ester analysis (Table BXI.γ.19; Yang et al., 1993d). Subsequent analysis of fatty acid data demonstrated a high degree of internal homogeneity within the species (Vauterin et al., 1996b). SDS-PAGE analysis of total proteins placed strains in Cluster 18 (Vauterin et al., 1991). Metabolic activity on carbon compounds, determined using Biolog GN microplates, clearly distinguishes this species from all others (Table BXII.γ.20).

The mol% G + C of the DNA is: 62.6–63.3 (Bd, T_m).

Type strain: ATCC 33239, CFBP 2157, ICMP 5715, LMG 708, NCPPB 1469.

GenBank accession number (16S rRNA): X95920.

10. **Xanthomonas hortorum** Vauterin, Hoste, Kersters and Swings 1995, 485[VP]

hor.to′rum. L. gen. n. *hortorum* of gardens.

The characteristics are as described for the genus and as listed in Tables BXII.γ.19 and BXII.γ.20. In general, pathovars are not distinguishable by phenotypic characterization and identification is reliant on knowledge of their hosts. Six strains isolated from necrotic spots on a variety of plants, mainly on leaves, were recovered in DNA hybridization group 2 in the study of Vauterin et al. (1995), with an average DNA binding value of 92% ± 10%. These strains were recovered in Cluster 10 by fatty acid methyl ester analysis (Table BXII.γ.19; Yang et al., 1993d), though *X. campestris* pathovar *vitians* type B (LMG 938), now included in this species, was found to be loosely associated with Cluster 1. Subsequent analysis of fatty acid data demonstrated a high degree of internal homogeneity within the species (Vauterin et al., 1996b). SDS-PAGE analysis of total proteins placed strains in Cluster 7e (*X. hortorum* pathovar *hederae*) and 15 (*X. hortorum* pathovar *pelargonii*; Vauterin et al., 1991). Metabolic activity on carbon compounds, determined using Biolog GN microplates, clearly distinguishes this species from 14 of the other 19 species, but not the remaining five *viz X. arboricola, X. cassavae, X. axonopodis, X. vesicatoria*, and *X. campestris* (Table BXII.γ.20). Laboratory identification of the species without complex apparatus and processes would seem to be uncertain.

In the following list of currently acceptable pathovars of *X. hortorum* and their pathotype strains (Vauterin et al., 1995; Young et al., 1996b), the synonyms, names of host plants, and their respective families are given. It should be noted that *X. campestris* pathovar *carotae* and *X. campestris* pathovar *taraxaci* were not studied by Vauterin et al. (1995) and their inclusion in this species was based on the work of Palleroni et al. (1993). The validity of this action could be questioned, as DNA from this pathovar was not hybridized against the full range of pathovars covered in the study of Vauterin et al. and operational differences between the analyses may have occurred.

The mol% G + C of the DNA is: 63.6–65.1 (T_m).

Type strain: ICMP 453, LMG 733, NCPPB 939.

GenBank accession number (16S rRNA): Y10759.

Additional Remarks: This is also the pathotype strain of *X. hortorum* pathovar *hederae* (there being no *X. hortorum* pathovar *hortorum*).

a. **Xanthomonas hortorum** *pathovar* **carotae** (Kendrick 1934) Vauterin, Hoste, Kersters and Swings 1995 (*Xanthomonas campestris* pathovar *carotae* (Kendrick 1934) Dye 1978.)

Host: *Daucus carota* (fam. Umbelliferae).

Type strain: ICMP 5723, LMG 8646, NCPPB 1422.

Additional Remarks: This strain has been reported to be unsuitable as a pathotype strain (Young et al. 1991a).

b. **Xanthomonas hortorum** *pathovar* **hederae** (Arnaud 1920) Vauterin, Hoste, Kersters and Swings 1995 (*Xanthomonas campestris* pathovar *hederae* (Arnaud 1920) Dye 1978.)

Host: *Hedera helix* (fam. Araliaceae).

Type strain: ICMP 453, LMG 733, NCPPB 939.

c. **Xanthomonas hortorum** *pathovar* **pelargonii** (Brown 1923) Vauterin, Hoste, Kersters and Swings 1995 (*Xanthomonas campestris* pathovar *pelargonii* (Brown 1923) Dye 1978.)

Host: *Geranium* spp., *Pelargonium* spp. (fam. Geraniaceae).

Type strain: CFBP 2533, ICMP 4321, LMG 7314, NCPPB 2985.

d. **Xanthomonas hortorum** *pathovar* **taraxaci** (Niederhauser 1943) Vauterin, Hoste, Kersters and Swings 1995 (*Xanthomonas campestris* pathovar *taraxaci* (Niederhauser 1943) Dye 1978.)

Host: *Taraxacum bicorne* (fam. Compositae).

Type strain: ATCC 19318, CFBP 410, ICMP 579, LMG 870, NCPPB 940.

11. **Xanthomonas hyacinthi** (ex Wakker 1883) Vauterin, Hoste, Kersters and Swings 1995, 486[VP]

hy.a.cin′thi. Gr. *hyakinthos* used as M.L. masc. n. *Hyacinthus* generic name; M.L. gen. n. *hyacinthi* of *Hyacinth*.

The characteristics are as described for the genus and as listed in Tables BXII.γ.19 and BXII.γ.20. The isolates cause yellow leaf stripes and wilt in *Hyacinthus orientalis* (fam. Liliaceae). Three strains of *X. campestris* pathovar *hyacinthi* studied were recovered in DNA hybridization group 17 of Vauterin et al. (1995), with an average DNA binding value of 98% ± 2%. They were considered sufficiently distinct from all other groups to be separated as a species. These strains were recovered in Cluster 19 by fatty acid methyl ester analysis (Table BXII.γ.19; Yang et al., 1993d). Subsequent analysis of fatty acid data demonstrated a very high degree of internal homogeneity within the species (Vauterin et al., 1996b). SDS-PAGE analysis of total proteins placed strains in Cluster 4 (Vauterin et al., 1991). Metabolic activity on carbon compounds, determined using Biolog GN microplates, clearly distinguishes this species from all others (Table BXII.γ.20).

The mol% G + C of the DNA is: 69.2–69.3 (T_m).

Type strain: ATCC 19314, ICMP 189, LMG 739, NCPPB 599.

GenBank accession number (16S rRNA): Y10754.

12. **Xanthomonas melonis** Vauterin, Hoste, Kersters and Swings 1995, 486[VP]

me.lo' nis. L. fem. gen. n. *melonis* of melon.

The characteristics are as described for the genus and as listed in Tables BXII.γ.19 and BXII.γ.20. The isolates cause a soft rot of fruit of *Cucumis melo* (fam. Cucurbitaceae). Two strains of *X. campestris* pathovar *melonis* studied were recovered in DNA hybridization group 13 of Vauterin et al. (1995), with an average DNA binding value of 88%. They were considered sufficiently distinct from all other groups to be separated as a species. These strains were recovered in Cluster 4 by fatty acid methyl ester analysis (Table BXII.γ.19; Yang et al., 1993d), alongside strains of *X. axonopodis* pathovar *citri*, *X. campestris* pathovar *lantanae*, and *X. campestris* pathovar *viticola*, the latter two pathovars were not examined by Vauterin et al. (1995) and their exact taxonomic position is at present uncertain. Subsequent analysis of fatty acid data demonstrated a very high degree of internal homogeneity within the species (Vauterin et al., 1996b). SDS-PAGE analysis of total proteins placed strains in Cluster 19 (Vauterin et al., 1991), alongside two strains of *X. campestris* pathovar *poinsettiicola*; the latter were not examined by Vauterin et al. (1995) and their exact taxonomic position is at present uncertain. Metabolic activity on carbon compounds, determined using Biolog GN microplates, clearly distinguishes this species from all others, except *X. axonopodis* (see Table BXII.γ.20).

The mol% G + C of the DNA is: 66.1 (T_m).

Type strain: ICMP 8682, LMG 8670, NCPPB 3434.

GenBank accession number (16S rRNA): Y10756.

13. **Xanthomonas oryzae** (ex Ishiyama 1922) Swings, van den Mooter, Vauterin, Hoste, Gillis, Mew and Kersters 1990, 309[VP] emend. van den Mooter and Swings 1990, 367.

o.ry' zae. M.L. gen. n. *Oryza* the generic name for rice; M.L. gen. n. *oryzae* of *Oryza*.

The characteristics are as described for the genus and as listed in Tables BXII.γ.18, BXII.γ.19, and BXII.γ.20. The isolates cause blight and leaf streak of *Oryza* spp. (fam. Gramineae). Six strains of *X. campestris* pathovar *oryzae* and *X. campestris* pathovar *oryzicola* studied were recovered in DNA hybridization group 10 of Vauterin et al. (1995), with an average DNA binding value of 91% ± 7%. These strains were recovered in Cluster 22 (*X. oryzae* pathovar *oryzicola*) and 24 (*X. oryzae* pathovar *oryzae*) by fatty acid methyl ester analysis (Table BXII.γ.19; Yang et al., 1993d). Subsequent analysis of fatty acid data demonstrated a high degree of internal homogeneity within the species (Vauterin et al., 1996b). SDS-PAGE analysis of total proteins placed strains in Cluster 14 (*X. oryzae* pathovar *oryzae*) and 15 (*X. oryzae* pathovar *oryzicola*) (Vauterin et al., 1991). Metabolic activity on carbon compounds, determined using Biolog GN microplates, clearly distinguishes this species from all others (see Table BXII.γ.20).

In the following list of currently acceptable pathovars of *X. oryzae* and their pathotype strains (Vauterin et al., 1995; Young et al., 1996b), the synonyms, names of host plants, and their respective families are given.

The mol% G + C of the DNA is: 64.6–65 (T_m).

Type strain: YK9, ATCC 35933, CFBP 2532, ICMP 3125, LMG 5047, NCPPB 3002.

GenBank accession number (16S rRNA): X95921.

a. **Xanthomonas oryzae** *pathovar* **oryzae** (Ishiyama 1922) Swings, Van den Mooter, Vauterin, Hoste, Gillis, Mew and Kersters 1990 (*Xanthomonas campestris* pathovar *oryzae* (Ishiyama 1922) Dye 1978.)

Hosts: *Isachne globosa*, *Leersia* spp., *Leptochola* spp., *Oryza* spp., *O. sativa*, *Phalaris arundinaceae*, *Phragmites communis*, *Zizania aquatica* (fam. Gramineae).

Type strain: ATCC 35933, CFBP 2532, ICMP 3125, LMG 5047, NCPPB 3002.

b. **Xanthomonas oryzae** *pathovar* **oryzicola** (Fang, Ren, Chen, Chu, Faan and Wu 1957) Swings, Van den Mooter, Vauterin, Hoste, Gillis, Mew and Kersters 1990 (*Xanthomonas campestris* pathovar *oryzicola* (Fang, Ren, Chen, Chu, Faan and Wu 1957) Dye 1978.)

Hosts: *Oryza sativa*, *Oryza* spp., *Leersia hexandra* (weak) (fam. Gramineae).

Type strain: ATCC 49072, CFBP 2286, ICMP 5743, LMG 797, NCPPB 1585.

14. **Xanthomonas pisi** (ex Goto and Okabe 1958) Vauterin, Hoste, Kersters and Swings 1995, 486[VP]

pi'si. M.L. gen. n. *Pisum* genus of peas; M.L. gen. n. *pisi* of *Pisum*.

The characteristics are as described for the genus and as listed in Tables BXII.γ.19 and BXII.γ.20. The isolates cause necrotic spots and rotting of *Pisum sativum* (fam. Leguminosae), and are strongly pectolytic, causing soft rot in vegetable tissue. The one strain of *X. campestris* pathovar *pisi* studied was recovered in DNA hybridization group 12 of Vauterin et al. (1995) and considered sufficiently distinct from all other groups to be separated as a species. This strain was distantly related to Cluster 1 by fatty acid methyl ester analysis (Table BXII.γ.19; Yang et al., 1993d), the latter encompassing many strains of *X. axonopodis*. As only one strain has been studied to date, no assessment of the homogeneity of this species can be made. SDS-PAGE analysis of total proteins could not cluster this strain with any other xanthomonad examined in this study (Vauterin et al., 1991). Metabolic activity on carbon compounds, determined using Biolog GN microplates, clearly distinguishes this species from all others (see Table BXII.γ.20). This is by far the most reactive species of *Xanthomonas*, being metabolically active on 70 out of the 95 carbon compounds on microplates (Vauterin et al., 1995). Especially useful diagnostically are the activities shown on *N*-acetyl-D-galactosamine, formate, D-galacturonate, *p*-hydroxyphenylacetate, glucuronamide, and thymidine, which are positive for *X. pisi* and negative for all other species studied.

The mol% G + C of the DNA is: 64.6 (T_m).

Type strain: ATCC 35939, ICMP 570, LMG 847, NCPPB 762.

GenBank accession number (16S rRNA): Y10758.

15. **Xanthomonas populi** (ex Ridé 1958) van den Mooter and Swings 1990, 367[VP]

po'pu.li. M.L. masc. n. *Populus* generic name of the poplar tree; M.L. n. *populi* of *Populus*.

The characteristics are as described for the genus and as listed in Tables BXII.γ.18, BXII.γ.19, and BXII.γ.20. In Dye's Medium C, acid was produced within 21 days from glucose, sucrose, mannose, galactose, trehalose, and fructose (Ridé and Ridé, 1978), but not from arabinose and cellobiose, which agrees with the results obtained from GN plates. Sodium chloride tolerance is low at 0.4–0.6%, and maximum growth temperature is the lowest in all xanthomonads at 27.5°C. Few of the cells in culture have a flagellum, but it can be detected by immunofluorescence. The isolates cause a weeping canker of poplar (*Populus* spp.) and all species in the sections *Aigeros*, *Deltoides*, *Tacamahaca*, *Leuce* (subsections *Albidae* and *Trepidae*) and their hybrids are susceptible (fam. Salicaceae). Three strains of *X. populi* studied were recovered in DNA hybridization group 3 of Vauterin et al. (1995), with an average DNA binding value of 96% ± 1%. These strains were recovered in Cluster 30 by fatty acid methyl ester analysis (Table BXII.γ.19; Yang et al., 1993d). Subsequent analysis of fatty acid data demonstrated a low internal homogeneity within the species, which may be due to poor or non-reproducible growth (Vauterin et al., 1996b). SDS-PAGE analysis of total proteins placed strains in Cluster 13 (Vauterin et al., 1991). Metabolic activity on carbon compounds, determined using Biolog GN microplates, clearly distinguishes this species from all others. One of the least reactive species, it is unequivocally positive for only 10 compounds; these are glycogen, sucrose, D-trehalose, D-fructose, α-D-glucose, and maltose, as well as the four shown in Table BXII.γ.20.

The mol% G + C of the DNA is: 62–65 (T_m).

Type strain: ATCC 51165, ICMP 5816, LMG 5743, NCPPB 2959.

GenBank accession number (16S rRNA): X95922.

16. **Xanthomonas sacchari** Vauterin, Hoste, Kersters and Swings 1995, 486[VP]

sac'cha.ri. L. gen. n. *sacchari* from sugar.

The characteristics are as described for the genus and as listed in Tables BXII.γ.19 and BXII.γ.20. The isolates, all of which have been isolated from Guadeloupe, cause a disease of sugarcane (*Saccharum officinarum*; fam. Gramineae) very like leaf scald. Two strains studied were recovered in DNA hybridization group 19 of Vauterin et al. (1995), with an average DNA binding value of 98%. They were considered sufficiently distinct from *X. albilineans* and all other groups to be separated as a species. All six strains were recovered in Cluster 27 by fatty acid methyl ester analysis (Table BXII.γ.19; Yang et al., 1993d). Subsequent analysis of fatty acid data demonstrated a very high degree of internal homogeneity within the species (Vauterin et al., 1996b). All Guadeloupe isolates were recovered in Cluster 6 by protein SDS-PAGE analysis (Vauterin et al., 1992), distinct from other xanthomonads from cereals and grasses, except for *X. campestris* pathovar *vasculorum* LMG 8285; the latter was not examined by Vauterin et al. (1995) and its exact taxonomic position is at present uncertain. Metabolic activity on carbon compounds, determined using Biolog GN microplates, clearly distinguishes this species from all others (Table BXII.γ.20). Other clear results not shown in the table include metabolic activity on quinate and L-phenylalanine, which also distinguish this species from most others.

The mol% G + C of the DNA is: 68.5 (T_m).

Type strain: LMG 471.

GenBank accession number (16S rRNA): Y10766.

17. **Xanthomonas theicola** Vauterin, Hoste, Kersters and Swings 1995, 486[VP]

the.i' co.la. L. fem. gen. n. *Thea* the generic name for tea; L.v. *colere* to inhabit; M.L. adj. *theicola* living in tea.

The characteristics are as described for the genus and as listed in Tables BXII.γ.19 and BXII.γ.20. The isolates cause bacterial canker of tea (*Camellia sinensis*; fam. Theaceae). A strain of *X. campestris* pathovar *theicola* was recovered in DNA hybridization group 18 in the study of Vauterin et al. (1995), and considered sufficiently distinct from all other groups to be separated as a species. Three strains of the species were recovered in Cluster 20 by fatty acid methyl ester analysis (Table BXII.γ.19; Yang et al., 1993d). Subsequent analysis of fatty acid data demonstrated a very high degree of internal homogeneity within the species (Vauterin et al., 1996b), though only three strains were studied. SDS-PAGE analysis of total proteins placed strains in Cluster 16 (Vauterin et al., 1991). Metabolic activity on carbon compounds, determined using Biolog GN microplates, clearly distinguishes this species from all others (see Table BXII.γ.20). *X. theicola* differs for three or more compounds from 15 species, differs clearly for two compounds from *X. axonopodis*, and only for α-ketovalerate from *X. vasicola* and *X. translucens*.

The mol% G + C of the DNA is: 69.2 (T_m).

Type strain: ICMP 6774, LMG 8684.

GenBank accession number (16S rRNA): Y10763.

18. **Xanthomonas translucens** (ex Jones, Johnson and Reddy 1917) Vauterin, Hoste, Kersters and Swings 1995, 487[VP]

trans.lu' cens. L. v. *transluceo* to be translucent; L. part. adj. *translucens* being translucent.

The characteristics are as described for the genus and as listed in Tables BXII.γ.19 and BXII.γ.20. The isolates cause disease in various members of the family Gramineae. Twenty-three strains were recovered in DNA hybridization group 16 in the study of Vauterin et al. (1995), with an average DNA binding value of 78% ± 11%. These strains were recovered in three closely related clusters, Clusters 7, 8, and 9, by fatty acid methyl ester analysis (Table BXII.γ.19; Yang et al. 1993d). Subsequent analysis of fatty acid data demonstrated a relatively low level of internal homogeneity within the species (Vauterin et al., 1996b). All isolates were recovered in Cluster 1 by protein SDS-PAGE analysis of total proteins (Vauterin et al., 1992), distinct from other xanthomonads isolated from cereals and grasses. Metabolic activity on carbon compounds, determined using Biolog GN microplates, enables clear distinction from 14 of the species, although seven of these used only one C-compound for clear differentiation (see Table BXII.γ.20). The remaining five species, not clearly differentiated by any of the 95 compounds, were *X. cucurbitae*, *X. axonopodis*, *X. oryzae*, *X. vasicola*, and *X. vesicatoria*.

In the following list of currently acceptable pathovars of *X. translucens* and their pathotype strains (Vauterin et al., 1995; Young et al., 1996b), the synonyms, names of host plants, and their respective families are given.

The mol% G + C of the DNA is: 69.1–69.7 (T_m).

Type strain: ATCC 19319, ICMP 5752, LMG 876, NCPPB 973.

GenBank accession number (16S rRNA): X99299.

a. **Xanthomonas translucens** *pathovar* **translucens** (Jones, Johnson and Reddy 1917) Vauterin, Hoste, Kersters and Swings 1995 (*Xanthomonas campestris* pathovar *hordei* (Hagborg 1942) Dye 1978) *Xanthomonas campestris* pathovar *translucens* (Jones, Johnson and Reddy 1917) Dye 1978.)

Hosts: *Hordeum* spp. and other cereals (fam. Gramineae).

Type strain: ATCC 19319, CFBP 2054, ICMP 5752, LMG 876, NCPPB 973.

b. **Xanthomonas translucens** *pathovar* **arrhenatheri** (Egli and Schmidt 1982) Vauterin, Hoste, Kersters and Swings 1995 (*Xanthomonas campestris* pathovar *arrhenatheri* Egli and Schmidt 1982.)

Host: *Arrhenatherum elatius* (fam. Poaceae).

Type strain: ATCC 33803, CFBP 2056, ICMP 7727, LMG 727, NCPPB 3229.

c. **Xanthomonas translucens** *pathovar* **cerealis** (Hagborg 1942) Vauterin, Hoste, Kersters and Swings 1995 (*Xanthomonas campestris* pathovar *cerealis* (Hagborg 1942) Dye 1978.)

Hosts: *Agropyron* spp., *Avena* spp., *Bromus* spp., *Hordeum* spp., *Secate cereale*, *Triticum* spp. (fam. Gramineae).

Type strain: CFBP 2541, ICMP 1409, LMG 679, NCPPB 1944.

d. **Xanthomonas translucens** *pathovar* **graminis** (Egli, Goto and Schmidt 1975) Vauterin, Hoste, Kersters and Swings 1995 (*Xanthomonas campestris* pathovar *graminis* (Egli, Goto and Schmidt 1975) Dye 1978.)

Hosts: *Alopecurus pratensis*, *Dactylis glomerata*, *Festuca* spp., *Lolium multiflorum*, *L. perenne*, *Phleum pratense* (fam. Gramineae).

Type strain: ATCC 29091, CFBP 2053, ICMP 5733, LMG 726, NCPPB 2700.

e. **Xanthomonas translucens** *pathovar* **phlei** (Egli and Schmidt 1982) Vauterin, Hoste, Kersters and Swings 1995 (*Xanthomonas campestris* pathovar *phlei* Egli and Schmidt 1982.)

Host: *Phleum pratense* (fam. Gramineae).

Type strain: ATCC 33805, CFBP 2062, ICMP 7725, LMG 730, NCPPB 3231.

f. **Xanthomonas translucens** *pathovar* **phleipratensis** (Wallin and Reddy 1945) Vauterin, Hoste, Kersters and Swings 1995 (*Xanthomonas campestris* pathovar *phleipratensis* (Wallin and Reddy 1945) Dye 1978.)

Host: *Phleum pratense* (fam. Gramineae).

Type strain: CFBP 2540, ICMP 5744, LMG 843, NCPPB 1837.

g. **Xanthomonas translucens** *pathovar* **poae** (Egli and Schmidt 1982) Vauterin, Hoste, Kersters and Swings 1995 (*Xanthomonas campestris* pathovar *poae* Egli and Schmidt 1982.)

Host: *Poa trivialis* (fam. Gramineae).

Type strain: ATCC 33804, CFBP 2057, ICMP 7726, LMG 728, NCPPB 3230.

h. **Xanthomonas translucens** *pathovar* **secalis** (Reddy, Godkin and Johnson 1924) Vauterin, Hoste, Kersters and Swings 1995 (*Xanthomonas campestris* pathovar *secalis* (Reddy, Godkin and Johnson 1924) Dye 1978.)

Hosts: *Secale cereale*, *Hordeum* and *Triticurn* spp. may become artificially infected (fam. Gramineae).

Type strain: CFBP 2539, ICMP 5749, LMG 883, NCPPB 2822.

i. **Xanthomonas translucens** *pathovar* **undulosa** (Smith, Jones and Reddy 1919) Vauterin, Hoste, Kersters and Swings 1995 (*Xanthomonas campestris* pathovar *undulosa* (Smith, Jones and Reddy 1919) Dye 1978.)

Hosts: *Triticum* spp. (fam. Gramineae), *Secale cereale*, *Hordeum* spp. by inoculation.

Type strain: ATCC 35935, CFBP 2055, ICMP 5755, LMG 892, NCPPB 2821.

19. **Xanthomonas vasicola** Vauterin, Hoste, Kersters and Swings 1995, 487[VP]

vas.i' co.la. L. neut. n. *vas* a vessel; M.L. a duct or vessel as in a vascular bundle; *cola* dwelling in; M.L. adj. (indecl.) *vasicola* living in vascular bundles.

The characteristics are as described for the genus and as listed in Tables BXII.γ.19 and BXII.γ.20. The isolates cause leaf streak of *Sorghum* spp. (fam. Gramineae), which may cover large areas and produce much bacterial exudate. Strains have also been found to cause a gummosis of sugar cane (*Saccharum officinarum*; fam. Gramineae). The three strains of *X. campestris* pathovar *holcicola* studied and three of the seven strains originally assigned to *X. campestris* pathovar *vasculorum* were recovered in Group 11, with an average DNA binding value of 90% ± 4%. As the remaining four strains of *X. campestris* pathovar *vasculorum*, including the pathotype strain, were quite different, Group 11 was considered a separate species and named *X. vasicola* (Vauterin et al., 1995). These strains were recovered in Cluster 21 by fatty acid methyl ester analysis (Table BXII.γ.19; Yang et al., 1993d). Subsequent analysis of fatty acid data demonstrated a very high degree of internal homogeneity within the species (Vauterin et al., 1996b). All isolates were recovered in Cluster 8 by protein SDS-PAGE analysis (Vauterin et al., 1992), distinct from other xanthomonads isolated from cereals and grasses. Metabolic activity on carbon compounds, determined using Biolog GN microplates, clearly distinguishes this species from 13 other *Xanthomonas* species, but no clear results distinguish it from the remaining six species *viz X. fragariae*, *X. arboricola*, *X. axonopodis*, *X. oryzae*, *X. vesicatoria*, and *X. translucens* (see Table BXII.γ.20).

Vauterin et al. (1995) proposed two pathovars, pathovar *holcicola* attacking *Sorghum* spp. and pathovar *vasculorum* attacking *Saccharum* sp.; however as no direct comparisons of their pathogenicity have been conducted, their status is uncertain.

The mol% G + C of the DNA is: 64.2 (T_m).

Type strain: ICMP 3103, LMG 736, NCPPB 2417.

GenBank accession number (16S rRNA): Y10755.

20. **Xanthomonas vesicatoria** (ex Doidge 1920) Vauterin, Hoste, Kersters and Swings 1995, 487[VP]

ve.si.ca.to' ri.a. L. *vesica* a blister; M.L. adj. *vesicatorius* causing a blister.

The characteristics are as described for the genus and as listed in Tables BXII.γ.19 and BXII.γ.20. The isolates cause necrotic lesions on leaves, stems, fruit and other parts of *Capsicum* spp. and *Lycopersicon* spp. (fam. Solanaceae). The two strains of *X. campestris* pathovar *vesicatoria* type B studied were recovered in DNA hybridization group 14 of Vauterin et al. (1995), with an average DNA binding value of 99%, and were quite different from three pathovar *ve-*

sicatoria type A strains that fell into *X. axonopodis*. They were considered sufficiently distinct from all other groups to be separated as a species. These strains were recovered in Cluster 2 by fatty acid methyl ester analysis (Table BXII.γ.19; Yang et al. 1993d), alongside strains of *X. arboricola*, *X. axonopodis*, and *X. campestris*. Subsequent analysis of fatty acid data demonstrated a very high degree of internal homogeneity within the species (Vauterin et al., 1996b). SDS-PAGE analysis of total proteins placed strains in Cluster 7 (Vauterin et al., 1991), alongside strains of *X. arboricola* and *X. hortorum*. Metabolic activity on carbon compounds, determined using Biolog GN microplates, clearly distinguishes this species from 13 of the other *Xanthomonas* spp., but not from *X. hortorum*, *X. cassavae*, *X. axonopodis*, *X. vasicola*, or *X. translucens* (Table BXII.γ.20). The only clear methods to distinguish this species from *X. axonopodis* seem to be SDS-PAGE of cell proteins and DNA hybridization. More suitable phenotypic tests are needed. Because of the re-arrangement of the original taxon into two species, the host ranges of both need re-examination.

The mol% G + C of the DNA is: 65.6 (T_m).

Type strain: ICMP 63, LMG 911, NCPPB 422.

GenBank accession number (16S rRNA): Y10761.

Species Incertae Sedis*

1. **Xanthomonas campestris** *pathovar* **alangii** (Padhya and Patel 1962) Dye 1978.
   Host: *Alangium lamarckii* (fam. Alangiaceae).
   Deposited strain: ICMP 5717, LMG 470; NCPPB 1336.

2. **Xanthomonas campestris** *pathovar* **amaranthicola** (Patel, Wankar and Kulkarni 1952c) Dye 1978.
   Host: *Amaranthus* spp. (fam. Amaranthaceae).
   Deposited strain: ATCC 11645, ICMP 441, LMG 498, NCPPB 570.

3. **Xanthomonas campestris** *pathovar* **amorphophalli** (Jindal, Patel and Singh 1972) Dye 1978.
   Host: *Amorphophallus campanulatus* (fam. Araceae).
   Deposited strain: ICMP 3033, LMG 499, NCPPB 2371.

4. **Xanthomonas campestris** *pathovar* **aracearum** (Berniac 1974) Dye 1978.
   Host: *Xanthosoma sagittifolium* (fam. Araceae).
   Deposited strain: ICMP 5381, LMG 532, NCPPB 2832.

5. **Xanthomonas campestris** *pathovar* **arecae** (Rao and Mohan 1970) Dye 1978.
   Host: *Areca catechu* (fam. Palmaceae).
   Deposited strain: ICMP 5719, LMG 533, NCPPB 2649.

6. **Xanthomonas campestris** *pathovar* **argemones** (Srinivasan, Patel and Thirumalachar 1961b) Dye 1978.
   Host: *Argemone mexicana* (fam. Papaveraceae).
   Deposited strain: ICMP 1617, LMG 534, NCPPB 1593.

7. **Xanthomonas campestris** *pathovar* **arracaciae** (Pereira, Paradella and Zagatto 1971) Dye 1978.
   Host: *Arracacia xanthorrhiza* (fam. Umbelliferae).
   Deposited strain: ICMP 3158, LMG 536, NCPPB 2436.

8. **Xanthomonas campestris** *pathovar* **asclepiadis** Flynn and Vidaver 1995.
   Host: *Asclepias* spp. (fam. Asclepiadaceae).
   Deposited strain: ICMP 10007.

9. **Xanthomonas campestris** *pathovar* **azadirachtae** (Desai, Gandhi, Patel and Kotasthane 1966) Dye 1978.
   Host: *Azadirachta indica* (fam. Meliaceae).
   Deposited strain: ICMP 3102, LMG 543, NCPPB 2388.

10. **Xanthomonas campestris** *pathovar* **badrii** (Patel, Kulkarni and Dhande 1950) Dye 1978.
    Hosts: *Xanthium strumarium* (fam. Compositae), *Pisum sativum* (fam. Leguminosae).
    Deposited strain: ATCC 11672, ICMP 571, LMG 546, NCPPB 571.

11. **Xanthomonas campestris** *pathovar* **betae** Robbs, Kimura and Ribeiro 1981.
    Host: *Beta vulgaris* (fam. Chenopodiaceae).
    Deposited strain: ICMP 8917, LMG 9040, NCPPB 2592.

12. **Xanthomonas campestris** *pathovar* **bilvae** Chakravarti, Sarma, Jain and Prasad 1984.
    Host: *Aegle marmelos* (fam. Rutaceae).
    Deposited strain: ICMP 8918, NCPPB 3213.

13. **Xanthomonas campestris** *pathovar* **blepharidis** (Srinivasan and Patel 1956) Dye 1978.
    Hosts: *Blepharis boerhaavifolia*, *B. molluginifolia* (fam. Acanthaceae).
    Deposited strain: ATCC 17995, ICMP 5722, LMG 557, NCPPB 1757.

14. **Xanthomonas campestris** *pathovar* **boerhaaviae** (Mathur, Swarup and Sinha 1964) Bradbury 1986.
    Host: *Boerhaavia repens* (fam. Nyctaginaceae).
    Deposited strain: ICMP 9423, LMG 9041, NCPPB 1612.

15. **Xanthomonas campestris** *pathovar* **brunneivaginae** (Luo, Liao and Chen 1988) Young, Saddler, Takikawa, De Boer, Vauterin, Gardan, Gvozdyak and Stead 1996b.
    Host: *Oryza sativa* (fam. Gramineae).
    Deposited strain: ICMP 9991.

16. **Xanthomonas campestris** *pathovar* **cannabis** Severin 1978.
    Host: *Cannabis sativa* (fam. Moraceae).
    Deposited strain: ICMP 6570, LMG 9042, NCPPB 2877.

17. **Xanthomonas campestris** *pathovar* **cannae** Easwaramurthy, Kaviyarasan and Gnanamanickam 1984.
    Host: *Canna generalis* (fam. Cannaceae).
    Deposited strain: ICMP 8306, LMG 9043.

18. **Xanthomonas campestris** *pathovar* **carissae** (Moniz, Sabley and More 1964) Dye 1978.
    Hosts: *Carissa congesta*, *C. carandas*, *Thevetia nerifolia* (fam. Apocynaceae), *Cestrum nocturnum* (fam. Solanaceae).
    Deposited strain: ICMP 3034, LMG 669, NCPPB 2373.

19. **Xanthomonas campestris** *pathovar* **centellae** Basnyat and Kulkarni 1979.
    Host: *Centella asiatica* (fam. Umbelliferae).
    Deposited strain: ICMP 6746, LMG 9044, NCPPB 3245.

20. **Xanthomonas campestris** *pathovar* **clerodendri** (Patel, Kulkarni and Dhande 1952b) Dye 1978.
    Host: *Clerodendron phlomoides* (fam. Verbenaceae).

Editorial Note: The following pathovars of *X. campestris* were not examined by Vauterin et al. (1995), and their exact taxonomic position is uncertain.

Deposited strain: ATCC 11676, ICMP 445, LMG 684, NCPPB 575.

21. **Xanthomonas campestris** *pathovar* **convolvuli** (Nagarkoti, Banerjee and Swarup 1973) Dye 1978.
    Host: *Convolvulus arvensis* (fam. Convolvulaceae).
    Deposited strain: ICMP 5380, LMG 685, NCPPB 2498.

22. **Xanthomonas campestris** *pathovar* **coriandri** (Srinivasan, Patel and Thirumalachar 1961a) Dye 1978.
    Hosts: *Coriandrum sativum, Foeniculum vulgare* (fam. Umbelliferae).
    Deposited strain: ATCC 17996, ICMP 5725, LMG 687, NCPPB 1758.

23. **Xanthomonas campestris** *pathovar* **daturae** (Jain, Dange and Siradhana 1975) Bradbury 1986.
    Host: *Datura metel* (fam. Solanaceae).
    Deposited strain: ICMP 12546, NCPPB 2932.

24. **Xanthomonas campestris** *pathovar* **durantae** (Srinivasan and Patel 1957) Dye 1978.
    Host: *Duranta repens* (fam. Verbenaceae).
    Deposited strain: ICMP 5728, LMG 696, NCPPB 1456.

25. **Xanthomonas campestris** *pathovar* **esculenti** (Rangaswami and Easwaran 1962) Dye 1978.
    Host: *Hibiscus esculentus* (fam. Malvaceae).
    Deposited strain: ICMP 5729, LMG 699, NCPPB 2190.

26. **Xanthomonas campestris** *pathovar* **eucalypti** (Truman 1974) Dye 1978.
    Hosts: *Eucalyptus citriodora, E. maculata* (fam. Myrtaceae).
    Deposited strain: ICMP 5382, LMG 700, NCPPB 2337.

27. **Xanthomonas campestris** *pathovar* **euphorbiae** (Sabet, Ishag and Khalil 1969) Dye 1978.
    Host: *Eurphorbia acalyphoides* (fam. Euphorbiaceae).
    Deposited strain: ICMP 5730, LMG 863, NCPPB 1828.

28. **Xanthomonas campestris** *pathovar* **fici** (Cavara 1905) Dye 1978.
    Host: *Ficus religiosa* (fam. Moraceae).
    Deposited strain: ICMP 3036, LMG 701, NCPPB 2372.

29. **Xanthomonas campestris** *pathovar* **guizotiae** (Yirgou 1964) Dye 1978.
    Host: *Guizotia abyssinica* (fam. Compositae).
    Deposited strain: ICMP 5734, LMG 731, NCPPB 1932.

30. **Xanthomonas campestris** *pathovar* **gummisudans** (McCulloch 1924) Dye 1978.
    Host: *Gladiolus* sp. (fam. Iridaceae).
    Deposited strain: ICMP 5780, LMG 732, NCPPB 2182.

31. **Xanthomonas campestris** *pathovar* **heliotropii** (Sabet, Ishag and Khalil 1969) Dye 1978.
    Hosts: *Heliotropium aegypticum, H. sudanicum* (fam. Boraginaceae).
    Deposited strain: ICMP 5778, LMG 735, NCPPB 2057.

32. **Xanthomonas campestris** *pathovar* **ionidii** (Padhya and Patel 1963a) Dye 1978.
    Host: *Ionidium heterophyllum* (fam. Violaceae).
    Deposited strain: ICMP 5736, LMG 744, NCPPB 1334.

33. **Xanthomonas campestris** *pathovar* **lantanae** (Srinivasan and Patel 1957) Dye 1978.
    Host: *Lantana camara* var. *aculeata* (fam. Verbenaceae).
    Deposited strain: ICMP 5737, LMG 754, NCPPB 1455.

34. **Xanthomonas campestris** *pathovar* **laureliae** (Dye 1963a) Dye 1978.
    Host: *Laurelia novae-zelandiae* (fam. Monimiaceae).
    Deposited strain: ICMP 84, LMG 755, NCPPB 1155.

35. **Xanthomonas campestris** *pathovar* **lawsoniae** (Patel, Bhatt and Kulkarni 1951a) Dye 1978.
    Host: *Lawsonia alba* (fam. Lythraceae).
    Deposited strain: ATCC 11674, ICMP 319, LMG 756, NCPPB 579.

36. **Xanthomonas campestris** *pathovar* **leeana** (Patel and Kotasthane 1969a) Dye 1978.
    Host: *Leea edgeworthii* (fam. Vitaceae).
    Deposited strain: ICMP 5738, LMG 9048, NCPPB 2229.

37. **Xanthomonas campestris** *pathovar* **leersiae** (ex Fang, Ren, Chen, Chu, Faan and Wu 1957) Young, Bradbury, Davis, Dickey, Ercolani, Hayward and Vidaver 1991a.
    Host: *Leersia hexandra* (fam. Gramineae).
    Deposited strain: ICMP 8788.

38. **Xanthomonas campestris** *pathovar* **malloti** Goto 1993.
    Host: *Mallotus japonicus* (fam. Euphorbiaceae).
    Deposited strain: ATCC 51262, ICMP 11536.

39. **Xanthomonas campestris** *pathovar* **mangiferaeindicae** (Patel, Moniz and Kulkarni 1948) Robbs, Ribeiro and Kimura 1974.
    Hosts: *Anacardium occidentale, Mangifera indica, Spondias mangifera* (fam. Anacardiaceae).
    Deposited strain: ATCC 11637, CFBP 1716, ICMP 5740, LMG 941, NCPPB 490.

40. **Xanthomonas campestris** *pathovar* **merremiae** (Pant and Kulkarni 1976b) Dye, Bradbury, Goto, Hayward, Lelliott and Schroth 1980.
    Host: *Merremia gangetica* (fam. Convolvulaceae).
    Deposited strain: ICMP 6747, LMG 9051, NCPPB 3114.

41. **Xanthomonas campestris** *pathovar* **mirabilis** (Durgapal and Trivedi 1976) Young, Bradbury, Davis, Dickey, Ercolani, Hayward and Vidaver 1991a.
    Host: *Mirabilis jalapa* (fam. Nyctaginaceae).
    Deposited strain: ICMP 8949.

42. **Xanthomonas campestris** *pathovar* **musacearum** (Yirgou and Bradbury 1968) Dye 1978.
    Hosts: *Ensete ventricosum, Musa* spp. (fam. Musaceae).
    Deposited strain: ATCC 49084, ICMP 2870, LMG 785, NCPPB 2005.

43. **Xanthomonas campestris** *pathovar* **nigromaculans** (Takimoto 1927) Dye 1978.
    Host: *Arctium lappa* (fam. Compositae).
    Deposited strain: ATCC 23390, ICMP 80, LMG 787, NCPPB 1935.

44. **Xanthomonas campestris** *pathovar* **obscurae** Chand and Singh 1994.
    Host: *Ipomoea obscura* (fam. Convolvulaceae).
    Deposited strain: ICMP 12547, NCPPB 3759 (incorrectly reported as NCPPB 3359 in Chand and Singh 1994).

45. **Xanthomonas campestris** *pathovar* **olitorii** (Sabet 1957) Dye 1978.
    Host: *Corchorus olitorius* (fam. Tiliaceae).
    Deposited strain: ICMP 359, LMG 9052, NCPPB 464.

46. **Xanthomonas campestris** *pathovar* **papavericola** (Bryan and McWhorter 1930) Dye 1978.

    Hosts: *Argemone mexicana, Eschscholzia californica, Meconopsis baileyi, Papaver* spp. (fam. Papaveraceae).

    Deposited strain: ATCC 14179, ICMP 220, LMG 809, NCPPB 2970.

47. **Xanthomonas campestris** *pathovar* **parthenii** Chand, Singh, Singh and Singh 1995.

    Host: *Parthenium hysteriphorus* (fam. Compositae).

    Deposited strain: ICMP 12476, NCPPB 3888.

48. **Xanthomonas campestris** *pathovar* **passiflorae** (Pereira 1969) Dye 1978.

    Host: *Passiflora edulis* (fam. Passifloraceae).

    Deposited strain: ICMP 3151, LMG 810, NCPPB 2346.

49. **Xanthomonas campestris** *pathovar* **paulliniae** Robbs, Medeiros and Kimura 1982.

    Host: *Paullinia cupana* var. *sorbilis* (fam. Sapindaceae).

    Deposited strain: ICMP 8919, LMG 9053, NCPPB 3079.

50. **Xanthomonas campestris** *pathovar* **pennamericanum** Qhobela and Claflin 1988.

    Hosts: *Panicum milaceum, Pennisetum americanum* (fam. Gramineae).

    Deposited strain: ATCC 49152, ICMP 9627.

51. **Xanthomonas campestris** *pathovar* **phormiicola** (Takimoto 1933) Dye 1978.

    Host: *Phormium tenax* (fam. Agavaceae).

    Deposited strain: ICMP 4294, LMG 702, NCPPB 2983.

52. **Xanthomonas campestris** *pathovar* **physalidis** (Srinivasan, Patel and Thirumalachar 1962) Dye 1978.

    Hosts: *Physalis minima, P. peruviana* (fam. Solanaceae).

    Deposited strain: ATCC 17994, ICMP 5746, LMG 846, NCPPB 1756.

53. **Xanthomonas campestris** *pathovar* **sesami** (Sabet and Dowson 1960) Dye 1978.

    Host: *Sesamum orientale* (fam. Pedaliaceae).

    Deposited strain: ICMP 621, LMG 865, NCPPB 631.

54. **Xanthomonas campestris** *pathovar* **spermacoces** (Srinivasan and Patel 1956) Dye 1978.

    Host: *Spermacoce hispida* (fam. Rubiaceae).

    Deposited strain: ATCC 17998, ICMP 5751, LMG 868, NCPPB 1760.

55. **Xanthomonas campestris** *pathovar* **syngonii** Dickey and Zumoff 1987.

    Host: *Syngonium podophyllum* (fam. Araceae).

    Deposited strain: ICMP 9154, LMG 9055, NCPPB 3586.

56. **Xanthomonas campestris** *pathovar* **tardicrescens** (McCulloch 1937) Dye 1978.

    Hosts: *Belamcanda* sp., *Iris* spp. (fam. Iridaceae).

    Deposited strain: ICMP 4295, LMG 9056, NCPPB 2984.

57. **Xanthomonas campestris** *pathovar* **thespesiae** Patil and Kulkarni 1981.

    Host: *Thespesia populnea* (fam. Malvaceae).

    Deposited strain: ICMP 7466, LMG 9057.

58. **Xanthomonas campestris** *pathovar* **thirumalacharii** (Padhya and Patel 1964) Dye 1978.

    Host: *Triumfetta pilosa* (fam. Tiliaceae).

    Deposited strain: ATCC 23577, ICMP 5852, LMG 872, NCPPB 1452.

59. **Xanthomonas campestris** *pathovar* **tribuli** (Srinivasan and Patel 1956) Dye 1978.

    Host: *Tribulus terrestris* (fam. Zygophyllaceae).

    Deposited strain: ICMP 5753, LMG 873, NCPPB 1454.

60. **Xanthomonas campestris** *pathovar* **trichodesmae** (Patel, Kulkarni and Dhande 1952a) Dye 1978.

    Host: *Trichodesma zeylanicum* (fam. Boraginaceae).

    Deposited strain: ATCC 11678, ICMP 5754, LMG 874, NCPPB 585.

61. **Xanthomonas campestris** *pathovar* **uppalii** (Patel 1948) Dye 1978.

    Hosts: *Ipomoea muricata* (fam. Convolvulaceae), *Tropaeolum majus* (fam. Tropaeolaceae).

    Deposited strain: ATCC 11641, ICMP 5756, LMG 893, NCPPB 586.

62. **Xanthomonas campestris** *pathovar* **vernoniae** (Patel, Desai and Patel 1968) Dye 1978.

    Host: *Vernonia cinerea* (fam. Compositae).

    Deposited strain: ICMP 5758, LMG 9058, NCPPB 1787.

63. **Xanthomonas campestris** *pathovar* **viegasii** Robbs, Neto, Malavolta and Kimura 1989.

    Host: *Pachystachys lutea* (fam. Acanthaceae).

    Deposited strain: ICMP 9261.

64. **Xanthomonas campestris** *pathovar* **viticola** (Nayudu 1972) Dye 1978.

    Hosts: *Azadirachta indica* (fam. Meliaceae), *Phyllanthus maderaspatensis* (fam. Euphorbiaceae), *Vitis vinifera* (fam. Vitaceae).

    Deposited strain: ICMP 3867, LMG 965, NCPPB 2475.

65. **Xanthomonas campestris** *pathovar* **vitiscarnosae** (Moniz and Patel 1958) Dye 1978.

    Host: *Vitis carnosa* (fam. Vitaceae).

    Deposited strain: ICMP 90, LMG 939, NCPPB 1149.

66. **Xanthomonas campestris** *pathovar* **vitistrifoliae** (Padhya, Patel and Kotasthane 1965b) Dye 1978.

    Host: *Vitis trifolia* (fam. Vitaceae).

    Deposited strain: ICMP 5761, LMG 940, NCPPB 1451.

67. **Xanthomonas campestris** *pathovar* **vitiswoodrowii** (Patel and Kulkarni 1951a) Dye 1978.

    Host: *Vitis woodrowii* (fam. Vitaceae).

    Deposited strain: ATCC 11636, ICMP 3965, LMG 954, NCPPB 1014.

68. **Xanthomonas campestris** *pathovar* **zantedeschiae** (Joubert and Truter 1972) Dye 1978.

    Host: *Zantedeschia aethiopica* (fam. Araceae).

    Deposited strain: ICMP 2372, LMG 9059, NCPPB 2978.

69. **Xanthomonas campestris** *pathovar* **zingibericola** (Ren and Fang 1981) Bradbury 1986.

    Host: *Zingiber mioga* (fam. Zingiberaceae).

    Deposited strain: ICMP 8787, LMG 9060.

70. **Xanthomonas campestris** *pathovar* **zinniae** (Hopkins and Dowson 1949) Dye 1978.

    Host: *Zinnia elegans* (fam. Compositae).

    Deposited strain: ICMP 5762, LMG 8692, NCPPB 2439.

Genus II. *Frateuria* Swings, Gillis, Kersters, De Vos, Gosselé and De Ley 1980, 555[VP]

JEAN SWINGS AND MARTIN SIEVERS

Fra.teur' i.a. M.L. fem. n. *Frateuria* named after the late Joseph Frateur (1903–1974), eminent Belgian microbiologist.

Regular straight rods, 0.5–0.7 × 0.7–3.5 µm, occurring singly or in pairs. Gram negative. **Generally motile by polar flagella or nonmotile. Obligately aerobic.** Optimum temperature for growth 25–30°C. Colonies on mannitol–yeast extract–peptone (MYP) agar[1] are yellow to orange. On glucose–yeast extract–CaCO₃ (GYC) agar,[2] most strains produce a typical brown water-soluble pigment. Oxidase negative. Grows at pH 3.6. No nitrate reduction. Starch and gelatin are not hydrolyzed. H₂S is produced. Chemoorganotrophic. Acid is produced from ethanol and a number of carbon sources. On D-glucose and D-xylose, the pH drops below 4.0. From D-glucose, 2-keto-, and 2,5-diketogluconic acids are formed, but not 5-ketogluconic acid. No requirements for growth factors. Isolated from *Lilium auratum* and from the fruit of *Rubus parvifolius* (raspberry) in Japan.

The mol% G + C of the DNA is: 62–64 (T_m).

Type species: **Frateuria aurantia** Swings, Gillis, Kersters, De Vos, Gosselé and De Ley 1980, 555.

FURTHER DESCRIPTIVE INFORMATION

On GYC agar, cells of *Frateuria* appear mostly as rods, rarely as ovoids, never in chains. Some strains form filaments. Frateuria grows luxuriantly on this medium and forms typical dark coffee-brown colonies, which measure 2–5 mm in diameter with a dark center, and are glistening or rough, flat, or raised, and circular with a regular edge. The pigment also diffuses into the agar. Colonies on MYP agar are yellow to orange, glistening or rough, regular or irregular, highly convex, measuring 1–3 mm in diameter. All strains can be grown in a defined Hoyer medium[3] with ammonium sulfate as the sole nitrogen source and mannitol as the carbon source. Luxuriant growth is obtained when mixtures of amino acids are supplied instead of NH₄⁺. Some strains grow in yeast extract broth or in peptone broth, but for routine cultivation of *Frateuria*, D-mannitol and D-glucose are incorporated in the media as carbon sources. In a mannitol–salts–vitamins medium[4], several single amino acids are used by *Frateuria* as sole nitrogen sources. L-Alanine, L-aspartic acid, L-glutamic acid, and L-proline are not utilized as carbon and nitrogen sources by *Frateuria*.

The following physiological tests are positive for all the strains: catalase, production of H₂S, ketogenesis on glycerol and on D-mannitol, and oxidation of DL-lactate to CO₂ and water. The following tests are universally negative: oxidase, reduction of nitrates, indole formation, gelatin hydrolysis, and oxidation of acetate.

Injected into apples or pears, *Frateuria* causes browning and rotting symptoms (Vanden Abeele et al., 1980).

Frateuria is able to grow and to degrade aniline at pH 4.0–7.0 (Aoki et al., 1984).

For growth on natural gas, co-cultures of *Methylococcus capsulatus* with *Frateuria aurantia* have been proposed. *Frateuria aurantia* consumes ethanol and acetate, both exometabolites of the underoxidation of the ethane (present in natural gas) by *Methylococcus capsulatus*. Both exometabolites inhibit the growth of *M. capsulatus* (Romanovskaya et al., 1992).

Frateuria strains produce the polyene antibiotics enacyloxin IIa, IIIa and IVa (Watanabe et al., 1992a, b, c).

---

1. Mannitol–yeast extract–peptone (MYP) agar has the following composition (g/l): mannitol, 25.0; yeast extract, 5.0; peptone, 3.0; and agar, 25.0.

2. Glucose–yeast extract–CaCO₃ (GYC) agar consists of (g/l): D-glucose, 50.0; yeast extract, 10.0; CaCO₃, 30.0; and agar, 25.0.

3. The Hoyer medium consists of (g/l): mannitol, 30.0; (NH₄)₂SO₄, 1.0; K₄HPO₄, 1.0; K₂HPO₄, 0.1; KH₂PO₄, 0.9; MgSO₄·7H₂O, 0.25; and FeCl₃, 0.005.

4. The medium for testing utilization of L-amino acids as sole nitrogen sources consists of D-mannitol, 30.0 g; d-biotin, 0.001 g; Ca-D-pantothenate, 0.001 g; thiamine, 0.001 g; pyridoxal hydrochloride, 0.0015 g; niacin, 0.0015 g; riboflavin, 0.0015 g; p-aminobenzoic acid, 0.001 g; vitamin B₁₂, 0.001 g; folic acid, 0.001 g; salt solution A, 5 ml; salt solution B, 5 ml; L-amino acid, 1 g; in 1 liter of 0.2 M Tris-maleate buffer, pH 5.4. Salt solution A contains (g/l distilled water): KH₂PO₄, 10.0; and K₂HPO₄, 10.0. Salt solution B contains (g/l distilled water): MgSO₄·7H₂O, 4.0; NaCl, 0.2; FeSO₄·7H₂O, 2.0; MnSO₄·H₂O, 0.15.

TABLE BXII.γ.21. Characteristics differentiating the genera *Frateuria*, *Gluconobacter*, and *Acetobacter*[a]

Characteristics	Frateuria	Gluconobacter	Acetobacter
Flagellar arrangement in motile strains:			
Polar	+	+	−
Peritrichous	−	−	+
Overoxidation of ethanol	−	−	+
Oxidation of DL-lactate to CO₂ and H₂O	+	−	+
Oxidation of acetate to CO₂ and H₂O	−	−	+
Formation of brown water-soluble pigments on GYC agar	+[b]	−	−
Formation of H₂S	+	−	−
Type of ubiquinone formed:			
Q₈	+	−	−
Q₉	−	−	D
Q₁₀	−	+	D
Growth on Frateur's Hoyer mannitol medium	+	−	−
Carbon sources for growth:			
D-Mannose, L-arabinose, D-xylose, L-xylose	+	−	−
Mol% G + C of DNA	62–64	57–64	51–65

[a]For symbols see standard definitions.

[b]Some strains are negative.

TABLE BXII.γ.22. Other characteristics of *Frateuria aurantia*[a]

Characteristic	Result or Reaction
Growth in 0.5% yeast extract broth or 0.5% peptone broth	d
Growth in Frateur's Hoyer medium with:	
Ethanol	−
Glucose	d
Mannitol	+
Growth in SM[b] containing 0.5–2% NaCl	d
Growth in SM at pH 3.6–8.1	+
Temperature range, growth in SM at:	
4°C	−
30–37°C	+
Ethanol tolerance, growth in SM containing:	
1% Ethanol	+
2–5% Ethanol	d
10% Ethanol	−
Glucose tolerance, growth in 0.5% yeast extract containing:	
20% Glucose	+
25–30% Glucose	d
35% Glucose	−
Tolerance to metals and dyes, growth in presence of:	
Malachite green, 0.001%; crystal violet, 0.0001%; brilliant green, 0.001%	+
Cd(CH$_3$COO)$_2$·2H$_2$O, 0.01%; CoSO$_4$·7H$_2$O, 0.01%	−
HgCl$_2$, 0.001%	+
CH$_3$COOTl	d
Vitamin requirements: *p*-aminobenzoic acid, thiamine, niacin, pantothenate	−
Final pH in D-glucose, D-xylose, or D-galactose media	4,5
Acid produced from: [c]	
Ethanol, D-ribose, D-mannose, *n*-propanol, *i*-inositol	+
m-Erythritol, D-arabinose[d], D-fructose, sucrose, *n*-butanol	d
Glycerol[d], L-rhamnose[d], L-sorbose, D-mannitol, sorbitol, D-cellobiose[d], D-lactose[d], maltose, raffinose, dextrin, starch, *n*-amyl alcohol	−
Carbon sources for growth: [e]	
Glycerol, *m*-erythritol, D-ribose, D-fructose, D-glucose, D-mannose, D-mannitol, D-sorbitol, L-arabinose, D-lyxose, L-lyxose, adonitol, *i*-inositol, acetate, glycerate, lactate	+
Ethanol, D-arabinose, D-xylose, L-rhamnose, D-galactose, L-sorbose, D-cellobiose, D-lactose, maltose, sucrose, raffinose, starch, methanol, ethanediol, L-arabitol, *m*-xylitol, oxalate, malonate, tartrate, malate, gluconate, dextrin	d
n-Propanol, dulcitol, formate, citrate	−
Amino acids as sole nitrogen source:	
Arginine, asparagine, glutamine, glutamate, isoleucine, leucine, phenylalanine	+
Alanine, aspartate, histidine, methionine, proline, serine, valine	d
Cysteine, glycine, lysine, threonine, tryptophan	−
Amino acids as sole carbon and nitrogen sources: [f]	
Alanine, aspartate, glutamate, proline	−
Antimicrobial agents (amount per disk):	
Bacitracin, 10 U; chloramphenicol, 30 μg; colistin sulfate, 10 μg; erythromycin, 10 μg; fusidic acid, 10 μg; gentamicin, 10 μg; lincomycin, 2 μg; methicillin, 10 μg; neomycin, 30 μg; nitrofurantoin, 200 μg; penicillin G, 10 U; polymyxin B, 300 U; sulfafurazole, 100 μg	R
Kanamycin, 30 μg; novobiocin, 30 μg; streptomycin, 10 μg; tetracycline, 30 μg	S
Ampicillin, 10 μg; cephaloridine, 25 μg; nalidixic acid, 30 μg	d
Catalase	+
Oxidase	−
Nitrate reduction, indole production, gelatin hydrolysis, change in litmus milk	−
Ferric chloride reaction on media containing:	
D-Glucose	d
D-Fructose	+
D-Galactose	−

[a]For symbols see standard definitions; R, resistant; S, susceptible.

[b]SM medium contains (g/l): D-glucose, 50.0; yeast extract, 5.0.

[c]Acid formation was tested in the following medium (g/l): yeast extract, 5.0; bromocresol purple, 0.02; carbon source, 10.0. A final pH below 5.9 was considered as positive for acid formation.

[d]After retesting, some corrections of our previous results (Swings et al., 1980) were necessary.

[e]Growth was tested on the following medium (g/l): carbon source, 3.0; yeast extract, 0.5; vitamin-free Casamino acids, 3.0; agar (Oxoid No. 1), 25.0.

[f]Frateur's Hoyer medium was employed, with D-mannitol omitted.

All known authentic *Frateuria* strains have been isolated in Japan.

ENRICHMENT AND ISOLATION PROCEDURES

Yamada et al. (1976) used the following medium for the enrichment of *Frateuria* from raspberries (per liter of 10% potato extract): D-glucose, 10.0 g; ethanol, 5.0 ml; yeast extract, 5.0 g; peptone, 3.0 g; and acetic acid, 0.3 ml. The pH was 4.5 and the incubation temperature 30°C. Isolation of *Frateuria* is possible on a CaCO$_3$-containing medium, on which the colonies dissolve the CaCO$_3$ and thus are surrounded by a clear zone.

MAINTENANCE PROCEDURES

Originally *Frateuria* was grown and maintained on a medium containing (g/l of 20% potato extract): D-glucose, 5.0; glycerol, 15.0; yeast extract, 30.0; peptone, 20.0; and CaCO$_3$, 10.0. *Frateuria* can be maintained for five months at 4°C on slants of this medium contained in screw-capped vials. *Frateuria* survives lyophilization for many years.

DIFFERENTIATION OF THE GENUS *FRATEURIA* FROM OTHER GENERA

Frateuria resembles the genera of acetic acid bacteria in its ability to oxidize ethanol to acetic acid and to grow at pH 3.6. It resembles mostly *Gluconobacter* because of its polar flagellation; however, it oxidizes lactate, requires no vitamins, and does not produce 5-ketogluconic acid. *Acetobacter* differs from *Frateuria* by its peritrichous flagellation and its ability to oxidize acetate. *Frateuria* makes the ubiquinone Q8, whereas *Gluconobacter* possesses ubiquinone Q10 and *Acetobacter* possesses Q9 or Q10.

Frateuria can be easily distinguished from its phylogenetically nearest neighbors within the *Gammaproteobacteria* (*viz, Lysobacter, Stenotrophomonas, Xanthomonas*) using FAME analysis, in which it characteristically possesses C$_{12:0\ 2OH}$ and C$_{17:0\ cyclo}$ fatty acids, which are absent in *Lysobacter, Stenotrophomonas,* and *Xanthomonas*. The latter three genera typically display C$_{11:0\ iso}$, C$_{11:0\ iso\ 3OH}$, and C$_{15:0\ anteiso}$ fatty acids, which are absent in *Frateuria*. The presence of more than 50% C$_{15:0\ iso}$ fatty acid is characteristic for *Frateuria*.

TAXONOMIC COMMENTS

When either phenotypic features or protein gel electrophoregrams of *Frateuria* strains are compared, a tight cluster is obtained, suggesting a high genetic resemblance among them. In the past, *"Acetobacter aurantius"* strains were considered as polarly flagellated acetic acid bacteria that were intermediate between *Acetobacter* and *Gluconobacter* (Asai, 1968). Yamada et al. (1976) already suspected that these strains form a separate new genus, mainly because of their typical ubiquinone Q8. The reclassification of these strains was performed by Swings et al. (1980).

DNA–rRNA hybridizations had already shown that *Frateuria* was not related to *Acetobacter* or to *Gluconobacter*, and did not belong in rRNA superfamily IV (*sensu* De Ley, 1978), of which the family *Acetobacteraceae* constitutes a separate branch (De Vos, 1980; Gillis and De Ley, 1980; Swings et al., 1980). 16S rDNA gene sequencing confirmed that *Frateuria* belongs in the *Gammaproteobacteria* (the former rRNA superfamily II *sensu* De Ley 1978). *Frateuria* is closely related to *Stenotrophomonas, Xanthomonas, Xylella,* and *Lysobacter*. 16S rRNA gene sequence similarity values of *F. aurantia* with the type strains of *Stenotrophomonas maltophilia, Xanthomonas campestris, Xylella fastidiosa,* and *Lysobacter enzymogenes* are 89.6%; 89.6%; 88.8%, and 88.8%, respectively. *Frateuria, Stenotrophomonas,* and *Xanthomonas* were shown to contain the polyamine spermidine (Hamana and Matsuzaki, 1993; Yang et al., 1993b).

FURTHER READING

Swings, J., M. Gillis, K. Kersters, P. De Vos, F. Gosselé and J. De Ley. 1980. *Frateuria*, a new genus for *"Acetobacter aurantius"*. Int. J. Syst. Bacteriol. *30*: 547–556.

List of species of the genus Frateuria

1. **Fratcuria aurantia** Swings, Gillis, Kersters, De Vos, Gosselé and De Ley 1980, 555^{VP}.

   au.ran' tia. L. v. *aurare* to overlay with gold; M.L. adj. *aurantius* gold colored, refers to the gold-yellow color of the strains on MYP agar.

   The description of the species is as for the genus. See Tables BXII.γ.21 and BXII.γ.22 for additional characteristics.

   Isolated from *Lilium auratum* and from the fruit of *Rubus parvifolius* in Japan.

   The mol% G + C of the DNA is: 62–64 (T_m).

   Type strain: IFO 3245, ATCC 33424, DSM 6220.

   GenBank accession number (16S rRNA): AJ010481.

Genus III. **Luteimonas** *Finkmann, Altendorf, Stackebrandt and Lipski 2000, 280^{VP}*

ANDRÉ LIPSKI AND ERKO S. STACKEBRANDT

Lu.te.i.mo' nas. M.L. adj. *luteus* yellow; Gr. n. *monas* a unit; M.L. fem. n. *Luteimonas* a yellow unit.

Rod-shaped cells 0.4–0.6 × 0.8–1.8 μm. Gram negative. **Aerobic**, having a strictly respiratory type of metabolism with oxygen as the terminal electron acceptor. Nitrite is reduced to nitrous oxide, but nitrate is not reduced. Colonies are yellow. Heterotrophic. Ubiquinone with eight isoprene units. **Iso-/anteiso-type fatty acid profile** with C$_{15:0\ iso}$ and C$_{17:1\ iso\ 9c}$ predominating. Isolated from several experimental biofilters supplied with the waste gas of an animal rendering plant.

The mol% G + C of the DNA is: 67.

Type species: **Luteimonas mephitis** Finkmann, Altendorf, Stackebrandt and Lipski 2000, 280.

ENRICHMENT AND ISOLATION PROCEDURES

All strains of the genus were isolated on thiosulfate agar from cultures for the enrichment of methyl-sulfide-degrading bacteria (Bendinger and Reichert, personal communication). However, the capability for sulfide degradation was not stable for the isolates of this genus and was lost during subcultivation (Reichert, personal communication).

MAINTENANCE PROCEDURES

Short-term maintenance on agar plates for several weeks at room temperature is acceptable. For long-term storage, preservation at below −70°C with 5% dimethylsulfoxide is recommended. Freezing at higher preservation temperatures results in rapid loss of viability.

DIFFERENTIATION OF THE GENUS *LUTEIMONAS* FROM OTHER GENERA

The genus can be separated from the genera *Xanthomonas* and *Stenotrophomonas* by the absence of the fatty acid 3-hydroxy-iso-tridecanoic acid, and from the species *Xylella fastidiosa* by the presence of branched-chain fatty acids. No gliding motility, which is a characteristic property of the genus *Lysobacter* (Christensen and Cook, 1978), was observed for the genus *Luteimonas*. Denitrification activity is absent in the genera *Xanthomonas*, *Stenotrophomonas*, *Xylella*, and *Lysobacter* (one *Lysobacter antibioticus* strain is positive), but present in the genus *Luteimonas*.

TAXONOMIC COMMENTS

The genus is a deep-branching member of the *Xanthomonas* group of the *Gammaproteobacteria*. *Luteimonas mephitis* is remotely and approximately equidistantly related to members of *Xanthomonas*, *Pseudoxanthomonas*, *Xylella*, *Lysobacter*, and *Stenotrophomonas* (93–95% 16S rDNA similarity); members of *Schineria*, *Rhodanobacter*, *Frateuria*, and *Nevskia* are even more distantly related.

Using members of *Stenotrophomonas* as a root, analyses by RDP

dendrogram (Maidak et al., 1997; De Soete, 1983), maximumlikelihood, and neighbor-joining (Felsenstein, 1993) place *Luteimonas mephitis* between *Lysobacter antibioticus* and the species *Rhodanobacter lindaniclasticus*, *Frateuria aurantia*, and *Nevskia ramosa*, which share a common ancestry with *L. mephitis*. *N. ramosa* is characterized by a significantly longer branch length than the other members of this phylogenetic cluster. Using the 16S rDNA sequence of *N. ramosa* as a root, *Frateuria aurantia* and *Rhodanobacter lindaniclasticus* are separated from *Luteimonas mephitis*, *Lysobacter antibioticus*, and members of *Xylella*, *Xanthomonas*, *Stenotrophomonas*, and *Pseudoxanthomonas*. Hence, the branching order of genera constituting the deeply branching lineages of the *Gammaproteobacteria* changes with the selection and number of sequences included in the analysis, and at present cannot be considered stable.

List of species of the genus Luteimonas

1. **Luteimonas mephitis** Finkmann, Altendorf, Stackebrandt and Lipski 2000, 280[VP]

me.phi′tis. L. gen. fem. n. *mephitis* of harmful odor.

The characteristics are the same as those given for the genus and as listed in Table BXII.γ.23. The predominant compounds of the fatty acid profile are $C_{15:0\ iso}$, $C_{17:1\ iso\ \omega9c}$, and $C_{17:0\ iso}$. The major hydroxy fatty acid is $C_{11:0\ iso\ 3OH}$. The fatty acids $C_{17:0\ cyclo}$ and $C_{13:0\ iso\ 3OH}$ are absent. At the beginning of the stationary phase in liquid media, cells tended to clump, showing "trembling"-like behavior. Stain-

ing of flagella (Heimbrook et al., 1989) was negative. As cells from this growth phase also showed no evidence of flagella by electron microscopic observation, the type strain of *L. mephitis* is considered nonmotile. The strain showed no growth under anaerobic conditions on Caso agar (Becton Dickinson and Company, Cockeysville, MD, USA). No pathogenic activity toward humans, animals, or plants is known.

The mol% G + C of the DNA is: 67 (HPLC).
Type strain: B1953/27.1, DSM 12574.
GenBank accession number (16S rRNA): AJ012228.

TABLE BXII.γ.23. Characteristics of *Luteimonas mephitis*[a]

Characteristic	Reaction
Nitrate reduction to nitrite or N_2	−
Nitrite reduction to N_2O	+
Tween 80 hydrolysis	+
Urease, lecithinase, arginine dihydrolase	−
Citrate alkalinization	d
Malonate alkalinization	−
Growth at 37°C	−
Acid from:	
Glucose	+
Rhamnose, sucrose, xylose, fructose, lactose, maltose	−
Hydrolysis of:	
pNP-*N*-acetyl-β-glucosaminide, pNP-β-galactopyranoside, pNP-β-glucuronide, pNP-β-xyloside, bis-pNP-phosphate, pNP-phosphocholine	−
pNP-β-glucopyranoside, pNP-α-D-maltoside, thymidine-5′-monophosphate-pNP-ester	d
L-alanine-pNA, L-arginine-pNA, glycine-pNA, L-leucine-pNA, L-lysine-pNA, L-proline-pNA, L-valine-pNA	−
γ-L-glutamate-pNA	d
Growth in the presence of:	
NaCl (3%), ZnCl$_2$ (0.01%)	−
Cadmium acetate (0.001%), lead acetate (0.01%)	d
Resistant to:	
Erythromycin (10 µg/disk), streptomycin (10 µg/disk), nalidixic acid (30 µg/disk), ampicillin (10 µg/disk), penicillin G (10 U)	+
Kanamycin (30 µg/disk)	−
Gentamicin (10 µg/disk), fucidin (10 µg/disk), tetracycline (30 µg/disk), novobiocin (30 µg/disk), neomycin (30 µg/disk)	d

[a]Symbols: see standard definitions; pNP, *para*-nitrophenyl; pNA, *para*-nitroanilide. Data from Lipski et al., 1992; Finkmann et al. 2000.

Genus IV. **Lysobacter** Christensen and Cook 1978, 372[AL]

PENELOPE CHRISTENSEN

Lys.o.bac' ter. Gr. adj. *lysis* loosing; M.L. n. *bacter* equivalent of Gr. neut. n. *bactrum* a rod; M.L. masc. n. *Lysobacter* the loosing rod; intended to mean the lysing rod.

Thin rods, 0.2–0.5 × 1.0–15 (sometimes up to 70) μm, which are Gram negative, nonflagellated, **gliding, flexing,** and aerobic. **Colonies are highly mucoid, cream-colored, pink, or yellow-brown;** many strains also produce a brown, water-soluble **pigment. Nonfruiting; no microcysts** produced. Growth in broth culture is silky. Most strains are resistant to actinomycin D. This organism **degrades chitin** and often other polysaccharides, but it does not degrade filter-paper cellulose, and it infrequently degrades agar. Strongly proteolytic, characteristically **lysing a variety of microorganisms** (both Gram-negative and Gram-positive bacteria, including actinomycetes, blue-green and green algae, yeasts, and filamentous fungi), as well as nematodes. Habitat: soil and freshwater.

The mol% G + C of the DNA is: 65.4–70.1.

Type species: **Lysobacter enzymogenes** Christensen and Cook 1978, 374.

FURTHER DESCRIPTIVE INFORMATION

Thin, flexible, gliding, Gram-negative rods, 0.2–0.5 × 1.0–15 (sometimes up to 70) μm (Figs. BXII.γ.27, BXII.γ.28, and BXII.γ.29). Colonies are slimy or mucoid, with the rubbery type of *L. gummosus* being particularly difficult to handle, and may be white, cream-colored, yellow, pink, or brown; many strains also produce a brown, water-soluble pigment, especially in older cultures. The nature of *Lysobacter* pigments is not known, but they are not carotenoids or flexirubins (Reichenbach et al., 1981). Colonies of *L. brunescens* spread in the typical *"Cytophaga"* fashion, but those of *L. antibioticus* and *L. enzymogenes* are more mucoid, and it is believed that this quality conceals the typical spreading. All strains of the latter two species have been observed to produce a thin, advancing fringe of cells at colony edges on one or more media. However, *L. gummosus* has never been observed to spread, most likely due to both its shorter cells and the large amount of gum produced on every medium on which it has been grown.

Liquid cultures show a characteristic silkiness when gently tapped, a property correlated with the presence of longer cells. Silkiness is not often seen with *L. gummosus* cultures, which the heavy gum seems to preclude lengthy cells. Liquid cultures of all four species are somewhat viscous, and *L. gummosus* is so much so that a 2-d-old tube culture may be inverted without spillage.

Aerobic. Mesophilic. The optimum temperature for growth of these organisms is relatively high (20–40°C) as compared to most other soil and water organisms. pH range for growth: 5–10. Growth of most strains is reduced by 1% NaCl.

Chemoorganotrophic. Metabolism is usually respiratory; molecular oxygen is used as the terminal electron acceptor. NO_3^-, NH_4^+, glutamate, and asparaginate are used as sole nitrogen sources, but urea is used by only a few strains. In general, the addition of yeast extract to a salts–glucose–nitrate agar or to a chitin-only medium stimulates growth. Chitin, but not filter-paper cellulose or agar, is hydrolyzed; some species degrade alginate, pectate, starch, and/or carboxymethyl cellulose (CMC). One of the extracellular enzymes of *L. enzymogenes* strain ATCC 27796, previously known as "myxobacter Al-1", was the first enzyme with both β-1,4-gluconase and chitosanase activities to be described (Hedges and Wolfe, 1974). This enzyme attacks amorphous forms of cellulose such as CMC but does not attack crystalline cellulose. It was the first enzyme with chitosanase activity to be purified to homogeneity and could be useful for the characterization of fungal cell walls and as a tool for the classification of fungi. It has been shown to be capable of attacking some fungal cell walls to release glucosamine. The enzyme has a molecular weight of 28,900–31,000 Da and contains large amounts of the basic amino acids lysine and arginine among its 230 residues.

Strongly proteolytic. The enzymes of two strains of *L. enzymogenes* have been the subjects of intensive study. One strain, ATCC 27796, known formerly as "myxobacter AL-1," excretes a bacteriolytic enzyme, designated "AL-1 protease I", with a molecular weight of 14,300 and containing 136 amino acids. In contrast to many exocellular enzymes, this protein contains a disulfide bond, 1 mol hexose per mol protein, and a relatively high content of aromatic amino acids; the E_{280} of a 1% solution in a 1 cm cuvette is 15.8 (Jackson and Wolfe, 1968). This enzyme has been found to cleave the pentaglycine bridge in the cell wall of *Staphylococcus* and other bacteria and to remove peptide moieties from the peptidoglycan (Katz and Strominger, 1967; Tipper et al., 1967). ATCC 27796 also produces another extracellular protease ("AL-1 protease II"), which crystallizes as fine needles and has a molecular weight of 17,000 Da, with 157 amino acid residues (Wingard et al., 1972). This enzyme does not possess cell-wall lytic activity, and it has a unique specificity for lysine residues, exhibiting peptide bond cleavage on the amino side of lysine.

It was not realized at the time that the AL-1 protease reported by Jackson and Matsueda (1970) was from the same species of bacterium as the α-lytic protease described in the paper immediately following it in that publication (Whitaker, 1970). The "myxobacterium *Sorangium*"/"myxobacter 495" (now ATCC 29487) has since been classified as *L. enzymogenes*, and it was with this strain that Whitaker's group did the initial enzymological characterization of α- and β-lytic proteases (Whitaker, 1965, 1967; Whitaker et al., 1965a, b). The proteases of "495" (now ATCC 29487) have provided much fruitful work for biochemists interested in the three-dimensional structure of microbial proteases as compared with mammalian examples. Although the classification of the zinc-containing α-lytic protease is uncertain, β-lytic protease has been shown to be a serine protease with the active site amino acid sequence Asp–Ser–Gly (Whitaker et al., 1966; Whitaker and Roy, 1967). This is the same sequence as that of the mammalian pancreatic serine protease, and this enzyme was the first microbial serine protease isolated that is not a member of the Thr–Ser–Met family. The subsequent elucidation of the complete polypeptide sequence has shown 198 amino acids in a single polypeptide chain, which contains three intrachain disulfide bridges and has an overall molecular weight of 19,869 (Olson et al., 1970). In a 2.8-Å resolution crystallographic study of α-lytic protease, the three-dimensional structure of the enzyme has been determined, facilitating a detailed description of the active site. Comparisons with other microbial serine proteases (viz *Streptomyces griseus* proteases A and B) support the possibility that they arose from a common ancestral gene (Brayer et al., 1979). Sub-

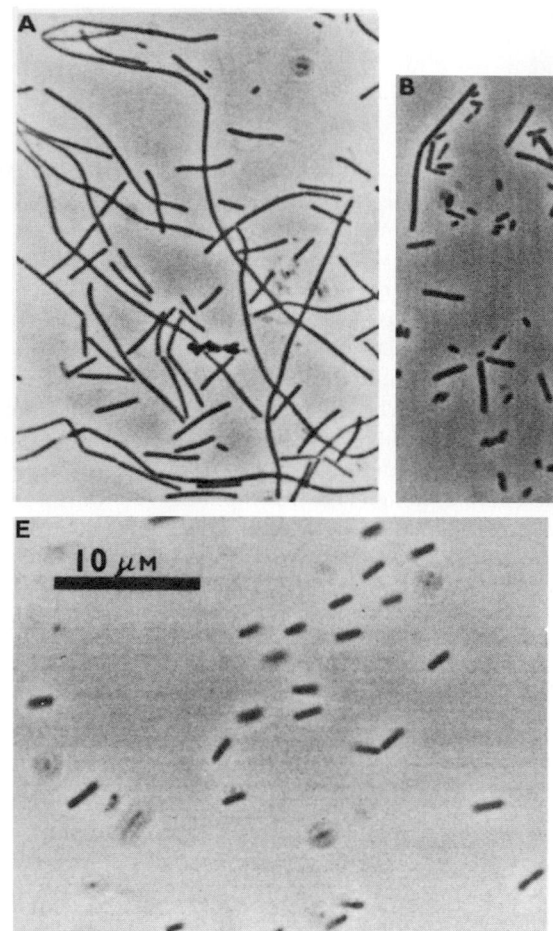

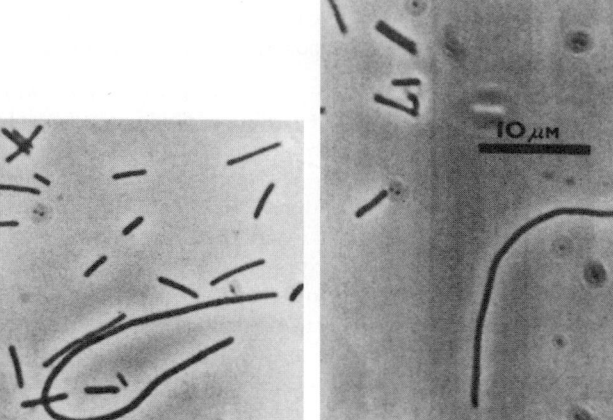

FIGURE BXII.γ.27. *(A–D):* cells of *L. enzymogenes* from skim–acetate broth cultures. *A*, ATCC 29487, 44 h. *B*, ATCC 21123, 44 h. *C*, ATCC 29485, 44 h. *D*, ATCC 29488, 60 h. All micrographs are at the same magnification. (Reproduced with permission from P.J. Christensen and F.D. Cook, International Journal of Systematic Bacteriology *28:* 383, ©1978, International Union of Microbiological Societies.). *E*, cells of *L. gummosus* from skim–acetate broth culture, ATCC 29489, 44 h. (Reproduced with permission from P.J. Christensen and F.D. Cook, International Journal of Systematic Bacteriology *28:* 390, 1978, ©International Union of Microbiological Societies.)

sequent work has included further study of the active site and comparison with mammalian elastase (Bauer et al., 1981), investigation of the molecular structure at 1.8-Å resolution (M.N.G. James, personal communication), and D. Agard's isolation of the α-lytic protease gene, sequencing, and point-specific mutations (M.N.G. James, personal communication).

The nucleases of *L. enzymogenes* strain ATCC 29487 (formerly "myxobacter 495") have been characterized by von Tigerstrom (1980, 1981, 1983), with the two major ones being extracellular RNase and DNase. The RNase consists of one polypeptide chain with a molecular weight of 46,000–47,000 Da that is most active at pH 8.0–8.5 and in the presence of Mg^{2+}. The nonspecific nuclease has a molecular weight of 22,000–28,000 Da, is most active at pH 8.0, and requires Mg^{2+} or Mn^{2+}.

Catalase, oxidase, and phosphatase are produced; work is in progress on the phosphatases of *L. enzymogenes* (R.G. von Tigerstrom, personal communication).* The indole, methyl-red, and Voges–Proskauer tests are negative.

Lyses Gram-negative and Gram-positive bacteria, including actinomycetes, yeasts and filamentous fungi, blue-green and green algae, and nematodes. Lysobacters do not attack Gram-negative bacteria as vigorously as they do Gram-positive species; this may be a self-protective feature.

All strains known at present have been isolated from soil or freshwater habitats; no studies have been aimed at finding them elsewhere, but this would probably be a fruitful area for investigation. Their polysaccharolytic and proteolytic abilities suggest a role in the degradation of biological structural and storage materials. The significance of their lytic abilities toward other organisms has yet to be fully assessed. It is quite possible that marine members of the genus will be found or that these organisms may have medical significance.* *Lysobacter* spp. have probably been confused with "eubacteria" in the past when encountered on regular plating media; nutritionally dilute media are needed in order for lysobacters and other gliding bacteria to be recognized, since they do not exhibit spreading on plate count agar (PCA).

One species, *L. antibioticus,* elaborates a potent, wide-spectrum antibiotic named "myxin," which is a 1-hydroxy-6-methoxyphenazine (Cook et al., 1971; Behki and Lesley, 1972). It is manufactured by Hoffman and Roche, but there are two important drawbacks to its widespread medical usage: it has too wide a spectrum,

---

Editorial Note: Since the publication of this chapter in the 1st edition of *Bergey's Manual of Systematic Bacteriology,* several papers on this topic have been published. See von Tigerstrom (1984); von Tigerstrom and Stelmaschuk (1985, 1986, 1987); Au et al. (1991).

Editorial Note: Since the publication of this chapter in the 1st edition of *Bergey's Manual of Systematic Bacteriology,* several papers on this topic have been published. See Bonner et al. (1988), O'Sullivan et al. (1988), Harada et al. (1988), and Tymiak et al. (1989).

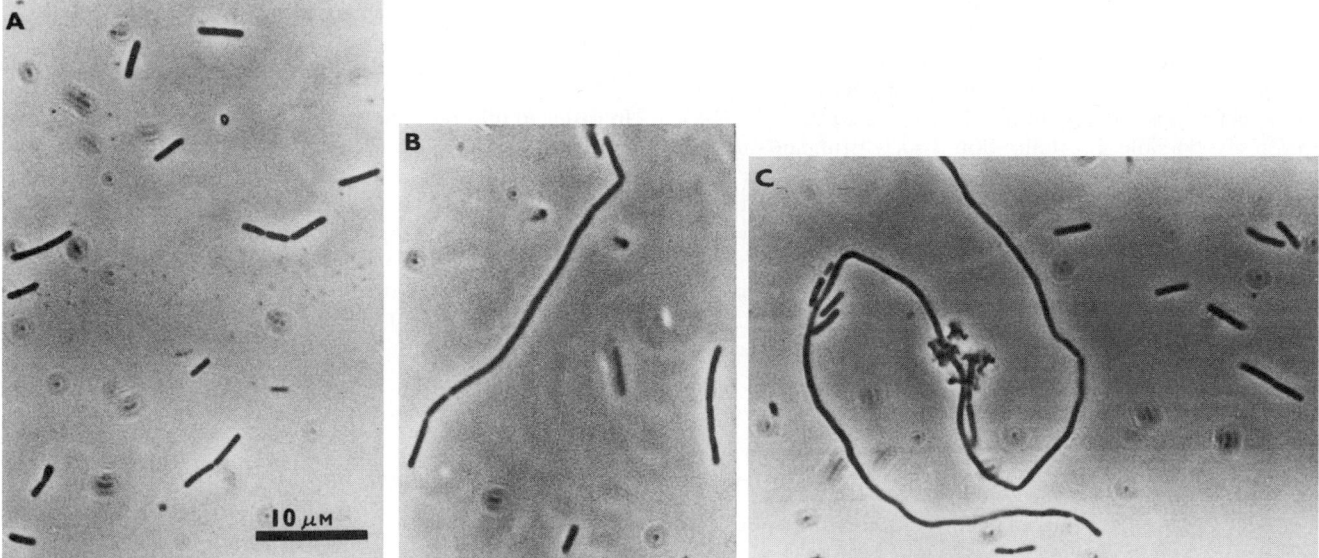

FIGURE BXII.γ.28. Cells of *L. antibioticus* from skim–acetate broth cultures. *A*, ATCC 29479, 66 h. *B*, UASM 4593, 32 h. *C*, UASM 4578, 72 h. All micrographs are at the same magnification. (Reproduced with permission from P.J. Christensen and F.D. Cook, International Journal of Systematic Bacteriology *28:* 389, 1978, ©International Union of Microbiological Societies)

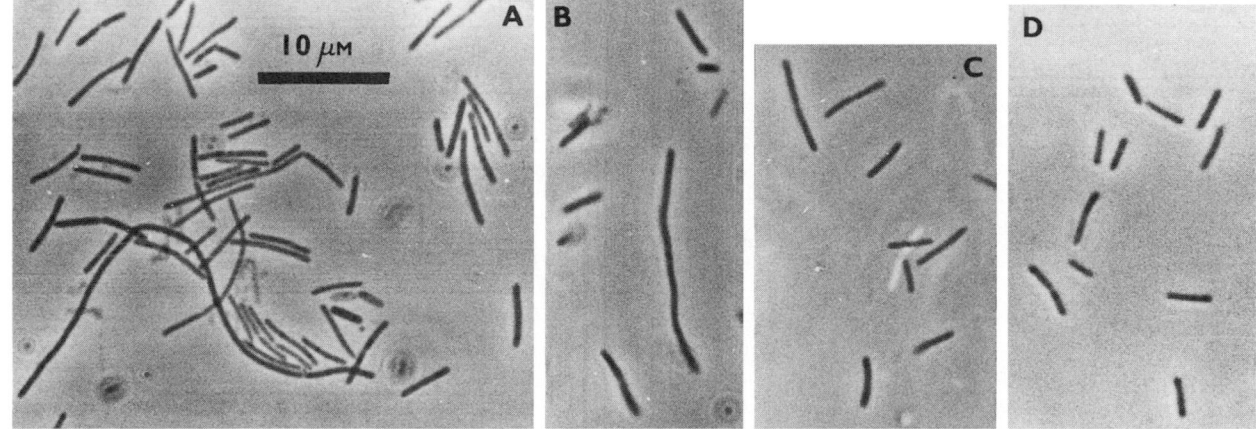

FIGURE BXII.γ.29. Cells of *L. brunescens* from skim–acetate broth cultures. *A*, ATCC 29483, 60 h. *B*, ATCC 29482, 60 h. *C*, UASM 4541, 44 h. *D*, UASM CB 5, 30 h. All micrographs are at the same magnification. (Reproduced with permission from P.J. Christensen and F.D. Cook, International Journal of Systematic Bacteriology *28:* 391, 1978, ©International Union of Microbiological Societies.)

especially for internal use, and it is explosive when dry. Its use is restricted to topical application for resistant fungal skin problems.

ENRICHMENT AND ISOLATION PROCEDURES

Lysobacters have the unusual ability to hydrolyze chitin, and their isolation is facilitated by providing chitin as a suspension, in ground mushrooms, or in autoclaved yeast cells. The simplest method is to enrich soil samples with one of the above sources of chitin, incubate it for at least 1 month, then plate it out onto yeast cell agar (Smit and Clark, 1971), which consists of 0.5% baker's yeast in a 1.5% agar medium. The number of colonies capable of lysing autoclaved yeast cells increases dramatically with

such enrichment procedures (Christensen and Cook, 1978). Water samples can be plated directly onto yeast cell agar.

MAINTENANCE PROCEDURES

Lysobacters grow profusely on PCA (plate count agar, from, for example, Difco), SAA (skim–acetate agar: 0.5% skim milk, 0.05% yeast extract, and 0.02% sodium acetate in 1.5% agar) or broth, or CCA (Cook's *Cytophaga* agar: 0.2% tryptone in 1.0% agar) (Christensen and Cook, 1972). The more dilute the medium, the greater the expression of the spreading morphology, thus CCA is recommended for this purpose and because on this medium, organisms do not appear to lose their polysaccharolytic and proteolytic potentials. On PCA, colonies of lysobacters are often indistinguishable from nongliding bacteria. Ordinary plating techniques are sufficient with incubation at room temperature for the species so far known; however, difficulty can occur with *L. gummosus* because of its intensely rubbery colonies, and, for this species, it is recommended that young cultures (<24 h) be used when possible. Lyophilization is successful, and cultures are readily obtainable from these preparations, which are recommended for long-term storage. Caution is advised on two aspects, however: (a) colony morphology may change slightly after freeze-drying, and (b) strains of *L. enzymogenes* do not exhibit as strong a microbial lytic action and do not maintain as high a level of certain proteolytic enzymes, after they have been maintained in the laboratory on skim milk media for some time, as

do cells that have not been freeze-dried, and this may also apply to lyophilized cultures.

PROCEDURES FOR TESTING SPECIAL CHARACTERS

In order to differentiate lysobacters from the myxobacteria, it is necessary to demonstrate the absence of fruiting bodies. At least two methods should be attempted, such as inoculation onto sterile wild herbivore pellets in water agar (Smit and Clark, 1971) and induction of myxospores following addition of 0.5 M glycerol during exponential growth (Dworkin and Gibson, 1964).

Suitable methods for demonstration of polysaccharide degradation, antimicrobial lytic action, and susceptibility to actinomycin D have been published by Christensen (1977).

DIFFERENTIATION OF THE SPECIES OF THE GENUS *LYSOBACTER*

Characteristics useful for distinguishing among the species of *Lysobacter* are given in Tables BXII.γ.24 and BXII.γ.25.

TABLE BXII.γ.24. Differential characteristics of the species of the genus *Lysobacter*[a,b,c]

Characteristics	1. *L. enzymogenes*	2. *L. antibioticus*	3. *L. brunescens*	4. *L. gummosus*
Broth culture viscous	+	+	−	+ (heavy)
Colonies:				
Type of growth	Sloppy, mucoid	Sloppy, mucoid	Spreading, thin	Pulvinate, gummy
Surface smooth	+ or −	+	+ or −	+
Opacity	Tp, Tl or Op	Tl or Op	Tp or Tl	Tl or Op
Color	Dark C to deep Y-C	Pi to Br-Pi or O-Br	Y to Ch	Pale Y-G to Y-G
Brown, water-soluble pigment produced	None to moderate	Weak to heavy	Weak to heavy	−
Myxin crystals in old cultures	−	Usually +	−	−
Urea as nitrogen source	+ or −	−	−	+
Acid from:				
Cellobiose	+	+	−	+
Sucrose	+	−	−	+
Lactose	+	d	−	+
Lipase–Tween 20	+	d	−	+
Hydrolysis of:				
Alginate	+ or −	d	−	−
CMC	+	+	−	+
Pectate	+	−[d]	+	+
SYS or NBS	−	−[e]	+	−
Potato starch	+ or −	−[f]	+	−
Sheep erythrocytes	α, β, or γ	β or α	γ or −	α
H$_2$S produced	d	−	+	−
Citrate as sole carbon source	+	+	−	+
Growth on:				
MacConkey's	+ (colorless or −)	d (colorless)	−	−
EMB	+ (Pi)	+ (Pi)	−	+ (Pi)
Completely inhibited by 0.1% SLS	+ or −	+	+	−
Susceptible to:				
Chloramphenicol (30 μg)	d	−	+	−
Penicillin G (10 U)	−	−	d	−
Actinomycin D	d	−	+	d
Habitat	Soil	Soil	Freshwater	Soil

[a]Data are from Christensen and Cook (1978), to which the reader is referred for details of tests.

[b]Symbols: +, 90% or more of strains are positive; −, 90% or more of strains are negative; and d, 11–89% of strains are positive.

[c]Abbreviations: Br, brown(ish); C, cream-colored; Ch, chocolate; EMB, eosin methylene blue; G, gray; NBS, nutrient broth–starch; O, orange: Op, opaque; Pi, Pink(ish); SLS, sodium lauryl sulfate; SYS, salts–yeast extract–starch; Tl, translucent; Tp, transparent; and Y, yellow(ish).

[d]Growth but not liquefaction.

[e]Only one strain recorded on SYS.

[f]Only one strain recorded.

TABLE BXII.γ.25. Other characteristics of the species of the genus *Lysobacter*[a,b,c]

Characteristics	1. *L. enzymogenes*	3. *L. antibioticus*	4. *L. brunescens*	2. *L. gummosus*
Cell dimensions (µm)	0.3–0.5 × 4–50	0.4 × 4–40	0.2–0.5 × 7–70	0.4 × 2
Flexing[d]	+	+	+	−
Gliding[d]	+	+	+	+ (short jerks)
Silkiness in broth culture[d]	+	+	+	+
Fruiting bodies induced by:				
Glycerol	−	−	−	−
Pellets	−	−	−	−
Partial inhibition by 1% NaCl	d	+	+	−
% NaCl causing complete inhibition	≤3	3	≤2	3
Preferred atmosphere	Air or 10% O₂	Air or 10% O₂	Air or 10% O₂	Air
Growth range temperature (°C)	5–40	2–40	4–50	10–40
Growth optimum temperature (°C)	25–35	25–33	30–40	20
Initial pH for growth	5 to >10	5 to >10	5 to >10	6 to >10
Nitrogen sources:				
Nitrate	+	+	+	+
Ammonia	+	+	+	+
Glutamate	+	+	+	+
Asparaginate	+	+	d	+
OF test (glucose):				
Oxidation	+	+	+	+
Fermentation	−	d	d	
Acid from:				
Glucose	+	+	+	+
Glycerol	d	−	−	−
Mannitol	d	−	−	−
Lipase–Tween 40	+	+	d	+
Lipase–Tween 60	+	+	+	+
Lipase–Tween 80	+	+	d	+
Hydrolysis of:				
Agar (gelase)	+ or −	−[e]	+	−
Agar tubes	−	−	−	−
Cellulose (filter paper)	−	−	−	−
Chitin	+	+	+	+
Gelatin liquefaction	+	+	+	+
Complete peptonization of milk (days)	1–3	1	1	1
Casein—growth and NH₃ produced	+,+	+,+	+,+	+,+
Penassay—growth	+	+[e]	+	+
Casitone—growth and NH₃ produced	+,+	+,+	+,+	+,+
Casamino acids–salts—growth and NH₃ produced	+,+	+,−[e]	+,−	+,−
Grows well on 0.2% tryptone	+	+	+	+
Catalase	+	+	+	+
Oxidase	+	+	+	nd
Phosphatase	+	+	+	+
Indole	−	−	−	−
Methyl red	−	−	−	−
Voges–Proskauer	−	−	−	−
Reduction of NO₃⁻ to NO₂⁻	−	d	−	−
Reduction of NO₂⁻ to gas	−	−	−	−
Growth reduced by 0.01% SLS	+ or −	−	d	
Susceptible to:				
10 µg streptomycin	−	d	d	−
300 U polymyxin B (Kirby–Bauer)	d	−	d	d
300 U polymyxin B (Christensen)	+	+	+	+
Antimicrobial lytic action:				
Gram-negative bacteria:				
Escherichia coli	d	d	d	−
Pseudomonas aeruginosa	d	−	d	−
Serratia marcescens	d	d	d	−
Gram-positive bacteria:				
Arthrobacter spp.	+	d	d	+
Bacillus subtilis	+	d	d	+
Actinomycetes:				
UASM 4432	+	−	−	+
UASM 4441	+	d	d	+
Fungi:				
Rhizopus spp.	+	d	d	d
Penicillium notatum	+	d	d	+
Sclerotinia sclerotiorum	d	+	d	+
Baker's yeast	+	d	+	+
Alga:				
Chlorella spp.	d	+	+	+

[a]Data are from Christensen and Cook (1978), to which the reader is referred for details of tests.

[b]Symbols: +, 90% or more of strains are positive; −, 90% or more of strains are negative; and d, 11–89% of strains are positive.

[c]Abbreviations: OF, oxidation/fermentation; and SLS, sodium lauryl sulfate.

[d]Flexing, gliding, and silkiness are more readily observed with longer cells.

[e]Only one strain recorded on nutrient broth.

List of species of the genus Lysobacter

1. **Lysobacter enzymogenes** Christensen and Cook 1978, 374[AL]
en.zy.mo'ge.nes. Gr. n. *zyme* leaven; M.L. n. *enzymum* enzyme; Gr. v. *gennaio* to produce; M.L. adj. *enzymogenes* enzyme-producing.

Two distinct colony forms are known: a dirty-white, mucoid colony and a yellowish, nonmucoid one. The mucoid colony produces nonmucoid mutants, but the yellowish, nonmucoid form does not produce revertants to the dirty-white, mucoid form. The colony descriptions given here and in Table BXII.γ.25 cover the whole range of colony forms observed. The forms are identical in other properties.

On CCA, 5-d-old colonies are dark cream-colored, with Munsell notation 2.5Y 6–8/4–8 and 5–10YR 6–8/4–8, and circular to irregular. Colonies usually have a smooth surface, but occasionally have a rough surface. Edges may be entire, undulate, lobate or erose; elevation is typically effuse with a raised or convex center, but occasionally flat. Colonies are transparent or translucent; no brown, water-soluble pigment is produced.

On SAA, 5-d-old colonies are dark cream-colored, with Munsell notation 2.5–7.5Y 6.5–8.5/4–6 and 7.5–10YR 5.5–7/5–7, and circular to irregular, with a smooth or rough surface. Edges may be entire, undulate or erose; elevation is typically effuse with a raised, convex or umbonate center, but occasionally flat. Colonies are transparent, translucent or opaque; no brown, water-soluble pigment is produced.

On PCA, 5-d-old colonies are deep yellow, cream-colored, or creamy brown, with Munsell notation 1.5–5Y 3–7/4–8, and more or less circular, usually with a smooth surface. Edges may be entire, undulate or erose; elevation is usually convex, but may be effuse with a convex or raised center or may be umbonate. Colonies are translucent or opaque; some brown, water-soluble pigment may be produced.

Older cultures of the majority of strains produce copious, dark-brown, water-soluble pigment on most media.

The mol% G + C of the DNA is: 65.4–70.1 (T_m).

Type strain: ATCC 29487, UASM 495.

Additional Remarks: Other strains: ATCC 27796 (UASM AL-1), ATCC 29488 (UASM 13B), and ATCC 29485/6 (colony types of UASM 18L).

2. **Lysobacter antibioticus** Christensen and Cook 1978, 387[AL]
an.ti.bi.o'ti.cus. Gr. pref. *anti* against; Gr. n. *bios* life; M.L. adj. *antibioticus* against life, antibiotic.

On CCA, 5-d-old colonies are pinkish brown, with Munsell notation 1–10YR 2.5–7/4–8, and circular to irregular, with a smooth surface. Edges are usually erose, but may be entire, undulate, or filamentous. Elevation is usually flat or raised, but sometimes effuse with a convex center. Translucent or opaque; some or much brown, water-soluble pigment is produced.

On SAA, 5-d-old colonies are pinkish brown, with Munsell notation 2.5Y 7/5 and 1.5–7.5YR 3–6/4–8, and more or less circular, with a smooth surface. Edges are usually erose, but may be entire or filamentous. Elevation is usually raised, but may be effuse, flat or convex. Translucent or opaque; some or much brown, water-soluble pigment is produced or may be absent.

On PCA, 5-d-old colonies are orange-brown, with Mun-

sell notation 1–2.5Y 4–6/3–4 and 5–10YR 2–5/2–6, and more or less circular, with a smooth surface. Edges are usually entire, but may be erose or undulate. Elevation is convex or raised. Colonies are usually opaque; weak to heavy, brown, water-soluble pigment is produced.

Older cultures on most media produce copious brown, water-soluble pigment, and deep-red crystals of the antibiotic myxin may be observed within the highly mucoid colonies.

The mol% G + C of the DNA is: 66.2–69.2 (T_m).

Type strain: ATCC 29479, UASM 3C.

GenBank accession number (16S rRNA): AB019582.

Additional Remarks: Other strains: ATCC 29480 (UASM L17) and ATCC 29481 (UASM 4045).

3. **Lysobacter brunescens** Christensen and Cook 1978, 390[AL]
bru.nes'cens. L. v. *brunesco* to become dark brown; L. part. adj. *brunescens* becoming dark brown.

On CCA, 5-d-old colonies are brownish yellow, with Munsell notation 2.5–5Y 3–4/4–8 and 10YR 4–5/6–8, and irregular, with a rough surface. Edges are usually filamentous, but may be lobate, undulate, or erose. Elevation is effuse with an umbonate or convex center. Colonies are transparent; amounts of brown, water-soluble pigment vary from moderate to absent.

On SAA, 5-d-old colonies are yellow-brown, with Munsell notation 2.5Y 4.5/6 and 7.5–10YR 4–6/6–8. Usually irregular, but occasionally circular. Colonies usually have a rough surface. Edges are typically erose or filamentous, but may be entire, undulate, or lobate. Elevation is effuse or raised, with convex or umbonate centers. Transparent or translucent; weak to moderate amounts of brown, water-soluble pigment are produced.

On PCA, 5-d-old colonies are deep yellow-brown, with Munsell notation 1–2.5Y 3–4/4–6 and 10YR 3–4/4–6. Circular to irregular, with a smooth or rough surface. Edges are usually entire or undulate, but sometimes erose or filamentous. Elevation is either effuse with convex or umbonate centers, raised, or convex. Translucent; production of brown, water-soluble pigment is moderate to heavy.

Older cultures on all media produce copious, dark-brown, water-soluble pigment.

The mol% G + C of the DNA is: 67.6–67.8 (T_m).

Type strain: ATCC 29482, UASM D.

Additional Remarks: Other strains: ATCC 29483 (UASM 2) and ATCC 29484 (UASM 6).

4. **Lysobacter gummosus** Christensen and Cook 1978, 388[AL]
gum.mo'sus. L. adj. *gummosus* slime (gum)-producing.

On CCA, 5-day-old colonies are pale yellowish-gray, with Munsell notation 5Y 8/2. Circular, with a smooth surface. Entire edge. Elevation pulvinate. Translucent; no water-soluble pigment produced. Gummy.

On SAA, 5-d-old colonies are pale yellowish-gray, with Munsell notation 5Y 8/2. Circular, with a smooth surface. Entire edge. Elevation pulvinate. Opaque; no water-soluble pigment produced. Very gummy.

On PCA, 5-d-old colonies are yellow-gray, with Munsell notation 5Y 7/4. Circular, with a smooth surface. Entire edge. Elevation pulvinate. Opaque; no water-soluble pigment produced. Very gummy.

Older cultures are intensely gummy and do not produce a water-soluble pigment. In very old cultures, the viscosity of the colony may decrease dramatically, although viable cells can still be recovered from this thin slime.

The mol% G + C of the DNA is: 65.7 (T_m).

Type strain: ATCC 29489, UASM 402.

Genus V. **Nevskia** *Famintzin 1892, 484*[AL]

HANS-DIETRICH BABENZIEN AND HERIBERT CYPIONKA

Nev′ski.a. M.L. fem. n. *Nevskia* from the Neva, a river in St. Petersburg.

Cells elongated, rod shaped, usually slightly bent with **acellular hyaline stalks** (Fig. BXII.γ.30). The **stalks branch dichotomously** because of binary fission of mature cells and the preference for laterally excreted slime material (Fig. BXII.γ.31). Undisturbed and slowly growing cells form **flat, rosette-like to bush-like microcolonies** (Fig. BXII.γ.32), up to 80 μm in diameter, **on the surface of freshwater**. Faster growth results in the formation of a thin surface pellicle without distinct stalks (Fig. BXII.γ.33A) and with

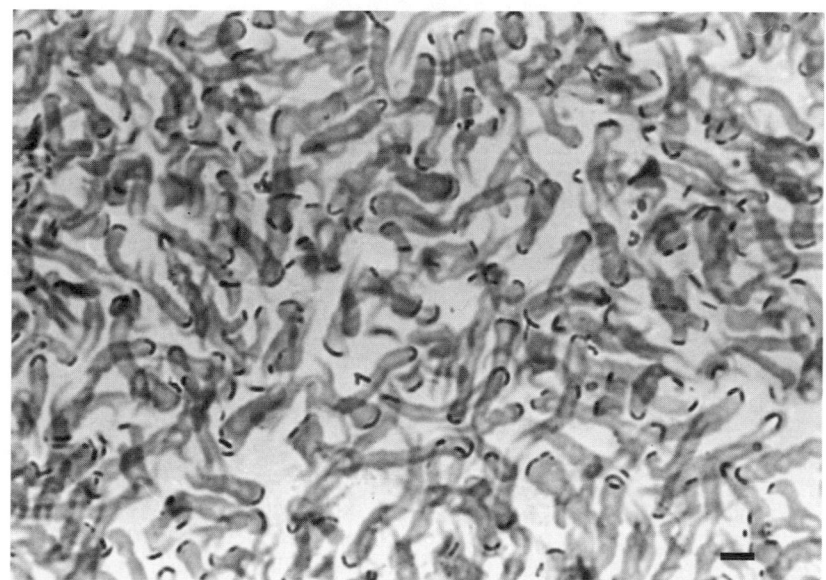

FIGURE BXII.γ.30. Enrichment culture of *N. ramosa*, thin bent rods at the tip of slime stalks. Formalin-toluidine blue. Bar = 10 μm.

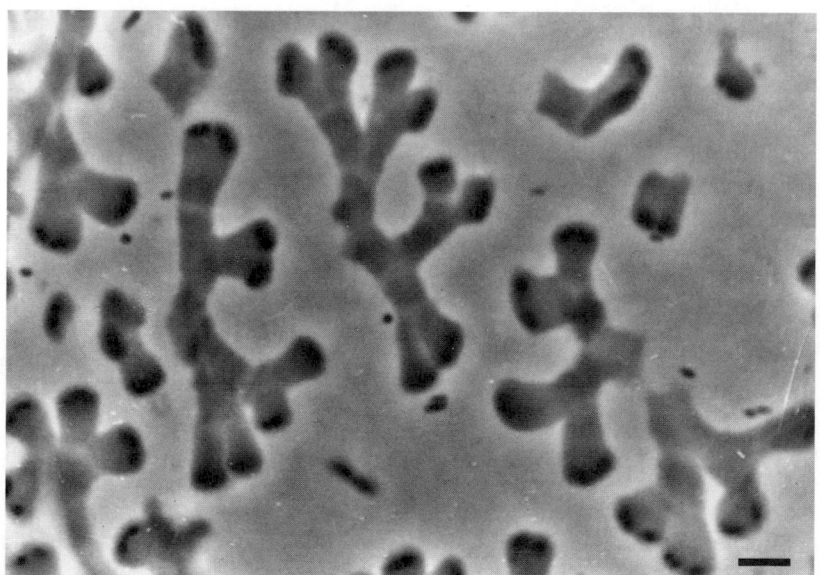

FIGURE BXII.γ.31. *Nevskia ramosa*, strain Soe 1 (DSM 11499). Dichotomously branched slime stalks with *Nevskia ramosa* cells at the tip. Phase-contrast micrograph. Bar = 10 μm.

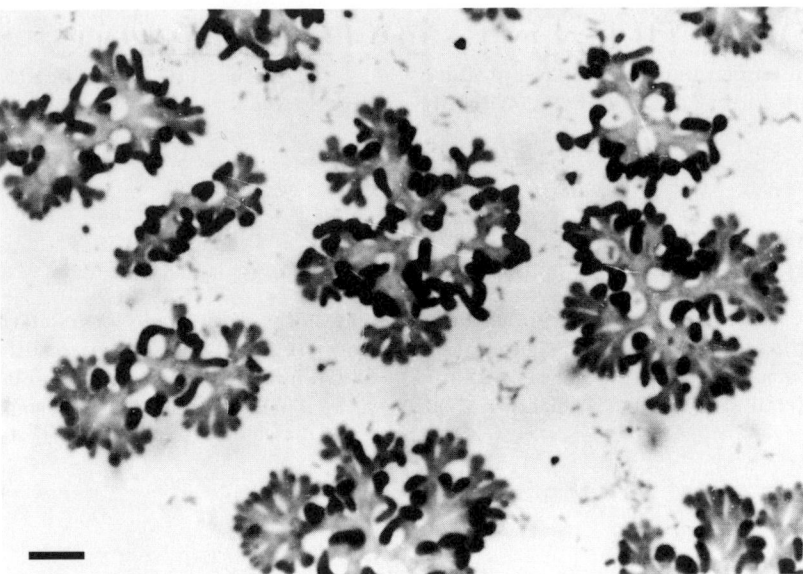

FIGURE BXII.γ.32. Habitat form of *N. ramosa*, showing the typical bush-like microcolony structure. Formalin-carbol-fuchsin. Bar = 20 μm.

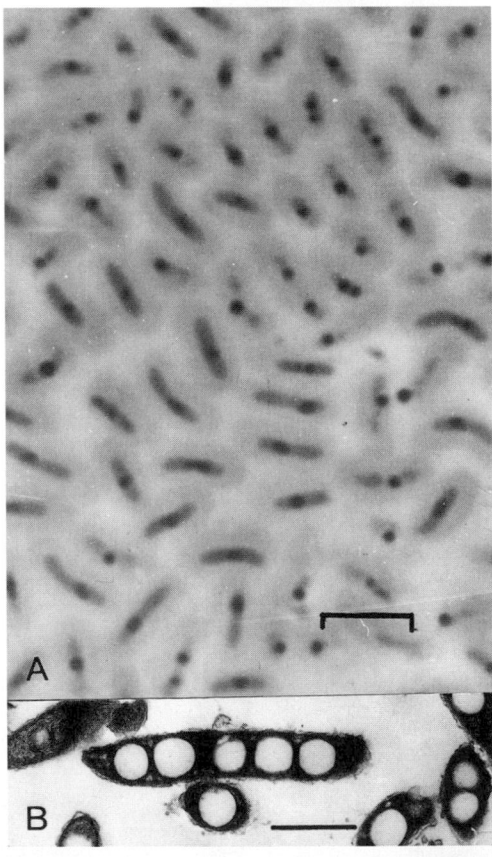

FIGURE BXII.γ.33. *Nevskia ramosa*, strain Soe 1 (DSM 11499). *A*, without distinct stalks in this growth phase. Phase-contrast micrograph. Bar = 10 μm. *B*, ultrastructure of organisms grown with Na-lactate. Cells filled with poly-β-hydroxybutyrate granules (white areas). Bar = 1 μm.

slime all around the cells. Cells filled with numerous refractile globules (Fig. BXII.γ.33B) which consist mainly of polyhydrox-yalkanoates. Resting stages are not known. **Gram negative**. In young cultures, some cells are **motile** by means of a polar fla-

gellum: they are set free from the microcolonies and can start the development of new colonies (Fig. BXII.γ.34). **Strictly aerobic.** Optimum growth temperature: 20–25°C. **Chemoorganotrophic**, with a broad range of organic compounds, having an oxidative type of metabolism. **Colorless colonies** on nutrient agar. Weak catalase activity, no cytochrome oxidase activity. Cytochromes of the *b* and *c* types are present. Originally found in the surface pellicle of an aquarium. Occur mainly on the surface of freshwater environments (neuston community).

Phylogenetic analysis of 16S rDNA sequences revealed that *N. ramosa* is a member of the phylum *Proteobacteria* and represents a deep branch of the *Gammaproteobacteria* with no known close relatives (Fig. BXII.γ.35).

The original strain described by Famintzin (1892) as the type strain was not isolated in pure culture.

The mol% G + C of the DNA is: 67.8–69.0 (HPLC), neotype strain Soe 1 (DSM 11499); 69.0, strain OL 1 (DSM 11500).

Type species: **Nevskia ramosa** Famintzin 1892, 484.

FURTHER DESCRIPTIVE INFORMATION

The two 16S rDNA sequences obtained have been deposited with EMBL (accession No. AJ001010 and AJ001011 for strains Soe 1 and OL 1, respectively). The 16S rDNA sequences of the two strains revealed 99.2% similarity; genomic fingerprinting by ERIC-PCR showed different banding patterns on agarose gel (Stürmeyer et al., 1998).

For *in situ* identification by hybridization, two oligonucleotide probes targeting positions 656–674 (NEV 656, 5'-CGCCTCCCTCTACCGTTT-3') and 177–195 (NEV 177, 5'-GCTCTTGCGAGATCATGC-3') of the 16S rRNA gene are available (Glöckner et al., 1998).

The type species was described by Famintzin (1892) as being 2–6 × 12 μm in size, but he also mentioned organisms identical to *N. ramosa* of "much smaller sizes". Henrici and Johnson (1935) described an organism practically identical to *N. ramosa* in morphology but of smaller dimensions, i.e., 1 × 3–4 μm. Babenzien (1965, 1967) isolated some strains from the surface of swamp ditches and ponds and the average size was 0.7 × 2.4–2.7 μm.

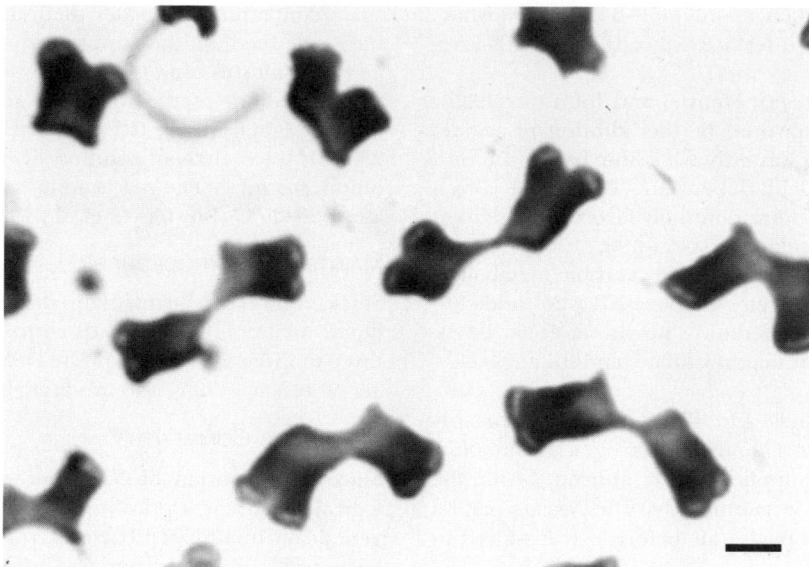

FIGURE BXII.γ.34. *Nevskia ramosa*, the beginning of colony growth, cells at the tip of the slime stalks. Formalin-toluidine blue. Bar = 10 μm.

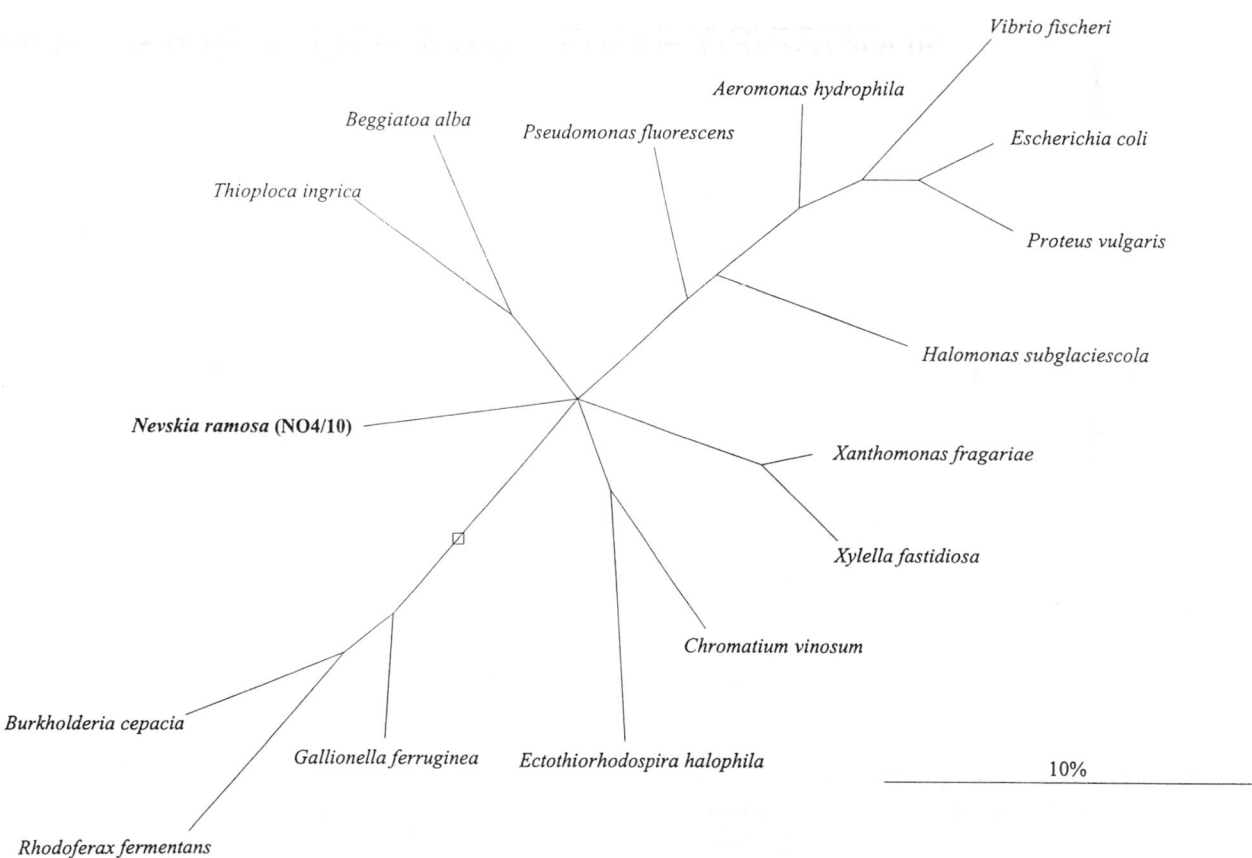

FIGURE BXII.γ.35. Phylogenetic tree, showing phylogenetic relationships between *Nevskia ramosa*, *Nevskia* clones NO4 and NO10, and closest relatives among the *Gammaproteobacteria*, based on comparative analysis of 16S rDNA. Bar = 10% estimated divergence. The open box indicates the root. (Reproduced with permission from Glöckner et al., Applied and Environmental Microbiology *64*: 1895–1901, 1998, ©American Society for Microbiology, Washington, D.C..)

Hirsch (1992) reported other *Nevskia*-like organisms also in this size range. The neotype strain DSM 11499 has a size of 0.7–1.1 × 1.4–2.3 μm during the logarithmic phase of growth and a tendency to form more elongated cells within the lag phase.

The globular bodies were thought to be composed of "etherical oils" and could be dissolved with 70% ethanol (Famintzin, 1892), whereas it was assumed by Henrici and Johnson (1935) that the vacuoles contained sulfur. According to Babenzien and

Hirsch (1974), globules contain mainly poly-β-hydroxybutyrate. Cells grown at the air-water interface typically contain 2–5 refractile globules.

In the enrichment cultures of Henrici and Johnson (1935), the bacteria were distinctly favored by the addition of sulfide, which indicated that *Nevskia* was either a sulfur bacterium or a microaerophile. As reported by Babenzien (1965) and Hirsch (1981), enrichment cultures were definitely favored by addition of 0.1% sodium lactate or acetate to lake water.

Pure cultures of *N. ramosa* have a strictly aerobic metabolism and grow with numerous substrates (Table BXII.γ.26) added to lake water or simple synthetic medium (Stürmeyer et al., 1998). No anaerobic growth with nitrate, thiosulfate, and sulfate as electron donors.

The development of a pellicle and stalk formation is primarily dependent on the availability of combined nitrogen. If ammonia, nitrate, or amino acids are supplied to the nutrient broth, the cultures grow submersed. It is assumed that the *Nevskia* biofilm in nature traps ammonia from the air before it is dissolved in the water.

By cultivating *N. ramosa* as a surface film, the fluid medium remains clear.

The slime stalks consist mainly of polysaccharides with rhamnose and small amounts of glucose and mannose, but other hydrophobic components may be present.

Nitrogenase activity was not detected.

According to numerous reports (Babenzien, 1966; Babenzien and Hirsch, 1974; Hirsch, 1981; Heldal and Tumyr, 1986; Kangatharalingam and Priscu, 1993; Stürmeyer et al., 1998), *Nevskia ramosa* is widely distributed in shallow aquatic habitats; it has never been reported from soils. Most findings have been from the water surface and, in fact, *Nevskia* appears to be a typical neuston microorganism. As indicated by the dull appearance and hydrophobicity of the biofilm, the flat rosettes live essentially outside of the water phase. While *Nevskia* cells are sensitive to UV radiation if kept in the dark after UV exposure, the isolates overcome UV damage by a very effective photorepair mechanism if incubated subsequently under light.

ENRICHMENT AND ISOLATION PROCEDURES

Surface films of swamp ditches, ponds, greenhouse pools, etc. are collected with sterile glass slides, needle loops, paper strips, or foils. The samples are transferred to Petri dishes with sterile water from the same sites, supplemented with 5 mM sodium lactate, and incubated in the dark at room temperature for about 10 days. Under microscopic control, bacterial microcolonies from the water surface are then picked up with a loop and transferred repeatedly into sterile Erlenmeyer flasks containing the same medium. For dilution purposes and to obtain pure cultures, multiwell chambers are also suitable. Pure cultures were obtained by repeated streaking of diluted samples on agar plates (1% Difco

agar) containing lake water medium. To obtain submersed cultures, the medium was supplemented with 5 mM NH₄Cl.

Pure cultures can also be cultivated as surface films in a synthetic medium of the following composition: sodium lactate, 5 mM; MgSO₄·7H₂O, 0.2 mM; CaCl₂·2H₂O, 0.1 mM; KH₂PO₄, 25 mM; trace element solution SL 9, 0.5 ml/l; and vitamin solution, 0.5 ml/l. The pH is adjusted to 7.0. Incubation temperature 20–25°C (Stürmeyer et al., 1998).

MAINTENANCE PROCEDURES

Stock cultures can be maintained in Erlenmeyer flasks containing liquid medium, covered with cotton plugs, with transfers every three months. The strains in the DSM (Braunschweig, Germany) are stored at −70°C and are available in lyophilized form.

TAXONOMIC COMMENTS

Since the taxonomy of *N. ramosa* was based solely on morphological data, it was previously found in the section on "Phylogenetically unaffiliated bacteria" (*The Prokaryotes*, 2nd Ed.) and Section 21 "Budding and/or appendaged bacteria" (*Bergey's Manual of Systematic Bacteriology*, 1st Ed.).

It is now clear that *N. ramosa* should not be affiliated with *Caulobacterales*, *Gallionella*, *Thiobacterium*, or *Beggiatoa*.

TABLE BXII.γ.26. Utilization of organic substrates by *Nevskia ramosa*

Component	Concentration (mM)	Growth	Rosette formation
D-Ribose	5	+	+
D-Glucose	5	+	+
D-Fructose	5	−	−
Sucrose	5	+	+
Starch	0.20%	+	−
Cellulose	0.20%	+	−
Formate	5	+	+
Acetate	5	+	+
Lactate	5	+	+
Lactate/NH₄Cl	5	+	−
Butyrate	8	−	−
Palmitate	2	−	−
L-Malate	10	−	−
Pyruvate	10	+	+
Citrate	5	−	−
Fumarate	10	+	+
Succinate	10	−	−
Ethanol	5	+	+
Glycerol	5	+	+
L-Alanine	5	−	−
L-Arginine	5	+	−
L-Cysteine	5	−	−
L-Glutamate	5	+	−
Benzoate	2	+	−
Tween 20	0.001%	+	−
Tween 80	0.001%	+	−

List of species of the genus Nevskia

1. **Nevskia ramosa** Famintzin 1892, 484[AL]

ra.mo′sa. L. adj. *ramosus* branched.

The characteristics are the same as those described for the genus. A spectrum of substrates and the mode of growth are shown in Table BXII.γ.26. The neotype strain Soe 1 (DSM 11499) was isolated from a small bog lake (L. Soel-

kensee, Germany) and formerly designated as strain IMET 10965.

The mol% G + C of the DNA is: 67.8 (HPLC), strain Soe 1 (DSM 11499); 69.0 (HPLC), strain OL 1 (DSM 11500).

Type strain: Soe 1, DSM 11499.

GenBank accession number (16S rRNA): AJ001010.

Genus VI. **Pseudoxanthomonas** Finkmann, Altendorf, Stackebrandt and Lipski 2000, 280[VP]

ANDRÉ LIPSKI AND ERKO S. STACKEBRANDT

Pseu.do.xan.tho.mo′nas. Gr. pseudes false; M.L. n. Xanthomonas a bacterial generic name; M.L. fem. n. Pseudoxanthomonas false Xanthomonas.

Rod-shaped cells. Gram negative. **Aerobic. Heterotrophic.** Colonies are yellow. Possess a ubiquinone with eight isoprene units. **The predominant fatty acids in the iso-/anteiso-type fatty acid profile are $C_{15:0 \text{ iso}}$ and $C_{15:0 \text{ anteiso}}$.**

Type species: **Pseudoxanthomonas broegbernensis** Finkmann, Altendorf, Stackebrandt and Lipski 2000, 280.

DIFFERENTIATION OF THE GENUS *PSEUDOXANTHOMONAS* FROM OTHER GENERA

The genus can be separated from the genera *Xanthomonas* and *Stenotrophomonas* by the absence of the fatty acid 3-hydroxy-iso-tridecanoic acid and from the species *Xylella fastidiosa* by the presence of branched-chain fatty acids.

TAXONOMIC COMMENTS

Pseudoxanthomonas broegbernensis is clearly a member of the family *Xanthomonadaceae*. The phylogenetic position within this family is not yet settled. Depending on the selection and number of reference sequences, the genus *Xanthomonas* or *Xylella* is the most closely related genus. The more precise phylogenetic relatedness of these genera must await analyses using more sequences of strains of the genus *Pseudoxanthomonas*, which is presently represented by a single sequence only.

At present, the 16S rDNA sequence of *P. broegbernensis* shares the closest relationship (97% similarity) to the sequence of an undescribed ultramicrobacterium strain 12-3 (Iizuka et al., 1998) whose accession number is AB008507.

List of species of the genus Pseudoxanthomonas

1. **Pseudoxanthomonas broegbernensis** Finkmann, Altendorf, Stackebrandt and Lipski 2000, 280[VP]

   broeg.ber′ nen.sis. M.L. fem. adj. *broegbernensis* pertaining to Brögbern (location near Lingen/Germany, from where the organism was isolated).

   The characteristics are as described for the genus and as listed in Table BXII.γ.27. Nitrate is not reduced, but nitrite is reduced to nitrous oxide (Finkmann et al., 2000). Predominant compounds of the fatty acid profile are $C_{15:0 \text{ iso}}$, $C_{15:0 \text{ anteiso}}$, $C_{16:0}$, $C_{16:1 \omega 7c}$, and $C_{17:1 \text{ iso } \omega 8c}$. The major hydroxy fatty acid is $C_{11:0 \text{ iso } 3OH}$. The fatty acids $C_{17:0 \text{ cyclo}}$ and $C_{13:0 \text{ iso } 3OH}$ are absent. No pathogenic potential for humans, animals, or plants known.

TABLE BXII.γ.27. Characteristics of *Pseudoxanthomonas broegbernensis*[a,b]

Characteristic	Reaction
Nitrate reduction to nitrite or N_2	−
Nitrite reduction to N_2O	+
Tween 80 hydrolysis	+
Urease, arginine dihydrolase	−
Alkalinization of citrate and malonate media	−
Lecithinase	−
Growth at 37°C	−
Acid from:	
Glucose, xylose, fructose, lactose, maltose	+
Rhamnose, sucrose	−
Hydrolysis of:[c]	
pNP-N-acetyl-β-glucosaminide, pNP-β-galactopyranoside	+
pNP-β-glucopyranoside, pNP-α-D-maltoside, pNP-β-xyloside	+
Bis-pNP-phosphate, pNP-phosphocholine, thymidine-5′-mono-phosphate-pNP-ester	−
Hydrolysis of:[c]	
L-Alanine-pNA, L-arginine-pNA, γ-L-glutamate-pNA, L-leucine-pNA, L-lysine-pNA	+
Glycine-pNA, L-proline-pNA, L-valine-pNA	−
Growth with:	
N-Acetylglucosamine, D-cellobiose, D-galactose, D-maltose, D-mannose, D-sucrose, D-trehalose	+
L-Arabinose, D-fructose, gluconate, D-ribose, i-inosite, D-mannite, sorbite, propionate, T-aconitate, adipate, DL-3-hydroxybutyrate, DL-lactate, L-malate, pyruvate, L-histidine, L-hydroxyproline, L-ornithine, L-proline, putrescine	−
Growth in the presence of:	
NaCl, 3%	−
Cadmium acetate, 0.001%; $ZnCl_2$, 0.01%; lead acetate, 0.01%	−
Growth in presence of antimicrobial agents (μg or U per disk):	
Erythromycin (10 μg); streptomycin (10 μg); nalidixic acid (30 μg); kanamycin (30 μg); ampicillin (10 μg); penicillin G (10 U); gentamicin (10 μg); fucidin (10 μg); tetracycline (30 μg); novobiocin (30 μg)	+
Neomycin, 30 μg	−

[a] Symbols: see standard definitions.

[b] Data from Lipski et al. (1992); Finkmann et al. (2000).

[c] pNP, *para*-nitrophenyl; pNA: *para*-nitroanilide.

The only strain of this genus was isolated from an experimental biofilter supplied with the waste gas of an animal rendering plant.

The mol% G + C of the DNA is: not known.
Type strain: B1616/1, DSM 12573.
GenBank accession number (16S rRNA): AJ012231.

Genus VII. **Rhodanobacter** *Nalin, Simonet, Vogel and Normand 1999, 22*[VP]

THE EDITORIAL BOARD

Rho.da′ no.bac.ter. L. masc. n. *Rhodanus* River Rhône; M.L. masc. n. *bacter* equivalent of *bacterium* a small rod; M.L. masc. n. *Rhodanobacter* rod isolated close to the River Rhône.

Gram-negative rods. **Nonmotile**. Colonies on SMA agar plates are yellow with clean edges, 0.2–1 mm in diameter after 3 d at 30°C. Do not form spores. Branching does not occur. Aerobic; chemoorganotrophic. **Catalase and oxidase positive**. Optimum growth temperature 30°C. Based on analysis of 16S rDNA sequences, *Rhodanobacter* belongs in the class *Gammaproteobacteria*, order *Xanthomonadales*, and family *Xanthomonadaceae*.

The mol% G + C of the DNA is: 63.

Type species: **Rhodanobacter lindaniclasticus** Nalin, Simonet, Vogel and Normand 1999, 22.

FURTHER DESCRIPTIVE INFORMATION

Rhodanobacter lindaniclasticus can degrade γ-hexachlorocyclohexane (lindane; γ-HCH) under aerobic conditions. The first gene in the lindane degradation pathway, *linA*, is highly conserved (99.4%) between *R. lindaniclasticus* and the other known aerobic lindane degrading bacterium, *Sphingomonas paucimobilis* (Thomas et al., 1996b).

ENRICHMENT AND ISOLATION PROCEDURES

Thomas et al. (1996b) collected γ-HCH-contaminated soil from several wood treatment facilities and added 5 g to a mixture of 10 g each Biolyte CX80 and CX85 in 100 ml of Ringer (1/4) solution. Following 1 h on a rotary shaker and 10 min of ultrasonication, 1 ml of the resulting supernatant was used as an inoculum for the enrichment culture and added to 3 l of a mineral medium containing γ-HCH as the sole carbon and energy source. Over 18 months, γ-HCH was added and the pH adjusted as necessary, using 1 N NaOH. Aliquots were periodically plated onto SMA agar and tested for ability to utilize γ-HCH.

DIFFERENTIATION OF THE GENUS *RHODANOBACTER* FROM OTHER GENERA

Rhodanobacter can be distinguished from other members of the *Xanthomonadaceae* by its ability to utilize γ-HCH as a growth substrate. Other distinguishing characteristics are listed in Table BXII.γ.17 in the chapter on *Xanthomonas*.

List of species of the genus Rhodanobacter

1. **Rhodanobacter lindaniclasticus** Nalin, Simonet, Vogel and Normand 1999, 22[VP]
   lin.da′ ni.clas.ti.cus. Fr. masc. n. *lindane* commercial name of γ-HCH; Gr. adj. *clasticus* breaking, from G. part. perf. *klastos* broken; M.L. masc. adj. *lindaniclasticus* lindane-breaking.

   Colonies on nutrient agar are yellow (nothoxanthin), circular, and convex, 0.2–1 mm in diameter. Growth on Milieu Base Salin (MBS) with γ-HCH as sole carbon source. Catalase, oxidase, and Tween 80 esterase positive. Urease,

   protease, tryptophanase, nitrate reductase, amylase, and DNase negative. Utilizes the following carbon sources: D-glucose, esculin, cellobiose, trehalose, D-xylose, lactate, caprate, malate, and citrate. Optimal growth at 30°C; also grows at 41°C; no growth at 4°C.

   The type strain was isolated from soil at a wood-treatment site.

   The mol% G + C of the DNA is: 63 (HPLC).
   Type strain: RP5557, LMG 18385.
   GenBank accession number (16S rRNA): AF039167.

Genus VIII. **Schineria** *Tóth, Kovács, Schumann, Kovács, Steiner, Halbritter and Máriageti 2001, 406*[VP]

THE EDITORIAL BOARD

Schi′ ner.i.a. N.L. fem. n. *Schineria* pertaining to Schiner who first described the fly *Wohlfahrtia magnifica* in 1862.

Gram-negative nonmotile aerobic rods. Catalase positive. Major fatty acids, $C_{14:0}$, $C_{16:0}$, and $C_{18:1}$. Major polar lipids include phosphatidylethanolamine, phosphatidylglycerol, and phosphatidylserine.

The mol% G + C of the DNA is: 42 (HPLC).

Type species: **Schineria larvae** Tóth, Kovács, Schumann, Kovács, Steiner, Halbritter and Máriageti 2001, 406.

FURTHER DESCRIPTIVE INFORMATION

The inner cell membrane forms invaginations that protrude into the cytoplasm (Tóth et al., 2001).

The *Schineria* strains studied were isolated from larvae of *Wohlfahrtia magnifica*, a dipteran that is an obligate parasite and that infects many vertebrates, including domestic livestock; it is possible that the strong chitin-degrading capability of *Schineria* strains plays a role in molting of the host larvae (Tóth et al., 2001).

Analysis of 16S rDNA sequences placed *Schineria larvae* in the *Gammaproteobacteria* (Tóth et al., 2001).

ENRICHMENT AND ISOLATION PROCEDURES

Schineria larvae strains were isolated on King's B medium from homogenates of washed *Wohlfahrtia magnifica* stage 1 and stage 2 larvae (Tóth et al., 2001).

MAINTENANCE PROCEDURES

Schineria larvae strains were maintained on King's B medium (Tóth et al., 2001).

List of species of the genus Schineria

1. **Schineria larvae** Tóth, Kovács, Schumann, Kovács, Steiner, Halbritter and Máriageti 2001, 406[VP]

   lar' vae. L. gen. fem. *larvae* pertaining to the origin of the type strain isolated from *Wohlfahrtia magnifica* maggots.

   Single straight encapsulated rods (0.8–0.9 × 2.0–3.0 μm). Produce H_2S. Hydrolyze chitin and urea; do not hydrolyze casein, esculin, or gelatin. Nonhemolytic. Do not utilize citrate. Voges–Proskauer test negative. No acid produced from carbohydrates in bioMéreiux API CH50 test strips.

   The mol% G + C of the DNA is: 42 (HPLC).
   Type strain: L1/68, CIP 107108, DSM 13226.
   GenBank accession number (16S rRNA): AF252143.

Genus IX. **Stenotrophomonas** *Palleroni and Bradbury 1993, 608[VP]**

NORBERTO J. PALLERONI

Ste.no.tro.pho.mo' nas. Gr. adj. *stenus* narrow; Gr. n. *trophus* one who feeds; Gr. n. *monas* a unit, monad; M.L. fem. n. *Stenotrophomonas* a unit feeding on few substrates.

Straight or curved rods but not helical, 0.5 × 1.5 μm, occur singly or in pairs. Do not accumulate poly-β-hydroxybutyrate granules as intracellular reserve material. No resting stages are known. Gram negative. **Motile by two or more polar flagella. Aerobic, having a strictly respiratory type of metabolism with oxygen as electron acceptor. Nitrate is reduced but it is not used as a nitrogen source for growth.** The colonies are yellow, greenish, or gray. The yellow color is not due to carotenoids or xanthomonadins. The colonies may turn a dark brown color with age. **The major polyamines are spermidine and cadaverine.** Aside from fatty acids in common with *Xanthomonas* species, one of the species (*S. maltophilia*) **is characterized by the presence of $C_{17:0}$ cyclopropane fatty acid and ubiquinone Q8.** No growth occurs at 4 or 41°C; the optimum growth temperature is ~35°C. **Oxidase negative.** Gelatinase and catalase positive. Strong lipolytic activity, as judged by the hydrolysis of Tween 80. **Chemoorganotrophic.** One of the species (*S. maltophilia*) is widely distributed in nature and is commonly isolated from clinical materials and nosocomial infections.

The mol% G + C of the DNA is: 66.9 ± 0.8.

Type species: **Stenotrophomonas maltophilia** (Hugh 1981) Palleroni and Bradbury 1993, 608 (*Pseudomonas maltophilia* Hugh 1981, 195; *Xanthomonas maltophilia* Swings, De Vos, Van Den Mooter and De Ley 1983, 412.)

FURTHER DESCRIPTIVE INFORMATION

Morphological characteristics *Stenotrophomonas* cells are morphologically uniform and appear as slender straight or curved rods, single or in pairs (Fig. BXII.γ.36). No granules of reserve materials are visible under the phase contrast microscope in cells grown under various conditions. No studies on fine structure of thin sections have been reported in the literature.

The cells have two or more polar flagella (Fig. BXII.γ.37),

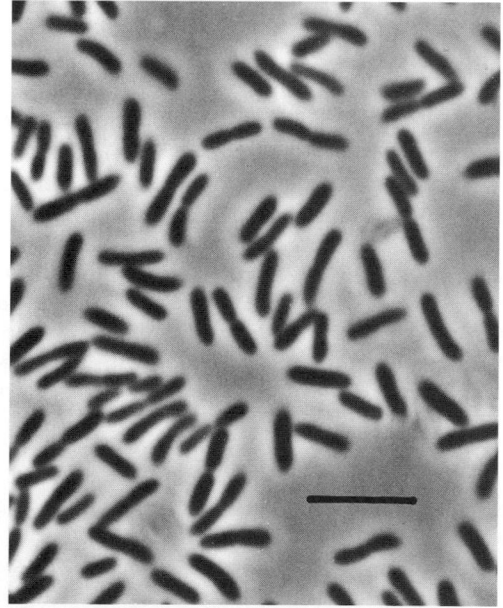

FIGURE BXII.γ.36. Cells of *Stenotrophomonas maltophilia* strain ATCC 13637[T] grown in nutrient broth. Phase contrast. Bar = 5 μm.

---

*Editorial Note: The literature search for the chapter on *Stenotrophomonas* was completed in January, 2000. During the course of unavoidable publication delays, a number of new species were described or reclassified after the chapter was completed. It was not possible to include these species in the text or to include their characteristics in the comparative tables. The reader is encouraged to consult the studies listed in the Further Reading section.

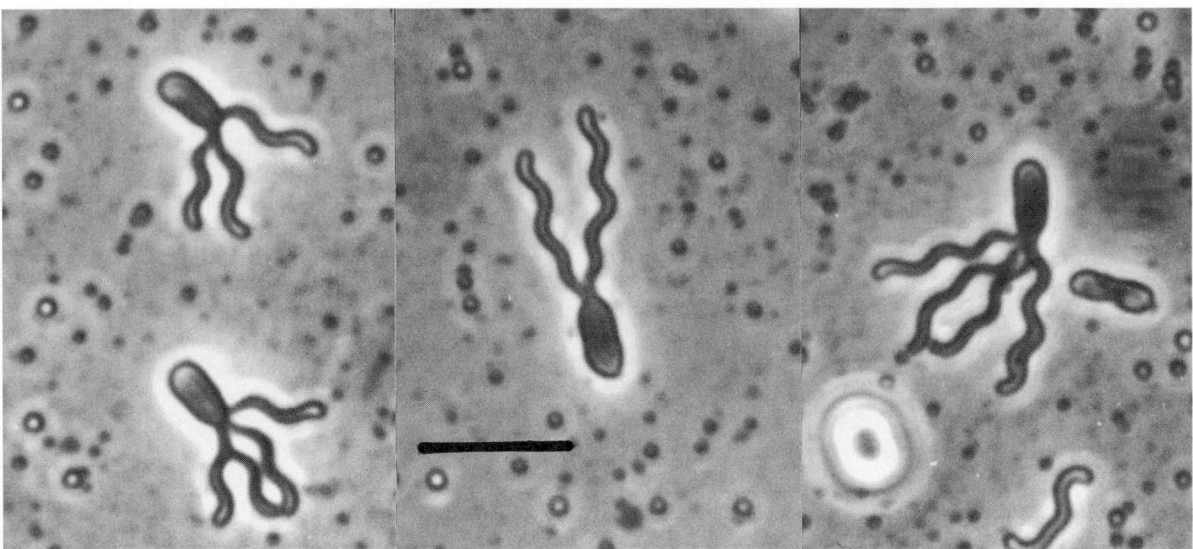

FIGURE BXII.γ.37. *Stenotrophomonas maltophilia* strain ATCC 13637^T. Leifson flagella stain. Bar = 5 μm.

whereas *Xanthomonas* species typically have only one polar flagellum per cell. A flagellin of molecular weight 33 kDa has been isolated (Montie and Stover, 1983).

Polar fimbriae occur in *S. maltophilia*, as well as in several other pseudomonads (Fuerst and Hayward, 1969). These elements may be necessary for attachment to host cells and other solid surfaces, and may also be involved in the twitching motility of the *Stenotrophomonas* cells on solid surfaces (Henrichsen, 1975).

Chemical composition Most studies on cell composition in this genus refer to *S. maltophilia*. This species is resistant to EDTA treatment, whereas *P. aeruginosa* and other species of this genus were found to be particularly sensitive, due to removal of components (metal ions, proteins, lipopolysaccharides, lipids) from the outer membrane of these organisms by the chelator (Wilkinson, 1970; Wilkinson et al., 1973).

An early report on the lipid composition of several species of *Xanthomonas* (Rietschel et al., 1975) shows a marked similarity to the lipid composition of whole cells of *S. maltophilia*, and clear differences from the composition of pseudomonads of other phenotypic groups (Moss et al., 1972, 1973), in agreement with the taxonomic relationships shown by rRNA–DNA hybridization experiments (Palleroni et al., 1973). Both *S. maltophilia* and *Xanthomonas* have a lipid A composition based on glucosamine (Hase and Rietschel, 1976), and similarities in the LPS composition include a high content of rhamnose, the presence of D-galacturonic acid, the absence of a heptose, and a low KDO content (Neal and Wilkinson, 1979, 1982). The major fatty acids of the outer membrane of *S. maltophilia* include 9-methyl-decanoic, 2-hydroxy-9-methyldecanoic, 3-hydroxy-9-methyldecanoic, 3-hydroxydodecanoic, and 3-hydroxy-11-methyldecanoic acids (Wilkinson et al., 1983b).

An interesting feature of the LPS of *S. maltophilia* is the presence of a pentose derivative identified as 3-*O*-methyl-L-xylose, which is also found in unrelated species such as *Rhodopseudomonas viridis* (Weckesser et al., 1974). Other observations on the LPS structure and a possible basis for the serological cross-reaction between *Stenotrophomonas* and *Brucella* have been reported (Di Fabio et al., 1987), as well as details of the synthesis of the tetrasaccharide repeating unit of *Stenotrophomonas* LPS (Liptak et

al., 1986). A specific extraction procedure has been described for the study of outer membrane proteins (Chin and Dai, 1990), which participate in the antibiotic susceptibility of *S. maltophilia* strains isolated from clinical species (Cullmann, 1991; Lecso-Bornet et al., 1992; Wilcox et al., 1994).

By transfer to a medium containing carbenicillin and polyvinyl pyrrolidone, Pease et al. (1989) were able to induce the formation of cell-wall-deficient forms in various stages, which were demonstrated using scanning electron microscopy. Induction of rapid reversion to normal bacterial forms has been attributed to plasmids by the authors. This rapid reversion makes it difficult (although not impossible) to maintain cell-wall-deficient strains in media containing carbenicillin.

The efficiency of solute diffusion through the outer membrane of *S. maltophilia* is roughly 3–5% of that of *E. coli*. This low value in *S. maltophilia* is attributed to the low copy number of porins rather than to differences in the size of the channels (Yamazaki et al., 1989).

No hopanes have been found in *S. maltophilia* cells (Rohmer et al., 1979). *Stenotrophomonas maltophilia* contains spermidine and cadaverine as the major polyamine components (Auling et al., 1991; Yang et al., 1993b).

The whole cell fatty acid composition of *Stenotrophomonas* will be discussed under Taxonomic Comments.

Pigmentation The colonies of *S. maltophilia* strains on various complex and chemically defined media are often yellowish. The color, which does not diffuse into the medium, is not due to carotenoid pigments but perhaps to flavins (Stanier et al., 1966; Palleroni and Bradbury, 1993). Although characterization of the pigment is still tentative, it is probably not a xanthomonadin (the pigment characteristic of *Xanthomonas*) or a closely related compound (Starr et al., 1977).

According to a published description (Hugh and Gilardi, 1980), the colonies of *S. maltophilia* characteristically appear lavender-green in color on blood agar media. A greenish discoloration of erythrocytes develops in the area of confluent growth, but there are no signs of hemolysis around isolated colonies. *S. africana* forms faint fluorescent green colonies on Trypticase soy agar and gray colonies on sheep blood agar and on MacConkey agar at both 30 and 37°C. (Drancourt et al., 1997). No information on the chemical nature of the pigment(s) is available.

Some strains of *S. maltophilia* develop a brown discoloration in agar media—in fact this was the basis for the name *"Pseudomonas melanogena"* (Iizuka and Komagata, 1963a; Komagata et al., 1974)—which may be due not to water-soluble melanin but to a secondary chemical reaction on some extracellular products (Hugh and Gilardi, 1980). The intensity of this brown color is enhanced after growth at 42°C.

Nutrition and growth conditions Many strains of *Stenotrophomonas* require the addition of growth factors (methionine or cystine plus glycine) for growth on chemically defined media (Iizuka and Komagata, 1964b; Hugh and Gilardi, 1980). Methionine-independent strains, however, have been known for many years (Ikemoto et al., 1980). As suggested elsewhere in this chapter, they may represent a different species of *Stenotrophomonas*.

Even though *Stenotrophomonas* species are able to reduce nitrate, this anion is not a nitrogen source for growth. This lack of growth is a useful characteristic for the differentiation of this genus from most other groups of aerobic pseudomonads (Stanier et al., 1966).

A strong ammonia odor develops after growth on blood agar medium (Hugh and Gilardi, 1980). This property is absent from *S. africana* (Drancourt et al., 1997).

Studies on the nutrition of *S. maltophilia* have shown that this species is not as nutritionally versatile as other species of aerobic pseudomonads of the genus *Pseudomonas* and other species formerly assigned to this genus (Stanier et al., 1966). This restricted nutritional capacity was the basis for the proposed new generic name (*Stenotrophomonas*) (Palleroni and Bradbury, 1993). Typically, strains of the two species of this genus utilize some disaccharides (maltose, cellobiose, lactose) as sole carbon and energy sources (Hugh and Ryschenkow, 1961; Stanier et al., 1966; Drancourt et al., 1997). These substrates are rarely utilized by other pseudomonads. Growth on lactose is poor, perhaps because of the fact that the cells do not use galactose.

The strong lipolytic activity of *S. maltophilia* is due, at least in part, to the presence of an esterase in the outer membrane. The esterase may play a role in the utilization of esters as substrates for growth (Debette and Prensier, 1989). The membrane-bound non-specific esterases of many strains have been analyzed by polyacrylamide gel electrophoresis and isoelectric focusing, and the relationship of these observations to the distribution of selected phenotypic properties of the strains supports the idea that species names such as *Pseudomonas beteli* and *P. hibiscicola* should be considered synonyms of *S. maltophilia* (Singer et al., 1994), as suggested by Van Den Mooter and Swings (1990).

A micromethod for the nutritional screening of strains of *S. maltophilia* has been described (Freney et al., 1983). In the earlier study by Stanier et al. (1966), 23 strains of *S. maltophilia* were subjected to an extensive nutritional screening (see the description of this species later in this chapter). No simple alcohols, polyalcohols or glycols, aromatic compounds, amines, various nitrogenous compounds, or paraffin hydrocarbons were utilized by the great majority of strains. Of the 145 substrates tested, only 23 supported growth of all or most strains. The strains tested in this study—provided by Dr. P. Thibault of the Pasteur Institute—had been isolated from various materials of clinical origin. However, later isolations from different natural habitats have resulted in the identification of strains of *S. maltophilia* capable of degrading sulfonic-aromatic compounds (Lee and Clark, 1993), the

herbicide Dicamba (3,6-dichloro-2-methoxybenzoic acid) (Yang et al., 1994; Wang et al., 1997), toluene and xylenes (Su and Kafkewitz, 1994), benzoate (Liu et al., 1996b), or chitin (Andreeva et al., 1996; Wiwat et al., 1996). *S. maltophilia* has been isolated from enrichments with kerosene and crude oil, inoculated with soil from petroleum regions in Japan. Although the isolates used hydrocarbons, they rapidly lost this capacity after subculture under laboratory conditions (Iizuka and Komagata, 1964b).

Some of these properties place *Stenotrophomonas* species among those of interest for applications in biotechnology and for bioremediation of contaminated environments. However, it is important to keep in mind that organisms reputed to be opportunistic human pathogens should be used under strictly controlled conditions in large-scale biotechnological projects.

Metabolism Studies on many basic aspects of metabolism and metabolic pathways in this group of bacteria are rather fragmentary. The strains of *Stenotrophomonas* species give a negative oxidase reaction, which is correlated with the lack of cytochrome c (Stanier et al., 1966). Neither an α-band at 551–554 nm nor a β-band at 522–524 nm was seen in the spectra. The oxidase reaction was mistakenly recorded as positive by Palleroni and Bradbury (1993), following the report of Van Den Mooter and Swings (1990) who recorded a positive reaction for the majority of their strains. Some strains gave at most a faint positive oxidase reaction.

The metabolism of fructose involves the participation of a phosphoenol-pyruvate (PEP) phosphotransferase system (Sawyer et al., 1977b). *S. maltophilia* is able to use lactose for growth, a property very rare in other aerobic pseudomonads. As mentioned above when dealing with the nutritional aspects of this species, growth on this disaccharide, however, is rather poor, which may be related to the fact that the cells do not use galactose as a carbon source.

In the tyrosine biosynthetic pathway, both *Pseudomonas* and *Stenotrophomonas* strains have a prephenate activity linked to NAD; this activity is sensitive to feedback inhibition in *Stenotrophomonas* but not in *Pseudomonas*. The latter genus also has an arogenate dehydrogenase linked to NAD, an enzyme that is lacking in *Stenotrophomonas* (Byng et al., 1980). Studies on the regulatory pattern of prephenate and arogenate dehydratase activities in the phenylalanine biosynthetic pathways have suggested that *Stenotrophomonas* be classified in the same group as the fluorescent species of *Pseudomonas* (Whitaker et al., 1981b).

In spite of the fragmentary biochemical information that is available, various reports on the isolation and properties of a number of enzymes are found in the literature. These refer to proteases (Debette, 1991; Margesin and Schinner, 1991; Singer and Debette, 1993), peptidases (Stevens et al., 1990; Suga et al., 1995; Kabashima et al., 1996), cocaine esterase (Britt et al., 1992a), a secondary alcohol dehydrogenase that has little activity on primary and aromatic alcohols (Britt et al., 1992b), deoxyribonuclease (Scally and Winder, 1991), and enzymes involved in ADP ribosylation (Edmonds et al., 1989), a mechanism of action found in important bacterial toxins. The peptidyl dipeptidase activity of *S. maltophilia* appeared to be the most active among those of 24 different bacterial species examined (Stevens et al., 1990).

Studies on the *S. maltophilia* β-lactamases will be mentioned in the section on antibiotic susceptibility.

As stated elsewhere, *S. maltophilia* is unable to use nitrate as a nitrogen source, but it can reduce this anion to nitrite (Iizuka and Komagata, 1964b; Stanier et al., 1966). The nitrate reductase of some strains has been characterized (Woodward et al., 1990; Ketchum and Payne, 1992). A membrane-bound methyl viologen nitrate reductase and a soluble molybdopterin have been described in a study on 23 strains of *S. maltophilia* (Woodward et al., 1990).

The above studies point to the fact that in the reduction of nitrate to nitrite, the former can act as an electron sink when the organisms are growing under semiaerobic conditions. In a study of utilization of toluene and xylenes by a strain of *S. maltophilia*, these aromatic substrates were used under low oxygen and anoxic conditions in the presence of nitrate (Su and Kafkewitz, 1994). For strains with nitrite and nitrate reductase activity, see the note at the end of the list of species.

Phages and plasmids The information available on these subjects is scanty and fragmentary. The sensitivity of *Xanthomonas* showed a major overlap with the sensitivity of *S. maltophilia* among the various species then assigned to the genus *Pseudomonas* (Starr, 1981). This finding is a further confirmation of the phylogenetic relationship between *Xanthomonas* and *Stenotrophomonas* as members of the *Gammaproteobacteria*.

It has to be admitted, however, that the practical application of phage sensitivity as a diagnostic tool frequently fails to fulfill expectations. As stated by Billing and Garrett (1980) with reference to problems of identification of plant pathogenic bacteria, "phages potentially useful in diagnosis have been found in some cases, but all too often in other cases they have proved nonspecific or too specific." However, there are instances in which phage sensitivity may become an important piece of evidence pointing to basic phylogenetic relationships. For example, a phage lytic for *Pseudomonas* species, both fluorescent and nonfluorescent, but not for *S. maltophilia*, was isolated from sewage and described in detail (Kelln and Warren, 1971). On the other hand, the isolation of a temperate phage capable of plaque formation on both *P. aeruginosa* and *S. maltophilia* has been described (Moillo, 1973).

Some factors required for phage nucleic acid replication may be conserved through evolution in some groups of Gram-negative bacteria. One of the elements required for *in vitro* replication of sex-specific single stranded RNA coliphage Qβ is the so-called host factor (HF), which is a heat-resistant RNA-binding protein of molecular weight 12,000, usually present in *E. coli* as a hexamer (Franze de Fernández et al., 1972). A component able to cross-react with an *E. coli* HF antiserum was detected in the extracts of a number of species of aerobic pseudomonads, among them, *S. maltophilia* (DuBow and Ryan, 1977). Cell extracts of members of rRNA similarity groups I and V (*Pseudomonas*, *Xanthomonas*, and *Stenotrophomonas*) had HF activity after heat treatment. Heat-treated extracts of species of pseudomonads belonging to different branches of *Proteobacteria* did not stimulate replication of Qβ RNA and, in addition, they inhibited the activity of systems containing saturating amounts of *E. coli* HF, in other words, they contained a heat-resistant inhibitor.

The promiscuous IncP-1 plasmid R18 was analyzed by means of Tn7 insertion mutations in the Tra1 region, involved in conjugational transmissibility from *P. aeruginosa* to other species, including *S. maltophilia* (Schilf and Krishnapillai, 1986).

From a collection of 18 strains of *S. maltophilia*, 5 were shown to harbor plasmids, of which one was studied in some detail. Its molecular weight is 4.4×10^6 daltons, and the presence of several single restriction sites makes it promising as a cloning vector (Shen et al., 1992).

Antigenic properties A 30 kDa immunoglobulin binding protein was isolated from *S. maltophilia* (Grover et al., 1991a) and later partially characterized (Grover and Odell, 1992). The protein binds IgGs of various subclasses and IgA at their Fc regions. A substantial number of reports on the composition of several antigens in the outer membrane of *S. maltophilia* by Prof. Wilkinson and collaborators can be found in the literature (Winn et al., 1995, 1996; Winn and Wilkinson, 1995, 1996, 1997).

A serological classification of *S. maltophilia* strains based on the heat-stable O antigens has been proposed, extending the 15 serotypes defined by antisera in previous investigations (Hugh and Ryschenkow, 1961) to a total of 26 (Schable et al., 1989).

The isolation of a membrane protein capable of cross-reacting with the β subunit and carboxyl tail of human pregnancy chorionic gonadotropin (hCG), was described (Grover et al., 1991b, 1993b; Carrell et al., 1993). The partial sequence of the receptor soon followed (Grover et al., 1993a). A complete sequence of the gene encoding the *S. maltophilia* gonadotropin-like protein is now available (Grover et al., 1995). Exposed sites on *S. maltophilia* cells bind both hCG and the native hCG-like ligand with equal affinity (Carrell and Odell, 1992).

Antibiotic sensitivity The high frequency of detection of *S. maltophilia* as an opportunistic pathogen and isolation from materials of clinical origin, has prompted numerous studies on the susceptibility of the strains to various antibiotics, as well as the intrinsic and exogenous conditions that affect such sensitivity (Cullmann, 1991; Lecso-Bornet et al., 1992; Sader et al., 1994; Wilcox et al., 1994; Poulos et al., 1995; Rahmati-Bahram et al., 1995, 1996; Vanhoof et al., 1995; Alonso and Martínez, 1997). The conditions for laboratory tests have also been examined (Yao et al., 1995).

S. maltophilia is resistant to most β-lactam antibiotics, and the aminoglycosides and quinolones have a modest activity (García-Rodríguez et al., 1991). However, some quinolones have been found to be quite active (Furet and Pechere, 1991), in particular, clinafloxacin (Visalli et al., 1997). In a Japanese study on *S. maltophilia* strains isolated from clinical materials, high sensitivity to minocycline was observed. The sensitivity did not show the yearly changes recorded for *P. aeruginosa* (Tabe and Igari, 1994). *S. africana* has an antibiotic susceptibility restricted to netilmicin, ciprofloxacin, trimethoprim-sulfamethoxazole, and colimycin (Drancourt et al., 1997).

A report on the transmission of antibiotic resistance from *S. maltophilia* to *Pseudomonas aeruginosa* calls attention to the combined danger that these two species represent as the most common Gram-negative opportunistic pathogens in the hospital environment (Babalova et al., 1995). Laboratory experiments with a strain of *S. maltophilia* have shown the possibility of transferring to other species its multiple resistance to drugs (which included amikacin and cephalosporins) coded for by genes in an R plasmid (Krcmery et al., 1985). Interestingly, antibiotic and heavy metal resistance genes of Gram-positive origin, also detected in *Staphylococcus aureus*, have been identified in a *S. maltophilia* strain of clinical origin (Alonso et al., 2000).

Among the resistance factors of *S. maltophilia*, β-lactamases have received much attention (Paton et al., 1994; Felici and Amicosante, 1995; Bonfiglio et al., 1995; Kelly and Mortensen, 1995). Sequence studies on the L1 metallo-β-lactamase have discovered significant differences from similar enzymes present in *Aeromonas hydrophila* and *Bacillus cereus*, which resulted in the addition of a new subclass of this type of enzymes to the ones already known (Walsh et al., 1994). The L1 penicillinase *blaS* gene of *S. malto-*

philia was cloned in *E. coli*, where it could express its capacity for imipem hydrolysis (Dufresne et al., 1988).

A recent study of the L2 serine β-lactamase from *S. maltophilia* has demonstrated that this enzyme is a clavulanic acid-sensitive cephalosporinase related to a β-lactamase from *Yersinia enterocolitica* (Walsh et al., 1997a).

A number of antibiotics have been tested against *S. africana* using the disk diffusion technique (Drancourt et al., 1997). This species appeared to be susceptible to netilmicin, ciprofloxacin, trimethoprim-sulfamethoxazole, and colimycin.

An antifungal compound, maltophilin, is produced by *S. maltophilia*. It is active against human and plant pathogenic fungi (Jakobi et al., 1996).

Relationships to plants A predominance of *S. maltophilia* over other pseudomonads has been observed in the rhizosphere of several cultivated plants, such as cabbage, rape, mustard, corn, and beet (Debette and Blondeau, 1980). Some of these plants have proteins of high sulfur content, and the abundance of *S. maltophilia*, which requires S-containing amino acids like methionine, may be due to excretion of such compounds by the roots. However, methionine could not be detected in significant amounts in the rhizosphere.

In the rhizosphere of oilseed rape, *S. maltophilia* is able to interfere with growth of plant pathogenic fungi (Berg et al., 1996). The antifungal compound maltophilin (Jakobi et al., 1996) may be involved in this activity. Use of chitinolytic strains of *S. maltophilia* has been proposed for biological control of summer patch on turf grass (Kobayashi et al., 1995).

S. maltophilia has been found on the surface of leaves of the grass *Lolium perenne*, although at a low frequency in comparison to other aerobic pseudomonads (Austin et al., 1978).

Pathogenicity for humans *S. maltophilia* has been implicated as a direct agent of disease and as the cause of secondary infections that aggravate various human pathological conditions. It has been identified in cystic fibrosis (Gladman et al., 1992; Denton et al., 1996), cancer (Khardori et al., 1990; Elsner et al., 1997), meningitis (Nguyen and Muder, 1994; Papadakis et al., 1997), soft tissue infections (Vartivarian et al., 1994), septic bursitis (Papadakis et al., 1996), urinary tract infections (Vartivarian et al., 1996), bacteremia (Muder et al., 1996), endophthalmitis following surgical procedures (Kaiser et al., 1997), and pneumonia (Elsner et al., 1997). The literature records concern for the emergence of the species as a dangerous opportunistic pathogen (Marshall et al., 1989), and for the role of anti-pseudomonadal antibiotics in this process.

A method based on the PCR technique for the detection of *Pseudomonas aeruginosa*, *Burkholderia cepacia*, and *S. maltophilia* in sputum of cystic fibrosis patients has been described, but the specificity is low for *S. maltophilia*, which may be due to improper PCR conditions (Karpati and Jonasson, 1996). It is interesting to mention here that strains of *S. maltophilia* do not produce alginate, and that genetic studies have demonstrated failure to transfer a plasmid carrying the gene *algT* from *P. aeruginosa* to *S. maltophilia*. The studies also failed to identify sequences homologous to this gene in samples examined by Southern analysis (Goldberg et al., 1993).

The type strain of *S. africana* was isolated from an HIV patient who exhibited primary meningoencephalitis (Drancourt et al., 1997).

Ecology The variety of materials from which *S. maltophilia* can be isolated is truly remarkable. A list compiled by Hugh and

Gilardi includes raw and pasteurized milk; well, stagnant, and river waters; sewage; frozen fish; feces of snakes, lizards, frogs, and rabbits; rotten eggs; and soil in petroleum zones. It has been isolated from the internal tissue of banana pseudostem, decaying banana sucker, cotton seed, bean pod, and tobacco seedling (Hugh and Gilardi, 1980). Strains similar to *Xanthomonas* and *Stenotrophomonas* have been frequently isolated from biofilms for waste gas treatment (Lipski et al., 1992; Lipski and Altendorf, 1997). Some strains have been assigned to a new species, *Stenotrophomonas nitritireducens* (Finkmann et al., 2000) (see note at the end of the list of species). From a soda lake, an alkaliphilic strain was studied by Duckworth et al. (1996).

For years, the species has been associated with a variety of infections in humans (Sutter, 1968; Gilardi, 1971), and it could be found in "contaminated tissue culture, streptomycin solution, distilled water, incubator reservoirs, respirators, nebulizers, and evacuated blood collection tubes" (Hugh and Gilardi, 1980). As mentioned above, strains of the species have been isolated from the rhizosphere and aerial parts of plants. These organisms may represent a very high proportion of the non-fermentative, Gram-negative bacteria in the soil microflora (Debette and Blondeau, 1977), and they have been isolated in large numbers from manured soils in Japan (Katoh and Suzuki, 1979) (cited by Ikemoto et al., 1980). The isolation of three strains similar to *S. maltophilia*, from gut and feces of the arthropod *Folsomia candida* was reported by Duckworth et al. (1996).

The distribution of *S. africana* in nature is not known.

ENRICHMENT AND ISOLATION PROCEDURES

Strains of *S. maltophilia* have been isolated from a bewildering variety of materials (Hugh and Gilardi, 1980) and they are second only to *P. aeruginosa* in the frequency of isolation in hospitals and clinical laboratories. Because the nutritional spectrum of the species is so restricted, it is not possible to design effective enrichment procedures based on carbon sources specifically used by strains of *S. maltophilia*. As mentioned elsewhere (Table BXII.γ.28), some carbon sources (maltose, lactose, cellobiose) are used by *S. maltophilia* strains and not by strains of other species of aerobic pseudomonads. However many other organisms found in natural samples are capable of using these substrates, which conspires against the effectiveness of enrichment procedures based on their use by *S. maltophilia*. The fact that so many strains have been isolated using traditional bacteriological procedures is indicative of their wide occurrence in appreciable numbers in many natural materials. In view of the fact that many strains require growth factors (methionine or other amino acids), direct isolation is best performed by streaking plates of any of several

TABLE BXII.γ.28. General characteristics of *S. maltophilia*

Characteristics	Typical phenotype
Methionine requirement	+
Oxidase reaction	−
Utilization of nitrate as nitrogen source	−
Growth on:	
Lactose	+
Maltose	+
L-Aspartate	−
β-Hydroxybutyrate	−
Glutarate	−
Glycerol	−
Cellobiose	+

complex media (Difco Nutrient Agar, Trypticase Soy, Brain Heart Infusion, etc).

MAINTENANCE PROCEDURES

As with most aerobic pseudomonads, *S. maltophilia* can be maintained in slants of common complex culture media, with transfers every one or two months. Although there is no information about *S. africana*, the above considerations might also apply to this species. *S. maltophilia* tolerates freeze-drying well, although freezing of cell suspensions at $-20°C$ or lower after addition of a cryo-protecting agent like glycerol (5–10% final concentration) may be the most reliable method for long-term preservation, since some strains do not revive from freeze-dried preparations.

PROCEDURES FOR TESTING SPECIAL CHARACTERS

Screening for nutritional properties Details of the methodology for determination of the nutritional spectra of aerobic pseudomonads are available (Stanier et al., 1966; Palleroni and Doudoroff, 1972; Palleroni, 1984). A simple minimal medium that does not include strong chelating compounds (Palleroni and Doudoroff, 1972)[1] gives better results than that of earlier formulations (Terry et al., 1991). Methionine (50 µg/ml) should be added from a sterile stock solution after the medium has been autoclaved. Good heterotrophic growth can be obtained by adding to the medium a single organic compound as carbon and energy source to a final concentration of 0.1%. Some compounds may be toxic at this concentration, but even if this is reduced to 0.0125%, visible growth above the level of the control without substrate can be obtained. Precautions and useful suggestions can be found in the above-mentioned papers.

Acid production from carbohydrates The recommended medium (Hugh and Leifson, 1953)[2] gives clear and reproducible results. It contains a small amount of peptone, and ammonification does not interfere with the acid production. As mentioned by Palleroni and Doudoroff (1972), the results do not necessarily correlate with utilization of the substrates for growth. There may be redundancy in some of the results, since a single dehydrogenase may cause acid production from different sugars.

Hydrolysis of Tween 80 The test devised by Sierra (1957) for the detection of lipolytic activity is very simple and the results are clear. Sierra's medium[3] contains Tween 80 and calcium chloride, and opacity formation around the patches of growth or the individual colonies indicates precipitation of calcium salts of the fatty acids liberated by lipases (Sierra, 1957).

Nitrate reduction Reduction of nitrate to nitrite can be tested according to Lelliott et al. (1966).

Oxidase reaction A description of this test can be found in the paper by Stanier and collaborators (Stanier et al., 1966).

Gelatinase Of the several techniques that have been proposed, the one described by Skerman (1967) is most convenient.

Egg yolk reaction The egg yolk test has been used for many years to detect lecithinase production (Esselman and Liu, 1961). However, the reaction is quite complex, although the results are in most cases clear and reproducible. The test can be performed according to published procedures (Lelliott et al., 1966; Stanier et al., 1966).

Catalase reaction The presence of catalase is simply tested by placing a drop of a hydrogen peroxide solution on top of a colony and observing the formation of bubbles of free O_2. The observation may be facilitated by using a dissecting microscope. In cases where the reaction is very weak, the use of an oxygen electrode may be advisable (Auling et al., 1978).

Nucleic acid hybridization Several methods are in use. In some of them, one of the DNA partners is immobilized on a membrane (Johnson, 1994a; Ballard et al., 1970; Palleroni et al., 1972). More stringent conditions are obtained by hybridization in solution followed by hydrolysis of single strands using S1 nuclease (Johnson and Palleroni, 1989; Johnson, 1994a). rRNA–DNA hybridization, as originally applied (Palleroni et al., 1973), has been largely replaced by sequence studies of PCR-amplified genes (Johnson, 1994b).

Fatty acid analysis Although several analytical procedures based on similar principles are in use, the method described by Stead (1992) can be recommended for the pseudomonads.

DIFFERENTIATION OF THE GENUS *STENOTROPHOMONAS* FROM OTHER GENERA

Stenotrophomonas can be easily differentiated from other genera of aerobic pseudomonads. The salient characteristics of *Stenotrophomonas* include the requirement for growth factors (mainly methionine), the inability to use nitrate as a nitrogen source, and the restricted range of substrates that can be used as sole carbon and energy sources. A clear differentiation of *S. maltophilia* from species of various genera of aerobic pseudomonads is possible using the small group of selected characteristics shown in Table BXII.γ.28. When the occurrence of these characters is scored for various other groups of pseudomonads, a clear separation is evident (Table BXII.γ.29). As stated elsewhere (Stanier et al., 1966), these organisms constitute the most isolated group among the aerobic pseudomonads. Useful phenotypic characteristics for the differentiation of *S. maltophilia* and *S. africana* from some species of aerobic pseudomonads are given in Table BXII.γ.30. *S. nitritireducens* could not be included in the table since no pertinent phenotypic details are given in the characterization of this species.

TAXONOMIC COMMENTS

Strains of *S. maltophilia* were included in the phenotypic study of Stanier et al. (1966), and their phenotypic characteristics were precisely defined. Later studies by Palleroni et al. (1973) separated this species from all other species of *Pseudomonas* and allocated it to rRNA similarity group V, together with the species of the genus *Xanthomonas*. Based on these results, Swings et al. (1983) proposed the transfer of *P. maltophilia* to the genus *Xanthomonas*, a decision that did not receive universal approval. In a study on numerical taxonomy and protein gel electrophoresis patterns as tools for the differentiation of pseudomonads, van Zyl and Steyn (1990) showed that *S. maltophilia* and *Pseudomonas hibiscicola* constituted a cluster that was not significantly close to

---

1. Medium of Palleroni and Doudoroff (1972) (g/l of 0.33 M Na-K phosphate buffer, pH 6.8): NH_4 Cl, 1.0; $MgSO_4$·$7H_2O$, 0.5; ferric ammonium citrate, 0.05; and $CaCl_2$, 0.005. The first two ingredients are added to the buffer and sterilized by autoclaving. The ferric ammonium citrate and $CaCl_2$ are added aseptically from a single stock solution that has been sterilized by filtration.

2. Hugh and Leifson medium (g/l): peptone 2.0; NaCl, 5.0; K_2HPO_4, 0.3; agar, 3.0; brom thymol blue, 0.03; carbohydrate, 10.0; pH 7.1.

3. Sierra's medium (g/l): Bacto-peptone (Difco), 10.0; NaCl, 5.0; $CaCl_2$·H_2O, 0.1; and agar, 17.0. Tween 80 is sterilized separately and added to the medium after autoclaving to give a final concentration of 10 g/l.

TABLE BXII.γ.29. Distribution of ten selected characters among various aerobic pseudomonad species[a]

Number of characters from table 1 that are shared by various species	Organisms			
	S. maltophilia (23 strains)	*Pseudomonas* spp. (202 strains)	*Comamonas* spp. (26 strains)	*Burkholderia cepacia* and *B. pseudomallei* (60 strains)
10	21	0	0	0
9	2	0	0	0
8	0	0	0	0
7	0	0	0	0
6	0	0	0	0
5	0	0	0	0
4	0	1	0	0
3	0	1	0	7
2	0	11	0	35
1	0	23	21	18
0	0	166	5	0

[a]Each figure in the table represents the number of strains positive for a given number of the characteristics listed in Table BXII.γ.28.

the *Xanthomonas* group, and the authors found it reasonable to accept the suggestion of Palleroni (1984) for the creation of a new genus name for the strains, which were at the time labeled *Xanthomonas maltophilia*. In fact, a later paper by the same authors (van Zyl and Steyn, 1992) clearly highlighted the differences between *X. maltophilia* and the xanthomonads, and presented compelling evidence that culminated in the proposal of the new genus *Stenotrophomonas* by Palleroni and Bradbury (1993).

Xanthomonas has been classically defined as a group of plant-pathogenic organisms and, from what we know about *Stenotrophomonas*, the pathogenic propensities of the two species assigned to the genus seem to be directed to human hosts. We know very little about the relationship of *S. maltophilia* with plants, but, as indicated in an earlier section of this chapter, the association seems to be a beneficial one. But even if we discard the value of these qualifications as part of a genus definition, various other characteristics have been invoked elsewhere to support allocating *(Xanthomonas) maltophilia* to a different genus (Palleroni and Bradbury, 1993). Additional evidence supporting this criterion is summarized in the following paragraphs.

Restriction analysis of PCR-amplified 16S rRNA genes, performed on a collection of *Xanthomonas* and *Stenotrophomonas* strains, resulted in the definition of 19 restriction fragment length polymorphism (RFLP) groups, which subsequently could be subdivided into two main clusters, one containing all the *Xanthomonas* species, and the other, *S. maltophilia*. An examination of the relationship between the DNA similarities as determined by hybridization methods, and the nucleotide substitution rate in the small rRNA subunit, confirmed the differentiation of the two genera (Nesme et al., 1995). A rapid method of classification based on PCR has demonstrated the production of DNA fragments in *S. maltophilia* extracts different from those obtained from the *Xanthomonas* strains tested (Maes, 1993).

S. maltophilia was included in a study using the AFLP (amplified fragment length polymorphism) technique (Janssen et al., 1996). As in the case of RFLP, this promising method has resulted in dendrograms that confirm relationships derived from other molecular approaches (Janssen et al., 1996; Rademaker et al., 2000).

Analysis of fatty acid composition for a large collection of strains of the genus *Xanthomonas* by Yang et al. (1993d) revealed the presence of nine fatty acids in more than 99% of the *Xanthomonas* strains. Three of the acids ($C_{11:0\ iso}$, $C_{11:0\ iso\ 3OH}$, and

$C_{13:0\ iso\ 3OH}$), are characteristic for the genus and therefore are taxonomically important to differentiate it from related genera. *S. maltophilia* strains contained the nine typical fatty acids, but, in addition, they were characterized by the presence of the fatty acid $C_{17:0}$ cyclopropane, which was absent from *Xanthomonas* species, with the exception of *X. campestris* pathovar *hyacinthi*. There was great uniformity in the fatty acid composition of the *S. maltophilia* strains, although they had been isolated from very diverse materials, but the DNA homology values among them ranged from 37 to 100% according to Ikemoto et al. (1980).

In a study of a large collection of *S. maltophilia* strains isolated from environmental and clinical sources, Hauben et al. (1999b) found a striking phenotypic uniformity but considerable genomic heterogeneity, and tcn genomic groups were observed. However, the lack of conventional phenotypic traits by which the various groups could be identified has prevented the authors from proposing a subdivision of *S. maltophilia* into different species.

Ikemoto et al. (1980) reported the results of studies on *Pseudomonas maltophilia* strains that do not require methionine. Strains of this new biovar (biovar II) differ from *S. maltophilia* (biovar I) in phenotypic properties, fatty acid composition, DNA composition, and requirement for methionine as an organic growth factor. DNA–DNA hybridization studies also point to DNA sequence differences between the two groups. Subsequently, Ha and Komagata (1984) extended the studies to the electrophoretic behavior of six enzymes, with results confirming the differences between the biovars.

Singer et al. (1994), in a comparison between phenotypic properties and esterase electrophoretic polymorphism, included another group of strains not requiring methionine, and confirmed the differences from *S. maltophilia* observed by the Japanese workers. At this point, sufficient evidence is available to suggest that biovar II of *S. maltophilia*, defined by Ikemoto et al. (1980), should be assigned to a new species of the genus. The observed differences in general properties and in DNA–DNA hybridization values clearly support this suggestion.

As previously mentioned, the phage isolated from sewage by Kelln and Warren (1971) was lytic for species of *Pseudomonas* but did not lyse *S. maltophilia* or strains of species of homology groups other than group I. The phylogenetic significance of this finding became apparent after demonstration of the genomic heterogeneity of the genus *Pseudomonas*, as classically defined.

TABLE BXII.γ.30. Phenotypic properties that differentiate *S. maltophilia* and *S. africana* from species of other genera of aerobic pseudomonads

Characteristics	S. maltophilia	S. africana	Pseudomonas aeruginosa	P. fluorescens	P. putida	P. alcaligenes	P. pseudoalcaligenes	P. stutzeri	P. mendocina	Burkholderia cepacia	B. pseudomallei	B. mallei	Ralstonia pickettii	R. solanacearum	Comamonas acidovorans	C. testosteroni
Flagellar number	>1	>1	1	>1	>1	1	1	1	1	>1	1	0	>1	1	>1	>1
PHB accumulation	−	−	−	−	−	−	v	−	−	+	+	+	+	+	+	+
Oxidase reaction	−	−	+	+	+	+	+	+	+	+	+	+	+	+	+	+
Gelatin liquefaction	+	+	+	+	−	+	v	−	−	+	+	+	−	−	−	−
Methionine requirement	+	+	−	−	−	−	−	−	−	−	−	−	−	−	−	−
Arginine dihydrolase	−	+	+	+	+	+	v	+	+	−	+	+	−	−	−	−
Denitrification	−	−	+	v	−	−	−	+	+	−	+	+	+	+	−	−
Growth on:																
Maltose	+	+	−	−	−	−	−	+	−	v	+	v	−	−	−	−
Cellobiose	+	+	−	−	−	−	−	−	−	+	+	+	−	−	−	−
Lactose	+	+	−	−	−	−	−	−	−	−	−	−	−	−	−	−
Trehalose	+		v	v	v	−	−	−	−	+	+	+	+	−	−	−
Arginine	−	+	+	+	+	+	+	+	+	+	+	+	−	−	−	−

FURTHER READING

Vauterin, L., J. Swings, K. Kersters, M. Gillis, T.W. Mew, M.N. Schroth, N.J. Palleroni, D.C. Hildebrand, D.E. Stead, E.L. Civerolo, A.C. Hay-ward, H. Maraîte, R.E. Stall, A.K. Vidaver and J.F. Bradbury. 1990. Towards an improved taxonomy of *Xanthomonas*. Int. J. Syst. Bacteriol. *40*: 312–316.

List of species of the genus Stenotrophomonas

1. **Stenotrophomonas maltophilia** (Hugh 1981) Palleroni and Bradbury 1993, 608^VP (*Pseudomonas maltophilia* Hugh 1981,195; *Xanthomonas maltophilia* Swings, De Vos, Van Den Mooter and De Ley 1983, 412.)

mal.to.phi' li.a. Anglo-Saxon n. *malt* Gr. n. *philia* friend; M.L. fem. n. *maltophilia* friend of malt.

Straight or slightly curved rods, 0.5 × 1.5 μm, occurring singly or in pairs. Motile by means of polar multitrichous flagella. Do not accumulate PHB as an intracellular reserve material. Colonies may be yellow or greenish yellow. The yellow color is not due to carotenoid pigments or xanthomonadins. Reduction of nitrate to nitrite is positive, but denitrification does not occur. The oxidase reaction is negative. Strongly lipolytic. Gelatinase reaction positive. Ammonia odor can be produced by colonies on blood agar. Methionine or cystine is required for growth, although strains lacking this requirement may be isolated (Ikemoto et al., 1980).

Limited in its nutritional spectrum. Individual strains used between 24 and 28 of 146 organic compounds tested as main carbon and energy sources (Stanier et al., 1966). The utilizable compounds were: glucose, mannose, sucrose, trehalose, maltose, cellobiose, lactose, salicin, acetate, propionate, valerate, malonate, succinate, fumarate, L-malate, lactate, citrate, α-ketoglutarate, pyruvate, L-alanine, D-alanine, L-glutamate, L-histidine, and L-proline. A variable number of strains also used fructose, isobutyrate, aconitate, and *n*-propanol. Under aerobic conditions, acid is readily produced in complex media with maltose but not with glucose (Hugh and Ryschenkow, 1960). Nitrate is not used as a nitrogen source.

No growth at 4°C or 41°C. Optimum growth temperature, 35°C. Many strains can be isolated from various natural materials, and frequently from clinical specimens. According to Hugh and Gilardi (1980), *S. maltophilia* is the second most frequently isolated aerobic pseudomonad, after *P. aeruginosa*, in the clinical laboratory. Strains of the species appear to be opportunistic human pathogens. Also found in water, milk, frozen foods, and many other sources. Debette and Blondeau (1980) were able to isolate more strains of *S. maltophilia* than of any other pseudomonad from the rhizosphere of some cultivated plants. Ikemoto et al. (1980) defined two biovars of this species and further studies perhaps will demonstrate that one of them (biotype II) deserves independent species status within the genus *Stenotrophomonas*.

The mol% G + C of the DNA is: 66.9 ± 0.8 (Bd) (Mandel, 1966).

Type strain: ATCC 13637, Hugh 810-2, DSM 50170, ICPB 2648-67, IMET 10402, NCIB 9203, NCTC 10257, NRC 729, RH 1168.

GenBank accession number (16S rRNA): M59158, X95923, AB008509.

2. **Stenotrophomonas africana** Drancourt, Niel and Raoult 1997, 162^VP

a.fri.ca' na. L. adj. *africana* from Africa.

Curved rods, 0.5 × 1.5 μm. Motile by means of two polar flagella. Colonies are gray to green. No ammonia odor is produced on blood plates. Indole and oxidase reactions are negative. Catalase positive. Basic phenotypic properties are similar to those of *S. maltophilia*, among them the basic biochemical characteristics, except for the failure to assimilate *cis*-aconitate. A total of 24 of 1377 16S rDNA base positions are different from those of *S. maltophilia*, and the DNA–DNA similarity between the two species is 35%. Isolated from the cerebrospinal fluid of a patient testing positive for HIV antibodies and exhibiting primary meningoencephalitis (Drancourt et al., 1995, 1997).

The mol% G + C of the DNA is: not known.

Type strain: MGB, CIP 104854.

GenBank accession number (16S rRNA): U62646.

3. **Stenotrophomonas nitritireducens** Finkmann, Altendorf, Stackebrandt and Lipski 2000, 281^VP

ni.tri.ti.re.du' cens. M.L. *nitritis* nitrite; L. *reducere* to reduce; L. part. pres. *reducens* reducing; M.L. part. pres. *nitritireducens* reducing nitrite.

The following description is transcribed in its entirety from the original publication by Finkmann et al. (2000). Predominant fatty acids are $C_{15:0 \text{ iso}}$, $C_{17:0 \text{ iso } \omega 7c}$, $C_{16:0 \text{ iso}}$, $C_{14:0 \text{ iso}}$, and $C_{15:0 \text{ anteiso}}$. $C_{17:0 \text{ cyclo}}$ and $C_{13:0 \text{ iso } 3OH}$ are present. The strains reduce nitrite to nitrous oxide, nitrate is not reduced, and nitrous oxide is the end product. Esculin is not hydrolysed and lecithin is not degraded. The strains grow in the presence of 0.01% $ZnCl_2$. The strains are resistant to fucidin, tetracycline, and novobiocin, but susceptible to streptomycin, nalidixic acid, and kanamycin. The strains show strong hydrolytic activity for *p*-nitrophenyl N-acetyl-β-D-glucosaminide. The strains were isolated from experimental biofilters supplied with artificial waste gases containing ammonia or ammonia and dimethyl disulfide. One strain (B1910/29.1) was isolated from an experimental biofilter supplied with the waste gas of an animal-rendering plant.

The mol% G + C of the DNA is: not known.

Type strain: L2, DSM 12575.

GenBank accession number (16S rRNA): AJ012229.

An examination of the properties reported in the above description clearly shows that a comparison of *S. nitritireducens* with *S. maltophilia* and *S. africana* is not possible, in view of the disparity of criteria used for characterization. Therefore, with the exception of nucleic acid similarity data, there is no phenotypic information that may account for assignment of *S. nitritireducens* to the genus *Stenotrophomonas*, and this species has been excluded from Table BXII.γ.30.

Genus X. **Thermomonas** Busse, Kämpfer, Moore, Nuutinen, Tsitko, Denner, Vauterin, Valens, Rosselló-Mora and Salkinoja-Salonen 2002, 480[VP]

EWALD B.M. DENNER, PETER KÄMPFER AND HANS-JÜRGEN BUSSE

Ther.mo.mo′ nas. Gr. n. *therme* heat; *monas* a unit, monad; M.L. fem. n. *Thermomonas* a thermophilic monad.

Gram-negative filamentous rods (1.0–12.5 × 0.5–0.75 μm). Motile by means of polar monotrichous flagellation or nonmotile. **Aerobic**. Endospores are not formed. **Moderately thermophilic**; temperature range for growth is 18–50°C, with an **optimum between 37 and 50°C. Catalase and oxidase positive**. Chemoorganotrophic. **Carbohydrates are usually not utilized**. Indole, Voges–Proskauer, methyl red, and Simmons citrate tests are negative. Urea is hydrolyzed. **Nitrate is not reduced to nitrite or nitrogen**; H_2S is not produced. **Ubiquinone Q-8** is the major respiratory lipoquinone system. The main component in the cellular polyamine content is **spermidine**. Polar lipid fingerprints are characterized by the presence of diphosphatidylglycerol, phosphatidylethanolamine, and phosphatidylglycerol. Cellular fatty acids are predominantly **iso-branched** with $C_{15:0\ iso}$ and $C_{16:0\ iso}$ as major components; other characteristic fatty acids are $C_{11:0\ iso}$, $C_{11:0\ iso\ 3OH}$, and $C_{17:1\ iso\ \omega9c}$. Based on 16S rRNA gene sequence data, *Thermomonas haemolytica* currently forms **a separate line of descent within the Xanthomonadaceae related to the genera Xanthomonas, Pseudoxanthomonas, Luteimonas, Stenotrophomonas, and Xylella**.

The mol% G + C of the DNA is: 67.1–68.7.

Type species: **Thermomonas haemolytica** Busse, Kämpfer, Moore, Nuutinen, Tsitko, Denner, Vauterin, Valens, Rosselló-Mora and Salkinoja-Salonen 2002, 480.

FURTHER DESCRIPTIVE INFORMATION

Currently, the genus contains only a single species, *T. haemolytica*. The four strains (A50-7-3[T], B50-7-1, B50-8-1, and D50-7-1) of the only species yet described characteristically show a marked β-hemolysis on blood agar. Cells of *T. haemolytica* are rod-shaped with a typical Gram-negative cell wall structure and have a tendency to form filaments (Fig. BXII.γ.38). *T. haemolytica* is motile by means of a polar flagellum. While tumbling can be frequently noticed, directed movement is only rarely observable. Motility of the cells is most pronounced at 37°C and higher. The flagellation of the cells can be visualized by the staining procedure of Kodaka et al. (1982), but has not yet been observed by electron microscopy.

T. haemolytica grows well on standard bacteriological media such as Tryptone soy agar (TSA), Luria–Bertani (LB) medium, or nutrient agar. Colonies on standard media are circular, smooth, and convex with an entire edge. At 50°C colonies are whitish/translucent, and at 37°C, transparent. After prolonged incubation (2–3 d) on TSA at 50°C, *T. haemolytica* produces a diffusible brownish discoloration of the medium and a characteristic ammonium-like odor. The temperature range for growth is 18–50°C; no growth occurs at 14 or 53°C.

Cellular fatty acid analyses indicate that with increasing growth temperature, over a range of 23–50°C, the proportion of unsaturated fatty acids in *Thermomonas haemolytica* decreases from 24% to 2.7% (mean of the four strains) of total fatty acids, mainly due to a decrease in the content of $C_{17:1\ iso\ \omega9c}$. The relative amount of $C_{17:0\ iso}$ remains constant in cells growing over the whole temperature range. This may indicate that the adaptation to growth at high temperatures by *Thermomonas haemolytica* does not occur by increasing the ratio of $C_{17:0\ iso}$ to $C_{17:1\ iso\ \omega9c}$. Instead, the relative proportion of the major fatty acid, $C_{15:0\ iso}$, increases

with increasing growth temperature. Interestingly, the relative amounts of short chain fatty acids, $C_{11:0\ iso}$ and $C_{11:0\ iso\ 3OH}$, also increase slightly at 50°C (Busse et al., 2002).

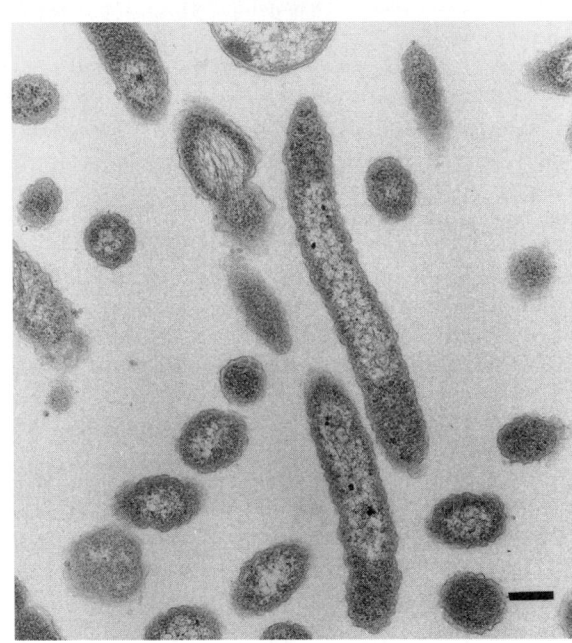

A

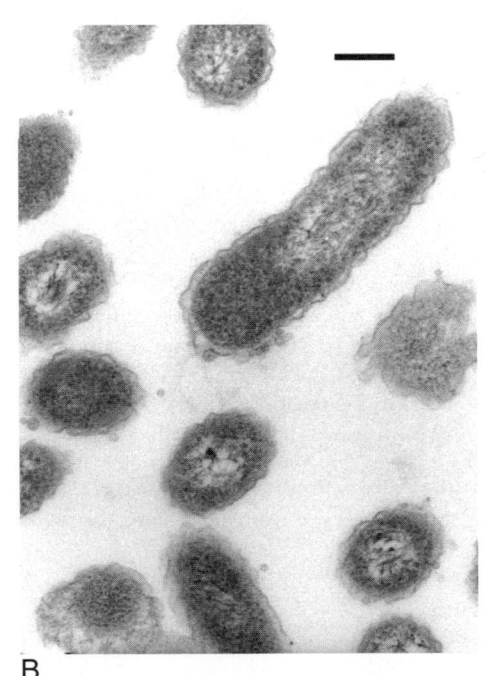

B

FIGURE BXII.γ.38. Electron micrographs of ultrathin sections of cells of *T. haemolytica* A50-7-3[T] showing (*A*) the filamentous and (*B*) rod-shaped forms. Bars = 200 nm. (Reprinted with permission from H.-J. Busse et al., International Journal of Systematic and Evolutionary Microbiology *52*: 473–483, 2002, ©International Union of Microbiological Societies.)

T. haemolytica also grows well on R2A agar, a low-nutrient complex medium developed previously for the enumeration and subculture of heterotrophic bacteria in potable water (Reasoner and Geldreich, 1985). *T. haemolytica* does not grow on the chemically defined yet simpler Czapek–Dox medium, which contains only sodium nitrate as inorganic source of nitrogen and sucrose as sole source of carbon.

As tested on selective/differential media, *T. haemolytica* is not able to grow in the presence of bile salt–neutral red (MacConkey agar) or in the presence of the cationic detergent cetyl trimethylammonium bromide (cetrimide agar). *T. haemolytica* is susceptible to amoxycillin/clavulanic acid, ampicillin, cephalexin, chloramphenicol, colistin sulfate, erythromycin, fusidic acid, gentamicin, kanamycin, neomycin, penicillin G, polymyxin B, streptomycin, and tetracycline, but is resistant to lincomycin, nitrofurantoin, and sulfamethoxazole/trimethroprim.

T. haemolytica characteristically uses only a limited number of organic acids and amino acids as a single source of carbon, whereas *T. haemolytica* A7-3-1[T] is the most versatile strain and B50-8-1 the most inactive strain of the four *T. haemolytica* strains (Table BXII.γ.31). Strain B50-8-1 is also somewhat distinct from the other three *T. haemolytica* strains, on the basis of mol% G + C content of DNA (67.1); however, by DNA–DNA similarity, riboprint patterns, and AFLP profiles strain B50-8-1 is univocally a strain of *T. haemolytica* (Busse et al., 2002).

ENRICHMENT AND ISOLATION PROCEDURES

The four strains of *T. haemolytica* were isolated at different times from papermaking chemicals (kaolin and pigment slurries) that were plated onto Plate Count Agar (Difco) with subsequent incubation at 50°C (Väisänen et al., 1998).

MAINTENANCE PROCEDURES

Slant cultures grown on standard bacteriological media should not be held for longer than 1 week at 4°C. Cells may also stored at −70°C in broth media, e.g., nutrient or LB broth containing 15–20% glycerol as cryoprotectant or under liquid nitrogen.

DIFFERENTIATION OF THE GENUS *THERMOMONAS* FROM OTHER GENERA

Thermomonas can be clearly differentiated from other genera of the *Xanthomonadaceae* on the basis of 16S rRNA gene sequence, cellular fatty acid composition, and several physiological and bio-

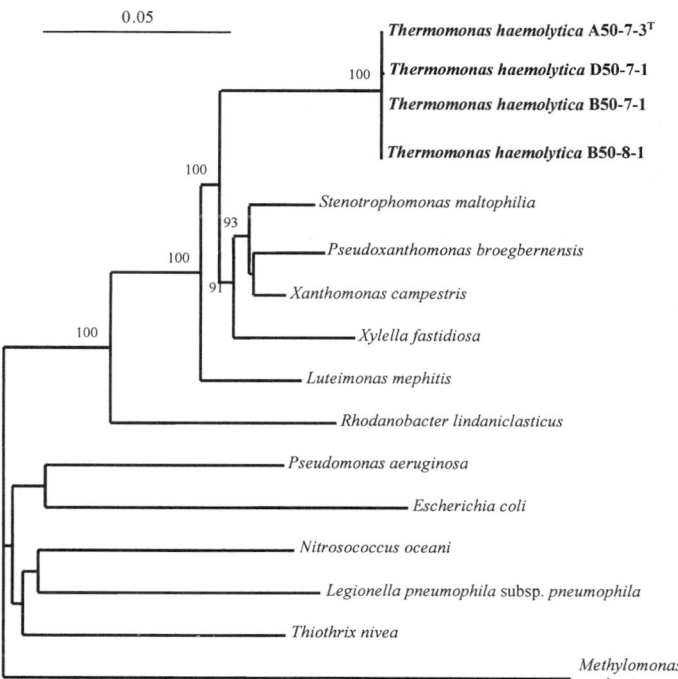

FIGURE BXII.γ.39. Phylogenetic position of *Thermomonas haemolytica* to related genera of the *Xanthomonadaceae*, and species representing distinct lineages within the *Gammaproteobacteria*. The dendrogram was generated from evolutionary distances, calculated from pair-wise dissimilarities, using the FITCH program contained in the PHYLIP package (Felsenstein, 1993). Bootstrap estimates of confidence, derived from 1000 resamplings, are shown for branchings with values greater than 80%, within the *Stenotrophomonas–Xanthomonas–Xylella* phylogenetic lineage. The topography of the branching order of the lineage within the dendrogram was supported by identical branching orders generated using an alternative treeing algorithm of maximum-likelihood (FastDNAml, version 1.0) (data not shown) and by estimated high bootstrap proportions of confidence for the respective branchings. GenBank/ EMBL/DDBJ accession numbers: *Escherichia coli*, M24836; *Legionella pneumophila* subsp. *pneumophila*, M59157; *Luteimonas mephitis*, AJ012228; *Methylomonas methanica*, M95660; *Nitrosococcus oceani*, M96395; *Pseudomonas aeruginosa*, Z76651; *Pseudoxanthomonas broegbernensis*, AJ012231; *Rhodanobacter lindaniclasticus*, AF039167; *Stenotrophomonas maltophilia*, X95923; *Thermomonas haemolytica*, A50-3-7[T] (AJ300185); *Thermomonas haemolytica*, B50-7-1, AJ300186 (partial sequence, see Busse et al., 2002); *Thermomonas haemolytica* B50-8-1, AJ300188 (partial sequence, see Busse et al., 2002); *Thermomonas haemolytica* D50-7-1, AJ300189 (partial sequence, see Busse et al., 2002); *Thiothrix nivea*, L40993; *Xanthomonas campestris*, X95917; *Xylella fastidiosa*, AF192343. Bar = 5 substitutions at any nucleotide position per 100 nucleotide positions (estimated evolutionary distance from the point of divergence of the 16S rRNA gene sequences).

TABLE BXII.γ.31. Physiological and biochemical characteristics of the *Thermomonas haemolytica* strains[a]

Characteristic	A50-7-3[T]	B50-7-1	D50-7-1	B50-8-1
Assimilation of:				
Acetate	+	−	+	−
Propionate	+	+	−	−
cis-Aconitate	+	+	−	−
L-Malate	+	−	−	−
Mesaconate	+	+	−	−
Oxoglutarate	+	−	−	−
Pyruvate	+	+	+	−
L-Alanine	+	+	−	−
L-Aspartate	+	+	−	−
L-Phenylalanine	+	−	−	−
L-Proline	−	+	−	−
L-Serine	+	−	−	−
Hydrolysis of:				
Gelatin	+	+	−	+

[a]All *Thermomonas haemolytica* strains assimilate DL-3-hyroxybutyrate, and L-leucine, whereas none of the strains assimilate N-acetyl-D-glucosamine, L-arabinose, *p*-arbutin, D-cellobiose, D-fructose, D-galactose, gluconate, D-glucose, D-mannose, D-maltose, α-D-melibiose, L-rhamnose, D-ribose, sucrose, salicin, D-trehalose, D-xylose, adonitol, *i*-inositol, maltitol, D-mannitol, D-sorbitol, putrescine, *trans*-aconitate, adipate, 4-aminobutyrate, azelate, citrate, fumarate, glutarate, itaconate, DL-lactate, suberate, β-alanine, L-histidine, L-ornithine, L-tryptophan, 3-hydroxybenzoate, 4-hydroxybenzoate, or phenylacetate. *p*NP-α-D-glucopyranoside, *p*NP-phenylphosphonate, 2-deoxythymidine-5'-*p*NP-phosphate, L-alanine-*p*NA, L-glutamate-γ-3-carboxy-*p*NA, and L-proline-*p*NA are hydrolyzed, but esculin, *p*NP-β-D-galactopyranoside, *p*NP-β-D-glucuronide, *p*NP-β-D-glucopyranoside, bis-*p*NP-phosphate, and *p*NP-phosphoryl-choline are not. Enzymatic tests for urease, alkaline phosphatase, esterase (C4), esterase lipase (C8), lipase (C14), leucine arylamidase, valine arylamidase, chymotrypsin, phosphatase acid, and naphthol-AS-BI-phosphohydrolase are uniformly positive.

chemical characteristics (Table BXII.γ.32). In particular, *Thermomonas* species exhibit a higher temperature optimum of 37–50°C.

TAXONOMIC COMMENTS

Comparative evolutionary distance analysis of the 16S rDNA sequence shows that the genus *Thermomonas* represents a separate line of descent within the *Gammaproteobacteria*. The nearest phylogenetic relatives, albeit relatively distant (16S rDNA sequence similarities <95%), are species of the genera *Xanthomonas*, *Pseudoxanthomonas*, *Stenotrophomonas*, *Luteimonas*, and *Xylella* (Fig. BXII.γ.39).

Very recently, an isolate SGM-6 was obtained from a hot spring at Sao Gemil in central Portugal, showing a 16S rDNA gene sequence similarity of 97.4% to *T. haemolytica* A50-7-3 [T] and able to grow at temperatures as high as 55°C. This isolate may represent another species of the genus *Thermomonas*, or at least a novel strain of *T. haemolytica* (M. Da Costa and F. Rainey, personal communication). Meanwhile, a second *Thermomonas* species has been described Alves et al., 2003a.

TABLE BXII.γ.32. Characteristics differentiating *Thermomonas haemolytica* from other related taxa within the *Xanthomonadaceae* [a]

Characteristic	*Thermomonas haemolytica*[b]	*Xanthomonas campestris*[c]	*Pseudoxanthomonas broegbernensis*[d]	*Stenotrophomonas maltophilia*[e]	*Luteimonas mephitis*[f]	*Xylella fastidiosa*[g]
Temperature optimum (°C)	37–50	28	28	35	28	26–28
Nitrate reduction	−	−	−	+	−	−
Esculin hydrolysis	−	+	−	+	−	−
Susceptibility to:						
Ampicillin	+	−	−	−	−	+
Penicillin G	+	−	−	−	−	−
Erythromycin	+	−	−	−	−	
Kanamycin	+	+	−	+	V+	+
Neomycin	+	−	+	−	V−	
Streptomycin	+	+	−	−	−	−
Predominant fatty acid:						
$C_{15:0\ iso}$	+	+	+	+	+	
$C_{15:0\ anteiso}$		+	+	+		+
$C_{16:0}$				+		
$C_{16:0\ iso}$	+					
$C_{16:1}$		+		+		+
$C_{17:0}$					+	+
$C_{17:0\ iso}$						
$C_{17:1}$					+	
$C_{17:1\ iso}$			+			
Hydroxy fatty acid:						
$C_{10:0\ 2OH}$						+
$C_{11:0\ iso\ 2OH}$			+			
$C_{11:0\ iso\ 3OH}$	+	+	+	+	+	
$C_{12:0\ 3OH}$		+		+		
$C_{13:0\ 2OH}$				+		
$C_{13:0\ iso\ 3OH}$		+		+		
Major polar lipids:[h]						
Diphosphatidylglycerol	+	+		+		
Phosphatidylethanolamine	+	+		+		
Phosphatidylglycerol	+	+		+		
Phosphatidylmonomethylethanolamine		+				
Unidentified phospholipid		+				
Quinone system	Q-8	Q-8	Q-8	Q-8	Q-8	
Major polyamines:[i]						
Cadaverine				+		
Spermidine	+	+		+		

[a]Symbols: +, positive for all strains; −, negative for all strains; nd, no data; V+, most strains are resistant; V−, most strains are susceptible.

[b]Data from Busse et al. (2002).

[c]Data from Oyaizu and Komagata (1983), Busse and Auling (1988), Auling et al. (1991), Yang et al. (1993b, d).

[d]Data from Finkmann et al. (2000).

[e]Data from Oyaizu and Komagata (1983), Palleroni (1984), Busse and Auling (1988), Stead (1992), Yang et al. (1993b, d), Palleroni and Bradbury (1993), Vauterin et al. (1995, 1996b), Finkmann et al. (2000).

[f]Data from Finkmann et al. (2000).

[g]Data from Wells et al. (1987).

[h]Data on polar lipids for *Xanthomonas campestris* and *Stenotrophomonas maltophilia* from Busse et al. (2002).

[i]No data for *Pseudoxanthomonas broegbernensis*, *Luteimonas mephitis*, and *Xylella fastidiosa*.

DIFFERENTIATION OF THE SPECIES OF THE GENUS *THERMOMONAS*

Only one species, *Thermomonas haemolytica*, is currently known.

List of species of the genus Thermomonas

1. **Thermomonas haemolytica** Busse, Kämpfer, Moore, Nuutinen, Tsitko, Denner, Vauterin, Valens, Rosselló-Mora and Salkinoja-Salonen 2002, 480[VP]

   hae.mo.ly' ti.ca. Gr. n. *haema* blood; Gr. adj. *lyticus* dissolving; M.L. fem. adj. *haemolytica* blood dissolving.

   Morphological, physiological, biochemical, and chemotaxonomic characteristics are the same as those described for the genus. Other characteristics are listed in Tables BXII.γ.31 and BXII.γ.32. Isolated from kaolin slurry used in the papermaking industry.

   The mol% G + C of the DNA is: 67.1–68.7 (HPLC); type strain, 68.5.

   Type strain: A50-7-3, DSM 13605, LMG 19653.

   GenBank accession number (16S rRNA): AJ300185.

   Additional Remarks: Reference strains include B50-71 (DSM 13598, LMG 19655; GenBank accession number AJ300186), B50-8-1 (DSM 13599, LMG 19654; GenBank accession number AJ296318), D50-7-1 (DSM 13610, LMG 19656; GenBank accession number AJ296319).

Genus XI. **Xylella** *Wells, Raju, Hung, Weisburg, Mandelco-Paul and Brenner 1987, 141*[VP]

GERARD S. SADDLER AND JOHN F. BRADBURY

Xy.lel' la. Gr. n. *xylon* wood; M.L. dim. ending *-ella*; M.L. fem. n. *Xylella* a small wood.

Single, straight rods, 0.25–0.35 × 0.9–3.5 μm, that can form long filamentous strands under some cultural conditions. Gram negative. **Nonmotile**, lacking flagella. Endospores not produced. **Aerobic**, having a strictly respiratory type of metabolism with oxygen as the terminal electron acceptor. **Colonies are cream to white.** Two colonial variants are frequently encountered: convex to pulvinate, smooth, opalescent with entire margins; and umbonate, rough with finely undulated margins. **Oxidase negative and catalase positive.** Optimum temperature for growth, 26–28°C; optimum pH for growth, 6.5–6.9. Nutritionally fastidious. Found mainly in the xylem of plant tissue.

The mol% G + C of the DNA is: 51.0–52.4 (T_m); 52–53.1 (Bd).

Type species: **Xylella fastidiosa** Wells, Raju, Hung, Weisburg, Mandelco-Paul and Brenner 1987, 141.

FURTHER DESCRIPTIVE INFORMATION

Cell morphology Long filamentous strands can be observed in many strains, but filaments are encountered more frequently when sampling from smooth colonies; rough colonial variants tend to exhibit only single rods (Wells et al., 1987). Flagella are not present and motility has not been observed. Cells exhibit rounded ends, though tapering can be observed at one end only. Irregular ridges or folds in the cell surface can be seen. Fibril-like structures are evident and, although peritrichous, predominate at the poles. There is some evidence of morphological heterogeneity, because strains of *X. fastidiosa* causing phony peach disease are thought to be slightly shorter (0.35 × 2.3 μm) and exhibit more pronounced ridges, or rippling of the cell wall when compared against strains causing Pierce's disease (Nyland et al., 1973). Staining with methylene blue is effective for the visualization of citrus variegated chlorosis strains *in planta* (De Lima et al., 1997).

Cell wall composition The major fatty acid components present are: pentadecanoic acid ($C_{15:0}$), hexadecanoic acid ($C_{16:0}$), hexadecenoic acid ($C_{16:1}$) and heptadecanoic acid ($C_{17:0}$) (Wells and Raju, 1984; Wells et al., 1987). The hydroxy fatty acids 2-hydroxydecanoic ($C_{10:0\ 2OH}$) and 3-hydroxydodecanoic ($C_{12:0\ 3OH}$) acids are reported to be present (Wells et al., 1987).

Fine structure The cell wall shows periodic, spiral, or annulated structural arrangement with the thickness varying from 25 to 35 nm in the furrows to 45–55 nm around the ridges (Nyland et al., 1973; Chagas, et al., 1992). The wall consists of an outer, an inner (each comprised of 3-layered unit membrane structure), and a middle peptidoglycan layer. Progressive changes in the cell wall occur with age; initially the inner and outer layer are thin, becoming denser with time, the outer layer increasing in density first (Mollenhauer and Hopkins, 1974). As the cells age, the cell wall structure can be more difficult to visualize and this is particularly true when examining senescent or dead cells. Cell content consists of DNA threads, ribosomes, and osmophilic granules of unknown function (Chagas et al., 1992). The distribution of cell constituents alters with age, with ribosomes aggregating along the periphery of aging cells. Fibril-like structures can also be observed.

Colonial characteristics Two colony types can be observed: smooth and rough, both are circular and opalescent. When grown on BCYE[1] medium, colonies are small, 0.6 mm in diameter after 10 days at 27°C, expanding to 1.5 mm after 30 days. On this medium, rough colonies exhibit green or red margins when viewed under reflected light. Cell replication is by binary fission (Wells et al., 1987; Chagas et al., 1992).

Nutrition, metabolism, and metabolic pathways Strains are aerobic, chemoorganotrophic, and use oxygen as the terminal electron acceptor. Many strains are nutritionally fastidious and require specialized media such as BCYE and PW[2] for growth (Wells et al., 1987). Little information on enzymic activity exists; however, protease activity has been determined by growing both virulent and weakly virulent strains of *X. fastidiosa* on PD3[3] medium (Davis et al., 1981c) amended with gelatin (Fry et al., 1994). Detailed study revealed the presence of at least two proteases, designated P1 and P2, of 54 and 50 kDa, respectively. Production reached a maximum with onset of the stationary phase, P2 becoming more, and P1 becoming less, predominant over time.

Genetics, mutants and plasmids The genetics of this genus have attracted limited interest thus far. The production of a cosmid clone bank (Goodwin, 1989), and evidence for the lack of homology between the *hrp* gene cluster of *Xanthomonas vesicatoria* and DNA from *Xylella fastidiosa*, (Leite et al., 1994b) have contributed to what limited information exists. The reader is directed towards the Bioinformatics Laboratory at the University of Campinas* for details of a project to sequence the entire genome of *X. fastidiosa* (Setubal, personal communication).

Plasmids have been observed in some strains of *X. fastidiosa* (Chen et al., 1992a). A restriction map has been generated for plasmid pXFPW1, which is 1.3 kb in size and was isolated from an *X. fastidiosa* strain causing periwinkle wilt disease. A copy number of 60 was estimated, and restriction analysis demonstrated a high degree of homology with two other plasmids from strains of Pierce's disease and mulberry leaf scorch disease. In contrast, a cryptic plasmid from *X. fastidiosa* ATCC 35868 was found in only one other strain of *X. fastidiosa*—ATCC strain 35878—and further searches revealed some nucleotide sequence homology with plasmid pNKH43 from *Stenotrophomonas maltophilia* (Pooler et al., 1997a).

Molecular data Restriction fragment length polymorphism (RFLP) analysis has been used to study diversity within *X. fastidiosa* by examining 24 strains from eight different hosts (Chen et al., 1992b). Two genomic libraries were constructed in pUC18 from a PD strain and a plum leaf scald strain. By studying 24

probe-enzyme combinations, 67 characters could be determined and PD strains, in particular, exhibited a high degree of uniformity. Subsequent studies produced similar results, this time using RAPD fingerprinting with 14 random primers on 17 different strains of *X. fastidiosa* (Chen et al., 1995); again, a high degree of similarity among the PD strains was evident, although clear differences between this group and *Quercus* leaf scorch, plum leaf scald strains, and a periwinkle wilt strain indicates a high degree of heterogeneity within the species overall. Analogous results were obtained by Albibi et al. (1998), using two random primers; OPA-03 and OPA-11 (Operon Technologies Inc.). With this method, PD strains could be distinguished from other strains of *X. fastidiosa* (citrus variegated chlorosis, mulberry leaf scorch, periwinkle wilt, plum leaf scald, and phony peach), as well as strains of *Xanthomonas campestris* pathovar *vesicatoria* and *Escherichia coli*. The use of RAPD markers has been further refined to develop a detection method, as demonstrated by Pooler and Hartung (1995b). Primers were developed for the specific detection of strains of citrus variegated chlorosis (CVC). A CVC-specific region of the *X. fastidiosa* genome that contains a 28-nucleotide insertion was identified, and single base changes that distinguish CVC from other *X. fastidiosa* strains. This method was used to develop a sensitive and specific assay for detecting *X. fastidiosa* in potential insect vectors, by the prior selection of the bacterium using immunomagnetic separation (Pooler et al., 1997b).

Infraspecific heterogeneity within *X. fastidiosa* has also been observed using other markers, such as PCR amplification with tRNA consensus primers (Beretta et al., 1997). With this method, it is possible to distinguish between CVC strains and mulberry leaf scorch strains. Further, a 7.4-kb *Eco*R1 fragment of genomic DNA of the type strain of *X. fastidiosa*, ATCC 35879, has been used as a probe and found to be conserved in 18 other strains (Minsavage et al., 1994). PCR primers developed from a 1.0-kb internal *Eco*RV portion of the fragment allowed the differentiation of the species into two groups or "pathovars". A PCR assay has been developed for the specific detection of *X. fastidiosa* strains from a wide variety of hosts, through the amplification of a 400-bp product for the variable regions V1 and V6 of the 16S rDNA sequence (Firrao and Bazzi, 1994).

Antigenic structure Polyclonal antibodies have been used in conjunction with ELISA and immunofluorescence for the detection of *X. fastidiosa* strains (Sherald and Lei, 1991; Zaccardelli et al., 1993; Hartman et al., 1995). However, not all strains in the species can be detected using this method. In a further refinement, polyclonals were successfully obtained by screening for unique protein bands of PD strains of *X. fastidiosa* (Bazzi et al., 1994b). The antibodies directed against the two best-suited protein bands (20.7 kDa and 19.8 kDa) did not show any cross-reaction in a Western blot, a solid phase ELISA, or using indirect fluorescent antibody staining, when tested against other bacteria commonly found on grapevine. Using these "monospecific" antibodies in conjunction with indirect ELISA, the detection limit was found to be 8.3×10^5 cells/ml. Similar levels of detection (3×10^5 cells/ml) using a series of 10 monoclonal antibodies specific to pear leaf scorch (PLS) strains of *X. fastidiosa*, also by indirect ELISA (Leu et al., 1998), have been reported. The monoclonals proved highly specific: 9 of the 10 did not cross-react with 14 other bacterial strains belonging to 9 genera. In Western blot analysis, all the monoclonal antibodies recognized a 46.9-kDa polypeptide in all 12 *X. fastidiosa* strains, and a specific 21.5 kDa polypeptide in PLS strains.

---

1. BCYE medium (Wells et al., 1981) (g/l distilled water): yeast extract, 10.0; activated charcoal, 2.0; L-cysteine HCl·H$_2$O, 0.4 (filter sterilized [0.2 μm pore size] and added after autoclaving); ferric pyrophosphate (soluble), 0.25 (filter sterilized [0.2 μm pore size] and added after autoclaving); ACES buffer, 10.0; agar, 17.0.

2. PW medium (Davis et al., 1981b) (g/l distilled water): Phytone peptone (BBL), 4.0; trypticase peptone (BBL), 1.0; hemin chloride, 0.01; MgSO$_4$·7H$_2$O, 0.4; K$_2$HPO$_4$, 1.2; KH$_2$PO$_4$, 1.0; phenol red, 0.02; L-glutamine, 4.0 (filter sterilized [0.2 μm pore size] and added after autoclaving); bovine serum albumin fraction V, 6.0 (filter sterilized [0.2 μm pore size] and added after autoclaving); agar, 12.0.

3. PD3 medium (Davis et al., 1981c) (g/l distilled water) Tryptone (Difco), 4.0; Soytone (Difco), 2.0; trisodium citrate, 1.0; disodium succinate, 1.0; hemin chloride, 0.01 (10 ml of 0.1% w/v solution in 0.05N NaOH); potato starch (soluble) 2.0; MgSO$_4$·7H$_2$O, 1.0; K$_2$HPO$_4$, 1.5; KH$_2$PO$_4$, 1.0; Bacto agar (Difco), 15.0.

Editorial Note: The website for the Bioinformatics Laboratory at the University of Campinas is http://www.lbi.dcc.unicamp.br/.

Antibiotic and drug sensitivity Data from *in vitro* studies have shown that different disease strains of *X. fastidiosa* exhibit a differential response to carbenicillin, penicillin G, and gentamycin when these antibiotics are added to CS-20[4] medium (Hopkins, 1988b). Data from *in vivo* studies of leaf scorch of pear trees (*Pyrus pyrifolia*) in the low-altitude areas of central Taiwan (Leu and Su, 1993) have shown that symptom remission can be achieved by injection of oxytetracycline.

Pathogenicity and ecology Symptoms produced in host plants vary, the most characteristic being a necrosis of interveinal leaf tissue, producing a characteristic scorching effect in a wide range of woody plants. Of primary importance is Pierce's disease of grapevine, which can cause 100% destruction in some vineyards, but which is normally not a serious pathogen (Goodwin and Meredith, 1988). The bacteria block xylem vessels and cause water stress, leading to leaf scorch symptoms, but no wilting is observed as leaf stomata close. Although transpiration is reduced, diseased leaves remain water stressed and various functions are impaired, resulting in carbohydrate imbalance, loss of chlorophyll and disruption of cell membranes. Physiological changes associated with water stress and senescence have been compared between leaves of healthy grapevines (*Vitis vinifera* cv. Chardonnay) and grapevines infected with the PD strains (Goodwin et al., 1988a, b). Vegetative growth of diseased vines was less than that of healthy vines under natural vineyard conditions. Although stomatal resistance of diseased leaves with marginal necrosis was similar to that of healthy leaves during early morning, the stomatal resistance of diseased leaves increased greatly later in the day, while that of healthy leaves remained relatively constant. The diurnal patterns of transpiration and photosynthesis demonstrated that these processes were inhibited by PD strains. The reduced photosynthesis of diseased leaves was at least partly due to inhibition of nonstomatal-related aspects of CO_2 fixation. Compared with healthy leaves, diseased leaves had higher concentrations of abscisic acid, glucose, fructose, Ca^{2+} and Mg^{2+}, but a lower concentration of K^+. Sucrose accumulated in green tissue of a diseased leaf, but the concentrations of sucrose and starch were lower in diseased chlorotic tissue than in diseased green or healthy leaf tissue. Chlorophyll reduction in diseased leaves was also associated with increases in electrolyte leakage, lipid peroxidation, and superoxide anion. Based on these physiological changes, it was proposed that the symptoms of PD were the result of relatively mild but prolonged water stress accelerating leaf senescence. Subsequent studies have demonstrated that PD strains were responsible for the production of occlusions in xylem vessels of susceptible hosts (Fry and Milholland, 1990).

Pierce's disease of grapevine can be transmitted by xylem-feeding leafhoppers (Hill and Purcell, 1997). Acquisition of PD strains has been demonstrated in the blue-green sharpshooter (*Graphocephala atropunctata*) after four days' incubation with infected grapes. Subsequent transmission to healthy plants can be achieved after 10 days. Previous studies have demonstrated that harvested fruit clusters from diseased grapes did not serve as inoculum sources for *G. atropunctata* (Purcell and Saunders,

1995). Only when *G. atropunctata* was given access to diseased grapevines did it subsequently transmit PD strains to healthy grapevine. Transmission efficiency by *G. atropunctata* remained high at seven days' post-acquisition and beyond (Hill and Purcell, 1995a). Bacterial population size in the heads of the inoculative vectors also increased up to day seven, but did not consistently increase further. Studies on grapevine (*Vitis vinifera*) alongside other *G. atropunctata* host plants (*Rubus discolor, Artemisia douglasiana, Echinochloa crusgalli,* and *Cynodon dactylon*) showed that distal movement from the inoculation site could only be detected in grapevine and *Rubus discolor* (Hill and Purcell, 1995b). These findings highlight the epidemiological importance of the diseased host as a reservoir for further dissemination and, in addition, agree with earlier work, which showed that strains isolated from natural hosts such as *Rubus* spp. could induce symptoms in grapevine (Hopkins, 1988a).

Similar results were obtained by studying CVC strains of *X. fastidiosa* in citrus nurseries in Brazil (Garcia et al., 1997) where the increases in vector populations of *Dilobopterus costalimai, Acrogonia terminalis,* and *Oncometopia facialis* and disease incidence were found to be linked. CVC strains, as well as PD strains, appear to be systemic, because xylem vessels of diseased plant petioles harbor large numbers of cells (Hartung et al., 1994).

The significance of disease reservoirs, such as weeds, has also been studied. Weeds collected from plum orchards infested with leaf scald were examined, and the presence of the pathogen was confirmed by isolation and by DAS-ELISA in *Brachiaria plantaginea, Digitaria* sp., *Facelis retusa, Hypochoeris brasiliensis, Leonurus sibiricus, Lolium multiflorum, Paspalum urvillei, Pennisetum clandestinum, Richardia* sp., *Taraxacum officinale,* and *Vernonia* sp. (Leite et al., 1997). Phony peach disease strains can be detected in symptomatic or asymptomatic hosts (Aldrich et al., 1992). Populations of *X. fastidiosa* in the root xylem fluid of symptomatic peach trees were high and the distribution of these bacteria was uniform both along the length of roots and among roots within the root ball; in contrast, distribution of the bacteria within asymptomatic trees was discontinuous.

ENRICHMENT AND ISOLATION PROCEDURES

It has long been considered that *X. fastidiosa* is fastidious and unable to grow on standard bacteriological media, despite reports that Nutrient Agar can support growth (Chang et al., 1990; Fry et al., 1990). More commonly, a range of media have been used successfully to culture *X. fastidiosa*, namely CS20, PD1,[5] PD2,[6] PD3, and PW.

MAINTENANCE PROCEDURES

Lyophilization and storage on silica gel at $-20°C$ are effective long-term storage methods (Sleesman and Leben, 1978; Hopkins, 1988b). Long-term culturing of *X. fastidiosa* frequently re-

---

4. CS-20 medium (Chang and Schaad in Hopkins, 1988b) (g/l distilled water): Tryptone, 2.0; Soytone, 2.0; $(NH_4)_2HPO_4$, 0.85; hemin chloride, 15 ml (0.1% w/v solution in 0.05N NaOH); KH_2PO_4, 1.0; $MgSO_4·7H_2O$, 0.4; L-glutamine, 6.0; dextrose, 1.0; potato starch (soluble), 2.0; L-histidine, 1.0; phenol red, 5.0 ml (0.2% solution prepared by dissolving 0.2 g in 30 drops 5N NaOH then making the volume to 100 ml in H_2O); agar, 12.0.

5. PD1 medium (Davis et al., 1980) (g/l distilled water): Tryptone (Difco), 7.0; Soytone (Difco), 3.0; disodium succinate, 1.0; hemin chloride, 0.01 (10 ml of 0.1% w/v solution in 0.05N NaOH); $MgSO_4·7H_2O$, 1.0; $FeCl_3·6H_2O$, 0.002; $MnSO_4·4H_2O$, 0.002; $(NH_4)_2SO_4$, 1.0; Bacto agar (Difco), 15.0; bovine serum albumin fraction V, 2.0 (10 ml of 20% v/v, added after autoclaving to medium held at 45–50°C).

6. PD2 medium (Davis et al., 1980) (g/l distilled water): Tryptone (Difco), 4.0; Soytone (Difco), 2.0; trisodium citrate, 1.0; disodium succinate, 1.0; hemin chloride, 0.01 (10 ml of 0.1% w/v solution in 0.05N NaOH); $MgSO_4·7H_2O$, 1.0; K_2HPO_4, 1.5; KH_2PO_4, 1.0; Bacto agar (Difco), 15.0; bovine serum albumin fraction V, 2.0 (10 ml of 20% v/v added after autoclaving to medium held at 45–50°C).

sults in the loss of virulence, e.g., after 18 months of weekly culturing on PD2 medium (Hopkins, 1984).

DIFFERENTIATION OF THE GENUS *XYLELLA* FROM OTHER GENERA

Characteristics useful for differentiating *Xylella* from other members of the family *Xanthomonadaceae* are listed in Table BXII.γ.17 in Genus *Xanthomonas*. In addition, a technique for identifying *X. fastidiosa* involving direct isolation of endophytic bacteria from propagation material, combined with one-dimensional SDS-PAGE of the total soluble proteins of the bacterial cell envelope, is described by Bazzi et al. (1994a). The gel electrophoretic pattern of this pathogen is characterized by a constant distribution of 14 major diagnostic bands.

TAXONOMIC COMMENTS

Analysis of 16S rRNA gene sequence data placed this genus in the *Gammaproteobacteria*, related to *Xanthomonas campestris* and *Stenotrophomonas maltophilia* (Wells et al., 1987). Similar findings have been presented by Kersters et al. (1996). In a subsequent study, an average sequence similarity of 94.7% was reported between *X. fastidiosa* and seven type strains of the genus *Xanthomonas*; the sequence similarity with *S. maltophilia* was 94.1% (Moore et al., 1997). Results obtained from restriction maps of PCR-amplified 16S rRNA genes of *Xylella*, *Xanthomonas*, and *Stenotrophomonas* strains highlighted relationships between the three

genera (Nesme et al., 1995). Nineteen restriction fragment length polymorphisms (RFLPs) were distinguished and strains were recovered in two main clusters, one of which contained *Xanthomonas* spp., the other *Stenotrophomonas* spp. The genera *Xanthomonas* and *Stenotrophomonas* formed a coherent group, with the genus *Xylella* recovered as a single member cluster and clearly more distantly related.

X. fastidiosa is a heterogeneous species producing a variety of symptoms on many hosts of economic significance: Pierce's disease of grapevine, citrus variegated chlorosis, phony peach, periwinkle wilt; alfalfa dwarf; ragweed stunt; and almond, coffee, elm, maple, mulberry, oak, plum, and sycamore leaf scorch. Differences between pathogenic populations can be observed using traditional methods (Table BXII.γ.33), RAPDs, (Chen et al., 1995; Pooler and Hartung, 1995a, b; Albibi et al., 1998), RFLP (Chen et al., 1992b), and monoclonal antibodies (Leu et al., 1998). Evidence from these data suggests that a pathovar structure may be appropriate to circumscribe infraspecific variation within *X. fastidiosa*, but clearly further work is required.

FURTHER READING

Hopkins, D.L. 1989. *Xylella fastidiosa*: xylem-limited bacterial pathogens of plants. Ann. Rev. Phytopathol. *27*: 271–290.

Purcell, A.H.1997. *Xylella fastidiosa*, a regional problem or global threat? J. Plant Pathol. *79*: 99–105.

Purcell, A.H. and D.L. Hopkins. 1996. Fastidious xylem-limited bacterial plant pathogens. Ann. Rev. Phytopathol. *34*: 131–151.

TABLE BXII.γ.33. Differentiation of some pathogenic strains of *Xylella fastidiosa*[a, b]

Characteristics	Disease-associated strains					
	Elm leaf scorch	Periwinkle wilt	Pierce's disease	Phony peach	Plum leaf scald	Ragweed stunt
Growth on:						
BCYE, PW, CS-20	+	+	+	+	+	+
PD3	+	−	+	−	−	−
Antibiotic resistance[c]:						
Carbenicillin, 12.5 µg/ml	+	+	−	+	+	nd
Penicillin, 6.25 µg/ml	+	+	+	−	−	nd
Gentamicin, 80 µg/ml	−	+	−	−	−	nd

[a]For symbols see standard definitions; nd, not determined.

[b]Adapted from Chang and Schaad, unpublished in Hopkins, 1988b.

[c]CS-20 used as basal medium.

List of species of the genus Xylella

1. **Xylella fastidiosa** Wells, Raju, Hung, Weisburg, Mandelco-Paul and Brenner 1987, 141[VP]

   fas.tid.i.o' sa. M.L. fem. adj. *fastidiosa* highly critical; referring to the nutritional fastidiousness of the organism, particularly on primary isolation.

   The characteristics are those described for the genus. Gelatin is hydrolyzed and gelatinase is produced. Many strains produce β-lactamase. β-galactosidase, coagulase, lip-

ase, amylase, and phosphatase negative; indole and H_2S are not produced.

   Differential characteristics of some pathogenic strains of *Xylella fastidiosa* are listed in Table BXII.γ.33.

   The mol% G + C of the DNA is: 51.0–52.4 (T_m) and 52–53.1 (Bd).

   Type strain: PCE-RR, ATCC 35879, DSM 10026, ICMP 11140.

   GenBank accession number (16S rRNA): AF192343.

Order IV. **Cardiobacteriales** *ord. nov.*

GEORGE M. GARRITY, JULIA A. BELL AND TIMOTHY LILBURN

Car.di.o.bac.te.ri.a' les. M.L. neut. n. *Cardiobacterium* type genus of the order; *-ales* ending
to denote order; M.L. fem. n. *Cardiobacteriales* the *Cardiobacterium* order.

The order *Cardiobacteriales* was circumscribed for this volume on the basis of phylogenetic analysis of 16S rDNA sequences; the order contains the family *Cardiobacteriaceae.*

Description is the same as for the family *Cardiobacteriaceae.*

Type genus: **Cardiobacterium** Slotnick and Dougherty 1964, 271[AL].

Family I. **Cardiobacteriaceae** Dewhirst, Paster, La Fontaine and Rood 1990, 431[VP]

FLOYD E. DEWHIRST AND BRUCE J. PASTER

Car.di.o.bac.te.ri.a' ce.ae. M.L. neut. n. *Cardiobacterium* type genus of the family; *-aceae* ending
to denote a family; M.L. fem. pl. n. *Cardiobacteriaceae* the family of *Cardiobacterium.*

Straight rods, 0.5–1.7 × 1–6 µm, with rounded ends. Cells occasionally occur in pair, chains, or rosettes. Gram negative but with a tendency to resist Gram decolorization. Nonmotile, as cells do not possess flagella, but may possess type IV pili and exhibit twitching motility. Aerobic and oxidase positive except for members of the anaerobic genus *Dichelobacter.* Catalase negative. Urease negative. Fermentative type of chemoorganotrophic metabolism except for members of the genus *Dichelobacter,* which are arginine chemoorganotrophs. Acid but not gas is produced from glucose, sucrose, maltose, mannose, and fructose in the fermentative species. Nitrates are not reduced. H_2S is produced (as determined by the lead acetate method). No growth occurs on MacConkey agar. Species in the family possess the following rare or unique signatures in their 16S rRNAs: uracil at positions 18, 1051, 1400; cytosine at positions 1124 and 1253; and adenine at position 917 (*E. coli* numbering). The family is composed of three genera—the type genus *Cardiobacterium* and the genera *Suttonella* and *Dichelobacter.*

The mol% G + C of the DNA is: 45–60.

Type genus: **Cardiobacterium** Slotnick and Dougherty 1964, 271.[AL]

Differentiation from other taxa   Based on 16S rRNA sequence analysis, it was recognized that *Kingella indologenes* and *Bacteroides nodosus* were taxonomically misplaced and that they were most closely related to *Cardiobacterium hominis* (Dewhirst et al., 1990). A neighbor-joining phylogenetic tree based on 16S rRNA sequence comparisons is shown in Fig. BXII.γ.40. The three organisms form a monophyletic group in the *Gammaproteobacteria. Bacteroides nodosus* clearly differs from *C. hominis* and *K. indologenes* in that it is an asaccharolytic anaerobe. It was therefore placed in its own genus as *Dichelobacter nodosus. Cardiobacterium hominis* and *K. indologenes* are quite similar phenotypically, but the mol% G + C content of their DNA differs by 11%, and *K. indologenes* was placed in its own genus as *Suttonella indologenes.* Strains for two additional *Cardiobacterium* species have been isolated, but not formally named (see Fig. BXII.γ.40). These three genera were placed in the family *Cardiobacteriaceae* based upon their forming a closely related (92% similarity) monophyletic group. The *Cardiobacteriaceae* cluster is supported in 100% of 200 bootstrap resamplings. The family branches deeply in the *Gammaproteobacteria,* being less than 87% similar to any other taxa. The most closely related organisms are *Thiobacillus neapolitanus, Halothiobacillus hydrothermalis* (shown on tree) and *T. halophilus.* However, these species clearly belong in a separate family because they have a chemolithotrophic metabolism, possess none of the 16S rRNA base signatures of the *Cardiobacteriaceae,* and have less than 87% 16S rRNA sequence similarity. Table BXII.γ.34 shows phenotypic traits that differentiate *Cardiobacteriaceae* genera from other genera.

Genus I. **Cardiobacterium** *Slotnick and Dougherty 1964, 271*[AL]

BRUCE J. PASTER AND FLOYD E. DEWHIRST

Car.di.o.bac.te' ri.um. Gr. n. *cardia* heart; Gr. n. *bakterion* small rod; M.L. neut. n. *Cardiobacterium* bacterium
of the heart.

Straight rods 0.5–0.75 × 1.0–3.0 µm, with rounded ends. Occasional long filaments, 7.0–35.0 µm may occur. **Pleomorphic.** Cells are arranged singly, in pairs, in short chains, and in rosette clusters. Gram negative, but retention of crystal violet may occur in the swollen ends or central portions of cells. Nonmotile. **Facultatively anaerobic.** CO_2 is required by some strains on isolation. Aerobic growth is scant unless humidity is elevated. Growth in candle jars or under anaerobic conditions is not dependent on an elevated humidity. Optimum temperature, 30–37°C. Colonies on blood agar are smooth, convex, and opaque. **Chemoorgano-** **trophic, having a strictly fermentative type of metabolism.** Acid but not gas is produced from fructose, glucose, mannose, sorbitol, mannitol, and sucrose. Lactic acid is the major product of glucose fermentation; smaller amounts of pyruvate, formate, and propionate are formed. Oxidase positive. **Catalase negative.** Small amounts of **indole** are formed. **Nitrates are not reduced.** No growth occurs on MacConkey agar. Urease negative. Ornithine decarboxylase (ODC) negative. Occur in nasal flora of humans; isolated from the blood of humans with bacterial endocarditis.

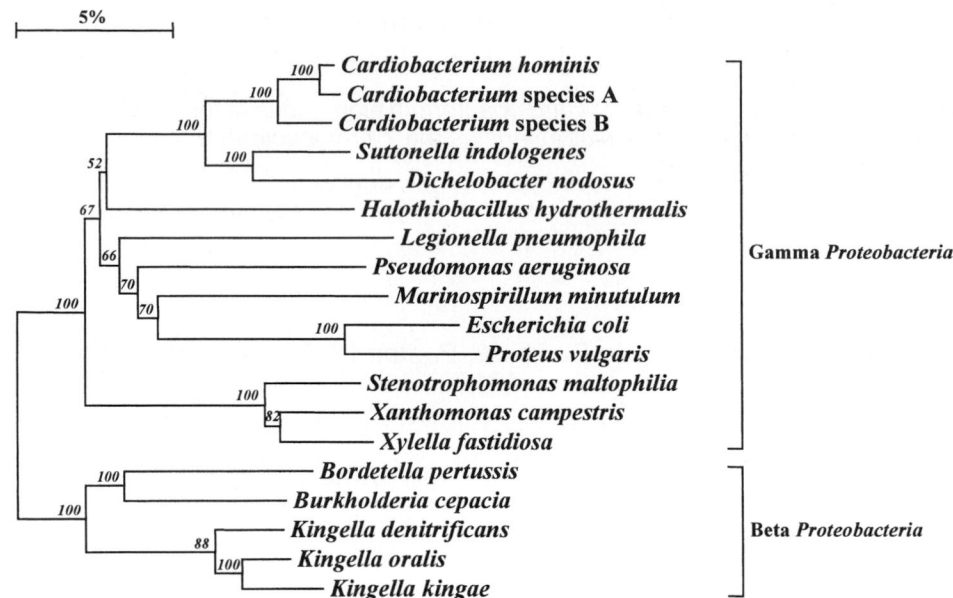

FIGURE BXII.γ.40. Phylogeny of *Cardiobacteriaceae*. The phylogenetic relationships of the three genera, namely *Cardiobacterium*, *Suttonella*, and *Dichelobacter*, of the family *Cardiobacteriaceae*, compared to other members of the *Betaproteobacteria* and *Gammaproteobacteria*. Bar = 5% difference in nucleotide sequences. The neighbor-joining method was used for tree construction. Two hundred bootstrap trees were generated, and bootstrap confidence levels (shown as percentages above the nodes) were determined. GenBank accession numbers for sequences of members of the *Cardiobacteriaceae*: *Cardiobacterium hominis*, M35014; *Cardiobacterium* sp. A, AF144697; *Cardiobacterium* sp. B, AF144696; *Suttonella indologenes*, M35015; *Dichelobacter nodosus*, M35016. Accession numbers for the sequences of the other species (all type strains) are available in GenBank.

TABLE BXII.γ.34. Phenotypic traits that differentiate *Cardiobacteriaceae* genera from other genera[a]

Characteristic	*Cardiobacterium*	*Dichelobacter*	*Suttonella*	*Bacteroides*	*Eikenella*	*Kingella*	*Moraxella*	*Pasteurella*	*Porphyromonas*
Aerobic	+	−	+	−	−	+	+	+	−
Oxidase	+	−	+	−	+	+	+	+	−
Catalase	−	−	−	D	−	−	D	+	D
Indole production	+	−	+	−	−	−	−	D	+
Nitrate reduction	−	−	−	nd	+	D	D	+	nd
Acidifies glucose	+	−	+	+	−	+	−	+	−
Acidifies D-sorbitol	D	−	−	D	−	−	−	D	−
γ-Glutamyltransferase activity	+	nd	+	nd	−	−	D	D	nd
Ornithine decarboxylase	−	+	−	−	+	−	D	D	−
Branched chain fatty acids	−	−	−	+	−	−	−	−	+

[a]For symbols see standard definitions; nd, not determined.

The mol% G + C of the DNA is: 59–60.

Type species: **Cardiobacterium hominis** Slotnick and Dougherty 1964, 271.

FURTHER DESCRIPTIVE INFORMATION

Cells of *Cardiobacterium* grown on tryptose blood agar (containing 5% human blood) are pleomorphic and have rounded ends. The usual cell dimensions are ~0.5–0.6 × 1.0–2.2 μm. Occasionally filaments are formed, which vary in length from 7.0–35.0 μm. Teardrop-shaped cells may occur. One or both ends of *Cardiobacterium* cells are frequently enlarged, and the crystal violet of the Gram stain tends to be retained in these areas. Other cells may show crystal violet retention in the central portions (Slotnick and Dougherty, 1964).

The degree of pleomorphism is apparently influenced by the medium on which *Cardiobacterium* is grown. Pleomorphism has been reported in cultures grown on agar media, which do not contain yeast extract. On yeast extract-containing media, the cells have been found to be generally uniform Gram-negative rods that are 0.5 × 2.0 μm (Savage et al., 1977).

Sudanophilic bodies and metachromatic granules have been reported with the use of the Sudan Black B and Albert's stain, respectively (Slotnick and Dougherty, 1964). However, Midgley et al. (1970) were unable to confirm these observations.

The cell wall is of the Gram-negative type with distinct layers— a dense outer layer, a unit membrane, and a dense inner layer. The substructure of the crystalline surface layer (S-layer) consists of closely packed units, which are nearly spherical and have an average diameter of about 3.5 nm (Reyn et al., 1971). Cells have 20–40 nm thick polar caps consisting of a dense, tuft-like material. In the periphery of the cytoplasm, numerous unit membrane profiles have been observed and are thought to be intrusions of the plasma membrane (Reyn et al., 1971).

After incubation for 24 h at 37°C on tryptose blood agar (containing 5% human blood), colonies are punctiform. After 48 h the maximum size is 1–2 mm. The colonies are circular, convex, smooth, moist, glistening, opaque, and butyrous, and have an entire edge. No clear zones of hemolysis are formed around the colonies (Slotnick and Dougherty, 1964).

Strains of *C. hominis* may be grown in a defined medium[1] described by Slotnick and Dougherty (1965). Essential components of the medium are pantothenate, niacinamide, thiamine, threonine, proline, leucine, histidine, glycine, and arginine. Biotin, pyridoxine, tyrosine, valine, and glutamate are not required but must be included for optimum growth. Glucose is not an essential carbon source or the sole energy source, but it is fermented and does increase the growth yield. The major end product of glucose fermentation is lactic acid (Slotnick and Dougherty, 1964). Maltose, sucrose, fructose, and mannose are also fermented and can be substituted effectively for glucose in the medium. Lactose, arabinose, and galactose are not fermented or utilized for growth. Low concentrations ($\sim$1.5–2.0 μg/ml) of riboflavin or flavin mononucleotide inhibit the growth of *C. hominis*. This inhibitory effect is neutralized by the presence of excess leucine in the medium.

Growth at 30°C and 37°C is equally good. Growth at 25°C is sporadic and light. No growth occurs above 42°C or at 22°C (Slotnick and Dougherty, 1964; Midgley et al., 1970).

The optimal pH range for growth is 7.0–7.2. *C. hominis* grows poorly in air unless the humidity is increased by a method such as placing filter paper strips, saturated with water, in a closed container. In a candle jar or under anaerobic conditions, it is not necessary to increase the humidity. The optimal atmosphere contains 3–5% CO_2 (Slotnick and Dougherty, 1964). Midgley et al. (1970) found that three of four strains grew equally well in a candle jar and in a sealed jar without increased CO_2. The fourth strain grew less well in the sealed jar than in the candle jar.

Slotnick and Dougherty (1964) produced antisera to four strains. All of the strains reacted to a high titer in each of the unabsorbed sera; however, by absorption studies the four strains were shown not to be antigenically identical. No antigenic relationship was demonstrated to members of the genera *Brucella*, *Pasteurella*, *Bordetella*, *Moraxella*, *Haemophilus*, *Streptobacillus*, *Corynebacterium*, *Bacteroides*, *Neisseria*, *Escherichia*, or *Lactobacillus*.

Savage et al. (1977) tested five strains and found the minimum inhibitory concentrations for all strains to be less than 2 μg/ml for ampicillin, carbenicillin, cephalothin, chloramphenicol, penicillin, tetracycline, streptomycin, kanamycin, gentamicin, and colistin.

In humans, *C. hominis* is a relatively rare cause of bacterial endocarditis. Dental caries and periodontal disease have been suggested as predisposing factors (Berbari et al., 1997; el Khizzi et al., 1997). A single case of abdominal abscess and bacteremia due to *Cardiobacterium hominis* and *Clostridium bifermentans* has been reported (Rechtman and Nadler, 1991). Tucker et al. (1962) found no evidence of infection in mice, guinea pigs, rabbits, hamsters, or pigeons after injection of 10^7–10^8 viable organisms intravenously, intraperitoneally, or subcutaneously.

C. hominis is apparently part of the normal nasal and pharyngeal flora of humans. It was isolated from the nose or throat of 68 of 100 individuals examined by Slotnick et al. (1964). The

organism was also isolated from cervical and vaginal cultures obtained from 2 of 159 subjects examined (Slotnick, 1968).

ENRICHMENT AND ISOLATION PROCEDURES

Cardiobacterium hominis has been isolated from blood in several different media under both aerobic and anaerobic conditions. Media that have been used successfully are brain heart infusion broth containing sodium polyanetholsulfonate (SPS) and *p*-aminobenzoic acid, casein soy broth with SPS, thioglycolate broth with SPS, and glucose broth (Midgley et al., 1970; Savage et al., 1977).

Slotnick et al. (1964) isolated *C. hominis* from the nose and throat of humans on trypticase soy agar plates containing 5% human blood. The plates were incubated at 37°C for 48–72 h in candle jars. Because of difficulty in recognizing the colonies of *C. hominis*, smears of selected areas of the plates were stained with fluorescent antibody. Isolation of strains was accomplished by subculturing from the areas that showed positive fluorescent staining reactions.

MAINTENANCE PROCEDURES

Strains of *C. hominis* may be preserved by lyophilization. Freezing also may preserve them. Cells from 18–24 h old agar cultures suspended in defibrinated rabbit blood have survived for more than 15 years at −50°C.

PROCEDURES FOR TESTING SPECIAL CHARACTERS

Indole formation is an important characteristic for the identification of *C. hominis*. Indole production, however, may be weak and may not be detected by test procedures that do not first concentrate the indole by xylene extraction.

Tryptone broth and heart infusion broth are suitable media for the indole test. A portion of broth culture should be tested after incubation of 18–24 h. If the reaction is negative, the remaining portion of the broth culture should be tested after incubation for an additional 24 h. To perform the test, add approximately 0.5–1 ml of xylene to 3–4 ml of broth culture. Shake vigorously to extract the indole. After the xylene layer has formed at the surface, add either Ehrlich's or Kovacs' reagent, allowing the reagent to flow down the side of the slightly tilted tube.

DIFFERENTIATION OF THE GENUS *CARDIOBACTERIUM* FROM OTHER GENERA

Characteristics by which *C. hominis* can be distinguished from other Gram-negative rod-shaped organisms that have similar physiological characteristics have been described by Bruun et al. (1984b) and are shown in Table BXII.γ.34 of the chapter *Cardiobacteriaceae*. The cellular fatty acid composition of *Cardiobacterium hominis* differs from *Eikenella corrodens*, *Kingella denitrificans*, and *Kingella kingae* by the absence of 3-hydroxylauric ($C_{12:0\ 3OH}$) acid, from *Suttonella indologenes* by the presence of 3-hydroxypalmitic ($C_{16:0\ 3OH}$) acid, and from *E. corrodens* by the presence of 3-hydroxymyristic ($C_{14:0\ 3OH}$) acid (Wallace et al., 1988). *Kingella denitrificans* and *Kingella kingae* contained myristic ($C_{14:0}$) and palmitic ($C_{16:0}$) acids as major acids, whereas *cis*-vaccenic ($C_{18:1\ \omega7c}$) and palmitic acids were the major acids in *Suttonella indologenes*, *C. hominis*, and *E. corrodens*.

TAXONOMIC COMMENTS

Cardiobacterium hominis is the only named species in the genus; however, two unnamed *C. hominis*-like species have been described (see Other Organisms). Based on 16S rRNA sequence

---

1. Defined medium has the following composition (per liter of distilled water): $MgSO_4 \cdot 7H_2O$, 0.4 g; $MnSO_4 \cdot H_2O$, 3 mg; NaCl, 50 mg; Na_2HPO_4, 2.84 g; KH_2PO_4, 2.72 g; $ZnSO_4 \cdot 7H_2O$, 4.43 mg; $FeSO_4 \cdot 7H_2O$, 4 mg; $CuSO_4 \cdot 5H_2O$, 0.5 mg; glucose, 5.0 g; arginine, 160 mg; glutamic acid, 200 mg; glycine, 176 mg; histidine, 126 mg; leucine, 426 mg; proline, 100 mg; threonine, 276 mg; valine, 185 mg; tyrosine, 35 mg; calcium pantothenate, 10 mg; niacinamide, 10 mg; pyridoxine hydrochloride, 20 mg; thiamine hydrochloride 10 mg; and biotin, 1 mg. Final pH is 7.0.

analysis, the genera *Cardiobacterium, Suttonella,* and *Dichelobacter* compose the family *Cardiobacteriaceae* (Dewhirst et al., 1990). The family branches deeply in the *Gammaproteobacteria* at the juncture of the *Gammaproteobacteria* and *Betaproteobacteria.* Neighboring families include "*Xanthomonadaceae*", "*Piscirickettsiaceae*", and "*Francisellaceae*". A phylogenetic tree including *Cardiobacterium hominis* and related bacteria is shown in Fig. BXII.γ.40 in the *Cardiobacteriaceae* chapter.

ACKNOWLEDGMENTS

Bruce J. Paster and Floyd E. Dewhirst were supported by NIH grants DE10374 and DE11443.

FURTHER READING

Dewhirst, F.E., B.J. Paster, S. La Fontaine and J.I. Rood. 1990. Transfer of *Kingella indologenes* (Snell and Lapage 1976) to the genus *Suttonella* gen. nov. as *Suttonella indologenes* comb. nov.; transfer of *Bacteroides nodosus* (Beveridge 1941) to the genus *Dichelobacter* gen. nov. as *D. nodosus* comb. nov.; and assignment of the genera *Cardiobacterium, Dichelobacter,* and *Suttonella* to *Cardiobacteriaceae* fam. nov. in the gamma division of *Proteobacteria* on the basis of 16S rRNA sequence comparisons. Int. J. Syst. Bacteriol. 40: 426–433.

Snell, J.J.S. and S.P. Lapage. 1976. Transfer of some saccharolytic *Moraxella* species to *Kingella* Henriksen and Bøvre 1976, with descriptions of *Kingella indologenes* sp. nov. and *Kingella denitrificans* sp. nov. Int. J. Syst. Bacteriol. 26: 451–458.

List of species of the genus Cardiobacterium

1. **Cardiobacterium hominis** Slotnick and Dougherty 1964, 271[AL]

   ho′ mi.nis. L. gen. n. *hominis* of man.

   The characteristics are as described for the genus.

   Occurs in nasal and oral flora of humans; isolated from the blood of humans with endocarditis.

   The mol% G + C of the DNA is: 59–60 (T_m).

   Type strain: 6573, ATCC 15826, DSM 8339, NCTC 10426.

   GenBank accession number (16S rRNA): M35014.

Other Organisms

Strains potentially representing two additional *Cardiobacterium* species have been described (Rothenpieler et al., 1986). These species are shown as *Cardiobacterium* sp. A and *Cardiobacterium* sp. B in Fig. BXII.γ.40 in the *Cardiobacteriaceae* chapter. *Cardiobacterium* sp. A is represented by the singe strain SSI AB2167 (FsK 3118; M. Kilian, human dental plaque; Denmark, 1985). This strain has been called *Suttonella indologenes*-like and has 97% 16S rRNA sequence similarity with *C. hominis. Cardiobacterium* sp. B is represented by at least two strains: CCUG 13150 (SSI AB 2102; G. Dahlen; human periodontitis; Sweden, 1982) and CCUG 31207 (SSI AB 2168; FsK3120; M. Kilian; human dental plaque; Denmark, 1985). These strains have also been called *Cardiobacterium hominis*-like and were designated by Enevold Falsen as EF Group 18 strains (Culture Collection, University of Gøteborg). There are additional EF Group 18 strains in the CCUG collection. *Cardiobacterium* sp. B has 95% 16S rRNA sequence similarity with *C. hominis.* The phenotypic traits differentiating these strains from *C. hominis* are shown in Table BXII.γ.35. Further characterization of strains in the genus *Cardiobacterium* is needed prior to naming strains these additional species. Inclusion of *Cardiobacterium* spp. A and B would necessitate modification of genus description to include indole-negative and alkaline phosphatase-positive species.

TABLE BXII.γ.35. Differentiation of unnamed *Cardiobacterium* species from *C. hominis*

Characteristic	*Cardiobacterium hominis*	*Cardiobacterium* sp. A	*Cardiobacterium* sp. B	*Suttonella indologenes*
Ferments D-mannitol or D-sorbitol	+	−	+	−
Indole	+	+	−	+
Alkaline phosphatase	−	−	+	+

Genus II. **Dichelobacter** *Dewhirst, Paster, La Fontaine and Rood 1990, 430*[VP]

JULIAN I. ROOD, DAVID J. STEWART, JILL A. VAUGHAN AND FLOYD E. DEWHIRST

Di.che′ lo.bac.ter. Gr. adj. *Dichelos* cloven hoofed; N.L. n. *bacter* a rod; N.L. masc. n. *Dichelobacter* cloven-hoofed rod, because this organism is the rod-shaped bacterium that causes footrot in sheep, goats, and cattle.

Large straight or slightly curved rods, 3–6 × 1.0–1.7 μm, with rounded ends. Freshly isolated cells often have terminal swellings, this feature is less pronounced after subculture. **Gram negative** but with a tendency to resist decolorization. Organisms stained with Loeffler methylene blue have prominent polychromatic granules toward the poles and at intermediate sites within the cells. **Grows slowly only under strictly anaerobic conditions but is not rapidly killed by exposure to oxygen.** The cells have large numbers of thin polar fimbriae or pili that vary in number according to changes in colony morphology. The pili belong to the type IV or *N*-methylphenylalanine family; consequently, **the cells exhibit the twitching motility and spreading of colonies that is characteristic of cells with type IV pili. The organism also produces extracellular serine proteases that are relatively heat stable.** Ammonia is produced from arginine, asparagine, serine, and threonine. Positive for ornithine decarboxylase, phosphatase (weak), H_2S production, selenite reduction, and proteolytic activity on gelatin, casein, and albumin. **Negative for acid or gas production from carbohydrates;** ammonia production from phenylalanine, cysteine, citrulline, and ornithine; starch hydrolysis; esculin hydrolysis; indole production; nitrate reduction; growth in 0.1% bile; hemolysis; arginine decarboxylase; catalase; oxidase; urease; DNase; coagulase; lipase; lecithinase; and hyaluronidase. The genus contains a single species, *Dichelobacter nodosus*. Further details on the properties of the genus are available elsewhere (Beveridge, 1941, Cato et al., 1979, Holdeman et al., 1984a, Skerman, 1989; Dewhirst et al., 1990).

The mol% G + C of the DNA is: 45.

Type species: **Dichelobacter nodosus** (Beveridge 1941) Dewhirst, Paster, La Fontaine and Rood 1990, 426 (*Bacteroides nodosus* (Beveridge 1941) Mráz 1963, 85; *Fusiformis nodosus* Beveridge 1941, 23.)

FURTHER DESCRIPTIVE INFORMATION

Dichelobacter is one of three genera that compose the family *Cardiobacteriaceae*. Based on 16S rRNA sequence data the members of this family are placed as deeply branching members of the *Gammaproteobacteria*. The 16S rRNA sequences of these genera, *Cardiobacterium*, *Suttonella*, and *Dichelobacter*, have approximately 93% identity (Dewhirst et al., 1990).

Little is known about the composition of the cell wall. The lipopolysaccharide (LPS) of *D. nodosus* is similar to enterobacterial LPS in that it contains 2-keto-3-deoxyoctonic acid (KDO), glucose, galactose, and heptose. The LPS is not highly toxic, although it does exhibit some characteristic endotoxin properties (Stewart, 1977). It is not immunoprotective (Stewart, 1978). Virtually nothing is known about the genes involved in LPS biosynthesis, although one gene, *lpsA*, whose product appears to interact with the acidic inner LPS core, has been identified (Billington et al., 1995).

The predominant fine structural feature is the presence of polar fimbriae or pili. These long filamentous structures are ~6 nm in diameter and up to 10 μm in length (Stewart, 1973; Walker et al., 1973). The fimbriae consist of identical subunits of a 17-kDa protein that has the typical sequence and structure of pilins belonging to the type IV pilin family (Elleman, 1988; Mattick, 1989). Several genes that appear to be involved in fimbrial biogenesis have been cloned and analyzed, including the gene encoding the prepilin peptidase (Johnston et al., 1995, 1998).

The organism is relatively slow growing and after 2–6 d of anaerobic growth on agar media produces flat spreading colonies 1–3 mm in diameter that are circular, concave, semiopaque, and slightly mottled, with a shiny, bumpy surface, and a pitted edge. Large colonies have concentric circles with the inner circle pitting the agar (Cato et al., 1979). However, there is considerable variation in colonial morphology, depending on both the isolate and the growth medium. Colonies may have a highly papilliated or beaded central zone overlying a pit-like depression etched into the agar surface, producing a sunken appearance, a pale, granular mid-zone surrounded by a peripheral zone with a ground glass texture, and a fimbriate edge. Changes in colony type within a strain can occur with subculturing (Skerman, 1989). Variations in the basic features of colony morphology by different strains are described (Stewart et al., 1986).

D. nodosus does not ferment carbohydrates but catabolizes arginine, a property that has been used in the formulation of the growth medium TAS (Skerman, 1989). Maximum growth requires trypticase and 0.02–0.05 M arginine. Serine also enhances growth, and in some strains so does lysine (Skerman, 1975). The organism is routinely grown in rich medium and little is known about its specific growth requirements or its metabolism. Growth is most rapid at 37°C and within a pH range of 6.8–

7.8 (optimal range 7.2–7.8). In broth cultures, growth is evenly dispersed with a light, turbid, or silky texture and is usually visible within 24–48 h. The organism hydrolyses gelatin, produces ammonia from amino acids, and has ornithine decarboxylase activity (Skerman, 1989). *D. nodosus* isolates are susceptible *in vitro* to a wide range of antimicrobial agents.

The organism is naturally transformable with 7 of 17 strains that have been tested able to be transformed with plasmid DNA preparations. Natural transformation can be used to make chromosomal mutants by allelic exchange (Kennan et al., 1998, 2001). The genome of *D. nodosus* strain VCS1703A is currently being determined (see http://www.tigr.org/). The genome of the reference *D. nodosus* strain A198 (198A, VCS1001, or ATCC 27521) has been mapped; it consists of a 1.54-Mb circular chromosome that has three rRNA operons (La Fontaine and Rood, 1996). Key features are that the putative virulence genes are not localized to one region of the genome and that the genome contains a large 27-kb virulence-related locus, the *vrl* region (Billington et al., 1999), which has a markedly different distribution of *Eag*I and *Stu*I restriction sites. Only one extrachromosomal element, the 10-kb plasmid pJIR896, has been identified. This plasmid carries a copy of another virulence-associated locus, the *vap* region, as well as the insertion sequence IS*1253*, which is also found on the *D. nodosus* chromosome (Billington et al., 1996b).

The primary antigenic determinants are the type IV fimbriae. *D. nodosus* isolates can be divided into nine major serogroups (A to I) based on their ability to react with antisera prepared against purified fimbriae (Claxton, 1989). The serogroups can be further divided into two major classes based on their fimbrial subunit structure and the genetic organization of the fimbrial subunit gene region (Elleman, 1988; Hobbs et al., 1991; Mattick et al., 1991). Class I isolates (serogroups A, B, C, E, F, G, and I) have a single gene *fimB* located between the fimbrial subunit gene *fimA* and the *clpB* gene. By contrast, in the same genomic position, class II isolates (serogroups D and H) have three genes, *fimC*, *fimD*, and *fimZ*. Mutation of the fimbrial subunit gene, *fimA*, has shown that production of fimbrial subunits is essential for fimbrial biogenesis, twitching motility, natural competence, protease secretion, and virulence in sheep (Kennan et al., 2001). These studies also showed that the *fimB* gene was not required for fimbrial biogenesis or virulence.

D. nodosus is the essential causative agent of footrot in ruminants, especially in sheep, goats, cattle, and deer. The ovine disease manifests as a continuum from virulent through intermediate to benign footrot. The severity of the disease depends on both the climatic conditions, being optimal under warm, moist conditions, and the virulence of the *D. nodosus* isolates. Clinical isolates are therefore often referred to as being virulent, intermediate, or benign. The only virulence factor that has been shown to be essential for virulence is the fimbrial subunit (Kennan et al., 2001). However, the production of extracellular serine proteases and the presence of the *vrl* genomic island have also been postulated to be involved in virulence (Billington et al., 1996a). The role of these and other virulence factors in the disease process remains to be determined.

Confirmed isolates of *D. nodosus* have been obtained only from ruminants such as sheep, goats, cattle, and deer (Egerton, 1989; Stewart, 1989). The organism does not survive for longer than 14 d in feces, soil, or pasture (Stewart, 1989). Although there have been instances of *D. nodosus* supposedly being isolated from other sources, such as human specimens, these strains were not confirmed as *D. nodosus* (Cato et al., 1979; Holdeman et al., 1984a).

ENRICHMENT AND ISOLATION PROCEDURES

Specimens are collected with a cotton-tipped swab, wooden applicator stick, or scalpel blade, by scraping active footrot lesions containing exudate. The specimens are then deposited into a variety of transport media (Skerman, 1989) and promptly sent to the laboratory within 24–48 h of collection where they are inoculated onto hoof agar (Thomas, 1958), trypticase arginine serine (TAS) agar (Skerman, 1975), or modified TAS agar incorporating hemoglobin-HEPES solution (Depiazzi et al., 1990) containing 4% agar. The plates are incubated at 37°C for 2–4 d in anaerobic jars containing CO_2/H_2 (10:90). A selective blood Eugonagar medium, using lincomycin, also has been devised (Gradin and Schmitz, 1977). Primary cultures can be inoculated by heavy surface streaking with lines 10 mm apart onto well-dried plates (Skerman, 1989) or by stab inoculation (Depiazzi et al., 1990). Surface colonies grow as translucent flat granular colonies, often with concentric zones with a fimbriate edge (Thorley, 1976). In the streaking method, colonies are picked off each primary plate with a sterile straight wire and subcultured onto a 4% hoof agar, TAS agar, or blood Eugonagar plate to obtain pure colonies and then incubated anaerobically for 3–4 d at 37°C. In the stab inoculation method, an agar section is cut from the isolation plate with the aid of a stereo microscope and using a nichrome bent, spatula-shaped wire. The agar then is placed colony side down onto the surface of a modified TAS maintenance agar plate and incubated anaerobically for 2 d at 37°C. The addition of hoof particles to modified TAS agar enhances the growth of spreading colonies and facilitates the isolation of *D. nodosus*. Small Petri dishes (55 mm) are used instead of conventional plates (90 mm), which enables less culture medium to be used and three times as many agar plates to be incubated in each anaerobic jar. If a pure colony is not obtained after incubation of the primary isolation plate, then with a sterile spatula the agar is flipped upside down into the lid so that the underside colony can be sampled and subcultured (Depiazzi et al., 1990). Alternatively, a sterile microscope cover glass, dipped into a bacterial suspension of a contaminated bacterial colony and rinsed to remove nonadherent bacteria, can be used to purify *D. nodosus* colonies. The cover glass is moved in a zigzag manner over the agar surface to inoculate the adherent *D. nodosus* cells. These modifications enable the time interval between submission and final reporting to be reduced to 8 d.

MAINTENANCE PROCEDURES

For routine maintenance of *D. nodosus* cultures under anaerobic conditions, either hoof agar, TAS agar, modified TAS agar, or blood Eugonagar (Gradin and Schmitz, 1977) can be used. For anaerobic broth cultures, TAS broth, HEPES TAS broth, or modified Eugonbroth (BBL) is recommended. To facilitate prompt growth in liquid medium from agar cultures, the broth can be inoculated with an aqueous vortexed suspension of pure colonies or a small portion of agar. Although *D. nodosus* is an anaerobe, fully grown plate cultures can tolerate exposure to air for up to 12 h on the laboratory bench (Skerman, 1989). Cultures will remain viable in anaerobic jars for several weeks at room temperature (Skerman, 1975). For long-term storage of cultures, *D. nodosus* may be preserved by lyophilization of 3–4-d-old cells suspended in horse serum containing 7.5% (w/v) glucose, skim milk, 0.25 M sucrose, or TAS broth, or alternatively by adding 20% glycerol to TAS broth suspensions and freezing at −70°C. Cryopreserved preparations are known to survive for more than 20 years.

PROCEDURES FOR TESTING SPECIAL CHARACTERS

Ovine footrot lesions differ in severity according to the virulence of the strain of *D. nodosus* involved; the two major clinical forms are termed virulent and benign footrot (Stewart, 1989). There are also strains of *D. nodosus* that are intermediate in virulence. However, for purposes of uniformity in footrot diagnosis, footrot is described as virulent or benign, with strains of intermediate virulence being regarded as effectively virulent. Because the extent of disease expression is also influenced by the host (e.g., the breed of sheep), the season, and the environment, laboratory diagnosis is increasingly being used to assist the clinical diagnosis of virulent or benign footrot. Although there is a correlation between virulence of wild-type isolates and colony type, this distinction is not sufficiently reliable as a laboratory test for differentiating virulent and benign isolates. The gelatin gel test is the recommended laboratory diagnostic test for differentiating virulent and benign footrot (Palmer, 1993); it measures the thermostability of proteases in 2–4-d-old broth cultures. In general, isolates of *D. nodosus* capable of producing virulent footrot produce proteolytic enzymes that are stable to heating at 67°C, whereas proteases from benign isolates are inactivated at this temperature. Proteolytic activity is detected by clearing of a gelatin substrate incorporated into agarose plates. The gelatin gel test has now superseded the elastase test because the latter may take 7–28 d for a definitive result. Virulent isolates cause digestion of elastin particles beneath the streak line in elastin TAS agar plates, whereas benign isolates do not produce clearing (Stewart, 1979). In the alternative zymogram protease test, the electrophoretic pattern of the protease isoenzymes is determined in a nondenaturing polyacrylamide slab gel electrophoresis system followed by detection with a gelatin/agarose contact overlay assay. Benign and virulent strains have different and characteristic profiles (Every, 1982; Kortt et al., 1983; Palmer, 1993). There is a range of different zymogram patterns for isolates with either stable (S) proteases or unstable (U) proteases, with the predominant pattern being the S1 and U1 pattern, respectively (Palmer, 1993). The zymogram test may be useful for classifying isolates where there are anomalous results with the gelatin gel test and virulence assessment.

More recently, two new technologies have become available for footrot diagnosis. These are the protease monoclonal antibody (Mab) ELISA (Stewart et al., unpublished) and the gene probe/PCR-based assay (Katz et al., 1991; Liu, 1994; Rood et al., 1996). The virulent and benign Mabs are able to distinguish antigenic differences between the proteases produced by virulent and benign strains, and there is a high level of agreement between the results for the gelatin gel test and the protease Mab ELISA. This difference in epitope specificity may result from a single amino acid change in one of the virulent and benign acidic proteases (Riffkin et al., 1995). The gene probe or PCR tests detect virulence-associated genomic DNA unrelated to the protease genes. The DNA probes, which specifically detect the *vrl* and *vap* gene regions, divide *D. nodosus* isolates into three major categories (1, 2, and 3). The PCR test is based on specific oligonucleotide primers for the *vrl* and *vap* specific plasmids, pJIR314B and pJIR318, respectively (Rood et al., 1996). The PCR test further divides isolates in gene probe categories 1 and 2 into groups L and H. Isolates in all three categories can belong to probe category + or − depending on the presence or absence of an additional gene region. It is recommended that rapid Mab tests for the proteases be used for differentiation of virulent and benign footrot and that the PCR assay should be used for molecular epidemiological studies.

DIFFERENTIATION OF THE GENUS *DICHELOBACTER* FROM OTHER GENERA

Dichelobacter nodosus, previously *Bacteroides nodosus*, can be differentiated from members of the genera *Bacteroides*, *Prevotella*, and *Porphyromonas* by fatty acid analysis. *D. nodosus* possesses $C_{16:1\ \omega7c}$, $C_{18:1\ \omega7c}$, and $C_{16:0}$ fatty acids, while the genera in the phylum "*Bacteroidetes*" contain large amounts of the methyl branched-chain fatty acids $C_{15:0\ iso}$ and $C_{15:0\ anteiso}$. Genera in the "*Bacteroidetes*" also possess sphingolipids. *D. nodosus* contains type IV pili, which have been found only in *Proteobacteria*. *D. nodosus* can be differentiated from other genera in the family *Cardiobacteriaceae* in that it is anaerobic, oxidase negative, indole negative, and ornithine decarboxylase positive.

TAXONOMIC COMMENTS

Dichelobacter nodosus is the only species in the genus *Dichelobacter*. The genera *Dichelobacter*, *Suttonella*, and *Cardiobacterium* compose the family *Cardiobacteriaceae*. The family branches deeply in the *Gammaproteobacteria* at the juncture of the *Gammaproteobacteria* and *Betaproteobacteria*. Neighboring families include "*Xanthomonadaceae*", "*Piscirickettsiaceae*", and "*Francisellaceae*". A phylogenetic tree including *Dichelobacter nodosus* is shown in the *Cardiobacteriaceae* chapter.

FURTHER READING

Billington, S.J., J.L. Johnston and J.I. Rood. 1996. Virulence regions and virulence factors of the ovine footrot pathogen, *Dichelobacter nodosus*. FEMS Microbiol. Lett. *145*: 147–156.

Dewhirst, F.E., B.J. Paster, S. La Fontaine and J.I. Rood. 1990. Transfer of *Kingella indologenes* (Snell and Lapage 1976) to the genus *Suttonella* gen. nov. as *Suttonella indologenes* comb. nov.; transfer of *Bacteroides nodosus* (Beveridge 1941) to the genus *Dichelobacter* gen. nov. as *D. nodosus* comb. nov.; and assignment of the genera *Cardiobacterium*, *Dichelobacter*, and *Suttonella* to *Cardiobacteriaceae* fam. nov. in the gamma division of *Proteobacteria* on the basis of 16S rRNA sequence comparisons. Int. J. Syst. Bacteriol. *40*: 426–433.

Kennen, R.M., O.P. Dhungyel, R.J. Whittington, J.R. Egerton and J.I. Rood. 2001. The type IV fimbrial subunit gene (*fimA* of *Dichelobacter nodosus* is essential for virulence, protease secretion and natural competence. J. Bacteriol. *183*: 4451–4458.

Skerman, T.M. 1989. Isolation and identification of *Bacteriodes nodosus*. *In* Egerton, Yong and Riffkin (Editors), Footrot and Foot Abscess in Ruminants, CRC Press, Boca Raton. 85–104.

Stewart, D.J. 1989. Footrot of sheep. *In* Egerton, Yong and Riffkin (Editors), Footrot and Foot Abscess in Ruminants, CRC Press, Boca Raton. 5–45.

List of species of the genus Dichelobacter

1. **Dichelobacter nodosus** (Beveridge 1941) Dewhirst, Paster, La Fontaine and Rood 1990, 426^{VP} (*Bacteroides nodosus* (Beveridge 1941) Mráz 1963, 85; *Fusiformis nodosus* Beveridge 1941, 23.)

   no.do′ sus. L. adj. *nodosus* full of knots, refers to the shape of the cells.

   The characteristics are as described for the genus. The organism is found on the hooves and interdigital skin of ruminants, especially sheep and cattle. The growth of most strains on an appropriate medium, such as TAS medium, is more luxurious than that of ATCC 25549. The genome of strain A198 is circular, 1.54 Mb in size, and has three rRNA operons (La Fontaine and Rood, 1996). The GenBank accession number of the 16S rRNA molecule from the reference strain A198 is M35016 (Dewhirst et al., 1990). There is no 16S rDNA sequence available from the type strain.

   The organism is the essential causative agent of footrot in sheep and other ruminants (Stewart, 1989). The fimbriae are the major protective antigens; immunization of sheep with purified recombinant fimbriae will protect against infection but only by members of the homologous serogroup (Egerton et al., 1987; Stewart and Elleman, 1987). The major virulence antigens appear to be the type IV fimbriae and the extracellular serine proteases produced by the organism. Virulent isolates produce four extracellular acidic serine proteases (V1–V3, V5), pI 5.2–5.6, and a basic protease (BprV), pI.~9.5. The benign protease isoenzymes, B1–B4, have distinctly different electrophoretic mobilities and pIs, elastolytic activity, and heat stability, but are very similar to V1–V3 from virulent strains in amino acid composition, and peptide profiles, and apparent M_r proteases V5 and B5 are identical. Benign strains secrete a distinct basic protease BprB), pI 8.6, that is very closely related to the virulent basic protease. In summary, there are three distinct types of *D. nodosus* proteases, V5/B5, V1–V3/B1–B4, and the basic protease. The gene sequences for the two acidic protease genes, *apr*V5 and *apr*V2/B2, and the basic protease genes, *bpr*V/bprB, have been determined. These serine proteases are members of the subtilisin family of proteases and presumably arose by gene duplication from an ancestral precursor gene. They show significant sequence similarity to the extracellular serine protease of *Xanthomonas campestris*, a plant pathogen. In addition, there is a 27-kb genomic region, the *vrl* locus, that appears to be preferentially associated with more virulent isolates. An additional integrated region, the *vap* locus, is found in all isolates that carry the *vrl* region, and in other isolates (Billington et al., 1996a). The genome of *D. nodosus* strain VCS1703A is currently being sequenced by the Institute for Genomic Research (TIGR). The genome has been assembled into 264 contigs and can be searched using BLAST at TIGR (http://www.tigr.org/) or NCBI (http://www.ncbi.nlm.nih.gov/blast/).

   The mol% G + C of the DNA is: 45.

   Type strain: ATCC 25549.

   Additional Remarks: The reference strain is ATCC 27521 (A198, 198A, or VPI5731-1).

Genus III. **Suttonella** *Dewhirst, Paster, La Fontaine and Rood 1990, 429*[VP]

Floyd E. Dewhirst and Bruce J. Paster

Sut.ton.el' la. L. dim. ending *-ella;* N.L. fem. n. *Suttonella* named after R.G.A. Sutton.

Large straight rods, 1.0×2–3 μm, with rounded ends. Cells occasionally occur in pair, chains, or rosettes. Gram negative but with a tendency to resist Gram decolorization. Nonmotile as determined by normal tests, but may be fimbriated (i.e., have pili) and exhibit twitching motility. Aerobic. Aerobic growth is enhanced by high humidity and CO_2, which makes the organism appear to be facultatively anaerobic. **Oxidase positive** (when tested with tetramethyl-*p*-phenylene diamine). **Catalase negative.** **Indole positive** (activity may be weak). Urease negative. DNase negative. Ornithine and lysine decarboxylase negative. **Alkaline phosphatase positive.** Trypsin (*N*-benzoyl-DL-arginine-2-naphthylamide) negative. Phosphohydrolase (naphthyl-AS-B1-phosphate) negative. Chemoorganotrophic, having a fermentative type of metabolism. Acid but not gas is produced from fructose, glucose, maltose, D-mannose, and sucrose. The maltose reaction is weak and delayed (positive at 28 d), but is negative as determined by rapid (4–24 h) protocols. **Nitrates are not reduced.** Nitrites are reduced by some strains. Resazurin is reduced. H_2S is produced (as determined by the lead acetate method, but not by the triple sugar iron method). No growth occurs on Mac-Conkey agar.

The mol% G + C of the DNA is: 49.

Type species: **Suttonella indologenes** (Snell and Lapage 1976) Dewhirst, Paster, La Fontaine, and Rood 1990, 430 (*Kingella indologenes* Snell and Lapage 1976, 456.)

Further descriptive information

Initial isolates of *Suttonella indologenes* were considered novel *Moraxella* strains (van Bijsterveld, 1970; Sutton et al., 1972; Bøvre et al., 1974). Snell and Lapage (1976) placed these saccharolytic and indole producing strains in the genus *Kingella* as the novel species *Kingella indologenes.* DNA–DNA hybridizations (Rothenpieler et al., 1986) and 16S rRNA sequence analysis (Dewhirst et al., 1990) demonstrated that *Kingella indologenes* was not related to other *Kingella* species in the family *Neisseriaceae,* but rather branched between the *Betaproteobacteria* and *Gammaproteobacteria* (rRNA homology groups I and II) with *Cardiobacterium hominis.* Subsequent 16S rRNA sequence analysis demonstrated that *Bacteroides nodosus* also clustered with these species. These three species were placed in the family *Cardiobacteriaceae,* and *Kingella indologenes* and *Bacteroides nodosus* were renamed *Suttonella indologenes* and *Dichelobacter nodosus,* respectively (Dewhirst et al., 1990).

Suttonella grows slowly and after 24 h colonies are 0.1–0.5 mm in diameter. After 72 h, colonies are 1–1.5 mm. Pitting of media or spreading edges has been seen in fresh colonies, but may be lost on storage (Snell and Lapage, 1976).

The major cellular fatty acids are as follows: $C_{18:1 \, \omega7c}$ (27%), $C_{16:0}$ (26%), $C_{16:1 \, \omega7c}$ (19%), $C_{14:0}$ (11%), $C_{12:0}$ (5%), and $C_{12:0 \, 3OH}$ (2%) (Wallace et al., 1988).

Ultrastructure has not been examined, but related genera *Cardiobacterium* and *Dichelobacter* both possess crystalline surface layers and intrusive membranes (Reyn et al., 1971; Stewart and Egerton, 1979). *Suttonella indologenes* possesses *N*-methylphenylalanine, or type IV pili.

Growth and biochemical characteristics from several references were summarized by Bruun et al. (1984b) and by Weyant et al. (1996). Growth at 22, 30, and 37°C. Acid is produced from dextrin, fructose, glucose, maltose, D-mannose, and sucrose. Growth in the presence of 4% NaCl. Production of acid phosphatase, γ-glutamyltransferase, indole, and hydrogen sulfide (by lead acetate paper). Hydrolysis of casein, Tween 20, and Tween 40. No growth at 5, 40, 42, or 45°C, or in the presence of 6% NaCl, 10% or 40% bile. No acid is produced from D-adonitol, L-arabinose, cellobiose, dulcitol, ethanol, D-galactose, glycerol, *i*-inositol, lactose, D-mannitol, melibiose, raffinose, L-rhamnose, salicin, starch, D-sorbitol, trehalose, or D-xylose. No liquefaction of gelatin or serum. No production of β-galactosidase, urease, ornithine decarboxylase, lysine decarboxylase, or trypsin. No citrate utilization or growth on hydroxybutyrate. No hydrolysis of esculin, starch, Tween 80, or arginine. No reduction of nitrate.

Currently, the only genetic information in GenBank for *Suttonella indologenes* is the 16S rRNA sequence under accession number M35015. The closely related organism *Dichelobacter nodosus* has been much more extensively studied, and over 60 sequences are in GenBank (see the chapter on the genus *Dichelobacter*). This sequence information should be extremely useful in seeking homologous genes in *Suttonella indologenes.*

Little is known regarding the pathogenicity of *Suttonella indologenes.* It has been isolated from patients with eye infections and endocarditis (van Bijsterveld, 1970; Sutton et al., 1972; Jenny et al., 1987).

Enrichment and Isolation Procedures

No particular methods have been developed specifically for isolation of *S. indologenes.* Strains have been recovered on standard blood plates grown under microaerophilic conditions.

Maintenance Procedures

Strains of *S. indologenes* may be preserved by lyophilization.

Procedures for Testing Special Characters

Indole formation may be weak and should be determined using extraction with xylene or chloroform and detection with Ehrlich's reagent (*p*-dimethylaminobenzaldehyde). The spot indole reagent *p*-aminocinnamaldehyde may fail to detect weak indole production.

Differentiation of the genus *Suttonella* from other genera

See Table BXII.γ.34 in the chapter on the family *Cardiobacteriaceae.* *Suttonella* may be distinguished from *Cardiobacterium* by a positive reaction for alkaline phosphatase, and negative reactions for D-sorbitol and D-mannitol fermentation. *Suttonella* may be distinguished from *Eikenella* and *Kingella* by indole production and γ-glutamyltransferase activity. *Suttonella* may be distinguished from many other genera by the unusual combination of being oxidase and indole positive and catalase and nitrate reduction negative.

Taxonomic Comments

Suttonella indologenes is the only species in the genus *Suttonella.* The genera *Suttonella, Dichelobacter,* and *Cardiobacterium* compose the family *Cardiobacteriaceae.* The family branches deeply in the *Gammaproteobacteria* at the juncture of the *Gammaproteobacteria* and *Betaproteobacteria*. Neighboring families include "*Xanthomonadaceae*", "*Piscirickettsiaceae*", and "*Francisellaceae*". A phylogenetic tree including *Suttonella indologenes* can be seen in Fig. BXII.γ.40.

ACKNOWLEDGMENTS

Floyd E. Dewhirst and Bruce J. Paster were supported by NIH grants DE10374 and DE11443.

FURTHER READING

Dewhirst, F.E., B.J. Paster, S. La Fontaine and J.I. Rood. 1990. Transfer of *Kingella indologenes* (Snell and Lapage 1976) to the genus *Suttonella* gen. nov. as *Suttonella indologenes* comb. nov.; transfer of *Bacteroides nodosus* (Beveridge 1941) to the genus *Dichelobacter* gen. nov. as *D. nodosus* comb. nov.; and assignment of the genera *Cardiobacterium, Dichelobacter*, and *Suttonella* to *Cardiobacteriaceae* fam. nov. in the gamma division of *Proteobacteria* on the basis of 16S rRNA sequence comparisons. Int. J. Syst. Bacteriol. *40*: 426–433.

Jenny, D.B., P.W. Letendre and G. Iverson. 1987. Endocarditis caused by *Kingella indologenes*. Rev. Infect. Dis. *9*: 787–789.

Snell, J.J.S. and S.P. LaPage. 1976. Transfer of some saccharolytic *Moraxella* species to *Kingella* Henriksen and Bøvre 1976, with descriptions of *Kingella indologenes* sp. nov. and *Kingella denitrificans* sp. nov. Int. J. Syst. Bacteriol. *26*: 451–458.

List of species of the genus Suttonella

1. **Suttonella indologenes** (Snell and Lapage 1976) Dewhirst, Paster, La Fontaine, and Rood 1990, 430[VP] (*Kingella indologenes* Snell and Lapage 1976, 456.)

in.dol.o'ge.nes. M.L. n. *indolum* indole; M.L. suff. *genes* producing; M.L. adj. *indologenes* indole-producing.

The characteristics are as described for the genus. Previously referred to as the "Bijsterveld/Sutton strains".
The mol% G + C of the DNA is: 49 (T_m).
Type strain: ATCC 25869, DSM 8309, NCTC 10717.
GenBank accession number (16S rRNA): M35015.

Order V. **Thiotrichales** *ord. nov.*

GEORGE M. GARRITY, JULIA A. BELL AND TIMOTHY LILBURN

Thi.o.tri.cha'les. M.L. fem. n. *Thiothrix* type genus of the order; *-ales* suffix to denote order; M.L. fem. n. *Thiotrichales* the *Thiothrix* order.

The order *Thiotrichales* was circumscribed for this volume on the basis of phylogenetic analysis of 16S rRNA sequences; the order contains the families *Thiotrichaceae, Piscirickettsiaceae*, and *Francisellaceae*.

Order is morphologically, metabolically and ecologically diverse. Includes obligate intracellular parasites of fish (*Piscirickettsia*) and of insects, arachnids, mammals, and man (*Francisella*). Other genera include chemoorganotrophs, methylotrophs, and chemolithotrophic sulfur-oxidizing organisms.

Type genus: **Thiothrix** Winogradsky 1888, 39[AL] emend. Howarth, Unz, Seviour, Blackall, Pickup, Jones, Yaguchi and Head 1999, 1824.

Family I. **Thiotrichaceae** *fam. nov.*

GEORGE M. GARRITY, JULIA A. BELL AND TIMOTHY LILBURN

Thi.o.tri.cha'ce.ae. M.L. fem. n. *Thiothrix* type genus of the family; *-aceae* ending to denote family; M.L. fem. pl. n. *Thiotrichaceae* the *Thiothrix* family.

The family *Thiotrichaceae* was circumscribed for this volume on the basis of phylogenetic analysis of 16S rRNA sequences; the family contains the genera *Thiothrix* (type genus), *Achromatium, Beggiatoa, Leucothrix, Thiobacterium, Thiomargarita, Thioploca*, and *Thiospira*.

Most genera deposit sulfur. Several form filaments. Motility, if present, is by gliding, except for *Thiospira*, which possesses flagella.

Type genus: **Thiothrix** Winogradsky 1888, 39[AL] emend. Howarth, Unz, Seviour, Blackall, Pickup, Jones, Yaguchi and Head 1999, 1824.

Genus I. **Thiothrix** Winogradsky 1888, 39[AL] emend. Howarth, Unz, Seviour, Blackall, Pickup, Jones, Yaguchi and Head 1999, 1824

RICHARD F. UNZ AND IAN M. HEAD

Thi'.o.thrix. Gr. n. *thium* sulfur; Gr. n. *thrix* hair; M.L. fem. n. *Thiothrix* sulfur hair.

Rods, 0.7–1.5 × 1.2–2.5 µm, depending on species, seriate in rigid, multicellular **filaments (trichomes)** of uniform (as long as 500 µm) or slightly tapering (base to tip and typically ranging from 100–200 µm in length with shorter trichomes present in agitated as opposed to static cultures) diameter. Gram negative or Gram variable. Capsules not produced. Trichomes clearly ensheathed in some but not all species. False branching never observed. **Rod-shaped gonidia with rounded ends formed at the apical region of the trichome. Gonidia motile by gliding.** The base of the trichome may have a holdfast. Rosettes of trichomes typically but not always produced. Only rosette-forming strains produce gonidia. Resting stages not known. Flagella absent, but a tuft of monopolar fimbriae may be present.

Aerobic or microaerophilic. Facultatively autotrophic, chemoorganotrophic, and mixotrophic species; the latter requiring any of several small organic compounds, e.g., acetate, lactate, propionate, pyruvate, succinate, fumarate, and oxalacetate, as well as a reduced inorganic sulfur source. **Sulfur globules deposited** within invaginations of the cytoplasmic membrane of cells grown in the presence of a reduced inorganic sulfur compound. **By light microscopy, sulfur globules appear to be internal.**

Found in sulfide-containing flowing water and in activated-sludge wastewater treatment systems.

The mol% G + C of the DNA is: 44–52.

Type species: **Thiothrix nivea** (Rabenhorst 1865) Winogradsky 1888, 39 (*Beggiatoa nivea* Rabenhorst 1865, 94.)

FURTHER DESCRIPTIVE INFORMATION

Advances have occurred in the taxonomy of the genus *Thiothrix* since the last edition of the *Manual* and it is no longer possible to rely on morphology alone to positively identify *Thiothrix* spp., especially, the polymorphic *T. eikelboomii* and *T. defluvii*. Notwithstanding dimensional differences, cells of *Thiothrix* spp. exhibit geometries that are typically cylindrical (Fig. BXII.γ.41a) as for *T. nivea*, *T. unzii*, and *T. fructosivorans*, and variously cuboidal, barrel-shaped, discoid, and bead-like (Fig. BXII.γ.41d and e) with respect to *T. eikelboomii* and *T. defluvii*. Trichomes are rigid in most *Thiothrix* species; however, *T. eikelboomii* and *T. defluvii* trichomes can form distinct loops or knots (Fig. BXII.γ.42), suggesting flexibility in the filaments. Several species do not contain ensheathed trichomes; however, when present, the sheath (Fig. BXII.γ.43) appears to be made up of several layers (Fig. BXII.γ.44a) and extends the full length of the trichome (Fig. BXII.γ.41c). Base to tip differentiation in cells of trichomes (Fig. BXII.γ.45) is prominent in some species. The cell wall is typically Gram negative (Fig. BXII.γ.44b), although, *T. eikelboomii* may demonstrate Gram stain variability and the formation of transverse septa in the process of cell division (Fig. BXII.γ.41e). Sulfur globules are deposited within invaginations of the cytoplasmic membrane (Fig. BXII.γ.46) and are therefore external to the cytoplasm, as in *Beggiatoa*. Poly-β-hydroxybutyrate (PHB; Fig. BXII.γ.41b and f) and polyphosphate intracellular inclusions (Fig. BXII.γ.47) may be produced. Larkin (1989) reported unconfirmed nitrogen fixation by several JP strains, including the neotype strain of *T. nivea*. Rosettes of trichomes are observed in all freshly isolated species but the property may not be stable upon repeated subculture in the laboratory. Rosettes may form from outgrowth of gonidia collected at the tip of a preexisting trichome or aggregated in the suspending fluid (Fig. BXII.γ.48). Initial attachment of the gonidia involves fimbriae (Fig. BXII.γ.49), with holdfast material being produced after attachment (Fig. BXII.γ.47). A life cycle is proposed by Larkin and Shinabarger (1983) and Larkin (1989).

Isolates of *Thiothrix* spp. develop visible colonies on LT agar after 2–3 d incubation at 28°C. Colonies are less than 2 mm in diameter, white to off-white, hard to the touch, flat to slightly raised with rough to slightly wavy or thread-like margins (Fig. BXII.γ.50). Liquid cultures are turbid to flocculent depending on species and contain freely dispersed gonidia and small rosettes. In addition, length of trichomes appears affected by culture conditions with shorter (<200 μm) trichomes occurring in shaken cultures and long (500 μm) trichomes present in static cultures.

Winogradsky (1888) first described *Thiothrix nivea* and sug-

gested the possible presence of autotrophy within the genus *Thiothrix*. Present evidence for autotrophic potential in *Thiothrix* spp. includes ribulose biphosphate carboxylase/oxygenase (RubisCo) activity (Unz and Williams, 1989; Odintsova et al., 1993) and carbon dioxide fixation (Strohl and Schmidt, 1984; McGlannan and Makemson, 1990). Chemolithoautotrophic growth is reported for "*Thiothrix ramosa*" (Odintsova et al., 1993) and *Thiothrix* strain CT3 (Tandoi et al., 1994) and *in situ* measurements of the physiology of *Thiothrix* cells in activated sludge further implicates autotrophy for these organisms (Nielsen et al., 2000). *T. nivea* has not been cultured chemolithoautotrophically or shown to fix carbon dioxide although DNA probes have detected genes encoding for RubisCo in the organism (Shively et al., 1986).

Thiothrix spp. are found in flowing sulfide-containing waters of circumneutral pH (Bahr and Schwartz, 1956; Brigmon et al., 1994) and low oxygen tension (Bahr and Schwartz, 1956; Bland and Staley, 1978) and in activated sludge systems treating septic or sulfide-bearing wastewaters (Farquhar and Boyle, 1971; Williams and Unz, 1985b; Unz and Williams, 1988). It may be the dominant filamentous sulfide-oxidizing organism in suitable waters (Lackey et al., 1965; McGlannan and Makemson, 1990). *Thiothrix* spp. have been observed in symbiotic (Brigmon and De Ridder, 1998) and epiphytic (Larkin et al., 1990) association with aquatic animals. The ability of *Thiothrix* spp. to attach to objects and to itself undoubtedly allows it to remain situated in a suitable location in flowing water.

ENRICHMENT AND ISOLATION PROCEDURES

Several methods have been reported for the enrichment of *Thiothrix* spp. In the procedure of Larkin (1989), fine-tipped forceps may be used to select tufts of suspected *Thiothrix* trichomes from *Thiothrix*-containing material contained in a Petri dish while under observation with a dissecting microscope and transmitted light. Bundles of trichomes are transferred to a second Petri dish containing about 5–10 ml of a salt solution (SS)[1] (Strohl and Larkin, 1978b), agitated with the forceps, and then transferred to another Petri dish containing the SS solution. The procedure is repeated through four or five transfers. A few drops from each dilution are transferred with a Pasteur pipette to either MP agar (SS plus 0.0001% sodium acetate, 0.03% Na₂S and 1.5% agar; Strohl and Larkin, 1978b) or MY agar (SS plus 0.01% each of sodium acetate, nutrient broth powder, and yeast extract, plus 0.03% Na₂S, and 1.5% agar; Larkin, 1980) in separate Petri dishes. Each dish is held on an angle so that the drops will flow across the agar surface. The excess is then withdrawn from the other side with the pipette. The dishes are incubated at about 25–30°C and are examined daily under transmitted light with the aid of a dissecting microscope. Colonies with a hairy or filamentous edge are transferred with sterile toothpicks to fresh media and are restreaked until pure.

Williams and Unz (1985a) transferred 1 ml amounts of activated sludge mixed liquor to standard test tubes containing 5 ml of solid media (0.15% glucose and/or 0.06% Na₂S·9H₂O plus 2.0% Bacto-agar) and an overlay of 15 ml of a mineral salts-vitamin solution (MSV).[2] Tubes are incubated at 22–25°C. Sur-

---

1. Salt solution contains: NH₄Cl, 0.02%; K₂HPO₄, 0.001%; MgSO₄, 0.001%; CaSO₄, 20 ml/l of a saturated solution; and trace element solution, 5 ml/l. pH is about 7.5. (Trace element solution contains: ZnSO₄·7H₂O, 0.001%; MnSO₄·4H₂O, 0.002%; CuSO₄·5H₂O, 0.00000005%; H₃BO₃, 0.001%; Co(NO₃)₂, 0.0001%; Na-molybdate, 0.0001%; EDTA, 0.02%; and FeSO₄·7H₂O, 0.07%.)

2. Mineral salts solution contains (per liter): (NH₄)₂SO₄, 0.5 g; MgSO₄·7H₂O, 0.1 g; CaCl₂·2H₂O, 0.05 g; K₂HPO₄, 0.11 g; KH₂PO₄, 0.085 g; FeCl₃·6H₂O, 0.002 g; EDTA, 0.003 g; and vitamin mix (Eikelboom, 1975), 1.0 ml. pH is about 7.5.

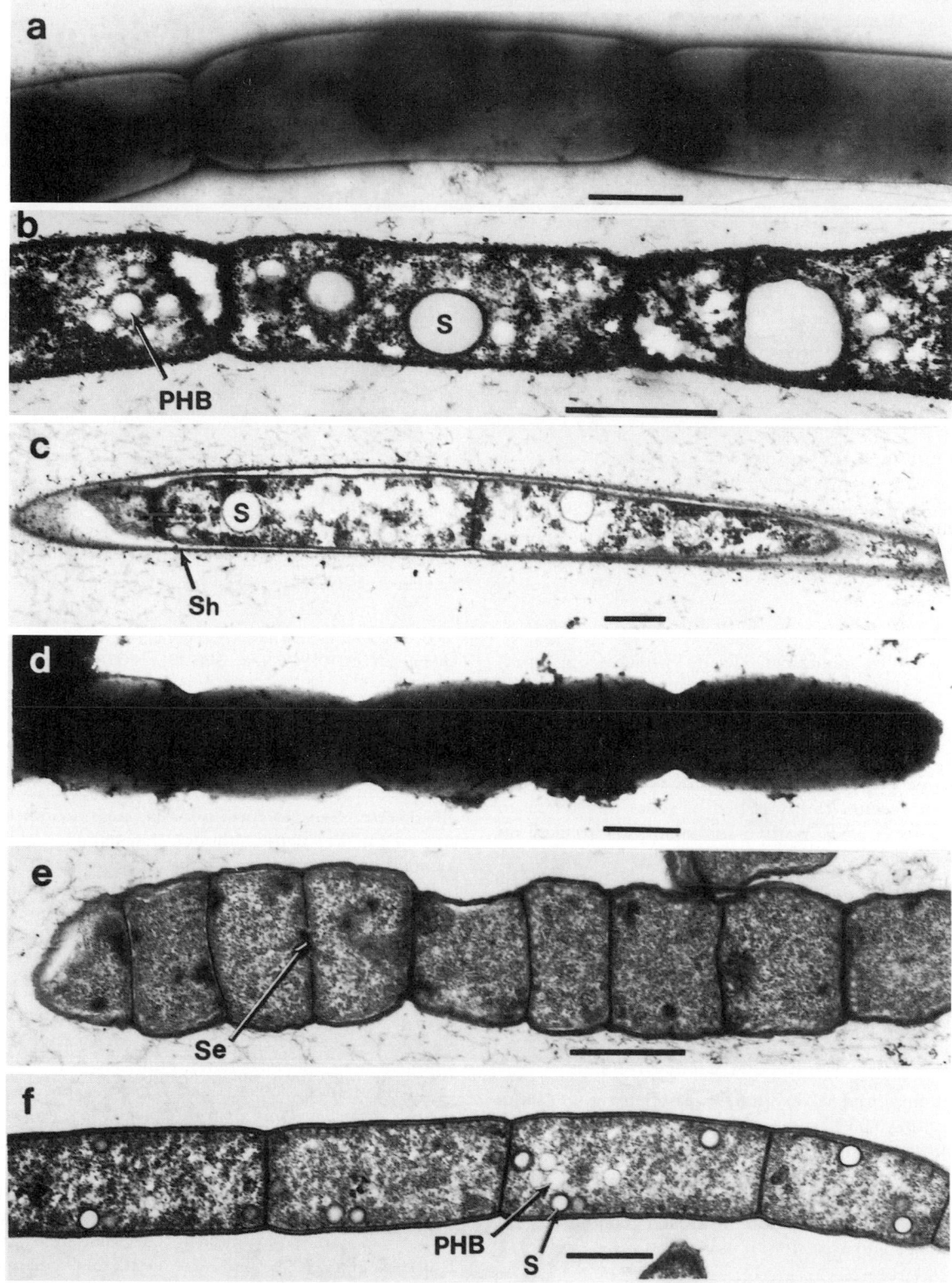

FIGURE BXII.γ.41. General ultrastructural characteristics of trichomes of *Thiothrix* species. *Thiothrix unzii* (*a*, whole cell; *b*, thin section); *Thiothrix fructosivorans* (*c*, thin section); *Thiothrix eikelboomii* (*d*, whole cell; *e, f*, thin section). Abbreviations: *Sh*, sheath; *Se*, septum; *S*, sulfur inclusion; *PHB*, poly-β-hydroxybutyrate. Bars = 1.0 μm. (Reproduced with permission from T.M. Williams et al., Applied and Environmental Microbiology *53*: 1560–1570, 1987, ©American Society for Microbiology.)

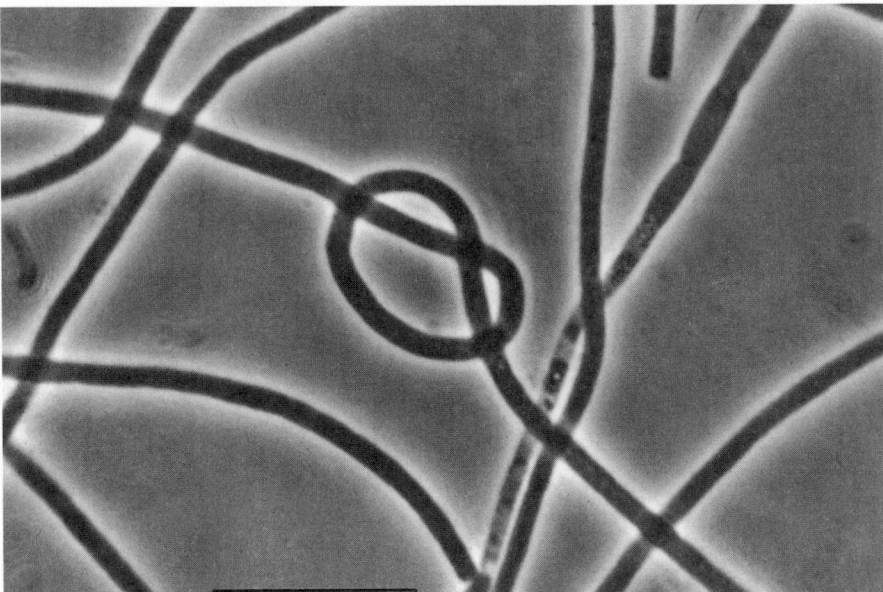

FIGURE BXII.γ.42. Knot produced in trichome by *Thiothrix eikelboomii*. Bar = 10 μm. (Reproduced with permission from T.M. Williams and R.F. Unz, Applied and Environmental Microbiology *49*: 887–898, 1985, ©American Society for Microbiology.)

face specimens are examined periodically by microscopy for evidence of the development of *Thiothrix* trichomes. Isolation of *Thiothrix* spp. is initiated by placing several loopfuls of enrichment culture films or dilute (10^{-1}) activated sludge in Petri dishes containing 7 ml of sterile MSV and, with the aid of a stereomicroscope (15–45× magnification), selecting 40–50 single filaments or rosettes to be washed by serial transfer six to seven times in fresh MSV. Washed trichomes are streaked on glucose–sulfide[3] or LT[4] medium.

MAINTENANCE PROCEDURES

The type species can be maintained in semisolid (0.15% agar) deeps of either MP or MY medium, with transfers at intervals of about 3–4 weeks. Other *Thiothrix* spp. survive monthly transfers in LTH[5] medium (sodium lactate supplied at 1.0 g/l) when stored at 10°C and cryopreservation at −83 to −90°C.

PROCEDURES FOR TESTING SPECIAL CHARACTERS

Larkin (1989) employed MP broth or appropriate modifications to test sole carbon (0.05%), nitrogen (0.02 or 0.05%), and sulfur (0.03%) sources in support of sustained growth of *T. nivea* strains through three successive transfers. Other *Thiothrix* species were nutritionally evaluated in MSV employing as a positive result the criterion of optical density equal to or greater than 4× that of controls (Williams and Unz, 1989).

---

3. Glucose–sulfide medium contains (per liter): glucose, 0.15 g; $(NH_4)_2SO_4$, 0.5 g; $Ca(NO_3)_2$, 0.01 g; $K_2HPO_4 \cdot 7H_2O$, 0.05 g; KCl, 0.05 g; $CaCO_3$, 0.1 g; $Na_2S \cdot 9H_2O$, 0.187 g; vitamin mix, 1.0 ml; and agar, 15.0 g, when needed. pH is about 7.5.

4. LT medium contains per liter of MSV: sodium lactate, 0.5 g; $Na_2S_2O_3 \cdot 5H_2O$, 0.5 g; and agar, 12.0 g. pH is about 7.5.

5. LTH medium contains per liter of MSV: sodium lactate, 0.5 g; $Na_2S_2O_3 \cdot 5H_2O$, 0.5 g; and HEPES buffer (*N*-2-hydroxyethylpiperazine-*N'*-2-ethanesulfonic acid), 0.01 M. pH is about 7.5.

DIFFERENTIATION OF THE GENUS *THIOTHRIX* FROM OTHER GENERA

Recent inclusion of new species within *Thiothrix* has exacerbated the problem of differentiation of the genus from other trichome-forming bacteria on morphological and physiological grounds. *Thiothrix* differs from *Beggiatoa* in that the latter is less nutritionally diverse, does not form gonidia, and trichomes are not ensheathed or in rosettes although they exhibit a gliding motility. *Thiothrix* is separated from *Thioploca* by the production of rosettes and the lack of gliding trichomes in the former and by the presence of multiple filaments within a single sheath and a multilayered cell wall in *Thioploca*. The unavailability of axenic cultures of *Thioploca* precludes further physiological comparison with *Thiothrix*. Analyses of 16S rDNA sequences recovered from purified *Thioploca* biomass collected from natural environments has shown that *Thioploca* spp. are distantly related to *Thiothrix* spp., although both genera belong to the *Gammaproteobacteria* (Teske et al., 1995; Ahmad et al., 1999). *Thioploca* spp. are more closely related to *Beggiatoa* spp. than to *Thiothrix* spp. (Teske et al., 1999; Ahmad et al., 1999). *Thiothrix* may be differentiated from *Leucothrix*, which it resembles morphologically, by the presence of a sheath in some species of the former and the unequivocal ability to deposit sulfur globules. It should be noted, however, that some *Leucothrix* strains may deposit intracellular sulfur (Brock, 1992; Dul'tseva et al., 1996; Grabovich et al., 1999); hence, this property alone may be insufficient to distinguish *Leucothrix* from *Thiothrix*. Although both genera exhibit gonidia formation, *Thiothrix* produces gonidia from only the end of a trichome, whereas the entire trichome of *Leucothrix* may fragment into gonidia.

TAXONOMIC COMMENTS

The type species of *Thiothrix* is *Thiothrix nivea*. The original specimen for which the type species was named is nonexistent and *T. nivea* JP2 (DSM 5202) is the approved neotype strain. Sub-

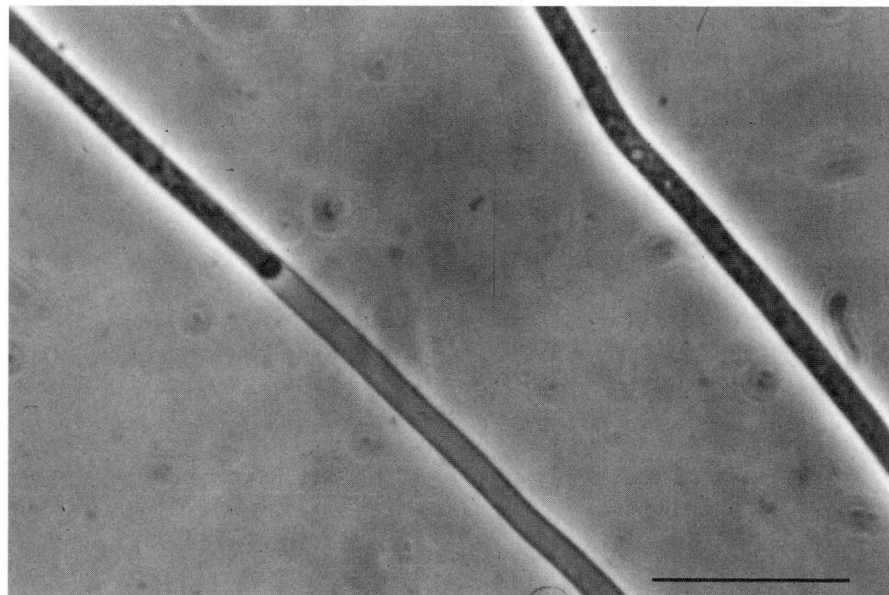

FIGURE BXII.γ.43. Sheath produced by trichome of *Thiothrix fructosivorans*. Bar = 10 μm. (Reproduced with permission from T.M. Williams and R.F. Unz, Applied and Environmental Microbiology *49:* 887–898, 1985, ©American Society for Microbiology.)

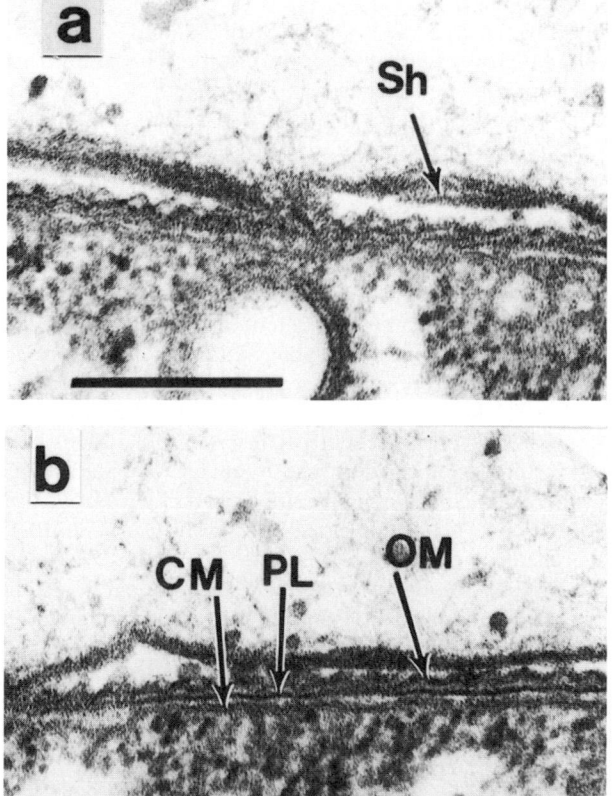

FIGURE BXII.γ.44. *a*, Ultrastructure of sheath surrounding trichome of *Thiothrix fructosivorans*. Abbreviation: *Sh*, sheath. Bar = 250 nm. *b*, Ultrastructure of cell in trichome of *Thiothrix fructosivorans*. Abbreviations: *CM*, cytoplasmic membrane; *PL*, peptidoglycan layer; *OM*, outer membrane. (Reproduced with permission from T.M. Williams et al., Applied and Environmental Microbiology *53:* 1560–1570, 1987, ©American Society for Microbiology.)

sequent descriptions of *Thiothrix* spp. have largely been based on morphological criteria that have not progressed beyond descriptions at the genus level.

Morphology has been a primary criterion in the taxonomy of many filamentous bacteria and it has been suggested that this may be sufficient for the characterization of *Thiothrix* spp. (Polz et al., 1996). However, features considered diagnostic for *Thiothrix* spp. (holdfast and rosette formation and the production of gliding gonidia) are not exclusive as may be noted for *Leucothrix mucor* (Williams and Unz, 1985b; Brock, 1992) and the Eikelboom type 021N bacteria associated with bulking activated sludge. Woese (1987) has stressed that morphological characteristics are of little value as indicators of phylogenetic relationships among a majority of bacteria and axenic cultures of *Thiothrix* species have been observed to undergo change in morphology of trichomes (Shuttleworth and Unz, 1991, 1993; Brigmon et al., 1995). Despite extensive use of *in situ* morphology and staining reactions to identify filamentous bacteria associated with activated sludge bulking and foaming (Eikelboom, 1975; Jenkins et al., 1993), these characteristics can vary depending on environmental conditions and are singularly of limited diagnostic value. Furthermore, bacterial morphotypes termed *"Leucothrix cohaerens"* (Pringsheim, 1957; Cyrus and Sladka, 1970) and *Thiothrix* sp. forms I and II (Farquhar and Boyle, 1971) have been equated with Eikelboom type 021N bacteria (Eikelboom, 1975). The taxonomic position of the polymorphic Eikelboom type 021N bacteria and their relationship to the genus *Thiothrix* has been questioned (Williams and Unz, 1985a, 1989; Ziegler et al., 1990). In recognition of the limitations of morphological characterization in the taxonomy of many filamentous bacteria, molecular biological approaches based on comparative 16S rRNA analyses have been used in the attempt to clarify relationships among such bacterial morphotypes.

Analysis of 16S rRNA sequences derived from six *Thiothrix* spp. (including one *Thiothrix* sp. form I) and 8 Eikelboom type 021N bacteria revealed that they fell within a single monophyletic

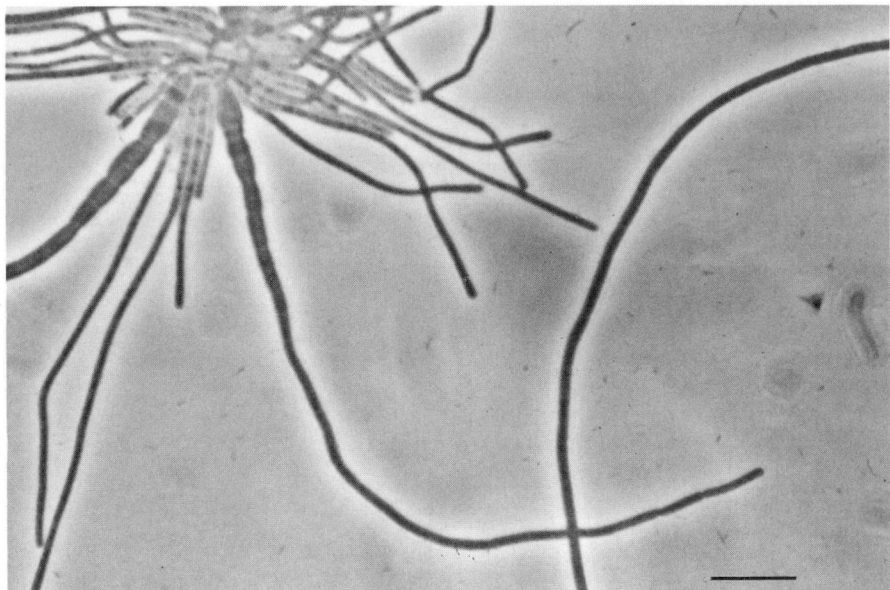

FIGURE BXII.γ.45. Base to tip differentiation in cell dimensions in a trichome of *Thiothrix fructosivorans*. Bar = 10 μm. (Reproduced with permission from T.M. Williams and R.F. Unz, Applied and Environmental Microbiology *49*: 887–898, 1985, ©American Society for Microbiology.)

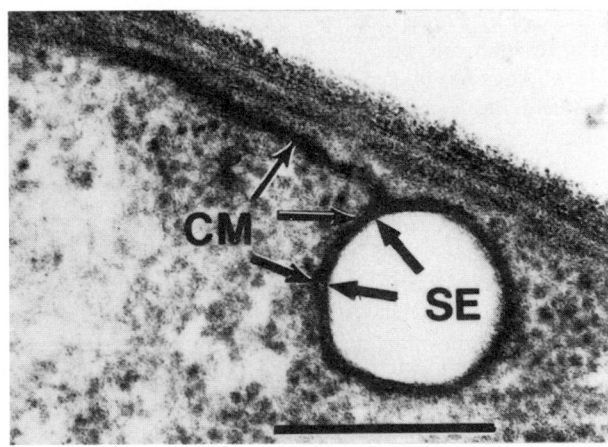

FIGURE BXII.γ.46. Sulfur inclusion in invaginated cytoplasmic membrane of a cell of *Thiothrix eikelboomii*. Abbreviations: *CM*, cytoplasmic membrane; *SE*, sulfur inclusion envelope. Bar = 250 nm. (Reproduced with permission from T.M. Williams et al., Applied and Environmental Microbiology *53*: 1560–1570, 1987, © American Society for Microbiology.)

group suggesting a clear relationship between Eikelboom type 021N and *Thiothrix* spp. (Howarth et al., 1999). Consequently, four new *Thiothrix* species have been validly described with sequence identities below 97.5% (Howarth et al., 1999) and include strains categorized as *incertae sedis* in the previous edition of the *Manual*. Strains previously categorized as Eikelboom type 021N are now accommodated within the species, *T. eikelboomii*, of which strain AP3 (ATCC 49788) is the type strain and a single *Thiothrix* sp. form I is named *T. defluvii* (type strain Ben57). *T. unzii* is represented by type strain A1 (ATCC 49747) that, like *T. nivea*, requires a source of reduced sulfur and organic carbon for growth. *T. fructosivorans* is a chemoorganotrophic species rep-

resented by the type strain Q (ATCC 49748) and strain I (ATCC 49749), which is essentially identical to the type strain. Despite the close phylogenetic relationship between *T. fructosivorans*, "*T. ramosa*", and *Thiothrix* sp. CT3 (>97% 16S rRNA sequence identity), *T. fructosivorans* is not known to be capable of autotrophic growth, notwithstanding evidence of RubisCo activity (Unz and Williams, 1989), whereas "*T. ramosa*" and *Thiothrix* sp. CT3 both grow autotrophically (Odintsova et al., 1993; Tandoi et al., 1994). Analysis of 16S rDNA sequences from a further 15 isolates, identified morphologically as Eikelboom type 021N, revealed the organisms to be most closely related to *Thiothrix* species (Fig. BXII.γ.51), although considerable additional phylogenetic diversity was discovered (Kanagawa et al., 2000). In addition to organisms related to *T. nivea* (group 0), several isolates related to *T. eikelboomii* (group II sensu Kanagawa et al., 2000), a completely distinct group (group I sensu Kanagawa et al., 2000), and a group related to, but significantly divergent from, *T. defluvii* (group III sensu Kanagawa et al., 2000) were resolved. Although these isolates showed a considerable degree of divergence in their 16S rRNA sequences, they had very homogeneous mol% G + C content (43.9–46.1) and all shared a characteristic deletion in helix 18 of their ribosomal RNA (positions 455–477 by *E. coli* numbering), which further supports their relationship to named *Thiothrix* spp. (Howarth et al., 1999; Kanagawa et al., 2000). Accordingly, groups I and III (Kanagawa et al., 2000) should be designated as new species.

During the preparation of this chapter the group of Takahiro Kanagawa described groups I and III as two new *Thiothrix* species: *Thiothrix disciformis* and *Thiothrix flexilis* (Aruga et al., 2002). *Thiothrix disciformis* corresponds to group I in Fig. BXII.γ.51. B3-1 is the type strain. *Thiothrix flexilis* corresponds to group III in Fig. BXII.γ.51 and EJ2M-B is the type strain. These organisms are morphologically recognizable as Eikelboom Type 021N and filaments exhibit considerable variability in cell morphology with discoid and ovoid cells often being present, as are elongate and

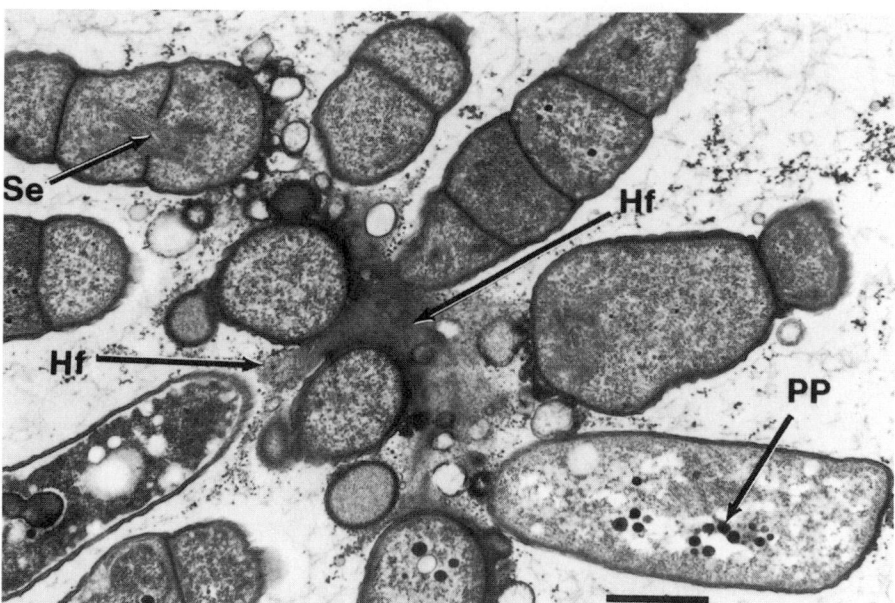

FIGURE BXII.γ.47. Thin section through rosette of *Thiothrix eikelboomii*. Abbreviations: *Hf*, holdfast material; *PP*, polyphosphate; *Se*, cell septum. Bar = 1.0 μm. (Reproduced with permission from T.M. Williams et al., Applied and Environmental Microbiology *53:* 1560–1570, 1987, ©American Society for Microbiology.)

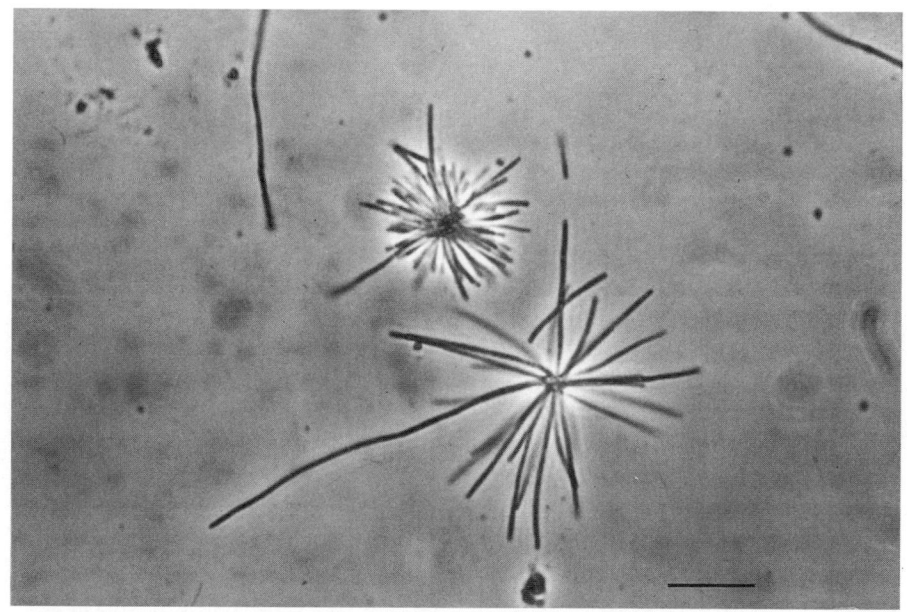

FIGURE BXII.γ.48. Rosettes of trichomes representative of *Thiothrix* spp. *Thiothrix eikelboomii*. Bar = 10 μm. (Reproduced with permission from T.M. Williams and R.F. Unz, Applied and Environmental Microbiology *49:* 887–898, 1985, ©American Society for Microbiology.)

swollen cells. Although elemental sulfur is deposited, reduced sulfur is not required for growth. Phenotypically the species are quite similar but show differences in carbon source utilization, the extent of sulfur deposition, nitrate reducing ability, and cell dimensions. They are, nevertheless, phylogenetically very distinct (Fig. BXII.γ.51). Readers are directed to the original publication for details of the phenotypic and genotypic analyses of these newly described species (Aruga et al., 2002).

The different phylogenetic groupings within the genus *Thiothrix* cannot be distinguished morphologically (Kanagawa et al., 2000). This further calls into question the validity of identification of trichome-forming microorganisms solely on morphological criteria. This is consistent with observations made using fluorescence *in situ* hybridization (FISH) where not all organisms morphologically identified as *Thiothrix* spp. hybridized with probes targeting *T. nivea* or Eikelboom type 021N bacteria (Per-

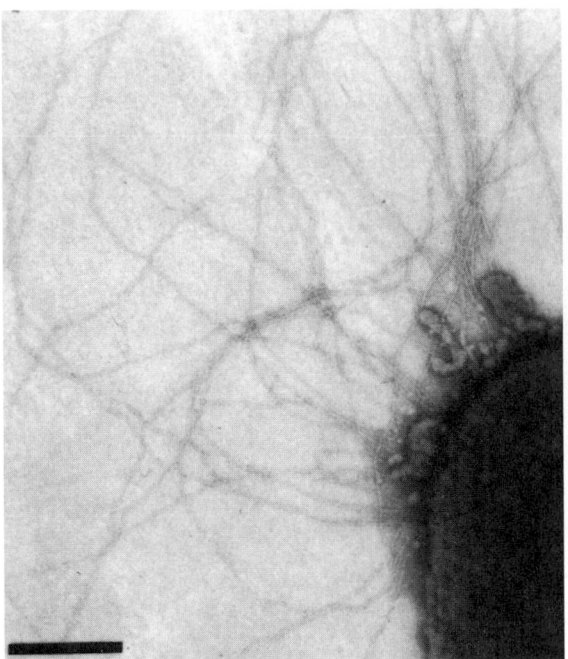

FIGURE BXII.γ.49. Fimbriae of a gonidium of *Thiothrix unzii*. Negative stained preparation. Bar = 0.5 μm. (Reproduced with permission from T.M. Williams et al., Applied and Environmental Microbiology *53:* 1560–1570, 1987, ©American Society for Microbiology.)

nelle et al., 1998). Also, trichomes morphologically identified as Eikelboom type 021N may bind with TNI probe specific for *T. nivea* but not an Eikelboom type 021N-specific probe (Nielsen et al., 1998), although results may, in part, be due to the poor specificity of the probes that were designed based on limited available sequence data. Nevertheless, some organisms with *Thiothrix*/Eikelboom type 021N-like morphology have proven to be phylogenetically very distantly related to *Thiothrix* spp., e.g., strain Ben47, which is most closely related to the *Alphaproteobacteria* (Howarth et al., 1999). A series of probes, designed for use in FISH or other rRNA hybridization assays to identify different groups of *Thiothrix* in environmental samples, are listed with their specificity indicated in Table BXII.γ.36. Readers are directed to the original publication for detailed information on the application of these diagnostic oligonucleotide probes (Kanagawa et al., 2000). Characteristics distinguishing *Thiothrix* from morphologically similar bacteria are listed in Table BXII.γ.37.

ACKNOWLEDGMENTS

The authors acknowledge the original contributions of John Larkin contained in the contents of this writing and are most grateful to Takahiro Kanagawa for providing data on *Thiothrix disciformis* and *Thiothrix flexilis* prior to publication.

FURTHER READING

Larkin, J.M. and W.R. Strohl. 1983. *Beggiatoa, Thiothrix,* and *Thioploca.* Annu. Rev. Microbiol. *37:* 341–367.

List of species of the genus Thiothrix

1. **Thiothrix nivea** (Rabenhorst 1865) Winogradsky 1888, 39[AL]
   (*Beggiatoa nivea* Rabenhorst 1865, 94.)
   ni′ve.a. L. adj. *nivea* snow-white.

Rods, about 1.0–1.5 μm in diameter, seriate in multicellular filaments (trichomes) of uniform diameter. Gram negative. Trichomes clearly ensheathed. Optimum temperature

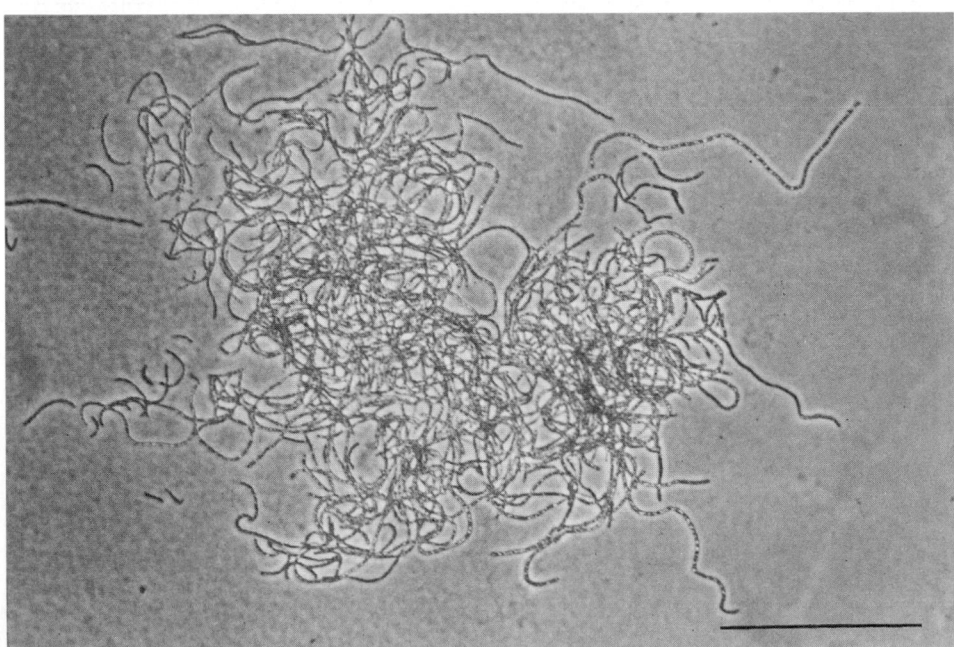

FIGURE BXII.γ.50. Colony morphology exemplified by *T. unzii*. LT medium. Bar = 60 μm. (Reproduced with permission from T.M. Williams and R.F. Unz, Applied and Environmental Microbiology *49:* 887–898, 1985, ©American Society for Microbiology.)

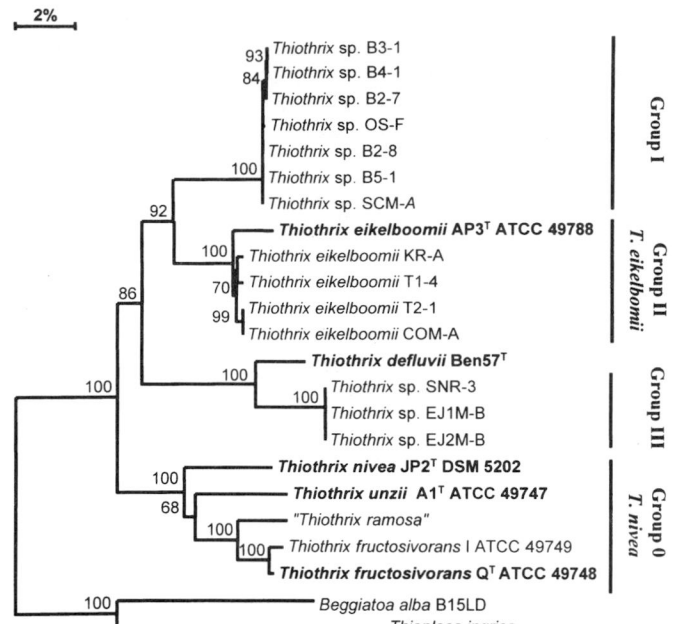

2%

93 Thiothrix sp. B3-1
84 Thiothrix sp. B4-1
Thiothrix sp. B2-7
Thiothrix sp. OS-F
Thiothrix sp. B2-8
100 Thiothrix sp. B5-1
Thiothrix sp. SCM-A
Group I

92
100 **Thiothrix eikelboomii AP3ᵀ ATCC 49788**
Thiothrix eikelboomii KR-A
70 Thiothrix eikelboomii T1-4
99 Thiothrix eikelboomii T2-1
Thiothrix eikelboomii COM-A
Group II
T. eikelboomii

86
100 **Thiothrix defluvii Ben57ᵀ**
100 Thiothrix sp. SNR-3
Thiothrix sp. EJ1M-B
Thiothrix sp. EJ2M-B
Group III

100
100 **Thiothrix nivea JP2ᵀ DSM 5202**
68 **Thiothrix unzii A1ᵀ ATCC 49747**
100 "Thiothrix ramosa"
100 Thiothrix fructosivorans I ATCC 49749
Thiothrix fructosivorans Qᵀ ATCC 49748
Group 0
T. nivea

100
100 Beggiatoa alba B15LD
Thioploca ingrica

FIGURE BXII.γ.51. Phylogenetic distance tree derived from 16S rRNA sequences of *Thiothrix* spp. The tree was generated from an alignment comprising 1148 unambiguously aligned nucleotides. Figures at nodes represent percentage bootstrap values. Bootstrap values below 50% have been omitted. Sequences from *Beggiatoa alba* B15LD and *Thioploca ingrica* were used as an outgroup to root the tree. The type strains of all validly named species are shown in bold typeface. Bar = 2 nucleotide changes per 100 nucleotides in the 16S rRNA sequence.

is 25–30°C; maximum about 32–34°C; minimum about 6–8°C. Mixotrophic. Additional characteristics of the neotype strain, which are representative of the species, appear in the description of the genus and the data in Table BXII.γ.38.

The mol% G + C of the DNA is: 52 (T_m).

Type strain: JP2, DSM 5205.

GenBank accession number (16S rRNA): L40993.

2. **Thiothrix defluvii** Howarth, Unz, Seviour, Blackall, Pickup, Jones, Yaguchi and Head 1999, 1824^VP

de.flu′vi.i. L. neut. n. *defluvium* sewage; L. gen. n. *defluvii* of sewage.

Rods, variable in shape (cylindrical, barrel-shaped, frequently elongate and swollen) in multicellular trichomes with base to tip differentiation; apical cells are 1.0–2.0 μm in diameter and 5.0–10.0 μm in length, whereas cells at the base of trichomes are 2.0–4.0 μm in diameter and 0.5–10 μm in length. Trichomes have no sheath but may form knots; rosettes, holdfasts, and gliding gonidia present in some strains. Gram negative or Gram variable. Growth occurs in the temperature range of 10–30°C but not at 4°C or 37°C. Isolates are extremely slow growing and biochemical properties of this organism have not been determined.

Isolated from activated sludge treatment plant located in Australia

The mol% G + C of the DNA is: unknown.

Type strain: Ben57.

GenBank accession number (16S rRNA): AF127020.

3. **Thiothrix disciformis** Aruga, Kamagata, Kohno, Hanada, Nakamura and Kanagawa 2002, 1315^VP

dis.ci.for′mis. L. masc. n. *discus* a disc; L. n. *forma* shape; N.L. adj. *disciformis* disc-shaped, the main cell form.

Readers are referred to the original publication for a detailed description of this species (Aruga et al., 2002).

The mol% G + C of the DNA is: 43.9–44.7 (HPLC).

TABLE BXII.γ.36. Diagnostic oligonucleotide probes for characterization of *Thiothrix* spp.[a]

Probe	Target Group	Sequence (5′ → 3′)	Target site (16S rRNA, *E. coli* numbering)
G1B	Group I	TGTGTTCGAGTTCCTTGC	1029–1046
G2M	Group II	GCACCACCGACCCCTTAG	842–859
G3M	Group III[b]	CTCAGGGATTCCTGCCAT	996–1013
G123T	Group 0-III	CCTTCCGATCTCTATGCA	697–714

[a]Data from Kanagawa et al., 2000.

[b]In addition to group III organisms *T. eikelboomii* AP3 also contains the target site for this probe.

TABLE BXII.γ.37. Characteristics used to differentiate *Thiothrix* spp. from related, morphologically similar genera[a,b]

Characteristic	*Thiothrix*	*Beggiatoa*	*Leucothrix*	*Thioploca*
Colorless	+	+	+	+
Strains deposit elemental sulfur	+	+	–[c]	+
Sheath	+	–	–	+
Holdfast	+	–	+	–
Gliding gonidia	+	–	+	–
Filament formation	+	+	+	+
Rosette formation	+	–	+	–
Swollen cells produced in filaments	+	–	+	–
Knots produced	+	–	+	–
mol% G + C	43–53	35–39	46–51	nd
Deletion in helix 18 of 16S rRNA	+	–	–	–

[a]Note that these characteristics are illustrative only and many isolates may not exhibit all of these characteristics.

[b]For symbols see standard definitions; nd, not determined.

[c]Most reports suggest that *Leucothrix* spp. do not deposit intracellular sulfur; an organism described as "*Leucothrix thiophila*" has been shown to grow lithoheterotrophically and deposit elemental sulfur (Dul′tseva et al., 1996).

TABLE BXII.γ.38. Characteristics of the type strains of species of the genus *Thiothrix* [a]

Characteristic	1. *T. nivea*[b] DSM 5205 (JP2)	2. *T. eikelboomii*[c] ATCC 49788 (AP3)	3. *T. fructosivorans*[c] ATCC 49748 (Q)	4. *T. unzii*[c] ATCC 49747 (A1)
Utilization as sole carbon source: [d,e]				
Acetate, malate, oxalacetate, pyruvate	+	+	+	+ (w, malate)
Formate, fumarate, lactate, propionate, succinate	−	+	+	+
α-Ketoglutarate, butyrate, hydroxybutyrate, valerate, benzoate		+	−	−
D-Fructose	−	+	+	−
D-Glucose		+	−	−
D-Galactose		+	−	−
D-Mannose	−	w	−	−
Sucrose	−	+	+	−
D-Maltose	−	+	−	−
D-Trehalose		+	−	−
D-Melezitose		+	−	−
D-Raffinose		w	−	−
Amygdalin, salicin		+	−	−
D-Mannitol,		+	−	−
i-Inositol,		w	−	−
D-Sorbitol		w	−	−
Utilization as carbon and nitrogen source: [d,f]				
L-Glutamate	−	+	−	−
L-Tyrosine, L-histidine, DL-alanine, Casamino acids		+	−	−
Utilization as sole nitrogen source: [d,g]				
DL-Tryptophan, DL-aspartate, DL-glutamine, L-asparagine, L-glutamate, Casamino acids		+	+	+
Hydroxy-L-proline		−	−	+
DL-Methionine, DL-valine, DL-isoleucine		+	−	+
L-Tyrosine, L-histidine		+	−	w
L-Cysteine, L-proline		+	+	−
L-Arginine, L-leucine, DL-phenylalanine		+	−	−
Glucosamine		−	+	+
Urea		+	+	+
Ammonia	+	+	+	+
Nitrate	+	−	+	+
NO$_3^-$ → NO$_2^-$ only	+	+	+	+
Utilitization of orthophosphate and Na$_5$P$_3$O$_{10}$		+	+	+
Polyphosphate stored		+	+	+
Sudanophilic granules formed		+	+	+
Oxidase	+	+	+	+
Catalase	−	−	+	−
Caseinase		−	−	−
Gelatinase		+	+	+
Sulfide or thiosulfate required sulfur source	+	−	−	+

[a]Symbols: See standard definitions; w, weak.

[b]Data obtained from Larkin (1989).

[c]Data obtained from Williams (1985).

[d]With sulfide or thiosulfate present in the case of *T. nivea* and *T. unzii.*

[e]Carbon sources not utilized by type strains in addition to those indicated in table: *T. nivea* (citrate, *cis*-aconitate, phosphoenolpyruvate, carbonate, glyoxylate, methylamine, D-ribose, D-arabinose, D-cellobiose, L-rhamnose, D-raffinose, D-sorbose, D-melebiose, D-xylose, ethanol, methanol, propanol, glycerol, i-erythritol); *T. unzii, T. fructosivorans,* and *T. eikelboomii* (citrate, L-rhamnose, D-xylose, L-fucose, D-ribose, mannoheptulose, L-sorbose, D-arabinose, glyceraldehyde phosphate, D-melibiose, D-gentiobiose, D-cellobiose, starch, inulin, esculin, glycerol, erythritol, galacturonic acid, glucuronic acid, gluconic acid, tributyrin, Tween 80, ethanol, *n*-propanol, *n*-butanol, *n*-amyl alcohol, methanol, isoamyl alcohol, phenol, *m*-toluate, glycolate, oleate, palmitate, myristate).

[f]Carbon and nitrogen sources not utilized by type strains in addition to those indicated in table: *T. unzii, T. fructosivorans,* and *T. eikelboomii* (DL-lysine, DL-cystine, DL-threonine, hydroxy-L-proline, glycine, L-arginine, DL-tryptophan, L-cysteine, L-proline, DL-phenylalanine, DL-serine, DL-methionine, DL-valine, DL-aspartate, DL-glutamine, L-asparagine, DL-isoleucine, L-leucine).

[g]Nitrogen sources not utilized by type strains in addition to those indicated in table: *T. nivea* (nitrite, dinitrogen); *T. unzii, T. fructosivorans,* and *T. eikelboomii* (nitrite, L-lysine, L-cystine, DL-threonine, glycine, DL-serine, DL-alanine).

Type strain: B3-1, DSM14473, JCM11364.
GenBank accession number (16S rRNA): AB042532.

4. **Thiothrix eikelboomii** Howarth, Unz, Seviour, Blackall, Pickup, Jones, Yaguchi and Head 1999, 1825[VP]

ei.kel.boom' i.i. M.L. gen. n. *eikelboomii* of Eikelboom; named for D.H. Eikelboom, who pioneered morphological identification of filamentous bacteria in wastewater.

Rods may vary in shape (cuboidal, barrel-shaped, cylindrical, discoid, bead-like) and size depending on location in multicellular trichomes; apical cells are 0.6–0.8 × 1.0–1.5 μm and frequently bead-like, whereas cells at the base of trichomes are 1.0–3.0 μm in diameter and 0.4–0.7 μm in length and discoid. Trichomes have no sheath but may form knots; rosettes and a holdfast present in some strains.

The type strain does not form rosettes. Gliding gonidia occur only in rosette-forming strains and have a tuft of monopolar fimbriae but lack flagella. Gram negative or Gram variable. Growth occurs in the temperature range of 10–33°C but not at 37°C and in the pH range of 6.5–8.5. Chemoorganotrophic. Additional characteristics of the type strain, which are representative of the species, are given in Table BXII.γ.38.

Isolated from activated sludge treating domestic wastewater.

The mol% G + C of the DNA is: 44.1–46.1 (HPLC).
Type strain: AP3, ATCC 49788.
GenBank accession number (16S rRNA): AB042819.

5. **Thiothrix flexilis** Aruga, Kamagata, Kohno, Hanada, Nakamura and Kanagawa 2002, 1315[VP]

fle.xi' lis. L. adj. *flexilis* pliable.

Readers are referred to the original publication for a detailed description of this species (Aruga et al., 2002).
The mol% G + C of the DNA is: 44.0–44.4 (HPLC).
Type strain: EJ2M-B, DSM14609, JCM11135.
GenBank accession number (16S rRNA): AB042545.

6. **Thiothrix fructosivorans** Howarth, Unz, Seviour, Blackall, Pickup, Jones, Yaguchi and Head 1999, 1824[VP]

fruc.to.si.vo' rans. M.L. neut. n. *fructosum* fructose; L. part. pres. *vorans* eating; M.L. adj. *fructosivorans* fructose-eating.

Rods, 1.2–2.5 × 2.7–4.5 µm, in multicellular trichomes. Gram negative, trichomes ensheathed, rosettes formed, holdfast present, gliding gonidia with monopolar fimbriae but no flagella. Growth occurs in the temperature range of 4–28°C but not at 33°C and in the pH range of 6.5–8.5. Chemoorganotrophic. Additional characteristics of the type strain, which are representative of the species, appear in Table BXII.γ.38.

Isolated from activated sludge treating domestic wastewater.

The mol% G + C of the DNA is: unknown.
Type strain: Q, ATCC 49748.
GenBank accession number (16S rRNA): L79962.

7. **Thiothrix unzii** Howarth, Unz, Seviour, Blackall, Pickup, Jones, Yaguchi and Head 1999, 1824[VP]

un' zi.i. M.L. gen. n. *unzii* of Unz; named for R.F. Unz.

Rods, 0.7–1.5 × 1.5–3.0 µm, in multicellular trichomes. Gram negative, trichomes without sheaths, rosettes formed, holdfast present, gliding gonidia with monopolar fimbriae but no flagella. Growth occurs in the temperature range of 4–33°C but not at 37°C and in the pH range of 6.5–8.5. Mixotrophic. Additional characteristics of the type strain, which are representative of the species, appear in Table BXII.γ.38.

Isolated from activated sludge treating domestic wastewater.

The mol% G + C of the DNA is: unknown.
Type strain: A1, ATCC 49747.
GenBank accession number (16S rRNA): L79961.

Other Organisms

Two species of *Thiothrix*, "*T. ramosa*" (Odintsova and Dubinina, 1990) and "*T. arctophila*" (Dul' tseva and Dubinina, 1994) are proposed but invalidly published. The general morphology of the illegitimate species befits the traditional description of the genus *Thiothrix* with respect to presence of nonmotile, multicellular trichomes and the ability to form gonidia and rosettes. However, in addition to gonidia, both species form gliding hormogonia through fragmentation of trichomes, and the hormogonia of "*T. ramosa*" may develop into pseudobranching filaments. Further, "*T. ramosa*" exhibits chemolithoautotrophic growth with thiosulfate or carbon disulfide as sole electron donors, mixotrophic growth on lactate and any of several organic sulfur compounds, including carbon disulfide, and chemoorganotrophic growth on lactate alone (Odintsova et al., 1993). Additional information on these organisms and other *Thiothrix* isolates is given in the following paragraphs.

1. "*T. ramosa*"

ra.mo' sa. L. n. *ramosum* branch.

Three strains, namely TA, T1, and T2, are known. Cells 0.7–2.5 × 2.5–5.0 µm in ensheathed trichomes up to 100 µm in length. Gram negative. Intracellular sulfur inclusions formed in media containing thiosulfate or sulfide; however, reduced sulfur is not required for growth. Strictly aerobic. Optimum growth temperature is 28°C with sparse growth at 20°C and no growth at 4°C. Optimum growth pH is 7.2–7.3. All strains utilize acetate, malate, succinate, fumarate, and oxalacetate, whereas isocitrate, glycolate, lactate, α-ketoglutarate, and mannitol utilization varies among strains. Sugars, alcohols, and amino acids are not utilized with the exception of mannitol by strain T2. No growth on peptone, gelatin, casein, or yeast extract. Vitamins are required for growth. Nitrogen sources for all strains include ammonium salts, peptone, and glutamine; asparagine and aspartate are utilized by strain T1. Nitrates are not reduced to nitrite. Catalase and oxidase positive. Obtained from a hydrogen sulfide spring in Latvia.

The mol% G + C of the DNA is: 48–52.4.

2. "*Thiothrix arctophila*"

arcto' phi' la. L. adj. *arcto* cool; Gr. adj. *philus* loving; M.L. *arctophila* cool loving.

Only strain IN is known. Cells 0.7–2.6 × 3.0–5.0 µm in trichomes of variable length covered with a protein–polysaccharide sheath. Gram negative. Intracellular sulfur inclusions formed in media containing thiosulfate, sulfide, tetrathionate, or elemental sulfur. Strictly aerobic. Optimum growth temperature is 15°C with no growth at 3°C and 34°C. Optimum growth pH is 7.2. Carbon sources supporting growth are acetate, malate, pyruvate, succinate, fumarate, lactate, oxalacetate, propionate, maltose, melezitose, and butanol. Compounds not supporting growth are malonate, citrate, glyoxylate, formate, benzoate, caproate, glycolate, α-ketoglutarate, carbonate, glucose, lactose, fructose, sucrose, ribose, mannose, arabinose, cellobiose, rhamnose, raffinose, xylose, galactose, trehalose, gentiobiose, melibiose, sorbitol, methanol, ethanol, propanol, asparagine, methionine, serine, glutamate, alanine, aspartate, methylamine, glycerol, cysteine, mannitol, peptone, yeast extract, starch, Tween 80, urea, and phenol. Indole not produced from tryptophan. Nitrogen sources utilized are ammonia, nitrate, nitrite, glutamine, casein hydrolysate, and peptone. Nitrogen sources not utilized are valine, gly-cine, phenylalanine, lysine, leucine, methionine, proline, cysteine, asparagine, aspartate, glutamate, urea, and dinitro-gen. Nitrate is reduced to nitrite. Vitamins are required for growth. Nitrogen sources for all strains include ammonium salts, peptone, and glutamine; asparagine and aspartate are utilized by strain T1. Nitrates are not reduced to nitrite. Catalase negative and oxidase positive. Obtained from a combined wastewater treatment plant.

The mol% G + C of the DNA is: 55.0 (T_m).

3. *Thiothrix* CT3.

Thiothrix CT3 is a poorly characterized isolate recovered from an activated sludge in Italy and reported to be fac-ultatively chemolithoautotrophic on thiosulfate, mixo-trophic on acetate and thiosulfate, and chemoorgano-trophic on acetate (Tandoi et al., 1994).

Several isolates of *Thiothrix* and Eikelboom type 021N, in ad-dition to the type strains, were isolated and studied in the course of investigations leading to the naming of the species *T. unzii*, *T. fructosivorans*, and *T. eikelboomii*. Some of these may no longer be extant; however, certain ones are referred to in publications by other workers and are described in the last edi-tion of the *Manual*. Strain A3 is essentially identical to the type strain of *T. unzii* and strain I (ATCC 49749) is essentially identical to the type strain of *T. fructosivorans*. Isolates TH1 and TH3 share many features of *T. fructosivorans* but are distinguished by the inability of the latter strain to utilize as many organic carbon compounds or any carbohydrates, ability to utilize several more amino acids and nitrate as nitrogen sources, display cell dimen-sions similar to *T. unzii*, require sodium ion for growth, and tolerate relatively high (3–4%) sodium chloride concentrations. A large number of Eikelboom type 021 N isolates were included in studies on the type strain of *T. eikelboomii* (Williams, 1985; Williams and Unz, 1985a, 1989), and notwithstanding minor morphological and nutritional differences, appear to befit the description of *T. eikelboomii*.

Genus II. **Achromatium** *Schewiakoff 1893, 1*[AL]

HANS-DIETRICH BABENZIEN, FRANK OLIVER GLÖCKNER AND IAN M. HEAD

A.chro.ma' ti.um. Gr. pref. *a* not; Gr. n. *chromatium* color, paint; M.L. neut. n. *Achromatium* that which is not colored.

Spherical to ovoid or cylindrical unicells with hemispherical ends. In natural populations, cells of widely different sizes appear. Cells vary in size from spheres of about 10 μm in diameter to giant cylindrical forms up to 30 × 125 μm (extremes are dividing stages); **cells are on average 15 × 50 μm.** Many size classes may be present within a sample. However, genetically different sub-populations that overlap in their dimensions can occur in a single habitat. **Division of cells by binary constriction. Gram negative. Calcium carbonate (calcite) is deposited intracellularly.** Addi-tionally, the **cells contain globules of S⁰.**

Movement, if any, of a peculiar slow, rolling, jerky type or gliding on solid surfaces. Not motile by means of flagella. The **cell surface** is covered by a glycocalyx. **Catalase negative.** Pure cultures not available.

The **natural habitat is the oxic–anoxic boundary** in the top 0–5 cm of sediments, suggesting that the cells may use **oxygen as the electron acceptor.** The habitats for *Achromatium* include different aquatic ecosystems around the world, e.g., lakes, ponds, wetlands, bog lakes, rivers, saline environments, brackish waters, and marine environments. **Resting stages are not known.**

Phylogenetic analyses including sequences from different ge-ographical locations confirm that *Achromatium* (e.g., *A. oxaliferum* **populations) forms a distinct, deeply branching monophyletic clade in the** *Gammaproteobacteria* (Fig. BXII.γ.52).

The topology of the clade indicates that beyond the two main clusters, A and B, at least three genetically distinct lineages exist (Fig. BXII.γ.52), and it is likely that several *Achromatium* spp. remain to be described. Several 16S rRNA-targeted oligonucle-otide probes are available for *in situ* identification and differ-entiation of co-existing *Achromatium* spp. by techniques like flu-orescence *in situ* hybridization (FISH).

The type species was originally found in the mud of the Neu-hofer Altrhein, Germany.

The mol% G + C of the DNA is: not known.

Type species: **Achromatium oxaliferum** Schewiakoff 1893, 1.

FURTHER DESCRIPTIVE INFORMATION

From experiments with cell enrichments, it is now apparent that members of the genus encompass more than one physiological type. Some of them are capable of oxidizing reduced sulfur spe-cies to sulfate, suggesting that they obtain energy from this pro-cess, in which S⁰ is deposited as an intracellular intermediate. The substrate spectrum appears to vary in different populations, and substrate utilization may involve a chemolithoautotrophic potential (assimilation of ¹⁴C-labeled bicarbonate) or the uptake of organic C-compounds (e.g., acetate, protein hydrolysate) by a chemoorganoheterotrophic, chemolithoheterotrophic, or mix-otrophic type of metabolism. Like the genus *Beggiatoa*, to which it is related, *Achromatium* appears to consist of a nutritionally heterogeneous group of bacteria. Without pure cultures, it is not possible to determine whether this physiological diversity exists within a single population, with induction or repression of dif-ferent metabolic pathways under different environmental con-ditions, or if it instead corresponds with genetic heterogeneity. Twenty nearly full-length 16S rRNA sequences and three partial sequences, which were amplified from DNA extracted from highly purified cell enrichments from fresh water sediments of various habitats, have been deposited in public databases. The accession numbers are L42543 and L48222–L48227 (Head et al., 1996), AJ010593–AJ010596 (Glöckner et al., 1999) and AF129548–AF129559 (Gray et al., 1999).

The prevailing temperature at their natural habitats is around 10°C.

The use of comparative 16S rRNA analysis has clarified the position of *Achromatium* and has shown that natural communities of *Achromatium oxaliferum* comprise genetically distinct subpop-ulations. With this information, the physiological diversity of *Achromatium* spp. becomes understandable.

The *Achromatium oxaliferum*-derived sequences are consistently recovered with the *Chromatium* assemblage but are without strong

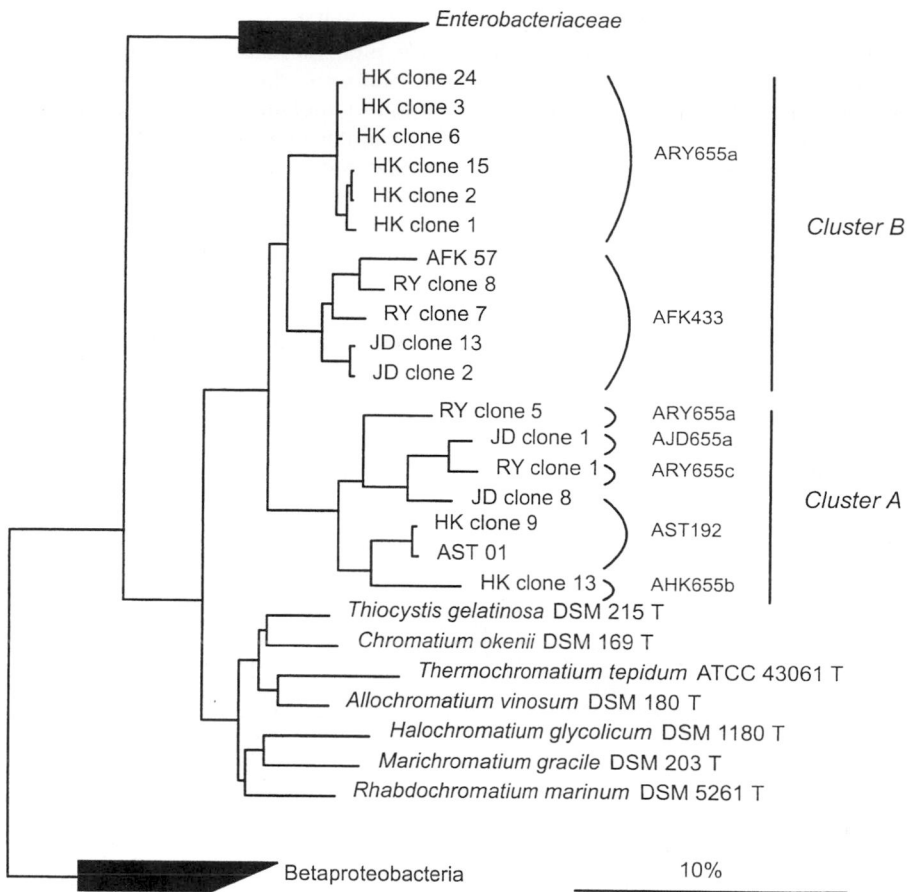

FIGURE BXII.γ.52. Phylogenetic tree based on comparative analysis of the 16S rRNA genes from all currently available full-length sequences from *Achromatium* and from selected sequences among the *Betaproteobacteria* and *Gammaproteobacteria*. The tree is a consensus tree based on maximum likelihood analysis, with highly variable positions excluded from the calculations. Sequences were obtained from *Achromatium* cells purified from different locations, including Hell Kettles (Co. Durham, UK), Rydal Water (Cumbria, UK), Jenny Dam (Cumbria, UK), Lake Fuchskuhle (Brandenburg, Germany), and Lake Stechlin (Brandenburg, Germany). These locations are designated with the suffixes HK, RY, JD, AFK, and AST, respectively. It has been proposed that the *Achromatium* spp. obtained from Lake Fuchskuhle (sequence AFK 57) is a new species *Candidatus* Achromatium minus. *T*, type strains. Bar = 10% estimated sequence divergence.

affiliation with any of the recognized lineages within this group. *A. oxaliferum* represents a novel clade within the *Gammaproteobacteria* and is therefore unrelated to the gliding bacteria of the *Cytophaga–Flavobacterium* group, as had been inferred prior to recent phylogenetic studies (La Riviere and Schmidt, 1989). 16S rRNA-sequences recovered from *Achromatium* cells from several locations have demonstrated that within the clade, two distinct phylogenetic clusters, termed cluster A and cluster B, are present. This bifurcation is strongly supported by distance analysis and maximum likelihood analysis (P <0.01). Support for cluster A by parsimony analysis is also high, while support for cluster B is less pronounced. The two clusters can also be distinguished based on secondary structure elements (helical motifs) in the V6 region of the 16S rRNA molecule. Sequences from cluster A have a characteristic deletion and lack helix 38 (1024–1036, *E. coli* numbering), as do sequences from a number of *Archaea* and *Eucarya*. In contrast, sequences in cluster B have a V6 region that is in accordance with that of the majority of the domain *Bacteria* (Gray et al., 1999; Glöckner et al., 1999). Cluster B comprises two genetically different subgroups.

Whole-cell hybridization with fluorescently-labeled oligonu-

cleotide probes specific for the different *Achromatium*-derived sequences has clearly demonstrated the co-existence of distinct subpopulations. In Lake Stechlin, Germany, a ratio of about 2:1:1 has been found for three different *Achromatium* populations identified by FISH, with the organism represented by 16S rDNA clone AST 01 accounting for nearly 50% of the total population.

In the habitats investigated using FISH, Lake Fuchskuhle (an acidic bog lake in Germany) was the only lake found to harbor a morphologically and genetically homogeneous population of *Achromatium*. Phylogenetic analysis of the 16S rDNA has assigned this population to cluster B of the *Achromatium* clade. Based on the unique features of the habitat, the homogeneous morphology and cell size, and the phylogenetic position of the population, a new species *Candidatus* Achromatium minus has been proposed (Figs. BXII.γ.52 and BXII.γ.54).

Unfortunately, until now there have been no probes available that cover the whole *Achromatium* clade or allow an exact separation of the two clusters by FISH. With the six probes shown in Fig. BXII.γ.52, it is possible to distinguish several subclusters and clones or, when probes are used in combination, to detect all currently available *Achromatium* sequences. Furthermore, with probes ARY655a and AFK433, it should be possible to distinguish cluster B from cluster A, excluding clone 5 derived from Rydal Water, UK. In addition, the combination of probes AJD655a, ARY655c, AST192, and AHK655b is specific for cluster A, again with the exception of clone 5 from Rydal Water. All probes should work under the same conditions and must be applied with appropriate competitor probes as described in Glöckner et al. (1999) and Gray et al. (1999).

Morphological heterogeneity with extreme differences in cell size (Figs. BXII.γ.53 and BXII.γ.54) has been noted by all investigators, but evidence that differently sized cells represent genetically different populations has been obtained only recently. Different sub-populations coexist within the same habitat (Head et al., 2000b).

The cell volume can reach 50,000–80,000 μm^3, and the abundance of *Achromatium* cells in sediments is in the range of 10^3–10^5 cells/cm^3. Because of the large size of individual cells, *Achromatium* populations can be the major component of the bacterial community in some surficial sediments in terms of biomass. Cells divide by constriction in the middle, and before separation takes place, the daughter cells are connected by a hyaline "bridge" (Fig. BXII.γ.55). Often, *Achromatium* populations share the same habitat with other sulfide-oxidizing bacteria, like *Beggiatoa* spp. (Babenzien, 1991).

The most striking feature of the cells is the intracellular deposition of calcium carbonate crystals (calcite), a property unique in the *Bacteria*. These inclusions can reach 5 μm in size (Fig. BXII.γ.56) and the mean calcium content is approximately

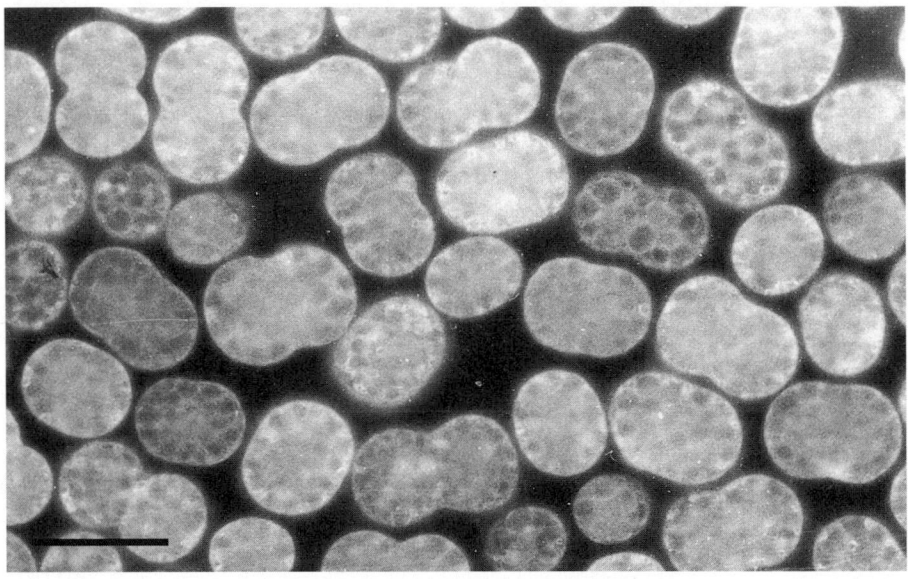

FIGURE BXII.γ.53. *Achromatium* cells collected from sediment of Lake Fuchskuhle (acidic bog lake, Brandenburg, Germany). In contrast to all other habitats this lake harbors a population that is homogeneous in morphology and size. Dark-field micrograph. Bar = 25 μm.

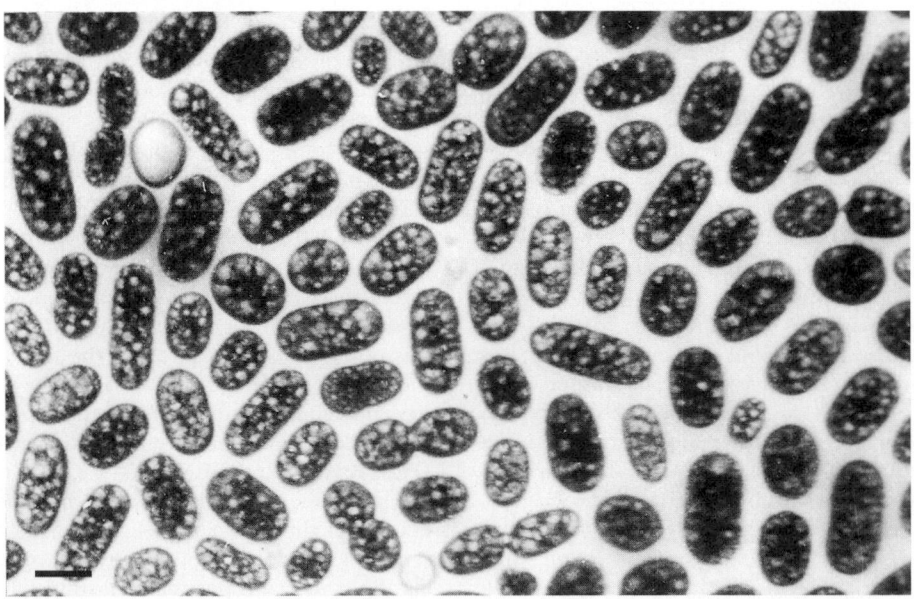

FIGURE BXII.γ.54. *Achromatium* cells collected from sediment of Lake Stechlin (Brandenburg, Germany). Considerable variation in size can be seen. Light micrograph. Bar = 50 μm.

2 ng/cell. In SEM preparations, the inclusions below the cell surface become visible due to shrinkage of the cell envelope (Fig. BXII.γ.57).

The function of the calcite inclusions is unknown, but it seems likely that the intracellular precipitation fulfils a metabolic role. The hypothesis that *Achromatium* possesses an active mechanism for concentrating calcium from the environment is consistent with the observation that cells in an acidic pond with very low levels of calcium in the water column contain calcite inclusions. In starvation experiments with physical cell enrichments, the inclusions tend to disappear. They can be easily removed by treating the cells with dilute acetic acid.

The presence of sulfur inclusions within the cells and the predominance of *Achromatium* in the micro-oxic zone of sediments suggest that *Achromatium* spp. may be sulfur-oxidizing chemolithoautotrophs. Microautoradiographic studies of the uptake of ^{14}C-labeled substrates have shown that *Achromatium* cells are able to assimilate ^{14}C from bicarbonate, acetate, and protein hydrolysate; however, glucose is not assimilated. The autotrophic potential for at least a proportion of cell enrichments is evident not only from inorganic carbon fixation, but also from the presence of genes similar to that encoding the large subunit of ribulose-1,5-bisphosphate carboxylase/oxygenase (RubisCo) (Head et al., 2000b).

From experiments incubating *Achromatium*-bearing sediment cores with ^{35}S-labeled sulfate in the presence or absence of sodium molybdate, it was deduced that the population was capable of oxidizing sulfide to intracellular S^0 and ultimately to sulfate. Moreover, sequences homologous to APS reductase genes (*aprBA*), were detected in purified DNA extracted from physically enriched cell preparations (Head et al., 2000b; Gray et al., 1997).

However, the pattern of organic substrate uptake within a single population suggests that physiological diversity occurs in natural *Achromatium* communities (Head et al., 2000a). In contrast to the response of some sulfide-oxidizing bacteria, e.g. *Beggiatoa* spp., sulfur droplets do not diminish in *Achromatium* enrichments under starvation conditions. It is possible that putatively heterotrophic members of the population can oxidize sulfide and deposit S^0 intracellularly as a mechanism for the detoxification of metabolically produced hydrogen peroxide in the

manner previously suggested for heterotrophic *Beggiatoa* spp. (Burton and Morita, 1964).

The cells are surrounded by a specific glycocalyx. SEM-preparations of cells show a typical network that covers the whole cell surface (Fig. BXII.γ.58), which is consistent with results of TEM-preparations (Fig. BXII.γ.59). Presumably, these structures play an important role in the colonization of solid surfaces (adhesion) and possibly in the peculiar type of cell movement exhibited by *Achromatium* cells.

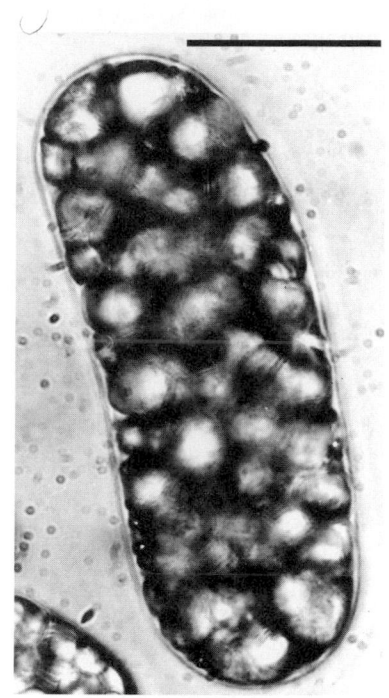

A

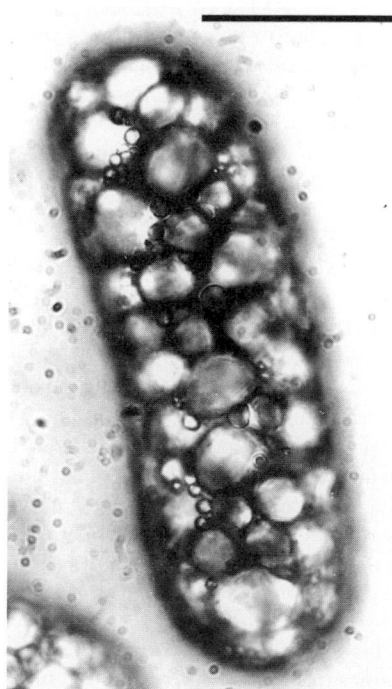

B

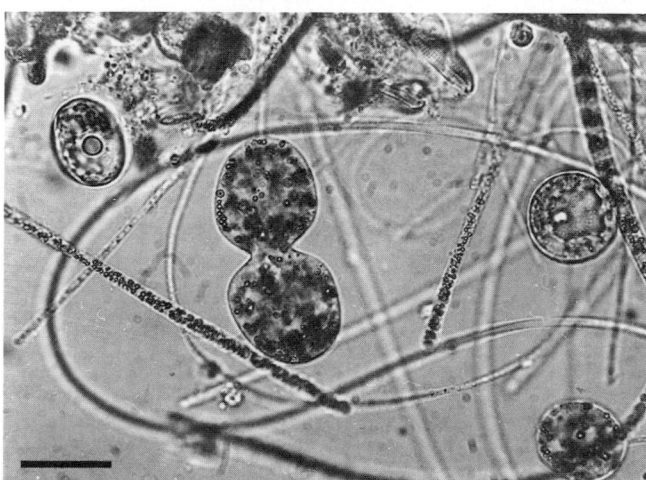

FIGURE BXII.γ.55. *Achromatium oxaliferum,* dividing stage of one cell. Sediment sample from Lake Dagow (Brandenburg, Germany). Filaments of *Beggiatoa* spp. and some diatoms are visible. Light micrograph. Bar = 25 μm.

FIGURE BXII.γ.56. Light micrographs of an identical *Achromatium* cell, focused at various levels. The whole cell volume is filled with calcite crystals and small sulfur droplets. (*A*) surface of the cell, (*B*) center of the cell. Bars = 25 μm.

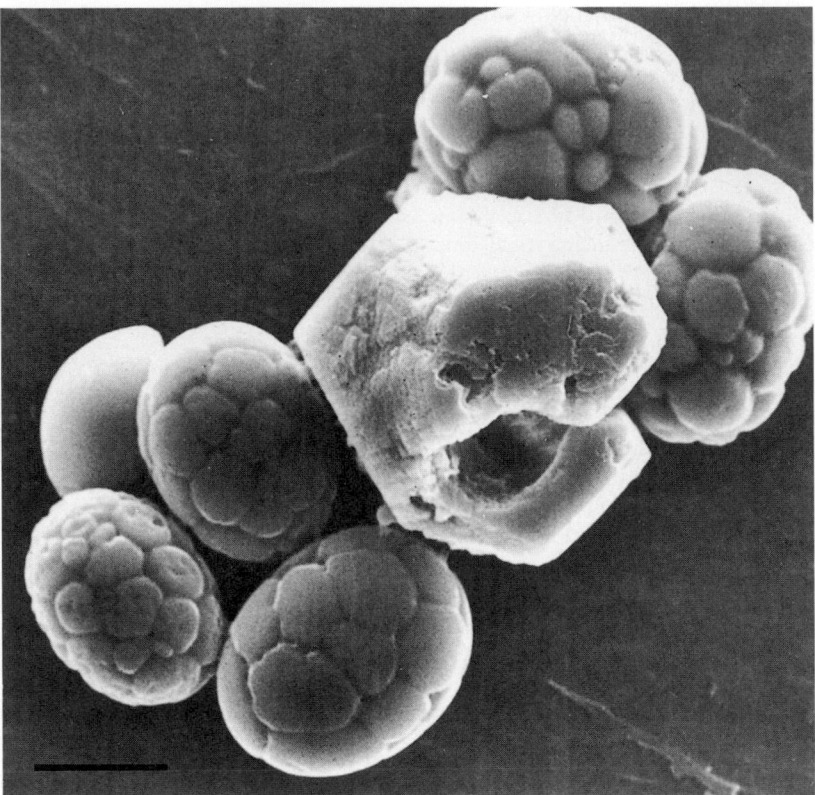

FIGURE BXII.γ.57. Scanning electron micrograph (SEM) showing some cells stuck to a calcite crystal from the sediment (Lake Stechlin, Brandenburg, Germany). The inclusions below the cell surface can be seen. Bar = 25 μm (Reproduced with permission from H.–D. Babenzien and H. Sass, Ergebnisse der Limnologie, *48:* 247–251, 1996, ©E. Schweizerbart'sche, Stuttgart.)

ENRICHMENT AND ISOLATION PROCEDURES

No enrichment or isolation procedures are available. For experiments, cells must be collected physically from sediments by a swirling technique ("gold panning method"; De Boer et al., 1971). These cell enrichments may be washed carefully in sterile water and transferred with a Pasteur pipette several times to obtain highly purified preparations of cells.

TAXONOMIC COMMENTS

There are two remarkable misinterpretations in the history of the genus *Achromatium*. The most striking inclusions present within cells were interpreted by Schewiakoff (1893) to be Ca-oxalate (hence the name of the species originally described); however, they consist of calcium carbonate. This was correctly described by Bersa (1920).

From the electron micrographs shown from De Boer et al. (1971), the glycocalyx was interpreted as peritrichously arranged appendages (flagella), and it was noted that "gliding motility should be caused by peritrichous flagella beating in a slime layer that surrounds the cell". The assumption of "gliding movement" was the main reason for inclusion of the organism in Section 23—Nonphotosynthetic, nonfruiting gliding bacteria—of the first edition of *Bergey's Manual of Systematic Bacteriology*. Today the means of movement of *Achromatium* remains unclear, but it has no flagella.

Although only one species (*A. oxaliferum*) is included in the Approved List of Bacterial Names, a second, *"A. volutans"*, was described earlier by Hinze (1903) under the proposed name *"Thiophysa volutans"*. The organism was found only in saline/marine waters, where it normally contains sulfur inclusions but lacks calcium carbonate crystals (La Riviere and Schmidt, 1992). More recently, Dando et al. (1998) described *"A. volutans"* at hydrothermal vents around the Greek island of Milos. The cells (diameter 5–90 μm) appeared with high density in the upper 1–15 mm of the sediment, forming "white sand" due to the reflective sulfur globules within the cells. Enzyme activity measurements of cell extracts (ribulosebisphosphate carboxylase, adenylphosphosulfate reductase, sulfate adenyltransferase) were indicative of autotrophic sulfur oxidizing activity.

The phylogenetic position of these populations is unclear and until further investigations are available, the position of *"Achromatium volutans"* as a genuine member of the genus *Achromatium* remains doubtful. Although only one species is validly published, the occurrence of genetically distinct subpopulations in *Achromatium* communities with greater than 2.5% divergence in their 16S rRNA sequences provides evidence to support the delineation of *Achromatium* into several different species.

ACKNOWLEDGMENTS

The authors would like to thank Dr. Renate Hanschke (University of Greifswald) for kindly providing SEM and TEM micrographs, and Dr. W. Ludwig (Technical University of Munich) for the excellent program package ARB. Prof. Dr. N. Pfennig and Dr. R. Amann are gratefully acknowledged for helpful discussions. The studies have been funded by the Deutsche Forschungsgemeinschaft, FRG, and the Natural Environmental Research Council, UK.

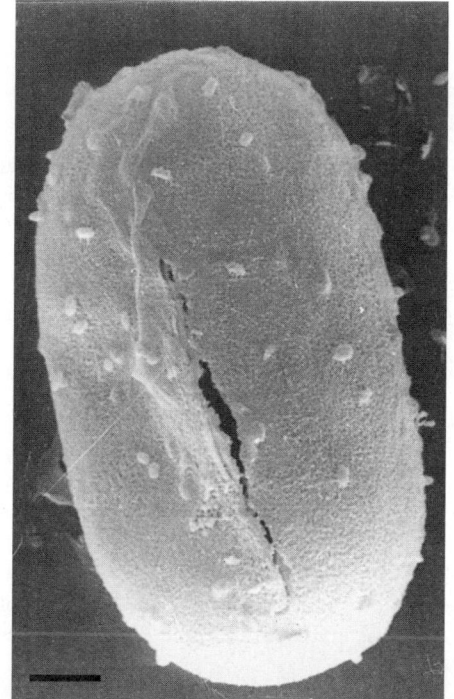

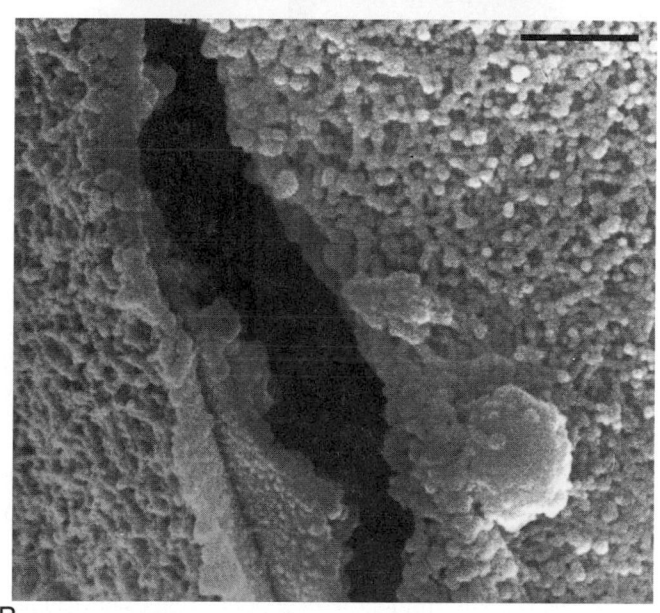

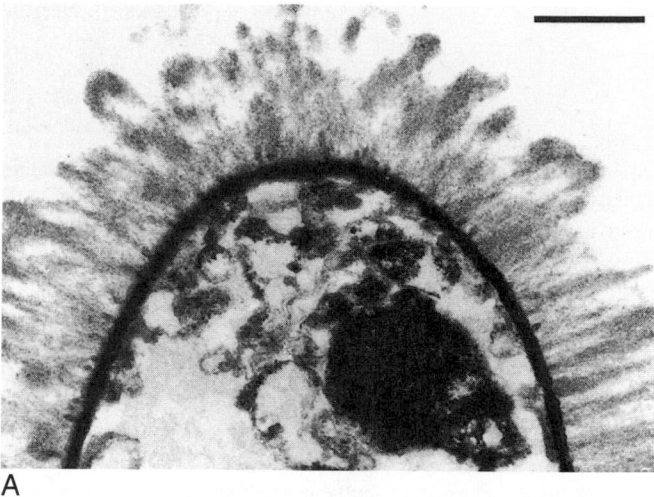

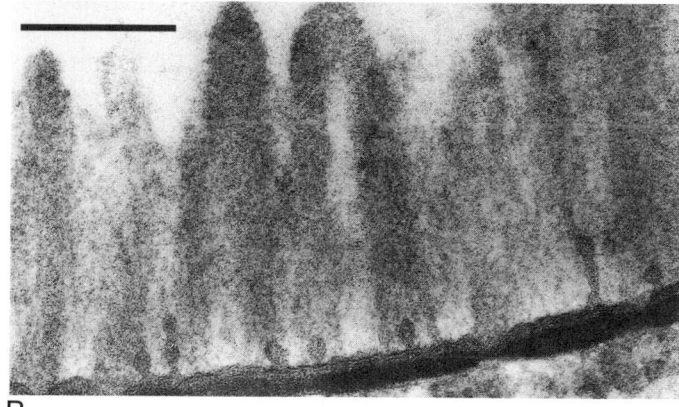

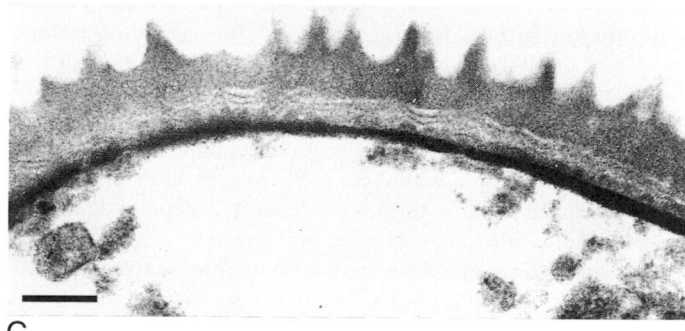

FIGURE BXII.γ.58. Scanning electron micrographs (SEM) from an *Achromatium* cell. The ruptured area of the cell is shown under a different magnification. A typical network can be seen. (*A*) whole cell. Bar = 5 µm. (*B*) Rupture of the cell, showing the thick glycocalyx. Bar = 1 µm.

FIGURE BXII.γ.59. Transmission electron micrographs (TEM) of ultrathin sections of *Achromatium* cells showing the extent of the glycocalyx covering the cell and its lobular structure. Calcite inclusions are solubilized. (*A*) Bar = 1 µm. (*B*), Bar = 0.5 µm. (*C*) Bar = 1.0 µm.

FURTHER READING

Head, I.M., N.D. Gray, R. Howarth, R.W. Pickup, K.J. Clarke and J.G. Jones. 2000. *Achromatium oxaliferum*. Understanding the unmistakable. Adv. Microb. Ecol. *16*: 1–40.

Head, I.M., N.D. Gray, H.D. Babenzien and F.O. Glöckner. 2000. Uncultured giant sulfur bacteria of the genus *Achromatium*. FEMS Microbiol. Ecol. *33*: 171–180.

List of species of the genus Achromatium

1. **Achromatium oxaliferum** Schewiakoff 1893, 1[AL]

   ox.al.i' fe.rum. M.L. n. *oxalatum* oxalate; Gr. n. *oxalis* sorrel, a sour plant; L. v. *fero* to carry; M.L. adj. *oxaliferum* oxalate-containing.

   The characteristics are as the same as those described for the genus.

   The mol% G + C of the DNA is: not known.

   Type strain: No culture isolated.

Genus III. **Beggiatoa** Trevisan 1842, 56[AL]

WILLIAM R. STROHL

Beg.gi.a.to'a. M.L. fem. n. *Beggiatoa* a genus of bacteria; named in remembrance of F.S. Beggiato, a physician of Vicenza, (g.b. 1807), who authored the *Delle Terme Euganea*.

Colorless cells, ~1.0–200 × 2.0–10.0 µm, occurring in filaments 5.0–10 cm in length. Filaments usually have a consistent width over the entire length. Freshwater strains mostly, if not universally, possess filament diameters less than 5.0 µm, whereas marine beggiatoas with filament diameters ranging from 1.0 to 200 µm have been observed microscopically. No strain wider than 5 µm has been isolated in axenic culture from any source. The filaments may contain up to 100 or more cells, and in rare cases, up to several thousand. Cells in filaments are cylindrical and are longer than they are wide in the thinner strains (i.e., those less than ~5 µm in diameter). In wider strains (i.e., those greater than ~7 µm in diameter), cells are usually disk-shaped and typically are wider than they are long. Filaments occur singly, in sheets of cottony masses on sediment surfaces, or in mats in which each filament retains its individuality. Reproduction is by transverse binary fission of cells within filaments; division occurs by septation, in which the peptidoglycan and cytoplasmic membranes close like the iris of a diaphragm. Filament dispersion is via sacrificial cell death (necridial cells), filament breakage, or simple disintegration. In some strains, the disintegration of filaments occurs until mostly single or double cell units (hormogonia) exist; a hormogonium then can grow to become a filament. **Cells contain inclusions of sulfur when grown in the presence of H_2S and, for some strains, thiosulfate.** Intracellular inclusions of poly-β-hydroxybutyrate or polyphosphate may be present, particularly in freshwater strains. **Marine or brackish water strains having a filament diameter of >10 µm contain a large (~80% of the cell volume), central liquid vacuole that contains concentrated nitrate.** Resting stages are not known. **Attachment holdfasts or sheaths are not present.** Capsules are not formed, but filaments usually produce a slime matrix. Gram negative. **Filaments and hormogonia have gliding motility.** No motility organelles are known. **Aerobic or microaerophilic, having a strictly respiratory type of metabolism,** with oxygen and, in some instance, nitrate as the terminal electron acceptor. Anaerobic growth is not proven with pure cultures, but may occur with nitrate as the terminal electron acceptor for some strains studied in nature. Internally stored sulfur may also serve as an electron acceptor by freshwater strains for short-term maintenance in the absence of oxygen. Sulfate does not substitute as the terminal electron acceptor for anaerobic growth in strains thus far studied. **Chemoorganotrophic (freshwater strains), and either facultatively or constitutively chemolithoautotrophic (marine strains). H_2S or thiosulfate may be used as the electron donor for chemolithotrophic metabolism.** Acetate is oxidized to CO_2 by all freshwater strains tested thus far. **Several C_2, C_3, and C_4 organic acids and, sometimes, their amino acid equivalents are utilized as sole carbon and energy sources for heterotrophic growth.** Growth factors are not required by most strains, although some strains may require vitamin B_{12}. Gelatin and starch are not hydrolyzed. **N_2 is fixed by all freshwater and marine strains tested thus far.** Nitrate, nitrite, ammonium, N_2, or certain amino acids may be used as the sole nitrogen source. **Oxidase positive. Catalase negative.** Freshwater, estuarine, and marine strains are known. **Beggiatoas are inherently gradient organisms, growing in horizontal layers in sediments at the interface between the un-** derlying anoxic sulfide-liberating zone and the overlying oxic zone. Growth can occur in the range of 0–40°C. Obligately thermophilic strains have not been characterized, although some beggiatoas have been observed in high temperature runoffs associated with thermal springs, or in mats adjacent to thermal vents in the ocean floor.

The mol% G + C of the DNA is: 35–39.

Type species: **Beggiatoa alba** (Vaucher 1803) Trevisan 1842, 56.

FURTHER DESCRIPTIVE INFORMATION

Three subgroups of *Beggiatoa* can be discerned at present. The type and only named species is *Beggiatoa alba*, which is a member of the first subgroup. The three subgroups are: (i) thin, heterotrophic freshwater strains; (ii) thin, obligately or facultatively chemolithoautotrophic marine strains; and (iii) wide, autotrophic, vacuolated marine strains. Characteristics of the three subgroups *Beggiatoa* are listed in Table BXII.γ.39. Other features are described below.

Motility Gliding motility of beggiatoas is relatively rapid (1.0–8.0 µm/sec) and is often accompanied by flexing, bending, and/or rotation of the filaments. Gliding motility determines the nature of growth and colony formation; on agar media containing relatively few nutrients, spiral colonies are usually produced. The filaments are capable of gliding on agar surfaces to form various patterns of waves and concentric rings (Fig. BXII.γ.60).

Cell structure and life cycle of thin freshwater strains Cells within the filaments are separated by the membranes and the first cell wall layer (Fig. BXII.γ.61; Strohl and Larkin, 1978a). Cell division occurs by separation of those two inside layers (Fig. BXII.γ.61). Nearly one fourth of the cells in a trichome may be dividing simultaneously (Hinze, 1901). The outer cell wall layers do not take part in septation (Strohl et al., 1982) and are longitudinally continuous with the length of the filament. Beggiatoas do not contain a classical sheath but do exude large quantities of extracellular neutral polysaccharide slime (Strohl and Larkin, 1978b; Larkin and Strohl, 1983).

Filament dispersion in the thin freshwater strains occurs via sacrificial cell death (necridia formation) and filament breakage (Strohl and Larkin, 1978a) in a manner reminiscent to that observed in *Oscillatoria* spp. Filament dispersion may also occur via simple separation of an end cell to produce a hormogonium (Wiessner, 1981), or the filaments may divide by simple fragmentation (Pringsheim, 1949). Fragmentation, however, is often associated with filament death.

Although all beggiatoas stain Gram-negative (Strohl and Larkin, 1978b), they have a cell envelope which is much more complex than that of typical Gram-negative bacteria (Fig. BXII.γ.62; Maier and Murray, 1965; Strohl et al., 1982, 1986b). The A layer of the cell wall is probably a peptidoglycan layer, and the B layer is a unit membrane-like layer that is positionally similar to the lipopolysaccharide layer of Gram-negative bacteria (Fig. BXII.γ.62; Strohl et al., 1982). Depending on the strain and the method of fixation, two or three further external layers may be observed (Maier and Murray, 1965; Strohl et al., 1981a, 1982), the most external of which usually has a fibrillar pattern (Strohl et al., 1982).

TABLE BXII.γ.39. Characteristics of the three subgroups of the genus *Beggiatoa*[a,b]

Characteristic	Thin, heterotrophic freshwater strains (including *B. alba*)	Thin, obligately or facultatively autotrophic marine strains	Wide, autotrophic, vacuolated marine strains
Typical filament diameters, mm	~1–5	~1–7	~10–200
Sulfide oxidized to sulfur	+	+	+
Sulfur oxidized to sulfate	−	+	+
Sulfur reduced to sulfide	+	?	?
Heterotrophic growth	+	−	−
Acetate assimilated	+	+	?
Acetate oxidized to CO_2	+	?	?
Autotrophic growth	−	+	+
Mixotrophic growth	?	−	−
CO_2 fixed by RubisCo	w	+	+
Large, central vacuoles	−	−	+
Nitrate accumulated	−	−	+
Nitrate reduced	+	+	+
N_2 product	+	?	+
NH_3 product	+	?	?
N_2 fixed	+	+	?
Anaerobic growth suggested	−	−	+
Microaerophilic	+	+	+
c-type cytochromes	+	+	+
CO-binding *c*-type cytochromes	+	?	−
Growth in distinct mats	+	+	+
Photophobic response	+	+	+ (?)
O_2-phobic response	+	+	+
H_2S-phobic response	+	+	+
Phylogenetic placement:			
Gammaproteobacteria	+	+	+
Closest relatives	*Vitreoscilla beggiatoides*	NA	wide, marine thioplocas
mol% G + C of the DNA	35–39[c]	NA	NA
Locations found:			
Marine sediments		+	+
Sediments	+		
Warm mineral springs			+
Representative strains	B18LD, OH-75-2a	MS-81-1c, MS-81-6	Monterey Canyon[d], Guaymas Basin[d], Warm Mineral Springs[d]

[a]Symbols: +, >90% positive; −, <90% negative; ?, not known; w, weak, insignificant levels; NA, not available; RubisCo, ribulose 1,5-bisphosphate carboxylase/oxygenase.

[b]The data presented in this table are not meant to be all-inclusive, but rather representative of the data, as they are known currently. There are likely to be exceptions to the given data based on individual strains or experimental conditions used.

[c]The mol% G + C content of the DNA of the freshwater *B. alba* strains B18LD (ATCC 33555), B15LD (ATCC 33556), and OH-75-2a are 37.1 mol%, 35.5 mol%, and 38.5 mol% (0.3 in duplicate and triplicate experiments), respectively, by T_m analysis (D.C. Nelson, personal communication).

[d]For the wide, marine example strains, there are no isolates. Thus, locations from which these organisms have been studied are given rather than strain names or numbers. It cannot be assumed, however, that the beggiatoas at these locations represent single species.

All *Beggiatoa* cells are known to contain at least three types of inclusions: poly-β-hydroxybutyrate (PHB; Pringsheim and Wiessner, 1963; Strohl and Larkin, 1978b; Strohl et al., 1982), polyphosphate (Maier and Murray, 1965; Strohl and Larkin, 1978b), and sulfur (Lawry et al., 1981; Strohl et al., 1981b). It is the latter inclusion that has traditionally separated organisms of the genus *Beggiatoa* from those in *Vitreoscilla* (Pringsheim, 1951; Leadbetter, 1974a; Strohl et al., 1986a). Several investigators have observed sulfur inclusions in thin sections of fixed and dehydrated *Beggiatoa* trichomes (Morita and Stave, 1963; Maier and Murray, 1965; Lawry et al., 1981; Strohl et al., 1981b; Larkin and Henk, 1996). The sulfur is dissolved away during dehydration with ethanol or acetone, leaving empty, electron-translucent spaces (Strohl et al., 1981b). Sulfur inclusions within these cytoplasmic membrane invaginations may be bound by envelopes typically composed of a single dense layer (Fig. BXII.γ.63; Strohl et al., 1981b). Unusual pentilaminar sulfur inclusion envelopes have been observed in cells of *B. alba* strain B15LD (Strohl et al., 1981b, 1982). Rudimentary pentilaminar envelopes have been observed in cells of strain B15LD grown in the total absence of reduced sulfur, indicating that a "primer" envelope was always present (Strohl et al., 1982). This may partially explain the rapidity with which sulfur inclusions appear after exposure of *Beggiatoa* cells to sulfide (Winogradsky, 1887; Burton and Morita, 1964; Schmidt et al., 1986). It may be physiologically significant that the sulfur inclusions of freshwater *Beggiatoa* strains are found within invaginations of the cytoplasmic membrane and are therefore periplasmic (i.e., external to the cytoplasmic membrane) in nature (Strohl et al., 1982).

Under certain growth conditions, PHB comprises over 50% of the cell dry weight of thin freshwater, heterotrophic *B. alba* strains (Güde et al., 1981; Strohl et al., 1981a). Furthermore, in one experiment, ~50% of the [14C]acetate assimilated by heterotrophically grown cells was incorporated into PHB (Strohl et

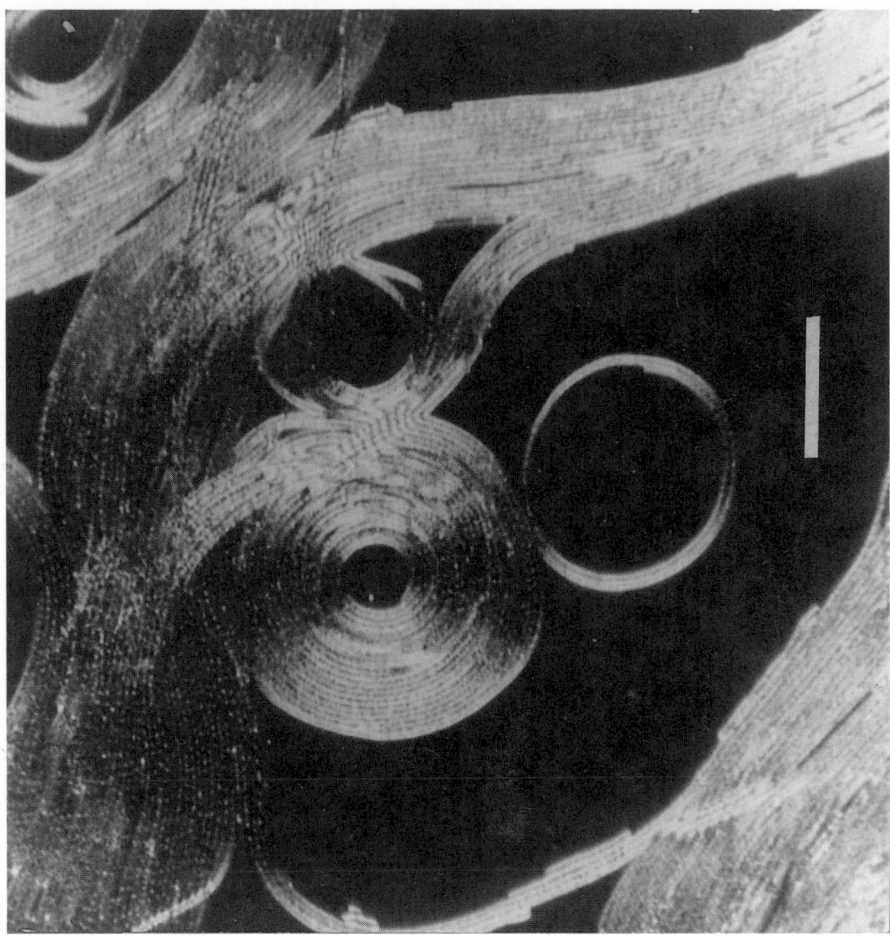

FIGURE BXII.γ.60. Light micrograph of a *Beggiatoa alba* colony on a 1.4% agar surface, taken through a dissecting microscope. Bar = 50 μm.

al., 1981a). The amount of PHB in the cells is usually proportional to the amount of acetate in the growth medium (Kowallik and Pringsheim, 1966). Cultures containing PHB have been shown to survive in media containing sulfide but lacking organic nutrients (Kowallik and Pringsheim, 1966), suggesting a survival role for the endogenous metabolism of PHB during carbon starvation. Furthermore, a freshwater *B. alba* strain containing massive PHB inclusions has been shown to be able to survive up to 7 days of starvation in a mineral salts medium lacking a carbon and an energy source (W.R. Strohl, unpublished observation).

Cell structure of wide, vacuolated, marine strains Larkin and Henk (1989) and Nelson et al. (1989b) simultaneously recognized that the large marine beggiatoas with diameters greater than 10 μm contain centrally located vacuoles (Figs. BXII.γ.64 and BXII.γ.65), similar to those previously observed in wide, marine *Thioploca* strains (Maier and Gallardo, 1984b). Larkin and Henk (1989) proposed that the vacuoles serve the purpose of generating a high surface to volume ratio, allowing for the large size of the trichomes. Later, Fossing et al. (1995) demonstrated that the vacuoles in *Thioploca* contain high concentrations (100–500 mM) of nitrate; similar analyses of the marine *Beggiatoa* central vacuoles also found high concentrations (130–150 mM) of nitrate (McHatton et al., 1996).

The cytoplasm, sulfur inclusions, and other cytoplasmic membrane intrusions are all found in a very narrow band at the outer edge of these large, vacuolated cells (Fig. BXII.γ.64 and BXII.γ.65). Although no high-magnification data are yet available, lower-magnification photographs of large, vacuolated beggiatoas (Nelson et al., 1989b; Larkin and Henk, 1989; Larkin et al., 1994) suggest a less complex cell wall with perhaps fewer layers than observed in the heterotrophic freshwater strains. Larkin and his colleagues also observed what appear to be symbiotic bacteria ultrastructurally similar to type II methanotrophic bacteria living within the cytoplasm of the wide, vacuolated marine beggiatoas (Larkin and Henk, 1996). Larkin and Henk (1996) proposed that the large, marine beggiatoas possessing internal symbionts might in fact be "dual" organisms much like the tubeworms and mussels found in the same habitat. This finding and its potential implications (e.g., potential symbiotic relationship) may have important implications for the potential ability to culture large beggiatoas axenically.

Sulfur metabolism of beggiatoas Oxidation of sulfide to sulfur by the heterotrophic freshwater strain *B. alba* is constitutive (Schmidt et al., 1986) and does not promote an increase in growth rate (Güde et al., 1981) or in the rate of protein synthesis (Schmidt et al., 1986). Sulfide oxidation by *B. alba* B18LD is oxygen-dependent (Vargas and Strohl, 1985b; Schmidt et al., 1987) and is inhibited by several specific electron transport inhibitors (Strohl and Schmidt, 1984; Schmidt et al., 1987), suggesting that sulfide oxidation is coupled with oxygen reduction

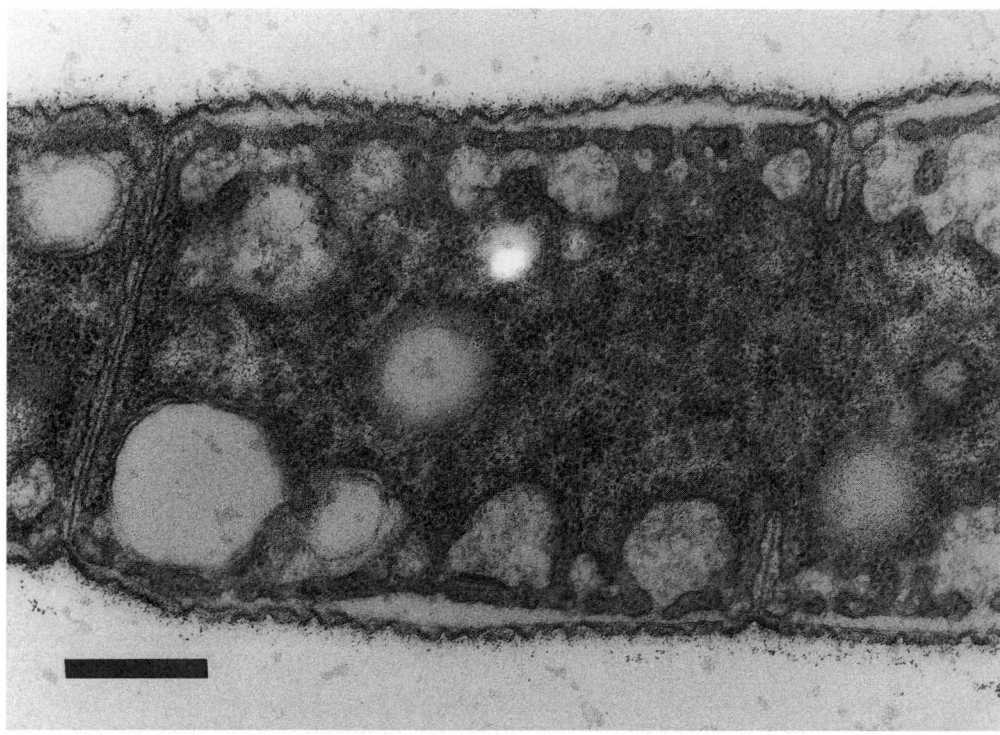

FIGURE BXII.γ.61. Thin section micrograph of *B. alba* B18LD showing the separation of cells in the filament by the cytoplasmic membranes and the peptidoglycan layer. A newly formed septation is also visible at the right. The outer walls, not well preserved by the modified Ryter–Kellenberger fixation technique (Strohl and Larkin, 1978b), are continuous with the length of the filament. Bar = 50 μm.

via the electron transport system. Nelson and Castenholz (1981b) have demonstrated that sulfur may comprise as much as 30% of the cell dry weight of the freshwater *Beggiatoa* strain OH-75-2a. Nelson et al. (1989b) have studied sulfide oxidation by the wide marine beggiatoas found near a hydrothermal vent at Guaymas Basin. They have shown that the sulfur, present mostly in the S_8 form, comprises ~14% of the cell dry weight (Nelson et al., 1989b). Although the form of sulfur in freshwater strains has not been determined, it is assumed to be in the S_8 form, as this is consistent with the sulfur found in other sulfur bacteria (Schmidt et al., 1971).

The significance of oxidation of the sulfur formed by freshwater strains such as *B. alba* to sulfate is not clear. Sulfur oxidation by strain OH-75-2a has been reported to be very slow (50 μmol sulfate produced/75 h/~30 mg cells; Nelson and Castenholz, 1981b); this rate—assuming 50% of cell dry weight is protein—calculates to a rate of ~0.4 nmol/min/mg protein. Hagen and Nelson (1997) have reported ATP sulfurylase activity in extracts of OH-75-2a to be about 1 nmol/min/mg of protein—slower than the rates of 23–30 nmol/min/mg protein and 1523–1710 nmol/min/mg protein for the facultatively autotrophic marine strain, MS-81-6, and the chemolithoautotrophic strain, MS-81-1c, respectively. In other experiments, Schmidt et al. (1987) did not observe significant *in vivo* oxidation of sulfur to sulfate (~5–10% maximum conversion) over periods of 22–72 hours, as evidenced by isotopic and chemical studies; a control strain, *Thiothrix nivea* strain JP3, produced considerable sulfate from sulfur inclusions in similar experiments (Schmidt et al., 1987). Similarly, another freshwater strain showed no (sulfuric) acid production from sulfur inclusions, whereas *Thiothrix* strains isolated from similar

sources did so (Williams and Unz, 1985a). On the other hand, Hagen and Nelson (1997) have recently demonstrated the presence, albeit low, in a heterotrophic, freshwater strain (OH-75-2a) of an enzyme activity (sulfite-dependent, AMP-independent ferricyanide reduction) that is normally considered to be involved in oxidation of sulfur to sulfate. This observation supports the contention that sulfur can be oxidized to sulfate by freshwater strains of *Beggiatoa*, albeit rather slowly as compared to the same process by marine strains (Hagen and Nelson, 1997).

Although sulfate appears to be a slowly produced product of sulfide oxidation by the heterotrophic, freshwater strains, the sulfur accumulated by these strains can be reduced back to sulfide under anaerobic conditions (Nelson and Castenholz, 1981b; Schmidt et al., 1987). For *B. alba* B18LD, this process occurs at a rate of ~6–7 nmol/min/mg protein (Schmidt et al., 1987), a 5–10-fold greater rate than the sulfur oxidation rates observed by Nelson and colleagues (Nelson and Castenholz, 1981a; Hagen and Nelson (1997)), and to be coupled with the oxidation of H_2 (Schmidt et al., 1987). Whether this process is ecologically relevant for these strains is as yet unknown, but it is possible that the heterotrophic beggiatoas utilize sulfur reduction as a temporary form of anaerobic respiration under anaerobic conditions, allowing them a form of energy to fuel the movement back to favorable conditions. Thus, sulfur reduction by the freshwater strains may provide an ecological role analogous to that proposed for anaerobic nitrate respiration in the wide, marine beggiatoas and thioplocas (Jørgensen and Gallardo, 1999).

Using cultures with vertical, convergent oxygen and sulfide gradients, Nelson et al. (1986a) have demonstrated in the thin, marine strain MS-81-6 a molar growth yield (Y_{H_2S}) on sulfide of

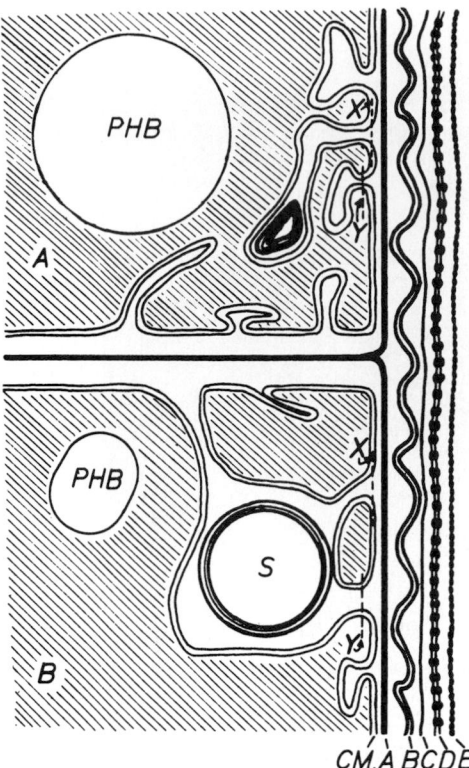

FIGURE BXII.γ.62. Model of a freshwater *Beggiatoa* cell wall (strain B15LD; Strohl et al., 1982). Note the pentilaminar sulfur (*S*) inclusion (which is single-layered in other *Beggiatoa alba* strains; Strohl et al., 1981b) external to the cytoplasmic membrane (*CM*); poly-beta-hydroxybutyrate, (*PHB*); cell wall layers from inside to outside (*A–E*); plane through which freeze-fracture shows invaginations (*X*); and plane through which thin sections have been made that give a different view of the invagination of membranes around sulfur inclusions (*Y*).

~8.4 g/mol. The marine strain grew linearly in a horizontal plate with a doubling time of ~11 h at the position at which the oxygen and sulfide gradients converged. Hagen and Nelson (1997) have compared the sulfide oxidation of two marine strains, the obligately chemolithoautotrophic marine *Beggiatoa* strain MS-81-1c and the facultatively autotrophic strain MS-81-6. They demonstrated that MS-81-1c stores more elemental sulfur and has a higher yield for sulfide oxidation than does strain MS-81-6 (Y_{H_2S} for MS-81-1c on sulfide was 15.9 g/mol, compared with 8.3–8.4 g/mol for MS-81-6). Similarly, O_2 consumption rates by strain MS-81-1c were observed to be consistently 2–3-fold greater than those of the facultatively autotrophic strain, MS-81-6. Moreover, strain MS-81-1c oxidizes sulfur to sulfate by either the adenosine phosphosulfate pathway or a highly regulated sulfite: acceptor oxidoreductase, whereas strain MS-81-6 uses only the sulfite: acceptor oxidoreductase (Hagen and Nelson, 1997).

Electron transport in beggiatoas   Although Burton and Morita (1964) stated that *Beggiatoa* strains lacked cytochrome systems, all beggiatoas thus far analyzed have been found to contain cytochromes (Cannon et al., 1979; Strohl et al., 1986b; Prince et al., 1988). Room temperature spectra have indicated that *B. alba* and other strains might contain several *b* type, *c* type, *a* type, and CO-binding cytochromes, as well as ubiquinone 8 and NAD(P)H dehydrogenases; however, the spectra obtained in those studies

were not of high quality (Strohl et al., 1986b). Schmidt and Dispirito (1990) have reinvestigated the cytochromes of *B. alba* B18LD and found four *c*-type cytochromes, three of which bind CO. One of the cytochromes was found to be a flavocytochrome *c*, somewhat analogous to cytochromes present in photosynthetic sulfide oxidizing bacteria (Schmidt and Dispirito, 1990).

Prince et al. (1988) determined that a sample of marine *Beggiatoa* filaments obtained from a marine sediment (suspension 1615B, which contained organisms of ~30 μm diameter; Nelson et al., 1989b) contained only *c*-type cytochromes. These *c*-type cytochromes from marine *Beggiatoa* filaments did not bind CO, nor were flavocytochromes detected (Prince et al., 1988), thus making them substantially different from the *c*-type cytochromes isolated from the freshwater strain, *B. alba* B18LD (Schmidt and Dispirito, 1990).

Carbon metabolism of beggiatoas   Freshwater strains of *Beggiatoa* grow heterotrophically (Pringsheim, 1964; Strohl and Larkin, 1978b; Nelson and Castenholz, 1981a; Strohl et al., 1981a; Larkin and Strohl, 1983). They can utilize acetate as the sole carbon and energy source (Larkin and Strohl, 1983; Strohl et al., 1986b). Moreover, several *Beggiatoa* strains have been shown to grow on C_2, C_3, and C_4 organic acids, although none of the beggiatoas described thus far has grown on C_5 or C_6 organic acids or on hexoses (Burton and Morita, 1964; Burton et al., 1966; Nelson and Castenholz, 1981a; Mezzino et al., 1984; Williams and Unz, 1985a). Although Burton et al. (1966) have suggested a type of reverse tricarboxylic acid cycle for CO_2 fixation and acetate utilization by a freshwater *Beggiatoa* strain, the accumulated data indicate that freshwater, heterotrophic *Beggiatoa* strains likely utilize the glyoxylate and tricarboxylic acid cycles during growth on acetate (Nelson and Castenholz, 1981a; Larkin and Strohl, 1983).

A small amount of CO_2 is fixed by freshwater strains of *Beggiatoa*, but most of it is assimilated primarily via heterotrophic metabolism mechanisms, rather than by the Calvin–Benson cycle (Nelson and Castenholz, 1981a; Strohl et al., 1981a). Low levels of ribulose 1,5 bisphosphate carboxylase/oxygenase (RubisCo) activity have been observed in extracts of the freshwater strains OH-75-2a and *B. alba* B18LD (Nelson et al., 1989a). In addition, a probe derived from the gene encoding the large subunit of *Anacystis nidulans* RubisCo has been found to hybridize strongly to DNA isolated from strain OH-75-2a but not to that from strain B18LD (Nelson et al., 1989a). Thus, although the data strongly suggest that the freshwater beggiatoas possess RubisCo, the levels of activity demonstrated so far are not of physiological significance (Nelson et al., 1989a).

Thin, marine strains of *Beggiatoa* can grow autotrophically. During such growth, CO_2 is fixed by RubisCo at levels consistent with the support of autotrophic growth (Nelson and Jannasch, 1983; Hagen and Nelson, 1996). Strain MS-81-1c, an obligately chemolithoautotrophic strain, has been shown to lack 2-ketoglutarate dehydrogenase activity and does not regulate its RubisCo activity greatly. This strain accumulates acetate into biomass but does not oxidize acetate to CO_2—similar to what has been observed with the chemolithoautotrophic thiobacilli (Hagen and Nelson, 1996). Strain MS-81-6, on the other hand, regulates both 2-ketoglutarate dehydrogenase and RubisCo activities and appears to possess facultative chemolithoautotrophic metabolism (Hagen and Nelson, 1996).

Suspensions of wide, marine beggiatoas have been shown by Nelson et al. (1989b) to fix CO_2 by RubisCo at levels consistent with chemolithoautotrophic growth, and the CO_2 fixation can

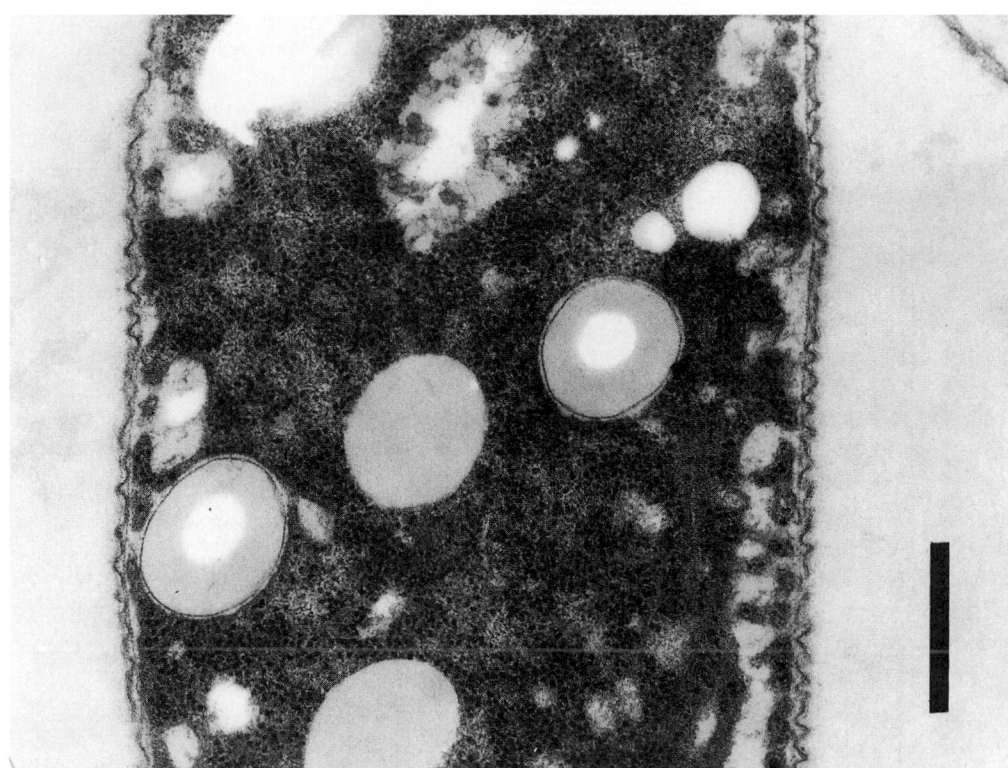

FIGURE BXII.γ.63. Thin section micrograph of *B. alba* B18LD showing the monolayered (4 nm thick envelope) sulfur inclusions external to the cytoplasmic membrane. Bar = 0.5 μm.

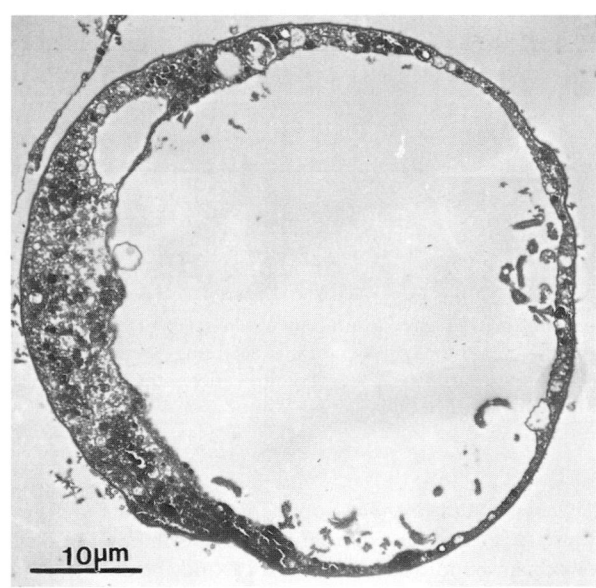

FIGURE BXII.γ.64. Thin section, cross-section of a wide, vacuolated, marine *Beggiatoa* strain showing an enormous central vacuole, with the cytoplasm and sulfur inclusions all squeezed into the thin outer "shell" surrounding the vacuole (Reproduced with permission from J.M. Larkin and M.C. Henk, Microscopy Research Technique *33:* 23–31, 1996, ©John Wiley & Sons, Inc.).

be stimulated approximately 7-fold by the addition of sulfide. Both RubisCo and phosphoribulokinase activities have been observed at significant levels in the suspensions of these marine beggiatoas (Nelson et al., 1989b), whereas 2-ketoglutarate dehydrogenase activities are below the limit of detection (McHatton et al., 1996). These data are supported by autoradiography, which has demonstrated extensive $^{14}CO_2$ fixation by intact strains (Nelson et al., 1989b). Interestingly, the vent beggiatoas from the Guaymas Basin exhibit 1.6–4.8-fold greater RubisCo activity at 50°C than at 30°C (whereas the thin, marine strain has approximately 4-fold more RubisCo activity at 30°C), suggesting a relatively high optimal temperature for their enzyme activity. Whole cell autoradiography of these organisms, however, demonstrates that CO_2 is not fixed above 40°C (Nelson et al., 1989b).

Steele et al. (1995) have reported that suspensions of the wide beggiatoas found in Warm Mineral Springs, FL, assimilate CO_2 at a level that would predict a 27-hr doubling time, suggesting that this organism also grows chemolithoautotrophically. This organism grows in an environment that is at approximately half the salinity of seawater (Lackey et al., 1965); thus, it is expected that it is more similar to the wide marine beggiatoas than to the freshwater strains.

Nitrogen metabolism Ammonia is a nitrogen source used by every *Beggiatoa* tested thus far (Larkin and Strohl, 1983). In studies by Vargas and Strohl (1985a), the glutamine synthetase (GS)–glutamate synthase (GOGAT) pathway was found to be the primary route for ammonia assimilation by *B. alba* B18LD. Glutamate dehydrogenase activity has not been observed and only low alanine dehydrogenase activity is observed under "high" ammonia conditions (Vargas and Strohl, 1985a). Other nitrogen sources that support growth of *B. alba* B18LD are nitrate, nitrite, urea, aspartate, and, to a lesser degree, asparagine, alanine, and thiourea (Vargas and Strohl, 1985a). Glutamate, aspartate, or

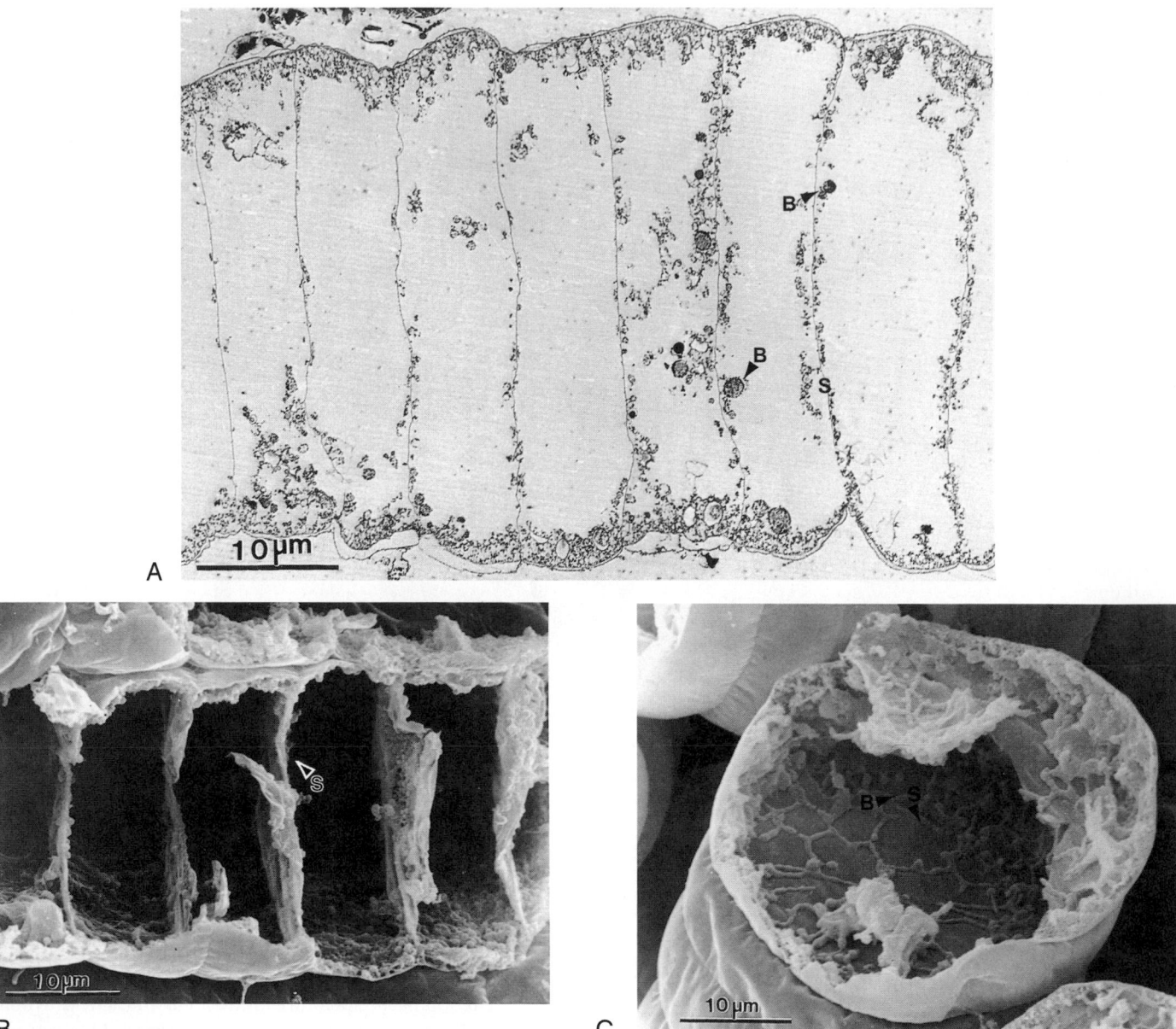

FIGURE BXII.γ.65. A typical thin section (*A*) of a *Beggiatoa* filament with a diameter of greater than 10 μm. Small blebs (*B*) of cytoplasm can be seen on the septa (*S*) and the remainder of the cytoplasm is restricted to a thin layer beneath the cell membrane. Scanning electron micrographs (*B, C*) of wide, marine, vacuolated *Beggiatoa* spp. The hollowed-out spaces where the nitrate vacuoles are located are easily observed. (Reproduced with permission from J.M. Larkin and M.C. Henk, Microscopy Research and Technique *33:* 23–31, 1996, ©John Wiley & Sons, Inc.)

asparagine support the growth of certain other beggiatoas, indicating that they can be used as sole nitrogen, carbon, and energy sources for those strains (Pringsheim, 1964).

Beggiatoas can utilize nitrate (Nelson et al., 1982; Vargas and Strohl, 1985b; Williams and Unz, 1985a) or nitrite (Vargas and Strohl, 1985b) as a sole nitrogen source. The product of nitrate reduction by *B. alba* is not N_2 or N_2O but ammonia, indicating that it has an assimilatory type nitrate reduction mechanism (Vargas and Strohl, 1985b). *B. alba* B18LD is unable to grow anaerobically by using nitrate as an electron acceptor, and nitrate reduction is not coupled with sulfide oxidation (Vargas and Strohl, 1985b). The nitrate reductase of *B. alba* B18LD has also been shown to be primarily in the soluble fraction (Vargas and Strohl, 1985b). Later experiments by Sweerts et al. (1990) showed that

a freshwater mat of *Beggiatoa* spp. denitrified nitrate to N_2 *in situ*. They also suggested, based on minimal experimental results, that under certain conditions *B. alba* B18LD may be able to denitrify nitrate to N_2 (Sweerts et al., 1990).

Perhaps the most significant recent discovery is that the wide, vacuolated marine strains of *Beggiatoa* (i.e., those strains greater than 10 μm in diameter) have the ability to assimilate and accumulate nitrate in their large, central liquid vacuoles (McHatton et al., 1996), in a manner similar to that of large, marine *Thioploca* strains (Fossing et al., 1995). These large, marine organisms concentrate nitrate to 130–160 mM, or ~3000–4000-fold greater than the concentration of ~40 μM measured in the sea water in which they are found (Fossing et al., 1995; McHatton et al., 1996; Otte et al., 1999). On the other hand, the nonvacuolated thin, marine strains MS-81-1c and MS-81-6 do not accumulate nitrate, indicating that the ability to accumulate and concentrate nitrate is correlated with the presence of the large central vacuoles (McHatton et al., 1996). Because the nitrate reductase activities of vacuolated, wide, marine beggiatoas are largely membrane bound—thus differing from the soluble nitrate reductase found in *B. alba* B18LD (Vargas and Strohl, 1985b) and the two thin, marine strains (McHatton et al., 1996)—the nitrate stored in these large vacuoles was originally thought to be denitrified to N_2 (McHatton et al., 1996). More recent evidence, however, suggests that ammonia is the major end product of nitrate reduction (D.C. Nelson, personal communication). Based on the thickness of the mats found at Guaymas Basin, the low level of oxygen present in the sea water, and the inadequacy of oxygen diffusion to support the sulfide oxidation present in such biomass, McHatton et al. (1996) made a strong case for the coupling of sulfide oxidation to nitrate reduction in those marine *Beggiatoa* mats. They calculated that enough nitrate was stored to support the organism for approximately 4.8 h of anaerobic growth without input of fresh nitrate (McHatton et al., 1996). Otte et al. (1999) have recently demonstrated that wide, marine *Thioploca* enrichments reduce most of their stored nitrate to NH_4^+, with a minor fraction (~15% of the NH_4^+ produced) going to N_2. Based on the rate of nitrate reduction to ammonia and the rate of sulfide oxidation, Otte et al. (1999) have suggested that the nitrate present in the vacuoles of marine *Thioploca* spp. might be sufficient for 200 hours of maintenance without additional input of nitrate. It has been calculated that the nitrate accumulation and reduction by these large marine beggiatoas and thioplocas should have a major impact on the nitrogen cycling in the oceanic sediments in which these organisms are found (Fossing et al., 1995; McHatton et al., 1996; Otte et al., 1999). Clearly, the fate of the nitrate reduction products (N_2 versus NH_4^+) by these organisms ought to be significant in the balance of nitrogen in the marine niches they occupy.

Nine strains of *Beggiatoa*—freshwater and marine, all with diameters less than 5 μm—have been shown to fix N_2 at rates of 3–12 nmol/min/mg protein when grown in tubes of semisolid agar in which the redox, sulfide, and oxygen gradients had been established (Nelson et al., 1982). The gradient cultures are required to provide proper redox conditions for N_2 fixation by both the freshwater and marine strains (Nelson et al., 1982). Nitrate or ammonia inhibits the fixation of nitrogen, and sulfide is required by some of the beggiatoas to fix nitrogen (Nelson et al., 1982). Polman and Larkin (1988) later found that ammonia and urea inhibit the activity of *B. alba* B18LD nitrogenase *in vitro*, whereas nitrate, nitrite, glutamine, and asparagine do not. Nitrite and nitrate, however, were found to inhibit *in vivo* nitrogenase induction. On the other hand, glutamine stimulates nitrogenase activity of *B. alba* B18LD (Polman and Larkin, 1988).

Autotrophy and sulfide oxidation Winogradsky (1887) developed his theory of chemoautotrophy based on his observations of sulfide oxidation by beggiatoas in slide cultures. The beggiatoas were considered autotrophs until the mid-1900s when Cataldi (1940), Faust and Wolfe (1961), Scotten and Stokes (1962), Burton and Morita (1964), Pringsheim (1964), and Burton et al. (1966) described heterotrophic growth for various freshwater strains of *Beggiatoa*. At that point, controversy began between various investigators about the role of sulfide oxidation in *Beggiatoa* metabolism. Pringsheim and his colleagues claimed that *Beggiatoa* was a facultative autotroph with the ability to grow autotrophically with sulfide used as the sole energy source or

heterotrophically with acetate used as the sole carbon and energy source (Kowallik and Pringsheim, 1966). Burton and Morita (1964) claimed that sulfide oxidation was responsible for the detoxification of peroxides produced during *Beggiatoa* growth. Because sulfide was oxidized in the presence of organic carbon, Pringsheim (1967, 1970) later rescinded his claim of autotrophy and suggested that the beggiatoas were capable of "mixotrophic" growth, obtaining their carbon from acetate and CO_2 and their energy from sulfide oxidation (and perhaps from acetate oxidation). *B. alba* B18LD was postulated to grow mixotrophically, with sulfide or thiosulfate serving as a supplemental energy source (Güde et al., 1981; Strohl and Schmidt, 1984).

On the other hand, very careful experiments carried out by Nelson and Castenholz (1981b) demonstrated that the increased increments in dry weights (~30% increases) brought about by the oxidation of sulfide could be entirely accounted for by the weight of the sulfur itself, indicating that no yield could be accounted for by energy conserved by the sulfide oxidation. Moreover, the continuous culture experiments carried out by Güde et al. (1981) have been brought into question by others (c.f., Kuenen and Beudeker, 1982; Nelson, 1989, 1992), leading most investigators in the field to conclude that mixotrophic growth could not be inferred by the outcome of those experiments. Finally, the oxidation of sulfur to sulfate by freshwater strains is, at best, very inefficient (at least under the laboratory conditions used in the various experiments; Nelson and Castenholz, 1981b; Schmidt et al., 1987; Hagen and Nelson, 1997), indicating that the most favorable energy conservation steps, i.e., oxidation of sulfur to sulfate, were not being utilized efficiently by these freshwater *Beggiatoa* strains. Thus, the overwhelming bulk of data available currently suggest that the thin, freshwater beggiatoas are solely heterotrophic and that the oxidation of sulfide must serve a physiological purpose other than energy conservation.

What purpose does sulfide oxidation serve for these strains, if not for the conservation of energy? As stated above, Burton and Morita (1964) suggested that this process might be coupled to the detoxification of peroxides. Experiments by Nelson and Castenholz (1981b) support that contention. On the other hand, both Nelson and Castenholz (1981b) and Schmidt et al. (1987) have demonstrated that sulfur can be reduced back to sulfide anaerobically. Schmidt et al. (1987) showed that this reduction can be coupled with the oxidation of H_2, an energetically favorable reaction; thus, perhaps sulfur is utilized as a short-term electron acceptor for maintenance metabolism under anaerobic growth conditions; this is somewhat analogous to the role suggested for nitrate metabolism by the wide, vacuolated marine beggiatoas (McHatton et al., 1996).

Although all of the experiments to date have fallen short in proving mixotrophic growth for freshwater beggiatoas, there are three bodies of evidence that are intriguingly suggestive as to its occurrence. First, experiments by Strohl and Schmidt (1984) and Schmidt et al. (1987) have demonstrated clearly that specific electron transport inhibitors inhibit not only the oxidation of acetate to CO_2, but also the oxidation of H_2S to internally stored sulfur, suggesting a linkage of sulfide oxidation with electron transport. *B. alba* B18LD has also been shown to possess a flavocytochrome *c*, not unlike those linking sulfide oxidation and electron transport in photosynthetic sulfur bacteria (Schmidt and Dispirito, 1990). Second, both strains OH-75-2a and B18LD were shown by Nelson et al. (1989a) to possess RubisCo activity, albeit at insignificant levels with respect to growth. Nevertheless, it is possible that conditions have not yet been found in the laboratory to optimize this activity. Finally, Hagen and Nelson (1997) have recently shown that the freshwater *B. alba* strain OH-75-2a possesses a regulated, sulfide-inducible sulfite:acceptor oxidoreductase, suggesting that perhaps under certain conditions, these strains might be capable of oxidizing sulfur to sulfate, a distinctly "lithotrophic-like" metabolism. While the rate of sulfur oxidation is very slow under the conditions tested thus far, oxidation does appear to occur. These three lines of data are still the most critical evidence to date for the potential conservation of energy from H_2S oxidation by heterotrophic freshwater strains. Nevertheless, without convincing orthogonal data to substantiate the mixotrophy claims, the freshwater strains should still be considered to be (only) heterotrophic. Perhaps the evolutionary progenitors of these freshwater strains possessed the ability to couple sulfide and sulfur oxidation with energy conservation mechanisms, but lost it, retaining the vestigial biochemical remains that we see today.

There is no controversy about the autotrophic nature of the marine strains. Careful experiments carried out by Nelson and colleagues (Nelson and Jannasch, 1983; Nelson et al., 1986b; Hagen and Nelson, 1996, 1997) have shown definitively that two thin, marine *Beggiatoa* strains, MS-81-1c and MS-81-6, can grow autotrophically in gradient cultures. Further experiments have in fact shown that strain MS-81-1c is an obligately chemolithoautotrophic marine *Beggiatoa* strain and that MS-81-6 is a facultatively autotrophic strain, capable of increasing growth yields by the assimilation of organic carbon. The data accumulated on the wide, marine beggiatoas all support chemolithoautotrophic growth with either oxygen or nitrate as the terminal electron acceptor (Nelson et al., 1989b; McHatton et al., 1996).

Ecological characteristics Beggiatoas exhibit four physiological characteristics that describe their ecological niche. First, they appear to be oligotrophic organisms that generally do not tolerate or utilize high concentrations of organic (W.R. Strohl, unpublished observations) or nitrogenous (Vargas and Strohl, 1985b) nutrients. Low concentrations (e.g., ≤ 20 mM) of organic acids and, in some cases, amino acids are apparently optimal for the growth and metabolism of freshwater, heterotrophic beggiatoas. Autotrophic, marine beggiatoas are even more fastidious and may or may not utilize organic nutrients.

Second, beggiatoas are respiratory organisms that couple the oxidation of organic acids (heterotrophic strains; Scotten and Stokes, 1962; Nelson and Castenholz, 1981a; Strohl et al., 1986b) and/or sulfide (autotrophic strains; Hagen and Nelson, 1997), with the reduction of oxygen (Schmidt et al., 1987; Hagen and Nelson, 1996, 1997) or nitrate (McHatton et al., 1996) as the terminal electron-accepting step. It appears that stored S^0 can be utilized, at least by the heterotrophic freshwater strains, as an alternative electron acceptor under strictly anaerobic conditions (Nelson and Castenholz, 1981b; Schmidt et al., 1987). Because the pool of internally stored sulfur would be limiting, however, the sulfur is probably used only as an electron acceptor *in situ* for maintenance purposes while the filaments glide to a new sulfide–oxygen interface. It is not yet known whether autotrophic strains can also utilize sulfur as an electron acceptor.

Third, beggiatoas are microaerophilic bacteria (Larkin and Strohl, 1983) that exhibit a phobic response to oxygen (Møller et al., 1985). Microaerophilic bacteria have been described as "aerobic or facultatively anaerobic bacteria, which under aerobic conditions use O_2 as hydrogen acceptor in a respiratory way of energy production and show optimal growth at low oxygen ten-

sions" (Stouthamer et al., 1979). Heterotrophic beggiatoas apparently produce hydrogen peroxide in potentially autolytic quantities (Burton and Morita, 1964); however, peroxide accumulation has not been quantified. High levels of nutrients usually induce post-growth autolysis of *Beggiatoa* (Faust and Wolfe, 1961; Scotten and Stokes, 1962; Strohl and Larkin, 1978b), which may be directly related to the amount of peroxide produced. Catalase is not produced to degrade the potentially toxic peroxides (Burton and Morita, 1964; Joshi and Hollis, 1976; Strohl and Larkin, 1978b; Nelson and Castenholz, 1981b). Addition of catalase to growth medium may enhance growth yields (Burton and Morita, 1964) or survival of certain *Beggiatoa* strains (Strohl and Larkin, 1978b). In a carefully performed set of experiments, however, Nelson and Castenholz (1981a) showed that catalase does not increase growth rates or yields significantly over cultures lacking catalase. Moreover, their experiments indicated that peroxides are not formed during exponential growth phase (Nelson and Castenholz, 1981b).

Finally, and most importantly, beggiatoas are gradient organisms that live in sediments, in which they occupy the horizontal interface between sulfide emanating upward from below and O_2 diffusing downward from above (Jørgensen and Revsbech, 1983; Nelson et al. 1986b). Jørgensen and his colleagues (Jørgensen and Revsbech, 1983; Møller et al., 1985; Nelson et al., 1986b) have described the sulfide/oxygen interface location of *Beggiatoa* mats and plates both *in situ* and *in vivo*. In marine *Beggiatoa* mats, gradients of O_2 and sulfide are created by the beggiatoas above and below the horizontal mats, respectively (Jørgensen and Revsbech, 1983). In creating such concentration gradients, the beggiatoas apparently do not produce high concentrations of most nutrients at their surface, a phenomenon probably preferential to their oligotrophic nature. O_2 in high concentrations, especially if coinciding with concentrations of carbon nutrients, is apparently toxic to the beggiatoas, probably due to formation of peroxide or other toxic product. Thus, the vertical gradient effect, which permits only minute amounts of oxygen to the organisms, is seemingly ideal for them (Nelson et al., 1986b). Stimulation of certain enzymes, such as nitrogenase (Nelson et al., 1982) and RubisCo (Nelson and Jannasch, 1983), and even the ability to grow (Nelson and Jannasch, 1983) may depend on the attributes of the gradient effect (i.e., gradients of oxygen and sulfide in which the organism selects its best position).

Beggiatoas possess unique physiological mechanisms to maintain their proper position in nutrient gradients. Nelson and his colleagues have elaborated a series of phobic responses by beggiatoas that guide their mat-forming behavior. First, the beggiatoas were found to possess a photophobic response that results in a daily migration of beggiatoas to the surface at night and down into the sediments during the sunlight hours (Nelson and Castenholz, 1982). It was later discovered that beggiatoas also possess phobic responses to both O_2 (Møller et al., 1985) and sulfide (Jørgensen and Postgate, 1982), resulting in the movement to the proper position in sediments at which the two substrates overlap (Jørgensen and Revsbech, 1983; Nelson et al., 1986b). A significant difference is observed between the freshwater, heterotrophic strains and the marine, chemolithoautotrophic strains in gradient cultures. The marine strains utilize sulfide much more efficiently, with the sulfide and oxygen gradients overlapping by only ~50–100 µm (D. Nelson, personal communication), whereas sulfide and oxygen overlap significantly in gradients of the less efficiently sulfide-oxidizing freshwater strains (Nelson et al., 1986b). The rapid gliding motility exhibited by beggiatoas allows them to position themselves at the optimal nutrient concentrations and to move quickly as environmental conditions change.

The interwoven, horizontal mats formed by wide, marine beggiatoas found near the hydrothermal vents in the Guaymas Basin can be up to 30 cm in thickness and are virtual monocultures of *Beggiatoa* filaments (Nelson et al., 1989b; McHatton et al., 1996). This is in striking contrast to mats formed by freshwater filaments, which typically reach only 1–2 mm in thickness at most, as well as to mats of ~1–2 cm formed by marine beggiatoas in other locations (Larkin et al., 1994). Marine *Thioploca* spp., on the other hand, form vertical columns of filaments reaching up out of the sediments, indicating that they possess an ecology that is somewhat unique when compared to the beggiatoas found in similar environments (McHatton et al., 1996; Jørgensen and Gallardo, 1999).

Beggiatoas of various morphological and physiological types have been found in environments as diverse as hydrothermal vents at 2000 m of depth (McHatton et al., 1996), desert hypersaline pools (Lackey et al., 1965), freshwater and marine marsh sediments, mineral and sulfur springs (Lackey et al., 1965), run-offs from thermal springs and geysers, activated sludge in sewage treatment plants (particularly those utilizing rotating biological contractors; Chung and Strom, 1997), microbial mats, and sediments ranging from sandy and pristine to polluted and silty. In marine microbial mats, a succession of colorless sulfur bacteria occur, starting with *Thiospira* and *Macromonas*, followed by *Thiovulum*, and then finally *Beggiatoa* spp. (Bernard and Fenchel, 1995).

An environment in which *Beggiatoa* strains appear to play a significant "economic" role is the rice rhizosphere, another gradient environment, set up by oxygen transport by the rice plants to their roots and the surrounding submerged, anoxic sediments (Pitts et al., 1972; Joshi and Hollis, 1977). A symbiotic relationship between the rice plants and *Beggiatoa* has been postulated in which the rice plants provide not only the gradient environment, but also catalase to the beggiatoas to protect them from toxic oxygen, while the beggiatoas protect the plants from sulfide by oxidizing it to harmless elemental sulfur (Joshi and Hollis, 1977).

Finally, a marine *Beggiatoa* spp., along with the cyanobacterium, *Phormidium corallyticum*, and sulfate-reducing bacteria, has been observed to be associated with black band disease of coral (Carlton and Richardson, 1995; Richardson, 1996). While not directly implicated in causation of the disease itself, the *Beggiatoa* is hypothesized to be a stable and contributing member to the consortium of bacteria associated with the presence of the disease (which is most likely caused by sulfide generated by the sulfate reducers) (Carlton and Richardson, 1995; Richardson, 1996).

Genetic characteristics Little has been done to characterize the beggiatoas genetically. The molecular mass of the *B. alba* genome, as characterized by C_0t analysis, is 2.02×10^9 (3.03×10^3 kilobase pairs; Genthner et al., 1985). Thus, the *B. alba* genome is approximately the same size as that of *Escherichia coli* (Genthner et al., 1985). Three *B. alba* strains contain plasmids with molecular masses of $12.3–12.8 \times 10^6$ (18.9–19.7 kb; Minges et al., 1983); however, no function has been ascribed to these plasmids. No phages specific for infection of *Beggiatoa* species have been found, either in sediment samples or endogenously in the genome of four strains tested (W.R. Strohl, unpublished data). No mutants or strains containing auxotrophic markers have yet been described for any *Beggiatoa* species.

ENRICHMENT AND ISOLATION PROCEDURES

Freshwater strains Keil (1912) isolated *Beggiatoa* by washing the filaments repeatedly with sterile stream water. The washed filaments were placed in Petri dishes containing stream water and were incubated under a bell jar containing the gases H_2S (7.8 Pa), O_2 (147 Pa), CO_2 (245 Pa), and H_2. It has been suggested that Keil's methods have been the only isolation procedures to be specifically designed to enrich for and isolate autotrophic strains.

Cataldi first observed *Beggiatoa* during her studies with iron bacteria (Cataldi, 1940), and she used the extracted hay technique that Winogradsky had developed for enrichment of iron bacteria. Recent investigators (Joshi and Hollis, 1976; Strohl and Larkin, 1978b) have successfully used various modifications of Cataldi's boiled hay procedure. Burton and Lee (1978) used an alternative procedure in which gallon jars were filled with raw sewage and aerated for 3 days, after which bundles of *Beggiatoa* filaments were observed on the surface.

For enrichment of salt marsh or estuarine beggiatoas, a marine salt mix can be added to a typical hay enrichment to approximate 50% seawater (Strohl and Larkin, 1978b). An excellent enrichment procedure for marine beggiatoas uses the sulfuretum concept of Baas-Becking (1925). A 1–2-inch layer of sea sand is covered by a 1-inch layer of sulfide-emanating mud and a few inches of 50–80% seawater. Decaying leaves (or boiled hay) and $CaSO_4$ (~1 g/l) are then added. Two thirds of the tank is darkened by wrapping in aluminum foil, and a light is shined on the remaining third so that a natural sulfur cycling takes place between the photosynthetic and the dark zones, as well as between the aerobic and reduced zones. Combined cyanobacterial, photosynthetic bacterial, and *Beggiatoa* mats develop in about 2 weeks and remain stable for months.

Depending on the nutritive conditions of the hay enrichments and the agar plates upon which the beggiatoas are isolated, it is likely that a wide variety of different nutritional types of beggiatoas can be obtained (Strohl and Larkin, 1978b). Cataldi (1940) used the gliding motility of *Beggiatoa* strains to isolate them. She placed washed filaments from enrichments into the center of a Petri dish containing 0.1% yeast extract, 0.6% ethanol, and 1.2% agar. After 48 h, the contaminant-free trichomes at the edge of the plates were picked up by a Pasteur pipette and were placed into the liquid medium (Cataldi, 1940). Most investigators have used various modifications of Cataldi's isolation procedures (Faust and Wolfe, 1961; Scotten and Stokes, 1962; Pringsheim, 1964; Burton and Lee, 1978; Strohl and Larkin, 1978b). The following is a general isolation procedure for *Beggiatoa* strains. Filaments from enrichments are washed 5 times with sterile basal salts containing 5 mM neutralized Na_2S. Tufts of filaments are transferred from one wash bath to the next using microforceps. The tufts of filaments are teased and prodded to remove as much macroscopic contaminating material as possible. The washed tufts are placed onto a dry 1.6% agar plate for about 1 min to absorb away as much of the excess fluid as possible. They can be touched with a corner of filter paper to further remove excess fluid (Burton and Lee, 1978). The tufts of filaments are then removed from the "drying plate" and placed onto pre-scored (Burton and Lee, 1978), freshly prepared 1.4% agar plates containing 2 mM Na_2S, 0.01% each of yeast extract and sodium acetate, and basal salts (Vargas and Strohl, 1985a). After 5–48 h, the plates may be observed with a dissecting microscope to detect any contaminant-free filaments, which, along with a small amount of the agar beneath them, are removed using flame-sterilized 23–26-gauge needles. It is usually best to wait for at least 24 h, so that the contaminated filaments are overgrown, and to make sure that the pure filaments are far away from the contaminants, making them easy to isolate.

For growth of most freshwater beggiatoas, a defined medium can be constructed that contains the following nutrients: 0.2–10.0 mM sodium acetate, succinate, or malate as carbon source, ~2.0 mM ammonium or nitrate as nitrogen source (Vargas and Strohl, 1985b), and standard basal salts (Vargas and Strohl, 1985a). An apparent requirement for a high concentration of calcium and a low concentration of phosphate is notable. A reduced sulfur source, such as sulfide or thiosulfate, can be added at concentrations of 1–2 mM to the heterotrophic medium above. Increasing carbon nutrients shortens the survival time of the organism, perhaps due to higher levels of the peroxides produced. A good general purpose medium for growth, isolation, and maintenance of most freshwater beggiatoas is one modified from the MY medium described by Larkin (1980). It contains 0.01% each of sodium acetate, yeast extract, and nutrient broth (Difco, Detroit, Michigan), 2 mM neutralized $Na_2S \cdot 9H_2O$, and basal salts (Vargas and Strohl, 1985a). For routine maintenance, Na_2S can be added prior to autoclaving, after which the medium should appear gray due to production of FeS. About 90% of the sulfide is lost upon autoclaving (Vargas and Strohl, 1985a). The medium should be inoculated immediately after cooling so that some of the sulfide can be used before it is chemically oxidized or volatilized. For critical experiments, the sulfide should be neutralized and then autoclaved separately under a nitrogen atmosphere in sealed Wheaton serum bottles. It may then be added to the medium just prior to inoculation.

Marine strains To date, the only marine strains that have been successfully cultured in the laboratory are two relatively thin, marine, chemolithoautotrophic strains—one obligately autotrophic and the other facultatively autotrophic (Nelson and Jannasch, 1983; Hagen and Nelson, 1996, 1997). One of the keys to the study of these thin, marine beggiatoas was the development and use of gradient cultures using gel-stabilized sulfide/oxygen gradients (Nelson et al., 1982, 1986a; Nelson and Jannasch, 1983). Wide, marine strains have been studied, however, by carefully washing the trichomes from thick marine mats in order to reduce the contamination to virtually background noise levels. These types of studies have generated significant physiological information about these strains, allowing investigators to determine their overall physiological and phylogenetic parameters.

Additionally, Lackey et al. (1965) reported that they were able to culture wide, brackish water beggiatoas in mixed-culture environmental chambers, such as the sulfuretum described above, for several months, although those of intermediate size ranges (described as *"Beggiatoa arachnoidea"* [5–14 µm diameter] and *"Beggiatoa mirabilis"* [15–21 µm diameter]) did not survive under those conditions.

MAINTENANCE PROCEDURES

The best media for general maintenance of freshwater beggiatoas are usually low nutrient media with adequate sulfide or thiosulfate concentrations and relatively low oxygen concentrations. The modified MY medium described above is a good general medium for the maintenance of freshwater beggiatoas. Biphasic cultures can be constructed by overlaying a 1.6% agar plug of MY medium, modified by increasing the sulfide concentration to 4 mM, with either MY liquid medium or MY and 0.2% semi-solid agar medium. Normally, cultures will survive about 1 month under any of these conditions. For growth and maintenance of marine cultures, refer to methods described by Nelson et al. (1982) and Nelson and Jannasch (1983).

Beggiatoa strains have been preserved for long term viability by freezing them in the presence of 20–30% glycerol at $-70°C$ or $-196°C$. Survival at $-196°C$ is at least several months for the strains tested.

DIFFERENTIATION OF THE GENUS *BEGGIATOA* FROM OTHER GENERA

Although the beggiatoas make up a relatively heterogeneous group, considering the major differences observed between the thin, freshwater strains and the wide, vacuolated, marine strains, they can still be differentiated as a group from members of other genera (Table BXII.γ.40). The primary distinguishing characteristics include their abilities to glide and deposit sulfur when grown on sulfide, their inability to produce sheaths or holdfasts, and their absence of pigmentation.

TAXONOMIC COMMENTS

Much has been learned about the physiology, ecology, and phylogeny of beggiatoas since the last edition of *Bergey's Manual of Determinative Bacteriology* (Strohl, 1989)—particularly in reference to the large, marine strains—that bear on *Beggiatoa* taxonomy. The three general groupings listed in Table BXII.γ.39 are heterogeneous and should not be construed as "taxonomic" designations; nevertheless, they do represent the three major physiological, morphological, ecological, and phylogenetic groups of

TABLE BXII.γ.40. Characteristics differentiating the genus *Beggiatoa* from other phylogenetically or phenotypically similar genera[a]

Characteristic	*Beggiatoa*	*Thioploca*	*Thiothrix*	*Vitreoscilla*
Colorless	+[a]	+	+	+
Strains deposit sulfur	+	+	+	−
Ensheathed	−	+	+	−
Holdfast	−	−	−	+
Gliding motility	+	+	+	+
Filament formation	+	+	+	±[b]
Individual cells observable within filaments	−	−	−	±
Class:				
Betaproteobacteria				+
Gammaproteobacteria	+	+	+	−

[a]Symbols: +, >90% positive; −, >90% negative.

[b]Different strains of this species give different results, and some individual strains do not give consistent results.

beggiatoas studied to date. As exemplified by the notable exceptions below, discovery and characterization of additional strains in various niches may significantly alter this profile in the future.

Currently, there are three possible exceptions to the generalized groupings given above. Both the marine strain L1401-14 ("strain 14") and a Pringsheim marine strain isolated from a marine estuary near Plymouth exhibit a requirement for saltwater and grow heterotrophically (Pringsheim, 1964). Also, Uphof (1927) reported the presence of wide (i.e., >10–20 μm diameter) beggiatoas in Florida freshwater sulfur springs; Lackey and Lackey (1961) later showed that many of the Florida sulfur springs are brackish, and suggested that Uphof may have observed large beggiatoas similar to those seen at Warm Mineral Springs, Florida, which apparently require salt for growth (Lackey et al., 1965; Larkin et al., 1994). The third possible exception is the most intriguing—the finding of a natural population of freshwater *Beggiatoa* filaments from Hunters Hot Springs that shows activities of RubisCo comparable to those of autotrophic marine strains (D.C. Nelson, personal communication). Further physiological and phylogenetic investigations of this group of organisms could clarify the taxonomy of the members of the genus *Beggiatoa*.

Phylogenetic analyses have greatly clarified the taxonomy of these organisms, but they have also opened new questions about taxonomic relationships, particularly between the wide, marine beggiatoas and the thioplocas (see the chapter on *Beggiatoa* and the chapter on *Thioploca* in this volume). All of the beggiaoas thus far analyzed fall into the *Gammaproteobacteria*, which is most closely aligned with the thioplocas (see the chapter on *Thioploca* in this volume and Teske et al., 1995, 1999; Ahmad et al., 1999; Jørgensen and Gallardo, 1999). Teske et al. (1995) have pointed out that positioning of the beggiatoas and thioplocas in the *Gammaproteobacteria* places them developing in the Proterozoic Period, suggesting that they are relatively "young" organisms phylogenetically.

While the resemblance of *Beggiatoa* filaments to those of cyanobacteria caused Pringsheim to label them as colorless forms of cyanobacteria (Pringsheim, 1949), none of the *Beggiatoa* species is phylogenetically related to any cyanobacterium (Reichenbach et al., 1986; Stahl et al., 1987). In addition, as pointed out by Stahl et al. (1987), phylogenetic analysis of beggiatoas and similar organisms has demonstrated that gliding motility is not a particularly useful taxonomic or phylogenetic marker. Additionally, the separation of *Beggiatoa* spp. from *Thioploca* spp. has traditionally been dependent on the formation of sheaths by the thioplocas. Several phylogenetic studies have now shown that wide, marine, autotrophic beggiatoas are more closely related to both the wide, marine and narrow, freshwater thioplocas than they are to the narrow, freshwater heterotrophic *Beggiatoa* strains. Although *Thiothrix* spp. are members of the *Gammaproteobacteria* (Polz et al., 1996), they are not closely related to either *Beggiatoa* or *Thioploca* spp. (Stahl et al., 1987; Lane et al., 1992; Polz et al., 1996; Teske et al., 1999; see chapter on *Thioploca* in this volume).

Recent 16S rRNA sequence analyses have shown that the freshwater *Beggiatoa* strains (OH-75-2a, B15LD, B18LD) cluster very tightly (see again chapter on *Thioploca*), making up the type species (which now can be more narrowly defined as freshwater only). Two other freshwater strains, L1401-13 (Pringsheim, 1964) and AA5A (unpublished strain from Azeem Ahmad; D.C. Nelson, personal communication; (also see *Beggiatoa* and *Thioploca* in this volume) which cluster near the three *B. alba* strains but are clearly different from them.

Sequence analysis of the 5S rRNA placed *Beggiatoa alba* B18LD (neotype strain) and *Vitreoscilla beggiatoides* strain B23SS (neotype strain) in the *Gammaproteobacteria* (Stahl et al., 1987; Woese, 1987; also see chapter on *Beggiatoa* in this volume). Stahl et al. (1987) stated, however, that *V. beggiatoides* was "specifically, but distantly, related to *B. alba*" strain B18LD. Based on the 5S rRNA similarity, it is likely that *Vitreoscilla beggiatoides*—a nonsulfur-depositing species that is similar in other features to the *B. alba* species—would fit within this cluster. However, the 16S rRNA of *V. beggiatoides* has not yet been sequenced. Subsequent analyses may show that deposition of sulfur—the key physiological criterion separating *V. beggiatoides* from freshwater, heterotrophic *Beggiatoa* spp.—may not be a valid criterion for separation of the two genera (just as gliding motility [see again chapter on *Beggiatoa* in this volume] and sheath formation by wide, marine thioplocas [see chapter on *Thioploca* in this volume] have proven to be invalid as taxonomic separators).

The two thin, chemolithoautotrophic marine strains of *Beggiatoa* (i.e., MS-81-1c and MS-81-6) cluster with the *Beggiatoa–Thioploca* group, but clearly are more closely related to the wide, marine beggiatoas than to the thin, freshwater, heterotrophic *B. alba* types. These two strains, however, are clearly separated from each other, both physiologically (Hagen and Nelson, 1996, 1997) and phylogenetically (Jørgensen and Gallardo, 1999(also see again chapter on *Thioploca* in this volume).

Finally, based on 16S rRNA sequence analyses, the wide marine "strains" of *Beggiatoa* from Bay of Concepción and Monterey Canyon form a discrete cluster with the wide, marine, autotrophic thioplocas (*Thioploca araucae* and *T. chileae*), and the freshwater, thin *Thioploca* strain, *T. ingrica* (Teske et al., 1995, 1999; Ahmad et al., 1999; Jørgensen and Gallardo, 1999; see chapter on *Thioploca* in this volume), as well as the newly described marine sulfur microorganism, *Thiomargarita namibiensis* (Schulz et al., 1999a). These phylogenetic data suggest that taxonomic changes are due for the genera *Beggiatoa* and *Thioploca*. While no specific changes can be made at this time, it is clear that rearrangements within at least these two genera, as well as within *Vitreoscilla* (see chapter on *Vitreoscilla* in this this volume), will need to be made based on the phylogenetic data. Additional 16S rRNA sequence analyses and strain–strain nucleic acid hybridizations with a greater variety of beggiatoas, thioplocas, and similar non-sulfur-depositing strains are clearly needed to place these organisms into their proper taxonomic locations with respect to related organisms.

Historical information *Beggiatoa alba* was the only species of *Beggiatoa* recognized in the ninth edition of *Bergey's Manual of Determinative Bacteriology* (Strohl, 1989). This remains true for the present edition, although evidence is accumulating that should allow additional species descriptions by the next version. In the seventh (Buchanan, 1957) and eighth (Leadbetter, 1974a) editions of *Bergey's Manual of Determinative Bacteriology*, six species were recognized, each differentiated on the basis of filament diameter alone: *B. alba* (type species; 2.5–5 μm diameter), "*Beggiatoa arachnoidea*" (5–14 μm diameter), "*Beggiatoa gigantea*" (26–55 μm diameter), "*Beggiatoa leptomitiformis*" (1–2.5 μm diameter), "*Beggiatoa minima*" (<1 μm diameter), and "*Beggiatoa mirabilis*" (15–21 μm diameter). In the eighth version, an additional 18 species names were listed as *species incertae sedis*, differentiated based on filament diameter, habitats and other general characteristics. Until strain B18LD was proposed as the neotype strain of the *B. alba* (Mezzino et al., 1984), species designations in *Beggiatoa* had been made solely based on filament diameter, general morphology, or habitat.

B. alba was originally described in 1803 as *"Oscillatoria alba"*, an organism characterized by "white filaments, of diameter of 1/800th of the length, distance between septa equal to diameter, extremities rounded and not pointed" (Vaucher, 1803). The length of *Beggiatoa* filaments is variable (Strohl and Larkin, 1978a, b), however; thus, the diameter of the first observed *"O. alba"* cannot be calculated. Trevisan (1842) described the genus *Beggiatoa*, in which *"O. alba"* (Vaucher) was included as a species. The two original species of *Beggiatoa*, *B. alba* and *"Beggiatoa leptomitiformis"*, were differentiated by their filament tips being rounded and pointed, respectively (Trevisan, 1842, 1845), not by filament diameter as in more recent years. Strohl (1989) has described in detail how the use of filament diameters has been both arbitrary and inconsistent (i.e., different authors using different filament diameters for the same species) over the years, especially for the thinner strains, making that criterion virtually useless in defining species. For example, historically, *"B. mirabilis"* has been described by different investigators as having a filament diameter (in µm) of 16 (original description), 20–40, 20–38, 20–45, 30–42, 16–45, and 15–45 (Klas, 1936). This interesting, but confusing, state of affairs in *Beggiatoa* taxonomy prompted the deletion of all but *B. alba*, the type species, from acceptance (Skerman et al., 1980), with the hope that the move would induce future descriptions of new *Beggiatoa* species to include physiological and phylogenetic data as well as morphological data. In subsequent descriptions, filament width should be considered as a single characteristic among several that differentiate the various species.

On the other hand, good historical logic and experimental evidence exist that support using filament diameter as a differentiation tool for certain beggiatoas. In various field investigations, beggiatoas have been observed ranging from <1 to >50 µm in diameter. Klas (1936) found that two size ranges of marine beggiatoas, 14.8–21.4 µm (18.4 µm average) and 26.4–49.2 µm (34.4 µm average), existed among trichomes from environmental samples. Likewise, Jørgensen (1978) found clusters in the size classes of 3–5 µm, 8–17 µm, and >23 µm among the marine beggiatoas he observed. McHatton et al. (1996) reported size clusters of 65–85 µm diameter from Clam Field Seep and 24–32 µm, 40–42 µm, and 116–122 µm diameter from Guaymas Basin. Some of these filaments were reported to reach 10 cm in length. Based on the maximum cell length of 36 µm reported in that study (McHatton et al., 1996), this would suggest that some filaments may contain as many as 270,000 cells, which would clearly qualify them as the most "multicellular" procaryotic organisms ever observed. Larkin and Henk (1996) similarly observed clusters of marine beggiatoas with diameters of 20–30 µm, 40–50 µm, and ~110–130 µm. They also observed marine *Beggiatoa* filaments with diameters of nearly 200 µm (Fig. BXII.γ.66) and lengths of over 2 cm (Larkin and Henk, 1996), making these among the largest procaryotes in existence, along with the recently discovered *Thiomargarita namibiensis* (Schulz et al., 1999a). Whereas a continuum of sizes is not observed from individual sites in nature, a compilation of all wide beggiatoas observed by nine different groups of investigators results in a broad continuum of reported sizes of ~7–50 µm, with much larger filament diameters also observed (Fig. BXII.γ.67). Clearly, more phylogenetic data on filaments of different sizes will be required to determine how to differentiate species among the larger marine beggiatoas.

ACKNOWLEDGMENTS

I sincerely thank Bo B. Jørgensen for sharing a preliminary version of the *Thioploca* chapter, which helped to stimulate thoughts about the taxonomy of these organisms, and Douglas C. Nelson for sharing his thoughts and expertise on *Beggiatoa* taxonomy and physiology, as well as for unpublished data and updates on his current research efforts in the area.

FURTHER READING

Ahmad, A.A., J.P. Barry and D.C. Nelson. 1999. Phylogenetic affinity of a wide, vacuolate, nitrate-accumulating *Beggiatoa* sp. from Monterey Canyon, California, with *Thioploca* spp. Appl. Environ. Microbiol. *65:* 270–277.

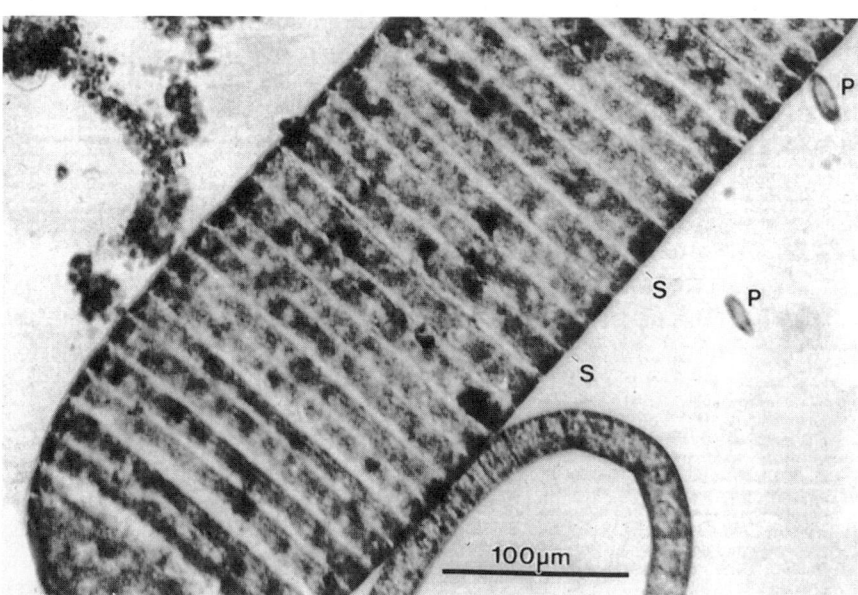

FIGURE BXII.γ.66. A light micrograph of a wide, marine *Beggiatoa* filament with a diameter of ~100 µm. Note how the cells, divided by visible septa (*S*) are the width of the filament but are only ~20 µm in length. Ciliated protozoa (*P*) are present, giving a comparison in size. (Reproduced with permission from J.M. Larkin and M.C. Henk, Microscopy Research and Technique *33:* 23–31, 1996, ©John Wiley & Sons, Inc.)

Marine *Beggiatoa*

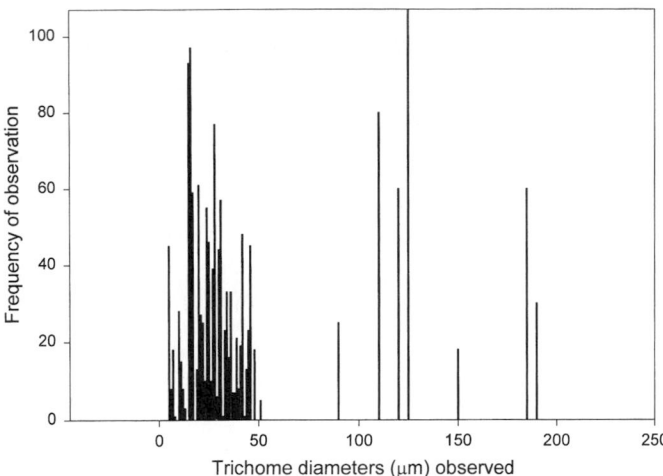

FIGURE BXII.γ.67. A bar graph depicting observations of wide, marine beggiatoas in nature at nine different locations. The Y-axis gives an arbitrary frequency of observation figure, as mentioned by several investigators, as a function of filament diameter (in µm).

Hagen, K.D. and D.C. Nelson. 1996. Organic carbon utilization by obligately and facultatively autotrophic *Beggiatoa* strains in homogeneous and gradient cultures. Appl. Environ. Microbiol. *62*: 947–953.

Hagen, K.D. and D.C. Nelson. 1997. Use of reduced sulfur compounds by *Beggiatoa* spp.: enzymology and physiology of marine and freshwater strains in homogeneous and gradient cultures. Appl. Environ. Microbiol. *63*: 3957–3964.

Larkin, J.M. and W.R. Strohl. 1983. *Beggiatoa, Thiothrix,* and *Thioploca.* Annu. Rev. Microbiol. *37*: 341–367.

McHatton, S.C., J.P. Barry, H.W. Jannasch and D.C. Nelson. 1996. High nitrate concentrations in vacuolate, autotrophic marine *Beggiatoa* spp. Appl. Environ. Microbiol. *62*: 954–958.

Nelson, D.C. 1989. Physiology and biochemistry of filamentous bacteria. *In* Schlegel and Bowien (Editors), Autotrophic Bacteria, Science Tech Publishers, Madison. 219–238.

Nelson, D.C. 1992. The Genus *Beggiatoa. In* Balows, Trüper, Dworkin, Harder and Schleifer (Editors), The Prokaryotes. A Handbook on the Biology of Bacteria: Ecophysiology, Isolation, Identification, Applications, 2nd Ed., Vol. 4, Springer-Verlag, New York. 3171–3180.

Nelson, D.C. and H.W. Jannasch. 1983. Chemoautotrophic growth of a marine *Beggiatoa* in sulfide-gradient cultures. Arch. Microbiol. *136*: 262–269.

Nelson, D.C., C.O. Wirsen and H.W. Jannasch. 1989. Characterization of large, autotrophic *Beggiatoa* spp. abundant at hydrothermal vents of the Guaymas Basin, (Gulf of California, USA). Appl. Environ. Microbiol. *55*: 2909–2917.

Teske, A., M.L. Sogin, L.P. Nielsen and H.W. Jannasch. 1999. Phylogenetic relationships of a large marine *Beggiatoa.* Syst. Appl. Microbiol. *22*: 39–44.

List of species of the genus Beggiatoa

1. **Beggiatoa alba** (Vaucher 1803) Trevisan 1842, 56.[AL]
al' ba. L. adj. *albus* white.

The characteristics are as described for the genus, with the following additional features. The filaments measure about 1.5–4.0 µm in diameter and are of uniform width; the filament diameter may vary with growth conditions. The cells are usually 3.0–9.0 µm long, with filament lengths averaging 60–120 µm. Ends of the filaments are rounded. Necridia and hormogonia may be produced. Circuitans type colonies are usually formed. Sulfur is deposited as inclusions external to the cytoplasmic membrane, but within the cell wall, when the organism is grown in the presence of sulfide or thiosulfate. PHB and polyphosphate deposits are often present.

Chemoorganotrophic growth is obtained by using acetate, fumarate, lactate, malate, pyruvate, succinate, or ethanol as sole carbon and energy sources. The oxidation of sulfide, thiosulfate, or H_2 has been proposed to be coupled with energy conservation for chemolithotrophic metabolism, although this has yet to be proven. Ammonia, nitrate, nitrite, or urea is used as the sole source of nitrogen.

Organisms grow by a respiratory metabolism, with molecular oxygen used as the terminal electron acceptor. Microaerophilic to aerobic. Maintenance under anaerobic conditions, with sulfur used as the terminal electron acceptor, may occur.

Gelatin, starch, and casein are not hydrolyzed. Filaments are catalase-negative, and *N,N,N',N'*-tetramethyl-*p*-phenylenediamine (TMPD) cytochrome oxidase-positive. No growth is observed in the presence of 0.05% sodium dodecyl sulfate or 1.5% NaCl. Strains thus far have been isolated only from freshwater sediments.

The neotype strain, B18LD, was isolated from an enrichment obtained from a rice paddy in Lacassine, Louisiana, U.S.A. This strain, along with other *B. alba* strains, was described in detail by Mezzino et al. (1984).

The mol% G + C of the DNA is: 35–39 (T_m).

Type strain: LSU B18LD, ATCC 33555.

Additional Remarks: Based on phylogenetic and physiological data, other well-characterized strains that now should be considered as *B. alba* strains are OH-75-2a (Nelson and Castenholz, 1981a, b) and B15LD (ATCC 35556; Strohl et al., 1982; Ahmad et al., 1999).

Other Organisms

Among the thin, heterotrophic, freshwater strains, a strain not presently assigned to the species *Beggiatoa alba* is L1401-13 ("strain 13" as described by Pringsheim, 1964; Kowallik and Pringsheim, 1966). The thin, marine strains include the obligately chemolithoautotrophic strain MS-81-1c and the facultatively chemolithoautotrophic strain MS-81-6 (Nelson and Jannasch, 1983; Nelson et al., 1986a; Hagen and Nelson, 1996, 1997)

The wide, autotrophic, vacuolated, marine beggiatoas include, e.g., Monterey Canyon, Guaymas Basin, and Bay of Concepcion suspensions, which are phylogenetically related to marine thioplocas (Table BXII.γ.39) (Jannasch et al., 1989; Nelson et al., 1989b; Ahmad et al., 1999; Jørgensen and Gallardo, 1999; Teske et al., 1999).

Genus IV. **Leucothrix** *Oersted 1844, 44*[AL]

JUDITH A. BLAND AND THOMAS D. BROCK

Leu'co.thrix. Gr. adj. *leucus* clear, light; Gr. n. *thrix, trichis* hair; M.L. fem. n. *Leucothrix* colorless hair.

Long filaments composed of short, cylindrical or ovoid cells (**gonidia**), cross-walls clearly visible, **colorless, unbranched**; typically uniform filaments may taper from base to apex, showing under some conditions an apical beady chain of gonidia connected from end to end. Filaments may attain a length of over 100 μm, and lengths of several hundred micrometers are not unusual (Harold and Stanier, 1955; Pringsheim, 1957). The filaments, particularly when attached to a substrate, may be somewhat tapered, with the apex being 2–3 μm and the base 5–6 μm in diameter (Harold and Stanier, 1955; Pringsheim, 1957; Brock, 1969). In nature, **filaments are usually attached to solid substrates by means of inconspicuous holdfasts; stalks** and **sheaths absent. Filaments do not glide**, but may wave sporadically from side to side. **Dispersal by means of gonidia** (single cells arising from cells of the filaments by rounding up, often released primarily from apices, but they may also be formed in an intercalary fashion); **gonidia often**, but not **always, show jerky gliding motion** on solid substrates. **Rosette formation** is a key diagnostic characteristic of the genus that is seen frequently in laboratory culture but less commonly in nature. The rosettes may be formed of gonidia or, after gonidial growth, of several or more filaments attached at their bases. Filaments in a laboratory culture often form true knots. These structures also occur in nature, but rarely. The organism morphologically resembles some filamentous cyanobacteria but does not form photosynthetic pigments. **Gram negative. Strictly aerobic, heterotrophic. Aquatic; typically marine**, although one freshwater strain has been reported. Strains require NaCl for growth; optimum concentration: about 1.5% NaCl; grows at concentrations of 0.3–6.0% NaCl. Most strains do not require growth factors. Optimum temperature: 25°C; maximum: 30–35°C; minimum: 0°C, forming visible colonies within 1–2 weeks. Strains from tropical waters are more stenothermal, not growing below 15°C. Catalase positive. Oxidase positive.

According to Woese et al. (1985), the "definition" of the *Leucothrix* group (two strains tested—*L. mucor* ATCC 25707 and a "*Thiomicrospira*-like isolate") is the set of 16S rRNA signature sequences consisting of AACCUUAUCCAUCCCUUG, CACUUUCAAUUG, UUAACUUUAG, and AUCUAUAG. The most specific 16S rRNA probe sequence published is CCCCTCTCCCAAACTCTA, positions 652–669 (a non-ATCC culture; Wagner et al., 1994a). Sequences of 16S rRNA sequences are also available for two other isolates (Reichenbach et al., 1986; Nielsen et al., 1998). Also defined is a 23S rRNA sequence (Ludwig et al., 1995).

The mol% G + C of the DNA is: 46.9–51 (Brock and Mandel, 1966; Kelly and Brock, 1969a, b).

Type species: **Leucothrix mucor** Oersted 1844, 44.

FURTHER DESCRIPTIVE INFORMATION

Phylogenetic treatment *Leucothrix* was first isolated in pure culture from a marine source by Harold and Stanier (1955), who characterized the organism "as a chemoheterotrophic counterpart of the colorless sulfur-oxidizing organism *Thiothrix*." The two organisms have many features in common, and the life cycles of both organisms can be superimposed. Both genera were placed in the family *Leucotrichaceae* (Buchanan and Gibbons, 1974). *Leucothrix* and *Thiothrix* were further described as colorless

evolutionary descendants of cyanobacteria (Harold and Stanier, 1955; Pringsheim, 1957; Raj, 1977), since each have a gliding form, the gonidium, and are loosely associated with other gliding, heterotrophic bacteria. Focus on the presence of a sheath in *Thiothrix*, its chemoautotrophy, and its sulfur metabolism eventually distanced this species from *Leucothrix*. Currently, as a result of 16S rRNA studies, *Leucothrix* is classified as a eubacterium belonging to the *Gammaproteobacteria* (Woese et al., 1985). Studies of the 23S rRNA support this classification (Ludwig et al. 1995). The findings of both Woese et al. (1985) and Ludwig et al. (1995) place *Leucothrix* by itself with its relationship to the other *Gammaproteobacteria*.

Cell morphology *Leucothrix* is a complex marine bacterium with diverse morphologies. A commonly seen form, especially in nature, is the filament (Bland and Brock, 1973; Harper and Talbot, 1984; Hansen and Olafsen, 1989), which is composed of individual cells that appear slightly beaded under the light microscope (Fig. BXII.γ.68). Filaments in nature extend perpendicularly outward from the surface to which they are attached via holdfast (Fig. BXII.γ.68). Cell division occurs throughout the filament (Brock, 1967). Filaments are of variable length, often much greater than 100 μm, with a diameter of 2–3 μm or more. They are colorless, unbranched, nonmotile (although they occasionally wave back and forth), and lack a sheath. Storage granules are often visible (Fig. BXII.γ.69). Filaments frequently grow intertwined or in dense tangles. One of the intriguing characteristics of *Leucothrix* is the ability of most strains to form true knots (Brock, 1964). The function of knot formation is unknown, and it may actually be an accidental process, but it seems to be characteristic of most *Leucothrix* strains as part of the growth process. Swollen cells often form apparently at random along filaments. Larger structures (bulbs) usually form in knotty cultures, possibly as the result of fusion of cells in the region of knots.

Gonidia (Figs. BXII.γ.69 and BXII.γ.70) are formed by rounding of cells in the filaments. At a low concentration of an organic energy source, filaments do not grow very long, and gonidia are formed more readily. Gonidial formation is also increased if growth rate is slowed by anaerobiosis, starvation, or reduction of the temperature of incubation (Pringsheim, 1957; Bland and Brock, 1973). Gonidia move by a jerky sort of gliding motility, which can only be seen well when observations are made in slide cultures. Single gonidia may move for extended periods before attaching and dividing to form filaments. If the gonidial density is high enough, gonidia appear to be attracted to each other by homotaxis, and they form aggregates, a process that ultimately results in rosette formation (Fig. BXII.γ.69). Rosettes are also seen in nature (Fig. BXII.γ.70; Solangi et al., 1979; Hansen and Olafsen, 1989). Free gonidia appear to lack holdfast material; however, once a gonidium has attached to a surface or to another gonidium (after elaborating holdfast), it remains attached indefinitely. Gonidia grow into filaments if conditions are favorable. Holdfasts can also be visualized by fluorescence microscopy. The fluorochrome dye primulin, when viewed in blue light, causes holdfast material to fluoresce red and the cell wall to fluoresce yellow.

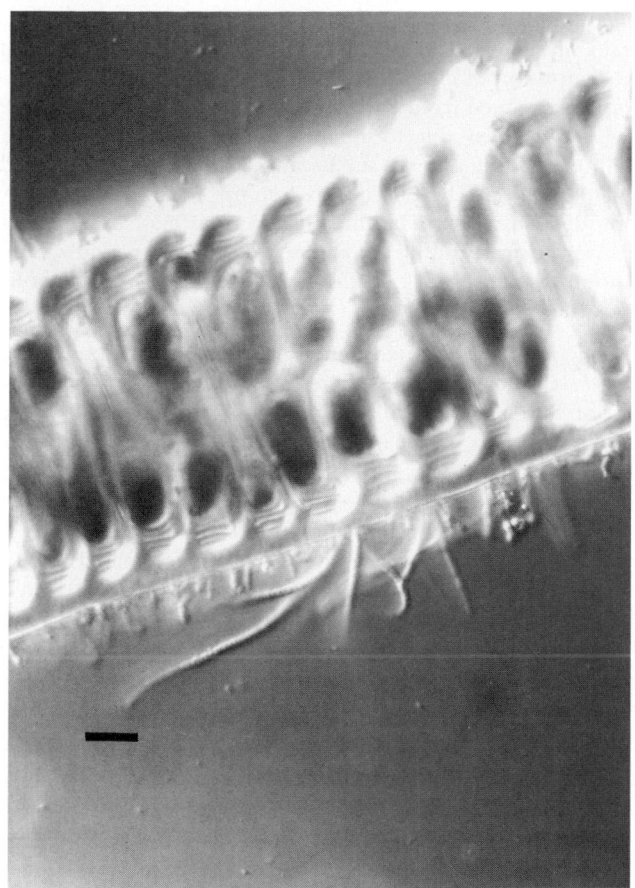

FIGURE BXII.γ.68. Filaments of *Leucothrix* attached to the red alga *Bangia fuscopurpurea*. Bar = 10 μm (Nomarsky interference optics).

FIGURE BXII.γ.69. Rosette of *L. mucor* in laboratory culture, with storage granules visible in the filaments. Free gonidia (*G*) are visible. Bar = 15 μm (Phase microscopy).

Cell wall composition To date, no chemistry for the cell walls of *Leucothrix* has been published.

Fine structure Electron microscopic studies of *Leucothrix* have been presented by Brock and Conti (1969), Snellen and Raj (1970), Raj (1977), Couch (1978), Solangi et al. (1979), and Hansen and Olafsen (1989). The filaments of *Leucothrix* are multicellular, with the individual, somewhat square cells being separated by well-defined cross-walls (Fig. BXII.γ.71). Membranous and storage vesicles are obvious in both filament cells and gonidia (Figs. BXII.γ.71 and BXII.γ.72). The cell wall shows structural features characteristic of the Gram-negative bacterial envelope. No sheath is apparent (Figs. BXII.γ.71 and BXII.γ.72). In gonidia formation, the vegetative cells in the filament become rounded with no apparent internal fine-structural changes (Brock and Conti, 1969) and disassociate from each other (Harold and Stanier, 1955). Released gonidia may either attach singly to a surface via holdfast or aggregate into rosettes and elaborate holdfast at the aggregated cell ends. Figure BXII.γ.72 depicts aggregated gonidia with the associated holdfast material easily seen in the electron microscope as an electron-dense material, possibly a polysaccharide, peripheral to the outer wall layer. A micrograph by Solangi et al. (1979) shows holdfast material external to a brine shrimp structure. A similar finding is demonstrated by Couch (1978). His pictures reveal *Leucothrix* attached via holdfast to the gill cuticle of pink shrimp. The holdfast adheres to the

cuticle (e.g., Fig. BXII.γ.73) and does not penetrate into the organism. Snellen and Raj (1970) have demonstrated that many bizarre forms of *Leucothrix*, including knots and bulbs, can be induced by calcium deficiency, and they have shown the resultant fine structure.

Life cycle The life cycle of *Leucothrix* is shown in Fig. BXII.γ.74. Rosette and gonidia formation are not obligatory; the organism can grow indefinitely in the filamentous form. Gonidia formation occurs in nature, ensuring the dispersal of the organism.

Nutrition and growth conditions *Leucothrix* is a versatile heterotroph that is able to grow on a variety of organic energy sources, including sugars, amino acids, sugar alcohols, alcohols, lipids, and organic acids (Raj, 1977; Williams and Unz, 1985a, 1989; Brock, 1992). Walker et al. (1975) reported utilization of a mixed hydrocarbon substrate by the type strain, ATCC 25107. Strains exhibit variable use of both inorganic and organic nitrogen sources (Williams and Unz, 1989), but *Leucothrix* is unable to fix N_2 (Mague and Lewin, 1974). Many strains do not need growth factors; however, a stenothermal strain required vitamin B_{12} for growth (Kelly and Brock, 1969b).

Metabolism and metabolic pathways The physiology and the metabolism of *Leucothrix* have been extensively reviewed by Raj (1977). *Leucothrix* is a strict aerobe, with an optimum temperature

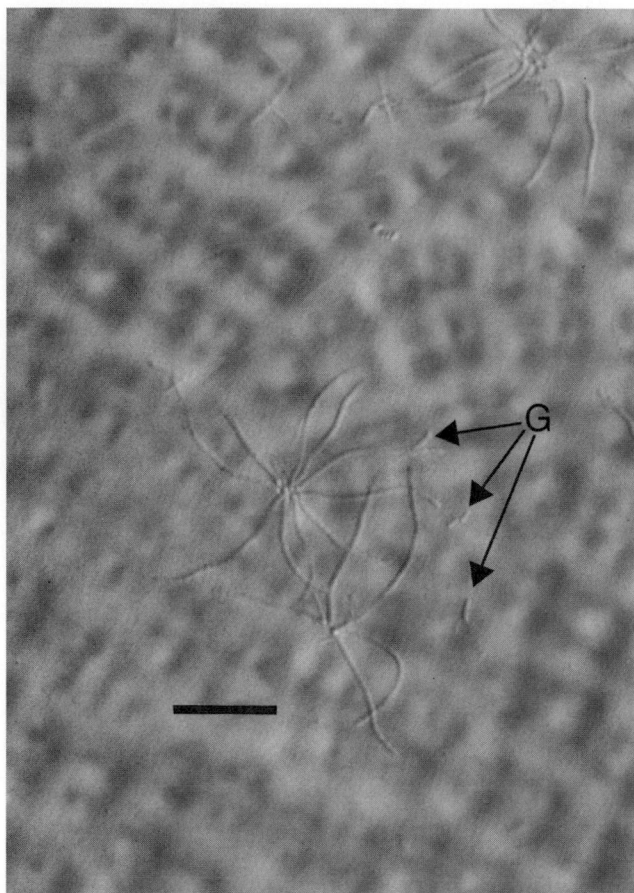

FIGURE BXII.γ.70. Rosette of *Leucothrix* on the surface of the red alga *Porphrya.* Gonidia (*G*) are visible. Bar = 20 μm (Nomarsky interference optics).

for growth of 25–30°C, an optimum pH for growth of 7.6, and an optimum salinity of 1.5%. Bland and Brock (1973) have shown that *Leucothrix* is able to withstand several hours of drying without any apparent reduction in viability, which is consistent with the fact that the organism extensively colonizes *Bangia fuscopurpurea*, a red algal species found high in the intertidal zone and frequently subject to drying during low tides. Radiorespirometric studies have shown that *Leucothrix* oxidizes carbohydrate substrates primarily via the Entner–Doudoroff pathway, with contributions from the pentose phosphate and the Embden–Meyerhoff pathways and the TCA cycle (Raj, 1977). Operation of the TCA cycle is consistent with the presence of isocitrate and malate dehydrogenases (Kelly and Brock, 1969a). The organism is catalase and oxidase positive (Poffe et al., 1979; Williams and Unz, 1985a). The electron transport pathway has been studied by Biggins and Dietrich (1968) using spectrophotometric methods. The terminal electron transport system is a particulate cytochrome chain containing cytochrome b_{562}, cytochrome c_{552}, and cytochrome b_{558}. Cytochrome b_{558} has been proposed to be functionally equivalent to cytochrome *o*, and it operates as the terminal oxidase. Cytochrome b_{558} has high CO-binding properties. The activity of the oxidase system is markedly affected by the degree of aeration during growth. Although several strains of *Leucothrix* have been tested for ability to oxidize hydrogen sulfide, these tests have all been negative (Harold and Stanier, 1955;

Williams and Unz, 1985a). However, K. Eimhjellen of the Technical University in Trondheim has found sulfide-oxidizing ability and deposition of elemental sulfur in a pure marine culture of *Leucothrix* isolated from Pacific Grove, California (Brock, 1992). He has also been able to demonstrate oxidation of thiosulfate to either tetrathionate or sulfate and followed this latter oxidation manometrically. Eimhjellen's strain is obviously different from those studied by Harold and Stanier (1955). In light of the uncertainty surrounding the ability of *Leucothrix* to oxidize sulfur compounds, more work on this aspect of *Leucothrix* metabolism is in order.

Genetics Very limited genetic information is available for *Leucothrix*. Buoyant density determinations of DNA base composition have been determined by Brock and Mandel (1966) and by Kelly and Brock (1969a) for 46 strains isolated from various locations around the world. The mol% G + C content for all strains is in the range of 46.9–51.0, indicating great homogeneity at the DNA level. However, enzyme mobility data for 35 of the strains tested by Kelly and Brock (1969a) has revealed more diversity. Electrophoresis of malate dehydrogenase (MDH) and isocitrate dehydrogenase results in 17 different banding patterns, with MDH showing the most variability. Strains from the same geographical area vary as much as those from different geographical locations.

Antibiotic sensitivity Limited antibiotic sensitivity data exist for *Leucothrix*. According to Raj (1977), growth of *Leucothrix* can be inhibited by penicillin (0.1 mg/l), streptomycin (5.0 mg/l), or chloromycetin (0.7–0.9 mg/l). An isolate tested by Williams and Unz (1985a), *Leucothrix* sp. N11, was found to be completely inhibited by <0.125 mg/l of streptomycin, gentamicin, tetracycline, ampicillin, and penicillin G. The same strain is resistant to sulfanilamide and lincomycin, at >64 mg/l for both antibiotics, and is somewhat sensitive to chloramphenicol (0.5 mg/l) and bacitracin (4.0 mg/l).

Pathogenicity *Leucothrix* is an opportunistic pathogen of marine organisms reared in aquaculture or aquaria. Johnson et al. (1971) showed that *Leucothrix* was able to grow extensively on benthic crustacea, invertebrates, and fish eggs (cod and winter flounder) in laboratory spawning tanks and aquaria where they were infected with *Leucothrix* introduced by the seawater system. Since that report, *Leucothrix* has been found to cause infestations of crustaceans and fish eggs under cultivation in high-density aquaculture situations (Shelton et al., 1975; Couch, 1978; Fisher, 1978; Solangi et al., 1979; Harper and Talbot, 1984; Dale and Blom, 1987; Hansen and Olafsen, 1989; Carr, 1996). Harper and Talbot (1984) were unable to show differences in mortalities of wild and laboratory spawned lobster eggs (*Homarus gammarus* and *H. americanus*) in cases where *Leucothrix* was one of the epiphytes observed. In contrast, Lightner et al. (1975) theorized that penaeid shrimp heavily infested with *Leucothrix* died as a result of oxygen depravation. In a similar vein, Couch (1978) reported that mortality of penaeid shrimp was proportionate to the extent of growth of *Leucothrix* on their gills. He suggested that the mortality may have been due to the massive growth of the bacterium over the gill cuticle, blocking normal gas diffusion across the gill surfaces. Solangi et al. (1979) noted that *Leucothrix* attached to all available surfaces on cultured brine shrimp, *Artemia salina*, and caused mortality. They found that exposure of shrimp to terramycin (100 ppm) for two days eliminated *Leucothrix*, while treatment with various chemicals proved toxic to

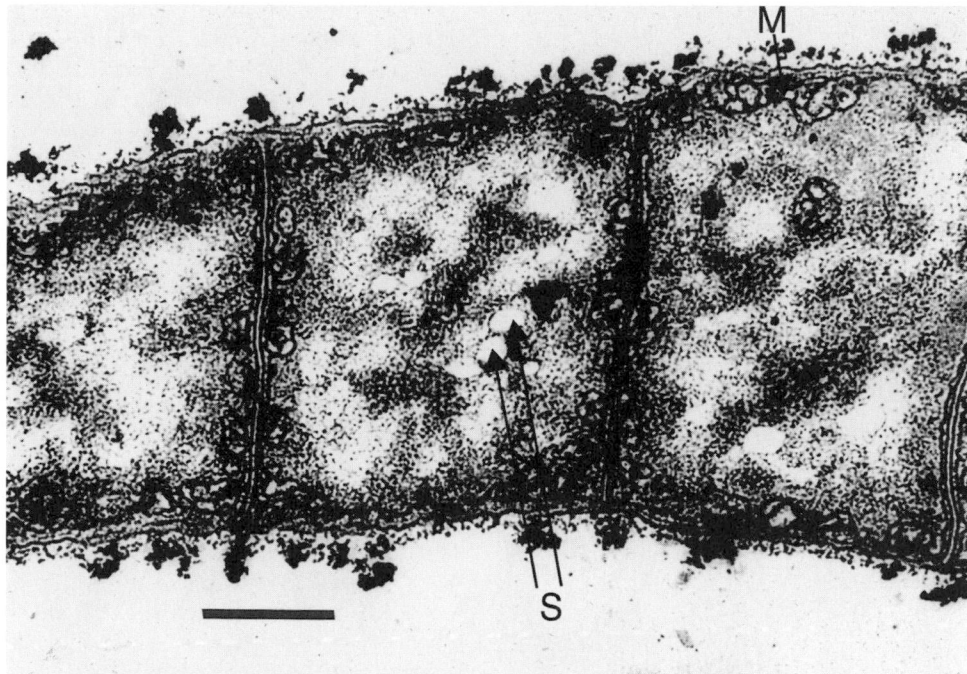

FIGURE BXII.γ.71. Electron micrograph of a thin section of a portion of a filament of *L. mucor*. Membranous vesicles (*M*) and storage vesicles (*S*) are visible. Bar = 0.5 μm. (Reproduced with permission from T.D. Brock and S.F. Conti, Archives of Microbiology *66:* 76–90, 1969, ©Springer-Verlag.)

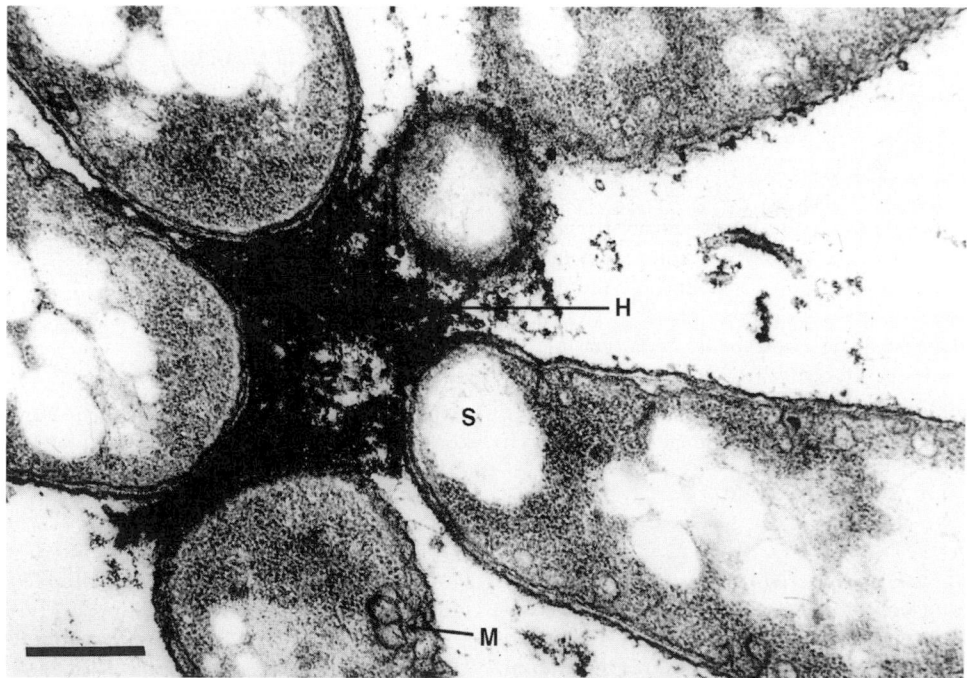

FIGURE BXII.γ.72. Electron micrograph of an *L. mucor* rosette, with the electron-dense holdfast material (*H*) shown. Membranous (*M*) and storage (*S*) vesicles are obvious in both filament cells and gonidia. Bar = 1 μm. (Reproduced with permission from T.D. Brock and S.F. Conti, Archives of Microbiology *66:* 76–90, 1969, ©Springer-Verlag.)

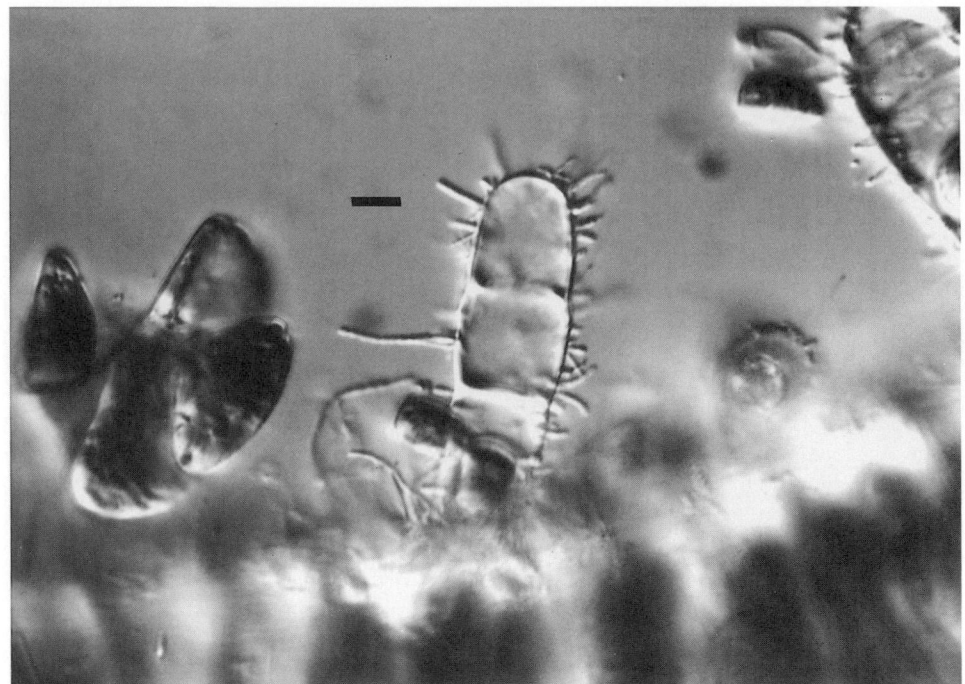

FIGURE BXII.γ.73. *Leucothrix* filaments attached to the cuticle of *B. fuscopurpurea*. Bar = 10 μm (Nomarsky interference optics).

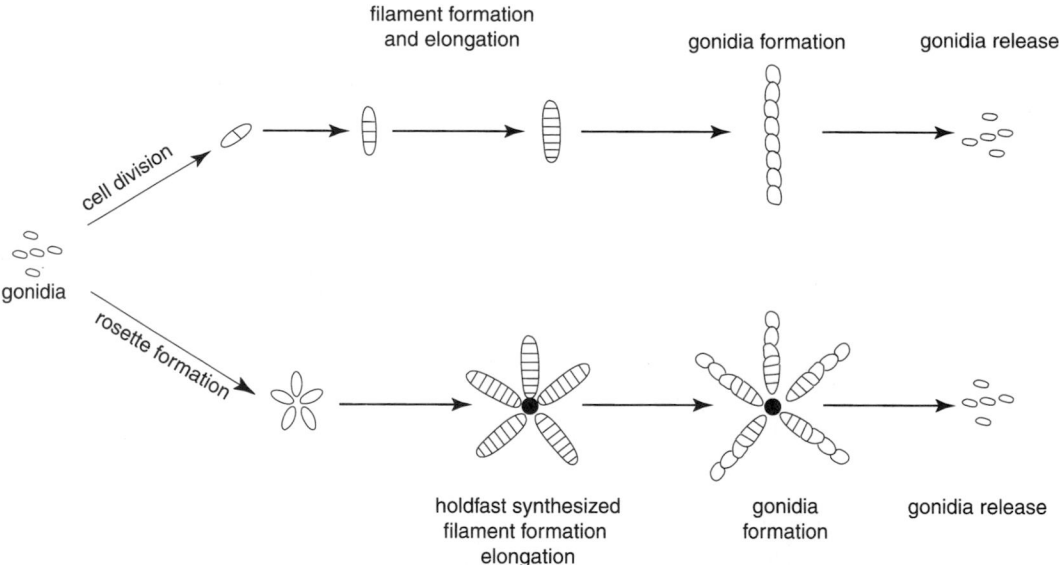

FIGURE BXII.γ.74. Life cycle of *Leucothrix*. Gonidia may attach to a substrate or to each other (rosette formation) with subsequent cell division. Ultimately, filament cells become gonidia that are released to repeat the cycle.

the shrimp. Colonization of uninfected organisms or eggs is apparently rapid. Cod and halibut eggs in aquaculture were colonized within 2 h of fertilization by a variety of bacteria including *Leucothrix* according to Hansen and Olafsen (1989). The authors further report heavier *Leucothrix* growth on cod eggs, giving them a "hairy" appearance that could be mistaken for fungal growth. Antibiotic treatment may be used to prevent such overgrowth. In addition to the antibiotics mentioned above, *Leucothrix* infec-

tions have been treated with penicillin, streptomycin, and neomycin (Johnson et al., 1971; Raj, 1977; Fisher, 1978). Also, Sadusky and Bullis (1994) report that dilute bleach solutions are effective in controlling *Leucothrix* infections in the lobster.

Ecology   *Leucothrix* is found widely in the littoral zone in marine environments worldwide, growing primarily as an epiphyte of micro- and macroscopic algae or animals. The dominant form of *Leucothrix* in the marine environment is the filament, suggesting an advantage to this morphology that might involve entrapment of available nutrients from seawater, better exposure to available oxygen, and access to any materials released by the host. The dispersive, or colonizing, form of *Leucothrix* is the gonidium. *Leucothrix* has been reported to colonize almost every type of marine algae, including species of diatoms and red, green, and brown algae (Brock, 1966; Kelly and Brock, 1969a; Johnson et al., 1971; Bland and Brock, 1973). Some algae are routinely more heavily colonized than others. Bland and Brock (1973) have reported that the red alga *Bangia fuscopurpurea*, a filamentous species living in the high intertidal region, has *Leucothrix* populations 10–30 times larger than those on other algal species. The cuticle of this alga contains mannose (Frei and Preston, 1964), a sugar readily utilized by *Leucothrix* (Bland and Brock, 1973). Johnson et al. (1971) have also reported a preference for red algal species. Brock (1966) has shown that *Leucothrix* can grow on nutrients produced or liberated from algae. However, algae also produce inhibitory chemicals, e.g., sulfuric acid, acrylic acid, and tannins (Bland and Brock, 1973). Furthermore, extracts of 5 of 39 species of corals tested were shown to inhibit a *Leucothrix* culture (Jensen et al., 1996). Although *Leucothrix* is also able to colonize artificial substrates (plastic strips) in the marine environment, Bland and Brock (1973) have shown that growth is much poorer on these strips than on seaweeds, a result that further supports the notion a living host provides a nutritional advantage for *Leucothrix*.

Johnson et al. (1971) have observed *Leucothrix* on many kinds of free-living benthic and pelagic crustaceans, including crabs, lobster, prawns, shrimp, and copepods. Filaments of the organism attach to antennae, pleopods, uropods, and egg masses. *Leucothrix* grows on cultured fish eggs (see section on Pathogenicity), but no data about its occurrence on wild fish has been found. *Leucothrix* may serve as a source of food for some crustaceans. The grass shrimp *Palaemonetes pugio* has been observed to graze on heavy growths of *Leucothrix* on its pleopods while grooming (Johnson et al., 1971). Bauer (1987) has experimentally determined that the growth of *Leucothrix* on the antennae and gills of the stomatopod *Gonodactylus oerstedii* is far heavier on organisms unable to groom than those that were. He has also observed that *Leucothrix* does not colonize the eyes of the organism or most other parts of the body, indicating that some mechanism restricts attachment or growth. McAllen and Hannah (1999) have observed that *Leucothrix* is a major biofoulant of the marine copepod *Tigriopus brevicornus*, a species that has no grooming structure. McAllen and Scott (2000) have additionally reported that the biofouling interferes with swimming behavior.

There are controversial reports about the presence of *Leucothrix* in sewage treatment processes. *Leucothrix*-like organisms have reportedly been associated with activated sludge systems, where they have been implicated in the phenomenon of bulking, and they have been observed in self-purification and other biological waste treatment processes (Eikelboom, 1975, 1977; Cyrus and Sladka, 1970). In most such cases, brackish water was involved (Eikelboom and van Buijsen, 1981), although one strictly freshwater strain, said to be *Leucothrix*, has been isolated from petrochemical wastewater undergoing activated sludge treatment (Poffe et al., 1979). Even this strain grows well at seawater salinity and thus may be considered simply more euryhaline than are other isolates. However, more recent data (Wagner et al., 1994a; Nielsen et al., 1998) from a *Leucothrix mucor* rRNA-targeted oligonucleotide probe did not identify *Leucothrix* in activated sludge samples where organisms morphologically similar to *Leucothrix* were seen. All in all, investigators claiming to have isolated *Leucothrix* from freshwater environments should be encouraged to confirm the culture identity with molecular genetic tools, such as probes.

ENRICHMENT AND ISOLATION PROCEDURES

Although any marine-type seawater salts base may be used in the formation of a culture medium, pH control and the avoidance of metal precipitation are easier in a Provasoli-type culture medium (Provasoli, 1964). The salts formulation (Brock, 1966) in *Leucothrix* medium[1] has proven to be quite effective. The low phosphate concentration in this salts base is critical, as media of more typical phosphate concentrations are frequently inhibitory. Most strains have no vitamin requirements and are able to use glutamate as their sole source of carbon, nitrogen, and energy. If a richer culture medium is desired, 0.1% tryptone plus 0.1% yeast extract can be used. Other media have been described by Raj (1977) and Williams and Unz (1985a, 1989).

In the initial isolation step, it is best to keep the concentration of organic materials low to avoid problems with overgrowth by unicellular bacteria. Although Harold and Stanier (1955) describe a procedure for enrichment of *Leucothrix* from rotting seaweeds, direct isolation from fresh seaweeds is preferable. Rapidly growing, highly motile, unicellular bacteria frequently present problems when isolation of *Leucothrix* from rotted materials is attempted. An effective way of obtaining cultures of *Leucothrix* is to place relatively clean seaweeds taken directly from the sea onto an agar medium that will support the growth of *Leucothrix*. The plates are incubated overnight at 20–25°C and examined within 12–18 h for the presence of *Leucothrix* colonies. This examination is best done by using ×125 magnification (×12.5 eyepiece and ×10 objective) with a long working-distance condenser. It is important to use short incubation periods in order to find *Leucothrix* colonies before they become overgrown by motile contaminants. A *Leucothrix* colony may be recognized by its coiled rope or thumbprint morphology. Colonies are immediately picked by touching them with a sterile insect pin and transferring them in patches to fresh agar plates of the same composition. One advantage of isolating *Leucothrix* in this manner is that the precise habitat of the organism can be recognized; such information is of considerable value in studies on the molecular evolution of the organism (Kelly and Brock, 1969a). Transfer of agar cultures to liquid medium frequently presents problems. It has been observed that if a small inoculum is used in a large volume of liquid medium, growth often does not occur, whereas if the inoculum is placed in a small volume of medium, i.e., 1–2 ml, heavy growth occurs overnight. Once satisfactory growth has been obtained in an initial small-volume liquid culture, it is possible to make transfers to large volumes of liquid medium. In liquid medium, growth is best when flasks are shaken gently, as on a wrist-action shaker or slowly on a rotary shaker. When a rotary shaker is used, growth rate is increased if the flasks contain small internal baffles, made by pushing in the sides of the flasks while heating with an oxygen flame.

---

1. *Leucothrix* medium consists of (g/l deionized water, pH 7.6): Basal salts: NaCl, 11.75; MgCl$_2$·6H$_2$O, 5.35; Na$_2$SO$_4$, 2.0; CaCl$_2$·2H$_2$O, 0.75, KCl. 0.35; Tris(hydroxymethyl)aminomethane, 0.5; Na$_2$HPO$_4$, 0.05; Organic ingredients: monosodium glutamate, 1.0 or tryptone + yeast extract, 1.0 + 1.0.

Identification While an experienced researcher can identify *Leucothrix* simply based on morphological examination with the light microscope, a novice might initially want to work with a known isolate. Other than its distinct morphology and usual occurrence in the marine habitat, there are no conventional diagnostic tests specific for *Leucothrix*. In the extensive isolations of Kelly and Brock (1969a), all of the isolates recognized morphologically had virtually the same physiological properties. The morphological properties described at the beginning of this article are generally found in any *Leucothrix* strain that is isolated. The availability of 16S and 23S rRNA (or other molecular) probes make confirmation possible when identification is critical, and these should especially be used for putative freshwater isolates.

MAINTENANCE PROCEDURES

Once rapidly growing liquid cultures have been obtained (see above), there are no special maintenance problems. Cultures in the early stages of growth can be stored in the refrigerator for several months, after which they may consist mostly of gonidia. Cultures can also be lyophilized, with the best suspension medium being skim milk. When a lyophilized culture is rehydrated, it is best if the dry plug is placed in 1–2 ml of liquid medium and 0.1 ml of this suspension placed in another tube of 1–2 ml of liquid medium. Both tubes are then incubated for 1 or 2 days. If there is no visible growth in the dilution, 0.1–0.2 ml of the tube containing the original rehydrated culture should be transferred to another tube containing 1–2 ml of medium, followed by another incubation period. Only after good growth has been obtained in a small volume of medium should large-volume cultures be prepared. Contamination is easy to recognize, since *Leucothrix* cultures never show uniform turbidity. Microscopic examination of a culture should permit contamination to be readily determined. Streaking onto a seawater agar medium containing tryptone and yeast extract is also a satisfactory way of recognizing many contaminants. If contaminants do become established, they will generally grow better than the *Leucothrix* and as a result will quickly take over the culture.

DIFFERENTIATION OF THE GENUS *LEUCOTHRIX* FROM OTHER GENERA

Molecular techniques using both 16S rRNA (Woese et al., 1985) and 23S rRNA (Ludwig et al., 1995) place *Leucothrix* in a group by itself in the *Gammaproteobacteria*. Woese et al. (1985) consider *Leucothrix* to be a peripheral member of the *Gammaproteobacteria* core and specifically related to the enterics, vibrios, and oceanospirilla. Ludwig et al. (1995) define the position of *Leucothrix* as "somewhat unstable . . . This organism may represent the currently deepest branch of the gamma-subclass of the proteobacteria or a branch at or even below the level of gamma- and beta-subclass relatedness." In short, it is clear what *Leucothrix* is not, but not what it is. *Thiothrix*, the genus with which *Leucothrix* was once paired, is also classed within the *Gammaproteobacteria* along with other sulfur-oxidizing, gliding bacteria (*Thioploca, Beggiatoa*, and *Achromatium*). A head-to-head use of 16S rRNA probes for *Leucothrix* and *Thiothrix* in activated sludge has shown no relatedness, as judged by the lack of cross-reaction (Nielsen et al., 1998). The gliding characteristic is now recognized as not being unique to any group of bacteria; gliders are located widely across the spectrum of the proteobacteria. Relatedness of *Leucothrix* to the cyanobacteria has also been ruled out; Reichenbach et al. (1986), using 16S rRNA techniques, have demonstrated that *Leucothrix* is not a colorless cyanobacterium.

TAXONOMIC COMMENTS

To date, the only recognized species of *Leucothrix* is *Leucothrix mucor*, with nine marine isolates available in the American Type Culture Collection. While the genus has many phenotypic features in common with other organisms, its 16S and 23S rRNA profiles put it in a class by itself. The only characteristic in dispute is the ability of *Leucothrix* strains to oxidize sulfide. Brock (1992) cites work by Eimhjellen that demonstrated the ability of an isolate to oxidize sulfide, sulfur, and thiosulfate. The sulfide oxidation resulted in the internal deposition of sulfur granules, causing *Leucothrix* to resemble *Thiothrix*. Unfortunately, this culture is not available. One of the two organisms used by Woese et al. (1985) for the *Leucothrix* group in his classic 16S rRNA study is identified as a "*Thiomicrospira*-like isolate," a label that suggests sulfide-oxidizing ability. Overall, the *Leucothrix* genus would benefit from new experimental attention to better characterize this possible ability.

FURTHER READING

Brock, T.D. 1992. The genus *Leucothrix*. *In* Balows, Trüper, Dworkin, Harder and Schleifer (Editors), The Prokaryotes-A Handbook on the Biology of Bacteria: Ecophysiology, Isolation, Identification, Applications., 2nd Ed., Springer-Verlag, New York. pp. 3247–3255.

Ludwig, W., R. Rossello-Mora, R. Aznar, S. Klugbauer, S. Spring, K. Reetz, C. Beimfohr, E. Brockmann, G. Kirchhof, S. Dorn, M. Bachleitner, N. Klugbauer, N. Springer, D. Lane, R. Nietupsky, M. Weiznegger and K.-H. Schleifer. 1995. Comparative sequence analysis of 23S rRNA from *Proteobacteria*. Syst. Appl. Microbiol. *18*: 164–188.

Raj, H.D. 1977. *Leucothrix*. Crit. Rev. Microbiol. *5*: 270–304.

Woese, C.R., W.G. Weisburg, C.M. Hahn, B.J. Paster, L.B. Zablen, B.J. Lewis, T.J. Macke, W. Ludwig and E. Stackebrandt. 1985. The phylogeny of purple bacteria: the gamma subdivision. Syst. Appl. Microbiol. *6*: 25–33.

List of species of the genus Leucothrix

1. **Leucothrix mucor** Oersted 1844, 44[AL]

    mu' cor. L. n. *mucor* mold; M.L. n. *mucor* a genus of molds.

    The characteristics are the same as those described for the genus.

    Occurs in marine habitats.

    The mol% G + C of the DNA is: 46.9–51 (T_m).

    Type strain: ATCC 25107, DSM 2157.

    GenBank accession number (16S rRNA): X87277.

Genus V. **Thiobacterium** *(ex Janke 1924) la Rivière and Kuenen 1989b, 496^(VP) (Effective publication: la Rivière and Kuenen 1989a, 1838)*

J. GIJS KUENEN

Thi'o.bac.te' ri.um. Gr. n. *thios* sulfur; Gr. dim. n. *bakterion* a small rod; M.L. neut. n. *Thiobacterium* small sulfur rod.

Rod-shaped cells, each containing one or more sulfur globules. Cells embedded in **gelatinous masses**, which are spherical when free-floating or are dendroid when attached to a solid substrate. **Nonmotile.** No resting stages known. Has not been grown in pure culture.

Type species: **Thiobacterium bovista** (ex Janke 1924) la Rivière and Kuenen 1989b, 496 (Effective publication: la Rivière and Kuenen 1989a, 1838.)

FURTHER DESCRIPTIVE INFORMATION

Rod-shaped cells 0.4–1.5 × 2.5–9 μm with up to nine sulfur inclusions. When these are present, cell masses are white in reflected light and black or bluish in transmitted light. Gram negative (Scheminzky et al., 1972). In the spherical colonies, the cells are embedded in the gelatinous walls of a bladder-like structure filled with water and up to 4 mm in diameter. Such colonies occur near the water surface and have the appearance of groups of puff balls of different sizes. Dendroid colonies show extensive branching and may reach 2–3 mm.

The only species in this genus, *T. bovista*, has been found in marine, brackish, and freshwater environments containing hydrogen sulfide and at temperatures of up to 45°C in thermal springs.

ENRICHMENT AND ISOLATION PROCEDURES

Molisch (1912) reported enrichment of the spherical colony type at the surface of Winogradsky columns prepared with decaying algae, mud, and seawater from the Gulf of Trieste. The dendroid form could not be enriched for but could, after being taken from nature, be kept alive for 3 months in the laboratory at 20°C in jars with water from the original habitat (Lackey and Lackey, 1961). Further details are given by la Rivière and Schmidt (1981). For micrographs, see articles by Lackey and Lackey (1961) and Scheminzky et al. (1972). No pure cultures are known.

TAXONOMIC COMMENTS

Although this genus and its type species do not appear in the Approved Lists of Bacterial Names, maintaining their descriptions in the present edition of the *Manual* and reviving the name appears justified because reports of *Thiobacterium* have occasionally been published (Scheminzky et al., 1972; Caldwell and Caldwell, 1974; Naganuma and Seki, 1991; Hedoin et al., 1996).

List of species of the genus Thiobacterium

1. **Thiobacterium bovista** (ex Janke 1924) la Rivière and Kuenen 1989b, 496^(VP) (Effective publication: la Rivière and Kuenen 1989a, 1838.)
   bo.vis' ta. M.L. fem. n. *bovista* puff ball.

For a description, see that of the genus.
The mol% G + C of the DNA is: unknown.
Type strain: no culture available.

Genus VI. **Thiomargarita** *Schulz, Brinkhoff, Ferdelman, Hernández Mariné, Teske and Jørgensen 1999b, 1325^(VP) (Effective publication: Schulz, Brinkhoff, Ferdelman, Hernández Mariné, Teske and Jørgensen 1999a, 493)*

HEIDE N. SCHULZ AND BO BARKER JØRGENSEN

Thi'o.mar.ga.ri' ta. Gr. n. *thion (theion)* sulfur; L. n. *margarita* pearl; M.L. fem. n. *Thiomargarita* sulfur pearl.

Very large **spherical cells held together** like a string of pearls **by a common gelatinous mucus.** The cells are not joined to each other, which is the main difference to the closely related filamentous genera, *Thioploca* and *Beggiatoa.* **Numerous sulfur inclusions** occur in a thin outer layer of cytoplasm. The cells appear hollow due to a **large central vacuole in which nitrate is stored.** Most strings of cells are linear and typically contain 20–60 cells, but some are branched or coiled. Both single cells and strings of more than 100 cells occur. **Nonmotile.** Gram negative. **Oxygen tolerant.** *Thiomargarita* appear to have an autotrophic or mixotrophic, sulfide-oxidizing, nitrate-reducing metabolism, comparable to the large marine forms of *Thioploca* and *Beggiatoa*, which are its closest relatives according to partial 16S rDNA sequence analysis.

Type species: **Thiomargarita namibiensis** Schulz, Brinkhoff, Ferdelman, Hernández Mariné, Teske and Jørgensen 1999b, 1325 (Effective publication: Schulz, Brinkhoff, Ferdelman, Hernández Mariné, Teske and Jørgensen 1999a, 493.)

FURTHER DESCRIPTIVE INFORMATION

The cell diameters of the only described species, *Thiomargarita namibiensis*, vary between 50 and 750 μm, with the majority of cells being between 100 and 300 μm in diameter. Although most cells are spherical, the shape of the cell is variable. Occasionally, more irregular or barrel-shaped cells (Fig. BXII.γ.75C) occur. The mucus sheath surrounds the single cells completely, which is apparent when cells disintegrate and leave empty mucus pockets (Fig. BXII.γ.75B, left side). The strings of cells seem to grow by lateral division (Fig. BXII.γ.75B, cells in division). Bacteria, including filamentous forms resembling *Thiothrix* and *Desulfonema* (Fig. BXII.γ.75A), are frequently found living attached to the outside of the mucus.

16S rRNA phylogeny and systematics   Partial 16S rDNA sequence analysis shows that *Thiomargarita* belongs to the *Gammaproteobacteria* and is the closest relative to the vacuolated marine *Thioploca* and *Beggiatoa* species. Details of the phylogeny are discussed in the chapter on the genus *Thioploca*.

Morphology and cell structure   Electron micrographs have shown that, similar to the nitrate-storing *Thioploca* and *Beggiatoa* species, the cytoplasm of *Thiomargarita* is restricted to a thin outer

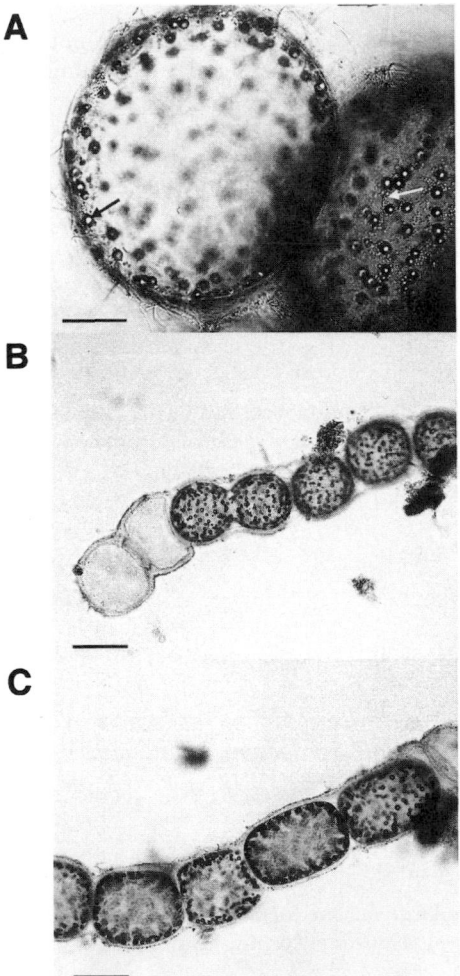

FIGURE BXII.γ.75. *Thiomargarita namibiensis* as it appears under light microscopy. (*A*) The terminal cell of a string. The black arrow points to a sulfur inclusion, the white arrow to a smaller unidentified type of inclusion. The inclusions are restricted to the periphery of the cell. Outside the cell there are various filamentous bacteria, resembling *Desulfonema*, attached to the mucus sheath. Bar = 50 μm. (*B*) To the left, two empty sheaths; in the middle, a dividing cell. Bar = 100 μm. (*C*) A chain of barrel-shaped *Thiomargarita* cells. Bar = 50 μm.

layer of 1–2 μm thickness. About 98% of the cell is comprised of a large liquid vacuole. Apart from the sulfur inclusions along the periphery of the cells (Fig. BXII.γ.75A, black arrow), numerous smaller inclusions which have not been identified can be observed (Fig. BXII.γ.75A, white arrow).

Metabolism   *Thiomargarita* stores nitrate at 0.1–0.8 M concentration and elemental sulfur at a concentration equivalent to 0.4–1.7 M. Growth of *Thiomargarita* can be induced by adding nitrate and sulfide to the natural sediment. Therefore, an autotrophic or mixotrophic metabolism is assumed, by which sulfide is oxidized with nitrate, comparable to the metabolism of the large marine *Thioploca* and *Beggiatoa* species. In contrast to these, however, *Thiomargarita* tolerates prolonged exposure to oxygen up to air saturation, as well as to high sulfide concentrations (~0.8 mM) (Schulz and de Beer, 2002). *Thiomargarita* has a remarkable capability to survive starvation. Sediment samples maintained at 5°C without addition of sulfide or nitrate contained intact cells after more than two years. Without addition of nitrate or sulfide, no dividing cells were observed and cell numbers decreased.

Ecology   The only known species, *Thiomargarita namibiensis*, was found in shelf sediments off the Namibian coast. The upwelling area off Namibia is characterized by high plankton productivity, which leads to sediments rich in organic material and with high sulfate reduction rates. In contrast to the upwelling area off the Pacific coast of South America, where the related *Thioploca* species thrive, the upwelling intensity off Namibia fluctuates more. *Thiomargarita* occurs in high density (about 50 g/m²) in the area of Walvis Bay. The population seems to be associated with an unusual type of sediment, a fluid, sulfidic diatom ooze. The highest density of *Thiomargarita* was observed directly at the sediment surface, declining exponentially with depth. As *Thiomargarita* cells are not motile, but are highly resistant towards nitrate starvation and high oxygen concentrations, it is assumed that they take up nitrate only occasionally, when nitrate-rich bottom water is introduced into the sediment, possibly co-occurring with elevated oxygen concentrations. Between such events, *Thiomargarita* is able to survive high concentrations of sulfide accumulating in the sediment. Due to its large volume (50–750 μm in diameter), which is mainly due to a central vacuole with nitrate concentrations up to 800 mM, *Thiomargarita* has a large storage capacity for its electron acceptor, nitrate. Sulfide, produced by sulfate-reducing bacteria, is permanently available as electron donor in the organic-rich sediment. Thus, *Thiomargarita* can continuously oxidize sulfide, although the electron acceptor, nitrate, is only intermittently available.

MAINTENANCE PROCEDURES

Thiomargarita may be grown in a 2–3 cm thick layer of natural sediment on agar containing 0.5 mM sulfide and covered by oxic or anoxic seawater containing 1 mM nitrate. A frequent resuspension of the sediment enhances growth, whereas addition of organic substrates such as acetate or glucose has no clear effect.

DIFFERENTIATION OF THE GENUS *THIOMARGARITA* FROM OTHER GENERA

In contrast to the closely related genera *Thioploca* and *Beggiatoa*, *Thiomargarita* forms filaments, but does not form trichomes. The extremely large, nonmotile cells are surrounded by a mucus that keeps them attached to each other in a string. The ability to survive under atmospheric oxygen concentration is unique among the vacuolated, nitrate-storing sulfur *Bacteria*.

1. **Thiomargarita namibiensis** Schulz, Brinkhoff, Ferdelman, Hernández Mariné, Teske and Jørgensen 1999b, 1325[VP] (Effective publication: Schulz, Brinkhoff, Ferdelman, Hernández Mariné, Teske and Jørgensen 1999a, 493.)

na.mi.bi.en' sis. M.L. gen. n. *namibiensis* of Namibia.

For a description, see that of the genus.
The mol% G + C of the DNA is: unknown.
Type strain: No pure culture has been isolated.

Genus VII. **Thioploca** Lauterborn 1907, 242[AL]

BO BARKER JØRGENSEN, ANDREAS TESKE AND AZEEM AHMAD

Thi.o.plo' ca. Gr. neut. n. *thion, theion* sulfur; Gr. fem. n. *ploke* a braid, a twist; M.L. fem. n. *Thioploca* sulfur braid.

Flexible, uniseriate, colorless filaments made up of numerous cells and enclosed by a **common gelatinous sheath**. Gram negative. Cells are disk-shaped or cylindrical, generally harbor **sulfur inclusions** in their cytoplasm, and are separated by distinct cross-walls. Cells of large marine *Thioploca* species **appear hollow due to central vacuoles in which nitrate is stored**. Individual filaments are of uniform diameter, often with **tapered terminal cells**. Filaments fall into distinct classes of **different diameters, which represent different species**. Filaments show **independent gliding movement** within a sheath. They can emerge from the end or from breaks in the sheath and may be found outside sheaths. **Chemotactic.** May be attracted by nitrate but avoid high oxygen and sulfide concentrations. *Thioploca* sheaths are mostly vertically orientated to allow the filaments to move between spatially separated pools of electron donor (sulfide) and acceptor (nitrate, possibly oxygen). The sheath surrounding several filaments is the main feature distinguishing the genus *Thioploca* from the similar, closely related genus *Beggiatoa*. Not in pure culture. Large, marine *Thioploca* species have an autotrophic or mixotrophic, sulfide-oxidizing, nitrate-reducing metabolism. By 16S rRNA, *Thioploca* is a member of the *Gammaproteobacteria* **and is the closest relative of the genera *Beggiatoa* and *Thiomargarita*.**

Type species: **Thioploca schmidlei** Lauterborn 1907, 242.

FURTHER DESCRIPTIVE INFORMATION

Thioploca filaments fall into distinct diameter classes, which are the traditional basis for species designation (Table BXII.γ.41, Fig. BXII.γ.76). At least two different groups of *Thioploca* species are apparent, the smaller freshwater and brackish water species, of which *T. ingrica* is the best-known representative, and the large marine species, which include *T. araucae* and *T. chileae*. Diameters of accepted species range from 0.8–43 µm, but occasionally marine sheathed *Thioploca* of up to 125 µm diameter have been observed (H.N. Schulz and B.B. Jørgensen, unpublished observations). The large, marine filaments are easily visible with the unaided eye.

Thioploca typically grow in bundles surrounded by a common sheath (Figs. BXII.γ.77 and BXII.γ.78A). The number of fila-

ments in a sheath ranges from a few to tens but may also reach a hundred. When bundles are observed (and slightly squeezed) in a microscope preparation, intertwining filaments may have the appearance of a braid, giving rise to the name *Thioploca* (Fig. BXII.γ.78C). Each bundle presumably represents a clone, and neighboring bundles may, accordingly, show small but significant differences in mean filament diameter (Schulz et al., 1996). Inspection of many sheaths of the marine *Thioploca* has revealed that ~20% of the sheaths have mixed bundles of two or three of the species *T. araucae*, *T. chileae*, and "*T. marina*". *Thioploca* filaments may leave their own sheath under environmental stress, such as sulfide accumulation in the sediment. They can also survive as free-living filaments and may perhaps subsequently enter another sheath or form a common sheath with another species. Such mixed bundles also occur in the freshwater species; these gave rise to the name "*T. mixta*" (Koppe, 1924), which is no longer in use. The transparent sheath of young and growing bundles is thin and tough but it may become very wide and loose around older and senescent bundles.

16S rRNA phylogeny and systematics Based on near-complete 16S rRNA sequences, filamentous sulfur-oxidizing bacteria of the genera *Thioploca* and *Beggiatoa* form a monophyletic, highly diversified cluster within the *Gammaproteobacteria* that currently consists of three distinct groups (Teske et al., 1995, 1999; Ahmad et al., 1999): *Thioploca* and large, marine *Beggiatoa*; heterotrophic, freshwater *Beggiatoa*; and autotropic, marine *Beggiatoa* (Tree 1 of Fig. BXII.γ.79). The clade of marine, vacuolated, nitrate-accumulating *Thioploca* spp. (*T. araucae* and *T. chileae*), large, marine *Beggiatoa* spp., and *Thioploca ingrica* has 82–100% bootstrap support for near-complete sequences. For partial sequences (Tree 2 of Fig. BXII.γ.79), this clade includes the species *T. araucae* and *T. chileae* as well as the unicellular *Thiomargarita namibiensis* (Schulz et al., 1999a; Schulz and Jørgensen (2001)). The heterotrophic, freshwater *Beggiatoa* and the autotrophic, marine *Beggiatoa* clades do not contain *Thioploca* species (Tree 1 of Fig. BXII.γ.79). The heterotrophic, freshwater *Beggiatoa* clade contains a tight cluster of the *Beggiatoa alba* type strain B18LD (Mezzino et al., 1984) and closely related strains, always supported by

TABLE BXII.γ.41. Differential characteristics of the species of the genus *Thioploca*

Characteristic	1. *T. schmidlei*	2. *T. araucae*	3. *T. chileae*	4. *T. ingrica*	5. "*T. marina*"	6. "*T. minima*"
Size (µm)	5–9	30–43	12–20	2.0–4.5	2.5-5.0	0.8–1.5
Freshwater	+	−	−	+	−	+
Marine	−	+	+	−	+	−

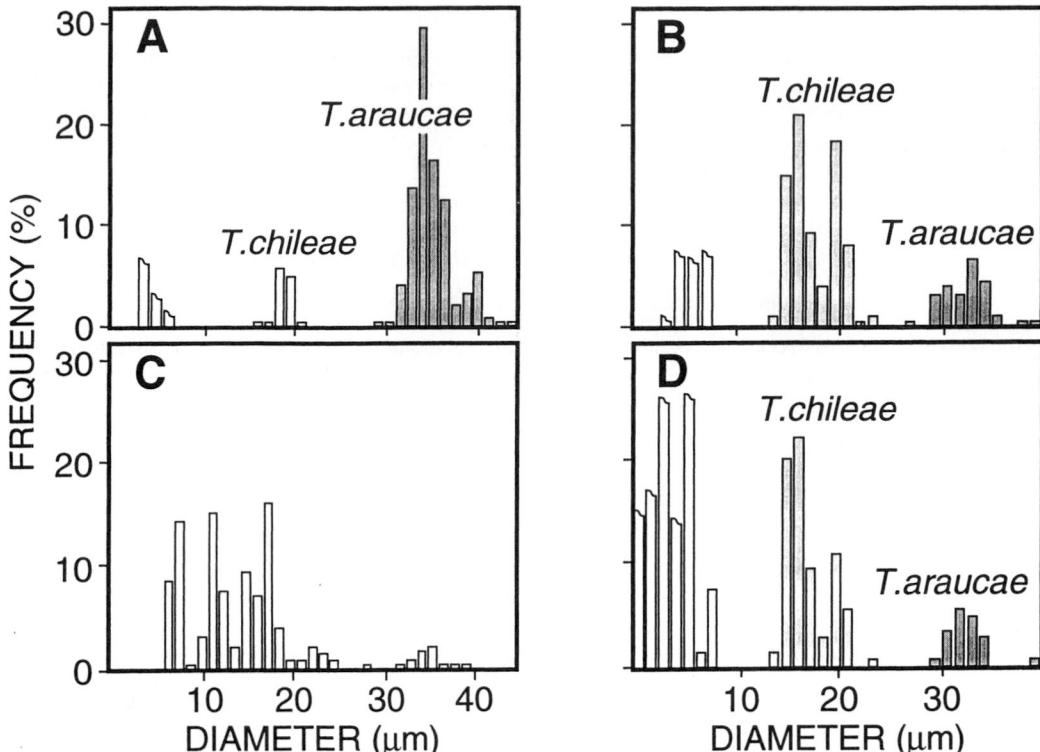

FIGURE BXII.γ.76. The presence of a common sheath and the diameters of trichomes are used to define the genus and species of *Thioploca*. The graphs illustrate the problems of this classification with regard to two marine populations from the surface layer of shelf sediments off the coast of Chile. Several hundred trichomes were randomly picked for each diagram, and the relative frequency distribution of their diameters is shown at ~1 μm resolution. Frames (*A*) and (*C*) are from a station at 40 m water depth and frames (*B*) and (*D*) from a station at 80 m water depth. Frames (*A*) and (*B*) represent trichomes picked from bundles surrounded by a common sheath. Frames (*C*) and (*D*) represent trichomes picked from the abundant free trichomes without visible sheaths. The two species, *T. chileae* and *T. araucae*, are clearly distinguished, as are the filaments of a narrower species which partially corresponds to *T. marina* (the narrow species occurred in larger numbers than indicated). At the shallow station (*A* + *C*), many free trichomes were of diameters not represented by the neighboring thioplocas in sheaths (they should consequently be called *Beggiatoa* spp.). At the deep station (*B* + *D*), there was a good correspondence between diameter distributions of sheathed and free trichomes, thus indicating that the latter were thioplocas outside sheaths. As indicated by the available 16S rRNA sequence database (see Fig. BXII.γ.79) the vacuolated marine thioplocas and beggiatoas are phylogenetically intertwined and, therefore, a distinction based on morphological characteristics (sheath, diameter) is probably not possible or meaningful. (Data from Schulz et al., 1996).

100% bootstrap. The sequence data demonstrate that the physiological and morphological similarities within and among these groups result from evolutionary relatedness.

Within the *Thioploca/Beggiatoa* clade, representatives of both genera are phylogenetically intertwined. The current distinction between the genera *Thioploca* and *Beggiatoa* does not follow phylogenetic lines but is instead determined by a morphological characteristic, the formation of a sheath around a filament bundle. This point is illustrated by a large, marine *Beggiatoa* from the Bay of Concepcion on the Chilean shelf that resembles *T. araucae* in filament diameter, 16S rRNA sequence, nitrate content, and habitat, but that occurs in individual filaments instead of sheathed filament bundles (Teske et al., 1999). The large *Beggiatoa* spp. from Monterey Canyon with 60–80 μm filament diameter (Ahmad et al., 1999) could represent a similar case, since ensheathed *Thioploca* with the same filament diameters are

also known from this habitat (K. Buck, personal communication). Representing a possible challenge to the generality of this interpretation are massive assemblages of vacuolated *Beggiatoa* collected from the Guaymas Basin vents (Nelson et al., 1989b) that show no indication of forming sheathed filament bundles (A. Teske, unpublished data; D. C. Nelson, unpublished data). Parallel observations come from extensive *Beggiatoa* mats in the Gulf of Mexico (D.C. Nelson, unpublished data).

The 16S rRNA data support the classification of *T. araucae* and *T. chileae*, previously defined only by filament diameter, as not mere morphotypes, but separate species. However, filament diameter classes do not always correspond to genospecies or metabolic types; they may instead cover up unexpected genetic and metabolic diversity, as illustrated by the two clades of heterotrophic, freshwater *Beggiatoa* and marine, autotrophic *Beggiatoa* (Tree 1 of Fig. BXII.γ.79). With the exception of strain MS-81-

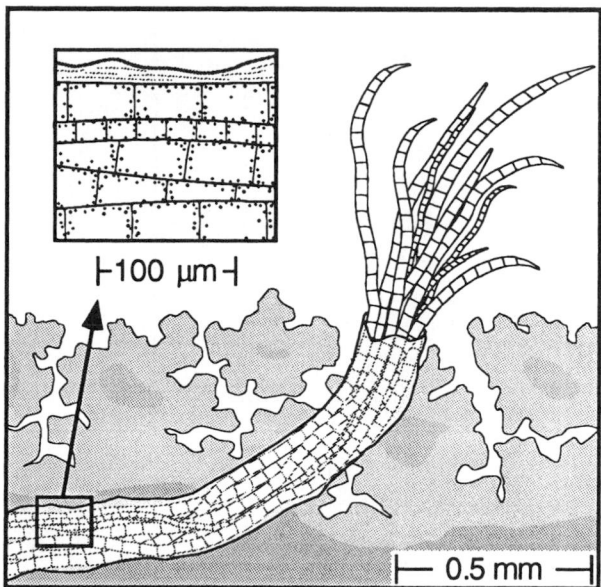

FIGURE BXII.γ.77. Schematic presentation of a bundle of marine thioplocas extending from their common sheath within the sediment and up into the overlying seawater. The insert shows the light-microscopic appearance of the trichomes, with crosswalls and fine sulfur globules along the cell periphery. (Drawing by Heide N. Schulz).

6, these *Beggiatoa* strains all share filament diameters of close to 2 μm. The genetic coherence of morphologically defined, widely distributed *Thioploca* species, for example *T. ingrica*, which is represented only by 16S rRNA sequences of specimens from a Danish brackish water habitat, is therefore not proven.

The thioplocas show some phenotypic similarities with cyanobacteria such as *Microcoleus*, both in morphology and in the formation of sheaths around bundles of filaments. The phylogenetic data clearly show, however, that the filamentous sulfide-oxidizing bacteria are unrelated to the cyanobacteria, which are not members of the *Proteobacteria*, but rather a separate monophyletic bacterial group.

Morphology and cell structure Each filament consists of a single row of cylindrical or barrel-shaped cells separated by septa. The septa may be difficult to observe microscopically in filaments rich in light-diffracting sulfur globules, but in those cases, their position is often indicated by the distribution pattern of sulfur globules (Fig. BXII.γ.78, D and F). For the smaller species with filament widths of <10 μm, the cell height is greater than the diameter, whereas the large marine species have more disk-shaped cells. Interestingly, the two diameter classes of *T. chileae* and *T. araucae* have been found to occur as two distinct populations of morphotypes: the normal type with a height-to-diameter ratio of 0.5:1.5 and a form with more rounded and flat cells having a height-to-diameter ratio of 0.3:1–0.5:1 (Schulz et al., 2000). The large, undescribed, marine thioplocas with diameters >50 μm all belong to this short-cell type. Each trichome contains a hundred to several thousand cells. The large, marine species are generally 1–2 cm long and may reach >5 cm in length (Huettel et al., 1996).

There is a distinct difference in cell structure between the small, freshwater species and the large, marine species of *Thioploca*. Each group shows many similarities to the small and large

Beggiatoa, respectively (Maier and Murray, 1965; Strohl et al., 1982; Maier et al., 1990). In both groups, the 90–130 nm thick cell wall has a complex, four-layered structure, of which only the innermost layer and the cytoplasmic membrane continue across the septum wall separating the cells. Intracytoplasmic membranes and numerous cell inclusions form complex structures, which are apparent from TEM micrographs, and which have functions related to transport and storage. Most of these inclusions are situated in intrusions of the cytoplasmic membrane and are, therefore, extracytoplasmic (Maier et al., 1990). Most prominent are globules of S^0, which are also visible in the light microscope. Membrane-bound vacuoles, which in *Thioploca* contain electron-dense material, are distributed throughout the cytoplasm in *T. ingrica* as well as in the 4 μm wide *Beggiatoa* (Morita and Stave, 1963; Maier and Murray, 1965).

In *T. araucae* and *T. chileae*, the cytoplasm, with its sulfur inclusions, is confined to a 0.5-1.0 μm thick layer along the periphery of each cell (Maier et al., 1990). The whole inner cell volume is filled by a single liquid vacuole surrounded by a cytoplasmic membrane without apparent connection to the cell surface (Fig. BXII.γ.78B). Such a vacuole, which comprises >90% of the cell volume, is also found in the large, marine *Beggiatoa* spp. of >15 μm diameter (Lackey et al., 1965; Nelson et al., 1989b). It serves as a reservoir for nitrate, which the organisms use as an electron acceptor when oxidizing sulfide within the sediment (Fossing et al., 1995; McHatton et al., 1996; Zopfi et al., 2001). The peripheral distribution of cytoplasm, furthermore, provides short transport distances to the external medium and, thus, counteracts the diffusion limitation of these very large procaryotes.

Metabolism Since none of the thioplocas has been isolated in pure culture, information on their physiology is based mostly on observations of whole communities and on experiments with whole bundles of filaments. However, since the organisms are large and conspicuous, it has also been possible to physically isolate individual filaments and to observe their metabolic activity, e.g., by feeding them radiolabeled substrates and subsequently performing microautoradiography.

Thioploca oxidizes H_2S to sulfate and stores S^0 as an intermediate oxidation product in the form of globules in the cells. The S^0 serves as a transportable energy source that makes the organism independent of a continuous source of H_2S. All large, marine thioplocas and beggiatoas that have been examined so far contain high (100–500 mM) intracellular concentrations of nitrate, an accumulation four orders of magnitude over nitrate concentrations in the ambient sea water (Fossing et al., 1995; McHatton et al., 1996). The nitrate is stored in the large central vacuole, which functions as an anaerobic "lung" as the organisms move around in the sulfidic sediment. *Thioploca* uses nitrate as a respiratory electron acceptor for sulfide oxidation and reduces the nitrate principally to ammonium, as does *Beggiatoa* (Vargas and Strohl, 1985b; McHatton et al., 1996; Otte et al., 1999). It is unknown to what extent the smaller, freshwater or brackish water *Thioploca* species utilize nitrate as electron acceptor.

T. araucae and *T. chileae* are able to fix CO_2 and may be able to grow autotrophically but probably mostly grow mixotrophically, similar to *Beggiatoa* and *Thiothrix* (Nelson and Jannasch, 1983; Strohl and Schmidt, 1984; Otte et al., 1999). They can incorporate acetate, mixed amino acids, glycine, glucose, and thymidine in the presence of sulfide, as shown by autoradiog-

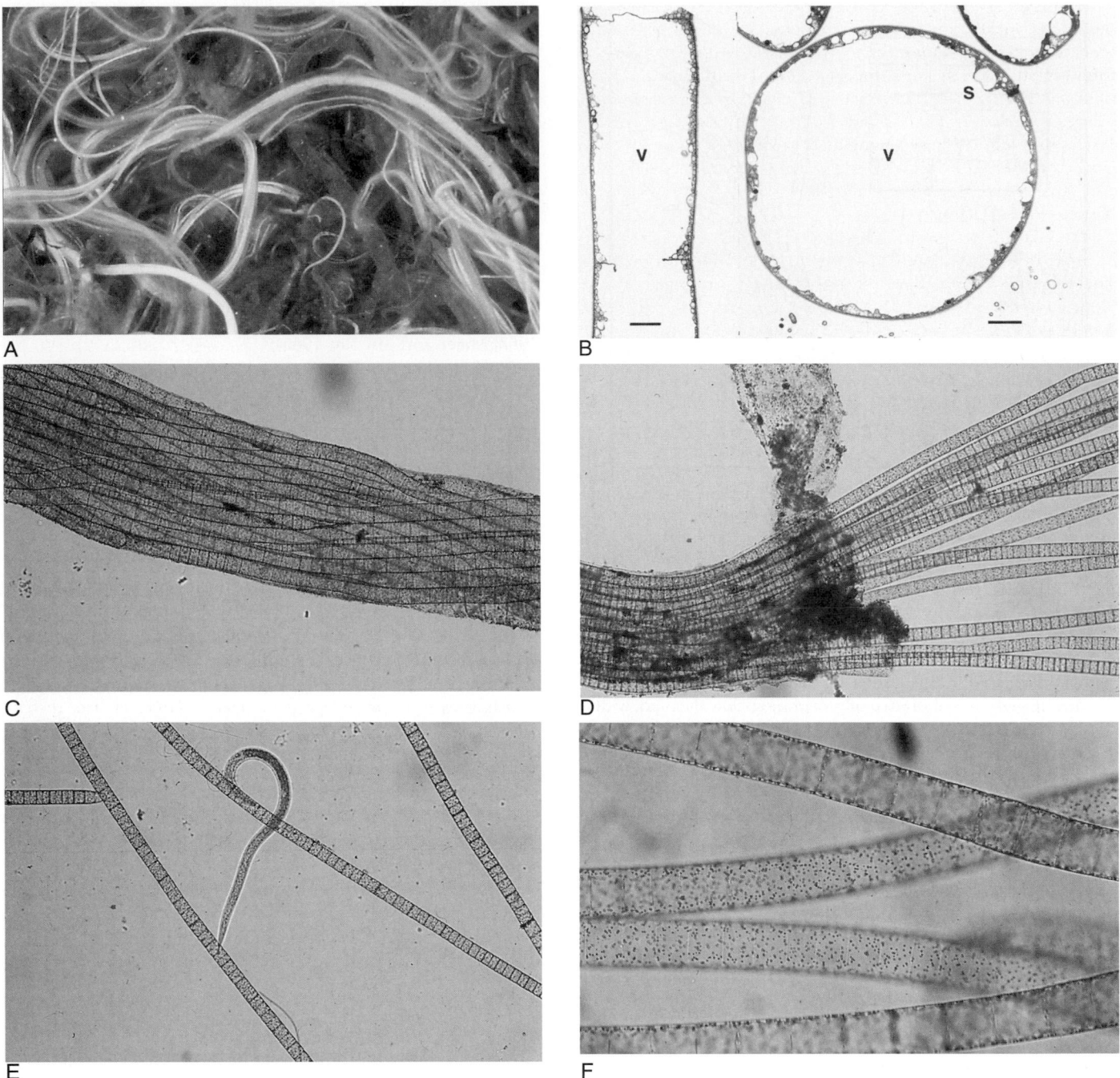

FIGURE BXII.γ.78. Marine thioplocas from the shelf off the coast of central Chile. (*A*) Thiplocas in gelatinous sheaths which were sieved out of the sediment. The frame is 8 mm wide. (*B*) Transmission electron micrographs of *Thioploca chileae* in cross-section (*left*) and longitudinal section (*right*). The cells appear empty due to the large central vacuole. Sulfur globules are stored in the thin peripheral layer of cytoplasm. The filaments are 20 μm in diameter. (*C*) *Thioploca* bundle densely packed in a sheath (not visible). The bundle has the appearance of a braid where the filaments cross over, hence the name *Thioploca*, meaning sulfur braid. (*D*) Bundle of *T. araucae* filaments extending out of their sheath. (*E*) Trichomes of *T. araucae* together with a marine nematode, illustrating the large size of these multicellular prokaryotes. (*F*) Light micrograph of *T. araucae* trichomes showing the dense sulfur globules as fine dots. In the central trichomes, the focus is on the surface, whereas in the outer trichomes, the focus is in the center, thus showing the hollow appearance with sulfur globules only along the periphery. (Part B reproduced with permission from S. Maier et al., Canadian Journal of Microbiology, *36:* 438–448, 1990, ©National Research Council of Canada; A courtesy of M. Hüttel; C–F courtesy of B.B. Jørgensen.)

raphy (Maier and Gallardo, 1984a). Methylotrophy has also been suggested (Morita et al., 1981), but this theory appears to be incorrect since *Thioploca* cannot assimilate methane or methanol.

Furthermore, in the marine habitats of *Thioploca*, methane is produced in insufficient amounts to play a role as a carbon source (Ferdelman et al., 1997).

Ecology *Thioploca* occurs in freshwater, brackish, and marine sediments, mostly under oxygen-depleted water. The organisms seem to prefer low sulfide concentrations, although they are also found in sulfide-rich sediments along the Peruvian coast (Henrichs and Farrington, 1984). Classical localities of the freshwater species are lakes in central and northern Europe (Lauterborn, 1907; Wislouch, 1912; Koppe, 1924; Kolkwitz, 1955), but *Thioploca* is also reported from large lakes in North America, central Russia, and Japan (Maier and Preissner, 1979; Namsaraev et al., 1994; Nishino et al., 1998).

The marine species form dense and extensive communities in areas of coastal upwelling (Jørgensen and Gallardo, 1999). The largest *Thioploca* community is found associated with the Humboldt current on the Pacific continental shelf of South America (Gallardo, 1977). Biomasses here reach 120 g fresh weight/m^2, or up to 1 kg wet weight/m^2 when sheaths are included (Schulz et al., 1996). There is a seasonal variation in the population size, with the highest biomasses occurring during the summer, when the sedimentation of organic matter is highest and the overlying water is strongly depleted in oxygen (Schulz et al., 2000). The population size is strongly depressed during "El Niño," when upwelling is reduced and the overlying water remains oxygenated (V.A. Gallardo and H.N. Schulz, personal communication). Other reported marine habitats include the continental shelf underlying the Benguela current along the South Atlantic coast of Africa (Gallardo et al., 1998; H.N. Schulz, personal communication), the monsoon-driven upwelling area of the northwestern Arabian Sea (Levin et al., 1997), and hydrothermal vent sites in the eastern Mediterranean Sea (Dando and Hooper, 1996).

The upwelling habitats are characterized by very high plankton productivities, high nitrate and low oxygen concentrations in the bottom water, high sedimentation rates of organic matter, and high rates of sulfide production in the sediment (Thamdrup and Canfield, 1996; Ferdelman et al., 1997). The bacteria contribute significantly to the oxidation of sulfide in these sediments and are able to maintain very low H$_2$S concentrations. The winding sheaths have a mostly vertical position in the muddy sediment and may reach 2–5 cm depth for the freshwater species and 10–15 cm for the large marine species (Schulz et al., 1996; Nishino et al., 1998). The sheaths form vertical tunnels through which the filaments can migrate between the sediment-water interface and the bulk sediment (Fig. BXII.γ.80). The filaments glide with a speed of 1–3 μm/s and may thus move ~10 cm/day. Although they are not the only sulfide oxidizers in the sediment, their ability to transport intracellular nitrate reserves down into the sulfide production zone allows them to compete efficiently with other sulfide-oxidizing bacteria that require overlapping or frequently fluctuating gradients of sulfide and an appropriate electron acceptor (Jørgensen and Revsbech, 1983; Nelson et al., 1986a; Gundersen et al., 1992). Dense populations of *Thioploca* may form loose, gelatinous mats of 1–2 cm thickness at the sediment surface. Inhabited sheaths have been found to be densely covered by filamentous sulfate reducing bacteria of the genus *Desulfonema* (M. Fukui et al., 1999), thereby providing a short diffusion path between the H$_2$S producers and consumers.

MAINTENANCE PROCEDURES

Although thioplocas have not been isolated in pure culture, they may be maintained in undisturbed sediment in the laboratory for months or even years. Maier (1984) described the following procedures for freshwater *Thioploca*: They may be maintained in jars overlaid with tap water at 8–20°C in the dark. At approximately yearly intervals, a few stems of extracted grass (Scotten and Stokes, 1962) may be stuck into the sediment; *Thioploca* often colonizes these stems. Alternatively, 0.2–0.3 g pulverized extracted hay is autoclaved in 60 ml tap water in 125-ml Erlenmeyer flasks and inoculated with 4–10 ml sediment. After a month of undisturbed incubation at room temperature, *Thioploca* bundles are added and incubation continues for many weeks, with intermittent inspection.

Attempts to enrich marine *Thioploca* have met little success. They may be maintained for months in undisturbed cores sampled from the natural populations and kept near the *in situ* temperature of 13°C in a basin of anoxic seawater with nitrate added (H.N. Schulz, personal communication).

DIFFERENTIATION OF THE GENUS *THIOPLOCA* FROM OTHER GENERA

Due to their conspicuous morphology and generally large size, members of the genus *Thioploca* can be easily distinguished from most other genera. However, they show a strong similarity and overlap in characteristics with *Beggiatoa* with respect to morphology, ultrastructure, physiology, ecology, and phylogeny. Tapered filament ends are characteristic of *Thioploca*, but these may also have rounded terminal cells with no tapering. Beggiatoas occasionally have a tapered terminal cell, though tapering is less than that for *Thioploca*. The best criterion of differentiation is the formation of bundles surrounded by a common sheath, which has so far been attributed only to the genus *Thioploca*. Sheath-inhabiting thioplocas tend to have a more vertical orientation in sediments than do beggiatoas, which often form whitish films covering solid–water interfaces. However, the reverse orientation may also be found for both species. When triggered by environmental stress, *Thioploca* filaments often glide partly or completely out of their sheaths, and they may perhaps even establish communities without apparent sheath formation, in which case they would be recognized as *Beggiatoa* (Schulz et al., 1996). *Thioploca* can be clearly differentiated from *Thiothrix* by their gliding movement and by the lack of a terminal holdfast or rosette formation. The closely related *Thiomargarita namibiensis* has separated, nonmotile, spherical cells.

TAXONOMIC COMMENTS

Future changes in classification of *Thioploca* and *Beggiatoa* are likely. The range of strains over which the genus designation *Beggiatoa* is used is overly broad. More importantly, the differentiation between *Thioploca* and *Beggiatoa* is currently based on the formation of a common sheath surrounding filament bundles, a characteristic that might vary in response to environmental conditions. In the absence of pure cultures, it may be impossible to prove or disprove whether any natural population of vacuolated *Beggiatoa* will form sheath bundles in some specific environment. The clade (Tree 2 of Fig. BXII.γ.79) comprised of three *Thioploca* strains, two *Beggiatoa* strains, and a *Thiomargarita* strain is united by the possession of a large central vacuole. This feature currently appears to be the best morphological candidate to replace sheath formation as a marker in a revised taxonomy of the group *Beggiatoa–Thioploca*. This marker, in addition to being consistent with 16S rRNA phylogeny, appears to be universally connected to intracellular nitrate accumulation, presumably in the vacuole, for nitrate respiration enabling sustained anaerobic metabolism. This generalization regarding metabolic capacity remains to be confirmed for *T. ingrica*. A future revision of the

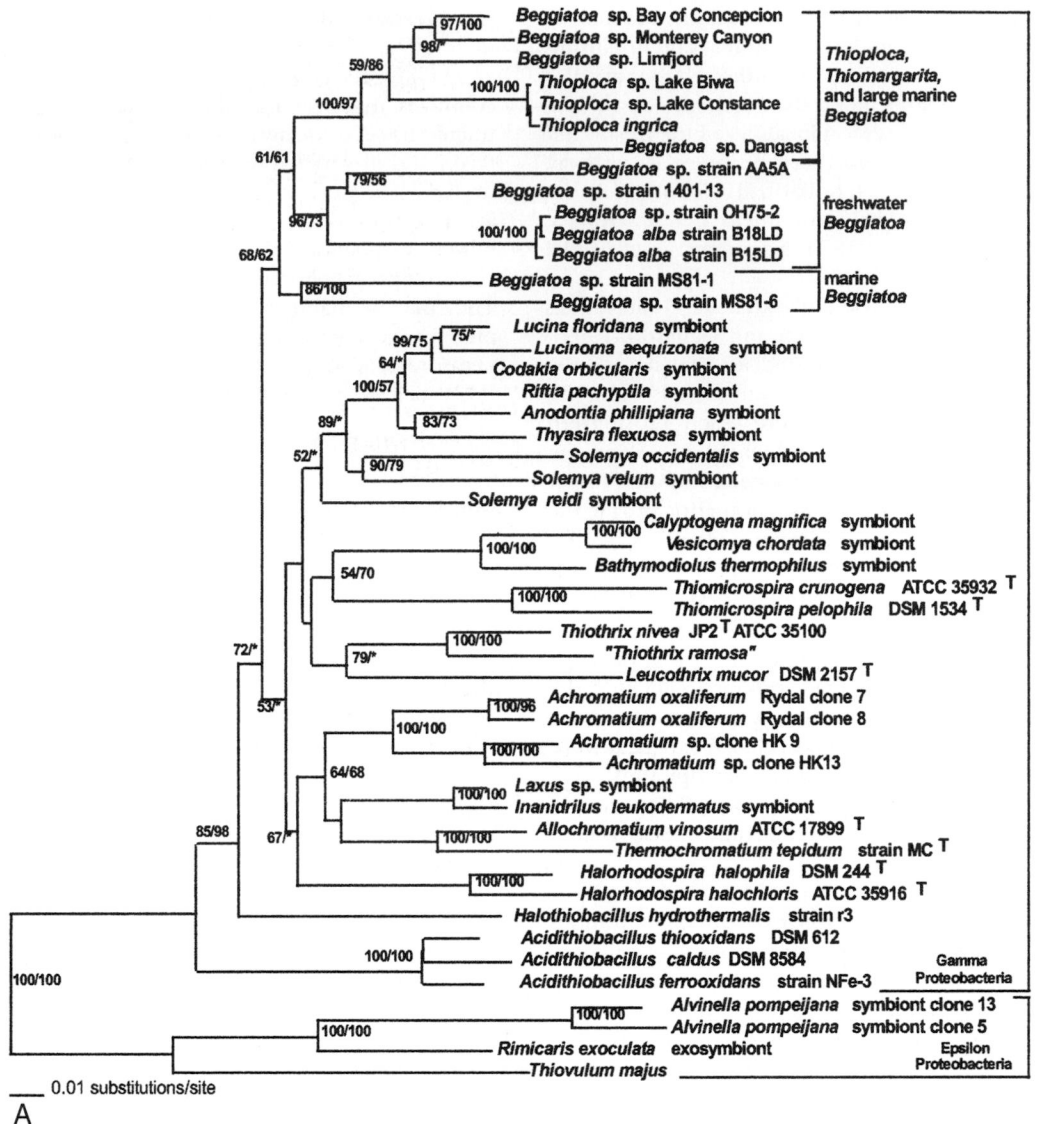

FIGURE BXII.γ.79. Phylogenetic relationships of *Thioploca*, *Beggiatoa*, and *Thiomargarita*, and representative sulfur-oxidizing bacteria of the *Gammaproteobacteria*. The trees are based on nearly complete 16S rDNA sequences (*E. coli* positions 28–1487), (*A*) and on partial sequences (*E. coli* positions 341–941), including those of marine *Thioploca* species and *Thiomargarita namibiensis*, for which only partial sequences are available. (*B*) The trees were routed

genus *Thioploca*, based on the vacuolated, nitrate-respiring phenotype and corresponding 16S rRNA clade, might include these gliding filaments regardless of whether they occur in sheathed bundles. For now, further investigations of natural populations of vacuolated *Beggiatoa* and *Thioploca*, as well as *Thiomargarita*, are required to check the consistency of this concept.

ACKNOWLEDGMENTS

Bo Barker Jørgensen was supported by the Max Planck Society and the Fonds der Chemischen Industrie. Andreas Teske was supported by a DFG postdoctoral stipend and a subsequent WHOI postdoctoral fellowship. Azeem Ahmad, as well as a portion of the phylogenetic analyses reported here, was supported by National Science Foundation award IBN-9513962 to Douglas C. Nelson. We thank Douglas C. Nelson for important contributions to the phylogeny of *Thioploca–Beggiatoa* and the following colleagues for illustrative material: Markus Hüttel, Siegfried Maier, Heide N. Schulz, and Horst Völker.

FURTHER READING

Jørgensen, B.B. and V.A. Gallardo. 1999. *Thioploca* spp: filamentous sulfur bacteria with nitrate vacuoles. FEMS Microbiol. Ecol. *28*: 301–313.

Schulz, H.N. and B.B. Jørgensen. 2001. Big bacteria. Ann. Rev. Microbiol. *55*: 105–137.

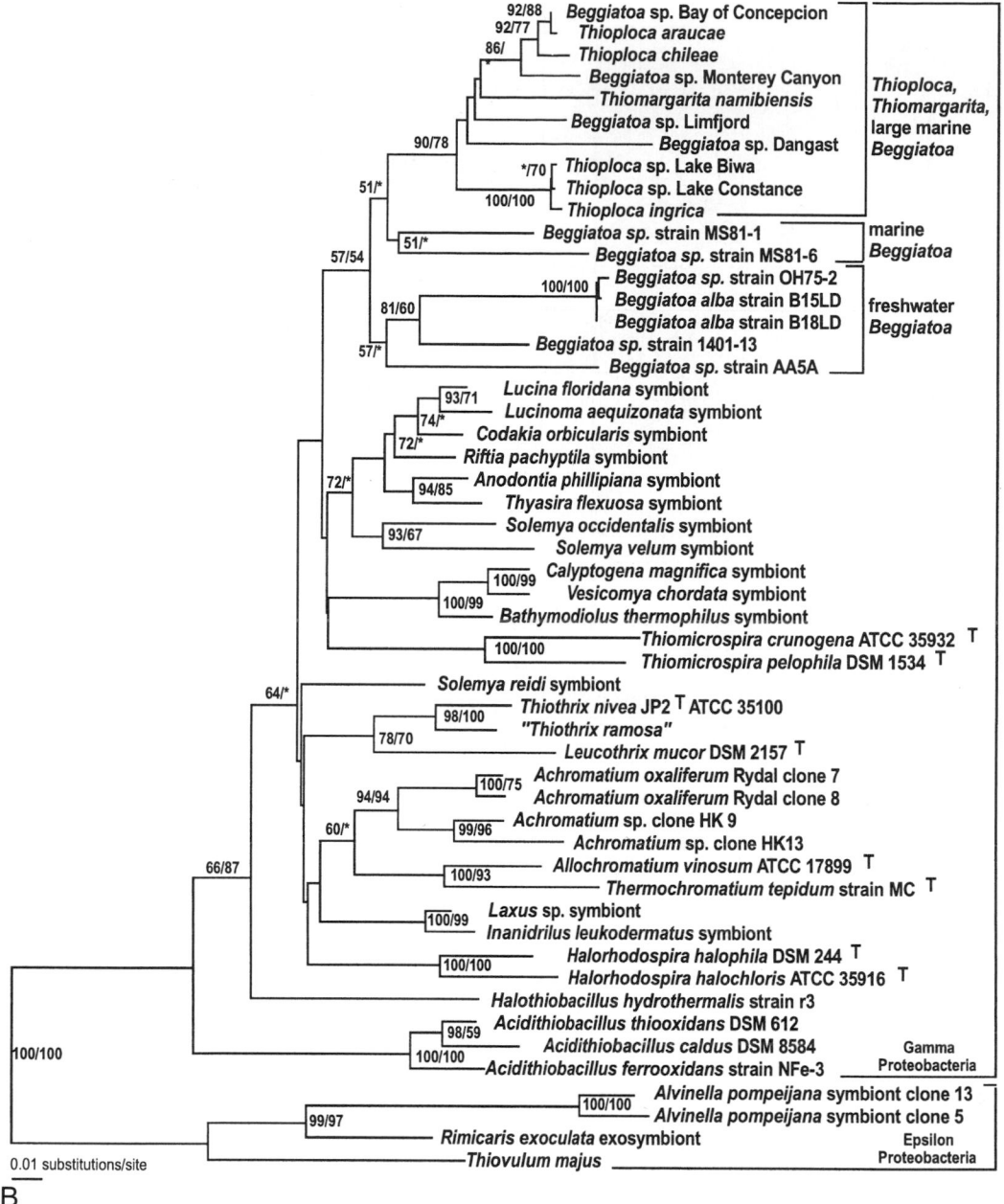

with *Thiovulum majus* of the *Epsilonproteobacteria* as the outgroup. Bootstrap values minimum evolution/parsimony are indicated at the nodes with more than 50% bootstrap support. Bar = 0.01 Jukes-Cantor substitutions per nucleotide.

List of species of the genus Thioploca

1. **Thioploca schmidlei** Lauterborn 1907, 242[AL]
   schmid'le.i. M.L. gen. n. *schmidlei* of Schmidle.

   Identified from sediments of fresh and brackish water localities in Europe and from Lake Baikal, Russia. Originally found in Lake Constance, southern Germany.

   The mol% G + C of the DNA is: not available.
   Type strain: none isolated.

2. **Thioploca araucae** Maier and Gallardo 1984b, 417[VP]
   a.rau'cae. M.L. gen. n. *araucae* of Arauco in Central Chile.

   Identified from oxygen-poor upwelling areas of the

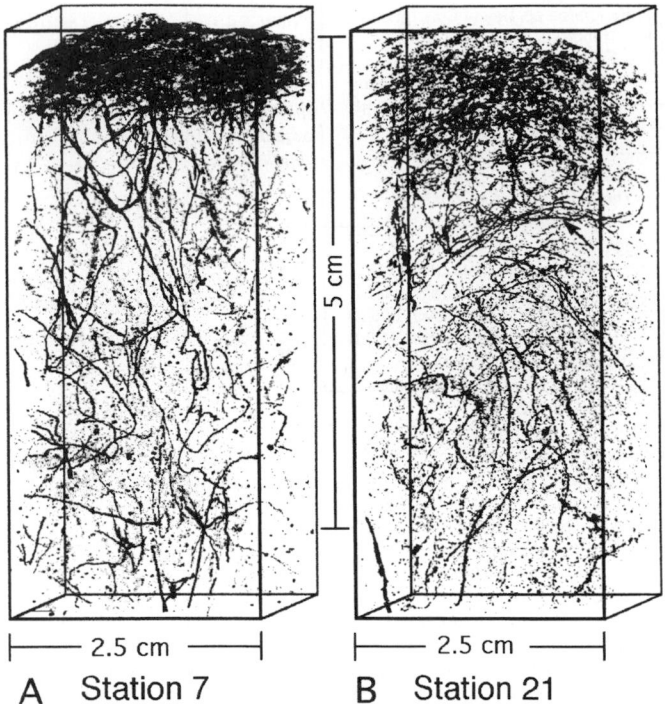

A   Station 7   B   Station 21

FIGURE BXII.γ.80. Three-dimensional reconstruction of marine *Thioploca* sheaths (black lines) in blocks of sediment from the Chilean coast at 40 m (*A*) and 100 m (*B*) water depth. The sediment blocks of 5 × 2.5 × 1 cm were rapidly frozen, cut vertically at 100 μm increments with a cryomicrotome, and photographed in polarized light. The 3-D image was developed from 100 sequential photographs. At the sediment surface, a grayish color shows a dense mat of more horizontally oriented sheaths, plus many free filaments of either *Beggiatoa* or *Thioploca* outside their sheaths. Below the mat, dark strands of *Thioploca* sheaths extend down through the sediment with a mostly vertical orientation. (Reproduced with permission from H.N. Schulz et al., Applied and Environmental Microbiology, *62:* 1855–1862, 1996, ©American Society for Microbiology.)

ocean, e.g., the Pacific continental shelf of South America, the Atlantic coast of southern Africa, and the northwest Arabian Sea.

The mol% G + C of the DNA is: not available.

Type strain: none isolated.

3. **Thioploca chileae** Maier and Gallardo 1984b, 417[VP]

chi' le' ae. M.L. gen. n. *chileae* of Chile.

Identified from oxygen-poor upwelling areas of the ocean, e.g., the Pacific continental shelf of South America, the Atlantic coast of southern Africa, and the northwest Arabian Sea.

The mol% G + C of the DNA is: not available.

Type strain: none isolated.

4. **Thioploca ingrica** Maier 1984, 344[VP]

in' gri.ca. M.L. adj. *ingrica* pertaining to Ingria, ancient district of Leningrad, Russia.

Identified from sediments of fresh and brackish water localities in central Europe and from Lake Erie, USA, and Lake Biwa, Japan.

The mol% G + C of the DNA is: not available.

Type strain: none isolated.

GenBank accession number (16S rRNA): L40998.

Other Organisms

1. *"Thioploca marina"*

ma.ri' na. L. adj. *marina* of marine.

Identified from oxygen-poor upwelling areas of the ocean, e.g., the Pacific continental shelf of South America (Maier and Gallardo, unpublished data).

2. *"Thioploca minima"* Koppe 1924, 630.

Identified from sediments of freshwater localities in Central Europe and Lake Erie.

Genus VIII. **Thiospira** *Visloukh 1914, 48*[AL]

J. GIJS KUENEN AND GALINA A. DUBININA

Thi'o.spi' ra. Gr. n. *thios* sulfur; Gr. n. *spira* a coil; M.L. fem. *Thiospira* sulfur coil or spiral.

Spirilla, usually with pointed ends, containing **sulfur inclusions**. **Motile** by monotrichous or polytrichous **polar flagella**. No resting stages known. Isolates conforming to the descriptions of the two "species" have now been obtained (Dubinina et al., 1993).

Type species: **Thiospira winogradskyi** (Omelianski 1905) Visloukh 1914, 48.

FURTHER DESCRIPTIVE INFORMATION

Cells are colorless spirilla, pointed at the ends and containing sulfur globules. Polar flagella, in some cases united in a tuft, are visible in the light microscope. *Thiospira winogradskyi* is 2–2.5 μm

wide and up to 50 μm long. *"Thiospira bipunctata"* is 1.7–2.4 × 6.6–14 μm.

Thiospira spp. are microaerophilic and vigorously chemotactic with respect to oxygen and possibly H_2S.

They are found in sulfurous marine and freshwater environments, including wastewater treatment systems.

The validity of the genus *Thiospira* has, in recent years, been called into question. The primary diagnostic feature of the genus was the accumulation of sulfur within spiral cells, and the separation of the two species was primarily morphological. Now that some strains have been grown in axenic cultures, it seems likely that members of the genus should actually be included within the genus *Aquaspirillum*, especially since *Aquaspirillum* species have been shown to accumulate sulfur in their cells (Dubinina et al., 1993). However, because the taxonomic position of the genus *Aquaspirillum* is itself under investigation (see, for example, Willems et al., 1991c; Pot et al., 1992), the genus *Thiospira* has not yet been formally removed, and because many authors still report observations of *Thiospira* species (e.g., Bernard and Fenchel, 1995), the description of the genus *Thiospira* will be maintained until the situation is clarified. What follows is adapted from the chapter by la Riviére and Kuenen in the last edition of *Bergey's Manual*.

ENRICHMENT AND ISOLATION PROCEDURES

T. winogradskyi was enriched by Omelianski (1905) in Winogradsky columns kept for some months at room temperature in the dark. The cell mass was in the form of "bacterial plates", i.e., discrete dense layers of bacteria, and could be kept alive in the laboratory for 2 years. Molisch (1912) enriched *"T. bipunctata"* from Black Sea mud which, when mixed with decaying algae, was used as an H_2S-generating sediment in a seawater-filled cylinder 20–30 cm high.

Recent successful isolations (Dubinina et al., 1993) have been made from H_2S or thiosulfate-rich samples on a medium containing (g/l): $(NH_4)_2SO_4$, 1.0; $MgSO_4 \cdot 7H_2O$, 1.0; casein hydrolysate, 3.0; sodium succinate, 1.0; vitamins and trace elements (Pfennig and Lippert, 1966). The pH is 7.0, and the necessary redox gradient is generated by adding a FeS suspension, which also serves as a source of sulfide.

TAXONOMIC COMMENTS

Although one of the species (*"T. bipunctata"*) does not appear on the Approved Lists, it is retained in the present description of the genus in view of its marked differences from the type species. It should be recalled that Bavendamm (1924) recognized five species.

Dubinina and Grabovich (1983) have described the isolation of heterotrophic spirilla morphologically similar to *Thiospira* and capable of forming internal sulfur granules in sulfide-containing media. Experiments indicate that the sulfur is formed through oxidation of sulfide by hydrogen peroxide.

Since other heterotrophs capable of forming sulfur inclusions in sulfide-containing media have also been found (Skerman et al., 1957), there is a distinct possibility that, on further examination, the spirilla isolated by Dubinina and Grabovich may turn out to be classifiable within the existing species of the genus *Spirillum* or, as now seems likely, *Aquaspirillum* (see above).

In spite of its present uncertain taxonomic position, the importance of the genus may prove to be considerable when further studies have been made of the population of sulfide oxidizers encountered at the deep-sea bottom around H_2S-emitting hydrothermal vents (Jannasch, 1984).

FURTHER READING

Bavendamm, W. 1924. Die farblosen und roten Schwefelbakterien des Süss- und Salz-wassers. *In* Kolkwitz (Editor), Pflanzenforschung, Gustav Fischer Verlag, Jena. pp. 7–156.

Dubinina, G.A. and M.Y. Grabovich. 1983. Isolation of pure *Thiospira* cultures and investigation of their physiology and sulfur metabolism. Mikrobiologiya *52*: 5–12.

Dubinina, G.A., M.Y. Grabovich, A.M. Lysenko, N.A. Chernykh and V.V. Churikova. 1993. Revision of taxonomic position of colorless sulfur spirilla of the genus *Thiospira* and description of a new species *Aquaspirillum bipunctata* comb. nov. Microbiology *62*: 368–644.

Jannasch, H.W. 1984. Microbial processes at deep sea hydrothermal vents. *In* Rona, Bostrom, Laubier and Smith (Editors), Hydrothermal Processes at Seafloor Spreading Centers, Plenum Publishing, New York. pp. 667–709.

List of species of the genus Thiospira

1. **Thiospira winogradskyi** (Omelianski 1905) Visloukh 1914, 48[VP]

   wi.no.grad'sky.i. M.L. gen. n. *winogradskyi* of Winogradsky; named for S.N. Winogradsky, a Russian microbiologist.

   Found in freshwater and marine environments overlaying sulfurous muds. See Table BXII.γ.42 for differentiation of *T. winogradskyi* and *"T. bipunctata"*.

   The mol% G + C of the DNA is: 38 (T_m).

   Type strain: No culture available.

2. **"Thiospira bipunctata"** (Molisch 1912) Visloukh 1914, 48.

   bi.punc.ta'ta. L. *bis* twice; L. part. adj. *pinctatus* punctate, dotted; M.L. fem. adj. *bipunctata* twice punctate.

   Found in sulfurous marine and brackish waters (see Table BXII.γ.42).

TABLE BXII.γ.42. Differentiation of the species of the genus *Thiospira*

Characteristic	*T. winogradskyi*	*"T. bipunctata"*
Size	2–2.5 × up to 50 μm	1.7–2.4 × 6.6–14 μm
Number of sulfur inclusions	Numerous	Few
Volutin granules at both ends	−	+

Family II. **Piscirickettsiaceae** *fam. nov.*

JOHN L. FRYER AND CATHARINE N. LANNAN

Pis′ci.ric.kett.si.a′ce.ae. M.L. fem. n. *Piscirickettsia* type genus of the family; *-aceae* ending to denote family, M.L. fem. pl. n. *Piscirickettsiaceae* the *Piscirickettsia* family.

Gram-negative, aerobic, coccoid, rod- or spiral-shaped bacteria, occasionally pleomorphic. Isolated predominantly from the marine environment. All belong to the *Gammaproteobacteria* with 16S rRNA similarities as the major unifying factor.

Type genus: **Piscirickettsia** Fryer, Lannan, Giovannoni and Wood 1992, 123.

Further Comments   *"Piscirickettsiaceae"* includes six genera, *Piscirickettsia, Cycloclasticus, Hydrogenovibrio, Methylophaga, Thioalkalimicrobium,* and *Thiomicrospira*. These groups are related phylogenetically, but share a limited number of phenotypic characters. *Piscirickettsia* is a Gram-negative, nonmotile, highly fastidious, intracellular rickettsia-like pathogen of fish. The bacterium replicates within membrane-bound cytoplasmic vacuoles in cells of infected hosts and does not grow on any known artificial medium. *Piscirickettsia* contains a single species that is known to cause serious mortality among salmonids and other species of fish in the marine environment.

The remaining five genera are composed of rod-, comma-, or spiral-shaped bacteria. They are all motile by means of a single polar flagellum. All can be cultured on artificial media. None is pathogenic or replicates within the cells of hosts. Optimum temperatures are higher than those of *Piscirickettsia* and all were isolated from aquatic environments. *Cycloclasticus* utilizes aromatic hydrocarbons as the sole source of organic carbon; *Methylophaga* utilizes C_1 compounds; and *Hydrogenovibrio* and *Thiomicrospira* both use CO_2. Members of the genus *Thiomicrospira* are obligately autotrophic, sulfur bacteria and have been isolated from marine hydrothermal vents and mud flats. The genus *Thioalkalimicrobium* contains chemolithoautotrophs that are isolated from alkaline lakes and that oxidize sulfide and thiosulfate to sulfate.

Genus I. **Piscirickettsia** *Fryer, Lannan, Giovannoni and Wood 1992, 123*[VP]

JOHN L. FRYER AND CATHARINE N. LANNAN

Pis′ci.ric.kett′si.a. L. n. *piscis* fish; generic name *Rickettsia*; N.L. fem. n. *Piscirickettsia* rickettsia-like organism affecting fish.

Pleomorphic, but predominantly coccoid organisms, ~0.5–1.5 μm in diameter. Gram negative, Gimenez negative, and stain dark blue with Giemsa's stain. Nonmotile. Aerobic. Characteristically replicate within cytoplasmic vacuoles in susceptible fish host cells. Cultivable in fish cell cultures but not on any known host-cell-free medium. Direct host-to-host transmission occurs in the aquatic environment without the requirement of a vector. The 16S rRNA gene sequence conforms to the secondary structural models for the *Gammaproteobacteria*. No signature sequence has been identified. **The etiological agent of the disease piscirickettsiosis in fishes.**

Type species: **Piscirickettsia salmonis** Fryer, Lannan, Giovannoni and Wood 1992, 123.

FURTHER DESCRIPTIVE INFORMATION

At present, *Piscirickettsia salmonis* is the only described species of the genus *Piscirickettsia*. This organism has a widespread geographic distribution, and causes disease in numerous salmonid host species (Table BXII.γ.43). *Piscirickettsia salmonis*-like bacteria have also been isolated from diseased sea bass, *Atractosion nobilis*, in California, USA (Chen et al., 2000a); and in Taiwan, from grouper, *Epinephelus melanostigma* (Chen et al., 2000b), and five species of cultured tilapia (Chern and Chao, 1994).

By electron microscopy, the organisms display the typical procytoplasmic structure of a procaryote and the cell wall of a Gram-negative bacterium (Fig. BXII.γ.81). Although primarily coccoid, the organisms also appear as pairs of curved rods or rings (Fig. BXII.γ.82). They replicate by binary fission within membrane-bound cytoplasmic vacuoles in cells of susceptible fish hosts or fish cell lines (Fig. BXII.γ.83) where they produce a cytopathic effect (Fig. BXII.γ.84). *In vitro* replication is optimal at 15–18°C, retarded above 20°C and below 10°C, and does not occur above 25°C (Fryer et al., 1990).

Little genetic variation has been observed in *P. salmonis* isolated from differing salmonid host species or originating from diverse geographic locations. 16S rDNA sequence data are available for five *P. salmonis* isolates representing three salmonid host species and three widely separated geographic regions (North America, South America, and Europe; Fryer and Mauel, 1997). With 1527 bases compared, the five isolates form a tight monophyletic cluster. Four of the isolates show sequence similarities >99.4%. One of the five is less closely related with similarities ranging from 98.5–98.9% (Mauel et al., 1999). The genetic differences between this isolate and others in the study do not correlate with either host species or geographic region as these characters are shared among them.

Hyperimmune rabbit serum generated against purified *P. sal-*

TABLE BXII.γ.43. Geographic region and host species of *Piscirickettsia salmonis*

Geographic region	Host species	Reference
Canada, Atlantic coast	*Salmo salar*	Cusack et al. (1997)
Canada, Pacific coast	*Oncorhynchus gorbuscha, O. tshawytscha, S. salar*	Evelyn and Kent (1992)
Chile	*Oncorhynchus kisutch*	Bravo and Campos (1989)
	Oncorhynchus mykiss, O. tshawytscha, S. salar	Garcés et al. (1991)
	Oncorhynchus masou	Bravo (1994)
Ireland	*S. salar*	Rodger and Drinan (1993)
Norway	*S. salar*	Olsen et al. (1997)

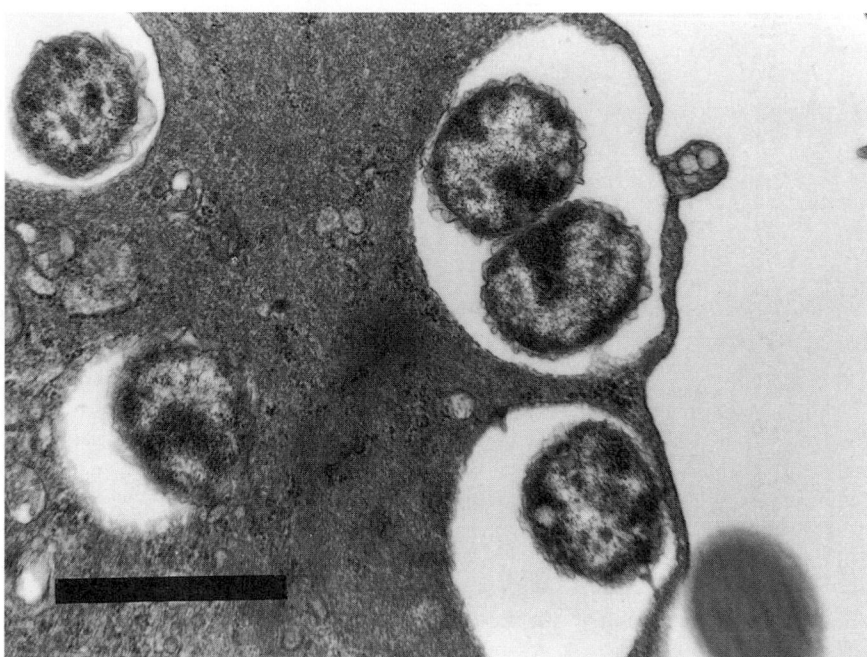

FIGURE BXII.γ.81. Ultrathin section of a cultured chinook salmon embryo (CHSE-214) cell containing *Pisciric-kettsia salmonis* organisms within cytoplasmic vacuoles. Bar = 1 μm. (Reprinted with permission from J.L. Fryer et al., Fish Pathology *25:* 107–114, 1990, ©The Japanese Society of Fish Pathology, Tokyo.)

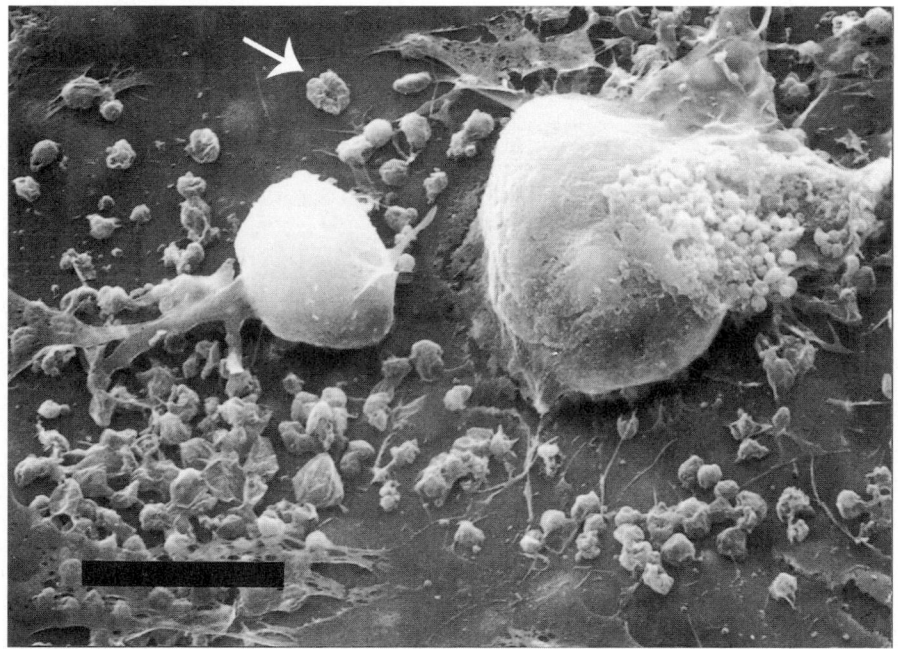

FIGURE BXII.γ.82. Scanning electron micrograph of *Piscirickettsia salmonis* organisms free or being released from cultured chinook salmon embryo (CHSE-214) cells. Note the ring form (*arrow*) and the varied sizes of the coccoid forms. Bar = 10 μm. (Reprinted with permission from J.L. Fryer et al., Fish Pathology *25:* 107–114, 1990, ©The Japanese Society of Fish Pathology, Tokyo.)

monis recognizes four surface-exposed protein antigens with relative molecular sizes of 65, 60, 54, and 51 kDa and two low molecular weight carbohydrate antigens, the lipopolysaccharide and a component lipo-oligosaccharide with relative molecular weights of 16 kDa and ~11 kDa, respectively (Kuzyk et al., 1996). Nevertheless, the humoral response elicited by *P. salmonis* infec-

tion in fish is apparently low. In the study by Kuzyk et al. (1996) convalescent serum from infected salmonid fish reacted with several minor immunoreactive protein antigens between 10 and 70 kDa in size as well as with a carbohydrate antigen with a relative molecular weight of ~11 kDa, but the antigens recognized by the convalescent salmonid fish serum differed greatly from those

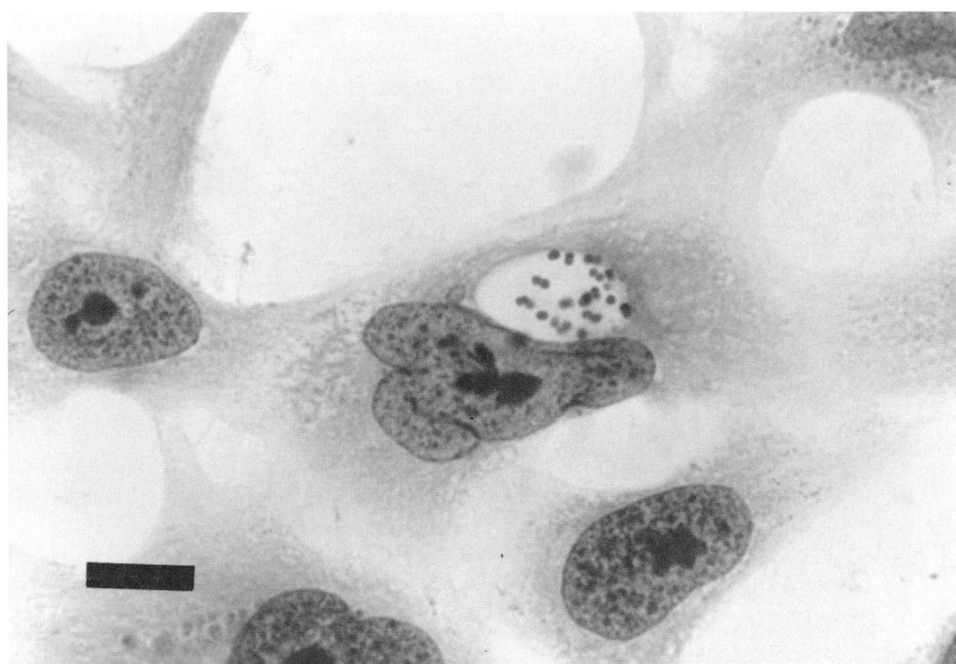

FIGURE BXII.γ.83. *Piscirickettsia salmonis* organisms within a cytoplasmic vacuole in a cultured chinook salmon embryo (CHSE-214) cell 3 d postinoculation. May-Grünwald-Giemsa stain. Bar = 10 μm. (Reprinted with permission from J.L. Fryer et al., Fish Pathology *25:* 107–114, 1990, ©The Japanese Society of Fish Pathology, Tokyo.)

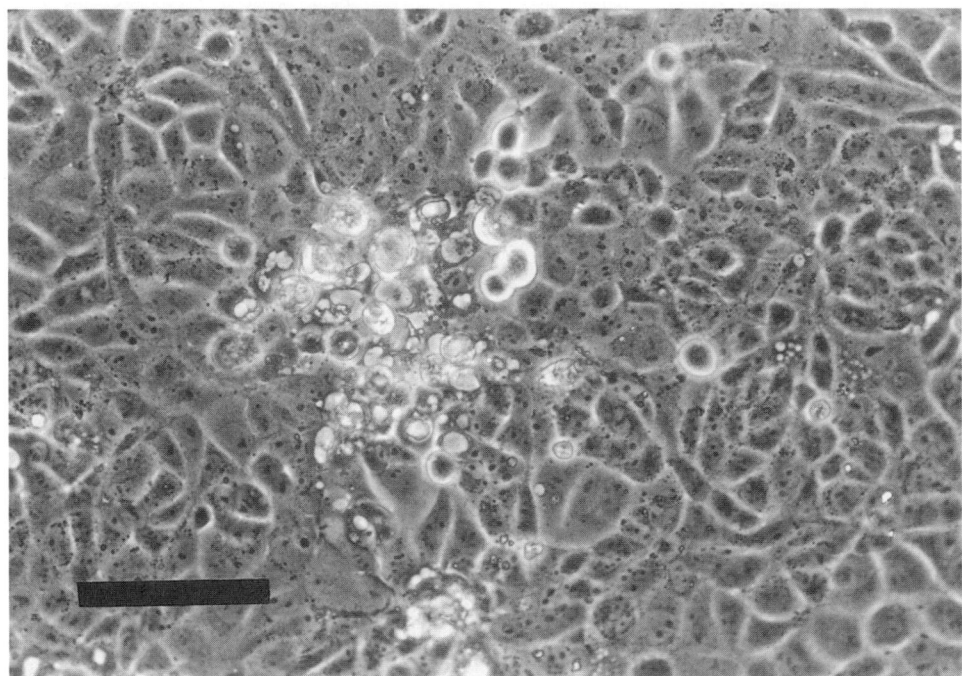

FIGURE BXII.γ.84. Cytopathic effect produced by *Piscirickettsia salmonis* in cultured chinook salmon embryo (CHSE-214) cells. Phase contrast microscopy. Bar = 100 μm. (Reprinted with permission from J.L. Fryer et al., Fish Pathology *25:* 107–114, 1990, ©The Japanese Society of Fish Pathology, Tokyo.)

identified by rabbit antiserum, suggesting a modest humoral response in the fish.

Piscirickettsia salmonis replication is inhibited *in vitro* by numerous antibiotics normally used in cell culture or in human and veterinary medicine. These include gentamicin, streptomycin, erythromycin, rifampin, tetracycline, doxycycline, oxolinic acid, and chloramphenicol (Fryer et al., 1992). Penicillin appears to have little effect. In spite of the broad antibiotic sensitivity of this organism, *in vivo* antibiotic therapy has had limited success. Intracellular concentrations of these preparations may not reach lethal levels *in vivo*, permitting replication to continue within host cells despite therapy.

FIGURE BXII.γ.85. *Oncorhynchus kisutch* infected with *Piscirickettsia salmonis*. Note the diagnostic, ring-shaped lesions on the liver. (Reprinted with permission from J.L. Fryer and M.J. Mauel, Emerging Infectious Diseases *3:* 137–144, 1997, ©Centers for Disease Control and Prevention, Atlanta, Georgia, USA.)

Piscirickettsial infection results in a systemic, epizootic fish disease, piscirickettsiosis, which can occur in chronic or acute form. Mortality may range from low to extremely high (≥90% of a naturally exposed population.) Marked difference in virulence has been reported among three isolates of *P. salmonis* each obtained from a different geographic location (House et al., 1999). Frequently, skin hemorrhages and lesions ranging from small areas of raised scales to shallow ulcers up to 2 cm in diameter are the first disease signs to appear (Branson and Diaz-Munoz, 1991). Pathological changes that include necrosis and inflammation, frequently found adjacent to blood vessels, may be observed in most internal organs of infected fish, especially in kidney, liver, spleen, and intestine (Cvitanich et al., 1991). At times, epithelial hyperplasia results in lamellar fusion of the gills. Organisms are found within macrophages and in the cytoplasm of other host cells. Diagnostic, ring-shaped, cream-colored lesions are present on the livers of chronically infected fish (Fig. BXII.γ.85), but in acute cases, mortality may be the only gross sign of disease observed.

The means of transmission of *P. salmonis* in natural infections is by direct fish-to-fish passage and occurs without the need of a vector. The reservoir for the pathogen has not been determined but may include one or more species of fish or other aquatic animals. Almendras et al. (1997) demonstrated direct horizontal transmission of *P. salmonis* under experimental conditions in both fresh and salt water. Vertical transmission from parent to offspring may also occur. The organisms can be found in the gonad of infected fish, and Larenas et al. (1996) reported *P. salmonis* organisms in the fertilized ova from experimentally infected parents. Despite this, piscirickettsiosis is less frequently observed in anadromous salmonids during the freshwater stage of their life cycle than after entry into the marine environment, suggesting that vertical transmission may be an uncommon occurrence.

ENRICHMENT AND ISOLATION PROCEDURES

Piscirickettsia salmonis can be isolated from the internal organs of infected fish by inoculation of homogenized tissues onto cell cultures derived from susceptible fish hosts. The cultures are incubated at 15–18°C for up to 28 d and microscopically observed for development of the characteristic cytopathic effect. However, isolation of these agents from clinical specimens presents a considerable challenge. Aseptic collection of diagnostic specimens in the field is difficult, enhancing the likelihood of adventitious bacterial contamination of the inoculum. In spite of this, the broad antibiotic sensitivity of *P. salmonis* dictates that the cultures be maintained in antibiotic-free medium. For this reason, preliminary diagnosis of piscirickettsiosis is normally made by examination of Gram stain, Giemsa stain, methylene blue, or acridine orange-stained kidney imprints or smears, with confirmation of identity provided by serological methods, e.g., immunofluorescence (Lannan et al., 1991) or immunohistochemistry (Alday-Sanz et al., 1994).

MAINTENANCE PROCEDURES

Piscirickettsia salmonis can be subcultured with the inoculation of spent growth medium from infected cultures onto fresh fish cell cultures in antibiotic-free medium. Infectious organisms can be recovered from spent culture medium stored for up to 3 weeks at 5°C (Lannan and Fryer, 1994). Long-term survival in culture medium requires the addition of 10% DMSO as a cryopreservative followed by storage at −70°C or in liquid nitrogen.

DIFFERENTIATION OF THE GENUS *PISCIRICKETTSIA* FROM OTHER GENERA

Piscirickettsia salmonis might be confused with *Neorickettsia helminthoeca*, the coccoid, Gram-negative, intracellular bacterium that is the etiologic agent of salmon-poisoning disease in canids. Although *N. helminthoeca* is not a parasite of salmonid fishes, salmonids are one host in the complex life cycle of *Nanophyetus salmincola*, the trematode vector of *N. helminthoeca*. Unlike *P. salmonis*, geographic distribution of *N. helminthoeca* is restricted to a portion of the Pacific coast of the United States. The differential characteristics of these two genera are listed in Table BXII.γ.44.

TAXONOMIC COMMENTS

The genus *Piscirickettsia* currently contains one species. Additional microorganisms assigned to this genus must be intracel-

TABLE BXII.γ.44. Differential characteristics of *Piscirickettsia salmonis* and *Neorickettsia helminthoeca*

Characteristic	P. salmonis	N. helminthoeca
Replicates within fish cells	+	−
Replicates at 37°C	−	+
Widespread geographic distribution	+	−
Pathogenic for *Canidae*	−	+
Reacts with convalescent serum from canine salmon poisoning disease	−	+
Alphaproteobacteria	−	+
Gammaproteobacteria	+	−

lular, Gram-negative, procaryotic parasites of fishes and perhaps other aquatic animals. These bacteria must be specifically related to *P. salmonis*. Members of the species *P. salmonis* are those recognized by specific antiserum against the type strain, LF-89.

Based on the intracellular location of *P. salmonis* and phenotypic similarities between this organism and the genus *Ehrlichia*, it was originally proposed that *Piscirickettsia* should be placed with *Ehrlichia* in the family *Rickettsiaceae* (Fryer et al., 1992). However, microorganisms historically included in the *Rickettsiaceae* appear to be polyphyletic (Weisburg et al., 1989). The majority fall in the *Alphaproteobacteria*, but certain of these intracellular bacteria, including *P. salmonis*, *Rickettsiella grylli* (Roux et al., 1997), *Coxiella burnetii*, and the bacterium designated *Wolbachia persica* (Weisburg et al., 1989), belong to the *Gammaproteobacteria*. Therefore, these organisms are phylogenetically remote from others in the family. *Piscirickettsia* appears to group with the genera *Cycloclasticus*, *Hydrogenovibrio*, *Methylophaga*, and *Thiomicrospira*, and is somewhat more distantly related to the *Legionella*, *Francisella*, and *Coxiella*. *Piscirickettsia* is now assigned to a new family, the "*Piscirickettsiaceae*".

Gram-negative obligate intracellular bacteria were not recognized as pathogens of fish before 1989, when *P. salmonis* was isolated from diseased salmonid fish in Chile (Fryer et al., 1990). Since that time, bacteria of this type have been observed or isolated from other species of salmonid and nonsalmonid fish at numerous locations (Brocklebank et al., 1992; Rodger and Drinan, 1993; Chern and Chao, 1994; Olsen et al., 1997; Chen et al., 2000a, b). Many of the isolates from salmonid fish have been serologically identified as *P. salmonis*, but the taxonomic placement of those found in other fishes has not been determined. Whether these organisms can be classified as *P. salmonis*, or as additional species in the genus *Piscirickettsia*, or whether they more properly belong to other genera as yet undescribed remains unresolved.

ACKNOWLEDGMENTS

We thank Sandra Bravo of Puerto Montt, Chile, who first reported the disease that was later identified as piscirickettsiosis, and L. H. Garcés, J. J. Larenas, and P. A. Smith of the Veterinary Sciences Faculty of the University of Chile, Santiago, all of whom have provided specimens, counsel, and technical help in the identification and characterization of *P. salmonis*. In addition, we thank Ena Urbach of the Department of Microbiology, Oregon State University, for aid in the phylogenetic description of this organism. This is Oregon Agricultural Experiment Station technical paper no. 11325.

FURTHER READING

Cvitanich, J.D., O.N. Garate and C.E. Smith. 1991. The isolation of a rickettsia-like organism causing disease and mortality in Chilean salmonids and its confirmation by Koch postulate. J. Fish Dis. *14*: 121–145.

Fryer, J.L. and R.P. Hedrick. 2003. *Piscirickettsia salmonis*: a Gram-negative intracellular bacterial pathogen of fish. J. Fish Dis. *26*: 251–262.

Fryer, J.L., C.N. Lannan, L.H. Garcés, J.J. Larenas and P.A. Smith. 1990. Isolation of a rickettsiales-like organism from diseased coho salmon (*Oncorhynchus kisutch*) in Chile. Fish Pathology *25*: 107–114.

Fryer, J.L., C.N. Lannan, J. Giovannoni and N.D. Wood. 1992. *Piscirickettsia salmonis* gen. nov., sp. nov., the causative agent of an epizootic disease in salmonid fishes. Int. J. Syst. Bacteriol. *42*: 120–126.

List of species of the genus *Piscirickettsia*

1. **Piscirickettsia salmonis** Fryer, Lannan, Giovannoni and Wood 1992, 123[VP]

   sal.mo'nis. L. gen. n. *salmonis* of salmon.

The characteristics are as described for the genus.

The mol% G + C of the DNA is: unknown.

Type strain: LF-89, ATCC(R) VR 1361.

GenBank accession number (16S rRNA): U36941.

Genus II. **Cycloclasticus** Dyksterhouse, Gray, Herwig, Lara and Staley 1995, 120[VP]

ALLISON D. GEISELBRECHT

Cy.clo.clas'ti.cus. Gr. n. *kyklos* circle or rings; Gr. adj. *klastos* broken; M.L. masc. n. *cycloclasticus* ring-breaker.

Unicellular, rod-shaped bacterium, 0.5 × 1.0–2.0 μm. **Gram negative.** Motile by means of a single polar flagellum. **Obligately aerobic**, nonfermentative, **heterotrophic. Grows poorly on complex bacteriological media containing no aromatic compounds.** Requires at least 1% salinity for growth. Aromatic compounds, including biphenyl, naphthalene, phenanthrene, anthracene, and toluene, used as sole or principal carbon sources for growth.

In addition, strains use selected fatty acids and amino acids, including acetate, propionate, and glutamate. **Oxidase** and **catalase** positive.

The mol% G + C of the DNA is: 37.

Type species: **Cycloclasticus pugetii** Dyksterhouse, Gray, Herwig, Lara and Staley 1995, 120.

TABLE BXII.γ.45. Characteristics of the genus *Cycloclasticus*[a]

Characteristic	*Cycloclasticus pugetii*
Cell shape	Straight rods
Motility	+
Flagellar arrangement	Polar, single
Poly-β-hydroxybutyrate accumulation	−
Na$^+$ required for growth	+
NO$_3^-$ reduced to NO$_2^-$	+
Oxidase	+
Catalase	+
Lipase	+
Amylase	−
Gelatinase	−
Intracellular membranes	−
Utilization of:	
Methane	−
Methanol	−

[a]From Dyksterhouse et al. (1995).

TABLE BXII.γ.46. Predominant whole-cell fatty acids and percent composition of *Cycloclasticus* strains[a]

Fatty acid	% total fatty acids in:	
	C. pugetii PS-1[b]	*Cycloclasticus* sp. N3[c]
Unknown[d]	6	6
C$_{12:0}$	2.5	3
C$_{12:0}$ 3OH	2.8	3
C$_{16:0}$ iso	nd[e]	3
C$_{16:1}$ ω7c	61.5	53
C$_{16:0}$	23	21
Total identified	95.7	89

[a]Only fatty acids representing 2.5% or more of the total fatty acid composition of each strain are listed. Cellular fatty acids were extracted, methylated, and detected with a Hewlett Packard model 5890 series II gas chromatograph as previously described (MIDI Inc., 1993).

[b]Data from Dyksterhouse et al. (1995). Strains were grown in liquid culture on 50 ml of Bacto Marine 2216 broth containing biphenyl crystals for 3 d with agitation, centrifuged, washed twice with ONR7a. 40 mg (wet weight) of the cell pellet was used for the analysis.

[c]Data from Geiselbrecht et al. (1998). Strains were grown on ONR7a plus Casamino acids (Difco) and salicylate (0.01% each) plates for 2 weeks at 20°C. Cells were harvested by scraping the entire plate surface.

[d]Unidentified peak at an equivalent carbon length of 11:798.

[e]nd, not detected.

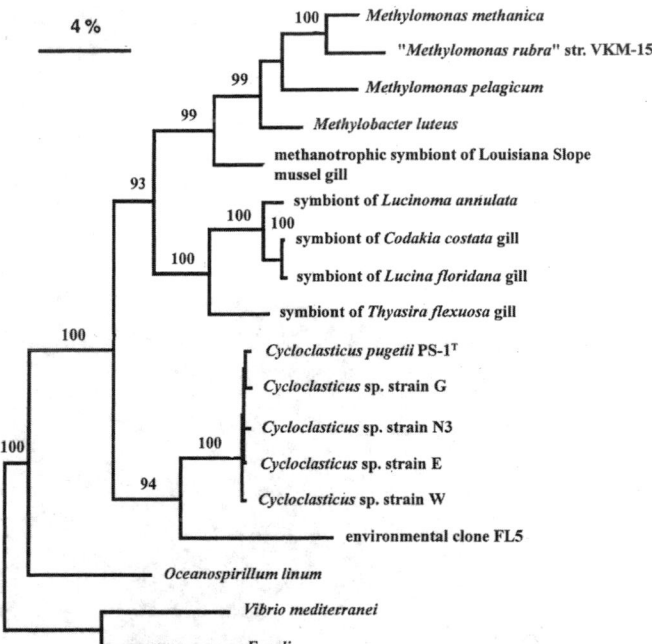

FIGURE BXII.γ.86. Phylogenetic relationships based on 16S rDNA sequences of *Cycloclasticus* and closely related species found in the Ribosomal Database Project, release 6.0 (Maidak et al., 1999). Environmental clone FL5 is an rDNA clone amplified from free-living bacteria in the marine water column (DeLong et al., 1993). This tree was the most parsimonious tree generated with PAUP, version 3.0s (Swofford, 1993). Heuristic unweighted bootstrap values are shown near the clades; only values greater than 50% are shown. Bar = ~0.04 change per nucleotide position.

FURTHER DESCRIPTIVE INFORMATION

The genus *Cycloclasticus* comprises a collection of marine strains with one validly named species. Phylogenetically, the closest relatives of these organisms are members of the methane-oxidizing genera *Methylomonas* and *Methylobacter*, a group of uncultured sulfur-oxidizing symbionts found in marine invertebrates, and a 16S rRNA environmental clone designated FL5. However, the mol% G + C DNA base composition of *Cycloclasticus* and *Methylomonas methanica* differ substantially (37 and 52, respectively), and *Cycloclasticus* cannot utilize methane or other C$_1$ compounds

as carbon sources. In addition, *Cycloclasticus* does not utilize thiosulfate as an energy source. Furthermore, existing strains are not phylogenetically closely related to other marine genera of aerobic, heterotrophic, motile, rod-shaped bacteria belonging to the *Gammaproteobacteria* (see Table BXII.γ.45 and Fig. BXII.γ.86) (Dyksterhouse et al., 1995).

Transmission electron microscopy of whole cells shows a single polar flagellum with several fimbriae originating from the cell surface and a cell envelope structure typical of Gram-negative bacteria, with a cell size of 0.5 × 1.0–2.0 µm. Colonies are small, smooth, round, and raised, with slow growth on solid media, and in some cases requiring up to 2 weeks for visible colony formation. Cultures have been shown to grow at temperatures ranging from 4–28°C, in the presence of salinity ranging from 1–7% (Dyksterhouse et al., 1995).

Whole cell fatty acid analyses show similar compositions between strains (Table BXII.γ.46). The predominant fatty acids in these and two additional strains tested were similar; C$_{16:1}$ ω7c was the most prevalent fatty acid.

A characteristic of this genus is an apparent preference for aromatic compounds as carbon sources. Carbon source utilization studies indicate that only a few additional carbon sources, including selected fatty acids and amino acids, can be used as sole carbon sources (Table BXII.γ.47). The concentrations of the carbon sources are crucial in determining carbon source utilization: strains responded differently to 0.01% and 0.1% w/v amounts of carbon source (Geiselbrecht, 1998).

Polyaromatic hydrocarbon (PAH) degradation studies performed with *Cycloclasticus* strains indicate similar, broad PAH-substrate ranges including naphthalene, substituted naphthalenes, phenanthrene, fluorene, biphenyl, and acenaphthene (Table BXII.γ.48) (Geiselbrecht et al., 1998). "*Cycloclasticus oligotrophus*" RB1 is capable of using a variety of aromatic compounds, such as toluene and *o*-, *m*-, and *p*-xylenes, as sole carbon sources (Wang et al., 1996).

Wang et al. (1996) cloned and sequenced a 5.7-kb DNA fragment from "*C. oligotrophus*" RB1. This DNA region contains genes that encode proteins somewhat homologous to the large-subunit (XcylC1) and small-subunit (XylC2) biphenyl dioxygenase of ter-

TABLE BXII.γ.47. Nonaromatic carbon sources utilized by selected *Cycloclasticus* strains

Carbon source	*C. pugetii*[a]	*Cycloclasticus* strain N3[b]	*Cycloclasticus* strain G[b]
D-Glucose	−	−	nd
D-Fructose	−	−	nd
D-Mannitol	−	nd	nd
DL-Malate	−	nd	nd
α-Ketoglutarate	−	nd	nd
Salicylate	+	−	−
Benzoate	+	nd	nd
Acetate	+	+[c]	+[d]
Propionate	+	+[d]	+[d]
Glutamate	+	+[c]	−
Succinate	nd	+[c]	+[d]
Pyruvate	−	−	+

[a]Data from Dyksterhouse et al. (1995).

[b]Data from Geiselbrecht (1998). Strain N3 was isolated from Puget Sound sediments; strain G from Gulf of Mexico sediments.

[c]Growth occurred when carbon source was provided at 0.1% but not at 0.01% w/v.

[d]Growth occurred when carbon source was provided at 0.01% but not at 0.1% w/v.

TABLE BXII.γ.48. PAHs used as sole carbon sources by *Cycloclasticus* strains

PAH	*C. pugetii*[a]	*Cycloclasticus* strain N3[b]	*Cycloclasticus* strain W[b]
Naphthalene	+	+	+
1-Methylnaphthalene	nd	+	+
2-Methylnaphthalene	nd	+	+
Acenaphthene	nd	+[c]	+[c]
2,6-Dimethylnaphthalene	nd	+	+
Biphenyl	+	+[c]	+
Phenanthrene	+	+	+
Anthracene	+	+	nd
Fluorine	nd	+	+

[a]PAH provided in excess of its saturation level as the sole carbon source in ONR7a broth. Growth indicates use of PAH as sole carbon source. From Dyksterhouse et al. (1995).

[b]PAH was provided as sole carbon source at or below its saturation level in ONR7a broth. The disappearance of PAH was monitored via gas chromatography after 7 d.

[c]PAH was only partially removed.

restrial bacteria, although these sequences are quite divergent from the sequences of known dioxygenases. By utilizing degenerate PCR primers, partial ISP sequences were obtained from *C. pugetii* and *Cycloclasticus* strain W that indicated the presence of *bph*-type dioxygenase ISP sequences similar to the region cloned in "*C. oligotrophus*" RB1, as well as a sequence corresponding to an additional PAH dioxygenase type more closely related to naphthalene type dioxygenases (Geiselbrecht et al., 1998). Although this study did not examine the functionality of these sequences, these data may indicate that *Cycloclasticus* contains multiple dioxygenases.

16S rDNA sequence analysis performed on three strains of *Cycloclasticus pugetii* revealed that 16S rDNA sequences from strain PS-1[T], -2, and -3 are nearly identical. Furthermore, *Cycloclasticus* sp. N3 contains only 3 base pair differences from *C. pugetii*[T] (Geiselbrecht et al., 1996). "*C. oligotrophus*" RB1 is also phylogenetically very closely related to *C. pugetii* [T]. All of these isolates share a distinctive 16S rDNA sequence at approximately base pair position 825, relative to *E. coli* (Gutell, 1993) of 5′-CUAACUG-UUGGGCGGGUUUCCGC-3′. No DNA–DNA reassociation data exist for strains of the genus.

Cycloclasticus strains have been isolated from coastal marine sediments and waters in Puget Sound, the Gulf of Mexico, and Resurrection Bay, Alaska (Button et al., 1993, 1998; Dyksterhouse et al., 1995; Geiselbrecht et al., 1996, 1998). In Puget Sound, *Cycloclasticus* is widespread in both contaminated and uncontaminated sediments (Geiselbrecht et al., 1996). Furthermore, they occur in the Puget Sound water column (Geiselbrecht, 1998). "*C. oligotrophus*" RB1 was isolated from the water column in Resurrection Bay, Alaska, following a 10^8 dilution, indicating that they can potentially be numerous in the marine water column (Button et al., 1993). In the Gulf of Mexico, they have been isolated from offshore and near-shore sites at sediment dilution numbers ranging from 2×10^3 to 2×10^5, indicating that they are somewhat numerous in these sediments (Geiselbrecht et al., 1998).

ENRICHMENT AND ISOLATION PROCEDURES

An artificial seawater mineral salts medium called ONR7a has been used for most *Cycloclasticus* isolations. This medium contains all of the major cations and anions that are present in concentrations greater than 1 mg/l in seawater. ONR7a contains (per liter of distilled or deionized water): NaCl, 22.79 g; $MgCl_2 \cdot 6H_2O$, 11.18 g; Na_2SO_4, 3.98 g; $CaCl_2 \cdot 2H_2O$, 1.46 g; TAPSO (3-[N-tris(hydroxymethyl)methylamino]-2-hydroxypropanesulfonic acid), 1.3 g; KCl, 0.72; NH_4Cl, 0.27 g; $Na_2HPO_4 \cdot 7H_2O$, 89 mg; NaBr, 83 mg; $NaHCO_3$, 31 mg; H_3BO_3, 27 mg; $SrCl_2 \cdot 6H_2O$, 24 mg; NaF, 2.6 mg; $FeCl_2 \cdot 4H_2O$, 2.0 mg. To prevent precipitation of ONR7a during autoclaving, three separate solutions are prepared and then mixed together after autoclaving when the solutions are cooled to at least 50°C: one solution contains NaCl, Na_2SO_4, KCl, NaBr, Na_2HPO_4, H_3BO_3, NaF, NH_4Cl, Na_2HPO_4, and TAPSO (pH adjusted to 7.6 with NaOH, then autoclaved); the second solution contains $MgCl_2$, $CaCl_2$, and $SrCl_2$ (divalent cation salts made as a $50\times$ stock and autoclaved); and the third solution contains $FeCl_2$ (made as a $200\times$ stock and then preferably filter-sterilized). For solid media, Bacto Agar (Difco) or agarose (Sigma) is used as a solidifying agent. ONR8a contains all the compounds listed above for the ONR7a plus 10 ml/l of a vitamin solution (Staley, 1968) and 10 ml/l of a trace element solution (Dyksterhouse et al., 1995). The type strain of the genus *Cycloclasticus*, *C. pugetii*, was isolated by blending Puget Sound sediments, diluting these sediments 1:5, and spreading the dilution onto ONR8a containing 0.2 g of biphenyl on the Petri dish lid. Colonies surrounded by a diffusible yellow zone indicative of PAH degradation were selected for further isolation (Dyksterhouse et al., 1995). The vitamin and trace element solutions were later found to be unnecessary for growth, and were eliminated.

Cycloclasticus has also been isolated following enrichment in dilute most probable number (MPN) tubes specific for PAH-degrading marine bacteria. Surficial (aerobic) 1-cm portions of sediments were used as the inocula for these MPN preparations. Briefly, this technique involved serially diluting wet sediments with ONR7a containing PAHs as sole carbon sources. Tubes were scored positive or negative based on PAH degradation. After incubation for approximately 5 months at 15°C, the most dilute tubes that showed color changes indicative of PAH degradation were used for isolation. Samples from positive MPN tubes were streaked onto plates containing ONR7a solidified with 0.8% agarose. Phenanthrene or naphthalene were used as the sole carbon sources; these PAHs are provided in the vapor phase by placing approximately 1 mg of PAH in the lid of each plate and storing the plates inverted and wrapped in Parafilm™. Colonies that were capable of growing with either of these PAHs as sole carbon sources are picked and inoculated into an ONR7a solution containing the appropriate PAH in crystal form at a concentration of approximately 1 mg/ml (Geiselbrecht et al., 1998). *Cycloclasticus* spp. have also been isolated from enrichments made by inoculating ONR7a + PAHs with fresh sediments, incubating for 2 weeks at room temperature, and plating onto ONR7a solidified with 0.8% agarose (Geiselbrecht et al., 1996).

"*C. oligotrophus*" RB1 was isolated following a 10^8 dilution of Resurrection Bay seawater into unamended seawater prepared by filtration through fired Gelman 1/E 47-mm-diameter filters, autoclaving, refiltering through other fired Gelman filters, and aseptically siphoning the sterile filtrate into incubation chambers (Button et al., 1993).

MAINTENANCE PROCEDURES

Although initial isolations were at 15°C, subsequent studies indicated a temperature growth range of 4–28°C, with strong growth at room temperature. Routine growth occurs in either ONR7a with PAHs as a sole carbon source, or in a seawater medium amended with acetate (Button et al., 1998). Strains have also been grown on Bacto Marine 2216 Petri plates with biphenyl

in the vapor phase (Dyksterhouse et al., 1995). Cultures can be frozen and stored at $-80°C$ using either DMSO or glycerol.

PROCEDURES FOR TESTING SPECIAL CHARACTERS

The use of PAHs as a sole carbon source is a characteristic of this genus. PAH degradation can be confirmed by several methods, but a simple method involves growing cultures with individual PAHs as sole carbon sources and monitoring the disappearance of the PAHs with a gas chromatograph equipped with a flame ionization detector in comparison to sterile controls. Briefly, 20-ml Balch tubes fitted with Teflon-lined stoppers can be used, and PAHs added as concentrated solutions in methylene chloride. After the methylene chloride is evaporated, 5 ml of a seawater medium (such as ONR7a) is added. Tubes are vortexed and shaken to allow the PAHs to solubilize. The tubes are inoculated with an exponential-phase culture of *Cycloclasticus*, grown on either acetate or a different PAH. Control tubes receive no inoculum. The tubes are incubated in the dark with shaking (for aeration), and then the cultures are extracted with hexane. Each extract can be analyzed with a gas chromatograph equipped with a 30-m DB-5 column (diameter, 0.5 mm) or comparable column, and the recovery of parent compound recovered versus controls indicate PAH degradation (Geiselbrecht et al., 1998).

DIFFERENTIATION OF THE GENUS *CYCLOCLASTICUS* FROM OTHER GENERA

The ability to utilize PAHs is not unique to this genus; other marine PAH-degrading, Gram-negative rods have been isolated and include members of the genus *Pseudomonas* (Shiaris and Cooney, 1983; Tagger et al., 1990), *Flavobacterium* (Shiaris and Cooney, 1983; Okpokwasili et al., 1984), *Moraxella* (Tagger et al., 1990), *Marinobacter* (Gauthier et al., 1992), *Vibrio* (West et al., 1984), *Sphingomonas* (Zylstra et al., 1997), and a new genus *Neptunomonas* (Hedlund et al., 1999a). *Cycloclasticus* can be distinguished from some of these genera by its 1% salt requirement. *Cycloclasticus* can be distinguished from all of these genera as well as other aerobic, heterotrophic, motile, rod-shaped marine bacteria by its distinct phylogenetic location.

TAXONOMIC COMMENTS

As noted, several strains of *Cycloclasticus* have been isolated. "*Cycloclasticus oligotrophus*" RB1 has been described as a distinct species based on its small size and inability to utilize polar substrates such as amino acids (Button et al., 1993, 1998). Although "*C. oligotrophus*" RB1 is phylogenetically closely related to *C. pugetii*[T] (Button et al., 1998), it has not yet been validly named.

FURTHER READING

Dyksterhouse, S.E., J.P. Gray, R.P. Herwig, J.C. Lara and J.T. Staley. 1995. *Cycloclasticus pugetii* gen. nov., sp. nov., an aromatic hydrocarbon-degrading bacterium from marine sediments. Int. J. Syst. Bacteriol. *45*: 116–123.

List of species of the genus Cycloclasticus

1. **Cycloclasticus pugetii** Dyksterhouse, Gray, Herwig, Lara and Staley 1995, 120[VP]

   pu.get' i.i. M.L. gen. n. *pugetii* of Puget, named in honor of Peter Puget, a British naval officer who participated in the Vancouver Expedition and for whom Puget Sound was named.

   The characteristics are as described for the genus.

   Isolated from surficial sediments of Sinclair Inlet, which is located on Puget Sound near the city of Bremerton, Washington.

   The mol% $G + C$ of the DNA is: 37 (HPLC).

   Type strain: PS-1, ATTC 51542.

   GenBank accession number (16S rRNA): U12624, L34955.

Other Organisms

A collection of 15 strains of *Cycloclasticus* was isolated from several Puget Sound locations subsequent to the isolation of *Cycloclasticus pugetii*. These were grouped on the basis of fatty acid analyses and a representative strain's 16S rDNA was sequenced, *Cycloclasticus* sp. strain N3 (GenBank U57920) (Geiselbrecht et al., 1996). This strain differs from *C. pugetii* at only 3 base pair positions in the 16S rDNA. Twenty-three strains of *Cycloclasticus* spp. were isolated from offshore and near-shore Gulf of Mexico sediments (Geiselbrecht et al., 1998). Partial 16S rDNA sequencing revealed that all 23 were members of the *Cycloclasticus* genus, and the 16S rDNA was sequenced in entirety from three strains: *Cycloclasticus* sp. strain G (GenBank AF093002), *Cycloclasticus* sp. strain E (GenBank AF093003), and *Cycloclasticus* sp. strain W (GenBank AF093004). All three are closely related to *C. pugetii*. At this time, it is unclear whether any of these strains are potentially new species of *Cycloclasticus*.

Genus III. **Hydrogenovibrio** Nishihara, Igarashi and Kodama 1991b, 132[VP]

HIROFUMI NISHIHARA

Hy.dro.ge.no.vib' ri.o. Gr. n. *hydro* water; Gr. n. *genus* offspring; M.L. masc. n. *hydrogenium* hydrogen, that which produces water; M.L. masc. n. *vibrio* that which vibrates; M.L. masc. n. *Hydrogenovibrio* hydrogen vibrio.

Cells are **comma-shaped rods** (0.2–0.5 × 1–2 μm) that occur singly (Fig. BXII.γ.87). **Gram negative.** Motile by means of a polar flagellum. No resting stages are known. **Aerobic. Chemolithoautotrophic, using H$_2$ or reduced inorganic sulfur compounds**, such as S^0, thiosulfate, and tetrathionate, as the electron donor **and CO$_2$ as the carbon source.** CO$_2$ is fixed via the **Calvin-Benson cycle.** Chemoorganotrophic growth has not been observed. The optimum temperature for growth is ~37°C. The optimum pH for growth is ~6.5. NaCl is required for growth with an optimum concentration of 0.5 M.

Straight chain saturated C$_{16:0}$ and C$_{18:0}$ acids and a straight-chain unsaturated C$_{16:1}$ acid are the major components of the cellular fatty acids. Ubiquinone-8 is the major component of the quinone system.

The mol% G + C of the DNA is: 44.1.

Type species: **Hydrogenovibrio marinus** Nishihara, Igarashi and Kodama 1991b, 132.

FURTHER DESCRIPTIVE INFORMATION

Reddish-brown, smooth colonies are formed on inorganic basal medium solidified with agar in an atmosphere of H$_2$/O$_2$/CO$_2$ (85:5:10) (1 atm at room temperature). Colonies are not formed when the partial pressure of oxygen is more than 10%. The colonies grow up to approximately 1 mm in diameter after a week of incubation at 26.5°C. No colony is formed heterotrophically on nutrient broth medium supplemented with 0.5 M NaCl.

H. marinus grows rapidly and vigorously in the presence of

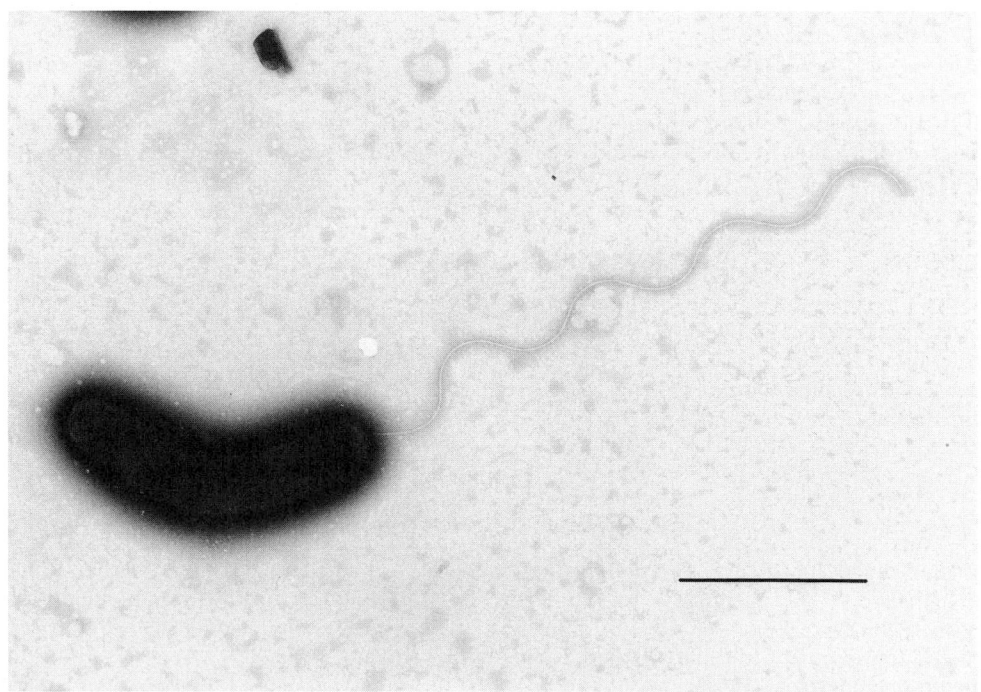

FIGURE BXII.γ.87. Electron micrograph of *H. marinus* strain MH-110 fixed with glutaraldehyde and negatively stained. Bar = 1 μm.

H_2 and CO_2 as the sole energy and carbon sources, respectively. Lag time of growth is not observed at any tension of oxygen (1–40%), nevertheless hydrogen-oxidizing bacteria are usually microaerophilic. Specific growth rate in the initial stage of growth is very high (0.60–0.67 h^{-1} at 37, pH 6.5) irrespective of any oxygen tension. The cell density reaches more than 20 g/l dry weight after 24 h cultivation under a continuous gas flow of $H_2/O_2/CO_2$ in a jar fermentor (Nishihara et al., 1991a). A reduced inorganic sulfur compound such as S^0, thiosulfate, or tetrathionate is also utilized as the sole energy source under both aerobic (shaking, $pO_2 = 20\%$, $pCO_2 = 10\%$) and microaerobic (standing, $pO_2 = 5\%$, $pCO_2 = 10\%$) conditions in flask cultures, but the growth is very poor. A decrease in pH value occurs when *H. marinus* is cultivated on these sulfur compounds. S^0 is deposited extracellularly when cultivated with thiosulfate or tetrathionate. Ferrous, manganous, ammonium, or nitrite ions are not utilized as an energy source under both aerobic and microaerobic conditions (Nishihara et al., 1989).

Chemoorganotrophic growth is not observed on the following organic compounds (0.1%, w/v) or media: acetate, propionate, oxalate, succinate, maleate, fumarate, DL-malate, α-ketoglutarate, DL-lactate, citrate, pyruvate, formate (5–100 mM), glycollate, DL-β-hydroxybutyrate, *p*-hydroxybenzoate, gluconate, D-xylose, D-galactose, sucrose, D-glucose, D-mannose, D-fructose, maltose, L-rhamnose, trehalose, D-raffinose, soluble starch, methanol, ethanol, mannitol, glycine, L-glutamate, L-aspartate, L-serine, L-leucine, L-valine, L-lysine, L-tryptophan, L-histidine, L-alanine, L-proline, L-arginine, L-threonine, L-phenylalanine, nutrient broth medium, yeast extract-peptone-glucose medium, brain heart infusion medium, and ZoBell 2216E medium (Nishihara et al., 1989).

Ammonium ions and urea are utilized as sole nitrogen sources, but nitrate ions, nitrite ions, and gaseous nitrogen are not. Nitrite inhibits growth. No growth is observed in the absence of NaCl, and Na^+ is not replaceable by K^+ or Li^+. The optimum NaCl concentration is 0.5 M. No growth factors are required (Nishihara et al., 1989).

Glycogen is accumulated when *H. marinus* is autotrophically cultivated on hydrogen and CO_2 under nitrogen or magnesium starvation or oxygen limitation. The maximum absorption of the iodine complex is observed at 453 nm (Nishihara et al., 1993).

The major hydrogenase is membrane-bound type. The natural electron carrier has not been identified. Pyridine nucleotides reducing activity is not observed in either the membrane or the soluble fraction. Type *b*, *c*, and *o* cytochromes have been found. The membrane-bound hydrogenase is characterized as highly oxygen-tolerant and thermostable. The optimum temperature for hydrogen-oxidation activity using methylviologen as an electron acceptor is 90°C (Nishihara et al., 1997).

CO_2 is fixed via the Calvin-Benson cycle. Three different molecules of ribulose-1,5-bisphosphate carboxylase/oxygenase (RuBisCo) have been found in *H. marinus* cells. Two of them are form I RuBisCos (CbbLS-1 and CbbLS-2), which are composed of eight large and eight small subunits, and the other is form II

RuBisCo (CbbM), which is composed of only large subunits (Chung et al., 1993a). The specific factors (τ values) of CbbLS-1 and CbbLS-2 are higher than that of CbbM, but lower than those of other form I RuBisCos from various organisms (Hayashi et al., 1998). The gene sequences of CbbLS-1, CbbLS-2, and CbbM from *H. marinus* are deposited in the DDBJ database under accession numbers D43621, D43622, and D28135, respectively. The deduced amino acid sequences of both CbbLS-1 and CbbLS-2 exhibited particularly high similarities to those of *Chromatium vinosum* and sulfur-oxidizing bacteria (*Thiobacillus*) (Nishihara et al., 1998). Large subunits of form I RuBisCos (CbbLs) from these bacteria formed a monophyletic group in the maximum parsimony tree of *cbbL* amino acids (Delwiche and Palmer, 1996).

The 16S rRNA gene sequence of *H. marinus* is deposited in the DDBJ database under accession number D86374. According to the sequence, *H. marinus* is affiliated with purple sulfur bacteria and sulfur-oxidizing bacteria belonging to the *Gammaproteobacteria*, members of *Thiomicrospira* being the closest relatives (Nishihara et al., 1998). Sequence similarities of the 16S rRNA genes and of the *cbbL* genes among these *Gammaproteobacteria* suggest a common autotrophic ancestry.

H. marinus was isolated from seawater (surface or shallow water) collected at a coastal region in Japan.

ENRICHMENT AND ISOLATION PROCEDURES

The following inorganic medium was used for the enrichment of *H. marinus*: K_2HPO_4, 2.5 g/l; KH_2PO_4, 0.5 g/l; $(NH_4)_2SO_4$, 2.0 g/l; NaCl, 29.3 g/l; $MgSO_4\cdot7H_2O$, 0.2 g/l; $CaCl_2$, 10.0 mg/l; $FeSO_4\cdot7H_2O$, 10.0 mg/l; and $NiSO_4\cdot7H_2O$, 0.6 mg/l; plus trace elements (pH 7.0) (Nishihara et al., 1989). The flasks were gassed with a gas mixture of $H_2/O_2/CO_2$ (75:15:10) (1 atm at room temperature) and shaken at 25°C. Single colony isolation was performed on plate medium solidified with 1.5% agar. The inoculated plates were placed in a desiccator and the gas phase was replaced by a gas mixture of $H_2/O_2/CO_2$ (85:5:10) (1 atm at room temperature).

MAINTENANCE PROCEDURES

H. marinus can be preserved by a freeze-drying method. 50 mM phosphate buffer (pH 7.0) containing 2% sodium glutamate was used as the suspending medium. The cultures can also be stored in 10–15% (v/v) glycerol at −80°C or in liquid nitrogen.

DIFFERENTIATION OF THE GENUS *HYDROGENOVIBRIO* FROM OTHER GENERA

Hydrogenovibrio is the only taxon for mesophilic and obligately chemolithoautotrophic hydrogen bacteria. *Thiomicrospira* is the closest relative of *Hydrogenovibrio* according to the 16S rRNA gene sequence analysis. Hydrogen autotrophy is the main feature that differentiates these two genera.. *H. marinus* grows best on hydrogen as an energy source. On the other hand, growth of *Thiomicrospira* on hydrogen has not been observed (Nishihara et al. 1991b) (see also the table in the description of the genus *Thiomicrospira*).

List of species of the genus Hydrogenovibrio

1. **Hydrogenovibrio marinus** Nishihara, Igarashi and Kodama 1991b, 132[VP]

    ma.ri'nus. L. adj. *marinus* marine, of the sea.

The description of this species is the same as that of the genus.

The mol% G + C of the DNA is: 44.1 (HPLC).

Type strain: MH-110, DSM 11271; JCM 7688.

GenBank accession number (16S rRNA): D86374.

Genus IV. **Methylophaga** Janvier, Frehel, Grimont and Gasser 1985, 138[VP]

JOHN P. BOWMAN

Me.thyl.o'pha.ga. Fr. n. *methyl* methyl radical; Gr. v. *phagein* to eat; M.L. n. *Methylophaga* methyl eating.

Rod-shaped cells 0.2 × 0.9-1.0 µm, occurring singly or in pairs. **Motile by means of a single polar flagellum. Gram negative.** Spores and other resting stages are not formed. **Chemoheterotrophic, having a strictly respiratory type of metabolism with oxygen as the terminal electron acceptor. Restricted methylotroph**, able to utilize only a limited range of substrates, mostly C_1 substrates including methanol, methylamine, dimethylamine, and trimethylamine. The only C_{2+} compounds used for growth are D-fructose and sucrose. The **ribulose monophosphate pathway** is used for assimilation of C_1 compounds. One species can utilize dimethylsulfide and grow mixotrophically by oxidizing H_2S to thiosulfate. Catalase and oxidase positive. **Na^+ and Mg^{2+} are required for optimum growth**. Requires vitamin B_{12} for growth. Major fatty acids are $C_{16:0}$ and $C_{16:1\ \omega7c}$. The major coenzyme Q is ubiquinone-8 (Q-8). Belongs to the *Gammaproteobacteria*. Isolated from seawater, brackish waters and marine benthic mats.

The mol% G + C of the DNA is: 42–45.

Type species: **Methylophaga marina** Janvier, Frehel, Grimont and Gasser 1985, 138.

FURTHER DESCRIPTIVE INFORMATION

Morphology Cells lack differentiated resting cells, prostheca, and other ultrastructural peculiarities except for having a somewhat wide periplasmic space with a width of 20–30 nm (Janvier et al., 1985). Cell walls are of the usual type for Gram-negative cells.

Cultural characteristics *Methylophaga* strains grow well on artificial seawater (ASW)-methanol medium[1] but do not grow on most standard complex marine media such as marine agar (ZoBell, 1941). The colonies of *Methylophaga* strains are circular and convex, possess entire edges and a butyrous consistency, and have a white to pale pink color.

Methylophaga strains require Na^+ and Mg^{2+} ions (20 mM and 5 mM, respectively) for optimal growth (Janvier et al., 1985), though some strains still grow weakly in their absence (Urakami and Komagata, 1987). Most *Methylophaga* strains are able to tolerate up to 9–10% NaCl. "*M. limanica*" exhibits optimum growth with 0.5–1.0 M NaCl (Doronina and Trotsenko, 1997). Strains usually possess a requirement for cyanocobalamin (vitamin B_{12}). The optimal pH for growth is equivalent to that of seawater (7.4–7.8) and the optimal growth temperature is 30–37°C. *M. thalassica* is slightly more thermotolerant than *M. marina* and can grow at 42°C (Urakami and Komagata, 1987).

Carbon and energy sources Only a few substrates are utilized for carbon and energy. These are mainly C_1 compounds including methanol and methylamine, which are utilized by all *Methylophaga* strains (Table BXII.γ.49). No strains grow on methane or methanesulfonate. Among a large number of C_{2+} compounds tested, only D-fructose supports growth, except for "*M. limanica*", which can also use sucrose (Janvier et al., 1985; de Zwart et al., 1996; Doronina et al., 1997).

M. sulfidovorans has the unusual ability to grow on dimethylsulfide (DMS) and oxidize it to thiosulfate. The doubling time is approximately 15 h compared with 2–3 h on methanol. This species can also grow mixotrophically by oxidizing H_2S to thiosulfate when in the presence of a carbon substrate such as DMS, which acts as a source of sulfide and carbon (de Zwart and Kuenen, 1997). Sulfide concentrations of 200 µmol/l reduce DMS oxidation by about 50%. The sulfide oxidation rate of *M. sulfidovorans* is equivalent to that in other sulfide-oxidizers such as *Thiobacillus thioparus* (de Zwart et al., 1997).

C_1 metabolism *Methylophaga* strains possess a dissimilatory pathway for methanol oxidation like that found in other methylotrophic bacteria. In this pathway, methanol is oxidized to formaldehyde by methanol dehydrogenase; then the formaldehyde is either oxidized to formate (and then to CO_2) or is fixed as cellular carbon. The methanol dehydrogenase in *Methylophaga marina* has a structure similar to that found in other methylotrophic bacteria (Janvier and Gasser, 1987). Mutagenesis studies indicate that apo-methanol dehydrogenase and pyrroloquinol quinone—the cofactor typically associated with methanol dehydrogenase—are insufficient to constitute an active enzyme and that another unknown factor is required (Janvier et al., 1992). *Methylophaga* strains use the 2-keto-3-deoxy-6-phosphogluconate aldolase variant of the ribulose monophosphate (RuMP) pathway for formaldehyde fixation (Doronina et al., 1997). The assimilation of methylamine, however, uses a variety of enzymes that differ between strains. Methylamine dehydrogenase is the primary methylamine-oxidizing enzyme in *M. marina* and "*M. limanica*" (Doronina et al., 1997), whereas in *M. thalassica* the predominant enzyme is *N*-methylglutamate dehydrogenase (Janvier et al., 1985). This differentiation is not absolute, however, because "*M. limanica*" and some strains of *M. marina* also exhibit *N*-methylglutamate dehydrogenase activity (Janvier et al., 1985; Doronina et al., 1997). Methylamine dehydrogenase converts methylamine directly to formaldehyde and ammonia using cytochrome *c* as an electron acceptor (Mehta, 1977). Methylamine can also be converted to γ-glutamylmethylamide (γ-GMA) by the addition of glutamate via γ-GMA synthetase. γ-GMA is then cleaved by γ-GMA lyase to yield ammonia and *N*-methylglutamate, and the latter is oxidized by *N*-methylglutamate dehydrogenase to produce formaldehyde and glutamate (Hou, 1984; Doronina et al., 1997). An alternative pathway has been discovered for *Methylophaga* strain AA-30 in which γ-GMA is cleaved into ammonia, formaldehyde, and 2-oxoglutarate; the latter compound is then transaminated to form glutamate (Kimura et al., 1990b, 1995). As in other methylotrophs possessing the RuMP pathway, the tricarboxylic acid cycle lacks 2-oxoglutarate dehydrogenase; moreover, strains lack pyruvate kinase and glyoxylate cycle enzymes (Doronina et al., 1997).

---

1. Artificial seawater (ASW)-methanol medium contains (g/l of natural or artificial seawater) KH_2PO_4, 0.14; Bis-Tris, 2.0; and ferric ammonium citrate, 0.06. The pH is adjusted to ~7.5 and 1.6% agar is added (optional). After autoclaving and cooling to ~50°C, methanol and vitamin B_{12} are added to final concentrations of 0.2% and 1 µg/ml, respectively. Artificial seawater consists of (g/l distilled water) NaCl, 24; $MgCl_2\cdot6H_2O$, 3; $MgSO_4\cdot7H_2O$, 2; KCl, 0.5; $CaCl_2\cdot2H_2O$, 1; Bis-Tris, 0.5; and 1 ml of trace element solution (Widdel and Pfennig, 1981) consists of (per liter of distilled water) 25% HCl, 10 ml; $FeCl_2\cdot4H_2O$, 1.5 g; $CoCl_2\cdot6H_2O$, 0.19 g; $MnCl_2\cdot4H_2O$, 0.1 g; $ZnCl_2$, 0.07 g; H_3BO_3, 0.062 g; $Na_2MoO_4\cdot2H_2O$, 0.036; $NiCl_2\cdot6H_2O$, 0.024 g; and $CuCl_2\cdot2H_2O$, 0.017 g. The ferrous chloride is dissolved in HCl and then added to the distilled water with the other reagents.

TABLE BXII.γ.49. Phenotypic characteristics of the species of the genus *Methylophaga*[a]

Characteristic	1. *M. marina*	2. *M. sulfidovorans*	3. *M. thalassica*	4. *"M. limanica"*
Nitrate reduction	−	−	d (weak)	−
Mixotrophy (H_2S oxidation)	−	+	−	nd
Catalase, oxidase	+	+	+	+
Optimal temperature (°C)	30–37	22–24	30–37	30–37
Optimal pH	7.5	7.4–7.8	7.5	6.7–7.2
Growth at 42°C	−	−	+	−
Growth with 9% NaCl	+	−	d	+
Methylamine dehydrogenase	+	nd	−	+
N-Methylglutamate dehydrogenase	d	nd	+	+
Utilizable carbon sources:				
Methanol, methylamine	+	+	+	+
Dimethylamine	d	+	+	−
Trimethylamine	−	−	d	−
Dimethylsulfide	−	+	−	nd
D-Fructose	+	−	d	+
Sucrose	−	−	−	+
Mol% G + C of DNA (T_m)	43–44	42	44–45	43–44

[a]For symbols see standard definitions; nd, no data.

Nitrogen metabolism *Methylophaga* strains utilize a variety of nitrogen sources including ammonia, nitrate, short chain amines (Kimura et al., 1990a), amino acids, and yeast extract. In general, strains cannot reduce nitrate to nitrite, though some *Methylophaga thalassica* strains have been reported to produce weak positive results in this test (Urakami and Komagata, 1987). Ammonia is assimilated by the glutamate cycle (glutamate synthase and glutamine synthetase) (Doronina et al., 1997).

DNA composition The mol% G + C of the DNA has been analyzed for several *Methylophaga* strains. In *M. marina* the values range from 43–44. *M. thalassica* has slightly higher values of 44–45. Additional *Methylophaga*-like strains analyzed by Urakami and Komagata (1987) have values as high as 48.

Fatty acids and quinones Chemotaxonomically the strains of *Methylophaga* form a homogeneous group. The major fatty acids include $C_{12:0}$ (4–5%), $C_{14:0}$ (3–4%), $C_{16:0}$ (33–35%), $C_{16:1\ \omega7c}$ (37–44%), $C_{18:1\ \omega7c}$ (4–11%) and $C_{10:0\ 3OH}$ (3–5%). The major isoprenoid quinone is ubiquinone-8 and all strains contain significant levels of squalene (Urakami and Komagata, 1986c, 1987).

Ecology Limited information is available on the distribution and ecology of *Methylophaga*. Studies using indirect immunofluorescence have shown that *M. marina* strains can be most frequently isolated from open seawater rather than estuarine waters (Sieburth et al., 1993). The ability of *M. sulfidovorans* to utilize dimethylsulfide and use H_2S for energy suggests the species has a significant role in cycling of DMS and sulfide in marine ecosystems (de Zwart et al., 1997). DMS is the breakdown product of dimethylsulfoniopropionate (DMSP), an important osmolyte produced copiously by various phytoplankton species (Ledyard et al., 1993; Visscher and Taylor, 1994). Other bacteria that can metabolize DMS, such as *Hyphomicrobium* and *Thiobacillus* spp. (Kanagawa and Mikami, 1989; Visscher and Taylor, 1994), have been isolated from freshwater sites. *"M. limanica"* was isolated from the hypersaline waters of a liman adjoining the Black Sea (Doronina and Trotsenko, 1997).

ENRICHMENT AND ISOLATION PROCEDURES

M. marina and *M. thalassica* can be isolated following initial enrichment of marine samples in ASW-methanol broth for 3–6 d at about 30°C (Janvier et al., 1985). Enrichment cultures are then serially diluted onto ASW-methanol agar and incubated at

30°C for about 1 week. The appearance of pale pink colonies 0.5–2 mm in size are usually indicative of *Methylophaga* strains. A similar approach can be taken for *"M. limanica"* except NaCl levels used should match sample salinity levels (from 1.0–2.0 M). *M. sulfidovorans* can be isolated from benthic mats and sediment by enriching samples in ASW-dimethylsulfide (DMS)[2] broth at 20–25°C (de Zwart et al., 1996). A sample slurry is serially diluted in ASW-DMS broth and growth from the highest dilution is then transferred to fresh medium. Isolation must be performed in liquid medium because no growth occurs on DSM agar media. The DMS-utilizing culture is passaged until it becomes morphologically homogeneous. Alternatively, purification can be achieved on ASW-methanol agar, with checks made to ensure DMS utilization is maintained.

MAINTENANCE PROCEDURES

Methylophaga can be preserved at −80°C in ASW-methanol media supplemented with 30% DMSO. Strains can also be freeze-dried using 20% skim milk as a cryoprotectant.

DIFFERENTIATION OF THE GENUS *METHYLOPHAGA* FROM OTHER GENERA

To date, *Methylophaga* is the only known restricted methylotroph isolated from marine habitats. A requirement for Na^+ ions differentiates this genus from other metabolically similar restricted methylotrophs belonging to the *Betaproteobacteria*, i.e., *Methylophilus*, *Methylobacillus*, and *Methylovorus*.

TAXONOMIC COMMENTS

16S rRNA sequences are available for *Methylophaga marina*, *Methylophaga thalassica*, and *Methylophaga sulfidovorans*, and these species form a tightly knit group with an overall sequence divergence of approximately 4–5%. *Methylophaga* forms a distinct lineage within the *Gammaproteobacteria* (Ando et al., 1989; Janvier and

2. ASW-dimethylsulfide (DMS) broth consists of (g/l) NaCl, 15–25; NH_4SO_4, 0.5; $CaCl_2 \cdot 6H_2O$, 0.33; KCl, 0.2; $MgSO_4 \cdot 7H_2O$, 1; KH_2PO_4, 0.02; Na_2CO_3, 2; $FeSO_4 \cdot 7H_2O$, 1 mg; trace element solution (similar to that given for ASW-methanol medium except that it is supplemented with 6 mM Na_2SeO_4), 1 ml. The pH is adjusted to ~7.5 and 10- or 20-ml portions of the medium are dispensed into 60- or 100-ml serum vials, sealed by butyl-rubber stoppers, and autoclaved. After the medium is cooled, DMS and vitamin B_{12} are added to final concentrations of 1 mM and 1 μg/ml, respectively.

Grimont, 1995; de Zwart et al., 1996). The closest relatives include the marine genus *Cycloclasticus* and various type I methanotrophs including *Methylomonas* and *Methylobacter*. *Methylophaga* are unrelated to other restricted methylotrophic bacteria, including *Methylophilus* (Jenkins et al., 1987), *Methylobacillus* (Urakami and Komagata, 1986b), and *Methylovorus* (Govorukhina and Trotsenko, 1991), which belong to the *Betaproteobacteria* and which are nonhalophilic.

"*Alteromonas thalassomethanolica*" (e.g., ATCC 33145 and NCIMB 2160) and "*Methylomonas thalassica*" (e.g., NCIMB 2162, ATCC 33146) described originally by Yamamoto et al. (1980) were renamed *Methylophaga thalassica* by Janvier et al. (1985). Further phenotypic analysis by Urakami and Komagata (1987)

indicates that *M. marina* and *M. thalassica* are very similar phenotypically and that they can only be identified definitively by a limited number of biochemical characteristics (Table BXII.γ.49), 16S rRNA analysis, and DNA–DNA hybridization. DNA–DNA hybridization studies indicate that *M. marina* strains share 81 to >100% hybridization, *M. thalassica* strains share 61–74% hybridization, and that no significant DNA hybridization occurs between the species.

The putative species "*M. limanica*" has yet to be validated. Doronina et al. (1997) reported that "*M. limanica*" strain hal-6 exhibited 30–40% DNA hybridization with *M. marina* and *M. thalassica*.

DIFFERENTIATION OF THE SPECIES OF THE GENUS *METHYLOPHAGA*

Phenotypic characteristics differentiating *Methylophaga* species are shown in Table BXII.γ.49. Electrophoresis of enzymes has also been used to distinguish *Methylophaga* species. For example, the electrophoretic mobility of glucose-6-phosphate dehydrogen-

ase is slightly greater for *M. thalassica* strains vs. *M. marina* strains. The methanol dehydrogenases of these species also differ in terms of mobility (Janvier et al., 1985).

List of species of the genus Methylophaga

1. **Methylophaga marina** Janvier, Frehel, Grimont and Gasser 1985, 138[VP]

ma.ri′na. L. adj. *marinus* of the sea.

Short, slender, Gram-negative rods 0.2 × 1 μm. Colonies on artificial seawater-methanol agar are white to pale pink, convex, and circular with an entire edge and a creamy consistency. Na⁺ ions, Mg²⁺ ions, and vitamin B_{12} are required for growth.

Optimal growth temperature, 30–37°C; range, 10–40°C. Optimal pH, ~7.5.

Other characteristics are as given in the genus description and are presented in Table BXII.γ.49.

Isolated from seawater and marine sediment.

The mol% G + C of the DNA is: 43–44 (T_m).

Type strain: 222, ATCC 35842, DSM 5989, NCIMB 2244.

GenBank accession number (16S rRNA): X87338, X95459.

2. **Methylophaga sulfidovorans** de Zwart, Nelisse and Kuenen 1998, 1083[VP] (Effective publication: de Zwart, Nelisse and Kuenen 1996, 266.)

sul.fi.do′vo.rans. L. n. *sulfur* sulfur; L. prefix *sulf-* + *ide* probably originally formed from German *sulfid*; L. v. *voro* to devour; M.L. part. adj. *sulfidovorans* sulfide-devouring.

Short, slender, Gram-negative motile rods, approximately 0.9 × 0.2 μm. Na⁺ ions, Mg²⁺ ions, and vitamin B_{12} are required for growth. Other characteristics are as given in the genus description and are presented in Table BXII.γ.49.

Isolated from a marine benthic mat off the coast of the Netherlands.

The mol% G + C of the DNA is: 42 (T_m).

Type strain: YK-4015, ATCC 33146, DSM 5690, IAM 12458, LMG 4055, NCIMB 2163.

GenBank accession number (16S rRNA): X87339, X95460.

3. **Methylophaga thalassica** Janvier, Frehel, Grimont and Gasser 1985, 138[VP]

tha.las′si.ca. Gr. n. *thalassica* of the sea.

Short, slender, Gram-negative rods, 0.2 × 1 μm. Colonies on artificial seawater-methanol agar are white to pale pink, convex, and circular with an entire edge and a creamy consistency. Na⁺ ions, Mg²⁺ ions, and vitamin B_{12} are required for growth.

Optimal temperature, 30–37°C; range, 10–42°C. Optimal pH, ~7.5.

Other characteristics are as given in the genus description and are presented in Table BXII.γ.49.

Isolated from seawater.

The mol% G + C of the DNA is: 44–45 (T_m).

Type strain: YK-4015, ATCC 33146, DSM 5690, IAM 12458, LMG 4055, NCIMB 2163.

GenBank accession number (16S rRNA): X87339, X95460.

Species Incertae Sedis

1. "**Methylophaga limanica**" Doronina, Krauzova and Trotsenko 1997, 438.

li.ma′ni.ca. M.L. n. *limanica* of a liman (a salty lagoon-like water body).

Short Gram-negative rods possessing a single polar flagellum. Do not form spores or prostheca.

Colonies on artificial seawater salts–methanol agar appear opaque, white, and circular, with an entire edge and a creamy consistency. No growth occurs on nutrient agar.

Requires vitamin B_{12} and Na⁺ ions for growth (0.5–1.0 M is optimal). Optimal temperature and pH for growth are 30–37°C and pH 6.7–7.2, respectively.

Isolated from a Black Sea liman.

Other characteristics are as given in the genus description and are presented in Table BXII.γ.49.

The mol% G + C of the DNA is: 43.5 (T_m).

Deposited strain: hal 6.

Genus V. **Thioalkalimicrobium** *Sorokin, Lysenko, Mityushina, Tourova, Jones, Rainey, Robertson and Kuenen 2001a, 576*[VP]

THE EDITORIAL BOARD

Thi.o.al.kal.i.mi.cro' bi.um. Gr. n. *thios* sulfur; N.L. n. *alkali* Arabic *al* the; *qaliy* soda ash; Gr. adj. *micros* small; Gr. masc. n. *bios* life; M.L. n. *Thioalkalimicrobium* sulfur alkaline microbe.

Motile Gram-negative rods, curved rods, or spirilla (0.4–0.5 × 0.8–1.5 µm). 1–3 monopolar flagella. Possess **carboxysomes. Chemolithoautotrophs; obligate aerobes. Oxidize sulfide and thiosulfate to sulfate**; do not oxidize sulfite. Carbon assimilation via Calvin cycle. Halotolerant up to 1.2–1.5 M Na^+; require at least 0.2–0.3 M NaCl.

The mol% G + C of the DNA is: 48–51.2.

Type species: **Thioalkalimicrobium aerophilum** Sorokin, Lysenko, Mityushina, Tourova, Jones, Rainey, Robertson and Kuenen 2001a, 578 (**Thioalkalimicrobium aerophilum** (sic) Sorokin, Lysenko, Mityushina, Tourova, Jones, Rainey, Robertson and Kuenen 2001a, 578.)

FURTHER DESCRIPTIVE INFORMATION

These organisms were found in the water and sediments of alkaline lakes in which sulfide is produced in the underlying sediments. Growth in the laboratory was studied in carbonate/bicarbonate buffered medium containing sulfide, thiosulfate, and polysulfide. Tetrathionate and elemental sulfur were not used by most strains. The pH optimum was 9–10. Ammonia, nitrite, nitrate, and thiocyanate were used as nitrogen sources (Sorokin et al., 2001a).

Analysis of the 16S rDNA sequence of the type strains of *T. aerophilum* and *T. sibiricum* revealed a close relationship with the genus *Thiomicrospira*, particularly with *Thiomicrospira pelophila*. Thus, the genus *Thioalkalimicrobium* is a member of the family "*Piscirickettsiaceae*" in the *Gammaproteobacteria* (Sorokin et al., 2001a).

ENRICHMENT AND ISOLATION PROCEDURES

These bacteria were obtained by enrichment and isolation in a carbonate/bicarbonate buffered liquid salts medium containing thiosulfate. For some strains, the amount of NaCl in the base medium had to be increased (Sorokin et al., 2001a).

MAINTENANCE PROCEDURES

Thioalkalimicrobium spp. can be stored at 4°C for 2 months between transfers in liquid salts medium (pH 10) containing 5mM $MgCl_2$ to stabilize the cell walls. They can also be stored in 10% glycerol (v/v) at –80°C (Sorokin et al., 2001a).

List of species of the genus Thioalkalimicrobium

1. **Thioalkalimicrobium aerophilum** Sorokin, Lysenko, Mityushina, Tourova, Jones, Rainey, Robertson and Kuenen 2001a, 578[VP] (*Thioalkalimicrobium aerophilum* (sic) Sorokin, Lysenko, Mityushina, Tourova, Jones, Rainey, Robertson and Kuenen 2001a, 578.)

ae.ro.phi' lum. Gr. masc. n. *aër* gas Gr. adj. *philus* loving; M.L. n. *aerophilum* air-loving.

Description as for the genus with the following additional characteristics. Grow preferentially in atmospheric O_2 levels. Tetrathionate oxidized by most strains. Colonies flat and spreading.

The mol% G + C of the DNA is: 48–51.2 (T_m).

Type strain: AL 3, DSM 13739.

GenBank accession number (16S rRNA): AF126548.

2. **Thioalkalimicrobium sibiricum** Sorokin, Lysenko, Mityushina, Tourova, Jones, Rainey, Robertson and Kuenen 2001a, 578[VP] (*Thioalkalimicrobium sibiricum* (sic) Sorokin, Lysenko, Mityushina, Tourova, Jones, Rainey, Robertson and Kuenen 2001a, 578.)

si.be' ri.cum. M.L. masc. adj. *sibericus* from Siberia.

Description as for the genus with the following additional characteristics. Grow preferentially at less than atmospheric O_2 levels. Tetrathionate not oxidized by most strains. Colonies compact.

The mol% G + C of the DNA is: 48.2–49.5 (T_m).

Type strain: AL 7, DSM 13740.

GenBank accession number (16S rRNA): AF126549.

Genus VI. **Thiomicrospira** *Kuenen and Veldkamp 1972, 253*[AL]

THORSTEN BRINKHOFF, JAN KUEVER, GERARD MUYZER AND HOLGER W. JANNASCH

Thi.o.mi.cro.spi' ra. Gr. n. *thios* sulfur; Gr. adj. *micros* small, little; Gr. n. *spira* spiral; M.L. fem. n. *Thiomicrospira* small sulfur spiral.

Small spiral-, comma-, or rod-shaped cells (0.2–0.5 × 0.8–3.0 µm) that occur singly. Gram negative. Motile by means of a polar flagellum or nonmotile. No resting stages are known. **Aerobic. Chemolithoautotrophic, using reduced inorganic sulfur compounds, such as S^0, sulfide, tetrathionate, thiosulfate** (but not sulfite or thiocyanate), **and CO_2 as carbon source.** CO_2 is fixed via the **Calvin-Benson cycle** (ribulose 1,5-bisphosphate carboxylase is present). The final oxidation product is sulfate, but S^0 may accumulate in the medium. During growth acid is produced. Ammonium salts serve as nitrogen source. Na^+ ions are required for growth. The temperature optimum lies between 28°C and 40°C. The optimal pH varies from 6.0 to 8.0. Metabolic properties are very similar to those of the genus *Thiobacillus*.

Q-8 is the major ubiquinone. Members of a **monophyletic group** within the *Gammaproteobacteria*.

The mol% G + C of the DNA is: 39.6–49.9.

Type species: **Thiomicrospira pelophila** Kuenen and Veldkamp 1972, 253.

FURTHER DESCRIPTIVE INFORMATION

All described *Thiomicrospira* spp. are obligate chemolithotrophs, and thus use CO_2 as their main carbon source. They do not have functional citric acid or glyoxylate cycles. However, *Thiomicrospira pelophila* can assimilate small amounts of organic substrates, such as acetate, while growing autotrophically under thiosulfate limitation (Kuenen and Veldkamp, 1972). Heterotrophic growth, as reported for *Thiomicrospira thyasirae* (Wood and Kelly, 1989) could not be confirmed with a culture obtained from the Deutsche Sammlung für Mikroorganismen und Zellkulturen (DSMZ) (Kuever, unpublished results). Strains of *Thiomicrospira crunogena* can use polymetal sulfides as electron donors (Eberhard et al., 1995; Wirsen et al., 1998). Members of the genus *Thiomicrospira* require oxygen as terminal electron acceptor for respiration. Good growth occurs at low to saturated oxygen conditions.

Temperature ranges of species of this genus are from 3.5 to 40°C. At optimal pH the maximum growth rates on TP or T-ASW medium[1] at 30°C ranged from 0.3 to 0.8 h^{-1}. The optimal pH values are between 6.0 and 8.0; however, growth of particular species was reported down to values of 4.0 or up to 9.0.

On 1% (w/v) agar plates with TP medium small, smooth, whitish colonies are first obtained. Later, bigger yellowish colored colonies with a rough surface are formed due to intensive sulfur precipitation. Production and consumption of S^0 globules is both pH and thiosulfate dependent (Javor et al., 1990). Under stress conditions, pleomorphism and spirals up to 5–15 μm length were observed for *T. thyasirae* (Wood and Kelly, 1993).

Although *Thiomicrospira* was present in one freshwater habitat (Brinkhoff and Muyzer, 1997), so far, all described species and strains were isolated from marine environments. Na$^+$ ions are required, with an optimal concentration of around 470 mM. The named species were obtained from intertidal mud flats, a continental shelf, and deep-sea hydrothermal vents. Deep-sea strains are barotolerant, but not barophilic, and are known to excrete approximately 10% of their fixed carbon as low molecular weight organic compounds (Ruby and Jannasch, 1982; Jannasch et al., 1985; Wirsen et al., 1998).

Thiomicrospira appears to be ecologically very significant in hydrothermal vent systems (e.g., Muyzer et al., 1995; Brinkhoff et al., 1999c), while in an intertidal mud flat habitat the genus was found in much lower densities than other sulfur-oxidizing bacteria (Brinkhoff et al., 1998).

ENRICHMENT AND ISOLATION PROCEDURES

Procedures for the enrichment and isolation of these organisms were described several times (Ruby et al., 1981; Kuenen et al., 1992; Brinkhoff and Muyzer, 1997). *Thiomicrospira* can be obtained after inoculation of a general marine medium for chemolithoautotrophic bacteria containing thiosulfate as the sole electron donor with a water or sediment sample. Incubation is aerobic at room temperature. After growth is obtained, indicated by a decreasing pH value, and often sulfur precipitation, an aliquot should be transferred to a 1–1.5% (w/v) agar plate, containing the same medium, to promote formation of the distinctive colony types. For some isolates enrichment by filtration of the sample before inoculation of the medium was described. A pore size of 0.22 μm was recommended for *T. pelophila* (Kuenen and Veldkamp, 1972) and one of 0.45 μm for *T. chilensis* (Brinkhoff and Muyzer, 1997).

MAINTENANCE PROCEDURES

Active cultures can be stored at 4°C for 2 weeks, but should then be transferred in fresh medium, to prevent acid formation that might easily result in irreversible inhibition of growth. For long-term storage, members of this genus are best maintained in liquid nitrogen or at −80°C in TP medium containing 20% (v/v) glycerol or 5% (v/v) DMSO (final concentration). They can be subcultured on mineral salts medium with sulfide, thiosulfate, or S^0.

PROCEDURES FOR TESTING SPECIAL CHARACTERS

Two oligonucleotide probes specific for the genus *Thiomicrospira* were designed targeting the 16S rRNA gene (Brinkhoff and Muyzer, 1997). The probes are located at positions 128–145 (TMS128F: 5′-GAA TCT RCC CTT TAG TTG-3′) and 830–849 (TMS849R: 5′-CTT TTT AAT AAG RCC AAC AG-3′) according to *Escherichia coli* numbering (Brosius et al., 1981). They were checked against the small-subunit rRNA sequences stored in the Ribosomal Database Project (RDP) using the Check Probe software program (Maidak et al., 1997) and contained at least two mismatches to all non-*Thiomicrospira* sequences. The probes require a hybridization temperature of 52°C for specific dot-blot hybridization with nearly complete PCR-amplified 16S rRNA gene fragments. If both probes are used combined as PCR primers, an annealing temperature of 44°C is required for specific amplification of *Thiomicrospira* 16S rRNA genes (Brinkhoff and Muyzer, 1997).

Two slightly different versions of probe TMS849R, which differ by an A/G mismatch, were developed for use in rRNA slot-blot hybridization (TMS849A: 5′-CTT TTT AAT AAG ACC AAC A-3′; TMS849G: 5′-CTT TTT AAT AAG GCC AAC A-3′) (Brinkhoff et al., 1998). Both probes contain one mismatch to several *Thiomicrospira* sequences. The combination of both probes matches all known species. Specificity was tested with rRNA isolated from *Thiomicrospira* species having either no mismatch or one. Both probes have a T_d of 41°C. Hybridization is done at 30°C, and washing at 41°C (Brinkhoff et al., 1998).

DIFFERENTIATION OF THE GENUS *THIOMICROSPIRA* FROM OTHER GENERA

Thiomicrospira species can be differentiated from the related genus *Thiobacillus* by their lower mol% G + C content and their higher sulfide tolerance (800–1000 μM) (Kuenen and Veldkamp, 1972; Ruby and Jannasch, 1982; Jannasch et al., 1985). Not all *Thiomicrospira* species have a spiral cell form; *T. chilensis* and *T. frisia* are rod-shaped. Therefore, morphology should not be used for distinguishing *Thiomicrospira* from thiobacilli. The most reliable method for a clear identification is the use of the genus-specific probes, as described above, or sequencing of the 16S rRNA gene followed by phylogenetic analysis.

---

1. TP (*Thiomicrospira pelophila*) medium (Kuenen and Veldkamp, 1972, modified by Brinkhoff et al., 1999b): synthetic seawater medium contained (g/l deionized water) NaCl, 25.0; (NH$_4$)$_2$SO$_4$, 1.0; MgSO$_4$·7H$_2$O, 1.5; CaCl$_2$·2H$_2$O, 0.42; K$_2$HPO$_4$, 0.5; Na$_2$ S$_2$O$_3$·5H$_2$O, 5.0. One ml of a trace element solution containing EDTA (Widdel and Bak, 1992) was used. Vitamin B$_{12}$ is added after sterile filtration (final concentration 50 ng/l). Bromothymol blue (4 mg/l) is used as pH indicator. K$_2$HPO$_4$ and Na$_2$S$_2$O$_3$·5H$_2$O are autoclaved separately, each in 10% of the final volume. After autoclaving the pH is adjusted to 7.2 with sterile 0.4% (w/v) Na$_2$CO$_3$ solution.

T-ASW (thiosulfate artificial seawater) medium (Jannasch et al., 1985): synthetic seawater medium contained (g/l deionized water) NaCl, 25.0; (NH$_4$)$_2$SO$_4$, 1.0; MgSO$_4$·7H$_2$O, 1.5; CaCl$_2$·2H$_2$O, 0.3; KH$_2$PO$_4$, 0.42; NaHCO$_3$, 0.2; Trizma hydrochloride, 3.0; Na$_2$S$_2$O$_3$, 1.58. One ml trace element solution SL-8 (Biebl and Pfennig, 1978) was used. Phenol red (2 ml of a 0.5% [w/v] solution) is used as pH indicator. KH$_2$PO$_4$ and Na$_2$S$_2$O$_3$ are autoclaved separately, each in 10% of the final volume. After autoclaving the pH is adjusted to 7.5 with sterile 0.4% (w/v) Na$_2$CO$_3$ solution.

Taxonomic Comments

Phylogenetic analysis of nearly complete 16S rRNA gene sequences showed that all *Thiomicrospira* spp., with the exception of *Thiomicrospira denitrificans* (Timmer-ten Hoor, 1975), form a monophyletic group (bootstrap value of 100%) within the *Gammaproteobacteria* (Fig. BXII.γ.88). *T. denitrificans* belongs to the *Epsilonproteobacteria* (Muyzer et al., 1995). This organism requires reclassification as a new genus. Even though *T. denitrificans*, based on phylogenetic and physiological analysis, does not belong to the genus *Thiomicrospira*, it is listed here as a species. The same description of this organism as given by Kuenen and Robertson (1989) was used. However, its characteristics are not included in the general description of the genus *Thiomicrospira* and the specific oligonucleotides do not match the 16S rRNA gene of *T. denitrificans*.

The *Thiomicrospira* group includes *Hydrogenovibrio marinus*, which differs from *Thiomicrospira* spp. only in the ability to use hydrogen as electron donor (Table BXII.γ.50) (Nishihara et al., 1991b, 1998). Since hydrogen utilization is a ubiquitous trait among many phylogenetically diverse bacteria, hydrogen utilization alone may be insufficient to separate this organism as a separate genus.

The rod-shaped *Thiomicrospira* species and strains, i.e., *T. frisia*, *T. chilensis*, *Thiomicrospira* spp. str. Art-3, and *Thiomicrospira* spp. str. Milos T-2, form a subcluster within the *Thiomicrospira* branch. This indicates a common ancestor of these organisms (Kuever, unpublished).

The reliable differentiation of *Thiomicrospira* species requires integration of physiological and genomic data. In cases of high 16S rRNA gene homology among closely related strains, DNA–

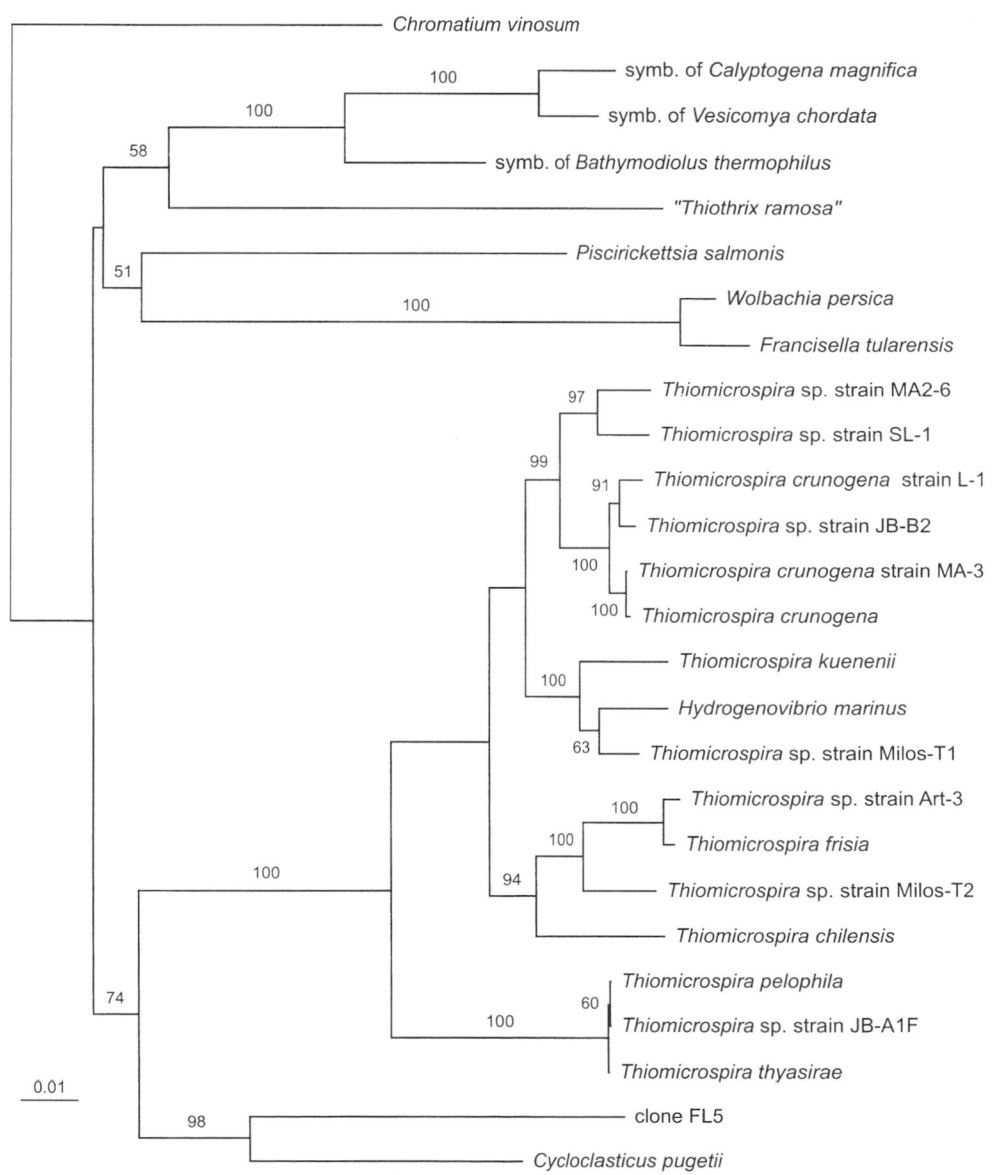

FIGURE BXII.γ.88. Neighbor joining tree based on nearly complete 16S rRNA gene sequences, showing the phylogenetic relationships between different *Thiomicrospira* spp. and closely related bacteria. The *Thiomicrospira* sequences were aligned to prealigned sequences taken from the Ribosomal Database Project (RDP; Maidak et al., 1997). *Chromatium vinosum* was taken as the outgroup. The numbers on the branches refer to bootstrap values; only values of 50% and higher are shown. Bar = 0.01 estimated nucleotide change per sequence position.

TABLE BXII.γ.50. Morphological and physiological characteristics of the species composing the genus *Thiomicrospira*, *Hydrogenovibrio marinus*, and *Thiomicrospira denitrificans*[a]

Characteristic	1. *T. pelophila*	2. *T. chilensis*	3. *T. crunogena*	4. *T. frisia*	5. *T. kuenenii*	6. *T. thyasirae*	7. *H. marinus*	8. *T. denitrificans*
Class:								
Gammaproteobacteria	+	+	+	+	+	+	+	
Epsilonproteobacteria								+
Shape	Vibrio	Rod	Vibrio	Rod	Vibrio	Vibrio	Vibrio	Spiral
Width (μm)	0.2–0.3	0.3–0.5	0.4–0.5	0.3–0.5	0.3–0.4	0.3	0.2–0.5	0.3
Length (μm)	1.0–2.0	0.8–2.0	1.5–3.0	1.0–2.7	1.0–2.5	0.8–1.2	1.0–2.0	variable
Motility	+	+	+	+	+	−	+	−
Mol% G + C content	45.7 (44)[b]	49.9	44.2 (42)[b]	39.6	42.4	45.6[c] (52)[b]	44.1	36.0
Major ubiquinone	Q-8	Q-8	Q-8	Q-8	Q-8	Q-8[d]	Q-8	nd
Maximum growth rate with thiosulfate, h^{-1}	0.3	0.4	0.8	0.45	0.35	0.07[e]	0.6	0.06
Optimal pH	7.0[f]	7.0	7.5–8.0	6.5	6.0	7.5	6.5	7.0
pH range	5.6–9.0[f]	5.3–8.5	5.0–8.5	4.2–8.5	4.0–7.5	7.0–8.4	nd	nd
Optimal temperature (°C)	28–30	32–37	28–32	32–35	29–33.5	35–40	37	22
Temperature range (°C)	3.5–42	3.5–42	4–38.5	3.5–39	3.5–42	10–45	nd	nd
Na$^+$ requirement	+	+	+	+	+	+	+	−
Optimal Na$^+$-concentration (mM)	470	470	nd	470	470	430	500	nd
Na$^+$-concentration range (mM)	40–1240[f]	100–1240	at least 45	100–1240	100–640	250–3000	nd	nd
Vitamin B$_{12}$-dependent	+	−	−	−	−	−	−	nd
Calvin Benson cycle	+	+	+	+	+	+	+	+
Formation of sulfur from thiosulfate at pH 7.0 in liquid medium	+	+	+	−	−	+	+	−
Growth on H$_2$ as the sole electron donor	−[c]	−	−[c]	−	−	−[c]	+	nd
Fully aerobic growth	+	+	+	+	+	+	+	−
Denitrification	−	−	−	−	−	−	−	+
Heterotrophic growth	−	−	−	−	−	−[c]	−	nd

[a]Symbols: +, 90% or more strains are positive; −, 90% or more strains are negative; nd, not determined; data from Kuenen and Veldkamp, 1972, 1973; Timmer-ten Hoor, 1975; Ruby and Jannasch, 1982; Jannasch et al., 1985; Kuenen and Robertson, 1989; Wood and Kelly, 1989, 1993; Nishihara et al., 1991b, 1998; Brinkhoff et al. 1999a, 1999b; Kuever, unpublished data.

[b]The values in brackets are from the original description, determined by thermal denaturation.

[c]J. Kuever, unpublished results.

[d]In the original description Q-10 was found (Wood and Kelly, 1989); we could only detect Q-8 and traces of Q-7 (J. Kuever, unpublished results).

[e]In the original description the maximum growth rate on acetate was 0.53 (h^{-1}) (Wood and Kelly, 1989). Growth rates on thiosulfate seemed to be similar to *T. pelophila* in the range of 0.2 0.3 (Kuever, unpublished results).

[f]Brinkhoff et al., 1999a.

DNA hybridization was necessary to decide whether these strains belong to one species or not. This method revealed that *Thiomicrospira* strains, which were obtained from different hydrothermal vent systems, and which showed specific metabolic adaptations to their environments, belong to one species, *T. crunogena* (Wirsen et al., 1998). DNA–DNA hybridization also confirmed *T. pelophila*, *T. crunogena*, *T. kuenenii*, *T. frisia*, and *T. chilensis* as distinct species (Brinkhoff et al., 1999a, 1999b).

Acknowledgments

The authors wish to thank Andreas Teske, Carl O. Wirsen, and Stefan Sievert for help and advice. The help of Hirofumi Nishihara and Yasuo Igarashi is greatly appreciated, because they supplied us with unpublished results and submitted manuscripts. We are grateful to Gijs Kuenen for providing the photograph of *Thiomicrospira pelophila*.

Further Reading

Brinkhoff, T. and G. Muyzer. 1997. Increased species diversity and extended habitat range of sulfur-oxidizing *Thiomicrospira* spp. Appl. Environ. Microbiol. *63*: 3789–3796.

Kuenen, J.G., L.A. Robertson and O.H. Tuovinen. 1992. The genera *Thiobacillus*, *Thiomicrospira*, and *Thiosphaera*. *In* Balows, Trüper, Dworkin, Harder and Schleifer (Editors), The Prokaryotes. A Handbook on the Biology of Bacteria: Ecophysiology, Isolation, Identification, Applications, 2nd Ed., Vol. III, Springer Verlag, New York. pp. 2638–2657.

List of species of the genus Thiomicrospira

1. **Thiomicrospira pelophila** Kuenen and Veldkamp 1972, 253[AL].

   pe.lo'phi.la. Gr. n. *pelos* mud; Gr. adj. *phila* loving; N.L. adj. *pelophila* mud-loving.

   Colonies on thiosulfate agar are small, white, and often surrounded by a halo of sulfur. On soft thiosulfate agar, colonies may be up to 1 cm in diameter and grow down into the agar as large disks. These disks are white from precipitated sulfur and only sparsely populated with bacteria. In media containing more than 0.2–0.3 mM thiosulfate, sulfur is formed. The type strain requires vitamin B$_{12}$; others do not. Sulfide tolerance is high in this species compared with *Thiobacillus* species (Kuenen and Veldkamp, 1972). Even though CO$_2$, fixed via the Calvin-Benson cycle, is the primary carbon source, small amounts of organic compounds such as acetate and succinate may serve as a secondary source under energy-limited conditions (Kuenen and Veldkamp, 1973). See Table BXII.γ.50 for further physiological and biochemical details; Fig. BXII.γ.89 illustrates morphology and flagellation.

Isolated from a marine intertidal mud flat of the Dutch Wadden Sea.

The mol% G + C of the DNA is: 45.7 (HPLC), 44.0 (T_m).

Type strain: ATCC 27801, DSM 1534.

GenBank accession number (16S rRNA): L40809.

2. **Thiomicrospira chilensis** Brinkhoff, Muyzer, Wirsen and Kuever 1999a, 878[VP]

chi.len' sis. M.L. adj. *chilensis* from Chile, a country of South America.

Cells are rod-shaped and motile, but during late exponential growth, cells have a strong tendency to clump and to form aggregates. Strong sulfur precipitation in liquid batch cultures as *T. crunogena*. See Table BXII.γ.50 for further details. Found in sediment samples of *Thioploca* mats obtained from the Chilean coast shelf. Isolated from a sample that was filtered through a 0.45-μm pore-size filter before inoculation (Brinkhoff and Muyzer, 1997).

The mol% G + C of the DNA is: 49.9 (HPLC).

Type strain: ATCC 700858, DSM 12352.

GenBank accession number (16S rRNA): AF013975.

3. **Thiomicrospira crunogena** Jannasch, Wirsen, Nelson, and Robertson 1985, 422[VP]

cru.no.ge' na. Gr. n. *crunos* spring; Gr. v. *genomai* to be generated; Gr. part. adj. *crunogena* spring-born.

T. crunogena can withstand relatively high concentrations of sulfide, continuing to fix CO_2 at concentrations up to 800 μM. Sulfide is toxic at 2000 μM. Strong precipitation of sulfur in liquid batch cultures. Barotolerant; at 250 atm, the CO_2 fixation rate is 80% of that obtained at 1 atm. Growth does not occur at 400 atm or above. Cells are motile and vibrio shaped. For further physiological and biochemical characteristics see Table BXII.γ.50.

Isolated from a marine hydrothermal vent system (21°N, 109°W, East Pacific Rise) at a depth of 2600 m.

The mol% G + C of the DNA is: 44.2. (HPLC), 42.0 (T_m).

Type strain: ATCC 35932, DSM 12353, LMD 84.00.

GenBank accession number (16S rRNA): L40810.

Additional Remarks: Two further described *Thiomicrospira* strains belong to this species as determined by 16S rRNA gene sequence analysis and DNA–DNA hybridization (Wirsen et al., 1998): *Thiomicrospira* sp. str. L-12 (Ruby and Jannasch, 1982) and *Thiomicrospira* sp. str. MA-3 (Wirsen et al., 1993). All three strains of *T. crunogena*, but especially strain MA-3, can grow on sulfidic minerals (Eberhard et al., 1995). They also show robust growth properties compared to the other species of the genus *Thiomicrospira*. *Thiomicrospira crunogena* str. L-12 was isolated from mussel periostracum from the Galapagos Rift hydrothermal vent site and deposited under DSM 12346. The 16S rRNA gene sequence is available from GenBank under accession number L01576. *Thiomicrospira crunogena* str. MA-3 was isolated from a polymetal sulfide rock from the Mid-Atlantic Ridge TAG hydrothermal vent site and deposited under ATCC 700270 (DSM 12354). The 16S rRNA gene sequence is available from GenBank under accession number AF069959.

4. **Thiomicrospira frisia** Brinkhoff, Muyzer, Wirsen, and Kuever 1999b, 391[VP]

fri' si.a. M.L. fem. adj. *frisia* pertaining to Frisia, a coastal region in northwestern Germany and the Netherlands from where the organism was obtained.

When thiosulfate is used as the primary energy source very small amounts of sulfur and sulfite are produced. Compared to most other *Thiomicrospira* spp., sulfur precipitation in liquid medium is less obvious. Cells are rod-shaped and motile, but during late exponential growth cells have a strong tendency to clump and to form aggregates. See Table BXII.γ.50 for more details on the physiology. See Fig. BXII.γ.90A for details of the morphology.

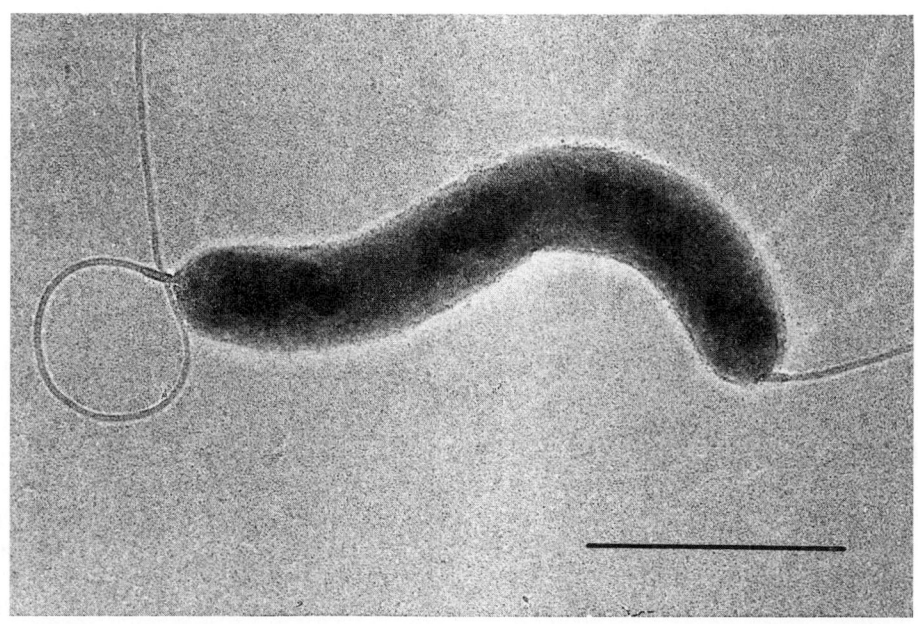

FIGURE BXII.γ.89. Electron micrograph of a *T. pelophila* cell, which was platinum-shadowed to show the shape and the flagellation. Bar = 0.5 μm.

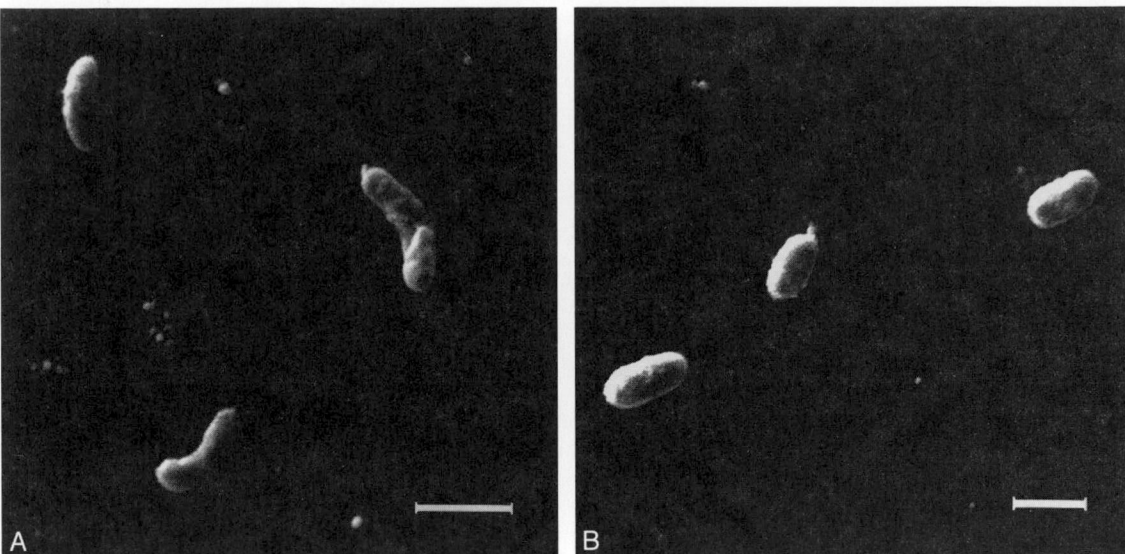

FIGURE BXII.γ.90. Electron micrographs of *T. frisia* and *T. kuenenii* to demonstrate the two morphologies found for *Thiomicrospira* spp. *A*, morphology of *T. kuenenii*, which is typical for most *Thiomicrospira* spp.; *B*, rod-like morphology of *T. frisia*. Bars = 1 µm. (Reproduced with permission from T. Brinkhoff et al., International Journal of Systematic Bacteriology *49*: 385–392, 1999, ©International Union of Microbiological Societies.)

Isolated from a marine intertidal mud flat of the German Wadden Sea.

The mol% G + C of the DNA is: 39.6 (HPLC).
Type strain: ATCC 700878, DSM 12351.
GenBank accession number (16S rRNA): AF013974.

5. **Thiomicrospira kuenenii** Brinkhoff, Muyzer, Wirsen, and Kuever 1999b, 390[VP]

kue.nen' i.i. M.L. gen. n. *kuenenii* of Kuenen; named after J.G. Kuenen, a Dutch microbiologist of the Delft School of Microbiology, who proposed the genus *Thiomicrospira*.

When thiosulfate is used as the primary energy source very small amounts of sulfur and sulfite are produced. Compared to most other *Thiomicrospira* spp., sulfur precipitation in liquid medium is less obvious. See also Table BXII.γ.50 and Fig. BXII.γ.90B. Isolated from a marine intertidal mud flat of the German Wadden Sea.

The mol% G + C of the DNA is: 42.4 (HPLC).
Type strain: ATCC 700877, JB-A1, DSM 12350.
GenBank accession number (16S rRNA): AF013978.

6. **Thiomicrospira thyasirae** (Wood and Kelly 1989) Wood and Kelly 1995, 418[VP] (Effective publication: Wood and Kelly 1993, 47) (*Thiobacillus thyasiris* Wood and Kelly 1989, 165.) *thy.a.si' rae.* M.L. adj. *Thyasira* a genus of the bivalve mollusk family of the *Thyasiridae*.

Morphologically and physiologically nearly identical to *T. pelophila*. 16S rRNA gene sequences of *T. thyasirae* and *T. pelophila* are almost identical (one mismatch in 1424 bp) (Brinkhoff and Muyzer, 1997). DNA–DNA hybridization data will be required to decide whether *T. thyasirae* can be retained as a separate species. Some biochemical and physiological properties of a culture obtained from the DSMZ vary to the original description (Table BXII.γ.50). Cells exhibit pleomorphism in response to environmental conditions.

Isolated from the gills of the mussel *Thyasira flexuosa* collected from Plymouth Sound, U.K.

The mol% G + C of the DNA is: 45.6 (HPLC), 52 (T_m).
Type strain: ATCC 51452, DSM 5322.
GenBank accession number (16S rRNA): AF016046.

Other Organisms

1. *Thiomicrospira* sp. strain MA2–6 Muyzer, Teske, Wirsen and Jannasch 1995.

Vibrio-shaped isolate from a marine hydrothermal vent site at the Mid-Atlantic Ridge, (23°2″ N, 44°57″ W).
Deposited strain: not publicly available
GenBank accession number (16S rRNA): L40811.
Additional Remarks:

2. *Thiomicrospira* sp. strain SL-1 Brinkhoff and Muyzer 1997.

Vibrio-shaped cells isolated from a microbial mat of the hypersaline Solar Lake, Sinai, Egypt. Physiological properties similar to *T. crunogena*, but probably a different species

(Kuever, unpublished). Isolated from a sample that was filtrated through a 0.45-µm pore-size filter before inoculation.
Deposited strain: not publicly available
GenBank accession number (16S rRNA): AF013971.
Additional Remarks:

3. *Thiomicrospira* sp. strain JB-B2 Brinkhoff and Muyzer 1997.

Vibrio-shaped cells, isolated from a water sample overlying a marine intertidal mud flat of the Jadebusen Bay, Germany. High similarity (99%) of the 16S rRNA gene sequence to the sequence of *T. crunogena* and therefore probably a strain of this species (Wirsen et al., 1998).

Deposited strain: not publicly available
GenBank accession number (16S rRNA): AF013972.
Additional Remarks:

4. *Thiomicrospira* sp. strain Milos T-1 Brinkhoff, Sievert, Kuever and Muyzer 1999c.

Vibrio-shaped isolate obtained from sediment of a marine shallow-water hydrothermal vent off the island of Milos, Greece.

Deposited strain: not publicly available
GenBank accession number (16S rRNA): AF082327.
Additional Remarks:

5. *Thiomicrospira* sp. strain Art-3 Brinkhoff and Muyzer 1997.

Rod-shaped isolate obtained from sediment of a saline spring at the city of Artern, Germany. This organism is morphologically and physiologically similar to *T. frisia* (Kuever, unpublished).

Deposited strain: not publicly available
GenBank accession number (16S rRNA): AF013973.
Additional Remarks:

6. *Thiomicrospira* sp. strain Milos T-2 Brinkhoff, Sievert, Kuever and Muyzer 1999c.

Rod-shaped isolate obtained from the sediment surface of a marine shallow water hydrothermal vent off the island of Milos, Greece.

Deposited strain: not publicly available
GenBank accession number (16S rRNA): AF082328.
Additional Remarks:

7. *Thiomicrospira* sp. strain JB-AF1 Brinkhoff and Muyzer 1997.

Isolated from a sample obtained from a marine intertidal mud flat of the Jadebusen Bay, Germany. The sample was filtrated through a 0.22-μm pore-size filter before inoculation. Morphologically and by 16S rRNA gene sequence identical to *T. pelophila*.

Deposited strain: not publicly available
GenBank accession number (16S rRNA): AF013976.
Additional Remarks:

Species Incertae Sedis

1. **Thiomicrospira denitrificans** Timmer-ten Hoor 1975, 344[AL]

de.ni.tri'fi.cans. L. prep. *de* away from; L. n. *nitrum* soda; M.L. n. *nitrum* nitrate; M.L. v. *denitrifico* denitrify; M.L. part. adj. *denitrificans* denitrifying.

Colonies are brown, 1–2 mm in diameter. Sulfur is not deposited. Chemolithotrophic, obtaining energy from the oxidation of reduced sulfur compounds such as sulfide and thiosulfate. CO_2, fixed via the Calvin cycle, is the primary source of carbon. Ammonium salts serve as nitrogen source. This species is microaerophilic, only able to use oxygen as an electron acceptor when pO_2 is <0.5%. In practice, aerobic growth has been attained only in an oxygen-limited chemostat. Nitrate or nitrite can serve as electron acceptors, being reduced to molecular nitrogen. The dissimilatory nitrate reductase is constitutive. Although isolated from marine samples, this species does not have a requirement for NaCl. See Table BXII.γ.50 for further physiological details.

Found in mud of marine tidal flats where sulfide is produced.

The mol% G + C of the DNA is: 36.0 (T_m).
Type strain: ATCC 33889, DSM 1251.
GenBank accession number (16S rRNA): L40808.

Family III. **Francisellaceae** *fam. nov.*

ANDERS B. SJÖSTEDT

Fran.ci.sel.la'ce.ae. M.L. fem. n. *Francisella* type genus of the family; *-aceae* ending for the family *Francisella*.

Short, rod-shaped or coccoid cells 0.2–0.7 × 0.2–1.7 μm. Gram negative, faintly staining. Endospores are not produced. **Nonmotile. Aerobic.** Produce acid from D-glucose. **Cysteine is required for, or enhances, growth.** Weakly catalase positive. **H_2S is produced in cysteine-supplemented medium. Unique fatty acid composition**; long-chain saturated and monounsaturated C_{18} to C_{26} acids, relatively large amounts of saturated even-chain acids ($C_{10:0}$, $C_{14:0}$, and $C_{16:0}$), and two long-chain hydroxy acids ($C_{16:0\ 3OH}$ and $C_{18:0\ 3OH}$). Susceptible to tetracycline and chloramphenicol. Intracellular parasites, transmissible to a wide range of animal species, including man, and the etiological agents of the zoonotic disease tularemia. Most commonly isolated from rodents, lagomorphs, blood-sucking insects, and ticks. The family belongs to the *Gammaproteobacteria*. Signatures of *Francisella* 16S rRNA gene sequences are listed in Table BXII.γ.51. A gene encoding a 17-kDa lipoprotein of *F. tularensis* is conserved in all francisellae investigated.

The mol% G + C of the DNA is: 30–36.
Type genus: **Francisella** Dorofe'ev 1947, 176.

FURTHER DESCRIPTIVE INFORMATION

The family comprises closely related organisms within the single genus *Francisella*. The two recognized species, *F. tularensis* and *F. philomiragia*, show a 16S rRNA gene sequence similarity of ≥98.3% (Forsman et al., 1994). The family is circumscribed and readily distinguishable from other families based on a unique set of phenotypic characteristics that include a coccoidal morphology, Gram negativity, acid but no gas production from a limited number of carbohydrates, requirement for cysteine, and a unique fatty acid composition. There is no DNA relatedness between francisellae and members of other genera showing minor serological cross-reactivity (Ritter and Gerloff, 1966). This has been substantiated by phylogenetic analysis based on comparison of 16S rRNA gene sequences, demonstrating that the family is a member of the *Gammaproteobacteria* (Forsman et al., 1994). It is most closely related to, but still relatively distant from, the family *Piscirickettsiaceae*.

F. tularensis shows slow growth and usually requires cysteine

TABLE BXII.γ.51.   16S rRNA sequence signatures for the genus *Francisella*[a]

Position[b]	*Francisella*[c]	*Bacteria*[d]
151	C	A (G)
336	A	G
402	U	C
849	C	U (G)
889	C	G
1280	A	U (C)
1290	G	C (U)
1453	A	G
1455	U	A (G)

[a]Adapted from Forsman et al. (1994).

[b]Positions numbered according to the nomenclature of *E. coli*.

[c]Bases found in the species of *Francisella*, including *W. persica* and the symbionts isolated from *D. andersoni* and *O. moubata*.

[d]Bases found in other *Bacteria*. Letter in parentheses denote bases found in less than one-fourth of the aligned strains of *Bacteria*.

or cystine for growth. The supplements enhance, but are not required for growth of isolates of *F. tularensis* subsp. *novicida* and *F. philomiragia*. Cystine heart agar supplemented with hemoglobin, glucose cystine heart agar, or Mueller–Hinton agar supplemented with hemoglobin and IsoVitaleX are commonly used for cultivation of *F. tularensis*. Although cultivation from clinical samples of tularemia patients is often avoided due to the contagiousness of *F. tularensis*, isolation is usually possible from most types of relevant clinical specimens, e.g., wound samples, lymph node aspirates, and blood, after prolonged cultivation.

Colony-type variants of *F. tularensis*, including those associated with virulence and immunogenicity, have been described (Eigelsbach et al., 1951). Possibly, these variants can be attributed to a reversible phase variation in their lipopolysaccharide (LPS) (Cowley et al., 1996). The LPS of *F. tularensis* has been characterized and contains a unique somatic antigen (Vinogradov et al., 1991). Extracts of *F. tularensis* strains contain phosphoantigens, pyrophosphoesters bound to derivatives of uracil monophosphate or thymidine monophosphate (Poquet et al., 1998). A gene encoding a 17-kDa lipoprotein of the live vaccine strain, *F. tularensis* strain LVS, is conserved in *Francisella* species (Sjöstedt et al., 1992) and has been used both for diagnostic and taxonomic purposes (Sjöstedt et al., 1997; Niebylski et al., 1997). Francisellae are widely distributed in nature. Members of *F. tularensis* are the etiological agents of tularemia, a zoonotic disease of humans. Tularemia is transmitted by direct contact with infected animals, through contaminated water or food, or by vectors such as biting insects or ticks within endemic regions over the northern hemisphere. The disease shows epidemic outbreaks, both in humans and in animals. The bacterium is highly virulent for humans and numerous animal hosts, in particular, rodents, hares, and rabbits. Clinical manifestations in humans depend on the means of transmission. *F. tularensis* subsp. *novicida* and *F. philomiragia* have been linked to water-borne transmission and can cause disease in immunocompromised individuals.

Genus I. **Francisella** *Dorofe'ev 1947, 176*[AL]

ANDERS B. SJÖSTEDT

Fran.ci.sel'la. M.L. dim. ending *-ella*; M.L. fem. n. *Francisella* named after Edward Francis, American bacteriologist, who extensively studied the etiologic agent and pathogenesis of tularemia and is credited with naming the disease.

Short, rod-shaped or **coccoid** cells, 0.2–0.7 × 0.2 μm (*Francisella tularensis* subsp. *tularensis*, *F. tularensis* subsp. *holarctica*, and *F. tularensis* subsp. *mediasiatica*) or 0.7 × 1.7 μm (*F. philomiragia* and *F. tularensis* subsp. *novicida*) when examined during active growth, pleomorphism occurs thereafter. Gram negative, faintly staining. Endospores are not produced. **Nonmotile. Aerobic**, growth enhanced under microaerophilic conditions. Colonies are distinct, convex, pale white, and reach maximum size in 3–4 days. **Cysteine (or cystine) is required for growth** (*F. tularensis* subsp. *tularensis*, *F. tularensis* subsp. *holarctica*, and *F. tularensis* subsp. *mediasiatica*) **or enhances growth** (*F. philomiragia* and *F. tularensis* subsp. *novicida*). Weakly catalase positive. *F. tularensis* is oxidase negative and *F. philomiragia* oxidase positive (Kovacs modification). **H₂S is produced** in cysteine-supplemented medium. Strains from both species, with the exception of *F. tularensis* subsp. *mediasiatica*, produce acid but no gas from D-glucose and maltose. Urease negative and no reduction of nitrate. **Strains are characterized by a unique fatty acid composition**; long-chain saturated and monosaturated C_{18} to C_{26} acids, relatively large amounts of saturated even-chain acids ($C_{10:0}$, $C_{14:0}$, and $C_{16:0}$), and two long-chain hydroxy acids ($C_{16:0\ 3OH}$ and $C_{18:0\ 3OH}$).

Strains are susceptible to quinolones, aminoglycosides, tetracycline, and chloramphenicol. β-Lactamase positive. *F. tularensis* is a **facultative intracellular bacterium** and *F. tularensis* subsp. *tularensis*, *F. tularensis* subsp. *holarctica*, and *F. tularensis* subsp. *mediasiatica* **cause the zoonotic disease tularemia** in humans and animals, predominantly rodents and lagomorphs. Also *F. tularensis* subsp. *novicida* may cause a tularemia-like disease. *F. philomiragia* has been associated with septicemic and pneumonic disease in humans with immunocompromised conditions.

16S rDNA sequence analysis has demonstrated that the genus belongs to the *Gammaproteobacteria*. *Francisella* strains show a high degree of 16S rRNA gene sequence similarity, ≥98.3%, and *F. tularensis* strains >99.8%. 16S rRNA gene sequence signatures of *Francisella* are shown in Table BXII.γ.51. Phylogenetically related to *Wolbachia persica* and two other endosymbionts of ticks. DNA relatedness of *F. tularensis* strains, including isolates of *F. tularensis* subsp. *novicida*, is 75–98% (65°C, hydroxyapatite method) to the type strain *F. tularensis* ATCC 6223[T]. Members of *F. philomiragia* show 72–83% relatedness to the type strain *F. philomiragia* ATCC 25015[T].

Type species: **Francisella tularensis** (McCoy and Chapin 1912) Dorofe'ev 1947, 176 (*Bacterium tularense* McCoy and Chapin 1912, 61.)

FURTHER DESCRIPTIVE INFORMATION

Francisella, the only recognized genus in "*Francisellaceae*", belongs to the *Gammaproteobacteria*. The most closely related organisms are the obligate intracellular bacterium *W. persica*, first isolated from the tick *Argas persicus* (*arboreus*), and two other endosymbionts of ticks (Fig. BXII.γ.91). *Francisella* shows the closest relationship, albeit a rather distant one, to *Piscirickettsiaceae*. The

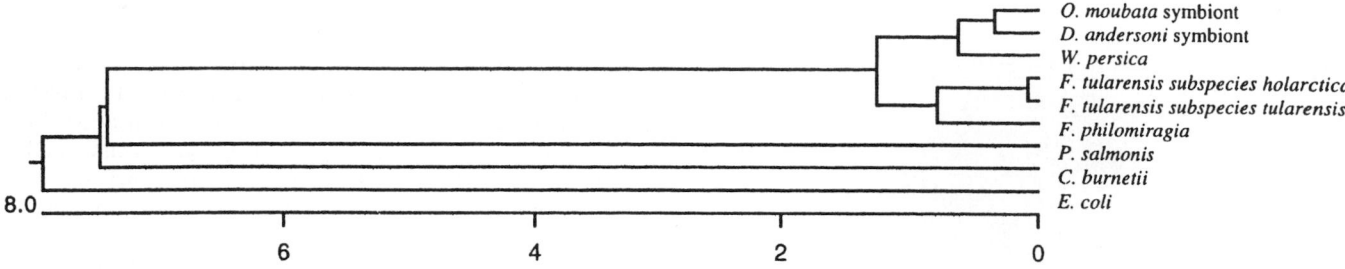

FIGURE BXII.γ.91. Evolutionary distance tree showing relationships among the recognized members of *Francisellaceae*, *Wolbachia persica*, and symbionts from the ticks *Dermacentor andersoni* and *Ornithodoros moubata*. *Escherichia coli* and *Coxiella burnetii* are included for further comparison to other members of the *Gammaproteobacteria*. The scale represents the number of nucleotide substitutions calculated against the number of compared nucleotide sites.

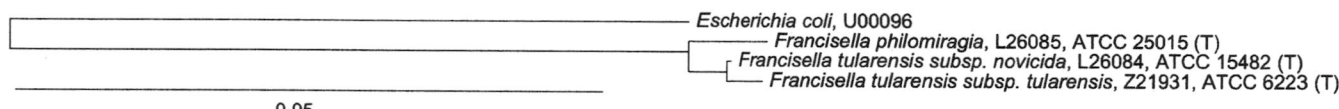

FIGURE BXII.γ.92. Phylogenetic relationships of the genus *Francisella*. The distances in the tree were calculated using 1101 positions (the least-squares method, Jukes-Cantor model). (Courtesy T. Lilburn of the Ribosomal Database Project.)

type species, *F. tularensis*, is the source of virtually all knowledge of the genus (see Fig. BXII.γ.92).

Cells of *Francisella* typically appear as short, rod-shaped or coccoid cells. They are small, singly occurring, nonmotile, and nonsporulating. Cells of *F. tularensis* are 0.2–0.7 × 0.2 μm and *F. tularensis* subsp. *novicida* and *F. philomiragia* are approximately 0.7 × 1.7 μm (Fig. BXII.γ.93). This morphology occurs typically during the logarithmic growth phase in a liquid medium. In infected tissue and during stationary growth phase, the organisms show pleomorphism. In those situations, cells often elongate, filamentous cells form, filament fragmentation occurs, and a small number of cells not larger than 300 nm are demonstrable (Ribi and Shepard, 1955). Minute forms have been shown to be infectious for mice and pass through membranes with 600 nm average pore diameter (Foshay and Hesselbrock, 1945). *Francisella* cells are faintly stained in tissue. The characteristic morphology is most readily demonstrable with a polychrome (Giemsa) stain.

Virulent strains of *F. tularensis* are surrounded by a relatively thick electron-transparent capsule that can be removed in hypertonic solutions; loss of capsule is accompanied by loss of virulence but viability is unaffected (Hood, 1977). Underlying the capsule is an outer layer surrounding the cell wall and a delicate double-layered cell wall. The lipid concentration of the capsule and cell wall is unusually high for Gram-negative bacteria. The composition is also characteristic for the genus, long-chain saturated and monosaturated C_{18} to C_{26} acids, relatively large amounts of saturated even-chain acids ($C_{10:0}$, $C_{14:0}$, and $C_{16:0}$), and two long-chain hydroxy acids ($C_{16:0\ 3OH}$ and $C_{18:0\ 3OH}$) (Jantzen et al., 1979; Hollis et al., 1989). Members of the genus also contain a ubiquinone with eight isoprene units as the major isoprenologs (Hollis et al., 1989). The live vaccine strain of *F. tularensis*, LVS, grown in a chemically defined medium, was found to have a lipid content of 21% (dry wt) and to contain phosphatidylethanolamine (76%) and phosphatidylglycerol (24%) as its two major phospholipids (Anderson and Bhatti, 1986). As opposed to other Gram-negative bacteria, all members of the genus seem to lack monounsaturated 16-carbon acid and most

strains lack dodecanoic acid but possess small amounts (1–2%) of a 16-carbon aldehyde (Hollis et al., 1989).

Francisella species are strictly aerobic. *F. tularensis* subsp. *holarctica*, *F. tularensis* subsp. *mediasiatica*, and *F. tularensis* subsp. *tularensis* grow more slowly and less abundantly than do isolates of *F. tularensis* subsp. *novicida* and *F. philomiragia*. *F. tularensis* is usually described as a fastidious organism since clinical isolates grow slowly and usually require cysteine or cystine for growth. These supplements enhance but are not required for growth of isolates of *F. tularensis* subsp. *novicida* and *F. philomiragia*. Cystine heart blood agar supplemented with 1% hemoglobin[1] provides excellent conditions for isolation of *F. tularensis* (Payne and Morton, 1992). Other suitable media are Mueller–Hinton agar (Baker et al., 1985)[2] supplemented with 2% IsoVitaleX and 0.025% ferric pyrophosphate, and glucose cysteine heart blood agar[3]. Glucose cysteine agar (BBL) and cystine heart agar (Difco) are two commercially produced solid media intended for cultivation of *F. tularensis*.

---

1. Cystine heart blood agar with hemoglobin. Suspend 102 g of cystine heart agar (Difco) in 1.0 liter of distilled water, then mix thoroughly. Heat with frequent agitation and boil for 1 min. Dispense into tubes and sterilize by autoclaving at 118–121°C for 15 min. For larger quantities, autoclave at the same temperature for 30 min. Cool to 45–48°C. Aseptically add 10 ml of hemoglobin and 50 ml of citrated bovine blood. Mix thoroughly and pour into plates. Incubate at 37°C for 24 h before use to decrease surface moisture and to test for sterility. Cystine heart agar without supplements also supports growth of *F. tularensis*.

2. Modified Mueller–Hinton II agar. Suspend 38 g of Mueller–Hinton II agar (BBL) in 1.0 liter of distilled water and mix thoroughly. Heat with frequent agitation and boil for 1 min to completely dissolve the powder. Sterilize by autoclaving at 121°C for 15 min. Aseptically add 0.1% glucose, 2% IsoVitaleX (BBL) which contains cysteine, and 0.025% ferric pyrophosphate. The pH is adjusted to 7.0 with sterile 1 N NaOH. Incubate at 37°C for 24 h before use to decrease surface moisture and to test for sterility.

3. Glucose cysteine heart blood agar. Suspend 58 g of glucose cysteine heart agar (BBL) in 1.0 liter of distilled water, then proceed as indicated in the preceding footnote. After autoclaving and cooling, aseptically add 50 ml of citrated bovine, rabbit, or sheep blood. Glucose cysteine heart agar without supplements also supports growth of *F. tularensis*.

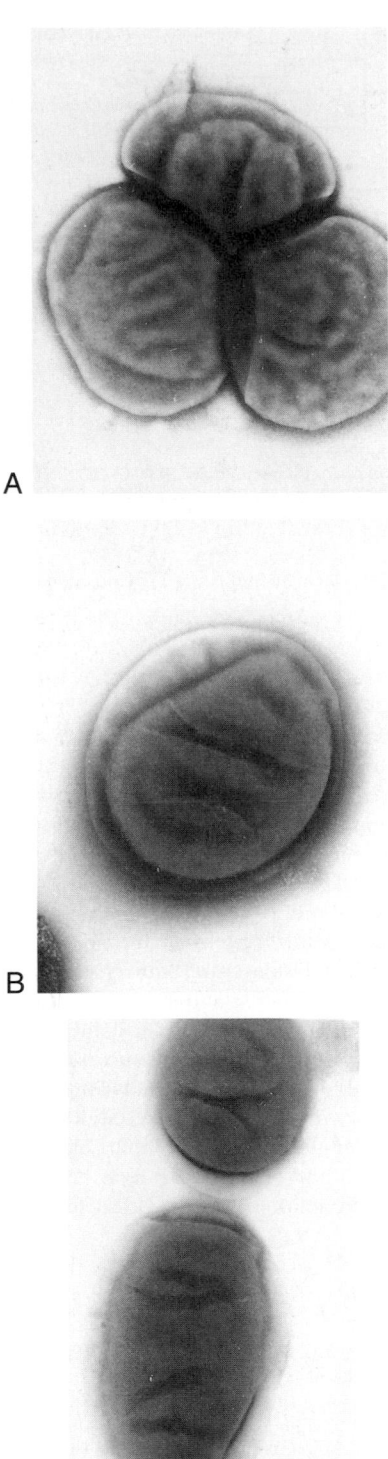

A

B

C

FIGURE BXII.γ.93. Electron micrographs of (*A*) the live vaccine strain (ATCC 29684) of *F. tularensis*, (*B*) the type strain of *F. tularensis* subsp. *novicida* (ATCC 15482), and (*C*) the type strain of *F. philomiragia* (ATCC 25015) (× 40,000).

There are also several reports documenting isolation of the organism on media lacking cysteine, e.g., enriched chocolate agar, Thayer–Martin agar, and Mueller–Hinton agar and in blood culture bottles (Provenza et al., 1986; Reary and Klotz, 1988; Bernard et al., 1994). In fact, cysteine enhanced, but was not

critically required for, growth of a number of field isolates on agar bases supplemented with 5% blood (Payne and Morton, 1992). The low-virulence *F. tularensis* type strain, ATCC 6223, is distinctly more fastidious than virulent strains (Eigelsbach and McGann, 1984) and, in contrast to virulent strains, it may not grow on trypticase soy blood agar or brain–heart infusion blood agar (Payne and Morton, 1992). Modified Mueller–Hinton broth (Baker et al., 1985), T-medium (Tresselt and Ward, 1964), or Chamberlain's chemically defined medium (Chamberlain, 1965) can be used for liquid cultivation of *Francisella*. Provided safety precautions can be met with cultivation from samples of tularemia patients, isolation is usually possible from most types of relevant clinical specimens, e.g., wound samples, lymph node aspirates, and blood. However, the success of cultivation is dependent on rapid transportation and only some 20% of wound samples from patients with ulceroglandular tularemia yielded growth of *F. tularensis* after an average transportation time of 3–4 days (Sjöstedt et al., 1997).

Degradation of a limited number of carbohydrates occurs under suitable conditions. Acid but no gas is produced. Reactions are characteristically slow. All subspecies of *F. tularensis* except *F. tularensis* subsp. *mediasiatica* form acid from D-glucose and maltose whereas degradation of sucrose is a distinguishing feature of *F. tularensis* subsp. *novicida*. *F. tularensis* subsp. *tularensis* differs from subsp. *holarctica* by its capacity to utilize glycerol and the possession of a citrulline ureidase pathway (Marchette and Nicholes, 1961). In addition, the subsp. *mediasiatica* ferments glycerol and citrulline (Olsufjev and Meshcheryakova, 1983). *Francisella* isolates produce H_2S in media supplemented with cysteine whereas the capability to form H_2S in triple sugar iron agar is confined to *F. philomiragia*. Most isolates of *F. philomiragia* and *F. tularensis* subsp. *novicida*, but not of other subspecies of *F. tularensis*, grow in nutrient broth containing 6% NaCl. Generally, strains within each subspecies of *F. tularensis*, as well as *F. philomiragia*, show a high degree of homogeneity in their biochemical profile. However, strains of *F. tularensis* subsp. *novicida* vary in degradation of maltose and glycerol. Moreover, in contrast to an absolute majority of other strains of *F. tularensis*, *F. tularensis* subsp. *novicida* strains lack a growth requirement for cysteine. *F. philomiragia* also lacks a cysteine requirement and degrades sucrose and, in addition, is distinguished by its clearing of blood agar and by being oxidase positive (Kovacs modification). Optimal growth occurs at 37°C for *F. tularensis* and at 25°C or 37°C for *F. philomiragia* strains.

Analysis of DNA relatedness strongly indicated the presence of two species, *F. tularensis* and *F. philomiragia*, within the genus *Francisella* (Hollis et al., 1989). All data were based on relative DNA reassociation at 65°C. The type strain *F. tularensis* ATCC 6223[T] showed close to 90% DNA relatedness to strains of *F. tularensis* subsp. *holarctica*, *F. tularensis* subsp. *tularensis*, and *F. tularensis* subsp. *novicida*, thereby supporting the assignment of the former species *F. novicida* as a subspecies of *F. tularensis*. The calculated divergence was <2.5%. The DNA relatedness of the type strain of *F. philomiragia* ATCC 25015[T] was, on average, 77% to 13 strains showing the phenotypic characteristics of the species *F. philomiragia*, with a maximum divergence of 3.5%. The type strain of *F. philomiragia* showed 39% average DNA relatedness to the *F. tularensis* type strain and seven other *F. tularensis* strains, including the reference strain of *F. tularensis* subsp. *novicida* and two *novicida*-like strains.

The 16S rDNA sequence similarities of strains within the genus *Francisella* (Forsman et al., 1994) corroborated the differentiation of the genus into the species *F. tularensis* and *F. philomiragia*. In spite of clear phenotypic distinctions between the two species, the 16S rDNA sequence similarity between *F. philomiragia* and *F. tularensis* strains is high, ≥98.3%. The 16S rDNA sequence analysis strengthened the argument for assigning subsp. *novicida* as a subspecies of *F. tularensis*, the degree of 16S rDNA similarity being ≥99.8% (Forsman et al., 1994).

By PCR amplification of the rRNA gene cluster from a large number of *Francisella* strains, combined with endonuclease digestion of the resulting amplicons with *Hae*III, four patterns were demonstrable: one for *F. philomiragia*, one for *F. tularensis* subsp. *novicida*, and two for other strains of *F. tularensis* (Ibrahim et al., 1996).

A spontaneous mutant of *F. tularensis* strain LVS expressing an antigenically distinct LPS has been described (Cowley et al., 1996). This phase variation has been linked to a capacity to switch to increased induction of nitric oxide (NO) and suppression of microbial intracellular growth in rat macrophages. When the LPS returns to its original antigenic form, NO production is reduced and intracellular growth is restored (Cowley et al., 1996). A capsule-deficient mutant of *F. tularensis* strain LVS has been described. In contrast to the encapsulated vaccine strain, the capsule-deficient variant is avirulent in mice and susceptible to the bactericidal effect of nonimmune human serum (Sandström et al., 1988). A cryptic plasmid has been isolated and characterized from *F. tularensis* subsp. *novicida* (Pavlov et al., 1994) and another from *F. tularensis* subsp. *holarctica* (Zakharenko et al., 1985).

Conventional serologic procedures indicate that most *F. tularensis* strains are qualitatively similar in antigenic composition, with the exception of some strains of *F. tularensis* subsp. *novicida*. *F. tularensis* shows a minor antigenic relationship to *Brucella* (Saslaw and Carlisle, 1961) and *Yersinia pestis* (Larson et al., 1951). Immunization with the attenuated vaccine strain *F. tularensis* strain LVS protects mammals against all subspecies of *F. tularensis*. A live *F. tularensis* subsp. *novicida* vaccine protects against homologous infection and against *F. tularensis* strains of moderate virulence (Claflin and Larson, 1972). Commercial *F. tularensis* antiserum agglutinates *F. tularensis* strains of *F. tularensis* subsp. *tularensis* and *F. tularensis* subsp. *holarctica*, but not all strains of *F. tularensis* subsp. *novicida* and not strains of *F. philomiragia* (Clarridge et al., 1996).

The lipopolysaccharide (LPS) of *F. tularensis* contains a unique O-antigen, though related to that of some species of *Pseudomonas aeruginosa* and *Shigella flexneri* (Vinogradov et al., 1991). It induced production of tumor necrosis factor-α, interleukin-1, and NO, but only at concentrations 100–1000 times higher than *E. coli* LPS (Sandström et al., 1992; Ancuta et al., 1996). Colony type variants of *F. tularensis*, including those associated with virulence and immunogenicity, have been described (Eigelsbach et al., 1951). Possibly, these variants can be attributed to phase variation in LPS as described above (Cowley et al., 1996).

A number of *F. tularensis* proteins have been cloned and characterized, among them a heat-labile outer membrane protein, a 17-kDa outer membrane lipoprotein, a cytoplasmic 23-kDa protein, an acid phosphatase, a triosephosphate isomerase, a putative ABC transporter protein, and the GroE and DnaK homologs (Nano, 1988; Sjöstedt et al., 1990; Mdluli et al., 1994; Zuber et al., 1995b; Reilly et al., 1996; Ericsson et al., 1997; Golovliov et al., 1997; Keeling and Doolittle, 1997). The gene encoding the 17-kDa lipoprotein of *F. tularensis* strain LVS is conserved in *Francisella* spp. although the immunoreactive proteins in *F. philomiragia* and *F. tularensis* subsp. *novicida* strains display a slightly dif-

ferent M_r (Sjöstedt et al., 1992). Primers specific to this gene can be used to amplify a DNA fragment from the tick endosymbiont designated DAS (Niebylski et al., 1997) and from *W. persica* (Forsman et al., 1994).

Extracts of *F. tularensis* strains have been found to contain so-called phosphoantigens, a pyrophosphoester bound to derivatives of uracil monophosphate or thymidine monophosphate, and potent inducers of Vγ9Vδ2 T cells in humans. Tularemia, but not vaccination with *F. tularensis* strain LVS, resulted in prominent expansion of Vγ9Vδ2 cells (Poquet et al., 1998).

F. tularensis shows pronounced sensitivity to aminoglycosides; streptomycin, gentamicin, and kanamycin, and also tetracyclines, chloramphenicol, and quinolones. Aminoglycosides and tetracyclines have a well-documented effect on tularemia. Notably, administration of doxycycline is recommended for at least 3 weeks (Tärnvik et al., 1997). β-Lactamase activity has been demonstrated in all *F. tularensis* isolates tested, except those from mid-Asia (Pavlovich and Mishankin, 1992). Antibiotic testing of 15 Scandinavian strains showed that all were resistant to β-lactams but had a mixed susceptibility/resistance pattern to macrolides. All strains were susceptible to gentamicin, chloramphenicol, doxycycline, and four quinolones (Scheel et al., 1993). Similar antimicrobial susceptibility was found among *F. philomiragia* strains (Hollis et al., 1989). Since quinolones show low MIC values, have good intracellular penetration, and proved to afford complete recovery in a small number of tularemia patients (Syrjälä et al., 1991; Scheel et al., 1992), they are promising agents for treatment.

F. tularensis is widely distributed in nature and is found throughout the Northern Hemisphere. Large outbreaks, especially in parts of the continental USA, the southern part of the former USSR, and northern Scandinavia, have been reported. Four subspecies, *F. tularensis* subsp. *tularensis*, *F. tularensis* subsp. *holarctica*, *F. tularensis* subsp. *mediasiatica*, and *F. tularensis* subsp. *novicida* (Olsufjev and Meshcheryakova, 1983; Hollis et al., 1989), are recognized. All have been associated with human disease, although they differ in virulence for humans and rabbits. Tularemia is transmitted by direct contact with infected animals, through contaminated water or food, or by vectors such as biting insects or ticks. Airborne transmission also occurs, especially during processing of agricultural products. The disease is often epidemic, both in humans and in animals, and clinical manifestations depend on the type of reservoir involved and the means of transmission. The bacterium has been isolated from approximately 250 wildlife species, many of which allow transmission of tularemia to man (Olsufjev, 1974).

F. tularensis subsp. *tularensis* has been isolated only in North America and is the predominant form on the continent. It is highly virulent for humans and numerous animal hosts, including domestic rabbits. Most isolates derive from ticks and rabbits. *F. tularensis* subsp. *holarctica* is spread throughout the Northern Hemisphere and causes a milder form of disease. The subspecies seems to be predominantly associated with aquatic rodents, e.g., muskrats and beaver in North America, and ground voles in the former USSR. In Europe, tularemia is most frequently observed in hares and rodents. *F. tularensis* subsp. *mediasiatica*, which has only been isolated in the Central Asian republics of the former USSR, is often found in species of *Lepus* and *Gerbillinae* and in ticks. *F. tularensis* subsp. *holarctica* and *F. tularensis* subsp. *mediasiatica* are significantly less virulent for rabbits than is *F. tularensis* subsp. *tularensis*. *F. tularensis* subsp. *tularensis* is highly infectious and even in humans, the infectious dose is 10 bacteria after subcutaneous injection and 25 organisms when given by aerosol (McCrumb, 1961; Saslaw et al., 1961). Human-to-human transmission is extremely rare.

F. tularensis subsp. *novicida* and *F. philomiragia* are often closely linked to water-borne transmission. The reference strain of subsp. *novicida* was isolated from water in Utah (Larson et al., 1955; Owen et al., 1964) and in some cases of human disease, the suspected source has been natural water (Clarridge et al., 1996). Five isolates of *F. philomiragia*, including the type strain, were isolated from a marshy area in Utah (Jensen et al., 1969). Five out of 13 patients with a disease caused by *F. philomiragia* had a history of near-drowning (Wenger et al., 1989).

Human tularemia is an acute, febrile, zoonotic disease. The clinical picture and severity vary considerably depending on the route of infection and the virulence of the etiological organism. The incubation period is usually 3–6 days but may vary from 1 to 14 days. Before the advent of antibiotic therapy, tularemia in untreated patients in North America was associated with a mortality ranging from 5 to 30% (Dienst, 1963; Evans et al., 1985). Due to the inherently lower virulence of *F. tularensis* subsp. *holarctica* and *F. tularensis* subsp. *mediasiatica*, virtually no mortality has been recorded in Eurasia.

F. tularensis subsp. *novicida* is generally less virulent than the other subspecies and the type strain is pathogenic only for mice, guinea pigs, and hamsters, while rabbits and rats are resistant. A few cases of human disease caused by *novicida*-like strains have been reported, including one case resembling glandular tularemia, one case of typhoidal tularemia, and two cases of pneumonia in compromised patients (Hollis et al., 1989; Clarridge et al., 1996). *F. philomiragia* may cause severe disease with pneumonia and/or septicemia. A majority of cases has occurred in immunocompromised individuals with a history of granulomatous disease or near drowning (Wenger et al., 1989).

The laboratory diagnosis of tularemia relies mainly on serology. Agglutination techniques or enzyme-linked immunoassays are routinely used (Carlsson et al., 1975; Viljanen et al., 1983; Koskela and Salminen, 1984). The most common antigen is purified LPS (Carlsson et al., 1975). Seroconversion occurs normally 6–10 days after onset of disease (Koskela and Salminen, 1984). Lymphocyte stimulation tests have also been performed for diagnostic purposes (Tärnvik and Löfgren, 1975; Syrjälä et al., 1984) although not on a routine basis. In a study of 200 cases of tularemia, more than 20% of the patients showed a positive lymphocyte stimulation test during the first week of infection and virtually 100% within 2 weeks (Syrjälä et al., 1984). Cultivation can be used for identification, but due to the contagiousness of the organism, it is usually performed only at reference laboratories. Identification of *F. tularensis* in clinical specimens has been based on direct or indirect fluorescent antibody techniques (Prochazka, 1966; Karlsson and Söderlind, 1973), although not generally included as part of routine diagnostic procedures on human specimens. It is a preferred method for establishing the diagnosis on postmortem material from deceased animals (Mörner, 1981).

Antigen detection in urine and RNA hybridization on a wound specimen have been attempted as diagnostic measures (Sandström et al., 1986; Forsman et al., 1990). PCR has been applied for sensitive and specific detection of *F. tularensis* in blood from infected mice (Long et al., 1993) and for wound specimens from tularemia patients (Sjöstedt et al., 1997). The latter study showed that PCR-based detection is a promising alternative for early diagnosis of ulceroglandular tularemia. Based on the specific signatures of 16S rRNA gene sequences of *Francisella* (Table BXII.γ.51), PCR-based identification of *Francisella* strains at the genus level, and differentiation at the species level, have been developed (Forsman et al., 1994).

ENRICHMENT AND ISOLATION PROCEDURES

Isolation of *F. tularensis* may be attempted from any tissue or specimen. Due to the contagiousness of the organism, all isolation procedures should be performed by trained staff working under biosafety level three conditions using efficient exhaust cabinets. Many laboratories rely only on serologic diagnosis of suspected cases of tularemia.

Detailed procedures for collection and storage of clinical specimens from patients with suspected tularemia have been described (Mollaret, 1991). Procedures for testing water samples or collection fluids from impinged air samples for the presence of *F. tularensis* have been described by Eigelsbach and McGann (1981). Successful isolation will depend on precautions taken to maintain the viability of *F. tularensis* organisms and to prevent overgrowth by normal flora. All specimens should be kept at a temperature of 10°C or lower and procedures for isolation performed as soon as possible.

Recommended media for isolation of *F. tularensis* include cystine heart blood agar supplemented with 1% hemoglobin, Mueller–Hinton agar supplemented with 2% IsoVitalex and 0.025% ferric pyrophosphate, and glucose cysteine heart blood agar. In addition, the commercially available media, glucose cysteine agar (BBL) and cystine heart agar (Difco), allow isolation of the organism from most types of clinical specimens.

Direct culture on any of the recommended media has the advantage of being more rapid and less hazardous than inoculation of susceptible laboratory animals. If specimens are from normally nonsterile sites, incorporation of antibiotics (penicillin, polymyxin B, and cycloheximide) in the medium is recommended (Berdal and Söderlund, 1977; Pavlovich and Mishankin, 1987). If adequate facilities are available for inoculating and housing infectious animals and isolation by direct culture is unsuccessful, materials can be inoculated intraperitoneally into guinea pigs or mice. Inoculation of only a few viable cells of *F. tularensis* usually results in death within 1 week. Moribund animals display gross enlargement of spleen, often 4–5 times normal size, and numerous minute, gray foci of necrosis. The presence of *F. tularensis* is readily confirmed from spleen, liver, and lung of the animals.

MAINTENANCE PROCEDURES

The organism can be maintained in the laboratory by transfer on the same media used for isolation. Attenuation has been observed after repeated passage and for preservation of original phenotypes of strains, plating from original isolates is recommended. Viability of isolates during long-term storage is improved by the addition of stabilizers. Recommended conditions are storage at −70°C or lyophilization in the presence of stabilizers such as 20% sucrose, 2.6% gelatin, 0.2% agar (Faibich, 1959), or 50% skim milk.

PROCEDURES FOR TESTING SPECIAL CHARACTERS

General characteristics Francisellae typically appear as small, coccobacillary, faintly stained Gram-negative organisms demonstrable after 2–4 days of cultivation of clinical specimens.

Media *Francisella* subspecies grow on the commercially available media, glucose cysteine agar, cystine heart agar, and

Mueller–Hinton agar, the latter with IsoVitaleX and ferric pyrophosphate as supplements. The etiological agents of tularemia, *F. tularensis* subsp. *tularensis*, *F. tularensis* subsp. *holarctica*, and *F. tularensis* subsp. *mediasiatica*, do not grow on standard blood agar, nutrient agar, or other commonly used media lacking cysteine.

Direct identification   Direct fluorescent microscopy is rarely used for diagnosis of tularemia in human specimens nowadays, but is a preferred method for establishing the diagnosis on postmortem material from deceased animals and has shown to be sensitive and specific (Mörner, 1981). A slide agglutination test based on commercial antiserum (Difco) can be used for rapid, tentative identification of *F. tularensis*.

Serology   ELISA (enzyme-linked immunosorbant assay) is the standard serological test used to diagnose tularemia. Seroconversion commonly occurs within 6–10 days after onset of disease (Koskela and Salminen, 1984).

Pathology   The postmortem findings of tularemia in hares and rodents are characterized by focal necrosis in liver, spleen, and bone marrow (Mörner et al., 1988). In humans, abscess necrosis of lymph nodes with or without epithelioid cell reaction is observed within one to two weeks of infection, and later caseous necrosis occurs (Sutinen and Syrjälä, 1986). These changes closely resemble those observed in tuberculosis.

Gas-liquid chromatography (GLC)   GLC analysis of cellular fatty acids identifies the unique fatty acid profile characteristic of *Francisella* strains (Jantzen et al., 1979; Hollis et al., 1989). The method has been, thus far, completely specific in identifying isolates of *Francisella* at the genus level.

Molecular diagnosis   Based on *Francisella*-specific 16S rRNA gene sequence signatures, PCR-based methods allowing identification of *Francisella* strains at the genus level and differentiation at the species level have been developed (Forsman et al., 1994). Another molecular target is a gene encoding a 17-kDa lipoprotein from *F. tularensis* strain LVS (Sjöstedt et al., 1990). This gene is conserved in *F. tularensis* and *F. philomiragia* isolates (Sjöstedt et al., 1992). Primer sequences are indicated in Table BXII.γ.52. Multiplex PCR-based identification of the 17-kDa lipoprotein gene and the *Francisella* 16S rRNA gene sequence has been applied on wound specimens from tularemia patients (Sjöstedt et al., 1997).

DIFFERENTIATION OF THE GENUS *FRANCISELLA* FROM OTHER GENERA

Table BXII.γ.53 indicates the characteristics useful to distinguish *Francisella* from other genera of clinically relevant, small, Gram-negative, nonsporeforming, coccobacillary bacteria pathogenic for man and other mammals. Based on DNA relatedness and 16S rRNA gene sequence analysis, no close relatives of *Francisella* are recognized (Ritter and Gerloff, 1966; Forsman et al., 1994).

The requirement of *F. tularensis* strains for cysteine in growth media is a unique characteristic that distinguishes it from virtually all other clinically relevant species. It grows on media developed for isolation of legionellae. Unlike *Legionella* spp., francisellae produce acid from carbohydrates, and do not contain branched-chain fatty acids. The less fastidious species of *Francisella*, *F. philomiragia* and *F. tularensis* subsp. *novicida*, lack the requirement for cysteine and may be more difficult to differentiate from the Gram-negative genera *Pasteurella*, *Brucella*, and *Yersinia*. The latter two genera may show weak agglutination with *F. tularensis* antiserum. In contrast to the case of *Pasteurella*, *Brucella*, and *Yersinia*, a capsule is not easily demonstrable in isolates of *Francisella* and they show a weak Gram stain (Table BXII.γ.53). Other characteristic, distinguishing features of *Francisella* include weak catalase production and no production of NH_3 in liquid media. The primers indicated in Table BXII.γ.52 only give amplicons from francisellae after PCR-based amplification.

TAXONOMIC COMMENTS

Circumscription   As indicated in the previous edition of this *Manual*, the taxonomic position of the genus *Francisella* long remained uncertain (Eigelsbach and McGann, 1984). The highly infectious nature of these bacteria has been one of the reasons for slow development of taxonomic classification. After originally being designated *"Bacterium"*, it was subsequently placed in the genus *Pasteurella*, based on serological studies. However, DNA hybridization studies indicated that the genus was not closely related to *Pasteurella*, *Yersinia*, or the coliforms (Ritter and Gerloff, 1966). These early findings using DNA hybridization were corroborated when 16S rDNA sequence analysis demonstrated that although the genus groups within the *Gammaproteobacteria*, it shows no close relationship to other characterized genera (Forsman et al., 1994). *Francisella* is the only genus described in *"Francisellaceae"*. The closest relationship is to the intracellular bacterium *W. persica* and, outside *"Francisellaceae"*, to members of *Piscirickettsiaceae*, in particular the intracellular bacterium *Piscirickettsia salmonis*, the etiological agent of a septicemic disease in fish (Fryer et al., 1992). The cellular fatty acid composition of *Francisella* and the unusually high lipid content of the cell wall are traits distinctly different from those of other Gram-negative bacteria (Hood, 1977).

In 1959, Olsufiev et al. (1959) recognized the presence of two etiological agents of tularemia varying in virulence. The designation *F. tularensis* biovar tularensis was proposed for the organism responsible for the severe form of human illness in North America, and *F. tularensis* biovar palaearctica for the less virulent organism encountered in Europe, Asia, and North America. In 1970, Olusfiev proposed that the biovars adopt the status of subspecies and suggested the designation *F. tularensis* subsp. *holarctica* for the less virulent form (Olsufjev, 1970). Moreover, the designation *F. tularensis* subsp. *mediasiatica* for strains from the Cen-

TABLE BXII.γ.52.   Nucleic acid composition of primers used in the PCR-based identification of *Francisella* isolates at the genus and species level[a]

Designation	Sequence (5′–3′)	Direction	Fragment size
F11[b]	TACCAGTTGGAAACGACTGT	Forward	1141 bp
F5	CCTTTTTGAGTTTCGCTCC	Reverse	
TUL4-435[b]	GCTGTATCATCATTTAATAAACTGCTG	Forward	428 bp
TUL4-863	TTGGGAAGCTTGTATCATGGCACT	Reverse	

[a]Adapted from Forsman et al. (1994) and Sjöstedt et al. (1997).

[b]F11 and F5 target the 16S rRNA gene. TUL4-435 and TUL4-863 target a 17-kDa lipoprotein gene. All primers anneal at 65°C.

TABLE BXII.γ.53. Differential characteristics of the genus *Francisella* and other genera of clinically relevant, small, Gram-negative, nonsporeforming coccobacillary bacteria[a]

Characteristic	*Francisella*	*Brucella*	*Pasteurella*	*Yersinia*
Mol% G + C of DNA	33–36	55–58	40–45	46–50
Cells: 0.25–0.5 μm	+[b]	−	d	−
Capsule easily demonstrated	−	−	+	+
Gram stain: weak counterstain	+	+	−	−
Strictly aerobic	+	+	−	−
CO_2 enhances growth	d	+	d	−
Optimum temperature, °C	37	37	37	25
Cysteine/cystine enhance growth	d[c]	−	−	−
Chains occur in liquid media	−	+	d	d
Carbohydrates are dissimilated	+[d]	−	+	+
Acid production from sucrose	−	−	+	d
Catalase	+[e]	+	+	+
Oxidase	d[f]	d	+	−
NH_3 produced in liquid media	−	+	+[e]	+
Sodium ricinoleate solubility	+	−	+	+
Penicillin-sensitive *in vitro*	−	−	d	d
Arthropod vectors can occur	+	−	−	+
Agglutination of *F. tularensis* antiserum	d[g]	d, weak	−	d, weak
Amplicons by use of *Francisella*-specific primers[h]	+	−	−	−

[a]Adapted from Eigelsbach and McGann (1984). For symbols see standard definitions.

[b]Cell size of tularemia-causing species in infected tissue.

[c]Absolute growth requirement for *F. tularensis* subsp. *tularensis*, *F. tularensis* subsp. *holarctica*, and *F. tularensis* subsp. *mediasiatica* and enhance growth of *F. tularensis* subsp. *novicida* and *F. philomiragia*.

[d]Acid but no gas produced. Weak reaction and some also delayed, 3–7 d.

[e]Weak reaction.

[f]*F. tularensis* negative, *F. philomiragia* positive (Kovacs modification).

[g]No agglutination of strains of *F. philomiragia* and some of *F. tularensis* subsp. *novicida*.

[h]Primer sequences specific to 16S rDNA and a gene encoding a 17-kDa lipoprotein of *F. tularensis* are given in Table BXII.γ.52.

tral Asian republics of the former USSR was proposed. Biochemical properties, virulence, and geographical origin of strains formed the basis for this differentiation. Subdivisions of *F. tularensis* subsp. *holarctica* were also proposed (Olsufjev and Meshcheryakova, 1983). Strains were distinguished with regard to susceptibility to erythromycin and designated biovar I Ery[S] and biovar II Ery[R], and Japanese strains were designated *F. tularensis* subsp. *holarctica* biovar japonica Rodionova. However, no, or very few, other phenotypic attributes support the subdivision of the subspecies. The designations type A and type B have been proposed for *F. tularensis* subsp. *tularensis* and *F. tularensis* subsp. *holarctica*, respectively (Jellison, 1974).

The original strain of *F. tularensis* subsp. *novicida* was isolated from a water sample in Utah in 1951, classified in the genus *Pasteurella* in 1955 (Larson et al., 1955), and transferred to the genus *Francisella*, as *F. novicida*, in 1959 (Olsufiev et al., 1959). In 1989, it was suggested as a biogroup of *F. tularensis*, based on DNA relatedness and biochemical characteristics (Hollis et al., 1989); the 16S rDNA sequences of *F. tularensis* subsp. *novicida* and *F. tularensis* subsp. *tularensis* are also very similar (Fig. BXII.γ.92). Members of *F. tularensis* subsp. *novicida* have less fastidious metabolic requirements than members of the other three subspecies. Despite differences in growth requirements and some metabolic reactions, the placement of *F. novicida* as a subspecies of *F. tularensis* is strongly supported. Analysis of fatty acid profiles showed *F. novicida* to be similar to members of the species *F. tularensis*; the two species also shared more than 75% DNA relatedness (Hollis et al., 1989). Moreover, an almost complete identity in 16S rDNA sequences between the two species has been demonstrated (Forsman et al., 1994). Although the type strain is considered nonpathogenic for humans and shows low virulence for mice, *novicida*-like isolates have been identified as the etiological agents of severe diseases in man (Hollis et al., 1989; Clarridge et al., 1996). The relationship between *F. tularensis* subsp. *novicida* strains and other members of *F. tularensis* is also supported by the fact that *F. tularensis* subsp. *novicida* could be transformed with DNA from *F. tularensis* strain LVS (Anthony et al., 1991). In summary, *F. tularensis* subsp. *novicida* demonstrates heterogeneity with regard to virulence and metabolic characteristics but displays the characteristic fatty acid profile of *F. tularensis*, >70% DNA relatedness, and almost complete 16S rRNA gene sequence similarity to other members of the species.

F. philomiragia was originally designated *Yersinia philomiragia* (Jensen et al., 1969) but subsequent analysis showed no relationship to other members of the genus *Yersinia* (Ursing et al., 1980b) and in the latest version of the *Manual*, it was considered a species *incertae sedis* misclassified in that genus (Bercovier and Mollaret, 1984). In 1989, transfer to the genus *Francisella* based on DNA relatedness and fatty acid analysis was proposed, resulting in the designation *F. philomiragia* comb. nov. (Hollis et al., 1989). The species shows a high 16S rRNA gene sequence similarity to, and the unique fatty acid profile typical of, strains of *F. tularensis*, but is distinguished from most *F. tularensis* strains by its ability to degrade certain carbohydrates and by having no requirement of cysteine for growth.

Anticipated changes The species *W. persica* described under the section "*species incertae sedis*" and the two other symbionts of ticks described in the section "Other organisms" all show the unique 16S rRNA gene sequence signatures of *Francisella*. Thus, they should be included as members of *Francisella* in the future (Fig. BXII.γ.91). At present, insufficient data for determining their relationship at the species level exist.

ACKNOWLEDGMENTS

I thank Arne Tärnvik and Gunnar Sandström for critically reviewing this manuscript, Thomas Svensson for help with sequence analyses, Lenore Johansson for taking electron micrographs, and Stig Granström for advice on agar media. The literature search for this chapter was concluded on 30 April, 1998.

FURTHER READING

Jellison, W.L. 1974. Tularemia in North America 1930–1974, University of Montana Foundation, Missoula.

Penn, R. L. 1995. *Francisella tularensis* (Tularemia). *In* Mandell, Bennett and Dolin (Editors), Mandell, Douglas and Bennett's Principles and Practice of Infectious Diseases, Churchill Livingstone, New York. 2060–2068.

Pollitzer, R. 1967. History and incidence of tularemia in the Soviet Union. A review, The Institute of Contemporary Russian Studies, Fordham University, New York.

Tärnvik, A. 1989. Nature of protective immunity to *Francisella tularensis*. Rev. Infect. Dis. *11*: 440–451.

Tärnvik, A., M. Ericsson, G. Sandström and A. Sjöstedt. 1992. *Francisella tularensis* — a model for studies of the immune response to intracellular bacteria in man [editorial]. Immunology. *76*: 349–354.

DIFFERENTIATION OF THE SPECIES OF THE GENUS *FRANCISELLA*

Differential characteristics of the species and subspecies of *Francisella* are presented in Tables BXII.γ.54 and BXII.γ.55.

List of species of the genus Francisella

1. **Francisella tularensis** (McCoy and Chapin 1912) Dorofe'ev 1947, 176[AL] (*Bacterium tularense* McCoy and Chapin 1912, 61.)

   tu.la.ren' sis. M.L. adj. *tularensis* tularensis pertaining to Tulare County, California, where the disease was first described in rodents.

   The characteristics of relevance for identification are as listed in Table BXII.γ.54 and other characteristics are given in Table BXII.γ.55. Four subspecies, *F. tularensis* subsp. *tularensis*, *F. tularensis* subsp. *holarctica*, *F. tularensis* subsp. *mediasiatica*, and *F. tularensis* subsp. *novicida*, are recognized; these are differentiated by the characteristics listed in Tables BXII.γ.54 and BXII.γ.55. Other differences are related to geographical origin of strains of the subspecies.

   The mol% G + C of the DNA is: 34 (T_m, Bd).

   Type strain: B38, ATCC 6223.

   GenBank accession number (16S rRNA): Z21931, Z21932.

   Additional Remarks: Strain B38 is an avirulent strain with more fastidious growth requirements and a greater tendency towards variation in colony type than is found in virulent strains.

   a. **Francisella tularensis** *subsp.* **tularensis** (McCoy and Chapin 1912) Dorofe'ev 1947, 176[AL] (*Bacterium tularense* McCoy and Chapin 1912, 61.)

      Characteristics distinguishing this subspecies from the other subspecies are indicated in Tables BXII.γ.54 and BXII.γ.55. Highest virulence of all subspecies, most readily demonstrable in higher mammals and man. Produce acid but no gas from maltose, D-glucose, and glycerol. No degradation of lactose and sucrose. Possession of a citrulline ureidase pathway. Isolates have been found only in North America.

TABLE BXII.γ.54. Characteristics of diagnostic value in identifying the genus *Francisella* and its species and subspecies[a,b]

Characteristics	F. tularensis subsp. tularensis	F. tularensis subsp. holarctica	F. tularensis subsp. mediasiatica	F. tularensis subsp. novicida	F. philomiragia
Size	<0.5 μm	<0.5 μm	<0.5 μm	<1.5 μm	<1.5 μm
Capsule[c]	y	y	y	n	nt
Gram stain	w[d]	w	w	w	w
Growth on MacConkey agar	−	−	−	d[e]	d
Cysteine required for growth[f]	+	+	+	−	−
H₂S production in cysteine-supplemented medium	+	+	+	+	+
β-lactamase	+	+	−	+	+
Acid production from:[g]					
Maltose	+	+	−	d	+
Lactose	−	−	−	−	−
Sucrose	−	−	−	+	+
D-Glucose	+	+	−	+	d
Glycerol	+	−	+	d	−
Citrulline ureidase pathway	+	−	+	+	nt
Agglutination of *F. tularensis* antiserum	+	+	+	D	−
% 16S rRNA gene sequence similarity to ATCC 6223[h]	≥99.8	≥99.8	≥99.8	≥99.8	≥98.3
Presence of *F. tularensis* 17-kDa lipoprotein gene	+	+	+	+	+

[a]For symbols see standard definitions; nt, not tested.

[b]Data from Eigelsbach and McGann (1984), Hollis et al. (1989), and Clarridge et al. (1996).

[c]Presence of capsule associated with virulence; y, yes; n, no.

[d]w, weak Gram-negative reaction.

[e]A majority of strains reportedly grow under these conditions.

[f]A few strains of *F. tularensis* subsp. *tularensis* and *F. tularensis* subsp. *holarctica* lack the cysteine requirement.

[g]Cysteine agar (carbohydrate base) used for assaying for *F. tularensis* subsp. *tularensis*, *F. tularensis* subsp. *holarctica*, and *F. tularensis* subsp. *mediasiatica* and O/F medium for *F. tularensis* subsp. *novicida* and *F. philomiragia*. Strains show weak reactions and some also delayed, 3–7 d.

[h]Refers to similarities of partial sequences, 390–450 bp in length.

TABLE BXII.γ.55. Other characteristics of *F. philomiragia* and the subspecies of *F. tularensis*[a,b]

Characteristics	*F. tularensis* subsp. *tularensis*	*F. tularensis* subsp. *holarctica*	*F. tularensis* subsp. *mediasiatica*	*F. tularensis* subsp. *novicida*	*F. philomiragia*
Mol% G + C of DNA	33–36	33–36	33–36	34	33–34
Aerobic, microaerophilic	+	+	+	+	+
CO$_2$ enhances growth	+	+	+	+	+
Optimum temperature, °C	37	37	37	37	25 or 37
Growth in nutrient broth, 0% NaCl	−	−	−	−	−
Growth in nutrient broth, 6% NaCl	−	−	−	d[c]	d[c]
Catalase	+w	+w	+w	+w	+w
Oxidase (Kovacs modification)	−	−	−	−	+
Indole	−	−	−	−	+
Urease	−	−	−	−	−
Nitrate reduction	−	−	−	−	−
H$_2$S slant, Triple Sugar Iron	−	−	−	−	+
Gelatin hydrolysis	−	−	−	−	+
Motility	−	−	−	−	−
LD$_{50}$ rabbits	<10^1	>10^6	>10^6	>10^6	nt
Median infectious dose for mice, <10^3	+	+	+	−	nt

[a]For symbols, see standard definitions; nt, not tested.

[b]Data from Eigelsbach and McGann (1984), Hollis et al., (1989), and Clarridge et al. (1996).

[c]A majority of strains reportedly grow under these conditions.

The mol% G + C of the DNA is: 34.(T_m).

Type strain: GIEM Schu, ATCC 6223.

GenBank accession number (16S rRNA): Z21931, Z21932.

Additional Remarks: Virulent strains, such as SCHU, are available only from individual investigators.

b. **Francisella tularensis** *subsp.* **holarctica** (ex Olsufiev, Emelyanova and Dunaeva 1959) Olsufjev and Meshcheryakova 1983, 872[VP]

hol.arc′ti.ca. Gr. prep. *holos* whole; Gr. n. *actos* the great and little bear constellations; M.L. adj. *holarctica* generally distributed in the arctic regions.

Characteristics distinguishing this subspecies from the other subspecies are indicated in Tables BXII.γ.54 and BXII.γ.55. Isolates show lower virulence than isolates of *F. tularensis* subsp. *tularensis*. Minimum infectious dose <10^3 for mice and LD$_{50}$ >10^6 for rabbits. Produce acid but no gas from maltose and D-glucose. No degradation of lactose, glycerol, and sucrose. Isolates have been found on all continents throughout the Northern Hemisphere.

The mol% G + C of the DNA is: 34 (T_m).

Type strain: GIEM 503.

Additional Remarks: Strain GIEM 503 is the attenuated live vaccine strain.

c. **Francisella tularensis** *subsp.* **mediasiatica** (Aikimbaev 1966) Olsufjev and Meshcheryakova 1983, 872[VP]

med.i.as.i.a′ti.ca. Gr. adj. *media* middle; Gr. adj. *asiatica* Asian; M.L. adj. *mediasiatica* pertaining to mid-Asia.

Characteristics distinguishing this subspecies from the other subspecies are indicated in Tables BXII.γ.54 and BXII.γ.55. Isolates show lower virulence than isolates of *F. tularensis* subsp. *tularensis*. Minimum infectious dose <10^3 for mice and LD$_{50}$ >10^6 for rabbits. Produce acid but no gas from glycerol. No degradation of lactose, sucrose, maltose, and D-glucose. Isolates have been found only in the Central Asian republics of the former USSR, in particular in the valleys and deltas of the rivers Amu-Darja, Ili, and Tshu.

The mol% G + C of the DNA is: unknown.

Type strain: 543.

Additional Remarks: Strain 543 is available from the Gamaleya Institute of Epidemiology and Microbiology, Academy of Medical Sciences, Moscow, Russia (alternative designation FSC147 is available at the Institute of Microbiology, FOA, Umeå, Sweden). It was isolated in 1965 in the Alma-Ata region, Kazakhstan.

d. **Francisella tularensis** *subsp.* **novicida** *subsp. nov.* (Larson, Wicht and Jellison 1955) (*Francisella novicida* (Larson, Wicht and Jellison 1955) Olsufiev, Emelyanova and Dunaeva 1959, 146[AL]; *Pasteurella novicida* Larson, Wicht and Jellison 1955, 253.)

no.vi′ci.da. L. adj. *novus* new; L. v. suff. *-cida* from L. v. *caedo* to cut, kill; M. L.n. *novicida* new killer.

Characteristics distinguishing this subspecies from the other subspecies are indicated in Tables BXII.γ.54 and BXII.γ.55. It was originally considered a separate species, *F. novicida* (Olsufiev et al. 1959). However, Hollis et al. (1989) showed by DNA relatedness studies that isolates of this species were indistinguishable from the type strain of *F. tularensis* and proposed that it be designated a biogroup of *F. tularensis*. It displays the characteristic fatty acid profile of *F. tularensis*, >70% DNA relatedness, and almost complete 16S rRNA gene sequence similarity to other subspecies of *F. tularensis*. The assignment of *F. novicida* as a distinct subspecies is supported by its less fastidious growth requirements, including lack of cysteine requirement, degradation of sucrose, unique morphology, and lower virulence as compared to the other subspecies of *F. tularensis*.

F. tularensis subsp. *novicida* is often closely linked to water-borne transmission. The type strain of *F. tularensis* subsp. *novicida* was isolated from water in Utah (Larson et al., 1955) and in some cases of human disease, the suspected source has been natural water (Clarridge et al., 1996). Isolates show lower virulence than those of other subspecies and there are only a few documented cases of human disease caused by isolates of the subspecies (Hollis et al., 1989; Clarridge et al., 1996). Unlike other subspecies, growth is enhanced by, but not absolutely dependent on, the presence of cysteine in media. Another characteristic distinguishing the subspecies is growth in nutrient broth containing 6% NaCl. Produce acid but no gas from sucrose and D-glucose, but no degradation of lactose. Some isolates produce acid from glycerol and maltose. Not all isolates are agglutinated by commercially available antiserum against *F. tularensis*. Isolates have been found only in North America.

The mol% G + C of the DNA is: 34 (T_m).

Type strain: ATCC 15482 (type strain).

GenBank accession number (16S rRNA): L26084.

Additional Remarks: The type strain was isolated from a water sample from Ogden Bay, Utah in 1951 (patient died Larson et al., 1955).

2. **Francisella philomiragia** (Jensen, Owen and Jellison 1969, 1237) Hollis, Weaver, Steigerwalt, Wenger, Moss, and Brenner 1990, 105[VP] (Effective publication: Hollis, Weaver, Steigerwalt, Wenger, Moss and Brenner 1989, 1601) (*Yersinia philomiragia* Jensen, Owen and Jellison 1969, 1237.)

phi.lo.mi.ra' gi.a. Gr. adj. *philos* loving; M.L. n. *miragia* plural of Latinized English word mirage; *philomiragia* loving mirages, because of the mirages observed in the area where the first isolations of this species were made.

The characteristics are listed in Tables BXII.γ.54 and BXII.γ.55. The name *Y. philomiragia* was proposed by Jensen et al. (1969) for a small Gram-negative bacillus isolated in 1959 from a dying muskrat found in a marshy area in Utah. The species description was based on this strain and four other strains isolated from water samples in the same area in 1960. Because of morphological resemblance to *Y. pestis* and some degree of DNA relatedness to *Y. pestis* (Ritter and Gerloff, 1966), the organism was assigned to the genus *Yersinia*. Later studies indicated, however, no significant DNA relatedness between *Y. philomiragia* and other *Yersinia* species, other members of the *Enterobacteriaceae*, or *Pasteurella multocida* (Ursing et al., 1980b). Hollis et al. (1989) reviewed data on 14 *Y. philomiragia* strains isolated from human specimens. The biochemical characteristics, fatty acid composition, and DNA relatedness of these strains and the five original strains demonstrated a close relationship at the species level and supported placement of this species in the genus *Francisella*.

F. philomiragia has been isolated from water, muskrats, and humans in North America and Europe. It rarely causes human disease, but serious infections may occur. Wenger et al. (1989) reviewed case reports of 14 patients infected with *F. philomiragia*. A majority of patients belonged to two risk groups. Patients with chronic granulomatous disease and near-drowning victims. Pneumonia and fever-bacteremia were reported in most patients and one patient died.

The mol% G + C of the DNA is: 34 (T_m).

Type strain: ATCC 25015, DSM 7535.

Additional Remarks: The type strain is the original isolate from a muskrat in Utah (Jensen et al., 1969).

Other Organisms

1. *Dermacentor andersoni* symbiont Niebylski, Peacock, Fischer, Porcella, and Schwan 1997, 3933 and *Ornithodoros moubata* symbiont Noda, Munderloh, and Kurtti 1997, 3926.

By analysis of their 16S rDNA sequences, two endosymbionts of ticks have been found to be phylogenetically related to members of the genus *Francisella*. An endosymbiont (DAS) present in >90% of isolates of the wood tick *D. andersoni* from Montana (Niebylski et al., 1997) and one found in the ovaries and Malpighian tubules from the African soft tick *O. moubata* (Noda et al., 1997) show ≥97.2% 16S rRNA gene sequence similarity to *F. tularensis* and *F. philomiragia*, and ≥99.4% similarity to *W. persica*. The DAS organism has been shown to cross-react in fluorescent antibody tests with *W. persica* (Burgdorfer et al., 1973). It can be cultivated in chicken embryo cell cultures and in Vero cells (Niebylski et al., 1997), and grows well in guinea pigs and golden hamsters, killing the animals when high inocula are given (Burgdorfer et al., 1973). The data generated so far showing the immunological cross-reactivity and almost complete identity with regard to 16S rRNA gene sequence (Fig. BXII.γ.91) demonstrate that the *D. andersoni* symbiont and *W. persica* are closely related. However, due to lack of DNA relatedness data, it cannot be concluded that they belong to the same species. Moreover, the 16S rRNA gene sequences of the *D. andersoni* and *O. moubata* endosymbionts show high similarity, 99.3%, to those of *Francisella* isolates and a parsimony analysis demonstrates a close relationship (Fig. BXII.γ.91). The two endosymbionts and *W. persica* all harbor the 16S rRNA sequence signatures of *Francisella* (Table BXII.γ.51). Thus, the data available at present support their membership in *Francisella* though, due to the lack of DNA relatedness data, their exact relationship at the species level cannot be determined.

Species Incertae Sedis

1. **Wolbachia persica** Suitor and Weiss 1961, 105[AL]

per' si.ca. L. fem. adj. *persica* from the specific epithet of the reputed host tick, *Argas persicus* (later reclassified as *A. arboreus*).

The name *W. persica* was proposed by Suitor and Weiss (1961) for a coccoid, Gram-negative bacillus first isolated from the tick *Argas persicus* feeding on the buff-backed heron, *Bubulcus ibis*, in Egypt. Because of morphological resemblance, Gram-negative staining, and an arthropod association, the organism was assigned to the genus *Wolbachia*. In 1989, an analysis of 16S rRNA gene sequences of members of the *Rickettsiae* demonstrated that *W. persica* had been misclassified and instead was a member of the *Gammaproteobacteria* (Weisburg et al., 1989). Subsequent analysis indicated that it has a close relationship to members of the genus *Francisella* (Forsman et al., 1994).

W. persica can be cultivated in the yolk sac of chicken embryos, where it grows profusely and kills the embryos in 5–10 days, depending on the dose of inoculum (Burgdorfer et al., 1973). It also grows well in several types of mammalian cells or insect cell cultures. An antigenically related organism was isolated from the hard tick *Dermacentor andersoni* collected in Montana (Burgdorfer et al., 1973) and further characterized by Niebylski et al. (1997) (see Other Organisms).

W. persica has been isolated by removing and triturating

the Malpighian tubules from ticks (Suitor and Weiss, 1961). After resuspension in a salt solution, this mixture was inoculated into chicken embryos via the yolk sac. Dying embryos are examined for the presence of coccoid bodies, which are most clearly visualized by Giemsa staining following Carnoy's fixation. For the production of seed and other tests, the isolates are passaged in eggs. *W. persica* metabolizes D-glucose, as well as serine, glutamine, and other substrates, including glutamate (Weiss et al., 1962). Erythromycin, chloramphenicol, tetracycline, and *p*-aminobenzoic acid inhibit growth in chicken embryos. The viability of a suspension of *W. persica* in sucrose, phosphate buffer, glutamate solution ([Bovarnick et al., 1950]) was retained for a decade at $-70°C$.

W. persica displays the unique 16S rRNA sequence sig- natures of *Francisella* (Table BXII.γ.51), supporting its placement in the genus *Francisella*. The 16S rRNA gene sequence shows high similarity to those of two symbionts of *D. andersoni* and *Ornithodoros moubata*, and the three organisms are located on the same branch of the phylogenetic tree, showing their close relationship (Fig. BXII.γ.51). However, the lack of DNA relatedness data hinders determination of the relationship between *W. persica*, the two tick endosymbionts, and *Francisella* spp. at the species level.

The mol% G + C of the DNA is: 30 (T_m, Bd) (Kingsbury and Weiss, 1968).

Type strain: ATCC VR-331.

GenBank accession number (16S rRNA): M21292.

ATCC VR-331 was the first *Argas persicus* (*A. arboretus*) isolate (Suitor and Weiss, 1961).

Order VI. **Legionellales** *ord. nov.*

George M. Garrity, Julia A. Bell and Timothy Lilburn

Le.gi.on.el.la′ les. M.L. n. *Legionella* type genus of the order; *-ales* ending to denote order; M.L. fem. n. *Legionellales* the *Legionella* order.

The order *Legionellales* was circumscribed for this volume on the basis of phylogenetic analysis of 16S rRNA sequences; the order contains the families *Legionellaceae* and *Coxiellaceae*.

Facultative (*Legionellaceae*) and obligate (*Coxiellaceae*) intra- cellular parasites infecting protozoa, invertebrates, animals, and man.

Type genus: **Legionella** Brenner, Steigerwalt and McDade 1979, 658.

Family I. **Legionellaceae** Brenner, Steigerwalt and McDade 1979, 658[AL]

Washington C. Winn, Jr.

Le.gi.on.el.la′ ce.ae. M.L. n. *Legionella* type genus of the family; *-aceae* ending to denote family; M.L. fem. pl. n. *Legionellaceae* the *Legionella* family.

Rods 0.3–0.9 × 2–20 μm or more. Do not form endospores or microcysts. Not encapsulated. Not acid-fast in culture. **Gram negative.** Most species are motile by one, two, or more straight or curved polar or lateral flagella; nonmotile strains of motile species are occasionally seen and motility may be lost *in vitro*. **Aerobic. L-cysteine-HCl and iron salts are required for growth.** Catalase positive; oxidase negative or weakly positive. Nitrates are not reduced. Urease negative. Gelatin is liquefied by most species. **Chemoorganotrophic,** using amino acids as carbon and energy sources. **Carbohydrates are neither fermented nor oxidized** (Tables BXII.γ.56 and BXII.γ.57). **Branched chain fatty acids** predominate in whole cell hydrolysates (Table BXII.γ.58). All species contain ubiquinones with 9–14 isoprene units in the side chains (Table BXII.γ.59). Isolated from surface water, mud or moist soil, thermally polluted lakes and streams, and potable water systems. Associated with free-living environmental amoebae. There is no known soil or animal source. Pathogenic or potentially pathogenic for humans. Isolated from sputum, bronchoalveolar lavage washings, lung tissue, pleural fluid, pericardial fluid, blood, heart valves, abscesses, and prosthetic devices.

Growth of stock cultures obtained from humans occurs at temperatures between 25 and 43°C, but not at 50°C. The optimum temperature is 36° ± 1°C (Weaver and Feeley, 1979; Thacker et al., 1981); however, in nature, the optimum growth temperature may be 45°C or higher (Fliermans et al., 1981; Or- rison et al., 1981). The optimum pH for growth of legionellae is 6.8–7.0 (Feeley et al., 1978).

The mol% G + C of the DNA is: 38–52.

Type genus: **Legionella** Brenner, Steigerwalt and McDade 1979, 658.

Further descriptive information

Analysis of highly conserved sequences of 16S RNA has established that the family *Legionellaceae* forms a subline within the *Gammaproteobacteria* (Fry et al., 1991b; Hookey et al., 1995, 1996; Birtles et al., 1996; Riffard et al., 1998); the two closest relatives are *Coxiella burnetii* and *Wolbachia persica*. The relationships among strains, as determined by 16S rRNA analysis, support the proposal that a single genus, *Legionella*, be included in the family, *Legionellaceae*. Over the succeeding years, usage among the microbiology community has established the precedence of the single genus approach.

The family contains one genus, *Legionella*, with more than 40 species. The number will undoubtedly continue to expand, because a number of unnamed species have been detected by DNA–DNA hybridization (D.J. Brenner, personal communication). All species tested grow in free-living amoebae, and at least one species has been isolated only in amoebae (Rowbotham, 1983). The family is circumscribed and readily distinguishable from other families on the basis of a unique set of phenotypic characteristics

TABLE BXII.γ.56. Selected characteristics of *Legionella* species isolated from humans[a]

Characteristic	*L. anisa*	*L. birminghamensis*	*L. bozemanii*	*L. cincinnatiensis*	*L. dumoffii*	*L. feeleii*	*L. gormanii*	*L. hackeliae*	*L. jordanis*	*L. lansingensis*	*L. longbeachae*
Gram reaction	−	−	−	−	−	−	−	−	−	−	−
Acid-fastness[c]	−	−	−	−	−	−	−	−	−	−	−
Growth on:											
BCYE-α agar	+	+	+	+	+	+	+	+	+	+	+
Blood agar with trypticase soy base	−	−	−	−	−	−	−	−	−	−	−
Growth requirement for L-cysteine	+	+	+	+	+	+	+	+	+	+	+
Browning of tyrosine-containing media	+	−	+	+	+[f]	+ (w)	+	+	+	−	+
Acid from carbohydrates	−	−	−	−	−	−	−	−	−	−	−
Nitrate to nitrite	−	−	−	−	−	−	−	−	−	−	−
Hippurate hydrolysis	−	−	−	−	−	±	−	−	−	−	−
Gelatin liquefaction	+	+	+	+	+	±	+	+	+	−	+
β-lactamase production	+	+	±	+	+	−	+	+	+	−	±
Catalase	+	+	+	+	+	+	+	+ (w)	+	+	+
Oxidase	+	±	±	−	−	−	−	+	+	+	+
Motility[i]	+	+	+	+	+	+	+	+	+	+	+
Autofluorescence[j]	BW	−	BW	−	BW	−	BW	−	−	−	−

[a]Characteristics based on varying numbers of strains. NA, not available; w, weak.

[b]Amoebal-associated pathogens.

[c]Modified decolorization

[d]May be acid-fast in tissue.

[e]On initial isolation; may adapt to growth without cysteine.

[f]One strain negative.

[g]On initial isolation; may be + after several transfers on BCYE agar.

[h]A few strains negative.

[i]One polar flagellum

[j]BW, blue-white; R, red.

(*continued*)

that include Gram negativity, nonfermentative metabolism, requirement for L-cysteine and iron salts, and predominantly branched-chain cellular fatty acids. Relatively large amounts of branched chain and hydroxy-substituted branched chain fatty acids are found in flavobacteria, but these organisms were shown by DNA–DNA hybridization to be unrelated to legionellae. DNA–DNA hybridization was also used to rule out relatedness between legionellae and members of other families that resemble *Legionellaceae* in some phenotypic characteristics (Brenner et al., 1979).

Legionellae in formalin-fixed, paraffin-embedded tissue may not be demonstrable with the staining techniques usually employed histologically. The simplest means for demonstrating legionellae in tissue is to prepare imprint smears of the fresh cut surface of the tissue, after which the smears are stained by Gram's method. If sufficient numbers of organisms are present, they may be seen in tissue sections using one of the variations of the tissue Gram stain; e.g., the Brown and Hopps stain, or even hematoxylin and eosin. One of the silver impregnation stains (Warthin-Starry, Steiner, or Dieterle), designed originally for spirochetes, provides a more sensitive means to demonstrate legionellae in paraffin-fixed tissue (Chandler et al., 1977; van Orden and Greer, 1977). The Gimenez stain, designed for rickettsiae, has been used to color *Legionella* cells in secretions and imprint smears from unfixed tissue (Greer et al., 1980). The Gimenez method employs carbol fuchsin, and the sensitivity of Gram's stain may be improved without overstaining inflammatory cells by adding 0.05%

carbol fuchsin to the saffranin counterstain (W.C. Winn, Jr., personal observation). All of these histochemical stains, which have varying degrees of sensitivity for detecting bacteria, are completely nonspecific; they provide information about the morphology of the bacterial cells at the light microscopic level.

Bacterial cells have vacuoles that stain with Sudan black B and may contain poly-β-hydroxybutyrate (Chandler et al., 1979). Diaminopimelic acid has been demonstrated in the cell wall (Guerrant et al., 1979). Some species show a blue-white or red autofluorescence under long wavelength (365 nm) ultraviolet light (Tables BXII.γ.56 and BXII.γ.57), and many strains of approximately half of the recognized species produce a water-soluble, brown pigment on tyrosine-containing media (Baine et al., 1978).

The term legionellosis is used by many to indicate all infections caused by legionellae. Strains of approximately 50% of the recognized species have been implicated in human disease, usually pneumonia. All species should be considered potential pathogens if the immunologic, inflammatory, and/or nonspecific defenses of the patient are sufficiently compromised. Pneumonia caused by *L. pneumophila* is commonly referred to as Legionnaires' disease (Fraser et al., 1977) and pneumonia due to *L. micdadei* has been referred to as Pittsburgh pneumonia (Pasculle et al., 1980). Several *Legionella* species, including *L. pneumophila*, *L. micdadei*, *L. feeleii*, and *L. anisa*, can also cause a mild, nonpneumonic, febrile disease termed Pontiac fever (Glick et al., 1978; Herwaldt et al., 1984a; Goldberg et al., 1989; Fields et al., 1990).

TABLE BXII.γ.56. *(cont.)*

Characteristic	*L. lytica*[b]	*L. maceachernii*	*L. micdadei*	*L. oakridgensis*	*L. parisiensis*	*L. pneumophila pneumophila*	*L. pneumophila fraseri*	*L. pneumophila pascullei*	*L. sainthelensi*	*L. tucsonensis*	*L. wadsworthii*
Gram reaction	NA	−	−	−	−	−	−	−	−	−	−
Acid-fastness[c]	NA	−	−[d]	−	−	−	−	−	−	−	−
Growth on:											
BCYE-α agar	NA	+	+	+	+	+	+	+	+	+	+
Blood agar with trypticase soy base	NA	−	−	−	−	−	−	−	−	−	−
Growth requirement for L-cysteine	NA	+	+	+[e]	+	+	+	+	+	+	+
Browning of tyrosine-containing media		+	−[g]	+	+	+	+	+	+	−	−
Acid from carbohydrates		−	−	−	−	−	−	−	−	−	−
Nitrate to nitrite		−	−	−	−	−	−	−	−	−	−
Hippurate hydrolysis		−	−	−	−	+[h]	±	+	−	−	−
Gelatin liquefaction		+	−	+	+	+	+	+	+	+	+
β-lactamase production		−	−	+ (w)	+	+	+	+	+	+	+
Catalase		+	+	+	+	+	+	+	+	+	+
Oxidase		+	+	±	+	+ or ±	+ or ±	+ or ±	+	−	−
Motility[i]		+	+	−	+	+	+	+	+	+	−
Autofluorescence[j]		−	−	−	BW	−	−	−	−	BW	−

[a]Characteristics based on varying numbers of strains. NA, not available; w, weak.

[b]Amoebal-associated pathogens.

[c]Modified decolorization

[d]May be acid-fast in tissue.

[e]On initial isolation; may adapt to growth without cysteine.

[f]One strain negative.

[g]On initial isolation; may be + after several transfers on BCYE agar.

[h]A few strains negative.

[i]One polar flagellum

[j]BW, blue-white; R, red.

Genus I. **Legionella** *Brenner, Steigerwalt and McDade 1979, 658*[AL]

WASHINGTON C. WINN, JR.

Le.gi.on.el′la. M.L. n. *legio* legion or army; M.L. dim. ending *-ella*; M.L. fem. n. *Legionella* small legion or army.

As the only genus in the family, the definition of *Legionella* is identical to that of *Legionellaceae*. **Aerobic. L-cysteine-HCl and iron salts are required for growth.** Oxidase negative or weakly positive. Nitrates are not reduced. Urease negative. Gelatin is liquefied by most species. Chemoorganotrophic, using amino acids as carbon and energy sources. **Carbohydrates are neither fermented nor oxidized** (Tables BXII.γ.56 and BXII.γ.57). **Branched chain fatty acids** predominate in whole cell hydrolysate (Table BXII.γ.58). All species contain ubiquinones with 9–14 isoprene units in the side chains (Table BXII.γ.59).

The mol% G + C of the DNA is: 38–52.

Type species: **Legionella pneumophila** Brenner, Steigerwalt and McDade 1979, 658.

FURTHER DESCRIPTIVE INFORMATION

Analysis of highly conserved sequences of 16S rRNA has established that the genus *Legionella* forms a subline within the *Gammaproteobacteria* (Fry et al., 1991b; Birtles et al., 1996); the two closest relatives are *Coxiella burnetii* and *Wolbachia persica*. The relationships among strains, as determined by 16S rRNA analysis, support the proposal that a single genus, *Legionella*, be included in the family *Legionellaceae* (Fry et al., 1991a). Genetic relationships among the species of *Legionella* indicate subgroups within some species (Fry et al., 1991b; Hookey et al., 1995, 1996; Birtles et al., 1996; Riffard et al., 1998) and subspecies within *L. pneumophila* (Brenner et al., 1988c) As a generalization, the species that exhibit blue-white autofluorescence under long-wavelength ultraviolet light (Wood's Lamp) tend to be most closely related to each other (Hookey et al., 1996), and the two species that fluoresce red are close relatives. Interestingly, *L. wadsworthii* appears in the blue-white fluorescent clade, although it does not exhibit autofluorescence. Hookey et al. (1996) suggest that *L. wadsworthii* may be descended from an autofluorescent ancestor. By analysis of 16S rRNA, Hookey et al. (1995) were able to distinguish all but seven *Legionella* species from each other. The indistinguishable species were *L. erythra* from *L. rubrilucens* (red-

TABLE BXII.γ.57. Selected characteristics of *Legionella* species that have been isolated only from the environment[a]

Characteristic	L. brunensis	L. cherrii	L. erythra	L. fairfieldensis	L. geestiana	L. gratiana	L. israelensis	L. jamestowniensis	L. londiniensis	L. moravica	L. nautarum	L. parisiensis
Gram reaction	−	−	−	−	−	−	−	−	−	−	−	−
Acid-fastness[b]	−	−	−	−	−	−	−	−	−	−	−	−
Growth on:												
BCYE-α agar	+	+	+	+	+	+	+	+	+	+	+	+
Blood agar with trypticase soy base	−	−	−	−	−	−	−	−	−	−	−	−
Growth requirement for L-cysteine	+	+	+	+	+	+	+	+	+	+	+	+
Browning of tyrosine-containing media	w	+	+	−	w	−	−	+	+	w	−	+
Acid from carbohydrates	−	−	−	−	−	−	−	−	−	−	−	−
Nitrate to nitrite	−	−	−	−	−	−	−	−	−	−	−	−
Hippurate hydrolysis	−	−	−	−	w	−	−	−	− or w	−	−	−
Gelatin liquefaction	+	+	+	−	+	+	w	+	+	+	−	+
β-lactamase production	+	+	+	−	−	+	NA	+	+	+	+	+
Catalase	+	+	+	+	+	+	+	+	+	+	+	+
Oxidase	− or w	−	+	w	−	+	−	−	−	− or w	+	+
Motility[c]	+	+	+	+	+	+	+	+	+	+	+	+
Autofluorescence[d]	−	BW	R	−	−	−	−	−	−	−	−	BW

[a]Characteristics based on varying numbers of strains. In some cases a species is represented by a single isolate. NA, not available; w, weak.

[b]Modified decolorization.

[c]One polar flagellum.

[d]BW, blue-white; R, red.

[e]Autofluorescence may be lost on subculture.

(*continued*)

fluorescing species), *L. anisa* or *L. cherrii* from *L. tucsonensis* (blue-white fluorescing species), and *L. quateirensis* from *L. shakespearei*. Riffard et al. (1998) have grouped closely related species into three phylogenetic groups, based on intergenic 16S-23S ribosomal spacer PCR analysis (Table BXII.γ.60).

Legionellae are rod shaped or filamentous, 0.3–0.9 × 2–20 μm or more in length. The filaments are found after growth on agar media, less commonly in yolk sac material, and rarely in human lung or guinea pig tissue (Chandler et al., 1979; Weaver and Feeley, 1979). Electron microscopic examination reveals filamentous nucleoids, ribosomes, and Sudan black B-staining vacuoles thought to contain poly-β-hydroxybutyrate granules (Chandler et al., 1979). Cells are enclosed by a double envelope, composed of two three-layered unit membranes; a peptidoglycan layer has been demonstrated for *L. pneumophila*, and for *L. micdadei* (Flesher et al., 1979; Pasculle et al., 1979; Gress et al., 1980). Cell division occurs by a pinching process (Chandler et al., 1979; Keel et al., 1979).

One, two, or occasionally more flagella occur per cell. The flagella are curved or straight and have a polar or lateral arrangement (Chandler et al., 1980).

Legionellae are not acid-fast by the Ziehl-Neelsen procedure for mycobacteria (Hébert et al., 1980a) after they have been isolated on agar. *L. micdadei*, however, may appear acid-fast in tissue; modified acid-fast stains that use weak acids as decolorizing agents may be required to demonstrate the acid-fastness (Pasculle et al., 1979).

Legionellae were first isolated in guinea pigs and in embryonated hens' eggs (Tatlock, 1944; Jackson et al., 1952; Bozeman et al., 1968; McDade et al., 1979). With the advent of supportive culture media, inoculation of animals and eggs has become obsolete. It has been suggested, however, that preincubation of environmental samples with protozoa increases the recovery of isolates (Rowbotham, 1983; Sanden et al., 1992). Legionellae do not grow on commonly employed media such as standard blood agar or nutrient agar. Mueller-Hinton agar, supplemented with hemoglobin and IsoVitaleX or GC base with similar supplements, was used to cultivate the first isolates of *L. pneumophila* (Centers for Disease Control, 1977). In one case, the agar was a commercially prepared chocolate agar (Dumoff, 1979). A buffered charcoal yeast extract (BCYE-α) agar supplemented with 0.1% α-ketoglutarate is now preferred for isolation of *L. pneumophila* (Feeley et al., 1979; Pasculle et al., 1980). All species grow on BCYE-α agar. The medium is buffered at pH 6.9, the optimum for growth of *Legionella* spp., by ACES buffer. The activated charcoal serves as a scavenger for toxic free radicals that are generated by the interaction of light with the medium (Hoffman et al., 1983). Yeast extract serves as a rich growth medium. Variations on BCYE-α agar have been developed (Wadowsky and Yee, 1981), and some have proven particularly useful for isolating legionellae from environmental specimens (Edelstein, 1982). Some investigators have found that addition of 1.0% bovine serum albumin enhances the growth of *L. micdadei* and some strains of *L. bozemanii* (Morrill et al., 1990). Antibiotics may be added to suppress the growth of indigenous human or environmental flora (Edelstein and Finegold, 1979; Edelstein, 1981). It is essential to include a nonselective BCYE-α agar, because both vancomycin and cefamandole, two commonly used agents for suppressing Gram-positive bacteria, will inhibit various *Legionella* species (Lee et al., 1993b). *L. pneumophila* grew better than other species. Strains of 25 of the 28 species tested were recovered on media that contained vancomycin. In contrast, cefamandole completely inhibited strains of 11 species and inhibited growth of strains from an additional 8 species. Several complex and semisynthetic broth

TABLE BXII.γ.57. *(cont.)*

Characteristic	*L. parisiensis*	*L. quateirensis*	*L. quinlivanii*	*L. rubrilucens*	*L. santicrucis*	*L. shakespearei*	*L. spiritensis*	*L. steigerwaltii*	*L. taurinensis*	*L. waltersii*	*L. worsleiensis*	Genomospecies 1
Gram reaction	−	−	−	−	−	−	−	−	−	−	−	−
Acid-fastness[b]	−	−	−	−	−	−	−	−	NA	−	−	NA
Growth on:												
BCYE-α agar	+	+	+	+	+	+	+	+	+	+	+	+
Blood agar with trypticase soy base	−	−	−	−	−	−	−	−	−	−	−	−
Growth requirement for L-cysteine	+	+	+	+	+	+	+	+	+	+	+	+
Browning of tyrosine-containing media	+	+	+	+	+	−	+	+	− (96%)	−	+	+
Acid from carbohydrates	−	−	−	−	−	−	−	−	−	−	−	−
Nitrate to nitrite	−	−	−	−	−	−	−	−	−	−	−	−
Hippurate hydrolysis	−	−	−	−	−	−	+ (w)	−	− (90%)	+	−	−
Gelatin liquefaction	+	+	+	+	+	+	+	+	+	+	+	+
β-lactamase production	+	+	−	+	+	+	+	+	+	+	+	−
Catalase	+	+	+	+	+	+	+	+	+	+	−	+
Oxidase	+	−		+		w	+		+	+		−
Motility[c]	+	+	+	+	+	+	+		+	+	+	+
Autofluorescence[d]	BW	−	−	R	−	−	−	BW	R (67%)/−[e]	−	−	−

[a]Characteristics based on varying numbers of strains. In some cases a species is represented by a single isolate. NA, not available; w, weak.

[b]Modified decolorization.

[c]One polar flagellum.

[d]BW, blue-white; R, red.

[e]Autofluorescence may be lost on subculture.

media are available for physiologic studies of *L. pneumophila* (Saito et al., 1981), but have not proven useful for isolation of clinical or environmental strains.

Media containing dyes that have some differential value have been described (Vickers et al., 1981), based on the appearance of isolated colonies. Data are not available for more recently isolated species.

Several species exhibit a blue-white or red autofluorescence, and many species produce a diffusible, brown pigment on tyrosine-containing agar (Tables BXII.γ.56 and BXII.γ.57).

Growth of stock cultures obtained from humans occurs at temperatures between 25°C and 43°C, but not at 50°C. The optimum temperature is 36° ± 1°C (Weaver and Feeley, 1979; Thacker et al., 1981); however, in nature, the optimum growth temperature may be 45°C or higher (Fliermans et al., 1981; Orrison et al., 1981). The optimum pH for growth of legionellae is 6.8–7.0 (Feeley et al., 1978).

CO_2 does not appear to stimulate the growth of legionellae, with the exception of *L. gormanii*. Because of this exception, and also the buffering effect of CO_2, incubation in an atmosphere of air + 2.5% CO_2, or in a candle jar, has been recommended for all legionellae; however, CO_2 may slightly inhibit growth in BCYE (Pasculle et al., 1980).

Pinpoint colonies appear on BCYE-α agar[1] in about 3 d after primary culture; the colony diameter reaches 3–4 mm after 5–7 d of incubation at 36° ± 1°C. Colonies are gray, glistening, convex, and circular with an entire edge. With a dissecting microscope one can visualize iridescence in varying colors of red, blue, and green; there is often a faceted or "cut-glass" appearance of the colonies when viewed with reflected light.

Carbohydrates are neither fermented nor oxidized by legionellae (Weaver and Feeley, 1979; Riley and Weaver, personal communication). Amino acids can presumably serve as the carbon source (Warren and Miller, 1979; George et al., 1980).

Legionellae are nutritionally fastidious. They require iron and L-cysteine for growth. As indicated previously, soluble ferric pyrophosphate, ferric nitrate, or hemoglobin can provide the iron, and IsoVitaleX (BBL) can serve as a source of the L-cysteine. The following amino acids are essential for growth: cysteine, serine, methionine, arginine, valine, leucine, isoleucine, and threonine; no vitamin or cofactor requirements have been demonstrated (George et al., 1980). Chemically defined and semisynthetic liquid media have been developed for legionellae (Pine et al., 1979; Warren and Miller, 1979; Ristroph et al., 1980).

One or more cryptic plasmids are present in many, but not all, strains of many species (Knudson and Mikesell, 1980). Neither pathogenicity nor antibiotic resistance, nor any other cell

---

1. BCYE agar: add 10.0 g of ACES buffer (*N*-2-acetamido-2-aminoethane-sulfonic acid; pKa = 6.9 at 20°C; available from Sigma Chemical Co., St. Louis, MO) to 500 ml of distilled water and dissolve by heating in a water bath at 45–50°C. Mix this solution with 440 ml of distilled water to which 40 ml of 1.0 N KOH has been added. Add the following ingredients: activated charcoal (Norit SG, available through Sigma Chemical Co., Cat. No. C5510 [acid-washed with phosphoric and sulfuric acids]), 2.0 g; yeast extract, 10.0 g; and agar (Difco or Oxoid), 17.0 g. Dissolve by boiling. Autoclave 15 min at 121°C. Cool to 50°C. Aseptically add L-cysteine-HCl·H₂O solution (0.4 g in 10 ml of distilled water, sterilized by filtration), followed by ferric pyrophosphate solution (0.25 g in 10 ml of distilled water, sterilized by filtration). (Soluble ferric pyrophosphate is available on request from

Director, Biological Products Div., Centers for Disease Control and Prevention, Atlanta, GA 30333. It must be kept dry and stored in the dark until used. It is not usable if the color changes from green to yellow or brown. Do not heat over 60°C to dissolve the ferric pyrophosphate; a 50°C water bath is satisfactory.) The pH of the final solid medium should be 6.9 ± 0.05 at room temperature. Since reagents vary, each laboratory must determine the amount of KOH required. Do this by holding the bulk medium at 50°C while pouring one plate and checking its pH. When necessary, adjust the bulk medium with either 1.0 N KOH or 1.0 N HCl. Note the pKa of ACES buffer is influenced by temperature (0.02/°C); consequently, this must be considered with all pH determinations. Dispense 20 ml portions of the complete medium into 10- × 100-mm plastic Petri dishes; swirl the medium between pouring plates to keep charcoal particles suspended.

TABLE BXII.γ.58. Major fatty acids of *Legionella* species (% total fatty acids)[a,b]

Fatty acids	L. adelaidensis*	L. anisa	L. birminghamensis*	L. bozemanii	L. brunensis	L. cherrii	L. cincinnatiensis*	L. dumoffii	L. erythra (ATCC 33303)	L. erythra	L. fairfieldensis*	L. feeleii	L. geestiana	L. gormanii	L. gratiana	L. hackeliae	L. israelensis	L. jamestowniensis	L. jordanis	L. lansingensis*	L. londiniensis	L. longbeachae	L. lytica[c]
C14:0 iso	7	6	—	3	2	5	34	2	tr	2	14	3	tr	4	2	2	2	2	2	NA	tr	5	NA
C15:0 iso	6	1	—	tr	4	tr	—	tr	1	1	—	tr	16	—	NA	4	1	1	2	5	tr	—	NA
C15:0 ante	48	24	100	29	100	27	—	30	8	12	15	18	9	22	NA	33	27	30	46	NA	5	15	NA
C16:1 iso A	100	NA	—	NA	NA	NA	—	NA	—	—	NA	2	0	NA	NA	NA	NA	NA	NA	NA	NA	—	NA
C16:1 iso B	—	—	—	—	—	—	—	—	—	—	—	—	—	—	—	—	—	—	—	—	—	—	NA
C16:0	44	23	46	14	14	29	100	14	4	24	100	16	2	17	14	15	22	21	18	40	11	25	NA
C16:1 ω7c	—	9	—	9	NA	9	—	8	38	27	NA	22	—	11	NA	13	7	4	4	NA	NA	24	NA
Un16	—	1	—	—	NA	tr	—	1	—	—	NA	—	—	1	NA	—	—	tr	—	NA	—	1	NA
C16:0	NA	8	NA	11	NA	6	—	7	26	14	NA	16	18	11	30	6	7	3	1	NA	15	9	NA
C16:1	—	NA	—	NA	NA	NA	—	NA	NA	NA	NA	NA	NA	NA	NA	NA	NA	NA	NA	NA	19	NA	NA
C17:1 ante	3	1	54	1	3	1	36	17	1	1	—	tr	9	1	3	2	1	3	1	12	—	—	NA
C17:0 ante	14	7	—	11	62	12	—	8	10	10	9	7	7	9	—	14	17	21	19	9	0	tr	NA
C17:0 cyc	—	9	—	9	0	3	—	NA	—	—	—	tr	—	10	—	3	5	5	2	100	23	8	NA
C18:0 iso	—	NA	—	NA	—	NA	—	1	4	2	—	tr	4	NA	4	NA	NA	NA	NA	—	—	4	NA
C19:0 ante	—	1	—	2	—	1	—	—	—	—	—	1	—	2	—	1	1	1	tr	—	8	tr	NA
C21:0	—	NA	—	—	—	NA	—	—	tr	tr	—	—	—	NA	—	—	NA	NA	—	5	—	1	NA
C14:1	4		—		0		—				2						2	1	—			—	
C14:0	12		10		3		10				5											—	
C15:0	16		—		3		8				6									5			
C15:1	9		—		1		—				33												
C16:0	44		43		30		—				51									20			
C16:1	100		37		22		—				—	1				—				12			
C17:0	3				1		7				4												
C18:0	12				3		7				8												
C20:0	12				5		5																
Lambert and Moss group (Lambert, Moss, 1989)	—	A15/16C	—	A15	—	A15/16C	—	A15	16C	16C	—	16C		A15/16C		A15	A15/16C	A15/16C	A15	—	—	16C	—

[a]Relative quantities among nonhydroxy fatty acids.

[b]NA, not available; tr, trace.

[c]Amoeba-associated strains.

(continued)

TABLE BXII.γ.58. *(cont.)*

Fatty acids	*L. maceachernii*	*L. micdadei*	*L. moravica**	*L. nautarum*	*L. oakridgensis*	*L. parisiensis*	*L. pneumophila pneumophila*	*L. pneumophila fraseri*	*L. pneumophila pascullei*	*L. quateirensis*	*L. quinlivanii**	*L. rubrilucens*	*L. sainthelensi*	*L. santicrucis*	*L. shakespearei*	*L. spiritensis*	*L. steigerwaltii*	*L. taurinensis*	*L. tucsonensis**	*L. wadsworthii*	*L. waltersii*	*L. worsleiensis*	Genomospecies 1
$C_{14:0}$ iso	tr	tr	10	2	tr	3	5	NA	NA	3	8	2	5	6	9	2	2	3	8	1	13	4	2
$C_{15:0}$ iso	–	1	1	0	–	tr	–	NA	NA	tr	NA	tr	tr	–	2	tr	tr	–	NA	tr	tr	tr	1
$C_{15:0}$ ante	28	32	45	19	2	23	13	NA	NA	8	100	12	20	10	10	20	23	1–20	100	37	9	12	23
$C_{16:1}$ iso A	NA	NA	NA	tr	2	NA	–	NA	NA	tr	NA	–	–	–	NA	–	NA	NA	NA	NA	NA	1	18
$C_{16:1}$ iso B	3	2	NA	9	26	17	3	NA	NA	20	94	23	22	30	29	5	15	>20	77	10	17	–	NA
$C_{16:0}$ iso	11	12	39	NA	14	7	32	NA	NA	NA	NA	27	25	22	NA	28	8	NA	NA	8	NA	9	21
$C_{16:1}$ ω7c	14	8	NA	NA	–	2	16	NA	NA	NA	NA	–	–	–	NA	15	1	NA	NA	1	NA	NA	NA
Un16	–	–	NA	NA	10	9	–	NA	NA	NA	NA	15	7	10	NA	–	18	NA	NA	6	NA	NA	NA
$C_{16:0}$	6	6	NA	19	NA	NA	7	NA	NA	7	NA	NA	NA	NA	NA	5	NA	10–20	NA	NA	NA	16	NA
$C_{16:1}$	NA	NA	NA	27	1	–	NA	NA	NA	35	NA	–	–	tr	NA	NA	–	10–20	NA	–	NA	35	NA
$C_{17:1}$ ante	5	5	0	0	–	1	tr	NA	NA	0	5	tr	10	4	1	3	8	10	49	1	NA	0	–
$C_{17:0}$ iso	22	tr	12	13	5	9	8	NA	NA	4		9	1	8	3	tr	14	–		19	2	5	1
$C_{17:0}$ ante	1	21			12	10	2	NA	NA			–	–	–		16	NA	–		6			12
$C_{17:0}$ cyc	NA	1			2	NA	–	NA	NA			–	tr	1		1	3	–		NA			
$C_{18:0}$ iso	tr	NA	3	2	13	2	3	NA	NA	10		tr	1	–		–	NA	tr		1		9	
$C_{18:0}$	–	tr			1	NA	–	NA	NA			2	–	–		tr	–	–		2			
$C_{19:0}$ ante	–	–			–	–	1	NA	NA			–	–			1		–		–			
$C_{21:0}$	2	2						NA	NA									–					
$C_{14:1}$			1					NA	NA		6							–			2		tr
$C_{14:0}$			8					NA	NA		9							NA			6		3
$C_{15:0}$			7					NA	NA		10				4			NA			6		4
$C_{15:1}$			2					NA	NA		79				5			NA	41		16		10
$C_{16:0}$			85					NA	NA		95				11			NA	78		19		
$C_{16:1}$			100					NA	NA		16				16				3		2		2
$C_{17:0}$			3					NA	NA		7				4				4		2		tr
$C_{18:0}$			35					NA	NA		6				1						3		1
$C_{20:0}$			11					NA	NA														
Lambert and Moss group (Lambert, Moss, 1989)	A15	A15	–	–	16C	A15/16C	16C	–	–	–	–	16C	16C	16C	–	16C	A15/16C	–	–	A15	–	–	–

[a] Relative quantities among nonhydroxy fatty acids.

[b] NA, not available; tr, trace.

[c] Amoeba-associated strains.

TABLE BXII.γ.59. Ubiquinones of *Legionella* species[a,b]

Legionella species	Q-9	Q-10	Q-11	Q-12	Q-13	Q-14	Ubiquinone group
L. adelaidensis	2	40	100	12	–	–	–
L. anisa	2–3	4	2–3	3	tr–1	–	B
L. birminghamensis	NA	NA	NA	NA	NA	NA	–
L. bozemanii	2–3	4	3–4	4	tr–1	–	B
L. brunensis	NA	NA	NA	NA	NA	NA	–
L. cherrii	2	3	3–4	4	1	–	B
L. cincinnatiensis	NA	NA	NA	NA	NA	NA	–
L. dumoffii	2–3	3–4	3–4	4	1	–	B
L. erythra	–	–	1–2	4	1–2	–	A
L. fairfieldensis	NA	NA	NA	NA	NA	NA	–
L. feeleii	–	–	–	tr	4	2–3	E
L. geestiana	NA	NA	1	4	39	52	–
L. gormanii	3	4	4	4	–	–	B
L. gratiana	1.7	4	1.9	2.2	tr	NA	–
L. hackeliae	–	–	–	2	4	tr–1	D
L. israelensis	–	tr	1–2	4	3–4	–	D
L. jamestowniensis	–	tr	1–2	4	4	tr	D
L. jordanis	–	tr	tr	3–4	4	tr	D
L. lansingensis	–	–	1X	1X	4X	1X	–
L. londiniensis	2	24	55	16	2	tr	–
L. longbeachae	3–4	4	3–4	3–4	tr	–	B
L. lytica; amoeba-associated strains	NA	NA	NA	NA	NA	NA	NA
L. maceachernii	–	–	–	tr–1	4	1	D
L. micdadei	–	–	tr–1	3	4	tr	D
L. moravica	NA	NA	NA	NA	NA	NA	–
L. nautarum	NA	NA	NA	18	76	6	–
L. oakridgensis	2	4	1–2	tr	–	–	C
L. parisiensis	1–2	3–4	3	4	1	–	B
L. pneumophila subsp. pneumophila	–	–	1–2	4	1–2	–	A
L. pneumophila subsp. fraseri	NA	NA	NA	NA	NA	NA	–
L. pneumophila subsp. pascullei	NA	NA	NA	NA	NA	NA	–
L. quateirensis	NA	tr	2	21	69	8	–
L. quinlivanii	NA	NA	NA	NA	NA	NA	–
L. rubrilucens	–	–	1–2	4	1–2	–	A
L. sainthelensi	3	4	3–4	3–4	tr	–	B
L. santicrucis	1–2	4	3–4	3–4	tr	1	B
L. shakespearei	–	1X	1X	3X	1X	–	–
L. spiritensis	–	–	–	tr–1	4	1	D
L. steigerwaltii	1	2	2–3	4	–	–	B
L. tucsonensis	NA	NA	NA	NA	NA	NA	–
L. taurinensis	–	–	1X	5X	1X	–	–
L. wadsworthii	2	4	2–3	tr–1	–	–	C
L. waltersii	NA	NA	NA	Major	NA	NA	–
L. worsleiensis	NA	1	4	26	62	4	–
Genomospecies 1	–	–	–	–	Major	–	–

[a]Relative proportion of ubiquinones

[b]Symbols: tr , trace; nX,relative proportion of ubiquinone, –, not present. .

function was correlated with the presence of any plasmid (Mikesell et al., 1981; Aye, Wachsmuth, and Feeley, personal communication).

Motile strains of all *Legionella* species possess cross-reacting H antigens (Lewallen et al., 1979; Hébert et al., 1980d). O antigens, which define the serogroups, are primarily defined by the characteristics of cellular lipopolysaccharide (Ciesielski et al., 1986; Otten et al., 1986). They are largely unique for each species, although some cross-reactions have been demonstrated (Table BXII.γ.60). More than one O-antigen group is present in several species (Table BXII.γ.61).

Legionella species are susceptible *in vitro* to a variety of antimicrobial agents, including erythromycin, rifampin, and newer macrolides, such as azithromycin and clarithromycin, which are effective *in vivo*, as measured in experimental animals or in clinical studies (Hamedani et al., 1991; Donowitz and Earnhardt, 1993; Stamler et al., 1994). Quinolones have been used successfully to treat *Legionella* infections (Saito et al., 1986; Unertl et al., 1989), but clinical failures have also been described (Kurz et al., 1988). Tetracycline, doxycycline, and minocycline are also effective *in vitro* and possibly *in vivo*. There are no prospective, random, placebo-controlled studies of antimicrobial therapy for *Legionella* infection. Retrospective analysis of two early epidemics (Philadelphia in 1976 and Burlington, Vermont, in 1977) documented the efficacy of erythromycin and, less conclusively, tetracycline (Fraser et al., 1977; Broome et al., 1979). Other antibiotics that are not effective against *Legionella* include aminoglycosides, sulfamethoxazole-trimethoprim, chloramphenicol, and a variety of cephalosporins (Thornsberry et al., 1978).

Many species of *Legionella* produce some level of β-lactamase activity when assayed with a chromogenic cephalosporin test (Tables BXII.γ.56 and BXII.γ.57) (Hébert et al., 1980b). The β-lactamase is assumed to be chromosomal rather than of plasmid origin and is more active on cephalosporins than on penicillins (Thornsberry and Kirven, 1978). The correlation between *in vitro* susceptibility and clinical efficacy of antibiotics for legi-

TABLE BXII.γ.60. Cross-reactions among *Legionella* species and with other bacteria

Species (antigen)	Method	Cross-reacts with antiserum to:	Absorbed by:	Reference
L. adelaidensis	Agglutination (unabsorbed neat)	*L. hackeliae*, serogroup 1 and *L. dumoffii* (1 +); no reaction with *L. adelaidensis* antiserum at optimal dilution	None	Benson et al., 1991a
L. anisa	Direct immunofluorescence	*L. longbeachae*, serogroup 2 and *L. bozemanii*, serogroup 2	None	Bornstein et al., 1989a
L. anisa	Direct immunofluorescence	*L. gratiana* (+ to + +)	NA	Bornstein et al., 1989a
L. anisa	Slide agglutination	*L. gratiana* (+ + +)	Removed by absorption with *L. anisa* cells	Bornstein et al., 1989a
L. anisa	Slide agglutination	*L. tucsonensis* (+ + +)	Cross-reaction removed by absorption of antiserum with *L. santicrucis* and *L. anisa* cells	Thacker et al., 1989a
L. bozemanii	Agglutination	*L. longbeachae*, serogroup 2; *L. jordanis*	NA	Pelaz et al., 1987
L. bozemanii	Microagglutination	*L. jordanis*	NA	Pelaz et al., 1987
L. bozemanii	Immunofluorescence	*L. jordanis*	NA	Pelaz et al., 1987
L. bozemanii, serogroup 2	Direct immunofluorescence	*L. gratiana* (+ to + +)	NA	Bornstein et al., 1989a
L. bozemanii, serogroup 2	Slide agglutination	*L. gratiana* (+ + +)	Removed by absorption with *L. anisa* cells	Bornstein et al., 1989a
L. cincinnatiensis	Slide agglutination	*L. sainthelensi* (2 +)	NA	Thacker et al., 1988a
L. cincinnatiensis	Direct immunofluorescence	*L. longbeachae*, *L. dumoffii*	NA	Jernigan et al., 1994
L. dumoffii	Agglutination	*L. pneumophila*, serogroup 7	NA	Pelaz et al., 1987
L. erythra, serogroup 2	Indirect immunofluorescence	*L. rubrilucens*	Cross-reaction partially removed by absorption with *L. erythra*, serogroup 2	Saunders et al., 1992
L. geestiana	Slide agglutination	*L. jamestowniensis* (+ + +)	Cross-reaction removed by absorption with *L. jamestowniensis* cells	Dennis et al., 1993
L. gratiana	Direct immunofluorescence	*L. bozemanii*, serogroup 2 (+ +); *L. longbeachae*, serogroup 2 (+ +); *L. anisa* (+ +); *L. longbeachae*, serogroup 1 (+); *L. bozemanii*, serogroup 1 (+)	NA	Bornstein et al., 1989a
L. israelensis	Slide agglutination	*L. wadsworthii* (+ + + to + + + +)	Removed by absorption with *L. wadsworthii* cells	Bercovier et al., 1986
L. jordanis	Agglutination, microagglutination, and immunofluorescence	*L. bozemanii*	NA	Pelaz et al., 1987
L. jordanis	Slide agglutination	*L. bozemanii*, serogroups 1 and 2 (+ + + +), *L. parisiensis* (+ + + +), *L. anisa* (+ + + +), *L. longbeachae*, serogroups 1 (+ +) and 2 (+).	Adsorption of antisera with *L. jordanis* removed cross-reactions	Thacker et al., 1988b
L. jordanis	Slide agglutination	*L. cincinnatiensis* (1 +)	Removed by *L. santicrucis* and *L. sainthelensi* cells	Thacker et al., 1988b
L. jordanis	Direct immunofluorescence	*L. bozemanii*	Reduced by adsorption with *L. bozemanii* cells	Cherry et al., 1982a
L. jordanis	Slide agglutination	*L. cincinnatiensis* (+)	Cross-reaction removed by absorption with *L. santicrucis* and *L. sainthelensi* cells	Thacker et al., 1988b
L. lansingensis	Slide agglutination	*L. pneumophila*, serogroup 3	Cross-reaction absorbed with *L. pneumophila*, serogroup 6 cells	Thacker et al., 1992
L. longbeachae, serogroup 1	Agglutination	*L. longbeachae*, serogroup 2	NA	Pelaz et al., 1987
L. longbeachae, serogroup 1	Microagglutination	*L. longbeachae*, serogroup 2	NA	Pelaz et al., 1987
L. longbeachae, serogroup 1	Immunofluorescence	*L. longbeachae*, serogroup 2	NA	Pelaz et al., 1987

(continued)

TABLE BXII.γ.60. *(cont.)*

Species (antigen)	Method	Cross-reacts with antiserum to:	Absorbed by:	Reference
L. longbeachae, serogroup 2	Agglutination, microagglutination, and immunofluorescence	*L. longbeachae*, serogroup 1; *L. jordanis*	NA	Pelaz et al., 1987
L. longbeachae, serogroup 2	Immunodiffusion	*L. longbeachae*, serogroup 1	NA	Pelaz et al., 1987
L. longbeachae, serogroups 1 and 2	Direct immunofluorescence	*L. gratiana* (+ to ++)	NA	Bornstein et al., 1989a
L. maceachernii	Direct immunofluorescence	Undiluted antiserum *L. micdadei* (+++)	NA	Brenner et al., 1985
L. oakridgensis	Direct immunofluorescence	At 1:8 dilution *L. sainthelensi* (+++ to ++++); at working dilution *L. sainthelensi* (++)	Absorption of antisera to *L. sainthelensi* with *L. longbeachae* or *L. oakridgensis* cells removed reactivity with the absorbing antigen	Campbell et al., 1984a
L. parisiensis	Direct immunofluorescence	*L. bozemanii*, serogroups 1 and 2	Not removed by absorption with s	Lo Presti et al., 1997
L. parisiensis	Direct immunofluorescence	*L. bozemanii* (+); *L. micdadei* (+); *L. jordanis* (++); with undilute antiserum *L. bozemanii* and *L. jordanis* (++++)	NA	Brenner et al., 1985
L. pneumophila, serogroup 1	Agglutination / immunofluorescence	*L. pneumophila*, serogroup 7	NA	Pelaz et al., 1987
L. pneumophila, serogroup 2	Agglutination	*L. pneumophila*, serogroup 3	NA	Pelaz et al., 1987
L. pneumophila, serogroup 2	Microagglutination	*L. jordanis*	NA	Pelaz et al., 1987
L. pneumophila, serogroup 2	Immunofluorescence	*L. pneumophila*, serogroups 3 and 6	NA	Pelaz et al., 1987
L. pneumophila, serogroup 3	Agglutination	*L. pneumophila*, serogroup 4	NA	Pelaz et al., 1987
L. pneumophila, serogroup 3	Microagglutination	*L. pneumophila*, serogroups 2 and 6	NA	Pelaz et al., 1987
L. pneumophila, serogroup 3	Immunodiffusion	*L. pneumophila*, serogroup 2	NA	Pelaz et al., 1987
L. pneumophila, serogroup 3	Immunofluorescence	*L. pneumophila*, serogroups 2 and 6	NA	Pelaz et al., 1987
L. pneumophila, serogroup 4	Slide agglutination	*L. worsleiensis* (++++)	Cross-reaction to serogroup 4 removed by absorption with serogroup 4 cells	Dennis et al., 1993
L. pneumophila, serogroup 4 (Los Angeles 1 strain)	Agglutination	*L. pneumophila*, serogroup 8	NA	Pelaz et al., 1987
s, serogroup 4 (Los Angeles 1 strain)	Immunodiffusion	*L. pneumophila*, serogroups 5 and 8	NA	Pelaz et al., 1987
L. pneumophila, serogroup 4 (Los Angeles 1 strain)	Immunofluorescence	*L. pneumophila*, serogroups 5 and 8	NA	Pelaz et al., 1987
L. pneumophila, serogroup 5 (Dallas 1-E strain)	Immunodiffusion	*L. pneumophila*, serogroup 8	NA	Pelaz et al., 1987
L. pneumophila, serogroup 6	Agglutination	*L. pneumophila*, serogroup 2	NA	Pelaz et al., 1987
L. pneumophila, serogroup 6	Immunodiffusion and immunofluorescence	serogroup 3	NA	Pelaz et al., 1987
L. pneumophila, serogroup 7	Agglutination	*L. pneumophila*, serogroup 1	NA	Pelaz et al., 1987
L. pneumophila, serogroup 7	Immunofluorescence	*L. dumoffii*	NA	Pelaz et al., 1987
L. pneumophila, serogroup 8	Agglutination	*L. pneumophila*, serogroups 3 and 4	NA	Pelaz et al., 1987
L. pneumophila, serogroup 8	Microagglutination	*L. pneumophila*, serogroup 5	NA	Pelaz et al., 1987
L. pneumophila, serogroup 8	Immunodiffusion	*L. pneumophila*, serogroup 4	NA	Pelaz et al., 1987
L. pneumophila, serogroup 8	Immunofluorescence	*L. pneumophila*, serogroups 4 and 5	Na	Pelaz et al., 1987
L. pneumophila, serogroup 8	Slide agglutination	*L. worsleiensis* (++)	NA	Dennis et al., 1993

(continued)

TABLE BXII.γ.60. *(cont.)*

Species (antigen)	Method	Cross-reacts with antiserum to:	Absorbed by:	Reference
L. pneumophila, serogroups 4, 10, 13	Slide agglutination	*L. quateirensis* (+ + + +)	Cross-reactions removed by absorption with *L. pneumophila*, serogroup 13	Dennis et al., 1993
L. quateirensis	Slide agglutination	*L. pneumophila*, serogroups 4, 5, 8, 10, 13, 14; *L. worsleiensis*	Cross-reactions removed by absorption with *L. quateirensis* cells	Dennis et al., 1993
L. quateirensis	Slide agglutination	*L. worsleiensis* (+ + + +)	Cross-reaction removed by absorption with *L. quateirensis* cells	Dennis et al., 1993
L. quinlivanii, serogroup 2	Slide agglutination	Unnamed genomospecies 1	Cross-reaction not removed by absorption	Benson et al., 1996
L. rubrilucens	Indirect immunofluorescence	*L. erythra*, serogroup 2	Cross-reaction removed by absorption with *L. rubrilucens* or *L. erythra*, serogroup 2 cells	Saunders et al., 1992
L. sainthelensi	Slide agglutination	*L. cincinnatiensis* (2+)	Removed by *L. santicrucis* and *L. sainthelensi* cells	Thacker et al., 1988a
L. sainthelensi	Direct immunofluorescence	At 1:8 dilution: *L. longbeachae*, serogroups 1 (+ + + to + + + +) and 2 (+ +); *L. oakridgensis* (+ + + to + + + +); at working dilution: *L. longbeachae*, serogroup 2 (+ +) and *L. oakridgensis* (+ +)	Absorption of antisera to *L. longbeachae*, serogroups 1 and 2 or *L. oakridgensis* with *L. sainthelensi* abolishes or reduces reactivity with *L. sainthelensi* and *L. longbeachae*, serogroups 1 and 2	Campbell et al., 1984a
L. sainthelensi	Slide agglutination	*L. cincinnatiensis* (+ +)	Cross-reactions removed by absorption with *L. sainthelensi* and *L. santicrucis* cells	Thacker et al., 1988a
L. santicrucis	Slide agglutination	*L. cincinnatiensis* (4+)	Removed by *L. santicrucis* and *L. sainthelensi* cells	Thacker et al., 1988a
L. santicrucis	Direct immunofluorescence	*L. longbeachae*, serogroup 1 (+ + +); with undiluted antiserum *L. sainthelensi* (+ +)	NA	Brenner et al., 1985
L. santicrucis	Slide agglutination	*L. tucsonensis* (+ + + +)	Cross-reaction removed by absorption of antiserum with *L. santicrucis* and *L. anisa* cells	Thacker et al., 1989a
L. taurinensis	Direct immunofluorescence	*L. spiritensis*, serogroup 1 (+ + + to + + + +)	NA	Lo Presti et al., 1999
L. tucsonensis	Slide agglutination	*L. gratiana* (+ +)	Removed by absorption with *L. anisa* cells	Bornstein et al., 1989a
L. tucsonensis	Slide agglutination	*L. longbeachae*, serogroup 1 (+ + + +); *L. longbeachae*, serogroup 2 (+); *L. cincinnatiensis* (+)	Cross-reactions with *L. longbeachae*, serogroup 1 removed by absorption	Thacker et al., 1989a
L. wadsworthii	Agglutination, microagglutination	*L. oakridgensis*	NA	Pelaz et al., 1987
L. worsleiensis	Slide agglutination	*L. pneumophila*, serogroups 4, 5, 8, 10; *L. quateirensis*	Cross-reactions with *L. pneumophila*, serogroup 4 and *L. quateirensis* not removed by absorption	Dennis et al., 1993
L. worsleiensis	Slide agglutination	*L. quateirensis* (+ + + +)	Cross-reactions removed by absorption with *L. worsleiensis* cells	Dennis et al., 1993
Genomospecies 1	Slide agglutination	*L. quinlivanii*, serogroup 2 (+)	NA	Benson et al., 1996

onellae is poor (Kirby et al., 1978). Clinical response and survival, however, do correlate with the timely administration of effective antibiotics (Heath et al., 1996). Resistance to penicillins and cephalosporins may be partially related to β-lactamase activity, although infections caused by *L. micdadei*, which does not produce β-lactamase, do not respond to penicillins or cephalosporins. The most likely explanation for the poor correlation between *in vitro* and *in vivo* results is the accumulation of antibiotics

TABLE BXII.γ.61. Serogroups of *Legionella* species[a]

Species	Serogroup	Type strain	Source of isolate	Reference
L. pneumophila	1	Philadelphia 1	Human lung	McKinney et al., 1979a
	2	Togus 1	Human lung	McKinney et al., 1979a, b
	3	Bloomington 2	Creek water	McKinney et al., 1979a
	4	Los Angeles 1	Human lung	McKinney et al., 1979a
	5	Dallas 1-E	Cooling tower water	England et al., 1980
	6	Chicago 2	Human lung	McKinney et al., 1979b
	7	Chicago 8	Environment	Bibb et al., 1983
	8	Concord 3	Human lung	Bissett et al., 1983
	9	N-23-G1-C2	Human lung	Edelstein et al., 1984
	10	Leiden 1	Human lung	Meenhorst et al., 1985
	11	797-PA-H	Human lung	Thacker et al., 1986
	12	570-CO-H	Human lung	Thacker et al., 1987
	13	82A3105	Human lung	Lindquist et al., 1988
	14		Human lung	Benson et al., 1988
	Lansing 3	Lansing 3	Human lung	Brenner et al., 1988c
	JENA-1	JENA-1	Hot water system	Luck et al., 1995
L. bozemanii	1	WIGA	Human lung	Brenner et al., 1980c
	2	Toronto 3	Human lung	Tang et al., 1984
L. erythra	1	SE-32A-C8	Cooling tower water	Brenner et al., 1985
	2	LC217	Water	Saunders et al., 1992
L. feeleii	1	WO-44C	Coolant water	Herwaldt et al., 1984a
	2	691XWI-H	Human lung	Thacker et al., 1985
L. hackeliae	1	Lansing 2	Human lung	Brenner et al., 1985
	2	798-PA-H	Human lung	Wilkinson et al., 1985b
L. longbeachae	1	Long Beach 4	Respiratory secretions	McKinney et al., 1981
	2	Tucker 1	Human lung	Bibb et al., 1981
L. quinlivanii	1	1442-AUS-E	Air Conditioning System	Benson et al., 1989
	2	LC870	Cooling tower water	Birtles et al., 1991
L. sainthelensi	1	Mt. St. Helens 4	Water	Campbell et al., 1984a
	2	1489-CA-H	Human bronchial washing	Benson et al., 1990a
L. spiritensis	1	Mt. St. Helens 9	Water	Brenner et al., 1985
	2	ML76	Environment	Harrison et al., 1988

[a]NA, not available.

within phagocytic cells. The most effective antibiotics are those that are effective *in vitro* and are also concentrated in macrophages.

Erythromycin is the drug of choice in treating human infection and should be given for a 3-week course (Kirby et al., 1978). Although there are few data, it has been suggested that rifampin should be used in combination with erythromycin when patients respond poorly to erythromycin alone or when serious disease is present (Cordes and Fraser, 1980).

All *Legionella* species are potentially pathogenic for humans. Antibodies to *Legionella* spp. are found in several species of domestic animals, but definitive evidence of active infection is lacking. Isolation of the bacterium has been accomplished only rarely and in a setting where autopsy contamination by water cannot be excluded completely (Boldur et al., 1987). Guinea pigs are easily infected by *Legionella* spp., and lethal infection often results (McDade et al., 1977; Winn et al., 1982). Rhesus monkeys and marmosets develop a fibrinopurulent pneumonia after infection with an aerosol of *L. pneumophila* (Baskerville et al., 1983). Mice and rats are variably resistant to the development of active infection, depending on the strain of rodent (Yamamoto et al., 1991b).

Disease caused by any *Legionella* species is often termed legionellosis. The two main types of legionellosis caused by *Legionella* are *Legionella* pneumonia (LP) (Fraser et al., 1977) and Pontiac fever (Glick et al., 1978). Most clinical, epidemiologic, and diagnostic data pertain to the most commonly isolated species, *L. pneumophila*; the limited data available for other species suggest that they follow the same pattern. The pneumonia has a 2- to 10-d incubation period after a presumed airborne ex-

posure to the organism. There is no proof of person-to-person spread. The earliest symptoms include malaise, myalgia, headache, and often a nonproductive cough. Rapidly rising fever and chills usually appear within a day, at which time there are definite clinical and radiologic indications of pneumonia. Other common symptoms are chest pain, abdominal pain, vomiting, diarrhea, and mental confusion. Pathologically there is focal, lobular pneumonia, which may coalesce and even produce a lobar distribution (Winn and Myerowitz, 1981). The inflammatory response usually consists of a mixture of polymorphonuclear neutrophils and macrophages. Treatment with antibiotics is necessary; untreated, the fatality rate is 15% or higher.

Subclinical and mild cases may not be recognized unless they are a part of a larger epidemic. Smoking, preexisting respiratory illness, and conditions where the patient is immunocompromised are predisposing factors. LP occurs most often in men above the age of 50. It has been estimated that there are 8000–18,000 cases of LP in the United States each year, many not recognized and/or reported (Marston et al., 1994). The disease occurs worldwide. LP has been caused by approximately half of the recognized species, and could potentially be caused by any if the patient is sufficiently immunocompromised.

Legionella pneumonia occurs as sporadic or epidemic disease, which may be acquired in the community (Storch et al., 1979) or as a nosocomial infection (England and Fraser, 1981). Remarkable differences in the incidence of LP in similar patient populations presumably relate to the frequency and quantity of virulent strains in the local environment (Goldstein et al., 1982). Subclinical disease occurs, even in severely immunocompromised patients (Jacobs et al., 1990). Mixed infections with multiple

Legionella species (Dowling et al., 1983) occur. Differing serologic types have also been isolated from human lung (Joly and Winn, 1984), although it has been suggested that strains with differing antigenic characteristics may derive from a common parent strain (Harrison et al., 1990).

Legionella spp. also produce acute, nonpneumonic, febrile illness accompanied by chills, headache, and myalgia (Pontiac fever). It has a high attack rate and a short incubation period of hours to several days. It is, by definition, an epidemic disease. Pontiac fever is not accompanied by clinically detectable disease of the lower respiratory tract. It is self-limiting and not fatal. Pontiac fever has been caused by *L. pneumophila* (Glick et al., 1978), *L. feeleii* (Herwaldt et al., 1984a), *L. micdadei* (Goldberg et al., 1989), and *L. anisa* (Fenstersheib et al., 1990). Severe *Legionella* pneumonia and nonpneumonic disease (differing from Pontiac fever in the sporadic nature of the infection) have occurred after exposure to the same environmental source (Girod et al., 1982).

All species of *Legionella* have been isolated from the environment. Natural environmental isolates have come from streams or lakes, sometimes heated by natural or industrial thermal source. *L. pneumophila* has been identified from moist soil adjacent to a body of water (Fliermans et al., 1981), and *L. longbeachae* has been recovered from moist potting soil (Steele et al., 1990). Although a variety of environmental cofactors have been described, it is becoming increasingly clear that the association of legionellae with a variety of free-living protozoa is the most important factor (Rowbotham, 1986; Fields, 1993). Isolation of strains may be accomplished only by coculture with amoebae, both for human (Rowbotham, 1983) and environmental (Sanden et al., 1992) sources.

There have been no isolates from dry soil. Urban environmental isolates were made from water collected from air conditioning cooling towers and evaporative condensers (Orrison et al., 1981). More recently, colonization of potable water systems by *Legionella* spp. has been frequently demonstrated., with isolates being more common on the hot water than the cold water side. The sediment of hot water heaters are a particularly productive source for isolates, especially if the water temperature has been lowered to minimize scalding and reduce heating costs. Water heaters in which the heating elements are remote from the bottom of the tank where the sediment accumulates are more likely to yield isolates than units with heating elements in the bottom of the tanks (Alary and Joly, 1991). It is interesting that the portions of municipal water systems that have been colonized are all peripheral; contamination of central water distribution facilities, where chlorination is performed, has not been demonstrated. Dissipation of chlorine as water flows through the system may explain this phenomenon.

Not surprisingly, epidemic disease can usually be traced to exposure to a contaminated water source. Sporadic cases are likely related to similar exposures, but investigation is more difficult. The contaminated water may be either environmental surface water or potable water. Factors that favor production of disease include a host who is compromised by immunosuppression or chronic cardiopulmonary disease, a vehicle for generating aerosols (which may be quite local), and undoubtedly the inoculum and virulence of the bacteria. The most vulnerable patients are in health care centers, where epidemic disease has been concentrated. Epidemics have also been associated, however, with hotels, cruise ships, and even grocery stores.

The primary hazard to researchers working with *Legionella*

species is exposure to aerosols, although self-inflicted wounds are a potential hazard. In the absence of large-scale aerosols, as might be produced in an experimental situation, the usual precautions for class II pathogens are sufficient. Organisms isolated on agar plates can be safely studied at the bacteriology bench. There are no documented occupational infections in standard diagnostic bacteriology laboratories.

ENRICHMENT AND ISOLATION PROCEDURES

The diagnosis of *Legionella* infection can be made by three methods, which are in decreasing order of importance, isolation of the organism in culture, direct identification of antigens or nucleic acid in tissues or body fluids, and demonstration of an immune response (Zuravleff et al., 1983).

Human specimens Lung tissue, lung aspirates, pleural fluid, transtracheal aspirate or sputum, and bronchoalveolar lavage fluid are the specimens usually cultured when *Legionella* is suspected (Edelstein et al., 1980). Samples that cannot be processed within 2 h of collection can be stored on wet ice or refrigerated for 2 d. Samples that will not be cultured within a reasonable time (2–5 d) should be frozen at −70°C. Ten percent tissue suspensions in distilled water or bacteriologic broth, body fluids, or sputum should be inoculated onto media designed to grow legionellae, such as BCYE-α. If indigenous flora is likely to be present, inoculate media with and without addition of antibiotics for selective inhibition of nonlegionellae. Media with selective antibiotics should never be inoculated without an accompanying noninhibitory plate.

Inhibition of *Legionella* growth, presumably by antibiotics or toxic compounds in the specimen, may be documented as a zone of no growth where the specimen was placed on the agar, in conjunction with heavy growth in adjacent areas, where the inoculum had been spread over the surface of the agar. Homogenization and acid treatment of respiratory secretions increased the number of isolates from sputum, but not bronchoalveolar lavage fluid in one study (Buesching et al., 1983), but this finding has not been reported by other investigators. Brief heat treatment of specimens also selectively inhibits contaminating flora, but *Legionella* spp. may also be inhibited (Edelstein et al., 1982b).

Blood cultures may be positive in severely ill patients who have high-level bacteremia (Rihs et al., 1985), but the sensitivity in mild to moderate cases is not known. Continuous-reading blood culture instruments do not register growth when *Legionella* spp. are present; staining of fluid with acridine orange or blind subculture onto BCYE-α agar may be required to document the presence of bacteria. *In vitro* experiments with spiked blood samples suggest that the lysis-centrifugation method, a convenient means of culturing blood for fastidious pathogens, may be used to recover legionellae from blood efficiently, but clinical confirmation is lacking.

Each batch of media must be tested for pH and sensitivity for primary isolation of legionellae. The pH of solid media is determined with a surface electrode or by emulsifying the agar from one plate in distilled water. The quality of media from various manufacturers can vary (Lee et al., 1993b). If media are commercially prepared, the manufacturer should supply documentation of successful quality control. If the media are prepared locally or if the sensitivity of media for primary isolation is in doubt, a control should be prepared from a freshly isolated strain that has been minimally passed on BCYE-α agar. After *in vitro* passage on non-charcoal-containing agar, strains with altered growth characteristics emerge. These strains will grow well on batches of media that do not support the growth of bacteria in fluid or tissue; a false sense of security as to the adequacy of the media may result (Keathley and Winn, 1985). Media without antibiotics, stored refrigerated in the dark, in sealed, plastic containers remain stable for at least 2 months. Users should follow manufacturers' instructions for storage and outdates of commercially prepared media.

Media for isolation of *Legionella* spp. should be incubated at 35–37°C. Incubation in an atmosphere of 2–5% CO_2 provides marginal enhancement of isolation for some *Legionella* spp., but higher concentrations of CO_2 may be inhibitory. Plates should be held for at least 5–7 d before discarding. Colonies are usually evident after incubation for 48–72 h. Use of a stereomicroscope will facilitate detection of colonies and allow assessment of fine features of colonial morphology. *Legionella* spp. produce smooth, entire colonies that exhibit a mottled, cut-glass appearance when viewed with a stereomicroscope under reflected light. Red, blue, or green iridescence may be visible early with a stereomicroscope and later with the naked eye in areas of heavy growth. *L. pneumophila* produces sticky colonies that "string" when a portion of the colony is lifted with an inoculating needle or loop.

Environmental samples   A comprehensive manual for culture of environmental specimens is available from the Centers for Disease Control and Prevention (CDC) (Centers for Disease Control and Prevention, 1992). Environmental, nonpotable water samples do not require concentration. Potable water samples must be concentrated either by filtration, as recommended by CDC, or by centrifugation (Brindle et al., 1987). Collection of at least 1 l of potable water is desirable, but as many as 10 l may be required if small numbers of legionellae are present. For filtration the water should be passed through a 0.2-μm polycarbonate filter, after which the filter is folded to the outside, placed in 10 ml of water, vortexed, and subcultured onto solid media. Serial dilutions may be performed to quantify the bacterial load. Plates are incubated for 7 d in 2.5–5.0% CO_2. Samples from faucets and showers may be taken with polyester swabs that are immersed in 3–5 ml of water; the faucet aerator or shower head should be removed if possible. Samples of sediment from hot water heaters should be plated on solid agar. Detailed instructions for recovering legionellae from air samples have been provided by CDC (Centers for Disease Control and Prevention, 1992).

Environmental samples, especially those from nonpotable water, are frequently contaminated with a variety of bacterial species. A variation on BCYE agar that includes glycine has proved useful for culturing environmental specimens (Wadowsky and Yee, 1981). Addition of selective antibiotics provides an additional competitive advantage for *Legionella* spp. Antibiotic combinations have included cefamandole or vancomycin for inhibition of Gram-positive flora, polymyxin B for Gram-negative flora, and anisomycin for yeast. Cefamandole may inhibit many isolates of *Legionella*, so selective plates should not be used alone. As has been recommended for sputum, brief acidification of the sample to pH 2.0 for 5 min followed by neutralization may provide additional suppression of contaminating environmental flora (Centers for Disease Control and Prevention, 1992). Heat enrichment of environmental specimens that were negative for legionellae on primary culture may yield as many as 30% additional isolates, putatively because the bacteria multiply in free-living amoebae in the samples. Tubes of water are incubated in closed tubes at 35°C, followed by periodic subculture onto agar.

If necessary, the aliquots for subculture may be treated with acid before testing.

MAINTENANCE PROCEDURES

Stock cultures on BCYE-α agar slants are recommended for maintenance of stock cultures. Inoculated slants are incubated for 2 d at 35°C and then stored in the dark at room temperature. Under these conditions, *L. pneumophila* cultures remain viable for 1–3 months (Hébert, 1980). Other species remain viable for variable periods from 4 months to 1 year. Lyophilization or storage at −50°C or lower is recommended for long-term maintenance (Feeley and Weaver, personal communication); a convenient and effective method for storage of isolates is to excise a piece of agar containing isolated colonies, after which the agar is frozen in a sterile container at −50°C or less (W.C. Winn, Jr., personal observation). Virulence and plasmids are often lost on lyophilization or subculture on non-charcoal-containing agar (McDade and Shepard, 1979). This conversion to avirulence is a one-way phenomenon and does not appear to occur if one of the charcoal agars, such as BCYE-α agar, is used for maintenance (Catrenich and Johnson, 1988).

PROCEDURES FOR TESTING SPECIAL CHARACTERS

Direct Detection of Antigen in Specimens

Human specimens   The first method used to detect *Legionella* antigen in clinical specimens was direct immunofluorescence, using polyclonal antisera. Commercially available polyclonal conjugates are available for some species and serotypes. Cross-reactions have been demonstrated with *Pseudomonas aeruginosa* and *Bacteroides fragilis* (Edelstein and Edelstein, 1989), *Bordetella pertussis* (Benson et al., 1987), *Francisella tularensis* (Roy et al., 1989), and with spores of *Bacillus cereus* (Flournoy et al., 1988). A commercially available monoclonal antibody to an antigen that is specific for the species *L. pneumophila* provides cleaner staining and fewer cross-reactions than the polyclonal antisera, but monoclonal reagents are not available for other species (Edelstein et al., 1985; Tenover et al., 1985). The direct immunofluorescence test has been discontinued in many laboratories because of low sensitivity (Edelstein et al., 1980; Buesching et al., 1983). Although the specificity is high, false-positive results cause unacceptably low predictive values if the prevalence of disease is very low. In addition, legionellae in water used to make reagents may cause clusters of false-positive tests. Radioimmunoassays and enzyme immunoassays have been developed for detection of *L. pneumophila* serogroup 1 antigen in urine. A commercially available enzyme immunoassay, using polyclonal antiserum, (BINAX, Portland, Maine) has a sensitivity of 70–80% and a specificity of 90% (Kazandjian et al., 1997). *L. pneumophila* antigen may be excreted for weeks or months, so that detection of antigen does not always indicate current disease (Kohler et al., 1984).

Environmental specimens   Initially direct immunofluorescence was commonly used for detection of legionellae in water samples. Many specimens that were positive by immunofluorescence could not be confirmed by culture, probably reflecting both insensitivity of early culture techniques, fastidious or nonreplicating legionellae, and/or immunologic crossreactions. The preferred method for testing environmental samples is culture.

Direct Detection of Nucleic Acid in Specimens

Human specimens   A direct nucleic acid probe for *L. pneumophila* was marketed briefly, but withdrawn because of unacceptably frequent false-positive reactions. There are no commercially available molecular amplification tests for clinical specimens, but amplification procedures that are approved for environmental specimens have been used successfully to detect legionellae in respiratory secretions (Kessler et al., 1993).

Environmental specimens Nucleic acid amplification, using the polymerase chain reaction, has been used successfully (Jaulhac et al., 1992), and an amplification kit is commercially available, although not approved for use with human specimens (Kessler et al., 1993; Oshiro et al., 1994).

Serologic Diagnosis

Human infections The indirect immunofluorescent antibody (IFA) test is the test of choice for serologic diagnosis of human infections. Although other methods have been used, appreciable experience has been accumulated only with the IFA test. Either heat-killed or formalin-inactivated bacterial antigen may be employed (Wilkinson and Brake, 1982). It is important to use antiglobulin conjugates that are active against both IgG and IgM (Wilkinson et al., 1979a). If the criterion for a positive test is a fourfold increase in paired sera to a titer of at least 1:128, the sensitivity of the test is 70–80% (Edelstein et al., 1980), and the specificity is approximately 99% (Wilkinson et al., 1981). The details of IFA are given by Wilkinson et al. (1979b). Sera should be examined for at least 6 weeks to demonstrate seroconversion; some patients do not seroconvert for 1 year or longer (Edelstein et al., 1980). A titer of 1:256 in a single serum has been used as presumptive evidence of *Legionella* infection during an epidemic, but the diagnosis of sporadic cases cannot be made with a single serum specimen. An elevated IgM titer suggests recent infection (Zimmerman et al., 1982), but these titers may persist for as long as a year in some patients (Zimmerman et al., 1982). Antisera are available for certain serogroups of *L. pneumophila* and for selected other species. Cross-reactions among *Legionella* serogroups and species make it impossible to assign a precise diagnosis with certainty (Wilkinson et al., 1983a). In addition, cross-reactions have been described with other bacterial species, including *Campylobacter* spp. (Boswell and Kudesia, 1992), *Pseudomonas* spp. (Edelstein et al., 1980; Klein, 1980; Collins et al., 1984), and *Bacteroides fragilis* (Edelstein et al., 1980). Under the best of circumstances, therefore, serologic diagnosis is only presumptive.

Animals Serologic studies have demonstrated anti-*Legionella* antibodies in a variety of animals (Collins et al., 1982). In the absence of bacterial isolates, it is difficult to interpret the serologic data.

Molecular Epidemiology of *Legionella* Infections It is not possible to ascribe an etiologic role to a *Legionella* isolate, simply because the same species has been recovered from human and environmental sources. An exhaustive investigation of all potential environmental sites must be undertaken. The likelihood of an etiologic association is increased by application of a variety of subtyping tools: monoclonal antibody typing (Joly and Winn, 1984; Gomez-Lus et al., 1993), ribotyping (Schoonmaker et al., 1992; Gomez-Lus et al., 1993), restriction endonuclease treatment with pulsed field electrophoresis (Saunders et al., 1991; Schoonmaker et al., 1992), polymerase chain reaction, arbitrarily primed polymerase chain reaction (Gomez-Lus et al., 1993; Ledesma et al., 1995), and alloenzyme analysis (Edelstein et al., 1986; Tompkins et al., 1987). The greatest discrimination comes from applying multiple typing tools (Barbaree, 1993; Mamolen

et al., 1993). It has been suggested that certain monoclonal antibody types of *L. pneumophila* are associated with increased virulence (Dournon et al., 1988; Helbig et al., 1995).

Media Legionellae grow on BCYE agars. They do not grow on standard blood agar, nutrient agar, or other commonly employed laboratory media that do not contain both L-cysteine and iron salts. To document the nutritional pattern of a fresh isolate, parallel subcultures should be made to BCYE-α and either sheep blood agar or BCYE-α agar without L-cysteine. Growth only on BCYE-α agar with L-cysteine suggests the possibility of *Legionella* spp. Legionellae must be differentiated phenotypically from two groups of organisms that grow on some modified charcoal media that contain L-cysteine, but not on common laboratory media. The first, and most significant, for clinical diagnosis and laboratory safety is *F. tularensis* (Westerman and McDonald, 1983; K. McLeod, C.P. Patton, and J.C. Feeley, unpublished data), and possibly *F. novicida*. *Francisella* spp. are easily distinguished from *Legionella* spp. because they produce acid from carbohydrates and do not contain branched chain fatty acids (Jantzen et al., 1979). *F. tularensis* constitutes a major biohazard for unsuspecting laboratories, so this possibility must be kept in mind. A number of thermophilic, spore-forming bacilli that stain Gram-negative mimic legionellae on BCYE agars and do not grow on common laboratory media. They also have branched chain fatty acid profiles that superficially resemble those of one or more *Legionella* species (Thacker et al., 1981). These strains are distinguished from legionellae by their ability to grow at 50°C or higher, to form spores, and to grow on certain media that lack cysteine, such as tryptone broth (Thacker et al., 1981; Weaver, unpublished data). These environmental organisms rarely, if ever, produce human disease, but may cause confusion if they contaminate reagents or specimens.

Pigments Pigments are of minimal value in identification of legionellae. Brown pigment is formed on supportive clear agars that contain tyrosine (Baine et al., 1978). Yellow, blue-white, or red autofluorescence may be seen with some species if the colonies are examined in the dark with a long-wavelength ultraviolet lamp (Wood's light) (Tables BXII.γ.56 and BXII.γ.57).

Serologic identification Serologic characterization of isolates, most commonly by the direct immunofluorescent antibody (DFA) test, is the most common method of identifying isolated strains. Polyclonal antisera, probably directed primarily against a lipopolysaccharide antigen, provide a serotype-specific diagnosis. Reagents are available commercially for selected serogroups of *L. pneumophila*, *L. micdadei*, and many other species (Hookey et al., 1995). A commercially available monoclonal antibody to a species-specific antigen of *L. pneumophila* provides more consistent and easily interpreted results (Edelstein et al., 1985; Tenover et al., 1985, 1986). Although serologic identification is sufficiently accurate to use in clinical and environmental laboratories, cross-reactions among serotypes and species do occur occasionally. The only certain way to establish the nature of an isolate is by nucleic acid analysis.

A slide agglutination test can also be used for the rapid identification of legionellae (Wilkinson and Fikes, 1980), especially in laboratories that are unable to perform immunofluorescence exams. Unfortunately, the agglutination test is not commercially available in the United States.

Gas-liquid chromatography (GLC) GLC analysis of cellular fatty acids is a method used, along with DFA, to test suspect isolates and to confirm the identification of known legionellae, new serogroups, and new species (Moss and Dees, 1979; Lambert and Moss, 1989) (Table BXII.γ.58). The procedure is, thus far, totally specific in identifying species of *Legionella* at the genus level and can identify most, but not all, legionellae to the species level. The procedure involves saponification of whole cells, methylation of the fatty acids, extraction of the resultant methyl esters with diethyl ether and then hexane, followed by GLC analysis on a nonpolar stationary-phase column. The retention time of the resultant elution peaks is recorded, and tentative identification is made by comparison to reference standards. Confirmation and quantitation of the fatty acids are obtained by subsequent procedures, including mass spectrometry, use of polar columns, hydrogenation, and acetylation.

Characterization of cellular ubiquinones is also useful in characterizing *Legionella* species (Table BXII.γ.59). The genus contains varying amounts of ubiquinones with 9-14 isoprene units in the side chains (Lambert and Moss, 1989).

DNA relatedness This is the only method whereby one can unequivocally identify legionellae to the level of species, confirm the existence of new serogroups in existing species, and identify new species. DNA relatedness, however, is not adaptable to routine laboratory use and should be used only for isolates that cannot be characterized with certainty by biochemical, antigenic, and fatty acid analysis.

DIFFERENTIATION OF THE GENUS *LEGIONELLA* FROM OTHER GENERA

Legionella pneumophila has no close relatives in other genera based on DNA relatedness (Brenner et al., 1979). The original four strains that we now know to represent three separate *Legionella* species were isolated between 1943 and 1959 (Tatlock, 1944; Jackson et al., 1952; Bozeman et al., 1968). At the time of their discovery, however, none of these strains was cultivatable on bacteriologic media. They were treated as rickettsiae and not further studied until 1978 (McDade et al., 1979; Hébert et al., 1980a,d). The "modern" study of what was to become the genus *Legionella* began in 1976 with the investigation of a large outbreak of respiratory disease at an American Legion Convention in Philadelphia, Pennsylvania (Fraser et al., 1977), resulting in the isolation of the causative agent, *L. pneumophila*, by Joseph McDade (McDade et al., 1977). Shortly thereafter, the Flint 1 strain was isolated as an unknown on bacteriologic media by Dumoff (Dumoff, 1979) and the Philadelphia strains were isolated on bacteriologic media by Weaver (Feeley et al., 1978). The first species, *L. pneumophila*, was named in 1979 (Brenner et al., 1979).

The named species of *Legionella* have similar phenotypic characteristics (Tables BXII.γ.56 and BXII.γ.57). These include growth on media designed for legionellae, optimum pH and temperature for growth, common flagellar antigens, pathogenicity for guinea pigs and embryonated hen's eggs, nonfermentative metabolism, presence of catalase, absence of urease, and *in vitro* antibiotic susceptibility patterns. The cellular fatty acid compositions of all species are characterized by relatively large amounts of branched chain acids. Approximately half of the recognized species have been directly or indirectly implicated as causative agents of human pneumonia. Available microbiological and biochemical tests are insufficient to separate all species with certainty. Once a species has been defined, it is quite simple to identify species and serogroups by DFA. As the number of recognized species has expanded, so has the number of immunological cross-reactions (Table BXII.γ.60). It is still appropriate to

identify clinical isolates by DFA, particularly if the species is *L. pneumophila*, but the possibility of cross-reactions should be recognized.

TAXONOMIC COMMENTS

A frequently used definition of a "genetic" species is "a group of strains whose DNAs are 70% or more related at optimal reassociation conditions and have 5% or less divergence in related sequences". Stackebrandt and Goebel (1994) consider that strains that have at least 70% relatedness will also have at least 96% sequence identity when conserved portions of 16S rRNA are analyzed. The technology of DNA hybridization is costly and time-consuming, whereas genetic sequences can now be determined quickly and relatively cheaply. Analysis of 16S rRNA, however, does not have the discriminatory ability of DNA–DNA hybridization at the species level. Stackebrandt and Goebel suggest that analysis of 16S rRNA genes be used as a screening tool to select likely candidates for hybridization studies.

Legionella species fulfill the current definition of bacterial species. DNA relatedness cannot be used to generate good genetic definitions for a genus or family. The most common method for establishing relationships at the genus level is now sequencing of 16S rDNA (Wayne et al., 1987). Some of the amoeba-associated pathogens, which are not available as isolated bacteria, are closely related to other *Legionella* species by analysis of 16S rDNA; others, representing as yet unnamed species, are more distantly related (Fry et al., 1991a, b).

New species and serogroups have been confirmed by DNA relatedness. Relatedness between strains of the same species is 75% or higher for all species except *L. bozemanii*, for which relatedness is 56–77%. The related sequences of all species, including *L. bozemanii*, exhibit 3% or less divergence (unpaired bases within heteroduplex DNA), and there is only a small decrease in relatedness when reactions are done at a supraoptimal incubation temperature. Using another approach to molecular taxonomy, Ludwig and Stackebrandt (1983) studied the phylogeny of four *Legionella* species that were closely related. Analysis of 16S rDNA from seven *Legionella* species demonstrated that they were closely related (>95%) to each other and that they fell into the *Gammaproteobacteria* (Fry et al., 1991b). Using several technical variations for detection of rDNA, several investigators have confirmed and extended these observations (Hookey et al., 1995, 1996; Birtles et al., 1996; Riffard et al., 1998).

A proposal was made to divide the family *Legionellaceae* into three genera: *Legionella*, *Tatlockia*, and *Fluoribacter* (Garrity et al., 1980; Brown et al., 1981). Other investigators have not supported the proposal and have recommended that a single genus, *Legionella*, be maintained (Brenner et al., 1985; Brenner, 1986). Analysis of 16S ribosomal RNA supported existence of a single genus (Fry et al., 1991b). Most importantly, the single-genus concept has been uniformly accepted within the community of microbiologists.

FURTHER READING

Dowling, J.N., Saha, A.K. and Glew, R.H.. 1992. Virulence factors of the family *Legionellaceae*. Microbiol. Rev. *56*: 32–60.

Edelstein, P.H., Meyer, R.D. and Finegold, S.M.. 1980. Laboratory diagnosis of Legionnaires' disease. Am. Rev. Respir. Dis. *121*: 317–327.

Edlestein, P.H. 1993. Legionnaires' disease. Clin. Infect. Dis. *16*: 741–747.

Eisenstein, B.I. and Engleberg, N.C.. 1986. Applied molecular genetics: new tools for microbiologists and clinicians. J. Infect. Dis. *153*: 416–430.

Kirby, B.D., Snyder, K.M., Meyer, R.D. and Finegold, S.M.. 1980. Legionnaires' disease: report of sixty-five noscomially acquired cases and review of the literature. Medicine. *59*: 188–205.

Stout, J.E. and Yu, V.L.. 1997. Current concepts: Legionellosis. N. Engl. J. Med. *337*: 682–687.

Winn, W.C. 1988. Legionnaires' disease: historical perspective. Clin. Microbiol. Rev. *1*: 60–81.

Winn, W.C 1999. *Legionella. In* Murray, Baron, Pfaller, Tenover and Yolken (Editors), Manual of Clinical Microbiology, ASM Press, Washington D.C. p.572–585.

DIFFERENTIATION OF THE SPECIES OF THE GENUS *LEGIONELLA*

Selected characteristics useful for differentiating legionellae from other organisms and from one another are presented in Tables BXII.γ.56 and BXII.γ.57. Cellular fatty acids of *Legionella* species are shown in Table BXII.γ.58, ubiquinone content in Table BXII.γ.59, and DNA relatedness data in Table BXII.γ.62.

Immunologic cross-reactions are detailed in Table BXII.γ.60, and the serogroups of Legionella species are described in Table BXII.γ.61. Morphology is shown in Figs. BXII.γ.94, BXII.γ.95, and BXII.γ.96.

List of species of the genus Legionella

1. **Legionella pneumophila** Brenner, Steigerwalt and McDade 1979, 658.[AL]

pneu.mo' phi.la. Gr. n. *pneumo* lung; Gr. adj. *philos* loving; M.L. adj. *pneumophila* lung-loving.

The characteristics are as described for the genus and as listed in Tables BXII.γ.56, BXII.γ.57, BXII.γ.58, and BXII.γ.59. Hydrolysis of hippurate is not positive for all strains of *L. pneumophila* (Hébert, 1981). Serologic cross-reactions are detailed in Table BXII.γ.60. The species contains 15 serogroups (Table BXII.γ.61). Causative agent of legionellosis, presenting either as a pneumonia with a high fatality rate in untreated patients (Legionnaires' disease) or as an acute, self-limited, febrile illness (Pontiac fever). Isolated from human lung, bronchoalveolar lavage fluid, sputum, blood, intravascular devices, peritoneal fluid, abscess fluids, and potable and surface water. DNA relatedness between strains of *L. pneumophila* is 75–100%. *L. pneumophila* is 4–21% related to other *Legionella* species (Table BXII.γ.62). The fatty acid composition of *L. pneumophila* closely resembles that of *L. longbeachae*, and the two species cannot be distinguished by this technique (Bibb et al., 1981). The phenotype of strains, derived by serogrouping, does not always match their genotype. Strains within the same serogroup may be genetically disparate, whereas strains in different serogroups may be closely related genetically. *L. pneumophila* was divided into three DNA groups, based on a study of 60 strains (Brenner et al., 1988c). Subspecies designations were proposed for these nucleic acid groups.

The mol% G + C of the DNA is: 39 (T_m, Bd).

Type strain: Philadelphia 1, ATCC 33152, DSM 7513.

GenBank accession number (16S rRNA): M59157, M36023.

Additional Remarks: The type strain was isolated from human lung tissue by McDade and Weaver in 1977. The oldest isolate is strain OLDA isolated by E. Jackson in 1947 (Jackson et al., 1952).

a. **Legionella pneumophila** *subsp.* **pneumophila** (Brenner, Steigerwalt and McDade 1979) Brenner, Steigerwalt, Epple, Bibb, McKinney, Starns, Colville, Selander, Edelstein and Moss 1989, 205[VP] (Effective publication: Brenner, Steigerwalt, Epple, Bibb, McKinney, Starns, Selander, Edelstein and Moss 1988c, 1701) (*Legionella pneumophila* Brenner, Steigerwalt and McDade 1979, 658.)

pneu.mo' phi.la. Gr. n. *pneumo* lung; Gr. adj. *philos* loving; M.L. adj. *pneumophila* lung-loving.

This subspecies represents DNA group I and contains strains in serogroups 1–10. All strains hydrolyze sodium hippurate. Strains of *L. pneumophila* subsp. *pneumophila* are >70% related to each other under optimal conditions (60°C) with ≤3.0% divergence in related sequences; they are >60% related under stringent conditions (75°C). DNA relatedness of these strains to other subspecies is below the species level in at least one and usually two of these hybridization parameters. Isolated from human lung tissue.

The mol% G + C of the DNA is: 39–41 (T_m); type strain: 39 (T_m).

Type strain: Philadelphia 1, ATCC 33152, DSM 7513.

GenBank accession number (16S rRNA): M59157, M36023.

b. **Legionella pneumophila** *subsp.* **fraseri** Brenner, Steigerwalt, Epple, Bibb, McKinney, Starns, Colville, Selander, Edelstein and Moss 1989, 205[VP] (Effective publication: Brenner, Steigerwalt, Epple, Bibb, McKinney, Starns, Colville, Selander, Edelstein and Moss 1988c, 1702.)

fra' ser.i. M.L. gen. n. *fraseri* in honor of D.W. Fraser, who headed the investigation of the Legionnaires' disease outbreak in Philadelphia that led to the recognition of legionellosis and isolation of *L. pneumophila*.

This subspecies represents DNA group II and contains strains from serogroups 1, 4, 5, and Lansing 3. Strains from serogroups 1 and 5 hydrolyze sodium hippurate; strains from serogroups 4 and Lansing 3 do not hydrolyze hippurate. Strains of *L. pneumophila* subsp. *fraseri* are >65% related to each other under optimal conditions with <3% divergence and >60% related under stringent conditions. DNA relatedness of these strains to other subspecies is below the species level in at least one and usually two of these hybridization parameters. Isolated from human lung tissue.

The mol% G + C of the DNA is: 38–39 (T_m); type strain: 39 (T_m).

Type strain: Los Angeles 2, ATCC 33156, DSM 7514.

GenBank accession number (16S rRNA): M36025.

c. **Legionella pneumophila** *subsp.* **pascullei** Brenner, Steigerwalt, Epple, Bibb, McKinney, Starns, Colville, Selander, Edelstein and Moss 1989, 205[VP] (Effective publication: Brenner, Steigerwalt, Epple, Bibb, McKinney, Starns, Colville, Selander, Edelstein and Moss 1988c, 1702.)

pas.cul' le.i. M.L. gen. n. *pascullei* in honor of A.W. Pasculle for his pioneering studies on media, isolation, and nosocomial spread of *L. pneumophila* and *L. micdadei*.

This subspecies represents DNA group III that hydrolyze sodium hippurate. Strains of subsp. *pascullei* are

TABLE BXII.γ.62. Base composition and DNA relatedness within and among species under optimal conditions

Legionella species	Mol% G + C Content	% Relatedness within species under optimal conditions	% Relatedness with other species under optimal conditions	Closest relative(s)	16S-23S rDNA Group[a]
L. adelaidensis	40	NA[b]	≤31	None	Distinctive
L. anisa	42	84–99	5–57	L. bozemanii	A
L. birminghamensis	42.9	NA	1–3	None	Distinctive
L. bozemanii	41–43	56–73	3–65	L. parisiensis	A
L. brunensis	41.5	NA	1–8	None	Distinctive
L. cherrii	39–40	94–100	6–67	L. steigerwaltii, L. bozemanii, L. dumoffii, L. gormanii, L. anisa, L. parisiensis	A
L. cincinnatiensis	36.3	NA	<44	L. santicrucis, L. longbeachae, L. sainthelensi	Distinctive
L. dumoffii	41–42	90	3–57	L. steigerwaltii, L. cherrii, L. parisiensis	A
L. erythra	51	NA	5–59	L. rubrilucens	B
L. fairfieldensis	42	87–94	1–17	None	Distinctive
L. feeleii	46	78–100 (stringent conditions)	≤15	None	Distinctive
L. geestiana	52	NA	≤11	None	Distinctive
L. gormanii	41	>90	6–47	L. cherrii, L. parisiensis, L. steigerwaltii	A
L. gratiana	38	NA	3–46	L. longbeachae, L. cincinnatiensis, L. sainthelensi, parisiensis	Distinctive
L. hackeliae	40	NA	3–22	L. cherrii, L. jamestowniensis	Distinctive
L. israelensis	40.9	95–100	≤12	None	Distinctive
L. jamestowniensis	42	NA	≤17	None	Distinctive
L. jordanis	45	93–100	2–20	L. spiritensis	Distinctive
L. lansingensis	42.6	NA	1–12	None	Distinctive
L. londiniensis	43	94–100	0–36	None	Distinctive
L. longbeachae	40	97–100	2–22	L. cherrii	Distinctive
L. lytica; amoeba-associated strains	NA	NA	NA	NA	NA
L. maceachernii	41	100	3–23	L. micdadei	Distinctive
L. micdadei	41–44	>90	1–23	L. maceachernii	Distinctive
L. moravica	41.5	NA	1–9 (named species); 24–36 (unnamed species)	None	D
L. nautarum	41	70–100	2–24	L. micdadei	Distinctive
L. oakridgensis	43–45	96–100	2–22	L. spiritensis	Distinctive
L. parisiensis	42	NA	NA	L. bozemanii, L. dumoffii, L. gormanii, L. cherrii, L. steigerwaltii, L. wadsworthii (37)	A
L. pneumophila	38–41	>70	4–21	L. cherrii	C
L. pneumophila subsp. subsp. pneumophila[c]	39–41	72–100		—	C
L. pneumophila subsp. fraseri	38–39	>70		—	C
L. pneumophila subsp. subsp. pascullei	39	96–100		—	C
L. quateirensis	39	NA	2–36	L. moravica	D
L. quinlivanii	43	92–100	1–30	L. birminghamensis	Distinctive
L. rubrilucens	52	NA	3–59	L. erythra	B
L. sainthelensi	41	91–100	3–58	L. longbeachae, L. santicrucis	Distinctive
L. santicrucis	38	NA	7–64	L. sainthelensi	Distinctive
L. shakespearei	45.5	NA	1–25	L. jordanis	Distinctive
L. spiritensis	46	NA	4–34	L. rubrilucens	Distinctive
L. steigerwaltii	40	NA	6–54	L. cherrii, L. bozemanii, L. dumoffii, L. gormanii, L. parisiensis	A

(continued)

TABLE BXII.γ.62. *(cont.)*

Legionella species	Mol% G + C Content	% Relatedness within species under optimal conditions	% Relatedness with other species under optimal conditions	Closest relative(s)	16S-23S rDNA Group[a]
L. tucsonensis	44	NA	1–66	*L. bozemanii, L. anisa, L. parisiensis, L. gormanii, L. wadsworthii, L. longbeachae, L. micdadei, L. dumoffii*	A
L. taurinensis	46	78–100	<3–64	*L. rubrilucens, L. erythra, L. spiritensis*	NA
L. wadsworthii	42	NA	3–36	*L. steigerwaltii, L. tucsonensis*	Distinctive
L. waltersii	NA	NA	1–8	None	Distinctive
L. worsleiensis	41	NA	1–24	*L. moravica*	D
Genomospecies 1	NA	NA	1–69	*L. quinlivanii*	Distinctive

[a]Riffard et al., 1998

[b]NA, not available.

[c]The DNA relatedness among the three subspecies of *L. pneumophila* is below the species level in at least one and usually two hybridization parameters.

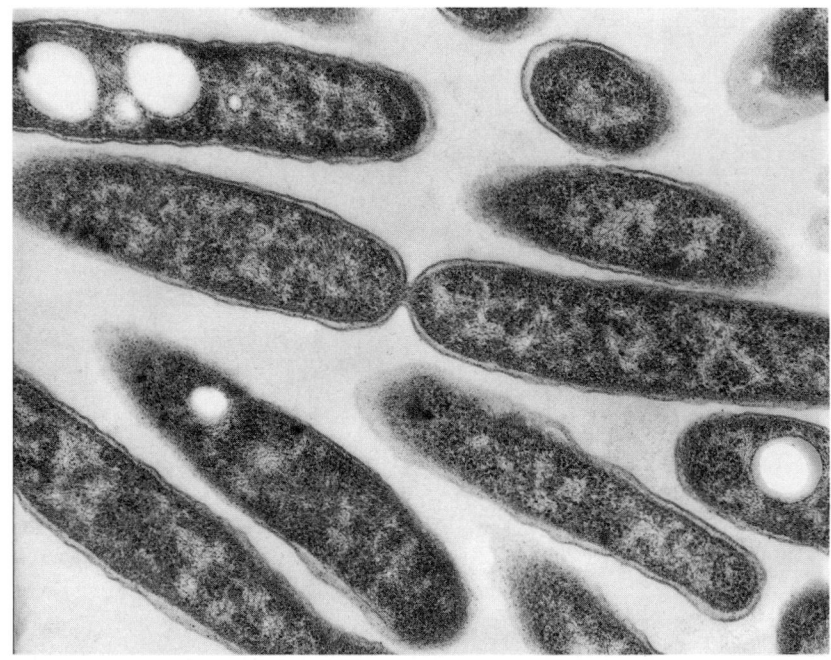

FIGURE BXII.γ.94. Electron micrograph of *Legionella pneumophila* grown in broth (× 90,000).

>95% related to each other at optimal conditions with a divergence of 1% and >95% related at stringent conditions. DNA relatedness of these strains to other subspecies is below the species level in at least one and usually two of these hybridization parameters.

The type strain was isolated from a hospital water supply, but has not been isolated from humans.

The mol% G + C of the DNA is: 39 (T_m).

Type strain: U8W, ATCC 33737, DSM 7515.

2. **Legionella adelaidensis** Benson, Thacker, Lanser, Sangster, Mayberry and Brenner 1991b, 580[VP] (Effective publication: Benson, Thacker, Lanser, Sangster, Mayberry and Brenner 1991a, 1005.)

a.del.aid.en′sis. M.L. fem. adj. *adelaidensis* coming from Adelaide, Australia.

The characteristics are as described for the genus and as listed in Tables BXII.γ.56, BXII.γ.57, BXII.γ.58, and BXII.γ.59. Serologic cross-reactions are detailed in Table BXII.γ.60. Not demonstrated to cause human disease. Isolated from surface water. *L. adelaidensis* was >31% related to 28 other *Legionella* species tested.

The mol% G + C of the DNA is: 40 (T_m).

Type strain: 1762-AUS-E, ATCC 49625.

Additional Remarks: The type strain was isolated from cooling tower water in Adelaide, Australia in 1987.

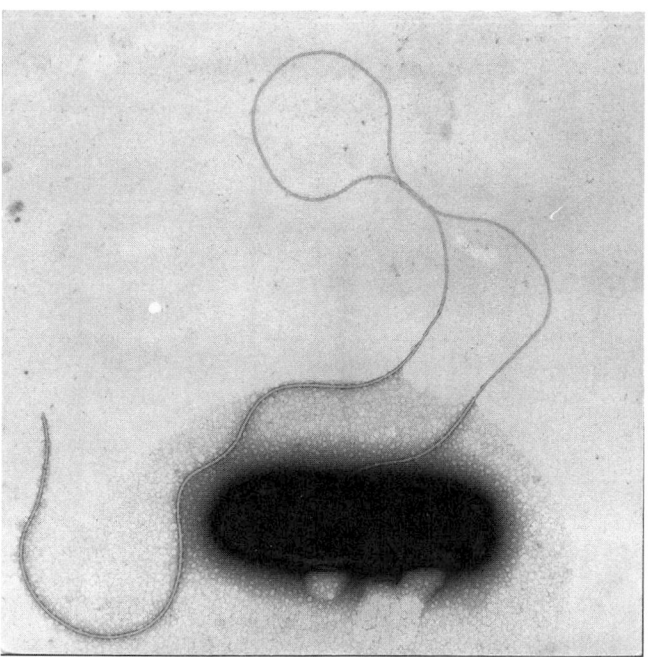

FIGURE BXII.γ.95. Transmission electron micrograph of *Legionella pneumophila* grown on CYE agar, showing a single, probably lateral, flagellum (× 61,700).

3. **Legionella anisa** Gorman, Feeley, Steigerwalt, Edelstein, Moss and Brenner 1985b, 375[VP] (Effective publication: Gorman, Feeley, Steigerwalt, Edelstein, Moss and Brenner 1985a, 308.)

a'ni.sa. M.L. gen. n. *anisa* from Gr. adj. *anisos* unequal, referring to the fact that blue-white autofluorescence is exhibited by most, but not all, strains.

The characteristics are as described for the genus and as listed in Tables BXII.γ.56, BXII.γ.57, BXII.γ.58, and BXII.γ.59. One of the original five strains did not exhibit autofluorescence (Gorman et al., 1985a). Serologic cross-reactions are detailed in Table BXII.γ.60. Causative agent of human pneumonia and empyema and of acute, nonpneumonic, febrile illness (Pontiac fever) (Gorman et al., 1985a; Bornstein et al., 1989b; Fenstersheib et al., 1990; Thacker et al., 1990a). Isolated from human bronchial lavage and pleural fluid and from surface and potable water. The strain that was associated with acute, nonpneumonic, febrile illness was isolated from a public fountain (Fenstersheib et al., 1990). The DNA relatedness among strains of *L. anisa* is 84–99% with divergence of 0.5–1.0% at 60°C ; relatedness at 75°C is 84–92%. *L. anisa* is 5–57% related to other *Legionella* species with divergence of 7.0–14% at 60°C and 1–27% at 75°C. It is most closely related to other species that exhibit blue-white autofluorescence (Table BXII.γ.62).

The mol% G + C of the DNA is: 42 (T_m).

Type strain: WA-316-C3, ATCC 35292.

GenBank accession number (16S rRNA): X73394, Z32635.

Additional Remarks: The type strain was isolated in Los Angeles, California, from hot water in a sink by George W. Gorman in 1981.

4. **Legionella birminghamensis** Wilkinson, Thacker, Benson, Polt, Brookings, Mayberry, Brenner, Gilley and Kirklin 1988b, 220[VP] (Effective publication: Wilkinson, Thacker, Benson, Polt, Brookings, Mayberry, Brenner, Gilley and Kirklin 1987, 2121.)

bir.ming.ham.en'sis. N.L. fem. adj. *birminghamensis* from Birmingham, Alabama, where the original isolate was made.

The characteristics are as described for the genus and as listed in Tables BXII.γ.56, BXII.γ.57, BXII.γ.58, and BXII.γ.59. Causative agent of human pneumonia (Wilkinson et al., 1987). Isolated from human lung biopsy. *L. birminghamensis* was 1–3% related to 18 other *Legionella* species tested (Table BXII.γ.62).

The mol% G + C of the DNA is: 42.9 (T_m).

Type strain: 1407-AL-H, ATCC 43702.

Additional Remarks: The type strain was isolated from a human lung biopsy in Birmingham, Alabama in 1986.

5. **Legionella bozemanii** (sic) Brenner, Steigerwalt, Gorman, Weaver, Feeley, Cordes, Wilkinson, Patton, Thomason and Lewallen Sasserville 1980d, 676[VP] (Effective publication: Brenner, Steigerwalt, Gorman, Weaver, Feeley, Cordes, Wilkinson, Patton, Thomason and Lewallen Sasserville 1980c, 114.)

boze.ma'ni.i. M.L. gen. n. *bozemanii* named after F. Marilyn Bozeman, the microbiologist who isolated and first studied the organism.

The characteristics are as described for the genus and as listed in Tables BXII.γ.56, BXII.γ.57, BXII.γ.58, and BXII.γ.59. Serologic cross-reactions are detailed in Table BXII.γ.60. *L. bozemanii* cannot be distinguished reliably from *L. micdadei*, *L. dumoffii*, or *L. gormanii* by analysis of cellular fatty acids (Karr et al., 1982). There are two serogroups of *L. bozemanii* (Table BXII.γ.61) (Tang et al., 1984). Causative agent of human pneumonia and empyema. Isolated from human lung, pleural fluid, and potable water. DNA relatedness between strains of *L. bozemanii* is 56–77%. *L. bozemanii* is 3–65% related to other *Legionella* species (Table BXII.γ.62).

The mol% G + C of the DNA is: 43 (T_m).

Type strain: WIGA, ATCC 33217.

GenBank accession number (16S rRNA): M36031.

Additional Remarks: The type strain was isolated in 1959 by Bozeman from human lung tissue.

6. **Legionella brunensis** Wilkinson, Drasar, Thacker, Benson, Schindler, Potuznkiova, Mayberry and Brenner 1989, 205[VP] (Effective publication: Wilkinson, Drasar, Thacker, Benson, Schindler, Potuznkiova, Mayberry and Brenner 1988a, 399.)

bru.nen'sis. N.L. gen. n. *brunensis* coming from Brno, Czechoslovakia.

The characteristics are as described for the genus and as listed in Tables BXII.γ.56, BXII.γ.57, BXII.γ.58, and BXII.γ.59. Not demonstrated to cause human disease. Isolated from surface water. *L. brunensis* is 1–8% related to 32 other described *Legionella* species (Table BXII.γ.62).

The mol% G + C of the DNA is: 41.5 (T_m).

Type strain: 441-1, ATCC 43878, CDC 1635-CZK-E.

GenBank accession number (16S rRNA): Z32636.

Additional Remarks: The type strain was isolated from cooling tower water in Brno, Czechoslovakia; first reported in 1986.

7. **Legionella cherrii** Brenner, Steigerwalt, Gorman, Wilkinson, Bibb, Hackel, Tyndall, Campbell, Feeley, Thacker, Ska-

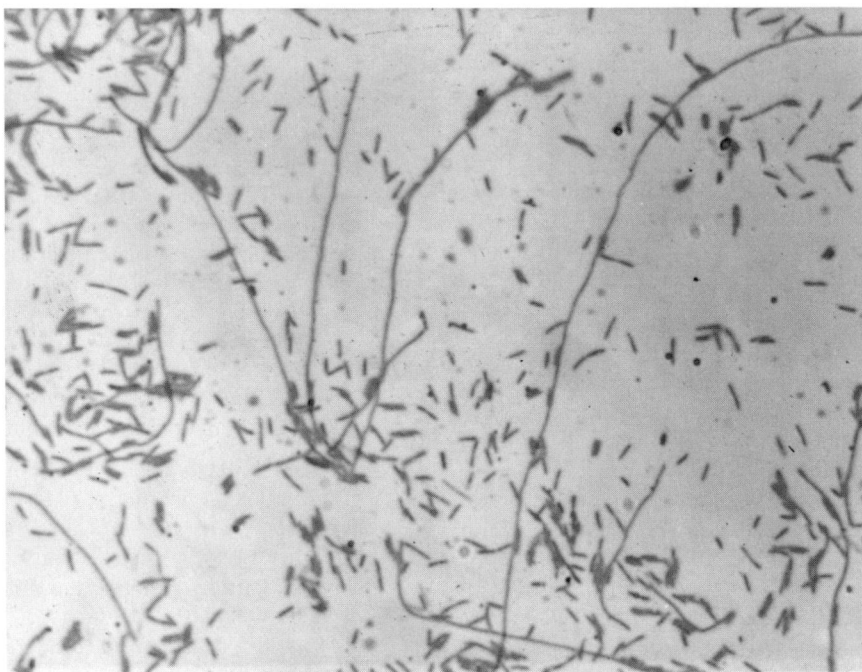

FIGURE BXII.γ.96. Carbol fuchsin-stained *Legionella pneumophila* grown on CYE agar, showing both short rods and filaments (× 1000).

liy, Martin, Brake, Fields, McEachern and Corcoran 1985, 57[VP]

cher′ ri.i. N.L. gen. n. *cherrii* in honor of William B. Cherry for his many pioneering contributions to our knowledge of legionellae.

The characteristics are as described for the genus and as listed in Tables BXII.γ.56, BXII.γ.57, BXII.γ.58, and BXII.γ.59. Not demonstrated to cause human disease. Isolated from surface water. *L. cherrii* strains are ≥94% related to each other at both 60°C and 75°C with 0.0–0.5% divergence. It is highly related to the following species that also exhibit blue-white autofluorescence: *L. steigerwaltii* (67%), *L. dumoffii* (57%), *L. bozemanii* (51%), and *L. gormanii* (47%). It is 24% related to another species that produces blue-white autofluorescence, *L. parisiensis*, and is 6–35% related to other *Legionella* species (Table BXII.γ.62).

The mol% G + C of the DNA is: 39–40 (T_m).

Type strain: ORW, ATCC 35252.

GenBank accession number (16S rRNA): X73404.

Additional Remarks: The type strain was isolated from thermally polluted water in Minnesota by R.L. Tyndall and C.B. Duncan in 1982.

8. **Legionella cincinnatiensis** Thacker, Benson, Staneck, Vincent, Mayberry, Brenner and Wilkinson 1989b, 205[VP] (Effective publication: Thacker, Benson, Staneck, Vincent, Mayberry, Brenner and Wilkinson 1988a, 419.)

cin.cin.na.ti.en′ sis. N.L. fem. adj. *cincinnatiensis* from Cincinnati, Ohio, where the species was first isolated.

The characteristics are as described for the genus and as listed in Tables BXII.γ.56, BXII.γ.57, BXII.γ.58, and BXII.γ.59. Serologic cross-reactions are detailed in Table BXII.γ.60. Isolates from two clinical cases of *L. cincinnatiensis* cross-reacted with *L. longbeachae*, serogroup 1, and *L. du-*

moffii in the direct immunofluorescence test and were initially misidentified (Thacker et al., 1988a; Jernigan et al., 1994). Causative agent of human pneumonia. Isolated from an open lung biopsy. At 60°C *L. cincinnatiensis* was 44% related to *L. longbeachae*, 39% related to *L. santicrucis,* and 36% related to *L. sainthelensi.* At 75°C it is 5–10% related to these species (Table BXII.γ.62).

The mol% G + C of the DNA is: 36 (T_m).

Type strain: 72-OH-H, ATCC 43753.

GenBank accession number (16S rRNA): X73407.

Additional Remarks: The type strain was isolated from a human open lung biopsy in Cincinnati, Ohio in 1982.

9. **Legionella dumoffii** Brenner, Steigerwalt, Gorman, Weaver, Feeley, Cordes, Wilkinson, Patton, Thomason and Lewallen Sasseville 1980d, 676[VP] (Effective publication: Brenner, Steigerwalt, Gorman, Weaver, Feeley, Cordes, Wilkinson, Patton, Thomason and Lewallen Sasseville 1980c, 115.)

du.mof fi.i. M.L. gen. n. *dumoffii* of Dumoff, named after Morris Dumoff, who first isolated *L. pneumophila* directly on bacteriologic media.

The characteristics are as described for the genus and as listed in Tables BXII.γ.56, BXII.γ.57, BXII.γ.58, and BXII.γ.59. One strain of *L. dumoffii* failed to produce brown pigment on tyrosine-containing media (Edelstein and Pryor, 1985). Serologic cross-reactions are detailed in Table BXII.γ.60. *L. dumoffii* cannot be distinguished reliably from *L. bozemanii,* *L. micdadei,* or *L. gormanii* by analysis of cellular fatty acids (Karr et al., 1982). Causative agent of human pneumonia (Dumoff, 1979). Isolated from human lung tissue, prosthetic heart valves (Tompkins et al., 1988), and from water. DNA relatedness between strains of *L. dumoffii* is 90%. *L. dumoffii* is 3–57% related to other *Legionella* species (Table BXII.γ.62).

The mol% G + C of the DNA is: 41–42 (T_m).

Type strain: NY-23, ATCC 33279.

GenBank accession number (16S rRNA): Z32637, X73405.

Additional Remarks: The type strain was isolated from water by G. W. Gorman in 1978.

10. **Legionella erythra** Brenner, Steigerwalt, Gorman, Wilkinson, Bibb, Hackel, Tyndall, Campbell, Feeley, Thacker, Skaliy, Martin, Brake, Fields, McEachern and Corcoran 1985, 55.[VP]

e.ryth'ra. Gr. fem. adj. *erythra* red after the red autofluorescence exhibited by the species.

The characteristics are as described for the genus and as listed in Tables BXII.γ.56, BXII.γ.57, BXII.γ.58, and BXII.γ.59. Serologic cross-reactions are detailed in Table BXII.γ.60. A second serogroup of *L. erythra* cannot be distinguished serologically from *L. rubrilucens* (Table BXII.γ.60) (Saunders et al., 1992). The species contains two serogroups (Table BXII.γ.61) (Saunders et al., 1992). Not yet demonstrated to cause human disease. Isolated from surface water. *L. rubrilucens* and *L. erythra* are 59–62% interrelated, with 9.0–9.5% divergence in related sequences. Relatedness to other *Legionella* species ranges from 5–59%. The closest other species was *L. spiritensis* with 20% relatedness. At 75°C *L. rubrilucens* and *L. erythra* are 27–34% related (Table BXII.γ.62).

The mol% G + C of the DNA is: 51 (T_m).

Type strain: SE-32A-C8, ATCC 35303.

GenBank accession number (16S rRNA): Z32638.

Additional Remarks: The type strain was isolated from cooling tower water in Seattle, Washington, by George W. Gorman in 1981.

11. **Legionella fairfieldensis** Thacker, Benson, Hawes, Gidding, Dwyer, Mayberry and Brenner 1991b, 580[VP] (Effective publication: Thacker, Benson, Hawes, Gidding, Dwyer, Mayberry and Brenner 1991a, 477.)

fair.field.en'sis. N.L. fem. adj. *fairfieldensis* coming from Fairfield, Victoria, Australia.

The characteristics are as described for the genus and as listed in Tables BXII.γ.56, BXII.γ.57, BXII.γ.58, and BXII.γ.59. Not demonstrated to cause human disease. Isolated from surface water. Strains of *L. fairfieldensis* are 87–94% related at 60°C, with 0.0–0.5% divergence. *L. fairfieldensis* was ≤17% related to 30 other *Legionella* species tested (Table BXII.γ.62).

The mol% G + C of the DNA is: 42 (T_m).

Type strain: 1725-AUS-E, ATCC 49588.

Additional Remarks: The type strain was isolated from cooling tower water in Fairfield, Victoria, Australia, in 1987.

12. **Legionella feeleii** Herwaldt, Gorman, McGrath, Toma, Brake, Hightower, Jones, Reingold, Boxer, Tang, Moss, Wilkinson, Brenner, Steigerwalt and Broome 1984b, 355[VP] (Effective publication: Herwaldt, Gorman, McGrath, Toma, Brake, Hightower, Jones, Reingold, Boxer, Tang, Moss, Wilkinson, Brenner, Steigerwalt and Broome 1984a, 338.)

fee.le'i.i. M.L. gen. n. *feeleii* of the late James Feeley, who contributed greatly to early work on *Legionella* and developed effective agar media for primary isolation.

The characteristics are as described for the genus and as listed in Tables BXII.γ.56, BXII.γ.57, BXII.γ.58, and BXII.γ.59. Hydrolysis of hippurate by *L. feeleii* has been reported by some investigators, but not others (Brenner et al., 1985). Serologic cross-reactions are detailed in Table BXII.γ.60. The species contains two serogroups (Thacker et al., 1985). Causative agent of human pneumonia (Thacker et al., 1985) and acute, nonpneumonic, febrile illness (Pontiac fever) (Herwaldt et al., 1984a). Isolated from human sputum and lung tissue and from surface water. The DNA relatedness among strains of *L. feeleii* is 78–100% at 75°C; *L. feeleii* is 1–15% related to other *Legionella* species (Table BXII.γ.60).

The mol% G + C of the DNA is: 46 (T_m).

Type strain: WO-44C, ATCC 35072.

GenBank accession number (16S rRNA): X73395.

Additional Remarks: The type strain was isolated from industrial coolant by George W. Gorman in 1981.

13. **Legionella geestiana** Dennis, Brenner, Thacker, Wait, Vesey, Steigerwalt and Benson 1993, 335.[VP]

gees.ti.a'na. M.L. gen. n. *geestiana* pertaining to the Geest office building.

The characteristics are as described for the genus and as listed in Tables BXII.γ.56, BXII.γ.57, BXII.γ.58, and BXII.γ.59. Serologic cross-reactions are detailed in Table BXII.γ.60. Not demonstrated to cause human disease. Isolated from potable water. *L. geestiana* is 0–10% related to other *Legionella* species tested (Table BXII.γ.62).

The mol% G + C of the DNA is: 52 (T_m).

Type strain: 1308ATCC 49504.

Additional Remarks: The type strain was isolated from a domestic hot water tap in the Geest Office Building in London, United Kingdom, between 1982 and 1984.

14. **Legionella gormanii** Morris, Steigerwalt, Feeley, Wong, Martin, Patton and Brenner 1980b, 676[VP] (Effective publication: Morris, Steigerwalt, Feeley, Wong, Martin, Patton and Brenner 1980a, 720.)

gor.man'i.i. M.L. gen. n. *gormanii* of Gorman; named after G.W. Gorman, who isolated and first studied the organism and pioneered in the isolation of legionellae from environmental and clinical sources.

The characteristics are as described for the genus and as listed in Tables BXII.γ.56, BXII.γ.57, BXII.γ.58, and BXII.γ.59. *L. gormanii* cannot be distinguished reliably from *L. bozemanii*, *L. micdadei*, or *L. dumoffii* by analysis of cellular fatty acids (Karr et al., 1982). Causative agent of human pneumonia (Griffith et al., 1988). Isolated from bronchial lavage fluid and from water. *L. gormanii* is 1–47% related to other *Legionella* species. The relationship of strains within the species is >90%.

The mol% G + C of the DNA is: 41 (T_m).

Type strain: LS-13, ATCC 33297.

GenBank accession number (16S rRNA): Z32639.

Additional Remarks: The type strain was isolated by G.W. Gorman from creek bank soil in 1978.

15. **Legionella gratiana** Bornstein, Marmet, Surgot, Nowicki, Meugnier, Fleurette, Ageron, Grimont, Grimont, Thacker, Benson and Brenner 1991, 580[VP] (Effective publication: Bornstein, Marmet, Surgot, Nowicki, Meugnier, Fleurette, Ageron, Grimont, Grimont, Thacker, Benson and Brenner 1989a, 550.)

gra'ti.an.a. M.L. fem. adj. *gratiana* of Gratianus, the Roman emperor who bathed in the hot springs where the species was first isolated.

The characteristics are as described for the genus and as listed in Tables BXII.γ.56, BXII.γ.57, BXII.γ.58, and BXII.γ.59. Serologic cross-reactions are detailed in Table BXII.γ.60. Not demonstrated to cause human disease. Isolated from surface and potable water. Using the hydroxyapatite method at 60°C, *L. gratiana* was most closely related to *L. longbeachae* (46%), *L. cincinnatiensis* (41%), *L. sainthelensi* (38%), and *L. parisiensis* (32%). The relationship to 23 other species was <16% (Table BXII.γ.62).

The mol% G + C of the DNA is: 38 (T_m).

Type strain: Lyon 8420412, ATCC 49413, CDC 1242.

Additional Remarks: The type species was isolated from spring water at a spa in the Savoy region of France, first reported in 1989.

16. **Legionella hackeliae** Brenner, Steigerwalt, Gorman, Wilkinson, Bibb, Hackel, Tyndall, Campbell, Feeley, Thacker, Skaliy, Martin, Brake, Fields, McEachern and Corcoran 1985, 55.[VP]

hack.el′i.ae. M.L. gen. n. *hackeliae* after Meredith Hackel, who first isolated the organism.

The characteristics are as described for the genus and as listed in Tables BXII.γ.56, BXII.γ.57, BXII.γ.58, and BXII.γ.59. Causative agent of human pneumonia (Brenner et al., 1985; Wilkinson et al., 1985b). Isolated from human bronchial biopsy tissue (Brenner et al., 1985; Wilkinson et al., 1985b). *L. hackeliae* is 3–22% related to other *Legionella* species under optimal conditions; it is 14–17% related to *L. jamestowniensis* in reciprocal studies and is 22% related to *L. cherrii* (Table BXII.γ.62).

The mol% G + C of the DNA is: 40 (T_m).

Type strain: Lansing 2, ATCC 35250.

GenBank accession number (16S rRNA): M36028.

Additional Remarks: The type strain was isolated from a bronchial biopsy specimen by Meredith Hackel in 1981.

17. **Legionella israelensis** Bercovier, Steigerwalt, Derhi-Cochin, Moss, Wilkinson, Benson and Brenner 1986, 369.[VP]

is.ra.el.en′sis. M.L. fem. adj. *israelensis* coming from Israel.

The characteristics are as described for the genus and as listed in Tables BXII.γ.56, BXII.γ.57, BXII.γ.58, and BXII.γ.59. Serologic cross-reactions are detailed in Table BXII.γ.60. Not demonstrated to cause human disease. Isolated from surface water. Strains of *L. israelensis* are 95–100% related to each other under optimal conditions. The relationship to other species tested was 3–12% (Table BXII.γ.62).

The mol% G + C of the DNA is: 40 (T_m).

Type strain: Bercovier 4, ATCC 43119.

GenBank accession number (16S rRNA): X73408, Z32640.

Additional Remarks: The type strain was isolated from an oxidation pond in Gaash, Israel, by Hervé Bercovier.

18. **Legionella jamestowniensis** Brenner, Steigerwalt, Gorman, Wilkinson, Bibb, Hackel, Tyndall, Campbell, Feeley, Thacker, Skaliy, Martin, Brake, Fields, McEachern and Corcoran 1985, 55.[VP]

james′town.i.en.sis. M.L. gen. n. *jamestowniensis* named after Jamestown, New York, where the organism was first isolated.

The characteristics are as described for the genus and as listed in Tables BXII.γ.56, BXII.γ.57, BXII.γ.58, and BXII.γ.59. Serologic cross-reactions are detailed in Table BXII.γ.60. Not yet demonstrated to cause human disease.

Isolated from wet soil. When tested under stringent conditions at 75°C *L. jamestowniensis* is 1–17% related to other *Legionella* species (Table BXII.γ.62).

The mol% G +˙ C of the DNA is: 42 (T_m).

Type strain: JA-26-G1-E2, ATCC 35298.

GenBank accession number (16S rRNA): X73409.

Additional Remarks: The type strain was isolated by G.W. Gorman from wet soil in 1979.

19. **Legionella jordanis** Cherry, Gorman, Orrison, Moss, Steigerwalt, Wilkinson, Johnson, McKinnery and Brenner 1982b, 384[VP] (Effective publication: Cherry, Gorman, Orrison, Moss, Steigerwalt, Wilkinson, Johnson, McKinnery and Brenner 1982a, 294.)

jor.da′nis. M.L. gen. n. *jordanis* named after the Jordan River in Bloomington, Indiana, source of the first isolate.

The characteristics are as described for the genus and as listed in Tables BXII.γ.56, BXII.γ.57, BXII.γ.58, and BXII.γ.59. Serologic cross-reactions are detailed in Table BXII.γ.60. Causative agent of human pneumonia (Thacker et al., 1988b). Isolated from human lung tissue and from water and sewage. DNA relatedness among strains of *L. jordanis* is >90%. *L. jordanis* is 2–20% related to other *Legionella* species (Table BXII.γ.62).

The mol% G + C of the DNA is: 45 (T_m).

Type strain: BL-540, ATCC 33623.

GenBank accession number (16S rRNA): Z32667, X73396.

Additional Remarks: The type strain was isolated from surface water by G.W. Gorman in 1978.

20. **Legionella lansingensis** Thacker, Dyke, Benson, Havlichek, Robinson-Dunn, Stiefel, Schneider, Moss, Mayberry and Brenner 1994, 595[VP] (Effective publication: Thacker, Dyke, Benson, Havlichek, Robinson-Dunn, Stiefel, Schneider, Moss, Mayberry and Brenner 1992, 2400.)

lan.sing.en′sis. M.L. fem. adj. *lansingensis* from Lansing, Michigan, where the species was first isolated.

The characteristics are as described for the genus and as listed in Tables BXII.γ.56, BXII.γ.57, BXII.γ.58, and BXII.γ.59. Serologic cross-reactions are detailed in Table BXII.γ.60. Causative agent of human pneumonia (Thacker et al., 1992). Isolated from human bronchoscopic washings. *L. lansingensis* was 1–12% related to the 28 other *Legionella* species tested (Table BXII.γ.62).

The mol% G + C of the DNA is: 42.6 (T_m).

Type strain: 1677-MI-H, ATCC 49751.

Additional Remarks: The type strain was isolated from human bronchoscopic washings in Lansing, Michigan, in 1986.

21. **Legionella londiniensis** Dennis, Brenner, Thacker, Wait, Vesey, Steigerwalt and Benson 1993, 335.[VP]

lon.din.i.en′sis. M.L. gen. n. *londiniensis* coming from London where the first isolate was made.

The characteristics are as described for the genus and as listed in Tables BXII.γ.56, BXII.γ.57, BXII.γ.58, and BXII.γ.59. Weak hydrolysis of hippurate may be observed. Not demonstrated to cause human disease. Isolated from surface water. Strains of *L. londiniensis* are >90% related to each other. They are 0–36% related to other *Legionella* species (Table BXII.γ.62).

The mol% G + C of the DNA is: 43 (T_m).

Type strain: 1477, ATCC 49505.

Additional Remarks: The type strain was isolated between 1982 and 1984 from cooling tower pond water in London, England.

22. **Legionella longbeachae** McKinney, Porschen, Edelstein, Bissett, Harris, Bondell, Steigerwalt, Weaver, Ein, Lindquist, Kops and Brenner 1982, 266[VP] (Effective publication: McKinney, Porschen, Edelstein, Bissett, Harris, Bondell, Steigerwalt, Weaver, Ein, Lindquist, Kops and Brenner 1981, 743.)

long.beach' ae. M.L. gen. n. *longbeachae* of Long Beach (California) where the organism was first isolated.

The characteristics are as described for the genus and as listed in Tables BXII.γ.56, BXII.γ.57, BXII.γ.58, and BXII.γ.59. Serologic cross-reactions are detailed in Table BXII.γ.60. *L. longbeachae* cannot be distinguished from *L. pneumophila* by analysis of fatty acid composition (Table BXII.γ.58) (Bibb et al., 1981). The species contains two serogroups (Table BXII.γ.61) (Bibb et al., 1981). Causative agent of human pneumonia. Isolated from human lung tissue and transtracheal aspirates and from surface water and moist potting soils. The DNA relatedness among strains of *L. longbeachae* is 97–100% *L. longbeachae* is 2–22% related to other *Legionella* species (Table BXII.γ.62).

The mol% G + C of the DNA is: 40 (T_m).

Type strain: Long Beach 4, ATCC 33462, DSM 10572, NCTC 11477.

GenBank accession number (16S rRNA): M36029.

Additional Remarks: The type strain was isolated from a transtracheal aspirate by R. Porschen in 1980.

23. **Legionella lytica** (Drozanski 1991) Hookey, Saunders, Fry, Birtles and Harrison 1996, 529[VP] (*Sarcobium lyticum* Drozanski 1991, 86.)

ly' ti.ca. M.L. neut. adj. *lyticum* that which causes lysis.

Identity was established between an amoebal pathogen, *Legionella lytica*, and a previously described amoebal pathogen, *Sarcobium lyticum* (Drozanski, 1991). The new nomenclature is proposed, because the genus *Legionella* has priority over *Sarcobium*. This species is an obligate intracellular parasite of amoebae with limited ability to grow on agar. The few known characteristics are listed in Table BXII.γ.56. Etiologic agent of human disease (Fry et al., 1991a). Isolated from human sputum and lung with amoebae. Isolated bacteria are not available for DNA hybridization experiments. The genetic sequence of *L. lytica* is <97% related to other *Legionella* spp. and thus likely to represent a new species (Fry et al., 1991aM).

The mol% G + C of the DNA is: 43 (T_m).

Type strain: L2, PCM 2298.

GenBank accession number (16S rRNA): X97364.

24. **Legionella maceachernii** Brenner, Steigerwalt, Gorman, Wilkinson, Bibb, Hackel, Tyndall, Campbell, Feeley, Thacker, Skaliy, Martin, Brake, Fields, McEachern and Corcoran 1985, 55.[VP]

mac.kath' er.ni.i. M.L. gen. n. *maceachernii* named after Harold V. McEachern, who first isolated the organism.

The characteristics are as described for the genus and as listed in Tables BXII.γ.56, BXII.γ.57, BXII.γ.58, and BXII.γ.59. Serologic cross-reactions are detailed in Table BXII.γ.60. Causative agent of human pneumonia (Wilkin-

son et al., 1985a). Isolated from human bronchial biopsy tissue, water from a home evaporative condenser, and a potable water cistern. When tested under stringent conditions at 75°C *L. maceachernii* is 3–23% related to other *Legionella* species; it is 14–17% related to *L. jamestowniensis* in reciprocal studies and is 22% related to *L. cherrii* (Table BXII.γ.62).

The mol% G + C of the DNA is: 41 (T_m).

Type strain: PX-1-G2-E2, ATCC 35300.

GenBank accession number (16S rRNA): Z32641.

Additional Remarks: The type species was isolated by George W. Gorman from water from a home evaporative condenser in 1979.

25. **Legionella micdadei** Hébert, Steigerwalt and Brenner 1980c, 676[VP] (Effective publication: Hébert, Steigerwalt and Brenner 1980b, 257) (*Legionella pittsburghensis* Pasculle, Feeley, Gibson, Cordes, Myerowitz, Patton, Gorman, Carmack, Ezzell and Dowling 1980, 731.)

mic.da' de.i. M.L. gen. n. *micdadei* of McDade, named after Joseph E. McDade, who isolated the etiologic agent of the 1976 Legionnaires' disease outbreak in Philadelphia.

Selected characteristics are as described for the genus and as listed in Tables BXII.γ.56, BXII.γ.57, BXII.γ.58, and BXII.γ.59. Serologic cross-reactions are detailed in Table BXII.γ.60. Causative agent of pneumonia. Originally referred to as the Pittsburgh pneumonia agent. Not acid-fast by the Ziehl-Neelsen method (Hébert et al., 1980d), but may appear acid-fast in tissue preparations stained by other methods (Pasculle et al., 1979). Browning was demonstrated on F-G agar only after some strains were well adapted to and growing well on this medium. This took 12 or more subcultures (Hébert et al., 1980d). Without adaptation or in the absence of heavy growth, no browning occurs on F-G agar (Morris et al., 1980a). Isolated from human lung, blood, soft tissue, and from potable water. DNA relatedness between strains of *L. micdadei* is 90–100%. *L. micdadei* is 1–23% related to other *Legionella* species (Table BXII.γ.62). *L. micdadei* cannot be distinguished reliably from *L. bozemanii*, *L. dumoffii*, or *L. gormanii* by analysis of cellular fatty acids (Karr et al., 1982).

The mol% G + C of the DNA is: 41–44 (T_m).

Type strain: TATLOCK, ATCC 33218.

GenBank accession number (16S rRNA): M36032.

Additional Remarks: The type strain was isolated by H. Tatlock from a guinea pig inoculated with human blood in 1943.

26. **Legionella moravica** Wilkinson, Drasar, Thacker, Benson, Schindler, Potuznkiova, Mayberry and Brenner 1989, 205[VP] (Effective publication: Wilkinson, Drasar, Thacker, Benson, Schindler, Potuznkiova, Mayberry and Brenner 1988a, 399.)

mo.ra' vi.ca. M.L. gen. n. *moravica* coming from Moravia, Czechoslovakia.

The characteristics are as described for the genus and as listed in Tables BXII.γ.56, BXII.γ.57, BXII.γ.58, and BXII.γ.59. Not demonstrated to cause human disease. Isolated from surface water. *L. moravica* was 1–9% related to 31 other named or described *Legionella* species and 24–36% related to two undescribed species (Table BXII.γ.62).

The mol% G + C of the DNA is: 41.5 (T_m).

Type strain: 316-36, ATCC 43877, CDC 1634-CZK-E.

Additional Remarks: The type strain was isolated from cooling tower water in Moravia, where the species was first isolated; first reported in 1987.

27. **Legionella nautarum** Dennis, Brenner, Thacker, Wait, Vesey, Steigerwalt and Benson 1993, 335.[VP]

nau.ta' rum. M.L. gen. n. *nautarum* of sailors.

The characteristics are as described for the genus and as listed in Tables BXII.γ.56, BXII.γ.57, BXII.γ.58, and BXII.γ.59. Not demonstrated to cause human disease. Isolated from potable water. Two strains of *L. nautarum* were 70% related to each other at 60°C and 78% at 75°C. Relatedness with other *Legionella* species ranges from 2–24% at 60°C (Table BXII.γ.62).

The mol% G + C of the DNA is: 41 (T_m).

Type strain: 1224, ATCC 49506.

Additional Remarks: The type strain was isolated from a domestic water system in London, England.

28. **Legionella oakridgensis** Orrison, Cherry, Tyndall, Fliermans, Gough, Lambert, McDougal, Bibb and Brenner 1983b, 672[VP] (Effective publication: Orrison, Cherry, Tyndall, Fliermans, Gough, Lambert, McDougal, Bibb and Brenner 1983a, 544.)

oak.ridg.en' sis. M.L. gen. n. *oakridgensis* named after Oak Ridge, Tennessee, where the species was first isolated.

The characteristics are as described for the genus and as listed in Tables BXII.γ.56, BXII.γ.57, BXII.γ.58, and BXII.γ.59. Serologic cross-reactions are detailed in Table BXII.γ.60. Causative agent of human pneumonia (Tang et al., 1985). Isolated from human lung (Tang et al., 1985) and from cooling tower water (Orrison et al., 1983a). The DNA relatedness among strains of *L. oakridgensis* is 90–100%; *L. oakridgensis* is 2–22% related to other *Legionella* species (Table BXII.γ.62).

The mol% G + C of the DNA is: 43–45 (T_m).

Type strain: Oak Ridge 10, ATCC 33761.

GenBank accession number (16S rRNA): X73397, Z32642.

Additional Remarks: The type strain was isolated from cooling tower water by R.L. Tyndall, C.B. Duncan, and E.L. Domingue in 1981.

29. **Legionella parisiensis** Brenner, Steigerwalt, Gorman, Wilkinson, Bibb, Hackel, Tyndall, Campbell, Feeley, Thacker, Skaliy, Martin, Brake, Fields, McEachern and Corcoran 1985, 55.[VP]

pa.ri.si.en' sis. M.L. fem. adj. *parisiensis* coming from Paris, France.

The characteristics are as described for the genus and as listed in Tables BXII.γ.56, BXII.γ.57, BXII.γ.58, and BXII.γ.59. Serologic cross-reactions are detailed in Table BXII.γ.60. Causative agent of human pneumonia (Lo Presti et al., 1997). Isolated from surface water and human tracheal aspirate. *L. parisiensis* is substantially related to the other *Legionella* species that exhibit blue-white autofluorescence: *L. bozemanii, L. dumoffii, L. gormanii, L. cherrii,* and *L. steigerwaltii.* Of the species that do not exhibit blue-white autofluorescence, *L. parisiensis* is most closely related to *L. wadsworthii* (37%) (Table BXII.γ.62).

The mol% G + C of the DNA is: 42 (T_m).

Type strain: PF-209C-C2, ATCC 35299.

Additional Remarks: The type strain was isolated from cooling tower water by George W. Gorman in 1981.

30. **Legionella quateirensis** Dennis, Brenner, Thacker, Wait, Vesey, Steigerwalt and Benson 1993, 336.[VP]

qua.teir.en' sis. M.L. gen. n. *quateirensis* coming from Quateira.

The characteristics are as described for the genus and as listed in Tables BXII.γ.56, BXII.γ.57, BXII.γ.58, and BXII.γ.59. Serologic cross-reactions are detailed in Table BXII.γ.60. *L. quateirensis* cannot be distinguished serologically from *L. worsleiensis* or *L. pneumophila* serogroup 4. Not demonstrated to cause human disease. Isolated from potable water. *L. quateirensis* is 2–36% related at 60°C to other *Legionella* species tested. It is most closely related to *L. moravica* (36%) (Table BXII.γ.62).

The mol% G + C of the DNA is: 39 (T_m).

Type strain: 1335, ATCC 49507.

Additional Remarks: The type strain was isolated from a shower in a bathroom of a hotel in Quateira, Portugal, between 1982 and 1984.

31. **Legionella quinlivanii** Benson, Thacker, Waters, Quinlivan, Mayberry, Brenner and Wilkinson 1990b, 105[VP] (Effective publication: Benson, Thacker, Waters, Quinlivan, Mayberry, Brenner and Wilkinson 1989, 196.)

quin.li.va' ni.i. N.L. gen. n. *quinlivanii* named after P.A. Quinlivan.

The characteristics are as described for the genus and as listed in Tables BXII.γ.56, BXII.γ.57, BXII.γ.58, and BXII.γ.59. Serologic cross-reactions are detailed in Table BXII.γ.60. Not demonstrated to cause human disease. There are two serogroups of *L. quinlivanii* (Birtles et al., 1991). Isolated from surface water. The relationship among strains of *L. quinlivanii* is 92–100%. *L. quinlivanii* is 1–30% related to other *Legionella* species; under optimum conditions the most closely related *Legionella* sp. was *L. birminghamensis* (23–30%) (Table BXII.γ.62).

The mol% G + C of the DNA is: 43 (T_m).

Type strain: 14442-AUS-E, ATCC 43830, NCTC 12433.

Additional Remarks: The type strain was isolated from water in Australia in 1986.

32. **Legionella rubrilucens** Brenner, Steigerwalt, Gorman, Wilkinson, Bibb, Hackel, Tyndall, Campbell, Feeley, Thacker, Skaliy, Martin, Brake, Fields, McEachern and Corcoran 1985, 55.[VP]

ru.bri.lu'sens. M.L. gen. n. *rubrilucens* "red shining" after the red autofluorescence exhibited by the species.

The characteristics are as described for the genus and as listed in Tables BXII.γ.56, BXII.γ.57, BXII.γ.58, and BXII.γ.59. Serologic cross-reactions are detailed in Table BXII.γ.60. A second serogroup of *L. erythra* cannot be distinguished serologically from *L. rubrilucens* (Table BXII.γ.60) (Saunders et al., 1992) Not yet demonstrated to cause human disease. Isolated from tap water. *L. rubrilucens* and *L. erythra* are 59–62% interrelated, with 9.0–9.5% divergence in related sequences. Relatedness to other *Legionella* species ranges from 3–59%. The most closely related other species was *L. spiritensis* at 34%. At 75°C *L. rubrilucens* and *L. erythra* are 27–34% related (Table BXII.γ.62).

The mol% G + C of the DNA is: 52 (T_m).

Type strain: WA-270A-C2, ATCC 35304, DSMZ 11884.

GenBank accession number (16S rRNA): X73398, Z32643.

Additional Remarks: The type strain was isolated from tap water in Los Angeles, California, by Giorge W. Gorman in 1980.

33. **Legionella sainthelensi** Campbell, Bibb, Lambert, Eng, Steigerwalt, Allard, Moss and Brenner 1984b, 355[VP] (Effective publication: Campbell, Bibb, Lambert, Eng, Steigerwalt, Allard, Moss and Brenner 1984a, 372.)

saint.he' len.si. M.L. gen. n. *sainthelensi* from Mount St. Helens, where the original isolate was made from surface water in the devastated area after a volcanic eruption.

The characteristics are as described for the genus and as listed in Tables BXII.γ.56, BXII.γ.57, BXII.γ.58, and BXII.γ.59. Serologic cross-reactions are detailed in Table BXII.γ.60. The species contains two serogroups (Benson et al., 1990a). Causative agent of human pneumonia (Benson et al., 1990a). Isolated from human respiratory secretions (Benson et al., 1990a) and from surface water (Campbell et al., 1984a). The DNA relatedness among strains of *L. sainthelensi* is 91–100%; the two species most closely related to *L. sainthelensi* are *L. santicrucis* (58% related with a divergence of 9% at 60°C and 11% related at 75°C) and *L. longbeachae* (37% related with 8.5% divergence at 60°C and 4% related at 75°C). *L. spiritensis* was 3–14% related to the other *Legionella* species tested (Table BXII.γ.62).

The mol% G + C of the DNA is: 41 (T_m).

Type strain: Mt. St. Helens 4, ATCC 35248.

GenBank accession number (16S rRNA): X73399.

Additional Remarks: The type strain was isolated from spring water by J. Campbell and S. Eng in 1981.

34. **Legionella santicrucis** Brenner, Steigerwalt, Gorman, Wilkinson, Bibb, Hackel, Tyndall, Campbell, Feeley, Thacker, Skaliy, Martin, Brake, Fields, McEachern and Corcoran 1985, 58.[VP]

san.ti.cru' cis. M.L. gen. n. *santicrucis* named after Santa Crux, the Latin name for St. Croix, Virgin Islands.

The characteristics are as described for the genus and as listed in Tables BXII.γ.56, BXII.γ.57, BXII.γ.58, and BXII.γ.59. Serologic cross-reactions are detailed in Table BXII.γ.60. Not demonstrated to cause human disease. Isolated from potable water. *L. santicrucis* is 7–64% related to other *Legionella* species. It is closely related to *L. sainthelensi* (64% related at 60°C and 23% at 75°C) and is 38% related to *L. longbeachae* (Table BXII.γ.62).

The mol% G + C of the DNA is: 38 (T_m).

Type strain: SC-63-C7, ATCC 35301.

Additional Remarks: The type strain was isolated from tap water in St. Croix, Virgin Islands, by George W. Gorman in 1982.

35. **Legionella shakespearei** Verma, Brenner, Thacker, Benson, Vesey, Kurtz, Dennis, Steigerwalt, Robinson and Moss 1992, 406.[VP]

shakes.pea.re' i. N.L. gen. n. *shakespearei* named after the playwright William Shakespeare, because the organism was isolated in Stratford-upon-Avon.

The characteristics are as described for the genus and as listed in Tables BXII.γ.56, BXII.γ.57, BXII.γ.58, and BXII.γ.59. The species is distinctive in production of uniformly pink colonies on BCYE agar. Not demonstrated to cause human disease. Isolated from surface water. *L. shakespearei* was 25% related to *L. jordanis* at 60°C. It was <10%

related to 26 other *Legionella* species tested (Table BXII.γ.62).

The mol% G + C of the DNA is: 45.5 (T_m).

Type strain: 214, ATCC 49655.

Additional Remarks: The type strain was isolated from cooling tower water in Stratford-upon-Avon, United Kingdom, reported in 1992.

36. **Legionella spiritensis** Brenner, Steigerwalt, Gorman, Wilkinson, Bibb, Hackel, Tyndall, Campbell, Feeley, Thacker, Skaliy, Martin, Brake, Fields, McEachern and Corcoran 1985, 55.[VP]

spi.ri.ten' sis. M.L. fem. adj. *spiritensis* coming from Spirit Lake near Mount St. Helens in Oregon.

The characteristics are as described for the genus and as listed in Tables BXII.γ.56, BXII.γ.57, BXII.γ.58, and BXII.γ.59. Not yet associated with human disease. There are two serogroups of *L. spiritensis* (Harrison et al., 1988). Isolated from surface water. *L. spiritensis* is most closely related (34%) to *L. rubrilucens*, a red fluorescing species. It is 4–34% related to a number of species, including *L. erythra*, another red fluorescing species (Table BXII.γ.62).

The mol% G + C of the DNA is: 46 (T_m).

Type strain: Mt. St. Helens 9, ATCC 35249.

GenBank accession number (16S rRNA): M36030.

Additional Remarks: The type strain was isolated from lake water by J. Campbell in 1981.

37. **Legionella steigerwaltii** Brenner, Steigerwalt, Gorman, Wilkinson, Bibb, Hackel, Tyndall, Campbell, Feeley, Thacker, Skaliy, Martin, Brake, Fields, McEachern and Corcoran 1985, 57.[VP]

stei.ger.wal' ti.i. M.L. gen. n. *steigerwaltii* named in honor of Arnold G. Steigerwalt, who did the definitive experiments to classify the first 22 *Legionella* species.

The characteristics are as described for the genus and as listed in Tables BXII.γ.56, BXII.γ.57, BXII.γ.58, and BXII.γ.59. Not demonstrated to cause human disease. Isolated from potable water. *L. steigerwaltii* is 32–54% related to other species that exhibit blue-white autofluorescence. It is 31–34% related to three species that do not autofluoresce: *L. santicrucis*, *L. wadsworthii*, and *L. sainthelensi*. The relationship with other species is 6–33% (Table BXII.γ.62).

The mol% G + C of the DNA is: 40 (T_m).

Type strain: SC-18-C9, ATCC 35302.

GenBank accession number (16S rRNA): X73400.

Additional Remarks: The type strain was isolated from tap water in St. Croix, Virgin Islands, by George W. Gorman in 1982.

38. **Legionella taurinensis** Lo Presti, Riffard, Meugnier, Reyrolle, Lasne, Grimont, Grimont, Vandenesch, Etienne, Fleurette and Freney 1999, 401.[VP]

tau.ri.nen' sis. M.L. adj. *taurinensis* pertaining to Turin, Italy, whose classical Latin name was Augusta Taurinorum.

The characteristics are as described for the genus and as listed in Tables BXII.γ.56, BXII.γ.57, BXII.γ.58, and BXII.γ.59. Not demonstrated to cause human disease. Isolated from potable and surface water. Strains of *L. taurinensis* were 78–100% related to each other. *L. taurinensis* was >20% related to *L. rubrilucens*, *L. erythra*, and *L. spiritensis*. Its closest relative was *L. rubrilucens*, which was 54% related. In reciprocal reactions *L. taurinensis* was 64% related to *L. rubrilucens* with 7.5% divergence under optimal conditions tested (Table BXII.γ.62).

The mol% G + C of the DNA is: 46 (T_m, HPLC).

Type strain: Turin 1, ATCC 700508.

GenBank accession number (16S rRNA): AF037597.

Additional Remarks: The type strain was isolated from a humidifier in Turin, Italy.

39. **Legionella tucsonensis** Thacker, Benson, Schifman, Pugh, Steigerwalt, Mayberry, Brenner and Wilkinson 1990b, 105[VP] (Effective publication: Thacker, Benson, Schifman, Pugh, Steigerwalt, Mayberry, Brenner and Wilkinson 1989a, 1833.)

tuc.so.nen'sis. M.L. fem. adj. *tucsonensis* from Tucson, Arizona, where the species was first isolated.

The characteristics are as described for the genus and as listed in Tables BXII.γ.56, BXII.γ.57, BXII.γ.58, and BXII.γ.59. Serologic cross-reactions are detailed in Table BXII.γ.60. Causative agent of human pneumonia (Thacker et al., 1989a). Isolated from pleural fluid. Of the species tested, *L. tucsonensis* was most closely related to the following species that show blue-white autofluorescence: *L. bozemanii*, *L. gormanii*, and *L. parisiensis*. It is only 1–11% related to *L. birminghamensis*, *L. cincinnatiensis*, *L. brunensis*, *L. moravica*, and *L. quinlivanii* (Table BXII.γ.62).

The mol% G + C of the DNA is: 44 (T_m).

Type strain: 1087-AZ-H, ATCC 49180.

GenBank accession number (16S rRNA): Z32644.

Additional Remarks: The type strain was isolated from human pleural fluid in Tucson, Arizona in 1984.

40. **Legionella wadsworthii** Edelstein, Brenner, Moss, Steigerwalt, Francis and George 1983, 672[VP] (Effective publication: Edelstein, Brenner, Moss, Steigerwalt, Francis and George 1982a, 812.)

wads'wor.thi.i. M.L. gen. n. *wadsworthii* of V.A. Wadsworth Medical Center after James Wadsworth, and the location where the patient was hospitalized and the bacterium isolated.

The characteristics are as described for the genus and as listed in Tables BXII.γ.56, BXII.γ.57, BXII.γ.58, and BXII.γ.59. Serologic cross-reactions are detailed in Table BXII.γ.60. Causative agent of human pneumonia (Edelstein

et al., 1982a). Isolated from human sputum. *L. wadsworthii* is 3–36% related to other *Legionella* species with a divergence of 9.5–16% at 60°C. When tested under stringent conditions at 75°C *L. wadsworthii* is 2–15% related to other *Legionella* species (Table BXII.γ.62).

The mol% G + C of the DNA is: 42 (T_m).

Type strain: Wadsworth 81-716A, ATCC 33877.

GenBank accession number (16S rRNA): X73401.

Additional Remarks: The type strain was isolated from sputum by Paul H. Edelstein in 1981.

41. **Legionella waltersii** Benson, Thacker, Daneshvar and Brenner 1996, 633.[VP]

wal'ters.i.i. M.L. gen. n. *waltersii* of Walters, to honor R.P. Walters, who isolated the organism.

The characteristics are as described for the genus and as listed in Tables BXII.γ.56, BXII.γ.57, BXII.γ.58, and BXII.γ.59. Not demonstrated to cause human disease. Isolated from potable and surface water. It was <10% related to other *Legionella* spp. tested.

The mol% G + C of the DNA is: unknown.

Type strain: 2074-AUS-E, ATCC 51914.

Additional Remarks: The type strain was isolated from a drinking water distribution system in South Australia by R.P. Waters in 1987–88.

42. **Legionella worsleiensis** Dennis, Brenner, Thacker, Wait, Vesey, Steigerwalt and Benson 1993, 336.[VP]

wors.lei.en'sis. M.L. gen. n. *worsleiensis* coming from Worsley.

The characteristics are as described for the genus and as listed in Tables BXII.γ.56, BXII.γ.57, BXII.γ.58, and BXII.γ.59. Serologic cross-reactions are detailed in Table BXII.γ.60. It cannot be distinguished serologically from *L. pneumophila* serogroup 4 or from *L. quateirensis*. Not demonstrated to cause human disease. Isolated from surface water. *L. worsleiensis* is 1–24% related to other *Legionella* species tested at 60°C (Table BXII.γ.62).

The mol% G + C of the DNA is: 41 (T_m).

Type strain: 1347, ATCC 49508.

Additional Remarks: The type strain was isolated from cooling tower return flow at an industrial site in Worsley, United Kingdom, between 1982 and 1984.

Other Organisms

1. *Legionella* genomospecies 1 Benson, Thacker, Daneshvar, and Brenner 1996.

   This genetically distinct organism was reported as a genomospecies, because it could not be distinguished biochemically or serologically from *L. quinlivanii*, serogroup 2. The characteristics of the genomospecies are as described for the genus and as listed in Tables BXII.γ.56, BXII.γ.57, BXII.γ.58, and BXII.γ.59. Serologic cross-reactions are detailed in Table BXII.γ.60. Not demonstrated to cause human disease. Isolated from surface water. The relationship between *Legionella* genomospecies 1 and *L. quinlivanii* in reciprocal reactions was 53–69% with 4.5–6.5% under optimal conditions and 31–51% under stringent conditions. In the same laboratory the relationship between strains of *L. quinlivanii* was 79–99% with 0.5–3.0% under optimal con-

ditions and 85–100% under stringent conditions is unknown. It was <10% related to other *Legionella* spp. tested (Table BXII.γ.62).

The mol% G + C of the DNA is: unknown.

Deposited strain: 2055-AUS-E, ATCC 51913.

Additional Remarks: The type strain was isolated from cooling tower water in South Australia, Australia.

Bacterial isolates recovered by conventional techniques may represent additional, as yet unnamed *Legionella* species (Wilkinson et al., 1990; Benson et al., 1996). Strains that have been described as probably new species include LC878, "*L. donaldsonii*" (Riffard et al., 1998i) and strain "glasgow" (Hookey et al., 1996). At least six genomospecies distinct from all named species except *L. taurinensis* have been characterized (D.J. Brenner, personal communication).

Family II. **Coxiellaceae** *fam. nov.*

GEORGE M. GARRITY, JULIA A. BELL AND TIMOTHY LILBURN

Co.xi.el.la' ce.ae. M.L. fem. n. *Coxiella* type genus of the family; *-aceae* ending to denote family; M.L. fem. pl. n. *Coxiellaceae* the *Coxiella* family.

The family *Coxiellaceae* was circumscribed for this volume on the basis of phylogenetic analysis of 16S rRNA sequences; the family contains the genera *Coxiella* (type genus) and *Rickettsiella*.

Found only in vacuoles in cells of invertebrate or vertebrate hosts. Serial passage in cell-free medium has not been achieved.

Type genus: **Coxiella** (Philip 1943) Philip 1948, 58[AL] (**Rickettsia (Coxiella) burnetii** Philip 1943, 306.)

Genus I. **Coxiella** *(Philip 1943) Philip 1948, 58[AL] (Rickettsia (Coxiella) burnetii Philip 1943, 306)*

MICHEL DRANCOURT AND DIDIER RAOULT

Co.xi.el' la. M.L. fem. dim. *-ella* ending; M.L. fem. n. *Coxiella* named after Harold R. Cox, who in collaboration with G.E. Davis, first isolated this organism in the United States shortly after its discovery in Australia and who introduced the technique of yolk sac inoculation of the chick embryo, which greatly facilitated the study of this and other genera.

Strictly intracellular bacteria, usually 0.2–0.4 μm × 0.4–1.0 μm. Best stained by **Gimenez staining**. They have no flagella or capsule. They live in close natural association with **arthropod and vertebrate hosts**. The genus includes *Coxiella burnetii*—the **agent of Q fever**—and endosymbionts of ticks and aquatic invertebrates. *C. burnetii* grows well in cultured cell lines and in the yolk sac of chicken embryos, where it undergoes a cycle of development with formation of an endospore-like body. *Coxiella burnetii* possesses **two antigenic phases**: natural virulent phase I and attenuated phase II, which is obtained after passage in cultured cell lines and yolk sac. The organism **grows in vacuoles of the host cell** rather than in the cytoplasm or the nucleus. *C. burnetii* phase I vacuoles share characteristics of late-stage, immature phagosomes whereas *C. burnetii* phase II vacuoles are bactericidal phagolysosomes. Highly resistant to chemical agents and elevated temperature. Coxiellae occur worldwide in ticks and various vertebrates, including humans. Infection is particularly prevalent in cattle, sheep and goats. Extremely infectious; a single aerosol-borne organism can cause the human disease Q fever. *C. burnetii* is one of the ten most feared potential **bioterrorism agents**.

The mol% G + C of the DNA is: 42.7.

Type species: **Coxiella burnetii** (Derrick 1939) Philip 1948, 58[AL] (*Rickettsia burnetii* (sic) Derrick 1939, 14.)

FURTHER DESCRIPTIVE INFORMATION

Coxiellae are generally regarded as Gram-negative rods, but they are best stained by Gimenez staining (Gimenez, 1964) (Fig.BXII.γ.97). They have no flagella or capsule and are not always retained by ordinary bacteriological filters. Coxiellae have never been cultured axenically in laboratory media, although outside the host, the organisms metabolize glutamate, glucose and other substrates if the pH is low. Electron microscopy of bacteria separated from host cell components reveals great variation in morphology.

Cell morphology Two forms of *C. burnetii* cells are found in infected eucaryotic cells: a metabolically inactive sporelike "small-cell variant" (SCV; 204–450 nm) and a "large-cell variant" (LCV;

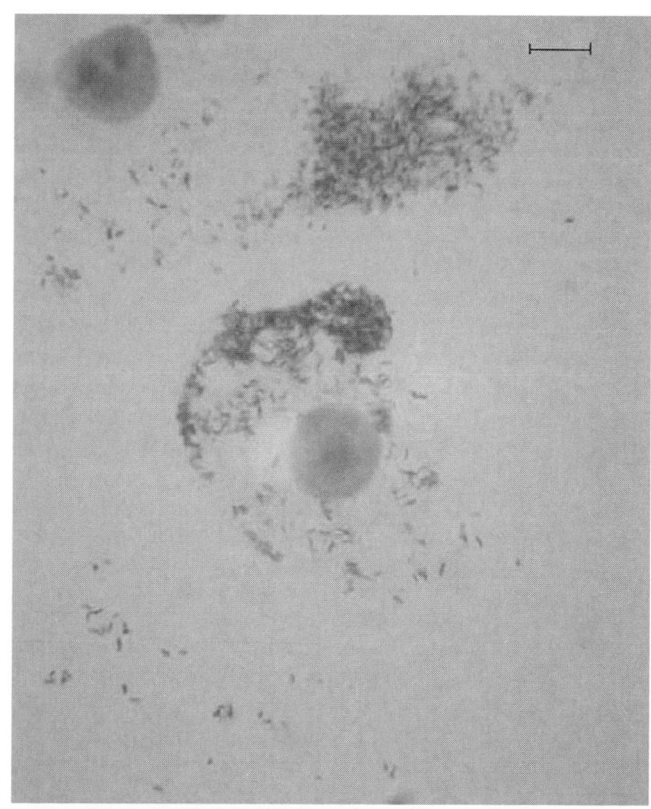

FIGURE BXII.γ.97. Gimenez-stained *Coxiella burnetii* inside a large vacuole of a L 929 cell. Bar = 5 μm.

up to 2 μm in length). SCVs develop from LCVs following an asymmetric cell division (McCaul, 1991) and are distinguished from LCVs by having a regular rod shape, a dense layer of peptidoglycan and protein between the two layers of the cell envelope, and condensed nucleoids. The more pleomorphic LCVs possess dispersed nucleoids and granules or sometimes fibrils in their cytoplasm. Condensation or dispersal of the nucleoid may be associated with the presence in SCVs of a 20-kDa protein—Hq1—that binds DNA; Hq1 is a homolog of histone H1 (Heinzen and Hackstadt, 1996).

SCVs also possess greater resistance than LCVs to drying, UV irradiation, acid or alkaline pH, disinfectants, and other chemicals (Ranson and Huebner, 1951; Babudieri, 1959; Maurin and Raoult, 1999). At 4°C, viability of SCVs is retained for 1 year or more in dried fomites such as tick feces and wool, as well as in sterile skim milk or unchlorinated water. Meats remain infected for at least 1 month. Complete inactivation is not always accomplished by exposure to 63°C for 30 min or 85–90°C for a few seconds. *C. burnetii* is rapidly inactivated by diethyl ether but not by ethanol (Ranson and Huebner, 1951; Babudieri, 1959).

Thus, the morphogenesis and properties of SCVs are comparable, although not identical, to cycles of cell differentiation and formation of resistant, inactive cell forms in other bacteria (McCaul and Williams, 1991). SCVs are released from host cells and bind to the membranes of phagocytic cells; hence, SCVs appear to be a resistant stage adapted to survive transmission between individual hosts. In addition, *C. burnetii* is highly infectious: a single organism can cause aerosol-borne disease. It has therefore been considered a potential weapon for biological warfare (Lederberg, 2000) and has been classified as a category B biological weapon by the Centers for Disease Control and Prevention, Atlanta, GA.

Antigenic structure Repeated serial passage (10–100 times) in cultured eucaryotic cells or in embryonated eggs produces attenuated "rough" or Phase II variants of naturally-occurring virulent "smooth" or Phase I *C. burnetii*. This phenomenon resembles the production of rough variants of smooth strains of enterobacteria (Stoker and Fiset, 1956; Brezina, 1978) and involves mutation in the genes encoding enzymes that synthesize the lipopolysaccharide (LPS) (Hackstadt et al., 1985; Hackstadt, 1988). Phase II *C. burnetii* strains exhibit changes in both the sugar composition and the length of the LPS (Amano and Williams, 1984). The sugars dihydrohydroxystreptose, galactosamine uronyl-α-(1,6)-glucosamine and L-virenose are present in Phase I but not Phase II LPS (Schramek and Mayer, 1982; Schramek et al., 1985; Amano et al., 1987; Hackstadt, 1990). The protein complement of the outer membrane also changes during the transition from Phase I to Phase II (Amano and Williams, 1984). Vodkin and Williams (1986) demonstrated large deletions in the genome of the Phase II Nine Mile strain.

Genetics The genome of *C. burnetii* consists of a chromosome plus a facultative 36–42-Kb plasmid whose function remains undetermined. The genome of the Nine Mile strain of *C. burnetii* is 1.99 Mb based on the complete genome sequence (GenBank accession number NC002971), but the genome size varies from 1.5 to 2.4 Mb among different *C. burnetii* strains (Willems et al., 1998). In addition, complete genome sequencing confirmed that the chromosome is circular, contrary to previous speculations of a linear geometry (Willems et al., 1996; 1998; Dr. J.E. Samuel,

personal communication). The *C. burnetii* Nine Mile strain genome comprises 6653 open reading frames (ORFs) that code for peptides of ≥50 residues. A physical macrorestriction map of *C. burnetii* Nine Mile phase I was constructed by Willems et al. (1996; 1998); 25 DNA fragments were distinguished by pulsed field gel electrophoresis (PFGE) after digestion of total DNA with *Not*I. Vodkin and Williams (1988) and Hoover and Williams (1990) have elucidated operon structures in *C. burnetii*.

Fifteen *C. burnetii* genes have been cloned to date in *Escherichia coli*, including genes involved in metabolism (*gltA* (Heinzen and Mallavia, 1987), *pyrB* (Hoover and Williams, 1990), and *icd* (Nguyen and Hirai, 1999)), transcription and translation (*rpoD* and *rpoS* (Seshadri and Samuel, 2001) and *serS* (Willems et al., 1996)); stress responses, (*sodB* (Heinzen et al., 1992), *htpA* and *htpB* (Vodkin and Williams, 1988), and *dnaJ* (Zuber et al., 1995a)), production of surface antigens (*omp* (Hendrix et al., 1990), *mucZ* (Zuber et al., 1995b), and *algC* (Willems et al., 1996)), and environmental sensing (*qrsA* (Mo and Mallavia, 1994)). In addition, a possible chromosomal origin of replication (*oriC*) has been cloned (Chen et al., 1990a), but Suhan et al. (1996) were unable to demonstrate the ability of this DNA fragment to initiate DNA replication in *C. burnetii*.

Plasmid DNA can be introduced into *C. burnetii* by electroporation; plasmids bearing both antibiotic resistance determinants (Suhan et al., 1994; 1996) and a green fluorescent protein gene (Lukacova et al., 1999) have been introduced and the genes expressed.

Four plasmids native to *C. burnetii* have been characterized: QpH1 (Samuel et al., 1983), QpRS (Samuel et al., 1985), QpDG (Mallavia, 1991), and QpDV (Valkova and Kazar, 1995). These plasmids are 33–42 Kb in size and share 30 kb of homology with each other (Minnick et al., 1990; Mallavia, 1991). The chromosomal DNAs of some plasmid-free genomic group V *C. burnetii* strains contain a region of homology to the DNA of plasmid QpRS (Savinelli and Mallavia, 1990); this region is derived from plasmid sequences (Lautenschlager et al., 2000). The entire DNA sequences of plasmids QpH1 and QpDV have been obtained (Thiele et al., 1994, GenBank accession number NC002118; and Valkova and Kazar (1995), GenBank accession number NC002131; respectively).

DNA–DNA hybridization studies showed that *C. burnetii* strains are homogenous (Vodkin and Williams, 1986). Early studies of genetic differences among strains revealed by RFLP analysis (Hendrix et al., 1991), plasmid analysis (Samuel et al., 1983, 1985; Mallavia, 1991), and immunoblotting of LPS (Hackstadt, 1986) suggested that six RFLP-defined groups differed in their association with "acute" or "chronic" cases of Q fever. However, recent studies on larger numbers of strains from a variety of locations and sources using improved techniques—genomic RFLP (Thiele and Willems, 1994; Jäger et al., 1998), PCR-based plasmid analysis (Stein and Raoult, 1993), immunoblotting of LPS with specific monoclonal antibodies (Yu and Raoult, 1994), and gene sequence analysis (Nguyen and Hirai, 1999; Sekeyova et al., 1999)—have shown that genetic variation in this pathogen is geographic in origin. (See also Thiele et al. (1993) and Willems et al. (1996).) Moreover, variation in host factors is more likely to explain whether individual cases of Q fever are "acute" or "chronic" than is genetic variation in the pathogen (La Scola et al., 1998b; Maurin and Raoult, 1999; Raoult et al., 2000; Fenollar et al., 2001).

Epidemiology In humans, contact between *C. burnetii* and a nonimmune individual results in a primary infection which can be either asymptomatic or symptomatic. The symptomatic primary infection is acute Q fever, which is usually followed by complete recovery of a normal host. In some hosts, *C. burnetii* can multiply despite an antibody response following primary infection, symptomatic or not. In these cases, chronic infection develops because the immune system is unable to control the infection. This distinction between acute and chronic infection is based on clinical expression, temporal course, and the serological profile of the patient, and is mainly determined by host factors. This hypothesis is supported by all available data from humans as well as from animal models.

The aerosol route is the principal mode of acquisition of *C. burnetii* infection (Babudieri, 1959; Dupuis et al., 1987; Tissot-Dupont et al., 1999). *C. burnetii* is highly infectious, and a single organism can cause the aerosol-borne Q fever. This fact and the great resistance of *C. burnetii* to environmental stresses have led to the classification of *C. burnetii* as a group B bioterrorism agent. Indeed, due to its sporelike form, *C. burnetii* may remain viable in soil and milk for several months following contamination. Cattle, sheep, and goats are the main reservoirs of *C. burnetii* infection for humans. High densities of *C. burnetii* are found in the placenta of infected parturient animals and are shed in the environment following labor. In Europe, most cases occur between January and June because of the heavily contaminated environment following lambing season. Cases among students following visits to farms as part of school activities and outbreaks following exposure to birth products of cats, dogs, and rabbits, have been reported. *C. burnetii* can also be transmitted by ingestion of unpasteurized dairy products (Fishbein and Raoult, 1992). Recently, cases were linked to ingestion of pasteurized milk in Canada (Hatchette et al., 2001). Human cases have been reported following contact with parturient products of an infected woman (in an obstetrician) and through transplacental, blood, and sexual transmission (Maurin and Raoult, 1999). Based on the results of animal experimentation (La Scola et al., 1997), it has been hypothesized that the route of acquisition of *C. burnetii* infection may influence the clinical presentation of the disease; i.e., infection via an aerosol results in pneumonia, whereas infection via intraperitoneal injection results in hepatitis.

Q fever has been described worldwide except in New Zealand. Estimates of the incidence of Q fever are complicated by the lack of systematic investigation of this illness worldwide. In regions of France where Q fever is endemic, the annual incidence of acute Q fever is estimated at 50/100,000 persons (Maurin and Raoult, 1999). The incidence of hospital admission for acute and chronic Q fever is estimated at 1.7 and 0.5/1,000,000 persons, respectively. Comparable figures have been reported from England.

Pathogenicity in humans In Q fever outbreaks, there is a high ratio of males to females in infected patients and a low ratio of children to adults (Maltezou and Raoult, 2002). Because the levels of exposure and seroprevalence are not different in these groups, it is suspected that host factors determine different clinical expression of acute disease. The incubation period for acute Q fever is 1–3 weeks. Data from the National Reference Center for Rickettsial Diseases, Marseille, France, and from large outbreaks show that up to 60% of infected patients remain asymptomatic. Among symptomatic patients, most (95%) develop a mild, non-specific flu-like illness, whereas the remaining 5% re-

quire hospitalization. Chronic Q fever accounts for about 10% of hospital cases and 0.2% of all cases (Dupuis et al., 1987; Maurin and Raoult, 1999). Acute Q fever has two primary clinical presentations: atypical pneumonia and hepatitis. Most infected patients experience a transient bacteremia with *C. burnetii*, usually late in the incubation period. The spread of the organisms to other organs can result in severe complications; endocarditis is the primary manifestation of chronic Q fever. Other aspects of the pathogenicity of *C. burnetii* in humans, the clinical aspects of Q fever, and current recommendations for treatment of Q fever have been reviewed in detail by Maurin and Raoult, 1999.

Animal infection and reservoirs Livestock—including cattle, sheep and goats—and pets are major reservoirs of *C. burnetii* (Langley et al., 1988; Laughlin et al., 1991; Buhariwalla et al., 1996). The pathogen can be found in urine, feces, milk, and birth products of these animals and transmitted to humans by these substances as well as by aerosols. Animals are usually not harmed by *C. burnetii*, although association of the organisms with infertility (Ho et al., 1995), abortion (Waldhalm et al., 1978), low birth weight (Ho et al., 1995), and death of newborns (Buhariwalla et al., 1996) in animal hosts has been reported. Serological studies of cattle herds have shown that large proportions of seronegative animals brought into a region where *C. burnetii* is endemic are exposed to *C. burnetii* (Huebner and Bell, 1951) and that the incidence of exposure may be increasing (Krauss, 1989). Some animals, including cattle and goats, may become chronically infected (Babudieri, 1959; Baca and Paretsky, 1983; Lang, 1990). There is a lack of current data on *C. burnetii* infection in livestock.

The pathogenicity of *C. burnetii* in various laboratory animals has been reviewed by Maurin and Raoult, 1999. Rats, mice, guinea pigs, rabbits, and monkeys have been used in laboratory studies; infection can produce no symptoms, can produce fever and granulomas, or can lead to death.

Interactions with host cells *C. burnetii* infects mammalian monocytes and macrophages *in vivo* (Dellacasagrande et al., 2000); the organism can be grown in embryonated eggs and in Vero, fibroblast, and mouse macrophagelike cells *in vitro*. *In vitro* cultures of *C. burnetii*-infected cells can be maintained for long periods (Burton et al., 1978; Baca et al., 1981, 1985; Akporiaye et al., 1983; Roman et al., 1986). Both the mechanisms of entry of Phase I and Phase II *C. burnetii* cells into mammalian cells and the adaptations that promote long-term survival and reproduction of *C. burnetii* cells inside mammalian cells specialized to kill them have been investigated extensively (Hackstadt and Williams, 1981; Akporiaye et al., 1983; Zuerner and Thompson, 1983; Hendrix and Mallavia, 1984; Roman et al., 1986; Chen et al., 1990b; Raoult et al., 1990; Maurin et al., 1992; Ghigo et al., 2002).

Antibiotic susceptibility Antibiotic susceptibility testing of *C. burnetii* has been carried out in cell cultures, in animals, and in embryonated eggs. These methods and the results obtained have been reviewed in detail by Maurin and Raoult (1999).

Vaccination Vaccination is a logical strategy for the prevention of Q fever both among humans at high risk for this infection and among animals. A killed Q fever vaccine was approved for use in Australia (O-Vax; Commonwealth Serum Laboratories). Further information about the effectiveness of this vaccine can be found in Ormsbee and Marmion (1990). In areas where Q fever is endemic, the following groups should be offered vaccination: dairy workers, abattoir workers; and others who work with cattle, sheep, or goats. Vaccination of cattle, sheep, and goats in these areas is also indicated, although few data are available to indicate the effectiveness and cost benefit of such a strategy in eliminating Q fever.

ENRICHMENT AND ISOLATION PROCEDURES

Isolation of *C. burnetii* requires a Biosafety Level 3 facility and poses a substantial risk to laboratory personnel. In the past, isolation was achieved by injecting potentially infected specimens into guinea pigs, mice, or embryonated eggs (Ormsbee, 1952; Perrin and Bengtson, 1942; Williams et al., 1986a); organisms are most readily recovered from the spleens of guinea pigs 5–8 days after infection. Tissue culture cells can be substituted for animals (Maurin and Raoult, 1999).

MAINTENANCE PROCEDURES

C. burnetii is maintained in liquid nitrogen but can be more conveniently maintained at −80°C. Phase I organisms are obtained by experimental passage in laboratory animals (guinea pig). Phase II organisms are obtained by passage in cultured cell systems or embryonated eggs.

DIFFERENTIATION OF THE GENUS *COXIELLA* FROM OTHER GENERA

Coxiella can be differentiated from *Rickettsiella* by its relatively high heat resistance and by its formation of an endospore-like structure. Although electron-dense forms have been observed in the developmental cycles of both *Rickettsiella* and *Coxiella burnetii* (Avakyan et al., 1983; Popov et al., 1991), *Rickettsiella* species do not develop intracellularly into larger particles that multiply and then condense to reform the smaller forms in a cycle that resembles that of *Chlamydia*.

TAXONOMIC COMMENTS

Phylogenetic analysis, chiefly based on 16S rRNA gene sequences, has shown that *Coxiella* belongs to the *Gammaproteobacteria* (Weisburg et al., 1985, 1989; Wilson et al., 1989b; Stein et al., 1993), with *Legionella*, *Francisella*, and *Rickettsiella* as its closest relatives (Fig. BXII.γ.98). Bacteria of the genus *Rickettsia* and of the family *Bartonellaceae* belong to the *Alphaproteobacteria* (Fig. BXII.γ.99). In addition, uncultured endosymbionts of ticks as diverse as *Rhipicephalus sanguineus*, *Haemaphysalis longicornis*, *Haemophysalis con-*

cinnae, and *Ornithodoros moubata* are closely related to *C. burnetii* by 16S rRNA gene sequence analysis (Noda et al., 1997). The 16S rDNA sequence of one rickettsia-like organism isolated in the yolk sac of chicken embryos using material from farm-reared *Cherax quadricarinatus* crayfish possessed 95.6% similarity to the *C. burnetii* 16S rRNA gene sequence; this organism has been proposed as *Coxiella cheraxi* sp. nov. (Tan and Owens, 2000) (Fig.BXII.γ.100), but the new name has not yet been validated.

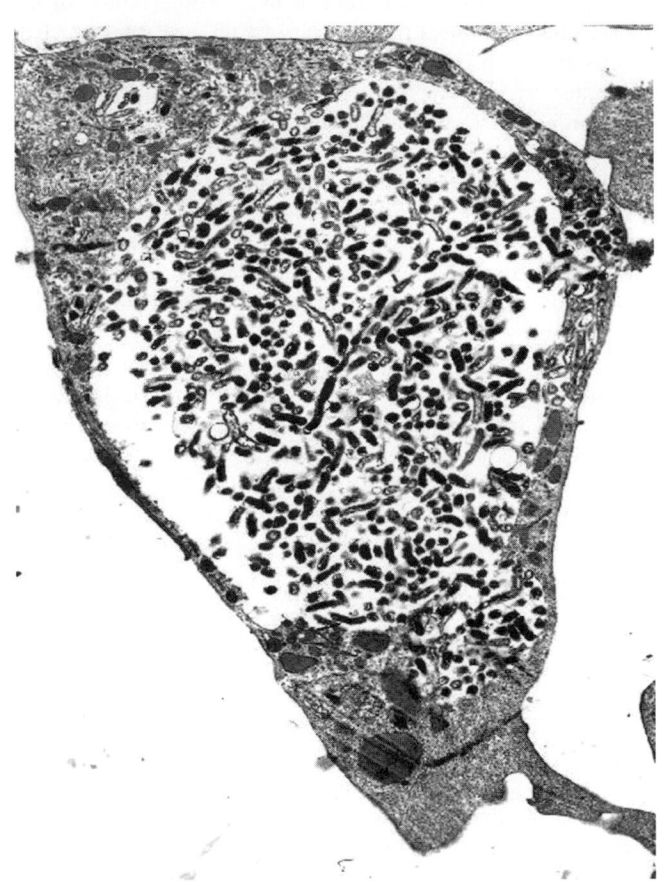

FIGURE BXII.γ.98. Transmission electron micrograph of *Coxiella burnetii* inside a vacuole of a Vero cell.

List of species of the genus Coxiella

1. **Coxiella burnetii** (Derrick 1939) Philip 1948, 58[AL] (*Rickettsia burneti* (sic) Derrick 1939, 14.)

   bur.ne′ ti.i. M.L. gen. n. *burnetii* of Burnet, named after Frank MacFarlane Burnet, who first studied the properties of this organism.

   The characteristics are as described for the genus. The complete genome sequence of strain RSA 493 is available (GenBank accession number NC002971).

   The mol% G + C of the DNA is: 42.6 (by sequencing).

   Type strain: ATCC VR 615 strain (Nine Mile Phase I).

   Reference strains include: (1) ATCC V5-145 [strain Henzerling; essentially in phase II]; (2) VR 616 [strain Nine Mile Q (Davis and Cox, 1938), essentially Phase II]; (3) VR 147 Strain Dyer, from blood from patient, USPHS Rocky Mountain Laboratory, 1938; (4) VR 542 Strain Ohio 314, from milk from cow; collected by Ohio Dept. Hlth., 1955; (5) VR 614 Strain California 76, from a dairy cow in California; and (6) VR 730 Strain Bangui, from a human in Central Africa.

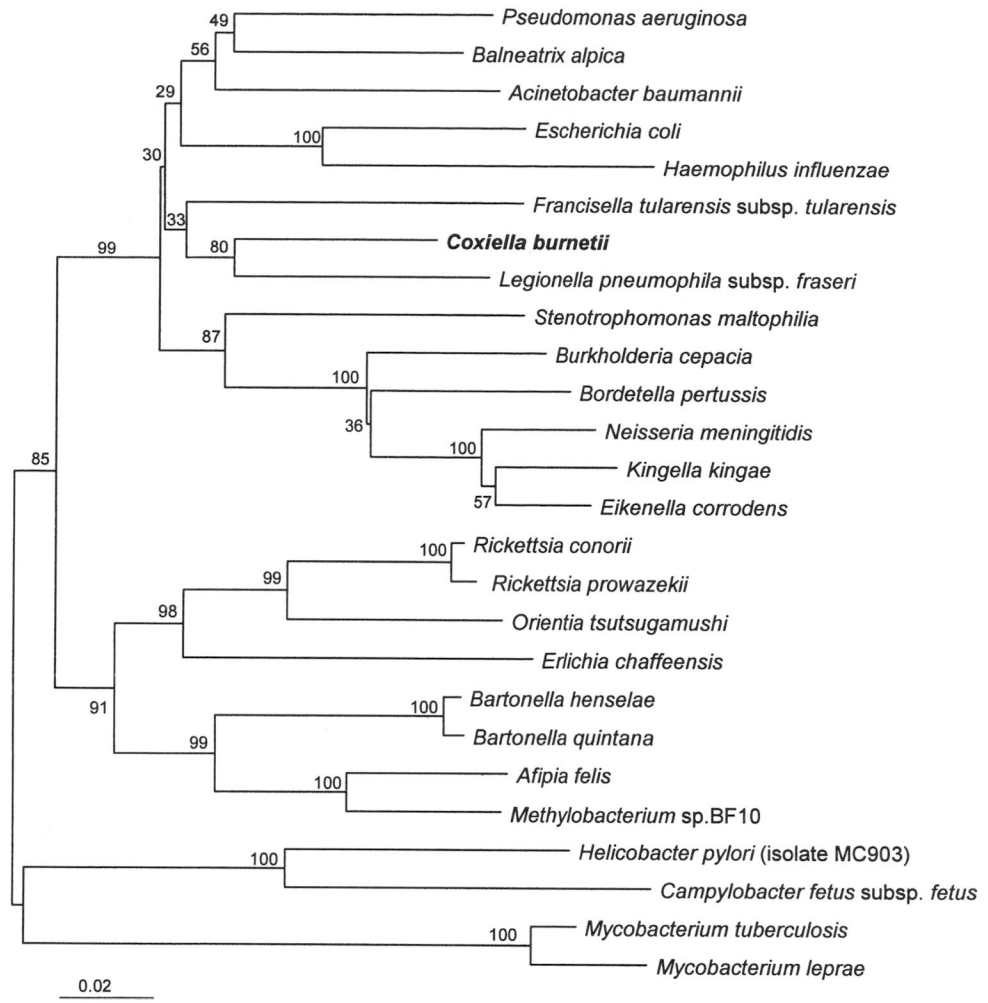

FIGURE BXII.γ.99. Dendogram based on 16S rRNA gene sequence comparison of *Coxiella burnetii* with other *Proteobacteria*. This dendogram includes *Coxiella* sequences derived from: *Coxiella burnetii* (strain 607, taxon 777), GenBank accession number D89792; *Francisella tularensis* (strain SCHU, isolate FSC 043, taxon:263), GenBank accession number Z21932; *Rickettsia conorii* (taxon 781), GenBank accession number L36107; *Rickettsia prowazekii* (taxon 782), GenBank accession number M21789; *Bartonella henselae* (strain BA-TF, taxon:38323, clone PL 84-1), GenBank accession number Z11684; *Bartonella quintana* (taxon 803), GenBank accession number AJ250247; *Mycobacterium tuberculosis* (taxon 1773), GenBank accession number X52917; *Mycobacterium leprae* (taxon 1769), GenBank accession number X55022; *Pseudomonas aeruginosa* (taxon 287), GenBank accession number M34133; *Haemophilus influenzae* (strain NCTC 11146, taxon:67859), GenBank accession number AF024530; *Orientia tsutsugamushi* (taxon 784), GenBank accession number AF062074; *Ehrlichia chaffeensis* (taxon 945), GenBank accession number AF147752; *Afipia felis* (taxon 1035), GenBank accession number M65248; *Methylobacterium* sp.BF10 (taxon 194432), GenBank accession number Z23156; *Bordetella pertussis* (ATCC 9797, taxon 520), GenBank accession number U094950; *Burkholderia cepacia* (taxon 292), GenBank accession number AB091761; *Kingella kingae* (ATCC 23330, taxon 504), GenBank accession number M22517; *Neisseria meningitidis* (strain NCTC 10025 [T], taxon 487), GenBank accession number X74900; *Helicobacter pylori* (isolate MC903, taxon 210), GenBank accession number U01332; *Campylobacter fetus* (taxon 196), GenBank accession number M65012; *Stenotrophomonas maltophilia* (strain N4-15, taxon:40324), GenBank accession number AF017749; *Legionella pneumophila* (strain Los Angeles 1, taxon:44), GenBank accession number M36025; *Balneatrix alpica* (taxon:75684), GenBank accession number Yl7112; *Acinetobacter baumannii* (isolate DSM 30008, taxon 470), GenBank accession number X81667; *Escherichia coli* (taxon 562), GenBank accession number J01859; *Eikenella corrodens* (taxon 539), GenBank accession number M22513.

Genus II. **Rickettsiella** *Philip 1956, 267[AL]*

PIERRE-EDOUARD FOURNIER AND DIDIER RAOULT

Rick.ett.si.el'la. M.L. dim. *-ella* ending; M.L. fem. n. *Rickettsia* genus of parasitic bacteria; M.L. fem. n. *Rickettsiella* small *Rickettsia*.

Rod- or disk-shaped organisms, developing intracellularly into larger particles that multiply and then condense to reform the smaller forms in a cycle that resembles that of *Chlamydia*. Sometimes produce or induce the formation of large crystalline bodies. Gram negative. **Grow in cell vacuoles of the fat body, he-patopancreas, and other organs of insects and other invertebrate hosts.** In some instances the organisms have been cultivated in invertebrates other than the host of origin, but such cultivation has occurred rarely and only for a few passages in vertebrate and invertebrate cell cultures and in a cell-free medium. **Pathogenic**

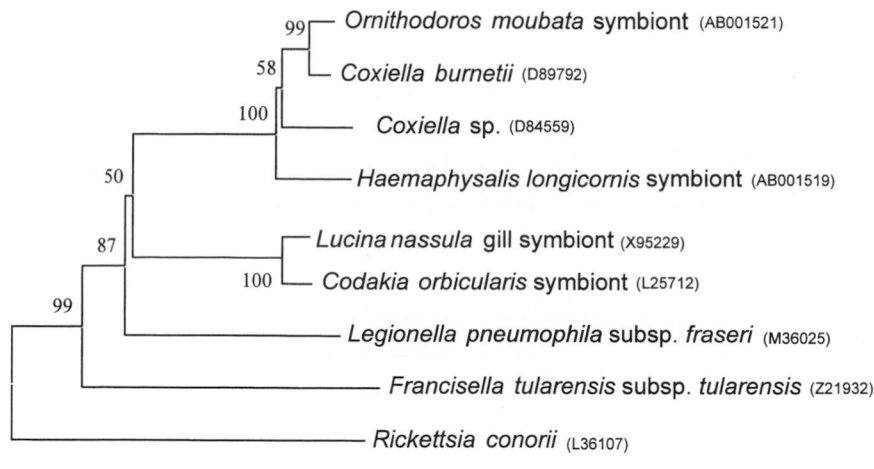

FIGURE BXII.γ.100. Dendrogram based on 16S rRNA gene sequence comparison of *Coxiella burnetii* to its close relatives. *Rickettsia conorii* (part of the *Alphaproteobacteria*) was used as an outgroup to root the tree. Bootstrap values are indicated for each node. Bar = 2 inferred nucleotide substitutions per 100 nucleotides.

for the larvae of the host species and for young and mature stages of other invertebrate hosts. Pathogenicity for vertebrate hosts demonstrated only in a mouse experimental model. Natural hosts include insects, crustaceans, and arachnids.

The mol% G + C of the DNA is: 36–41.

Type species: **Rickettsiella popilliae** (Dutky and Gooden 1952) Philip 1956, 267 (*Coxiella popilliae* Dutky and Gooden 1952, 749)

FURTHER DESCRIPTIVE INFORMATION

Most of the knowledge of this genus is based on light and electron microscopic observations. Bacteria were classified in the genus *Rickettsiella* mainly based on morphologic criteria, including their intracellular location, their oval, rod-like, or pleomorphic forms, the occurrence of a complex intravacuolar cycle, and the occurrence of crystals. Studies of host specificity, attempts at cultivation in cells or axenic media, or antigenic analyses have not been extensive and only minimally successful. Table BXII.γ.63 lists the species that, to the best of our limited knowledge, appear to belong to this genus.

Phylogenetic classification In the previous edition of the *Manual*, the genus *Rickettsiella* was classified in the order *Rickettsiales* based on a very few morphological data, including a close association with arthropods and intravacuolar location during multiplication. Recently, analysis of the 16S rDNA sequence of *Rickettsiella grylli* has revealed that this bacterium is most closely related to *Coxiella burnetii* (Philip, 1943) in the *Gammaproteobacteria*; therefore, it has been provisionally removed from the order *Rickettsiales* (Roux et al., 1997). An analysis of the 16S rDNA sequences from another two rickettsiellae—isolates from the psyllid *Cecidotrioza sozanica* (*Insecta, Psylloidea*) (Spaulding and Hennessy, 1960) and from the tick *Ixodes woodii* (*Arachnida, Acarina*) (Kurtti et al., 2002)—has demonstrated that these organisms cluster with *R. grylli* with a high statistical probability (Fig. BXII.γ.101). The 16S rDNA sequence of *R. grylli* has 98.2% and 94.3% similarity to that of the psyllid and the tick isolates, respectively, and 88.3% similarity to that of *Coxiella burnetii*. However, the 16S rDNA sequences of *Rickettsiella popilliae* and *Rick-*

ettsiella chironomi are not available, and thus their phylogenetic position is not clear.

Cell structure All strains studied in some detail have been shown to undergo a cycle of development. The infectious form is a small, dense particle (elementary body) that gains entrance into a vacuole of its host cell by phagocytosis. There it enlarges into intermediate and large forms of much lower electron density and which are capable of multiplication (reticulate body). As the host-cell vacuole becomes crowded with multiplying bacteria, the large forms condense to reform the small dense particles, which eventually escape from the vacuole to start another cycle.

There are striking differences between the cycles of *R. popilliae* and *R. chironomi*. The cycle of *R. grylli* resembles that of *R. popilliae*. *R. popilliae* cells, although displaying some plasticity, retain typical outer membranes and a rod-shaped appearance through most of their cycle (Devauchelle et al., 1972). Some large forms become giant round cells from which bipyramidal crystalline bodies (0.8–1.8 × 3.8 μm in size) arise. The roles of the bacteria and of the host cell in the production of the crystalline bodies are not entirely clear. Except for the absence of tyrosine, these bodies have approximately the same amino acid composition as the albuminoid spheres that are found in greatest numbers in late larval and pupal stages in normal insects. It has been postulated that the crystalline bodies are derived from the albuminoid reserve as the result of a disturbance of host cell metabolism (Huger, 1959; Krieg, 1959). Other evidence suggests a close association of the crystalline bodies with the bacterial cycle (Huger, 1962; Devauchelle et al., 1972) and host dependence. A *Cetonia* strain (Table BXII.γ.63) produces crystalline bodies in *Coleoptera* but not in *Orthoptera* or *Lepidoptera* (Meynadier and Monsarrat, 1969), and a *Melolontha* strain does not produce crystalline bodies in mammalian cell cultures (Pourquier et al., 1963).

In *R. chironomi*, the infectious particle is a disk-shaped elementary body 0.06 × 0.6 μm in size. This body enlarges in a newly infected cell into a spherical initial body 1 μm in diameter,

TABLE BXII.γ.63. Ecology of the species of the genus *Rickettsiella* and proposed nomenclature

Nomenclature used in the present edition of the *Manual*	Host zoological class	Host zoological order	Host common name (scientific name)	Location	Subjective synonyms	References
R. popilliae	Insecta	Coleoptera	Japanese beetle (*Popillia japonica*)	USA	"*Rickettsiella popilliae*"	Dutky and Gooden, 1952; Philip, 1956
			Other scarabeid and carabid beetles	USA, UK	Unnamed	Sutter and Kirk, 1968; Carter and Luff, 1977
			Cockchafer (*Melolontha melolontha*)[a]	Germany	"*Rickettsiella melolonthae*" ("*Rickettsia melolonthae*")	Krieg, 1955a; Philip, 1956
			Mealworm (*Tenebrio molitor*)	Germany	"*Rickettsiella tenebrionis*"	Krieg, 1965
			Cetonid beetle (*Cetonia* sp.)	Madagascar	"*Rickettsiella cetonidarum*"	Meynadier and Monsarrat, 1969
		Diptera	Crane fly (*Tipula paludosa*)	Germany	"*Rickettsiella tipulae*"	Müller-Kogler, 1958
		Dictyoptera	Cockroach (*Blatta orientalis*)	Germany	"*Rickettsiella blattae*"	Huger, 1964
	Crustacea	Isopoda	Isopode (*Armadillidium vulgare*)[a]	France	"*Rickettsiella armadillidii*"	Vago et al., 1970
R. grylli	Insecta	Orthoptera	Cricket (*Gryllus bimaculatus*)	France	*Rickettsiella grylli*	Vago and Martoja, 1963
			Desert locust (*Schistocerca gregaria*)	Jordan	"*Rickettsiella schistocercae*"	Vago and Meynadier, 1965
	Crustacea	Amphipoda	Amphipod (*Crangonyx floridanus*)	USA	Unnamed	Federici et al., 1974
R. chironomi	Insecta	Diptera	Midge (*Chironomus tentans*)[a]	Germany	*Rickettsiella chironomi*	Weiser, 1963
	Arachnida	Aranea	Spider (*Argyrodes gibbosus*)	Spain (Canary islands)	Unnamed	Meynadier et al., 1974
Organisms for which insufficient data exist to support assignment to a species	Insecta	Coleoptera	Spider (*Pisaura mirabilis*)	France	Unnamed	Morel, 1977
			Stethorus beetle (*Stethorus punctum*)	Morocco	*Rickettsiella stethorae*	Hall and Badgley, 1957
		Lepidoptera	Sturnid moth (*Samia cynthia*)	UK	Unnamed	Entwistle et al., 1968
			Navel orange worm (*Paramyelois transitella*)	USA	Unnamed	Kellen et al., 1972
		Psylloidea	Psyllid (*Cecidotrioza sozanica*)	USA	Unnamed	Spaulding and von Dohlen, 2001
		Dictyoptera	Cockroach (*Blatta orientalis*)	Germany	"*Rickettsiella crassifcans*"	Radek, 2000
	Arachnida	Acarina	Mite (*Phytoseiulus persimilis*)	USSR	"*Rickettsiella phytoseiuli*"	Sutakova and Ruttgen, 1978
			Tick (*Ixodes woodii*)	USA	*Rickettsiella* sp. Strain GSU	Kurtti et al., 2002
		Scorpionida	Scorpion (*Buthus occitanus*)	France	"*Rickettsiella buthi*" ("*Porochlamydia buthi*")	Morel, 1976
	Crustacea	Decapoda	Crab (*Carcinus mediterraneus*)	France (Mediterranean sea)	Unnamed	Bonami and Pappalardo, 1980
			Redclaw crayfish (*Cherax quadricarinatus*)	UK	Unnamed	Romero et al., 2000
		Isopoda	Isopode (*Porcellio scaber*)	Slovenia	Unnamed	Drobne et al., 1999

[a]And related species.

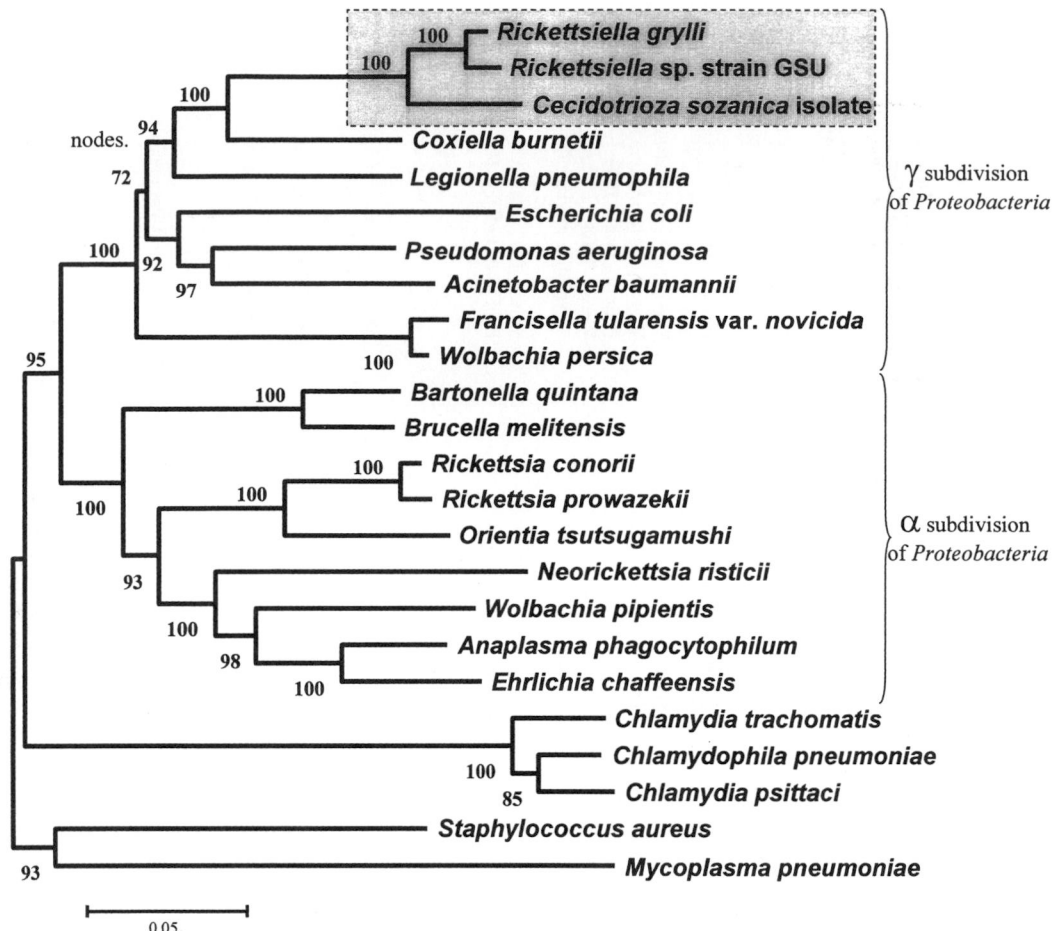

FIGURE BXII.γ.101. Unrooted 16S rDNA-based phylogenetic tree showing the position of *Rickettsiella grylli*. The tree was constructed using the neighbor-joining method, and the bootstrap values are indicated at the nodes. This tree includes *Rickettsiella* sequences derived from: *Rickettsiella grylli*, GenBank accession number U97547; *Rickettsiella* sp. (strain GSU), GenBank accession number AF383621; isolate from *Cecidotrioza sozanica*, GenBank accession number AF286124; *Coxiella burnetii* (strain Q177), GenBank accession number M21291; *Legionella pneumophila* (strain ATCC 33152), GenBank accession number M59157; *Escherichia coli*, GenBank accession number Z83204; *Pseudomonas aeruginosa*, GenBank accession number AB073312; *Acinetobacter baumannii* (ATCC 19606T), GenBank accession number Z93435; *Francisella tularensis* subsp. *novicida*,* GenBank accession number L26084; *"Francisella persica"* (= *Wolbachia persica*), GenBank accession number M21292; *Bartonella quintana* (strain Fuller), GenBank accession number M11927; *Brucella melitensis* biovar abortus, GenBank accession number X13695; *Rickettsia conorii* (strain Moroccan), GenBank accession number L36105; *Rickettsia prowazekii* (strain Breinl), GenBank accession number M21789; *Orientia tsutsugamushi* (strain Karp), GenBank accession number D38623; *Neorickettsia risticii* (strain SHSN-1), GenBank accession number AF037210; *Wolbachia pipientis*, GenBank accession number U23709; *Anaplasma phagocytophilum*, GenBank accession number U02521; *Ehrlichia chaffeensis*, GenBank accession number AF147752; *Chlamydia trachomatis* (strain:B/TW-5/OT), GenBank accession number D85719; *Chlamydia pneumoniae* (strain TW183), GenBank accession number L06108; *Chlamydia psittaci* (strain 6BC, ATCC VR-125T), GenBank accession number AB001778; *Staphylococcus aureus* (strain M2), GenBank accession number Y15585; *Mycoplasma pneumoniae* (strain FH, ATCC 15531), GenBank accession number M29061.

*Editorial Note: In this volume Sjöstedt proposes that *Francisella tularensis* var. *novidica* be regarded as a subspecies of *Francisella tularensis* (see page 208).

which divides by binary fission. The initial bodies are eventually reduced to intermediate forms that subsequently condense to reform the elementary bodies. Occasionally the initial bodies continue to grow to become particles 1.5 × 2.0 μm in diameter, which divide equally or unequally (Federici, 1980), or which may undergo multiple divisions that result in the formation of as many as 30 elementary bodies (Götz, 1972). Unlike those of *R. popilliae*, the giant cells in *R. chironomi* do not evolve into typical crystalline bodies (Weiser and Zizka, 1968).

Cultivation Most strains of *R. popilliae* have been grown in the larvae of *Melolontha*, the common cockchafer. The *R. popilliae* and *R. grylli* strains from the cetonid beetle have been grown in *Orthoptera*, *Coleoptera*, and *Lepidoptera* (Meynadier and Monsarrat, 1969). Intracoelomically inoculated female *Dermacentor reticulatus* ticks (*Arachnida, Acarina*) have been used to isolate and maintain *"Rickettsiella phytoseiuli"*, a *Rickettsiella* sp. isolated from the mite *Phytoseiulus persimilis* (*Arachnida, Acarina*) (Sutakova and Ruttgen, 1978).

Limited growth of *R. popilliae* has been obtained in chicken embryo endodermal cell cultures and in McCoy cells incubated at 28 or 32°C. Typical organisms were first detected after 1 week and their numbers increased during the following 2 weeks. Infectivity was reduced progressively during serial passage. No growth occurred after 4–5 passages (Suitor, 1964). *R. popilliae* has also been passed three times in other mammalian cell lines (Pourquier et al., 1963). *R. grylli* was grown in cricket cardiac cell cultures (Meynadier et al., 1967).

The only cultivation of a *Rickettsiella* in a cell-free medium has been reported for *"R. phytoseiuli"* in SM IMV-72 medium, which is used for cultivation of phytopathogenic mycoplasmas (Sutakova et al., 1991). The bacteria were maintained until day 66 in primary culture and until day 16 after the first passage. Further passages failed.

Drugs and antibiotic susceptibility Currently, there are no data on the antimicrobial susceptibility of rickettsiellae available.

Pathogenicity The chief interest in rickettsiellae stems from the effect of these organisms on laboratory insectaries and other animal collections as well as from the possible use of these organisms as biological control agents for agricultural pests. Rickettsiellae are not as virulent and persistent as some of the viral and mycotic pathogens of agricultural pest insects; however, rickettsiellae stay alive in the soil for years. Infection of offspring is effected through contamination of soil, rather than by transovarian passage (Hurpin and Robert, 1972, 1976, 1977). The wide distribution of rickettsiellae geographically and in arthropod taxa suggests an early appearance in the course of evolution.

R. popilliae causes blue disease of insect larvae, a deadly disease characterized by the discoloration of the infected larvae. In the cricket and locust, *R. grylli* causes swelling of the abdomen and, at later stages, turgidity of the intersegmentary membranes and ventral inclination of the head, leading to death. *R. grylli* induces a "behavioral fever" in experimentally infected crickets *Acheta domesticus* and *Gryllus bimaculatus* (Louis et al., 1986; Adamo, 1998). These infected invertebrates, when exposed to a temperature gradient, are more likely to survive at 32–33°C than at 26°C—their optimal temperature when uninfected. Infected insects move to areas that are at the higher temperature. This behavior is thought to be used by the insects to create a fever and protect them from infection.

Rickettsiella species have never been reported to be human pathogens. Virulence for vertebrate hosts has not extensively been studied. *R. popilliae* (*Melolontha* strain) is virulent for mice only when massive doses are inoculated intraperitoneally (Krieg, 1955b). Successful infection by the intranasal route has also been demonstrated (Giroud et al., 1958). When inoculated intraperitoneally or by inhalation, *R. grylli* multiplies in the mouse, but the severity of the infection depends on the inoculum size. With a low inoculum, the bacterium has either no effect or causes local inflammation; the mouse overcomes the infection and the microorganism disappears within 45 d (Croizier and Meynadier, 1972; Delmas and Timon-David, 1985).When injected with a high inoculum, mice develop a cirrhotic hepatitis and malignant tumors, including hematosarcomas.

ENRICHMENT AND ISOLATION PROCEDURES

Successful propagation of *Rickettsiella* has been achieved by the inoculation of healthy larvae of the same or compatible insect species with infected blood or other infectious material. Japanese beetles or cockchafers inoculated with *R. popilliae* are incubated under appropriate conditions of temperature, food and moisture for about 35 d. Following surface sterilization, the organisms are obtained from the surviving larvae by grinding. The bacteria can be separated from tissue components by cycles of adsorption and dilution from Celite and filtration through diatomaceous filters of medium porosity (Dutky, 1959) or by comparable more recently developed methods such as differential centrifugation (Morel, 1980).

Isolation in cell culture is useful for the study of antigenic cross-reactions by fluorescent antibody techniques or for other comparative investigations. However, unlimited growth in cell culture has not been demonstrated.

MAINTENANCE PROCEDURES

If *R. popilliae* suspensions are not contaminated with other bacteria, viability can be retained for 1 year at 4°C but not for 3 years. Viability is retained for least 3 years at −70°C.

DIFFERENTIATION OF THE GENUS *RICKETTSIELLA* FROM OTHER GENERA

These intracellular bacteria are pathogenic for, or symbiotic with, invertebrates and exhibit a developmental cycle similar to that of *Chlamydia* species. Rickettsiellae are phylogenetically related to *Coxiella burnetii* within the *Gammaproteobacteria*.

TAXONOMIC COMMENTS

Little of the sort of information that is generally used for bacterial classification is available for the classification of *Rickettsiella*.

Historically, based on the similarity in the ultrastructure and developmental cycles, many investigators suggested that rickettsiellae more closely resembled *Chlamydia* than *Rickettsia* and that they should be transferred to the order *Chlamydiales* (Yousfi et al., 1979; Morel, 1980; Avakyan and Popov, 1984). Points of similarity with *Chlamydia* are: (1) both *Chlamydia* and rickettsiellae multiply in the vacuoles of their host cells, and (2) both undergo a developmental cycle consisting of up to six stages that involve the formation of a small dense particle whose function is to infect and a much larger particle of lower density whose function is to multiply. Points of difference between *Chlamydia* and rickettsiellae are (1) the rod or disk shape of the infectious particles of rickettsiellae, (2) the formation of large crystalline bodies, and (3) the absence of relatedness between *Rickettsiella* and *Chlamydia* based on serological analyses (Croizier et al., 1975). Resemblances to other members of the *Rickettsiales* include a close association with arthropods and the intravacuolar location during multiplication—characters that are found in *Ehrlichia* and other genera in the order *Rickettsiales*. By pulsed field gel electrophoresis, the genome sizes of members of the genus *Rickettsiella* range from 1.55 Mb to 2.65 Mb (Frutos et al., 1989) (Table BXII.γ.64). On the basis of DNA–DNA hybridization, DNA restriction site polymorphism, DNA melting profiles, and mol% G + C data (Table BXII.γ.64), Frutos et al. (1994) concluded that the classification of members of the genus *Rickettsiella* was valid at the generic level but not at the species level. Three distinct groups were identified: (1) the *R. grylli* complex containing *R. grylli* and *R. popilliae*, with *"Rickettsiella armadillidii"* being recognized as a strain of *R. popilliae* rather than a synonym of *R. grylli*; (2) *R. chironomi*; and (3) *"R. buthi"*, a strain previously considered to

TABLE BXII.γ.64. Differential characteristics of the recognized species of the genus *Rickettsiella*[a]

Characteristics	*R. popilliae*	*R. chironomi*	*R. grylli*
Morphology of infectious particles:			
Rod	+	−	+
Disk	−	+	−
Crystalline body production	+	−	+
Serological reaction with anti-*R. popilliae* serum	+	−	−
Natural hosts:			
Insecta	+	+	+
Coleoptera	+	−	−
Diptera	+	+	−
Orthoptera	−	−	+
Crustacea	+	−	+
Arachnida	−	+	−
Genome size (Mb)	1.72	2.65	2.1
Mol% G + C of the DNA	38.9	36.7	36.3–38.4
DNA thermal denaturation midpoint (°C)			
In 0.1 × SSC	69.1	68.2	68.0–68.9
In 1 × SSC	85.3	84.4	84.2–85.1

[a]Symbols: +, positive feature; −, negative feature.

belong to the species *R. chironomi*, from the scorpion *Buthus occitanus* (*Arachnida, Scorpionida*) (Frutos et al., 1994). According to Frutos et al., *Rickettsiella, Coxiella, Chlamydia,* and *Rickettsia* are taxonomically distinct genera, and the classification of the genus *Rickettsiella* into a separate order might be justifiable (Frutos et al., 1994). However, based on 16S rDNA sequence analysis, *R. grylli* is most closely related to *Coxiella burnetii* and *Legionella* species in the *Gammaproteobacteria*, whereas members of the order *Rickettsiales* belong to the *Alphaproteobacteria* (Roux et al., 1997) (Fig. BXII.γ.101). This finding is supported by the observation of electron-dense forms in the developmental cycles of both *Rickettsiella* and *Coxiella burnetii* (Avakyan et al., 1983; Popov et al., 1991). Based on 16S rDNA analysis, *Rickettsiella* strain GSU—which is present in the ovarian tissues and Malpighian tubules of female *Ixodes woodii* ticks (*Acarina: Ixodidae*)—has also been classified in the *Gammaproteobacteria* (Kurtti et al., 2002) as has a *Rickettsiella*-like endosymbiont of the psyllid *Cecidotrioza sozanica* (*Insecta*) (Spaulding and von Dohlen, 2001). Both bacteria were closely related phylogenetically to *R. grylli*. However, because the 16S rDNA sequences of *R. popilliae* and *R. chironomi* have not yet been determined, it is not known whether all species of the genus *Rickettsiella* are misplaced in the order *Rickettsiales* and whether all current members of the genus are phylogenetically homologous (Roux et al., 1997).

The differential characteristics of the species of the genus *Rickettsiella* are presented in Table BXII.γ.64. Six specific names previously proposed (Table BXII.γ.63) are treated as synonyms of the first described species, *R. popilliae*, since no major differences have been encountered among them. The *Popillia, Melolontha, Tipula,* and *Cetonia* pathogens have common antigens (Krieg, 1958) (Croizier and Meynadier, 1971). The pathogens of *Tenebrio* and *Blatta* have not been extensively studied.

The separation of *R. popilliae* from *R. grylli* was initially based primarily on serological cross-reactions between the cricket and isopod pathogens and lack of reaction with *R. popilliae* antisera (Croizier and Meynadier, 1971). The developmental cycles of the cricket and isopod pathogens appear to be quite similar (Louis et al., 1977a, b), but the results of DNA hybridization studies indicate that *R. grylli* and *R. popilliae* are separate species, with "*R. armadillidii*"—the isopod isolate—showing greater DNA re-

latedness to *R. popilliae* than to *R. grylli* (Frutos et al., 1994). The hosts from which *R. grylli* have been isolated—*Orthoptera* and Crustacea—may not constitute a valid criterion for classification, but they might serve as a guide for the provisional placement of the less well-studied pathogens of the desert locust and amphipod (Table BXII.γ.63). *R. popilliae* and *R. grylli* have different genome sizes, but their DNAs have similar mol% G + C contents and DNA melting temperatures (Table BXII.γ.64).

The third species, *R. chironomi*, can be clearly separated from the other two by its unusual life cycle, in which flat disks are formed instead of rods. There is a remarkable similarity between the cycle of the midge pathogen and the cycles of the arachnid pathogens (Morel, 1977; Federici, 1980), although differences have been described (Louis et al., 1979). In 1976, the creation of the genus "*Porochlamydia*" was proposed to accommodate the phenotypically different isolates from the midge and the scorpion, with "*P. chironomi*" being present in the midge larvae and "*P. buthi*" infecting the scorpion *Buthus occitanus* (*Arachnida, Scorpionida*). This genus was not included in the Approved Lists or the 1984 issue of the *Manual*. The midge and scorpion isolates were later shown to have chromosomal DNAs markedly different in size (Frutos et al., 1989), to exhibit no DNA relatedness as demonstrated by the absence of cross-hybridization, and to have distinct DNA melting profiles. However, the differentiation of the scorpion isolate as a different species has not been officially proposed.

The stethorus beetle pathogen, for which the name *R. stethorae* has been assigned (Table BXII.γ.63), is clearly different in morphology from the above listed species of *Rickettsiella* (Hall and Badgley, 1957), but further studies are necessary to allow proper classification. The other isolates listed in Table BXII.γ.63 appear to be typical rickettsiellae, but additional information is required for appropriate placement in a species.

It should be noted that four of the species recognized in the present edition of the *Manual, R. popilliae, R. grylli, R. chironomi,* and *R. stethorae* were included on the Approved Lists of Bacterial Names. However, as indicated above, sufficient evidence is not available at present to warrant considering *R. stethorae* as a distinct species.

FURTHER READING

Devauchelle, G., G. Meynadier and V. C. Vago. 1972. Étude ultrastructurale du cycle de multiplication de *Rickettsiella melolonthae* (Krieg), Philip, dans les hémocytes de son hôte. J. Ultrastruc. Res. *38*: 134–148.

Dutky, S.R. 1959. Insect microbiology. Adv. Appl. Microbiol. *1*: 175–200.

Federici, B.A. 1980. Reproduction and morphogenesis of *Rickettsiella chironomi*, an unusual intracellular procaryotic parasite of midge larvae. J. Bacteriol. *143*: 995–1002.

Frutos, R., B.A. Federici, B. Revet and M. Bergoin. 1994. Taxonomic studies of *Rickettsiella, Rickettsia,* and *Chlamydia* using genomic DNA. J. Invertebr. Pathol. *63*: 294–300.

Götz, P. 1972. "*Rickettsiella chironomi*": an unusual bacterial pathogen which reproduces by multiple cell division. J. Invertebr. Pathol. *20*: 22–30.

Louis, C. , G. Morel, G. Nicolas and G. Kuhl. 1979. Étude d' une *Rickettsiella* (Rickettsie) se développant chez un arachnide, l' araignée *Pisaura mirabilis*. Ann. Microbiol. (Institut Pasteur). *128A*: 49–59.

Roux, V., M. Bergoin, N. Lamaze and D. Raoult. 1997. Reassessment of the taxonomic position of *Rickettsiella grylli*. Int. J. Syst. Bacteriol. *47*: 1255–1257.

Suitor, E.C., Jr. 1964. Propagation of *Rickettsiella popilliae* (Dutky and Gooden) Philip and *Rickettsiella melolonthae* (Krieg) Philip in cell cultures. J. Insect Pathol. *6*: 31–40.

DIFFERENTIATION OF THE SPECIES OF THE GENUS *RICKETTSIELLA*

Characteristics useful for the differentiation of *Rickettsiella* species are listed in Table BXII.γ.64.

List of species of the genus Rickettsiella

1. **Rickettsiella popilliae** (Dutky and Gooden 1952) Philip 1956, 267[AL] (*Coxiella popilliae* Dutky and Gooden 1952, 749.) *pop.il' li.ae.* M.L. gen. n. *popilliae* of *Popillia*, the generic name of the Japanese beetle, one of its hosts.

This species includes the subjective synonyms *"Rickettsiella melolonthae"*, *"Rickettsiella tenebrionis"*, *"Rickettsiella cetonidarum"*, *"Rickettsiella tipulae"*, *"Rickettsiella blattae"*, *"Rickettsiella armadillidii"*, and an unnamed isolate from various scarabaeid and carabid beetles (Table BXII.γ.63). The form usually obtained from infected host tissues by the separation procedures described above is a rod 0.2 × 0.6 μm or slightly larger; the rods may be oval, curved or kidney-shaped, with rounded edges. Satisfactorily stained by the Giemsa or Macchiavello methods (Macchiavello, 1937) and presumably by the Gimenez (1964) method. The developmental cycle, the formation of crystalline bodies, and attempts at cultivation outside the host have been discussed above in the section on Further Descriptive Information.

R. popilliae is the etiological agent of blue disease of insect larvae, a name reflecting the discoloration of the infected larvae. Other names reflecting either the locality where the disease was discovered or the host have also been used. Infection is most commonly encountered among larvae but is not confined to this stage. The infection starts in the fat body but ultimately spreads to the blood and the other organs. Larvae have been infected experimentally by injection, by feeding, or by holding them in soil inoculated with suspensions of the organism. By injection, fewer than six organisms are sufficient to infect a Japanese beetle larva with its naturally occurring pathogen. The time required for appearance of symptoms is dependent on dosage and on the temperature of incubation of the larvae (optimum 27°C). With an injection of 10^4 organisms per larva, the symptoms appear after 19 d and death occurs in 19–26 d after the first appearance of symptoms. Larvae can be protected by injection of streptomycin (20 μg/larva) or sulfadiazine (200 μg/larva) but not penicillin or chlortetracycline (Dutky, 1959).

There is considerable variation in the time of appearance of symptoms and death of the larvae infected by the various strains of *R. popilliae*. It is not known to what extent this variation reflects strain virulence or host susceptibility. The organisms can be inactivated by incubation at 60°C for 10 min. The genome size is 1.72 Mb (Frutos et al., 1989, 1994).

The mol% G + C of the DNA is: 36.9 (T_m).

Type strain: no culture has been isolated.

2. **Rickettsiella chironomi** (ex Weiser 1963) Weiss, Dasch and Chang 1984b, 356[VP] (Effective publication: Weiss, Dasch and Chang 1984a, 716.) *chi.ro.no' mi.* M.L. gen. n. *chironomi* of *Chironomus*, the generic name of the midge, one of its hosts.

This species includes unnamed isolates from the midge *Chironomus tentans* and the spider *Argyrodes gibbosus* (Table BXII.γ.63). The infectious stage has an unusual disk-shaped morphology, 0.06 × 0.6 μm, as indicated above in the section on Further Descriptive Information. There is probably a greater diversity among the strains previously included in this species than among the strains of the other two species of *Rickettsiella*. "R. buthi", which infects the scorpion *Buthus occitanus* (*Arachnida, Scorpionida*) has a markedly different chromosomal DNA size (Frutos et al., 1989), exhibits no DNA relatedness to *R. chironomi*, and has a DNA melting profile distinct from that of *R. chironomi* (Frutos et al., 1994) (Table BXII.γ.63). However, the differentiation of the scorpion isolate as a different species has not been officially proposed.

In the midge, the disease affects primarily the last two instars and the pupal stages. The infection starts in the fat body and progresses to lobes around the intestine, which eventually rupture and fill the whole body with white fluid. Experimental infection results in death in 7–14 d (Weiser and Zizka, 1968). In arachnids, the hepatopancreas and intestinal diverticula are the organs primarily affected. The genome size of *R. chironomi* is 2.65 Mb, whereas the genome size of "R. buthi" is 1.55 Mb and its mol % G + C of the DNA is 41 (T_m) (Frutos et al., 1989, 1994).

The mol% G + C of the DNA is: 36.7 (T_m).

Type strain: no culture has been isolated.

3. **Rickettsiella grylli** (ex Vago and Martoja 1963) Weiss, Dasch and Chang 1984b, 356[VP] (Effective publication: Weiss, Dasch and Chang 1984a, 716.) *gryl' li.* M.L. gen. n. *grylli* of *Gryllus*, the generic name of the cricket, one of its hosts.

This species includes the subjective synonym *"Rickettsiella schistocercae"* and an unnamed isolate from the amphipod *Crangonyx floridanus* (Table BXII.γ.63). The organism resembles *R. popilliae* in the morphology of its infectious particle and subsequent developmental stages and crystalline body production. It can be differentiated from *R. popilliae* by serological tests and, to a certain extent, by the natural hosts infected, although it does not have a high degree of host specificity (Table BXII.γ.63).

The disease in naturally infected hosts affects both larvae and adults and develops very slowly. In the cricket and locust, it is characterized by swelling of the abdomen and, at later stages, by turgidity of the intersegmentary membranes and by ventral inclination of the head. The time of death following first appearance of symptoms is quite variable. In the isopod, death is preceded by loss of weight and a whitish coloration of the intersegmentary membranes due to the accumulation of iridescent fluid in the body cavities. Diseased amphipods are also recognized by their opaque pale green iridescence. Most patently diseased amphipods die within six weeks.

The genome size is 2.1 Mb (Frutos et al., 1989, 1994). Based on the 16S rDNA sequence, *R. grylli* is most closely related to *Coxiella burnetii* within the *Gammaproteobacteria* (Roux et al., 1997).

The mol% G + C of the DNA is: 36.3–38.4 (T_m).

Type strain: no culture has been isolated.

Order VII. **Methylococcales** *ord. nov.*

JOHN P. BOWMAN

Me.thy.lo.coc.ca'les. M.L. fem. pl. n. *Methylococcus* type genus of the order; *-ales* ending to denote an order; M.L. fem. pl. n. *Methylococcus* order.

Morphology varies from **rods to cocci**. Cells are usually single or diploid, occasionally forming chains and tetrads. Do not form rosettes. If motile, possess single polar flagellum. **Gram negative.** The majority of species produce a polysaccharide **cyst-like resting stage.** Cysts may be desiccation resistant or sensitive. Dissimilatory methane oxidation is associated with **type I intracytoplasmic membranes,** which appear as stacks of vesicular disks derived from convolutions of the cytoplasmic membrane. Catalase and oxidase are usually produced. **Strictly aerobic respiratory metabolism** with oxygen as the electron acceptor. **Obligate utilizers of methane and other C$_1$ compounds for carbon and energy. Compounds with carbon–carbon bonds are not utilized.** High levels of organic solutes and ammonia ions inhibit growth. C$_1$ compounds are incorporated by the **ribulose monophosphate pathway.** The **tricarboxylic acid cycle is incomplete,** with 2-oxoglutarate dehydrogenase activity absent. Nitrogenase activity may be present. Contain mainly **C$_{16}$ fatty acids,** which are predominantly monounsaturated with several different C$_{16:1}$ isomers present. **Ubiquinone-8 or 18-methyleneubiquinone-8** are the predominant lipoquinones. Habitat: isolated from aerobic and microaerobic habitats that have an influx of methane (from methanogenesis or hydrocarbon seeps), including benthic sediments of fresh water and marine bodies of water; oxyclines and thermoclines of lakes and oceans; wetland muds; coal mine drainage waters and coal surfaces; sewage sludge and wastewater; most soils; subsurface sediment; groundwater; and natural gas reserves. Some members form endosymbiotic relationships with invertebrates, including mytilid mollusks and pogonophora. Genome sizes range from 2.5–5.5 Mb. Member of the *Gammaproteobacteria*.

The mol% G + C of the DNA is: 43–65.

Type genus: **Methylococcus** Foster and Davis 1966, 1929, emend. Bowman, Sly, Nichols and Hayward 1993b, 748.

FURTHER DESCRIPTIVE INFORMATION

Bacteria that have the capability to oxidize and utilize methane as a sole carbon and energy source are referred to as methanotrophic bacteria or methanotrophs (Whittenbury et al., 1970b). The original description of the family *Methylococcaceae* (Whittenbury and Krieg 1984a) included two distinct groups of methanotrophs referred to in the literature as "type I" and "type II" methanotrophs. In addition, there is a third group called the "type X" methanotrophs (Hanson and Hanson, 1996) that essentially represents a subgroup of type I methanotrophs. Type I and II methanotrophs possess extensive biological differences that are morphological, biochemical, chemotaxonomic, and phylogenetic in nature. As type II methanotrophs belong to the *Alphaproteobacteria*, they are dealt with in a separate section of this volume. Currently, the family *Methylococcaceae* includes all type I methanotrophs (Bowman et al., 1995) in five genera: *Methylococcus, Methylosphaera, Methylobacter, Methylomicrobium,* and *Methylomonas*. Phenotypic properties and fatty acid components differentiating these genera are shown in Tables BXII.γ.65 and BXII.γ.66, respectively.

Methane oxidation   The major characteristic that separates methanotrophs from other procaryotes is their ability to utilize methane as a sole carbon and energy source. Methanotrophs oxidize methane to CO$_2$ in a dissimilatory pathway that generates energy and allows access to metabolizable carbon units in the form of formaldehyde. The critical first step of this pathway is the oxidation of methane to methanol, which is performed by methane monooxygenase (MMO). Methanol is then oxidized to formaldehyde by a pyrrolquinol quinone-containing methanol dehydrogenase. The formaldehyde is then either used for cell carbon (see below) or oxidized to formate and then CO$_2$; the

TABLE BXII.γ.65. Phenotypic characteristics differentiating member genera of the Family *Methylococcaceae*[a,b]

Characteristics	Methylococcus	Methylobacter	Methylomicrobium	Methylomonas	Methylosphaera
Cell morphology:					
Cocci					+
Cocci-rods	+				
Cocci-ellipsoid		+			
Rods			+	+	
Motility	D	D	+	+	−
Cyst formation	+	+	−	+	ND
Desiccation resistant	−	+	−	−	−
Pigmentation:					
Carotenoids	−	−	−	+	−
Melanin-like	D	D	−	−	−
Other	−	D	−	−	−
Growth at 25°C	−	+	+	+	−
Growth at 45°C	+	−	−	−	−
Requires seawater or Na$^+$	−	D	D	−	+
RuBisCo[c]	+	−	−	−	−
Nitrogen fixation	+	−	−	D	+
Major quinone[d]	MQ-8	Q-8	Q-8	MQ-8	ND
Mol% G + C of DNA (T_m)	59–65	48–55	48–60	50–59	43–46

[a]For symbols, see standard definitions.

[b]Data from Bowman et al. (1993b) and Bowman et al. (1997c).

[c]RuBisCo, ribulose-1,5-bisphosphate carboxylase.

[d]MQ-8, 18-methylene-quinone-8; Q-8, quinone-8. Data from Collins and Green (1985).

TABLE BXII.γ.66. Fatty acids of member genera of the Family *Methylococcaceae*[a,b]

Fatty acids[c]	*Methylococcus*	*Methylobacter*	*Methylomicrobium*	*Methylomonas*	*Methylosphaera*
Saturated:					
$C_{14:0}$	1–6	7–10	1–2	19–25	2–3
$C_{15:0}$	1–13	0–4	–	0–1	1–2
$C_{16:0}$	33–56	8–9	11–24	4–9	14–15
$C_{18:0}$	tr–2	–	0–3	0–trc	tr
Monounsaturated:					
$C_{16:1\ \omega8c}$	–	–	12–19	19–41	37–41
$C_{16:1\ \omega7c}$	17–46	56–58	14–24	8–15	16–19
$C_{16:1\ \omega7t}$	–	–	–	–	2–3
$C_{16:1\ \omega6c}$	4–12	4–5	8–14	5–13	17–18
$C_{16:1\ \omega5c}$	3–9	6–8	6–8	2–6	0–tr
$C_{16:1\ \omega5t}$	0–6	10–11	8–28	8–17	–
$C_{17:1\ \omega8c}$	0–tr	–	–	–	tr–1
$C_{17:1\ \omega7c}$	0–2	–	–	0–1	tr–1
$C_{17:1\ \omega7t}$	0–2	–	–	0–tr	–
$C_{18:1\ \omega9c}$	0–3	–	–	0–tr	0–1
$C_{18:1\ \omega7c}$	tr–7	–	0–3	tr–2	1–2
Other fatty acids:					
$C_{15:0\ iso}$	0–1	0–tr	–	0–2	–
$C_{15:0\ ante}$	0–tr	–	–	0–2	–
$C_{16:0\ cyc}$	0–1	–	–	–	–
$C_{17:0\ cyc}$	0–15	–	–	0–2	–
$C_{19:0\ cyc}$	0–3	–	–	0–tr	–

[a]For symbols, see standard definitions; tr, trace (fatty acid present at a level of less than 0.5%).

[b]Data from Bowman et al. (1991a), Bowman et al. (1993b), and Bowman et al. (1997c).

[c]Fatty acids are designated as follows: the number of carbon atoms:number of double bonds. The number (n) of carbon atoms of the closest double bond from the methyl end (ω) of the molecule is given by ωn. The suffixes iso, ante, and cyc indicate iso-branched, anteiso-branched, and cyclopropane fatty acids, respectively. The suffixes c and t indicate *cis* and *trans* geometry, respectively

final steps form reducing equivalents needed to drive methane oxidation.

In all type I methanotrophs, MMO is present in a membrane-bound form and is referred to as particulate MMO (pMMO). This enzyme consists of three polypeptides (sizes 45, 35, and 26 kDa) and contains copper (Semrau et al., 1995). Synthesis of a cytoplasmic version of MMO, referred to as soluble MMO (sMMO) (Dalton, 1992; Murrell, 1992; Lipscomb, 1994), appears to be a strain-specific trait (Koh et al., 1993; Miguez et al., 1997; Shen et al., 1997) that is relatively rare in type I methanotrophs. This enzyme is copper-repressible and contains three components. The first includes a 245 kDa hydroxylase that contains an unusual non-heme iron–oxygen linked active site. The second component is a 15 kDa regulatory protein, and the third is a ferredoxin-based reductase. Particulate MMO has a high affinity for methane, allowing for more efficient growth yields compared to sMMO, which appears to be an adaptation to copper-limiting growth conditions (Hanson and Hanson, 1996). However, the substrate specificity of sMMO is exceptionally broad, and it can oxidize an extraordinarily wide range of aliphatic, aromatic, and heterocyclic compounds (Haber et al., 1984). This property has been exploited biotechnologically, with methanotrophs proposed for organic transformations in industrial applications (Lidstrom and Stirling, 1990) and as a bioremediation agent (Hanson and Hanson, 1996).

The amino acid sequence of the *pmo*A gene, which encodes a pMMO protein component in type I methanotrophs, has been used in phylogenetic analyses. This gene is very similar to that encoding the ammonia monooxygenase (*amo*A) in *Nitrosococcus oceani*, suggesting a common evolutionary link (Holmes et al., 1995a; Bodrossy et al., 1997). Several physiological comparisons, in fact, have been drawn between methanotrophs and ammonia-oxidizing bacteria, because both groups are capable of oxidizing methane and ammonia by similar means (Hanson and Hanson,

1996). The sequences of *pmo*A in type II methanotrophs are quite different from those of type I methanotrophs. The amino acid sequences of sMMO polypeptides (*mmo*A, *mmo*B, *mmo*C, and *mmo*X) show a high degree of conservation, however, and are similar between type I and II methanotrophs (Stainthorpe et al., 1991; Murrell, 1992). These genes have been used to develop oligonucleotide probes to specifically target methanotrophs (Holmes et al. 1995b; McDonald and Murrell, 1997; Miguez et al., 1997).

C₁ carbon assimilation Methanotrophs are specialized to utilize the C_1 compounds methane, methanol, and formaldehyde. Occasionally, other C_1 compounds can also be utilized, but this is the exception rather than the rule. Compounds with carbon–carbon bonds (acetate, D-glucose, etc.) are not utilized as energy sources by methanotrophs; however, they may be used as supplementary carbon sources when cells are grown in the presence of methane or other utilizable C_1 compounds (Hanson et al., 1992). So far, no evidence exists for the presence of facultative methanotrophic bacteria that are otherwise similar to the type I methanotrophs.

Formaldehyde is toxic to methanotrophic cells at relatively low concentrations and so is often not tested as a carbon source. In addition, freshly isolated methanotrophs often prove sensitive to methanol (due to accumulation of formaldehyde). Consequently, methanol utilization should be tested with only low methanol concentrations (10–30 mM) or with methanol provided as a vapor in a sealed container. Methanotrophs may be "trained" to tolerate higher concentrations of methanol (Lidstrom, 1988). Type I methanotrophs use the ribulose monophosphate (RuMP) pathway to incorporate formaldehyde into anabolic pathways. Type II methanotrophs, on the other hand, use the serine transhydroxymethylase (Serine) pathway to fix carbon (Strom et al., 1974). In the RuMP pathway, formaldehyde is combined with ribulose-5-phosphate to form D-erythro-L-glycero-3-hexulose-6-phosphate through the action of the enzyme hexulose-6-phosphate synthase. Subsequently, 1-glyceraldehyde-3-phosphate is formed via 1-fructose-6-phosphate and 3-fructose-6-phosphate. Ribulose-5-phosphate is regenerated via rearrangement reactions from 3-fructose-5-phosphate. Hexulose-6-phosphate synthase is often used as an indicator of the RuMP pathway and thus can be used as a direct way to distinguish type I and II methanotrophs. Some type I methanotrophs, mostly *Methylococcus* strains, possess a low level of activity for α-hydroxypyruvate reductase, a key enzyme for the Serine pathway. The tricarboxylic acid cycle is incomplete and lacks the enzyme 2-oxoglutarate dehydrogenase. The cofactor for isocitrate dehydrogenase differs among type I methanotrophs: in some strains the enzyme activity is dependent on either NAD or NADP; in others it is dependent only on NAD and is unresponsive to NADP, and in still others it is NADP-dependent and unresponsive to NAD (Davey et al., 1972).

Intracytoplasmic membranes When cells are grown in the presence of methane (and sometimes methanol), characteristic intracytoplasmic membranes (ICMs) are formed from convolutions of the cytoplasmic membrane. ICMs are about 80–90 nm thick, appear as typical lipid bilayers, and in type I methanotrophs, are arranged as layered bundles of vesicular disks (Proctor et al., 1969; Davies and Whittenbury, 1970; De Boer and Hazeu, 1972; Hyder et al., 1979) (Fig. BXII.γ.102). Type II methanotrophs, on the other hand, have ICMs that extend peripherally throughout the cell and are arranged parallel to the cell wall. The relative ICM content in cells is enhanced in conditions of limiting oxygen and methane tensions or elevated copper ion concentrations and when a stable and rapid rate of methane transfer is available (Scott et al., 1981; Stanley et al., 1983).

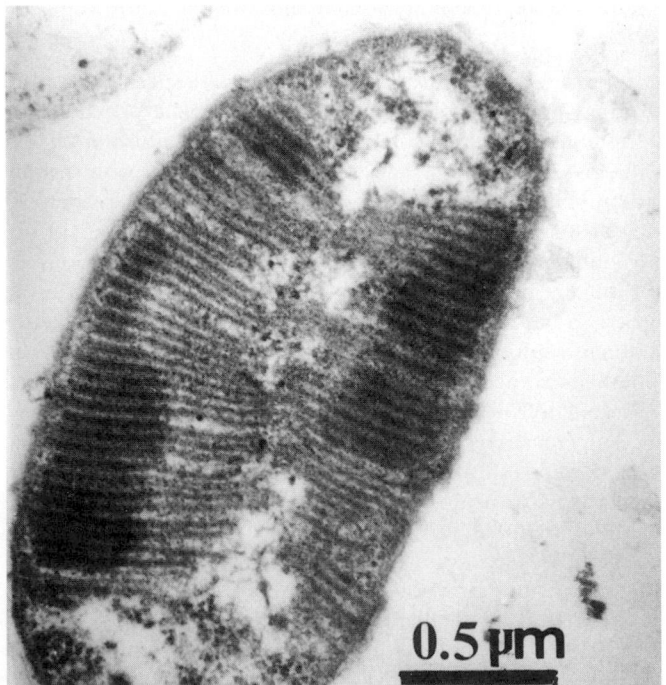

FIGURE BXII.γ.102. Electron micrograph of a ruthenium red-stained thin section of *Methylomonas fodinarum*, showing typical type I methanotrophic intracytoplasmic membrane ultrastructure.

Resting stages Type I methanotrophs often produce single- or multiple-bodied cyst-like inclusions often regarded as being similar to the cysts observed in the genus *Azotobacter*. Encystment usually increases in the stationary growth phase and when cells are grown at low methane tensions (Whittenbury et al., 1970a). Cyst formation depends on cell density, calcium ion concentration, pH, and the rate of cell autolysis (Mulyukin et al., 1997). Cysts may be readily observed by the *Azotobacter* cyst-staining procedure developed by Vela and Wyss (1964). Cysts appear to be made up of a glucan-like polysaccharide, unlike those of the type II methanotroph *Methylocystis*, which are composed mostly of lipid (Sutherland and MacKenzie, 1977). Cyst production appears to allow some methanotrophs to resist methane starvation. In other species, it can also confer desiccation resistance. Heat resistance, however, is not present; it requires more specialized structures such as the exospore formed by the type II methanotroph *Methylosinus*. Excellent photographic descriptions of both multi- and single-bodied cysts have been published by Hazeu et al. (1980a).

Fatty acid composition Members of the *Methylococcaceae* contain mainly C_{16} fatty acids. Although this is not unusual among Gram-negative bacteria, type I methanotrophs species contain abundant levels of several $C_{16:1}$ fatty acid isomers that are rarely encountered in other prokaryotes, including $C_{16:1\ \omega8c}$, $C_{16:1\ \omega6c}$, $C_{16:1\ \omega5c}$, and $C_{16:1\ \omega5t}$ (Makula, 1978; Nichols et al., 1985; Bowman et al., 1991a). The combinations of these fatty acids are diagnostic for the identification of type I methanotrophic genera (Andreev and Gal'chenko, 1978; Bowman et al., 1991a; Bowman et al., 1993b, 1997c) (Table BXII.γ.66). By comparison, type II methanotrophs contain mostly $C_{18:1}$ fatty acids; however, as in the type I methanotrophs, the double bond in most is an unusual position ($C_{18:1\ \omega8c}$) (Nichols et al., 1985). These unusual fatty acids are very useful as signatures for detection and quantitation of methanotroph populations in environmental samples (Nichols et al., 1987; Guckert et al., 1991; Jahnke et al., 1995; Sundh et al., 1997; Guezennec and Fiala-Medioni, 1996).

Lipopolysaccharide composition Studies of the carbohydrate moieties of the lipopolysaccharide of various type I methanotrophs have revealed a pattern typical of Gram-negative bacteria (i.e., D-glucose, L-fucose, and D-heptose) (Sutherland and Kennedy, 1986). The lipid moieties of lipopolysaccharide are composed mostly of $C_{16:0\ 3OH}$, with smaller quantities of β-hydroxylated and saturated components with carbon chain lengths of 10–15. The phospholipids found in type I methanotrophs are generally quite similar to one another and are composed mainly of phosphatidylethanolamine and phosphatidylglycerol, with lower levels of phosphatidylcholine and diphosphatidylglycerol (cardiolipin) (Makula, 1978; Andreev and Gal'chenko, 1983).

Quinones The isoprenoid lipoquinone (coenzyme Q) content of many type I methanotrophs consists mainly of ubiquinone (Q)-8 (94–99%), with small amounts of Q-7 also present. In several species, coenzyme Q is substituted with a methylene group identified as 18-methylene-Q-8 or MQ-8 (Collins and Green, 1985). The MQ-8 is accompanied by minor amounts of MQ-7.

Habitats Methane is an important greenhouse gas and the most abundant organic gas in the atmosphere (Crutzen, 1991). Methanotrophs are the largest global methane sink and, as a result, are ubiquitous in nature. They produce the highest and most active populations in environments where oxygen and methane are readily available (Reeburgh et al., 1993). Recent evidence suggests that methanotrophs make up a high proportion of the total bacterial biomass (up to 40%) in many aquatic environments and surface sediments (Ross et al., 1997; P. I. Boon, personal communication). Methane is highly ^{13}C-depleted, and stable isotopic analyses indicate that a considerable proportion of carbon found in aquatic life at different trophic levels has its origin from methanotrophic bacteria (Boschker et al., 1998). Habitats from which methanotrophs have been successfully isolated include marshes and swamps, rivers and streams, rice paddies, oceans, ponds, lakes, soils of meadows and forests, groundwater, sewage sludge, and coal mines (Hanson and Hanson, 1996). Psychrophilic methanotrophs have been isolated from tundra soils and Antarctic lakes (Omel'chenko et al., 1996; Bowman et al., 1997c). Methanotrophs may be halophilic, although so far, only a few have been described with this property (Lidstrom, 1988; Lees et al., 1991; Sieburth et al., 1993; Bowman et al., 1997c). Methanotrophs have also been isolated from the effluent of an underground hot spring found within a natural gas

field (Bodrossy et al., 1995). Methanotrophs have been detected and/or isolated from acidic peat bog soils (Dedysh et al., 1998) and deep sea sediments (Guezennec and Fiala-Medioni, 1996). Methanotrophs can be endosymbiotic in invertebrates associated with marine hydrocarbon seeps (see below). There is also good evidence that methanotrophs are closely associated with aquatic macrophytes (King, 1994; Calhoun and King, 1998).

Endosymbionts Methanotrophic endosymbionts, in coexistence with sulfur-oxidizing chemoautotrophic endosymbionts (Childress et al., 1986; Cavanaugh et al., 1987; Fisher et al., 1993), are the metabolic basis of unusual invertebrate communities associated with hydrocarbon seeps in the Pacific and Atlantic Oceans. The invertebrate hosts include deep-sea mytilid mussels (family *Mytlidae*) (Childress et al., 1986), and the pogonophoran tubeworm *Siboglinum poseidoni* (Dando et al., 1994). Cold-water reefs of algae-free coral found in waters north of Norway, associated with hydrocarbon seeps, are also believed to use methane

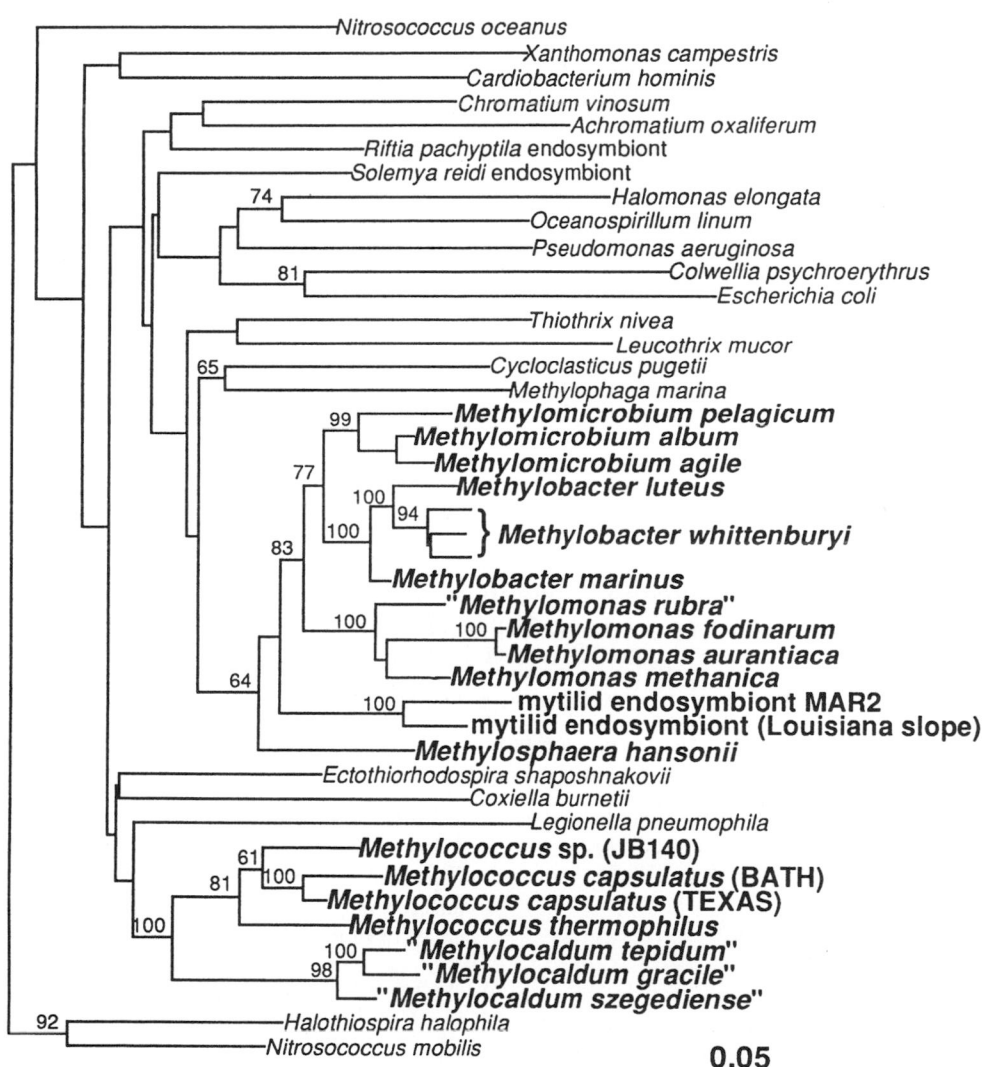

FIGURE BXII.γ.103. Unrooted phylogenetic tree based on 16S rRNA sequences, showing the position of the various genera belonging to the Order *Methylococcales*, Family *Methylococcaceae* (shown in bold type) in the *Gammaproteobacteria*. The tree is based on maximum likelihood values and joined by the neighborliness algorithm (outgrouped with the sequence for *Desulfovibrio desulfuricans*) using the PHYLIP v. 3.57c software package (Phylogenetic Inference Package, J. Felsenstein, University of Washington, Seattle). Values at the branch nodes are bootstrap percentage values generated from 250 replicated analyses. Bar = a taxonomic distance of 0.05.

as a major nutrient and thus presumably contain methanotroph symbionts (Hovland and Judd, 1988). The invertebrates can exist on methane as a primary carbon source, as shown by stable carbon isotope analyses (Southward et al., 1981; Cavanaugh, 1993). Methanotrophs growing in the gills of mytilids have been identified as type I methanotrophs by the presence of key enzymes of the RuMP pathway and characteristic fatty acids, as well as by oligonucleotide probes (Cavanaugh, 1993; Distel and Cavanaugh, 1994; Jahnke et al., 1995). Phylogenetic analyses indicate that mytilid symbionts form a distinct lineage within the *Methylococcaceae* (Distel and Cavanaugh, 1994). Cultivation of methanotroph symbionts has been unsuccessful to date.

TAXONOMIC COMMENTS

The use of 16S rRNA-based phylogenetic analysis has helped to resolve many of the previously identified nomenclatural problems affecting methanotrophs (Bratina et al., 1992; Brusseau et al., 1994; Bowman et al., 1995). On the basis of 16S rDNA sequencing, type I methanotrophs belong in the *Gammaproteobacteria* and in 1994-1995 were thought to comprise a single clade. However, in most phylogenetic trees based on currently available sequences, the *Methylococcales* is now made up of two separate clades (Fig. BXII.γ.103). The first cluster includes the "group X" methanotrophs—thermotolerant and thermophilic methanotrophs of *Methylococcus* and "*Methylocaldum*". This clade represents a relatively deep lineage within the *Gammaproteobacteria*, with *Ectothiorhodospira* and *Legionella* being among the closest relatives. The second cluster contains mesophilic and psychrophilic type I methanotrophs, including the genera *Methylosphaera*, *Methylobacter*, *Methylomicrobium*, and *Methylomonas*. This cluster is most closely related to *Methylophaga*, *Cycloclasticus*, and various chemoautotrophic endosymbionts. From this, it could be argued that because the *Methylococcales* does not have phylogenetic coherency,

it should be divided into two families. However, sequence analysis still indicates that the type I and group X methanotrophic clades are similar (evolutionary distance 0.10–0.13). More importantly, both groups possess a number of traits, not the least being methane oxidation, that sets them apart from other members of the *Gammaproteobacteria*. It can be safely assumed that many novel methanotrophic species remain to be isolated and that their 16S rDNA sequences could hypothetically introduce topological changes to current phylogenetic trees. Thus, until a broader knowledge of methanotroph phylogeny and evolution is obtained it seems prudent to retain the family *Methylococcaceae* as it currently stands.

FURTHER READING

Bowman, J.P., L.I. Sly, P.D. Nichols and A.C. Hayward. 1993 Revised taxonomy of the methanotrophs: description of *Methylobacter* gen. nov., emendation of *Methylococcus*, validation of *Methylosinus* and *Methylocystis* species, and a proposal that the family *Methylococcaceae* includes only the group I methanotrophs. Int. J. Syst. Bacteriol. *43*: 735–753.

Cavanaugh, C.M. 1993. Methanotroph–invertebrate symbioses in the marine environment: ultrastructural, biochemical and molecular studies. *In* Murrell and Kelley (Editors), Microbial Growth on C_1 Compounds, Intercept Press Ltd., Andover. pp. 315–328.

Hanson, R.S. and T.E. Hanson. 1996 Methanotrophic bacteria. Microbiol. Rev. *60*: 439–471.

Hanson, R.S., A.I. Netrusov and K. Tsuji. 1992. The obligate methanotrophic bacteria *Methylococcus*, *Methylomonas*, and *Methylosinus*. *In* Balows, Trüper, Dworkin, Harder and Schleifer (Editors), The Prokaryotes. A Handbook of Bacteria: Ecophysiology, Isolation, Identification, Applications, 2nd Ed., Vol. 3, Springer-Verlag, New York. pp. 2350–2364.

Whittenbury, R.A., K.C. Phillips and J.F. Wilkinson. 1970 Enrichment, isolation and some properties of methane-utilizing bacteria. J. Gen. Microbiol. *61*: 205–218.

Family I. **Methylococcaceae** Whittenbury and Krieg 1984b, 355[VP] (Effective publication: Whittenbury and Krieg 1984a, 256) emend. Bowman, Sly, Nichols and Hayward 1993b, 745.

JOHN P. BOWMAN

Me.thy.lo.coc.ca' ce.ae. M.L. masc. n. *Methylococcus* type genus of the family; -*aceae* ending to denote a family; M.L. fem. pl. n. *Methylococcaceae* the *Methylococcus* family.

The description of the family is the same as that for the order.

Type genus: **Methylococcus** Foster and Davis 1966, 1929 emend. Bowman, Sly, Nichols and Hayward 1993b, 748

Differentiation of the Genera of the Family *Methylococcaceae*
A key to the genera of the family *Methylococcaceae* is provided below. In addition, Table BXII.γ.67 lists phenotypic features that differentiate the genera and Table BXII.γ.68 indicates differences in fatty acid composition between the genera.

Key to the genera of the family Methylococcaceae

I. Can grow at at 45°C or higher. Form cysts and possess ribulose-1,5-diphosphate carboxylase activity. Cells are cocci or rods that may be motile. Colonies are light tan to dark brown. $C_{16:0}$ and $C_{16:1\ \omega7c}$ are the major fatty acid components in cells. The major lipoquinone is MQ-8. The mol% G + C content of the DNA ranges from 59–65 (T_m).

Genus *Methylococcus*

II. Mesophilic. Elliptical rod-like morphology. Usually motile. Colonies on agar are light tan to brown with either a light brown pigment diffusing into agar media or a yellow pigment that may or may not diffuse into agar media. Cells are desiccation resistant. Nitrogenase activity is absent. $C_{16:1\ \omega7c}$ is the major fatty acid component in cells. The major lipoquinone is Q-8. The mol% G + C content of the DNA ranges from 49–55 (T_m).

Genus *Methylobacter*

III. Growth at temperatures > 40°C. Cells coccoid to long rods. Form cysts. Motile or non-motile. Not resistant to desiccation. Colonies buff to brown. Nitrogenase absent. Mol% G + C 57–59 (Bd).

Genus *Methylocaldum*

IV. Mesophilic. Rod-like morphology. Motile. Grows at 25°C and may be slightly halophilic. Colonies on agar are non-pigmented. Nitrogenase activity is absent. Major fatty acids include $C_{16:0}$, $C_{16:1\ \omega8c}$, $C_{16:1\ \omega7c}$, and $C_{16:1\ \omega5c}/C_{16:1\ \omega5t}$. The major lipoquinone is Q-8. The mol% G + C content of the DNA ranges from 48–60 (T_m).

Genus *Methylomicrobium*

V. Mesophilic. Rod-like morphology. Motile. Colonies on agar contain red, pink, or orange non-diffusible carotenoid pigments. Cells are desiccation sensitive. Nitrogenase activity may be present. $C_{14:0}$ and $C_{16:1\ \omega8c}$ are the major fatty acids in cells. The major lipoquinone is MQ-8. The mol% G + C content of the DNA ranges from 50–59 (T_m).

Genus *Methylomonas*

VI. Growth at temperatures ≤37°C. Cells coccoid or fusiform; form packets. Motile. Form cysts. Colonies buff to brown. Nitrogenase absent. Major fatty acids $C_{16:0}$, $C_{16:1\ \omega5t}$, $C_{16:1\ \omega6c}$, $C_{16:1\ \omega7c}$, $C_{16:1\ \omega8c}$. Mol% G + C 54–55 (HPLC).

Genus *Methylosarcina*

VII. Psychrophilic, growing only at −2–20°C. Cells are coccoidal, lack cysts, form gas vesicles, and are nonmotile. No pigmentation. No growth on solid agar media. Oxygen-tolerant nitrogenase activity present. Contains high levels of $C_{16:1\ \omega8c}$ and lacks $C_{16:1\ \omega5c}/C_{16:1\ \omega5t}$. The mol% G + C content of the DNA ranges from 43–46 (T_m).

Genus *Methylosphaera*

TABLE BXII.γ.67. Phenotypic characteristics differentiating member genera of the Family *Methylococcaceae*[a,b]

Characteristics	Methylococcus	Methylosphaera	Methylobacter	Methylomicrobium	Methylomonas
Cell morphology	Cocci-rods	Cocci	Cocci-ellipse	Rods	Rods
Motility	D	−	D	+	+
Cyst formation	+	−	+	−	+
Desiccation resistance	−	−	+	−	−
Pigmentation:					
Carotenoids	−	−	−	−	+
Melanin-like	D	−	D	−	−
Other	−	−	D	−	−
Growth at 25°C	−	−	+	+	+
Growth at 45°C	+	−	−	−	−
Requires seawater or Na$^+$	−	+	D	D	−
RuBisCo[c]	+	−	−	−	−
Nitrogen fixation	+	+	−	−	D
Major quinone[d]	MQ-8	ND	Q-8	Q-8	MQ-8
Mol% G + C of DNA (T_m)	59–65	43–46	48–55	48–60	50–59

[a]Symbols, see standard definitions.

[b]Data from Bowman et al. 1993b and Bowman et al., 1997c.

[c]RuBisCo, ribulose-1,5-bisphosphate carboxylase.

[d]MQ-8, 18-methylene-quinone-8; Q-8, quinone-8. Data from Collins and Green (1985).

TABLE BXII.γ.68. Fatty acids of member genera of the Family *Methylococcaceae*[a, b]

Fatty acids	*Methylococcus*	*Methylosphaera*	*Methylobacter*	*Methylomicrobium*	*Methylomonas*
Saturated:					
$C_{14:0}$	1–6	2–3	7–10	1–2	19–25
$C_{15:0}$	1–13	1–2	0–4	–	0–1
$C_{16:0}$	33–56	14–15	8–9	11–24	4–9
$C_{18:0}$	tr-2	tr	–	0-3	0-tr
Monounsaturated:					
$C_{16:1}$ ω8c	–	37–41	–	12–19	19–41
$C_{16:1}$ ω7c	17–46	16–19	56–58	14–24	8–15
$C_{16:1}$ ω7t	–	2–3	–	–	–
$C_{16:1}$ ω6c	4–12	17–18	4–5	8–14	5–13
$C_{16:1}$ ω5c	3–9	0–tr	6–8	6–8	2–6
$C_{6:1}$ ω5t	0–6	–	10–11	8–28	8–17
$C_{7:1}$ ω8c	0–tr	tr–1	–	–	–
$C_{7:1}$ ω7c	0–2	tr–1	–	–	0–1
$C_{7:1}$ ω7t	0–2	–	–	–	0–tr
$C_{8:1}$ ω9c	0–3	0–1	–	–	0–tr
$C_{8:1}$ ω7c	tr–7	1–2	–	0–3	tr–2
Other fatty acids:					
$C_{15:0}$	0–1	–	0–tr	–	0–2
$C_{15:0}$ cyc	0–tr	–	–	–	0–2
$C_{16:0}$ cyc	0–1	–	–	–	–
$C_{17:0}$ cyc	0–15	–	–	–	0–2
$C_{19:0}$ cyc	0–3	–	–	–	0–tr

[a] Data from Bowman et al. (1995).

[b] These values are % of total fatty acids; symbols: tr, trace component (<0.1%); −, not detected.

Genus I. **Methylococcus** *Foster and Davis 1966, 1929*[AL] *emend. Bowman, Sly, Nichols and Hayward 1993b, 748*

JOHN P. BOWMAN

Me.thy.lo.coc'cus. Fr. n. *méthyle* the methyl radical; Gr. n. *coccus* a grain, berry; M.L. n. *Methylococcus* methyl coccus.

Cells appear as **cocci or rods**, 0.8–1.5 × 1.0–1.5 µm. Usually occur singly or in pairs. Form a polysaccharide-containing **cyst-like resting stage** that is desiccation sensitive. Poly-β-hydroxybutyrate granules present. **Gram negative.** If motile, cells are propelled by single polar flagellum. Catalase and oxidase positive. **Strictly aerobic respiratory metabolism** with oxygen as the electron acceptor. **Obligate utilizers of methane and methanol for carbon and energy.** Contain a **partially functional Benson–Calvin cycle** (phosphoribulokinase and ribulose-1,5-diphosphate carboxylase activity present); however, incorporation of CO_2 for cellular carbon occurs only in the presence of methane. **Fix atmospheric nitrogen** via an oxygen-sensitive nitrogenase. **Thermotolerant or moderately thermophilic** with optimal growth between 40–60°C. Major fatty acids include $C_{16:0}$ **and** $C_{16:1 \, \omega7c}$. **18-Methyleneubiquinone-8** (MQ-8) is the predominant lipoquinone. Isolated from sediments of freshwater lakes and rivers, wetland muds, activated sludge, and wastewater. Belongs to the *Gammaproteobacteria* as part of the Order *Methylococcales*, Family *Methylococcaceae*.

The mol% G + C of the DNA is: 59–65 (T_m).

Type species: **Methylococcus capsulatus** Foster and Davis 1966, 1929.

FURTHER DESCRIPTIVE INFORMATION

Morphology and ultrastructure Cells of *Methylococcus capsulatus* are usually spherical in shape; however, in stationary growth phase cultures, a significant proportion of cells may appear as coccobacilli or short rods. *Methylococcus thermophilus* has a more pleomorphic morphology, with cells ranging from cocci to rods. Cells usually occur singly, in pairs, and sometimes in chains; however, rosette formations are absent. Although *Methylococcus capsulatus* strains lack flagella, *Methylococcus thermophilus* cells are usu-

ally vigorously motile upon isolation. Motility is spontaneously lost in many isolates of *Methylococcus thermophilus* (including the type strain) after several serial transfers. Motility is also more pronounced in younger cultures and in cultures grown on methanol.

The cell envelope of *Methylococcus* species is typical of Gram-negative bacteria. Hydroxy fatty acids from the outer membrane lipopolysaccharide have been analyzed in detail in *Methylococcus capsulatus*, *Methylococcus thermophilus*, and *Methylococcus* sp. JB140 (Bowman et al., 1991a). The major components include $C_{10:0 \, 3OH}$, $C_{12:0 \, 3OH}$, $C_{14:0 \, 3OH}$, and $C_{16:0 \, 3OH}$. Cells are covered by an exopolysaccharide capsule (Whittenbury et al., 1970a), which often causes a degree of cell aggregation.

The cells of *Methylococcus* species possess the intracytoplasmic membrane (ICM) and cyst formation that commonly occur in many members of the Family *Methylococcaceae*. The ICM formation and ultrastructure have been examined most extensively in *Methylococcus capsulatus* ATCC 33009 (strain "Bath") (Hyder et al., 1979). ICMs are formed when *Methylococcus capsulatus* is grown on methane or methanol, and ICM levels increase in proportion to methane oxidation rates. Growth in the absence of copper (levels less than 1.0 µmol/g dry weight) suppresses formation of ICMs, as well as drastically reducing pMMO activity and derepressing sMMO activity (Prior and Dalton, 1985).

Cysts Single-bodied spherical cysts similar to those observed in *Azotobacter chroococcum*, but comparatively simpler and less defined in structure, are formed by *Methylococcus capsulatus* and *Methylococcus thermophilus*, usually in the stationary growth phase (Whittenbury et al., 1970a; Malashenko et al., 1975b). Increased cyst formation is usually associated with increasing cell refractility, but cyst formation is not as profuse as in species of *Methylobacter*. Cysts can be visualized by light microscopy after staining (Vela and Wyss, 1964). The cysts are mainly made up of glucan-type polysaccharides (Sutherland and MacKenzie, 1977), which are discarded as cells revert to a normal vegetative state. Production of cysts allow *Methylococcus capsulatus* to survive methane starvation for several weeks even though vegetative cells die rapidly in the absence of methane. The cysts, however, do not confer desiccation or heat resistance. In addition, *Methylococcus capsulatus* cells contain granules of poly-β-hydroxybutyrate, which also contribute to changes in cell refractility. Cells may also contain polyphosphate (volutin) inclusions.

Nutrition and growth conditions *Methylococcus* strains grow well in nitrate mineral salts (NMS) liquid media[1] *Methylococcus thermophilus* tends to produce dark brown colonies with an associated diffusible melanin-like brown pigment; the diffusible pigment is usually lost rapidly upon subculture. *Methylococcus capsulatus* colonies appear off-white to light tan in color and never produce diffusible pigments. Both *Methylococcus* species, depending on the strain, can readily form both rough and smooth colonial variants, with rough colonies possessing a somewhat cartilaginous consistency. This property appears to be particularly prevalent in isolates of *Methylococcus thermophilus* and phenotypically similar species (Bodrossy et al., 1997). With constant subculture, colonies tend to become consistently more butyrous and mucoid.

Methylococcus species are thermotolerant or moderately thermophilic, with optimal growth temperatures of 40–60°C. No strains have been found so far that can grow at temperatures higher than 65°C or below ~25°C.

No growth factors are required, and the strains are nonhalophilic. Some *Methylococcus capsulatus* strains can utilize methylamine, formate, and/or formamide as sole carbon and energy sources. However, *Methylococcus thermophilus* strains, the type strain of *Methylococcus capsulatus* ATCC 19069 (strain "Texas"), and the more heavily studied strain ATCC 33009 cannot utilize any of these compounds. No carbon–carbon bonded compounds can be used for growth.

Nitrogen sources are usually provided in the form of a nitrate or ammonium salt, although *Methylococcus* strains can also use yeast extract, Casamino acids (Difco), and amino acids. In addition, *Methylococcus capsulatus* and most strains of *Methylococcus thermophilus* can also fix atmospheric nitrogen by means of an oxygen-sensitive nitrogenase (Murrell and Dalton, 1983; Zhivotchenko et al., 1995b). A modified acetylene reduction procedure is used to detect nitrogenase activity in methanotrophic cultures (Takeda, 1988).

Some *Methylococcus capsulatus* strains can form sMMO, which allows these strains to grow fairly well in media lacking copper (Prior and Dalton, 1985). Evidence that a strain lacks sMMO is indicated by poor growth in copper-free mineral media; production of sMMO can also be rapidly detected by oxidation of naphthalene to α-naphthol (Brusseau et al., 1990). PCR detection using sMMO-specific primers or detection by filter or colony hybridization techniques can be used to definitively detect the presence of sMMO genes (Stainthorpe et al., 1991; Miguez et al., 1997).

Methylococcus species primarily uses the RuMP pathway to fix formaldehyde for cell carbon. However, the key enzyme for the serine pathway (α-hydroxypyruvate reductase) has also been detected (Strom et al., 1974). The tricarboxylic acid cycle is incomplete, lacking 2-oxoglutarate dehydrogenase activity, but an NAD-dependent isocitrate dehydrogenase is present. *Methylococcus* species also contain a partially functional Benson–Calvin pathway for CO_2 fixation, including activity of the enzymes ribulose-1,5-diphosphate carboxylase and phosphoribulokinase (Whittenbury and Krieg 1984a; Bowman et al., 1993b). The importance of this ability with respect to the cell carbon budget is still not understood, since autotrophic growth of *Methylococcus* species has not been obtained in the absence of methane.

Lipid composition The major fatty acids in *Methylococcus* species have been found to be $C_{16:0}$ and $C_{16:1\ \omega7c}$ (Table BXII.γ.68 of the chapter describing the family *Methylococcaceae*). The high level of $C_{16:0}$ is a characteristic that distinguishes this species from psychrophilic and mesophilic type I methanotrophs. The neutral lipid fraction of *Methylococcus capsulatus* is unusual among procaryotes as it contains squalene, methylated sterols, and hopanoids (Bird et al., 1971; Neunlist and Rohmer, 1985). The unusual distribution of cyclic triterpenes and cyclopropane fatty acids (which become more prevalent with growth under reduced oxygen tension) is believed to improve the stability of outer and intracytoplasmic membranes (Jahnke and Nichols, 1986; Jahnke et al., 1992). The major respiratory coenzyme Q in *Methylococcus capsulatus* and *Methylococcus thermophilus* is 18-methylene-ubiquinone-8 (MQ-8) (Collins and Green, 1985).

Ecology Little information exists on the specific ecology of *Methylococcus* species; however, they have been isolated from a range of aquatic and terrestrial habitats, including soils and sediments of rivers and ponds, sewage sludge, and in the vicinity of various thermal environments (Malashenko et al., 1975a).

ENRICHMENT AND ISOLATION PROCEDURES

Methylococcus species can be enriched and isolated from the microaerobic and aerobic sediments of freshwater lakes and rivers, soil, activated sludge, and near certain thermal environments. A small amount of a sample is added to NMS liquid mineral medium in serum vials or in cotton-stoppered flasks placed within airtight containers. Methane is added directly to the vials and containers, usually by first removing a portion of the headspace. The best methane/air ratio is equivocal but should be in a range of 1:10–1:1. Methane should be of high purity, as natural gas may contain acetylene, which is a suicide substrate of pMMO and sMMO and will hinder or prevent growth, even at low concentrations. Static incubation should then proceed at 40–60°C. Successful enrichments of *Methylococcus* strains develop a white to brown turbidity that is often accompanied by a surface pellicle. Growth from the enrichments can be plated onto mineral salts agar plates, which are then incubated under 1:1 methane/air at 45–60°C.

One of the most problematic areas of methanotroph study is that of obtaining pure cultures. In practically all situations, methanotroph enrichments are severely contaminated by nonmethanotrophic (often methylotrophic) bacteria, which can easily overgrow and/or predate cultures. Due to the relatively slow growth of *Methylococcus* species and the humid conditions in

---

1. Nitrate mineral salts medium is slightly modified from Whittenbury et al. (1970b) and consists of (per liter of tap water): $MgSO_4 \cdot 7H_2O$, 1.0 g; $CaCl_2 \cdot 6H_2O$, 0.2 g; ferric-EDTA, 10 mg; NH_4Cl, 0.5 g or KNO_3, 1.0 g; trace element solution, 1.0 ml. The trace element solution solution contains (per liter of distilled water): $CuSO_4 \cdot 5H_2O$, 0.2 g; $ZnSO_4 \cdot 7H_2O$, 10 mg; $MnCl_2 \cdot 4H_2O$, 3 mg; H_3BO_4, 30 mg; $CoCl_2 \cdot 6H_2O$, 10 mg; and $Na_2MoO_4 \cdot 7H_2O$, 10 mg; pH 4.0. The medium is adjusted to pH 6.8 before autoclaving. The autoclaved medium is allowed to cool, and 2 ml of 1.0 M sodium phosphate buffer (pH 6.8) is added. Addition of extra copper as $CuSO_4 \cdot 5H_2O$ promotes the growth of methanotrophs (Zhivotchenko et al.,1995a).

which plates are incubated, fungal contamination is also frequent unless containers are thoroughly cleaned with ethanol before each use. Fungicides, such as cycloheximide or nystatin, added to the medium are usually effective in reducing this problem.

A straightforward method for the isolation of *Methylococcus* and other methanotrophs uses mineral agar media that contain small amounts of yeast extract (0.05% w/v) and methanol (0.05% v/v) (Malashenko et al., 1975a), with incubation occurring under a methane/air atmosphere. The enrichment culture is then serially diluted onto the media to the point of extinction. Single colonies from these spread-plates are then transferred to liquid media. A number of passages from liquid media to spread-plates and back to liquid media are often required. The purity of the culture can be assessed by assuring that strains show no growth on nutrient media in the presence or absence of methane and that cells possess a reasonably consistent morphology. More details on enrichment, isolation, and potential problems have been detailed by Whittenbury et al. (1970b) and Hanson et al. (1992).

MAINTENANCE PROCEDURES

Cryopreservation (with cyroprotectants such as DMSO or glycerol) at −80°C and lyophilization are generally ineffective for storing *Methylococcus* strains, as dramatic losses in viability follow freezing (Nesterov et al., 1986). *Methylococcus* strains can be stored in liquid nitrogen (with 20–30% v/v glycerol) for up to 6 months before recoverability becomes problematic. The storage of cultures at low temperature (4°C) under a methane atmosphere has been found to be very effective for short-term storage of up to 6 months.

PROCEDURES FOR TESTING SPECIAL CHARACTERS

Nitrogen fixation in methanotrophs cannot be tested by the standard acetylene reduction technique because acetylene acts as a suicide substrate of methane monooxygenase. Instead, resting cultures are used. First, the methanotroph is cultured with methanol or methane in vials using nitrogen-free NMS media prepared with 0.2–0.3% (w/v) purified agar to make it semisolid, thereby reducing oxygen diffusion and creating a microaerobic environment that promotes nitrogen fixation. After adequate growth has occurred, acetylene is injected into the headspace (final concentration 2–5%, v/v) and the vials are incubated for several hours. During this time, samples are taken periodically and tested by gas chromatography-thermal conductivity to detect ethylene in the headspace (Murrell and Dalton, 1983; Takeda, 1988).

Putative soluble methane monooxygenase activity of methanotrophs can be rapidly assessed using the naphthalene oxidation technique (Brusseau et al., 1990). First, the methanotroph to be tested is cultured in copper-free NMS liquid medium. Glassware used for cultivation should be washed with 1% (v/v) nitric acid to strip off contaminating copper ions. An equal volume of a saturated naphthalene solution (~30 mg/l at 25°C) is added to a suspension of methanotrophic cells (degassed briefly of residual methane, using nitrogen). The mixture is incubated at room temperature for approximately one hour. To the suspension, 20 μl of a freshly prepared solution of 1% (w/v) tetraotized *o*-dianisidine (Fast Blue salt B) is then added and mixed. The presumptive presence of soluble methane monooxygenase is indicated by the immediate production of a reddish semi-soluble diazo compound, formed from α-naphthol, which can be measured by spectrophotometry at 540 nm.

DIFFERENTIATION OF THE GENUS *METHYLOCOCCUS* FROM OTHER GENERA

Phenotypic characteristics and fatty acid profiles distinguishing *Methylococcus* from other type I methanotrophic genera are shown in Tables BXII.γ.67 and BXII.γ.68 of the chapter describing the family *Methylococcaceae*.

TAXONOMIC COMMENTS

Numerical taxonomic analysis suggested that several of the methanotrophic groups described by Whittenbury et al. (1970b) were related to *Methylococcus*. These groups were subsequently published as new species of *Methylococcus* (Romanovskaya et al., 1978), and their names subsequently appeared in validation list No. 7 as *Methylococcus bovis*, *Methylococcus chroococcus*, *Methylococcus luteus*, *Methylococcus vinelandii*, and *Methylococcus whittenburyi* ("*Methylobacter capsulatus*"). Subsequent studies using immunological, protein electrophoresis, and fatty acid analysis indicated that this expanded version of the genus *Methylococcus* was clearly heterogeneous (Andreev and Gal'chenko, 1978; Gal'chenko and Nesterov, 1981; Bezrukova et al., 1983; Meyer et al., 1986; Bowman et al., 1991a). Moreover, genomic characteristics demonstrated that the genus was made up of two major groups (Bowman et al., 1991b). This nomenclatural problem was resolved only by retention of *M. capsulatus* and *M. thermophilus* in the genus *Methylococcus*, with the description of *Methylococcus* appropriately emended (Bowman et al., 1993b) and transfer of the other species to the genus *Methylobacter*, which was created to accommodate them.

"*Methylococcus fulvus*" Malashenko, Romanovskaya and Kvashnikov 1972, 877 possesses phenotypic traits highly similar to those of *Methylobacter luteus* (Romanovskaya et al., 1978) and lacks a type strain.

The single strain making up the species *Methylococcus mobilis* Hazeu et al., 1980b, 676 has been lost, and this species should be regarded as invalid until a replacement strain has been isolated and described. *Methylococcus* species form a deeply branching clade within the *Gammaproteobacteria* (Fig. BXII.γ.103). Outside of this clade, the closest bacterial relative in terms of 16S rRNA sequence similarity is the photosynthetic species *Ectothiorhodospira shaposhnikovii* (evolutionary distance 0.09–0.10). The *Methylococcus* clade groups separately from the remaining type I methanotrophs, but evolutionary distances are roughly equivalent to the interposing genera (evolutionary distance 0.10–0.14).

DNA–DNA hybridization analysis indicates that *Methylococcus capsulatus* ATCC 19069 possesses high levels of hybridization with strain ATCC 33009 (Bowman et al., 1993b). A DNA hybridization range of 40–43% is found with the non-extant strain JB140; however, only negligible levels of DNA hybridization (<25%) is found between strains of *Methylococcus capsulatus* and *Methylococcus thermophilus*. Moderately thermophilic methanotrophs were recently described by Bodrossy et al. (1997) as the genus *Methylocaldum*, which includes the species *Methylocaldum szegediense* (the type species), *Methylocaldum tepidum*, and *Methylocaldum gracile*. (The latter was renamed from "*Methylomonas gracilis*" NCIMB 11128.) Bodrossy and colleagues have noted phenotypic similarity between *Methylococcus thermophilus* and *Methylocaldum* species; moreover, high levels of DNA hybridization have been reported between *Methylococcus thermophilus* IMV-2Yu and "*Methylomonas gracilis*" NCIMB 11912 (Bowman et al., 1993b). The phenotypic differences between *Methylococcus* and *Methylocaldum* appear to be trivial in nature and, based on the current description of *Methylo-*

coccus, it is nearly impossible to differentiate *Methylococcus* from *Methylocaldum*. Both groups possess similar fatty acid profiles, mol% G + C contents of the DNA, genomic characteristics, and *pmo*A sequences. On the other hand, 16S rDNA analysis suggests that the two groups are phylogenetically distinct (evolutionary distance ~0.08). This is in contrast to the above-mentioned similarities between the two groups, particularly the hybridization data for *Methylococcus thermophilus* IMV-2Yu and "*Methylomonas gra-*

cilis" NCIMB 11128 (Bowman et al., 1993b), and this anomaly suggests that the 16S rDNA sequence of *Methylococcus thermophilus* needs to be verified. It is certainly clear that, based on DNA hybridization data, *Methylocaldum gracile* is a subjective synonym of *Methylococcus thermophilus* and is thus misnamed. Further DNA hybridization analysis is needed to confirm or deny the genotypic distinctiveness of other *Methylocaldum* species from *Methylococcus thermophilus*.

DIFFERENTIATION OF THE SPECIES OF THE GENUS *METHYLOCOCCUS*

Phenotypic and genotypic characteristics differentiating the species of *Methylococcus* are shown in Table BXII.γ.69, while additional characteristics are shown in Table BXII.γ.70. The phe-

notypic characteristics possessed by strains of *Methylococcus thermophilus* are essentially the same as those possessed by *Methylocaldum* strains.

List of species of the genus Methylococcus

1. **Methylococcus capsulatus** Foster and Davis 1966, 1929[AL]

    cap.su.la' tus. M.L. adj. *capsulatus* encapsulated.

    Cells are coccoidal and nonmotile. Colonies are white to light tan, round, convex, entire- or lobate-edged, with a cartilaginous or butyrous consistency. Other characteristics are as described for the genus and as listed in Tables BXII.γ.69 and BXII.γ.70. For further descriptive information, see Foster and Davis (1966) and Bowman et al. (1993b).

    Optimum temperature for growth, ~45°C. pH range for growth, 5.5–8.5; optimum, 6.5–7.0.

    Isolated from lakes and river mud, wetland muds, activated sludge, and wastewater.

    The mol% G + C of the DNA is: 62.5 (Bd), type strain. Analysis of several other strains using the thermal denaturation method found a range of 62–65 mol% (Bowman et al., 1991b).

    Type strain: Texas, ACM 1292, ATCC 19069 NCIMB 11853.

    GenBank accession number (16S rRNA): X72770.

    Additional Remarks: Another important strain is Bath (ATCC 33009, NCIMB 11132, VKM-TRMC, IMET 10544).

2. **Methylococcus thermophilus** Malashenko, Romanovskaya, Bogachenko and Shved 1975b, 779[AL]

    ther.mo' phil.us. Gr. n. *therme* heat; Gr. adj. *philus* loving; M.L. adj. *thermophilus* heat loving.

    Cells are coccoidal and usually motile on initial isolation. Motile strains usually lose motility after several subcultures. Colonies are light to dark brown, round, raised convex, with entire to irregular edges, and have a viscid or cartilaginous consistency. Produce a diffusible brown pigment, which is usually lost during the strain purification process. Other

characteristics are as described for the genus and as listed in Tables BXII.γ.69 and BXII.γ.70. For further descriptive information, see Bowman et al. (1993b).

Optimum growth temperature, 50–55°C. pH range for growth, 5.5–9.0; optimum, 7.0.

Isolated from therapeutic mud, moist subtropical–tropical soils, creek mud, and pond sediment.

TABLE BXII.γ.69. Characteristics differentiating the species of the genus *Methylococcus*[a,b]

Characteristic	1. *M. capsulatus*	2. *M. thermophilus*
Motile cells	−	+
Melanin-like pigment production	−	+
Growth at 55°C	−	+
Mol% G + C of DNA (T_m)	62.5–65.2	59.1–60.8

[a]For symbols, see standard definitions.

[b]Phenotypic data from Bowman (1992) and Bowman et al. (1993b).

TABLE BXII.γ.70. Other phenotypic characteristics of the species of the genus *Methylococcus*[a,b]

Characteristics	1. *M. capsulatus*	2. *M. thermophilus*
Rod-like cells formed	d	+
Viscid–cartilaginous colonies	d	+
Poly-β-hydroxybutyrate inclusions	+	+
Polyphosphate inclusions	d	d
Capsular exopolysaccharide	+	+
Heat resistance (80°C, 20 min)	−	−
Desiccation resistance	−	−
Growth in the presence of:		
0.0075% KCN	+	+
0.02% sodium azide	−	−
0.001% crystal violet	−	−
0.001% malachite green	−	−
0.01% SDS	−	−
3.0% NaCl	−	d
Lysed by 0.2% SDS	−	−
Reduction of 0.01% triphenyltetrazolium chloride	−	−
Esculin hydrolysis	−	−
Catalase activity	+	+
Oxidase activity	+	d
Soluble methane monooxygenase activity	d	nd
Ribulose-1,5-diphosphate carboxylase activity	+	+
Phosphatase activity	d	−
Urease activity	−	d
Nitrate reduction (to nitrite)	d	+
Nitrogen fixation	+	d
Sole carbon and energy sources:		
Methane, methanol	+	+
Formate, methylamine, formamide, dimethylamine, trimethylamine, trimethylamine *N*-oxide	d	−

[a]For symbols see standard definitions; nd, not determined.

[b]Phenotypic data from Bowman (1992) and Bowman et al. (1993b).

The species *Methylocaldum gracile* is a synonym of *Methylococcus thermophilus*, as indicated by DNA–DNA hybridization data (Bowman et al., 1993b). The phylogenetic discrepancy between strain IMV-2Yu and NCIM 11912 is attributed to a potentially erroneous *Methylococcus thermophilus* 16S rRNA sequence.

The mol% G + C of the DNA is: 59–61 (T_m).

Type strain: VKM-2Yu, V 2YU, ACM 3585.

GenBank accession number (16S rRNA): X73819.

Additional Remarks: GenBank accession number for *Methylocaldum gracile* is U89298.

Species Incertae Sedis

1. **"Methylococcus fulvus"** Malashenko, Romanovskaya and Kvashnikov 1972, 877.

    ful' vus. L. adj. *fulvus* tawny, yellowish brown.

    See Taxonomic Comments.

2. **Methylococcus mobilis** Hazeu, Batenburg-van der Vegte and de Bruyn 1980b, 676[VP] (Effective publication: Hazeu Batenburg-van der Vegte and de Bruyn 1980a, 211.)

    mo' bi.lis. L. masc. adj. *mobilis* motile.

    See Taxonomic Comments

    The mol% G + C of the DNA is: unknown.

Genus II. Methylobacter Bowman, Sly, Nichols and Hayward 1993b, 749[VP] emend. Bowman, Sly and Stackebrandt 1995, 184.

JOHN P. BOWMAN

Me.thy.lo.bac' ter. Fr. n. *méthyle* the methyl radical; Gr. n. *bacter* rod; M.L. n. *Methylobacter* methyl rod.

Cells are **spherical or elliptical**, 0.8–1.5 × 1.2–3.0 μm. Usually occur singly, in pairs, and occasionally in chains. **Gram negative.** When motile, form single polar flagellum. Form a **glucan-containing cyst-like resting stage,** which usually confers **desiccation resistance.** Poly-β-hydroxybutyrate granules present. **Strictly aerobic respiratory metabolism** with oxygen as the electron acceptor. **Obligate utilizers of methane for carbon and energy.** Carbon–carbon bonded compounds are not utilized. Oxidase and catalase positive. Uses the **ribulose monophosphate pathway** to fix formaldehyde for cell carbon. Key enzyme activity of the serine transhydroxymethylase pathway is absent. **Nitrogen fixation is absent.** Not susceptible to lysis by 0.2% sodium dodecyl sulfate. **Mesophilic,** with optimal growth between 30°C and 35°C. One species requires sodium ions and other factors for growth. The major fatty acid is $C_{16:1 \, \omega7c}$. **Ubiquinone-8 (Q-8)** is the predominant lipoquinone. Habitat: Isolated from sediments of freshwater lakes and rivers, wetland muds, activated sludge, and wastewater. Belongs to the *Gammaproteobacteria* as part of the Order *Methylococcales,* Family *Methylococcaceae.*

Type species: **Methylobacter luteus** (Romanovskaya, Malashenko and Bogachenko 1978) Bowman, Sly, Nichols and Hayward 1993b, 749 (*Methylococcus luteus* Romanovskaya, Malashenko and Bogachenko 1978, 124; *Methylococcus bovis* Romanovskaya, Malashenko and Bogachenko 1978, 124.)

FURTHER DESCRIPTIVE INFORMATION

Morphology and ultrastructure Cells of *Methylobacter* species possess a characteristic elliptical, rod-like morphology, with a width of 0.8–1.5 × 1.2–3.0 μm. They occur mostly singly or in pairs, although chain formation is prevalent in some strains in the late exponential growth phase. When motile, the cells are propelled by a single polar flagellum. *Methylobacter luteus* is nonmotile, while *Methylobacter whittenburyi* strains are usually motile when first isolated but can spontaneously lose this ability after extensive subculture. Motility appears most pronounced in young cultures of *M. whittenburyi* and *Methylobacter marinus;* older cultures are often devoid of motile cells (Romanovskaya et al., 1978; Bowman et al., 1993b). Cell walls have an ultrastructure that is typical of Gram-negative bacteria. Cells are surrounded by capsular material that is detectable by India ink staining.

Cells contain type I intracytoplasmic membranes, typical for other members of the family *Methylococcaceae. Methylobacter*

species form well-defined *Azotobacter*-type cysts, which give cells a refractile appearance (Whittenbury et al., 1970a). Cyst formation is most pronounced in stationary-growth-phase cultures and appears to be stimulated when oxygen becomes limiting for growth. The cysts are readily detectable by the *Azotobacter* cyst-staining procedure of Vela and Wyss (1964). The cysts are not heat resistant, but they are relatively desiccation resistant. Spontaneous loss of cysts has been noted in one study and usually leads to problems in maintaining culture viability (Lidstrom, 1988). Though the cysts may be a mechanism for surviving times of oxygen deprivation and desiccation, *Methylobacter* strains are still susceptible to methane starvation and eventually lose viability within 1–3 weeks without methane or methanol substrate. Poly-β-hydroxybutyrate granules tend to be formed by *Methylobacter* strains, particularly in younger cultures.

Cultural features *Methylobacter* strains form an even turbidity in NMS liquid medium, with little or no pellicle development. On agar, colonies are uniformly circular and convex with entire edges, possess a creamy consistency, and reach a diameter of 1–2 mm after 7–14 days under methane/air atmosphere (1:1, v/v). *M. whittenburyi* and *M. marinus* strains form colonies that are initially an off-white color but become increasingly tan to brown after prolonged incubation. At the same time, a tan diffusible melanin-like pigment is slowly released into the medium. *M. luteus* differs by forming lemon-yellow colonies, with some strains also releasing a water-soluble yellow pigment. Detailed chemical analysis of the yellow pigment has yet to be performed, but preliminary spectrophotometric scans indicated that it is not carotenoid in nature, nor does it undergo a bathochromic shift if exposed to strong alkaline solutions.

Methylobacter species are mesophilic, and most strains grow in the range of 15–40°C, with optimal growth occurring at about 30°C. The pH range for growth ranges from 5.5–9.0, with optimal growth occurring at about pH 7.0. No *Methylobacter* strains require growth factors, and all are nonhalophilic. The estuarine species *M. marinus* grows optimally with about 0.1 M NaCl in tap water or with half-strength seawater salts. One strain of *Methylobacter marinus* requires nicotinic acid for growth (Lidstrom, 1988).

Nutrition and metabolism *Methylobacter* strains are strictly aerobic obligate methanotrophs, with carbon and energy substrates limited to methane and methanol. A few strains, however, can also utilize methylamine. No strains have been demonstrated to grow on carbon–carbon bonded compounds. Upon isolation, most *Methylobacter* strains may be sensitive to methanol and do not grow on media containing 0.1% (v/v) methanol. However, good growth occurs when methanol is provided only as a vapor. It is possible to acclimate strains to methanol by slowly adding increasingly higher concentrations of methanol to media while growing the strain on methane. Methane appears to be oxidized by only the particulate form of methane monooxygenase. The presence of soluble methane monooxygenase has not been demonstrated in *Methylobacter* strains so far (Stainthorpe et al., 1991; Smith et al., 1997c). Formaldehyde is fixed by the ribulose monophosphate pathway, and key enzymes for the serine pathway are absent. Ribulose-1,5-diphosphate carboxylase activity is also absent. *Methylobacter* strains can utilize nitrate and ammonia salts, yeast extract, Casamino acids, and various amino acids as nitrogen sources, but high levels (>0.5% w/v) of complex organic compounds are inhibitory to growth. Some strains, particularly those of *M. luteus*, can produce a urease and can use urea as a sole nitrogen source. No *Methylobacter* strain has been found to be capable of fixing atmospheric nitrogen.

Fatty acid composition The outer membrane lipopolysaccharide hydroxy fatty acids have been analyzed in detail in *M. luteus* and *M. whittenburyi*. The major components in *M. luteus* include $C_{10:0\ 3OH}$ and $C_{16:0\ 3OH}$, with smaller quantities of $C_{12:0\ 2OH}$, $C_{14:0\ 3OH}$, and $C_{15:0\ 3OH}$. *M. whittenburyi* contains almost entirely $C_{16:0\ 3OH}$ (Bowman et al., 1991a). The fatty acid profiles of *Methylobacter* species have been found to be very similar (Table BXII.γ.68 of the chapter describing the family *Methylococcaceae*; Andreev and Gal'chenko, 1978). The predominant fatty acid is $C_{16:1\ \omega7c}$, accompanied by lower levels of $C_{14:0}$, $C_{16:1\ \omega6c}$, $C_{16:1\ \omega5c}$, and $C_{16:0}$ (Bowman et al., 1991a). The lack of $C_{16:1\ \omega8c}$ and relatively low levels of $C_{14:0}$ and $C_{16:0}$ distinguish *Methylobacter* from other type I methanotrophs.

Ecology The specific ecology of *Methylobacter* species is generally unknown. Indirect immunofluoresence studies, however, indicate that they are common in surface sediments and in the water columns of various freshwater and brackish water environments making up as much as 10–40% of the total bacterial population (Malashenko et al., 1987; Gal'chenko et al., 1988; Gal'chenko, 1994; Saralov et al., 1984).

ENRICHMENT AND ISOLATION PROCEDURES

The enrichment and isolation approach described for the genus *Methylococcus* can also be applied for *Methylobacter* strains. However, the incubation temperature is in this case lowered to 25–30°C. Salt composition and pH of the media all should closely reflect the habitat being investigated.

MAINTENANCE PROCEDURES

Methods for maintenance are the same as given for the genus *Methylococcus*.

DIFFERENTIATION OF THE GENUS *METHYLOBACTER* FROM OTHER GENERA

Phenotypic, chemotaxonomic, and genotypic properties differentiating *Methylobacter* from other type I methanotrophs are

shown in Tables BXII.γ.67 and BXII.γ.68 of the chapter describing the family *Methylococcaceae*.

TAXONOMIC COMMENTS

The genus *Methylobacter* was formed from the genus *Methylococcus* (Bowman et al., 1993b). *Methylobacter* species are equivalent to a similarly named group first coined by Whittenbury et al. (1970b). DNA–DNA hybridization, fatty acid analyses, and phylogenetic analyses have further refined the genus, so that species lacking cyst formation and different fatty acid profiles have been moved into the genus *Methylomicrobium*. This has necessitated an emendation of the description of *Methylobacter* (Bowman et al., 1995).

Phylogenetic analysis using 16S rDNA sequences indicates that *Methylobacter* forms a relatively tight cluster within the *Gammaproteobacteria*. *Methylobacter* is part of a larger phylogenetic cluster that includes *Methylosphaera*, *Methylomicrobium*, *Methylomonas*, and methanotrophic mytilid endosymbionts. The genus is most closely related to the genera *Methylomicrobium* and *Methylomonas* (evolutionary distance 0.04–0.06) (Bratina et al., 1992; Bowman et al., 1995).

The mol% G + C of the DNA of *Methylobacter* ranges from 48–55 (T_m). (Malashenko et al., 1975b; Meyer et al., 1986; Lysenko et al., 1988; Bowman et al., 1991b) (Table BXII.γ.71); this value is similar to those for *Methylomonas* and *Methylomicrobium*. DNA–DNA hybridization has been found to be useful in sorting out some of the taxonomic problems affecting the genus *Methylococcus*. Various *Methylococcus* species, including *Methylococcus vinelandii*, *Methylococcus chroococcus*, "*Methylococcus ucrainicus*", *Methylococcus bovis* IMET 10593 (a non-pigmented strain), *Methylococcus whittenburyi*, and phenotypically similar isolates, all exhibit high levels of DNA–DNA hybridization (54–93% similarity) with each other indicating that they were a single genospecies (Lysenko et al., 1988; Bowman et al., 1993b). This argument is supported by inherent phenotypic similarity, similar protein electrophore patterns (Gal'chenko and Nesterov, 1981), immunological cross-reactivity (Bezrukova et al., 1983), and close phylogenetic affiliation. (See Fig. BXII.γ.103 in the description of the order *Methylococcales*.). In addition, *Methylococcus luteus* shares very high levels of DNA–DNA hybridization (98% similarity) with the yellow-pigmented strain *Methylococcus bovis* IMV-B-3098 (Bowman et al., 1993b). Overall DNA–DNA hybridization levels among the three valid species of *Methylobacter* are 20–40%.

Recently, two species, *Methylobacter psychrophilus* (Omel'chenko et al., 1996) and "*Methylobacter alkaliphilus*" (Khmelenina et al., 1997), were proposed for strains isolated from tundra soil and soda lake habitats, respectively. *Methylobacter psychrophilus* possesses phenotypic properties and DNA base composition similar to those of *Methylosphaera hansonii*; both species consist of spherical nonmotile cells, lack cysts, possess gas vesicles, have psychrophilic growth requirements, and possess relatively low mol% G + C contents of their DNA (approximately 45–46). "*Methylobacter alkaliphilus*" possesses morphological traits typical of *Methylobacter* species but has a different fatty acid profile that includes high levels of $C_{16:0}$ and, depending on the growth conditions, $C_{16:0\ iso}$ (Khmelenina et al., 1997); $C_{16:1}$ fatty acid is only a relatively minor component. A 16S rDNA sequence for this species would be useful in confirming that it belongs to the genus *Methylobacter*.

DIFFERENTIATION OF THE SPECIES OF THE GENUS *METHYLOBACTER*

Phenotypic and genotypic characteristics differentiating the species of *Methylobacter* are shown in Table BXII.γ.71. Additional characteristics are shown in Table BXII.γ.72.

List of species of the genus Methylobacter

1. **Methylobacter luteus** (Romanovskaya, Malashenko and Bogachenko 1978) Bowman, Sly, Nichols and Hayward 1993b, 749[VP] (*Methylococcus luteus* Romanovskaya, Malashenko and Bogachenko 1978, 124 *Methylococcus bovis* Romanovskaya, Malashenko and Bogachenko 1978, 124.)

lu' te.us. L. adj. *luteus* golden-yellow.

Cells are nonmotile cocci or elliptical rods, 1.2–1.5 × 1.5–2.0 μm. Other characteristics are as described for the genus and as listed in Tables BXII.γ.71 and BXII.γ.72. For further descriptive information, see Bowman et al. (1993b).

Optimum temperature for growth, ~30°C; range, 15–40°C. pH optimum, ~7.0; range, 5.5–9.0.

Isolated from sediments of freshwater lakes and ponds.

The mol% G + C of the DNA is: 50.3–51.1 (T_m).

Type strain: ACM 3304, IMET 10584, NCIMB 11914, VKM-53B.

GenBank accession number (16S rRNA): M95657.

Additional Remarks: "*Methylococcus fulvus*" Malashenko, Romanovskaya and Kvashnikov 1972, 877 is phenotypically indistinguishable from *Methylobacter luteus* (Romanovskaya et al., 1978); moreover, it lacks an available strain, thereby making genotypic comparisons impossible.

2. **Methylobacter marinus** Bowman, Sly, Nichols and Hayward 1993b, 749[VP]

mar' i.nus. L. adj. *marinus* from the sea.

Cells are motile elliptical rods, 0.8–1.2 × 1.5–2.0 μm. Requires 0.1 M NaCl and tap water (or half-strength natural or artificial seawater salts) for growth. May require nicotinic acid.

Other characteristics are as described for the genus and as listed in Tables BXII.γ.71 and BXII.γ.72. For further descriptive information, see Lidstrom (1988) and Bowman et al. (1993b).

Optimum temperature for growth, ~35°C; range, 15–40°C. pH optimum, ~7.0; range, 5.5–9.0.

Isolated from near a coastal sewage outfall in Pasadena, California.

TABLE BXII.γ.71. Characteristics differentiating the species of the genus *Methylobacter*[a,b]

Characteristic	1. *M. luteus*	2. *M. marinus*	3. *M. whittenburyi*
Motility	−	+	d
Pigmentation	Y	Bb	Wb
NaCl and growth factor(s) required	−	+	−
Urease	+	−	d
Phosphatase	+	−	−
Mol% G + C of DNA (T_m)	50.3–51.1	54.3–55.0	48.7–53.7

[a]For symbols, see standard definitions; Y, yellow; Bb, buff-brown; Wb, white-brown.

[b]Phenotypic data from Bowman et al. (1993b).

TABLE BXII.γ.72. Other phenotypic characteristics of the species of the genus *Methylobacter*[a,b]

Characteristic	1. *M. luteus*	2. *M. marinus*	3. *M. whittenburyi*
Chains of cells	d	−	−
Poly-β-hydroxybutyrate present	+	+	+
Polyphosphate present	−	−	−
Capsule present	d	+	+
Cyst formation	+	d	+
Diffusible pigment formed	d	+	+
Growth at 37°C	d	+	d
Growth in the presence of:			
0.0075% KCN	d	d	d
0.02% sodium azide	−	d	d
0.001% crystal violet	−	d	−
0.001% malachite green	+	d	d
0.01% SDS	+	d	d
3% NaCl	d	+	−
Desiccation resistance	+	d	+
Heat resistance	−	−	−
Catalase	+	+	+
Oxidase	+	+	d
Serine pathway enzymes	−	−	−
Ribulose-1,5-diphosphate carboxylase	−	−	−
Nitrate (to nitrite) reduction	−	−	d
Nitrogen fixation	−	−	−
Reduction of 0.01% TTC[c]	d	−	d
Carbon and energy sources:			
Methanol as vapor	+	+	+
Methanol as solution (0.1% v/v)	d	−	d
Other C$_1$ compounds [d]	−	−	−

[a]For symbols, see standard definitions.

[b]Phenotypic data from Bowman et al. (1993b).

[c]TTC, triphenyltetrazolium chloride.

[d]Compounds tested (at 0.1%) included formate, formamide, methylamine, dimethylamine, and trimethylamine.

The mol% G + C of the DNA is: 54.3–55.0 (T_m).
Type strain: A4, ACM 4717.
GenBank accession number (16S rRNA): M95658.

3. **Methylobacter whittenburyi** (Romanovskaya, Malashenko and Bogachenko 1978) Bowman, Sly, Nichols and Hayward 1993b, 750^{VP} (*Methylococcus whittenburyi* Romanovskaya, Malashenko and Bogachenko 1978, 125)

whit.ten.bur' y.i. L. n. *whittenburyi* after Whittenbury, in honor of British microbiologist Roger Whittenbury for his seminal research on methanotrophic bacteria.

Cells are motile elliptical rods, 1.2–1.5 × 1.5–3.0 μm. Other characteristics are as described for the genus and as listed in Tables BXII.γ.71 and BXII.γ.72. For further descriptive information, see Bowman et al. (1993b).

Optimum temperature for growth, ~30–35°C; range, 15–40°C. pH optimum, ~7.0; range, 5.5–9.0.

Isolated from sediments of freshwater lakes and ponds.
The mol% G + C of the DNA is: 48.7–53.7 (T_m).
Type strain: Y, 1521, ACM 3310, NCIMB 11128.
GenBank accession number (16S rRNA): L20843, X72773.

Species Incertae Sedis

1. **"Methylobacter alkaliphilus"** Khmelenina, Kalyuzhnaya, Starostina, Suzina and Trotsenko 1997, 260.

al.ka.li.phi' lus. M.L. n. *alcali* (from Arabic *al* the end; *qaliy* soda ash); Gr. adj. *philus* loving; M.L. adj. *alkaliphilus* loving alkaline conditions.

Coccoidal to elliptical cells, 1.2–1.3 × 2.0–3.0 μm. Motile by a single polar flagellum. Multiply by binary division. *Azotobacter*-type cysts and type I intracytoplasmic membranes are formed. Possess a protein surface layer (cup-shaped units).

Colonies on agar are white to cream, circular, with entire margins, and 1–2 mm in diameter after sufficient incubation in methane/air. No water-soluble pigments are released.

Halophilic and alkaliphilic. Grows at 0.15–10% NaCl, optimum growth at 2–4% NaCl. Optimum temperature, 30°C; range, 8–37°C. Optimum pH, 9.0–9.5; range, 7.0–10.5.

Strictly aerobic. Obligate methanotroph: growth occurs only on methane and methanol. Carbon is assimilated by the ribulose monophosphate pathway. The serine-pathway enzyme α-hydroxypyruvate reductase is absent. The tricarboxylic-acid-cycle enzyme 2-oxoglutarate dehydrogenase is absent. Nitrate and ammonia are utilized as nitrogen sources via the glutamine and glutamate synthetase system.

Major fatty acids (when grown under optimal growth conditions, 0.7 M NaCl, pH 9.0) include $C_{16:0}$ (50.1%), $C_{16:0\ cyc}$ (11.6%), $C_{16:0\ iso}$ (8.7%), $C_{14:0}$ (8.0%), and $C_{16:1}$ (7.4%). At suboptimal growth conditions (low salt and neutral pH), cells produce more $C_{16:0\ iso}$ than $C_{16:0}$. Major phospholipids (under optimal growth conditions) include phosphatidylcholine, phosphatidylglycerol, and diphosphatidyglycerol (cardiolipin).

Isolated from soda lakes (pH 9.5, 1–3% salinity) in the Tuva region of Central Asia.
The mol% G + C of the DNA is: 47.6–47.9 (T_m).
Deposited strain: VKM B-2133.

2. **Methylobacter psychrophilus** Omel' chenko, Vasil' eva, Zavarzin, Savel' eva, Lysenko, Mityushina, Khmelenina and Trotsenko 2000, 423^{VP} (Effective publication: Omel' chenko, Vasil' eva, Zavarzin, Savel' eva, Lysenko, Mityushina, Khmelenina and Trotsenko 1996, 342.)

psy.chro' phi.lus. Gr. n. *psychros* cold; Gr. adj. *philus* loving; L. adj. *psychrophilus* cold loving.

Cells are coccoidal, 1.0–1.7 μm in diameter, occurring singly or in pairs. Possess a thin capsule. Multiply by binary division. Do not form cysts or other resting stages. Possess type I intracytoplasmic membranes and gas vesicles. Gas vesicles are more common in cells grown at higher cultivation temperatures and are cigar-shaped, 80–110 nm wide and up to 300 nm long.

Colonies are cream, circular, and convex, with an entire edge and a smooth shiny surface and possessing an aqueous consistency. Water-soluble pigments are not formed.

Psychrophilic. Optimum temperature, 3.5–10°C; range, 1.0–20°C. Optimum pH, 6.8–7.2; range, 5.9–7.6. Nonhalophilic. Growth factors are not required for growth.

Strictly aerobic obligate methanotroph. Growth occurs only on methane and methanol. Carbon is assimilated by the ribulose monophosphate pathway. Isocitrate dehydrogenase is NAD-specific, and 2-oxoglutarate dehydrogenase activity is absent. Ribulose-1,5-diphosphate carboxylase activity is absent.

The predominant fatty acid is $C_{16:1}$ with lower levels of $C_{14:0}$ and $C_{16:0}$.

Isolated from swampy soil of dwarf birch tundra in Russia.
The mol% G + C of the DNA is: 45.6 (T_m).
Deposited strain: Z-0021, VKM B-2103.
GenBank accession number (16S rRNA): AF152597.

Genus III. **Methylocaldum** *Bodrossy, Holmes, Holmes, Kovács and Murrell 1998, 631^{VP}*
(Effective publication: Bodrossy, Holmes, Holmes, Kovács and Murrell 1997, 501)

THE EDITORIAL BOARD

Me.thy.lo.cal.dum. N.L. n. *methylo* the methyl radical; L. adj. *caldus* hot; M.L. n. *Methylocaldum* heat-loving methylotroph.

Gram-negative, **coccoidal to long rods**. Moderately thermophilic; growth at temperatures >40°C. **Utilize methane as sole carbon and energy source. No growth on compounds containing carbon–carbon bonds.** Enzymes for ribulose monophosphate and serine pathways of formaldehyde assimilation are present. No soluble methane monooxygenase. Produce large cysts. Habitat: soil, hot springs.

The mol% G + C of the DNA is: 57–59.

Type species: **Methylocaldum szegediense** Bodrossy, Holmes, Holmes, Kovács and Murrell 1998, 631 (Effective publication: Bodrossy, Holmes, Holmes, Kovács and Murrell 1997, 501.)

List of species of the genus Methylocaldum

1. **Methylocaldum szegediense** Bodrossy, Holmes, Holmes, Kovács and Murrell 1998, 631[VP] (Effective publication: Bodrossy, Holmes, Holmes, Kovács and Murrell 1997, 501.) *sze.ged.i.en.se.* Hung. n. *Szeged* the town of Szeged; M.L. adj. *szegediense* from the town of Szeged.

Gram-negative, coccoidal to long rods, 0.6–1.0 × 1.0–1.5 μm. Motile at some stages of growth. Chains may occur in late exponential phase. Pleomorphic in stationary phase. Nondiffusible pigments produced in later stages of growth. Colonies mucoid or cartilaginous, buff to light brown after 1 week. Older colonies may be shiny and smooth with entire margins or dry and rough with irregular margins. Utilizes methane as sole carbon and energy source. No growth on compounds containing carbon–carbon bonds. Sensitive to desiccation. Produces an extracellular polymeric substance in liquid culture. Optimum temperature 55°C (range 37–62°C). No growth on methanol, formate, formamide, or methylamine. Does not fix N_2. Resistant to heating at 80°C. Enzymes for ribulose monophosphate and serine pathways of formaldehyde assimilation are present. No soluble methane monooxygenase. Produces large cysts. Habitat: soil, hot springs.

The mol% G + C of the DNA is: 57.

Type strain: OR2, NCIMB.

GenBank accession number (16S rRNA): U89300.

2. **Methylocaldum gracile** (ex Romanovskaya, Malashenko and Bogachenko 1978) Bodrossy, Holmes, Holmes, Kovács and Murrell 1998, 631[VP] (Effective publication: Bodrossy, Holmes, Holmes, Kovács and Murrell 1997, 502.) *gra.cil.e.* L. adj. *gracile* slim, slender, thin.

Gram-negative, coccoidal to long rods, 0.4–0.5 × 1.0–1.5 μm. Motile. No chain formation. Pleomorphic in stationary phase. Nondiffusible pigments produced in later stages of growth. Colonies are pasty or cartilaginous, buff to light brown after 1 week. Older colonies can turn dark brown. Utilizes methane as sole carbon and energy source. No growth on compounds containing carbon–carbon bonds. Optimum temperature 42°C (range 20–47°C). No growth above 50°C. Sensitive to desiccation and heating to 80°C. Does not fix N_2. Enzymes for ribulose monophosphate and serine pathways of formaldehyde assimilation are present. No soluble methane monooxygenase. Produces large cysts. Habitat: soil, hot springs.

The mol% G + C of the DNA is: 59.

Type strain: VKM-14L, NCIMB 11912.

GenBank accession number (16S rRNA): U89298.

3. **Methylocaldum tepidum** Bodrossy, Holmes, Holmes, Kovács and Murrell 1998, 631[VP] (Effective publication: Bodrossy, Holmes, Holmes, Kovács and Murrell 1997, 502.) *te.pi.dum.* M.L. adj. *tepidum* lukewarm.

Gram-negative, coccoidal to long rods, 1.0–1.2 × 1.0–1.5 μm. Chains may occur in late exponential phase. Pleomorphic in stationary phase. Nondiffusible pigments produced in later stages of growth. Colonies are buff and butyrous or cartilaginous after 1 week. Older colonies can turn light brown. Utilize methane as sole carbon and energy source. No growth on compounds containing carbon–carbon bonds. Sensitive to desiccation. Optimum temperature 42°C (range 30–47°C). No growth above 50°C. Does not fix N_2. Enzymes for ribulose monophosphate and serine pathways of formaldehyde assimilation are present. No soluble methane monooxygenase. Produces large cysts. Habitat: soil, hot springs.

The mol% G + C of the DNA is: 59.

Type strain: LK6.

GenBank accession number (16S rRNA): U89297.

Genus IV. **Methylomicrobium** *Bowman, Sly and Stackebrandt 1995, 183*[VP]

JOHN P. BOWMAN

Me.thy.lo.mi.cro'bi.um. Fr. n. *méthyle* the methyl radical; Gr. adj. *micros* small; Gr. n. *bios* life; M.L. n. *Methylomicrobium* a small organism able to utilize methyl units.

Rods 0.5–1.0 × 1.5–2.5 μm. Usually occur singly or in pairs. **Gram negative. Motile**, propelled by a single, unsheathed, polar flagellum. **Lack cysts** or other differentiated resting stages. Desiccation and heat-sensitive. Colonies are **nonpigmented. Have a strictly aerobic respiratory metabolism,** with oxygen as the electron acceptor. **Obligate utilizers of methane and methanol for carbon and energy.** Carbon–carbon bonded compounds are not utilized. Oxidase and catalase positive. Uses the **ribulose monophosphate pathway** to fix formaldehyde for cell carbon. Key enzyme activity of the serine transhydroxymethylase pathway is absent. Nitrogen fixation is absent. Not susceptible to lysis by 0.2% SDS. **Mesophilic** with optimal growth at 20–30°C. One species requires seawater salts for growth. Major fatty acids are $C_{16:1\ \omega8c}$, $C_{16:1\ \omega7c}$, $C_{16:1\ \omega5c}$, $C_{16:1\ \omega5t}$, and $C_{16:0}$. **Ubiquinone-8 (Q-8)** is the predominant lipoquinone. Isolated from sediments of freshwater lakes and rivers, wetland muds, agricultural and swampy soils, upper mixing layers of oceans, and estuarine waters. Belongs to the *Gammaproteobacteria* as part of the Order *Methylococcales*, Family *Methylococcaceae*.

The mol% G + C of the DNA is: 48–60.

Type species: **Methylomicrobium agile** (Bowman, Sly, Nichols and Hayward 1993b) Bowman, Sly and Stackebrandt 1995, 183 (*Methylobacter agilis* Bowman, Sly, Nichols and Hayward 1993b, 749.)

FURTHER DESCRIPTIVE INFORMATION

Morphology and ultrastructure Cells of *Methylomicrobium* species are regular short rods, 0.5–1.5 × 1.5–2.5 μm, occurring singly or in pairs. There is no propensity to form chains or rosettes. All species are actively motile by means of single polar flagellum. Cell walls are of the standard Gram-negative type and are surrounded by a thin slime capsule. Cells contain type I intracytoplasmic membranes typical for other members of the family *Methylococcaceae* but do not form glucan-containing cysts. Most strains contain poly-β-hydroxybutyrate and polyphosphate granules. Cells are not heat or desiccation-resistant and are somewhat sensitive to methane starvation, losing viability in only a few days when exposed to a methane-free atmosphere.

Lipopolysaccharide   The carbohydrate fraction of the outer membrane lipopolysaccharide in *Methylomicrobium album* includes D-glucose, L-fucose, and D-heptose (Sutherland and Kennedy, 1986). The hydroxy fatty acids from the lipopolysaccharide fraction of cell material of *Methylomicrobium album* and *Methylomicrobium agile* (Bowman et al., 1991a) are predominantly composed of $C_{16:0\ 3OH}$.

Cultural features   *Methylomicrobium* strains form an even turbidity in NMS liquid media without pellicle formation—a trait that is much more prevalent in *Methylomonas* and *Methylococcus* species. On agar, colonies are uniformly circular and convex, with entire edges, possessing a creamy consistency, and reaching a diameter of 1–2 mm after 7–14 days under a methane/air (1:1, v/v) atmosphere. Colonies are nonpigmented, with a translucent to opaque appearance, and no diffusible pigments are formed. *Methylomicrobium* species are mesophilic, growing at 10–30°C. *Methylomicrobium agile* and *Methylomicrobium album* grow best at 25–30°C, while *Methylomicrobium pelagicum* grows optimally at about 20–25°C. The pH range for growth ranges from 6.0–9.0, with optimal growth occurring at about pH 7.0. Neither *Methylomicrobium agile* nor *Methylomicrobium album* requires growth factors, and both are nonhalophilic. The marine species *Methylomicrobium pelagicum* grows optimally in media containing either natural or artificial seawater (Sieburth et al., 1987).

Nutrition and metabolism   *Methylomicrobium* strains are strictly aerobic methanotrophs, with carbon and energy substrates limited to methane and some other C_1 compounds; however, one strain of *Methylomicrobium agile* has the capacity utilize methylamine, dimethylamine, and trimethylamine (the type strain of *M. agile* can use only methylamine). No strains have been demonstrated to grow on carbon–carbon bonded compounds. Upon isolation, most *Methylomicrobium* strains are able to tolerate methanol and grow quite well on it. Methane appears to be oxidized predominantly by a particulate methane monooxygenase. The presence of soluble methane monooxygenase has not been demonstrated in *Methylomicrobium* strains so far (Stainthorpe et al., 1991). Formaldehyde is fixed by the ribulose monophosphate pathway, and key enzymes for the serine pathway, as well as ribulose-1,5-diphosphate carboxylase activity, are absent. *Methylomicrobium* species can utilize nitrate and ammonia salts, yeast extract, Casamino acids, and various amino acids as nitrogen sources, although high levels (>0.5% w/v) of complex organic compounds are inhibitory to growth. Urease and nitrogen fixation activities are absent.

Fatty acid composition   Fatty acid profiles of *Methylomicrobium* species have been found to all be quite similar, with the predominant fatty acids including $C_{16:1\ \omega 5c}$, $C_{16:1\ \omega 5t}$, $C_{16:1\ \omega 8c}$, $C_{16:1\ \omega 7c}$, and $C_{16:0}$ being the most abundant components (Table BXII.γ.68 of the chapter describing the family *Methylococcaceae*) (Bowman et al., 1991a). The low level of $C_{14:0}$ and abundance of $C_{16:1\ \omega 5t}$ and $C_{16:1\ \omega 8c}$ distinguish *Methylomicrobium* from *Methylobacter* and *Methylomonas* species. The high level of the *trans* fatty acid $C_{16:1\ \omega 5t}$ is unusual and does not appear to be due to stressful cultivation conditions. The primary respiratory lipoquinone has been found to be ubiquinone-8 (Q-8) (Collins and Green, 1985).

Ecology   The ecology and distribution of *Methylomicrobium* species is generally unknown. Indirect immunofluorescence studies indicate that *Methylomicrobium pelagicum* appears to be the dominant methanotrophic inhabitant of the upper mixing layers of ocean waters but is significantly less common in estuarine waters (Sieburth et al., 1993).

ENRICHMENT AND ISOLATION PROCEDURES

The enrichment and isolation approach described for the genus *Methylomicrobium* can be applied for *Methylobacter* strains, provided that the incubation temperature is lowered to 20–30°C. The salt composition and pH of the media should closely reflect the habitat being investigated. For isolation of *Methylomicrobium pelagicum*, cycloheximide should be added to media to eliminate predation of cells by protozoa (Sieburth et al., 1987).

MAINTENANCE PROCEDURES

Methods for maintenance are the same as those described for the genus *Methylococcus*.

DIFFERENTIATION OF THE GENUS *METHYLOMICROBIUM* FROM OTHER GENERA

Phenotypic, chemotaxonomic, and genotypic properties differentiating *Methylomicrobium* from other type I methanotrophs are shown in Tables BXII.γ.67 and BXII.γ.68 of the chapter describing the family *Methylococcaceae*.

TAXONOMIC COMMENTS

Phylogenetic analysis using 16S rDNA sequences indicates that *Methylomicrobium* forms a tight cluster within the *Gammaproteobacteria* and is part of a larger phylogenetic cluster that includes *Methylosphaera*, *Methylobacter*, *Methylomonas*, and methanotrophic mytilid endosymbionts. The genus is most closely related to *Methylobacter* and *Methylomonas* (evolutionary distance 0.04–0.06) (Bowman et al., 1995).

DNA–DNA hybridization analysis indicates that the three known *Methylomicrobium* species are genotypically distinct, with only relatively low levels of hybridization occurring between the species (Bowman et al., 1993b).

DIFFERENTIATION OF THE SPECIES OF THE GENUS *METHYLOMICROBIUM*

Phenotypic and genotypic characteristics differentiating the species of *Methylomicrobium* are shown in Table BXII.γ.73 and additional characteristics are shown in Table BXII.γ.74.

List of species of the genus Methylomicrobium

1. **Methylomicrobium agile** Bowman, Sly, Nichols and Hayward 1993b) Bowman, Sly and Stackebrandt 1995, 183^{VP} (*Methylobacter agilis* Bowman, Sly, Nichols and Hayward 1993b, 749.)

   a'gi.le. L. adj. *agilis* agile, nimble.

   Characteristics are as described for the genus and as listed in Tables BXII.γ.73 and BXII.γ.74. For further descriptive information, see Bowman et al. (1993b).

   Optimum temperature, 25–30°C range, 10–37°C. pH optimum, ~7.0; range, 6.0–9.0.

   Isolated from freshwater sediments and swampy soil.

TABLE BXII.γ.73. Characteristics differentiating the species of the genus *Methylomicrobium*[a,b]

Characteristic	1. *M. agile*	2. *M. album*	3. *M. pelagicum*
Poly-β-hydroxybutyrate inclusions	d	+	−
Growth at 37°C	+	d	−
Growth occurs with 3.0% NaCl	−	−	+
NaCl (seawater) required for growth	−	−	+
Phosphatase activity	+	+	−
Nitrate reduction (to nitrite)	−	+	−
Methylamine used as a sole carbon and energy source	+	−	−
Mol% G + C of DNA (T_m)	58.1–59.6	54.4–56.1	48.5

[a]For symbols, see standard definitions.

[b]Data from Bowman et al. (1993b).

TABLE BXII.γ.74. Other phenotypic characteristics of the species of the genus *Methylomicrobium*[a,b]

Characteristics	1. *M. agile*	2. *M. album*	3. *M. pelagicum*
Polyphosphate inclusions	+[b]	+	+
Capsular exopolysaccharide	+	d	+
Cyst formation	−	−	−
Motility	+	+	+
Pigmentation	−	−	−
Desiccation resistance	−	−	−
Heat resistance (80°C, 20 min)	−	−	−
Growth occurs in the presence of:			
0.0075% KCN	+	+	+
0.02% sodium azide	+	+	+
0.001% crystal violet	−	+	−
0.001% malachite green	+	d	−
0.01% SDS	d	−	−
Reduction of 0.01% triphenyltetrazolium chloride (TTC)	d	−	−
Lysed by 0.2% SDS	−	−	−
Catalase	+	+	+
Oxidase	+	d	+
Urease	−	−	−
Nitrogen fixation	−	−	−
Sole carbon and energy sources:			
Methane, methanol	+	+	+
Dimethylamine, trimethylamine	d	−	−
Formate, formamide, trimethylamine-*N*-oxide	−	−	−

[a]for symbols, see standard definitions

[b]Phenotypic data from Bowman et al. (1993b).

The mol% G + C of the DNA is: 58.1–59.6 (T_m).

Type strain: A30, ACM 3308, ATCC 35068, NCIMB 11124.

GenBank accession number (16S rRNA): X72767.

2. **Methylomicrobium album** (Bowman, Sly, Nichols and Hayward 1993b) Bowman, Sly and Stackebrandt 1995, 184[VP] (*Methylobacter albus* Bowman, Sly, Nichols and Hayward 1993b, 749.)

al'bum. L. adj. *album* white.

Characteristics are as described for the genus and as listed in Tables BXII.γ.73 and BXII.γ.74. For further descriptive information, see Bowman et al. (1993b).

Optimum temperature, 25–30°C; range, 10–37°C. pH optimum, ~7.0; range, 6.0–9.0.

Isolated from freshwater sediments and swampy soil.

The mol% G + C of the DNA is: 54.4–56.1 (T_m).

Type strain: ACM 3314; IMET 10526, NCIMB 11123, VKM-BG8.

GenBank accession number (16S rRNA): X72777.

3. **Methylomicrobium pelagicum** (Sieburth, Johnson, Eberhardt, Sieracki, Lidstrom and Laux 1987) Bowman, Sly and Stackebrandt 1995, 184[VP] (*Methylomonas pelagica* Sieburth, Johnson, Eberhardt, Sieracki, Lidstrom and Laux 1987, 291; *Methylobacter pelagicus* Bowman, Sly, Nichols and Hayward 1993b, 749.)

pe.la'gi.cum. L. adj. *pelagicus* belonging to the sea.

Characteristics are as described for the genus and as listed in Tables BXII.γ.73 and BXII.γ.74. Requires seawater for optimal growth. For further descriptive information, see Sieburth et al. (1987) and Bowman et al. (1993b).

Optimum temperature, ~20–25°C; range, 10–30°C. pH optimum, ~7.0; range, 6.0–8.5. Isolated from the upper mixing zone of oceans and, to a lesser extent, from estuarine waters.

The mol% G + C of the DNA is: 48.5 (T_m).

Type strain: AA-23, ACM 3505, NCIMB 2265.

GenBank accession number (16S rRNA): X72775.

Genus V. **Methylomonas** *Whittenbury and Krieg 1984b, 355^(VP) (Effective publication: Whittenbury and Krieg 1984a, 260) emend. Bowman, Sly, Nichols and Hayward 1993b, 746.*

JOHN P. BOWMAN

Me.thy.lo.mo' nas. Fr. n. *méthyle* the methyl radical; Gr. n. *monas* unit; M.L. n. *Methylomonas* methyl monad or methyl unit.

Straight or slightly curved rods 0.5–1.0 × 1.0–4.0 μm, occurring singly or in pairs. Form a polysaccharide-containing **cyst** that is desiccation sensitive. Poly-β-hydroxybutyrate granules usually present. **Motile** by means of a single polar flagellum. **Gram negative. Have a strictly aerobic respiratory metabolism**, with oxygen as the electron acceptor. **Obligate utilizers of methane and methanol for carbon and energy.** No growth occurs on carbon–carbon bonded compounds. Fix formaldehyde by the **ribulose monophosphate pathway**. Ribulose-1,5-diphosphate carboxylase activity is absent. Some species fix atmospheric nitrogen via an oxygen-sensitive nitrogenase. Form **red, pink, and orange carotenoid non-water soluble pigments**. Mesophilic, growing between 10–40°C, optimal growth temperatures 25–30°C. Non-halophilic. Major fatty acids include $C_{14:0}$ and $C_{16:1\ \omega8c}$. **18-Methyleneubiquinone-8 (MQ-8)** is the predominant lipoquinone. Habitat: Isolated from sediments of freshwater lakes and rivers, wetland muds, activated sludge and wastewater, and coal mine drainage waters. Belongs to the *Gammaproteobacteria* as part of the Order *Methylococcales*, Family *Methylococcaceae*.

The mol% G + C of the DNA is: 50–59 (T_m).

Type species: **Methylomonas methanica** Whittenbury and Krieg 1984b, 355 (Effective publication: Whittenbury and Krieg 1984a, 260.)

FURTHER DESCRIPTIVE INFORMATION

Morphology and ultrastructure The cells of *Methylomonas* species are regularly shaped rods that are either straight or slightly curved and are occasionally branched. Cells occur singly, in pairs, and sometimes as chains. All species are motile by single, unsheathed, polar flagellum. Cell walls are of the standard Gram-negative type, and well-defined slime capsules occur. *Methylomonas* species contain standard type I intracytoplasmic membranes (see Fig. BXII.γ.102 in the description of the order *Methylococcales*), and cells contain simple, single-bodied cysts similar to, but less poorly defined than, those typically observed in *Azotobacter chroococcum*. The cysts are similar to those observed in *Methylococcus* species and do not confer either desiccation or heat resistance. Most strains possess short shelf lives, dying off within a matter of days when lacking methane.

Lipopolysaccharides The hydroxy fatty acids from the outer membrane lipopolysaccharide have been analyzed in detail in *Methylomonas methanica*, *Methylomonas fodinarum*, and *Methylomonas aurantiaca* (Bowman et al., 1991a). The major component hydroxy fatty acid detected was $C_{16:0\ 3OH}$. Smaller proportions of $C_{12:0\ 3OH}$, $C_{13:0\ iso\ 3OH}$, and $C_{14:0\ 2OH}$ are also present, though their levels vary considerably between the species.

Cultural features In static liquid cultures, growth is usually restricted to a surface pellicle, which is pink-red or orange in color. On agar, *Methylomonas methanica* forms bright pink colonies that are circular, entire or lobate-edged, and butyrous in consistency, and that reach a diameter of 1–2 mm in diameter after 2 weeks under a methane/air atmosphere. The colonies of *Methylomonas fodinarum* are similar except that they have a bright orange color. The colonies of *Methylomonas aurantiaca* are larger, orange-pigmented, circular to irregular in shape, and possess a

raised convex elevation and a mucoid–viscid or cartilaginous consistency. Strains of *Methylomonas methanica* and *Methylomonas aurantiaca* often segregate into rough and smooth colonial variants, and the rough colonies usually have a somewhat cartilaginous consistency. Pigments formed by *Methylomonas* species are carotenoid in nature, lacking solubility in water but readily extractable in chloroform and acetone (Bowman et al., 1990a). Preliminary HPLC analysis of the pigments of *Methylomonas fodinarum* ACM 3268 shows the presence of as many as 24 different carotenoid components (S. Liaaen-Jensen, personal communication). *Methylomonas methanica* has been reported to form blue diffusible pigments when grown on iron deficient agar media (Whittenbury et al., 1970b).

All *Methylomonas* species have a mesophilic and nonhalophilic ecophysiology and do not require growth factors. Growth occurs between 10°C and 42°C. *Methylomonas fodinarum* strains have a temperature optimum of approximately 25–30°C. The temperature optima of *Methylomonas methanica* and *Methylomonas aurantiaca* range from 30–35°C. The pH range is 5.5–8.5, and best growth occurs at approximately pH 7.0.

Nutrition and metabolism *Methylomonas* species are obligate methanotrophs, with their sole carbon and energy sources restricted to methane and methanol. *Methylomonas aurantiaca* strains, however, can also utilize methylamine and, in some cases, dimethylamine. Other C_1 compounds are not utilized; these include formate, trimethylamine, and trimethylamine N-oxide. Methane is oxidized by a particulate methane monooxygenase; however, two *Methylomonas methanica* strains have been found to contain soluble methane monooxygenase (Koh et al., 1993; Shen et al., 1997). The ribulose monophosphate pathway is clearly the major route for fixation of formaldehyde, although *Methylomonas* strains may possess weak activity for α-hydroxypyruvate reductase, the key enzyme of the serine pathway. Ribulose-1,5-diphosphate carboxylase activity is absent, and isocitrate dehydrogenase activity is usually NAD-specific (Bowman et al., 1990a).

Nitrogen sources include nitrate, ammonia, yeast extract, Casamino acids, and various amino acids. The addition of complex nutrient sources, such as yeast extract, generally does not lead to growth stimulation and can be growth inhibitory at concentrations greater than 0.25% (w/v). Most *Methylomonas* strains are urease-positive and can use urea as a nitrogen source. *Methylomonas fodinarum* and *Methylomonas aurantiaca* are also able to fix atmospheric nitrogen by an oxygen-sensitive nitrogenase, whereas nitrogenase activity has been detected in only a few *Methylomonas methanica* strains. However, southern blot analysis has detected in several *Methylomonas methanica* strains (including the type strain ATCC 35067) a gene homologous to *nifH*, with a size of approximately 9 kb (Oakley and Murrell, 1993).

Fatty acid composition Fatty acid profiles of *Methylomonas* species are similar to one another, with the fatty acids $C_{16:1\ \omega8c}$, $C_{14:0}$, $C_{16:1\ \omega7c}$, and $C_{16:1\ \omega5t}$ being the most abundant components. The relatively high levels of $C_{14:0}$ and $C_{16:1\ \omega8c}$ distinguish *Methylomonas* species from those of *Methylobacter* and *Methylomicrobium*. The primary respiratory lipoquinone has been found to be 18-methylene-ubiquinone-8 (MQ-8) (Collins and Green, 1985).

Ecology Knowledge of the specific ecology of *Methylomonas* species is currently limited. Indirect immunofluorescence studies suggest that *Methylomonas* species are common in surface sediments and the water columns of various fresh and brackish water environments and swampy soils (Reed and Dugan, 1978; Saralov et al., 1984; Malashenko et al., 1987; Gal'chenko et al., 1988). High populations of *Methylomonas methanica* are found in the littoral sediments of various subtropical rivers in Eastern Australia, with most probable number counts reaching 10^8–10^9 cells per gram of sediment. *Methylomonas aurantiaca* is extremely populous in activated sludge samples, while *Methylomonas fodinarum* is the most common methanotroph in Australian coal mine drainage water (Bowman, 1992).

ENRICHMENT AND ISOLATION PROCEDURES

The enrichment and isolation approach described for the genus *Methylococcus* can also be applied for *Methylomonas* strains, provided that the incubation temperature is lowered to 20–30°C. The salt composition and pH of the media should closely reflect the habitat being investigated.

MAINTENANCE PROCEDURES

Methods for maintenance are the same as those given for the genus *Methylococcus*.

DIFFERENTIATION OF THE GENUS *METHYLOMONAS* FROM OTHER GENERA

Phenotypic, chemotaxonomic, and genotypic properties differentiating *Methylomonas* from other type I methanotrophs are shown in Tables BXII.γ.67 and BXII.γ.68 of the chapter describing the family *Methylococcaceae*.

TAXONOMIC COMMENTS

The genus *Methylomonas* previously contained several species that were either obviously misclassified or invalid owing to lack of descriptions, validation, and/or type strains. A number of these nomenclatural and taxonomic problems have been resolved. Several invalid species are now recognized as being synonymous with *Methylobacillus glycogenes* (Urakami and Komagata, 1986b); these include "*Methylomonas methylovora*" (Kuono et al., 1973), "*Methylomonas methanolica*", "*Methylomonas espexii*", "*Methylomonas methanocatalesslica*", and "*Methylomonas methanofructolica*" (Urakami and Komagata, 1986a). "*Methylomonas clara*" (Faust et al., 1977) is synonymous with *Methylophilus methylotrophus* (Jenkins et al.,

1987). All of these species are restricted methylotrophs utilizing methanol and methylamine as primary carbon and energy sources but without the capacity to utilize methane. "*Methylomonas methaninitrificans*" (Davis et al., 1964) Leadbetter 1974b, 269 and "*Methylomonas methanooxidans*" (Brown and Strawinski, 1958) Leadbetter 1974b, 269 in all probability are synonymous with the genus *Methylosinus* (Whittenbury et al., 1970a, 1970b), based on nitrogen fixing ability and morphology; however, this cannot be verified because no representative strains are extant. The species "*Methylomonas margaritae*" Takeda, Motomatsu, Hachiya, Fukuoka and Takahara 1974, 795 and "*Methylomonas flagellata*" Morinaga, Yamanaha, Otsuka and Hirose 1976, 1544 possess traits similar to *Methylomicrobium agile* or *Methylomicrobium album*, but there are no extant cultures of either species. The species "*Methylomonas gracilis*" Romanovskaya, Malashenko and Bogachenko 1978, 122 is synonymous with *Methylococcus thermophilus* (see comments for the genus *Methylococcus*).

Phylogenetic analysis using 16S rDNA sequences indicates that *Methylomonas* forms a relatively tight cluster within the *Gammaproteobacteria* and is part of a larger phylogenetic cluster that includes *Methylosphaera*, *Methylobacter*, *Methylomicrobium*, and methanotrophic mytilid endosymbionts. The genus is most closely related to *Methylobacter* and *Methylomicrobium* (evolutionary distance 0.05–0.06) (Bratina et al., 1992; Bowman et al., 1995).

The mol% G + C of the DNA of *Methylomonas* cover a relatively broad range of 50–59 (T_m), much like those for *Methylomonas* and *Methylobacter* (Table BXII.γ.75). DNA–DNA hybridization analysis has found the three valid *Methylomonas* species to be genotypically distinct; however, significant DNA homology (30–50%) is present between *Methylomonas fodinarum* and *Methylomonas aurantiaca*. This corresponds with the high level of phylogenetic similarity shared by these species (See Fig. BXII.γ.103 in the description of the order *Methylococcales*.). *Methylomonas methanica* is more distinct, sharing less than 25% DNA homology with its sister species. Very high levels of homology are present between *Methylomonas methanica* and "*Methylomonas rubra*", indicating that "*Methylomonas rubra*" is a junior subjective synonym of *Methylomonas methanica*. This result agrees with the high levels of phenotypic similarity that these species share (Bowman et al., 1990a; 1993b). Phylogenetic analysis, however, indicates that *Methylomonas methanica* is somewhat distinct from other *Methylomonas* species (evolutionary distance 0.03–0.04) (Bratina et al., 1992). The genotypic relationship of "*Methylomonas rubra*" to *Methylomonas methanica* requires further verification.

DIFFERENTIATION OF THE SPECIES OF THE GENUS *METHYLOMONAS*

Phenotypic and genotypic characteristics differentiating the species of *Methylomonas* are shown in Table BXII.γ.75 and additional characteristics are shown in Table BXII.γ.76.

TABLE BXII.γ.75. Characteristics differentiating the species of the genus *Methylomonas*[a,b]

Characteristics	1. *M. methanica*	2. *M. aurantiaca*	3. *M. fodinarum*
Colony pigmentation	Pr	Or	Or
Growth at 37°C	d	+	−
Oxidase	+	−	−
Alkaline phosphatase	−	+	-
Nitrate reduction (to nitrite)	−	−	+
Methylamine as a sole carbon and energy source	+	+	+
Mol% G + C of DNA (T_m)	50.6–54.0	55.3–57.6	57.8–59.1

[a]For symbols, see standard definitions; Pr, pink-red; Or, orange.

[b]Data from Bowman et al. (1993b).

TABLE BXII.γ.76. Other phenotypic characteristics of the species of the genus *Methylomonas*[a,b]

Characteristics	1. *M. methanica*	3. *M. aurantiaca*	2. *M. fodinarum*
Chains of cells	d	−	−
Poly-β-hydroxybutyrate inclusions	+	+	+
Polyphosphate inclusions	+	d	+
Capsular exopolysaccharide	d	+	+
Cyst formation	+	+	+
Motility	+	+	+
Desiccation resistance	−	−	−
Heat resistance (80°C, 20 min)	−	−	−
Growth in the presence of:			
0.0075% KCN	+	+	+
0.02% sodium azide	d	−	+
0.001% crystal violet	d	d	d
0.001% malachite green	d	d	+
0.01% SDS	−	d	−
3.0% NaCl	−	−	−
Lysed by 0.2% SDS	+	+	+
Reduction of 0.01% triphenyltetrazolium chloride (TTC)	−	+	+
Esculin hydrolysis	+	−	d
Catalase activity	+	+	+
Urease activity	+	d	+
Nitrogen fixation	d	+	+
Sole carbon and energy sources:			
Methane, methanol	+	+	+
Dimethylamine	−	d	−
Formate, formamide, trimethylamine, trimethylamine-*N*-oxide	−	−	−

[a]For symbols, see standard definitions.

[b]Data from Bowman et al. (1993b).

List of species of the genus Methylomonas

1. **Methylomonas methanica** Whittenbury and Krieg 1984b, 355[VP] (Effective publication: Whittenbury and Krieg 1984a, 260.)

me.tha′ ni.ca. M.L. adj. *methanica* related to or associated with methane.

Cells are motile rods. Gram negative. Characteristics are as described for the genus and as listed in Tables BXII.γ.75 and BXII.γ.76. For further descriptive information, see Bowman et al. (1993b).

Optimum temperature, ~25–30°C; range, 10–37°C. pH optimum ~7.0; range, 5.5–9.0.

Isolated from freshwater sediment, lake and pond water, and swampy soil.

The mol% G + C of the DNA is: 50.6–54.0 (T_m).

Type strain: ACM 3307, ATCC 35067, IMET 10543, NCIMB 11130.

GenBank accession number (16S rRNA): AF304196.

2. **Methylomonas aurantiaca** Bowman, Cox, Sly and Hayward 1990b, 470[VP] (Effective publication: Bowman, Cox, Sly and Hayward 1990a, 285.)

au.ran.ti′ a.ca. M.L. adj. *aurantiacus* orange-colored.

Cells are motile rods. Gram negative. Characteristics are as described for the genus and as listed in Tables BXII.γ.75 and BXII.γ.76. For further descriptive information, see Bowman et al. (1993b).

Optimum temperature, ~30–35°C; range, 10–40°C. pH optimum, ~7.0; range, 5.0–9.0.

Isolated from activated sludge, wastewater and swampy soil.

The mol% G + C of the DNA is: 57.6–59.1 (T_m).

Type strain: JB103, UQM 3406.

GenBank accession number (16S rRNA): X72776.

3. **Methylomonas fodinarum** Bowman, Cox, Sly and Hayward 1990b, 470[VP] (Effective publication: Bowman, Cox, Sly and Hayward 1990a, 283.)

fo.di.na′ rum. M.L. n. *fodinarum* of coal mine.

Cells are motile rods. Gram negative. Characteristics are as described for the genus and as listed in Tables BXII.γ.75 and BXII.γ.76. For further descriptive information, see Bowman et al. (1993b).

Optimum temperature, ~25–30°C; range, 10–35°C. pH optimum ~7.0; range, 5.5–9.0.

Isolated from coal mine drainage water and freshwater sediments.

The mol% G + C of the DNA is: 55.3–57.6 (T_m).

Type strain: LD2, UQM 3268.

GenBank accession number (16S rRNA): X72778.

4. **Methylomonas scandinavica** Khalyuzhnaya, Khmelenina, Kotelnikova, Holmquist, Pedersen and Trotsenko 2000, 949[VP] (Effective publication: Khalyuzhnaya, Khmelenina, Kotelnikova, Holmquist, Pedersen and Trotsenko 1999, 571[VP]

Cells are motile rods which are single or which form short chains. Gram negative. Single polar flagellum. Colonies are pigmented pink. Carbon sources include only methane and methanol metabolized using the RuMP pathway. Use ammonia and nitrate as nitrogen sources via the glutamate cycle and by reductive amination of pyruvate and 2-oxoglutarate. Optimum temperature, ~17°C; range 5–30°C. pH optimum 6.8–7.6; range 5.0–9.0. Generation time for growth is about 15 h. Nonhalophilic. Cells contain phosphatidylglycerol and phosphatidylethanolamine but not cardiolipin. Major lipoquinone is MQ-8. Isolated from groundwater of the Äspö Hard Rock Laboratory tunnel, southeastern Sweden.

The mol% G + C of the DNA is: 53.8 (T_m).

Type strain: SR-5, VKM B-2140.

GenBank accession number (16S rRNA): AJ131369.

Species Incertae Sedis

1. **"Methylomonas margaritae"** Takeda, Motomatsu, Hachiya, Fukuoka and Takahara 1974, 795.

2. **"Methylomonas flagellata"** Morinaga, Yamanaha, Otsuka and Hirose 1976, 1544.

3. **"Methylomonas methaninitrificans"** (Davis, Coty and Stanley 1964) Leadbetter 1974b, 269 (*"Pseudomonas methaninitrificans"* Davis, Coty and Stanley 1964, 471.)

   "*Methylomonas methaninitrificans*" is probably a strain of the genus *Methylosinus*.

4. **"Methylomonas methanooxidans"** (Brown and Strawinski 1958) Leadbetter 1974b, 269 (*"Methanomonas methanooxidans"* Brown and Strawinski 1958.)

   "*Methylomonas methanooxidans*"is probably a strain of the genus *Methylosinus*.

Genus VI. **Methylosarcina** *Wise, McArthur and Shimkets 2001, 620*[VP]

THE EDITORIAL BOARD

Me.thyl.o.sar.ci′na. N.L. neut. n. *methylum* the methyl group; L. fem. n. *sarcina* pack, bundle; N.L. fem. n. *Methylosarcina* methane-utilizing bundle of cells.

Gram-negative strictly aerobic short, coccoid, or fusiform rods. Form aggregates in slime layer or capsule. Catalase positive; oxidase negative. **Form cysts. Only carbon sources methane and methanol.** Major fatty acids include $C_{16:0}$, $C_{16:1\,\omega8c}$, $C_{16:1\,\omega7c}$, $C_{16:1\,\omega5t}$.

The mol% G + C of the DNA is: 54.

Type species: **Methylosarcina fibrata** Wise, McArthur and Shimkets 2001, 620.

FURTHER DESCRIPTIVE INFORMATION

Cells of both species of *Methylosarcina* contain inclusion bodies and stacked membranes such as are found in other type I methanotrophs. *Methylosarcina fibrata* cells are often embedded in a fibrillar matrix. At the end of growth, cells of both species enlarge to produce cysts; these cysts do not survive pasteurization or drying for 7 d (Wise et al., 2001).

Both *Methylosarcina* species use only methane and methanol as carbon and energy sources; both use ammonia, nitrate, and a variety of organic nitrogen sources, and are unable to fix N_2. Both species reduce nitrate to nitrate but do not grow using nitrate as an electron acceptor and methane or methanol as the source of carbon and electrons under anaerobic conditions. Both possess the enzyme hexulose phosphate synthase of the ribulose monophosphate pathway for carbon assimilation and the *pmoA* gene for particulate methane monooxygenase (Wise et al., 2001).

Analysis of 16S rDNA sequences showed that the type strains of *Methylosarcina fibrata* and *Methylosarcina quisquiliarum* are most closely related to each other and form a cluster distinct from other type I methylotrophs. The *Methylosarcina* 16S sequences were most similar to those of *Methylobacter*. Analysis of partial *pmoA* sequences showed that the *Methylosarcina* sequences were most closely related to those of *Methylomicrobium album* (Wise et al., 1999, 2001).

ENRICHMENT AND ISOLATION PROCEDURES

These methanotrophs were obtained from liquid enrichments of landfill soil serially diluted in varying concentrations of a mineral salts medium adjusted to different pH values and incubated under atmospheres of differing air/CH_4/CO_2 mixtures. They were isolated by streaking onto agar of the same composition. *Methylosarcina fibrata* was more common in acidic (pH 4.8) enrichments; *Methylosarcina quisquiliarum*, in enrichments at pH 6.8 (Wise et al., 1999).

DIFFERENTIATION OF THE GENUS *METHYLOSARCINA* FROM OTHER GENERA

Wise et al. (2001) provide a comparison of the fatty acid profiles of the type strains of *Methylosarcina fibrata* and *Methylosarcina quisquiliarum* with those of members of the genera *Methylomonas*, *Methylobacter*, *Methylomicrobium*, and *Methylococcus*.

List of species of the genus Methylosarcina

1. **Methylosarcina fibrata** Wise, McArthur and Shimkets 2001, 620[VP]

   fi.bra′ ta. L. fem. adj. *fibra* a fiber or filament; M.L. fem. adj. *fibrata* covered with fibers or fibrils.

   Description as for the genus with the following additional characteristics. Cells coccoid (diameter 0.8–1.5 μm). Form packets. Single cells, diplococci, and tetrads may be motile. Fibrils on cell surface.

   The mol% G + C of the DNA is: 54.1 ± 0.2 (HPLC).

   Type strain: AML-C10, ATCC 700909, DSM 13736.

   GenBank accession number (16S rRNA): AF177296.

2. **Methylosarcina quisquiliarum** Wise, McArthur and Shimkets 2001, 620[VP]

   quis.qui.li.a′ rum. L. pl. fem. n. *quisquiliae* rubbish, trash, or refuse; L. gen. pl. fem. n. *quisquiliarum* of rubbish, trash, etc., denoting strains that were isolated from a landfill site.

   Description as for the genus with the following additional characteristics. Cells pleomorphic: ovoid, fusiform, or rods (1 × 1–6 μm). Coccoid cells occur in packets. No fibrils on cell surface.

   The mol% G + C of the DNA is: 54.3 ± 0.3 (HPLC).

   Type strain: AML-D4, ATCC 700908, DSM 13737.

   GenBank accession number (16S rRNA): AF177297.

Genus VII. **Methylosphaera** Bowman, McCammon and Skerratt 1998d, 327[AL] (Effective publication: Bowman, McCammon and Skerratt 1997c, 1457)

JOHN P. BOWMAN

Me.thy.lo.sphae'ra. Fr. n. *méthyle* the methyl radical; Gr. n. *spaira* sphere; M.L. n. *Methylosphaera* methyl sphere.

Cells are spherical, 1.5–2.0 μm in diameter. Occur singly or in pairs. Cysts or other differentiated **resting stages are not formed. Gas vesicles present.** Poly-β-hydroxybutyrate granules present. **Gram negative. Nonmotile. Nonpigmented. Strictly aerobic respiratory metabolism**, with oxygen as the electron acceptor. **Obligate utilizer of methane and methanol for carbon and energy.** Carbon–carbon bonded compounds are not utilized. Oxidase and catalase positive. The **ribulose monophosphate pathway** is used to fix formaldehyde for cell carbon. Key enzyme activity of the serine transhydroxymethylase pathway is absent. Nitrate, ammonium salts, and L-glutamine are used as nitrogen sources. Nitrate is reduced to nitrate. **Fix atmospheric nitrogen.** Ribulose-1,5-diphosphate carboxylase activity is not present. **Psychrophilic**, growing at −2–20°C with an optimal temperature range of 10–15°C. **No growth occurs on solid agar media. Seawater salts are required for optimal growth.** Major fatty acids include $C_{16:0}$, $C_{16:1 \omega 8c}$, $C_{16:1 \omega 7c}$, and $C_{16:1 \omega 6c}$. Habitat: Isolated from hypolimnetic and benthic zones of an Antarctic meromictic lake of marine salinity. Belongs to the *Gammaproteobacteria* as part of the Order *Methylococcales*, Family *Methylococcaceae*.

The mol% $G + C$ of the DNA is: 43–46 (T_m).

Type species: **Methylosphaera hansonii** Bowman, McCammon and Skerratt 1998d, 327 (Effective publication: Bowman, McCammon and Skerratt 1997c, 1457.)

FURTHER DESCRIPTIVE INFORMATION

Morphology and ultrastructure During all phases of growth *Methylosphaera hansonii* strains appear as featureless spherical cells that exhibit varying degrees of refractility under phase contrast microscopy. Although cells show signs of uneven binary division, evidence for budding division is still lacking. Cells possess cell walls of the standard Gram-negative type and possess type I intracytoplasmic membranes when grown under methane but not methanol. Strains are nonmotile and possess no flagella. They also do not form cysts or other types of resting stages typically found in other methanotrophic bacteria. Thus, *M. hansonii* cells have no desiccation or heat resistance and die rapidly in the absence of methane or methanol substrate. A strong accumulation of poly-β-hydroxybutyrate may occur in some cells. Preliminary electron microscopic examination indicates the occurrence of gas vesicles 0.05–0.1 × 0.2 μm (Bowman et al., 1997c; Bowman, unpublished data.).

Cultivation *M. hansonii* can be grown in nitrate mineral salts media supplemented with natural or artificial seawater salts. Strangely, strains exhibit no growth on solid agar media including those with high purity agars at relatively low concentrations. Agars tested included Bacto-agar (Difco), noble agar, agarose type V, and low melting point agarose at concentrations of 0.3–1.5% (w/v). This inability to grow on agar media has been previously observed with unidentified type I methylotrophs isolated from seawater off the British Isles (Lees et al., 1991). Those strains, however, were able to grow on semisolid noble agar whereas *M. hansonii* does not. When growing in statically incubated liquid media, *M. hansonii* forms a white turbidity with a distinct layer of sediment. It requires seawater for growth, with growth occurring only at 0.25–1.5× strength seawater salts. Poor growth occurs if only NaCl, rather than seawater, is provided, implying that strains require additional components found in seawater, although these components have yet been identified. All strains are psychrophilic, with optimum and maximum growth temperatures varying slightly between strains. Optimal growth occurs at 10–15°C; no growth occurs at 25°C. A doubling time of 20–24 h has been determined for strains growing at or close to their optimum temperature. In NMS–seawater liquid media, growth occurs between pH 6.0–8.0, and the optimum pH is approximately 7.5.

Nutritional and metabolic features *M. hansonii* is metabolically specialized, with only methane and methanol serving as suitable sources of carbon and energy. Methane monooxygenase activity appears to be restricted to the particulate (pMMO) form. No putative sMMO activity has been detected with the naphthalene oxidation assay. That this species lacks sMMO activity is supported by the fact that strains grow very poorly in media lacking copper. Growth on methanol (at 0.1%, v/v) is slower than that on methane but results in similar growth yields. However, no growth has been found to occur on a range of other C_1 compounds, including formate, formamide, methylamine, dimethylamine, trimethylamine, and trimethylamine N-oxide. Strains use the ribulose monophosphate pathway to fix formaldehyde for cell carbon, as demonstrated by the presence of hexulose-6-phosphate synthase activity. However, activity of α-hydroxypyruvate reductase, a key enzyme of the serine pathway, is absent. Ribulose-1,5-diphosphate carboxylase activity is also absent. Growth is strictly aerobic, and no growth occurs anaerobically in the presence of an alternative electron acceptor, such as nitrate or ferric chloride.

Strains utilize nitrate, ammonia and L-glutamine for nitrogen. L-Glutamine also stimulates growth yield and growth rate. More complex sources of nitrogen, such as yeast extract or Casamino acids (Difco), are less suitable, causing partial inhibition of growth when tested at 0.05–0.1% (w/v) and complete inhibition of growth at concentrations of 0.25–0.5% (w/v). Urease is not formed, and urea does not serve as a nitrogen source. Nitrogen fixation occurs and is not quite as oxygen sensitive as in other methanotrophs; at an oxygen level of about 10% (v/v), nitrogen fixation occurs in *Methylosphaera hansonii* but is substantially inhibited or abolished in *Methylosinus trichosporium* and *Methylococcus capsulatus* (Murrell and Dalton, 1983; Zhivotchenko et al., 1995b).

Genetic features The DNA base composition values of *Methylosphaera hansonii* are the lowest among the known methanotrophs, with a range of 43–46 mol% $G + C$. The type strain ACAM 549 (AM6) has a mol% $G + C$ of 44.8.

Ecology Little is known about the ecology of *Methylosphaera*. Populations are known to become concentrated in the surface benthic sediments and the oxygen gradient zones of sulfate-depleted, marine-salinity lakes (Ace Lake and Burton Lake) found in the Vestfold Hills area of Eastern Antarctica. The waters of these lakes have an average annual temperature of about 2°C.

The distribution of *Methylosphaera hansonii* outside of these lakes is unknown; however, the Vestfold Hills lakes are of recent origin (8000–10000 years old), formed from a marine uplift (Franzmann, 1996) that followed the retreat of the ice sheet after the last ice age. This leads to the hypothesis that *Methylosphaera hansonii* may also be common in Antarctic coastal sediments.

ENRICHMENT AND ISOLATION PROCEDURES

Low temperature marine sediments and hypolimnetic zones of polar marine-salinity lakes (and possibly coastal marine zones) are the best sites for isolation of *Methylosphaera* strains. Enrichment is performed in mineral salts media prepared in serum vials with either natural or artificial seawater. A headspace of 1:1 methane/air is created and the vials incubated at 2–4°C. After 3–4 weeks, a white turbidity forms that is accompanied by white sedimented material. *Methylosphaera hansonii* enrichments contain a high population of nonmethanotrophic, usually methylotrophic, co-contaminants. Known strains of *Methylosphaera* cannot be grown on solid agar and, consequently, the isolation procedure described for most methanotrophs (see enrichment and isolation details for the genus *Methylococcus*) is not possible. Isolation can be achieved by serially diluting enrichments to extinction in 96-well plastic microtiter trays (Bowman et al., 1997c) containing NMS seawater liquid media and incubated in containers containing 1:1 methane/air. Several strains may be purified in the same tray simultaneously. After sufficient incubation, the wells containing the highest dilutions that show growth are examined by microscopy. A number of separate transfers may be required to eventually obtain morphologically homogenous cultures.

MAINTENANCE PROCEDURES

Long term preservation of *Methylosphaera hansonii* is currently problematic. The best procedure is to keep strains in serum vials with NMS seawater media under a methane/air atmosphere held at about 2°C, periodically subculturing them. Viability is rapidly lost following freezing, including snap freezing in liquid nitrogen.

DIFFERENTIATION OF THE GENUS *METHYLOSPHAERA* FROM OTHER GENERA

Phenotypic, genotypic, and chemotaxonomic traits that differentiate *Methylosphaera* from other type I methanotrophs are shown in Tables BXII.γ.67 and BXII.γ.68 of the chapter describing the family *Methylococcaceae*.

TAXONOMIC COMMENTS

16S rDNA based phylogenetic analysis indicates that *Methylosphaera hansonii* forms a separate branch at the periphery of the cluster of type I methanotrophs that includes the genera *Methylobacter*, *Methylomicrobium*, and *Methylomonas*. The closest relatives are methanotrophic mytilid endosymbionts (evolutionary distance 0.09–0.10).

DNA–DNA hybridization analysis indicates that strains from different lake isolation sites are closely related, showing 65–100% hybridization with ACAM 549.

List of species of the genus Methylosphaera

1. **Methylosphaera hansonii** Bowman, McCammon and Skerratt 1998d, 327[VP] (Effective publication: Bowman, McCammon and Skerratt 1997c, 1457.)

   han.son'i.i. M.L. gen. n. *hansonii* of Hanson, named after American microbiologist R.S. Hanson.

   Cells are coccoidal and nonmotile. Colonies do not form on solid agar media. Liquid media cultures possess no pigment. Characteristics are as described for the genus and as listed in Tables BXII.γ.67 and BXII.γ.68 of the chapter describing the family *Methylococcaceae*.

   Optimum temperature for growth, ~10–15°C; range, −2–20°C. pH range for growth, 6.0–8.5; optimum, ~7.5.

   Known habitat includes certain Antarctic marine-salinity meromictic lakes.

   The mol% G + C of the DNA is: 43.5–45.9 (T_m).

   Type strain: AM6, ACAM 549.

   GenBank accession number (16S rRNA): U67929.

Order VIII. **Oceanospirillales** *ord. nov.*

GEORGE M. GARRITY, JULIA A. BELL AND TIMOTHY LILBURN

O.ce.an.o.spi.ril.la' les. M.L. neut. n. *Oceanospirillum* type genus of the order; *-ales* ending to denote order; M.L. fem. n. *Oceanospirillales* the *Oceanospirillum* order.

The order *Oceanospirillales* was circumscribed for this volume on the basis of phylogenetic analysis of 16S rDNA sequences; the order contains the families *Oceanospirillaceae*, *Alcanivoraceae*, *Hahellaceae*, *Halomonadaceae*, *Oleiphilaceae*, and *"Saccharospirillaceae"*.

Most genera halotolerant or halophilic. Motile except for *Al-canivorax*. Aerobic, microaerophilic, or facultatively anaerobic chemoorganotrophs.

Type genus: **Oceanospirillum** Hylemon, Wells, Krieg and Jannasch 1973, 361[AL].

Family I. **Oceanospirillaceae** *fam. nov.*

GEORGE M. GARRITY, JULIA A. BELL AND TIMOTHY LILBURN

O.ce.an.o.spi.ril.la' ce.ae. M.L. neut. n. *Oceanospirillum* type genus of the family; *-aceae* ending to denote family; M.L. fem. pl. n. *Oceanospirillaceae* the *Oceanospirillum* family.

The family *Oceanospirillaceae* was circumscribed for this volume on the basis of phylogenetic analysis of 16S rDNA sequences; the family contains the genera *Oceanospirillum* (type genus), *Balneatrix*, *Marinomonas*, *Marinospirillum*, *Neptunomonas*, *Oceanobacter*, *Oleispira*, and *Pseudospirillum*. *Oceanobacter*, *Pseudospirillum* and *Oleispira* were proposed after the cut-off date for inclusion in this volume (June 30, 2001) and are not described here (see Satomi et al. (2002) and Yakimov et al. (2003a), respectively).

Motile by polar flagella. Aerobic; strictly respiratory except for *Neptunomonas*, which gives weak fermentation reactions. Aquatic; *Balneatrix* is found in fresh water, whereas other genera are marine.

Type genus: **Oceanospirillum** Hylemon, Wells, Krieg and Jannasch 1973, 361[AL].

Genus I. **Oceanospirillum** *Hylemon, Wells, Krieg and Jannasch 1973, 361*[AL]*

BRUNO POT AND MONIQUE GILLIS

O.ce.an.o.spi.ril' lum. M.L. n. *oceanus* ocean; Gr. n. *spira* a spiral; M.L. dim. neut. n. *spirillum* spirillum a small spiral; *Oceanospirillum* a small spiral (organism) from the ocean (seawater).

Rigid, helical cells 0.4–1.2 μm in diameter. Motile by bipolar tufts of flagella. A polar membrane underlies the cytoplasmic membrane at the cell poles in all species so far examined by electron microscopy. **Intracellular poly-β-hydroxybutyrate is formed. All species form thin-walled coccoid bodies that predominate in old cultures.** Gram negative. **Aerobic**, having a strictly respiratory type of metabolism with oxygen as the terminal electron acceptor. Nitrate respiration does not occur. Nitrate can be reduced to nitrite in all oceanospirilla. Optimum temperature for growth, 25–32°C. **Oxidase positive.** Indole and aryl sulfatase negative. Casein, starch, hippurate, and esculin are not hydrolyzed. **Seawater is required for growth. Carbohydrates are neither oxidized nor fermented.** Amino acids or the salts of organic acids serve as carbon sources. Growth factors are not usually required. Isolated from coastal seawater, decaying seaweed, and putrid infusions of marine mussels.

The mol% G + C of the DNA is: 45–50.

Type species: **Oceanospirillum linum** (Williams and Rittenberg 1957) Hylemon, Wells, Krieg and Jannasch 1973, 374 (*Spirillum linum* Williams and Rittenberg 1957, 82.)

FURTHER DESCRIPTIVE INFORMATION

All species of *Oceanospirillum* consist of helical cells; however, variants having less curvature may arise after prolonged transfer. For example, the type strain of *O. japonicum* consisted initially of long, helical cells with several turns (Watanabe, 1959), but now consists of slightly curved or S-shaped cells. The cells have a constant and characteristic type of clockwise (right-handed) helix. Only one phylogenetically unrelated species (*O. pusillum*) has

a counterclockwise (left-handed) helix (Terasaki, 1972). Photographs showing the size and shape of various species of oceanospirilla and *O. minutulum* (now *Marinospirillum minutulum*) are presented in Fig. BXII.γ.104.

An unusual elaboration of the plasma membrane, the "polar membrane", occurs in all of the species so far examined (Beveridge and Murray, unpublished results). It is attached to the inside of the plasma membrane by bar-like links and is located, most commonly, in the region surrounding the polar flagella (Murray and Birch-Andersen, 1963). Such a membrane has been found mainly in genera of helical bacteria, such as *Spirillum*, *Campylobacter*, *Aquaspirillum*, *Ectothiorhodospira*, and *Rhodospirillum*.

All species have intracellular poly-β-hydroxybutyrate, but granules may not be evident in cells having a small diameter and chemical analysis may be required to demonstrate the polymer.

All species have bipolar tufts of flagella and all species show extensive formation of coccoid bodies (sometimes termed "microcysts") in old cultures. These bodies have thin walls and resemble spheroplasts; however, they are resistant to lysis in distilled water (Kelly, 1959). Whether coccoid bodies are resistant to desiccation is not known. Three main modes of formation of coccoid bodies were described by Williams and Rittenberg (1957), as follows: (a) two cells may entwine and apparently fuse. The cells become shorter and thicker and a protuberance develops at the point of fusion. This gradually enlarges and absorbs the organisms to form the coccoid body. More than one coccoid body may develop from a pair of entwined spirilla; (b) a spirillum may become shorter and thicker and a protuberance arises from the center of the cell or from each end of the cell. The protuberances enlarge and eventually merge into a single coccoid body as the helical cell is absorbed; (c) a spirillum may undergo a gradual shortening and rounding to form a coccoid body. The majority of coccoid bodies present in old cultures appears to be viable and can "germinate" when placed into a fresh medium (Williams and Rittenberg, 1956). Germination is by unipolar or bipolar growth of a helical cell from the coccoid body, with the latter being absorbed into the developing helical cell.

Seawater is required for the growth of all species. Media prepared with natural seawater or with 2.75% NaCl have been used for enrichment and isolation (Williams and Rittenberg, 1957; Terasaki, 1963, 1970, 1980). Commonly used culture media for

---

Editorial Note: The genus *Oceanospirillum* has recently undergone taxonomic re-evaluation. *O. minutulum* has been transferred to the genus *Marinospirillum* as *Marinospirillum minutulum* (Satomi et al., 1998); see the Taxonomic Comments section in this chapter. *O. commune* and *O. vagum* are homotypic synonyms of *Marinomonas communis* and *Marinomonas vaga*, respectively. After the cut-off date for inclusion of taxonomic changes in this volume of the *Systematics*, Satomi et al. (2002) emended the description of *Oceanospirillum* and proposed the transfer of *O. jannaschii* to the genus *Marinobacterium* as *Marinobacterium jannaschii*, the transfer of *O. japonicum* to the new genus *Pseudospirillum* as *Pseudospirillum japonicum*, the transfer of *O. kriegii* to the new genus *Oceanobacter* as *Oceanobacter kriegii*, and the transfer of *O. pusillum* to the new genus *Terasakiella* as *Terasakiella pusilla*.

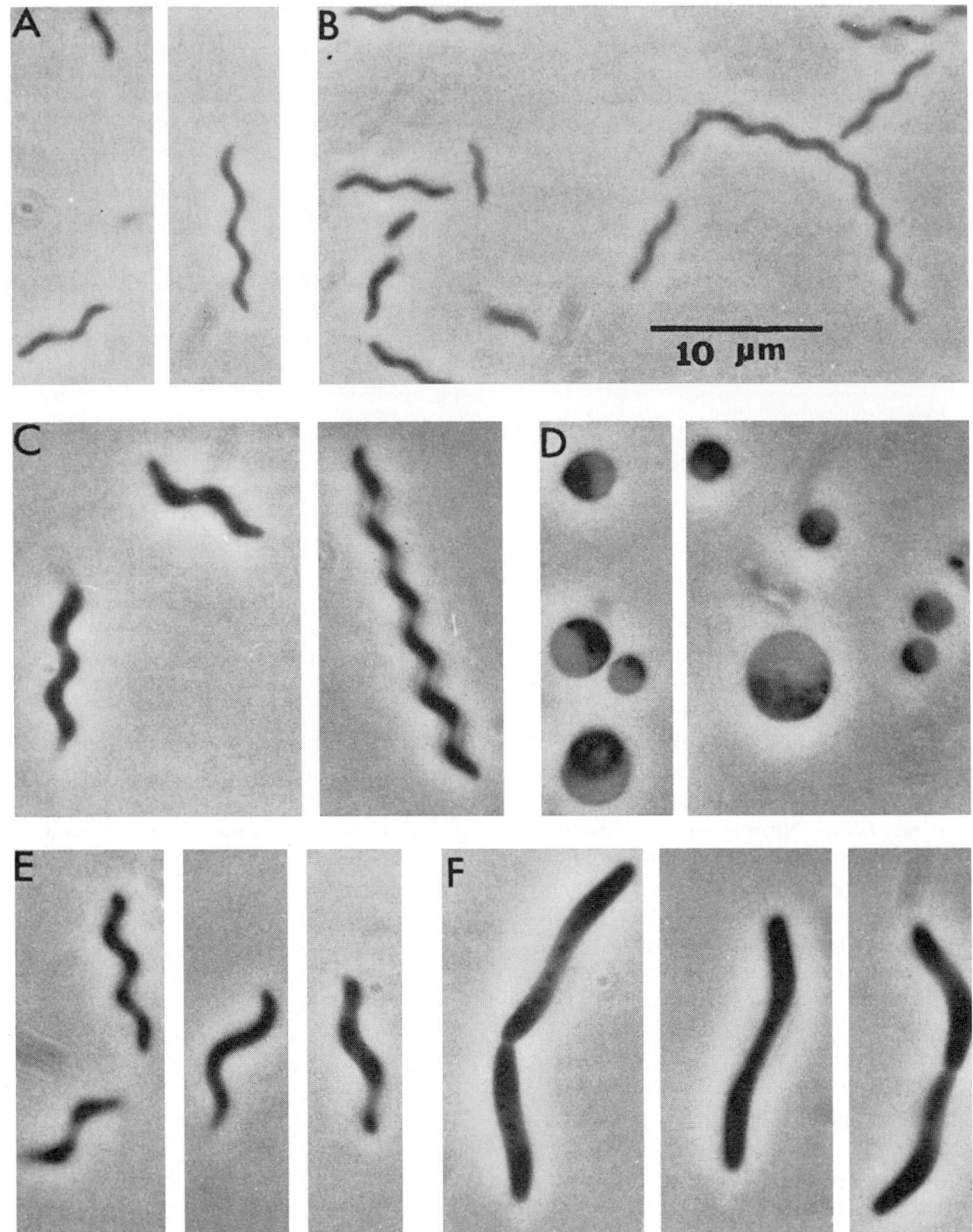

FIGURE BXII.γ.104. Phase contrast photomicrographs of several species of the genus *Oceanospirillum.* All photomicrographs were taken at the same magnification. *A, Marinospirillum minutulum* ATCC 19193. *B, O. linum* ATCC 11336. *C, O. maris* ATCC 27509. *D,* coccoid bodies of *O. maris* formed after 7 d of incubation. *E, O. beijerinckii* subsp. *beijerinckii* ATCC 12754. *F, O. japonicum* ATCC 19191. Reproduced with permission from N.R. Krieg, Bacteriological Reviews, *40:* 55–115, 1976, ©American Society for Microbiology.)

oceanospirilla are nutrient broth prepared with natural seawater and PSS, or MPSS broth[1] prepared with artificial seawater[2].

Oceanospirilla generally produce moderate to abundant, turbid growth in 2–3 d in PSS seawater broth (Hylemon et al., 1973). In seawater-nutrient broth, membranous masses are often formed at the surface and can be dispersed with shaking to yield turbid cultures (Terasaki, 1972).

Colonies of oceanospirilla generally develop within 2–3 d on PSS seawater agar and are usually white, circular, and convex, ranging from pinpoint to 1.5 mm in diameter (Hylemon et al., 1973). Colonies on seawater-nutrient agar are generally pinpoint in size at 48 h but become larger (up to 2.0 mm in diameter) at 7 d; they are usually convex or umbonate, glistening, opaque, pale yellow, and butyrous (Terasaki, 1972). Rough (R) colonies may arise on prolonged transfer; for example, the colonies of the type strain of *O. japonicum* are presently of the R type. Some species produce a water-soluble, yellow-green fluorescent pigment on PSS seawater agar.

Most species grow best at a temperature of 30–32°C; however, *O. maris* subsp. *hiroshimense* grows best at 25°C (Terasaki, 1972).

---

1. See the genus *Aquaspirillum* for recipes for these media.

2. Artificial sea water for use in PSS broth, g/l of distilled water: NaCl, 27.5; MgCl$_2$, 50; MgSO$_4$, 2.0; CaCl$_2$, 0.5; KCl, 1.0; and FeSO$_4$, 0.01.

The nutrition of oceanospirilla is generally simple. Most species grow in simple defined media with amino acids or the salts of organic acids as carbon sources and ammonium ions as the nitrogen source. However, *O. linum* is specifically stimulated by methionine in a medium containing succinate and malate as carbon sources, and *O. maris* subsp. *williamsae* has a growth factor requirement that has not yet been identified. A listing of the carbon sources for oceanospirilla is given below in Table BXII.γ.77. Some apparent contradictions occur between the results obtained from different laboratories, although the results within each laboratory are reproducible. These differences are likely attributable to differences in definitions of what constitutes a positive growth response, and in some cases to the use of different strains.

The use of antisera in agglutination tests with a limited number of strains has indicated that the species of *Oceanospirillum* can be distinguished serologically (McElroy and Krieg, 1972). The antisera were prepared against whole cells and adsorbed with heated cells, leaving only antibodies against thermolabile antigens.

Oceanospirilla have been isolated from coastal seawater (Williams and Rittenberg, 1957), decaying seaweed (Jannasch, 1963), and putrid infusions of marine mussels (Terasaki, 1963, 1970, 1980). By direct microscopic counts of the bacteria present in clear and turbid seawaters near Port Aransas, Texas, Oppenheimer and Jannasch (1962) found that spirilla comprised only 0.1–2.5% of the total bacteria present. Whether oceanospirilla occur in the open sea is not known. Based on chemostat experiments, Jannasch (1963) suggested that the growth of oceanospirilla might be restricted to environments of higher nutrient concentration than is found in ordinary seawater, such as in zones surrounding decaying particulate matter. With regard to occurrence of oceanospirilla in putrid infusions of marine mussels, the source is most likely marine mud adherent to the mussels (Terasaki, 1970).

ENRICHMENT AND ISOLATION PROCEDURES

The enrichment and isolation method used by Williams and Rittenberg (1957) is as follows. A seawater sample is mixed with an equal volume of Giesberger's base medium (NH$_4$Cl, 0.1%; K$_2$HPO$_4$, 0.05%; MgSO$_4$, 0.05%) plus 1.0% calcium lactate. After incubation and appearance of spirilla, a portion of the initial culture is sterilized and mixed with an equal volume of sterile Giesberger's medium lacking NH$_4$Cl. This mixture is then inoculated from the unsterilized portion of the initial culture. One to three subcultures done in this manner are sufficient to establish the spirilla as the predominant type. For isolation, the enrichment is diluted 1:100 to 1:100,000 with sterile seawater. The dilution bottles are shaken vigorously and allowed to stand at room temperature for 20 min to allow migration of spirilla to the surface of the dilution. Isolation is then accomplished by streaking the surface water onto a suitable agar medium such as nutrient agar prepared with seawater and containing 0.3% yeast autolysate. Plates are incubated at 30°C and after 24 h examined for distinctive, granular, umbonate or pulvinate colonies with a ground-glass appearance.

The method of Terasaki (1970) has yielded excellent results for the isolation of oceanospirilla from putrid infusions. Marine mussels are smashed with a hammer and placed in a Petri dish with a teaspoon of marine mud. Sterilized seawater is poured into the dish until the mussels sink completely in the solution. The infusion is incubated at 27–28°C and examined for the development of spirilla after 1, 2, 4, and 7 d. Isolation is accomplished by streaking dilutions onto suitable agar media.

For enrichment by use of continuous cultures, see Jannasch (1967).

MAINTENANCE PROCEDURES

Oceanospirilla may be maintained in semisolid PSS seawater medium (containing 0.15% agar to give a jelly-like consistency) at 30°C with weekly transfer (Hylemon et al., 1973). Cultures may also be maintained as stabs in seawater-nutrient agar at room temperature with monthly transfer (Terasaki, 1972).

Preservation is most easily accomplished by suspending a dense concentration of cells in seawater-nutrient broth containing 10% (v/v) dimethylsulfoxide, with subsequent freezing in liquid nitrogen. A method for freeze-drying oceanospirilla has been reported by Terasaki (1975).

PROCEDURES FOR TESTING SPECIAL CHARACTERS

Characterization methods for oceanospirilla have been described in detail by Terasaki (1972, 1979) and Hylemon et al. (1973). The comments given in this *Manual* for the genus *Aquaspirillum* also apply to the genus *Oceanospirillum*, except that media containing natural or artificial seawater must be used for all characterization tests.

DIFFERENTIATION OF THE GENUS OCEANOSPIRILLUM FROM OTHER GENERA

See the genus *Aquaspirillum*, in Volume 2 Part C in this *Manual*, for characteristics of *Oceanospirillum* that distinguish the genus from other morphologically or physiologically similar genera.

TAXONOMIC COMMENTS

In the eighth edition of *Bergey's Manual of Determinative Bacteriology* (Krieg, 1974), a single genus, *Spirillum*, contained all of the various aerobic and microaerophilic spirilla, including freshwater and marine species. However, the DNA base composition for the genus ranged from 38 to 65 mol% G + C and appeared to be unusually broad for a bacterial genus. Moreover, three groups were evident within the genus: (a) the aerobic, freshwater spirilla that could not tolerate 3% NaCl (mol% G + C 50–65); (b) the aerobic, marine spirilla that required seawater for growth (mol% G + C = 42–48); and (c) the large, microaerophilic spirilla that belong to the species *S. volutans* (mol% G + C = 38). Accordingly, Hylemon et al. (1973) divided the genus into the three genera *Spirillum*, *Aquaspirillum*, and *Oceanospirillum*, with the marine organisms comprising the latter genus. This subdivision was used in the first edition of *Bergey's Manual of Systematic Bacteriology* (Krieg, 1984a). Although this scheme proved useful for practical purposes, it was only gradually that the phylogenetic aspects of the three subdivisions were revealed.

In an analysis of the 16S rRNA oligonucleotide catalogs of the species *O. minutulum* (now *Marinospirillum minutulum*) and *Oceanospirillum maris*, Woese et al. (1982) found that both organisms belonged to group III of the phototrophic bacteria as defined by Gibson et al. (1979), but they were not closely related to each other. Later, Woese et al. (1985) studied three additional species of *Oceanospirillum*—*O. japonicum*, *O. linum*, and *O. beijerinckii*. These species, together with the families *Enterobacteriaceae* and *Vibrionaceae*, constituted the core of 'subgroup 3' of the *Gammaproteobacteria* (Stackebrandt et al., 1988).

An organism known as "*Spirillum lunatum*" (Williams and Rit-

TABLE BXII.γ.77.　Carbon sources used by *Oceanospirillum* species and *Marinospirillum minutulum*

Characteristic	O. linum		O. beijerinckii subsp. beijerinckii		O. beijerinckii subsp. pelagicum	O. maris subsp. maris	O. maris subsp. hiroshimense	O. multiglobuliferum	O. jannaschii	O. japonicum		O. kriegii	O. pusillum	M. minutulum	
Method[a],[b]	A[c]	B[d]	A	B	B[d]	A[e]	B	B	A[f]	A	B	A[f]	B	A	B
Carbon source:															
Citrate	−	−	−	−	d	−	−	+	+	−	−	+	+	−	−
Aconitate	−	nd	−	nd	nd	−	nd	nd	−	−	nd	−	nd	−	nd
Isocitrate	−	nd	−	nd	nd	−	nd	nd	nd	−	nd	nd	nd	−	nd
α-Ketoglutarate	−	nd	−	nd	nd	−	nd	nd	d	−	nd	+	nd	d	nd
Succinate	−	−	−	+	+	−	+	+	+	−	+	+	+	+	+
Fumarate	−	−	−	+	+	−	+	+	+	−	+	+	+	+	+
Malate	−	−	+	−	+	d	+	+	+	+	+	+	+	+	+
Oxaloacetate	−	nd	−	nd	nd	+	nd	nd	nd	+	nd	nd	nd	+	nd
Pyruvate	−	−	+	−	+	−	+	+	+	+	+	+	+	+	+
Lactate	−	−	−	−	+	−	+	+	+	+	+	+	+	+	+
Malonate	−	−	−	−	−	−	−	−	−	−	−	d	+	−	−
Tartrate	nd	−	nd	−	+	nd	+	−	−	nd	nd	−	−	d	−
Acetate	d	−	−	−	+	−	+	+	+	+	+	+	+	−	+
Propionate	−	−	−	−	+	−	+	+	+	−	+	+	+	+	+
Butyrate	nd	−	−	−	nd	nd	nd	+	+	nd	+	d	+	nd	+
Caproate	−	nd	−	nd	nd	−	nd	nd	d	−	nd	d	nd	−	nd
β-Hydroxybutyrate	−	nd	−	nd	nd	−	nd	nd	+	−	nd	+	nd	−	nd
p-Hydroxybenzoate	−	nd	−	nd	d	−	d	nd	−	−	nd	+	nd	−	nd
Ethanol	−	−	−	−	d	−	d	−	+	−	−	+	−	−	−
n-Propanol	−	−	−	−	d	−	d	−	+	−	−	+	−	−	−
n-Butanol	−	−	−	−	−	−	−	−	d	−	−	+	−	−	−
Glycerol	−	−	−	−	nd	−	nd	−	−	−	−	−	+	−	−
L-Histidine	−	nd	−	nd	nd	−	nd	nd	−	−	nd	−	nd	−	nd
L-Tyrosine	−	nd	−	nd	nd	−	nd	nd	−	−	nd	−	nd	−	nd
L-Phenylalanine	−	nd	−	nd	nd	−	nd	nd	d	−	nd	−	nd	−	nd
L-Alanine	−	nd	−	nd	nd	−	nd	nd	+	+	nd	+	nd	−	nd
L-Glutamate	−	nd	−	nd	nd	+	nd	nd	d	+	nd	d	nd	+	nd
L-Aspartate	−	nd	−	nd	nd	−	nd	nd	d	−	nd	−	nd	−	nd
L-Glutamine	−	nd	−	nd	nd	−	nd	nd	nd	+	nd	nd	nd	d	nd
Asparagine	−	nd	−	nd	nd	−	nd	nd	nd	−	nd	nd	nd	−	nd
L-Proline	−	nd	−	nd	nd	d	nd	nd	+	−	nd	+	nd	+	nd
L-Hydroxyproline	−	nd	−	nd	nd	−	nd	nd	nd	−	nd	nd	nd	−	nd
L-Ornithine	−	nd	−	nd	nd	−	nd	nd	+	−	nd	−	nd	−	nd
L-Citruline	−	nd	−	nd	nd	−	nd	nd	+	−	nd	−	nd	−	nd
L-Arginine	−	nd	−	nd	nd	−	nd	nd	+	−	nd	d	nd	−	nd
L-Lysine	−	nd	−	nd	nd	−	nd	nd	d	−	nd	−	nd	−	nd
Putrescine	−	nd	−	nd	nd	−	nd	nd	+	−	nd	+	nd	−	nd
L-Methionine	−	nd	−	nd	nd	−	nd	nd	nd	−	nd	nd	nd	−	nd
L-Serine	−	nd	−	nd	nd	−	nd	nd	d	−	nd	−	nd	−	nd
L-Cysteine	−	nd	−	nd	nd	−	nd	nd	nd	−	nd	nd	nd	−	nd
Glycine	−	nd	−	nd	nd	−	nd	nd	d	−	nd	−	nd	−	nd
L-Leucine	−	nd	−	nd	nd	−	nd	nd	d	−	nd	−	nd	−	nd
L-Isoleucine	−	nd	−	nd	nd	−	nd	nd	d	−	nd	−	nd	−	nd
L-Valine	−	nd	−	nd	nd	−	nd	nd	d	−	nd	−	nd	−	nd
L-Tryptophan	−	nd	−	nd	nd	−	nd	nd	nd	−	nd	−	nd	−	nd

[a]Method A (Hylemon et al., 1973): A turbidimetrically standardized cell suspension in synthetic seawater was inoculated into a defined, vitamin-free medium containing the carbon sources (0.1%) and ammonium sulfate as the nitrogen source. Growth responses were measured turbidimetrically after one 72-h serial transfer from the initial cultures, using a Klett colorimeter with the blue (420 nm) filter and 16-mm cuvettes. Symbols: +, 10 or more Klett units of turbidity for all strains tested; −, less than 10 Klett units of turbidity; d, differs among strains; nd, not determined.

[b]Method B (Terasaki, 1972, 1979): A cell suspension washed in basal, defined, vitamin-free medium (Williams and Rittenberg, 1957) containing natural seawater and lacking carbon sources. The cells were inoculated into similar media containing the test compounds (0.05%) and ammonium chloride as the nitrogen source. After 7 d, growth was estimated turbidimetrically. Symbols: +, a turbidity of 0.025 absorbance units or greater for all strains tested; −, a turbidity of less than 0.025; d, differs among strains; nd, not determined.

[c]Strain ATCC 12753 failed to grow with any sole carbon source, while strain ATCC 11336 grew only with acetate. Both strains grew abundantly when succinate plus malate were supplied as carbon sources and L-methionine as the nitrogen source.

[d]Strain OF3 (Terasaki, 1972, 1973) differs from the results given in the table in that it grows with a large veriety of sole carbon sources: citrate, succinate, fumarate, malate, pyruvate, lactate, acetate, propionate, and butyrate. Whether this strain should be included in the species O. linum is uncertain.

[e]The results are given for *O. maris* subsp. *maris*. *O. maris* subsp. *williamsae* fails to grow with any sole carbon (or sole nitrogen) source and, therefore, appears to have an auxotrophic growth requirement. This requirement has not yet been defined.

[f]As reported by Bowditch et al. (1984a).

tenberg, 1957) was included in the genus *Oceanospirillum* by Hylemon et al. (1973), but this posed taxonomic problems. The characteristics of the type strain (ATCC 11337 or NCMB 54) did not fit the original description of the species, and Linn and Krieg (1978) found that NCMB strain 54 consisted of a mixture of two dissimilar organisms. The first type was a short, vibrioid rod that possessed a single polar flagellum, grew in either the presence or absence of seawater, catabolized sugars, did not form coccoid bodies, and had a mol% G + C of 63–64. The second type was a larger, helical organism that possessed bipolar flagellar tufts, required seawater, failed to attack sugars, formed coccoid bodies, and had a mol% G + C of 45. The smaller organism did not appear to belong to either *Oceanospirillum* or *Aquaspirillum* and it remains unclassified. The larger organism had characteristics more in accord with the original description of "*S. lunatum*" but differed in certain respects; it has been classified as a new subspecies of *O. maris*: *O. maris* subsp. *williamsae*.

Bowditch et al. (1984a, b) added four species to the genus *Oceanospirillum*, mainly based on immunological relationships. These species were *Oceanospirillum commune*, for the organism previously named *Marinomonas communis* (Van Landschoot and De Ley, 1983, 1984), *Oceanospirillum vagum* for *Marinomonas vaga* (Van Landschoot and De Ley, 1983, 1984), and two species *Oceanospirillum jannaschii* and *Oceanospirillum kriegii* for two groups of unnamed marine bacteria I-1 and H-1, respectively. As a result, the genus definition of *Oceanospirillum* needed to be changed drastically, with the unfortunate loss of most of the readily determinable phenotypic features from the genus definition (Krieg, 1984a) and the extension of the upper mol% G + C limit for the genus from 51 to 57. By this extension, a considerable overlap of mol% G + C range was introduced between the genera *Aquaspirillum* (49–65 mol% G + C) and *Oceanospirillum* (42–51 mol% G + C). In this way, one of the most reliable genotypic features discriminating both genera was lost. Phylogenetic data (Pot et al. 1989, Pot, 1996; Satomi et al., 1998), however, have since shown that all four species cannot be regarded as members of the genus *Oceanospirillum*.

On the basis of a polyphasic approach including DNA–DNA and DNA–rRNA hybridizations, Pot et al. (1989) showed that only five species, including the type species, constituted a separate rRNA branch in the *Gammaproteobacteria* and redefined the genus *Oceanospirillum* to contain *O. linum*, *O. maris*, *O. beijerinckii*, *O. multiglobuliferum*, and, more distantly, *O. japonicum*. Based on

DNA–DNA hybridizations (as suggested by Krieg, 1984a) and numerical comparison of whole-cell proteins, *O. maris* subsp. *hiroshimense* and *O. beijerinckii* subsp. *pelagicum* were created for the former species *O. hiroshimense* and *O. pelagicum*. *O. pusillum* was shown to belong to the *Alphaproteobacteria*, and *O. commune* and *O. vagum* were relegated to their original generic positions as *Marinomonas communis* and *Marinomonas vaga*, respectively. The two species *O. jannaschii* and *O. kriegii* were shown to be phylogenetically too remote to be considered members of the genus *Oceanospirillum*, and, together with *O. minutulum*, they constituted separate rRNA branches in the *Gammaproteobacteria*.

Subsequently, this phylogenetic heterogeneity was confirmed by studies of fatty acid, quinone, and polyamine compositions (Hamana et al., 1994; Sakane and Yokota, 1994; Bertone et al., 1996). All species, except *O. pusillum*, contained ubiquinone-8 (Q-8) as a major respiratory quinone (Table BXII.γ.78). Like other spirilla from the *Alphaproteobacteria* (see the genus *Aquaspirillum* in this book), *O. pusillum* contained over 90% Q-10. The thirteen strains of *Oceanospirillum* that have been investigated for their fatty acid composition by Sakane and Yokota were divided into three groups (Table BXII.γ.79 and BXII.γ.80). Group I included the 10 strains belonging to *O. linum*, *O. maris* subsp. *hiroshimense*, *O. maris* subsp. *williamsae*, *O. beijerinckii* subsp. *beijerinckii*, *O. beijerinckii* subsp. *pelagicum*, *O. multiglobuliferum*, and *O. japonicum*, all of which have a low mol% G + C (42.5–48.4). Group II included the two type strains of *O. jannaschii* and *O. kriegii* and had a high mol% G + C content (54.8–54.9). Group III included only *O. pusillum* and could be clearly distinguished from other marine spirilla in having $C_{14:0\ 3OH}$ as the major 3-hydroxy fatty acid, besides Q-10 (Table BXII.γ.80). Bertone et al. (1996) confirmed the separate position of *O. japonicum*, *O. jannaschii*, and *O. kriegii*.

All *Oceanospirillum* species including *O. jannaschii* and *O. kriegii* contain both putrescine and spermidine. The relative content (Table BXII.γ.81) of putrescine is very small when compared with the level found in members of the *Alphaproteobacteria*. The relative concentration of putrescine for *O. pusillum* corresponds with that of other members of the *Alphaproteobacteria*. The absence of 2-hydroxy putrescine and homospermidine is a unifying character for the *Gammaproteobacteria*. The polyamine profile of *Oceanospirillum* I and II is not different, nor are their fatty acid profiles.

Later, 16S rDNA sequence analysis of all *Oceanospirillum* spe-

TABLE BXII.γ.78. Cellular quinone systems in *Oceanospirillum* species and *Marinospirillum minutulum*[a]

Species[b]	Strain	Group	Quinone system				
			Q-6	Q-7	Q-8	Q-9	Q-10
O. linum	IFO 15448[T]	Ic	3	4	91	2	
	IFO 15449	Ic	1	2	96	1	
O. beijerinckii subsp. *beijerinckii*	IFO 15445[T]	Id	1	12	83	4	
O. beijerinckii subsp. *pelagicum*	IFO 13612[T]	Id	2	7	91	1	
O. maris subsp. *hiroshimense*	IFO 13616[T]	Ic	1	4	94	1	
O. maris subsp. *williamsae*	IFO 15468[T]	Ic	12	7	80	1	
O. multiglobuliferum	IFO 13614[T]	Ic	1	4	94	1	
O. jannaschii	IFO 15466[T]	IIb	3	7	89	1	
O. japonicum	IFO 15446[T]	Ib	1	14	84	1	
	IFO 15447	Ib		9	88	3	
O. kriegii	IFO 15467[T]	IIa	2	5	89	4	
O. pusillum	IFO 13613[T]	III			1	6	93
Marinospirillum minutulum	IFO 15450[T]	Ia	1	1	97	1	

[a]After Sakane and Yokota (1994).

[b]*Oceanospirillum maris* subsp. *maris* has not been investigated.

TABLE BXII.γ.79. Cellular concentrations of non-polar fatty acids in *Oceanospirillum* species and *Marinospirillum minutulum*[a]

Species[b]	Strain	Non-polar fatty acid[c]											
		$C_{12:0}$	$C_{12:1}$	$C_{14:0}$	$C_{14:1}$	$C_{15:0}$	$C_{16:0}$	$C_{16:1}$	$C_{17:0}$	$C_{17:1}$	$C_{18:0}$	$C_{18:1}$	$C_{20:0}$
O. linum	IFO 15448[T]	3	2	1			16	48					30	
	IFO 15449	4	4	1			16	45					29	
O. beijerinckii subsp. *beijerinckii*	IFO 15445[T]	4		4		1	32	50					9	
O. beijerinckii subsp. *pelagicum*	IFO 13612[T]	4		2	1		22	46					23	3
O. maris subsp. *hiroshimense*	IFO 13616[T]	4	2	1			27	49				1	15	
O. maris subsp. *williamsae*	IFO 15468[T]	4	4	2			31	47				1	11	
O. multiglobuliferum	IFO 13614[T]	3	2	2			28	44					20	
O. jannaschii	IFO 15447	2		1			19	46				1	31	
O. japonicum	IFO 15466[T]	3		1			25	57					14	
	IFO 15446[T]	3		1			22	53					21	
O. kriegii	IFO 15467[T]	7	4	1	1	2	16	36	2	3	1		27	
O. pusillum	IFO 13613[T]	3		1	3		15	18				1	58	
Marinospirillum minutulum	IFO 15450[T]	2			4		35	26					32	

[a]After Sakane and Yokota (1994).

[b]*Oceanospirillum maris* subsp. *maris* has not been investigated.

[c]The percentage of the acid relative to the total non-polar acids.

TABLE BXII.γ.80. Cellular concentrations of 2- and 3-hydroxy fatty acids in *Oceanospirillum* species and *Marinospirillum minutulum*[a]

Species[b]	Strain	3-hydroxy fatty acids[c]						2-hydroxy fatty acid[d]
		$C_{10:0}$	$C_{12:0}$	$C_{14:0}$	$C_{14:1}$	$C_{16:0}$	$C_{18:0}$	
O. linum	IFO 15448[T]	100						−
	IFO 15449	100						−
O. beijerinckii subsp. *beijerinckii*	IFO 15445[T]	63		30		6		−
O. beijerinckii subsp. *pelagicum*	IFO 13612[T]	60		30		9		−
O. maris subsp. *hiroshimense*	IFO 13616[T]	100						−
O. maris subsp. *williamsae*	IFO 15468[T]	100						−
O. multiglobuliferum	IFO 13614[T]	100						−
O. jannaschii	IFO 15447	100						−
O. japonicum	IFO 15466[T]	4	96					−
	IFO 15446[T]	3	97					−
O. kriegii	IFO 15467[T]	19	54			27		−
O. pusillum	IFO 13613[T]			87		2	10	+ ($C_{18:1}$)
Marinospirillum minutulum	IFO 15450[T]		61	3	36			−

[a]After Sakane and Yokota (1994).

[b]*Oceanospirillum maris* subsp. *maris* has not been investigated.

[c]The percentage of the acid relative to the total 3-hydroxy acids.

[d]−, absent; +, present

cies confirmed the above findings. Kawasaki et al. (1997), in a phylogenetic study of helically shaped bacteria in the *Alphaproteobacteria*, showed that *O. pusillum* was not related to other taxa of spirilla and constituted a separate branch in the *Alphaproteobacteria*, also confirming previous findings of Woese et al. (1985) and Pot (1996). Therefore, *O. pusillum* cannot belong to the genus *Oceanospirillum* (Kawasaki et al., 1997). Phenotypic characteristics support the removal of *O. pusillum* from the genus: a single flagellum at each pole, a counterclockwise type of helix, and a mol% G + C of 51, which is slightly higher than the range of 42–48 for the rest of the genus. As a formal transfer has not been proposed, *O. pusillum* is therefore listed below as "Species assigned but phylogenetically not belonging in *Oceanospirillum*".

Satomi et al. (1998) determined and compared 16S rDNA sequences of all the *Oceanospirillum* species. They found that *O. linum, O. maris, O. beijerinckii,* and *O. multiglobuliferum* constituted a single rRNA cluster, separate from the branches formed by *Marinobacter, Marinobacterium,* and *Marinomonas*. Based on their findings, they also excluded *O. japonicum* from the genus *Oceanospirillum*. Phenotypically, *O. japonicum* is different from other *Oceanospirillum* species since it does not form coccoid bodies and its flagella appear to be crescent shaped with less than one helical

turn, whereas those of other species have one or more helical turns. Moreover, *O. japonicum* grows best at 35–37°C. As a formal new description has not been proposed, *O. japonicum* is therefore listed below as a "Species assigned but phylogenetically not belonging in *Oceanospirillum*".

In the same study, it was shown that *O. minutulum* clustered on a separate branch together with new isolates from kusaya gravy (Satomi et al., 1998). For this branch, a new genus *Marinospirillum* has been proposed, containing the two species *M. minutulum* and *M. megaterium* (Satomi et al., 1998).

Compared to the other *Oceanospirillum* species, *O. jannaschii* and *O. kriegii* both have a higher mol% G + C (54.8–54.9) and occupy a separate phylogenetic position (Satomi et al., 1998). Many phenotypic characteristics discriminate these species from the genus *Oceanospirillum* (Table BXII.γ.82). Although not discussed separately by the authors, *O. jannaschii* occurred on the same branch as *Marinobacterium* (González et al., 1997). Further taxonomic research, including DNA–DNA hybridizations, should be performed to substantiate the exact level of genotypic relationship between *O. jannaschii* and *Marinobacterium georgiense*. *O. kriegii* constituted a separate branch in the 16S rRNA dendrogram (Satomi et al., 1998).

TABLE BXII.γ.81. Cellular concentrations of polyamines in *Oceanospirillum* and *Marinospirillum minutulum*[a,b,c]

Species[d]	Strain	Medium[e]	Dap	H-Put	Put	Cad	Spd	HSpd
O. linum	IFO 15448[T]	199SW	−	−	0.01	−	0.65	−
	IFO 15449	199SW	−	−	0.02	−	0.80	−
O. beijerinckii subsp. *beijerinckii*	IFO 15445[T]	199SW	−	−	0.01	−	0.64	−
O. beijerinckii subsp. *pelagicum*	IFO 13612[T]	199S	−	−	0.06	−	0.48	−
		199SW	−	−	0.01	−	0.65	−
O. maris subsp. *hiroshimense*	IFO 13616[T]	199S	−	−	0.02	−	0.86	−
		199SW	−	−	0.03	−	0.90	−
O. maris subsp. *williamsae*	IFO 15468[T]	199SW	−	−	0.03	−	0.90	−
O. multiglobuliferum	IFO 13614[T]	199S	−	−	0.08	−	0.40	−
		199SW	−	−	0.01	−	0.45	−
O. jannaschii	IFO 15466[T]	199SW	−	−	0.02	−	0.80	−
O. japonicum	IFO 15446[T]	199SW	−	−	0.01	−	1.11	−
	IFO 15447	199SW	−	−	0.01	−	1.23	−
O. kriegii	IFO 15467[T]	199SW	−	−	0.03	−	0.84	−
O. pusillum	IFO 13613[T]	199S	−	−	0.15	−	1.40	−
		199SW	−	−	0.24	−	1.19	−
Marinospirillum minutulum	IFO 15450[T]	199SW	−	−	0.10	−	0.72	−

[a]Cells were harvested at stationary growth phase.

[b]After Hamana et al. (1994).

[c]Abbreviations: Dap, diaminopropane; H-Put, 2-hydroxyputrescine; Put, putrescine; Cad, cadaverine; Spd, spermidine; HSpd, homospermidine; −, not detectable (<0.005).

[d]*Oceanospirillum maris* subsp. *maris* has not been investigated.

[e]Media:199, polyamine-free growth medium from Flow Lab., Irvine, U.K., pH 7.0.; 199S, medium 199 dissolved in 70% synthetic seawater, pH 7.0; 199SW, medium 199 dissolved in seawater, pH 7.0.

Based on 16S rRNA gene sequence analysis, the genus *Oceanospirillum* should therefore be limited to *O. linum*, *O. maris* subsp. *maris*, *O. maris* subsp. *hiroshimense*, *O. maris* subsp. *williamsae*, *O. beijerinckii* subsp. *beijerinckii*, *O. beijerinckii* subsp. *pelagicum*, and *O. multiglobuliferum*. Consequently, *O. japonicum*, *O. jannaschii*, *O. kriegii*, and *O. pusillum* should be removed from the genus. The genus definition described above has been adapted accordingly. The precise taxonomic affiliation of the last four species needs to be further determined.

Note added in proof: Satomi et al. (2002) have formally revised the taxonomic status of the genus *Oceanospirillum* and proposed the formal transfer of the species assigned to but not belonging to the genus *Oceanospirillum* to new and existing genera. (See Kimura et al., 2002 in Further Reading section below.)

ACKNOWLEDGMENTS

We acknowledge Dr. N.R. Krieg for the template text, figures, and tables on which this chapter has been based.

FURTHER READING

Krieg, N.R. 1976. Biology of the chemoheterotrophic spirilla. Bacteriol. Rev. *40*: 55–115.

Krieg, N.R. and P.B. Hylemon. 1976. The taxonomy of the chemoheterotrophic spirilla. Annu. Rev. Microbiol. *30*: 303–325.

Pot, B., Gillis, M. and De Ley, J.. 1992. The genus *Oceanospirillum*. *In* Balows, Trüper, Dworkin, Harder and Schleifer (Editors), The Prokaryotes. A Handbook on the Biology of Bacteria, Ecophysiology, Isolation, Identification, Applications, 2nd ed., Springer-Verlag, New York. pp. 3230–3236.

Satomi, M., B. Kimura, T. Hamada, S, Harayama, and T. Fujii. 2002. Phylogenetic study of the genus *Oceanospirillum* based on 16S rRNA and *gyrB* genes: emended description of the genus *Oceanospirillum*, description of *Pseudospirillum* gen. nov., *Oceanobacter* gen. nov. and *Terasakiella* gen. nov. and transfer of *Oceanospirillum jannaschii* and *Pseudomonas stanieri* to *Marinobacterium* as *Marinobacterium jannaschii* comb. nov. and *Marinobacterium stanieri* comb. nov. Int J Syst Evol Microbiol. *52*: 739–747.

DIFFERENTIATION OF THE SPECIES OF THE GENUS *OCEANOSPIRILLUM*

Morphological and physiological characteristics of the species of *Oceanospirillum* are indicated in Tables BXII.γ.77 and BXII.γ.82.

Chemotaxonomic characteristics of the species are indicated in Tables BXII.γ.78, BXII.γ.79, BXII.γ.80, and BXII.γ.81.

List of species of the genus Oceanospirillum

1. **Oceanospirillum linum** (Williams and Rittenberg 1957) Hylemon, Wells, Krieg and Jannasch 1973, 374[AL] (*Spirillum linum* Williams and Rittenberg 1957, 82.)

li' num. L. n. *linum* flax, thread.

The morphological characters are depicted in Fig. BXII.γ.104 and listed in Table BXII.γ.82. The physiological and chemotaxonomic characters are indicated in Tables BXII.γ.78, BXII.γ.79, BXII.γ.80, BXII.γ.81, and BXII.γ.82.

Sole carbon sources are listed in Table BXII.γ.77. Growth in defined media is usually poor; however, abundant growth occurs in defined media containing malate plus succinate as carbon sources and methionine as the nitrogen source. Nitrate is not used.

Strain OF3, isolated by Terasaki (1972, 1973), differs from other strains of *O. linum* in that it can grow well in defined media with a variety of sole carbon sources and

TABLE BXII.γ.82. Morphological and physiological characteristics of the species of *Oceanospirillum* and of *Marinospirillum minutulum*[a]

Characteristics	*O. linum*	*O. beijerinckii*	*O. maris*	*O. multiglobuliferum*	*O. jannaschii*	*O. japonicum*	*O. kriegii*	*O. pusillum*	*Marinospirillum minutulum*
Type of helix[b]	C	C	C	C	SR	C	SR	CC	C
Wavelength of helix, μm	1.8–4.0	6.3–7.2	3.5–7.0	3.5–5.0	nd	7.0–20.0	nd	1.7–2.0	2.0–2.8
Helix diameter, μm	0.8–1.4	1.5–3.0	1.4–2.8	1.0–2.0	nd	2.0–5.0	nd	1.0–1.2	0.6–1.5
Length of helix, μm	4.0–30.0	2.0–15.5	2.5–40.0	2.0–10.0	nd	5.0–75.0	nd	1.2–4.0	3.0–8.0
Cell diameter, μm	0.4–0.6	0.6–1.2	0.6–1.1	0.5–0.9	nd	0.8–1.4	nd	0.3–0.5	0.3–0.4
Flagellar arrangement[c]	BT	BT	BT	BT	M1-2	BT	MS	BS	BT
Polar membrane present	nd	+[d]	+[d]	nd	+	+	nd	nd	+
Intracellular poly-β-hydroxybutyrate formed	+	+	+	+	+	+	+	+	+[e]
Coccoid bodies predominant:									
After 3–4 weeks	+	+	+	+	nd	–	nd	+	+
After 24–48 h	–	–	–	+	nd	–	nd	–	–
Range of NaCl (%) for growth in peptone water after 7 d[f]	0.5–8.0	0.5–8.0	nd	0.5–4.0	nd	0.5–8.0	nd	0.5–8.0	0.5–8.0
Growth in PSS-seawater broth containing[g]									
9.75% total NaCl	+	+	+	nd	nd	nd	nd	nd	+
12.75% total NaCl	–	–	–	nd	nd	nd	nd	nd	d
Temperature range for growth (°C)[h]	11–39	8–41	2–35	6–37	nd	10–43	nd	6–40	11–37
Optimum growth temperature 25°C rather than 30–32°C	+ or w	–	d[i]	+	+	–	+	–	–
Phosphatase	+	+	d[i]	+	nd	w	nd	w	+
Nitrate reduced to nitrite	–	+	d[i]	–	+	–	+	+	+
Oxidase	+	+	+	+	+	+	+	+	+
Catalase	+	+ or w	–	+	nd	w or –	nd	w or –	–
Anaerobic growth with nitrate	–	–	–	–	nd	–	nd	–	–
Denitrification	–	–	–	–	–	–	–	–	–
Acid produced from sugars	–	–	–	–	nd	–	nd	–	–
Hydrolysis of esculin, hippurate, starch, or casein	–	–	–	–	nd	–	nd	–	–
Indole test	–	–	–	nd	nd	–	nd	–	–
Sulfatase (0.1% phenolphthalein disulfate)[g]	–	–	–	nd	nd	nd	nd	nd	–
Gelatin liquefaction at 20°C after:[f,j]									
7 d	–	d	–	–	–	d	–	–	–
28 d	+	d	–	–	–	d	–	–	–
42 d	+	d	–	–	–	d	–	–	–

(continued)

TABLE BXII.γ.82. (cont.)

Characteristics	O. linum	O. beijerinckii	O. maris	O. multiglobuliferum	O. jannaschii	O. japonicum	O. kriegii	O. pusillum	Marinospirillum minutulum
Gelatin hydrolysis at 30°C after:									
4 d[d,g]	−	−	−	nd	−	−	−	nd	−
Reduction of 0.3% H_2SeO_3[g]	+	−	−	nd	nd	nd	nd	nd	+
Water-soluble brown pigment formed in the presence of:[g]									
0.1% tyrosine	+	+	d	nd	nd	nd	nd	nd	−
0.1% phenylalanine	−	−	−	nd	nd	nd	nd	nd	−
0.1% tryptophan	d	−	d	nd	nd	nd	nd	nd	d
Growth in the presence of:[g]									
0.1% oxgall	+	+	+	nd	nd	nd	nd	nd	+
0.1% glycine	+	−	+	nd	nd	nd	nd	nd	+
Growth on:[g]									
Eosin methylene blue agar	−	−	d	nd	nd	nd	nd	nd	−
MacConkey agar	−	−	−	nd	nd	nd	nd	nd	−
Triple-Sugar iron agar	+	−	+	nd	nd	nd	nd	nd	+
Sellers agar	+	−	−	nd	nd	nd	nd	nd	−
Methyl red-Voges-Proskauer broth	+	−	d	nd	nd	nd	nd	nd	−
Water-soluble yellow-green fluorescent pigment formed[g]	−	−	−	nd	−	−	−	nd	−
Deoxyribonuclease[g]	d	+	−	nd	nd	nd	nd	nd	−
Ribonuclease[g]	−	+	−	nd	nd	nd	−	nd	−
Urease[g]	+	−	−	nd	nd	−	nd	nd	−
Auxotrophic growth requirement	+[k]	−	d[k]	−	nd	−	nd	−	−
Mol% (G + C) of DNA	48–50	47–49	45–47	46	56–57	45	54–56	51	42–44

[a]Phenotypic data are from Krieg (1984a), Baumann et al. (1972), and Bowditch et al. (1984a); +, present in all strains; −, lacking in all strains; d, depending on the strain; w, weak reaction; nd, not determined.

[b]C, clockwise helix; CC, counterclockwise helix (determined by focusing on the bottom of the cells.

SR, straight rod; BT, bipolar tufts, BS, bipolar single; MI-2, 1–2 flagella at one pole; MS, monopolar single.

[c]Characteristic has not been tested in O. maris subsp. hiroshimense and O. beijerinckii subsp. pelagicum.

[d]No granules are visible microscopically in Marinospirillum minutulum but chemical analysis indicates presence of the polymer.

[f]Data from Terasaki (1972, 1979).

[g]Data from Hylemon et al. (1973).

[h]Oceanospirillum kriegii grows at 35°C but not at 40°C; Oceanospirillum jannaschii does not grow at either 35°C or 40°C.

[i]Oceanospirillum maris subsp. hiroshimense has an optimal growth temperature of 25°C; catalase reaction can be negative, weak, or positive; phosphatase reaction can be positive or negative. For detailed information: see Hylemon et al. (1973), Terasaki (1972, 1979), and Table BXII.γ.83.

[j]Gelatin liquefaction was not tested in O. maris subsp. maris. It was negative in O. maris subsp. hiroshimense, positive in O. beijerinckii subsp. beijerinckii; and differed depending on the strain in O. beijerinckii subsp. pelagicum.

[k]Oceanospirillum linum grows poorly or not at all in a defined medium with a single carbon source and ammonium ions as the nitrogen source; however, abundant growth occurs in a defined medium containing succinate plus malate as carbon source and methionine as the nitrogen source. (See also footnote [d] in Table BXII.γ.82.) Oceanospirillum maris subsp. williamsae ATCC 2954[T] does not grow in vitamin-free medium and requires an unidentified growth factor.

ammonium ions as the nitrogen source (see footnote d, Table BXII.γ.77). Other characteristics of this strain are similar to those of the type strain (Terasaki, 1973), but whether it should be included in this species is uncertain.

The species includes organisms previously assigned to the two species *Spirillum linum* and *Spirillum atlanticum* by Williams and Rittenberg (1957). The two species were combined into the single species *O. linum* by Hylemon et al. (1973), based on a high degree of similarity in phenotypic characters and in DNA base composition.

Isolated from coastal seawater.

The mol% G + C of the DNA is: 48–50 (T_m).

Type strain: ATCC 11336, DSM 6292, NCMB 56.

2. **Oceanospirillum beijerinckii** (Williams and Rittenberg 1957) Hylemon, Wells, Krieg and Jannasch 1973, 374[AL] (*Spirillum beijerinckii* Williams and Rittenberg 1957, 90.)

bei.jer.inck' i.i. M.L. gen. n. *beijerinckii* of Beijerinck; named after Prof M.W. Beijerinck of Delft, Holland.

The morphological characters are depicted in Fig. BXII.γ.104 and listed in Table BXII.γ.82. The physiological and chemotaxonomic characters are indicated in Tables BXII.γ.78, BXII.γ.79, BXII.γ.80, BXII.γ.81, and BXII.γ.82. Sole carbon sources are listed in Table BXII.γ.77. Ammonium ions can serve as a sole nitrogen source; nitrate is not used.

Isolated from coastal seawater.

The mol% G + C of the DNA is: 47 (T_m).

Type strain: ATCC 12754, DSM 7166, NCMB 52.

a. **Oceanospirillum beijerinckii** *subsp.* **beijerinckii** (Williams and Rittenberg 1957) Hylemon, Wells, Krieg and Jannasch 1973, 375[AL] (*Spirillum beijerinckii* Williams and Rittenberg 1957, 90.)

Differs from *O. beijerinckii* subsp. *pelagicum* as indicated in Table BXII.γ.83.

The mol% G + C of the DNA is: 47 (T_m).

Type strain: ATCC 12754 , DSM 7166, NCMB 52.

b. **Oceanospirillum beijerinckii** *subsp.* **pelagicum** (Terasaki 1973) Pot, Gillis, Hoste, Van de Velde, Bekaert, Kersters and De Ley 1989, 32[VP] (*Spirillum pelagicum* Terasaki 1973, 65.)

pe.la' gi.cum. L. neut. adj. *pelagicum* belonging to the sea.

Differs from the *O. beijerinckii* subsp. *beijerinckii* as indicated in Table BXII.γ.83.

Isolated from putrid infusions of marine mussels.

The mol% G + C of the DNA is: 49 (T_m).

Type strain: ATCC 33337, DSM 6288, IFO 13612, NCMB 2228.

GenBank accession number (16S rRNA): AB006761.

3. **Oceanospirillum maris** Hylemon, Wells, Krieg and Jannasch 1973, 376[AL]

ma' ris. L. n. *mare* the sea; L. gen. n. *maris* of the sea.

The morphological characters are depicted in Fig. BXII.γ.104 and listed in Table BXII.γ.82. The physiological and chemotaxonomic characters are indicated in Tables BXII.γ.78, BXII.γ.79, BXII.γ.80, BXII.γ.81, and BXII.γ.82. Sole carbon sources are listed in Table BXII.γ.77.

Isolated from coastal seawater (Jannasch, 1967).

The mol% G + C of the DNA is: 45–46 (T_m).

Type strain: ATCC 27509, DSM 6286, LMG 5213.

GenBank accession number (16S rRNA): AB006771.

TABLE BXII.γ.83. Differentiating characteristics for the subspecies of *Oceanospirillum beijerinckii*[a]

Characteristic	O. beijerinckii subsp. beijerinckii	O. beijerinckii subsp. pelagicum
Temperature range for growth (°C)	14–37	8–41
Growth with:		
Lactate, citrate, malate, pyruvate, acetate, propionate, tartrate	−	+
p-Hydroxybenzoate, ethanol, *n*-propanol	−	d
Gelatin liquified at 20°C	+	d
Growth after 7 d in peptone water with:		
0.5–6.0% NaCl	+	−
0.5–8.0% NaCl	−	+
Mol% G + C of DNA	47	49

[a]Phenotypic data from Terasaki (1972, 1979); +, present in all strains; −, lacking in all strains; d, depending on the strain

c. **Oceanospirillum maris** *subsp.* **maris** Hylemon, Wells, Krieg and Jannasch 1973, 376[AL]

Differs from the *O. maris* subsp. *williamsae* and the *O. maris* subsp. *hiroshimense* as indicated in Table BXII.γ.84.

The mol% G + C of the DNA is: 46 (T_m).

Type strain: ATCC 27509, DSM 6286, LMG 5213.

GenBank accession number (16S rRNA): AB006771.

d. **Oceanospirillum maris** *subsp.* **hiroshimense** (Terasaki 1973) Pot, Gillis, Hoste, Van de Velde, Bekaert, Kersters and De Ley 1989, 33[VP] (*Spirillum hiroshimense* Terasaki 1973, 62.)

hi.ro.shi.men' se. M.L. neut. adj. *hiroshimense* pertaining to Hiroshima. Japan.

Characters are as described for the species. Differs from the *O. maris* subsp. *maris* and the *O. maris* subsp. *williamsae* as indicated in Table BXII.γ.84.

Isolated from putrid infusions of marine mussels.

The mol% G + C of the DNA is: 47 (T_m).

Type strain: IFO 13616, DSM 9524.

e. **Oceanospirillum maris** *subsp.* **williamsae** Linn and Krieg 1984, 355[VP]

will' iam.sae. M.L. gen. n. *williamsae* of Williams; named after Marion A. Williams, who was the first to describe species of marine spirilla.

Characters are as described for the species. Differs from the *O. maris* subsp. *maris* and the *O. maris* subsp. *hiroshimense* as indicated in Table BXII.γ.84.

Isolated from a mixture of organisms comprising NCMB strain 54 by Linn and Krieg (1978).

The mol% G + C of the DNA is: 45 (T_m).

Type strain: ATCC 29547, IFO 15468.

GenBank accession number (16S rRNA): AB006763.

4. **Oceanospirillum multiglobuliferum** (Terasaki 1973) Terasaki 1979, 143[AL] (*Spirillum multiglobuliferum* Terasaki 1973, 69.)

mul.ti.glo.bu.li' fe.rum. L. adj. *multus* much, many; L. dim. n. *globulus* a small sphere, globule; L. v. *fero* to bear, carry; M.L. neut. adj. *multiglobuliferum* bearing many globules.

The morphological characters are listed in Table BXII.γ.82. The physiological and chemotaxonomic characters are indicated in Tables BXII.γ.78, BXII.γ.79, BXII.γ.80, BXII.γ.81, and BXII.γ.82. Sole carbon sources

TABLE BXII.γ.84. Differentiating characteristics for the subspecies of *Oceanospirillum maris*[a]

Characteristic	*O. maris* subsp. *maris*	*O. maris* subsp. *hiroshimense*	*O. maris* subsp. *williamsae*
Optimal growth temperature (°C)	30–32	25	30–32
Catalase reaction	strongly +	d	weakly +
Phosphatase activity	−	+	−
DNase activity	−		+
RNase activity	−		+
Growth with 1% glycine	+		−
Growth in vitamin-free, defined growth medium	+		−
Growth with:			
L-Glutamate, oxaloacetate	+		−
Succinate, pyruvate, lactate, tartrate, acetate, propionate	−	+	−
Mol% G + C of DNA	46	47	45

[a]Phenotypic data from Hylemon et al. (1973); +, present in all strains; −, lacking in all strains; d, depending on the strain

are listed in Table BXII.γ.77. Differs from other species by forming unusually large numbers of coccoid bodies even in 24- to 48-h-old broth cultures. Ammonium ions can serve as a sole nitrogen source; nitrate is not used.

Isolated from putrid infusions of marine mussels.

The mol% G + C of the DNA is: 46 (T_m).

Type strain: IFO 13614.

GenBank accession number (16S rRNA): AB006764.

Species assigned but phylogenetically not belonging in *Oceanospirillum*

1. **Oceanospirillum jannaschii** Bowditch, Baumann and Bauman 1984b, 503[VP] (Effective publication: Bowditch, Baumann and Bauman 1984a, 227.)

   jan.nasch'i.i. M.L. gen. n. *jannaschii* of Jannasch; named after H.W. Jannasch.

   Straight rods. Some morphological, physiological, and chemotaxonomic characters are listed in Tables BXII.γ.78, BXII.γ.79, BXII.γ.80, BXII.γ.81, and BXII.γ.82 (see also Baumann et al., 1972). Sole carbon sources are listed in Table BXII.γ.77. Utilize 39–46 organic compounds including fatty acids, tricarboxylic acid cycle intermediates, alcohols, amino acids, and amines. Utilizes γ-aminovalerate and histamine (Baumann et al., 1972).

   Isolated from seawater after enrichment (Baumann et al., 1972).

   The mol% G + C of the DNA is: 56–57 (T_m).

   Type strain: ATCC 27135, DSM 6295, IFO 15466.

   GenBank accession number (16S rRNA): AB006765.

2. **Oceanospirillum japonicum** (Watanabe 1959) Hylemon, Wells, Krieg and Jannasch 1973, 375[AL] (*Spirillum japonicum* Watanabe 1959, 78.)

   ja.pon'i.cum. M.L. neut. adj. *japonicum* pertaining to Japan.

   The morphological characters are depicted in Fig. BXII.γ.104 and listed in Table BXII.γ.82; however, the type strain presently has morphological features that differ from the original description, in that the cells are no longer helical with several waves but instead are curved, straight, or S-shaped; moreover, colonies of this strain are presently of the R (rough) type. Therefore, it is likely that the type strain has undergone alteration since its isolation in 1959. Three reference strains isolated by Terasaki (1972, 1973) have morphological features that more nearly correspond

to those given in the original description (strains IF4, IF8, and UF3), and also they form colonies of the S (smooth) type.

The physiological and chemotaxonomic characters are indicated in Tables BXII.γ.78, BXII.γ.79, BXII.γ.80, BXII.γ.81, and BXII.γ.82. Sole carbon sources are listed in Table BXII.γ.77.

Isolated from putrid infusions of marine mussels.

The mol% G + C of the DNA is: 45 (T_m).

Type strain: ATCC 19191, DSM 7165.

3. **Oceanospirillum kriegii** Bowditch, Baumann and Bauman 1984b, 503[VP] (Effective publication: Bowditch, Baumann and Bauman 1984a, 227.)

   krie'gi.i. M.L. gen. n. *kriegii* of Krieg; named after N.R. Krieg.

   Straight rods. Some morphological and physiological characters are listed in Tables BXII.γ.78, BXII.γ.79, BXII.γ.80, BXII.γ.81, and BXII.γ.82 (see also Baumann et al., 1972). Sole carbon sources are listed in Table BXII.γ.77. Utilize 29–33 organic compounds including D-glucose and D-fructose but no other pentose, hexose, or disaccharide; also utilize tricarboxylic acid cycle intermediates, alcohols, amino acids, and amines. Produces an extracellular lipase (Baumann et al., 1972).

   Isolated from see water after enrichment (Baumann et al., 1972).

   The mol% G + C of the DNA is: 54–56 (T_m).

   Type strain: ATCC 27133, DSM 6294, IFO 15467, NCMB 2042.

   GenBank accession number (16S rRNA): AB006767.

4. **Oceanospirillum pusillum** (Terasaki 1973) Terasaki 1979, 142[AL] (*Spirillum pusillum* Terasaki 1973, 67.)

   pu.sil'lum. L. dim. neut. adj. *pusillum* very small.

   The morphological characters are listed in Table BXII.γ.82. The physiological and chemotaxonomic characters are indicated in Tables BXII.γ.78, BXII.γ.79, BXII.γ.80, BXII.γ.81, and BXII.γ.82. Sole carbon sources are listed in Table BXII.γ.77. Ammonium ions can serve as a sole nitrogen source; nitrate is not used.

   Isolated from putrid infusions of marine mussels.

   Belongs to the *Alphaproteobacteria*.

   The mol% G + C of the DNA is: 51 (T_m).

   Type strain: ATCC 33338, DSM 6293, IAM 14442, IFO 13613, NCMB 2229.

   GenBank accession number (16S rRNA): AB006768.

Genus II. **Balneatrix** Dauga, Gillis, Vandamme, Ageron, Grimont, Kersters, De Mahenge, Peloux and Grimont 1993b, 624[VP] (Effective publication: Dauga, Gillis, Vandamme, Ageron, Grimont, Kersters, De Mahenge, Peloux and Grimont 1993a, 42)

CATHERINE DAUGA

Bal' ne.a.trix. L. fem. n. *balneatrix* bather.

Straight or curved rods, 0.5–0.7 × 2.8–5 μm, sometimes elongated and flexuous. Gram negative. **Motile by single polar flagellum. Aerobic**, having a strictly respiratory type of metabolism with oxygen as the terminal electron acceptor. Produce convex, round, smooth colonies on solid media. Optimum temperature, 30°C; range, 20–46°C. Optimum pH, 6.5–7.5. Growth factors are not required. **No growth occurs in medium containing more than 1% NaCl.** Chemoorganotrophic. Acid is produced from glucose and certain other carbohydrates. **Oxidase positive.** Weakly catalase positive. **Indole positive.** Urease negative. Reduce nitrate to nitrite. Freshwater organisms; occasionally pathogenic for humans.

The mol% G + C of the DNA is: 54.

Type species: **Balneatrix alpica** Dauga, Gillis, Vandamme, Ageron, Grimont, Kersters, De Mahenge, Peloux and Grimont 1993b, 624 (Effective publication: Dauga, Gillis, Vandamme, Ageron, Grimont, Kersters, De Mahenge, Peloux and Grimont 1993a, 43).

FURTHER DESCRIPTIVE INFORMATION

Balneatrix strains grow after a 2-d incubation at 30°C. The center of the colony is pale yellow after 2–3 d, and pale brown after 4–5 d.

Balneatrix strains are nonfermentative, as verified by oxidation–fermentation medium (Casalta et al., 1989). All strains tested are Voges–Proskauer negative by the Barrit reference method. Nutritional tests can be done using Biotype-100 strips (BioMerieux, Marcy l' Etoile, France) with M70 minimal medium (Véron, 1975) supplemented with 0.15% agar. Strains of *Balneatrix* are nutritionally diverse, and some strains can grow with citrate, D-alanine, ethanolamine, D-glucosamine, L-histidine, DL-lactate, D(+)-malate (see Table BXII.γ.85). Arginine dihydrolase, lysine and ornithine decarboxylases, acetamide, and starch hydrolysis are invariably negative. Tetrathionate is not reduced. Gelatin is weakly hydrolyzed. Citrate can serve sometimes as the sole carbon source for growth. Tween-80 hydrolysis also differs among strains. Tributyrin is hydrolyzed, and the egg yolk reaction (lecithinase) is positive. The gamma-glutamyl transferase test is positive. There is no ONPG hydrolysis and no extracellular DNase (Casalta et al., 1989).

Balneatrix strains are susceptible *in vitro* to a variety of antimicrobial agents, including β-lactam, macrolides, and aminoglycoside antibiotics, sulfamethoxazole–trimethoprim, chloramphenicol, deoxycycline, minocycline, ofloxacin, and nalidixic acid. They are resistant to clindamycin and vancomycin.

At this writing, *Balneatrix* strains have been isolated only from thermal water and clinical specimens at a spa therapy center in southern France. Thirty-five cases of pneumonia and two cases of meningitis have been caused by *Balneatrix* strains (Hubert et al., 1991). According to epidemiological data, the bacteria were present in the hot water spring spa, and favorable growing conditions were found only in vapor baths. After disinfection of water pipes by chlorination, no further cases of infection were observed.

ENRICHMENT AND ISOLATION PROCEDURES

Balneatrix strains can be isolated by traditional culture techniques and incubation under aerobic conditions at 20–41°C. Environ-

TABLE BXII.γ.85. Nutritional characteristics of *Balneatrix alpica*, as determined with Biotype-100 strips[a]

Substrates	Growth response[b]
N-acetyl-glucosamine, *cis*-aconitate, L-alanine, DL-γ-amino-*n*-butyrate, DL-δ-amino-*n*-valerate, L-aspartate, betaine, D-fructose, fumarate, D-gluconate, D-glucose, L-glutamate, glutarate, glycerol, DL-β-hydroxybutyrate, *myo*-inositol, α-ketoglutarate, L-malate, maltose, maltotriose, D-mannitol, D-mannose, L-proline, propionate, putrescine, D-ribose, L-serine, D-sorbitol, succinate, L-tyrosine	+
trans-Aconitate, adonitol, D-arabinose, D-arabitol, L-arabitol, benzoate, caprate, caprylate, D-cellobiose, *m*-coumarate, dulcitol, *i*-erythritol, esculin, L-fucose, D-galactose, D-galacturonate, gentisate, β-gentiobiose, D-glucuronate, DL-glycerate, *m*-hydroxybenzoate, *p*-hydroxybenzoate, histamine, itaconate, 2-keto-D-gluconate, 5-keto-D-gluconate, lactose, lactulose, D-lyxose, malonate, maltitol, D-melezitose, D-melibiose, 1-*O*-methyl-α-galactopyranoside, 1-*O*-methyl-β-galactopyranoside, 3-*O*-methyl-D-glucopyranose, 1-*O*-methyl-α-D-glucopyranoside, 1-*O*-methyl-β-D-glucopyranoside, mucate, palatinose, phenylacetate, 3-phenylpropionate, protocatechuate, quinate, D-raffinose, L-rhamnose, D-saccharate, saccharose, L-sorbose, D(−)-tartrate, L(+)-tartrate, *meso*-tartrate, D-tagatose, D-trehalose, tricarballylate, trigonelline, tryptamine, L-tryptophan, D-turanose, xylitol, D-xylose	−
D-Alanine	d (7)
Citrate	d (1)
Ethanolamine	d (2)
D-Glucosamine	d (2)
L-Histidine	d (6)
DL-Lactate	d (7)
D(+)-Malate	d (6)

[a]Symbols : +, all strains positive ; −, all strains negative ; d, differs among strains. Numbers in parentheses indicate the number of strains that are positive.

[b]Based on results for eight strains.

mental and clinical strains can grow in 24 h in trypto-casein soy broth (Sanofi Diagnostics Pasteur, Marnes-la-Coquette, France). Brain–heart infusion and IsoVitaleX chocolate agar (Sanofi Diagnostics Pasteur) can be used for blood and CSF (cerebrospinal fluid) samples, respectively. Strains can be subcultured on trypto-casein soy agar or nutrient agar. Antimicrobial disks can be tested on Mueller–Hinton agar (Sanofi Diagnostics Pasteur).

MAINTENANCE PROCEDURES

The organisms survive at 30°C in trypto-casein soy broth for several weeks. They can survive for several years when frozen at −80°C in brain–heart infusion containing 50% glycerol.

DIFFERENTIATION OF THE GENUS *BALNEATRIX* FROM OTHER GENERA

The characteristics which differentiate *Balneatrix* from other genera are listed in Table BXII.γ.86.

Because *Balneatrix* shares some properties with *Flavobacterium*, such as isolation requirements, pigmented colonies, and biochemical reactions, it was initially compared with this genus; how-

TABLE BXII.γ.86. Phenotypic characteristics differentiating the genera *Balneatrix*, *Oceanospirillum*, *Marinomonas*, *Halomonas*, and *Flavobacterium*[a]

Characteristic	*Balneatrix*	*Flavobacterium*[b]	*Halomonas*[c]	*Marinomonas*[d]	*Oceanospirillum*[e]
Cell diameter, μm	0.5–0.7	0.5–3	0.6–1.9	0.5–4	0.3–1.4
Shape:					
Curved rods	+			+	
Straight rods	+	+	+	+	
Helical					+
Flagellar arrangement:					
Number	1–2	none	1–2	single	single or tufts
Arrangement	polar		polar	bipolar	bipolar
Salt tolerance (%NaCl, w/v)	0–1	nd	0.1–32	0.7–3.5	2.75–9.75
Maximum temperature (°C)	46	42	45	40	39–43
Nitrate reduced to nitrite	+	D	+	−	D
Anaerobic growth with nitrate	−	−	+	−	−
Indole	+	D	d	nd	−
Urease	−	D	d	−	−
Esculin hydrolyzed	−	+	d	nd	−
Acid from carbohydrates	+	+	+	+	−
Mol% G + C of DNA	54	31–42	60–61	45–50	45–51

[a]Symbols: +, positive for all species; −, negative for all species; nd, not determined; D, differs among species; d, differs among strains.

[b]Data from Holmes et al., 1984.

[c]Characteristics of *Halomonas elongata*. Data from Vreeland, 1984.

[d]Characteristics of *Marinomonas vaga* and *Marinomonas communis*, which were misnamed *Oceanospirillum vagum* and *O. commune* (Pot et al., 1989). Data from Baumann et al., 1972.

[e]Data from Krieg, 1984b.

ever, the mol% G + C value of 54 for the DNA of *Balneatrix* does not correspond to that of *Flavobacterium*.

Despite its environmental habitat and oxidative ability, *Balneatrix* is easily differentiated from such genera as *Oceanospirillum*, *Marinomonas*, or *Halomonas* based on salt tolerance, shape, flagellar arrangement, and ability to catabolize several carbohydrates. *Balneatrix* can be quickly differentiated from these genera by the inability to grow with more than 1% NaCl.

TAXONOMIC COMMENTS

Strains of *Balneatrix* are closely related by DNA–DNA hybridization and belong to the same species. Strain 4-87, isolated from cerebrospinal fluid, has been selected as the type strain.

rRNA–DNA hybridization and 16S rDNA sequencing have shown that the genus *Balneatrix* belongs to the class *Gammaproteobacteria*. Phylogenetically, *Balneatrix* is located within a subline including many *Oceanospirillum* species (Fig. BXII.γ.105). In this group, bacteria are heterogeneous and difficult to classify. Like *Balneatrix*, *Halomonas halmophila* was first misidentified as *Flavobacterium halmophilum* (Franzmann et al., 1988). *Halomonas aquamarina*, *Halomonas halophila*, and *Halomonas marina* were first assigned to the genus *Deleya* (Dobson et al., 1993).

Phylogenetically, *Balneatrix* branches in the vicinity of *Oceanospirillum* species, but *O. vagum*, *O. commune*,* *O. jannaschii*, and *O. kriegii* are misnamed according to DNA–rRNA hybridization (Pot et al., 1989). This, in addition to the fact that *Balneatrix alpica* is the only freshwater pathogenic bacterium in the subline, argues that *Balneatrix* is a new genus.

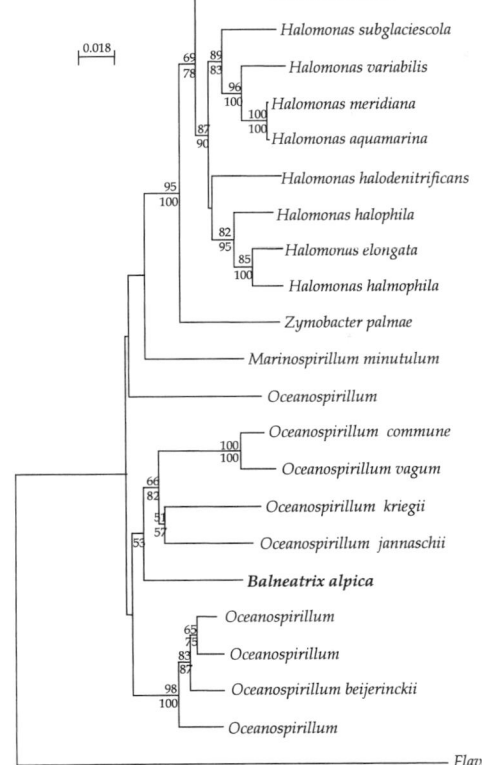

FIGURE BXII.γ.105. Relationships between the marine bacteria of the class *Gammaproteobacteria* and *Balneatrix*. *Flavobacterium aquatile*, which shares some phenotypic characteristics with *Balneatrix*, served as an outgroup. The tree shown is the result of distance analysis (NEIGHBOR routine in PHYLIP 3.5c, Kimura 2-parameters). The same tree topology was obtained with parsimony analysis (heuristic search option in PAUP 3.1.1 with 1000 replicates of the random-addition sequence). Percentages of bootstrap replicates supporting a branching pattern are given above and below the corresponding branches; the value above the branch is the parsimony analysis bootstrap percentage, and that below is from the distance analysis (100 bootstrap replicates).

*Editorial Note: Oceanospirillum vagum and O. commune are junior objective synonyms of Marinomonas vaga and M. communis.

List of species of the genus Balneatrix

1. **Balneatrix alpica** Dauga, Gillis, Vandamme, Ageron, Grimont, Kersters, De Mahenge, Peloux and Grimont 1993b, 624[VP] (Effective publication: Dauga, Gillis, Vandamme, Ageron, Grimont, Kersters, De Mahenge, Peloux and Grimont 1993a, 43.)

al'pi.ca. L. fem. adj. *alpica* pertaining to the Alps.

The description is the same as that given for the genus. Other characteristics are listed in Table BXII.γ.85. The type strain has all the properties given for the species and hydrolyses Tween-80.

The mol% G + C of the DNA is: 54 (HPLC).

Type strain: 4-87, CIP 103589.

GenBank accession number (16S rRNA): Y17112.

Genus III. **Marinomonas** *Van Landschoot and De Ley 1984, 91[VP] (Effective publication: Van Landschoot and De Ley 1983, 3071)*

ANTONIO SANCHEZ-AMAT AND FRANCISCO SOLANO

Ma.ri.no.mo'nas. L. adj. *marinus* pertaining to the sea; Gr. n. *monas* a unit, monad; M.L. *Marinomonas* sea monad.

Gram-negative, straight or curved rods. Motile by means of polar flagella at one or both poles. **Aerobic,** having a strictly respiratory type of metabolism. **Oxidase positive or negative. Na[+] is required for growth. Do not accumulate poly-β-hydroxybutyrate.** Do not require organic growth factors. **Do not produce extracellular amylase. Utilize acetate but not butyrate or valerate.** Utilize glutamate, sorbitol, and malate. Commonly isolated from seawater.

The mol% G + C of the DNA is: 45–50.

Type species: **Marinomonas communis** (Baumann, Baumann, Mandel and Allen 1972) Van Landschoot and De Ley 1984, 91 (Effective publication: Van Landschoot and De Ley 1983, 3071) (*Alteromonas communis* Baumann, Baumann, Mandel and Allen 1972, 420.)

FURTHER DESCRIPTIVE INFORMATION

Species of *Marinomonas* are rods, either curved such as *M. communis,* or straight such as *M. vaga* and *M. mediterranea* (Fig. BXII.γ.106). They are motile; *M. communis* and *M. vaga* show a single, unsheathed, polar flagellum inserted at one or both poles of the cell. *M. mediterranea* shows only a single flagellum at one pole of the cell (Fig. BXII.γ.106).

Marinomonas strains grow in typical marine media such as Marine Agar (Zobell, 1941). In this complex medium, only *M. mediterranea* shows pigmentation; it synthesizes a brown to black melanin pigment derived from L-tyrosine as main precursor. This pigment is observed both in the colonies and diffusing into the surrounding medium (Fig. BXII.γ.107). *Marinomonas* strains also grow in basal marine agar (BMA) containing 0.2% D-glucose as carbon and energy source (Baumann et al., 1972). This basal medium is used in studies of carbon and energy source utilization. No organic growth factor is required; however, growth of *M. mediterranea* is favored in media containing organic compounds as the nitrogen source. The most suitable chemically defined medium for growing this species is MMM[1]. All culture media for *Marinomonas* must contain a marine salts base composition. *Marinomonas* species show an absolute requirement for Na[+] ion.

The physiological characteristics of the *Marinomonas* species are listed in Tables BXII.γ.87 and BXII.γ.88. Thirty-three strains of *M. communis* and 17 *M. vaga* strains isolated from the Pacific Ocean have been phenotypically characterized (Baumann et al., 1972). In contrast, only a single *M. mediterranea* strain—isolated from the Mediterranean Sea as its species name indicates—has been described (Solano et al., 1997). Unless otherwise indicated, the characterization of this strain has been also performed according to the protocols of Baumann et al. (1972), described as well by Gauthier and Breittmayer (1992).

Members of the genus *Marinomonas* are aerobic and use oxygen as a universal electron acceptor. No *Marinomonas* species is able to denitrify, although *M. mediterranea* is able to reduce nitrate to nitrite. *M. communis* is oxidase positive whereas the other two species are oxidase negative. In some cases permeabilization of *M. vaga* strains with toluene will allow a positive oxidase reaction to occur. It has been suggested that this result is due to a low level of cytochrome *c* in *M. vaga* cells (Baumann et al., 1984d); however, *M. mediterranea* always gives a negative result in cytochrome *c* oxidase tests, even when the cells are toluene-treated (Solano et al., 1997).

In assays for extracellular degradative activities, *Marinomonas* species give negative results, with the exception of *M. mediterranea*, which shows lipase and gelatinase activities. Gelatin hydrolysis was detected by comparison of the solidification of Marine Broth, supplemented with 12% gelatin and inoculated with *M. mediterranea*, with an uninoculated control. Under these conditions, a weak gelatinase activity was detected (Solano et al., 1997).

Marinomonas species are able to utilize acetate, but not butyrate or valerate, as a sole carbon and energy source. All of the *M. communis* and *M. vaga* strains metabolize *m*-hydroxybenzoate, *p*-hydroxybenzoate, and quinate by means of *meta* cleavage of the intermediate protocatechuate (Baumann et al., 1972). In contrast, *M. mediterranea* is unable to metabolize *m*-hydroxybenzoate. All *Marinomonas* species can use D-glucose. *Marinomonas communis* and *M. vaga* metabolize D-glucose and D-fructose via the Entner–Doudoroff pathway (Sawyer et al., 1977a). Aspartate kinase activity has been detected in these two species (Baumann and Baumann, 1974). All species can use glutamate as a sole carbon and energy source, but none can use glycine or L-tyrosine. Data for other carbon sources are listed in Tables BXII.γ.87 and BXII.γ.88.

---

1. MMM is a chemically defined medium containing (per liter): NaCl, 20 g; MgSO$_4$·7H$_2$O, 7.0 g; MgCl$_2$·6H$_2$O, 5.3 g; KCl, 0.7 g; CaCl$_2$, 1.25 g; FeSO$_4$·7H$_2$O, 25 mg; CuSO$_4$·5H$_2$O, 5 mg; K$_2$HPO$_4$, 75 mg; sodium glutamate, 2 g; and Tris base, 6.1g. The pH of this medium is adjusted to 7.4.

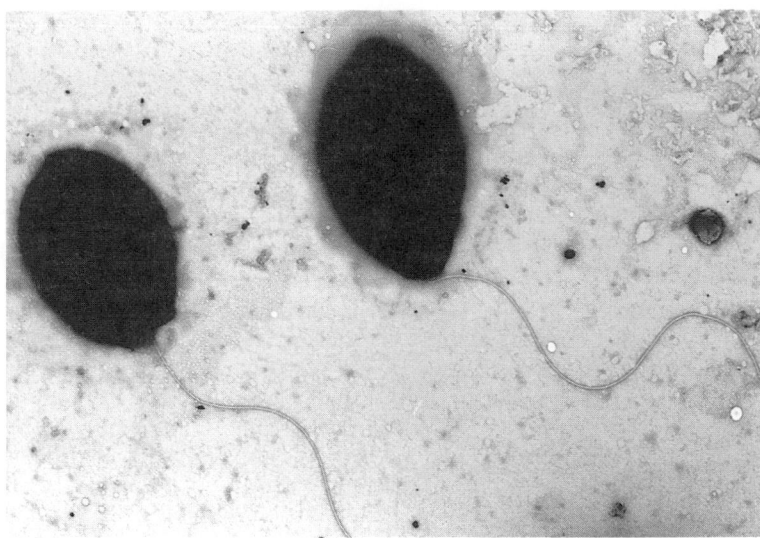

FIGURE BXII.γ.106. Electron micrograph of *Marinomonas mediterranea* MMB-1. Negatively stained with phospho-tungstic acid (× 20,000).

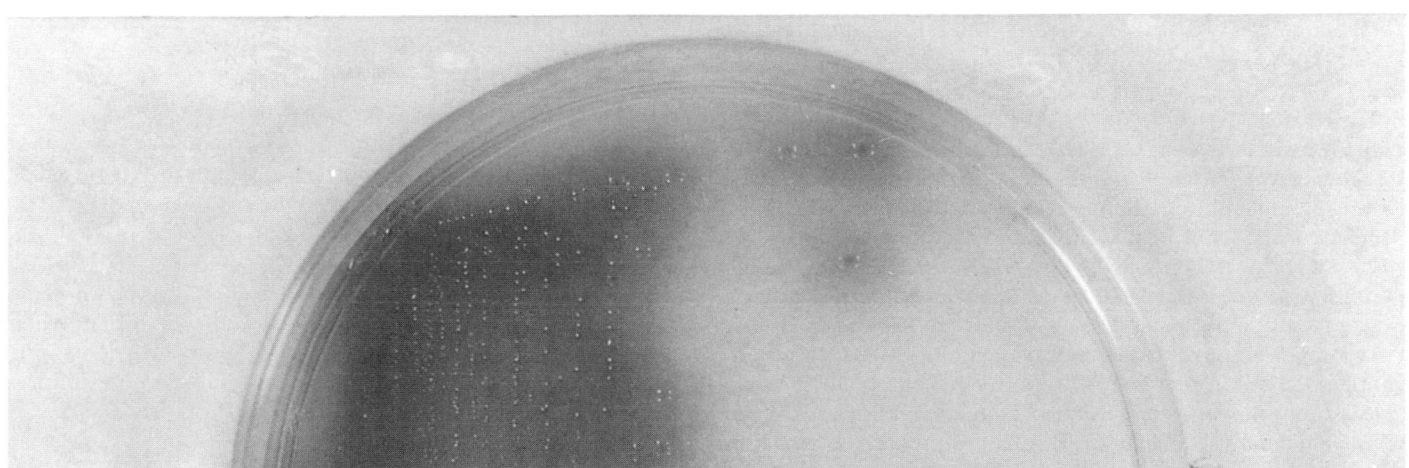

FIGURE BXII.γ.107. Colonies of *Marinomonas mediterranea* MMB-1 on Marine Agar after four days of incubation at 25°C (× 2.1).

Marinomonas communis and *M. vaga* can serve as prey for the growth of marine bdellovibrios (Taylor et al., 1974).

The mol% G + C content of the DNA of *M. communis* and *M. vaga* has been calculated using the buoyant density method. The mol% G + C of the DNA of *M. mediterranea* strain MMB-1 was determined by HPLC analysis after DNA hydrolysis and de-phosphorylation (Solano et al., 1997). In this strain, the chromatogram revealed two additional peaks besides those of the four expected standard deoxynucleosides. The base modifications in the compounds giving these peaks remain unidentified, although their UV adsorption spectra suggest that they are methylated derivatives of the purine deoxynucleosides. The method used to estimate the mol% G + C content in the other two species of *Marinomonas* does not allow one to know whether this feature is shared by all the *Marinomonas* species.

Resistance of *M. mediterranea* to some antimicrobial agents was tested in order to evaluate them as genetic markers. This species was sensitive to ampicillin (50 μg/ml), chloramphenicol (10 μg/ml), gentamicin (10 μg/ml), kanamycin (40 μg/ml), streptomycin (10 μg/ml), rifampicin (50 μg/ml), and tetracycline (10 μg/ml). Resistance to the latter compound was assayed in media lacking Mg^{2+}.

TABLE BXII.γ.87. Differential characteristics of the species of the genus *Marinomonas*[a]

Characteristic	1. *M. communis*	2. *M. mediterranea*	3. *M. vaga*
Cell morphology:			
Straight rod	−	+	+
Curved rod	+	−	−
Growth at:			
5°C	−	−	−
25°C	+	+	+
35°C	+	−	+
40°C	+	−	−
NO_3^- reduction to NO_2^-	−	+	−
Cytochrome *c* oxidase	+	−	−
Pigmentation	−	+	−
Lipase	−	+	−
Gelatinase	−	+	−
Utilization of:			
D-Fructose	+	−	+
m-Hydroxybenzoate	+	−	+
α-Ketoglutarate	+	−	+

[a]Symbols: see standard definitions.

Several molecular techniques have been applied to *M. mediterranea* (Solano et al., 2000). Vectors containing the RP4 *mob* site are mobilizable by conjugation. Conjugation can be performed at 25°C on the surface of plates containing the medium LB2216[2]. Plasmids, such as pKT230, with the p15 origin of replication are stable in this species and hence can be used as cloning vectors. On the contrary, plasmids containing the *oriR6K* behave as suicidal vectors in *M. mediterranea*, which makes them useful as transposon delivery vectors. Tn*5* and Tn*10* derivatives have been assayed for this purpose. Higher transposition frequencies were obtained with Tn*10* derivatives encoding kanamycin and gentamicin resistance than with Tn*10* transposons with a different marker, or with Tn*5* transposons. Other genetic markers such as ampicillin and chloramphenicol resistance were not properly expressed in *M. mediterranea* even if they were present in plasmids. *M. mediterranea* is normally lac(−), but after transposon mutagenesis with one mini-Tn10 derivative containing the *lacZ* reporter gene, approximately 55% of the colonies showed blue coloration when grown in media containing X-Gal, thus indicating the usefulness of *lacZ* as a reporter gene in *M. mediterranea*. Mutants of *M. mediterranea* can also be obtained by nitrosoguanidine mutagenesis (Solano et al., 1997). UV radiation is not an efficient mutagen for this strain, however.

A relevant feature in *M. mediterranea* is its capacity to synthesize melanins. These pigments are made from L-tyrosine as precursor and by the involvement of the enzyme tyrosinase (EC 1.14.18.1) (Solano et al., 1997). Tyrosinase is a copper protein that belongs to the group of polyphenol oxidases (PPOs). The other important copper enzyme in this group is laccase (EC 1.10.3.2). *M. mediterranea* also shows this activity, due to a multipotent enzyme showing both tyrosinase and laccase activities (Sanchez-Amat and Solano, 1997).

Some other aerobic, Gram-negative, marine proteobacteria are able to synthesize brown to black pigments in complex media. However, aside from *M. mediterranea*, actual PPO activity has been described only in the strain 2-40 (Kelley et al., 1990; Solano and Sanchez-Amat, 1999). In some cases, melanins result from spon-

taneous oxidation of intermediates of L-tyrosine catabolism, such as homogentisate in *Shewanella colwelliana* (Ruzafa et al., 1994; Kotob et al., 1995). In this case, melanin-like pigments are synthesized only in aerated media with a high amount of tyrosine. These pigments, designated pyomelanins, are very similar in their chemical properties to melanins obtained by activity of the enzyme tyrosinase. To differentiate between these two possibilities, actual PPO activity should be determined spectrophotometrically or revealed by staining after PAGE (Solano et al., 1997).

A high number of *M. communis* and *M. vaga* strains have been isolated using enrichment methods with different compounds as carbon sources (Baumann et al., 1972). These isolations led to the view that *Marinomonas* is a usual component of the bacterial flora in marine waters; however, there are no data about the ecological distribution of *M. mediterranea*. The single strain of this species was detected after screening thousands of colonies of marine microorganisms for the ability to synthesize melanins. No other strains with pigmentation identical to *M. mediterranea* were detected, which suggests that it may not be very abundant. It is important to bear in mind that it is not easy to differentiate between *Marinomonas* and other aerobic marine bacteria using only phenotypic characteristics (see below). Strains for which 16S rDNA sequence analyses clearly indicate they belong to the genus *Marinomonas* have been isolated from North Sea bacterioplankton (Eilers et al., 2000). In addition, a clone showing a close relationship to the genus *Marinomonas* was detected in a 16S rDNA sequence clone library obtained from bacterial communities associated with the seagrass *Halophila stipulacea* (Weidner et al., 2000).

ENRICHMENT AND ISOLATION PROCEDURES

Marinomonas species can be isolated by direct plating of seawater samples on a complex medium such as Marine Agar (Zobell, 1941). *M. communis* and *M. vaga* are able to use *m*-hydroxybenzoate, and this property can be used for an enrichment method. Seawater (500 ml) is transferred to a 2-liter Erlenmeyer flask and supplemented with 25 ml 1M Tris-HCl (pH 7.5), 0.5 g NH_4Cl, 38 mg K_2HPO_4, 14 mg $FeSO_4 \cdot 7H_2O$, and 0.5 g *m*-hydroxybenzoate. The culture is incubated at 20–25°C for up to 10 d. When growth is observed, the culture is streaked onto BMA containing 0.1% (w/v) *m*-hydroxybenzoate in order to obtain pure cultures (Baumann et al., 1984d).

MAINTENANCE PROCEDURES

Long-term preservation of *Marinomonas* species can be achieved by lyophilization. A suitable protocol for preparation of the cells has been described by Gauthier and Breittmayer (1992). The lyophilized cultures are reconstituted by adding 0.5 ml of Marine Broth. A few drops are streaked onto Marine Agar and the remaining is transferred to a tube containing 4 ml of Marine Broth. It is advisable to avoid high aeration of the culture during the first hours of incubation. Growth is observed after 2–3 d.

An alternative method of preservation is freezing. Glycerol is added to an overnight culture in Marine Broth to a final concentration of 20% and the culture is immediately stored at −75°C. Using this protocol, frozen *M. mediterranea* cells have remained viable for more than five years.

Strains can be maintained by serial transfer on Marine Agar for routine work in the lab. After 2–3 days of growth at 25°C, the plates can be preserved for 3–4 weeks at 15°C. It is not recommended to keep the cultures at 4°C, because viability is lost much faster than when the cultures are stored at 15°C.

2. LB2216 is obtained by mixing—after autoclaving—equal amounts of LB medium (containing 1.5% NaCl) and Marine Agar 2216.

TABLE BXII.γ.88. Other characteristics of the species of the genus *Marinomonas*[a]

Characteristics	1. *M. communis*[b]	2. *M. mediterranea*	3. *M. vaga*[b]
Motility	+	+	+
Flagellar arrangement:			
Polar	+	+	+
Peritrichous	−	−	−
Anaerobic growth	−	−	−
Na⁺ requirement	+	+	+
Poly-β-hydroxybutyrate accumulation	−	−	−
Denitrification	−	−	−
Luminescence	−	−	−
Amylase	−	−	−
Utilization of:			
D-Glucose, D-mannose, D-sorbitol, citrate, acetate, succinate, malate, glycerol	+	+	+
DL-β-Hydroxybutyrate	+	+	d
β-Alanine, L-serine, L-lysine, L-arginine, spermine	+	nd	d
Ethanol, propanol	+	nd	−
D-Galactose, L-rhamnose, L-arabinose	d	nd	+
Maltose, propionate	d	−	d
DL-Glycerate, ethanolamine	d	nd	−
Cellobiose, N-acetylglucosamine, erythritol, heptanoate	−	nd	+
Lactose, butyrate, valerate, methanol, L-glycine, L-tyrosine	−	−	−
D-Ribose, D-arabinose, caproate, caprylate, pelargonate, ribitol, trigonelline	−	nd	d
Butanol	−	nd	−
Mol% G + C of DNA	45.9–48	46.3 ± 0.9	46.4–49.3

[a]Symbols: see standard definitions; nd, not determined.

[b]*M. communis* and *M. vaga* are able to use the following compounds as principal sources of carbon and energy: D-gluconate, fumarate, DL-lactate, aconitate, D-mannitol, γ-aminobutyrate, sarcosine, putrescine, saccharate, pyruvate, m-inositol, L-alanine, D-alanine, L-ornithine, L-histidine, L-proline, betaine.

The following compounds are not used as sole or principal sources of carbon and energy by *Marinomonas communis* or *M. vaga*: sucrose, melibiose, salicin, D-fucose, inulin, cellulose, mucate, L-threonine, L-leucine, L-isoleucine, L-valine, L-tryptophan, D-tryptophan, L-norleucine, DL-α-aminobutyrate, DL-α-aminovalerate, δ-aminovalerate, formate, isobutyrate, isovalerate, oxalate, malonate, maleate, glutarate, adipate, pimelate, suberate, azelate, sebacate, D-tartrate, *meso*-tartrate, glycolate, levulinate, D-mandelate, L-mandelate, citraconate, itaconate, mesaconate, ethyleneglycol, 2,3-butanediol, methanol, isopropanol, isobutanol, n-butanol, propyleneglycol, benzoylformate, DL-kynurenine, kynurenate, anthranilate, m-aminobenzoate, p-aminobenzoate, methylamine, benzylamine, histamine, tryptamine, butylamine, α-amylamine, 2-amylamine, pentylamine, creatine, hippurate, pantothenate, acetamide, nicotinate, nicotinamide, allantoin, adenine, guanine, cytosine, thymine, uracil, n-dodecane, phenylethanediol, phenol, naphthalene.

11–89% of the strains included in *M. communis* and *M. vaga* are able to use the following compounds as principal sources of carbon and energy: D-xylose, D-trehalose, D-glucuronate, D-galacturonate, caprate, L-tartrate, L-aspartate, L-phenylalanine, L-citrulline.

DIFFERENTIATION OF THE GENUS *MARINOMONAS* FROM OTHER GENERA

The characteristics that differentiate *Marinomonas* species from other aerobic, marine, Gram-negative bacteria are listed in Table BXII.γ.89. Further characterization of more strains might help to establish other phenotypic criteria for differentiating these genera; however, all of these genera share many physiological and biochemical features and their differentiation can be problematic. In this regard, some chemotaxonomic markers such as isoprenoid quinone and fatty acid composition have been tested (Akagawa-Matsushita et al., 1992a; Ivanova et al., 2000c). Molecular analysis, such as rRNA–DNA hybridization and particularly rDNA sequencing, have shown that *Marinomonas* species form a well-defined cluster (Van Landschoot and De Ley 1983; Solano and Sanchez-Amat, 1999). 16S rDNA sequencing of additional *Marinomonas* strains could be useful for designing nucleic acid probes for the identification and quantification in natural habitats of microorganisms belonging to this genus.

TAXONOMIC COMMENTS

There are many Gram-negative, nonfermentative, heterotrophic, marine bacteria with polar flagella. Those with mol% G + C contents below 50 were temporarily assigned to the genus *Alteromonas*, and those with a mol% G + C of 57–64 were included in the genus *Pseudomonas* (Baumann et al., 1972). Further characterization of the genus *Alteromonas* revealed a great diversity

TABLE BXII.γ.89. Differential characteristics of the genus *Marinomonas* and other aerobic, Gram-negative, marine proteobacteria with polar flagellation[a]

Characteristic	*Marinomonas*	*Oceanospirillum*[b]	*Alteromonas*[c]	*Marinobacterium*[d]	*Marinobacter*[e]	*Neptunomonas*[f]	*Pseudoalteromonas*[c]	*Pseudomonas*[g]
Poly-β-hydroxybutyrate granules	−	+	−	−	−	+	−	D
Amylase	−	−	+	−	−	−	+[h]	D
Utilization of:								
Carbohydrates	+	−	+	+	−	+	+	+
Malate	+	D	−	+	−	−	−	+
D-Sorbitol	+		−		−	−	−	−
Mol% G + C of DNA	45–50	42–51	44–47	54.9	52.7	46	37–50	55–64

[a]For symbols see standard definitions.

[b]Data from Krieg (1984b).

[c]Data from Gauthier et al. (1995a).

[d]Data from González et al. (1997).

[e]Data from Gauthier and Brittmayer (1992).

[f]Data from Hedlund et al. (1999a).

[g]Data from Baumann et al. (1972, 1983b).

[h] *Pseudoalteromonas nigrifaciens* is negative.

(Van Landschoot and De Ley, 1983; Gauthier et al., 1995a). Two of the species originally included in the genus *Alteromonas* were *A. communis* and *A. vaga*. Immunological analysis of Fe-containing superoxide dismutase and glutamine synthetase established a relationship between these species and the genus *Oceanospirillum*, and hence their assignment to this genus under the species names *Oceanospirillum commune* and *O. vagum* was proposed (Bowditch et al., 1984a). DNA–rRNA hybridization studies applied to some Gram-negative bacteria, including those initially assigned to the genus *Alteromonas*, revealed that *A. communis* and *A. vaga* constituted a separate rRNA branch (Van Landschoot and De Ley, 1983). Taking into consideration this fact and also phenotypic traits, the genus *Marinomonas* was created to contain the two species as *M. communis* and *M. vaga*. Phylogenetic analysis based on 16S rDNA sequences of these two species and a new strain, MMB-1, revealed that they were closely related, with percentage identities greater than 95% (Fig. BXII.γ.108). These data supported creation of the genus *Marinomonas*, in which the strain MMB-1 constituted the third species, namely *M. mediterranea* (So-lano and Sanchez-Amat, 1999). The data also showed that these species were more closely related to members of the genus *Oceanospirillum* than to the species remaining in the modified genus *Alteromonas* or *Pseudoalteromonas*.

FURTHER READING

Baumann, L., P. Baumann, M. Mandel and R.D. Allen. 1972. Taxonomy of aerobic marine eubacteria. J. Bacteriol. *110*: 402–429.

Gauthier, M.J. and V.A. Breittmayer. 1992. The genera *Alteromonas* and *Marinomonas. In* Balows, Trüper, Dworkin, Harder and Schleifer (Editors), The Prokaryotes. A Handbook on the Biology of Bacteria: Ecophysiology, Isolation, Identification, Applications, 2nd Ed., Vol. 3, Springer-Verlag, New York. 3046–3070.

Solano, F. and A. Sanchez-Amat. 1999. Studies on the phylogenetic relationships of melanogenic marine bacteria: proposal of *Marinomonas mediterranea* sp. nov. Int. J. Syst. Bacteriol. *49*: 1241–1246.

Van Landschoot, A. and J. De Ley. 1983. Intra- and intergeneric similarities of the rRNA cistrons of *Alteromonas*, *Marinomonas* (gen. nov.) and some other Gram-negative bacteria. J. Gen. Microbiol. *129*: 3057–3074.

DIFFERENTIATION OF THE SPECIES OF THE GENUS *MARINOMONAS*

Table BXII.γ.87 shows the main differential characteristics of the three *Marinomonas* species described. Other characteristics are listed in Table BXII.γ.88.

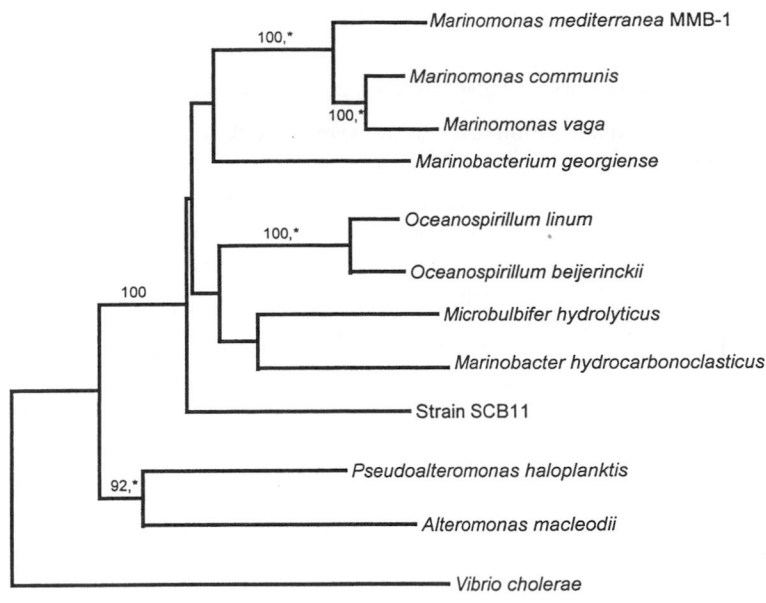

FIGURE BXII.γ.108. Phylogenetic position of *Marinomonas mediterranea* strain MMB-1 in the context of other marine bacteria. The tree was generated by the neighbor-joining method from complete 16S rDNA sequences. Numbers indicate the support for these branches by bootstrap analysis. Asterisks indicate branches also identified by maximum parsimony method. Only values higher than 90% are included (bar = 1% dissimilarity). (Reprinted with permission from F. Solano and A. Sanchez-Amat, *International Journal of Systematic Bacteriology 49:* 1241–1246, 1999 ©International Union of Microbiological Societies.)

List of species of the genus Marinomonas

1. **Marinomonas communis** (Baumann, Baumann, Mandel and Allen 1972) Van Landschoot and De Ley 1984, 91[VP] (Effective publication: Van Landschoot and De Ley 1983, 3071) (*Alteromonas communis* Baumann, Baumann, Mandel and Allen 1972, 420.)

   com.mu′ nis. L. adj. *communis* common.

   The characteristics are as described for the genus and as listed in Tables BXII.γ.87 and BXII.γ.88. Curved rods, 0.7–1.5 × 1.8–3.0 μm. Motile by means of polar flagella at one or both poles. Oxidase positive. Metabolize *m*-hydroxybenzoate, *p*-hydroxybenzoate, and quinate, but not benzoate or *o*-hydroxybenzoate, by means of *meta* cleavage of the intermediate protocatechuate.

   The mol% G + C of the DNA is: 45–48 (Bd).
   Type strain: ATCC 27118, DSM 5604, LMG 2864.

2. **Marinomonas mediterranea** Solano and Sanchez-Amat 1999, 1245[VP]

   med.i.terr.an′e.a. L. gen. n. *mediterranea* of the Mediterranean, referring to the isolation of the type strain from the Mediterranean Sea in the region of Murcia, on the southeastern Spanish coast.

   The characteristics are as described for the genus and as listed in Tables BXII.γ.87 and BXII.γ.88. Gram-negative rods, 0.5–0.9 × 1.1–1.8 μm. Motile by a single polar flagellum. Cytochrome oxidase negative. No denitrification occurs. Lipase positive; gelatinase weakly positive. Utilize D-glucose, D-mannose, D-sorbitol, citrate, β-hydroxybutyrate, succinate, glycerol, and malate. Produce melanin. Contain a pluripotent polyphenol oxidase and a tyrosinase activated by SDS.

   The mol% G + C of the DNA is: 46.3 ± 0.9 (HPLC).
   Type strain: MMB-1, ATCC 700492, CECT 4803.
   GenBank accession number (16S rRNA): AF063027.

3. **Marinomonas vaga** (Baumann, Baumann, Mandel and Allen 1972) Van Landschoot and De Ley 1984, 91[VP] (Effective publication: Van Landschoot and De Ley 1983, 3071) (*Alteromonas vaga* Baumann, Baumann, Mandel and Allen 1972, 420.)

   va′ ga. L. adj. *vaga* wandering.

   The characteristics are as described for the genus and as listed in Tables BXII.γ.87 and BXII.γ.88. Straight rods 0.7–1.5 × 1.8–3.0 μm. Motile by means of polar flagella at onc or both polcs. Oxidase negative. Metabolize *m*-hydroxybenzoate, *p*-hydroxybenzoate and quinate, but not benzoate or *o*-hydroxybenzoate, by means of *meta* cleavage of the intermediate protocatechuate.

   The mol% G + C of the DNA is: 46–50 (Bd).
   Type strain: ATCC 27119.
   GenBank accession number (16S rRNA): X67025.

Genus IV. **Marinospirillum** *Satomi, Kimura, Hayashi, Shouzen, Okuzumi and Fujii 1998, 1346*^{VP}

MASATAKA SATOMI, BON KIMURA AND TATEO FUJII

Ma.ri.no.spi.ril' lum. L. adj. *marinus* of the sea; Gr. n. *spira* a spiral; M.L. dim. neut. n. *spirillum* a small spiral; *Marinospirillum* a small spiral from the sea.

Helical cells. Motile by bipolar tufts of flagella. Thin-walled coccoid bodies are formed in aging cultures. Endospores are not formed. Aerobic, having a strictly respiratory type of metabolism with oxygen as the terminal electron acceptor. **Na$^+$ is required for growth.** Chemoheterotrophic. **Accumulate poly-β-hydroxybutyrate. Oxidase positive. Carbohydrates are neither oxidized nor fermented.** The major isoprenoid quinone type is Q-8. Belongs to the class *Gammaproteobacteria*. One species has been isolated from putrid infusions of marine mussels; another has been isolated from kusaya gravy.

The mol% G + C of the DNA is: 42–45.

Type species: **Marinospirillum minutulum** (Watanabe 1959) Satomi, Kimura, Hayashi, Shouzen, Okuzumi and Fujii 1998, 1346 (*Oceanospirillum minutulum* (Watanabe 1959) Hylemon, Wells, Krieg and Jannasch 1973, 373; "*Spirillum minutulum*" Watanabe 1959, 83.)

FURTHER DESCRIPTIVE INFORMATION

Because *Marinospirillum minutulum*—like most members of the genus *Oceanospirillum*—was isolated from putrid infusions of marine mussels, its source is most likely marine mud adherent to the mussels (Terasaki, 1970).

Marinospirillum megaterium was isolated from kusaya gravy (Fujii et al., 1990), which is used for producing Japanese traditional dried fish, and its true ecological niche is unknown. Under microscopic observation, this species is usually observed as the dominant microbial population in the gravy, characterized by large helical cells, but is unculturable on agar surfaces. The main chemical characteristics of kusaya gravy are as follows: salinity of approximately 3% NaCl concentration, low dissolved oxygen concentration, and large amounts of volatile basic nitrogen compounds (200–500 mg/100 g) (Satomi et al., 1997).

ENRICHMENT AND ISOLATION PROCEDURES

Terasaki (1970) enriched and isolated *M. minutulum* using the following methods. Marine mussels were smashed with a hammer and placed in a Petri dish with a teaspoon of marine mud. Sterilized seawater was poured into the dish until the mussels sank completely in the solution. The infusion was incubated at 27–28°C and examined for the development of spirilla after 1, 2, and 7 d. Isolation was accomplished by streaking dilutions onto suitable agar media.

Fujii et al. (1990) enriched and isolated *M. megaterium* as follows. A loopful of kusaya gravy was used to inoculate the inlet tube of a horizontal glass tube, in which the inlet and outlet ends were bent upwards. After an appropriate time, the isolates accumulated predominantly in the outlet because of their high motility. By repeating this selective accumulation and purification process, the isolates could almost be purified (by microscopic observation). Isolation was accomplished by streaking a large amount of cells onto the suitable agar medium, because the ability of *M. megaterium* to form colonies on an agar surface was low. The ability to form colonies on the agar surface was lost after subsequent transfers of the culture.

MAINTENANCE PROCEDURES

Marinospirillum species may be maintained on stab cultures in TSSY semi-solid agar, which has the following composition (g/l): Trypticase peptone (BBL), 17.0; Phytone peptone (BBL), 3.0; Yeast nitrogen base (Difco), 1.0; NaCl, 30.0; and agar, 2.0; pH 8.0. Cultures are incubated at 25°C with biweekly transfer. Long-term preservation of *M. minutulum* by freeze-drying has been reported by Terasaki (1975).

PROCEDURES FOR TESTING SPECIAL CHARACTERS

Characterization methods for marinospirilla have been described in detail by Terasaki (1972, 1973) and Satomi et al. (1998). Media containing natural or artificial seawater or 2.5–3.0% sodium chloride must be used for all characterization tests. Media for *M. megaterium* must be supplemented with 0.1–0.3% agar because the organism requires semisolid media for growth.

DIFFERENTIATION OF THE GENUS *MARINOSPIRILLUM* FROM OTHER GENERA

The genus *Marinospirillum* cannot be differentiated phenotypically from *Oceanospirillum*, but each *Marinospirillum* species can be differentiated from *Oceanospirillum*, by positive nitrate reduction (*M. minutulum*) and a prominent microaerophilic nature with no growth on agar surfaces (*M. megaterium*).

TAXONOMIC COMMENTS

Phylogenetic analysis based on 16S rDNA gene sequences of the genus *Marinospirillum*, consisting of two species *M. minutulum* and *M. megaterium*, showed that this genus was closely related to the genera *Oceanospirillum* and *Halomonas* in the *Gammaproteobacteria* group (Fig. BXII.γ.109). Although members of the genera *Marinospirillum* and *Oceanospirillum* share helical cell morphology, there is a large phylogenetic distance between *Marinospirillum* and *Oceanospirillum*.

DIFFERENTIATION OF THE SPECIES OF THE GENUS *MARINOSPIRILLUM*

Table BXII.γ.90 lists the characteristics that differentiate *M. minutulum* from *M. megaterium*.

List of species of the genus Marinospirillum

1. **Marinospirillum minutulum** (Watanabe 1959) Satomi, Kimura, Hayashi, Shouzen, Okuzumi and Fujii 1998, 1346^{VP} (*Oceanospirillum minutulum* (Watanabe 1959) Hylemon, Wells, Krieg and Jannasch 1973, 373; "*Spirillum minutulum*" Watanabe 1959, 83.)

   mi.nu' tu.lum. L. dim. neut. adj. *minutulum* very little.

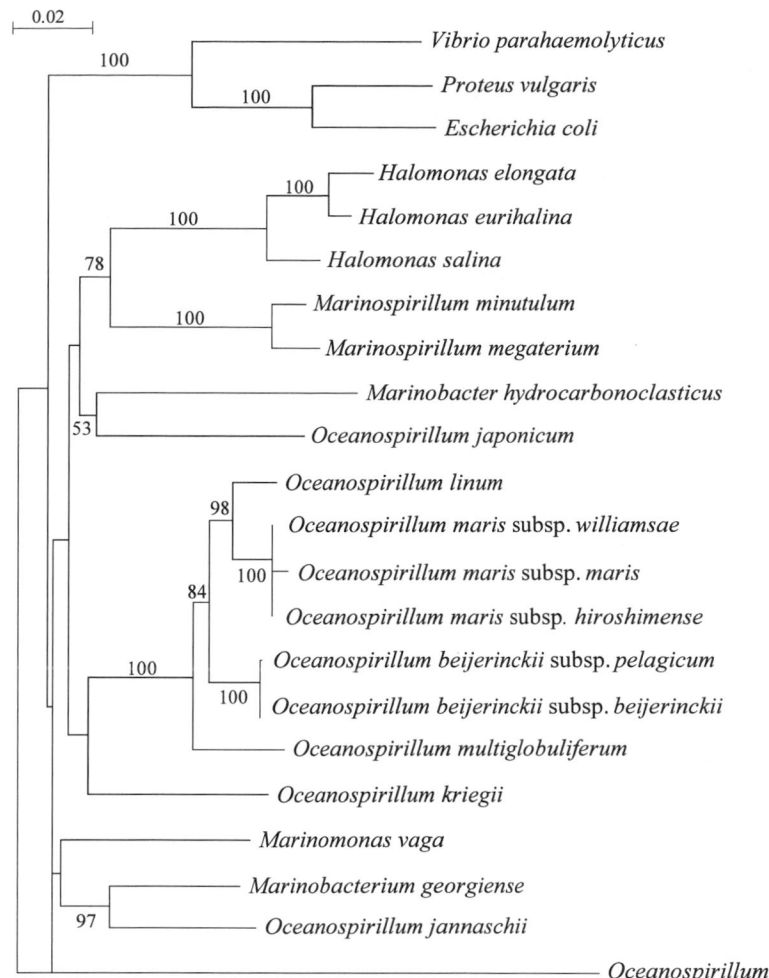

FIGURE BXII.γ.109. Phylogenetic trees of the genus *Marinospirillum* and related bacteria based on the nucleotide sequences of the 16S rRNA gene. The tree was constructed using the neighbor-joining method, and the nucleotide substitution rates (*K*nuc values) were computed by Kimura's 2-parameter model. The scale bar indicates the genetic distance of 0.02 *K*nuc. The numbers at the nodes indicate the percentages of occurrence in 1000 boot-strapped trees. Only values greater than 40% are shown. *Oceanospirillum pusillum* was used as an outgroup. Accession numbers for sequences used: *Marinospirillum minutulum* (AB006769), *M. megaterium* (AB006770), *Oceanospirillum linum* (M22365), *O. maris* subsp. *maris* (AB006771), *O. maris* subsp. *hiroshimense* (AB006762), *O. maris* subsp. *williamsae* (AB006763), *O. beijerinckii* subsp. *beijerinckii* (AB006760), *O. beijerinckii* subsp. *pelagicum* (AB006761), *O. multiglobuliferum* (AB006764), *O. japonicum* (AB006766), *O. kriegii* (AB006767), *O. jannaschii* (AB006765), *O. pusillum* (AB006768), *Halomonas elongata* (X67023), *Halomonas eurihalina* (X87218), *Halomonas salina* (X87217), *Marinomonas vaga* (X67025), *Marinobacter hydrocarbonoclasticus* (X67022), *Marinobacterium georgiense* (U58339), *Vibrio parahaemolyticus* (X56580), *Escherichia coli* (J01859), *Proteus vulgaris* (X07652).

TABLE BXII.γ.90. Characteristics differentiating the species of the genus *Marinospirillum*[a,b]

Characteristic	1. *M. minutulum*[c]	2. *M. megaterium*[d]
Cell diameter, μm	0.3–0.4	0.8–1.2
Microaerophilic	−	+
Catalase	+	− or W
Nitrate reduced to nitrite	+	−
Growth in the presence of:		
1% oxgall	+	−
1% glycine	+	−

[a]Symbols: see standard definitions.

[b]Data from Satomi et al. (1998).

[c]ATCC 19193.

[d]Strains H7 and Sp5.

Rigid, helical cells with clockwise turns (Terasaki, 1973). Cell diameter, 0.3–0.4 μm; wavelength, 2.0–2.8 μm; helix diameter, 0.6–1.5 μm; length of helix, 3.0–8.0 μm. Coccoid bodies are predominant at 3–4 weeks but not after 24–48 h growth. NaCl is required for growth: optimum concentration, 2.5–3.0%; range 0.2–10%. Optimum temperature, 25–30°C; range, 4–30°C. Optimum pH, 8.0; range, 7.0–10.5. Aerobic. Grow in liquid media under static or aerobic conditions with shaking, or on plates. Catalase positive. Do not hydrolyze gelatin, hippurate, or starch. Reduce nitrate to nitrite. Do not produce DNase, RNase, urease, or phosphatase. The type strain was isolated from putrid infusions of marine mussels.

The mol% G + C of the DNA is: 42–44 (HPLC).
Type strain: ATCC 19193, DSM 6287.
GenBank accession number (16S rRNA): AB006769.

2. **Marinospirillum megaterium** Satomi, Kimura, Hayashi, Shouzen, Okuzumi and Fujii 1998, 1346[VP]
me.ga.te′ ri.um. Gr. adj. *mega* large; Gr. n. *teras, teratis* monster, beast; M.L. n. *megaterium* big beast.

Rigid, helical cells, 0.8–1.2 × 5–15 μm, having clockwise turns; thin-walled coccoid bodies 2.0–2.5 μm in diameter. NaCl is required for growth: optimum concentration, 3%; range, 0.5–9.0%. Optimum temperature, 20–25°C; range, 4–25°C. Optimum pH, 8; range, 7.5–9.0. Microaerophilic: no growth occurs in liquid media under static or aerobic conditions with shaking, or on plates. Catalase weakly positive or negative. Do not hydrolyze gelatin, hippurate, or starch. Do not reduce nitrate. Do not produce DNase, RNase, urease, or phosphatase. The type strain was isolated from kusaya gravy.

The mol% G + C of the DNA is: 44–45 (HPLC).
Type strain: H7, JCM 10129.
GenBank accession number (16S rRNA): AB006770.

Genus V. **Neptunomonas** Hedlund, Geiselbrecht, Bair and Staley 1999b, 1325[VP] (Effective publication: Hedlund, Geiselbrecht, Bair and Staley 1999a, 258)

BRIAN P. HEDLUND

Nep.tu.no.mo′ nas. Rom. myth n. *Neptune* the Roman god of the sea; Gr. n. *monas* a unit, monad; M.L. n. *Neptunomonas* Neptune's monad.

Rod-shaped or slightly curved bacteria. Approximately 0.7–0.9 × 2.0–3.0 μm. (See Fig.BXII.γ.110). Cells may produce a capsule that is visible by India ink staining. Coccoid bodies may predominate in old cultures and are associated with a loss of viability. Poly-β-hydroxybutyrate is accumulated and may form small inclusions that are visible by phase-contrast microscopy. **Motile by a single polar flagellum. Aerobic; however, fermentation tests with some sugars and sugar alcohols are weakly positive. Acid is produced slowly from some carbohydrates. Na+ is required for growth. Oxidase and catalase positive.** Phosphatase positive. **Utilize some carbohydrates, polycyclic aromatic hydrocarbons**, amino acids, organic acids, and sugar alcohols as sole carbon sources for chemoorganotrophic growth. Temperature range for known strains, ≤4–30°C. Indigenous to coastal marine sediments. Belong to the *Oceanospirillum* group of the class *Gammaproteobacteria* based on phylogenetic analyses of 16S rDNA gene sequences.

The mol% G + C of the DNA is: 46.
Type species: **Neptunomonas naphthovorans** Hedlund, Geiselbrecht, Bair and Staley 1999b, 1325 (Effective publication: Hedlund, Geiselbrecht, Bair and Staley 1999a, 258.)

FURTHER DESCRIPTIVE INFORMATION

Only one species of *Neptunomonas* is currently known, *N. naphthovorans*.

Polycyclic aromatic hydrocarbon (PAH) catabolism by *N. naphthovorans* strains NAG-2N-113 and NAG-2N-126 has been studied in some detail (Hedlund et al., 1999a). Both strains grow on naphthalene or 2-methylnaphthalene as a sole carbon and energy source. Only strain NAG-2N-113 grows weakly on phenanthrene. Both strains degrade 1-methylnaphthalene and acenaphthene under certain conditions; however, the degradation of these molecules is not coupled to growth. Colonies produce indigo when

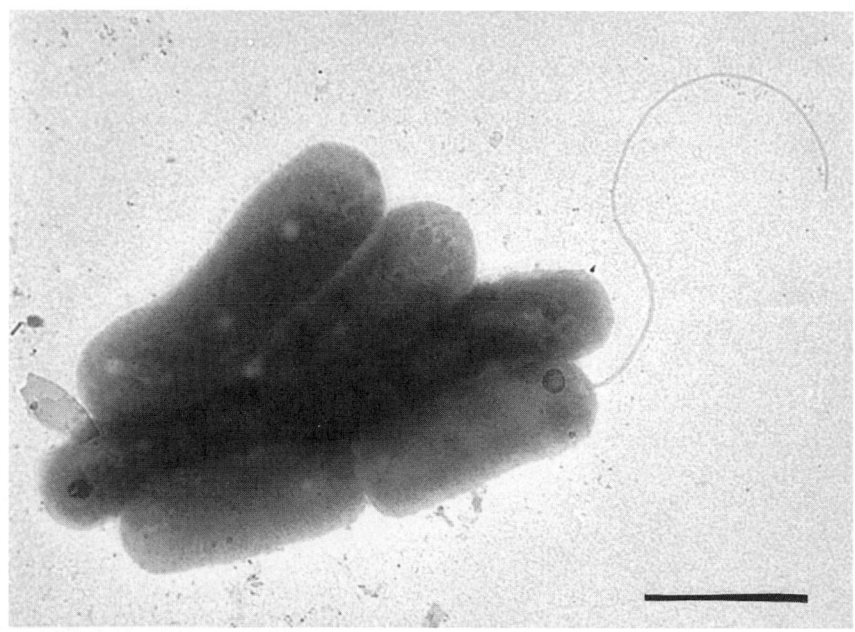

FIGURE BXII.γ.110. Electron micrograph of negatively stained *Neptunomonas naphthovorans* strain NAG-2N-126. Bar = 2 μm. (Reproduced with permission from B. Hedlund et al., Applied and Environmental Microbiology *65:* 251–259, 1999 ©American Society for Microbiology, Washington, D.C.)

naphthalene-induced cells are exposed to indole vapors, a characteristic *Neptunomonas* shares with naphthalene-degrading strains of *Pseudomonas*, *Burkholderia*, and *Marinobacter*, among others. This reaction is catalyzed by the naphthalene dioxygenase. A portion of the gene encoding the naphthalene dioxygenase large subunit has been sequenced from strains NAG-2N-113 and NAG-2N-126; they bear roughly 50–65% sequence identity to *Pseudomonas* and *Burkholderia* naphthalene dioxygenases. However, the low mol% G + C for the *Neptunomonas* dioxygenase gene indicates it is not newly acquired from organisms with a high mol% G + C by horizontal gene transfer. *Neptunomonas* does not produce colored intermediates characteristic of the *meta*-cleavage pathway when grown on naphthalene or other aromatic compounds.

A variety of carbon sources are used by *Neptunomonas* as sole carbon and energy sources, including naphthalene, 2-methyl-naphthalene, *p*-hydroxybenzoate, D-fructose, D-glucose, citrate, DL-β-hydroxybutyrate, glutarate, succinate, DL-lactate, pyruvate, L-arginine, L-serine, L-glutamate, acetate, L-proline, DL-alanine, glycerol, mannitol, and D-arabitol.

All strains of this species have been isolated from a single area in Puget Sound near Seattle, WA USA.

ENRICHMENT AND ISOLATION PROCEDURES

Neptunomonas strains have been isolated from coastal marine sediment contaminated with coal tar creosote, which consists mainly of polycyclic aromatic hydrocarbons. Two alternative methods have been used. The first method involves using a most probable number (MPN) approach to enumerate marine bacteria capable of growth on naphthalene and subsequently obtaining isolates from the positive MPN tubes. All manipulations have been done at 4°C with media that have been equilibrated to that temperature for several hours, but whether this is a necessary precaution is not yet known. Serial dilutions (10-fold) of the sediment are made into 5- or 10-ml portions of the artificial seawater solution ONR7a,[1] and naphthalene crystals are added as a sole carbon and energy source. The tubes are typically incubated near *in situ* temperature without shaking and monitored daily for turbidity. It should be noted that *Neptunomonas* strains do not remain viable for long periods in ONR7a with naphthalene; therefore, sampling the dilution tubes within days of the culture becoming turbid is recommended. Samples from positive tubes are diluted with ONR7a, and the dilutions are spread onto plates of ONR7a solidified with 0.8% agarose or onto a complex marine medium, such as Marine Broth 2216 solidified with 1.5% agar. Naphthalene crystals are added to the Petri dish lids, the plates are inverted, and the plates are sealed with Parafilm. If *Neptunomonas* is outnumbered by other naphthalene-degrading bacteria such

as *Cycloclasticus*, *Neptunomonas* strains can be selected based on their ability to grow well in complex media such as Difco Marine Broth 2216.

A second isolation method employs a direct plating approach. Sediment is diluted into ONR7a and spread onto plates of ONR7a solidified with 0.8% agarose. Naphthalene crystals are added to the Petri dish lids, the plates are inverted, and then sealed.

MAINTENANCE PROCEDURES

Strains of *Neptunomonas* can be maintained for up to 1 month at 4°C on Parafilm-sealed plates of solidified Difco Marine Broth 2216. Additionally, freezer stocks containing glycerol or DMSO can be stored at −80°C for at least 2 years. Lyophilized cultures have also been revived; however, their long-term viability has not yet been evaluated.

DIFFERENTIATION OF THE GENUS *NEPTUNOMONAS* FROM OTHER GENERA

The characteristics that differentiate *Neptunomonas* from phylogenetically similar genera are listed in Table BXII.γ.91.

Several other genera of naphthalene-degrading marine bacteria exist, including *Cycloclasticus* (Dyksterhouse et al., 1995), *Marinobacter* (Gauthier et al., 1992), *Vibrio* (Geiselbrecht et al., 1996), *Pseudoalteromonas* (Hedlund et al. 1996a), and *Sphingomonas* (Zylstra, personal communication). Differential characteristics of these bacteria are shown in Table BXII.γ.92. However, it is generally recommended that 16S rRNA genes be sequenced in order to confidently identify new isolates of marine PAH-degraders.

TAXONOMIC COMMENTS

Based on phylogenetic analyses using nearly complete 16S rRNA gene sequences, *Neptunomonas* is a member of the *Oceanospirillum* group of the class *Gammaproteobacteria*. However, the exact phylogenetic relationship among *Neptunomonas*, *Oceanospirillum*, *Marinomonas*, and *Marinobacterium* is uncertain, as evidenced by the low bootstrap value for the node connecting *Neptunomonas* to the tree (Fig. BXII.γ.111).

The phylogeny and taxonomy of the *Oceanospirillum* group is further complicated by nomenclatural problems associated with some members of the group. Four species, *O. linum*, *O. maris*, *O. beijerinckii*, and *O. multiglobuliferum* form a monophyletic cluster based on rRNA hybridization (Pot et al., 1989), multilocus enzyme electrophoresis (Pot et al., 1989), quinone analysis (Sakane and Yokota, 1994), fatty acid analysis (Sakane and Yokota, 1994), and 16S rRNA gene phylogenetic analysis (Hedlund et al., 1999a). Other taxa, such as *O. kriegii*, and *O. jannaschii*, are not as closely related by any measurement, and it is likely each will be assigned to a new genus. *O. japonicum* is phylogenetically distinct from the true members of the genus *Oceanospirillum*; however, it does share many phenotypic similarities with them. *O. pusillum* obviously should not be included in the genus *Oceanospirillum* since it belongs to the class *Alphaproteobacteria* (Satomi et al., 1998). Therefore, the taxonomy of the *Oceanospirillum* group will likely change dramatically in the near future.

FURTHER READING

Hedlund, B.P., A.D. Geiselbrecht, T.J. Bair and J.T. Staley. 1999. Polycyclic aromatic hydrocarbon degradation by a new marine bacterium, *Neptunomonas naphthovorans* gen. nov., sp. nov. Appl. Environ. Microbiol. 65: 251–259.

---

1. ONR7a consists of several solutions that are prepared separately and mixed when cool. First, a 10× seawater salts solution is prepared, consisting of (g/l distilled water): NaCl, 228; Na$_2$SO$_4$, 40; KCl, 7.2; NaBr, 0.83, NaHCO$_3$, 0.31; H$_3$BO$_3$, 0.27; and NaF, 0.026. This 10× solution is diluted to 1×, and the following are added (g/l 1× solution): Na$_2$HPO$_4$·7H$_2$O, 0.089; NH$_4$Cl, 0.27; and TAPSO buffer (3-[N-tris(hydroxymethyl)methylamino]-2-hydroxypropanesulfonic acid), 1.3. After the solution has been autoclaved and cooled to at least 50°C, 20.0 ml of a 50× divalent cation solution and 5.0 ml of a 200× Fe(II) solution are added per liter. The 50× divalent cations solution, which is prepared and autoclaved separately, contains (g/l): MgCl$_2$·6H$_2$O, 559.1; CaCl$_2$·2H$_2$O, 72.8; and SrCl$_2$·6H$_2$O, 1.21. The 200× solution of Fe(II) is prepared by dissolving 0.04 g of FeCl$_2$·4H$_2$O with a few drops of concentrated HCl; water is then added to 100 ml, and the solution is sterilized by filtration. Long-term storage of the complete ONR7a solution is not recommended.

List of species of the genus Neptunomonas

1. **Neptunomonas naphthovorans** Hedlund, Geiselbrecht, Bair and Staley 1999b, 1325[VP] (Effective publication: Hedlund, Geiselbrecht, Bair and Staley 1999a, 258.)

naph.tho.vo'rans. Chem. n. *naphtho-* combining form of *naphthalene*, a white, crystalline hydrocarbon; L. v. *voro* to devour; M.L. part. adj. *naphthovorans* naphthalene-devouring.

The characteristics are as described for the genus and as listed in Table BXII.γ.91. The type strain was isolated from Eagle Harbor, a creosote-contaminated EPA Superfund site in Puget Sound, Washington, USA.

The mol% G + C of the DNA is: 46.3 (T_m).

Type strain: NAG-2N-126, ATCC 700637.

GenBank accession number (16S rRNA): AF053734.

TABLE BXII.γ.91. Characteristics differentiating the genus *Neptunomonas* from related genera[a]

Characteristic	Neptunomonas	Marinobacterium	Oceanospirillum[b]	O. japonicum	O. kriegii
Cell shape:					
Curved rods	+				+
Straight rods	+	+			+
Spirilla			+	+	
Flagellar arrangement	single polar	single polar	bipolar tufts	bipolar or single tufts	single polar
Cell diameter, μm	0.7–0.9	0.5–0.7	0.3–1.2	0.8–1.4	0.8–1.2
Poly-β-hydroxybutyrate accumulation	+	−	+	+	+
NO_3^- reduced to NO_2^-	−	−	D	−	−
Catalase	+	+	D	w or −	NR
Amylase	−	−	−	−	NR
Gelatinase	−	−	D	d	NR
Lipase	−	+	NR	NR	+
Phosphatase	+	NR	D	w	NR
Acid from carbohydrates	+[c]	−	−	−	−
Carbohydrates fermented	+[c]	−	−	−	−
Temperature range, °C	≤4–30	4–41	2–43	10–43	20–35
Water-soluble, brown pigment	+	NR	D	−	NR
mol% G + C of DNA	46	54.9	45–50	45	54.8

[a]For symbols see standard definitions; NR, not reported; w, weakly positive.

[b]This includes the members of the genus *Oceanospirillum* as described in this book.

[c]Carbohydrate oxidation/fermentation tests were done in PSS media (Hylemon et al., 1973). Acidification took up to two weeks to occur.

TABLE BXII.γ.92. Differentiation of *Neptunomonas* from some other marine naphthalene-degrading bacteria[a,b]

Characteristic	Neptunomonas	Cycloclasticus	Marinobacter	Pseudoalteromonas	Vibrio
Indigo from indole[c]	+	−	+	d	−
Utilization of phenanthrene[c]	− or w	+	−[d]	d	+
Na[+] requirement	+	+	D	+	+[e]
Growth on marine broth 2216	+	− or w	+	+	+
Utilization of carbohydrates	+	−	−	+	+
D-glucose	+	−	−	+	+
D-fructose	+	−	−	d	D
Utilization of amino acids	+	+[f]	+[g]	+	+
Acid from D-glucose	+ or w	NG		−	+
mol% G + C	46	39	57.3–57.7[d]	37–50	38–51

[a]For symbols see standard definitions; NG, no growth in test medium; w, weakly positive.

[b]Marine bacteria are defined here as requiring Na[+] ions for growth. Taxa that have not been fully described are not considered here.

[c]The results for these phenotypes are only given for naphthalene-degrading strains. Otherwise, characteristics are generalized to encompass the whole genus.

[d]For these features we used the results of Spröer et al. (1998), which differ from those obtained previously (Gauthier et al., 1992).

[e]All *Vibrio* strains require Na[+] except *V. cholerae*.

[f]*Cycloclasticus* utilizes only L-glutamate.

[g]*Marinobacter* utilizes only L-proline and L-glutamate.

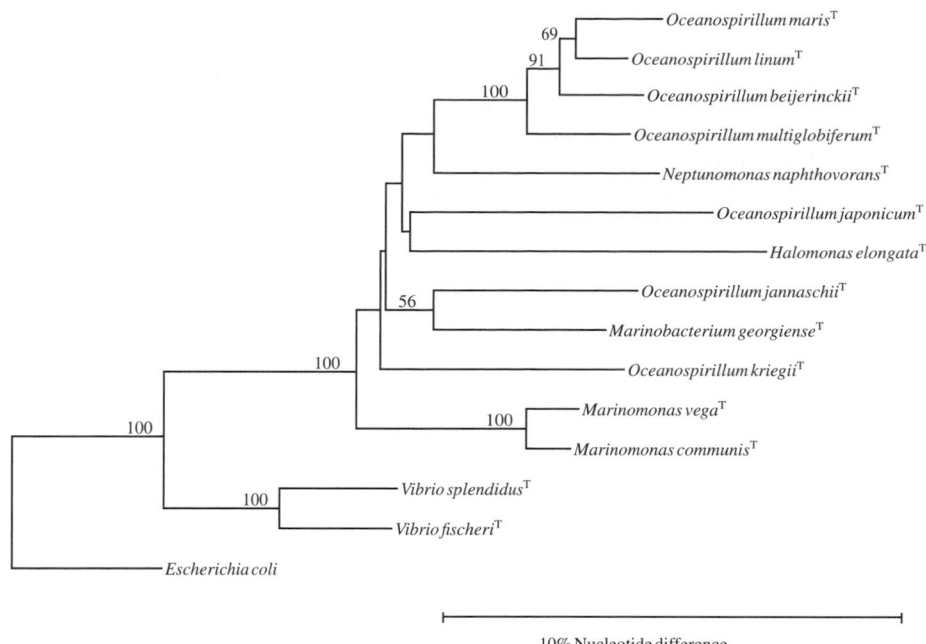

FIGURE BXII.γ.111. Dendrogram showing the relationship between *Neptunomonas* and its close relatives. The dendrogram was created in TreeCon using Jukes-Cantor distance with the Kimura 2–parameter correction. Numbers at branch nodes represent bootstrap values for 100 replications. Numbers below 50 are not shown. (Reproduced with permission from B. Hedlund et al., Applied and Environmental Microbiology *65:* 251–259, 1999 ©American Society for Microbiology, Washington, D.C.)

Family II. **Alcanivoraceae** *fam. nov.*

PETER N. GOLYSHIN, SHIGEAKI HARAYAMA, KENNETH N. TIMMIS AND MICHAIL M. YAKIMOV

Al.ca.ni.vo.ra'.ce.ae. M.L. masc. n. *Alcanivorax* type genus of the family; *-aceae* ending to denote family; M.L. masc. pl. n. *Alcanivoraceae* the *Alcanivorax* family.

Rods 0.6–0.8 × 1.6–2.5 μm in size, depending on the substrate used for growth. Some isolates are capable of anaerobic growth. Reduce nitrate to nitrite. **Principal carbon and energy sources are linear-chain alkanes and their derivatives**, with carbon chain length between 9 and 20. Only a few simple organic compounds are used as carbon and energy source (formate, acetate, propionate, methyl-pyruvate, and α-ketoglutarate). **Moderately halo-**

philic, NaCl content optimal for growth is between 3 and 10%. Strains of all recognized species and recent isolates are mesophilic, with the temperature optima about 25–30°C. Some isolates produce glucose lipid surfactants.

Type genus: **Alcanivorax** Yakimov, Golyshin, Lang, Moore, Abraham, Lünsdorf and Timmis 1998, 346.

Genus I. **Alcanivorax** Yakimov, Golyshin, Lang, Moore, Abraham, Lünsdorf and Timmis 1998, 346[VP]

PETER N. GOLYSHIN, SHIGEAKI HARAYAMA, KENNETH N. TIMMIS AND MICHAIL M. YAKIMOV

Al.ca.ni.vo'.rax. M.L. masc. n. *alcanum* alkane, aliphatic hydrocarbon; L. adj. *vorax* voracious, gluttonous; M.L. masc. n. *Alcanivorax* alkane-devouring.

Rods 0.6–0.8 × 1.6–2.5 μm when grown in pyruvate-supplemented medium and 0.6–0.8 × 1.0–1.5 μm when grown on *n*-alkanes. Nonmotile. Gram negative. Colonies 2–3 days old are circular, colorless, and transparent; in older cultures, they turn opaque and light yellow if grown on alkanes. **Aerobic and microaerophilic**, having a strictly respiratory type of metabolism with oxygen as the terminal electron acceptor. Nitrate is reduced to nitrite, but anaerobic growth does not occur on nitrate. **Oxidase and catalase positive. Growth occurs on aliphatic hydrocarbons as sole or principal carbon sources.** Some strains produce glucose lipid surfactants that exist in two forms. **Moderately**

halophilic; optimum NaCl concentration 3% (range 0.5–15%). Isolated from seawater and seawater sediments enriched with crude oil, petroleum hydrocarbons, or single aliphatic hydrocarbons.

The mol% G + C of the DNA is: 53–66.

Type species: **Alcanivorax borkumensis** Yakimov, Golyshin, Lang, Moore, Abraham, Lünsdorf and Timmis 1998, 347.

FURTHER DESCRIPTIVE INFORMATION

Some strains produce a glycine-containing biosurfactant precursor that is linked to the cell surface and increases the hydrophobicity of the cells and their affinity to oil droplets suspended in the water phase. Some strains produce another form—the mature extracellular glucose lipid biosurfactant, which lacks the terminal glycine residue, and which forms micelles with water-insoluble oil fractions, thereby increasing their bioavailability.

These organisms are distributed in natural marine environments around the world, and their predominance in microbial communities accidentally polluted, or experimentally spiked with oil, suggests on important role in natural petroleum biodegradation (Harayama et al., 1999).

Strains of "*Alcanivoraceae*" were isolated or detected in 16S rDNA clone libraries from different geographic locations (Table BXII.γ.93).

Genome format The genome size estimated for *Alcanivorax borkumensis* SK2T and AP1 is about 3 Mbp. Up to now, no plasmids have been detected in these bacteria.

Isolates of *Alcanivorax* are amenable to conjugative DNA transfer, and transposon mutagenesis (miniTn5-mutagenesis has been characterized by a random positioning of the inserts in the chromosome of *A. borkumensis* strain SK2T with an appropriate antibiotic marker (chloramphenicol, streptomycin/spectinomycin). Although isolates are sensitive to kanamycin, the relatively high salinity of common *Alcanivorax* growth media interferes with the use of this antibiotic marker for selection purposes.

The fluorescent *in situ* hybridization probe 5′-CCA AGA ATA CTA AGA TTC CC—complementary to rRNA between nucleotides 821-842 (*E. coli* numbering)—labeled with Cy3 to detect "*Alcanivoraceae*" in environmental samples has given satisfactory results under stringent conditions (35% (w/vol) formamide; hybridization at 46°C for 2 h). Probes that distinguish *Fundibacter* and *Alcanivorax* species hybridize to 16S rRNA between nucleo-

tides 203 and 223 (*E. coli* numbering); these probes are 5′-GCG AGC TCA TCC ATC **T**GC A and 5′-GCG AGC TCA TCC ATC **A**GC A, respectively (mismatches are shown in boldface type).

ENRICHMENT AND ISOLATION PROCEDURES

The principal sources of these organisms are seawater and seawater sediments, enriched with crude oil, petroleum hydrocarbons, or single aliphatic hydrocarbons. *n*-Hexadecane can serve as the sole carbon source for enrichment of the seawater/sediment samples; addition of nitrogen and phosphorus is required. The flat, colorless, transparent colonies of *Alcanivorax*-related strains develop on the agar surface within 3–5 d after plating. After incubation for longer than 1 week, colonies turn opaque and cream-colored.

DIFFERENTIATION OF THE GENUS *ALCANIVORAX* FROM OTHER GENERA

Characteristics that differentiate the genus from other related or phenotypically similar genera are given in Table BXII.γ.94.

TAXONOMIC COMMENTS

The analysis of 16S rRNA gene sequencing data of both recognized and recently isolated strains belonging to the genera *Alcanivorax* and *Fundibacter* (Yakimov et al., 1998; Bruns and Berthe-Corti, 1999) suggest these bacteria belong to a distinct taxonomic group, presumably a new family, "*Alcanivoraceae*", phylogenetically equidistant from the characterized lineages of the class *Gammaproteobacteria*, "*Alteromonadaceae*", *Halomonas* spp., belonging to *Halomonadaceae* and members of family *Oceanospirillaceae* (Fig. BXII.γ.112).

A controversy exists regarding the taxonomic position of the

TABLE BXII.γ.93. List of strains relevant to the genus *Alcanivorax* and clones derived from 16S rRNA gene libraries from different geographic locations

Strain/clone designation	Origin	Notes
Alcanivorax borkumensis SK2, DSM 11573[T]	Isle of Borkum, North Sea	Type and only species of *Alcanivorax*
Alcanivorax jadensis T9, DSM 12178[T]	Jadebusen, North Sea	Formerly known as *Fundibacter jadensis* T9
Alcanivorax borkumensis AP1, SK, SK1, SK4A, SK7, MM1[a]	Isle of Borkum, North Sea	Isolates obtained from the enrichments along with the type strain SK2
Alcanivorax sp. ME104[a]	Mediterranean Sea, Harbor of Messina, Sicily	98% 16S rRNA gene similarity with the type strain
Alcanivorax borkumensis ME103, ME106, ME111[a]	Mediterranean Sea, Harbor of Messina, Sicily	100% 16S rRNA gene similarity with the type strain
Alcanivorax borkumensis LE4	Singapore shore	100% 16S rRNA gene similarity with the type strain
Alcanivorax sp. SCB54	California (E. Pacific)	Partially sequenced environmental clone
Alcanivorax sp. ME105, ME112[a]	Mediterranean Sea, Harbor of Messina, Sicily	95–98% 16S rRNA gene similarity with the type strain
MBIC4324–4330, 4335–4439, 4346–4351[b]	Sea of Japan	
MBIC4322, 4323, 4332, 4334[b]	Palau (S. Pacific)	
MBIC4197–4100[b]	Canada (NE Pacific)	
MBIC4064, 4066, 4067[b]	Adriatic Sea, Croatia	
MBIC4331,4168[b]	Okinawa (W. Pacific)	
MBIC4089, 4090[b]	Miyagi (Central Pacific)	
MBIC4426, 4427[b]	Shizuoka (W. Pacific)	
MBIC4103, 4104[b]	Hawaii (Central Pacific)	
"*Alcanivorax profundus*" MRS3300[c]	Deep area (3300m) of Mediterranean Sea located between Straits of Messina and South Ionian Sea: 36°30′N-15°50′E	98% 16S rRNA gene similarity with *Alcanivorax jadensis* strain T9
Alcanivorax borkumensis 9B1, 7B[c]	Coral Sea (S. Pacific), harbor of Noumea, New Caledonia	100% 16S rRNA gene similarity with the type strain

[a]Available in the strain collection of the Department of Microbiology of the GBF (German Research Center for Biotechnology), Mascheroder Weg 1, 38124 Braunschweig, Germany (Email pgo@GBF.de).

[b]Available in the strain collection of Marine Biotechnology Institute, 3-75-1, Kamaishi Laboratory, Heita, Kamaishi, Iwate 026-0001, Japan (Email shigeaki.harayama@kamaishi.mbio.co.jp).

[c]Available in the strain collection of the Istituto Sperimentale Talassographico CNR, Spianato S. Raineri 86, 98122 Messina, Italy (Email iakimov@its.me.cnr.it).

TABLE BXII.γ.94. Comparison of key phenotypic characteristics of *Alcanivorax* spp. and other heterotrophic marine bacteria belonging to the *Gammaproteobacteria*[a,b]

Characteristic	*Alcanivorax* spp.	*Halomonas* spp.	*Marinobacter* spp.	*Marinomonas* spp.	*Oceanospirillum* spp.
Typical Gram-negative cell wall	+	+	+	+	+
Morphology	Rods	Rods	Rods	Rods	Helical
Number and arrangement of flagella	None	Peritrichous	None or single polar	Single polar	Bipolar tufts
Optimal temperature for growth (°C)	20–30	30	32	20–25	25–32
Growth at 4°C	±	+	−	−	±
Growth at 45°C	−	NR	+	+	−
Optimal NaCl concentration (%) for growth	3–5	3–8	3–6	0.7–3.5	0.5–8
Maximal NaCl concentration (%) for growth	15	32.5	20	NR	9.75
Nitrate reduced to nitrite	+	+	+	−	−
Oxidase test (Kovacs)	+	+	+	−	+
Arginine dihydrolase activity	−	−	−	−	NR
Ornithine decarboxylase	−	+	−	−	NR
Gelatin liquefaction	−	±	−	−	−
Starch hydrolysis	−	−	−	−	−
Lysine decarboxylase	−	+	−	−	NR
Urease activity	−	±	−	−	−
Utilization as sole carbon source:					
DL-Alanine	−	+	−	+	±
L-Arginine	−	+	−	±	−
Aspartate	−	+	−	NR	−
Cellobiose	−	+	−	±	−
D-Fructose	−	+	−	+	−
D-Gluconate	−	+	−	+	NR
D-Glucose	−	+	−	+	−
L-Glutamate	−	+	+	+	±
Glycerol	−	+	−	+	−
Hexadecane	+	+	+	+	NR
p-Hydroxybenzoate	−	−	+	+	±
D-Mannitol	−	+	−	+	−
D-Mannose	−	+	−	+	−
L-Serine	−	+	−	±	−
Succinate	−	+	+	+	+
Sucrose	−	+	−	−	−
Accumulation of poly-β-hydroxybutyrate	±	+	−	−	+
Mol% G+C of the DNA	53–66	60–61	53	44–48	42–51

[a]+, Positive reaction or growth; ±, variable reaction; negative reaction or growth; NR, not reported.

[b]Modified from Yakimov et al. (1998).

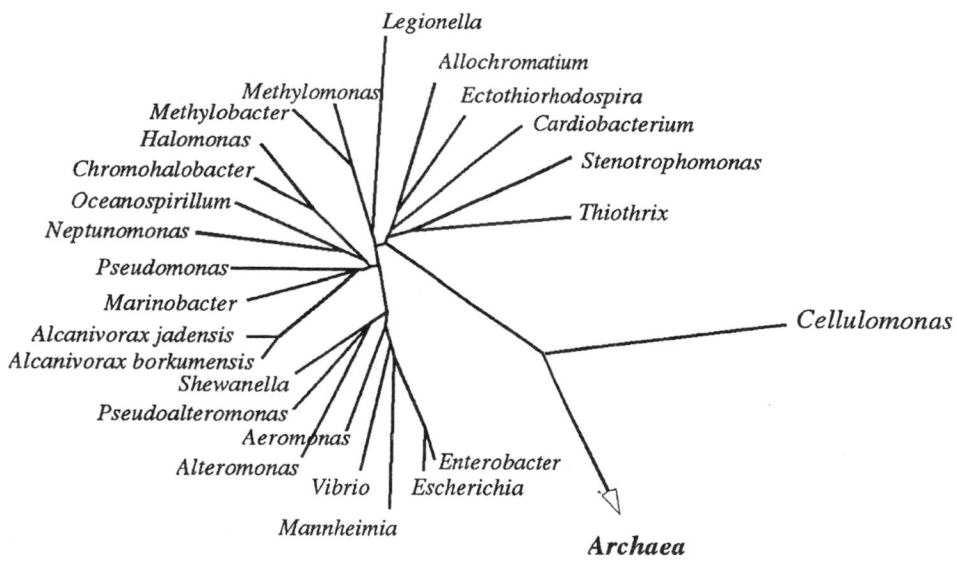

FIGURE BXII.γ.112. Phylogenetic position of species of genus *Alcanivorax* among principal genera within *Gammaproteobacteria*. Phylogenetic tree was rooted from an archaeon, *Pyrodictium occultum*, and *Cellulomonas* was used as an outgroup. The scale bar represents 0.1 fixed point mutation per sequence position. *Fundibacter jadensis* was transferred to the genus *Alcanivorax* as *Alcanivorax jadensis* (González and Whitman, 2002; Fernández-Martínez et al., 2003).

genus *Fundibacter* (see Other Organisms, below). Because two validly described genera—*Alcanivorax* (Yakimov, Golyshin, Lang, Moore, Abraham, Lünsdorf and Timmis, 1998) and *Fundibacter* (Bruns and Berthe-Corti, 1999)—share significant 16S rRNA homology (about 98%) and because no clear-cut phenotypic differentiation can be drawn between these genera, which were isolated from similar environmental sources (German North Sea, alkane-enriched aerobic pelagic sediments), it appears appropriate to assign the genus *Fundibacter* to the genus *Alcanivorax*.

DNA/DNA hybridization studies between *A. borkumensis* and *F. jadensis* clearly indicate, however, that these two organisms do not belong to the same species (Steiner and Tindall, personal communication).

The sequencing of *gyrB* genes from above type cultures and a large number of recently isolated—but not validly published—*Alcanivorax* strains suggest existence of at least five genotypically distinct groups, presumably at the level of species (Fig. BXII.γ.113).

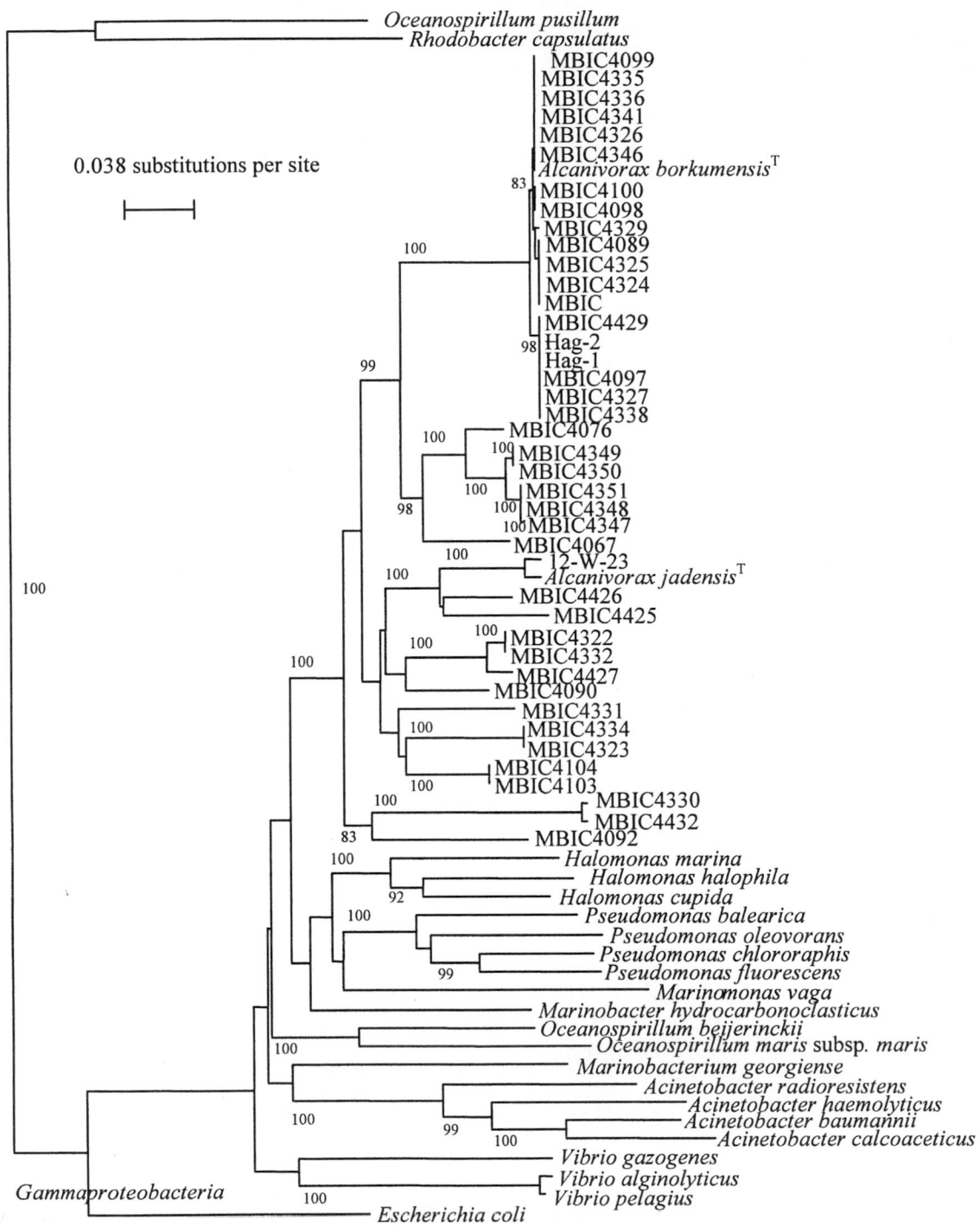

FIGURE BXII.γ.113. Phylogenetic tree derived from the similarities between the *gyrB* sequences. Bootstrap values over 80% are shown at the branch points. Strains are listed in Table BXII.γ.93. *Fundibacter jadensis* was transferred to the genus *Alcanivorax* as *Alcanivorax jadensis* (González and Whitman, 2002; Fernández-Martínez et al., 2003).

List of species of the genus Alcanivorax

1. **Alcanivorax borkumensis** Yakimov, Golyshin, Lang, Moore, Abraham, Lünsdorf and Timmis 1998, 347[VP]

bor.ku.me' n.sis. M.L. adj. *borkumensis* from the island of Borkum, a small island in Western-Elms harbor in the North Sea, located close to the German-Dutch border.

The characteristics are as described for the genus, with the following additional information. Optimum temperature 20–30°C; range, 10–35°C. Optimum NaCl concentration 3–10%; range, 1.0–12.5%. Reduce nitrate to nitrite, though no anaerobic growth occurs. Do not produce agarase, amylase, arginine dihydrolase, ornithine decarboxylase, lysine decarboxylase, gelatinase, or esculinase. Exhibit Tween 80 hydrolysis. Utilize aliphatic hydrocarbons (formate, acetate, propionate, methylpyruvate, α-ketoglutarate) and some short-chain fatty acids as sole or principal carbon source. Produce glucose lipids (Fig. BXII.γ.114) that exist in two forms: a glycine-containing biosurfactant precursor and the mature extracellular glucose lipid. Principal fatty acids are $C_{16:0}$, $C_{16:1}$, and $C_{18:1}$. Resistant to ampicillin, tetracycline, nalidixic acid; sensitive to streptomycin/spectinomycin, chloramphenicol, kanamycin.

The mol% G + C of the DNA is: 53.4 (HPLC).

Type strain: SK2, ATCC 700651, CIP 105606, DSM 11573.

GenBank accession number (16S rRNA): Y12579.

2. **Alcanivorax jadensis** Fernández-Martínez, Pujalte, García-Martínez, Mata, Garay, and Rodríguez-Valera 2003, 337[VP] (*Fundibacter jadensis* Bruns and Berthe-Corti 1999, 447.)

ja.den' sis. M. L. adj. *jadensis* referring to the region Jade, which forms part of the bay "Jadebusen", which belongs to the German North Sea coast.

Rods 0.3–0.7 μm wide and 0.8–1.8 μm long. Nonmotile. Pili are present. Gram negative. Colonies are circular, 0.25–1.25 mm in diameter and ivory-pigmented. The temperature spectrum ranges from 10–40°C, with optimal growth occurring at 30°C. Shows weak halotolerance and grows at NaCl concentrations ranging from 0.5–15% (w/v); opti-

mum, 3%. It is able to grow both aerobically and anaerobically, and reduces nitrate. It grows with the carbon sources pyruvate and acetate if a synthetic medium is used and also with tetradecane, hexadecane, and pristane if the medium is supplemented with vitamins and mineral salts. The cells are resistant to ampicillin, cefazolin, cephalotin, erythromycin, linomycin, nalidixic acid, novobiocin, ofloxacin, oxacillin, penicillin, and tetracycline. Oxidase and catalase positive. Growth occurs on aliphatic hydrocarbons as the sole or principal carbon source (tetradecane, hexadecane, and pristane). Pyruvate and acetate can be used. The organisms emulsify alkanes, but do not produce glucose lipid.

The mol% G + C of the DNA is: 63.6 (T_m).

Type strain: T9, ATCC 700854, DSM 12178.

GenBank accession number (16S rRNA): AJ001150.

FIGURE BXII.γ.114. Cell-associated glucose lipid from *A. borkumensis* SK2[T].

Family III. **Hahellaceae** *fam. nov.*

GEORGE M. GARRITY, JULIA A. BELL AND TIMOTHY LILBURN

Ha. hel.la' ce.ae. M.L. fem. n. *Hahella* type genus of the family; *-aceae* ending to denote family; M.L. fem. pl. n. *Hahellaceae* the *Hahella* family.

The family *Hahellaceae* was circumscribed for this volume on the basis of phylogenetic analysis of 16S rDNA sequences; the family contains the genera *Hahella* (type genus) and *Zooshikella*. *Zooshikella* was proposed after the cut-off date for inclusion in this volume (June 30, 2001) and is not described here (see Yi et al. (2003)).

Aerobic or facultatively anaerobic chemoorganotrophs. Require NaCl for growth.

Type genus: **Hahella** Lee, Chun, Moon, Ko, Lee, Lee and Bae 2001a, 664[VP].

Genus I. **Hahella** *Lee, Chun, Moon, Ko, Lee, Lee and Bae 2001a, 664*[VP]

THE EDITORIAL BOARD

Ha. hel' la. M.L. fem. n. named after Yung Chil Hah, a Korean bacteriologist who pioneered microbiological research in Korea.

Gram-negative motile rod. Facultatively anaerobic; produces acid from sugars. Reduces nitrate to nitrite. Hydrolyzes esculin and gelatin. **Produces extracellular polysaccharides and red pigment. Requires NaCl.**

The mol% G + C of the DNA is: 55.

Type species: **Hahella chejuensis** Lee, Chun, Moon, Ko, Lee, Lee and Bae 2001a, 665.

FURTHER DESCRIPTIVE INFORMATION

Analysis of 16S rDNA sequences showed that *Hahella chejuensis* is a member of the *Gammaproteobacteria* but is not closely related to any known genera (Lee et al., 2001a).

ENRICHMENT AND ISOLATION PROCEDURES

The strain was isolated directly from diluted marine sediment spread on ZoBell's medium (Lee et al., 2001a).

MAINTENANCE PROCEDURES

The strain was stored at −80° in 20% (w/v) glycerol (Lee et al., 2001a).

DIFFERENTIATION OF THE GENUS *HAHELLA* FROM OTHER GENERA

Lee et al. (2001a) provide a table of characteristics that differentiate *Hahella chejuensis* from other marine/halophilic bacteria in the *Gammaproteobacteria*.

List of species of the genus Hahella

1. **Hahella chejuensis** Lee, Chun, Moon, Ko, Lee, Lee and Bae 2001a, 665[VP]

   che.ju.en' sis. M.L. adj. *chejuensis* pertaining to Cheju Island, Republic of Korea, geographical origin of the type strain of the species.

   Young cells 0.5–0.7 × 1.6–9.0 μm; older cells 0.7–0.8 × 1.4–1.7 μm. Halophilic; produces red pigment soluble in methanol. Growth at 1–8% NaCl (optimum 2% NaCl), 10–45°C, and pH 6–10 (optimum pH 7). Grows on and produces acid from adonitol, D(+)-fructose, D(+)-glucose, in-

ositol, D(+)-maltose, D(−)-mannitol, D(+)-mannose, D(−)-sorbitol, sucrose, and D(+)-trehalose. Does not grow on L(+)-arabinose, citrate, D(+)-galactose, D(+)-lactose, malate, malonate, D(+)-melibiose, D(+)-raffinose, L(+)-rhamnose, and D(+)-xylose. Major fatty acids octadecenoic, *cis*-9-hexadecenoic/*iso*-2-hydroxypentadecanoic, hexadecanoic, and 3-hydroxydodecanoic.

   The mol% G + C of the DNA is: 55 (T_m).

   Type strain: 96CJ10356, KCTC 2396, IMSNU 11157.

   GenBank accession number (16S rRNA): AF195410.

Family IV. **Halomonadaceae** Franzmann, Wehmeyer and Stackebrandt 1989, 205[VP] emend. Dobson and Franzmann 1996, 558

GEORGE M. GARRITY, JULIA A. BELL AND TIMOTHY LILBURN

Ha.lo.mo.na.da' ce.ae. M.L. fem. n. *Halomonas* type genus of the family; *-aceae* ending to denote family; M.L. fem. pl. n. *Halomonadaceae* the *Halomonas* family.

The family *Halomonadaceae* was circumscribed for this volume on the basis of phylogenetic analysis of 16S rDNA sequences; the family contains the genera *Halomonas* (type genus), *Carnimonas*, *Chromohalobacter*, *Cobetia*, *Deleya*, and *Zymobacter*. *Cobetia* was proposed after the cut-off date for inclusion in this volume (June 30, 2001) and is not described here (see Arahal et al. (2002a)).

Halotolerant or halophilic, except for *Zymomonas*. Motile by means of flagella. Aerobic or facultatively anaerobic chemoorganotrophs.

Type genus: **Halomonas** Vreeland, Litchfield, Martin, and Elliot 1980, 494[VP] emend. Dobson and Franzmann 1996, 557.

Genus I. **Halomonas** *Vreeland, Litchfield, Martin, and Elliot 1980, 494*[VP] *emend. Dobson and Franzmann 1996, 557*

RUSSELL H. VREELAND

Ha.lo.mo' nas. Gr. n. *hals, halos* salt of the sea; Gr. n. *monas* a unit, monad; M.L. fem. n. *Halomonas* salt(-tolerant) monad.

Straight or curved rod shaped cells, generally 0.6–0.8 × 1.6–1.9 μm. One species (*H. halodenitrificans*) presents coccoid cells. Species may form elongated, flexuous filaments under certain conditions. Endospores are not formed. **Gram negative. Rod shaped species are motile** by lateral, polar, or peritrichous flagella. Colonies are white or yellow, turning light brown with age. **Possess**

a mainly respiratory type of metabolism with oxygen as the terminal electron acceptor. Some species are also capable of anaerobic growth in the presence of nitrate. Some species have been reported to grow under anaerobic conditions in the absence of nitrate if supplied with glucose (but not other carbohydrates or amino acids). Nitrate is reduced to nitrite; nitrogen gas is not formed. All species tested are catalase positive; 17 of the 21 known species are also oxidase positive. Chemoorganotrophic. Carbohydrates, amino acids, polyols, and hydrocarbons can serve as sole carbon sources in mineral media. Ammonium sulfate can serve as a sole nitrogen source. **Intracellular granules are not produced. Halotolerant** (also described as slight to moderate halophiles), **able to grow in NaCl concentrations ranging from 0.1–32.5% (w/v).** The major respiratory quinone is ubiquinone 9. The major fatty acids are $C_{16:1}$, $C_{17:0\ cyc}$, $C_{16:0}$, $C_{18:1}$, and $C_{19:0\ cyc}$. In addition to having the signature sequences found in the *Halomonadaceae*, *Halomonas* species contain the following four signature bases (by *E. coli* numbering): C at position 1424, U at position 1439, A at position 1462, and C at position 1464. **Isolated from saline environments around the world**, including solar salt facilities, intertidal estuaries, the open ocean, and hypersaline lakes (Dead Sea, Israel; Organic Lake, Antarctica).

The mol% G + C of the DNA is: 52–68.

Type species: **Halomonas elongata** Vreeland, Litchfield, Martin and Elliot 1980, 495.

FURTHER DESCRIPTIVE INFORMATION

During exponential growth, 20 of the 21 *Halomonas* species consist of mixtures of straight and curved rods. Upon entry into stationary phase, several *Halomonas* species also form elongated, flexuous filaments of various lengths. The percentage and length of these elongated cells depends upon the type of growth medium. In complex Casamino acids medium (CAS)[1] with 8% NaCl, all of the cells may be elongated, with many filaments containing irregular loops and bends. In a chemically defined mineral salts medium (MS)[2] containing 8% NaCl, as few as 25% of the cells may be elongated, and the filaments seldom produce irregular looping or extensive bending.

Electron microscopy of thin sections from optimum growth conditions reveals a typical Gram-negative type of cell wall (Vreeland and Martin, 1980).

The number and arrangement of flagella on short cells depends upon the species examined. Cells generally possess 4–7 flagella. In some strains, the flagella are arranged laterally, while others possess only polar flagella. One biovar of this genus has both lateral and polar flagellation. Motile *Halomonas* cells describe a helix when viewed by light microscopy. Motility is generally lost rapidly under low oxygen concentrations. Flagella have not been observed on elongated cells or on cells grown in low NaCl concentrations (<0.2%).

On solid CAS or MS medium containing 8% NaCl, most *Halomonas* colonies are white to cream-colored, smooth, glistening, opaque, and ~2 mm in diameter after 24 h at 30°C. The colonies

generally become yellow to light brown and spread following prolonged incubation.

Halomonas species require Na^+ for growth—at least 0.1% w/v NaCl added to CAS medium or 0.3% w/v NaCl added to MS medium—and will grow in a wide range of NaCl concentrations (0.1–32% NaCl w/v). When subjected to osmotic shock from either rapid dilution or increased NaCl concentration, *Halomonas* cells respond as osmometers and exhibit swelling or plasmolysis. Osmotically shocked *Halomonas* cells do not lose viability and soon reestablish osmotic balance and resume growth. This reestablishment of osmotic balance is mediated by an increased concentration of ectoine, which is the primary compatible solute found in *H. elongata*. The optimum NaCl concentration is 2.2–8.0% at 30°C. The Na^+ requirement of the type species can be satisfied by NaCl, $NaNO_3$, or NaBr (Vreeland and Martin, 1980). The NaCl requirement of the other species has not been tested in detail. NaCl tolerance is affected by growth temperature, but apparently not by the carbon source being used (Vreeland and Martin, 1980). In terms of the ability to promote salt tolerance, temperatures can be arranged from greatest degree of promotion to least as follows: 30°C, 23°C, 37°C, 15°C, 45°C, 4°C (Vreeland et al., 1980).

In CAS medium containing 8% NaCl, the *Halomonas* species that have been tested grow at pH values from 5.0–9.0. Alkaliphilic strains generally grow from pH 7.0–11.0 with an optimum at 9.5. The pH tolerance has not been tested on MS medium and has not been tested at different temperatures.

The physiological characteristics of all properly recognized *Halomonas* species are listed in Tables BXII.γ.95 and BXII.γ.96. Although considered mainly aerobic, some *Halomonas* species have been reported to grow on glucose under anaerobic conditions in the absence of nitrate or other terminal electron acceptors. If the glucose is replaced by other carbohydrates or an amino acid, nitrate is required for anaerobic growth.

Members of the genus are nutritionally versatile. The following compounds often serve as sole carbon sources for growth: glucose, gluconate, glycerol, fructose, mannose, sucrose, cellobiose, succinate, mannitol, alanine, glutamine, glutamate, asparagine, aspartate, lysine, histidine, phenylalanine, tyrosine, tryptophan, proline, arginine, leucine, isoleucine, valine, methionine, cysteine, serine, and threonine.

Halomonas species have been isolated from a solar salt facility on the island of Bonaire, Netherlands Antilles (Vreeland et al., 1980) from the Dead Sea and Canada (Huval et al., 1995), the Antarctic (Franzmann et al., 1987; James et al., 1990a), Great Salt Lake, Utah (Fendrich, 1988), estuaries (Hebert and Vreeland, 1987), and the Pacific Ocean (Baumann et al., 1983a). Representatives of the genus may be even more widespread in nature.

ENRICHMENT AND ISOLATION PROCEDURES

Halomonas species may be isolated using a wide variety of media. Brine samples can be spread onto the surface of the medium and incubated at close to the environmental temperature and under high humidity for 2–7 d. *Halomonas* colonies are white to cream colored and easily distinguished from red-pigmented halophiles. Alternatively, the brine samples can be added directly to filter pads saturated with medium. This method has the advantage of allowing the use of small (9 mm) Petri plates, facilitating shipment to distant field areas. A disadvantage of the technique is that the white *Halomonas* colonies can be difficult to see

---

1. CAS medium is a modification of the medium of Abram and Gibbons, as described by Gibbons (1969), and contains (g/l of distilled water): yeast extract, 1.0; casamino acids (Difco) (not "vitamin-free"), 7.5; Proteose peptone No. 3 (Difco), 5.0; sodium citrate, 3.0; $MgSO_4 \cdot 7H_2O$, 20.0; K_2HPO_4, 7.5; and NaCl (or solar salt), 80.0. The pH is adjusted to 8.0 ± 0.1 with NaOH prior to sterilization and is 7.5 ± 0.1 after autoclaving at 121°C for 20 minutes (Vreeland et al., 1980, 1984). CAS medium is stable for several weeks when stored in the dark. It should be discarded if any crystal formation is seen.

2. MS medium is similar to that described by Vreeland and Martin (1980). It contains (g/l): $MgCl_2 \cdot 6H_2O$, 5.3; KCl 0.75; and $(NH_4)_2SO_4$, 4.1. It is supplemented with a carbon source (10–50 mM) and NaCl (2.92–198.7 g/l). Phosphate ($K_2HPO_4 \cdot 3H_2O$) is added as a sterile 10× concentrate after the medium has been autoclaved to give a final concentration of 0.5 g/l. The medium may also be supplemented with $CaCl_2$ (0.11 g/l) to enhance growth. The pH of the medium is adjusted to 7.2 with 1 N KOH prior to autoclaving and is 7.0 after sterilization.

on white filter pads; also, the pad fiber makes colonial isolation tedious.

A more direct isolation technique is to take advantage of the halotolerance of all of the species by setting up a salt-mediated selection medium. High salt brines can be mixed with low salt or distilled water, then spread onto a low salt medium. Alternatively, low salt samples can be added to high salt media to select for salt-tolerant organisms.

MAINTENANCE PROCEDURES

Halomonas strains can be maintained on agar slants containing 8% (w/v) NaCl. Stock cultures are transferred every 6 months, allowed to grow at 30°C for 1–7 d, and then stored in the dark at 4°C (Vreeland et al., 1980). Some media that have proven useful for growing Halomonas are HSC medium[3], MH medium[4], and AOL medium[5].

Most species of *Halomonas* have been lyophilized, but survival rates may vary with the strain and species. *Halomonas elongata* strains ATCC 33173 and 33174 have survived for up to 2 years. The growth from a fresh CAS slant or broth is suspended in 0.5 ml of a solution containing Proteose peptone No. 3 (Difco) (0.5%), yeast extract (1.0%), and NaCl (0.2%). A small amount of the suspension (0.1–0.2 ml) is transferred to a lyophilization vial, dipped into liquid N_2 for 5–10 min, and lyophilized. Preserved cultures are reconstituted by suspending the cells in a small amount of CAS lacking salt. After 15–30 min, the cells are transferred to CAS broth containing NaCl and incubated at 30°C.

Halomonas strains may also be stored frozen on ceramic Protec® or Microbank® beads at −80°C. When using ceramic beads, it is best to grow the cultures under optimum growth conditions and then add the culture to fresh medium and place it onto the pre-sterilized beads. The cultures may then be reincubated at 37°C for several hours to overnight to allow time for the cells to attach to the beads. After this incubation, the fluid may be drawn off and the culture frozen. *Halomonas* cultures have been preserved for up to 10 years using this technique; however, the quality of the culture has been found to deteriorate slowly under these conditions (R. Vreeland, unpublished observations).

Halomonas may also be stored frozen on ceramic beads under liquid N_2. Due to the variation in survival rates for other storage methods, liquid N_2 storage is the method of choice for long-term preservation.

PROCEDURES FOR TESTING SPECIAL CHARACTERS

A common characteristic of all *Halomonas* species is their ability to grow at a wide variety of salt concentrations. Huval and Vreeland (unpublished studies) have determined that the phenotypic characteristics of these bacteria are often affected by the salt concentration of the growth and testing medium. These effects are particularly noticeable at the low and high salt extremes. This aspect has also been encountered by Hebert and Vreeland (1987) during their characterization of *Halomonas halodurans*. Consequently, prior to conducting any taxonomic analysis on isolates suspected of belonging to *Halomonas*, the optimum growth conditions with respect to NaCl concentration, temperature, and pH must be determined. These characteristics are best tested by use of complex media, each containing 0, 3.5, 8.0, 15.0, 20.0, or 32.0% (w/v) NaCl or solar salt. Temperature tolerance and qualitative effects of temperature on salt tolerance can be tested by incubating tubes of each salt concentration at temperatures from 4–45°C. After 2–3 weeks, all tubes without visible growth should be incubated for an additional 2–3 weeks at the optimum temperature to determine whether any apparent inhibitory conditions have actually been lethal (Vreeland, 1993).

DNA may be extracted from *Halomonas* by the method of Marmur (1961) with some modification. *Halomonas* suspensions and the sodium dodecyl sulfate (SDS) used for lysis must be heated to 50°C. Following SDS addition (2% final concentration), the lysate is held at 50°C for 15 min. Pronase is not added to the suspension. Water-saturated phenol (also at 50°C) is then added to fully denature the protein. The preparation may then be cooled to room temperature and the rest of the Marmur procedure followed. This modification is necessary since some *Halomonas* strains possess a nuclease, which, unless inactivated by heating, causes rapid breakdown of DNA following cell lysis.

All other characterization tests may be performed using conventional techniques (Holding and Collee, 1971; Vreeland, 1993). All media are supplemented with NaCl at the optimum concentration, as determined above. Uninoculated controls are necessary to ensure that any reactions detected are not artifacts caused by the NaCl.

DIFFERENTIATION OF THE GENUS *HALOMONAS* FROM OTHER GENERA

Halomonas is easily differentiated from such genera as *Zymomonas*, *Cellvibrio*, *Oceanospirillum*, or *Serpens* on the basis of salt tolerance, fermentative ability, flexibility, high mol% G + C value, and the ability to catabolize several carbohydrates. *Halomonas* can be quickly differentiated from other genera by its ability to survive exposure to, and to grow in, very high (20%) NaCl concentrations.

TAXONOMIC COMMENTS

The genus *Halomonas* currently contains 23 recognized, validly named species, which have been included because of the simi-

---

3. HSC medium is prepared as two separate solutions (A and B). Solution A contains NaCl or solar salt (25%, w/v); the pH is adjusted to 11 with 1 N NaOH, and "vitamin-free" casein (Difco) is added to a concentration of 15 g/l. This mixture is incubated at 30°C overnight to allow protein acidic groups to be exposed. The pH is then adjusted to 7.9 ± 0.1 with sterile NaOH or HCl. Solution B also contains 25% (w/v) solar salt or NaCl and is supplemented with yeast extract (0.2%), sodium citrate (0.6%), MgSO₄·7H₂O (4.0%), and ferric ammonium sulfate (0.01%). This solution is adjusted to pH 7.9 with 1 N NaOH and sterilized by autoclaving. Just before use, solutions A and B are mixed 1:1 (v/v). If a solid medium is to be used, agar (4.0%) is added to solution B prior to autoclaving.

4. MH medium contains 1% yeast extract, 0.5% Proteose Peptone no. 3 (Difco), and 0.1% glucose. It may be supplemented with synthetic sea salts. The concentrations of sea salts that have been used include: 0.5, 5.0, 10, 20, and 25% (w/v). The pH of this medium is usually adjusted to 7.2 prior to sterilization. Note that the salt concentrations listed for this medium do not represent the concentration of individual salts, such as NaCl, KCl, and MgCl₂; consequently, media made with sea salts obtained from different suppliers may differ in terms of the actual ion contents. Given the tolerance of the various *Halomonas* species, however, this should not be a problem.

5. AOL medium (Franzmann et al., 1987) is a synthetic medium that mimics the Antarctic hypersaline lake from which several species of *Halomonas* have been isolated. It consists of NaCl, 80.0 g; MgSO₄·7H₂O, 9.5 g; KCl, 5.0 g; CaCl₂·2H₂O, 0.2 g; (NH₄)₂SO₄, 0.1 g; KNO₃, 0.1 g; and yeast extract, 1.0 g; suspended in 960 ml of distilled water. The pH is adjusted to 7.0 and the medium is sterilized by autoclaving. When the solution cools to 60°C, 20 ml of Huntner mineral base solution (Atlas, 1997) is added, followed by 1.0 ml of a filter-sterilized vitamin solution containing (mg/100 ml): cyanocobalamin, 10.0; biotin, 2.0; thiamine hydrochloride, 10.0; calcium pantothenate, 5.0; folic acid, 2.0; nicotinamide, 5.0; and pyridoxine hydrochloride, 10.0. The medium is then completed by adding 20.0 ml of a sterile phosphate solution containing 0.25 g K₂HPO₄ and 0.25 g KH₂PO₄ per 100 ml of distilled water. The medium may be solidified with 15 g/l agar. An AOL-peptone medium may also be made by adding 5.0 g/l peptone to AOL medium.

TABLE BXII.γ.95. Carbon sources utilized by *Halomonas* species[a]

Characteristic	1. *H. elongata*	2. *H. aquamarina*	3. *H. campisalis*	5. *H. cupida*	6. *H. desiderata*	7. *H. eurihalina*	8. *H. halmophila*	9. *H. halodenitrificans*	10. *H. halodurans*	11. *H. halophila*	13. *H. magadii*	14. *H. marina*	15. *H. marisflavae*	16. *H. maura*	17. *H. meridiana*	18. *H. pacifica*	19. *H. pantelleriensis*	20. *H. salina*	21. *H. subglaciescola*	22. *H. variabilis*	23. *H. venusta*
L-Arabinose	+	−	−	+	+	+	−			+	+	−	+	+	d	−	−	d		−	−
Cellobiose	+	−		d	D	D				+	+		+		d	−	−	d		−	−
D-Glucose	+	+	+	+	+	+	+	+	+	+	+	+	+	+	+	+	+	d		−	+
Lactose	+	−	−	d		D	−	−	+	d	+	−	+		d	−	−	d		−	−
Maltose	+	d	+	−	+	+	+			+		−	+	+	+	+	+	d		−	d
Mannose	+	−	−	+	D	+	+		+	+	+	−	+	+	+	−	+	d		−	−
Sucrose	+	d	+	d	+	+	+			+	+	−	+	+	+	−	+	d		−	d
Acetate	+	+	+	+	+			−	+	+	+	−	+	+	+	+	+	+		+	+
Citrate	+	d		+	+			+	+	+	+	+	+		+	+	+	d		−	+
Gluconate	+	+		+	+	+		+	−	+	+	+	+	+	+	+	+	+		−	+
Lactate	+	+	+	+	+			+		+		+		+		+		d	+		+
Propionate	+	+		+		+	+			+		+		+	+	+	+	+	d		+
Succinate	+	+		+	+	+	−	+	+	+	+			+	+		+	d	+		+
L-Alanine	+	+		+	+			+	+	+	+	+		+				d	+	−	+
L-Arginine	+	+		+		+	+	−		+	+	d		+	+	+	+	d	+		+
L-Histidine	+	−		+		−	+	−	−	d	+	−		+	−	+		−	−		+
L-Isoleucine	+	+		d		+	+	+		d	+	d			+	−		d		−	d
L-Lysine	+	+		+		D	+	+		d	+	d			d	+		d	+		+
L-Serine	+	−		+		D	+	+	−	+	+	+		+	d	+	−	d	−		−
L-Tryptophan	−	−		−		+		−	+	d	+	−		−	+	−		d	−		d
Glycine	+	−		d		D	+	+			−				D	+			+	−	−
L-Leucine	+	d		+		+	+	−	+	d	+			+	+	+		d	−		+

[a]For symbols see standard definitions.

larity of their 16S rDNA sequences and the presence of respiratory ubiquinone-9. Although the genus is currently claimed to constitute a monophyletic group of organisms, it contains at least three main evolutionary branches. Subgroup 1 is composed of *H. halmophila*, *H. elongata* (the type species), *H. eurihalina*, *H. salina*, *H. halophila*, *H. pacifica*, *H. cupida*, and *H. halodenitrificans*. The 16S rDNA sequence similarities between pairs of these species range from 93.1 to 99.2%. The second subgroup contains both biovars of *H. subglaciescola* (ACAM 21 and ACAM 12), *H. halodurans*, *H. venusta*, *H. aquamarina*, *H. meridiana*, and *H. variabilis*, and the 16S rDNA sequence similarities among these species range from 94.4 to 100%. The third subgroup is composed of a single species, *H. marina*. When 16S rDNA sequences are used to distinguish the various phylogenetic subgroups within the genus, the situation improves only marginally. At the most restrictive similarity levels (98% and above), the genus seems to contain at least 10 individual phylogenetic groups (Fig. BXII.γ.115). Unfortunately, resolution of these subgroups deteriorates rapidly below this level. In fact, sequence homology levels existing between some members of the current genus are lower than those between members of the *Halomonas* and species currently listed in a different genus (Dobson and Franzmann, 1996).

Confusion about the current taxonomy of this group has arisen because the data that have been used to describe some of these individual species have been inconsistent. For instance, in the initial descriptions of *Halomonas subglaciescola* (Franzmann et al., 1987) and *Halomonas meridiana* (James et al., 1990a), data were used to compare these species to the type species *H. elongata*. Unfortunately, these two reports had only 55 kinds of characteristics in common, and within these common tests, 15 (27%) gave different results for *H. elongata*. Moreover, 17 of the 55 (31%) test results reported by James et al. (1990a) differed from

those reported by Franzmann et al. (1987) for the type strain of *H. subglaciescola*. There is presently no information on the degree to which such discrepancies would affect the taxonomic placement of the species involved.

A similar situation exists with the organisms that were originally part of genera such as *Deleya* and *Halovibrio*, and with the species "*Paracoccus halodenitrificans*", all of which were merged with the *Halomonas* based entirely upon 16S rDNA sequence similarities and a single chemotaxonomic feature. Very few of the phenetic characters originally used to describe these taxa are equivalent to those used to describe the rest of the *Halomonas* species. In fact, over 80 of the nearly 100 phenotypic descriptors for *Deleya* involve utilization of sole carbon sources, with virtually no information about the presence or absence of specific enzyme systems or abilities. Thus, it is difficult to provide a coherent and useable phenotypic description of the genus *Halomonas* and to provide reliable identifying features for the individual species.

At the chemotaxonomic level, the genus *Halomonas* represents a reasonably coherent group of bacteria; however, they do not possess specific or unique cell components that can be considered to be taxonomically distinct markers. All of the species possess ubiquinone-9 as their major respiratory quinone. When cells are grown on AOL with peptone, the major fatty acids of the *Halomonas* species that have been examined are $C_{16:1\ \omega7c}$, $C_{16:0}$, $C_{17:0\ cyclo}$, $C_{18:1}$, and $C_{19:0\ cyclo\ 11-12}$. The major polar lipids are phosphatidylethanolamine, phosphatidylglycerol, and an unidentified periodate–Schiff positive phospholipid (Franzmann and Tindall, 1990). Vreeland et al. (1984) have shown that the type strain *H. elongata* also contains cardiolipin. This particular lipid may be detected in increased concentration in cells grown in higher salt concentration.

Common molecular features among the species include the

TABLE BXII.γ.96. Additional characteristics of *Halomonas* species[a]

Characteristic	1. H. elongata	2. H. aquamarina	3. H. campisalis	5. H. cupida	6. H. desiderata	7. H. eurihalina	8. H. halmophila	9. H. halodenitrificans	10. H. halodurans	11. H. halophila	13. H. magadii	14. H. marina	15. H. marisflavae	16. H. maura	17. H. meridiana	18. H. pacifica	19. H. pantelleriensis	20. H. salina	21. H. subglaciescola	22. H. variabilis	23. H. venusta
Rods/curved rods	+	+	+	+	+	+	+		+	+	+	+	+	+	+	+	+	+	+	+	+
Cocci								+													
Gram reaction	−	−	−	−	−	−	−	−	−	−	−	−	−	−	−	−	−	−	−	−	−
Motile	+	+	+	+	+	+	−	−	+	+	+	+	+	−	+	+	+	−	+	+	+
Pigmentation[b]	C	C		C		C	Y	W	W	C	C	C	YO	C	W	C	CP	C	W	B	C
Growth conditions:																					
0% NaCl	+	−	−	−	+	D	−	−	+	−	+		−	−	d				−	−	
0.5% NaCl	+		+		+						+		+								
3.5% NaCl	+	+	+	+	+	+	+	+	+	+	+		+	+	+		+	+	+	+	+
8.0% NaCl	+	+	+	+	+	+	+	+	+	+	+		+	+	+		+	+	+	+	+
15% NaCl	+	+	+	+	+	+	+	+	+	+	+	+	+	+	+	−	+	+	+	+	+
25% NaCl	−	−	+	−	+	+	+		+	+	−		+		+		−	+	+	+	−
<10°C	−	−										+	+			+					
15°C	+		+	−	+	+		+	+	+	+	+	+	+	+	+	+	+	+	+	+
25°C	+	+	+	+	+	+	+	+	+	+	+	+	+	+	+	+	+	+	−	+	+
30°C	+	+	+	+	+	D	+	+	+	+	+	+	+	+	d	+	+	+		+	+
37°C	+	+	+	+	+	+			−	+	+	−	+	+	d	−	−	+	d	+	+
45°C	−	−	+	−	+				+		+	−	−	−	+	+	+	+		+	−
Oxidase present	+	+	+	+	+	+		+			+		−		+	−	−	d	+	+	+
Nitrate reduced to nitrite	+	+	+	+	−	−		+	+		−		−	−	−	+		+	−	−	+
Urease produced	d	d	−	−	−	+		−	−	d	−		−	−	+		+	+	−	+	+
Starch hydrolysis	−	+	+	−	−	d		−	−	−	d	−	−	−	d		−		−	−	−
mol% G + C content	60–61	57–58	66	60–63	66	59–65	62.9	64–66	63.2	66.7	62.2	62–64	59	64.1	58.8–59.1	67–68	65	60.7–64.2	60.9–62.9	57	52–54

[a]For symbols see standard definitions.

[b]B, light brown; C, cream-beige; CP, cream to pink; W, white; Y, yellow; YO, yellow-orange.

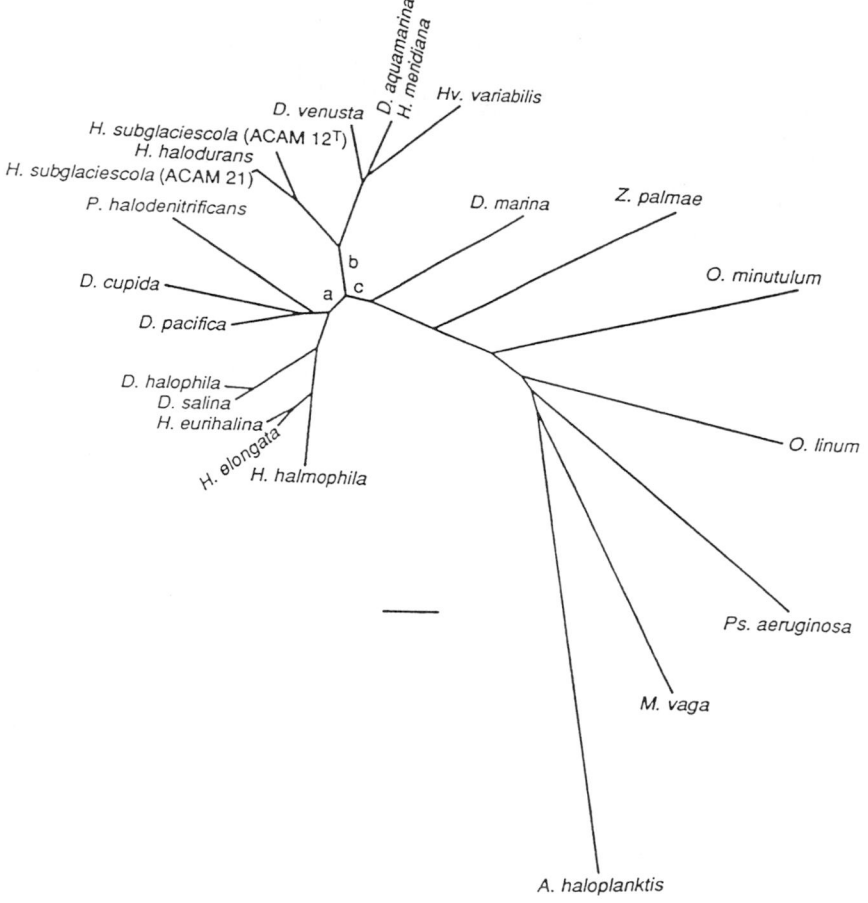

FIGURE BXII.γ.115. Distance tree containing the members of the genus *Halomonas* and other members of the class *Gammaproteobacteria*. Bar = 0.01 distance calculated by the Jukes–Cantor equation. Lower case letters a, b, and c refer to short internodes connecting the main branches of the monophyletic *Halomonas*. Abbreviations: *H.*, *Halomonas*; *O.*, *Oceanospirillum*; *Z.*, *Zymobacter*; *M.*, *Marinomonas*; *Ps.*, *Pseudomonas*; *A.*, *Alteromonas*. Redrawn and reprinted with permission from S. Dobson and P.D. Franzmann, International Journal of Systematic Bacteriology *46:* 550–558, 1996 ©International Union of Microbiological Societies.

presence of specific bases at 16 specific sites along the 16S rRNA molecule (Table BXII.γ.97). A taxonomic study of six species of *Halomonas* has recently been conducted by Huval et al. (1995). The DNA–DNA reassociation data presented in that article shows *Chromohalobacter israelensis* and *Chromohalobacter canadensis* possess 49% and 55% similarity, respectively, with the type species *Halomonas elongata*. *Halomonas halodurans* demonstrates DNA similarity of 28% with *H. elongata*. In contrast, *H. halophila* shows no detectable similarity with the type species of this genus (Table BXII.γ.98). Dobson et al. (1993) have shown that *H. aquamarina* shares a 41% DNA similarity with *H. venusta*, 16% similarity with *H. cupida*, and only 12% similarity with *H. pacifica* and *H. marina*. DNA–DNA hybridizations have not yet been conducted using any other species in the genus.

ACKNOWLEDGMENTS

The author wishes to express his appreciation to Drs. C.D. Litchfield and E.L. Martin for their continued support and assistance in the preparation of this manuscript. This review was prepared with the support of National Science Foundation grants EAR 9714203 and EAR0085371.

TABLE BXII.γ.97. 16S rRNA signatures of the genus *Halomonas*[a]

Position number	Sequence of nucleotide
76–93	6 bp stem
484	A
486	C
640	G
660-[b]	A-
745	U
668-	A-
738-	U-
669-	A-
737	U
776	U
1124	U
1297	U
1298	C
1423	A
1424	C
1439	U
1462	A
1464	C

[a]Data from Dobson and Franzmann (1996).

[b]Dashes indicate that the specified nucleotide base pairs with the next one listed.

TABLE BXII.γ.98. DNA–DNA similarities between the type strains of *Halomonas elongata, Chromohalobacter canadensis* and strains of related species[a]

Source target DNA	% Hybridization with radiolabeled DNA	
	H. elongata ATCC 33173[T]	*C. canadensis* NRCC 41227[T]
H. elongata ATCC 33173[T]	100	57
H. elongata ATCC 33174	72	49
H. halodurans ATCC 29868[T]	28	34
H. halophila CCM 662[T]	0	7
C. canadensis NRCC 41227[T]	55	100
C. israelensis ATCC 43985[T]	49	20

[a]Data from Huval et al. (1995).

FURTHER READING

Dobson, S.J. and P.D. Franzmann. 1996. Unification of the genera *Deleya* (Baumann et al. 1983), *Halomonas* (Vreeland et al. 1980), and *Halovibrio* (Fendrich 1988) and the species *Paracoccus halodenitrificans* (Robinson and Gibbons 1952) into a single genus, *Halomonas*, and placement of the genus *Zymobacter* in the family *Halomonadaceae*. Int. J. Syst. Bacteriol. *46*: 550–558.

Vreeland, R.H., C.D. Litchfield, E.L. Martin and E. Elliot. 1980. *Halomonas elongata*, a new genus and species of extremely salt-tolerant bacteria. Int. J. Syst. Bacteriol. *30*: 485–495.

Vreeland, R.H. 1991. The Family *Halomonadaceae. In* Balows, Trüper, Dworkin, Harder and Schleifer (Editors), The Prokaryotes, 2nd edition, Springer-Verlag, New York. pp. 3181–3188.

DIFFERENTIATION OF THE SPECIES OF THE GENUS *HALOMONAS*

Characteristics differentiating the species of the genus *Halomonas* are listed in Table BXII.γ.96.

List of species of the genus Halomonas

1. **Halomonas elongata** Vreeland, Litchfield, Martin and Elliot 1980, 495[VP]

e.lon′ga.ta. L. fem. part. adj. *elongata* elongated, stretched out.

The characteristics are as described for the genus and as listed in Tables BXII.γ.95 and BXII.γ.96. Morphological features are shown in Figs. BXII.γ.116 and BXII.γ.117. Additional characteristics of the species include the ability to grow at 0.5% and 30% NaCl. Grows at pH 5.0–9.0 in complex medium with 8% NaCl at 30–37°C. Grows at 15°C. Catalase, phosphatase, lysine decarboxylase, β-galactoside, and indole are produced. Some strains may be positive for esculin hydrolysis. Hydrogen sulfide negative. Does not possess DNase or phenylalanine deaminase. Hydrolyzes Tween 20, but not Tween 80. This species has been reported to produce acid from fructose, D-glucose, lactose, D-maltose, and sucrose, but not from D-galactose, inulin, or D-mannitol. In addition to those listed in Table BXII.γ.95, the following compounds serve as sole carbon sources: esculin, D-galactose, L-rhamnose, D-salicin, D-xylose, fumarate, pyruvate, β-hydroxybutyrate, L-aspartic acid, L-glutamic acid, L-ornithine, L-valine, proline, L-asparagine, and tyrosine. The species does not utilize inulin, raffinose, trehalose, malonate, L-tryptophan, L-threonine, or methionine as sole carbon sources. Sensitive to gentamicin (10 μg), mercuric chloride (1:5000), neomycin (30 μg), streptomycin (10 μg), and

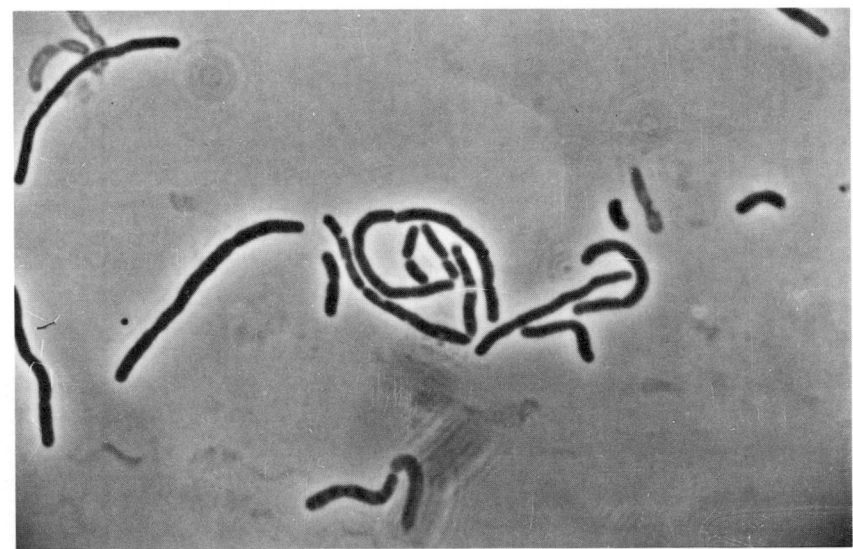

FIGURE BXII.γ.116. Phase contrast micrograph of *Halomonas elongata* strain 1H9 (ATCC 33173), showing both long and short cell forms (× 2000).

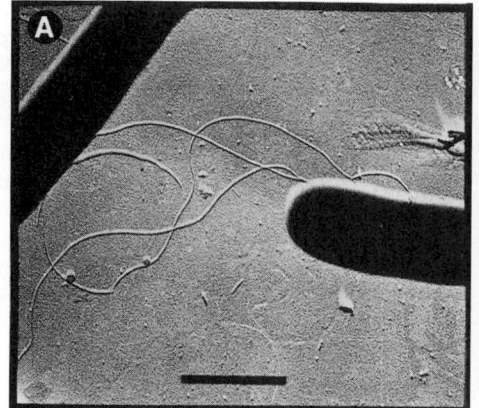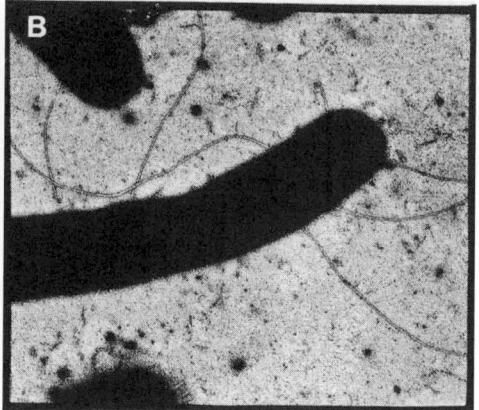

FIGURE BXII.γ.117. Electron micrographs of *Halomonas elongata*, showing flagellar arrangement. *A*, strain 1H9 (ATCC 33173); carbon shadowed. *B*, strain 1H15 (ATCC 33174); negatively stained (Bar = 1.0 μm).

chloramphenicol (30 μg). Resistant to ampicillin (10 μg), bacitracin (10 i.u.), cephalothin (30 μg), erythromycin (15 μg), nalidixic acid (30 μg), novobiocin (20 μg), vibriostat O/129 (10 μg), penicillin G (10 i.u.), and tetracycline (30 μg).

Sodium salts, such as sodium glutamate, $NaNO_3^-$, and NaBr, can substitute for NaCl. Na_2SO_3 and NaI cannot. The chloride salts of magnesium, lithium, potassium, and ammonia cannot substitute for NaCl in growth media. Possesses peptidoglycan. Based upon amino acid analyses, the peptidoglycan contains leucine in a crossbridge (Vreeland et al., 1984). Present in solar salterns, estuaries, salt marshes, and other environments containing NaCl. This species originally contained two biovars (ATCC 33174 = 1H15 [Vreeland et al., 1980] and DSM 3043 = 1H11 [Vreeland et al., 1980]). Both biovars have now been reclassified as *Chromohalobacter salexigens* (Arahal et al., 2001a).

The mol% G + C of the DNA is: 60.5 ± 0.5 (Bd, T_m).

Type strain: 1H9 of Vreeland et al., 1980, ATCC 33173, DSM 2581.

GenBank accession number (16S rRNA): M93355, X67023.

2. **Halomonas aquamarina** (ZoBell and Upham 1944) Dobson and Franzmann 1996, 556^{VP} (*Alcaligenes aquamarinus* (ZoBell and Upham 1944) Hendrie, Holding and Shewan 1974, 537; *Deleya aquamarina* (ZoBell and Upham 1944) Akagawa and Yamasato 1989, 466; "*Achromobacter aquamarinus*" ZoBell and Upham 1944, 264.)

a.qua.ma.ri'na. L. n. *aqua* water; L. adj. *marina* of the sea; L.adj. *aquamarina* pertaining to seawater.

The characteristics are as described for the genus and as listed in Tables BXII.γ.95 and BXII.γ.96. This species may or may not produce H$_2$S and phenylalanine deaminase. The species does not hydrolyze esculin or produce DNase or β-galactosidase. This species utilizes malate, pyruvate, β-hydroxybutyrate, caprylate, L-aspartic acid, L-glutamic acid, and proline as sole sources of carbon. The species also utilizes butyrate, isobutyrate, valerate, isovalerate, caproate, heptanoate, adipate, pimelate, azelate, sebacate, levulinate, erythritol, *meso*-inositol, *n*-propanol, *n*-butanol, isobutanol, cytosine, and uracil. This is the only *Halomonas* species known to hydrolyze starch (Table BXII.γ.96) and utilize suberate. Accumulates polyhydroxyalkanoates. Some strains

of the species may utilize D-galactose, trehalose, L-ornithine, and L-valine. The species does not grow on inulin, L-rhamnose, D-salicin, D-xylose, formate, malonate, oxalacetate, D-tartrate, benzoate, L-threonine, or tyrosine as the sole carbon source. There are no data available about the sensitivity of this species to antimicrobials. Isolated from the Pacific Ocean. The similarity of the 16S rRNA sequence of this species to the other *Halomonas* species ranges from 93.6–100%.

The mol% G + C of the DNA is: 60.5 ± 0.5 (Bd, T_m).

Type strain: ATCC 14400, DSM 30161, IAM 12550, NCMB 557.

GenBank accession number (16S rRNA): M93352.

3. **Halomonas campisalis** Mormile, Romine, Garcia, Ventosa, Bailey and Peyton 2000, 949^{VP} (Effective publication: Mormile, Romine, Garcia, Ventosa, Bailey and Peyton 1999, 556.)

cam.pi.sa'lis. L. masc. n. *campus* plain, field; L. masc. n. *sal* salt; gen. *salis* of salt; L. masc. n. *campisalis* of the plain of salt, of the salt plain.

General characteristics are as described for the genus and are presented in Tables BXII.γ.95 and BXII.γ.96. Additional characteristics include the ability to grow at temperatures from 4°C to 50°C. Optimal growth occurs at 30°C. This species reproduces at pH between 6.0 and 11.0 with optimal growth at pH 9.5. The optimal salt concentration is 1.5 M with a growth range from 0.2 to 4.5 M. In addition to those listed in Table BXII.γ.95, the species utilizes *n*-acetyl-glucosamine, ethanol, D(−)-fructose, D(+)-glucosamine, glycerol, pyruvate, and yeast extract. The species does not utilize benzoic acid, D(+)-galactose, D(+)-mannose, methanol, pectin, D(−)-ribose, L-rhamnose, L(−)-sorbose, whey, and D(+)-xylose. The organism does not grow fermentatively. The species gives negative tests for arginine dihydrolase, phenylalanine deaminase, indole production, and Voges–Proskauer.

This species is susceptible to ampicillin, chloramphenicol, and rifampin. It is resistant to kanamycin, neomycin, and streptomycin. The concentrations of antibiotics used were not provided in the original description.

The species was isolated from a soil sample just below the crystalline salt surface of a salt flat south of Alkali Lake in eastern Washington State, USA.

The mol% G + C of the DNA is: 66 (T_m).

Type strain: 4A, ATCC 700597.

GenBank accession number (16S rRNA): AFO54286.

4. **Halomonas cupida** (Baumann, Baumann, Mandel and Allen 1972) Dobson and Franzmann 1996, 556[VP] (*Deleya cupida* (Baumann, Baumann, Mandel and Allen 1972) Baumann, Bowditch and Baumann 1983a, 801; *Alcaligenes cupidus* Baumann, Baumann, Mandel and Allen 1972, 426.)

cu′pi.da. L. adj. *cupidus* desiring.

The characteristics are as described for the genus and as listed in Tables BXII.γ.95 and BXII.γ.96. This species possesses peritrichous flagella; it has been described as utilizing 69–85 organic compounds, possibly more than any other *Halomonas* species. In addition to those listed, the species also utilizes D-galactose, D-xylose, fumarate, malate, pyruvate, β-hydroxybutyrate, benzoate, caprylate, L-aspartic acid, L-glutamic acid, proline, tyrosine, melibiose, saccharate, galacturonate, glucuronate, N-acetylglucosamine, butyrate, isobutyrate, valerate, isovalerate, caproate, heptanoate, pelargonate, caprate, glycolate, DL-glycerate, aconitate, erythritol, *meso*-inositol, n-propanol, isobutanol, benzoylformate, p-hydroxybenzoate (but not the *ortho* or *meta* isomers), putrescine, spermine, betaine, sarcosine, hippurate, and allantoin. The species may or may not grow on L-rhamnose, D-salicin, D-trehalose, malonate, and L-ornithine. This organism does not use formate, oxalate, D-tartrate, L-valine, or L-threonine as the sole source of carbon. The similarity of the 16S rDNA of this species to other *Halomonas* species ranges from 92.6–96.5%. Isolated from marine habitats.

The mol% G + C of the DNA is: 60–63 (Bd).

Type strain: ATCC 27124, DSM 4740.

GenBank accession number (16S rRNA): L42615.

5. **Halomonas desiderata** Berendes, Gottschalk, Heine-Dobbernack, Moore and Tindall 1997, 242[VP] (Effective publication: Berendes, Gottschalk, Heine-Dobbernack, Moore and Tindall 1996,165.)

de.si.de.ra′ta. L. part. adj. *desiderata* wished for, the strain wished for.

Cells are motile with peritrichous flagella and are 0.4–0.6 × 1.0–2.6 μm. The colony type has not been described, but the species does not produce pigments. Requires sodium ions. Optimum pH is 9.7 and the optimum temperature for growth is 37–42°C. The characteristics are as described for the genus and as listed in Tables BXII.γ.95 and BXII.γ.96.

Facultatively anaerobic in the presence of nitrate. Obligately alkaliphilic; optimum pH 9.5. Grows at 45°C. Denitrifying. Does not grow below pH 7 or above pH 11. Does not grow at 50°C. Supplemental vitamins or amino acids are not needed. Catalase positive and phenylalanine deaminase positive. Arginine dihydrolase negative. Does not hydrolyze starch, gelatin, casein, Tween 80, Tween 60, hippurate, or pullulan. Produces poly-β-hydroxybutyrate as a reserve material but does not produce levan from sucrose. Grows on MacConkey agar, *Salmonella–Shigella* agar, and Cetrimide agar. At pH 9.7, the species is resistant to up to 1% bile salts.

Utilizes D-xylose, ribose, trehalose, glycerol, DL-β-hydroxybutyrate, and benzoate as sole sources of carbon. Does not utilize inositol. Not fermentative; indole and Voges–

Proskauer negative. The major respiratory quinone is ubiquinone-9. The major fatty acids are $C_{18:1}$ (52–65% depending on the strain), $C_{16:0}$ (13–23% depending on the strain), and $C_{16:1 \omega7c}$. Other fatty acids detected in various strains include $C_{10:0}$, $C_{10:0 3OH}$, $C_{12:0}$, $C_{12:0 3OH}$, $C_{14:0}$, $C_{17:0 cyclo}$, $C_{18:0}$, and $C_{19:0 cyclo}$. The major phospholipids are phosphatidylglycerol, phosphatidylethanolamine, diphosphatidylglycerol, and an uncharacterized phosphoglycolipid. The similarity of the 16S rDNA of this species to other *Halomonas* species ranges from 89.7–94.8%. DNA–DNA hybridization data show similarities below 60–70%. Isolated from municipal sewage.

The mol% G + C of the DNA is: 66 (HPLC).

Type strain: DSM 9502.

GenBank accession number (16S rRNA): X92417.

6. **Halomonas eurihalina** (Quesada, Valderrama, Bejar, Ventosa, Gutierrez, Ruiz-Berraquero and Ramos-Cormenzana 1990) Mellado, Moore, Nieto and Ventosa 1995b, 715[VP] (*Volcaniella eurihalina* Quesada, Valderrama, Bejar, Ventosa, Gutierrez, Ruiz-Berraquero and Ramos-Cormenzana 1990, 265.)

eu.ri.ha.li′ na. Gr. adj. *euris* wide, broad; Gr. adj. *halinos* salted; M.L. adj. *eurihalina* growing at a wide range of salt concentrations.

The characteristics are as described for the genus and as listed in Tables BXII.γ.95 and BXII.γ.96. Colonies grown on MH medium with 7.5% NaCl at 32°C for 72 h are 4–5 mm in diameter, circular, convex, smooth, opaque, cream-colored with entire margins, and very mucoid. This species requires Na$^+$ ions. Sodium can be supplied as NaCl, Na_2SO_4, or NaBr. Under optimum conditions, the species grows at pH 5.0–9.0. Catalase positive. Phosphatase, lysine decarboxylase, and β-galactosidase negative. Does not hydrolyze esculin. Produces acid from fructose and maltose. Some strains may also produce acid from D-galactose, D-glucose, lactose, mannitol, and sucrose. Utilizes fumarate, β-hydroxybutyrate, L-valine, L-threonine, proline, and tyrosine as sole carbon sources. Some strains also utilize D-galactose, raffinose, D-salicin, D-trehalose, L-ornithine, and methionine as sole carbon sources. This species hydrolyzes gelatin and grows on KCN, MacConkey, and Cetrimide agar. Some strains of this species produce H$_2$S and hydrolyze Tween 20. Sensitive to chloramphenicol (30 μg). Some strains may be sensitive to gentamicin (10 μg) and neomycin (30 μg). Resistant to most other antimicrobials. Isolated from hypersaline habitats (soils and salterns) and from ocean water.

The mol% G + C of the DNA is: 59.1–65.7 (T_m).

Type strain: ATCC 49336, DSM 5720.

GenBank accession number (16S rRNA): L42620, X87218.

7. **Halomonas halmophila** (Elazari-Volcani 1940) Franzmann, Wehmeyer and Stackebrandt 1989, 205[VP] (Effective publication: Franzmann, Wehmeyer and Stackebrandt 1988, 19) emend. Dobson, James, Franzmann and McMeekin 1990, 462 (*Flavobacterium halmophilum* Elazari-Volcani 1940, 85.)

hal.mo.phi′ lum. Gr. n. *halme* brine, seawater; Gr. adj. *philos* loving; M.L. n. adj. *halmophilum* seawater loving.

The characteristics are as described for the genus and as listed in Tables BXII.γ.95 and BXII.γ.96. Does not produce acid from sugars, D-mannitol, or inulin and does not produce indole. The species is sensitive to nalidixic acid

(30 μg) and resistant to ampicillin (10 μg), gentamicin (10 μg), neomycin (30 μg), streptomycin (10 μg), and tetracycline (30 μg). This species was transferred into the genus *Halomonas* when 16S rRNA cataloging showed an S_{AB} value of 0.66 to the type species of the genus *Halomonas*. The similarity of the 16S rDNA of this species to other *Halomonas* species ranges from 92.2–97.9%. The mol% G + C content of the DNA has been reported to be 49.7. The exact site from which this bacterium was isolated is not known, but was probably somewhere in the Dead Sea.

The mol% G + C of the DNA is: 49.7 (method unknown).
Type strain: ATCC 19717, DSM 5349.
GenBank accession number (16S rRNA): M59153.

The name of this species has been incorrectly spelled as *Flavobacterium halmephilum* (Weeks, 1974) and as *Flavobacterium halmephilium* (Weeks, 1974).

8. **Halomonas halodenitrificans** (Robinson and Gibbons 1952) Dobson and Franzmann 1996, 556^VP (*Paracoccus halodenitrificans* (Robinson and Gibbons 1952); *Micrococcus halodenitrificans* Robinson and Gibbons 1952, 154.)

ha.lo.de.ni.tri' fi.cans. Gr. n. *hals, halis* salt; M.L. v. *denitrifico* to denitrify; M.L. part. adj. *halodenitrificans* salt (requiring) denitrifying.

This is the only member of this genus that has a coccoid morphology. The characteristics are as described for the genus and as listed in Tables BXII.γ.95 and BXII.γ.96. Most of the characteristics have been taken from the description of Kocur (1984). When grown on nutrient agar containing 6% NaCl, the colonies of this species are described as circular, entire, convex, butyrous, glistening, cream-colored, and opaque. The optimum growth temperature is 25–30°C; the growth range is 5–32°C. The species cannot grow in media containing less than 3% NaCl. Optimum growth occurs with 4.4–8.8% NaCl; slow growth occurs with up to 20% NaCl in media. Catalase positive; phosphatase is present. Hydrogen sulfide is not produced. Esculin is not hydrolyzed. Lysine decarboxylase, phenylalanine deaminase, and β-galactosidase negative. Indole and Voges–Proskauer negative. Does not hydrolyze Tween 80. Acid is not produced from either D-glucose or sorbitol. Acid production from other sugars or polyols has not been tested. L-valine, L-threonine, proline, and L-asparagine serve as sole carbon sources. Does not utilize L-aspartic acid, L-glutamic acid, or methionine as sole carbon sources. Sensitive to mercuric chloride and polymyxin B, although no concentrations have been specified. Resistant to gentamicin (10 μg), vibriostat O/129 (10 μg), penicillin G (10 i.u.), streptomycin (10 μg), tetracycline (30 μg), and chloramphenicol (30 μg). The similarity of the 16S rDNA of this species to each member of the genus *Halomonas* ranges from 91.5–95.6%. Isolated from meat-curing brines.

The mol% G + C of the DNA is: 64–66 (T_m, Bd).
Type strain: ATCC 13511, CCM 286, DSM 735, NCMB 700.
GenBank accession number (16S rRNA): L04942.

9. **Halomonas halodurans** Hebert and Vreeland 1987, 350^VP
ha.lo.du' rans. Gr. n. *hals, halos*; the sea, salt; L. v. *durare* to last endure; M.L. masc. adj. *Halodurans* salt tolerating.

The characteristics are as described for the genus and as listed in Tables BXII.γ.95 and BXII.γ.96. The original description of this species was based on tests using media supplemented with 2.6% NaCl. Later descriptions of this species by Hebert and Vreeland (1987) were performed at the optimum NaCl concentration for growth of the species (8.0%). Hebert and Vreeland (1987) noted several differences in the results of the phenotypic testing and attributed them to the differences in the salt content of the media. The biochemical properties listed in Tables BXII.γ.95 and BXII.γ.96 for this species are those corresponding to tests conducted with media containing 8% NaCl.

The cells are motile by single polar flagellum that is 8–10 μm in length and 12–14 nm wide. Colonies are circular, convex, smooth, entire, 2–3 mm in diameter, and nonpigmented following 48 h of growth at 20–35°C. This species does not grow in 30% NaCl. Grows at pH 6.0–8.0 in 8% NaCl.

Catalase positive; phosphatase, and lysine decarboxylase are present. Phenylalanine deaminase and β-galactosidase are absent. Hydrolyzes esculin. Tolerant to KCN. Hydrolyzes Tween 80. Hydrogen sulfide, indole, and acetyl-methyl carbinol (Voges–Proskauer) are not produced. Produces acid from D-fructose, D-galactose, D-glucose, lactose, salicin, sucrose, and D-xylose. Does not produce acid from D-mannitol, D-mannose, or L-rhamnose. Utilizes D-galactose, inulin, ribose, malonate, pyruvate, benzoate, L-ornithine, L-threonine, proline, and L-asparagine as sole sources of carbon. The species cannot use L-rhamnose, malate, β-hydroxybutyrate, or methionine as sole carbon sources. Sensitive to erythromycin (15 μg), nalidixic acid, (30 μg), streptomycin (10 μg), tetracycline (30 μg), and polymixin B (concentration unknown). The species is resistant to vibriostat O/129 (10 μg), penicillin G (10 i.u.), and chloramphenicol (30 μg). The similarity of the 16S rDNA of this species to other species of *Halomonas* ranges from 93.7–99.9%. The level of DNA–DNA homology between this species *H. elongata* and *H. canadensis* is shown in Table BXII.γ.98. Isolated from Great Bay Estuary (New Hampshire, USA).

The mol% G + C of the DNA is: 63.2 ± 1.1 (T_m, Bd).
Type strain: ATCC 29686, DSM 5160.
GenBank accession number (16S rRNA): L42619.

10. **Halomonas halophila** (Quesada, Ventosa, Ruiz-Berraquero and Ramos-Cormenzana 1984) Dobson and Franzmann 1996, 556^VP (*Deleya halophila* Quesada, Ventosa, Ruiz-Berraquero and Ramos-Cormenzana 1984, 290.)

ha.lo.phi' la. Gr. n. *halos* salt; Gr. adj. *philus* loving; M.L. adj. *halophila* salt loving.

The characteristics are as described for the genus and as listed in Tables BXII.γ.95 and BXII.γ.96. When grown with 7.5% sea salt for 4 days at 32°C, the colonies are circular, low convex, smooth, opaque, and cream-colored. Grows in 30% NaCl and at pH 5.0–9.0. Tests positive for catalase and phosphatase. DNase and phenylalanine deaminase negative. Produces hydrogen sulfide and hydrolyzes esculin, but does not hydrolyze Tween 80. Indole and Voges–Proskauer negative. Produces acid from D-galactose, D-glucose, and D-maltose. Some strains may also produce acid from sucrose, D-trehalose, and D-xylose. Acid is not produced from lactose and mannitol. In addition to the carbon sources listed in Table BXII.γ.95, this species also utilizes D-galactose, D-salicin, D-trehalose, D-xylose, fumarate, malate, pyruvate, L-aspartic acid, and L-glutamic acid as sole carbon sources. Some strains may also use L-ornithine. The species does not utilize L-rhamnose, hippurate, D-tartrate, benzoate, caprylate, or L-valine as sole carbon sources. The species is sensitive to ampicillin (10 μg), cephalothin (30 μg), erythromycin (15 μg), chloramphenicol (30 μg), polymyxin B, and rifampin A (no concentrations have been specified for the latter two anti-microbials). Resistant to bacitracin (10 i.u.) and lincomycin (concentration unknown). The similarity of the 16S rDNA of this species to each member of the *Halomonas* ranges from 93.8–99.2%. Isolated from saline soils.

The mol% G + C of the DNA is: 66.7 (T_m).

Type strain: CCM3662, DSM 4770.

GenBank accession number (16S rRNA): L42619 and M93353.

11. **Halomonas magadii** Duckworth, Grant, Jones, Meijer, Márquez and Ventosa 2000b, 1415[VP] (Effective publication: Duckworth, Grant, Jones, Meijer, Márquez and Ventosa 2000a, 59.)

ma.ga' di.i. M.L. gen. n. *magadi* of Magadi, named for lake Magadi, a saline soda lake in Kenya.

The characteristics are as described for the genus and are listed in Tables BXII.γ.95 and BXII.γ.96. Other descriptive characters are listed below. Rods 4.0–6.0 × 0.6–0.8 μm. Catalase positive, produces circular, low convex opaque cream-colored colonies. Alkaliphilic, growth occurs between pH 7.0 and 11.0 with an optimum around 9.5. Grows at salt concentrations from 0% to 20% between 25°C and 40°C with an optimum at 37°C. Phosphatase and phenylalanine deaminase negative. Does not hydrolyze gelatin or casein. Indole negative. No anaerobic growth in the presence or absence of nitrate. Produces H_2S from cysteine. When grown in alkaline medium, the species utilizes amygdalin, esculin, D-fructose, D-fucose, D-galactose, D-melibiose, L-raffinose, D-ribose, salicin, starch, trehalose, D-xylose, adonitol, D-glucuronolctone, D-glucosamine, dulcitol, erythritol, ethanol, glycerol, D-mannitol, *m*-inositol, propanol, D-sorbitol, *N*-acetylglucosamine, propionide, quinate, D-saccharate, salicylate, suberate, tartrate, L-asparagine, DL-aspartic acid, DL-phenylalanine, L-glutamine, α-aminovalerate, aconitate, α-ketoglutarate, butyrate, caprylate, fumarate, DL-glycerate, D-glucuronate, glutamate, *p*-hydroxybenzoate, hippurate, DL-malate, malonate, oxalate, L-ornithine, L-leucine, L-methionine, L-proline, L-threonine, L-valine, and pyruvate. Sarcosine, ethionine, creatine, DL-aminobutyrate, Tween 80, inulin, and L-rhamnose are not used. Sensitive to erythromycin (5 mg), streptomycin (10 mg), polymyxin (300 IU), sulphafurazole (100 mg), chloramphenicol (25 mg), and oleandomycin (5 mg). Not sensitive to gentamicin (10 mg), kanamycin (30 mg), fusidic acid (10 mg), tetracycline (25 mg), rifampicin (2 mg), bacitracin (10 IU), neomycin (30 mg), novobiocin (5 mg), and vancomycin (30 mg). Listed as insensitive to sulphamethoxazole, nitrofurantoin, trimethoprim, ampicillin, penicillin G, and methicillin, but no concentrations were specified.

The major polar lipids are phosphatidylglycerol, phosphatidyl glycerol phosphate, diphosphatidylglycerol, phosphatidylethanolamine, and a small amount of an unidentified glycolipid. Cells also contain ubiquinones 6 and 9.

The mol% G + C of the DNA is: 62 (T_m).

Type strain: 21M1, NCIMB 13595.

GenBank accession number (16S rRNA): X92150.

12. **Halomonas marina** (Cobet, Wirsen and Jones 1970) Dobson and Franzmann 1996, 556[VP] (*Deleya marina* (Cobet, Wirsen and Jones 1970) Baumann, Bowditch and Baumann 1983a,801; *Pseudomonas marina* (Cobet, Wirsen and Jones 1970) Baumann, Baumann, Mandel and Allen 1972, 423; "*Arthrobacter marinus*" Cobet, Wirsen and Jones 1970, 159.) *ma.ri' na.* L. adj. *marina* of the sea.

The characteristics are as described for the genus and as listed in Tables BXII.γ.95 and BXII.γ.96. The cells possesses polar flagella. Baumann et al. (1983a) note that this species has flagella that tend to bend back toward the cell, giving arrangements that appear to be lateral or peritrichous. The species has been described as utilizing 35–47 organic compounds as sole sources of carbon. In addition to those listed in Table BXII.γ.95, this species utilizes D-galactose, fumarate, malate, pyruvate, β-hydroxybutyrate, L-aspartic acid, L-glutamic acid, L-proline, butyrate, valerate, isovalerate, caproate, heptanoate, pelargonate, caprate, DL-glycerate, α-ketoglutarate, aconitate, *n*-butanol, sarcosine, and creatine. Some strains may also utilize malonate and L-ornithine as sole carbon sources. The species does not use L-rhamnose, ribose, D-salicin, D-trehalose, D-xylose, hippurate, D-tartrate, benzoate, L-valine, or L-threonine. The similarity of the 16S rDNA of this species to other *Halomonas* species is only 92.3–94.8%. Isolated from marine habitats.

The mol% G + C of the DNA is: 63 (Bd).

Type strain: C 25374, DSM 4741.

GenBank accession number (16S rRNA): M93354.

13. **Halomonas marisflavae** Yoon, Choi, Lee, Kho, Kang, and Park 2001a, 1176[VP]

ma.ris.fla' vae. L. gen. neut. n. *maris* of the sea; L. neut. adj. *flavum* yellow; L. gen. neut. n. *marisflavae* of the Yellow Sea, Korea.

The characteristics are as described for the genus and are listed in Tables BXII.γ.95 and BXII.γ.96 The species produces rod or oval shaped cells that are 0.9–1.3 × 1.7–2.3 μm following 3 days growth at 28°C. Cells are motile with a single polar flagellum. Colonies are yellow-orange, smooth, glistening, circular, and slightly irregular on Tryptic Soy Agar (TSA) and on Marine Agar (MA). The colonies are described as being convex on TSA and concave on MA. Optimal growth occurs between 0.5% and 12% NaCl (w/v) and at 25–30°C. The optimal pH for growth is 7.0–8.0. Grows anaerobically on MA medium although the exact conditions for this characteristic were not specified. Catalase positive. Hydrolyzes esculin and gelatin, but not casein, hypoxanthine, Tween 80, tyrosine, or xanthine. Indole and arginine deaminase negative. Assimilates and produces acid from D-glucose, arabinose, mannose, mannitol, gluconate, malate, and citrate. The species is also described as producing acid from glycerol, erythritol, ribose, D-xylose, galactose, D-fructose, arbutin, salicin, maltose, melibiose, trehalose, gentiobiose, D-turanose, D-lyxose, and D-furanose. Also produces a weak acid reaction from adonitol, mannitol, and xylitol.

The predominant respiratory quinone is ubiquinone-9. The major fatty acids are $C_{18:1}$, $C_{16:0}$, $C_{16:1\ \omega7c}$ and/or $C_{15:0\ iso\ 2OH}$.

Isolated from the Yellow Sea, Korea.

The mol% G + C of the DNA is: 59 (HPLC).

Type strain: SW32, JCM 10873, KCCM 80003.

GenBank accession number (16S rRNA): AF251143.

14. **Halomonas maura** Bouchotroch, Quesada, Del Moral, Llamas and Bejar 2001, 1630[VP]

mau'ra. L. adj. *maurus* northwest African.

The characteristics are as described for the genus and are listed in Tables BXII.γ.95 and BXII.γ.96. This species produces one of the longest rods (6.0–9.0 × 0.5–0.7 μm) in the genus. It and *Halomonas eurihalina* are the only species known to produce large amounts of exopolysaccharide. Colonies are circular, convex, mucoid, cream-colored, and 2–3 mm in diameter after 24 hours growth at 32°C. Colonies expand to over 5 mm in diameter after 72 hours. Grows in salt concentrations between 1% and 15% with optimal growth occurring between 7.5 and 10% (w/v). The salt concentrations given actually reflect the percentage of total sea salts and not specifically NaCl. However, the original description of this species states that the organism is able to grow optimally in medium containing 1.2M NaCl (7.0% w/v) and 0.2M $MgSO_4 \cdot 7H_2O$ (4.5%). The description also states that $MgCl_2 \cdot 6H_2O$ may be substituted for magnesium sulfate. The species is capable of anaerobic growth in the presence of nitrate but not fumarate. Catalase positive. Does not produce acids from sugars. Reduces selenite and oxidizes gluconate. The species produces H_2S from peptones and hydrolyzes Tween 20. Indole, methyl-red, and Voges–Proskauer tests are negative. The species does not utilize esculin, tyrosine, DNA, gelatin, casein, lecithin, and Tween 80, nor does it hydrolyze blood, produce phenylalanine deaminase, grow on cetrimide agar or under anaerobic conditions on normal medium. In addition, the type strain of the species produces phosphatase and grows on MacConkey and KCN agars and is ONPG negative.

In addition to the compounds listed in Table BXII.γ.95, the species utilizes D-galactose, *myo*-inositol, and D-sorbitol as sole sources of carbon and energy. The species does not utilize formate, D-salicin, or L-sorbose. The type strain differs somewhat in these characters in that it also utilizes citrate, ethanol, fumarate, D-fructose, glycerol, D-rhamnose, and D-ribose. The type strain also does not utilize lactose or D-trehalose. The species also utilizes the amino acids listed in Table BXII.γ.95, as well as L-ornithine and L-valine, as sole sources of carbon and nitrogen. The organism does not utilize cysteine or L-tryptophan in this manner. As a whole, the species is susceptible to amoxycillin, ampicillin, cephalothin, cefoxitin, chloramphenicol, nalidixic acid, nitrofurantoin, polymixin, rifampicin, and trimethoprim-sulfamethoxazole. The species is resistant to tetracycline and Vibriostat O/129. The type strain is also sensitive to gentamicin, penicillin, sulfonamide, and tobramycin while being resistant to erythromycin. Unfortunately, no antibiotic concentrations were provided in the original species description. These data may be provided by Bouchotroch et al. (1999) which was cited by Bouchotroch et al. (2001).

The major fatty acids (up to 81% total) of the type strain are $C_{18:1 \ \omega7c}$, $C_{16:1 \ \omega7c}/C_{15:0 \ iso \ 2OH}$, and $C_{16:0}$. The bacterium produces exo-polysaccharide designated EPS S-31 which is composed of neutral sugars glucose, mannose, and galactose in a ratio of 1:4:2.5. EPS S-31 emulsifies crude oil.

The species was isolated from a solar saltern in Asilah Morocco.

The mol% G + C of the DNA is: 62.2–64.1 (T_m).

Type strain: S-31, CECT 5298, DSM13445.

GenBank accession number (16S rRNA): AJ271864.

15. **Halomonas meridiana** James, Dobson, Franzmann and McMeekin 1990b, 470[VP] (Effective publication: James, Dobson, Franzmann and McMeekin 1990a, 277.)

me.ri.di.a'na. L. adj. *meridiana* of the south.

The characteristics are as described for the genus and as listed in Tables BXII.γ.95 and BXII.γ.96. The colonies produced by this species are smooth, circular, convex, and white to off-white, with entire margins. Yellow colonies may be produced on media containing alanine or pyruvate. The cells may possess either lateral or polar flagella. The species grows in media of pH 6.0–9.0; some strains may grow at pH 5.0. Catalase positive. Test results for phosphatase, DNase, lysine decarboxylase, phenylalanine deaminase, and β-galactosidase are all negative. Most strains have been reported to hydrolyze Tween 80 and Tween 20. Produces acid from D-fructose, maltose, and sucrose. Some strains may produce acid from D-galactose, D-glucose, lactose, and mannitol. Utilizes D-xylose, fumarate, β-hydroxybutyrate, L-valine, proline, and L-asparagine as sole sources of carbon. Most strains also utilize D-galactose and D-rhamnose. Some strains may utilize raffinose, D-salicin, D-trehalose, L-ornithine, L-threonine, and methionine as sole carbon sources. Does not use malonate. Resistant to most antibiotics, including ampicillin (10 μg), bacitracin (10 i.u.), cephalothin (30 μg), erythromycin (10 μg), mercuric chloride 1:5000 (200 mg/l), nalidixic acid (30 μg), novobiocin (20 μg), vibriostat O/129 (10 μg), penicillin G (10 i.u.), streptomycin (10 μg), and tetracycline (30 μg). Some strains show sensitivity to gentamicin (10 μg), neomycin (30 μg), and chloramphenicol (30 μg). The similarity of the 16S rDNA of this species to each member of the *Halomonas* ranges from 93.6–100%. Isolated from Antarctic lakes.

The mol% G + C of the DNA is: 58.8–59.1 ± 0.8 (T_m).

Type strain: ACAM 246, ATCC 49692, DSM 5425, UQM 3352.

GenBank accession number (16S rRNA): M93356.

16. **Halomonas pacifica** (Baumann, Baumann, Mandel and Allen 1972) Dobson and Franzmann 1996, 556[VP] (*Deleya pacifica* (Baumann, Baumann, Mandel and Allen 1972) Baumann, Bowditch and Baumann 1983a, 801; *Alcaligenes pacificus* Baumann, Baumann, Mandel and Allen 1972, 426.)

pa.ci'fic.a. M.L. adj. *pacifica* pertaining to the Pacific ocean.

The characteristics are as described for the genus and as listed in Tables BXII.γ.95 and BXII.γ.96. In addition, the species has been shown to utilize fumarate, malate, pyruvate, β-hydroxybutyrate, benzoate, L-aspartic acid, L-glutamic acid, L-ornithine, proline, *N*-acetylglucosamine, butyrate, isobutyrate, valerate, isovalerate, caproate, heptanoate, glutarate, DL-glycerate, 2,3-butyleneglycol, benzoate, *p*-hydroxybenzoate (but not the *ortho* and *meta* isomers), phenylacetate, δ-aminovalerate, DL-kynurenine, kynurenate, anthranilate, benzylamine, putrescine, spermine, histamine, butylamine, betaine, sarcosine, and nicotinate as carbon sources. Some strains may use caprylate as well. The species does not use D-galactose, L-rhamnose, ribose, D-salicin, D-trehalose, D-xylose, malonate, D-tartrate, L-valine, or L-threonine as sole carbon sources. The similarity of the 16S rDNA of this species to each member of the *Halomonas* ranges from 93.2–96.6%. Isolated from marine habitats.

The mol% G + C of the DNA is: 67–68 (Bd).

Type strain: ATCC 27122, DSM 4742.

GenBank accession number (16S rRNA): L42616.

17. **Halomonas pantelleriensis** Romano, Nicolaus, Lama, Manca and Gambacorta 1997, 601[VP] (Effective publication: Romano, Nicolaus, Lama, Manca and Gambacorta 1996, 332.)

pan.tel.le.ri.en' sis. from the place of isolation, Pantelleria Island in the south of Sicily, Italy.

The characteristics are as described for the genus and as listed in Tables BXII.γ.95 and BXII.γ.96. Colonies are cream to pink, making this the only species of the genus with pink colonies. Colonies are smooth, granulate, entire, circular, raised, and 0.11 mm in diameter. The species produces PHB. In defined medium, the species requires biotin or thiamine (50 μg/l) in addition to the carbon sources listed in Table BXII.γ.95. Optimum growth occurs in media containing 10% NaCl, 33–35°C, pH 9.0. The species is alkalophilic and grows at a pH range of 7.5–11.0. Catalase and α-glucosidase positive; β-galactosidase negative. Decomposes tyrosine. Hydrolyzes hippurate, but does not attack casein, gelatin or L- or D-N'-benzoylarginine-p-nitroanilide. The species gives a weak reaction for phenylalanine deaminase and is not sensitive to lysozyme. In addition to the substrates listed in Table BXII.γ.95, the species also utilizes fructose, D-xylose, galactose, trehalose, salicin, sorbitol, glycerol, pyruvate, heptanoate, and valine. The organism cannot utilize raffinose, ethanol, or valerate. The species is sensitive to chloramphenicol (30 μg) and erythromycin (30 μg). It is insensitive to lincomycin (15 μg), gentamicin (10 μg), ampicillin (25 μg), tetracycline (50 μg), penicillin (10 μg), neomycin (30 μg), bacitracin (10 μg), and novobiocin (30 μg). The similarity of the 16S rDNA of this species to other members of the *Halomonas* ranges from 93.1–95.8%. This species represents a distinct phylogenetic line within the larger genus, and is most closely related to *H. marina* and *H. halodenitrificans*. Isolated from hard sand on the island of Pantelleria, south of Sicily, Italy.

The mol% G + C of the DNA is: 65.02 (HPLC).

Type strain: ATCC 700273, DSM 9661.

GenBank accession number (16S rRNA): X93493.

18. **Halomonas salina** (Valderrama, Quesada, Bejar, Ventosa, Gutierrez, Ruiz-Berraquero and Ramos-Cormenzana 1991) Dobson and Franzmann 1996, 556[VP] (*Delaya salina* Valderrama, Quesada, Bejar, Ventosa, Gutierrez, Ruiz-Berraquero and Ramos-Cormanzana 1991, 382.)

sa.li' na. L. adj. *salina* salted, saline.

The characteristics are as described for the genus and as listed in Tables BXII.γ.95 and BXII.γ.96. Of all of the *Halomonas* species, *H. salina* is currently the easiest to distinguish at the phenotypic level. The species has several characteristics that are not shared by the other halomonads. It forms coccoid cells when grown in low salts, but produces typical elongated filaments when grown in optimum (7.5% total sea salts) and higher salts, but not in 30% salts. The cells are nonmotile, produce a capsule, and accumulate poly-β-hydroxybutyrate. The colonies produced after 48 h at 32°C on MH medium are small (1–2 mm in diameter), convex, smooth, glossy, and translucent yellow to cream and have entire margins. This is also the only *Halomonas* species known to reduce selenite and to require Mg^{2+}, in addition to Na^+. Unlike other *Halomonas* species, *H. salina* can only use cations supplied as either Cl^- or SO_4^{2-} salts. Members of the species grow at pH 6.0–9.0. Some strains also grow

at pH 5.0. Catalase positive. Some strains also possess DNase, phenylalanine deaminase, and β-galactosidase. Some strains produce hydrogen sulfide, are KCN tolerant and hydrolyze Tween 80 and Tween 20. Some strains grow on MacConkey and cetrimide agars. Does not possess phosphatase, lysine decarboxylase, or lecithinase. Some strains may produce acid from D-mannitol and sucrose. Acid is not produced from other sugars or poly-alcohols. The members of this species utilize D-salicin, fumarate, malate, and L-aspartic acid as sole sources of carbon. Some strains may also utilize D-galactose, inulin, raffinose, L-rhamnose, sorbose, D-trehalose, D-xylose, formate, pyruvate, D-tartrate, β-hydroxybutyrate, and L-glutamic acid. The species does not utilize oxalate, benzoate, or caprylate as sole carbon sources. Sensitive to ampicillin (10 μg), cephalothin (30 μg), erythromycin (15 μg), nalidixic acid (15 μg), penicillin G (10 i.u.), and chloramphenicol (30 μg). Some strains are also sensitive to gentamicin (10 μg), streptomycin (10 μg), and tetracycline (30 μg). Isolated from hypersaline habitats such as soils, salt ponds, salt lakes, and from the sea.

The mol% G + C of the DNA is: 60.7–64 (T_m).

Type strain: ATCC 49509, DSM 5928.

GenBank accession number (16S rRNA): L42617, X87217.

19. **Halomonas subglaciescola** Franzmann, Burton and McMeekin 1987, 32[VP]

sub.gla' ci.es.co.la. L. pref. *sub* below; L. n. *glacies* ice; L. suff. *cola* to dwell; M.L. adj. *subglaciescola* dwelling below ice.

The characteristics are as described for the genus and as listed in Tables BXII.γ.95 and BXII.γ.96. When motile, the cells are peritrichously flagellated. On solid media, the colonies produced are white to cream. Yellow colonies may form on media enriched with alanine or hydroxy-L-proline. This species cannot be grown on simple media, such as CAS or MH. Grows at temperatures as low as −5°C. The species has two biovars. Biovar I is motile and produces long filamentous cells >10 μm long. Biovar II is nonmotile and produces cells <10 μm long. The type strain belongs to Biovar I. Some characteristics of the species are listed in Tables BXII.γ.95 and BXII.γ.96. The species grows at pH 5.0–9.0 and does not grow above 25°C. The species does not grow in the presence of 30% NaCl. Catalase and phenylalanine deaminase positive. Phosphatase, lysine decarboxylase, and β-galactosidase negative. Does not produce indole or hydrogen sulfide; does not hydrolyze esculin. The species does not produce acid from either adonitol or sorbitol. There is no data available for acid production from other compounds. This species has been reported to grow on L-threonine and proline as primary carbon sources. It does not utilize sucrose, L-aspartic acid, L-glutamic acid, methionine, or L-asparagine as sole carbon sources. Sensitive to mercuric chloride (200 mg/l). Some strains are sensitive to Vibriostat O/129 (10 μg), penicillin G (10 i.u.), and chloramphenicol (30 μg). Resistant to gentamicin (10 μg), neomycin (30 μg), streptomycin (10 μg), and tetracycline (30 μg). The similarity of the 16S rDNA of both biovars of this species to other *Halomonas* species ranges from 93.7–96.1% for Biovar I and from 93.6–99.9% for Biovar II. The 16S rDNA sequence of Biovar II shows only 98.2% similarity to that of Biovar I. However, Biovar II shows 99.9% sequence similarity to *Halomonas halodurans*, indicating that Biovar II is likely a variant of the later species. Isolated from Antarctic hypersaline lakes.

The mol% G + C of the DNA is: 60.9–62.9 ± 1.0 (method unknown).

Type strain: ATCC 43668, DSM 4683, UQM 2926.

GenBank accession number (16S rRNA): M93358.

20. **Halomonas variabilis** (Fendrich 1989) Dobson and Franzmann 1996, 556[VP] (*Halovibrio variabilis* Fendrich 1989, 205.) *va.ri.a' bi.lis.* L. adj. *variabilis* changeable, variable; referring to variation of the cell diameter with changing salt concentrations.

The characteristics are as described for the genus and as listed in Tables BXII.γ.95 and BXII.γ.96. The cells are curved rods (described as vibrion shaped). Coccoid bodies may form in older cultures. Cells have monotrichous flagellation. Colonies are circular, entire, smooth, slimy, raised, and produce a light brown pigment. The cells grow in NaCl concentrations from 1.2–4.9 M, with an optimum salt concentration of 1.0 M. The optimum pH for growth is 7.5; minimum, 6.5; maximum, 8.4. Temperature range for growth: 15–37°C; optimum, 33°C. Minimum doubling time, 7.5 h. The similarity of the 16S rDNA of this species to other *Halomonas* species ranges from 91.5–96.8%. Isolated from the North Arm of the Great Salt Lake Utah (USA).

The mol% G + C of the DNA is: 57.0 (method unknown).

Type strain: DSM 3051.

GenBank accession number (16S rRNA): M93357, X90483.

21. **Halomonas venusta** (Baumann, Baumann, Mandel and Allen 1972) Dobson and Franzmann 1996, 556[VP] (*Delaya venusta* (Baumann, Baumann, Mandel and Allen 1972) Baumann, Bowditch and Baumann 1983a, 801; *Alcaligenes venustus* Baumann, Baumann, Mandel and Allen 1972, 426.) *ve.nus' tus.* L. adj. *venustus* lovely, beautiful.

The characteristics are as described for the genus and as listed in Tables BXII.γ.95 and BXII.γ.96. The cells possess peritrichous flagella. The species utilizes 52–76 organic compounds. Also utilized, in addition to compounds listed in Table BXII.γ.95, are: galacturonate, glucuronate, *N*-acetylglucosamine, butyrate, isobutyrate, valerate, isovalerate, caproate, heptanoate, caprate, glutarate, glycolate, α-ketoglutarate, aconitate, *meso*-inositol, propyleneglycol, 2,3-butyleneglycol, *n*-propanol, *n*-butanol, isobutanol, L-mandelate, benzoylformate, benzoate, *o*-, *m*-, and *p*-hydroxybenzoate, phenylate, quinate, δ-aminovalerate, anthranilate, ethanolamine, putrescine, betaine, sarcosine, acetamide, and allantoin. The similarity of the 16S rDNA of this species to other *Halomonas* species ranges from 93.2–98%.

The mol% G + C of the DNA is: 58.7 (Bd).

Type strain: ATCC 27125, DSM 4743.

GenBank accession number (16S rRNA): L42618.

Genus II. **Carnimonas** Garriga, Ehrmann, Arnau, Hugas and Vogel 1998, 684[VP]

MARGARITA GARRIGA, MATTHIAS A. EHRMANN, JACINT ARNAU, MARTA HUGAS AND RUDI F. VOGEL

Car.ni' mo.nas. L. gen. n. *carnis* of meat; Gr. n. *monas* a unit, monad; *Carnimonas* a monad of meat.

Straight or slightly curved rods, 0.5–0.6 × 1.0–1.7 μm, occurring singly or in pairs. Gram negative. Does not form spores. **Nonmotile. Oxidase and catalase positive. Aerobic, having a strictly respiratory type of metabolism with oxygen as the terminal electron acceptor. Moderately halotolerant.** Optimum temperature for growth is 28–30°C. **No growth occurs at 5°C or 37°C.** Chemoorganotrophic. **Acid, but no gas, is produced from D-glucose, D-xylose, and various other carbohydrates. β-Galactosidase (ONPG) activity occurs.** The main respiratory quinone is ubiquinone-9. Main components in the polar lipid composition are diphosphatidylglycerol, phosphatidylglycerol, and phosphatidylethanolamine. Major fatty acids are $C_{16:0}$, $C_{16:1}$, $C_{18:1}$, and $C_{19:0cyc}$. Belongs to the class *Gammaproteobacteria*. Forms dark spots on the surface of raw, cured meat products.

The mol% G + C of the DNA is: 56.

Type species: **Carnimonas nigrificans** Garriga, Ehrmann, Arnau, Hugas and Vogel 1998, 685.

FURTHER DESCRIPTIVE INFORMATION

On tryptone soy agar, colonies appear white, convex, shiny, and circular. No pigmentation is visible in any culture media. Optimum temperature for growth is 28–30°C. No growth occurs at 5°C or 37°C. Growth occurs in the presence of 8% NaCl, but not at higher levels.

Indole production, reduction of nitrate, and the Voges–Proskauer reaction are negative. All strains produce acid from glucose, xylose, melibiose, maltose, and saccharose. No gas is produced from glucose. Additional physiological data are provided in Table BXII.γ.99.

Garriga et al. (1998) have reported that the main polar lipids are diphosphatidylglycerol, phosphatidylglycerol, and phosphatidylethanolamine. In addition, there are three unidentified components. The fatty acid profile contains major amounts of several saturated and unsaturated straight chain fatty acids and only minor fractions of 3-hydroxylated fatty acids. The main fatty acids are palmitic acid ($C_{16:0}$), which comprises 40% of total fatty acids, and the cyclopropanic acid $C_{19:0\ cyc}$, which comprises 21.07%. Oleic acid ($C_{18:1\ \omega9t}$) comprises 7.45% and a $C_{18:1\ \omega7c/\omega9t/\omega12t}$ group comprises 12.9%. Moreover, a $C_{16:1}$ fatty acid comprises 6.7%, whereas only minor amounts of $C_{12:1\ 3OH}$ (1.85%) and traces of $C_{16:0\ 3OH}$ (0.46%) are found. No match was found after comparison with the microbial ID, TSBA library (Newark, DE, USA).

No pathogenic effect on mice is detectable 8 d after intraperitoneal injection of up to 1.9 × 10^10 CFU of *Carnimonas nigrificans*. Under similar conditions, the bacteria used as controls (*Escherichia coli* HM-42, *Pseudomonas aeruginosa* HS-116, *Staphylococcus aureus* HS-93) show a concentration-dependent lethality.

The presence of dark spots on the surface of raw, cured meat products was first described by Hugas and Arnau (1987). In 1993, Arnau and Garriga identified a Gram-negative bacterium (now *C. nigrificans*) as being responsible for this defect. The defect, a rust-like color turning to black within hours, can be reproduced on comminuted pork meat containing salt (40 g/kg) and dextrose (20 g/kg) after inoculation of an overnight culture and storage in an aerobic environment at 30°C. The browning effect is increased by some amino acids, i.e., glycine, L-arginine, L-glutamine, and L-monosodium glutamate, when added (20 g/kg) to a meat mixture containing salt and dextrose. N-Acetyl-L-cysteine, L-cysteine, potassium metabisulfite, and propyl-3,4,5-trihydroxybenzoate (5 g/kg) are useful in the prevention of this defect (Arnau and Garriga, 2000). The browning produced by C. nigrificans with dextrose and amino acids and the inhibitory properties of some of the substances studied show similarities to the Maillard reaction. However, important differences exist in the temperature pattern, the oxygen effect, and carbohydrates involved.

ENRICHMENT AND ISOLATION PROCEDURES

The isolation of C. nigrificans from raw, cured meat products must be done at the beginning of browning; no recovery is possible later. The organism is capable of growing on Cetrimide agar and MacConkey agar; consequently, these selective media could be used for isolation procedures. Incubation should be performed under aerobic conditions at 30°C for 2–3 days. After isolation, C. nigrificans grows well on the usual nutrient media (e.g., tryptone soy broth, brain–heart infusion broth).

MAINTENANCE PROCEDURES

Cultures in routine use can be maintained in tryptone soy broth or agar at 4°C. Cultures may be preserved for long-term storage in tryptone soy broth with 20% glycerol at −80°C.

PROCEDURES FOR TESTING SPECIAL CHARACTERS

Comparative sequence analysis of 16S rRNA gene sequences has revealed a diagnostic sequence that can be used as a target site for specific amplification. The sequence of the oligonucleotide used as specific primer for C. nigrificans strain CTCBS1 is 5′-TAA CGT CCT TCA TGC CGG-3′ (binding position 469–486 in the E. coli numbering system). This primer (bs1) has been checked for its specificity against more than 10,000 16S rRNA sequences by using the probe-checking software provided by the Ribosomal Database Project (Maidak et al., 1996). PCR and cycle conditions for the species-specific reaction with primer bs1 and universal primer 616V are as follows: one initial cycle 94°C (120 s), followed by 32 cycles of 94°C (45 s), 50°C (90 s), and 72°C (120 s). Master mixes are prepared with reaction buffer containing 10 mM Tris-HCl, 1.5 mM $MgCl_2$, 50 mM KCl, pH 8.3, 200 nM dNTP, 0.5 pmol each primer, and Taq DNA polymerase (Boehringer, Mannheim, Germany).

Specific amplification of a 480 bp fragment using primer bs1 and 616V occurs only in C. nigrificans strains.

DIFFERENTIATION OF THE GENUS CARNIMONAS FROM OTHER GENERA

The closest relatives of Carnimonas nigrificans are Zymobacter palmae and the genus Halomonas (at least Chromohalobacter). Some physiological and morphological properties, e.g., relationship to oxygen, growth temperature, hydrolysis of starch, oxidase reaction, as well as flagellation and motility, are useful in differentiating Carnimonas from these organisms (Table BXII.γ.99 and Table BXII.γ.100).

The mol% G + C content of 56, the presence of ubiquinone 9 as the major respiratory lipoquinone, and the lack of $C_{19:0 \, cyc}$ in the non-polar lipid profile are sufficient to exclude these organisms from Oceanospirillum. The mol% G + C content of Carnimonas nigrificans CTCBS1 differs by at least 4.2% from those of Halomonas and Chromohalobacter, and there are distinct differences in the fatty acid profiles of these organisms; the levels of $C_{16:0}$ in the profiles of members of the genus Halomonas range from 15.5 to 32%, whereas the level in Carnimonas nigrificans CTCBS1 is 40%. Additionally, in contrast to Halomonas, no $C_{17:0 \, cyc}$ could be detected in Carnimonas nigrificans CTCBS1.

TAXONOMIC COMMENTS

The phylogenetic tree, based on 16S rDNA sequences, is shown in Fig. BXII.γ.118. The topology of all organisms illustrated in the tree is based on maximum likelihood analysis of 16S rDNA sequences (Fig. BXII.γ.118) and is consistent with a tree previously published by Dobson and Franzmann (1996). The closest relatives of strain Carnimonas nigrificans are Zymobacter palmae with 93.3% sequence similarity and, to a lesser extent, species of the genus Halomonas (at least 91.9%) and Chromohalobacter (91.5%).

The phylogenetic tree shows that the closest relationship is

TABLE BXII.γ.99. Features differentiating the genus Carnimonas from other genera[a]

Characteristic	Carnimonas	Chromohalobacter	Halomonas	Oceanospirillum	Zymobacter
Cells are rod-shaped	+	+	+	D	+
Cells are helical	−	−	−	D	−
Relationship to oxygen:					
Aerobic	+	+	+	+	−
Facultatively anaerobic	−	−	−	−	+
Acid production from hexoses	+	+	+	−	+
Motile	−	+	D	+	+
Oxidase	+	−	D	+	−
Growth at 37°C	−	+	+	+	+
Growth with 0% NaCl	+	−	D	−	−
Tolerates 15% NaCl	−	+	+	−	−
Hydrolysis of:					
Esculin	+	−	D	−	
Starch	+	−	−	−	−
Violet to brown colonies on culture media	−	+	−	−	−
Habitat:					
Raw, cured meat	+	−	−	−	−
Palm sap	−	−	−	−	+
Coastal sea water	−	−	−	+	−
Solar salt facilities, intertidal estuaries, and hypersaline lakes	−	+	+	−	−

[a] Symbols: see standard definitions.

TABLE BXII.γ.100. Other characteristics of the genus *Carnimonas*[a]

Characteristic	Reaction or result
α-Glucosidase (PNPG), phenylalanine deaminase	+
Arginine dihydrolase, urease, DNase, lecithinase	−
Nitrate reduction	−
Hydrolysis of gelatin, casein, Tween 80	−
Acid production from:	
Amygdalin, esculin, galactose, melibiose, sucrose, salicin	+
L-Arabinose	−
Fructose, mannose, mannitol, melezitose, raffinose, ribose, trehalose	d
Utilization of2:[b]	
Adipate, L-arabinose, caprate malate, mannitol, N-acetylglucosamine	−
Citrate, mannose, maltose, gluconate	+

[a]Symbols: see standard definitions.

[b]On media containing vitamins and yeast extract.

to *Zymobacter palmae* ATCC 51623[T]. *Zymobacter palmae* is the single species of the genus *Zymobacter* and the closest phylogenetic relative to strain CTCBS1 (Okamoto et al., 1993). The G + C content of its DNA is 55.8 mol%, in the range of that determined for CTCBS1. The profiles of the major fatty acids of the two strains are similar; however, a significant amount (5%) of a $C_{12:0}$ fatty acid that might be unique to *Zymobacter palmae* is not detected in strain CTCBS1 or the other closely related genera.

FURTHER READING

Arnau, J. and M. Garriga. 1993. Black spot in cured meat-products. Fleischwirtschaft *73*: 1412–1413.

Arnau, J. and M. Garriga. 2000. The effect of certain amino acids and browning inhibitors on the 'black spot' phenomenon produced by *Carnimonas nigrificans*. J. Sci. Food Agric. *80*: 1655–1658.

Garriga, M., M.A. Ehrmann, J. Arnau, M. Hugas and R.F. Vogel. 1998. *Carnimonas nigrificans* gen. nov., sp. nov., a bacterial causative agent for black spot formation on cured meat products. Int. J. Syst. Bacteriol. *48*: 677–686.

Hugas, M. and J. Arnau. 1987. Aparicíon de manchas de color en la corteza y grasa del jamón durante el post-salado. *In* Arnau, Hugas and Monfort (Editors), Jamón Curado, Aspectos Técnicos, Institut de Recerca i Tecnología Agroalimentáries, Monells, Spain. pp. 179–182.

List of species of the genus Carnimonas

1. **Carnimonas nigrificans** Garriga, Ehrmann, Arnau, Hugas and Vogel 1998, 685[VP]

   nig.rif i.cans. L. adj. *niger* black; L. v. *facere* to make; M.L. n. *nigrificans* black-making.

   Characteristics are the same as for the genus. Colonies are nonpigmented, white, convex, shiny, and circular. Produce acid from glucose, fructose, maltose, xylose, melibiose, and saccharose. Hydrolyze esculin and starch, but not gelatin, casein, or DNA. Voges–Proskauer negative. Arginine dihydrolase, urease, lecithinase, and phenylalanine deaminase negative. Do not produce indole.

   The mol% G + C of the DNA is: 56 ± 0.3 (T_m).

   Type strain: CTCBS1, CECT 4437, CIP 105703, NCIMB 13550.

   GenBank accession number (16S rRNA): Y13299.

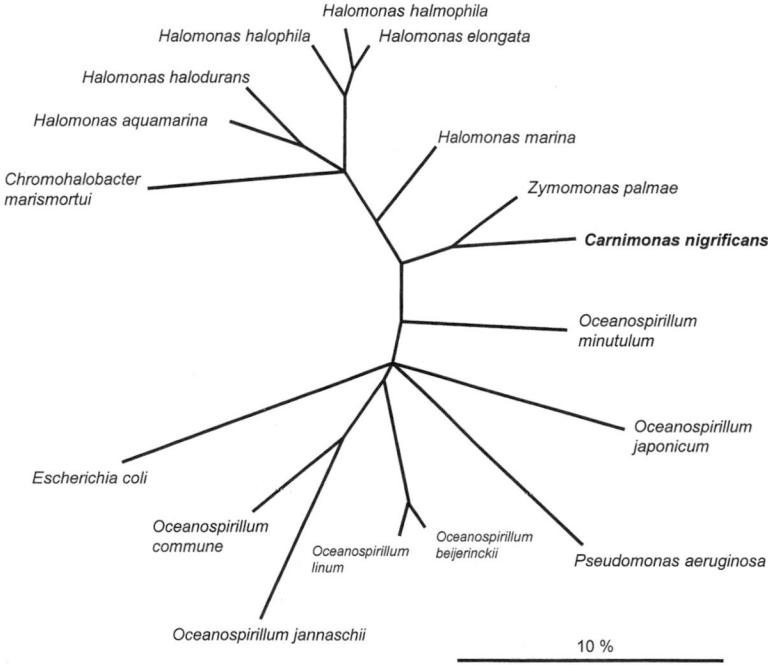

FIGURE BXII.γ.118. Phylogenetic relationships of *Carnimonas nigrificans* and selected representatives of the class *Gammaproteobacteria*.

Genus III. **Chromohalobacter** *Ventosa, Gutierrez, García and Ruiz-Berraquero 1989, 384^VP emend. Arahal, García, Ludwig, Schleifer and Ventosa 2001a, 1446)*

ANTONIO VENTOSA

Chro.mo.ha' lo.bac'ter. Gr. n. *chroma* color; Gr. n. *halos* the sea, salt; M.L. n. *bacter* rod; M.L. masc. n.*Chromohalobacter* colored salt rod.

Gram-negative, straight or sometimes slightly curved, rods (0.6–1.2 × 1.5–4.2 μm). Motile by polar or peritrichous flagella. Cells occur singly, in pairs, and in short chains. Spores are not formed. **Moderately halophilic.** Salt is required for growth. **The optimum salt concentration for growth is between 8 and 10%.** May grow at salt concentrations up to 30%. The broader ranges of temperature and pH observed for growth are 5–45°C (optimal 30–37°C) and pH 5.0–10.0 (optimal pH 7.5), respectively. **Aerobic.** Chemoorganotrophic. Catalase positive. **Oxidase negative. Most strains reduce nitrates.** Phenylalanine deaminase test is negative. Gelatin, starch, Tween 80, esculin, DNA, and tyrosine are not hydrolyzed. Acid is produced aerobically from D-glucose and other sugars. Carbohydrates, amino acids, and some polyols can serve as sole carbon sources. Colonies are cream to brown-yellow pigmented.

The mol% G + C of the DNA is: 62–66.

Type species: **Chromohalobacter marismortui** (ex Elazari-Volcani 1940) Ventosa, Gutierrez, García and Ruiz-Berraquero 1989, 384.

FURTHER DESCRIPTIVE INFORMATION

The genus *Chromohalobacter* was created by Ventosa et al. (1989) to accommodate seven moderately halophilic isolates obtained during the course of an extensive taxonomic study of bacteria isolated from salterns in the Mediterranean coast (Ventosa et al., 1982). These isolates showed phenotypic and chemotaxonomic features very similar to those of "*Chromobacterium marismortui*", a species originally described by Elazari-Volcani (1940) based on strains isolated from the Dead Sea. This species was described in the 7th edition of *Bergey's Manual of Determinative Bacteriology* (Breed et al., 1957), and in the 8th edition it was included in the genus *Chromobacterium* as a species *incertae sedis* since it did not produce violacein and lacked the typical flagellar arrangement of the genus *Chromobacterium* (Buchanan and Gibbons, 1974). It was not included in the Approved Lists of Bacterial Names (Skerman et al., 1980). Until recently, the genus *Chromohalobacter* was represented by a single species, *Chromohalobacter marismortui*, which was phylogenetically closely related to species of the genus *Halomonas* and included in the family *Halomonadaceae* (Mellado et al., 1995b). Very recent studies based on 16S rDNA gene sequence comparison and other phenotypic and molecular data supported the conclusion that two species previously described as members of the genus *Halomonas*, *Halomonas canadensis* and *Halomonas israelensis*, should be placed in the genus *Chromohalobacter*, as *Chromohalobacter canadensis* and *Chromohalobacter israelensis*, respectively (Arahal et al., 2001a). Finally, a study of two strains previously assigned to the species *Halomonas elongata*, strains DSM 3043 and ATCC 33174, showed that they represent a different species of the genus *Chromohalobacter*, and named them as *Chromohalobacter salexigens* (Arahal et al., 2001b).

All species currently described as members of the genus *Chromohalobacter* have a specific requirement for Na$^+$. They grow optimally in media containing 8–10% salt and are defined as moderately halophilic microorganisms (Ventosa et al., 1998). Few studies have been carried out with the type species *C. marismortui*;

however, *C. canadensis* and *C. israelensis* have been extensively used for physiological studies. These species were formerly designated as strain Ba$_1$ (Rafaeli-Eshkol, 1968), isolated from the Dead Sea (*C. canadensis*) and strain NRCC 41277, isolated by Matheson et al. (1976) as a medium contaminant (*C. israelensis*). More detailed information about their physiological and biochemical features can be found in a recent review (Ventosa et al., 1998).

To cope with the high external salinity, halophilic bacteria need to balance their cytoplasm with the osmotic pressure exerted by the external medium. However, in *C. canadensis* and *C. israelensis*, as in many other moderate halophiles, the sum of the apparent intracellular concentration of Na$^+$ and K$^+$ ions is much lower than the external concentration (Matheson et al., 1976; Goldberg and Gilboa, 1978). They accumulate intracellular organic compounds, called compatible solutes, such as glycine betaine (Rafaeli-Eshkol and Avi-Dor, 1968; Shkedy-Vinkler and Avi-Dor, 1975). However, the most wide spread compatible solutes found in halophilic bacteria, and in *Chromohalobacter* in particular, are ectoine and its β-hydroxy derivative, hydroxyectoine (Ventosa et al., 1998). During recent years, several studies have focused on the osmoregulatory mechanisms of *C. salexigens* DSM 3043. It was selected as an excellent model organism because it displays one of the widest ranges of salt tolerance found in nature. In order to maintain its internal osmolarity and generate turgor in environments with high salinities, *C. salexigens* accumulates compounds such as glycine betaine, choline or choline-*O*-sulfate, when present externally. However, the main osmoadaptation mechanism is the *de novo* synthesis of ectoine and hydroxyectoine (Cánovas et al., 1996, 1997, 1998b, 1999; Nieto et al., 2000). Genes involved in the biosynthesis of ectoine and the oxidation of choline to betaine have been recently characterized (Cánovas et al., 1998a, 2000). The ectoine synthesis genes of *C. salexigens* and *Halomonas elongata* have lower sequence homology values than was expected from two closely related halophilic microorganisms (Nieto et al., 2000).

The genome size, determined by pulsed-field gel electrophoresis, of three strains of *C. marismortui* (including the type strain) ranged from 1770 to 2295 kb and that of *C. israelensis* was estimated to be 2490 kb (Mellado et al., 1998). Several plasmids have also been reported (Ventosa et al., 1994; Mellado et al., 1995a; Vargas et al., 1995a). The basic replicon of the narrow-host-range plasmid pCM1 from *C. marismortui* has been sequenced and characterized in detail (Mellado et al., 1995a). Cloning and shuttle vectors useful for the genetic manipulation of these bacteria have been reported (Mellado et al., 1995c; Vargas et al., 1995a). Conjugation is the only genetic transfer mechanism that has been described for this genus (Vargas et al., 1997).

ENRICHMENT AND ISOLATION PROCEDURES

Selective procedures for the enrichment and isolation of species of the genus *Chromohalobacter* have not been reported. They can be isolated from hypersaline environments by direct plating on complex media supplemented with a salt mixture and incubation at 30–37°C for 7–10 d. The isolation medium described by Ventosa et al. (1982) can be used.

Maintenance Procedures

For short-term storage, the species of the genus *Chromohalobacter* can be maintained on complex medium containing 10% salts. An appropriate medium is described by Ventosa et al. (1989). Long-term storage of members of this genus can be carried out by freeze-drying, storage at −80°C, or storage in liquid nitrogen.

Differentiation of the genus *Chromohalobacter* from other genera

The genus *Chromohalobacter* belongs to the family *Halomonadaceae* and is closely related to the genus *Halomonas* (Fig. BXII.γ.119), as well as to *Zymobacter* and *Carnimonas*. Since the genus *Halomonas* is taxonomically heterogeneous and contains greater than 20 species, it is very difficult to differentiate these two genera based on phenotypic or chemotaxonomic characteristics. Typical features of species of the genus *Chromohalobacter* are their optimal growth in media containing 8–10% salt and characteristic results for the biochemical tests reported in the genus description. Species of the genus *Chromohalobacter* can be phylogenetically differentiated from other related genera by comparison of 16S rRNA gene sequences.

Differential features of the four currently recognized species of the genus *Chromohalobacter* are given in Table BXII.γ.101.

Acknowledgments

I am grateful to Drs. R.H. Vreeland and J.T. Staley for the critical reading of this article and to Dr. D.R. Arahal for help with the phylogenetic analysis.

Further Reading

Arahal, D.R., M.T. García, W. Ludwig, K.H. Schleifer and A. Ventosa. 2001. Transfer of *Halomonas canadensis* and *Halomonas israelensis* to the genus *Chromohalobacter* as *Chromohalobacter canadensis* comb. nov and *Chromohalobacter israelensis* comb. nov. Int. J. Syst. Evol. Microbiol. *51*: 1443–1448.

Mellado, E., E.R.B. Moore, J.J. Nieto and A. Ventosa. 1995. Phylogenetic inferences and taxonomic consequences of 16S ribosomal DNA-sequence comparison of *Chromohalobacter marismortui*, *Volcaniella eurihalina*, and *Deleya salina* and reclassification of *Volcaniella eurihalina* as *Halomonas eurihalina* comb. nov. Int. J. Syst. Bacteriol. *45*: 712–716.

Ventosa, A., M.C. Gutierrez, M.T. García and F. Ruiz-Berraquero. 1989. Classification of "*Chromobacterium marismortui*" in a new genus, *Chromohalobacter* gen. nov. as *Chromohalobacter marismortui*, comb. nov., nov. rev. Int. J. Syst. Bacteriol. *39*: 382–386.

Ventosa, A., J.J. Nieto and A. Oren. 1998. Biology of moderately halophilic aerobic bacteria. Microbiol. Mol. Biol. Rev. *62*: 504–544.

List of species of the genus Chromohalobacter

1. **Chromohalobacter marismortui** (ex Elazari-Volcani 1940) Ventosa, Gutierrez, García and Ruiz-Berraquero 1989, 384[VP] *ma.ris.mor′ tu.i.* L. gen. n. *maris* of the sea; L. adj. *mortuus* dead; M.L. gen. n. *marismortui* of the Dead Sea.

   See Table BXII.γ.101 and generic description for many features. Cells are Gram negative, rod shaped, and sometimes slightly curved, the length varies with the concentra-

tion of salt; at 10% salts, the cells are 0.6–1.0 × 1.5–4.0 μm; at higher and lower salt concentrations, the cells are longer. Cells occur singly or in pairs. Motile by means of peritrichous flagella. Spores are not formed. On solid, complex media containing 10% salts, colonies are circular, convex, smooth, entire, and concentrically ringed with dark brown centers followed by bluish brown, grayish brown, and

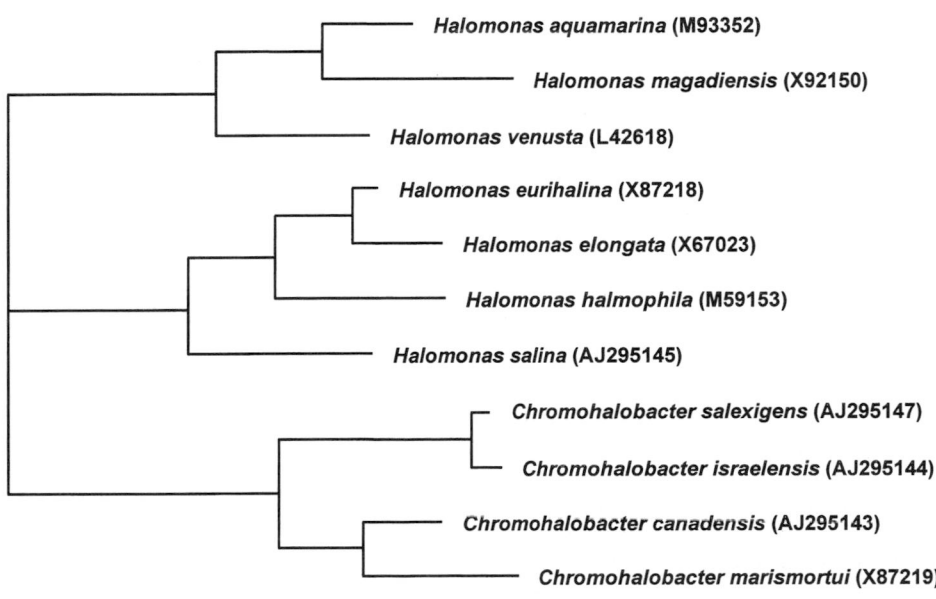

FIGURE BXII.γ.119. Phylogenetic tree derived from analysis of the 16S rRNA gene sequences of the type strain of species of the genus *Chromohalobacter* and some representative related species of the genus *Halomonas*. Figures in parentheses indicate the accession numbers of 16S rRNA gene sequence data. Bar = 5% sequence difference.

TABLE BXII.γ.101. Differential characteristics of species of the genus *Chromohalobacter*[a]

Characteristic	1. *C. marismortui*	2. *C. canadensis*	3. *C. israelensis*	4. *C. salexigens*
Indole production	−	+	+	−
H$_2$S production	−	−	−	+
Simmons citrate	+	−	−	+
Hydrolysis of casein	−	−	−	+
Urease	−	−	−	+
Lysine decarboxylase	−	+	+	−
Ornithine decarboxylase	−	+	+	−
Acid production from:				
Arabinose	−	+	+	+
Maltose	+	−	+	+
Sucrose	+	−	+	+
D-Trehalose	+	−	−	−
Growth on:				
D-Cellobiose	−	+	+	−
Sucrose	+	−	+	+

[a]For symbols see standard definitions.

yellow rings. They produce a yellow pigment and a violet-blue pigment which is not violacein. Production of pigment is favored by suboptimal growth temperatures, glycerol, and a salt concentration of 10%. In liquid medium containing 10% salts, turbidity and a pellicle are produced; usually brown-to-yellow pigments are also produced.

Grows in media containing 1–30% salts, with an optimum at about 10% salts. Growth occurs at 5–45°C (optimum temperature is 37°C) and pH 5–10 (optimal growth at pH 7.5) on solid media containing 10% salts. Acid is produced without gas under aerobic conditions in marine oxidation–fermentation medium supplemented with 10% salts and D-glucose, D-galactose, maltose, lactose, D-arabinose, D-xylose, sucrose, trehalose, glycerol, and D-mannitol. Generally, nitrate is reduced to nitrite (except the type strain). Gelatin, casein, starch, esculin, tyrosine, Tween 80, and DNA are not hydrolyzed. Phosphatase, Voges–Proskauer, β-galactosidase, phenylalanine deaminase, and arginine dehydrolase tests are negative.

The following compounds are utilized as sole carbon and energy sources: dulcitol, D-fucose, D-galactose, D-gluconate, D-glucose, glutamate, *meso*-inositol, maltose, D-mannitol, D-mannose, pyruvate, D-ribose, sucrose, D-sorbitol, and D-xylose. The following compounds are not utilized as sole carbon and energy sources: N-acetylglucosamine, amygdalin, DL-α-aminobutyrate, butyrate, cellobiose, citrate, esculin, p-hydroxybenzoate, hippurate, inulin, malonate, melibiose, oxalate, raffinose, salicylate, salicin, and D-tartrate. The following compounds are utilized as sole carbon, nitrogen, and energy sources: L-alanine, DL-arginine, L-glutamine, L-ornithine, L-proline, putrescine, and L-serine. The following compounds are not utilized as sole carbon, nitrogen, and energy sources: L-allantoin, betaine, creatine, ethionine, L-isoleucine, L-leucine, phenylalanine, sarcosine, L-threonine, and L-valine.

Isolated from the Dead Sea and marine salterns.

The mol% G + C of the DNA is: 62.1–64.9 (T_m).

Type strain: ATCC 17056, CCM 3518, DSM 6770, NCIMB 8731.

GenBank accession number (16S rRNA): X87219.

Additional Remarks: Other sequences: strain A-65 (X87220), strain A-100 (X87221), strain A-492 (X87222), and strain T1093 (U78719).

2. **Chromohalobacter canadensis** (Huval, Latta, Wallace, Kushner and Vreeland 1996) Arahal, García, Ludwig, Schleifer and Ventosa 2001a, 1447[VP] (*Halomonas canadensis* Huval, Latta, Wallace, Kushner and Vreeland 1995, 1130.)

ca.na.den'sis. M.L. n. *canadensis* pertaining to Canada, the country where the type strain was isolated.

See Table BXII.γ.101 and generic description for many features. Cells are Gram negative, straight or curved rods with rounded ends, 0.6–1.2 × 2.0–3.8 μm; cells occur singly, in pairs, or in short chains. Pleomorphic cells are not present. Both culture age and salt concentration affect chain length. Motile by a single polar flagellum. Spores are not formed. On solid complex media with 8% salts, colonies are translucent and convex with a smooth glistening surface, produce a nondiffusible white pigment, and have a diameter of 2–6 mm after 7 days of growth at 30°C.

Grows in media containing 3–25% NaCl from 15 to 30°C; grows in media with 8–32% NaCl at 45°C. Optimal growth at 7.5% NaCl. Growth occurs at 5–45°C (optimum temperature is 30°C) and from pH 5 to 9 on solid media containing 8% NaCl. Acid is produced from mannose, lactose, glycerol, and cellobiose but not from sucrose. Nitrate is reduced to nitrite. Phenylalanine deaminase and Voges–Proskauer tests are negative. Gelatin, starch, DNA, esculin, and agar are not hydrolyzed.

The following compounds are utilized as sole carbon and energy sources: acetate, citrate, DL-malate, succinate, lactate, D-glucose, xylose, L-arabinose, D-fructose, D-mannitol, *meso*-inositol, ethanol, mannose, esculin, glycerol, and gluconic acid. The following compounds are not utilized as sole carbon and energy sources: butyrate, isobutyrate, propionate, tartrate, benzoate, maltose, melibiose, L-sorbose, D-ribose, D-sorbitol, n-propanol, n-butanol, ethylene glycol, n-hexadecane, pyridine-1-oxide, formaldehyde, formamide, N,N-dimethylacetamide, and ethylenediamine. Isolated as a contaminant on Sehgal and Gibbons medium containing 25% NaCl.

The mol% G + C of the DNA is: 62 (T_m).

Type strain: ATCC 43984, DSM 6769, CECT 5385, CCM 4919, CIP 105571, NCIMB 13767, NRCC 41227.

GenBank accession number (16S rRNA): AJ295143.

3. **Chromohalobacter israelensis** (Huval, Latta, Wallace, Kushner and Vreeland 1996) Arahal, García, Ludwig, Schleifer

and Ventosa 2001a, 1447[VP] (*Halomonas israelensis* Huval, Latta, Wallace, Kushner and Vreeland 1996, 1189.) *is.ra.el.en'sis*. M.L. n. *israelensis* an inhabitant of Israel.

See Table BXII.γ.101 and generic description for many features. Gram negative. Cells are straight rods with rounded ends, 0.6–0.9 × 1.5–4.2 µm, occurring singly and occasionally in pairs or short chains. Pleomorphic cells are not present. On solid complex medium with 8% NaCl colonies are flat with entire margins, produce a nondiffusible cream pigment, and have 2–6 mm in diameter after 7 days.

Grows in media containing 3.5–20% NaCl (optimum at 8% NaCl). Growth occurs at 15–45°C (optimum temperature is 30°C) and from pH 5 to 9. Acid is produced without gas from mannose, lactose, glycerol, cellobiose, and sucrose. Nitrate is reduced to nitrite. Phenylalanine deaminase, Voges–Proskauer, and methyl red tests are negative. Gelatin, starch, esculin, and DNA are not hydrolyzed.

The following compounds are utilized as sole carbon and energy sources: cellobiose, lactose, mannose, esculin, glycerol, acetate, starch, gluconic acid, and sucrose.

The mol% G + C of the DNA is: 65 (T_m).

Type strain: ATCC 43985, DSM 6768, CECT 5287, CCM 4920, CIP 106853, NCIMB 13766.

GenBank accession number (16S rRNA): AJ295144.

4. **Chromohalobacter salexigens** Arahal, García, Vargas, Cánovas, Nieto and Ventosa 2001b, 1460[VP]

sal.ex' i.gens. L. n. *sal* salt; L. v. *exigo* to demand; M.L. part. adj. *salexigens* salt-demanding.

See Table BXII.γ.101 and generic description for many features. Cells are Gram-negative rods, 0.7–1.0 × 2.0–3.0 µm; cells occur singly or in pairs. Motile. Spores are not formed. On solid, complex media containing 10% salts, colonies are cream, opaque, and circular and less than 2 mm in diameter; spreading may occur after extended incubation. In liquid medium containing 10% salts, a homogeneous turbidity is produced.

Grows in media containing 0.9–25% salts, and optimal growth at 7.5–10% salts. Growth occurs at 15–45°C (optimum temperature is 37°C) and from pH 5 to 10 on liquid media containing 10% salts (optimal growth at pH 7.5). Acid is produced from L-arabinose, D-fructose, D-galactose, glycerol, D-glucose, lactose, maltose, D-mannose, sucrose, and D-xylose, but not from trehalose. Nitrate is reduced to nitrite but nitrite is not reduced. Gelatin, starch, esculin, DNA, and Tween 80 are not hydrolyzed. Methyl red positive. Indole and acetoin are not produced. Phenylalanine deaminase, lysine decarboxylase, and ornithine decarboxylase are not produced.

The following compounds are utilized as sole carbon and energy sources: L-arabinose, erythritol, D-fructose, D-galactose, D-glucose, maltose, D-mannitol, D-mannose, D-sorbitol, sucrose, D-trehalose, D-ribose, D-xylose, ethanol, glycerol, *meso*-inositol, dulcitol, acetate, α-aminovalerate, α-ketoglutarate, citrate, fumarate, DL-glycerate, glutamate, malate, malonate, propionate, D-saccharate, succinate, and D-tartrate. The following compounds are not utilized as sole carbon and energy sources: adonitol, cellobiose, L-fucose, α-lactose, D-melibiose, D-raffinose, L-rhamnose, galactosamine, gluconolactone, inulin, DL-α-aminobutyrate, butyrate, caprylate, lactate and oxalate. The following compounds are utilized as sole sources of carbon, nitrogen and energy: L-arginine, L-asparagine, betaine, glycine, L-glutamine, L-lysine, L-ornithine, L-proline, and L-serine. The following compounds are not utilized as sole sources of carbon, nitrogen and energy: L-alanine, creatine, L-methionine, putrescine, sarcosine, L-threonine, and L-valine.

Isolated from salterns.

The mol% G + C of the DNA is: 64.2–66.0 (T_m).

Type strain: ATCC BAA-138, CECT 5384, CCM 4921, CIP 106854, DSM 3043, NCIMB 13768.

GenBank accession number (16S rRNA): AJ295146.

Additional Remarks: Other sequence: AJ295147 (*C. salexigens* ATCC 33174).

Genus IV. Zymobacter Okamoto, Taguchi, Nakamura, Ikenaga, Kuraishi and Yamasato 1995, 418[VP] (Effective publication: Okamoto, Taguchi, Nakamura, Ikenaga, Kuraishi and Yamasato 1993, 336)

TOMOYUKI OKAMOTO, HIROSHI KURAISHI AND KAZUHIDE YAMASATO

Zy.mo.bac' ter. Gr. n. *zyme* leaven, ferment; M.L. n. *bacter* masc. equivalent of Gr. neut. n. *bakterion* rod♂sc. n. *Zymobacter* the fermenting rod.

Rod-shaped cells with rounded ends, 1.3–2.4 × 0.7–0.9 µm; usually single. **Motile by peritrichous flagella** that are nonsheathed. Gram negative. **Facultatively anaerobic.** Chemoorganotrophic. **Grow on and ferment 1 mol of glucose or the hexose moiety of maltose to produce approximately 2 mol each of ethanol and CO$_2$, with trace amounts of acids.** Ferment hexoses, α-linked di- and trisaccharides, and sugar alcohols. Growth occurs at pH 4.7–8.1. Catalase positive. **The predominant cellular fatty acids are oleic acid, cyclopropanic acid of C$_{19:0}$, and palmitic acid. The hydroxylated acid is characteristically C$_{12:0\ 3OH}$. The quinone system is ubiquinone-9.**

The mol% G + C of the DNA is: 55.4–56.2 (HPLC).

Type species: **Zymobacter palmae** Okamoto, Taguchi, Nakamura, Ikenaga, Kuraishi and Yamasato 1995, 418 (Effective publication: Okamoto, Taguchi, Nakamura, Ikenaga, Kuraishi and Yamasato 1993, 336.)

FURTHER DESCRIPTIVE INFORMATION

The morphology of the cells is depicted in Fig. BXII.γ.120. The cells can possess as many as 20 peritrichous flagella.

When grown on MY agar[1], the colonies are round, entire, smooth, opaque, and milky white. Colonies of similar size develop under aerobic and anaerobic growth conditions. When the organism is grown in MY broth, the medium is slightly turbid, becoming cloudy with flocculent cells during the late log phase growth, then transparent due to sedimentation of cells during the stationary phase. No pellicle is formed. Growth is better in static cultures than in shaken cultures. *Zymobacter* requires nic-

---

1. MY agar (per liter): yeast extract (Difco), 10.0; maltose, 20.0; KH$_2$PO$_4$, 2.0; NaCl, 5.0; agar (Difco), 15.0; pH 6.0.

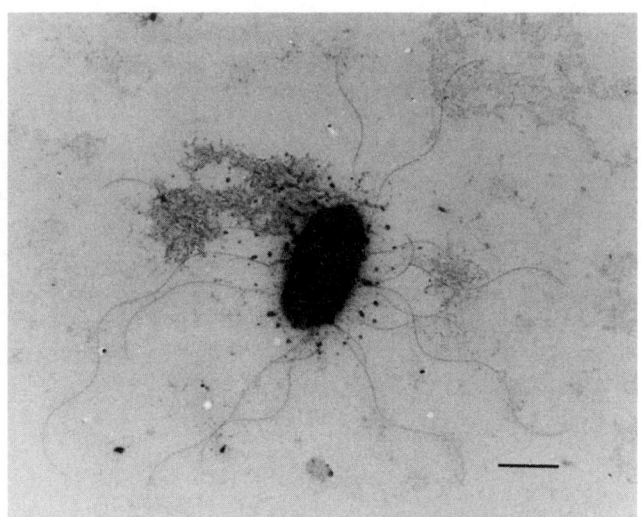

FIGURE BXII.γ.120. Electron micrograph of peritrichous flagella of *Zymobacter palmae* T109[T]. Bar = 1 μm.

otinic acid for growth. Growth does not occur in the absence of a sugar or sugar alcohol.

Growth occurs at 15–37°C (optimum, 30°C) and pH 4.7–8.1 (optimum, 6.0). The organisms are neither halophilic nor halotolerant, but they are considerably tolerant to ethanol and produce 5.8% ethanol from maltose after 6 d of fermentation.

The organism produces more than 0.5% (w/v) ethanol in the culture broth from glucose, fructose, maltose, sucrose, melibiose, raffinose, sorbitol, mannitol, and—depending on the strain— mannose and galactose. The specific growth rate on glucose is higher than that on maltose at sugar concentrations less than 7.5% (w/v). Growth is poor at glucose concentrations greater than 15% (w/v). The specific growth rate on maltose is substantially similar over a concentration range of 2.5–20% (w/v) (Table BXII.γ.102). Growth is initiated at 50% (w/v) maltose, but not at 25% (w/v) glucose; this divergence may be due to the differing osmotic pressures exerted by these sugars.

The genera *Zymomonas*, *Saccharobacter*, and *Zymobacter* are unique among bacteria in terms of their energy-yielding metabolism, which produces ethanol, CO_2, and small amounts of other products. *Zymomonas* has been studied as a potential bacterium for the industrial production of ethanol, but its spectrum of fermentable sugars is limited to glucose, fructose, and sucrose (Ingram et al., 1998). *Zymobacter* (and *Saccharobacter*) may be preferable to *Zymomonas* in this regard, as it has a wider profile of fermentable sugars, especially maltose and raffinose.

Palmitic acid ($C_{16:0}$) comprises 46–54% of the whole cell fatty acids. Oleic acid ($C_{18:1 \, \omega 9c}$) $C_{19:0 \, cyc}$ account for 30–42% (Table BXII.γ.103).

ENRICHMENT AND ISOLATION PROCEDURES

Okamoto et al. (1993) have used a two-step procedure to first isolate ethanol-tolerant bacteria and then select maltose-fermenting, ethanol-producing isolates to obtain *Zymobacter* strains. The composition of the isolation medium is (g/l): yeast extract (Difco), 10.0; glucose, 20.0; KH_2PO_4, 2.0; Bacto agar (Difco), 15.0 g; pH 6.0. Ethanol is added to a concentration of 5% (v/v). To prevent multiplication of fungi and yeasts, the antibiotic Kabicidine (Wako Pure Chemical Industries, Osaka, Japan) is added to a concentration of 100 mg/l. Diluted samples are plated onto the medium and incubated at 30°C for 4 d. Isolates are incubated at 30°C for 4 d in MY medium, and those that produce concentrations of ethanol in excess of 2% are selected. Four strains have been obtained from palm sap from among 500 samples of plant sap, fermented foods, and alcoholic beverages collected from tropical and subtropical regions of Brazil, Indonesia, Japan, Thailand, and other countries.

MAINTENANCE PROCEDURES

Zymobacter cultures can be maintained by serial transfer every 2 weeks on MY agar and stored at 4°C. *Zymobacter* can be preserved for many years by lyophilization, L-drying, or freezing with 50% glycerol at −80°C.

PROCEDURES FOR TESTING SPECIAL CHARACTERS

Characterization procedures for *Zymobacter* strains have been described by Okamoto et al. (1993). Conventional taxonomic features can be determined through standard methods. Test media that cannot support the growth must be supplemented with nicotinic acid (1 mg/l) and/or yeast extract (10.0 g/l), as well as with glucose or maltose. Flagellation is observed when cells are grown at 20°C for 12 h in brain–heart infusion broth (Difco) supplemented with nicotinic acid (1 mg /l), maltose (20.0 g/l), and liver infusion (10% (v/v); pH 6.0. The liver infusion is prepared by gently boiling 100 g of sliced cattle liver in 250 ml water for 30 min and filtering through cloth. Cells can be shadowed with platinum and visualized by transmission electron microscopy. The basal liquid medium for testing growth-factor requirement and fermentation of carbon compounds is composed of (g/l): K_2HPO_4, 7.0; KH_2PO_4, 2.0; $MgSO_4 \cdot 7H_2O$, 0.1; $(NH_4)_2SO_4$, 1.0; pH 6.0. Maltose is added when assessing growth-factor requirements, and nicotinic acid is added when evaluating the fermentation of carbon compounds.

DIFFERENTIATION OF THE GENUS *ZYMOBACTER* FROM OTHER GENERA

Table BXII.γ.104 indicates the salient features that differentiate *Zymobacter* from other phylogenetically related genera. Characteristics that distinguish *Zymobacter* from *Halomonas* and *Chromohalobacter* are ethanol and CO_2 production from glucose, growth in relationship to oxygen, halophilic property, and habitat.

TABLE BXII.γ.102. Specific growth rate of *Zymobacter palmae* T109[T] as affected by the initial concentration of glucose and maltose

| Sugar | Specific growth rate μ (h^{-1}) at a sugar concentration of (%): | | | | | | | |
	2.5	5	7.5	10	12.5	15	17.5	20
Glucose	0.47	0.43	0.38	0.26	0.19	0.08	0.07	0.05
Maltose	0.2	0.24	0.25	0.29	0.26	0.26	0.2	0.2

Phenotypically and ecologically, *Zymobacter* is similar to the ethanol-producing genera *Zymomonas* and *Saccharobacter*, both of which occur in sugar-containing niches, such as plant juices. The features that differentiate *Zymobacter* from these genera are highlighted in Table BXII.γ.105. *Zymobacter* is clearly differentiated from these two genera by flagellation, sugar fermentation, vitamin requirements for growth, osmotic tolerance, mol% G + C of DNA, and ubiquinone system.

Taxonomic Comments

Zymobacter belongs to the family *Halomonadaceae*. The 16S rRNA gene of *Zymobacter* palmae T109[T] (accession number D14555) is

TABLE BXII.γ.103. Cellular fatty acids of *Zymobacter palmae*[a]

Fatty Acid	Range, %	Mean, %
$C_{12:0}$	5–6	5
$C_{14:0}$	3–4	4
$C_{16:0}$	46–54	51
$C_{16:1}$	<0.5	<0.5
$C_{18:0}$	4–6	5
$C_{18:1\ \omega9c}$	8–21	11
$C_{19:0\ cyc}$	21–26	24
$C_{10:0\ 3OH}$	<0.5	<0.5
$C_{12:0\ 3OH}$	8–13	12

[a]From an examination of four strains. The percentages of fatty acid compositions were calculated on the basis of total non-hydroxylated acids.

TABLE BXII.γ.104. Differential characteristics of the genus *Zymobacter* and other genera in the family *Halomonadaceae*[a]

Characteristics	Zymobacter	Halomonas	Chromohalobacter
Motility	+	D	+
Flagella:			
Peritrichous	+	+	+
Polar	−	+	−
Mol% G + C of DNA	55.4–56.2	52–68	62.1–64.9
Growth in relationship to oxygen:			
Facultatively anaerobic	+	−	−
Aerobic	−	+[b]	+
Glucose fermented to ethanol/CO_2	+	−	−
Halophilic	−	+	+
Habitat:			
Palm sap	+	−	−
Saline environment	−	+	+

[a]Symbols: see standard definitions.

[b]Some strains can grow anaerobically in the presence of nitrate.

TABLE BXII.γ.105. Differential characteristics of *Zymobacter palmae* and other ethanol-producing bacteria[a]

Characteristics	Zymobacter palmae	Zymomonas mobilis	Saccharobacter fermentatus
Cell diameter, μm	0.7–0.9	1.0–1.4	0.5–0.9
Cell length, μm	1.3–2.4	2.0–6.0	1.0–1.9
Motility	+	− (usually)	+
Flagellar arrangement:			
Peritrichous	+	−	+
Polar, 1–4 flagella	−	+	−
Utilization of citrate	−	−	+
Arginine dihydrolase	−	−	+
Phenylalanine deaminase	−	−	+
β-galactosidase	−	−	+
Utilization of:			
L-Aarabinose	−	−	+
D-Xylose	−	−	+
Mannose	+	−	
Maltose	+	−	+
Sucrose	+	d	+
Melibiose	+	−	+
L-Rhamnose	−	−	+
Trehalose	−	−	+
Lactose	+	−	+
Starch	−	−	+
Growth factor requirement:			
Nicotinic acid	+	−	
Biotin	−	+	
Pantothenate	−	+	
Osmotic tolerance to:			
Glucose	<25%	40%	35%
Maltose	50%		
Mol% G + C of DNA	55.4–56.2	47.5–49.5	63.3–63.8
Ubiquinone system	Ubiquinone-9	Ubiquinone-10	
Characteristic non-hydroxylated acids, %			
$C_{16:0}$	51	12	
$C_{18:1}$ (oleic)	11		
$C_{18:1}$ (vaccenic)		69[b]	
$C_{19:0}$ cyc	24		

[a]Symbols: see standard definitions.

[b]Data from Carey and Ingram (1983).

TABLE BXII.γ.106. The signature characteristics of *Zymobacter palmae* and related bacteria

Position(s)	*Zymobacter palmae*[a]	*Halomonas*[b]	*Chromohalobacter marismortui*[c]	*Oceanospirillum linum*
76–93	6-bp stem	6-bp stem	6-bp stem	6-bp stem
484	A	A	A	G
486	C	C	C	U
640	G	G	G[d]	G
660[e]	A	A	A	A
745	U	U	U	U
668[f]	A	A	A	G
738	U	U	U	C
669[g]	A	A	A	U
737	U	U	U	A
776	U	U	U	G
1124	U	U	G[h]	U
1297	U	U	U	G
1298	C	C	C	U
1423	A	A	A	G
1424	U	C	C	U
1439	C	U	U	C
1462	G	A	A[i]	G
1464	U	C	C[j]	C

[a]*Zymobacter palmae* T109[T] (D14555).

[b]Data from Dobson and Franzmann, 1996.

[c]*Chromohalobacter marismortui* ATCC 17056[T] (X87219).

[d]G for strains T1093 (U78719), A-65 (X87220), and A-492 (X87222).

[e]Nucleotides at 660 and 745 form a base pair.

[f]Nucleotides at 668 and 738 form a base pair.

[g]Nucleotides at 669 and 737 form a base pair.

[h]G for strains ATCC 17056 (X87219), A-65 (X87220), A-100 (X87221), and A-492 (X87222); U for strain T1093 (U78719).

[i]Strains T1093 (U78719), A-65 (X87220), and A-492 (X87222).

[j]Strains T1093 (U78719), A-65 (X87220), and A-492 (X87222).

89.3–92.5% similar to that of *Halomonas* and 91.1% similar to that of *Chromohalobacter* (Dobson and Franzmann, 1996).

Zymobacter belongs to the family *Halomonadaceae*, which is in the class *Gammaproteobacteria*, and includes the marine genera *Halomonas* and *Chromohalobacter*. In contrast with these genera, *Zymobacter* is of terrestrial origin, is neither halophilic nor halotolerant, is facultatively anaerobic, and produces ethanol and CO_2. The nucleic acid similarity among the 16S rRNAs of these genera is about 90%, and, like *Halomonas* and *Chromohalobacter*, *Zymobacter* demonstrates the 15 signature sequences necessary to place it in the family *Halomonadaceae* (Dobson and Franzmann, 1996) (Table BXII.γ.106). *Zymobacter* (type strain of *Zymobacter palmae*) differs from *Chromohalobacter* and *Halomonas* at 16S rRNA positions 1424, 1439, 1462, and 1464 (*E. coli* numbering system), at which *Halomonas* has the sequences described as characteristic of its genera (Dobson and Franzmann, 1996) (Table BXII.γ.106). The 16S rRNA similarity of *Zymobacter* to another marine genus, *Oceanospirillum*, is 88.7–89.7%—similar to that between *Zymobacter* and *Halomonas* and *Chromohalobacter*. However, *Oceanospirillum* is phylogenetically distinct; it lacks the signature sequences of the *Halomonadaceae* (Dobson and Franzmann, 1996). Chemotaxonomically, the *Halomonadaceae* share the ubiquinone-9 respiratory system, whereas *Oceanospirillum* uses ubiquinone-8 (Sakane and Yokota, 1994). Because the genus *Zymobacter* and the species *Zymobacter palmae* were established in light of the taxonomic features of four strains isolated from a limited source (palm sap from the southern region of Japan), a precise circumscription of the attributes of this genus and species requires the isolation of strains from more varied sources.

The phenotypic features of *Zymobacter* are similar to those of the genera *Zymomonas* and *Saccharobacter*. All of these organisms are Gram negative, facultatively anaerobic, inhabit plant juice and the like, and produce ethanol and CO_2 from carbohydrates. Phylogenetically, *Zymobacter* is distinct from *Zymomonas*, which be-

TABLE BXII.γ.107. Other characteristics of *Zymobacter palmae*[a]

Characteristics	Reaction or result
Catalase	+
Oxidase	−
Nicotinic acid required for growth	+
Methyl red test	+
Voges–Proskauer test	+
Utilization of citrate (Koser, Christensen)	−
Nitrate reduction	−
Indole production	−
Gelatin liquefaction	−
Hydrolysis of starch	−
Phenylalanine deaminase	−
Arginine dihydrolase	−
Lysine decarboxylase	−
Ornithine decarboxylase	−
α-Glucosidase	+
β-Galactosidase	−
Utilization of carbon sources:	
L-Arabinose	−
D-Xylose	−
D-Glucose	+
D-Fructose	+
D-Mannose	d
D-Galactose	d
Maltose	+
Sucrose	+
Melibiose	+
L-Rhamnose	−
Trehalose	−
Cellobiose	−
Lactose	−
Raffinose	+
Sorbitol	+
Mannitol	+
Dextrin	−
Inulin	−

[a]Symbols: see standard definitions.

longs to the *Alphaproteobacteria*. Like *Zymobacter*, *Saccharobacter* is peritrichously flagellated, but it has a higher mol% G + C (63.5 versus 55.8). The generic description of *Saccharobacter* lacks a description of chemotaxonomic characteristics and phylogenetic

analysis. The phylogenetic relationship and exact differentiation of *Zymobacter* from *Saccharobacter* requires detailed comparative study.

List of species of the genus Zymobacter

1. **Zymobacter palmae** Okamoto, Taguchi, Nakamura, Ikenaga, Kuraishi and Yamasato 1995, 418[VP] (Effective publication: Okamoto, Taguchi, Nakamura, Ikenaga, Kuraishi and Yamasato 1993, 336.)

   pal' mae. L. gen. n. *palmae* of palm.

The characteristics are as given for the genus. See also Tables BXII.γ.104, BXII.γ.105, and BXII.γ.107. Isolated from palm sap in Okinawa Prefecture, Japan.

   The mol% G + C of the DNA is: 55.4–56.2 (HPLC).
   Type strain: T109, ATCC 51623, IAM 14233.
   GenBank accession number (16S rRNA): D14555, AF211871.

Order IX. **Pseudomonadales** Orla-Jensen 1921, 270[AL]

GEORGE M. GARRITY, JULIA A. BELL AND TIMOTHY LILBURN

Pseu.do.mon.a.da' les. M.L. fem. n. *Pseudomonas* type genus of the order; *-ales* suffix to denote order; M.L. fem. n. *Pseudomonadales* the *Pseudomonas* order.

The order *Pseudomonadales* was circumscribed for this volume on the basis of phylogenetic analysis of 16S rRNA sequences; the order contains the families *Pseudomonadaceae* and *Moraxellaceae*.

Aerobic chemoorganotrophs with respiratory metabolism. Most are motile by means of flagella. *Azomonas* spp. and *Azotobacter*

spp. fix nitrogen; *Azotobacter* spp. form cysts; *Moraxella* spp. inhabit the mucosa of animals and man.

Type genus: **Pseudomonas** Migula 1894, 237[AL] (Nom. Cons., Opin. 5 of the Jud. Comm. 1952, 121.)

Family I. **Pseudomonadaceae** Winslow, Broadhurst, Buchanan, Krumwiede, Rogers and Smith 1917, 555[AL]

GEORGE M. GARRITY, JULIA A. BELL AND TIMOTHY LILBURN

Pseu.do.mon.a.da' ce.ae. M.L. fem. n. *Pseudomonas* type genus of the family; *-aceae* ending to denote family; M.L. fem. pl. n. *Pseudomonadaceae* the *Pseudomonas* family.

The family *Pseudomonadaceae* was circumscribed for this volume on the basis of phylogenetic analysis of 16S rRNA sequences; the family contains the genera *Pseudomonas* (type genus), *Azomonas*, *Azotobacter*, *Cellvibrio*, *Mesophilobacter*, *Rhizobacter*, and *Rugamonas*. *Serpens* is also included.

Aerobic chemoorganotrophs with respiratory metabolism. Most are motile by means of flagella. *Azomonas* and *Azotobacter* fix nitrogen; *Azotobacter* forms cysts.

Type genus: **Pseudomonas** Migula 1894, 237[AL] (Nom. Cons., Opin. 5 of the Jud. Comm. 1952, 121.)

Genus I. **Pseudomonas** Migula 1894, 237[AL] (Nom. Cons., Opin. 5 of the Jud. Comm. 1952, 121)*

NORBERTO J. PALLERONI

Pseu.do' mo.nas or *Pseu.do.mo' nas.* Gr. adj. *pseudes* false; Gr. n. *monas* a unit, monad; M.L. fem. n. *Pseudomonas* false monad.

Straight or slightly curved rods but not helical, 0.5–1.0 × 1.5–5.0 μm. **Most of the species do not accumulate granules of polyhydroxybutyrate**, but accumulation of polyhydroxyalkanoates of monomer lengths higher than C_4 may occur when growing on alkanes or gluconate. Do not produce prosthecae and are not surrounded by sheaths. No resting stages are known. Gram negative.

ative. **Motile by one or several polar flagella**; rarely nonmotile. In some species lateral flagella of short wavelength may also be formed. **Aerobic, having a strictly respiratory type of metabolism with oxygen as the terminal electron acceptor; in some cases nitrate can be used as an alternate electron acceptor**, allowing growth to occur anaerobically. **Xanthomonadins are not produced.** Most, if not all, species fail to grow under acid conditions (pH 4.5 or lower). Most species do not require organic growth factors. Oxidase positive or negative. Catalase positive. Chemoorganotrophic. **Strains of the species include in their composition the hydroxylated fatty acids $C_{10:0\ 3OH}$ and $C_{12:0}$, and $C_{12:0\ 2OH}$, and ubiquinone Q-9.** Widely distributed in nature. Some species are pathogenic for humans, animals, or plants.

Editorial Note: The literature search for the chapter on *Pseudomonas* was completed in January, 2000. During the course of unavoidable publication delays, a number of new species were described or reclassified after the chapter was completed. It was not possible to include these species in the text of to include their characteristics in the comparative tables. The reader is encouraged to consult the studies listed in the Further Reading section and the *International Journal of Systematic and Evolutionary Microbiology* (2000–2003).

Note: The above-mentioned characteristics do not allow an absolute differentiation of this genus from other genera of aerobic pseudomonads[1] belonging to other ribosomal RNA groups. The following approaches add considerable solidity to the decision to assign newly isolated strains to *Pseudomonas*. One is the determination of sequence similarity in ribosomal RNA, which can be demonstrated either by hybridization techniques or, more practically, by nucleotide sequence determination. The second is determination of the fatty acid and ubiquinone composition. These two approaches will be discussed in separate sections of this chapter.

The mol% G + C of the DNA is: 58–69.

Type species: **Pseudomonas aeruginosa** (Schroeter 1872) Migula 1900, 884 (*Bacterium aeruginosum* Schroeter 1872, 126.)

FURTHER DESCRIPTIVE INFORMATION

General Comments The following sections provide further descriptive information on the properties of species assigned to the genus *Pseudomonas*, beyond the basic characteristics that have been highlighted in the above genus definition. Unfortunately, the comments that follow refer to only a few species, and the available information for these species is uneven. Thus, a literature survey shows that most references on *P. aeruginosa* are focused on its medical importance as an opportunistic pathogen; *P. putida* references are concentrated on its remarkable biochemical versatility; and those on *P. fluorescens* also highlight biochemical properties, some of which are directed to understanding its role in the promotion of plant growth. *P. stutzeri* and *P. mendocina* have attracted attention as denitrifiers and, in recent years, an infamous side of their reputation is their potential activities as opportunistic pathogens.

Many of the species listed at the end of this chapter or in the tables are little known beyond the details published in the original species descriptions. In addition to these unavoidable limitations, the details that are given for the better known species of the genus represent, by necessity, a radical simplification of the mass of information found in the literature since the date of publication of the first volume of the *Manual* (Palleroni, 1984).

Cell morphology and fine structure The cells of *Pseudomonas* strains occasionally differ substantially in size and shape from the general definition. For instance, the original description of "*P. ovalis*" (now considered to be a synonym of *P. putida*) highlighted the oval shape of the cells, while the cells of other *P. putida* strains and of some plant pathogenic fluorescent species may far exceed the 5 µm indicated in the definition.

A morphological character that was used to differentiate members of the genus *Pseudomonas* from other aerobic pseudomonads was the inability of the former to accumulate endocellular granules of poly-β-hydroxybutyrate (PHB) when growing in media of low nitrogen content on various carbon sources. More recently, the fluorescent members of *Pseudomonas* were found to be able to accumulate polyhydroxyalkanoates (PHAs) composed of monomers of medium chain lengths (C_6 to C_{12}) when grown on carbon sources such as gluconate, alkanes, and alkenes (Huisman et al., 1989; Anderson and Dawes, 1990; Steinbüchel and Valentin, 1995). Further comments on this subject may be found in the section Procedures for Testing Special Characters, below.

Thin sections of cells of *Pseudomonas* show cell walls and membranes characteristic of Gram-negative bacteria. In freeze-etched preparations of *P. aeruginosa* cells, as many as nine layers can be defined, accounting for all the electron-dense and electron-transparent layers that can be observed in thin sections (Lickfield et al., 1972; Gilleland et al., 1973). Gilleland and collaborators have shown that the outer membrane of the cell wall could be split down the middle when the cells were freeze-etched in the presence of a cryoprotective agent (Gilleland et al., 1973), thus confirming an earlier interpretation of similar experiments. The outer layer thus separated from the outer membrane had in its inner face numerous spherical units, which are made up of protein and in some cases were aggregated in the shape of small rods. *P. aeruginosa* is extremely sensitive to the action of EDTA, which provokes cell lysis. The spherical elements were removed by EDTA treatment but could be restored by addition of magnesium ions, which also restored cell stability.

Flagella and pili (fimbriae) Typically, *Pseudomonas* cells have polar flagella. The flagellar insertion in some instances is not exactly polar but subpolar, and occasionally it may be difficult to differentiate the latter from the so-called degenerately peritrichous type observed in members of other genera. In addition to the polar flagella, lateral flagella of short wavelength may be produced by strains of some species (*P. stutzeri*, *P. mendocina*) and are shed much more easily than the polar flagella. Growth on solid media favors the formation of lateral flagella, which suggests that they may be involved in swarming of the population on solid surfaces (Shinoda and Okamoto, 1977).

The number of flagella has taxonomic importance. Based on the work performed at the University of California in Berkeley, the number of flagella (usually expressed as "one" or "more than one") was highly characteristic for all the strains of a species that have been examined. To get good results, it is advisable to follow well-controlled growth conditions and to express the results on a statistical basis, as indicated by Lautrop and Jessen (1964). Some practical recommendations and basic details of the technique described by these authors will be summarized in the section Procedures for Testing Special Characters. Unfortunately, they have not been followed in the description of most species. In Jessen's classical study of *P. aeruginosa*, 97% of the cells with stained flagella had only one flagellum, 3% two flagella at one pole, and only three cells (about 0.05% of the collection) had three flagella. According to the definition proposed by Lautrop and Jessen (1964), the group can be characterized as polar monotrichous. Only one strain deviated from the limit defined for monotrichous strains in having two flagella on 15% of the flagellated cells, instead of the accepted maximum of 10%.

The amino acid sequence of the flagella of a given species is not always homogeneous. An analysis of the virulence-associated locus *fliC* of *P. aeruginosa* revealed two types of flagellin genes, one variable and the other conserved. Differences in the flagellar sequences amount to about 35%, but the two proteins fold into similar structures during polymerization (Spangenberg et al., 1996). In two strains of *P. putida*, flagellins of apparent molecular masses of 81 kDa and 50 kDa, respectively, were identified. Within the chromosomal fragment encoding the large component, two genes homologous to two flagellin genes of *Salmonella typhimurium* and one gene homologous to a *P. aeruginosa* flagellin gene were identified. The deduced molecular mass of the product of

---

[1]. The word "pseudomonad" will be used quite often in this chapter. Although it may be clear to many, the precise meaning is given in the paper "The aerobic pseudomonads: a taxonomic study" by Stanier et al. (1966). It comprises all the species assigned to the genus *Pseudomonas* as classically defined. Ever since this publication, a fairly accurate definition has also been included in the editions of Webster Collegiate Dictionary.

the first flagellin gene was 68 kDa instead of 81 kDa, suggesting posttranslational modification. Both the N- and C-terminal sequences are conservative, while the middle section is variable (Winstanley et al., 1994). PCR amplification of flagellar genes of *P. aeruginosa* gave products that were analyzed for restriction fragment length polymorphism, and 13 groups could be defined, suggesting that this approach may provide a useful genetic marker for the study of genetic variation among closely related species (Winstanley et al., 1996).

Fimbriae (pili) of polar insertion have been reported for *P. aeruginosa* and *P. alcaligenes* in the early studies of Fuerst and Hayward (1969). No fimbriae have been observed in the strains of *P. fluorescens*, *P. chlororaphis*, and *P. putida* that have been examined.

The *P. aeruginosa* pili are about 6 nm wide, thinner than those of enteric bacteria. They act as receptors for various phages and are retractile (Bradley, 1972a, b), but not all the pili of this species have the property of withdrawing into the cell (Bradley, 1974). Fig. BXII.γ.121 shows the polar fimbriae of *P. aeruginosa*, whose genetic determinants are mobilized by the FP plasmids (Bradley, 1980b)

P. aeruginosa produces type-4 pili, which are found in a number of pathogenic bacteria and are involved in cell adhesion to epithelial cells (Woods et al., 1980; Doig et al., 1988). The adhesive region is located at the tip of the pilus, but this mode of attachment is not universal among fimbriae of different species (Smyth et al., 1996). In the C-terminal region of pilin there is a disulfide loop of 12–17 amino acids, and even though pili of different prototypes have little sequence similarity, the loops in all cases are structurally similar. However, the loops do not seem to be essential in maintaining the functionality of the binding domain. In addition to the fimbriae, exoenzyme S and flagella are additional adhesins of importance in the attachment process (Hahn, 1997).

Type-4 pili not only determine attachment to epithelial cells, but also cause a form of surface translocation called twitching motility (Henrichsen, 1975; Bradley, 1980a). Adhesion of *P. aeruginosa* cells to mucin is pilus-independent. A gene *fliO* complemented mutations in motility and adhesion to mucin (Simpson et al., 1995). A gene homologous to the *fliF* gene of other species has been identified in *P. aeruginosa*. It is involved in both motility and adhesion to mucin. Further characterization of the markers involved showed that the basal body structures of flagella also are important for adhesion (Arora et al., 1996).

Pilus formation is genetically very complex and depends on the expression of more than 30 genes distributed in at least six regions of the *P. aeruginosa* chromosome (Mattick et al., 1996; Tan et al., 1996; Watson et al., 1996a). The locations of a number of the pili genes have now been determined. Some of the genes encode proteins highly homologous to factors involved in protein secretion (Alm and Mattick, 1995, 1997; Alm et al., 1996b; Hahn, 1997). At least four different mechanisms are known to participate in the assembly of fimbriae from subunits (Vu-Thien et al., 1996

Some years ago, it was found that the amino terminal sequence of *P. aeruginosa* pili had marked similarity with that of *Neisseria meningitidis* and *Moraxella nonliquefaciens* (Buchanan and Pearce, 1979). One of the gene clusters encodes proteins strikingly similar to chemotaxis components of enteric bacteria and the gliding *Myxococcus xanthus*. In *M. xanthus*, type-4 pili have a role in what has been called "social motility" (Darzins, 1994; Darzins and Russell, 1997; Wu et al., 1997b). Work done with

polyclonal antibodies raised against pilins expressed by *P. aeruginosa*, *Moraxella bovis*, *Neisseria gonorrhoeae*, *Dichelobacter nodosus*, and *Vibrio cholerae* showed that these polypeptides have conserved antigenic and, in most instances, immunogenic determinants, which are located in the highly similar amino-terminal domains (Patel et al., 1991).

AlgR is considered to be a regulator of alginate production in *P. aeruginosa*; in addition, it is required for twitching motility. Adjacent to the *algR* gene, a sensor gene (*fimS*) has been identified, and AlgR and FimS appear to represent another example of the two-component signal transduction systems that have been identified in bacteria. AlgU, an alternate sigma factor, also affects both alginate production and twitching motility, indicating that type-4 fimbriae and alginate, two important virulence determinants, are closely related (Whitchurch et al., 1996).

A gene region identified in nonfimbriated, phage-resistant transposon mutants of *P. syringae* pathovar *phaseolicola* (in this treatment described as a pathovar of *P. savastanoi*) has a high degree of similarity with a *P. aeruginosa* region required for pilus type-4 production (Alm and Mattick, 1995). Pili mutants that have lost the twitching motility characteristic of type-4 fimbriae, could be complemented in *trans* by homologous or heterologous subunits from the same strains (PAO or PAK) and from strains of other species such as *Dichelobacter nodosus* (Watson et al., 1996b).

Composition of cell envelope The cytoplasmic membrane separates the bacterial cytoplasm from other components of the cell envelopes, and its efficiency as a barrier is related to the lipid bilayer composition, which may undergo relatively rapid changes to adjust its fluidity to changes in the composition of the medium and the physical conditions of the environment.

In contrast to the cytoplasmic membrane, the structure of the outer membrane is asymmetrical. It is composed of a special lipid (lipopolysaccharide, LPS) with hydrocarbon chains giving low fluidity to the inner region of the LPS. Toward the outside region, chains of carbohydrates give to the whole cells a specific antigenic identity. The outer membrane is quite effective as a barrier against passage of molecules such as the antibiotics, which may have a deleterious effect on the cells, but at the same it must be permeable to nutrients. This discrimination is effected by the presence of porins.

The outer membrane proteins of pseudomonads carry the designation Opr followed by a letter or a letter and number (Hancock and Carey, 1980). One monomeric *Pseudomonas* protein, OprF, is similar to OmpA of *E. coli*; it is highly antigenic and not specific toward the solutes that enter the cells (Nikaido, 1992). In *Pseudomonas*, porins are composed of monomeric proteins, and not trimeric proteins as in *E. coli*. The outer membrane of *P. fluorescens* can change its permeability in direct relationship with growth temperature, even though no difference in composition could be found to correlate with this transition. The effect was attributed to a change in the structure of the porins that affects their function, since a decrease in the growth temperature induced a reduction in the conductance in the major component, OprF, suggesting a structural change (De et al., 1997).

Protein fractions representing the OprF of two *P. fluorescens* strains were purified and characterized. A sequence similarity of 94% was found among OprFs of strains from origins as different as milk and soil (De et al., 1995). OprF of *P. aeruginosa* is a major outer membrane component, and as such it is potentially inter-

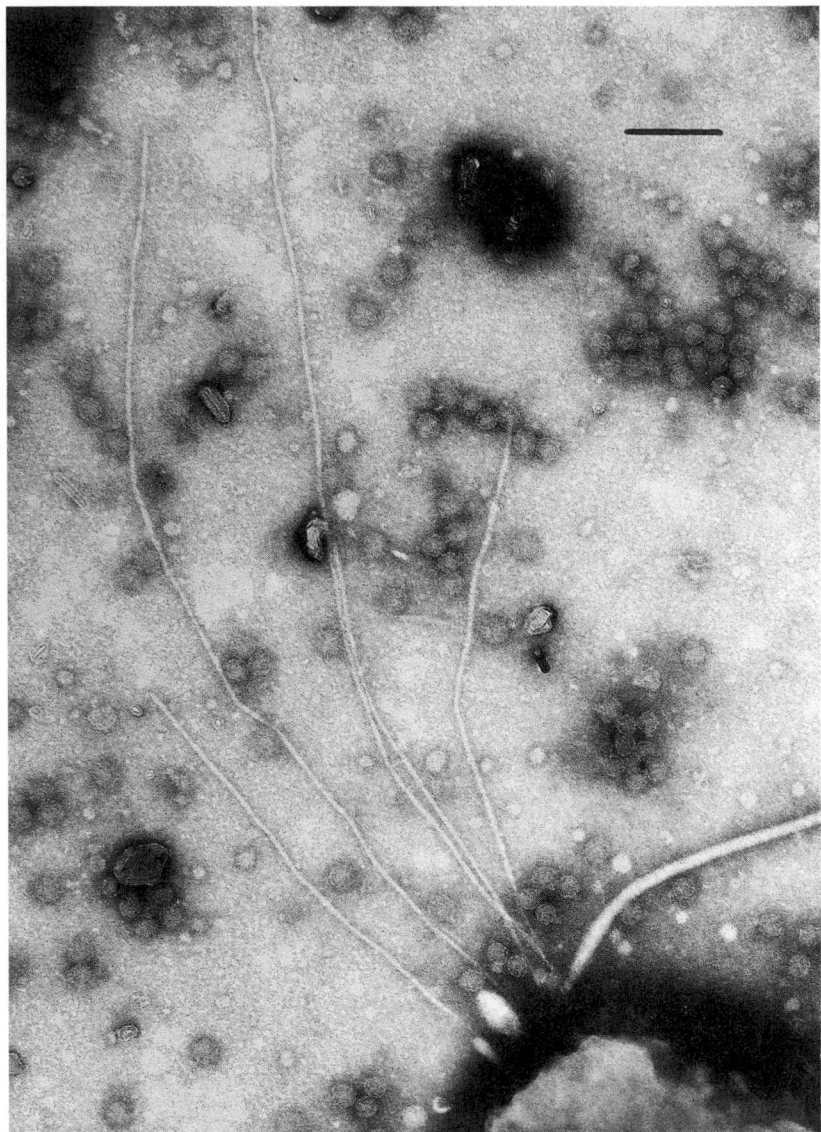

FIGURE BXII.γ.121. *Pseudomonas aeruginosa* strain PAC5 polar pili, which have been prevented from retracting by the adsorption of a pilus-specific bacteriophage; this can be seen scattered over the field. Bar = 0.1 μm. (Reproduced with permission from D.E. Bradley, Canadian Journal of Microbiology *26*:155–160, 1980, ©National Research Council of Canada.)

esting for the preparation of vaccines (Rawling et al., 1995). There are differences in the location of OprFs between the two species. Thus, while in *P. fluorescens* the protein is embedded in the outer membrane in such a way that it offers a surface-exposed region rich in proline, this region is absent in *P. aeruginosa* and in *P. syringae*. In addition, in *P. aeruginosa*, the entire protein is embedded, without an exposed region to the outside (De Mot et al., 1994).

An outer membrane protein of *P. fluorescens* that was inducible under conditions of phosphate limitation was purified and characterized (Leopold et al., 1997). It did not show similarity to any of the known outer membrane proteins. Its distribution among different strains of the species was uneven, i.e., it was not present in all strains, which simply may reflect the internal heterogeneity of *P. fluorescens*. A method of immunofluorescence microscopy involving cell permeability was developed to visualize the specific

expression of the protein in cells exposed to limiting phosphate concentrations, in order to use the system to measure phosphate availability (Leopold et al., 1997). In *P. aeruginosa*, exposure to limiting phosphate provoked the expression of porin OprP. Only one of three lysine residues of this protein was required to form the specific phosphate binding site in OprP (Sukhan and Hancock, 1996).

Resistance to polymyxin B, gentamicin, and EDTA is associated with the outer membrane protein OprH in *P. aeruginosa*, which is expressed under magnesium-limiting conditions (Rehm and Hancock, 1996).

P. aeruginosa is able to simultaneously express two different LPS types, which migrate to different positions in electrophoresis and have been named bands A and B. The B band corresponds to the component that confers to the cell its O-antigen serological specificity and the A band to a common antigen. Two of the eight genes that are part of the operon that controls band A biosynthesis and transport encode proteins that are highly similar to a number of proteins that are part of the ABC (acronym for "ATP-binding cassette") transport systems (Rocchetta and Lam, 1997). Of the two LPS O-polysaccharide species, one confers to the cells high hydrophobicity, and the other, hydrophilicity. The factors that provoke a change in the ratio between these two components affect both adhesion and survival (Makin and Beveridge, 1996).

Pigments and siderophores Early taxonomic treatments of the genus *Pseudomonas* included pigmentation as a generic character, but this is no longer valid. In fact, the colonies and other cell masses always display some colors due to normal cellular components, which, in some instances, become quite apparent. Thus, *P. stutzeri* is grouped with the nonpigmented species, even though the colonies of many strains become dark brown due to high concentration of cytochrome *c* in the cells.

In his comprehensive monograph on the taxonomy of *P. aeruginosa* and other fluorescent pseudomonads, Jessen (1965) listed six pigments produced by the type species: four phenazines (pyocyanin, pyorubin, chlororaphin, oxiphenazin), the *Pseudomonas* blue protein, and pyoverdine. Several phenazine pigments can be produced by a single strain (Chang and Blackwood, 1969). The best known and most characteristic of these pigments is the phenazine blue pigment pyocyanin, identified many years ago as the cause of the blue color of the pus of wounds infected with *P. aeruginosa*. Synthesis of pyocyanin by *P. aeruginosa* can be stimulated by growth in King A medium (King et al., 1954). It is inducible in a dose-dependent manner and its production is strongly enhanced by addition of L-N-(3-oxohexanoyl) homoserine lactone to the medium (Stead et al., 1996).

Phenazine pigments synthesized by other fluorescent pseudomonads are the green, almost insoluble, chlororaphin of *P. chlororaphis*, and the orange phenazine-monocarboxylic acid characteristic of *P. aureofaciens* (in the present treatment these taxa are considered to be subspecies of *P. chlororaphis*). The species "*P. lemonnieri*", which was considered by Stanier et al. (1966) as a biovar of *P. fluorescens*, produces an intracellular blue pigment that has been chemically characterized (Starr et al., 1960, 1967). Some strains of *P. aeruginosa* are able to produce melanin pigments (Mann, 1969).

Other important pigments from the physiological and taxonomic standpoints are the pyoverdines. They are the typical yellow-green pigments of the so-called fluorescent pseudomonads. Jessen (1965) found that his collection of strains of *P. aeruginosa* contained 328 fluorescent strains producing pyocyanin, 18 fluorescent strains not producing pyocyanin, five nonfluorescent strains with pyocyanin, and three nonpigmented strains. Strains of this species lacking one or more of the characteristic pigments, however, could be identified based on other phenotypic properties, and, in general, pigmentation is a most striking but not always dependable characteristic for species identification. In any case, identification of fluorescent *Pseudomonas* species by examination of the fluorescence profiles of pigments diffusing into the medium has been proposed by Shelly et al. (1980).

Pyoverdines are also physiologically important because they function as efficient siderophores (Meyer and Abdallah, 1978; Meyer and Hornsperger, 1978). Their production is enhanced under conditions of iron-starvation. All pyoverdines share a quinoleinic chromophore, which is linked to peptides of different compositions and sizes. This may vary from 6–12 D- and L-amino

acid residues. With minor exceptions, each fluorescent *Pseudomonas* type strain produces a pyoverdine with a specific amino acid composition (Budzikiewicz, 1993). However, within a single species there is a diversity of peptide structures.

The presence of D- and L-amino acid in pyoverdines as well as possible cyclic structures suggests a similarity the cyclic peptides of antibiotic properties produced by some Gram-positive bacteria and toxins synthesized by *Pseudomonas syringae* pathovars. These observations are indicative of a nonribosomal type of synthesis (Georges and Meyer, 1995; Merriman et al., 1995). In a search for components of such a system, high molecular weight cytoplasmic proteins have been identified. As expected, they are produced by cells of fluorescent pseudomonads in iron-deficient media, for which they have been named iron-repressed cytoplasmic proteins (IRCPs). They vary from M_r 180–600 kDa, and they give characteristic electrophoresis profiles for strains producing different pyoverdines. On the other hand, they are absent from mutants in pyoverdine synthesis and from cells of nonfluorescent pseudomonads (Georges and Meyer, 1995). As mentioned by Meyer et al. (1997), pyoverdines provide information of use in strain typing and epidemiological applications.

Using a system with a promotorless *lacZ* gene, 24 insertion mutants of *P. aeruginosa* unable to synthesize pyoverdine were isolated. All the mutations could be allocated within a 103-kb region ("pyoverdine region") at 47 min of the *P. aeruginosa* strain PAO genetic map, very near the catechol region (Tsuda et al., 1995). The genetic determinants of pyoverdine production occupy a region of at least 78 kb, but few genes have been characterized at present. *pvdE*, recently identified, codes for a protein that is a member of the ATP-binding cassette group of membrane transporters (McMorran et al., 1996). *pvdA* encodes an enzyme that catalyzes a key step in the synthesis. Three tightly iron-regulated regions are located in a fragment upstream of *pvdA*. Fur (ferric uptake repressor, a protein that controls expression of iron-repressible genes) indirectly controls *pvdA* transcription. When iron is not limiting, Fur blocks the *pvdA* promoter, thus inhibiting transcription of several pyoverdin genes (Leoni et al., 1996). Gene *pvdS* is also required for pyoverdine biosynthesis by *P. aeruginosa*. Under conditions of unlimited iron, Fur also acts here as a repressor, binding to the *pvdS* promoter and preventing expression (Cunliffe et al., 1995).

Because siderophores are iron-scavenging compounds, a large number of strains of *P. aeruginosa* have been analyzed by different methods for pyoverdine-mediated iron incorporation, and in all cases the collection could be subdivided into the same three groups, in spite of the inclusion of strains devoid of pyoverdine production in some of the groups (Meyer et al., 1997). The *P. aeruginosa* cells also are able to incorporate iron combined with siderophores of foreign origin. For instance, they can produce a receptor (PfeA) for the iron trapped by enterobactin, a siderophore synthesized by enteric bacteria. PfeA production depends on a regulator and a sensor, which are members of the family of two-component regulatory systems. The operon of regulator/sensor is probably regulated by iron, since the regulator gene in this system has a sequence similarity to the iron uptake regulator *fur*, the gene that codes for the above-mentioned Fur protein (Dean et al., 1996). Iron can be transported into *P. aeruginosa* not only by enterobactin but also by at least one of its breakdown products, 2,3-dihydroxybenzoyl-L-serine, in a process that is neither iron-repressible nor strongly energy-dependent (Spangenberg et al., 1995).

Attempts to complement a pyoverdine mutant from a *P. ae-*

ruginosa PAO cosmid bank resulted in the recovery of an apparent wild-type phenotype. Physical mapping indicated that the cloned fragment corresponded to a different region of the PAO chromosome and that the properties of the transconjugants were not the result of a true complementation. The yellow-green fluorescent compound was different in its properties from pyoverdine and it was named pseudoverdine. It lacks a peptide chain, but it resembles pyoverdine in its spectral properties (Stintzi et al., 1996).

Aside from pyoverdines, the fluorescent pseudomonads also produce other strain-specific, but chemically related, siderophores called pseudobactins. A method for isolation of pseudobactin from pseudomonads, and an assay for fluorescent siderophores based on reverse-phase HPLC, have been described (Nowak-Thompson and Gould, 1994). When grown in an iron-deficient medium, each of two strains of *P. putida* and *P. fluorescens* produced two different novel yellow-green fluorescent pseudobactins. All four compounds contained a dihydroxyquinoline-based chromophore. The receptor proteins of the two species are similar but not identical (Khalil-Rizvi et al., 1997). The composition of the corresponding peptides has been described, but the reason for the differences are not apparent, since the uptake system seems to consist of a single receptor in both organisms. Fragments of the peptide component of one pseudobactin have been synthesized. Antifungal peptides called pseudomycins have related structures, but the one mentioned here does not have activity against fungi (Koushik et al., 1997).

A *P. putida* strain WCS358 produces a fluorescent pseudobactin that has been described in detail (von der Hofstad et al., 1986). In spite of the results mentioned in the previous reference, the results here indicate that the ability of *P. putida* WCS358 to use different pseudobactins of various origins is related to the presence of multiple outer membrane receptor proteins (Koster et al., 1995).

In addition to the siderophores mentioned above, the fluorescent pseudomonads produce another one, pyochelin, which is not pigmented (that is, it does not have an absorption spectrum in the visible region), and appears to have low efficiency as an iron-scavenging compound (Liu and Shokrani, 1978). In the composition of pyochelin there is one molecule of salicylic acid (which is by itself a siderophore) condensed to two cysteinyl residues. In *P. aeruginosa*, the pathway of salicylate synthesis has been clarified: conversion of chorismate (which also acts as an intermediate of aromatic amino acid biosynthesis) to isochorismate, which is converted to salicylate plus pyruvate. These steps are catalyzed by two iron-repressible proteins, PchA and PchB (Serino et al., 1995).

The iron trapped by the siderophores is utilized by the producing organism presumably after reduction to ferrous ion, a step catalyzed by ferrisiderophore reductases. Ferripyoverdine reductase is ubiquitous among *Pseudomonas* species (Halle and Meyer, 1992), and since its activity is not under iron regulation, it is not part of the siderophore genetic system (J.-M. Meyer, personal communication).

There are earlier reports of other fluorescent compounds that have been isolated from fluorescent pseudomonads. These include four pteridine derivatives isolated from "*P. ovalis*" (presumably *P. putida*) (Suzuki and Goto, 1971); two fluorescent antibiotics isolated from *P. fluorescens*, which were named fluopsin C and fluopsin F (Shirahata et al., 1970); 6-hydroxymethylpteridine, an intermediate in folic acid synthesis (Viscontini and Fra-

ter-Schröder, 1968); and erithroneopterine of *P. putida* (Suzuki and Goto, 1972).

P. fragi, a nonpigmented member of the genus, does not produce siderophores in detectable amounts and it is very sensitive to iron-limiting conditions. Growth can be strongly stimulated by iron combined with siderophores of foreign origin. The siderophores include enterobactin, some pyoverdines, and siderophores of eucaryotic origin, such as transferrin, lactoferrin, and hemoglobin. Probably this property is related to the capacity of *P. fragi* to grow in milk, from which the first strain of this species was isolated many years ago (Gruber, 1905). Iron starvation in this species may induce the synthesis of the siderophore-mediated iron uptake system (Champomier-Vergès et al., 1996).

The following is a useful summary of siderophores produced by *Pseudomonas* species, kindly supplied by J.-M. Meyer from a manuscript submitted for publication. All strains of the fluorescent species *P. aeruginosa*, *P. fluorescens*, *P. chlororaphis*, and *P. putida* produce pyoverdines as main siderophores. Some strains also produce pyochelin and/or salicylic acid as secondary siderophores (Cox et al., 1981; Meyer et al., 1992; Visca et al., 1993). *P. stutzeri* ATCC 17588 produces desferrioxamines E and D2 (Meyer and Abdallah, 1980; Azelvandre, 1993); a different strain of *P. stutzeri* (RC7) gives a catechol-type siderophore (Chakraborty et al., 1990) and an unclassified *Pseudomonas* sp., aerobactin (Buyer et al., 1991). No siderophores have been detected in *P. stutzeri* YPL-1 (Lim et al., 1991), in *P. fragi* (Stintzi et al., 1996), and in *P. mendocina*. It is interesting to note here that the internal heterogeneity of the nonfluorescent species *P. stutzeri* is also reflected in the capacity for siderophore production.

P. alcaligenes, *P. mendocina*, and *P. flavescens* produce yellow to orange pigments that have not been chemically characterized. Strains of one of the biovars of *P. fluorescens*, which were originally assigned to the species "*P. lemonnieri*", produce an intracellular insoluble pigment of structure related to that of indigoidine, the purple pigment of "*Pseudomonas indigofera*". Both pigments are derivatives of 3,3'-bipyridyl (Kuhn et al., 1965). The pigment of "*P. lemonnieri*" has been reexamined, its chemical structure has been clarified, and it has received the name "lemonnierin" (Ferguson et al., 1980; Jain and Whalley, 1980). Pigments that are not soluble in water and remain associated with the cell mass are found in many former *Pseudomonas* species now classified in other genera.

Nutrition and growth conditions   Strains of *Pseudomonas* species can grow in minimal, chemically defined media with ammonium ions or nitrate as nitrogen source and a single organic compound as the sole carbon and energy source. Some of the species previously assigned to *Pseudomonas* have true growth factor requirements, but none of the species belonging to the rRNA group I (Palleroni, 1984) has an absolute dependence on these nutritional supplements. In media of minimal composition, strains of phytopathogenic *P. syringae* grow very slowly in comparison with strains of the main saprophytic species, and that growth is enhanced by addition of small amounts of complex organic materials (yeast extract, peptones). However, in most cases no true dependence on organic growth factors can be demonstrated. Pantothenate is required by strains of *P. syringae* pathovar *avellanae*, now considered as an independent species, *P. avellanae* (Janse et al., 1996) (see List of Species). In addition, organic growth factors are required by some species of uncertain phylogenetic position (*P. iners*, *P. lanceolata*, *P. spinosa*) (see List of Species). Occasionally, attempts have been made to improve the poor growth of phytopathogenic *Pseudomonas* in chemically defined media. In one such proposal, a defined medium for "*P. tomato*" (*P. syringae* pathovar *tomato*) includes L-asparagine, L-glutamine, or L-threonine as a nitrogen source, and D-galactose as carbon source. The organism grows in this medium as well as in complex media (Bashan et al., 1982).

From time to time the capacity for nitrogen fixation has been claimed for some species that later on proved to be unable to perform this function under strictly controlled conditions. In the first edition of this *Manual,* it was assumed that none of the *Pseudomonas* species could be considered legitimate nitrogen fixers; however, this ability is reported to occur in strains of the nonfluorescent species *P. stutzeri* (Krotzky and Werner, 1987; Puente and Bashan, 1994), which is also a vigorous denitrifier.

Requirement for sodium ions has been observed for *Pseudomonas elongata* which, according to Anzai et al. (2000) should be placed in the genus *Microbulbifer* following further taxonomic studies, and for *P. halophila* (described in the section Other Species). This requirement, which is one of the characteristics typical of so-called marine eucaryotes, has not been determined in the majority of *Pseudomonas* species.

P. aeruginosa can utilize one of many different sources of sulfur in the medium. The list includes inorganic and organic compounds, from which the aromatic sulfur compounds are excluded. If sulfur sources other than the preferred ones (sulfate, cysteine, or thiocyanate) are the only ones available, a set of 10 sulfate starvation induced (SSI) proteins are upregulated (Hummerjohann et al., 1998). One of these proteins is periplasmic and has high affinity for sulfate. Even though no similarity of the other nine SSI proteins to other proteins of known function has been detected, they also may represent scavenging elements for the S-sources preferred by the cells. Studies on the genetics of the starvation response have indicated the role of cysteine biosynthetic intermediates and the possibility that at least two independent co-repressors are operative in *P. aeruginosa* (Hummerjohann et al., 1998).

The ability to grow in very simple mineral media at the expense of many organic compounds has served as the basis for extensive nutritional characterization of a large number of strains, providing a mass of phenotypic data ideally suited for taxonomic studies by numerical methods. Aside from their taxonomic implications, nutritional investigations on the utilization of certain groups of compounds have served as the bases for many interesting studies on metabolic pathways, their regulatory mechanisms, and their phylogenetic significance.

The best growth temperature for growth of most strains is approximately 28°C. Some species grow at a substantial rate at 4°C and thus can be considered psychrotrophic. For others the maximum temperature is about 45°C, and therefore they are not true thermophiles. None of the members of the genus tolerates acidic conditions and growth is invariably negative at pH 4.5.

Metabolism and metabolic pathways The metabolism of *Pseudomonas* is typically respiratory with oxygen as the terminal electron acceptor, but some species also can use nitrate as an alternate electron acceptor and can carry out oxygen-repressible denitrification (dissimilatory reduction of nitrate to N_2O or N_2). In most cases, denitrification is the property of all members of a given species, and only a few strains may be unable to denitrify. Denitrification is not a species characteristic in *P. fluorescens*; only some biovars are able to denitrify (Stanier et al., 1966).

Some cytochromes are involved in denitrification through the

participation of a special cytochrome oxidase that probably is a remnant of a very primitive mechanism dating from the pre-aerobic era of the planet (Yamanaka, 1964). An excellent review on the cellular and molecular aspects of denitrification has been published by Zumft (1997). *Pseudomonas* is rich in denitrifying species. The core structures of the denitrification apparatus in these and in denitrifiers belonging to other genera demand the participation of around 50 genes.

Assimilation of nitrate occurs through the reduction of nitrate to ammonia. Although there may be common intermediates in the assimilatory and dissimilatory routes of nitrate reduction (Hartingsveldt et al., 1971), the pathways are encoded by different sets of genes. Some mutations with pleiotrophic effects are due to alterations in a gene involved in molybdenum incorporation into the nitrate reductases (Sias et al., 1980). A comprehensive review on the diversity of enzymes involved in denitrification in the pseudomonads and the genetic organization is available (Zumft and Körner, 1997), as well as a description of the localization of the genes in the *P. aeruginosa* PAO map (Vollack et al., 1998). The genes for nitrate reduction (*nir*), nitric oxide reduction (*nor*), and nitrous oxide reduction (*nos*) have been located in a 30-kb gene cluster in the chromosome of *P. stutzeri* (Braun and Zumft, 1992), and a sequence analysis of a 9.72-kb internal segment that includes the genes is now available (Glockner and Zumft, 1996). Very useful accounts of the enzyme diversity and gene organization of the denitrification genes in denitrifying pseudomonads, as well as the regulatory elements of the denitrification system of *P. stutzeri*, have been published by Zumft and collaborators (Cuypers and Zumft, 1992; Zumft and Körner, 1997). As mentioned before, a review by Zumft (1997) includes much useful information on this subject.

The oxidative degradation of some substrates (particularly the aromatic compounds) or their intermediates by *Pseudomonas* occasionally involves the participation of oxygenases. Both mono- and dioxygenases coupled to a variety of electron donors are well represented in species of the genus. Oxygenases acting on aliphatic compounds such as alkanes may be part of complex systems, and this is also true for the oxidation of some compounds like camphor by *P. putida*, with steps involving oxygenases of considerable complexity that include the iron-sulfur protein putidaredoxin and cytochrome P-450$_{CAM}$ (Gunsalus et al., 1971). Other systems include the iron protein rubredoxin (Lode and Coon, 1971) and cytochrome *o* (Peterson, 1970).

The classical reactions of the tricarboxylic acid cycle are found in all species of *Pseudomonas* that have been examined. A key reaction is the synthesis of citrate from oxaloacetate and acetyl-CoA, which is under a control system typical of absolute aerobic organisms (Weitzman and Jones, 1968). The control of peripheral catabolic enzymes (amidase, histidase, enzymes of aromatic compounds, and camphor metabolism) by intermediates of the tricarboxylic acid cycle are a manifestation of the central position that it occupies in metabolism. Those intermediates that are used for biosynthetic purposes can be replenished by carboxylation of pyruvate and by the action of the enzymes isocitrate lyase and malate synthase, which are part of the anaplerotic system known as the glyoxylate cycle (Kornberg and Madsen, 1958). A multienzyme complex of tricarboxylic acid cycle enzymes (fumarase, malate dehydrogenase, citrate synthase, aconitase, and isocitrate dehydrogenase) that catalyzes the reactions from fumarate to α-ketoglutarate has been identified in *P. aeruginosa* cells, from which it can be released by gentle osmotic lysis. The complex can be reconstituted from the individual enzymes provided that one of the two citrate synthase isoenzymes is present (Mitchell, 1996).

A variety of macromolecules can be degraded by some strains by means of extracellular enzymes. Hydrolytic enzymes that have been studied in detail include the proteases of *P. aeruginosa* (Morihara, 1964; Morihara et al., 1965), which are important in infections caused by this organism. In addition to these early studies, others will be discussed in relation to pathogenesis to humans and animals.

One of the differences between *P. fluorescens* and *P. putida* is the ability of the former species to produce extracellular proteases, to which the property of gelatin liquefaction can be attributed. A metalloprotease of *P. fluorescens* that is inhibited by EDTA and has a trypsin-like activity has been found to be highly homologous to zinc metalloproteases of diverse origins (Kim et al., 1997a).

Carbohydrates   The amylolytic activity of *P. stutzeri* is responsible for rapid starch hydrolysis, one of the characteristic phenotypic properties of the species. The enzymology of the exoamylase, which is responsible for the formation of maltotetraose as end product, has been examined at the molecular level, and the enzyme has been cloned (Morishita et al., 1997). Phytopathogenic pseudomonads have been found to hydrolyze pectin (Hildebrand, 1971; Ohuchi and Tominaga, 1973, 1975; Wilkie et al., 1973), xylan (Maino et al., 1974), and glycosides (Hayward, 1977).

Common monosaccharides (glucose, fructose, galactose, L-arabinose) are used by strains of most species of the genus, but growth of some of the species (*P. stutzeri*, *P. mendocina*, *P. syringae*) may be slow. Most hexoses are degraded by the Entner-Doudoroff pathway, which was discovered in studies on *P. saccharophila* (Entner and Doudoroff, 1952).

The fluorescent pseudomonads have multiple peripheral pathways for glucose oxidation that converge for the synthesis of 6-phosphogluconate, which is further degraded by the Entner-Doudoroff pathway (Eisenberg et al., 1974). Of these routes, one involves direct oxidation of the sugar (oxidative pathway), and either gluconate or 2-ketogluconate can serve as a precursor of 6-phosphogluconate. However, in *P. putida*, 6-phosphogluconate is synthesized preferentially from 2-ketogluconate (Vicente and Cánovas, 1973). Induction of the oxidative pathway in *P. aeruginosa* can only occur in the presence of oxygen. Under denitrifying conditions, only the so-called phosphorylative pathway (starting with the phosphorylation of glucose) is operative in this organism (Hunt and Phibbs, 1981).

A rather peculiar situation is represented by the oxidative degradation of glucose by *P. putida*, which has been shown to oxidize the sugar with a dehydrogenase peripherally located in the cells. This results in gluconate accumulation in the surrounding medium, from which it is taken back into the cells after induction of a specific energy-dependent transport system. The properties of this system are different from a second one specific for glucose uptake by the same cells (Schleissner et al., 1997).

The metabolism of fructose by several species of fluorescent and nonfluorescent pseudomonads occurs by means of a phosphoenol-pyruvate (PEP) phosphotransferase system. The product is fructose-1-phosphate, which may be further phosphorylated and cleaved by an aldolase. An isomerization of free mannose to fructose can be demonstrated in cell-free extracts of strains of many species (Alicia Palleroni and N.J. Palleroni, unpublished), although the significance of this conversion is obscure at present.

An excellent review on alternative pathways of carbohydrate metabolism by pseudomonads is available (Lessie and Phibbs, 1984).

Polyhydroxyalkanoates   At the time the main body of *Pseudomonas* taxonomy was being developed, the general ideas on reserve materials of the species of rRNA group I were vague. As mentioned before, although other groups of aerobic pseudomonads accumulate PHB in granules easily observable under phase microscopy, particularly when grown in media of low nitrogen and high carbon content, with few exceptions the species of rRNA similarity group I (Palleroni, 1984) now classified in the genus *Pseudomonas* did not seem to have the same capacity, and it was even assumed that the reserve material might be dispensable proteins, because of production of ammonia during respirometric studies on the endogenous metabolism of cells. Later it was discovered that the fluorescent pseudomonads (typical representative species of the genus *Pseudomonas*) were able to accumulate PHAs of medium chain length (C_6 to C_{12}) (de Smet et al., 1983; Lageveen et al., 1988). The studies confirmed the inability of these organisms to accumulate PHB, and this inability remains a reliable negative characteristic for differentiation from most other aerobic pseudomonads (Anderson and Dawes, 1990). It has been suggested that the ability to accumulate the medium chain length PHAs, which is not dependent on the presence of a plasmid, may be in itself of taxonomic value (Huisman et al., 1989). Additional comments on this subject will be found in the section Differentiation of the Genus *Pseudomonas* from Other Genera.

Hydrolysis of PHAs for the utilization of these carbon reserve materials can be catalyzed by extracellular depolymerases, of which one, produced by a strain of *P. fluorescens*, has been purified and characterized (Schirmer et al., 1993). The depolymerases are considered to belong to the family of serine hydrolases (Schirmer and Jendrossek, 1994; Schirmer et al., 1995). All lipases and an esterase of *P. fluorescens* had hydrolytic action on triolein, but this property was absent from the PHA depolymerases tested, although these enzymes have the lipase consensus sequence in their structure. The finding confirms the differences in function between lipases and PHA depolymerases (Jaeger et al., 1995).

Biotechnological applications of some pseudomonad lipases have been reported (Tan et al., 1996).

Polyalcohols   Genes involved in the conversion of 2,3-butylene glycol to central metabolites by *P. putida* have been characterized, and there is a high similarity between the genes that encode 2,3-butylene glycol dehydrogenase and those that encode alcohol dehydrogenases (Huang et al., 1994). Degradation of polyethylene glycol has been demonstrated with *P. stutzeri* (Obradors and Aguilar, 1991). The mannitol dehydrogenase of strains of *P. fluorescens* has a broad specificity, since it is also capable of oxidation of other polyalcohols such as sorbitol and arabitol (Brunker et al., 1997).

Acetamide and biochemical evolution   Acetamide is used for growth by several species of aerobic pseudomonads, among them *P. aeruginosa* (Stanier et al., 1966). It is hydrolyzed with liberation of ammonia, which is used then as a nitrogen source, but the enzyme can also catalyze acyl transfer reactions. Amides hydrolyzed by the wild-type enzyme include propionamide, acetamide, formamide, and butyramide, in order of decreasing rates of hydrolysis. Among the amides there are excellent inducers (lactamide) as well as anti-inducers (butyramide). Clarke and collaborators found this system to be admirably suited for studies of experimental evolution, during which it became possible to evolve amidases with altered substrate specificities, and the work of this group, performed during many years, stands out as a beautiful example of experimentation on evolution achieved in the laboratory (Brown et al., 1969; Betz and Clarke, 1972; Brown and Clarke, 1972; Betz et al., 1974). Among the many papers that followed these, a general article by Clarke and Drew (1988) gives a good account of the experimental evolution of the system.

The genetic organization and complete sequence of the amidase operon is now known (Drew and Wilson, 1992), as well as details of its transcription antitermination regulation (Wilson et al., 1996).

Aromatic compounds   For obvious reasons, as in the preceding sections, the short review that follows has been restricted to very few of the many valuable reports available in a well-populated literature on the degradation of aromatic compounds and some derivatives.

Members of the genus *Pseudomonas* are notorious for their capacity of aerobic degradation of a number of hydrocarbons, aromatic compounds, and their derivatives, of which there are natural compounds and final products or intermediates from industrial activities. A considerable number of these compounds are toxic and are found widely dispersed as environmental contaminants. Many of them can be used for growth by *Pseudomonas* species. Compounds as diverse as benzoate, *p*-hydroxybenzoate, mandelate, tryptophan, phthalate, salicylate, polycyclic compounds, and many derivatives may be metabolized by strains of the group following pathways that converge to a common intermediate, β-ketoadipate. This intermediate is formed soon after the last aromatic intermediate is cleaved by a 1,2-dioxygenase, in a type of ring cleavage frequently referred to as an *ortho* cleavage. The attack of the aromatic rings and the mode of cleavage are properties of taxonomic importance that have been of help in the circumscription of the rRNA similarity group I, which comprises the *Pseudomonas* species. In recent times, much information has been added to our basic knowledge of this pathway, but few details will be mentioned here. A transporter and chemoreceptor protein from *P. putida* that is part of the β-ketoadipate regulon, PcaK, was localized in the membrane when it was expressed in *E. coli*, adding to our understanding of active transport in aromatic compound metabolism (Nichols and Harwood, 1997).

The *ortho* ring cleavage, however, is not the only type of ring opening that can be catalyzed by these organisms, and the conditions adopted for the experimental protocols should be precisely stated. Thus, in the *P. putida* type strain, cleavage of the diphenolic intermediate catechol can be caused by either a 1,2-dioxygenase or a 2,3-dioxygenase (which causes a *meta* cleavage) depending on whether the substrate on which the cells had grown was benzoate or phenol, respectively (Feist and Hegeman, 1969). Often the genes encoding for *meta*-cleavage systems are located in plasmids (Austen and Dunn, 1980), as in the case of the well-known plasmid TOL, but this is not always so. Hewetson and collaborators have presented evidence of an *ortho* pathway for *p*-cresol degradation catalyzed by enzymes encoded by plasmid genes (Hewetson et al., 1978). The *ortho*-enzymes (1,2-dioxygenases, also known as pyrocatechases) in different *P. putida* strains have all a common ancestry (Nakai et al., 1995).

The pathways of aromatic compounds degradation are still being extensively analyzed. Regulatory mechanisms, genetic or-

ganization of the genes involved, immunological properties of the enzymes, and molecular studies as the bases for strain improvement in possible bioremediation applications are all subjects of great basic and practical importance. Detailed reports on the molecular aspects of the analysis of aromatic hydrocarbon degradation have been published (Zylstra and Gibson, 1991; Zylstra, 1994). Figs. BXII.γ.122 and BXII.γ.123 have been taken from one of these sources (Zylstra and Gibson, 1991). Fig. BXII.γ.122 is a graphical representation of the main differences among aerobic pseudomonads in the initial steps of pathways of toluene degradation, and illustrates differences among some members of rRNA groups I and II (represented by *Burkholderia cepacia* and *Ralstonia pickettii*). The ring fissions of the dihydroxy intermediates (catechol, protocatechuic acid, and 3-methyl-catechol) occur in different ways. For organisms carrying the TOL plasmid (A) and *P. putida* F1 (E), the cleavage of the catechols results from the action of a 2,3-dioxygenase (*meta* cleavage), while cleavage of the protocatechuate by *P. mendocina* is of the *ortho* type (1,2-dioxygenase). Further details of the initial steps of toluene degradation by *P. putida* F1 are shown in Fig. BXII.γ.123.

Tolerance to toluene can develop in cells that normally are killed by contact with solvents. A *P. putida* strain able to grow in the presence of high concentrations of aromatic hydrocarbons was isolated from enrichments. The tolerance was inducible in this strain. Electron micrographs showed that the cells had a periplasmic space wider than that in the nontolerant cells and

FIGURE BXII.γ.122. Pathways for the degradation of toluene by aerobic pseudomonads. The black dot indicates the position of the subsequent hydroxylation that precedes ring cleavage. For more details, see text and also Zylstra and Gibson (1991).

enzymes   **genetic markers**

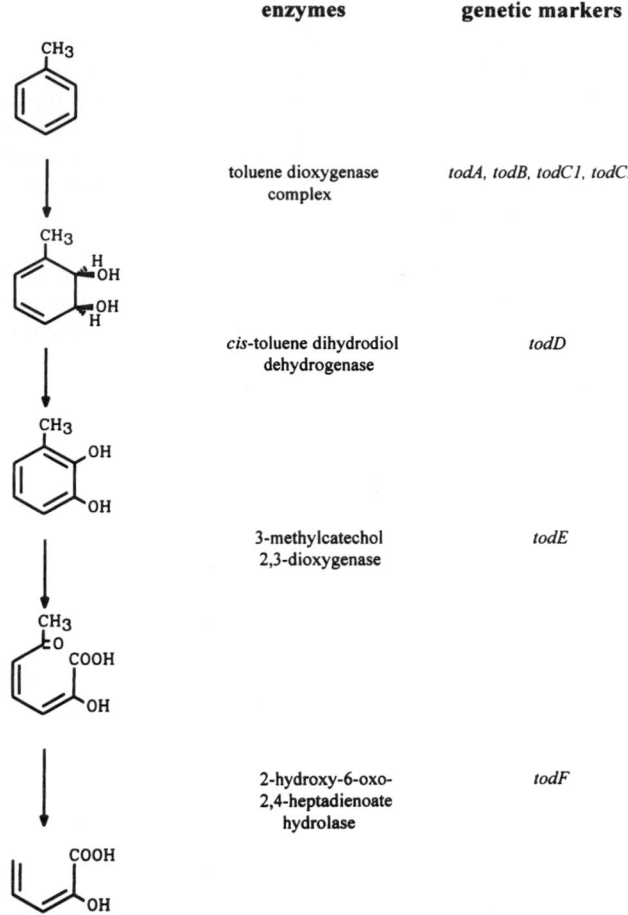

toluene dioxygenase   *todA, todB, todC1, todC2*
complex

cis-toluene dihydrodiol   *todD*
dehydrogenase

3-methylcatechol   *todE*
2,3-dioxygenase

2-hydroxy-6-oxo-   *todF*
2,4-heptadienoate
hydrolase

FIGURE BXII.γ.123. Initial steps in the pathway for the aerobic degradation of toluene by *Pseudomonas putida* strain F1. More details are found in Zylstra and Gibson (1991).

that the outer membrane integrity was preserved. Tolerance was also acquired to *m*- and *p*-xylene, and the strain gained the capacity of using these two compounds after receiving the plasmid TOL (Ramos et al., 1995). An energy-dependent efflux system may be part of the tolerance of this species to toluene (Isken and De Bont, 1996).

In a personal communication, J.A.M. de Bont has summarized as follows some of the mechanisms of tolerance to solvents: modification of the composition of the head groups of phospholipids (Ramos et al., 1997), of outer membrane proteins (Li et al., 1995; Ramos et al., 1997), of LPS (Pinkart et al., 1996), changes in the rate of turnover of membrane components (Pinkart and White, 1997), and most importantly, alterations in the composition of the fatty acids of phospholipids (Heipieper et al., 1994). This last is a very straightforward response, mostly due to the action of a *cis-trans* isomerase in *P. putida*. This enzyme has been cloned and expressed in *E. coli* (Holtwick et al., 1997). The isomerization is quite fast and does not demand an energy input. The presence of solvents triggers this activity, but other stress factors (heavy metals, high temperatures, low pH, and water stress due to salt) can act as inducers (Heipieper et al., 1996).

As expected, plasmid TOL, which was discovered in a strain now classified as *P. putida*, has continued receiving much attention. There are interesting relationships between the pathways encoded by TOL genes and those chromosomally located in the

host. Thus, when growing on unlimited amounts of toluene, a *P. putida* carrying TOL lost the capacity of using xylenes and toluates, but could grow on toluene by using initially the enzymes of the TOL upper pathway and the chromosomal system for benzoate metabolism (Brinkmann et al., 1994). Catabolite repression by succinate and glucose on the TOL enzymes has been examined (Duetz et al., 1994; Holtel et al., 1994).

The catabolism of toluene and *o*-xylene in *P. stutzeri* has been analyzed by gene cloning (Baggi et al., 1987; Bertoni et al., 1996).

In studies on the degradation of polycyclic aromatic hydrocarbons (PAHs), a mixed natural population in which *P. aeruginosa* and *P. putida* were represented was more efficient in the degradation of the poorly soluble components of the mixture than any of the pure cultures isolated from the population. The latter in turn were more active on the more water-soluble compounds (among them acenaphthene, fluorene, phenanthrene, and fluoranthene) (Trzesicka-Mlynarz and Ward, 1995).

Naphthalene degradation by *P. stutzeri* has been found to be catalyzed by enzymes encoded by highly homologous sets of catabolic genes, although the hosts had been allocated to four different genomovars (Rosselló-Mora et al., 1994c).

Many reports in the literature refer to the degradation by *Pseudomonas* species of a number of derivatives of simple aromatic compounds including *p*-cymene (Eaton, 1996, 1997), nitrophenols (Rani and Lalithakumari, 1994; Meulenberg et al., 1996; Michan et al., 1997), and styrene (O'Connor et al., 1995; Marconi et al., 1996; Beltrametti et al., 1997).

Due to their widespread use in a number of applications based on highly desirable physical and chemical characteristics, the halogenated derivatives of hydrocarbons and aromatic compounds occupy an outstanding position in the studies on environmental microbiology. Many of the multitude of halogenated compounds are toxic and, due to their remarkable stability, they tend to persist in the environment, where they are slowly degraded by a number of organisms, under both aerobic and anaerobic conditions. In the aerobic degradation of these recalcitrant compounds, the pseudomonads are among the most active organisms, and some of them also are involved in the anaerobic degradation when nitrate is present as an alternate electron acceptor. The biodegradation of many halogenated organic compounds has been adequately summarized in reviews (Häggblom, 1992; Fetzner and Lingens, 1994).

In view of the fact that anaerobic bacteria are more active in the initial degradation of haloaromatics that are reductively dehalogenated, attempts have been made to combine this property with the ability of aerobic bacteria for active degradation of the aromatic ring. A proper combination of genes in a *P. putida* strain resulted in the metabolism of polyhalogenated compounds by sequential reductive and oxidative reactions (Wackett et al., 1994).

In *P. putida* chloroaromatic and methylaromatic compounds are degraded via *meta*-cleavage pathways. The chlorocatechol produced in the first case usually inactivates the 2,3-dioxygenase, but one strain was developed with a resistant dioxygenase and had the ability to degrade both toluene and chlorobenzene via the *meta*-cleavage route (Mars et al., 1997).

Phenol, monochlorophenols, monochlorobenzoate, dichlorophenols, 2,3,5-trichlorophenol, and several alkylbenzenes, were oxidized by a strain of *P. putida* by means of an inducible toluene dioxygenase. A study of this property revealed not only the remarkable versatility of the species, but also the influence of other factors (e.g., growth rate) on the competence of the cells for use as bioremediation agents (Heald and Jenkins, 1996).

The degradative enzymes that convert these unusual compounds to intermediates of the central metabolism in distantly related hosts frequently show a high level of similarity. A rather extreme case is illustrated by a comparative study of the deduced amino acid sequences of components of the polychlorinated biphenyl degradation systems of the Gram-positive *Rhodococcus globerulus* and of Gram-negative bacteria, which showed a high similarity between the toluene dioxygenase of *P. putida* and the biphenyl dioxygenase of the Gram-positive organism, suggesting a possible transfer of genes crossing the Gram-positive/Gram-negative barrier (Asturias et al., 1995). This is certainly a reminder of the danger of drawing phylogenetic conclusions based on similarities found in metabolic systems involved in the degradation of these unusual organic substrates.

Amino acid catabolism   One aspect that has attracted the attention of biochemists and bacteriologists for many years is the utilization of arginine by the pseudomonads. Among the degradative pathways, the arginine deiminase ("dihydrolase") system has been used by taxonomists since the early 1960s for differentiation of species. The reactions catalyzed by the system are the conversion of arginine to citrulline, and of citrulline to ornithine, with liberation of ammonia (Slade et al., 1954). Motility of the cells can be activated by air or arginine (Sherris et al., 1957). In the latter case the required energy is provided by the ATP formed in the citrulline to ornithine step. This system is constitutively produced by fluorescent pseudomonads but it is inhibited under aerobic conditions.

A rise of pH in the medium with arginine, because of ammonia liberation, indicates the presence of the dihydrolase system, and this is the basis of one of the determinative methods (Thornley, 1960). This method is still widely used because it is simpler (although less reliable) than methods based on arginine disappearance. Presence of arginine deiminase is not diagnostic for species of the genus *Pseudomonas* (for instance, *P. stutzeri* is negative, while species now assigned to other genera of aerobic pseudomonads are positive).

Several other pathways of arginine degradation have been described for species of *Pseudomonas*. They are characterized by the following key reactions: arginine oxidase in *P. putida* and arginine decarboxylase in fluorescent *Pseudomonas* species (Stalon and Mercenier, 1984), arginine dehydrogenase in *P. aeruginosa* (Jann et al., 1988), and arginine succinyl transferase in fluorescent *Pseudomonas* species and in *P. mendocina* (Vander Wauven and Stalon, 1985; Stalon et al., 1987). The arginine decarboxylase/agmatine deiminase pathway appears to be characteristic of *Pseudomonas*, and it is the source of polyamines (Stalon and Mercenier, 1984). A brief discussion on the importance of these compounds as taxonomic tools will be given below, in the section on taxonomic considerations.

Pseudomonas species use only one of the various agmatine catabolic pathways, in which there is participation of an inducible agmatine deiminase system. Evidence for the presence of arginase (which converts arginine to ornithine and urea) has not been found in any *Pseudomonas* species. Further studies on arginine metabolism by organisms of this genus have included a number of ureido and guanido compounds, with findings that suggest that the corresponding screening may be of taxonomic value (Tricot et al., 1990).

A study of the distribution and induction of guanidinoacetate amidinohydrolase (GAH), guanidinopropionate amidinohydrolase (GPH), and guanidinobutyrate amidinohydrolase (GBH) among some fluorescent species of *Pseudomonas* has shown the presence of GBH in most strains, with arginine as an inducer (Yorifuji et al., 1983). GPH was detected in *P. aeruginosa* only, and the results further indicate that this species lacks the *P. putida* enzyme(s) required for degradation of L-arginine to 4-guanidinobutyrate.

The information collected on the metabolism of arginine and related compounds suggests an interesting phylogenetic scenario. The arginine deiminase pathway may be a primitive remnant from the time when anaerobic conditions prevailed on our planet, and amino acids may have provided both building blocks and energy (Prieto et al., 1992). If so, the appearance of arginine oxidase may have occurred in more recent times.

In summary, in spite of the variety of arginine catabolic pathways among the aerobic pseudomonads, members of the genus *Pseudomonas* are characterized by the presence of the arginine decarboxylase/agmatine deiminase system. The finding of arginine dehydrogenase in *P. aeruginosa* (Jann et al., 1988) would require a survey of other species not yet tested in order to assess its phylogenetic significance.

Lysine catabolism can occur in *Pseudomonas* by at least three different pathways eventually converging to glutarate, which generates acetyl-CoA. For short, these pathways may be called the "oxygenase", the "pipecolate", and the "cadaverine" routes (Chang and Adams, 1971; Miller and Rodwell, 1971; Fothergill and Guest, 1977), and their distribution in species of *Pseudomonas* is indicated in Table BXII.γ.108.

Amino acid biosynthesis   Catabolic pathways offer a larger variety of metabolic routes than the biosynthetic ones. A good example is the diversity of arginine degradative pathways discussed above. In contrast, similar anabolic routes can be used by widely different organisms. Biosynthetic pathways also are interconnected with other pathways and regulatory systems, and consequently there is a selective pressure favoring their conservation, thus preventing perturbations in pathways belonging to the same network. However, taken in combination with its regulatory mechanisms, a given biosynthetic pathway often can be used as a tool for the exploration of distant relationships among organisms.

The regulatory mechanisms of biosynthetic pathways of some aliphatic amino acids in the pseudomonads have interesting phylogenetic implications. The regulation of activity of aspartokinase and homoserine dehydrogenase is clearly different in organisms of the fluorescent group and species now assigned to other genera. Long ago it was found that *P. stutzeri* and *Burkholderia cepacia* (formerly *Pseudomonas cepacia*) resembled the fluorescent species (Cohen et al., 1969). In fact, this was one of the first findings suggesting a phylogenetic relationship between *P. stutzeri* and other species of *Pseudomonas*. However, *B. cepacia* also showed similarities with the fluorescent pseudomonads, in spite of its phylogenetic position in a different branch of *Proteobacteria*.

The multibranched pathway of biosynthesis of aromatic amino acids has offered not only more variations in biochemical details, but also a richer source of regulatory information. These are the reasons for which Jensen and his collaborators chose the interconnected pathways as a model for the study of phylogenetic relationships among the aerobic pseudomonads. The results of this work, which has direct taxonomic implications (Byng et al.,

TABLE BXII.γ.108. Distribution of lysine catabolic pathways in some fluorescent *Pseudomonas* species and *Burkholderia cepacia*[a]

Lysing catabolic pathways	P. aeruginosa	P. fluorescens	P. putida biovar A	P. putida biovar B	Burkholderia cepacia
Oxygenase	−	+	+	+	−
Pipecolate	+	+	+	−	+
Cadaverine	+	+	−	−	−

[a]For symbols see standard definitions. Taken from Palleroni (1984); data from Miller and Rodwell (1971), Chang and Adams (1971), and Fothergill and Guest (1977).

1980, 1983; Whitaker et al., 1981b) were in agreement with the subdivision of the pseudomonads into five so-called rRNA homology groups (Palleroni et al., 1973). In fact, the enzymatic assays are simpler than the hybridization methods of ribosomal similarity studies, which makes the former good determinative tools indeed.

Glutamine synthetase, an important enzyme involved in nitrogen metabolism, has been subjected to immunological comparative studies that were expressed in combination with the results of nucleic acid hybridization experiments (Baumann and Baumann, 1978). These studies, aside from representing further confirmation of the internal subdivision of the pseudomonads, have useful determinative applications.

Genetics Some of the *Pseudomonas* species have attracted a great deal of attention from bacterial geneticists because of their widespread occurrence, their biological and medical importance, their nutritional and biochemical versatility, and the simplicity of conditions required for their cultivation in the laboratory.

Of them, *P. aeruginosa* is by far the best known from the genetic point of view. Strain PAO ("*Pseudomonas aeruginosa* One") is the one that has been most extensively studied. A detailed genomic map including more than 360 genes is now available (Holloway et al., 1994). This map represents the last of a number of reports from the Department of Genetics of Monash University, Clayton, Australia (see below), in combination with work performed elsewhere on the construction of physical maps. Recent additions to the genomic map available for strain PAO include the physical mapping of 32 genes by application of modern molecular techniques. The approach used will increase the number of genes included in databases, where now nearly 40% of the genes of the species are represented (Liao et al., 1996).

A second strain of *P. aeruginosa*, PAT ("*Pseudomonas aeruginosa* Two"), is also familiar to geneticists, and its map closely resembles that of PAO. The strains have different geographical origin, and their similarity is a confirmation of the homogeneity of the species already noticed by many taxonomists.

The three best-known systems of genetic exchange, namely, conjugation, transduction, and transformation, have been observed in strains of *P. aeruginosa*. Conjugation and transduction have been most important in genetic studies of this species. In contrast, transformation occurs in *P. stutzeri*, *P. mendocina*, and other nonfluorescent members of the genus under natural conditions. Conjugation, which results in transfer of substantial chromosomal segments, has been the most effective source of information for mapping purposes in *P. aeruginosa* genetic studies. The transfer depends on the chromosomal mobilizing ability (cma) of some plasmids, of which FP2 has been used extensively, and to a lesser extent, FP5, FP39, and FP110. Each of these fertility plasmids has one predominant attachment site, and this was the main reason that delayed the demonstration of circularity of the map. This was finally achieved based on 2- and 3-factor crosses with cma plasmids FP2, FP5, FP110, and R68.45 (Royle et al., 1981).

An approach recently introduced in the study of bacterial genomes combines the action of endonucleases that infrequently cut the chromosome, with pulsed-field agarose gel electrophoresis (PFGE), which makes possible the resolution of large DNA fragments. A limited number of cuts per genome is desirable for genomic restriction mapping (McClelland et al., 1987), with clear advantages over frequently cutting endonucleases for the production of fingerprints characteristic of different taxa (Mielenz et al., 1979; Dobritsa, 1985; Sorensen et al., 1985). Application of this principle has resulted in a practical method for constructing and analyzing macrorestriction patterns of 234 strains of different species of aerobic pseudomonads (Grothues and Tümmler, 1991). The restriction nucleases *Asn*I, *Dra*I, *Spe*I, *Xba*I, and *Pac*I are among the most appropriate by being specific for AT-rich regions (which are less common than GC pairs in *Pseudomonas*), or for sites including the extremely uncommon tetranucleotide CTAG (McClelland et al., 1987).

The macrorestriction patterns are compared for the number and position of the bands, and use of appropriate equations often indicates a correlation between the estimated similarities of fingerprints and conventional taxonomic groupings. The results obtained by Grothues and Tümmler (1991) in general confirm the classification of pseudomonads based on extensive numbers of phenotypic properties and nucleic acid hybridization studies.

Comparisons involving mol% G + C content of the chromosome, codon usage, and genome size can be used for the calculation of similarity coefficients useful for determinative purposes. The genome restriction patterns also have the practical advantage of helping in epidemiological studies. Differences in these parameters were found between strains susceptible or resistant to multiple antibiotics (Yamashita et al., 1997).

Differences in restriction fragment size distribution may result from various chromosomal rearrangements and/or mutations at the restriction sites, which suggests that the methodology based on restriction fingerprints should be supplemented with other approaches to determine the degree of similarity of the fragments. Large genomic rearrangements have been observed in *P. aeruginosa* strains isolated from clinical samples or from the environment (Schmidt et al., 1996c). The former came mostly from cystic fibrosis cases. A 95-kb plasmid was detected in environmental strains, and it was integrated in the chromosome in cystic fibrosis strains. Exchange of DNA blocks and large DNA inversions led to divergence of clones in this species. The presence of inversions only in cystic fibrosis strains suggests that this niche causes or tolerates substantial changes in the genome (Römling et al., 1997).

A study on pathovars of fluorescent plant pathogenic pseudomonads (Grothues and Rudolph, 1991) indicated that two strains belonging to the same pathovar, but of different origins, can give almost identical restriction fingerprints. However, this is not always the case. Pathogens of wide host range give more diverse restriction patterns than those of restricted host range. The methodology was also used for the examination of representative strains of *P. stutzeri*, a species notorious for its heterogeneity (Stanier et al., 1966; Palleroni et al., 1970). The results show a marked correlation of genome structure with fatty acid composition, and with data of nucleic acid hybridization experiments (Rainey et al., 1994b). A high degree of heterogeneity in macrorestriction patterns was also observed by another group of workers, which did not correlate with the subdivision of the species in genomovars. The marked heterogeneity of this species is attributed, at least in part, to large chromosomal rearrangements (Ginard et al., 1997). The results of these two groups also indicate that the genome size of *P. stutzeri* ranges from 3.4–4.64 Mba for the strains subjected to their studies.

Pseudomonas species other than *P. aeruginosa* have been less rewarding subjects for chromosomal mapping. This is unfortunate in the case of a species like *P. putida*, which has been extensively investigated from the biochemical viewpoint. This species has presented serious problems for the development of a satisfactory chromosome transfer system, and, consequently, a less detailed chromosome map is available (Mylroie et al., 1977; Strom and Morgan, 1990). Circularity of the *P. putida* map strain PPN (related to strain ATCC 12633, which was the subject of many metabolic studies for many years) has been demonstrated (Dean and Morgan, 1983). A genetic map of *P. syringae* pathovar *syringae* is also available (Nordeen and Holloway, 1990). The situation is even more regrettable in the case of the very complex species *P. fluorescens*, with many strains extensively known for their phenotypic properties. A good starting point now is the report on a complete physical map of a strain of the species with a genome of 6.63 Mbp. A total of 139 restriction sites and 31 genes have been located in the map (Rainey and Bailey, 1996). Similarly, an efficient mutagenic procedure using electroporation for *P. fluorescens* and a transposon delivery vector has been described (Artiguenave et al., 1997). A *recA* mutant has been obtained, and sequence studies have shown that the chromosomal organization was very similar to that of *P. aeruginosa* and *Azotobacter vinelandii*. The regulatory region and the structural gene differed from those of *Burkholderia cepacia*. By insertion of a kanamycin cassette in the *recA* gene, the mutant obtained had an increased UV sensitivity and was much impaired in its recombinatorial activity (De Mot et al., 1993).

A *recA* mutant strain of *P. stutzeri* is also available (Vosman and Hellingwerf, 1991). It is completely deficient in natural transformation with chromosomal DNA and it is sensitive to UV and methyl-methane sulfonate. The wild-type gene complements an *E. coli recA* mutant (Bennasar et al., 1996).

Plasmids, phages, bacteriocins Plasmids are important components of the genetic makeup of *Pseudomonas*. Some of them act as fertility factors, some (R plasmids) may impart resistance to various agents, and others confer the capacity of degradation of unusual carbon sources, thus contributing to the nutritional versatility that is a striking feature of many members of the genus. A large number of *Pseudomonas* plasmids have been described in the past decades, and it is impossible to cover all references here. Some of the plasmids are mentioned in other sections of this chapter. Examples of the properties of *Pseudomonas* encoded by plasmid genes are:

1. Resistance characters to antibiotics and other antibacterial compounds, such as carbenicillin, chloramphenicol, gentamicin, kanamycin, streptomycin, tetracycline, tobramycin, sulfonamides;

2. Resistance to chemical and physical agents (borate, chromate, various metal ions, organomercurials, tellurite, ultraviolet radiation);

3. Resistance to bacteriophage propagation; interference with lysogenization by some temperate phages; DNA restriction and modification;

4. Resistance to bacteriocins.

Various other characteristics due to plasmid genetic determinants are the chromosome donor ability, donor-specific phage susceptibility, inhibition of bacteriocin production, fertility inhibition, incompatibility with other plasmids, and, most important for some species, utilization of various organic compounds not normally used for growth by species other than the aerobic pseudomonads. Occasionally plasmids are found that confer on the cells the capacity to use simple organic compounds. Thus, for instance, a *P. fluorescens* plasmid that carries genes of resistance to ampicillin, kanamycin, and streptomycin also allows the host to grow on malonate because of a plasmid gene encoding malonate decarboxylase. The plasmid (pPSF1) can be transferred to *E. coli*, where the ability to grow on malonate can also be expressed (Kim and Kim, 1994).

Plasmids can be classified most effectively by incompatibility, that is, by the incapacity of a given plasmid to coexist in the same cell with other plasmids of the same group. In *Pseudomonas*, at least 10 incompatibility groups have been defined (Jacoby, 1977; Korfhagen et al., 1978). Useful information on *P. aeruginosa* plasmids may be found in a chapter by Jacoby (Jacoby, 1979). In general it can be stated that most of the research with *Pseudomonas* plasmids has been focused on the fluorescent species, and the information on other species is fragmentary.

Plasmids of the various incompatibility groups have different host ranges. The widest range is that of the IncP-1 plasmids, while those of groups 2 and 5–9 are more specific. IncP-2 plasmids include R factors and plasmids carrying genes for the degradation of unusual carbon compounds. Some of them are among the largest of *Pseudomonas* plasmids, exceeding 300 MDa, while the majority range from 10–60 MDa. Outside of the IncP-2, degradative plasmids are also found in the P-9 group.

Degradative plasmids named CAM, OCT, SAL, NAH, TOL, and XYL are involved in the degradation of camphor, *n*-octanol, salicylate, naphthalene, toluates, and xylene, respectively, and have received much attention because they confer to the cells the capacity to degrade environmental pollutants. The naphthalene degradative enzymes of plasmids identified in *P. fluorescens* are also involved in the degradation of high molecular weight polyaromatic hydrocarbons other than naphthalene (Menn et al., 1993).

Of the above-mentioned plasmids, TOL, which was originally isolated from a strain now classified as *P. putida*, is the best known and it still receives a great deal of attention by bacteriologists. An excellent review of the properties of TOL and naturally occurring variants is available (Assinder and Williams, 1990). The upper and lower operons of this plasmid seem to have different origins. In support of this hypothesis, it has been found that their respective codon usages are different, but they are the same for the genes within each operon (Harayama, 1994). TOL can be transferred from cell to cell by conjugation, which occurs at a sufficiently high rate to maintain the plasmid in a dense microbial community without the help of selective pressures (Smets et al., 1993).

Comai and Kosuge (1980) have presented evidence for plasmid involvement in the oxidative degradation of tryptophan to the plant hormone indoleacetic acid (IAA) by *P. savastanoi*. Concentrations of IAA higher than those normally found in plant tissues result in gall formation (Gardan et al., 1992

The range of antibiotic resistance determined by plasmids and the mechanisms by which the resistance is manifested are similar to those found in *E. coli*. Some plasmids are transferred from cell to cell by conjugation; others lack this capacity, but some of them can be mobilized by other conjugative plasmids.

Many different lytic and temperate bacteriophages have been identified in *Pseudomonas*, particularly in *P. aeruginosa*. Their morphological diversity is at least as great as for phages of other bacterial genera. Lysogeny is a very common phenomenon in *P. aeruginosa*, and transducing phages have been very useful in linkage studies. Most *Pseudomonas* phages contain double-stranded DNA, but some are RNA phages, including one with double-stranded RNA (Semancik et al., 1973; Vidaver et al., 1973). The sequence of *P. aeruginosa* single-stranded RNA phage PP7 has been determined and the results suggest that the phage is related to coliphages but branched off before the coliphages diverged into separate groups (Olsthoorn et al., 1995).

An example of a rather unusual application of a *P. stutzeri* bacteriophage has been its use for the selection of denitrification-negative mutants. The procedure takes advantage of the fact that the phage (phi PS5) adsorbs to the outer membrane protein NosA. Mutants defective in NosA production do not grow with N_2O and are resistant to the phage (Clark et al., 1989).

Even though host specificity is the rule, some bacteriophages attack different but related species, with interesting taxonomic implications (M.P. Starr, personal communication). This approach to bacterial classification was attempted many years ago by Billing, in an effort to improve the methodology for the differentiation of phytopathogenic pseudomonads (Billing, 1963, 1970a). The sensitivity tests were not very useful in themselves, but they were valuable as a complement of the biochemical tests of identification. Pathovars of *P. syringae* isolated from pear could be distinguished by phage sensitivity but not by serological tests from the strains isolated from cherry. On the other hand, pathogenic and saprophytic pseudomonads could not be distinguished by their lysotypes (phage types or phagovars) (Crosse and Garrett, 1963).

Phage sensitivity can also be used successfully as a method of typing *P. aeruginosa* strains, but the procedure is not considered sufficiently reliable to be used by itself as an epidemiological tool without comparison to other typing procedures (Brokopp and Farmer, 1979).

Some factors required for phage nucleic acid replication appear to have been conserved through evolution in some groups of Gram-negative bacteria. One of the elements required for *in vitro* replication of sex-specific single-stranded RNA coliphage Qβ is the so-called host factor (HF). This is a heat-resistant RNA-binding protein of molecular weight 12,000, usually present in *E. coli* as an hexamer (Franze de Fernández et al., 1972). In *P. putida*, a polypeptide of molecular weight 11,000 gives an immunological cross-reaction with HF and allows Qβ replication *in vitro* (DuBow and Blumenthal, 1975). Material cross-reacting with *E. coli* HF antiserum was found in extracts of *P. aeruginosa*, *P. fluorescens*, *P. putida*, and in other species of different rRNA similarity groups (Palleroni, 1984)), but the material was heat stable only when it came from organisms of RNA similarity groups I and V (*Pseudomonas*, *Xanthomonas*, and *Stenotrophomonas*), which

are located in neighbor branches in the *Gammaproteobacteria* (DuBow and Ryan, 1977).

Bacteriocins are proteins produced by some bacterial strains, which have a lethal action on other strains of the same species. They have been frequently detected in *Pseudomonas*, and in *P. aeruginosa* they have been named pyocins (from the old name *P. pyocyanea*) and aeruginosins, although both names are improper. Pyocin can be mistaken for pyocyanin, the blue diffusible pigment characteristic of the species, and aeruginosin has been used for two red pigments produced by some strains of *P. aeruginosa* (aeruginosins A and B) (Holliman, 1957; Herbert and Holliman, 1964).

Different types of *P. aeruginosa* bacteriocins have been described. One resembles bacteriophage contractile tails, a second one has the appearance of slender, flexuous rods, and a third (S type) is amorphous and is sensitive to proteolytic attack (Govan, 1974a, b). A novel S-type pyocin has been characterized by molecular techniques (Duport et al., 1995).

Bacteriocin typing can be performed in two ways, which are based on production of these elements, or on sensitivity to them, respectively. The first approach is more commonly used, and it requires a collection of sensitive strains (Govan, 1978; Brokopp and Farmer, 1979). Bacteriocin production by species other than *P. aeruginosa* has been reported in a few instances (Vidaver et al., 1972; Smirnov et al., 1984). In a study on bacteriocin typing of unknown *P. syringae* strains, the specificity was uneven, and the highest level was shown for the case of the phaseolicins of *P. syringae* pathovar *phaseolicola* (in this treatment described as a pathovar of *P. savastanoi*) (Vidaver et al., 1972). In an extension of these studies, the correlation found between bacteriocin type and host plant of origin of phytotoxin production was rather poor (Vidaver and Buckner, 1978). The approach is probably susceptible to further refinements, since 86% of the strains examined were able to produce bacteriocins.

Antigenic structure   Again here, the literature is dominated by the number of papers dedicated to *P. aeruginosa*, reflecting in this case, as in others, the medical importance of this species.

Agglutination of intact *P. aeruginosa* cells can be caused by specific antibodies elicited in animals by specific components that are similar to the O-antigens of other Gram-negative bacteria. The heat-stable O-antigen, considered the most stable marker for *P. aeruginosa*, is represented by one of the components of the lipopolysaccharide (LPS). The specificity is related to the composition of the polysaccharide chains (the O antigens themselves) projecting to the outside of the cells. The LPS of *P. aeruginosa* has O-specific polysaccharide chains with unbranched oligosaccharides including amino sugars that are not acetylated (Wilkinson, 1983).

P. aeruginosa shows a marked serological diversity. The LPS composition determines at least 17 heat-stable O antigens, and antisera are adequate for serotype identification, in spite of variations in titer and specificity. In experiments using a panel of 48 monoclonal antibodies against eight of the serotypes, various degrees of activity were shown by some of the antibodies, one of which bound to all serotype strains and also to strains of *P. fluorescens* and *P. putida* (Gaston et al., 1986). A further refinement of this approach showed that the antigenic specificity of various parts of the LPS molecule could be clearly demonstrated by preparing monoclonal antibodies recognizing the inner core, outer core, and lipid A regions of LPS. Antibodies to the complex (lipid A + core + one repeat of the O-specific polysaccharide chain) reacted with a lower number of serotypes than the antibodies recognizing the outer core without the oligosaccharide. The specificity was even lower for the antibodies that recognized the inner core, since they reacted not only with the largest number of serotypes tested, but with all the other Gram-negative species included in the experiment (de Kievit and Lam, 1994).

Mild heating of a cell suspension reduces the agglutinability by specific antisera, but the property can be fully recovered by a more intense treatment (100–120°C for 2–2.5 h). The alginate of mucous strains that can be selected under appropriate conditions (Govan and Fyfe, 1978) or isolated from cystic fibrosis cases (Doggett, 1969) does not interfere with the O-antigenicity.

In addition to the immune response against *P. aeruginosa* LPS, which in its early manifestation consists of IgM antibodies (Høiby, 1979), there is a humoral response to cross-reactive antigens that are present in other, mostly Gram-negative, bacteria (Høiby, 1975). A heat labile antigen common to a wide range of bacteria has been isolated from *P. aeruginosa* and shown to be an acidic protein composed of subunits of molecular weight 62,000, present in the cytoplasmic fraction of the cells (Sompolinsky et al., 1980a, b).

Various other antigens have been identified in *P. aeruginosa*. Heat-labile surface antigens are represented by flagella and fimbriae (Bradley and Pitt, 1975; Pitt and Bradley, 1975). An extracellular slime can elicit the production of an agglutinin, and exoenzymes such as phosphatases, proteases, and phospholipases can also act as antigens. The outer membrane protein OprF was subjected to epitope mapping, and, as mentioned before, the fact that it is a major component suggests that it may be a good candidate for use as a vaccine and as a target for monoclonal antibodies for immunotherapeutic applications and for diagnosis (Rawling et al., 1995).

Isolation and characterization of monoclonal antibodies for outer membrane antigens have been described (Hancock et al., 1982). As a practical identification tool for the identification of glucose oxidizing *Pseudomonas* species, Mutharia and Hancock (1985) have proposed the use of a monoclonal antibody (MA1-6), specific for a single antigenic epitope on the outer membrane lipoprotein H2 of *P. aeruginosa*. The epitope was detected in all 17 serotype strains of *P. aeruginosa*, in numerous clinical isolates of the same species, and in other *Pseudomonas* species but not in species of other rRNA similarity groups (Palleroni, 1984). Two strains of *P. aeruginosa* (out of a total of 52) failed to give the reaction. In a *P. aeruginosa* collection that included 30 environmental isolates, no major differences in outer membrane proteins (other than quantitative variations in lipoprotein H2 content) or lipopolysaccharide patterns were observed when compared with those of previously studied clinical isolates (Hancock and Chan, 1988).

Similar conclusions on the advantages of using monoclonal antibodies against outer membrane proteins, in addition to other tools useful for determinative purposes, have been reached from experiments on the detection of the lipoprotein I gene in species of *Pseudomonas* (De Vos et al., 1993), as an extension of the original observations by Saint-Onge et al. (1992).

Many schemes of serotyping have been proposed, but the system of Habs (1957) has gained wide acceptance and, with some modifications, is in general use. Habs defined 12 somatic groups that could be identified by agglutination tests. Different additional proposals and modifications include the addition of 5 O-groups. Standard strains and typing sera are commercially available.

In the opinion of Brokopp and Farmer (1979), serological typing of *P. aeruginosa* based on the O-antigens produces more reliable evidence for relatedness than can be obtained by other, less specific typing methods. Thermolabile surface antigens can also be used for serotyping, and various methods based on flagellar (H) antigens have been proposed (Verder and Evans, 1961; Lányi, 1970). These approaches have not been widely accepted, the main reason being the difficulties encountered in the preparation of specific flagellar antisera (Brokopp and Farmer, 1979).

For epidemiological purposes, serotyping is the most important typing method and, as usually practiced, consists of the reaction of a cell suspension toward a standard set of antibody preparations. Serotyping can be supplemented by other typing methods, such as production and sensitivity to bacteriocins and lytic phages, biotyping (or characterization of strains by their biochemical and physiological properties) and antibiograms, and sensitivity to antibiotics.

To these procedures we now have to add modern techniques based on molecular concepts. One of them is the arbitrary PCR fingerprinting of strains, which provides a simple and practical typing approach considered to be more discriminatory than the traditional serotyping scheme, although the maximum discriminatory power is achieved by a combination of both methodologies (Hernández et al., 1997). Comparative typing of *P. aeruginosa* also has been carried out by random amplification of polymorphic DNA or pulsed-field gel electrophoresis of DNA macrorestriction fragments (Renders et al., 1996), to which we can add the fingerprinting of whole cell proteins (Khan et al., 1996), and studies on comparative ribotyping and genome fingerprinting (Bennekov et al., 1996).

In the field of fluorescent plant pathogenic pseudomonads, the serological approach has had variable success, and has not helped substantially in the circumscription of nomenspecies. The lack of a solid taxonomic frame of reference has often made it very difficult to interpret the resolving power of the serological techniques. Otta and English (1971) were unable to define a precise serological differentiation of virulent strains from non-virulent ones. On the other hand, species-specific antigens could be identified in *P. syringae* pathovar *phaseolicola* (described here as a pathovar of *P. savastanoi*) (Guthrie, 1968) and in *P. syringae* pathovar *lachrymans* (Lucas and Grogan, 1969a, b). *P. syringae* strains isolated from different groups of host plants could be distinguished immunologically (Otta, 1977). In general, however, the reproducibility of serological data by different authors using different strains of the same nomenspecies appears to be rather poor.

Susceptibility to antibiotics   Many aerobic pseudomonads are resistant to a number of antibacterial agents. This subject is of particular medical importance because members of this group are serious opportunistic human pathogens and are often isolated from patients and clinical materials. The capacity to resist antibiotics is included in some of the descriptions of new taxa.

As expected, most of the available information on *Pseudomonas* refers to *P. aeruginosa* because of its importance as a serious opportunistic human pathogen. The antibiotics that are most effective in the treatment of *P. aeruginosa* infections include some β-lactams, such as carbenicillin and ticarcillin, third-generation cephalosporins (cefsulodin, cephoperazone, ceftazidime), the synthetic monocyclic β-lactam aztreonam, carbapenems (among them the extremely broad-spectrum semisynthetic imipenem), the aminoglycosides (gentamicin, tobramycin), and the quinolones (ciprofloxacin). In practice, *P. aeruginosa* infections are treated with a mixture of tobramycin and some β-lactams active against *Pseudomonas*, such as piperacillin or azlocillin. Various quinolones are in use for the treatment of chronic or mild corneal or urinary infections.

P. aeruginosa and other fluorescent *Pseudomonas* species are in general resistant to β-lactams other than those indicated above. Carbenicillin is moderately effective and was extensively used in therapeutic applications either by itself or in mixtures with aminoglycosides. Carbenicillin produces cell enlargement and filament formation, but the sensitivity of the cells is not very high, and years ago the emergence of resistant *P. aeruginosa* mutants was reported (Lowbury et al., 1969; Gaman et al., 1976). The mucoid strains that are frequently found in infections in cystic fibrosis patients are somewhat more resistant to the antibiotic.

Resistance to carbenicillin may be due to β-lactamases coded for by genes carried by plasmids of the incompatibility group 1 (IncP-1) that can be transferred to *E. coli* (Lowbury et al., 1969), but similar genes may be found in host restricted IncP-2 plasmids (Bryan, 1979). Seven different kinds of β-lactamases have been identified in 24 *Pseudomonas* plasmids belonging to at least eight of the incompatibility groups (Jacoby and Matthew, 1979). β-lactamase genes are not invariably carried by plasmids and, in fact, there is in *P. aeruginosa* a constitutive penicillinase (Furth, 1975) and an inducible cephalosporinase (Sabath et al., 1965; Sykes and Matthew, 1976) whose genetic determinants are chromosomal.

Even though the third-generation cephalosporin ceftazidime is considered to be highly effective against many *P. putida* strains (Yang et al., 1996b), a metallo-β-lactamase gene *blaIMP* was detected by PCR in a number of Gram-negative organisms, among them resistant strains of *P. fluorescens* and *P. putida* (Senda et al., 1996). It is also interesting that the response to β-lactam antibiotics in *Pseudomonas* can result in their utilization as substrates for growth. A soil strain of *P. fluorescens* was able to grow at the expense of benzylpenicillin as carbon, nitrogen, and energy source (Johnsen, 1977). This property also is present in pseudomonads now assigned to different genera (for instance, *Burkholderia cepacia*) (Beckman and Lessie, 1979).

Aside from the production of enzymes such as the β-lactamases, the pseudomonads are able to resist many antibiotics by mechanisms such as a low cell wall permeability, the production of modifying enzymes, and efflux systems. The most abundant porin in the outer membrane, OprF, is probably very important in reducing the permeability to antibiotics, although other porins may be involved.

Enzymes capable of modification of aminoglycosides as a resistance mechanism have been described, and they correspond to the types already known for the enteric bacteria. These enzymes catalyze phosphorylation, adenylation, or acetylation of the antibiotics, although the resistance of *P. aeruginosa* to the aminoglycosides may be largely nonenzymatic.

A useful and comprehensive review on multidrug efflux systems present in both eucaryotes and bacteria has been published by Paulsen et al. (1996). Drug accumulation and efflux system studies suggest that *P. aeruginosa* has at least two different proton motive force-dependent efflux systems (Li et al., 1994, 1995). A chromosomal operon was discovered that confers multidrug resistance to this organism, with three genes, *mexA*, *mexB*, and *oprM*, encoding for components of a transport system originally shown to be able to correct a defect in iron metabolism (Poole et al.,

1993a, b; Ryley et al., 1995). Prior to this work, it was known that OprM was involved in conferring resistance to multiple drugs, and it was later confirmed as a component of the efflux system. The operon *mexAB/oprM* is inducible under iron-limiting conditions, and is co-regulated with components of the pyoverdin-mediated iron transport. Mutants lacking *mexA* or *mexB* are unable to grow under conditions of iron limitation in the medium. All *P. aeruginosa* strains that have been examined have the operon, although its components seem to operate with unequal efficiencies according to the strain.

Mutations in the operon result in intracellular accumulation of chloramphenicol, tetracycline, norfloxacin, benzylpenicillin, carbenicillin, and quinolones. A second efflux system was later identified (*mexC-mexD-oprJ*) encoding proteins that prevent the cytoplasmic accumulation of quinolones, tetracycline, chloramphenicol, and the cephems (Li et al., 1994, 1995; Liu et al., 1995b; Poole et al., 1996).

The most common mechanisms of gentamicin resistance in *P. aeruginosa* are enzymatic modifications of the antibiotic consisting of *N*-acetylation, O-adenylation, and, to a much lesser extent, phosphorylation. These properties are controlled by plasmid genes, mainly IncP-2 plasmids (Bryan et al., 1972, 1973, 1974; Jacoby, 1974a, b). Permeability factors may also be involved in the case of gentamicin and other aminoglycosides (Mathias et al., 1976; Bryan, 1979).

Polymyxins are very active against *P. aeruginosa* and other pseudomonads when tested *in vitro*, but their efficacy *in vivo* is limited. Resistance is unstable, and this condition seems to correlate with an increase in EDTA tolerance. Gilleland and Murray (1976) noted in these variant strains the disappearance of the particles characteristic of the concave cell wall layer separated by freeze-etching (see above in the section on fine structure), but the particles reappeared by growth in the absence of polymyxin, and acquisition of temporary resistance is accompanied by a decrease in the phosphorus content of the outer membrane.

Sensitivity to metals and metalloids Copper compounds are extensively used in agriculture, and since only minute amounts are needed for nutrition, the excess can affect higher organisms as well as procaryotes, among which the plant pathogens are of particular interest. In *P. syringae* strains that carry the operon *cop*, copper is excluded by combination with components of the periplasm and in outer membrane. A two-component sensory transduction mechanism similar to equivalent systems in other organisms operates in the regulation of expression of copper resistance, and it has strong similarity with a gene that seems to regulate the uptake or efflux of copper in "*Streptomyces lividans*" (Mills et al., 1993).

Inducible copper resistance in *P. aeruginosa* is encoded by chromosomal genes, although the strains that were examined had plasmids (Vargas et al., 1995b).

In *P. fluorescens* isolated from copper-contaminated soil, a chromosomal locus with determinants for copper resistance and competitive fitness was cloned, and genes responsible for conferring copper resistance and production of cytochrome *c* were identified (Yang et al., 1993a, 1996a). A *P. putida* strain isolated from electroplating effluent could accumulate cupric ions in a concentration as high as 6.5% of its dry weight. The capacity was highest when the culture was previously grown under sulfate-limiting conditions (Wong and So, 1993).

Acid-labile sulfide levels were found to be generally higher in silver-resistant *P. stutzeri* strains than in sensitive ones. The resistance to the metal may be due to formation of silver sulfide complexes, since no complex formation with polyphosphate or metal-binding proteins has been found to be the cause (Slawson et al., 1992).

Three plasmids were found in a silver-resistant strain isolated from a silver mine. The largest plasmid (MW 49.4×10^6), which specifies silver resistance, is nonconjugative, but it could be transferred to *P. putida* by mobilization with plasmid R68.45 (Haefeli et al., 1984).

P. fluorescens detoxifies aluminum by elaboration of a soluble, aluminum-complexing metabolite. When iron was present in the medium, the two trivalent metals were immobilized in a lipid-rich complex containing Al, Fe, and P, after an early stage in which aluminum was found to be associated with phosphatidyl-ethanolamine (Appanna et al., 1995; Appanna and Hamel, 1996). Workers from the same group examined the adaptation of *P. fluorescens* to stress caused by excess of cesium (Appanna et al., 1996), and were also involved in determining the basis for resistance of the species to various metals (aluminum, iron, zinc, calcium, and gallium). These metals were supplied as complexes with citrate, which was completely oxidized. As in the case of the resistance to aluminum mentioned above, the metals appeared associated to phosphatidylethanolamine, which later was found in the lipid-rich complex where the metals were located (Appanna and St. Pierre, 1996). In a study on the mechanism of metal-citrate complexes by *P. fluorescens*, the biodegradation depended on the nature of the complex. Thus, the bidentate ferric, nickel, and zinc citrate complexes were readily degraded, while the tridentate cadmium and copper citrates were not. The latter, similarly to uranium citrate, were neither transported inside the cells nor metabolized by cell-free extracts (Joshi-Tope and Francis, 1995).

About one-half of the *P. aeruginosa* known plasmids confer resistance to mercuric ions (Jacoby and Shapiro, 1977). A plasmid (pPB) was found to confer mercury and organomercurial resistance to a *P. stutzeri* strain. The plasmid had two regions of functional and independently regulated *mer* genes, probably transcribed from different promoters (Reniero et al., 1995). Resistance to chromium, boron, and tellurium was also determined by some *Pseudomonas* plasmids (Summers et al., 1978). After growth in the presence of increasing concentrations of tellurite, strains of *P. aeruginosa* and *P. putida* harboring certain plasmids that determine tellurite resistance accumulated crystalline structures containing tellurium in their periplasmic space. From there, these materials were released into the medium in vesicles that are pinched off the outer membrane (Suzina et al., 1995).

Operons of genes coding for arsenic resistance are usually carried in plasmids. Arsenate is reduced to arsenite, which is eliminated by an export system. A chromosomal operon was identified and cloned in *E. coli* and was able to hybridize with chromosomal of other enteric bacteria and *P. aeruginosa*. This chromosomal operon may be the evolutionary precursor of the plasmid operons, with the advantage of a multicopy system as a means of natural amplification (Diorio et al., 1995).

Antibiotic production Plant root colonization by certain strains of the species *P. fluorescens* and *P. putida* may result in an enhancement of plant growth, an effect that is accompanied by a selective inhibition of other bacterial species and fungi. In part this action is due to the production of siderophores, which remove iron required by organisms that lack systems for utilization of the ferric siderophores. An example is pseudobactin 358 (pro-

duced by *P. putida* WBS358) whose structure, resembling that of other siderophores, has been elucidated (Van der Hofstad et al., 1986). The molecular aspects of iron assimilation by the pseudobactin-producing organisms have been analyzed (Leong et al., 1992).

Fragments of the peptide component of pseudobactin do not have fungicidal activity, although they resemble the antifungal compounds pseudomycins (Okonya et al., 1995). However, the fluorescent organisms do produce antifungal antibiotics. Thus, a strain of *P. fluorescens* produces the antibiotics pyrrolnitrin, pyoluteorin, and 2,4-diacetylphloroglucinol, which are able to suppress root diseases due to fungal pathogens. Of the three compounds, the latter represents an important factor in the control of plant diseases. A 6.5-kb chromosomal fragment has been isolated that contains the genes that encode the biosynthetic enzymes (Bangera and Thomashow, 1996). An examination of this and other molecular aspects has led to identifying a sigma factor encoded by gene *rpoS* that has an influence on the production of the antibiotics, on the biological control activity, and on the survival capacity of the strain on plant surfaces (Fedi et al., 1996). Pyrrolnitrin requires the participation of four genes, the only determinants that appear to be absolutely necessary for the complete biosynthesis (Hammer et al., 1997).

In a *P. fluorescens* strain, the production of the above antifungal compounds and of an extracellular protease can be abolished by mutations of an antibiotic production gene (*apdA*). Interestingly, the sequence of this gene is strikingly similar to that of genes that encode sensor kinases required for the pathogenicity of *P. syringae* pathovar *syringae* and *P. viridiflava*, suggesting that *apdA* encodes a putative sensor kinase component of a classical two-component regulatory system that, in this case, is required for the synthesis of a secondary metabolite (Corbell and Loper, 1995).

A different strain of *P. fluorescens* was able to produce three antibiotics: 2,4-diacetyl-phloroglucinol, pyoluterin, and + (S)-dihydroaeruginoic acid. This last compound (detected for the first time in the reported investigations) inhibits phytopathogenic fungi—among them, *Septoria tritici* (Carmi et al., 1994). Antifungal compounds produced by *P. fluorescens* include chitinase, cyanide, and pyrrolnitrin, all of which have a protective action on a variety of seedlings against damping-off caused by *Rhizoctonia solani* (Gaffney et al., 1994). Brominated derivatives of pyrrolnitrin were synthesized by a strain of *P. aureofaciens* (in this treatment considered to be a subspecies of *P. chlororaphis*) (Van Pée et al., 1983). The characteristics of fluorescent species of the genus *Pseudomonas* that participate in the inhibition of plant root pathogens have been analyzed in a useful review (O'Sullivan and O'Gara, 1992).

Miscellaneous antibiotics produced by fluorescent pseudomonads include (1) karalicin, an antibiotic isolated from a strain in the *P. putida*/*P. fluorescens* cluster, with some inhibitory action on yeasts, and a weak antiviral activity against herpes simplex viruses (Lampis et al., 1996a, b); (2) fluviols, a group of antibiotics with antitumor activity that were isolated from *P. fluorescens* and described (Smirnov et al., 1997); (3) 2-alkyl-quinolones, which inhibit growth of *Helicobacter pylori*, were produced by *P. aeruginosa* strains (Lacey et al., 1995); (4) fluopsins C and F (Shirahata et al., 1970); and (5) obafluorin, a β-lactone produced by *P. fluorescens*, with activity against *Bacillus* species. In spite of its β-lactone structure, this compound is hydrolyzed by β-lactamases (Wells et al., 1984); fosfomycin (phosphonomycin), is produced by a strain of *P. syringae* (Shoji et al., 1986).

An interesting practical application has been found for alginate, which is notorious for causing complications in cystic fibrosis cases. The polysaccharide has been found to be a convenient carrier for seed inoculation with genetically modified *P. fluorescens* strains producing antifungal antibiotics (Russo et al., 1996).

Pathogenicity for plants and mushrooms Phytopathogenic pseudomonads are allocated to three of the five rRNA similarity groups (Palleroni, 1984). Only the ones in rRNA group I, the present genus *Pseudomonas*, will be discussed here. Various symptoms produced in plants by these organisms such as tumorous outgrowth, rot, blight or chlorosis, and necrosis are the consequence of alterations of the normal metabolism of plant cells by substances excreted by the pathogens. Among these excretions there are toxins, plant hormones, and enzymes capable of attacking various components of plant tissues.

Bibliographic sources of the original papers describing various hydrolytic enzymes produced by phytopathogens can be found in the first edition of the *Manual* (Palleroni, 1984). Various nomenspecies of plant-pathogenic bacteria described in this treatment as pathovars of *P. syringae* have been found to produce phytotoxins capable of producing disease symptoms in susceptible plants. Table BXII.γ.109, taken from published information (Durbin, 1992), presents a summary of well-known examples of toxins as well as their respective targets of mode of action. Several phytotoxins have an amino acid of peptide nature (Leisinger and Margraff, 1979), and one of them, tabtoxin, has a β-lactam structure, a rare example among secondary metabolites of *Pseudomonas* (Stewart, 1971; Durbin et al., 1978).

Toxins do not necessarily parallel the host specificity of the respective phytopathogen (Patil, 1974). Their importance as taxonomic tools is very limited but their practical significance is considerable. In certain instances their production correlates with the presence of plasmids, and this has provoked renewed interest in them (Leisinger and Margraff, 1979). Loss of a plasmid by treatment with acridine orange resulted in loss of syringomycin production (González and Vidaver, 1979), but no correlation was observed between phaseolotoxin production and the presence of plasmids in *P. syringae* pathovar *phaseolicola* (described

in this treatment as a pathovar of *P. savastanoi*) (Jamieson et al., 1981).

High concentrations of the plant hormone indole-3-acetic acid produced by *P. savastanoi* are thought to be the cause of tumors in plants of the family Oleaceae (Wilson and Magie, 1964).

The interpretations given to data on host–parasite interactions in plant pathology have been deeply influenced by experiments making use of the genetic variability of the host species and of the pathogenic organisms. The main ideas developed during the first three-quarters of the 20th century, and the consequent conceptual changes in the interpretation of the host-pathogen interactions have been reviewed (Ellingboe, 1981).

In recent times, work performed by H.H. Flor with a fungal pathogen of flax gave origin to the gene-for-gene hypothesis. According to this hypothesis, the development of disease or the production of a well-circumscribed hypersensitive reaction in plant tissues depends on matching genes of resistance in the host plants with avirulence genes in the pathogens. In genetic experiments, host resistance was found to be dominant over susceptibility, and the pathogen avirulence dominant over virulence. Matching host resistance with avirulence in the pathogen results in "incompatibility", manifested in a hypersensitive reaction in the host tissues. In the other three combinations (resistant host/virulent pathogen, susceptible host/avirulent pathogen, and susceptible host/virulent pathogen), a "compatible" association is established, which results in disease. These concepts are being subjected to scrutiny in the genetic study of a number of pathovars of the fluorescent species *P. syringae* (Vivian, 1992), which has opened new horizons in attempting to understand the role played by the avirulence genes in the gene-for-gene hypothesis.

Pathogenicity for humans and animals Some references on this subject have been cited on sections of this chapter dealing with outer membrane composition, genetics, pili, and flagella.

As stated by Mekalanos (1992) in an excellent review on environmental signals that control virulence in bacteria, an expanded view of virulence comprises not only the properties directly linked to the pathogenic ability of microorganisms (adherence to host tissues, production of toxins, invasion mechanisms, host defenses) but also other factors involved in "house-

TABLE BXII.γ.109. Some toxins produced by phytopathogenic *Pseudomonas* species and pathovars[a]

Species	Toxins	Target or mechanism
P. syringae		
P. syringae pathovar *atropurpurea*	Coronatine	
P. syringae pathovar *coronafaciens*	Tabtoxinine-β-lactam	Glutamine synthetase
P. syringae pathovar *garcae*	Tabtoxinine-β-lactam	Glutamine synthetase
P. syringae pathovar *glycinea*	Coronatine, polysaccharide	
P. syringae pathovar *lachrymans*	Polysaccharides	
P. syringae pathovar *maculicola*	Coronatine	
P. syringae pathovar *morsprunorum*	Coronatine	
P. syringae pathovar *phaseolicola*[b]	Phaseolotoxin	Ornithine transcarbamylase
P. syringae pathovar *savastanoi*[b]	Indole acetate, cytokinins	
P. syringae pathovar *syringae*	Syringomycins	Plasma membrane
	Syringopeptins	
	Syringotoxins	Plasma membrane
P. syringae pathovar *tabaci*	Tabtoxinine-β-lactam	Glutamine synthetase
P. syringae pathovar *tagetis*	Tagetitoxin	Chloroplastic RNA polymerase
P. syringae pathovar *tomato*	Coronatine	
P. tolaasii	Tolaasin	Plasma membrane

[a]For symbols see standard definitions. Modified from Durbin (1992). The nomenclature used is the one preceding the proposal by Gardan et al. (1992).

[b] *P. syringae* pathovar *savastanoi* is described in this treatment as an independent species, *P. savastanoi*, and pathovar *phaseolicola* is considered as a pathovar of *P. savastanoi*; see description of this species in the list of species).

keeping" functions. In regard to opportunistic pathogens such as the fluorescent pseudomonads and, in particular, *P. aeruginosa*, the functions in this "gray area" can lead to different interpretations. Thus, competition of *P. aeruginosa* for iron in the host tissues by production of a strong chelator such as pyochelin is on the whole comparable to the reaction of this organism in an iron-deficient artificial medium. However, the fact that pyoverdine, the main iron-scavenging compound produced by *P. aeruginosa*, allows normal growth of pyoverdine-negative mutants in a medium containing human iron-transferrin complex justifies considering pyoverdine production as a component of *P. aeruginosa* virulence complex (Meyer et al., 1996). In addition, iron-bound pyochelin acts as an efficient catalyst for hydroxyl radical (HO·) formation, and it contributes to endothelial cell damage from exposure to the superoxide radical (O_2^-) and H_2O_2 (Britigan et al., 1997).

P. aeruginosa seldom infects healthy individuals outside the hospital environment, and the condition of the host is essential in determining the clinical relevance of this opportunistic pathogen. Strains of *P. aeruginosa* can be isolated from a bewildering variety of sources, since perhaps it is one of the most, if not the most, common species in nature. Numbers are very low in human feces, which may be due to competition with other species, since the numbers increase significantly because of antibiotic treatments (Levison, 1977).

In chronically debilitated and immunocompromised patients, *P. aeruginosa* is capable of causing serious and even fatal infections. Individuals with extensive burns, or those who have been subjected to surgical procedures, catheterization, and treatment with broad-spectrum antibiotics, are particularly vulnerable targets. Factors involved in the transmission of *P. aeruginosa* in hospitals have been reviewed and analyzed, and preventive measures are being recommended (Doring et al., 1996).

Iron and osmolarity are mentioned by Mekalanos as environmental signals inducing in *P. aeruginosa* the expression of dissimilar virulence determinants. Control of these and other factors provoked by environmental signals usually act at the transcription level (Mekalanos, 1992). *P. aeruginosa* controls the expression of multiple genes coding for virulence factors by means of LasR, a transcriptional activator, a key component that acts in a cell-density manner ("quorum sensing"), with participation of a *Pseudomonas* autoinducer and an *N*-acylhomoserine lactone. Production of virulence factors (exotoxin A, elastase, alkaline protease, alginate, phospholipases, and rhamnolipids acting as extracellular surfactants) can be stimulated in mutants by gene transfer, or by addition of the synthetic *N*-acylhomoserine lactones (Ochsner and Reiser, 1995). Characteristically, quorum sensing becomes maximal at the time the culture is entering the stationary phase of growth (Rust et al., 1996; Albus et al., 1997).

A global activator, GacA, is responsible for the production of exoenzymes and secondary metabolites in *P. aeruginosa*. Its inactivation resulted in reduced formation of the cell density signal and of LasR. Amplification of the *gacA* gene carried on a multicopy plasmid causes early and enhanced production of LasR and of the lactone. GacA is important in the regulation of the synthesis of virulence factors (Reimmann et al., 1997).

P. aeruginosa exotoxin A is a virulence factor resembling diphtheria toxin in its mode of action. It is able to transfer an ADP-ribosyl group to elongation factor 2, causing translation termination, inhibition of protein synthesis, and cell death (Vasil et al., 1977). The toxin is a protein of 613 amino acid residues, of which residues 60–120 are of importance for excretion and

contain information for interaction with eucaryotic cells (Lu and Lory, 1996). The toxin is excreted when the cells grow under iron-limiting conditions, and its production is repressed when iron is not limiting, which suggests that the ferric uptake regulator Fur is involved, as in pyoverdine synthesis. However, the situation is more complicated, and accessory iron regulatory systems may be involved (Barton et al., 1996; Ochsner et al., 1996). The genetic picture is quite complicated. A regulatory gene (*ptxR*) is involved, and when introduced in a multicopy plasmid, it increases toxin production four- to fivefold, while other virulence factors remain unchanged.

P. aeruginosa is able to produce several proteases. In fact, the organism can cause serious corneal infections and conditions that complicate diseases of the respiratory tract and burn cases, and in all these instances proteases are responsible for the tissues alterations. The most notorious of the proteases as a virulence factor is elastase, with activity on elastin, which lines the blood vessels. The protease is produced in the cell as a precursor of 54 kDa, but it is excreted as a smaller molecule (39.5 kDa) (Morihara and Homma, 1985). Various other proteases produced by the same organism are the cause of lesions in different human and animal tissues. Their importance in pathogenesis is not easy to determine, but mutants defective in protease formation are known to be less virulent.

Alginate is an exopolysaccharide produced by several species of *Pseudomonas*. Various aspects of this extracellular product already have been mentioned, one of which refers to the relationship to fimbriae, which, like alginate, are involved in adherence to tissues (Baker, 1990). A recent review covers important aspects of alginate biosynthesis by bacteria (Gacesa, 1998). The effect that environmental sensory signals have on alginate production as a virulence factor has been examined by DeVault et al. (1989). Specific factors that have been examined include osmolarity, which affects alginate formation by various fluorescent pseudomonads (Singh et al., 1992), and metals (Kidambi et al., 1995).

The *P. aeruginosa* alginate has been the one most thoroughly studied, but production of the polysaccharide also has been detected and characterized in *P. mendocina* (Govan et al., 1981; Hacking et al., 1983; Anderson et al., 1987; Sengha et al., 1989), *P. putida* (Govan et al., 1981), *P. fluorescens* (Govan et al., 1981; Conti et al., 1994; Smit et al., 1996), and even *P. syringae* (Kidambi et al., 1995). Sequences of alginate genes have been detected in other alginate producers and nonproducers belonging to rRNA similarity group I and in species of other rRNA similarity groups (Palleroni, 1984; Fialho et al., 1990; Fett et al., 1992). Similarity of the genetic determinants of alginate in *P. aeruginosa* and *Azotobacter vinelandii* (Ertesvag et al., 1995; Rehm et al., 1996) suggests either a transfer of these determinants among related members of *Proteobacteria*, or else a relic of common ancestry.

Interest in the *P. aeruginosa* alginate, which was first identified in *Pseudomonas* more than three decades ago (Linker and Jones, 1966), derives mostly from its importance as a virulence factor causing respiratory tract infections that aggravate the condition of many cystic fibrosis patients.

Alginate is composed of alternating units of mannuronic and L-guluronic acids joined by a β (1,4) linkage, where part of the carboxyl groups of mannuronic acid are acetylated. The polymer does not show the repetition of a certain type of unit, but instead there are regions where one or the other uronic acid predominates.

The steps of alginate biosynthesis are formulated starting with fructose-6-phosphate, going through mannose-6-phosphate, mannose-1-phosphate, GDP-mannose, and GDP-mannuronic acid (steps catalyzed by enzymes encoded by the genes *algA, C, A,* and *D,* respectively), followed by polymerization, acetylation, epimerization, and final export of the alginate polymer (Chakrabarty, 1998). The genetic aspect of the process is quite complicated because it includes participation of many biosynthetic and regulatory proteins encoded by many more genes than the ones above indicated. The genes have been mapped in three regions of the *P. aeruginosa* chromosome (Goldberg, 1992).

The polymer is not used as a reserve material, and its synthesis demands energy that is not recovered by the cells. It is hard to think of a role other than protection of the pathogen against the host defenses. Secretion of alginate represents a detoxification mechanism to protect the bacterial cells against accumulation of toxic energy-rich nucleoside triphosphates under conditions that prevent active multiplication (Chakrabarty, 1998).

P. stutzeri is frequently isolated from clinical materials but rarely causes disease. In the few cases in which it was the cause of infections, the patients often had another serious underlying disease, and they responded to antibiotics (aminoglycosides, some of the β-lactams, or cephalosporins) (Noble and Overman, 1994).

Ecology   Numerous natural materials are good sources for the isolation of strains of *Pseudomonas,* as well as of strains of species of other rRNA groups now assigned to different genera. It is unfortunate that few procedures can be recommended for the selective isolation of *Pseudomonas* species with exclusion of these other organisms. In some of the habitats, strains of *Pseudomonas* species may represent a minority of the microbial flora; however, certain conditions—a pH close to neutrality, organic matter in solution, a temperature in the mesophilic range, a good supply of dissolved oxygen—together with a capacity for rapid growth in the absence of complex organic factors can favor their predominance. Even in media of extremely low nutrient content, pseudomonads occasionally multiply to a considerable extent. Thus, *P. aeruginosa* has been found capable of growth at the expense of minor impurities present in hospital distilled water (Favero et al., 1971).

Strains of many species are ubiquitous, and isolation data often throw little light on their ecology. When dealing with organisms of such versatility, ecological conclusions are particularly difficult to draw. Of the 57 strains of fluorescent *Pseudomonas* species isolated by den Dooren de Jong, 23 had their origin in soil, and with one exception all were classified as *P. putida,* while all other strains were gelatin-liquefiers isolated from water, and were assigned to *P. fluorescens* (den Dooren de Jong, 1926). However, the conclusion that *P. putida* is a soil organism whereas *P. fluorescens* predominates in water could not be supported by an analysis of isolation data of many other strains from the collections of the Department of Bacteriology at Berkeley, California, and of the Seruminstitut in Copenhagen (unpublished observations). Obviously, a decision on this point is not easy in view of the possibility of cross-contamination of materials from the two habitats.

In recent years, more precise data regarding habitats have become available. An early observation suggests that the predominant fluorescent pseudomonads in wheat rhizosphere comprise one of the biotypes of *P. fluorescens* (Sands and Rovira, 1971). Many other reports have confirmed the fact that these organisms proliferate in general in plant rhizospheres, where they seem to have a stimulating effect on plant growth. In part, this may be due to inhibition of plant pathogenic organisms by iron starvation (Kloepper et al., 1980; Sarniguet et al., 1995) and to the utilization of the iron-siderophore complex by the plants (Powell et al., 1980). An additional beneficial effect is suggested by studies on the rhizosphere of plants growing in the Canadian Arctic, in which *P. putida* can grow and stimulate root elongation of plants during spring and winter at 5°C, as shown by using a strain that can survive exposure to −20°C to −50°C. Following growth at 5°C, the strain secreted a protein with antifreeze activity (Sun et al., 1995). A numerical taxonomic analysis of strains of fluorescent pseudomonads associated with the roots of tomato plants has been published (Stenström et al., 1990).

A recent survey indicates that of the species of bacteria isolated from the rhizosphere of wheat, 40% stimulated plant growth, 40% were inhibitory, and 20% had no effect (Lugtenberg and De Weger, 1992). The properties of fluorescent *Pseudomonas* species that are involved in the suppression of pathogenic species in the root system of plants have been reviewed (O'Sullivan and O'Gara, 1992).

A gnotobiotic system for the study of colonization of plant rhizospheres by *Pseudomonas,* based on seed inoculation and subsequent estimation of different parameters, has shown that many strains were outcompeted by a chosen *P. fluorescens* strain. Slow-growing and autxotrophic mutants of this strain were at a disadvantage, indicating that growth rate and the capacity for multiplication in the absence of exogenous growth factors were important for the persistence of a strain in the rhizosphere community (Simons et al., 1996).

Many species of *Pseudomonas* have been isolated from *Lolium* leaves (Stout, 1960). *P. fluorescens* has been found frequently, and *Burkholderia cepacia* and *Stenotrophomonas maltophilia* less so (Austin et al., 1978). In their review of the literature, these authors mention isolation of *P. fluorescens* from leaves of *Phaseolus vulgaris, Fagus,* and *Pinus* by various workers, and consider that this species is indigenous to leaf surfaces. Whether the strains can become opportunistic plant pathogens is an open question, but some of the epiphytic pseudomonads have definite pathogenic potentialities, as was suggested by Billing (1970b) for *P. viridiflava.*

Plant pathogenic pseudomonads are normally isolated from lesions in plant hosts, which are natural enrichment cultures. Because of their ecological niches, these bacteria may offer some of the most interesting materials for the study of bacterial speciation (Palleroni and Doudoroff, 1972). In general, animals are not as good sources of *Pseudomonas* species, unless these are involved in infections as opportunistic pathogens.

In different areas, speciation of plant pathogens may proceed at different rates according to the selective forces imposed by various environmental factors. Some pathovars of pathogenic species may develop as distinctive ecotypes according the selective pressures predominating in a given area (Garrett et al., 1966).

Continued association with living hosts is important to the survival of members of the *P. syringae* group (Schroth et al., 1981). The association does not necessarily have to involve lesions, since the pathogen may be able to survive as an epiphyte on hosts as well as on nonhosts (Ercolani et al., 1974). Survival of the *P. syringae* pathovars in soil may not be long, in contrast with other fluorescent species (Schroth et al., 1981).

A brief discussion of a few points related to modern biotechnological applications of *Pseudomonas* strains may be pertinent to the present discussion. It has been reassuring to observe that the introduction of a plant growth-promoting strain of *P. fluorescens* and a genetically modified derivative did not appreciably disturb the indigenous microflora when introduced to the soil (Mahaffee and Kloepper, 1997), and that carbon-starved *P. fluorescens* had a good survival capacity and resistance to stress in this environment (Van Overbeek et al., 1995).

The attractive possibility of using genetic means to improve the metabolic versatility of *Pseudomonas* species for the degradation of environmental pollutants has raised questions about the convenience of keeping the lateral spread of genes in the environment under control. These questions have stimulated the development of ingenious methodologies to quantify the degree of horizontal transfer of genes (Jaenecke et al., 1996), to prevent such transfer (Diaz et al., 1994), and to contain the spread of inoculated strains (Jensen et al., 1993b). Modern biotechnological developments also have resulted in improvements for observation of cells in natural environments. These improvements include the combined use of an appropriate probe and scanning confocal laser microscopy (Moller et al., 1996b), as well as in the construction of a biosensor for monitoring metabolic activities *in situ* with a bioluminescent reporter *P. fluorescens* strain (Heitzer et al., 1994).

Pseudomonas species are important members of natural microbial communities. The capacity of these organisms to react to environmental changes is in part related to their capacity to exchange genetic material. An evaluation of this process may be attempted using a strategy that tries to avoid selecting for the characters acquired by the recipients during exchange. The core of the procedure is the use in donor cells of an engineered *lazZ* reporter gene whose expression is shut down by chromosomal repressors at the levels of transcription and translation. The gene will be freely expressed when it can escape repression in the new host, which, consequently, can be identified (Jaenecke et al., 1996).

The consequences of the high surface/volume ratio in procaryotes are manifested, among other things, in concentrating on their surface components of the liquid medium. Fine-grained minerals precipitate around the cell clusters as consequence of their activities when the cells are either freely suspended ("planktonic"), or attached to surfaces with formation of a biofilm. LPS is important in the attachment, which results in a biofilm. Membrane vesicles containing degradative enzymes may bleb off the cell surface and have a predatory action against cells of surrounding populations, with liberation of nutrients that help the growth of the biofilm, as suggested by results obtained with a *P. aeruginosa* model system (Beveridge et al., 1997).

ENRICHMENT AND ISOLATION PROCEDURES

Direct isolation of *Pseudomonas* species often can be achieved from many natural materials, especially soil and water. From these materials, direct isolation in solid complex or minimal media is feasible. When heavy contamination with fungi is expected, these media can be supplemented with antifungal compounds such as cycloheximide (Actidione), which is added to the medium after autoclaving (final concentration, 20–50 μg/ml), from a concentrated solution that does not need to be sterilized.

A medium that is frequently used for direct isolation especially by plant pathologists is medium B^2 of King et al. (1954), which enhances pyoverdin production by fluorescent organisms but is also satisfactory for nonpigmented strains as a general purpose medium. Some selective solid media have been developed based

---

2. Medium B (g/l): protease-peptone (Difco), 20.0; Bacto-agar (Difco), 15.0; glycerol, 10.0; K$_2$PO$_4$, 1.5; and MgSO$_4$·7H$_2$O, 1.5. The pH is adjusted to 7.2.

on medium B. An example is a medium that contains penicillin G, novobiocin, and cycloheximide (Sands and Rovira, 1970). These compounds do not inhibit the fluorescent pseudomonads, and their colonies can be identified on the plates by the characteristic diffusible, fluorescent pigment. A modification of this medium later was published for the isolation of *Pseudomonas* species with pectolytic properties (Sands et al., 1972). The medium contains sodium polypectate, and the pectolytic activity is detected by flooding the plates after growth with a solution of hexadecyltrimethylammonium bromide, which precipitates intact pectin. Once identified, the colonies are isolated as soon as possible since the compound is toxic to the cells.

Two selective media containing the detergent lauroyl sarcosine and the antibiotic trimethoprim, and differing in having either Casamino acids (medium S1) or L-asparagine (medium S2), gave good selectivity and detection of fluorescence on initial plating. Of the two media, S1 gave the highest recovery, although S2 was more selective (Gould et al., 1985).

Selective media for the oxidase-negative fluorescent phytopathogens have been described (Sands et al., 1980).

Some methods for the isolation of certain species may go through an initial enrichment step. The lesions produced by plant pathogenic pseudomonads on various plant organs represent a natural enrichment, from which direct isolation can be attempted. The lesions selected should not be too old, when other organisms may have gained access to the diseased tissues. Isolation of denitrifying pseudomonads (*P. aeruginosa*, *P. stutzeri*, *P. mendocina*, some biovars of *P. fluorescens*, etc.) after enrichment is often successful. The denitrification experiment can be performed using particular carbon sources as electron donors, and the incubation can be done at a temperature that favors multiplication of the desired organism. However, due to our ignorance of the proper conditions, or unpredicted transformations by other organisms in the microflora, these experiments may fail to give the expected results.

It is convenient to keep in mind that that species of aerobic pseudomonads other than the ones mentioned above may predominate in the enrichment cultures. In the chapter on *Burkholderia* in this *Manual*, a table has been included that summarizes the properties of *Pseudomonas* species that are able to denitrify, together with selected species of denitrifiers of other rRNA groups.

Methods of enrichment and isolation are dispersed in a number of publications, and some of them are very specific for species originally assigned to *Pseudomonas* but now allocated to newly created genera and phylogenetically belonging to other rRNA groups (Palleroni, 1984) than group I. Additional details may be found in the second edition of the treatise *The Prokaryotes* (Palleroni, 1992a), and in the first edition of this *Manual* (Palleroni, 1984).

MAINTENANCE PROCEDURES

Most *Pseudomonas* strains can be maintained on slants of common bacteriological media (nutrient agar or other standard complex media, or various chemically defined media with the addition of 0.5% yeast extract and 0.1% lactate or glycerol), with transfers every 1–2 months. Slants can be stored at 4–8°C. The collection examined by Stanier et al. (1966) included a number of strains of fluorescent and nonfluorescent *Pseudomonas* species that had been kept on slants of ordinary media with periodic transfers since the 1920s, and their phenotypic properties appeared to have remained essentially unchanged. In contrast, some strains of *Pseudomonas* species as well as members of other rRNA groups (Palleroni, 1984) are not so easy to maintain for long periods in slants in the refrigerator.

Lyophilization is generally reliable for long-term preservation, particularly when proteinaceous material (e.g., skim milk) is used as an ingredient of the suspending fluid. There are many variations of lyophilization procedures traditionally followed by various laboratories, with good results. A substantial proportion of the cell mass of cultures of *Pseudomonas* species does not survive even the most careful lyophilization procedure, and lyophilization has been replaced in many laboratories by freezing cultures at a temperature of $-20°C$ or lower (usually from about $-70°$ to $-80°C$ in mechanical freezers or at lower temperatures in liquid nitrogen) after addition of a cryoprotective agent to the liquid before freezing (for instance, 5–10% glycerol, final concentration).

PROCEDURES FOR TESTING SPECIAL CHARACTERS

Flagellar number and insertion The observation of these characteristics is best done on smears prepared by the method described by Jessen (1965), combined with results expressed on a statistical basis, as mentioned earlier (Lautrop and Jessen, 1964). On clean microscope slides, thin smears are prepared from dilute suspensions in distilled water of bacteria that have grown in broth agar cultures at 23°C or 30°C for about 18 h. Best results are obtained using freshly poured solid media whose surface has been dried in an incubator for no more than 1 h. The number of cells with flagella in the smears often varies with the degree of dryness of the agar medium in which the strains have been growing. The smears are stained for flagella by the method of Leifson (1960)[3]. Only slides with well-dispersed cells and well stained for flagella should be used. The number of flagella on each of the first 50–100 flagellated cells is noted. Cells with flagella at both ends are disregarded, since they probably represent cells in an early stage of their division. Some results of this technique have been mentioned in the early section on flagella and pili (fimbriae).

Screening of nutritional and other physiological properties The medium described by Stanier et al. (1966) has been used extensively for nutritional screenings and for testing various physiological properties of cultures of pseudomonads. One disadvantage of this medium is that not all cultures tolerate well the heavily chelated composition of the medium. More satisfactory results are obtained with a formulation that is both simpler and safer, such as the medium recommended by Palleroni and Doudoroff (1972)[4], which is most satisfactory for autotrophic and heterotrophic enrichments and as a general medium for cultivation and short-term preservation. Minimal media of similar

composition have been described (Schlegel and Lafferty, 1971; Zavarzin and Nozhevnikova, 1977). These last authors recommend supplementation of the medium with minor elements for best results with fastidious organisms, but such additions are unnecessary when dealing with *Pseudomonas* strains.

Good heterotrophic growth in all these minimal media can be obtained by addition of a single organic compound as a carbon and energy source (usually 0.1%, final concentration). Experimental details of the nutritional analysis have been extensively discussed elsewhere (Stanier et al., 1966; Palleroni and Doudoroff, 1972). Ideally, the analysis should be performed under conditions that prevent cross-feeding or competition among the various strains when the medium is solidified with agar and several strains are patched on a single plate. Organic impurities in the agar may also be a serious problem. For critical tests, liquid media may be preferable, although the amount of work involved and the facilities required may be considerably greater. Carbon compounds that are to be tested in different concentrations or that require special treatments for their use in nutritional screenings have been discussed by Palleroni and Doudoroff (1972). Description of many species of aerobic pseudomonads performed at Berkeley included nutritional studies using many organic compounds as substrates. The studies were extended to the plant pathogenic pseudomonads in two laboratories (Misaghi and Grogan, 1969; Sands et al., 1970).

At present, the studies on nutritional properties of various prokaryotic groups are more often performed using commercial kits designed to reduce the labor involved in traditional procedures. Some of them (API, Biotype-100 strips, Biolog GN Microplate System, Hayward, CA) have been used in studies on *Pseudomonas*, and the results will be briefly discussed below in the section on taxonomic comments.

Biosynthesis of medium-chain-length polyhydroxyalkanoic acids (mcl-PHAs) The following observations on the biosynthesis and detection of mcl-PHAs have been provided by Birgit Kessler, a member of B. Witholt's group at the Swiss Federal Institute of Technology in Zürich.

1. As with poly-β-hydroxybutyrate (PHB), accumulation of mcl-PHAs can be elicited in a medium of high C/N ratio, and the polymer can be detected after staining with Sudan Black or by direct observation under phase contrast microscopy. However, granules of the two polymers cannot be differentiated by either procedure, and the identification procedure may involve either gas chromatographic analysis (Timm and Steinbüchel, 1990) or a solubility test. The mcl-PHAs are soluble in acetone, while PHB is not.

2. Assimilation of medium- and long-chain-length fatty acids (saturated, unsaturated, or certain branched and substituted acids) enables the cells to accumulate mcl-PHAs. Some *Pseudomonas* strains (*P. aeruginosa* PAO and *P. putida* KT2442) accumulate mcl-PHAs when grown on carbohydrates. In contrast, *P. oleovorans* lacks this capacity, but it accumulates mcl-PHAs when grown on alkanes. Growth on butyrate and β-hydroxybutyrate induces production of small amounts of PHAs, and the granules can be seen under phase contrast directly or after staining with Sudan Black.

3. Several species of aerobic pseudomonads have been tested for their capacity of PHB and/or PHA accumulation (Timm and Steinbüchel, 1990). Production of both types of polymers by a single strain did not receive special attention, but recently a strain of an unidentified *Pseudomonas* sp. was found

---

3. Leifson's reagent for flagella staining (Leifson, 1960): 1.2% solution of basic fuchsin in 96% ethanol, 3% solution of tannic acid in water, and 1.5% solution of NaCl in water. Equal parts of the three reagents are mixed before use. The mixture can be left overnight in a refrigerator, where a precipitate may form. The clear stain is applied over the smear, and after about 2–3 min a fine precipitate will form. At this point, the staining is interrupted by washing the slides with water at room temperature. The final point may have to be ascertained by trial and error before optimal conditions are defined. Absolute cleanliness of the slides is essential for good results.

4. Medium of Palleroni and Doudoroff (1972) (g/l of 0.33 M Na-K phosphate buffer, pH 6.8): NH₄Cl, 1.0; MgSO₄·7H₂O, 0.5; ferric ammonium citrate, 0.05; and CaCl₂, 0.005. The first two ingredients are added to the buffer and sterilized by autoclaving. The ferric ammonium citrate and calcium chloride are added aseptically from a single stock solution that has been sterilized by filtration.

with such capacity when grown on fatty acids or carbohydrates.

Pigment production   Perhaps the most widely used medium for pyoverdine production is the medium B of King et al. (1954). Paton (1959) has described a medium whose preparation involves extensive treatment of the solution and the agar with iron-chelating agents, but the removal of iron is so complete that it often results in inhibition of bacterial growth (Garibaldi, 1967). Garibaldi recommends instead a medium that is based on the well-known iron-chelating capacity of conalbumin, a component of egg-white.[5] Luisetti and collaborators (1972) recommend a medium[6] for the enhancement of fluorescent pigment production by plant pathogenic and other pseudomonads that fail to fluoresce in the medium B of King and collaborators.

A point frequently incorrectly stated or disregarded altogether in descriptions of procedures for the detection of pyoverdins by fluorescent pseudomonads is the type of ultraviolet lamp to be used for observing fluorescence. Only the "true" fluorescent pigment will fluoresce under a source of short wavelength ultraviolet light (around 254 nm). Some species belonging to rRNA group II (Palleroni, 1984) (*Burkholderia cepacia, B. gladioli, B. caryophylli*) produce diffusible yellow-green pigments that are sometimes mistaken for the fluorescent siderophores.

Medium A of King et al. (1954) is generally recommended for the production of phenazine pigments to which pyocyanin belongs. In our experience, this medium is not always very effective, but unfortunately no alternative can be recommended. Phenazine pigment production, therefore, appears erratic, particularly with cultures that have been kept for a long time under ordinary laboratory conditions.

The blue pigment of *P. fluorescens* biovar IV ("*P. lemonnieri*") can be produced abundantly by fresh cultures in the potato medium[7] familiar to mycologists. As mentioned earlier, the pigment is related to indigo iodine (Starr et al., 1967). Interestingly, this same medium induces the production of indigo iodine by an unrelated organism, *Corynebacterium insidiosum* (Starr, 1958).

Ring fission mechanisms   The *ortho* or *meta* cleavage of two central intermediates in the metabolism of aromatic compounds (catechol and protocatechuate) can be tested by the method originally suggested by K. Hosokawa (Stanier et al., 1966). After growth in a chemically defined medium with an aromatic substrate, the cells are suspended in 0.02 M Tris buffer (pH 8). To each 2-ml portion of suspension are added a drop of toluene and 0.2 ml of either 0.1 M catechol or 0.1 M protocatechuate. A bright yellow color appearing in a short time (usually within

a few seconds) indicates a *meta* cleavage. The tubes are shaken for 1 h at 30°C, and then tested for the appearance of β-ketoadipate (indicative of an *ortho* cleavage) by the Rothera reaction, as follows. Solid ammonium sulfate is added to saturation and the pH is brought up to about 10 by the addition of two drops of 5 N ammonium hydroxide. One drop of freshly prepared 25% sodium nitroprusside in water is added. A deep purple color indicates a positive reaction.

Arginine dihydrolase reaction   The method giving the most unequivocal results is the direct one, based on an estimation of arginine disappearance under anaerobic conditions. A bacterial suspension (200 Klett units, as measured with green filter 54) is incubated for 2 h at 30°C in the presence of arginine (2.5 × 10^{-4} M) under anaerobic conditions. The tubes are immersed in a boiling water bath for 15 min and arginine is estimated by the method of Rosenberg in a sample of the supernatant after removing the cells. A control, without addition of arginine, is included in the experiment (Rosenberg et al., 1956).

The Rosenberg procedure is cumbersome and time-consuming, and other methods appear to be much simpler and sufficiently reliable. Among them, the method of Thornley (1960) is perhaps the most convenient. As recommended by Lelliott et al. (1966), Thornley's medium[8] can be conveniently dispensed in 3-ml portions into 5-ml screw cap vials. After autoclaving, the vials are inoculated by stabbing and sealed with melted paraffin. Anaerobic formation of ammonia from arginine can be detected by a change of color of the indicator within 3 d.

Acid production from carbohydrates   As pointed out by Palleroni and Doudoroff (1972), the reaction of acid production from carbohydrates is not necessarily correlated with nutritional data, since the acid produced by oxidation of a sugar may not always be a suitable carbon source for growth. However, this should not diminish the taxonomic value of the test. More serious, though, is the objection of redundancy, in that a single enzyme may affect the oxidation of several different sugars (Weimberg, 1962).

Acid production from carbohydrates has been used very extensively for descriptive purposes since the last century, and it is still being used in various laboratories. It is described here among the recommended procedures, provided that the results are interpreted with proper reservations. The method of choice is that described by Hugh and Leifson (1953), which gives clear and reproducible results. The main advantage of this method is the use of a medium containing a low concentration of peptone[9]. If the concentration is high, the deamination can neutralize a positive acid reaction from the sugar.

Hydrolysis of Tween 80   This reaction is indicative of lipolytic activity, and is carried out in the medium of Sierra (1957)[10]. The medium is usually dispensed in Petri dishes and is spot-inoculated with the cultures. During the incubation, the plates are observed daily for opacity around the patches, due to formation of insoluble calcium soaps (Sierra, 1957). If the reaction has been negative for 10 d, the bacterial mass may be scraped off to observe

---

5. Garibaldi's medium: sterile egg white is drawn aseptically from eggs that have been sterilized externally by immersion in 70% ethanol for 5 min, drained and flamed to remove the residual alcohol. The egg white is warmed to 45°C and added aseptically to give a final concentration of 10% (v/v) to any commercial complex solid medium that has been autoclaved and cooled to 45°C. After mixing, the medium is dispensed into Petri dishes. A solution of the protein conalbumin (available commercially) that has been sterilized by filtration can be used instead of egg white at a final concentration of 1.7 mg/ml of medium.

6. Medium of Luisetti et al. (1972) (g/l): vitamin-free Casamino acids, 10.0; K_2HPO_4, 1.0; $MgSO_4 \cdot 7H_2O$, 1.0; sucrose, 10.0; gelatin, 30.0; and agar, 20.0; pH 7.0.

7. A formula recommended by D.C. Hildebrand (personal communication), who has isolated a number of strains of these organisms, is as follows. Sliced potatoes (250 g) are steeped in 1 liter of water at 65°C for 1 h. The preparation is filtered through three layers of cheesecloth, and 2.0 g of glucose and 20.0 g of agar are added to the solution. The medium is sterilized at 20 psi for 30 min.

8. Thornley's medium (g/l): peptone, 1.0; NaCl, 5.0; K_2HPO_4, 0.3; agar, 3.0; phenol red, 0.01; and L-arginine-HCl, 10.0; pH 7.2.

9. Hugh-Leifson medium (g/l): peptone, 2.0; NaCl, 5.0; K_2HPO_4, 0.3; agar, 3.0; bromothymol blue, 0.03; and carbohydrate, 10.0; pH 7.1.

10. Sierra's medium (g/l): Bacto-peptone (Difco), 10.0; NaCl, 5.0; $CaCl_2 \cdot H_2O$, 0.1; and agar, 17.0. Tween 80 is sterilized separately and added to the medium after autoclaving to give a final concentration of 10.0 g/l.

the medium under the patch. The presence of a precipitate, however, may be due to release of endocellular lipases by lysis of part of the population, and therefore, for practical purposes, the reaction is considered negative.

Nitrate reduction and denitrification Reduction of nitrate to nitrite can be tested according to Lelliott et al. (1966), and denitrification according to Stanier et al. (1966). Some of the problems encountered in the denitrification test have been discussed by Palleroni and Doudoroff (1972).

Miscellaneous tests Additional tests that are performed in the identification of *Pseudomonas* species include the following, with references to bibliographic sources for the methods: oxidase test (Kovács, 1956; Stanier et al., 1966); levan formation from sucrose (Lelliott et al., 1966; Stanier et al., 1966); gelatinase (Skerman, 1967); and the egg yolk reaction (Lelliott et al., 1966; Stanier et al., 1966). In testing for catalase, the usual procedure of placing a drop of hydrogen peroxide solution on top of a colony and observing the formation of bubbles is usually satisfactory, but occasionally the reaction may be so weak as to require observation under a dissection microscope, or sensitive detection methods involving the use of an oxygen electrode (Auling et al., 1978).

A number of additional techniques have been added as taxonomic tools to supplement the phenotypic analysis of the aerobic pseudomonads. The results of many of these methods will be summarized below under general taxonomic comments, and the papers cited are sources of information on the methodologies followed in each case. The methods described in these bibliographic references cover various aspects of nucleic and protein studies, the analysis of the fatty acid composition, as well as the description of numerical procedures for data processing.

DIFFERENTIATION OF THE GENUS *PSEUDOMONAS* FROM OTHER GENERA

The first edition of the *Manual* included a discussion of the properties that distinguish the aerobic pseudomonads of the genus *Pseudomonas sensu lato* from other phenotypically similar Gram-negative, nonsporulating aerobic bacteria. The reader is referred to this source for comments on this subject in its broader sense (Palleroni, 1984).

Over a quarter of a century ago, strains of *Pseudomonas* species were subjected to experiments of ribosomal RNA–DNA hybridization. Based on the conservative nature of the ribosomal RNA cistrons that had been demonstrated in species of the genus *Bacillus* (Doi and Igarashi, 1965; Dubnau et al., 1965), these experiments attempted to disentangle the phylogenetic complexity that was suspected in the unrelated genus *Pseudomonas*. The results were indeed very striking. The rRNA–DNA hybridization experiments clearly suggested an internal subdivision of *Pseudomonas* into five rRNA groups that seemed "to be very distantly related to each other phylogenetically . . . [deserving] at least independent genus (and possibly family or order) assignment" (Palleroni et al., 1973). One of these groups, which included the type species, *P. aeruginosa* (rRNA group I), was to retain the genus name. It showed a closer relationship to *Xanthomonas* and to *E. coli* than to other groups of pseudomonads.

It was soon realized by other workers that extension of this approach to comparative studies of other bacterial taxa could yield results of use "as an index of general relatedness [to] provide a sound base on which to construct taxonomic or maybe identification schemes" (Johnson and Francis, 1975). This pro-

phetic statement found ample corroboration in the years that followed.

Pseudomonas species were examined again at Gent (Belgium) on a much broader basis, but using essentially the same experimental approach, nucleic acid hybridization, and the results, published 10 years later (De Vos and De Ley, 1983), confirmed the conclusions of the original observations. In view of the phylogenetic diversity demonstrated in the genus *Pseudomonas* as classically defined, one of the striking consequences from the taxonomic point of view was that the set of basic properties that had been considered characteristic for this group of organisms— i.e., rod shaped, Gram negative, nonsporeforming, motile by means of polar flagella, and aerobic—was inadequate for practical diagnosis of the genus *Pseudomonas*. The circumscription of *Pseudomonas* is now restricted to the members of rRNA similarity group I of Palleroni et al. (1973), and it is unfortunate that the similarities to members of other groups (which, in the first place, was the basic reason for their original assignment to the genus *Pseudomonas*) make it difficult to define discrete sets of differential phenotypic characters.

The hybridization methods originally designed for the evaluation of rRNA similarities (Palleroni et al., 1973; De Vos and De Ley, 1983) were eventually replaced by other procedures (oligonucleotide cataloging, rDNA nucleotide sequences). Their use, as expected, confirmed the results of the earlier hybridization experiments. Sequencing the 16S rDNA gene after amplification by PCR is now universally used as the basis for assignment of new species to the genus *Pseudomonas*. As shall be mentioned below, fatty acid analysis can also be effective in differentiating *Pseudomonas* from the other similarity groups.

Pseudomonas sensu stricto (rRNA similarity group I) is still a group of considerable heterogeneity. Internal subgroups can be defined based on plant pathogenicity or pigment production. Among the pigments, the fluorescent siderophores are characteristic of the so-called fluorescent pseudomonads, and production of these compounds is sufficient *per se* to allocate a strain in the genus *Pseudomonas*.

However, the genus also includes nonfluorescent species. It is interesting to note that during the early phenotypic studies carried out at Berkeley, in spite of the differences in pigmentation, a phylogenetic relationship was suspected between some pigmented and nonpigmented species. Obviously, the extensive phenotypic study to which the strains had been subjected implied the expression of a considerable proportion of genes, and it was observed that "the nutritional spectrum of the [alcaligenes group, that is represented by strains that are the least nutritionally versatile of all aerobic pseudomonads] recall[ed] in many respects that of the fluorescent pseudomonads, but it [was] much narrower, notably in the areas of sugars and sugar-acids, and of aromatic compounds" (Stanier et al., 1966). In agreement with the above inference, the numerical taxonomic studies of Sneath et al. (1981), based on part of the published data from the Berkeley group, resulted in a three-dimensional graphical representation, where the so-called fluorescent complex is placed on the same side of the horizontal plane as the nonfluorescent species *P. stutzeri* and *P. alcaligenes*. Although discrete sets of phenotypic properties appear to be inadequate for an absolute differentiation of *Pseudomonas* from other genera of aerobic pseudomonads, the wealth of information contained in extensive nutritional screenings can provide, after appropriate processing, a generally satisfactory graphical representation of the relationships between *Pseudomonas* and other rRNA similarity groups (Sneath et al.,

1981). The position of nonfluorescent species is uncertain in graphical representations of values obtained using the Gower or the pattern coefficients, but association of *P. stutzeri* and *P. alcaligenes* is more distinct when using three-dimensional representation by principal coordinate analysis of Euclidean distances (Sneath et al., 1981). Moreover, the phylogenetic representation of Fig. BXII.γ.124 shows that the nonfluorescent species *P. alcaligenes* and *P. pseudoalcaligenes*, are placed in the neighborhood of the versatile fluorescent species *Pseudomonas aeruginosa*.

The original descriptions of the various phenotypic groups of *Pseudomonas* species included the inability of species of group I to accumulate PHB granules as carbon reserve material, with the exception of some *P. pseudoalcaligenes* strains that appeared to accumulate small amounts of the polymer. The diagnostic value of this negative characteristic appeared to be rather limited: on the one hand, the species of group V (*Xanthomonas* and *Stenotrophomonas*) are also negative, and on the other, the present treatment includes some species (*P. corrugata*, *P. amygdali*, *P. ficuserectae*) that have been reported in the literature to be able to accumulate PHB. However, *P. corrugata* and *P. ficuserectae* actually do not synthesize PHB but instead accumulate PHAs of different monomer composition than PHB, thus buttressing the taxonomic value of inability to accumulate PHB as a generic character (Kessler and Palleroni, 2000). The lack of ability to form PHB is shared by species of group V (*Xanthomonas* and *Stenotrophomonas*), which is interesting in view of the relatively close relationship of this group with group I (*Pseudomonas*). (See also the chapter on *Stenotrophomonas* in this *Manual*.)

The main body of the present treatment of the genus *Pseudomonas* comprises the species listed in group I of the first edition of the *Manual*. In addition, there are species that were not included in the early experiments, although they were mentioned as probable members of group I, and were later assigned to *Pseudomonas* mainly from additional screenings performed at the University of Gent, Belgium (De Vos and De Ley, 1983; De Vos et al., 1985, 1989). An excellent source of information on the genuine *Pseudomonas* species as well as the present position of many species assigned to *Pseudomonas* but now transferred to rRNA groups other than group I (Palleroni, 1984) is the review article by Kersters et al. (1996). The list of species below will refer with special detail to species whose assignment to *Pseudomonas* has been decided based on phylogenetic criteria.

TAXONOMIC COMMENTS

The proposed division of the aerobic pseudomonads into five RNA similarity groups has received confirmation from numerous contributions by workers in many different laboratories. The investigations have followed various approaches, including investigations on metabolic pathways and their regulatory mechanisms, nucleic acid similarity studies, amino acid sequences of selected proteins, immunological studies, and cell wall composition. The results obtained by application of these approaches were extensively discussed in previous reviews (Palleroni, 1975, 1986, 1992a, b, 1993), and determinative keys have been proposed (Palleroni, 1977; Bergan, 1981; Stolp and Gadkari, 1981). The data add considerable precision to the circumscription of the genus *Pseudomonas*, and some of the methodologies can be of help for determinative purposes.

Interest in the taxonomic intricacies of this genus has not subsided, and of particular interest is a compilation of papers published by members of a consortium of European research groups in vol. 19, pp. 465–568 of *Systematic and Applied Microbiology*. These reports represent to a great extent further explorations into the formidable phenotypic and genomic diversity of the genus *Pseudomonas* as presently circumscribed using a number of experimental approaches.

In addition to the techniques for the determination of special properties already described, this section includes a discussion of other methods or variations of previous methodologies that have been particularly informative in taxonomic analysis of *Pseudomonas* species. Details of the experimental procedures may be found in the respective bibliographic sources.

Nutritional and metabolic studies   Even though the formidable body of phenotypic information collected at Berkeley constituted a very appropriate subject for numerical taxonomic studies, this type of data treatment was reported only for a fraction of the available strains. A considerable amount of work was done in this area, but it remained unpublished, except for a few selected instances (Sands et al., 1970; Palleroni et al., 1972; Champion et al., 1980). Numerical treatment of phenotypic data also is available in original Ph.D. theses from the Department of Bacteriology (Ballard, 1970; Barrett Ralston, 1972).

A more extensive numerical analysis of published nutritional data of the Berkeley collection has given results in very good agreement with the proposed groupings, as indicated by Sneath et al. (1981). Their paper includes an illustrative three-dimensional representation using the system of principal coordinate analysis. A different system of data processing not requiring a complete matrix with all pairwise comparisons also results in useful three-dimensional representations, and it has been applied to published data of the pseudomonads (Hildebrand et al., 1984).

In addition to the extensive phenotypic studies performed at Berkeley on many strains of aerobic pseudomonads following conventional bacteriological techniques, in recent times additional observations have been performed using available commercial kits. Many strains of *Pseudomonas* species were subjected to nutritional studies using API Biotype-100 strips (BioMérieux, La Balme les Grotes, France) for carbon assimilation and the Biolog GN MicroPlate System (Biolog Inc.), which is based on substrate oxidation as detected by the change of color of an indicator (Grimont et al., 1996). Each of these systems included a list of nearly 100 organic compounds, the data obtained were analyzed by computerized systems, and the results were represented graphically in dendrogram form. *P. aeruginosa*, *P. tolaasii*, *P. mendocina*, *P. cichorii*, *P. viridiflava*, *P. fragi*, *P. stutzeri*, *P. agarici*, *P. alcaligenes*, and *P. pseudoalcaligenes* appeared as homogeneous phenons. In contrast, some of the fluorescent species (*P. putida*, *P. fluorescens*, *P. marginalis*) did not form discrete clusters but were found distributed in several phenons, confirming the heterogeneity previously reported elsewhere.

In general, according to Grimont et al. (1996), there was a remarkable correlation between the Biotype-100 data and the results obtained over 30 years ago using a more conventional methodology (Stanier et al., 1966; Ogawa et al., 1992). This is in agreement with the basic idea behind the two approaches, which rely on the utilization of substrates for growth. The work of Stanier et al. (1966) did not include testing for the production of acid from carbohydrates, a technique widely used at the time. As discussed elsewhere (Palleroni and Doudoroff, 1972) and in a previous section of this chapter, in spite of their reproducibility the results of the oxidation tests may be redundant; i.e., one enzyme may produce acid from more than one substrate. On

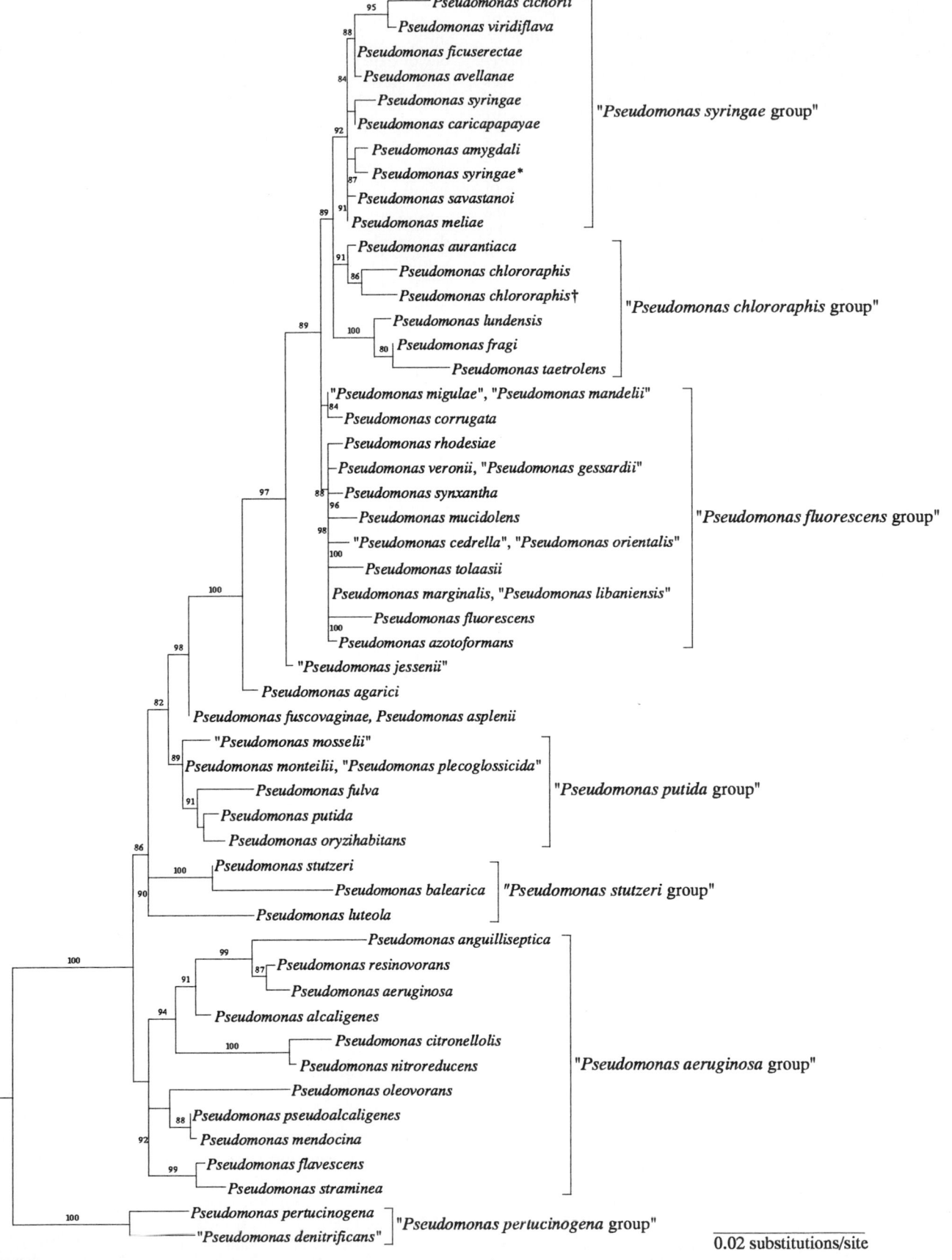

FIGURE BXII.γ.124. Phylogenetic tree of species of the genus *Pseudomonas* derived from the 16S rDNA sequences. Bootstrap values of 80% or higher are indicated at the branch points. *Escherichia coli* (V00348) was used as the root organism. (Reproduced with permission from Y. Anzai et al., International Journal of Systematic and Evolutionary Bacteriology *50:* 1563–1589, 2000, ©International Union of Microbiological Societies.)

the other hand, acid production does not necessarily correlate with utilization of a given substrate for growth. However, acid production from carbohydrates in some instances is useful and provides an additional metabolic test.

A study of strains of *P. putida* and other fluorescent pseudomonads was performed by a determination of the assimilation of carbon compounds of three chemical families (carbohydrates, acids, and amino acids); this was done using the API system (BioMérieux, La Balme les Grotes, France) and following a methodology described elsewhere for a numerical taxonomic study on nonfluorescent species of *Pseudomonas* (Gavini et al., 1989a). This determination was supplemented by an extensive enzymatic study using kits from the same manufacturer for testing 59 peptidase, 10 esterase, and 20 oxidase activities (Elomari et al., 1994). Five main phenotypic clusters were obtained by application of the unweighted pair group average linked method and the Dice similarity coefficient. Of these, one group (II) was identified as *P. putida* biotype A, and could be subdivided into four subclusters, IIa–IId. Strains of cluster II were characterized by ribotyping (see below) using the restriction enzyme *Pvu*II, and seven ribotype subclusters were obtained. Two phenotypic subclusters correlated completely with two ribotype clusters. Based on the results of these studies, the authors urged a revision of *P. putida* and its biovars (of which A represents the "genuine" *P. putida*). The studies indicated the convenience of creating a third biovar, in confirmation of a similar suggestion that emerged from the studies of Barrett et al. (1986), and they serve to illustrate once again the enormous complexity of the species of the fluorescent group. In future studies, the *P. putida* biovars may be circumscribed more precisely as different species (Elomari et al., 1994).

The phenetic taxonomy of a large number of fluorescent pseudomonads isolated from tomato roots was performed using Jaccard similarity coefficients (Stenström et al., 1990). The field strains belong to the *P. fluorescens*/*P. putida* complex, and a suggestion was made to convert the biovars of the first species to the rank of individual species. Once again, the *P. putida* strains included a group that differed from the two known biovars (Stenström et al., 1990).

The usefulness of 36 determinative tests was estimated for 32 pathovars of the phytopathogenic fluorescent species *P. syringae*. The results allowed differentiation of pathovars in some cases, but in others, no distinctive clusters could be detected (Young and Triggs, 1994).

A numerical taxonomic analysis of the nonfluorescent species of RNA similarity group I (*Pseudomonas*), including a large number of strains, suggested that there may be future subdivision in a larger number of species than those recognized at present (Gavini et al., 1989a). Since the analysis was restricted to phenotypic properties, further work using molecular approaches will be necessary to substantiate such suggestion. Other examples illustrate the power of the numerical analysis of extensive phenotypic data for defining clusters that, upon further analysis, may be segregated as biovars or even described as independent species within a population of related strains. One case in point is the circumscription of *P. lundensis*, which, as mentioned in the species description (see below in list of species), had been previously identified as a well-defined cluster within a collection of strains isolated from meat (Molin and Ternström, 1982, 1986). It was similarly defined independently by an analysis of a collection of fluorescent strains from the Berkeley collection (Barrett et al., 1986). The studies resulted in defining four clusters of strains, of which two were identified as *P. fragi*. The other two

clusters, of fluorescent pseudomonads, could not be identified as any of the described biovars of *P. fluorescens* or *P. putida*. The authors emphasized the value of substrate utilization tests for elucidating the relationships among *Pseudomonas* strains (Shaw and Latty, 1982).

The numerical analysis of a collection of strains isolated from tomato pith necrosis demonstrated that the strains of *P. corrugata* constituted a single phenon. This study served as the basis for an emended description of this species (Sutra et al., 1997). Numerical taxonomic studies of strains isolated from natural mineral waters led to the discovery of the new species *P. rhodesiae* (Coroler et al., 1996) and *P. veronii* (Elomari et al., 1996).

Mol% G + C studies   The original phenotypic studies by Stanier and collaborators in 1966 were supplemented with base composition determinations of the DNA of the strains (Mandel, 1966), which were later extended for inclusion in the description of new species under study in the same laboratory. In modern species descriptions, mol% G + C values of the DNAs are routinely included, although they are still missing for some of the species assigned to the genus.

DNA–DNA hybridization studies   The phenotypic studies on the aerobic pseudomonads were followed by DNA–DNA hybridization studies, and, by virtue of the wide range of similarity values obtained, the results provided the first clues of the profound genomic differences among the species of this group of organisms. The results eventually suggested the need for similarity studies involving molecules of a more conservative nature. Within the genus *Pseudomonas*, the DNA–DNA hybridization studies included experiments with strains of the *P. stutzeri* group (Wendt-Potthoff et al., 1992), with fluorescent organisms (Palleroni et al., 1972), and with miscellaneous members of other groups. In all these experiments, the hybridizations were done by the competition technique of Johnson and Ordal (1968), but in later experiments the S1 nuclease procedure described by Johnson (1994a) was chosen.

The most important taxonomic application of DNA–DNA hybridization techniques involving the whole genome is in defining relationships at the species level. When applied to strains of aerobic pseudomonads, the hybridization experiments confirmed the relationships among species of the same phenotypic groups; at the same time, as mentioned above, the values were very low or negligible for species of different groups, even among strains of species that unequivocally could be assigned to the same genus. In spite of these observations, in some instances the results of DNA hybridization experiments occasionally have been used to attempt the precise circumscription of genera. One example was the proposal of two new generic names (*Chryseomonas* and *Flavimonas*) (Holmes et al., 1987) for two species previously assigned to the genus *Pseudomonas* (Kodama et al., 1985); the low DNA similarity values obtained in hybridization with DNA of *Pseudomonas* species were taken as the basis for such a proposal. Later studies suggested that the newly created generic names should in fact be considered as junior subjective synonyms of *Pseudomonas* (Anzai et al., 1997), and they are described as such in the present treatment.

Restriction analysis of the whole genome   A useful approach adopted for the characterization of bacterial genomes combines the action of endonucleases that infrequently cut the chromosome with pulsed-field agarose gel electrophoresis for the separation of large DNA fragments. A limited number of cuts per genome is desirable for genomic restriction mapping (Mc-Clelland et al., 1987), with clear advantages over frequently cutting endonucleases for the production of fingerprints characteristic of different taxa (Mielenz et al., 1979; Dobritsa, 1985; Sorensen et al., 1985). Application of this principle has resulted in a practical method for constructing and analyzing macrorestriction patterns of 234 strains of different species of aerobic pseudomonads (Grothues and Tümmler, 1991). The restriction nucleases *AsnI*, *DraI*, *SpeI*, *XbaI*, and *PacI* were found to be most appropriate by being specific for AT-rich regions, or for sites including the extremely uncommon tetranucleotide CTAG (McClelland et al., 1987).

The macrorestriction patterns are compared for the number and position of the bands, and use of appropriate equations often indicates a correlation between the estimated similarities of fingerprints and conventional taxonomic groupings. The results obtained by Grothues and Tümmler (1991) in general confirm the classification of pseudomonads based on extensive numbers of phenotypic properties and nucleic acid hybridization studies.

Strains sharing the same mol% G + C contents in the chromosomal DNA, the same codon usage, and similar genome size, give similarity coefficients useful for determinative purposes. The genome patterns also have the practical advantage of helping in epidemiological studies. Differences have been found between *Pseudomonas* strains that are susceptible or multiply resistant to antibiotics (Yamashita et al., 1997).

Differences in restriction fragment size distribution may result from various chromosomal rearrangements and/or mutations at the restriction sites, which suggests that the methodology based on restriction fingerprints should be supplemented with other approaches to determine the degree of homology of the fragments. Large genomic rearrangements have been observed in *P. aeruginosa* strains isolated from clinical samples or from the environment (Schmidt et al., 1996c). The former strains came mostly from cystic fibrosis cases. A 95-kb plasmid was detected in environmental strains and it integrated into the chromosome in cystic fibrosis strains. Exchange of DNA blocks and large DNA inversions led to divergence of clones in this species. The presence of inversions only in cystic fibrosis strains suggests that this niche causes or selects for substantial changes in the genome (Römling et al., 1997).

A study of pathovars of fluorescent plant pathogenic pseudomonads (Grothues and Rudolph, 1991) indicated that two strains belonging to the same pathovar, but of different origins, can give almost identical restriction fingerprints. However, this is not always the case. Pathogens of wide host range give more diverse restriction patterns than those of restricted host range. The methodology was also used for the examination of representative strains of *P. stutzeri*, a species notorious for its heterogeneity (Stanier et al., 1966; Palleroni et al., 1970). The results show a marked correlation of genome structure with fatty acid composition, and with data of nucleic acid hybridization experiments (Rainey et al., 1994b). A high degree of heterogeneity in macrorestriction patterns that did not correlate with the subdivision of the species in genomovars was also observed by another group of workers, and the marked heterogeneity of *P. stutzeri* was attributed, at least in part, to large chromosomal rearrangements (Ginard et al., 1997). The results obtained by these two groups of workers also indicate that the genome size of *P. stutzeri* ranges from 3.4–4.64 Mb for the strains subjected to their studies.

Rotating field electrophoresis was used to separate fragments of DNA of common and of genetically modified strains of *Pseu-domonas* species after digestion with rare-cutting restriction endonucleases. The technique was adapted to the identification of *Pseudomonas* strains deliberately released in the environment. The fingerprints were different for strains of the same species, although they were identical for related strains, and the differences were not affected by the presence of natural or genetically modified plasmids (Claus et al., 1992).

Sequence determination of 16S/23S spacer regions A study of the genomic organization of the ribosomal RNA genes in the *P. aeruginosa* has been the basis for the hypothesis that this species carries at least four sets of genes, each containing the genes for 16S, 23S, and 5S (Hartmann et al., 1986). Much information is now available for the 16S rRNA genes, in contrast with the other components of the ribosomal RNA complex, and it has been suggested that the spacer regions between the respective genes may be interesting subjects for sequence studies applicable to identification and typing procedures. Working with *Pseudomonas* strains isolated from different types of soils, Gill et al. (1994) amplified the spacer regions by PCR and subjected them to sequencing studies. The results suggest a limited degree of variability among strains identified as *P. putida* and *P. fluorescens*. However, the data may be useful for the recognition of particular *Pseudomonas* environmental strains. A review with information on studies of the spacer regions in a number of bacterial species, including some aerobic pseudomonads, is now available (Gürtler and Stanisich, 1996).

Distribution of repetitive sequences in the genome These sequences have been taken as the basis for genomic studies based on amplification of the fragments separating the repeated sequences. Work on pseudomonads (Louws et al., 1994) is an extension of studies performed mostly on rhizobia and *Xanthomonas* (Judd et al., 1993), following an original report suggesting the use of the distribution of the repetitive sequences as a tool for the characterization of the genome (Versalovic et al., 1991).

Identification of special genetic markers Saint-Onge et al. (1992) succeeded in cloning and sequencing the lipoprotein I gene (*oprI*) of *P. aeruginosa* PAO. The sequence showed similarity to published sequences from other sources, and oligonucleotide primers were designed to amplify the gene by PCR. The amplified gene or the oligonucleotide primers are convenient identification tools in Southern blot analyses. DNA obtained from small amounts of material taken from colonies could be used for amplification, and when bands of the expected mobility were obtained, they could be further identified by the use of a probe prepared with the amplified gene from *P. aeruginosa*. The gene was identified in species of *Pseudomonas* (rRNA similarity group I) and in strains of the genera *Chryseomonas* and *Flavimonas*, whose genomic DNAs reacted with the probe in Southern blots. Saint-Onge and collaborators concluded that the *oprI* gene was found primarily in strains of group I, but they wisely added that it would be of interest to search for other indications leading to a reclassification of the organisms assigned to these newly created genera. Evidence supporting this prediction was eventually found, and, as mentioned above in the section on DNA–DNA hybridization, supported the conclusion that the genera *Chryseomonas* and *Flavimonas* should be considered as synonyms of *Pseudomonas* (Anzai et al., 1997). The main conclusions of the work of Anzai and collaborators are summarized graphically in Fig. BXII.γ.125. For the time being, the two genera *Chryseomonas* and *Flavimonas* should be considered as junior subjective synonyms of *Pseudomonas*.

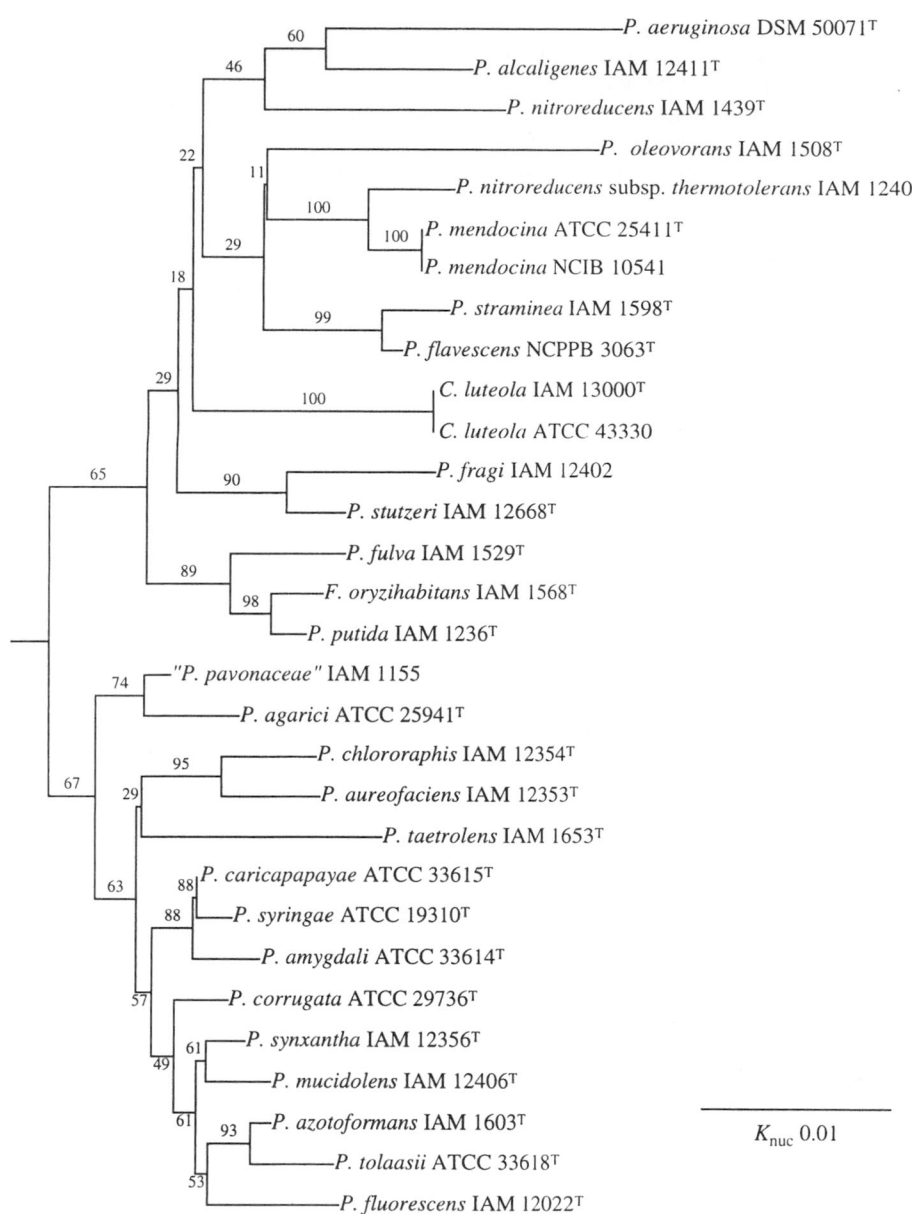

FIGURE BXII.γ.125. Phylogenetic tree for different species of the genus *Pseudomonas* derived from 16S rDNA sequence similarities to show the relative position of *Chryseomonas* and *Flavimonas*. Bootstrap values are indicated at the branch points. The confidence limits for the position of the branches has been estimated using the bootstrap analysis. *Escherichia coli* was chosen as root organism. (Reproduced with permission from Y. Anzai et al., International Journal of Systematic Bacteriology *47*: 249–251, 1997, ©International Union of Microbiological Societies.)

Additional confirmation of the relationship among species of *Pseudomonas* came from work demonstrating that four alginate production genes were conserved within the group (Fialho et al., 1990). Conservation also was observed in the *oprP* gene in various species of similarity group I (Sichnel et al., 1990).

rRNA–DNA hybridization This technique has had fundamental importance in *Pseudomonas* taxonomic and phylogenetic studies because its application resulted in the demonstration of the complexity of the aerobic pseudomonads as a bacterial group (Palleroni et al., 1973). Because of these studies, *Pseudomonas* is now restricted to one of five rRNA similarity groups. In these early studies, the competition technique was used (see Kilpper-Bälz, 1991, for a review of the methodology). Later, modifications

of the hybridization technique were used by other workers, confirming and extending the original results (De Vos and De Ley, 1983; De Vos et al., 1985, 1989). Since the development of practical methods for sequencing the components of the rRNA operon, rRNA–DNA hybridization methods are seldom being used for taxonomic purposes.

rRNA gene sequencing Following the initial rRNA–DNA hybridization studies on the aerobic pseudomonads, oligonucleotide catalogs (Uchida et al., 1974) were used for phylogenetic characterization of bacterial species. Comparisons among catalogs were expressed numerically by means of association coefficients (S_{AB}), and the values could be represented graphically in dendrograms (Fox et al., 1977). Cataloging eventually gave way to a more comprehensive approach directed to the determination of the total sequence of ribosomal RNA components. Initially this was achieved by the isolation of rRNA and DNA synthesis by transcriptase reaction, but the advent of the PCR technique has greatly simplified the experimental methodologies by avoiding RNA isolation, and it is widely used today. An article by Moore et al. (1996) specifically refers to determination and comparison of the 16S rRNA sequences of *Pseudomonas* species. A European compilation of pseudomonad sequences is available (van de Peer et al., 1996).

The results of the contribution by Moore et al. (1996) have been summarized in the creation of "lineages", which are grouped into two "clusters", the *P. aeruginosa* intrageneric cluster and the *P. fluorescens* intrageneric cluster. Within the first cluster there are four lineages: the *P. aeruginosa* lineage (*P. aeruginosa*, *P. citronellolis*, *P. alcaligenes*, *P. stutzeri*), the *P. resinovorans* lineage (*P. resinovorans*, *P. balearica*), the *P. mendocina* lineage (*P. mendocina*, *P. pseudoalcaligenes*, *P. oleovorans*), and the *P. flavescens* lineage (only *P. flavescens*). Within the *P. fluorescens* cluster there are located five lineages: the *P. fluorescens* lineage (*P. fluorescens*, *P. marginalis*, *P. tolaasii*, *P. chlororaphis*, *P. aureofaciens*, *P. viridiflava*), the *P. syringae* lineage (*P. syringae*, *P. amygdali*, "*P. coronafaciens*", *P. ficuserectae*), the *P. cichorii* lineage (*P. cichorii*), the *P. putida* lineage (*P. putida*, *P. asplenii*), and the *P. agarici* lineage (*P. agarici*) (Moore et al., 1996). It is, however, hard to draw any solid conclusions on the proposed classification. In the first place, since only type strains have been included in the published sequences, it is impossible to know the intraspecific variations, and although several strains of some of the species have been included, the selection has not taken into account the degree of heterogeneity in both DNA similarities and in phenotypic features characteristic of some of these species. Thus, while *P. aeruginosa* and *P. mendocina* are notoriously homogeneous in their phenotypic and genotypic properties, they are represented by a total of 12 strains. On the other hand, a very heterogeneous species in both phenotypic features and in DNA similarity such as *P. stutzeri* is represented by only one strain. The authors did not find obvious correlations of their molecular data with the results of standard phenotypic criteria. This is actually information from papers published by other groups; data corresponding to the strains subjected to the 16S rRNA sequence analysis are not reported.

The above-mentioned study included a comparison of sequences of the 16S rRNA variable region (*E. coli* positions 453–479) for the various lineages and for the species within each lineage. With respect to this approach, it may be interesting to mention here the findings on the *P. putida* biovars reported by Yamamoto and Harayama (1998). In a study of fluorescent pseudomonads done on the basis of the sequences of 16S rRNA, gyrase B subunit (*gyrB*) and RNA polymerase σ70 factor (*rpoD*), they found that in *P. putida* B the pairwise distances estimated from the variable regions of the 16S rRNA gene correlated rather poorly with the synonymous distances estimated from the *gyrB* and *rpoD* genes. On the other hand, the correlation of the nonvariable regions of these genes was highly significant, and a recommendation is formulated that only the nonvariable regions should be used for phylogenetic analysis.

Ribotyping The methodology allows comparisons of restriction patterns of ribosomal RNA genes (Stull et al., 1988), and it became widely used for epidemiological purposes because of its capacity for detection of prokaryotic diversity at the infraspecific level. When applied to the aerobic pseudomonads (*Pseudomonas*

sensu lato), using the restriction enzymes *Sma*I and *Hind*II, followed by hybridization with 16 and 23S rRNA, the method was capable of resolving 169 and 159 unique ribotyping patterns, respectively (Brosch et al., 1996). Of the two enzymes, *Sma*I was the most discriminative, except for some of the species. An analysis of *P. putida* biovar A strains in combination with a phenotypic analysis supported earlier observations, and suggested that a revision of the taxonomic status and internal subdivision of this species might be in order (Elomari et al., 1994). These considerations are mentioned in some detail in the section on taxonomic comments for this species. The methodology has limitations from the phylogenetic point of view, particularly because of significant differences that may be created by single mutations. However, aside from being a useful typing tool (Grimont and Grimont, 1986), it also appears to have taxonomic implications, and some of the observations by Brosch and collaborators confirm the marked heterogeneity of some of the species, particularly *P. fluorescens*.

Nucleic acid probes Based on the capacity of nucleic acid hybridization for the elucidation of genomic relatedness among bacteria, the development of nucleic acid probes allows for high-resolution, rapid, automated identification of specific microorganisms when they are mixed with other members of natural communities. The preparation and use of probes have been described in detail (Stahl and Amann, 1991; Schleifer et al., 1993).

Schleifer et al. (1992) have described a very effective use of the nucleic acid probes for the specific identification of pseudomonads in the environment. More recently, Amann et al. (1996a) have described the application of such probes to the detection of certain *Pseudomonas* species and pseudomonad species now assigned to other genera. In this report, the main goal was to identify the target organisms as either "former" or "genuine" pseudomonads. Unfortunately, the title of the paper carries this distinction, which is most confusing, since all pseudomonads, no matter what genera they have been assigned to, are equally "genuine".

A simplified cell blot technique combined with the use of 16S rRNA-directed probes was used for the identification of environmental isolates. The sequence of a group probe identifying *P. aeruginosa*, *P. mendocina*, *P. fluorescens*, *Comamonas acidovorans*, and "*Flavobacterium lutescens*" is described (Braun-Howland et al., 1993).

Low molecular weight (LMW) RNA profiles As an important addition to phylogenetic methodologies, Höfle (1991) has proposed a "rapid genotypic approach that bridges the gap between the solid systematics of type strains and the everyday need for relating new isolates and old reference strains". His technique is based on the examination of profiles of low-molecular-weight (LMW) RNA components in *Pseudomonas* species using high-resolution gel electrophoresis (Höfle, 1988). The reference strains of this genus, according to the author, showed very specific band patterns with discriminative power at the genus and species level (Höfle, 1990).

Whole-cell protein fingerprinting Protein banding patterns obtained by polyacrylamide gel electrophoresis, particularly when performed under denaturing conditions, are highly characteristic at the strain level, and have been used for many years for the purpose of typing and classification. An interesting application is the verification of strain authenticity, a sensitive and rapid control method that bypasses lengthy and tedious procedures for rechecking exchanged cultures (Kersters et al., 1994). At present, objective quantification of the band patterns, particularly when a substantial number of strains are analyzed, is being done by computer-assisted methods. Among the variants on the basic theme of protein separation in an electric field, electrophoresis in the presence of a denaturing agent such as sodium dodecyl ("lauryl") sulfate (SDS) (Laemmli, 1970) is the technique of choice. SDS polyacrylamide gel electrophoresis (SDS-PAGE) combines high resolution and good reproducibility, which from a practical point of view compensate for the fact that no specific identification of the bands is usually performed. A description of various methodologies and their evaluation has been conveniently summarized by Jackman (1985, 1987).

Of the various references available for descriptions of whole-cell protein fingerprinting methods, a recent publication of a one-dimensional electrophoretic analysis applied to *Pseudomonas* species is a convenient source of technical and bibliographic information (Vancanneyt et al., 1996a). Over 200 reference strains of various species were included. Aside from the few species for which only the type strain has been analyzed, uniform and distinct patterns were observed for many species of fluorescent and nonfluorescent organisms (*P. aeruginosa, P. agarici, P. alcaligenes, P. amygdali, P. caricapapayae, P. chlororaphis, P. chichorii, "P. coronafaciens", P. corrugata, P. ficuserectae, P. fragi, P. mendocina, P. pertucinogena, P. tolaasii,* and *P. viridiflava*). In contrast, *P. fluorescens, P. marginalis, P. pseudoalcaligenes, P. putida, P. stanieri,* and *P. stutzeri* were found to be particularly heterogeneous, in confirmation with earlier data on their phenotypic characteristics.

Ribosomal proteins These proteins, similarly to the ribonucleic acids with which they are associated in the ribosome, are of a highly conservative nature. Obviously, this should be reflected in a comparison of their sequences. In fact, partial amino acid sequences of preparations of protein L30 from species of various ribosomal RNA similarity groups give clusters similar to those that had originally been defined in rRNA–DNA hybridization studies (Ochi, 1995). Despite different levels of conservatism, other ribosomal proteins (S20, S21, L27, L20, L31, L32, and L33 protein families) give comparable results.

Other membrane proteins The discussion on this point will be limited to the identification of lipoprotein I in species of *Pseudomonas*. Work performed on the specificity of the *P. aeruginosa* lipoprotein I gene used as a probe for rapid identification purposes was an extension of the experiments of Saint-Onge et al. (1992). The work included more strains of species of RNA group I (*Pseudomonas*), as well as the related "*Azotomonas*" and *Azomonas* species and other species of Gram-negative organisms. All of the species of *Pseudomonas* that were tested gave positive reactions of variable intensities in PCR and dot blot tests using monoclonal antibodies for the lipoprotein I, whereas species of other similarity groups and of other Gram-negative genera gave negative reactions. These data, still considered to be of a preliminary nature, suggest that *P. aeruginosa* can be differentiated from the other species of the same genus that have been tested, with the exception of *P. mendocina, P. pseudoalcaligenes,* and *P. oleovorans,* which have similar restriction patterns in their lipoprotein I gene (De Vos et al., 1993).

Fatty acid analysis Fatty acid composition and the quinone system have been used for groupings of *Pseudomonas* species (Ikemoto et al., 1978; Yamada et al., 1982; Oyaizu and Komagata,

1983). A study of the fatty acid composition of 50 strains of various *Pseudomonas* species revealed the presence of the straight-chain saturated acid of $C_{16:0}$ and straight-chain unsaturated acids $C_{16:1}$ and $C_{18:1}$ in all strains (Ikemoto et al., 1978). Distribution of hydroxy acids, cyclopropane acids, and branched-chain acids follows quite closely the accepted subdivision of the aerobic pseudomonads into RNA similarity groups. These cellular components are very useful for identification purposes.

The results of fatty acid and quinones analyses, performed on a collection of 75 strains that included phytopathogenic pseudomonads, have been reported (Oyaizu and Komagata, 1983). The survey gave a basis for the classification of the collection into groups with particular reference to the presence of 3-OH fatty acids, and some of the groups coincided with the RNA similarity groups already defined (Palleroni et al., 1973).

The analysis of a large number of fatty acid profiles obtained from a collection of 340 strains of pseudomonads that includes plant pathogenic species has been published (Stead, 1992). The results on the distribution of 2-OH and 3-OH fatty acids could serve as one of the bases for a subdivision of the collection into six groups warranting independent generic designations.

The above comments on fatty acid composition studies refer to *Pseudomonas sensu lato* and coincide in supporting the division of these organisms into the various genera that later were to be proposed in various communications from different laboratories. A summary of the results of the above sources is presented in Table BXII.γ.110.

In a more recent communication (Vancanneyt et al., 1996b) the fatty acid composition of whole-cell hydrolysates and phospholipid fractions are reported. The results confirm some of the conclusions of the above-cited references. In general, *Pseudomonas* species are characterized by the presence of $C_{10:0\ 3OH}$ and $C_{12:0\ 3OH}$ acids. When the fatty acid contents of phospholipids were taken into account, no clear separation could be found between *Pseudomonas* RNA group I organisms and RNA similarity groups II and III.

Polyamine composition Polyamine patterns appear to be good chemotaxonomic markers for the Gram-negative bacteria in the phylum *Proteobacteria*, allowing ready differentiation of some of the pseudomonad groups (Busse and Auling, 1988). In addition to quinone analysis, polyamines are useful for rapid

TABLE BXII.γ.110. Fatty acid and ubiquinone composition of *Pseudomonas* and other rRNA similarity groups[a]

| Compound | Ribosomal RNA similarity groups | | | | |
	I	II	III	IV	V
Fatty acids:					
$C_{10:0\ 3OH}$	+		+		+
$C_{11:0\ 3OH}$					+
$C_{11:0\ iso\ 3OH}$					+
$C_{12:0\ 3OH}$	+			+	+
$C_{12:0\ iso\ 3OH}$					+
$C_{13:0\ iso\ 3OH}$					+
$C_{14:0\ 3OH}$		+		+	
$C_{16:1\ 3OH}$		+			
$C_{12:0\ 2OH}$	(+)				
$C_{16:0\ 2OH}$		(+)			
$C_{16:1\ 2OH}$		(+)			
$C_{18:1\ 2OH}$		+			
Ubiquinones	Q-9	Q-8	Q-8	Q-10	Q-8

[a]Data taken from Oyaizu and Komagata (1983) and Stead (1992). Data in parentheses: not all strains of the group are positive.

identification at the genus level (Busse et al., 1989). This last contribution includes a discussion of various other approaches applicable to species and strain identification.

In an attempt to relieve the dearth of simple and reliable tests for the classification of newly isolated strains of the genus *Xanthomonas*, a determination of polyamine composition of strains of this genus and of phytopathogenic *Pseudomonas* species was introduced as a rapid chemotaxonomic identification tool (Auling et al., 1991). Spermidine was found to be the main polyamine of *Xanthomonas*, whereas strains of *Pseudomonas* RNA group I were characterized by the presence of putrescine. The polyamine pattern of *Azotobacter* and *Azomonas* was the same as that of the fluorescent pseudomonads, which are placed in the same branch of *Proteobacteria*. The results obtained by Auling et al. (1991) were confirmed in a more recent study by Goris et al. (1998). Aside from the presence of putrescine and spermidine mentioned above, the species of *Pseudomonas* could be divided into two sublineages, of which only one contained cadaverine. *P. aeruginosa* (as well as *Azotobacter vinelandii*) is in the sublineage that contains this polyamine.

FURTHER READING

Achouak, W., L. Sutra, T. Heulin, J.-M. Meyer, N. Fromin, S. Degraeve, R. Christen and L. Gardan. 2000. *Pseudomonas brassicacearum* sp. nov. and *Pseudomonas thivervalensis* sp. nov., two root-associated bacteria isolated from *Brassica napus* and *Arabidopsis thaliana*. Int. J. Syst. Evol. Microbiol. *50*: 9–18.

Andersen, S.M., K. Johnsen, J. Sorensen, P. Nielsen and C.S. Jacobsen. 2000. *Pseudomonas frederiksbergensis* sp. nov., isolated from soil at a coal gasification site. Int. J. Syst. Evol. Microbiol. *50*: 1957–1964.

Anzai, Y. , H. Kim, J.Y. Park, H. Wakabayashi and H. Oyaizu. 2000. Phylogenetic affiliation of the pseudomonads based on 16S rRNA sequence. Int. J. Syst. Evol. Microbiol. *50*: 1563–1589.

Brown, G.R., I.C. Sutcliffe and S.P. Cummings. 2001. Reclassification of [*Pseudomonas*] *doudoroffii* (Baumann et al. 1983) in the genus *Oceanomonas* gen. nov. as *Oceanomonas doudoroffii* comb. nov., and description of a phenol-degrading bacterium from estuarine water as *Oceanomonas baumannii* sp. nov. Int. J. Syst. Evol. Microbiol. *51*: 67–72.

Guasp, C., E.R.B. Moore, J. Lalucat and A. Bennasar. 2000. Utility of internally transcribed 16S–23S rDNA spacer regions for the definition of *Pseudomonas stutzeri* genomovars and other *Pseudomonas* species. Int. J. Syst. Evol. Microbiol. *50*: 1629–1639.

Nishimori, E., K. Kita-Tsukamoto and H. Wakabayashi. 2000. *Pseudomonas plecoglossicida* sp. nov., the causative agent of bacterial haemorrhagic ascites of ayu, *Plecoglossus altivelis*. Int. J. Syst. Evol. Microbiol. *50*: 83–89.

Rowe, N.J., J. Tunstall, L. Galbraith and S.G. Wilkinson. 2000. Lipid composition and taxonomy of (*Pseudomonas*) *echinoides*: Transfer to the genus *Sphingomonas*. Microbiology (Read.) *146*: 3007–3012.

Uchino, M., Y. Kosako, T. Uchimura and K. Komagata. 2000. Emendation of *Pseudomonas straminea* Iizuka and Komagata 1963. Int. J. Syst. Evol. Microbiol. *50*: 1513–1519.

Yumoto, I., K. Yamazaki, M. Hishinuma, Y. Nodasaka, A. Suemori, K. Nakajima, N. Inoue and K. Kawasaki. 2001. *Pseudomonas alcaliphila* sp. nov., a novel faculatatively psychrophilic alkaliphile isolated from seawater. Int. J. Syst. Evol. Microbiol. *51*: 349–355.

List of species of the genus Pseudomonas

The following list includes those species whose assignment to the genus *Pseudomonas* has been definitely established. Part of the information for some of the species has been condensed in the form of tables representing updated versions of those of the first edition of this *Manual*.

The order is alphabetical, thus avoiding the problem of ordering the species on the basis of properties such as pathogenicity, pigmentation, or certain physiological peculiarities, which only seldom correlate with the order of groups based on 16S rRNA sequence similarities (Figs. BXII.γ.124 and BXII.γ.125). *Pseudomonas* can be subdivided into two big groups, the fluorescent and the nonfluorescent species and, with respect to plant pathogenicity, a subdivision of the genus into saprophytes and plant pathogenic species is possible. These subdivisions do not correlate with each other. Some correlations, however, are apparent. In Fig. BXII.γ.124 the plant pathogens congregate mostly in the so-called *P. syringae* group.

The members of some pairs of species (*P. migulae–P. mandelii*; *P. veronii–P. gessardii*; *P. cedrella–P. orientalis*; *P. marginalis–P. libanensis*; *P. fuscovaginae–P. asplenii*, and *P. monteilii–"P. ayucida"*) were found to have identical 16S rRNA sequences (see Fig. BXII.γ.124) (Anzai et al., 1997). It is to be hoped that the precise taxonomic allocation of these taxa may be further investigated. Some comments are included in the corresponding descriptions below.

A number of species assigned to *Pseudomonas* will not be treated here. As mentioned by Kersters et al. (1996), the phylogenetic position of the following species has not yet been determined. In the list that follows, references to the original descriptions are given in parentheses, and the number in italics refers to the page where the species is described in the first edition of this *Manual* (Palleroni, 1984): *P. antimicrobica* (Attafuah and Bradbury, 1989), *P. cissicola* (Goto and Makino, 1977), *P. flectens* (Johnson, 1956), *P. gelidicola* (Kadota, 1951), *P. halophila* (Fendrich, 1988), *P. indigofera* (Elazari-Volcani, 1939), *P. iners* (Iizuka and Komagata, 1964b), *P. lanceolata* (Leifson, 1962a), *P. mephitica* (Haynes and Burkholder, 1957), and *P. spinosa* (Leifson, 1962b). Since the descriptions of *P. antimicrobica*, *P. flectens*, and *P. halophila* are not given in the first edition of the *Manual*, they will be added at the end of the list of species, under the subtitle Other Species.

Table BXII.γ.111 lists a number of species previously assigned to *Pseudomonas sensu lato*, which have been transferred to various other genera or groups. The information has been taken mainly from Kersters et al. (1996), and also Behrendt et al. (1999) and Anzai et al. (1997). The paper by Kersters et al. (1996) includes a handy table with the current classification of the species that have been assigned to *Pseudomonas* prior to the subdivision of the genus into rRNA similarity groups, their present assignments, and references to the studies deciding phylogenetic allocations and/or revised taxonomic status for each species.

Recently, three new species have been added to the list of those requiring further confirmation—*Pseudomonas abietaniphila*, *P. multiresinivorans*, and *P. vancouverensis*, which are isolated from resin acids (Mohn et al., 1999a, b). Even though the reported catabolic activities of these organisms are unique, DNA–DNA hybridization data to confirm their genomic relationships vs. related organisms have not been reported. For this reason, the description of these species will be found at the end, in the section Other Species.

Finally, many new species of *Pseudomonas* have been described in the *International Journal of Systematic and Evolutionary Microbiology* since mid-2001, the cut-off point for inclusion of new genera and species in this edition of the *Manual*.

1. **Pseudomonas aeruginosa** (Schroeter 1872) Migula 1900, 884[AL] (*Bacterium aeruginosum* Schroeter 1872, 126.) *ae.ru.gi.no'sa.* L. fem. adj. *aeruginosa* full of copper rust or verdigris, hence green.

Characteristics of the species are summarized in Tables BXII.γ.108, BXII.γ.112, BXII.γ.113, and BXII.γ.114. Rods, motile by means of a single polar flagellum. Two main colony types can be observed on common solid media. One is large, smooth, with flat edges and elevated center ("fried egg" appearance), and the other is small, rough, convex. On the large colonies, silver-gray metallic shining patches may be observed, and pitting is fairly common. Clinical materials are, in general, good sources of the large colony type, whereas the small type is commonly obtained from

TABLE BXII.γ.111. *Pseudomonas sensu lato* species transferred to other taxa

Pseudomonas species	Previous description in *Bergey's Manuals*	New allocation	References
P. aminovorans	Palleroni (1984)	*Aminobacter*	Urakami et al. (1992)
P. beijerinckii	Palleroni (1984)	*Halomonas* rRNA lineage	De Vos et al. (1989)
P. beteli	Palleroni (1984)	*Xanthomonas* rRNA lineage	De Vos et al. (1985)
P. boreopolis	Palleroni (1984)	*Xanthomonas* rRNA lineage	De Vos et al. (1989)
P. carboxydohydrogena	Palleroni (1984)	*Bradyrhizobium-Rhodopseudomonas* rRNA lineage	Auling et al. (1988)
"*P. carboxydovorans*"	Palleroni (1984)	*Oligotropha*	Meyer et al. (1993)
"*P. compransoris*"	Palleroni (1984)	*Zavarzinia*	Meyer et al. (1993)
P. doudoroffii	Palleroni (1984)	*Aeromonadaceae*	De Vos et al. (1989)
P. echinoides	Palleroni (1984)	*Sphingomonas* rRNA lineage	De Vos et al. (1989)
P. elongata	Palleroni (1984)	*Oceanospirillum*	De Vos et al. (1989)
"*P. extorquens*"	Haynes and Burkholder (1957)	*Methylobacterium*	Bousfield and Green (1985)
P. geniculata	Haynes and Burkholder (1957)	*Xanthomonas* rRNA lineage	Byng et al. (1983)
P. hibiscicola	Palleroni (1984)	*Xanthomonas* rRNA lineage	De Vos et al. 1985)
P. huttiensis	Palleroni (1984)	*Herbaspirillum*	Kersters et al. (1996)
P. lemoignei	Palleroni (1984)	rRNA group II	De Vos and De Ley (1983), Mergaert et al. (1996)
P. marina	Palleroni (1984)	*Halomonas* rRNA lineage	De Vos et al. (1989)
P. mesophilica	Palleroni (1984)	*Methylobacterium*	Green and Bousfield (1983)
P. mixta		*Telluria*	Bowman et al. (1988, 1993a)
P. nautica	Palleroni (1984)	*Oceanospirillum*	De Vos et al. (1989)
P. paucimobilis	Palleroni (1984)	*Sphingomonas*	De Vos et al. (1989)
P. pictorum	Palleroni (1984)	*Xanthomonas* rRNA lineage	De Vos et al. (1989)
P. radiora	Palleroni (1984)	*Methylobacterium*	Green and Bousfield (1983)
P. rhodos	Palleroni (1984)	*Methylobacterium*	Green and Bousfield (1983)
"*P. riboflavina*"	Haynes and Burkholder (1957)	*Devosia*	Nakagawa et al. (1996)
"*P. rosea*"		*Methylobacterium*	Bousfield and Green (1985)
P. rubrisubalbicans	Palleroni (1984)	*Herbaspirillum*	Baldani et al. (1996)
P. saccharophila	Palleroni (1984)	*Comamonadaceae*	Palleroni et al. (1973), Willems et al. (1992)
P. stanieri		*Marinomonas*	Y. Ansai, personal communication

TABLE BXII.γ.112. General phenotypic characteristics of some fluorescent *Pseudomonas* species[a]

Characteristics	*P. aeruginosa*	*P. chlororaphis* subsp. *chlororaphis*	*P. chlororaphis* subsp. *aureofaciens*	*P. flavescens*	*P. fluorescens* bv. I	*P. fluorescens* bv. II	*P. fluorescens* bv. III	*P. fluorescens* bv. IV	*P. fluorescens* bv. V	*P. lundensis*	*P. monteilii*	*P. putida* biovar A	*P. putida* biovar B	*P. rhodesiae*	*P. veronii*
Number of flagella	1	>1	>1	1	>1	>1	>1	>1	>1	1		>1	>1	1	1
Nonfluorescent pigment, color:															
Green	−	+	−	−	−	−	−	−	−	−	−	−	−	−	−
Orange	−	−	+	−	−	−	−	−	−	−	−	−	−	−	−
Yellow	−	−	−	+	−	−	−	−	−	−	−	−	−	−	−
Blue	−	−	−	−	−	−	−	+	−	−	−	−	−	−	−
Gelatin liquefaction	+	+	+	−	+	+	+	+	+	+	−	−	−	−	+
Denitrification	+	+	−	−	−	+	+	+	−	−	−	−	−	−	+
Levan formation	−	+	+	−	+	+	−	+	−	−	−	−	−	+	−
Lecithinase	−	+	+	−	+	d	+	+	d	−	−	−	−	+	−
Lipase	+	+	+	−	+	−	d	d	d	−	−	−	−	−	−
Growth at 4°C	−	+	+	−	+	+	+	+	+	+	−	d	+	+	+
Growth at 41°C	+	−	−	−	−	−	−	−	−	−	+	−	−	−	−
Mol% G + C of the DNA	67	63	64	63	60	60	60	60	60	60	60	62	61	59	61

[a]For symbols see standard definitions. Data from Palleroni (1984), Molin et al. (1986), Hildebrand et al. (1994), Elomari et al. (1996, 1997), and Coroler et al. (1996).

TABLE BXII.γ.113. Nutritional characteristics of some fluorescent *Pseudomonas* species[a]

Substrates	P. aeruginosa	P. chlororaphis subsp. chlororaphis	P. chlororaphis subsp. aureofaciens	P. flavescens	P. fluorescens bv. I	P. fluorescens bv. II	P. fluorescens bv. III	P. fluorescens bv. IV	P. fluorescens bv. V	P. lundensis	P. monteilii[b]	P. putida biovar A	P. putida biovar B	P. rhodesiae[c]	P. veronii
Acetamide	+	–	–	–	–	–	–	–	–	–	–	d	d	–	–
Acetate, L-alanine[d], γ-aminobutyrate, L-asparagine[e], L-aspartate, betaine, caprate, caprylate, citrate, fumarate, L-glutamate[d], glutarate, glycerol, heptanoate, β-hydroxybutyrate, DL-lactate, L-malate[d], pelargonate[e], L-proline, putrescine, pyruvate[d], succinate[d]	+	+	+	+	+	+	+	+	+	+	+	+	+	+	
cis-Aconitate	+	d	+	+	+	+	d	+	d	–	+	+	+	+	+
Adipate	+	–	–	–	–	–	d	–	–	–	–	–	d	–	d
Adonitol	–	–	–	–	+	+	d	–	d	–	–	–	–	–	–
β-Alanine, L-arginine, spermine, L-tyrosine	+	+	+		+	+	+	+	+	+	+	+	+	+	+
D-Alanine	+	+	–	+	+	+	+	+	–	–	+	+	+	+	+
α-Aminobutyrate	–	–	–	–	–	–	–	–	–	–	–	d	–	+	+
α-Aminovalerate	–	–	–	–	–	–	d	–	–	–	–	d	–		
δ-Aminovalerate	+	d	d	–	d	+	+	+	+	–	+	+	+	+	
α-Amylamine	–	–	+	–	–	d	d	–	d	–	+	d	+	–	–
Anthranilate	+	+	+	–	d	d	d	–	d	–	–	–	+	–	d
D-Arabinose	–	–	–	–	–	–	–	–	–	–	+	–	–	–	–
L-Arabinose	–	–	+	+	+	+	d	+	d	+	+	d	+	+	+
Azelate	+	–	–	–	–	d	–	–	–	–	–	–	–	–	–
Benzoate	+	+	d	–	d	d	d	+	d	+	+	d	+	–	–
Benzoylformate	+	+	+	–	–	–	–	–	–	–	–	d	d		
Benzylamine	–	–	–	–	–	d	–	d	–	–	+	d	+		
Butanol	+	d	–	–	d	d	d	d	+	d	+	+	+		
Butylamine	–	–	d	–	–	–	–	–	d	–	+	+	+		
2,3-Butylene glycol	+	d	d	+	d	+	d	+	d	–		d	d		
Butyrate	+	+	d	–	–	d	d	+	d	+	+	+	+	–	+
Caproate	+	d	+	+	+	d	+	+	+	+	+	+	+	+	+
Cellobiose, ethylene glycol[e], D-fucose, inulin[f], isopropanol[e,f], lactose, maleate, maltose, methanol[e,f], oxalate, phthalate, poly-β-hydroxybutyrate[e,f], salicin[f], starch, L-threonine[e,f]	–	–	–	–	–	–	–	–	–	–	–	–	–	–	–
Citraconate	–	–	–	–	d	d	d	+	d	–	–	–	d	–	
L-Citrulline	d	d	d	+	d	d	d	–	d	–	–	d	d	–	d
Creatine	–	–	–	+	–	–	d	–	–	+	+	d	d	–	–
Dodecane, hexadecane	d	–	–	–	–	–	–	–	–	–	–	–			
Erythritol	–	–	–	–	d	d	+	–	d	–	–	–	–	–	+
Ethanol, propanol	+	d	–	–	–	+	d	–	d	–		d	d		
Ethanolamine	d	d	+	+	+	d	d	+	d	+	d	+	d	+	+
D-Fructose	+	d	d	–	+	d	+	+	+	+	+	+	+	+	+
D-Galactose	–	d	+	+	+	+	d	+	d	–	–	–	d	+	+
Geraniol	+	–	–	–	–	–	–	–	–	–	–	–			
D-Glucose	+	+	d	+	+	+	+	+	+	+	+	+	+	+	+
Gluconate	+	+	d	+	+	+	+	+	+	+	+	+	+	+	+
Glycerate	+	+	+	+	+	+	d	+	d	–	+	d	+	+	+
Glycine	d	–	–	–	–	–	d	–	d	–	–	d	d		
Glycolate	–	–	–	+	–	–	–	–	–	–	–	d	–		d
Hippurate	–	–	–	–	–	–	d	–	–	+	–	d	d		
Histamine	+	d	d	–	d	–	d	–	d	–	d	d	+	–	–
L-Histidine	+	+	+	+	+	d	+	+	d	+	+	+	+	+	d
o-Hydroxybenzoate	–	d	–	–	–	–	–	–	–	–	–	d	d	–	–
m-Hydroxybenzoate	–	d	d	–	–	–	–	–	–	–	–	d	d		
p-Hydroxybenzoate	+	+	+	+	+	+	d	+	d	–	+	+	+	+	+
Hydroxymethylglutarate	–	–	–	+	d	d	+	–	d	–	–	–			

[a]For symbols see standard definitions. Data from Palleroni (1984), Molin et al. (1986), Hildebrand et al. (1994), Elomari et al. (1996, 1997), and Coroler et al. (1996).

[b]*P. monteilii* was also reported to be unable to use the following carbon sources: N-acetylglucosamine, esculin, m-aminobenzoate, p-aminobenzoate, 3-aminobutyrate, amygdalin, D-arabitol, L-arabitol, arbutin, L-cysteine, dulcitol, ethylamine, L-fucose, β-gentiobiose, glucosamine, glycogen, isophthalate, 5-ketogluconate, D-lyxose, L-lyxose, melezitose, melibiose, L-methionine, α-methylglucoside, α-methyl-D-mannoside, α-methyl-xyloside, raffinose, L-sorbose, terephthalate, raffinose, tagatose, D-tryptophan, D-turanose, xylitol, and L-xylose.

[c]*P. rhodesiae* was also found to use acetylglucosamine and D-arabitol, and was unable to grow on esculin, 2-aminobenzoate, 3-aminobenzoate, 4-aminobenzoate, 3-aminobutyrate, amygdalin, L-arabitol, arbutin, L-cysteine, dulcitol, ethylamine, L-fucose, gentiobiose, glucosamine, glycogen, norvaline, raffinose, salicin, sorbose, D-tagatose, terephthalate, D-turanose, urea, xylitol, and L-xylose.

[d]Positive for *P. veronii*; results on other substrates not reported.

[e]*P. monteilii* not tested.

[f]No information on *P. veronii*.

(continued)

Substrates	*P. aeruginosa*	*P. chlororaphis* subsp. *chlororaphis*	*P. chlororaphis* subsp. *aureofaciens*	*P. flavescens*	*P. fluorescens* bv. I	*P. fluorescens* bv. II	*P. fluorescens* bv. III	*P. fluorescens* bv. IV	*P. fluorescens* bv. V	*P. lundensis*	*P. monteilii*[b]	*P. putida* biovar A	*P. putida* biovar B	*P. rhodesiae*[c]	*P. veronii*
m-Inositol	−	+	+	−	d	+	d	+	d	−	+	−	−	+	+
Isobutanol	+	−	−	−	d	d	−	d	−			d	d	d	
Isobutyrate	+	−	d	−	−	d	d	d	−	d		+	d	d	−	+
L-Isoleucine, L-valine	d	+	+	−	+	+	+	+	+	+	+	+	+	+	+
Isovalerate	+	+	+	−	d	d	d	−	d	−	+	+	+	d	−
Itaconate, mesaconate	+	+	+	+	+	d	d	−	d	−	−	d	−	−	
2-Ketogluconate	+	+	d	−	+	+	+	d	+		+	d	+	+	+
α-Ketoglutarate	+	+	+	+	+	+	+	+	+	+		+	+		
L-Kynurenine	+	+	+	−	d	d	d	d	−	d		−	−		
Kynurenate	d	d	d	−	d	−	d	−	−	−		−	d		
L-Leucine	+	+	+	+	+	d	+	+	+	+	+	+	+	+	d
Levulinate	+	d	d	−	d	−	d	−	d	−	−	d	d	−	d
L-Lysine	+	d	d	−	+	d	d	+	d	+	+	+	d	−	−
D-Malate	d	d	−	+	−	d	d	+	d			d	d	d	−	−
Malonate	+	+	+	+	+	+	d	+	d	−		d	d	+	+
D-Mandelate	−	−	−	−	−	−	−	−	−	−		−	d	d	−
L-Mandelate	+	−	−	−	−	−	−	−	d	−		−	−	d	−	−
Mannitol	+	+	+		+	+	d	+	d	−		−	d	d	+	+
D-Mannose	−	+	+	+	+	+	+	+	d	+		−	d	d	+	+
Mucate	−	+	+	+	+	+	d	+	+	−		−	d	+	+	d
Naphthalene	−	−	−	−	−	−	−	−	−	−		−	d		
Nicotinate	−	−	−	−	−	−	−	−	d	−		d	+	
DL-Norleucine	−	−	−	−	−	−	−	−	−		+	−	−		−
L-Ornithine	+	d	+	−	+	d	d	d	d	+	+	+	+		d
Pantothenate	−	−	−	−	−	−	d	−	−	−				
Phenol	−	−	−	−	−	d	−	−	−			d	d		
Phenylacetate	−	d	+	−	−	d	d	−	d	−		+	d	+	−	−
L-Phenylalanine	d	d	+	−	d	d	d	+	d	+		+	+	+	+
Phenylethanediol	−	−	−	−	−	−	−	−	−			d	−	
Pimelate, suberate	d	−	−	−	−	d	−	−				−	−	−	
Propionate	+	+	+	−	+	+	d	+	+	+	+	+	+	+	+
Propylene glycol	d	−	−	−	d	−	−					−	−	
Quinate	d	+	+	−	+	+	+	+	d	−		+	+		
L-Rhamnose	−	−	−	−	d	d	−	d	−			−	−	−	−
D-Ribose	+	+	+	+	+	+	d	+	d	+	+	d	d	+	+
Saccharate	−	+	+	+	+	+	+	+	−			−	−		
Sarcosine	d	d	−	+	d	d	d	+	+		+	+	+	+	+
Sebacate	+	−	−	−	−	d	−	−	−		−	−	d	−	
L-Serine	d	d	+	+	+	d	+	+	d		d	d	d	+	+
Sorbitol	−	−	−	−	+	+	d	+	d	−		−	−	d	+	+
Sucrose	−	+	d	+	+	+	−	+	d	−		−	−	d		+
D(−)-Tartrate	−	−	d	−	d	−	d		−			d	d	−	−
L(+)-Tartrate	−	d	−	−	−	+	−	d				d	d	−	−
m-Tartrate	−	−	+	−	d	−	d	−	d			−			
Testosterone	−	−	−	−	−	−	−	−	−			−	+		
Trehalose	−	+	d	+	+	+	d	+	d	−		−	−	+	d
Trigonelline	−	−	−	+	d	d	d	−	d	−		d	d	+	−	+
Tryptamine	−	−	−	−	d	d	−	−	−			d	+	−	+
D-Tryptophan	−	−	−	−	−	−	−	−	−			−	d	−	−
L-Tryptophan	d	+	+	−	+	d	d	−	d	−		−	+	−	+
Valerate	+	+	−		d	d	d	d	−	d		+	+	+
D-Xylose	−	−	−	+	+	d	d	d	d	−		−	d	d	+	+

[a] For symbols see standard definitions. Data from Palleroni (1984), Molin et al. (1986), Hildebrand et al. (1994), Elomari et al. (1996, 1997), and Coroler et al. (1996).

[b] *P. monteilii* was also reported to be unable to use the following carbon sources: N-acetylglucosamine, esculin, *m*-aminobenzoate, *p*-aminobenzoate, 3-aminobutyrate, amygdalin, D-arabitol, L-arabitol, arbutin, L-cysteine, dulcitol, ethylamine, L-fucose, β-gentiobiose, glucosamine, glycogen, isophthalate, 5-ketogluconate, D-lyxose, L-lyxose, melezitose, melibiose, L-methionine, α-methylglucoside, α-methyl-D-mannoside, α-methyl-xyloside, raffinose, L-sorbose, terephthalate, raffinose, tagatose, D-tryptophan, D-turanose, xylitol, and L-xylose.

[c] *P. rhodesiae* was also found to use acetylglucosamine and D-arabitol, and was unable to grow on esculin, 2-aminobenzoate, 3-aminobenzoate, 4-aminobenzoate, 3-aminobutyrate, amygdalin, L-arabitol, arbutin, L-cysteine, dulcitol, ethylamine, L-fucose, gentiobiose, glucosamine, glycogen, norvaline, raffinose, salicin, sorbose, D-tagatose, terephthalate, D-turanose, urea, xylitol, and L-xylose.

[d] Positive for *P. veronii*; results on other substrates not reported.

[e] *P. monteilii* not tested.

[f] No information on *P. veronii*.

TABLE BXII.γ.114. Characteristics differentiating *Pseudomonas aeruginosa, P. balearica, P. stutzeri,* and *P. putida*[a]

Characteristics	*P. aeruginosa*	*P. balearica*	*P. putida*	*P. stutzeri*
Type of colony:				
Smooth	+		+	
Wrinkled		+		+
Number of flagella	1	1	>1	1
Hydrolysis of:				
Gelatin	+	−	−	−
Starch	−	+	−	+
Utilization of:				
Maltose	−	+	d	+
Xylose	−	+	d	−
γ-Aminobutyrate	−	−	d	d
Malate	d	+	−	+
Suberate	d	−	−	d
Mannitol	+	−	−	d
Ethylene glycol	−	−	−	+
Denitrification	+	+	−	+
Growth at:				
42°C	+	+		d
46°C	−	+	−	d
Growth in media with 8.5% NaCl	−	+	−	−
Fatty acid content (%):				
$C_{17:0\ cyclo}$	0.8	4.71	>5	0.28–1.72
$C_{19:0\ cyclo}$	1.2	3.8	Traces	0.32-1.45
Mol% G + C of the DNA	67	64.1–64.4	60.7–62.5	60.9–64.9

[a]For symbols see standard definitions. Data from Bennasar et al. (1996) and Stanier et al. (1966).

natural sources (Véron and Berche, 1976). Variation of the large type to the small is rather common but the reverse variation is extremely rare. A third colony type (mucoid) often can be obtained from respiratory and urinary tract secretions and was first observed by Sonnenshein (1927). Mucoid mutants can be divided into two groups according to the capacity of forming mucus (alginate) in chemically defined media (Fyfe and Govan, 1980). Aside from pyoverdine and pyocyanin, other pigments may be produced by some strains, including a dark red pigment (see above section on pigments).

Denitrification and gelatin liquefaction is present in the great majority of strains. The hydrolysis of Tween 80 is weak and the egg-yolk reaction is negative. Optimum temperature, 37°C. *P. aeruginosa* is probably the most widespread of all bacterial species. It can be isolated from soil and water, particularly from enrichment cultures for denitrifying bacteria. Commonly isolated from clinical specimens (wound, burn, and urinary tract infections). Causative agent of "blue pus", which accounts for the origin of the synonym pyocyaneus. Occasionally pathogenic for plants. Strains isolated from leaf spot of tobacco, identical with or similar to *P. aeruginosa* have been named "*P. polycolor*" (Clara, 1930). The species can be internally divided into a number of subgroups (types) useful for epidemiological purposes.

The mol% G + C of the DNA is: 67.2 (Bd).

Type strain: ATCC 10145, DSM 50071, NCIB 8395, NCTC 10332.

GenBank accession number (16S rRNA): X06684, Z76651.

2. **Pseudomonas agarici** Young 1970, 985[AL]

a.ga*r*'i.ci. M.L. n. *Agaricus* a genus of fungi; M.L. gen. n. *agarici* of *Agaricus*.

The following description is a summary of the one presented by Young (1970).

Short rods (no dimensions given), motile by one or, rarely, two polar flagella. Green diffusible pigment with

weak fluorescence under ultraviolet light of unspecified wavelength (in our experience, the maximum intensity of fluorescence is observed at low wavelength UV, ~257 nm). Acid is produced from arabinose, glucose, and mannitol; little acid from fructose, galactose, and ribose, and no acid from rhamnose, xylose, mannose, lactose, sucrose, maltose, trehalose, melibiose, cellobiose, raffinose, starch, inulin, dextrin, glycogen, adonitol, sorbitol, inositol, and salicin.

Acetate, benzoate, citrate, formate, fumarate, gluconate, lactate, propionate, and succinate are utilized. Galacturonate, oxalate, and tartrate are not. Oxidase and catalase reactions positive; nitrate reduction, pectate liquefaction, starch hydrolysis, esculin hydrolysis, levan production, and growth factor requirements are all negative. Further extensive phenotypic studies on strains of this species in comparison with other fluorescent organisms have been published (Fahy, 1981). Some of the reported properties are summarized in Table BXII.γ.115. Details of its fatty acid composition are known (Stead, 1992).

The organism causes drippy gill of mushrooms, and one of the main differences with another mushroom pathogen, *P. tolaasii*, is in the utilization of benzoate. Catechol is oxidized to a black pigment diffusing into the medium, a character also present in *P. agarici*. The species was tentatively assigned to RNA group I by Byng et al. (1980), and this position was further confirmed by rRNA–DNA hybridization (De Vos et al., 1985) and by rDNA sequencing (Moore et al., 1996).

The mol% G + C of the DNA is: 58.8–61.1 (T_m).

Type strain: ATCC 25941, DSM 11810, LMG 2289.

GenBank accession number (16S rRNA): Z76652.

3. **Pseudomonas alcaligenes** Monias 1928, 332[AL]

al.ca.li'ge.nes. M.L. adj. *alcaligenes* alkali-producing.

Characteristics useful to differentiate the species from other *Pseudomonas* species are given in Tables BXII.γ.116 and BXII.γ.117. For further descriptive information see Ral-

TABLE BXII.γ.115. Characters distinguishing some fluorescent *Pseudomonas* species associated with mushroom culture[a]

Characteristics	P. agarici	P. cichorii	P. fluorescens biovar II	P. tolaasii
Levan formation from sucrose	−	−	+	−
Arginine dihydrolase	−	−	+	+
Denitrification	−	−	+	−
Gelatin hydrolysis	−	−	+	+
Egg yolk reaction	−	+	−	+
Growth at the expense of:				
Trehalose	−	−	+	d
2-Ketogluconate	−	−	+	d
meso-Inositol	−	+	+	+
L-Valine	d	−	+	+
β-Alanine	+	−	+	+
L-Arabinose	−	+	+	d
Sucrose	−	+	+	−
Sorbitol	−	−	+	+
Adonitol	−	−	d	d
Ethanol	−	−	+	d
meso-Tartrate	−	+	−	+
Nicotinate	−	−	−	+
Staining of mushroom caps	d			+
Pitting of mushroom caps	−			+

[a]For symbols see standard definitions. Data from Fahy (1981) and Stanier et al. (1966).

TABLE BXII.γ.116. General characteristics of some nonfluorescent *Pseudomonas* species[a]

Characteristics	P. alcaligenes	P. corrugata	P. luteola	P. mendocina	P. oryzihabitans	P. pseudoalcaligenes	P. stutzeri
Cell diameter, μm	0.5		0.8	0.7–0.8	0.8	0.7–0.8	0.7–0.8
Cell length, μm	2.0–3.0		2.5	1.4–2.8	2	1.2–2.5	1.4–2.8
Number of flagella	1	>1	>1	1b	1	1	1[b]
Yellow or orange pigments	d	+	+	+	+	−	−
Oxidase reaction	+	+	−	+	−	+	+
PHB accumulation	−	+	+	−	−	d	−
Gelatin liquefaction	d	d	+	−	−	d	−
Starch hydrolysis	−	−	−	−	−	−	+
Lecithinase	−	+		−		−	−
Lipase	d	d	−	+	−	−	+
Growth at 41°C	+	−	+	+	+	+	+
Denitrification	+	−	−	+	−	d	+
Arginine dihydrolase	+	d	+	+	−	d	+
Mol% G + C of the DNA	64–68	58–61	55.4	63–64	65.1	62–64	61–66

[a]For symbols see standard definitions. Data from Palleroni (1984), Kodama et al. (1985), and Sutra et al. (1997).

[b]Lateral flagella of short wavelength may be produced under certain conditions.

ston-Barrett et al. (1976) and Stanier et al. (1966). The nutritional spectrum is very narrow, resembling that of highly mutated fluorescent organisms. The gelatinase reaction is negative. The type strain was isolated from swimming pool water (Hugh and Ikari, 1964).

The mol% G + C of the DNA is: 64–68 (Bd).

Type strain: Stanier 142, ATCC 14909, LMG 1224NCIB 9945, NCTC 10367.

GenBank accession number (16S rRNA): Z76653.

4. **Pseudomonas amygdali** Psallidas and Panagopoulos 1975, 105[AL]

a.myg′ da.li. L. n. *amygdalum* almond; L. gen. n. *amygdali* of the almond.

The following description is taken from Psallidas and Panagopoulos (1975).

Rods, 0.7 × 1.7 μm or much longer (filaments can be 10–15 times the length of normal cells). Motile by means of one to six polar flagella. No PHB accumulated. Grows better in potato-dextrose medium than in nutrient agar. Growth range, 3–32°C. No growth below pH 5. No fluo-

rescent pigment produced. Acid is formed from D-ribose, L-arabinose, glucose, mannose, galactose, fructose, sucrose, mannitol, and sorbitol. No utilization of xylose, L-rhamnose, L-sorbose, cellobiose, lactose, maltose, melibiose, trehalose, raffinose, inulin, esculin, amygdalin, arbutin, salicin, dulcitol, erythritol, glycerol, inositol, dextrin and α-methyl-D-glucoside. Malate, citrate, succinate, and fumarate are utilized. Gluconate is slowly assimilated. Acetate, propionate, oxalate, maleate, malonate, tartrate, lactate, sulfanilic acid, picrate, hippurate, and benzoate are not utilized. Among the natural amino acids, serine, aspartate, glutamate, arginine, asparagine, proline, and histidine are utilized. Not used as carbon and/or nitrogen sources are glycine, β-alanine, leucine, isoleucine, valine, lysine, ornithine, tyrosine, phenylalanine, tryptophan, cystine, cysteine, methionine, and creatine.

Some isolates are urease positive. Tween 80 and tributyrin are rapidly hydrolyzed. Lechithinase and arginine dihydrolase negative. Gelatin, casein, esculin, arbutin, and starch are not hydrolyzed. Nitrates are not reduced. Rotting of potato slices does not occur, but the organism is positive

TABLE BXII.γ.117. Nutritional characteristics of some nonfluorescent species of *Pseudomonas* [a]

Substrates[b]	P. alcaligenes	P. corrugata[c]	P. mendocina	P. pseudoalcaligenes	P. stutzeri
Acetate, L-alanine, caprate, caprylate, fumarate, L-glutamate, α-ketoglutarate, lactate, L-proline, succinate	+	+	+	+	+
Aconitate, caproate, heptanoate, L-tyrosine, valerate	d	+	+	d	d
Adipate	−	−	−	−	d
β-Alanine	d		+	d	−
D-Alanine, citrate	d	+	+	d	+
α-Aminobutyrate	−	−	+	−	d
γ-Aminobutyrate	+	+	−	+	−
L-Arabinose, D-galactose, m-inositol, D-mannose, D-ribose, sucrose, trehalose, trigonelline, D-xylose	−	+	−	−	−
δ-Aminovalerate	d	+	−	d	d
α-Amylamine	d		−	−	−
L-Arginine	+	+	+	+	−
L-Aspartate, L-isoleucine, malonate, L-valine	−	+	+	−	d
Azelate, maltose, sebacate, starch	−	−	−	−	+
Benzoate	−	−	d	−	d
Betaine	−	+	+	+	−
Butanol, propanol, putrescine	d		+	+	d
Butylamine	d		−	−	−
2,3-Butylene glycol	−		−	−	d
Butyrate	d	d	+	+	+
Isobutyrate, citraconate	−	d	d	−	−
Creatine, sorbitol	−	−	−	d	−
Dodecane	d		−	−	−
Ethanolamine	−	d	−	+	d
Ethylene glycol	−		+	d	+
D-Fructose	−	+	d	+	d
Gluconate, glutarate, L-serine	−	+	+	d	d
D-Glucose	−	+	+	−	+
Glycerate	−	d	+	+	d
Glycerol	−	+	d	d	+
Glycine, D-malate	−	−	d	d	d
Glycolate	−	−	+	−	+
Histamine	d	d	−	d	−
L-Histidine	d	+	+	d	−
p-Hydroxybenzoate, mannitol	−	+	−	−	d
β-Hydroxybutyrate	−	+	+	+	+
Hydroxymethylglutarate	−		d	−	d
Isobutanol	d		+	−	d
Isovalerate, mucate, saccharate	−		+	−	d
Itaconate	−		+	d	+
L-Leucine	+	+	+	d	+
Levulinate	−	−	+	−	−
L-Lysine, DL-norleucine, L-ornithine, D(−)-tartrate, m-tartrate	−	d	−	−	−
L-Malate	+	+	−	+	+
Mesaconate	−	−	+	+	+
Pelargonate, propionate, spermine	+	+	+	+	d
L-Phenylalanine	−	+	d	d	d
Pyruvate	+	d	d	+	+
Sarcosine	−	+	+	d	−
L(+)-Tartrate, tryptamine	−	−	d	−	−

[a]For symbols see standard definitions. Data from Palleroni (1984) and Sutra et al. (1997).

[b]The following compounds were not used by any of the species: adonitol, anthranilate[d], D-arabinose, benzoyl-formate[d], benzylamine, cellobiose, citrulline, erythritol, hexadecane[d], hippurate[d], o-hydroxybenzoate, m-hydroxybenzoate, isopropanol, 2-ketogluconate, kynurenate[d], lactose, maleate, mandelate, D- and L-mandelate, naphthalene[d], nicotinate[d], oxalate, panthothenate[d], phenol[d], phenyl-acetate, phenyl-ethanediol[d], phthalate, pimelate, L-rhamnose, salicin, suberate, testosterone[d], and L-threonine.

[c]N-acetylglucosamine, D-arabitol, diaminobutane are all positive for *P. corrugata* (other species untested). D-lyxose was variable for *P. corrugata* (other species untested). Substrates not used by *P. corrugata* (other species not tested) are adonitol, 2, 3, and 4-aminobenzoates, amygdalin, L-arabitol, arbutin, creatine, cysteine, dulcitol, ethylamine, D- and L-fucose, gentiobiose, glycogen, inulin, 5-ketogluconate, melezitose, melibiose, methionine, methylglucoside, methylmannoside, methylxyloside, norvaline, isophthalate, terephthalate, raffinose, tagatose, turanose, urea, and xylitol.

[d]Compounds not been tested for *P. corrugata*.

in the hypersensitivity test on tobacco leaves. Further details are given in the original paper. Assignment to rRNA group I has been decided based on rRNA–DNA hybridization studies (De Vos et al., 1985), and later confirmed by rDNA sequence analysis (Moore et al., 1996; Anzai et al., 1997). The fatty acid composition places the species in one of the subgroups of group I (Stead, 1992). Pathogenic for the almond tree (*Prunus dulcis*, family Rosaceae), in which it

produces a hyperplasic bacterial canker. Not pathogenic for other fruit trees.

The mol% G + C of the DNA is: 57.7–58.5 (T_m); strain NCPPB 2610, 52.2.

Type strain: ATCC 33614, DSM 7298, LMG 2123 , NCPPB 2607.

GenBank accession number (16S rRNA): D84007, Z76654.

Additional Remarks: The DNA similarity studies of Gardan

et al. (1999) have included the type strains of this species as well as those of *P. ficuserectae*, *P. meliae*, and *P. savastanoi*. They indicate that these names should be considered as synonymous, and that *P. amygdali* is the correct one. Unfortunately, no decision on this point seems advisable in view of the similarity in known phenotypic properties among these organisms.

5. **Pseudomonas anguilliseptica** Wakabayashi and Egusa 1972, 584[AL]

an.guil.li.sep' ti.ca. L. n. *anguilla* eel; Gr. adj. *septica* putrefactive; M.L. adj. *anguilliseptica* pertaining to diseased eels.

The description below is taken from the original paper. Rods, 0.4×2.0 µm, with a tendency to become filamentous. Motile by a single polar flagellum; motility is better at 15°C than at 25°C. Catalase and oxidase reactions positive. Gelatin is liquefied. Nitrate reduction, urease, fluorescent pigment production, and starch hydrolysis are all negative. There is no acid production from arabinose, xylose, rhamnose, fructose, galactose, glucose, mannose, sorbose, lactose, maltose, raffinose, sucrose, starch, dextrin, salicin, glycerol, mannitol, or inositol. rDNA sequencing studies place this species close to *P. aeruginosa* (see Fig. BXII.γ.124). Isolated from diseased pond-cultured eels (*Anguilla japonica*).

The mol% G + C of the DNA is: unknown.

Type strain: ATCC 33660, DSM 12111, NCMB 1949.

GenBank accession number (16S rRNA): X99540.

6. **Pseudomonas asplenii** (Ark and Tompkins 1946) Săvulescu 1947, 11[AL] (*Phytomonas asplenii* Ark and Tompkins 1946, 760.)

a.sple' ni.i. M.L. neut. n. *Asplenium* genus of ferns, spleenworts; M.L. gen. n. *asplenii* of *Asplenium*.

A description is given by Haynes and Burkholder (1957) in the seventh edition of the *Manual*.

The species was tentatively placed in RNA group I by Byng et al. (1980), an assignment confirmed by nucleic acid hybridization studies (De Vos et al., 1985; Kersters et al., 1996). Pathogenic for the bird's-nest fern (*Asplenium nidus*), which has been the source of isolation.

The mol% G + C of the DNA is: unknown.

Type strain: ATCC 23835, LMG 2137.

GenBank accession number (16S rRNA): Z76655.

7. **Pseudomonas aurantiaca** Nakhimovskaya 1948, 64[AL]

au.ran.ti' a.ca. M.L. adj. *aurantiaca* orange colored.

The description is taken from the original paper.

Rods, $0.3–0.5 \times 0.8–2.0$ µm; motile, lophotrichous, with four to six flagella. Two main diffusible pigments, green and orange, are produced. The colony may remain orange, while the medium is stained green and fluoresces. Colorless colonies are also produced, presumably by mutation. Gelatin liquefaction is rapid. Starch is not hydrolyzed. Growth is good in complex and synthetic media. Nitrogen sources assimilated include ammonia, nitrate, amino acids, and peptone. The organism has been described as "oligonitrophilic", that is, its nitrogen requirements are satisfied by the traces of nitrogenous compounds that are contained in the "nitrogen-free" medium used in the experiments (Nakhimovskaya, 1948). Optimum temperature 25°C. Capable of growth and acid formation from arabinose, xylose, glucose, galactose, sucrose, raffinose, glycerol, and mannitol. Lactose, maltose, and starch are not utilized.

Studies on the type strain of this species have shown that it actually corresponds to *P. chlororaphis* subsp. *aureofaciens* (Kiprianova et al., 1985). These authors propose to retain the name *P. aurantiaca* for pigmented strains of biovar II of *P. fluorescens*.

The mol% G + C of the DNA is: unknown.

Type strain: NCIB 10068.

8. **Pseudomonas avellanae** Janse, Rossi, Angelucci, Scortichini, Derks, Akkermans, De Vrijer and Psallidas 1997, 601[VP] (Effective publication: Janse, Rossi, Angelucci, Scortichini, Derks, Akkermans, De Vrijer and Psallidas 1996, 594.)

a' vel.la.nae. L. adj. *avellana* the species name of the host, *Corylus avellana*.

The following description is taken from Janse et al. (1996), who transcribed the information from the original paper by Psallidas (1993).

Rods occurring single or in pairs. Motile by means of one to four polar flagella. PHB is not accumulated. Slow growth in nutrient agar. On sucrose-containing media they form circular, dome-shaped, glistening, semitranslucent, butyrous, radially striated, cream-white or pearl-white colonies, 2–2.5 mm and 3.5–4 mm in diameter after 3 and 7 d, respectively. A levan is formed on nutrient medium plus 5% sucrose. Weak, blue-green fluorescent diffusible pigment formed in King B medium. Pantothenate is required for growth. 4% NaCl in the medium is tolerated. Catalase and lipase (Tween 80 hydrolysis) are positive. Urease, oxidase reaction, tyrosinase, arginine and ornithine dihydrolase, amylase, gelatin liquefaction, hydrolysis of casein, and protopectinase reaction (rotting of potato slices) are all negative. Nitrates are not reduced. Hydrolysis of esculin and arbutin are negative. H_2S is not produced from peptone, cysteine, or thiosulfate. Ammonia and 2-ketogluconate are not produced.

The following carbon compounds are used for growth: L-arabinose, D-ribose, D-xylose, D-glucose, D-fructose, D-galactose, D-mannose, sucrose, glycerol, mannitol, *meso*-inositol, citrate, formate, fumarate, maleate, malate, malonate, mucate, nonanoate, propionate, and succinate. Some strains use D-alanine, L-alanine, L- and DL-aspartate, L-glutamate, glutamine, L-proline, and L- and DL-serine. Organic compounds not used for growth include adonitol, esculin, amygdalin, arbutin, cellobiose, dulcitol, inulin, lactose, maltose, melezitose, melibiose, *meso*-erythritol, α-methylglucoside, L-rhamnose, salicin, L-sorbose, sorbitol, trehalose, adipate, benzoate, cinnamate, hippurate, itaconate, lactate, oxalate, picrate, sulfanilate, D(−)-tartrate, L(+)-tartrate, DL-tartrate, L-aspartate, β-alanine, L-arginine, L- or DL-asparagine, L-citrulline, L-cysteine, L-cystine, glycine, L-histidine, DL-homoserine, L-hydroxyproline, L-leucine, L-lysine, L-methionine, DL-ornithine, DL-phenylalanine, L-sarcosine, DL-tyrosine, L-threonine, DL-tryptophan, and L-valine. Further details are mentioned in the original paper (Janse et al., 1996). Optimum growth temperature, 23–25°C. Maximum, 30°C. Strains contain plasmids in the range of 33–100 MDa. It causes hypersensitive reaction in tobacco leaves. Isolated from hazelnut trees (*Corylus avellana* L.), where it causes cankers and dieback on stems and branches.

The mol% G + C of the DNA is: unknown.

Type strain: F11, DSM 11809, ICMP 9746, NCPPB 3487.

Additional Remarks: Interesting additional information

on this species is found in the literature (Janse et al., 1996). Fourteen strains were analyzed for plasmid DNA, fatty acids of whole cells, whole-cell proteins, and restriction fragment patterns of 16S rRNA. In comparison with other fluorescent species of the genus *Pseudomonas*, *P. avellanae* strains are homogeneous in their properties, and the molecular approaches indicate that the species, until recently considered to be a pathovar of *P. syringae* (*Pseudomonas syringae* pathovar *avellanae*, Psallidas, 1993), warrants independent species allocation.

9. **Pseudomonas azotoformans** Iizuka and Komagata 1963b, 137[AL]

a.zo.to.for' mans. Fr. n. *azote* nitrogen; L. v. *formo* to fashion, form; M.L. part. adj. *azotoformans* nitrogen forming (by denitrification).

The description is taken from the paper by Iizuka and Komagata (1963b).

Rods, 0.6–0.8 × 1.4–2.0 µm, motile with polar flagella. Fluorescent pigment is produced. Oxidase, gelatin liquefaction, nitrate reduction, and denitrification are all positive. H_2S is not produced and starch is not hydrolyzed. Acid is produced from glucose, glycerol, xylose, sucrose, but not from lactose or starch. Glucose, gluconate, 2-ketogluconate, citrate, succinate, ethanol, *p*-hydroxybenzoate, protocatechuate, and anthranilate are assimilate, but not phenol, benzoate, salicylate, *m*-hydroxybenzoate, gentisate, *p*-aminobenzoate, or 5-ketogluconate. Additional details can be found in the original paper. Optimum growth temperature, 25–30°C. No growth at 37°C. The description suggests that this species is similar to some of the denitrifying biovars of *P. fluorescens*, but no comparative studies have been performed. Assignment to the genus *Pseudomonas* has been confirmed by 16S rDNA sequence analysis studies (Anzai et al., 1997). Isolated from Japanese rice paddies.

The mol% G + C of the DNA is: unknown.

Type strain: IAM 1603.

GenBank accession number (16S rRNA): D84009.

10. **Pseudomonas balearica** Bennasar, Rosselló-Mora, Lalucat and Moore 1996, 204[VP]

ba.le.a' ri.ca. L. fem. adj. *balearica* of the Balearic Islands, where the organism was isolated.

Description taken from the paper by Bennasar et al. (1996).

Short, straight rods, 0.3–0.5 × 1.5–3.0 µm. Motile by means of a single polar flagellum. It grows aerobically, or under anaerobic conditions in the presence of nitrate. Vigorous denitrifier, liberating copious amounts of nitrogen gas from nitrate. Nonpigmented. It shares with *P. stutzeri* some of the phenotypic and morphological characteristics, including starch hydrolysis, growth with maltose, negative arginine dihydrolase, and gelatinase reactions, and a typical wrinkled colony morphology when freshly isolated. Differences with *P. stutzeri* include the ability to grow at 46°C and to tolerate 8.5% NaCl in the medium. It is also able to grow on xylose, but it does not utilize ethylene glycol, mannitol, 4-aminobutyrate, and suberate. Table BXII.γ.114 summarizes differences with other species of *Pseudomonas*. Some characteristics that differentiate the species from other denitrifying pseudomonads are presented in the chapter on *Burkholderia*. The two strains that have been isolated are

able to degrade 2-methylnaphthalene at 40°C in wastewater or marine sediment samples.

The mol% G + C of the DNA is: 64.1–64.4 (HPLC).

Type strain: SP1402, DSM 6083.

GenBank accession number (16S rRNA): U26418.

11. **Pseudomonas caricapapayae** Robbs 1956, 74[AL]

ca.ri.ca.pa' pay.ae. M.L. gen. n. *caricapapayae* of *Carica papaya*, pawpaw.

The description is taken from Robbs (1956).

Straight or slightly curved rods, 0.9–1.1 × 1.3–3.0 µm. Motile by three to six polar flagella. Green fluorescent pigment produced. Gelatin is liquefied. Starch is not hydrolyzed. Nitrates are not reduced. Acid is produced from glucose, mannose, sucrose, glycerol, and mannitol, but not from lactose, maltose, salicin, or starch. Citrate and tartrate are utilized; lactate is not. rRNA–DNA hybridization studies have confirmed the allocation of this species among the fluorescent members of the genus *Pseudomonas* (De Vos et al., 1985). Optimum growth temperature: 23–29°C; no growth below 7°C and above 45°C. Isolated from water-soaked, angular spots on leaves of pawpaw.

The mol% G + C of the DNA is: 58.9–59.4 (T_m).

Type strain: ATCC 33615, NCPPB 1873.

GenBank accession number (16S rRNA): D84010.

12. **"Pseudomonas cedrella"** Dabboussi, Hamze, Elomari, Verhille, Baida, Izard and Leclerc 1999b, 303

ced.rel' la. M.L. fem. gen. n. *cedrella* of the cedar tree.

Rods, motile by means of a single polar flagellum. Colonies on nutrient agar are circular and smooth. On blood agar there are no signs of hemolysis. Fluorescent pigment is produced on King B agar. Oxidase, catalase, and arginine dihydrolase reactions are all negative. Some strains (2/7) liquefy gelatin. Levan is produced from sucrose. Two out of seven strains liquefied gelatin. Growth occurs at 4°C but not at 41°C. No growth occurs with 5% NaCl in the medium. Various enzymatic reactions are described in the paper (Dabboussi et al., 1999b), but gelatinase is not reported. Denitrification is a variable character. Growth at the expense of various carbon sources, particularly carbohydrates and derivatives, and various enzymatic activities are reported in the description. A comparison of growth characteristics and some general phenotypic characters, with those of other fluorescent species, is presented in the paper.

The results of DNA–DNA hybridization experiments suggest that perhaps this species should be considered as a genomovar of some of the biovars of *P. fluorescens*. Isolated from Lebanese spring water.

The mol% G + C of the DNA is: 59–60.

Type strain: CFML 96-198; CIP 105541.

GenBank accession number (16S rRNA): AF064461.

13. **Pseudomonas chlororaphis** (Guignard and Sauvageau 1894) Bergey, Harrison, Breed, Hammer and Huntoon 1930, 166[AL] (*Bacillus chlororaphis* Guignard and Sauvageau 1894, 841.)

chlo.ro.ra' phis. Gr. adj. *chlorus* green; Gr. n. *raphis* a needle; M.L. fem. n. *chlororaphis* a green needle.

Fluorescent pigment is produced. Even though the species name refers to the production of the green insoluble phenazine pigment chlororaphin, which often is excreted and crystallizes around the colonies, *P. chlororaphis* has been found to be closely related to *P. aureofaciens* Kluyver (which does not produce this pigment, but instead makes another phenazine pigment that is orange and freely diffuses into the medium. The relationship between the two species was demonstrated by a high DNA similarity as determined by DNA–DNA hybridization (Palleroni et al., 1972; Johnson and Palleroni, 1989). For this reason, *P. chlororaphis* may include two subspecies: subsp. *chlororaphis*, producing chlororaphin, and subsp. *aureofaciens*, which produces the water-soluble phenazine monocarboxylic acid. Other phenotypic properties that differentiate the two subspecies are the utilization of L-arabinose and α-amylamine by subsp. *aureofaciens*, and of D-alanine by subsp. *chlororaphis*. Optimum temperature, about 30°C. The type strain of *P. chlororaphis* subsp. *chlororaphis* was isolated from dead larvae of the cockchafer (a large European beetle). Strains of this subspecies and of subsp. *aureofaciens* have been isolated from various sources, including water. The first strain of subsp. *aureofaciens* (ATCC 13985, NCIB 9030) was isolated from clay suspended in kerosene for 3 weeks (Kluyver, 1956).

The mol% G + C of the DNA is: 63.5 (Bd).

Type strain: ATCC 9446, DSM 50083, IFO 3904, NRRL B-560, NCIB 9392.

GenBank accession number (16S rRNA): D86004, Z76657.

14. **Pseudomonas cichorii** (Swingle 1925) Stapp 1928, 291[AL] (*Phytomonas cichorii* Swingle 1925, 730.)
ci.cho' ri.i. Gr. pl. *cichora* succory, chicory; L. n. *cichorium* chicory; M.L. gen. n. *cichorii* of chicory.

Selected characteristics of the species are given in Tables BXII.γ.115 and BXII.γ.118. A fluorescent pigment is produced. The oxidase reaction is weak and slow and the gelatinase reaction is negative. Optimum temperature, ~30°C. Isolated from *Cichorium intybus* and *C. endivia*, for which it is pathogenic.

The mol% G + C of the DNA is: 59 (Bd).

Type strain: ATCC 10857, DSM 50259, LMG 2162, NCPPB 943, PDDCC 5707.

GenBank accession number (16S rRNA): Z76658.

15. **Pseudomonas citronellolis** Seubert 1960, 428[AL]
cit.ro.nel' lo.lis. M.L. gen. n. *citronellolis* of citronellol.

The following description is summarized from the original paper by Seubert (1960).

Rods, 0.5×1.0–1.5 μm. Motile by means of a single polar flagellum. Optimum growth temperature, approximately 31°C. Gelatin is not liquefied. Nitrates are reduced to nitrites. Growth can occur anaerobically in the presence of nitrate. H_2S is not produced. Besides citronellol, glucose, acetate, farnesol, and ionone can support growth. No growth was observed with squalene or camphoric acid. Ammonium salts, nitrate, peptone, or yeast extract can be used as nitrogen sources. Acid is produced from glycerol, but not from glucose, galactose, arabinose, fructose, sucrose, maltose, lactose, dulcitol, inositol, mannitol, inulin, or dextrin.

16S rDNA sequence analysis allocates this species in the genus *Pseudomonas* (Moore et al., 1996). Isolated from soil by enrichment with citronellol as carbon source.

The mol% G + C of the DNA is: unknown.

Type strain: ATCC 13674, DSM 50332.

GenBank accession number (16S rRNA): Z76659.

16. **Pseudomonas corrugata** Roberts and Scarlett 1981, 216[VP] (Effective publication: Roberts and Scarlett *in* Scarlett, Fletcher, Roberts and Lelliott 1978, 109) emend. Sutra, Siverio, Lopez, Hunault, Bollet and Gardan 1997, 1032.
cor.ru' ga.ta. L. v. *corrugare* to wrinkle up; part adj. *corrugatus* wrinkled up.

The following description is summarized from the original paper by Scarlett et al. (1978).

Rods motile by multitrichous polar flagella. According to the original description, the cells are reported to accumulate PHB as carbon reserve material; however, the granules are in fact poly-hydroxyalkanoates of different monomer composition (Kessler and Palleroni, 2000 . The colonies are wrinkled (the reason for the species name), yellowish, sometimes with a green center. With age, the color may change to khaki or fawn, depending on the medium. A yellow to yellow-green diffusible, nonfluorescent pigment is produced. Gelatin is hydrolyzed, but starch is not. Egg yolk (lecithinase) reaction positive. Levan production negative. There is growth at 37°C but not at 41°C. Among the characteristics that differentiate this species from *Burkholderia cepacia* and *B. gladioli* (*"Pseudomonas alliicola"*) are the absence of pectate hydrolysis and rot of onion slices, and the lack of utilization of D-arabinose, cellobiose, adipate, *meso*-tartrate, and citraconate. All these characteristics were positive for the strains of the last two species included in the study.

The above description has been emended recently (Sutra et al., 1997). This detailed study included a large number of strains of the species as well as strains of fluorescent pseudomonads also isolated from tomato pith necrosis, and the results of the analysis were studied by numerical methods, which showed that *P. corrugata* constituted a single phenon. The general phenotypic characters and nutritional properties of the strains, taken from this comprehensive report, are summarized in Tables BXII.γ.116 and BXII.γ.117. The cellular fatty acid composition places this species in a subgroup within the genus (Stead, 1992). The results of rRNA–DNA hybridization experiments allocate this species as a member of the genus *Pseudomonas* (De Vos et al., 1985), which was later confirmed by 16S rDNA sequence analysis (Anzai et al., 1997). Isolated from tomato pith necrosis.

The mol% G + C of the DNA is: 58.4–60.8 (T_m).

Type strain: ATCC 29736, DSM 7228, NCPPB 2445.

GenBank accession number (16S rRNA): D84012.

17. **Pseudomonas ficuserectae** Goto 1983a, 547[VP]
fi.cus.e.rec' tae. L. n. *ficuserectae* of *Ficus erecta*, the name of the host species.

The following description is taken from the paper by Goto (1983a)

Nonencapsulated straight rods, 0.5×2.0 μm. Motile by means of one to five polar flagella. The original description reports the accumulation of PHB granules by the cells. However, an examination of the type strain has shown that these granules are of poly-β-hydroxyalkanoates other than PHB (Kessler and Palleroni, 2000). Colonies on yeast extract-peptone agar at 28°C are white, transparent, circular, convex, 0.2–3.0 mm in diameter after 6 d. Growth on yeast extract-peptone agar slants becomes very viscid after several days. Denitrification, nitrate reduction, growth factor re-

TABLE BXII.γ.118. Nutritional properties of some of the fluorescent phytopathogenic *Pseudomonas* species[a]

Characteristics	*P. cichorii*[b]	*P. syringae* pathovars[c]	*P. viridiflava*[d]	*P. viridiflava*[e]
Utilization of[f]:				
Glucose, mucate, succinate, glycerol, L-aspartate, L-glutamate, L-glutamine, γ-aminobutyrate	+	+	+	+
D-Ribose, D-xylose, acetate, propionate, β-hydroxybutyrate	+	d	d	d
L-Arabinose, gluconate, L-malate, citrate, aconitate	+	d	d	+
D-Mannose, D-galactose, caproate, L-arginine, betaine	+	d	+	d
D-Fructose, caprylate, pelargonate, lactate, mannitol, *m*-inositol, *p*-hydroxybenzoate, quinate, L-serine, L-proline	+	d	+	+
Raffinose	d	d	−	−
Fumarate	+	+	+	−
Sucrose, glutarate	+	d	+	
Saccharate	+	d	d	+
Valerate	d	d	d	−
Caprate	d	+	+	d
Malonate, *m*-tartrate	+	d	d	
D-Malate, glycerate, trigonelline	+	d	+	
D(−)-Tartrate	−	−	+	+
L(+)-Tartrate, α-ketoglutarate	+	d	−	−
Hydroxymethylglutarate	−	d	d	
Pyruvate	+	+	+	d
Erythritol, sorbitol	−	d	+	+
L-Alanine	+	+	d	+
D-Alanine	d	d	+	
L-Leucine	−	−	−	+
L-Histidine	d	d	+	d
L-Tyrosine	+	−	d	−
L-Tryptophan	−	−	−	d
Putrescine	−	d	d	−
Sarcosine	d	d	−	d
Laurate				d
L-Sorbose, melezitose, amygdalin, dextrin, formate, dulcitol, isophthalate, L-methionine, *m*-aminobenzoate, *p*-aminobenzoate, methylamine				−
Linolenate	d	d	+	
Triacetin	+	d	+	
Tripropionin	+	d	+	
Tricaproin	d	d	+	
Ascorbate	−	d	−	
Isoascorbate	−	d	+	
Lecithin	+	d	d	
L-Asparagine	+	d	+	

[a]For symbols see standard definitions.

[b]Data from Sands et al., 1970.

[c]Data from Sands et al., 1970.

[d]Data from Sands et al., 1970.

[e]Data from Billing, 1970b.

[f]The following compounds were not utilized by any strains of the species in the table: D-arabinose, D-fucose[g], L-rhamnose, trehalose, maltose, cellobiose, lactose, melibiose, methylglucoside[g], starch, inulin, 2-ketogluconate[g], salicin[g], N-acetylglucosamine[g], isobutyrate, isovalerate, linoleate[g], laurylsulfate[g], tannate[g], oxalate, maleate, adipate, pimelate, suberate, azelate, sebacate, glycolate, thioglycolate[g], levulinate, citraconate, itaconate, mesaconate, 3-phosphoglycerate[g], hydroxymethylbutyrate[g], adonitol, ethylene glycol, propylene glycol, 2,3-butyleneglycol, methanol[g], ethanol, n-propanol, isopropanol, n-butanol, isobutanol, geraniol[g], D-mandelate, L-mandelate[g], benzoylformate[g], benzoate, *o*-hydroxybenzoate, *m*-hydroxybenzoate, phthalate, phenylacetate, phenylethanediol[g], eicosenedioate[g], naphthalene, phenol, testosterone, glycine, β-alanine, L-threonine, L-isoleucine, L-norleucine, L-valine, L-lysine, L-ornithine, L-citrulline, α-aminobutyrate, δ-aminovalerate[g], L-phenylalanine, L-hydroxyproline[g], D-tryptophan[g], indoleacetic acid[g], L-kynurenine[g], kynurenate[g], anthranilate, methylamine, ethanolamine, benzylamine, spermine, histamine[g], tryptamine[g], butylamine, α-amylamine, creatine, choline[g], hippurate, urate[g], pantothenate, acetamide, nicotinate, dodecane, hexadecane, poly-β-hydroxybutyrate[g], pectate[g], chlorogenate[g], and uridine[g].

[g]Compounds not been tried by Billing, 1970b.

quirement, arginine dihydrolase reaction, starch hydrolysis, decarboxylase reaction with lysine and ornithine, phenylalanine deaminase, growth in 5% NaCl, gelatin liquefaction, hydrolysis of Tween 80, urease, and growth at 41°C are all negative. The strains grow with L-arabinose, glucose, mannose, sucrose, raffinose, glycerol, and malonate, but do not grow at the expense of galactose, maltose, trehalose, erythritol, dulcitol, xylan, inulin, glycogen, salicin, carboxymethyl cellulose, α-methylglucoside, esculin, oxalate, glycolate, butyrate, sebacate, hippurate, arginine, betaine, valine, or ethylene glycol. Additional characteristics can be found in the original paper by Goto (1983a).

A numerical analysis that included strains of this species as well as many others of aerobic pseudomonads gave results showing that *P. ficuserectae* and *P. meliae* formed a single cluster, and suggested that they may be pathovars of *P. syringae* (Hu et al. 1991). This comment hinges on the observation that the strains under study did not accumulate PHB. For the moment, the position of *P. ficuserectae* remains unsettled, and more authentic strains will be needed for examination to confirm the correct taxonomic status. However, rDNA sequence analysis confirms its allocation to *Pseudomonas* (Moore et al., 1996). See also Additional Remarks at the end of the description of *P. amygdali* in this list. Pathogenic for *Ficus erecta*, in which it produces leaf spots that develop beside the thick veins.

The mol% G + C of the DNA is: 59 (T_m).

Type strain: ATCC 35104, DSM 6929, LMG 5694, PDDCC 7848.

GenBank accession number (16S rRNA): Z76661.

18. **Pseudomonas flavescens** Hildebrand, Palleroni, Hendson, Toth and Johnson 1994, 413[VP]

fla.ves' cens. L. v. *flavesco* to become golden yellow; L. part. adj. *flavescens* becoming golden yellow.

The information for the following description is taken from Hildebrand et al. (1994).

Characteristics useful for differentiation from some of *Pseudomonas* species are given in Tables BXII.γ.112 and BXII.γ.113. Slightly curved rods, 0.6–0.7 × 1.6–2.3 μm. Motile by means of a single polar flagellum. No PHB accumulated. The following characteristics are negative: arginine dihydrolase, ice nucleation, hypersensitivity reaction, denitrification, and hydrolysis of gelatin, starch, or Tween 80. It grows well in nutrient agar and other common complex media. The colonies are yellow on King medium B and other complex media. In King B, a diffusible fluorescent pigment is produced. The original report gives a list of more than 50 carbon compounds that can be used individually as carbon and energy sources, and a list of more than 80 compounds that do not support growth. Optimum temperature 28°C. No growth occurs at 37°C. Isolated from walnut blight cankers.

The mol% G + C of the DNA is: 63 (T_m).

Type strain: B62, DSM 12071, NCPPB 3063.

GenBank accession number (16S rRNA): U01916.

19. **Pseudomonas fluorescens** (Trevisan 1889b) Migula 1895a, 29[AL] (*Bacillus fluorescens* Trevisan 1889b, 18.)

flu.o.res' cens. L. n. *fluor* a flux; M.L. v. *fluoresce*; M.L. part. adj. *fluorescens* fluorescing.

Characteristics of the species are given in Tables BXII.γ.108, BXII.γ.112, BXII.γ.113, BXII.γ.115, and BXII.γ.119. Optimum temperature, 25–30°C. Found in soil and water, from which it can be isolated after enrichment in media containing various carbon sources, incubated aerobically; strains of the denitrifying biovars can be enriched in similar media containing nitrate, incubated under anaerobic conditions. Commonly associated with spoilage of foods (eggs, cured meats, fish, and milk). Often isolated from clinical specimens. Some strains assigned to this species (biovar II) have been isolated from diseased plants (e.g., lettuce), and identified as *Pseudomonas marginalis* (Brown, 1918) Stevens 1925[AL]. For more details on this species see the description below. Biovar I (biotype A of Stanier et al., 1966) is considered to by typical of *P. fluorescens*, and the type strain of the species belongs to this group. Aside from *P. marginalis* strains, biovar II (previously biotype B) also includes saprophytic organisms. In their studies on the species *P. aurantiaca* Nakhimovskaya 1948, Kiprianova et al. (1985) found that the type strain was actually *P. aureofaciens* (now a subspecies of *P. chlororaphis*), but they propose to retain the name *P. aurantiaca* for the pigmented strains of biovar II of *P. fluorescens*. Within biovar III (biotype C of Stanier et al., 1966) at least two subgroups, which differ from each other in their capacity for utilization of dicarboxylic acids, can be defined. Biovar IV (biotype F of Stanier

et al., 1966) contains the type strain of "*P. lemonnieri*" (Lasseur 1913) Breed 1948, 178. Several strains are known at present, and these can be grouped into at least two clusters.

The group of miscellaneous strains assigned to *P. fluorescens* biotype G by Stanier et al. (1966) constitutes the last (V) biovar of the species. The biovar is very heterogeneous in its nutritional properties, and it may consist of strains that have lost one or more of the properties considered to be of diagnostic importance in differentiating among the better characterized biovars. Among the non-authentic strains assigned to this biovar are strains labeled "*P. schuylkilliensis*" and *P. geniculata* (Wright 1895) Chester 1901, 313[AL] (see Stanier et al., 1966). The group of strains isolated from meat that eventually was described as a new species under the name *P. lundensis* (Molin et al., 1986) corresponded to a cluster identified within biotype G (Barrett et al., 1986) (see below). Strains of biovar V are common in various materials, and have been isolated from soils (Sands and Rovira, 1971). It is quite possible that isolation of more strains of each of the different biovars may result in the future in the proposal of new species with the general properties of *P. fluorescens*, but circumscribed based on genomic differences.

The mol% G + C of the DNA is: 59.4–61.3 (Bd).

Type strain: ATCC 13525, DSM 50090, NCIB 9046, NCTC 10038.

GenBank accession number (16S rRNA): Z76662.

20. **Pseudomonas fragi** (Eicholz 1902) Gruber 1905, 122[AL] (*Bacterium fragi* Eicholz 1902, 425.)

fra' gi. L. neut. n. *fragum* strawberry; L. gen. n. *fragi* of the strawberry.

The description of this species is given in the seventh edition of the *Manual* (Haynes and Burkholder, 1957).

Rods motile with a polar flagellum. The gelatinase reaction is positive. Nitrate is not reduced to nitrite. Ammonia is produced from peptone. Description of growth in various media and acid production from some sugars and sugaralcohols are given by Haynes and Burkholder (1957). Growth occurs from 10–30°C. No growth at 37°C. The organism is very sensitive to heat.

This species was tentatively placed in rRNA group I by Byng et al. (1980), a conclusion supported by rRNA–DNA hybridization studies (De Vos et al., 1989) and by 16S rDNA sequence analysis (Moore et al., 1996; Anzai et al., 1997). Isolated from milk, dairy products, water.

The mol% G + C of the DNA is: 60.6 (T_m).

Type strain: ATCC 4973, DSM 3456, LMG 2191.

21. **Pseudomonas fulva** Iizuka and Komagata 1963b, 138[AL]

ful' va. L. adj. *fulva* tawny, yellowish brown.

Rods, 0.6–0.8 × 1.2–1.8 μm. Motile with one to three polar flagella. Fluorescent pigment is produced. Oxidase reaction feebly positive. Gelatin is not liquefied. Nitrate not reduced to nitrite or denitrified. H_2S not produced. Starch is not hydrolyzed. Acid is produced from glucose. No acid from glycerol, xylose, sucrose, lactose, or starch. Glucose, gluconate, 2-ketogluconate, citrate, succinate, ethanol, *p*-hydroxybenzoate, and protocatechuate are assimilated. Phenol, benzoate, *m*-hydroxybenzoate, gentisate, *p*-aminobenzoate, and 5-ketogluconate are not assimilated.

Optimum temperature for growth, 25–30°C. Poor growth at 37°C and no growth at 42°C. Allocation to the

TABLE BXII.γ.119. Characteristics of *Pseudomonas chlororaphis* and *P. fluorescens* biovars[a]

Characteristics	*P. chlororaphis* subsp. *chlororaphis*	*P. chlororaphis* subsp. *aureofaciens*	*P. fluorescens* bv. I	*P. fluorescens* bv. II	*P. fluorescens* bv. III	*P. fluorescens* bv. IV	*P. fluorescens* bv. V
P. fluorescens biovars as designated by Stanier et al. (1966)	D	E	A	B	C	F	G
Nonfluorescent pigments, color:							
Green (chlororaphin)	+	−	−	−	−	−	−
Orange (phenazine-1-carboxylate)	−	+	−	−	−	−	−
Blue, nondiffusible	−	−	−	−	−	+	−
Levan formation from sucrose	+	+	+	+	−	+	−
Denitrification	+	d	−	+	+	+	−
Substrates used for growth:							
L-Arabinose	−	+	+	+	d	+	d
Sucrose	+	d	+	+	−	+	d
Saccharate	+	+	+	+	d	+	d
Propionate	+	+	+	−	d	+	+
Butyrate	+	d	−	d	d	+	d
Sorbitol	−	−	+	+	d	+	d
Adonitol	−	−	+	−	d	−	d
Propylene glycol	−	−	−	+	d	−	d
Ethanol	d	−	−	+	d	−	d

[a]For symbols see standard definitions. Data from Palleroni (1984).

genus *Pseudomonas* has been confirmed by rRNA sequence analysis (Anzai et al., 1997). Isolated from Japanese rice paddies.

The mol% G + C of the DNA is: 60.6 (T_m).

Type strain: IAM 1529.

GenBank accession number (16S rRNA): D84015.

22. **Pseudomonas fuscovaginae** (ex Tanii, Miyajima and Akita 1976) Miyajima, Tanii and Akita 1983, 656[VP]

fus.co.va.gi' nae. L. adj. *fuscus* fuscous; L. fem. n. *vagina* vagina, sheath; M.L. fem. n. *fuscovaginae* of a fuscous vagina.

The following description is taken from the original paper.

Rods with round ends, 0.5–0.8 × 2.0–3.5 μm. Cells occur singly or in pairs. Motile by one to four polar flagella. Moderate growth on nutrient agar after 4 or 5 d at 28°C; white to light brown colonies, smooth, glistening, raised, translucent, circular, butyrous. A green fluorescent, diffusible pigment is produced on King's medium B. No slime is produced on nutrient agar plus 5% sucrose. Catalase and Kováč's oxidase reaction are positive. Denitrification and nitrate reduction are negative. Hydrolysis of Tween 80 and of gelatin is positive; esculin and arbutin are not hydrolyzed. Arginine dihydrolase reaction positive, but phenylalanine deaminase and formation of 2-ketogluconate and H_2S are negative. No organic growth factors are required. Citrate, malonate, succinate, urate, acetate, β-alanine, L-valine, and L-lysine are utilized for growth, but not tartrate, hippurate, 2-ketogluconate, or polygalacturonic acid. Optimal growth temperature is approximately 28°C. No growth occurs at 37°C. Additional characteristics may be found in the paper by Miyajima et al. (1983).

Pathogenic for rice (*Oryza sativa*) and various other plants of the family Gramineae. In rice it produces first water-soaked, dark green spots, which later become brown to dark brown. Characteristics differentiating this species from other fluorescent pseudomonads are mentioned in the original paper. Assignment to the genus *Pseudomonas* has been confirmed by rRNA–DNA hybridization studies (De Vos et al., 1985). Isolated from diseased leaf sheaths of *O. sativa* in Japan.

The mol% G + C of the DNA is: unknown.

Type strain: 6801, DSM 7231, NCPPB 3085, PDDCC 5940.

23. **Pseudomonas gessardii** Verhille, Baïda, Dabboussi, Hamze, Izard and Leclerc 1999a, 1566[VP]

ges.sar' di.i. gessardii of C. Gessard, French chemist who isolated "*Bacterium aeruginosum*" for the first time in 1882 and studied its pigment.

The following description is taken from the original paper (Verhille et al., 1999a).

Rods, motile by means of a single flagellum. Fluorescent pigment is produced on King B; no phenazine pigments are observed on King A. Oxidase, catalase, and production of levan from sucrose are all positive. Many compounds were tested as carbon and energy sources for growth and a number of enzymatic reactions were determined with the cultures. The results are listed in the original paper. According to the authors, the utilization of L.(−)-arabitol, xylitol, *myo*-inositol, adonitol, and *i*-inositol and the absence of growth in the presence of L-arabinose, D-xylose, D-saccharate, *meso*-tartrate, tricarballylate, glucuronate, galacturonate phenylacetate, and histamine, differentiate this species from *P. migulae*, a closely related species. Differentiation from other *Pseudomonas* species is given in tables in the original report. A minority of the strains (1/13) had gelatinase activity. No growth is observed in the presence of 5% NaCl in the medium. The range of growth temperatures goes from 4°C to 35°C, and the optimum is around 30°C. Isolated from natural mineral waters.

The mol% G + C of the DNA is: 58.

Type strain: CIP 105469.

GenBank accession number (16S rRNA): AF074384.

24. **Pseudomonas graminis** Behrendt, Ulrich, Schumann, Erler, Burghardt and Seyfarth 1999, 306[VP]

gra' mi.nis. L. n. *gramen* grass; *graminis* of grass, the source of the organism.

Rods, 0.5–1.0 × 3.5–5.0 μm, motile by one polar flagellum. Colonies are yellow, glistening, moderately convex, circular, with smooth edges. Fluorescent pigment production, levan production from sucrose, starch hydrolysis, oxidase, nitrate reduction to nitrite, denitrification, arginine dihydrolase, lysine and ornithine decarboxylase, tryptophan deaminase, DNase, urease, β-hemolysis of sheep blood, and β-galactosidase reactions are all negative. Catalase and hydrolysis of Tweens 40 and 80 are positive. Utilization of a number of carbon compounds is listed in the original paper by Behrendt et al. (1999). Unfortunately, few of the substrates tested are the same as the ones that appear in the tables of this chapter, and the species has not been included for comparison. However, the paper gives a selection of properties that allow a differentiation of this species from others of similar characteristics. The strains use monosaccharides but not the disaccharides tested.

A comparison of rRNA sequences with those of other species of *Pseudomonas* and species of other RNA similarity groups indicates that the assignment to the genus *Pseudomonas* is correct. The creation of a new species name for the strains is supported by low DNA similarity with other species of the genus. The paper is an excellent source of taxonomic information on the allocation of this species within the group of fluorescent pseudomonads. The original description includes the presence of ubiquinone Q-9 and the hydroxy fatty acids 3-hydroxydodecanoic and 2-hydroxydodecanoic. Isolated from the phyllosphere of grasses.

The mol% G + C of the DNA is: 60–61.

Type strain: P 294/08, DSM 11363.

GenBank accession number (16S rRNA): Y11150.

25. **Pseudomonas jessenii** Verhille, Baïda, Dabboussi, Izard and Leclerc 1999b, 54[VP]

jes.sen' ni.i. M.L. masc. gen. n. *jessenii* of Jessen, named after O. Jessen, eminent Danish bacteriologist who contributed substantially to the knowledge of fluorescent pseudomonads.

Rods, motile by means of a single polar flagellum. Colonies on nutrient agar are circular, smooth, nonpigmented. Nonhemolytic when grown on blood agar. Fluorescent pigment is produced on King B agar medium. Catalase, oxidase, and arginine dihydrolase reactions are positive. Denitrification, gelatinase, and lecithinase are negative. Nitrate is reduced to nitrite. The strains grow at 4°C but not at 41°C (range: 4–35°C). Optimal: 30°C. No growth occurs in the presence of 5% NaCl. Other properties as well as the results of nutritional and enzymatic screenings are listed in the original paper. Isolated from mineral water.

The mol% G + C of the DNA is: 57–58.

Type strain: CIP 105274.

GenBank accession number (16S rRNA): AF068259.

26. **Pseudomonas libanensis** Dabboussi, Hamze, Elomari, Verhille, Baïda, Izard and Leclerc 1999a, 1099[VP]

li.ba.nen' sis. L. n. *Libanus* Lebanon, in southern Syria; L. adj. *libanensis* from or of Lebanon.

The description is taken from the original paper.

Rods, motile by means of a single polar flagellum. No poly-β-hydroxybutyrate is accumulated. Colonies on nutrient agar are smooth, circular, and nonpigmented. No hemolysis is observed on blood agar. A fluorescent pigment is produced on King B medium, and no phenazine pigments are produced on King A medium. Lipase, elastase, and tetrathionate reductase are negative. Arginine dihy-

drolase, lecithinase, catalase, levan formation from sucrose, and oxidase reaction are all positive. The strains do not denitrify. The majority of the strains (6/7) liquefy gelatin. A number of enzymatic reactions and the results of a nutritional screening on a number of compounds (mostly carbohydrates and derivatives) are reported in the original paper by Dabboussi et al. (1999a). The cultures tolerate 3% but not 7% of NaCl in the medium. Growth occurs between 4 and 36°C, with optimal growth at 30°C. The strains were isolated from Lebanese spring waters.

The authors of the original description (Dabboussi et al., 1999a) discuss the relationship of this species to other fluorescent organisms. *P. libanensis* seem to be closer to *P. fluorescens* biovar A, as suggested by the results of DNA–DNA hybridization experiments. This similarity extends to the main characteristics of biovar, namely, a negative denitrification test and the capacity of formation of levan from sucrose.

The mol% G + C of the DNA is: 58.

Type strain: CFML 96-195; CIP 105460.

GenBank accession number (16S rRNA): AF057645.

27. **Pseudomonas lundensis** Molin, Ternström and Ursing 1986, 339[VP]

lund.en' sis. M.L. adj. *lundensis* referring to the city of Lund, Sweden.

The following description is taken from the original paper.

Rods, 0.5–1.0 × 1.0–3.0 μm, motile by means of a single polar flagellum. Colonies on nutrient agar are 1–5 mm in diameter after 3 d at 25°C, circular, and smooth. No organic growth factors are required. Fluorescent pigment is produced in King B medium. Denitrification is negative. PHB is not accumulated in the cells. Arginine dihydrolase, catalase, and oxidase are positive. Gelatin liquefaction is variable. Only one strain has been observed to produce lipase against Tween 80.

All strains produce acid but not gas from L-arabinose, D-galactose, D-glucose, D-mannose, D-ribose, cellobiose, maltose, melibiose, and D-xylose. In assimilation tests, the last four sugars do not serve as carbon and energy sources. The following carbon compounds are assimilated (number of negative strains from a collection of 60 strains of the species, are given in parentheses): acetate, L-alanine, β-alanine, 4-aminobutyrate, 2-aminoethanol (1), 5-aminopantanoate, L-arabinose (1), L-arginine, L-asparagine, L-aspartate, betaine (2), citrate (1), fructose, fumarate, D-gluconate (4), D-glucose, L-glutamate, L-glutamine, glycerate (1), glycerol (1), heptanoate, L-histidine, *meso*-inositol (6), L-lactate (2), L-leucine (1), L-malate (2), L-ornithine, 2-oxoglutarate, pelargonate, L-phenylalanine, L-proline, propionate, putrescine, pyruvate, succinate, DL-tyrosine and DL-valine (1). Less than 90% of the strains can use D-arabinose, benzoate, carnosine, creatine, deoxycholate, D-galactose, DL-hydroxybutyrate, L-isoleucine, D-malate, D-ribose, sarcosine, and taurocholate. Slow utilization of butyrate, caproate, L-lysine, D-mannose, and trehalose. The main characteristics of the species are summarized in Tables BXII.γ.112 and BXII.γ.113. They refer to the type strain.

Growth occurs between 0 and 33°C, and one-third of the strains are able to grow at 37°C. Optimal growth is at 25°C. No growth at pH below 4.5. Isolated from meat. The strains correspond to cluster 2 of a previous numerical taxonomic study of psychrotrophic pseudomonads that had been isolated from meat, soil, and water (Molin and Ternström, 1982, 1986). Interestingly, results of numerical taxonomic studies performed independently at about the same time on the fluorescent pseudomonad collection at Berkeley, which included the biotype G of *P. fluorescens*, identified a tight cluster with the same properties as *P. lundensis* (Barrett et al., 1986).

The mol% G + C of the DNA is: 58–60.

Type strain: ATCC 49968 , CCM 3503, DSM 6252.

28. **Pseudomonas luteola** Kodama, Kimura and Komagata 1985, 473^{VP} (*Chryseomonas luteola* (Kodama, Kimura and Komagata 1985) Holmes, Steigerwalt, Weaver and Brenner 1987, 246.)

lu.te′o.la. L. dim. adj. *luteola* yellowish.

The following description is taken from the original paper.

Rods, 0.8 × 2.5 µm, with rounded ends, occurring singly, rarely in pairs. Motile by means of polar multitrichous flagella. A small number of PHB granules accumulate in the cells. Colonies on 0.5% glucose nutrient agar are smooth or wrinkled, entire or erose, flat or convex, light yellow or pale yellow, 3 mm in diameter after 2 d of growth at 30°C. A pellicle is formed on the surface of 0.5% glucose nutrient broth. A water-insoluble yellow pigment is produced. No water-soluble fluorescent pigment is produced. Catalase positive but oxidase negative. Nitrate is reduced to nitrite, but no denitrification occurs. Production of indole, hydrogen sulfide, and hydrolysis of starch, gelatin, Tween 80, esculin, and ONPG are all negative. No organic growth factors are required. Growth does not occur in a medium with 6.5% NaCl. Growth occurs at 42°C. Cleavage of protocatechuate is of the *ortho* type.

Acid is produced from L-arabinose, D-xylose, D-ribose, D-glucose, D-fructose, D-mannose, D-galactose, L-rhamnose, maltose, trehalose, mannitol, inositol, and salicin, but not from sucrose, lactose, cellobiose, adonitol, sorbitol, or inulin. L-arabinose, D-xylose, D-ribose, D-glucose, D-fructose, D-mannose, D-galactose, maltose, trehalose, mannitol, glycerol, acetate, pyruvate, malonate, DL-β-hydroxybutyrate, fumarate, 2-ketogluconate, gluconate, succinate, *p*-hydroxybenzoate, and glutamate are utilized, but sucrose, lactose, raffinose, inulin, starch, phenol, *o*-hydroxybenzoate, and *m*-hydroxybenzoate are not. Some properties are summarized in Table BXII.γ.116. The fatty acid and ubiquinone composition is typical of that of the genus *Pseudomonas*. Isolated from clinical specimens.

The mol% G + C of the DNA is: 55.4 (T_m).

Type strain: KS0921, DSM 6975 , IAM 13000, JCM 3352.

GenBank accession number (16S rRNA): D84002.

Additional Remarks: This species, originally assigned to the genus *Pseudomonas* (Kodama et al., 1985), was subsequently considered a member of a newly proposed genus, *Chryseomonas* (Holmes et al., 1987). Later studies (Anzai et al., 1997) demonstrated that *Pseudomonas* should be its correct generic allocation.

29. **Pseudomonas mandelii** Verhille, Baïda, Dabboussi, Izard and Leclerc 1999b, 56^{VP}

man.de′li.i. M.L. gen. n. *mandelii* of Mandel, dedicated to M. Mandel, American bacteriologist.

Colonies on nutrient agar are smooth, circular, and non-pigmented. Fluorescent pigment produced on King B agar medium. Catalase, oxidase, denitrification, and levan formation are positive. Gelatin and starch are not hydrolyzed. Sodium chloride inhibits growth at 5% concentration. Growth occurs at 4°C but not at 41°C. Optimal growth at 30°C. Growth at the expense of many organic compounds, as well as the result of many enzymatic reactions, are listed in the original paper (Verhille et al., 1999b).

DNA–DNA hybridization experiments indicate a relative high DNA similarity with that of biovar IV (*"P. lemonnieri"*) of *P. fluorescens*. Isolated from mineral water.

The mol% G + C of the DNA is: 57.

Type strain: CFML 95-303, CIP 105273.

GenBank accession number (16S rRNA): AF058286.

30. **Pseudomonas marginalis** (Brown 1918) Stevens 1925, 30^{AL} (*Bacterium marginale* Brown 1918, 386; *Phytomonas marginalis* (Brown 1918) Bergey, Harrison, Breed, Hammer and Huntoon 1923, 182; *Phytomonas intybi* Swingle 1925, 730; *Chlorobacter marginale* (Brown 1918) Patel and Kulkarni 1951b, 80.)

mar.gi.na′ lis. L. *margo, marginis* edge, margin; M.L. adj. *marginalis* marginal.

Description taken from Haynes and Burkholder (1957), who cite the original sources.

Rods, motile with one to three polar flagella. Green fluorescent pigment produced in culture. Agar colonies cream to yellowish. Gelatin is liquefied. There is a discrepancy among different authors on nitrate reduction. Feeble hydrolysis of starch. Acid but no gas from glucose, galactose, fructose, mannose, arabinose, xylose, rhamnose, mannitol, and glycerol. Alkali from (indicating utilization of) acetate, citrate, malate, formate, lactate, succinate, and tartrate. Sucrose, maltose, raffinose, and salicin are not degraded. Optimum growth temperature between 25 and 26°C. Minimum, 0°C; maximum, 38°C.

Isolated from marginal lesion on lettuce from Kansas; pathogenic on lettuce and related plants. Strains of this species have been included in biovar II of *P. fluorescens* (Stanier et al., 1966), and it appears under such designation in Tables BXII.γ.112, BXII.γ.113, and BXII.γ.119. Later, the taxon was subdivided by Young et al. (1978) into three pathovars. *P. marginalis* pathovar *marginalis*, which includes the type strain of the species, ATCC 10844; *P. marginalis* pathovar *alfalfa* Shinde and Lukezic 1974b, associated with discolored alfalfa (reference strain PDDCC 5708, NCPPB 2644), and *P. marginalis* pathovar *pastinacae* (Burkholder 1960a), a pathogen of cultivated parsnip (*Pastinaca sativa*) (reference strain ATCC 13889, PDDCC 5709, NCPPB 806). Fatty acid analysis data were taken as the basis for placing this species in a subgroup with other fluorescent phytopathogenic and saprophytic *Pseudomonas* (Stead, 1992). This subgroup also contains the nonfluorescent *P. meliae*.

The mol% G + C of the DNA is: 60 (Bd).

Type strain: ATCC 10844, LMG 2210, NCPPB 667, PDDCC 3553.

GenBank accession number (16S rRNA): Z76663.

31. **Pseudomonas meliae** Ogimi 1981, 382^{VP} (Effective publication: Ogimi 1977, 547.)

me′ li.ae. Gr. n. *Melia* Chinaberry tree, *Melia azedarach*; M.L. gen. n. *meliae* of Melia.

The following description is taken from the original paper.

Rods, 0.4–0.5 × 1.4–2.0 µm (average 0.5 × 1.8 µm). Motile by means of one to two polar flagella. Noncapsulated. Do not accumulate PHB as carbon reserve material. Colonies on nutrient agar very small after 48 h, and 0.3–1.0 mm in 72 h. Circular, smooth, convex, translucent, and white when viewed with reflected light. On potato-sucrose agar, the colonies have a rough surface and irregular shape. The rough colonies revert to the smooth type after transfer to nutrient agar. The following properties are negative: nitrate reduction, gelatin liquefaction, H_2S and indole formation, production of pyocyanine, fluorescein, and other pigments, levan formation on nutrient-sucrose agar, arginine and lysine dihydrase reactions, starch and esculin hydrolysis, production of gluconate, tyrosinase and urease reactions, egg-yolk and lipase (margarine hydrolysis), malonate utilization, phenylalanine deaminase, methyl red, and Voges–Proskauer reactions. The catalase and oxidase reactions are positive. Nicotinate is required as an organic growth factor.

The following carbon compounds are used for growth: ribose, glucose, mannose, galactose, fructose, sucrose, glycerol, citrate, succinate, and malate. No growth is observed at the expense of xylose, rhamnose, arabinose, lactose, maltose, cellobiose, melibiose, trehalose, dextrin, glycogen, starch, inulin, mannitol, sorbitol, inositol, adonitol, dulcitol, salicin, or tartrate. Growth occurs between 4 and 37°C (optimum, 27°C) and between pH 5 and 8 (optimum, pH 7–8). rRNA–DNA hybridization studies have confirmed the allocation of this species in the genus *Pseudomonas* (De Vos et al., 1985). Based on fatty acid analysis, *P. meliae* was placed in the same subgroup constituted by phytopathogenic and saprophytic fluorescent *Pseudomonas* species (Stead, 1992). Isolated from Chinaberry tree (*Melia azedarach* L.) (family Meliaceae), where this pathogen produces the bacterial galls. The galls are formed in the leaf petioles and in branches and trunk.

The mol% G + C of the DNA is: 57.9 (T_m).

Type strain: ATCC 33050, DSM 6759, LMG 2220, NCPPB 3033.

Additional Remarks: See the Additional Remarks at the end of the description of *P. amygdali* in this list.

32. **Pseudomonas mendocina** Palleroni *in* Palleroni, Doudoroff, Stanier, Solanes and Mandel 1970, 220[AL]

men.do.ci′na. Sp. fem. n. *mendocina* native of Mendoza (Argentina).

Characteristics of the species useful for differentiation from other species are given in Tables BXII.γ.116 and BXII.γ.117. Properties useful for distinguishing the species from some denitrifying pseudomonads are summarized in the chapter on the genus *Burkholderia*. Colonies are yellowish because of production of carotenoid pigment; not adherent or wrinkled in appearance. Optimum temperature, approximately 35°C. For further descriptive information, see Palleroni et al. (1970). Found in soil and water; isolated by enrichment in media with nitrate under anaerobic conditions, especially at 40°C. Ethanol and L(+)-tartrate can be used as carbon sources in the enrichments. It has also been isolated from clinical specimens (Hugh and Gilardi, 1980).

The mol% G + C of the DNA is: 62.8–64.3 (Bd).

Type strain: ATCC 25411, DSM 50017, LMG 1223.

GenBank accession number (16S rRNA): M59154, Z76664.

33. **Pseudomonas migulae** Verhille, Baïda, Dabboussi, Hamze, Izard and Leclerc 1999a, 1570[VP]

mi′gu.lae. M.L. gen. n. *migulae* of W. Migula, who created the generic name *Pseudomonas*.

The following description is taken from the original paper.

Rods, motile by means of a single polar flagellum. Colonies on nutrient agar are smooth and circular. Fluorescent pigment is produced on King B agar. No phenazine pigments are synthesized. Starch and gelatin are not hydrolyzed. Levan is produced from sucrose and the arginine dihydrolase system is present in the cells. Low concentrations of NaCl (up to 0.8%) are tolerated, and 5% concentration is inhibitory. Growth occurs at 4°C but not at 41°C. The original paper reports the results of a number of enzymatic tests, as well as a list of compounds that can serve as sources of carbon and energy for growth, and a selection of these characters is used for differentiation from other fluorescent species.

DNA–DNA hybridization experiments suggest that this species may be considered as a genomovar of some of the biovars of *P. fluorescens*. Isolated from natural mineral water.

The mol% G + C of the DNA is: unknown.

Type strain: CFML 95-321, CIP 105470.

GenBank accession number (16S rRNA): AF074383.

34. **Pseudomonas monteilii** Elomari, Coroler, Verhille, Izard and Leclerc 1997, 849[VP]

mon.tei′li.i. M.L. masc. gen. n. *monteilii* of Monteil, in honor of Henri Monteil, a French microbiologist.

The following description is taken from the original paper.

Motile rods. Colonies on nutrient agar are circular and nonpigmented. Fluorescent pigment is produced. No hemolysis on blood agar. The strains are lipase, elastase, lecithinase, and tetrathionate reductase negative. Arginine dihydrolase, catalase, and cytochrome oxidase are produced. Denitrification, formation of levan from sucrose, gelatinase, starch hydrolysis, hydrolysis of esculin and starch, deamination of phenylalanine, exonuclease activity, lysine and ornithine decarboxylase, and the tributyrin and fibrinolysis tests are all negative. Temperature range for growth is 10–36°C; optimal growth at 30°C. Growth occurs in the presence of 3% NaCl but not with 5% NaCl. The original paper by Elomari et al. (1997) includes an extensive nutritional and enzymatic characterization of the species. Part of the available information has been summarized in Tables BXII.γ.112 and BXII.γ.113. The strains have been isolated from clinical specimens.

The mol% G + C of the DNA is: 60 (T_m).

Type strain: CFML 90-60, CIP 104883.

GenBank accession number (16S rRNA): AF064458.

35. **Pseudomonas mucidolens** Levine and Anderson 1932, 344[AL]

mu.ci′do.lens. L. adj. *mucidus* musty; L. v. *oleo* to smell of; L. part. adj. *mucidolens* musty smelling.

The description below is taken from the original paper. Rods, rounded ends, occurring singly and in pairs, actively motile. Filamentous cells are frequently observed. Fluorescent pigment is produced. Gelatin is liquefied. Nitrates reduced to nitrites and gas. Optimum growth temperature 23–25°C; slight growth at 37°C and at 10°C. Acid produced from glucose, rhamnose, arabinose, erythritol, sorbitol, trehalose, galactose, fructose, and mannose. Moderate amount of acid from glycerol, mannitol, dulcitol, glycogen, inulin, maltose, melezitose, pectin, raffinose, salicin, starch, sucrose, and xylan. Lactate, citrate, and urate can be used for growth. Acetate, oxalate, sulfanilate, tartrate, salicylate, and formate do not support growth. A physiologically less active form of this species is described in the same paper under the name " *P. mucidolens* biovar tarda".

P. mucidolens is one of the species causing mustiness in eggs, which were the source for its isolation. The general characteristics as described above roughly conform to those of one of the denitrifying biovars of *P. fluorescens*, but extensive comparative studies of the phenotypic properties have not been done. However, DNA–rRNA hybridization data (De Vos et al., 1989) and rRNA sequence analysis (Anzai et al., 1997) confirm the allocation to the genus *Pseudomonas*.

The mol% G + C of the DNA is: 61.0 (T_m).

Type strain: ATCC 4685, LMG 2223.

36. **Pseudomonas nitroreducens** Iizuka and Komagata 1964a, 214[AL]

ni.tro.re.du' cens. L. n. *nitrum* nitate; L. v. *reduco* to draw backwards, bring back to a state or condition; M.L. part. adj. *nitroreducens* nitrate reducing.

The following description is taken from the original paper.

Rods, 0.4–0.6 × 1.4–1.8 µm, occurring singly, rarely in pairs; motile with polar flagella. Fluorescent pigment produced. Gelatin is not liquefied. Nitrates are reduced to nitrites and to gas. Oxidase reaction positive. H_2S not produced. Starch is not hydrolyzed. Acid is produced from glucose. Glucose, gluconate, 2-ketogluconate, citrate, succinate, ethanol, *p*-hydroxybenzoate, and protocatechuate can be utilized as sole carbon sources. 5-Ketogluconate, benzoate, salicylate, gentisate, anthranilate, and *p*-aminobenzoate are not utilized. The fresh isolate could utilize kerosene, but the activity was lost after subcultivation in the laboratory. Optimum growth temperature, 25–30°C. No growth at 37°C. The source of isolation was oil brine in Japan.

The species has two characteristics never found in association in any of the fluorescent species described in RNA group I, namely, the capacity for denitrification and the inability to utilize gelatin. Its assignment to the genus *Pseudomonas* has been confirmed by molecular studies (Anzai et al., 1997).

The mol% G + C of the DNA is: unknown.

Type strain: IAM 1439.

GenBank accession number (16S rRNA): D84021.

37. **Pseudomonas oleovorans** Lee and Chandler 1941, 377[AL]

o.le.o' vo.rans. L. n. *oleum* oil; L. v. *voro* to destroy, consume; M.L. part. adj. *oleovorans* oil consuming.

Description taken from the original paper.

When grown on agar, the cells are almost coccoid (0.5 × 0.8 µm), but the length increases to about 1.5 µm during the exponential phase in broth. The colonies have a typical fluorescence that is not imparted to the medium. Nitrate is reduced to nitrite. Gelatin not liquefied. Starch is hydrolyzed. Isolated from oil-water emulsions used as lubricants and cooling agents in the cutting and grinding of metals. Apparently the organism lives on some normal constituent of the cutting compound, probably the naphthenic acids, which act as emulsifying agents. Consequently, the name *oleovorans* may not be appropriate. Allocation to the genus *Pseudomonas* was confirmed by DNA–rRNA hybridization studies (De Vos et al., 1989) and by 16S rRNA sequence analysis (Moore et al., 1996; Anzai et al., 1997).

The mol% G + C of the DNA is: 63.5 (T_m).

Type strain: ATCC 8062, DSM 1045.

GenBank accession number (16S rRNA): Z76665.

38. **Pseudomonas orientalis** Dabboussi, Hamze, Elomari, Verhille, Baida, Izard and Leclerc 1999b, 303[VP]

or.i.en.tal' is. M.L. masc. adj. *orientalis* pertaining to the Orient.

Rods, motile by means of a single polar flagellum. Colonies on nutrient agar are circular, smooth, and nonpigmented. Fluorescent pigment is produced on King B agar. Catalase, oxidase, gelatinase, esterase-lipase, urease, and arginine dihydrolase reactions are all positive. Levan is produced from sucrose. Reduction of nitrate to nitrite is positive, but there is no denitrification. All strains liquefied gelatin. Various enzymatic reactions are described in the original paper, but not the ability to liquefy gelatin. Assimilation of various organic compounds, particularly carbohydrates, is described.

DNA–DNA hybridization experiments give low reannealing values with other fluorescent species and biovars. Growth occurs at 4°C. Isolated from Lebanese spring water.

The mol% G + C of the DNA is: 60.

Type strain: CFML 96-170, CIP 105540.

GenBank accession number (16S rRNA): AF064457.

39. **Pseudomonas oryzihabitans** Kodama, Kimura and Komagata 1985, 472[VP] (*Flavimonas oryzihabitans* (Kodama, Kimura and Komagata 1985) Holmes, Steigerwalt, Weaver and Brenner 1987, 245.)

o.ry.zi' ha.bi.tans. L. fem. n. *oryza* rice; L. fem. adj. *habitans* inhabiting; M.L. fem. adj. *oryzihabitans* rice.

The description is taken from the original paper.

Rods, 0.8 × 2.0 µm, with rounded ends, occurring singly, rarely in pairs. Motile by means of a monotrichous flagellum. Granules of PHB do not accumulate. Colonies on 0.5% glucose nutrient agar are smooth or wrinkled, entire or erose, flat to convex, light yellow, 2 mm in diameter after 2 d incubation at 30°C. A pellicle is formed on the surface of 0.5% glucose nutrient broth. A water-insoluble yellow pigment is produced. No water-soluble fluorescent pigment is produced. Catalase is produced; the oxidase reaction is negative. Nitrate reduced to nitrite, but no denitrification occurs. Urease is produced. Production of indole, hydrogen sulfide, and hydrolysis of starch, Tween 80, esculin, and ONPG are all negative.

Growth occurs in medium containing 6.5% NaCl. No organic growth factors are required. Cleavage of protocatechuate is of the *ortho* type. Acid is produced from L-arabinose, D-xylose, D-glucose, D-fructose, D-mannose, D-galactose, maltose, trehalose, mannitol, sorbitol, and inositol, but not from L-rhamnose, sucrose, lactose, cellobiose, adonitol, salicin or inulin. L-arabinose, D-xylose, D-ribose, D-glucose, D-mannose, D-galactose, maltose, trehalose, mannitol, glycerol, acetate, pyruvate, malonate, DL-β-n-hydroxybutyrate, fumarate, gluconate, 2-ketogluconate, succinate, *p*-hydroxybenzoate, and glutamate are utilized, but not sucrose, lactose, raffinose, inulin, starch, phenol, *o*-hydroxybenzoate, or *m*-hydroxybenzoate. Hydrogen is not utilized. Some properties are given in Table BXII.γ.116. The cellular fatty acid and ubiquinone composition is the typical type of the genus *Pseudomonas*. Isolated from a Japanese rice paddy.

The mol% G + C of the DNA is: 65.1 (T_m).

Type strain: KS0036, ATCC 43272, DSM 6835, IAM 1568, JCM 2592.

GenBank accession number (16S rRNA): D84004.

Additional Remarks: This species was originally described as a member of the genus *Pseudomonas* (Kodama et al., 1985), but was later assigned to the proposed new genus name *Flavimonas* (Holmes et al., 1987). Further studies (Anzai et al., 1997) demonstrated that the latter should be considered a synonym of *Pseudomonas*.

40. **Pseudomonas pertucinogena** Kawai and Yabuuchi 1975, 318[AL]

per.tu.ci.no′ge.na. pertucin (coined word), a bacteriocin active against *Bordetella pertussis*; L. v. *gigno* to produce; M.L. fem. adj. *pertucinogena* intended to mean pertucin producing.

The description is from the original paper, referring to two strains of the species.

Rods, 0.4 × 1.1 μm, motile by means of single polar flagellum. Do not accumulate PHB. Oxidase reaction positive. No pigments are produced. Hydrolysis of gelatin, starch, or Tween 80, and arginine dihydrolase reaction are all negative. Acid production from sugars is in general weak or negative (glucose, D-arabinose, and galactose are positive). Pyruvate, succinate, oxaloacetate, β-hydroxybutyrate, and L-alanine are utilized for growth. A number of amino acids, carbohydrates, and alcohols are not utilized. Further details are given in the original paper. The two known strains produce pertucin, a bacteriocin active against *Bordetella pertussis*, and were kept for many years in the American Type Culture Collection as members of this species.

The mol% G + C of the DNA is: ~60.

Type strain: ATCC 190.

41. **Pseudomonas pseudoalcaligenes** Stanier *in* Stanier, Palleroni and Doudoroff 1966, 247[AL]

pseu.do.al.ca.li′ge.nes. Gr. adj. *pseudes* false; M.L. adj. *alcaligenes* alkali-producing; M.L. adj. *pseudoalcaligenes* false alkali-producing.

Characteristics useful for differentiation from other *Pseudomonas* species are given in Tables BXII.γ.116 and BXII.γ.117, and the chapter on the genus *Burkholderia*, which summarizes information on denitrifying pseudomonads. Optimum temperature, 35°C. Further descriptive information can be found in Ralston-Barrett et al. (1976) and Stanier et al. (1966). The nutritional spectrum is quite narrow (see *P. alcaligenes*). The gelatinase reaction is negative. The species is rather heterogeneous. The collection examined by Stanier et al. (1966) included at least two groups. Strains of one of the groups were capable of PHB accumulation and gave a positive arginine dihydrolase re-

action, while the strains of the second group were negative for the two properties and also differed in some nutritional characteristics (Doudoroff and Palleroni, 1974).

Three subspecies were thought to integrate this species, namely, *P. pseudoalcaligenes* subsp. *pseudoalcaligenes* (corresponding to the original description by Stanier et al., 1966), *P. pseudoalcaligenes* subsp. *citrulli* Schaad et al. 1978 (a name given to strains isolated from water soaked lesions on cotyledons of watermelon, *Citrullus lanatus*, and also pathogenic for other plants of the same family) (Schaad et al., 1978), and *P. pseudoalcaligenes* subsp. *konjaci* (the causal agent of bacterial blight of konjac, *Amorphophallus konjaci* Koch) (Goto, 1983b). Nutritional characteristics that were found useful for rapid identification were the utilization of L-arabinose and galactose by the subspecies *citrullus* (subsp. *pseudoalcaligenes* and subsp. *konjaci*, and *P. alcaligenes* are negative), and the utilization of mannitol by subsp. *konjaci* but not by subsp. *pseudoalcaligenes* and subsp. *citrulli*, and *P. alcaligenes* (Goto, 1983b).

The fatty acid composition of *P. pseudoalcaligenes* and its subspecies placed this taxon in one of the subgroups of *Pseudomonas* together with other nonfluorescent species (Stead, 1992). In spite of the above considerations, the results of a numerical analysis supplemented by DNA–DNA reassociation experiments (Hu et al., 1991) were taken as the basis for the conclusion that subsp. *citrulli* and subsp. *konjaci* should be transferred to a different species, *P. avenae*, which had been allocated to rRNA similarity group III (Palleroni, 1984). Consequently, this species should include three subspecies: *P. avenae* subsp. *avenae* Manns 1909 (type strain ICMP 3183), *P. avenae* subsp. *citrulli* (Schaad et al., 1978) comb. nov. (type strain ICMP 7500), and *P. avenae* subsp. *konjaci* (Goto, 1983a) comb. nov. (type strain ICMP 7733) (Hu et al., 1991). Studies by Willems et al. (1992) support the conclusion that these various phytopathogenic pseudomonad taxa should be transferred to the genus *Acidovorax* and the new combinations *Acidovorax avenae* subsp. *citrulli* and *A. avenae* subsp. *cattleyae* as well as the new name *Acidovorax konjaci* have been proposed. *P. pseudoalcaligenes* has been isolated from various natural materials and from clinical specimens.

The mol% G + C of the DNA is: 62.2–63.2 (Bd).

Type strain: ATCC 17440, LMG 1225, NCIB 9946.

GenBank accession number (16S rRNA): Z76666.

42. **Pseudomonas putida** (Trevisan 1889b) Migula 1895a, 29[AL] (*Bacillus putidus* Trevisan 1889b, 18.)

pu′ti.da. L. fem. adj. *putida* stinking, fetid.

Characteristics of the species (103 strains of biovar A and 9 strains of biovar B) are given in Tables BXII.γ.112 and BXII.γ.113. Other properties of interest are summarized in Tables BXII.γ.108 and BXII.γ.114. The characteristics that are useful for a separation from *P. aeruginosa* and *P. fluorescens* include the inability to liquefy gelatin, to produce any phenazine pigments, to denitrify or to give an egg-yolk reaction, and to grow at 41°C. In this constellation of negative properties, the incapacity to hydrolyze gelatin is the one that has classically defined *P. putida*. Optimum temperature, 25–30°C.

Isolated from soil and water after enrichment in mineral media with various carbon sources. The majority of the strains have been assigned to biovar A (biotype A of Stanier et al., 1966), which is considered typical. Biovar B differs from biovar A only in a few phenotypic properties: all known strains of this biovar utilize L-tryptophan, kynurenine, and anthranilate, and most use D-galactose as carbon sources. None of the strains of biovar B uses nicotinate. Recent studies on many strains of this species, combining a numerical analysis of phenotypic data with rRNA gene restriction patterns (ribotyping), have indicated the convenience of a revision of its present internal subdivision into biovars (Elomari et al., 1994), as previously suggested by Barrett et al. (1986) and by Stenström et al. (1990) (see also the sections on taxonomic comments and on ribotyping, in particular the contribution by Brosch et al., 1996).

The mol% G + C of the DNA is: 62.5 (biovar A; thermal denaturation); 60.7 (biovar B).

Type strain: ATCC 12633, DSM 291, NCIB 9494.

GenBank accession number (16S rRNA): D37923.

43. **Pseudomonas resinovorans** Delaporte, Raynaud and Daste 1965, 1075[AL]

re.si.no′vor.ans. L. n. *resina* resin; L. v. *voro* to devour, digest; M.L. part. adj. *resinovorans* resin digesting.

The description is taken from the original paper.

Rods, 0.6–0.7 × 2.0–2.5 μm. Motile by means of a polar flagellum. Fluorescent pigment is produced. Gelatin is not liquefied. Nitrate reduction is weak, and denitrification is negative. Oxidase reaction positive. Optimum growth temperature, 28–30°C; no growth at 5°C or 42°C. No acid is produced from arabinose, xylose, rhamnose, glucose, fructose, galactose, mannose, lactose, maltose, sucrose, raffinose, inulin, salicin, dextrin, glycerol, mannitol, inositol, or dulcitol. Starch hydrolysis very weak. Growth occurs at the expense of colophony, Canada balsam, or abietic acid. Phenol, phenanthrene, salicylic acid, *m*-cresol, and naphthalene can also be used as carbon and energy sources for growth. Further information may be found in the original paper by Delaporte et al. (1965). Assignment to the genus *Pseudomonas* is confirmed by results on nucleic acid hybridization (De Vos et al., 1989) and by 16S rDNA sequence analysis (Moore et al., 1996).

The mol% G + C of the DNA is: 63.7 (T_m).

Type strain: ATCC 14235, CCUG 4439, LMG 2274.

GenBank accession number (16S rRNA): Z76668.

44. **Pseudomonas rhodesiae** Coroler, Elomari, Hoster, Izard and Leclerc 1996, 603[VP]

rho.de′ si.ae. M.L. fem. gen. n. *rhodesiae* of Rhodes, in honor of M.R. Rhodes, an English microbiologist.

Rods, motile by means of single polar flagella. Colonies on nutrient agar are smooth, circular, and nonpigmented. No hemolysis observed on blood agar. Fluorescent pigment is produced in King B medium. Arginine dihydrolase, catalase, cytochrome oxidase, lecithinase, nitrate reduction to nitrite, formation of levan from sucrose, growth on cetrimide agar, lecithinase reaction, and decomposition of L-tyrosine are all positive. PHB is not accumulated. Phenazine pigments are not produced. Gelatin liquefaction, lipase reaction, esculin and starch hydrolysis, deamination of phenylalanine, and exonuclease production are all negative. A large number of enzymatic properties are described in the original paper. A summary of properties of this species is presented in Tables BXII.γ.112 and BXII.γ.113. A number of additional characteristics can be found in the original paper and elsewhere (Elomari et al., 1995). Isolated from natural mineral water.

The mol% G + C of the DNA is: 59 ± 1 (T_m).

Type strain: CIP 104664.

GenBank accession number (16S rRNA): AF064459.

45. **Pseudomonas savastanoi** (Janse 1982) Gardan, Bollet, Abu Ghorrah, Grimont and Grimont 1992, 611[VP] (*Pseudomonas syringae* subsp. *savastanoi* Janse 1982, 168.)

sa.vas.ta′ no.i. L. gen. n.*savastanoi* of Savastano, the first worker who studied olive knot disease.

The description is taken from the above-mentioned paper.

Rods, 0.4–0.8 × 1.0–3.0 μm. Motile by means of one to four polar flagella. Rather slow growing. Colonies are white or cream, smooth, flat, and glistening with entire or erose margins. Strains produce an hypersensitive reaction on tobacco leaves. Other properties of interest are summarized in Table BXII.γ.109. Metabolism is respiratory. Oxidase negative. Nitrates are not reduced. Blue fluorescent pigment is produced on King B medium and can be observed under UV light. Arginine dihydrolase negative. Esculin, gelatin, and starch are not hydrolyzed. Strains assimilate sucrose, L-arabinose, gluconate, caprylate, fumarate, DL-glycerate, L-malate, pyruvate, citrate, D-alanine, and L-proline. They do not assimilate lactose, L-xylose, adonitol, 2-aminobutyrate, DL-lactate, D-β-hydroxybutyrate, D(−)-tartrate, L-cysteine, L-methionine, and L-valine.

Strains have been isolated for many years from members of the family Oleaceae (olive, oleander). These do not produce levan from sucrose. Strains isolated from another species of the same family (*Fraxinus excelsior* L.), *Phaseolus vulgaris*, and *Glycine max* (Leguminosae) do produce levan. The species is now divided into three pathovars. *P. savastanoi* pathovar *savastanoi* causes knots, galls, and cankers on plants of various genera of the family Oleaceae; *P. savastanoi* pathovar *glycinea* causes bacterial blight of soybean; and *P. savastanoi* pathovar *phaseolicola* causes halo blight of bean.

The mol% G + C of the DNA is: 60 (T_m).

Type strain: ATCC 13522, CFBP 1670, ICMP 4352, NCPPB 639.

Additional Remarks: The paper by Gardan et al. (1992) essentially is a proposal to revive the name *P. savastanoi*, which had been created by Stevens in 1913 in a valid publication. This name was omitted from the 1980 approved list of bacterial species, an omission that had its origin in the fact that the name had been included in a list of pathovars of the species *P. syringae*.

The restoration of *P. syringae* pathovar *savastanoi* to its original status as independent species (*P. savastanoi*) after two demotions (one to the subspecies category, and the other to pathovar), has been formulated based on a phenotypic analysis and DNA–DNA hybridization studies (Gardan et al., 1992). In fact the phenotypic properties that differentiate this taxon from other related fluorescent, oxidase negative plant pathogens, are very few. In addition, the conditions of the nucleic acid hybridization used by Gardan et al. (1992) apparently were those used in the original description of the adopted methodology (Crosa et al., 1973b), even to the point of the incubation at 60°C, which is low for organisms of mol% G + C content higher than *E. coli*. As a consequence, the estimated similarities reported in the paper tend to be high, although the figures can be grouped into blocks that reflect differences between *P. savastanoi* and other *P. syringae* pathovars, with the exception of the type strains of pathovars *phaseolicola* and *glycinea*. Consequently, the proposal subdivides *P. savastanoi* into pathovar *savastanoi*, pathovar *phaseolicola*, and pathovar *glycinea*. These last two names had also been used until now to designate pathovars of *P. syringae*

Recent studies on DNA similarity by hybridization methods indicate that *P. savastanoi*, *P. ficuserectae*, *P. meliae*, and *P. amygdali* should be considered as synonyms, in which case the latter is the correct name of the species (Gardan et al., 1999).

46. **Pseudomonas straminae** Iizuka and Komagata 1963b, 139[AL] emend. Uchino, Kosako, Uchimura and Komagata 2000, 1518.

stra.mi′ na.e. L. adj. *straminae* made of straw.

Slender rods, 0.3 × 3.0 μm. Motile with single polar flagellum. Colonies are yellow. Fluorescent pigment is produced. Oxidase reaction positive. Gelatin is liquefied. Nitrates are not reduced to nitrites. Denitrification is negative. H_2S is produced. Starch is not hydrolyzed. Acid produced from glucose. Glucose, gluconate, citrate, succinate, and ethanol are assimilated. Phenol, benzoate, salicylate, *m*-hydroxybenzoate, *p*-hydroxybenzoate, protocatechuate, gentisate, anthranilate, *p*-aminobenzoate, 2-ketogluconate, and 5-ketogluconate are not assimilated. Optimum growth temperature, 25–30°C. No growth at 37°C. Assignment to the genus *Pseudomonas* has been confirmed by 16S rDNA sequence analysis (Anzai et al., 1997). The organism has been isolated from Japanese rice paddies.

The mol% G + C of the DNA is: unknown.

Type strain: IAM 1598.

GenBank accession number (16S rRNA): D84023.

47. **Pseudomonas stutzeri** (Lehmann and Neumann 1896) Sijderius 1946, 115[AL] (*Bacterium stutzeri* Lehmann and Neumann 1896, 237.)

stut′ze.ri. *stutzeri* Stutzer patronymic, of Stutzer.

Characteristics differentiating the species from other species of *Pseudomonas* are given in Tables BXII.γ.116 and BXII.γ.117. Differentiation from a number of denitrifying pseudomonads can be found in Table BXII.γ.114 and in the chapter on the genus *Burkholderia* appearing in Volume 2 Part C in this *Manual*. Recently isolated colonies are adherent, dark brown, and have a characteristic wrinkled appearance, which usually is lost after repeated transfers in laboratory media. The colonies may eventually become smooth, butyrous, and pale in color. The gelatinase reaction is negative. Among the characteristics useful for its identification are the vigorous denitrification, the appearance of the colonies, and the use of starch as a carbon and energy source. Variations in some of the typical characteristics are observed in some populations, depending of the enrichment conditions. Some strains grow at 43°C. Optimum temperature, ~35°C.

The species is markedly heterogeneous in nutritional properties and in DNA base composition. Further descriptive information: van Niel and Allen (1952), Sijderius (1946), and Palleroni et al. (1970). Found in soil and water,

from which it can be isolated after enrichment in media with nitrate under anaerobic conditions at 30°C using various carbon sources. L(+)-Tartrate gives excellent results in the enrichments (van Niel and Allen, 1952) although, paradoxically, strains obtained in this manner may not grow with tartrate in pure culture. Many strains have been isolated from clinical specimens. Recently, the results of studies on the genomic structure and organization, as well as on the genotypic and phenotypic diversity of *P. stutzeri*, have been published (Rosselló et al., 1991, Rosselló–Mora et al.,1994b; Rainey et al., 1994b; Ginard et al., 1997). Seven DNA–DNA similarity groups (genomovars) have been defined in this species (Rosselló-Mora et al., 1994b). This subdivision correlated with clusters obtained by comparison of 16S rDNA gene sequences. One of the genomovars was raised to the category of species, for which the name *P. balearica* was created (see below). The name *P. stanieri* (Mandel 1966) has been proposed for strains with a mol% G + C of ~62. However, this species is not clearly differentiated from *P. stutzeri* based on phenotypic characteristics (Palleroni et al., 1970).

The mol% G + C of the DNA is: from 60.6–66.3 (Bd).

Type strain: AB 201, ATCC 17588, CCUG 11256, DSM 5190.

GenBank accession number (16S rRNA): U26262.

48. **Pseudomonas synxantha** (Ehrenberg 1840) Holland 1920, 220[AL] (*Vibrio synxanthus* Ehrenberg 1840, 202.)

syn.xan′ tha. Gr. pref. *syn* along with, together; Gr. adj. *xanthus* yellow; M.L. adj. *synxanthus* with yellow.

The description of this species is given by Haynes and Burkholder (1957) in the seventh edition of the *Manual*.

Gelatinase positive. Growth characteristics in various solid and liquid media are described by Haynes and Burkholder (1957). A distinctive characteristic is the production of an intense, diffusible, yellow to orange pigment in cream or in the cream layer in milk.

rRNA–DNA hybridization studies indicate that this is a species of the genus *Pseudomonas* (De Vos et al., 1989), which was later confirmed by 16S rDNA sequencing analysis (Anzai et al., 1997). Isolated from bitter milk.

The mol% G + C of the DNA is: 61. (T_m).

Type strain: ATCC 9890, LMG 2190.

49. **Pseudomonas syringae** Van Hall 1902141[AL] (*Phytomonas syringae* (van Hall 1902) Bergey, Harrison, Breed, Hammer and Huntoon 1930, 257; *Pseudomonas barkeri* (Berridge 1924) Clara 1934, 11; *Pseudomonas citrarefaciens* (Lee 1917) Stapp 1928, 190; *Pseudomonas citriputealis* (Smith 1913) Stapp 1928, 190; "*Pseudomonas hibisci*" (Nakada and Takimoto 1923) Stapp 1928; *Pseudomonas prunicola* Wormald 1930, 742; *Pseudomonas punctulans* (Bryan 1933) Săvulescu 1947, 12; *Pseudomonas rimaefaciens* Koning 1938, 11; *Pseudomonas spongiosa* (Leifson 1962b); *Pseudomonas tonelliana* (Ferraris 1926) Burkholder 1948b, 132; *Pseudomonas trifoliorum* (Jones, Williamson, Wolf and McCulloch 1923) Stapp 1928; *Pseudomonas utiformica* Clara 1932, 111; *Pseudomonas vignae* Gardner and Kendrick 1923, 275; *Pseudomonas viridifaciens* Tisdale and Williamson 1923, 150.)

sy.rin′ gae. M.L. fem. n. *Syringa* generic name of lilac; M.L. fem. gen. n. *syringae* of the lilac.

Characteristics of the species and of some of its pathovars are described in Tables BXII.γ.118 and BXII.γ.120. The

TABLE BXII.γ.120. Characteristics of some *Pseudomonas syringae* pathovars and of *Pseudomonas savastanoi* and its two biovars, biovar glycinea and biovar phaseolicola[a]

Characteristics	*P. savastanoi*	*P. savastanoi* pathovar *glycinea*	*P. savastanoi* pathovar *phaseolicola*	*P. syringae* pathovar *antirrhini*	*P. syringae* pathovar *aptata*	*P. syringae* pathovar *cannabina*	*P. syringae* pathovar *coronafaciens*	*P. syringae* pathovar *delphinii*	*P. syringae* pathovar *eriobotryae*	*P. syringae* pathovar *lachrymans*	*P. syringae* pathovar *mori*	*P. syringae* pathovar *morsprunorum*	*P. syringae* pathovar *passiflorae*	*P. syringae* pathovar *persicae*	*P. syringae* pathovar *pisi*	*P. syringae* pathovar *sesami*	*P. syringae* pathovar *striafaciens*	*P. syringae* pathovar *syringae*	*P. syringae* pathovar *tabaci*	*P. syringae* pathovar *tomato*
Levan formation	−	+	+	+	+	+	+	d	+	+	+	+	−	+	+	+	+	+	+	+
Pectate gel pitting[b]	4	4	4		−	4	4,8	4	4	4,8		4	4	4	−	4	−	−	4,8	4
β-Glucosidase	−	−	−	+	+	−	+	+	+	+	−	−	+	−	d	−	+	+	+	+
Growth on:																				
Mannitol	+	d	−	+	+		+	+	+	+	+	+	+	+	+	−	+	+	+	+
Betaine	+	−	d	+	+	+	+	+	+	−	−	+	+	−	+	+	+	+	+	+
Inositol	d	+	−	+	+	−	+	+	−	+	+	+	+	−	+	+	+	+	+	+
Sorbitol	d	−	−	+	+	−	+	+	+	+	d	+	+	+	+	−	+	+	+	+
Trigonelline	d	+	+	+	+	+	−	+	+	+	+	−	+	+	+	+	+	+	+	+
Quinate	−	+	+	+	+	−	+	+	+	+	+	−	+	+	+	+	+	+	+	+
Erythritol	−	−	−	−	+	−	+	+	+	+	−	d	+	−	d	+	+	+	d	−
L-Tartrate	d	−	−	−	−	−	−	+	+	−	+	−	+	−	+	−	−	+	+	−
D-Tartrate	−	−	−	+	+	−	−	−	−	−	−	−	−	−	−	−	−	d	−	+
L-Lactate	−	−	−	−	+	−	−	−	−	−	−	−	−	d	−	−	−	+	−	−
Anthranilate	d	−	−	−	−	−	−	−	−	−	−	+	−	−	−	−	−	−	−	−
Homoserine	−	−	−	+	−	−	−	−	−	−	−	−	−	−	+	−	−	−	−	−

[a]For symbols see standard definitions. Data from Sands et al. (1970) and from Gardan et al. (1992).

[b]Method of Hildebrand (1971); numbers represent pH values at which pitting occurs.

pattern of acid production from sugars will be found in Haynes and Burkholder (1957) and in the original papers describing the nomenspecies here included as pathovars.

Cytochrome *c* is not detectable. Fluorescent pigment is produced. The gelatinase reaction is positive. Rare strains may require organic growth factors. Growth of most strains is slow in mineral media with a single carbon source and is relatively slow in complex media. Nutritional spectrum is less extensive and more heterogeneous than that of the saprophytic fluorescent pseudomonads. Optimum temperature, 25–30°C. The original strain was isolated from lilac (*Syringa vulgaris*, family Oleaceae) but strains conforming to the original description are pathogenic for many unrelated plants.

The mol% G + C of the DNA is: 60.5.

Type strain: ATCC 19310, DSM 6693, LMG 1247, NCPPB 281, PDDCC 3023.

Additional Remarks: The following is a list of pathovars of *P. syringae* and of the respective pathotypes proposed by Dye et al. (1980). The names of the most important host plants and of their botanical families are also included. The list includes pathovar *mori*, although this name has priority over *syringae*. As recommended by Young et al. (1978), *P. syringae* is widely known and it should be conserved over "*P. mori*". Many of the pathovars in the list may not be distinguishable from *P. syringae* except for their host range. Few phenotypic characters are taxonomically useful.

Studies on the internal subdivision of *P. syringae* into pathovars using various molecular techniques, in particular DNA–DNA hybridization, have suggested that some of the pathovars should be elevated to independent species status.

At present, however, attempts in this direction have materialized only in a few instances. The difficulty is in finding sets of phenotypic properties that could help in the ready identification of such species without the need of nucleic acid similarity experiments. A recent study centered mainly on this approach and in ribotyping (Gardan et al., 1999) extended early investigations that defined clusters of genomically related strains within *P. syringae* (Pecknold and Grogan, 1973).

The studies of Gardan and collaborators (1999) defined nine genomospecies. The conclusions included the proposal for synonymy of the names *P. savastanoi*, *P. ficuserectae*, *P. meliae*, and *P. amygdali*, with *P. amygdali* as the earliest name in this group. In addition, genomospecies 3 and 7 may be named in the future "*P. tomato*" and "*P. tagetis*", respectively, once phenotypic data for differentiation become available. Finally, genomospecies 5 and 9, for which such data are available, are given the species names *P. tremae* and *P. cannabina*, respectively.

The genomospecies 5 (*P. tremae*) is represented by a single strain (Gardan et al., 1999). According to the description given by Gardan and collaborators, *P. tremae* is a nonfluorescent organism, but evaluation of its relationships with other species or pathovars or with nonfluorescent species is not clear at this stage. The description includes a nutritional analysis where the strain appears to be unable to use L-alanine, fructose, glycerol, or gluconate, and it is stated that it did not assimilate the 91 other carbon sources of the Biotype-100 strips (BioMérieux). Since there is only one strain available at this moment, it is not possible to ascertain the intraspecies diversity of the species. Experiments on the DNA relatedness of *P. tremae* and *P. cannabina* to other taxa have been limited mainly to *P. syringae* pathovars but have not included most other species of the genus. Descriptions of *P. tremae* and *P. cannabina*, as given in the paper by Gardan et al. (1999), are transcribed below in the section Other Species.

As a general comment, the taxonomic situation of the members of the conglomerate known under the name *P. syringae* still remains largely unsettled, but it is reasonable to expect that future studies will contribute to reach a satisfactory internal subdivision of the species and a more precise circumscription of pathovars segregated as individual species on the basis of genomic and phenotypic properties.

a. **Pseudomonas syringae** *pathovar* **syringae** van Hall 1902.
    Hosts: lilac (*Syringa vulgaris*, family Oleaceae) and several unrelated plants.
    Deposited strain: ATCC 19310, NCPPB 281, PDDCC 3023 (neopathotype strain).

b. **Pseudomonas syringae** *pathovar* **aceris** (Ark 1939) Young et al. 1978.
    Hosts: *Acer macrophyllum, Acer* spp. (family Aceraceae).
    The mol% G + C of the DNA is: 61.1.
    Deposited strain: ATCC 10853, NCPPB 958, PDDCC 2802 (Sneath and Skerman, 1966) (neopathotype strain).

c. **Pseudomonas syringae** *pathovar* **antirrhini** (Takimoto 1920) Young et al. 1978.
    Host: *Antirrhinum majus* (family Scrophulariaceae).
    The mol% G + C of the DNA is: 60.4.
    Deposited strain: NCPPB 1817, PDDCC 4303 (neopathotype strain).

d. **Pseudomonas syringae** *pathovar* **apii** (Jagger 1921) Young et al. 1978.
    Host: celery (*Apium graveolens*, family Umbelliferae).
    The mol% G + C of the DNA is: 60.8.
    Deposited strain: ATCC 9654, NCPPB 1626, PDDCC 2814 (neopathotype strain).

e. **Pseudomonas syringae** *pathovar* **aptata** (Brown and Jamieson 1913) Young et al. 1978.
    Hosts: sugar beet (*Beta vulgaris*, family Chenopodiaceae), *Nasturtium* sp. (family Cruciferae); lettuce (*Lactuca sativa*, family Compositae).
    Deposited strain: NCPPB 871, PDDCC 459 (neopathotype strain).

f. **Pseudomonas syringae** *pathovar* **atrofaciens** (McCulloch 1920) Young et al. 1978.
    Hosts: wheat (*Triticum* sp.) and other plants of the family Gramineae.
    Deposited strain: NCPPB 2612, PDDCC 4394 (neopathotype strain).

g. **Pseudomonas syringae** *pathovar* **atropurpurea** (Reddy and Godkin 1923) Young et al. 1978.
    Hosts: *Bromus inermis* and many other *Bromus* spp., *Agropyron repens* and many other plants of the family Gramineae.
    Deposited strain: NCPPB 2397, PDDCC 4457 (neopathotype strain).

h. **Pseudomonas syringae** *pathovar* **berberidis** (Thornberry and Anderson 1931a) Young et al. 1978.
    Hosts: barberry (*Berberis thunbergii, B. inermis*, family Berberidaceae).
    The mol% G + C of the DNA is: 59.7–60.1.
    Deposited strain: NCPPB 2724, PDDCC 4116 (neopathotype strain).

i. **Pseudomonas syringae** *pathovar* **cannabina** (Šutić and Dowson 1959) Young et al. 1978.
    Hosts: *Cannabis sativa* (family Moraceae); *Phaseolus vulgaris, Vicia sativa* (family Leguminosae).
    The mol% G + C of the DNA is: 59.3–60.3.
    Deposited strain: PDDCC 2823 (NCPPB 1437) (neopathotype strain).

j. **Pseudomonas syringae** *pathovar* **ciccaronei** (Ercolani and Caldarola 1972) Young et al. 1978.
    Host: carob (*Ceratonia siliqua*, family Leguminosae).
    Deposited strain: NCPPB 2355, PDDCC 5710 (pathotype strain).

k. **Pseudomonas syringae** *pathovar* **coronafaciens** (Elliott 1920) Young et al. 1978.
    Hosts: *Avena sativa, Bromus inermis, Agropyron repens* and various other wild and cultivated plants of the family Gramineae. Description as "*P. coronafaciens*" is transcribed by Haynes and Burkholder (1957) in the seventh edition of *Bergey's Manual of Determinative Bacteriology*.
    The mol% G + C of the DNA is: 59.3.
    Deposited strain: NCPPB 600, PDDCC 3113 (neopathotype strain).

l. **Pseudomonas syringae** *pathovar* **delphinii** (Smith 1904) Young et al. 1978.
    Host: *Delphinium* sp. (family Ranunculaceae).
    Deposited strain: NCPPB 1879, PDDCC 529 (neopathotype strain).

m. **Pseudomonas syringae** *pathovar* **dysoxyli** (Hutchinson 1949) Young et al. 1978.
    Host: *Dysoxylum spectabile* (family Meliaceae).
    Deposited strain: : ATCC 19863, NCPPB 225, PDDCC 545 (neopathotype strain).

n. **Pseudomonas syringae** *pathovar* **eriobotryae** (Takimoto 1931) Young et al. 1978.
    Host: loquat (*Eriobotrya japonica*, family Rosaceae).
    The mol% G + C of the DNA is: 58.3.
    Deposited strain: NCPPB 2331, PDDCC 4455 (neopathotype strain).

o. **Pseudomonas syringae** *pathovar* **garcae** (Amaral, Teixeira and Pinheiro 1956) Young et al. 1978.
    Host: coffee (*Coffea arabica*, family Rubiaceae) and many other unrelated plants.
    Deposited strain: ATCC 19864, NCPPB 588, PDDCC 4323 (pathotype strain).

p. **Pseudomonas syringae** *pathovar* **helianthi** (Kawamura 1934) Young et al. 1978.
    Host: sunflower (*Helianthus debilis*, family Compositae).
    Deposited strain: NCPPB 2640, PDDCC 4531 (neopathotype strain).

q. **Pseudomonas syringae** *pathovar* **japonica** (Mukoo 1955) Dye et al. 1980 (*Pseudomonas striafaciens* pathovar *japonica* Mukoo 1955.)

Hosts: rye (*Secale cereale*), barley (*Hordeum sativum*), wheat (*Triticum vulgare*), rice (*Oryza sativa*) and many other Gramineae of the genera *Setaria, Panicum, Bromus, Lolium, Andropogon, Alopecurus,* etc., as well as plants of other botanical families (Solanaceae, Chenopodiaceae, Oxalidaceae, etc.).

Deposited strain: NCPPB 3093, PDDCC 6305 (neopathotype strain).

r. **Pseudomonas syringae** *pathovar* **lachrymans** (Smith and Bryan 1915) Young et al. 1978.

Host: cucumber (*Cucumis sativus*, family Cucurbitaceae).

Deposited strain: ATCC 7386, NCPPB 537, PDDCC 3988 (neopathotype strain).

s. **Pseudomonas syringae** *pathovar* **lapsa** (Ark 1940) Young et al. 1978.

Hosts: corn (*Zea mays*), sugarcane (*Saccharum officinarum*) (family Gramineae).

The mol% G + C of the DNA is: 59.4.

Deposited strain: NCPPB 2096, PDDCC 3947 (neopathotype strain).

t. **Pseudomonas syringae** *pathovar* **maculicola** (McCulloch 1911) Young et al. 1978.

Hosts: cabbage, cauliflower (*Brassica oleracea*, family Cruciferae).

The mol% G + C of the DNA is: 61.3.

Deposited strain: NCPPB 2039, PDDCC 3935 (neopathotype strain).

u. **Pseudomonas syringae** *pathovar* **mellea** (Johnson 1923) Young et al. 1978.

Host: tobacco (*Nicotiana tabacum*, family Solanaceae).

Deposited strain: NCPPB 2356, PDDCC 5711 (neopathotype strain).

v. **Pseudomonas syringae** *pathovar* **mori** (Boyer and Lambert 1893) Young et al. 1978.

Host: mulberry (*Morus* spp., family Moraceae).

Deposited strain: ATCC 19873, NCPPB 1034, PDDCC 4331 (neopathotype strain).

w. **Pseudomonas syringae** *pathovar* **morsprunorum** (Wormald 1931) Young et al. 1978.

Host: *Prunus* spp. (family Rosaceae).

Deposited strain: ATCC 19322, NCPPB 2995, PDDCC 5795 (Sneath and Skerman, 1966) (neopathotype strain).

x. **Pseudomonas syringae** *pathovar* **panici** (Elliott 1923) Young et al. 1978.

Host: proso or broom-corn millet (*Panicum miliaceum*, family Gramineae).

The mol% G + C of the DNA is: 60.7.

Deposited strain: ATCC 19875, NCPPB 1498, PDDCC 3955 (neopathotype strain).

y. **Pseudomonas syringae** *pathovar* **papulans** (Rose 1917) Dhanvantari 1977.

Host: apple (*Pyrus malus*, family Rosaceae).

Deposited strain: NCPPB 2848, PDDCC 4048 (neopathotype strain).

z. **Pseudomonas syringae** *pathovar* **passiflorae** (Reid 1938) Young et al. 1978.

Host: *Passiflora edulis* (family Passifloraceae).

The mol% G + C of the DNA is: 59.4–60.3.

Deposited strain: NCPPB 1387, PDDCC 129 (neopathotype strain).

aa. **Pseudomonas syringae** *pathovar* **persicae** (Prunier et al. 1970) Young et al. 1978.

Host: peach (*Prunus persicae*, family Rosaceae).

Deposited strain: NCPPB 2761, PDDCC 5846 (neopathotype strain).

ab. **Pseudomonas syringae** *pathovar* **pisi** (Sackett 1916) Young et al. 1978.

Host: pea (*Pisum sativum*, family Leguminosae).

Deposited strain: NCPPB 2585, PDDCC 2452 (neopathotype strain).

ac. **Pseudomonas syringae** *pathovar* **primulae** (Ark and Gardner 1936) Young et al. 1978.

Host: *Primula polyantha* (family Primulaceae).

The mol% G + C of the DNA is: 60.4.

Deposited strain: ATCC 19306, NCPPB 133, PDDCC 3956 (neopathotype strain).

ad. **Pseudomonas syringae** *pathovar* **ribicola** (Bohn and Maloit 1946) Young et al. 1978.

Host: golden currant (*Ribes aureum*, family Saxifragaceae).

The mol% G + C of the DNA is: 60.6.

Deposited strain: ATCC 13456, NCPPB 963, PDDCC 3882 (neopathotype strain).

ae. **Pseudomonas syringae** *pathovar* **sesami** (Malkoff 1906) Young et al. 1978.

Host: sesame (*Sesamum indicum*, family Pedaliaceae).

The mol% G + C of the DNA is: 59.4.

Deposited strain: ATCC 19879, NCPPB 1016, PDDCC 763 (neopathotype strain).

af. **Pseudomonas syringae** *pathovar* **striafaciens** (Elliott 1927) Young et al. 1978.

Hosts: oats (*Avena sativa*), barley (*Hordeum sativum*) (family Gramineae).

The mol% G + C of the DNA is: 59.1.

Deposited strain: ATCC 10730, NCPPB 1898, PDDCC 3961 (pathotype strain); NCPPB 2394, PDDCC 4483 (pathogenic reference strain).

ag. **Pseudomonas syringae** *pathovar* **tabaci** (Wolf and Foster 1917) Young et al. 1978.

Host: tobacco (*Nicotiana tabacum*, family Solanaceae).

Deposited strain: NCPPB 1427, PDDCC 2835 (neopathotype strain).

ah. **Pseudomonas syringae** *pathovar* **theae** (Hori 1915) Young et al. 1978.

Host: tea plant (*Thea sinensis*, family Theaceae).

Deposited strain: NCPPB 2598, PDDCC 3923 (neopathotype strain).

ai. **Pseudomonas syringae** *pathovar* **tomato** (Okabe 1933) Young et al. 1978.

Host: tomato (*Lycopersicum esculentum*, family Solanaceae).

Deposited strain: NCPPB 1106, PDDCC 2844 (neopathotype strain).

aj. **Pseudomonas syringae** *pathovar* **ulmi** (Šutić and Tešić 1958) Young et al. 1978.

Host: elm (*Ulmus* sp., family Ulmaceae).
The mol% G + C of the DNA is: 58.7.
Deposited strain: ATCC 19883, NCPPB 632, PDDCC 3962 (neopathotype strain).

ak. **Pseudomonas syringae** *pathovar* **viburni** (Thornberry and Anderson 1931b) Young et al. 1978.

Host: elm (*Ulmus* sp., family Ulmaceae).
The mol% G + C of the DNA is: 60.1.
Deposited strain: ATCC 13458, NCPPB 1921, PDDCC 3963 (neopathotype strain).

50. **Pseudomonas taetrolens** Haynes 1957, 108[AL]

taet' ro.lens. L. adj. *taeter* offensive; L. part. adj. *olens* having an odor; M.L. part. adj. *taetrolens* foul smelling.

The description of this species is given by Haynes and Burkholder (1957) in the seventh edition of the *Manual.*

The cells are short rods with rounded ends, motile by means of one to five polar flagella. Gelatin and starch are not hydrolyzed. Nitrate is not reduced to nitrite. Acid production from a number of substrates is described by Haynes and Burkholder (1957). A decision to allocate the species in the genus *Pseudomonas* has been made based on rRNA–DNA hybridization studies (De Vos et al., 1989). Isolated from foods that have a musty odor.

The mol% G + C of the DNA is: 59.8 (T_m).
Type strain: ATCC 4683, IAM 1653, LMG 2336.
GenBank accession number (16S rRNA): D84027.

51. **Pseudomonas tolaasii** Paine 1919, 210[AL]

to.laa' si.i. Tolaas patronymic; M.L. gen. n. *tolaasii* of Tolaas.

Description is given by Haynes and Burkholder (1957) in the seventh edition of the *Manual.*

Rapid identification tests have been described (Wong and Preece, 1979). Results of a detailed phenotypic study of a collection of strains of this species have been published (Fahy, 1981). Some of the reported characteristics are summarized in Table BXII.γ.115. A gene cluster encoding proteins required for the synthesis of the toxin tolaasin has been identified and characterized (Rainey et al., 1993). Detailed composition of fatty acids is known (Stead, 1992). rRNA–DNA hybridization studies indicate allocation of this species to the genus *Pseudomonas* (De Vos et al., 1985). Fluorescent pigment is produced. Gelatin is hydrolyzed (Haynes and Burkholder, 1957). Pathogenic for cultivated mushrooms. Isolated from brown-spot of cultivated mushrooms.

The mol% G + C of the DNA is: 60.8–61.3 (T_m).
Type strain: ATCC 33618, LMG 2342, NCPPB 1873.
GenBank accession number (16S rRNA): D84028, Z76670.

52. **Pseudomonas veronii** Elomari, Coroler, Hoste, Gillis, Izard and Leclerc 1996, 1142.[VP]

ve.ro'ni.i. M.L. masc. gen. n. *veronii* of Véron, in honor of Professor M.M. Véron, a distinguished French microbiologist.

The following description is taken from the original paper.

Rods, motile by means of single polar flagella. PHB is not accumulated by the cells. The cultures produce a fluorescent pigment on King B medium. The colonies on nutrient agar are smooth, circular, and nonpigmented. They are nonhemolytic on blood agar. The oxidase, catalase, and arginine dihydrolase reactions are positive. Most strains liquefy gelatin. All the strains grow on α-aminobutyrate, D-xylose, L-arabinose, D-mannose, D-galactose, sucrose, butyrate, isobutyrate, erythritol, sorbitol, inositol, D-alanine, L-tryptophan, and trigonelline as the sole source of carbon and energy. No strain is able to use isovalerate, sebacate, azelate, L-mandelate, benzoate, L-kynurenine, histamine, or acetamide. The original paper (Elomari et al., 1996) includes extensive information on a variety of enzymatic activities. Additional properties are summarized in Tables BXII.γ.112 and BXII.γ.113 and in a previous communication by the same group of workers (Elomari et al., 1995).

All strains have been isolated from natural mineral waters.

The mol% G + C of the DNA is: 61–62.
Type strain: CFML 92-134, CIP 104663, DSM 11331.
GenBank accession number (16S rRNA): AF064460.

53. **Pseudomonas viridiflava** (Burkholder 1930) Dowson 1939, 177[AL] (*Phytomonas viridiflava* Burkholder 1930, 63.)

vi.ri.di.fla' va. L. *viridis* green; L. *flavus* yellow; M.L. adj. *viridiflavus* greenish yellow.

Characteristics that are useful for comparison with related taxa are given in Table BXII.γ.118. Further information is given by Haynes and Burkholder (1957), Clara (1934), and Billing (1970b). The following characteristics are mentioned in this last paper. The bacterial mass usually has a yellow tinge in media with 5% sucrose and olive to golden brown in media with yeast extract and glycerol. A blue-green insoluble pigment is produced by some strains. Aside from the characteristics described in the tables, the potato rot and esculin reactions are positive. Fluorescent pigment is produced, and gelatin is liquefied (Haynes and Burkholder, 1957).

From her extensive nutritional screening of strains of the species, Billing (1970b) has concluded that few substrates have diagnostic value. The only ones distinguishing *P. viridiflava* from most other oxidase-negative plant pathogens are the inability to use sucrose and the capacity for use of D(−)-tartrate. This substrate is used by *P. viridiflava*, *P. syringae* pathovar *tomato*, and rarely by other species and pathovars. In addition to these characteristics, Billing (1970b) mentions that the reaction in beans was similar to that produced by *P. syringae* and both were different from the water-soaked lesions produced by "*P. phaseolicola*".

Pathogenic on bean (*Phaseolus vulgaris*).
The mol% G + C of the DNA is: 59.9 (NCPPB 1810).
Type strain: ATCC 13223, DSM 6694, LMG 2352, NCPPB 635, PDDCC 2848.
GenBank accession number (16S rRNA): Z76671.

Other Species

Some putative *Pseudomonas* species have an uncertain phylogenetic position. The following descriptions are those that are not included in the first edition of the *Manual* (Palleroni, 1984) or in previous editions of *Bergey's Manual of Determinative Bacteriology.*

1. **Pseudomonas abietaniphila** Mohn, Wilson, Bicho and Moore 1999b, 935[VP] (Effective publication: Mohn, Wilson, Bicho and Moore 1999a, 76.)

a.bie.ta.ni'phi.la. M.L neut. n. *abietanum* abietane; Gr. adj. *philos* loving, friendly to; M.L. fem. adj. *abietaniphila* abietane-loving.

The following description is taken from the paper by Mohn et al. (1999a).

Rods. When grown on dehydroabietic acid, the cells are $0.7 \times 1.2–2.5$ μm. Motile. No information is given on the flagellar number and insertion. Colonies are clear, translucent, smooth, circular, and convex. Catalase positive and oxidase negative. Abietic, dehydroabietic, linoleic and pyruvic acids, as well as L-arabinose, D-galactose, D-glucose, D-xylose, and glycerol are used for growth; 12- plus 14-chlorodehydroabietic and palmitic acids are used poorly. Pimaric, isopimaric, and acetic acids are not used. Main cellular fatty acids are $C_{16:1 \omega 7c}$, $C_{16:0}$, and $C_{18:0}$; minor ones include $C_{10:0 3OH}$, $C_{12:0 2OH}$, and $C_{12:0 3OH}$. Isolated from bleached Kraft pulp mill effluent treatment system near Kamlops, British Columbia, Canada.

The mol% G + C of the DNA is: unknown.

Type strain: Strain BKME-9, ATCC 700689.

2. **Pseudomonas antimicrobica** Attafuah and Bradbury 1990, 320[VP] (Effective publication: Attafuah and Bradbury 1989, 571.)

an.ti.mi.cro'bi.ca. Gr. pref. *anti* against; Gr. adj. *micrus* small; Gr. n. *bios* life; *antimicrobica* against microbes (referring to a wide antimicrobial activity).

Description taken from the original paper (Attafuah and Bradbury, 1989).

Rods, $0.5 \times 0.8–1.5$ μm, occurring singly, in groups, and occasionally in pairs. Motile by means of one or two polar flagella. Two types of colonies were observed on nutrient agar. Those of one variant (NCIB 9897) were circular, raised, creamy-white, and opaque, rugose and butyrous when raised, with microundulate margins; the other type (NCIB 9898) was moist, smooth, glistening, and slightly viscid, with entire margins and occasionally an orange tinge. A yellowish-orange, water-soluble, nonfluorescent pigment was produced in King A and B media, tyrosine medium and others. No pigment was produced on potato dextrose agar. Gelatinase activity, levan formation, accumulation of PHB, starch hydrolysis, pectolytic activity, and rotting of plant tissue were all negative. Growth occurs with NaCl up to 4%, but it decreases to a trace with 6%.

Acid was produced from arabinose, dulcitol, fructose, galactose, glucose, glycerol, lactose, mannitol, *meso*-inositol, sorbitol, and xylose, but not from maltose, raffinose, salicin, or sucrose. Carbon sources for growth were *N*-acetylglucosamine, D-alanine, L-arabinose, caprate, citrate, gluconate, glucose, glutamate, *p*-hydroxybenzoate, *meso*-inositol, α-ketoglutarate, malonate, mannitol, D-mannose, phenylacetate, L-proline, propionate, and raffinose. No growth on adipate, D-iso-ascorbate, malate, *meso*-tartrate, starch, or ethanol. No organic growth factors were required. A wide spectrum of antibiotic activity was present. Good growth occurred between 15 and 37°C; optimum around 30°C. No growth at 41°C. Additional information may be found in the original paper (Attafuah and Bradbury, 1989) and some comments on the relationship of this species to those of other groups of aerobic pseudomonads. Isolated from the mealybug *Planococcoides njalensis*.

The mol% G + C of the DNA is: unknown.

Type strain: DSM 8361, NCIB 9898.

3. **Pseudomonas cannabina** (ex Šutić and Dowson 1959) Gardan, Shafik, Belouin, Broch, Grimont and Grimont 1999, 477[VP]

can.na'bi.na. L. fem. adj. *cannabina* pertaining to *Cannabis*, the generic name of the host plant, *Cannabis sativa* L.

Rods, $1.1–3.0 \times 3.0–4.0$ μm, motile by means of one to four polar flagella. Colonies are gray color and are slightly convex. Production of fluorescent pigment in King B medium. The results of LOPAT (levan production, oxidase test, potato rotting, arginine dihydrolase, and tomato hypersensitivity) are +, −, −, −, and +. Nitrate is not reduced. Negative for hydrolysis of starch, esculin, and gelatin. The strains assimilate D-glucose, glycerol, D-saccharate, mucate, citrate, D-gluconate, L-histidine, L-aspartate, L-glutamate, L-proline, L-alanine, and L-serine, but not the other 83 carbon sources of the Biotype-100 strips (BioMérieux). Pathogenic on *Cannabis sativa* L.

The mol% G + C of the DNA is: 60.2 (T_m).

Type strain: CFBP 2341; ICMP 2823; NCPPB 1437.

4. **Pseudomonas flectens** Johnson 1956, 144[AL.]

flec'tens. L. v. *flectere* to bend; *flectens* that bends.

The following description is summarized from that in the original paper.

Rods, $0.5–0.75 \times 1.4–2.0$ μm, with rounded ends. Motile by means of one to two polar flagella. Colonies on meat infusion agar, after 5 d at 27.5°C, circular, 1–1.5 mm in diameter, convex, amorphous, smooth, glistening, with an entire edge, grayish white. No fluorescent pigment is formed. Gelatin liquefaction does not occur. Acid is produced from glucose, mannose, and sucrose, but not from maltose, lactose, starch, glycerol, mannitol, sorbitol, or esculin. Starch hydrolysis is very slight. H_2S is not produced and nitrate is not reduced. The organism is the agent of the pod twist disease of French beans (*Phaseolus vulgaris* L.), where it produces large, diffuse water-soaked lesions on twisted young pods.

The mol% G + C of the DNA is: 31.8 (type strain).

Type strain: ATCC 12775.

Additional Remarks: Strain NCPPB 539, presumed to be the type strain, is not phylogenetically related to the genus *Pseudomonas*, as shown by rRNA–DNA hybridization experiments (De Vos et al., 1989). However, the authenticity of this strain may be questioned because of its unexpectedly low mol% G + C.

5. **Pseudomonas halophila** Fendrich 1989, 205[VP] (Effective publication: Fendrich 1988, 42.)

ha.lo'phi.la. Gr. n. *hals, halos* salt; Gr. v. *philein* to love; M.L. adj. *halophila* salt-loving.

The description is taken from the original paper (Fendrich, 1988).

Rod-shaped cells, $0.8–1.0 \times 1.5–5.0$ μm, single or in pairs. Motile by means of a polar flagellum. Colonies are circular with entire margins and smooth surfaces, slimy, raised, and reddish-brown colored. Chemoorganotrophic growth on arabinose, cellobiose, fructose, galactose, glucose, lactose, maltose, mannitol, sorbitol, sucrose, trehalose, and xylose with no acid production, and on acetate, caproate, citrate, lactate, pelargonate, propionate, pyruvate, succinate, alanine, glutamine, esculin, ethanol, and glycerol as carbon and energy sources. Catalase and cytochrome oxidase positive. Gelatinase and urease positive. Hydrolysis of starch or cellulose, production of indole or sulfide, arginine dihydrolase reaction, and lysine and ornithine decarboxylase activities are all negative. Strictly aerobic. Growth occurs between 0.02 and 3.3 ml/l NaCl, with optimum at 0.8 mol/l. The pH range is 4.5–9.6; optimum at 7. Temperature range, 4–37°C; optimum at 28°C. Susceptible to ampicillin, chloramphenicol, erythromycin, nalidixic acid, and penicillin G. Isolated from the north arm of Great Salt Lake, Utah, USA.

The mol% G + C of the DNA is: 57.

Type strain: DSM 3050.

6. **Pseudomonas multiresinivorans** Mohn, Wilson, Bicho and Moore 1999b, 935[VP] (Effective publication: Mohn, Wilson, Bicho and Moore 1999a, 77.)

mul.ti.re.si.ni.vo′ rans. L. adj. *multi* many; L. fem. n. *resina* resin; L. part. adj. *vorans* devouring; M.L. fem. adj. *multiresinivorans* devouring many resins.

Rods. When grown on isopimaric acid, cells are 1×1.5–2 μm, motile. No information is given about insertion and number of flagella. The cells form clumps. Colonies are clear, translucent, smooth, asymmetrical, and flat. Catalase and oxidase reactions are positive. Can live anaerobically in the presence of nitrate. Pimaric, isopimaric, palmitic, linoleic, benzoic, and acetic acids, as well as L-arabinose, D-glucose, citronellol, ethanol, glycerol, and n-hexadecane are used for growth. Poor utilization of abietic, dehydroabietic, and pyruvic acids, 12- plus 14-chlorodehydroabietic acids, D-galactose, or D-xylose. Main cellular fatty acids are $C_{18:1}$, $C_{16:0}$, and $C_{16:1 \omega 7c}$; minor ones include $C_{10:0 3OH}$, $C_{12:0 2OH}$, and $C_{12:0 3OH}$. Isolated from a laboratory sequencing batch reactor in Vancouver, British Columbia, Canada.

The mol% G + C of the DNA is: unknown.

Type strain: Strain IpA-1, ATCC 700690.

7. **Pseudomonas tremae** Gardan, Shafik, Belouin, Broch, Grimont and Grimont 1999, 477[VP]

tre′ ma.e. M.L. gen. fem. n. *tremae* of *Trema*; generic name of the host plant, *Trema orientalis* BL.

Rods, motile by means of one to four polar flagella. No fluorescent pigment is produced on King B medium. Of the LOPT tests (levan formation from sucrose, oxidase reaction, potato rotting ability, arginine dihydrolase reaction, and tobacco hypersensitive test), the first four are negative. Nitrate is not reduced. Negative for hydrolysis of starch, esculin, gelatin, and Tween 80. The strain assimilates D-saccharate, D- and L-malate, citrate, succinate, fumarate, L-aspartate, and L-glutamate, but not any of the 91 other carbon sources of the Biotype-100 strips (BioMérieux). Pathogenic for *Trema orientalis*.

The mol% G + C of the DNA is: 60.5 (T_m).

Type strain: CFBP3229; ICMP 9151; NCPPB 3465.

8. **Pseudomonas vancouverensis** Mohn, Wilson, Bicho and Moore 1999b, 935[VP] (Effective publication: Mohn, Wilson, Bicho and Moore 1999a, 76.)

van.cou.ver.en′ sis. M.L. adj. *vancouverensis* pertaining to the city of Vancouver, Canada.

Rods. When grown on dehydroabietic acid, the cells are 0.9×0.9–3.0 μm. Motile. No information is given about number and insertion of flagella. The cells form clumps. Colonies pale yellow, translucent, smooth, circular, and convex. Catalase and oxidase reactions are positive. Abietic, acetic, benzoic, dehydroabietic, linoleic, palmitic, and pyruvic acids, as well as β-citronellol, D-glucose, and glycerol, are used for growth. Acetic, isopimaric, and pimaric acids, and L-arabinose, D-galactose, and D-xylose are not used. Main cellular fatty acids are $C_{16:1 \omega 7c}$, $C_{16:0}$, $C_{17:0 cyc}$, and $C_{18:1}$; minor ones include $C_{10:0 3OH}$, $C_{12:0 2OH}$, and $C_{12:0 3OH}$. Isolated from forest soil in Vancouver, British Columbia, Canada.

The mol% G + C of the DNA is: unknown.

Type strain: Strain DhA-51, ATCC 700688.

Genus II. **Azomonas** *Winogradsky 1938, 391*[AL]

CHRISTINA KENNEDY AND PAUL RUDNICK

A.zo.mo′ nas. Gr. adj. *a* not; Gr. n. *zoê* life; Gr. n. *azoê* not sustaining life, nitrogen Gr. n. *monas* a unit, monad; M.L. fem. n. *Azomonas* nitrogen monad.

Cells are 2.0 μm or more in diameter, of variable length and shape, although usually 2.5–3.5 μm in length and generally ellipsoidal- to rod-shaped. They cannot easily be distinguished from cells of *Azotobacter*. Cells may occur singly, in pairs, or in clumps. Motile by peritrichous, lophotrichous, or polar flagella. Gram negative, but sometimes Gram variable, depending on culture age. **Cysts are not formed in aging cultures. All species fix atmospheric nitrogen under aerobic conditions.** Alternative nitrogenases containing vanadium (nitrogenase-2) or iron (nitrogenase-3) only may be synthesized in Mo-deficient media. Cultures can grow both aerobically and microaerobically. Chemoorganotrophic. Sugars, alcohols, and organic acids are used as carbon sources. Ammonium salts and sometimes nitrate (*A. insignis* only) are used as nitrogen sources; amino acids are not used. **Water-soluble and fluorescent pigments are produced by nearly all strains.** Species are catalase-positive. The optimum pH

for nitrogen fixation is close to neutrality, but certain strains can also fix nitrogen at a pH of 4.6–4.8. Species are isolated from water or soil.

The mol% G + C of the DNA is: 52–58.6.

Type species: **Azomonas agilis** (Beijerinck 1901) Winogradsky 1938, 400 (*Azotobacter agilis* Beijerinck 1901, 577.)

FURTHER DESCRIPTIVE INFORMATION

Please note that the large body of data collected by Thompson and Skerman (1979) forms the current basis for the description of the genus *Azomonas* and its species. Most of the information below is provided from a summary of the work of those authors. Exceptions and new data are noted.

The cell shape of *Azomonas* species is similar to that for *Azotobacter* species (Fig. BXII.γ.126) but may vary in aging cultures of *A. agilis*. In *A. insignis*, the cell shape is retained but the size

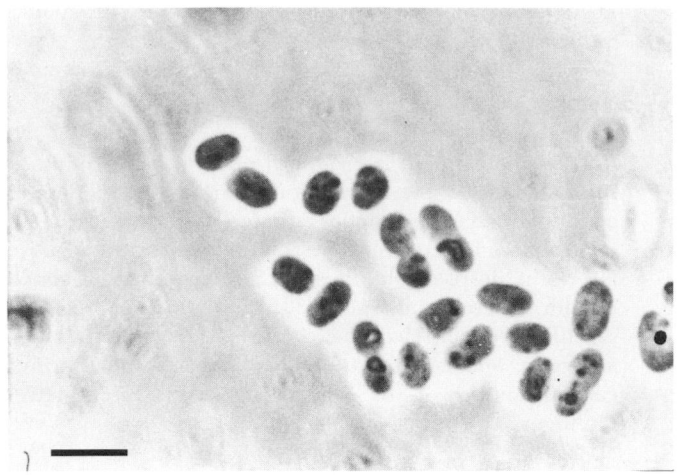

FIGURE BXII.γ.126. Cells of *Azomonas agilis* (phase-contrast microscopy). Bar = 10 μm.

can change. *A. macrocytogenes* produces very large cells and sometimes spindle-shaped or filamentous forms, especially when ethanol is used as the carbon source (Fig. BXII.γ.127). On peptone-yeast extract medium, *A. macrocytogenes* may show distortion of cells, with irregular, filamentous forms predominating. *A. insignis* does not show any distortion, but the cells appear thinner than when grown in glucose medium. *A. agilis* is less affected and does not show distortion.

No cysts are formed by any species; however, in *A. macrocytogenes* cells emerging from saclike capsular structures may occur and be mistakenly identified as germinating cysts (Jensen, 1955a; Tchan, 1968). Page and Collinson (1987) described five *A. macrocytogenes* isolates from Alberta soils, none of which formed cysts after prolonged growth on glucose or with butanol—an encystment inducer for *Azotobacter* species. Moreover, no 5-*n*-alkyl resorcinols—characteristic components of *Azotobacter vinelandii* cysts—could be identified in lipid extracts of the Alberta *A. mac-*

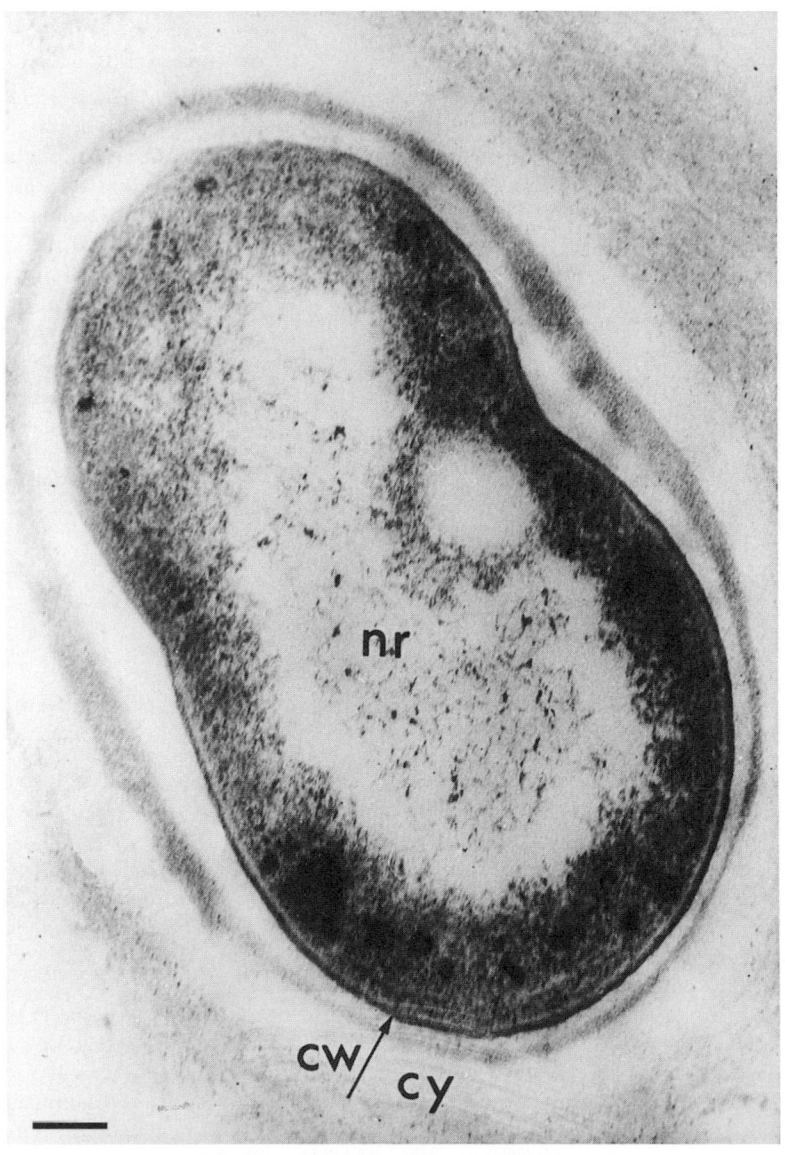

FIGURE BXII.γ.127. Electron micrograph of *Azomonas macrocytogenes*. The nuclear region (*nr*) and cell wall (*cw*) are readily visible. The structure *cy* probably consists of polysaccharide, but can be mistaken for the cyst coat of *Azotobacter*. Bar = 0.2 μm

rocytogenes strains. *A. agilis* and *A. insignis* stain uniformly Gram negative. *A. macrocytogenes* tends to be Gram variable.

All members of the genus are motile, with the flagellar arrangement varying with the species (Table BXII.γ.121).

All species are obligate aerobes, but nitrogen fixation can be enhanced by a decreased oxygen tension (0.04 atm) in some strains. On nitrogen-free agar medium with a sugar as the carbon source, *A. agilis* and *A. insignis* grow at a relatively rapid rate, forming colonies 2–6 mm in diameter after incubation for a few to several days at 30°C. The colonies are opaque, glistening, smooth and convex, and variants rarely appear. Colonies of *A. insignis* are usually smaller than those of *A. agilis*. *A. macrocytogenes* forms colonies ranging from 1–6 mm in diameter after 1 week of incubation; initially, they may be coarsely wrinkled but later they become smooth. On sucrose agar *A. macrocytogenes* produces voluminous and moist colonies due to the formation of colony-retained homopolysaccharides. No species of *Azomonas* produces diffusible homopolysaccharides, i.e., there is no formation of a diffuse halo around the colonies.

Azomonas macrocytogenes and *A. agilis* produce a yellow-green visible, blue-white fluorescent pigment when grown in iron-deficient medium. This siderophore, isolated from *A. macrocytogenes* ATCC 12334 and named azoverdin, is an isopyoverdin-like chromopeptide (Michalke et al., 1996). Azoverdin is produced in Fe-limited cultures grown at 28°C but not at 34°C (Collinson et al., 1990). This strain synthesizes two outer membrane proteins of 74 kDa and 70 kDa in response to Fe starvation. These proteins may act as azoverdin binding and/or transport sites (Collinson and Page, 1989). *A. macrocytogenes* also produces 3,4-dihydroxybenzoic acid, which can solubilize iron from minerals and may represent the red-violet diffusible pigment reported to be produced by all species of *Azomonas* (Tchan and New, 1984; Collinson et al., 1987). *A. insignis* differs from other azomonads by its production of a brown-black pigment in the presence of benzoate. *A. insignis* is also the only species that fails to form a fluorescent pigment.

Nitrogen fixation can occur by alternative nitrogenase enzymes in species of *Azomonas* (see the genus *Azotobacter*). The evidence for such alternative nitrogenases is physiological and genetic. Growth and nitrogen fixation can occur in molybdate-starved cultures of *A. macrocytogenes*, although at lower rates than in molybdenum-sufficient medium. DNA hybridization studies show that *A. agilis* (ATCC strain 7494) carries genes for all three nitrogenase enzymes encoded by *nif, vnf,* and *anf* genes (Fallik et al., 1991). A PCR amplification product specific for the Fe-nitrogenase *anfG* gene but not for the V-nitrogenase *vnfG* gene was detected using *A. macrocytogenes* (ATCC strain 12334) DNA (see *Azotobacter*, Fig. BXII.γ.138). This finding suggests that this species might have one alternative nitrogenase—the Fe enzyme—in addition to the Mo nitrogenase (Loveless and Bishop, 1999). If so, this would be the only member of the *Azotobacter–Azomonas* group with this combination of two nitrogenases. Other nitrogen sources for *Azomonas* species are ammonium salts. Only *A. insignis* can generally utilize nitrate. All species are negative for urease activity and can presumably not utilize urea as a nitrogen source—in contrast to *Azotobacter* species. No species can utilize glutamate as an N source.

The optimum temperature for growth is approximately 30°C, although some strains of *A. agilis* prefer 37°C. The minimum temperature for growth is generally close to 14°C; however, some strains, particularly of *A. insignis*, are able to grow at 9–10°C.

Compounds used as sole carbon and energy sources by all species of the genus are shown in Table BXII.γ.122 and are similar to those utilized by all species of *Azotobacter*. Only *A. macrocytogenes* can use D-galactose, melizitose, mannitol, and sorbitol. Maltose is used by *A. macrocytogenes* and most strains of *A. agilis*. Caproate, malonate, and mucate are used by *A. agilis* and *A. insignis* but not by *A. macrocytogenes*. *A. agilis* utilizes benzoate, catechol, 4-hydroxybenzoate, naphthalene, protocatechuate, and 4-toluate; the other species were not tested (Chen et al., 1993). A soil isolate identified as *A. macrocytogenes* degraded and used polyacrylamide as both a nitrogen and a carbon source (Nakamiya and Kinoshita, 1995). None of 17 amino acids tested can serve as a carbon source for *Azomonas* species, possibly due to a lack of ability to transport most amino acids—as in species of *Azotobacter*.

Catalase is present in all species. A positive oxidase reaction is given by the majority of *A. agilis* and *A. insignis* but rarely by *A. macrocytogenes*. Peroxidase is produced by the other species but not by *A. macrocytogenes*.

TABLE BXII.γ.121. Characteristics differentiating the species of the genus *Azomonas*[a]

Characteristic	1. *A. agilis*	2. *A. insignis*	3. *A. macrocytogenes*
Presence of enlarged cells in media with ethanol	−	−	+
Flagellar arrangement:			
Peritrichous	+	−	−
Lophotrichous	−	+	−
Monotrichous	−	−	+[b]
Diffusible pigments:			
Brown-black, on benzoate medium	−	d	−
Blue-white fluorescent, on iron-deficient medium	+	−	d
Utilization as carbon sources:			
Sucrose	+	−	+
Mannitol	−	−	+
Maltose	d	−	+
Malonate	+	+	−
Raffinose	−	−	+
Utilization of nitrate as a nitrogen source	−	+	−

[a]Symbols: see standard definitions.

[b]Rarely, two flagella may occur at one pole.

TABLE BXII.γ.122. Other characteristics of species of the genus *Azomonas*[a]

Characteristic	1. *A. agilis*	2. *A. insignis*	3. *A. macrocytogenes*
Nitrogen fixation occurs at pH:			
5.5	−	−	+
6	d	−	+
6.5–10.0	+	+	+
Nitrogen fixation genes:[b]			
nif	+		+
vnf	+		−
anf	+		+
Growth at a temperature of:			
9°C	d	d	d
14–28°C	+	+	+
32°C	+	+	d
37°C	+	−	−
Pigmentation:			
Diffusible red-violet	d	+	+
Diffusible red-purple	−	d	−
Diffusible yellow-green on iron-deficient media[b]	+	d	d
Diffusible blue-white fluorescent on iron-deficient media[b]	+	−	d
Utilization as sole carbon source:[c]			
Fructose, glucose, acetate, pyruvate, fumarate, malate, succinate, α-oxoglutarate, lactate, DL-gluconate, acetyl methyl carbinol	+	+	+
D-Galactose, sorbitol	−	−	+
Trehalose	−	−	d
Glutarate, benzoate	−	−	−
Habitat:			
Soil	−	−	+
Water	+	+	−
Susceptibility to antimicrobial agents:			
Streptomycin, 0.2 µg/ml	S	S	S
Neomycin, kanamycin, 2 µg/ml	d	S	d
Erythromycin, 2 µg/ml	d	R	S
Chloramphenicol, 25µg/ml	R	R	R
Sulfanilamide, 25 µg/ml	d	d	d
Penicillin G, 5 U/ml	d	S	d
Phenol, 0.95%	R	d	R
Sodium benzoate, 0.5%	R	d	R
Sodium benzoate, 1.0%	d	S	R
Sodium fluoride, 0.01 M	d	R	d
Mercuric chloride, 10µg/ml	R	d	d
Iodoacetate, 1mM	d	S	S

[a]For symbols see standard definitions; S, susceptible; R, resistant; blank space, not reported.

[b]See Procedures for Testing Special Characters.

[c]See also Table BXII.γ.121 for additional compounds.

The genome of *Azomonas macrocytogenes* (ATCC 12334) is much smaller than *Azotobacter* species genomes (2.4 versus 3.9–5.3 Mb) (see *Azotobacter* chapter, Table BXII.γ.125). There are no reports of indigenous plasmids within or transfer of plasmids from other bacteria into *Azomonas* species.

No bacteriophages from *Azomonas* species have been reported. Hegazi and Jensen (1973) reported that bacteriophages isolated with *A. chroococcum* and *A. vinelandii* as hosts were unable to infect or form plaques on *A. agilis*, *A. insignis*, and *A. macrocytogenes*.

Tchan and New (1984) indicated that antisera produced against *A. agilis* and *A. insignis* do not cross-agglutinate with members of *Azotobacter*. Norris and Kingham (1968), however, reported that antisera against *A. agilis* do cross-react with *A. vinelandii*. This discrepancy may be due to strain differences or to variability in immune response between animals. Within the genus *Azomonas*, antiserum against *A. insignis* cross-agglutinates with *A. agilis* (Tchan and New, 1984). Immunodiffusion produces numerous cross-precipitation lines. Rocket-line immunoelectrophoresis indicated that *Azomonas* and *Azotobacter* can be differentiated by the unshared precipitation peaks using an appropriate antiserum (Tchan et al., 1983), although at least one thermoresistant antigen is shared by *Azomonas* and *Azotobacter* (Tchan et al., 1980).

All species of *Azomonas* are sensitive to streptomycin (0.2 µg/ml), as are *Azotobacter* species. Sensitivities to chlortetracycline, oxytetracycline, polymyxin, sulfanilamide, penicillin, and neomycin vary depending on the strain. Only *A. insignis* is entirely sensitive to penicillin (Thompson and Skerman, 1979). As for *Azotobacter* species, those of *Azomonas* are somewhat resistant to chloramphenicol.

Azomonas species are resistant to phenol, benzoate, and mercuric chloride to varying degrees. Most strains tolerate sodium fluoride up to 0.01 M. Cadmium tolerance and the ability to remove this metal from the culture medium was reported for *A. agilis* PYO (You and Park, 1998). The resistance of *A. agilis* to iodoacetate can be used for selective isolation of this species (Thompson and Skerman, 1979).

A. agilis and *A. insignis* are aquatic bacteria. *A. macrocytogenes* is found in soil, although no information is available as to whether it also has an aquatic habitat (for further details, see Becking, 1992). Thus, it appears that the genus is ecologically heterogeneous. Only *A. macrocytogenes* is resistant to desiccation: five of seven strains survived desiccation for at least 1 month (Thompson and Skerman, 1979). All species tolerate a salt concentration of up to 1%, suggesting that the aquatic members are capable of living in contaminated water where concentrations of organic matter and mineral salts can be relatively high.

ENRICHMENT AND ISOLATION PROCEDURES

The enrichment and isolation techniques depend on the ability of *Azomonas* to grow by fixing nitrogen under aerobic conditions, using organic substrate as energy source. Nonselective enrichment methods are similar to those described in the chapter on *Azotobacter*.

Selective enrichment methods (Thompson and Skerman, 1979) Media containing either 1% sodium benzoate (Derx, 1951; Jensen, 1955a) or 1 mM iodoacetate are selective for *A. agilis*. The growth of *A. macrocytogenes* is favored by incorporating 1% benzoate into nitrogen-free sucrose medium (pH of 6.0) and incubating the cultures at a decreased incubation temperature (10–12°C). No selective enrichment method is available for *A. insignis*.

Isolation methods These are similar to those described previously for *Azotobacter* species. The enrichment cultures are inspected daily. When macroscopic growth becomes apparent, the culture is examined by phase-contrast microscopy to detect the presence of *Azomonas* cells (ovoid cells 2.5 × 5.0 μm, or sometimes filamentous forms). Growth will usually occur within 2–5 days. Positive cultures are streaked onto nitrogen-free agar medium with glucose as the carbon source. For selective isolation, nitrogen-free agar media with appropriate selective substances should also be used. Sometimes nitrogen-free iron-deficient agar can be used to encourage pigment production by some species.

MAINTENANCE PROCEDURES

For routine maintenance, *Azomonas* strains should be subcultured at monthly or bimonthly intervals on Winogradsky's agar medium with glucose. Other maintenance methods are similar to those described in the chapter on *Azotobacter*.

PROCEDURES FOR TESTING SPECIAL CHARACTERS

Methods for determining presence of genes for alternative nitrogenases, for testing ability of strains to grow diazotrophically under Mo-deprivation conditions, cell morphology, and pigment production are the same as those described in the chapter on the genus *Azotobacter*.

DIFFERENTIATION OF THE GENUS *AZOMONAS* FROM OTHER GENERA

The single most distinguishing feature between *Azomonas* and *Azotobacter* is the failure of the former to form cysts in aging cultures or after induction with butanol. For other features that distinguish *Azomonas* from other diazotrophs and from pseudomonads, to which they are most related, see the chapter on *Azotobacter*.

TAXONOMIC COMMENTS

Analysis of 16s rRNA reveals the close but separate relationship of *Azomonas macrocytogenes* ATCC 12334 to the four *Azotobacter* species characterized (MacDonald and Melton, unpublished results) (see *Azotobacter* chapter, Figs. BXII.γ.139 and BXII.γ.140). Sequences are not available for strains of *A. agilis* and *A. insignis*. Nevertheless, the justification of a single genus with three species remains sound, based on DNA base composition, hybridization analysis, and rRNA cistron similarities (De Smedt et al., 1980). General physiological characteristics, polyamine patterning, distribution of alginate biosynthetic genes, and aromatic biosynthetic pathway enzyme distribution support the common lineage of *Azotobacter* and *Azomonas* species with group I pseudomonads (Byng et al., 1986; Fialho et al., 1990; Goris et al., 1998). While the genus is more heterogeneous than *Azotobacter*, further molecular characterization of *Azomonas* species and strains would be required to justify separation of the group into more than one genus.

DIFFERENTIATION OF THE SPECIES OF THE GENUS *AZOMONAS*

The differential characteristics of the three species of *Azomonas* are presented in Table BXII.γ.121.

List of species of the genus Azomonas

1. **Azomonas agilis** (Beijerinck 1901) Winogradsky 1938, 400[AL] (*Azotobacter agilis* Beijerinck 1901, 577.)

   a'gi.lis. L. adj. *agilis* quick, agile.

   The characteristics are as described for the genus and as listed in Tables BXII.γ.121 and BXII.γ.122. The morphology is as shown in Fig. BXII.γ.126. The cells are typically 2–4 × 1.6–3.0 μm. Cells are motile by means of numerous peritrichous flagella—a distinguishing characteristic of this species. The flagella have a wavelength of 2.0–3.0 μm and an amplitude of 0.40–0.59 μm.

   Colonies are opaque, low convex, butyrous, and usually mucoid, glistening, and smooth.

   Physiological and nutritional characteristics are presented in Tables BXII.γ.121 and BXII.γ.122. Starch is not hydrolyzed. Ammonium salts can be used as a sole source of nitrogen for growth. Neither nitrate nor glutamate can be used as a nitrogen source.

   The type strain was isolated by Kluyver and van den Bout (1936).

   The mol% G + C of the DNA is: 52–53.2 (T_m).

   Type strain: ATCC 7494, DSM 375, WR 83.

2. **Azomonas insignis** (Derx 1951) Jensen 1955b, 156[AL] (*Azotobacter insignis* Derx 1951, 344.)

   in.sig'nis. L. adj. *insignis* distinguished by a mark.

   The characteristics are as described for the genus and as listed in Tables BXII.γ.121 and BXII.γ.122. Motile by lophotrichous flagella, whose full wavelength may exceed 4 μm). The numerous flagella tend to bind together to form thick, whip-like structures.

   Colonies are translucent, low convex, butyrous, and smooth. They are usually small, probably due to low production of extracellular polysaccharides.

   Physiological and nutritional characteristics are presented in Tables BXII.γ.121 and BXII.γ.122. Starch is not hydrolyzed. Ammonium or nitrate, but not glutamate, can be utilized as a sole nitrogen source for growth. The ability to use nitrate as a nitrogen source and the inability to use sucrose as a carbon source are characteristics distinctive for this species.

   The mol% G + C of the DNA is: 55.1–58.3 (T_m).

   Type strain: UQM 1966.

3. **Azomonas macrocytogenes** (ex Baillie, Hodgkiss and Norris 1962, 118) New and Tchan 1982, 381[VP] (*Azotobacter macrocytogenes* Jensen 1955a, 280; *Azotobacter macrocytogenes* (Jensen 1955a) Thompson and Skerman 1981, 215; *Azotobacter agilis* subsp. *jakutiae* Krasil'nikov 1949, 508.)

mac.ro.cy.to'ge.nes. Gr. adj. *macrus* large; Gr. n. *kytos* cell; Gr. v. *gennaio* produce; M.L. part. adj. *macrocytogenes* large cell producing.

The characteristics are as described for the genus and as listed in Tables BXII.γ.121 and BXII.γ.122. The morphology is shown in Fig. BXII.γ.127. Cells are typical of species of *Azomonas* except that motility occurs by polar rather than peritrichous flagella. Usually one or, less frequently, two flagella per cell are observed. The flagellar wavelength is 1.3–1.7 μm and the amplitude is 0.25–0.52 μm. The cells are often surrounded by large capsules.

Colonies on sucrose media are mucoid and large (15 mm in diameter in 7 d) due to the production of colony-retained homopolysaccharides. Starch is not hydrolyzed. Ammonium, but not nitrate or glutamate, can serve as a sole nitrogen source for growth.

The gene *anfG*—but not *vnfG*—is present in strain ATCC 123334, suggesting that this species contains the alternative nitrogenase-3 in addition to nitrogenase-1. The possession of *anfG* is possibly a unique characteristic of this species. The genome size of *A. macrocytogenes* ATCC strain 12334 is 2.4 Mb—smaller than the genomes of other *Azomonas* and *Azotobacter* species (see Table BXII.γ.123 in the chapter on *Azotobacter*). The significance of this small size with respect to the diversity of this organism is not known.

The mol% G + C of the DNA is: 58.2–58.6 (T_m).

Type strain: Jensen O, ATCC 12335, DSM 721, NCIB 8700, WR-111.

Additional Remarks: Other strains include strain M (ATCC 12336), a mutant isolated by Jensen, and strain WR-140, cotype of "*A. agilis* subsp. *jakutiae*" Krasil'nikov 1949.

Genus III. **Azotobacter** Beijerinck 1901, 567[AL]

CHRISTINA KENNEDY, PAUL RUDNICK, MELANIE L. MACDONALD AND THOYD MELTON

A.zo.to.bac'ter. Fr. n. *azote* nitrogen; M.L. masc. n. *bacter* the equivalent of Gr. neut. n. *bactrum* a rod or staff; M.L. masc. n. *Azotobacter* a nitrogen rod.

Cells range from straight rods with rounded ends to more ellipsoidal or coccoid, depending on the culture medium and age. **Cells are up to 2 μm or more in diameter and 4 μm in length**. *A. paspali* cells are usually longer, 5–10 μm in length, and can be filamentous, up to 60 μm long. Cells are usually single but may occur in pairs, irregular clumps (especially with *A. paspali*), or, more rarely, in chains of varying length. **Encystment** occurs during late stationary phase at low frequency or at high frequency after culturing on butanol. **Motile with peritrichous flagella or nonmotile. Aerobic, having a strictly respiratory type of metabolism with oxygen as the terminal electron acceptor. Nitrogen is fixed** under microaerobic conditions (2% oxygen), under full aerobiosis, or after adaptation in hyperbaric oxygen. N_2 fixation uses Mo-, V-, or Fe-containing nitrogenase enzymes, depending on the environmental metal supply. **Water-soluble and water-insoluble pigments are produced by some strains of all species.** Growth is heterotrophic; sugars, alcohols, and salts of organic acids are used as carbon sources. Ammonium salts, nitrate, and urea are used as sources of fixed nitrogen. Very few amino acids are used, probably due to a general deficiency in amino acid transport. The minimum pH for growth in the presence of fixed nitrogen sources ranges from 4.8 to 6.0 with maximum pH 8.5. The optimum pH for diazotrophic growth is 7.0–7.5. **Most isolates are from soil, but a few are from water.** One species (*A. paspali*) has been isolated only from roots of the tropical grass *Paspalum notatum*.

The mol% G + C of the DNA is: 63.2–67.5.

Type species: **Azotobacter chroococcum** Beijerinck 1901, 567.

FURTHER DESCRIPTIVE INFORMATION

Preliminary note The immense amount of data accumulated by Thompson and Skerman (1979) remains the foundation for the basic morphological and physiological characterization of both *Azotobacter* and *Azomonas*. The data resulted from an analysis of some 90 strains representing the several species of *Azotobacter* and *Azomonas*—including established strains found in international culture collections and their own isolates—for 230 different morphological and physiological attributes. The data were summarized by Tchan and New (1984) in the previous edition of the *Manual*. More current knowledge of *Azotobacter* species is very uneven, particularly with respect to genetic, physiological, and biochemical studies, with *A. vinelandii* being the most studied followed by *A. chroococcum*. Relatively little information has been reported for the other five species, especially for *A. nigricans* and *A. armeniacus*, in each of which only four isolates were previously studied (Thompson and Skerman, 1979). The present authors have drawn on those contributions and on updates from the literature since 1984 to provide a current overview of the characteristics of the genus *Azotobacter*.

Cell morphology In nitrogen-free Winogradsky medium[1] with glucose as the carbon source, the young cells of different species are remarkably similar in appearance (Figs. BXII.γ.128, BXII.γ.129, and BXII.γ.130): rods with rounded ends, 1.6–2.7 × 3.0–7.0 μm. In older cultures, the cells of some species may tend to be ellipsoidal to coccoid. In other species, chains and filamentous forms become more common, and dark-colored granules are observed. *A. paspali* produces long filamentous forms even in young cultures (Fig. BXII.γ.131), and this characteristic—together with the inability of this species to use several carbon sources—differentiates it from the others. In peptone–yeast ex-

---

1. Winogradsky's nitrogen-free mineral medium is prepared from a 200× stock solution which has the following composition (g/l): KH₂PO₄, 50.0; MgSO₄·7H₂O, 25.0; NaCl, 25.0; FeSO₄·7H₂O, 1.0; Na₂MoO₄·2H₂O, 1.0; and MnSO₄·4H₂O, 1.0; the pH is adjusted to 7.2 with NaOH. The medium is prepared by using 5.0 ml of the stock solution and 0.1 g/l of CaCO₃ and then sterilized at 120°C for 20 min. To make nitrogen-free organic medium, a suitable quantity of organic substrate is added to the mineral medium. (Certain sugars, including glucose, must be sterilized separately before addition to the sterilized mineral medium.) For a solid medium, 15.0 g/l of agar is added.

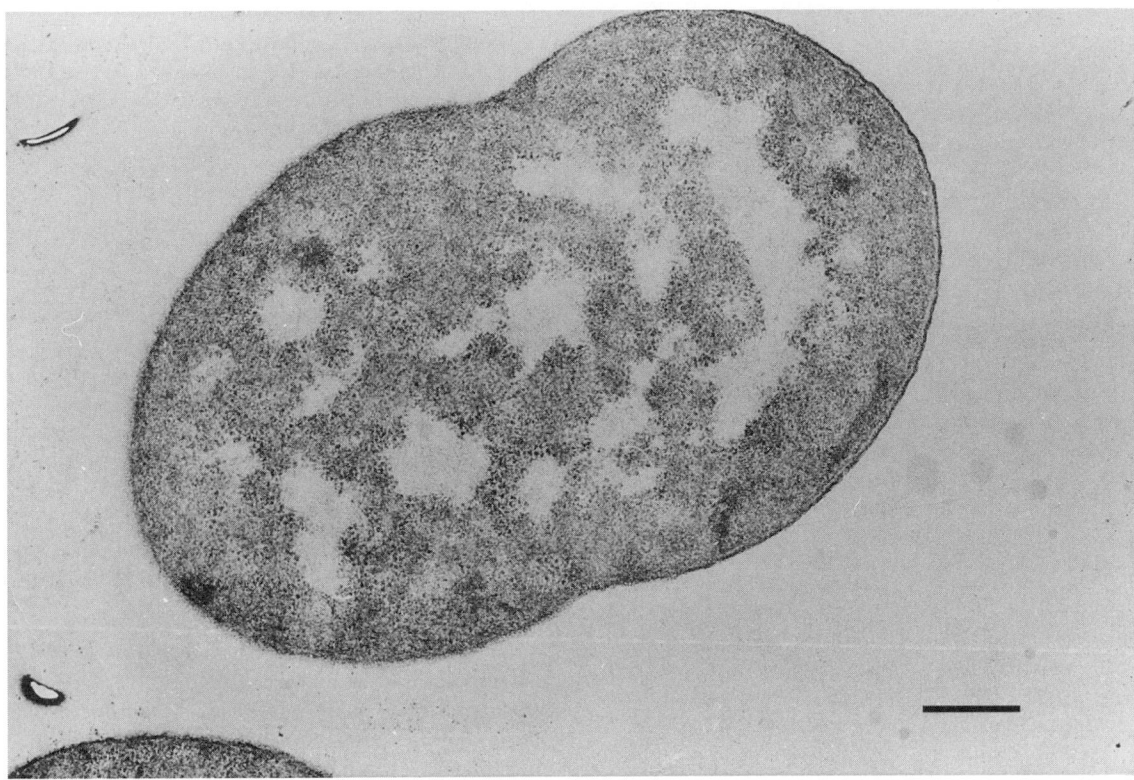

FIGURE BXII.γ.128. Electron micrograph of a vegetative cell of *Azotobacter chroococcum*. Bar = 0.2 μm.

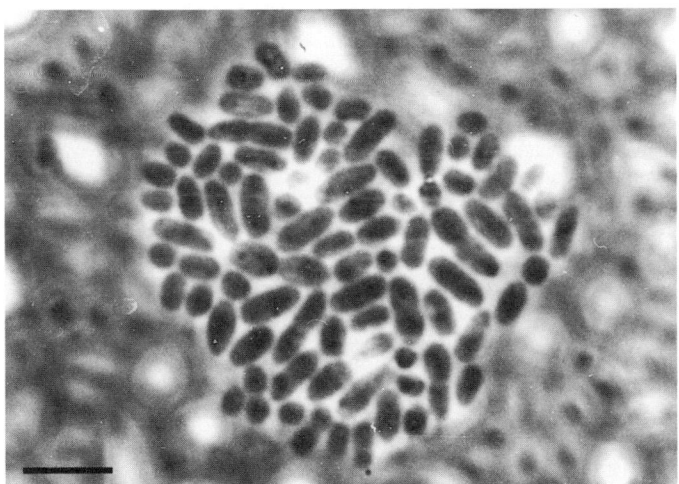

FIGURE BXII.γ.129. Vegetative cells of *Azotobacter chroococcum* (phase-contrast microscopy). Bar = 10 μm.

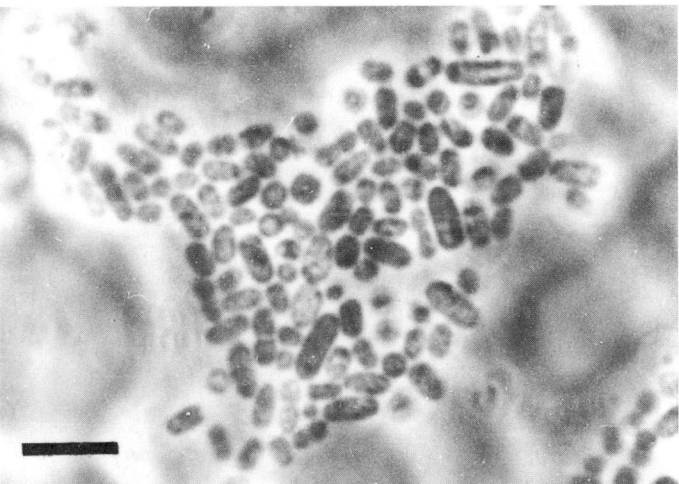

FIGURE BXII.γ.130. Vegetative cells of *Azotobacter nigricans* (phase-contrast microscopy). Bar = 10 μm.

tract agar[2], all members of the genus produce distorted cells (Tchan and New, 1984). The widely-used, nongummy strain AvOP of *A. vinelandii* (see below) cannot grow at all on Luria–Bertani medium (C. Kennedy, unpublished results), a medium with peptone and yeast extract (Sambrook et al., 1989). Cells grown under nitrogen-fixing conditions are smaller than ammonia-grown cells. *A. paspali* cells aggregate in late-log or sta-

tionary phase, especially with an increasing C:N ratio (Abbass and Okon, 1993a). The large size of *A. vinelandii* cells correlates with their high DNA content (see later), and their volume was estimated to be 16 times greater than that of *Escherichia coli* cells (Efuet et al., 1996) (Fig. BXII.γ.132).

Cell wall composition *Azotobacter* species form a proteinaceous S-layer as the outermost cell surface component. The layer is structurally similar to that of *Aeromonas salmonicida* and two other *Pseudomonas* species (Bingle et al., 1987). Unlike the S-layer in these organisms, however, the S-layer of *A. vinelandii* can be extracted from whole cells or outer membrane fractions with

2. Peptone-yeast extract agar (g/l): Bacto peptone (Difco), 1.0; yeast extract (Difco), 0.5; NaCl, 0.5; Bacto agar (Difco), 15. The pH is adjusted to 7.2 with NaOH prior to autoclaving, giving a final pH of 7.0.

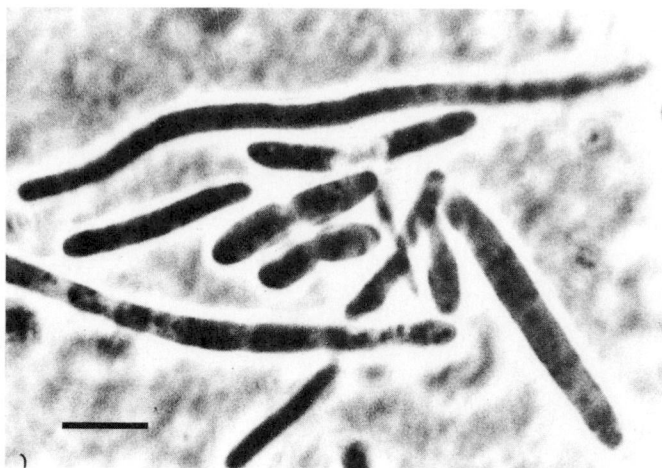

FIGURE BXII.γ.131. Vegetative cells of *A. paspali* (phase-contrast microscopy). Bar = 10 μm.

water. The S-layer of *A. vinelandii* is composed of a ~60 kDa protein, which assembles into tetramers every 12–13 nm. N_2-fixing cells form an elaborate inner membrane network (Oppenheim and Marcus, 1970). Such networks are formed concomitantly with the synthesis of nitrogenase and an increase in the rate of respiration, suggesting an involvement in protecting nitrogenase from oxygen damage. Ammonia-grown cells of *A. vinelandii* develop large deposits of poly-β-hydroxybutyrate (PHB), and the cytoplasm appears less dense. PHB deposits are not seen with cells grown with nitrate or N_2. PHB biosynthesis seems not to be influenced by the fixed nitrogen or carbon supply in *A. chroococcum* (Martinez-Toledo et al., 1995). In *A. beijerinckii*, PHB synthesis is variable and dependent on media and specific culture conditions (Bormann et al., 1998).

The fatty acid composition of vegetative cells differs from that of cysts. In vegetative cells analyzed during mid-exponential phase, 75% of the total fatty acids are C_{16} derivatives, 4% are C_{14}, and 21% are C_{18} (Su et al., 1979). Seventy percent of the total lipids are replaced with 5-*n*-alkylresorcinols and 6-*n*-alkylpyrones during encystment in *A. vinelandii* (Su and Sadoff, 1981; Reusch and Sadoff, 1983a) and in *A. chroococcum* (Kozubek et al., 1996). These molecules—thought previously to be strictly components of plant membranes—are unique to the cysts and probably contribute to desiccation resistance in both *A. vinelandii* (Reusch and Sadoff, 1983b) and in several strains of *A. chroococcum* (Kozubek et al., 1996).

Motility *A. beijerinckii* and *A. nigricans* are nonmotile. Other species are motile by peritrichous flagella, which are often present in large numbers (Fig. BXII.γ.133). *A. vinelandii* OP is a fast-moving bacterium, with a measured average velocity of 74 μm/sec (Haneline et al., 1991). Chemotaxis occurs and requires oxygen; the speed of the cells increases as a function of attractant concentration. Attraction to fructose, glucose, xylitol, and mannitol occurs, especially after pregrowth in these compounds, indicating that chemotaxis is often inducible. Chemotaxis to other compounds, including hexoses, hexitols, pentitols, pentoses, disaccharides, and amino sugars, occurs.

Cultural characteristics When grown at 30°C with sugar—usually sucrose—as the carbon source at 1–2% (w/v) on either Winogradsky's nitrogen-free agar medium or on Burk's medium[3], colonies appear within 48 h at 30°C and reach a diameter of 2–6 mm in a week, depending on the numbers (crowding) per plate. Burk's medium is now more commonly used than Winogradsky's medium for growth of laboratory strains of *A.*

---

3. Burk's medium has the following composition (g/l): K_2HPO_4, 0.64; KH_2PO_4, 0.20; $MgSO_4 \cdot 7H_2O$, 0.2; NaCl, 0.2; $CaSO_4 \cdot 2H_2O$, 0.05. Sucrose or other carbon sources should be added separately after autoclaving from sterile 20% solutions to give a final concentration of 20 g/l. Na_2MoO_4 and $FeSO_4$ should be added from sterile stock solutions to give a final concentration of 0.001 g/l and 0.003 g/l Fe. Agar is added to a final concentration of 1.2–1.5% (w/v).

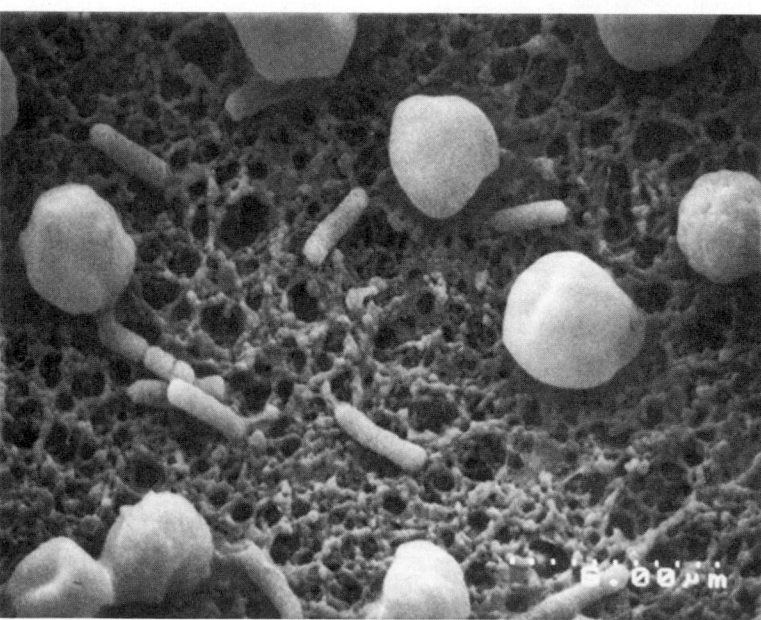

FIGURE BXII.γ.132. Scanning electron micrograph of *E. coli* and *A. vinelandii* vegetative cells, × 1000 magnification. (Reproduced with permission from E.T. Efuet et al., Journal of Basic Microbiology *36*: 229–234, 1996, ©Wiley-VCH.)

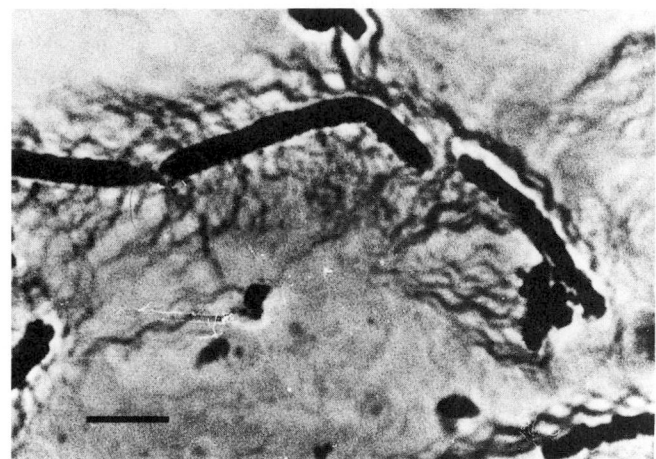

FIGURE BXII.γ.133. Flagella stain of *A. paspali*. Bar = 10 μm.

FIGURE BXII.γ.134. Violamine-stained cysts. The central body is surrounded by the cyst coat. Bar = 5 μm.

vinelandii and *A. chroococcum.* Brown's medium[4] with glucose results in higher plate counts of *A. chroococcum* colonies (Thompson, 1989). Colonies of most species are generally smooth, glistening, opaque, low convex, and mucoid; however, variations may occur. For *A. vinelandii*, smaller variant colonies may appear due to decreased production of extracellular polysaccharide. *A. armeniacus* sometimes produces translucent colonies and *A. paspali* forms undulate edged and unevenly convex colonies with a dull or rough surface. On sucrose or raffinose agar, the production of diffusible homopolysaccharides—which results in a diffuse halo around the colony—is species dependent. *A. chroococcum* and some strains of *A. beijerinckii* and *A. nigricans* produce diffusible homopolysaccharides from both sugars, whereas *A. armeniacus* and some other strains of *A. beijerinckii* and *A. nigricans* produce them only from sucrose (Thompson and Skerman, 1979).

Cysts Cysts are formed in fewer than 1% of late stationary cells of *Azotobacter* species grown on glucose or mannitol. Encystment can reach >90% 5 d after the addition of butan-1-ol or β-hydroxybutyrate to growing cultures (Lin and Sadoff, 1968) (Fig. BXII.γ.134). Isopropanol appears to be a better inducer for strains of *A. chroococcum* (Marengo, 1983). During encystment, there is a cessation of most metabolic activities. Cysts are significantly more resistant than vegetative cells to desiccation and damage due to physical and chemical agents (Socolofsky and Wyss, 1962). In ultrastructure, they are composed of a central body containing β-hydroxybutyrate surrounded by capsule-like exine and intine layers, of which alginate is a major component (Fig. BXII.γ.135). Deposition of capsular materials and membrane components in the absence of cell division produces the layers of the exine coat (Hitchins and Sadoff, 1970). Intine vesicles migrate towards the exine where they form a layer internal to the exine. These vesicles have also been visualized during vegetative growth (Vela et al., 1970). This observation provides ev-

idence that cysts are simply resting vegetative cells and are not analogous to bacterial spores, which are more differentiated. The composition of the exine coat in *A. vinelandii* is 32% carbohydrate, 28% protein, and 30% lipid (Lin and Sadoff, 1969).

Alginate synthesis The alginate biosynthetic pathway in *A. vinelandii* has been characterized by Pindar and Bucke (1975). Fructose-6-phosphate is converted to GDP-mannuronic acid, which is polymerized and secreted. The polymer is modified by an O-acetylase and an extracellular C-5 epimerase to form alginate (Rehm et al., 1996). Alginate biosynthesis is apparently essential for encystment in *A. vinelandii* because biosynthetic mutants do not form cysts (Campos et al., 1996). Alginate production is remarkably similar to that in the pathogen *Pseudomonas aeruginosa* (reviewed by May and Chakrabarty, 1994), and genes from each organism can rescue mutations in the other (Moreno et al., 1998). There is industrial interest in the exopolysaccharides synthesized by *A. vinelandii* and *A. chroococcum*, and these polysaccharides can be produced in high concentrations depending on cultural conditions (De la Vega et al., 1991; Pena et al., 2000). The best medium for production of alginate by *A. vinelandii* is that described by Sabry et al. (1996).[5] In *A. vinelandii*, alginate is made to a concentration of 4.5 g/l when cells are incubated at 300 rpm on a shaker under 5% O_2. Alginate production increases with agitation speeds upwards to 700 rpm; however, increasing the speed leads to a decrease in the polymer size from 680,000 Da at 300 rpm to 352,000 Da at 700 rpm. Reduction in size of the polymer is correlated with alginase activity, detected in the more vigorously aerated cultures. The alginate produced by *A. vinelandii* differs from other bacterial sources in that guluronic acid residues can be introduced after polymerization by conversion of a mannuronic residue, and there are apparently eight enzymes which code for this particular C-5 epimerase activity (Ertesvag et al., 1994, 1995; Rehm et al., 1996; Svanem et al., 1999). The resultant alginate formed by the activity of these enzymes gives it industrially favorable gel-forming properties.

---

4. Brown's medium has the following composition (g/l): glucose or mannitol, 5.0; K_2HPO_4, 0.8; $MgSO_4 \cdot 7H_2O$, 0.2; $FeSO_4 \cdot 7H_2O$, 0.04; Na_2MoO_4, 0.005; $CaCl_2$ (anhydrous), 0.15; agar, 15. The phosphate is prepared as a separate 8% (w/v) solution and added to the agar after sterilization by autoclaving (standard conditions) and cooling to 42°C just before pouring plates.

5. Medium for alginate production (per liter): $MgSO_4 \cdot 7H_2O$, 0.3 g; NaCl, 0.4 g; $CaCl_2 \cdot 2H_2O$, 42 mg; KH_2PO_4, 4 mg; K_2HPO_4, 16 mg; $FeSO_4 \cdot 7H_2O$, 2.5 mg; H_3BO_3, 2.9 mg; $ZnSO_4 \cdot 7H_2O$, 2 mg; $Na_2MoO_4 \cdot 2H_2O$, 2 mg; $CuSO_4 \cdot 5H_2O$, 0.3 mg; and $MnCl_2 \cdot 4H_2O$, 0.2 mg.

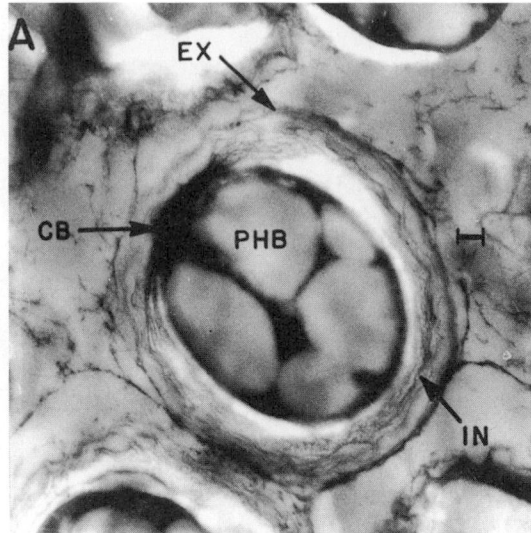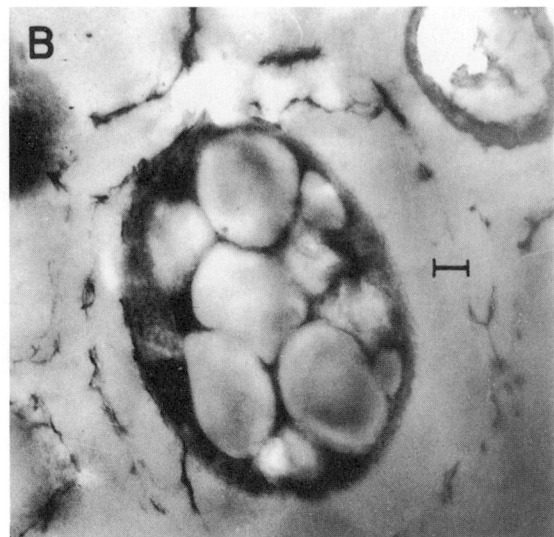

FIGURE BXII.γ.135. Electron micrographs of the cysts formed by strains ATCC 9046 (*A*) and ATR8 (*algR⁻*) (*B*). Abbreviations: EX, exine; IN, intine; CB, central body; PHB, poly-hydroxybutyrate. Bars = 0.4 μm. (Reprinted with permission from C. Nuñez et al., *Journal of Bacteriology, 181:* 141–148, 1999, ©American Society for Microbiology.)

The particular exopolysaccharides produced by *A. chroococcum* vary among strains. Cote and Krull (1988) reported the existence of two distinct exopolysaccharides from strain NRRL B-14341 (ATTC 7491), one of which has been identified as an alginate. *A. chroococcum* ATCC 4412 apparently produces only one type of exopolysaccharide, which is alginate-like. Production of this molecule varies with carbon source and air-flow rate, and the best yields occur at 8% sucrose (De la Vega et al., 1991). Since alginate production varies among strains, it may not be possible to differentiate between species based on the production of alginate. *A. beijerinckii* B-1615 produces two different exopolysaccharides in a 9:1 ratio (Likhosherstov et al., 1991). The major polysaccharide consists of D-galactose, L-rhamnose, and pyruvic acid in a 2:1:1 ratio. The minor polysaccharide contains mannuronic and guluronic acid in a 2:3:1 ratio and is an alginate.

Genes encoding products involved in alginate synthesis are found in all members of *Azotobacter* and *Azomonas* and throughout most of *Pseudomonas* group I. (Fialho et al., 1990). The alginate biosynthetic genes of *A. vinelandii* are clustered as operons (Martinez-Salazar et al., 1996; Nuñez et al., 1999), and a similar arrangement is predicted for *A. chroococcum* (Peciña et al., 1999). Although not all of the *alg* genes have been identified in any member of *Azotobacter* or *Azomonas*, their arrangement is probably similar to that in *P. aeruginosa*, in which three large clusters at distinct locations on the chromosome comprise all functions of alginate biosynthesis and regulation (for review, see Gacesa, 1998). It should be noted that not all *A. vinelandii* strains produce alginate. The commonly used nonmucoid laboratory strain, UW136 (a rifampicin resistant derivative of AvOP/UW), may have arisen from insertion of small IS element in *algU* (Martinez-Salazar et al., 1996).

Siderophores   As noted earlier, some *Azotobacter* species produce the yellow-green fluorescent peptide siderophore, azobactin, in response to iron limitation. *A. vinelandii* also produces the three nonfluorescent catecholate siderophores, azotochelin, protochelin, and aminochelin, under similar growth conditions (Cornish and Page, 1998). Iron uptake by each of the sidero-

phores is biphasic and energy-dependent. Fe^{+3} uptake mediated by each of the siderophores is sensitive to metal salts: the addition of 40 mM Na$^+$, K$^+$, Li$^+$, or Mg^{+2} inhibits 60% of total uptake. NH$_4$$^+$ also inhibits uptake but to a lesser extent (Knosp et al., 1984). The production of both types of siderophores is repressed by iron compounds in the medium. However, repression of azotobactin synthesis is sensitive to lower concentrations of Fe than is production of the catecholate siderophores. The promoter of the gene encoding the first enzyme in catecholate siderophore synthesis has been characterized (Tindale et al., 2000), and within this promoter is a Fur-box as well as sequences that match the Sox-boxes of *E. coli*. The appearance of these regulatory features, as well as an increase in expression in response to superoxide, suggests that azotochelin may provide protection against oxidative stress under nitrogen-fixing conditions. *A. chroococcum* produces Fe-binding hydroxamates in response to Fe limitation as well as an 85 kDa outer member protein (Fekete et al., 1989). *A. paspali* produces 3,4-dihydroxybenzoic acid (protocatechuate) in both Fe-limited and Fe-sufficient media; this compound solubilizes Fe from minerals (Collinson et al., 1987).

A. salinestris produces a catechol melanin pigment described as an "iron trap" and may function to protect cells from oxidative damage mediated by H$_2$O$_2$ and the Fenton reaction (Page and Shivprasad, 1995). The siderophore-like acquisition of iron mediated by this pigment is proposed to occur in a two-step process. In the first step, iron is concentrated in the periplasm by an energy-dependent process and is promoted by a hydroxamate compound produced in conjunction with the pigment. In the second step, the iron is brought into the cytoplasm; this process is sensitive to inhibitors of Na$^+$-dependent activities of the organism.

Nutrition   The ability of species of *Azotobacter* to utilize various carbon and nitrogen sources is important for experimental and applied work, particularly considering how carbon sources can influence the efficiency of nitrogen fixation and how various nitrogen sources can be used in the analysis of Nif$^-$ and other mutants.

Utilization of carbon sources by *Azotobacter* species is shown in Tables BXII.γ.123 and BXII.γ.124. Some carbon sources are utilized by all species, whereas others differentiate the species. Several other less commonly used compounds can be utilized by some species (Thompson and Skerman, 1979). Few reports that further differentiate the species with respect to carbon source utilization have appeared since the work of Thompson and Skerman (1979) and the last edition of the *Manual*. De La Vega et al. (1991) reported that sucrose is degraded by an extracellular invertase in *A. chroococcum* ATCC 4412, whereas it is degraded by an intracellular invertase in *A. vinelandii* UW (ATCC 13705). Chen et al. (1993) found that *A. chroococcum* (ATCC 9043) and *A. vinelandii* UW utilize not only the aromatic compounds benzoate and 4-OH-benzoate but also catechol, naphthalene, protocatechuate, and 4-toluate, which confirmed and extended the findings of Thompson and Skerman (1979). *A. chroococcum* expressed only *ortho* cleavage dioxygenases during growth on naphthalene and 4-toluate, and only *meta* cleavages of other aromatics. *A. vinelandii* expressed both *ortho* and *meta* cleavage enzymes (Chen et al., 1993). Interestingly, growth rates and nitrogenase activities were nearly as high or higher on certain aromatic compounds—benzoate and naphthalene, in particular—for both *A. chroococcum* and *A. vinelandii*. Among isolates enriched from British soils near Leicester for growth on monochloroacetate, one strain—RC26—had many cultural and physiological characteristics that placed it in the genus *Azotobacter* (Diez et al., 1995). This strain was able to dehalogenate a number of different compounds used as pesticides at rates similar to those determined for species of *Pseudomonas*. While this strain has not been characterized with respect to 16S rDNA sequence, it may represent a soil-inhabiting species of *Azotobacter* with capability for bioremediation of agrochemicals.

Nitrogen fixation Nitrogen fixation under aerobic conditions is a principle characteristic of the genus. The major and minor clusters of genes encoding the enzyme synthesized under Mo-sufficient conditions in *A. vinelandii*, sequenced and extensively characterized, are shown in Fig. BXII.γ.136. *A. chroococcum* MCD-1 (derived from ATCC 4412) has a similar major cluster, but has been less well characterized in terms of mutational analysis. Some specific functions of the *nif* gene products are mentioned below; a more extensive review of them can be found in Rangaraj et al. (2000). The *nifHDK* genes encode the basic structural proteins of the nitrogenase enzyme. The other gene products encoded by this *nif* cluster are required for (1) synthesis of the Fe-Mo-cofactor, an essential component of the enzyme for substrate (N$_2$) binding and reduction (*nifV, nifEN, nifH*); (2) formation of Fe-S clusters required at other sites in the enzyme which are involved in electron transfer to the cofactor (*nifU, nifS*); and (3) participation in other aspects of enzyme maturation (*nifM, nifW*). In addition to this is another cluster, located elsewhere in the *A. vinelandii* genome, which includes *nifL* and *nifA* in an operon encoding the regulatory proteins NifL and NifA, and other proteins involved in nitrogenase cofactor synthesis and Mo processing (products of the *nifB fdxN nifO nifQ* operon). NifL prevents expression of the other *nif* genes if high levels of ammonium or oxygen are present, conditions that would either make nitrogenase unnecessary for growth or become inactive, respectively. The NifL protein, which carries a flavin cofactor, inhibits the activity of NifA, which is a transcriptional activator of the other *nif* genes and operons under these conditions (Dixon, 1998). The presence of a *nifL* gene is characteristic of diazotrophs in the *Gammaproteobacteria*, which includes another important diazotroph of this taxon, *Klebsiella pneumoniae*, as well as *Azotobacter vinelandii*—the two species most studied. The *nifL* gene has not been identified in the *Alphaproteobacteria*, *Betaproteobacteria*, or *Deltaproteobacteria* groups of diazotrophs, with the possible exception of *Azoarcus* species in the class *Betaproteobacteria* (B. Reinhold-Hurek, personal communication). Muta-

TABLE BXII.γ.123. Differential characteristics of the species of the genus *Azotobacter*[a]

Characteristic	1. A. chroococcum	2. A. armeniacus	3. A. beijerinckii	4. A. nigricans	5. A. paspali	6. A. salinestris[b]	7. A. vinelandii
Motility	+	+	−	−	+	+	+
Water soluble pigments:							
Yellow-green fluorescent[c]	−	−	−	−	+	−	+
Green	−	−	−	−	−	−	d
Brown-black	−	−	−	d	−	+	−
Brown-black to red-violet	−	+	−	+	−	−	−
Red-violet	−	+	−	d	+	−	d
Nitrate reduced to nitrite	+	−	+	+	−	+	+
Carbon sources:[d]							
D-Glucuronate	−	−	+	−	−	+	−
D-Galacturonate	−	+	+	−	−	nd	−
Glutarate	−	d	−	−	−	nd	+
Glycolate	−	−	−	−	−	nd	+
Phenol	−	−	d	−	−	nd	+
Rhamnose	−	−	−	−	−	−	+
Caproate	+	−	−	−	−	nd	+
Caprylate	−	+	−	−	−	nd	+
meso-Inositol	−	+	+	−	−	nd	+
Malonate	+	−	−	d	−	nd	+
Propan-1-ol	−	−	d	d	−	nd	+
Trehalose	+	+	d	+	−	nd	−
Starch hydrolyzed	+	+	d	d	−	nd	−

[a]For symbols see standard definitions; nd, not determined.

[b]Although there are limited data for *A. salinestris*, many of the carbon sources used by *A. chroococcum* also support growth of this species (Page and Shivprasad, 1991a).

[c]On iron-deficient medium. See Procedures for Testing Special Characters.

[d]The following carbon sources are not used by any species: lactose, mannose, xylose, arabinose, ribose, and fucose.

TABLE BXII.γ.124. Other characteristics of the species of the genus *Azotobacter*[a]

Characteristic	1. A. chroococcum	2. A. armeniacus	3. A. beijerinckii	4. A. nigricans	5. A. paspali	6. A. salinestris	7. A. vinelandii
Peritrichous flagella	+	+	−	−	+	+	+
Nitrogen fixation occurs at pH:[b]							
5.0–5.5	−	−	d	d	−	−	−
6	−	−	+	d	d	d	+
6.5–9.5	+	+	+	+	+	+	+
10	+	d	+	+	+	−	+
Nitrogen fixation genes:[c]							
nif	+	d	+	+	+	d	+
vnf	+	nd	+	+	+	d	+
anf	−	nd	−	−	+	nd	+
Growth at a temperature of:							
9°C	−	−	d	d	−	−	−
14°C	d	−	+	d	+	−	+
18°C	+	D	+	+	+	−	+
32°C	+	+	+	+	+	d	+
37°C	d	d	−	−	+	+	+
Diffusible exopolysaccharides produced	+	d	d	+	−	d	d
Peroxidase	d	d	d	d	−	+	+
Urease	+	+	+	+	+	+	+
Oxidase	+	d	+	+	+	nd	+
Production of H₂S from:							
Thiosulfate	d	−	d	d	+	nd	+
Cysteine	−	−	−	−	d	nd	d
Utilization as sole carbon source:[e]							
Fructose, glucose, acetate, pyruvate, fumarate, malate, succinate, α-oxoglutarate, lactate, DL-gluconate, or acetyl methylcarbinol	+	+	+	+	+	+	+
Sucrose	+	d	+	+	+	+	+
Propionate	+	d	+	−	−	nd	+
β-Phenylpropionate	d	−	−	−	−	nd	+
n-Butyrate	+	+	+	d	−	nd	+
Glutarate	−	d	−	−	−	nd	+
DL-β-Hydroxybutyrate	d	+	+	d	−	nd	d
Benzoate	d	−	+	−	−	+	+
Trehalose	d	+	d	+	−	nd	−
Melibiose	+	+	+	d	−	+	+
Maltose	+	+	d	d	−	nd	+
Raffinose	+	d	d	d	−	nd	d
Propan-1-ol	d	−	d	d	d	nd	+
Butan-1-ol	+	d	+	d	−	nd	+
Glycerol	d	−	d	−	−	−	+
Mannitol	+	+	d	d	−	nd	+
Sorbitol	+	+	d	d	−	nd	+
Susceptibility to antimicrobial agents:							
Streptomycin, 0.2 μg/ml	S	S	S	S	S	S	S
Tetracycline, 2 μg/ml	d	R	d	d	R	nd	d
Chloramphenicol, 25 μg/ml	d	d	d	R	d	R	R
Sulfanilamide, 25 μg/ml	d	R	d	d	S	nd	d
Penicillin G, 5 U/ml	d	S	d	R	R	nd	d
Phenol, 0.05%	R	S	d	d	d	nd	R
Sodium benzoate, 0.5%	d	R	d	R	d	nd	R
Sodium fluoride, 0.01 M	d	R	R	d	R	nd	R
Mercuric chloride, 10 μg/ml	d	R	d	d	R	nd	d
Iodoacetate, 1mM	S	S	S	S	S	nd	S
Erythromycin, 2 μg/ml	S	R	d	R	d	nd	R
Neomycin or kanamycin, 1 μg/ml	S	S	S	S	S	S	S

[a]For symbols see standard definitions; S, susceptible; R, resistant; nd, not determined.

[b]Results of Thompson and Skerman (1979) on solid media. Machado and Döbereiner (1969) found no growth of *A. paspali* above pH 8.0 in their liquid medium.

[c]See Procedures for Testing Special Characters.

[d]Although there are no reports of PCR or hybridization results to confirm the presence of these genes, they can be inferred to be present on the basis of tests for diazotrophic growth and/or nitrogenase activity.

[e]See Table BXII.γ.123 for additional compounds; also see Thompson and Skerman (1979).

tions in *nifL* can result in constitutive expression of *nif* genes leading to unregulated nitrogenase enzyme overexpression and excretion of ammonium in *A. vinelandii* (Bali et al., 1992). Unlike in most other diazotrophs in the *Proteobacteria*, ammonium inhibition of nitrogen fixation in *A. vinelandii* occurs at only one target, namely, inactivation of NifA by NifL. Posttranslational control of nitrogenase activity by ammonium might occur in *A. chroococcum* (Munoz-Centeno et al., 1996).

In addition to using the classic Mo-containing enzyme for nitrogen fixation, all *Azotobacter* species examined have the potential to synthesize one or more alternative nitrogenases that do not contain Mo. Evidence for such alternative nitrogenases first came from studies with *A. vinelandii* mutants UW10 and UW9—derivatives of AvOP (UW) (ATCC 13705) carrying mutations in the *nifH* and *nifD* genes which encode protein subunits of nitrogenase. Bishop et al. (1980) found that these mutants gave rise to Nif$^+$ colonies at low frequency and that these strains were resistant to tungstate, an antagonizer of Mo transport. In addition, both UW10 and UW91 were able to grow and fix nitrogen if Mo was eliminated from the growth medium, and it was proposed that *A. vinelandii* could synthesize a Mo-independent form of nitrogenase. Subsequently, by genetics, biochemistry, and gene sequencing, *A. vinelandii* was found to encode three different nitrogenase enzymes with different structural subunits: (1) the traditional Mo nitrogenase (nitrogenase-1, the structural subunits of the enzyme encoded by the *nifHDK* genes), (2) a V-containing enzyme (nitrogenase-2, encoded by the *vnfH vnfDGK* genes), and (3) a Fe-containing nitrogenase (nitrogenase-3, encoded by the *anfHDGK* gene cluster) (see Fig. BXII.γ.136 for *nif, vnf*, and *anf* gene cluster arrangements). These alternate nitrogenase systems have been reviewed by Bishop and Joerger (1990) and Rangaraj et al. (2000). Parallel studies with *A. chroococcum* strain MCD1 have indicated the presence of the Nif and Vnf enzymes only. They are encoded by *nif* and *vnf* genes, which have sequences and arrangements similar to those in *A. vinelandii* (see references in legend to Fig. BXII.γ.136). In *A. vinelandii*, all three enzymes require the activity of five other *nif*

gene products found within the major and minor clusters of *nif* genes, including *nifB, nifM, nifS, nifU,* and *nifV* (Joerger and Bishop, 1988; Kennedy and Dean, 1992). Some of these *nifB* and *nifV* genes are required for Fe–V cofactor biosynthesis (Ruttimann-Johnson et al., 1999) and presumably also for Fe–Fe cofactor formation. The molybdate or vanadate concentration in the growth medium (usually Burk's sucrose medium treated to extract metals; see Procedures for testing Special Characters) with 30 µM Fe determines which nitrogenase is available (Bishop and Joerger, 1990). Thus, Mo (whether or not V is present) at a concentration of about 50 nM represses *vnf* and *anf* gene expression. In cultures grown at 30°C, molybdate deficiency (less than approximately 50 nM) prevents expression of the nitrogenase structural genes, *nifHDK* (Jacobson et al., 1986), and when vanadate is present (approximately 50 nM), the *vnfHDGK* genes but not the *anfHDGK* genes are expressed. Repression by Mo of *vnf* and *anf* gene expression occurs far less severely in cultures grown at lower temperatures (14°C or 21°C) (Walmsley and Kennedy, 1991), probably because Mo is less well transported at lower temperatures. This characteristic may have ecological significance, possibly correlating with the better efficiency of the V-nitrogenase for reduction of acetylene at lower temperatures (Dilworth et al., 1988).

Three strains of *A. chroococcum* [MCD1 (ATCC 4412), ATCC 480, and ATCC 9043] have been analyzed for alternative nitrogenase genes by Southern hybridization. All were found to have *nif* and *vnf* genes, but not *anf* genes (Fallik et al., 1991). Con-

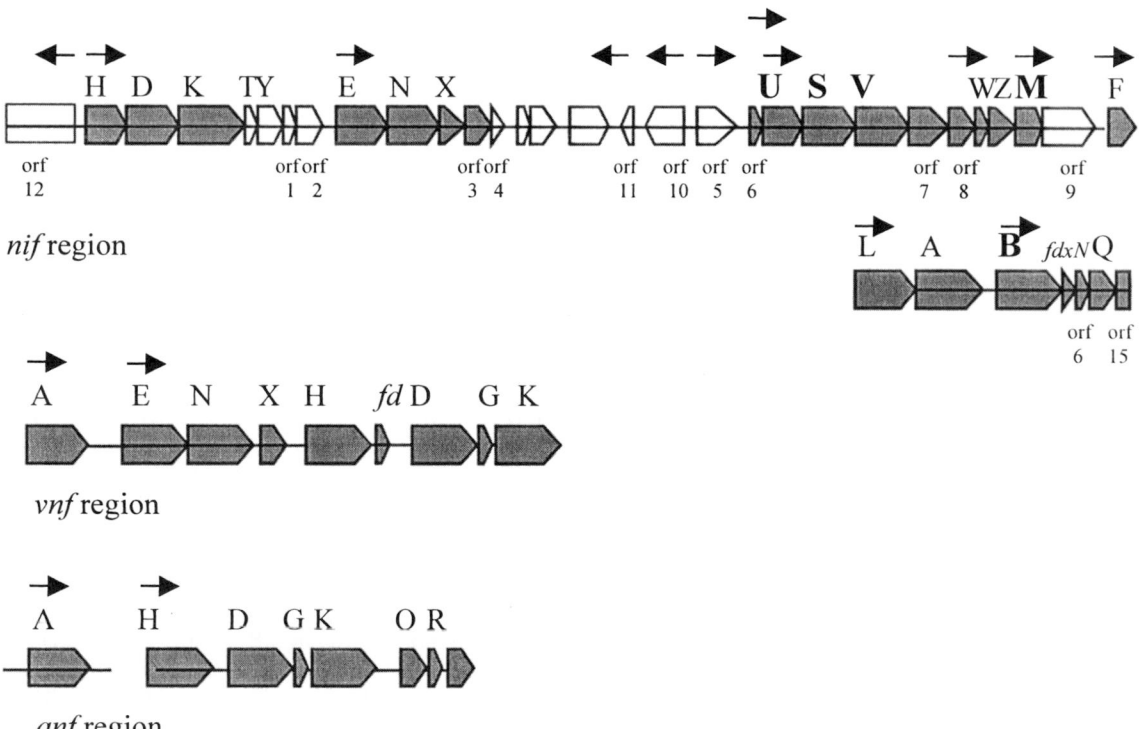

FIGURE BXII.γ.136. Physical map of the *A. vinelandii* nitrogen-fixation genes. *nif* genes encode products involved in the synthesis of the primary Mo-containing enzyme (Jacobson et al., 1989; Joerger and Bishop, 1988; Joerger et al., 1990) while the *vnf* (Joerger et al., 1990) and *anf* clusters encode alternative nitrogenases (see text) (Joerger et al., 1989; Mylona et al., 1996). Bold letters in larger font indicate genes which are essential for all three nitrogenase systems. *Open arrows* depict genes that are not essential for nitrogen fixation (Jacobson et al., 1989). *Small arrows* indicate locations of promoters and direction of transcription. *A. chroococcum nif* and *vnf* genes are arranged similarly (Evans et al., 1988; Fallik and Robson, 1990).

sequently, this property may characterize the species (Nif$^+$, Vnf$^+$, Anf$^-$). *A. beijerinckii* (ATCC 19360), *A. nigricans* (ATCC 35009), and *A. salinestris* (ATCC 49674) also apparently contain *nif* and *vnf* genes only (Fallik et al., 1991; Loveless and Bishop, 1999). Only *A. vinelandii* and *A. paspali* (strains ATCC 23367 and ATCC 23833) among the *Azotobacter* species (and also *Azomonas agilis*) carry *nif*, *vnf*, and *anf* genes and can synthesize any of the three nitrogenases, depending on environmental conditions of Mo or V supply. Consequently, their phenotype is Nif$^+$, Vnf$^+$, Anf$^-$.

Oxygen control of nitrogenase activity in Azotobacters Nitrogen fixation is a process that is oxygen-sensitive due to the O_2-lability of the nitrogenase component proteins. However, in *Azotobacter* species, N_2 fixation can occur under fully aerobic conditions in all species examined except *A. salinestris*. Even the latter species can adapt to aerobic diazotrophy under conditions of Fe limitation, which increases production of a melanin catechol that may trap Fe and be involved in oxygen tolerance (Page and Shivprasad, 1995). In *A. chroococcum*, a shift of cultures to hyperbaric oxygen conditions ($> 20\%$) is followed by a period of cessation of nitrogenase activity which then returns after further incubation, probably because of conformational protection by the FeSII protein, discussed below and in Robson and Postgate (1980). Aerotolerance in *A. vinelandii* is due mainly to a low O_2 affinity, high K_m cytochrome *d* terminal oxidase encoded by the *cydAB* genes. Mutation in either of these genes results in mutant strains unable to grow aerobically on N_2, which represents a loss of respiratory protection. Under low (2%) O_2, the mutants can fix nitrogen and grow (Kelly et al., 1990). The CydR protein represses transcription of *cydAB* under conditions of low O_2 (Wu et al., 1997a) during which the cytochrome *c* pathway provides respiration (Rey and Maier, 1997). A supply of ATP and reducing equivalents are additional factors that may be important to prevent oxygen damage to nitrogenase (Oelze, 2000). Another factor of significance is the FeSII or Shethna protein (Dervartanian et al., 1969). This protein binds to and protects Mo nitrogenase from oxygen damage—but apparently does not bind to or protect either of the alternative Vnf or Anf nitrogenases—during periods of O_2 stress (Dervartanian et al., 1969; Moshiri et al., 1994). *A. vinelandii* mutant strains lacking the FeSII Shethna protein show reduced viability under carbon deprivation if O_2 is present (Maier and Moshiri, 2000). This protein is also found in *A. chroococcum* where it provides "conformational protection" from O_2 damage to nitrogenase (Robson, 1979). An FeSII DNA probe from *A. vinelandii* hybridized only to *A. chroococcum* genomic DNA among several diazotrophs tested (Moshiri et al., 1994). The FeSII protein may therefore be found only in species of *Azotobacter* and thus may be a distinguishing characteristic. Production of alginates and an alginate capsule may form a barrier to oxygen transfer (Sabra et al., 2000), adding to the battery of mechanisms elaborated by azotobacters for protecting nitrogenase from damage by oxygen in an obligate aerobe.

Other nitrogen sources Most *Azotobacter* species can utilize nitrate as a nitrogen source in an assimilatory fashion by reduction to nitrite and then ammonium. Genes for nitrate and nitrite reductases, *nasB* and *nasA*, have been isolated from *A. vinelandii* OP. The *nasAB* operon, regulated from a σ^{54}, NtrC-dependent promoter, is repressed by ammonium and requires nitrate for full expression (Ramos et al., 1993). Soluble nitrate reductases have been purified from *A. vinelandii* and *A. chroococcum* (Guerrero et al., 1973; Gangeswaran et al., 1993). Nitrate transport and reduction have been studied in *A. chroococcum*. Although the

genes have not been characterized in this organism, the physiology of nitrate assimilation is similar to that in *A. vinelandii*. Ammonium prevents full synthesis/activity of the enzyme, and nitrate is required for full activity.

All *Azotobacter* species are urease positive in physiological tests—a feature that distinguishes *Azotobacter* from *Azomonas* species (Thompson and Skerman, 1979). This feature is consistent with the fact that urea is a good nitrogen source in both *A. vinelandii* and *A. chroococcum*. It is also a useful trait, because urea added to media at 5–10 mM does not prevent nitrogenase synthesis, in contrast to the effects of ammonium. Consequently, urea allows studies of expression of these genes in liquid cultures (e.g., Walmsley and Kennedy, 1991).

Other aspects of metabolism H_2 uptake systems are similar in both *A. vinelandii* and *A. chroococcum*, having been best characterized physiologically in the latter species. H_2 recycling occurs in N_2-fixing cultures: H_2, a byproduct of the reduction of N_2 to NH_4^+, is re-oxidized to H^+ by a Ni-dependent hydrogenase encoded by the 16-gene *hup* cluster in *A. chroococcum* (Du and Tibelius, 1994; Du et al., 1994) and by a similar 16-gene *hox/hyp* cluster in *A. vinelandii* (Menon et al., 1992; Garg et al., 1994). Hydrogenase activity is apparently beneficial in *A. chroococcum*, as shown by the ability of the Hup$^+$ strain MCD-1 to outcompete a Hup$^-$ mutant for survival under controlled carbon-limited chemostat conditions (Yates and Campbell, 1989). In *A. vinelandii*, both Hox$^+$ and Hox$^-$ strains yielded very similar amounts of protein in chemostat cultures (Linkerhagner and Oelze, 1995). In addition, both had similar respiratory activities, and, in the wild-type, only a small fraction of total respiratory activity was due to H_2-dependent O_2 consumption. This finding led the authors to conclude that hydrogenase does not benefit *A. vinelandii* during carbon-limited growth under N_2-fixing conditions.

In regard to ferredoxins, it is remarkable that *A. vinelandii* can synthesize at least 12 different ferredoxin-like proteins (Jung et al., 1999), some of which—Fe-SI, FixX, FdxN, FixFd—have been implicated in electron transfer to nitrogenase. However, no mutation in a single ferredoxin-encoding gene prevents electron transfer to nitrogenase; thus, there is an apparent redundancy of function in this respect. The VnfFd ferredoxin is required for activity of the V-containing nitrogenase (Fig. BXII.γ.136) (Raina et al., 1993). Another ferredoxin is the Fe-SII Shethna protein, described above, which protects nitrogenase from inactivation by O_2. Known ferredoxins in *A. chroococcum* include those encoded by a gene adjacent to *vnfH*, and FdI and FdII, corresponding to ferredoxins characterized in *A. vinelandii* (George et al., 1984; Robson et al., 1986; Thomson, 1991).

Genetic Characteristics *A. vinelandii* can be readily studied genetically because the widely-used, non-gummy strain AvOP (ATCC 13705), derived from AvO (ATCC 12518), is easy to handle. AvOP was also named strain UW in studies at the University of Wisconsin (Bishop and Brill, 1977) and CA in studies at North Carolina State University (P. Bishop laboratory). This strain harbors no indigenous plasmids, is a high-frequency recipient during conjugation, and can be transformed with either linear or plasmid DNA. It becomes competent for DNA uptake after growth on Fe- and Mo-free medium (Page, 1985). Another widely-used derivative of UW is UW136, wild-type for nitrogen fixation but carrying a mutation for rifampicin resistance (Bishop and Brill, 1977). The features of UW and UW136 allow easy genetic studies and manipulation by introduction of wide-host-range plasmids and isolation of mutant strains, either by transposition or gene replacement techniques (see later). In addition, nitrogenase and other important enzymes have been isolated and analyzed from strain AvOP (UW) because this strain is relatively easy to grow on either a small or large scale for preparation of extracts for protein purification. The strain has more recently become a focus of study for alginate biosynthesis. To date, sequences are available for more than 200 of its genes.

Azotobacter chroococcum has also been well-studied genetically with respect to a number of traits, and although there are no reports of successful transformation in strains of this species, the cells can receive plasmids by conjugation with *Escherichia coli* donors. Strain MCD1, a non-gummy derivative of ATCC 4412 (NCIB 8033) cured of three of five indigenous plasmids, has been used for most genetic studies (Robson et al., 1984). The sequences of approximately 50 genes in *A. chroococcum* have been reported to date.

The genome sizes of several *Azotobacter* species have been determined by pulsed-field gel electrophoresis of large chromosomal DNA fragments generated by rare-cutting restriction endonucleases. The sizes range from 3.10 Mb for *A. chroococcum* NCIB 9043 to 4.57 Mb for *A. vinelandii* UW (MacDonald and Melton, manuscript submitted) (Table BXII.γ.125). In another study, Manna and Das (1994) found the M4 strain of *A. chroococcum* to have a genome size of 5.3 Mb. This size disparity supports previous findings of genomic variation within *A. chroococcum* (Becking, 1992). The presence of a single circular chromosome in *A. vinelandii* is indicated from genetic analysis (Blanco et al., 1990). The size of the *A. paspali* genome is about the same as that of *A. vinelandii*, ~4.5 Mb, even though it might be expected that the *A. paspali* genome would be smaller because the organism uses fewer compounds as carbon sources and has a more restricted rhizosphere habitat.

The origin of replication of *A. vinelandii*, including a 200 bp segment capable of replicating plasmid pBR322 in *E. coli* polA, has been identified and sequenced. A larger fragment of 1652 bp, which includes the 200 bp ori sequence, contains 14 putative DnaA protein binding sites, ten putative binding sites for the IHF protein, and 19 GATC boxes (potential sites for Dam protein methylation), all of which are involved in initiation of DNA replication in *E. coli* and other organisms (Singh et al., 2000). Genes involved in general recombination in *A. vinelandii* include *recA*

and *recF* (Venkatesh and Das, 1992; Badran et al., 1999). The former is more closely related to the *recA* gene of *P. aeruginosa* than to that of *E. coli*, which supports other phylogenetic data discussed below.

Multiple chromosomes are common in the two species examined, *A. vinelandii* and *A. chroococcum*. In *A. vinelandii*, the chromosome number is variable and can reach about 80 copies per cell in late-exponential phase cells; in old stationary-phase cultures the amount of DNA is drastically reduced, possibly in preparation for cyst formation (Maldonado et al., 1994). In exponential cultures of *A. chroococcum* MCD-1, cells contained 20–25 genome equivalents (Robson et al., 1984).

No indigenous plasmids in *A. vinelandii* AvOP (UW) wild-type and derived mutant strains have been detected in the author's laboratory or by several other investigators who study this organism. One strain of *A. paspali* did not harbor any plasmids (Do Nascimento and Tavares, 1987). Robson et al. (1984) reported finding between two and six plasmids in each of eight strains of *A. chroococcum* examined; these plasmids ranged in size from 10.5 kb to 300 kb. The *A. chroococcum* strains included NCIB 8003 (ATCC 4412); strain MCC-1, a Nalr Smr and non-gummy derivative of NCIB 8003; and six new isolates from soils in Sussex, UK. One plasmid in NCIB 8003 appeared to be of the IncP class. Plasmids were cured by growing strains for several generations using the agents ethidium bromide, acriflavin, or mitomycin. The cured strains showed no altered phenotypes that are usually associated with plasmids, except for possibly one trait, the production of agar-diffusible exopolysaccharides. Evidence for a plasmid encoding a gene or genes for degradation of 2,4-dichlorophenoxyacetic acid (2,4-D) in strain MSB1 of *A. chroococcum* was reported (Balajee and Mahadevan, 1989). The presence of single or multiple plasmids might be a distinguishing feature of *A. chroococcum*. Reports of plasmids in other species of *Azotobacter* are not available.

Wide-host-range plasmids of the IncQ and IncP classes can be transferred from *E. coli* by conjugation to *A. chroococcum* strain MCD1 and its derivatives (Kennedy and Robson, 1983; Ramos and Robson, 1987) and by either conjugation or transformation to *A. vinelandii* strain AvOP and mutant derivatives (Kennedy and Robson, 1983; Doran et al., 1987; Kennedy and Toukdarian, 1987; Glick et al., 1989). This genetic manipulation allows for the isolation of numerous genes from IncP cosmid libraries by mutant complementation, and analysis of gene function by specific genes cloned into either IncP or IncQ plasmids. Introducing wide-host-range plasmids of the IncQ compatibility group can result in reduced growth rates in *A. vinelandii* (Glick et al., 1986).

The state of competence for transformation of strains of *A. vinelandii* is associated with siderophore production and can be induced by growth on Fe-deficient or Fe- and Mo-deficient media (Page, 1985; Page and Grant, 1987). Competent cells can be transformed with either plasmid or linear DNA (Doran et al., 1987). Interestingly, the transformation frequency was enhanced 10- to 50-fold when pKT210 (IncQ) carried an insert fragment of *A. vinelandii nif* DNA, suggesting that *A. vinelandii* possesses a homology-facilitated transformation system. Mutated genes can be introduced directly on plasmids unable to replicate or on linear chromosomal DNA prepared from mutant strains; recombination with the chromosome at sites flanking the mutation leads to replacement of the wild-type genes (e.g., as described by Jacobson et al., 1989, and in many other reports).

The rRNA genes of several aerotolerant diazotrophs were an-

TABLE BXII.γ.125. Comparison of genome size and rRNA (*rrn*) copy number among isolates of the family *Azotobacter* and *Azomonas*

Strain	Genome size[a] (Mb)	Copies of *rrn* genes
Azotobacter chroococcum (ATCC 7493)	3.9 ± 0.02	8
Azotobacter chroococcum (ATCC 9043)	3.1 ± 0.32	7
Azotobacter chroococcum[b] M4	5.3	NR[c]
Azotobacter paspali (ATCC 23822)	4.3 ± 0.24	6
Azotobacter paspali AP1	4.5 ± 0.05	6
Azotobacter paspali (ATCC 11562)	4.6 ± 0.01	6
Azotobacter paspali (ATCC 12095)	4.6 ± 0.01	6
Azotobacter vinelandii UW (ATCC 1218)	4.6 ± 0.35[d]	6
Azomonas macrocytogenes	2.4 ± 0.01	7

[a]Genome size determined by summing restriction fragments generated by rare-cutting enzyme digestion. Values are derived from at least three data sets of two separate enzyme digestions (MacDonald and Melton, manuscript submitted).

[b]Determined by Manna and Das (1994).

[c]NR, not reported.

[d]Manna and Das (1993) reported a genome size of 4.5 Mb for this strain.

alyzed in hybridization experiments by De Smedt et al. (1980) to determine their phylogenetic relationship. The results indicated the validity of the grouping of all the *Azotobacter* species identified in this description and the species described below (with the exception of *A. salinestris*, which was not available at the time of their analysis). In a further and more recent analysis, the *Azotobacter* species *A. vinelandii* (ATCC 13705 derivative, CA), *A. chroococcum* (ATCC 9043), and *A. salinestris* (ATCC 49674) all carry an arrangement in the rRNA operon that consists of the 16S rRNA gene (*rrs*), a spacer region containing the tandem tRNA gene sequences for tRNAIle and tRNAAla, the 23S rRNA gene (*rrl*), a second spacer region, and the 5S rRNA gene (MacDonald and Melton, manuscript submitted). *A. paspali* (ATCC 23833) also possesses the 16S–23S–5S gene arrangement; however, the tRNAs within the 16S–23S spacer region were not identified. All the *Azotobacter* species examined here contain six, seven, or eight duplicated copies of the rRNA gene cluster (Table BXII.γ.125).

Not surprisingly, the codon usage of *A. vinelandii* is similar to that of *A. chroococcum* (Rudnick and Kennedy, unpublished analysis). The mol% G + C content of the DNA of the former species was earlier estimated from T_m experiments to be 64.9–66.5 and the latter to be 65.8–67.5% (De Ley, 1968). For genes that encode proteins, the sequences of individual genes available for each species numbering about 200 and 50, respectively, indicated a mol% G + C content of 62.5 for *A. vinelandii* and 66.7% for *A. chroococcum*. Interestingly, the *anf* genes of *A. vinelandii* have a mol% G + C value (58.3) lower than that of the *nif* and *vnf* genes (64.4% each). The *anf* cluster of *A. paspali* is also lower in mol% G + C (54.3%) than that of the genome (63.2–64.6%, from T_m experiments). In addition, the clusters of genes encoding uptake hydrogenases, *hox/hyp* in *A. vinelandii* and *hup* in *A. chroococcum*, had a higher mol% G + C value than the total DNA in these species—70.7 and 70.2, respectively. This may reflect the acquisition of these genes—both *anf* and *hup/hox*—by lateral transfer from other organisms during the course of evolution of the genus *Azotobacter*.

Isolation of mutants Transposon mutagenesis is easily achieved in *A. vinelandii* and *A. chroococcum* using Tn5 carried on suicide vectors unable to replicate in these host organisms (for examples, see Kennedy et al., 1986; Joerger et al., 1989; Contreras et al., 1991; Luque et al., 1993; Tibelius et al., 1993; Wu et al., 1997a). In addition, Tn10 was successfully used to isolate mutants of *A. vinelandii*, including methionine auxotrophs (Contreras and Casadesus, 1987). When Tn5 was introduced into *A. beijerinckii* (NCIB 11292), the cells of most of the isolates had abnormal morphologies and rapidly lost viability when subcultured. In contrast, Tn76 was successfully used to make mutants in this strain (Owen and Ward, 1985). Contreras et al. (1991) developed a method to "freeze" Tn5 mutations in *A. vinelandii* by replacing the original transposon with a defective Tn5, resulting in highly stable mutants.

Isolation of auxotrophic mutants of *A. vinelandii*, attempted in many laboratories during the 1970s and 1980s, was largely unsuccessful. It was believed that high DNA content (originally reported by Sadoff et al., 1979), and hence high chromosome copy numbers, leads to difficulty in segregation of mutant phenotypes. However, the successful isolation of many other types of mutations over the past 15 years by transposon mutagenesis or antibiotic resistant cassette insertion in cloned genes, followed by transfer into strains and subsequent gene replacement, has

shown that isolation of mutant strains of *Azotobacter* is not inherently difficult. The variable DNA content of this organism, described above, does indicate that there are growth phases and conditions when the chromosome number is low, thereby allowing for "normal" segregation of introduced, selectable mutations in nonessential genes.

Although isolation of auxotrophs remains difficult, success has been achieved in the construction of Ade⁻, Met⁻, and Leu⁻ mutants (Mishra and Wyss, 1968; Contreras and Casadesus, 1987; Manna and Das, 1997). *A. vinelandii* is unable to transport many amino acids (Toukdarian et al., 1990; D. Dean, personal communication). Although Mishra et al. (1991) reported that threonine was transported at a rate similar to that found for *E. coli* and methionine at a rate more than twice that in *E. coli*, five others tested were transported at a fraction of the rate in *E. coli*. In *A. beijerinckii*, RP4::Tn76 was successfully used to isolate Nif⁻ mutants and auxotrophs requiring adenine or leucine (Owen and Ward, 1985). D. Dean and coworkers (personal communication) found that tryptophan auxotrophs could be isolated in *A. vinelandii* only if the strain harbored a wide-host-range plasmid carrying the tryptophan transport genes from *Escherichia coli*. Also consistent with the idea that amino acids are not easily transported into species of *Azotobacter* and *Azomonas* are the data of Thompson and Skerman (1979), who found that none of the 17 amino acids tested could serve as carbon sources in any of the species of these genera. Thus, limited capability of amino acid transport is probably a general characteristic of *Azotobacter* and *Azomonas*.

Mutations have now been introduced into many genes of *A. vinelandii* and *A. chroococcum*, resulting in gene replacement and the construction of mutant strains. A Nif⁻ mutant of *A. paspali* was constructed by the conjugation and recombination of a kanʳ cassette with the *nifHDK* operon, leading to gene replacement (Fallik et al., 1993). Some mutations are "lethal" in *A. vinelandii*, due to the essential nature of the genes under study. These genes include *glnA*, which encodes glutamine synthetase—the sole pathway for assimilation of ammonium (Kennedy and Toukdarian, 1987; Toukdarian et al., 1990); the *isc* and *fdxD* genes, which encode enzymes and a ferredoxin for Fe–S cluster formation (Zheng et al., 1998; Jung et al., 1999); and the *glnK* and *glnD* genes, which encode proteins that sense and signal the status of the nitrogen supply to glutamine synthetase to activate or inactivate the enzyme (Meletzus et al., 1998; Colnaghi et al., 2001). The ability of *A. vinelandii* to harbor high copy numbers of the chromosome is an advantage in that this feature allows identification of essential genes. Both wild-type and mutated copies marked by antibiotic resistance gene insertion can be maintained and identified by hybridization or PCR analysis; when selective pressure is removed by growth in medium without antibiotics, the chromosomes carrying the mutated essential genes are rapidly lost along with antibiotic resistance.

Bacteriophages Bacteriophages isolated from soil using *A. vinelandii* and *A. chroococcum* as indicator strains are capable of lysing cultures of *A. chroococcum*, *A. vinelandii*, and, to a lesser degree, *A. beijerinckii*; however, they do not affect *Azomonas* (Hegazi and Jensen, 1973). Four serologically distinguishable phages have been identified that can apparently lysogenize *A. vinelandii* strain O (ATCC 12518), resulting in a pseudolysogenic state in which the bacteria are converted to an unencapsulated form (Thompson et al., 1980). Bishop et al. (1977) described a general technique for isolating phage able to infect *A. vinelandii* from soil samples.[6] No information on phages is available for *A. nigricans*, *A. armeniacus*, and *A. paspali*.

Antigenic structure The most recent and informative study is that of Tchan et al. (1983), which, by a method of immuno-electrophoresis, related cross-reacting antigenic determinants between and within the species of *Azotobacter* and of *Azomonas*. In these studies, species of *A. vinelandii* and *A. paspali* were found to be immunologically more homogeneous than strains of *A. chroococcum*—probably because of the wider geographical distribution of isolates of the latter. In addition, *A. vinelandii* and *A. paspali* were more similar to each other than to other *Azotobacter* species, but all were related. *Azomonas* species were related at about the same level to each other as to *Azotobacter* species.

Antibiotic sensitivity Table BXII.γ.124 summarizes the data from Thompson and Skerman (1979), from other researchers, and from one of the authors' laboratory. Of special note is that all isolates of *Azotobacter* are very sensitive to streptomycin and kanamycin/neomycin: the minimal inhibitory concentration (MIC) for streptomycin is <0.2 µg/ml, and for kanamycin/neomycin it is <1.0 µg/ml. This was also observed by Page and Shivprasad (1991a) for a more recently distinguished *Azotobacter* species, *A. salinestris*. These concentrations are important for constructing mutants that carry cassettes encoding genes for resistance to these antibiotics. Most strains are also relatively insensitive to chloramphenicol (MIC being 50 µg/ml or more), useful in counterselection of *E. coli* donor strains in conjugation experiments.

Ecology and plant growth stimulation *Azotobacter* species are generally found in soils of slightly acid to alkaline pH (reviewed in Becking, 1992). While the numbers are generally fairly low (<10^4 cells per g of soil), they are found throughout the world, typically in 30–80% of soils sampled, but rarely in polar regions. Because of their relatively high requirement for phosphorus, they are more commonly found in fertile soils than in sand. The numbers are generally not higher in rhizospheres than in open locations (Jensen, 1965). There are several reports of isolation of *Azotobacter* species from plant leaves and other surfaces. Most isolates are of *A. chroococcum*, which has also been isolated from freshwater canals and watercourses. *A. vinelandii* has been isolated far less frequently and may be more abundant in tropical regions; it has occasionally been isolated from freshwater (Thompson and Skerman, 1979). *A. beijerinckii* is found more often in acid soils than the other *Azotobacter* species, which reflects its laboratory tolerance to lower pH. *A. salinestris*, a Na-dependent species, has been isolated from slightly saline soils in western Canada, in Egypt, and in Australia (Page and Shivprasad, 1991a; Wang et al., 1993). The other species are rarely isolated, and may be restricted regionally. Little information is available for *A.*

nigricans and *A. armeniacus*, and the distribution of these two species in their habitats remains unknown. Only *A. paspali* is known to have a constant association with the root system of a plant, *Paspalum notatum* cv Batatais (Döbereiner, 1966), where availability of organic substances and a more suitable pH is conditioned by the plant in its rhizosphere. The restriction of *A. paspali* to the rhizosphere of *Paspalum* may be related to its reduced ability to utilize many organic substances that can be utilized by other species of *Azotobacter* that are more widely distributed in soils. In addition, *A. paspali* is the only species antagonistic to Gram-positive bacteria, a property that may be advantageous for life in the rhizosphere (Thompson and Skerman, 1979).

The ability of *Azotobacter* species, mainly *A. chroococcum*, to enhance the growth of various crop plants was widely studied in Egypt and Russia in 1950–1970. Although beneficial effects were often observed, on the order of 10–20% better growth, the practice of inoculation of crops has been largely abandoned due to inconsistency of results. The benefits sometimes observed are likely due to the production of plant growth-stimulating substances by azotobacters, rather than to transfer of bacterially fixed nitrogen to the plants. *A. paspali* synthesizes and excretes indole acetic acid, three gibberellins, and two cytokinins (Barea and Brown, 1974). Stimulation of growth of inoculated plants (both tomato and *Paspalum notatum*) was observed by these authors, even though no acetylene reduction activity could be measured. *A. paspali* also enhanced growth of several crop seedlings, particularly in the root surface area (Abbass and Okon, 1993b) (Fig. BXII.γ.137). Although *A. paspali* initially colonized the roots in both sets of experiments, cell survival was low over time (Barea and Brown, 1974; Abbass and Okon, 1993b). Two strains of *A. chroococcum* that were able to stimulate growth of wheat seedlings produced indoleacetic acid and gibberellins (Pati et al., 1995). In another report, growth of *A. chroococcum* was enhanced in the presence of maize root exudates, which also stimulated the production of auxins, gibberellins, and cytokinins by this organism (Martinez-Toledo et al., 1988). These results should prompt a reexamination of whether numbers of *Azotobacter* might be higher in rhizospheres than in nonrhizosphere soils.

ENRICHMENT AND ISOLATION PROCEDURES

The enrichment techniques are based on the ability of *Azotobacter* to grow by fixing nitrogen aerobically or microaerobically and to use organic substrates as an energy source. Soil, macerated roots or leaves, or other samples are suspended in water; then aliquots of diluted suspension are incubated in liquid or semisoft N-free agar or on the surface of N-free agar plates using Winogradsky's or Burk's medium with 1% glucose or sucrose. All species except *A. salinestris* will grow diazotrophically on agar plates incubated in air, a feature useful for their enrichment and isolation because of the particularly high tolerance to oxygen during diazotrophic growth.

A specific enrichment method for *Azotobacter* species is that of Döbereiner (1995). Plates containing LG growth medium (Burk's medium with 2 ml/l bromothymol blue [0.5% solution in ethanol]) are inoculated with about 100 mg of finely sieved soil suspended in 1 ml of sterile water, or with appropriate dilutions of soil. Colonies of *A. chroococcum* appear after 24 h of incubation at 30°C as moist white colonies that become dark brown after 3–5 d. *A. vinelandii* colonies are similar but do not turn dark brown. *A. paspali* colonies appear only after 48 h; they become yellowish in the center due to assimilation of bromothymol blue and acidification of the medium.

---

6. Sites selected for enrichment were high in decomposed organic material. Inoculum was prepared by growing a 15-liter culture of *A. vinelandii* to about 5×10^8 cells/ml in Burk's medium. After inoculation, each site was irrigated with 1 liter 2% sucrose every other day. Soil sample extracts for phage assay were prepared by suspending 5.0 g soil in 15 ml phosphate buffer. Soil particles were removed by centrifugation. Extracts were filter-sterilized (45 µm pore size), then 1 ml of the sterile extract was added to 19 ml of a culture of *A. vinelandii* (~2.5×10^7 cells/ml). After incubation overnight with vigorous shaking, filter-sterilized samples (0.1 ml) of cultures were tested for the presence of phage by a soft agar overlay method. Four ml of molten 0.5% agar were mixed with 0.1 ml freshly grown *A. vinelandii* cells (~10^8) and with aliquots of extracts and dilutions. The mixture was poured onto a fresh plate of Burk's agar medium and incubated for 24–48 h at 30°C and observed for the presence of phage plaques.

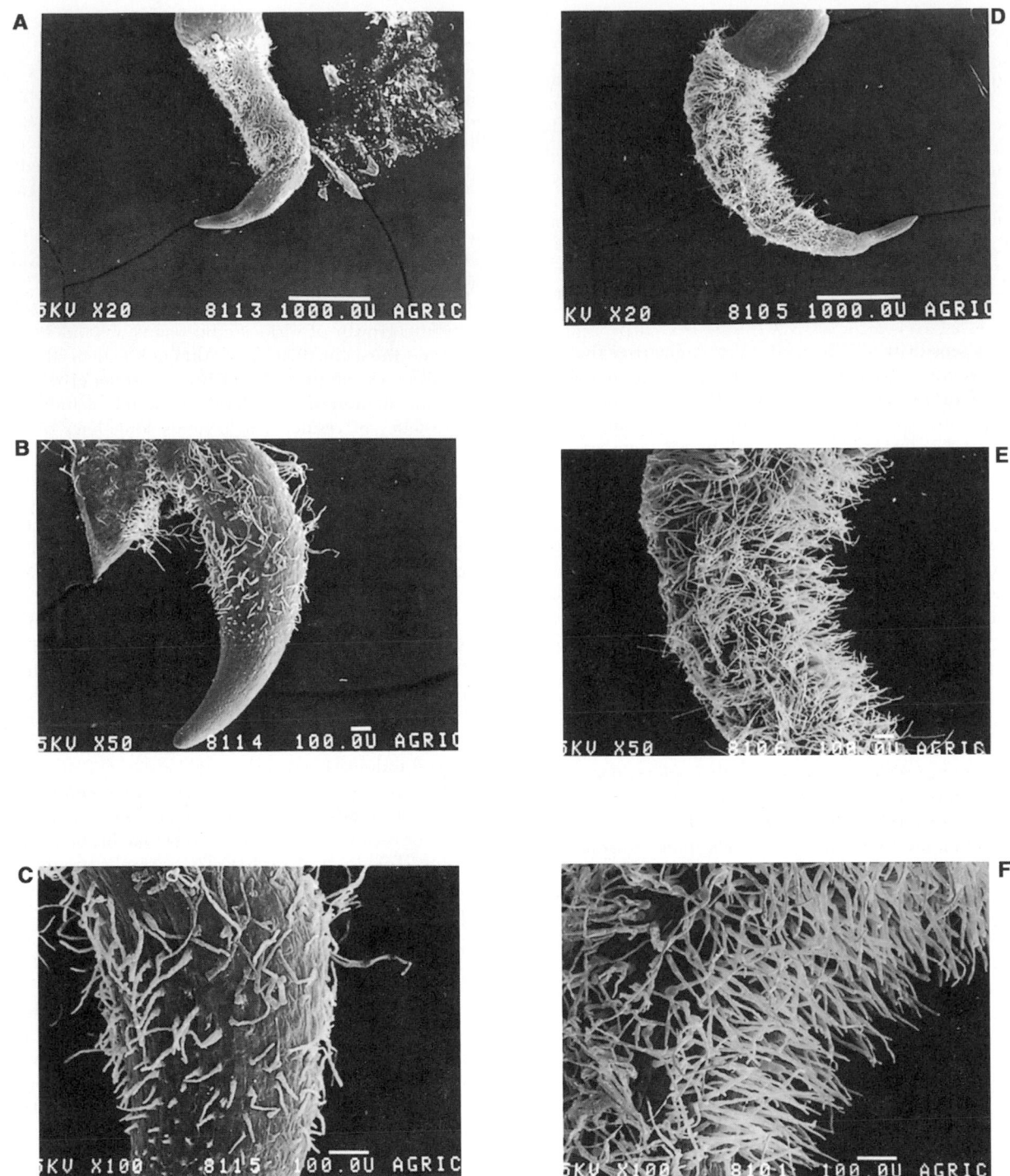

FIGURE BXII.γ.137. Scanning electron microphotographs of the inoculation effects with *A. paspali* (5×10^8 CFU/ml) 48 h after inoculation on the morphology of root hair development of canola seedlings. *Right column*, inoculated seedling; *left column*, uninoculated seedlings.

Thompson and Skerman (1979) described selective enrichment for various *Azotobacter* species based upon the addition to Winogradsky's medium of substances selectively stimulatory or inhibitory to species of *Azotobacter*. *A. vinelandii* can be selectively enriched by using L-rhamnose, ethylene glycol, erythritol, or D-arabitol as the carbon source. Alternatively, 1.0% sodium benzoate or 0.1% phenol can be used to inhibit the growth of other *Azotobacter* species. Incubation at 37°C favors the development of

A. vinelandii. A. beijerinckii can be selectively enriched using L-tartrate, *o*-hydroxybenzoate, D-glucuronate, or D-galacturonate. A pH of 6 or slightly less favors the growth of this species. *A. chroococcum* is the most common species and needs no special enrichment for its isolation. *A. armeniacus* can be enriched with caprylate but *A. vinelandii* will also grow in this medium; however, further isolation can differentiate these two species. *A. paspali* can be isolated from the rhizosphere of *Paspalum notatum*. There is no selective medium for this species but an incubation temperature of 35–37°C will favor its growth. There is no specific method to isolate *A. nigricans*.

For isolation of *Azotobacter* strains, liquid enrichment cultures are inspected daily. When macroscopic growth becomes apparent, the culture is examined with phase-contrast microscopy to detect the presence of *Azotobacter* cells, which are recognizable by their general morphology. Growth is usually obtained within 2–5 d. Positive cultures are streaked onto nitrogen-free agar medium with glucose or sucrose as the carbon source. For selective isolation, nitrogen-free agar media with appropriate selective substances should also be used. Sometimes nitrogen-free iron-deficient agar media can be used to encourage pigment production by some species. For soils containing sufficiently high numbers of *Azotobacter*, the different species can be isolated directly on agar plates: 0.1–0.5 g dry soil or an appropriate dilution is spread on the surface of nitrogen-free agar medium as described above.

MAINTENANCE PROCEDURES

For routine maintenance, *Azotobacter* should be subcultured at monthly or bimonthly intervals on Winogradsky's or Burk's agar medium with sucrose, except that glucose should be substituted for sucrose in the case of *A. armeniacus* and certain other strains which prefer glucose.

Thompson (1989) preserved many strains by suspending the cells from agar-grown cultures (10^9–10^{10} cells/ml) in 0.5 ml of 10% (v/v) glycerol. The suspensions were placed in glass ampules, rapidly frozen, and stored in liquid nitrogen. On revival, ampules were rapidly warmed by agitation in a water bath at 33°C until thawing occurred. All cultures remained viable for up to 7.5 years, grew with short lag phases, and suffered no apparent phenotypic changes. The following method of preservation has been used in the Kennedy lab: high density cell suspensions of *A. vinelandii* in glycerol (10–50%) have been stored in cryogenic microfuge tubes at −20°C or −80°C, with successful revival after four or more years.

Dimethyl sulfoxide (DMSO) has been used in the laboratory of Dr. Dennis Dean, Virginia Polytechnic Institute and State University, with great success for preservation of *A. vinelandii* wild-type and mutant derivatives of strain AvOP (personal communication). One-ml portions of freshly prepared 7% DMSO in 0.1 M sterile phosphate buffer (pH 7) are transferred to a sterile cryotube and a few loopfuls of fresh cells grown on Burk's medium are suspended in the DMSO solution. The suspension is vortexed to ensure a homogenous suspension without aggregates, then frozen at −80°C. The application of this method to other *Azotobacter* species has not been tested.

A paraffin oil preservation method has been used by Tchan and New (1984). Ten milliliters of a suitable nitrogen-free agar medium is poured into a 28 ml capacity McCartney bottle and allowed to solidify to form a slope approximately 3 cm in length with a base 2 cm in height. The culture is inoculated in the middle of the surface of the slope with a single stroke. When the growth reaches about 1 mm in thickness, the culture is checked for purity. Then 5–7 ml of sterile paraffin (mineral) oil is poured into the bottle to completely cover the slope. (It is important that no agar medium should be in direct contact with air, otherwise the culture can dry out and some cultures may die.) Usually *Azotobacter* survives for many months (even years) by this method. This method is particularly advantageous for frequently used stock cultures.

For preservation by lyophilization, dense culture suspensions in 10% sucrose solution (Bascombe and Jackson, 1965) can be freeze-dried and stored at 5°C. This method may not be suitable for all species.

PROCEDURES FOR TESTING SPECIAL CHARACTERS

Identification of alternative nitrogenase genes Although the characterization of species with respect to their ability to grow on Mo-deficient medium, with or without the addition of V, is incomplete in terms of the spectrum of isolates examined, the results of Fallik et al. (1991) and Loveless and Bishop (1999) suggest that (1) *A. chroococcum*, *A. beijerinckii*, *A. nigricans*, and *A. salinestris* carry the *nif* and *vnf* genes; (2) *A. vinelandii*, *A. paspali*, and *Azomonas agilis* have genes for all three nitrogenases, *nif*, *vnf*, and *anf*; and (3) *Azomonas macrocytogenes* has the *nif* and *anf* genes only. In addition, Loveless et al. (1999) described seven new isolates from aquatic samples or wood-chip mulch in North Carolina by their ability to grow on Mo-deficient medium. These organisms, by 16S rDNA analysis, are taxonomically within the *Gammaproteobacteria*, and most related to the fluorescent pseudomonads. All had genes encoding the Mo nitrogenase (nitrogenase-1) and the V nitrogenase (nitrogenase-2). Four of the isolates also encoded the Fe-only nitrogenase (nitrogenase-3) These isolates may represent a new taxon of diazotrophs. The procedures for identification of these genes by hybridization and/or PCR analysis are therefore important in characterizing isolates in terms of the ability to synthesize the alternative nitrogenases that contain V (nitrogenase-2) or Fe only (nitrogenase-3), in addition to the classical Mo nitrogenase (nitrogenase-1) found in all other diazotrophs characterized. The methods for determining the presence of alternative nitrogenases are as follows:

1. Method of Loveless et al. (1999): blots of genomic DNA isolated from test strains restricted with enzymes such as *Eco*RI are hybridized to specific gene probes isolated from *A. vinelandii* and radiolabeled. The probes are *nifD* (from pTMR18; Bishop et al., 1985), *vnfD* (from pVDSJ1; Jacobitz and Bishop, 1992), and *anfD* (from pPJD3A2; Premakumar et al., 1992).

2. Method of Fallik et al. (1991): hybridization of Southern blots of restriction enzyme-digested DNA from test strains to radiolabeled gene probes. The probes consist of *A. chroococcum nifDK* (from pER4; Jones et al., 1984), *vnfDGK* (from pSEQ10; Robson et al., 1989), and *A. vinelandii anfDGK* (from pJWD3; Joerger et al., 1989).

3. Method of Loveless and Bishop (1999): PCR analysis for the presence of the genes *vnfG* or *anfG*, which indicate the presence of nitrogenase-2 and nitrogenase-3, respectively. *vnfG* is located between the *vnfD* and *vnfK* genes, and *anfG* between the *anfD* and *anfK* genes (see Fig. BXII.γ.138). The PCR primers flanking the *vnfG* and *anfG* genes—which anneal to conserved sequences within *vnfD* and *vnfK* or within *anfD* and *anfK*—will amplify the *vnfG* or *anfG* genes if they are present. The PCR products of these reactions can be analyzed by determination of their size on agarose gels and subsequent sequence determination. Sequences of specific primers used for identification of *anfG* and *vnfG* in the *Proteobacteria* and the expected product sizes obtained after PCR amplification are shown in Table BXII.γ.126.

Diazotrophic growth in the absence of molybdenum This procedure is carried out to help reveal the presence of alternative nitrogenase-2 (with the addition of V) or of alternative nitro-

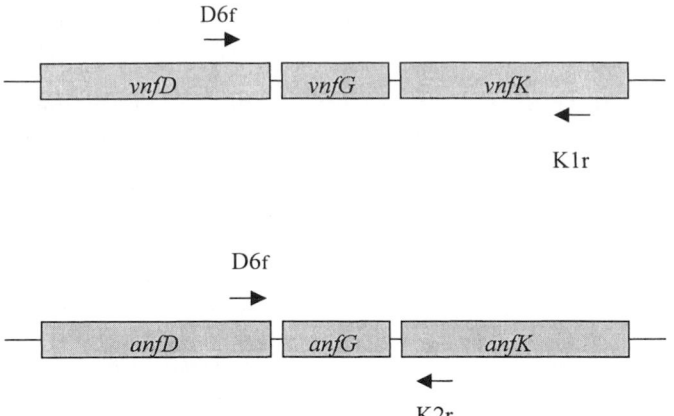

FIGURE BXII.γ.138. PCR primer annealing sites for the amplification of *vnf*G or *anf*G which can be used for the identification of alternative nitrogenases. (Reproduced with permission from T.M. Loveless and P.E. Bishop, Canadian Journal of Microbiology *45*: 312–317, 1999, ©National Research Council of Canada.)

genase-3, traits characteristic of species of *Azotobacter* and *Azomonas*. The preparation of media to test the growth of diazotrophs in the absence of Mo is critical because Mo can contaminate many sources of chemicals used to prepare media and the water supply. Alternative nitrogenase synthesis can be inhibited by as little as 20 nM Mo. Two methods are described here that have been successfully used to extract Mo from laboratory-prepared media. The first, modified from Eady and Robson (1984), uses 8-hydroxyquinoline (a chelator of Mo) and dichloromethane to remove 8-hydroxyquinoline from the medium. Chemicals for Burk's medium are dissolved in double-glass-distilled water, phosphates and other chemicals prepared separately as two different 20× stock solutions (20 fold more concentrated than in the final medium). The 20× solution of chemicals (but not the 20× phosphate buffer) is adjusted to pH 3.5 with 5M HCl. 8-Hydroxyquinoline (0.2 g/l) and dichloromethane (25 ml/l) are added to both 20× stock solutions. The stock solutions are stirred using a Teflon-coated magnetic bar for 2 h. The dichloromethane is allowed to settle and then siphoned off. This extraction procedure is repeated six times. The pH is then adjusted to 7.1 with pellets of NaOH, and the excess 8-hydroxyquinoline removed by repeated extractions with dichloromethane until the medium is colorless. An alternative is to use 1,2-dichloroethane instead of dichloromethane for extraction (Luque et al., 1993).

The second method employs activated charcoal to extract molybdenum from concentrated stock solutions (Schneider et al., 1991). In this method, 50 g of pulverized activated charcoal

(Merck) is suspended in 1 liter of doubly distilled water and stirred for 2 d at room temperature. This suspension is degassed by boiling for 10 min then cooled to room temperature, and 50-ml aliquots are placed on a 5 cm round filter paper disc (e.g., Whatman #1). The water is suctioned through the filter, resulting in a 5-mm layer of activated carbon. After washing the filter with 100 ml doubly distilled water, the stock solutions (50 ml) are suctioned through the filter. The method is more efficient if concentrated single-nutrient-component solutions, adjusted to pH 2–4, are filtered through the activated charcoal then mixed together by dilution into doubly distilled water, followed by pH adjustment.

Regardless of the method used to extract Mo from media, it is essential to use either commercially supplied plasticware, which is usually free of contaminating Mo, or glassware that has been soaked for 24 h or more in a solution of 5% Decon (or similar commercial product) and 0.5% EDTA in pure water (as purified through a MonoQ column, Millipore) for 24 h, followed by rinsing in pure water. Additions of Mo or V compounds are made as necessary, usually $Na_2MoO_4·2H_2O$ or V_2O_5 at 1 μM.

Cell morphology Cultures grown for 24–48 h on both agar and liquid media should be used for the study of general morphology by phase contrast microscopy. Motility may be observed in wet mounts, but it is advisable to incorporate some air bubbles in the mount to insure an adequate supply of oxygen. Motile cells are usually found near the edge of an air bubble.

Cysts These resting forms are best studied with cultures at least 2 weeks old that have been grown on nitrogen-free medium containing 0.2% butanol. If the culture will not grow on butanol (e.g., *A. paspali*), nitrogen-free medium with 0.2% glucose can be used. The cysts may be stained with violamine[7] (Winogradsky, 1938), acridine orange[8] (Tchan, unpublished data), or with a mixture of neutral red and light green SF yellowish (Vela and Wyss, 1964). Examination by phase-contrast microscopy is sometimes helpful when violamine stain is used.

Pigment production The production of diffusible pigments in the presence or absence of benzoate is not constant on iron-

---

7. Violamine stain consists of violamine "2 R", 1.0 g; phenol, 5.0 g; and water, 100 ml. An air-dried, fixed smear is treated with violamine stain for 45 sec, washed, and counterstained with dilute crystal violet (0.05% or less) for 1 min. After rinsing with water, the preparation is mounted in water for microscopic examination.

8. Acridine orange stain: Cells are suspended in a 0.01% aqueous solution of acridine orange. One drop is deposited on a slide, covered with 2 drops of mineral oil. Drops are then spread and a cover glass applied. A monolayer of cell suspension can be found in the preparation where the depth of liquid is such that motile cells are immobilized to facilitate microscopic examination. When examined by fluorescence microscopy, the vegetative cells are green and the cysts have a green central body and a red coat.

TABLE BXII.γ.126. PCR primer sequences for identification of *vnf*G and *anf*G, which indicate the presence of genes for nitrogenase-2 or nitrogenase-3[a]

Primer	Anneals to	Sequence[b]	Expected size of PCR product[c]
D6f	*vnf*D	5'-CGGGATCCGAAGACTTYGARAAGGTCAT-3'	1700 bp (*vnf*G)
K1r	*vnf*K	5'-GGGGGTACCACTTCYAWGCAGAACT-3'	
D7f	*anf*D	5'-GCTCTAGACGCSATCTAYTCGCCGA-3'	750 bp (*anf*G)
K2R	*anf*K	5'-CGGAATTCCGATGCAATCCTTGAT-3'	

[a]From Loveless and Bishop (1999).

[b]Single letter code: I, inosine; Y, C or T; R, A or G; S, C or G; W, A or T.

[c]Use of primers D6f and K1r with test strain DNA should amplify a fragment carrying *vnf*G; primers D7f and K2R should amplify a fragment carrying *anf*G (see Fig. BXII.γ.139). For PCR amplification conditions, see Loveless and Bishop (1999).

containing media, and is unreliable for strain differentiation. Diffusible pigments are readily produced on the following iron-deficient media (Thompson and Skerman, 1979). Any one of the three basal agar media[9] is modified by omission of $FeSO_4 \cdot 7H_2O$ and reduction of $Na_2MoO_4 \cdot 2H_2O$ to 1 µg/ml. Glucose is added to give a concentration of 10 g/l. The medium is inoculated by depositing a small drop of a suspension of culture on the surface of the agar. The plates are examined in daylight for diffusible pigment and under ultraviolet light (wavelength 364 nm) for fluorescent pigments. The medium of Stainer and Scholte (1970) can also be used for the production of nondiffusible pigments or, for those strains that will not grow on this medium, the basal medium of Thompson and Skerman can be used when enriched with sodium gluconate (2.0 g/l).

DIFFERENTIATION OF THE GENUS *AZOTOBACTER* FROM OTHER GENERA

Phenotypic characteristics can differentiate the genus *Azotobacter* from others. Genera most closely related taxonomically, i.e., found in the *Gammaproteobacteria* class, order *Pseudomonadales*, family *Pseudomonadaceae*, are *Azomonas* and several others for which no nitrogen-fixing species/strain has been identified, except for certain isolates of *Pseudomonas stutzeri*. Like *Azotobacter*, *Azomonas* species can fix nitrogen under aerobic conditions while *P. stutzeri* cannot. *Azotobacter* species can utilize urea as nitrogen source and form cysts in older cultures or after growth in butanol (or propanol) while *Azomonas* species cannot, although general cultural and morphological characteristics are similar. In addition, among genera able to grow aerobically on dinitrogen, which includes not only *Azotobacter* and *Azomonas*, but also *Beijerinckia* and *Derxia*, members of the *Alphaproteobacteria* class, the latter two genera are morphologically and physiologically different from *Azotobacter*. Distinguishing characteristics are summarized in Table BXII.γ.127.

TAXONOMIC COMMENTS

One new species, *A. salinestris*, has been added to the genus since the last edition of the *Manual*. Several isolates of this species were previously classified as strains of *A. chroococcum*, and then distinguished on the basis of Na-dependent growth under most conditions and decreased aerotolerance during diazotrophic growth (Page and Shivprasad, 1991a).

In this edition of the *Manual*, *Azotobacter* and *Azomonas* are classified in the phylum *Proteobacteria*, class *Gammaproteobacteria*,

order *Pseudomonadales*, and family *Pseudomonadaceae*. In the previous edition of the *Manual*, these two genera comprised the family *Azotobacteraceae* (Tchan, 1984). 16S rDNA sequencing of species of the two genera indicated that the family was distinct from, but closely related to, the fluorescent pseudomonads in *Pseudomonas* Superfamily I within the gamma group of *Proteobacteria* (MacDonald and Melton, unpublished studies). By the least squares distance matrix method, *Azotobacter* strains formed an assemblage into one group with a close association to *Azomonas macrocytogenes*, and clusters of pseudomonads in another group; both groups were stemmed to *E. coli* with an evolutionary distance of 0.01 or less (Fig. BXII.γ.139). The maximum parsimony tree exhibited a limited outcropping of organisms (Fig. BXII.γ.140). This can be interpreted as a hierarchy of phylogeny, which begins with the stem organism *E. coli*, progressively branches off the pseudomonads and *Azomonas macrocytogenes*, then clusters the *Azotobacter* species. The parsimony tree had a higher average bootstrap confidence level than the least squares distance tree, 89.5% vs. 75.8%, respectively (MacDonald and Melton, manuscript submitted).

Phylogenetic analysis using other genes or features have placed *A. vinelandii* and *Azomonas macrocytogenes* in the *Pseudomonas* group I lineage with *P. aeruginosa*. These studies included a comparison of four *P. aeruginosa* alginate (*alg*) genes to those from species of the *Pseudomonas* group I, group V, and enteric lineages (Fialho et al., 1990). The comparison indicated the greatest homology with the species of *Pseudomonas* group I, which included *A. vinelandii* but not *P. stutzeri*—which did not have an *algD* gene—or *Azomonas macrocytogenes*—which also did not show homology at this locus. A comparison of aromatic amino acid biosynthetic pathways also indicated the same evolutionary relationship between the group I pseudomonads and the species of *Azotobacteraceae* (Byng et al., 1986); this was consistent with groupings previously defined by rRNA–DNA hybridization studies. As with the rDNA analysis (De Smedt et al., 1980), immunoelectrophoresis—performed with antisera prepared to each of several strains and tested against one another (Tchan et al., 1983)—indicated that the six *Azotobacter* species were more similar to each other than to any *Azomonas* species and that the three *Azomonas* species were as distant from each other as each is from *Azotobacter* (Tchan et al., 1983).

A proposed reclassification of *Azotobacter paspali* to a new genus, *Azorhizophilus* (Thompson and Skerman, 1979), is not justified because this species clearly belongs to the *Azotobacter* genus based on 16S rDNA relatedness (Figs. BXII.γ.139 and BXII.γ.140). Moreover, as in *A. vinelandii*, *A. paspali* carries genes for the synthesis of all three types of nitrogenase. Another genus, *Azomonotrichon*, was suggested for the single species *A. macrocytogenes*, which was originally classified as an *Azomonas* species based on numerical analysis of common features (Thompson and Skerman, 1979). Reclassification as "*Azomonotrichon macrocytogenes*" is not supported by 16S rRNA gene sequence analysis, im-

---

9. Basal agar media (as cited in Thompson and Skerman, 1979): The medium of Norris and Jensen (1958) consists of (g/l): K_2HPO_4, 1.0; $CaCl_2 \cdot 2H_2O$, 0.1; $MgSO_4 \cdot 7H_2O$, 0.2; $FeSO_4 \cdot 7H_2O$, 0.05; $Na_2MoO_4 \cdot 2H_2O$, 0.005; and Ionagar (Oxoid), 10.0; pH 7.3. An acidic modification of this medium consists of (g/l): K_2HPO_4, 0.3; KH_2PO_4, 0.7; $MgSO_4 \cdot 7H_2O$, 0.2; $FeSO_4 \cdot 7H_2O$, 0.05; $Na_2MoO_4 \cdot 2H_2O$, 0.005; and Ionagar (Oxoid), 10.0; pH 6.2. An enriched modification of the medium consists of Norris and Jensen's medium containing 0.5% yeast extract (Difco).

TABLE BXII.γ.127. Differential characteristics of the genus *Azotobacter* and other aerobic diazotrophic genera[a]

Characteristic	Azotobacter	Azomonas	Beijerinckia	Derxia
Dumbbell-shaped cells with polar granules	−	−	+	−
Monotrichous flagellum	−	D	−	+
Cyst formation	+	−	−	−
Autotrophic growth on N_2	−	−	−	+
Mol% G + C of DNA	63.2–67.5	52–58.6	55–61	69–73

[a]For symbols see standard definitions.

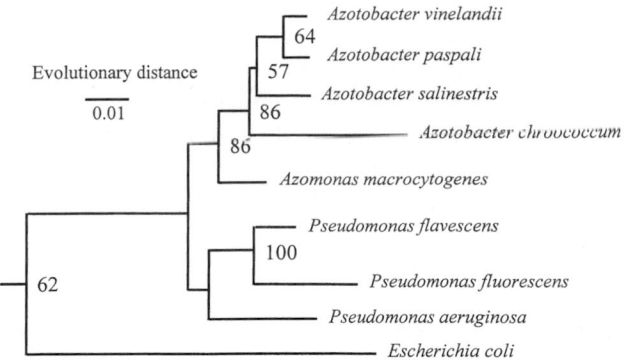

FIGURE BXII.γ.139. Least squares distance matrix tree showing phylogenetic relationships of *Azotobacter* and *Azomonas*. The tree is representative of the best fit of 100 data sets by the least squares distance algorithm and 100 bootstrapped pseudosamples. The evolutionary distance is 0.01. Bootstrap confidence values of 50% or greater are represented at each node of the tree (Courtesy of M.L. MacDonald and T.T. Melton, unpublished studies).

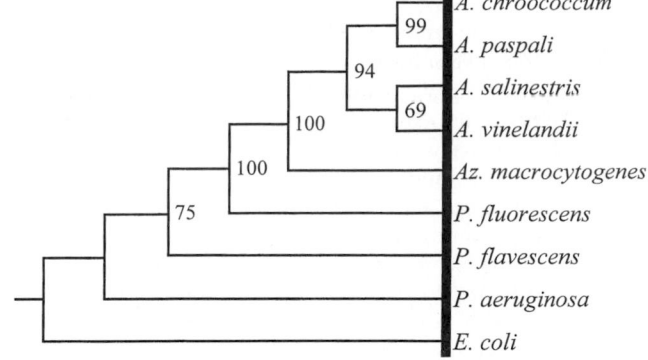

FIGURE BXII.γ.140. Maximum parsimony tree showing phylogenetic relationships of *Azotobacter* and *Azomonas*. The tree is representative of the best fit of 100 data sets by the maximum parsimony algorithm and 100 bootstrapped pseudosamples. Tree branch length is not significant for this algorithm. Bootstrap confidence values of 50% or greater are represented at each node of the tree (Courtesy of M.L. MacDonald and T.T. Melton, unpublished studies).

munological analysis (Tchan et al., 1983), and comparison of protein profiles (Jana et al., 1992). Previous inclusion of the nitrogen-fixing genera *Beijerinckia* and *Derxia* in the family *Azotobacteraceae* is now fully discounted by 16S rRNA gene sequence analysis. In this edition of the *Manual*, *Beijerinckia* is included in the class *Alphaproteobacteria*, and *Derxia* is placed in the class *Betaproteobacteria*, but distant from *Azotobacter* or *Azomonas* species.

Most species of *Azotobacter* and *Azomonas* were not extensively studied after 1979 with the exception of *A. vinelandii** and *A. chroococcum*, as reviewed in this chapter. A detailed history of the species, together with cultural, morphological, and physiological

Editorial Note: During 2002 and 2003, a rough draft of the sequence of the genome of *A. vinelandii* became available through the U.S. Department of Energy Joint Genome Institute. Automated annotation of this genome has been carried out at Oak Ridge National Laboratories by Frank Larrimer and associates and at The Institute for Genome Research. A fully annotated finished genome may be available in 2004.

characteristics of approximately 90 isolates, can be found in Thompson and Skerman (1979) and references therein. See also Becking (1992) for further descriptive information.

ACKNOWLEDGMENTS

We thank Adam Field and Angela Smedke for assistance in compiling DNA sequences for analyzing codon usage and mol% G + C content. We also thank Mali Gunatilaka for help at many stages of preparation of the manuscript. Finally, we would like to dedicate this chapter to Thoyd Melton who passed away before its publication. His contributions to the studies of the genetics and physiology of *Azotobacter vinelandii* were significant. He was a devoted scientist who was ardent about research, teaching, and his fellow colleagues.

FURTHER READING

Becking, J.H. 1992. The family *Azotobacteraceae. In* Balows, Trüper, Dworkin, Harder and Schleifer (Editors), The Prokaryotes, 2nd ed., Vol. 3, Springer-Verlag, New York. 3144–3170.

DIFFERENTIATION OF THE SPECIES OF THE GENUS AZOTOBACTER

The differential characteristics of the species of *Azotobacter* are presented in Table BXII.γ.123. Other characteristics of the species are indicated in Table BXII.γ.124.

List of species of the genus Azotobacter

1. **Azotobacter chroococcum** Beijerinck 1901, 567[AL]

   chro.o.coc′ cum. Gr. n. *chroa* color; Gr. n. *coccus* a grain; M.L. neut. n. *chroococcum* colored coccus.

   The morphology is shown in Figs. BXII.γ.128 and BXII.γ.129. Cells appear as blunt rods to ellipsoid forms, 1.6–2.5 µm in diameter and 3–5 µm in length. Chains of cells are occasionally observed in some strains. Cysts (sometimes referred to as "microcysts") are formed in old cultures or after addition of isopropanol to exponential cultures. Cells have randomly placed granules, described as metachromatic or sudanophilic (Thompson and Skerman, 1979). Motile cells from 1- to 2-d-old cultures have peritrichous flagella with a wavelength of 2.5–3.3 µm and an amplitude of 0.37–0.76 µm. Colonies are opaque and convex, and, for most strains, are mucoid, glistening, and

smooth. A brown-black nondiffusible pigment is produced in aging colonies. Many organic compounds can be used as carbon sources. Distinguishing physiological and nutritional characteristics of the species are presented in Tables BXII.γ.123 and BXII.γ.124. N$_2$, ammonium, nitrate, and urea are used as nitrogen sources. N$_2$ is fixed under aerobic or microaerobic conditions. Ammonium prevents N$_2$ fixation by inhibition of expression of *nif* genes and possibly by inactivation of Mo nitrogenase (nitrogenase-1). In three strains tested, one "alternative" nitrogenase is synthesized when Mo supply is very low (nitrogenase-2, carrying the Fe-V cofactor). This correlates with the presence of *vnf* but not *anf* genes in these strains. For other characteristics, less distinguishing, see Thompson and Skerman (1979). The genome size varies between 3.1 and 5.3 Mb. There are 6 or 7 copies of the *rrn* operon.

The mol% G + C of the DNA is: 65.8–67.5 (T_m).

Type strain: R. L. Starkey's strain 43, ATCC 9043, DSM 2286, WR-88.

2. **Azotobacter armeniacus** Thompson and Skerman 1981, 215[AL] (Effective publication: Thompson and Skerman 1979, 272 (*Azotobacter agilis* subsp. *armeniae* Kirakosyan and Melkonyan 1964, 41; *Azotobacter vitreus* subsp. *armeniae* Kirakosyan and Melkonyan 1964, 41.)

ar.me.ni.a' cus. M.L. neut. n. *armeniacus* named after Armenia, U.S.S.R., where the species was isolated.

The morphology is as described for the genus and most other species. Cells are motile with peritrichous flagella having a wavelength of 2.0–2.9 μm and an amplitude of 0.40–0.59 μm. Colonies are opaque, mucoid, and more highly convex than those of other species. Brown-black to red-violet water-soluble pigments are produced. No strain has been tested for the presence of alternative nitrogenases. Distinctive features include inability to utilize nitrate or ammonium for growth, but strains are urease positive. Physiological and nutritional characteristics of the species are presented in Tables BXII.γ.123 and BXII.γ.124.

The mol% G + C of the DNA is: 63.5–65.0 (T_m).

Type strain: N28, DSM 2284, WR-136.

3. **Azotobacter beijerinckii** Lipman 1904, 248[AL]

bei.jer.inck' i.i. M.L. gen. n. *beijerinckii* of Beijerinck; named after M.W. Beijerinck of Delft, the Dutch microbiologist.

The morphology of cells is similar to that of *A. chroococcum* and can vary depending on culture conditions. The cells are nonmotile and lack flagella. Colonies are smooth, but variants may arise due to the quantity of extracellular polysaccharide produced. Growth at 37°C is poor or does not occur. A yellowish or cinnamon-colored pigment is produced in aging cultures. Many isolates are acid tolerant. Physiological and nutritional characteristics of the species are presented in Tables BXII.γ.123 and BXII.γ.124. Ammonium and nitrate are used as nitrogen sources. Type strain ATCC 193600 carries both *nif* and *vnf* genes but not *anf*, indicating that *A. beijerinckii* can synthesize nitrogenase-1 and nitrogenase-2.

The mol% G + C of the DNA is: 66.0–66.2 (T_m).

Type strain: ATCC 19360, DSM 378, NCIB 8948, WR-135.

Two subgroups of *A. beijerinckii* have been delineated. The acid-tolerant subgroup corresponds to "*A. beijerinckii* subsp. *acidotolerans*" Tchan. The other subgroup is distinguished by its sensitivity to 0.05% phenol or 40 μg/ml of diamond fuchsin, its inability to use sorbitol or aconitate, and its failure to produce diffusible homopolysaccharides.

4. **Azotobacter nigricans** Krasil' nikov 1949, 506[AL]

ni' gri.cans. L. adj. *nigricans* black.

The cells (Fig. BXII.γ.130) are round-ended rods, as are typical of other species of *Azotobacter*. The cells are nonmotile and lack flagella. Colonies are smooth, but variants occur due to the quantity of extracellular polysaccharide produced. Most colonies are opaque but can be translucent. Pigments are produced and are usually red-violet. Growth at 37°C is poor or does not occur. Strains have been isolated from soils in Eastern Europe. Physiological and nutritional characteristics are presented in Tables BXII.γ.123 and BXII.γ.124. Ammonium and nitrate are used as nitrogen source in preference to nitrogen. The type strain carries

nif and *vnf* but not *anf* genes, indicating an ability to synthesize nitrogenase-1 and nitrogenase-2.

The mol% G + C of the DNA is: 64.5 (T_m).

Type strain: ATCC 35009, DSM 2288, UQM 1967.

a. **Azotobacter nigricans** *subsp.* **nigricans** Krasil' nikov 1949, 506[AL]

In aging colonies, a brown-black water-soluble pigment is usually produced.

The mol% G + C of the DNA is: 64.5 (T_m).

Type strain: ATCC 35009, DSM 2288, UQM 1967.

b. **Azotobacter nigricans** *subsp.* **achromogenes** Thompson and Skerman 1981, 215[VP] (Effective publication: Thompson and Skerman 1979, 12.)

a.chro.mo' ge.nes. Gr. pref. *a* not; Gr. n. *chroma* color; Gr. suff. *-genes* born of; N.L. neut. adj. *chromogenes* not producing color.

Aging colonies produces only a yellow nondiffusible pigment within the colony.

The mol% G + C of the DNA is: 64.5 (T_m).

Type strain: A-2 of Jensen (WR-41).

5. **Azotobacter paspali** Döbereiner 1966, 364[AL] (*Azorhizophilus paspali* (Döbereiner 1966) Thompson and Skerman 1981, 215.)

pas.pal' i. M.L. gen. n. *paspali* of Paspalum; named after *Paspalum*, the generic name of a grass.

The general morphology is similar to that of other *Azotobacter* species, but the cells are longer than usual, 5–10 μm in the exponential phase. Long filamentous cells up to 60 μm can occur (Fig. BXII.γ.131). Cells are motile by numerous peritrichous flagella having a wavelength of 2.3–2.8 μm and an amplitude of 0.40–0.59 μm (Fig. BXII.γ.133). Some strains have "curly" flagella with a wavelength of 1.1–1.4 μm and an amplitude of 0.25–0.34 μm. Colonies are opaque with an entire or undulate edge, raised, usually not evenly convex, and butyrous with a dull or rough surface. Physiological and nutritional characteristics are presented in Tables BXII.γ.123 and BXII.γ.124. Ammonium but not glutamate can be used as the sole nitrogen source for growth. Thompson and Skerman (1979) reported the ability of *A. paspali* isolates to utilize nitrate as nitrogen source but failed to detect nitrite production from nitrate or nitrite disappearance in standard tests. This conclusion should be reexamined because nitrate assimilation is not known to occur other than by nitrate reduction to nitrite and nitrite reduction to ammonium. Urea is hydrolyzed. Genes for both alternative nitrogenases (nitrogenase-2 *vnf* and nitrogenase-3 *anf*) were detected in two strains (the type strain ATCC 23833 and also ATCC 23367). Distinctive features include (1) production of a yellow-green fluorescent pigment (as for *A. vinelandii*); (2) utilization of fewer compounds as carbon sources than other *Azotobacter* species. The carbon sources used include maltose, trehalose, melibiose, raffinose, mannitol, and others (Thompson and Skerman, 1979). The organism has never been isolated from environments other than the rhizosphere of the grass *Paspalum notatum* cv. Batatais. Genome size is uniform, 4.3–4.6 Mb in four strains tested, as is number of copies (six) of the *rrn* operon.

The mol% G + C of the DNA is: 63.2–64.6 (T_m).

Type strain: Ax-8, ATCC 23833, DSM 2283, WR-129.

6. **Azotobacter salinestris** Page and Shivprasad 1991b, 374[VP]
sal.in.es' tris. N.L. adj. *salinus* saline; L. suff. -*estris* belonging
to, living in; N.L. adj. *salinestris* living in saline environments.

The morphology is described as coccobacilli (oval with
pointed ends); cells are approximately $2 \times 3–4$ μm in young
or exponential cultures Older cells are round with a diameter of 3–5 μm, pleomorphic, and are often found in
pairs and chains of six-to-eight cells. Cysts are formed. Cells
are motile by means of peritrichous flagella 9–10 μm long,
with a wavelength of 2.7–2.9 μm and an amplitude of 0.36–
0.55 μm. Cells from old cultures are nonmotile. Physiological characteristics include growth at temperatures between
30° and 36°C. The optimal temperature for N_2 fixation is
35°C. Cells and colonies grow at pH 6.2–8.0. The optimal
pH for growth on N_2 is 7.2; on nitrate it is pH 7.3, and on
ammonium it is pH 7.5. N_2 fixation occurs best in microaerophilic cultures, but cells can be adapted to aerobic diazotrophy. Most strains, except the type strain, produce a
capsule and synthesize poly-β-hydroxybutyrate. The type
strain, ATCC 49674, and strain 253 contain *nif* and *vnf*
genes and can fix nitrogen in the absence of Mo with V.
Growth under most conditions requires the presence of 1
mM Na^+. Strains are urease positive, amylase positive, and
catalase positive (with Na^+). No fluorescent compounds are
synthesized, even when cultures are Fe limited. Carbon
sources include fructose, galactose, glucose, sucrose, mannitol, melibiose, 0.25% sodium benzoate, and starch. Nitrogen sources are N_2, ammonium, and nitrate. A brown-
black to tan-brown pigment, allomelanin (a catechol melanin), is produced by most strains when grown at high
aeration under N-fixing conditions. Growth is inhibited by
K^+ and Rb^+, with Rb^+ having the greatest effect. For other
characteristics, see Tables BXII.γ.123 and BXII.γ.124.[10]

The mol% G + C of the DNA is: 65.19–65.98 (T_m).

---

10. Not all traits tested by Thompson and Skerman (1979) for the other *Azotobacter*
species have been tested in *A. salinestris*.

Type strain: 184, ATCC 49674, DSM 11553.
GenBank accession number (16S rRNA): AF035213.

7. **Azotobacter vinelandii** Lipman 1903, 238[AL.] (*Azotobacter mis-*
cellum Pshenin 1964, 617.)
vine.lan' di.i. M.L. gen. n. *vinelandii* of Vineland; named after
Vineland, New Jersey, where the species was first isolated.

The morphology is similar to that of *A. chroococcum* and
can vary from rods to more coccoid forms. The cells typically are $1.6–2.5 \times 3–5$ μm. Ammonium-grown cells are
more coccoid in stationary phase (Fig. BXII.γ.133). Cysts
are formed in old cultures or after addition of butanol to
exponential cultures. Motile cells from 1- to 2-d-old cultures
have peritrichous flagella with a wavelength of 2.0–3.0 μm
and an amplitude of 0.40–0.59 μm. The velocity of motile
cells can be 74 μm/sec. A few strains are nonmotile. Colonies are opaque and convex, and, for most strains, are
mucoid, glistening, and smooth. Distinguishing physiological and nutritional characteristics of the species are presented in Tables BXII.γ.123 and BXII.γ.124; utilization of
rhamnose and phenol as carbon sources is distinctive. A
yellow-green fluorescent pigment is produced, especially
under conditions of Fe limitation. Other diffusible pigments can also be produced. N_2, ammonium, nitrate, and
urea, but not glutamate, are used as nitrogen sources. N_2
is fixed under aerobic or microaerobic conditions. Ammonium prevents N_2 fixation by inhibition of expression of
nif genes and possibly by inactivation of Mo nitrogenase
(nitrogenase-1, carrying the Fe–Mo cofactor). In strain
AvOP (ATCC 13705), two "alternative" nitrogenases can be
synthesized when Mo supply is very low (nitrogenase-2, carrying the Fe–V cofactor and nitrogenase-3, with a Fe–Fe
cofactor, synthesized in the absence of Mo and V). This
correlates with the presence of *vnf* and *anf* genes. For other
characteristics, less distinguishing, see Thompson and Skerman (1979). The genome size is 4.6 Mb. Six copies of the
rrn operon are present.

The mol% G + C of the DNA is: 64.9–66.5 (T_m).
Type strain: ATCC 478, DSM 2289, WR-100.

Genus IV. **Cellvibrio** Blackall, Hayward and Sly 1986, 354[VP] (Effective publication: Blackall, Hayward and Sly 1985, 95)

THE EDITORIAL BOARD

Cell.vib' ri.o. M.L. n. *cell* abbreviation of M.L. n. *cellulosum* cellulose, which is degraded by the organism;
M.L. masc. n. *Vibrio* a generic name; M.L. masc. n. *Cellvibrio* cellulose (degrading) vibrio.

Aerobic Gram-negative **slender curved rods** ($0.2–0.5 \times 1.0–1.3$
μm). Motile with **mixed flagellation. Oxidize glucose and hydrolyze cellulose. Produce curdlan polysaccharide from glucose**. Catalase and oxidase positive. Do not utilize organic acids. No growth
factors required.

The mol% G + C of the DNA is: 46.1–54.1.

Type species: **Cellvibrio mixtus** Blackall, Hayward and Sly 1986,
354 (Effective publication: Blackall, Hayward and Sly 1985, 95.)

FURTHER DESCRIPTIVE INFORMATION

Cells possess mixed flagellation, with up to 11 lateral flagella; the
lateral flagella, which are not always present, have a smaller diameter and a shorter wavelength than the single polar flagellum.

The enzymes involved in the degradation of polysaccharides and
in the metabolism of the degradation products have been studied
in *Cellvibrio mixtus*, "*Cellvibrio vulgaris*", "*Cellvibrio fulvus*", and
"*Cellvibrio gilvus*", but the taxonomic position of the latter three
strains has not been determined (see Taxonomic Comments).

ENRICHMENT AND ISOLATION PROCEDURES

Enrichment and isolation can be carried out by incubation of
rhizosphere soil samples in liquid mineral salts medium containing cellulose or dextran as a carbon source followed by streaking
onto agar medium containing the same carbon source (Blackall
et al., 1985).

MAINTENANCE PROCEDURES

Strains can be stored on slants of sucrose peptone or peptone yeast extract agar at 7°C (Blackall et al., 1985).

TAXONOMIC COMMENTS

The genus *Cellvibrio* was first described by Winogradsky (1929), but did not appear on the Approved Lists (Skerman et al., 1980). Blackall et al. (1985) revived the genus based on the description of a number of newly isolated cellulose- and dextran-hydrolyzing strains which they compared to selected *Pseudomonas*, *Aeromonas*, *Vibrio*, *Xanthomonas*, *Cytophaga*, and *Cellulomonas* strains as well as to strains of "*Cellvibrio vulgaris*" and "*Cellvibrio fulvus*". In a numerical taxonomic analysis of phenotypic characters the new isolates clustered with each other and with "*Cellvibrio vulgaris*" and "*Cellvibrio fulvus*" but separately from all the other organisms. However, the new isolates differed from the "*Cellvibrio vulgaris*" and "*Cellvibrio fulvus*" strains and from the organisms described by Winogradsky in failing to produce pigment on cellulose agar; they also differed from "*Cellvibrio vulgaris*" and "*Cellvibrio fulvus*" in the mol% G + C of the DNA (50.1 ± 4.0 for the new isolates and 44.9 and 44.6 for the latter two strains).

List of species of the genus Cellvibrio

1. **Cellvibrio mixtus** Blackall, Hayward and Sly 1986, 354[VP] (Effective publication: Blackall, Hayward and Sly 1985, 95.) *mix′ tus.* M.L. masc. adj. *mixtus* mixed referring to the type of flagellation.

   On semi-opaque mineral salts–cellulose medium all strains form circular white colonies with entire edges surrounded by a zone of clearing. Produce acid from L-arabinose, dextrin, galactose, glucose, lactose, maltose, mannose, melibiose, raffinose, sucrose, trehalose, and xylose. Hydrolyze chitin, esculin, pectate (at neutral and alkaline pH only), and starch, but not casein, DNA, gelatin, or Tween 80. Do not produce gas from nitrate, H_2S from cysteine or indole from tryptophan. Voges–Proskauer and methyl red tests negative. Produce α-glucosidase, β-glucosidase, and β-xylosidase.

   The mol% G + C of the DNA is: 53–55 (T_m).
   Type strain: UQM 2601.

   a. **Cellvibrio mixtus** *subsp.* **mixtus** Blackall, Hayward and Sly 1986, 354[VP] (Effective publication: Blackall, Hayward and Sly 1985, 95.)

   Properties are as for the species with the following additions: produce acid from cellobiose and melezitose. Produce β-glucuronidase and β-galactosidase. Do not hydrolyze dextran. Gram-negative; cells soluble in 3% KOH.

   The mol% G + C of the DNA is: 52.6 (T_m).
   Type strain: UQM 2601.
   GenBank accession number (16S rRNA): AF448515.

   b. **Cellvibrio mixtus** *subsp.* **dextranolyticus** Blackall, Hayward and Sly 1986, 354[VP] (Effective publication: Blackall, Hayward and Sly 1985, 95.)
   dex.trano.lyticus. dextran dextran, a chemical; Gr. adj. *lyticus* dissolving; M.L. masc. adj. *dextranolyticus* dextran dissolving.

   Properties are as for the species with the following additions: hydrolyze dextran and produce acid from fructose. Gram-negative, but cells not soluble in 3% KOH.

   The mol% G + C of the DNA is: 55.2 (T_m).
   Type strain: UQM 1666.

Genus V. **Mesophilobacter** Nishimura, Kinpara and Iizuka 1989, 380[VP]

THE EDITORIAL BOARD

Me.so.phi.lo.bac′ ter. Gr. n. *mesos* middle; Gr. adj. *philus* loving; M.L. *bacter* rod, staff; M.L. masc. n. *Mesophilobacter* mesophilic rod.

Gram-negative nonmotile aerobic pleomorphic rods. (0.5–0.6 × 1.0–2.0 μm). Catalase positive; oxidase negative. Nonsporeforming, non–acid-fast, nonfermentative, nonencapsulated. Resistant to penicillin. **Optimum growth at 33–37°C.**

The mol% G + C of the DNA is: 44.0–46.9.
Type species: **Mesophilobacter marinus** Nishimura, Kinpara and Iizuka 1989, 380.

FURTHER DESCRIPTIVE INFORMATION

The strains were isolated from seawater (Nishimura et al., 1986b, 1989).

Cells of young cultures may be stout (0.8–1.0 × 1.5–3.0 μm) or may be elongated up to 15 μm in length. Cells in stationary cultures are coccoid with diameters 0.6–0.9 μm (Nishimura et al., 1989).

ENRICHMENT AND ISOLATION PROCEDURES

Strains of this species were isolated from seawater on seawater agar or modified nutrient agar as described by Nishimura et al. (1986b).

DIFFERENTIATION OF THE GENUS *MESOPHILOBACTER* FROM OTHER GENERA

Nishimura et al. (1989) provide a table of characteristics that differentiate *Mesophilobacter* from the genera *Moraxella*, *Acinetobacter*, and *Psychrobacter*.

List of species of the genus Mesophilobacter

1. **Mesophilobacter marinus** Nishimura, Kinpara and Iizuka 1989, 380[VP]
   ma.ri′ nus. L. adj. *marinus* of the sea, marine.

   Description as for the genus with the following additional characteristics. No change in litmus milk. No H_2S produced. Voges-Proskauer test negative. Methyl red test positive. Reduces nitrate to nitrite. Assimilates acetate, citrate, D-fructose, fumarate, gluconate, D-glucose, malate, maltose, D-ribose, succinate, sucrose, and trehalose. Produces acid from cellobiose, D-fructose, D-glucose, D-mannitol, and D-ribose. Grows in presence of 7% NaCl. Major fatty acids $C_{16:0}$, $C_{16:1}$, and $C_{18:1}$.

The mol% G + C of the DNA is: 44.0–46.9 (HPLC).

Type strain: 14S-4, DSM 9142, IAM 13185, NCIMB 13183.

Genus VI. **Rhizobacter** *Goto and Kuwata 1988, 238*[VP]

Masao Goto

Rhi.zo.bac'ter. Gr. n. *rhiza* a root; M.L. masc. n. *bacter* masc. equivalent of Gr. neut. n. *bacterion* rod, staff; M.L. masc. n. *Rhizobacter* root rod.

Straight to slightly curved, nonsporeforming, encapsulated rods 0.9–1.3 × 2.1–2.5 μm. Poly-β-hydroxybutyrate granules are formed. Cells stain Gram negative. **Motile by polar flagella or lateral flagella or both** but motile cells are rare in the populations. **Aerobic** with respiratory metabolism of glucose. Good growth is obtained at 28–30°C; no growth at 35°C. Growth occurs at pH between 5.0 and 9.0. Utilizes D-glucose as a sole source of carbon and energy. **White or yellowish white, plicated, tough, or viscid colonies on agar plates**; if yellow, the pigment differs from xanthomonadines. **Floccular growth consisting of globular units in liquid media; fingerlike projections never occur. Oxidase and catalase positive. The following are positive**: H₂S from cysteine and auxin, reduction of nitrate to nitrite, *o*-nitrophenyl-β-D-galactopyranoside (ONPG) test, and growth inhibition by KCN. Negative for denitrification, methyl red, and production of arginine dihydrolase, nitrogenase, fluorescent pigment, indole, and acetoin formation. Susceptible to 10 μg of vibriostatic agent O/129 phosphate. **Hydrolyzes starch, dextrin, and glycogen.** Cannot utilize benzene derivatives as carbon sources. Does not require NaCl or growth factors. Contains ubiquinone Q8. The type species occurs in soil and is a plant pathogen, causing galls on carrot roots in nature.

The mol% G + C of the DNA is: 67–71.

Type species: **Rhizobacter dauci** *corrig.* Goto and Kuwata 1988, 238 (*Rhizobacter daucus* (sic) Goto and Kuwata 1988, 238.)

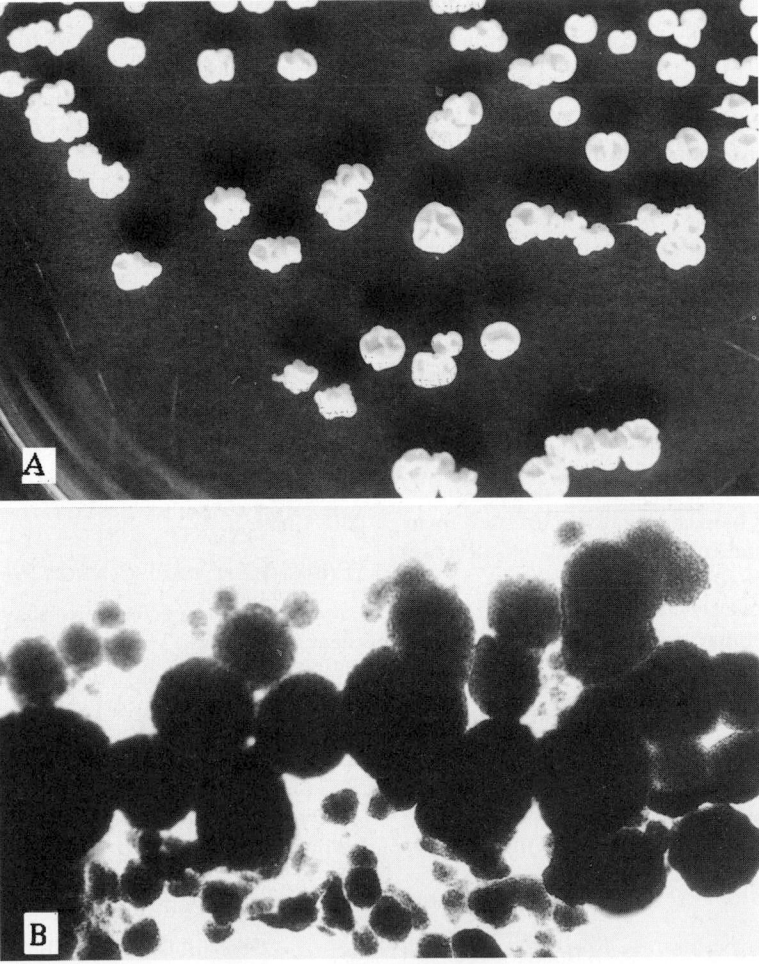

FIGURE BXII.γ.141. Colony morphology of *R. dauci* H6[T]. (*A*) Plicated colonies of wild type on DPPG agar. (*B*) Globular aggregates of bacterial cells in DPPG liquid medium. (× 100)

FURTHER DESCRIPTIVE INFORMATION

The growth on agar plates depends on the concentrations of ingredients: growth is best on a 10-fold dilution of potato-peptone-glucose (DPPG) agar, a 10-fold dilution of yeast extract-peptone-glucose (DYPG) agar, or Casitone-yeast extract agar[1]. This oligotrophic trait seems to be due to the presence of peptone, which significantly retards the growth of the bacterium.

Although this bacterium is aerobic, it can grow slowly under anaerobic conditions. After several days, the colonies on DPPG agar plates are 1.5–2.0 mm in diameter, white, opaque, tough, and pulvinate or capitate with entire margins. Later the colonies develop highly plicated surfaces and undulating margins (Fig. BXII.γ.141A).

Plication is less distinct or absent when glucose is omitted from the medium. The colonies on DYPG agar plates are yellowish white but otherwise identical to those on DPPG agar plates. The colonies are so tough and adhere so firmly to the agar surfaces that they are difficult to remove with a platinum loop. The agar surface is slightly depressed for about 1–2 mm around a colony.

Growth in liquid media is floccular, and the medium never clouds uniformly. The flocs are composed of globular aggregates of bacterial cells 60–200 μm in diameter (average, 120 μm) (Fig. BXII.γ.141B). The flocs adhere to the test tube wall in the early stages of growth and then develop into a film or pellicle, which eventually fall to the tube bottom.

The results of the oxidation-fermentation test are not clearcut. Because of the slow growth, as well as the considerable amount of growth and acid production that take place in the tubes sealed with mineral oil, the oxidation-fermentation test may be misread in the early stages of growth. After several days of incubation, however, growth and acid production in the open tubes become conspicuously greater than growth and acid production in the sealed tubes, making determination of oxidative metabolism easier.

This bacterium uses diverse kinds of compounds as a sole source of carbon: L-arabinose, D-arabinose, xylose, fructose, galactose, glucose, lactose, maltose, melibiose, sucrose, raffinose, carboxymethylcellulose, dextrin, glycogen, adonitol, inositol, mannitol, sorbitol, salicin, gluconate, mucate, quinate, and tartrate. The following are used weakly: isoleucine, leucine, methanol, and propanol.

Other physiological and biochemical characteristics include positive reactions for gelatinase, lecithinase, urease, pit formation on pectate gel, hydrolysis of Tween 80 and esculin, production of H_2S, and indoleacetic acid and/or its derivatives, and negative reactions for casein hydrolysis, tyrosinase, deoxyribonuclease, and phenylalanine deaminase. No detectable change occurs in litmus milk cultures.

Slimy colony mutants are often detected during the storage of agar slant cultures at room temperature. These mutants grow faster than the parents and form abundant viscous slime on agar media.

The bacterium causes bacterial carrot gall that is a typical soilborne disease. In natural infection, galls develop along the entire length of the storage roots from the crown to the root tip. Small galls may also be found on the fibrous roots. Galls vary in size from a few millimeters to approximately 1 cm in diameter (average, several millimeters). Large galls may be formed by the union of several smaller galls. Galls are often distributed along lenticles, in severe cases forming a ring around the root (Fig. BXII.γ.142A). The galls are light brown and have very rough surfaces. Another characteristic of the disease is the development of numerous adventitious fibrous roots.

On artificially inoculated carrot seedlings, hypertrophied areas become apparent on the wounded roots 7–14 d after inoculation, and outgrowths like those on naturally infected plants develop about 1 month after inoculation. The disease is particularly severe on the roots, which are injured by scratching the epidermis with a sterilized sharp needle. The bacterial cells occupy the narrow fissures formed in gall tissues, and deposit heav-

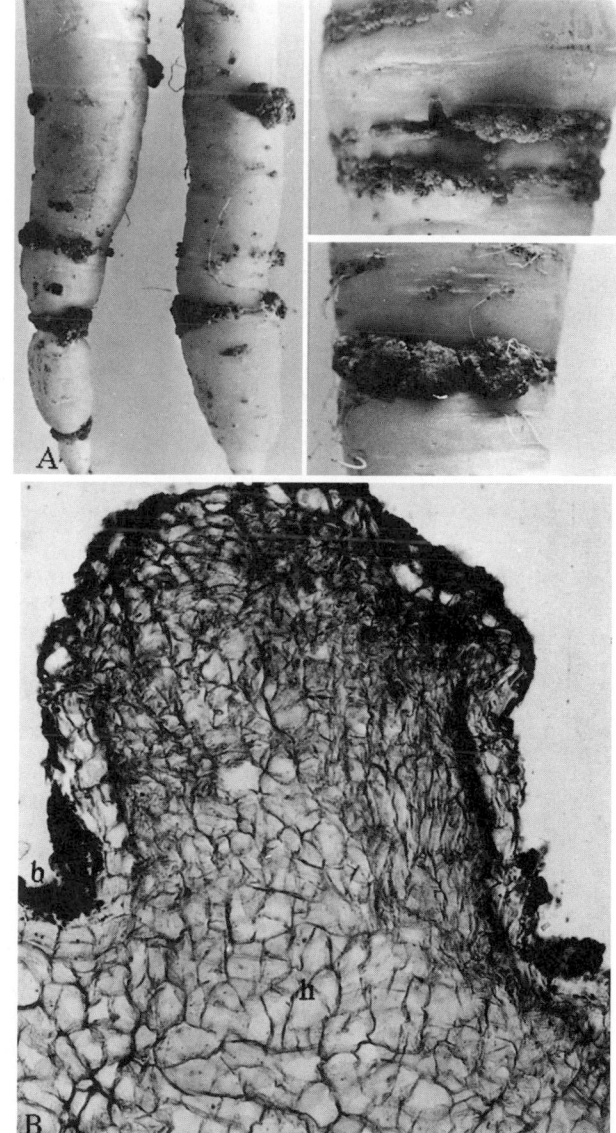

FIGURE BXII.γ.142. Characteristics of carrot bacterial gall. (*A*) Symptoms of galls formed on carrot roots. (*B*) Transverse section of an outgrowth showing hyperplastic tissues (*h*) and bacterial cells found in bacterial fissures and on gall surface (*b*).

---

1. DPPG agar consists of potato extract (20 g of potato in 1000 ml of water), 1000 ml; peptone, 1 g; glucose, 10 g; and agar, 15 g; pH 6.8. DYPG agar consists of peptone, 1 g; yeast extract, 0.5 g; glucose, 10 g; agar, 15 g; and water, 1000 ml; pH 6.8. Casitone-yeast extract agar consists of Casitone (Difco), 5 g; yeast extract, 2.5 g; agar, 15 g; and water, 1000 ml; pH 6.8.

ily on the surface of galls (Fig. BXII.γ.142B). When tomato seeds are immersed for several minutes in bacterial suspension and planted in sterilized soil, the conspicuous outgrowths develop along the main root, giving a thick club-shaped appearance and causing a remarkable growth inhibition of whole plant parts.

ENRICHMENT AND ISOLATION PROCEDURES

Isolation of *Rhizobacter dauci* is difficult because of its preference for oligotrophic media and its slow growth on agar plates, which can be overgrown by soil bacteria that contaminate the habitat. In practice, a small piece (approximately a 2-mm cube) of outgrowth including the surface is cut out of a fresh gall, washed briefly, macerated in 10 ml of sterilized distilled water, and subjected to 10-fold serial dilution with sterilized distilled water. The suspensions are plated onto a DPPG agar. The plates are incubated aerobically at 28°C. The bacterium is easily recognizable after 1 week or so because of its slow-growing, white, plicated, tough colonies. It may be further purified by single-colony isolation procedures.

MAINTENANCE PROCEDURES

For routine work, the bacterium may be stored either as the suspensions in sterilized distilled water at room temperature or as the stab cultures on DPPG agar medium at 4°C. Lyophilization is preferable for long-term preservation. For reviving the lyophilized cultures, freeze-dried cells are suspended in a few milliliters of sterilized distilled water and plated immediately on DPPG agar, and again after storing the suspension overnight at room temperature. The typical wild-type colonies with plicated surface are often formed on the latter plates.

DIFFERENTIATION OF THE GENUS *RHIZOBACTER* FROM OTHER GENERA

See Table BXII.γ.128.

TAXONOMIC COMMENTS

This bacterium is characterized by the lateral flagellation and the susceptibility to vibriostatic agent 0/129. These features have been known in the members of the family *Vibrionaceae*, facultative anaerobes (Baumann and Schubert, 1984), but are not common in the aerobes (Goto, 1988). Until the phylogenetic data became available, the bacterium was most appropriately placed in the family *Pseudomonadaceae* based on various other phenotypic characteristics.

Rhizobacter dauci resembles *Rhizomonas suberifaciens* in its preference for oligotrophic media, wrinkled colony formation, and polar or lateral flagellation. However, the former is clearly distinguished from the latter by the higher mol% G + C of the DNA, a different major ubiquinone, the large number of carbon sources utilized, and significantly larger dimensions of the cells (van Bruggen et al. 1990).

Rhizobacter dauci resembles *Zoogloea* spp. in formation of zoogloeae but is differentiated by its plant pathogenicity, shape of the zoogloeae, susceptibility to the vibriostatic agent 0/129 phosphate, lower tolerance to NaCl, lack of growth at 35°C, positive lecithinase reaction, lack of a litmus milk reaction, and a wider spectrum of carbon sources used (Crabtree and McCoy, 1967; Friedman and Dugan, 1968).

Colonies of *Rhizobacter dauci* resembles those of *Beijerinckia* spp. grown on nitrogen-free glucose mineral agar (Becking, 1984). However, *Beijerinckia* spp. are distinct from *R. dauci* in plant pathogenicity, positive nitrogenase activity, and the significantly lower mol% G + C.

List of species of the genus Rhizobacter

1. **Rhizobacter dauci** *corrig.* Goto and Kuwata 1988, 238[VP] (*Rhizobacter daucus* (sic) Goto and Kuwata 1988, 238.) *dau' ci.* M.L. n. *Daucus* generic name of carrot, host plant; M.L. masc. gen. n. *dauci* of the carrot.

The characteristics are as described for the genus and as listed in Table BXII.γ.128. The morphology is as depicted as Fig. BXII.γ.141.

The mol% G + C of the DNA is: 67–71 (T_m).

Type strain: H6, ATCC 43778, ICMP 9400, LMG 9036.

TABLE BXII.γ.128. Differential characteristics of the genus *Rhizobacter* and other genera of the family *Pseudomonadaceae*

Characteristics	*Rhizobacter*[a]	*Rhizomonas*[b]	*Zoogloea*[c]	*Beijerinckia*[c]
Cell diameter, >1 μm	+	−	+	+
Lateral flagellation	+	+	−	+
Susceptibility to 0/157	+	nd	−	nd
Zoogloea	+[e]	−	+[f]	−
Acid production from glucose	+	+	−	+
Denitrification to N_2	−	−	+	nd
Inhibition by peptone	+	+	+	+
Lecithinase activity	+	nd	−[b]	nd
Nitrogenase activity	−	−	nd	+
Carbon sources utilized	Wide	Narrow	Narrow	Wide
Isoprenoid quinone	Q8	Q10	nd	nd
Mol% G + C content of DNA	66.9–70.6	59	65.3	55–61
Plant Pathogenicity	+	+	−	nd

[a]Data from Goto and Kuwata (1988).

[b]Data from Van Bruggen et al. (1990).

[c]Data from Unz (1984).

[d]Data from Becking (1984).

[e]Globular zoogloea.

[f]Fingerlike zoogloea.

Genus VII. **Rugamonas** Austin and Moss 1987, 179[AL] (Effective publication: Austin and Moss 1986, 1907)

MAURICE O. MOSS AND DAWN A. AUSTIN

Ru.gamo' nas. L. fem. n. *ruga* wrinkle; Gr. fem. n. *monas* unit, monad; M.L. n. *Rugamonas* wrinkled unit.

Rods 0.8–0.9 × 2.4–4.0 μm, in young cultures. **Intracellular granules of prodigiosenes develop in cells more than 7 d old.** Cells in cultures less than 7 d are motile by one or more polar or subpolar flagella. **Colonies on Bennett's agar**[1] **after 4 d at 20°C are pink to deep red, wrinkled, and rubbery in consistency after 5 d**, with a diameter of approximately 3 mm. Oxidase and catalase positive, strictly aerobic and oxidative metabolism, unable to metabolize carbohydrates fermentatively. Nitrates are reduced to N_2 anaerobically. Growth occurs in the presence of 0.0–0.5% (w/v) NaCl. Arginine decarboxylase negative. Lysine and ornithine decarboxylases are produced by some strains. Esculin, chitin, DNA, RNA, gelatin, lecithin, Tween 20, 40, 60, and 80, tyrosine, and urea are degraded but not allantoin, cellulose, elastin, hypoxanthine, starch, and xanthine. H_2S is not produced on triple sugar iron agar. The gluconate and Koser's citrate tests are positive; Voges–Proskauer, methyl red, and melonate tests are negative. β-Galactosidase (ONPG method) positive. Has been isolated from river water.

The mol% G + C of the DNA is: 66–67.

Type species: **Rugamonas rubra** Austin and Moss 1987, 179 (Effective publication: Austin and Moss 1986, 1908.)

FURTHER DESCRIPTIVE INFORMATION

As cultures age, cells show reduced motility and become increasingly pleomorphic. Cultures grown in Bennett's broth form pink/red flocs after 7 d. Colonies on Bennett's agar at 20°C are white, shiny, and circular after 2–3 d, pink after 4 d, and deep red, wrinkled, and rubbery in consistency after 5 d, with a diameter of approximately 3 mm. The tough rubbery consistency is characteristic of all isolates of *R. rubra* obtained so far, as is the production of red pigments of the prodigiosene family. The type strain was shown to produce 6-methoxy-2-methyl-3-pentyl-prodigeosene (prodigiosin) but the predominant pigment was the 3-heptyl analogue (Fig. BXII.γ.143).

On rich media such as Bennett's agar, the initially pillar-box red colonies become a deep maroon color and colonies often have a green metallic sheen. These color changes reflect both the reduction in pH and overproduction of the prodigiosenes, which can be seen as dark granular inclusions in wet mounts prepared for motility studies. At this stage viability drops rapidly and no viable organisms can usually be recovered from colonies after 14 d incubation on Bennett's agar.

Temperature range for growth after 168 h is 1.2–30.6°C; optimum, 20°C. No growth at 37°C. The cardinal temperatures are sensitive to the period of incubation and after 48 h the following have been recorded: T_{min}, 5.6°C; T_{opt}, 26.4°C; T_{max}, 30.6°C (Moss, 1983).

R. rubra has only been reported from a lowland, freshwater river site where it occurred in low but persistent numbers during a 5-month period (January to May 1984), especially after rainfall 24–48 h before collection of samples (Austin and Moss, 1986).

Although it is not possible to assign this organism to a particular ecological niche, the utilization of a wide range of sugars, amino acids, and organic acids as sole carbon sources suggests that it may have the ability to exploit a range of environments.

ENRICHMENT AND ISOLATION PROCEDURES

Maximum recovery of isolates from river water was achieved by making fivefold serial dilutions up to 1/125 using ¼ strength Ringer's buffer and plating 0.5-ml samples onto large (12-cm) plates of modified Bennett's agar supplemented with neomycin hydrochloride, cycloheximide, and nystatin, all at 50 μg/ml. Plates are incubated at 20°C for up to 14 d, and tough membranous colonies are streaked onto plates of Bennett's agar to ensure purity (Austin and Moss, 1986).

MAINTENANCE PROCEDURES

For medium-term storage, cultures are inoculated into dilute peptone water, incubated to allow turbidity to develop, and then stored at ~4°C. For long-term storage, freeze drying is not successful but storage in liquid nitrogen has given adequate survival. The type strain has been successfully recovered after 15 years' storage in liquid nitrogen using the following double-strength cryoprotectant (composition per liter of distilled water): K_2HPO_4, 12.6 g; sodium citrate, 0.9 g; $MgSO_4 \cdot 7H_2O$, 0.18 g; $(NH_4)_2SO_4$, 1.8 g; KH_2PO_4, 3.6 g; glycerol, 44 ml; sterilized by filtration. A 1-ml sample of a 3-d-old culture of the organism in Bennett's broth is thoroughly mixed with 1 ml of cryoprotectant in a freeze vial (Nunc) and stored in liquid nitrogen.

DIFFERENTIATION OF THE GENUS *RUGAMONAS* FROM OTHER GENERA

One of the most striking features of *Rugamonas* is the production of flocs in liquid culture and tough rubbery colonies on solid media. Table BXII.γ.129 shows some of the characteristics that distinguish three well-characterized Gram-negative, motile, prodigiosin-producing species of *Rugamonas*, *Alteromonas*, and *Serratia*.

FIGURE BXII.γ.143. The structure of 6-methoxy-2-methyl-3-heptylprodigeosene, which is produced by *Rugamonas* strains.

---

1. Bennett's broth (Keeble and Cross, 1977) contains (g/l distilled water): yeast extract (Difco), 1.0; beef extract (Lab Lemco), 1.0; Bacto casitone (Difco), 2.0; and glucose, 10.0; pH 7.3. For a solid medium, 10.8 g/l of agar (LabM no. 2 or Oxoid no. 1) are added.

TABLE BXII.γ.129. Characteristics differentiating three Gram-negative, motile, prodigiosin-producing species

Characteristic	Rugamonas rubra	Alteromonas rubra	Serratia rubidaea
Oxidase	+	+	−
Fermentative metabolism	−	−	+
Growth at 35°C		+	+
Anaerobic growth	−	−	+
Utilizes D-sorbitol	+	−	
Flagella	1–2 Polar or subpolar	1 Polar	Peritrichous
G + C DNA	66.7	46–48	53–59

TAXONOMIC COMMENTS

The general characteristics of *Rugamonas* isolates suggest an affinity with the *Pseudomonadaceae* but they cannot be comfortably placed in any existing genera. They are distinguished from *Pseudomonas* and *Alteromonas* by their floc-forming ability in liquid cultures and from *Zoogloea* by the mol% G + C of the DNA.

List of species of the genus Rugamonas

1. **Rugamonas rubra** Austin and Moss 1987, 179[AL] (Effective publication: Austin and Moss 1986, 1908.)

   rub'ra. L. fem. adj. *rubra* red.

   The characteristics are as given for the genus. In addition, the following enzymes are produced: alkaline phosphatase, acid phosphatase, esterase-lipase, leucine arylamidase, and phosphoamidase. Esterase is produced by some strains. Phenylalanine deaminase is not produced. Sole carbon sources include D-alanine, L-arabinose, DL-arginine, *meso*-inositol, sodium DL-lactate, cellobiose, D-fructose, D-galactose, L-histidine, maltose, D-mannose, D-raffinose, D-ribose, DL-serine, sodium acetate, trisodium citrate, D-sorbitol, sucrose, and trehalose. The following are not used: adonitol, *i*-erythritol, sodium malonate, D-melibiose, D-mel-izitose, D-rhamnose, DL-valine, and D-xylose. Lactose is used by some strains.

   By disk assay, strains are sensitive to chloramphenicol, chlortetracycline, erythromycin, gentamicin, oxytetracycline, streptomycin, and tetracycline (each at 10 µg/disk); furazolidone (50 µg), kanamycin (30 µg), novobiocin (5 µg), sulfafurazole (500 µg), and cotrimoxazole (25 µg). Resistant to ampicillin (2 µg), cloxacillin (5 µg), colistin sulfate (10 µg), neomycin (10 µg), nitrofurantoin (200 µg), and penicillin G (1.5 U).

   Isolated from river water.

   The mol% G + C of the DNA is: 66.7 ± 0.1 (T_m).

   Type strain: Strain MOM 28/2/79, ATCC 43154, CCM 3730, NCIMB 12552.

Genus VIII. **Serpens** Hespell 1977, 380[AL]*

ROBERT B. HESPELL

Ser'pens. L. fem. n. *serpens* snake, serpent.

Rod-shaped cells, 0.3–0.4 µm wide by 8–12 µm long. Occur singly in pairs. Cysts or coccoid bodies not formed, but cells in the stationary phase of growth are longer (16–25 µm) and often possess blebs or spherical protuberances. Gram negative. **Extremely flexible and capable of serpentine-like motility in agar gels. Possess bipolar tufts of 4–10 flagella and also a few lateral flagella.** Poly-β-hydroxybutyrate or other internal granules not formed. Have a strictly respiratory type of metabolism with oxygen as the sole electron acceptor. **Grow aerobically but prefer oxygen concentrations less than that of an air atmosphere. Catalase and oxidase positive.** Chemoorganotrophic. **Lactate is the only effective carbon and energy source,** although very slight growth occurs with acetate or α-ketoglutarate. Carbohydrates, fatty acids, and sugar alcohols are not catabolized. Casein hydrolysate, peptone, yeast extract, and, for most strains, ammonium chloride can serve as nitrogen sources; nitrates and nitrites are not used. Vitamins are stimulatory but not required. Optimum temperature, 28–30°C. On media containing 1.8–2.0% agar, colonies are cream colored, round, 3–6 mm in diameter, and have a filamentous edge. **On media with less than 1.5% agar,** **only subsurface spreading colonies occur.** Found in the sediments of eutrophic freshwater ponds.

The mol% G + C of the DNA is: 66.

Type species: **Serpens flexibilis** Hespell 1977, 381.

FURTHER DESCRIPTIVE INFORMATION

The most striking morphological feature is the complex, serpentine-like movement of the cells when observed in slides of agar pieces removed from the edge of a spreading colony (Fig. BXII.γ.144). The organisms move very rapidly and display a furious lashing and bending of the cells; coiling into a knot-like configuration is also common. In liquids, the movement is less violent and the cells move rapidly with reversals along a straight line axis with a gentle flexing of the body, and the distal or trailing tip of the cell vibrates intensely (Fig. BXII.γ.144E–H). No tumbling type of movement is evident. Motility occurs in buffer solutions of polyvinylpyrrolidine having a viscosity greater than 1000 centipoise (Greenberg and Canale-Parola, 1977). In comparison, viscosities of 60 and 1000 centipoise prevent the motility of *Escherichia coli* and *Spirochaeta halophila*, respectively. The structural basis permitting such cell flexibility is unknown, but the organisms possess a thin peptidoglycan layer and a morphologically typical Gram-negative outer membrane.

---

*Editorial Note: The chapter on *Serpens* is reprinted from the 1st edition of *Bergey's Manual of Systematic Bacteriology*.

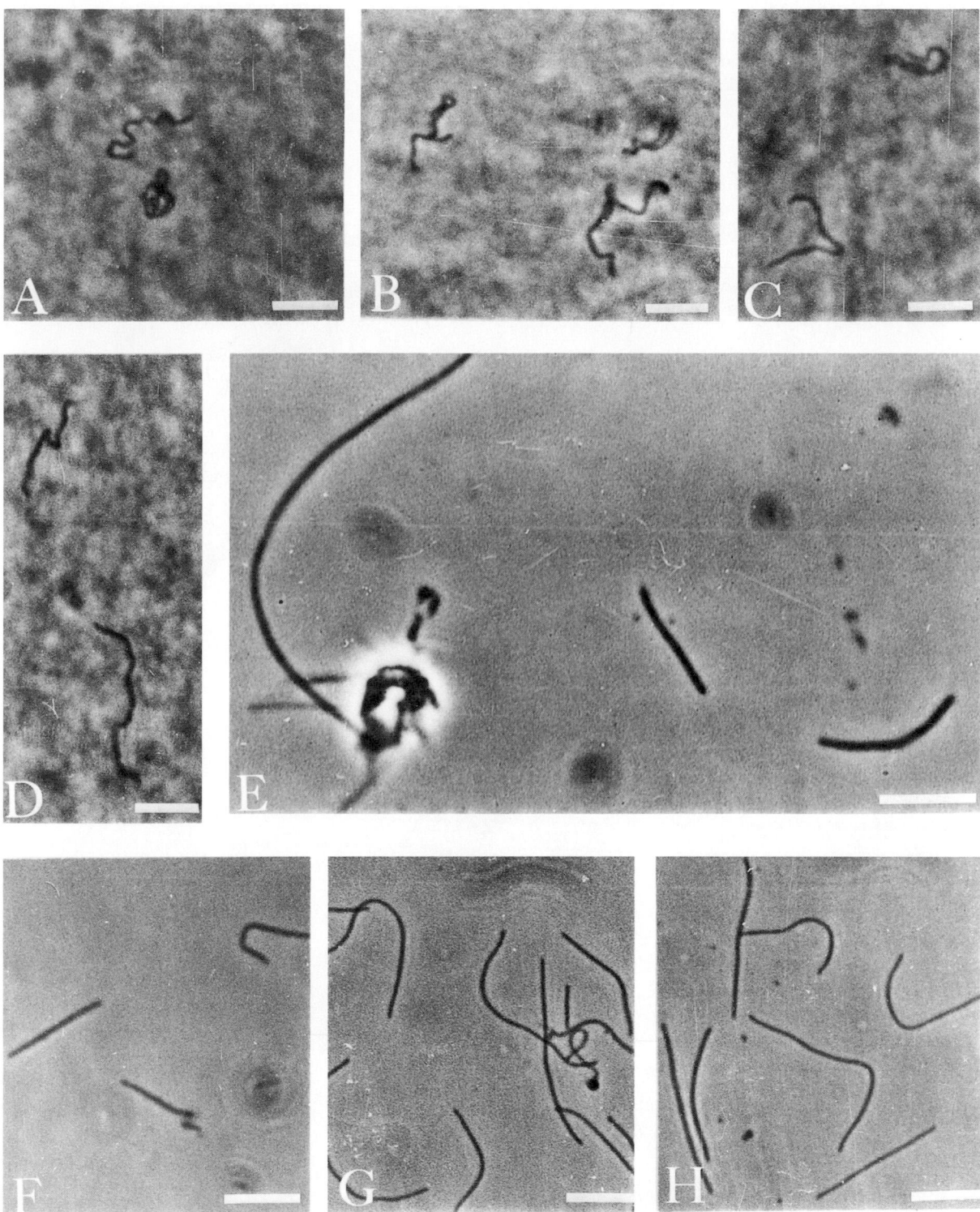

FIGURE BXII.γ.144. Living cells of *Serpens flexibilis* PFR-1. Wet-mount preparations, phase contrast. Cells in agar pieces excised from subsurface, spreading growth (*A–D*) displayed rapid motion with bending, lashing, and serpentine-like movements. Cells rapidly reversed themselves between coiled and uncoiled forms (*A*). In liquid menstrua (*E– H*) logarithmic-phase cells (*F* and *H*) showed straight-line movements with flexing of the cell body. Stationary cells often clumped (*E*) and had protuberances (*G*). Bars = 3.0 μm. (Reproduced with permission from R.B. Hespell, International Journal of Systematic Bacteriology *27*: 371–381, 1977, ©International Union of Microbiological Societies.)

Although growth occurs aerobically, *S. flexibilis* prefers oxygen tensions less than that of an air atmosphere, growing subsurface (1–3 mm deep) in stationary liquid media or in media with agar concentrations of less than 1.5%. The type strain and all other strains examined can use only lactate as an effective oxidizable carbon and energy source. Cell extracts contain only trace levels of the key enzymes of the Embden–Meyerhof–Parnas or hexose monophosphate pathways, but the tricarboxylic acid cycle enzymes are in high levels. Vitamins are not required for growth, but yeast extract is often stimulatory. Most strains can use ammonium chloride as a sole nitrogen source, but increased growth rates and cell yields occur with media containing peptone or casein hydrolysate.

ENRICHMENT AND ISOLATION PROCEDURES

S. flexibilis can be readily isolated from the upper few centimeters of the sediments of eutrophic freshwater ponds. Traditional enrichment culture techniques employing lactate as the carbon source are not useful for isolation as the organisms often become

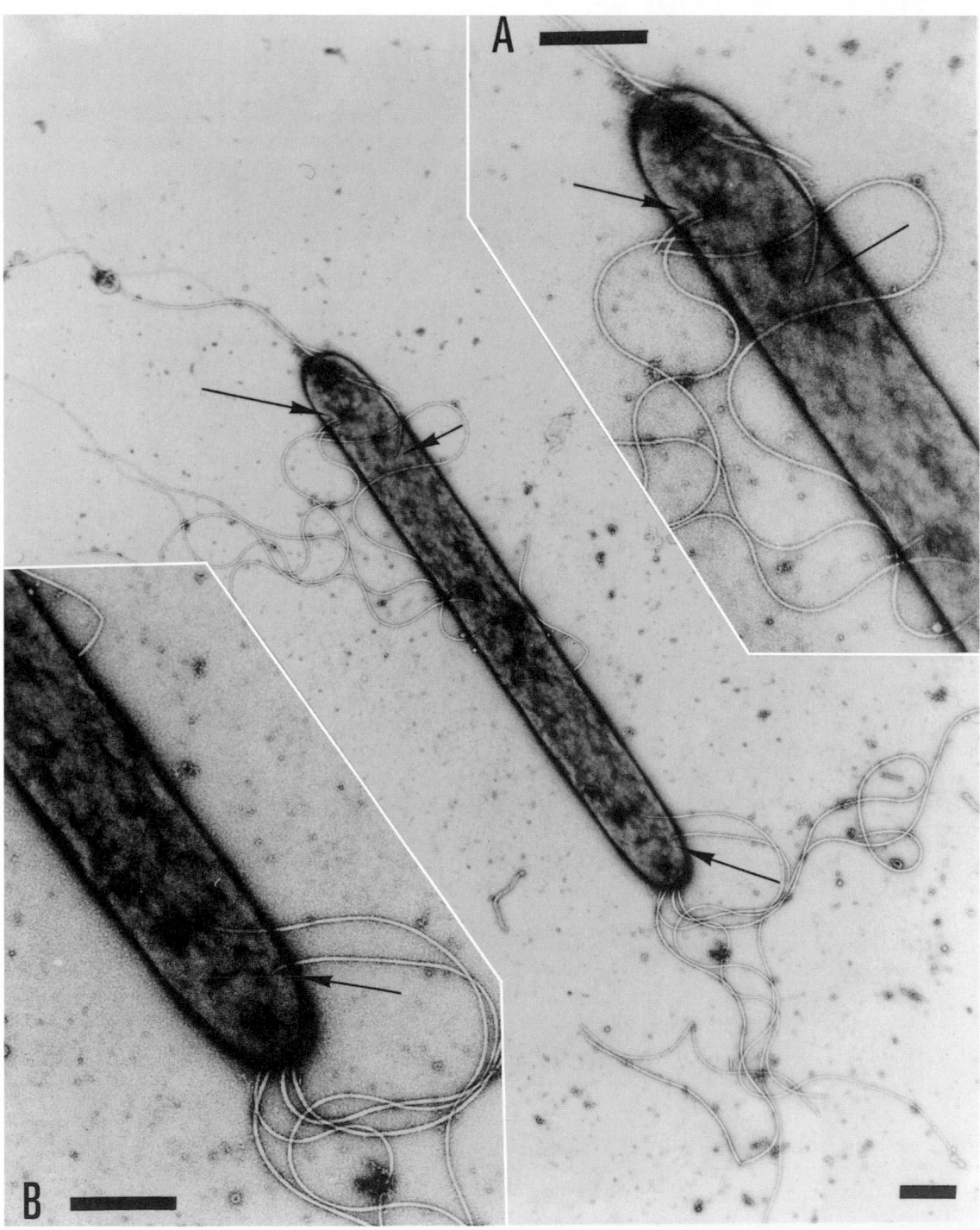

FIGURE BXII.γ.145. Transmission electron micrograph of *S. flexibilis* PFR-1 cells. The insets (*A* and *B*) are higher magnifications of the ends of the same cell stained with phosphotungstic acid. Although the overall cell shape has been preserved, some flagella have detached, and an aggregate of four flagella remains at one end of the cell. Several laterally inserted flagella are present, and their insertion points (*arrows*) are more clearly evident in the insets. Bars = 0.25 μm. (Reproduced with permission from R.B. Hespell, International Journal of Systematic Bacteriology *27*: 371–381, 1977, ©International Union of Microbiological Societies.)

overgrown by other heterotrophic bacteria. However, selective isolation can be accomplished by the use of membrane filters overlayed on a Petri dish of appropriate media, a technique that has also been employed for the isolation of thin spirilla and spirochetes (Canale-Parola et al., 1966). A small inoculum of pond water/mud slurry is deposited in the center of a sterile cellulose filter disk (0.3–0.45 μm pore diameter) that has been placed on the surface of a plate of isolation medium.[1] After incubation for 6–12 h at 30°C, the disk is removed and the plate is incubated for 2–4 d. The organisms grow as a subsurface, whitish veil that diffuses from the center of the plate. By picking from the edge of the veil and streaking onto a second plate, cloned cultures may be obtained. Although aerobic spirilla and spirochetes can also be obtained in this manner, these organisms can usually be distinguished from *S. flexibilis* by light microscopic

observations, but definitive confirmation should be done by electron microscopy.

MAINTENANCE PROCEDURES

S. flexibilis is grown routinely in LYPP broth[2] and can be maintained by biweekly transfers on slants of LYPP agar. The organism can be preserved indefinitely by lyophilization or by storage in liquid nitrogen.

TAXONOMIC COMMENTS

The genus presently consists of a single species, *S. flexibilis.**

---

1. Isolation medium (per 90 ml water): yeast extract, 0.2 g; peptone, 0.1 g; hay extract, 10 ml; and agar 1.0 g. The hay extract is prepared by boiling 10 g of hay in 100 ml of water for 15 min and clarifying the mixture by centrifugation. The pH of the isolation medium is adjusted to 7.0 with KOH before sterilization.

---

2. LYPP broth (per 100 ml water): 60% sodium lactate syrup, 1.0 ml; yeast extract, 0.3 g; peptone, 0.2 g; and K$_2$HPO$_4$, 0.35 g. The pH of the medium is adjusted to 7.2 prior to autoclaving. LYPP agar is prepared by adding 1.5 g of agar.

Editorial Note: Woese et al. (1982) reported that *S. flexibilis* is closely related to *Pseudomonas pseudoalcaligenes* (SAB = 0.9) and might be considered as a variant pseudomonad that had initially developed a defect in its system for synthesizing septa. Ahmad and Jensen (1987) found that *S. flexibilis* lacked the "overflow" pathway in the biosynthetic pathway for aromatic amino acids and shared this trait with most members of the *Pseudomonas aeruginosa* group.

List of species of the genus Serpens

1. **Serpens flexibilis** Hespell 1977, 381[AL]

flex.i.bi′ lis. L. adj. *flexibilis* flexible, pliant.

The description of the species is the same as that for

the genus. The morphological features of the type strain are depicted in Figs. BXII.γ.144 and BXII.γ.145.

The mol% G + C of the DNA is: 66 (Bd).

Type strain: PFR-1, ATCC 29606.

Family II. **Moraxellaceae** Rossau, Van Landschoot, Gillis and De Ley 1991, 317[VP]

ELLIOT JUNI AND KJELL BØVRE

Mo.ra.xel.la′ ce.ae. M.L. fem. pl. n. *Moraxella* type genus of the family; *-aceae* ending to denote a family; M.L. fem. pl. n. *Moraxellaceae* the *Moraxella* family.

Organisms are **rod shaped, coccoid,** or coccal or may exhibit a characteristic multicellular micromorphology. Cells usually single but may also occur in pairs or short chains. Endospores are not formed. Gram negative, but there may be a tendency to resist decolorization. The cells are **nonmotile** in liquid media but pilus-associated surface-bound motility ("twitching motility") may be observed. Psychrotrophic or mesophilic. Strains of all recognized species usually have an optimum growth temperature between 14 and 36°C. Some species have capsules. Pili (fimbriae), pilus-associated colony morphology, and **competence for natural transformation** are frequently detected. Colonies are not pigmented. Chemoorganotrophic and aerobic. Several species have complex growth factor requirements while some species grow readily in simple defined media containing a single organic carbon and energy source. Except for acinetobacters, all strains are oxidase positive. **Usually catalase positive. Indole is not produced.** Except for some strains of *Acinetobacter* and *Psychrobacter*, **no acid is produced from carbohydrates**. True waxes may be present. Fatty acid profiles show the presence of mainly unbranched, saturated and mono- or di-unsaturated fatty acids composed of 16 and 18 carbon atoms. Most strains reside indigenously on mucosal membranes of humans and animals. Except for *Moraxella catarrhalis*, *Moraxella nonliquefaciens*, *Moraxella bovis*, and *Moraxella osloensis*, most strains are rarely involved in disease of humans or other

animals. Strains of *Acinetobacter* and *Psychrobacter* are found in soil and water or in association with food products and have been isolated from the skin or mucous membranes of humans and other animals. The *Moraxellaceae* belong to the *Gammaproteobacteria* and to rRNA superfamily II.

The mol% G + C of the DNA is: 38–50.

Type genus: **Moraxella** Lwoff 1939, 173 emend. Henriksen and Bøvre 1968, 391.

Further comments With the exception of *M. caviae, M. caprae, M. phenylpyruvica,* and *M. lincolnii,* strains competent for natural transformation have been found for the other genera of the *Moraxellaceae,* thus making it possible to establish genetic relationships among all members of the family. Genetic studies involving the highly conserved streptomycin-resistance locus (resulting from mutation of S12, a protein component of the small ribosomal subunit) have made it possible to detect intergeneric, as well as interspecies, relationships. Because of the definitive, although frequently weak, genetic interactions involving the streptomycin-resistance marker, it is possible to recognize newly discovered and previously uncharacterized members of the *Moraxellaceae.* Using genus-specific transformation of nutritionally deficient mutants of certain conserved genes, it has been possible to assign strains of all species of a genus to that particular genus in a few of the genera of the *Moraxellaceae.*

TAXONOMIC COMMENTS

The first description of a rod-shaped bacterium involved in eye infection in humans was made independently by Morax (1896) and by Axenfeld (1897). This so-called Morax–Axenfeld bacillus was Gram negative, nonmotile, occurred frequently as diploid pairs, and required serum or other complex media for growth. In subsequent years, organisms with similar properties were isolated from patients with conjunctivitis. In 1939, Lwoff proposed that these organisms, as well as some others, be placed in a new genus, *Moraxella*. Seeking to find a family that included the genus *Moraxella*, Henriksen (1952) suggested, and it was later formally proposed (Henriksen and Bøvre, 1968), that based on similarities in morphology, Gram stain and oxidase reaction, as well as other properties, the family *Neisseriaceae* was a likely candidate. However, from results of studies involving genetic transformation (Bøvre and Hagen, 1981), DNA–DNA hybridization (Tønjum et al., 1989), DNA–rRNA hybridization (Rossau et al., 1986, 1989), and 16S rDNA sequencing (Enright et al., 1994), it can be concluded that there is presently no basis for including the moraxellae in the family *Neisseriaceae*.

In 1962, Bøvre and Henriksen demonstrated transformation of streptomycin resistance among the oxidase-positive moraxellas. In independent studies by Bøvre (1963) and Catlin (1964), it was shown that interspecies transformation of the high-level streptomycin resistance marker could take place between *Neisseria catarrhalis* and strains of moraxellae. It was proposed that organisms previously called *Neisseria catarrhalis* be renamed as *Moraxella catarrhalis* (Henriksen and Bøvre, 1968). In 1970, Catlin proposed that organisms named *Neisseria catarrhalis* be transferred to the genus *Branhamella*. For pedagogic reasons Bøvre (1979) suggested that the genus *Moraxella* be divided into the subgenus *Moraxella* for the six rod-shaped species (*Moraxella (Moraxella) lacunata*, *M. (M.) bovis*, *M. (M.) nonliquefaciens*, *M. (M.) osloensis*, *M. (M.) phenylpyruvica*, and *M. (M.) atlantae* and the subgenus *Branhamella* for the coccal species (*Moraxella (Branhamella) catarrhalis*, *M. (B.) caviae*, *M. (B.) ovis*, and *M. (B.) cuniculi*: the latter three coccal groups frequently being referred to as the "false neisseriae"). In the light of recent findings, use of the subgeneric terms *Moraxella* and *Branhamella* should be discontinued. Transformation studies (Bøvre and Hagen, 1981) have shown that all the coccal species are members of the genus *Moraxella*. Other groups considered to have been species of the genus *Moraxella* are, in fact, separate genera of the family *Moraxellaceae*.

Early studies of the nutritional and physiological properties of the moraxellae (Baumann et al., 1968a) revealed that one group (*M. osloensis*) was nutritionally unexacting and three groups (*M. lacunata*, *M. nonliquefaciens*, and *M. bovis*) displayed various levels of nutritional requirements.

For many years, bacteria now known to be strains of *Acinetobacter* were associated with the genus *Achromobacter*. In 1923, the Bergey's Manual Committee proposed a new genus, *Achromobacter*, to include all nonpigmented, Gram-negative, and aerobic saprophytes (see Ingram and Shewan, 1960). In 1954, Brisou and Prévot proposed that nonmotile achromobacters be considered species of the newly defined genus *Acinetobacter*. With the introduction of the oxidase test in diagnostic bacteriology, it was noted that organisms previously classified as strains of *Acinetobacter* included both oxidase-positive and oxidase-negative species (Henriksen, 1952; Piéchaud, 1961). In 1967, Thornley published the results of an extensive investigation of both oxidase-positive and oxidase-negative bacteria, all considered strains of *Acinetobacter*. Acinetobacters have been difficult to classify in the past because they do not possess a sufficient number of unique phenotypic properties that enable them to be differentiated with certainty from other bacteria of similar appearance. In addition to being referred to as strains of *Achromobacter*, organisms subsequently demonstrated to be acinetobacters had been assigned at least 15 different names; those used most frequently include "*Bacterium anitratum*", "*Herellea vaginicola*", and "*Mima polymorpha*" (Baumann et al., 1968b). In a resolution based on classical taxonomic and genetic studies, the Subcommittee on the Taxonomy of *Moraxella* and Allied Bacteria suggested that the genus *Acinetobacter* should include only the oxidase-negative strains (Lessel, 1971). It was also proposed that the generic name *Achromobacter* be rejected (Hendrie et al., 1974). The strongest evidence that acinetobacters and moraxellas belong to the same family is that DNAs from streptomycin-resistant acinetobacters showed a weak, but consistently positive, transformation of *Moraxella catarrhalis* and *Moraxella osloensis* (Bøvre, 1967), and that relatively strong DNA–rRNA hybridization between the oxidase-positive and oxidase-negative organisms (Johnson et al., 1970) has been demonstrated. These results were supported by 16S rDNA sequence analysis (Pettersson et al., 1998).

The genus *Psychrobacter* contains most of the oxidase-positive strains of the original *Achromobacter* group. *Moraxella*-like psychrotrophic bacteria have been isolated from fish, poultry, and meat products (Gennari et al., 1992). Unlike the moraxellae and similar to some acinetobacters, some psychrobacters are able to form acid from glucose and other sugars. The demonstration of weak transformation of *M. osloensis* and other *Moraxella* species with DNA from streptomycin-resistant psychrotrophic strains (Bøvre and Hagen, 1981) established the relationship of these bacteria to the moraxellae. These organisms were subsequently classified as members of a new genus, *Psychrobacter* (Juni and Heym, 1986b).

Except for strains of *Acinetobacter* and *Psychrobacter*, all other members of the *Moraxellaceae* have been isolated as parasites of humans and other mammals. Several such organisms from sheep, cattle, and pigs have been studied (Bøvre and Hagen, 1981) and two of these, *M. caprae* (Kodjo et al., 1995) and *M. boevrei* (Kodjo et al., 1997), both isolated from goats, have been named. It seems most likely that there are many more species of the *Moraxellaceae* remaining to be discovered, particularly in less well-studied animals.

From studies of DNA–rRNA hybridization a new family, the *Moraxellaceae*, was proposed to accommodate the genera *Moraxella*, *Acinetobacter*, and *Psychrobacter* (Rossau et al., 1991). The family name *Moraxellaceae* is a most appropriate designation since it includes all organisms that have been shown to have genetic interactions with each other using the high-level-streptomycin resistance marker. The family name *Branhamaceae* was proposed to accommodate the spherical and rod-shaped species of *Moraxella* (Catlin, 1991). Unlike the family name *Moraxellaceae*, the proposed family name *Branhamaceae* fails to include *Acinetobacter* and *Psychrobacter*, genera that are clearly members of the same family. A summary of properties of individual members of some of the *Moraxellaceae* is given in Table BXII.γ.130.

Criteria for establishing membership in the *Moraxellaceae*
Strains are considered to belong to the family *Moraxellaceae*, provided that preferably all, but at least the first, of the criteria listed below are fulfilled:

TABLE BXII.γ.130. Phenotypic characteristics of the moraxellae[a,b,c]

Characteristics	M. lacunata	M. nonliquefaciens	M. bovis	M. caprae	M. ovis	M. caviae	M. catarrhalis	M. canis	M. cuniculi	M. osloensis	M. phenylpyruvica	M. atlantae	M. boevrei	M. lincolnii
Morphology	R	R	R	R	C	C	C	C	C	R	R	R	SR	C, R
Motility	−	−	−	−	−	−	−	−	−	−	−	−	−	
Catalase activity	+	+	(+)	+	+	+	+	+	+	+	+	+	+	+
Oxidase activity	+	+	+	+	+	+	+	+	+	+	+	+	+	+
Growth on MacConkey agar	−	−	−	−	−	−	−	−	−	v	nd	+	−	nd
Acids produced from glucose	−	−	−	−	−	−	−	−	−	−	−	−	−	−
Growth on minimal medium[d]	−	−	−	−	−	−	−	(+)	−	+	−	−	−	nd
Hemolysis	−	−	+	+	(+)	w	−	+	−	−	−	−	+	−
Nitrate reduction	+	+	(−)	+	+	+	(+)	+	−	(−)	(+)	−	+	−
Liquefaction of gelatin	+	−	+	−	−	−	−	−	−	−	−	−	+	−
DNase activity	−	−	−	−	(−)	(−)	+	(+)	−	−	−	−	−	−
Proteolysis on Löffler slants	+	−	I	−	−	−	−	−	−	−	−	−	+	
Indole	−	−	−	−	−	−	−	−	−	−	−	−	−	nd
Phenylalanine deaminase activity	−	−	−	−	−	−	−	−	−	−	+	nd	−	−
Hydrolysis of Tween 80	+	−	+	+	−	−	−	−	−	−	nd	−	+	−
Alkaline phosphatase activity	+	−	−	−	+	+	+	+	+	+	nd	+	−	−
Esterase activity	+	+	+	−	+	+	+	+	+	+	nd	+	+	+
Acid phosphatase activity	w	−	w	−	−	−	−	−	w	+	nd	+	−	−
Mol% G + C content	40–44.5	40–44	41–44.5	40–41.5	44.5–46.5	44.5–47.5	40–43	45.5–49.6	44.5	43–46	42.5–43.5	46.5–47.5	41–41.5	44

[a]Reproduced with permission from A. Kodjo et al., International Journal of Systematic Bacteriology 47: 115–121, 1997, ©International Union of Microbiological Societies.

[b]C, coccus; R, rod; SR, short rod.

[c]−, negative reaction; +, positive reaction; w, weak reaction; (+), most strains are positive; (−), most strains are negative; nd, not tested.

[d]Minimal medium containing ammonium and acetate.

1. All strains of the family must show either a weak or a strong genetic interaction of the conserved high-level streptomycin resistance marker with at least one established *Moraxellaceae* strain. Since competence for natural transformation is widespread in the *Moraxellaceae*, there should be no difficulty in satisfying this criterion.
2. 16S rRNA gene sequence similarity of all strains within the family should be significantly greater than is found between a given strain and any other bacterium that does not satisfy criterion 1.
3. In DNA–rRNA hybridizations, using DNA from any member of the *Moraxellaceae* and rRNA from a strain being tested, $T_{m(e)}$ values should be significantly greater for rRNAs derived from all members of the family than the corresponding value obtained using rRNA from any other bacterial strain.

At the present time, available data support criteria 2 and 3. However, because of the heterogeneity of *Acinetobacter* and *Psychrobacter*, as demonstrated by wide variation in DNA base composition, large numbers of species (as determined by DNA–DNA homology), and fairly widespread 16S rRNA gene sequence differences within the *Moraxellaceae*, it is possible that strains belonging to the family that do not fulfill one, or both, of criteria 2 and 3 may be found.

Criteria for distinguishing genera in the *Moraxellaceae* A genus of the *Moraxellaceae* can be distinguished from other genera in the family if it can be determined that one or more of the properties listed below apply to all members of the genus:

1. Distinct phenotypic properties. (Example: only members of the genus *Acinetobacter* are oxidase negative.)
2. Very similar rRNAs, as determined by sequence similarities or by DNA–rRNA relatedness. (Examples: high 16S rRNA gene sequence similarity for strains of the species *M. lacunata*, *M. bovis*, and *M. nonliquefaciens* (Pettersson et al., 1998), all of which are members of the genus *Moraxella*, and unique $T_{m(e)}$ range for strains of the genus *Acinetobacter* and also for strains of the genus *M. osloensis* [Rossau et al., 1991]).
3. In quantitative transformation of the high-level streptomycin resistance marker, demonstration of a ratio of interstrain to intrastrain transformation equal to or greater than 10^{-4}, for a strain being tested and compared with at least one member strain of the genus. (Examples: strains of all species of *Moraxella*, all strains of *M. osloensis*, and all strains of *M. atlantae* [Bøvre and Hagen, 1981; Kodjo et al., 1997]).
4. Ability of DNAs from all strains of a genus to transform a conserved nutritional marker of a competent member of the genus. (Examples: transformation of a *trpE*⁻ auxotroph of *Acinetobacter*, strain BD413 [Juni, 1972], transformation of a *trpE*⁻ auxotroph of a competent strain of *M. osloensis* [Juni, 1974], and transformation of a hypoxanthine and thiamin auxotroph of a competent strain of *Psychrobacter* [Juni and Heym, 1980]).

Criteria for establishing named species in the *Moraxellaceae*

1. Strains of the same species should show 70% or greater DNA–DNA relatedness with 5°C or less ΔT_m (Wayne et al., 1987).
2. Strains of the same species may have one or more phenotypic properties that are unique for members of the species, as defined by DNA–DNA hybridization. (Example: selected phenotypic properties of the named species of *Acinetobacter*.)

Genera and species of *Moraxellaceae*

I. Genus *Moraxella*
  A. *M. lacunata*
    1. "*M. lacunata* subsp. *lacunata*"
    2. "*M. lacunata* subsp. *liquefaciens*"
  B. *M. bovis*
  C. *M. canis*
  D. *M. caprae*
  E. *M. catarrhalis*
  F. *M. caviae*
  G. *M. cuniculi*
  H. *M. equi*
  I. *M. nonliquefaciens*
  J. *M. ovis*

  Species Incertae Sedis

  A. *M. atlantae*
  B. *M. boevrei*
  C. *M. lincolnii*
  D. *M. osloensis*
  E. *M. phenylpyruvica*
II. Genus *Acinetobacter*
  A. *A. baumannii*
  B. *A. calcoaceticus*
  C. *A. haemolyticus*
  D. *A. johnsonii*
  E. *A. junii*
  F. *A. lwoffii*
  G. *A. radioresistens*
  H. *A. venetianus*
  I. Other species: 14 as yet unnamed species (DNA groups 3, 6, 9, 10, 11, TU13, TU14 (= BJ13), TU15, CTTU13, 1-3, BJ14, BJ15, BJ16, and BJ17)
III. Genus *Psychrobacter*
  A. *P. faecalis*
  B. *P. frigidicola*
  C. *P. glacincola*
  D. *P. immobilis*
  E. *P. marincola*
  F. *P. pacificensis*
  G. *P. proteolyticus*
  H. *P. submarinus*
  I. *P. urativorans*

Note concerning the significance of natural transformation in bacterial taxonomy Natural transformation, when it occurs, provides a unique mechanism for recombination of bacterial genes. Most early transformation studies made use of antibiotic resistance markers. Streptomycin resistance, a frequently employed genetic marker, can be expected to give somewhat different results compared with those obtained using other chromosomal markers. Single step mutation to high levels of streptomycin resistance (or dependence) results from specific base changes in

DNA coding for the S12 (formerly P10) protein of the small ribosomal subunit (Ozaki et al., 1969). Since genes coding for ribosomal components (proteins and RNAs) have been shown to be highly conserved (Dubnau et al., 1965), transformation of a competent strain with a ribosomal marker DNA from any member of the same bacterial family is possible. Intergeneric transformation of such a marker is generally considerably weaker than intrageneric (interspecies) transformation. It has also been shown, using the streptomycin-resistance marker, that interspecies transformation is usually weaker than intraspecies transformation (Bøvre and Hagen, 1981). Extremely weak intergeneric transformation of the high-level streptomycin-resistance marker has, in the past, led to the conclusion that the interacting strains are members of the same genus. This view is no longer tenable in the light of more recent findings for the existence of several independent genera in the *Moraxellaceae*.

In contrast to ribosomal markers, nutritional markers appear to be limited to intrageneric (interspecies) transformation of competent strains. Studies of a variety of nutritional markers in competent strains have revealed that all of them can undergo intrastrain transformation. Only a relatively small number of them, however, are conserved in the sense that they can be transformed by DNAs from strains of all species of a particular genus. The reason for the conserved behavior of this latter class of nutritional markers is currently not understood. It has been proposed that the ability of DNA from one strain to transform another competent strain signifies that both strains are members of the same genus (Ravin, 1963). Using conserved nutritional markers this hypothesis has been verified for several genera of the *Moraxellaceae*: *Acinetobacter* (*trpE⁻* marker) (Juni, 1972), *Psychrobacter* (a hypoxanthine and thiamine marker) (Juni and Heym, 1980), and *M. osloensis* (*trpE⁻* marker) (Juni, 1974). The ability of DNAs from a large number of strains of *Oligella* (formally *Moraxella urethralis*) to transform a *trp⁻* marker has also been demonstrated (Juni, 1977). In these examples, transformation of the appropriate nutritional mutant by DNA from any strain of the genus has never failed. Furthermore, a false positive reaction with DNA from an organism that is not a member of the same genus has never been observed in these transformation assays. It would thus appear that the ability to transform conserved nutritional markers (but not conserved ribosomal markers) in competent strains of a bacterial genus might, in fact, be the best way to establish membership in that genus.

Using nonconserved nutritional markers, it has been shown that interstrain transformation takes place with high efficiency for strains of the *Moraxella* species *M. catarrhalis* (Juni, 1990), *M. nonliquefaciens* (Juni et al., 1987), and *M. bovis* (Juni et al., 1988). DNAs from other species of *Moraxella* failed to transform either of two *M. catarrhalis* auxotrophs (Juni, 1990). Auxotrophic mutants of *M. nonliquefaciens* and *M. bovis* were also transformed with lower efficiencies by DNAs from some, but not all, species of *Moraxella* (Juni et al., 1987, 1988).

It is possible to transform competent strains of any genus in the family for the ability to grow in the presence of streptomycin, using DNA from a streptomycin-resistant (or dependent) mutant of any strain of the family. Otherwise, the high-level streptomycin-resistance marker resembles conserved nutritional markers in its ability to transform any competent member of the genus from which the streptomycin-resistance marker was obtained. As with the case of conserved nutritional markers, interstrain transformation of the streptomycin-resistance marker can be less efficient than intrastrain transformation (Bøvre and Hagen, 1981). Intergeneric transformation of the high-level streptomycin-resistance marker probably occurs because of the conserved nature of all ribosomal genes, which are even more similar in sequence than conserved nutritional markers.

For most genetic markers, interfamily transformation has not been observed. In a rare exception, however, it has been shown that 23S rDNA from a variety of unrelated bacteria can stably transform any of the seven rDNA operons in competent *Acinetobacter* because of the highly conserved nature of rRNA genes (Strätz et al., 1996). It was also shown that the further the phylogenic distance (as determined from 16S rRNA gene sequence data), the lower the frequency of such interfamily transformation (Strätz et al., 1996).

Phylogenetic relationships of the *Moraxellaceae*

DNA–rRNA hybridization analysis DNA–rRNA hybridization and 16S rRNA (or rDNA) nucleotide sequencing of 1000 or more bases are two of the main procedures that have been used to obtain data for establishing phylogenetic relationships (Murray et al., 1990b). The (family) relationships of all genera of the *Moraxellaceae* were originally established by genetic transformation experiments (Bøvre and Hagen, 1981).

DNA–rRNA hybridization studies of the family *Neisseriaceae*, as this family was previously constituted, led to the conclusion that organisms presently considered to be members of the *Moraxellaceae* are not related to the *Neisseriaceae* (Rossau et al., 1986, 1989). Hybridization of rRNA from a strain of *Moraxella lacunata* and a strain of *Psychrobacter immobilis*, to DNAs from strains of all

members of the *Moraxellaceae* resulted in a distinct cluster of $T_{m(e)}$ values ($T_{m(e)}$ is the temperature at which one-half of a DNA–rRNA duplex is denatured) ranging from 69.9 to 80.0°C. Significantly lower $T_{m(e)}$ values were obtained for DNAs from strains that were members of other families of rRNA superfamily I, of rRNA superfamily II, and of rRNA superfamilies V and VI (Rossau et al., 1991). $T_{m(e)}$ values for acinetobacters formed a distinct group. The other strains were separated into subgroups of a so-called *Moraxella* group (Rossau et al., 1991). The *Moraxella lacunata* subgroup included strains of the classical "*Moraxella lacunata* group" as well as the coccal species of *Moraxella*, which could not be distinguished from each other by this procedure.

Only three strains of *M. osloensis* were examined. They formed a distinct subgroup when their DNAs were hybridized to "*M. lacunata* subsp. *liquefaciens*" rRNA but were clustered together with the *M. lacunata* subgroup when these DNAs were hybridized to *Psychrobacter immobilis* rRNA (Rossau et al., 1991). Using *M. lacunata* rRNA and DNA from other strains, *M. phenylpyruvica* clustered with strains of *Psychrobacter*. When *Psychrobacter* rRNA was used for hybridization, however, strains of *M. phenylpyruvica* appeared to reside in a unique subcluster. The one strain of *M. atlantae* examined could be assigned to either the *Psychrobacter* or the *Moraxella lacunata* subgroup, depending upon which rRNA was used for hybridization (Rossau et al., 1991).

Although DNA–rRNA hybridization is a useful procedure for establishing membership in the *Moraxellaceae*, results from such studies do not necessarily provide the means for distinguishing genera, or species of a particular genus, within the family.

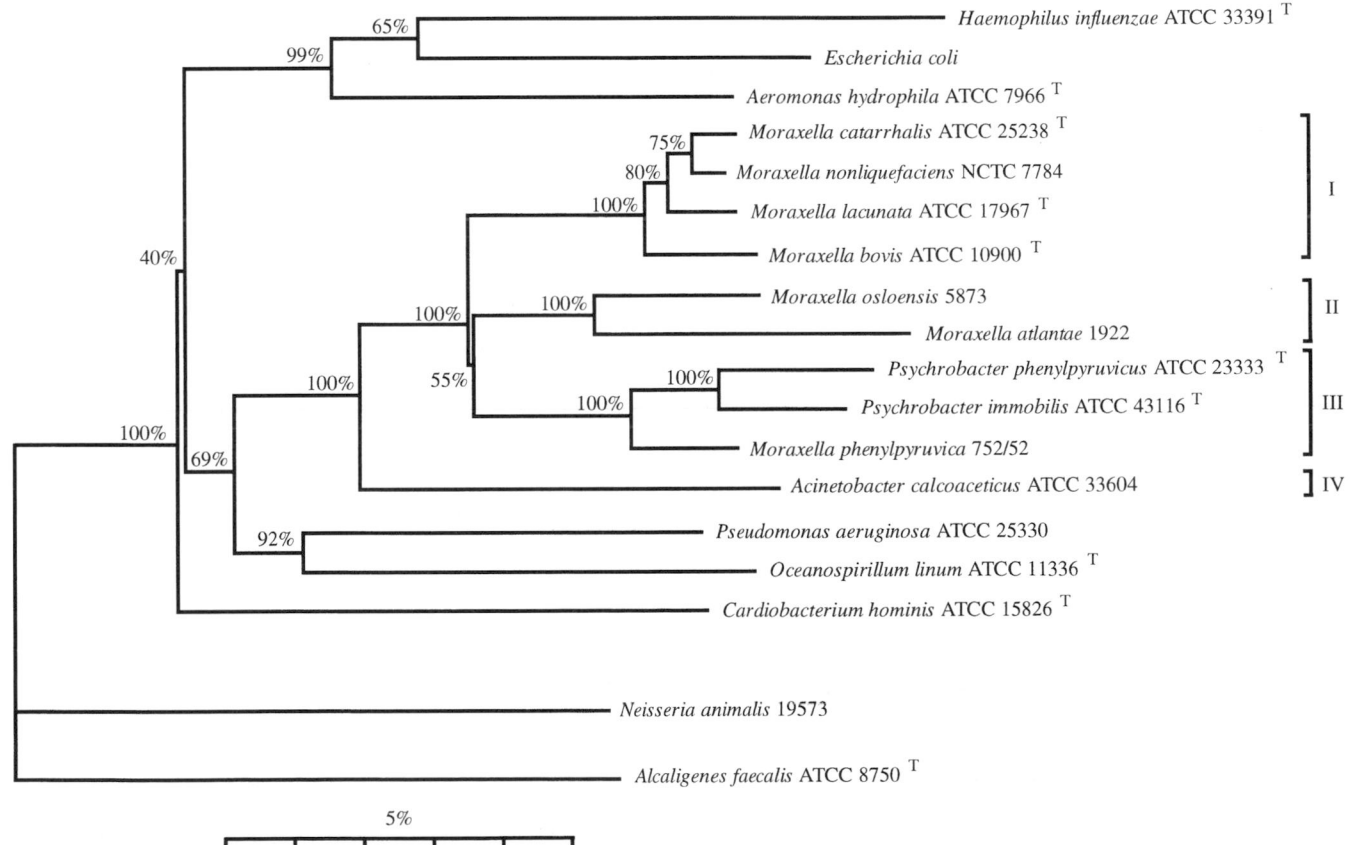

FIGURE BXII.γ.146. Phylogenetic tree based on distance matrix analysis of 1355 positions in the 16S rRNA genes for selected members of *Moraxellaceae*. (Reproduced with permission from B. Pettersson et al., International Journal of Systematic Bacteriology *48:* 75–89, 1998, ©International Union of Microbiological Societies.)

16S rRNA gene sequence analysis 16S rRNA gene sequence analysis is similar to DNA–rRNA hybridization in that it enables isolated strains to be recognized as members of the *Moraxellaceae*. With both procedures, it was possible to demonstrate that the genera *Acinetobacter*, *Moraxella*, and *Psychrobacter* are not members of the *Neisseriaceae* (Rossau et al., 1991; Enright et al., 1994).

A phylogenetic study of 33 strains, including 11 (previously designated) species of the family *Moraxellaceae*, by 16S rDNA sequence analysis (Pettersson et al., 1998) showed that the family *Moraxellaceae* formed a distinct clade consisting of four phylogenetic groups, as determined from branch lengths, bootstrap values, and signature nucleotides (Fig. BXII.γ.146). Group I contained the classical and coccal *Moraxella* species. Group II consisted of *Moraxella osloensis* and *Moraxella atlantae*. Group III contained strains of *Psychrobacter* and a single strain of *Moraxella phenylpyruvica*. Group IV contained two strains of *Acinetobacter*. The phylogenetic tree showed that the members of the *Moraxellaceae* constituted a monophyletic taxon forming a distinct line of descent within the *Gammaproteobacteria*. A phylogenetic tree for organisms belonging to group I is shown in Fig. BXII.γ.147.

16S rRNA nucleotides U and A in positions 514 and 537, respectively, flank a highly conserved oligonucleotide that should be useful for the development of diagnostic probes for detection of members of the *Moraxellaceae* (Pettersson et al., 1998). Phylogenetic conclusions regarding individual genera of the *Moraxellaceae* will be discussed below, along with the other features of each genus. Studies of other bacterial families have shown that information obtained through phylogenetic analysis of 16S rRNA

does not necessarily parallel results obtained by DNA–DNA hybridization at the species level (Stackebrandt and Goebel, 1994).

As this volume was going to press, it was noticed that the genus *Enhydrobacter* (Staley et al., 1987) resembles a genus of the *Moraxellaceae*, based on 16S rRNA sequence similarity (sequence similarity was determined using the RDP Sequence Match program and comparing *Enhydrobacter aerosaccus* strain LMG 21877 = ATCC 27094[T] (AJ550856) to the NCBI database). The nearest-neighbors of *Enhydrobacter aerosaccus* were *Moraxella osloensis* strains Ben 58 (X95304), 5873 (AF005190), PIV-10-1 (AJ505859), and AU1220 (AY043376) with S_{AB} values ranging from 0.956–0.973 and *M. osloensis* strain 3 (Y15855), S_{AB} = 0.893.

It should be noted, however, that *Enhydrobacter* DNA has a mol% G + C of 66, whereas the mol% G + C values of other genera in the family *Moraxellaceae* range from 38–50. In addition, *Enhydrobacter*, unlike all members of the *Moraxellaceae*, has several other unique properties, such as an extremely slow growth rate, growth at an optimum temperature of 37–39°C, ability to carry out sugar fermentations, and being gas-vacuolated. Although the relationship of *Enhydrobacter* 16S rRNA to that of *M. osloensis* suggests that *Enhydrobacter* may be a member of the *Moraxellaceae*, the several startlingly unusual properties of *Enhydrobacter* cast doubt on this conclusion.

It might be considered that 16S rDNA in *Enhydrobacter* may have been derived from *M. osloensis* by horizontal gene transfer or by mutation of 16S rDNA in an unrelated strain over eons of time resulting in the rRNA sequence in *Enhydrobacter* that is presently observed. It is also possible that a strain of *M. osloensis* may

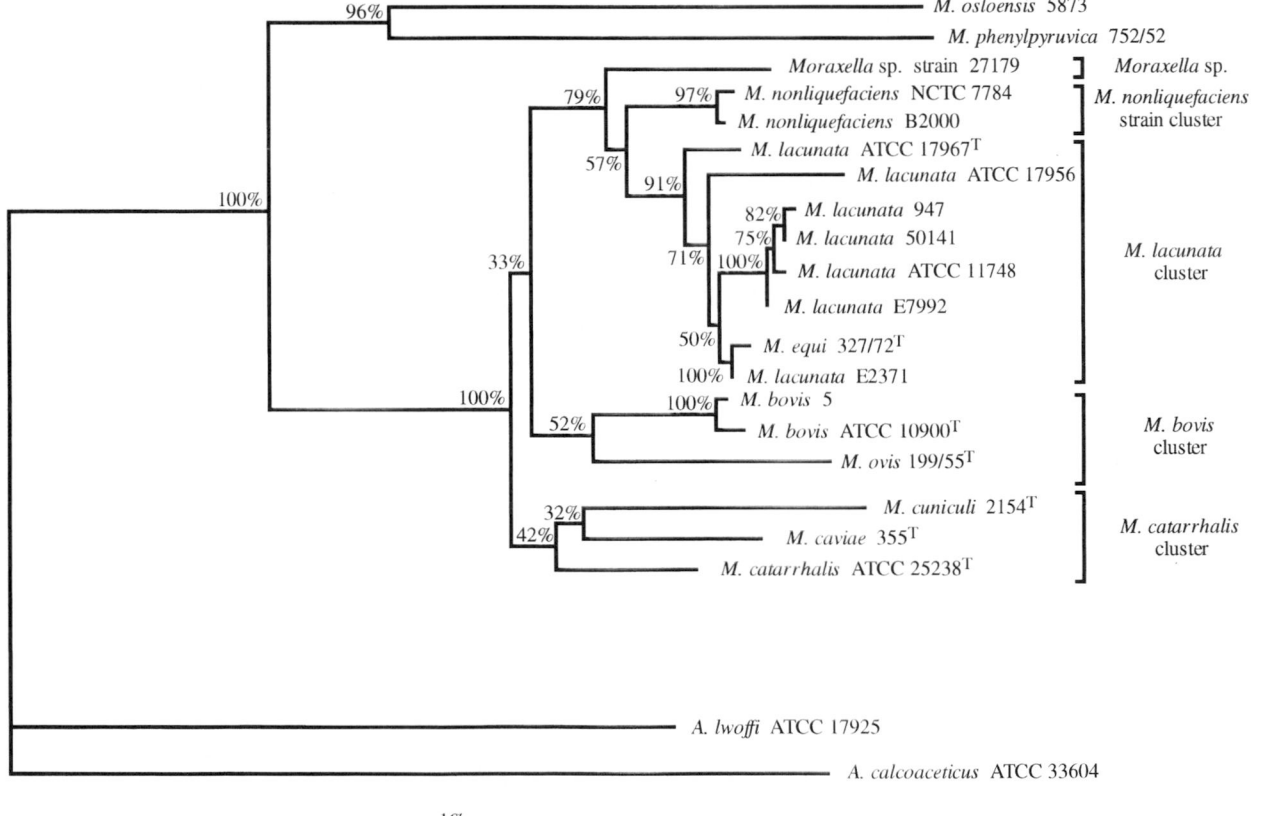

FIGURE BXII.γ.147. Phylogenetic tree based on distance matrix analysis of 1440 positions in the 16S rRNA genes of the moraxellae. (Reproduced with permission from B. Pettersson et al., International Journal of Systematic Bacteriology *48:* 75–89, 1998, ©International Union of Microbiological Societies.)

have undergone very extensive genetic modification of the many genes responsible for the unique phenotypes described above resulting in the organism now referred to as *Enhydrobacter*. In any case, this dilemma can be resolved readily by checking for ability of DNA from a high level streptomycin-resistant mutant of *Enhydrobacter* to transform competent strains of *M. osloensis*, or any of several other competent strains of several genera of the *Moraxellaceae* where even a weak, but reproducible, transformation would prove that *Enhydrobacter* is indeed a member of the *Moraxellaceae*. Also, *Enhydrobacter* DNA should be tested for ability to transform the genus-specific *trpE⁻* mutant (Juni, 1974) where, once again, a positive result would prove that *Enhydrobacter* is a member of the *Moraxellaceae*. Until such tests are performed, it is premature to consider that *Enhydrobacter* is a member of the *Moraxellaceae* based solely on 16S rRNA data.

ACKNOWLEDGMENTS

This manuscript is dedicated to the memory of Kjell Bøvre, friend and colleague. It is largely because of his critical studies that we now have a much better understanding of the moraxellae.

FURTHER READING

Bøvre, K. 1984. Genus II. *Moraxella* Lwoff 1939. *In* Krieg and Holt (Editors), Bergey's Manual of Systematic Bacteriology, 1st Ed., Vol. 1, The Williams & Wilkins Co., Baltimore. 296–303.

Bøvre, K. and N. Hagen. 1981. The family *Neisseriaceae*: rod-shaped species of the genera *Moraxella*, *Acinetobacter*, *Kingella*, and *Neisseria*, and the *Branhamella* group of cocci. *In* Starr, Stolp, Trüper, Balows and Schlegel (Editors), The Prokaryotes. A Handbook on Habitats, Isolation and Identification of Bacteria, 1st Ed., Vol. 2, Springer-Verlag, New York. pp. 1506–1529.

Henriksen, S.D. 1973. *Moraxella, Acinetobacter*, and the *Mimeae*. Bacteriol. Rev. *37*: 522–561.

Henriksen, S.D. 1976. *Moraxella, Neisseria, Branhamella*, and *Acinetobacter*. Annu. Rev. Microbiol. *30*: 63–83.

Juni, E. 1978. Genetics and physiology of *Acinetobacter*. Annu. Rev. Microbiol. *32*: 349–371.

Juni. E. 1984. Genus III. *Acinetobacter* Brisou and Prévot 1954. *In* Krieg and Holt (Editors), Bergey's Manual of Systematic Bacteriology, 1st Ed., Vol. 1, The Williams & Wilkins Co., Baltimore. 303–307.

Juni, E. 1992. The genus *Psychrobacter*. *In* Balows, Trüper, Dworkin, Harder and Schleifer (Editors), The Prokaryotes, 2nd Ed., Vol. IV, Springer-Verlag, New York. 3241–3246.

Genus I. **Moraxella** *Lwoff 1939, 173, emend. Henriksen and Bøvre 1968, 391*[AL]

ELLIOT JUNI AND KJELL BØVRE

Mo.rax.el'la. M.L. dim. *-ella* ending; M.L. fem. n. *Moraxella* named after V. Morax, a Swiss ophthalmologist who pioneered the recognition of the type species.

Rods or cocci. The rods are often very short and plump, frequently approaching a coccus shape (1.0–1.5 × 1.5–2.5 µm); they usually occur in pairs and short chains (one plane of division). Variation in cell size, shape, and filament or chain formation is often seen in cultures, the pleomorphism being enhanced by lack of oxygen and by incubation temperatures above the optimum. **The cocci are usually smaller** (0.6–1.0 µm in diameter) and occur as single cells or in pairs with the adjacent sides flattened (differing planes of division); division in two planes at right angles to each other sometimes results in the formation of tetrads. Gram negative, but often with a tendency to resist Gram decolorization. Flagella are absent. Both rod-shaped and coccal species may be fimbriated. Swimming motility is absent, but surface-bound "twitching motility" has been observed in some rod-shaped species. **Aerobic, but some strains may grow weakly under anaerobic conditions.** May be encapsulated. Chemoorganotrophic. Most species are nutritionally fastidious and all grow on complex media; some are stimulated significantly by fatty acids (bile salts, Tween 80). Optimum temperature for growth, 33–35°C. **Colonies are not pigmented. Oxidase positive** (with either tetra- or dimethyl-*p*-phenylenediamine reagent). **Usually catalase positive. No acid is produced from carbohydrates.** Usually highly sensitive to penicillin. Parasitic on the mucous membranes of humans and other warm-blooded animals. Strains of the several species of this genus have been shown to be genetically closely related by transformation of the high-level streptomycin resistance marker (Bøvre and Hagen, 1981) as well as by transformation of nutritional markers (Juni et al., 1987, 1988; Juni, 1990). DNA–DNA hybridization studies have shown high relative interspecies binding ratios for DNAs from strains of the genus *Moraxella* (Tønjum et al., 1989). Species of *Moraxella* share 95% similarity in their 16S rRNA gene sequences.

The mol% G + C of the DNA is: 40.0–47.5 (T_m, Bd, HPLC).

Type species: **Moraxella lacunata** (Eyre 1900) Lwoff 1939, 173 (*Bacillus lacunatus* Eyre 1900, 5; (*Moraxella (Moraxella) lacunata* (Eyre 1900) Bøvre 1979, 404.)

List of species of the genus Moraxella

1. **Moraxella lacunata** (Eyre 1900) Lwoff 1939, 173[AL] (*Bacillus lacunatus* Eyre 1900, 5; (*Moraxella (Moraxella) lacunata* (Eyre 1900) Bøvre 1979, 404.)*

   la.cu.na' ta. L. n. *lacuna* a shallow depression; M.L. fem. adj. *lacunata* pitted.

   Two subspecies of *Moraxella lacunata* have been recognized (Henriksen, 1969).

   The mol% G + C of the DNA is: 40.0–44.5.

   Type strain: ATCC 17967, NCTC 11011.

   GenBank accession number (16S rRNA): AF005160, AF005170, D64049.

   a. **"Moraxella lacunata subsp. lacunata"**

   Medium thick to plump rods, 0.8–1.2 µm in diameter, coccoid to distinctly bacillary, occurring predominantly in pairs and short chains. Frequently pleomorphic. May form narrow capsules. Colonies on blood agar are small (0.1–0.3 mm in diameter). Colonies are translucent to semiopaque. Pitting of the agar may be observed. No hemolysis occurs, but on heated blood (chocolate) agar, large dark zones around the colonies usually occur.

   Strains may show improved growth on chocolate agar medium and may not grow on certain rich media without serum or oleic acid. They are unable to grow on medium MB, a defined medium that supports growth of strains of *M. bovis* (Juni et al., 1988). Other characteristics of

---

Editorial Note: Neither "*Moraxella lacunata* subspecies *lacunata*" nor "*Moraxella lacunata* subspecies *liquefaciens*" are validly published names.

the species are given in Table BXII.γ.130 of the description of the family *Moraxellaceae*. Human strains have been derived mainly from inflamed, as well as healthy, conjunctiva and from sites in the upper respiratory tract (Graham et al., 1990). They have occasionally been isolated from the blood. Attachment to epithelial tissues is mediated by type 4 pili (Ruehl et al., 1988), which are regulated by the *piv* gene. This gene encodes a site-specific DNA invertase (Marrs et al., 1990), the transcription of which is initiated at *pivp*, a σ^{70}-dependent promoter (Heinrich and Glasgow, 1997). The organism appears to be a significant causative agent of human conjunctivitis (and keratitis) and was more frequently isolated in the past. It is only rarely isolated at the present time.

The mol% G + C of the DNA is: 40.0–44.5 (Bd).

Type strain: none.

GenBank accession number (16S rRNA): AF005170.

Additional Remarks: There is presently no type strain of this subspecies; the previously listed type strain, ATCC 17967 (NCTC 11011), is actually a strain of "*M. lacunata* subsp. *liquefaciens*". Strain ATCC 11748 (NCTC 10748) should be proposed as the neotype strain since it has typical properties of the species, as well as being competent for natural transformation.

b. **"Moraxella lacunata** subsp. **liquefaciens"**

li.que.fa'ciens. M.L. part. adj. *liquefaciens* dissolving.

This subspecies is very similar to "*M. lacunata* subsp. *lacunata*" in morphology and in its response to most conventional tests, but differs in being able to grow on media without serum and to liquefy the usual gelatin media. "*M. lacunata* subsp. *liquefaciens*" was first described as the cause of keratitis. It has also been isolated from cases of endocarditis and sinusitis. Less fastidious than "*M. lacunata* subsp. *lacunata*" (Bøvre, 1965c); can grow in a defined medium (medium MB) containing 11 growth factors (Juni et al., 1988). Colonies on blood agar medium have a greater diameter (~3 mm) than colonies of "*M. lacunata* subsp. *lacunata*". Properties enabling further distinction between strains of "*M. lacunata* subsp. *liquefaciens*" and "*M. lacunata* subsp. *lacunata*" are discussed below. A number of isolates from the conjunctiva of healthy guinea pigs were shown to have the same morphological, cultural and biochemical reactions as "*M. lacunata* subsp. *liquefaciens*" (Ryan, 1964). Furthermore, quantitative transformation studies, using high-level streptomycin-resistance markers, revealed that two of these guinea pig isolates were very closely related to authentic strains of "*M. lacunata* subsp. *liquefaciens*" (Bøvre, 1965c). Because there are antigenic differences between guinea pig isolates and an authentic strain of "*M. lacunata* subsp. *liquefaciens*" (Ryan, 1964), it is possible that the animal strains represent another distinct subspecies of *M. lacunata*.

The mol% G + C of the DNA is: 41.5–43 (Bd).

Type strain: ATCC 17952, NCTC 7911 should be proposed as the neotype strain.

GenBank accession number (16S rRNA): AF005169.

Further comments The occurrence of genetically competent strains permits identification by quantitative genetic transformation (Bøvre and Hagen, 1981; Tøn-jum et al., 1992). Although "*M. lacunata* subsp. *lacunata*" and "*M. lacunata* subsp. *liquefaciens*" have many phenotypic similarities and cannot be differentiated readily based on 16S rRNA gene sequence analysis (Pettersson et al., 1998), it is possible to accomplish this by comparison of certain phenotypic properties. It has long been known that strains of "*M. lacunata* subsp. *lacunata*" are more nutritionally fastidious and generally have smaller colonies on blood agar medium than strains of "*M. lacunata* subsp. *liquefaciens*" (Bøvre, 1965c; Bøvre and Hagen, 1981). Studies of a collection of bacteria, received as strains of *M. lacunata*, showed that not all of these strains could grow on a defined medium (medium MB) suitable for growth of strains of *M. bovis* (Juni et al., 1986; 1988); one group of these strains (group 1) being able to grow on medium MB whereas strains of a second group (group 2) being unable to grow on this medium. It was subsequently shown that group 1 consisted of strains of "*M. lacunata* subsp. *liquefaciens*" whereas strains in group 2 consisted of strains of "*M. lacunata* subsp. *lacunata*" (Tønjum et al., 1992).

Furthermore, other studies of a collection of strains identified as *M. lacunata* based on results of classical phenotypic tests disclosed that these organisms fell into two classes based on their respective cellular fatty acid profiles (Moss et al., 1988). Profiles I and II are characteristic for strains of "*M. lacunata* subsp. *liquefaciens*" and "*M. lacunata* subsp. *lacunata*", respectively. Many of the same strains were used in both the nutritional (Juni et al., 1988) and fatty acid studies (Moss et al., 1988).

Although there was some overlap, the clustering of strains of "*M. lacunata* subsp. *lacunata*" and "*M. lacunata* subsp. *liquefaciens*" into distinct groups was also apparent from total genomic DNA–DNA hybridization, (Tønjum et al., 1992). These two subspecies were also distinguished based on genetic variation of 12 enzyme loci (Tønjum et al., 1992). In quantitative transformation studies using the high-level streptomycin-resistant marker, it was shown that "*M. lacunata* subsp. *lacunata*" and "*M. lacunata* subsp. *liquefaciens*" were closely related but could, nevertheless, be distinguished from each other (Tønjum et al., 1992).

The different nutritional requirements and different cellular fatty acid patterns, as well as variation of enzyme loci of the subspecies, suggest that these two organisms normally occupy different ecological niches. "*M. lacunata* subsp. *lacunata*" may normally reside at a site so well supplied with nutrients that it has, over eons of time, evolved and lost considerable biosynthetic ability. By contrast, "*M. lacunata* subsp. *liquefaciens*" may reside at a site deficient in certain nutrients and has evolved to synthesize those growth factors not supplied at its normal host site, while losing the ability to synthesize growth factors not required at this site. This accounts for its relatively increased biosynthetic ability. Evolution in membrane structure, as indicated by different fatty acid compositions of the two subspecies, may result in membranes that, in some manner, contribute to the ability to compete successfully with other bacteria for access to a particular site. If these properties do, in fact, contribute to the ability to preferentially occupy certain sites, it would seem prudent to compare phenotypic properties, such as those described above, when attempting to allocate bacteria to species since these may be the very factors that determine speciation. The subspecies nomenclature is retained pending a more precise definition of the terms species/subspecies as well as further possible findings regarding other different characteristics of the entities.

Phylogeny of the *Moraxella lacunata* cluster A phylogenetic study of the *Moraxellaceae* by 16S rDNA sequence analysis revealed four distinct groups (Pettersson et al., 1998) (Fig. BXII.γ.146). Strains of *Moraxella* were shown to reside in three clusters of group 1 (Fig. BXII.γ.147). A so-called *M. lacunata* cluster included 13 strains of "*M. lacunata* subsp. *lacunata*", "*M. lacunata* subsp. *liquefaciens*", and *M. equi*. Eight strains of this cluster had unique 16S rRNA gene sequences, and interstrain similarity values within the cluster ranged from 98.4 to 99.9% (Pettersson et al., 1998). Five strains of "*M. lacunata* subsp. *lacunata*", obtained from different geographic regions, clustered tightly together, and shared 99.8% 16S rRNA gene nucleotide similarity (Pettersson et al., 1998).

2. **Moraxella bovis** (Hauduroy, Ehringer, Urbain, Guillot and Magrou 1937) Murray 1948, 591[AL] (*Haemophilus bovis* Hauduroy, Ehringer, Urbain, Guillot and Magrou 1937, 247; *Moraxella (Moraxella) bovis* (Hauduroy, Ehringer, Urbain, Guillot and Magrou 1937) Bøvre 1979, 404.)

bo' vis. L. gen. n. *bovis* of cattle.

Rods, showing variation in shape, size and arrangement. The cells are often fimbriated and show "twitching motility." A distinct variation in colony type occurs. Freshly isolated strains form hemispherical to flat colonies (~1 mm in diameter in 48 h) on blood agar; they corrode the agar and often show surface spreading. These characteristics are associated with fimbriation of the cells and "twitching motility" (Bøvre and Frøholm, 1972b; Henrichsen et al., 1972; Pedersen et al., 1972). The texture of such colonies is often friable. Nonfimbriated variants occur spontaneously in cultures and form noncorroding, nonspreading colonies which are convex and usually larger than their corroding counterparts (often 3.0 mm after 48 h), and have a butyrous consistency. Fimbriated cells of collection strains may also form large colonies, but these may be less corroding than the colonies of fresh isolates. With few exceptions, bovine strains produce distinct hemolytic zones around colonies. Dark zones are produced on chocolate agar. Nutritionally fastidious. Other characteristics are listed in Table BXII.γ.130 of the description of the family *Moraxellaceae*.

Most frequently isolated from bovine eyes in cases of infectious keratoconjunctivitis, but has also been isolated from unaffected eyes and the nasal cavity of cattle, the eyes of sheep and mice, but not the eyes of rats, rabbits, or guinea pigs. Considered as potentially pathogenic for animals, dependent on accessory factors to elicit disease. Conjunctival colonization and disease development in UV-irradiated bovine eyes have been found associated with fimbriation of the bacteria (Pedersen et al., 1972).

Further comments Growth of most strains of *M. bovis* occurs in a defined medium containing 11 growth factors (Juni and Heym, 1986a). The cells are frequently transformable, genetic competence being associated with fim-

briation (piliation) and corrosion (Bøvre and Frøholm, 1972a). A *M. bovis* pilin gene has been cloned and sequenced (Marrs et al., 1985; Rozsa and Marrs, 1991). Strains may be identified by quantitative genetic transformation of the high-level streptomycin resistance marker (Bøvre, 1965a, c), or qualitatively by transformation of an auxotrophic mutant of a competent strain (Juni et al., 1988). The species is slightly genetically heterogeneous, with interstrain transformation affinities as low as 10% of intrastrain transformation affinities (Bøvre, unpublished results). A so-called *M. bovis* cluster consisting of strains of *M. bovis* and *M. ovis* has been observed in phylogenetic studies involving 16S rRNA gene sequence analysis (Fig. BXII.γ.147); the 16S rRNA genes of *M. bovis* and *M. ovis* were 97.1–97.4% similar (Pettersson et al., 1998).

Two sequence-specific endonucleases (*Mbo*I and *Mbo*II) have been demonstrated in *M. bovis*, ATCC 10900 (Gelinas et al., 1977). The genes encoding the *Mbo*I restriction-modification enzymes are adjacent to each other with this gene cluster encoding two distinct methyl transferases (Ueno et al., 1993) and have been cloned and expressed in *E. coli*. An enzyme from *M. bovis* that methylates the *Mbo*II recognition sequence has been purified (McClelland et al., 1985). The two genes encoding this class II restriction-modification system are located adjacent to each other and have each been cloned separately in *E. coli* (Bocklage et al., 1991). Details of the interaction of the *Mbo*II restriction endonuclease with its DNA target site have been investigated (Sektas et al., 1995).

The mol% G + C of the DNA is: 42.5–43 (Bd).

Type strain: ATCC 10900.

GenBank accession number (16S rRNA): AF005182.

3. **Moraxella canis** Jannes, Vaneechoutte, Lannoo, Gillis, Vancanneyt, Vandamme, Verschraegen, Van Heuverswyn and Rossau 1993, 448[VP]

ca' nis. L. gen. n. *canis* of a dog.

Cells are cocci (diameter ~0.8–1.3 μm) and usually occur in pairs or short chains. Gram negative, but there may be a tendency to resist decolorization. A brown pigment may be observed on Mueller–Hinton agar. Cells are nonmotile in liquid medium. Chemoorganotrophic and aerobic. Mesophilic. Cell growth is usually observed on a medium containing mineral salts and ammonium acetate, on lactose– bromothymol blue agar, on eosine–methylene blue agar, on Mueller–Hinton medium, and on MacConkey agar at 37°C. Growth occurs on Mueller–Hinton agar at room temperature. No growth occurs in the presence of 6% NaCl or at 5°C. Usually indifferent to bile salts. Oxidase and catalase positive. Extracellular DNase, butyrase, esterase, and γ-glutamyl-aminopeptidase are produced. Phenylalanine deaminase, indole, urease, and carbonic anhydrase are not produced. No acid is produced from glucose. Tween 80 and starch are not hydrolyzed. Autoagglutination in physiological solution is usually not observed. Gelatin and serum are not liquefied. No hemolysis of human and sheep blood occurs. The major fatty acids are oleic acid ($C_{18:1\ \omega9c}$) and palmitoleic acid ($C_{16:1\ \omega7c}$). Isoheptadecanoic acid ($C_{17:0\ iso}$) is present in moderate amounts, and myristic acid ($C_{14:0}$) is absent. Typical characteristics are listed in Table BXII.γ.130 of the description of the family *Moraxellaceae*.

Isolated from the saliva of dogs or cats and occasionally

from humans (dog bite wounds or blood). Pathogenicity is not known.

Further comments Strains are frequently competent for genetic transformation. Transformation studies of strain NCTC 4103 (ATCC 23246), originally considered a strain of *M. catarrhalis*, showed that this strain interacted weakly with authentic strains of *M. catarrhalis* (Catlin and Cunningham, 1964). In an investigation involving transformation of nutritional mutants of *M. catarrhalis* and NCTC 4103, it was shown that strain NCTC 4103 is more closely related to *M. canis* than to *M. catarrhalis* (Juni, 1990). Significant levels of DNA–DNA relatedness (52–61%) were found between strain NCTC 4103 and strains of *M. canis*, and it has been suggested that strain NCTC 4103 occupies a position somewhere between *M. catarrhalis* and *M. canis* (Jannes et al., 1993). From the results of a polyphasic taxonomic investigation, it was concluded that strains of *M. canis* constitute a distinct cluster within the genus *Moraxella* (Jannes et al., 1993).

A review of the properties of *M. catarrhalis* appeared in 1992 (Doern, 1992).

The mol% G + C of the DNA is: 45.9–47.5 (T_m).

Type strain: N7, CCUG 8415A, LMG 11194.

4. **Moraxella caprae** Kodjo, Tønjum, Richard and Bøvre 1995, 471[VP]

ca′prae. L. gen. n. *caprae* of a goat.

Gram negative. Medium to large straight rods with a tendency to grow as diplobacilli or in short chains. Nonmotile, aerobic, and grow well at temperatures between 30 and 37°C. Colonies are grayish white, smooth, and convex, vary in size from 1 to 1.5 mm, and are surrounded by a wide area of hemolysis after 36–48 h of aerobic incubation on 5% blood agar plates. Grow poorly on nutrient agar. Oxidase and catalase positive. All strains hydrolyze Tween 80 and tributyrin and reduce nitrate. No acid is produced from carbohydrates. All strains are nonproteolytic (Loeffler slant and gelatin negative). Indole and urease reactions are also negative. The main features that distinguish *M. caprae* from *M. bovis* are reduction of nitrate and absence of proteolysis. Other characteristics are listed in Table BXII.γ.130 of the description of the family *Moraxellaceae*. The clinical significance of this organism is not known. All strains were isolated from the nasal cavity of healthy goats.

Further comments Strains of *M. caprae* have a high degree of genetic affinity to *M. bovis*, as shown by quantitative and qualitative genetic transformation assays, and exhibit high DNA–DNA relative binding ratios to each other, but lower levels of DNA relatedness to all other species of *Moraxella* (Kodjo et al., 1995). None of the eight strains of *M. caprae* tested was competent for genetic transformation.

The mol% G + C of the DNA is: 40–41.5 (HPLC).

Type strain: CCUG 33296, NCTC 12877.

5. **Moraxella catarrhalis** (Frosch and Kolle 1896) Henriksen and Bøvre 1968, 391[AL] (*Mikrokokkus catarrhalis* Frosch and Kolle *in* Flügge 1896, 15; *Branhamella catarrhalis* (Frosch and Kolle 1896) Catlin 1970, 157; *Moraxella* (*Branhamella*) *catarrhalis* Bøvre 1979, 404.)

ca.tarrh.a′lis. Gr. adj. *catarrhus* downflowing, catarrh; M.L.adj. *catarrhalis* of catarrh.

Cocci; division in two planes at right angles to each other and tetrad formation may be observed. May be fimbriated. Colonies are approximately 2.0 mm in diameter in 48 h, opaque, hemispherical, becoming considerably larger and convex, almost flat, on prolonged incubation. They usually have a friable texture and can be shifted around and picked from the agar without losing their semi-convex shape and without adherence to the agar. They are nonhemolytic. Extracellular DNase is produced. Other characteristics are listed in Table BXII.γ.130 of the description of the family *Moraxellaceae*.

Previously, as a rule, highly sensitive to penicillin, but strains resistant because of β-lactamase production are now generally encountered. Frequently isolated from the human nasopharynx, which appears to be its main natural habitat (Murphy, 1996). Has also been isolated from inflammatory secretions of the middle ear and maxillary sinus, from bronchial aspirate in bronchitis and pneumonia, and occasionally from systemic infections. The organism is most often considered to be a well-adapted parasite, but it is also a significant cause of disease in both the upper and lower respiratory tracts. It is responsible for otitis media and sinusitis in children and infants (Murphy, 1996; Enright and McKenzie, 1997; Catlin, 1990).

Further comments *M. catarrhalis* has been shown to possess several outer membrane proteins; the outer membrane protein profiles of this organism are highly conserved (Murphy, 1996). It produces a rough-type lipopolysaccharide attached to a lipid A moiety, and lacks high-molecular-mass O-polysaccharide chains (Johnson et al., 1976). There are at least three serotypes of *M. catarrhalis* lipopolysaccharide (Vaneechoutte et al., 1990).

The cells are frequently competent for genetic transformation and are easily identified by transformation procedures. In a qualitative transformation assay, in which nutritional mutants of a competent strain of *M. catarrhalis* were transformed to prototrophy with homologous wild-type DNA, DNAs from 20 independently isolated strains of *M. catarrhalis* all transformed these mutants with the same efficiency (Juni, 1990). In quantitative transformations, comparing ratios of interstrain to intrastrain transformation of the high-level streptomycin-resistance marker, a competent strain of *M. catarrhalis* showed extremely poor transformation (ratios $<5 \times 10^{-5}$) with DNAs from representative strains of most species of *Moraxella*. The exception was *M. cuniculi*, where a ratio of 1×10^{-3} was obtained (Bøvre and Hagen, 1981). In another study of transformation of the high-level streptomycin-resistance marker, an affinity ratio $>10^{-4}$ for *M. catarrhalis* and *M. bovis* was observed (Kodjo et al., 1997).

The amino acids glycine and arginine are the principal growth factors in a defined medium for all strains of *M. catarrhalis* (Juni et al., 1986).

The species appears genetically homogeneous, but one deviating strain, NCTC 4103 (ATCC 23246), has been detected (Catlin and Cunningham, 1964; Bøvre, 1965b, 1980; Bøvre and Hagen, 1981). Strain NCTC 4103 was originally reported to show a significant level (≤59%) of DNA–DNA hybridization with *M. catarrhalis* (Rossau et al., 1991), but these results could not be reproduced in a subsequent study (Jannes et al., 1993).

Phylogenetic investigations based on 16S rRNA gene sequence analysis place *M. catarrhalis*, *M. cuniculi*, and *M. caviae* together in a somewhat diverse cluster (Pettersson et al., 1998). A review of the properties of *M. catarrhalis* appeared in 1992 (Doern, 1992).

The mol% G + C of the DNA is: 40–43 (Bd).
Type strain: Ne 11, ATCC 25238, NCTC 11020.
GenBank accession number (16S rRNA): AF005185.

6. **Moraxella caviae** (Pelczar 1953) Henriksen and Bøvre 1968, 391[AL] (*Neisseria caviae* Pelczar 1953, 744; *Moraxella (Branhamella) caviae* (Pelczar 1953) Bøvre 1979, 404.)
ca'vi.ae. M.L. fem. n. *Cavia* generic name of the guinea pig; M.L. gen. n. *caviae* of *Cavia*.

Cocci commonly arranged in pairs with adjacent sides flattened; division in two planes at right angles to each other. Tetrad formation may be observed. The colonies are low, conical, and have a diameter of approximately 2.5 mm in 48 h. They are semiopaque and have a butyrous consistency. Weak hemolysis may be observed, and in this respect, the colonies may resemble those of *M. ovis*.

Sensitive to penicillin, but in some studies the organism has been shown to resist 0.1 U/ml penicillin. Other characteristics are listed in Table BXII.γ.130 of the description of the family *Moraxellaceae*. Isolated from the pharynx and mouth of healthy guinea pigs.

Further comments All strains examined are noncompetent for genetic transformation. In quantitative transformation of the high-level streptomycin-resistance marker to competent strains of *M. catarrhalis* and *M. cuniculi*, the ratios of interstrain to intrastrain transformation were extremely low (2×10^{-5} and 5×10^{-5}, respectively), but a higher ratio of 5×10^{-4} for transformation of *M. ovis* (Bøvre and Hagen, 1981) was found. Similar transformation studies involving DNA from streptomycin-resistant *M. caviae* resulted in extremely low interactions with competent strains of *M. bovis* (Henriksen, 1973). Phylogenetic studies of 16S rRNA gene sequences place one strain of *M. caviae* in the same, somewhat unstable, cluster, which includes *M. catarrhalis* and *M. cuniculi* (Pettersson et al., 1998). It would appear that *M. caviae* might be the most distantly related species of *Moraxella*.

The mol% G + C of the DNA is: 44.5–47.5 (Bd).
Type strain: ATCC 14659, CCUG 355, NCTC 10293.

7. **Moraxella cuniculi** (Berger 1962) Bøvre and Hagen 1984, 355[VP] (Effective publication: Bøvre and Hagen 1981, 1522) (*Neisseria cuniculi* Berger 1962, 455.)
cun.i' cu.li. L. gen. n. *cuniculi* of the rabbit.

Cocci commonly arranged in pairs with adjacent sides flattened; division in two planes at right angles to each other. Tetrad formation may be observed. The colony size in 48 h is similar to, or slightly smaller than, that of *M. catarrhalis*. The colonies are conical, opaque, and often friable. Nonhemolytic. Isolated from the mouths of healthy rabbits. No evidence of pathogenicity. Other characteristics are listed in Table BXII.γ.130 of the description of the family *Moraxellaceae*.

Further comments Using the high-level streptomycin-resistance marker, ratios of interstrain to intrastrain transformation of 2×10^{-3}, 2×10^{-4}, and 1×10^{-4} for interaction with *M. catarrhalis*, *M. ovis*, and *M. lacunata*, respectively, were obtained (Bøvre and Hagen, 1981); these results verify that *M. cuniculi* is a species of the genus *Moraxella*. Only low level DNA–DNA hybridization between *M. cuniculi* and either *M. ovis* or *M. caviae* could be demonstrated (Véron et al., 1993). The type strain is competent for genetic transformation, making genetic identification of new isolates possible.

The mol% G + C of the DNA is: 44.5 (Bd).
Type strain: ATCC 14688, NCTC 10297.

8. **Moraxella equi** Hughes and Pugh 1970, 457[AL]
e'qui. L. gen. n. *equi* of the horse.

Nonmotile diplococcus. Colonies on blood agar are 0.5–1.0 mm in diameter, resembling *M. bovis*, but are nonhemolytic and butyrous. Other properties resemble those of *M. bovis*. Oxidase and catalase positive. Produces an alkaline reaction in several carbohydrate fermentation media, grows well on Herellea agar, unlike *M. bovis* which reportedly fails to grow on Herellea agar (Hughes and Pugh, 1970). Liquefies gelatin at 37°C, does not produce ornithine or lysine decarboxylases, does not reduce nitrate to nitrite, does not produce H_2S or indole, and does not grow on Simmons citrate medium, KCN medium, or urea broth. Precipitin lines indistinguishable from those with control homologous antiserum were produced when *M. equi* cells were tested against three *M. bovis* antisera (Hughes and Pugh, 1970).

Isolated from the eyes of two groups of horses and ponies that had signs of conjunctivitis and erosion of the epithelium of their eyelid margins. Becomes established when introduced into equine eyes but not into bovine eyes, in contrast to *M. bovis*, which does not survive in equine eyes (Hughes and Pugh, 1970).

Further comments *M. equi* has shown overlapping affinities in streptomycin-resistance transformation with some bovine isolates of *M. bovis* (Bøvre, unpublished results); an interstrain to intrastrain transformation ratio of 8×10^{-2} was found for interaction with the type strain of *M. bovis* (Kodjo et al., 1995). *M. equi* was indistinguishable from strains of "*M. lacunata* subsp. *liquefaciens*" in ability to grow on a defined medium containing 11 growth factors, and in transformation of nutritional mutants of *M. bovis* (Juni et al., 1988), such transformations being significantly less efficient than those observed with DNAs from strains of *M. bovis* (Juni et al., 1988). Furthermore, the 16S rRNA gene of *M. equi* has a sequence similarity of 99.9% to that of a strain of "*M. lacunata* subsp. *liquefaciens*" (Pettersson et al., 1998). Nevertheless, in DNA–DNA hybridization and multilocus enzyme electrophoresis studies *M. equi* and *M. bovis* are virtually indistinguishable (Tønjum et al., 1992). *M. equi* and *M. bovis* do not show a close relationship to "*M. lacunata* subsp. *liquefaciens*" in DNA–DNA hybridization (Tønjum et al., 1992). These findings confirm that *M. equi* is a unique *Moraxella* species. It might be considered that *M. equi* first appeared when DNA from either *M. bovis* or "*M. lacunata* subsp. *liquefaciens*" was transferred from one to the other and integrated into the chromosome of the recipient strain, resulting in a hybrid strain. Such a strain may have survived because it is better suited to occupy certain specific sites in the horse. Currently only a single strain of *M. equi* (ATCC 25576) is available. Further studies, involving isolation of moraxellae from horses in different geographical regions, are required to determine whether *M. equi* is universally distributed.

The mol% G + C of the DNA is: not reported.
Type strain: ATCC 25576.

9. **Moraxella nonliquefaciens** (Scarlett 1916) Lwoff 1939, 175[AL] (*Bacillus duplex non liquefaciens* Scarlett 1916, 107; *Moraxella (Moraxella) nonliquefaciens* Bøvre and Henriksen 1967a, 134.)

non.li.que.fa' ci.ens. L. pref. *non* not; L. part. adj. *liquefaciens-*dissolving; L. part. adj. *nonliquefaciens* not dissolving.

Rods, showing variation in shape, size, and arrangement. The colony size and colony type variation in relation to fimbriation and "twitching motility" are similar to that described for *M. bovis*, but the corroding colonies of freshly isolated strains may often be larger and more spreading (Bøvre et al., 1970; Bøvre and Frøholm, 1972b; Henrichsen et al., 1972). Colonies are translucent to semiopaque. Some strains are very mucoid. Cell walls of nonmucoid strains often contain polysaccharide that is chemically and immunologically identical to capsular polysaccharide of serogroup B meningococci (Bøvre et al., 1983). No hemolysis occurs on blood agar. Other characteristics are listed in Table BXII.γ.130 of the description of the family *Moraxellaceae.*

M. nonliquefaciens is nutritionally fastidious; 22 independently isolated strains were able to grow in a defined medium containing nine amino acids and three vitamins (Juni et al., 1984). It fails to grow on Hugh and Leifson's OF medium. Previously, as a rule, highly sensitive to penicillin, but strains resistant because of β-lactamase production are now often encountered. Most frequently isolated from the nasal cavity, which is probably the main natural habitat. Also found in other samples from the respiratory tract, including bronchial aspirates in chronic bronchitis and upper respiratory disease (Graham et al., 1990), the latter in small children. It is considered a well-established parasite of humans with good adaptation to the host and, although rarely reported, may play a role, similar to *M. catarrhalis*, as an agent of disease. It has been found in individuals with endophthalmitis (Ebright et al., 1982; Lobue et al., 1985) and occasionally in blood in systemic disease (Brorson et al., 1983; Tønjum et al., 1992).

Further comments The cells are frequently transformable, with competence for genetic transformation being associated with fimbriation of cells and with corroding, spreading colonies (Bøvre and Frøholm, 1972a). Strains of *M. nonliquefaciens* have type 4 pilin genes (*tfa*) (Tønjum et al., 1991). *M. nonliquefaciens* strains can be identified by quantitative (Bøvre, 1964, 1965a, c) and qualitative (Juni et al., 1987) genetic transformation. The species is genetically homogeneous (Bøvre and Henriksen, 1967a). Strains of *M. nonliquefaciens* have high interstrain 16S rRNA gene sequence similarities (99.8%) and can be distinguished from other strains of the *M. lacunata* cluster by such analysis (Pettersson et al., 1998).

The mol% G + C of the DNA is: 40–44 (Bd).

Type strain: 4663/62, ATCC 19975, NCTC 10464.

10. **Moraxella ovis** (Lindqvist 1960) Henriksen and Bøvre 1968, 391[AL] (*Neisseria ovis* Lindqvist 1960, 165; *Moraxella* (*Branhamella*) *ovis* (Lindqvist 1960) Bøvre 1979, 404.)

o' vis. L. gen. n. *ovis* of the sheep.

Cocci commonly arranged in pairs with adjacent sides flattened; division in two planes at right angles to each other. Tetrad formation may be observed. Colonies on blood agar are ~2.5 mm in diameter in 48 h, low convex, becoming almost flat after longer incubation. They are grayish white and slightly smaller and more opaque than the largest colony forms of *M. bovis*. The colonies are friable or soft, and are usually surrounded by a narrow zone of clear hemolysis. Nonhemolytic variants may arise spontaneously in cultures, and primarily nonhemolytic strains (genetically confirmed as belonging to the species) have been isolated directly from sheep and cattle (and a horse). Other characteristics are listed in Table BXII.γ.130 of the description of the family *Moraxellaceae*

Isolated from the conjunctiva of sheep and cattle and from upper respiratory sites in sheep (and a horse). Frequently found in cultures from eyes in infectious keratoconjunctivitis of sheep, and often found together with *M. bovis* in cases of such disease in cattle. Considered to be of low pathogenicity, most frequently occurring as a harmless parasite.

Further comments The strains are frequently competent for genetic transformation and can be easily identified by quantitative transformation (Bøvre and Hagen, 1981). Demonstrated to be a species of *Moraxella* by virtue of transformation of the high-level streptomycin-resistance marker (Bøvre and Hagen, 1981) as well as by transformation of an auxotrophic mutant of *M. bovis* (Juni et al., 1988). Other closely related coccal and rod-shaped moraxellae, not yet described, have been isolated from sheep and cattle and these must be taken into consideration when identifying *M. ovis*, as well as *M. bovis*, in these animals (Bøvre, 1980; Bøvre and Hagen, 1981). Strains of *M. ovis* grow on a defined medium (medium MB) suitable for growth of strains of *M. bovis* (Juni et al., 1988). There is a significant amount of DNA–DNA hybridization between *M. ovis* and *M. bovis* (Tønjum et al., 1989). The 16S rRNA genes of *M. ovis* and *M. bovis* have been shown to be 97.1–97.4% similar (Pettersson et al., 1998).

The mol% G + C of the DNA is: 44.5–46.5 (Bd).

Type strain: 199/55, ATCC 33078, NCTC 11227.

GenBank accession number (16S rRNA): AF005186.

Species Incertae Sedis[1]

1. **Moraxella atlantae** Bøvre, Fuglesang, Hagen, Jantzen and Frøholm 1976, 520[AL] (*Moraxella (Moraxella) atlantae* Bøvre 1979, 404.)

at.lan' tae. M.L. gen. n. *atlantae* of Atlanta, the American city where strains of the species were first recognized as a distinct group by Elizabeth O. King.

This group is comprised of strains originally designated *Moraxella atlantae*. Rods, showing variation in shape and size. There is little tendency to form chains. Cells are often fimbriated and may show "twitching motility." Colonies are small (0.2–0.5 mm in diameter, occasionally up to 1 mm) in 48 h on either blood agar or chocolate agar. Agar corrosion and spreading of colonies may be seen. Permanent dissociation of such colonies formed by fimbriated cells into nonfimbriated, noncorroding variants has not been observed. Colonies are semiopaque and may appear slightly pink, probably due to blood pigment accumulation. Nonhemolytic. Bile salts (or Tween 80) stimulation of growth occurs with levels up to 1%. Penicillin sensitive, but usually resists 0.05 U/ml penicillin. Other characteristics are listed in Table BXII.γ.130 of the description of the family *Moraxellaceae.*

---

1. Although originally considered to be species of *Moraxella*, each of the five species listed in this section appears to represent an independent genus of the *Moraxellaceae.*

Only rarely isolated. Found in human blood, cerebrospinal fluid, and spleen. The natural habitats and pathogenicity are not yet defined.

Further comments The strains are often competent for genetic transformation and can be identified by transformation methods (Bøvre et al., 1976). The cellular lipid composition closely resembles that of *M. phenylpyruvica*; however, *M. atlantae* differs by containing true waxes (Bøvre et al., 1976; Bryn et al., 1977). Although previously considered a species of *Moraxella*, this organism constitutes a unique genus in the family *Moraxellaceae*, based on the following evidence: 1. Ratios of interstrain to intrastrain transformation of the high-level streptomycin-resistance marker for all other strains in the family tested are $\leq 2 \times 10^{-5}$ (Bøvre and Hagen, 1981). 2. There is only weak DNA–DNA hybridization with *M. phenylpyruvica* and no significant level with other members of the family (Tønjum et al., 1989). 3. Can be differentiated from other *Moraxellaceae* species by the presence of mannose in its whole-cell monosaccharide pattern (Jantzen et al., 1974). 4. 16S rRNA gene sequence similarity with other members of the *Moraxellaceae* is highest for *M. osloensis* (93%) and is lower for other members of the family (Pettersson et al., 1998).

M. atlantae should be given a new genus designation.

The mol% G + C of the DNA is: 46.5–47.5 (Bd).

Type strain: ATCC 29525, CDC 5118, NCTC 11091.

2. **Moraxella boevrei** Kodjo, Richard, and Tønjum 1997, 120^{VP}

boev' re.i. L. gen. n. *boevrei* of Bøvre, in honor of Kjell Bøvre.

This group is comprised of strains originally designated *Moraxella boevrei*. Cells are thin rods to coccus-like structures and have a tendency to occur as diplobacilli or in short chains. Colonies on blood agar are gray-white, rough, and small (diameter after 48 h, 0.5 mm) and are surrounded by a wide area of hemolysis. Isolates are nonmotile, aerobic, and mesophilic. Growth occurs in an aerobic atmosphere (preferably containing 5% CO_2) on 5% sheep or human blood agar plates or chocolate agar plates. Growth does not occur or occurs weakly on nutrient agar. The oxidase and catalase reactions are strongly positive. All strains hydrolyze Tween 80 and reduce nitrate. No acid is produced from carbohydrates. All strains are proteolytic (Loeffler slants and gelatin positive). Indole and urease reactions are negative. The only conventional characteristic that distinguishes *M. boevrei* from *M. bovis* is the reduction of nitrate. Otherwise, all other conventional and hydrolytic enzymes of *M. boevrei* and *M. bovis* are identical. Other characteristics are listed in Table BXII.γ.130 of the description of the family *Moraxellaceae*.

Isolated from the nasal flora of goats. Pathogenicity has not been demonstrated.

Further comments The strains are competent for genetic transformation and can be identified by quantitative transformation of the high-level streptomycin-resistance marker. The basis for considering *M. boevrei* to be a new genus includes the following: 1. Using *M. boevrei* DNA with *M. bovis*, *M. lacunata*, *M. atlantae*, and *M. catarrhalis* as recipients, ratios of interstrain to intrastrain transformation to streptomycin resistance were less than 5×10^{-5} to 6.5×10^{-5}. 2. There is little or no DNA–DNA hybridization of *M. boevrei* with most other species of *Moraxellaceae* and other Gram-negative bacteria tested. 3. The results of rDNA restriction

pattern analysis showed that the strains of *M. boevrei* were similar to each other and different from 10 other member strains of the *Moraxellaceae* (Kodjo et al., 1997).

It should be noted, however, that DNA–DNA hybridization resulted in relative binding ratios of 12–23 between *M. boevrei* and strains of *M. bovis*, *M. nonliquefaciens*, and *M. ovis* as compared with ratios of 58 to >100 among *M. boevrei* isolates (Kodjo et al., 1997). In view of these findings, *M. boevrei* may actually be a distantly related species of the genus *Moraxella* (*sensu stricto*). Pending further studies to establish the precise taxonomic position of *M. boevrei*, this entity is presently considered a possibly unique genus of the *Moraxellaceae*, based on the criteria used to classify members of this family.

The mol% G + C of the DNA is: 41–41.5 (HPLC).

Type strain: 88365, ATTC 700022, CIP 104716, CCUG 35435, NCTC 12925.

3. **Moraxella lincolnii** Vandamme, Gillis, Vancanneyt, Hoste, Kersters, and Falsen 1993, 479^{VP}

lin' col.ni.i. N.L. gen. n. *lincolnii* of K. Lincoln, a Swedish microbiologist.

This group is comprised of strains originally designated *Moraxella lincolnii*. Cells are nonmotile, coccus-like to plump rods 1–1.5 × 1.5–2.5 μm. They often occur in pairs and may form short chains. After 2 d incubation, colonies are whitish, smooth, convex, and circular and have a diameter of 1–3 mm. The colonies of some strains may have a flattened edge. No hemolysis and no production of pigment or odor. Aerobic. Optimal growth occurs at 28–33°C. Growth also occurs at 36–37°C but not at 42°C. Growth occurs in the absence of NaCl; no growth occurs in the presence of 1.5% NaCl. Oxidase and catalase activities are present. All strains are susceptible to penicillin (10 μg discs). Most strains reduce nitrite but fail to reduce nitrate. Acid is not produced from D-glucose, D-fructose, maltose, or sucrose. Lactate cannot be used as a carbon source. No denitrification, liquefaction of gelatin, proteolysis on Loeffler slants, hydrolysis of Tween 80, or indole production. No urease, DNase, or β-galactosidase activity. The major fatty acids are $C_{10:0}$, $C_{12:0\,3OH}$, $C_{16:1\,\omega7c}$, $C_{16:0}$, $C_{18:1\,\omega9c}$, and $C_{18:0}$. Other characteristics are listed in Table BXII.γ.130 of the description of the family *Moraxellaceae*.

Has been isolated mainly from the respiratory tract of humans. Its clinical significance, if any, is unknown.

Further comments Assignment of these organisms to *Moraxellaceae* is based on the results of DNA–rRNA hybridization studies in which DNAs from strains of *M. lincolnii* were hybridized with radioactively labeled rRNA from *M. lacunata* (Vandamme et al., 1993). $T_{m(e)}$ values of 74.4–74.8°C obtained are well within the range for all members of the *Moraxellaceae* ($T_{m(e)}$ range: ~70–80.0°C [Rossau et al., 1991]). Further studies are required to establish possible relationships to other members of the family.

M. lincolnii will require a new genus designation should further investigations verify that it is, indeed, a distinct genus of the *Moraxellaceae*.

The mol% G + C of the DNA is: 44 (T_m).

Type strain: CCUG 9405, LMG 5127.

4. **Moraxella osloensis** Bøvre and Henriksen 1967a, 131^{AL} (*Moraxella (Moraxella) osloensis* Bøvre 1979, 404.)

os.lo.en' sis. M.L. *adj. osloensis* pertaining to Oslo, Norway, where the species was first recognized.

This group is comprised of strains originally designated *Moraxella osloensis.* Rods, showing variation in size, shape, and arrangement similar to that of the genus *Moraxella* (*sensu stricto*). The cells may be indistinguishable from cocci in some specimens or when freshly isolated. A distinct rod shape occurs with pronounced formation of fusiform cells when cultured in sublethal concentrations of penicillin; diplocells may sometimes appear like "waterbuffalo horns" (Bøvre, unpublished). Intracellular inclusions of poly-β-hydroxybutyrate are usually formed, especially when the nitrogen source is limiting.

Colonies are 2.0–2.5 mm in diameter in 48 h and are semiopaque. They do not exhibit colony type variation.

Strains usually grow in mineral media with ammonium ions and acetate, but one genetically verified exception has been found. Highly or, most often, moderately sensitive to penicillin, but resistant strains which possess β-lactamase have been found. Other characteristics are listed in Table BXII.γ.130 of the description of the family *Moraxellaceae.* Isolated from the upper respiratory tract, genitourethral specimens, blood, cerebrospinal fluid, and pyogenic manifestations in joints, bursae, and other sites from humans. Not yet isolated with certainty from nonhuman sources. Usually considered a harmless parasite, but significant pathogenicity is possible.

Further comments Although originally designated as members of *M. nonliquefaciens,* two strains (19116/51 and 752/52) of the 22 studied were considered to represent two new taxonomic groups, based on quantitative transformation studies using the high-level streptomycin-resistance marker (Bøvre, 1964). Further studies of strains similar to strain 19116/51 led to the designation of *M. osloensis* (Bøvre, 1965d; Bøvre and Henriksen, 1967a). Strains are frequently competent for genetic transformation and procedures for their identification by transformation are readily performed (Bøvre, 1965d; Bøvre et al., 1977; Juni, 1974). Although previously considered to be a species of *Moraxella,* this organism constitutes a unique genus in the family *Moraxellaceae,* based on the following evidence: 1. Ratios of interstrain to intrastrain transformation of the high-level streptomycin-resistance marker for all other strains in the family tested (Bøvre and Hagen, 1981) are $\leq 4 \times 10^{-5}$. 2. Except for a weak reaction with DNA from a strain of *M. phenylpyruvica* (752/52), there is no detectable DNA–DNA hybridization with DNA from most members of the family (Tønjum et al., 1991). 3. DNAs from strains of *M. osloensis* have a unique average intergroup $T_{m(e)}$ value (75.6°C) when hybridized with rRNA from "*M. lacunata* subsp. *liquefaciens*", ATCC 17952 (Rossau et al., 1991). 4. A *trpE*⁻ mutant of *M. osloensis* was transformed to prototrophy only by DNAs from strains of *M. osloensis* (Juni, 1974). 5. Unique cellular fatty acid composition (Jantzen et al., 1974; Bøvre et al., 1977; Moss et al., 1988). 6. Ability to grow on a defined medium containing a single organic carbon and energy source with an ammonium salt as the nitrogen source (Bøvre and Henriksen, 1967a; Juni, 1974). A specific endonuclease, *Mos*I, has been purified from *M. osloensis,* ATCC 19976, and shown to recognize the same sequence as *Mbo*I (Gelinas et al., 1977).

Phylogenetic analysis of 16S rRNA gene sequences has shown that *M. osloensis* and *M. atlantae* form a common and distinct clade of the *Moraxellaceae* (Pettersson et al., 1998) (Fig. BXII.γ.147). Strains of *M. osloensis* and *M. atlantae* were shown to be distantly related to each other, their 16S rRNA genes displaying 93.4% similarity and ≤93% similarity, respectively, to other members of the *Moraxellaceae* (Pettersson et al., 1998).

M. osloensis should be given a new genus designation.
The mol% G + C of the DNA is: 43–46 (Bd).
Type strain: ATCC 19976, CDC A1920, NCTC 10465.
Additional Remarks: A frequently used (transformable) reference strain is CDC 5873.

5. **Moraxella phenylpyruvica** Bøvre and Henriksen 1967b, 344[AL] Epit. spec. cons. Opin. 42, Jud. Comm. 1971, 107 (*Moraxella (Moraxella) phenylpyruvica* Bøvre 1979, 404.)
phe.nyl.py.ru' vi.ca. M.L. n. *acidum phenylpyruvicum* phenylpyruvic acid; M.L. fem. adj. *phenylpyruvica* pertaining to phenylpyruvic acid, the product of deamination of phenylalanine by this organism.

This group is comprised of strains originally designated *Moraxella phenylpyruvica.* Rods, showing variation in shape, size, and arrangement. The colonies are relatively small (0.9–1.0 mm in diameter) in 48 h on blood agar. They often appear semiopaque with a very slight pink hue, probably due to blood pigment accumulation. Nonhemolytic. Phenylalanine deaminase and urease positive, but strains negative for one or, very rarely, both activities may occur. Grows distinctly, but slowly, at 4–10°C. Usually grows at high salt or bile concentrations; the minimum inhibitory concentrations are usually 7.5–9.0% NaCl and 5% bile salts (Oxoid). Bile salt (Tween 80) stimulation of growth occurs, with levels up to 4% (Snell et al., 1972; Bøvre et al., 1976). Although usually highly sensitive to penicillin, several strains of human and animal origin have been found to be penicillin resistant due to β-lactamase production. Other characteristics are listed in Table BXII.γ.130 of the description of the family *Moraxellaceae.*

Isolated from human blood and cerebrospinal fluid, the genitourethral tract, and from other sites and specimens of humans; also isolated from the genital tract and brain of sheep and cattle, the intestine of a goat, and the genital tract of pigs. Pathogenicity for humans, and perhaps animals, has not been assessed adequately, although judging from its sources of isolation this species may be a significant potential pathogen.

Further comments Strains competent for genetic transformation have not yet been found, but genetic relationships to other taxa of the family *Moraxellaceae* have been detected by the use of strains as DNA donors in transformation. The cellular lipid composition is unique among recognized *Moraxellaceae* species in that true waxes are absent; in other respects, the lipid composition resembles that of *M. atlantae* (Jantzen et al., 1974; Bøvre et al., 1976; Bryn et al., 1977; Moss et al., 1988). Although previously considered a species of *Moraxella,* this organism constitutes a unique genus in the family *Moraxellaceae,* based on the following evidence: 1. Ratios of interstrain to intrastrain transformation of the high-level streptomycin-resistance marker for all other strains in the family tested are equal to, or less than, 2×10^{-5} (Bøvre and Hagen, 1981). 2. There is only slight DNA–DNA hybridization with *M. osloensis* and *M. at-*

lantae, and nondetectable levels with other members of the family (Tønjum et al., 1989). 3. A streptomycin-dependent marker of *M. phenylpyruvica*, which had been transformed into the competent strain of *Acinetobacter* (a rare occurrence), was transformed to streptomycin independence by DNAs from 35 strains of *M. phenylpyruvica;* DNAs from other members of the *Moraxellaceae* were inactive (Juni, 1990).

M. phenylpyruvica may be heterogeneous since strains 752/ 52 and ATCC 23333 ^{T} have a 16S rRNA gene sequence similarity of only 95.9% (Pettersson et al., 1998). Both of these strains react in the same positive manner in trans-

formation of an *M. phenylpyruvica* streptomycin-dependent marker located in the competent strain of *Acinetobacter*, indicating a strong relationship (Juni, 1990).

M. phenylpyruvica should be given a new genus designation.*

The mol% G + C of the DNA is: 42.5–43.5 (Bd).

Type strain: ATCC 23333, CDC 2863, NCTC 10526.

Editorial Note: Bowman et al. (1996) have proposed the transfer of *Moraxella phenylpyruvica* to the genus *Psychrobacter* as *Psychrobacter phenylpyruvicus*.

Genus II. **Acinetobacter** Brisou and Prévot 1954, 727[AL]

ELLIOT JUNI

A.ci.ne′ to.bac.ter. Gr. adj. *akinetos* unable to move; M.L. n. *bacter* the masculine form of the Gr. neut. n. *bactrum* a rod; M.L. masc. n. *Acinetobacter* nonmotile rod.

Rods 0.9–1.6 × 1.5–2.5 μm, becoming spherical in the stationary phase of growth. Colonies are generally nonpigmented and are mucoid when the cells are encapsulated. Cells commonly occur in pairs and in chains of variable length. Do not form spores. Gram negative but occasionally difficult to destain. Swimming motility does not occur but the cells display "twitching motility", presumably because of the presence of fimbriae. Aerobic, having a strictly respiratory type of metabolism with oxygen as the terminal electron acceptor. Most strains do not reduce nitrate to nitrite. Most strains grow between 20 and 37°C, having temperature optima of 33–35°C. Some strains cannot grow at 37°C. **Oxidase negative. Catalase positive.** Grow well on most complex media. Most strains grow in defined media containing a single carbon and energy source, such as acetate or lactate, using ammonium or nitrate salts, or one of several common amino acids, as a supply of nitrogen. Frequently amino acids such as glutamic acid or aspartic acid can serve as a single source of carbon, energy, and nitrogen in a defined mineral medium. With rare exceptions, they display no growth factor requirements. Most frequently saprophytic, occurring naturally in soil, water, sewage, and foods such as raw vegetables. Can also reside, possibly indigenously, on the human skin and in the human respiratory tract. Can cause nosocomial infections such as bacteremia, secondary meningitis, pneumonia, and urinary tract infections in humans.

The mol% G + C of the DNA is: 38–47.

Type species: **Acinetobacter calcoaceticus** (Beijerinck 1911) Baumann, Doudoroff and Stanier 1968b, 1538 emend. Bouvet and Grimont 1986, 238 (*Micrococcus calcoaceticus* Beijerinck 1911,1067.)

FURTHER DESCRIPTIVE INFORMATION

Habitat Acinetobacters occur naturally in soil and water (Baumann, 1968) and are present in sewage (Warskow and Juni, 1972). It has been estimated that at least 0.001% of the total culturable, heterotrophic, aerobic population in soil and water are acinetobacters (Baumann, 1968). Acinetobacters have been demonstrated to associate with soybean root nodules in relatively large numbers (Wong et al., 1986). They have been shown to occur in raw, washed, and frozen vegetables (Gennari and Stegagno, 1986), in fresh, frozen, and stored fish products (Gennari and Stegagno, 1985), and in spoiled meat, milk, and cheese (Gennari et al., 1992). Body sites, such as the skin and respiratory

tract, also appear to harbor certain acinetobacters (Bergogne-Bérézin and Towner, 1996).

General Physiology

CELL STRUCTURES Rapidly growing cells tend to be plump rods, whereas cells in the stationary phase of growth are spherical and have a somewhat smaller diameter than the rods (Baumann et al., 1968b). An occasional rod-shaped cell can frequently be observed in a microscopic field of stationary phase cells. Cells occur typically in pairs, and occasionally in chains. Many strains are encapsulated and capsules may be seen readily in India ink wet mounts. Colonies are generally nonpigmented and are mucoid when the cells are encapsulated.

Electron microscopy of thin sections of cells has revealed a cell wall ultrastructure that is typical of Gram-negative bacteria (Breuil et al., 1975; Scott et al., 1976). The peptidoglycan contains muramic acid, glucosamine, alanine, D-glutamic acid, and *meso*-diaminopimelic acid (Martin et al., 1973; Horisberger, 1977). Unlike the cell walls of several Gram-negative bacteria, the murein of one strain of *Acinetobacter* had an unusually high proportion of cross-linked peptide side chains (Quintela et al., 1995). It is known that antibodies against *Chlamydia* cross-react with *Acinetobacter* (Brade and Brunner, 1979). Nurminen et al. (1984) present evidence that this cross reaction involves a lipopolysaccharide component; other findings suggest that *Chlamydia* antiserum reacts with the cell wall of *Acinetobacter* (Espinosa et al., 1996).

When originally investigated, one strain of *Acinetobacter* was shown to possess a lipopolysaccharide containing D-glucose, glucosamine, galactosamine, lipid A, ethanolamine, fatty acids, phosphate, and protein (Adams et al., 1970). Lipopolysaccharide composition is not the same in all strains studied (Borneleit and Kleber, 1991). Numerous investigations have elucidated the structures of O-antigen lipopolysaccharides from a variety of *Acinetobacter* strains (Vinogradov et al., 1997; Haseley and Wilkinson, 1997; Haseley et al., 1997, 1998; and other references to be found in these papers). It has been suggested that because of the presence of deoxy sugars, amino acids, and branched polymers, *Acinetobacter* O-antigens display a marked hydrophobicity, which is consistent with the ability of acinetobacters to grow on hydrophobic substrates (Haseley et al., 1997).

Evolution of acinetobacters has resulted in a large variety of surface antigens, many of them capsular, on different strains. A serologic system for identification of strains by immunofluorescent staining revealed 28 serovars among strains capable of forming acid aerobically from glucose, as well as a series of other serovars among acinetobacters that do not form acid from glucose (Marcus et al., 1969). The capsular polysaccharide of one strain of *Acinetobacter* has been shown to interact with antisera prepared against group B and group G streptococci, as well as with antipneumococcal type XX serum (Heidelberger et al., 1969).

Outer membranes of *A. calcoaceticus* and *A. radioresistens* have been analyzed and found to contain several unique proteins ranging in molecular weight from 12,000 to 50,000 daltons (Fischer et al., 1984; Nishimura et al., 1986a). The major outer membrane protein of *A. calcoaceticus* has a molecular weight of 38,000 daltons, in contrast to that of *A. radioresistens*, which has two major outer membrane proteins with molecular weights of 39,000 and 42,000 daltons, respectively (Nishimura et al., 1986a). Two other studies, which examined larger numbers of strains, showed a good correlation between outer membrane cell envelope protein patterns and DNA–DNA hybridization groups (Ino and Nishimura, 1989; Dijkshoorn et al., 1990).

METABOLISM Nitrite and nitrate can serve as nitrogen sources for acinetobacters, and these organisms possess an assimilatory nitrate reductase (Jyssum and Joner, 1965). Although most acinetobacters are not able to reduce nitrate to nitrite in the conventional nitrate reduction assay, a few strains can do so (Riley and Weaver, 1974); these strains are unable to grow anaerobically with nitrate as the terminal electron acceptor (Juni, 1978). Since all acinetobacters are oxidase negative, they lack cytochrome *c* (Baumann et al., 1968a). However, they do contain cytochromes *b*, *o*, and occasionally *d* and P-450, as well as flavoproteins, and appear to have a normal electron transport system (Asperger and Kleber, 1991).

Enzymes of the tricarboxylic acid cycle have been shown to be present in acinetobacter cell-free extracts. The presence of NADP-isocitrate dehydrogenase has been demonstrated (Weitzman, 1991). The absence of an NAD-isocitrate dehydrogenase, which is typical for most bacteria, and the relatively slow growth rate with acetate as the sole carbon source, indicate that for acinetobacters, as well as for the vast majority of bacteria, the tricarboxylic acid cycle is more important in the synthesis of amino acids (specifically aspartate, glutamate, and proline) and porphyrin than it is in general cellular respiration. Enzymes of the glyoxylate cycle have been demonstrated in *Acinetobacter* and all evidence verifies the obligate requirement of this cycle for growth with two-carbon compounds such as acetate as sole carbon and energy source (Juni, 1978). The enzymes of the Embden–Meyerhof pathway, for conversion of phosphoenolpyruvate to hexose phosphates, are present in *Acinetobacter* and are used for gluconeogenesis from C_4-dicarboxylic acids, but not for the degradation of sugars (Taylor and Juni, 1961).

Relatively few strains can use D-glucose as a carbon source for growth (Baumann et al., 1968b; Juni, 1972). One strain that can use glucose as a carbon source has been shown to catabolize it via the Entner–Doudoroff pathway, involving the key enzymes 6-phosphogluconate dehydratase and 2-keto-3-deoxy-6-phosphogluconate aldolase (Taylor and Juni, 1961). Many acinetobacters that are unable to grow on glucose contain an aldose dehydrogenase and are able to acidify glucose media rapidly (gluconic acid being formed), as well as media containing other sugars such as D-xylose, L-arabinose, D-galactose, D-mannose, L-rhamnose, maltose, lactose, and cellobiose (Hauge, 1960); this enzyme is responsible for the positive diagnostic test for the production of acid from glucose. Strains of *Acinetobacter* able to form acid from glucose have been demonstrated to produce two glucose dehydrogenases, a soluble and a membrane-bound enzyme (Matsushita et al., 1989; Duine et al., 1982). The two glucose dehydrogenases are distinct enzymes; the membrane-bound form appears to be the only enzyme used by intact cells for the oxidation of glucose (Duine, 1991). Both enzymes use pyrroloquinoline quinone (PQQ, methoxatin) as a coenzyme (Duine, 1991). Genes involved in synthesis of PQQ and coding for membrane-integrated quinoprotein glucose dehydrogenase in *Acinetobacter* have been cloned in *E. coli* (Goosen et al., 1987, 1989a, b; Cleton-Jansen et al., 1989; Dewanti and Duine, 1998).

GROWTH AND NUTRITION Most strains of *Acinetobacter* can grow in a simple mineral medium containing a single carbon and energy source such as ethanol, acetate, lactate, pyruvate, malate, or α-ketoglutarate. Ammonium and nitrate salts serve as nitrogen sources. Some organic compounds, such as amino acids, can serve as single sources of carbon, energy, and nitrogen. Rare strains have growth factor requirements; such strains frequently revert spontaneously to prototrophy when cultured in the laboratory (Warskow and Juni, 1972). Acinetobacters resemble saprophytic pseudomonads in being able to use a wide variety of organic compounds as carbon, energy, and/or nitrogen source for growth in mineral media, and generally oxidize such compounds to completion. Utilizable organic nutrients include certain sugars, fatty acids, aliphatic alcohols, unbranched hydrocarbons, dicarboxylic acids, amino acids, and many aromatic and alicyclic compounds (Baumann et al., 1968b); the spectrum of compounds utilized is frequently species specific (Bouvet and Grimont, 1986). Acinetobacters, normally associated in relatively large numbers with soybean root nodules, have also been shown to obtain energy for growth by oxidizing H_2, which is released by the nodules (Wong et al., 1986). Some strains can utilize a single hexose, D-glucose, and the pentoses D-ribose, D-xylose, and L-arabinose. Other carbohydrates, disaccharides, and sugar alcohols (including glycerol) are not used. There is no evidence that acinetobacters can obtain energy for growth by fermentation. Since acinetobacters are nonmotile, growth in liquid media requires vigorous aeration. Growth can take place from pH 5.0 to 8.0, with a pH optimum for growth of approximately 6.5 (Baumann, 1968; Juni, 1982).

Biochemistry Because of their ability to metabolize many organic compounds, acinetobacters normally synthesize a wide variety of degradative enzymes. Most metabolic and genetic studies in *Acinetobacter* have been directed towards elucidating the pathways involved, as well as regulation of expression of the relevant enzymes. Since aromatic compounds are available through normal breakdown of dead plant and animal tissues, it is not surprising to find that acinetobacters can degrade organic compounds such as benzoate, *p*-hydroxybenzoate, mandelate, quinate, and tryptophan (Fewson, 1991).

The normal soil and water habitats of acinetobacters, coupled with their ability to degrade a wide variety of organic compounds, make it likely that acinetobacters can be used for bioremediation. Although toxic to most microorganisms, aromatic compounds such as salicylate, halogenated aromatics, and phenol (Schirmer et al., 1997) can be degraded by acinetobacters, occasionally by strains harboring specific degradative plasmids (Fewson, 1991).

Because of their ability to utilize hydrocarbons, acinetobacters are capable of playing a role in the degradation of crude oil. It has been demonstrated that certain acinetobacters elaborate and excrete polymers that emulsify hydrocarbons, such as oils, thus making these substances available for degradation in an aqueous environment (Gutnick et al., 1991; Navon-Venezia et al., 1995). Acinetobacters that degrade alkanes are frequently isolated from areas contaminated with petroleum. Various aspects of alkane metabolism have been investigated (Asperger and Kleber, 1991; Maeng et al., 1996; Parche et al., 1997; Ratajczak et al., 1998). Many acinetobacters possess the ability to degrade alkanes by pathways that, in some cases, involve rubredoxin, rubredoxin reductase, and cytochrome P-450 (Asperger and Kleber, 1991; Geissdörfer et al., 1995).

Some acinetobacters elaborate an extracellular lipase, as determined by hydrolysis of Tween 80 (Baumann et al., 1968b) and other lipids (Breuil and Kushner, 1975; Fischer, 1986; Kok et al., 1995a,b, 1996). Strains that show hemolysis on blood agar plates excrete phospholipase (Lehmann, 1971, 1973).

Early studies of pathways for degradation of aromatic compounds in *Acinetobacter* demonstrated that these compounds are generally converted to β-ketoadipate, and that succinate and acetyl CoA are the products of degradation of β-ketoadipate (Stanier and Ornston, 1973). Evolution of genes for the β-ketoadipate pathway has been reviewed (Ornston and Neidle, 1991). There is evidence for supraoperonic clustering on the chromosome of certain genes involved in aromatic catabolism, where the product of one gene is further metabolized by a second gene in an independent and closely located transcriptional operon (Ornston and Neidle, 1991; Averhoff et al., 1992). An apparent DNA slippage structure has been proposed to explain the extensive evolutionary divergence of *pcaD* and *catD*, genes that encode identical catalytic activities in *Acinetobacter* but are located in different regions of the chromosome (Hartnett and Ornston, 1994). Evidence has been presented for gene conversion in the nonreciprocal genetic exchange involved in repair of defective *pcaIJF* by nucleotide sequences of a functional *catIJF* gene in *Acinetobacter* (Kowalchuk et al., 1995), a process that requires a functional *recA* gene (Gregg-Jolly and Ornston, 1994). Transcriptional activation of aromatic genes, similar to *LysR* type activation (Collier et al., 1998), as well as other transcriptional activators, has been reported (DiMarco and Ornston, 1994; Gerischer et al., 1998). Using a mutant of *Acinetobacter* sp. strain ADP1 that was unable to grow on benzene as the sole carbon source, since it lacked both transcriptional activators of the *cat* genes, it was demonstrated that spontaneous mutants able to grow with benzoate carried 10–20 amplified chromosomal regions (amplicons), each encompassing the *cat* genes (Reams and Neidle, 2003).

It has been demonstrated that certain features of aromatic amino acid biosynthesis are unique and characteristic for all acinetobacters (Byng et al., 1985).

Acinetobacter chromosomes have been reported to encode genes for synthesis of several restriction endonucleases and, most probably, their corresponding modification enzymes. These include *AccI* and *AccII* (Roberts, 1985), *AccIII* (Kita et al., 1985), *AlwNI* (Morgan et al., 1987), *AclI* (Degtyarev et al., 1992), *AjoI* (Nowak et al., 1994), *AspM* (Zelinskaia et al., 1996), and *AccBSI* (Abdurashitov et al., 1997). The genes of the *AccI* restriction-modification system have been sequenced following cloning and expression in *E. coli* (Kawakami et al., 1991).

Polyphosphate has been shown to be accumulated in wastewater by some acinetobacters (Gutnick et al., 1991; Kortstee et al., 1994). These organisms appear to be responsible for removing inorganic phosphate from sewage through their association with activated sludge (Fuhs and Chen, 1975; Buchan, 1983); the presence of K^+ seems to be essential for phosphate uptake (Van Groenestijn et al., 1988). A strain from activated sludge was demonstrated to accumulate large amounts of inorganic polyphosphate (3.7–5.7 mg P/100 mg dry weight of cells) and has been identified as *A. johnsonii* (Bonting et al., 1992). It has been suggested that a polyphosphate-synthesizing system is encoded on a 20-kb plasmid (Bayly et al., 1991). A strain of *A. johnsonii* that accumulates inorganic phosphate under aerobic conditions was shown to contain a high-affinity phosphate-uptake system that is induced in the absence of inorganic phosphate (Van Veen et al., 1993; Kortstee et al., 1994). A polyphosphate kinase gene from strain ADP1 (BD413) has been isolated and shown to be induced by phosphate starvation (Geissdörfer et al., 1998). Inorganic polyphosphate can serve as an energy reserve, as can poly-β-hydroxybutyrate and simple wax esters, all of which have been shown to accumulate in many strains of *Acinetobacter* (Fixter and Sherwani, 1991).

Studies of amino acid transport in *Acinetobacter* have revealed that alanine, lysine, and proline are taken up and concentrated by high affinity systems that are coupled to membrane-bound glucose dehydrogenase (Van Veen et al., 1994).

A phosphotyrosine-protein phosphatase from *A. johnsonii*, which specifically dephosphorylates an endogenous protein kinase known to autophosphorylate at multiple tyrosine residues in the inner membranes of this organism, has been studied and shown to be a member of the phosphotyrosine-protein phosphatase family (Grangeasse et al., 1998).

Mutarotase, an enzyme that catalyzes the anomeric interconversion of D-glucose and other aldoses, has been demonstrated in *Acinetobacter*; its gene has been cloned and expressed in *E. coli* (Gatz et al., 1986), and the nucleotide sequence of the gene directing the synthesis of this enzyme has been determined (Gatz and Hillen, 1986).

Genetics It is to be expected that a relatively large part of the *Acinetobacter* chromosome will code for enzymes involved in degradative pathways, because of the large number of organic compounds that can be degraded by these organisms. A measure of the percentage of the chromosome devoted to these activities comes from the studies of Gutnick et al. (1969), where it was shown that for about 600 organic compounds screened as possible carbon and nitrogen sources for *Salmonella typhimurium* LT-2, an organism previously considered to have only a limited number of possible carbon and energy sources, about 100 utilizable compounds were found. It might be expected that the portion of the chromosomes of *Acinetobacter* strains devoted to the genes controlling synthesis of enzymes involved in degrading organic compounds must be considerably greater than the 10% estimated for *S. typhimurium* (Gutnick et al., 1969).

Using plasmids, it has been possible to mobilize the *Acinetobacter* chromosome and conjugally transfer chromosomal genes to a recipient strain (Towner and Vivian, 1976b). In this manner, it has been demonstrated, using a series of genetic markers, that the *Acinetobacter* chromosome is circular (Towner and Vivian, 1976a) and loci of 23 different mutations were mapped on a circular linkage group (Towner, 1978). The orientation and precise location of more than 40 genes have been physically mapped on the circular chromosome of *Acinetobacter* sp. strain EBF/65, which has been shown to consist of 3780 ± 191 kb (Gralton et al., 1997). Analysis of codon usage in *Acinetobacter* structural genes has revealed that all possible triplet codons are used, some preferentially (White et al., 1991).

Transformation Genetic exchange in *Acinetobacter* can be mediated by transformation, transduction, and conjugation (Vivian, 1991). Although rare, competence for genetic transformation in *Acinetobacter* has been observed in two strains (Juni and Janik, 1969; Ahlquist et al., 1980). Two clinically isolated strains of *Acinetobacter* were reported to be weakly competent (Bergan and Vaksvik, 1983). Nutritional, as well as antibiotic resistance, markers are readily transformed in competent strain BD413, and transformation of a *trpE⁻* auxotroph of this strain to prototrophy can be accomplished specifically by DNAs from acinetobacters (Juni, 1972). Genes involved in capsule biosynthesis of a competent strain have been studied by transformation (Juni and Janik, 1969). A partial mapping of genes directing synthesis of enzymes for the tryptophan biosynthetic pathway in *Acinetobacter* has been accomplished using transformation (Sawula and Crawford, 1972). A genetic analysis of the genes for proline biosynthesis has also been reported (Ginther, 1978). Natural transformation has been useful in the analysis of catabolic pathways in *Acinetobacter* (Kloos et al., 1995).

Using natural transformation in *Acinetobacter*, it has been possible to introduce randomly mutated genes, resulting from errors produced during PCR replication, into the chromosome of a competent recipient strain without prior cloning of PCR fragments (Kok et al., 1997). It has been demonstrated that plant DNA can be taken up and recombinationally incorporated into suitable plasmids resident in competent *Acinetobacter* (de Vries and Wackernagel, 1998; Gebhard and Smalla, 1998). A general method, where randomly mutated PCR products are transformed into competent *Acinetobacter* and captured by marker-replacement recombination in a plasmid resident in the transformation recipient, has been described (Melnikov and Youngman, 1999).

Physiological characterization of natural transformation in *Acinetobacter* has been reported (Palmen et al., 1993). The DNA uptake process in *Acinetobacter* does not make use of specific sequence recognition for binding and uptake of DNA (Lorenz et al., 1992; Palmen et al., 1993). Investigations of the molecular basis for DNA uptake in competent strains of *Acinetobacter* have implicated elements in the assembly of the type IV pilus (Palmen and Hellingwerf, 1997; Link et al., 1998), findings that are consistent with earlier observations correlating competence and piliation in other bacteria (Bøvre and Frøholm, 1971).

Phages Lytic phages for acinetobacters are isolated readily from sewage (Blouse and Twarog, 1966; Twarog and Blouse, 1968; Herman and Juni, 1974). A generalized transducing phage (P78) specific for one strain of *Acinetobacter* has been isolated and demonstrated to lysogenize the host strain (Herman and Juni, 1974). Weigle reactivation has been demonstrated for UV-irradiated temperate phage P78, suggesting that *Acinetobacter* possesses an inducible DNA repair pathway (Berenstein, 1982). Ultraviolet light was shown to induce temperate prophage P78 (Berenstein, 1986). Several *Acinetobacter* phages have been classified according to structural characteristics of phage heads and tails (Ackermann et al., 1994).

Plasmids More than 80% of all acinetobacters isolated harbor multiple plasmids (Hartstein et al., 1988; Gerner-Smidt, 1989; Seifert et al., 1994b), some of which are cryptic (Hunger et al., 1990; Minas and Gutnick, 1993; Seifert et al., 1994b); others carry antibiotic resistance markers (Murray and Moellering, 1979; Hinchliffe and Vivian, 1980; Gerner-Smidt, 1989; Garcia et al., 1996). Transmissible antibiotic resistance in *Acinetobacter* has been shown to be associated with plasmids that belong to broad host-range incompatibility groups (Towner, 1991). Plasmids from many incompatibility groups have been transferred to *Acinetobacter*, by conjugation or mobilization, where most of them are not completely stable (Towner, 1991). By making use of the ability of P1 incompatibility plasmid RP4 to mobilize the *Acinetobacter* chromosome, it was first demonstrated that the chromosomal DNA is circular (Towner and Vivian, 1976a; Towner, 1978). Some *Acinetobacter* plasmids encode genes involved in degradation of certain organic compounds (Winstanley et al., 1987, 1991; Fujii et al., 1997). Plasmid-mediated heavy metal resistance in *Acinetobacter* has been reported (Towner, 1991). A shuttle plasmid for *Acinetobacter* and *Escherichia coli* was constructed by ligation of a cryptic *Acinetobacter* plasmid and pBR322; the *trpE* gene of *Acinetobacter* was cloned in this vector (Hunger et al., 1990). The same shuttle vector was used to construct a gene bank of *Acinetobacter* chromosomal DNA (Hunger et al., 1990). Three cryptic plasmids have been observed in competent strain BD413 and two of these, when ligated with pUC19, resulted in shuttle vectors that replicated in both *Acinetobacter* and in *Escherichia coli* (Minas and Gutnick, 1993). Plasmids have been introduced into strain BD413 by natural transformation (Singer et al., 1986; Hunger et al., 1990). The means by which resident plasmids were originally introduced into isolated strains are not known.

Transposable elements Some acinetobacters carry transposons on their chromosomes and these encode antibiotic resistance genes (Towner, 1991). Studies of the gene for chloramphenicol acetyltransferase (CAT) in a strain of *A. baumannii* revealed that this chromosomal gene is part of a transposon similar to Tn*2670* from plasmid NR1; the 5′ end of the cloned CAT gene is linked to IS *1* and the 3′ end is linked to components of Tn *21* (Elisha and Steyn, 1991). A newly discovered insertion sequence, IS*1236*, which resembles IS*3*, has been discovered in *Acinetobacter* sp. strain BD413 (ADP1) where it exists in seven copies (Gerischer et al., 1996). Transposon Tn *3171*, which may be identical to Tn*7*, was shown to be incorporated in the chromosome of *Acinetobacter* in a nonrandom fashion (Towner, 1983). When introduced into *Acinetobacter* on a suicide plasmid, Tn*5* was shown to insert nonrandomly into the chromosome at a limited number of sites (Singer and Finnerty, 1984). However, Tn *5* has been inserted into a variety of cloned fragments of *Acinetobacter* DNA that can be transformed into strain BD413 (Doten et al., 1987; Goosen et al., 1987; Reddy et al., 1989; Gregg-Jolly and Ornston, 1994). When present in *Acinetobacter*, Tn *5* confers low-level resistance to streptomycin due to the presence of a Tn*5*-encoded streptomycin phosphotransferase (O'Neill et al., 1984). Low transposition frequencies for Tn*7* and Tn*10* to the *Acinetobacter* chromosome have been reported (Ely, 1985). Several different kinds of mutations in an *Acinetobacter* strain were produced by introduction of a derivative of Tn *10* (mini-Tn*10* PttKm) (Leahy et al., 1993).

Integrons, genetic elements that integrate cassettes encoding antibiotic resistance, have been demonstrated to be present in the majority (48 of 59 strains) of clinical isolates of *A. baumannii* biotype 9 (Gonzalez et al., 1998).

Pathogenesis and antibiotic susceptibility Although considered normally nonpathogenic, acinetobacters are causative agents of nosocomial infections, particularly in debilitated individuals and those residing in intensive care units (Bergogne-

Bérézin and Towner, 1996). From 1971 to 1981, species of *Acinetobacter* accounted for 1.4% of all nosocomial infections in a university hospital in the United States (Larson, 1984). Infection with *Acinetobacter* has been shown to result in septicemia, meningitis, endocarditis, brain abscess, lung abscess, pneumonia, empyema, urinary tract infections, skin and wound infections, as well as other clinical manifestations (Glew et al., 1977; Bergogne-Bérézin and Towner, 1996). There appears to be a clear association between hospital instrumentation, such as the use of mechanical ventilators, endotracheal and gastric tubes, vascular catherization, lumbar punctures, myelography, ventriculography, indwelling urinary catheterization, and penetrating keratoplasty, as well as the fitting of contact lenses and subsequent infection with *Acinetobacter*. Evidence has been presented indicating that nosocomial infection with *Acinetobacter* has a distinct seasonal variation (Retailliau et al., 1979).

Although acinetobacters are frequently isolated from several areas of the human body, there is some uncertainty as to whether they are present as contaminants rather than as commensals. Bacteria able to survive and multiply in human tissue must have a mechanism for sequestering iron. Several different siderophores are synthesized by acinetobacters (Smith et al., 1990; Echenique et al., 1992; Actis et al., 1993; Okujo et al., 1994; Yamamoto et al., 1994); a high affinity iron transport system has been demonstrated in a strain of *A. baumannii* (Echenique et al., 1992).

Virtually all acinetobacters can grow in simple mineral media containing one of a large number of organic carbon and energy sources (Warskow and Juni, 1972; Baumann, 1968), a characteristic not usually observed for bacteria whose normal habitat is an animal site. It therefore seems likely that, although some of these bacteria can act as opportunistic pathogens, they should probably not be considered as part of the indigenous microflora. The finding that acinetobacters are frequently isolated from the skin of hospital inpatients (Al-Khoja and Darrell, 1979; Seifert et al., 1997) can account for the occurrence of nosocomial infections with strains of this organism. Although isolated readily from the hands of hospital personnel, there is no evidence for persistent colonization with *Acinetobacter* (Buxton et al., 1978). Of the 20 or more recognized *Acinetobacter* genomic species (DNA groups) and named species, *A. baumannii* and the unnamed groups 3 and 13 *sensu* Tjernberg and Ursing (TU) are the ones most frequently isolated from hospitalized patients (Dijkshoorn et al., 1993; Seifert et al., 1993b).

High rates of antibiotic resistance, frequently to multiple drugs, are found in *Acinetobacter* spp. (Bergogne-Bérézin and Towner, 1996). Unlike most of the *Moraxellaceae*, acinetobacters are generally resistant to penicillin. Growth of most strains is not inhibited by 1 U/ml of penicillin G and the majority of strains are resistant to 100 U/ml (Baumann et al., 1968a). Since the 1970s, there has been an increase in the number of antibiotics, particularly the older antibiotics, to which hospital-isolated acinetobacters have become resistant. Most strains of *A. baumannii* now show resistance to aminopenicillins, ureidopenicillins, cephalosporins, cephamycins, many aminoglycosides-aminocyclitols, chloramphenicol, and tetracyclines (Bergogne-Bérézin and Towner, 1996). Currently, imipenem appears to be the most effective drug, although imipenem-resistant strains have been isolated already (Tankovic et al., 1994; Lytikäinen et al., 1995).

Taxonomic relationships of acinetobacters

***Assignment of isolated strains to the genus* Acinetobacter** The following phenotypic properties are generally used to assign iso-

lated strains to the genus *Acinetobacter*: Gram-negative coccobacilli, oxidase negative, aerobic (nonfermenting), nonmotile, and, for most strains, negative nitrate reduction.

The ability of DNA from strains of *Acinetobacter* to transform a conserved *trpE⁻* mutant of the naturally competent BD413 strain to prototrophy (Juni, 1972) has never failed to be demonstrated during the many years since the introduction of this test. Several hundred independently isolated strains have already been identified with this assay. Transformation of *trpE⁻* has been shown to be genus specific and no exceptions have been reported. The specificity of this assay has been confirmed in a clinical setting (Brooks and Sodeman, 1974).

A genus-specific 16S rRNA-targeted oligonucleotide probe has been used to detect acinetobacters (Wagner et al., 1994b).

DNA–rRNA hybridization The highly conserved nature of 16S rRNA makes it useful for recognition of generic relationships among isolated bacterial strains. In DNA–rRNA hybridization studies, employing rRNA from *Acinetobacter calcoaceticus* ATCC 23055^T and filter-fixed DNAs from 23 strains of *Acinetobacter*, $T_{m(e)}$ (midpoint of thermal elution) values from 75.5 to 79.5°C were obtained (Van Landschoot et al., 1986). Similar studies, where DNAs from five established strains of *Acinetobacter* were hybridized with rRNA from *Moraxella lacunata*, resulted in hybrids having $T_{m(e)}$ values ranging from 71 to 73°C; corresponding values ranging from 73.1 to 80°C were obtained using DNAs from strains of all other genera in the *Moraxellaceae* (Rossau et al., 1991). When DNA–rRNA hybridizations were performed using the same five established strains of *Acinetobacter* and rRNA from *Psychrobacter immobilis*, $T_{m(e)}$ values ranging from 69.9 to 71.7°C were obtained; this range being unique for the five strains of *Acinetobacter* (Rossau et al., 1991). In both of these studies, corresponding DNA–rRNA hybridizations using DNA from bacteria not in *Moraxellaceae* resulted in $T_{m(e)}$ values from 53.5 to 67.5°C.

In studies employing rRNA from *A. calcoaceticus* ATCC 23055^T, it was proposed that two strains previously designated as *Pseudomonas cruciviae* and *Pseudomonas pavonacea* were "probably not members of the genus *Acinetobacter*", since although $T_{m(e)}$s of their hybrids (75.5°C and 76.0°C, respectively) indicated that they belong to the *Acinetobacter* rRNA branch, these strains differed significantly in phenotypic properties from some typical *Acinetobacter* species (Van Landschoot et al., 1986). The finding that DNA from each of these atypical strains transformed *Acinetobacter* strain BD413 *trpE⁻* to prototrophy (Juni, unpublished results) proves that these strains are indeed acinetobacters, and serves to confirm the significance of $T_{m(e)}$, as well as transformation, for identification of acinetobacters.

***Acinetobacter* species and DNA groups delineated by DNA–DNA hybridization** It has been suggested, and generally accepted, that bacterial species be defined primarily on the basis of DNA reassociation, wherein members of a particular species show approximately 70% or greater DNA–DNA relatedness and with 5°C or less ΔT_m (Wayne et al., 1987). It was also recommended that genomic species that cannot be identified by phenotypic properties should not be given a species name. Early DNA–DNA hybridization studies of 45 strains of *Acinetobacter*, using a nitrocellulose filter method, distinguished groups with members that showed 50%, or greater, relatedness to one of five reference strains, and a sixth group consisted of strains showing less than 50% relatedness to any of the reference strains (Johnson et al., 1970). The results of this investigation have been correlated with hybridizations studies using the S1 nuclease method where, within species, the levels of DNA relatedness to reference strains ranged from 62 to 95%, with ΔT_m values of less than 3.5°C (Bouvet and Grimont, 1986). To date, extensive DNA–DNA hybridization investigations have revealed the existence of approximately 21 species of *Acinetobacter* (Bouvet and Grimont, 1986; Bouvet and Jeanjean, 1989; Tjernberg and Ursing, 1989; Gerner-Smidt and Tjernberg, 1993). Six species were named (and assigned species numbers) in one study (Bouvet and Grimont, 1986), and a seventh species, *A. radioresistens*, was named by another laboratory (Nishimura et al., 1988). The other genomic (DNA) groups, which could not be differentiated by phenotypic methods, or contained fewer than 10 strains, were assigned a number. It should be noted that genomic species 13–15 of Bouvet and Jeanjean (1989) and Tjernberg and Ursing (1989) were described independently, and in order to avoid confusion, it is current practice to add the suffices BJ or TU to denote the groups of the respective studies. Genomic species 13 BJ and DNA group 14 TU have one reference strain in common, and have been considered one genomic species, while the other groups are apparently distinct.

Numerous reports refer to strains that cannot be allocated to any of the described (genomic) species. Since almost all of the acinetobacters studied were derived from hospital sources and because the most common habitats of these organisms are soil and water, it seems likely that many naturally occurring strains have yet to be isolated and that the current listing is probably not complete.

A distinct correlation between the PAGE-types of the outer membrane proteins and several *Acinetobacter* DNA relatedness groups has been reported (Ino and Nishimura, 1989; Dijkshoorn et al., 1990).

Although the highly competent strain BD413, a soil derivative, was originally reported to be a strain of *A. calcoaceticus*, phenotypic characterization (Bouvet and Grimont, 1986), studies of outer membrane protein patterns (Ino and Nishimura, 1989), and 16S rDNA sequence analysis (Strätz et al., 1996) suggest that this strain may, in fact, belong to a unique species. Furthermore, phylogenetic analysis of nucleotide sequences of *gyrB* genes revealed that strain BD413 was not strongly linked to any currently recognized genomic species (Yamamoto and Harayama, 1996) and may, in fact, represent a new genomic species (Yamamoto et al., 1999).

A group of *n*-alkane-degrading *Acinetobacter* strains have been described as members of a new species, "*A. venetianus*"; these strains harbored plasmids containing genes homologous to *alkBFGH* of *Pseudomonas oleovorans*, which encode the enzymatic steps of *n*-alkane oxidation (Di Cello et al., 1997). It has been shown that the oil-degrading *Acinetobacter* strain RAG-1 is a member of this species and that all strains of "*A. venetianus*" have a unique restriction pattern of amplified 16S rDNA and share similarities in DNA–DNA hybridization and biochemical characteristics (Vaneechoutte et al., 1999).

It is most likely that *Acinetobacter* species, particularly those which include many virtually identical independently isolated strains, consist of organisms that have evolved to survive in some particular ecological niche(s). In each instance, the particular genetic constitution of strains of a species may represent an arrangement that is especially advantageous to the survival of those strains in their natural niche; the similarity of such arrangements in isolated strains may account for the observed high levels of DNA–DNA hybridization. It is possible that mutations in 16S

rDNA may not interfere with the ability of strains of a particular species to retain a selective advantage for occupation of their normal niche, thus accounting for frequent lack of correlation between data from DNA–DNA hybridization and 16S rDNA sequence analysis, as described below.

16S rDNA analysis 16S rRNA gene sequences from single strains of 21 reported *Acinetobacter* (genomic) species have been shown to have binary similarities of more than 94% and to constitute a phylogenetically coherent group of organisms within the *Gammaproteobacteria* (Ibrahim et al., 1997; Rainey et al., 1994a) (Figs. BXII.γ.148 and BXII.γ.149). Most strains of *Acinetobacter* were found to belong to one of five clusters (Fig. BXII.γ.149). Cluster I included DNA groups 2 and TU13; cluster II included DNA groups 3, CTTU13, and 1-3; cluster III included DNA groups 6, 8, and 9; cluster IV included DNA groups BJ15 through BJ17; and cluster V included DNA groups 1, BJ14, 10, and 11 (Ibrahim et al., 1997). Several DNA groups (4, 5, 7, 12, TU14, and TU15) are not associated significantly with any of the five clusters. Although reassociation of DNA from group 10 to DNA from group 11 is low (Bouvet and Grimont, 1986), strains from these groups have almost identical 16S rDNA sequences (Ibrahim et al., 1997). Neither study (Ibrahim et al., 1997; Rainey et al., 1994a) had complete correspondence between DNA–DNA hybridization values and clustering of 16S rDNA sequences, similar to observations for other groups of bacteria (Stackebrandt and Goebel, 1994). The finding that two authentic strains of *A. calcoaceticus* have 16S rRNAs with a similarity coefficient (S_{AB}) of only 0.66 (Byng et al., 1985) is yet another demonstration of the lack of correspondence of data obtained by DNA–DNA hybridization and 16S rRNA gene sequences.

Although there appear to be unambiguous sequence differences in the highly variable regions of the 16S rDNAs of each of the 21 DNA groups examined (Ibrahim et al., 1997), it seems premature to use these signatures for absolute species identification until more strains from each DNA group are analyzed in a similar manner. In addition, since the *Acinetobacter* chromosome encodes at least seven rRNA operons (Strätz et al., 1996; Gralton et al., 1997) it is possible that one, or more, of the rDNA sequences might have experienced mutational and recombinational changes during evolution of individual members of a species (Sneath, 1993).

Phylogenetic studies involving GYRB, RPOD, and RECA genes It has been suggested that the resolution of 16S rRNA-based phylogenetic analysis is not sufficient to distinguish closely related genomic species, because of the extremely slow rate of base substitution in 16S rDNAs (Yamamoto and Harayama, 1996). Genes that code for proteins evolve more rapidly than the genes coding for rRNA. It has been proposed that the highly conserved nature of type II DNA topoisomerase genes makes them excellent candidates for studying bacterial diversity (Huang, 1996). The DNA topoisomerase *gyrB* of acinetobacters has been subjected to phylogenetic analysis based on nucleotide sequences of these genes and the corresponding amino acid sequences of the gene products (Yamamoto and Harayama, 1996). It was shown that the percentage of nucleotide substitutions in the *gyrB* genes of 15 strains of *Acinetobacter* varied from 0.3 to 30.4%, but that the percentage of substitutions in the corresponding amino acid sequences was between 0.5 and 13.3%, indicating that nucleotide substitution which results in amino acid substitution is less frequent than synonymous substitution (Yamamoto and Harayama,

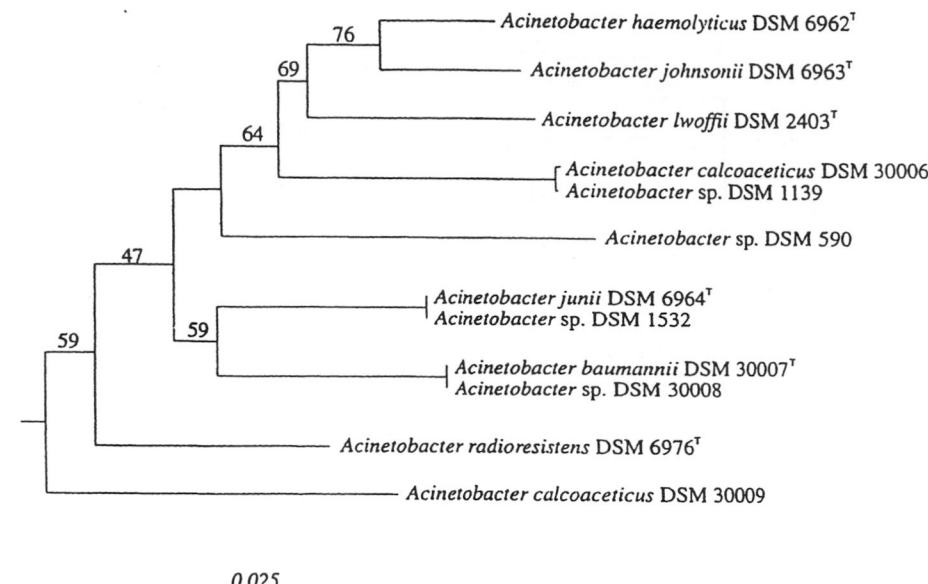

0.025

FIGURE BXII.γ.148. Intrageneric relationships of the genus *Acinetobacter* as determined by 16S rDNA sequence analysis. (Reproduced with permission from F.A. Rainey et al., FEMS Microbiology Letters. *124:* 349–354, 1994, ©Elsevier Science B.V.)

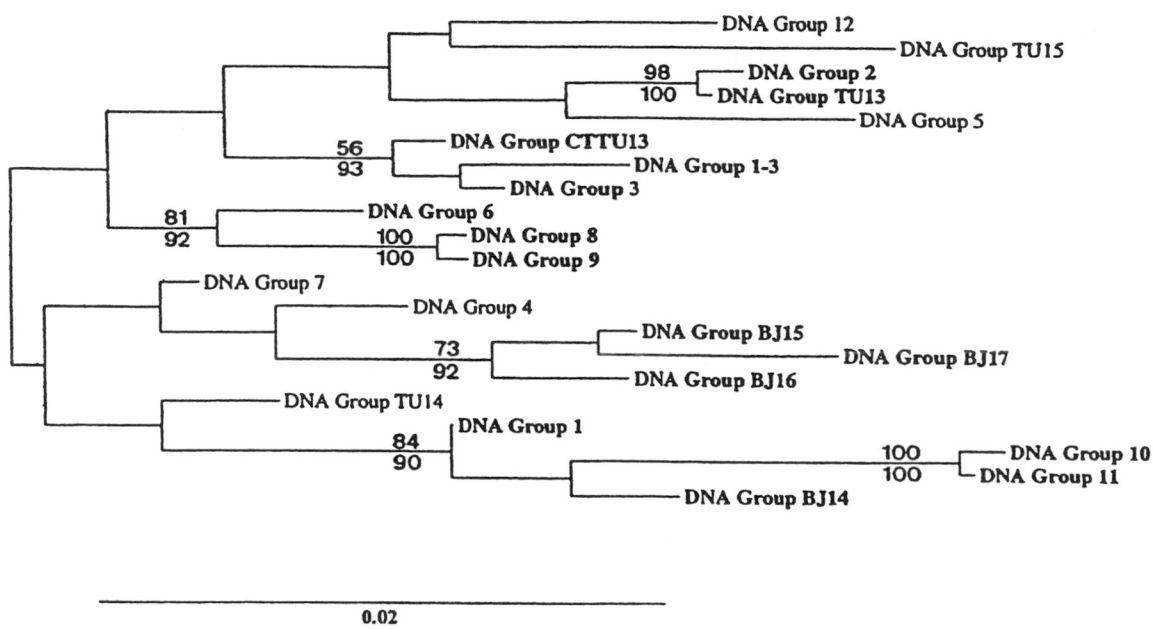

0.02

FIGURE BXII.γ.149. Unrooted phylogenetic dendrogram constructed for 21 strains representing all known DNA groups described for the genus *Acinetobacter.* (Reproduced with permission from A. Ibrahim et al., International Journal of Systematic Bacteriology *47:* 837–841, 1997, ©International Union of Microbiological Societies.)

1996). The results of phylogenetic analysis using *gyrB* nucleotide sequences was consistent with those using DNA–DNA hybridization and phenotypic comparisons.

In a more extensive investigation using 49 *Acinetobacter* strains, 46 of which had previously been classified into 18 genomic species using the results of DNA–DNA hybridization, it was shown that the phylogenetic grouping of such strains based on *gyrB* genes was almost congruent with that based on DNA–DNA hybridization studies (Yamamoto et al., 1999) (Fig. BXII.γ.150). Three strains that had not been characterized previously by

DNA–DNA hybridization appeared to belong to two new genomic species (Fig. BXII.γ.150). Since minor discrepancies existed in the grouping of strains of genomic species 8, 9, and BJ17, the phylogenetic tree for these strains was reconstructed from the sequence of *rpoD*, the structural gene for RNA polymerase σ⁷⁰ factor (Yamamoto et al., 1999); this tree was completely congruent with the grouping based on DNA–DNA hybridization (Fig. BXII.γ.150).

Using *Acinetobacter recA* obtained by polymerase chain reaction amplification, it was possible to distinguish 17 genomic species

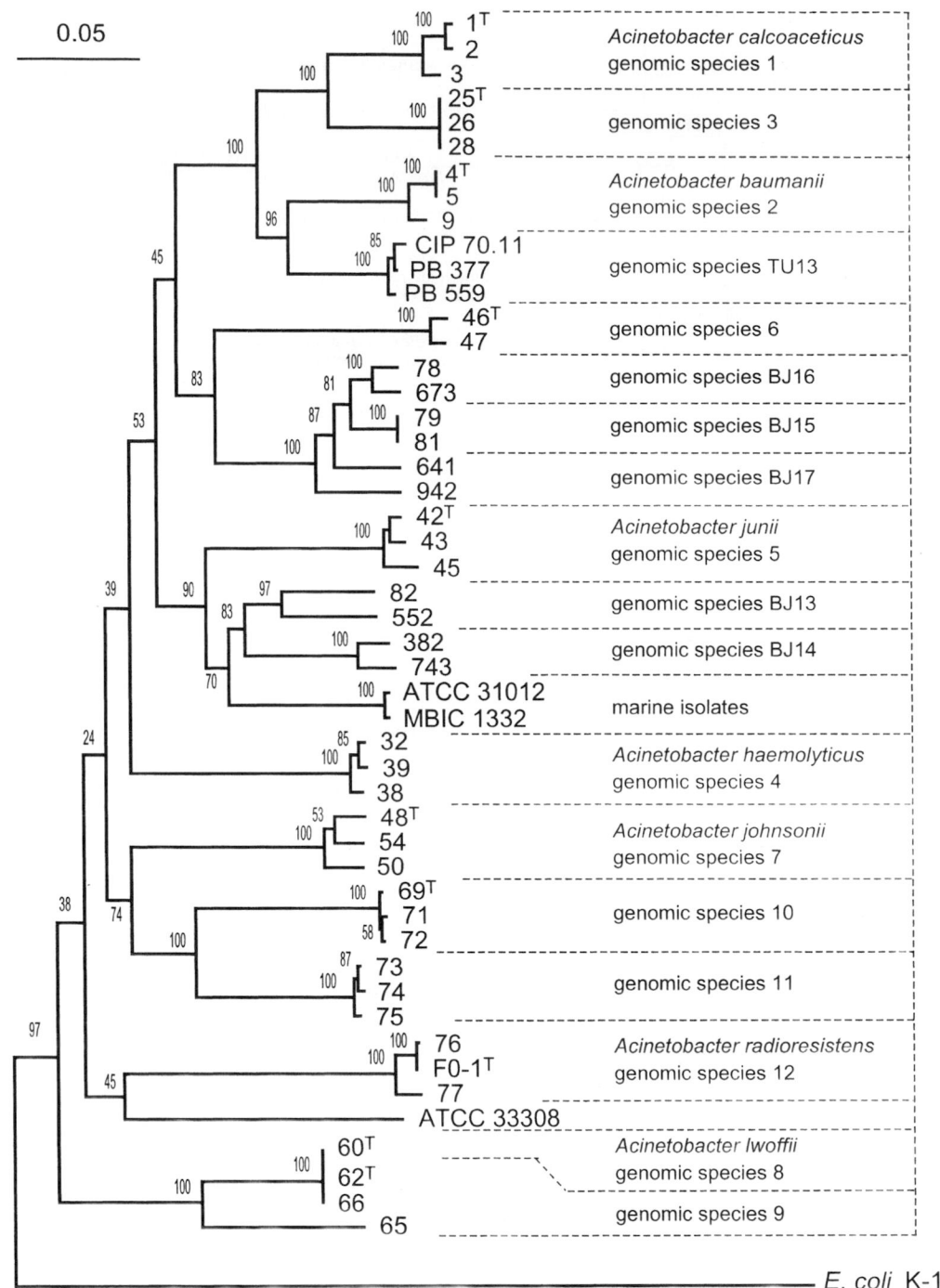

FIGURE BXII.γ.150. Phylogenetic trees of 49 *Acinetobacter* strains from the nucleotide sequences of the *gyrB* genes. Reproduced with permission from S. Yamamoto et al., International Journal of Systematic Bacteriology *49*: 87–95, 1999, ©International Union of Microbiological Societies.)

by restriction fragment length polymorphism analysis with two restriction enzymes (Nowak et al., 1995).

Methods used for* Acinetobacter *(genomic) species identification The recognition of distinct DNA relatedness groups within the genus *Acinetobacter* has raised interest in the ecology and pathology of the different groups. A great many approaches to identification of these (genomic) species by methods easier to perform than DNA–DNA hybridization have been taken. Table BXII.γ.131 presents phenotypic properties that frequently make

it possible to distinguish strains from most of 19 DNA groups. It should be noted though, that phenotypic systems, including nutritional and physiological tests (Bouvet and Grimont, 1986, 1987b; Gerner-Smidt et al., 1991; Kämpfer et al., 1993) or test galleries of commercial systems (Soddell et al., 1993; Bernards et al., 1995, 1996), do not allow for identification of all DNA groups. In particular, using phenotypic properties, it is difficult to discriminate between the closely related (genomic) species *A. calcoaceticus* (DNA group 1), *A. baumannii* (DNA group 2), and

TABLE BXII.γ.131. Differentiation of genomic species of *Acinetobacter* by phenotypic tests[a,b,c]

Characteristics	1,2,3, and 13 TU[d]	4 and 6	5	7–15 TU[d]	8–9	10	11	12	13 BJ–14 TU	14 BJ	15 BJ	16BJ	17BJ
Most discriminatory tests (n = 0):													
Utilization of:													
Azalate	[+]	−	−	[−]	D	D	+	[+]	−	+	−	(+)[e]	−
β-Alanine	[+]	[ ]	−	[−]	−	+	+	−	D	+	+	−	+
L-Leucine	[+]	[+]	D	[−]	−	−	−	+	[−]	+	(+)	(+)	(+)
L-Phenylalanine	[+]	−	−	−	D	−	−	+	+	+	+	+	+
L-Tryptophan	[+]	−	−	−	−	−	−	[v]	+	+	+	−	+
4-Hydroxybenzoate	[+]	[+]	−	[−]	−	D	[+]	[−]	+	+	+	−	+
Phenylacetate	[+]	−	−	−	[+]	−	D	[+]	+	+	+	+	+
Growth at 37°C	[+]	[+]	+	−	D	+	−	+	[−]	+	+	+	+
Acid from glucose	[+]	D	−	−	D	+	−	[−]	+	+	−	−	−
Gelatinase	−	[+]	−	−	−	−	−	−	+	+	+	+	+
Additional tests (n = 4):													
Hemolysis of:													
Sheep blood	−	[+]	D	−	−	−	−	−	+	+	+	+	+
Human blood	−	[+]	D	−	−	−	−	−	+	+	+	+	+
Utilization of:													
Citrate	[+]	D	D	D	[−]	+	D	−	+	+	−	+	+
Glutarate	[+]	[−]	[−]	D	D	+	+	+	D	+	−	(+)	(+)
L-Arginine	+	[ǀ]	[+]	D	[−]	−	−	[+]	[+]	+	+	+	+
L-Aspartate	[+]	D	D	D	−	+	+	D	D	−	−	(+)	−
L-Histidine	+	+	[+]	−	−	+	+	−	+	+	+	+	+
Phenylpyruvate	[+]	−	−	−	D	−	−	[+]	+	+	+	+	+
Protocatechuate	+	+	[−]	D	−	+	+	−	+	+	+	+	+
Vanillate	[+]	[+]	−	[−]	−	D	D	−	−	+	−	+	−
Hydrolysis of:[f]													
pNP-β-D-Xylopranoside [g]	[+]	D	−	−	−	D	−	−	−	−	−	+	+
γ-L-Glutamate-pNA [h]	[+]	D	−	−	−	−	−	−	D	−	−	+	+
Susceptibility to:													
Chloramphenicol	[−]	D	D	D	+	−	D	[−]	−	nd	nd	nd	nd
Penicillin G	[−]	−	D	D	[+]	−	−	D	−	nd	nd	nd	nd

[a]Table reproduced with permission from Towner, K.J., 1996. Biology of *Acinetobacter* spp. *In Acinetobacter*: Microbiology, Epidemiology, Infections, Management, edited by Bergogne-Bérézin, Joly-Guillou, and Towner , © CRC Press, Boca Raton, pp. 44–45.

[b]Results for genomic species 4–17BJ are based on only one strain and should be verified by testing more authentic strains.

[c]Adapted from Kämpfer et al. (1993), except for susceptibility tests which are according to Gerner-Smidt and Frederiksen (1993). Symbols: +, all strains positive; [+], 80% or more positive; D, 20–79% positive; [−], 20% or less positive; −, all strains negative. The following genomic species have been named: 1, *A. calcoaceticus*; 2, *A. baumannii*; 4, *A. haemolyticus*; 5, *A. junii*; 7, *A. johnsonii*; 8, *A. lwoffii*; 12, *A. radioresistens*.

[d]Genomic species grouped according to common characteristics.

[e][+] Weak positive.

[f]Only one strain investigated.

[g]pNp, *p*-nitrophenyl; pNA, *p*-nitroanilide.

[h]Weak reaction.

the unnamed groups 3 and 13TU, sometimes referred to as the *A. calcoaceticus-A. baumannii* (Acb) complex. Table BXII.γ.132 lists phenotypic properties that permit some differentiation of the species of this complex.

Other methods used for species identification of acinetobacters (Dijkshoorn, 1996) are based on structural characteristics of strains and include serotyping (Traub and Leonhard, 1994), fatty acid compositional analysis (Kämpfer, 1993), outer membrane protein patterns (Ino and Nishimura, 1989; Dijkshoorn et al., 1990), and enzyme electrophoresis (Bouvet and Jeanjean, 1995). None of these methods has been completely successful for assignation of isolated strains to the currently described (genomic) species.

Due to the difficulties in phenotypic identification, and given the clinical relevance of some genomic species, a variety of molecular methods has been explored for their usefulness in identification of *Acinetobacter* species. These methods include ribotyping (Gerner-Smidt, 1992; Dijkshoorn et al., 1993), amplified rDNA restriction analysis (ARDRA) (Vaneechoutte et al., 1995; Dijkshoorn et al., 1998; Jawad et al., 1998) and AFLP™ (amplified fragment length polymorphism) (Janssen et al., 1997), 16S-23S rDNA spacer analysis (Dolzani et al., 1995; Nowak et al., 1995), and tRNA spacer fingerprinting (Ehrenstein et al., 1996).

Typing methods Primarily because of an interest in tracing strains of *Acinetobacter* involved in nosocomial infections and epidemic spread, a great effort has been made to distinguish strains. Early phenotypic methods used to allocate strains to definitive types include phage typing (Bouvet et al., 1990) and bacteriocin typing (Andrews, 1986). Other phenotypic methods include biotyping (Bouvet and Grimont, 1987b), multilocus enzyme electrophoretic typing (Picard et al., 1989; Thurm and Ritter, 1993), cell envelope protein typing (Dijkshoorn et al., 1987), antibiotic susceptibility typing (Dijkshoorn et al., 1993; Seifert et al., 1993a; Marcos et al., 1994; Lytikäinen et al., 1995), and serotyping (Traub and Leonhard, 1994; Pantophlet et al., 1998). Antibiogram typing is widely used in screening for strain identity, although gain or loss of resistance genes may lead to false conclusions. Biotyping has been shown to be a practical method for allocating strains of the Acb complex to as many as 19 biotypes (Bouvet and Grimont, 1987b; Bouvet et al., 1990) and has ena-

TABLE BXII.γ.132. Phenotypic differentiation of genomic species of the *A. calcoaceticus–A. baumannii* complex[a,b]

Characteristic	Genomic species			
	1[c]	2[d]	3	13 TU[e]
Growth at 41°C	[−]	+	d	+
Growth at 44°C	−	+	−	d
Growth with:				
L-Arabinose	d	[+]	d	+
D-Glucarate	d	[+]	[+]	[−]
D-Ribose	d	[+]	d	[+]
Sorbinic acid	[−]	d	[−]	[+]
Glycerate	[+]	[−]	d	[−]
Levulinate	d	d	−	[−]
L-Tartrate	[−]	d	[+]	−]
Acetyl-L-glutamine	[−]	d	d	d
D-Asparagine	+	−		[−]
L-Hydroxyproline	[−]	[+]	[−]	d

[a]Table reproduced with permission from Towner, K.J., 1996. Biology of *Acinetobacter* spp. *In Acinetobacter*: Microbiology, Epidemiology, Infections, Management, edited by Bergogne-Bérézin, Joly-Guillou, and Towner , © CRC Press, Boca Raton, pp. 46.

[b]Adapted from Kämpfer et al. (1993). For symbols, see Table BXII.γ.131.

[c]*A. calcoaceticus.*

[d]*A. baumannii.*

[e]DNA–DNA hybridization group 13 (Tjernberg and Ursing, 1989).

bled the recognition of specific types from different countries (Marcos et al., 1994; Dominguez et al., 1995; Ratto et al., 1995).

Rapid developments in molecular biology have led to the application of numerous DNA-based typing methods for *Acinetobacter* strain identification, including discrimination of strains of the Acb complex. These include plasmid DNA fingerprinting (Gerner-Smidt, 1989; Seifert et al., 1994b), ribotyping (Gerner-Smidt, 1992), PCR fingerprinting methods with detection of amplification products by agarose gel electrophoresis (Reboli et al., 1994) or automated laser fluorescence (Webster et al., 1996), pulsed field gel electrophoresis (Tankovic et al., 1994), and selective amplification of restriction fragments by AFLP™ (Janssen and Dijkshoorn, 1996; Janssen et al., 1997). Nosocomial acinetobacters can be very similar; therefore, it is current practice to use a combination of methods for strain identification (Dijkshoorn et al., 1993; Struelens et al., 1993; Seifert et al., 1994a; Lytikäinen et al., 1995). By application of such a polyphasic strategy, including ribotyping, protein electrophoretic typing, AFLP™, biotyping and antibiogram typing, it was possible to recognize and differentiate two highly similar strains found in different NW European hospitals and countries (Dijkshoorn et al., 1996). It was shown recently that multiply-resistant strains with features similar to these two clones are also common in the Czech Republic (Nemec et al., 1999). It is not yet known whether these strains are also responsible for outbreaks in other parts of the world.

Analysis of 170 acinetobacter strains isolated from various foods revealed that although strains belonging to almost every species were represented, 46% were identified as *A. johnsonii* and 15% as *A. lwoffii* (Gennari and Lombardi, 1993). A majority (78%) of *Acinetobacter* strains isolated from activated sludge, in a pilot-scale treatment of sewage for enhanced removal of phosphate, was identified as *A. johnsonii* (Duncan et al., 1988).

Study of quantitative transformation of the high-level streptomycin-resistance marker Since quantitative transformation of the high-level streptomycin-resistance marker was instrumental in elucidation of the taxonomic relationships within the *Moraxellaceae* (Bøvre and Hagen, 1981), similar studies with 54 clinical isolates

and 17 reference strains of *Acinetobacter*, using competent strain BD4 as the recipient, were initiated (Bergan and Vaksvik, 1983). Ratios of interstrain to intrastrain transformation of 0.06–6.4% were obtained, DNAs from streptomycin-resistant mutants of 13 of the clinical isolates failing to transform the competent streptomycin-sensitive strain. Although these studies were limited, the results obtained suggest that evolutionary change in ribosomal protein S12 (*StrA*) in the genus *Acinetobacter* may be as extensive as that found in the genus *Moraxella* (Bøvre and Hagen, 1981).

ENRICHMENT AND ISOLATION PROCEDURES

Isolation of acinetobacters can be accomplished using ordinary laboratory media such as heart infusion agar, brain heart infusion agar, or trypticase soy agar. Although many acinetobacters grow well at 37°C, the optimum growth temperature for most strains is 33–35°C. Acinetobacters having lower optimum temperatures and unable to grow at 37°C have been reported (Breuil et al., 1975). A general enrichment procedure for isolation of acinetobacters from soil and water involves inoculation of 20 ml of an acetate-mineral medium with 5 ml of a water sample, or of a filtered 10% soil suspension, followed by vigorous aeration during incubation at 30°C, or at room temperature (Baumann, 1968). Aeration is especially favorable for enrichment of nonmotile acinetobacters since, in a nonaerated culture, motile oxygen-consuming pseudomonads tend to move to the surface layer, thereby decreasing the amount of oxygen diffusing into the interior of the liquid volume. Acinetobacters have a slightly acid pH optimum for growth, and aeration at a pH from 5.5 to 6.0 favors enrichment of these organisms from soil and water samples (Baumann, 1968; Juni, 1982). Presumptive identification of colonies of *Acinetobacter* growing on any medium involves a demonstration of nonmotile coccobacilli in a wet mount, an oxidase-negative reaction, and a Gram-negative stain (Warskow and Juni, 1972). When these criteria are satisfied, definitive identification can be achieved through use of the transformation assay (Juni, 1972).

Selective media for isolation of acinetobacters from environmental and clinical sources are available and include Herellea

Agar (Difco) (Mandel et al., 1964) and Leeds *Acinetobacter* Medium (Jawad et al., 1994).

MAINTENANCE PROCEDURES

Cultures of *Acinetobacter* can be maintained on heart infusion agar plates (Difco) stored in plastic bags (to delay evaporation and drying) at room temperature for periods as long as six months before being transferred (Marcus et al., 1969). This procedure is not recommended, however, since it has been observed that the ability to use certain carbon sources is lost, probably because of spontaneous mutation, when cultures are transferred infrequently in this manner (Juni, unpublished). Storage at low temperatures, or by lyophilization, appears to be best for preservation

of properties initially observed for a particular strain. Cultures can be conveniently stored by preparing a heavy suspension of organisms in buffer, or in broth, and adding 0.5 ml of this suspension to 1.0 ml of sterile glycerol in a screw-capped test tube. After mixing, this suspension is stored at approximately $-65°C$. Since 67% glycerol remains liquid at temperatures above $-46°C$, a stored culture can be sampled conveniently by withdrawing a loopful, or 0.1 ml, of the glycerol stock and streaking this material directly on a heart infusion, or other suitable medium, plate without significant warming of the glycerol suspension.

ACKNOWLEDGMENTS

I would like to thank Lenie Dijkshoorn for invaluable assistance in preparation of the section dealing with taxonomy of acinetobacters.

List of species of the genus Acinetobacter

1. **Acinetobacter calcoaceticus** (Beijerinck 1911) Baumann, Doudoroff and Stanier 1968b, 1538[AL] emend. Bouvet and Grimont 1986, 238 (*Micrococcus calcoaceticus* Beijerinck 1911, 1067.)

   cal.co.a.ce′ti.cus. L. n. *calx* chalk; L. n. *acetum* acetic acid; M.L. n. *calcoaceticus* calcium acetate, which was used by Beijerinck in the enrichment medium from which he isolated the organism.

   Characteristics as given for the genus. Colonies on tryptocase soy agar are circular, convex, smooth, and slightly opaque with entire margins; colonies are 0.5–1.5 mm in diameter after 24 h and 2.5–3.5 mm in diameter after 48 h at 30°C. Good growth occurs at 15–37°C in one day. No growth occurs at 41°C and above. Nitrates are not reduced to nitrites in complex medium. Grows on citrate (Simmons). Acid is produced from D-glucose. γ-Glutamyltransferase is produced, but β-xylosidase is not produced. Horse or sheep blood is not hemolyzed; gelatin is not hydrolyzed.

   The following compounds are utilized as sole sources of carbon and energy: L-phenylalanine, L-histidine, phenylacetate, malonate, azelate, β-alanine, *trans*-aconitate, L-arginine, L-ornithine, DL-lactate, 2,3-butanediol, DL-4-aminobutyrate, glutarate, L-aspartate, and L-tyrosine. D-malate and histamine are not utilized. L-Leucine is utilized by some strains.

   Isolated from soil.
   Genomic species 1.
   The mol% G + C of the DNA is: 40–42 (T_m).
   Type strain: ATCC 23055, CIP 81.08, DSM 30006.
   GenBank accession number (16S rRNA): X81661, Z93434.

2. **Acinetobacter baumannii** Bouvet and Grimont 1986, 239[VP]
   bau.ma′ni.i. M.L. gen. *baumannii* in honor of Paul and Linda Baumann.

   Characteristics as given for the genus. Colonies on tryptocase soy agar are circular, convex, smooth, and slightly opaque with entire margins and sometimes have a butyrous aspect. Colonies are 1.5–2.0 mm in diameter after 24 h and 3.0–4.0 mm in diameter after 48 h at 30°C. Good growth occurs at 15–44°C. Occasional strains may require methionine for growth on minimal medium. Nitrates are not reduced to nitrites in complex medium (exceptions occur). Acid is produced from D-glucose by most strains. Strains which produce acid from D-glucose also produce β-xylosidase. γ-Glutamyltransferase is produced by most strains. Horse or sheep blood is not hemolyzed; gelatin is not hy-

   drolyzed. Citrate (Simmons) is utilized by prototrophic strains. The following compounds are utilized: DL-lactate, glutarate, L-aspartate, L-tyrosine, ethanol, 2,3-butanediol, and DL-4-aminobutyrate. The following compounds are utilized by most strains: L-phenylalanine and phenylacetate (growth on L-phenylalanine is associated with growth on phenylacetate in this species), L-histidine, azelate, D-malate, L-leucine, β-alanine, *trans*-aconitate, L-arginine, and L-ornithine. Histamine is not utilized.

   Isolated from human clinical specimens or the natural environment. Most *Acinetobacter* strains isolated from nosocomial infections belong to this species.
   Genomic species 2.
   The mol% G + C of the DNA is: 40–43 (T_m).
   Type strain: ATCC 19606, CIP 70.34, DSM 30007.
   GenBank accession number (16S rRNA): X81660.

3. **Acinetobacter haemolyticus** Bouvet and Grimont 1986, 239[VP]
   hae.mo.ly′ti.cus. Gr. n. *haema* blood, dissolving.

   Characteristics as given for the genus. Colonies on tryptocase soy agar are circular, convex, smooth, and slightly opaque with entire margins and sometimes have a sticky consistency. Colonies are 1.5–2.0 mm in diameter after 48 h at 30°C. Good growth occurs at 15–37°C. No growth occurs at 41°C. Nitrates are not reduced to nitrites in complex medium (exceptions occur). On horse or sheep blood agar, colonies are surrounded by a clear zone which is limited by a dark red ring after 24 h at 37°C or after 48 h at 30°C. Gelatin is hydrolyzed (exceptions occur). Some strains produce acid from D-glucose and produce β-xylosidase. Some strains produce γ-glutamyltransferase. Most strains grow on Simmons citrate agar.

   DL-4-Aminobutyrate is utilized by all strains. The following compounds are utilized by most strains: L-histidine, D-malate, L-leucine, ethanol, and L-arginine. The following compounds are not utilized: DL-lactate, azelate, histamine, β-alanine, 2,3-butanediol, and L-ornithine.

   Isolated from clinical specimens and the environment.
   Genomic species 4.
   The mol% G + C of the DNA is: 40–43 (T_m).
   Type strain: Strain Mannheim 2446/60, B40, ATCC 17906, CIP 64.3, DSM 6962, NCTC 10305.
   GenBank accession number (16S rRNA): X81662.

4. **Acinetobacter johnsonii** Bouvet and Grimont 1986, 239[VP]
   john′so.ni.i. M.L. gen. *johnsonii* in honor of John L. Johnson.

Characteristics as given for the genus. Colonies on tryptocase soy agar are circular, convex, smooth, and slightly opaque with entire margins; colonies are 1.0–1.5 mm in diameter after 24 h and 2.0–3.0 mm in diameter after 48 h at 30°C. Good growth occurs at 15 to 30°C; no growth occurs at 37°C. Nitrates are not reduced to nitrites in complex medium (exceptions occur). Acid is not produced from D-glucose; β-xylosidase and γ-glutamyltransferase are not produced. Horse or sheep blood is not hemolyzed; gelatin is not hydrolyzed. No growth factor is required (in a few cases methionine is required). Grows on Simmons citrate. DL-Lactate and ethanol are utilized as sole sources of carbon and energy.

The following compounds are utilized by many strains: malonate, D-malate, L-aspartate, L-tyrosine, 2,3-butanediol, L-arginine, L-ornithine, and DL-4-aminobutyrate. The following compounds are not utilized: glutarate, L-phenylalanine, phenylacetate, L-histidine, azelate, L-leucine, histamine, β-alanine, and trans-aconitate.

Isolated from clinical specimens, activated sludge, foods, and eviscerated chickens.

Genomic species 7.

The mol% G + C of the DNA is: 44–45 (T_m).

Type strain: ATCC 17909, CIP 64.6, DSM 6963, NCTC 10308.

GenBank accession number (16S rRNA): X81663.

5. **Acinetobacter junii** Bouvet and Grimont 1986, 239[VP]
ju'ni.i. M.L. gen. *junii* in honor of Elliot Juni.

Characteristics as given for the genus. Colonies on tryptocase soy agar are circular, convex, and smooth with entire margins. Colonies are 1.0–1.5 mm in diameter and translucent after 24 h and 2.0–2.5 mm in diameter and slightly opaque after 48 h at 30°C. Good growth occurs at 15–37°C; occasionally growth occurs at 41°C. No growth occurs at 44°C. No growth factor is required. Nitrates are not reduced to nitrites in complex media. Acid is not produced from D-glucose. β-Xylosidase and γ-glutamyltransferase are not produced. Horse or sheep blood is not hemolyzed; gelatin is not hydrolyzed.

Citrate (Simmons) is occasionally utilized. DL-Lactate, L-histidine, D-malate, and ethanol are utilized as sole sources of carbon and energy. The following compounds are utilized by many strains: L-aspartate, L-leucine, L-tyrosine, L-arginine, and DL-4-aminobutyrate. The following compounds are not utilized: glutarate, L-phenylalanine, phenylacetate, malonate, azelate, histamine, β-alanine, 2,3-butanediol, trans-aconitate, and L-ornithine.

Isolated from human clinical specimens.

Genomic species 5.

The mol% G + C of the DNA is: 42 (T_m).

Type strain: Mannheim 2723/59, ATCC 17908, CIP 64.5, DSM 6964, NCTC 10307.

GenBank accession number (16S rRNA): X81664.

6. **Acinetobacter lwoffii** (Audureau 1940) Brisou and Prévot 1954, 727[AL] emend. Bouvet and Grimont 1986, 238.
lwof.fi.i. M.L. gen. *lwoffii* in honor of Andre Lwoff.

Characteristics as given for the genus. Colonies on tryptocase soy agar are circular, convex, smooth, and slightly opaque with entire margins; colonies are 1.0–1.5 mm in diameter after 24 h and 3.0–4.0 mm in diameter after 48 h at 30°C. Good growth occurs at 15–37°C in one day. No growth occurs at 41°C and above. Nitrates are not reduced to nitrites in complex medium. No growth occurs on citrate (Simmons). Acid is not produced from D-glucose; β-xylosidase and γ-glutamyltransferase is not produced. Horse or sheep blood is not hemolyzed; gelatin is not hydrolyzed.

Azelate and DL-lactate are utilized as sole sources of carbon and energy. The following compounds are not utilized: L-phenylalanine, L-histidine, malonate, L-aspartate, L-leucine, histamine, β-alanine, 2,3-butanediol, glutarate, trans-aconitate, and L-arginine. Some strains can utilize phenylacetate, L-tyrosine, ethanol, L-ornithine, and DL-4-aminobutyrate.

Habitat unknown.

Genomic species 8.

The mol% G + C of the DNA is: 46 (T_m).

Type strain: ATCC 15309, CIP 64.10, DSM 2403, NCTC 5866.

GenBank accession number (16S rRNA): X81665.

7. **Acinetobacter radioresistens** Nishimura, Ino, and Iizuka 1988, 210[VP]
ra.di.o.re.sis'tens. L. n. *radius* ray, beam; L. part. adj. *resistens* resisting; M.L. part. adj. *radioresistens* ray resisting, because of high resistance to gamma-ray irradiation.

Cells are rods 0.6–0.7 × 0.8–1.0 μm. Pleomorphic. A plump form is 0.8–0.9 ×1.2–1.8 μm; elongate cells (15 μm or more long) occur in young cultures; and coccoid cells (0.6–0.9 μm in diameter) occur in old cultures. The cells occur singly or in pairs. Nonsporeforming, nonencapsulated, nonmotile, non-acid fast, Gram negative. Colonies on nutrient agar smooth, entire, convex, glistening, opaque, and yellowish white to pale yellow. A pellicle is formed on the surface of nutrient broth.

Metabolism is strictly respiratory. Catalase is produced. Cytochrome oxidase is not produced. Nitrate is reduced to nitrite in succinate-nitrate medium but not in Casitone-nitrate medium. Indole and hydrogen sulfide are not produced. Tween 80 is hydrolyzed. Lysine and ornithine decarboxidase, pheylalanine deaminase, arginine dihydrolase, gelatinase, and urease are not produced. Hemolysis on sheep blood agar is negative. Growth occurs in media containing 5% NaCl. Good growth occurs at 27–31°C. Growth occurs at 42°C. L-Arabinose, D-ribose, D-xylose, D-galactose, D-glucose, D-fructose, D-mannose, L-rhamnose, D-glucosamine, D-cellobiose, lactose, maltose, melibiose, sucrose, and raffinose are not oxidized and are not assimilated. Acetate, fumarate, DL-lactate, L-malate, malonate, succinate, hexadecane, heptadecane, octadecane, eicosane, ethanol, *n*-butanol, L-alanine, L-glutamate, L-leucine, and L-proline are assimilated. Citrate, gluconate, decane, nonadecane, methanol, *n*-propanol, isopropanol, ethanolamine, β-alanine, DL-phenylalanine, L-histidine, L-isoleucine, DL-norleucine, DL-serine, L-threonine, dulcitol, *i*-inositol, D-mannitol, and D-sorbitol are not assimilated.

Highly resistant to gamma-ray irradiation (D_{10} value, 125 to 220 krads under air equilibrium in 0.067 M phosphate buffer). Resistant to benzylpenicillin (100 IU).

The major cellular fatty acids are the even-numbered straight-chain acids $C_{18:1}$, $C_{16:1}$, and $C_{16:0}$. The ubiquinone system is Q-9.

Isolated from cotton and from soils.

Assigned to genomic species 12 (Tjernberg and Ursing, 1989).

The mol% G + C of the DNA is: 44.1–44.8 (T_m).

Type strain: Strain FO-1, DSM 6976, IAM 13186.

GenBank accession number (16S rRNA): X81666.

8. **"Acinetobacter venetianus"** Di Cello, Pepi, Baldi and Fani 1997, 245.

venetianus. venetianus because it was isolated from the Venice lagoon

Characteristics are as given for the genus. Strains tested (2) are β-hemolytic on sheep blood agar. Growth occurs at 37°C but not at 41°C or 44°C. Aerobic acidification of glucose on OF medium is negative. β-Xylosidase is not produced. Strains test positive for gelatinase production; caprate, malate, and citrate hydrolysis; and assimilation of acetate, alanine, caprate, valerate, citrate, histidine, and 3-hydroxybutyrate. There is no growth on rhamnose, *N*-acetyl-glucosamine, ribose, inositol, fucose, sorbitol, arabinose, propionate, 5-ketogluconate, glycogen, *m*-hydroxybutyrate, and 2-ketogluconate.

Strains grow on fuel oil and the *n*-alkanes (C_{10}, C_{14}, and C_{20}), and their respective oxidation products as sole carbon sources. Strains harbor plasmids that encode the enzyme activities required for oxidation of alkanes and dehydrogenation of aliphatic alcohols.

Isolated from lagoons and other waters that are contaminated with fuel oil or other products derived from petroleum.

The mol% G + C of the DNA is: 43.6–43.9 (T_m).

Deposited strain:

GenBank accession number (16S rRNA): X80825 (for strain RAG-1).

Additional Remarks: No type strain has been designated. Strains investigated include VE-C3, RAG-1, and T4.

Other Organisms

The following genomic groups have been recognized by virtue of possessing unique DNA–DNA hybridization groups. They are not yet named because none of them has been shown to display a set of phenotypic properties that distinguish them from the other species of *Acinetobacter*. Some of the phenotypic properties of these genomic species are shown in Tables BXII.γ.131 and BXII.γ.132.

Unnamed genomic species of *Acinetobacter*: 3, 6, 9, 10, 11, TU13, TU14 (= BJ13), TU15, CTTU13, 1-3, BJ14, BJ15, BJ16, and BJ17.

Genus III. **Psychrobacter** *Juni and Heym 1986b, 389*[VP]

ELLIOT JUNI

Psy.chro'bac.ter. Gr. adj. *psychros* cold; M.L. masc. n. *bacter* rod; M.L. masc. n. *Psychrobacter* a rod that grows at low temperatures.

Rods or cocci, 0.9–1.3 × 1.5–3.8 μm. Rods can vary in length from extremely short (coccobacilli) to relatively long. For some strains the rods tend to be somewhat swollen. Also observed to grow as cocci or coccobacilli. Nonpigmented. Colonies on heart infusion agar are usually cream-colored, smooth, and opaque. Gram negative; the stain frequently tends to be retained. Nonmotile, and aerobic. **Most strains are psychrotrophic; they are able to grow at 5°C, have temperature optima near 20°C, and generally are unable to grow at 35–37°C.** Strains able to grow well at 35–37°C usually cannot grow at 5°C. Grow well on most common complex media. Most strains can grow on a mineral-agar medium containing a single carbon and energy source and ammonium salts as the nitrogen source. Some strains form acid aerobically from D-glucose, D-mannose, D-galactose, L-arabinose, D-xylose, and L-rhamnose. They are **halotolerant and grow in the presence of 6.5% or greater NaCl.** Many strains are radiation resistant. **Oxidase positive. Catalase positive.** Many strains are **competent for natural genetic transformation.** Have been isolated from the open sea, the deep sea, sea ice, ornithogenic soils, skins and gills of fish, skins of poultry, various food products, irradiated foods, air contaminants, and from a variety of human sources. Not known to be clinically significant. Cultures of *Psychrobacter* can be maintained in the same manner as for cultures of *Acinetobacter*, as described above.

The mol% G + C of the DNA is: 41–50.7.

Type species: **Psychrobacter immobilis** Juni and Heym 1986b, 388.

FURTHER DESCRIPTIVE INFORMATION

Bacterial strains that were oxidase positive and psychrotrophic (subsequently shown to be psychrobacters) were proposed as members of the *Moraxellaceae* because of weak but distinct genetic interactions with authentic *Moraxella* species, involving the high-level streptomycin-resistance marker (Bøvre and Hagen, 1981). Based on 16S rRNA gene sequence information, *Psychrobacter immobilis* has been placed in group III (of four groups) of the family *Moraxellaceae* (See Fig. BXII.γ.146 of the description of the family *Moraxellaceae*.) (Pettersson et al., 1998). Nine distinct species of *Psychrobacter* have been revealed from analysis of phenotypic properties, DNA–DNA hybridization, and 16S rRNA sequences (see Fig. BXII.γ.151) (Bowman et al., 1996, 1997a, 1997d; Maruyama et al., 2000; Denner et al., 2001a; Kämpfer et al., 2002a; Romanenko et al., 2002a). 16S rRNA sequences having similarities of 93.9–99.9% have been reported for strains of all species of *Psychrobacter* (Bowman et al., 1996, 1997d; Denner et al., 2001a; Kämpfer et al., 2002a; Romanenko et al., 2002a). Based on the extent of 16S rDNA similarity, psychrobacters are more closely related to the moraxellae than they are to the acinetobacters (Bowman et al., 1996; Pettersson et al., 1998) (Fig. BXII.γ.146). The wide range of DNA composition of psychrobacters (mol% G + C 41–50.7) provides evidence for heterogeneity of the genus.

It has been reported (Romanenko et al., 2002a) that strains of *P. submarinus* and *P. marincola* have 99.9% similarity in 16S rRNA sequences but showed DNA–DNA reassociation of only 15% similarity. Furthermore, the mol% G + C of the DNAs of these strains differed by 3.9%. These data indicate that *P. submarinus* and *P. marincola* must have a long history of independent evolution, from a presumed common ancestor, in spite of conservation of the sequences of their respective 16S rRNAs. By contrast, it was demonstrated (Denner et al., 2001a) that 16S rRNA similarities between *P. proteolyticus* and those of *P. glacincola*

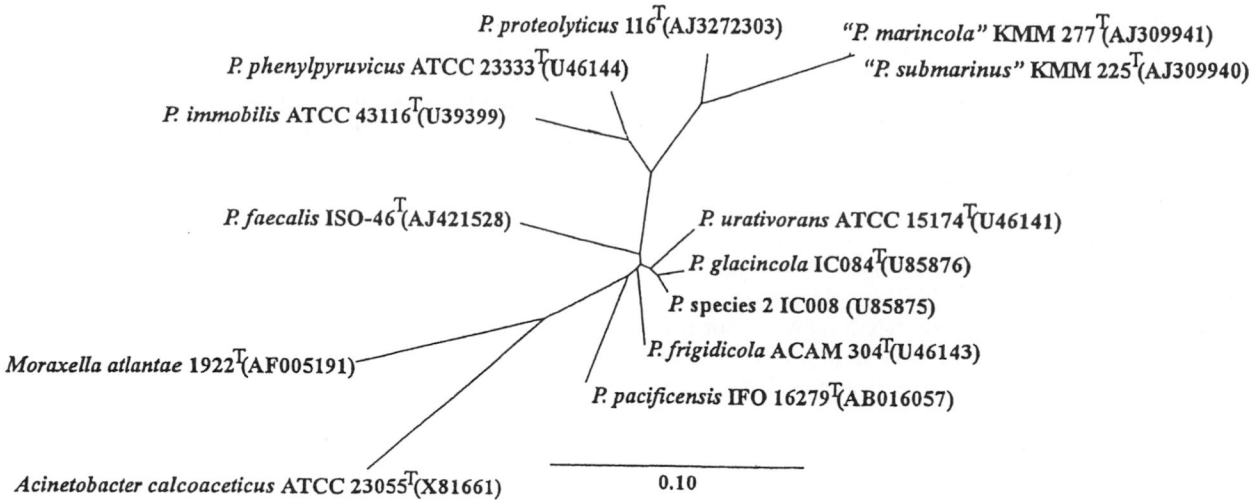

FIGURE BXII.γ.151. Phylogenetic tree for the species of the genus *Psychrobacter*. (Reproduced with permisson from P. Kämpfer et al., Systematic and Applied Microbiology *25:* 31–36, 2002, ©Urban & Fischer.)

and *P. urativorans* were 97.4 and 96.9%, respectively, whereas the corresponding DNA–DNA reassociation values were 62.2 and 35.9%, respectively. It is clear, as already noted (Stackebrandt and Goebel, 1994), that the relative evolutionary stability of 16S rRNA does not mirror corresponding stability of chromosomal DNA.

The organism originally designated as *"Micrococcus cryophilus"* (McLean et al., 1951) has been shown to be a psychrobacter (Juni and Heym, 1980), *P. urativorans* (Bowman et al., 1996).

The relatively large individual cells of many *Psychrobacter* strains are in contrast to the smaller sizes of most other Gram-negative bacteria. Major fatty acid profiles include $C_{16:1 \, \omega7c}$, 35–47%; $C_{17:1 \, \omega8c}$, 3–7%; and $C_{18:1 \, \omega9c}$, 37–70% (Nishimura et al., 1979; Moss et al., 1988; Bowman et al., 1996). A novel fatty acid, $C_{12:1 \, \omega9c}$, has been shown to be present in *P. immobilis* (Moss and Daneshvar, 1992). Contain wax esters (Gallagher, 1971; Russell and Volkman, 1980; Bowman et al., 1996, 1997d). The major ubiquinone of *Psychrobacter* contains eight isoprene units (Q8) (Moss et al., 1988). Cell membranes resemble those of other Gram-negative bacteria (Thorne, et al., 1973).

Most strains grow well at room temperature on plates of a variety of complex media. Strains of *P. frigidicola* grow best at 14–16°C; some fail to grow at room temperature and may be considered psychrophilic. Growth in liquid media requires aeration. Most psychrobacters can grow on mineral media containing sodium lactate and/or monosodium glutamate and an ammonium salt as the source of nitrogen. Many organic compounds can serve as sole carbon and energy sources for growth of psychrobacters (Shaw and Latty, 1988). The growth lag phase in such mineral media is decreased by addition of amino acids and, frequently, Tween 80 (Juni, 1992). Psychrobacters are aerobic and appear to contain all the enzymes of the tricarboxylic acid cycle (Tai and Jackson, 1969). They have been isolated from foods following gamma-ray irradiation at 300 and 500 krad (Ito et al., 1976), but strains isolated from nonirradiated foods are frequently radiation sensitive (Ito and Iizuka, 1983). They have also been isolated from sea life and the deep sea (Georgala, 1958; Maruyama et al., 2000; Denner et al., 2001a; Romanenko et al., 2002a), the skin of chickens (Lahellec et al., 1975), food products (Shaw and

Latty, 1988; Gennari et al., 1992), Antarctic ornithogenic soils (Bowman et al., 1996), Antarctic sea ice (Bowman et al., 1997d), and a bioaerosol originating from pigeon feces (Kämpfer et al., 2002a).

DNAs from strains of all species of *Psychrobacter* tested are able to transform a hypoxanthine and thiamine auxotroph (ATCC 43117), of a competent *Psychrobacter immobilis* strain (ATCC 43116), for ability to grow on a defined medium (Juni and Heym, 1980). DNAs from strains of all other members of the *Moraxellaceae* fail to transform this auxotroph to prototrophy; the transformation assay appears to be genus specific. Approximately one-third of strains isolated from humans and previously classified as CDC group EO-2 were shown to be psychrobacters, based on their phenotypic properties and the results of the transformation assay (Hudson et al., 1987).

DNA–rRNA hybridization studies have shown that *Moraxella phenylpyruvica* is more closely related to psychrobacters than to the other members of the *Moraxellaceae* (Rossau et al., 1991). The overall levels of 16S rDNA sequence similarity between one strain of *M. phenylpyruvica* and the several species of *Psychrobacter* was shown to be 93–96% (Bowman et al., 1996, 1997d; Denner et al., 2001a; Kämpfer et al., 2002a). *M. phenylpyruvica* also resembles *Psychrobacter* in being able to grow (slowly) at 4°C and tolerate NaCl concentrations up to 9%, and it has been proposed that *M. phenylpyruvica* be transferred to the genus *Psychrobacter* as *Psychrobacter phenylpyruvicus* comb. nov. (Bowman et al., 1996). There are three lines of evidence suggesting that *M. phenylpyruvica* is, in fact, not a species of *Psychrobacter*. 1. DNAs from strains of *M. phenylpyruvica* fail to transform the hypoxanthine and thiamine auxotroph of *Psychrobacter* to prototrophy (Juni and Heym, 1980), this assay being genus specific for all strains of *Psychrobacter* tested. 2. In a transformation assay that makes use of a streptomycin-dependent marker from a strain of *M. phenylpyruvica* that was transformed into the competent *Acinetobacter* strain BD413 (a rare transformant), it was shown that DNAs from 34 independently isolated strains of *M. phenylpyruvica* (including strain ATCC 23333^T, the type strain of *P. phenylpyruvicus*) all transformed the streptomycin-dependent *Acinetobacter* to streptomycin-independence (Juni, 1990). DNAs from strains of *Psychrobacter* were inactive (Juni, unpublished results). 3. There are characteristic differences in fatty acid composition between strains of *Psychrobacter* and strains of *M. phenylpyruvica* (Moss et al., 1988; Bowman et al., 1996, 1997d) and, unlike *Psychrobacter* and other members of the *Moraxellaceae*, *M. phenylpyruvica* does not possess true waxes (Bowman et al., 1996). In the description of *M. phenylpyruvica* (listed under species of the genus *Moraxella*),

it is suggested that strains bearing this designation are members of a unique genus of the family *Moraxellaceae*.

A *Psp*PI type-II restriction-modification system of a strain of *Psychrobacter* has been investigated; the gene encoding the *Psp*PI methyltransferase has been cloned, expressed in *E. coli*, and its nucleotide sequence has been determined (Rina et al., 1997).

List of species of the genus Psychrobacter

1. **Psychrobacter immobilis** Juni and Heym 1986b, 388^{VP}.
   im.mo' bi.lis. L. adj. *immobilis* motionless.

   Plump rods or coccobacilli frequently showing diploforms, 0.9–1.3 µm in diameter. Nonmotile. Aerobic. Gram positive. Oxidase and catalase positive. Most strains grow on a defined medium with a single carbon and energy source and a mineral or organic source of nitrogen (i.e., an amino acid such as glutamate). Many organic compounds can serve as carbon and energy sources. Most strains form acid aerobically from D-glucose, D-mannose, D-galactose, L-arabinose, D-xylose, and L-rhamnose and use some of these compounds as carbon and energy sources for growth. Phenylalanine is deaminated. Other phenotypic characteristics appear in Table BXII.γ.133.

   The mol% G + C of the DNA is: 44–47 (T_m).

   Type strain: ATCC 43116, CCUG 9708, DSM 7229, LMG 7091, LMG 7203.

   GenBank accession number (16S rRNA): U39399.

2. **Psychrobacter faecalis** Kämpfer, Albrecht, Buczolits and Busse 2002b, 1438^{VP} (Effective publication: Kämpfer, Albrecht, Buczolits and Busse 2002a, 35.)
   fae.ca' lis. L. n. *faex, faecis* dregs; M.L. adj. *faecalis* fecal.

   Aerobic, cells are straight rods 0.8–1.2 × 1–2 µm and occur singly; they are nonmotile, Gram negative, oxidase- and catalase-positive. Colonies are circular, slightly raised, and beige with entire margins on nutrient agar. Grows well at 4–36°C on different complex media; no growth was found at 45 and 55°C on nutrient agar. No growth occurs on Salmonella-Shigella Agar. H$_2$S is not produced. Indole production, growth on Simmons citrate, urease, arginine dihydrolase, lysine decarboxylase, and ornithine decarboxylase, were all negative. No acid production from the following sugars could be detected: glucose, lactose, sucrose, D-mannitol, dulcitol, salicin, adonitol, inositol, sorbitol, L-arabinose, raffinose, L-rhamnose, maltose, D-xylose, trehalose, cellobiose, methyl-D-glucoside, erythritol, melibiose, D-arabitol, and D-mannose. The following compounds are not utilized as sole sources of carbon: L-arbutin, cellobiose, D-galactose, gluconate, D-maltose, D-mannose, α-D-melibiose, L-rhamnose, D-ribose, sucrose, salicin, D-trehalose, D-xylose, adonitol, *i*-inositol, D-sorbitol, glycerol, D-fructose, D-mannitol, maltitol, putrescine, *trans*-aconitate, mesaconate, adipate, azelate, itaconate, suberate, β-alanine, L-phenylalanine, L-serine, L-proline, L-histidine, L-tryptophan, 3-hydroxybenzoate, and phenylacetate. Ubiquinone Q-8 is the major quinone type, in addition to small amounts of ubiquinone Q-9 and Q-7; spermidine is the major polyamine. The polar lipid profile is characterized by the presence of phosphatidylethanolamine, phosphatidylglycerol, and diphophotidylglycerol. The fatty acids C$_{18:1 \omega 9c}$, C$_{17:1 \omega 8c}$, C$_{16:1 \omega 7c}$, C$_{15:0 iso 2OH}$, and C$_{12:0 3OH}$ are produced. Isolated from an aerosol originating from pigeon feces. Other phenotypic characteristics appear in Table BXII.γ.133.

   The mol% G + C of the DNA is: not determined.

   Type strain: Iso-46, DSM 14664, CIP 107288.

   GenBank accession number (16S rRNA): AJ421528.

3. **Psychrobacter frigidicola** Bowman, Cavanagh, Austin and Sanderson 1996, 847^{VP}.
   fri' gid.i.col.a. L. adj. *frigidus* cold; L. n. *cola* dweller; M.L. n. *frigidicola* cold dweller.

   Cells are coccoid, 1.5–2.0 µm in diameter; occur in pairs, tetrads, or rarely in short chains. Aerobic. Nonmotile. No growth factors are required. Able to grow in the absence of NaCl but growth is stimulated by 0.25–0.30 M NaCl. Can tolerate up to 9% NaCl. Optimal temperature for growth 14–16°C. Maximum temperature for growth 19–22°C. Phenylalanine and tryptophan are deaminated. Urease, nitrate reduction, and production of acid from carbohydrates are negative. Isolated from ornithogenic soils from Antarctica. Other phenotypic characteristics appear in Table BXII.γ.133.

   The mol% G + C of the DNA is: 41–42 (T_m).

   Type strain: ACAM 304, CCUG 34377.

   GenBank accession number (16S rRNA): U46143.

4. **Psychrobacter glacincola** Bowman, Nichols and McMeekin 1997e, 1274^{VP} (Effective publication: Bowman, Nichols and McMeekin 1997d, 214.)
   gla.cin.co' la. L. n. *glacies* ice; L. *incola* inhabitant; M.L. n. *glacincola* the ice inhabitant.

   Cells are coccoidal with a diameter of 1.0–1.5 µm. Nonmotile. Colonies are up to 4 mm in diameter, circular, have an entire edge, low convex elevation, butyrous consistency, and an off-white color. No fluorescent pigments formed. Seawater is required for optimal growth but most strains will grow at between 0 and 2.1 M NaCl. Optimal growth occurs at 19–22°C. Growth factors are not required. Fail to utilize or form acid from carbohydrates. Utilize nitrate and ammonium salts as nitrogen sources. Some strains produce urease and reduce nitrate to nitrite. Tween 40 and Tween 80 are hydrolyzed. Phenylalanine is not deaminated. Isolated from Antarctic ice. Other phenotypic characteristics appear in Table BXII.γ.133.

   The mol% G + C of the DNA is: 43–44 (T_m).

   Type strain: ACAM 483, DSM 12194.

   GenBank accession number (16S rRNA): U46145.

5. **Psychrobacter marincola** Romanenko, Schumann, Rohde, Lysenko, Mikhailov and Stackebrandt 2002a, 1296^{VP}
   ma.rin.co' la. L. gen. n. *maris* of the sea; L. n. *incola* inhabitant; N.L. n. *marincola* inhabitant of the sea.

   Aerobic, Gram negative, nonmotile nonpigmented, nonsporeforming, coccoidal cells, 0.7–1.0 µm in diameter. Oxidase- and catalase-positive. Colonies on nutrient agar are whitish, circular with an entire edge, convex, 1–2 mm in diameter. No fluorescent pigments are formed. Moderately halophilic. Seawater is required for growth; can grow at 0.5–

TABLE BXII.γ.133. Properties of *Psychrobacter* species[a]

Characteristic	*P. immobilis*[b]	*P. faecalis*[c]	*P. frigidicola*[b]	*P. glacincola*[b]	*P. marincola*[d]	*P. pacificensis*[e]	*P. proteolyticus*[f]	*P. submarinus*[d]	*P. urativorans*[b]
Maximum growth temperature, °C	35–39	36	19–22	19–22	35	33–38	<35	35	25–27
Optimum growth temperature, °C	27–31	15–30	14–16	13–15	25–28	25–28	ND	25–28	17–19
Tolerates 2.0 M NaCl	v+	+	−	+	+	−	+	+	−
Tolerates 5% bile salts	+	+	−	+	+	ND	−	+	−
Stimulated by bile salts	−	ND	−	−	ND	ND	ND	ND	−
Acid from carbohydrates	+	−	−	−	−	+	+	+	−
Arginine arylamidase	+	ND	v+	−	ND	ND	ND	ND	+
Phenylalanine arylamidase	v+	ND	+	+	ND	ND	ND	ND	−
Phenylalanine deaminase	+	ND	−	−	−	−	ND	−	+
Nitrate reduction	v−	ND	−	v+	−	−	−	−	v−
Urease	v+	−	−	v-	−	+	+	−	v+
Lecithinase (egg yolk reaction)	v+	ND	−	+	ND	ND	ND	ND	−
Tween 40 hydrolysis	+	ND	−	+	−	ND	ND	+	−
Tween 80 hydrolysis	+	ND	−	+	+	ND	+	ND	−
Carbon sources:									
3-Hydroxybutyrate	+	+	−	v+	ND	v	+	ND	+
Pyruvate	+	+	−	+	ND	ND	+	ND	+
L-Malate	+	+	+	−	−	+	−	−	v+
Oxaloacetate	+	ND	−	+	ND	ND	ND	ND	−
DL-Lactate	+	+	−	v+	−	v	−	−	v+
Propionate	v+	(+)	−	+	−	−	+	−	−
Isovalerate	v−	ND	−	v+	ND	ND	ND	ND	−
Heptanoate	v+	ND	−	+	ND	ND	ND	ND	v−
Pimelate	−	ND	−	v+	ND	ND	ND	ND	−
Azelate	v−	−	+	+	ND	ND	+	ND	−
L-Alanine	+	+	−	v+	−	v	+	−	−
Hydroxy-L-proline	v−	ND	−	+	ND	ND	ND	ND	−
L-Phenylalanine	+	−	+	−	ND	ND	−	ND	v+
Urate	v+	ND	+	+	ND	ND	ND	ND	+
Mol% G + C (T_m) of the DNA	44–47	ND	41–42	43–44	50.7	44–45	43.6	46.7	44–46

[a]Symbols: +, positive for 90–100% of strains; v, positive for 11–89% of strains; v+, positive for 11–89% of strains but type strain is positive; v−, positive for 11–89% of strains but type strain is negative; −, positive for 0–10% of strains; ND, not determined.

[b]Data for *P. immobilis*, *Psychrobacter frigidicola*, *P. glacincola*, and *P. urativorans* from Bowman et al., 1997d.

[c]Data for *P. faecalis* from Kämpfer et al., 2002a.

[d]Data for *P. marincola* and *P. submarinus* from Romanenko et al., 2002a.

[e]Data for *P. pacificensis* from Maruyama et al., 2000.

[f]Data for *P. proteolyticus* from Denner et al., 2001a.

15.5% (w/v) NaCl. Growth occurs at 7–35°C, with an optimum at 25–28°C. Does not grow at 4, 37, or 40°C. The pH range for growth is 5.5–9.5, with optimum growth at 6.5–8.5. Negative for Simmons citrate test, Voges–Proskauer reaction, ONPG test, H₂S production and denitrification. Susceptible to ampicillin (10 µg), benzylpenicillin (10 U), gentamycin (10 µg), kanamycin (30 µg), carbenicillin (25 µg), lincomycin (15 µg), oleandomycin (15 µg), polymyxin (300 U), streptomycin (30 µg), tetracycline (30 µg) and neomycin (15 µg). The major fatty acid is $C_{18:1\ \omega9c}$. The phospholipids are phosphatidylethanolamine, phosphatidylglycerol, phosphatidylinositol and phosphatidylserine. Isolated from internal tissues of the ascidian *Polysyncraton* sp., collected in the Indian Ocean. Other characteristics appear in Table BXII.γ.133.

The mol% G + C of the DNA is: 50.7 (T_m).

Type strain: KMM277, DSM 14160.

GenBank accession number (16S rRNA): AJ309941.

6. **Psychrobacter pacificensis** Maruyama, Honda, Yamamoto, Kitamura and Higashihara 2000, 845[VP]

pa.ci.fi.cen'sis. M.L. adj. *pacificensis* pertaining to the Pacific Ocean.

Strains are aerobic, Gram negative, nonmotile, nonpigmented, nonsporulating, oxidase-positive, catalase-positive coccobacilli 1.0–1.5 µm long and about 1 µm wide. Strains produce many fimbriae as appendages, but no flagella. Circular, convex, off-white colonies with a complete edge occur on agar plates containing polypeptone and yeast extract. No fluorescent pigments formed. Seawater or NaCl at about 3% is required for optimal growth, with most strains not growing at 0 or 8% NaCl. It takes 1–2 weeks to reach stationary phase at 4°C, but isolates give growth yields at 4°C comparable to those at 20°C. Optimal growth at about 25°C, with the maximum growth temperature being 38°C. Acid is formed aerobically from glucose, xylose and arabinose. Urease is positive, but phenylalanine deaminase and tryptophan deaminase are negative. The species is negative for the following biochemical tests: glucose fermentation, indole production, esculin hydrolysis, gelatin liquefaction, and arginine dihydrolase. Strains use L-histidine and DL-malate as sole carbon and energy sources. Some strains use acetate, L-alanine, 3-hydroxybutyrate, lactate, malonate and suberate, but none use p-hydroxybutyrate, citrate, gluconate, propionate, L-serine or n-valerate. The major fatty acid is $C_{18:1\ \omega9c}$ and Q-8 is the major quinone. Isolated from 6,000 m deep seawater in the Japan Trench off Hachijo Island, Japan. Other characteristics appear in Table BXII.γ.133.

The mol% G + C of the DNA is: 43–44 (HPLC).

Type strain: NIHB P2K6.

GenBank accession number (16S rRNA): AB016057.

7. **Psychrobacter proteolyticus** Denner, Mark, Busse, Turkiewicz and Lubitz 2001b, 1619.[VP] (Effective publication: Denner, Mark, Busse, Turkiewicz and Lubitz 2001a, 51.)

pro.te.o.ly' ti.cus. M.L. n. adj. *proteolyticus,* proteolytic.

Cells are Gram negative, nonpigmented, nonmotile, nonsporeforming, rod-shaped to coccoidal, and 0.5–1.25 μm wide by 1.0–2.5μm long. Colonies on complex standard media are cream-colored, circular, entire, convex, and of butyrous consistency; diameters are as large as 1.0 mm after 2 d incubation at room temperature. The following biochemical tests are positive: oxidase, catalase, urease, alkaline phosphatase, C4 esterase, C8 esterase, lipase, leucine arylamidase, valine arylamidase, acid phosphatase, and naphthol-AS-BI-phosphohydrolase. Gelatin, casein, and tributyrin are hydrolyzed whereas esculin, starch, and DNA are not. The following biochemical tests are negative: arginine dihydrolase, lysine decarboxylase, ornithine decarboxylase, C14 lipase, cystine arylamidase, trypsin, chymotrypsin, α-galactosidase, β-galactosidase, β-glucuronidase, α-glucosidase, β-glucosidase, α-mannosidase, α-fucosidase, H_2S production, nitrate reduction, and indole production. Citrate is used as a single carbon source. Acid is produced from the carbohydrates: glucose, rhamnose, melibiose, amygdalin, and arabinose. No acid is produced from the carbohydrates: mannitol, inositol, sorbitol, and saccharose. Ubiquinone-8 is the main respiratory lipoquinone. Phosphatidylethanolamine, phosphatidylglycerol, and diphosphatidylglycerol are the major polar lipids. The predominant nonpolar fatty acids are: $C_{18:1\,\omega9c}$ and $C_{16:1\,c}$; $C_{17:1\,\omega8c}$, $C_{16:0}$, $C_{18:0}$, and $C_{12:0}$ are present in minor amounts. The predominant 2-hydroxy fatty acid is $C_{12:0\,2OH}$. Isolated from a stomach specimen of an Antarctic krill. Other phenotypic characteristics appear in Table BXII.γ.133.

The mol% G + C of the DNA is: 43.6 (HPLC).

Type strain: 116, CIP 106830, DSM 13887.

GenBank accession number (16S rRNA): AJ272303.

8. **Psychrobacter submarinus** Romanenko, Schumann, Rohde, Lysenko, Mikhailov and Stackebrandt 2002a, 1296[VP]

sub.ma.ri' nus. L. prep. *sub* under; L. adj. *marinus* of the sea; N.L. adj. *submarinus* from under the sea.

Aerobic, Gram negative, nonmotile, nonpigmented, nonsporeforming, ovoid cells, 1.6–1.9 μm long and 0.7–1.0 μm in diameter. Oxidase- and catalase-positive. Colonies on nutrient agar are whitish, circular with an entire margin, convex, 2–3 mm in diameter. No fluorescent pigments are formed. Moderately halophilic. Seawater is required for growth; growth occurs in the presence of 0.5–15.5% (w/v) NaCl. Temperature range for growth is 4–35°C, with an optimum at 25–28°C. Does not grow at 37 or 40°C. pH range for growth is 5.5–9.5 with optimum growth at 6.5–8.5. The strain is negative for Simmons citate test, ONPG test, acetoin production, H_2S production, and denitrification. Susceptible to ampicillin (10 μg), gentamicin (10 μg), kanamycin (30 μg), carbenicillin (25 μg), polymyxin (300 U), streptomycin (30 μg), and neomycin (15 μg). Resistant to benzylpenicillin (10 U), lincomycin (15 μg), oleandomycin (15 μg), tetracycline (15 μg). The major fatty acid is $C_{18:1\,\omega9c}$. The phospholipids are phosphatidylethanolamine, phosphatidylglycerol, diphosphatidylglycerol, phosphatidylinositol, and phosphatidylserine. Other characteristics appear in Table BXII.γ.133.

The mol% G + C of the DNA is: 46.7 (T_m).

Type strain: KMM 225, DSM 14161.

GenBank accession number (16S rRNA): AJ309940.

9. **Psychrobacter urativorans** Bowman, Cavanagh, Austin and Sanderson 1996, 847[VP]

u.ra.ti.vor' ans. M.L. n. *uratum* salt of uric acid; M.L. part. adj. *vorans* devouring; M.L. part. adj. *urativorans* uric acid devouring.

Cells are coccoid, 1.5–2.0 μm in diameter, and occur in pairs and tetrads. Nonmotile. Aerobic. No growth factors are required. Growth is inhibited by 1% bile. Tolerates 9% NaCl. Growth is not stimulated by NaCl. Optimal temperature for growth 18–20°C. Maximum temperature for growth 25–27°C. Does not deaminate phenylalanine or tryptophan. Acid is not produced from carbohydrates. Strains may produce urease and reduce nitrate to nitrite. Other phenotypic characteristics appear in Table BXII.γ.133.

The mol% G + C of the DNA is: 44–46 (T_m).

Type strain: ACAM 534, ATCC 15174, CCUG 4982, DSM 20429.

GenBank accession number (16S rRNA): U46141.

Family III. *Incertae Sedis*

Genus I. **Enhydrobacter** *Staley, Irgens and Brenner 1987, 290*[VP]

JAMES T. STALEY AND DON J. BRENNER

En.hy' dro.bac.ter. Gr. adj. *enhydros* aquatic Gr. neut. n. *bacter* masc. form of Gr. neut. n. *bacterion* a rod; N.L. masc. n. *Enhydrobacter* aquatic rod.

Coccobacillary to rod-shaped cells, 0.5–0.7 × 1.0–5.0 μm. **Unicellular; pairs and short chains occur.** No resting stages known. **Gram negative. Facultative aerobic.** Chemoheterotrophic. **Oligotrophic. Grow very slowly.** Sugars are fermented anaerobically, and organic acids are used aerobically. Growth occurs on mineral medium with D-glucose as the sole carbon source and inorganic ammonium compounds as the sole nitrogen sources. No growth on nutrient agar, Clark–Lubs medium or EMB agar. Vitamins are required by the type strain when grown on defined media. Oxidase and catalase positive. Type strain isolated from the oxygen-depleted zone of a eutrophic lake.

The mol% G + C of the DNA is: 66.

Type species: **Enhydrobacter aerosaccus** Staley, Irgens and Brenner 1987, 290.

FURTHER DESCRIPTIVE INFORMATION

A recent analysis of 16S rRNA sequences indicated that *Enhydrobacter aerosaccus* is closely related to *Moraxella osloensis* (Thompson et al., 2003a). The genus *Enhydrobacter* has therefore been placed in the family *Moraxellaceae* as a *genus incertae sedis*; no further testing has been done to relate it to or to differentiate it from other genera in this family.

Unlike other described members of aeromonads, *Enhydrobacter aerosaccus* is a true oligotrophic bacterium that grows very slowly on typical media used for cultivation of *Proteobacteria*. Biochemical reactions on Hugh–Leifson medium take 30–60 days at room temperature.

This monospecific genus most closely resembles members of the genus *Aeromonas* in that it grows on a defined medium with D-glucose as the sole carbon source and uses ammonium salts as the sole nitrogen sources for growth. Like *Aeromonas* spp., *Enhydrobacter* can use some amino acids such as L-alanine, L-serine, and L-arginine as carbon sources for growth. The type strain of *E. aerosaccus* is nonmotile, a feature that is also found in some species of *Aeromonas*, such as *A. salmonicida*.

However, *E. aerosaccus* differs from *Aeromonas* spp. in several ways. The mol% G + C content of the DNA is 66, considerably higher than the 57–63% found in *Aeromonas* spp. Furthermore, it does not grow on typical media used for cultivation of *Aeromonas* spp., such as methyl red–Voges–Proskauer (Clark–Lubs) medium, EMB medium, nutrient agar, and King A and King B media. Furthermore, *E. aerosaccus* is gas vacuolate, a property not found in any *Aeromonas* spp.

Finally, DNA–DNA reassociation experiments indicate that *Enhydrobacter* is sufficiently distantly related to *Aeromonas* to justify the creation of a separate genus (Staley et al., 1987).

ENRICHMENT AND ISOLATION PROCEDURES

The sole strain of this genus has been isolated from a eutrophic lake, Wintergreen Lake, near Kalamazoo, Michigan (Van Ert and Staley, 1971). A dilute peptone enrichment culture is made using 100 ml of lake water from the thermocline of the oxygen-depleted zone (< 1 mg/l dissolved O_2), supplemented with 10 mg of sterile Bacto Peptone (Difco laboratories, Detroit, MI). After incubation for 18 days at room temperature, 0.1 ml of a 1:100 dilution is spread on plates containing 0.01% peptone with mineral salts and vitamins (Van Ert and Staley, 1971). After incubation for 27 d, material from the plate, containing colonies of planctomycetes, is restreaked on PYG medium (0.025% peptone, 0.025% yeast extract, and 0.025% glucose with vitamins and mineral salts; see genus *Ancalomicrobium* description, this volume). A small, chalky white colony containing gas-vacuolated, rod-shaped cells appears after one month. This is streaked to purity on a medium containing 0.1% glucose, 0.1% Difco Vitamin Free Casamino acids, mineral salts, and vitamins.

MAINTENANCE PROCEDURES

The organism has been maintained by lyophilization for 30 years.

DIFFERENTIATION OF THE GENUS *ENHYDROBACTER* FROM OTHER GENERA

Table BXII.γ.134 shows the differences between *Enhydrobacter* and *Aeromonas*. Although the genera resemble one another in morphology, the type strain and species of *Enhydrobacter* is gas vacuolate. *Enhydrobacter aerosaccus* does not grow on many of the typical media used for cultivation of *Aeromonas* spp., and the time for growth in biochemical tests is 30–60 d.

List of species of the genus Enhydrobacter

1. **Enhydrobacter aerosaccus** Staley, Irgens and Brenner 1987, 290.

   aer′o.sac.cus. Gr. n. *aer* air; Gr. n. *sakkos* sack; N.L. *aerosaccus* gas vacuolate.

   Nonmotile, gas-vacuolated, nonflagellated rods. Gas vacuoles are not produced when growing on sugars, but are produced when cells are grown on certain organic acids, such as acetate, pyruvate, and succinate. Facultatively aerobic, but best growth aerobically is under microaerophilic conditions. No growth at 7 or 43°C; growth occurs at 20–41°C, with optimum growth at 37–39°C. Grows at pH 5.0–9.5. Nitrate is reduced to nitrite. Folic acid and biotin are required for growth. Lysine, ornithine, and arginine are decarboxylated after 5 weeks at room temperature. The following carbon sources are utilized after incubation at 30°C for 60 d: L-arabinose, D-xylose, L-fucose, L-rhamnose, D-ribose, D-glucose, D-fructose, D-mannose, L-sorbose, D-galactose, D-xylose, sucrose, maltose, acetate, citrate, formate, fumarate, glycerol, succinate, inulin, ethanol, pyruvate, lactate, *meso*-malate, oxaloacetate, and *meso*-tartrate. Colonies, which adhere to the agar, are colorless with an entire margin and convex elevation. The type strain was isolated from the oxygen-depleted zone of a eutrophic lake.

   The mol% G + C of the DNA is: 66 (Bd).

   Type strain: ATCC 27094, DSM 8914, LMG 81277.

   GenBank accession number (16S rRNA): AJ550856.

TABLE BXII.γ.134. Differentiation of the genera *Enhydrobacter* and *Aeromonas* [a]

Characteristic	Enhydrobacter	Aeromonas
Growth on:		
Nutrient agar	−	+
EMB agar	−	+
Clark–Lubs medium	−	+
Average time for biochemical reactions (d)	30–60	1–2
Growth rate	very slow	rapid
Gas vacuole formation	+[b]	−
Mol% G + C of DNA	66	57–63

[a]For symbols, see standard definitions.

[b]Only one strain has been described, and it is gas vacuolate.

Order X. **Alteromonadales** ord. nov.

JOHN P. BOWMAN AND THOMAS A. MCMEEKIN

Al.ter.o.mon.a.da' les. M.L. fem. n. *Alteromonas* the type genus of the order; L. suff. *-ales* ending which denotes an order; M.L. fem. n. *Alteromonadales* the *Alteromonas* order.

Gram negative, straight or curved rods. Motile by means of a **single polar flagellum.** Do not form endospores or cysts. **Chemoheterotrophs** which are **either facultative anaerobes or strictly aerobic.** If able to grow anaerobically, growth occurs by respiration and, less commonly, by fermentation. Oxygen is used universally as an electron acceptor. **Most strains are catalase and oxidase positive.** Possess psychrophilic, psychrotolerant, and mesophilic growth regimes. Most species **require Na$^+$ ions** for growth and most species grow optimally on seawater-based media. Most species are able to utilize ammonia as a nitrogen source. Some species have a requirement for amino acids. **Primarily marine inhabitants** found in a wide range of oceanic ecosystems ranging from surface waters and sea ice to abyssal sediments. Usually nonpathogenic. The major coenzyme Q type is ubiquinone-8.

The mol% G + C of the DNA is: 36–54 (T_m, Bd).

Type genus: **Alteromonas** Baumann, Baumann, Mandel and Allen 1972, 418 emend. Gauthier, Gauthier and Christen 1995a, 760

FURTHER DESCRIPTIVE INFORMATION

The order *Alteromonadales* and family *Alteromonadaceae* are created here to accommodate an assemblage of marine Gram-negative bacteria that form a phylogenetic clade in the *Gammaproteobacteria*, positioned between the families *Pseudomonadaceae* and *Enterobacteriaceae*. Members of this group include the genera *Colwellia*, *Moritella* (*Vibrio marinus*), *Shewanella*, *Ferrimonas*, *Alteromonas*, and *Pseudoalteromonas*. These genera are characteristically rod-like in morphology, motile by a single polar flagellum, oxidase positive, require or stimulated by sodium ions, and possess DNA base compositions ranging from 36 to 54 mol% G + C. Cells generally lack inclusions such as poly-β-hydroxybutyrate, resting stages such as spores and microcysts, and cellular extensions such as prostheca, buds, and spinae. The metabolism of the members of the order *Alteromonadales* is quite varied. Some members are facultative anaerobes able to grow anaerobically by fermentation and respiration (*Colwellia*, *Moritella*, and some *Shewanella* species) or by respiration alone (*Ferrimonas* and most *Shewanella* species). Still others are strictly aerobic (*Alteromonas* and *Pseudoalteromonas*).

Overall, there are few if any common properties (except those stated) which the genera in the order *Alteromonadales* have in common. Based on the molecular clock theory (Ochman and Wilson, 1987), the divergence in 16S rRNA gene sequences suggests that members of the order *Alteromonadales* have been evolving in the oceans for at least 500–600 million years (assuming a 1% sequence difference equals ~50 million years). During that time, a multiplicity of marine (and some terrestrial) habitats have been colonized, with the development of a wide range of ecophysiological specialization and versatility. For example, the order *Alteromonadales* is noteworthy for containing all known psychrophilic bacteria belonging to the *Gammaproteobacteria*, the majority of barophilic species, and most bacterial species capable of forming omega-3 polyunsaturated fatty acids.

TAXONOMIC COMMENTS

Before phylogenetic analysis became commonplace, members of the order *Alteromonadales* were grouped either within the family *Vibrionaceae* (e.g., *Shewanella*; MacDonell and Colwell, 1985; Farmer, 1992) or were simply ungrouped (e.g., *Alteromonas*, Baumann et al., 1984d; Gauthier and Breittmayer, 1992). The advent of 16S rRNA-based phylogenetic analysis has resulted in considerable reorganization of the species in this group. This includes major reassignment of species of *Alteromonas* to *Shewanella* (MacDonell and Colwell, 1985; Coyne et al., 1989), *Pseudoalteromonas* (Gauthier et al., 1995a), and *Marinomonas* (van Landschoot and De Ley, 1983). *Colwellia* (Deming et al., 1988a) was created to accommodate a psychrophilic *Vibrio* species (D'Aoust and Kushner, 1972) which, on the basis of 5S and 16S rRNA gene sequence analysis, was clearly distinct from other *Vibrio* species. *Vibrio marinus* (Baumann et al., 1984b) also groups phylogenetically within the family *Alteromonadaceae* and has been recently transferred to a new genus, *Moritella*, as *Moritella marina* (Urakawa et al., 1998). *Ferrimonas*, a genus containing iron-reducing species, is similar in many respects to *Shewanella* and is represented by only a single strain (Rosselló-Mora et al., 1995a).

Family I. **Alteromonadaceae** Ivanova and Mikhailov 2001b, 1229[VP] (Effective publication: Ivanova and Mikhailov 2001a, 15.)

JOHN P. BOWMAN AND THOMAS A. MCMEEKIN

Al.ter.o.mo.na.da' ce.ae. M.L. fem. n. *Alteromonas* the type genus of the family; L. suff. *-aceae* which denotes a family; M.L. fem. n. *Alteromonadaceae* the *Alteromonas* family.

The description is the same as for the order *Alteromonadales*. See Table BXII.γ.135 for characteristics differentiating some genera of the family *Alteromonadaceae.**

Type genus: **Alteromonas** Baumann, Baumann, Mandel and Allen 1972, 418 emend. Gauthier, Gauthier and Christen 1995a, 760.

---

*Editorial Note: Only the genera *Alteromonas, Colwellia, Idiomarina,* and *Pseudoalteromonas* were included in the family by Ivanova and Mikhailov (2001a). The genera *Alishewanella, Ferrimonas, Glaciecola, Marinobacter, Marinobacterium, Microbulbifer, Moritella, Psychromonas, Shewanella,* and *Thalassomonas* have been added to this family based on 16S rDNA sequence analysis.

TABLE BXII.γ.135. Characteristics differentiating some genera belonging to the family *Alteromonadaceae*[a]

Characteristics	*Alteromonas*	*Colwellia*	*Ferrimonas*	*Moritella* (*Vibrio marinus*)	*Pseudoalteromonas*	*Shewanella*
Growth in presence of:						
0% NaCl	−	−	−	−	−	D
10% NaCl	+	−	−	−	D	D
Growth at:						
4°C	−	+	−	+	D	D
25°C	+	−	+	D	+	D
35°C	+	−	+	−	D	D
Barophilic	−	D	−	D	−	D
Nitrate reduction	−	+	+	+	D	+
Denitrification	−	−	+	−	D	D
Anaerobic growth:						
Fermentative	−	+	−	+	−	D
Respiratory	−	+	+	+	−	+
Hydrolysis of:						
Gelatin	+	+	−	+	+	+
Starch	+	+	−	−	D	D
Utilization of:						
D-glucose, maltose	+	+	nd	+	+	D
Salicin, galacturonate, DL-glycerate	+	−	nd	−	−	−
PUFA	−	DHA	−	DHA	−	EPA
Mol% G + C of DNA	44–47	35–46	54	40–42	36–48	38–54

[a]For symbols, see standard definitions; nd, not determined; PUFA, polyunsaturated fatty acid; EPA, eicosapentaenoic acid (20:5 ω3c); DHA, docosahexaenoic acid (C$_{20:6\ \omega3c}$).

Key to Some Genera of the Family Alteromonadaceae

I. Straight to curved rod-shaped cells, motile, that require or are stimulated by NaCl and are oxidase positive. Strictly aerobic; no anaerobic growth with nitrate as the electron acceptor.

   A. Utilizes D-turanose, salicin, D-glucuronate, and/or DL-glycerate.

Genus Alteromonas

   B. Does not utilize D-turanose, salicin, D-galacturonate, and/or DL-glycerate.

Genus Pseudoalteromonas

II. Straight to curved rod-shaped cells, motile, that require or are stimulated by NaCl and are oxidase positive. Anaerobic growth using nitrate as an electron acceptor.

   A. Gelatinase positive.

     1. Dissimilatory iron-reduction positive and/or form eicosapentaenoic acid (EPA); most species amylase negative.

Genus Shewanella

     2. Dissimilatory iron-reduction negative; form docosahexaenoic acid (DHA).

       a. Amylase positive.

Genus Colwellia

       b. Amylase negative.

Moritella (Vibrio marinus)

   B. Gelatinase negative.

Genus Ferrimonas

Genus I. **Alteromonas** Baumann, Baumann, Mandel and Allen 1972, 418, emend. Gauthier, Gauthier and Christen 1995a, 760

JOHN P. BOWMAN AND THOMAS A. MCMEEKIN

Al.te.ro.mo' nas. L. adj. *alter* another; Gr. n. *monas* a unit, monad; M.L. fem. n. *Alteromonas* another monad.

Straight rods, approximately 1 × 2-3 μm, occurring singly or in pairs. Do not form spores or cysts. Do not accumulate poly-β-hydroxybutyrate. **Gram negative. Motile**, propelled by a single polar unsheathed flagellum. Colonies are nonpigmented. **Chemoheterotrophic. Aerobic, having a strictly respiratory type of metabolism** with oxygen as the terminal electron acceptor. Catalase and oxidase positive. **Nitrate is not reduced, nor can it be used as an alternative electron acceptor.** Do not grow by hydrogen autotrophy. A constitutive arginine dihydrolase system is absent. A range of carbon substrates can be used, including carbohydrates, alcohols, organic acids, and amino acids. Ammonia serves as a nitrogen source. Growth factors are not required. **Require Na$^+$ ions for growth.** Growth occurs between 10 and 40°C; **no growth at 4°C** or 45°C. Major quinone is ubiquinone-8 (Q-8). Major fatty acids are C$_{16:0}$, C$_{16:1\ \omega7c}$, C$_{17:1\ \omega8c}$, and C$_{18:1\ \omega7c}$. Isolated from seawater. Member of the order *Alteromonadales*, family *Alteromonadaceae*, class *Gammaproteobacteria*.

The mol% G + C of the DNA is: 44–48 (T_m).

Type species: **Alteromonas macleodii** Baumann, Baumann, Mandel and Allen 1972, 418.

FURTHER DESCRIPTIVE INFORMATION

Alteromonas macleodii is the only validly published species included in the genus at present.

When growing on various complex or defined seawater or NaCl-containing media such as Difco marine 2216 agar (ZoBell, 1941), basal medium agar[1] (BMA), or yeast extract agar[2] (YEA) (Baumann et al., 1972), the colonies possess an off-white color, smooth texture, circular convex shape, entire edge, and creamy consistency.

All strains grow best on media containing a seawater salts base and fail to grow if Na$^+$ ions are absent or if Na$^+$ ions are completely replaced by K$^+$ ions. *A. macleodii* is relatively halotolerant and able to grow fairly well on media containing 10% NaCl (Ortigosa et al., 1994). Strains are also characteristically mesophilic, growing between 10 and 40°C, with an optimal temperature of 30–35°C. *A. macleodii* and the closely related "*Alteromonas infernus*" are chemoheterotrophs able to utilize a wide range of carbohydrates for growth (Baumann et al., 1984a; Raguénès et al., 1997b). In addition, *A. macleodii* can utilize a range of organic acids, alcohols, and amino acids, but is unable to grow on aromatic compounds (Table BXII.γ.136). *A. macleodii* produces several extracellular enzymes and is able to hydrolyze gelatin, lipids, and starch. A few strains can also attack alginate, but are unable to degrade agar, chitin, or cellulose.

A. macleodii possesses an inducible Entner–Doudoroff pathway (Baumann and Baumann, 1973), which is used for catabolism of D-fructose and D-glucose. Activity for aspartokinase, which is involved in the initial stages of the synthesis of the aspartate family of amino acids, is absent (Baumann and Baumann, 1974).

A. macleodii is not susceptible to predation by *Bdellovibrio* (strains of which were initially isolated by using *Vibrio* species as hosts) (Taylor et al., 1974). *Alteromonas*-like strains have been found to form proteinaceous substances, referred to as marinostatins, which are potent inhibitors of a range of proteases, including subtilisin, papain, ficin, and α-chymotrypsin (Imada et al., 1985a, b, 1986). Several *Alteromonas* strains have been reported to form antibiotic-like substances; however, it is likely that these strains actually belong to the genus *Pseudoalteromonas* (see section on the genus *Pseudoalteromonas*).

The mol% G + C values for the DNA of *A. macleodii* range from 44.9–46.4 (T_m). The mol% G + C contents of "*A. macleodii* subsp. *fijiensis*" (Raguénès et al., 1996) and "*A. infernus*" (Raguénès et al., 1997b) have slightly higher values, of 47–48 (T_m).

Ubiquinone-8 (Q-8) is the major quinone (97% of total), with traces of Q-9 (2%) and Q-10 (1%) present (Akagawa-Matsushita et al., 1992b). The major whole-cell fatty acids of *A. macleodii* (determined only for strain ATCC 27126) include $C_{15:0}$ (3%), $C_{15:1\ \omega 8c}$ (2%), $C_{16:1\ \omega 7c}$ (37%), $C_{16:0}$ (24%), $C_{17:1\ \omega 8c}$ (7%), $C_{17:0}$ (5%), and $C_{18:1\ \omega 7}$ (9%) (Svetashev et al., 1995). Analysis using neural networks is not able to distinguish the fatty acid pattern of *A. macleodii* from those of some other members of the genus *Pseudoalteromonas* (Bertone et al., 1996). Hydroxy fatty acids comprise only a small amount (~1%) of the whole cell fatty acids. The major polyamines in *A. macleodii* include putrescine, 2-hy-

droxyputrescine, cadaverine, 5-hydroxyspermidine, and spermidine (Hamana, 1997). *Alteromonas* species form acidic polysaccharides. For example, "*A. macleodii* subsp. *fijiensis*" produces a uronic acid-rich polysaccharide with a viscosity similar to that of xanthan gum (Raguénès et al., 1996). "*A. infernus*" secretes two unusual polysaccharides, the first of which is a water-soluble exopolymer made up of monomers of glucose, galactose, galacturonate, and glucuronate and an insoluble gel-forming exopolymer (Raguénès et al., 1997b). Several other *Alteromonas* strains produce a range of novel acidic exopolysaccharides (Ivanova et al., 1994; Vincent et al., 1994; Nazarenko et al., 1993; Samain et al., 1997).

Little is known about the specific ecology of *Alteromonas*. Stud-

TABLE BXII.γ.136. Characteristics of *Alteromonas macleodii*[a]

Characteristics	A. macleodii	ATCC 27126^T
	Reaction or result for	
Accumulation of poly-β-hydroxybutyrate	−	−
Catalase, oxidase	+	+
Hydrogen autotrophy	−	−
Arginine dihydrolase	−	−
Nitrate reduction	−	−
Requirement for Na$^+$ ions for growth	+	+
Growth at:		
4°C	−	−
30°C	+	+
40°C	d	+
45°C	−	−
Lipase, gelatinase, amylase	+	+
Alginase	(−)	+
Chitinase	−	−
Utilization of:[b]		
D-Fructose, D-galactose, D-glucose, cellobiose, lactose, maltose, melibiose, sucrose, trehalose, salicin, glycerol, acetate, propionate, butyrate, isobutyrate, valerate, caproate, caprylate, pelargonate, caprate, pyruvate, DL-glycerate, L-alanine	+	+
D-Gluconate, D-galacturonate, ethanol	(+)	+
L-Tyrosine	(+)	−
N-Acetylglucosamine, D-mannitol, propanol, heptanoate, DL-lactate, glycine	d	+
Propane-1,2-diol, isovalerate, DL-3-hydroxybutyrate, L-arginine, L-aspartate, L-glutamate, L-isoleucine, L-leucine, L-serine, L-valine	d	−
D-Xylose	(−)	+
Mol% G + C of DNA (T_m)	45–46	46

[a]Symbols: +, positive for 90% or more strains; (+), positive for 75–89% of strains; d, positive for 26–74% of strains; (−), positive for 11–25% of strains; −, positive for 0–10% of strains.

[b]The following substrates are not utilized: D-ribose, D-arabinose, L-arabinose, D-fucose, L-rhamnose, D-mannose, inulin, *i*-erythritol, D-sorbitol, *meso*-inositol, adonitol, ethane-1,2-diol, butane-2,3-diol, methanol, isopropanol, isobutanol, saccharate, mucate, D-glucuronate, formate, oxalate, malonate, succinate, adipate, pimelate, suberate, azelate, sebacate, DL-malate, D(−)-tartrate, L(+)-tartrate, *meso*-tartrate, 2-oxoglutarate, fumarate, citrate, aconitate, maleate, levulinate, citraconate, itaconate, mesaconate, D-mandelate, L-mandelate, benzoyl formate, benzoate, *o*-hydroxybenzoate, *m*-hydroxybenzoate, *p*-hydroxybenzoate, phenylacetate, quinate, D-alanine, β-alanine, L-citrulline, L-histidine, L-lysine, L-norleucine, L-ornithine, L-phenylalanine, L-proline, L-threonine, D-tryptophan, L-tryptophan, DL-kynurenine, kynurenate, anthranilate, *m*-aminobenzoate, *p*-aminobenzoate, γ-aminobutyrate, δ-aminovalerate, hippurate, methylamine, ethanolamine, benzylamine, putrescine, spermine, histamine, tryptamine, butylamine, 2-amylamine, pentylamine, betaine, sarcosine, creatine, pantothenate, acetamide, nicotinate, nicotinamide, trigollenine, allantoin, adenine, guanine, cytosine, thymine, uracil, and *n*-hexadecane.

---

1. BMA consists of (per liter): Tris-HCl, 6.1 g (to give a pH of 7.5); NH$_4$Cl, 1.0 g; K$_2$HPO$_4$·3H$_2$O, 0.075 g; FeSO$_4$·7H$_2$O, 0.028 g; artificial seawater (ASW), 500 ml; distilled water, 500 ml; and agar, 20 g. ASW (MacLeod, 1968) consists of (g/l distilled water): NaCl, 23.4; MgSO$_4$·7H$_2$O, 24.6; KCl, 1.5; and CaCl$_2$·H$_2$O, 2.9.

2. YEA medium consists of BMA supplemented with 0.5% (w/v) yeast extract.

ies suggest the species is a common inhabitant of waters of the Mediterranean Sea (Ortigosa et al., 1994) and is probably common in subtropical and tropical waters, owing to its mesophilic temperature requirement. Many *Alteromonas* strains have been isolated from tropical oceanic regions. The type strain of *A. macleodii* was isolated from seawater in the Hawaiian archipelago area (Baumann et al., 1972). "*A. infernus*" and "*A. macleodii* subsp. *fijiensis*" were isolated from a field of hydrothermal vents populated by dense stands of the tubeworm *Rifta pachyptila* where the water temperature was 9–18°C (Raguénès et al., 1996, 1997b).

ENRICHMENT AND ISOLATION PROCEDURES

Alteromonas macleodii strains can be isolated directly from seawater. After filtering seawater samples through membrane filters having a pore diameter of 0.22 or 0.45 μm, the filter is placed on BMA agar containing 0.2% (w/v) lactose. After several days of incubation at approximately 25°C, colonies are transferred and purified on the same medium (Gauthier and Breittmayer, 1992). "*Alteromonas infernus*" can be isolated from hydrothermal vent fluid by enrichment in marine 2216 broth buffered with 50 mM MOPS buffer at pH 7.5, with incubation at 25°C for 2 d. Isolation from the enrichment broths then proceeds by serially diluting the enrichment broth onto marine 2216 agar (Raguénès et al., 1997b).

MAINTENANCE PROCEDURES

Alteromonas strains can be maintained at approximately 15°C on marine agar slants and transferred on a monthly basis. Storage at 15°C is preferred, because survival at 4°C seems to be poor. Cryopreservation and storage in liquid nitrogen using glycerol as a cryoprotectant or lyophilization is effective for long-term storage.

DIFFERENTIATION OF THE GENUS *ALTEROMONAS* FROM OTHER GENERA

Differentiation of the *Alteromonas* species from other marine genera appears to be problematic. The most phenotypically similar genus, *Pseudoalteromonas*, is chemotaxomically and metabolically very similar. Both genera possess similar fatty acid profiles (Svetashev et al., 1995), quinones (Akagawa-Matsushita et al., 1992b), and polyamine profiles, although *Pseudoalteromonas* species lack 2-hydroxyputrescine and 5-hydroxyspermidine (Hamana, 1997). The major traits distinguishing these genera are growth at 4°C and utilization of a variety of carbon compounds (see Key to Some Genera of the Family *Alteromonadaceae* and Table BXII.γ.135 of the chapter describing the family). Beyond these limited phenotypic criteria, only 16S rRNA sequencing can provide a definitive identification. *Alteromonas* differs from the marine genera *Marinobacterium*, *Microbulbifer*, *Marinobacter*, and *Halomonas* primarily by possessing lower mol% G + C values for the DNA. In addition, *Alteromonas* can be distinguished from the genera *Oceanospirillum* and *Marinomonas* based on cellular morphology.

TAXONOMIC COMMENTS

In the last decade or so, following rRNA–DNA hybridization and 5S and 16S rRNA sequence analyses, the genus *Alteromonas* has undergone a reclassification of most of its species and has been reduced to a single validly described species, *A. macleodii*. A 1983 study using rRNA–DNA hybridization was the first to indicate that *A. macleodii* should be classified separately from other species that were originally placed in *Alteromonas* (van Landschoot and De Ley, 1983). The 23S rRNA–DNA hybridization values range from 68.5–71.5 $T_{m(e)}$°C, which indicate *A. macleodii* is genotypically distinct at the genus level. Subsequent phylogenetic analysis of 16S rRNA sequences has also demonstrated that *A. macleodii* is distinct from the other *Alteromonas* species (Kita-Tsukamoto et al., 1993; Gauthier et al., 1995a). This resulted in nomenclatural restructuring of the genus *Alteromonas*, with *Alteromonas macleodii* being retained as the sole species of the genus because it was the type species of the genus (as stipulated by rule 39b of the Nomenclatural Code (Lapage et al., 1992). Former *Alteromonas* species having a predominantly straight rod morphology, including pigmented and nonpigmented strains, have been classified in the genus *Pseudoalteromonas* (Gauthier et al., 1995a). The bioluminescent species *Alteromonas hanedai* (Jensen et al., 1980) and the oyster species *Alteromonas colwelliana* (Weiner et al., 1988) were renamed *Shewanella hanedai* (MacDonell and Colwell, 1985) and *Shewanella colwelliana* (Coyne et al., 1989), respectively. *Alteromonas vaga* and *Alteromonas communis* have been found to form a distinct branch in the vicinity of the genus *Pseudomonas* on the basis of rRNA–DNA homology (van Landschoot and De Ley, 1983) and 16S rRNA sequences (Gauthier et al., 1992; Kita-Tsukamoto et al., 1993) and have been renamed *Marinomonas vaga* and *Marinomonas communis* (van Landschoot and De Ley, 1983). Methanol and methylamine-utilizing marine strains initially described as "*Alteromonas thalassomethanica*" (Yamamoto et al., 1980) have been reclassified as *Methylophaga thalassica* (Janvier et al., 1985; Urakami and Komagata, 1987).

Overall, the genus *Alteromonas* is most closely related to the genus *Pseudoalteromonas* and to various pigmented psychrophilic isolates from Antarctic sea ice (Bowman et al., 1997a). DNA–DNA hybridization studies indicate *A. macleodii* shares no significant levels of hybridization with phenotypically similar, nonpigmented species of the genus *Pseudoalteromonas* (Akagawa-Matsushita et al., 1993).

Isolates forming unusual polysaccharides ("*A. macleodii* subsp. *fijiensis*" and "*A. infernus*") have been shown to be closely related to *A. macleodii* on the basis of 16S rRNA sequences, forming a tight cluster (sequence divergence ~2%) in the *Gammaproteobacteria* (Raguénès et al., 1996, 1997b). A significant level of DNA hybridization (48%) occurs between "*A. infernus*" and the type strain of *A. macleodii* (ATCC 27126), establishing that "*A. infernus*" is a distinct but closely related genospecies (Raguénès et al., 1997b). The names of these organisms have not yet been validated.

List of species of the genus Alteromonas

1. **Alteromonas macleodii** Baumann, Baumann, Mandel and Allen 1972, 418[AL]

   mac.leod' i.i. M.L. gen. n. *macleodii* of MacLeod, named after R.A. MacLeod, a Canadian microbiologist.

   Cells are straight rods, about 1 × 2-3 μm. Motile by a single polar flagellum.

   Colonies on marine agar are off-white, smooth, circular, convex, with an entire edge, and butyrous in consistency.

   Physiological and nutritional traits are presented in Table BXII.γ.136.

   Isolated from seawater.

   The mol% G + C of the DNA is: 45–46 (T_m).BNL-1

Type strain: 107, ATCC 27126, DSM 6062, IAM 12920, NCMB 1963.

GenBank accession number (16S rRNA): X82145.

Species Incertae Sedis

1. **"Alteromonas infernus"** Raguénès, Peres, Ruimy, Pignet, Christen, Loaec, Rougeaux, Barbier and Guezennec 1997b, 428.

    in.fer' nus. Gr. n. *infernus* from infernal regions, deep-sea hydrothermal vents.

    Cells are straight rods, 0.6-.8 × 1.4-2.0 µm. Encapsulated and motile by a single polar flagellum. Poly-β-hydroxybutyrate is accumulated. Strictly aerobic chemoheterotroph. Oxidase and catalase positive. Does not denitrify. Grows between 20–35°C, with no growth at 4°C or 45°C. Na⁺ ions are required for growth. Does not require growth factors. Hydrolyzes starch, esculin, and gelatin. Positive for β-galactosidase, alkaline phosphatase, C_8 esterase, and leucine arylamidase. Acid phosphatase and C_4 esterase activity is weak.

    Utilizes the following carbon substrates: D-xylose, D-fructose, D-galactose, D-gentiobiose, D-glucose, D-turanose, cellobiose, lactose, maltose, melibiose, sucrose, amygdalin, esculin, salicin, D-mannitol, and D-gluconate. The following carbon substrates are not utilized: D-melezitose, trehalose, arbutin, N-acetylglucosamine, and glycogen.

    Isolated from seawater from a deep-sea hydrothermal field at 2000 m depth in a rift system of the Guaymas basin.

    The mol% G + C of the DNA is: 48.1 (T_m).

    Deposited strain: GY785, CNCM I-1628.

2. **"Alteromonas macleodii** *subsp.* **fijiensis"** Raguénès, Pignet, Gauthier, Peres, Rougeaux, Christen, Barbier and Guezennec 1996, 72.

    Phenotypic characteristics are similar to those for *Alteromonas macleodii* ATCC 27126; however, this subspecies can be distinguished from *Alteromonas macleodii* ATCC 27126 in that it is able to utilize D-ribose, α-methyl-D-glucoside, and raffinose, but not D-xylose, mannitol, N-acetylglucosamine or melezitose.

    Isolated from seawater from a deep-sea hydrothermal field near Fiji, Pacific Ocean.

    The mol% G + C of the DNA is: 47.6 (T_m).

    Deposited strain: ST716, CNCM I-1627.

Genus II. **Alishewanella** Fonnesbech Vogel, Venkateswaran, Christensen, Falsen, Christiansen and Gram 2000, 1140[VP]

THE EDITORIAL BOARD

A.li.she.wa.nel' la. L. pron. *alius* the other; M.L. fem. n. *Alishewanella* the other *Shewanella*.

Gram-negative, facultatively anaerobic, nonmotile rods. Electron acceptors are O_2, nitrate, thiosulphate, and TMAO. **Nonfermative.** One polar flagellum. Catalase and oxidase positive. **Requires NaCl.** Hydrolyzes esculin and gelatin. Does not produce arginine dihydrolase, β-galactosidase, urease, indole, or H_2S.

The mol% G + C of the DNA is: 51.

Type species: **Alishewanella fetalis** Fonnesbech Vogel, Venkateswaran, Christensen, Falsen, Christiansen and Gram 2000, 1141.

FURTHER DESCRIPTIVE INFORMATION

Alishewanella fetalis R2422019 was isolated from a human fetus at autopsy and originally thought to be related to *Shewanella*.

Analysis of 16S rDNA sequences showed that strain R2422019 was related to but distinct from members of the genera *Shewanella*, *Aeromonas*, *Vibrio*, *Photobacterium*, *Erwinia*, *Klebsiella*, and *Shigella*.

DIFFERENTIATION OF THE GENUS *ALISHEWANELLA* FROM OTHER GENERA

Fonnesbech Vogel et al. (2000) provide tables of characteristics that distinguish *Alishewanella fetalis* R2422019 from *Shewanella putrefaciens* and *S. algae* (potential human pathogens having phenotypes similar to R2422019) as well as from the genera *Shewanella*, *Aeromonas*, *Vibrio*, *Photobacterium*, *Erwinia*, *Klebsiella*, and *Shigella*.

List of species of the genus Alishewanella

1. **Alishewanella fetalis** Fonnesbech Vogel, Venkateswaran, Christensen, Falsen, Christiansen and Gram 2000, 1141[VP]

    fe.ta' lis. L. adj. *fetalis* pertaining to the fetus, from which the organism was isolated.

    Description as for the genus with the following additional characteristics. Cells 0.5–1.0 × 2.0 µm. Grows in medium containing up to 8% in NaCl.

    The mol% G + C of the DNA is: 51 (HPLC).

    Type strain: R2422019, ATCC BAA-284, CCUG 30811, CIP 106648.

    GenBank accession number (16S rRNA): AF144407.

Genus III. **Colwellia** Deming, Somers, Straube, Swartz and MacDonell 1988b, 328[VP]
(Effective publication: Deming, Somers, Straube, Swartz and MacDonell 1988a, 159)

JODY W. DEMING AND KAREN JUNGE

Col.wel' li.a. M.L. ending *-ia*; M.L. fem. n. *Colwellia* of Colwell, named in honor of the American microbiologist Professor Rita R. Colwell.

Curved or straight rods 0.4–1.0 × 1.5–5.0 µm. Generally motile by polar flagella. Gram negative. Asporogenous. Colonies are off-white, convex, and mucoid, except for the type strain, *Colwellia* *psychrerythraea*, which produces red pigment. **Facultatively anaerobic. Chemoorganotrophic**. Able to grow by fermentation of a suitable carbohydrate as indicated by acid (but no gas) production. Generally able to utilize amino acids and organic acids, but not many of the sugars or alcohols. Capable of **nitrate reduction** but not iron or trimethylamine N-oxide reduction or indole or H_2S production. Generally **chitinolytic**, able to degrade starch, and positive for oxidase, catalase, and alkaline phosphatase. Stimulated by, but not dependent on, growth factors in yeast extract and vitamin solutions. Require **sea salts**, with no growth occurring below 25% sea-salt concentration. **Psychrophilic**. Able to grow on solid media in the temperature range of 0° to <20°C (optimum usually <15°C), with slightly higher maximal and lower minimal temperatures in liquid media. **Sometimes barophilic**, requiring high hydrostatic pressure for growth. Produce docosahexaenoic acid ($C_{22:6 \, \omega 3}$ or DHA) and other **polyunsaturated fatty acids** to help maintain homeoviscosity of cellular membranes under low temperature and high pressure. Cultured strains derived from **permanently cold**, and often deep, marine habitats.

The mol% G + C of the DNA is: 35–46.

Type species: **Colwellia psychrerythraea** (ex D'Aoust and Kushner 1972) Deming, Somers, Straube, Swartz and MacDonell 1988b, 328 (Effective publication: Deming, Somers, Straube, Swartz and MacDonell 1988a, 159 (*"Vibrio psychroerythrus"* D'Aoust and Kushner 1972, 342.)

FURTHER DESCRIPTIVE INFORMATION

Cell morphology varies from spherical to curved to straight rod-shaped cells with occasional short filaments. The fine structure appears typical of asporogenous Gram-negative bacteria, with no obviously unique structural features, even in the obligate barophiles (see electron micrographs in Deming et al., 1988a, and Deming and Baross, 2000). *C. rossensis*, however, is unique among *Colwellia* species for its ability to form gas vesicles (Gosink and Staley, 1995). Potentially unique cell-membrane composition is implied by the documented production of polyunsaturated fatty acids.

The most distinctive features of the genus *Colwellia* derive from its habitation of permanently cold marine environments. All characterized species are cold adapted, if not strictly psychrophilic, as defined by Morita (1975) (T_{opt} for growth of 15°C or less, T_{max} of ≤20°C). *Colwellia* (along with *Moritella*) thus appears to be unusual among all genera for encompassing exclusively cold-adapted organisms. Cardinal growth temperatures of individual species tend to be several degrees higher in liquid media than in solid media, causing at least one of the type species, *C. psychrotropica*, to be considered psychrotolerant (T_{max} of 25°C in liquid) (Bowman et al., 1998a). *C. psychrerythraea* strain 34H grows at −5°C in liquid media (Huston, 2003) and swims at subzero temperatures to −10°C (Junge et al., 2003). Two of the known strains, *C. hadaliensis* (Deming et al., 1988a) and *Colwellia* strain MT41 (DeLong et al., 1997), are not only strictly psychrophilic but also obligately barophilic, derived from the deepest portions of the cold Atlantic and Pacific Oceans, respectively. The production of polyunsaturated fatty acids (PUFAs), a feature common to members of the genus so analyzed, is attributed to the need to maintain membrane fluidity for metabolism and growth at the constantly cold temperatures (<4°C) and often high pressures of their native habitats, e.g., polar waters, sea ice, and hadal depths in the cold ocean. Their common growth requirement for sea salts and preference for full-strength sea salt concentration also reflect their native marine habitats.

The type species, *C. psychrerythraea*, is unique in the genus for its production of red pigment during colony formation. All other species produce chalk-white, off-white, or colorless colonies, typically of convex, circular, raised, and mucoid morphology, when grown on solid media. The obligate barophiles do not grow on standard agar plates at atmospheric pressure, but rather form spherical off-white colonies when grown within a solid medium under high hydrostatic pressure. *C. psychrotropica* produces a diffusible brown pigment during L-tyrosine hydrolysis in liquid media.

All *Colwellia* species are chemoorganotrophic facultative anaerobes able to grow by fermentation of a suitable carbohydrate, as indicated by acid production in Leifson oxidation–fermentation medium. In general, no gas is detected during fermentation. Carbohydrate oxidation and fermentation vary between species, as does the ability to hydrolyze various substrates (Tables BXII.γ.137 and BXII.γ.138). Utilization of amino acids and organic acids is favored over sugars and alcohols. Most species are able to utilize L-glutamate, L-proline, acetate, butyrate, fumarate, pyruvate, succinate, N-acetylglucosamine, oxalacetate, DL-lactate, valerate, caproate, and γ-aminobutyrate, but not lactose, maltose, D-melibiose, L-raffinose, sucrose, trehalose, D-xylose, D-adonitol, D-arabitol, m-inositol, D-mannitol, D-sorbitol, L-threonine, L-valine, D-gluconate, D-glucuronate, isovalerate, nonanoate, pimelate, saccharate, α-glycerophosphate, or putrescine. Individual species are distinguished by specific metabolic or hydrolytic traits: e.g., only *C. demingiae* utilizes L-serine; only *C. hornerae* produces acid from L-rhamnose; only *C. rossensis* produces acid from D-galactose, L-arabinose, D-fructose, and D-gluconate; and only *C. psychrotropica* degrades uric acid. Some strains produce extracellular hydrolytic enzymes with unique properties; e.g., *Colwellia* sp. strain 34H produces the most cold-adapted extracellular proteases yet discovered (Huston et al., 2000).

Only *C. hornerae* and *C. maris* are susceptible to vibriostatic agent O/129. Antigenic structure and other antibiotic or drug sensitivities, along with mutants and plasmids, remain to be documented or developed, although *Colwellia*-specific phages have been reported (Borriss et al., 2003). None of the *Colwellia* species are believed to be pathogenic, in keeping with their growth requirement for cold temperatures.

In direct phylogenetic analyses of environmental samples, 16S rRNA sequences specific to *Colwellia* have been detected in cold marine samples from the deep sea (e.g., Moyer et al., 1995), in keeping with what is known about cultured species, but also in coastal waters at temperate latitudes (DeLong et al., 1997). Absent any culturing information, however, the temperature adaptations of such organisms remain unknown.

ENRICHMENT AND ISOLATION PROCEDURES

Members of the genus *Colwellia* are isolated from cold marine samples, using liquid enrichment media containing complex carbon sources, such as peptone, and full-strength sea-salt concentration. The most common medium used is 2216 marine broth (Difco). An initial enrichment period of days to weeks in liquid media, before transfer and cultivation on agar plates, appears to favor the recovery of these psychrophilic strains, although *C. maris* was first enriched on agar medium (Takada et al., 1979). Initial colony formation may require up to 2 months of incubation time at cold temperature. Attention to low-temperature (0° to <4°C) storage of the original sample and of subsequent enrichment incubations is important.

Sea-ice samples are first melted in seawater or brine solutions prior to enrichment procedures to avoid hypotonic shock to salt-requiring strains. Deep-sea samples can be recovered at atmo-

TABLE BXII.γ.137. Characteristics differentiating the species of the genus *Colwellia*[a]

Characteristic	1. *C. psychrerythraea*	2. *C. demingiae*	3. *C. hadaliensis*	4. *C. hornerae*	5. *C. maris*	6. *C. psychrotropica*	7. *C. rossensis*
Growth at 25°C	−	−	−	−	−	+	−
Barophilic growth	−		+				
Prodigiosin-like pigments	d[b]	−	−	−	−	−	−
Motility	+	+	+	+	+	+	−
Gas vesicles	−	−	−	−	−	−	+
Susceptible to vibriostatic agent O/129	−	−		+	+	−	−
Hydrolysis of:							
Chitin	+	+	+	−	−	+	+
Starch	+	+		+	−	−	+
Uric acid	−	−		−		+	−
Urea	+	−		−		+	+
Tween 80	+	−		+	+	+	NG
Gelatin	+	−		−	+	−	−
Casein	+	+		+	−	+	−
DNA	−	−		−	+	−	−
Production of acid from:							
D-Glucose	+	−	+	−	+	−	+
Maltose	+	−		−	−	−	−
Glycerol	−	−		+	−	−	+
Mol% G + C of DNA	35–40	37	46	39	39	42	38

[a]Symbols: see standard definitions; blank space, no data available; NG, no growth occurred on test medium.

[b]Type strain is positive.

spheric pressure, but barophilic *Colwellia* strains are subsequently enriched and isolated under hydrostatic pressures simulating those encountered *in situ*. Methodologies for cultivation of bacteria, in either liquid or solid media, under high hydrostatic pressure are described in detail by Deming (2002).

MAINTENANCE PROCEDURES

Colwellia spp. are routinely maintained by periodic subcultivation on 2216 marine agar or broth (Difco) at low temperatures: <4°C for strict psychrophiles, but also up to 10°C. Obligate barophiles must be subcultivated at low temperatures under high hydrostatic pressure (Deming, 2002). Cryopreservation is the preferred form of long-term storage, freezing pelleted cultures in glycerol (15% glycerol is effective at −20°C to −80°C; 30% glycerol at −80°C) or using lyophilization. Cryopreservation is advised even for short-term storage of the type strain, *C. psychrerythraea*, since it is not recovered reliably from plates, and for obligate barophiles, which have been problematic to store (the availability of recoverable *C. hadaliensis* is currently uncertain). All *Colwellia* cultures should be shipped frozen and packed on ice to maximize recoverability.

PROCEDURES FOR TESTING SPECIAL CHARACTERS

Special characteristics of *Colwellia* include growth requirements at low temperatures and sometimes at high pressures and the production of PUFA. Standard incubation devices, from refrigerators to water baths, are used to achieve and maintain low temperatures, but special equipment is required for testing growth and other parameters under high pressure, as described by Deming (2002). Many standard tests for physiological and nutritional features have not yet been developed for high-pressure applications, making molecular assessments essential to determining phylogenetic status. Analysis of 16S rRNA sequences proceeds normally for *Colwellia* spp., if attention is paid to low-temperature requirements in first preparing the cultures. Whole-cell fatty acid analyses rely on standard GC and GC-MS procedures (Bowman et al., 1998a).

DIFFERENTIATION OF THE GENUS *COLWELLIA* FROM OTHER GENERA

The genus *Colwellia* is differentiated from other closely related genera by the combination of being cold adapted, facultatively anaerobic, and able to produce chitinase. Nearest neighbors at the genus level, *Idiomarina*, *Alteromonas*, and *Pseudoalteromonas*, are not usually adapted to growth at low temperatures, are strictly aerobic, and do not typically produce chitinase (some species of *Alteromonas* and *Pseudoalteromonas* do produce the enzyme). The production of PUFA also appears to be an important distinguishing trait for *Colwellia*. Use of 16S rRNA sequence data, however, remains the essential means to distinguish *Colwellia* definitively from other genera. The main features for distinguishing one *Colwellia* species from another are provided in Table BXII.γ.138.

TAXONOMIC COMMENTS

A phylogenetic treatment based on 16S rRNA sequence data from organisms in culture indicates that the closest relatives to members of the genus *Colwellia* are species of the newly described genus *Idiomarina* (Ivanova et al., 2000b), as represented by the deep-sea isolate *I. abyssalis* in Fig. BXII.γ.152. Sequence similarities between *I. abyssalis* and *Colwellia* spp. range from 85.2% (*Colwellia* sp. strain MT41) to 89.6% (type species *C. psychrerythraea*). *Colwellia* also has close phylogenetic affinities, as evidenced by similar degrees of sequence similarity, to members of the genera *Alteromonas* and *Pseudoalteromonas*, as described previously by Bowman et al. (1998a) and depicted in Fig. BXII.γ.152.

Analysis of 16S rRNA sequences for *Colwellia* species in culture (Fig. BXII.γ.152) indicates sequence similarities that range from 92.4% to 98.7%. The low end of the range derives from a comparison of the deep-sea obligate barophile, *Colwellia* sp. strain MT41, with sea-ice isolate *C. demingiae*. Strain MT41 in general has the lowest sequence similarity values to the other described *Colwellia* species (92.4–94.6%), with the exception of *C. maris* (96.3%). Only 5S (not 16S) rRNA sequence data are available for the other obligate barophile, *C. hadaliensis*, precluding comparisons of the two with each other or with other species. A

TABLE BXII.γ.138. Other characteristics of the species of the genus *Colwellia*[a]

Characteristic	1. *C. psychrerythraea*	2. *C. demingiae*	3. *C. hadaliensis*	4. *C. hornerae*	5. *C. maris*	6. *C. psychrotropica*	7. *C. rossensis*
Temperature range for growth in liquid media, °C	0–19	0–18	2–10	0–23	0–22	0–25	0–15
Single polar flagellum	+		+		+		
Gas vesicles	−	−	−	−	−	−	+
Toleration of 200% sea salts	d[b]	−		+		+	−
Growth on 0–25% sea salts	−	−		−	−[c]	−	−
Growth on 400% sea salts	−	−		−		−	−
Growth on NaCl	2.75%				3–6.5%		
Susceptibility to vibriostatic agent O/129	d[b]	−		+	+	−	−
Growth factor requirement (yeast extract)	−	−		+		−	−
Catalase, nitrate reduction	+	+		+	+	+	+
Oxidase	+	+	+	+	+	+	+
Alkaline phosphatase	+	+	+	+		+	+
Acid phosphatase, trypsin			+				
Fe(III) reduction, trimethylamine *N*-oxide reduction	−	−		−		−	−
Indole production	−	−		−	−	−	−
H₂S production	−	−		−		−	−
Hydrolysis of:							
Alginic acid					−		
Casein	+	+		+	−	+	−
Chitin	+[e]	+[e]	+	−	−	+[e]	+[e]
Dextran	−	−		−		−	−
DNA	−	−		−	+	−	−
Esculin	+	+		+		−	NG
Gelatin	d[b]	−		−	+	−	−
ONPG	−	−		−		−	
Starch	+	+		+	−	+	+
Tween 80	+	−		+	+	+	NG
Tyrosine	d[d]	−		−		+	−
Urea	d[b]	−		−		+	+
Uric acid	−	−		−		+	−
Arginine dihydrolase, lysine decarboxylase, ornithine decarboxylase	−	−		−		−	−
Production of acid from:							
N-Acetylglucosamine	+[e]	+[e]		−		+[e]	+[e]
L-Arabinose, D-fructose, D-mannose, melibiose, sucrose	−	−		−	−	−	−
D-Arabinose, inositol, mannitol, sorbitol, trehalose, xylose, lactose	−	−		−		−	−
Cellobiose	d[b]	−		−		−	−
D-Galactose	−	−		−		−	+
D-Glucose	+[e]	−	+	−	−	−	+[e]
Maltose	+[e]	−		−	−	−	−
L-Rhamnose	−	−		+[e]		−	−
Glycerol	−	−		+		−	+
Utilization of:							
L-Proline, acetate, butyrate, fumarate, L-glutamate, succinate, pyruvate	+	+		+		+	+
Oxalacetate	+	+		+		+	−
Valerate, caproate, γ-aminobutyrate	−	+		+		+	+
N-Acetylglucosamine	+	+		−		+	+
3-Hydroxybutyrate	+	−		−		+	+
Citrate	−	+		+		−	+
Glycogen	+	d[d]		−		−	+
DL-Lactate	d[b]	+		+		+	−
D-Glucose	d[b]	−		−	+	−	+
Propionate	d[d]	+		+		−	−
L-Asparagine	−	+		−		+	+
L-Alanine, L-aspartate	−	d[b]		−		+	+
Hydroxy-L-proline	−	d[b]		+		−	−
L-Phenylalanine	−	d[b]		−		−	−
L-Serine	−	+		−		−	
Azelate	−	d[b]		+		−	−
Caprylate	−	d[b]		−		−	
Glutarate	−	+		+		−	−
Heptanoate	−	db		+		−	−
Isobutyrate	−	db		−		+	−
Malonate	−	db		−		−	+
D-Mannose, D-raffinose					−		
D-Melibiose							
Glycerol	−	−		+		−	+
L-Malate	−	−		−		+	+

(*continued*)

TABLE BXII.γ.138. (*cont.*)

Characteristic	1. *C. psychrerythraea*	2. *C. demingiae*	3. *C. hadaliensis*	4. *C. hornerae*	5. *C. maris*	6. *C. psychrotropica*	7. *C. rossensis*
2-Oxoglutarate	−	−		−		+	−
L-Arabinose, D-fructose, D-gluconate	−	−		−	−	−	+
L-Raffinose, trehalose, D-adonitol, D-arabitol, *m*-inositol, D-mannitol, D-sorbitol, L-threonine, L-valine, aconitate, adipate, D-gluconate, D-glucuronate, nonanoate, saccharate, α-glycerophosphate, putrescine, isovalerate, pimelate	−	−		−		−	−
Lactose, maltose, sucrose, D-xylose	−	−		−	−	−	−
Fatty acid composition (% of total):							
$C_{14:0}$	5.1–7.8	7.6–8.0		3		0.8	4.6
$C_{14:1\ \omega7c}$	5.1–7.3	9.1–9.3		2.8		2	2.8
$C_{15:1\ \omega8c}$	0–2.3	1.9–2.6		20.3		4.2	4.1
$C_{15:1\ \omega6c}$	0-0.4	−		1.1		0.1	−
$C_{15:0}$	1.7–11.0	0.9–1.4		14.3		2.7	2.9
$C_{16:0\ iso}$	0–0.2	−		10.3		−	−
$C_{16:1\ \omega9c}$	6.2–8.8	9.5–11.8		2	20	−	1.8
$C_{16:1\ \omega7c}$	31.4–36.3	37.5–37.8		15.4	6	56.8	43.4
$C_{16:1\ \omega9t}$					20		
$C_{16:0}$	26.8–33.2	21.9–23.6		13.5	18	21.9	27.1
$C_{17:1\ \omega8c}$	0–1.3	tr^c		5.6		4.5	0.5
$C_{17:1\ \omega6c}$	0–0.9	−		1.9		0.9	−
$C_{17:0}$	0–1.3	tr		2.5	5	1.5	0.1
$C_{18:1\ \omega11c}$					6		
$C_{18:1\ \omega9c}$	0–1.7	0.2		1.4		0.3	0.8
$C_{18:1\ \omega7c}$	0.3–2.1	1.3–1.4		−		1.9	4.2
$C_{18:0}$	0.1–2.4	0.2–0.5		2		0.4	tr
$C_{20:5\ \omega3}$	0–1.5	−		−		0.1	tr
$C_{22:6\ \omega3}$	5.5–8.0	1.7–2.2		2.1		0.7	6
Isoprenoid quinone					ubiquinone-8 (Q-8)		

^{a}Symbols: see standard definitions; blank space, no data available; NG, no growth occurred on test medium; tr, trace fatty acid component making up 0.1% or less of total fatty acid content.

^{b}Reaction differs among strains; type strain is positive.

^{c}No growth occurs without NaCl.

^{d}Reaction differs among strains; type strain is negative.

^{e}Acid (but no gas) was formed from this substrates in Leifson oxidation/fermentation medium.

possible close relationship between phylogeny and obligate barophily within the genus thus remains elusive (DeLong et al., 1997). The high end of the sequence similarity range for *Colwellia* species (98.7%) derives from comparison of the type strain *C. psychrerythraea* with *Colwellia* sp. strain 34H, suggesting the latter is also *C. psychrerythraea*. Confirmation of this suggestion comes from DNA–DNA hybridization tests (Huston, 2003). The availability of the whole genome sequence for strain 34H from The Institute for Genomic Research* thus well represents the genus *Colwellia* and enables a first-order genetic analysis of psychrophily and other traits specific to *Colwellia* or unique to the type species. Existing DNA–DNA hybridization tests show clear separation at the species level, with homologies of 12–35% among the type species (Table BXII.γ.139).

The list of *Colwellia* type species has expanded from the first two, described in 1988, to seven only in the last few years. This expansion reflects renewed ecological and biotechnological interests in cold-adapted organisms and their enzymes (Bowman et al., 1998a; Deming and Baross, 2000, 2002). Characterizations of numerous additional strains and species can be expected in the near future, including *Colwellia* sp. strain 34H and possibly strain MT41. Current culture collections of psychrophilic bacteria likely contain many undescribed *Colwellia* spp. Recent acquisition of 16S rRNA sequence data from our collection reveals at least three species, awaiting further taxonomic characterization (Junge et al., 2002).

ACKNOWLEDGMENTS

We gratefully acknowledge John Bowman for helpful communications and his substantial contributions to the discovery and characterization of *Colwellia* species.

FURTHER READING

Bowman, J.P., J.J. Gosink, S.A. McCammon, T.E. Lewis, D.S. Nichols, P.D. Nichols, J.H. Skerratt, J.T. Staley and T.A. McMeekin. 1998. *Colwellia demingiae* sp. nov., *Colwellia hornerae* sp. nov., *Colwellia rossensis* sp. nov., and *Colwellia psychrotropica* sp. nov.: psychrophilic Antarctic species with the ability to synthesize docosahexaenoic acid (22:6ω3). Int. J. Syst. Bacteriol. *48*: 1171–1180.

Deming, J.W., L.K. Somers, W.L. Staube, D.G. Swartz and M.T. MacDonell. 1988. Isolation of an obligately barophilic bacterium and description of a new genus, *Colwellia* gen. nov. Syst. Appl. Microbiol. *10*: 152–160.

Yumoto, I., K. Kawasaki, H. Iwata, H. Matsuyama and H. Okuyama. 1998. Assignment of *Vibrio* sp. strain ABE-1 to *Colwellia maris* sp. nov., a new psychrophilic bacterium. Int. J. Syst. Bacteriol. *48*: 1357–1362.

Editorial Note: At the time this volume went to press, the website for the Institute for Genomic Research was http://www.tigr.org.

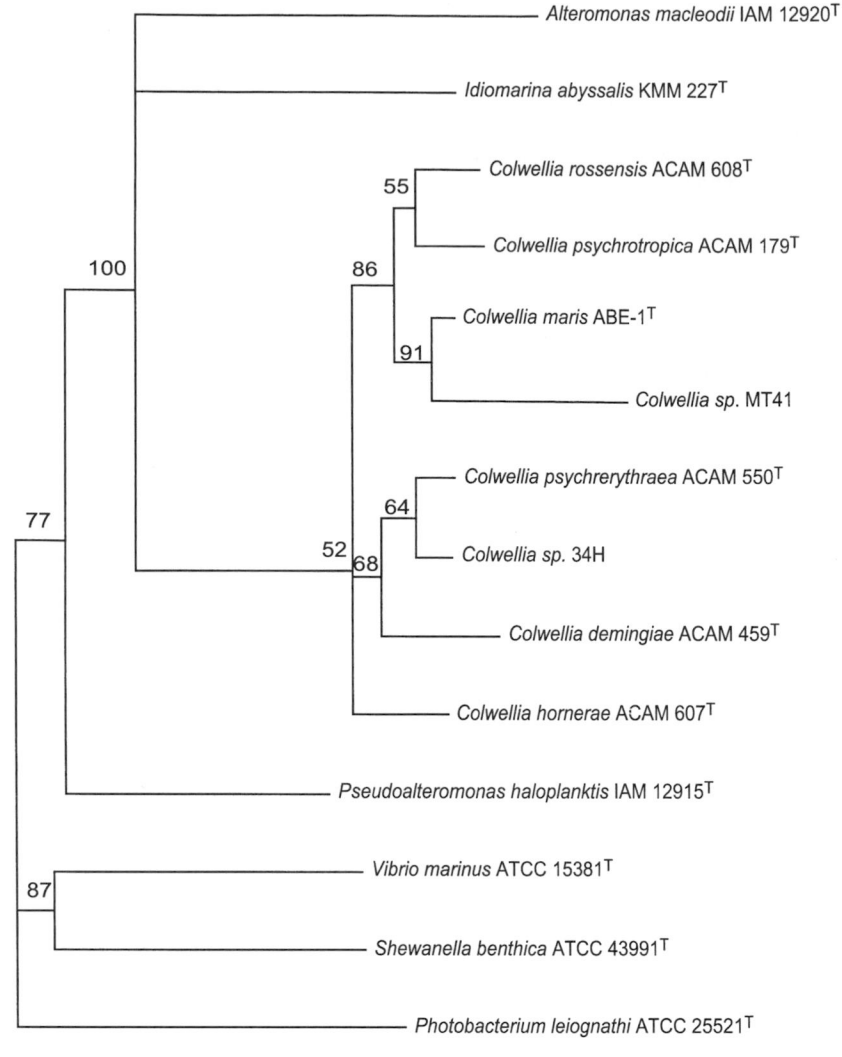

1%

FIGURE BXII.γ.152. Phylogenetic tree based on 16S rRNA sequence comparisons showing the positions of members of the genus *Colwellia* within the *Colwellia* assemblage radiation of the *Gammaproteobacteria*. Phylogenetic analysis was conducted with reference sequences obtained from GenBank. Preliminary alignments were made using the Sequence Align program of the Ribosomal Database Project Web site (RDP) and requesting that common gaps be preserved. Complete sequence alignments were prepared with the sequence alignment editor SeqApp (D.G. Gilbert, 11 July 1994, posting date, Seq-App program, version 1.9a169, http://ftp.bio.indiana.edu/molbio/seqapp) and by manual comparison to secondary structures provided by RDP. For further phylogenetic analysis, aligned sequences corresponding to *E. coli* positions 30–1452 were used. Phylogenetic relationships were inferred using the PAUP program (4.0b8, D.L. Swofford, Smithsonian Institution, 2001). Tree topology was inferred from neighbor-joining analysis determined with the Kimura two-parameter model. *Photobacterium leiognathi* was used as an outgroup. Results from neighbor-joining bootstrap analysis (1000 replicates) are indicated by numbers (as percentages) near each node. Only nodes supported in greater than 50% of the bootstrap replicates are indicated. The scale bar corresponds to 1 fixed mutation per 100 sequence positions. The following 16S rDNA sequences (with their GenBank accession numbers in parentheses) were obtained for the phylogenetic analysis: *Alteromonas macleodii* (X82145), *Colwellia demingiae* (U85845), *Colwellia hornerae* (U85847), *Colwellia maris* (AB002630), *Colwellia psychrerythraea* (AF001375), *Colwellia psychrotropica* (U85846), *Colwellia rossensis* (U14581), *Colwellia* sp. MT41 (U91595), *Colwellia* sp. 34H (AF396670), *Idiomarina abyssalis* (AF05270), *Photobacterium leiognathi* (X74686), *Pseudoalteromonas haloplanktis* (X67024), *Shewanella benthica* (U91594), and *Vibrio marinus* (X82142).

List of species of the genus Colwellia

1. **Colwellia psychrerythraea** (ex D'Aoust and Kushner 1972) Deming, Somers, Straube, Swartz and MacDonell 1988b, 328[VP] (Effective publication: Deming, Somers, Straube, Swartz and MacDonell 1988a, 159 (*"Vibrio psychroerythrus"* D'Aoust and Kushner 1972, 342.)

   psy.chre.ryth' rae.a. Gr. adj. *psychros* cold; L. adj. *erythraeus, -a, -um* red.

TABLE BXII.γ.139. DNA–DNA homologies among the species of the genus *Colwellia*[a]

Percentage DNA–DNA hybridization	1. *C. psychrerythraea*	2. *C. demingiae*	3. *C. hadaliensis*	4. *C. hornerae*	5. *C. maris*	6. *C. psychrotropica*	7. *C. rossensis*
1. *C. psychrerythraea*	100	35		12	22	12	13
2. *C. demingiae*	35	100		21			21
3. *C. hadaliensis*							
4. *C. hornerae*	12	21		100		18	22
5. *C. maris*	22						
6. *C. psychrotropica*	12			18		100	21
7. *C. rossensis*	13	21		22		24	100

[a]Blank space, not determined.

The characteristics are as described for the genus and as listed in Tables BXII.γ.137 and BXII.γ.138, with the following additional information. Cell length, 2.5–3.5 μm; width, 0.5 μm. Gas vesicles not formed. Motile by means of a single, long polar flagellum. Colonies are white at first appearance (7–8 d) becoming red within 10–12 d, opaque, raised, 3–5 mm in diameter, with entire edge. Psychrophilic. Growth in liquid media occurs at 0–19°C, but not at temperatures higher than 20°C when cell lysis occurs. Requires 2.75% NaCl for growth. Growth factors are not required. Resistant to vibriostatic agent O/129 (100 μg/ml), although growth is slowed by it. Acid (with no gas) is formed fermentatively and oxidatively from glucose and maltose, chitin, and *N*-acetylglucosamine. Casein, gelatin, esculin, chitin, starch, urea, and Tween 80 are hydrolyzed. The following compounds are utilized as sole sources of carbon and energy: D-glucose, glycogen, *N*-acetylglucosamine, 3-hydroxybutyrate, acetate, butyrate, succinate, fumarate, pyruvate, DL-lactate, oxalacetate, L-glutamate, and L-proline. Some strains can also utilize propionate. Isolated from flounder eggs, Norway, and fast sea ice, Antarctica.

The mol% G + C of the DNA is: 35–40 (T_m).

Type strain: ATCC 27364, DSM 8813.

GenBank accession number (16S rRNA): AF001375.

2. **Colwellia demingiae** Bowman, Gosink, McCammon, Lewis, Nichols, Nichols, Skerrat, Staley, and McMeekin 1998a, 1178[VP]

de.ming'i.ae. L. adj. *demingiae* in honor of Jody W. Deming, an American microbiologist who has expanded the knowledge of deep-sea bacteria.

Individual cells are straight to curved rods, with spherical and short filamentous cells occasionally occurring. Cell length, 1.5–4.5 μm; width, 0.4–0.6 μm. Gas vesicles not formed. Motile. Colonies are off-white, have a mucoid consistency, raised elevation, and convex circular shape with entire to lobate edges. Psychrophilic. In liquid media T_{opt} is about 10–12°C and T_{max} is about 18°C. Requires sea salts for growth. Growth factors are not required. Resistant to vibriostatic agent O/129 (100 μg/ml). Acid (with no gas) is formed fermentatively and oxidatively from chitin and *N*-acetylglucosamine. Casein, esculin, chitin, and starch are hydrolyzed. The following compounds are utilized as sole sources of carbon and energy: *N*-acetylglucosamine, acetate, propionate, butyrate, valerate, caproate, succinate, glutarate, citrate, fumarate, pyruvate, DL-lactate, oxalacetate, L-asparagine, L-glutamate, L-proline, L-serine, and γ-amino-

butyrate. Some strains can also utilize glycogen, isobutyrate, heptanoate, caprylate, malonate, azelate, L-alanine, L-aspartate, L-phenylalanine, and hydroxy-L-proline. Isolated from fast sea ice of the Prydz Bay coast, Antarctica.

The mol% G + C of the DNA is: 37 (T_m).

Type strain: ACAM 459.

GenBank accession number (16S rRNA): U85845.

3. **Colwellia hadaliensis** Deming, Somers, Straube, Swartz and MacDonell 1988b, 328[VP] (Effective publication: Deming, Somers, Straube, Swartz and MacDonell 1988a, 159.)

ha'dal.i.en'sis. M.L. adj. ending *-iensis;* Eng. adj. *hadal* used in oceanographic terminology; of, or pertaining to, the greatest depths of the ocean; derived from Gr. n. *Hades* god of the underworld.

Individual cells are curved rods. Cell length, 3.0–5.0 μm; width, 0.8 μm. Gas vesicles not formed. Motile by means of polar flagella. Colonies are off-white but form only within solid media under hydrostatic pressure. Psychrophilic. Growth occurs in liquid media at 2–10°C, but not at temperatures of 15°C or higher. Obligately barophilic. Requires hydrostatic pressures of 300–1020 atm for growth at 2°C and 500–1020 atm at 10°C. Cell lysis occurs at <200 atm and 2°C and at <400 atm and 10°C; absolute upper pressure limits for growth and survival unknown. Acid is formed fermentatively and oxidatively from glucose. Chitin is hydrolyzed. Associated with cold (<4°C), hadal (depth of 6000–11,000 m) marine waters and sinking particles.

The mol% G + C of the DNA is: 46 (T_m).

Type strain: BNL-1.

Additional Remarks: The 16S rDNA sequence is not available for *C. hadaliensis*, but see Deming et al. (1988a) for 5S rRNA sequence.

4. **Colwellia hornerae** Bowman, Gosink, McCammon, Lewis, Nichols, Nichols, Skerrat, Staley and McMeekin 1998a, 1178[VP]

horn.ner'ae. L. adj. *hornerae* in honor of Rita Horner, an American biologist who pioneered studies on sea-ice microbiota.

Individual cells are straight to curved rods, with spherical cells occasionally occurring. Cell length, 1.5–3.0 μm; width, 0.4–0.8 μm. Gas vesicles not formed. Motile. Colonies are off-white, have a mucoid consistency, raised elevation, and convex circular shape with entire to lobate edges. Psychrophilic. In liquid media T_{opt} is about 12°C and T_{max} is about 23°C. Requires sea salts for growth. Growth occurs in the presence of sea salts at three times normal concentration and is sensitive to vibriostatic agent O/129 (100 µg/ml). Growth factors are not required. Acid (with no gas) is formed fermentatively and oxidatively from L-rhamnose. Acid is formed oxidatively from glycerol. Tween 80, casein, esculin, and starch are hydrolyzed. The following compounds are utilized as sole sources of carbon and energy: glycerol, acetate, propionate, butyrate, valerate, caproate, heptanoate, succinate, glutarate, azelate, citrate, fumarate, pyruvate, DL-lactate, oxaloacetate, L-glutamate, L-proline, hydroxy-L-proline, and γ-aminobutyrate. Isolated from fast sea ice of the Prydz Bay coast, Antarctica.

The mol% $G + C$ of the DNA is: 39 (T_m).

Type strain: ACAM 607, CIP 105821.

GenBank accession number (16S rRNA): U85847.

5. **Colwellia maris** Yumoto, Kawasaki, Iwata, Matsuyama and Okuyama 1998, 1361[VP]

mar' is. L. gen. n. maris of the sea.

Individual cells are curved rods. Cell length, 2–4 µm; width, 0.6–1.0 µm. Gas vesicles not formed. Motile by means of a single polar flagellum. Colonies are colorless. Catalase and oxidase reactions are positive. Psychrophilic. Growth occurs in liquid media at 0–22°C, with T_{opt} of 15°C. Growth factors not required. Growth occurs in media supplemented with 3% or 4% NaCl, but not in the absence of NaCl or at a salinity higher than 6.5%. Sensitive to vibriostatic agent O/129 (10 and 150 µg/ml). Acids are produced from D-glucose under aerobic, but not anaerobic, conditions. No acids are produced from L-arabinose, D-fructose, maltose, D-mannose, melibiose, and sucrose either aerobically or anaerobically. Gelatin, DNA and Tweens 20, 40, 60, and 80 are hydrolyzed, but not casein, chitin, starch, or alginic acid. D-glucose is utilized as a sole source of carbon and energy. Does not utilize D-mannose, raffinose, D-xylose, L-arabinose, D-fructose, glycerol, lactose, maltose, melibiose, or sucrose. The major isoprenoid quinone is Q-8. The whole-cell fatty acids contain saturated and monounsaturated fatty acids with 10–18 C atoms; saturated and monounsaturated C_{16} fatty acids are predominant in cells grown at 15°C. The strain contains a unique trans-unsaturated fatty acid (9-trans-hexadecanoic acid, $C_{16:1 \omega 9t}$). Isolated from ice-impacted seawater in the Okhotsuku Sea, Japan.

The mol% $G + C$ of the DNA is: 39 (HPLC).

Type strain: ABE-1, CIP 106458, JCM 10085.

GenBank accession number (16S rRNA): AB002630.

6. **Colwellia psychrotropica** Bowman, Gosink, McCammon, Lewis, Nichols, Nichols, Skerrat, Staley and McMeekin 1998a, 1179[VP]

psy.chro.tro.pi' ca. Gr. adj. psychros cold; Gr. n. tropica circle; M.L. fem. adj. psychrotropica having an affinity for cold.

Individual cells are straight to curved rods, with spherical cells occasionally occurring. Cell length, 1.5–3.0 µm; width, 0.4–0.8 µm. Gas vesicles not formed. Motile. Colonies are off-white, have a mucoid consistency, raised elevation, and convex circular shape with entire to lobate edges. Psychrophilic. In liquid media T_{opt} is about 18°C and T_{max} is about 25°C. Requires sea salts for growth. Growth factors not required. Growth occurs in the presence of sea salts at three times normal concentration and is sensitive to vibriostatic agent O/129 (100 µg/ml). Acid (with no gas) is formed fermentatively and oxidatively from chitin and N-acetylglucosamine. Urea, uric acid, Tween 80, casein, chitin, and L-tyrosine are hydrolyzed. The following compounds are utilized as sole sources of carbon and energy: N-acetylglucosamine, acetate, butyrate, valerate, caproate, succinate, 2-oxoglutarate, 3-hydroxybutyrate, L-malate, fumarate, pyruvate, DL-lactate, oxaloacetate, L-alanine, L-aspartate, L-asparagine, L-glutamate, L-proline, and γ-aminobutyrate. Isolated from the pycnocline of an Antarctic marine-salinity meromictic lake (Burton Lake).

The mol% $G + C$ of the DNA is: 42 (T_m).

Type strain: ACAM 179.

GenBank accession number (16S rRNA): U85846.

7. **Colwellia rossensis** Bowman, Gosink, McCammon, Lewis, Nichols, Nichols, Skerrat, Staley and McMeekin 1998a, 1179[VP]

ross.en' sis. M.L. fem. adj. rossensis for the Ross Sea, Antarctica.

Individual cells are straight to curved rods, with spherical cells occasionally occurring. Cell length, 1.5–3.0 µm; width, 0.4–0.8 µm. Gas vesicles are formed. Nonmotile. Colonies are chalky-white, have a raised elevation and convex circular shape with entire edges. Psychrophilic. In liquid media T_{opt} is about 10°C and T_{max} is about 15°C. Requires sea salts for growth. Yeast extract required for growth. Resistant to vibriostatic agent O/129 (100 µg/ml). Acid (with no gas) is formed fermentatively and oxidatively from chitin, N-acetylglucosamine, and D-glucose. Also produces acid oxidatively from D-galactose and glycerol. Urea, chitin, and starch are hydrolyzed. The following compounds are utilized as sole sources of carbon and energy: glycogen, L-arabinose, N-acetylglucosamine, D-fructose, D-glucose, glycerol, acetate, butyrate, valerate, caproate, malonate, succinate, citrate, 3-hydroxybutyrate, L-malate, fumarate, pyruvate, L-alanine, L-aspartate, L-asparagine, L-glutamate, L-proline, and γ-aminobutyrate. Isolated from the sea/ice interface of fast sea ice McMurdo Sound, Antarctica (Gosink and Staley, 1995).

The mol% $G + C$ of the DNA is: 38 (T_m).

Type strain: S51-W(gv)1, ACAM 608.

GenBank accession number (16S rRNA): U14581.

Genus IV. **Ferrimonas** Rosselló-Mora, Ludwig, Kämpfer, Amann and Schleifer 1996, 362[VP] (Effective publication: Rosselló-Mora, Ludwig, Kämpfer, Amann and Schleifer 1995a, 200)

RAMON A. ROSSELLÓ-MORA

Fer' ri.mo.nas. L. neut. n. ferrum iron; Gr. n. monas unit; M.L. fem. n. Ferrimonas iron (III)-reducing cell.

Straight rods 0.3–0.5 × 1.2–1.5 µm with rounded ends. Occur singly, occasionally with some pairs or short chains. No resting stages are known. Definite capsules are not evident. Gram negative. **Motile by means of a single polar flagellum.** No intracellular inclusions have been observed. **Chemoorganotroph, having a strictly respiratory metabolism with oxygen, nitrate, Fe(III)-oxy-**

hydroxide, Fe(III)-citrate, and MnO₂ used as electron acceptors. Lactate is oxidized to CO_2 but not to acetate. Sodium dependent; a minimum of 0.5% NaCl is required in the medium. Catalase, oxidase, phenylalanine deaminase, DNAse, and lipase (Tween 80 and Tween 20) positive. Gelatinase, urease, amylase, Simmons citrate, arginine dihydrolase, and lysine decarboxylase negative. Growth at 42°C. Member of the *Gammaproteobacteria*.

The mol% G + C of the DNA is: 54 (T_m).

Type species: **Ferrimonas balearica** Rosselló-Mora, Ludwig, Kämpfer, Amann and Schleifer 1996, 362 (Effective publication: Rosselló-Mora, Ludwig, Kämpfer, Amann and Schleifer 1995a, 200.)

FURTHER DESCRIPTIVE INFORMATION

Fresh isolates may not form colonies on PYG[1] agar medium, but after several subcultivations in enrichment medium colony-forming units may appear. Colonies on TSI-agar[2] medium produce a black iron precipitate. Aerobic growth on solid medium is often brown and mucous. *Ferrimonas* undergoes autolysis within 5 d under aerobic conditions. Reduction of Fe(III)-oxyhydroxide with lactate as sole electron donor yields magnetite under anaerobic conditions. The range of NaCl tolerance is 0.5–7.5%. *Ferrimonas* does not exhibit growth at 5°C and 44°C but at 42°C. The pH range for growth is 6–9.

ENRICHMENT AND ISOLATION PROCEDURES

Only one strain of *Ferrimonas* has been isolated; it was simultaneously obtained from the same sample by using two different electron donors. Inocula were obtained from the upper few centimeters of a marine sediment of the Palma de Mallorca harbor (Spain). Sediments were inoculated in amorphous Fe(III)-oxyhydroxide medium[3] containing either 0.4% (w/v) sodium acetate or 0.25% (w/v) sodium tartrate as sole electron donors, and in both instances 2% NaCl was added to the medium. The anaerobic atmosphere was composed of 97% N_2 and 3% H_2. Incubations were done at 20°C without shaking and in the dark. Enrichments yielded magnetite after 1 month of incubation. After three subcultivations in enrichment medium the isolation of the organism was carried out by streaking the enrichment culture

---

1. PYG broth: 1% peptone, 0.5% yeast extract, 0.5% glucose, and 0.8% NaCl.

2. TSI-agar: Triple Sugar Iron agar, Difco.

3. Anaerobic basal mineral medium of Lovley and Phillips (1986): (in grams per liter deionized water) NaHCO₃, 2.5; CaCl₂.2H₂O, 1.0; KCl, 0.1, NH₄Cl, 1.5; NaH₂PO₄, 0.6; NaCl, 20.0; MgCl₂.6H₂O, 5.3; MgSO₄.7H₂O, 0.1; MnCl₂.4H₂O, 0.005; NaMoO₄.2H₂O, 0.001; and yeast extract, 0.05. Media is adjusted to pH 7. Electron acceptors were used at 250 mmol/l Fe(III)-oxyhydroxide (for preparation see Lovley and Phillips, 1986) or 20 mmol/l Fe(III)-citrate.

on Fe(III)-citrate agar. Pure culture was obtained by continuous restreaking under anaerobic incubations for at least 2 weeks.

MAINTENANCE PROCEDURES

Aerobic cultures should be reinoculated often due to the rapid autolysis of this bacterium; it does not survive more than 5 days on either solid or liquid complex medium incubated aerobically. However, anaerobic growth on Fe(III)-oxyhydroxide lactate agar shakes can be maintained for several months. The organism survives for several years frozen at −80°C in nutrient PYG broth containing 15% glycerol. It also survives freeze-drying. However, in cultures maintained by both methods viability may be low due to the tendency of the cells to lyse.

DIFFERENTIATION OF THE GENUS *FERRIMONAS* FROM OTHER GENERA

Ferrimonas can be distinguished from other strictly respiratory Gram-negative genera of the *Gammaproteobacteria* able to undergo Fe(III) reduction (i.e. *Shewanella putrefaciens*, *S. baltica*, and *S. algae*) as well as from *Alteromonas* species by the following characters: its ability to denitrify, to grow at 42°C and its phenylalanine deaminase activity, its inability to grow in medium containing 0% NaCl, its lack of gelatinase and urease, and its negative Simmons citrate reaction.

An oligonucleotide probe specific for *Ferrimonas* has been designed (5'-ACCCCCCTCTCAAGGACT-3' complementary to the position 654-672 in *E. coli* 16S rRNA numbering; Brosius et al., 1981). A comparison with 5300 complete or almost complete accessible 16S rRNA sequences present in the RDP database (Maidak et al., 1997) did not reveal any identical complementary sequences other than that of the target organism. T_m, derived from the thermal denaturation curve, is 53.2°C. Optimal stringency conditions for dot blot hybridization with extracted nucleic acids are at 62°C. This probe is also optimized for whole cell *in situ* hybridization. Specificity of the probe is obtained by using 35% formamide with the standard protocol (Manz et al., 1992).

TAXONOMIC COMMENTS

The genus *Ferrimonas* is phylogenetically affiliated with the *Gammaproteobacteria* as illustrated in Fig. BXII.γ.153. The genus presently consists of a single species, *F. balearica*; to date, no similar strains have been reported.

ACKNOWLEDGMENTS

K.-H. Schleifer and W. Ludwig are gratefully acknowledged for critically reviewing this chapter.

FURTHER READING

Rosselló-Mora, R.A., W. Ludwig, P. Kämpfer, R. Amann and K.-H. Schleifer. 1995. *Ferrimonas balearica* gen. nov., sp. nov., a new marine facultative Fe(III)-reducing bacterium. Syst. Appl. Microbiol. *18*: 196–202.

List of species of the genus Ferrimonas

1. **Ferrimonas balearica** Rosselló-Mora, Ludwig, Kämpfer, Amann and Schleifer 1996, 362[VP] (Effective publication: Rosselló-Mora, Ludwig, Kämpfer, Amann and Schleifer 1995a, 200.)

   ba.le.a'ri.ca. L. fem. adj. *balearica* of the Balearaic Islands where the organism was isolated.

The description is the same as that given for the genus.
The mol% G + C of the DNA is: 54 (T_m).
Type strain: PAT, DSM 9799.
GenBank accession number (16S rRNA): X93021.

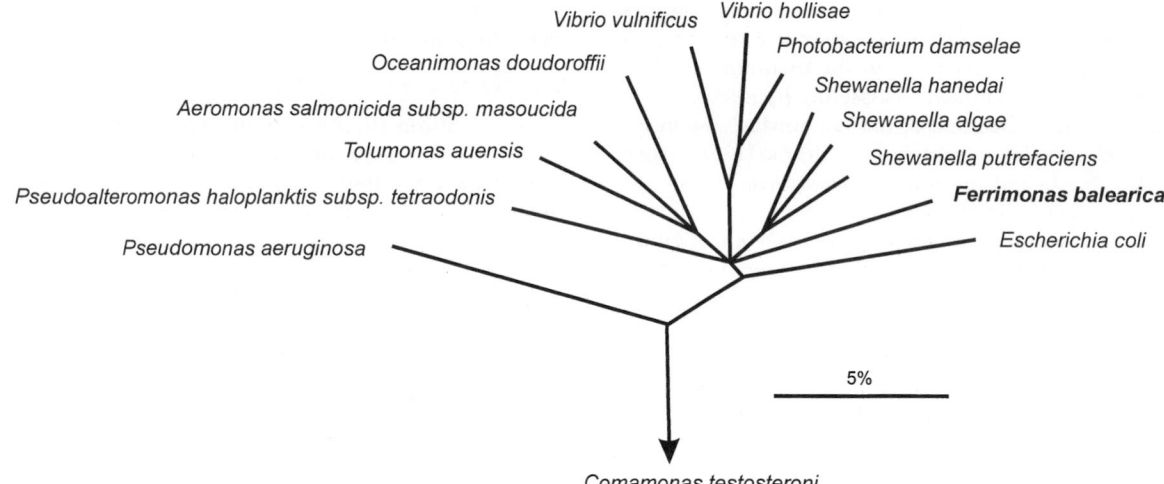

FIGURE BXII.γ.153. 16S rRNA based tree reflecting the phylogenetic affiliation of *Ferrimonas balearica* and a selection of *Proteobacteria*. The tree is based on the results of a distance matrix analysis including only sequence positions that share at least 50% of all available complete or almost complete 16S rRNA sequences from bacteria of this class. Tree topology was evaluated and corrected according to the results of distance matrix, maximum parsimony, and maximum likelihood analyses of various data sets. Accession numbers of shown sequences in the tree are the following: AB021371, M11224, M34133, X74680, X74700, X74707, X76333, X80725, X81623, X82132, X82139, X92889, X93021, and Z95486. Multifurcations indicate topologies that could not be unambiguously resolved. The bar indicates 5% estimated sequence divergence.

Genus V. **Glaciecola** *Bowman, McCammon, Brown and McMeekin 1998b, 1220*[VP]

JOHN P. BOWMAN

Gla.ci.e′co.la. L. fem. n. *glacies* ice; L. gen. n. *incola* an inhabitant; M.L. fem. n. *Glaciecola* inhabitant of ice.

Slender rod-shaped cells, 0.2–0.4 × 1.0–5.0 μm. May be slightly curved, curved, or spiral. Gram negative. **Motile.** Nonsporeforming. **Aerobic,** having a strictly respiratory type of metabolism with oxygen as the only known terminal electron acceptor. Chemoheterotrophic. **Psychrophilic,** growing best at temperatures well below room temperature. In liquid media, the maximal growth temperature for most strains is at about 25–26°C; on agar media, no growth occurs above 20°C. **Seawater is required for growth. Oxidase and catalase positive.** Major fatty acids are hexadecaenoic acid ($C_{16:1 \omega7c}$), hexadecanoic acid ($C_{16:0}$), and octadecaenoic acid ($C_{18:1 \omega7c}$). Belongs to the *Gammaproteobacteria*. Habitat: Antarctic sea-ice diatom assemblages.

The mol% G + C of the DNA is: 40–46.

Type species: **Glaciecola punicea** Bowman, McCammon, Brown and McMeekin 1998b, 1220.

FURTHER DESCRIPTIVE INFORMATION

Glaciecola cells have a generally rod-shaped morphology (0.2–0.4 × 1.0–5.0 μm) and are motile. At the optimal growth temperature, ~15°C, cells are very slender and appear spiral or curved. At supraoptimal (>20°C) and suboptimal (<10°C) temperatures, the cells appear as straight rods. Short filaments also occur, especially in stationary growth cultures. All strains are motile, although information on the type and position of flagella present is still lacking.

The colonies of *G. punicea* are circular and convex in shape, with lobate edges and a butyrous consistency. After 14 d incubation, colony diameter is about 3–5 mm. *G. pallidula* strains form even-edged colonies that have a mucoid consistency and are 3–5 mm after 14 d incubation. Various *Glaciecola* strains are pigmented with unknown compounds that are not extractable with either polar or nonpolar solvents. In the presence of a strong base (20% KOH), the pigment is bleached white. Pigmentation is most obvious in strains of *G. punicea*, which appear musk pink to magenta in color. By comparison, *G. pallidula* colonies have a pale pink coloration.

Strains grow well on marine media, such as Marine 2216 agar (Difco Laboratories). All strains require seawater salts for optimal growth. No growth occurs if NaCl is added alone to the medium; Mg^{2+} and Ca^{2+} ions are also required for growth. Strains of *G. pallidula* are nutritionally non-exacting and will grow in a defined mineral salts medium, such as that used for testing carbon and energy sources of *Shewanella*, *Alteromonas*, and *Pseudoalteromonas* (Baumann et al., 1984a). Most strains of *G. punicea* require yeast extract for growth, and growth of this species can be achieved in a defined mineral salts medium supplemented with 0.02% yeast extract. In any case, growth in minimal media by most *Glaciecola* strains is relatively slow and scant, but it can be boosted somewhat by the addition of a vitamin solution, such as that formulated by Balch et al. (1979).

All strains are psychrophilic. In liquid media, the optimal growth temperature is about 15°C and the maximal growth temperature for most strains is about 25–26°C. Strains will grow at −2°C in liquid marine 2216 media. On agar media, cardinal growth temperature points are reduced by up to 5°C, with no growth occurring at temperatures above 20°C. A detailed study of the salt and temperature optima of *Glaciecola punicea* ACAM 611[T] has been performed by Nichols et al. (1999).

All *Glaciecola* strains are strictly aerobic chemoheterotrophs and strongly produce catalase and cytochrome *c* oxidase. None of the strains grows anaerobically by either fermentation or by anaerobic respiration. Oxidative acid production from carbohydrates is weak and slow on Leifson's oxidation–fermentation medium (Leifson, 1963) and is restricted to a small range of carbohydrates (Table BXII.γ.140). Up to 10 d incubation at 15°C is required before acid production is detectable. The hydrolytic capacity of *Glaciecola* strains is relatively limited. Some strains of *G. punicea* can degrade uric acid (slowly) and esculin, whereas most strains of *G. pallidula* can degrade Tween 80 and produce an amylase. All strains so far tested are able to form alkaline phosphatase, α-galactosidase and β-galactosidase. *G. punicea* strains also produce a 6-phospho-β-galactosidase. Strains are able to utilize a limited range of organic acids, amino acids, and other compounds as sole carbon sources (Table BXII.γ.140). The metabolic pathways and nutrient transport mechanisms in *Glaciecola* strains have yet to be studied.

G. punicea strains possess a DNA base composition of 44–46 mol% G + C—significantly higher than that of *G. pallidula* strains, which have an average value of 40 mol% G + C.

Whole cell fatty acid profiles of *Glaciecola* species are very similar to one another. The major fatty acids include hexadecaenoic acid ($C_{16:1\ \omega7c}$; 52–62%), octadecaenoic acid ($C_{18:1\ \omega7c}$; 12–21%), hexadecanoic acid ($C_{16:0}$; 9–12%), and heptadecaenoic acid ($C_{17:1\ \omega8c}$; 4–5%). Strains also contain low levels of saturated iso-branched-chain fatty acids ($C_{13:0\ iso}$, $C_{14:0\ iso}$, $C_{15:0\ iso}$, $C_{16:0\ iso}$, and $C_{17:0\ iso}$; total 2.5–3.3%). Polyunsaturated fatty acids and hydroxylated fatty acids (usually associated with the cell envelope) are either absent or are present only at trace levels. The fatty acids present are similar to those of *Alteromonas macleodii* (Svetashev et al., 1995); however, monounsaturated fatty acids predominate in *Glaciecola* strains, whereas saturated fatty acids are the most significant components in *Alteromonas* strains.

So far, *Glaciecola* strains have only been isolated from sea-ice algal assemblages. None have been isolated from under-ice seawater below the same sea-ice collection points. The genus is present at both poles; strains have been isolated from both Antarctic and Arctic (Barrow, Alaska) sea ice (Staley and Gosink, 1999).

ENRICHMENT AND ISOLATION PROCEDURES

Glaciecola strains have been obtained only from algal assemblages present within coastally attached sea ice (fast sea ice). Algal assemblages in fast ice usually occur perennially as a bottom assemblage. To obtain this bottom assemblage, care must be taken to avoid losing the layer when drilling out the ice core. The ice samples are melted in seawater at 4°C or less to avoid hypotonic shock to the bacteria present. The melted sample is then added to Marine 2216 liquid media for 1–2 d at 2°C before plating onto Marine 2216 agar. This enrichment phase encourages the growth of isolates that will not grow readily when directly plated onto a solid agar surface, such as *Glaciecola* spp. Incubation of marine agar plates proceeds for at least 4 weeks at 2°C. The unusual pigments produced by *Glaciecola* strains, particularly those of *G. punicea*, make them distinctive on primary isolation plates and easy to select for further purification. Some strains (all so far undescribed) that are related to the genera *Psychrobacter* and *Marinobacter* can also appear pale pink and so can be potentially confused with *G. pallidula* strains.

MAINTENANCE PROCEDURES

Glaciecola strains can be stored as active cultures on Marine 2216 agar plates or slants at 2°C for at least 12 months. The medium should contain an antifungal agent (nystatin or cycloheximide) to prevent contamination. Strains can also be cryopreserved. For cryopreservation, a heavy suspension of cells from a fresh culture is made in Marine 2216 liquid media supplemented with 30% glycerol. The suspension is then frozen at −20°C and then stored at −70°C. The frozen suspension is then checked within 2 weeks of preparation and every 6 months thereafter. Multiple small tubes of frozen suspension should be prepared to avoid loss of viability from repeated freezing and thawing.

DIFFERENTIATION OF THE GENUS *GLACIECOLA* FROM OTHER GENERA

The sea-ice habitat, distinctive pigmentation and psychrophilic properties of the genus *Glaciecola* make them readily identifiable. The genera that are most closely related to *Glaciecola* include

TABLE BXII.γ.140. Characteristics differentiating the species of the genus *Glaciecola*[a]

Characteristic[b,c]	1. *G. punicea*	2. *G. pallidula*
Pigmentation	Pink-red	Pale pink
Yeast extract requirement	d	−
β-Galacto(6-phosphate)sidase	+	−
Acid from maltose	−	(+)
Hydrolysis of:		
Urate	d	−
Esculin	d	−
Starch	−	d
Tween 80	d	+
Utilization of:		
α-Glycerophosphate	d	+
Succinate, L-malate, fumarate, L-proline	+	−
Glycerol, glycogen, acetate, pyruvate, DL-lactate, L-glutamate	−	+
L-Tyrosine	d	−
Butyrate	−	d
Mol% G + C of the DNA	44–46	40

[a]Symbols: +, test positive for all strains; d, variable (10–90% of strains) results between strains; −, strains negative for test; (+), the test exhibited a weak and delayed reaction.

[b]All *Glaciecola* strains tested are positive for the following tests: growth on Marine 2216, R2A (Oxoid Pty. Ltd.) + 2.5% NaCl, and ZoBells marine medium (5 g bacteriological peptone, 2 g yeast extract dissolved in 1000 ml seawater); growth at <15°C on agar media; catalase, oxidase, α-galactosidase, β-galactosidase; utilization of oxaloacetate; and acid production (but weak and delayed) from D-glucose, D-galactose, D-melibiose, and glycerol.

[c]All *Glaciecola* strains tested are negative for the following tests: growth at 25°C or higher on agar media; growth with 3× or 4× strength seawater; tolerance to 5% ox-bile salts; nitrate as a nitrogen source; nitrate reduction, denitrification, and nitrogen fixation; hydrolysis of urea, xanthine, egg yolk, casein, gelatin, elastin, chitin, dextran, and DNA; production of L-phenylalanine deaminase, L-tryptophan deaminase, arginine dihydrolase, glutamate decarboxylase, lysine decarboxylase, ornithine decarboxylase, α-glucosidase, β-glucosidase, α-fucosidase, α-arabinosidase, β-glucuronidase, β-N-acetylglucosaminidase; production of indole from L-tryptophan; production of hydrogen sulfide from thiosulfate or L-cysteine; acid production from L-arabinose, D-mannose, D-fructose, L-rhamnose, D-xylose, D-mannitol, N-acetylglucosamine, sucrose, lactose, cellobiose, trehalose, D-raffinose, dextran, adonitol, L-arabitol, D-sorbitol, and m-inositol; utilization of N-acetylglucosamine, L-arabinose, cellobiose, and D-xylose

Alteromonas and *Pseudoalteromonas*. *Alteromonas* strains have not been isolated from polar environments and seem to be more adapted to tropical and temperate marine waters, such as the Mediterranean Sea or mid-Pacific Ocean. *Pseudoalteromonas* species are common in southern ocean seawater and have also been isolated from sea ice (mostly strains of *P. antarctica*, *P. nigrifaciens*, and *P. prydzensis*) (Bowman, 1998); however these species are easy to distinguish from *Glaciecola* spp. because they are nonpigmented, able to grow at temperatures up to 30–35°C on agar, and nutritionally versatile, with strong saccharolytic activity.

TAXONOMIC COMMENTS

The 16S rDNA sequence analysis of *Glaciecola* strains indicates that they form a lineage within the order *Alteromonadales*, which includes other marine genera, such as *Colwellia*, *Shewanella*, *Moritella*, *Ferrimonas*, *Pseudoalteromonas*, and *Alteromonas*. The genus *Alteromonas* is by far the closest related taxon, with a sequence similarity of 0.90–0.92. *Glaciecola punicea* and *Glaciecola pallidula* strains both form individually distinct branches and share a sequence similarity of about 0.93.

The levels of DNA hybridization between strains of the different species have been examined using spectrophotometric-based renaturation rate kinetics (Huss et al., 1983). The levels are equivalent to background levels of reassociation.

DIFFERENTIATION OF THE SPECIES OF THE GENUS *GLACIECOLA*

Characteristics differentiating *G. punicea* and *G. pallidula* are listed in Table BXII.γ.140.

List of species of the genus Glaciecola

1. **Glaciecola punicea** Bowman, McCammon, Brown and McMeekin 1998b, 1220[VP]

   pu.nice' a. L. fem. adj. *punicea* pink-red, referring to the species pigmentation.

   The characteristics are as described for the genus and as listed in Table BXII.γ.140. Other characteristics are as follows. Optimum temperature, 15°C; maximum 20–25°C. Colonies are bright pink-red, circular, and convex, with an entire edge and butyrous consistency. Yeast extract is required by some strains; growth is stimulated by vitamin growth factors. Oxidative acid production from carbohydrates is weak and slow; acid may be formed from D-glucose, D-galactose, D-melibiose, and glycerol. Some strains slowly degrade uric acid, esculin, and L-tyrosine. The following enzymes are produced: α-galactosidase, β-galactosidase, 6-phospho-β-galactosidase, and alkaline phosphatase. The following compounds are used as carbon and energy sources: succinate, L-malate, fumarate, oxaloacetate, and L-proline. Some strains utilize α-glycerophosphate. The major fatty acids are hexadecaenoic acid ($C_{16:1 \, \omega 7c}$), hexadecanoic acid ($C_{16:0}$), and octadecaenoic acid ($C_{18:1 \, \omega 7c}$). The type strain was isolated from a sea-ice core from the O'Gorman Rocks area of Prydz Bay, Antarctica.

   The mol% G + C of the DNA is: 44–46 (T_m).

   Type strain: IC067, ACAM 611.

   GenBank accession number (16S rRNA): U85853.

2. **Glaciecola pallidula** Bowman, McCammon, Brown and McMeekin 1998b, 1221[VP]

   pal.lid' u.la. L. adj. *pallidula* pallidula somewhat pale, referring to the weak pigmentation of the species.

   The characteristics are as described for the genus and as listed in Table BXII.γ.140. Other characteristics are as follows. Optimum temperature, 10–15°C; maximum, 18–20°C. Colonies are pale pink, circular, and raised convex, with entire edges and a butyrous–mucoid consistency. Growth is stimulated by vitamin growth factors. Oxidative acid production from carbohydrates is weak and slow; acid may be formed from D-glucose, D-galactose, D-melibiose, and glycerol. Tween 80 is hydrolyzed. Produce the following enzymes: α-galactosidase, β-galactosidase, and alkaline phosphatase. Some strains slowly degrade starch. Utilize the following compounds as carbon and energy sources: glycogen, glycerol, α-glycerophosphate, acetate, pyruvate, DL-lactate, oxaloacetate, and L-glutamate. Some strains utilize butyrate. Major fatty acids are hexadecaenoic acid ($C_{16:1 \, \omega 7c}$), hexadecanoic acid ($C_{16:0}$), and octadecaenoic acid ($C_{18:1 \, \omega 7c}$). The type strain was isolated from sea-ice cores from Taynaya Bay, Antarctica.

   The mol% G + C of the DNA is: 40 (T_m).

   Type strain: IC079, ACAM 615.

   GenBank accession number (16S rRNA): U85854.

Genus VI. **Idiomarina** *Ivanova, Romanenko, Chun, Matte, Matte, Mikhailov, Svetashev, Huq, Maugel and Colwell 2000b, 906*[VP]

THE EDITORIAL BOARD

I.di.o.ma.ri' na. Gr. adj. *idios* original, true; L. fem. adj. *marina* of the sea, marine; M.L. fem. n. *Idiomarina* pertaining to the peculiar, true marine nature of microorganisms from the ocean (seawater).

Chemoorganotrophic Gram-negative obligately aerobic rods. Motile; one polar flagellum. **Require Na$^+$ for growth**; no required vitamins or amino acids. Temperature optimum 20–22°C; pH optimum 7.5–8.0. **Oxidase, catalase, gelatinase, lipase, and DNase positive.** Do not hydrolyze starch or agar. Major cellular fatty acids are branched *iso* forms having odd numbers of carbon atoms.

The mol% G + C of the DNA is: 48.9–50.4.

Type species: **Idiomarina abyssalis** Ivanova, Romanenko, Chun, Matte, Matte, Mikhailov, Svetashev, Huq, Maugel and Colwell 2000b, 906.

FURTHER DESCRIPTIVE INFORMATION

The type strains of both *Idiomarina* species utilized alaninamide, L-alanine, L-alanyl-glycine, L-arginine, glycyl-L-glutamic acid, α-ketobutyric acid, α-ketovaleric acid, monomethylsuccinate, and L-tyrosine in the Biolog™ test system. Neither utilized D-arabinose, glycerol, lactose, L-lysine, maltose, mannitol, D-mannose, melibiose, L-phenylalanine, D-rhamnose, or sucrose (Ivanova et al., 2000b).

The type strains of both *Idiomarina* species were resistant to ampicillin, benzylpenicilllin, kanamycin, lincomycin, oleandomycin, oxacillin, tetracycline, and vancomycin; both were susceptible to erythromycin (Ivanova et al., 2000b).

Analysis of 16S rDNA sequences of *I. abyssalis* and *I. zobellii* placed them together in a clade in the *Gammaproteobacteria* with *Alteromonas*, *Pseudoalteromonas*, and *Colwellia* species (Ivanova et al., 2000b).

ENRICHMENT AND ISOLATION PROCEDURES

Strains were isolated from seawater that was collected at 4000–5000 m, spread on seawater agar amended as described by Ivanova et al. (2000b), and incubated at atmospheric pressure.

MAINTENANCE PROCEDURES

Strains can be maintained at 4°C on semisolid medium B covered with mineral oil (Ivanova et al., 2000b) or at −80°C suspended in Marine Broth containing 30% (v/v) glycerol (Ivanova et al., 2000b).

DIFFERENTIATION OF THE GENUS *IDIOMARINA* FROM OTHER GENERA

Characteristics that differentiate the genus *Idiomarina* from other members of the family *Alteromonadaceae* (*Alteromonas*, *Pseudoalteromonas*, and *Colwellia*) are given by Ivanova and Mikhailov (2001a); characteristics that differentiate *Idiomarina* from those three genera and from members of the genera *Halomonas*, *Marinomonas*, and *Oceanospirillum* are given by Ivanova et al. (2000b). A comparison of fatty acid profiles of members of the genera *Idiomarina*, *Marinomonas*, *Alteromonas*, *Pseudoalteromonas*, and *Glaciecola* was presented by Ivanova et al. (2000c).

DIFFERENTIATION OF THE SPECIES OF THE GENUS *IDIOMARINA*

I. abyssalis KMM 227 grows in the presence of NaCl up to 15% at 4°C; *I. zobellii* KMM 231, up to 10%. *I. zobellii* KMM 231 possesses fimbriae and produces chitinase; *I. abyssalis* KMM 227 does not.

I. abyssalis KMM 227 oxidizes acetic acid, cyclodextrin, dextrin, glucose-6-phosphate, glycerol, glycogen, methylpyruvate, and proline. *I. zobellii* KMM 231 utilizes L-ornithine.

List of species of the genus Idiomarina

1. **Idiomarina abyssalis** Ivanova, Romanenko, Chun, Matte, Matte, Mikhailov, Svetashev, Huq, Maugel and Colwell 2000b, 906[VP]

   a.bys.sal'is. L. fem. masc. adj. *abyssalis* deep sea; M.L. fem. adj. *abyssalis* of abssyal depths of the ocean (1000–6000 m) from which the organism was isolated.

   Description as for the genus with the following additional characteristics. Grows in the presence of 0.6–15% NaCl. Produces proteinase.

   The mol% G + C of the DNA is: 50.4 (T_m).

   Type strain: KMM 227, ATCC BAA-312.

   GenBank accession number (16S rRNA): AF052740.

2. **Idiomarina zobellii** Ivanova, Romanenko, Chun, Matte, Matte, Mikhailov, Svetashev, Huq, Maugel and Colwell 2000b, 906[VP]

   zo.bell'i.i. M.L. fem. adj. *zobelli* of Zobell; named after C.E. Zobell, a pioneer marine microbiologist.

   Description as for the genus with the following additional characteristics. Grows in the presence of 1–10% NaCl. Produces chitinase.

   The mol% G + C of the DNA is: 48 (T_m).

   Type strain: KMM 231, ATCC BAA-313.

   GenBank accession number (16S rRNA): AF052741.

Genus VII. **Marinobacter** *Gauthier, Lafay, Christen, Fernandez, Acquaviva, Bonin and Bertrand 1992, 574*[VP]

JOHN P. BOWMAN AND THOMAS A. MCMEEKIN

Mar.i' no.bac.ter. L. adj. *marinus* of the sea; Gr. n. *bacterion* rod or staff; M.L. masc. n. *Marinobacter* rod of the sea.

Rod-shaped cells 0.6–0.8 × 1.6–2.0 μm. Form surface blebs when grown on *n*-eicosane and other hydrocarbons. **Motile by a single polar flagellum. Gram negative.** Nonpigmented colonies. Aerobic with a strictly respiratory type of metabolism. Oxidase and catalase positive. **Can grow anaerobically by denitrification** coupled to the oxidation of a suitable donor carbon substrate. Chemoheterotrophic. Constitutive arginine dihydrolase system is absent. **Na$^+$ is required for growth. Halotolerant**, growing in 0.08–3.5 M NaCl. **Mesophilic** growing at temperatures 10–45°C. Organic acids, alcohols, amino acids, and hydrocarbons are used as carbon and energy sources. **Carbohydrates are not utilized.** Polysaccharides are not degraded. Type species produces a nondialyzable bioemulsifier when grown on hydrocarbons and is resistant to vibriostatic agent O/129. The major polyamine is spermidine. Isolated from oil brine. **Member of the *Gammaproteobacteria*.**

The mol% G + C of the DNA is: 53 (T_m).

Type species: **Marinobacter hydrocarbonoclasticus** Gauthier, Lafay, Christen, Fernandez, Acquaviva, Bonin and Bertrand 1992, 574.

FURTHER DESCRIPTIVE INFORMATION

Marinobacter hydrocarbonoclasticus is represented by a single strain (ATCC 49840). The cells do not accumulate poly-β-hydroxybutyrate (PHB) inclusions or form spores or cysts. When cultivated on synthetic salt (SM) medium[1] containing the C_{20} alkane eicosane as a sole carbon source (SM-alkane medium), cells are covered by bleb-like extensions of the cell wall. Flagellation is lost when the strain is grown at Na^+ concentrations well above and below the optimal level - i.e., >1.0 M and <0.2 M NaCl.

On marine agar and SM-peptone (0.5% w/v) agar young colonies are white, circular, and convex, possessing entire edges and a creamy consistency. Colonies usually reach 2–4 mm after 7 d of incubation at 32°C. Older colonies develop a pinkish pigment probably due to accumulation of cytochromes. Diffusible and fluorescent pigments are not produced. The cells are not bioluminescent.

M. hydrocarbonoclasticus has an absolute requirement for Na^+ ions with an optimal growth level of 0.6 M NaCl. The requirement for Na^+ cannot be reversed by replacement of K^+ and Li^+ ions. The strain also exhibits considerable halotolerance and grows at NaCl levels of 0.08–3.5 M NaCl. When suspended in a solution containing 0.5 M NaCl and in the absence of Mg^{2+} ions, cells undergo severe hypotonic shock with the suspension losing about 60% of its optical density. When suspended or washed in the presence of Mg^{2+}, cell walls are considerably stabilized and show only limited, if any, hypotonic shock even if subsequently suspended in distilled water. Stability appears to be due to linking of Mg^{2+} ions between the anionic groups of membrane components (Rayman and MacLeod, 1975).

The optimal growth temperature is 32°C; range 10–45°C. The optimal pH is 7.0–7.5; range 6.0–9.5. The strain is nutritionally nonexacting and does not require growth factors. Nitrate and ammonia ions serve as nitrogen sources.

Cytochrome *c* oxidase and catalase are readily detectable in cell masses; however, cytochrome *aa3* oxidase and cytochrome P-450 appear to be completely absent. Cells can grow anaerobically with nitrate as electron acceptor and a suitable carbon source/electron donor such as citrate, acetate, or succinate. A variety of organic acids and hydrocarbons, such as eicosane and hexadecane, and a few amino acids can be used as carbon and energy sources. Lipase and lecithinase occur but the cells do not hydrolyze polysaccharides, gelatin, or DNA. A constitutive arginine dihydrolase system is absent.

The cells form a nondialyzable bioemulsifier when grown on hydrocarbons as sole sources of carbon and energy (Fernandez-Linares et al., 1996a). When grown on eicosane at various NaCl concentrations, ectoine is the major compatible solute in cells; however, ectoine is readily replaced by glycine betaine (Fernandez-Linares et al., 1996b). Glycine betaine reduces the growth rate at high Na^+ ion concentrations, possibly by interfering with eicosane metabolism as a consequence of decreased ectoine synthesis (Gauthier et al., 1997). *M. hydrocarbonoclasticus* has been shown to degrade 6,10,14-trimethylpentadecan-2-one, a common isoprenoid ketone found in marine sediment (Rontani et al., 1997).

Spermidine is the predominant polyamine constituent (Hamana, 1997), while the major phospholipids of cells include phosphatidylethanolamine, phosphatidylglycerol, and diphospatidylglycerol (Gerin and Goutx, 1993).

---

1. SM medium consists of (per liter): 30 g NaCl, 12.3 g Tris, 3.7 g NH_4Cl, 6.2 g $MgSO_4 \cdot 7H_2O$, 1.5 g $CaCl_2$, and 0.75 g KCl. The pH of the medium is adjusted with 10 M HCl to 7.5. For solid media, agar (15 g/liter) can be added. SM-alkane medium consists of SM medium supplemented with 0.1% (w/v) alkane (eicosane, hexadecane etc.). SM-peptone medium consists of SM medium supplemented with 0.5% peptone.

Little is known about the broader ecology of the genus. The ability to degrade alkane hydrocarbons and broad halotolerance suggests the organisms are commonly associated with oil brines. A gradient diffusion chamber containing increasing levels of NaCl and *p*-toluate resulted in the enrichment of *M. hydrocarbonoclasticus*-like strains (identified as *Pseudomonas nautica*) (Emerson and Breznak, 1997). *M. aquaeolei* was isolated from the head of an offshore oil/gas platform (Nguyen et al., 1999). Several isolates from sea ice and seawater (Bowman et al., 1997c) have been found to be closely related to *M. hydrocarbonoclasticus* and include a variety of psychrophilic and psychrotrophic strains including a strain possessing gas vesicles (Gosink and Staley, 1995).

ENRICHMENT AND ISOLATION PROCEDURES

M. hydrocarbonoclasticus can be isolated by enriching oil brine, seawater, and oil brine contaminated soil samples with SM-alkane liquid broth, which contains a long-chain alkane such as eicosane or hexadecane as the sole carbon and energy source. Enrichment flasks are shaken for several days at approximately 35°C. Growth from these cultures is serially diluted onto SM-alkane agar or marine agar for further purification. Subsequent testing of the traits presented in Table BXII.γ.141 should follow in order to confirm identity of the isolates.

M. aquaeolei VT8 was obtained from an enrichment culture incubated in a complex medium containing 10% NaCl (Nguyen et al., 1999).

MAINTENANCE PROCEDURES

M. hydrocarbonoclasticus can be maintained at −80°C in Marine 2216 broth containing a cyroprotectant agent such as 20–30% glycerol or dimethylsulfoxide. Strains can also be freeze-dried using 20% skim milk as a cyroprotectant. *M. aquaeolei* can be stored in 25% glycerol, 5% NaCl at −70°C (Nguyen et al., 1999).

PROCEDURES FOR TESTING SPECIAL CHARACTERS

Denitrification in *M. hydrocarbonoclasticus* can be determined by measuring accumulation of nitrous oxide during anaerobic growth with nitrate. The testing medium is SM supplemented with 0.1% acetate, 0.1% succinate, and 4 mM nitrate, and is prepared anaerobically using 100% nitrogen. Denitrification is halted by the addition of acetylene (10 kPa). Nitrous oxide levels are measured by gas chromatography (Bonin et al., 1987).

DIFFERENTIATION OF THE GENUS *MARINOBACTER* FROM OTHER GENERA

Marinobacter is most likely to be confused with nonfermentative Gram-negative, rod-shaped halophilic marine genera including *Halomonas*, *Marinomonas*, *Marinobacterium*, *Microbulbifer*, *Alteromonas*, *Pseudoalteromonas*, and *Shewanella*. The most useful traits available for differentiation of *Marinobacter* strains from these genera include the inability to utilize carbohydrates, or to degrade polysaccharides or gelatinase, the ability to denitrify, and broad halotolerance. Traits useful for differentiation of the genus from some other Gram-negative marine bacteria are presented in Table BXII.γ.141.

TAXONOMIC COMMENTS

Only limited chemotaxonomic data are available for *M. hydrocarbonoclasticus*. However, the phenotypic traits of *Marinobacter* are not particularly distinct and it possesses some similarity to species of several families of the *Gammaproteobacteria* including *Halomonadaceae* and *Alteromonadaceae*.

A phylogenetic analysis of *M. hydrocarbonoclasticus* ATCC 49840 using 16S rDNA sequence data indicated that its closest named relative is *Microbulbifer hydrolyticus*, a polysaccharide-degrading marine species (González et al., 1997). Spröer et al. (1998) later showed that *Pseudomonas nautica* is also very closely related to ATCC 49840 and proposed the transfer of *P. nautica* to *M. hydrocarbonoclasticus* (see below). The 6,10,14-trimethylpentadecan-2-one-degrading strain CAB and an isolate from an oil well (strain VT8) were also very closely related to ATCC 49840; Nguyen et al. (1999) have described strain VT8 as *Marinobacter aquaeolei*. Other strains whose 16S rDNA sequences were closely related to *Marinobacter* include a variety of isolates from Antarctic sea ice and under-ice seawater (Gosink and Staley, 1995; Bowman et al., 1997c) (Fig. BXII.γ.154). These strains share the inability to catabolize carbohydrates with *M. hydrocarbonoclasticus* but have other traits different from the current definition of *Marinobacter*: they are unable to reduce nitrate and are psychrophilic and stenohaline. These differences could be sufficient to support the description of a separate genus, closely related to *Marinobacter*.

16S rDNA sequence data (>99% similarity) provided compelling evidence that the species *Pseudomonas nautica* (Baumann et al., 1983b) is a subjective synonym of *M. hydrocarbonoclasticus*. In addition, Spröer et al. (1998) reported 100.8% hybridization of DNA from the type strains of *P. nautica* and *M. hydrocarbonoclasticus* in DNA–DNA hybridization experiments. Finally, the phenotypic similarities of *P. nautica* and *M. hydrocarbonoclasticus* (Table BXII.γ.142), including the ability to utilize a range of hydrocarbon alkanes (Monpert, 1996; Husain et al., 1997a, b) and active denitrification, provide a strong argument to support the transfer of *P. nautica* to *M. hydrocarbonoclasticus*. One initial discrepancy was a difference in the reported DNA base compositions. Reported values for the mol% G + C of the DNA of *P. nautica* were 57.8–61.1 (determined by buoyant density centrifugation; Baumann et al., 1972); these values were higher than the value of 52.7 reported for *M. hydrocarbonoclasticus* (determined by thermal denaturation; Gauthier et al., 1992). However, Spröer et al. (1998) analyzed DNA from the type strain of each species by HPLC and obtained values of 57.7 mol% G + C for *P. nautica* and 57.3% for *M. hydrocarbonoclasticus*. The mol% G + C of *M. aquaeolei* is 55.7 ± 0.1 (determined by HPLC; Nguyen et al., 1999).

TABLE BXII.γ.141. Phenotypic characterisitics differentiating the genus *Marinobacter* from other marine bacteria

Characteristic	*Marinobacter*	*Marinobacterium*	*Microbulbifer*	*Marinomonas*	*Oceanospirillum*
Morphology	Rod	Rod	Rod	Curved rods	Curved, helical
Na$^+$ requirement	+[a]	+	+	+	+
Gelatinase activity	−	−	+	−	D
Denitrification	+	−	−	−	−
Carbohydrate utilization	−	+	+	+	−
Mol% G+C	52.7[b]; 57.3[c]	55	58	45–50	45–51

[a]Symbols: +, positive; D, test results vary between species; − negative.

[b]Gauthier et al., 1992.

[c]Spröer et al., 1998.

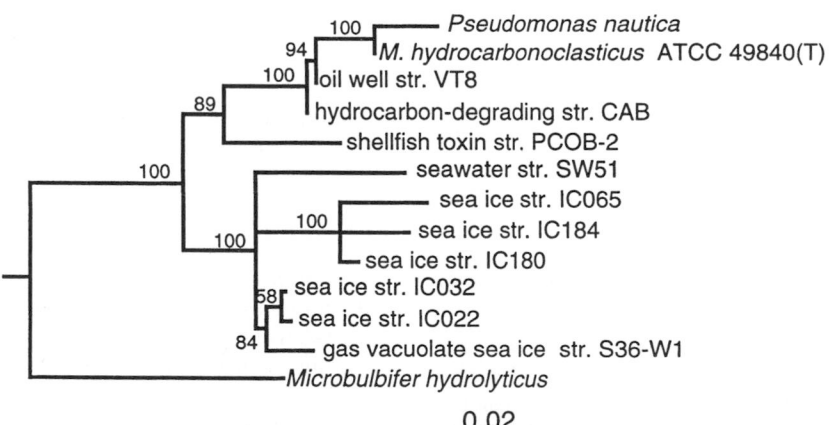

0.02

FIGURE BXII.γ.154. Phylogenetic tree of *Marinobacter hydrocarbonoclasticus* ATCC 49840, *Pseudomonas nautica* ATCC 27132, and close relatives based on nearly complete 16S rRNA sequences. The tree was determined using maximum likelihood distances and clustered by neighbor-joining. Strain VT8 has been described as *M. aquaeolei* (Nguyen et al., 1999; GenBank accession no. AJ000726). Various other strains on the tree remain unclassified, including hydrocarbon-degrading strain CAB (Rontani et al., 1997; U61848), shellfish toxin strain PCOB-2 (Maas et al., unpublished; AJ000647); under-ice seawater and sea ice strains SW51, IC065, IC180, IC184, IC032 and IC022 (Bowman et al., 1997c; AF001374, U85863-7); and a gas vacuolate sea ice strain S36-W1 (Gosink and Staley, 1995; U14584). Numbers at branch nodes are bootstrap values (calculated from 250 replicate analyses). Only bootstrap values over 50% are shown. Bar equals a distance of 0.02.

TABLE BXII.γ.142. Phenotypic characteristics of *Marinobacter hydrocarbonoclasticus* and *Pseudomonas nautica*[a,b,c]

Characteristic	Marinobacter hydrocarbonoclasticus	Pseudomonas nautica
Morphology	Rods	Rods
Flagella	Single polar or none	Single polar or none
Accumulation of poly-β-hydroxybutyrate	−	−
Optimal temperature for growth	32°C	35°C
Growth at 4°C	−	−
Growth at 45°C	+	d
NaCl requirement	+	+
Tolerates 20% NaCl	+	ND
Catalase, oxidase, denitrification	+	+
Hydrogen autotrophy	−	−
Arginine dihydrolase, lysine decarboxylase, ornithine decarboxylase	−	−
Gelatin liquefaction	−	−
Starch hydrolysis	−	−
Lipase (Tween 80 hydrolysis)	+	+
Lecithinase (egg yolk reaction)	+	+
Urease activity	−	−
Acid from glucose	−	−
Utilization of:		
Acetate, butyrate, caproate, succinate, fumarate, DL-lactate, *p*-hydroxybenzoate, L-proline, hexadecane, eicosane	+	+
Adipate, citrate, L-glutamate	+	d
DL-Alanine, benzoate	−	d
Propionate, isobutyrate, valerate, heptanoate, caprylate, pelargonate, caprate, malonate, pyruvate, DL-malate, DL-3-hydroxybutyrate, L-histidine	ND	+
Isovalerate, sebacate, propyleneglycol, ethanol, propanol, butanol, isobutanol	ND	d
Tetradecane, heneicosane, pristane, phenyldecane, phenanthrene	+	ND

[a]Symbols: +, 80–100% of strains are positive; d, 20–80% of strains are positive; −, 0–20% of strains are positive; ND, no data.

[b]Phenotypic data from Gauthier et al. (1992) and Baumann et al. (1983b). *M. hydrocarbonoclasticus* and *P. nautica* do not utilize DL-arabinose, D-ribose, D-fructose, D-glucose, D-mannose, lactose, sucrose, cellobiose, D-mannitol, D-sorbitol, glycerol, D-gluconate, *N*-acetylglucosamine, glycolate, 2-oxoglutarate, *m*-hydroxybenzoate, *o*-hydroxybenzoate, DL-mandelate, propionate, L-arginine, L-asparagine, L-aspartate, glycine, L-lysine, L-methionine, L-ornithine, L-serine, DL-tryptophan, and sarcosine.

[c]In addition most strains of *P. nautica* (<20% positive) do not utilize D-xylose, D-fucose, L-rhamnose, D-galactose, trehalose, maltose, melibiose, saccharate, mucate, D-galacturonate, D-glucuronate, salicin, inulin, cellulose, agar, adonitol, *i*-erythritol, *m*-inositol, ethyleneglycol, methanol, isopropanol, geraniol, formate, oxalate, glutarate, pimelate, suberate, azelate, L(+)-tartrate, D(−)-tartrate, meso-tartrate, glycerate, aconitate, maleate, levulinate, citraconate, itaconate, mesaconate, benzoyl formate, phenylacetate, phenylethanediol, naphthalene, quinate, β-alanine, L-citrulline, L-isoleucine, L-leucine, L-norleucine, L-phenylalanine, L-threonine, L-tyrosine, L-valine, kynurenate, kynurenine, trigollenine, anthranilate, *m*-aminobenzoate, *p*-aminobenzoate, α-aminobutyrate, γ-aminobutyrate, δ-aminovalerate, methylamine, ethanolamine, benzylamine, tryptamine, butylamine, 1-amylamine, 2-amylamine, putrescine, spermine, betaine, creatine, histamine, hippurate, acetamide, pantothenate, nicotinate, nicotinamide.

List of species of the genus Marinobacter

1. **Marinobacter hydrocarbonoclasticus** Gauthier, Lafay, Christen, Fernandez, Acquaviva, Bonin and Bertrand 1992, 574[VP]

hy′ dro.car.bo.no.clas.ti.cus. M.L. part. adj. *hydrocarbonoclasticus* hydrocarbonoclastic, hydrocarbon dismantler.

Phenotypic characteristics are described in the text and presented in Tables BXII.γ.141 and BXII.γ.142. Isolated from seawater of the Mediterranean Sea, near a petroleum refinery outlet, Gulf of Fos, France.

The mol% G + C of the DNA is: 53 (T_m); 57.3 (HPLC).

Type strain: SP.17, ATCC 49840, DSM 8798.

GenBank accession number (16S rRNA): X67022 (ATCC 49840).

2. **Marinobacter aquaeolei** Nguyen, Denner, Dang, Wanner and Stan-Lotter 1999, 373[VP]

a.quae.o′le.i. L. n. *aqua* water; L. n. *oleum* oil; M.L. gen. n. *aquaeolei* from water of oil, isolated from an oil-field brine.

Gram-negative motile rods (0.4–0.5 × 1.4–1.6 μm). Single polar flagellum. Nonsporeforming. Growth at 13–50°C, optimum 30°C; pH 5–10, optimum 7.3; and NaCl concentrations 0–20%; optimum 5%. Catalase and oxidase positive. Grows aerobically. Also grows anaerobically on acetate, citrate, and succinate when nitrate is present. Carbon and energy sources include acetate, L-alanine, butyrate, citrate, fumarate, L-glutamate, L-glutamine, L-isoleucine, DL-lactate, L-leucine, L-proline, and succinate. Degrades *n*-hexadecane, pristine, and some compounds in crude oil. Isolated from the head of an offshore oil/gas platform in southern Vietnam.

The mol% G + C of the DNA is: 55.7 ± 0.1 (HPLC).

Type strain: VT8, ATCC 700491, DSM 11845.

GenBank accession number (16S rRNA): AJ000726.

TABLE BXII.γ.143. Differential characteristics of *Marinobacterium* and some aerobic, rod-shaped, marine bacteria that require NaCl to grow[a]

Characteristic	*Marinobacterium*	*Alteromonas*,[b] *Pseudoalteromonas*,[c] and *Shewanella*[d]	*Halomonas marina*	*Halomonas* spp.[e]	*Marinobacter hydrocarbonoclasticus*	*Marinomonas*[f]	*Mesophilobacter*	*Methylomicrobium pelagicum*[g]	*Methylophaga marina*	*Oceanospirillum jannaschii*[h]	*Oceanospirillum kriegii*	*Pseudomonas doudoroffii*	*Pseudomonas stanieri*	*Pseudomonas stutzeri*[j]
Cell shape:														
Straight rod	+	+[l]	+	+	+	D	+[k]	+	+	+	+	+	+	+
Curved rod	−	−[l]	−	−	−	D[m]	−	−	−	−	−	−	−	−[n]
Motility	+	+	+	+	+	+	−	+	+	+	+	+	+	+
Flagellar arrangement:														
Polar	+	+	+	−	+	+	nd	+	+	+	+[i]	+[i]	+[i]	+
Lateral	−	−	+	+	−	−	nd	−	−	−	−	−	−	−
Cell diameter, μm	0.5–0.7	0.5–1.5	0.8–1.1	0.8–1.1	0.3–0.5	0.7–1.5	0.5–0.6	0.8–1.2	0.2	1.0–1.4	0.8–1.2	0.7–1.2	0.6–0.8	0.5–0.7
Poly-β-hydroxybutyrate accumulation	−	+	+	+	−	−	−	+	nd	+	+	+	+	−
Utilization of:														
DL-Malate	+	−	+	D	d	−	+	−	−	d	+	+	d	+
α-Ketoglutarate	nd	−[o]	+	D	−	+	nd	−	−	nd	+	+	+	+
β-Hydroxybutyrate	+	+	nd	nd	d	+	nd	+	−	−	+	+	+	+
Methane	nd	−	−	−	−	−	nd	+	−	−	−	−	−	−
Methanol	+	−	−	−	−	+	nd	+	+	−	−	−	−	−
Methylamine	−	−	−	−	−	−	nd	+	+	−	−	−	−	−
NO₃⁻ reduced to NO₂⁻	−	−[p]	−	D	d	D	+	nd	D	−	−	d	−	+
Oxidase	+	+	−	D	+	−	+	nd	+	+	+	+	+	+
Lipase	+	+	nd	nd	d	−	nd	nd	nd	−	−	−	−	−
Gelatinase	−	+	−	−	+	+	nd	−	nd	−	−	−	−	−
D-Glucose	+	+[q]	+	+	+	+	+	−	−	−	−	+	+	+
D-Fructose	+	D	+	D	−	+	+	−	−	−	−	+	+	+
Saccharate	nd	nd	−	D	+	+	nd	nd	nd	−	+	+	+	+
Glycolate	nd	nd	−	D	−	−	nd	nd	nd	−	−	+	+	+
Mannitol	nd	nd	−	D	−	−	+	nd	nd	−	−	+	+	+
Isobutanol	nd	nd	−	D	−	−	nd	nd	nd	−	−	−	−	+
Quinate	nd	nd	−	D	+	+	nd	nd	nd	+	+	−	+	+
δ-Aminovalerate	nd	nd	+	D	−	−	nd	nd	nd	−	−	+	d	+
L-Tyrosine	nd	D	+	D	−	+	nd	nd	nd	+	+	+	d	+
Sarcosine	nd	nd	d	D	−	+	nd	nd	nd	+	−	+	−	−

[a] For symbols, see standard definitions; nd, not determined.
[b] The species considered is *A. macleodii*.
[c] The species considered are *P. haloplanktis*, *P. espejiana*, *P. undina*, *P. rubra*, *P. luteoviolacea*, *P. citrea*, *P. aurantia*, *P. nigrifaciens*, and *P. denitrificans*, all of which were previously included in the genus *Alteromonas*.
[d] The species considered are *S. hanedai* and *S. colwelliana*, which were previously included in the genus *Alteromonas*.
[e] The species considered are *H. aquamarina*, *H. cupida*, *H. halophila*, *H. pacifica*, and *H. venusta*, all of which were previously assigned to the genus *Deleya*.
[f] The species considered are *M. communis* and *M. vaga*, which were formerly assigned to the genus *Alteromonas*.
[g] *Methylomicrobium pelagicum* was previously included in the genera *Methylomonas* and *Methylobacter*.
[h] The two species considered here are straight rods. All other species of *Oceanospirillum* are helical.
[i] Marine pseudomonads differ from terrestrial species by requiring over 75 mM NaCl for optimum growth rate and cell yield.
[j] Refers to strains previously classified as *P. perfectomarina* but shown to belong to *P. stutzeri* by DNA hybridization experiments.
[k] Coccoid cells occur in old cultures.
[l] Only one species, *P. undina*, is a curved rod. Long, filamentous helices, often exceeding 20 μm, occur when *S. colwelliana* grows on surfaces, in nutrient-poor media, or during late growth phases.
[m] *M. communis* is curved; *M. vaga* is straight.
[n] Some cells of *P. stutzeri* (*P. perfectomarina*) may be slightly curved.
[o] *S. colwelliana* utilizes α-ketoglutarate.
[p] *S. hanedai*, *S. colwelliana*, and some strains of *P. haloplanktis* are positive. *P. denitrificans* carries out denitrification with gas formation.
[q] *S. colwelliana* and some strains of *S. hanedai* are unable to use D-glucose.

Genus VIII. **Marinobacterium** *González, Mayer, Moran, Hodson and Whitman 1997, 375*[VP]

JOSÉ M. GONZÁLEZ

Ma.ri.no.bac.te′ri.um. L. adj. *marinus* of the sea; Gr. n. *bakterion* a small rod; M.L. neut. n. *Marinobacterium* small marine rod.

Rod-shaped cells, 0.5–0.7 × 1.6–2.3 μm. Motile by means of a single polar flagellum. Gram negative. Nonsporeforming and noncystforming. **Have a strictly aerobic type of metabolism with O$_2$ as the terminal electron acceptor.** Denitrification does not occur. Nonpigmented. **Oxidase and catalase positive. Sea salt-based medium is required for growth.** Chemoorganotrophic, growing on various sugars, fatty acids, aromatic compounds, and amino acids. C$_1$ compounds are utilized, such as methanol or formate. Vitamins, in the form of 0.005% yeast extract or a complex vitamin solution, are required for optimal growth.

The mol% G + C of the DNA is: 54.9.

Type species: **Marinobacterium georgiense** González, Mayer, Moran, Hodson and Whitman 1997, 375.

FURTHER DESCRIPTIVE INFORMATION

When the type strain is observed by electron microscopy and negative staining, the cells exhibit numerous blebs and vesicles on their surface, and an S layer composed of repeating units covers the surfaces of the blebs and vesicles. Poly-β-hydroxybutyrate is not accumulated.

Temperature range for growth, 4–41°C; optimum, 37°C. The NaCl range for growth is 0.01–2 M; optimum, 0.1–0.5 M.

The type strain contains the following major fatty acids: C$_{16:0}$, C$_{10:0\ 3OH}$, C$_{10:0}$, C$_{12:0}$, C$_{18:0}$, C$_{18:1\ \omega7c}$, C$_{18:1\ \omega9t}$, C$_{18:1\ \omega12t}$, C$_{16:1\ \omega7c}$, and C$_{15\ iso\ 2OH}$.

ENRICHMENT AND ISOLATION PROCEDURES

The type strain was isolated from a marine enrichment community growing on the high molecular weight fraction of a black liquor sample from pulp mill effluent, which is rich in lignin, lignin byproducts, and other plant polymers in a smaller proportion (González et al., 1997). The original inoculum was from a salt marsh on the coast of Georgia, U.S.A. The enrichment medium consists of filter-sterilized seawater containing 5 μM NH$_4$NO$_3$, 1 μM KH$_2$PO$_4$, and the liquor fraction at a concentration of 20 mg/l. This medium is inoculated with cloudy, brown-green seawater from the salt marsh. Flasks are incubated aerobically with shaking. After enrichment, the organism is isolated on YTSS[1] agar plates incubated at room temperature.

MAINTENANCE PROCEDURES

The type strain is maintained in YTSS broth with 15% glycerol and 15% DMSO at −70°C. Lyophilized cultures are also used.

DIFFERENTIATION OF THE GENUS *MARINOBACTERIUM* FROM OTHER GENERA

Table BXII.γ.143 (see previous page) lists characteristics differentiating *Marinobacterium* from other aerobic marine bacteria. The NaCl requirement for growth excludes *Marinobacterium* from the family *Halomonadaceae* and the genus *Marinobacter*. The mol% G + C is different from values of members of the genera *Alcaligenes*, *Oceanospirillum*, *Marinomonas*, and *Microbulbifer*.

TAXONOMIC COMMENTS

Based on 16S rRNA sequence analysis, *Marinobacterium* is related to *Pseudomonas*, *Oceanospirillum*, *Marinomonas*, *Microbulbifer*, *Halomonas*, and *Marinobacter*. However, the percent similarity with any of these genera is less than 90%, indicating that *Marinobacterium* is a distinct genus.

---

1. YTSS agar (g/l distilled water): yeast extract (Difco Laboratories, Detroit, MI, USA), 4.0; tryptone (Difco), 2.5; sea salts (Sigma Chemical Co., St. Louis, MO), 20; and agar, 18. YTSS broth is the same except that agar is omitted.

List of species of the genus Marinobacterium

1. **Marinobacterium georgiense** González, Mayer, Moran, Hodson and Whitman 1997, 375[VP]

geor.gi.en′ se. L. adj. *georgiense* from Georgia, U.S.A., referring to the place where it was first isolated.

The description is as given for the genus and as given in Table BXII.γ.143. In addition, fimbriae are present and small capsules are produced.

Colonies on Marine Agar 2216 (Difco Laboratories, Detroit, Mich.) are translucent.

Growth occurs between pH 5.5 and 9.5; optimum, 7.5. The NaCl range for growth is 0.01–2.0 M, optimum, 0.1–0.5 M.

Growth occurs on the C$_1$ compounds methanol and formate, organic acids, monosaccharides, disaccharides, amino acids, and aromatic compounds. Included among the aromatic compounds are lignin-related compounds, such as *p*-coumarate, cinnamate, ferulate, and vanillate. Phenol, *p*-hydroxybenzoate, and benzoate are also metabolized.

Cellulose, chitin, gelatin, starch, agar, and xylan are not hydrolyzed.

The mol% G + C of the DNA is: 54.9 (HPLC).

Type strain: KW-40, ATCC 700074, CIP 105236, DSM 11526.

GenBank accession number (16S rRNA): U58339.

Genus IX. **Microbulbifer** *González, Mayer, Moran, Hodson and Whitman 1997, 375*[VP]

JOSÉ M. GONZÁLEZ

Mi.cro.bul′ bi.fer. Gr. adj. *micro* small; L. masc. n. *bulbus* bulb; suff. *-fer* carrying, bearing; L. masc. n. *Microbulbifer* small bearer of bulbs.

Rod-shaped cells, 0.3–0.5 × 1.1–1.7 μm. Gram negative. Capsules are produced. Nonsporeforming and noncystforming. **Nonmotile.** Poly-β-hydroxybutyrate is not produced. Colonies are nonpigmented. **Possess a strictly aerobic type of metabolism with oxygen as the terminal electron acceptor.** Denitrification does not occur. **Oxidase and catalase positive. A sea salt-based medium**

is required for growth. Chemoorganotrophic, growing on sugars, fatty acids, aromatic compounds, and amino acids. Cellulose, xylan, chitin, gelatin, starch, and Tween 80 are hydrolyzed. Related to but outside the group of true *Pseudomonas* in the *Gammaproteobacteria*.

The mol% G + C of the DNA is: 57.6.

Type species: **Microbulbifer hydrolyticus** González, Mayer, Moran, Hodson and Whitman 1997, 375.

FURTHER DESCRIPTIVE INFORMATION

The cell envelope has numerous surface vesicles derived from the outer membrane.

Microbulbifer contains the following major fatty acids: $C_{15:0\ iso}$, $C_{17:1\ iso\ \omega9c}$, $C_{11:0\ iso\ 3OH}$, $C_{11:0\ iso}$, $C_{17:0\ iso}$, $C_{15:0}$, $C_{17:1\ \omega8c}$, $C_{17:0}$, $C_{18:1\ \omega7c}$, $C_{18:1\ \omega9t}$, $C_{18:1\ \omega12t}$, $C_{16:1\ \omega7c}$, and $C_{15\ iso\ 2OH}$.

ENRICHMENT AND ISOLATION PROCEDURES

The type strain was isolated from a marine enrichment community growing on the high-molecular-weight fraction of a black liquor sample from pulp mill effluent, which is rich in lignin, lignin byproducts, and other plant polymers in a smaller proportion (González et al., 1997). The original inoculum was from a salt marsh on the coast of Georgia. The enrichment medium consists of filter-sterilized seawater containing 5 μM NH_4NO_3, 1 μM KH_2PO_4, and the liquor fraction at a concentration of 20 mg/l. This medium is inoculated with cloudy, brown-green sea water from the salt marsh. Flasks are incubated aerobically with shaking. After enrichment, the organism is isolated on YTSS[1] agar plates incubated at room temperature.

MAINTENANCE PROCEDURES

The type strain is maintained in YTSS broth with 15% glycerol and 15% DMSO at −70°C. Lyophilized cultures are also used.

DIFFERENTIATION OF THE GENUS *MICROBULBIFER* FROM OTHER GENERA

The NaCl requirement for growth excludes *Microbulbifer* from the family *Halomonadaceae* and the genus *Marinobacter*. The mol% G + C content of the DNA is different from that of members of the genera *Oceanospirillum*, *Marinomonas*, and *Marinobacter*. *Microbulbifer* is able to use a limited range of carbohydrates, in contrast to most *Pseudomonas* species. The fatty acid profile also differs from those of other bacteria. Among other characteristics, the fatty analysis reveals a low resemblance with *Marinobacterium*. Although the characteristics of *Microbulbifer* are very similar to those of *Alteromonas*, the mol% G + C is outside the range of this genus.

TAXONOMIC COMMENTS

Based on its 16S rRNA sequence, besides *Pseudomonas*, other closely related bacteria are the marine genera *Oceanospirillum*, *Marinomonas*, *Marinobacterium*, *Halomonas*, and *Marinobacter*.

---

1. YTSS agar (g/l distilled water): yeast extract (Difco Laboratories, Detroit, MI, USA), 4.0; tryptone (Difco), 2.5; sea salts (Sigma Chemical Co., St. Louis, MO), 20; and agar, 18. YTSS broth is the same except that agar is omitted.

List of species of the genus Microbulbifer

1. **Microbulbifer hydrolyticus** González, Mayer, Moran, Hodson and Whitman 1997, 375[VP]

   hy.dro.ly′ ti.cus. Gr. n. *hydor* water; Gr. adj. *lyticus* dissolving, splitting; M.L. adj. *hydrolyticus* splitting with water, referring to the hydrolytic activity of the bacterium.

   Cells are rod shaped, 0.3–0.5 × 1.1–1.7 μm. Cells appear as single cells or short chains. In medium with xylan or viscous medium, they form long chains. They do not possess flagella. An S layer is present. Capsules are produced in liquid media. Polyhydroxybutyrate is not produced.

   Colonies on marine agar 2216 are circular, convex with entire margins, and cream colored.

   Growth occurs between pH 6.5 and 8.5; optimum, 7.5. The temperature range for growth is 10–41°C; optimum, 37°C. Sea salt-based medium is required for growth. The NaCl range for growth is 0.1–1.0 M; optimum, 0.1–0.5 M. Vitamins are required for optimal growth.

   Growth occurs on organic acids, amino acids, and a few monosaccharides, disaccharides, and aromatic compounds (glucose, xylose, *N*-acetyl-D-glucosamine, cellobiose, vanillate, and ferulate). Cellulose, xylan, chitin, gelatin, starch, and Tween 80 are hydrolyzed. This species has been found by oligonucleotide probes to be numerically dominant in a pulp mill effluent enrichment community containing only bacteria.

   The mol% G + C of the DNA is: 57.6 (HPLC).

   Type strain: IRE-31, ATCC 700072; CIP 105235, DSM 11525.

   GenBank accession number (16S rRNA): U58338.

Genus X. **Moritella** Urakawa, Kita-Tsukamoto, Steven, Ohwada and Colwell 1999b, 341[VP]
(Effective publication: Urakawa, Kita-Tsukamoto, Steven, Ohwada and Colwell 1998, 376)

THE EDITORIAL BOARD

Mo.ri.tel′ la. M.L. dim. *-ella* ending; M.L. fem. n. *Moritella* named after Richard Y. Morita to honor his work in marine microbiology.

Gram-negative, chemoorganotrophic, facultatively anaerobic motile rods. Polar flagella. **Halophilic.** Catalase and oxidase positive. Voges–Proskauer negative. Negative for production of indole and H_2S. Do not produce arginine dihydrolase or ornithine decarboxylase. Produce acid from D-glucose. Utilize *N*-acetyl glucosamine.

The mol% G + C of the DNA is: 40–45.

Type species: **Moritella marina** (Baumann, Furniss and Lee 1984c) Urakawa, Kita-Tsukamoto, Steven, Ohwada and Colwell

1999b, 341 (Effective publication: Urakawa, Kita-Tsukamoto, Steven, Ohwada and Colwell 1998, 376) (*Vibrio marinus* Baumann, Furniss and Lee 1984c, 356.)

FURTHER DESCRIPTIVE INFORMATION

The genus *Moritella* includes four species; all inhabitants of marine environments. *M. marina* was isolated from fish skin, seawater, and marine sediment (Colwell and Morita, 1964; Urakawa et al., 1998). *Moritella japonica*, which is barotolerant, was obtained from sediments from the Japan Trench at a depth of 6,356 m (Nogi et al., 1998a). *M. yayanosii*, which requires pressures of 50 MPa or greater, was obtained from sediments from the Mariana Trench at a depth of 10,898 m (Nogi and Kato, 1999a).

M. viscosa is an etiologic agent of the disease "winter ulcer" in Atlantic salmon (*Salmo salar*) and may cause skin disease in other species of fish (Lunder et al., 2000).

Analysis of 16S rDNA sequences placed the genus *Moritella* in the *Gammaproteobacteria* (Urakawa et al., 1998).

ENRICHMENT AND ISOLATION PROCEDURES

Sampling and isolation procedures for *Vibrio marinus* (*Moritella marina*) from seawater are described by Colwell and Morita (1964).

Moritella japonica and *Moritella yayanosii* were isolated by diluting bottom mud sediments in Marine Broth 2216™ (Difco)

and incubating at 4–15°C either at atmospheric pressure or in a pressurized vessel (Kato et al., 1995; Nogi et al., 1998a; Nogi and Kato, 1999a).

Moritella viscosa can be isolated from tissue samples on blood agar made 0.5–2.0% in NaCl as described by Lunder et al. (1995).

MAINTENANCE PROCEDURES

Media and conditions for maintenance of *Vibrio marinus* (*Moritella marina*) are described by Colwell and Morita (1964).

Moritella japonica and *Moritella yayanosii* can be maintained on Marine Agar 2216™ (Difco) with incubation at 4–15°C either at atmospheric pressure or in a pressurized vessel as appropriate (Kato et al., 1995; Nogi et al., 1998a; Nogi and Kato, 1999a).

Moritella viscosa strains can be maintained at 4°C on Difco brain–heart infusion agar made 2% in NaCl or on Difco Marine Agar 2216™; long-term storage can be achieved at –80°C in brain–heart infusion broth made 2% in NaCl and 30% in glycerol or in Marine Agar 2216™ made 10% in glycerol (Benediktsdøttir et al., 2000; Lunder et al., 2000).

DIFFERENTIATION OF THE GENUS *MORITELLA* FROM OTHER GENERA

Tables of phenotypic characteristics that differentiate *Moritella* from species of the genera *Shewanella* and *Vibrio* are available (Nogi et al., 1998a; Nogi and Kato, 1999a; Lunder et al., 2000).

DIFFERENTIATION OF THE SPECIES OF THE GENUS *MORITELLA*

Nogi et al. (1998a), Nogi and Kato (1999a), and Lunder et al. (2000), provide tables of phenotypic characteristics that differentiate *Moritella* species from each other.

List of species of the genus Moritella

1. **Moritella marina** (Baumann, Furniss and Lee 1984c) Urakawa, Kita-Tsukamoto, Steven, Ohwada and Colwell 1999b, 341[VP] (Effective publication: Urakawa, Kita-Tsukamoto, Steven, Ohwada and Colwell 1998, 376) (*Vibrio marinus* Baumann, Furniss and Lee 1984c, 356)

   ma.ri′na. L. fem. adj. *marina* marine, of the sea.

   Description as for the genus with the following additional characteristics. Nitrate reduced to nitrite. Optimum growth temperature 18°C. Major fatty acids $C_{14:0}$, $C_{16:0}$, $C_{16:1}$, $C_{22:6}$.

   The mol% G + C of the DNA is: 42.2 (T_m).

   Type strain: ATCC 15381, CIP 102861, NCCB 79030, NCIMB 1144.

   GenBank accession number (16S rRNA): AB038033, X82142.

2. **Moritella japonica** Nogi, Kato and Horikoshi 1999, 341[VP] (Effective publication: Nogi, Kato and Horikoshi 1998a, 294.)

   ja.pon.i′ca. L. gen. n. *japonica* named after the Japan Trench, where this strain originated.

   Description as for the genus with the following additional characteristics. Psychrophilic (growth optimum 10–15°C) and halophilic (growth optimum 3% NaCl). Able to grow both at atmospheric pressure and at pressures up to 70 MPa. Nitrate reduced to nitrite. Lysine decarboxylase produced. Acid produced from glycerol. Major fatty acids $C_{14:0}$, $C_{16:0}$, $C_{16:1}$, $C_{22:6}$.

   The mol% G + C of the DNA is: 45 (HPLC).

   Type strain: DSK1, CIP 106291, JCM 10249.

   GenBank accession number (16S rRNA): D21224.

3. **Moritella viscosa** (Lunder, Sørum, Holstad, Steigerwalt, Mowinckel and Brenner 2000) Benediktsdøttir, Verdonck, Sproer, Helgasön and Swings 2000, 487[VP] (*Vibrio viscosus* Lunder, Sørum, Holstad, Steigerwalt, Mowinckel and Brenner 2000, 447.)

   vis.co′sa. L. fem. adj. *viscosa* viscous, because of its thread-forming, adherent colonies.

   Description as for the genus with the following additional characteristics. Requires 1–4% NaCl. Lysine decarboxylase produced. Hydrolysis of casein, DNA, gelatin, lecithin, starch (19 of 20 strains), Tween 80, and urea. Acid produced from D-galactose.

   The mol% G + C of the DNA is: 42.5 (T_m).

   Type strain: NVI 88/478, ATCC BAA-105, NCIMB 13584.

   GenBank accession number (16S rRNA): Y17574.

4. **Moritella yayanosii** Nogi and Kato 1999b, 1325[VP] (Effective publication: Nogi and Kato 1999a, 76.)

   ya.ya.nos.i′i. M.L. gen. n. *yayanosii* in honor of American deep-sea biologist Aristides Yayanos.

   Description as for the genus with the following additional characteristics. Cannot grow at pressures less than 50 MPa; able to grow at pressures up to 100 MPa. Requires NaCl. No reduction of nitrate or nitrite. Major fatty acids $C_{14:0}$, $C_{16:0}$, $C_{16:1}$, $C_{22:6}$.

   The mol% G + C of the DNA is: 44.6 (HPLC).

   Type strain: DB21MT-5, JCM 10263.

   GenBank accession number (16S rRNA): AB008797.

Genus XI. **Pseudoalteromonas** Gauthier, Gauthier and Christen 1995a, 759[VP]

JOHN P. BOWMAN AND THOMAS A. MCMEEKIN

Pseu.do.al.te.ro.mo'nas. Gr. adj. *pseudes* false; M.L. n. *Alteromonas* the genus *Alteromonas*; M.L. n. *Pseudoalteromonas* false *Alteromonas.*

Straight rods, approximately 1×2–3 µm, occurring singly or in pairs. Do not form spores or cysts. Do not accumulate poly-β-hydroxybutyrate. Nonbioluminescent. **Gram negative. Motile by means of a single polar flagellum**, either unsheathed or sheathed. Colonies are nonpigmented or pigmented. **Chemoheterotrophic. Aerobic, possessing a strictly respiratory type of metabolism** with oxygen as the terminal electron acceptor. No growth occurs anaerobically. A constitutive arginine dihydrolase system is usually absent. One species (*P. denitrificans*) is able to denitrify. Utilize a range of carbon substrates, including carbohydrates, alcohols, organic acids, and amino acids. Form gelatinase and lipase. Ammonia serves as a nitrogen source. Amino acids may be required by some strains for growth. **Requires Na$^+$ ions for growth.** Mostly psychrotolerant, with growth occurring between 0 and 40°C. Several species produce autotoxic antibiotic compounds. Isolated from seawater, sediment, sea ice, surfaces of stones, marine algae, marine invertebrates, and salted foods. The major quinone is ubiquinone-8 (Q-8). Major fatty acids are $C_{16:0}$, $C_{16:1}\,\omega 7c$, $C_{17:1}\,\omega 8c$, and $C_{18:1}\,\omega 7c$. Member of the order *Alteromonadales*, family *Alteromonadaceae*, in the *Gammaproteobacteria*.

The mol% G + C of the DNA is: 38–48 (T_m).

Type species: **Pseudoalteromonas haloplanktis** (Zobell and Upham 1944) Gauthier, Gauthier and Christen 1995a, 759 (*Alteromonas haloplanktis* (Zobell and Upham 1944) Reichelt and Baumann 1973a, 438; "*Vibrio haloplanktis*" Zobell and Upham 1944, 261.)

FURTHER DESCRIPTIVE INFORMATION

Motility Sheathed flagella occur on cells of *P. denitrificans* (Enger et al., 1987), *P. luteoviolacea* (Novick and Tyler, 1985), and *P. piscicida* (Hansen et al., 1963, 1965). Cells occasionally form three or more polar flagella. Older cultures often exhibit significantly reduced levels of motile cells, and filamentous cells and involution bodies often occur.

Cultural characteristics Colonies on Difco Marine 2216 agar are typically circular and convex, possessing an entire edge and a creamy or mucoid consistency. Colonies are nonpigmented or pigmented with noncarotenoid type pigments. When growing on tyrosine-containing media, some species form black-brown, melanin-like pigments (Table BXII.γ.144). In some instances, such as in the type strain of *P. nigrifaciens* (ATCC 27126), melanin production occurs constitutively on complex media. *P. citrea*, *P. aurantia*, and *P. piscicida* form yellow or orange pigments. Spectral and preliminary chemical analyses of the pigments of *P. piscicida* suggest the pigments are pteridine derivatives (Weeks et al., 1962). Both *P. rubra* and *P. denitrificans* form bright red prodigiosin-type pigments. When *P. denitrificans* is grown on protamine or L-glutamine, the color of cultures changes from red to bright purple or blue (Enger et al. 1987). Strains of *P. luteoviolacea* form violet colonies owing to accumulation of violacein and can simultaneously form yellow pigments, giving colonies an unusual and characteristic appearance (Gauthier, 1982).

Pseudoalteromonas are either psychrotrophic or mesophilic. No psychrophilic strains, however, have been isolated, even from the cold-water and sea-ice environments of the Antarctic (Bozal et al., 1997; Bowman, 1998).

Antibiotic production Most pigmented and a few nonpigmented *Pseudoalteromonas* strains have the ability to form antibiotic substances. The red pigment of *P. denitrificans*, identified as cycloprodigiosin hydrochloride, has been shown to suppress T-cell proliferation and has been proposed as a immunosuppressant therapeutic agent. Cycloprodigiosin hydrochloride apparently selectively uncouples proton translocation from the ATPase reaction (Kawauchi et al., 1997). Strains of *P. citrea*, *P. aurantia*, *P. rubra*, and *P. luteoviolacea* form autotoxic, high-molecular-weight polyanionic substances (70–100 kDa in size), which can inhibit Gram-negative and Gram-positive bacteria (Gauthier, 1976b; Gauthier and Flatau, 1976; Ballester et al., 1977). Analysis of the agent produced by *P. rubra* indicates it is an outer membrane glycoprotein of complex composition (Ballester et al., 1977). *P. citrea* produces two different proteinaceous components, as determined by electrophoretic profiles (Gauthier, 1977). The mode of action of these agents involves inducing an increase in the rate of oxygen consumption in susceptible bacteria, leading to lethal cell damage due to accumulation of oxygen free radicals (Gauthier, 1976b). In addition to this, *P. luteoviolacea* can form a variety of low-molecular-weight brominated compounds, including pentabromopseudilin, which are bacteriolytic (Gauthier and Flatau, 1976; Sakata et al., 1982, 1986; Hanefeld et al., 1994). *P. haloplanktis* forms an unusually bioactive iron siderophore, bisucaberin, which is able to block DNA synthesis in tumor cells and induce macrophage-mediated cytolysis (Kameyama et al., 1987; Takahashi et al., 1987). Siderophore production by *Pseudoalteromonas* strains is thought to give an advantage in the marine environment, where iron often is very scarce (Reid and Butler, 1991; Reid et al., 1993). The ability of *Pseudoalteromonas* to produce antibiotics may have some unexpected advantages; for example, some strains of *P. haloplanktis* produce a heat-labile, proteinaceous compound able to inhibit various pathogenic *Vibrio* species, which has been proposed as a probiotic agent to protect reared scallops and oyster larvae (Riquelme et al., 1996).

The exact scope, ecological role, mechanisms, and ecological impact of antibiotic production by *Pseudoalteromonas* are under increasing investigation. Obviously, antibiotic production would help *Pseudoalteromonas* species effectively compete for potentially scarce nutrients by inhibiting other marine bacteria. Another possible role is that they act as chemical barriers to predation; for example, various marine biota have been observed to possess coats of pigmented, aerobic, rod-shaped bacteria, which produce bioactive substances (Fenical, 1993). *Pseudoalteromonas* strains, however, also have an aggressive role in the marine food chain and are able to kill various types of toxic "red-tide" dinoflagellates and raphidiophytes (such as *Gymnodinium* and *Chatonella* species) using uncharacterized, proteinaceous, secreted substances (Imai et al., 1995; Lovejoy et al., 1998). *P. piscicida* has been claimed to produce a neuromuscular toxin, which is able to kill a variety of fish and crab species (Bein, 1954; Meyers et al., 1959; Hansen et al., 1965). In addition, *P. haloplanktis* subsp. *tetraodonis* can apparently form tetrodotoxin, an exceptionally potent neurotoxin that has been isolated from the puffer fish (*Fugu poecilonotus*), from which this subspecies was first described (Yasumoto

TABLE BXII.γ.144. Characteristics differentiating the species of the genus *Pseudoalteromonas*[a]

Characteristics	*P. haloplanktis*	*P. antarctica*	*P. atlantica*	*P. aurantia*	*P. carragenovora*	*P. citrea*	*P. denitrificans*	*P. espejiana*	*P. luteoviolacea*	*P. nigrifaciens*	*P. piscicida*	*P. prydzensis*	*P. rubra*	*P. tetraodonis*	*P. undina*
Sheathed polar flagellum	−	−	−	−	−	−	+	−	d		+	−	−	−	−
Pigments:															
Yellow-orange	−	−	−	+	−	d	−	−	d	−	+	−	−	−	−
Red (prodigiosin)	−	−	−	−	−	−	+	−	−	−	−	−	+	−	−
Violet (violacein)	−	−	−	−	−	−	−	−	+	−	−	−	−	−	−
Melanin-like	+	−	+	+	d	d	nd	−	d	+	+	−	nd	−	−
NO₃→NO₂	d	−	d	−	−	−	+	−	−	−	−	−	−	+	−
Denitrification	−	−	−	−	−	−	+	−	−	−	−	−	−	−	−
Catalase	+	+	+	+	+	d	+	+	−	+	d	+	−	+	+
Agarase	−	−	+	−	−	d	−	−	−	−	−	−	−	−	−
Chitinase	d	−	−	−	−	−	+	−	−	+	+	+	−	+	+
Alginase	−		+	+	+	+	+	+	+		+	+	+	+	−
Amylase	d	d	+	−	+	+	+	+	nd	−	−	−	nd	+	d
Growth at:															
4°C	−	+	+	+	+	d	+	−	d	d	+	+	−	+	d
35°C	+	−	+	−	+	d	−	d	+	d	+	−	+	+	+
Tolerates 9% NaCl	+	+	+	−	+	−	−	+	−	+	+	+	−	nd	+
Amino acid requirement	d	−	−	+	−	d	+	+	+	−	+	−	+	+	+
Antibiotic production	d	nd	nd	+	nd	d	+	−	+	nd	+	nd	+	+	−
Utilization of:															
D-Ribose	−	−	−	−	−	+	nd	−	−	+	−	nd	−	−	−
D-Fructose	d	−	+	+	+	+	−	d	−	+	d	−	−	−	−
D-Galactose	d	+	+	−	+	−	−	+	−	+	d	d	−	−	−
D-Mannose	+	d	+	+	−	d	nd	d	−	d	+	+	+	+	−
Cellobiose	−	d	+	−	+	d	−	d	−	d	−	d	−	−	−
Lactose	−	−	+	−	+	d	−	+	−	d	−	−	−	−	−
Maltose	+	+	+	d	+	−	+	+	+	+	+	+	−	+	+
Melibiose	−	−	+	−	+	nd	nd	+	−	+c	−	−	nd	−	−
Sucrose	+	−	+	−	+	d	−	+	−	+	+	+	−	+	+
Trehalose	d	−	+	+	−	d	nd	+	+	+	+	+	+	+	d
Mannitol	d	+	+	−	+	d	−	+	−	+	−	+	−	−	−
D-Gluconate	−	d	−	−	−	d	−	−	−	d	−	+	−	−	−
Glycerol	−	+	d	−	+	d	−	−	−	d	−	+	−	−	−
Acetate	+	−	+	−	+	d	+	−	−	d	+	+	−	+	+
Succinate	+	+	d	−	+	+	−	−	−	d	+	+	−	+	+
Pyruvate	+	nd	nd	−	nd	+	+	+	d	+	+	+	−	+	+
Fumarate	+	nd	d	−	+	d	−	−	−	d	+	+	−	+	+
DL-Malate	−	−	−	−	+	−	nd	−	−	−	+	+	−	+	−
DL-Lactate	−	−	−	−	−	d	−	−	d	+	−	−	−	−	−
Citrate	+	d	+	−	+	d	−	+	−	−	+	+	−	−	−
2-Oxoglutarate	−	nd	−	−	+	−	nd	−	−	−	−	−	−	−	−
Aconitate	+	nd	+	−	−	+	nd	+	nd	+	−	+	nd	nd	−
p-Hydroxybenzoate	−	−	−	−	+	+	−	−	−	+	−	nd	nd	−	−
DL-α-Alanine	+	+	−	−	−	d	+	+	+	+	−	−	nd	+	+
L-Glutamate	d	+	−	−	−	+	+	−	−	+	−	+	nd	+	−
L-Ornithine	−	nd	−	−	−	+	−	d	−	+	−	−	nd	−	d
L-Proline	d	+	−	−	−	+	+	+	d	+	d	+	nd	+	d
Mol% G + C (T_m, Bd)	42–44	41–42	41–42	38–43	39–40	39–45	36–37	43–44	40–43	39–41	44–45	38–39	46–48	41	43–44

[a]For symbols see standard definitions; nd, not determined.

et al., 1986; Simidu et al., 1987, 1990; Gallacher and Birkbeck, 1993). *Pseudoalteromonas* spp. are generally not pathogenic; however, *P. haloplanktis* may cause disease in oysters (Colwell and Sparks, 1967) and the red-spot disease of the seaweed *Laminaria japonica* (Ezura et al., 1988a, b; Sawabe et al., unpublished).

Sodium ion requirement All *Pseudoalteromonas* species require Na⁺ ions for growth; the minimal levels vary somewhat, but are usually in the range of 20 mM. Studies have shown that the Na⁺ requirement cannot be replaced by supplying K⁺ ions but can be partially reduced in some strains by supplying seawater levels of Ca^{2+}, Na^{2+}, and Mg^{2+} (Thompson et al., 1970; Thompson and MacLeod, 1973). Na^+ is necessary for the activity of cellular uptake systems in *Pseudoalteromonas*, including the transport of amino acids, tricarboxylic acid intermediates, galactose, orthophosphate, and K^+ (MacLeod, 1968; Fein and MacLeod, 1975; Droniuk et al., 1987; MacLeod and MacLeod, 1992). Na^+ ions are also necessary for maintaining the integrity of the cell wall (Forsberg et al., 1970). Most nonpigmented *Pseudoaltero-*

monas species demonstrate some degree of halotolerance and can grow at NaCl concentrations of 9–12%. Strains of *P. prydzensis* can even tolerate NaCl levels of 15–20% (Bowman, 1998). Pigmented species appear to be more stenohaline in nature, with the exception of *P. piscicida*.

Metabolism *Pseudoalteromonas* species are strictly aerobic chemoheterotrophs using a respiratory metabolism with oxygen as the electron acceptor. *P. denitrificans* is able to grow microaerobically by denitrification (Enger et al., 1987), while some strains of *P. haloplanktis* and *P. atlantica* can reduce nitrate to nitrite. A constitutive arginine dihydrolase system is generally absent. Glucose and fructose are assimilated by an inducible Entner–Doudoroff pathway (Sawyer et al., 1977a). All strains possess cytochrome *c* oxidase activity. Catalase activity is usually present, though it may be weakly expressed or absent in certain pigmented species. *P. luteoviolacea* and *P. rubra* possess peroxidase (H_2O_2 oxidoreductase) instead of catalase (Gauthier, 1976a, 1982). The following tests have been found to be negative for all tested species: lysine decarboxylase, ornithine decarboxylase, tryptophan deaminase, phenylalanine deaminase, indole production from tryptophan, fermentation of D-glucose, methyl red test, anaerobic growth with nitrate as the electron acceptor, and sensitivity to vibriostat agent O/129.

Nutrition *Pseudoalteromonas* strains grow well on marine 2216 agar and similar seawater media. Though most strains can use ammonia as a sole nitrogen source, some species, in particular the pigmented species, require amino acids for growth. For these strains, growth in defined media can be achieved by the addition of individual amino acids (Baumann et al., 1984a, c) or by addition of low levels of yeast extract or casein hydrolysate (0.1–0.5% w/v). Media useful for growing *Pseudoalteromonas* species are given in the section on the genus *Alteromonas*.

Pseudoalteromonas species can utilize a range of organic compounds as sole carbon and energy sources, including carbohydrates, alcohols, and amino acids. Pigmented species in general tend to be less nutritionally versatile than the nonpigmented species. *Pseudoalteromonas* species also have the ability to degrade a range of macromolecules. Most species can degrade gelatin, casein, starch, and DNA, and can form a lipase. Several species possess chitinolytic and alginolytic activity. *P. atlantica* possesses a type I β-agarase, the gene for which has been cloned, sequenced, and overexpressed in *Escherichia coli* (Belas, 1989). *P. citrea* may also form agarases (Ivanova et al., 1998), and *P. carrageenovora* possesses the unusual ability to degrade κ-carrageenan (Yaphe and Baxter, 1955) by means of a β-1,3-1,4-glucanase-like κ-carrageenase (Barbeyron et al., 1994). The amylase from *P. haloplanktis* strain A23 is cold adapted and thermolabile (Feller et al., 1992). *P. undina* synthesizes an efficient organophosphorus acid anhydrolase (Cheng et al., 1993).

Genetic characteristics The genome structure of *P. haloplanktis* is complex. Analysis by restriction digestion and pulsed-field gel electrophoresis has detected the presence of two genetic units, one with a length of 2.7 Mb and a second replicon with a length of 0.8 Mb (Lanoil et al., 1996).

Lipids, quinones, polyamines The major fatty acids in *Pseudoalteromonas* include $C_{15:0}$ (2–9%), $C_{15:1\ \omega8c}$ (1–7%), $C_{16:0}$ (18–34%), $C_{16:1\ \omega7c}$ (35–50%), $C_{17:1\ \omega8c}$ (1–13%), and $C_{18:1\ \omega7c}$ (1–9%) (Monteoliva-Sanchez and Ramos-Cormenzana, 1987; Matsui et al., 1991; Svetashev et al., 1995; Bozal et al., 1997; Bowman, 1998). So far, fatty acid characterization has been done only for some of the nonpigmented species. In any case, the profiles are so similar to that of *Alteromonas macleodii* that neural network analysis cannot tell the difference (Bertone et al., 1996). The major isoprenoid quinone is ubiquinone-8 (Q-8, 84–100%), with smaller amounts of Q-6 (0–9%), Q-7 (0–6%), and Q-9 (0–15%) present (Akagawa-Matsushita et al., 1992b). The major cellular polyamines in *Pseudoalteromonas* include putrescine, cadaverine, and spermidine (Hamana, 1997). Some species of *Pseudoalteromonas* synthesize unusual acidic exopolysaccharides (Gorshkova et al., 1993, 1997; Bozal et al., 1994).

Ecology Due to the relatively limited scope of individual studies and often inadequate identification of isolates, it is difficult to obtain a precise idea of how widely distributed each *Pseudoalteromonas* species is. However, strong evidence indicates the genus is common and widespread in the marine ecosystem, with strains isolated from coastal and estuarine seawater, organic detritus, deep water, benthic sediments, sea ice, surfaces of marine fauna and flora, aquaculture systems, and salterns (Kakimoto et al., 1980; Gavrilovic et al., 1982; Sakata et al., 1982; Austin, 1982; Kameyama et al., 1987; Lemos et al., 1985; Nair and Simidu, 1987; Nissen, 1987; Oliver, 1987; Ezura et al., 1988a; del Moral et al., 1988; Nair et al., 1988; Noble et al., 1990; Gauthier and Breittmayer, 1992; Bozal et al., 1994, 1997; Ortigosa et al., 1994; Uchida et al., 1997; Yoshikawa et al., 1997; Bowman, 1998). In addition, strains of *Pseudoalteromonas nigrifaciens* have also been isolated from various salted food products, including butter, from which it was originally described (White, 1940). *P. haloplanktis* seems to have a very wide distribution, having been isolated from all major oceans, save for the polar regions (Gauthier and Breittmayer, 1992). *P. antarctica* appears to be very common in the Antarctic marine environment and is one of the most common isolates from seawater and sea ice (Bozal et al., 1997; Bowman et al., 1997a; Bowman, 1998). Pigmented species are often very sensitive to UV irradiation, thus affecting potential isolation sites (Gauthier and Breittmayer, 1992; Enger et al., 1987).

ENRICHMENT AND ISOLATION PROCEDURES

The isolation of specific *Pseudoalteromonas* species is dependent on recognition of certain obvious phenotypic properties, especially pigmentation (Gauthier and Breittmayer, 1992). In most instances, special enrichment or selection conditions for *Pseudoalteromonas* species are not available, and the organisms usually can only be directly isolated from suitable source material onto marine agar or BMA supplemented with 0.1–0.2% concentration of a particular carbon source that is utilized by the species in question. Certain species have been isolated by enrichment, including *P. denitrificans*, which was isolated from a medium containing L-arginine and L-glutamine as sole carbon and nitrogen sources (Enger et al., 1987). Antibiotic production of colonies on isolation plates can be used to screen for antibiotic-producing *Pseudoalteromonas* species by means of the replica plating procedure (Smibert and Krieg, 1994), as described by Gauthier and Breittmayer (1992). For more details on methods useful in isolating *Pseudoalteromonas* species, see Gauthier and Breittmayer (1992).

MAINTENANCE PROCEDURES

Most nonpigmented strains can be stored on marine agar plates or slants at 2–5°C for several months before subculturing is necessary. Pigmented strains, however, often produce autotoxic antibiotic compounds and so need to be transferred at least once per week. *Pseudoalteromonas* strains can be successfully stored for several years as dilute suspensions in sterile, unsupplemented seawater at 7–8°C (Enger et al., 1987). They can also be preserved by freeze-drying with 20% skim milk as a cryoprotectant or by freezing in liquid nitrogen or in a −80°C freezer with 20–30% glycerol as a cryoprotectant.

DIFFERENTIATION OF THE GENUS *PSEUDOALTEROMONAS* FROM OTHER GENERA

Characteristics differentiating the genus *Pseudoalteromonas* from other members of the family *Alteromonadaceae* are shown in Table BXII.γ.135 of the chapter describing the family. Nonpigmented *Pseudoalteromonas* species are most likely to be confused with the genus *Alteromonas*, with which they share many phenotypic and chemotaxonomic similarities. In broad terms, *Pseudoalteromonas* species can be distinguished by their ability to grow at 4°C and inability to utilize a number of carbon substrates. These few phenotypic traits seem inadequate, but until a wider knowledge of the phenotypic differences between these genera are available, the only other effective methods for the conclusive identification are 16S rDNA sequencing, DNA–DNA hybridization, and rRNA–DNA hybridization.

TAXONOMIC COMMENTS

Pseudoalteromonas was created to accommodate most of the former *Alteromonas* species that comprise a phylogenetic clade distinct from *Alteromonas macleodii* (van Landschoot and De Ley, 1983; Kita-Tsukamoto et al., 1993; Gauthier et al., 1995a). *Pseudoalteromonas piscicida*, previously classified as "*Pseudomonas piscicida*" (Buck et al., 1963), was also included in the new genus (Gauthier et al., 1995a).

More specifically, both rRNA–DNA hybridization and 16S rDNA sequencing studies (Kita-Tsukamoto et al., 1993; Gauthier et al., 1995a) indicate *Pseudoalteromonas* species comprise a relatively tight clade adjacent to the genus *Alteromonas*. Fig. BXII.γ.155 shows a phylogenetic tree based on 16S rDNA sequences for *Pseudoalteromonas*. One interesting feature is the high sequence similarity shared by the nonpigmented species; indeed, six species possess sequences that diverge less than 2%—well within the potential estimated range of sequence divergence for a single species (Stackebrandt and Goebel, 1994). *P. prydzensis* is an outlier of this group. The pigmented species show a greater divergence in sequence than do the nonpigmented species, except for *P. citrea* and *P. aurantia*, which are very closely related.

DNA–DNA (Akagawa-Matsushita et al., 1993) and rDNA–DNA hybridization (van Landschoot and De Ley, 1983) results confirm the very close relatedness of the nonpigmented strains. Overall $T_{m(e)}$ values range from 74.0–77.5°C, while DNA–DNA hybridization values range from 30–60%. *P. piscicida* has a $T_{m(e)}$ of 71.5°C, which corresponds with its more distinct phylogenetic position. On the basis of DNA–DNA hybridization data, Ivanova et al. (1998) have indicated that the species *Alteromonas distincta* (Romanenko et al., 1995b) and *Alteromonas elyakovii* (Ivanova et al., 1996b) are closely related to *P. nigrifaciens* (sharing 40–50% DNA hybridization) while *Alteromonas fuliginea* (Romanenko et al., 1994) is a junior subjective synonym of *P. citrea*.

Several strains are novel, yet to be described, species, based on their phylogenetic positions (Fig. BXII.γ.155). Various sea-ice and under-ice seawater strains (strains IC006, IC013, MB6-03, MB6-05, MB8-02, SW08, and SW29) are distributed throughout the nonpigmented *Pseudoalteromonas* clade; however, based on phenotypic data and DNA hybridization, they are closely related to *P. antarctica* (Bowman, 1998; unpublished data).

Several melanin-producing strains isolated from marine fauna from the eastern Russian-Pacific area have been described as *Alteromonas* species. These include *A. fuliginea* (Romanenko et al., 1994), *Alteromonas distincta* (Romanenko et al., 1995b), and *A. elyakovii* (Ivanova et al., 1996b). Subsequent analysis by DNA–DNA hybridization now indicates that these species have a closer allegiance to the genus *Pseudoalteromonas*, with *A. fuliginea* being a junior subjective synonym of the species *P. citrea* (Ivanova et al., 1998). Because these species do not have their generic positions officially established yet, they have been listed as *Species Incertae Sedis*.

DIFFERENTIATION OF THE SPECIES OF THE GENUS *PSEUDOALTEROMONAS*

Characteristics distinguishing *Pseudoalteromonas* species are presented in Table BXII.γ.144. Considerable variation in carbon-source utilization patterns is evident within some species. This may be due to differences in testing procedures, because carbon source results sometimes differ significantly among different studies; examples include the results for the type strains of *P.* *haloplanktis* subsp. *tetraodonis* (Simidu et al., 1990; Akagawa-Matsushita et al., 1992b) and *P. nigrifaciens* (Baumann et al., 1984a; Akagawa-Matsushita et al., 1992b; Ivanova et al., 1996a). The carbon source data shown in Tables BXII.γ.144 and BXII.γ.145 should therefore be treated with some degree of caution.

List of species of the genus Pseudoalteromonas

1. **Pseudoalteromonas haloplanktis** (Zobell and Upham 1944) Gauthier, Gauthier and Christen 1995a, 759[VP] (*Alteromonas haloplanktis* (Zobell and Upham 1944) Reichelt and Baumann 1973a, 438; *Vibrio haloplanktis* Zobell and Upham 1944, 261.)

   ha.lo.plank' tis. Gr. n. *halos* sea; Gr. adj. *planktos* wandering; M.L. adj. *haloplanktis* sea-wandering.

   Cells are rod-shaped, 0.8–1.2 × 1.8–3.0 μm. Colonies are nonpigmented, smooth, and convex, with an entire edge. Sodium ions are required for growth. Growth factors (amino acids) are usually not required for growth. Grows between 10°C and 35°C, with optimum growth occurring at approximately 30°C. Some strains may grow at 4°C. Other characteristics are shown in the genus description and in Tables BXII.γ.144 and BXII.γ.145. Strains utilize D-mannose, citrate, and aconitate, but do not hydrolyze alginate or utilize L-malate, isovalerate, caproate, or L-lysine.

   Isolated from seawater.

   The mol% G + C of the DNA is: 41–45 (T_m).

   Type strain: 545, ATCC 14393, DSM 6060, NCIMB 2084.

   GenBank accession number (16S rRNA): X67024; X82143.

2. **Pseudoalteromonas antarctica** Bozal, Tudela, Rosselló-Mora, Lalucat and Guinea 1997, 350.[VP]

   ant.arc' ti.ca. L. adj. *antarcticus* southern (in this instance pertaining to the Antarctic environment, where the organism was isolated).

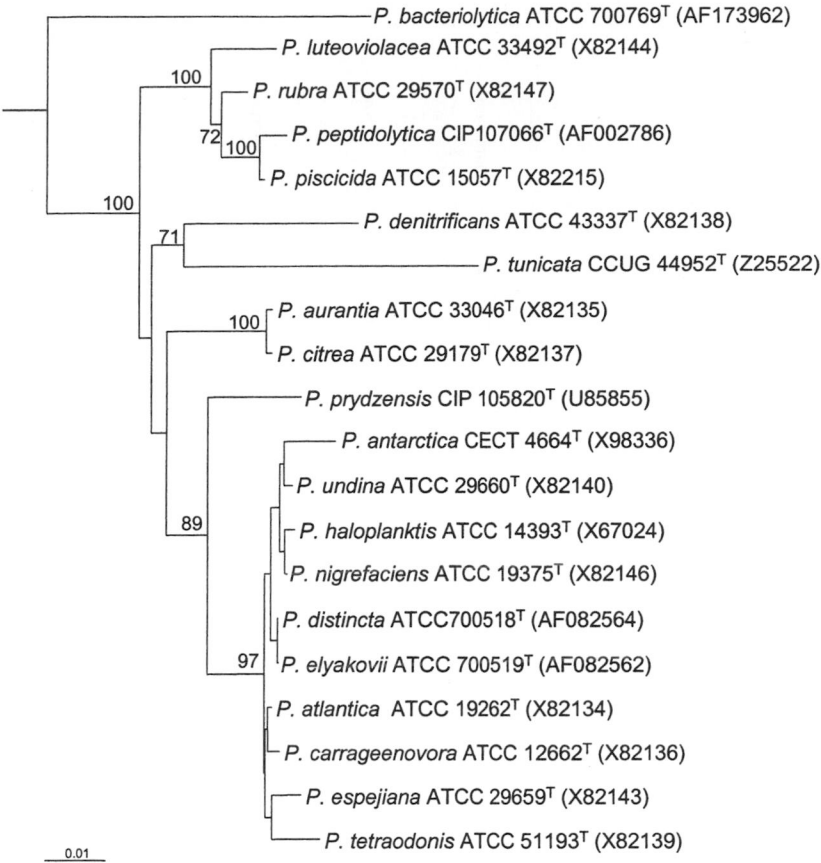

FIGURE BXII.γ.155. Phylogenetic tree of the genus *Pseudoalteromonas* based on nearly complete 16S rRNA sequences, including species validly published as of 2000–2001. The tree was determined using maximum-likelihood distances and clustered by neighbor-joining. Numbers at branch nodes are bootstrap values (calculated from 250 replicate analyses). Only bootstrap values over 50% are shown. *T* denotes a type strain. Bar = distance of 0.01. Numbers in brackets denote GenBank accession codes.

Cells are straight and rod-like, 0.5–0.9 × 0.7–3 μm. Colonies are beige, smooth, convex, and mucoid, with entire edges. Require approximately 17 mM NaCl for growth, but do not require amino acids or other growth factors. Grow between 4 and 30°C; optimum temperature is 25–30°C. Other characteristics are as given in the genus description and in Tables BXII.γ.144 and BXII.γ.145. In addition, strains are lecithinase negative, and acid is formed oxidatively from D-glucose, maltose, starch, D-galactose, and D-mannitol, but not from L-arabinose, D-fructose, rhamnose, sucrose, or D-sorbitol. Acid may be formed from D-mannose or glycerol. Can grow on MacConkey agar, but not on Simmons citrate agar. May grow on cetrimide agar. See Bozal et al. (1994, 1997) for additional characteristics.

Isolated from muddy soils and sediments collected from Antarctic coastal areas.

The mol% G + C of the DNA is: 41–42 (*T*ₘ).

Type strain: NF3, CECT 4664.

GenBank accession number (16S rRNA): X98336.

3. **Pseudoalteromonas atlantica** (Akagawa-Matsushita, Matsuo, Koga and Yamasoto 1992b) Gauthier, Gauthier and Christen 1995a, 760^VP (*Alteromonas atlantica* Akagawa-Matsushita, Matsuo, Koga and Yamasoto 1992b, 624.)

at.lan′ti.ca. L. adj. *atlanticus* from Gr. adj. *Atlantikos* pertaining to the Atlantic Ocean.

Cells are rod-shaped, 0.5–0.8 × 1.5–2.3 μm. Colonies are sunken by agarase activity. Colonies are beige to pale orange, smooth, and convex, with an entire edge. May produce a melanin-like pigment on tyrosine-containing media. Sodium ions are required for growth, but strains do not require amino acids or other growth factors. Grow between 5 and 35°C, with optimum growth at about 30°C. Grows between pH 5.5 and 8.5. Other characteristics are as given in the genus description and in Tables BXII.γ.144 and BXII.γ.145. In addition, strains can reduce methylene blue and coagulate litmus milk; gluconate oxidation, 3-ketolactose production from lactose, and Congo red adsorption tests are negative; strains are unable to degrade carrageenan or xanthine; acid is formed from D-glucose and D-mannitol; and no growth occurs on media containing 0.4% phenethyl alcohol.

Isolated from marine macroalgae and seawater.

The mol% G + C of the DNA is: 40.6–41.7 (*T*ₘ).

Type strain: ATCC 19262, IAM 12927, LMG 2138, NCIMB 301.

GenBank accession number (16S rRNA): X82134.

4. **Pseudoalteromonas aurantia** (Gauthier and Breittmayer 1979) Gauthier, Gauthier and Christen 1995a, 760^VP (*Alteromonas aurantia* Gauthier and Breittmayer 1979, 371.)

TABLE BXII.γ.145. Other characteristics of selected species of the genus *Pseudoalteromonas*[a]

Characteristics	*P. haloplanktis*	*P. atlantica*	*P. antarctica*	*P. carrageenovora*	*P. espejiana*	*P. nigrifaciens*	*P. tetraodonis*	*P. undina*
Oxidase, gelatinase, caseinase	+	+	+	+	+	+	+	+
Lipase	+	+	d	+	+	+	+	+
Urease	+	+	d	−	+	+	−	d
β-Galactosidase	−	+	−	−	+	+	−	+
β-Glucosidase (esculin)	nd	+	d	nd	nd	nd	nd	+
Deoxyribonuclease	nd	+	d	nd	nd	nd	nd	+
Phosphatase	nd	+	+	nd	nd	nd	nd	+
H₂S production	nd	+	d	nd	nd	nd	nd	+
Voges–Proskauer Test	nd	−	d	nd	nd	nd	nd	d
Decomposition of L-tyrosine	+	+	−	+	+	+	−	+
Utilization of:[b]								
DL-Arabinose	d	−	−	d	−	d	−	−
D-Xylose	−	−	−	−	−	d	−	−
D-Gentiobiose	+	+	−	+	+	−	+	+
D-Glucose	+	+	+	+	+	+	+	+
L-Rhamnose	−	−	−	−	−	d	−	−
D-Raffinose	−	+	−	−	−	−	−	−
Adonitol	−	−	−	−	−	d	nd	−
meso-Inositol	−	−	−	−	−	+	nd	−
D-Sorbitol	−	−	−	−	−	+	−	−
Amygdalin	−	+	d	−	−	−	−	+
Salicin	−	−	−	−	−	−	−	−
α-Methyl-D-mannoside	nd	−	nd	nd	nd	nd	nd	−
D-Glucosamine	−	−	nd	+	−	−	+	−
N-Acetylglucosamine	+	−	d	+	−	−	+	−
Glycogen	−	−	d	−	+	+	−	+
Ethanol	d	nd	nd	+	−	+	nd	nd
n-Propanol	d	nd	nd	+	−	+	nd	nd
n-Butanol	−	nd	nd	+	d	−	nd	nd
Propionate	+	+	−	+	+	+	nd	+
Butyrate	d	−	−	d	+	+	+	d
Isobutyrate	d	−	nd	+	+	+	+	d
Valerate	d	+	−	+	d	+	+	d
Isovalerate	−	−	nd	−	+	−	+	d
Caproate	−	−	nd	−	−	+	+	d
Heptanoate	d	−	nd	−	−	−	−	d
Caprylate	d	+	nd	−	−	d	−	d
Pelargonate	d	+	nd	d	+	−	nd	+
Caprate	+	+	−	+	+	−	nd	+
Oxalate	−	+	nd	−	−	−	nd	+
Malonate	−	−	d	−	−	−	nd	−
Azelate	−	−	nd	−	−	−	nd	−
DL-3-Hydroxybutyrate	d	−	d	d	d	+	+	d
DL-Tartrate	−	−	nd	−	−	−	nd	−
Mesaconate	−	−	nd	−	−	−	−	d
o-Hydroxybenzoate	−	+	nd	−	−	−	−	d
Phenylacetate	−	−	−	−	−	+	nd	−
L-Arginine	d	−	nd	d	d	+	+	−
L-Aspartate	d	−	nd	−	+	−	nd	−
L-Citrulline	−	−	nd	d	+	−	nd	−
L-Cysteine		−	nd	nd	nd	nd	nd	−
Glycine	d	−	nd	d	+	−	nd	−
L-Histidine	d	−	−	−	−	d	nd	−
L-Isoleucine	−	−	nd	d	d	−	nd	−
L-Leucine	d	−	d	+	d	−	nd	−
L-Lysine	−	−	−	−	−	+	+	−
L-Methionine	nd	−	nd	nd	nd	nd	nd	−
L-Phenylalanine	−	−	−	d	d	d	−	−
L-Serine	+	−	d	+	+	+	nd	−
L-Threonine	d	−	nd	d	+	+	nd	−
L-Valine	d	−	nd	−	−	−	nd	−
DL-γ-Aminobutyrate	−	−	nd	−	−	−	nd	−
L-Amylamine	−	−	nd	−	−	−	nd	−
Adipate, pimelate	−	−	nd	−	−	−	nd	−
Glutarate	−	−	nd	−	−	−	nd	−

(continued)

TABLE BXII.γ.145. *(cont.)*

Characteristics	*P. haloplanktis*	*P. atlantica*	*P. antarctica*	*P. carrageenovora*	*P. espejiana*	*P. nigrifaciens*	*P. tetraodonis*	*P. undina*
Putrescine	–	–	nd	–	–	–	nd	–
m-Hydroxybenzoate	–	–	–	–	–	–	nd	–
DL-Fucose	–	–	–	–	–	–	nd	–
Suberate, sebacate, *meso*-tartrate, itaconate	–	–	–	–	–	–	nd	–
α-Methyl-D-glucoside, D-turanose, 5-keto-D-gluconate	–	–	nd	–	–	–	–	–
D-Melezitose	–	–	nd	–	–	–	–	–
DL-Tryptophan	–	–	nd	–	–	–	nd	–
Inulin, creatine	–	–	nd	–	–	–	nd	–
i-Erythritol, DL-glycerate, L-norleucine, DL-kynurenine, DL-α-aminobutyrate, DL-α-aminovalerate, DL-δ-aminovalerate, ethanolamine, histamine, tryptamine, spermine, trigollenine, betaine, sarcosine, acetamide, butylamine, benzylamine, anthranilate, *m*-aminobenzoate, *p*-aminobenzoate	–	–	nd	–	–	–	nd	–
Glycolate	–	–	nd	–	–	–	nd	–
Citraconate, levulinate, maleate, benzoate, DL-maleate, β-alanine	–	–	nd	–	–	–	nd	–
Arbutin	nd	–	nd	–	–	–	nd	–
D-Glucuronate	–	nd	nd	–	–	–	nd	nd
Saccharate	–	nd	nd	–	–	–	nd	nd
Mucate, methanol, isopropanol, isobutanol, ethylene glycol, propyleneglycol, butylene glycol, formate, quinate, benzoylformate, phenylthanediol, phenol, naphthaene, kynurenate, methylamine, pentylamine, 2-amylamine, hippurate, pantothenate, nicotinate, nicotinamide, allantoin, adenine, cytosine, guanine, thymine, uracil, dodecane	–	nd	nd	–	–	–	nd	nd
DL-Arabitol	nd	–	nd	nd	nd	nd	nd	–
Dulcitol, urea	nd	–	nd	nd	nd	nd	nd	–
D-Lyxose, D-tagatose, sorose, xylitol, β-methyl-D-glucoside, 2-keto-D-gluconate, L-norvaline, ethylamine	nd	–	nd	nd	nd	nd	nd	–
Phthalate, isophthalate, *tert*-phthalate	nd	–	nd	nd	nd	nd	nd	–
D-Galacturonate	–	nd	nd	–	–	nd	nd	nd

[a]For symbols see standard definitions; nd, not determined.

[b]No information is available concerning ascorbate, oxaloacetate, hydroxy-L-proline, L-asparagine, L-glutamine, protamine, 2-deoxyribose, deoxycholate, galacitol, D-galactosamine, α-glycerophosphate, urate, and glycylglycine.

au.ran' ti.a. M.L. n. *aurantium* orange; M.L. fem. adj. *aurantia* orange-colored.

Cells are rod-shaped, 0.5–1.5 × 1.5–4.0 µm. Colonies on agar media are orange, smooth, and convex, with an entire edge. Sodium ions (optimal levels 0.5–0.6 M) and amino acids are required for growth. Grows between 1 and 8% NaCl. Produces a polyanionic, autotoxic antibiotic on complex media. Grows between 4 and 30°C, with optimum growth occuring at about 25°C. Other characteristics are as shown in the genus description and in Tables BXII.γ.144 and BXII.γ.146. In addition, strains can form phospholipase C, collagenase, leucine aminopeptidase, valine aminopeptidase, and phosphoamidase. Strains may also form α-glucosidase, β-glucosaminidase, and α-chymotrypsin and reduce methylene blue. Tryptophanase, α-galactosidase, β-galactosidase, β-glucuronidase, α-mannosidase, α-fucosidase, and C14 lipase are absent. Trimethylamine *N*-oxide is not reduced. See Gauthier and Breittmayer (1979) for additional characteristics.

Isolated from surface seawater.

The mol% G + C of the DNA is: 38.8–42.5 (*T*$_m$).

Type strain: 208, ATCC 33046, DSM 6057, NCIMB 2052.

GenBank accession number (16S rRNA): X82135.

5. **Pseudoalteromonas bacteriolytica** Sawabe, Makino, Tatsumi, Nakano, Tajima, Iqbal, Yumoto, Ezura and Christen 1998a, 773[VP]

bac.te.ri.o.ly' ti.ca. Gr. n. *baktron* rod or staff; Gr. adj. *lytica* dissolving; M.L. fem. adj. *bacteriolytica* bacteria-dissolving.

Gram negative. Cells are rod-shaped, 1.9–2.5 × 0.6–0.9 µm, occurring singly or in pairs. Motile by polar flagella. Does not accumulate poly-β-hydroxybutyrate. Colonies are red-pigmented or non-pigmented, circular, smooth, and convex with entire edges. Requires NaCl and amino acids for growth. Grows from 15–35°C; no growth at 37°C.

Strictly aerobic chemoheterotroph. Oxidase positive. Catalase activity is weak. Produces acid oxidatively from D-glucose. Produces amylase. Digests casein and Tween 80. Does not produce chitinase, alginase, agarase, or κ-carragennase. Does not reduce nitrate to nitrite. Utilizes D-glucose, D-mannose, D-fructose, sucrose, D-galactose, D-glucosamine, fumarate, succinate, acetate, and propionate. Does not utilize xylose, lactose, melibiose, D-gluconate, D-glucuronate, *N*-acetylglucosamine, D-sorbitol, erythritol, glycerol, citrate, DL-malate, aconitate, α-ketoglutarate, *m*-hydroxybenzoate, putrescine, δ-aminovalerate, γ-aminobutyrate, or L-proline.

TABLE BXII.γ.146. Other characteristics of selected species of the genus *Pseudoalteromonas*[a]

Characteristics	P. aurantia	P. citrea	P. denitrificans	P. luteoviolacea	P. piscicida	P. prydzensis
Oxidase, gelatinase, caseinase	+	+	+	+	+	+
Lipase	+	+	+	+	+	+
Urease	−	nd	nd	−	−	d
β-Galactosidase	−	−	−	−	−	−
β-Glucosidase (esculin)	−	nd	nd	d	d	d
Deoxyribonuclease	+	+	+	+	+	+
Phosphatase	+	+	+	+	+	+
H₂S production	−	−	nd	−	d	−
Voges–Proskauer Test	−	−	nd	nd	−	−
Decomposition of L-Tyrosine	+	+	+	d	+	−
Utilization of:[b]						
DL-Arabinose	−	−	−	−	−	+
D-Xylose	−	−	nd	−	−	nd
D-Glucose	+	d	+	d	+	+
L-Rhamnose	−	+	nd	−	−	−
D-Raffinose	−		nd	−	−	−
Adonitol	−	+	nd	−	−	−
meso-Inositol	−	−	nd	−	−	nd
D-Sorbitol	−	d	−	−	−	−
Amygdalin	−	nd	nd	nd	−	nd
Salicin	−	d	nd	−	−	nd
α-Methyl-D-mannoside	−	nd	nd	nd	+	nd
D-Glucosamine	+	nd	nd	+	+	nd
N-Acetylglucosamine	+	nd	−	+	+	+
Glycogen	−	+	nd	nd	+	+
Ascorbate	d	nd	nd	d	−	nd
Ethanol	nd	+	−	nd	nd	nd
Propionate	nd	+	+	nd	nd	+
n-Butyrate	nd	+	+	nd	nd	+
Isobutyrate	nd	nd	nd	nd	nd	+
n-Valerate	nd	−	nd	nd	nd	d
Isovalerate	nd	nd	nd	nd	nd	d
Caproate	nd	+	nd	nd	nd	−
Heptanoate	nd	nd	nd	nd	nd	−
Caprylate	nd	−	nd	nd	nd	d
Pelargonate	nd	−	−	nd	nd	−
Caprate						
Oxalate	−	nd	nd	d	−	nd
Malonate	−	nd	nd	−	−	d
Azelate	nd	nd	nd	nd	nd	d
Oxaloacetate, hydroxy-L-proline	nd	nd	nd	nd	nd	+
DL-3-Hydroxybutyrate	nd	nd	−	−	nd	+
DL-Tartrate	−	nd	nd	d	−	nd
o-Hydroxybenzoate	−	nd	nd	−	−	nd
Phenylacetate	nd	+	nd	nd	nd	nd
L-Arginine	+	+	+	d	+	nd
L-Asparagine	nd	nd	+	+	nd	−
L-Aspartate	−	+	nd	nd	−	−
L-Citrulline	d	nd	−	nd	+	nd
L-Cysteine	−	nd	−	d	−	nd
Glycine	+	nd	nd	nd	−	nd
L-Glutamine, protamine	nd	nd	+	nd	nd	nd
L-Histidine	d	−	nd	d	d	−
L-Isoleucine	−	nd	nd	nd	+	nd
L-Leucine	d	nd	nd	−	+	−
L-Lysine	−	−	nd	d	nd	nd
L-Methionine	−	nd	nd	nd	d	nd
L-Phenylalanine	−	−	nd	d	d	nd
L-Serine	−	nd	nd	nd	−	−
L-Threonine	d	d	nd	+	−	−
L-Valine	−	nd	nd	nd	−	nd
γ-Aminobutyrate	−	nd	nd	nd	−	+
L-Amylamine	−	nd	nd	nd	d	nd
Adipate, pimelate	nd	nd	nd	nd	nd	−
Glutarate	nd	−	nd	nd	nd	−
Putrescine	−	nd	−	nd	−	−
m-Hydroxybenzoate	nd	nd	nd	−	nd	nd
DL-Fucose	−	nd	nd	nd	−	nd
α-Methyl-D-glucoside,	−	nd	nd	nd	−	nd
D-Turanose, 5-keto-D-gluconate	nd	nd	nd	nd	nd	nd
DL-Tryptophan	−	nd	−	−	−	nd

(*continued*)

TABLE BXII.γ.146. *(cont.)*

Characteristics	P. aurantia	P. citrea	P. denitrificans	P. luteoviolacea	P. piscicida	P. prydzensis
Inulin, creatine	−	nd	nd	−	−	nd
i-Erythritol, DL-glycerate	nd	nd	nd	−	nd	nd
L-Norleucine, DL-kynurenine, DL-α-aminobutyrate, DL-α-aminovalerate, DL-δ-aminovalerate, ethanolamine, histamine, tryptamine, spermine, trigollenine, betaine, sarcosine, acetamide, butylamine, benzylamine, anthranilate, *m*-aminobenzoate, *p*-aminobenzoate	−	nd	nd	nd	−	nd
Glycolate	nd	nd	−	nd	nd	nd
Arbutin	−	nd	nd	nd	−	nd
D-Glucuronate	nd	nd	nd	−	nd	−
Saccharate	nd	nd	nd	nd	nd	−
DL-Arabitol	−	nd	nd	nd	−	−
Dulcitol, urea	nd	nd	nd	−	nd	nd
D-Lyxose, D-tagatose, sorbose, xylitol, β-methyl-D-glucoside, 2-keto-D-gluconate, L-norvaline, ethylamine	−	nd	nd	nd	−	nd
D-Galacturonate	nd	nd	nd	nd	nd	−
2-Keto-D-gluconate	−	nd	nd	nd	−	nd
α-Glycerophosphate, urate	nd	nd	nd	nd	nd	−
Deoxycholate	nd	nd	nd	−	nd	nd
Galacitol, D-galactosamine, glycylglycine	−	nd	nd	nd	−	nd

[a]For symbols see standard definitions; nd, not determined.

[b]No information is available concerning the following substrates: caprate, D-gentiobiose, *n*-propanol, *n*-butanol, mesaconate, suberate, sebacate, *meso*-tartrate, itaconate, D-melizitose, citraconate, levulinate, maleate, benzoate, DL-maleate, β-alanine, mucate, methanol, isopropanol, isobutanol, ethyleneglycol, propyleneglycol, butyleneglycol, formate, quinate, benzoylformate, phenylethanediol, phenol, napthalene, kynurenate, methylamine, pentylamine, 2-amylamine, hippurate, pantothenate, nicotinate, nicotinamide, allantoin, adenine, cytosine, guanine, thymine, uracil, dodecane, phthalate, isophthalate, and *tert*-phthalate.

Isolated from, and causative agent of, red spot disease in beds of *Laminaria japonica*.

The mol% G + C of the DNA is: 44–46 (HPLC).

Type strain: IAM 14595.

GenBank accession number (16S rRNA): D89929.

6. **Pseudoalteromonas carrageenovora** (Akagawa-Matsushita, Matsuo, Koga and Yamasoto 1992b) Gauthier, Gauthier and Christen 1995a, 760[VP] (*Alteromonas carrageenovora* Akagawa-Matsushita, Matsuo, Koga and Yamasoto 1992b, 626.)

car.ra.gee.no'vo.ra. M.L. n. *carrageenum* carrageen, another name for carrageenan; L. v. *vorare*, to devour; M.L. fem. adj. *carrageenovora* carrageenan decomposing.

Cells are rod-shaped, 0.7–0.8 × 1.9–2.5 μm. Colonies are beige to pale yellow, smooth, and convex, with an entire edge. Produces diffusible melanin-like pigments on tyrosine-containing media. Sodium ions are required for growth, but amino acids and other growth factors are not required. Grow between 5 and 35°C, with optimum growth at about 30°C. Grow between pH 5.5 and 9.0. Other characteristics are as given in the genus description and in Tables BXII.γ.144 and BXII.γ.145. In addition, strains can reduce methylene blue and coagulate litmus milk. Gluconate oxidation, 3-ketolactose production from lactose, and Congo red adsorption tests are negative. Degrade carrageenan, but not xanthine. Acid is formed from D-glucose and D-mannitol. No growth occurs on media containing 0.4% phenethyl alcohol (Akagawa-Matsushita et al., 1992b).

Isolated from a mixture of seawater and marine macroalgae (Yaphe and Baxter, 1955).

The mol% G + C of the DNA is: 39.5 (T_m).

Type strain: ATCC 43555, DSM 6820, IAM 12662, IFO 12985, NCIMB 302.

GenBank accession number (16S rRNA): X82136.

7. **Pseudoalteromonas citrea** (Gauthier 1977) Gauthier, Gauthier and Christen 1995a, 760, emend. Ivanova, Kiprianova, Mikhailov, Levanova, Garagulya, Gorshkova, Vysotskii, Nicolau, Yumoto, Taguchi and Yoshikawa 1998, 255[VP] (*Alteromonas citrea* Gauthier 1977, 354; *Alteromonas fuliginea* Romanenko, Lysenko, Mikhailov and Kurika 1995a, 879.)

ci'tre.a. L. adj. *citreus* lemon-yellow.

Cells are straight rods, 0.5–0.8 × 1.0–1.5 μm. Colonies on agar media are smooth, convex with an entire edge, and either nonpigmented, pigmented yellow-orange, or containing diffusible blue-black or brown, melanin-like pigments. Sodium ions are required for growth; optimal growth occurs with about 2–3% NaCl. May require amino acids for growth. Some strains produce a polyanionic, autotoxic antibiotic on complex media. Grows between 4 and 35°C, with optimal growth at about 30°C. Other characteristics are as shown in the genus description and in Tables BXII.γ.144 and BXII.γ.146. Positive for levan production. Strains may hydrolyze pustulan and laminarin. Strains are resistant to rhistomycin, ampicillin, neomycin, kanamycin, streptomycin, benzylpenicillin, lincomycin, and cephalexin, but are sensitive to rifampicin, gentamicin, oxacillin, polymyxin, erythromycin, and ofloxacin.

Isolated from surface seawater, mussels, ascidians, and sponges.

The mol% G + C of the DNA is: 38.9–44.7 (T_m).

Type strain: ATCC 29719, DSM 6058, NCIMB 1889.

GenBank accession number (16S rRNA): X82137.

8. **Pseudoalteromonas denitrificans** (Enger, Nygaard, Solberg, Schei, Nielsen and Dundas 1987) Gauthier, Gauthier and Christen 1995a, 760[VP] (*Alteromonas denitrificans* Enger, Nygaard, Solberg, Schei, Nielsen and Dundas 1987, 421.)

de.ni.tri'fi.cans. M.L. v. *denitrifico* to denitrify; M.L. part. adj. *denitrificans* denitrifying.

Cells are rod-shaped, 0.5–0.7 × 2–4 μm. Motile by a single, sheathed, polar flagellum; rarely, up to 3 polar flagella may be observed. Cells produce a number of prodigiosin-type pigments, with a red pigment predominant and identified as cycloprodigiosin. Colonies on agar are initially pink, becoming more intensely red with age, and are smooth and convex with an entire edge. Produce bright blue, purple, or red pigments in liquid media. Produce a polyanionic, autotoxic antibiotic on complex media. Sodium ions are required for growth; the optimal level is about 2.5% NaCl; no growth occurs in the presence of 6% NaCl. Require amino acids for growth. Grow between 2 and 30°C, with optimum growth at about 25°C. Other characteristics are as given in the genus description and in Tables BXII.γ.144 and BXII.γ.146. Strains are also resistant to tetracycline and penicillin, but are susceptible to chloramphenicol. Strains possess a high sensitivity to ultraviolet radiation.

Isolated from seawater collected from fjords of western Norway.

The mol% G + C of the DNA is: 36.5–37.1 (T_m).

Type strain: Nygaard 1977, ATCC 43337, DSM 6059, IAM 14445, NCIMB 13179.

GenBank accession number (16S rRNA): X82138.

9. **Pseudoalteromonas distincta** (Romanenko, Mikhailov, Lysenko and Stapanenko 1995a) Ivanova, Chun, Romanenko, Matte, Mikhailov, Frolova, Huq and Colwell 2000a, 143^{VP} (*Alteromonas distincta* Romanenko, Mikhailov, Lysenko and Stapanenko 1995a, 879.)

di.stinc'ta. M.L. adj. *distincta* distinct.

Cells are rods, about 0.9 × 2.5 μm. Motile by a single polar flagellum, very rarely by 2–4 polar flagella with 5–7 lateral flagella. Cells do not form capsules. Colonies on marine agar are black-brown, smooth, convex, and circular with entire edges. Produce a diffusible, black-brown, melanin-like pigment that is formed most strongly at low incubation temperatures and on tyrosine and arginine-containing media. Sodium ions are required for growth. Growth factors are not required. Grow between 4 and 35°C; no growth occurs at 40°C. Grow between pH 6.0 and 9.0. Strictly aerobic with a respiratory metabolism. Oxidase and catalase positive. Do not denitrify. Produce lipase and gelatinase, but do not hydrolyze starch, chitin, alginate, carrageenan, or agar. Decompose L-tyrosine and L-arginine. Utilize citrate, but not D-arabinose, D-xylose, D-glucose, D-mannose, L-rhamnose, cellobiose, lactose, maltose, sucrose, D-mannitol, or N-acetylglucosamine.

Isolated from a marine sponge near the Komandorskiye Islands, Russian Pacific region.

The mol% G + C of the DNA is: 44 (T_m).

Type strain: ATCC 700518, KMM 638.

GenBank accession number (16S rRNA): AF043742, AF082564.

10. **Pseudoalteromonas elyakovii** (Ivanova, Mikhailov, Kiprianova, Levanova, Garagulya, Frolova and Swetashev 1997) Sawabe, Tanaka, Iqbal, Tajima, Ezura, Ivanova and Christen 2000, 270^{VP} (*Alteromonas elyakovii* Ivanova, Mikhailov, Kiprianova, Levanova, Garagulya, Frolova and Swetashev 1997, 601.)

el.ya.kov'i.i. M.L. gen. n. *elyakovii* of Elyakov, named after G.B. Elyakov for his work in microbial biotechnology.

Cells are rods, about 0.5–0.7 × 1.0–1.2 μm. Motile by a single polar flagellum. Cells do not form spores or capsules.

Colonies on marine agar are nonpigmented, smooth, convex, and circular, with entire edges. Melanin-like pigments are not formed. Nonbioluminescent. Sodium ions are required for growth. Growth factors are not required. Grow between 4 and 35°C; no growth occurs at 41°C. Grow between pH 7 and 8.5. Strictly aerobic with a respiratory metabolism. Oxidase and catalase positive. Do not denitrify. Produce lipase, amylase, and gelatinase, but do not hydrolyse chitin. Decompose L-tyrosine and L-arginine. Utilize D-ribose, D-xylose, D-fructose, D-galactose, D-glucose, cellobiose, lactose, salicin, adonitol, D-sorbitol, m-inositol, ethanol, propionate, butyrate, succinate, lactate, 2-oxoglutarate, glycolate, p-hydroxybenzoate, phenylacetate, L-alanine, L-histidine, L-lysine, L-ornithine, L-phenylalanine, and L-threonine, but not arabinose, rhamnose D-mannose, cellobiose, lactose, maltose, sucrose, trehalose, D-gluconate, D-mannitol, acetate, caprylate, valerate, caproate, pelargonate, glutarate, citrate, fumarate, L-malate, pyruvate, glycerate, aconitate, L-asparagine, L-glutamate, or L-proline. Other characteristics as are described by Ivanova et al. (1996b).

Isolated from the coelomic fluid of the mussel *Crenomytilus grayanus.*

The mol% G + C of the DNA is: 39 (T_m).

Type strain: 40MC, ATCC 700519, KMM 162, VKPM B3905.

GenBank accession number (16S rRNA): AF082562.

11. **Pseudoalteromonas espejiana** (Chan, Baumann, Garza and Baumann 1978) Gauthier, Gauthier and Christen 1995a, 760^{VP} (*Alteromonas espejiana* Chan, Baumann, Garza and Baumann 1978, 220.)

es.pe.ji.a'na. M.L. fem. adj. *espejiana* of Espejo, a Chilean microbiologist who isolated one of the first lipid-containing bacteriophages.

Cells are rod-shaped, 0.2–1.0 × 2.0–3.5 μm. Colonies are nonpigmented, smooth, and convex, with an entire edge. Sodium ions (0.2–0.3 M concentration optimal) required for growth. Amino acids are required for growth. Grow between 10 and 35°C, with optimal growth at about 30°C. No growth at 4°C. Other characteristics are as shown in the genus description and in Tables BXII.γ.144 and BXII.γ.145.

Isolated from seawater.

The mol% G + C of the DNA is: 43–44 (T_m).

Type strain: 261, ATCC 29659, CCUG 16147, DSM 9414, IAM 12640, LMG 2866, NCIMB 2127.

GenBank accession number (16S rRNA): X82143.

12. **Pseudoalteromonas luteoviolacea** (Gauthier 1982) Gauthier, Gauthier and Christen 1995a, 760^{VP} (*Alteromonas luteoviolacea* Gauthier 1982, 85; *Alteromonas luteo-violaceus* [sic] Gauthier 1976c, 147.)

lu.te.o.vi.o.la'ce.a. L. adj. *luteus* yellow; L. adj. *viola* violet; M.L. adj. *luteoviolacea* yellow-violet.

Cells are rod-shaped, 0.8–1.5 × 2–4 μm. Motile by single polar flagellum, which may or may not be sheathed. Colonies on agar are smooth or waxy and circular convex, with a more or less violet pigment (violacein); edges of colonies may be pigmented pale violet or yellow. Some strains form diffusible brown pigments on complex media. Most strains produce a polyanionic, autotoxic antibiotic on complex media. Sodium ions are required for growth; the optimal level is about 2–3% NaCl; no growth occurs in the presence of 8% NaCl. May require amino acids for growth. Grow between 10 and 30°C, with optimal growth at about 25°C. Other characteristics are as given in the genus description and in Tables BXII.γ.144 and BXII.γ.146. In addition, strains can form lecithinase and produce acid from trehalose and maltose. Acid from glucose varies between strains. Negative for the following tests: cellulose hydrolysis, methylene blue reduction, β-galactosidase, glycine decarboxylase, asparagine decarboxylase, and alanine decarboxylase. Acid is not produced from D-mannose, D-xylose, D-fructose, lactose, sucrose, D-mannitol, or glycerol. Susceptible to polymyxin B, chloramphenicol, and oleandomycin.

Isolated from surface seawater.

The mol% G + C of the DNA is: 40.9–42.2 (T_m).

Type strain: CH130, ATCC 33492, DSM 6061, NCIMB 1893.

GenBank accession number (16S rRNA): X82144.

13. **Pseudoalteromonas nigrifaciens** (Baumann, Baumann, Bowditch and Beaman 1984a) Gauthier, Gauthier and Christen 1995a, 760[VP] emend. Ivanova, Kiprianova, Mikhailov, Levanova, Garagulya, Gorshkova, Yumoto and Yoshikawa 1996b, 227 (Alteromonas nigrifaciens Baumann, Baumann, Bowditch and Beaman 1984a, 146.)

ni.gri.fa' ci.ens. L. adj. niger black; L. v. facio to make; M.L. part. adj. nigrifaciens blackening.

Cells are straight rods, 0.8–1.2 × 1.8–2.3 μm. Cells may be encapsulated. Colonies are smooth, circular, convex with an entire edge, and nonpigmented. May form a brown-black, melanin-like pigment constitutively at low temperatures. On tyrosine-containing media, melanin-like pigments are formed by all strains. Sodium ions are required for growth. Most strains do not require growth factors. Grow between 4 and 35°C, with optimal growth at about 30°C. Other characteristics are as given in the genus description and in Tables BXII.γ.144 and BXII.γ.145. See Ivanova et al. (1996a) for additional phenotypic information.

Isolated from seawater, mussels and salted food.

The mol% G + C of the DNA is: 38.5–40.2 (T_m).

Type strain: ATCC 19375, DSM 6063, NCIMB 8614.

GenBank accession number (16S rRNA): X82146.

14. **Pseudoalteromonas peptidolytica** Venkateswaran and Kohmoto 2000, 572[VP]

pep.ti.do.lyt' i.ca. Gr. n. peptos a class of organic compounds; Gr. adj. lytica dissolving; M.L. fem. adj. peptidolytica peptide dissolving.

Gram negative. Cells are rods, 2.0–3.0 × 0.7–0.8 μm. Motile with a polar unsheathed flagellum. Colonies on marine agar are circular, opaque, and yellow-pigmented. Requires Na⁺ ions for growth. Grows between 1–10% NaCl. Growth occurs at 15–40°C. Optimum growth temperature is 30°C.

Chemoheterotrophic and strictly aerobic. Catalase and oxidase positive. Produces gelatinase, amylase, lipase (Tween 80), and alginase. Arginine dihydrolase, lysine decarboxylase, ornithine decarboxylase, chitinase, and β-galactosidase are not produced. Does not produce H_2S from thiosulfate. Indole and Voges–Proskauer tests are negative. Does not denitrify. Utilizes glucose, maltose, N-acetylglucosamine, succinate, and fumarate as sole carbon sources.

Does not utilize mannose, fructose, galactose, lactose, sucrose, salicin, mannitol, glycerol, citrate, DL-lactate, γ-aminobutyrate, sarcosine, putrescine, α-ketoglutarate, sorbitol, DL-malate, acetate, m-hydroxybenzoate, or methanol.

Isolated from surface seawater, Yamoto Islands, Sea of Japan.

The mol% G + C of the DNA is: 41–42 (HPLC).

Type strain: F12–50–A1, MBICC F1250A1.

GenBank accession number (16S rRNA): AF007286.

15. **Pseudoalteromonas piscicida** (ex Bein 1954) Gauthier, Gauthier and Christen 1995a, 760 ("Pseudomonas piscicida" (Bein 1954) Buck, Meyers and Leifson 1963, 1125; "Flavobacterium piscicida" Bein 1954, 115.)

pi.sci.ci' da. L. n. piscis fish; L. suff. -cida from L. v. caedo to cut or kill; M.L. n. piscicida fish killer.

Cells are straight rods, 0.2–0.6 × 1.2–2.0 μm. Fimbriae are present on cells. Possess single, sheathed, polar flagella. Colonies are yellow-orange, translucent, smooth, circular, and convex, with an entire edge and a butyrous consistency. Liquid cultures possess even, moderate turbidity, with a pigmented pellicle.

Cell cultures and filtrates produce an antibiotic active against yeasts. Filtrates and cells also induce neuromuscular toxicity syndromes in a variety of fish and crab species, including minnow (Gambusia sp.), schoolmaster (Lutjans apodus), sandperch (Eucinostomus pseudosula), killifish (Fundulus similis), mollie (Mollieresia latipinna), and fiddler crabs (Uca pugnas and Uca pugilator).

Sodium ions required for growth, with cultures growing between 0.5% and 10% NaCl. Use ammonia, but not nitrate as a nitrogen source; require amino acids for growth. Grow between 10 and 37°C; optimal growth is at about 30°C; no growth occurs at 4°C or 40°C. Grow between pH 6.0 and 8.0. Other characteristics are as given in the genus description and in Tables BXII.γ.144 and BXII.γ.146. Strains can also reduce trimethylamine-N-oxide to trimethylamine, oxidize gluconate to 2-keto-D-gluconate, coagulate litmus milk, and produce lecithinase. Acid is formed from D-glucose, sucrose, and maltose, but not from lactose, D-xylose, or D-mannitol.

Isolated from an area of dead fish.

The mol% G + C of the DNA is: 43.5–45.5 (Bd).

Type strain: MB-1, ATCC 15057, NCMB1142.

GenBank accession number (16S rRNA): X82215.

16. **Pseudoalteromonas prydzensis** Bowman 1998, 1040[VP]

prydz.en' sis. M.L. fem. adj. prydzensis pertaining to Prydz Bay, Antarctica, the site of isolation.

Cells are rod-shaped, 0.5–0.7 × 1–2.5 μm. Colonies on agar are nonpigmented, translucent, smooth, circular, and convex, with an entire or lobate edge; diffusible pigments are not formed on complex or tyrosine-containing media. Sodium ions are required for growth; the optimal level is about 2–3% and cultures grow at NaCl concentrations up to about 15%. Growth factors are not required. Grow between −2 and 30°C, with optimal growth at about 22–25°C. Other characteristics are as shown in the genus description and in Tables BXII.γ.144 and BXII.γ.146. In addition, the strains can produce acid from L-arabinose, D-glucose, D-mannose, maltose, sucrose, trehalose, and N-acetylglucosamine. Acid from cellobiose varies between strains. Do not hydrolyze dextran, urate or xanthine. Acid is not produced from D-xylose, L-rhamnose, D-fructose, D-galactose, lactose, melibiose, D-raffinose, dextran, adonitol, D-mannitol, *meso*-inositol, D-sorbitol, or glycerol. Resistant to ampicillin and bile salts.

Isolated from sea ice.

The mol% G + C of the DNA is: 38–39 (T_m).

Type strain: MB8–11, ACAM 620.

17. **Pseudoalteromonas rubra** (Gauthier 1976a) Gauthier, Gauthier and Christen 1995a, 760[VP] (*Alteromonas rubra* Gauthier 1976a, 464.)

rub' ra. L. fem. adj. *rubra* red.

Cells are rod-shaped, 1.0–1.5 × 2–4 μm. Colonies on agar media are smooth, convex with an entire edge, and pigmented bright red. Sodium ions are required for growth. Require amino acids for growth. Grow between 10 and 35°C; optimum growth occurs at about 30°C, and no growth occurs at ≥40°C. Produce a polyanionic, autotoxic antibiotic on complex media. Other characteristics are shown in the genus description and Tables BXII.γ.144 and BXII.γ.146.

Isolated from surface seawater.

The mol% G + C of the DNA is: 46–48 (T_m).

Type strain: ATCC 29570, DSM 6064.

GenBank accession number (16S rRNA): X82147.

18. **Pseudoalteromonas tetraodonis** (Simidu, Kita-Tsukamoto, Yasumoto and Yotsu 1990) Ivanova, Romanenko, Matté, Matté, Lysenko, Simidu, Kita-Tsukamoto, Sawabe, Vysotskii, Frolova, Mikhailov, Christen and Colwell 2001b, 1077[VP] (*Pseudoalteromonas haloplanktis* subsp. *tetraodonis* Gauthier, Gauthier and Christen 1995a, 760; *Alteromonas tetraodonis* Simidu, Kita-Tsukamoto, Yasumoto and Yotsu 1990, 336.)

tet' ra.o.do.nis. M.L. *tetraodonis* of *Tetraodontidae*, a family of plectognathic fishes.

Cells are rod-shaped, 2.0–3.0 × 1.0 μm. Colonies are non-pigmented, smooth, and convex with an entire edge. Sodium ions are required for growth. Grows between 1–10% NaCl. Growth factors (amino acids) are not required for growth. Grows between 10–35°C, with optimum growth occurring at about 30°C. Optimum pH for growth is 7.5–8.0. Growth occurs between pH 5.5–9.5.

Utilizes glucose, sucrose, maltose, galactose, DL-lactate, valerate, caproate, butyrate, citrate, fumarate, succinate, caprylate, L-glutamate, acetate, pyruvate, ethanol, and L-tyrosine. Does not utilize mannose, fructose, mannitol, L-histidine, L-proline, or DL-malate (Ivanova et al., 2001b). Strains are susceptible to benzylpenicillin, carbenicillin, vancomycin, and tetracycline. Other characteristics are

shown in the genus description and in Tables BXII.γ.144 and BXII.γ.145.

The mol% G + C of the DNA is: 41.5 (T_m).

Type strain: GFC, ATCC 51193, DSM 9166, IAM 14160, KMM 458, NCIMB 13177.

GenBank accession number (16S rRNA): X82139; AF214729.

19. **Pseudoalteromonas tunicata** Holström, James, Neilan, White and Kjelleberg 1998, 1210[VP]

tu.ni.ca' ta. L. fem. n. *tunicata* clothed with a tunic

Gram negative. Cells are rods, 2.0–3.4 μm in length. Motile with a polar sheathed flagellum. Colonies on marine agar are circular, opaque, and dark green. Colonies on trypticase soya agar are non-pigmented. Requires Na+ ions for growth with biomass produced optimally with 1–2% NaCl. Growth occurs at 4–30°C; no growth at 37°C. Optimum growth temperature is 25°C. Optimum pH for growth is 7–8; no growth at pH 5 or 11.

Chemoheterotrophic and strictly aerobic. Catalase, oxidase, and gelatinase positive. Utilizes trehalose, glucose, maltose, and Tween 20 as sole carbon sources. Does not utilize galactose, fructose, xylose, sucrose, lactose, raffinose, melibiose, xylitol, erythritol, glycine, threonine, arabinose, L-asparagine, or L-histidine. Susceptible to erythromycin, rifampicin, gentamycin, tetracycline, ampicillin, neomycin, kanamycin, and nalidixic acid at 100 μg/ml. Susceptible to the vibriostatic agent O/129 at 150 μg/ml.

Isolated from an adult of the tunicate species *Ciona intestinalis* collected from the coast of western Sweden.

The mol% G + C of the DNA is: 42.2 (T_m).

Type strain: Strain DS, CUG 26757.

GenBank accession number (16S rRNA): Z25522.

20. **Pseudoalteromonas undina** (Chan, Baumann, Garza and Baumann 1978) Gauthier, Gauthier and Christen 1995a, 760[VP] (*Alteromonas undina* Chan, Baumann, Garza and Baumann 1978, 220.)

un.di' na. L. fem. n. *undina* water nymph.

Cells are rod-shaped, 0.7–0.9 × 1.8–3.0 μm. Colonies are nonpigmented, smooth, and convex with an entire edge. Sodium ions are required for growth. Require amino acids for growth. Grow between 4 and 30°C, with optimum growth occurring at about 25°C. No growth at ≥35°C. Other characteristics are as shown in the genus description and in Tables BXII.γ.144 and BXII.γ.145.

Isolated from seawater.

The mol% G + C of the DNA is: 43–44 (T_m).

Type strain: ATCC 29690, DSM 6065, NCIMB 2128.

GenBank accession number (16S rRNA): X82140.

Genus XII. **Psychromonas** *Mountfort, Rainey, Burghardt, Kaspar and Stackebrandt 1998b, 631[VP] emend. Nogi, Kato and Horikoshi 2002, 1531 (Effective publication: Mountfort, Rainey, Burghardt, Kaspar and Stackebrandt 1998a, 237)*

THE EDITORIAL BOARD

Psy.chro.mo' nas. Gr. adj. *psychros* cold; Gr. fem. n. *monas* a unit; N.L. n. *Psychromonas* a cold monad.

Motile Gram-negative rod-shaped or oval cells; polar flagellum. **Grow at temperatures <22°C.** Catalase and oxidase positive. Major quinone Q-8; major fatty acids are $C_{14:1}$, $C_{16:0}$, and $C_{16:1}$. **Facultative anaerobes;** use carbohydrates for energy for growth. **Found in marine or high salinity environments.**

The mol% G + C of the DNA is: 38.1–43.8.

Type species: **Psychromonas antarctica** Mountfort, Rainey, Burghardt, Kaspar and Stackebrandt 1998b, 631 (Effective publication: Mountfort, Rainey, Burghardt, Kaspar and Stackebrandt 1998a, 237.)

FURTHER DESCRIPTIVE INFORMATION

The genus *Psychromonas* includes organisms adapted to a variety of conditions in marine or other saline aquatic environments, including an Antarctic saline pond (*Psychromonas antarctica*), seawater (*Psychromonas arctica* and *Psychromonas marina*), and deep sea sediments (*Psychromonas kaikoae* and *Psychromonas profunda*). All species are able to grow at minimum temperatures of 0–4°C and exhibit both fermentative and respiratory metabolism (Mountfort et al., 1998a; Kawasaki et al., 2002; Nogi et al., 2002; Groudieva et al., 2003; Xu et al., 2003). Increased pressure enhances the growth of *Psychromonas profunda* (Xu et al., 2003) and is required for the growth of *Psychromonas kaikoae* (Nogi et al., 2002).

ENRICHMENT AND ISOLATION PROCEDURES

Psychromonas antarctica was isolated under anaerobic conditions and maintained in a basal medium emended with 3% NaCl (w/v) and filtered water from the Antarctic salt pond from which the organism was isolated (5%, v/v) (Mountfort et al., 1998a). *Psychromonas arctica* was enriched and isolated on a complex marine medium (Groudieva et al., 2003) containing starch. High pressure incubation was used for the isolation of *Psychromonas kaikoae* (Nogi et al., 2002). Difco Marine Medium 2216 was used for *Psychromonas kaikoae*, *Psychromonas marina*, and *Psychromonas profunda* (Kawasaki et al., 2002; Nogi et al., 2002; Xu et al., 2003).

DIFFERENTIATION OF THE GENUS *PSYCHROMONAS* FROM OTHER GENERA

Psychromonas species are facultatively anaerobic, whereas members of the genera *Alteromonas*, *Glaciecola*, *Idiomarina*, *Marinobacterium*, *Microbulbifer*, and *Pseudoalteromonas* are strictly aerobic. *Psychromonas* species exhibit both fermentative and respiratory metabolism, whereas *Ferrimonas balearica* and *Alishewanella* species have a strictly respiratory metabolism; in addition, *Alishewanella* species are nonmotile. *Psychromonas* species utilize carbohydrates as energy sources, whereas *Marinobacter* species do not. *Psychromonas* species can be distinguished from *Shewanella* species by the unusual fatty acid profiles of the latter. *Psychromonas* species are ecologically and phenotypically similar to members of the genera *Colwellia* and *Moritella*.

DIFFERENTIATION OF THE SPECIES OF THE GENUS *PSYCHROMONAS*

Differential characteristics of *Psychromonas* species are given in Table BXII.γ.147.

List of species of the genus Psychromonas

1. **Psychromonas antarctica** Mountfort, Rainey, Burghardt, Kaspar and Stackebrandt 1998b, 631[VP] (Effective publication: Mountfort, Rainey, Burghardt, Kaspar and Stackebrandt 1998a, 237.)

ant.arc.ti.ca. M.L. gen. n. *antarctica* of Antarctica.

Rods, 1.3 × 2.5–6.0 μm. Grows at NaCl concentrations 0–4% (w/v); 3% NaCl is optimum and protects against killing by freeze-thaw. Killed quickly by temperatures >50°C or by temperatures >30°C for >20 h. Catalase positive; oxidase positive. Ferments glycogen, starch, *N*-acetyl glucosamine, and hexose monosaccharides and disaccharides with the production of CO_2, formate, acetate, ethanol, lactate, and butyrate. Does not hydrolyze esculin. Does not grow on *N*-acetyl muramic acid, alanine, casamino acids, esculin, formate, fumarate, glycerol, isoleucine, lactate, lactose, malate, mannose, putrescine, rhamnose, sorbitol, valine, yeast extract, xylan, or xylose.

TABLE BXII.γ.147. Differential characteristics of *Psychromonas* species[a,b]

Characteristic	*Psychromonas antarctica*	*Psychromonas arctica*	*Psychromonas kaikoae*	*Psychromonas marina*	*Psychromonas profunda*
Temperature range (optimum) (°C)	2–17 (12)	0–25 (20)	4–15 (10)	0–25 (14–16)	2–13 (3–4)
Growth at atmospheric pressure	+	+	−	+	+
Indole production	−	nd	−	−	+
H₂S production	−	nd	−	+	+
Acid produced from:					
Inositol	−	nd	−	−	+
Lactose	−	nd	−	+	+
Rhamnose	−	nd	−	−	+
Trehalose	+	nd	+	−	+
Growth on:					
Alanine	−	+	nd	nd	+[c]
Cellobiose	−	nd	+	+	+[c]
Fumarate	−	+	nd	nd	+
Glycerol	−	+	−	+	+
Mannose	−	+	+	−	−
Xylose	−	−	−	+	+[c]
Hydrolysis of:					
Gelatin	+	−	+	−	−
Starch	+	+	−	+	w

[a]Abbreviations: +, positive; −, negative; nd, not determined; w, weak.

[b]Data from Mountfort et al.(1998a), Kawasaki et al. (2002), Nogi et al. (2002), Groudieva et al. (2003), and Xu et al.(2003).

[c]Extended incubation (up to 4 w) may be required.

Isolated from the anaerobic sediment of a saline pond on the McMurdo Ice Shelf, Antarctica.

The mol% G + C of the DNA is: 42.8 (HPLC).

Type strain: star-1, DSM 10704.

GenBank accession number (16S rRNA): Y14697.

2. **Psychromonas arctica** Groudieva, Grote and Antranikian 2003, 544[VP]

arc' ti.ca. L. fem. adj. *arctica* from the Arctic, referring to the site where the type strain was isolated.

Rods 0.7–1.7 × 1.3–2.6 µm. Aging cultures contain coccoid cells 1.3–1.7 µm in diameter. Forms biofilms by secreting fibrous exopolysaccharides. Optimal NaCl concentration for growth, 2% (w/v); range, 1–7%. Temperature optimum, 20°C; range, 0–25°C. Killed by temperatures >27°C for >4 h. Carbon sources include acetate, alanine, fructose, fumarate, glucose, glycerol, lactose, maltose, mannitol, mannose, pyruvate, succinate, and sucrose. Fermentation products are CO_2, formate, acetate, ethanol, and lactate. No growth on chitin, citrate, gelatin, lactate, malate, propionate, xylan, or xylose. No vitamin requirement.

Isolated from seawater collected off Svalbard, Spitzbergen, Norway.

The mol% G + C of the DNA is: 40.1 (HPLC).

Type strain: Pull 5.3, DSM 14288, KCTC 12111.

GenBank accession number (16S rRNA): AF374385.

3. **Psychromonas kaikoae** Nogi, Kato and Horikoshi 2002, 1531[VP]

kai.ko' ae. N.L. fem. gen. n. *kaikoae* of KAIKO, the unmanned submersible that collected the samples from which the organism was isolated.

Motile rods, 0.8–1.0 × 2.0–4.0 µm; one polar flagellum (unsheathed). NaCl required for growth; optimal concentration 3% (w/v). Catalase and oxidase positive. Optimum growth at 10°C and 50 Mpa; does not grow at atmospheric pressure. Possesses both fermentative and respiratory pathways. Hydrolyzes gelatin. Reduces nitrate to nitrite; nitrite not further reduced. No H_2S, amylase, or indole production. Major quinone Q-8; major fatty acids are $C_{14:0}$, $C_{14:1}$, $C_{16:0}$, and $C_{16:1}$.

Isolated from sediment from the Japan Trench at a depth of 7434 m.

The mol% G + C of the DNA is: 43.8 (method not given).

Type strain: JT7304, ATCC BAA-353, JCM 11054.

GenBank accession number (16S rRNA): AB052160.

4. **Psychromonas marina** Kawasaki, Nogi, Hishinuma, Nodasaka, Matsuyama and Yumoto 2002, 1458[VP]

ma.ri' na. L. adj. *marina* of the sea.

Motile rods, 0.8–1.2 × 1.5–2.0 µm; one polar flagellum. Catalase and oxidase positive. NaCl required for growth; optimum, 3–5% (w/v); maximum, 7%. Temperature range for growth, 0–25°C; no growth at or above 26°C. Ferments D-fructose, D-galactose, D-glucose, D-maltose, D-mannitol, sucrose, and D-xylose with no production of gas; in addition, utilizes N-acetyl glucosamine, D-cellobiose, D-galactose, glycerol, and lactose. Does not utilize L-arabinose, *myo*-inositol, D-mannose, melibiose, raffinose, rhamnose, raffinose, sorbitol, or trehalose. No vitamins, amino acids, or organic nitrogen sources required. Reduces nitrate to nitrite; produces H_2S but not indole. Produces β-galactosidase. Hydrolyzes alginic acid, DNA, starch, tributyrin, Tween 20, Tween 40, Tween 60, and Tween 80. Does not hydrolyze gelatin, casein, or chitin. Susceptible to vibriostatic agent O/129. Major fatty acids are $C_{14:1}$, $C_{16:0}$, $C_{16:0\ iso}$, $C_{16:1}$, $C_{18:1}$; also contains $C_{22:6}$.

Isolated from seawater collected in the Okhotsk Sea, Japan.

The mol% G + C of the DNA is: 43.5 (HPLC).

Type strain: 4-22, ATCC BAA-724, JCM 10501, KCTC 12105.

GenBank accession number (16S rRNA): AB023378.

5. **Psychromonas profunda** Xu, Nogi, Kato, Liang, Rüger, Kegel and Glansdorff 2003, 530[VP]

pro.fun' da. L. fem. adj. *profunda* from the deep.

Motile rods, 0.9–1.2 × 2.0–5.5 µm; one polar flagellum (unsheathed). NaCl required for growth. Growth occurs at 2–14°C. Greater than atmospheric pressure enhances growth. Possesses both fermentative and respiratory pathways. Produces acid from both fermentation and oxidation of cellobiose, dulcitol, fructose, galactose, glucose, glycerol, inositol, lactose, maltose, mannitol, mannose, rhamnose, salicin, sucrose, trehalose, and xylose. Hydrolyzes esculin and DNA. Weak hydrolysis of starch; no gelatin hydrolysis. Catalase and oxidase positive. Reduces nitrate to nitrite. Produces indole, H_2S, and β-galactosidase. Susceptible to chloramphenicol, furazolidone, penicillin, polymyxin B, tetracycline, and vibriostatic agent O/129. Major quinone Q-8. Major fatty acids are $C_{14:1}$, $C_{16:0}$, and $C_{16:1}$.

Isolated from Atlantic sediment samples taken at 2770 m depth off the coast of West Africa.

The mol% G + C of the DNA is: 38.1 (HPLC).

Type strain: 2825, JCM 11437.

GenBank accession number (16S rRNA): AJ416756.

Genus XIII. **Shewanella** *MacDonell and Colwell 1986, 355[VP] (Effective publication: MacDonell and Colwell 1985, 180)*

JOHN P. BOWMAN

Shew.a.nel' l.a. M.L. dim. ending *-ella*; *Shewanella* named after James Shewan for his work in fisheries microbiology.

Straight or curved rods, 0.5–0.8 × 0.7–2.0 µm. Endospores and microcysts are not formed. **Motile by a single, unsheathed, polar flagellum.** Gram negative. Colonies are often pale tan to pink-orange, due to cytochrome accumulation. **Oxidase and catalase positive.** Chemoheterotrophic. **Facultatively anaerobic. Oxygen** is used as the electron acceptor during aerobic growth. **Anaerobic growth is predominantly respiratory;** the oxidation of organic carbon compounds or H_2 is coupled to the reduction of various inorganic and organic electron acceptors, including NO_3^-, NO_2^-, Fe^{3+}, trimethylamine-N-oxide, fumarate, various sulfur compounds, and Mn^{4+}. **May also be fermentative**, producing acid (but usually no gas) from carbohydrates, such as D-glucose and N-acetylglucosamine. Constitutive arginine dihydrolase is usually absent. Most strains can form H_2S from thiosulfate. **May require Na^+ ions** for growth. **Most species can grow at 4°C.** Some species are psychrophilic. One species is psychrophilic and barophilic. Major fatty acids are $C_{13:0\ iso}$, $C_{14:0}$, $C_{15:0}$, $C_{15:0\ iso}$, $C_{16:1\ \omega7c}$, $C_{16:0}$, and $C_{17:1\ \omega8c}$. Several species can also form significant quantities of the omega-3 fatty acid eicosapentaenoic acid ($C_{20:5\ \omega3c}$). Isoprenoid quinone components include ubiquinone-8, ubiquinone-7, and menaquinone-7, with smaller amounts of methylmenaquinone-7. Major polyamines are putrescine and cadaverine. Member of the order *Alteromonadales*, family *Alteromonadaceae*, class *Gammaproteobacteria*. Isolated from clinical samples, chilled meats, butter, cutting oils, fresh water, freshwater sediment, estuaries, salt marshes, marine algae, seawater, marine sediment, fish, marine invertebrates, sea ice, marine snow, and abyssal ocean waters.

The mol% G + C of the DNA is: 38–54.

Type species: **Shewanella putrefaciens** (Lee, Gibson and Shewan 1977) MacDonell and Colwell 1986, 355 (Effective publication: MacDonell and Colwell 1985, 180) (*Alteromonas putrefaciens* Lee, Gibson and Shewan 1977, 449.)

FURTHER DESCRIPTIVE INFORMATION

Morphology Cells are pleomorphic rods, possessing either straight or slightly curved axes. In nutrient media, short filaments and, more rarely, helical forms may develop in older cultures. Cells do not contain poly-β-hydroxybutyrate inclusions or gas vesicles, form endospores or microcysts, or develop cellular appendages, such as prosthecae or spinae. Inclusion bodies of unknown function, possibly consisting of protein, have been found in *S. putrefaciens* (Krause et al., 1996).

Cultural characteristics Colonies on complex nutrient media typically have a pale tan to pink-orange or salmon color, which is due to strong accumulation of cytochrome proteins. This tendency is less pronounced in some species, such as *S. hanedai* and *S. benthica*. *S. colwelliana* and some *S. hanedai* strains form dark brown, melanin-like pigments when growing on complex media or media containing L-tyrosine. Studies on *S. colwelliana* indicate that this pigmentation derives from L-tyrosine, which is catabolized to homogentisate, which in turn is polymerized to form the pigment pyomelanin (Fuqua and Weiner, 1993; Coon et al., 1994; Ruzafa et al., 1994; Kotob et al., 1995).

Several *Shewanella* species require Na^+ for growth, and best growth occurs on seawater media such as marine 2216 agar (Difco Laboratories, Detroit, Michigan). *S. putrefaciens*, *S. baltica*, *S. algae*, and *S. frigidimarina* can grow well on marine agar and most commonly used nutrient media, including nutrient agar (Oxoid Ltd., Basingstoke, UK), plate count agar (Oxoid), triple sugar iron agar (Oxoid), etc. *S. colwelliana* requires amino acids, including L-aspartate and L-glutamate, for growth (Weiner et al., 1988). *S. hanedai* and *S. benthica* strains may also require amino acids and/or vitamins for growth (Baumann et al., 1984d; Bowman et al., 1997b). In defined media, such as basal medium agar (see *Alteromonas*), these requirements can be met by adding yeast extract at concentrations of 0.05–0.1% (w/v). *S. putrefaciens*, *S. baltica*, *S. frigidimarina*, *S. algae*, and *S. gelidimarina* can grow well on defined seawater media containing ammonia as a nitrogen source. The growth requirements for *S. woodyi* are unknown. Some *S. algae* strains and most *S. frigidimarina* strains do not require Na^+ but

are stimulated by its presence. Growth occurs between 0.05–1.0 M NaCl, with optimal NaCl concentrations being 0.4–0.5 M. Strains of *S. algae* and *S. frigidimarina* exhibit moderate halotolerance and are able to tolerate Na^+ ion concentrations of up to 1.7–2.0 M.

The physiological diversity of *Shewanella* species is reflected by the wide range of habitats from which they can be isolated. Species may be mesophilic (*S. algae*, *S. colwelliana*, *S. putrefaciens* DNA hybridization group III), psychrotolerant (*S. putrefaciens* group I, *S. baltica*, *S. frigidimarina*, *S. woodyi*), psychrophilic (*S. gelidimarina*, *S. hanedai*), and both psychrophilic and barophilic (*S. benthica*). The majority of *S. algae* strains can grow at 41°C, but no *Shewanella* strains can grow at 45°C or higher.

Energy metabolism *Shewanella* species are chemoheterotrophic facultative anaerobes, with anaerobic growth typically of a respiratory nature; however, some species can also grow fermentatively. These include *S. frigidimarina* and *S. benthica*, both of which can ferment D-glucose, and *S. gelidimarina*, which can ferment N-acetylglucosamine and chitin, but not D-glucose (MacDonell and Colwell, 1985; Bowman et al., 1997b). Gas is not produced by these species during fermentation; however, *S. violacea*, a barotolerant species closely related to *S. benthica*, can ferment D-glucose and produce both acid and gas (Nogi et al., 1998b). The end products and relevant biochemistry of this process has not been studied.

Most studies on the diverse modes of anaerobic respiration in *Shewanella* have concentrated on strains of *S. putrefaciens* and *S. algae*; however, all *Shewanella* strains can use nitrate as an electron acceptor for growth (although denitrification is a strain-specific characteristic). Trimethylamine-N-oxide (TMAO) is also a common terminal electron acceptor among *Shewanella* species, and its reduction to trimethylamine is usually responsible for the odors associated with *Shewanella* food spoilage (Shewan, 1971). *Shewanella* strains can grow anaerobically by reduction of various sulfur compounds to H_2S, including S^0, thiosulfate, and sulfite (Semple and Westlake, 1987; Perry et al., 1993; Moser and Nealson, 1996), thiosulfate, and sulfite. Anaerobic growth can also occur via reduction of fumarate to succinate, coupled to the oxidation of formate. The fumarate reductase of *S. frigidimarina* NCIMB strain 400 is similar to a fumarate reductase flavoprotein found in *Wolinella succinogenes* (Simon et al., 1998).

Extensive studies have concentrated on the relatively unusual ability of *Shewanella* to facultatively reduce ferric iron, manganese, and other metals. The majority of *Shewanella* strains tested can grow anaerobically by coupling the oxidation of carbon compounds or H_2 (in the presence of a utilizable carbon source) to the reduction of Fe^{3+} to Fe^{2+} (Semple and Westlake, 1987; Lovley and Phillips, 1986; Lovley, 1993) or of Mn^{4+} to insoluble Mn^{3+} (Myers and Nealson, 1988; Caccavo et al., 1996b). Mn^{4+} and Fe^{3+} have redox potentials higher than that of sulfate and are capable of out-competing electron acceptors of lower potential, such as SO_4^{2-} (used for sulfate reduction) and CO_2 (used for methanogenesis) (Nealson and Saffarini, 1994). Dissimilatory metal reduction is believed to be important in terms of metal cycling and mobilization, as well as for its contribution to geomagnetism and sedimentary diagenesis (Nealson and Saffarini, 1994; Lovley, 1997; Pedersen, 1997). Owing to the broad specificity of the *Shewanella* anaerobic reductase enzyme system, *S. putrefaciens* strains can reduce and mobilize toxic and radioactive metallic pollutants, including Co, Cr, Se, Tc, and U (Lovley, 1993; Caccavo et al., 1994, 1996b; Lloyd and McCaskie, 1996; Ganesh et al., 1997; Gorby et al., 1998). In addition, strains can reductively dehalogenate and transform organic pollutants, such as tetrachloroethane and carbon tetrachloride (Petrovskis et al., 1994; Picardal et al., 1995; Backhus et al., 1997). Studies indicate that the different modes of anaerobic respiration utilize different sets of enzymes. For example, a transposon mutant of *S. putrefaciens* MR-1 lacks fumarate reduction ability, but other reductive pathways are not compromised (Myers and Myers, 1997). Genes for anaerobic respiration by *S. putrefaciens* strain MR-1 appear to be both chromosomal and borne on megaplasmids (Saffarini et al., 1994).

During aerobic and anaerobic growth, *Shewanella* species can utilize a range of organic acids as sole sources of carbon and energy. In general, aromatic and heterocyclic compounds are not utilized, while the degree of utilization of carbohydrates varies considerably among the species, with some strains unable to use any carbohydrates. Glucose is catabolized via the Entner–Doudoroff pathway (Abu et al., 1994; Scott and Nealson, 1994). Most strains can degrade gelatin and form lipases, while certain species form chitinases, but in general do not attack starch, except for *S. colwelliana* and some strains of *S. woodyi* and *S. hanedai*. *S. gelidimarina* appears to be adapted almost specifically to chitin degradation, given its ability to utilize both aerobically and anaerobically only *N*-acetylglucosamine and chitin, out of a wide range of carbohydrates tested.

S. putrefaciens produces a potent iron chelator, putrebactin, that is structurally similar to chelators formed by *Pseudoalteromonas haloplanktis* (bisucaberin) and *Bordetella* spp. (alcaligin) (Ledyard and Butler, 1997). *S. algae* strains have been shown to form 4-epitetrodotoxin, a derivative of tetrodotoxin, which is a potent neurotoxin (Simidu et al., 1987, 1990; Nozue et al., 1992).

Genetic characteristics DNA base composition values for *Shewanella* vary from 38 to 54 mol% (T_m, Bd), and the ranges for individual species are shown in Table BXII.γ.148.

Temperature and pressure-regulated genes have been studied in a number of barotolerant and barophilic *S. benthica* strains (Kato et al., 1997; Chilukuri and Bartlett, 1997) and include a series of cytochrome proteins, including a cytochrome *c* oxidase. A set of PCR-amplified, pressure-regulated gene products have been found to be universal in *S. benthica* strains and have been proposed as a rapid means to identify this particular barophilic species (Li et al., 1998).

S. hanedai and *S. woodyi* have the distinction of being the only known nonfermentative bioluminescent species. Studies of the *luxA* sequence of *S. woodyi* have shown it to have only little homology with the *luxA* nucleotide sequence of *S. hanedai*. Overall, the *S. hanedai luxA* gene shows closer similarity to sequences found in *Vibrio logei* and *Vibrio fischeri* (*Photobacterium fischeri*) (Makemson et al., 1997).

Fatty acid profiles *Shewanella* species have a distinctive fatty acid pattern, as compared with other bacteria of the *Gammaproteobacteria*, in that they are rich in branched and odd-chain-length fatty acids (Wilkinson and Caudwell, 1980; Moule and Wilkinson, 1987; Rosselló-Mora et al., 1995a; Bowman et al., 1997b). Major fatty acids (and their ranges of % composition) include: $C_{13:0\ iso}$ (1–16%), $C_{14:0}$ (1–12%), $C_{15:0\ iso}$ (4–24%), $C_{15:0}$ (1–8%), $C_{16:1\ \omega7c}$ (16–55%), $C_{16:0}$ (5–31%), $C_{17:1\ \omega8c}$ (1–15%), and $C_{18:1\ \omega7c}$ (0–8%). The quantitative proportions of fatty acid components vary considerably among species, but it is difficult to judge with the data available whether these differences would be useful in species differentiation or whether they are merely due to variations in cultivation conditions. *Shewanella* possess both aerobic and anaerobic desaturatase pathways for fatty acid synthesis. *Shewanella* strains grown anaerobically thus form a different fatty acid profile that those grown aerobically (Nichols et al., 1997b; Venkateswaran et al., 1999).

Psychrotolerant and psychrophilic *Shewanella* species have the unusual ability to synthesize eicosapentaenoic acid (EPA, $C_{20:5\ \omega3}$), the levels of which range from 2–22% (Bowman et al., 1997b). Few other bacterial species known to possess EPA, including *Flexibacter polymorphus* (Johns and Perry, 1977) and *Psychroflexus torquis* (Nichols et al., 1997a; Bowman et al. 1998c), which are members of the flavobacteria division. Several studies have detected EPA in unidentified marine bacteria with fatty acid profiles very similar to *Shewanella* (Iwanami et al., 1995, Watanabe et al., 1996; Yazawa, 1996; Jostensen and Landfald, 1997; Yano et al., 1997). Compelling evidence is available that links production of EPA synthesis with cold adaptation. In *S. gelidimarina* ACAM 456, EPA levels decrease markedly when incubation temperatures are above its optimal growth temperature (Nichols et al., 1997b), and there seems to be a correlation between the optimal growth temperatures of *Shewanella* species and their inherent EPA levels (Bowman et al., 1997b). In *S. benthica*, EPA levels increase until the hydrostatic pressure reaches the particular strain's growth pressure optimum (DeLong and Yayanos, 1986). Increased levels of EPA decrease the homeoviscosity of cellular membranes (Nichols et al., 1995), an adaptation important for organisms living in perpetually cold and high-pressure environments. Because omega-3 fatty acid precursors cannot be synthesized *de novo* by most metazoa, it is thought that bacteria and microalgae act as a dietary supply of these lipids (DeLong and Yayanos, 1986). The EPA synthesis gene cluster of *Shewanella* sp. SCRC-2738 has been cloned and sequenced, and weak expression of the genes with EPA production has been conferred to a *Synechococcus* strain (Takeyama et al. 1997). The synthetic pathway of EPA has been partially characterized in the same strain and appears to be derived from an aerobic desaturation pathway (Watanabe et al., 1997). Mesophilic species do not appear to produce EPA, and their fatty acid profiles are similar to those of *Ferrimonas balearica*, another mesophilic, iron-reducing halophilic species isolated from sediment (Rosselló-Mora et al., 1995a). The major polar lipids have been determined in *S. putrefaciens*, *S. baltica*, and *S. algae* and include mostly phosphatidylglycerol and phosphatidylethanolamine, a pattern typical of Gram-negative bacteria (Moule and Wilkinson, 1987).

Lipopolysaccharides (LPS) *S. putrefaciens* possesses R-type lipopolysaccharides, rather than the ladder-type LPS typically found in many other Gram-negative bacteria (Sledjeski and Weiner, 1991). The chemical composition of the LPS varies among various species of *Shewanella* (Moule and Wilkinson, 1989). The lipid fraction of the LPS in *S. putrefaciens* (strains ATCC 8071 and 8073), *S. baltica* (strain ATCC 8072), and *S. algae* (NCIMB 11157) consists of (% range) $C_{13:0\ iso}$ (4–9%), $C_{13:0}$ (6–13%), $C_{10:0\ 3OH}$ (8–13%), $C_{12:0\ 3OH}$ (11–16%), $C_{13:0\ iso\ 3OH}$ (8–14%), $C_{13:0\ 3OH}$ (13–21%), and $C_{14:0\ 3OH}$ (7–9%). In addition, *S. baltica* contains high levels of $C_{12:0\ iso\ 3OH}$ (10%) and $C_{14:0\ iso\ 3OH}$ (8%); these components are much less abundant in the other strains tested (0–2%). The neutral and amino sugar content of the LPS core oligosaccharide of all strains tested contained of galactose and heptose. In addition *S. putrefaciens* ATCC 8071 contains glucose and 3-amino-3,6-dideoxyglucose; both *S. putrefaciens* ATCC 8073 and *S. baltica* contain galactosamine; and *S. algae* contains quinovosamine.

TABLE BXII.γ.148. Phenotypic characteristics which differentiate the species of the genus *Shewanella*[a]

Characteristics	*S. putrefaciens*	*S. algae*	*S. amazonensis*	*S. baltica*	*S. benthica*	*S. colwelliana*	*S. frigidimarina*	*S. gelidimarina*	*S. hanedai*	*S. oneidensis*	*S. violacea*	*S. woodyi*
Bioluminescence	−	−	−	−	−	−	−	−	+	−	−	+
Violet pigment	−	−	−	−	−	−	−	−	−	−	+	−
Growth at 4°C	d	−	+	+	+	−	+	+	+	d	+	+
Growth at 25°C	+	+	+	+	−	+	−	−	d	+	−	−
Growth at 37°C	+	+	+	−	−	−	−	−	−	+	−	−
Barophilic	−	−	−	−	+	−	−	−	−	−	+	−
Requires Na$^+$ for growth	−	S	+	−	+	+	S	+	+	−	+	+
Tolerates 6% NaCl	−	+	−	−	−	−	+	−	−	−	nd	nd
Fermentation of:												
D-Glucose	−	−	−	−	+	−	+	−	−	−	+	
D-Glucose (with gas)	−	−	−	−	−	−	−	−	−	−	+	
N-Acetylglucosamine	−	−	nd	−	+	−	−	+	−	nd	nd	nd
Chitinase	−	−	−	−	+	−	−	+	+	nd	nd	−
Amylase	−	−	−	−	−	+	−	−	−	−	−	d
Lipase	+	+	nd	+	+	−	+	+	+	nd	nd	+
Hemolysis	−	+	+	−	−	−	−	−	−	−	nd	−
Denitrification	d	−	+	−	−	−	d	−	d	−	−	+
Ornithine decarboxylase	+	+	−	+	−	−	d	−	−	−	−	nd
EPA synthesis[b]	−	−	−	−	+	nd	+	+	+	−	+	nd
H$_2$S production	+	+	+	+	+	−	d	+	+	+	nd	nd
Iron reduction	+	+	+	+	d	nd	+	+	−	+	nd	nd
Acid from D-glucose	d	d (weak)	nd	+	+	−	+	−	+	nd	+	nd
Utilization of:												
D-Glucose	d	−	nd	+	+	−	+	−	d	nd	nd	+
Cellobiose	−	−	nd	+	+	−	+	−	−	nd	nd	+
Maltose, sucrose	d	−	nd	+	−	−	+	−	−	−	nd	−
N-Acetylglucosamine	d	+	nd	+	+	−	−	+	+	nd	nd	−
D-Gluconate	−	−	nd	+	−	+	d	−	d	nd	nd	nd
Mol% G + C	43–49	54	52	46–47	46–47	46	40–43	48	44–47	45	47	46

[a]For symbols see standard definitions; S, growth is stimulated by the presence of Na$^+$ ions.

[b]EPA, eicosapentaenoic acid.

Lipids A number of additional substituted lipids have been found in *S. putrefaciens* ATCC 8071, including two glycolipids, identified as β-D-glucopyranosyldiacylglycerol and β-D-glucopyranuronosyldiacylglycerol, and an ornithine amide lipid (Wilkinson, 1968a, 1972; Wilkinson et al., 1973).

Quinones The quinones found in *S. putrefaciens* ("*Pseudomonas rubescens*"), *S. algae*, *S. baltica*, and *S. frigidimarina* are ubiquinone-7 (11–51%), ubiquinone-8 (31–58%), and menaquinone-7 (5–52%), with small quantities of methylmenaquinone-7 (thermoplasmoquinone-7, 0–6%) also present (Itoh et al., 1985; Moule and Wilkinson, 1987; Akagawa-Matsushita et al., 1992a). This pattern distinguishes *Shewanella* spp. from other related bacteria, such as *Alteromonas* and *Pseudoalteromonas*, which contain mostly Q-8 (Akagawa-Matsushita et al., 1992a). Menaquinones appear to play an important role in anaerobic respiratory activity in *Shewanella* (Myers and Myers, 1993).

Polyamines The major polyamines found in *Shewanella* species are putrescine and cadaverine. This pattern differs from those of *Pseudoalteromonas* and *Alteromonas*, which, in addition to putrescine and cadaverine possess hydroxylated polyamine derivatives. *Moritella marina*, by comparison, possesses only cadaverine (Hamana, 1997).

Exopolysaccharides *S. putrefaciens* strain S29 possesses a phosphorylated exopolysaccharide made up of tetrasaccharide–phosphate repeating units consisting of 2-acetamido-2,6-dideoxyglucose and 4-acetamido-4,6-dideoxyglucose (Shashkov et al., 1997).

Pathogenicity *S. algae* and, less often, *S. putrefaciens* have been implicated occasionally in polymicrobial bacteremia and septicemia (Vogel et al., 1997) and are associated with a wide spectrum of clinical syndromes, such as surface and soft-tissue infections, eye infections, pneumonia, arthritis, peritonitis, and empyema (von Graevenitz and Simon, 1970; Debois et al., 1975; Holmes et al., 1975; Kim et al., 1989; Heller et al., 1990; Chen et al., 1991, 1997; Brink et al., 1995; Butt et al., 1997a; Dominguez et al., 1996; Yohe et al., 1997; Levy and Tessier, 1998). Hemolysin production by *S. algae* is believed to be an important factor contributing to the pathogenicity of this species (Khashe and Janda, 1998), but the exact details are yet to be resolved. *Shewanella*-like isolates have been proposed as a possible cause of lesions on some fish species (Subasinghe and Shariff, 1992; Mashima et al., 1997), although goldfish directly inoculated with a number of *Shewanella* strains have failed to exhibit any infection (Decostere et al., 1996).

Ecology *Shewanella* is a ubiquitous group of bacteria that has been isolated from a wide range of habitats. *S. putrefaciens* and its close relative *S. baltica* are among the major spoilers of chilled foods and are frequently isolated from a wide range of dairy, poultry, beef, and seafood products (Shewan, 1971; Lee et al., 1977; Molin and Ternström, 1982, Stenström and Molin, 1990;

Borch et al., 1996; Gram and Huss, 1996; Leroi et al., 1998). *Shewanella* has also been isolated from cutting and lubricating oils, oil brine, seawater, sediment, soil, ponds and lakes, sewage, and subsurface groundwater (Pivnick, 1955; Iizuka and Komagata, 1964b; Semple and Westlake, 1987; Aznar et al., 1992; DiChristina and DeLong, 1993; Höfle and Brettar, 1996; Pedersen et al., 1996a; Ziemke et al., 1997; Bowman et al., 1997a).

The exact isolation sites of *S. putrefaciens senso stricto* are unclear, owing to the species heterogeneity. The most common sites for isolation include various lightly salted food products, the coastal marine environment, freshwater sediments and overlying waters, and water and soil mixed with crude oil and other petroleum products (Semple and Westlake, 1987). A close relative of *S. putrefaciens*, *S. oneidensis* (Venkateswaran et al., 1998) has been isolated from lake sediment. Another close relative, *S. baltica*, appears to occur frequently in coastal marine environments. Genetic analysis of *S. baltica* (described in the study as *S. putrefaciens*) in Baltic Sea waters indicates the population is stable at the clonal level (Ziemke et al., 1997). Populations spike in correspondence with increased sulfur levels in the Baltic Sea and other locations (Nealson et al., 1991; Brettar and Höfle, 1993). *S. algae* comprises the bulk of strains isolated from clinical specimens (Nozue et al., 1992, Vogel et al., 1997), and this species has also been isolated from various saline habitats, including salt marshes (Rosselló-Mora et al., 1994a), oil brine (Semple and Westlake, 1987), surfaces of macroalgae (Simidu et al., 1990), seawater, and occasionally salted food products (Vogel et al., 1997). Most *Shewanella* species appear to be strictly marine in origin. *S. amazonensis* has so far been isolated only from coastal marine muds off the Brazilian coast. The bioluminescent species *S. woodyi* (Makemson et al., 1997), as well as other novel *Shewanella* strains, has been shown to be associated with the ink and reproductive organs of certain squid (*Loligo* spp.) (Leonardo et al., 1999). *S. woodyi* has also been isolated from seawater and organic detritus collected at depths of 200–300 meters in parts of the Mediterranean Sea. *S. colwelliana* has been isolated from cultured oysters and their vicinity in the Chesapeake Bay, United States (Weiner et al., 1988). Studies suggest that *S. colwelliana* can enter a symbiotic arrangement with a host oyster (Bonar et al., 1986).

A host of cold-adapted *Shewanella* species has been isolated from marine ecosystems. *S. frigidimarina* appears to be ubiquitous in the Antarctic marine environment and has been isolated from seawater, ice, sediments, and cyanobacterial mat communities of the southern ocean and continental, marine-derived lakes (Bowman et al., 1997a, b; Rea et al., unpublished). It has also been isolated from the deep sea (DeLong et al., 1997) and the North Sea (Reid and Gordon, 1999), as indicated by 16S rDNA sequencing of isolates. *S. gelidimarina*, by comparison, has a much more restricted distribution and inhabits diatom-rich communities found in coastal Antarctic sea ice (Bowman et al., 1997a). *S. hanedai* has also been isolated from Antarctic sea ice diatom communities, but was first isolated from sediments of the Arctic and Southern Oceans and from coastal areas of Canada (Jensen et al., 1980). Strains of *S. benthica* as well as *S. violacea* have been isolated from several deep-sea sites, including the Marianas Trench at a depth of nearly 11,000 m (Kato et al., 1995, 1996, 1998). This species has been found in water and sediment samples and organic detritus, but is most frequently associated with decaying deep sea invertebrates, including amphipods and holothurians (Deming et al., 1988a; DeLong et al., 1997).

ENRICHMENT AND ISOLATION PROCEDURES

Most *Shewanella* species can be directly isolated from source material onto marine 2216 (Difco Laboratories) or nutrient agar (Oxoid) without prior enrichment. *Shewanella* colonies are often identifiable by their pale tan or pink-salmon color. *S. putrefaciens* and *S. algae* can be isolated semi-selectively from clinical and environmental samples on a medium containing 1% NaCl, 0.1% ox bile salts, and 1% peptone (Nozue et al., 1992). Nonfermentative, H_2S-producing isolates (which are usually positive for ornithine decarboxylase) can be further selected on deoxycholate–hydrogen sulfide–lactose medium (Eiken Chemical Co., Ltd., Tokyo, Japan). Additional selection for *S. algae* includes incubation at 40°C on nutrient agar containing 6% NaCl. The identification of *S. algae* strains can be confirmed by their α-hemolytic activity on sheep blood agar; strains can also be selected and isolated by plating clinical samples onto *Salmonella–Shigella* agar (Nozue et al., 1992). *S. putrefaciens* and *S. algae* strains and several other *Shewanella* species can also be enriched and isolated with iron-, nitrate- or sulfur-reducing media (see Procedures for Testing Special Characters). Alternatively, direct isolation can be performed anaerobically on S^0 reduction agar (Moser and Nealson, 1996) with transfer of growth to nutrient or marine agar.

A brief enrichment of samples in marine 2216 broth at 2°C for 1–2 d is advantageous for the isolation of psychrophilic species such as *S. gelidimarina* (Bowman et al., 1997a). *S. woodyi* and *S. hanedai* produce bioluminescence most effectively on marine agar, rather than on the standard glycerol-containing luminous medium normally used to detect bioluminescent *Vibrio* or *Photobacterium* spp. (Ruby et al., 1980; Hastings and Nealson, 1977).

Isolation of *S. benthica* involves a relatively straightforward method that can also be used to isolate barophilic species belonging to other genera (Sakiyama and Ohwada, 1997); it is similar to methods employed in other studies (DeLong and Yayanos, 1986; DeLong et al., 1997; Kato et al., 1995, 1996, 1998). Small portions of sediment or seawater filters are added to 0.5× marine 2216 broth in small, sterile polyethylene bags (Whirl-Pak, Nasco, USA) and heat sealed. The bags are then incubated at low temperatures (2–4°C) in stainless steel pressure vessels under a hydrostatic pressure equivalent to that at the isolation site. For further details on the use of high-hydrostatic-pressure culture equipment, see Yayanos et al. (1979, 1982). Strains are isolated and purified using marine 2216 broth solidified with 2% SeaPrep low-melting-point agarose (FMW Inc., Maryland, USA) or silica gel (Dietz and Yayanos, 1978) in tubes sealed with silicone rubber double stoppers and pressurized to about 30–70 mPa (300–700 atmospheres). Individual colonies are then cut out of the medium with a scalpel and transferred to fresh marine 2216 broth.

MAINTENANCE PROCEDURES

In general, *Shewanella* species are relatively robust and can be preserved and maintained in various ways in a carbohydrate-free medium in which they grow well, such as nutrient or marine agar. *Shewanella* species can be lyophilized using 20% skim milk as a cryoprotectant. In addition, strains can sustain viability for over 6 months when frozen in liquid nitrogen or cryopreserved at −80°C in broth containing 20–30% glycerol or DMSO. *S. algae* and *S. putrefaciens* strains can be maintained at 15–20°C in semisolid agar medium containing 0.1% Proteose peptone no. 3 (Difco Laboratories), 0.1% yeast extract, 0.05% phytone, 0.02% sodium thiosulfate, 0.005% sodium sulfite, 0.004% ferric citrate, and 0.3% agar dissolved in 3:1 aged seawater (or artificial sea salts) and water; pH 7.6. (Simidu et al., 1990). *S. frigidimarina* and *S. gelidimarina* can be stored on marine agar slants or as heavy suspensions in sterile seawater at 2°C for 12 months or longer.

PROCEDURES FOR TESTING SPECIAL CHARACTERS

Dissimilatory iron reduction is a common property of *Shewanella* species and can be used in their isolation. Iron-reducing medium[1] Lovley and Phillips, 1988) supplemented with 0.05% yeast extract to boost growth can be used to test this property and to enrich for iron reducing strains. A variety of iron electron acceptors can be used, including amorphic ferric oxide, ferric citrate, and ferric pyrophosphate (Lovley and Phillips, 1986, 1988). These electron acceptors can be added separately or in combination. Electron donors that are typically utilized include organic acids, such as lactate, citrate, malate, etc. The addition of H_2 to the headspace can substantially stimulate growth (Caccavo et al., 1992). Samples are suspended in phosphate-buffered saline buffer or seawater and then sparged with nitrogen. The diluted sample is then added to the media by syringe or added under a stream of nitrogen, and vials or tubes are crimp sealed. Reduction of Fe^{3+} to Fe^{2+} is indicated by the change of the rust-colored medium to a clear solution that contains white (smectite or vivianite) or black (FeS_2, pyrite) precipitates. A similar approach can be used with other electron acceptors, such as nitrates[2] and MnO_2[3] (Lovley and Phillips, 1988).

Anaerobic sulfur reduction can be tested using a plate assay developed by Moser and Nealson (1996), which contains polysulfide[4]. Colonies within clearing zones on the sulfur plates can then be transferred to nutrient or marine agar for further purification with aerobic incubation.

---

1. Iron-reducing medium consists (per liter distilled water): $NaHCO_3$, 2.5 g; $CaCl_2 \cdot 2H_2O$, 0.1 g; KCl, 0.1 g; NH_4Cl, 1.5 g; NaH_2PO_4, 0.6 g; carbon source (acetate, lactate, citrate, etc.), 20 mM; trace element solution, 10 ml; and 200 mM amorphic ferric oxide. The medium is prepared with vigorous boiling and gassing with nitrogen to remove oxygen and dispensed into vials or tubes sealed with thick butyl rubber stoppers. Following autoclaving, the medium pH is about 6.7. The trace element solution (Balch et al., 1979) contains (per liter distilled water) nitriloacetate, 1.5 g; $MgSO_4 \cdot 7H_2O$, 3 g; $MnSO_4 \cdot H_2O$, 0.5 g; NaCl, 1 g; $FeSO_4 \cdot 7H_2O$, 0.1 g; $CoCl_2 \cdot 6H_2O$, 0.1 g; $CaCl_2 \cdot 2H_2O$, 0.1 g; $ZnSO_4 \cdot 7H_2O$, 0.1 g; $CuSO_4 \cdot 5H_2O$, 0.01 g; $AlK(SO_4)_2 \cdot 12H_2O$, 0.01 g; H_3BO_3, 0.01 g; $Na_2MoO_4 \cdot 2H_2O$, 0.01 g; $NiSO_4 \cdot 6H_2O$, 0.03 g; Na_2SeO_3, 0.02 g; and $Na_2WO_4 \cdot 2H_2O$, 0.02 g. The nitriloacetate is dissolved in 500 ml distilled water and the pH is adjusted to 6.5 by KOH. The remaining salts are added one at a time, and the volume is brought up to 1 liter. Samples are added using Hungate techniques or in an anaerobic chamber. N_2/CO_2 (80:20) or $N_2/CO_2/H_2$ (80:10:10) can be used as the headspace atmosphere. For enrichment of marine samples, the medium is supplemented with 27 g/l NaCl and 3.7 g/l $MgCl_2$. Amorphic ferric oxide is prepared by neutralizing a 0.4 M $FeCl_3$ solution to pH 7 with NaOH. Ferric citrate can be used instead of amorphic ferric oxide and is added at 20 mM (with $CaCl_2 \cdot 2H_2O$ omitted to prevent precipitation).

2. Nitrate-reducing medium: similar to iron-reducing medium, but with amorphic ferric oxide replaced by 20 mM KNO_3 ($CaCl_2 \cdot 2H_2O$ is omitted to prevent precipitation).

3. Manganese-reducing medium: similar to iron-reducing medium, but with amorphic ferric oxide replaced by 15 mM MnO_2. The latter is prepared by slowly adding 30 mM $MnCl_2$ to 20 mM $KMnO_4$ with constant stirring.

4. For sulfur-reduction plates, the basal medium used is similar to iron-reducing medium, but the amorphic ferric oxide is omitted and the medium is supplemented with 0.5% Casamino acids and 1.5% agar. Immediately before dispensing the medium into plates, 40 mM polysulfide is added from a polysulfide stock (2.25 M total sulfur) prepared by adding 7.2 g sulfur flowers and 24 g $Na_2S \cdot 9H_2O$ to 100 ml boiling water and stirring for 15 min. After adding the polysulfide, the medium is left in air overnight to allow for sufficient precipitation of sulfur globules and to shift the pH back to neutrality. The medium is then inoculated and incubated anaerobically.

The ability to ferment carbohydrates is tested most effectively in the oxidation/fermentation medium of Leifson (1963), to which carbohydrates are added at a concentration of 0.5% (w/v). This medium is more sensitive than the usual Hugh and Leifson medium because it contains phenol red instead of bromothymol blue. Other biochemical and nutritional traits can be tested using standard procedures, provided that the incubation-temperature and Na^+-ion requirements are met.

DIFFERENTIATION OF THE GENUS *SHEWANELLA* FROM OTHER GENERA

Characteristics that differentiate *Shewanella* from other members of the *Alteromonadaceae* are shown in Table BXII.γ.135 of the chapter describing the family. Due to the diversity of species, there are only a few properties shared by all the members of the genus, thus resulting in a potential for misidentification. Individually, most species possess sufficiently distinct properties (H_2S production, ornithine decarboxylase activity, psychrophily, etc.) to make them relatively easy to identify, assuming enough tests are employed. The generally nonfermentative nature of *Shewanella* species distinguishes them from members of the *Vibrionaceae*. *S. frigidimarina* is an exception; however, it lacks sheathed flagella and possesses a mol% G + C value lower than those of most *Vibrio* species. There is a possibility that *S. algae* can be confused with *Ferrimonas balearica*, but, unlike the latter, it can lyse sheep's blood, degrade gelatin, and tolerate 10% NaCl and is unable to produce phenylalanine deaminase (Rosselló-Mora et al., 1995a). In ambiguous cases, fatty acid analysis and/or partial 16S rRNA sequencing provides direct identification.

TAXONOMIC COMMENTS

Phylogenetic trees based on 16S rRNA sequences of several *Shewanella* strains are shown in Fig. BXII.γ.156. Overall, two major groups are delineated within the genus, with *S. algae* and *S. amazonensis* (Venkateswaran et al., 1998), forming relatively distinct branches. One large group includes the psychrotolerant, nonhalophilic species *S. putrefaciens*, *S. baltica*, and *S. frigidimarina*. The second group includes the psychrotolerant and psychrophilic Na^+-requiring species *S. benthica*, *S. hanedai*, *S. gelidimarina*, *S. woodyi*, and *S. pealeana* (Leonardo et al., 1999). Overall, *Shewanella* is phylogenetically most closely related to the genera *Pseudoalteromonas*, *Alteromonas*, *Moritella*, *Ferrimonas*, and *Colwellia*, all of which are members of the family *Alteromonadaceae*.

DNA–DNA hybridization (Owen et al., 1978; Semple et al., 1989), ribotyping, and protein electrophoretic pattern analysis (Vogel et al., 1997) have shown the *S. putrefaciens* complex (not including *S. algae*) to be made up of at least three groups. DNA hybridization data for a variety of *S. putrefaciens* and related strains available in major culture collections are shown in Table BXII.γ.149. Other species, such as *S. algae* (Nozue et al., 1992; Vogel et al., 1997), *S. frigidimarina*, *S. gelidimarina*, and *S. hanedai* (Bowman et al., 1997b), each form genotypically distinct groups (Table BXII.γ.149).

S. putrefaciens was first described in 1931 as an *Achromobacter* species (Derby and Hammer, 1931), but was transferred to *Pseudomonas* because of its rod-like morphology, motility, and nonfermentative metabolism (Long and Hammer, 1941). Because its mol% G + C values are considerably lower than those of other pseudomonads (Levin, 1972), the species was reclassified as a species of *Alteromonas* (Lee et al., 1981a). Finally, based on 5S rRNA sequences, the species was renamed *S. putrefaciens* and designated the type species of the genus *Shewanella*. The description

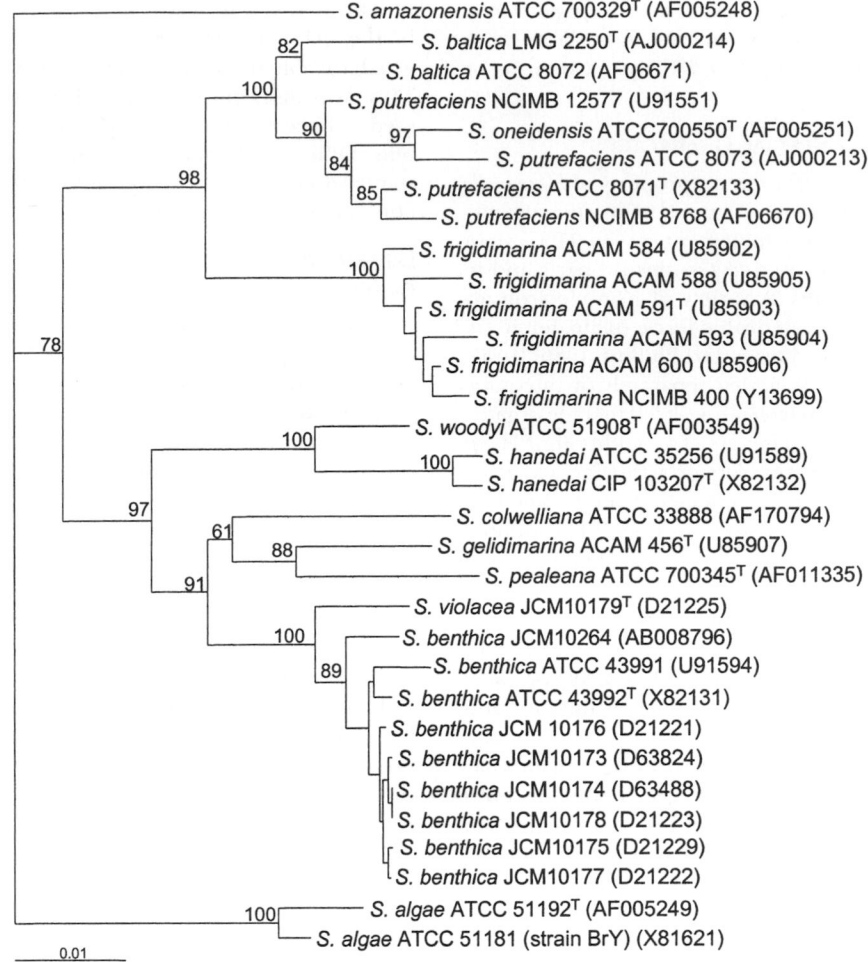

FIGURE BXII.γ.156. Phylogenetic tree of the genus *Shewanella* based on nearly complete 16S rRNA sequences. Species included are those validly published as of 2000. The tree was determined using maximum-likelihood distances and clustered by neighbor-joining. Numbers at branch nodes are bootstrap values (calculated from 250 replicate analyses). Only bootstrap values over 50% are shown. *T* denotes a type strain. Bar = distance of 0.01.

TABLE BXII.γ.149. DNA–DNA relatedness (% hybridization) between *Shewanella putrefaciens* Owen DNA-hybridization groups[a]

	S. putrefaciens				S. baltica		S. algae
	Group I		Group III		Group II		Group IV
Strain (source)	ATCC 8071[T]	ATCC 8073	NCTC 10695	CL256/73	NCTC 10735[T]	ATCC 8072	NCIMB 11157
S. putrefaciens - Group I							
ATCC 8071[T] (butter)	100		15	34	11–26	27	5
ATCC 19857 (unknown)	86	100	9		41	37	5
NCTC 10695 (cutting oil)		100		32	47		27
S. putrefaciens - Group III							
ATCC 8073 (butter)	10	100		42	38	22	6
CL256/73 (spinal fluid)		29		100	32		40
S. putrefaciens (ungrouped)							
NCIMB 1733 (fish)	9	7				21	6
S. baltica - Group II							
NCTC 10735[T] (oil brine)		5	39	28	100		13
ATCC 8072 (butter)	12	7		21	58	100	4
NCTC 10736 (fish)			30	31	83		
NCTC 10737 (squid)			32	26	57		
S. algae - Group IV							
NCTC 10738 (feces)	5	4	25	44	13	9	100
NCTC 10762 (pus)			24		9		
NCTC 10763 (bottled blood)			11		27		

[a]Data from Owen et al. (1978); Semple et al. (1989); Bowman et al. (1997b); Ziemke et al. (1998).

of this genus also included descriptions of *S. hanedai* (previously *Alteromonas hanedai*; Jensen et al., 1980) and *S. benthica* (MacDonell and Colwell, 1985). Although *Shewanella* was initially thought to be a member of the *Vibrionaceae* (MacDonell and Colwell, 1985; Farmer, 1992), 16S rRNA sequences have indicated that *Shewanella* is more closely related to the genus *Alteromonas* and its relatives (Kita-Tsukamoto et al., 1993; Gauthier et al., 1995a). *Shewanella colwelliana* was added to the genus following reassessment of its taxonomy using 5S rRNA sequences (Coyne et al., 1989). *S. algae* was created by Simidu et al. (1990) and has been subsequently emended by Nozue et al. (1992). *S. woodyi* (Makemson et al., 1997), *S. frigidimarina*, *S. gelidimarina* (Bowman et al., 1997b), and *S. baltica* (Ziemke et al., 1998) have only recently been described.

The nomenclature of *S. putrefaciens* is still evolving, but at the time of its renaming as *S. putrefaciens* it was a heterogeneous complex of strains. Owen et al. (1978) have demonstrated that *S. putrefaciens* comprises four DNA hybridization groups. In earlier phenotypic studies of *S. putrefaciens*, at least two groups, or "Gilardi" biovars, were observed. Gilardi biovar 2 includes mostly mesophilic clinical strains, which can grow at 42°C and tolerate 6% NaCl, but cannot utilize glucose or maltose. Gilardi biovars 1 and 3 include mostly food and environmental strains, which are psychrotrophic, saccharolytic, and unable to grow in the presence of 6% NaCl (Gilardi, 1972; Riley et al., 1972; Richard et al., 1985; Khashe and Janda, 1998). Subsequently, Nozue et al. (1992) have determined Gilardi biovar 2 to be comprised mainly of strains of *S. algae*, a species that is equivalent to Owen's DNA hybridization group IV. Additional data in the form of ribotypes and protein electrophoretic patterns indicate that strains belonging to DNA groups I, II, and III can be clearly distinguished from *S. algae* (Vogel et al., 1997). This corresponds well with the distinct phylogeny of *S. algae* and *S. putrefaciens* strains (Fig. BXII.γ.156). However, the resolution among DNA groups I, II, and III using the same procedures is quite poor (Vogel et al., 1997). These methods tend to suggest that *S. putrefaciens* is composed of a continuum of a series of genetically different strains. In their study of fish isolates, Stenström and Molin (1990) have claimed that numerical taxonomic analysis of phenotypic data does not correlate well with the hybridization groups of Owen et al. (1978); however, in the light of more recent data, this conclusion may be biased by their isolation of predominantly group II *S. putrefaciens* strains. Phenotypic analyses of *S. putrefaciens* by Semple and Westlake (1987) and Ziemke et al. (1998) suggest that phenotypic resolution of these groups may be possible, although DNA hybridization results (Semple et al., 1989) suggest that they are not necessarily definitive. For the purpose of completeness, phenotypic data differentiating the various DNA groups of *S. putrefaciens* are presented in Table BXII.γ.150. One limitation is that relatively few strains identified as belonging to group III have been examined. Based on 16S rDNA hybridization and the aforementioned phenotypic properties, group II strains have recently been given species status and named *S. baltica* (Ziemke et al., 1998). Overall, *S. baltica* represents a large proportion of isolates conforming to Gilardi biovars 1 and 3. From phylogenetic studies (Fig. BXII.γ.156), it is evident that heterogeneity remains in the vicinity of the type strain of *S. putrefaciens*, particularly among *S. putrefaciens* DNA group III strains (Venkateswaran et al., 1999).

The taxonomy of the remaining species is more clear-cut. Phenotypic differences can readily distinguish these species (Table BXII.γ.148), although data for *S. woodyi* are somewhat limited. DNA hybridization analysis has also not been performed between *S. hanedai* and *S. woodyi*.

DIFFERENTIATION OF THE SPECIES OF THE GENUS *SHEWANELLA*

Phenotypic characteristics that distinguish the different species and genomospecies of *Shewanella* are presented in Tables BXII.γ.148 and BXII.γ.150.

List of species of the genus Shewanella

1. **Shewanella putrefaciens** (Lee, Gibson and Shewan 1977) MacDonell and Colwell 1986, 355[VP] (Effective publication: MacDonell and Colwell 1985, 180) (*Alteromonas putrefaciens* Lee, Gibson and Shewan 1977, 449.)

   pu.tre.fa' ci.ens. L. v. *putrefacere* to make rotten; L. part. adj. *putrefaciens* making rotten or putrefying.

   Cells are rod-shaped, 0.5–1.0 × 1.5–2.0 μm. Colonies on nutrient agar are light tan to salmon pink, opaque, circular, convex with entire edges, and butyrous in consistency. NaCl is usually not required for growth. Can tolerate up to 6% NaCl. Does not require growth factors.

   Mesophilic. Growth occurs between 10 and 40°C; optimum, 30–35°C.

   Other characteristics are as given in the genus description and in Tables BXII.γ.148 and BXII.γ.150.

   Isolated from fish, poultry, chilled meats, machine cutting oils, and freshwater and marine samples.

   The current description of *Shewanella putrefaciens* incorporates at least two distinct genomospecies, including group I (*Shewanella putrefaciens stricto sensu*), group III, and probably other groups. Refer to Table BXII.γ.150 for phenotypic differentiation of groups I and III and to Fig. BXII.γ.156 for phylogeny.

   The mol% G + C of the DNA is: 44–47 (T_m, Bd).

   Type strain: 95, ATCC 8071, DSM 6067, ICPB 352, LMG 2268, NCIMB 10471.

   GenBank accession number (16S rRNA): X82133, X81623.

2. **Shewanella algae** Simidu, Kita-Tsukomoto, Yasumoto and Yotsu 1990, 335[VP] emend. Nozue, Hayashi, Hashimoto, Ezaki, Hamasaki, Ohwada and Terawaki 1992, 633.

   al' gae. L. gen. n. *algae* of an alga.

   Cells are straight rods, about 0.8 × 1.6 μm. On marine agar, colonies are pale tan to salmon, circular, convex with an entire edge, and butyrous in consistency. Growth is stimulated by Na$^+$ ions, and NaCl concentrations of up to 12% are tolerated. Most strains can grow on *Salmonella–Shigella* agar.

   Mesophilic. Growth occurs between 10 and 42°C.

   Other characteristics are as given in the genus description and presented in Tables BXII.γ.148 and BXII.γ.150. In addition, strains also produce indophenol oxidase, lecithinase, deoxyribonuclease, and acid from D-ribose. Some

TABLE BXII.γ.150. Characteristics differentiating members of the *Shewanella putrefaciens* species complex, *S. baltica*, *S. algae*, and DNA hybridization Group III[a]

Characteristic	S. putrefaciens	S. algae	S. baltica	DNA hybridization Group III
Owen DNA Hybridization Group	I	IV	II	III
Growth at:				
4°C	+	−	+	d
37°C	+	+	−	+
42°C	−	+	−	d
Tolerates 6% NaCl	−	+	−	−
Reduction of sulfite (to H$_2$S)	+	d	−	−
Acid from:				
D-Glucose	−	d[b]	+	+
Maltose, sucrose	−	−	+	+
Arabinose	+	−	+	+
Growth on *Shigella–Salmonella* agar	+	+	−	−
Christensen's citrate (alkali)	−	−	+	−
Chymotrypsin	+	+	d[b]	+
N-Acetyl-β-glucosaminidase	−	+[b]	d	−
Urease	−	d	d	−
Utilization of:				
α-Cyclodextrin, glucose, maltose	−	−	+	+
Dextrin	−	−	d	+
L-Arabinose	d	−	−	+
Gentobiose, cellobiose, D-gluconate, citrate	−	−	+	−
Glycogen, adipate	−	−	d	−
Tween 80	+	+	d	−
Sucrose	−	−	+	d
N-Acetylgalactosamine	−	+	−	d
N-Acetylglucosamine	d	+	+	+
α-Hydroxybutyrate	−	+	−	−
DL-Malate	d	+	d	d
Caprate	−	+	d	d
Mol% G + C (T_m, Bd)	43–45	52–54	44–48	46–49

[a]For symbols see standard definitions.

[b]Weak reaction.

strains can produce urease and give a positive Simmons citrate test. Alginate is not hydrolyzed. Acid is not formed from D-arabinose, L-arabinose, D-mannose, D-mannitol, *meso*-inositol, L-rhamnose, D-xylose, sucrose, maltose, or lactose.

Several environmental and clinical strains can produce 4-epi-tetrodotoxin.

Isolated from red algae (*Jainia* spp.), salt marsh surface sediments, and human clinical samples, including blood, pus, sputum, feces, and tissue samples.

The species is equivalent to *Shewanella putrefaciens* DNA hybridization group IV of Owen et al. (1978).

The mol% G + C of the DNA is: 52–54 (T_m).

Type strain: OK-1, ACAM 541, ATCC 51192, IAM14159.

GenBank accession number (16S rRNA): AF005249, U91546.

3. **Shewanella amazonensis** Venkateswaran, Dollhopf, Aller, Stackebrandt, Nealson 1998, 971[VP]

a.ma.zo.nen' sis. M.L. gen. n. *amazonensis* named after the area from the species was isolated.

Cells are rods, 0.4–0.7 × 2–3 μm. Colonies on marine agar are beige to pinkish, circular, convex with entire edges, and butyrous in consistency. No diffusible pigments are formed. NaCl is required for good growth, but cultures grow slowly in the absence of Na$^+$ ions. For optimal growth, about 1–2% NaCl is required. No growth occurs above 3% NaCl.

Psychrotolerant. Grows between 4 and 40°C; optimum temperature is approximately 37°C.

Other characteristics are as given in the genus descrip-

tion and in Tables BXII.γ.148 and BXII.γ.151. Can couple growth to reduction of S^0 and manganese oxide.

Isolated from oysters.

The mol% G + C of the DNA is: 52 (T_m).

Type strain: SB2B, ATCC 700329.

GenBank accession number (16S rRNA): AF005248.

4. **Shewanella baltica** Ziemke, Höfle, Lalucat and Rosselló-Mora 1998, 184.[VP]

bal' ti.ca. M.L. fem. adj. *baltica* of the Baltic Sea.

Cells are short, straight rods. On nutrient agar, colonies are pale tan, circular, convex with an entire edge, and butyrous in consistency. Do not require NaCl or organic factors for growth.

Psychrotrophic. Growth occurs between 4 and 30°C. No growth at 37°C.

Other characteristics are as given in the genus description and presented in Tables BXII.γ.148 and BXII.γ.150.

Isolated from brackish water, seawater, oil brine, fish and seafood products, and other food products.

The species is equivalent to *Shewanella putrefaciens* DNA hybridization group II of Owen et al. (1978).

The mol% G + C of the DNA is: 44–47 (T_m).

Type strain: CECT 323, DSM 9439, IAM 1477, LMG 2250, NCTC 10735.

GenBank accession number (16S rRNA): AJ000214.

5. **Shewanella benthica** MacDonell and Colwell 1986, 355[VP] (Effective publication: MacDonell and Colwell 1985, 180.)

ben' thi.ca. Gr. n. *benthos* deep sea; L. suff. *-ica* of or pertaining to; M.L. adj. *benthica* pertaining to the deep sea.

TABLE BXII.γ.151. Additional phenotypic characteristics of marine *Shewanella* species[a]

Characteristics	S. amazonensis	S. colwelliana	S. frigidimarina[b]	S. gelidimarina[b]	S. hanedai[b]	S. woodyi
Utilization of:						
D-Ribose	nd	nd	nd	nd	d	nd
D-Mannose	+	nd	−	−	−	−
D-Fructose	+	−	d	−	−	nd
D-Xylose	nd	−	d	−	−	nd
D-Galactose	nd	−	d	−	d	+
Trehalose, isobutyrate	nd	−	+	−	−	nd
D-Mannitol, DL-malate	−	nd	+	−	−	−
Lactose, glycerol, sorbitol	−	nd	−	−	−	−
D-Glucuronate	nd	nd	−	−	−	+
Glycogen, L-phenylalanine	nd	nd	d	−	nd	nd
Acetate	nd	−	+	+	+	+
Propionate	nd	−	+	−	+	+
Butyrate	nd	−	+	+	+	nd
Valerate	nd	−	+	+	−	nd
Caproate, pelargonate	nd	nd	d	−	d	nd
Hepatanoate	nd	nd	−	−	d	nd
Caprate, L-isoleucine, spermine	nd	nd	nd	nd	+	nd
Succinate	+	−	+	−	−	+
DL-Glycerate, glycine	nd	−	nd	nd	+	nd
DL-Lactate	+	−	+	+	−	nd
Pyruvate	nd	nd	+	+	+	nd
Citrate	+	−	d	−	−	−
2-Ketoglutarate	nd	+	−	−	−	+
Fumarate	+	+	+	−	−	nd
Oxaloacetate	nd	nd	+	−	nd	nd
DL-Alanine	nd	−	−	−	+	+
L-Aspartate	nd	nd	−	−	+	nd
L-Glutamate	nd	nd	d	+	+	nd
L-Histidine	−	+	−	−	−	nd
L-Leucine	nd	nd	+	−	d	+
Hydroxy-L-proline, γ-aminobutyrate	nd	nd	d	−	nd	nd
L-Proline	nd	nd	d	−	−	nd
L-Serine	+	nd	−	−	+	+
L-Threonine	nd	−	−	−	−	+
L-Tyrosine	nd	−	−	−	d	nd
Putrescine	−	nd	−	−	+	+

[a]Symbols: +, 90% of strains are positive; d, 10–90% of strains are positive; −, 0–10% of strains are positive; nd, not determined.

[b] *S. frigidimarina*, *S. gelidimarina*, and *S. hanedai* do not utilize DL-arabinose, L-rhamnose, lactose, melibiose, adonitol, *meso*-inositol, D-sorbitol, saccharate, isovalerate, caprylate, adipate, pimelate, azelate, DL-3-hydroxybutyrate, or aconitate. In addition, *S. hanedai* does not utilize D-galacturonate, *i*-erythritol, salicin, ethanol, *n*-propanol, *n*-butanol, propyleneglycol, malonate, glutarate, L-tartrate, benzoate, *m*-hydroxybenzoate, *p*-hydroxybenzoate, phenylacetate, quinate, β-alanine, L-arginine, L-citrulline, L-lysine, L-ornithine, L-valine, DL-δ-aminovalerate, ethanolamine, betaine, hippurate, sarcosine, or trigollenine.

Cells are slightly curved rods, 0.5 × 1.5–3.0 μm. Colonies on marine agar are off-white to pale tan, translucent, circular, convex with an entire edge, and butyrous in consistency. NaCl is required for growth. No growth occurs with 6% NaCl. Barophilic, growing optimally at 400–600 MPa. Grows slowly at atmospheric pressure.

Psychrophilic. Growth occurs at 0–10°C; optimum temperature, ~5°C at atmospheric pressure.

Other characteristics are given in the genus description and in Table BXII.γ.148.

Isolated from abyssal ocean water, sediments, organic detritus, and deep-sea invertebrates.

The mol% G + C of the DNA is: 47 (*T*$_m$).

Type strain: W 145, ATCC 43992, DSM 8812.

GenBank accession number (16S rRNA): X82131.

6. **Shewanella colwelliana** (Weiner, Coyne, Brayton, West and Raiken 1988) Coyne, Pillidge, Sledjeski, Hori, Ortiz-Conde, Muir, Weiner and Colwell 1990, 320[VP] (Effective publication: Coyne, Pillidge, Sledjeski, Hori, Ortiz-Conde, Muir, Weiner and Colwell 1989, 278[VP] (*Alteromonas colwelliana* Weiner, Coyne, Brayton, West and Raiken 1988, 242.)

col.well' i.a.na. L. gen. n. *colwelliana* named after Rita Colwell for her contributions to marine microbiology.

Cells are straight or slightly curved rods, 0.7–1.5 × 1.0–3.0 μm long. Filamentous and helical cells occur in nutrient-poor media. Colonies on marine agar are circular to irregular, convex with undulate edges, and butyrous in consistency. A dark brown, melanin-like, diffusible pigment may be produced on complex growth media by some strains. NaCl is required for growth; optimal NaCl levels for growth are 2–4%. No growth with 6% NaCl. Amino acids are required for growth.

Mesophilic. Grows between 8 and 30°C; optimum temperature, approximately 25°C. Optimal pH for growth is 7.4–7.8.

Other characteristics are as given in the genus description and in Tables BXII.γ.148 and BXII.γ.151. In addition, phosphatase and deoxyribonuclease are produced. Elastin, casein, albumin, L-tyrosine, alginate, and chondroitin are not degraded. No activity detected for lecithinase or sulfatase. Gluconate oxidation, Voges–Proskauer, and chicken-cell agglutination tests are negative. Acid is not produced from carbohydrates.

Isolated from oysters.

The mol% G + C of the DNA is: 46 (T_m).

Type strain: LST-W, ATCC 39565.

Additional Remarks: ATCC 39565 is not currently available; however, a similar strain (ATCC 33888) is available from Dr. R. Weiner, University of Maryland, USA.

7. **Shewanella frigidimarina** Bowman, McCammon, Nichols, Skerratt, Rea, Nichols and McMeekin 1997b, 1045.[VP]

fri.gid.i.ma.ri′ na. L. adj. *frigidus* cold; L. fem. adj. *marina* belonging to the sea; M.L. adj. *frigidimarina* belonging to the cold sea.

Cells are curved or straight rods, 0.5–0.8 × 1.0–2.5 μm. Colonies on marine agar are tan, translucent, circular, convex with entire edges, and butyrous in consistency, but becoming increasingly mucoid with prolonged incubation. NaCl is not required for growth. Can tolerate up to 9% NaCl. Growth factors are not required.

Psychrotrophic. Growth occurs between <0 and 30°C; optimum temperature, ~20–22°C.

Other characteristics are as given in the genus description and in Tables BXII.γ.148 and BXII.γ.151. In addition, strains produce deoxyribonuclease and lecithinase and hydrolyze casein. Strains may degrade L-tyrosine and give a positive Simmons citrate test. Do not hydrolyze alginate, esculin, urea, urate, or xanthine. Acid is produced from cellobiose, maltose, D-mannitol, and sucrose. Strains may produce acid from trehalose, D-galactose, D-mannose, and D-xylose. Acid is not formed from L-arabinose, L-rhamnose, lactose, D-melibiose, D-melezitose, D-raffinose, dextran, N-acetylglucosamine, adonitol, *meso*-inositol, D-sorbitol, or glycerol. Resistant to 0.0075% KCN and 5% ox-bile salts.

Isolated from sea ice, seawater, marine sediments, and waters and sediments of saline meromictic lakes.

The mol% G + C of the DNA is: 40–43 (T_m).

Type strain: IC-P1, ACAM 591, DSM 12253.

GenBank accession number (16S rRNA): U85903.

8. **Shewanella gelidimarina** Bowman, McCammon, Nichols, Skerratt, Rea, Nichols and McMeekin 1997b, 1045[VP]

ge.li.di.ma.ri′ na. L. adj. *gelidus* ice-cold; L. fem. adj. *marina* belonging to the sea; M.L. adj. *gelidimarina* belonging to the icy sea.

Cells are curved or straight rods, 0.5–0.8 × 1.0–2.5 μm. Colonies on marine agar are light tan, opaque, circular, convex with entire edges, and butyrous in consistency. NaCl is required for growth. Can tolerate up to 6% NaCl. Growth factors are not required.

Psychrophilic. Growth occurs between <0 and 23°C in liquid media; no growth occurs at 20°C on agar media. Optimal temperature, approximately 15–17°C.

Other characteristics are as given in the genus description and in Tables BXII.γ.148 and BXII.γ.151. In addition, strains produce lecithinase and can hydrolyze casein. Strains may produce deoxyribonuclease. Simmons citrate and urease tests are negative. Unable to hydrolyze alginate, esculin, urate, or xanthine. Acid is produced oxidatively and fermentatively from N-acetylglucosamine and chitin, but not from other carbohydrates. Resistant to 0.0075% KCN and 5% ox-bile salts.

Isolated from sea-ice diatom blooms.

The mol% G + C of the DNA is: 40–43 (T_m).

Type strain: IC-P6, ACAM 456.

GenBank accession number (16S rRNA): U85907.

9. **Shewanella hanedai** (Jensen, Tebo, Baumann, Mandel and Nealson 1981) MacDonell and Colwell 1986, 355[VP] (Effective publication: MacDonell and Colwell 1985, 180) (*Alteromonas hanedai* Jensen, Tebo, Baumann, Mandel and Nealson 1981, 382.)

han.e′ dai. M.L. gen. n. *hanedai* of Haneda; named after Y. Haneda, a Japanese biologist who pioneered studies on bioluminescence.

Cells are rod-shaped, 0.5–1.0 × 1.5–2.0 μm. Colonies on marine agar are light tan, translucent, circular, convex with entire edges, and butyrous in consistency. NaCl is required for growth. Can tolerate up to 6% NaCl. Strains may require yeast extract or vitamins for growth.

Psychrophilic. Growth occurs between 0 and 25°C; optimum temperature, approximately 15°C.

Other characteristics are as given in the genus description and in Tables BXII.γ.148 and BXII.γ.151. In addition, alginate is not hydrolyzed.

Isolated from sediments and seawater of the Arctic Ocean, Southern Ocean, and western coast of Canada.

The mol% G + C of the DNA is: 44–47 (T_m).

Type strain: ATCC 33224, CIP 103207, DSM 6066.

GenBank accession number (16S rRNA): X82132, U91590.

10. **Shewanella oneidensis** Venkateswaran, Moser, Dollhopf, Lies, Saffarini, MacGregor, Ringleberg, White, Nishijima, Sano, Burghardt, Stackebrandt and Nealson 1999, 721[VP]

o.nei.den′ sis. M.L. gen. n. *oneidensis* named after Lake Oneida where the species was isolated.

Cells are rods, 0.5–0.6 × 2–3 μm. Colonies on nutrient agar are pinkish beige, circular, smooth, convex with entire edges, and butyrous in consistency. No diffusible pigments are formed. NaCl is not required for growth.

Grows between 4 and 40°C; optimum temperature, ~30°C.

Other characteristics are as given in the genus description and in Table BXII.γ.148. Can couple growth to the reduction of manganese oxide (Nealson and Saffarini, 1994) and S^0 (Moser and Nealson, 1996). Does not form indole, acetoin, alginase, arginine dihydrolase, or lysine decarboxylase. Utilizes D-galactose, but not D-fructose, citrate, D-mannitol, glycerol, D-sorbitol, DL-malate, or DL-lactate. Some strains can utilize lactose, succinate, and citrate.

Isolated from water and sediment of Lake Oneida, New York, USA and from clinical specimens.

The mol% G + C of the DNA is: 45 (T_m).

Type strain: MR-1, ATCC 700550.

GenBank accession number (16S rRNA): AF005251.

11. **Shewanella pealeana** Leonardo, Moser, Barbieri, Branstner, MacGregor, Paster, Stackebrandt and Nealson 1999, 1349[VP]

peal′ le.ana. M.L. adj. *pealeana* from *peale or pealei* the species name of the squid; *Loligo pealei* from which the bacterial species was isolated.

Gram negative. Cells are rods, 2.0–3.0 × 0.4–0.6 μm. Motile with a single polar flagellum. Colonies on marine agar are circular, opaque, and salmon-colored. Requires Na⁺ ions for growth with biomass produced optimally at 0.5 M NaCl. Also requires choline chloride as an essential growth factor in minimal media. Grows between 0.125–0.75

M NaCl. Growth occurs at 4–30°C. Optimum growth temperature is 25°C. Grows between pH 6–8. Optimum pH for growth is 7.0.

Chemoheterotrophic facultative anaerobe. Can grow anaerobically using nitrate, fumarate, iron, manganese, TMAO, thiosulfate, and elemental sulfur as alternative electron acceptors with lactate acting as the carbon source. Catalase, oxidase, and lipase positive. Amylase and gelatinase negative. Glucose, galactose, lactate, acetate, pyruvate, citrate, succinate, glutamate, Casamino acids, yeast extract, and peptone are used aerobically as energy sources. Fructose, glycerol, sorbitol, arabinose, formate, and ethanol are not utilized.

Isolated from the accessory nidamental glands of female adults of the squid species *Loligo pealei*.

The mol% G + C of the DNA is: 45.0 (HPLC).

Type strain: ANG-SQ1, ATCC 700345.

GenBank accession number (16S rRNA): AF011335.

12. **Shewanella violacea** Nogi, Kato and Horikoshi 1999, 341[VP] (Effective publication: Nogi, Kato and Horikoshi 1998b, 337.)

vi.o.la' ce.a. L. gen. n. *violacea* of violet.

Cells are straight or slightly curved rods, 0.8–1.0 × 2–4 μm. Colonies on marine agar are circular, smooth, convex with entire edges, and butyrous in consistency. After 2–3 d, colonies are nonpigmented; after more than 7 d, colonies appear violet. NaCl is required for growth; optimal levels for growth are 2–3%. No growth with 6% NaCl.

Psychrophilic. Grows optimally between 4–10°C. Barophilic. Optimal pressure for growth is 30 MPa.

Other characteristics are as given in the genus description and in Table BXII.γ.148. Acid is produced from cellobiose and D-galactose, but not from DL-arabinose, D-fructose, glycerol, inositol, lactose, maltose, D-mannitol, D-mannose, D-raffinose, L-rhamnose, D-sorbitol, sucrose, D-trehalose, or D-xylose.

Isolated from the Ryukyu Trench, northwest Pacific Ocean, at a depth of 5110 m.

The mol% G + C of the DNA is: 47 (T_m).

Type strain: DSS12, JCM 10179.

GenBank accession number (16S rRNA): D21225.

13. **Shewanella woodyi** Makemson, Fulayfil, Landry, Van Ert, Wimpee, Widder and Case 1997, 1039[VP]

wood' y.i. M.L. gen. n. *woodyi* of Woody, in honor of the American biologist J. Woodland Hastings.

Cells are rod-shaped, 0.4–1.0 × 1.4–2.0 μm. Colonies on marine agar are pink-orange due to the accumulation of cytochromes. NaCl is required for growth. Growth factors are not required.

Psychrophilic. Growth occurs between 4 and 25°C; optimum temperature, 20–25°C; no growth at 30°C.

Other characteristics are as given in the genus description and in Tables BXII.γ.148 and BXII.γ.151.

Isolated from squid ink, seawater and marine snow (collected from the Alboran Sea).

The mol% G + C of the DNA is: 39 (by measurement of the relative binding of DNA-binding-fluorescent dyes bisbenzimide and chromomycin A3).

Type strain: MS32, ATCC 51908, DSM 12036.

GenBank accession number (16S rRNA): AF003549.

Order XI. "Vibrionales"

Vib.ri.o.na' les. M.L. masc. n. *Vibrio* type genus of the order; *-ales* ending to denote order; M.L. fem. pl. n. *Vibrionales* the order of bacteria whose circumscription is based on the genus *Vibrio*.

Description is the same as for the family *Vibrionaceae*.

Type genus: **Vibrio** Pacini 1854, 411.

Family I. **Vibrionaceae** Véron 1965, 5245[AL]

J.J. FARMER III AND J. MICHAEL JANDA

Vib.ri.o.na' ce.ae. M.L. masc. n. *Vibrio* type genus of the family; *-aceae* ending to denote family; M.L. fem. pl. n. *Vibrionaceae* the family of bacteria whose circumscription is based on the genus *Vibrio*.

Gram-negative straight or curved rods. Motile by means of polar flagella. Additional lateral flagella may be produced when grown on solid media; these differ in wavelength and antigenicity from the polar flagellum and may number from a few to over 100 flagella/cell. Do not form endospores or microcysts. Chemoorganotrophs. Facultative anaerobes, having both a respiratory and a fermentative metabolism. Oxygen is a universal electron acceptor. Do not denitrify. Most strains: are oxidase positive, reduce nitrate to nitrite, ferment D-glucose and utilize it as a sole or principal source of carbon and energy, grow in minimal media with D-glucose or other compounds as the sole source of carbon and energy and use NH_4^+ as the sole nitrogen source. A few species require vitamins and amino acids. Ferment and utilize a wide variety of simple and complex carbohydrates and utilize a wide variety of other carbon sources. Most species require Na^+ or a seawater base for growth and require 0.5–3% NaCl for optimum growth. Several species are bioluminescent; other species include a few bioluminescent strains. Primarily aquatic. Found in fresh, brackish, and sea water, often in association with aquatic animals and plants. Several species are pathogenic for humans.

Other species are pathogenic for fish, eels, and other aquatic animals. The mol% G + C of the DNA is 38–51%. The family is classified in the phylum *Proteobacteria* in the class *Gammaproteobacteria*.

Type genus: **Vibrio** Pacini 1854, 411.

Historical overview A history of the family *Vibrionaceae* as it has appeared in *Bergey's Manual* is given in Table BXII.γ.152. Related families include *Enterobacteriaceae*, *Aeromonadaceae*, and *Pasteurellaceae*. The family *Vibrionaceae* has undergone intense study since the first edition of *Bergey's Manual of Systematic Bacteriology* (Krieg and Holt, 1984) was published in 1984. In their chapter on the family *Vibrionaceae* in that edition, Baumann and Schubert (1984) included the genera *Vibrio*, *Photobacterium*, *Aeromonas*, and *Plesiomonas*. These are the same four genera included in the original classification of the family proposed by Véron almost twenty years earlier (Véron, 1965). In this *Manual Aeromonas* and *Plesiomonas* are classified in other families (Table BXII.γ.152). For practical identification schemes, it is still useful to consider *Aeromonas* and *Plesiomonas* together with other oxidase positive genera of fermentative bacteria such as *Vibrio* and *Photobacterium* (Table BXII.γ.153). A detailed history of changes in the classification of *Vibrio* and related genera that occurred as new methods were introduced has been given by Farmer (1992). These methods include examinations of the structure, function, and regulation of proteins; comparison of mol% G + C content; DNA–DNA hybridization; rRNA–DNA hybridization; 5S rRNA cataloging and sequence comparisons; and 16S rRNA gene sequence comparisons (Fig. BXII.γ.157).

The family *Vibrionaceae* presently includes three genera:

Genus 1. *Vibrio* (the type genus)
Genus II. *Photobacterium*
Genus III. *Salinivibrio*

The type strain of the type and only species of the genus *Allomonas*, *Allomonas enterica*, is very closely related to *Vibrio fluvialis* by DNA–DNA hybridization studies and phenotypic analysis (Kalina et al., 1984). *Allomonas* and *Allomonas enterica* are not described separately in this edition of the *Manual*; the reader is referred to the description of *V. fluvialis*. The two species of the genus *Listonella*, *Listonella anguillarum* (the type species) and *Listonella pelagia*, are included in the genus *Vibrio* in this edition of the *Manual* as *Vibrio anguillarum* and *Vibrio pelagius*.*

FURTHER DESCRIPTIVE INFORMATION

Habitats The ecological niches of members of the family *Vibrionaceae* have been described by Campbell (1957); Baumann and Baumann (1981a); Sakazaki and Balows (1981); and Simidu and Tsukamoto (1985)). In humans, some vibrios cause diarrhea, wound infections, and occasionally other extraintestinal infections. In aquatic animals, vibrios cause wound and generalized infections. Many vibrios and related organisms are also widely distributed in aquatic environments. Many factors govern the distribution of these organisms, but the most important probably include: particular human, animal or plant hosts; inorganic nutrients and carbon sources available; temperature; salinity; dissolved oxygen; and depth below the surface for the species that are found in the ocean (Simidu and Tsukamoto, 1985). A few species are adapted to particular hosts. For example, *Vibrio chol-*

erae serogroup O1 is adapted to humans and is the cause of cholera, a life-threatening diarrheal disease. Recent studies have shown that the ecology of this organism is more complex than originally thought. *Photobacterium leiognathi* is usually isolated from fish in shallow tropical water, and *P. phosphoreum* is usually found in the luminous organs of fish that live at depths of 200–1200 meters (Hastings and Nealson, 1981).

Isolation Most members of the *Vibrionaceae* grow well on ordinary complex media. Samples are spread onto solid medium or diluted in an enrichment broth. NaCl concentrations of 0.5–0.85% satisfy the requirements of most species, although a few require greater concentrations of NaCl. Incubation temperatures are also important. A few species grow only at temperatures <25°C; others grow at 25°C but not at 35–37°C. General and selective media for *Vibrionaceae* are described in the chapter on the genus *Vibrio*.

Identification Methods for the isolation and identification of *Vibrio* spp. from clinical specimens and non-clinical samples are discussed in detail in the chapter describing the genus *Vibrio*. Assignment of non-clinical isolates to a species can be problematic because over 50 species of *Vibrio* and *Photobacterium* must be considered and because comparative data for these organisms are sparse relative to data available for clinically important species.

The US Centers for Disease Control and Prevention maintain computer programs and databases for the identification of isolates subjected to a battery of 45–60 phenotypic tests; for details contact the *Vibrio* Laboratory at the CDC.

These alternatives to phenotypic methods are now being used routinely and have proven extremely useful in a research setting. It will be important to evaluate the sensitivity and specificity, and to understand the advantages and disadvantages of these methods. In the United States, the reporting of cultures from human specimens is subject to specific government regulations (the Clinical Laboratory Improvement Amendments of 1988), which has limited the application of these approaches in clinical and public health laboratories.

ACKNOWLEDGMENTS

We dedicate this chapter to M. Véron for giving us the name *Vibrionaceae* and for all his contributions to our understanding of the family, its organisms, and their close and distant relatives.

FURTHER READING

Baumann, P. and L. Baumann. 1977. Biology of the marine enterobacteria: genera *Beneckea* and *Photobacterium*. Ann. Rev. Microbiol. *31*: 39–61.

Baumann, P. and L. Baumann. 1981. The marine Gram-negative eubacteria. *In* Starr, Stolp, Trüper, Balows and Schlegel (Editors), The Prokaryotes, a Handbook on Habitats, Isolation and Identification of Bacteria, 1st Ed., Springer-Verlag, New York. pp. 1352–1394.

Baumann, P., L. Baumann, S.S. Bang and M.J. Woolkalis. 1980. Reevaluation of the taxonomy of *Vibrio*, *Beneckea*, and *Photobacterium*: abolition of the genus *Beneckea*. Curr. Microbiol. *4*: 127–132.

Baumann, P., L. Baumann and M. Mandel. 1971. Taxonomy of marine bacteria: the genus *Beneckea*. J. Bacteriol. *107*: 268–294.

Baumann, P., A.L. Furniss and J.V. Lee. 1984. Genus I. *Vibrio*. *In* Krieg and Holt (Editors), Bergey's Manual of Systematic Bacteriology, 1st Ed., Vol. 1, The Williams & Wilkins Co., Baltimore. pp. 518–538.

Baumann, P. and R.H.W. Schubert. 1984. Family II. *Vibrionaceae*. *In* Krieg and Holt (Editors), Bergey's Manual of Systematic Bacteriology, 1st Ed., Vol. 1, The Williams & Wilkins Co., Baltimore. pp. 516–517.

Brenner, D.J., G.R. Fanning, F.W. Hickmann-Brenner, J.V. Lee, A.G. Stei-

---

Editorial Note: Readers are advised that *Listonella pelagia* and *Listonella anguillarum* are valid names currently in use in the literature regarding fish pathology.

TABLE BXII.γ.152. Different classifications of *Vibrio*, *Photobacterium*, *Aeromonas*, *Plesiomonas*, and related organisms in five editions of *Bergey's Manual*[a] (1957–2001)

Edition	Classification

2004 *Bergey's Manual of Systematic Bacteriology*, 2nd Ed. (See the Revised Road Map by Garrity et al., in Volume 2(A) of this *Manual*.)	Phylum BXII. *Proteobacteria* phyl. nov.
	  Class III. *Gammaproteobacteria* class. nov.
	    Order VIII. *Oceanospirillales* ord. nov.
	      Family IV. *Halomonadaceae*
	        Genus I. *Halomonas* (19 species)
	    Order X. *Alteromonadales* ord. nov.
	      Family I. *Alteromonadaceae*
	        Genus III. *Colwellia* (7 species)
	          Genus X. *Moritella* (4 species)
	        Genus XII. *Shewanella* (13 species)
	    Order XI. "*Vibrionales*"
	      Family I. *Vibrionaceae*
	        Genus I. *Vibrio* (Type genus; 44 species)
	        Genus II. *Photobacterium* (6 species)
	        Genus III. *Salinivibrio* (1 species)
	    Order XII. *Aeromonadales* ord. nov.
	      Family I. *Aeromonadaceae*
	        Genus I. *Aeromonas* (Type genus) (14 species)
	    Order XIII. "*Enterobacteriales*"
	      Family I. *Enterobacteriaceae*
	        Genus XXVII. *Plesiomonas* (1 species)

1994 *Bergey's Manual of Determinative Bacteriology*, 9th Ed. (Holt et al., 1994)	Group 4: Gram-negative aerobic/microaerophilic rods and cocci
	  Genus *Halomonas*
	Group 5: Facultative anaerobic Gram-negative rods
	Subgroup 2, Family *Vibrionaceae*
	  Genus *Vibrio* (Type genus)
	  Genus *Aeromonas*
	  Genus *Enhydrobacter*
	  Genus *Photobacterium*
	  Genus *Plesiomonas*

1984 *Bergey's Manual of Systematic Bacteriology*, 1st Ed. (Krieg and Holt, 1984)	Section 5: Facultative anaerobic Gram-negative rods
	  Family II. *Vibrionaceae*
	    Genus I. *Vibrio* (Type genus)
	    Genus II. *Photobacterium*
	    Genus III. *Aeromonas*
	    Genus IV. *Plesiomonas*

1974 *Bergey's Manual of Determinative Bacteriology*, 8th Ed. (Buchanan and Gibbons, 1974)	Part VIII. Gram-negative facultatively anaerobic rods
	  Family II. *Vibrionaceae*
	    Genus I. *Vibrio* (Type genus)
	    Genus II. *Aeromonas*
	    Genus III. *Plesiomonas*
	    Genus IV. *Photobacterium*
	    Genus V. *Lucibacterium*

1957 *Bergey's Manual of Determinative Bacteriology*, 7th Ed. (Breed et al., 1957)	Class II. *Schizomycetes*
	  Order I. *Pseudomonadales*
	    Family IV. *Pseudomonadaceae*
	      Genus IV. *Aeromonas*[a]
	      Genus V. *Photobacterium*[a]
	      Genus IX. *Alginomonas*[a]
	    Family VII. *Spirillaceae*
	      Genus I. *Vibrio*[a]
	  Order IV. *Eubacteriales*
	    Family III. *Achromobacteriaceae*
	      Genus IV. *Agarbacterium*[a]
	      Genus V. *Beneckea*[a]
	    Family IV. *Enterobacteriaceae*
	      Genus V. *Alginobacter*[a]

[a]These seven genera contain species or strains that (probably) would be classified in the family *Vibrionaceae* as defined by Véron (1965). The genus *Plesiomonas* is not listed because it was not named until 1962 (Habs and Schubert, 1962). "*Pseudomonas shigelloides*" Bader 1954 (now *Plesiomonas shigelloides*) was not listed in the 1957 edition.

gerwalt, B.R. Davis and J.J. Farmer. 1983a. DNA relatedness among *Vibrionaceae*, with emphasis on the *Vibrio* species associated with human infection. INSERM Colloq. *114*: 175–184.

Chakraborty, S., G.B. Nair and S. Shinoda. 1997. Pathogenic vibrios in the natural aquatic environment. Rev. Environ. Health. *12*: 63–80.

Farmer, J.J. 1992. The family *Vibrionaceae*. *In* Balows, Trüper, Dworkin, Harder and Schleifer (Editors), The Prokaryotes. A Handbook on the Biology of Bacteria: Ecophysiology, Isolation, Identification, Applications, 2nd Ed., Vol. 3, Springer-Verlag, New York. pp. 2938–2951.

Farmer, J.J., M.J. Arduino and F.W. Hickman-Brenner. 1992. *Aeromonas* and *Plesiomonas*. *In* Balows, Trüper, Dworkin, Harder and Schleifer (Editors), The Prokaryotes. A Handbook on the Biology of Bacteria:

TABLE BXII.γ.153. Properties of the four genera originally classified in the family *Vibrionaceae* by Véron (1965); comparison with *Enterobacteriaceae*

Test or property[a]	Vibrio	Photobacterium	Aeromonas	Plesiomonas	Enterobacteriaceae
Associated with diarrhea and extra-intestinal infections in humans	+		+	+	+
Mol% G + C content of DNA	38–51	40–44	57–63	51	38–60
Sheathed polar flagellum	+				
Peritrichous flagella when grown in liquid media					+
Accumulate poly-β-hydroxybutyrate, but do not utilize β-hydroxybutyrate		+			
Na$^+$ is required for growth or stimulates growth	+	+			
Sensitive to the vibriostatic compound 0/129	+	+		+	
Lipase production	+	d	+		d
D-Mannitol fermentation	+		+		+
Enterobacterial common antigen (eca)				+	+

[a]Symbols: +, trait is present; d, trait varies among species; blank, data not available. These are properties of most species of the genera and most strains of the species, but there are exceptions. Adapted from Baumann and Schubert (1984).

Ecophysiology, Isolation, Identification, Applications, 2nd Ed., Vol. 3, Springer-Verlag, New York. pp. 3012–3045.

Farmer, J.J. and F.W. Hickman-Brenner. 1992. The genera *Vibrio* and *Photobacterium*. *In* Balows, Trüper, Dworkin, Harder and Schleifer (Editors), The Prokaryotes. A Handbook on the Biology of Bacteria: Ecophysiology, Isolation, Identification, Applications, 2nd Ed., Vol. 3, Springer-Verlag, New York. pp. 2952–3011.

Janda, J.M. 1998. *Vibrio, Aeromonas* and *Plesiomonas*. *In* Balows and Duerden (Editors), Topley & Wilson's Microbiology and Microbial Infections, 9th Ed., Vol. 2, Arnold, London. pp. 1065–1089.

Janda, J.M., C. Powers, R.G. Bryant and S.L. Abbott. 1988. Current per-

spectives on the epidemiology and pathogenesis of clinically significant *Vibrio* spp. Clin. Microbiol. Rev. *1*: 245–267.

Ruimy, R., V. Breittmayer, P. Elbaze, B. Lafay, O. Boussemart, M. Gauthier and R. Christen.. 1994. Phylogenetic analysis and assessment of the genera *Vibrio, Photobacterium, Aeromonas,* and *Plesiomonas* deduced from small-subunit rRNA sequences. Int. J. Syst. Bacteriol. . *44*: 416–426.

Sakazaki, R. and A. Balows. 1981. The genera *Vibrio, Plesiomonas,* and *Aeromonas*. *In* Starr, Stolp, Trüper and Schlegel (Editors), The Prokaryotes. A Handbook on Habitats, Isolation, and Identification of Bacteria, Vol. 2, Springer-Verlag, New York. pp. 1272–1301.

Genus I. **Vibrio** Pacini 1854, 411[AL]

J.J. FARMER III, J. MICHAEL JANDA, FRANCES W. BRENNER, DANIEL N. CAMERON AND KAREN M. BIRKHEAD

Vib' ri.o. L. v. *vibrio* move rapidly back and forth, vibrate; M.L. masc. n. *Vibrio* the vibrating, darting organism.

Small, straight, slightly curved, curved, or comma-shaped rods, 0.5–0.8 × 1.4–2.6 μm. Involution forms often occur in old cultures and are formed under adverse growth conditions. Do not form endospores or microcysts. Gram negative. In liquid media, motile by monotrichous or multitrichous **polar flagella enclosed in a sheath** continuous with the outer membrane of the cell wall. **On solid media, some species synthesize numerous lateral flagella** with a wavelength shorter than that of the sheathed polar flagellum. **Facultative anaerobes** capable of both **fermentative and respiratory metabolism.** Molecular oxygen is a universal electron acceptor. Most do not denitrify or fix molecular nitrogen. All are chemoorganotrophs; most are able to **grow in a mineral medium containing D-glucose as the sole carbon source and NH$^+$ as the sole nitrogen source.** A few strains have organic growth factor requirements. **Na$_4^+$ stimulates growth of all species and is an absolute requirement for most**; the minimal concentration necessary for optimal growth ranges from 5 to 700 mM (0.029–4.1%). Most species grow well in media containing a seawater base. **All ferment D-glucose producing acid but rarely gas; several species produce acetoin and acetyl methyl carbinol (positive Voges–Proskauer test).** Most ferment and utilize D-fructose, maltose, and glycerol; are **oxidase positive and reduce nitrate to nitrite.** Several grow at 4°C; all grow at 20°C; most grow at 30°C; many grow at 35–37°C. **A few species are bioluminescent,** as are a few strains of normally nonluminescent species. **Primarily aquatic**; species distribution is usually dependent on Na$^+$ and nutrient content of the water as well as its temperature. Very common in marine and estuarine environments and on the surfaces and in the intestinal contents of marine animals. Species with a low Na$^+$ requirement are also found in freshwater habitats.

Twelve species occur in human clinical specimens; **11 of these are apparently pathogenic for humans, causing diarrhea or extraintestinal infections. Several species cause diseases of other vertebrates and invertebrates**.

The mol% G + C of the DNA is: 38–51.

Type species: **Vibrio cholerae** Pacini 1854, 411.

FURTHER DESCRIPTIVE INFORMATION

Cell morphology The cell morphology of *Vibrio cholerae* and related organisms has been studied extensively since *V. cholerae* was first depicted by Robert Koch. Examples of cell morphology, including cytoplasmic inclusions and cell surface features such as tubular appendages and "blebs", are shown in the section below describing flagella and in figures accompanying the species descriptions. Vibrios commonly appear as small, straight, slightly curved, curved or comma-shaped rods 0.5–0.8 × 1.4–2.6 μm in size. The amount of curvature is generally more pronounced in early stationary phase in liquid media than during exponential growth (Baumann et al., 1984b). However, considerable pleomorphism can occur. Some marine vibrios naturally have significant morphologic variation when grown in broth, and irregularly shaped bacilli co-exist with "involuted cells" that resemble L-forms (Onarheim et al., 1994). This phenomenon often occurs in stationary phase cultures (Farmer and Hickman-Brenner, 1992). Other halophilic species, including those associated with human disease, may form spherical or thin elongated forms in response to growth conditions containing suboptimal electrolyte (NaCl) concentrations. When vibrios are deprived of nutrients, a series of morphologic changes occurs over 1–7 days. Early events in this morphogenesis include a decreasing cell diameter

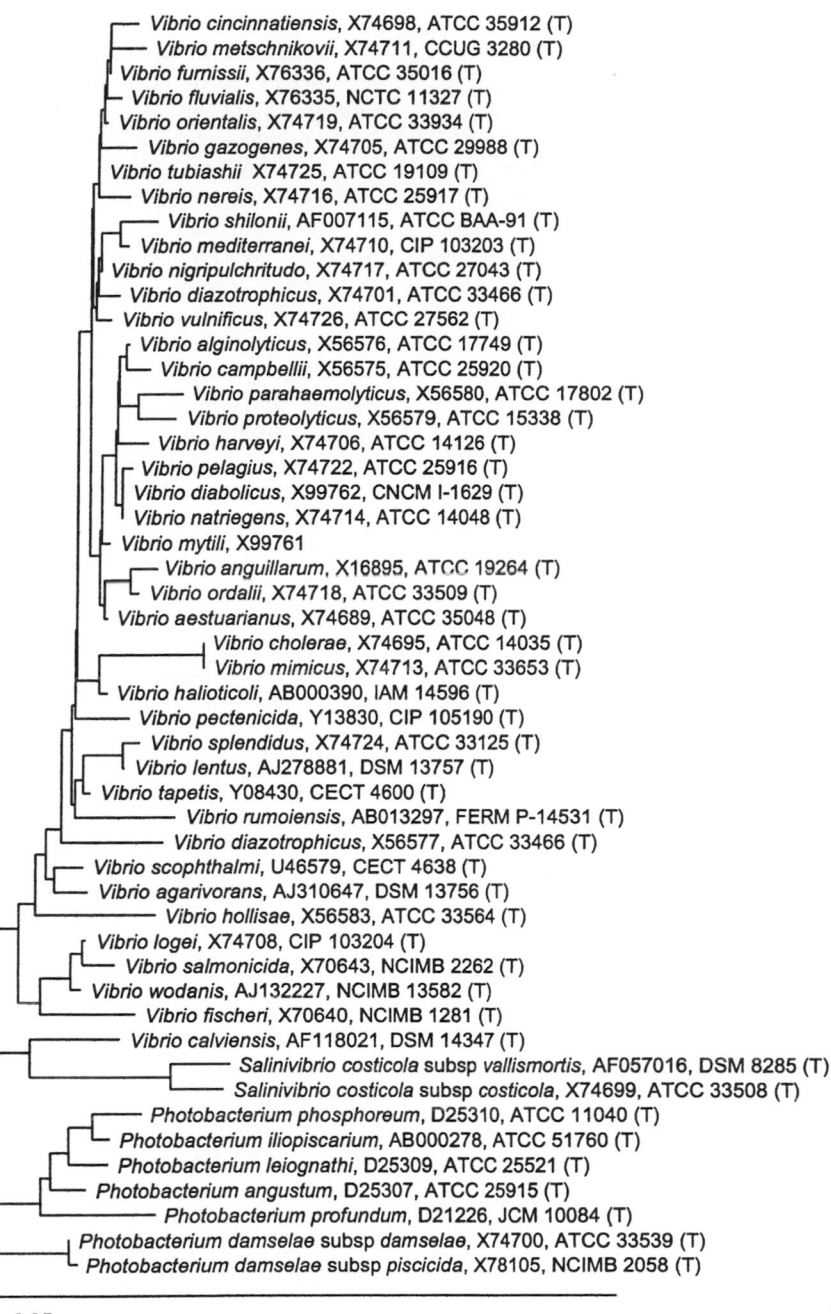

0.05

FIGURE BXII.γ.157. Relationship of most of the species of *Vibrio* and relatives based on 16S rRNA gene sequences.* Only the type strain of each species was included. The distances in the tree were calculated using 1101 positions (the least-squares method, Jukes-Cantor model). (Courtesy T. Lilburn of the Ribosomal Database Project.)

*Editorial Note: Photobacterium damselae subsp. damselae is a junior objective synonym of Vibrio damsela. Vibrio pelagius *and* Vibrio anguillarum *are synonyms of* Listonella pelagia *and* Listonella anguillarum, *respectively.*

and volume (15–67%) and a conversion of rods into coccal forms called spherical ultramicrocells (Holmquist and Kjelleberg, 1993; Kondo et al., 1994; Nelson et al., 1997). As the length of nutrient starvation increases, cytoplasmic inclusions and granules disappear, cell cultivability decreases, and the nuclear region becomes compressed (Hood et al., 1986). There are also noticeable differences in the integrity of the outer membrane and cell wall. Some changes may be linked to specific nutrient starvation (for

example, nitrogen starvation produces long filaments and phosphorus starvation produces swollen large rods), whereas others occur regardless of the type of nutritional stress (Holmquist and Kjelleberg, 1993). "Non-culturable" *V. cholerae* O1 strains produced in response to nutrient deprivation display a number of ultrastructural changes, which include an undulating outer membrane, a surface layer of fine fibers, and a thicker peptidoglycan layer (Kondo et al., 1994).

Poly-β-hydroxybutyrate granules (PHB) can be found in a number of *Vibrio* species, including *V. cholerae* O1 and O139 and *V. harveyi* (Hood et al., 1986; Sun et al., 1994; Finkelstein et al., 1997). In *V. cholerae*, accumulation of PHB appears to be related to colonial opacity and growth on glycerol-containing media (Finkelstein et al., 1997). In *V. harveyi*, PHB accumulation is dependent on cell density and is controlled by the autoinducer, *N*-(3-hydroxybutanoyl) homoserine lactone (Sun et al., 1994). Other kinds of granules can be found in vibrios, including electron dense lipoid particles and electron translucent inclusions of unknown composition (Sun et al., 1994; Finkelstein et al., 1997).

Cell wall composition   Vibrios contain the same three lipopolysaccharide (LPS) elements found in other Gram-negative bacteria: lipid A, core polysaccharide, and an O polysaccharide side chain that determines serological specificity. The most extensive work on biochemical characterization of *Vibrio* LPS has been done on *V. cholerae*. The lipid A portion consists of a $\beta(1'-6)$-linked glucosamine disaccharide backbone with two phosphoryl groups (Janda, 1998). Pyrophosphorylethanolamine is linked to one of these phosphoryl groups at the C-1 position of the reducing sugar, and a phosphate group ester is bound to the nonreducing glucosamine residue (Manning et al., 1994). Three fatty acids are ester linked at hydroxyl positions to this disaccharide backbone: tetradecanoic acid ($C_{14:0}$), hexadecanoic acid ($C_{16:0}$), and 3-hydroxydodecanoic acid ($C_{12:0\ 3OH}$). A fourth, 3-hydroxytetradecanoic acid ($C_{14:0\ 3OH}$), is connected to the backbone by an amide bond.

The core oligosaccharide region of *V. cholerae* contains KDO (keto-3-deoxy-D-mannose-octulosonic acid), D-glucose, heptose (L-glycero-D-manno-heptose), D-fructose, and ethanolamine phosphate (Manning et al., 1994). KDO, a normal constituent of the core oligosaccharide of enteric LPS, was originally thought to be absent in *Vibrio* species. However, when conventional periodate-thiobarbituric acid tests were replaced by strong acid hydrolysates, KDO was detected in *Vibrio* (Janda, 1998). The KDO molecule of *V. cholerae* differs in several aspects from those of enteric bacteria such as *Escherichia coli* and the genus *Salmonella*: only a single KDO molecule has been detected in the core oligosaccharide of *V. cholerae*, and the KDO moiety is phosphorylated at the C4 position (Kondo et al., 1990; Manning et al., 1994). The C5 position binds to a distal portion of the core region (heptose) similar to the KDO-C5 binding of L-glycero-D-manno-heptose (Janda, 1998). The other sugars form the remaining portion of the core oligosaccharide region and often contain additional sugar substitutions at various positions.

The O-polysaccharide side chain of *V. cholerae* O1 is a homopolymer of D-perosamine (4,6-dideoxy-D-mannose) approximately 17–18 units in length (Manning et al., 1994; Knirel et al., 1997). The amino groups of perosamine units are commonly acetylated with 3-deoxy-L-glycero-tetronic acid. Another compound, quinovosamine, is thought to be a "capping sugar" on either the distal or the proximal end of the O antigen (Manning et al., 1994). An unusual sugar, 4-amino-4,6-dideoxy-2-O-methyl-mannose is present only in the LPS of serogroup Ogawa and may have a role in serological specificity (Itoh et al., 1994).

The LPS composition of *V. cholerae* O139—a second serotype capable of causing pandemic cholera—is remarkably similar to that of O1 (Hisatsune et al., 1993; Isshiki et al., 1996). The lipid A moieties of O1 and O139, including fatty acid substitutions, appear to be identical (Hisatsune et al., 1993). The core oligo-

saccharide region contains two subtle differences: the presence of 2-aminoethyl phosphate, which is the O-acetyl group, and the presence of a second fructose molecule (Knirel et al., 1997). The most profound differences between O-groups 1 and 139 occur in the O-polysaccharide side chain. Unlike serogroup O1, which has long O-polysaccharide side chains, *V. cholerae* O139 has a short chain LPS similar to "SR strains" (Knirel et al., 1997). These truncated side chains migrate with the core oligosaccharide-lipid A fraction in LPS SDS-PAGE gels (Waldor et al., 1994). Classic "ladder-like" profiles of silver stained LPS side chains in SDS-PAGE gels are absent in O139 strains (Hisatsune et al., 1993; Nandy et al., 1995). Perosamine, the main component of the O1 side chain, is also absent in O139 strains (Hisatsune et al., 1993); instead, the unique sugar colitose (3,6-dideoxy-L-galactose)—which is not found in any other *Vibrio* species—is the main side chain subunit in O139 strains (Hisatsune et al., 1993). The abbreviated O-polysaccharide side chain of *V. cholerae* O139 appears to be a hexasaccharide containing colitose residues and a cyclic phosphate group (Knirel et al., 1997). The LPS of other *Vibrio* species is similar in many aspects to that of O139. KDO-phosphate has been detected in *V. parahaemolyticus* by gas chromatography-mass spectrometry analysis (Janda, 1998). The O-polysaccharide side chains of *Vibrio* species produce only a single fast-migrating band on silver-stained SDS-PAGE gels (Amaro et al., 1992; Iguchi et al., 1995). This result suggests that the side chains are short; a chain length of ≤10 monosaccharides has been proposed for *V. parahaemolyticus* (Iguchi et al., 1995). Some species however (e.g., *V. vulnificus*) may exhibit ladder-like patterns by immunoblotting with whole cell antisera; this result suggests that the LPS O-polysaccharide side chains are acidic. Lambert et al. (1983) studied the cellular fatty acids of most of the *Vibrionaceae* and postulated that differences among the *Vibrio* species might prove useful for identification.

Flagella   Two types of flagella are synthesized by vibrios in different environments. In liquid culture, swimmer cells predominate due to production of a single sheathed polar flagellum in most species (Figs. BXII.γ.158 and BXII.γ.159). The sheath is an extension of the outer membrane (Fig. BXII.γ.160). The polar flagella are 24–30 nm in diameter with a central core 14–16 nm in thickness with a wavelength of 1.4–1.8 μm (Baumann et al., 1984b; Janda, 1998). Some *Vibrio* species (e.g., *V. harveyi*, *V. fischeri*, *V. logei*, and *V. salmonicida*) produce tufts (3–12) of polar flagella (Fig. BXII.γ.161) with a wavelength of approximately 3.6 μm (Baumann et al., 1984b; Ishimaru et al., 1996). Polar flagella provide chemotactic motility in liquid media and derive their energy from the sodium membrane potential (McCarter, 1995). In some marine vibrios (e. g., *V. anguillarum*), the polar flagellum appears critical for disease production in estuarine fish (Milton et al., 1996; O' Toole et al., 1996). When vibrios come into contact with solid surfaces, a series of morphogenetic changes are initiated that result in the conversion of swimmer cells into swarmer cells in some marine species such as *V. parahaemolyticus*, *V. alginolyticus*, *V. diabolicus*, and *V. pectenicida* (Raguénès et al., 1997a; Lambert et al., 1998). During this process, cell septation ceases, the cells elongate from 1 to 30 μm, and numerous lateral flagella are formed (Fig. BXII.γ.162) (McCarter and Silverman, 1990). These lateral flagella, 14–15 nm in diameter with a wavelength of 0.9 μm, are distinct from polar flagella. They are unsheathed, have a different protein subunit composition, and are internally driven by the protonmotive force (Baumann et al., 1984b; McCarter, 1995). Formation of lateral flagella permits swarmer

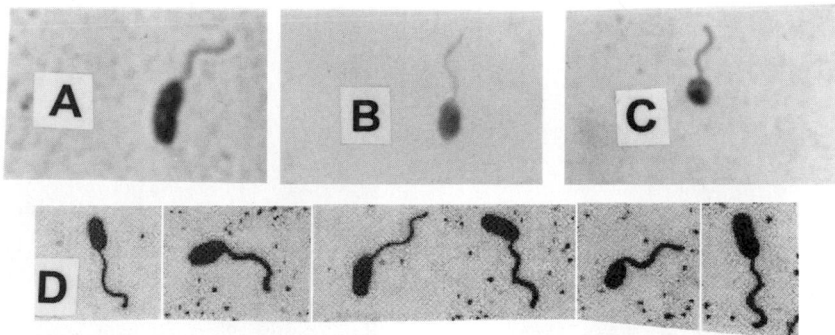

FIGURE BXII.γ.158. Leifson flagella stain of *Vibrio cholerae.* (Source: CDC archive, courtesy of Ed Ewing.)

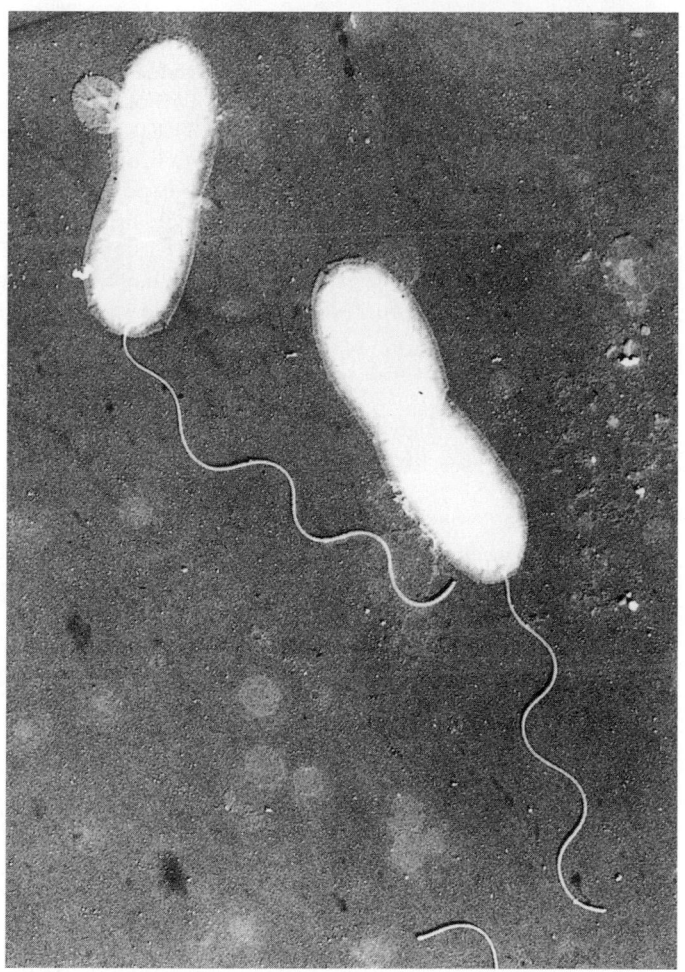

FIGURE BXII.γ.159. Electron micrograph of *Vibrio alginolyticus* grown in liquid medium. Note the sheathed polar flagellum and absence of peritrichous flagella. Shadowed preparation. × 13,000. (Reproduced with permission from C. Golten and W.A. Scheffers, Netherlands Journal of Sea Research *9:* 351–364, 1975, ©Netherlands Institute for Sea Research.)

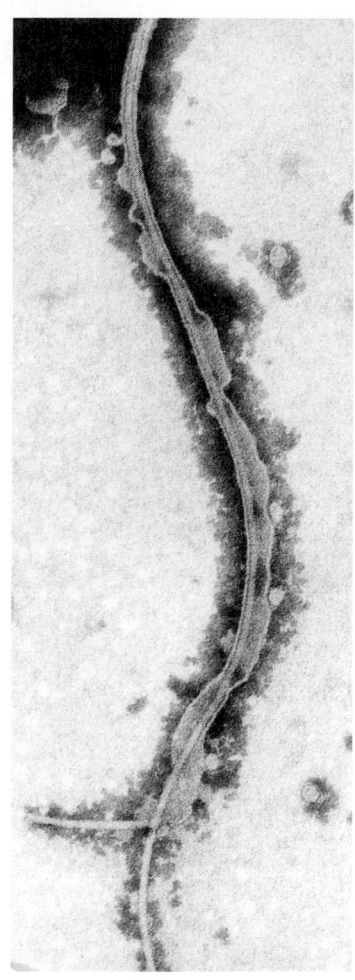

FIGURE BXII.γ.160. Electron micrograph of a polar flagellum of *Vibrio alginolyticus.* Note that the sheath has partially disintegrated exposing the inner core. Negatively stained preparation. × 30,000. (Courtesy of R.D. Allen.)

migration across solid surfaces and results in progressive spreading of the bacterial colony (McCarter and Silverman, 1990), a phenomenon called swarming. Swarming in many vibrio species is dependent upon a number of factors including agar concentration, media composition, iron availability, temperature, and relative viscosity (Baumann et al., 1984b; McCarter and Silver-

man, 1990). The microscopic morphology of vibrio cells removed from different concentric zones of swarming has been studied in some *Vibrio* strains (Sar and Rosenberg, 1989). Innermost zones consist of irregular cells including bent rods that progressively evolve into short rods and then into large rods with bundles of detached flagella (Fig.BXII.γ.163), whereas cells in

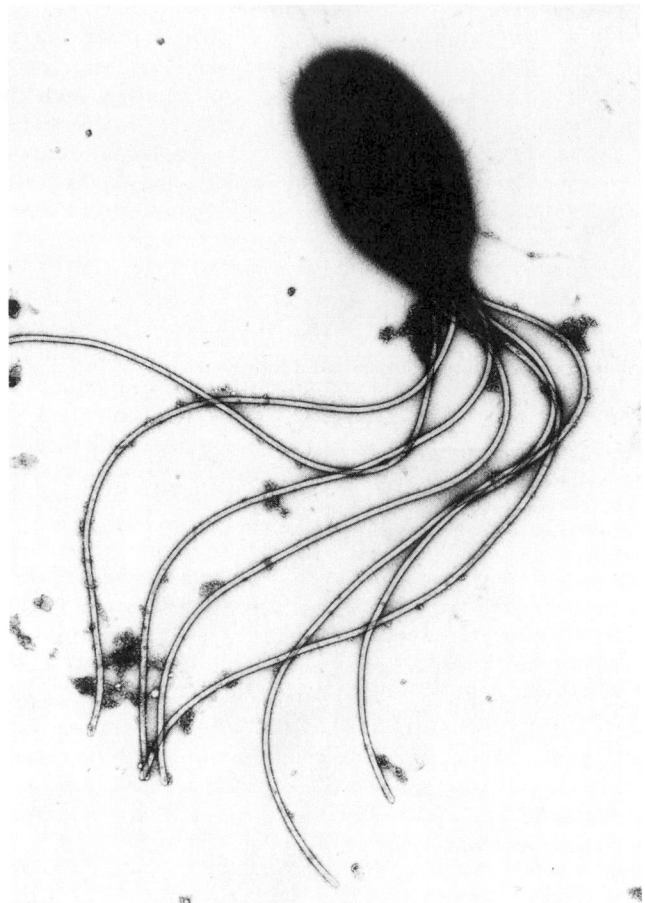

FIGURE BXII.γ.161. Electron micrograph of *Vibrio fischeri*. Note the tufts of sheathed polar flagella. Negatively stained preparation. × 23,000. (Reproduced with permission from: J.L. Reichelt and P. Baumann, Archives of Mikrobiology *94:* 283–330, 1973, ©Springer-Verlag, Berlin.)

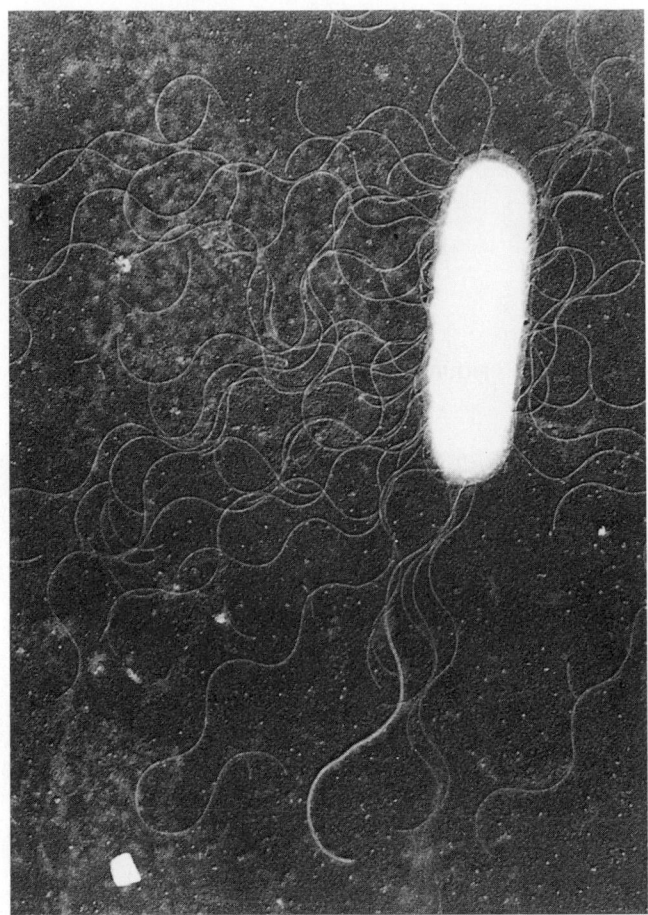

FIGURE BXII.γ.162. Electron micrograph of *Vibrio alginolyticus* grown on solid medium. Note the thick, sheathed flagellum and numerous unsheathed lateral flagella. Shadowed preparation. × 18,000. (Reproduced with permission from W.E. de Boer et al., Netherlands Journal of Sea Research *9:* 197–213, 1975, ©Netherlands Institute for Sea Research.)

the outermost circles of swarming colonies consist of long filamentous forms.

Fimbriae Fimbriae are produced by a number of pathogenic vibrios such as *V. cholerae* O1 and non-O1, *V. parahaemolyticus*, and *V. vulnificus* (Hall et al., 1988; Honda et al., 1988; Gander and LaRocco, 1989; Nakasone and Iwanaga, 1990). Several different morphologic types of fimbriae have been described in *V. cholerae* O1. These include both wavy pili 3 nm in diameter and rigid filaments 5–6 nm wide and 180–800 nm in length (Hall et al., 1988). The most important of these pili is composed of the protein TcpA; these pili are 5–6 nm wide and form bundles of parallel undulating filaments up to 15 μm long (Hall et al., 1988). TcpA formation is coregulated with cholera toxin expression and is a key determinant of *in vivo* colonization. The gene encoding TcpA appears to reside on a pathogenicity island.

Capsules Capsules have been detected surrounding cells of strains of *V. cholerae* O139 and *V. vulnificus* strains with a variety of staining techniques such as uranyl acetate, polycationic ferritin, and ruthenium red (Janda, 1998). The *V. vulnificus* polysaccharide capsule is 60 nm thick and has a low electron density (Amako et al., 1984; Hayat et al., 1993). The carbohydrate composition of the capsule of *V. vulnificus* varies from strain to strain. Sugars detected in different isolates include α-*N*-acetyl quino-

vosamine, α-*N*-acetyl galactosamine uronic acid, rhamnosamine, and fucosamine (Hayat et al., 1993).

Colonial morphology Most vibrios grow well on a variety of media, including protein-based agars and marine and seawater media, if sufficient Na$^+$ is present (Baumann et al., 1984b; Farmer and Hickman-Brenner, 1992). On most selective media, vibrios appear as smooth, buff-to-cream-colored colonies 2–5 mm in diameter, with an entire margin after overnight incubation (Baumann et al., 1984b; Janda, 1998). Some species tend to produce grayish colonies, particularly on blood agar. Considerable variation in colonial morphology has been reported for some species, and this is best demonstrated by observing colonies with a dissecting microscope at 10–25 magnification with oblique lighting. *V. cholerae* strains can have several different colonial morphologies (smooth, rough, and rugose forms) in response to different growth conditions. Rugose colonies are often chlorine-resistant. They are usually found in older cultures and are composed of an amorphous intercellular matrix of aggregated bacteria and exopolysaccharide material (Morris et al., 1996). Formation of rugose colonies can be enhanced by growth in enrichment broths such as alkaline peptone water (APW) or by picking the growth that has migrated up the sides of a culture tube. They can largely be avoided by picking a smooth colony and freezing it. In addition to these colony types, several path-

has been proposed (Hood et al., 1984). Depending on nutrient scarcity and the density/availability of particulate matter, *V. cholerae* can exist in several states. An epibiotic form attached to plankton predominates during periods of relatively high nutrient/particulate matter concentrations; this form changes to a microvibrio form (small rounded cells) during times of nutrient and particulate deprivation (Hood et al., 1984; Janda, 1998). This latter form may be analogous to the "viable but nonculturable state" (Colwell, 1984) that has been described for many *Vibrio* species including *V. cholerae*, *V. parahaemolyticus*, *V. vulnificus*, *V. anguillarum*, *V. campbellii*, *V. harveyi*, and *V. fischeri* during colder seasons of the year (Oliver, 1995). The microvibrio and "nonculturable" stages may be dormant phases for vibrios in winter from which subsequent blooms are triggered in response to increasing temperatures in spring and summer. However, it is still unclear whether any dormant state actually exists and whether blooms are due to the growth of a small number of cultivable cells (present at all times) or the actual resuscitation of dormant cells (Ravel et al., 1995; Bogosian et al., 1998).

Nutrition and growth conditions *Vibrio* species vary in their nutrition and growth requirements. The most important feature is that Na^+ is required for or stimulates growth. Minimum concentrations of Na^+ required for optimal growth (Fig. BXII.γ.164) range from 5–15 mM (0.029–0.087%) for *V. cholerae* and *V. metschnikovii* to 600–700 mM (3.5–4.1%) for *Salinivibrio costicola* (Baumann et al., 1984b). Most species grow well in solid or liquid media containing 0.5–2% NaCl. Some species (*Photobacterium iliopiscarium*) form bacterial aggregates in broth culture containing 2% NaCl (Onarheim et al., 1994).

The "salt requirement" of a strain will often depend on the test conditions. The main variables are temperature, the growth medium used prior to testing, the suspending medium, and the testing medium. All *Vibrio* species except *V. cholerae* and *V. mimicus* have an absolute requirement for Na^+ (Fig. BXII.γ.164). In some instances, this requirement may be partially offset by concentrations of Mg^{2+} or Ca^{2+} similar to those normally present in seawater (Baumann et al., 1984b). However, most species exhibit a specific requirement for Na^+ (Pujalte and Garay, 1986; Borrego et al., 1996). The range and optimum concentrations of NaCl supporting growth of some of the more recently described *Vibrio* species are listed in Table BXII.γ.154. No single medium or NaCl concentration is optimal for the recovery or growth of all *Vibrio* species. Many vibrios will grow in mildly alkaline conditions. Although most species prefer a pH range of 7–8 (Raguénès et al., 1997a), some, including *V. cholerae* and *V. metschnikovii*, will even grow at a pH of 10 (Baumann et al., 1984b).

Vibrios also vary in their temperature requirements for growth. Almost all *Vibrio* species grow well at 18–22°C. Some will grow at 0–4°C, whereas others can grow at temperatures up to 45°C. The temperature at which vibrios can grow is also dependent upon other factors including the composition of the medium and the NaCl concentration (Onarheim et al., 1994).

Most *Vibrio* species do not require specific organic growth factors such as vitamins or amino acids, although amino acid supplementation may be required to revive some strains stored for prolonged periods (Baumann et al., 1984b). Complex nutrients are required to induce growth of some species (Baumann et al., 1984b; Raguénès et al., 1997a). Such required supplements include yeast extract (for *V. anguillarum*, *Moritella marina*, and *V. logei*) and a seawater base (for *V. diabolicus*).

Vibrios use a variety of compounds as carbon and energy

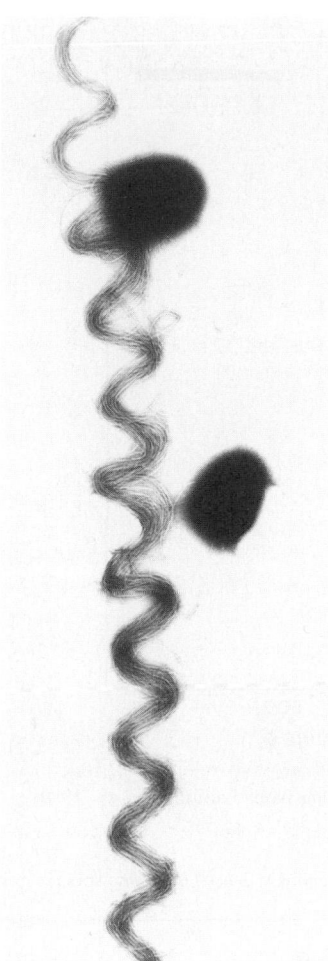

FIGURE BXII.γ.163. Large bundle of flagella in a culture of *Vibrio harveyi*. These bundles are frequently observed in cultures grown on solid medium. Negatively stained preparation. × 13,000. (Courtesy of R.D. Allen.)

ogenic vibrios, including *V. cholerae* and *V. vulnificus*, produce opaque and translucent varieties of smooth colonies on common media such as heart infusion and meat extract agars (Simpson et al., 1987; Finkelstein et al., 1992, 1997). Cells from colonies of these different morphologies differ from each other in a number of characteristics, including encapsulation, cell surface composition, cellular metabolism, and ability to survive under adverse conditions.

Pigmentation Several *Vibrio* species produce pigmented colonies. *V. nigripulchritudo* produces an insoluble blue-black pigment that accumulates in a crystalline form within the colonies (Baumann et al., 1984b). Other *Vibrio* species also produce blue-black crystals under various growth conditions, but typically do not produce the characteristic blue-black colonies of *V. nigripulchritudo*. Similar blue-black colonies are produced by a few strains of *Kluyvera* (Farmer et al., 1981a). Other pigmented species include *V. gazogenes* (red) and *V. fischeri* and *V. logei* (yellow-orange). A few strains of *V. cholerae* produce a brown diffusible melanin-like pigment (Ivins and Holmes, 1980).

Life cycles The marine environment is the natural habitat of vibrios, and the life cycle of these organisms is probably quite complex. A model for the life of *V. cholerae* in Gulf Coast estuaries

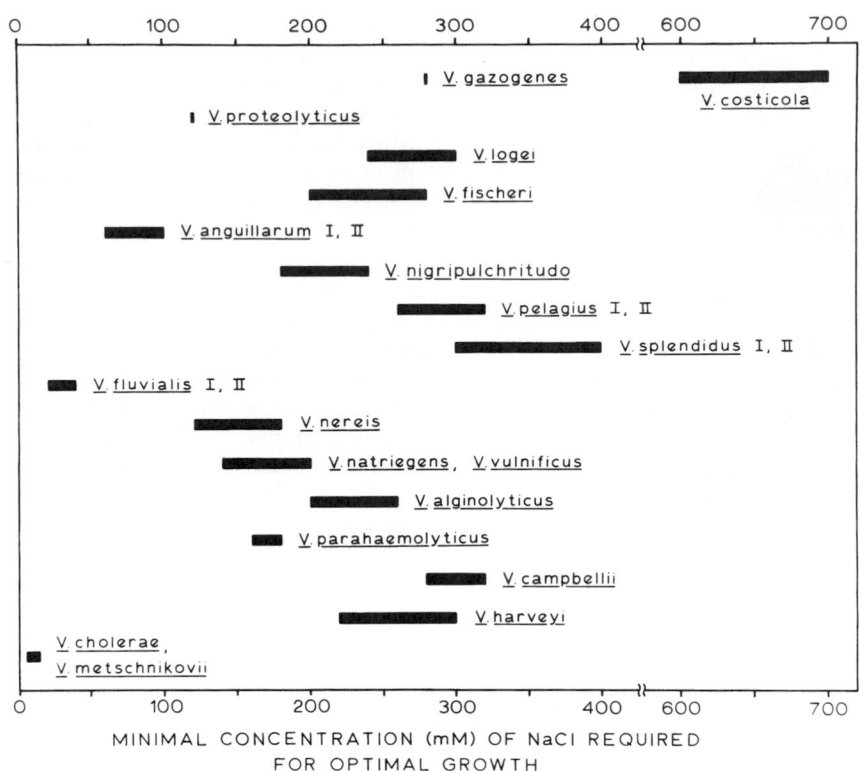

FIGURE BXII.γ.164. Effect of Na$^+$ (NaCl) on the growth of *Vibrio* species. Data from Baumann et al., 1980b.

TABLE BXII.γ.154. Growth temperatures, electrolyte optima, and nutritional versatility of 10 recently described *Vibrio* species and *Photobacterium iliopiscarium*[a]

Characteristic	V. diabolicus	V. halioticoli	V. ichthyoenteri	V. mediterranei	V. mytili	V. navarrensis	V. penaeicida	V. scophthalmi	V. tapetis	V. trachuri	Photobacterium iliopiscarium
Temperature range, °C	20–45	10–30	15–30	18–35	10-37	10–42	20–30	22–35	4–22		4–22
Temperature optima, °C	35–40				~28	30–37		22–30	18	~30	~20
NaCl range (%)	1–5		1–6	1.5–6	1–10	0.5-7	1–5	0.5–5	1–3	3–7	0.5–4
NaCl optima (%)	2.5–3.0	3						1–3			2
Nutritional versatility[b]	15/50	7/35	1/13	29/47	43/95	24/87	6/15	5/49	8/47		6/40

[a]Blank space, data not available. Data compiled from Urdaci et al. (1991); Pujalte et al. (1993); Onarheim et al. (1994); Iwamoto et al. (1995); Borrego et al. (1996); Ishimaru et al. (1996); Cerdà-Cuéllar et al. (1997); Sawabe et al. (1998b).

[b]Nutritional versatility is the number of carbon compounds used as a sole energy source/number of compounds tested.

sources. Some of the newly described environmental species (*V. ichthyoenteri*, *V. scophthalmi*) and related organisms (*Photobacterium iliopiscarium*) can only utilize a small number of substrates (<10 compounds) as sole carbon or energy sources (Table BXII.γ.154), while other species utilize over 40 compounds. The carbohydrates utilized include polysaccharides, monosaccharides, disaccharides, and polyhydroxyl alcohols. Vibrios produce a variety of extracellular enzymes that degrade DNA, RNA, proteins, lipids, lecithin, mucin, chitin, agar, alginic acid, starch, and related compounds. They also produce biologically active molecules such as hemolysins, and other toxins.

Physiology and metabolism Vibrios and photobacteria are chemoorganotrophic, facultatively anaerobic bacteria that produce several different acids during fermentation. Common metabolic end products of D-glucose metabolism include formic, acetic, lactic, succinic, and pyruvic acids (Baumann et al., 1984b).

Most vibrios ferment sugars without visible gas production ("anaerogenic" fermentation), although some species such as *V. furnissii*, *V. gazogenes*, *V. mytili*, *Photobacterium iliopiscarium*, and a few strains of *Vibrio damsela* produce CO_2 and H_2 as fermentation products. D-Glucose is transported into cells and phosphorylated to glucose-6-phosphate by the phosphoenolpyruvate:carbohydrate phosphotransferase (PTS) system (Postma et al., 1993). A number of other sugars are transported by the glucose–PTS system in *Vibrio* species including trehalose, D-fructose, D-mannose, and D-mannitol (Postma et al., 1993; Sarker et al., 1994). In *V. parahaemolyticus* the glucose–PTS is constitutive or partially inducible. A second mechanism for glucose transport in *V. parahaemolyticus* has been recently identified (Sarker et al., 1994). This Na$^+$/glucose symport system is stimulated by Na$^+$ and to a lesser extent by Li$^+$. D-Galactose, α-D-fucose, and methyl-α-glucosidase also appear to be transported by this mechanism, whereas D-glucose partially represses this Na$^+$/glucose symport system. A number of aromatic compounds are degraded by the *meta*-cleavage of the intermediate protocatechuate by the α-ketoacid pathway (Baumann et al., 1984b). All *Vibrio* species except *V. metschnikovii* and *V. gazogenes* are oxidase positive; possession of this trait is correlated with the presence of cytochrome *c* systems (Baumann et al., 1984b). Cytochromes *b*, *c*, *d*, *o*, and *a₁* occur in *V. cholerae*, *V. anguillarum*, *V. natriegens*, and *V. alginolyticus*. A number of different *c*-type cytochromes (*c4*, *c5*) have been shown to be present in some species based upon data from N-terminal amino acid sequencing of proteins and spectral analysis (Petushkov and Lee, 1997). *V. metschnikovii*, which is oxidase negative, lacks *c*-type cytochromes (Baumann et al., 1984b). *V. alginolyticus* possesses an ubiquinol oxidase of the cytochrome *bo*-type, which may function as a proton pump in the respiratory chain (Miyoshi-Akiyama et al., 1993).

Vibrios can acquire iron for growth by producing iron chelating molecules termed siderophores. Many *Vibrio* species including *V. cholerae*, *V. anguillarum*, *V. parahaemolyticus*, *V. fluvialis*, and *V. mediterranei* produce catechol-like siderophores called vibriobactins (Amaro et al., 1990; Guerinot, 1994). Less often, some *V. cholerae* non-O1, *V. parahaemolyticus*, and *V. fluvialis* strains produce a second hydroxymate-type siderophore. In *V. mimicus* and *V. hollisae* this hydroxymate-type siderophore has been identified as aerobactin (Okujo and Yamamoto, 1994). In *V. anguillarum*, a siderophore, called anguibactin is produced; anguibactin has both catechol and hydroxymate moieties (Guerinot, 1994). In addition to these iron chelators, vibrios can acquire iron by non-siderophore-dependent mechanisms. Ferric citrate can be actively transported into the cell by some vibrios via a non-inducible energy-dependent system (Mazoy et al., 1997). A number of other compounds can also serve as sole iron sources; these compounds include ferrichrome, hemoglobin, and saturated ferritin and transferrin (Simpson and Oliver, 1987; Fouz et al., 1994; Guerinot, 1994).

Several *Vibrio* species have a constitutive arginine-dihydrolase system. All *Vibrio* species produce a single iron-containing superoxide dismutase (Baumann et al., 1984b). The first cysteine protease in a *Vibrio* species has recently been identified in a luminous strain of *V. harveyi* isolated from a diseased tiger prawn (Liu et al., 1997c).

Bioluminescence A striking property of some *Vibrio* strains is the ability to emit yellow to blue-green light in response to certain environmental triggers. Bioluminescent marine species include *V. fischeri*, *V. harveyi*, *V. logei*, and *V. splendidus*. Light is produced as a consequence of the oxidation of the reduced flavin mononucleotide FMNH₂ and long-chain aliphatic aldehydes (Hastings and Nealson, 1977). The enzyme catalyzing this reaction is luciferase, an 80,000 MW heterodimer composed of an α (42,000 MW) and a β (38,000 MW) subunit; the α subunit contains the site of catalytic activity (Baumann et al., 1984b). Considerable N-terminal amino acid sequence homology exists (57–71%) between the α and β subunits of different luminescent species. At least 35–40% homology also exists between heterologous subunits; this fact suggests that these sequences originally arose from the duplication of a common ancestral gene (Baumann et al., 1984b).

Bioluminescence in marine *Vibrio* species is dependent upon quorum sensing (Stevens and Greenberg, 1997). Quorum sensing involves the ability of bacterial species to recognize relative population densities in their immediate environment and to induce specific cell functions once certain bacterial concentrations

have been achieved (Stevens and Greenberg, 1997). In seawater, the concentrations of *V. fischeri* are <10² cells/ml and bioluminescence is not induced (Kolibachuk and Greenberg, 1993). However, *V. fischeri* also exists in symbiotic associations with marine animals such as the bobtail sepiolid squid. This symbiotic association may be regulated by ADP-ribosyltransferases halovibrin α and β, which could act as signaling molecules in the symbiosis between *V. fischeri* and the squid (Reich et al., 1997). In the light organ of the squid (Fig. BXII.γ.165), population sizes of *V. fischeri* may reach 10¹⁰–10¹¹ cells/ml and luminescence is induced (Kolibachuk and Greenberg, 1993; Ruby and Lee, 1998).

In vivo, luminescence is triggered when autoinducers reach critical intracellular levels. *N*-(3-oxohexanoyl) homoserine lactone and *N*-(3-hydroxybutanoyl) homoserine lactone are the autoinducers for *V. fischeri* and *V. harveyi* respectively (Sun et al., 1994). The cells are freely permeable to these compounds. At high cell densities, greater concentrations of autoinducers accumulate intracellularly and trigger the expression of specific luminescence genes (Kolibachuk and Greenberg, 1993). Two genes activated in this process are *luxI* and *luxR* (Stevens and Greenberg, 1997). *luxI* encodes an autoinducer synthetase and *luxR* encodes an autoinducer-dependent activator of luminescence genes (Stevens and Greenberg, 1997). *luxR* activates transcription from the promoter of the *lux* operon, which contains seven genes. LuxR is a 250-amino acid polypeptide containing two separate domains. The N-terminal region of LuxR binds the autoinducer; this binding then allows the C-terminal region to activate transcription of the *lux* operon (Stevens and Greenberg, 1997).

Genetics Recent studies employing pulsed-field gel electrophoresis (PFGE), representational difference analysis, and restriction fragment length polymorphism indicate that there are major differences in genome size and composition between *V. cholerae* El Tor, O1, and O139 (Choudhury et al., 1994; Calia et al., 1998). The estimated genome size of *V. cholerae* O1 (3.0 Mb) is at least 500 Kb larger than that of the El Tor biotype (Choudhury et al., 1994). Recent construction of the physical map of a classical biotype strain, *V. cholerae* 395 (Ogawa), by PFGE of genomic DNA digested with I-*CeuI*, *SfiI*, and *NotI* indicates a substantially larger chromosome (4.0 Mb) than previously estimated for this biotype (Trucksis et al., 1998).

Furthermore, it appears that both the classical and El Tor biotypes of *V. cholerae* O1 contain two unique and circular megareplicons rather than a single chromosome (Trucksis et al., 1998). Generation of novel or emerging pandemic strains of *V. cholerae* may occur *in vivo* by transduction. The chromosomal genes for cholera toxin are located on the pathogenicity island CTXΦ, which actually represents the genome of a transducing filamentous bacteriophage (Waldor and Mekalanos, 1996; Mekalanos et al., 1997). It is postulated that new strains may arise as a result of recombination in the intestinal tract of humans simultaneously coinfected with both toxigenic and nontoxigenic strains of *V. cholerae*. A second bacteriophage, K139, belongs to the kappa family of bacteriophages and can be found in a number of *V. cholerae* isolates, including O139 (Mekalanos et al., 1997; Calia et al., 1998). This bacteriophage carries a *glo* gene that encodes for a G-related protein of unknown function. Other large segments of *V. cholerae* chromosomal DNA may be involved in genetic exchange between strains. The pathogenicity island VPI, which houses the structural gene for the toxin coregulated pilus, may have arisen in *V. cholerae* by horizontal gene transfer

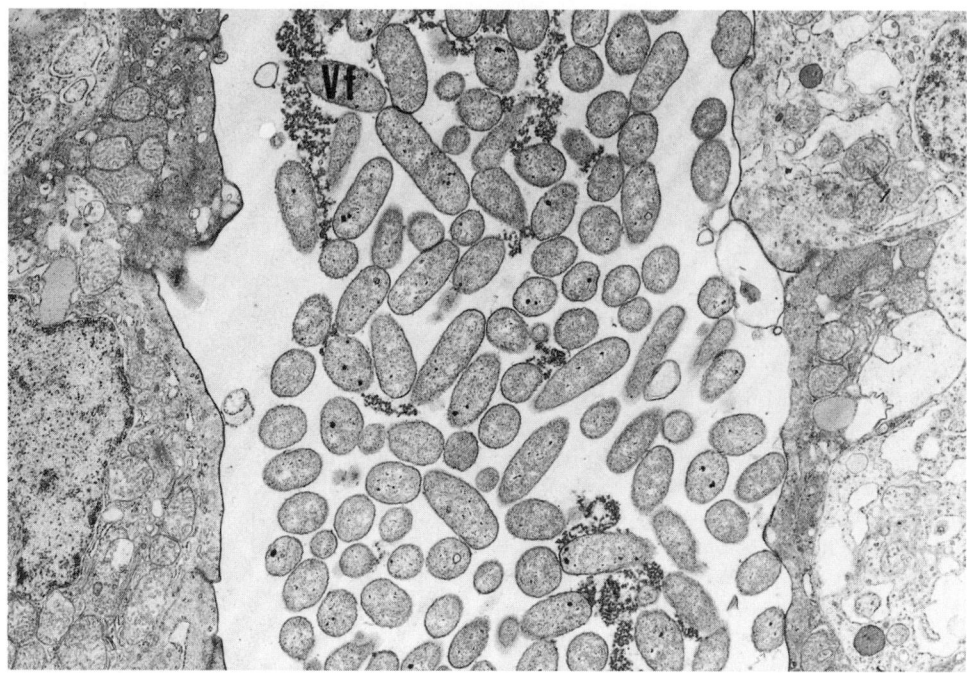

FIGURE BXII.γ.165. Ultrathin section through the luminous organ of *Monocentris japonicus* showing many cells of *Vibrio fischeri*. (× 9000). (Courtesy of K.H. Nealson and B.M. Tebo.)

from a heterologous species and may be transmissible (Mekalanos et al., 1997). A filamentous bacteriophage, designated VPIφ, carries the VPI pathogenicity island (Karaolis et al., 1999). VPIφ can be transferred between some *V. cholerae* strains, and the toxin coregulated pilin subunit of the pathogenicity island, TcpA, appears to be a coat protein of VPIφ. Thus, it appears that the VPIφ encodes for a protein (toxin coregulated pilus) that serves as a surface receptor for entry of a second phage, CTXΦ, into the bacterial cell. These findings suggest a sequential infection process of the bacterial cells by the two bacteriophages, a scenario which may lead to the emergence of new pathogenic strains of *V. cholerae*.

A 62-Kb chromosomal region, the SXT element, specifies resistance to streptomycin, trimethoprim, and sulfamethoxazole and can be transferred between *V. cholerae* strains by an unknown cell-dependent mechanism (Mekalanos et al., 1997). Transformation of *V. cholerae* strains with plasmids is very inefficient, presumably due to the large amounts of extracellular deoxyribonuclease (DNase) produced. However, transformation of DNase-negative strains using electroporation has been successful (Marcus et al., 1990). Conjugal mating in *V. cholerae*, which is mediated by sex factor P, is another potential mechanism of genetic exchange (Baumann et al., 1984b).

Filamentous phages have been isolated from a number of *V. cholerae* O139 strains (Honma et al., 1997; Shimodori et al., 1997; Jouravleva et al., 1998). These phages contain either single- or double-stranded DNA molecules ranging in size from 6.3 to 9.3 Kb. An open reading frame in one of these filamentous phages codes for a potential protein of 384 amino acid residues that has significant homology to the *zot* gene of *V. cholerae* (Honma et al., 1997). This finding suggests that such filamentous phage might play important roles in the horizontal transmission of genetic information between O1 and O139 strains (Jouravleva et al., 1998).

Current information on the genetics of other *Vibrio* species, although limited, suggests that gene transfer occurs among phylogenetically distant species. The thermostable direct hemolysin of *V. parahaemolyticus* has been detected in a number of vibrios including *V. cholerae* non-O1, *V. mimicus*, and *V. hollisae* (Nishibuchi et al., 1996). In the case of *V. cholerae* and *V. mimicus*, horizontal gene transfer from *V. parahaemolyticus* to these species appears to have occurred recently and to have been mediated by insertion sequence-like elements flanking the *tdh* genes. The *V. hollisae tdh* gene, exhibits significant sequence divergence from *tdh* genes of other *Vibrio* species; this finding suggests an earlier stable transfer of this hemolysin to the former group. Analysis of six *Salinivibrio costicola* strains indicates that the genome size varies from 2.1 to 2.6 Mb and that the DNA is highly methylated (Mellado et al., 1997). DNA methylation in *Salinivibrio costicola* might be affected by salt concentrations present in growth medium.

Mutants Chemical mutagens have been used *in vivo* and *in vitro* to generate mutants of both pathogenic and non-pathogenic *Vibrio* species. Compounds used include *N*-methyl-*N'*-nitro-*N*-nitrosoguanidine, ethylmethane sulfonate, and hydroxylamine (Baumann et al., 1984b; Iida et al., 1995). The Kappa-type lysogenic bacteriophages VcA-1, VcA-2, and VcA-3 have been exploited to generate "random" mutations in the chromosome of *V. cholerae* (Guidolin and Manning, 1987). Insertion of these bacteriophages into bacterial DNA inactivates the genes into which the prophage has been inserted. This process is analogous to that seen with bacteriophage phage Mu. Transposons have been successfully used to construct isogenic mutants in vibrios. The Tn*phoA* system has been used to produce *V. cholerae* mutants in chromosomal iron-regulated genes expressed on the cell surface and gene fusions for important colonization factors such as TcpA (Taylor et al., 1987; Goldberg et al., 1990). However, many strains of some species such as *V. vulnificus* are refractory to transposon mutagenesis.

Plasmids P plasmids are conjugative extrachromosomal elements in *V. cholerae* that are capable of self-transfer from P^+ to P^- cells (Guidolin and Manning, 1987). Stable integration of P into the host cell DNA does not occur. P plasmids are approximately 68 Kb in size; the exact function of these plasmids is presently unknown. Possession of P plasmids has been linked to attenuated virulence in *V. cholerae* and is associated with decreased TCP pilus production (Bartowsky et al., 1990).

In a study of over 100 *V. cholerae* O1 strains, only 2% were found to contain plasmids (Newland et al., 1984). In contrast, 21–33% of the strains of *Vibrio cholerae* non-O1 examined contained extrachromosomal elements (Newland et al., 1984; Barja et al., 1990; Dalsgaard et al., 1995). Most plasmids are of low molecular weight (2.5 to 22 Kb) (Barja et al., 1990; Dalsgaard et al., 1995). The function of these small plasmids is not known, although some large plasmids (>170 Kb) have been detected which carry genes encoding antibiotic resistance determinants and genes for cytotoxin production (Barja et al., 1990). *V. parahaemolyticus* and *V. vulnificus* infrequently harbor plasmids (Twedt et al., 1981; Davidson and Oliver, 1986). Strains of *Vibrio damsela*, *V. metschnikovii*, and *V. furnissii* commonly contain plasmids (Dalsgaard et al., 1996; 1997; Pedersen et al., 1997). Large plasmids of 190–200 kb have been detected in both *Vibrio damsela* and *V. metschnikovii*, but their function is unknown. Plasmids have also been detected in *V. fischeri*, *V. salmonicida*, *V. ordalii*, and *V. anguillarum* (Giles et al., 1995; Pedersen et al., 1996c). Most strains of *V. anguillarum* serogroup O1 have a 65–67-Kb plasmid named pJM1; this plasmid is a major virulence factor in hemorrhagic septicemia disease in salmonid fish (Litwin and Calderwood, 1993). pJM1 encodes several proteins associated with iron acquisition, including the siderophore anguibactin and an outer membrane receptor protein OM2. Plasmidless derivatives have reduced virulence. A 32-Kb cryptic plasmid is found in most *V. ordalii* strains, although some restriction fragment length polymorphism exists (Pedersen et al., 1996b). A few *Salinivibrio costicola* strains carry multiple plasmids ranging in size from 3–21 Kb up to a megaplasmid of undetermined mass; however, most strains of *S. costicola* do not harbor extrachromosomal elements (Mellado et al., 1997).

Phages and phage typing Temperate bacteriophages active against many *V. cholerae* strains serve as the basis for several phage-typing schemes. Only the scheme of Murkerjee has been used widely to type the classical biovar of *V. cholerae* (Baumann et al., 1984b). A typing system based on lytic phages originally developed by Murkerjee and Basu for *Vibrio cholerae* El Tor has been expanded to provide greater sensitivity (Chattopadhyay et al., 1993); the modified scheme incorporates five new phages as well as modifications of the protocol. Of 1000 *V. cholerae* O1 biogroup El Tor strains tested, >99% could be placed in one of 146 different phage types.

V. cholerae O139 isolates have a capsule and thus they are not usually lysed by choleraphages active against *Vibrio cholerae* O1. Albert et al. (1996) isolated an O139-specific bacteriophage named JA1 that lysed 119 of 122 *V. cholerae* O139 strains. Two Tn*phoA* mutants of *V. cholerae* O139 that lacked the O antigen and capsular polysaccharide were not lysed by phage JA1; this result suggests that the capsule is the receptor for this cholera-phage (Albert et al., 1996). A typing scheme for *V. cholerae* O139 strains has been devised based upon the responses of strains to five lytic phages (Chakrabarti et al., 1997); over 60% of the isolates fell into only one of eight phage patterns.

Bacteriocins and bdellovibrios Little recent work has been done on vibriocins. *Enterococcus faecalis*, *Enterococcus faecium*, and *Enterococcus durans* produce substances that are active *in vitro* against *V. cholerae* O1 and non-O1 (Simonetta et al., 1997). The activity is abolished by pre-treatment with proteases, but is resistant to 100°C for 30 min.

A number of *Vibrio* species serve as hosts for predatory marine *Bdellovibrio* strains (Baumann et al., 1984b; see the chapter describing the genus *Bdellovibrio*). The ability of bdellovibrios to lyse *V. parahaemolyticus* strains is temperature dependent; this finding suggests one possible explanation for the seasonal life cycle of this halophile (Joseph et al., 1982).

Antigenic structure As in the *Enterobacteriaceae*, the antigenic structure of *Vibrio* species is defined on the major classes of bacterial antigens: somatic (O), capsular (K), and flagellar (H), which are associated with the bacterial lipopolysaccharide, extracellular layer or capsule, and flagellum respectively (Janda, 1998). Within *V. cholerae* there are over 180 defined O antigens, but serotypes O1 and O139 contain all strains that have caused all large epidemics and pandemics of cholera (Shimada et al., 1994b; Yamai et al., 1997). Strains of *Vibrio cholerae* O1 can be further differentiated into two subtypes designated Ogawa and Inaba, which have antigenic formulas of the AB and AC type, respectively (Table BXII.γ.155; Shimada et al., 1994b). A third, unstable, antigenic type, Hikojima, with serofactors A, B, and C, has also been described (Janda, 1998). The "Hakata" group of *V. cholerae* possesses the C (Inaba) factor but not B (Ogawa) or A (common major O1 antigen) factors and has been designated as serogroup O140 by Shimada et al. (1994b). O serogroups 22 and 155 also share antigenic factors with O139 strains (Shimada et al., 1994b). The flagellar antigens of *V. mimicus* are identical or very similar to those of *V. cholerae*, and most *V. mimicus* strains fall into recognized O-groups of *V. cholerae*.

V. parahaemolyticus strains are differentiated serologically based on O and K (capsular) antigens. At present, 13 O antigens and more than 60 K antigens are recognized (Hisatsune et al., 1993; Janda, 1998). Compositional analysis of the lipopolysaccharide component of *V. parahaemolyticus* serogroups suggests that the side chain consists of no more than 10 monosaccharides (Iguchi et al., 1995). All serogroups produce R-type lipopolysaccharide. Carbohydrate analysis of Sephadex G-50 fractions indicates that glucose, L-glycero-D-manno-heptose, and D-glucuronic acid are common to all 13 serogroups (Iguchi et al., 1995). Serogroup-

TABLE BXII.γ.155. Antigenic serofactors of *Vibrio cholerae* O1: subtypes Ogawa, Inaba, and Hikojima

Serotype O1 subtype	O factors present in culture	Agglutination in absorbed serum:	
		Ogawa[a]	Inaba[a]
Ogawa	A, B	+	−
Inaba	A, C	−	+
Hikojima[b]	A, B, C	+	+

[a]The specific O factor sera are prepared by absorption. For example, an Ogawa antiserum is prepared by injecting an Ogawa culture, and then absorbing the resulting antiserum with an Inaba culture, which removes the antibodies to O serofactor A, leaving antibodies to O serofactor B.

[b]Most authors do not recognize subtype Hikojima and report cultures as either Inaba or Ogawa, based on which serum causes the fastest and strongest agglutination.

related sugars include fucose, arabinose, D-glycero-D-manno-heptose, D-glucosamine, D-galactosamine, 3-amino-3,6-dideoxyglucose, 3-amino-3,6-dideoxygalactose, 2-keto-3-deoxy-D-threo-hexonic acid, and 2-keto-3-deoxyoctonic acid (Iguchi et al., 1995).

Serologic typing schemes have been developed by the National Institutes of Health in Tokyo, Japan for several of the more commonly isolated *Vibrio* species. *V. vulnificus* strains can be subdivided into one of 7 O antigen groups (Shimada and Sakazaki, 1984). Most *V. vulnificus* (90%) were typable by this system, with O groups 1 and 4 predominating. Some strains have a heat-labile masking antigen that inhibits O agglutination reactions (Shimada and Sakazaki, 1984). All O antisera contain R antibody. The flagellar antigens of all *V. vulnificus* strains examined are identical. An international typing scheme composed of 35 serotypes has been developed for *V. fluvialis* and *V. furnissii* (Shimada et al., 1991). Some serotypes are apparently identical to those of *V. cholerae* O6, O39, and O41. Serogroup O19 possesses the C (Inaba) factor of *V. cholerae* O1 but not A or B (Shimada et al., 1987). At least 16 serogroups of *V. anguillarum* are presently recognized; O1 and O2 occur worldwide (Grisez and Ollevier, 1995). All strains of *V. ordalii* belong to O-group 2 of *V. anguillarum*. Flagellins from the polar flagella of many *Vibrio* species share common antigenic determinant(s): these species include *V. cholerae*, *V. mimicus*, *V. metschnikovii*, *V. harveyi*, *V. campbellii*, *V. parahaemolyticus*, *V. alginolyticus*, *V. anguillarum*, *V. fischeri*, *V. splendidus*, *V. nigripulchritudo*, and *Salinivibrio costicola* (Baumann et al., 1984b).

Antibiotic sensitivity There are surprisingly few detailed descriptions of the *in vitro* antibiotic susceptibility of halophilic and nonhalophilic *Vibrio* species. Most recent reports highlight only the susceptibility patterns of clinically significant groups (Morris et al., 1985; Clark, 1992; Farmer and Hickman-Brenner, 1992) although some other species, notably *V. harveyi* (now known to be same as the clinically important species *V. carchariae*), have been studied to some extent (French et al., 1989; Liu et al., 1997b). Most studies indicate that most vibrio isolates, regardless of species designation, are susceptible to tetracyclines, aminoglycosides (gentamicin), chloramphenicol, monobactams, carbapenems, quinolones, and nalidixic acid (Table BXII.γ.156) (Clark, 1992; Farmer and Hickman-Brenner, 1992; Janda, 1998). Widespread resistance to sulfa-containing antimicrobials, trimethoprim, penicillin, carbenicillin, ampicillin, and older cephalosporins has been reported and is often linked to specific species or groups (French et al., 1989; Farmer and Hickman-Brenner, 1992). Resistance to newer cephalosporins such as cefotaxime and ceftazidime is less common (French et al., 1989). Cefotaxime and minocycline show a synergistic activity that inhibits the growth of one clinical isolate of *V. vulnificus in vitro* (Chuang et al., 1997). Discrepancies in the literature regarding the susceptibility of *Vibrio* species to various chemotherapeutic agents may be related to test methodology or geographical variation. In a study of 240 strains representing nine *Vibrio* species, about 60% of all isolates expressed a β-lactamase based upon the nitrocefin test; >95% of all *V. alginolyticus* and *V. splendidus* were β-lactamase-positive, whereas <10% of *V. fluvialis* produced this enzyme (French et al., 1989).

Since the late 1970s, increasing drug resistance in *V. cholerae* has been reported. Outbreaks of disease in Bangladesh and Tanzania caused by multiresistant strains of *V. cholerae* O1 El Tor have been reported (Morris et al., 1985; Janda, 1998). These strains carry R factors encoding resistance to many common antimicro-

bial agents including ampicillin, chloramphenicol, and tetracycline. Drug resistance in these strains has been linked to possession of an IncC plasmid (Janda, 1998). Many isolates (>30%) of *V. cholerae* O1 from an outbreak of cholera in the western hemisphere were resistant to chloramphenicol, doxycycline, tetracycline, and trimethoprim-sulfamethoxazole. Large conjugative plasmids of the IncC group were detected in these strains. An unusual strain of *V. cholerae* O1 El Tor isolated during the same outbreak was resistant to cefotaxime and could transfer this resistance to *E. coli* by conjugation. After an outbreak of cholera in India in the 1990's due to *V. cholerae* O139, resistance of *V. cholerae* O1 strains to co-trimoxazole (trimethoprim-sulfamethoxazole), nalidixic acid, and chloramphenicol increased dramatically compared to O1 strains isolated prior to this epidemic (Mukhopadhyay et al., 1995). These findings suggest rapid emergence of drug resistance in *V. cholerae* facilitated by mobile genetic elements. However, studies have not shown a link between drug resistance in *V. cholerae* (non-O1) and plasmid carriage (Dalsgaard et al., 1995).

Early isolates of *V. cholerae* O139 from India and Bangladesh had different antibiotic susceptibilities (Yamamoto et al., 1995). Most strains were resistant to trimethoprim, sulfamethoxazole, and streptomycin. Six strains (3.5%) carried R factors mediating resistance to tetracycline, ampicillin, chloramphenicol, kanamycin, and gentamicin. Resistance markers were carried on a 200-kb self-transmissible plasmid of incompatibility group C. All plasmids were identical by restriction endonuclease analysis using *Eco*RI and *Sal*I. Most *V. cholerae* O139 strains isolated between 1993 and 1994 carried resistance to co-trimoxazole, furazolidone, and ampicillin (Mukhopadhyay et al., 1995; Albert et al., 1997; Vijayalakshmi et al., 1997). However, beginning in 1995 and continuing into 1996 with the resurgence of *V. cholerae* O139 in India, a change in susceptibility to antimicrobial agents was detected. Unlike previous isolates, strains isolated beginning in 1995 were susceptible to co-trimoxazole but still resistant to ampicillin and furazolidone (Mitra et al., 1996; Mukhopadhyay et al., 1996). These later strains were also resistant to neomycin (95%), streptomycin (85%), and to tetracycline (25%). These data indicate continued temporal shifts in phenotypes and resistance patterns associated with *V. cholerae* O139.

Sensitivity to pteridine compounds, particularly 2,4-diamino-6,7-diisopropylpteridine (commonly known as vibriostatic agent O/129, or simply as O129), varies among vibrios. Intrinsic resistance to O/129 occurs in a number of *Vibrio* species including *V. furnissii*, and *V. fluvialis* (Farmer and Hickman-Brenner, 1992). Recently acquired resistance to pteridine compounds has risen dramatically in previously susceptible species. Studies from Japan and India on *V. cholerae* O1 and non-O1 isolates recovered in Southeast Asia between 1973 and 1983 found only rare strains (1.3–1.8%) resistant to both the 10 µg and 150 µg concentrations of O/129 available in commercial discs (Sundaram and Murthy, 1983; Matsushita et al., 1984). By the early 1990s, widespread resistance (63–83%) to O/129 at both concentrations had been found in water, plankton, and fecal isolates of *V. cholerae* O1 from India, Bangladesh, and Malaysia (Huq et al., 1992; Ramamurthy et al., 1992; Mahalingam et al., 1993). Resistance to O/129 has been associated with concomitant resistance to co-trimoxazole and/or trimethoprim and has been found to be associated with transmissible R plasmids (Matsushita et al., 1984; Ramamurthy et al., 1992; Mahalingam et al., 1993).

TABLE BXII.γ.156. Antibiotic susceptibility patterns of the 12 *Vibrio* species that occur in human clinical specimens*

	Percentage of strains susceptible:[a]											
Antibiotic[b]	*V. cholerae*[c] (480)	*V. alginolyticus* (69)	*V. cincinnatiensis* (14)	*Vibrio damsela* (21)	*V. fluvialis* (25)	*V. furnissii* (9)	*V. harveyi* (2)	*V. hollisae* (34)	*V. metschnikovii* (22)	*V. mimicus* (75)	*V. parahaemolyticus* (144)	*V. vulnificus* (130)
Penicillin G (12–21)	2	0	0	0	0	0	0	97	9	3	0	2
Ampicillin (12–13)	87	0	36	52	32	11	0	100	31	97	12	99
Carbenicillin (18–22)	64	0	7	14	16	0	0	100	27	8	1	54
Cephalothin (15–17)	98	32	100	76	40	0	100	100	100	100	17	65
Colistin (9–10)	4	25	93	76	100	100	0	100	91	61	11	2
Tetracycline (15–18)	98	94	93	86	88	89	100	97	73	100	98	99
Sulfadiazine (13–16)	26	16	36	71	36	11	50	56	5	17	3	28
Chloramphenicol (13–17)	99	100	100	10	88	100	100	100	100	100	100	100
Streptomycin (12–14)	60	54	86	24	84	100	50	100	32	61	17	42
Kanamycin (14–17)	92	62	79	43	88	100	100	100	14	89	37	53
Gentamicin (13–14)	98	100	100	100	100	100	100	100	100	99	97	100
Nalidixic acid (14–18)	99	97	100	100	100	100	100	100	100	99	99	99

[a]Studied at the CDC *Vibrio* Laboratory and done on Mueller–Hinton agar (with no added NaCl) at 35–37°C.

[b]The numbers in parentheses give the zone size range for the category intermediate. For example, (12–21) means that resistant strains have 6- to 11-mm zones, strains of intermediate susceptibility have zones that are 12- to 21-mm, and susceptible strains have zones of 22 mm or larger. These particular break points are the ones established in the early 1970s for each antibiotic, and they have been used in the CDC *Vibrio* Reference Laboratory for almost 30 years for taxonomic studies. They may differ slightly from current break points for use in treating human infections. Therefore, these data should only be used as an aid in identifying cultures.

[c]The number of strains studied is given in parentheses.

Editorial Note: Photobacterium damselae subsp. *damselae* is a junior objective synonym of *Vibrio damsela*.

Pathogenicity Several *Vibrio* species cause diarrhea or infections of the gastrointestinal tract (Table BXII.γ.157) and a number of reviews are available (Morris and Black, 1985; Janda et al., 1988; Kaper et al., 1995). Symptoms can range from mild bouts of enteritis to a fulminant secretory diarrhea with vomiting such as cholera gravis, which is associated with some *V. cholerae* O1 infections (Kaper et al., 1995). The two most frequently isolated enteric pathogens of this group are *V. cholerae*, the etiologic agent of pandemic cholera, and *V. parahaemolyticus*, a major cause of foodborne disease, particularly in Japan and South East Asia (Joseph et al., 1982). Epidemiologic and other data strongly suggest that *V. mimicus*, *V. hollisae*, and *V. fluvialis* can cause diarrhea or infections of the gastrointestinal tract (Abbott and Janda, 1994b; Hlady and Klontz, 1996). *Vibrio furnissii* and occasionally other species have been isolated from feces of people with diarrhea, but there is no strong evidence that these other species can actually cause intestinal infections.

V. cholerae O1 is the causative agent of pandemic cholera. Of the seven pandemics recorded since 1817, both the fifth (1881–1896) and sixth (1899–1923) pandemics were attributed to the classical biovar (Kaper et al., 1995). The seventh pandemic, which is ongoing (1961–present), is caused by *V. cholerae* O1 biogroup El Tor. It has been suggested that a second serogroup, O139, has pandemic potential, but O139 is currently restricted to epidemic disease in the Indian subcontinent and South East Asia (Swerdlow and Ries, 1993; Albert, 1994). Despite their similarities in disease presentation, significant evolutionary divergence exists between these groups. Differences in chromosomal structure and DNA sequences of housekeeping genes indicate that the classical and El Tor biotypes are only distantly related to each other on a short-term evolutionary basis (Mekalanos et al., 1997). Representational difference analysis further indicates that *V. cholerae* O139 is more closely related to the El Tor biogroup than the

TABLE BXII.γ.157. Worldwide occurrence of 12 *Vibrio* species in human clinical specimens*

Species	Occurrence in human clinical specimens:[a]	
	Intestinal	Extraintestinal
V. cholerae		
Serogroups O1 and O139	+ + + +	+
Serogroup non-O1	+ +	+ +
V. alginolyticus	+	+ +
V. cincinnatiensis	−	+
V. damsela	−	+ +
V. fluvialis	+ +	−
V. furnissii	+ +	−
V. harveyi (*V. carchariae*)	−	+
V. hollisae	+ +	−
V. metschnikovii	−	+
V. mimicus	+ +	+
V. parahaemolyticus	+ + + +	+
V. vulnificus	+	+ + +

[a]The symbols +, + +, + + +, and + + + + give the relative frequency of each organism in specimens, and apply to the world, rather than to a particular country.

Editorial Note: Photobacterium damselae subsp. *damselae* is a junior objective synonym of *Vibrio damsela*.

classical and El Tor biogroups are related to each other (Calia et al., 1998).

Infection with *V. cholerae* O1 and O139 can result in gastrointestinal disease of varying severity (mild diarrhea, enteritis, cholera) although most infected persons (60–75%) are clinically asymptomatic (Kaper et al., 1995). Detection and control of such unapparent illnesses can be critical in preventing the spread of sporadic cases to produce outbreaks or epidemics. Two critical regions of DNA on the chromosomes of both *V. cholerae* O1 and O139 are responsible for the pandemic potential of strains of these biotypes as well as the ability to cause cholera in susceptible populations. Both regions are located in pathogenicity islands (Groisman and Ochman, 1996; Mekalanos et al., 1997). The first island, designated CTXΦ (cholera toxin), is a 7000–9700 bp region encoding at least six genes (Waldor and Mekalanos, 1996; Mekalanos et al., 1997). The most important of these gene products is cholera toxin, an oligomeric protein (MW 84,000) composed of five B subunits (*ctxB*) and one A subunit (*ctxA*). The B subunit binds holotoxin to the cell receptor, whereas the A subunit provides toxigenic activity intracellularly after proteolytic cleavage into 2 peptides, A_1 and A_2. Internal activation of the A_1 peptide results in ADP-ribosyltransferase activity leading to altered ion transport and hypersecretion of water and Cl^- into the lumen of the intestine. A number of excellent reviews on various aspects of the cholera toxin have been published (Spangler, 1992; Kaper et al., 1995; Sears and Kaper, 1996; Scott et al., 1996). The structural genes for the CTX element have recently been shown to reside on a filamentous phage, CTXΦ (Waldor and Mekalanos, 1996). This indicates that toxigenic *V. cholerae* can arise *de novo* by horizontal gene transfer, which presumably occurs in the gastrointestinal tract. Other genes located in this virulence cassette region include an accessory cholera toxin (*ace*), a zonula occludens toxin (*zot*), core encoded pilin (*cep*), and an open reading frame of unknown function (Janda, 1998). An RTX toxin gene cluster (hemolysins/leukotoxins) in *V. cholerae* El Tor is tightly linked to the cholera toxin prophage but is enzymatically independent of the CTX element (Lin et al., 1999). The toxin RtxA exhibits cytotoxic activity against HEp-2 cells. Classic strains that lack this cytotoxic activity contain a deletion in the RTX gene cluster involving *rtxA* (toxin) and *rtxC* (activator).

The second pathogenicity island is designated VPI and is associated with epidemic and pandemic strains of *V. cholerae*. VPI is 39.5 Kb in size and contains two ToxR-regulated genes: a regulator of virulence genes (ToxT) and a gene cluster containing essential colonization factors, including the toxin co-regulated pilus (TCP) (Karaolis et al., 1998). There is some genetic evidence that this island can be transferred from *V. cholerae* O1 to non-O1 strains. The *tcp* gene encodes for a 20.5-kDa protein that forms bundles of long filamentous pili 6–7 nm in diameter (Kaper et al., 1995; Janda, 1998). Both *in vitro* and *in vivo* experimentation indicates that TCP is essential for colonization and, therefore, for infectivity. Volunteers fed mutant strains lacking TCP failed to become colonized with *V. cholerae* O1, and no diarrhea was observed. Codon usage within this pathogenicity island differs from that of the rest of the *V. cholerae* chromosome; this finding suggests that the island may have been acquired by *V. cholerae* by horizontal transfer from an unrelated microorganism (Mekalanos et al., 1997).

V. cholerae O139, an emerging agent of epidemic cholera, is another important cause of diarrhea (Morris, 1995). Like serogroup O1 strains, O139 strains carry the structural genes encoded by the CTX operon and TCP (Hall et al., 1994; Janda, 1998). However, these two groups differ in several ways. The number of copies of *ctxA* gene(s) and their arrangement in O139 strains differ from that of serotype O1. In addition, a maltose-inducible outer-membrane protein, OmpS, is constitutively expressed (Janda, 1998) and O139 strains are encapsulated and lack specific genes (*rfbR*, *rfbS*) involved in O1 antigen synthesis (Johnson et al., 1994). Serotype O139 appears to have resulted from a number of genetic rearrangements in an O1 strain including deletion of the O1 *rfb* region and acquisition by horizontal transfer from a non-O1 strain of a 35 Kb DNA region that encodes the surface polysaccharide (Bik et al., 1995; Comstock et al., 1996). The recent reemergence of *V. cholerae* O139 in late 1996 has been associated with a number of chromosomal (genomic) rearrangements. When compared with O139 strains isolated in 1992, reemergent strains harbor three copies of the CTX genetic element connected by a single direct repeat sequence and carry an additional restriction site in the ribosomal RNA operon (Khetawat et al., 1999).

Isolates of *Vibrio cholerae* belonging to serogroups other than O1 and O139 normally lack cholera toxin genes and have never been found to carry TCP. These isolates can cause sporadic cases of gastroenteritis but have not spread as epidemics or pandemics of diarrheal disease. Pathogenic mechanisms include the production of "non-cholera enterotoxins", most of which are poorly characterized. However, unlike O1 strains, non-O1 strains of *V. cholerae* are commonly involved in invasive disease such as septicemia in immunocompromised hosts (Safrin et al., 1988; Lin et al., 1996).

V. parahaemolyticus has caused numerous cases of gastroenteritis, including many outbreaks. Cases are associated with the consumption of raw or undercooked shellfish such as oysters, shrimp, crabs, and lobster. Early studies suggested an association between human strains causing diarrhea and production of a positive hemolytic reaction on a special medium, Watgatsuma agar. This reaction, termed the Kanagawa phenomenon (KP), was positive for virtually all clinical isolates, whereas most non-clinical strains were "Kanagawa negative". Subsequent studies demonstrated a direct correlation between "Kanagawa positive" strains and the elaboration of a thermostable direct hemolysin (TDH). This hemolysin is enterotoxigenic in rabbit ileal loops and appears to alter ion flux in intestinal cells, thereby causing a secretory response and gastroenteritis (Nishibuchi and Kaper, 1995). Strains classified as "Kanagawa strong positive" have two copies of the TDH gene (*tdh1*, *tdh2x*), whereas strains classified as "Kanagawa weak" have only a single copy. "Kanagawa negative" strains lack this gene (Nishibuchi and Kaper, 1995). In the mid-1980s however, "Kanagawa negative" strains associated with gastroenteritis began to appear from many different geographic areas, including the Maldives (Honda et al., 1987). It was later found that these "Kanagawa negative" strains produced a different hemolysin named "TDH-related hemolysin" and abbreviated as "TRH". The gene for the TDH-related hemolysin, *trh*, has 70–80% homology to *tdh*, and appears to be closely associated with urease expression (Nishibuchi and Kaper, 1995; Suthienkul et al., 1995). Thus, "Kanagawa negative" strains of *V. parahaemolyticus* can apparently also cause gastroenteritis if they have the *trh* gene. The *tdh* gene is also found in several unrelated vibrios, including *V. hollisae*, *V. mimicus*, and *V. cholerae* non-O1 (Nishibuchi et al., 1996). This finding suggests that horizontal transfer of *tdh* genes to other species is a possible common enteropathogenic mechanism.

V. vulnificus is an important pathogen that causes wound infections and septicemia (Hollis et al., 1976; Blake et al., 1979; Tacket et al., 1984; Klontz et al., 1988). Wound infections result from direct inoculation of the organism into traumatized cutaneous surfaces after contact with marine animals or the marine or estuary environment. Septicemia commonly develops in immunodeficient persons with hepatic disease who consume raw or improperly cooked oysters. On rare occasions, a classic secretory diarrhea, typically observed with other vibrio species, has been reported in patients with *V. vulnificus* in their feces (Klontz et al., 1988), but the etiological role of *V. vulnificus* as a cause of diarrhea is not proved. Other rare complications of *V. vulnificus* infection include meningitis, myositis, endometritis, peritonitis, and ocular disease. Despite intense investigation, the pathogenic mechanisms of *V. vulnificus* are poorly defined. A unique *V. vulnificus* cytotoxin-hemolysin apparently does not play a major role in microbial pathogenesis (Wright and Morris, 1991), although it is a highly sensitive and specific genetic marker for the species *V. vulnificus*. Other postulated virulence factors include a capsule, iron-scavenging systems, and resistance to complement-mediated lysis.

Two distinct biogroups of *Vibrio vulnificus* have also been described, and both can infect humans. Strains of *Vibrio vulnificus* biogroup 2 have caused outbreaks in eels in Taiwan, Japan, and Spain and apparently cause human wound infections (Tison et al., 1982; Amaro and Biosca, 1996). Strains of *Vibrio vulnificus* biogroup 3 have caused outbreaks of severe wound infection in Israel (Bisharat et al., 1999). The infections occurred when people purchased living, pond-raised, "Saint Peter's fish" (*Talapia* spp.) from local vendors and then were inoculated with the organism through an existing skin break or trauma incurred while handling the fish. Interestingly, the halophilic strains of *V. vulnificus* biogroup 3 were able to grow in the "fresh water" inland fishponds. However, these ponds had a high salt content because of the source of the water and evaporation during the hot dry summer months.

Vibrio damsela is a rare but important pathogen of fish and mammals, including humans. Originally described as the causative agent of ulcerative disease in damselfish (Love et al., 1981), *Vibrio damsela* causes wound infections in bottlenose dolphins, leatherback turtles, yellowtail, sea bream, barramundi, turbot, brown shark, and rainbow trout (Pedersen et al., 1997). Damselfish infected with *Vibrio damsela* after scarification of the dermis develop large ulcers within three days, and death follows shortly thereafter (Love et al., 1981). In humans, wound infections due to *Vibrio damsela* are occasionally reported. Illnesses commonly result from traumas sustained in contact with objects in contaminated waters (e.g., a reef), and from penetrating injuries resulting from catfish barbs or stings from stingrays (Morris et al., 1982). A less frequent but serious complication is fulminant *Vibrio damsela* septicemia, which mimics *V. vulnificus* septicemia (Clarridge and Zighelboim-Daum, 1985; Lang, 1992; Shin et al., 1996). A hemolytic cytolysin with phospholipase D activity is postulated to be a virulence factor in wound pathology and subsequent invasive infections (Kreger et al., 1987).

Many non-clinical *Vibrio* species cause serious infections in a variety of marine animals including shellfish and fish. The best studied of these pathogens is *V. anguillarum*, the causative agent of epizootic disease in marine fish and shellfish (Austin and Austin, 1993). The general term vibriosis is used to describe fulminant septicemic infections produced by a number of vibrio species in marine fish. The etiologic agent of the disease may be specific for a single type of aquatic life or may have a very broad host range. Table BXII.γ.158 summarizes the important pathogens of marine animals (Hada et al., 1984; Ishimaru et al., 1995, 1996; Iwamoto et al., 1995; Borrego et al., 1996; Liu et al., 1996a; Yii et al., 1997).

Ecology Water (and its associated microorganisms, animals and plants) is the natural habitat of most of the nonhalophilic and marine vibrios. Environmental factors regulating vibrio concentrations and distribution in both freshwater and seawater ecosystems include the concentration of organic and inorganic

TABLE BXII.γ.158. Diseases in marine fish and invertebrates caused by or associated with *Vibrio* species

Disease(s)	Susceptible Animals	Species
Whitespot disease	Kuruma prawns, tiger prawns	*V. alginolyticus*
Terminal hemorrhagic septicemia	Eels, ayu, rainbow trout, salmonids	*V. anguillarum*
Gastroenteritis	Groupers	*V. harveyi*
Vibriosis (larvae and juveniles)	Kuruma prawns	*V. harveyi*
Intestinal necrosis (larvae)	Flounders	*V. ichthyoenteri*
Brown spot disease	Kuruma prawns	*V. penaeicida*
Brown ring disease	Manila and fine clams	*V. tapetis*
Intestinal hemorrhage, skin discoloration	Japanese horse mackerel	*V. trachuri*
Diseased larvae	Clams, oysters, scallops	*V. pectenicida, V. tubiashii*

chemicals, pH, temperature, salinity, oxygen tension, and exposure to UV light (Chakraborty et al., 1997). However, salinity (specifically, the concentration of Na^+) is the most important factor governing the environmental distribution of vibrios. Na^+ concentrations required for optimal growth of *Vibrio* spp (Fig. BXII.γ.164) range from a low of 5–15 mM (0.029–0.087%) for *V. cholerae* and *V. metschnikovii* to a high of 600–700 mM (3.5–4.1%) for the halophile *Salinivibrio costicola* (Baumann et al., 1984b). This wide range partially explains why *V. cholerae*, *V. mimicus*, *V. fluvialis*, and *V. furnissii* are recovered from freshwater rivers, streams, and ponds; other species such as *V. vulnificus* prefer moderate salinity and grow in shellfish beds in coastal areas (Kaysner et al., 1987; Chakraborty et al., 1997; Motes et al., 1998). Although salinity is a critical parameter, it does not completely explain the environmental distribution of all vibrios because halophilic species such as *V. parahaemolyticus* can survive in suboptimal Na^+ concentrations. Other *Vibrio* species have been implicated in human infections resulting from apparent contact with freshwater catfish (Lowry et al., 1986; Chakraborty et al., 1997). *V. mimicus*, a nonhalophilic species, is found in aquatic environments where the average salinity is less than 1% (Chakraborty et al., 1997).

Independent of salinity, vibrio densities tend to increase in warmer waters when enough inorganic and organic nutrients are present for growth. Vibrios are typically more common in geographic regions having temperate or tropical climates. Studies of northern Gulf of Mexico and Atlantic coast oysters indicate that large populations of *V. vulnificus* occur in shellfish meats from May through October; populations then decline until late March (Motes et al., 1998). This seasonality is apparent in most *Vibrio* species, with larger populations being observed from late spring to early fall. In one study, *V. fluvialis*, *V. furnissii*, *V. mimicus*, and *V. parahaemolyticus* survived in sea water at 20°C for 15 days without a detectable reduction in cell viability (Munro et al., 1994). Although *V. metschnikovii* and *V. cholerae* numbers initially declined for 2–6 days after inoculation into seawater, a regrowth phase occurred, and populations then rose to initial values. In contrast, *V. vulnificus* survived poorly under identical conditions, with viability being reduced >10,000-fold within the same interval.

The microbial ecology of vibrios is best exemplified by *V. chol-*

erae. V. cholerae can survive in a "free living state" in both freshwater and saline environments. It is widely distributed in habitats such as sewage, brackish water, estuaries, and coastal inlets, as well as in polluted streams, rivers, ponds, and lakes. *V. cholerae* can also persist in an epibiotic form associated with various microscopic life (Hood et al., 1984). Common microorganisms associated with *V. cholerae* include cyanobacteria, phytoplankton (diatoms, fresh water algae), and zooplankton (Huq et al., 1995; Chakraborty et al., 1997). *V. cholerae* may attach to the tissues or chitinous exoskeleton of crustaceans and production of the enzyme chitinase might be important in this process. All these combined niches provide a continuous environmental source for the maintenance and dissemination of *V. cholerae* throughout the world.

Transmission of *V. cholerae* O1 and O139 to humans usually occurs through consumption of contaminated water or foods. An outbreak of cholera in South America caused by *V. cholerae* O1 is a grim reminder of the rapidity with which cholera can spread in naive (susceptible) populations. The initial source of this epidemic, which began in 1991, is unknown, but the disease rapidly became waterborne and was quickly disseminated throughout the continent by introduction into water, unwashed fruits and vegetables, raw seafood, and cooked crab (Mujica et al., 1994; Weber et al., 1994). Cholera can be introduced into industrialized countries through the importation of contaminated foods brought by travelers from affected areas (Finelli et al., 1992).

Strains of non-O1 *V. cholerae* greatly outnumber O1 strains in the environment. *V. cholerae* non-O1 has been recovered from birds, amphibians, herbivores, and freshwater fish (Baumann et al., 1984b; Rhodes et al., 1985). A majority of these isolates lacks the classic virulence factors that are present in O1 strains, such as cholera toxin and the toxin co-regulated pilus (Baumann et al., 1984b; Kaysner et al., 1987).

V. parahaemolyticus inhabits inshore coastal areas and estuaries and has only rarely been recovered from pelagic (open ocean) regions (Joseph et al., 1982). It has been isolated from various parts of the water column, sediment, zooplankton, shellfish, and fish. Occasionally, *V. parahaemolyticus* has been recovered from fresh water in Indonesia and India (Chakraborty et al., 1997). It has also been recovered from inland bodies of water in the United States where the Na$^+$ content is high. Like *V. cholerae*, growth of *V. parahaemolyticus* in the marine environment is favored by warmer temperatures and saline concentrations approximating 12 parts per thousand (ppt) (Joseph et al., 1982). *V. parahaemolyticus* has been isolated from a variety of marine animals including clam, oyster, lobster, scallop, sardine, squid, eel, crab, and shrimp (Joseph et al., 1982). Most outbreaks of gastroenteritis caused by *V. parahaemolyticus* have been linked to the consumption of crabs, shrimp, oysters, and lobsters. In Japan, *V. parahaemolyticus* is a major cause of food poisoning and is associated with the ingestion of raw fish such as sashimi and sushi (Chakraborty et al., 1997). Most strains of *V. parahaemolyticus* isolated from human infections are "Kanagawa positive," and most environmental isolates are "Kanagawa negative." This distinction was originally thought to indicate that "Kanagawa negative" strains do not cause human disease; however, as mentioned previously, it is now known that some strains produce a thermostable-related hemolysin, a virulence factor that has significant homology to the thermostable direct hemolysin responsible for a positive Kanagawa reaction.

V. vulnificus has been isolated from various locales around the world, including Europe and the Pacific and Atlantic coasts of the United States as far north as Washington and Maine, respectively (O'Neill et al., 1990; Veenstra et al., 1994; Chakraborty et al., 1997). *V. vulnificus* is typically found in waters having intermediate salinities (5–25 ppt) and temperatures up to 26°C (Motes et al., 1998); it does not grow at temperatures less than 10°C (Chakraborty et al., 1997). DePaola et al. (1994) isolated *V. vulnificus* from seawater, crustacea, and estuarine fish from U.S. waters in the Gulf of Mexico. The highest concentration of *V. vulnificus* in one study was found in the intestinal contents of bottom-feeding estuarine fish (sea catfish, sheepshead, Atlantic croaker) that consume mollusks and crustacea (DePaola et al., 1994); it is rarely recovered from offshore fish. *V. vulnificus* populations decline as coastal and estuary water temperatures cool in late fall. The presence of *V. vulnificus* in shellfish may result from the constant filtering by these organisms of seawater containing vibrios rather than the active multiplication of *V. vulnificus* in shellfish tissues (Kelly and Dinuzzo, 1985). However, *V. vulnificus* can survive in oysters for 10–14 d when they are held at 2–10°C (Kaysner et al., 1989). In humans, *V. vulnificus* septicemia is almost exclusively associated with the consumption of raw oysters (Klontz et al., 1988). Some indirect evidence suggests that counts of approximately 10^3 bacteria/gm oyster meat are associated with human disease (Jackson et al., 1997). Strains of *Vibrio vulnificus* biotype 2 strains are pathogenic for eels (Amaro et al., 1995).

Few studies have addressed the environmental distribution of other *Vibrio* species. *V. fluvialis* was identified in 40% of 177 samples collected from rivers, ponds, swamp, estuaries, sewage, sediment, and crabs in Louisiana (Nishibuchi et al., 1983). Data on the association of many pathogenic and marine vibrios with aquatic microbiota suggest that they are widely dispersed in the marine environment. *V. fischeri* has developed a symbiotic association with the bobtail sepiolid squid, *Euprymna scolopes* (Ruby and Lee, 1998). *Vibrio damsela* causes ulcerative disease in the damselfish, *Chromis punctipinnis*, as well as in rainbow trout and in a number of other marine animals (Love et al., 1981; Pedersen et al., 1997). Many other vibrios cause disease in crustacea and fish (Table BXII.γ.158).

ENRICHMENT AND ISOLATION PROCEDURES

Much has been written about the isolation of *Vibrio* cultures, and several good reviews and manuals are available (see Further Reading). These topics have also been covered in detail by Farmer and Hickman-Brenner (1992). Two different sets of methods have been used, depending on whether the sample was taken from a human clinical specimen or from the marine environment. Species of *Vibrio* and *Photobacterium* will grow in many microbiological media; however, the low Na$^+$ content of some media (some have no added NaCl, others have a final concentration of 0.5–0.85%) will prevent growth of some marine species that require a higher concentration (Fig. BXII.γ.164). Only limited data exist on the growth of the species isolated from human diseases on many of the media traditionally used by marine microbiologists. In spite of the differences in Na$^+$ requirement/tolerance, several media are useful for all types of vibrio work.

There are several good general media for isolation of *Vibrio* and *Photobacterium* (Table BXII.γ.159). Marine agar is a nonselective medium, and almost all *Vibrio* strains will grow on it. Dozens of other media been described and are useful in particular instances, but are not in general use.

Marine agar is probably the most useful general medium for

TABLE BXII.γ.159. General and selective media for the genera *Vibrio, Photobacterium, Aeromonas,* and *Plesiomonas*[a]

Medium	Vibrio	Photobacterium	Aeromonas	Plesiomonas
Media and enrichments often used in clinical microbiology laboratories:				
Sheep blood agar	+ + + +	+ +	+ + + +	+ + + +
MacConkey agar	+ +		+ + + +	+ + + +
Alkaline peptone water	+ + + +	+ +	+ + +	+ + +
Nonselective "marine" media:				
Marine broth	+ + + +	+ + + +	+ +	+ +
Marine agar	+ + + +	+ + + +	+ +	+ +
Photobacterium agar	+ + + +	+ + + +	+ +	+ +
Photobacterium broth	+ + + +	+ + + +	+ +	+ +
Selective media for Vibrio:[b]				
TCBS agar	+ + + +	+ +	no growth	no growth
Vibrio agar	+ + + +			
Selective media for Aeromonas and Plesiomonas:				
Bile salts-brilliant green agar			+ + + +	+ + + +
Dextrin-fuchsin-sulfite medium			+ + + +	
Rimler-Shotts medium			+ + + +	
Aeromonas agar:			+ + + +	
Blood agar plus ampicillin			+ + + +	
DNase agar plus ampicillin and toluidine blue			+ + + +	
MacConkey agar plus ampicillin and Tween 80			+ + + +	
Pril–xylose–ampicillin agar			+ + + +	
Starch–ampicillin agar			+ + + +	

[a]The symbols "+ + + +" to "+" and "−" indicate the general usefulness of the medium and apply to most species in the genera; however, some species and strains may be exceptions and may differ on a particular medium. For example, most strains of *V. hollisae* do not grow well on TCBS agar (see the following chapter). A blank space indicates that data are not available or are contradictory.

[b]Most of the media formulated for *Vibrio* isolation were specifically designed to isolate *V. cholerae.*

Vibrio and *Photobacterium*. It is also useful for the isolation and growth of organisms whose natural habitat is marine or other environments with a high content of Na⁺. It is particularly useful for the species of *Vibrio* and *Photobacterium* (Fig. BXII.γ.164) that require Na⁺ in amounts higher than the usual content of bacteriological media (0.5–1.0%).

Marine semisolid medium (marine broth plus 0.4% agar) is a convenient semisolid medium for maintaining "working cultures" of *Vibrio, Photobacterium,* or other marine organisms that require added Na⁺ and other ions for growth.

TCBS (thiosulfate–citrate–bile salts–sucrose) agar is useful for the isolation of *V. cholerae* and *V. parahaemolyticus* from fecal specimens. It is probably the medium most widely used as a general isolation medium for *Vibrio* strains from human clinical specimens and from the environment (Fig. BXII.γ.166). Some *Vibrio* strains do not grow on TCBS agar or have a low plating efficiency; others are inhibited and grow as small colonies. Details regarding the growth characteristics of *Vibrio* species on TCBS agar are given in Table BXII.γ.160. Enterococci and some other bacteria will sometimes grow on TCBS agar, but most other bacteria are inhibited. Dehydrated TCBS agar is available from several commercial companies.. The exact composition of the medium varies from manufacturer to manufacturer and from lot to lot of the same manufacturer. This variation will influence the medium's selectivity, thus affecting both sensitivity and specificity.

The routine use of TCBS medium for isolation is too expensive for many laboratories that do not isolate vibrios frequently. Kelly (described in Farmer et al., 1985d) found that *Vibrio* isolates obtained on TCBS medium were always obtained on other media also, so long as oxidase-positive colonies were screened.

Alkaline peptone water is probably the most useful liquid medium for isolating and growing vibrios (Colwell 1984; Furniss et al., 1978). This medium is used as an enrichment for *Vibrio cholerae* and other *Vibrio* species, which often grow better than un-

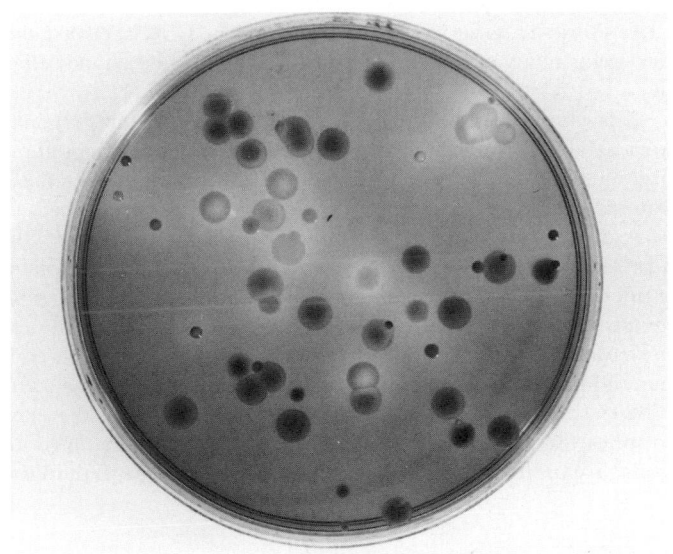

FIGURE BXII.γ.166. Growth of vibrios on TCBS agar. 0.01 ml of seawater collected about 10 m from shore was spread on the agar surface and incubated for 3 days at ambient temperature (about 25°C). Note the many different sizes and shapes of the colonies, which also very considerably in color depending on whether they ferment sucrose. (Reproduced with permission from J.J. Farmer III and F.W. Hickman-Brenner, *In* Balows, Trüper, Dworkin, Harder and Schleifer (Editors), The Prokaryotes: A Handbook on the Biology of Bacteria: Ecophysiology, Isolation, Identification, Applications, 2nd Edition, Volume III, ©Springer-Verlag, New York, 1992, pp. 2952–3011.)

desired organisms at high pH. Vibrios also tend to grow better than most other organisms at the meniscus. Enrichment in alkaline peptone water is usually followed by spreading onto TCBS

TABLE BXII.γ.160. Growth of *Vibrio* species on TCBS agar*

	Colony appearance on TCBS (%)		Growth-plating efficiency
Organism	Green	Yellow	
V. cholerae	0[a]	100[a]	Good
V. alginolyticus	0	100	Good
V. cincinnatiensis	0	100	Very poor
V. damsela	95	5	Reduced at 36°C
V. fluvialis	0	100	Good
V. furnissii	0	100	Good
V. harveyi	0	100	Good
V. hollisae	100	0	Very poor
V. metschnikovii	0	100	May be reduced
V. mimicus	100	0	Good
V. parahaemolyticus	99	1	Good
V. vulnificus	90[b]	10[b]	Good
"Marine vibrios"	Variable	Variable	Variable

[a]Percentage of strains that produce green colonies and yellow colonies, respectively.

[b]The original report describing this species gave the percentage positive for sucrose fermentation as 3%. At the CDC *Vibrio* laboratory, about 15% of the strains have been sucrose positive.

Editorial Note: Photobacterium damselae subsp. *damselae* is a junior objective synonym of *Vibrio damsela.*

agar or onto a similar plating medium selective for *Vibrio*. The NaCl content of alkaline peptone water is not standardized but is usually 0.5–1%. It can be increased to 2% to enhance the growth of marine vibrios. If the NaCl is omitted ("alkaline peptone water-saltless"), the medium becomes much more selective for *V. cholerae* and *V. mimicus*. The type of peptone used in the medium has also varied depending on several factors including local availability. The formula given below (Formula 1494[2] of the CDC Vibrio Laboratory) is the one used by CDC's *Vibrio* Laboratory for many years. Other formulations may be equally effective.

It is often desirable to use a growth or suspending medium with all of the inorganic constituents of seawater, but without organic compounds or toxic contaminants found in natural sea water. Many artificial types of seawater have been described.

Artificial Sea Water #1 (MacLeod's formula)[3] does not contain several of the ingredients found in formulas for "complete" artificial seawater. Some of the missing components are trace metals. This medium is satisfactory for most strains of *Vibrio* and *Photobacterium*. Artificial Sea Water #2, "Instant Ocean"[4], is a commercial packaged salt mixture available from Aquarium Systems (33208 Lakeland Blvd., Eastlake, Ohio 44094). Stores that have supplies for marine aquariums also sell it. Marine cation solutions[5] are useful alternatives to artificial sea waters, and the *Vibrio*

laboratory uses them for making "marine media" used for biochemical testing.

Standard techniques used in clinical microbiology laboratories are suitable for isolating *Vibrio* species from human clinical specimens. Benenson et al. (1964), Hugh and Sakazaki (1972), Morris et al. (1979), World Health Organization (1983), Wachsmuth (1984), and Barua and Greenough (1992)) should be consulted for specialized information. It is important to note that some *Vibrio* cultures do not grow readily on the selective media used for the isolation of enteric pathogens and that most *Vibrio* species require added NaCl for growth (Baumann et al., 1984b; Fig. BXII.γ.164). Many laboratory media have amounts of Na$^+$ less than 0.5% NaCl.

There are no special procedures for *Vibrio* in regard to the collection, transport, and storage of specimens. Fecal specimens should be obtained within 24 h of the onset of illness and prior to antibiotic treatment. Specimens should be spread on an isolation medium as soon as possible. Vibrios survive longer in the alkaline rice-water stool of cholera patients than in formed stools. *Vibrio* strains are easily killed by desiccation. Samples can be transported in alkaline peptone water or in Cary Blair transport medium, which maintains viability of *Vibrio* cultures for up to 4 weeks. "Buffered glycerol–saline" medium is unsatisfactory. Tellurite–taurocholate–peptone broth has also been used as a transport medium. *Vibrio* strains can survive up to 5 weeks on strips of blotting paper soaked in liquid stool and stored at ambient temperature in airtight plastic bags.

Medically important *Vibrio* strains from fecal specimens grow readily on blood agar. Most strains of *V. cholerae* have a characteristic appearance on blood agar (Fig. BXII.γ.167) and are strongly hemolytic, except for strains of the classical biogroup of *V. cholerae*, which are nonhemolytic. Inclusion of oxidase testing and/or inclusion of TCBS medium in the isolation scheme are recommended for more thorough searches.

Medically important *Vibrio* strains from fecal specimens can grow on blood agar where they may be beta-hemolytic (*V. cholerae* non-O1 and some *V. cholerae* O1 El Tor), alpha-hemolytic (*V. vulnificus* and many others), or nonhemolytic. Vibrios also usually grow on MacConkey agar and produce colorless (lactose-negative) colonies. Colonies grown on blood agar and lactose-negative colonies grown on MacConkey medium should be tested for oxidase activity.

2. Formula 1494 for alkaline peptone water: dissolve peptone (10 g) and sodium chloride (5 g) in 994 ml water. Adjust pH to 8.4 with 1N NaOH (about 6 ml will be required). Dispense into tubes and autoclave at 121°C for 15 min. The final medium will be clear and amber colored.

3. Artificial Sea Water #1, MacLeod's formula, contains (g/l) NaCl (23.38 g), MgSO$_4$·7H$_2$O (24.65 g), KCl (1.49 g), and CaCl$_2$·2H$_2$O (2.94 g). A clear, colorless solution with a pH of about 6.6 will result. Dispense into screw cap bottles or tubes and autoclave at 121°C for 15 min. Store at room temperature.

4. Artificial Sea Water #2, "Instant Ocean", is prepared by adding 40 g Instant Ocean to 1 liter of water and stirring until all the salt dissolves. Dispense and autoclave at 121°C for 15 min. Store at room temperature.

5. Marine cations supplement 1558 (Na$^+$, K$^+$, Mg^{2+}, and Ca^{2+} (10 X) was modified by the CDC *Vibrio* Laboratory from the "electrolyte supplement" of Furniss et al. (1978)). In 912 ml water, dissolve (in the order listed) NaCl (150 g), KCl (3.7 g), MgCl$_2$·6H$_2$O (51 g), and CaCl$_2$·2H$_2$O (7.4 g). Aseptically add one volume to 9 volumes of sterile medium to be supplemented and mix thoroughly. Marine cations supplement 1559 (Na$^+$, K$^+$, and Mg^{2+} (10X)) is the same as marine cations supplement 1558 except that it lacks the calcium chloride.

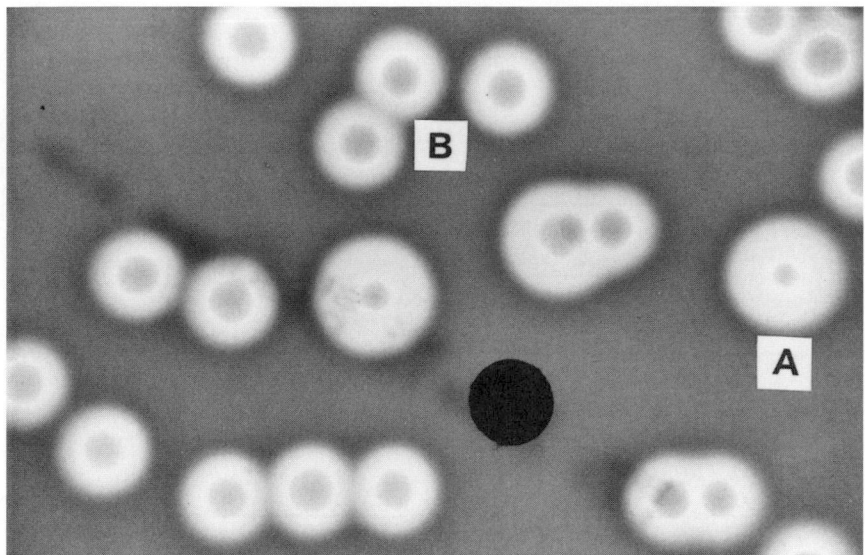

FIGURE BXII.γ.167. Strong hemolytic reaction of the Gulf Coast strain of *Vibrio cholerae* incubated overnight at 36°C on sheep blood agar. Note that the small colony type has a much larger zone of hemolysis *(A)* than the large colony type *(B)*. (Reproduced with permission from J.J. Farmer III and F.W. Hickman-Brenner, *In* Balows, Trüper, Dworkin, Harder and Schleifer (Editors), The Prokaryotes: A Handbook on the Biology of Bacteria: Ecophysiology, Isolation, Identification, Applications, 2nd Edition, Volume III, ©Springer-Verlag, New York, 1992, pp. 2952–3011.)

Fig. BXII.γ.168 gives a plan for isolating vibrios. Simply testing the oxidase reaction of hemolytic colonies on sheep blood agar plates will detect many *Vibrio* strains as well as many *Aeromonas* strains. Nonhemolytic colonies can also be tested. Procedures for oxidase testing are given in Farmer and Hickman-Brenner (1992).

A protocol adapted from Wachsmuth (1980) and Furniss et al. (1978) for the isolation of *V. cholerae* from stool specimens is shown in Fig. BXII.γ.168. This protocol is also applicable to the isolation of vibrios from other specimens. The use of TCBS agar as the primary plating medium and use of an enrichment step in alkaline peptone water would also enhance recovery of vibrios.

Rapid detection of *Vibrio cholerae* O1 in feces can be accomplished by using a latex bead agglutination test that detects the O1 antigen of *V. cholerae*. Antisera to *V. cholerae* O139 could also be used. A confirmatory culture is needed for the definitive diagnosis of cholera.

Direct detection of *Vibrio* species in feces can be accomplished by microscopic examination. In 1883, Koch described small curved rods that could be observed in the stools of cholera patients but not in the feces of patients without cholera. Later it was observed that sera from convalescent cholera patients stopped the characteristic motility of *V. cholerae* in rice water stools. Later still, antiserum produced in animals was used in this "microscopic vibriocidal assay". The cellular morphology of *V. cholerae* cells varies and typical curved rods, straight rods, short noncurved rods, and "involution forms" can all be seen in the same culture.

Methods for the enrichment of vibrios from seawater have been described by Baumann et al. (1984b). Many of the marine species can be obtained from enrichments involving the addition of 500-ml seawater samples to sterile 2-liter Erlenmeyer flasks containing 25 ml 1 M Tris-HCl (pH 7.5), 0.5 g NH_4Cl, 0.38 g $K_2HPO_4 \cdot 7H_2O$, 14 mg $FeSO_4 \cdot 7H_2O$, and 0.5 g or 0.5 ml of one of a variety of different organic carbon and energy sources. Ma-

terial from flasks with visible turbidity (generally within 10 days at 20–25°C) is spread onto basal medium agar (BMA; Baumann and Baumann, 1981a) containing 0.1–0.2% (w/v for solids and v/v for liquids) of the homologous carbon and energy source. Suitable organic compounds include sugars, sugar alcohols, sugar acids, tricarboxylic acid cycle intermediates, fatty acids, and amino acids. Strains of marine vibrios have also been obtained by direct isolation from seawater. Seawater is passed through nitrocellulose filters having a pore size of 0.22 or 0.45 μm; the filters are then placed on BMA containing 0.1 or 0.2% of an appropriate carbon and energy source or on yeast extract agar or marine agar. After 2–10 days incubation at 20–25°C, colonies are transferred to homologous medium. Vibrios have also been isolated by plating onto nonselective seawater-based media. For additional information on the isolation of marine vibrios, see Baumann and Baumann (1981a) and Baumann et al. (1984b).

Some strains of *Vibrio* swarm on complex solid media, and isolated colonies cannot be obtained when this occurs. Swarming can be inhibited or greatly reduced by increasing the concentration of agar in complex media to 4% (w/v) (Reichelt and Baumann, 1973b) or by streaking the organism onto a minimal medium such as BMA containing 0.2% (v/v) glycerol (Baumann et al., 1971a).

The isolation of bioluminescent strains of *Vibrio* from seawater does not involve enrichment procedures. Instead, samples are placed directly onto a suitable complex medium and luminescence is detected visually. A solid medium (luminous medium or LM) that promotes luminescence of marine isolates consists of BMA modified by the addition of 0.3% (v/v) glycerol and (in g/l) 5.0 g yeast extract, 5.0 g tryptone, 1.0 g $CaCO_3$, and 20.0 g agar). Because there is strain variation with respect to the intensity and duration of light emission and because some strains will only luminesce as isolated colonies, it is best to examine cultures periodically within 12–36 h of inoculation. A suitable incubation temperature is 15°C; many strains do not produce luminescence

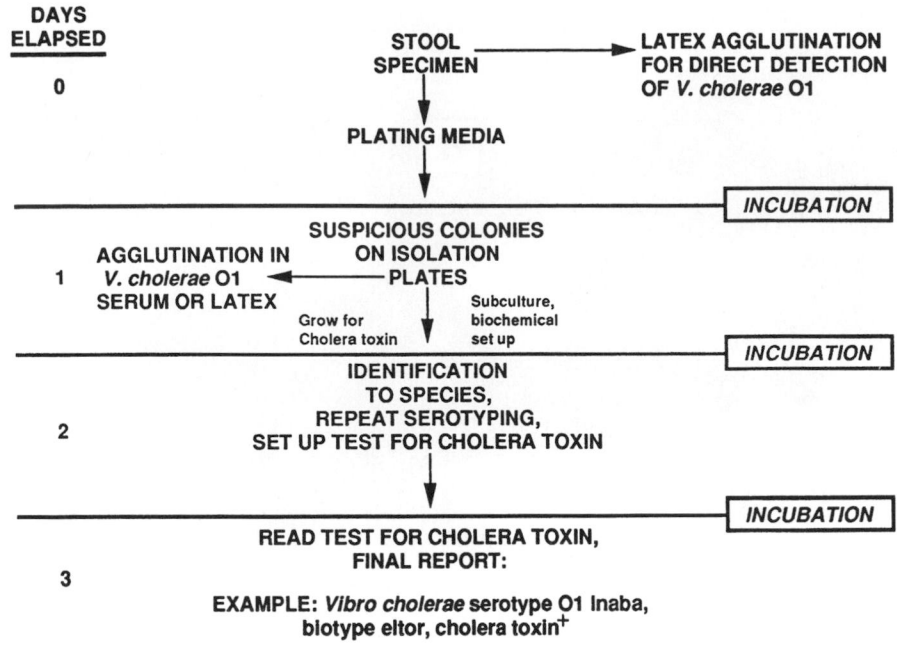

FIGURE BXII.γ.168. Overall plan and special steps for the isolation and identification of *Vibrio cholerae* and other *Vibrio* cultures from feces. (Reproduced with permission from J.J. Farmer III and F.W. Hickman-Brenner, *In* Balows, Trüper, Dworkin, Harder and Schleifer (Editors), The Prokaryotes: A Handbook on the Biology of Bacteria: Ecophysiology, Isolation, Identification, Applications, 2nd Edition, Volume III, ©Springer-Verlag, New York, 1992, pp. 2952–3011.)

at 20°C or 25°C. Sources of luminous isolates include seawater, the surfaces and intestinal contents of marine animals, luminous organs and, more rarely, freshwater sources. Additional details for the isolation of luminous bacteria are given by Cosenza and Buck (1966), Baumann and Baumann (1981a), Hastings and Nealson (1981), Ruby et al. (1980), and Desmarchelier and Reichelt (1981).

MAINTENANCE PROCEDURES

The following procedures have been used for many years in the Enteric Reference Laboratory, US Centers for Disease Control and Prevention. The methods of two other laboratories with many years of experience in *Vibrio* work have been described in considerable detail and are highly recommended (Furniss et al., 1978; Baumann et al., 1984b).

Strains of *Vibrio* are easy to grow and maintain compared to many other organisms. However, strains will die if allowed to dry on agar plates, in agar slants, or in agar "deeps." Freezing and storage at low temperature is the method of choice for long-term preservation and maintenance. Lyophilization and storage at 4°C are used routinely by the American Type Culture Collection with good results.

If liquid nitrogen storage is not available, *Vibrio* cultures can be stored in a laboratory freezer at −70°C, or even at −10 to −20°C. The lowest temperature available should be used for storage. The culture to be preserved is first inoculated heavily and grown on a marine agar plate for at least 24 h. Cultures of *V. cholerae* and *V. mimicus* can be grown on sheep blood agar instead of marine agar because of their low salt requirement. Growth is removed with a cotton swab and a dense suspension is made in 10% skim milk or other suspending medium. One ml of this suspension is then placed in a sterile plastic freezer vial. The vial is frozen in an alcohol bath that has been kept in the freezer. The vial is stored a divided box that holds 81–100 vials and kept at the lowest temperature available.

For preparation of a "working stock" culture, growth from the top of a single, isolated colony is touched and stabbed to the bottom of 13 × 100-mm screw cap tube of marine semisolid medium. This medium can also be used for *V. cholerae* and *V. mimicus*. Marine semisolid medium is used for the halophilic vibrios. The culture is grown overnight and checked to insure good growth. The tube is sealed by closing the cap tightly. To avoid the problem of drying, two alternative methods are used. First, a small volume (about 0.3 ml) of sterile mineral oil can be added to cover the surface to a depth of about 3 mm above the interface. A second sealing method is to insert a number "000" white rubber stopper into the neck of the screw-cap tube to seal it. After they are sealed, cultures are conveniently stored at room temperature in the dark. Many, but not necessarily all, species of *Vibrio* survive for years in such sealed tubes.

PROCEDURES FOR TESTING SPECIAL CHARACTERS

Many different procedures have been described for testing the phenotypic characteristics of *Vibrio* strains. The three most important variables are (1) medium composition, (2) addition of Na^+, K^+, Mg^+, and other cations as supplements, and (3) the temperature of incubation. Most clinical and public health laboratories use media and tests designed to identify *Enterobacteriaceae* for *Vibrio* strains. These methods are successful for *V. cholerae* and *V. mimicus* because these species do not require added Na^+. However, most halophilic vibrios require additional Na^+ for growth and expression of various metabolic traits (Fig. BXII.γ.164). In Table BXII.γ.161, results obtained using standard enteric test media are listed first and the percentage of strains positive for each test is given for each species; these data are followed by the percentage of strains positive for the same strains

TABLE BXII.γ.161. The effect of media (especially Na% content) and test conditions on biochemical test results for 10 species of *Vibrio**

Test or property	Percentage positive for:[a]									
	V. cholerae	*V. alginolyticus*	*V. damsela*	*V. fluvialis*	*V. furnissii*	*V. hollisae*	*V. metschnikovii*	*V. mimicus*	*V. parahaemolyticus*	*V. vulnificus*
Indole production:										
Peptone water	86	14	0	0	0	30	22	94	35	39
Peptone water + 1% NaCl	86	24	0	0	0	93	26	98	80	65
Heart infusion + 1% NaCl	97	42	0	13	11	97	17	95	89	94
Methyl red:										
Standard	25	NG	NG	0	0	NG	17	14	NG	NG
Standard + 1% NaCl	99	77	100	96	100	0	96	99	78	79
Voges–Proskauer:										
Standard	74	NG	NG	0	0	NG	26	2	NG	NG
Standard + 1% NaCl	73	8	33	0	0	0	24	1	0	0
Standard + 1% NaCl; Barrit method	93	83	95	0	0	0	96	9	0	0
Arginine dihydrolase:										
Moellers	0	0	81	52	0	0	0	0	0	0
Moellers + 1% NaCl	0	0	95	93	100	0	59	0	0	0
Lysine decarboxylase:										
Moellers	99+	29	0	0	0	0	13	99	63	85
Moellers + 1% NaCl	99+	99	52	0	0	0	36	100	100	99
Ornithine decarboxylase:										
Moellers	99+	3	0	0	0	0	0	99	71	47
Moellers + 1% NaCl	99+	53	0	0	0	0	0	99	89	53
Gelatin hydrolysis:										
Standard	52	45	5	41	56	0	65	60	66	55
Standard + 1% NaCl	62	76	6	85	86	0	38	63	89	75
Esculin hydrolysis:										
Standard	0	0	0	0	0	0	4	0	0	0
Standard + 1% NaCl	2	2	0	8	0	0	59	0	1	39
Nitrate reduction to nitrite:										
Standard	99+	NG	NG	59	33	NG	0	99+	0	2
Standard + 1% NaCl	99+	100	100	100	100	100	0	100	100	100

[a]The number indicates the percentage positive after 48 h of incubation at 36°C; NG, most cultures do not grow in the medium.

Editorial Note: Photobacterium damselae subsp. *damselae* is a junior objective synonym of *Vibrio damsela.*

tested in a modified medium with 1% NaCl added. The table also shows other effects of different test conditions; for example, more *Vibrio* species are indole positive when grown in heart infusion broth than when grown in peptone water. More strains are positive for the Voges–Proskauer test when the reagent for detecting acetylmethylcarbinol contains α-naphthol (Barritt method).

Bioluminescence test A bioluminescence test has been described by Baumann et al. (1984b). Several *Vibrio* and *Photobacterium* species usually produce bioluminescence (*V. fischeri*, *V. logei*, *V. orientalis*, *V. splendidus* biogroup 1, *Photobacterium phosphoreum*, and *P. leiognathi*), in contrast to most other *Vibrio* species. Some strains of *V. harveyi* and rare strains of *V. cholerae* and *V. vulnificus* are luminescent (Oliver et al., 1986). Details concerning growth conditions and observation of cultures for bioluminescence have been described by Farmer and Hickman-Brenner (1992).

Carbon assimilation tests Methods for carbon assimilation tests were described by Baumann and Baumann (1981a), who used artificial sea water #1 as a basal medium in their extensive studies to determine the growth of marine bacteria on different substrates. The carbon sources are added to the basal medium at a final concentration of 0.1 to 0.2%.

Susceptibility to vibriostatic compound O/129 Susceptibility to 2,4-diamino-6,7-diisopropyl-pteridine phosphate (vibriostatic compound O/129) is used to differentiate *Vibrio* isolates (usually susceptible) from *Aeromonas* isolates (usually resistant) by means of a disc diffusion assay. This test can also be used to differentiate some *Vibrio* species from each other. Details concerning growth

conditions and interpretation of results have been given by Farmer and Hickman-Brenner (1992).

Salt requirement (Na⁺ requirement) and tolerance Salt requirement (Na⁺ requirement) and tolerance form the basis of useful differential tests for the species of *Vibrio* and *Photobacterium*. In contrast to the halophilic *Vibrio* species, *V. cholerae* and *V. mimicus* are able to grow in 0% NaCl. The data are based on the media and methods given by Farmer and Hickman-Brenner (1992); use of other media may lead to different results.

String test Cells grown on agar medium are suspended in 0.5% sodium deoxycholate. Lysis of the cells is shown by the formation of "strings" of DNA that can be drawn from the suspension with a loop. Isolates that lyse within 60 sec are considered positive. Many *Vibrio* species are positive; *Aeromonas hydrophila* is negative.

Serotyping Serotyping of *Vibrio* was discussed in the section on antigenic structure. Simonson and Siebeling (1988) developed a latex slide agglutination test based on antisera to the flagellin portion of the flagella for the routine identification of several of the *Vibrio* species. *V. cholerae* O1 and O139 antisera are used to confirm isolation of *V. cholerae* O1 or O139 for reporting to health authorities.

Antibiotic susceptibility Clinically important *Vibrio* species can be tested for antibiotic susceptibility using a disk diffusion susceptibility test on Mueller–Hinton agar. However, some marine vibrio species grow poorly on Mueller–Hinton agar, presumably because of their higher requirement for Na⁺. Broth dilution susceptibility tests for most *Vibrio* species can usually be done in unmodified Mueller–Hinton broth, because the broth contains sufficient Na$^+$ (Hollis et al., 1976).

Antibiotic resistance is relatively rare in *Vibrio* species (Table BXII.γ.156). The term "resistant" is used with the note that the zone size breakpoints that define resistance and susceptibility have been based primarily on isolates of *Enterobacteriaceae* rather than *Vibrio*. In most cases, "resistance" appears to be an intrinsic property of the species rather than a trait acquired through plasmid transfer or mutation. An exception to this generalization is the antibiotic resistance found in some outbreak or imported strains of *V. cholerae* O1 that have become resistant through the acquisition of R factors. Resistance to polymyxin B and colistin can be useful in identifying *V. cholerae* El Tor and *V. vulnificus*. Most other *Vibrio* species are more susceptible.

DIFFERENTIATION OF THE GENUS *VIBRIO* FROM OTHER GENERA

In this section of the last edition of this *Manual*, Baumann et al. (1984b) stated: "While the number of readily determinable diagnostic traits between these genera is limited, there is little difficulty in their differentiation since the constituent species are generally well defined." We believe that this statement is no longer true and that it is now very difficult to write operational definitions for the genera *Vibrio* and *Photobacterium* that apply to all the species included, and to devise a useful but accurate differential table.

Table BXII.γ.153 in the chapter on *Vibrionaceae* lists the characteristics that distinguish *Vibrio* from *Photobacterium*, *Aeromonas*, *Plesiomonas*, and *Enterobacteriaceae*. However, the reader is warned that some of the *Vibrio* and *Photobacterium* species that have been named recently are incompletely described, and some characteristics have not been determined for all species. Additional data and the eventual subdivision of these two genera into more logical phylogenetic units should make it easier to devise an accurate differential table. Full descriptions and additional information on *Photobacterium*, *Moritella* and *Salinivibrio* can be found in the chapters describing those genera.

TAXONOMIC COMMENTS

Over the years there has been much speculation on the phylogeny of the genus *Vibrio*. However, only recently has it been possible to do systematic studies that operationally define the evolutionary relatedness of the species in *Vibrio*, *Photobacterium*, and related genera. This topic was introduced in the chapter describing the family *Vibrionaceae*. Many phylogenetic analyses based on the 16S rDNA sequences of the genera *Vibrio* and *Photobacterium* and their relatives have appeared in recent years (Aznar et al., 1994; Cerdà-Cuéllar et al., 1997; Raguénès et al., 1997a; Lambert et al., 1998; Nogi et al., 1998a, b; Sawabe et al., 1998b; Urakawa et al., 1998, 1999a; Yumoto et al., 1999a).

The concept of the genus *Vibrio* itself has changed drastically over the years. For many years organisms were classified in *Vibrio* based on a single morphological property, "curved rods". However, over the years many of these species have been reclassified in other genera with closer phylogenetic relatives. For example, "*Vibrio fetus*" (Smith and Taylor 1919, 301) is now classified in the genus *Campylobacter* (Sebald and Véron, 1963) as *C. fetus* and is the type species for that genus of curved rods, which differ from true *Vibrio* species in being microaerophilic. Similarly, the pink-pigmented methanol-oxidizing bacterium "*Vibrio extorquens*" (Stocks and McCleskey 1964) is now usually classified with other methanol-oxidizing bacteria in the genus *Methylobacterium*. Another species *Vibrio succinogenes* is a microaerophile and is now

the type species of the genus *Wolinella*. Today, the trend is to include the term "vibrio" as part of a new genus name to indicate morphologies that include "comma-shaped", "curved", or "slightly curved". Examples include *Bdellovibrio*, *Cellvibrio*, *Halovibrio*, *Micavibrio*, *Vampirovibrio*, *Acetivibrio*, *Anaerovibrio*, *Butyrivibrio*, *Succinivibrio*, *Desulfovibrio*, "*Nitrosovibrio*", and *Falcivibrio*. These organisms have many different types of metabolism and are found throughout the phylogenetic tree. Clearly, the circumscription of the genus *Vibrio* has evolved over the decades.

Three techniques introduced in the 1960s and 1970s greatly aided our understanding of the phylogeny of the genus *Vibrio*: DNA–DNA hybridization, rRNA–DNA hybridization, and the immunological relationship of enzymes and other proteins (Farmer and Hickman-Brenner, 1992; Baumann et al., 1984b). There is good agreement on many aspects of the classification and phylogeny of *Vibrio*, and the main problem today is how it should be subdivided. Before this issue is discussed, it is helpful to consider the relationships among the species of the genus *Vibrio*.

rRNA–DNA hybridization studies by Baumann and Baumann (1981a), Baumann et al. (1984b), and Fox et al. (1980) of the relationships among *Vibrio*, *Photobacterium*, *Aeromonas*, *Plesiomonas*, and *Enterobacteriaceae* indicated that the genus most closely related to *Vibrio* was *Photobacterium*, and that species of the genera *Pasteurella* and *Aeromonas*, and of the family *Enterobacteriaceae* are more distantly related.

Fig. BXII.γ.169 in this chapter shows that *Vibrio cholerae* is highly related to *Vibrio mimicus* by DNA–DNA hybridization but is much less closely related to the other *Vibrio* species. All other techniques agree that *V. cholerae* and *V. mimicus* are extremely closely related, except that the results of one study that employed 5S rRNA sequencing suggested that *V. cholerae* and *V. mimicus* are separated by a considerable distance and that other *Vibrio* species are closer relatives of *V. cholerae* than *V. mimicus* MacDonell and Colwell (1985).

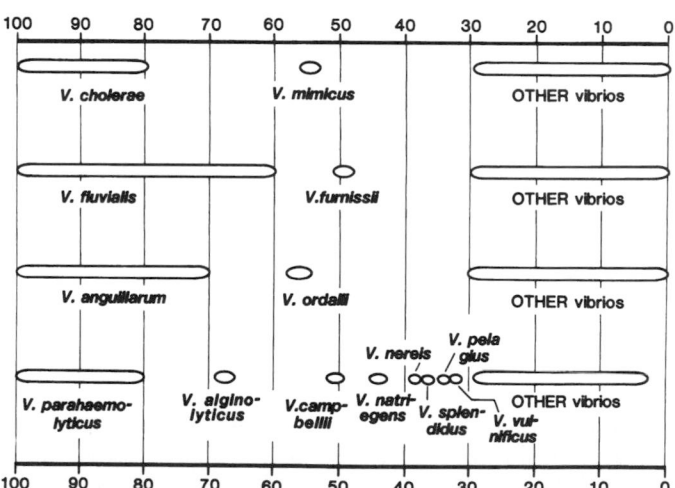

% RELATEDNESS BY DNA – DNA HYBRIDIZATION

FIGURE BXII.γ.169. Relatedness of several *Vibrio* species by DNA–DNA hybridization. (Conditions: hydroxyapatite, ^{32}P, 60°C). The figure was redrawn from published and unpublished data summarized by Brenner et al. (1983b). (Reproduced with permission from J.J. Farmer III and F.W. Hickman-Brenner, *In* Balows, Trüper, Dworkin, Harder and Schleifer (Editors), The Prokaryotes: A Handbook on the Biology of Bacteria: Ecophysiology, Isolation, Identification, Applications, 2nd Edition, Volume III, ©Springer-Verlag, New York, 1992, pp. 2952–3011.)

Most of the phylogenetic analyses of the genus *Vibrio* currently being published are based solely on 16S rRNA gene sequences. However, phylogenetic trees rarely agree with each other *in toto*. Inconsistencies can be due to a number of factors, such as failure to use type strains; use of incorrectly identified or mislabeled strains; use of incorrect sequences, including ones obtained from databases; use of incorrect sequence alignments; and use of different methods of calculation in data analysis. For all these reasons data from 16S rRNA sequencing should be examined critically before being used as the basis for any taxonomic proposal.

Subdivision of the genus Over the years reclassifications of the organisms included in the genus *Vibrio* have resulted in proposals of at least 11 new genera (Table BXII.γ.162). Six of these have standing in nomenclature: *Allomonas* (Kalina et al., 1984), *Beneckea* (Campbell, 1957), *Listonella* (MacDonell and Colwell, 1985), *Lucibacterium* (Hendrie et al., 1970), *Moritella* (Urakawa et al., 1998), and *Salinivibrio* (Mellado et al., 1996). Five of the proposed genera are older, do not have standing in nomenclature, and so are written in quotation marks. These are *"Agarbacterium"* (Angst, 1929), *"Alginomonas"* (Thjøtta and Kåss, 1945), *"Alginobacter"* (Thjøtta and Kåss, 1945), *"Alginovibrio"* (Thjøtta and Kåss, 1945) and *"Oceanomonas"* (Miyamoto, et al., 1961).

The genus *Allomonas* was proposed by Kalina et al. (1984) and contained oxidase positive, fermentative organisms that had some properties of *Aeromonas* and some of *Vibrio*. Based on DNA–DNA hybridization experiments, the strains did not appear to be closely related to *Aeromonas hydrophila*, *Plesiomonas shigelloides*, *V.*

TABLE BXII.γ.162. Alternative classifications that have been proposed that would subdivide the genus *Vibrio*[a]

Genus	Species
Names with standing in the nomenclature:	
Allomonas	*A. enterica*[T]
Beneckea	*B. alginolytica*
	B. campbellii[T]
	B. gazogenes
	B. harveyi
	B. natriegens
	B. nereis (*B. nereida*)[b]
	B. nigripulchritudo (*B. nigripulchritudo*)
	B. parahaemolytica
	B. pelagia
	B. splendida
	B. vulnifica
Listonella	*L. anguillarum*[T] (*L. anguillara*)
	L. damsela
	L. pelagia
Lucibacterium	*L. harveyi*[T]
Moritella	*M. japonica*
	M. marina[T]
	M. viscosa
	M. yayanosii
Photobacterium[c]	
Salinivibrio	*S. costicola*[T]
	S. costicola subsp. *vallismortis*
Names without standing in nomenclature:	
"Listonella ordalii"	
"Agarbacterium"	
"Alginobacter"	
"Alginomonas"	
"Alginovibrio"	
"Oceanomonas"	

[a]The merits and deficiencies of these classifications are discussed in the text.

[b]Names in parentheses are alternative spellings.

[c]See the chapter describing the genus *Photobacterium*.

cholerae, or eight other species in the family *Vibrionaceae*. However, the type strain of *Allomonas enterica* was later found to be highly related to *V. fluvialis* by DNA hybridization and phenotypic analysis (Subcommittee, 1975).

The genus *Beneckea* was proposed as a new genus in the family *"Achromobacteriaceae"* by Campbell (1957) and was defined as a genus of rod-shaped bacteria that were Gram negative, had peritrichous flagella, fermented carbohydrates, and digested chitin. Baumann et al. (1971a) studied a large group of "marine vibrios" and classified them in a redefined genus *Beneckea* along with *Vibrio alginolyticus* and *Vibrio parahaemolyticus*. However, as additional species of *Beneckea* were described the genus began to become heterogeneous, and Baumann et al. (1980b) concluded that all the *Beneckea* species should be classified in the genus *Vibrio*.

The genus *Listonella* was proposed by MacDonell and Colwell (1985) as new genus of *Vibrionaceae* that included three *Vibrio* species (Table BXII.γ.162): the type species *Listonella anguillarum* (*Vibrio anguillarum*), *Listonella damsela* (*Vibrio damsela*; *Photobacterium damselae*) and *Listonella pelagia* (*Vibrio pelagius*). *Listonella* included organisms fitting the following description: "Curved rods, Gram negative, motile by monotrichous or peritrichous flagella. Chemoorganotrophic, oxidase positive, associated with marine environments. Generally pathogenic for fish or eels. Do not grow at 40°C. Require NaCl for growth. G + C mol% 43–46." *V. anguillarum* is closely related to *V. ordalii* by DNA hybridization (Fig. BXII.γ.169).

The genus *Lucibacterium* was proposed in 1970 by Hendrie et al. (1970); this genus contains a single species, *Lucibacterium harveyi* (= *Vibrio harveyi*). The names *Lucibacterium* and *Lucibacterium harveyi* have standing in nomenclature but are rarely used.

The genus *Moritella* was proposed by Urakawa et al. (1998) based on 16S rRNA gene sequence divergence of 11 strains of *Moritella marina* from *Vibrio cholerae* and other *Vibrio* species. Three additional species, *Moritella japonica* (Nogi et al., 1998a), *Moritella yayanosii* (Nogi and Kato, 1999a), and *M. viscosa* (Benediktsdóttir et al., 2000) have been described.

The genus *Salinivibrio* was proposed by Mellado et al. (1996) based on 16S rRNA gene sequence divergence of six *S. costicola* (= *Vibrio costicola*) strains from *Vibrio cholerae* and other *Vibrio* species; these results were also in agreement with previous phenotypic and genotypic data.

Table BXII.γ.162 provides a summary of previous proposals that subdivided the *Vibrio* as it existed at the time each proposal was made. A successful classification should also be reconcilable with phenotypic data. It should be possible to classify newly discovered and described species into new genera that are limited so as to include only close relatives of the type species. The ideal grouping is illustrated by the pair *Vibrio cholerae* and *Vibrio mimicus* (Fig. BXII.γ.169).

Future of the genus The genus *Vibrio* and many other large heterogeneous genera such as *Pseudomonas* and *Bacillus* will change drastically in circumscription. These genera will eventually be limited to the type species and closely related species. Based on all data available we believe the following changes are likely to occur in the genus *Vibrio*, but it is difficult to determine a time frame for the changes and their acceptance by the scientific community.

The species *Vibrio cholerae* and *Vibrio mimicus* will be the basis of a more limited genus *Vibrio*; any redefinition of the genus will include these two species.

The *Vibrio* species not closely related to *Vibrio cholerae* will form the bases of many new genera, such as the genera *Salinivibrio* and *Moritella*. The figures in this chapter, the chapter describing the family *Vibrionaceae*, and the chapter on *Photobacterium* suggest other possible new genera.

We feel it is logical to include only *Listonella anguillarum* and *V. ordalii* in the genus *Listonella*. These two organisms are closely related by DNA–DNA hybridization (Fig. BXII.γ.169), the immunological relationship of superoxide dismutase (Fig. BXII.γ.170), 16S rDNA sequencing (Fig. BXII.γ.171), and phenotype. *Listonella damsela* and *Listonella pelagia* do not appear to be closely related to *Listonella anguillarum* and *"L. ordalii"*.

Vibrio fluvialis and *Vibrio furnissii* are closely related by DNA–DNA hybridization (Fig. BXII.γ.169), 16S rDNA sequencing (Fig. BXII.γ.171), the immunological relationship of superoxide dismutase (Fig. BXII.γ.170), and phenotype. These two closely related species could be used to define another new genus.

The marine species *Vibrio parahaemolyticus* and *Vibrio alginolyticus* are closely related by DNA–DNA hybridization (Fig. BXII.γ.169), DNA–RNA hybridization, the immunological relationship of superoxide dismutase (Fig. BXII.γ.170), and phenotype. However, data from 16S rRNA gene sequencing (Fig. BXII.γ.171) indicate that *Vibrio parahaemolyticus* is more closely related to other *Vibrio* species than to *Vibrio alginolyticus*. Data from DNA–DNA hybridization studies suggest that *Vibrio campbellii* is the closest relative of this pair. Gradations in digress of relatedness in the "marine vibrio group" will make it difficult to define a new genus.

As originally noted by Baumann and Baumann (1981a) *Vibrio pelagius* biogroups 1 and 2 are closely related and could form the basis of another genus. This close relationship is shown by DNA–DNA hybridization (Brenner et al., unpublished) and by the immunological relationship of superoxide dismutases (Fig. BXII.γ.170).

Vibrio metschnikovii and *Vibrio gazogenes* are unique among *Vibrio* species because they are nitrate negative and oxidase negative. They are distinct from other *Vibrio* species based on the immunological relationship of their superoxide dismutases (Fig. BXII.γ.170) and DNA–DNA hybridization (Fig. BXII.γ.172). In Figure 1 of Cerdà-Cuéllar et al. (1997) they form a tight cluster (B2) based on 16S rDNA sequencing; this cluster also includes *V. cincinnatiensis*. However, *V. metschnikovii* and *V. gazogenes* are less than 30% related by DNA–DNA hybridization (Brenner et al., unpublished). Further data are needed to determine whether *Vibrio metschnikovii* and *Vibrio gazogenes* form the basis of another new genus.

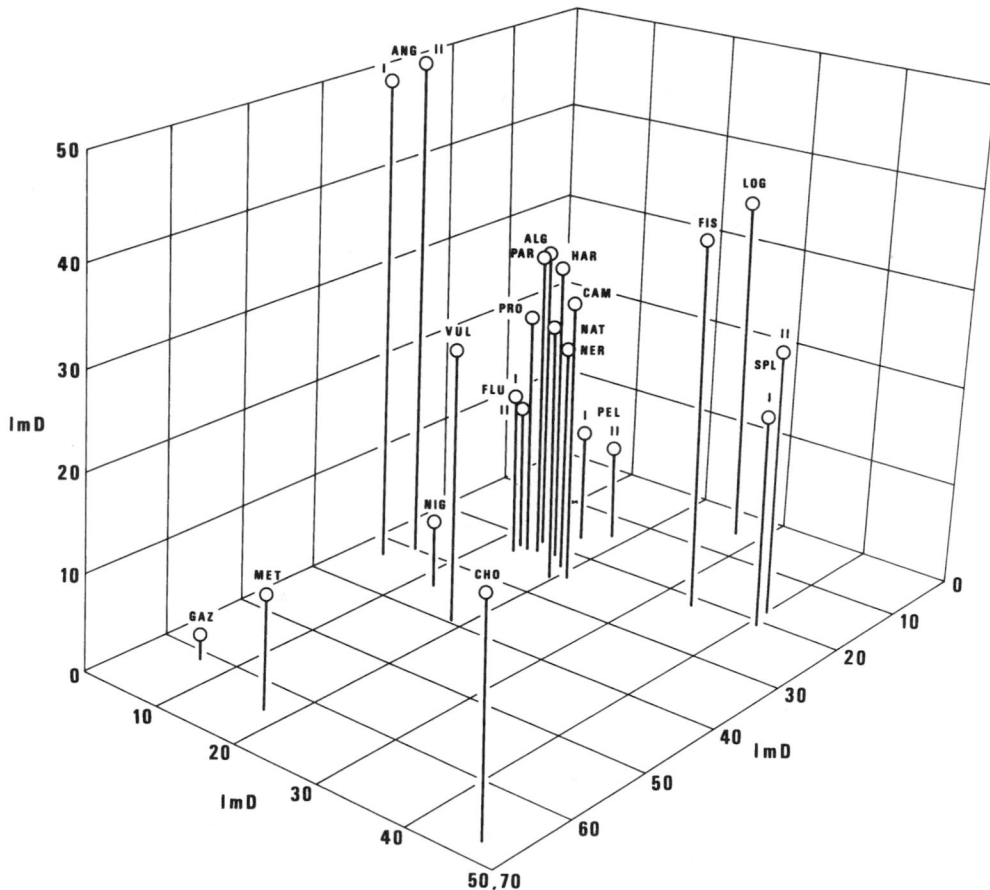

FIGURE BXII.γ.170. Relatedness of *Vibrio* species based on a three-dimensional representation of the immunological relationships of their superoxide dismutase. The spheres represent the relative positions of the species and biogroups. ImD, immunological distance, is a parameter related to percent amino acid sequence difference. *Vibrio* species are designated by the first three letters of their species name. "SPL I" and "SPL II" are *Vibrio splendidus* biogroups 1 and 2. "PEL I" and "PEL II" are *Vibrio pelagius* biogroups 1 and 2. "FLU I" and "FLU II" are *Vibrio fluvialis* biogroups 1 and 2; the latter is now usually classified as *Vibrio furnissii*. "ANG I" and "ANG II" are for *Vibrio anguillarum* biogroups 1 and 2; the latter is now usually classified as *Vibrio ordalii*. Three or four strains from each species or biogroup were used in the analysis; the maximal standard deviation was ± 1.2 ImD. *Salinivibrio costicola* was distant from all other species, and its position is not shown in this figure. (Source: Baumann et al., 1984b; the data were taken from Baumann et al. 1980a, and from S.S. Bang, L. Baumann, and P. Baumann, unpublished).

FIGURE BXII.γ.171. Relatedness of *Vibrio* and related species based on 16S rRNA gene sequences.* The distances in the tree were calculated using 1101 positions (the least-squares method, Jukes-Cantor model). (Courtesy T. Lilburn of the Ribosomal Database Project.)

Editorial Note: Photobacterium damselae subsp. *damselae* is a junior objective synonym of *Vibrio damsela. Vibrio pelagius* and *Vibrio anguillarum* are synonyms of *Listonella pelagia* and *Listonella anguillarum,* respectively.

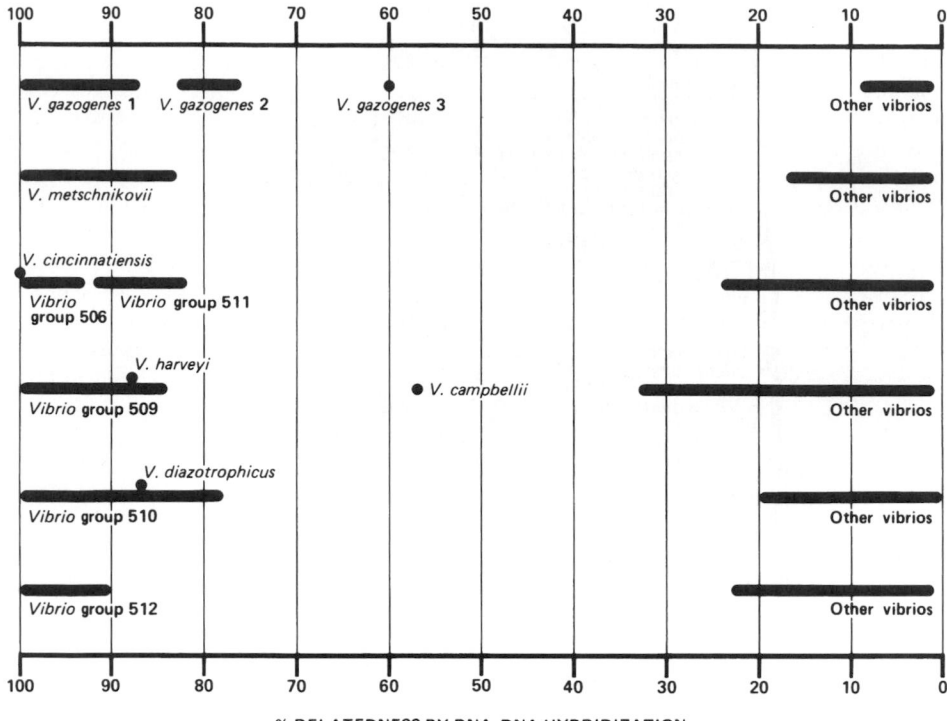

FIGURE BXII.γ.172. DNA relatedness of *Vibrio gazogenes*, *V. metschnikovii*, and five "*Vibrio* groups".

ACKNOWLEDGMENTS

We are grateful to Paul Baumann, Linda Baumann, A.L. Furniss and John Lee and to the Bergey's Manual Trust for allowing us to use some of the data, text, tables, and figures from the *Vibrio* chapter in the first edition of *Bergey's Manual of Systematic Bacteriology*. We acknowledged this in our text with the citation Baumann et al. (1984b). We believe these contributions greatly enhance our chapter. We thank Frances Brenner and Don J. Brenner and their collaborators for the detailed information on the relatedness of all the *Vibrio* species by DNA–DNA hybridization; these data are cited in the text as Brenner et al., unpublished. We thank Ed Ewing, George Garrity, and I. Yumoto for figures; and R.D. Allen, S.S. Bang, W.E. de Boer, C. Golten, K.H. Nealson, J.L. Reichelt, W.A. Schaffers, B.M. Tebo, I. Yumoto and S.W. Watson for figures previously used in the first edition; and Don Brenner for a very helpful review of the manuscript.

We dedicate this chapter to Linda Baumann, Paul Baumann, Don J. Brenner, G. Richard Fanning, J.L. Reichelt, and Arnold J. Steigerwalt (and to their many collaborators). Their pioneering studies using nucleic hybridization methods have formed the basis of our current understanding of the circumscription and classification of the genera *Vibrio* and *Photobacterium* and their relatives.

FURTHER READING

Baumann, P. and L. Baumann. 1977. Biology of the marine enterobacteria: genera *Beneckea* and *Photobacterium*. Annu. Rev. Microbiol. *31*: 39–61.

Baumann, P.. and L. Baumann. 1984. Genus *Photobacterium*. *In* Krieg and Holt (Editors), Bergey's Manual of Systematic Bacteriology, Vol. 1, The Williams & Wilkins Co., Baltimore. pp. 539–545.

Baumann, P. and L. Baumann. 1981. The marine Gram-negative eubacteria. *In* Starr, Stolp, Trüper, Balows and Schlegel (Editors), The Prokaryotes, a Handbook on Habitats, Isolation and Identification of Bacteria, 1st Ed., Springer-Verlag, New York. pp. 1352–1394.

Baumann, P., L. Baumann and M. Mandel. 1971. Taxonomy of marine bacteria: the genus *Beneckea*. J. Bacteriol. *107*: 268–294.

Baumann, P., L. Baumann, M.J. Woolkalis and S.S. Bang. 1983. Evolutionary relationships in *Vibrio* and *Photobacterium*: a basis for a natural classification. Ann. Rev. Microbiol. *37*: 369–398.

Baumann, P., A.L. Furniss and J.V. Lee. 1984. Genus I. *Vibrio*. *In* Krieg and Holt (Editors), Bergey's Manual of Systematic Bacteriology, 1st Ed., Vol. 1, The Williams & Wilkins Co., Baltimore. pp. 518–538.

Blake, P.A.. , R.E. Weaver and D.G. Hollis. 1980. Diseases of humans (other than cholera) caused by vibrios. Annu. Rev. Microbiol. *34*: 341–367.

Centers for Disease Control and Prevention 1999. Laboratory Methods for the Diagnosis of Epidemic Dysentery and Cholera, Centers for Disease Control and Prevention, Atlanta, Georgia..

Colwell, R.R. (Editor). 1984. Vibrios in the Environment, John Wiley and Sons, New York. 634 pp.

Farmer, J.J. and F.W. Hickman-Brenner. 1992. The genera *Vibrio* and *Photobacterium*. *In* Balows, Trüper, Dworkin, Harder and Schleifer (Editors), The Prokaryotes. A Handbook on the Biology of Bacteria: Ecophysiology, Isolation, Identification, Applications, 2nd Ed., Vol. 3, Springer-Verlag, New York. pp. 2952–3011.

Farmer, J.J., F.W. Hickman-Brenner and M.T. Kelly. 1985. *Vibrio*. *In* Lemmette, Balows, Hausler and Shadomy (Editors), Manual of Clinical Microbiology, American Society for Microbiology, Washington, D.C. pp. 282–310.

Fujino, T., G. Sakaguchi, R. Sakazaki and Y. Takeda. 1974. International Symposium on *Vibrio parahaemolyticus*, Saikon Publishing, Tokyo.

Furniss, A.L., J.V. Lee and T.J. Donovan. 1978. The Vibrios. Public Health Laboratory Service Monograph Series, Her Majesty's Stationery Office, London.

Hastings, J.W. and K.H. Nealson. 1981. The symbiotic luminous bacteria. *In* Starr, Stolp, Trüper, Balows and Schlegel (Editors), The Prokaryotes, a Handbook on Habitats, Isolation and Identification of Bacteria, Springer-Verlag, New York. pp. 1332–1345.

Joseph, S.W., R.R. Colwell and J.B. Kaper. 1982. *Vibrio parahaemolyticus* and related halophilic Vibrios. Crit. Rev. Microbiol. *10*: 77–124.

Miwatani, T. and Y. Takeda. 1976. *Vibrio parahaemolyticus*: a causative bacterium of food poisoning, Saikon Publishing, Tokyo.

DIFFERENTIATION OF THE SPECIES OF THE GENUS *VIBRIO**

Tables BXII.γ.163 and BXII.γ.164 summarize the characteristics of *Vibrio* spp. and include data on *Salinivibrio, Photobacterium,* and *Moritella* spp. for purposes of comparison. Phenotypic characteristics useful for the differentiation of the 12 *Vibrio* species that occur in human clinical specimens are given in Tables BXII.γ.156, BXII.γ.165, and BXII.γ.166. Table BXII.γ.161 gives data on the effects of such factors as basal medium, NaCl concentration, and test method on the results of nine biochemical tests for ten *Vibrio* species.

Species-specific probes have been developed for a number of vibrios. A 17-base pair oligonucleotide probe based on 5S rRNA sequence data is highly specific for the fish pathogens *V. anguillarum* and *V. ordalii* (Ito et al., 1995). However, the limit of detection is 1.5×10^5 viable cells. 16S rRNA gene sequences have also been used to detect *V. anguillarum* (Martínez-Picado et al., 1994). A number of probes designed to detect virulence-associated genes have been described for pathogenic vibrios. A 3.2-Kb probe containing the structural gene for the cytotoxin-hemolysin of *V. vulnificus* is species-specific (Morris et al., 1987). It is highly sensitive (60 viable cells can be detected) and detects

all three *Vibrio vulnificus* biogroups. A 0.69-kb probe for the dam-selysin gene reacts with all hemolytic strains of *Vibrio damsela* but not with nonhemolytic strains or with hemolytic isolates of other vibrios such as *V. cholerae* and *V. fluvialis* (Cutter and Kreger, 1990). A 19-base-pair oligonucleotide probe has been developed that spans an 11-base-pair deletion in the *hlyA* gene in classical biotype strains of *V. cholerae* (Alm and Manning, 1990). This probe detects all O1 El Tor and non-O1 *V. cholerae* strains but not strains of the classical biotype. Two DNA probes (100 and 1000 bp) specific for the chromosomal region that encodes genes for the synthesis of the lipopolysaccharide side chain region of all *V. cholerae* O139 strains detects >99% of O139 strains but not those belonging to other serogroups or species (Nair et al., 1995).

Molecular techniques have also been used to subtype strains of *V. cholerae, V. parahaemolyticus,* and some environmental species such as *V. ordalii.* These techniques include pulsed-field gel electrophoresis, restriction fragment length polymorphism, ribotyping, multilocus enzyme electrophoresis, and arbitrarily primed polymerase chain reaction (Koblavi et al., 1990; Faruque et al., 1994; Mahalingam et al., 1994; Evins et al., 1995; Popovic et al., 1995; Kurazono et al., 1996; Pedersen et al., 1996b, c; Okuda et al., 1997).

---

*Editorial Note: Photobacterium damselae subsp. damselae is a junior objective synonym of Vibrio damsela.

List of species of the genus Vibrio

1. **Vibrio cholerae** Pacini 1854, 411[AL]

   chol' er.ae. Gr. n. *cholera* cholera; M.L. gen. n. *cholerae* of cholera, a severe diarrheal disease.

   The characteristics are as described for the genus and as listed in Tables BXII.γ.163 and BXII.γ.164. Fig. BXII.γ.158 shows the polar flagellum; Fig. BXII.γ.173 shows colonies of *V. cholerae* on TCBS agar. The antigenic types of *V. cholerae* are listed in Table BXII.γ.155; the division of *V. cholerae* strains into serogroups is given in Table BXII.γ.167. Characters that differentiate the classical and El Tor biotypes of *V. cholerae* are given in Table BXII.γ.168. Characters that distinguish *V. cholerae* from *V. mimicus* are given in Table BXII.γ.169. Characters that distinguish *V. cholerae* from other *Vibrio* spp. found in human clinical specimens are given in Tables BXII.γ.156, BXII.γ.157, BXII.γ.160, BXII.γ.161, BXII.γ.165 and BXII.γ.166. A few *V. cholerae* strains are bioluminescent, and were formerly classified as *Vibrio albensis* or *Vibrio cholerae* biovar albensis.

   V. cholerae causes a diarrhea that has a wide range of severity and symptoms. In severe cholera (cholera gravis), there is severe diarrhea with rice water stools, devastating dehydration and electrolyte imbalance. There can also be cholera that is less severe and even a mild diarrhea with rapid recovery. Current strains that cause pandemic cholera produce cholera toxin and belong to serogroup O1 and to serogroup O139, which emerged recently. Epidemic strains of *V. cholerae* O1 are usually the El Tor biotype, but in some areas of the world, the classical biotype still occurs. The original El Tor strains hemolyzed sheep erythrocytes strongly, but "pandemic" strains isolated in recent years have been very poorly hemolytic or even nonhemolytic. Nontoxigenic strains of *V. cholerae* O1 may be isolated from cases of diarrhea and from the environment in areas where

   cholera is absent. These strains are usually hemolytic. Fig. BXII.γ.168 shows a procedure for the isolation of *V. cholerae* from feces.

   The mol% G + C of the DNA is: 47–49 (T_m, Bd).

   Type strain: ATCC 14035, CDC 9061-79, NCTC 8021.

   GenBank accession number (16S rRNA): X74695, Z21856.

   Additional Remarks: Reference strain, ATCC 14033 (NCTC 8457, CDC 9061-79), is a representative of the El Tor biotype, and was isolated from pilgrims in the El Tor quarantine camp. It was used in the extensive DNA–DNA hybridization studies of Davis et al. (1981a). It is characterized as: serogroup: O1 Inaba, biogroup: El Tor, pathogen classification: Class III, and is listed as nontoxigenic in the 1996 ATCC Catalog.

2. **Vibrio aerogenes** Shieh, Chen and Chiu 2000, 327[VP]

   a.e.ro' ge.nes. Gr. masc. n. *aer* air; Gr. v. *gennanio* to produce; M. L. adj. *aerogenes* gas-producing.

   The properties of *V. aerogenes* are given in Table BXII.γ.164. Rods or curved rods 0.6–0.8 × 2–3 μm. Motile by double-sheathed polar flagella. Does not swarm. Not bioluminescent. Facultatively anaerobic; produces organic acids, H_2 and CO_2 during fermentation of carbohydrates. No growth factors required. No PHB granules formed. Grows in media containing 1–7% NaCl; does not grow in 1% or 10% NaCl. The major fatty acid is $C_{12:0}$.

   Isolated from sediment in a stand of seagrass, Nanwan Bay, Kenting National Park, Taiwan.

   A 16S rDNA sequence analysis that included 35 *Vibrio* species showed that *V. aerogenes* was most closely related to *V. mytili* (Shieh et al., 2000).

   The mol% G + C of the DNA is: 46

   Type strain: FG1, ATCC 700797, CCRC 17041.

   GenBank accession number (16S rRNA): AF124055.

TABLE BXII.γ.163. Biochemical reactions of the named species and unnamed groups of the genera *Vibrio*, *Moritella*, and *Salinivibrio*.*

Columns V. cholerae through V. vulnificus biogroup 3 are grouped under Vibrio (clinical); *columns V. aestuarianus through V. natriegens are grouped under* Vibrio (non-clinical).

Characteristic	V. cholerae	V. alginolyticus	V. cincinnatiensis	Vibrio damsela	V. fluvialis[a]	V. furnissii[a]	V. harveyi	V. hollisae	V. metschnikovii	V. mimicus	V. parahaemolyticus	V. vulnificus	V. vulnificus biogroup 2	V. vulnificus biogroup 3	V. aestuarianus	V. anguillarum	V. campbellii	V. diazotrophicus	V. fischeri	V. gazogenes	V. ichthyoenteri	V. logei[b]	V. mediterranei	V. mytili	V. natriegens
Number of strains studied	5	5	2	5	5	5	6	5	5	5	5	5	3	2	3	4	5	5	4	24	2	3	1	1	5
Oxidase (Kovacs)	100	100	100	100	100	100	100	100	0	100	100	100	100	100	100	100	100	100	100	4	100	100	100	100	100
Nitrate reduced to nitrite	100	100	100	100	100	100	100	100	0	100	100	100	0	100	100	100	80	100	75	4	100	100	100	100	100
Indole production	100	80	0	0	60	0	100	100	100	100	100	100	0	100	100	75	80	100	0	0	0	0	100	0	0
Voges–Proskauer	100	80	0	80	0	0	0	0	100	0	0	0	0	0	33	100	100	0	50	67	0	0	67	0	20
Lysine decarboxylase (2 d)	100	100	100	60	0	0	83	0	60	100	100	100	100	0	100	75	0	0	0	0	0	0	67	100	0
Arginine dihydrolase (2 d)	0	0	0	0	100	100	0	0	100	0	0	0	0	100	0	0	20	0	0	0	0	33	67	0	0
Ornithine decarboxylase (2 d)	100	40	0	0	0	0	83	0	0	100	100	100	0	100	33	0	0	0	0	0	0	0	0	0	0
Lysine decarboxylase (7 d)	100	100	100	60	0	0	83	0	60	100	100	100	100	100	100	0	100	100	100	0	0	100	0	100	100
Arginine dihydrolase (7 d)	0	0	0	100	100	100	0	0	100	0	0	0	0	0	0	100	0	100	0	0	0	0	67	100	0
Ornithine decarboxylase (7 d)	100	0	0	0	0	0	83	0	0	100	100	100	0	100	100	0	20	100	0	0	0	0	0	0	0
Motility	100	100	100	100	100	100	100	0	60	100	100	100	100	100	100	0	80	100	100	100	100	100	100	100	100
D-Glucose, acid	100	100	100	100	100	100	100	100	100	100	100	100	100	100	100	100	100	100	75	100	100	100	100	100	100
D-Glucose, gas	0	0	0	40	0	100	83	0	0	0	0	0	0	0	0	0	0	0	0	96	0	0	0	100	0
Fermentation of:																									
Adonitol	0	0	0	0	0	0	0	0	0	0	0	0	0	0	0	0	0	0	0	0	0	0	0	0	0
L-Arabinose	0	0	100	0	100	100	0	100	0	0	80	0	0	0	0	75	0	100	0	100	0	0	0	100	100
D-Arabitol	0	0	0	0	80	100	0	0	0	0	0	0	0	0	33	0	0	0	0	0	0	0	0	0	100
Cellobiose	0	100	100	100	20	0	100	0	0	0	0	100	100	100	100	75	100	100	100	100	100	100	100	100	80
Dulcitol	0	0	0	0	0	0	0	0	0	0	0	0	0	0	0	0	0	0	0	0	0	0	0	0	0
Erythritol	0	0	0	0	0	0	0	0	0	0	0	0	0	0	0	0	0	0	0	0	0	0	0	0	0
D-Galactose	60	20	100	100	100	100	67	100	20	40	60	100	100	100	100	25	0	100	100	88	0	100	100	100	60
D-Galacturonate	0	0	0	0	100	100	0	0	0	0	0	0	0	0	0	0	0	100	0	0	0	0	0	0	0

[a]The strains for *V. fluvialis* and *V. furnissii* grew in 0% NaCl when incubated at 25°C but not at 36°C.

[b]*V. logei* was tested at 15°C.

[c]Each number is the % positive for the test after 48 h of incubation, except for oxidase and reduction of nitrate to nitrite, which were done only at 24 h. All species except *V. logei* were tested at 25°C. All testing was done in the CDC *Vibrio* Reference Laboratory using standardized "enteric" media supplemented with marine cations (Na+, K+, and Mg++). Results in this table may differ from those in the original descriptions of organisms due to variables in media, incubation temperature, reading time, and salt content.

Editorial Note: Photobacterium damselae subsp. *damselae* is a junior objective synonym of *Vibrio damsela.*

(continued)

TABLE BXII.γ.163. (*cont.*)

Characteristic	Vibrio (clinical) V. cholerae	V. alginolyticus	V. cincinnatiensis	Vibrio damsela	V. fluvialis[a]	V. furnissii[a]	V. harveyi	V. hollisae	V. metschnikovii	V. mimicus	V. parahaemolyticus	V. vulnificus	V. vulnificus biogroup 2	V. vulnificus biogroup 3	Vibrio (non-clinical) V. aestuarianus	V. anguillarum	V. campbellii	V. diazotrophicus	V. fischeri	V. gazogenes	V. ichthyoenteri	V. logei[b]	V. mediterranei	V. mytili	V. natriegens
myo-Inositol	0	0	100	0	0	0	0	0	40	0	0	0	0	0	0	0	0	0	0	0	0	0	0	0	0
Lactose	0	0	0	0	0	0	0	0	20	100	0	100	33	100	67	75	100	60	25	88	0	0	0	100	0
Maltose	100	100	100	100	100	100	100	0	80	80	100	100	100	100	100	75	40	100	100	100	100	100	100	100	100
D-Mannitol	100	100	100	0	100	100	100	0	100	100	80	80	0	0	100	75	100	100	50	96	100	100	100	100	100
D-Mannose	80	100	100	100	80	60	100	100	0	100	100	100	67	100	100	100	100	20	100	100	100	100	100	0	40
Melibiose	0	0	0	0	0	0	40	0	0	0	0	0	0	0	0	0	0	0	0	71	0	0	67	0	60
α-Methyl-D-glucoside	0	0	50	20	0	0	0	0	0	0	0	0	0	0	0	0	0	0	0	0	0	0	0	0	60
Raffinose	0	0	0	0	0	0	0	0	20	0	0	0	0	0	0	0	0	0	0	0	0	0	0	0	40
L-Rhamnose	0	0	0	0	0	0	0	0	40	0	0	0	0	0	0	0	0	0	0	0	0	0	0	100	100
Salicin	0	0	100	0	80	0	0	0	100	0	0	20	0	0	0	100	0	100	75	100	0	0	0	100	0
L-Sorbitol	0	0	0	20	0	0	0	0	100	0	0	0	0	0	0	100	0	100	0	71	0	0	100	0	100
Sucrose	100	100	100	20	100	100	83	0	100	0	0	100	100	100	100	100	0	100	0	100	50	67	100	100	100
Trehalose	80	100	100	40	100	100	100	0	100	100	100	100	100	100	100	75	100	100	0	100	100	0	100	100	100
D-Xylose	0	0	100	0	0	0	0	0	0	0	0	0	0	0	0	0	0	100	0	100	0	0	0	0	0
Growth in:																									
0% NaCl	100	0	0	0	100[c]	100	0	0	0	100	0	0	0	0	0	0	0	0	0	0	0	0	0	0	0
0.1% NaCl	100	0	100	0	100	100	0	0	80	100	0	40	0	0	100	100	0	100	0	0	0	0	0	0	0
0.2% NaCl	100	0	100	0	100	100	0	0	100	100	0	60	100	50	100	100	0	100	0	0	0	0	0	0	0
0.3% NaCl	100	80	100	40	100	100	0	0	100	100	0	80	100	100	100	100	0	100	0	0	0	0	50	0	0
0.4% NaCl	100	80	100	80	100	100	83	20	100	100	80	100	100	100	100	100	0	100	0	8	0	0	50	100	40
0.5% NaCl	100	100	100	80	100	100	100	100	100	100	80	100	100	100	100	100	60	100	75	29	50	0	67	100	100
1.0% NaCl	100	100	100	100	100	100	100	100	100	100	100	100	100	100	100	100	100	100	100	88	100	100	100	100	100
2.0% NaCl	100	100	100	100	100	100	100	100	100	100	100	100	100	100	100	100	100	100	100	100	100	0	100	100	100
3.5% NaCl	100	100	100	100	100	100	100	100	100	100	100	100	100	100	100	100	100	100	100	100	100	100	100	100	100
6.0% NaCl	100	100	100	40	100	100	100	100	100	80	100	100	100	100	100	100	100	100	75	100	100	0	100	100	100
8.0% NaCl	20	100	100	0	80	100	67	20	40	0	0	0	0	0	67	0	60	40	0	92	0	0	33	0	20
10.0% NaCl	0	100	0	0	80	20	0	20	0	0	0	0	0	0	0	0	0	0	0	42	0	0	0	100	0
12.0% NaCl	0	100	0	0	0	0	0	0	0	0	0	0	0	0	0	0	0	0	0	0	0	0	0	100	0
Swarming	0	100	0	0	0	33	33	0	0	0	80	0	0	0	0	0	0	0	0	0	0	0	100	0	0

[a]The strains for *V. fluvialis* and *V. furnissii* grew in 0% NaCl when incubated at 25°C but not at 36°C.

[b]*V. logei* was tested at 15°C.

[c]Each number is the % positive for the test after 48 h of incubation, except for oxidase and reduction of nitrate to nitrite, which were done only at 24 h. All testing was done in the CDC *Vibrio* Reference Laboratory using standardized "enteric" media supplemented with marine cations (Na+, K+, and Mg++). Results in this table may differ from those in the original descriptions of organisms due to variables in media, incubation temperature, reading time, and salt content.

Editorial Note: Photobacterium damselae subsp. *damselae* is a junior objective synonym of *Vibrio damsela*.

(continued)

TABLE BXII.γ.163. *(cont.)*

Column group headers: "Unnamed *Vibrio* group" spans *Vibrio* Group 509 through Baumann Group E3; "*Moritella*" spans *Moritella marina*; "*Photobacterium*" spans *Photobacterium angustum* through *P. phosphoreum*; "*Salinivibrio*" spans *Salinivibrio costicola*.

Characteristic	*V. nereis*	*V. nigripulchritudo*	*V. ordalii*	*V. orientalis*	*V. pectenicida*	*V. pelagius* biogroup 1	*V. pelagius* biogroup 2	*V. penaeicida*	*V. proteolyticus*	*V. scophthalmi*	*V. splendidus* biogroup 1	*V. splendidus* biogroup 2	*V. tapetis*	*V. trachuri*	*V. tubiashii*	*Vibrio* Group 509	*Vibrio* Group 510	*Vibrio* Group 511	*Vibrio* Group 512	Baumann Group E3	*Moritella marina*	*Photobacterium angustum*	*P. iliopiscarium*	*P. leiognathi*	*P. phosphoreum*	*Salinivibrio costicola*
Number of strains studied	5	3	1	4	2	4	2	1	1	4	4	2	1	1	4	4	3	2	6	1	1	5	2	5	3	5
Oxidase (Kovacs)	100	100	100	100	100	100	100	100	100	100	100	100	100	100	100	100	100	100	100	100	100	20	0	60	0[c]	100
Nitrate reduced to nitrite	60	100	0	100	100	100	100	0	100	0	100	100	0	100	100	100	100	100	0	100	100	40	100	100	33	60
Indole production	60	100	0	100	0	25	100	0	0	0	100	50	0	100	100	100	100	0	0	0	0	0	100	20	0	0
Voges–Proskauer	0	0	0	0	0	0	0	0	0	0	0	0	0	0	0	0	0	0	17	0	0	60	100	60	67	80
Lysine decarboxylase (2 d)	0	0	100	100	0	0	0	0	100	0	0	0	0	100	0	100	0	0	0	0	0	0	100	80	67	0
Arginine dihydrolase (2 d)	80	0	0	25	0	0	0	0	100	0	25	0	0	0	100	0	0	0	83	100	0	40	0	0	0	40
Ornithine decarboxylase (2 d)	0	0	0	0	0	0	0	0	0	0	0	0	0	100	0	75	0	0	0	0	0	0	100	60	0	0
Lysine decarboxylase (7 d)	0	0	100	100	0	0	0	0	100	0	0	0	0	100	0	100	0	100	100	0	0	0	100	80	67	0
Arginine dihydrolase (7 d)	100	0	0	50	0	0	0	0	100	0	75	50	0	0	100	0	0	0	100	100	0	60	0	0	0	60
Ornithine decarboxylase (7 d)	0	0	0	0	0	0	0	0	100	0	0	0	0	100	0	100	0	0	0	0	0	0	100	60	0	0
Motility	100	100	100	100	100	100	100	100	100	100	100	100	100	100	100	100	100	100	100	100	100	100	50	100	33	80
D-Glucose, acid	100	100	100	100	100	100	100	0	100	100	100	100	100	100	100	100	100	100	100	100	100	100	100	100	100	100
D-Glucose, gas	0	0	0	0	0	0	0	0	0	0	0	0	0	0	0	0	0	0	0	0	0	0	100	20	100	0
Fermentation of:																										
Adonitol	0	0	0	0	0	0	0	0	0	25	0	0	0	0	0	0	0	0	0	0	0	0	0	0	0	0
L-Arabinose	0	0	0	0	0	0	0	0	0	0	0	0	0	0	0	0	100	100	0	0	0	0	0	0	0	0
D-Arabitol	0	0	0	100	0	0	0	0	0	0	0	0	0	0	0	0	0	0	0	0	0	0	0	0	0	0
Cellobiose	0	100	0	0	0	0	0	100	100	0	100	100	0	100	100	100	100	50	67	0	0	0	0	0	0	20
Dulcitol	0	0	0	0	0	0	0	0	0	0	0	0	0	0	0	0	0	0	0	0	0	0	0	0	0	0
Erythritol	0	0	0	0	0	0	0	0	0	0	0	0	0	0	0	0	0	0	0	0	0	0	0	0	0	0
D-Galactose	0	33	0	0	0	0	33	0	0	25	0	0	0	100	50	100	100	100	100	0	100	0	100	80	100	20
D-Galacturonate	0	0	0	0	0	0	0	0	0	0	0	0	0	0	0	100	100	100	100	0	0	0	0	0	0	0

[a]The strains for *V. fluvialis* and *V. furnissii* grew in 0% NaCl when incubated at 25°C but not at 36°C.

[b]*V. logei* was tested at 15°C.

[c]Each number is the % positive for the test after 48 h of incubation, except for oxidase and reduction of nitrate to nitrite, which were done only at 24 h. All testing was done in the CDC *Vibrio* Reference Laboratory using "enteric" media supplemented with marine cations (Na$^+$, K$^+$, and Mg^{++}). Results in this table may differ from those in the original descriptions of organisms due to variables in media, incubation temperature, reading time, and salt content.

Editorial Note: Photobacterium damselae subsp. *damselae* is a junior objective synonym of *Vibrio damsela.*

(continued)

TABLE BXII.γ.163. (cont.)

Group headers: **Unnamed Vibrio group** (Vibrio Group 509, 510, 511, 512; Baumann Group E3) · **Moritella** (Moritella marina) · **Photobacterium** (Photobacterium angustum, P. iliopiscarium, P. leiognathi, P. phosphoreum) · **Salinivibrio** (Salinivibrio costicola)

Characteristic	V. nereis	V. nigripulchritudo	V. ordalii	V. orientalis	V. pectenicida	V. pelagius biogroup 1	V. pelagius biogroup 2	V. penaeicida	V. proteolyticus	V. scophthalmi	V. splendidus biogroup 1	V. splendidus biogroup 2	V. tapetis	V. trachuri	V. tubiashii	Vibrio Group 509	Vibrio Group 510	Vibrio Group 511	Vibrio Group 512	Baumann Group E3	Moritella marina	Photobacterium angustum	P. iliopiscarium	P. leiognathi	P. phosphoreum	Salinivibrio costicola
myo-Inositol	0	0	0	0	0	0	0	0	0	0	0	0	0	0	0	0	0	100	100	100	0	0	0	0	0	0
Lactose	0	67	0	0	0	25	0	0	0	0	0	0	0	0	0	0	67	0	100	100	0	0	0	0	0	0
Maltose	100	0	100	100	100	100	100	0	100	100	100	100	100	100	100	100	100	100	0	100	100	60	100	0	100	80
D-Mannitol	0	33	0	100	0	0	0	0	100	100	100	100	0	100	100	100	100	100	100	100	0	20	100	0	100	60
D-Mannose	20	0	0	100	0	0	100	0	0	100	100	100	100	0	100	100	67	100	100	0	0	100	0	100	0	60
Melibiose	0	0	0	0	0	0	0	0	0	100	0	0	0	0	25	50	0	0	100	0	0	0	0	0	0	0
α-Methyl-D-glucoside	0	0	0	0	0	0	0	0	0	0	0	0	0	0	0	0	0	50	0	0	0	0	0	0	0	0
Raffinose	0	0	0	0	0	0	0	0	0	0	0	0	0	0	0	0	0	0	0	0	0	0	0	0	0	0
L-Rhamnose	0	0	0	0	0	0	0	0	0	0	0	0	0	50	0	0	100	0	0	0	0	0	0	0	0	0
Salicin	0	0	0	0	0	0	0	0	100	0	0	0	0	100	0	0	0	100	67	0	0	0	0	0	0	0
L-Sorbitol	0	0	0	0	0	0	0	0	0	0	0	0	0	0	100	100	100	0	0	0	0	60	0	0	0	60
Sucrose	100	0	100	100	0	100	67	0	0	100	75	0	0	100	100	100	100	100	100	100	0	40	50	0	0	80
Trehalose	100	100	0	100	100	100	100	0	100	100	75	100	100	100	100	100	100	100	100	100	0	100	0	0	0	0
D-Xylose	0	0	0	0	0	0	0	0	0	0	0	0	0	0	0	100	100	0	0	100	0	0	0	0	0	0
Growth in:																										
0% NaCl	0	0	0	0	0	0	0	0	0	0	0	0	0	0	0	0	33	0	0	0	0	0	0	0	0	0
0.1% NaCl	0	0	0	0	0	0	0	0	0	0	0	0	0	0	0	0	100	100	17	0	0	0	0	0	0	0
0.2% NaCl	0	0	0	0	0	0	0	0	100	0	0	0	0	0	0	0	100	100	100	0	0	0	0	0	0	0
0.3% NaCl	0	0	0	0	0	0	0	0	100	0	0	0	0	0	50	0	100	100	100	0	0	0	0	0	0	0
0.4% NaCl	0	0	0	0	0	0	0	0	100	0	0	0	0	50	50	50	100	100	100	0	0	0	0	0	0	0
0.5% NaCl	0	0	100	25	0	25	0	0	100	0	0	0	0	100	75	100	100	100	100	0	0	0	0	60	33	40
1.0% NaCl	100	100	100	100	0	100	100	100	100	25	75	100	100	100	100	100	100	100	100	100	0	40	50	100	67	80
2.0% NaCl	100	100	100	100	0	100	100	100	100	100	100	100	100	100	100	100	100	100	100	100	0	100	100	100	100	100
3.5% NaCl	100	100	100	100	100	100	100	100	100	100	100	100	100	100	100	100	100	100	100	100	100	100	100	100	100	100
6.0% NaCl	100	100	0	100	50	0	100	0	100	100	100	50	0	100	100	100	100	100	100	100	0	100	100	100	67	100
8.0% NaCl	100	0	0	100	0	0	0	0	100	0	50	50	0	0	50	100	100	100	100	100	0	0	0	0	0	80
10.0% NaCl	20	0	0	0	0	0	0	0	100	0	0	0	0	0	0	25	67	100	67	0	0	0	0	0	0	60
12.0% NaCl	0	0	0	0	0	0	0	0	100	0	0	0	0	0	0	0	0	0	17	0	0	0	0	0	0	0
Swarming	0	0	0	0	0	0	0	0	100	0	0	0	0	0	0	100	0	0	0	0	0	0	0	0	0	0

a The strains for V. fluvialis and V. furnissii grew in 0% NaCl when incubated at 25°C but not at 36°C.

b V. logei was tested at 15°C.

Each number is the % positive for the test after 48 h of incubation, except for oxidase and reduction of nitrate to nitrite, which were done only at 24 h. All species except V. logei were tested at 25°C. All testing was done in the CDC Vibrio Reference Laboratory using "enteric" media supplemented with marine cations (Na+, K+, and Mg++). Results in this table may differ from those in the original descriptions of organisms due to variables in media, incubation temperature, reading time, and salt content.

*Editorial Note: Photobacterium damselae subsp. damselae is a junior objective synonym of Vibrio damsela.

TABLE BXII.γ.164. Properties of the species and biogroups of *Vibrio*[a]*

Characteristics	*V. cholerae*	*V. aerogenes*	*V. aestuarianus*	*V. alginolyticus*	*V. anguillarum*	*V. campbellii*	*V. cincinnatiensis*	*V. cyclitrophicus*	*V. damsela*	*V. diabolicus*	*V. diazotrophicus*	*V. fischeri*	*V. fluvialis*	*V. furnissii*	*V. gazogenes*	*V. halioticoli*	*V. harveyi*	*V. hollisae*	*V. ichthyoenteri*	*V. logei*	*V. mediterranei*	*V. metschnikovii*	*V. mimicus*	*V. mytili*	*V. natriegens*
Number of strains tested	161	1	9	38	20	44	1	1	21	1	13	12	15	7	1	6	91	16	6	11	4	6	51	5	8
3–12 polar flagella	−	−		+	−	+		−			+	+	−	−	−	−	−			+	−	−		−	−
Lateral flagella/solid media	−		+	+	−	−		−			−	−	d	d	−	−	+	−	−	−	−	d	−	+	−
Swarming on complex media	d	−	−	+	−	−		d				+	d	+	−	−	+			+		d		−	+
Straight rods[b]	d			−	−	−		d	+		+	+	d	d	−	−	+			+		d		+	+
Curved rods	d			−	+	−		d	+		+	−	d	+	+	+	−					−			+
PHB accumulation	−	−	+	−	+	−		d	+		−	−	−	d	−	−	−			−		−		−	+
Pigment:																									
Yellow-orange	−	−	+	−	−	−						+								+					
Blue-black	−																								
Red	−							w	+		+		+	+	+					+				+	+
Arginine dihydrolase[c]	+	+	+	−	+	−	−	−	+	−	+	+	+	+	−	−	−	−	−	+	−	−	−	+	−
Oxidase	+	−	+	+	+	+	+	+	+	+	+	+	+	+	+	+	+	+	+	+	+	−	+	+	+
Reduction of NO₃ to NO₂	+	+	+	+	+	+	+	−	+	+	+	d	+	+	+	+	+	+	+	+	d	+	+	+	+
Luminescence	−	−		−	−	−	−	−	−	−	−	+	−	−	−	−	d	−	−	−	−	−	−	−	−
Gas from D-glucose	+	+	−	−	−	−	+	−	+	−	−	−	−	d	+	−	−	−	−	−	−	d	−	−	−
Voges-Proskauer	+	+	+	+	+	+	+	−	+	−	+	−	+	+	+	+	+	+	+	+	d	+	−	+	+
Na⁺ required for growth	−		+	−	−	−	+	−	+	+	+	+	−	−	−	+	+	+	+	+	+	d	−	+	+
Organic growth factors required	d			−	−	−	+	−	−		+	+	−	−	−	+	+		+	d	−	d	−	+	−
Growth at:																									
4°C	−	+	+	−	−	−	−	+	−		+	+	−	−	−	−	−	−	−	+	−	−	−	−	−
30°C	+	+	+	+	+	+	+	+	+	+	+	−	+	+	+	+	+	+	+	+	+	+	+	+	+
35°C	+	+	+	+	+	+	+	+	+	+	+	d	+	+	+	+	+	+	+	+	−	+	+	+	+
40°C	+	−	−	+	+	+	+	+	−		d	+	+	+	+	−	d	−	−	−	−	+	−	−	−
Production of:																									
Alginase	−	+	−	−	−	−	−	−	−		+	−	−	−	−	−	d	−		+	−	−		−	−
Amylase	+	−	+	+	+	+	+	+	−		−	d	+	d	+	+	+	+	+	+	+	d	−	+	d
Chitinase	+	+	+	+	+	+	+	+	−		−	−	+	+	+	−	+		+	−	d	+	−		−
Gelatinase	+	+	+	+	+	+	+	+	−		−	+	+	d	+	−	+		+	d	+	d	+	−	d
Lipase	+			+	−	+	+	+	−		−	−	+	−	+	−	+		+	−	+	d	d	+	+
Utilization of:																									
β-Alanine	−			−	−	−	−	+	−			−	−	+	−	−	−			−		−			d
D-α-Alanine	d	+		+	d	+	+	+	−			d	+	+	−	−	+		−	+	+	d		−	
L-α-Alanine	−			+	d	d	+	+	−			+	+	+	+	+	d		−	−	d	+		+	d
Aminobutyrate	−			−	−	−	−	−	−			−	+	+	+	−	−			−	+	−			+
Aminovalerate	−	+		+	+	d	+	+	−			+	+	d	+	−	+	+		d	−	d	+	−	+
L-Arabinose	−		d	−	−	−	+	−	−			−	d	d	+	−	d	+		d		d	−	−	+
Cellobiose	−			+	+	d	+	+	−			+	+	d	+	−	+	−		+	−	+	−	−	+

Citrate	d	+	+	+	+	d	+		−	+		d	+	+	+		+	−		−		d	+		
Ethanol	−			d	−	−	−					−	+	+	−		−			−		−			
D-Galactose	+	+		d	d	−		+		+		+	+	+	+		d	+	−	+		d	+		d
D-Galacturonate	−			−	−	−						−	+	+	−		−			−		−			
D-Gluconate	+			+	+	−	+			+		−	+	+	−	−	+		−	+	+	+		−	+
D-Glucuronate	−			−	−	−	−	+				−	+	−	−		+			−		−			+
L-Glutamate	+	+		+	+	−	+	+			+	d	+	+	+	−	+			+	+	+		+	+
Glutarate	−			−	−	−						−	−	+			−			−		−			
Heptanoate	−			+	−	−						−	+	+	−		+			−		d			
L-Histidine	d			+	+	−						−	+	d		−	d			−		d		−	+
p-Hydroxybenzoate	−			−	−	−	−	−				−	+	d	−		−			−		−			
Hydroxybutyrate	−	−		−	−	−	−					−	+	d	−		−			−		−			
myo-Inositol	−	+		−	d	−	+		−			−	−	−	−		−	−		−		d	−		
Ketoglutarate	+			+	+	+	+					−	+	+	+		+			−		−			
DL-Lactate	+			+	+	+		+				−	+	+	+		+			−		+			
Lactose	−	−	+	−	−	−	−		−		+	−	−	−	−		−[e]	−		−	+	d	d	+	+
L-Leucine	−			+	−	−	−					−	−	−	−		−		−	−		−			d
DL-Malate	+	−		+	+	d						−	+	+	+		d			−		+			
D-Mannitol	+	+	+	+	+	d	−		−	+		+	+	+	+		+	−		+		+	+		
D-Mannose	d	+	+	d	+	d	+	−	+	+	d	+	+	d	+		+	+		d	+	d	+	+	+
Melibiose	−	−	d	−	−	−			−			−	−	−		−	−	−		−		−	−	−	+
L-Proline	+			+	+	+	+				+	+	+	+	+		+			+	+	+		+	+
Propionate	+			+	+	+		−				−	+	+	+		+			−		−			
Putrescine	−			d	−	−	−					−	d	+	−	−	−		−	−	+	−		−	+
Pyruvate	+			+	+	+		+				−	+	+	+		+			−		+			
Salicin	−		−	−	−	−	+	−	−	−		d	+	−	+	−	d	−				−	−	+	d
L-Serine	d	−		+	+	d		+			d	−	+	+	+		+			d	+	+		+	+
D-Sorbitol	−	−	+	−	+	−	−		−			−	−	−	−		−	−		−		d	−		
Sucrose	+	+	+	+	+	−	+	+	−	+	+	−	+	+	+	−	d	−		−	+	+	−	+	+
Trehalose	+	−	+	+	+	+	+	+	−	+		d	+	+		−	+	−	−	−	−	+	+	−	+
L-Tyrosine	−			+	−	d	−	+				−	+	+	−	−	+	−		−		−	d	+	d
Valerate	−			+	−	−						−	−	−	−		−			−		−			
D-Xylose	−	+	−	−	−	−	+	+	−	−	+	−	−	−	+	−	−	−	−	−	−	−	−	+	−
Benzoate, betaine, hippurate, malonate, L-rhamnose, sarcosine, spermine	−			−	−	−						−	−	−	−		−			−		−			+

[a]Symbols: see standard definitions; w, weak; blank, data not determined or not applicable. Precautionary note: limitations of this table for identifying "unknown" cultures. This table is an attempt to revise and update Table 5.58 (pp. 532–533) of the *Vibrio* chapter in the first edition of this *Manual*, in which Baumann et al. studied each of the 20 *Vibrio* species. The data in their Table 5.58 were based on over 500 cultures tested under essentially identical conditions. Authors who described new *Vibrio* species since 1984 did not perform all of these tests and may have used different sets of media and methods. The following are probably the most important variables in the new data: media and methods used, incubation temperature, incubation time (final reading at 1–2 days vs 28 days or more), NaCl or "artificial seawater" added to media to speed growth and enhance metabolism, inoculation method and size, and "endpoint definition" (criteria for coding a test as positive or negative).

[b]Straight rods in exponential phase of growth becoming curved in stationary phase.

[c]Originally defined by Baumann et al. (1984b) as "the anaerobic production of ornithine from arginine." These authors have described the problems associated with the Thornley method, which measures alkali production from arginine.

[d]Luminous strains of this species have been found by Desmarchelier and Reichelt (1981).

[e]Wild-type strains are unable to utilize lactose, but many strains acquire this property by mutation. Most strains ferment lactose in "peptone type" media.

**Editorial Note: Photobacterium damselae* subsp. *damselae* is a junior objective synonym of *Vibrio damsela*.

TABLE BXII.γ.164. *(cont.)*

Characteristics	*V. navarrensis*	*V. nereis*	*V. nigripulchritudo*	*V. ordalii*	*V. orientalis*	*V. parahaemolyticus*	*V. pelagius* biogroup 1	*V. pelagius* biogroup 2	*V. pectenicida*	*V. penaeicida*	*V. proteolyticus*	*V. rumoiensis*	*V. scophthalmi*	*V. splendidus* biogroup 1	*V. splendidus* biogroup 2	*V. tapetis*	*V. trachuri*	*V. tubiashii*	*V. vulnificus* biogroup 1	*V. vulnificus* biogroup 2	*V. vulnificus* biogroup 3	*V. wodanis*	*Moritella marina*	*M. viscosa*	*P. iliopiscarium*	*Salinivibrio costicola*
Number of strains tested	11	6	14	11	5	134	7	4	5	6	1	1	6	4	15	24	3	6	15	3	19	35	1	20	37	2
3–12 polar flagella	−	−	−			+	−	−	+	−	+	−	−	−	d	−	−	−	−	−	−	+	−	−		−
Lateral flagella/solid media	−	−	−			−	−	−	+	−	+	−		−	−				−	−		−		d		−
Swarming on complex media	−	+	+[b]			+	−	−		−	+				d							d	+	d		−
Straight rods	−	+	+	−	−	+	+	−	+					−	d				+			d	+	d	+	+
Curved rods	−	+	d	+	+	−	+	+						+	d				+			d	+	d	+	−
PHB accumulation	−	−	−	−	+	−	−	+		−	−		−	−	−			−	−			−			+	−
Pigment: Yellow-orange	−	−	−	−	−	−	−	−	−	−	−	−	−	−	−	−	−	−	−	−	−	−	−	−	−	−
Pigment: Blue-black	−	−	+	−	−	−	−	−	−	−	+	−	−	−	−	−	−	−	−	−	−	−	−	−	−	−
Pigment: Red	−	−	+	+	+	−	+	+	−	+	+	+	+	+	+	−	+	+	−	+	−	+	−	−	+	+
Arginine dihydrolase[c]	+	−	+	+	+	−	+	+	+	+	+	+	+	d	+	+	+	d	−	−	−	+	+	−	−	+
Oxidase	+	+	+	+	+	+	+	+	+	+	+	+	+	+	+	+	+	+	+	+	+	d	+	−	+	+
Reduction of NO₃ to NO₂	+	+	+	+	+	+	+	+	+	+	+	+	+	−	+	+	+	+	+	−	+	d	+	d	+	d
Luminescence	−	−	−	−	−	−	−	−	−	−	−	−	−	−	−	−	−	−	−	−	−	−	−	−	−	−
Gas from D-glucose	−	−	−	−	−	−	−	−	−	−	−	−	−	−	−	−	−	−	−	−	−	−	−	−	−	−
Voges–Proskauer	−	−	+	−	+	−	+	+	+	+	+	+	+	+	+	+	+	+	+	+	+	+	+	+	+	+
Na⁺ required for growth	d	+	+	+	d	+	+	−	+	+	+	+	+	+	+	+	+	+	+	+	+	+	+	+	+	d
Organic growth factors required	−	−	−	−	−	−	−	−	−	−	−	−	−	d	−	−	−	−	−	−	−	−	−	−	−	−
Growth at: 4°C	+	d	−	−	+	−	d	d	−	−	−	+	−	d	−	+	−	−	−	−	−	+	+	+	−	−
Growth at: 30°C	+	+	+	+	+	+	+	+	+	+	+	+	+	+	+	−	+	+	+	+	+	−	+	−	+	+
Growth at: 35°C	+	+	+	−	+	+	+	+	−	+	+	+	−	d	+	−	+	−	+	−	−	+	+	−	−	+
Growth at: 40°C	−	d	−	−	−	−	−	−		−	−	+	−	−	−	−	−	−	+	+	−	+	+	+	−	−
Production of: Alginase	+	−	−	−	−	−	d	d	+	d	+	+	−	d	−	+			−	+		d	−	−	+	−
Production of: Amylase	−	d	+	−	+	+	+	+	+	d	+	+	+	+	+	−	+	+	+	+		+	+	+	−	−
Production of: Chitinase	+	d	+	+	+	+	d	+	+	+	+	+	−	+	+	−		+	+	+		−	+	−	−	−
Production of: Gelatinase	−	d	+	−	+	+	+	+	−	+	+	+	−	+	+	+		+	+	+		+	+	+	+	−
Production of: Lipase	−	−	+	+	+	+	+	+	+	+	+	+	+	+	+	+	+	+	+	+		+	+	+	+	+
Utilization of: β-Alanine	−	+	d	−	−	+	−	d	−	d	+	−	−	d	−	−		−	−	+		d	−	−	−	+
Utilization of: D-α-Alanine	−	+	−	−	+	+	−	+	−	d	+	+	+	+	+	+		+	+	+		+	+	+	−	+
Utilization of: L-α-Alanine	+	+	−	+	+	d	d	+	+	+	+	+	−	+	+	−		+	+	+		+	+	+	−	−
Utilization of: Aminobutyrate	+	−	−	−	−	−	−	d	−	d	+	+	−	−	−	−		−	+	+		−	+	−	−	+
Utilization of: Aminovalerate	+	+	d	−	−	d	−	+	−	−	−	−	−	−	−	−		−	+	+		+	−	−	−	+
Utilization of: L-Arabinose	−	+	d	−	−	d	−	d	−	d	+	−	−	−	−	+		−	−	−	−	+	+	−	−	−
Utilization of: Cellobiose	−	+	−	−	−	−	−	+	−	−	+	+	−	−	+	+		−	+	+	−	+	−	+	−	−

Citrate

Ethanol

D-Galactose

D-Galacturonate

D-Gluconate

D-Glucuronate

L-Glutamate

Glutarate

Heptanoate

L-Histidine

p-Hydroxybenzoate

Hydroxybutyrate

myo-Inositol

Ketoglutarate

DL-Lactate

Lactose

L-Leucine

DL-Malate

D-Mannitol

D-Mannose

Melibiose

L-Proline

Propionate

Putrescine

Pyruvate

Salicin

L-Serine

D-Sorbitol

Sucrose

Trehalose

L-Tyrosine

Valerate

D-Xylose

Benzoate, betaine, hippurate, malonate, L-rhamnose, sarcosine, spermine

[a]Symbols: see standard definitions; w, weak; blank, data not determined or not applicable. Precautionary note: limitations of this table for identifying "unknown" cultures. This table is an attempt to revise and update Table 5.58 (pp. 532–533) of the *Vibrio* chapter in the first edition of this *Manual*, in which Baumann et al. studied each of the 20 *Vibrio* species. The data in their Table 5.58 were based on over 500 cultures tested under essentially identical conditions. Authors who described new *Vibrio* species since 1984 did not perform all of these tests and may have used different sets of media and methods. The following are probably the most important variables in the new data: media and methods used, incubation temperature, incubation time (final reading at 1–2 days vs 28 days or more), NaCl or "artificial seawater" added to media to speed growth and enhance metabolism, inoculation method and size, and "endpoint definition" (criteria for coding a test as positive or negative).

[b]Straight rods in exponential phase of growth becoming curved in stationary phase.

[c]Originally defined by Baumann et al. (1984b) as "the anaerobic production of ornithine from arginine." These authors have described the problems associated with the Thornley method, which measures alkali production from arginine.

[d]Luminous strains of this species have been found by Desmarchelier and Reichelt (1981).

[e]Wild-type strains are unable to utilize lactose, but many strains acquire this property by mutation. Most strains ferment lactose in "peptone type" media.

*Editorial Note: Photobacterium damselae subsp. damselae is a junior objective synonym of Vibrio damsela.

TABLE BXII.γ.165. Biochemical test results and other properties of the 12 *Vibrio* species that occur in human clinical specimens[a]*

Test	V. cholerae	V. alginolyticus	V. cincinnatiensis	V. damsela	V. fluvialis	V. furnissii	V. harveyi	V. hollisae	V. metschnikovii	V. mimicus	V. parahaemolyticus	V. vulnificus
						Percentage positive[b]						
Indole production (HIB, 1% NaCl)[c]	99	85	8	0	13	11	100	97	20	98	98	97
Methyl red (1% NaCl)	99	75	93	100	96	100	100	0	96	99	80	80
Voges–Proskauer (1% NaCl; Barritt)[c]	75	95	0	95	0	0	50	0	96	9	0	0
Citrate, Simmons	97	1	21	0	93	100	0	0	75	99	3	75
H₂S on TSI	0	0	0	0	0	0	0	0	0	0	0	0
Urea hydrolysis	0	0	0	0	0	0	0	0	0	1	15	1
Phenylalanine deaminase	0	1	0	0	0	0	NG	0	0	0	1	35
Arginine, Moellers, (1% NaCl)[c]	0	0	0	95	93	100	0	0	60	0	0	0
Lysine, Moellers, (1% NaCl)[c]	99	99	57	50	0	0	100	0	35	100	100	99
Ornithine, Moellers, (1% NaCl)[c]	99	50	0	0	0	0	0	0	0	99	95	55
Motility, (36°C)	99	99	86	25	70	89	0	0	74	98	99	99
Gelatin hydrolysis, (1% NaCl, 22°C)	90	90	0	6	85	86	0	0	65	65	95	75
KCN test (percentage that grow)	10	15	0	5	65	89	0	0	0	2	20	1
Malonate utilization	1	0	0	0	0	11	0	0	0	0	0	0
D-Glucose, acid production [c]	100	100	100	100	100	100	50	100	100	100	100	100
D-Glucose, gas production [c]	0	0	0	10	0	100	0	0	0	0	0	0
Acid production from:												
D-Adonitol	0	1	0	0	0	0	0	0	0	0	0	0
L-Arabinose [c]	0	1	100	0	93	100	0	97	0	1	80	0
D-Arabitol [c]	0	0	0	0	65	89	0	0	0	0	0	0
Cellobiose[c]	8	3	100	0	30	11	50	0	9	0	5	99
Dulcitol	0	0	0	0	0	0	0	0	0	0	3	0
Erythritol	0	0	0	0	0	0	0	0	0	0	0	0
D-Galactose	90	20	100	90	96	100	0	100	45	82	92	96
Glycerol	30	80	100	0	7	55	0	0	100	13	50	1
myo-Inositol	0	0	100	0	0	0	0	0	40	0	0	0
Lactose[c]	7	0	0	0	3	0	0	0	50	21	1	85
Maltose[c]	99	100	100	100	100	100	100	0	100	99	99	100
D-Mannitol [c]	99	100	100	0	97	100	50	0	96	99	100	45
D-Mannose	78	99	100	100	100	100	50	100	100	99	100	98
Melibiose	1	1	7	0	3	11	0	0	0	0	1	40
α-Methyl-D-glucoside	0	1	57	5	0	0	0	0	25	0	0	0
Raffinose	0	0	0	0	0	11	0	0	0	0	0	0
L-Rhamnose	0	0	0	0	0	45	0	0	0	0	1	0
Salicin[c]	1	4	100	0	0	0	0	0	9	0	1	95
D-Sorbitol	1	1	0	0	3	0	0	0	45	0	1	0
Sucrose[c]	100	99	100	5	100	100	50	0	100	0	1	15
Trehalose	99	100	100	86	100	100	50	0	100	94	99	100
D-Xylose	0	0	43	0	0	0	0	0	0	0	0	0
Mucate-acid production	1	0	0	0	0	0	0	0	0	0	0	0
Tartrate-Jordan	75	95	0	0	35	22	50	65	35	12	93	84
Esculin hydrolysis	0	3	0	0	8	0	0	0	60	0	1	40
Acetate utilization	92	0	14	0	70	65	0	0	25	78	1	7
Nitrate reduced to nitrate[c]	99	100	100	100	100	100	100	100	0	100	100	100
Oxidase[c]	100	100	100	95	100	100	100	100	0	100	100	100
DNase (25°C)	93	95	79	75	100	100	100	0	50	55	92	50
Lipase[c]	92	85	36	0	90	89	0	0	100	17	90	92
ONPG test[c]	94	0	86	0	40	35	0	0	50	90	5	75
Yellow pigment at 25°C	0	0	0	0	0	0	0	0	0	0	0	0
Tyrosine clearing	13	70	0	0	65	45	0	3	5	30	77	75
Growth in nutrient broth with:												
0% NaCl[c]	100	0	0	0	0	0	0	0	0	100	0	0
1% NaCl[c]	100	99	100	100	99	99	100	99	100	100	100	99
6% NaCl[c]	53	100	100	95	96	100	100	83	78	49	99	65
8% NaCl[c]	1	94	62	0	71	78	0	0	44	0	80	0
10% NaCl[c]	0	69	0	0	4	0	0	0	4	0	2	0
12% NaCl[c]	0	17	0	0	0	0	0	0	0	0	1	0
Swarming (marine agar, 25°C)	−	+	+	−	−	−	100	−	−	−	+	−
String test	100	91	80	80	100	100	100	100	100	100	64	100
O/129, zone of inhibitionc	99	19	25	90	31	0	100	40	90	95	20	98
Polymyxin B, % with a zone of inhibition	22	63	92	85	100	89	100	100	100	88	54	3

[a]Symbols: +, most strains (generally about 90–100%) positive; −, most strains negative (generally about 0–10% positive); 1% NaCl in parentheses indicates 1% NaCl has been added to the standard media to enhance growth; HIB, heart infusion broth; TSI, triple sugar iron agar; ONPG, *o*-nitrophenyl-β-D-galactopyranoside.

[b]The number gives the percentage positive after 48 h of incubation at 36°C (unless other conditions are indicated). Most of the positive reactions occur during the first 24 hours. NG (no growth) means that the organism does not grow, probably because the NaCl concentration is too low.

[c]Tests recommended as part of the routine set for *Vibrio* identification.

Editorial Note: Photobacterium damselae subsp. *damselae* is a junior objective synonym of *Vibrio damsela*.

TABLE BXII.γ.166. Differential tests that divide the 12 *Vibrio* species found in clinical specimens into six groups[a,b*]

Test	Group 1		Group 2	Group 3	Group 4	Group 5			Group 6			
	V. cholerae	*V. mimicus*	*V. metschnikovii*	*V. cincinnatiensis*	*V. hollisae*	*V. damsela*	*V. fluvialis*	*V. furnissii*	*V. alginolyticus*	*V. harveyi*	*V. parahaemolyticus*	*V. vulnificus*
Growth in nutrient broth with no NaCl added	+	+	−	−	−	−	−	−	−	−	−	−
Oxidase			−	+	+	+	+	+	+	+	+	+
Nitrate reduced to nitrite			−	+	+	+	+	+	+	+	+	+
myo-Inositol fermentation	−	−	d	+	−	−	−	−	−	−	−	−
Arginine dihydrolase					−	+	+	+	−	−	−	−
Lysine decarboxylase					−				+	+	+	+

[a]All data are for reactions within 2 days at 35–37°C, unless otherwise specified.

[b]Symbols: +, most strains (generally about 90–100%) positive; d, strain-to-strain variation (generally about 25–75% positive); −, most strains negative (generally about 0–10% positive).

Editorial Note: Photobacterium damselae subsp. damselae is a junior objective synonym of Vibrio damsela.

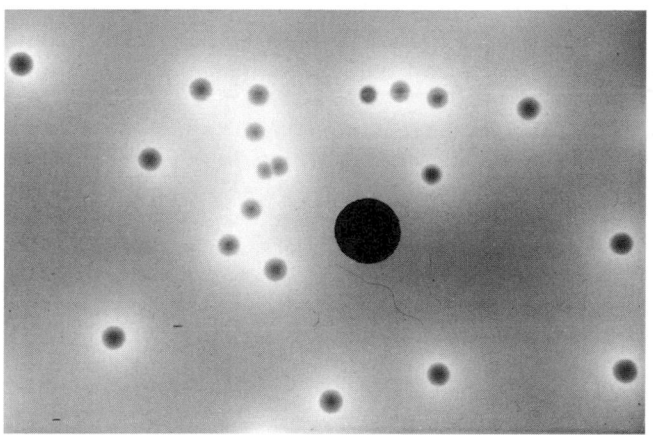

FIGURE BXII.γ.173. Colonies of *Vibrio cholerae* on TCBS agar. Yellow colonies, 2- to 3-mm in diameter are present after overnight incubation at 36°C. (Reproduced with permission from J.J. Farmer III and F.W. Hickman-Brenner, *In* Balows, Trüper, Dworkin, Harder and Schleifer (Editors), The Prokaryotes: A Handbook on the Biology of Bacteria: Ecophysiology, Isolation, Identification, Applications, 2nd Edition, Volume III, ©Springer-Verlag, New York, 1992, pp. 2952–3011.)

TABLE BXII.γ.167. Subdivision of *Vibrio cholerae* below the level of species

Serogroup	Serotype	Biogroup
V. cholerae serogroup O1[a]	Inaba	Classical
	Inaba	El Tor
	Ogawa	Classical
	Ogawa	El Tor
V. cholerae serogroup O139		
V. cholerae "non-O1, non-O139"		
(≈150 serogroups)		

[a]A third serotype of serogroup O1 named Hikojima is recognized by a few authors.

3. **Vibrio aestuarianus** Tison and Seidler 1983, 699[VP]

aes.tu.a.ri.a′ nus. L. n. *aestuarium* estuary; N.L. masc. adj. *aestuarianus* pertaining to an estuary.

The characteristics are as described for the genus and as listed in Tables BXII.γ.163 and BXII.γ.164. Straight or curved rods 0.5 × 1.5–2.0 μm; motile by a single polar flagellum. Colonies are β-hemolytic on sheep blood agar.

TABLE BXII.γ.168. Properties and differentiation of the classical and El Tor biogroups of *V. cholerae* serogroup O1[a]

Tests or property	Biogroup	
	Classical	El Tor
Frequency of isolation:		
On the Indian subcontinent	Occasional[b]	Common
In the rest of the world	Very rare	Common
Differential tests:		
Hemolysis of red blood cells	−	+
Voges–Proskauer test	−	+
Inhibition by polymyxin B (50-unit disks)	+	−
Agglutination of chicken red blood cells	−	+
Lysis by bacteriophage:		
Classical IV	+	−
FK	+	−
El Tor 5	−	+

[a]Symbols +, most strains (generally about 90–100%) positive; −, most strains negative (generally about 0–10 % positive).

[b]The classical biotype has reappeared on the Indian subcontinent, and has been isolated in some locations there.

TABLE BXII.γ.169. Key characteristics of *Vibrio cholerae* and *V. mimicus* and tests for their differentiation[a]

Tests or property	*V. cholerae*	*V. mimicus*
Frequency of isolation	Very common	Occasional
Oxidase	+	+
Growth in nutrient broth with:		
No added NaCl	+	+
1% NaCl	+	+
Lysine decarboxylase	+	+
Arginine dihydrolase	−	
Ornithine decarboxylase	+	+
Differentiation of the species:		
Sucrose fermentation	+[b]	−
Lipase (corn oil)	+	−
Voges–Proskauer	+[c]	−
Lactose fermentation (1–2 days)	−	d
Lactose fermentation (3–7 days)	d	d

[a]All data are for reactions within 2 days at 35–37°C unless otherwise specified. Symbols: +, 90–100% positive; d, 11–89% positive; −, 0–10% positive.

[b]Most of the positive reactions occur during the first 24 h of incubation

[c]Almost all strains of *V. cholerae* non-O1 and most strains of *V. cholerae* O1 currently isolated are of the El Tor biogroup, which is almost always positive for the Voges–Proskauer test. The classical biogroup is negative for the Voges–Proskauer test.

Isolated from estuary water, shellfish, and crabs off the coast of Oregon, USA. The type strain was isolated from an oyster.

Based on the results of DNA–DNA hybridization experiments, strains of *Vibrio aestuarianus* are highly related to the type strain (Tison and Seidler, 1983; Brenner et al., unpublished) and less than 25% related to other *Vibrio* species, including *V. anguillarum*, *V. nereis*, and *V. splendidus* which have similar phenotypic characteristics. The closest relative is *Vibrio anguillarum* to which it is 30–41% related (Brenner et al., personal communication). Fig. BXII.γ.171 illustrates the relationship of *Vibrio aestuarianus* to other *Vibrio* species, based on 16S rRNA gene sequence analysis (Brenner, personal communication).

The mol% G + C of the DNA is: 43–44 (T_m).

Type strain: OY-0-002, ATCC 35048.

GenBank accession number (16S rRNA): X74689.

4. **Vibrio alginolyticus** (Miyamoto, Nakamura and Takizawa 1961) Sakazaki 1968, 360[AL] (*"Oceanomonas alginolytica"* Miyamoto, Nakamura and Takizawa 1961, 481; *Beneckea alginolytica* (Miyamoto, Nakamura and Takizawa 1961, 481) Baumann, Baumann and Mandel 1971a, 289.)

al.gi.no.ly′ti.cus. L. fem. n. *alga* seaweed; M.L. adj. *alginolyticus* pertaining to alginic acid from seaweed; Gr. adj. *lyticus* dissolving; M.L. adj. *alginolyticus* alginic acid-dissolving.

The characteristics are as described for the genus and as listed in Tables BXII.γ.163 and BXII.γ.164. The morphology is illustrated in Figs. BXII.γ.159 and BXII.γ.162. Characters that differentiate *V. alginolyticus* from *V. harveyi*, *V. parahaemolyticus*, and *V. vulnificus* are given in Table BXII.γ.170. Characters that distinguish *V. alginolyticus* from other *Vibrio* spp. found in human clinical specimens are given in Tables BXII.γ.156, BXII.γ.157, BXII.γ.160, BXII.γ.161, BXII.γ.165 and BXII.γ.166.

Occurs in human clinical specimens, particularly in soft tissue infections (Table BXII.γ.157) and in the marine environment.

Many reports describe the isolation of *V. alginolyticus* from soft tissue infections (Rubin and Tilton, 1975; Pien et al., 1977). Wound infections and ear infections are usually mentioned, with eye infections mentioned much less fre-quently. There is a clear association of *V. alginolyticus* with infection at these sites; however, the etiological role of *V. alginolyticus* has rarely been shown conclusively and little has been reported about the pathogenesis of these infections. Most authors list *V. alginolyticus* as a pathogen, particularly of wound and ear infections. Antibiotic treatment has been used in most cases; surgical debridement has been used in some.

The mol% G + C of the DNA is: 45–47 (T_m, Bd).

Type strain: 118, ATCC 17749, CCM 2578, DSM 2171, IMET 11295.

GenBank accession number (16S rRNA): X56576, X74690.

5. **Vibrio anguillarum** Bergman 1909, 28[AL] (*Listonella anguillarum* (Bergman 1909) MacDonell and Colwell 1986, 354; *Listonella anguillara* (sic) (Bergman 1909) MacDonell and Colwell 1985, 180.)*

an.guil.la′rum. L. n. *anguilla* eel; L. gen. pl. n. *anguillarum* of eels.

The characteristics are as described for the genus and as listed in Tables BXII.γ.163, BXII.γ.164, and BXII.γ.171. In contrast to *V. ordalii*, whose strains belong only to O antigen group 2, *V. anguillarum* strains belong to O antigen groups 1–16.

Based on the results of DNA–DNA hybridization experiments, the type strain of *V. anguillarum* is closely related to other strains of this species. The closest relative is *V. ordalii* (Schiewe et al., 1981; also see Fig. BXII.γ.169, based on Brenner et al., 1983b) to which it is 58–76% related with T_m values of 2.5–4.0 (Brenner et al., unpublished). This close relationship has been confirmed by the structural similarity of superoxide dismutase (Fig. BXII.γ.170) and by 16S rRNA gene sequencing (Fig. BXII.γ.171). Although *V. an-*

---

*Editorial Note: In the last edition of this *Manual*, Baumann et al. (1984b) used the names *V. anguillarum* biovar I and *V. anguillarum* biovar II. In this edition, these organisms are classified as *V. anguillarum* and *V. ordalii*, respectively. *V. anguillarum* has also been known as *Vibrio anguillarum* biotype 1 (Schiewe et al., 1977), *Beneckea anguillara* biotype I (Baumann et al., 1978), *Vibrio anguillarum* biotype I (Baumann et al., 1980b), and *Vibrio anguillarum* biovar I (Baumann et al., 1984b). In addition, the name *Listonella anguillarum* has standing in the nomenclature (MacDonell and Colwell, 1985); see description of *V. pelagius*.

TABLE BXII.γ.170. Differentiation of the arginine-negative, lysine-positive species *V. alginolyticus*, *V. parahaemolyticus*, *V. vulnificus*, and *V. harveyi* [a]

Test or property	V. alginolyticus	V. harveyi	V. parahaemolyticus	V. vulnificus
Voges–Proskauer test	+	−	−	−
Growth in nutrient broth with:				
8% NaCl	+	−	(+)	−
10% NaCl	d	−	−	−
Fermentation of:				
Sucrose	+	+	−	(−)
Salicin	−	−	−	+
Cellobiose	−	−	−	+
Lactose	−	−	−	(+)
L-Arabinose	−	−	(+)	−
Swarming (marine agar, 25°C)	+	+	+	−
Size of zone of inhibition around:				
Colistin	Large	Small	Large	Small
Ampicillin	Small	Small	Small	Large
Carbenicillin	Small	Small	Small	Large

[a]All data for biochemical reactions are within 2 days at 35–37°C. Symbols: +, 90–100% positive, (+), 75–89.9% positive; d, 25.1–74.9% positive; (−), 10.1–25% positive; −, 0–10% positive.

guillarum and *V. ordalii* are closely related in a phylogenetic sense, they differ considerably in phenotypic properties (Table BXII.γ.171). *Vibrio anguillarum* is more versatile biochemically, has more diverse O antigens, and is less fastidious.

The mol% G + C of the DNA is: 44–46 (T_m, Bd).

Type strain: ATCC 19264, IFO 13266, LMG 4437, NCCB 72050.

GenBank accession number (16S rRNA): X16895.

6. **Vibrio campbellii** (Baumann, Baumann and Mandel 1971a) Baumann, Baumann, Bang and Woolkalis 1981, 217[VP] (Effective publication: Baumann, Baumann, Bang and Woolkalis 1980b, 128) (*Beneckea campbellii* Baumann, Baumann and Mandel 1971a, 288.)

camp.bel' li.i. M.L. gen. n. *campbelli* named to honor L.L. Campbell, an American bacteriologist.

The characteristics are as described for the genus and as listed in Tables BXII.γ.163 and BXII.γ.164. The cell morphology is illustrated in Fig. BXII.γ.174.

Occurs in ocean water some distance from the cost of Hawaii (Baumann et al., 1971a) and in open waters of the Atlantic ocean (Grimes et al., 1986).

Phylogenetically, *Vibrio campbellii* is one of the marine *Vibrio* species (Figs. BXII.γ.170 and BXII.γ.171). By DNA–DNA hybridization, it is most closely related to *V. harveyi* (61–74%); this result agrees with the close relationship indicated by 16S rRNA gene sequence analysis (Fig. BXII.γ.171). Based on the results of DNA–DNA hybridization experiments *V. campbellii* is 40–57% related to several of the marine *Vibrio* species (Reichelt et al., 1976; Brenner et al., 1983b).

TABLE BXII.γ.171. Properties and differentiation of *Vibrio anguillarum* and *V. ordalii*[a]

Test or property	V. anguillarum	V. ordalii
Number of strains studied	12	11
O antigens represented	O1–O16	O2 only
Colony size on TSA after 48 hours	3–5 mm	Barely visible
Growth at 37°C	92	0
Growth at 25°C	100	100
Growth at 22°C	100	100
Growth at 15°C	100	100
Growth in 0% NaCl	0	0
Growth in 0.5% NaCl	100	100
Growth in 3% NaCl	100	100
Growth in 7% NaCl	0	0
Indole production	58	0
Voges–Proskauer	100	0
Arginine dihydrolase	100	0
Nitrate reduction to nitrite	100	66
Citrate utilization (Simmons)	75	0
Citrate utilization (Christensen)	100	0
Starch hydrolysis	100	0
ONPG test	100	0
Lipase	100	0
L-Arabinose fermentation	58	0
Cellobiose fermentation	92	0
D-Galactose fermentation	100	55
Glycerol fermentation	100	0
D-Sorbitol fermentation	100	0
Trehalose fermentation	100	0

[a]Data are summarized from Table 2 of Schiewe et al. (1981). Conditions: incubation at 22°C; media were supplemented with NaCl to final concentration of 1.5%; final reading for negative tests was 14 days. Each number gives the percentage of strains that were positive for the test.

The mol% G + C of the DNA is: 46–48 (T_m, Bd).

Type strain: ATCC 25920, IFO 15631, LMG 11216, NCCB 73002.

GenBank accession number (16S rRNA): X56575.

7. **Vibrio cincinnatiensis** Brayton, Bode, Colwell, MacDonell, Hall, Grimes, West and Bryant 1986a, 355[VP] (Effective publication: Brayton, Bode, Colwell, MacDonell, Hall, Grimes, West and Bryant 1986b, 107.)

cin.cin.nat.i.en' sis. L. masc. adj. *cincinnatiensis* derived from Society of Cincinnati, from which the city of Cincinnati, Ohio, was named.

The characteristics are as described for the genus and as listed in Tables BXII.γ.163 and BXII.γ.164. Characters that distinguish *V. cincinnatiensis* from other *Vibrio* spp. found in human clinical specimens are given in Tables BXII.γ.156, BXII.γ.157, BXII.γ.160, BXII.γ.165 and BXII.γ.166. Small rods approximately 0.7–2.0 µm; motile; single polar flagellum. Colonies on nutrient agar are 1–2 mm in diameter, round, smooth, glossy, and cream colored. Grows poorly on TCBS medium (Table BXII.γ.160), but isolates from extraintestinal clinical specimens should grow on one of the non-selective plating media such as blood agar. Requires NaCl for growth (Tables BXII.γ.163 and BXII.γ.165). Ferments *myo*-inositol (Table BXII.γ.163).

Rarely isolated from human clinical specimens.

This organism was isolated by Brayton et al. (1986b) from blood and cerebral spinal fluid of a 70-year-old man with no known contact with seafood or saltwater. Biochemical testing and the 5S rRNA sequence led to the conclusion that this was a new species, which they named *V. cincinnatiensis*. Comparative 5S rRNA sequence analysis indicated that *V. cincinnatiensis* shared a recent ancestor with *V. gazogenes* (98% homology), but this postulated relationship has not been confirmed by other techniques including 16S rRNA gene sequencing (Fig. BXII.γ.171).

The mol% G + C of the DNA is: 45.0.

Type strain: ATCC 35912, CIP 104173, LMG 789.

GenBank accession number (16S rRNA): X74698.

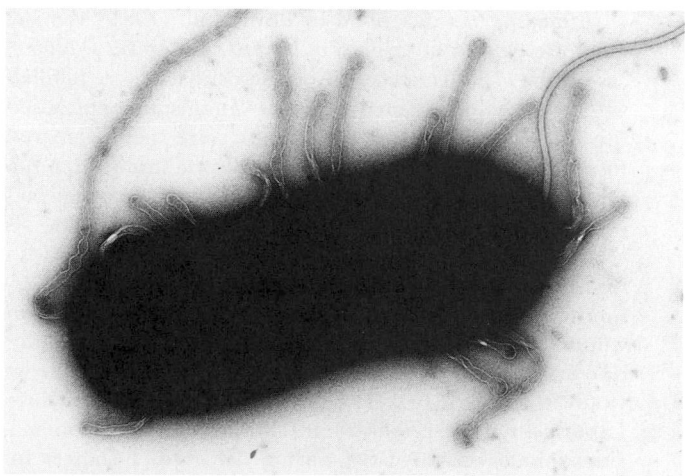

FIGURE BXII.γ.174. Electron micrograph of *Vibrio campbellii* illustrating the appearance of the tubular appendages that are also observed in some other marine vibrios. Negatively stained preparation. × 36,000. (Reproduced with permission from R.D. Allen and P. Baumann, *Journal of Bacteriology 107*: 295–302, 1971, ©American Society for Microbiology.)

8. **Vibrio cyclitrophicus** Hedlund and Staley 2001, 65[VP] (*Vibrio cyclotrophicus* (sic) Hedlund and Staley 2001, 65.)

cy′cli.tro′phi.cus. Gr. n. *kyklos* circle or ring; Gr. adj. *trophikos* pertaining to food.

Properties of *V. cyclitrophicus* are given in Table BXII.γ.164. Rods 0.6 × 1.5–2.5 µm. Motile by 1–2 polar or subpolar flagella. Grows in media containing ≥ 1.75% NaCl. Utilizes some polycyclic aromatic hydrocarbons as carbon and energy sources.

The type strain was isolated from creosote-contaminated marine sediment, Eagle Harbor, Puget Sound, Washington, USA, in 1993.

An analysis of the 16S rDNA sequences of 10 *Vibrio* species indicated that the closest relative of *V. cyclitrophicus* was *V. splendidus* (Hedlund and Staley, 2001). Fig. BXII.γ.171 illustrates the relationship of *V. cyclitrophicus* to other species of *Vibrio*.

The mol% G + C of the DNA is: 39 (T_m).

Type strain: P-2P44, ATCC 700982, CIP 106644.

GenBank accession number (16S rRNA): U57919.

9. **Vibrio damsela** Love, Teebken-Fisher, Hose, Farmer, Hickman, and Fanning 1982, 267[VP] (Effective publication: Love, Teebken-Fisher, Hose, Farmer, Hickman, and Fanning 1981, 1140.)*

dam.sel′a. M.L. n. *damsela* from the modern zoological term damselfish.

Classified as *Listonella damsela* by MacDonell and Colwell (1985, 1986) and as *Photobacterium damselae* by Smith et al. (1991); Trüper and De′ Clari (1997) corrected spelling of the species epithet of *Photobacterium damselae* to *damselae*.

Morphological, growth, physiological and biochemical characteristics are listed in Table BXII.γ.163. Biochemical and other phenotypic characteristics useful for identification are listed in Table BXII.γ.164. Additional characteristics are listed in Tables BXII.γ.156, BXII.γ.157, BXII.γ.160, BXII.γ.161, BXII.γ.165, BXII.γ.166, and BXII.γ.173. Causes skin ulcers on the blacksmith *Chromis punctipinnis* (a damselfish) and causes human wound infections. Isolated from marine algae, which may be important in disease transmission to fish.

V. damsela was described by Love et al. (1981) who isolated the type strain from skin lesions on *Chromis punctipinnis* off the California coast. Koch's postulates were fulfilled when it was documented that this organism alone caused the fish lesions. Strains of *V. damsela* were then compared to a large collection of unidentified *Vibrio* strains from human clinical specimens, which revealed that *V. damsela* had occurred in human infections (Love et al., 1981). Morris et al. (1982) reviewed the case histories of six patients with wound infections positive for *V. damsela*. All had been acquired in coastal areas. *V. damsela* appears to cause human wound infections, but its causal role needs to be further documented. Most of the clinical isolates have been from wounds, and 9 of 10 were from leg or foot wounds (Vibrio Laboratory, CDC, unpublished). Strains were also from marine animals, particularly marine fish. Other sources include: fish from Senegal, sewage, oysters, and a wound cul-

ture from a raccoon (Vibrio Laboratory, CDC, unpublished). By DNA–DNA hybridization, strains of *V. damsela* are 76–97% related to the type strain and only 0–26% related to all other *Vibrio* species (Brenner et al., personal communication). Based on 16S rDNA sequencing *V. damsela* is more closely related to several *Photobacterium* species than to *Vibrio cholerae*.

The mol% G + C of the DNA is: 43 (Bd).

Type strain: CDC 2588-80, ATCC 33539, DSM 7482.

GenBank accession number (16S rRNA): X74700.

10. **Vibrio diabolicus** Raguénès, Christen, Guezennec, Pignet and Barbier 1997a, 994[VP]

di.a.bol′i.cus. L. adj. *diabolicus* devilish, diabolic.

The morphological, growth, physiological and biochemical characteristics are listed in Tables BXII.γ.154 and BXII.γ.164. Small cells 0.8 × 2.2 µm, motile via a single polar flagellum in liquid media. Produce swarming colonies on solid media; the cells from these colonies have lateral flagella. The organisms synthesize a novel exopolysaccharide that contains high levels of uronic acids and hexosamines when grown in batch culture with glucose.

Only a single strain of this species has been described. This strain was isolated from the polychaete annelid *Alvinella pompejana* from a deep-sea (2600 m) hydrothermal vent in the East Pacific Rise and was found during a search for novel exopolysaccharides of marine origin.

Based on DNA–DNA hybridization experiments Raguénès et al., 1997a found that the closest relatives of *V. diabolicus* were *V. mytili* (27% relatedness) and *V. nereis* (15%). The closest relatives indicated by the results of 16S rRNA gene sequence analysis were *V. mytili* (98.9%), *V. alginolyticus* (98.9%), *V. pelagius* (98.9%), *V. parahaemolyticus* (98.8%), *V. natriegens* (98.7%), *V. nereis* (98.5%), *V. campbellii* (98.5%), and *V. tubiashii* (98.4%). The authors concluded that "*Vibrio diabolicus* belongs to a well-defined monophyletic unit that also includes *V. tubiashii*, *V. nereis*, and *V. mytili*." Fig. BXII.γ.171 illustrates the relationship of *V. diabolicus* to other *Vibrio* species based on 16S rRNA gene sequence analysis.

The mol% G + C of the DNA is: 49.6 (T_m).

Type strain: HE800, CNCM I-1629.

GenBank accession number (16S rRNA): X99762.

11. **Vibrio diazotrophicus** Guerinot, West, Lee and Colwell 1982, 356[VP]

di.a.zo.tro′phi.cus. Gr. prefix *di* two, double; N.L. n. *azotum* nitrogen; Gr. n. *trophus* one that feeds; L. suffix *icus* relating to; M.L. adj. *diazotrophicus* one that feeds on dinitrogen, or less literally, nitrogen fixing.

The characteristics are as described for the genus and as listed in Tables BXII.γ.163 and BXII.γ.164. Small cells 0.5 × 1.5–2.0 µm. Motile by means of a sheathed polar flagellum when grown in broth. Cells grown on solid media produce unsheathed lateral flagella with a short wavelength. The organisms fix nitrogen.

Distributed throughout estuarine and marine environments. Isolated from the intestinal tract of sea urchins, river water, Chesapeake Bay water, ditch water and sediment, and the surface of reeds growing in a drainage ditch.

This organism may occur in human clinical specimens.

Editorial Note: Photobacterium damselae subsp. *damselae* is a junior objective synonym of *Vibrio damsela*.

Culture 2550-90 sent to the *Vibrio* Reference Laboratory was isolated from human blood and was reported as "*Vibrio diazotrophicus* group" (*Vibrio* Reference Laboratory, CDC, unpublished). If confirmed, it would be the first clinical isolate of this species.

Based on DNA–DNA hybridization experiments, Guerinot et al. (1982) found that four strains were 88–100% related to the type strain; however, one strain was only 62% related to the type strain. Also based on DNA–DNA hybridization experiments, Brenner et al. (unpublished) found that strains of *V. diazotrophicus* were highly related to the type strain (77–84% related), but distantly related to all other named *Vibrio* species. Unnamed "*Vibrio* group 510" was the closest relative (65–74% related).

The mol% G + C of the DNA is: 45.9–47.2 (T_m, Bd).

Type strain: NS1, ATCC 33466, DSM 2604, IAM 14402.

GenBank accession number (16S rRNA): X56577, X74701.

12. **Vibrio fischeri** (Beijerinck 1889) Lehmann and Neumann 1896, 342[AL] (*Photobacterium fischeri* Beijerinck 1889, 402.)

fisch'er.i. M.L. gen. n. *fischeri* named to honor Bernhard Fischer, one of the earliest microbiologists to study bioluminescent bacteria.

The characteristics are as described for the genus and as listed in Tables BXII.γ.163 and BXII.γ.164. The cell morphology is as depicted in Figs. BXII.γ.161, BXII.γ.165, and BXII.γ.175. Produces a cell-associated yellow-orange pigment.

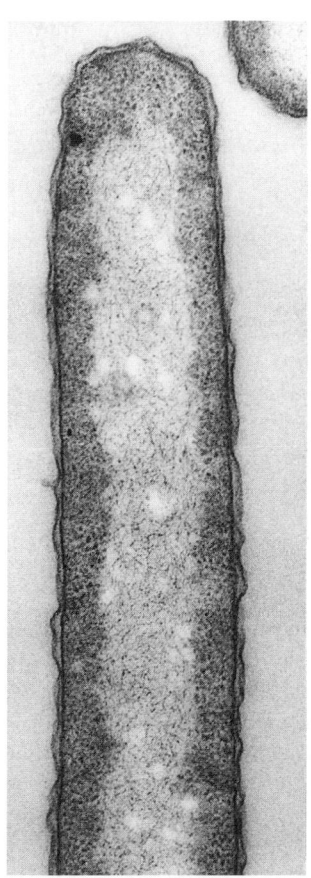

FIGURE BXII.γ.175. Ultrathin section of *Vibrio fischeri*. Exponential growth phase in yeast extract broth. × 40,000. (Courtesy of S.W. Watson.)

Isolated from seawater and marine animals.

Phenotypically and phylogenetically, *Vibrio fischeri* is closely related to *Vibrio logei* (Tables BXII.γ.163 and BXII.γ.164; Figs. BXII.γ.170 and BXII.γ.171). Based on DNA–DNA hybridization experiments, the two species are 36% related (Brenner et al., unpublished)

The mol% G + C of the DNA is: 39–41 (T_m, Bd).

Type strain: ATCC 7744 (strain 398, DSM 507.

GenBank accession number (16S rRNA): X70640.

13. **Vibrio fluvialis** Lee, Shread, Furniss and Bryant 1981c, 217[VP] (Effective publication: Lee, Shread, Furniss and Bryant 1981b, 92.)

flu.vi.a' lis. L. adj. *fluvialis* of or belonging to a river.

The characteristics are as described for the genus and as listed in Tables BXII.γ.163 and BXII.γ.164. Characters that distinguish *V. fluvialis* from *V. furnissii* are given in Tables BXII.γ.172 and BXII.γ.173. Characters that distinguish *V. fluvialis* from *Vibrio damsela* and *Aeromonas* spp. are given in Table BXII.γ.173. Characters that distinguish *V. fluvialis* from other *Vibrio* spp. found in human clinical specimens are given in Tables BXII.γ.156, BXII.γ.157, BXII.γ.160, BXII.γ.161, BXII.γ.165 and BXII.γ.166.

Causes sporadic cases of human diarrhea and occasional outbreaks. Also isolated from natural waters (often from estuaries) and animals that live there.

Lee et al. (1981b) gave the name *Vibrio fluvialis* to a group of halophilic vibrios that had been previously been known as "marine aeromonads" and "Group F" in the United Kingdom (Furniss et al., 1977) and as "Group EF6" at the Special Bacteriology Laboratory, CDC (Huq et al., 1980). These organisms had been isolated from a number of sources throughout the world and from humans with diarrhea. Lee et al. (1981b) defined two biogroups in this species based on the source of isolation and the results of certain biochemical tests. *V. fluvialis* biogroup I strains did not produce gas during fermentation and were isolated from cases of human diarrhea as well as from the environment. *V. fluvialis* biogroup II strains produced gas during fermentation and were isolated from the environment, but not from humans with diarrhea. Brenner et al. (1983c) later used DNA–DNA hybridization to show that strains of *V.*

TABLE BXII.γ.172. Differentiation of *Vibrio fluvialis* and *V. furnissii*

Test or property	Percentage positive for:	
	V. fluvialis	*V. furnissii*
Simple tests:[a]		
Gas production during fermentation	0	99
Esculin hydrolysis	72	0
Carbon source utilization:[b]		
Citrulline	97	4
D-Glucuronic acid	94	4
Putrescine	31	100
δ-Aminovalerate	0	63
Cellobiose	63	4
Glutaric acid[c]	−	+

[a]For the two simple tests, each number gives the percentage positive after 48h incubation at 35–37°C. Most of the positive reactions occur during the first 24 hours.

[b]Data from Lee et al. (1981b) except for glutaric acid, which is from Baumann et al. (1984b). These carbon-source utilization tests are usually done in research laboratories rather than in public health or clinical laboratories.

[c]Percentage data not available.

TABLE BXII.γ.173. Differentiation of the arginine-positive species *V. fluvialis*, *V. furnissii*, and *V. damsela* and comparison with *Aeromonas*[a]

Test or property	V. damsela	V. fluvialis	V. furnissii	Aeromonas
Growth in nutrient broth with:				
No added NaCl	−	−	−	+
1% NaCl	+	+	+	+
6% NaCl	+	+	+	−
Voges–Proskauer test	+	−	−	d
Citrate, Simmons	−	+	+	d
Fermentation of:				
D-Galacturonate	−	+	+	−
L-Arabinose	−	+	+	d
D-Mannitol	−	+	+	+
Sucrose	−	+	+	(+)
Gas production during fermentation	d	−	+	d

[a]All data are for reactions within 2 days at 35–37°C unless otherwise specified. Symbols: +, 90–100% positive; (+), 75–89.9% positive; d, 25.1–74.9 positive; −, 0–10% positive.

fluvialis biogroup II were related to strains of *V. fluvialis* biogroup I, but they were sufficiently different to be a new species, which was named *V. furnissii*.

The sources of early clinical isolates of *V. fluvialis* showed a marked association with diarrhea (Lee et al., 1981b), and further studies have strengthened its causative role. One difficulty in assessing the clinical symptoms and epidemiology of *V. fluvialis* has been the presence of other possible pathogens; most reports have come from geographical areas where several possible pathogens are often present in feces. Gastroenteritis associated with *V. fluvialis* is usually described as "cholera-like". Symptoms include watery diarrhea with vomiting (97% of patients), abdominal pain (75%), moderate to severe dehydration (67%), and fever (35%). Usually infants, children, and young adults are affected. Frank blood is found in a small percentage of stool samples, but red or white blood cells are found in most cases (75%). The pathogenesis of *V. fluvialis* diarrhea is beginning to be studied. Assays for heat-labile and heat-stable enterotoxin (LT and ST) and invasiveness have generally been negative. About 20% of isolates give a positive rabbit ileal loop test for enterotoxin. Recent studies have provided additional evidence for enterotoxin or for "enterotoxin-like" molecules (Lockwood et al., 1982; Nishibuchi et al., 1983).

The mol% G + C of the DNA is: 49.3–50.3 (T_m, Bd).

Type strain: VL 5125, ATCC 33809, IMET 11293, NCTC 11327.

GenBank accession number (16S rRNA): X74703, X76335.

14. **Vibrio furnissii** Brenner, Hickman-Brenner, Lee, Steigerwalt, Fanning, Hollis, Farmer, Weaver, Joseph and Seidler 1984b, 91[VP] (Effective publication: Brenner, Hickman-Brenner, Lee, Steigerwalt, Fanning, Hollis, Farmer, Weaver, Joseph and Seidler 1983c, 823.)

fur.niss'i.i. L. gen. n. *furnissii* of Furniss; named to honor A.L. Furniss, Maidstone Public Health Laboratories, Maidstone, England who was involved in the original discovery and naming of the "group F" vibrios.

The characteristics are as described for the genus and as listed in Tables BXII.γ.163 and BXII.γ.164. Characters that distinguish *V. furnissii* from *V. fluvialis* are given in Tables BXII.γ.172 and BXII.γ.173. Characters that distinguish *V. furnissii* from *Vibrio damsela* and *Aeromonas* spp. are

given in Table BXII.γ.173. Characters that distinguish *V. furnissii* from other *Vibrio* spp. found in human clinical specimens are given in Tables BXII.γ.156, BXII.γ.157, BXII.γ.160, BXII.γ.161, BXII.γ.165 and BXII.γ.166.

Widely distributed in natural waters (often in estuaries) and animals that live there; occasionally isolated from animal feces. Although *V. furnissii* strains were isolated from two outbreaks of acute gastroenteritis, other pathogens or possible pathogens were also isolated (*V. parahaemolyticus*, *V. cholerae* non-O1, *V. fluvialis*, *Salmonella*, and *Plesiomonas*). Therefore, the role of *V. furnissii* as the actual cause of the diarrhea was only speculative.

V. furnissii was established for strains that resembled *V. fluvialis* but produced gas during fermentation. Based on DNA–DNA hybridization experiments, three strains of *V. furnissii* were shown to be 79% or more related to the type strain of *V. furnissii* but only 50% related to the type strain of *V. fluvialis* (Brenner et al., 1984b).

The mol% G + C of the DNA is: 50.4 (T_m).

Type strain: ATCC 35016, CDC B3215, CIP 102972, LMG 7910.

GenBank accession number (16S rRNA): X74704, X76336.

15. **Vibrio gazogenes** (Harwood, Bang, Baumann and Nealson 1980) Baumann, Baumann, Bang and Woolkalis 1981, 217[VP] (Effective publication: Baumann, Baumann, Bang and Woolkalis 1980b, 128) (*Beneckea gazogenes* (ex Harwood 1978) Harwood, Bang, Baumann and Nealson 1980, 655.) *gaz.og'e.nes.* Fr. n. *gaz* gas; L. n. *genesis* birth; M.L. adj. *gazogenes* gas-producing.

The characteristics are as described for the genus and as listed in Tables BXII.γ.163 and BXII.γ.164. Relatedness of strains based on DNA–DNA hybridization experiments is given in Figs. BXII.γ.172 and BXII.γ.176. Three characters distinguish *V. gazogenes* from all other *Vibrio* species: production of a non-diffusible red ("prodigiosin-like") pigment, a negative reaction for oxidase, and an inability to reduce NO_3^- to NO_2^-. The determination of the oxidase reaction is complicated by the red pigmentation of colonies; this difficulty can be overcome by testing spontaneous white colonial mutants, which are produced by most cultures held for long periods in agar slants or deeps.

Isolated from estuary water and salt marsh mud off the eastern coast of the United States (Farmer et al., 1988), and from a hypersaline lake in Baja California (Giovannoni and Margulis, 1981).

DNA–DNA hybridization experiments (Figs. BXII.γ.172 and BXII.γ.176) showed that strains of *V. gazogenes* fall into three highly related groups (Farmer et al., 1988). However, the three groups cannot be completely differentiated by phenotype. The type strain is less than 15% related to other *Vibrio* species tested (Brenner et al., unpublished). Analysis of 16S rRNA gene sequences showed that *V. gazogenes* is not closely related to other *Vibrio* species (Fig. BXII.γ.171).

The mol% G + C of the DNA is: 47.1 (T_m).
Type strain: PB1, ATCC 29988.
GenBank accession number (16S rRNA): X74705.

16. **Vibrio halioticoli** Sawabe, Sugimura, Ohtsuka, Nakano, Tajima, Ezura and Christen 1998b, 578[VP]

hal.i.o.ti' co.li. M.L. n. *Haliotis* genus name for abalone, a rock-climbing gastropod mollusk; Gr. n. *colon* gut; M.L. gen. n. *halioticoli* pertaining to the gut of *Haliotis*.

The morphological, growth, physiological and biochemical characteristics are listed in Tables BXII.γ.154 and BXII.γ.164. Cells are rod shaped with rounded ends when grown in Zobell's medium 2216A but are spherical when grown in this medium with added alginate. Nonmotile. Cells bind to and utilize alginic acid.

Isolated from gut of the abalone *Haliotis discus hannai*, where this bacterium may aid in the digestion of algae that contain alginic acid.

Sawabe et al. (1998b) concluded that *Vibrio halioticoli* is a separate *Vibrio* species with no close relatives based on DNA–DNA hybridization experiments and 16S rRNA gene sequence analysis (also see Fig. BXII.γ.171).

The mol% G + C of the DNA is: 41.6–43.1 (HPLC).
Type strain: A431, ATCC 700680, IAM 14596.
GenBank accession number (16S rRNA): AB000390.

17. **Vibrio harveyi** (Johnson and Shunk 1936) Baumann, Baumann, Bang and Woolkalis 1981, 217[VP] (Effective publication: Baumann, Baumann, Bang and Woolkalis 1980b, 128) (*"Achromobacter harveyi"* Johnson and Shunk 1936, 587; *Lucibacterium harveyi* (Johnson and Shunk 1936) Hendrie, Hodgkiss and Shewan 1970, 166; *Beneckea harveyi* (Johnson and Shunk 1936) Reichelt and Baumann 1973b, 320; *Vibrio carchariae* Grimes, Stemmler, Hada, May, Maneval, Hetrick, Jones, Stoskopf and Colwell 1985, 224.)

har' vey.i. M.L. gen. n. *harveyi* of Harvey; named to honor E.N. Harvey, a biologist who was a pioneer in the systematic study of bioluminescence.

The characteristics are as described for the genus and as listed in Tables BXII.γ.163 and BXII.γ.164. Characters that differentiate *V. harveyi* from *V. alginolyticus*, *V. parahaemolyticus*, and *V. vulnificus* are given in Table BXII.γ.170. Characters that distinguish *V. harveyi* from other *Vibrio* spp. found in human clinical specimens are given in Tables BXII.γ.156, BXII.γ.157, BXII.γ.160, BXII.γ.165 and BXII.γ.166. The morphology is as illustrated in Figs. BXII.γ.163 and BXII.γ.177.

FIGURE BXII.γ.176. Subgroups in the species *Vibrio gazogenes* based on divergence values in DNA–DNA hybridization experiments. Divergence values reflect non-hybridizing DNA sequences. The numbers above or beside the dots are the strain numbers given in Farmer et al. (1988). (Reproduced with permission from J.J. Farmer III et al., Journal of Clinical Microbiology *2l*: 46–76, 1988, ©American Society for Microbiology.)

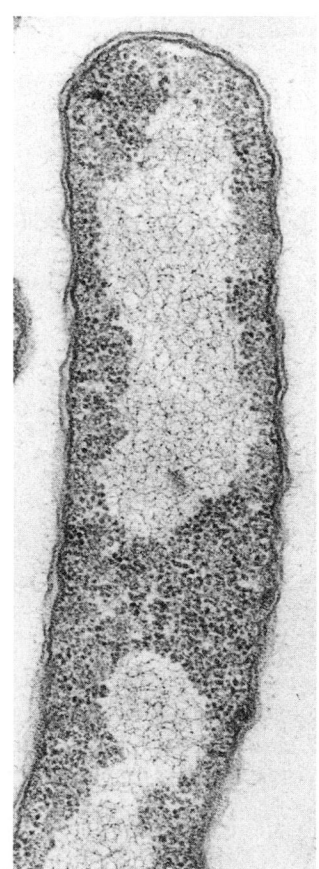

FIGURE BXII.γ.177. Ultrathin section of *Vibrio harveyi*. Exponential growth phase in yeast extract broth. × 43,000. (Courtesy of S.W. Watson.)

Occurs in marine animals, the mouth of sharks (reported as *Vibrio carchariae*), and the marine environment. Isolated from a human wound infection (reported as *Vibrio carchariae*) associated with a shark bite. The type strain was isolated from an amphipod (*Talorchestia* sp.).

Grimes et al. (1984), studied two urease-positive halophilic vibrios isolated from a brown shark (*Carcharhinus plumbeus*) that had died in captivity in a large aquarium. One of the organisms (strain 1116a) was identified as *V. damsela*, but the other (strain 1116b) could not be identified. After further phenotypic testing and DNA–DNA hybridization, Grimes et al. (1984) concluded that this was a new species and named it *V. carchariae*. *V. carchariae* has been isolated from other sharks since the original report, but it was recently shown to occur in human clinical specimens. Pavia et al. (1989) described a case in which the organism was isolated from a wound following a shark bite.

The name *Vibrio carchariae* was used for this organism until the identity of *V. carchariae* with *V. harveyi* was shown conclusively (Farmer and Hickman-Brenner, 1992; Pedersen et al., 1998; Brenner et al., unpublished; see Fig. BXII.γ.171). *Vibrio carchariae* and *Vibrio harveyi* are objective synonyms.

The mol% G + C of the DNA is: 46–48 (T_m).

Type strain: 384, ATCC 14126, CCUG 28584, CIP 103192, IFO 15634, LMG 4044, NCMB 1280.

GenBank accession number (16S rRNA): X56578, X74706.

18. **Vibrio hollisae** Hickman, Farmer, Hollis, Fanning, Steigerwalt Weaver, and Brenner 1982b, 384[VP] (Effective publication: Hickman, Farmer, Hollis, Fanning, Steigerwalt, Weaver and Brenner 1982a, 398.)

hol'lis.ae. M.L. gen. n. *hollisae* of Hollis; named to honor Dannie G. Hollis, who first recognized this organism as a new and distinct vibrio.

The characteristics are as described for the genus and as listed in Tables BXII.γ.163 and BXII.γ.164. Characters that distinguish *V. hollisae* from other *Vibrio* spp. found in human clinical specimens are given in Tables BXII.γ.156, BXII.γ.157, BXII.γ.160, BXII.γ.161, BXII.γ.165 and BXII.γ.166. Small cells, 0.5 × 1.5–2 μm with a single polar flagellum when grown in liquid media. On sheep blood agar colonies are 1–2 mm in diameter and are weakly hemolytic. Does not grow on TCBS or MacConkey agar.

Appears to cause sporadic cases of diarrhea in humans and has been isolated from blood. Rarely isolated from other human or environmental sources. The type strain was isolated from human feces.

V. hollisae is strongly associated with human diarrhea, but additional evidence is needed to further document the causal role and pathogenic mechanisms of this organism. Fifteen of the original sixteen strains were from feces; many of those patients had diarrhea (Hickman et al., 1982a). Morris et al. (1982) described the clinical and epidemiological features of 11 cases of diarrhea from which *V. hollisae* was isolated.

Based on DNA–DNA hybridization experiments, strains of *V. hollisae* are 76–97% related to the type strain and only 0–4% related to all other *Vibrio* species (Brenner et al. unpublished). 16S rRNA gene sequence analysis indicates that

V. hollisae is not closely related to other *Vibrio* species (Fig. BXII.γ.171).

V. hollisae has also been known as "Group EF-13" of the Special Bacteriology Laboratory, CDC, unpublished (see Hickman et al., 1982a).

The mol% G + C of the DNA is: 49.3–51.0.

Type strain: ATCC 33564, CDC 0075-80, IMET 12291.

GenBank accession number (16S rRNA): X56583, X74707.

19. **Vibrio ichthyoenteri** Ishimaru, Akagawa-Matsushita and Muroga 1996, 159[VP]

ich.thy.o.er' te.ri. Gr. n. *ichthys* fish; Gr. n. *enteron* gut; Gr. gen. n. *ichthyoenteri* of fish gut.

The characteristics are as described for the genus and as listed in Tables BXII.γ.154, BXII.γ.163 and BXII.γ.164. Short rods, straight or slightly curved, 1.6–2.5 × 0.6–0.8 μm. Motile by a single polar flagellum. No lateral flagella are produced when grown on solid media. Most strains produce yellow colonies on TCBS agar, but the type strain is a weak sucrose fermenter and produces green colonies.

Causes opaque intestines, intestinal necrosis, and enteritis with high mortality rates in Japanese flounder larvae (*Paralichthys olivaceus*) reared in marine hatcheries.

V. ichthyoenteri was previously known as "*Vibrio* species INFL" (Masumura et al., 1989). Ishimaru et al. (1996) found that five strains were 89–100% related to the type strain based on DNA–DNA hybridization experiments. There was only one close relative; strain P-8706 of "*Vibrio* species INFL-2" from a diseased flounder larva, which was 37% related. Other *Vibrio* species were only 3–18% related.

The mol% G + C of the DNA is: 43.4–44.3 (HPLC).

Type strain: : F-2, ATCC 700023, IFO 15847.

20. **Vibrio logei** (Harwood, Bang, Baumann and Nealson 1980) Baumann, Baumann, Bang and Woolkalis 1981, 217[VP] (Effective publication: Baumann, Baumann, Bang and Woolkalis 1980b, 128) (*Photobacterium logei* (ex Bang, Baumann and Nealson 1978) Harwood, Bang, Baumann and Nealson 1980, 655; *Vibrio salmonicida* Egidius, Wiik, Anderson, Hoff and Hjeltnes 1986, 519.)

log' e.i. M.L. gen. n. *logei* of Loge; from the German Loge, Norse god of fire and mischief.

The characteristics are as described for the genus and as listed in Tables BXII.γ.163 and BXII.γ.164. Produces a characteristic yellow-orange pigment that is cell-associated.

Causes Hitra disease ("hemorrhagic syndrome")—a disease originally thought to be caused by *Vibrio salmonicida*—in salmonid fish raised in ponds in Norway. Isolated from scallops, intestinal contents of fish, marine sediments, and lesions of the exoskeleton of tanner crabs (Baumann and Baumann, 1981a). The type strain was isolated from the intestine of arctic mussel.

The mol% G + C of the DNA is: 40–42 (T_m, Bd).

Type strain: 584, ATCC 29985, CIP 104991.

21. **Vibrio mediterranei** Pujalte and Garay 1986, 279[VP]

me.di.ter.ra' ne.i. L. gen. n. *mediterranei* of the Mediterranean Sea.

The characteristics are as described for the genus and as listed in Tables BXII.γ.154, BXII.γ.163, and BXII.γ.164. Cells are 0.5 × 1.0–2.0 μm and possess polar flagella when grown in liquid medium. Colonies are circular, translucent, nonswarming, and nonpigmented on marine agar and yellow on TCBS agar.

Isolated from plankton, sediments, and seawater off the coast of Valencia, Spain.

Based on DNA–DNA hybridization experiments, the type strain of *V. mediterranei* is highly related (75–100%) to other strains included in the species but is not closely related to other *Vibrio* species. *V. harveyi* was the closest relative (11%) in the original study and *V. alginolyticus* was the closest relative (25%) in the DNA–DNA hybridization study of Brenner et al. (unpublished). Fig. BXII.γ.171 illustrates its relationships to other *Vibrio* species based on 16S rRNA gene sequence analysis.

The mol% G + C of the DNA is: 42–43 (T_m).

Type strain: ATCC 43341, CECT 621, CIP 103203, NCTC 11946.

GenBank accession number (16S rRNA): X74710.

22. **Vibrio metschnikovii** Gaméléia 1888, 485[AL] (*Vibrio cholerae* biovar proteus Shewan and Véron 1974, 344.)

metsch.ni.kov'i.i. M.L. masc. gen. n. *metschnikovii* of Metschnikoff; named to honor E. Metschnikoff, a Russian biologist.

The characteristics are as described for the genus and as listed in Tables BXII.γ.163 and BXII.γ.164. Characters that distinguish *V. metschnikovii* from other *Vibrio* spp. found in human clinical specimens are given in Tables BXII.γ.156, BXII.γ.157, BXII.γ.160, BXII.γ.161, BXII.γ.165 and BXII.γ.166. Oxidase negative. Does not reduce nitrate to nitrite.

Frequently isolated from fresh, brackish, and marine waters, but rarely from human clinical specimens (Lee et al., 1978). The type strain was isolated from fowl.

In 1888, Gaméléia reported the isolation of a new organism, *Vibrio metschnikovii*, from a fowl that had died of a "cholera-like" disease (Lee et al., 1978). Little was written about the organism until Lee et al. (1978) began isolating similar organisms from marine and fresh water environments in the United Kingdom. These strains were oxidase negative and did not reduce nitrate to nitrite but otherwise were typical halophilic vibrios. Lee et al. (1978) proposed that the strains be classified as *V. metschnikovii*, and emended the description of this organism. Lee et al. (1978) described 40 strains of *V. metschnikovii*, but none were from human clinical specimens. Jean-Jacques et al. (1981) reported the first clinically significant isolate of this newly redefined species. A blood culture from an 82-year-old woman with peritonitis and an inflamed gallbladder yielded *V. metschnikovii*. Nonhuman sources include rivers, sewage, cockles, shrimp, lobster, crab, and fowl (Lee et al., 1978; *Vibrio* Laboratory, CDC, unpublished). These data suggest that *V. metschnikovii* is widely distributed in the environment and that humans may acquire the organism from these sources.

The mol% G + C of the DNA is: 44–46 (T_m, Bd).

Type strain: ATCC 700040, LMG 11664, NCTC 8443.

23. **Vibrio mimicus** Davis, Fanning, Madden, Steigerwalt, Bradford, Smith and Brenner 1982, 267[VP] (Effective publication: Davis, Fanning, Madden, Steigerwalt, Bradford, Smith and Brenner 1981a, 636.)

mim'i.cus. M.L. adj. *mimicus* pertaining to a mimic; named to indicate its close phenotypic similarity to *V. cholerae*.

The characteristics are as described for the genus and as listed in Tables BXII.γ.163 and BXII.γ.164. Characters that distinguish *V. mimicus* from *V. cholerae* are given in Table BXII.γ.169. Characters that distinguish *V. mimicus* from other *Vibrio* spp. found in human clinical specimens are given in Tables BXII.γ.156, BXII.γ.157, BXII.γ.160, BXII.γ.161, BXII.γ.165 and BXII.γ.166. *V. mimicus* is similar to *Vibrio cholerae* non-O1 in its ecology, distribution, and pathogenicity. Does not require Na+.

Occurs in aquatic environments. Some strains apparently cause diarrhea and human intestinal infections, and some virulence factors have been postulated. Occasionally causes extraintestinal human infections. The type strain was isolated from a human ear culture.

V. mimicus was discovered during a DNA–DNA hybridization study of "biochemically atypical" strains of *V. cholerae* (Davis et al., 1982). Strains that were originally characterized as "*Vibrio cholerae* sucrose+" were highly related to, but distinct from, *V. cholerae* by DNA–DNA hybridization; the new species *V. mimicus* was based on these results (Davis et al., 1982).

The mol% G + C of the DNA is: unknown.

Type strain: ATCC 33653, CDC 1721-77.

GenBank accession number (16S rRNA): X74713.

24. **Vibrio mytili** Pujalte, Ortigosa, Urdaci, Garay and Grimont 1993, 360[VP]

my.ti'li. L. gen. n. *mytili* of mussels, from the genus name *Mytilus*.

The characteristics are as described for the genus and as listed in Tables BXII.γ.154, BXII.γ.163, and BXII.γ.164. Coccobacilli. Motile by a single polar flagellum. Colonies are nonpigmented on marine agar and yellow on TCBS agar. No growth occurs on MacConkey agar.

Isolated from mussels (*Mytilus edulis*) off the Atlantic coast of Spain.

Based on DNA–DNA hybridization experiments Pujalte et al. (1993) concluded that strains of *V. mytili* were highly related (94–100%) to the type strain but not closely related to 23 other species of *Vibrio* and *Photobacterium*. *Vibrio parahaemolyticus* was the closest relative (25%). Brenner et al. (unpublished) found 25–40% relatedness of *V. mytili* to *V. alginolyticus* and other marine *Vibrio* species. Fig. BXII.γ.171 illustrates the relationships of *V. mytili* to other *Vibrio* species based on 16S rRNA gene sequence analysis.

The mol% G + C of the DNA is: 45–46 (T_m).

Type strain: 165, ATCC 51288, CECT 632.

25. **Vibrio natriegens** (Payne, Eagon and Williams 1961) Baumann, Baumann, Bang and Woolkalis 1981, 217[VP] (Effective publication: Baumann, Baumann, Bang and Woolkalis 1980b, 128) ("*Pseudomonas natriegens*" Payne, Eagon and Williams 1961, 125; "*Vibrio natriegens*" (Payne, Eagon and Williams 1961) Webb and Payne 1971, 1080; *Beneckea natriegens* (Payne, Eagon and Williams 1961) Baumann, Baumann and Mandel 1971a, 291.)

na.tri.e'gens. M.L. *natrium* sodium; L. *egens* pres. part. *egere* to be in need; M.L. adj. *natriegens* sodium-requiring.

The characteristics are as described for the genus and as listed in Tables BXII.γ.163 and BXII.γ.164. Cells usually have a single, long, polar flagellum and occasionally 2–3 flagella. Cells often contain poly-β-hydroxybutyrate granules (Fig. BXII.γ.178). Doubling time (generation time) is 9.8 minutes (Eagon, 1962). *V. natriegens* is differentiated from other species of *Vibrio* by its short generation time and its ability to utilize a much wider variety of carbon sources than most other species (Table BXII.γ.163).

Isolated from salt marsh mud and coastal seawater (Payne et al., 1961; Furniss et al., 1978; Baumann and Baumann, 1981a). The type strain was isolated from a salt marsh mud, Sapelo Island, Georgia, USA.

Based on DNA–DNA hybridization experiments, strains of *V. natriegens* are highly related (83% or greater) to the type strain and 33–53% related (Fig. BXII.γ.169) to other marine *Vibrio* species (Reichelt and Baumann, 1973b). These results were confirmed by Brenner et al. (unpublished) who found the closest relatives of *V. natriegens* according to DNA–DNA hybridization data to be *V. alginolyticus*, *V. campbellii*, *V. harveyi*, and *V. parahaemolyticus* (31–45% related). Based on 16S rRNA gene sequence analysis (Fig. BXII.γ.171 and analysis of the immunological relationships of superoxide dismutase (Fig. BXII.γ.170), *Vibrio natriegens* falls into the marine *Vibrio* group.

The mol% G + C of the DNA is: 45.1–47 (T_m, Bd).

Type strain: ATCC 14048, DSM 759, NCMB 857.

GenBank accession number (16S rRNA): X74714.

26. **Vibrio navarrensis** Urdaci, Marchand, Ageron, Arcos, Sesma and Grimont 1991, 293^{VP}

na.var.ren′ sis. M. L. masc. adj. *navarrensis* pertaining to Navarra, Spain, the Spanish province where the organism was isolated.

Morphological, growth, physiological, and biochemical characteristics are listed in Tables BXII.γ.154 and BXII.γ.164. Cells are 0.8–1.0 × 1.0–2.0 µm and are motile by a single polar flagellum when grown on solid and liquid media. Colonies are 2–3 mm in diameter on nutrient agar with 2% NaCl after overnight incubation at 30°C and are round, opaque, and nonpigmented. Colonies on TCBS agar are yellow. Slight requirement for Na$^+$. Seven of ten strains studied grew weakly in peptone water without added NaCl. Synthesizes a unique fatty acid of undetermined structure not found in 22 other *Vibrio* species (Urdaci et al., 1991) that accounts for 1–3 % of the total fatty acids of *V. navarrensis*.

Isolated from sewage, irrigation water and river water in Spain.

Based on DNA–DNA hybridization experiments, Urdaci et al. (1991) found that eight strains of *V. navarrensis* were highly related (82–98% with T_m values of 0–2.0) to the type strain; another strain (1381-1) was 74% related.. The closest relative was *Vibrio natriegens*, which was 39% related. The other *Vibrio* species tested were only 3–15% related. Based on DNA–DNA hybridization experiments, Brenner et al. (unpublished) found 14–22% relatedness of *V. navarrensis* to *V. natriegens* and 28–36% relatedness of *V. navarrensis* to *V. harveyi*, *V. parahaemolyticus*, *V. pelagius*, and *V. vulnificus*. Fig. BXII.γ.171 illustrates the relationship of *V. navarrensis* to other species of *Vibrio*.

The mol% G + C of the DNA is: 45–47 (T_m).

Type strain: ATCC 51183, CIP 103381.

GenBank accession number (16S rRNA): X74715.

27. **Vibrio nereis** (Harwood, Bang, Baumann and Nealson 1980) Baumann, Baumann, Bang and Woolkalis 1981, 217^{VP} (Effective publication: Baumann, Baumann, Bang and Woolkalis 1980b, 128) (*Beneckea nereida* (ex Baumann, Baumann and Mandel 1971a) Harwood, Bang, Baumann and Nealson 1980, 655.)

ne′ re.is. L. n. *nereis* a sea nymph.

The characteristics are as described for the genus and as listed in Tables BXII.γ.163 and BXII.γ.164. The morphology is illustrated in Figs. BXII.γ.179 and BXII.γ.180.

Isolated from seawater off the coast of Hawaii. Kusuda et al. (1986) isolated *V. nereis* from an outbreak of fatal infections of sea bream (*Acanthopagrus schlegeli*) at hatcheries in western Japan, but did not think this organism was causing disease in the fish. The type strain was isolated from seawater off the coast of Hawaii.

Based on DNA–DNA hybridization experiments, Reichelt et al. (1976) found that strains of *Vibrio nereis* were

FIGURE BXII.γ.178. Ultrathin section of *Vibrio natriegens* containing granules of poly-β-hydroxybutyrate that accumulate during stationary phase of growth. Grown in basal medium containing limiting nitrogen and excess β-hydroxybutyrate. × 32,000. (Courtesy of R.D. Allen.)

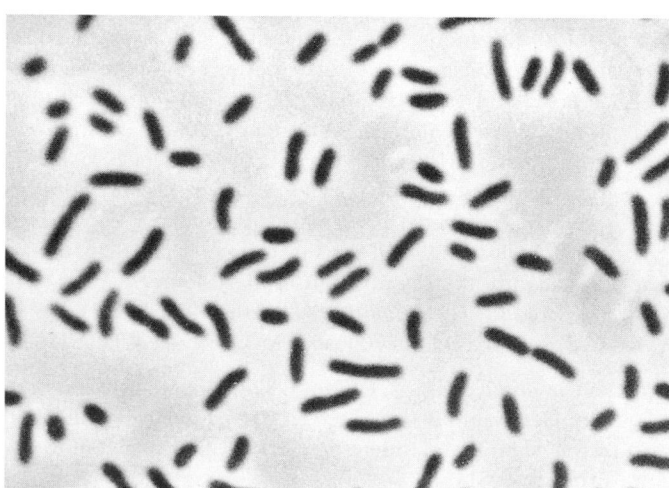

FIGURE BXII.γ.179. Phase contrast micrograph of *Vibrio nereis.* Exponential growth phase in yeast extract broth. × 3000.

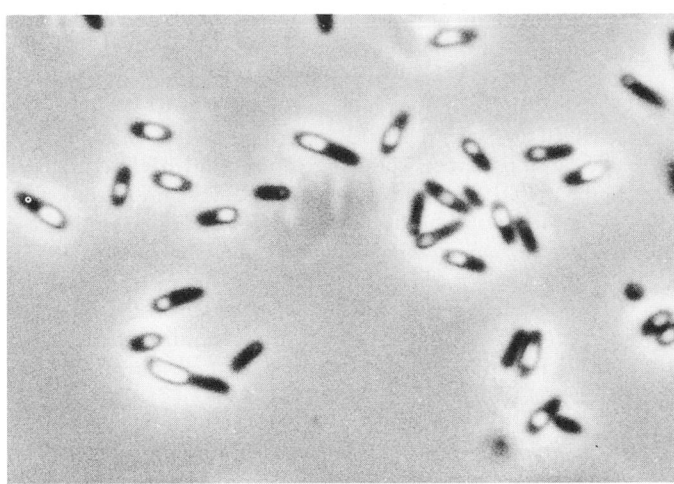

FIGURE BXII.γ.180. Phase contrast micrograph of *Vibrio nereis* containing refractile granules of poly-β-hydroxybutyrate that accumulate during stationary phase. The organisms were grown in basal medium containing limiting nitrogen and excess β-hydroxybutyrate. × 3000.

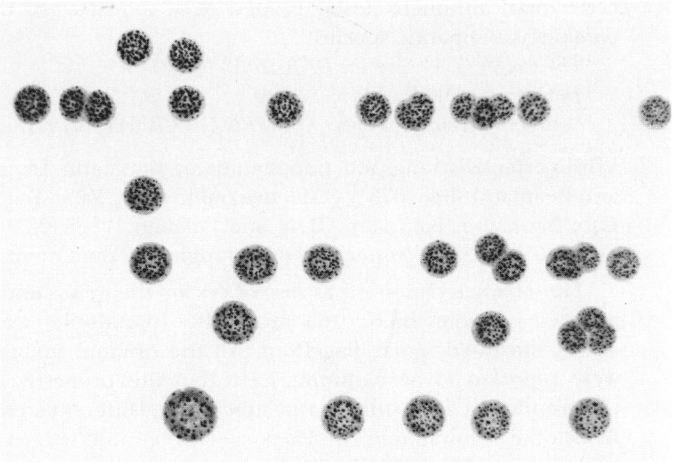

FIGURE BXII.γ.181. Colonies of *Vibrio nigripulchritudo* containing crystals of a blue-black pigment. Incubation for 2 days on basal medium agar with 0.2% [v/v] glycerol. × 4.5. (Reproduced with permission from P. Baumann et al., Journal of Bacteriology *108:* 1380–1383, 1971, ©American Society for Microbiology.)

highly related (87–94%) to the type strain and less than 30% related to 13 other *Vibrio* species. These results were confirmed by Brenner et al. (unpublished), who found the closest relatives by DNA–DNA hybridization to be *V. alginolyticus, V. parahaemolyticus, V. harveyi, V. proteolyticus,* and two unnamed *Vibrio* groups (all 22–40% related). By DNA–DNA hybridization (Fig. BXII.γ.169), 16S rRNA sequencing (Fig. BXII.γ.171), and the immunological relationship of superoxide dismutase (Fig. BXII.γ.170), *Vibrio nereis* falls into the marine *Vibrio* group.

The mol% G + C of the DNA is: 46–47 (T_m, Bd).

Type strain: ATCC 25917, CIP 103194, IFO 15637, LMG 3895, NCCB 73020.

GenBank accession number (16S rRNA): X74716.

28. **Vibrio nigripulchritudo** (Baumann, Baumann, Mandel and Allen 1971a) Baumann, Baumann, Bang and Woolkalis 1981, 217[VP] (Effective publication: Baumann, Baumann, Bang and Woolkalis 1980b, 128) (*Beneckea nigripulchritudo* Baumann, Baumann, Mandel and Allen 1971b, 1383.) *ni.gri.pul.chri.tu' do.* L. *niger* black; L. *pulchritudo* beauty; M.L. adj. *nigripulchritudo* black beauty, which refers to its striking blue-black colonies.

The characteristics are as described for the genus and as listed in Tables BXII.γ.163 and BXII.γ.164. Instantly recognizable and differentiated from other *Vibrio* species by its formation of striking blue-black colonies (Fig. BXII.γ.181).

Isolated from sea water off the coast of Hawaii.

Based on DNA–DNA hybridization experiments, Reichelt et al. (1976) found that strains of *Vibrio nigripulchritudo* were highly related (91–96%) to the type strain and less than 20% related to 15 other *Vibrio* species. These results were essentially confirmed by Brenner et al. (unpublished) who found the closest relative by DNA–DNA hybridization to be *V. penaeicida* (36% related). By 16S rRNA gene sequence analysis (Fig. BXII.γ.171 and the immunological relationship of superoxide dismutase (Fig. BXII.γ.170), *Vibrio nigripulchritudo* falls into the marine *Vibrio* group, but may have diverged slightly from that group.

The mol% G + C of the DNA is: 46–47 (T_m, Bd).

Type strain: ATCC 27043, CCUG 28586, CIP 103195, LMG 3896.

GenBank accession number (16S rRNA): X74717.

29. **Vibrio ordalii** Schiewe, Trust and Crosa 1982, 384[VP] (Effective publication: Schiewe, Trust and Crosa 1981, 347.) *or.dal' i.i.* M.L. gen. n. *ordalii* of Ordal; named to honor Erling J. Ordal for his many contributions to the current understanding of the biology of bacterial fish pathogens

The characteristics are as described for the genus and as listed in Tables BXII.γ.163 and BXII.γ.164. The cells are 2.5–3.0 × 1.0 μm in size and are motile by a single polar flagellum. Grows very slowly, forming visible colonies on trypticase soy agar only after 4–6 days (Table BXII.γ.171). Growth is enhanced by the addition of NaCl. All isolates of *V. ordalii* belong to O antigen group 2, in contrast to *V. anguillarum* (O antigen groups 1–16). Closely related phylogenetically to *V. anguillarum* (Figs. BXII.γ.169, BXII.γ.170, and BXII.γ.171), but less biochemically versatile and more likely to have additional nutritional requirements (Table BXII.γ.171).

Important fish pathogen isolated from diseased salmonid fish in Washington, Oregon, British Columbia, and Japan (Schiewe et al., 1981), and from ayu (*Plecoglossus altivelis*) and rockfish fingerlings (*Sebastes schlegeli*) in Japan (Muroga et al., 1986). The type strain was isolated from the kidney of a moribund Coho salmon (*Oncorhynchus kisutch*) at a commercial salmon farm in Puget Sound, Washington, U.S.A.

V. ordalii is a halophilic *Vibrio* species that is very closely related to *Vibrio anguillarum.* Previously, other investigators had studied vibriosis of marine fish and isolated a halophilic vibrio similar to, but less biochemically versatile than *V. anguillarum.* This organism was referred to as "*Vibrio* sp. 1669 group", "*Vibrio* sp. RT group", "*Beneckea anguillara*" biotype I or *Vibrio anguillarum* biotype 2. However, DNA–DNA hybridization studies by Schiewe et al. (1981) indicated that, although closely related to *V. anguillarum*, *V. ordalii* was a separate species.

The mol% G + C of the DNA is: 43–44 (T_m).

Type strain: DF3K, ATCC 33509.

GenBank accession number (16S rRNA): X70641, X74718.

30. **Vibrio orientalis** Yang, Yeh, Cao, Baumann, Baumann, Tang and Beaman 1983b, 673[VP] (Effective publication: Yang, Yeh, Cao, Baumann, Baumann, Tang and Beaman 1983a, 98.) *or.i.en.ta' lis*. L. adj. *orientalis* of or belonging to the Orient.

The characteristics are as described for the genus and as listed in Tables BXII.γ.163 and BXII.γ.164. Motile by a single, sheathed, polar flagellum. All the original strains were reported to be bioluminescent, but this property is apparently lost after storage and subculture. Differentiated from other bioluminescent *Vibrio* species because it accumulates poly-β-hydroxybutyrate and grows at 4°C.

Isolated from seawater and shrimp from the Yellow Sea, China.

V. orientalis was originally defined and named because it was distinct from other species of *Vibrio* and *Photobacterium* in phenotype and in the relationship of its superoxide dismutase (Yang et al., 1983a). Based on DNA–DNA hybridization experiments, the type strain of *Vibrio orientalis* was highly related (75% or greater with 0.5% divergence) to three other strains (Brenner et al., unpublished). The closest relatives were *V. parahaemolyticus*, *V. pelagius*, *V. tubiashii*, *Vibrio* group 515 and *Vibrio* group 9084-82 (19–37% related) (Brenner et al., unpublished). Fig. BXII.γ.171 illustrates the relationship of *V. orientalis* to other *Vibrio* species based on 16S rRNA gene sequence analysis. Based on the immunological relationship of its superoxide dismutase *V. orientalis* falls into the marine *Vibrio* group (see Fig. 3 of Yang et al., 1983a).

The mol% G + C of the DNA is: 45.3–45.8 (Bd).

Type strain: ATCC 33934, CCUG 16389, CIP 102891, IFO 15638, LMG 7897.

GenBank accession number (16S rRNA): X74719.

31. **Vibrio parahaemolyticus** (Fujino, Okuno, Nakada, Aoyama, Fukai, Mukai and Ueho 1951) Sakazaki, Iwanami and Fukumi 1963, 181[AL] ("*Pasteurella parahaemolytica*" Fujino, Okuno, Nakada, Aoyama, Fukai, Mukai and Ueho 1951, 11; "*Oceanomonas parahaemolytica*" (Fujino, Okuno, Nakada, Aoyama, Fukai, Mukai and Ueho 1951) Miyamoto, Nakamura and Takizawa 1961, 477; *Beneckea parahaemolytica* (Fujino, Okuno, Nakada, Aoyama, Fukai, Mukai and Ueho 1951) Baumann, Baumann, and Mandel 1971a, 291.) *para.hae.mo.ly' ti.cus*. Gr. prep. *para* by the side of, beside; Gr. n. *haema* blood; Gr. adj. *lyticus* dissolving; M.L. adj. *parahaemolyticus* dissolving blood.

The characteristics are as described for the genus and as listed in Tables BXII.γ.163 and BXII.γ.164. Characters that differentiate *V. parahaemolyticus* from *V. alginolyticus*, *V. harveyi*, and *V. vulnificus* are listed in Table BXII.γ.170. Characters that distinguish *V. parahaemolyticus* from other *Vibrio* spp. found in human clinical specimens are listed in Tables BXII.γ.156, BXII.γ.157, BXII.γ.160, BXII.γ.161, BXII.γ.165 and BXII.γ.166.

Occurs in human clinical specimens (Table BXII.γ.157) and is an important cause of diarrhea. Also occurs in marine environments.

Outbreaks of gastroenteritis caused by *V. parahaemolyticus* occur worldwide. In Japan, *V. parahaemolyticus* causes 50–70% of foodborne enteritis cases (Sakazaki and Balows, 1981), which are invariably associated with seafood. In the early 1970s, several articles described *V. parahaemolyticus* as the cause of severe wound infections and bacteremia (Roland, 1970, 1971; Zide et al., 1974; Weaver and Ehrenkranz, 1975). It is now known that the infectious agent in the case was really *Vibrio vulnificus* (Weaver and Ehrenkranz, 1975), not *V. parahaemolyticus*. Similarly, many ecological and environmental reports about *V. parahaemolyticus* may be incorrect because of our limited understanding of the large number of *Vibrio* species that inhabit these environments.

The mol% G + C of the DNA is: 46–47 (T_m, Bd).

Type strain: 113, ATCC 17802, DSM 30189, NCMB 1326.

GenBank accession number (16S rRNA): M59161, X56580, X74720.

32. **Vibrio pectenicida** Lambert, Nicolas, Cilia and Corre 1998, 486[VP]

pec.ten.i.ci' da. M.L. n. *Pecten* genus name of scallops; M.L. v. *caedo* to kill; *pectenicida* scallop-killer.

The characteristics are as described for the genus and as listed in Tables BXII.γ.163 and BXII.γ.164. Cells are curved rods with a large single polar flagellum when grown in liquid medium. Smaller, unsheathed, lateral flagella are produced during growth on solid medium. Colonies after 48 h of growth are circular, smooth, and unpigmented and exhibit waves of swarming. Isolated from moribund scallop (*Pecten maximus*) larvae during outbreaks of disease in a hatchery in France.

Lambert et al. (1998) compared the type strain of *Vibrio penaeicida* to the type strains of 30 other *Vibrio* and *Photobacterium* species by 16S rRNA gene sequence analysis. The two closest relatives were *Vibrio tapetis* (97.2% sequence similarity) and *V. splendidus* (95.5% sequence similarity). DNA–DNA hybridization was used to show that the type strain of *V. pectenicida* was 86–100% related to four other strains and less than 5% related to the type strains of *V. tapetis* and *V. splendidus*. Brenner et al. (unpublished) showed that the type strain of *V. pectenicida* was 98% related to one other strain, and 1–25% related to all other *Vibrio* species tested based on the results of DNA–DNA hybridization experiments. Fig. BXII.γ.171 shows the relationship of *Vibrio pectenicida* to other *Vibrio* species based on 16S rRNA gene sequence analysis. *V. logei* is the closest relative of *V. pectenicida* based on phenotypic similarities.

The mol% G + C of the DNA is: 39–41(T_m).

Type strain: A365, ATCC 700783, CIP 105190.

GenBank accession number (16S rRNA): Y13830.

33. **Vibrio pelagius** (Baumann, Baumann and Mandel 1971a) Baumann, Baumann, Bang and Woolkalis 1981, 217[VP] (Effective publication: Baumann, Baumann, Bang and Woolkalis 1980b, 128) (*Beneckea pelagia* Baumann, Baumann and Mandel 1971a, 291.)*

pe.la' gi.us. L. adj. *pelagius* of the sea.

Editorial Note: The name *Listonella pelagia* also has standing in the nomenclature (MacDonell and Colwell, 1985).

TABLE BXII.γ.174. Nomenclature, classification and differentiation of the two groups of *Vibrio pelagius* that are distinct by DNA hybridization and phenotype

Test or property	*Vibrio pelagius* group 1	*Vibrio pelagius* group 2
Type or reference strain	ATCC 25916^T (strain 99)	ATCC 33786 (strain 106)[a]
Other strains in ATCC[b]	33781 (96)	33504 (105)
	33782 (100)	33784 (103)
	33783 (101)	33785 (104)
	33808 (97)	33786 (106)
Nomenclature and classification according to:		
Baumann et al. (1971a)	Group F, cluster 1	Group F, cluster 2
Baumann et al. (1971a)	*Vibrio pelagius*	*Vibrio pelagius*
Reichelt et al. (1976)	*Vibrio pelagius* biotype I	*Vibrio pelagius* biotype II
Baumann et al. (1984b)	*Vibrio pelagius* biotype I	*Vibrio pelagius* biotype II
MacDonell and Colwell (1985)	*Listonella pelagia*	*Listonella pelagia*
Differential tests:		
Cellular morphology	Straight rods	Curved rods
Amylase production	0	100
Gelatinase production	0	100
γ-Aminobutyrate utilization	0	100
δ-Aminovalerate utilization	0	100

[a]This strain was the labeled strain in the DNA hybridization study of Reichelt et al. (1976).

[b]The ATCC number is listed first. The strain number from Baumann et al. (1971a) is listed in parentheses.

The characteristics are as described for the genus and as listed in Tables BXII.γ.163 and BXII.γ.164. Nomenclature, classification, and differential phenotypic traits of the two groups of *Vibrio pelagius*, which are also distinct based on the results of DNA hybridization experiments (see Taxonomic Comments), are given in Table BXII.γ.174.

Isolated from seawater of the coast of Hawaii. Furniss et al. (1978) found that *V. pelagius* is one of the most common vibrios isolated along the British coasts.

Beneckea pelagia was first described and named by Baumann et al. (1971a). It was defined as a single species based on phenotypic properties of 11 strains isolated from waters off the coast of Hawaii; however, the authors noted that the strains "clustered in two groups." Reichelt et al. (1976) confirmed the existence of two groups by DNA–DNA hybridization experiments. The two groups were classified as biotypes and referred to as *Beneckea pelagia* I and II. Baumann et al. (1980b) "abolished" the genus *Beneckea*, and proposed that *Beneckea pelagia* be classified in the genus *Vibrio* as *V. pelagius*.

Brenner et al. (1983b) showed by DNA–DNA hybridization experiments that the type strain of *V. pelagius* was 30–45% related to type strains of *V. campbellii*, *V. harveyi*, *V. parahaemolyticus*, and *V. splendidus*. Recently Brenner et al. (unpublished) used DNA–DNA hybridization experiments to compare strains of the two groups of *V. pelagius* and other strains that had been identified as belonging to this species. They confirmed the results of Reichelt et al. (1976) that showed that *V. pelagius* II was distinct from *V. pelagius* I and suggested that the two groups be classified as subspecies of *V. pelagius*, rather than biogroups. They also showed that the most closely related species (20–44%) are "*V. algosus*", *V. harveyi* biogroup 2, *V. mytili*, *V. orientalis*, *V. parahaemolyticus*, *V. splendidus* biogroup 2, and *Vibrio* groups 514, 515, and 516. Fig. BXII.γ.171 shows the relationship of *Vibrio pelagius* to other *Vibrio* species based on 16S rRNA gene sequence analysis. Based on the immunological relationship of their superoxide dismutases (Fig. BXII.γ.170) the two

groups of *Vibrio pelagius* are distinct, and both cluster in the marine *Vibrio* group (Baumann et al., 1984b).

Based on similarity of 5S rRNA sequences, MacDonell and Colwell (1985) proposed that *V. pelagius* should be united with *Vibrio anguillarum* and *Moritella marina* species in a new genus *Listonella*, and they proposed the name *Listonella pelagia*. Thus, *V. pelagius* currently has standing in nomenclature in three different genera as *Beneckea pelagia*, *Vibrio pelagius*, and *Listonella pelagia* (Table BXII.γ.162). All three names are based on the same type strain.

The mol% G + C of the DNA is: 45–47 (T_m, Bd).

Type strain: ATCC 25916, CECT 4202.

GenBank accession number (16S rRNA): X74722.

34. **Vibrio penaeicida** Ishimaru, Akagawa-Matsushita and Muroga 1995, 138^VP

pe.nae.i.ci'da. L. n. *Penaeus* genus of kuruma prawns; L. v. *caedo* to kill; L. adj. *penaeicida* *Penaeus* killer.

The characteristics are as described for the genus and as listed in Tables BXII.γ.154, BXII.γ.163, and BXII.γ.164. Small rods, straight or slightly curved, 0.5–0.8 × 1.5–2.0 μm. Motile by means of a single polar flagellum. No lateral flagella are produced during growth on solid media. Colonies are low, cream colored, and translucent with entire margins. All reported isolates have the same O antigen.

Causes vibriosis of kuruma prawns (*Penaeus japonicus*) in Japan. Isolated from diseased and apparently healthy kuruma prawns and from water samples from ponds where prawns are grown.

V. penaeicida was formerly known as "*Vibrio species PJ*" (de la Peña et al., 1992). Ishimaru et al. (1995) showed by DNA–DNA hybridization experiments that the type strain of *V. penaeicida* was highly related (87–99%) to five other strains but was only 4–18% related to 29 other *Vibrio* and *Photobacterium* species tested. On the basis of DNA–DNA hybridization experiments, Brenner et al. (unpublished) also confirmed that *V. penaeicida* is a distinct species and found that its closest relative is *V. nigripulchritudo* (36% related).

The mol% G + C of the DNA is: 46.2–47 (HPLC).

Type strain: KH-1, ATCC 51841, IFO 15640, JCM 9123.

35. **Vibrio proteolyticus** (Merkel, Traganza, Mukherjee, Griffin and Prescott 1964) Baumann, Baumann, Bang and Woolkalis 1982, 267[VP] (Effective publication: Baumann, Baumann, Bang and Woolkalis 1980b, 128) ("*Aeromonas proteolytica*" Merkel, Traganza, Mukheriee, Griffin and Prescott 1964, 1230; *Aeromonas hydrophila* subsp. *proteolytica* (Merkel, Traganza, Mukheriee, Griffin and Prescott 1964) Schubert 1969, 412.)

pro.te.o.ly' ti.cus. Ger. *protein* from Gr. *protos* first; Gr. adj. *lyticus* dissolving; M.L. adj. *proteolyticus* protein-dissolving.

The characteristics are as described for the genus and as listed in Tables BXII.γ.163 and BXII.γ.164. Grows on solid media as swarming waves. Isolated from *Limnoria tripunctata* taken from wood pilings Charleston, South Carolina, U.S.A.

The type strain was isolated from the alimentary canal of the marine isopod crustacean.

Reichelt et al. (1976) showed by DNA–DNA hybridization that the type strain of *V. proteolyticus* was only 0–25% related to 11 other *Vibrio* species. These results were confirmed by Brenner et al. (1983b) who showed that the type strain was less than 5% related to type strains of 16 other *Vibrio* species. Brenner et al. (unpublished) studied almost all newly described *Vibrio* species and concluded that *V. proteolyticus* is a distinct species whose closest relatives are CDC *Vibrio* groups 515 and 516 (33% relatedness), which have not yet been described in the literature. Fig. BXII.γ.171 shows the relationship of *V. proteolyticus* to other *Vibrio* species based on 16S rRNA gene sequence analysis.

The mol% G + C of the DNA is: 50.5 (Bd).

Type strain: ATCC 15338, DSM 30189, IFO 13287, NCMB 1326.

GenBank accession number (16S rRNA): X56579, X74723.

36. **Vibrio rumoiensis** Yumoto, Iwata, Sawabe, Ueno, Ichise, Matsuyama, Okuyama and Kawasaki 1999b, 935[VP] (Effective publication: Yumoto, Iwata, Sawabe, Ueno, Ichise, Matsuyama, Okuyama and Kawasaki 1999a, 71.)

ru.moi.en' sis. L. adj. *rumoiensis* from Rumoi, location of isolation.

Morphological, growth, physiological, and biochemical characteristics are listed in Table BXII.γ.164. Cells 0.5–0.9 × 0.7–2.1 μm. No flagella. Numerous blebs occur on the cell surface (Fig. BXII.γ.182). Colonies are white and circular. Produces high level of catalase.

Isolated from a drainage pool at a fish product processing plant where hydrogen peroxide was used as a bleaching agent.

Based on DNA–DNA hybridization experiments carried out using the fluorometric method, Yumoto et al. (1999a) found that the type strain of *V. rumoiensis* was only 3.2–9.0% related to 11 other *Vibrio* species. Based on the results of 16S rRNA gene sequence analysis, the type strain of *V. rumoiensis* was 92.4–95.5% similar to *Vibrio* and *Photobacterium* species. Fig. BXII.γ.171 illustrates the relationship of *V. rumoiensis* to other species of *Vibrio*.

The mol% G + C of the DNA is: 43.2 (HPLC).

Type strain: S-1, FERM-P 14531.

GenBank accession number (16S rRNA): AB013297.

37. **Vibrio salmonicida** Egidius, Wiik, Andersen, Hoff and Hjeltnes 1986, 519[VP]

sal.mon.i.ci' da N. L. n. *salmon* salmon; L. v. *caedo* to kill; N. L. adj. *salmonicida* salmon killer.

See description of *V. logei*.

By DNA–DNA hybridization experiments, Brenner et al. (unpublished) showed that the type strain of *Vibrio salmonicida* was highly related to the type strain of *Vibrio logei* (see also Fig. BXII.γ.171) and concluded that these two organisms should be classified in the single species, *Vibrio logei*.

The mol% G + C of the DNA is: 42 (T_m).

Type strain: HI 7751, ATCC 43839, CIP 103166, LMG 14010, NCIMB 2262.

38. **Vibrio scophthalmi** Cerdà-Cuéllar, Rosselló-Mora, Lalucat, Jorfe and Blanch 1997, 60[VP]

scoph.thal' mi. L. gen. n. *scophthalmi* derived from the genus name *Scophthalmus* because all the original isolates were from the turbot (fish) species *Scophthalmus maximus*.

The characteristics are as described for the genus and as listed in Tables BXII.γ.154, BXII.γ.163, and BXII.γ.164. Halophilic; grows poorly in media containing ≤ 0.5% NaCl. Colonies are unpigmented on TSA and yellow on TCBS agar.

Isolated from the intestines of juvenile turbot (*Scophthalmus maximus*) in a hatchery in Spain.

Based on DNA–DNA hybridization experiments (dot hybridization method) Cerdà-Cuéllar et al. (1997) found that the type strain was highly related to five other strains. Brenner et al. (unpublished) confirmed these results and concluded that *V. scophthalmi* is a distinct species. They also found that the type strain of *Vibrio scophthalmi* was 99% related to the reference strain of *Vibrio* INFL-2. The closest other relative was *V. ichthyoenteri*, which was 32% related. Based on the results of 16S rRNA gene sequence analysis, Cerdà-Cuéllar et al. (1997) found that the closest relative was *V. aestuarianus*, which had a sequence similarity of 97.8%. Fig. BXII.γ.171 shows the relationship of *Vibrio scophthalmi* to other *Vibrio* species based on 16S rRNA gene sequence analysis.

The mol% G + C of the DNA is: 44–44.5 (HPLC).

Type strain: A089, CECT 4638.

GenBank accession number (16S rRNA): U46579.

39. **Vibrio splendidus** (Beijerinck 1900a) Baumann, Baumann, Bang and Woolkalis 1981, 217[VP] (Effective publication: Baumann, Baumann, Bang and Woolkalis 1980b, 128) ("*Photobacter splendidum*" Beijerinck 1900a, 362; *Beneckea splendida* (Beijerinck 1900a) Reichelt, Baumann and Baumann 1979, 80.)

splen' di.dus. L. adj. *splendidus* brilliant.

The characteristics are as described for the genus and as listed in Tables BXII.γ.163 and BXII.γ.164. The cell morphology is shown in Fig. BXII.γ.183. Properties and differentiation of the three biogroups described by Reichelt et al. (1976) are given in Table BXII.γ.175.

Isolated from marine fish and seawater.

Reichelt et al. (1976) determined that *V. splendidus* consisted of three distinct groups that could be defined on the basis of both phenotypic characteristics and DNA–DNA hybridization experiments. These groups were referred to as

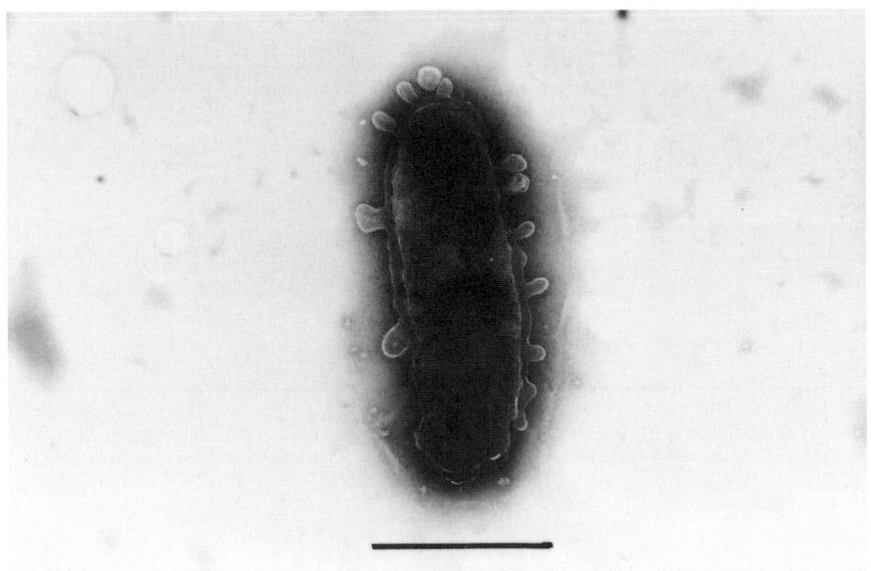

FIGURE BXII.γ.182. Electron micrograph of *Vibrio rumoiensis* illustrating the appearance of "blebs" on the cell surface. Negatively stained preparation. Bar = 1 μm. (Reproduced with permission from Yumoto et al., Applied and Environmental Microbiology, *65:* 67–72, 1999a ©American Society for Microbiology.)

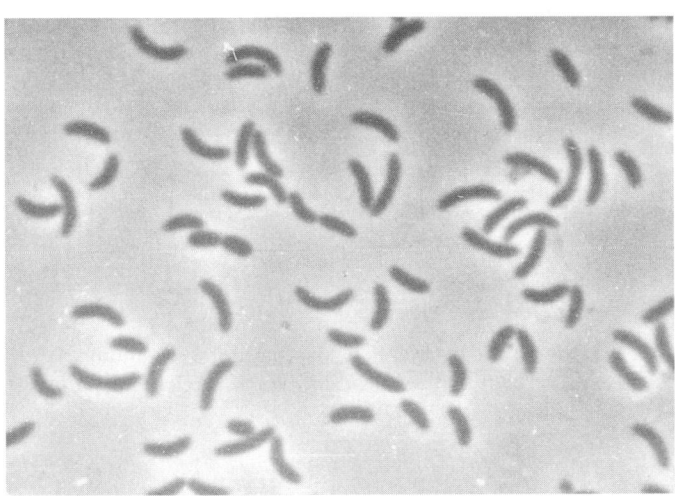

FIGURE BXII.γ.183. Phase contrast micrograph of *Vibrio splendidus* biogroup II. Exponential growth phase in yeast extract broth. × 3000.

biotypes I, II, and a third unnamed biotype (Table BXII.γ.175). Brenner et al. (unpublished) also confirmed that *Vibrio splendidus* is a distinct *Vibrio* species based on DNA–DNA hybridization experiments and showed that the type strain of *V. splendidus* was 30–45% related to the type strains of *V. campbellii*, *V. harveyi*, *V. parahaemolyticus*, and *V. pelagius*. Based on the immunological relationship of superoxide dismutases (Fig. BXII.γ.170). Baumann et al. (1984b) concluded that *V. splendidus* biogroups I and II are closely related, and (along with *V. fischeri* and *V. logei*) are more distantly related to the marine *Vibrio* group. Fig. BXII.γ.171 shows the relationship of the type strain of *V. splendidus*, which belongs to biogroup 1, to other *Vibrio* species based on 16S rRNA gene sequence analysis.

The mol% G + C of the DNA is: 45–46 (T_m, Bd).
Type strain: ATCC 33125, NCMB 1.
GenBank accession number (16S rRNA): X74724.

40. **Vibrio tapetis** Borrego, Castro, Luque, Paillard, Maes, Garcia and Ventosa 1996, 483[VP]

ta.pe′ tis. L. gen. n. *tapetis* of clams (the genus *Tapes*).

The characteristics are as described for the genus and as listed in Tables BXII.γ.154, BXII.γ.163, and BXII.γ.164. Coccobacilli, 0.5 × 1.0–1.5 μm. Motile by a single polar flagellum. Grows at 4–22°C; optimum growth at 22°C. Colonies on marine agar circular, translucent, and nonpigmented. Colonies on TCBS agar are green.

Causes brown ring disease in several species of wild and cultured clams. The type strain was isolated from cultured manila clam (*Tapes philippinarum*), Landeda, France.

Based on DNA–DNA hybridization experiments, Borrego et al. (1996) found that the type strain of *Vibrio tapetis* was highly related (100%) to eight other strains and less related to seven other *Vibrio* species. *Vibrio vulnificus* (58%), *V. alginolyticus* (53%), *V. fischeri* (48%), and *V. splendidus* (46%) were the closest relatives. Based on DNA–DNA hybridization experiments, Brenner et al., (unpublished) confirmed that *Vibrio tapetis* is a distinct species, and found that *V. wodanis* was the closest relative (19%). Results of this study indicated that *Vibrio tapetis* was 15% or less related to other *Vibrio* species, including the species listed above. Fig. BXII.γ.171 shows the relationship of *V. tapetis* to other *Vibrio* species based on 16S rRNA gene sequence analysis.

The mol% G + C of the DNA is: 42.9–45.5 (T_m).
Type strain: B1090, CECT 4600.
GenBank accession number (16S rRNA): Y08430.

41. **Vibrio trachuri** Iwamoto, Suzuki, Kurita, Watanabe, Shimizu, Ohgami and Yanagihara 1996, 625[VP] (Effective publication: Iwamoto, Suzuki, Kurita, Watanabe, Shimizu, Ohgami and Yanagihara 1995, 836.)

TABLE BXII.γ.175. The three biotypes (subgroups) of *Vibrio splendidus*

Test or property	Biotype I	Biotype II	Unnamed biotype
Type[a] or reference strain	ATCC 33125[T] = NCMB 1	ATCC 33789 (strain 2)[b]	Strain 16 (not in ATCC)
Other strains	ATCC 33869 (strain 378);	ATCC 25914 (strain 8)	
	ATCC 33870 (strain 379);		
	ATCC 33871 (strain 380)		
Differential tests:[c]			
Bioluminescence	+	−	−
Arginine dihydrolase	+	−	
D-Mannose utilization	+	−	
D-Galactose utilization	+	−	
D-Glucuronate utilization	+	−	
L-Arginine utilization	+	−	
L-Citrulline utilization	+	−	
Heptonate utilization	+	−	

[a]The type strain for the species *Vibrio splendidus* is ATCC 33125.

[b]The strain numbers of Reichelt et al. (1976) are given in parentheses. Reichelt et al. (1976) labeled strain ATCC 33789 for DNA–DNA hybridization experiments and showed it was 82–91% related to the other four strains included in biotype II.

[c]Symbols: +, positive; −, negative.

tra' chu.ri. L. gen. masc. n. *trachuri* derived from *Trachurus japonicus*, Japanese horse mackerel (fish).

The characteristics are as described for the genus and as listed in Tables BXII.γ.154, BXII.γ.163, and BXII.γ.164.

Causes vibriosis of "ma-aji", Japanese horse mackerel (*Trachurus japonicus*). The type strain was isolated from Japanese horse mackerel grown at Uchiura Bay, Numazu, Japan.

Based on DNA–DNA hybridization experiments (fluorometric assay in microdilution plates), Iwamoto et al. (1995) found that the type strain of *Vibrio trachuri* was highly related (95–98%) to two other strains, but less related (1–40%) to 15 other *Vibrio* species tested. The closest relative was *V. harveyi*, to which *Vibrio trachuri* was 40% related. *Vibrio trachuri* was 24–25% related to *V. alginolyticus* and *V. campbellii*, and 10% or less related to 12 other *Vibrio* species. Based on DNA–DNA hybridization experiments, Brenner et al. (unpublished) also concluded that *Vibrio harveyi* is the closest relative of *V. trachuri*. However, they found "species level relatedness" between the type strain of *V. trachuri* and the type strain of *V. harveyi*. Since these two species have different type strains, they concluded that *V. trachuri* would probably become a junior objective synonym of *V. harveyi*. Different methods were used in the two DNA–DNA hybridization studies; this fact may account for the differences in the results.

The mol% G + C of the DNA is: 45.4–47.8 (HPLC after nuclease P1 digestion).

Type strain: T9210, JCM 9677.

42. **Vibrio tubiashii** Hada, West, Lee, Stemmler and Colwell 1984, 2[VP]

tu.bi.ash' i.i. L. gen. n. *tubiashii* of Tubiash; named to honor H.S. Tubiash who first isolated the organism.

The characteristics are as described for the genus and as listed in Tables BXII.γ.163 and BXII.γ.164. Short rods, 0.5×1.5 μm. Motile by means of a single polar flagellum when grown in liquid medium. Lateral short-wavelength flagella are synthesized when cultures are grown on solid medium. Colonies on marine agar smooth, circular, off-white and sometimes mucoid; yellow on TCBS agar.

Causes disease, with resulting economic losses, in larvae and juveniles of bivalve mollusks. Isolated from larvae and juveniles of the hard clam (*Mercenria mercenaria*), oyster spat (*Crassostrea virginica*), and adult oyster.

Based on DNA–DNA hybridization experiments (nitrocellulose method), Hada et al. (1984) found that the type strain of *V. tubiashii* was highly related (68–96%) to five other strains, but was only 28% related to *Vibrio anguillarum*. They also found that *V. tubiashii* strain NCMB 2166 was only 13–26% related to 11 other *Vibrio* species. Brenner et al. (unpublished) agreed that *V. tubiashii* is a distinct species, and found the closest relatives of *V. tubiashii* to be *V. orientalis* and *Vibrio* groups 515 and 516 (29–33% related). Fig. BXII.γ.171 illustrates the relationship of *V. tubiashii* to other *Vibrio* species based on 16S rRNA gene sequence analysis.

The mol% G + C of the DNA is: 43–45% (T_m).

Type strain: ATCC 19109, CCUG 38428, CIP 102760, IFO 15644, LMG 10936.

GenBank accession number (16S rRNA): X74725.

43. **Vibrio vulnificus** (Reichelt, Baumann and Baumann 1979) Farmer 1980, 656[VP] (*Beneckea vulnifica* Reichelt, Baumann and Baumann 1979, 80.)

vul.ni' fi.cus. L. adj. *vulnificus* inflicting wounds.

The characteristics are as described for the genus and as listed in Tables BXII.γ.163 and BXII.γ.164. Characters that differentiate *V. vulnificus* from *V. alginolyticus*, *V. harveyi*, and *V. parahaemolyticus* are given in Table BXII.γ.170. Characters that distinguish *V. vulnificus* from other *Vibrio* spp. found in human clinical specimens are given in Tables BXII.γ.156, BXII.γ.157, BXII.γ.160, BXII.γ.161, BXII.γ.165 and BXII.γ.166. Differential characteristics of the three biogroups of *V. vulnificus* are given in Table BXII.γ.176.

Occurs in human clinical specimens (Table BXII.γ.157) and the marine environment. Cause of wound infections, bacteremia, and septicemia. The type strain was isolated from human blood.

The type stain and most other strains belong to biogroup 1. All information given below on mol% G + C of the DNA, type strain accession numbers, and GenBank accession numbers pertains to biogroup 1. Biogroups 2 and 3 differ in phenotype, host range, ecology, and several other ways.

V. vulnificus biogroup 2 was described by Tison et al. (1982). Deposited strains include ATCC 33147, 33148, and

TABLE BXII.γ.176. Differentiation of three *Vibrio vulnificus* biogroups[a]

Test	Biogroup:		
	1	2	3
Citrate (Simmons)	+	+	−
Indole production	+	−	+
ONPG test[b]	+	+	−
Ornithine decarboxylase	+	−	+
Fermentation of:			
Cellobiose	+	+	−
Lactose	+	+	−
D-Mannitol	+	−	−
Salicin	+	+	−
D-Sorbitol	−	+	−

[a]Symbols: +, 75–100% of strains positive; −, 75–100% of strains negative. Data from Bisharat et al., 1999.

[b]ONPG, *o*-nitrophenyl-β-D-galactopyranoside.

33149. Strains of biogroup 2 were originally shown to be pathogenic for eels (Tison et al., 1982). Biogroup 2 appears to be a cause of human wound infections as well (Amaro and Biosca, 1996).

Biogroup 3 strains were described as a separate biogroup by Bisharat et al. (1999). Biogroup 3 strains were originally identified as a variety of different *Vibrio* species (Bisharat and Raz, 1996); all isolates were associated with human wound infections and exposure to cultured tilapia. There are no deposited strains.

The mol% G + C of the DNA is: 46–48 (T_m, Bd) (biogroup 1).

Type strain: B9629, ATCC 27562, DSM 10143, IMET 11292 (biogroup 1).

GenBank accession number (16S rRNA): X56582, X74726, X76333 (biogroup 1).

44. **Vibrio wodanis** Lunder, Sørum, Holstad, Steigerwalt, Mowinckel and Brenner 2000, 446[VP]

wo.da′ nis. M. L. gen. n. *wodanis* from Wodan, the Norse god of art, culture, war, and the dead.

Morphological, growth, physiological, and biochemical characteristics are listed in Table BXII.γ.164. This is the second of the two new *Vibrio* species from Atlantic salmon with "winter ulcer" that were described by Lunder et al. (2000). Fig. BXII.γ.171 illustrates the relationship of *V. wodanis* to other species of *Vibrio*.

The mol% G + C of the DNA is: 40 (T_m).

Type strain: NVI 88/441, ATCC BAA-104, NCIMB 13582.

GenBank accession number (16S rRNA): AJ132227.

Other Organisms

1. Baumann Group E-3[1]

Properties are given in Table BXII.γ.163. This group was originally described and given the vernacular name "E-3" by Baumann et al. (1971a). The group included Baumann strains 94 and 95. Based on the immunological relationships of the enzyme superoxide dismutase from different *Vibrio* and *Photobacterium* species, Bang et al. (1981) showed that Group E-3 was more closely related to three *Photobacterium* species than to *V. cholerae, V. alginolyticus,* and other *Vibrio* species (see Fig. 4 of Bang et al., 1981). Group E-3 has been referred to as *Vibrio* group 513 by Farmer (unpublished).

Deposited strain: ATCC 33523 from seawater, Oahu, Hawaii (strain 95 of Baumann et al. [1971]; isolated with acetate enrichment).

2. *Vibrio* Group 509[2]

Properties are given in Table BXII.γ.163. The name "Vibrio Group 509" includes a group of four Na⁺-requiring

Vibrio strains isolated from blood cultures of Kemp's Ridley sea turtles during an outbreak at the National Marine Fisheries Services facilities, Galveston, Texas. The organisms were seen in stained sections of lung tissue and the animals were observed to vomit blood. This condition was thought to resemble "hemorrhagic septicemia". Based on DNA–DNA hybridization experiments, Brenner et al. (unpublished) found that the strains were 85% related under stringent hybridization conditions.. The strains of *Vibrio* Group 509 strains were later found to be closely related to the type strain of *V. harveyi* (Brenner et al., personal communication).

3. *Vibrio* group 510

Properties are given in Table BXII.γ.163. This group includes the reference strain, which was isolated from an infected human leg wound, and five additional strains. Based on DNA–DNA hybridization experiments (Brenner et al., personal communication), these strains had an average relatedness to the reference strain of 78% under stringent hybridization conditions. *Vibrio* group 510 is closely related to *V. diazotrophicus* and may eventually become a named biogroup (Table BXII.γ.164) or subspecies of *V. diazotrophicus. Vibrio* group 510 can be distinguished phenotypically from *V. diazotrophicus* by being arginine negative at 7 days, fermenting D-mannose, and failing to ferment salicin. *V. diazotrophicus* strains usually have the opposite reactions.

4. *Vibrio* group 511

Properties are given in Table BXII.γ.163. *Vibrio* group 511 appears closely related to or perhaps a synonym of *V. cincinnatiensis* (see Table BXII.γ.164, and *V. cincinnatiensis*).

5. *Vibrio* group 512

Properties are given in Table BXII.γ.163. Strains of *Vibrio* group 512 were isolated from seawater and mud in the

---

1. "Baumann Groups"—In the 1970s, many different vibrios were studied in the laboratories of Paul and Linda Baumann. Scientific names were assigned to many of their distinct groups. However, some groups were never given scientific names. Also, in the Other Organisms Section of the *Vibrio* chapter in the last edition of this *Manual,* Baumann et al. (1984b) discussed four strains or groups of strains that they considered were distinct from all the species and named groups known at that time. These four vibrio groups should be studied with the techniques now available.

2. For many years vernacular names such as "Vibrio group 509" were assigned by one of us (JJF) to a new or unusual strain or group of strains. This nomenclature is similar to the one that assigned a vernacular name "Enteric Group" to an organism thought to belong in the family *Enterobacteriaceae* (Farmer et al., 1985a). The "Vibrio group names" have been included for many years in the databases (see Table BXII.γ.163) and computer identification programs distributed by the *Vibrio* Reference Laboratory. Many have been included in various DNA–DNA hybridization studies (Brenner et al., 1983b; Brenner et al., unpublished). Four of the groups are described below. Additional groups have been defined but require further study and will be described in the future. The relationships of these groups to other *Vibrio* species as shown by DNA–DNA hybridization experiments are given in Fig. BXII.γ.172.

United Kingdom. Based on DNA–DNA hybridization experiments, five strains were found to be 89% related to the reference strain under stringent hybridization conditions (Brenner et al., unpublished). The closest relative of group 512 is *V. cincinnatiensis* (including *Vibrio* group 511), which is 15% related under stringent hybridization conditions.

Genus II. **Photobacterium** *Beijerinck 1889, 401*[AL]

AN THYSSEN AND FRANS OLLEVIER

Pho.to.bac.te′ ri.um. Gr. n. *phos* light; Gr. neut. dim. n. *bakterion* a small rod; M.L. neut. n. *Photobacterium* light (-producing) bacterium.

Plump, straight rods, 0.8–1.3 × 1.8–2.4 μm; some coccobacillary and bipolar. Involution forms are usually seen in old cultures or under adverse conditions of cultivation. Gram negative. **Motile by 1–3 unsheathed polar flagella**; some species nonmotile. **Facultative anaerobic.** Chemoorganotrophic, having **both a respiratory and a fermentative type of metabolism**. Optimal growth temperature range appears to be 18–25°C, except for the psychrophilic species *P. profundum*. **Sodium ions are required for growth.** D-Glucose and D-mannose are catabolized with the production of acid and, in the case of *P. damselae* subsp. *damselae*, *P. iliopiscarium*, and *P. phosphoreum*, with gas. Oxidase reaction is variable. Most strains are lysine decarboxylase and arginine dihydrolase positive. Ornithine decarboxylase negative. **Accumulate poly-β-hydroxybutyrate** (PBH) under certain conditions of cultivation; **do not use the exogenous monomer β-hydroxybutyrate**. Most strains **grow in a minimal medium containing a seawater base, D-glucose, and NH₄Cl**; other strains also require L-methionine. Utilize D-glucose, D-fructose, glycerol, and D-mannose. Some species are bioluminescent.

The mol% G + C of the DNA is: 39–44 (Baumann and Baumann, 1977).

Type species: **Photobacterium phosphoreum** (Cohn 1878) Beijerinck 1889, 401 (*Micrococcus phosphoreus* Cohn 1878, 126.)

FURTHER DESCRIPTIVE INFORMATION

The genus *Photobacterium* is situated in the family *Vibrionaceae* and belongs to the class *Gammaproteobacteria*. Within the genus *Photobacterium*, the group of *P. damselae* subsp. *piscicida*, *P. damselae* subsp. *damselae*, and *P. leiognathi* forms a distinct subcluster (Nogi et al., 1998c). *P. phosphoreum*, *P. angustum*, and *P. profundum* are also closely related. High DNA–DNA hybridization values (which vary with the hybridization techniques used) have been reported between *P. damselae* subsp. *piscicida* and *P. damselae* subsp. *damselae* (Gauthier et al., 1995b; Thyssen et al., 2000). Low DNA–DNA hybridization values have been reported between (i) *P. damselae* subsp. *piscicida* and *P. leiognathi* (Thyssen et al., 2000), (ii) *P. damselae* subsp. *damselae* and *P. leiognathi* (Nogi et al., 1998c; Thyssen et al., 2000), (iii) *P. angustum* and *P. phosphoreum* (Baumann and Baumann, 1981b; Nogi et al., 1998c), (iv) *P. angustum* and *P. profundum* (Nogi et al., 1998c), (v) *P. phosphoreum* and *P. profundum* (Nogi et al., 1998c), (vi) *P. phosphoreum* and *P. leiognathi* (Baumann and Baumann, 1981b).

Photobacterium spp. display morphologic characteristics typical of members of the *Vibrionaceae*, appearing as straight rods; *P. damselae* subsp. *damselae*, and *P. leiognathi* appear as coccobacilli.

In the early stationary phase of growth, PHB, which is stored by the cells as a reserve product, can be detected as discrete, refractile, intracellular granules (Reichelt and Baumann, 1973b). When a substantial amount of PHB is accumulated, the entire cell appears refractile. PHB is only formed when cells are grown in a basal medium with glucose as the sole carbon source; PHB is not formed when cells are grown in rich media containing peptone. Except for *P. damselae* subsp. *piscicida*, all *Photobacterium* species are motile. One to three polar flagella have been described (Reichelt and Baumann, 1973b). These flagella, unlike those of *Vibrio*, are not enclosed in a sheath.

The fatty acid profile of *Photobacterium* species has been recently determined by Nogi et al. (1998c), who found high concentrations of $C_{16:1}$ and $C_{16:0}$ fatty acids. Moreover, $C_{12:0}$, $C_{12:0\ 3OH}$, $C_{18:1}$, and to a lesser extent $C_{14:0}$, $C_{14:0\ 3OH}$, and $C_{14:1}$ fatty acids are detected. Q-8 and Q-7 have been reported as major isoprenoid quinones in *Photobacterium* species (Nogi et al., 1998c).

Colonies of *Photobacterium* are entire, smooth, semitranslucent, convex, and ivory. The color of the colonies is somewhat whiter than that of many other Gram-negative marine bacteria, probably due to a relatively low content of cytochromes.

Photobacterium species are unable to grow in the absence of Na⁺. For optimal growth, 160–280 mM Na⁺ is required. Most *Photobacterium* strains have no organic growth factor requirement; some *P. phosphoreum* strains, however, require L-methionine, either alone or in combination with other amino acids (Reichelt and Baumann, 1973b; Ruby et al., 1980). All species except *P. profundum* are able to grow at 20°C; *P. iliopiscarium*, *P. phosphoreum*, *P. profundum*, and some *P. angustum* strains are able to grow at 4°C.

The nutritional versatility of *Photobacterium* species is relatively limited; only 7–22 carbon compounds can be utilized as sole or principal sources of carbon and energy. These compounds include hexoses and a few pentoses, disaccharides, sugar acids, tricarboxylic acid cycle intermediates, and amino acids. Almost none of the *Photobacterium* species can utilize D-mannitol as the sole source of carbon energy; most species, however, except *P. damselae* subsp. *piscicida* and *P. leiognathi*, can utilize maltose. Most species are able to use *N*-acetyl-D-galactosamine, *N*-acetyl-D-glucosamine, D-fructose, glycerol, glycogen, DL-lactate, D-glucose, D-galactose, maltose, D-mannose, and turanose as sole carbon sources. Adonitol, L-arabinose, cellobiose, *myo*-inositol, α-D-lactose, raffinose, L-rhamnose, D-sorbitol, and sucrose are not used. Acid with gas from D-glucose fermentation is detected with some of the *Photobacterium* species. None of the species have an extra-

TABLE BXII.γ.177. Differential characteristics of species of *Photobacterium*[a]

Characteristic	1. *P. phosphoreum*	2. *P. angustum*	3a. *P. damselae* subsp. *damselae*	3b. *P. damselae* subsp. *piscicida*	4. *P. iliopiscarium*	5. *P. leiognathi*	6. *P. profundum*
Luminescence	+	−	−	−	−	+	−
Motility	+	+	+	−	+	+	+
Growth, 4°C	+	d	−	−	+	−	+
Growth, 37°C	−	+	d	d	−	+	−
Gas production from D-glucose	+	−	+	−	+	−	−
Indole production	−	−	−	−	−	−	+
Nitrate reduction	+	−	+	−	+	+	+
Gelatinase	−	d	d	−	−	−	nd
Lipase	−	d	d	+	nd	+	nd
Acetate utilization	−	+	−	nd	nd	+	nd
DL-Glycerate utilization	+	−	−	nd	nd	d	nd
Maltose utilization	+	d	+	−	nd	−	+
L-Proline utilization	−	−	nd	nd	nd	+	nd
Pyruvate utilization	−	+	nd	−	nd	+	nd
D-Xylose utilization	−	+	−	−	nd	−	nd

[a]For symbols, see standard definitions; nd, not determined.

cellular gelatinase or alginase; some strains have an extracellular chitinase, lipase, or amylase. The methyl red and Voges–Proskauer tests are positive. Oxidase reaction is variable; catalase reaction is positive. Indole, hemolysin, and H2S are not produced; except in the case of *P. profundum*, which is able to produce indole. Most species reduce nitrate, except *P. angustum* and *P. damselae* subsp. *piscicida* (Table BXII.γ.177).

Phylogenetic analyses based on the 5S rRNA, 16S rRNA, and 16S rDNA sequences have been carried out to clarify the taxonomic and phylogenetic position of members of the *Vibrionaceae* and species belonging to the genus *Photobacterium* (MacDonell and Colwell, 1985; Kita-Tsukamoto et al., 1993; Gauthier et al., 1995b; Nogi et al., 1998c). Moreover, the levels of homology of the chromosomal DNA of the different *Photobacterium* species have been intensively studied by Nogi et al. (1998c) and Thyssen et al. (2000).

Several serological typing systems have been developed, but these focus primarily on *P. damselae* subsp. *piscicida* and *P. damselae* subsp. *damselae* (Fouz et al., 1992; Magariños et al., 1992; Bakopoulos et al., 1997; Kawahara et al., 1998). In addition, plasmid profiling and a variety of DNA-based typing methods, such as amplified fragment length polymorphism (AFLP), restriction fragment length polymorphism (RFLP), and ribotyping, have been used in epidemiological studies (Magariños et al., 1992, 1997; Pedersen et al., 1997; Thyssen et al., 2000). Plasmids have been described in a variety of species, including *P. damselae* subsp. *piscicida* and *P. damselae* subsp. *damselae* (Fouz et al., 1992; Kim and Aoki, 1993; Pedersen et al., 1997). Chloramphenicol, tetracycline, sulfamonomethoxine, and kanamycin resistances have been shown to be plasmid mediated and transferable (Kim and Aoki, 1993).

Photobacterium species are susceptible to vibriostatic agent O/129 (10 μg and 150 μg). Drug resistance has been reported for many species, especially for *P. damselae* subsp. *piscicida*, which is resistant to a variety of antimicrobial agents (Aoki, 1992).

P. damselae subsp. *piscicida* and *P. damselae* subsp. *damselae* are pathogenic for several marine fish species. The polysaccharide capsular layer, iron uptake system, phopholipase activity of *P.*

damselae subsp. *piscicida*, and extracellular products have been recognized as virulence factors (Romalde and Magariños, 1997). A phospholipase toxin with hemolytic and cytotoxic activities has been documented in *P. damselae* subsp. *damselae* (Toranzo and Barja, 1993).

Photobacterium strains are widespread in the marine environment and have been isolated from seawater, sediment, the surface and intestines of diseased marine animals, and the specialized light organs from marine fish. Some strains are found as symbionts in specialized luminous organs of marine fish.

Strains of *P. leiognathi* and *P. phosphoreum* are able to emit light due to the presence of luciferase. The reaction catalyzed by luciferase involves the luminescent oxidation of reduced flavin mononucleotide (FMNH2) and a long chain aliphatic aldehyde by molecular oxygen. The reaction can be expressed as $FMNH_2 + O_2 + RCHO \rightarrow hv + FMN + RCOOH$.

ENRICHMENT AND ISOLATION PROCEDURES

Photobacterium species may be recovered from the marine environment and on the surfaces and the intestinal contents of marine animals using blood agar, brain–heart infusion agar, tryptic soy agar (supplemented with 1% (w/v) NaCl), thiosulfate citrate bile sucrose agar, and marine agar. Details of the isolation of *Photobacterium* in luminous organs of fish are given by Baumann and Baumann (1981b) and Hastings and Nealson (1979). Nutrient agents affecting the growth, luminescence intensity, and luciferase synthesis are described by Rodicheva et al. (1993).

MAINTENANCE PROCEDURES

Photobacterium strains may be maintained at room temperature on long-term preservation medium (West and Colwell, 1984) for several months and reactivated on brain–heart infusion agar (Difco) to which 1% (w/v) NaCl is added. For long-term storage, cultures should be frozen and maintained at −70°C. A method for long-term preservation of luminous bacteria, such as *P. leiognathi* and *P. phosphoreum*, is described by Janda and Opekarova (1989).

PROCEDURES FOR TESTING SPECIAL CHARACTERS

A somewhat unusual combination of properties of *Photobacterium* is the ability to accumulate PHB as an intracellular reserve product, coupled with the inability to utilize the exogenous monomer, β-hydroxybutyrate, as a sole or principal source of carbon and energy. Accumulation of refractile PHB granules can be readily detected by phase microscopy in early stationary-phase cultures grown in basal medium containing 0.2% (w/v) D-glucose. In the case of organisms requiring growth factors, the medium is supplemented with 0.05–0.1% (w/v) yeast extract.

DIFFERENTIATION OF THE GENUS *PHOTOBACTERIUM* FROM OTHER GENERA

Classical biochemical differentiation of the genus *Photobacterium* from related genera, such as *Vibrio*, *Aeromonas*, and *Plesiomonas*, is primarily achieved via the identification of the individual species. *Photobacterium* can be differentiated from *Vibrio*, *Aeromonas*, and *Plesiomonas* by its production of arginine dihydrolase, Voges–Proskauer reaction, inability to produce alginase, gelatinase, and indole, fermentation of sucrose, utilization of D-mannitol, and requirement of over 100 mM Na$^+$ for optimal growth. Moreover, *Photobacterium* can be distinguished from *Vibrio* by the presence of unsheathed polar flagella. Analysis of poly-β-hydroxybutyrate enables clear differentiation of *Photobacterium* strains from *Vibrio* strains, but is of limited value in a routine diagnostic laboratory.

The decay kinetics of light emission by luciferase in the presence of dodecanal subdivides luminous bacteria into two major groups having either "slow" or "fast" kinetics (Jensen et al., 1980; Hastings and Nealson, 1981). "Slow" decay kinetics are characteristic of the enzyme from *Vibrio harveyi*, *V. splendidus* biovar I, *V. cholerae*, and *Xenorhabdus luminescens*, while "fast" decay kinetics occur with the luciferases of *Photobacterium*, *V. fischeri*, *V. logei*, and *Alteromonas hanedai*. Although *Photobacterium* is not unique in having "fast" decay kinetics, this property still has diagnostic value, since the luminous species of *Vibrio* and *Alteromonas* are readily distinguished from *Photobacterium* by a few easily determined properties; *V. fischeri* and *V. logei* have a yellow-orange, cell-associated pigment, while *Photobacterium* does not, and *A. hanedai* is a strict aerobe, whereas *Photobacterium* is facultative anaerobe.

TAXONOMIC COMMENTS

Since its creation in 1889, a variety of Gram-negative, facultatively anaerobic, PHB-producing bacteria have been included in the genus *Photobacterium*. Originally, the genus was composed of 3 species: *P. angustum*, *P. leiognathi*, and *P. phosphoreum*. Later phylogenetic work based on ribosomal RNA and DNA sequence analysis and/or hybridization experiments has revealed the presence of other species in this genus, such as *P. fischeri* (formerly *V. fischeri*) (Reichelt and Baumann, 1973b), *P. logei* (Harwood et al., 1980), *P. histaminum* (Okuzumi et al., 1994), *P. damselae* subsp. *piscicida* (formerly *Pasteurella piscicida*) (Gauthier et al., 1995b), *P. damselae* subsp. *damselae* (formerly *Vibrio damsela* or *Listonella damsela*) (Gauthier et al., 1995b), *P. profundum* (Nogi et al., 1998c), and *P. iliopiscarium* (formerly *V. iliopiscarius*) (Urakawa et al., 1999a). *P. logei*, however, has been removed from *Photobacterium* and classified in the genus *Vibrio*. The taxonomic position and classification of *P. fischeri* (*Vibrio fischeri*) remains uncertain. More details are given in the chapter on *Vibrionaceae*.

There are some taxonomic problems at the species level. Although the names *P. damselae* subsp. *piscicida* and *P. damselae* subsp. *damselae* have been validly published, some doubt exists whether or not these organisms should be classified as two separate species. They are phenotypically and genotypically dissimilar, develop different diseases and have different ecological niches (Thyssen et al., 1998; Thyssen et al., 2000). DNA–DNA hybridization and 16S rRNA sequence analysis, however, indicate that they are closely related (Gauthier et al., 1995b).

Another problem concerns the taxonomic position of *P. iliopiscarium* and *P. histaminum*. *P. iliopiscarium* was originally placed in the genus *Vibrio* (Onarheim et al., 1994), but has recently been reclassified as *P. iliopiscarium* based on 16S rRNA gene sequences (Urakawa et al., 1999a). *P. histaminum* (Okuzumi et al., 1994) is considered to be a later heterotypic synonym of *P. damselae* subsp. *damselae*, based on 16S rDNA sequencing and DNA–DNA hybridization (Kimura et al., 2000).

List of species of the genus Photobacterium

1. **Photobacterium phosphoreum** (Cohn 1878) Beijerinck 1889, 401^{AL} (*Micrococcus phosphoreus* Cohn 1878, 126.)
   phos.pho' re.um. Gr. v. *phosphoreo* bring light; M.L. neut. adj. *phosphoreum* light-bearing.

   Straight rod. Motile by means of unsheathed polar flagella. Luminescent. Contains high concentrations of $C_{16:1}$, $C_{16:0}$, and $C_{14:0}$ fatty acids, as well as $C_{12:0}$ and $C_{12:0\ 3OH}$, and to a lesser extent $C_{14:1}$, $C_{14:0\ 3OH}$, $C_{18:0}$, and $C_{18:1}$ (Nogi et al., 1998c). Isoprenoid quinone Q-8 accounts for 100% of the total isoprenoid quinones (Nogi et al., 1998c). May require L-methionine, either alone or in combination with other amino acids. Able to grow at 4°C, but not at 35°C or 40°C. Optimum growth at 160–280 mM Na$^+$ and 20–25°C. The description is the same as that for the genus. Gas production. Oxidase, caseinase, and lipase are not produced. L-arabinose, cellobiose, *i*-inositol, D-mannitol, salicin, D-sorbitol, sucrose, trehalose, and D-xylose are not fermented. Acetate, alanine, arginine, citrate, ethanol, glycine, leucine, putrescine, proline, pyruvate, and threonine are not used as sole carbon source. See Table BXII.γ.178 for other characteristics.

   DNA–DNA hybridization experiments have revealed a DNA relatedness of 20.5% between *P. phosphoreum* and *P. profundum* (Nogi et al., 1998c). Moreover, there was a DNA-rRNA relatedness of 98.7% to *P. leiognathi* (Baumann and Baumann, 1981b). Bacteriophages active against *P. phosphoreum* have been isolated from the marine environment (Spencer, 1963). Isolated from the light organs of five fish families (*Macrouridae*, *Merluccidae*, *Opisthoproctidae*, *Trachichthydae*, and *Moridae*) that are usually found in deep, cold waters. *P. phosphoreum* is a psychrophilic, deepwater species found at depths of 50–500 m.

   The mol% G + C of the DNA is: 40.8 (HPLC).

   Type strain: ATCC 11040, LMG 4233.

   GenBank accession number (16S rRNA): D25310, X74687, Z19107.

2. **Photobacterium angustum** Reichelt, Baumann and Baumann 1979, 79^{AL}

TABLE BXII.γ.178. Other biochemical characteristics of *Photobacterium* species[a]

Characteristic	1. *P. phosphoreum*	2. *P. angustum*	3a. *P. damselae* subsp. *damselae*	3b. *P. damselae* subsp. *piscicida*	4. *P. iliopiscarium*	5. *P. leiognathi*	6. *P. profundum*
Growth at 20°C	+	+	+	+	+	+	−
Growth at 30°C	d	+	+	+	−	+	−
Growth at 40°C	−	−	−	−	−	−	−
Oxidase	−	d	d	+	nd	d	+
Catalase	+	+	+	+	+	d	+
Arginine dihydrolase	+	+	+	+	+	+	+
Lysine decarboxylase	+	−	d	−	+	−	−
Ornithine decarboxylase	−	−	−	−	−	−	−
Hemolysin production	nd	−	d	−	−	−	nd
Amylase	−	−	d	d	−	−	nd
Alginase	−	−	−	−	nd	−	nd
Production H₂S	−	−	−	−	−	−	−
Utilization of:							
N-Acetyl-D-galactosamine	−	+	+	nd	nd	+	−
N-Acetyl-D-glucosamine	+	+	+	+	+	+	−
Adonitol	−	−	−	−	−	−	−
L-Arabinose	−	−	−	−	−	−	−
Cellobiose	−	−	+	−	−	−	−
D-Fructose	+	+	+	+	+	+	−
D-Galactose	+	+	+	+	+	+	+
Glycogen	+	+	+	−	nd	+	+
D-Gluconate	d	+	−	−	d	+	nd
D-Glucose	+	+	+	+	+	+	+
L-Glutamate	d	−	nd	+	nd	d	nd
Glycerol	+	+	+	−	d	+	+
myo-Inositol	−	−	−	−	−	−	−
DL-Lactate	d	+	+	d	nd	+	nd
α-D-Lactose	−	−	−	−	nd	−	−
Maltose	+	+	+	+	+	+	+
D-Mannitol	−	−	−	−	−	−	+
D-Mannose	+	+	+	+	+	+	+
Raffinose	−	−	−	−	−	−	−
L-Rhamnose	−	−	−	−	−	−	−
D-Ribose	d	+	+	d	+	+	nd
D-Sorbitol	−	−	−	−	−	−	−
Sucrose	−	d	−	−	nd	−	−
Trehalose	−	d	d	−	d	−	+
Turanose	−	+	+	nd	−	+	+
Tween 40	+	−	+	nd	nd	−	+
Tween 80	−	−	−	+	nd	−	+

[a]For symbols, see standard definitions; nd, not determined.

an.gus' tum. L. neut. adj. *angustum* limited, with respect to nutritional versatility.

Rod. Motile by means of polar flagella. Not luminescent. The fatty acid profile has recently been determined by Nogi et al. (1998c), who have found that the bacterium contains high concentrations of $C_{16:1}$, $C_{16:0}$, and $C_{18:1}$ fatty acids. Moreover, $C_{12:0}$ and $C_{12:0\ 3OH}$, and to a lesser extent $C_{14:0}$, $C_{14:0\ 3OH}$, $C_{14:1}$, $C_{15:0}$, $C_{17:0}$, $C_{17:1}$, and $C_{18:0}$ fatty acid have been detected. The major isoprenoid quinone is Q-8 (Nogi et al., 1998c). Optimal growth at 20–25°C and 160–280 mM Na^+. No growth in absence of NaCl, or at NaCl concentration higher than 8%. Not able to grow at 42°C. No gas production from D-glucose. Oxidase and DNase positive. Urease, lysine decarboxylase, and phenylalanine deaminase are not produced. D-adonitol, L-arabinose, cellobiose, esculin, *i*-inositol, D-mannitol, L-rhamnose, salicin, and D-sorbitol are not fermented. Arginine, citrate, citrulline, ethanol, glycine, leucine, malate, methionine, proline, and pu-

trescine are not used as sole carbon sources. Some strains are gelatinase positive. See Table BXII.γ.178 for other characteristics.

DNA–DNA hybridization experiments have indicated relatedness between *P. angustum* and *P. damselae* subsp. *piscicida*, *P. damselae* subsp. *damselae*, *P. leiognathi*, *P. phosphoreum*, and *P. profundum* of, respectively, 21%, 18.5%, 43.5%, 33.5%, and 24.5% (Nogi et al., 1998c; Thyssen et al., 2000). Susceptible to vibriostatic agent O/129 (2,4-diamino-6,7-diisopropyl pteridine) (10 μg and 150 μg). Resistant to penicillin G, oxacillin, streptomycin, and amoxicillin. Isolated from the open ocean and the light organs of several fish species.

The mol% G + C of the DNA is: 41.0 (HPLC).

Type strain: ATCC 25915.

GenBank accession number (16S rRNA): D25307, X74685.

3. **Photobacterium damselae** (Love, Teebken-Fisher, Hose, Farmer, Hickman and Fanning 1981) Smith, Sutton, Fuerst

and Reichelt 1991, 533^{VP} (*Vibrio damsela* Love, Teebken-Fisher, Hose, Farmer, Hickman and Fanning 1981, 1140.) *dam.se' la.e.* L. *damselae* named after damselfish.

Not luminescent. Na$^+$ required for growth. Does not produce phenylalanine deaminase. D-Glucose and D-mannose are fermented. No fermentation of L-arabinose, lactose, D-mannitol, melibiose, or trehalose. Several molecular techniques, such as AFLP, ribotyping, and RFLP, have been used to characterize and differentiate between the two subspecies of *Photobacterium damselae*, *P. damselae* subsp. *piscicida* and *P. damselae* subsp. *damselae* (Magariños et al., 1992, 1997; Pedersen et al., 1997; Thyssen et al., 2000). Plasmids have been reported. Isolated from fish.

The mol% G + C of the DNA is: 40.6–41.4 (enzymatic hydrolysis + HPLC).

Type strain: ATCC 33539.

a. **Photobacterium damselae** *subsp.* **damselae** (Love, Teebken-Fisher, Hose, Farmer, Hickman and Fanning 1981) Gauthier, Lafay, Ruimy, Breittmayer, Nicolas, Gauthier and Christen 1995b, 142^{VP} emend. Kimura, Hokimoto, Takahashi and Fujii 2000, 1341 (*Photobacterium damselae* Smith, Sutton, Fuerst and Reichelt 1991, 533; *Vibrio damsela* Love, Teebken-Fisher, Hose, Farmer, Hickman and Fanning 1981, 1140.)

Analyses of lipopolysaccharides and outer membranes have shown that different patterns can be seen, although some bands were common (Fouz et al., 1992). The fatty acid profile of the bacterium contains high concentrations of $C_{16:1}$, $C_{16:0}$, and $C_{18:1}$ fatty acids; $C_{12:0}$ and $C_{12:0\ 3OH}$, and, to a lesser extent, $C_{14:0}$, $C_{14:0\ 3OH}$, $C_{14:1}$, $C_{15:0}$, $C_{17:0}$, $C_{17:1}$, and $C_{18:0}$ fatty acids are also present (Nogi et al., 1998c) . The major isoprenoid quinone is Q-8, and a minor quinone is Q-7; these account for 90% and 10% of the total isoprenoid quinones, respectively (Nogi et al., 1998c). Round to slightly irregular, off-white, opaque, flat, shiny colonies. Able to grow at 1–6% NaCl, with an optimum of 3% NaCl. Produces urease, phosphatase, DNase, and arginine dihydrolase, but not gelatinase or ornithine decarboxylase. Production of amylase and lysine decarboxylase is variable. Some strains produce β-hemolysis. Able to grow on TCBS and MacConkey agar. Gas is produced during fermentation of carbohydrates. Citrate is not used as a sole source of carbon and energy. Maltose is fermented. No fermentation of D-adonitol, D-arabitol, dulcitol, erythritol, *i*-inositol, raffinose, L-rhamnose, salicin, D-sorbitol, or sucrose. See Table BXII.γ.178 for other characteristics.

DNA–DNA hybridization experiments show a DNA relatedness of 21%, 21%, and 19.5% between *P. damselae* subsp. *damselae* and *P. leiognathi*, *P. phosphoreum*, and *P. profundum*, respectively (Nogi et al., 1998c). Plasmids have been reported to be common in *P. damselae* subsp. *damselae* (Fouz et al., 1992). Plasmids can be heterogeneous in size, varying from very small (3.0 kb) to very large (~190 kb) (Pedersen et al., 1997). A high-molecular-weight plasmid band can be detected in most strains. Antigenic differences exist among different *P. damselae* subsp. *damselae* strains, and at least 4 distinct groups are recognized. The serological results are supported by lipopolysaccharide profiles and the outer membrane protein patterns (Fouz et al., 1992).

Resistant to penicillin G, ampicillin, erythromycin, streptomycin, and kanamycin. Susceptible to vibriostatic agent O/129 (150 μg). Virulence tests have shown that *P. damselae* subsp. *damselae* strains are pathogenic for turbot and rainbow trout (Fouz et al., 1992). Pathogenicity is due to production of toxins. *P. damselae* subsp. *damselae* strains have been shown to possess similar virulence determinants for poikilothermic and homeothermic hosts, secreting a potent, lethal phospholipase toxin with hemolytic and cytotoxic activities (Toranzo and Barja, 1993). Isolated from several marine fish (damselfish, yellowtail, seabream, brown shark, lemon shark, etc.), mammals (dolphin), reptiles (turtle), mollusks (octopus), and uninfected fish. Also associated with wound infections in humans.

The mol% G + C of the DNA is: 40.6–41.4 (enzymatic hydrolysis, HPLC).

Type strain: ATCC 33539, DSM 7482.

GenBank accession number (16S rRNA): X74700.

b. **Photobacterium damselae** *subsp.* **piscicida** (ex Janssen and Surgalla 1968) Gauthier, Lafay, Ruimy, Breittmayer, Nicolas, Gauthier and Christen 1995b, 142^{VP} *pis.ci.ci' da.* L. n. *piscis* a fish; L. suff. *-cida* from L. v. *caedo* to cut or to kill; M.L. n. *piscicida* fish killer.

Nonflagellated, straight rods with bipolar staining, measuring 1.5 × 1.0 μm. Pleomorphic, changing from coccoidal to long rods under different culture conditions. The toxicity and chemistry of the lipopolysaccharides (LPS) has been studied by Salati et al. (1989). LPS structures are similar for different isolates (Nomura and Aoki, 1985), possessing O-side chains consisting of high-molecular-mass bands in a ladder-like pattern (Magariños et al., 1992). Outer membrane protein patterns (OMP) are similar for different strains (Magariños et al., 1992); when grown under iron-restricted conditions, strains are able to produce iron-regulated outer membrane proteins (IROMPs) (Bakopoulos et al., 1997). Capsular proteins (CP) are not produced by all strains, but production can be induced by growth under special conditions (Magariños et al., 1996). The chemical composition of the capsular polysaccharides has been studied by Bonet et al. (1994). Fatty acid analysis has revealed that saturated and unsaturated fatty acids of 16 carbon atoms are the predominant fatty acids (Romalde et al., 1995). Colonies are regular, uniformly round, 1–2 mm in diameter, glistening, convex, opaque, and viscid. Capsules may be produced. Cells are not able to grow at NaCl concentrations lower than 0.02 M. Optimal growth at 0.3 M NaCl; osmotic requirement of 0.5–3% NaCl. Cells are able to grow at 25°C, but not at 4°C. Growth at 37°C is variable. Cells are able to grow at pH 6–9; optimal growth is at pH 6.8. Oxidase and catalase variable. Lecithinase and lipase positive. Does not produce urease, elastinase, caseinase, phospholipase, and lysine decarboxylase. The following carbon compounds are utilized as sole carbon and energy sources: glutamate, adipate, and L-proline. Fructose, D-galactose, sucrose, and raffinose are fermented anaerobically. L-rhamnose, D-glucitol, D-xylose, maltose, glycerol, galactitol, and cellobiose are not fermented. Gas is not produced. See Table BXII.γ.178 for other characteristics.

By DNA–DNA hybridization, DNA relatedness between *P. damselae* subsp. *piscicida* and *P. damselae* subsp. *damselae* is 77%. DNA relatedness between *P. damselae* subsp. *piscicida* and both *P. leiognathi* and *P. angustum* is much lower, at 23% and 21%, respectively (Thyssen et al., 2000). Direct analysis of DNA sequences has revealed that the rRNA sequences of *P. damselae* subsp. *piscicida* and *P. damselae* subsp. *damselae* differ only by one nucleotide (Gauthier et al., 1995b). Considerable variation in the prevalence, number, and size of plasmids found in *P. damselae* subsp. *piscicida* has been described. R-plasmids encode resistance for chloramphenicol, tetracycline, sulfamonomethoxine, and kanamycin (Kim and Aoki, 1993). *P. damselae* subsp. *piscicida* was formerly thought to be serologically homogeneous (Magariños et al., 1992), but serological differences have been recently detected (Bakopoulos et al., 1997; Kawahara et al., 1998). The median values for minimum inhibitory concentrations of antibiotics are (µg/ml): oxytetracycline, 6.4; penicillin, 0.1; tetracycline, 0.8; amoxicillin, 0.1; ampicillin, 0.05; chloramphenicol, 1.6; flumequin, 0.2; gentamycin, 12.8; neomycin, 12.8; nitrofurantoin, 25.6; sulfonamide, 51.2; streptomycin, 25.6; and trimethoprim, 6.4. Can be extremely virulent for fish. The presence of capsules, the iron uptake system, and the phospholipase are thought to play an important role in the virulence (Romalde and Magariños, 1997). Isolated from diseased fish (mainly *Seriola quinqueradiata*, *Sparus aurata*, and *Dicentrarchus labrax*). Restricted to the East Coast of North America, Japan, and the Mediterranean area.

The mol% G + C of the DNA is: 41.1–41.3 (enzymatic hydrolysis and HPLC).

Type strain: NCMB 2058.

GenBank accession number (16S rRNA): X78105.

4. **Photobacterium iliopiscarium** (Onarheim, Wiik, Burghardt and Stackebrandt 1994) Urakawa, Kita-Tsukamoto and Ohwada 1999a, 260[VP] (*Vibrio iliopiscarius* Onarheim, Wiik, Burghardt and Stackebrandt 1994, 377.)

i.li.o.pis.ca' ri.um. L. n. pl. *ilia* intestines, guts; *ilio* pertaining to intestines; L. adj. *piscarius* belonging to fish *iliopiscarium*, belonging to intestines of fish.

Cells are highly pleomorphic. Morphology of both the colonies and the cells varies, even under apparently identical growth conditions. Bacteria growing in broth cultures at salt levels higher than 2% tend to aggregate. Other factors that influence the morphology are the presence of glucose, temperature, and age of the culture. In broth medium with 2% NaCl and lacking glucose, the dominant cell morphology is rods and curved rods. Motile. Not luminescent. The dominant colony morphology of young cultures comprises small, colorless to grayish, opaque colonies with entire edges. At least 0.5% NaCl is required for growth; this varies with the temperature. No growth occurs at NaCl concentrations of ≥5%. Optimum growth in media with 2% NaCl. The ability to grow at different temperatures varies with the medium composition. No growth is observed at 30°C; 20°C is the optimal temperature when grown in peptone water containing 2% salt. Bacteria are able to grow at 4°C. Growth is better under aerobic conditions rather than anaerobic conditions.

Produce gas from glucose. Cells possess lysine decarboxylase. Negative with respect to urease, and tryptophan deaminase. Citrate, *i*-inositol, D-sorbitol, L-rhamnose, sucrose, melibiose, amygdaline, erythritol, D-xylose, L-xylose, adonitol, β-methylxyloside, L-sorbose, dulcitol, α-methyl-D-mannoside, α-methyl-D-glucoside, arbutin, salicin, melibiose, inulin, melezitose, xylitol, D-turanose, D-lyxose, D-tagatose, D-fucose, L-fucose, L-arabitol, 2-cetogluconate and 5-cetogluconate are not utilized. See Table BXII.γ.178 for other characteristics. Maximum-likelihood, maximum-parsimony, and neighbor-joining methods of nearly complete 16S rRNA gene sequences have shown that the nearest phylogenetic neighbor is *P. phosphoreum* (Urakawa et al., 1999a). Susceptible to vibriostatic agent O/129 at 150 µg/disc, but only slightly sensitive at 10 µg/disc. Resistant to penicillin-G, but sensitive to novobiocin, chloramphenicol, sulfamethoxazole, tetracycline, nitrofurantoin, oxytetracycline, and trimethoprim. Slightly sensitive to ampicillin and streptomycin. Isolated from the intestines of fish (herring, coal fish, cod, and salmon) living in cold seawater.

The mol% G + C of the DNA is: 39 ± 1 (T_m).

Type strain: PS1, ATCC 51760, DSM 9896.

GenBank accession number (16S rRNA): AB000278.

5. **Photobacterium leiognathi** (Hendrie, Hodgkiss and Shewan 1970) Boisvert, Chatelain and Bassot 1967, 521[AL] (*Photobacterium mandapamensis* Hendrie, Hodgkiss and Shewan 1970, 165.)

lei.o.gna' thi. M.L. gen. n. *leiognathi* named after fish of the family *Leiognathidae*.

Coccobacillus, approximately 1.6 × 3.2 µm. Motile. Luminescent. The fatty acid profile of the bacterium contains high concentrations of $C_{16:1}$, $C_{16:0}$, and $C_{18:1}$ fatty acids; as well as $C_{12:0}$, $C_{12:0\ 3OH}$, $C_{14:0}$, and, to a lesser extent, $C_{14:1}$, $C_{14:0\ 3OH}$, $C_{15:0}$, $C_{17:0}$, $C_{17:1}$, and $C_{18:0}$ fatty acids (Makemson et al., 1997; Nogi et al., 1998c). The major isoprenoid quinone is Q-8, and a minor quinone is Q-7; these account for 90% and 10% of the total isoprenoid quinones, respectively (Nogi et al., 1998c). Colonies are 1 mm in diameter, regular, convex, smooth, transparent, and yellow. The optimal temperature is 30°C. Growth can be observed at 37°C, but not at 4°C. Cells can grow at a salt concentration of 0.5–6%. Optimal NaCl concentration is 1.5%. Negative for the production of oxidase and for tryptophan deaminase. No production of lysine decarboxylase. Esculin is not hydrolysed. Fermentation occurs without gas production from fructose, D-glucose, D-galactose, glycerin, D-mannose, *N*-acetylglucosamine, and ribose. Cells are not able to ferment L-arabinose, cellobiose, dulcitol, *i*-inositol, lactose, maltose, D-sorbitol, saccharose, trehalose, or D-xylose. Citrate, citruline, glucuronate, leucine, malonate, methionine, putrescine, and tartrate are not utilized. See Table BXII.γ.178 for other characteristics.

DNA relatedness of *P. leiognathi* to *P. phosphoreum* and *P. profundum* is 26% and 19%, respectively (Nogi et al., 1998c). DNA–rRNA hybridization experiments have shown a relatedness of >98% between *P. leiognathi* and *P. phosphoreum* (Baumann and Baumann, 1981b). Susceptible to vibriostatic agent O/129, chloramphenicol, kanamycin, thiamphenicol, tetracycline, erythromycin, novobiocin, nitrofurantoin, polymyxin, trimethoprim, ampicillin, flumequin, gentamycin, and neomycin. Resistant to penicillin G, oxacillin, and amoxicillin. Isolated from the light organs of Leiognathidae and Apogonidae, which are usually found in warm, shallow, tropical water.

The *mol% G + C of the DNA is*: 41.6 (enzymatic hydrolysis, HPLC).

Type strain: ATCC 25521.

GenBank accession number (16S rRNA): D25309, X74686.

6. **Photobacterium profundum** Nogi, Masui and Kato 1998d, 631[VP] (Effective publication: Nogi, Masui and Kato 1998c, 6.)

pro.fun' dum. L. adj. n. *profundum* deep, living within the depth of the oceans.

Cells are rod shaped; 0.8–1.0 × 2–4 μm. Motile by means of single, unsheathed polar flagella. Nonluminescent. The fatty acid profile is very distinct from those of other *Photobacterium* species and is particularly characterized by substantial amounts of $C_{16:0 \ iso}$ and $C_{20:5}$. The dominant cellular fatty acids are $C_{16:1}$, $C_{16:0 \ iso}$, $C_{16:0}$, $C_{18:1}$, and $C_{20:5 \ \omega 3c}$ (IPA). The major isoprenoid quinone is Q-8, and a minor quinone is Q-7; these account for 95% and 5% of the total isoprenoid

quinones, respectively (Nogi et al., 1998c). Colonies are entire, smooth, semitranslucent, and ivory. They are 0.7–1.0 mm in diameter after 48 h of incubation at 10°C. Growth occurs between 4°C and 18°C at atmospheric pressure, with optimum growth displayed between 8°C and 12°C. No growth occurs at 0°C or 20°C. Growth occurs at pressures between 0.1 MPa and 70 MPa at 10°C, with optimum growth displayed at 10 MPa. Best growth occurs at a NaCl concentration of ~3%. No growth occurs in the absence of NaCl. Moderately barophilic and psychrophilic.

Acid is produced from D-glucose, but gas is not produced. Production of indole. D-Mannitol, but not *N*-acetyl-D-galactosamine, *N*-acetyl-D-glucosamine, or D-fructose, can be utilized as a sole carbon and energy source. See Table BXII.γ.178 for other characteristics. The level of DNA relatedness of *P. profundum* to *P. angustum*, *P. damselae* subsp. *damselae*, *P. leiognathi*, and *P. phosphoreum* is 24.5%, 19.5%, 19%, and 20.5% respectively (Nogi et al., 1998c). Resistant to vibriostatic agent O/129 (10 μg and 150 μg). Isolated from sediment at a depth of 5110 m.

The *mol% G + C of the DNA is*: 42 (reverse-phase HPLC).

Type strain: DSJ4, JCM 10084.

GenBank accession number (16S rRNA): D21226 for Strain DSJ4.

Genus III. **Salinivibrio** Mellado, Moore, Nieto and Ventosa 1996, 820[VP]

ANTONIO VENTOSA

Sa.li.ni.vib' ri.o. L. adj. *salinus* saline; L. v. *vibrio* move rapidly to and from, vibrate; M.L. masc. n. *Salinivibrio* saline organism which vibrates.

Gram-negative curved rods (0.5–06 × 1.0–3.2 μm). Cells occur singly, in pairs, or occasionally united by S shapes or spirals. **Motile by one polar flagellum.** Spores are not formed.

Moderately halophilic. The optimum NaCl concentration for growth is between 2.5 and 10% at 37°C; grows between 0 and 20% NaCl. Temperature range for growth: 5–50°C; optimum temperature: 37°C. pH range for growth: 5–10; optimum pH: 7.3–7.5. No growth factors are required for growth.

Facultatively anaerobic. Chemoorganotrophic. Catalase and oxidase are produced. Acid is produced from D-glucose. Gelatin is hydrolyzed. Voges–Proskauer and arginine decarboxylase tests are positive. Indole, β-galactosidase, lysine, and ornithine decarboxylase tests are negative. Colonies are circular, convex, opaque, smooth, and cream colored. Broth cultures are uniformly turbid.

The *mol% G + C of the DNA is*: 49.4–50.5.

Type species: **Salinivibrio costicola** (Smith 1938) Mellado, Moore, Nieto and Ventosa 1996, 820 emend. Huang, Garcia, Patel, Cayol, Baresi and Mah 2000, 620 (*Vibrio costicola* Smith 1938, 29.)

FURTHER DESCRIPTIVE INFORMATION

The genus is currently represented by a single species, *S. costicola*, with two subspecies: *S. costicola* subsp. *costicola* and *S. costicola* subsp. *vallismortis*. The genus was established to separate *Vibrio costicola* from the other species of the genus *Vibrio*, based on their phenotypic features and the lack of a close phylogenetic relationship between *S. costicola* and other *Vibrio* species (Mellado et al., 1996). The species *S. costicola* was isolated from hypersaline

environments (salterns, saline soils) and from salted food (Garcia et al., 1987b; Mellado et al., 1996). Recently, Huang et al. (2000) isolated a single strain from a hypersaline pond located in Death Valley, California, USA, with phenotypic and genotypic features typical of *S. costicola*. Based on 16S rRNA gene sequence comparison, it was closely related to *S. costicola* (mean sequence similarity 97.7%) and DNA–DNA hybridization studies showed a high degree of relatedness (93%). However, there were some phenotypic differences and it was proposed to place the new strain in a subspecies of *S. costicola*, designated *S. costicola* subsp. *vallismortis*, which automatically created *S. costicola* subsp. *costicola* (Huang et al., 2000).

The most peculiar feature of the species *S. costicola* is its ability to grow optimally at salt concentrations ranging from 2.5 to 10%, with an upper limit for growth at 20–25% salts. For that reason it is classified as a moderately halophilic bacterium (Ventosa et al., 1998) and especially with respect to its physiology and biochemistry, has been considered a model organism representative of that physiological group (Ventosa et al., 1998). Several studies have determined that *S. costicola* has a specific requirement for Na$^+$. Addition of high concentrations of compounds such as glucose or glycerol lowered the NaCl requirement from 0.5 M to 0.3 M NaCl (Adams et al., 1987). Temperature and medium composition affect the salt requirement (Kushner, 1978; Adams and Russell, 1992).

S. costicola strains have simple growth requirements. The minimum growth requirements have been determined, and two synthetic media have been described. One contained D-glucose, L-cysteine or cystine, glutamate, L-arginine, L-valine, L-isoleucine, and salts (Flannery and Kennedy, 1962). Kamekura et al. (1985) reported a minimal medium based on D-glucose, glutamate, biotin, thiamine, choline, and salts. Complex media stimulate growth at high salt concentrations. The effect may be due to the presence in the medium of compatible solutes or their precursors, or to the fact that other growth factors may be synthesized more slowly under the high salt conditions (Ventosa et al., 1998).

Strains of *S. costicola* show a heterogeneous response to antibiotics; they are very sensitive to chloramphenicol and rifampin (Nieto et al., 1993). The salt concentration may influence the susceptibility of *S. costicola* to antimicrobial compounds. However, *S. costicola* shows a high sensitivity to rifampin and trimethoprim, regardless of the salt concentration (Coronado et al., 1995). The susceptibility of *S. costicola* to several heavy metals has also been studied (Garcia et al., 1987a). All 58 strains tested were susceptible to cadmium, copper, silver, zinc, and mercury, but showed tolerance to lead (Garcia et al., 1987a).

The genome organization of *S. costicola* has been investigated by pulsed field gel electrophoresis. The genome sizes of six strains ranged from 2100 to 2600 Kb. (Mellado et al., 1997). Several plasmids and one megaplasmid have also been reported (Fernandez-Castillo et al., 1992; Mellado et al., 1997). Conjugation is the only genetic transfer mechanism that has been described for this organism. A bacteriophage, phage UTAK, that infected and lysed *S. costicola* was isolated from salterns in Alicante (Spain), and propagated optimally at 1–2 M NaCl (Goel et al., 1996). However, genetic transfer by transduction has not been studied.

Besides its presence in salted foods, the natural environment of *S. costicola* is hypersaline habitats such as salt lakes and salterns. This organism is present in ponds of salterns with 10–25% salts, and is one of the predominant bacteria in ponds with less that 15% salts, as determined by isolation of *S. costicola* from these habitats in culture media (Rodriguez-Valera et al., 1985; Márquez et al., 1987). However, its presence in hypersaline soils is less evident (Quesada et al., 1982).

Since *S. costicola* has served as a model organism in different studies of moderately halophilic bacteria, there is much information concerning its ecology, physiology, and biochemistry (internal ions concentration and ion pumps, compatible solutes, cytoplasmic and membrane-bound enzymes, polar lipid and fatty acid composition, etc.). Further information can be found in a review on moderately halophilic bacteria (Ventosa et al., 1998).

ENRICHMENT AND ISOLATION PROCEDURES

Selective procedures for the enrichment and isolation of *S. costicola* have not been reported. It can be isolated in complex media with the addition of a salt mixture and incubated at 37°C under aerobic or preferentially anaerobic (to avoid the growth of strictly aerobic species) conditions. The isolation medium described by Ventosa et al. (1982) can be used. Another isolation procedure has been described by Huang et al. (2000). Enrichments are prepared by inoculating samples into serum bottles containing 25 ml anaerobic enrichment medium (12% NaCl and 0.5% D-glucose) and incubation at 37°C without shaking.

MAINTENANCE PROCEDURES

Salinivibrio costicola may be maintained on complex medium containing 10% salts. An appropriate medium is described by Garcia et al. (1987b). Long-term storage of members of this genus can

be carried out by freeze-drying, storage at −80°C, or by storing in liquid nitrogen.

DIFFERENTIATION OF THE GENUS *SALINIVIBRIO* FROM OTHER GENERA

The genus *Salinivibrio* can be differentiated from other related genera, except *Vibrio*, based on morphology, since it contains curved rods. Typical features of *Salinivibrio costicola* are its optimal growth in media containing 10% salts, and its ability to produce arginine decarboxylase. The genus *Salinivibrio* can also be differentiated from *Vibrio* and other related genera of the family *Vibrionaceae* by comparative analysis of 16S rRNA gene sequences. Besides, in comparison with species of the genus *Vibrio*, *Salinivibrio costicola* has two unique helical sequences and secondary structures at positions 178–197 and 197–219 of the 16S rRNA (*E. coli* 16S rRNA gene sequence numbering) (Huang et al., 2000).

TAXONOMIC COMMENTS

The single species of the genus *Salinivibrio* was isolated from rib bones in Australian bacon and described originally by Smith (1938), as *"Vibrio costicolus"*. In *Bergey's Manual of Determinative Bacteriology* (7th ed.), this species was included as a member of the genus *Vibrio* (Breed et al., 1957) and in the 8th edition it was named correctly as *Vibrio costicola* (Shewan and Véron, 1974). Very few taxonomic studies were performed on this species and they were carried out on isolates from cured meats (Gardner, 1980). Garcia et al. (1987b) amended the description of this species, including strains isolated from cured meats as well as 54 isolates from several salterns located in different areas of Spain. Further DNA–DNA hybridization studies determined that isolates from cured meats and salterns have a high level of DNA relatedness and should be considered members of the same species (Gutiérrez et al., 1989). The analysis of complete 16S rRNA gene sequences of six strains showed that phylogenetically they form a monophyletic branch that is distinct from other *Vibrio* species and related genera (Fig. BXII.γ.184). This evidence, together with several phenotypic differences, permitted the establishment of a different genus for this species, *Salinivibrio costicola* (Mellado et al., 1996).

ACKNOWLEDGMENTS

I am grateful to J.T. Staley for the critical reading of this article and to D.R. Arahal for his help with the phylogenetic analysis.

FURTHER READING

Garcia, M.T., A. Ventosa, F. Ruiz-Berraquero and M. Kocur. 1987. Taxonomic study and amended description of *Vibrio costicola*. Int. J. Syst. Bacteriol. *37*: 251–256.

Gutiérrez, M.C., M.T. Garcia, A. Ventosa and F. Ruiz-Berraquero. 1989. Relationships among *Vibrio costicola* strains assessed by DNA–DNA hybridization. FEMS. Microbiol. Lett. *61*: 37–40.

Huang, C.Y., J.L. Garcia, B.K.C. Patel, J.L. Cayol, L. Baresi and R.A. Mah. 2000. *Salinivibrio costicola vallismortis* subsp. nov., a halotolerant facultative anaerobe from Death Valley, and emended description of *Salinivibrio costicola*. Int. J. Syst. Evol. Microbiol. *50*: 615–622.

Mellado, E., , E.R.B. Moore, J.J. Nieto and A. Ventosa.. 1996. Analysis of 16S rRNA gene sequences of *Vibrio costicola* strains: description of *Salinivibrio costicola* gen. nov., comb. nov. Int. J. Syst. Bacteriol. *46*: 817–821.

Ventosa, A., J.J. Nieto and A. Oren. 1998. Biology of moderately halophilic aerobic bacteria. Microbiol. Mol. Biol. Rev. *62*: 504–544.

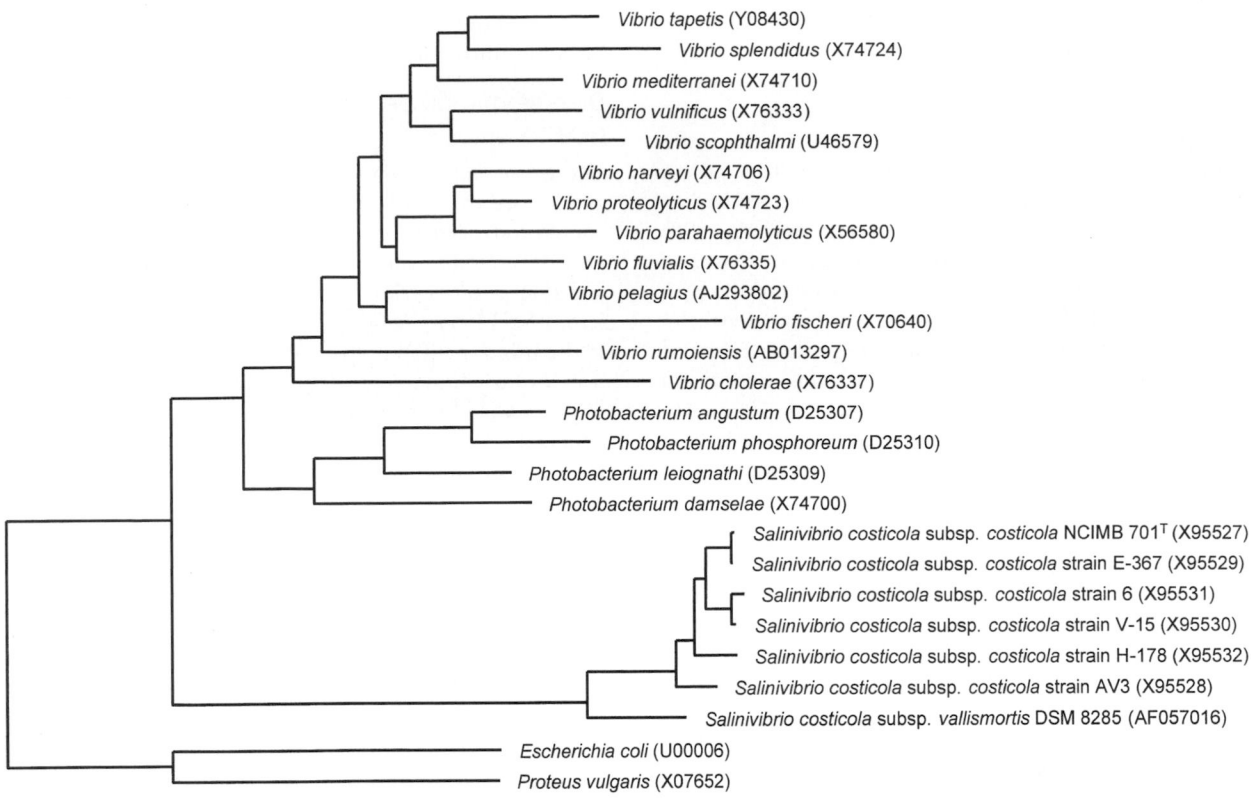

FIGURE BXII.γ.184. Phylogenetic tree derived from the analysis of 16S rRNA gene sequences of strains of *Salinivibrio costicola* and related species of the family *Vibrionaceae*, as well as two outgroup representatives. Figures in parenthesis indicate the accession numbers of 16S rRNA gene sequence data. Bar = 5% sequence difference.

List of species of the genus Salinivibrio

1. **Salinivibrio costicola** (Smith 1938) Mellado, Moore, Nieto and Ventosa 1996, 820[VP] emend. Huang, Garcia, Patel, Cayol, Baresi and Mah 2000, 620 (*Vibrio costicola* Smith 1938, 29.)

cos.ti' co.la. L. n. *costa* rib; L. subst. *cola* dweller; M.L. n. *costicola* rib dweller.

Cells are Gram-negative, curved rods, 0.5–0.6 × 1.0–3.2 μm and occur singly, in pairs, or rarely in chains. Motile by one, or rarely two, polar or subpolar flagella. Spores are not formed. Colonies are circular, convex, opaque, smooth, and cream colored. Broth cultures are uniformly turbid. Moderately halophilic. Grows at NaCl concentrations between 0 and 20%; the optimum NaCl concentration for growth is between 2.5 and 10% at 37°C. Growth occurs at 5–50°C (optimal growth at 37°C) and at pH 5–10 (optimal growth at pH 7.5). No growth factors are required, but yeast extract enhances growth. Facultatively anaerobic. Chemoorganotrophic. Catalase and oxidase are produced. Acid is produced from D-glucose, maltose, sucrose, and D-trehalose. Gelatin is hydrolyzed. Voges–Proskauer and arginine decarboxylase tests are positive. Indole, β-galactosidase, lysine, and ornithine decarboxylase tests are negative. Nitrates usually not reduced to nitrites. Nitrites are not reduced. Isolated from hypersaline environments and from salted food.

The mol% G + C of the DNA is: 49.4–50.5 (*T_m*).

Type strain: ATCC 33508, CCM 3575, DSM 11403, NCIMB 701.

GenBank accession number (16S rRNA): X95527, X74699.

a. **Salinivibrio costicola** *subsp.* **costicola** (Smith 1938) Mellado, Moore, Nieto and Ventosa 1996, 820[VP] Huang, Garcia, Patel, Cayol, Baresi and Mah 2000, 620 (*Vibrio costicola* Smith 1938, 29.)

Cells are Gram-negative, curved rods, 0.5 × 1.5–3.2 μm, motile by one polar flagellum (Fig. BXII.γ.185). The optimum salts concentration for growth is 10% at 37°C. Grows in the presence of 0.5–20% salts. No growth in the absence of NaCl. Other features are reported in the species description. Some characteristics differentiating this subspecies from *S. costicola* subsp. *vallismortis* are shown in Table BXII.γ.179. The following compounds are utilized as sole carbon and energy sources: D-trehalose, glycerol, acetate, *N*-acetylglucosamine, fumarate, glutamate, lactate, DL-malate, pyruvate, propionate, and succinate. The following compounds are not utilized as sole carbon and energy sources: amygdalin, D-arabinose, D-cellobiose, esculin, D-galactosamine, D-gluconolactone, lactose, D-melibiose, L-rhamnose, D-xylose, erythritol, *cis*-aconitate, benzoate, *p*-hydroxybenzoate, hip-

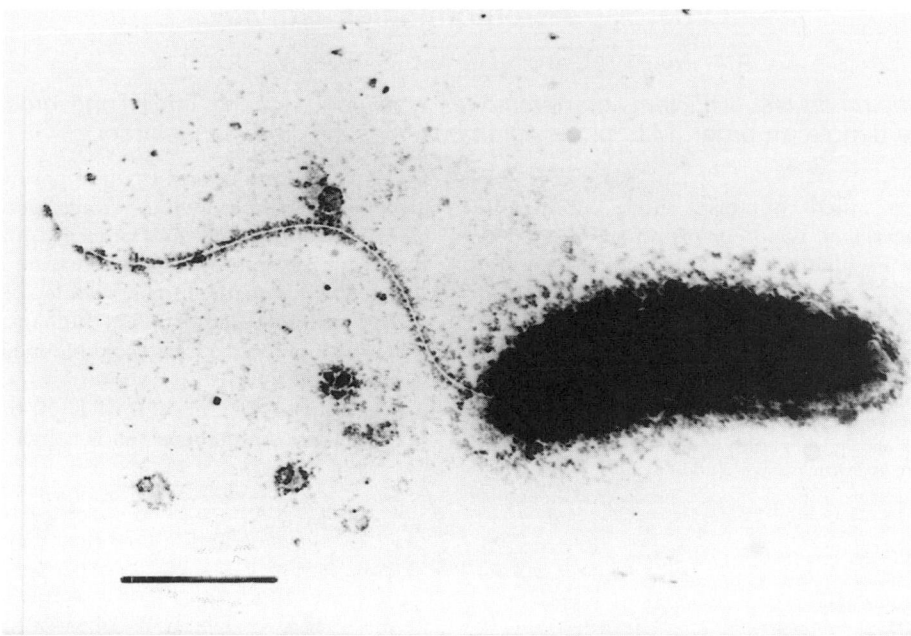

FIGURE BXII.γ.185. Electron micrograph of polar monotrichously flagellated cell of *Salinivibrio costicola*. Bar = 1 µm. (Reprinted with permission from M.T. Garcia et al., International Journal of Systematic Bacteriology *37:* 251–256, 1987, ©American Society for Microbiology.)

purate, malonate, quinate, salicylate, suberate, and D-tartrate. The following compounds are utilized as sole carbon, nitrogen, and energy sources: L-alanine, L-asparagine, L-glutamine, L-ornithine, L-proline, and L-serine. The following compounds are not utilized as sole carbon, nitrogen, and energy sources: L-allantoin, L-aspartic acid, betaine, creatine, ethionine, L-isoleucine, L-methionine, putrescine, sarcosine, L-tryptophan, and L-valine. Isolated from hypersaline habitats and from salted food.

The mol% G + C of the DNA is: 49.4–50.5 (T_m).

Type strain: ATCC 33508, CCM 3575, DSM 11403, NCIMB 701.

GenBank accession number (16S rRNA): X95527.

b. **Salinivibrio costicola** *subsp.* **vallismortis** Huang, Garcia, Patel, Cayol, Baresi and Mah 2000, 620[VP]

val.lis.mor′ tis. L. gen. n. *vallis* of the valley; L. gen. n. *mortis* of death; M.L. fem. n. *vallismortis* of the valley of death, named after Death Valley, California, USA.

Cells are Gram-negative, short, curved rods, 0.5–0.6 × 1.0–1.8 µm, highly motile by means of one, or rarely two, subpolar flagella. Optimum growth at 2.5% NaCl. Growth occurs at NaCl concentrations between 0 and 12.5%. Other features are reported in the species description. Some characteristics differentiating this subspecies from *S. costicola* subsp. *costicola* are shown in Table BXII.γ.179. The following compounds are used aerobically: D-glucose, D-mannose, sucrose, trehalose, starch, adipate, caproate, D-gluconate, pyruvate, L-glutamic acid, and yeast extract. The following compounds are not used: L-arabinose, cellobiose, D-galactose, D-galacturonate, lactose, maltose, L-rhamnose, salicin, citrate, 2-oxoglutarate, DL-lactate, L-malate, malonate, phenylacetate, propionate, valerate, ethanol, D-mannitol, *m*-inositol, D-sorbitol, DL-α-alanine, β-alanine, L-leucine, L-serine, L-tyrosine, benzoate, *p*-hydroxybenzoate, betaine, sarcosine and spermine. Isolated from Death Valley, California, USA.

The mol% G + C of the DNA is: 50 (HPLC).

Type strain: DV, DSM 8285.

GenBank accession number (16S rRNA): AF057016.

TABLE BXII.γ.179. Differential characteristics of the subspecies of *Salinivibrio costicola*[a]

Characteristic	*S. costicola* subsp. *costicola*	*S. costicola* subsp. *vallismortis*
Growth without NaCl	−	+
Growth in the presence of 20% NaCl	+	−
Anaerobic gas production	−	+
Hydrolysis of:		
Starch	−	+
Esculin	+	−
Utilization of:		
L-Alanine	+	−
Lactate	+	−
D-Mannose	−	+
D-Mannitol	+	−
L-Serine	+	−

[a]For symbols see standard definitions.

Order XII. **Aeromonadales** *ord. nov.**

AMY MARTIN-CARNAHAN AND SAMUEL W. JOSEPH

Ae.ro.mo.na.da′ les. M.L. fem. n. *Aeromonas* type genus of the family and order; suff. *-ales* to denote an order; M.L. pl. n. *Aeromonadales* the *Aeromonas* order.

The order *Aeromonadales*, which contains a single family, *Aeromonadaceae*, with the *Aeromonas*, is herein proposed. Gram-negative, straight, rigid rods. Facultatively anaerobic; oxidase positive

---

**Editorial Note:* The family *Succinivibrionaceae* has been placed in this chapter by the editors as *familia incertae sedis* based on a phylogenetic analysis that indicated distant relationships between the 16S rDNA sequences of the type species of the four genera included in the *Succinivibrionaceae* and the 16S rDNA sequences of some members of the genus *Aeromonas*. This placement should not be taken to imply that members of the *Succinivibrionaceae* are or should be included in the definition of the order *Aeromonadales* as proposed here by Martin-Carnahan and Joseph.

with rare exceptions and catalase positive. Includes motile (single, polar flagellum) and nonmotile species, as well as mesophilic and psychrophilic species. Primarily aquatic inhabitants of freshwater, marine, and especially estuarine environments, often associated with aquatic animals. Some species are either primary or opportunistic pathogens in humans, as well as a variety of other warm-blooded and cold-blooded animals and invertebrates.

The mol% G + C of the DNA is: 57–63.

Type genus: **Aeromonas** Stanier 1943, 213.

Family I. **Aeromonadaceae** Colwell, MacDonell and De Ley 1986, 474[VP]

AMY MARTIN-CARNAHAN AND SAMUEL W. JOSEPH

Ae.ro.mo.na.da′ ce.ae. M.L. fem. n. *Aeromonas* type genus of the family; suff. *-aceae* to denote a family; M.L. fem. pl. n. *Aeromonadaceae* the *Aeromonas* family.

Gram-negative, straight, rigid, nonsporeforming rods. Motility by means of polar flagella with some evidence of lateral flagella; some species are considered nonmotile. Occur singly, occasionally in pairs, or in short chains.

Facultatively anaerobic and chemoorganotrophic. Fermentative or respiratory with oxygen as a universal electron acceptor. Reduce nitrates, but do not denitrify. Ammonium salts utilized by most isolates as a sole source of nitrogen. Oxidase positive, except for *Tolumonas*. Catalase positive. Most species are mesophilic, but psychrophilic species do exist. *Oceanimonas* requires NaCl.

Primarily aquatic, most frequently isolated from fresh and estuarine waters and in association with aquatic animals; also found in sewage, surface waters, sediments, and biofilms; several species are pathogenic for humans, other warm-blooded animals, fish, eels, frogs, as well as other vertebrates and invertebrates, such as leeches.

The mol% G + C of the DNA is: 57–63.

Type genus: **Aeromonas** Stanier 1943, 213.

TAXONOMIC COMMENTS

The analysis of rRNA–DNA hybridization studies of the members of the family *Vibrionaceae*, in which *Aeromonas* was then included, was summarized in the first edition of *Bergey's Manual of Systematic Bacteriology* (Baumann and Schubert, 1984). The evidence from those studies suggested that the genus *Aeromonas* was sufficiently different from other genera in *Vibrionaceae* to merit its placement in a new family.

Subsequent analyses by Colwell et al. (1986) using 16S rRNA cataloging, 5S rRNA gene sequence comparisons, and rRNA–DNA hybridization data revealed that aeromonads demonstrated an evolutionary divergence that was approximately equidistant

from the *Enterobacteriaceae* and the *Vibrionaceae*, thereby justifying classification of the genus *Aeromonas* in its own family. These authors formally proposed the creation of the family *Aeromonadaceae* to incorporate the single genus, *Aeromonas*, which was later supported in 1992 by analyses using 16S rDNA sequencing and dendrograms showing that members of the genus *Aeromonas* form a distinct line within the *Gammaproteobacteria* (Martinez-Murcia et al., 1992a).

Additional evidence for the separation of the family *Aeromonadaceae* from the family *Vibrionaceae* soon followed with the analysis of 50 reference strains from 10 genera on the basis of 600 16S rRNA nucleotides by using reverse transcriptase sequencing (Kita-Tsukamoto et al., 1993). Shortly thereafter, the sequencing of nearly complete 16S rRNAs of 54 reference strains from the genera *Vibrio*, *Photobacterium*, *Aeromonas*, and *Plesiomonas* was added to all other known sequences for bacteria belonging to the *Gammaproteobacteria* to generate a database of 70 sequences for subsequent comparison (Ruimy et al., 1994). Their results showed that the family *Vibrionaceae* should include only *Vibrio* and *Photobacterium*, and that the genus *Aeromonas* should be moved to a separate family, *Aeromonadaceae*. The genus *Oceanimonas* has provisionally been placed in the family *Aeromonadaceae*.

Genus *Incertae Sedis* Although the genus *Tolumonas* is included as a member of the family *Aeromonadaceae* in this edition of *Bergey's Manual*, there is evidence to question the validity of this placement. In the genus description of *Tolumonas* in this edition, Fischer-Romero and Tindall have described the genus and single species, *Tolumonas auensis*, as either an anaerobic or aerobic Gram-negative rod that is oxidase negative, with a mol% G + C content of 49 (T_m). This description also differs from that given herein for the family *Aeromonadaceae*.

Genus I. Aeromonas Stanier 1943, 213[AL]

AMY MARTIN-CARNAHAN AND SAMUEL W. JOSEPH

Ae.ro.mo'nas. Gr. n. *aer* air, gas; Gr. n. *monas* unit, monad; M.L. fem. n. *Aeromonas* gas(-producing) monad.

Cells straight, coccobacillary to bacillary with rounded ends, 0.3–1.0 × 1.0–3.5 µm. Occur singly, in pairs, or rarely in short chains. Gram negative. **Most species are motile by a single, polar flagellum** of 1.7 µm wavelength; peritrichous flagella may be formed on solid media in young cultures and lateral flagella occur in some species. Facultatively anaerobic. Chemoorganotrophic, displaying oxidative and fermentative metabolism of D-glucose. Acid and often acid with gas produced from many carbohydrates, especially D-glucose. Nitrate is reduced to nitrite. A variety of exoenzymes such as arylamidases, amylase, DNase, esterases, peptidases, and other hydrolytic enzymes are produced. Main cellular fatty acids are hexadecanoic acid ($C_{16:0}$), hexadecenoic acid ($C_{16:1}$), and octadecenoic acid ($C_{18:1}$). **Usually oxidase positive and catalase positive.** Optimum growth temperature varies between 22°C and 37°C; growth temperature can range from 0 to 45°C, and some species do not grow at 35°C. **Generally resistant to 150 µg of the vibriostatic agent 2,4 diamino-6, 7-diisopropyl-pteridine (0/129).** Occur in fresh, brackish, tap, well, and chlorinated water, as well as biosolids and sewage. Some of the species have been associated with disease in a wide variety of warm-blooded and cold-blooded animals, including humans, domestic animals, frogs, fresh and salt water fish, and invertebrates. The phylogenetic position of *Aeromonas*, as determined by 16S rRNA gene sequence analysis, is in the *Gammaproteobacteria*, with its closest relatives in the families *Vibrionaceae* and *Enterobacteriaceae*. 16S rDNA sequences (signature sequences) have been determined for nearly all validly named species and are deposited in GenBank, EMBL, or RDP databases (Table BXII.γ.180).

The mol% G + C of the DNA is: 57–63.

Type species: **Aeromonas hydrophila** (Chester 1901) Stanier 1943, 213 (*Bacillus hydrophilus* Chester 1901, 235.)

FURTHER DESCRIPTIVE INFORMATION

The genus *Aeromonas* has traditionally been divided into two major groups of species, i.e., motile versus nonmotile species and mesophilic versus psychrophilic species. This general approach will be used here with the caveat that there are isolated instances where species considered to be nonmotile (*Aeromonas salmonicida*) do harbor flagellin genes (Umelo and Trust, 1997) and although *A. salmonicida* was generally considered not to grow above 30°C, it may actually be able to grow at 37°C (Austin, 1993).

The larger of these main categories includes the mesophilic, motile aeromonads which include 14 phenospecies that correspond to at least 17 genomospecies (DNA hybridization groups or HGs) (Table BXII.γ.181). The majority of human clinical isolates to date have been isolated from only six genomospecies, namely 1, 4, 8/10, 9, 12, and 14 (Holmberg et al., 1986a; Kuijper et al., 1989a; Altwegg, 1990; Altwegg et al., 1990; Abbott et al., 1992; Carnahan and Joseph, 1993). The category of psychrophilic, nonmotile aeromonads contains *A. salmonicida* (genomospecies 3), which includes four subspecies and is a major cause of fish disease worldwide (Austin et al., 1989a; Hänninen and Siitonen, 1995; Austin and Adams, 1996).

Phylogenetic treatment In the early 1980s, DNA relatedness studies of the motile aeromonads resulted in the establishment of three phenotypic species, namely *Aeromonas hydrophila*, *Aeromonas sobria*, and *Aeromonas caviae*, from among the eight reported genomospecies or DNA–DNA hybridization groups (Popoff et al., 1981). The number of aeromonad hybridization groups was subsequently expanded to 12 (Fanning et al., 1985; Farmer et al., 1986), and there are at present at least 17 hybridization groups, of which only two are unnamed: DNA hybridization Group 11 and *Aeromonas* Group 501 (HG 13) (Table BXII.γ.181).

Current recommendations define a genetic species as strains with approximately 70% or greater DNA–DNA relatedness and with 5°C or less ΔT_m to enable clear-cut differentiation of genomospecies (Wayne et al., 1987). 16S rDNA sequence analysis, which is routinely used to distinguish and establish relationships between genera, is often not as clear cut at the species level and has a lower discriminatory value when sequence analysis reveals a similarity of 97% or higher (Stackebrandt and Goebel, 1994).

TABLE BXII.γ.180. Nucleotide composition at variable positions of the 16S rDNA sequences determined from all known species of *Aeromonas*[a]

Species	77	129	131–132	154–156	165–167	199	218	230–232	250	258	264	457–464	469–476
A. hydrophila	A	A	A U	A G U	A C U	C	G	A U G	A	A	U	U G A U G C C U	C G U A U C A A
A. allosaccharophila	–	G	U C	– – –	– – –	–	–	G A A	–	–	–	– – G – A G – G	A C – G C – – G
A. bestiarum	–	G	U C	– – –	– – –	G	U	G A A	–	G	–	– – G C – – – –	– – – G – – – –
A. caviae	–	–	– –	– – –	– – –	–	–	– – –	–	–	–	C A G – A G – –	U C – G C U G G
A. encheleia	–	G	U C	– – –	– – –	G	U	G A A	–	G	–	– – – – – – – –	– – – – – – – –
A. eucrenophila	–	G	U C	– – –	G – –	–	–	G A A	–	G	–	– – – – – – – –	– – C – – – G
A. jandaei	–	G	U C	U A C	G U A	–	–	G A A	–	–	C	C A G – A G – –	U C – G C U G G
A. media	–	–	– –	– – –	– – –	–	–	– – –	U	–	–	– – – – – – – –	– C – – – G
A. popoffii	–	G	U C	– – –	– – –	G	U	G A A	–	G	–	– – – U – – – –	– C C – G – – –
A. salmonicida	–	G	U C	– – –	– – –	G	U	G A A	–	G	–	– – G C – – – –	– – – G – – – –
A. schubertii	G	G	U C	U A C	G U A	–	–	G A A	U	–	C	– – G – – G U –	– C – G C – – G
A. sobria	–	G	U C	– – –	– – –	G	U	G A A	–	G	–	– – G C A G – –	U C – G – – – G
A. trota	–	–	– –	– – –	– – –	–	–	– – –	–	–	–	C A G – A G – –	U C – G C U G G
A. veronii	–	G	U C	U A C	G U A	–	–	G A A	–	–	–	– – G – A G – –	A C – G C – – G
Aeromonas sp. DHG 11	–	G	U C	– – –	– – –	G	U	G A A	–	G	–	– – G – C G – –	A C G G C – – –
Aeromonas sp. G501	–	G	U C	U A C	G U A	–	–	G A A	U	–	C	– – G – A G – –	A C – G C – – G

[a]Numbers are homologous positions following the numbering system in *E. coli* (Brosius et al., 1978). Dashes indicate the same nucleotide as that of *A. hydrophila*. The two triplets characteristic of species from the "Schubertii" branch are underlined. Martínez-Murcia, unpublished.

TABLE BXII.γ.180.   *(cont.)*

Species	647	649–650	650	707	834	839	847	1009–1011	1018–1019	1153	1285	1308	1329	1355	1367
A. hydrophila	C	A G	G	C	C	U	A	U G C	G C	G	A	C	G	A	U
A. allosaccharophila	–	– –	–	–	–	–	–	C U A	U G	–	–	–	–	–	–
A. bestiarum	–	– –	–	–	–	–	–	– – –	– –	–	–	–	–	G	C
A. caviae	–	– –	–	–	–	–	–	– – –	– –	–	–	U	A	–	–
A. encheleia	–	– –	–	–	–	–	–	– – U	A –	–	–	–	–	–	–
A. eucrenophila	–	– –	–	G	–	–	–	– – U	A –	–	–	–	–	–	–
A. jandaei	–	– –	–	–	–	–	–	– – –	– –	–	–	–	–	–	–
A. media	–	– –	–	–	–	–	–	– – –	– –	–	–	–	–	–	–
A. popoffii	–	– –	–	–	–	–	–	– – –	– –	–	–	–	–	–	–
A. salmonicida	–	– –	–	–	–	–	–	– – U	A –	–	–	–	–	G	C
A. schubertii	–	G A	A	–	U	C	G	– – U	A –	A	U	U	A	–	–
A. sobria	U	– –	–	–	–	–	–	– – –	A –	–	–	–	–	–	–
A. trota	–	G –	–	–	–	–	–	– – –	– –	–	–	U	A	–	–
A. veronii	–	– –	–	–	–	–	–	– – U	A –	–	–	–	–	–	–
Aeromonas sp. DHG 11	–	– –	–	–	–	–	–	– – U	A –	–	–	–	–	–	–
Aeromonas sp. G501	–	G A	A	–	U	C	G	– – –	– –	A	U	U	A	–	–

[a]Numbers are homologous positions following the numbering system in *E. coli* (Brosius et al., 1978). Dashes indicate the same nucleotide as that of *A. hydrophila*. The two triplets characteristic of species from the "Schubertii" branch are underlined. Martínez-Murcia, unpublished.

The comparative analysis of the 16S rRNA gene sequences for *Aeromonas* species generally correlates with species designations derived from DNA–DNA hybridization studies (Martinez-Murcia et al., 1992a). While there is some lack of congruence between DNA–DNA hybridization studies and 16S rDNA sequencing results, the overall differentiation between groups is very similar. Table BXII.γ.180 provides the nucleotide composition at variable positions of the 16S rDNA sequences (signature sequences) determined from all currently recognized *Aeromonas* species. Fig. BXII.γ.186 is a phylogenetic tree showing the relationships between described *Aeromonas* species (type strains and reference strains) as determined by a continuous 1502-nucleotide 16S rDNA sequence comparison using the neighbor-joining method (Martinez-Murcia, 1999). Within the genus *Aeromonas*, the species are closely related and high levels of sequence similarity (~97.8–100%, which corresponds to 0–33 base differences) are observed. For example, the 16S rDNA of *A. hydrophila* ATCC 7966[T] (HG 1) exhibits only three nucleotide differences with the sequence of *A. media* ATCC 33907[T] (HG 5B). *A. jandaei* and *A. schubertii* represent quite separate lines, and together with *A. veronii*, seem to form a separate subbranch or affiliation within the genus (Fig. BXII.γ.187). Support for this loose association comes from the presence of paired signature triplets at the V2 region (positions 154–156 and 165–167) seen in Table BXII.γ.180 (Martinez-Murcia, unpublished), which distinguishes this "Schubertii" cluster from all other aeromonads. *A. schubertii* reveals the highest values of nucleotide differences (up to 33) within the genus and shows the highest number of sites with "unique" nucleotide composition.

Quite presciently, Sneath suggested that there was evidence for genetic crossing-over or recombination based on his analysis of the ribosomal sequences of Martinez-Murcia et al. (1992a) (Sneath, 1993). Although crossing-over may obliterate true phylogenetic relationships, Sneath reconstructed a speculative evolutionary phenogram from the right and left parts of the ribosomal sequence that seemed to suggest that *A. schubertii* might be the basal lineage species of the genus.

Further problems in using 16S rDNA for species identification of aeromonads were revealed when multiple strains from each species were sequenced. This subsequent research used Multilocus Sequence Typing (MLST) to examine a large collection of 94 well-characterized aeromonads, including type and reference strains, for four gene loci: 16S rDNA, *recA*, *chiA*, and *gyrB* (Carnahan, 2001). 16S rDNA exhibited less than 1% variable nucleotides within a species, which means that 16S rDNA sequencing can clearly define an isolate at the genus level only. In contrast, the other three loci were all more variable than 16S and could be more helpful in species identification. The phylogenetic analyses from these three loci arranged the aeromonads into their expected species clusters based on previous phenotypic and genotypic data. However, there were several instances where a single strain would cluster with its proper species designation for 2 of 3 loci, but for the other locus that strain would fall into an entirely different species cluster, again suggesting that recombination or lateral gene transfer is occurring within *Aeromonas*. For each locus examined, *A. schubertii* appeared to be the basal lineage species, consistent with Sneath's earlier assertion.

These observations of recombination across species boundaries and the recent touting of MLST as a "method of great promise" for the re-evaluation of the species definition in bacteriology (Stackebrandt et al., 2002) strongly suggest that the analysis of multiple gene loci should undoubtedly be pursued in future taxonomic studies of the genus *Aeromonas*.

Cell Morphology   The members of the genus *Aeromonas* are Gram-negative, straight rods with rounded ends, but sometimes can appear as coccobacilli or with filamentous forms. Cells are 0.3–1.0 × 1.0–3.5 μm and can occur singly, in pairs, or even as short chains (Altwegg, 1999).

Cell Wall Composition   Early studies of the composition of core oligosaccharides from motile aeromonads, with respect to combinations of hexose and heptose monosaccharide residues present in the core region of the cell wall lipopolysaccharides, revealed three general groups (Shaw and Hodder, 1978). These three groups generally matched later species designations of *A. hydrophila*, *A. caviae*, and *A. sobria* (Popoff et al., 1981). Analysis of the electrophoretic mobilities of purified or partially purified (proteinase K-treated) lipopolysaccharide (LPS) components by sodium dodecyl sulfate-polyacrylamide gel electrophoresis (SDS-PAGE) or either autoradiography or silver staining techniques indicated at least three major resolvable fractions (Chart et al., 1984; Dooley et al., 1985). Three major patterns among LPS O polysaccharide side chain profiles include a common heterogeneous profile that appears as a series of ladder-like bands ex-

TABLE BXII.γ.181. Current genomospecies and phenospecies within the genus *Aeromonas*

DNA hybridization group (HG)	Type strain[T] and/or HG definition strains or reference strains	Genomospecies	Phenospecies	Remarks
1	ATCC 7966[T]	*A. hydrophila*	*A. hydrophila*	Isolated from clinical specimens
2	ATCC 14715[T]	*A. bestiarum*	*A. hydrophila*-like	Isolated from clinical specimens
3	ATCC 33658[T]	*A. salmonicida*	*A. salmonicida* subsp. *salmonicida*[a]	
3	ATCC 33659[T]	*A. salmonicida*	*A. salmonicida* subsp. *achromogenes*[a]	
3	ATCC 27013[T]	*A. salmonicida*	*A. salmonicida* subsp. *masoucida*[a]	
3	ATCC 49393[T]	*A. salmonicida*	*A. salmonicida* subsp. *smithia*[a]	
3	CDC 0434-84 (DS), Popoff C316	unnamed	*A. hydrophila*-like[b]	Isolated from clinical specimens
4	ATCC 15468[T]	*A. caviae*	*A. caviae*	Isolated from clinical specimens
5A	CDC 0862-83(DS)	*A. media*	*A. caviae*-like	Isolated from clinical specimens
5B	CDC 0435-84 (DS)	*A. media*	*A. media*	
5B	ATCC 33907[T]	*A. media*	*A. media*	
6	ATCC 23309[T], NCMB 74[T]	*A. eucrenophila*	*A. eucrenophila*[c]	
7	CIP 7433[T], NCMB 12065[T]	*A. sobria*	*A. sobria*	
8X	CDC 0437-84 (DS)	*A. veronii*	*A. sobria*	
8Y	ATCC 9071 (RS)	*A. veronii*	*A. veronii* biovar sobria[d]	Isolated from clinical specimens
9	ATCC 49568[T]	*A. jandaei*	*A. jandaei*	Isolated from clinical specimens;
10	ATCC 35624[T]	*A. veronii* biovar veronii	*A. veronii* biovar veronii	Isolated from clinical specimens; ornithine decarboxylase positive
11	ATCC 35941 (RS)	unnamed	*Aeromonas* sp. (ornithine positive)[e]	
12	ATCC 43700[T]	*A. schubertii*	*A. schubertii*	Isolated from clinical specimens
13	ATCC 43946 (RS)	*Aeromonas* Group 501	*A. schubertii*-like	Isolated from clinical specimens
14	ATCC 49657[T]	*A. trota*	*A. trota*[f]	Isolated from clinical specimens; ampicillin susceptible
15	ATCC 51208[T], CECT 4199[T]	*A. allosaccharophila*	*A. allosaccharophila*	
16	ATCC 51929[T], CECT 4342[T]	*A. encheleia*	*A. encheleia*[g]	
17	LMG 17541[T]	*A. popoffii*	*A. popoffii*	

[a]This includes nonmotile, psychrophilic strains.

[b]This includes motile, mesophilic strains.

[c]It has been proposed that *A. eucrenophila* Subgroup II may reside within *A. encheleia* (Huys et al., 1996a, 1997a).

[d]*A. ichthiosmia* appears to be synonymous with *A. veronii* biovar sobria (Martinez-Murcia et al., 1992a; Collins et al., 1993; Huys et al., 2001).

[e]It has been proposed that members of HG Group 11 may reside within *A. encheleia* (Huys et al., 1996a, 1997a).

[f]*A. enteropelogenes* appears to be synonymous with *A. trota* (Martinez-Murcia et al., 1992a; Collins et al., 1993).

[g]It has been proposed to emend the description of *A. encheleia* to include members of HG 11 and Subgroup II of *A. eucrenophila* (Huys et al., 1997a).

tending throughout the gel lane; a second homogeneous pattern that appears as a small number of prominent O side chain bands, sometimes arranged as a doublet; and a third pattern often associated with serologically rough strains and characterized by a lack of detectable side chains (Dooley et al., 1985; Kokka et al., 1990, 1991a).

The major fatty cellular fatty acids produced are hexadecanoic acid ($C_{16:0}$), hexadecenoic acid ($C_{16:1}$), and octadecenoic acid ($C_{18:1}$) (Lambert et al., 1983; Huys et al., 1994). Gas–liquid chromatographic analysis of cellular fatty acid methyl esters (FAMEs) can be used to separate the basic phenotypic clusters of *A. hydrophila*, *A. caviae*, and *A. sobria* (Canonica and Pisano, 1988; Huys et al., 1995). However, the growth medium used can affect the cellular fatty acid composition, and hence the chemotaxonomic differentiation ability of FAME analysis (Huys et al., 1997c).

There is limited general information on *Aeromonas* outer membrane profiles (OMPs) other than studies of particular subsets of virulent aeromonads (Aoki and Holland, 1985; Kuijper et al., 1989b).

Fine structure *Aeromonas* species exhibit many fine structures outside the cell membrane and cell wall such as S-layer, flagella, pili, and capsule. A review of these features can be found in the chapter on "Pathogenic mechanisms" in *The Genus Aeromonas* (Gosling, 1996b). Further discussion of these individual features can be found in the pathogenicity section of this chapter *vide infra*.

Colonial or cultural characteristics On standard laboratory media, colonies of motile mesophilic aeromonads are 1–3 mm in diameter, smooth, circumscribed, circular, convex, translucent,

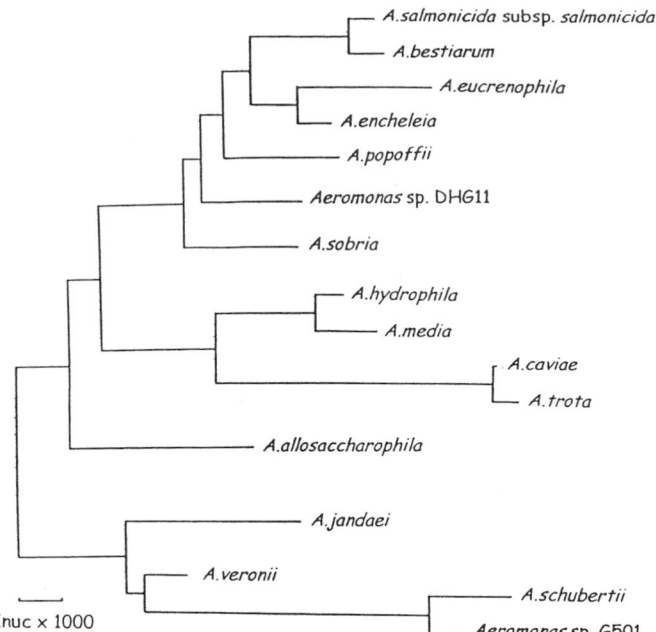

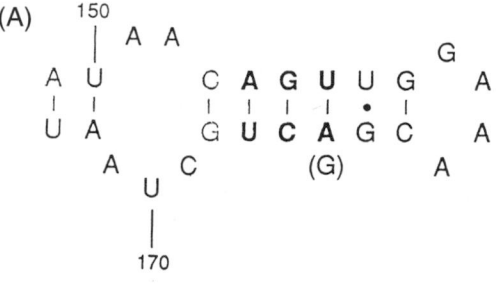

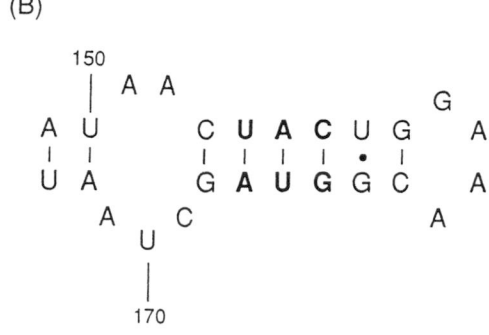

FIGURE BXII.γ.186. Phylogenetic relationships of described *Aeromonas* genomospecies as determined by a continuous 1502-nucleotide 16S rDNA sequence comparison using the neighbor-joining method (Saitou and Nei, 1987) and the MEGA Program version 1.01. Strain designations from different culture collections (in parentheses) and 16S rDNA sequence accession numbers [in brackets] are: *Aeromonas hydrophila* (HG 1) ATCC 7966^T [X60404]; *A. bestiarum* (HG 2) ATCC 51108^T (CIP 7430^T) [X60406]; *A. salmonicida* (HG 3) ATCC 33658^T (NCIMB 1102^T) [X60405]; *A. caviae* (HG 4) ATCC 15468^T (NCIMB 13016^T) [X60408]; *A. media* (HG 5) ATCC 33907^T [X60410]; *A. eucrenophila* (HG 6) NCIMB 74^T [X60411]; *A. sobria* (HG 7) CIP 7433^T (NCIMB 12065^T) [X60412]; *A. veronii* (HGs 8/10) ATCC 35624^T [X60414]; *A. jandaei* (HG 9) ATCC 49368^T [X60413]; *Aeromonas* sp. (HG 11) ATCC 35941 [X60417], *A. schubertii* (HG 12) ATCC 43700^T [X60416]; *Aeromonas* Group 501 (HG 13) ATCC 43946 (CECT 4254) [U88663]; *Aeromonas trota* (HG 14) ATCC 49657^T [X60415], *Aeromonas allosaccharophila* (HG 15) CECT 4199^T (S39232); *Aeromonas encheleia* (HG 16) CECT 4342^T [AJ224309]; and *Aeromonas popoffii* (HG 17) LMG 17541^T [AJ224308]. The sequences that start with the letter "X" are assigned with EMBL accession numbers and the other prefix letters, such as AJ, S, and U, can be found on the NCBI Website at http://www.ncbi.nlm.nih.gov (Reprinted with permission from A.J. Martinez-Murcia, International Journal of Systematic Bacteriology 49: 1403–1408, 1999, ©International Union of Microbiological Societies.)

FIGURE BXII.γ.187. Primary structures of the diagnostic sequences at the V2 regions of the 16S rDNA (position ~160) which distinguish the *Aeromonas* "type" groups (*A*) from the "Schubertii" cluster (*B*). The nucleotide in parentheses occurs only in *Aeromonas eucrenophila*. (Reprinted with permission from A.J. Martinez-Murcia et al., International Journal of Systematic Bacteriology 42: 412–421, 1992a, ©International Union of Microbiological Societies.)

and grayish-white to buff with a sometimes buttery consistency after 24–48 h of incubation at 35°C. Older colonies can develop a greenish hue, similar to that seen with certain *Vibrio* species, and a somewhat strong odor. There can be variation in colony size in some instances, either on original isolation or after subculture. *A. media* strains can produce a brown diffusible pigment on trypticase soy agar (TSA) (Allen et al., 1983). Colonies of the psychrophilic, nonmotile *A. salmonicida* subspecies are pinpoint in size after 18–24 h at 20–22°C, but after 4 d of incubation are circular, convex, entire, friable, and 1–2 mm in diameter (Griffin et al., 1953). Several of the *A. salmonicida* subspecies produce a brown, diffusible pigment after 5 d, especially on media containing tyrosine (Bernoth and Artzt, 1989). The presence of conspicuous encapsulation among several of the species results in the appearance of glistening to mucoid colonies.

There is variability in hemolysin production both within and between the species, as well as differences in the presence and/or type of hemolysis displayed depending on the type of red cells used. In one study, out of erythrocytes from nine animal species tested, the most sensitive blood agar medium contained mouse red cells, and sheep red cells were among the least sensitive for detection of β-hemolysis (Brenden and Janda, 1987). *A. hydrophila* and *A. veronii* are two of the more commonly isolated species that are strongly β-hemolytic on sheep blood agar, with less commonly isolated species often exhibiting β-hemolysis as well (Table BXII.γ.182). This β-hemolysis can be a broad zone, a double zone of partial hemolysis, or a narrow zone of β-hemolysis just under the colony edge. Many *A. caviae* isolates are partially hemolytic, displaying α-hemolysis, but can display narrow zones of β-hemolysis as well (Carnahan and Joseph, 1993).

On MacConkey agar, the majority of the mesophilic species grow as nonlactose fermenters, but a fair number of *A. caviae* isolates can ferment lactose. This can lead to confusion when testing water sources for coliforms, which are also lactose fermenters.

Nutrition and growth conditions The aeromonads grow over a wide temperature range (0–45°C), with the mesophilic strains growing between 10 and 42°C (Hänninen, 1994). The optimum temperature range is 22–37°C depending on the strain. While earlier studies have reported 28°C as the optimum temperature for motile aeromonads (Popoff, 1984a), at least one study on the effect of incubation temperature on growth and soluble protein profiles suggests that in some cases 37°C may be the optimum growth temperature (Statner et al., 1988). The psychrophilic

Characteristic	A. hydrophila	A. allosaccharophila (1)	A. bestiarum	A. caviae	A. encheleia (4)	A. eucrenophila	A. jandaei	A. media	A. popoffii	A. salmonicida[b]	A. schubertii	A. sobria	A. trota	A. veronii biovar veronii	A. veronii biovar sobria
Motility	+	+	+	d	+	+	+	d	+	d	+	−	+	+	+
Indole	+	+	+	+	+	+	+	+	d	+	−	+	+	+	+
Urea	−	−	−	−	−	−	−	−	−	−	−	−	−	−	−
ONPG	+	+	+	+	+	+	+	+	d	+	d	+	+	+	+
Citrate	d	d	d	+	−	−	+	d	+	d	d	+	d	+	d
Acetate	+	+	+	+	+	d	+	d	d	+	d	+	+	+	+
Malonate	−	−	−	−	−	−	−	−	−	−	−	−	−	−	−
KCN	+	−	d	+	+	+	d	d	d	d	−	−	d	−	d
Voges–Proskauer	+	−	d	−	−	−	d	−	d	d	d	−	−	d	d
Gelatin	+	+	d	+	d	d	+	d	+	+	d	+	+	+	+
Lysine decarboxylase	+	+	d	−	−	−	+	−	−	d	d	+	+	+	+
Arginine dihydrolase	+	d	d	+	d	d	+	d	+	d	+	−	+	−	+
Ornithine decarboxylase	−	d	−	−	−	−	−	−	−	−	−	−	−	+	−
Phenylpyruvic acid production	d	+	d	+	d	d	+	d	d	d	d	+	+	+	+
DNase	+	−	d	+	+	d	+	d	d	d	d	−	+	d	d
Corn oil lipase	+	+	d	d	+	d	+	d	+	+	+	+	−	+	+
D-Glucose (gas)	+	+	d	−	d	d	+	−	+	d	−	d	d	d	d
L-Arabinose	d	d	+	+	−	d	−	+	d	+	−	−	−	−	d
L-Rhamnose	d	d	d	−	d	d	−	−	−	−	−	−	−	−	−
D-Xylose	−	−	−	−	−	−	−	−	−	−	−	−	−	−	−
Cellobiose	−	+	d	+	−	d	d	+	−	d	−	+	+	d	d
Lactose	d	−	d	d	−	d	d	d	−	+	−	−	−	d	d
Maltose	+	+	+	+	+	+	+	+	+	+	+	+	+	+	+
Sucrose	+	+	+	+	d	d	−	+	−	+	−	+	d	+	+
D-Trehalose	+	+	+	+	+	+	+	+	+	+	+	+	+	+	+
D-Raffinose	−	d	−	−	−	−	−	−	−	−	−	−	−	−	−
Adonitol	−	−	−	−	−	−	−	−	−	−	−	−	−	−	−
Dulcitol	−	−	−	−	−	−	−	−	−	−	−	−	−	−	−
Erythritol	−	−	−	−	−	−	−	−	−	−	−	−	−	−	−
Glycerol	+	+	+	d	+	d	+	d	+	+	−	+	d	+	+
Inositol	−	−	−	−	−	−	−	−	−	−	−	−	−	−	−
D-Mannitol	+	+	+	+	+	+	+	+	+	+	−	+	d	+	+
D-Sorbitol	−	−	−	−	−	−	−	−	−	d	−	−	−	−	−
α-Methyl-D-glucoside	d	−	d	−	−	−	d	−	+	d	−	d	−	+	d
Salicin	d	−	d	+	d	d	−	d	−	d	−	−	−	+	−
Melibiose	−	−	−	−	−	−	d	−	−	−	−	−	−	−	−
D-Amygdalin	−	−	−	d	−	−	−	−	−	−	−	−	−	−	−
D-Arabitol															
H₂S from gelatin cysteine throsulfate	+	d	d	−	d	d	d	−	+	d	−	+	−	d	d
Gluconate oxidation[c]	d	−	d	−	−	−	d	−	−	−	−	−	−	d	d
Elastase	d	−	d	−	−	−	−	−	d	−	−	−	−	−	−
Ascorbate	nd	nd	d	nd	nd	−	d	−	nd	−	−	nd	−	−	nd
Esculin hydrolysis	+	d	+	+	+	d	−	d	−	d	−	−	−	+	−
Growth in 0% NaCl	+	+	+	+	+	+	+	+	+	+	+	+	+	+	+
Growth in 3% NaCl	+	+	+	+	+	d	+	+	+	+	+	+	+	+	+
Hemolysis[d]	+	d	+	−	d	d	+	d	−	d	d	−	d	+	+
Ampicillin (S)[e]	−	−	−	−	nd	−	−	d	−	d	−	d	+	−	−
Cephalothin (S)[f]	d	−	d	−	d	d	d	d	−	d	d	+	−	d	+
O/129 (R)	+	+	+	+	+	d	+	+	+	+	+	+	+	+	+
Pyrazinamidase[g]	−	nd	−	d	nd	+	−	d	−	d	−	−	d	−	d
Mannose	+	nd	+	d	nd	+	+	+	+	+	+	+	+	+	+
Stapholysin	d	−	d	−	nd	−	−	−	−	d	−	−	−	+	−
Oxidase	+	+	+	+	+	+	+	+	+	+	+	+	+	+	+
Catalase	+	+	+	+	+	+	+	+	+	+	+	−	+	+	+
Nitrate reduction	+	+	+	+	+	+	+	+	+	+	+	+	+	+	+
Polypectate (25°C)	−	−	−	−	−	−	−	−	−	d	−	−	−	−	−
Mucate[h]	−	−	−	−	−	−	−	−	−	−	−	−	−	−	−

[a]Symbols: +, >90%; −, <10%; d, 11 89% positive with incubation at 35°C for 7 d except for *A. popoffii* and *A. sobria*, which were incubated at 25°C (data from Abbott et al., 2003); nd, not determined.

[b]The *A. salmonicida* results are for the motile, mesophilic strains residing in HG 3 that are a currently unnamed subspecies.

[c]For determination of gluconate oxidation, the culture was incubated for only 48 h.

[d]Hemolysis detected on TSA on 5% sheep blood agar.

[e]Ampicillin susceptibility performed with 10 μg/disk and the culture incubated for only 24 h.

[f]Cephalothin susceptibility performed with 30 μg/disk and the culture incubated for only 24 h.

[g]Pyrazinamidase activity slants were incubated for only 48 h.

[h]Utilization as a sole carbon source.

strains in the *A. salmonicida* subspecies grow at temperatures generally ranging from 2 to 30°C.

Clinical isolates are generally incubated at 37°C, and environmental isolates are quite often incubated at 22–30°C, so that interlaboratory comparisons of biochemical reactions are difficult to make. Some mesophilic strains appear to be biochemically more active at 22°C than at 37°C. There are even cases of some biochemical tests having quite different results at the two different temperatures for the same strain (Ali et al., 1996a). In brain heart infusion medium at 28°C, the growth range for *A. hydrophila* is between pH 4.5 and 9.0 and 0–4% NaCl (Palumbo et al., 1985b).

Metabolism Aeromonads are facultative anaerobes that ferment D-glucose to acid or acid with gas and are oxidase and catalase positive, reduce nitrates to nitrites, and are enzymatically very active. They have been reported to produce amylase, DNase, chitinase, elastase, esterases, peptidases, arylamidases, and other hydrolytic enzymes (Hsu et al., 1981; Waltman et al., 1982; Janda, 1985; Carnahan et al., 1988; Hänninen, 1994). Recently, an adenylyl cyclase 2 was isolated from *A. hydrophila* with thermophilic properties and sequence similarities to proteins from hyperthermophilic *Archaea* (Sismeiro et al., 1998). There is new evidence that *Aeromonas* may exhibit anaerobic respiration and dissimilatory metal reduction (Knight and Blakemore, 1998).

Numerous large-scale phenotypic biochemical studies on *Aeromonas* have been published (Eddy, 1960, 1962; Ewing and Johnson, 1960; Ewing et al., 1961; Eddy and Carpenter, 1964; Popoff, 1969; McCarthy, 1975; Popoff and Veron, 1976; Lee, 1987). Both motile and nonmotile aeromonads utilize a number of other carbohydrates in addition to glucose (Popoff, 1984a; Arduino et al., 1988; Renaud et al., 1988; Carnahan et al., 1989a; Kuijper et al., 1989a; Altwegg et al., 1990; Abbott et al., 1992; Kämpfer and Altwegg, 1992; Carnahan and Joseph, 1993; Hänninen and Siitonen, 1995; Noterdaeme et al., 1996; Oakey et al., 1996a). In two recent major biochemical studies examining representatives of all known hybridization groups, excluding *A. encheleia* (HG 16), *A. popoffii* (HG 17), and nonmotile psychrophilic *A. salmonicida* subspecies (HG 3), it was determined that members of HGs 1–15 (including motile, mesophilic HG 3 *A. hydrophila*-like strains) shared a number of common features other than the aforementioned production of acid from glucose, oxidase and catalase activity, and reduction of nitrate (Abbott et al., 1992, 2003). These included the production of acid from D-trehalose; failure to utilize malonate or mucate as the sole carbon source; inability to ferment adonitol, dulcitol, erythritol, inositol, and D-xylose; and growth in nutrient broth containing 0 and 3% NaCl. Test results for several other phenotypic properties were nearly always positive (98–99%) and included motility, β-galactosidase activity, and resistance to O/129; fewer than 2% of the strains tested hydrolyzed urea, degraded pectin, or produced acid from arabitol, D-raffinose, or D-amygdalin. Variable results were obtained for 36 tests which included production of acetyl methyl carbinol (Voges–Proskauer test), indole, phenylpyruvic acid, acetate, citrate, and ascorbate utilization; elaboration of elastase, hemolysin, staphylosin, deoxyribonuclease, and corn oil lipase; esculin and gelatin hydrolysis; growth in KCN (potassium cyanide) broth; ornithine and lysine decarboxylase, arginine dihydrolase, and pyrazinamidase activities; H$_2$S production in GCF (gelatin–cysteine–thiosulfate) medium, pigment production at 25°C, gluconate oxidation; susceptibility to cephalothin and ampicillin; and acid production from L-arabinose, cellobiose, glyc-

erol, lactose, α-methylglucoside, maltose, D-mannitol, D-mannose, melibiose, L-rhamnose, salicin, and sorbitol fermentation (Abbott et al., 1992, 2003) (Table BXII.γ.182). Nine tests were determined by these authors as well as others to be discriminatory enough to use as an initial primary battery of tests to separate aeromonads from the more commonly isolated species in the clinical laboratory (Carnahan et al., 1991a; Abbott et al., 1992, 2003; Furuwatari et al., 1994). However, these test results may not be reliable for the identification of HG 3 nonmotile, psychrophilic *A. salmonicida* subspecies strains, since the most recent studies noted above were conducted at 35–37°C and generally evaluated only motile aeromonad isolates. Therefore, the HG 3 strains in the Abbott et al. (1992) study were members of the unnamed, motile subspecies within *A. salmonicida* (Hänninen, 1994), which are genetically identical with nonmotile, psychrophilic fish pathogens (HG 3), but phenotypically resemble HG 1 *A. hydrophila* strains (Table BXII.γ.183). It should be noted that rapid identification systems to date are not very accurate at identifying motile *Aeromonas* species other than *A. hydrophila*, *A. caviae*, and *A. veronii* biovar sobria (HG 8; formerly termed *A. sobria* HG 7) from clinical specimens. Often the rarer clinical species are misidentified by these systems (Hickman-Brenner et al., 1987; Carnahan et al., 1991a; Janda, 1991; Abbott et al., 1992, 1998; Janda and Abbott, 1998b; Altwegg, 1999).

Over the years, there have likewise been extensive biochemical studies of large numbers of nonmotile *A. salmonicida* strains (both typical and atypical) (Griffin et al., 1953; Smith, 1963; Schubert, 1974; Popoff, 1984a; Austin et al., 1989a, 1998a; Dalsgaard et al., 1994, 1998; Hänninen et al., 1995, 1997). A recent review and examination of the results from these studies revealed several discrepancies for key reactions previously published as useful for separating the various subspecies (Millership, 1996). This is most likely due to a combination of reasons, namely that only a limited number of strains were available in older studies, and that test methods and reagents varied, particularly incubation temperature and length of incubation. Recent extensive biochemical studies of the *A. salmonicida* subspecies, including the increasing numbers of "atypical" strains, are those by Hänninen et al. (1995) and Hänninen and Hirvelä-Koski (1997), who studied Scandinavian fish isolates (Table BXII.γ.183); Austin et al. (1998a); and a review article by Wiklund and Dalsgaard (1998). There is an urgent need for standardization of methods both within and between laboratories (Dalsgaard et al., 1998) (*vide infra* Taxonomic Issues).

Genetic classification of aeromonads has been accomplished by mol% G + C composition (Popoff, 1984a), DNA–DNA relatedness studies (Popoff et al., 1981; Fanning et al., 1985; Farmer et al., 1986), 16S rDNA sequence analysis (Martinez-Murcia et al., 1992a; Martinez-Murcia, 1999), and multilocus sequence typing (MLST) analysis (Carnahan, 2001). Popoff et al. (1981) originally designated eight DNA–DNA hybridization groups using the DNA S1 nuclease method. The number of HGs was expanded to 12 (Fanning et al., 1985; Farmer et al., 1986) using the hydroxyapatite method and the recommended specified values of >70% relatedness at the optimum temperature of 60°C, >55% relatedness at the stringent temperature of 75°C, and divergence of 5% or less.

A variety of other molecular methods has been employed for both taxonomic purposes and/or epidemiological typing of aeromonads. An excellent review of these can be found in the articles by Carnahan and Altwegg (1996) and Altwegg (1996), and include plasmid analysis, restriction enzyme analysis (REA), ribo-

TABLE BXII.γ.183. Differential characteristics for psychrophilic nonmotile *Aeromonas* species and subspecies (HG 3)

Characteristic	*A. salmonicida* subsp. *salmonicida*	*A. salmonicida* subsp. *salmonicida*	*A. salmonicida* subsp. *achromogenes*	*A. salmonicida* subsp. *masoucida*	*A. salmonicida* subsp. *smithia*	Atypical *A. salmonicida*
Strain	ATCC 33658[T]	47 strains[a]	ATCC 33659[T]	ATCC 27013[T]	ATCC 49393[T]	34 strains[a]
Motility	−	− (100%)	−	−	−	− (100%)
β-Hemolysis[b]	+	+ (100%)	−	+	−	− (100%)
Oxidase	+	+ (100%)[c]	+	+	+	+ (100%)
Brown diffusible pigment[d]	+	+ (100%)	+	−	−	+ (68%)
Esculin hydrolysis	+	+ (100%)	−	+	−	− (97%)
Indole	−	− (100%)	+	+	−	+ (18%)
Arginine dihydrolase	−	− (100%)	+	+	−	nd
Voges–Proskauer	−	nd	−	+	−	nd
Fermentation of D-glucose	+	+ (100%)	+	+	+	+ (100%)
Gas from D-glucose	+	+ (100%)	−	+	−	− (97%)
Fermentation of:						
Sucrose	−	− (100%)	+	+	+	+ (100%)
Maltose	+	+ (100%)	+	+	+	+ (100%)
Salicin	+	+ (100%)	−	−	−	− (100%)
Arbutin[e]	+	+ (91%)	−	−	−	− (100%)
D-Galactose[e]	+	+ (100%)	+	+	−	+ (76%)
D-Mannitol	+	+ (100%)	−	+	−	+ (94%)
L-Arabinose[e]	+	+ (100%)	−	+	−	− (100%)
i-Inositol[e]	−	− (100%)	−	−	−	− (100%)
N-Acetyl glucosamine assimilation[e]	+	+ (100%)	−	−	+	+ (37%)

[a]Data from Hirvelä-Koski et al. (1994) (deals with both typical *A. salmonicida* subsp. *salmonicida* strains (n 47), as well as atypical strains (n = 34), Hänninen and Hirvelä-Koski (1997) (atypical strains).

[b]β-Hemolysis detected on ox blood agar.

[c]Oxidase-negative strains have been isolated from turbots.

[d]Pigment production for all strains in this table were evaluated using the medium containing tyrosine (Bernoth and Artzt, 1989) (see text for medium formulation).

[e]Tested with API 50 CH.

typing, restriction fragment length polymorphism (RFLP), and amplified fragment length polymorphism (AFLP). In addition, the use of pulsed-field gel electrophoresis (PFGE) (Hänninen and Hirvelä-Koski, 1997), species-specific probes, polymerase chain reaction (PCR) tests, and RFLP-PCR of 16S rDNA has been proposed (Husslein et al., 1992; Ash et al., 1993a, 1993b; Dorsch et al., 1994; Cascon et al., 1996; Oakey et al., 1996b, 1998; Borrell et al., 1997; Hoie et al., 1997; Khan and Cerniglia, 1997; Graf, 1999a; Figueras et al., 2000). These findings reaffirm the opinion of some investigators that the separation of species and accurate identification requires a polyphasic approach (Altwegg, 1993; Carnahan and Joseph, 1993).

In addition, a variety of phenotypic methods outside of conventional or rapid identification kits have been used for subtyping of *Aeromonas* isolates. These include phage typing (DeMarta and Peduzzi, 1984); resistotyping (determining antimicrobial susceptibility patterns) (Motyl et al., 1985; Overman and Janda, 1999); protein analysis using SDS-PAGE protein profiles (Stephenson et al., 1987); SDS-PAGE of outer membrane proteins (OMPs) (Kuijper et al., 1989b); immunoblots of SDS-PAGE gels (Mulla and Millership, 1993); analysis of quinone, polyamine, and fatty acid patterns (Kämpfer et al., 1994); FAME (gas liquid chromatographic analysis of cellular fatty acid methyl esters) (Hansen et al., 1991; Huys et al., 1994, 1995); and enzyme analyses (Altwegg, 1996). Methods to detect the presence or absence of a particular enzyme using chromogenic substrates have already been discussed, but a second method of enzyme analysis involves the analysis of electrophoretic mobilities of enzymes using nondenaturing gel electrophoresis followed by staining for particular enzyme activities. Electrophoretic typing of esterases in an *Aeromonas* hospital outbreak (Picard and Goullet, 1985) was followed by the development of multilocus enzyme electrophoresis (MEE) based on the relative frequencies of four alleles at only four loci to separate HGs 1–12 (Altwegg et al., 1991).

Antigenic structure There are several serotyping schema for aeromonads that are based on the presence of unique heat-stable somatic (O) determinants (Sakazaki and Shimada, 1984; Fricker, 1987; Guinee and Jansen, 1987; Cheasty et al., 1989; Thomas et al., 1990). Only one comparison study of two of these schema has been published (Shimada and Kosako, 1991). The original Sakazaki and Shimada scheme recognizes 44 serogroups with an additional 52 provisional serogroups (Albert et al., 1995b). Using the Sakazaki and Shimada scheme on 268 strains, *Aeromonas* species were found to be serologically heterogeneous, with individual serogroups present in more than one species (Janda et al., 1996a). Most type and reference strains were not serologically representative of the genomospecies at large. Further, three particular serogroups—O:11 (24%), O:16 (14%), and O:34 (10%)—were predominant. They were not necessarily associated with any particular genomospecies, but did appear to be associated with clinical infections (48%). Following the report of Albert et al. (1995b) that some strains of *Aeromonas trota* cross-reacted with *Vibrio cholerae* O:139 antisera, it was determined that most of the *A. trota* strains examined by another investigator did not express the O139 antigen of *Vibrio cholerae* (Janda et al., 1996a).

Antimicrobial susceptibility Most motile *Aeromonas* species are generally resistant to penicillin, ampicillin, carbenicillin, and ticarcillin, but continue to be susceptible to second- and third-generation cephalosporins, aminoglycosides, carbapenems, chloramphenicol, tetracyclines, trimethoprim-sulfamethoxazole, and the quinolones (Koehler and Ashdown, 1993; Janda and Abbott, 1998b; Altwegg, 1999). In the three earliest studies on antimicrobial susceptibility, all clinical isolates were considered to be *A. hydrophila*, so that species-related differences in susceptibility could not be inferred (Overman, 1980; Fass and Barnishan, 1981; Fainstein et al., 1982b). A study comparing the "*in vitro*" susceptibilities among *A. hydrophila*, *A. veronii*s biovar sobria (formerly considered to be *A. sobria*), and *A. caviae* found *A. hydrophila* generally more resistant than the other species (Motyl et al., 1985). Most aeromonad species produce "inducible" chromosomal β-lactamases, and at least three distinct types have been noted (Bakken et al., 1988; Iaconis and Sanders, 1990; Morita et al., 1994; Walsh et al., 1997b). There have been reports of rapid commercial susceptibility systems that were unable to accurately detect this inducible resistance factor (Schadow et al., 1993). It appears that cephalothin can be a potential marker for clinical strains of *A. veronii* biovar sobria from clinical sources (Janda and Motyl, 1985) and that a certain minimum inhibitory concentration (MIC) of colistin (4 µg/ml) could be a marker for *A. jandaei* (Carnahan et al., 1991c). *Aeromonas trota* has a unique susceptibility to ampicillin, with up to 30% of some *A. caviae* isolates being susceptible as well (Kilpatrick et al., 1987; Carnahan et al., 1991b). Therefore, the use of ampicillin in selective media may lead to underestimation of these species. Examination of the more rarely isolated clinical species for antimicrobial resistance patterns included 17 *A. jandaei*, 12 *A. schubertii*, 15 *A. trota*, and 12 *A. veronii* biovar veronii strains (Overman and Janda, 1999). All of the *A. trota* isolates were susceptible to all of the agents tested except for 20% resistance to cefazolin and 13% to cefoxitin. All of the *A. schubertii*, *A. veronii* biovar veronii, and 88% of the *A. jandaei* isolates were resistant to ampicillin. Overall, *A. veronii* biovar veronii and *A. schubertii* had markedly increased resistance to tobramycin; *A. veronii* biovar veronii and *A. jandaei* were generally resistant to imipenem; and *A. schubertii* and *A. trota* were less susceptible to cefoxitin, an expanded-spectrum cephalosporin, than to broad-spectrum cephalosporins. These species-related antibiograms may be useful adjuncts to species identification of clinical isolates.

Antimicrobial resistance to tetracycline, trimethoprim-sulfamethoxazole, some extended-spectrum cephalosporins, and aminoglycosides seems to be increasing among clinical aeromonad isolates in Taiwan, as compared to isolates from Australia and the United States (Ko et al., 1996). A study of the spectrum of extraintestinal disease due to *Aeromonas* species in Queensland, Australia found that in nine cases, the empirical antibiotic regimen prescribed did not adequately cover infection due to *Aeromonas* (Kelly et al., 1993). This suggests that identification to the species level may play a role in the selection of appropriate antimicrobial therapy for infection with motile *Aeromonas* species.

Reports from several countries indicate increasing resistance of *A. salmonicida* isolates from fish. As with the motile species, there is a report of at least three β-lactamases among isolates of *A. salmonicida* (Hayes et al., 1994). Farmed Atlantic salmon in Scotland were assessed for antimicrobial disk diffusion values and compared with MIC results for amoxycillin, oxolinic acid, oxytetracycline, and potentiated sulfonamide (Grant and Laidler, 1993). Of the 65 isolates tested, 23 were susceptible to all four antimicrobials, and 12, 13, and 16 isolates showed resistance to 1, 2, or 3 antimicrobials, respectively. Among 35 Finnish isolates, resistance was found only to oxytetracycline in strains from 9 of the 35 farms (Hirvelä-Koski et al., 1994). A total of 130 strains originating from farmed salmonid fish isolated in Denmark, Norway, Scotland, Canada, and the United States were tested for susceptibility to 22 antimicrobial agents (Dalsgaard et al., 1994). Antibiograms revealed increasing resistance to oxytetracycline

and the quinolones, with multiresistant strains found in several countries.

Plasmids carrying resistance to streptomycin, chloramphenicol, tetracycline, and sulfathiazole as R factors have been isolated from *A. salmonicida* (Aoki et al., 1971; Toranzo et al., 1983). They have likewise been detected in motile aeromonads (Chang and Bolton, 1987). There is widespread plasmid-mediated resistance by means of 20–30 MDa plasmids in fish farms and eel ponds, and this most likely affects the potential benefit of antimicrobial compounds (Toranzo et al., 1983; Aoki, 1988). Broth conjugation experiments showed the transfer of the oxytetracycline (OT)-resistant phenotype from *A. salmonicida* to *Escherichia coli* and found this resistant phenotype to be encoded by high-molecular-weight R-plasmids that could be characterized by restriction digest profiles (Adams et al., 1998). One study of *Aeromonas* from clinical and environmental sources, representing seven species, found single or multiple resistance in 107/108 strains examined, with the highest incidence shown for β-lactam antibiotics other than cefotaxime. Thirty-five strains were found to contain transferable resistance plasmids, encoding resistance to ampicillin, cephalexin, cefoxitin, erythromycin, and furazolidone, either alone or in combination (Chaudhury et al., 1996).

Pathogenicity The early interest of von Graevenitz in *Aeromonas* stimulated others to begin to study this aquatic-borne organism, which seemed to be more common in clinical samples than previously realized (von Graevenitz and Mensch, 1968). While initially believed to be an opportunistic organism capable of infecting only immunocompromised individuals, a body of evidence now indicates that *Aeromonas* is a primary cause of extraintestinal illness and is strongly associated with gastrointestinal disease (Kelly et al., 1993; Janda and Abbott, 1996, 1998b; Joseph, 1996). Isolation rates have ranged from <1% to more than 60% in diarrheic populations in various geographic locations. The first well-documented extraintestinal infection was a case of septicemia in a menstruating woman who had been swimming in Jamaican waters. The organism was identified as "*Vibrio jamaicensis*" at the time but was later shown to be an *Aeromonas* sp. (Hill et al., 1954; Caselitz, 1955). The lack of final proof that *Aeromonas* is an etiologic agent of gastrointestinal disease lies in the absence of significantly large outbreaks and in one unsuccessful human volunteer study (Morgan et al., 1985). However, there is significant circumstantial evidence in the literature of sporadic cases and small outbreaks (Joseph, 1996; Janda and Abbott, 1998b).

Before the aeromonads were recognized as human pathogens, they were commonly isolated from poikilothermic animals, particularly amphibians, reptiles, and fish (Marcus, 1971, 1981; Shotts et al., 1972; Brackee et al., 1992; Burns et al., 1996; Gosling, 1996a; Schillinger, 1997). One of the first diseases attributed to *Aeromonas* was "red-leg" disease in frogs (Sanarelli, 1891). Aeromonads are also found with some frequency to cause various types of diseases in birds and domestic animals including pneumonia, peritonitis, and various localized infections (Gray, 1984; Shane et al., 1984; Shane and Gifford, 1985; Gosling, 1996a). Various species of fish develop hemorrhagic disease, ulcerative disease, furunculosis, red sore disease, and septicemia, resulting from infections with both motile and nonmotile *Aeromonas* species (Joseph and Carnahan, 1994; Austin and Adams, 1996). Humans can suffer from a variety of extraintestinal infections including wound infections, septicemia, meningitis, ophthalmitis, and infections associated with leech therapy, as well as gastroenteritis, especially in children <5 years (Altwegg, 1985; Janda and Duffey, 1988; Altwegg and Geiss, 1989; Janda, 1991; Fedorko, 1993; Janda and Abbott, 1996, 1998b; Graf, 1999b). The species most frequently isolated from clinical specimens are *A. hydrophila* (HG 1), *A. veronii* biovar sobria (HG 8), *A. caviae* (HG 4), *A. jandaei* (HG 9), *A. schubertii* (HG 12), and *A. trota* (HG 14) (Altwegg, 1990; Altwegg et al., 1990; Carnahan et al., 1991a; Janda, 1991; Carnahan and Joseph, 1993; Janda and Abbott, 1998b). There is some indication from experimental mouse lethality studies that some species are more virulent than others (Janda and Kokka, 1991).

Current theory suggests that the virulence of *Aeromonas* species may be multifactorial. Possible virulence-associated substances and components described for *Aeromonas* spp. include toxins (cytotoxic and cytotonic), proteases, hemolysins, lipases, adhesins, agglutinins, pili, enterotoxins, various enzymes, and outer membrane arrays (Cahill, 1990; Janda and Abbott, 1996; Gosling, 1996a, b; Howard et al., 1996). Unfortunately, with some of the early work, the species identification of the organisms was not consistent with present day criteria, so it is difficult to determine to which and to how many aeromonads these results apply.

Filamentous structures on aeromonads are diverse. There are two broad morphological types of pili (short, rigid and long, wavy), as well as polar and lateral flagella. Short, rigid pili are the main pilus type expressed on heavily piliated aeromonads. Amino acid sequencing of their pilin showed they are homologous to *E. coli* type 1 and Pap pili (Ho et al., 1992). Long, wavy pili on fecal isolates belong predominantly to a family of a type IV pili which has been designated Bfp for "bundle-forming pilus" (Kirov and Sanderson, 1996). Their pilins have a molecular mass of 19–23 kDa. Despite the tendency to form bundles, the pilins of Bfp pili have closer N-terminal homology to the type IVA pilins, such as MSHA of *Vibrio cholerae*, than they do to the type IVB pilins of enteropathogenic *E. coli* and *V. cholerae*. There is compelling evidence that this pilus type mediates adherence to enterocytes as revealed by observations made with EM and immunoelectron microscopy, reduction of adhesion when pili are removed from the organism, adhesions to Henle 407 cells, and blocking experiments with purified pili or antibody to Bfp (Honma and Nakasone, 1990; Iwanaga and Hokama, 1992; Kirov et al., 1999). Bfp pili have not yet been genetically characterized. However, a type IV pilus gene cluster (*tapABCD*) has been identified in *Aeromonas* spp. It was originally cloned from a strain of *A. bestiarum* AH65 (previously published as *A. hydrophila*), and the product was designated Tap for type IV *Aeromonas* pilus (Pepe et al., 1996). *TapD* encodes a type IV leader peptidase/methyltransferase, which in addition to processing Tap pilin, is responsible for the extracellular secretion of aerolysin and other enzymes via the type II secretion pathway (Strom and Lory, 1993; Strom and Pepe, 1999). The *tap* cluster was subsequently cloned from a strain of *A. veronii* biovar sobria from which Bfp had been purified and it was clearly established that Tap pili was a second and distinct family of pili from the Bfp family (Barnett et al., 1997). Tap pili have a predicted molecular mass of approximately 17 kDa.

To date, there is little known regarding the function(s) of Tap pili. If assembled on the cell surface, they represent a minor pilus species on fecal isolates grown under standard laboratory conditions. Studies with Tap pilin mutants in *in vitro* adhesion assays and in animal models have found no evidence that Tap pili are important enterocyte adhesins or colonization factors. The widespread conservation of the *tap* gene cluster among all

Aeromonas spp., however, suggests that the cluster does play an important function (Barnett and Kirov, 1999a, b).

Genetic characterization of the polar (Pof) and lateral (Laf) flagella of *Aeromonas* spp. has been described very recently (Shaw and Kirov, 1999). Evidence was presented that Pof are adhesins for *Aeromonas* spp. The Laf, when purified from a strain of *A. caviae*, showed considerable homology to the Laf of *Vibrio parahaemolyticus*. Laf genes are present in approximately 50% of *Aeromonas* isolates and lateral flagella are optimally expressed when bacteria are grown on solid media for <8 h at 37°C or 22°C. Preliminary evidence suggests that Laf serve as accessory colonization factors at the cell surface (Shaw and Kirov, 1999).

Enterotoxic (aerolysin), cytotoxic (Act), and two cytotonic toxins (Alt and Ast) have been detected in *A. hydrophila* (Chopra et al., 1996; Xu et al., 1998; Chopra and Houston, 1999). It appears that there may be a family of related hemolytic, cytotoxic enterotoxins. Because presumed "aerolysin" has been described in different species and/or strains, it is difficult to determine if these are all the same protein owing to some variance in the description of this toxin. Comparisons of Act of *A. hydrophila* SSU with the aerolysin of *A. trota* AH-2 (Chakraborty et al., 1987) and that of *A. bestiarum* (formerly *A. hydrophila*) AH-65 (Buckley et al., 1981) show differences in restriction maps, divergent flanking sequences, possibly different host cell receptors, and variability in the role of selected amino acid residues in biological function, e.g., in post-cleavage activity of the pro-toxin and eventual folding. Act appears to increase levels of tumor necrosis factor α (TNF-α) and interleukin (IL)-1 β in macrophage cell line RAW 264.7, as well as other pro-inflammatory cytokines, which could mediate inflammation and tissue damage during *Aeromonas* infections. There was also an increase in prostaglandin activity (PG₁ or PG₂). Use of PG inhibitors such as NS 398 and Celebrex eliminated activity or reduced the time of production of PG including INOS and SsPLA₂ (GrpV) levels in Chinese hamster ovary cells (Chopra and Houston, 1999).

Aerolysin from *A. trota* AH2 (formerly *A. hydrophila/A. sobria*) was first shown to be an important virulence feature by marker exchange mutagenesis, where the mutants were found to be much less toxic to mice (Chakraborty et al., 1987). It is first released as a proaerolysin, which is inactive until it is cleaved proteolytically and a C-terminal fragment of approximately 40 amino acids is removed.

A type II secretion pathway was described for a different "aerolysin" from *A. bestiarum* AH 65 that contained two operons, *exeAB* and *exeC–N*, forming the central components of this pathway in *Aeromonas*. The products of these genes form an apparatus that appears to culminate in an outer membrane secretion port. Phosphate bond hydrolysis and proton motive force are required for secretion through this port. The two operons appear to form an inner membrane complex, and *in vitro* mutagenesis of *exeA* suggests that this complex functions in the energy-dependent gating of the port (Howard et al., 1993, 1996; Howard, 1999). Their proaerolysin is a dimer in crystalline form and in solution (van der Goot et al., 1993). Murine erythrocytes, which have a glycoprotein receptor, are particularly susceptible to aerolysin activity and are easily lysed. The mode of action of this "aerolysin" involves channel formation in the target cell, and it functions in similar fashion to the action of membrane porins. The heptameric toxin inserts into the membrane of the cell and forms a 1–2 nm channel, causing the cell to lose its stability because of increased permeability and it is eventually destroyed (Howard, 1999).

Aeromonas also produces a lipase, named glycerophospholipid:cholesterol acyltransferase (GCAT), which can produce cholesteryl esters and can act as a phospholipase by digesting plasma membranes (Buckley, 1983). Two distinct types of proteases have been isolated from aeromonads including metallo- and serine proteases (Rodriguez et al., 1992). While considered accessory toxins in most instances, some studies with *A. salmonicida* suggest that they are primary virulence features. These enzymes probably function to protect the organism in its various environments, including humans, by inducing tissue destruction and assisting in the degradation of substrates for catabolic metabolism. Other miscellaneous toxins include amylase, chitinase, elastase, lecithinase, and nucleases (Gosling, 1996a).

The presence of a polysaccharide capsule has been reported for *A. salmonicida* and *A. hydrophila* serogroups O:11 and O:34 (Garrote et al., 1992; Martinez et al., 1995). Functionally, it is presumed that the capsule is a virulence factor that aids in complement resistance and/or in adherence to and invasion of fish cell lines (Merino et al., 1996, 1997).

S-layers of *Aeromonas* spp. are paracrystalline structures composed of identical protein subunits that are translocated across the cytoplasmic membrane, periplasm, and outer membrane to the cell surface, where they are assembled and tethered to the cell via an interaction with the O-polysaccharide side chains of the lipopolysaccharide. The S-Layer was originally referred to as an "A-Layer" first described in *A. salmonicida* (Udey and Fryer, 1978). It was considered a virulence feature because strains with this feature were more virulent than those without it. Specifically, strains with the S-layer could cause furunculosis in fish, while those without S-layers were avirulent. S-layer mutants have an altered ability to produce disease (Noonan and Trust, 1997). The significance of the S-layer in mesophilic aeromonads such as *A. hydrophila* and *A. veronii* biovar sobria appears to be less significant. Studies to correlate the presence of an S-layer on human isolates of *Aeromonas* through an autoagglutination phenomenon yielded inconclusive evidence for a direct involvement in mouse pathogenicity (Janda et al., 1987; Paula et al., 1988; Kokka et al., 1990, 1991a, b).

Aeromonads also possess more than one type of siderophore. In *A. salmonicida* there are at least two different mechanisms for iron acquisition (Chart and Trust, 1983). Mesophilic aeromonads usually produce either a unique siderophore, amonabactin (a four-peptide-based *bis*-catecholate siderophore), or an enterobactin-like siderophore (Telford and Raymond, 1997). Interestingly, *A. sobria* (HG 7), a species not isolated from humans thus far, does not produce siderophores (Zywno et al., 1992). *Aeromonas* spp. display significantly increased siderophore production under iron limiting conditions, and clinical strains are generally more productive than environmental strains (Naidu and Yadav, 1997).

Ecology Mesophilic aeromonads are indigenous to and have been isolated from various aquatic environments worldwide including fresh, estuarine (brackish), surface water (especially recreational), drinking water (including treated, well water, and bottled water), and polluted waters (Zimmermann, 1890; Hazen et al., 1978; Rippey and Cabelli, 1980, 1985; Seidler et al., 1980; Kaper et al., 1981; LeChevallier et al., 1982; Van der Kooj, 1988; Araujo et al., 1989; Alonso et al., 1994; Holmes and Nicolls, 1995; Holmes et al., 1996), as well as waste water effluent sludge (Schubert, 1975; Montfort and Baleux, 1991). They are not generally considered marine organisms, but can be found naturally in marine systems that interface with fresh waters, and found at all salinities, except the most extreme (Hazen et al., 1978). However, they are usually not a part of the autochthonous bacterial population of groundwater, which is normally poor in nutrients (Havelaar et al., 1990). However, in nutrient-rich waters, *Aeromonas* species can grow to large numbers, generally reaching a peak in the warmer temperatures of the summer months in both temperate freshwater lakes and chlorinated drinking water, showing a seasonal distribution (Burke et al., 1984a, b). While not considered to be of fecal origin, they seem to have a capacity to tolerate polluted environments including those of chemical origin (Seidler et al., 1980).

There is increasing interest in the incidence of *Aeromonas*, both during water treatment and distribution, particularly with regard to public health interest and to their role as indicators of chlorine resistance, disinfection efficacy, regrowth potential and biofilm development, even at 4°C (Holmes et al., 1996; Kuhn et al., 1997a, b; Sisti et al., 1998; Chamorey et al., 1999; Massa et al., 1999). Following the detection of increased numbers of aeromonads in the Netherlands in 1985, the public health authorities defined "maximum values" for *Aeromonas* densities in drinking water. These values are currently 20 CFU/100 ml of water for water leaving the production plant and 200 CFU/100 ml for drinking water in distribution (Van der Kooj, 1988). Examination of biofilms from exhumed pipe lengths revealed that 30% of the pipes examined contained an average population of aeromonads of 118 CFU/g dry weight (LeChevallier et al., 1987). After disinfection with 1 mg/l of chlorine, they could still isolate *Aeromonas* from 10% of the pipe lengths, with an average population of 51 CFU/g wet weight, suggesting that even though free cells of *Aeromonas* may be relatively susceptible to disinfection, populations may survive high chlorine dosing when associated with biofilms (LeChevallier et al., 1987; Holmes and Nicolls, 1995). Recent research revealed the apparent role of "quorum sensing" in biofilm formation on stainless steel by *A. hydrophila* (Lynch et al., 1999).

Aeromonas spp. have a capacity to infect fishes, reptiles, amphibians, humans, and other animals after exposure to such aquatic environments (Austin and Adams, 1996; Gosling, 1996a; Janda and Abbott, 1996, 1998b). There is good evidence that the species associated with clinical infections in humans are different hybridization groups than those isolated from drinking water and environmental sources, although putative virulence factors can be found in both categories (Holmberg et al., 1986a; Millership et al., 1986; Cahill, 1990; Havelaar et al., 1992; Hänninen, 1994; Kirov et al., 1994; Hänninen and Siitonen, 1995; Holmes and Nicolls, 1995; Kuhn et al., 1997a, b). Given the fact that the presence of aeromonads in water supplies may be connected with the increased incidence of *Aeromonas*-related gastroenteritis, the significance of which HGs are isolated from water as opposed to clinical isolates needs to be studied in greater detail (Holmes et al., 1996; Joseph, 1996).

Aeromonas species have been isolated from various food sources, including raw meats and fresh grocery produce, thereby establishing a possible source for *Aeromonas* human consumption (Kirov, 1993; Palumbo, 1996). These include raw chicken (Kirov et al., 1990; Akan et al., 1998); milk (Santos et al., 1994, 1996; Eneroth et al., 1998); cheese (Santos et al., 1994, 1996); ground meats (Okrend ct al., 1987; Singh, 1997); seafood (Tsai and Chen, 1996; Lipp and Rose, 1997; Pianetti et al., 1997); poultry eggs (Yadav and Verma, 1998); fish, fish eggs, and shrimp (Hänninen et al., 1997); and a swine slaughter plant (Palumbo et al.,

1999). Produce sources include vegetables (Callister and Agger, 1987; Pedroso et al., 1997).

The presence of *Aeromonas* species in foods most likely reflects contact of these foods with water, as reflected in the name of the type species, *A. hydrophila*, which means "water-loving". The factors controlling the numbers of these organisms in these products include temperature, the trophic state of water, pH, NaCl, nitrite, atmosphere, and miscellaneous inhibitors such as essential oils of clove, coriander, nutmeg, and pepper (Palumbo et al., 1985b; Stecchini et al., 1993; Palumbo, 1996). However, a conclusive link between the consumption of *Aeromonas*-containing food and diarrheal disease has not yet been identified. As with the water isolates discussed above, putative virulence factors are found in both clinical and food isolates (Palumbo, 1996; Pedroso et al., 1997; Pin et al., 1997). It may be that temperature-related expression of virulence features, as yet undiscovered virulence features, and/or the existence of unique virulence "subsets" within several *Aeromonas* species can help to explain the absence of disease when *Aeromonas* is so ubiquitous (Daily et al., 1981; Palumbo, 1993; Joseph, 1996).

A. salmonicida is an inhabitant of natural waters and especially aquaculture ponds. It has been long considered a chief cause of furunculosis in various species of fishes (Shotts et al., 1972; Austin and Adams, 1996; Wiklund and Dalsgaard, 1998). Surprisingly, the organism has been difficult to isolate from the environment but is isolated from infected fish, particularly in the kidney. However, this organism was recently isolated in association with purple-pigmented bacteria in sediment from a Scottish loch (Austin et al., 1998b). Another possible reason for its cryptic character may be found in the recovery of cell wall deficient forms or L forms (McIntosh and Austin, 1990).

ENRICHMENT AND ISOLATION PROCEDURES

The use of enrichments for the increased recovery of *Aeromonas* isolates from specimens is somewhat controversial. Ecological studies attempting to collect environmental samples should not use enrichments since the predominant strain(s) will quickly overgrow other strains that may be present. Enrichment broth can be useful as a presence–absence test for aeromonads in drinking water and food, or for monitoring marine populations in shellfish harvesting areas (Moyer, 1996). Among fish samples, both typical and atypical *A. salmonicida* have been recovered using an enrichment procedure with tryptone soy broth (TSB) (Austin and Adams, 1996). Food samples may be placed in enrichment broths when small numbers of aeromonads are expected and if there is the probability of cell injury from environmental pressures. The most frequently used broths are tryptic soy ampicillin (30 mg/l) broth (TSAB) and alkaline peptone water (APW). There are various formulations for these broths, as APW has been used with or without ampicillin for enrichment and enumeration of aeromonads from foods by the most probable number (MPN) technique.

The use of enrichment broths in clinical laboratories for the detection of low numbers of *Aeromonas* spp. has been exclusively used for the recovery from stool (fecal) samples. Early comparative studies found that nutrient broth gave the best recovery; but the authors contended that enrichment led to isolation of aeromonads in low concentrations that might be expected in convalescent patients, carriers, and those with subclinical infections. They further suggested that enrichment not be routinely used since it would interfere with interpretation of epidemiological studies trying to interpret the relationship between *Ae-*

romonas spp. and acute diarrhea (Robinson et al., 1984, 1986). The currently recommended broth for enrichment from fecal samples is APW adjusted to pH 8.6–9.8 (Shread et al., 1981; Millership et al., 1983; Moulsdale, 1983; von Graevenitz and Bucher, 1983; Moyer et al., 1991). The basic formulation contains 1% peptone in water at pH 8.6, but some investigators have increased the pH to 9.8 and added ampicillin (10–40 mg/l) and/or 0.5% desoxycholate to increase the selectivity of the medium (Khardori and Fainstein, 1988). The matter of the proper incubation temperature for APW enrichments is controversial as well, with some researchers finding better growth at 25°C than at 37°C (Millership and Chattopadhyay, 1984), but others finding no difference at all between these two temperatures (Price and Hunt, 1986). The use of GN (Gram-negative) broth for enrichment of aeromonads is only half as efficient as using APW (Millership et al., 1983), and Moyer et al. (1991) found that enrichment with APW increased the number of *A. caviae* isolates, suggesting that APW may be detecting not bona fide pathogens, but merely transient colonizers. The question of clinical relevance of aeromonads detected from enrichment broths only is clearly open to interpretation at this date and enrichment broths are generally not recommended for routine cultures.

Several studies have been conducted on the use of Cary–Blair transport medium for the transport of *Aeromonas* isolates from a variety of simulated clinical specimens (Siitonen and Mattila, 1990; Altwegg and Lüthy-Hottenstein, 1992; Koehler and Ashdown, 1993). Comparisons were made with other transport media, such as Amies medium, modified Stuart's medium, and glycerol-buffered saline. The consensus was that Cary–Blair transport medium was the best overall choice for transport of viable *Aeromonas*. Altwegg and Lüthy-Hottenstein (1992) also compared the behavior of *Aeromonas* species in Cary–Blair transport medium tubes incubated at 4°C, 25°C, and 37°C and found that the strains behaved similarly at both 25°C and 37°C, but very differently at 4°C with an initial drop in the colony count of all six strains examined during the first 24 h followed by a rapid increase for only the two *A. hydrophila* strains. The colony count of both *A. caviae* strains steadily dropped over the next few weeks, producing results consistent with those reported by Siitonen and Mattila (1990) in that storage at 4°C resulted in higher numbers of organisms only after several days.

Aeromonas species can be isolated from fish, food, water, and clinical specimens (most notably fecal) on a variety of primary plating media. They grow routinely on eugonic media such as blood and chocolate as well as on most enteric differential agar, although somewhat inhibited (Desmond and Janda, 1986). The following references include tables listing differential and selective media, their abbreviations, the application for which they are used, and the literature citation containing their formulation (von Graevenitz and Bucher, 1983; Joseph et al., 1988; Farmer et al., 1992; Moyer, 1996).

Aside from the enrichment above for *A. salmonicida* from fish, nearly all *A. salmonicida* strains can be isolated on trypticase soy agar (TSA). They require incubation at 22–25°C for 48 h and one should try to detect the presence of brown pigmentation, although furunculosis agar (recipe given below) seems superior (Bernoth and Artzt, 1989). There is only sparse growth of *A. salmonicida* on enteric agars and thiosulfate citrate bile salts sucrose (TCBS) medium, which is generally used for the isolation of *Vibrio* species. Rimler–Schotts (RM) medium (Shotts and Rimler, 1973) and a starch–glutamate–ampicillin–penicillin-based medium (SGAP-10C) (Jenkins and Taylor, 1995), both solid media, have been recommended for isolating *Aeromonas* species in general from piscine sources.

Methods for isolation of aeromonads from foods have been somewhat standardized (Palumbo et al., 1992). Some of the current recommended media used for the detection of motile aeromonads from food are starch–ampicillin agar (SA), blood ampicillin agar (BA, 30 mg/l), bile salts irgasan brilliant green agar (BIBG), Ryan Aeromonas agar (Oxoid, Basingstoke, England), and APW (Palumbo et al., 1985a; Moyer, 1996). Several comparison studies of media for food isolation resulted in slightly different recommendations for the preferred medium (Fricker and Tompsett, 1989; Ciufecu et al., 1990; Gobat and Jemmi, 1995). Trypticase soy ampicillin broth (TSBA, 30 mg/l) and MacConkey agar are currently considered the best combination for the isolation of *Aeromonas* from oysters (Abeyta et al., 1986).

An early recommendation for the detection of aeromonads from environmental samples was the use of dextrin fuchsin sulfite (DFS) in conjunction with membrane filtration for enumeration of aeromonads from clear waters (Schubert 1967a, 1987). This was followed by a selective and differential Rimler Shotts (RS) agar, which was subsequently modified into MRSM (Shotts and Rimler, 1973; Seidler et al., 1980). Rippey and Cabelli (1979) formulated mAeromonas medium or mA (a medium with trehalose, ampicillin, and ethanol) for use with a membrane filter. A comparison study of 11 media for recovery of aeromonads from polluted waters found mA to be the preferred medium followed by DNase–toluidine blue–ampicillin agar (DNTA), MacConkey Tween-80 (MACT), and starch bile (SB) agar (Arcos et al., 1988). A later study also found mA agar to be useful (Poffe and Op de Beeck, 1991). Analysis of surface waters using three media (mA, RS, and Ryan Aeromonas agar) found mA to be the most suitable (Bernagozzi et al., 1994). Ampicillin dextrin agar (ADA), mA, starch ampicillin (SA), Pril xylose ampicillin agar (PXA), and SGAP-10 appear to be the most widely used plating media (Moyer, 1996). However, there is no published comparison of these media nor does there appear to be a single optimal recovery medium.

Isolation of aeromonads from drinking water sources (chlorinated, unchlorinated, and well water) has been accomplished using a variety of media and methods, including anaerobic incubation (Burke et al., 1984a, b; Millership and Chattopadhyay, 1985; Cunliffe and Adcock, 1989; Havelaar et al., 1990; Moyer et al., 1992). In a study comparing ADA, xylose ampicillin agar (XAA), Ryan Aeromonas agar, and Aeromonas agar (Difco), Ryan's medium was found superior (Holmes and Sartory, 1993). The presence of lactose-fermenting aeromonad species, usually *A. caviae* phenotypes, has interfered with coliform determination by MF, MPN, and single tube coliform detection methods (Moyer, 1996; Landre et al., 1998). Aeromonads have been isolated from bottled waters by investigators in Canada, Saudi Arabia, Spain, and elsewhere (Gonzalez et al., 1987; Warburton et al., 1992; Hunter, 1993).

The recovery of *Aeromonas* from clinical specimens can be achieved with a number of media, many of which have already been discussed. Isolates from normally sterile sites should be easily detected on either blood or MacConkey agar. Blood agar has the advantage of allowing the simultaneous detection of hemolysis, oxidase, and indole production from a single primary medium, as well as the presence of more than one *Aeromonas* species (Janda et al., 1984). However, detecting fecal isolates of *Aeromonas*, particularly lactose-fermenting *A. caviae*-like strains on MacConkey, could be problematic as they would generally not be selected for further identification. Clinically significant *Aeromonas* species are also generally inhibited on TCBS, the selective medium for vibrios. Nearly all *Aeromonas* species can be detected on a modification of CIN (cefsulodin–irgasan–novobiocin) agar, which is normally used for the selection of *Yersinia enterocolitica* (Altorfer et al., 1985). This modified CIN, known as Yersinia selective agar (YSA, Difco), has a lower concentration of cefsulodin (4 mg/l) than the original CIN I formulation (15 mg/l) which tended to inhibit some species (Alonzo et al., 1996). This allows for simultaneous detection of both *Aeromonas* and *Yersinia* by streaking the plates with fecal matter and incubating at 25°C for 28 h. The incorporation of varying concentrations of ampicillin (10, 15, 20, and 30 mg/l) into blood agar for the purpose of inhibiting normal enteric flora has presented a few problems. Since several species of clinically significant *Aeromonas* are susceptible to ampicillin, most notably *A. trota* and a fair percentage of *A. caviae* (the most commonly isolated species from pediatric fecal samples), the incorporation of ampicillin in the media is controversial (Kilpatrick et al., 1987; Carnahan et al., 1991a; Singh and Sanyal, 1994). Most investigators use the lowest ampicillin concentration of 10 mg/l, if at all. A comparison of five selective media for detection of *Aeromonas* species from human and animal feces found that sheep blood agar with 30 mg/l of ampicillin (ASBA30) yielded a higher percentage of positive specimens, and that the addition of DNase–toluidine blue agar allowed them to detect 98% of all isolates (Mishra et al., 1987). Another comparison of five media included blood agar, blood agar with 20 mg/l of ampicillin, MacConkey, ampicillin Tween-80 agar (MAT), and YSA (modified CIN) for isolation of aeromonads from stools transported in Cary–Blair medium. Optimal recovery required the use of both BA20 and YSA (Kelly et al., 1988). An international multilaboratory study to establish the optimal culture media, incubation time, and incubation temperature found that APW enrichment streaked to blood agar with 10 mg/l of ampicillin gave the highest recovery of any single plate medium tested (Moyer et al., 1991). Ryan Aeromonas Medium Base with an ampicillin-selective supplement can also be used for clinical species isolation.

It should be noted that a modified MIO semisolid agar known as AH medium (Kaper's multitest medium) was proposed for screening presumptive *Aeromonas* colonies taken from selective medium (Kaper et al., 1979). It allowed for the determination of D-mannitol and *i*-inositol fermentation, ornithine decarboxylation, indole production, motility, H_2S, and gas production. It was evaluated by other investigators using an incubation temperature of 25°C for 24 h, rather than the originally described 30°C for 24 h (Toranzo et al., 1986). They detected false-negative indole reactions and the production of gas was difficult to determine. Furthermore, with species that are D-mannitol and indole negative (*A. schubertii*) and ornithine decarboxylase positive (*A. veronii* biovar veronii), the investigator would be better served to use one of several well-developed schema employing conventional biochemical tests (Carnahan et al., 1991a; Abbott et al., 1992; Altwegg, 1999).

MAINTENANCE PROCEDURES

The maintenance of stock cultures of *Aeromonas* can be accomplished by lyophilization, freezing in trypticase soy broth (TSB) with 15% glycerol at −20°C for short periods (months), TSB with 20% glycerol at −70°C for long periods (several years), or freezing in serum inositol broth (containing 25 g of *i*-inositol dissolved in 50 ml of distilled water, filter sterilized, and aseptically added to 450 ml of sterile calf serum) at −70°C for long-term storage. For short-term storage, working cultures of *Aeromonas* spp. grown on nutrient media that do not contain fermentable carbohydrates can be placed at 5°C and subcultured every 2–3 weeks (Moyer, 1996). There is a medium for maintenance of marine strains that contains 0.5% peptone, 0.1% yeast extract, 2.4% NaCl, 0.7% Mg₂SO₄, 0.075% KCl, and 1.5% agar (Effendi and Austin, 1991). There is also a long-term room temperature maintenance medium (up to 15 years) from H.M. Atkinson that was originally developed for salmonellae, but has been adapted for use in storing aeromonads (Crowder, 1974).

PROCEDURES FOR TESTING SPECIAL CHARACTERS

H₂S from GCF (gelatin–cysteine–thiosulfate) (Véron and Gasser, 1963) Aeromonads were assayed for production of H₂S from cysteine using a medium containing Lab Lemco powder (Oxoid, 3 g), tryptone (Oxoid, 10 g), KCl (4 g), L-cysteine hydrochloride (0.1 g), sodium thiosulfate (0.1 g), ferric ammonium citrate (0.4 g), and agar (Difco Bacto, 12 g), w/v in 1000 ml of distilled water and steamed to dissolve all ingredients. The pH was adjusted to 7.4 and the media dispensed in 3-ml volumes (as broths). Special autoclave instructions included 10-lb pressure for 10 min and then the small agar butts were cooled and refrigerated until used. Inoculation involved stabbing the semisolid agar tubes with growth from an 18–24-h TSA culture and then incubating at 35–37°C. A positive reaction is indicated by a diffuse blackening of the medium radiating from the stab line after 72 h of incubation.

Pyrazinamidase Activity (Carnahan et al., 1990) Aeromonads were assayed using a medium obtained from Remel (Lenexa, KS) that contained tryptic soy agar (15 g), yeast extract (1.5 g), Tris-maleate (0.2 M, pH 6, 500 ml) buffer, and pyrazinamide (0.5 g), and was dispensed in 5-ml aliquots into screw-capped tubes (16 × 150 mm), autoclaved at 121°C for 15 min, and slanted. Slants were inoculated with 18–24 h bacterial growth taken from tryptic soy agar and incubated for 48 h at 36°C. One millimeter of 1% (w/v) freshly prepared ferrous ammonium sulfate (aqueous) solution was flooded over each slant, and a positive (pinkish rust color) or negative reaction (colorless) was recorded after 15 min. Positive pyrazinamidase activity indicated the presence of pyrazinoic acid resulting from the action of the enzyme pyrazinamidase.

DL-Lactate utilization (Altwegg et al., 1990; Janda et al., 1996a) Aeromonads were tested for utilization of DL-lactate as a carbon source using slants prepared according to the following formula, per liter: DL-lactic acid (60% w/v; Sigma, St. Louis, MO.), 2.5 ml; NaCl, 5 g; NH₄H₂PO₄, 1 g; K₂HPO₄, 1 g; MgSO₄·7H₂O, 0.2 g; agar, 15 g; and 0.2% bromothymol blue, 40 ml, adjusted to a pH of 6.8. The medium was then autoclaved, dispensed in 5-ml slants, and inoculated with growth from an 18–24-h culture. Incubation was carried out at 25°C, 30°C, or 35°C for 48 h, 72 h, or 5 d with a positive reaction indicated by growth and blue color that begins to develop from the top of the slant.

Urocanic acid utilization (Hänninen, 1994) Aeromonads were tested for utilization of urocanic acid with a medium containing NaCl (5 g), NH₄H₂PO₄ (1 g), K₂HPO₄ (1 g), MgSO₄·7H₂O (0.2 g), agar (9 g), distilled water (1000 ml), and bromothymol blue (0.2% solution) 40 ml, adjusted to a pH of 6.8. The urocanic acid carbon source was added after the autoclaving of the basic medium as a filter-sterilized solution

(2 g/l). The medium was dispensed in 5-ml amounts and slanted. Slants were inoculated and incubated as described above for DL-lactate testing and a positive reaction was recorded as growth and a bright blue color on the slant.

Furunculosis agar (Bernoth and Artzt, 1989 Hänninen and Hirvelä-Koski, 1997) The most sensitive method for brown pigment production among *A. salmonicida* isolates is furunculosis agar (tryptone, 10 g; yeast extract, 5 g; L-tyrosine, 1 g; NaCl, 2.5 g; agar, 15 g; 1000 ml water, pH 7.3). This medium was incubated at 22°C for 3, 7, and 18 d.

DIFFERENTIATION OF THE GENUS *AEROMONAS* FROM OTHER GENERA

Aeromonas is most closely related to *Vibrio* and *Plesiomonas*. Key tests to differentiate these genera are given in Table BXII.γ.184. It should be noted that the classic test of resistance to 150 mg of O/129 bears the caveat that there are reports of increasing resistance to O/129 among isolates of *Vibrio cholerae* from India and Bangladesh (Ramamurthy et al., 1992).

TAXONOMIC COMMENTS

Great strides have been made over the past 20 years in clarifying the taxonomic order of the aeromonads. Prior to 1985, only three species were recognized. Now there are at least 14 species and a new family, *Aeromonadaceae*, has been proposed. Though much of the earlier confusion has been dispelled, some concerns remain including the taxa HG 3 (motile, mesophilic subspecies of *A. salmonicida* only), HG 5A, and HG 5B; the biovars in *A. veronii*; the status of *A. ichthiosmia* and *A. enteropelogenes*; and whether to consider *A. trota* as HG 13 or HG 14. Several discrepancies can occur when trying to group isolates of *A. salmonicida* into one of four subspecies, subsp. *salmonicida*, subsp. *achromogenes*, subsp. *masoucida*, and subsp. *smithia*, thus creating a number of atypical groups caused by lack of congruence of phenotypic and molecular data (Table BXII.γ.183). These atypical groups require attention because members of these groups are increasingly being isolated, especially from marine fish (Hänninen et al., 1995; Hänninen and Hirvelä-Koski, 1997; Austin et al., 1998a; Wiklund and Dalsgaard, 1998).

In HG 3 (*A. salmonicida*) there is an orphan group that is motile, grows well at 37°C, and has been isolated from humans.

TABLE BXII.γ.184. Differential characteristics for the genera *Aeromonas*, *Plesiomonas*, and *Vibrio*[a]

Characteristic	Aeromonas	Plesiomonas	Vibrio
Resistance to O/129[b] (150 μg)	R	S	S[c]
String Test[d]	−	−	+
Growth in 6.5% NaCl	−	−	(+)
Ornithine decarboxylase	(−)	+	+
Fermentation of i-inositol	−	+	(−)
Fermentation of D-mannitol	(+)	−	+
Fermentation of sucrose	(+)	−	(+)
Gelatin liquefaction	+	−	+
Growth on TCBS[e]	(−)	−	(+)

[a]Symbols: +, positive for the majority of species within the genus; −, negative for the majority of species within the genus; (+), positive for most species, but a few species may be negative; (−), negative for most species, but a few may be positive; R, resistant; S, susceptible.

[b]Vibriostatic agent (2,4-diamino-6,7-diisopropylpteridine) (150 μg/disk).

[c]Vibrios are generally susceptible to O/129 with increased reports of resistance in *V. cholerae* strains from India and Bangladesh (Ramamurthy et al., 1992).

[d]Performed using a 0.5% solution of sodium desoxycholate.

[e]Thiosulfate citrate bile salts sucrose agar.

TABLE BXII.γ.185. Identification of genomospecies within the phenospecies for *Aeromonas hydrophila*[a,b]

Characteristic	HG 1 *A. hydrophila*		HG 2 *A. bestiarum*		HG 3[c] Motile, mesophilic *A. salmonicida* subspecies	
	25°C	35°C	25°C	35°C	25°C	35°C
Acid from:						
D-Rhamnose	−	−	+	+	−	−
D-Sorbitol	−	−	−	−	+	+
Lactose	−	d	−	d	d	+
Utilization of:						
DL-Lactate	d	+	−	−	−	−
Urocanic acid	−	−	+	+	+	+
Elastase	+	d	+	−	+	d
Gluconate oxidation	nd	d	−	−	nd	−
Lysine decarboxylation	+	+	+	−	+	d
Maximum growth Temperature	41°C		38–39°C		38–39°C	

[a]Symbols: +, >75% of strains positive; −, <25% of strains positive; d, 26–74% of strains positive; nd, not determined.

[b]Data from Abbott et al. (1992); Hänninen (1994); Ali et al. (1996a); and Altwegg (1999); note that results for some tests vary in relation to the temperature of incubation.

[c]These are motile, nonpigmented, indole-positive strains that grow at 35°C.

It is unlike the other members of the group but exhibits subspecies level DNA relatedness to HG 3. See Tables BXII.γ.182 and BXII.γ.185 for differentiation of this motile *A. hydrophila*-like group from HG 1 (*A. hydrophila*) and HG 2 (*A. bestiarum*). It remains undesignated at present, but it is possible that a subspecies for this group could be proposed in the future.

For many years, the remaining subspecies members of the nonmotile *A. salmonicida* group were thought to be phenotypically distinct, but after additional taxonomic studies of the group by several researchers, it became evident that there are many more atypical strains in the environment than previously realized. An *Aeromonas* Taxonomy working group within the Subcommittee on the Taxonomy of *Vibrionaceae* has been established and is presently studying *A. salmonicida* in an attempt to clarify the position of atypical organisms and to reevaluate the presently accepted taxonomic designations for this group. It is questionable whether the currently recognized subspecies, as well as the possible incorporation of *Haemophilus piscium*, should be retained in the future (Austin et al., 1989a, 1998a; Thornton et al., 1999).

HG 5 is currently subdivided into HG 5A and HG 5B. HG 5A genetically resembles *A. media* but phenotypically resembles *A. caviae*. HG 5B genetically and phenotypically is *A. media*. It has been suggested that the two groups should be considered as either two subspecies of *A. media*, since HG 5B includes the type strain for *A. media* (Altwegg, 1990), or as biovars of the genomospecies *A. media* (Huys et al., 1996b).

HG 7, *A. sobria*, has not been isolated from human clinical specimens. However, there is a continuing use of this species name for clinical isolates, which resemble *A. sobria* but are actually *A. veronii* biovar sobria (HG 8) as opposed to *A. veronii* biovar veronii (HG 10). The terminology *A. veronii* biovar sobria for HG 8 and *A. veronii* biovar veronii HG 10 was suggested (Joseph et al., 1991) to clarify the situation and has been generally accepted based on usage in current literature and inclusion in the databases of most rapid identification systems. Although this suggestion to use the term "biovar sobria" for the biovar designation was to clarify that these HG 8 clinical isolates were genomically *A. veronii*, but phenotypically most similar to what we had heretofore called *A. sobria*, some concern has been expressed over the possibility that the term *sobria*, being used as both a species designation for HG 7 and the biovar designation for HG 8, might lead to confusion.

The proposed species *Aeromonas ichthiosmia* (Schubert et al.,

1990b) has now been shown to actually be *A. veronii* biovar sobria (HG 8) by 16S ribosomal DNA sequencing (Collins et al., 1993), AFLP (Huys et al., 1996b, and DNA–DNA hybridization studies (Huys et al., 2001). Therefore, it is a junior synonym of *A. veronii* biovar sobria.

There is a proposal to divide *A. eucrenophila* into two groups, *A. eucrenophila* I and *A. eucrenophila* II based on ribotyping, AFLP fingerprinting, electrophoretic profiling of soluble cellular proteins, and cellular fatty acid methyl ester profiles (Huys et al., 1996a), with *A. encheleia* and HG 11 strains also residing in Group II. DNA hybridization studies (optical renaturation) (De Ley et al., 1970) of Group I and II *A. eucrenophila* and strains of *A. encheleia* and HG 11 grouped them into two distinct hybridization groups, with the first group composed only of *A. eucrenophila* I. The other group comprised *A. eucrenophila* subgroup II, *A. encheleia*, and HG 11 strains. DNA relatedness between strains in the *A. eucrenophila* I group versus strains in the second group, which included *A. encheleia*, *A. eucrenophila*, subgroup II, and HG 11, ranged from 42 to 52%. These findings suggest that members of this second group, which includes *A. encheleia*, *A. eucrenophila* subgroup II, and HG 11 strains, are genetically (DNA relatedness of values of 74–105%) and phenotypically very similar and should therefore all reside in *A. encheleia* (Huys et al., 1997a). However, a recent rebuttal to this suggestion was published with equally compelling arguments for their continued separation based on 16S rDNA sequence data and conflicting DNA relatedness values below the species level using the competitive nitrocellulose filter method (Martinez-Murcia, 1999). It appears that additional investigation is required to resolve this matter.

Aeromonas Group 501 was initially designated by Hickman-Brenner et al. (1988a) to identify three strains, that were phenotypically close to the newly proposed *Aeromonas schubertii*, and genotypically close to each other, but not related at the species level to any of the other HG groups, including *A. schubertii*. The group of unnamed organisms currently resides in HG 13 (Carnahan et al., 1991a) (Table BXII.γ.181). 16S rDNA sequencing of one Group 501 strain shows that there is a four-nucleotide difference between this strain and *A. schubertii* (Table BXII.γ.180).

Aeromonas trota (HG 14) and *A. enteropelogenes* are synonymous (Collins et al., 1993). While the frequency of terminology usage in the literature shows a marked preference for *A. trota*, the name

A. enteropelogenes was validated earlier (Schubert et al., 1990a). A request for a Judicial Opinion will eventually be made to decide which validated name will be accepted. Although they do not share the same type strain, phenotypic, typing, and molecular studies suggest they are the same organism (Carnahan, 1992; Collins et al., 1993; Huys et al., 2002a).

There is a discrepancy between Opinion 48 of the Judicial Commission in response to a request from Schubert (1971) and the Approved List of Bacterial Names (Amended Edition, 1989). Opinion 48 states that Kluyver and van Niel (1936) are no longer to be associated with the genus designation of *Aeromonas* (Editorial Board, 1973). In the future, Stanier (1943) should be associated with both the genus *Aeromonas* and the type species, *A. hydrophila*. A recent request for an Opinion from the Judicial Commission to correct the Kluyver and Van Niel citation in the 1980 Approved Lists for the genus *Aeromonas* has been submitted.

A. punctata and *A. caviae* have been considered objective synonyms and share the same type strain, ATCC 15468. This is because the type strain of *A. punctata* is incorrectly listed on the 1980 Approved List of Bacterial Names as ATCC 15468, when it should have been NCMB 74 (ATCC 23309). As part of the same request for an Opinion from the Judicial Commission on the Kluyver and van Niel issue mentioned above, it has also been registered to correct this *A. punctata/A. caviae* mistake and make *A. caviae* the single legitimate name for the type strain ATCC 15468. This would, however, cause *A. punctata* and *A. eucrenophila* to then become objective synonyms since both species would then share the same type strain of NCMB 74 (ATCC 23309). Although

A. punctata is older and would have priority (it has a confusing history), it has been reported to have differing phenotypes and is not being used by the scientific community. Therefore, a second request for an Opinion from the Judicial Commission has been made to conserve the name *A. eucrenophila* Schubert and Hegazi, 1988a as the correct name for HG 6 with the type strain of NCMB 74 (ATCC 23309).

ACKNOWLEDGMENTS

The authors wish to acknowledge the following persons for their invaluable support in supplying material for use in this chapter as well as their comments and suggestions: Martin A. Altwegg, Brian Austin, Geert Huys, Marja-Liisa Hänninen, J. Michael Janda, Sylvia M. Kirov, Antonio J. Martinez-Murcia, Hans G. Trüper, and Michael Waddington.

FURTHER READING

Altwegg, M. and H.K. Geiss. 1989. *Aeromonas* as a human pathogen. Crit. Rev. Microbiol. *16*: 253–286.

Austin, B., M. Altwegg, P.J. Gosling and S. Joseph (Editors). 1986. The Genus *Aeromonas*, John Wiley & Sons, Ltd., Chichester.

Altwegg, M. 1999. *In* Murray, Baron, Pfaller, Tenover and Yolken (Editors), Manual of Clinical Microbiology, ASM Press, Washington, D.C. 507–516.

Janda, J.M. and S.L. Abbott. 1998. Evolving concepts regarding the genus *Aeromonas*: an expanding panorama of species, disease presentations, and unanswered questions. Clin. Infect. Dis. *27*: 332–344.

Joseph, S.W. and A.M. Carnahan. 1994. The isolation, identification, and systematics of the motile *Aeromonas* species. Annu. Rev. Fish Dis. *4*: 315–343.

DIFFERENTIATION OF THE SPECIES OF THE GENUS *AEROMONAS*

Each validly named species *vide infra* has a particular HG (hybridization group) designation and can be found in either Tables BXII.γ.182 and BXII.γ.186 (motile aeromonad HGs) or in Table BXII.γ.183 (nonmotile aeromonad HGs).

List of species of the genus Aeromonas*

1. **Aeromonas hydrophila** (Chester 1901) Stanier 1943, 213^AL (*Bacillus hydrophilus* Chester 1901, 235.)
   hy.dro' phi.la. Gr. n. *hydro* water; Gr. adj. *philos* loving; M.L. adj. *hydrophila* water-loving.

   Hybridization group 1 of 17 (Popoff et al., 1981; Farmer et al., 1986).

   At 37°C hydrolyzes esculin; produces acetoin from D-glucose (positive Voges-Proskauer test); displays pyrazinamidase activity; produces acid from D-mannitol and sucrose and variably from arabinose; resistant to ampicillin and cephalothin. Decarboxylates lysine but not ornithine; produces indole, H_2S (from GCF medium), and gas from D-glucose; β-hemolysis on TSA with 5% sheep blood agar. Differential tests to aid in the identification of *A. hydrophila* from all validly named motile species are listed in Tables BXII.γ.185 and BXII.γ.186. Descriptive tests for all the validly named motile species are listed in Table BXII.γ.182. Isolated from fresh and marine waters, diseased fish and poikilothermic aquatic animals (e.g., frogs with red leg disease), and warm-blooded animals; associated with both extraintestinal and diarrheal disease in humans (Austin and Adams, 1996; Gosling, 1996a ; Holmes et al., 1996; Janda and Abbott, 1996; Joseph, 1996).

   The mol% G + C of the DNA is: 58–62 (Bd, T_m).
   Type strain: ATCC 7966, DSM 30187.
   GenBank accession number (16S rRNA): X60404 (16S rDNA).

2. **Aeromonas allosaccharophila** Martinez-Murcia, Esteve, Garay and Collins 1992c, 511^VP (Effective publication: Martinez-Murcia, Esteve, Garay and Collins 1992b, 203.)
   al.lo.sa.ca.ro' phi.la. Gr. adj. *allos* different; Gr. n. *saccharo* sugar; Gr. adj. *philos* loving; M.L. adj. *allosaccharophila* different sugar loving.

   Hybridization Group 15 of 17 (Martinez-Murcia et al., 1992b; Esteve et al., 1995b).

   Motile. Diffusible brown pigment is not produced. Growth occurs in 0–3% (w/v) NaCl. Resistant to 2,4-diamino-6,7-diisopropylpteridine (vibriostatic agent 0/129). Acid and gas are produced from D-glucose. Lysine is decarboxylated (Moeller's medium); indole, and β-galactosidase are positive. H_2S production from thiosulfate and Voges–Proskauer test are negative. Growth occurs at 4–42°C and at pH 9.0, but not at pH 4.5. Acid is produced from sucrose, D-cellobiose, maltose, D-trehalose, D-galactose, D-mannose, glycerol, and D-mannitol, but not from adonitol, arbutin, D-xylose, lactose, *myo*-erythritol, dulcitol, *myo*-inositol, or D-sorbitol. Sole carbon and energy sources include L-arabinose, sucrose, D-cellobiose, maltose, D-trehalose, D-galactose, D-mannose, glycerol, D-mannitol, L-histidine, L-arginine, L-proline, L-glutamate, D-gluconate, fumarate, and

---

Editorial Note: Since the submission and review of this chapter, two new subspecies and one new species have been proposed and validly published. These are *Aeromonas hydrophila* subsp. *dhakensis* (Huys et al., 2002b), *Aeromonas hydrophila* subsp. *ranae* (Huys et al., 2003) and *Aeromonas culicicola* (Pidiyar et al., 2002).

TABLE BXII.γ.186. Differential characteristics among motile *Aeromonas* species[a]

Characteristic	*A. hydrophila*[b] HG 1	*A. allosaccharophila*[c] HG 15	*A. bestiarum*[d] HG 2	*A. caviae*[b] HG 4	*A. encheleia*[e] HG 16	*A. eucrenophila*[f] HG 6	*A. jandaei*[b] HG 9	*A. media*[g] HG 5	*A. popoffii*[h] HG 17	*A. schubertii*[b] HG 12	*A. sobria*[i] HG 7	*A. trota*[b] HG 14	*A. veronii* biovar *veronii*[b] HG 10	*A. veronii* biovar *sobria*[b] HG 8
Esculin hydrolysis	+	d	+	+	+	+	−	+	−	−	−	−	+	
Gas from D-glucose	+	+	+	−	+	+	+	−	+	−	+	+	+	+
Voges–Proskauer	+	−	+	−	−	−	+	−	+	d	weak+	−	+	+
Indole production	+	+	+	+	+	+	+	d	d	−	+	+	+	+
Pyrazinamidase[j]	+	nd	+		nd	+	−	d	nd	−	nd	—	—	—
Acid from:														
L-Arabinose	d	d	+	+	−	+	−	+	nd	−	−	−	+	−
D-Mannitol	+	+	+	+	+	+	+	+	+	−	+	+	+	+
Sucrose	+	+	+	+	+	d	−	+	−	−	+	−	+	+
Decarboxylase:														
Lysine	+	+	−	−	−	−	+	−	−	+	weak+	+	+	+
Ornithine	−	d	−	−	−	−	−	−	−	−	−	−	+	−
Arginine dihydrolase	+	d	+	+	d	+	+	+	+	+	−	+	−	+
Arbutin hydrolysis	+	nd	+	+	−	+	−	+	nd	−	nd	d	+	−
H₂S from GCF	+	nd	+	−	nd	+	+	−	nd	−	nd	+	+	+
Hemolysis on SBA	+	nd	+	d	−	+	−	−	+	−	+	−	+	+
Susceptibility:[k]														
Ampicillin (10 µg)	R	R	R	R	R	S	R	S	R	R	R	S	R	R
Carbenicillin (30 µg)	R	nd	R	R	R	nd	R	nd	nd	R	S	S	R	R
Cephalothin (30 µg)	R	nd	R	R	nd	S	S	d	nd	S	S	R	d	S
Colistin (4 µg/ml)	d	nd	d	S	nd	S	R	S	nd	S	nd	S	S	S

[a]Symbols: +, ≥75% of the strains are positive; −, ≤25% of the strains are positive; d, 26–74% of the strains are positive; nd, not determined.

[b]Data from Carnahan et al. (1991a); all analyses were performed at 36°C ± 1°C for 3 d.

[c]Data from Martinez-Murcia et al. (1992b); analyses were performed at 35°C.

[d]Data from Ali et al. (1996a) and Carnahan and Joseph (1993); analyses were performed at 35°C for 7 d (see Table BXII.γ.185 for results at 25°C).

[e]Data from Esteve et al. (1995a) and Huys et al. (1996a, 1997a); analyses were performed at 28°C unless otherwise indicated.

[f]Data from Schubert and Hegazi (1988a); analyses were performed at 30°C and 20°C; Carnahan et al. (1991a) (36°C ± 1°C); and Abbott et al. (1992) (35°C for 7 d).

[g]Data from Allen et al. (1983); analyses were performed at 22°C for 7 d; Carnahan et al. (1991a) (36°C ± 1°C); and Abbott et al. (1992) (35°C for 7 d). Although originally described as nonmotile, later work by Austin and Austin (1990) reported on the motility of this species at 37°C; original description by Allen et al. (1983) also included production of a brown diffusible pigment on TSA (trypticase soy agar).

[h]Data from Huys et al. (1997b); analyses were performed at 28°C and 37°C for 7 d unless otherwise indicated.

[i]Data from Carnahan (unpublished data); analyses were performed at 25°C for 7 d incubation, unless otherwise indicated.

[j]Data from Carnahan et al. (1990); analysis performed at 35°C for 48 h.

[k]Antimicrobial susceptibility results were incubated at 35°C for 24 h; R, resistance; S, susceptibility; d, 26–74% susceptibility.

succinate. There is no growth on lactose, *myo*-erythritol, dulcitol, *myo*-inositol, ethanol, L-citrulline, L-leucine, L-alanine, glycine, L-serine, L-glutamine, DL-3-hydroxybutyrate, propionate, γ-aminobutyrate, D-glucuronate, α-ketoglutarate, or putrescine. Gelatin, casein, egg yolk, Tween 80, starch, and DNA are degraded but not urea, sodium dodecyl sulfate, or elastin (Martinez-Murcia et al., 1992b). Differential tests to aid in the identification of *A. allosaccharophila* from all motile species are listed in Table BXII.γ.186. Descriptive tests for all the motile species are listed in Table BXII.γ.182. Found in diseased elvers and diarrheic stools.

The mol% G + C of the DNA is: 59.5 (method unknown).

Type strain: ATCC 51208, CECT 4199, DSM 11576.

GenBank accession number (16S rRNA): S39232.

3. **Aeromonas bestiarum** Ali, Carnahan, Altwegg, Lüthy-Hottenstein, and Joseph 1996b, 1189[VP] (Effective publication: Ali, Carnahan, Altwegg, Lüthy-Hottenstein, and Joseph 1996a, 163.)

bes.ti.a′ rum. L. n. *bestiarum* of beasts (whether wild or domestic).

Hybridization Group 2 of 17 (Popoff et al., 1981; Fanning et al., 1985; Farmer et al., 1986; Ali et al., 1996a).

At 25°C and 35°C it produces acid from L-rhamnose; does not usually produce acid from lactose or D-sorbitol; utilizes urocanic acid, but does not utilize DL-lactate. At 25°C it hydrolyzes elastin; utilizes phenylpyruvate and decarboxylates lysine. Differential tests to aid in the identification of *A. bestiarum* from all motile species are listed in Table BXII.γ.186 and tests for separation from closely related species are listed in Table BXII.γ.185. Descriptive tests for all the motile species are listed in Table BXII.γ.182. Primarily found in the environment in both wild and domestic animals and birds and fish. Isolated from human feces (Figueras et al., 1999).

The mol% G + C of the DNA is: not reported.

Type strain: ATCC 51108, CDC 9533-76.

GenBank accession number (16S rRNA): X60406 (16S rDNA).

Additional Remarks: X60406 is identical to *A. salmonicida* subsp. *achromogenes* X60407.

4. **Aeromonas caviae** (ex Eddy 1962) Popoff 1984b, 355^VP (Effective publication: Popoff 1984a, 548.)

ca' vi.ae. M.L. fem. n. *cavia* generic name of a guinea pig; M.L. gen. n. *caviae* of a guinea pig

Hybridization Group 4 of 17 (Popoff et al., 1981; Farmer et al., 1986).

Does not produce gas during D-glucose fermentation, Voges–Proskauer (acetoin) negative, lysine decarboxylase negative, H₂S not produced from GCF media. Utilizes DL-lactate and citrate as sole source of carbon. Differential tests to aid in the identification of *A. caviae* from all motile species are listed in Table BXII.γ.186 with supplementary tests to separate *A. caviae* given in Table BXII.γ.187. Descriptive tests for all the motile species are in Table BXII.γ.182. Found in fresh water, sewage, and on domestic and wild animals, birds, and fish. May be associated with gastrointestinal disease in adults (Holmberg et al., 1986a; Kuijper et al., 1989a; Altwegg et al., 1990; Joseph, 1996). Etiological agent of gastroenteritis in young children (Altwegg, 1985; Namdari and Bottone, 1990). Causes extraintestinal disease primarily in immunocompromised humans (Janda and Duffey, 1988; Janda, 1991; Janda and Abbott, 1998b).

The mol% G + C of the DNA is: 61–63 (Bd, T_m).

Type strain: ATCC 15468, DSM 7323, NCIMB 13016.

GenBank accession number (16S rRNA): X74674.

5. **Aeromonas encheleia** Esteve, Gutiérrez and Ventosa 1995a, 464^VP emend. Huys, Kämpfer, Altwegg, Coopman, Janssen, Gillis and Kersters 1997a, 1162.

en.che' le.ia. Gr. n. *encheyls* eel; M.L. adj. *encheleia* from eels.

Hybridization Group 16 of 17 (Martinez-Murcia et al., 1992b).

Motile. Colonies develop within 24 h at 28°C on TSA and are not pigmented. Old cultures (10–15 d) may have colonies with light brown pigmented colonies. No brown water-soluble pigment is produced. Growth occurs on MacConkey agar but not on thiosulfate–citrate–bile salts–sucrose agar. Acid is produced from D-glucose; gas production is variable. Resistant to vibriostatic agent 0/129. Growth in the presence of 0–3% NaCl at 4 and 37°C and under

alkaline conditions (pH 9.0). Arginine dihydrolase variable; indole positive. H₂S variable, and lysine and ornithine decarboxylase negative. Voges–Proskauer reactions are usually negative. Acid is produced from D-mannose, D-trehalose, D-galactose, and D-mannitol, but not from L-arabinose, D-cellobiose, lactose, D-xylose, D-melibiose, D-raffinose, and *myo*-erythritol, dulcitol, *myo*-inositol, or D-sorbitol. Acid is sometimes produced from sucrose, salicin, maltose, L-rhamnose, and glycerol. Hydrolyzes esculin, arbutin, casein, collagen, chitin, Tween 80, egg yolk, and DNA, but not chondroitin sulfate, elastin, keratin, or urea. SDS-alkyl sulfatase negative. Fibrinogen and starch, but not mucin, are sometimes hydrolyzed. Human erythrocytes are hemolyzed. Sole carbon and energy sources include sucrose, salicin, maltose, D-mannose, D-trehalose, L-proline, L-serine, L-malate, fumarate, and D-mannitol. L-Histidine is sometimes utilized. The following are not used as sole carbon or energy sources: L-arabinose, D-cellobiose, L-rhamnose, lactose, D-raffinose, L-arginine, L-citrulline, L-leucine, L-alanine, glycine, L-glutamine, L-tyrosine, citrate, L-aspartate, DL-3-hydroxybutyrate, propionate, D-gluconate, D-glucuronate, α-ketoglutarate, *myo*-inositol, *myo*-erythritol, dulcitol, ethanol, and putrescine. L-Glutamate and D-galactose are sometimes utilized (Esteve et al., 1995a) See Table BXII.γ.182 for tests to differentiate *A. encheleia* from other *Aeromonas* spp. Differential tests to aid in the identification of *A. encheleia* from all motile species are listed in Table BXII.γ.186. Descriptive tests for all the motile species are listed in Table BXII.γ.182. Found in healthy European eels reared in a freshwater farm; not pathogenic for mice.

The mol% G + C of the DNA is: 59.4–60.8 (T_m).

Type strain: ATCC 51929, CECT 4342, DSM 11577.

6. **Aeromonas eucrenophila** Schubert and Hegazi 1988b, 449^VP (Effective publication: Schubert and Hegazi 1988a, 34) emend. Huys, Kämpfer, Altwegg, Coopman, Janssen, Gillis and Kersters 1997a, 1162.

eu.cre.no.phi' la. Gr. fem. n. *eu* good; *krene* well; Gr. adj. *philos* loving; *eucrenophila* good well water.

Hybridization Group 6 of 17 (Popoff et al., 1981; Farmer et al., 1986).

Rod-shaped with rounded ends to coccoid 1.0–4.4 μm in diameter, occasionally forming filaments up to 8 μm long; occurs singly, in pairs, or chains. Motile by polar flagella, generally monotrichous. Voges–Proskauer negative. Facultatively anaerobic. Grows on a mineral medium with ammonia as the sole source of nitrogen and one of the following as the sole source of carbon: D-glucose, arginine, asparagine, or histidine. No growth at 40°C; maximum growth temperature is 37°C; minimum growth temperature is 0–5°C; optimal growth temperature 30°C, although incubation at 20°C is best for some tests; growth at pH range 5.5–9.0. Resistant to vibriostatic agent 0/129. Produces gas from D-glucose, but not from glycerol. Fructose, maltose, trehalose, and esculin are usually utilized; D-adonitol, dulcitol, i-inositol, inulin, melezitose, D-sorbose, and D-xylose are not fermented. Starch and dextrin are hydrolyzed; gelatin is liquefied; deoxyribonuclease, arginine dihydrolase, and phosphatase are usually produced; glutamic acid is not decarboxylated; urea is not hydrolyzed. Differential tests to aid in the identification of *A. eucrenophila* from all motile species are listed in Table BXII.γ.186. Descriptive tests for

TABLE BXII.γ.187. Identification of the genomospecies within the phenospecies *Aeromonas caviae*[a,b]

Characteristic	HG 4 *A. caviae*	HG 5A (unnamed)	HG 5B *A. media*[c]
Acid from:			
Lactose	d	+	+
Cellobiose	d	+	+
Utilization of:			
DL-Lactate	+	−	+
Citrate	+	−	−

[a]Symbols: +, >75% strains positive; −, <25% strains positive; d, 25–75% strains positive.

[b]Results derived from API50 and API 32GN strips at an incubation temperature of 30°C for 48 hr incubation. (Reprinted with permission from M. Altwegg. *In* Murry, Baron, Pfaller, Tenover and Yolken (Editors), *Manual of Clinical Microbiology*, 7th. Ed., American Society for Microbiology, Washington D.C., pp. 507 516, 1999.)

[c]HGs 5A and 5B may constitute two subspecies with HG 5B consisting of two biovars.

all the motile species are listed in Table BXII.γ.182. Found in unpolluted surface and groundwater such as oligosaprobic (living in an environment rich in organic matter and relatively free of oxygen) streams and clean well waters.

The mol% G + C of the DNA is: 59.8–62.6 (Bd, T_m).

Type strain: ATCC 23309, NCMB 74.

GenBank accession number (16S rRNA): X74675.

7. **Aeromonas jandaei** Carnahan, Fanning and Joseph 1992b, 191[VP] (Effective publication: Carnahan, Fanning and Joseph 1991c, 562.)

jan.dae' i. M.L. gen. n. *jandaei* named after J. Michael Janda.

Hybridization Group 9 of 17 (Farmer et al., 1986).

Its unique biochemical profile includes negative reactions for esculin hydrolysis, fermentation of sucrose and cellobiose, and resistance to colistin (4 μg/ml). It differs from *A. schubertii* (HG 12), because *A. jandaei* is indole positive, ferments D-mannitol, and produces gas from D-glucose fermentation. A sucrose-negative reaction usually differentiates *A. jandaei* from other *Aeromonas* spp., although sucrose-negative strains that are not *A. jandaei* can occur. Differential tests to aid in the identification of *A. jandaei* from all motile species are listed in Table BXII.γ.186. Descriptive tests for all the motile species are listed in Table BXII.γ.182. Isolation has been from specimens of diverse geographic origin including wounds, blood, and feces of humans, aquatic sources, fish, and prawns (Carnahan et al., 1991c; Joseph et al., 1991; Janda and Abbott, 1996).

The mol% G + C of the DNA is: not determined.

Type strain: ATCC 49568, DSM 7311.

GenBank accession number (16S rRNA): X74678.

8. **Aeromonas media** Allen, Austin and Colwell 1983, 603[VP]

me' di.a. M.L. fem. gen. n. *media* in the middle.

Hybridization Group 5 of 17 (Popoff et al., 1981; Farmer et al., 1986).

Colonies on TSA are creamy, shiny, smooth, round, raised, entire, and 2 mm in diameter after incubation for 2 d at 22°C. Originally described as nonmotile, but motility can be expressed under certain laboratory conditions using appropriate media, temperature, and incubation periods (Austin and Austin, 1990). A diffusible, brown, nonfluorescent pigment is produced. Cultures in peptone broth are uniformly turbid. Chemoorganotrophic; metabolism is fermentative. Temperature range of 4–37°C; no growth at 42°C. Growth occurs in 0–3% NaCl. β-Galactosidase and arginine dihydrolase are usually produced. Phosphatase is not produced. Lysine and ornithine are not decarboxylated. Blood, gelatin, starch, ribonucleic acid, Tween 20, Tween 40, Tween 60, and Tween 80 are degraded, but cellulose, elastin, lecithin, xanthine, and urea are not. Methyl red test positive. Voges–Proskauer, gluconate oxidation, and H_2S production are negative. Utilizes DL-lactate, L-arabinose, glycerol, D-mannitol, L-proline, L-serine, sodium acetate, sodium glutamate, sodium pyruvate, and sodium succinate as sole carbon sources for energy and growth. Does not utilize either adonitol, D-cellobiose, *meso*-erythritol, ethanol, D-fructose, *meso*-inositol, inulin, L-leucine, D-melezitose, *p*-hydroxybenzoic acid, D-raffinose, sodium benzoate, sodium formate, sodium malonate, D-sorbitol, sucrose, D-trehalose, L-valine, or D-xylose (Allen et al., 1983). Differential tests to aid in the identification of *A. media* from all motile species

are listed in Tables BXII.γ.186 and BXII.γ.187. Descriptive tests for all the motile species are listed in Table BXII.γ.182. Strains isolated from fishponds and other aquatic sources (Allen et al., 1983). One rare report of association with human diarrheal disease (Rautelin et al., 1995a).

The mol% G + C of the DNA is: 62.3 (Bd, T_m).

Type strain: ATCC 33907, DSM 4881.

GenBank accession number (16S rRNA): X74679.

9. **Aeromonas popoffii** Huys, Kämpfer, Altwegg, Kersters, Lamb, Coopman, Lüthy-Hottenstein, Vancanneyt, Janssen and Kersters 1997b, 1170[VP]

po.pof'fi.i. M.L. gen. n. *popoffii* of Popoff, named after Michel Popoff.

Hybridization Group 17 of 17 (Huys et al., 1997b).

Motile. Reduces nitrate to nitrite; resistant to vibriostatic agent 0/129. Optimal growth after 24 h at 28°C on TSA medium; acid and gas produced from D-glucose and glycerol. Growth in KCN broth but not in 3, 6, 8, and 10% NaCl; positive for arginine dihydrolase, DNase, and Voges–Proskauer test; variable production for indole; no production of urease, tryptophan deaminase, ornithine and lysine decarboxylase; H_2S is sometimes produced; citrate and malonate utilization is variable. Acid is uniformly produced from D-galactose, D-mannitol, D-mannose, methyl-D-glucoside, and D-trehalose, but not from adonitol, D-arabitol, D-cellobiose, dulcitol, erythritol, *myo*-inositol, lactose, α-D-melibiose, D-raffinose, L-rhamnose, salicin, D-sorbitol, D-sucrose, and D-xylose. The following substrates are used as carbon and energy sources: *N*-acetyl-D-glucosamine, L-aspartate, fumarate, D-galactose, D-gluconate, v-glucose, L-glutamate, L-glutamine, glycerol, L-histidine, L-malate, D-mannitol, D-mannose, D-maltose, putrescine, pyruvate, D-ribose, succinate, L-serine, D-trehalose, and L-tyrosine. The following are not utilized: acetate, *cis*-aconitate, *trans*-aconitate, adipate, adonitol, β-alanine, 4-aminobutyrate, arbutin, azelate, D-cellobiose, citrulline, dulcitol, ethanol, erythritol, D-glucuronate, DL-3-hydroxybutyrate, *myo*-inositol, itaconate, L-leucine, maltitol, α-D-melibiose, mesaconate, L-ornithine, oxoglutarate, phenylacetate, L-phenylalanine, L-proline, propionate, D-raffinose, L-rhamnose, salicin, D-sorbitol, suberate, D-sucrose, L-tryptophan, and D-xylose.

It hydrolyzes the following substrates: L-alanine-*p*NA, casein, 2-deoxythymidine-5'-*p*NP-phosphate, gelatin, lecithin, *bis-p*NP-phenylphosphonate, *p*NP-phosphoryl choline, L-proline-*p*NA, starch, and Tween 80. It does not hydrolyze chitin, esculin, L-glutamate-γ-3-carboxy-*p*NA, *p*NP-α-D-glucopyranoside, *p*NP-β-D-glucuronide, and *p*NP-β-D-xyloside (Huys et al., 1997b). Differential tests to aid in the identification of *A. popoffii* from all motile species are listed in Table BXII.γ.186. Descriptive tests for all the motile species are listed in Table BXII.γ.182. Found in drinking water production plants and reservoirs.

The mol% G + C of the DNA is: 57.7–59.6 (T_m).

Type strain: LMG 17541.

GenBank accession number (16S rRNA): AJ224308.

10. **Aeromonas salmonicida** (Lehmann and Neumann 1896) Griffin, Snieszko and Friddle 1953, 138[AL] (*Bacterium salmonicida* Lehmann and Neumann 1896, 240.)

sal.mon.ic' i.da. L. n. *salmo, salmonis* salmon; L. suff. *-cida* from L. v. *caedo* cut or kill; M.L. n. *salmonicida* salmon-killer.

Hybridization group 3 of 17 (Popoff et al., 1981; Fanning et al., 1985; Farmer et al., 1986).

Short Gram-negative rods, some pleomorphism and coccobacillary forms. In nutrient broth, pairs, chains, and clumps are usually observed in phase-contrast preparations. Nonencapsulated (Popoff, 1984a). Rarely motile. Usually possesses an S-layer (Noonan and Trust, 1997). Optimum growth temperature is 22–25°C. Maximum growth temperature in nutrient broth is 34.5°C (Griffin et al., 1953). Colonies on agar after 24 h are pinpoint. After 48–72 h colonies are circular, raised, convex, translucent, entire, and friable. On blood agar there is similar morphology and some subspecies produce β-hemolysis after 2–4 d (Millership, 1996). Colony pigmentation, when present, is usually yellowish; absence is indicated by grayish white colony. A brown, diffusible pigment appears after 24 h and reaches maximum color within 48–72 h. Differential and descriptive tests to aid in the identification of all nonmotile *A. salmonicida* subspecies, as well as atypical strains, are listed in Table BXII.γ.183. Differential and descriptive tests for the motile *A. salmonicida* strains within HG 3, which may be either a new subspecies or a biovar, are listed in Tables BXII.γ.182 and BXII.γ.185. Various diseases, including swelling at the vent and in the kidney, external and internal hemorrhages, ascitic fluid, and especially furunculosis are caused in a variety of fish. Interestingly, though not usually isolated from related aquatic locations, *A. salmonicida* was recently recovered in co-culture with purple-pigmented bacteria from fresh water lake sediment (Austin et al., 1998b).

The mol% G + C of the DNA is: 57–59 (Bd, T_m).

Type strain: ATCC 33658, NCMB 1102.

Additional Remarks: Neotype proposed Schubert, 1967b.

a. **Aeromonas salmonicida** *subsp.* **salmonicida** (Lehmann and Neumann 1896) Griffin, Snieszko and Friddle 1953, 138[AL] (*Bacterium salmonicida* Lehmann and Neumann 1896, 240.)

RBrown water-soluble pigment production on agar containing 0.1% tyrosine or phenylalanine. β-Hemolytic on ox blood agar, hydrolyzes esculin, indole negative, gas produced during D-glucose fermentation, ferments salicin and L-arabinose. Differential and descriptive tests to aid in the identification of this nonmotile *A. salmonicida* subspecies, as well as atypical strains, are listed in Table BXII.γ.183.

The mol% G + C of the DNA is: 57–59 (Bd, T_m).

Type strain: ATCC 33658, NCMB 1102.

GenBank accession number (16S rRNA): X60405.

Additional Remarks: Neotype proposed Schubert, 1967b.

b. **Aeromonas salmonicida** *subsp.* **achromogenes** (Smith 1963) Schubert 1967c, 278[AL] (*Necromonas achromogenes* Smith 1963, 273.)

a.chro.mo.ge' nes. Gr. adj. *achromos* colorless; Gr. v. *gennaio* produce; M.L. adj. *achromogenes* not producing color.

Generally do not produce brown, water-soluble pigment unless grown on specific media such as furunculosis agar. May produce indole, arginine dihydrolase, ferment D-mannitol, and assimilate *N*-acetyl glucosamine. Gas is not produced during glucose fermentation; es-

culin is not hydrolyzed; no β-hemolysis on ox blood agar. Differential and descriptive tests to aid in the identification of all nonmotile *A. salmonicida* subspecies, as well as atypical strains, are listed in Table BXII.γ.183.

The mol% G + C of the DNA is: 57–59 (Bd, T_m).

Type strain: ATCC 33659, NCMB 1110.

GenBank accession number (16S rRNA): X60407.

c. **Aeromonas salmonicida** *subsp.* **masoucida** Kimura 1969, 52[AL]

ma.sou.ci' da. Japan. n. *masou* specific epithet of *Onchorhynchus masou*, (fish); L. v. suff. *-cida* from L. v. *caedo* cut or kill; M.L. fem. n. *masoucida Onchorhyncus masou*-killer.

Does not produce brown, water-soluble pigment on agar; β-hemolytic on ox blood agar; may hydrolyze esculin; arginine dihydrolase positive; gas produced during glucose fermentation; L-arabinose fermented; *N*-acetyl glucosamine not assimilated. Differential and descriptive tests to aid in the identification of this nonmotile *A. salmonicida* subspecies, as well as atypical strains, are listed in Table BXII.γ.183.

The mol% G + C of the DNA is: not available.

Type strain: ATCC 27013.

GenBank accession number (16S rRNA): X74680.

d. **Aeromonas salmonicida** *subsp.* **pectinolytica** Pavan, Abbott, Zorzópulos and Janda 2000, 1123[VP]

Isolated from heavily polluted river.

The mol% G + C of the DNA is: not known.

Type strain: 34mel, DSM 12609.

GenBank accession number (16S rRNA): AF134065.

e. **Aeromonas salmonicida** *subsp.* **smithia** Austin, McIntosh and Austin 1989b, 495[VP] (Effective publication: Austin, McIntosh and Austin 1989a, 288.)

smi' thi.a. M.L. fem. gen. n. *smithia* of Smith, named after Isabel W. Smith.

Nonmotile, approximately 1–2 μm in size, with rounded ends. Cultures on trypticase soy agar (TSA) dissociate into "rough", "smooth", and "G-phase" colonies (Duff, 1937). Does not readily produce brown diffusible pigment. In peptone broth, cultures are uniformly turbid. Metabolism is fermentative. Growth occurs at 4–25°C but not at 30°C and in 0–2%, but not 3% NaCl. No growth on MacConkey agar. Catalase, β-galactosidase, H_2S, oxidase, phosphatase, and phosphoamidase are produced, but not indole, arginine dihydrolase, lysine or ornithine decarboxylase, or phenylalanine deaminase. DNA, gelatin, RNA, and starch are degraded, but not aesculin, blood, chitin, elastins, guanine, hypoxanthine, lecithin, Tween 20, 40, 60, 80, or xanthine. Carbon-containing compounds are not readily utilized (Austin et al., 1989a). Differential and descriptive tests to aid in the identification of this nonmotile *A. salmonicida* subspecies, as well as atypical strains, are listed in Table BXII.γ.183. Pathogenic for fish, with mortality within 24–72 h. External hemorrhages are evident in the vent and on the fins, as is tail rot. Accumulation of ascitic fluid occurs in the peritoneal cavity, along with extensive muscle hemorrhaging, liquefaction, and gastroenteritis.

The mol% G + C of the DNA is: 55.9 (T_m).

Type strain: 138, ATCC 49393, CCM 4103.

GenBank accession number (16S rRNA): AJ009859.

11. **Aeromonas schubertii** Hickman-Brenner, Fanning, Arduino, Brenner and Farmer 1989, 205^VP (Effective publication: Hickman-Brenner, Fanning, Arduino, Brenner and Farmer 1988a, 1563.)

schu.ber' ti.i. M.L. masc. gen. n.*schubertii,* named after Ralph H.W. Schubert.

Hybridization Group 12 of 17 (Farmer et al., 1986).

At 36°C, positive reactions for methyl red, lysine decarboxylase, arginine dihydrolase, motility, lipase, DNase, reduction of nitrate to nitrite, oxidase, and growth in nutrient broth with 0 and 1% NaCl. No growth in 6% NaCl. Fermentation of D-glucose, D-galactose, maltose, D-mannose, and trehalose. No fermentation of adonitol. L-arabinose, D-arabitol, cellobiose, dulcitol, erythritol, *myo*-inositol, lactose, D-mannitol, melibiose, α-methyl-D-glucoside, raffinose, L-rhamnose, salicin, D-sorbitol, sucrose, and D-xylose. Esculin is not hydrolyzed. String test negative. Differential reactions include failure to ferment D-mannitol, sucrose, L-arabinose, and salicin; failure to produce indole, or hydrolyze esculin, and lack of gas production from glucose (Hickman-Brenner et al., 1988a; Carnahan et al., 1991a). A D-mannitol negative reaction usually differentiates *A. schubertii* from other *Aeromonas* spp. although mannitol-negative strains that are not *A. schubertii* can occur. Extracellular amylase, protease, and DNases produced. High cell surface charge but low hydrophobicity. Major cell wall fatty acids are hexadecenoic ($C_{16:1}$), hexadecanoic ($C_{16:0}$), and octadecanoic acid ($C_{18:1}$) (Kokka et al., 1992). Differential tests to aid in the identification of *A. schubertii* from all motile species are listed in Table BXII.γ.186. Descriptive tests for all the motile species are listed in Table BXII.γ.182. Strains isolated from wounds, abscesses, pleural fluid, and blood. There have been no isolates from human stools (Hickman-Brenner et al., 1988a; Carnahan et al., 1989b).

The mol% G + C of the DNA is: undetermined.
Type strain: ATCC 43700, DSM 4882.
GenBank accession number (16S rRNA): X74682.

12. **Aeromonas sobria** Popoff and Véron 1981, 215^VP (Effective publication: Popoff and Véron 1976, 20.)

so.bri' a. M.L. fem. adj. *sobria* moderate.

Hybridization Group 7 of 17 (Popoff et al., 1981; Farmer et al., 1986).

Biochemical results obtained at 25°C after 7 d of incubation. Produces acid and gas from glucose, acid from mannitol and sucrose, and is weakly positive Voges–Proskauer reaction. Lysine decarboxylase is weakly positive after 7 d; arginine dihydrolase and ornithine decarboxylase are not produced. Does not produce acid from L-arabinose. Does not hydrolyze esculin. Does not produce β-hemolysis on sheep blood agar (SBA) (5%). There is no growth at 42°C. Produces gas from glucose at 25°C and no gas at 37°C after 48 h. This species closely resembles *A. veronii* biovar sobria (HG 8) biochemically, although they differ in their arginine dihydrolase and hemolytic reaction (SBA) (Table BXII.γ.186) and their ability to grow at 42°C. Other differences include the ability of *A. sobria* to produce a positive string test (with 0.5% sodium desoxycholate) and susceptibility to O129 (150 µg). Differential tests to aid in the identification of *A. sobria* from all motile species are listed in Table BXII.γ.186. Descriptive tests for all the motile species are listed in Table BXII.γ.182.

The few known strains were isolated from fish, sewage, and water. These strains generally grow optimally at 25–30°C with few exceptions. Thus far, there are no reported human or animal clinical isolates. (There is frequent incorrect reference to this species as a human clinical isolate; see Taxonomic Comments.)

The mol% G + C of the DNA is: 58–60 (T_m).
Type strain: ATCC 43979, CIP 7433, NCIMB 12065.
GenBank accession number (16S rRNA): X74683.

13. **Aeromonas trota** Carnahan, Chakraborty, Fanning, Verma, Ali, Janda and Joseph 1992a, 191^VP (Effective publication: Carnahan, Chakraborty, Fanning, Verma, Ali, Janda and Joseph 1991b, 1207.)

tro' ta. Gr. adj. *trotos* vulnerable; M.L. fem. adj. *trota* vulnerable.

Hybridization Group 14 of 17 (Carnahan et al., 1991b).

Its unique biochemical profile includes negative reactions for esculin hydrolysis, L-arabinose fermentation, Voges–Proskauer test, and gluconate oxidation. Positive for cellobiose fermentation, lysine decarboxylase, and citrate utilization. Susceptible to ampicillin (in contrast to almost all other aeromonads) and carbenicillin (Tables BXII.γ.182 and BXII.γ.183). Differential tests to aid in the identification of *A. trota* from all motile species are listed in Table BXII.γ.186. Descriptive tests for all the motile species are listed in Table BXII.γ.182. Strains isolated from human stools, aquatic sources, and extraintestinal human infections. First thought to be found primarily in Southern Asia. Now considered to have worldwide distribution (Carnahan and Joseph, 1993). Documented case of pediatric diarrhea (Reina and Lopez, 1996b).

The mol% G + C of the DNA is: undetermined.
Type strain: AH2, ATCC 49657, DSM 7312.

The species name *Aeromonas enteropelogenes* has been validly published (Schubert et al., 1990a). It appears to be identical to *A. trota*, a more frequently used name (Carnahan et al., 1991b; Collins et al., 1993; Huys et al., 2002a) (see Taxonomic Comments).

14. **Aeromonas veronii** Hickman-Brenner, MacDonald, Steigerwalt, Fanning, Brenner and Farmer 1988b, 220^VP (Effective publication: Hickman-Brenner, MacDonald, Steigerwalt, Fanning, Brenner and Farmer 1987, 901.)

ve.ro' ni.i. M.L. gen. n. *veronii* of Véron, named after M.M. Véron.

Hybridization Groups 8/10 of 17 (Popoff et al., 1981; Farmer et al., 1986).

Subsequent to the proposal for this species (formerly termed Enteric Group 77 and HG 10), it was learned that although they had different biochemical reactions, HG 8 was genetically identical to HG 10. Thus, *A. veronii* contained two biogroups (Kuijper et al., 1989a; Altwegg et al., 1990). The first was the originally proposed ornithine decarboxylase positive species, *A. veronii* (HG 10) (Hickman-Brenner et al., 1987). The second group was lysine decarboxylase positive, arginine dihydrolase positive, and ornithine decarboxylase negative, and contained the so-called "clinical" *A. sobria* (HG 8). While the HG 7 environmental strains of *A. sobria* and the former HG 8 *A. sobria* are phenotypically very similar, they are genetically and clinically quite different. For this reason, it was decided to place the two biogroups of *A. veronii* into two separate biovars, namely biovar veronii and biovar sobria (Joseph et al., 1991).

The mol% G + C of the DNA is: undetermined.

Type strain: ATCC 35624, DSM 7386.

GenBank accession number (16S rRNA): X74684.

Note that the species name *Aeromonas ichthiosmia* (Schubert et al., 1990b) has been validly published, but has been unarguably shown to be identical to *A. veronii* (Carnahan, 1992; Collins et al., 1993, Huys et al., 2001) (see Taxonomic Comments).

a. **Aeromonas veronii** *biovar* **veronii**

String test negative, produces gas from D-glucose, is resistant to O/129, and grows in the absence of added NaCl. This biovar is resistant to ampicillin (10 µg) and carbenicillin (30 µg), and sensitive to cephalothin (30 µg) and colistin (4 µg/ml). Indole positive and Voges–Proskauer positive, produces acid from D-mannitol and sucrose, does not produce acid from L-arabinose, and is pyrazinamidase negative. β-Hemolytic on sheep blood agar (5%).

Aeromonas veronii biovar veronii is characteristically ornithine decarboxylase positive, arginine dihydrolase negative, and esculin and arbutin hydrolysis positive. This biovar can be misidentified as *Vibrio cholerae* in some rapid identification systems (Hickman-Brenner et al., 1987). Differential tests to aid in the identification of *Aeromonas veronii* biovar veronii from all motile species are listed in Table BXII.γ.186. Descriptive tests for all the motile species are listed in Table BXII.γ.182. Isolated from aquatic environments, including leeches (Graf, 1999b) and specimens from human clinical disease including diarrhea, wounds, pulmonary complications, sinusitis (Hickman-Brenner et al., 1987), wounds (Joseph

et al., 1991), and bacteremia (Abbott et al., 1994; Janda and Abbott, 1996).

The mol% G + C of the DNA is: 57.6–58.2 (Bd, T_m).

Type strain: ATCC 35624, DSM 7386.

GenBank accession number (16S rRNA): X74684.

b. **Aeromonas veronii** *biovar* **sobria**

String test negative, produces gas from D-glucose, is resistant to O/129, and grows in the absence of added NaCl. This biovar is resistant to ampicillin (10 µg) and carbenicillin (30 µg), and sensitive to cephalothin (30 µg) and colistin (4 µg/ml). Indole positive and Voges–Proskauer positive, produces acid from D-mannitol and sucrose, does not produce acid from L-arabinose, and is pyrazinamidase negative. β-Hemolytic on sheep blood agar (5%).

Aeromonas veronii biovar sobria is characteristically ornithine decarboxylase negative; lysine decarboxylase and arginine dihydrolase positive, and esculin and arbutin hydrolysis negative; β-hemolytic on 5% sheep blood agar and able to grow at 42°C. Differential tests to aid in the identification of *Aeromonas veronii* biovar sobria from all motile species are listed in Table BXII.γ.186. Descriptive tests for all the motile species are listed in Table BXII.γ.182. Associated with gastrointestinal disease and a wide array of extraintestinal and systemic disease including septicemia, wound infections, meningitis, peritonitis, septic arthritis, and hepatobiliary disease (Steinfeld et al., 1998; Janda and Abbott, 1996, 1998b). The reference strain for these clinical isolates, which were formerly considered to be HG 7 *A. sobria*, but in actuality are HG 8 *Aeromonas veronii* biovar sobria, is ATCC 9071.

The mol% G + C of the DNA is: undetermined.

Type strain: ATCC 9071.

Other Organisms

1. *"Aeromonas arequipensis"*

Isolated from a lake in the Arequipa region of Peru; may have diarrheagenic potential for humans (Matte et al., 1999).

2. *"Aeromonas dechromatica"*

This organism was proposed by Kvasnikov et al. (1985). Isolated from chromium-containing industrial wastewater; has basic characteristics of *Aeromonas* spp. Exceptions are requirement for sodium chloride (optimal 3%), mixed flagellation, Gram variable, absence of catalase activity, presence of a crystalline microcapsule; so named because it reduces highly toxic hexavalent chromium ions.

3. *"Aeromonas guangheii"*

This organism was isolated from eels. Unpublished; only one strain.

Deposited strain: 95-173.

GenBank accession number (16S rRNA): AB028881.

4. *"Aeromonas pastoria"*

This organism was proposed (Torres, 1990; Torres et al., 1991) and later included in a large phenotypic study of *Aeromonas* isolates (Noterdaeme et al., 1996). Two strains of *"Aeromonas pastoria"* were examined, one of which fell into their Cluster 2A, which was characterized by nonfermentation of L-arabinose and α-methyl-D-glucoside and by assimilation of citrate. Another strain of *"A. pastoria"*, which

was received separately, fell into Cluster 7, which was characterized by the following reactions: negative for production of ornithine decarboxylase; hydrolysis of esculin; assimilation and fermentation of L-arabinose; fermentation of α-methyl-D-glucoside, arbutin, salicin, and lactose; and positive for the assimilation of citrate.

5. *"Aeromonas* Group 501" (formerly Enteric Group 501).

A designation given to two strains in the CDC collection by Hickman-Brenner et al. (1988a), which closely resembled *A. schubertii* but were genotypically and phenotypically different and currently reside in HG 13; see Taxonomic Comments above for further discussion.

Deposited strain: ATCC 43946.

GenBank accession number (16S rRNA): U88663.

6. *"Aeromonas* species ornithine positive" (formerly Enteric Group 77)

Currently resides in HG 11 (Fanning et al., 1985; Farmer et al., 1986; Hickman-Brenner et al., 1987); see Taxonomic Comments above for further discussion about these strains.

Deposited strain: ATCC 35941.

GenBank accession number (16S rRNA): X60417.

7. "Halophilic *Aeromonas* species"

There is a report of a strictly halophilic *Aeromonas* sp. from Atlantic salmon in association skin lesions leading to severe widespread lesions elsewhere, a condition that suggested production of a potent toxin(s). On primary culture, the organism was slow growing (best in the presence of horse blood, 10% w/v) and had a strict requirement for at least 1.0% NaCl. Optimum growth temperature is between 15 and 20°C. No growth at 37°C. Growth at 30°C after 72 h yielded small (1–2 mm), round, low, convex, mucoid, translucent, β-hemolytic colonies. Individual cells were motile, straight, Gram-negative rods. Resistant to vibriostatic agent 0/129 (10 and 150 μg) and susceptible to oxytetracycline, oxalinic acid, furazolidone, and penicillin. Oxidase positive, reduced nitrates, and used D-glucose fermentatively. Majority of biochemical tests were negative, except for gelatinase and D-glucose. Voges–Proskauer test was variable. The organism (NCIMB 2263) was at and was deposited in the National Collection of Industrial and Marine Bacteria Ltd., Aberdeen (Cox et al., 1986).

Species Incertae Sedis

1. **Haemophilus piscium** Snieszko, Griffin and Friddle 1950, 699[AL]

pis.ci' um. L. n. *piscis* a fish; L. gen. pl. *piscium* of fishes.

This microorganism was removed from the genus *Haemophilus* in the first edition of *Bergey's Manual of Systematic Bacteriology* (Kilian and Biberstein, 1984). This exclusion was made on the basis of its DNA base composition, growth factor requirements, cell wall structure, respiratory requirements, respiratory quinones, antigenic composition, and phage susceptibility. It was later concluded that it represents an atypical form of *Aeromonas salmonicida* (Paterson et al., 1980; Austin et al., 1989a). 16S rRNA encoding genes for the reference strains of *A. salmonicida* subsp. *smithia* and *H. piscium*, when amplified by PCR, revealed that these organisms showed 99.4 and 99.6% 16S rRNA gene sequence identity, respectively, with *A. salmonicida* subsp. *salmonicida* (Thornton et al., 1999).

It should be noted that an authentic type strain of *Haemophilus piscium* was not available in any culture collection, so the *H. piscium* strain (NCIMB 1952) was obtained for use from the National Collection of Industrial and Marine Bacteria, Aberdeen, Scotland (Thornton et al., 1999).

The mol% G + C of the DNA is: 55.1 (T_m).

Type strain: ATCC 10801.

GenBank accession number (16S rRNA): AJ009860.

Genus II. **Oceanimonas** Brown, Sutcliffe and Cummings 2001, 71[VP]

THE EDITORIAL BOARD

O.ce.a.ni.mo' nas. Gr. n. *okeanus* ocean; Gr. fem. n. *monas* monad unit; M.L. fem. n. *Oceanimonas* ocean monad.

Gram-negative motile rods (0.7–1.2 × 2.0–2.5 mm). Up to four polar flagella. **Respiratory metabolism** using O_2. Phosphatidylethanolamine, phosphatidylglyerol, and diphosphatidylglycerol are the main phospholipids. **Require Na$^+$ for growth**

The mol% G + C of the DNA is: 54 (HPLC).

Type species: **Oceanimonas doudoroffii** (Baumann, Bowditch, Baumann and Beamann 1983b) Brown, Sutcliffe and Cummings 2001, 71 (*Pseudomonas doudoroffii* Baumann, Bowditch, Baumann and Beamann 1983b, 863.)

FURTHER DESCRIPTIVE INFORMATION

These organisms are chemoorganotrophs found in marine environments.

ENRICHMENT AND ISOLATION PROCEDURES

O. doudoroffii strains were first isolated from Pacific Ocean surface water by enrichment in liquid medium containing allantoin, benzoate, creatine, caprylate, or D-aminovalerate followed by streaking on the same medium (Baumann et al., 1972). *O. baumannii* was isolated from estuary mud by enrichment in liquid minimal salts medium containing phenol as a carbon source and 5% NaCl, followed by streaking onto the same medium solidified with agar (Brown et al., 2001).

MAINTENANCE PROCEDURES

O. doudoroffii strains were maintained on Marine Agar (Difco) slopes with monthly transfer (Baumann et al., 1972).

DIFFERENTIATION OF THE GENUS *OCEANIMONAS* FROM OTHER GENERA

O. doudoroffii DSM 7028^T and *O. baumannii* GB6^T can be distinguished from the type strains of both *Aeromonas hydrophila* and *Tolumonas auensis* by the inability of the former two strains to utilize arabinose, gluconate, glucose, maltose, mannitol, mannose, and sucrose, and by the inability of the latter two strains to utilize malate. *O. doudoroffii* DSM 7028^T and *O. baumannii* GB6^T can be further distinguished from *Tolumonas auensis* by the inability of the latter to grow above 25°C and from *A. hydrophila* by the inability of the latter to utilize citrate and by its ability to produce indole, to produce acid from glucose, to produce arginine hydrolase, to hydrolyze esculin, and to utilize *N*-acetylglucosamine.

TAXONOMIC COMMENTS

The major reevaluation of the taxonomy of the genus *Pseudomonas* and the reassignment of species in RNA groups II–V to other genera are discussed elsewhere in this volume. In the first edition of the *Manual*, Palleroni (1984) placed *Pseudomonas doudoroffii* in a separate section devoted to organisms whose relationships to those in the *Pseudomonas* RNA groups I–V were not known. Subsequent studies (De Vos et al., 1989; Kersters et al., 1996; Anzai et al., 2000) did not clarify the taxonomic relationships of *Pseudomonas doudoroffii*. Brown et al. (2001) place the genus *Oceanimonas* in the *Gammaproteobacteria* but do not assign it to a family.

List of species of the genus Oceanimonas

1. **Oceanimonas doudoroffii** (Baumann, Bowditch, Baumann and Beamann 1983b) Brown, Sutcliffe and Cummings 2001, 71[VP] (*Pseudomonas doudoroffii* Baumann, Bowditch, Baumann and Beamann 1983b, 863.)

dou.do.rof fi.i. M.L. gen. n. *doudoroffii* of Doudoroff; named after M. Doudoroff.

Gram-negative rods (0.7–1.2 × 2.0–2.5 μm). One to three bipolar flagella. Utilizes acetate, aconitate, D-α-alanine, L-α-alanine, β-alanine, allantoin, γ-aminobutyrate, δ-aminovalerate, L-arginine, L-aspartate, benzoate, betaine, butylamine, caprate, citrate, creatine, ethanol, D-fructose, fumarate, L-glutamate, glutarate, DL-glycerate, glycine, glycolate, histamine, L-histidine, DL-β-hydroxybutyrate, α-ketoglutarate, DL-lactate, DL-malate, L-ornithine, phenol, L-proline, putrescine, pyruvate, L-serine, sarcosine, spermine, and succinate. Does not utilize galactose or glycerol. Oxidase positive. Accumulates poly-β-hydroxybutyrate. Fatty acids include dodecanoic, hexdecanoic, cis-hexadecenoic, and cis-octadecenoic.

The mol% G + C of the DNA is: 54 (HPLC).
Type strain: ATCC 27123, DSM 7028.
GenBank accession number (16S rRNA): AB019390.

2. **Oceanimonas baumannii** Brown, Sutcliffe and Cummings 2001, 71^{VP}

bau.man' ni.i. M.L. gen. n. *baumannii* of Baumann; named after Paul and Linda Baumann, who first studied these organisms.

Gram-negative rods (1 × 2 μm). Aerobic growth on L-alanine, betaine, caprate, citrate, ethanol, L-glutamate, malate, L-proline, phenol, sarcosine, and succinate. Grows in NaCl up to 7% (w/v). Growth temperatures 10–30°C. Utilizes galactose and glycerol. Fatty acids include dodecanoic, hexdecanoic, cis-hexadecenoic, and cis-octadecenoic.

The mol% G + C of the DNA is: 54 (HPLC).
Type strain: GB6, ATCC 700832, NICMB 13685.
GenBank accession number (16S rRNA): AF168367.

Genus incertae sedis III. **Tolumonas** *Fischer-Romero, Tindall and Jüttner 1996, 187^{VP}*

CARMEN FISCHER-ROMERO AND BRIAN J. TINDALL

To.lu.mo' nas. L. n. *toluolum* toluol (German for toluene); Gr. n. *monas* unit; M.L. fem. n. *Tolumonas* toluene-producing unit.

Cells are **nonmotile**, Gram-negative rods (2.5–3.2 × 0.9–1.2 μm) that occur singly or in pairs. **Growth** occurs under **aerobic and anaerobic** conditions. Anaerobic cultures are catalase and oxidase negative; aerobic cultures are **catalase positive** and **oxidase negative**. **Toluene is produced from phenylalanine, phenylacetate, phenyllactate, and phenylpyruvate** only in the presence of an additional carbon source. Phenol is produced when phenylalanine is replaced by tyrosine. The optimum conditions for anaerobic growth are **22°C** and pH 7.2. The major fermentation products in peptone–yeast extract–glucose cultures are ethanol, acetic acid, and formic acid. Under anaerobic conditions D-arabinose, D-cellobiose, D-fructose, D-glucose, glycogen, inulin, maltose, D-mannose, D-melecitose, melibiose, D-raffinose, L-rhamnose, D-ribose, salicin, sucrose, D-trehalose, D-mannitol, and D-sorbitol are used as carbon sources. In the presence of air the following organic substances are used: Tween 40, Tween 80, L-arabinose, D-fructose, D-galactose, D-glucose, methyl-pyruvate, monomethyl succinate, acetic acid, cis-aconitic acid, D-galacturonic acid, D-gluconic acid, D-glucuronic acid, β-hydroxybutyric acid, α-ketoglutaric acid, DL-lactic acid, malonic acid, quinic acid, sebacic acid, succinic acid, glucuronamide, L-alanine, L-asparagine, L-aspartic acid, L-glutamic acid, L-histidine, L-threonine, urocanic acid, and inosine. The enzymes lipase, lecithinase, urease, and exoprotease are not produced. Indole, lipase, and H$_2$S are not formed. According to the results of 16S rDNA sequence analysis this genus belongs to the *Gammaproteobacteria*.

The mol% G + C of the DNA is: 49.

Type species: **Tolumonas auensis** Fischer-Romero, Tindall and Jüttner 1996, 187.

FURTHER DESCRIPTIVE INFORMATION

Ecological studies (Fischer-Romero, unpublished results) on the distribution of *Tolumonas auensis* in different lakes and streams of central Europe suggest that this bacterium is not ubiquitously distributed. Cells were not found in surface samples, whereas in the water column the number of *Tolumonas auensis* increased with the depth and reached a maximum in the sediment and the overlying water. In the course of one year the maximum number of *Tolumonas auensis* cells was counted in samples taken in autumn, before the autumnal circulation, concurring with the highest seasonal toluene concentration in the hypolimnion.

ENRICHMENT AND ISOLATION PROCEDURES

Members of the genus *Tolumonas* can be enriched in TP medium inoculated with anoxic freshwater sediment and isolated by agar deep dilutions (Fischer-Romero et al., 1996).

MAINTENANCE PROCEDURES

Strains of the genus *Tolumonas* can be maintained in TP medium at 4°C, after incubation to allow good growth, with monthly transfers or preserved by lyophilization or by storage in liquid nitrogen.

PROCEDURES FOR TESTING SPECIAL CHARACTERS

To determine the concentration of toluene in the culture media, the method described by Jüttner (1988) may be used. The medium (0.1–75 ml) was supplied with perdeutero-toluene (1 μl perdeutero-toluene/ethanol; 1:1000 v/v) as internal standard and stripped for 30 min in a closed-loop stripping apparatus. Toluene was adsorbed on a Tenax-filled cartridge (150 g Tenax TA) and transferred from these by thermodesorption into a combined GC–MS. Toluene was separated by gas–liquid chromatography on a DB 1301 fused silica capillary column (30 m × 0.32 mm, J & W Scientific) and detected by single ion monitoring (*m/z* 91 and *m/z* 98) on a Hewlett Packard 5970A mass spectrometer.

DIFFERENTIATION OF THE GENUS *TOLUMONAS* FROM OTHER GENERA

Tolumonas transforms phenylalanine and other phenyl precursors into toluene. A few bacteria that are able to transform amino acids into aromatic hydrocarbons have been described. "*Clostridium aerofoetidum*" transforms phenylalanine in a Stickland reaction with methionine into toluene (Pons et al., 1984). *Lacto-*

TABLE BXII.γ.188. Differentiation of the genus *Tolumonas* from the closest phylogenetic relatives[a]

Characteristics	*Tolumonas*	*Aeromonas*	*Citrobacter*	*Escherichia*	*Hafnia*	*Plesiomonas*	*Serratia*	*Vibrio*
Mol% G + C	49	57–63	50–52	48–52	48–49	51	52–60	38–51
Motility	−	+	+	D	D	+	+	+
Optimal temperature, °C	22	22–28	37	37	30–37	37	30–37	20
Glycerol utilization	−	d	+	D	+	+	+	+
Fucose utilization	−	nd	nd	nd	nd	nd	+	nd
Catalase	+	+	+	+	+	+	+	nd
Oxidase	−	+	−	−	−	+	nd	+

[a]For symbols see standard definitions; nd, not determined.

bacillus sp. (Yokoyama and Carlson, 1981) and *Clostridium* sp. strains (Elsden et al., 1976) liberate *p*-cresol and phenol from tyrosine. These organisms differ from *Tolumonas* in their cytophysiological features and in the biochemistry of toluene production.

Other characteristics that differentiate members of the genus *Tolumonas* from their closest phylogenetic relatives, *Escherichia*, *Citrobacter*, *Hafnia*, *Plesiomonas*, *Serratia*, *Vibrio*, and *Aeromonas* (Holt et al., 1994) are shown in Table BXII.γ.188.

Based on the 16S rDNA sequence analysis *Tolumonas auensis* shows about 10% sequence divergence from its nearest neighbors. These include members of the genera *Escherichia*, *Serratia*, *Enterobacter* (i.e., members of the family *Enterobacteriaceae*), *Vibrio*, *Aeromonas*, *Shewanella*, *Alteromonas*, and *Pseudoalteromonas*. A greater degree of dissimilarity is observed between *Tolumonas* and members of the genera *Haemophilus*, *Actinobacillus*, *Pasteurella* (i.e., members of the family *Pasteurellaceae*) *Succinivibrio*, *Succinimonas*, *Anaerobiospirillum*, and *Ruminobacter* (Hippe et al., 1999).

T. auensis produces both ubiquinone 8 and menaquinone 8 (Fischer-Romero et al., 1996). Ubiquinone 8 is found in all members of the genera *Escherichia*, *Serratia*, *Enterobacter* (i.e., members of the family *Enterobacteriaceae*), *Vibrio*, *Aeromonas*, *Shewanella*, *Alteromonas*, and *Pseudoalteromonas* (Wilkinson, 1988; Akagawa-Matsushita et al., 1992a). In some cases, menaquinones with the same isoprenoid chain length are also found, and may comprise both menaquinones and demethyl menaquinones. Members of the genera *Alteromonas* and *Pseudoalteromonas* do not produce menaquinones, while those members of the genus *Shewanella* examined to date produce ubiquinone 7 and 8, but menaquinone 7, and usually, but not always, methylated menaquinone 7 (Akagawa-Matsushita et al., 1992a). Menaquinones are absent from *Ruminobacter amylophilus* (Stackebrandt and Hippe, 1986), while members of the family *Pasteurellaceae* produce either menaquin-

ones and/or demethylnaphthoquinones with seven or eight isoprene units; ubiquinones with the corresponding isoprenoid side chain length may be found, although they apparently are not always present (Kroppenstedt and Mannheim, 1989; Engelhard et al., 1991).

The fatty acid composition of *T. auensis* is dominated by $C_{16:0}$ and $C_{16:1}$ fatty acids (Fischer-Romero et al., 1996), as are the fatty acids of members of the majority of genera listed above. The most notable exceptions are members of the genus *Shewanella* where *iso-* and *anteiso-*branched fatty acids are present in significant quantities. A $C_{14:0\ 3OH}$ fatty acid is also present in *Tolumonas auensis*, approximately half of which appears to be amide linked, probably to the lipopolysaccharide. In members of the family *Enterobacteriaceae* it is well documented that the lipopolysaccharides contain 3-hydroxy fatty acids, in which $C_{14:0\ 3OH}$ predominates, and is present in both an ester- and an amide-linked form (Wilkinson, 1988). It is interesting to note that, in addition to members of the genera *Vibrio*, *Aeromonas*, *Haemophilus*, *Actinobacillus*, *Pasteurella*, *Succinivibrio*, *Succinimonas*, *Anaerobiospirillum*, and *Ruminobacter* also contain a $C_{14:0\ 3OH}$ fatty acid (Miyagawa et al., 1979; Moore et al., 1994a; Urdaci et al., 1990; Engelhard et al., 1991; Huys et al., 1994; Weyant et al., 1996). Depending upon the genus, additional hydroxylated and nonhydroxylated fatty acids may be present, allowing some degree of differentiation between the genera; 2-hydroxy fatty acids are absent in *T. auensis*, and all of the genera listed above.

Phosphatidylethanolamine and phosphatidylglycerol are the major phospholipids present in *T. auensis* (Fischer-Romero et al., 1996). Based on the scant data currently available it appears the same phospholipids predominate in the closest evolutionary relatives, although differentiation between the genera may also be possible based on the presence of additional polar lipids (Wilkinson, 1988; Engelhard et al., 1991).

List of species of the genus Tolumonas

1. **Tolumonas auensis** Fischer-Romero, Tindall and Jüttner 1996, 187[VP]

   au.en' sis. M.L. gen. n. *auensis* of Lake Au, the location of the first isolation of this organism.

   The characteristics are those of the genus, as described in the text and in Table BXII.γ.189. The major lipoquinones present are ubiquinone 8 and menaquinone 8 under both oxic and anoxic growth conditions. The fatty acid profile of the cells is: $C_{12:0}$, $C_{14:0}$, $C_{14:0\ 3OH}$, $C_{16:1\ \omega7c}$, $C_{16:0}$, and $C_{18:1\ \omega7c}$. Isolated from the anoxic sediment of a freshwater lake.

   The mol% G + C of the DNA is: 49 (T_m).
   Type strain: TA 4, DSM 9187.
   GenBank accession number (16S rRNA): X92889.

TABLE BXII.γ.189. Biochemical characteristics of the genus *Tolumonas*

Characteristics	Reaction/Result
Gram stain	−
Motility	−
Toluene production	+
Catalase	+
Oxidase	−
D-Glucose acid production	+
Hydrogen sulfide production	−
Indole production	−
Lipase	−

Family *incertae sedis* II. **Succinivibrionaceae** Hippe, Hagelstein, Kramer, Swiderski and Stackebrandt 1999, 782ᵛᴾ*

GEORGE M. GARRITY, JULIA A. BELL AND TIMOTHY LILBURN

Suc.ci.ni.vib.ri.o.na' ce.ae M. L. masc. n. *Succinivibrio* type genus of the family; *-aceae* ending to denote a family; M. L. fem. pl. n. *Succinivibrionaceae* the *Succinivibrio* family.

Gram-negative, short to long, **straight, curved, or helical rods.** Nonsporeforming. **Strict anaerobes.** Catalase negative. **Ferment carbohydrates to succinate and acetate.** No production of gas or reduction of nitrate. Take up CO_2.

The mol% G + C of the DNA is: 39–44.

Type genus: **Succinivibrio** Bryant and Small 1956, 22.

Editorial Note: This family was proposed based on 16S rDNA sequence analysis (Hippe et al., 1999).

Genus I. **Succinivibrio** Bryant and Small 1956, 22ᴬᴸ*

MARVIN P. BRYANT

Suc.ci.ni.vib' ri.o. M.L. n. *acidum succinicum* succinic acid; M.L. masc. n. *vibrio* that which vibrates, a generic name; M.L. masc. n. *Succinivibrio* the succinic acid vibrio.

Curved rods, 0.4–0.6 × 1.0–7.0 µm, with pointed ends. The cells are helically twisted with less than one coil to three or more coils per cell. The cells may become straight or only slightly curved after maintenance on artificial media. Gram negative. **Possess a progressive vibrating type of motility by means of a single polar flagellum.** Strictly anaerobic. Chemoorganotrophic, having a fermentative type of metabolism with carbohydrates being the main fermentable substrates. **The major products of glucose fermentation are succinate, acetate, formate, and sometimes lactate.** Butyrate and H_2 are not formed. A large net uptake of CO_2 may occur. Occur in the rumen of cattle and sheep.

The mol% G + C of the DNA is: unknown.

Type species: **Succinivibrio dextrinosolvens** Bryant and Small 1956, 22.

FURTHER DESCRIPTIVE INFORMATION

Cells from young cultures on RGCA slants[1] are short with one or less than one complete coil, but longer cells containing two or three coils are commonly present also. Cells are mainly single but a few short chains may occur. In aging cultures swollen, giant spirillar forms and round bodies may be present.

Surface colonies on RGCA agar in roll tubes are 1–2 mm in diameter, entire, translucent, slightly convex, and light tan in color after 3 d incubation at 37°C. Deep colonies are lenticular. In liquid media with glucose as the energy source growth occurs as a heavy, flocculent sediment that is easily dispersed; light turbidity also occurs.

Good growth of *S. dextrinosolvens* occurs in a chemically defined medium containing glucose, minerals, *p*-aminobenzoic acid, 1,4-naphthoquinone, ammonium ions, cysteine, methionine, leucine, and serine. The use of a CO_2/HCO_3^- buffer system (pH 6.7) is necessary for good growth (Gomez-Alarcon et al., 1982). Growth occurs at 30–39°C but not at 22° or 45°C.

The final pH in lightly buffered glucose medium is 4.8–5.2. *S. dextrinosolvens* requires a fermentable carbohydrate for growth and does not ferment amino acids. Succinic and acetic acids, and often formic and lactic acids, are produced from fermentation of glucose. Gas is not produced. Xylose, galactose, glucose, maltose, and dextrin can serve as fermentable substrates. Arabinose, fructose, cellobiose, sucrose, esculin, salicin, and mannitol are fermented by some strains. No fermentation of lactose, trehalose, cellulose, xylan, inulin, glycerol, or inositol occurs. Starch is partially fermented but not completely hydrolyzed. Nitrate is not reduced. Gelatin is not hydrolyzed. Acetoin, indole, and H_2S are not produced. The catalase test is negative.

Urease is produced by many strains but is strongly repressed in media containing large amounts of ammonia or other utilizable nitrogen sources (Wozny et al., 1977).

S. dextrinosolvens is not pathogenic for animals as far as is known. Isolates corresponding to the description of the organism have been isolated from a few cases of human bacteremia (Southern, 1975; Porschen and Chan, 1977).

S. dextrinosolvens is found in the rumen of cattle and sheep, especially when high-grain diets containing large amounts of starch are fed (Bryant and Small, 1956; Bryant et al., 1961; Wozny et al., 1977). The organism appears to be a major fermenter of dextrins under these conditions.

ENRICHMENT AND ISOLATION PROCEDURES

S. dextrinosolvens is sometimes a predominant organism from the rumen of cattle fed high-grain diets, and can be isolated nonselectively in anaerobic roll tubes of RGCA medium incubated for 3 d or longer at 37°C.

MAINTENANCE PROCEDURES

S. dextrinosolvens can be maintained on stab-inoculated RGCA slants at −70°C for 1 year or longer. The organisms can also be preserved indefinitely by lyophilization.

DIFFERENTIATION OF THE GENUS *SUCCINIVIBRIO* FROM OTHER GENERA

Succinivibrio differs from *Anaerobiospirillum* by having monotrichous rather than lophotrichous flagella. It differs from the genus *Anaerovibrio* by not producing a large amount of propionate, by not fermenting glycerol, and by its ability to ferment dextrin.

Editorial Note: The chapter on *Succinivibrio* is reprinted from the 1st edition of *Bergey's Manual of Systematic Bacteriology.*

1. RGCA slants (g/l): K_2HPO_4, 0.23 g; KH_2PO_4, 0.45; $(NH_4)_2SO_4$, 0.45; NaCl, 0.45; $MgSO_4$, 0.023; $CaCl_2$, 0.023; resazurin, 0.001; glucose, 0.5, cellobiose, 0.5; soluble starch, 0.5; rumen fluid (centrifuged), 400; cysteine·HCl, 0.25; $Na_2S·9H_2O$, 0.25; Na_2CO_3, 4.0, agar (Difco), 10.0; pH 6.7; gas phase, 100% CO_2. Prepared and used with the Hungate technique (Bryant, 1972).

FURTHER READING

Bryant, M.P. and N. Small. 1956. Characteristics of two new genera of anaerobic curved rods isolated from the rumen of cattle. J. Bacteriol. 72: 22–26.

Porschen, R.K. and P. Chan. 1977. Anaerobic vibrio-like organisms cultured from blood: *Desulfovibrio desulfuricans* and *Succinivibrio* species. J. Clin. Microbiol. 5: 444–447.

List of species of the genus Succinivibrio

1. **Succinivibrio dextrinosolvens** Bryant and Small 1956, 22.[AL]
   dex.tri.no.sol'vens. M.L. n. *dextrinosum* dextrin; L. part. adj. *solvens* dissolving; M.L. part. adj. *dextrinosolvens* dextrin-dissolving.

   The characteristics are as described for the genus.

Found in the bovine and ovine rumen.

The mol% G + C of the DNA is: unknown.

Type strain: 554, 24 of Bryant and Small (1956), ATCC 19716, DSM 3072.

GenBank accession number (16S rRNA): Y17600.

Genus II. **Anaerobiospirillum** Davis, Cleven, Brown and Balish 1976, 503[AL] emend. Malnick 1997, 383

HENRY MALNICK

An.ae.ro.bi.o.spi.ril'lum. Gr. prefix *an* not; Gr. n. *aer* air; Gr. n. *bios* life; M.L. dim. neut. n. *spirillum* a small spiral; M.L. neut. n. *Anaerobiospirillum* anaerobic small spiral.

Helical rods 0.6–0.8 × 3–15 µm, usually occurring singly. Some cells are up to 32 µm in length. **Motile by bipolar tufts of flagella.** Do not form endospores. Gram negative. **Anaerobic, having a strictly fermentative type of metabolism.** Catalase negative or weakly positive. Oxidase negative. Do not hydrolyze esculin, gelatin, hippurate, or urea. Do not reduce nitrate. Lipase activity does not occur. Indole is not produced and meat is not digested. **Ferment carbohydrates; produce succinic and acetic acids from glucose.** May also produce traces of lactic and formic acids. Optimal temperature 37–44°C. **Isolated from feces of dogs and cats. Pathogenic for humans, causing septicemia and/or diarrhea.** Belongs to the class *Gammaproteobacteria*, order *Aeromonadales*, and family *Succinivibrionaceae*.

The mol% G + C of the DNA is: 39–44.

Type species: **Anaerobiospirillum succiniciproducens** Davis, Cleven, Brown and Balish 1976, 503.

FURTHER DESCRIPTIVE INFORMATION

The diameter of the cell helix is 0.9–1.1 µm. Straight rods and spherical forms may occur, especially in liquid cultures. Some strains of *Anaerobiospirillum succiniciproducens*, including the type strain, can be pleomorphic, often producing bizarre forms. This feature seems to be rare in *A. thomasii*. The cells have a corkscrew-like motility. The flagella are 14 nm in diameter and are arranged as bipolar tufts of ~16 flagella.

Ultrastructural examination of *A. succiniciproducens* has shown that the cells possess fibrils arranged along the longitudinal axis within the cell. So far, this has not been described in any other organism. The flagellar base within the cell is disc shaped (Wecke and Horbach, 1999).

Surface colonies on most types of blood agar are 0.5–1.0 mm in diameter, circular, convex, and translucent after 2–3 days anaerobic incubation at 37°C. Colonies may have raised centers. Spreading or feathery edges from heavily inoculated areas is not uncommon on some media. Growth is most rapid at 37–40°C and does not occur below 33°C or above 44°C. Cells do not survive 80°C for 10 min.

Although requiring an anaerobic atmosphere for growth, *Anaerobiospirillum* spp. will survive in an atmosphere of reduced oxygen within a jar until other bacteria present have reduced the atmosphere. It is in this way that fecal isolates of *A. thomasii* were isolated on Skirrow's *Campylobacter* isolation medium.

At usual therapeutic levels, isolates of *Anaerobiospirillum* spp. are susceptible to amoxicillin-clavulanate, azithromycin, cefoxitin, ciprofloxacin, chloramphenicol, and trovafloxacin, and are resistant to cephalexin, clarithromycin, clindamycin, lincomycin, nalidixic acid, polymyxin B, sulfamethoxazole, and vancomycin. Susceptibility to ampicillin, erythromycin, penicillin G, and metronidazole is variable. *A. succiniciproducens* is mezlocillin resistant and gemifloxacin variable, and has an MIC for trimethoprim of 4 µg/ml, whereas *A. thomasii* is mezlocillin and gemifloxacin sensitive and has an MIC for trimethoprim of >256 µg/ml (McNeil et al., 1987b; Goldstein et al., 1999, Malnick, unpublished results).

Under favorable conditions of pH 6.2 and high CO_2 concentrations, *A. succiniciproducens* can produce industrially significant levels of succinate (up to 35 g/l) (Datta, 1989). The phosphoenolpyruvate-carboxykinase-encoding gene (*pckA*) involved has been sequenced, and the amino acid sequence—deduced from the data—encodes a 532-residue polypeptide. Phosphoenolpyruvate carboxykinase is a CO_2-fixing enzyme (Laivenieks et al., 1997).

Anaerobiospirillum species have been reported as causing septicemia and diarrhea in man. There appears to be a difference in pathogenicity between the two species. *A. succiniciproducens* has been isolated from blood cultures of patients suffering from septicemia, and in a number of cases it was reported either as the cause of death or as a contributory factor. Most of the patients involved had underlying disorders. So far, the isolation of *A. thomasii* has only been reported from cases of diarrhea (McNeil et al., 1987b; Malnick et al., 1990).

Of three isolates of *A. succiniciproducens* described by Davis et al. (1976), one came from the throat and two from cecal homogenate of dogs. Further studies have shown both species are found as part of the fecal flora in cats and dogs. Although the organisms seem to be part of the normal intestinal flora of these animals, no definite ecological function is known (Malnick et al., 1990).

ENRICHMENT AND ISOLATION PROCEDURES

An enrichment and serial dilution method for obtaining isolates from throat and cecal contents of dogs was used without the aid of a selective medium by Davis et al. (1976). A selective medium for the isolation of *Anaerobiospirillum* spp. has been described (Malnick et al., 1990). It consists of a suitable anaerobic agar medium supplemented with (per liter): polymyxin, 2500 IU; vancomycin, 20 mg; sulfamethoxazole, 100 mg; Victoria blue B (Cl 44045 Sigma), 250 mg; and horse blood lysed with saponin, 50 ml. The Victoria blue was initially dissolved in 4 ml of ethanol. After 48–72 h of anaerobic incubation, *Anaerobiospirillum* species produce dark blue colonies, depending on the basal medium. Other organisms produce white or pale blue colonies.

MAINTENANCE PROCEDURES

Anaerobiospirillum spp. can be maintained at $-70°C$ or in liquid nitrogen on beads, with any of a variety of cryoprotectants. The organisms can also be preserved by lyophilization.

DIFFERENTIATION OF THE GENUS *ANAEROBIOSPIRILLUM* FROM OTHER GENERA

The genus is distinguished from *Succinivibrio* and *Campylobacter* by its lophotrichous flagellar arrangement. In clinical situations, *Anaerobiospirillum* species have most often been confused with *Campylobacter*; however, *Campylobacter* species are oxidase positive and do not ferment carbohydrates. Production of succinic and acetic acids from glucose may help differentiate *Anaerobiospirillum* from other curved and spiral anaerobic bacteria.

TAXONOMIC COMMENTS

The phylogenetic relationship between *Anaerobiospirillum* and other Gram-negative, strictly anaerobic bacteria has been determined by 16S rDNA sequence analysis, placing these organisms in the family *Succinivibrionaceae* (Hippe et al., 1999).

The DNA–DNA hybridization values between the type strain of *A. succiniciproducens* and the type strain of *A. thomasii* are less than 6% (Malnick, 1997).

DIFFERENTIATION OF THE SPECIES OF THE GENUS *ANAEROBIOSPIRILLUM*

Characteristics useful in differentiating the two species of the genus are listed in Table BXII.γ.190.

List of species of the genus Anaerobiospirillum

1. **Anaerobiospirillum succiniciproducens** Davis, Cleven, Brown and Balish 1976, 503[AL]

    suc.ci.ni.ci.pro.du' cens. M.L. n. *acidum succinicum* succinic acid; L. pres. part. *producens* producing; M.L. part. adj. *succiniciproducens* producing succinic acid.

    The characteristics are as described for the genus and in Table BXII.γ.190, with the following additional information. Helical rods $0.6–0.8 \times 3–15$ μm. Produces weak β-hemolysis. Fixes CO_2.

    The mol% G + C of the DNA is: 42–44 (T_m).
    Type strain: S411, ATCC 29305, NCTC 11536.
    GenBank accession number (16S rRNA): U96412.

2. **Anaerobiospirillum thomasii** Malnick 1997, 383[VP]

    thom' as.i.i. N.L. gen. n. *thomasii* of Thomas, named after M.E.M. Thomas (Public Health Laboratory Service, 1947–1984), whose interest in and enthusiasm for uncommon medical bacterial infections led to the initiation of this study.

    The characteristics are as described for the genus and in Table BXII.γ.190, with the following additional information. Helical rods, $0.5–0.7 \times 5–32$ μm. Optimum temperature 37°C; range, 33–43°C. Colonies are flat with even, spreading edges, 1–2 mm in diameter after 24–48 h on horse blood agar. Hemolysis does not occur. Isolated from cat and dog feces and as a pathogen in human diarrheal feces.

    The mol% G + C of the DNA is: 39–41 (T_m).
    Type strain: NCTC 12467.
    GenBank accession number (16S rRNA): AJ420985.

TABLE BXII.γ.190. Differential characteristics of *A. succiniciproducens* and *A. thomasii*[a,b]

Characteristics	1. *A. succiniciproducens*	2. *A. thomasii*
Acid production from:		
Adonitol	−	+
Fructose	+	−
Inulin	+	−
Lactose	+	−
Raffinose	+	−
Sucrose	+	−
Enzyme production:[c]		
β-D-Galactosidase	+	−
α-Glucosidase	+	−
α-Maltosidase	d	−
L-Lysine arylamidase	−	+
Glycine arylamidase	d	+
Methionine arylamidase	−	+

[a]Symbols: see standard definitions.

[b]Data from Davis et al. (1976), Malnick (1997), and Malnick (unpublished results).

[c]API-ZYM enzyme test kits (API Biomerieux, Basingstoke, Hants, UK).

Genus III. **Ruminobacter** *Stackebrandt and Hippe 1987, 179[VP] (Effective publication: Stackebrandt and Hippe 1986, 205)*

ERKO S. STACKEBRANDT

Ru.mi.no.bac' ter. L. adj. *ruminalis* of the rumen; M.L. masc. n. *bacter* equivalent of bacterium, a small rod; M.L. masc. n. *Ruminobacter* small rod of the rumen.

Oval to long rods. Gram negative. Nonsporeforming. Nonmotile. Chemoorganotroph. **Metabolize carbohydrates; fermentation products include succinate, acetate, and formate; trace amounts of lactate and ethanol may be formed. No rumen fluid requirement for growth. Obligately anaerobic.** CO_2 required, which is incorporated into succinic acid. NH_3 is essential as a nitrogen source. No cytochromes. Peptidoglycan contains *meso*-diaminopimelic acid. Sphingophospholipids absent. Straight chain saturated $C_{16:0}$ fatty acid predominates, while $C_{14:0}$ acid and the monounsaturated $C_{16:1}$ and $C_{18:1}$ acids occur in smaller amount. Low levels of iso- and anteiso-methyl branched fatty acids. Analysis of 16S rDNA places the genus into the family *Succinivibrionaceae* of the *Gammaproteobacteria*.

The mol% G + C of the DNA is: 40–42.

Type species: **Ruminobacter amylophilus** (Hamlin and Hungate 1956) Stackebrandt and Hippe 1987, 179 (Effective publication: Stackebrandt and Hippe 1986, 205) (*Bacteroides amylophilus* Hamlin and Hungate 1956, 552.)

FURTHER DESCRIPTIVE INFORMATION

As the genus currently contains a single species, the description of the genus is that of the species. The presence of an A–U base pair at positions 19 and 916 indicates that *R. amylophilus* belongs to a broad phylogenetic cluster of organisms within the *Gammaproteobacteria*, all of which are defined by the presence of this signature, such as members of the *Succinivibrionaceae, Aeromonadaceae, Enterobacteriaceae, Vibrionaceae, Pasteurellaceae, Methylococcaceae, Halomonadaceae,* and *Oceanospirillaceae*.

The colonial and cultural properties are based on the descriptions of Hamlin and Hungate (1956), Holdeman and Moore (1974), Cato et al. (1978), and Holdeman et al. (1984b). Cells of the type strain of *R. amylophilus* are pleomorphic after incubation for 2 d in peptone–yeast extract-maltose broth. Gramnegative, oval to long rods with tapered or round ends, and include some swollen forms and irregularly curved cells, 0.9–1.2 × 1.1–8.0 µm (Cato et al., 1978). Surface colonies on rumen fluid-glucose-cellobiose agar roll tubes are 1 mm in diameter, circular, entire, slightly convex, translucent, smooth, glistening, white to tan. Colonies on blood roll streak tubes are pinpoint to 2 mm in diameter, circular, entire to slightly erose, convex to slightly erose, convex to umbonate, translucent, colorless, shiny, and smooth after incubation for 4 d. Growth is sparse on freshly prepared blood agar plates incubated anaerobically. Surface colonies on rumen fluid–glucose–cellobiose agar roll tubes are 1 mm in diameter, circular, entire, slightly convex, translucent, smooth, glistening, white to tan. Colonies in deep agar are 0.8–1 mm in diameter, lenticular, entire or irregular, white, and soft butyrous. Growth is not stimulated by amino acids, volatile fatty acids (Miura et al., 1980), vitamins, or hemin (Macy and Probst, 1979). No colonies appear on egg yolk agar plates and on the surface of plates incubated in an aerobic jar.

The organism is strictly anaerobic. Only bicarbonate and/or CO_2, a fermentable carbohydrate, and minerals are required for growth. Rumen fluid is not required (Bryant and Robinson, 1962). Ammonia is essential as a nitrogen source. CO_2 is fixed and ammonia is assimilated. In the presence of CO_2 maltose is fermented to succinate: 2 maltose + $2CO_2$ → 2 succinate + 2 acetate + 2 formate (Kühn, 1979). Production of succinic acid has been patented (U.S. Pat. 5,143,833). The organism is able to grow well in basal medium supplemented with starch. Starch broth cultures are turbid and have a final pH of 5.3–5.5. Amylase production is stimulated by Tween 80 (McWethy and Hartman, 1977). Autolysis occurs after growth ceases and the degradation of bacterial protein is accompanied by the production of branched-chain amino acids (Miura et al., 1980). The organism is strongly proteolytic (Abou-Akkada and Blackburn, 1963), and protease is produced in the exponential growth phase (Blackburn and Hallah, 1974).

Optimum temperature for growth is near 37°C. Hydrogen is not produced; no growth in 20% bile; esculin not digested; milk and meat reaction negative. Cellulose is not digested. Casein is not utilized but is digested when cells are grown in ammonia-containing media (Caldwell et al., 1973). Protease is produced in the exponential growth phase (Blackburn, 1968; Blackburn and Hallah, 1974). Acid is produced from dextrin, glycogen, maltose, and starch. No acid is produced from amygdalin, arabinose, cellobiose, esculin, fructose, D-galactose, D-glucose, glycerol, inositol, inulin, lactose, D-mannitol, raffinose, L-rhamnose, ribose, salicin, sucrose, trehalose, and D-xylose. Strains are very sensitive to inhibition by heavy metals and trace elements (Forsberg, 1978).

An almost complete *rrn* operon, consisting of the 5′ upstream promoter region, the 16S rDNA gene of 1539 nucleotides (accession number Y15992), the intergenic spacer, including a tRNA (*gln*) gene (Martens et al., 1987), and the 23S rDNA (Spiegl et al., 1988) (accession number X06765) has been sequenced from a BgII-Kpn1 fragment of strain DSM 1361[T] (plasmid number DSM 4825). A partial 16S rDNA sequence, consisting of 1429 nucleotides, of a *R. amylophilus* strain with no assigned strain number has recently been released under the accession number AB004908. The similarity of the common stretch of the two 16S rDNA sequences is 99.7%.

Cytochromes (Reddy and Bryant, 1977) and sphingolipids have not been detected (Kunsman and Caldwell, 1974; Miyagawa et al., 1978). Menaquinones (R. Kroppenstedt, personal communication) are absent. Peptidoglycan contains *meso*-diaminopimelic acid (Miyagawa et al., 1981).

Ruminobacter amylophilus is one of the predominant anaerobic bacteria obtained in pure culture from ruminal contents of a Holstein cow. The bacterium thrives on basal medium supplemented with maltose and ammonia. A casein hydrolysate plus volatile fatty acids is only slightly stimulatory and glucose and cellobiose cannot substitute for maltose (Bryant and Robinson, 1962). *R. amylophilus* occurs sporadically in the rumen contents of cattle but, when present, may be the predominant starch digester and may constitute as much as 10% of the bacterial population of the rumen. It also occurs also in the ovine rumen (Blackburn and Hobson, 1962; Bryant and Robinson, 1962).

ENRICHMENT AND ISOLATION PROCEDURES

None of the media described in the literature on the isolation of *Ruminobacter (Bacteroides) amylophilus* was designed to specifically enrich this organism. The media used are those to survey proteolytic bacteria present in the rumen of sheep (Blackburn and Hobson, 1962) and cows (Hamlin and Hungate, 1956; Bryant and Robinson, 1962). In some of these studies, the animals were fed different diets. As *R. amylophilus* does not require rumen fluid, a nonrumen fluid medium is described that has been used to isolate *R. amylophilus* (Bryant and Robinson, 1962). The basal medium contains 0.3% maltose, 0.0001% resazurin, 0.4% Na_2CO_3, 0.025% (each) cysteine-$HCl \cdot H_2O$ and $Na_2S \cdot 9H_2O$, 0.09% (each) KH_2PO_4 and NaCl, 0.002% (each) $CaCl_2$ and $MgCl_2 \cdot 6H_2O$, 0.001% $MnCl_2 \cdot H_2O$, 0.001% $CoCl_2 \cdot 6H_2O$, 0.2 mg per 100 ml (each) of thiamine-HCl, Ca-D-pantothenate, nicotinamide, riboflavine, and pyridoxal, 0.01 mg per 100 ml of *p*-aminobenzoic acid, 0.05 mg per 100 ml (each) of biotin, folic acid, and DL-thioctic acid, 0.02 mg per 100 ml of cobalamin, and is equilibrated with oxygen-free CO_2. All ingredients except cysteine, Na_2S, and Na_2CO_3 were mixed, adjusted to pH 6.5 with H_2SO_4, and autoclaved at 15 psi for 15 min. Sterile, CO_2-equilibrated Na_2CO_3 solution is added and the medium is added to tubes (13 × 100 mm) in 2.8 ml amounts, using anaerobic methods and CO_2 (Bryant and Robinson, 1961). Two milliliters each of 0.09% $(NH_4)_2SO_4$, and 2 mg/100 ml hemin plus 0.2 ml of a solution of cysteine, and Na_2S brings the medium constituents to final concentration.

According to Blackburn and Hobson (1962), samples are taken via a rumen cannula and rapidly strained to remove coarse debris. Tenfold dilutions of rumen fluid are made under CO_2 in a solution prepared exactly like the medium. Bacteria are cultured in 50-ml flasks each containing 25 ml of medium. For dilutions, the role tube method can be used as described by Bryant and Burkey (1953).

MAINTENANCE PROCEDURES

R. amylophilus is routinely maintained in the DSM-German Collection of Microorganisms and Cell Cultures in medium DSM 147 (DSM catalog of strains, 2001), containing (per 1000 ml distilled water): KH_2PO_4, 0.45 g; K_2HPO_4, 0.45 g; $(NH_4)_2SO_4$, 0.9g; NaCl, 0.9g; $MgSO_4 \cdot 7H_2O$, 0.18g; soluble starch, 5.0 g; casitone (Difco), 10.0 g; resazurin, 1.0 mg; cysteine hydrochloride, 0.5 g; $NaHCO_3$, 6.0 g; pH 7.0, at a gas atmosphere of 100 CO_2. For short-term storage, stab cultures in a modified medium DSM 147 is recommended, in which the starch content is reduced to 0.1%. For long-term conservation, freezing in liquid nitrogen is recommended.

DIFFERENTIATION OF THE GENUS *RUMINOBACTER* FROM OTHER GENERA

Recent phylogenetic analyses on some Gram-negative strictly anaerobic, nonsporeforming bacteria indicated *Ruminobacter amylophilus* to be a member of a separate line of descent that includes *Succinimonas amylolytica*, *Succinivibrio dextrinosolvens*, and members of the genus *Anaerobiospirillum*. This phylogenetic group has been described as *Succinivibrionaceae* (Hippe et al., 1999) (Fig. BXII.γ.188).

TAXONOMIC COMMENTS

The first culture of *Bacteroides amylophilus*, isolated and described by Hamlin and Hungate (1956), was lost. Since then, organisms conforming to the original description of *B. amylophilus* have been isolated frequently from the rumen (Blackburn and Hobson, 1962; Bryant and Robinson, 1962; Caldwell et al., 1969; Holdeman et al., 1977). Cato et al. (1978) designated strain H 18 isolated by Blackburn and Hobson (1962), as the neotype strain of *B. amylophilus*. This species differs from the majority of *Bacteroides* species in a number of chemotaxonomic properties, e.g., fatty acid composition (Miyagawa et al., 1979) and the lack of sphingophospholipids (Kunsman and Caldwell, 1974; Miyagawa et al., 1978). These differences led Miyagawa et al. (1979) and Shah and Collins (1983) to the conclusion that *B. amylophilus* should be removed from *Bacteroides sensu stricto*. These conclusions were supported by the finding that *B. amylophilus* and authentic members of *Bacteroides* belong to different main phylogenetic lines of descent. The results of 16S ribosomal RNA cataloging (Woese et al., 1985) and later by 16S rDNA analysis (Martens et al., 1987) revealed that *B. amylophilus* groups loosely with the "core" organisms (e.g., enterobacteria, vibrios, oceanospirilla, alteromonads) of the *Gammaproteobacteria*, while authentic *Bacteroides* species (e.g., *B. fragilis*, *B. asaccharolyticus*, *B. distasonis*, *B. melaninogenicus*, *B. ovatus*, *B. ruminicola*, *B. thetaiotaomicron*, *B. uniformis*, and *B. vulgatus*) form a coherent cluster within the *Bacteroides-Cytophaga-Flavobacterium* line of descent (Paster et al., 1985). Consequently, *Bacteroides amylophilus* was reclassified as *Ruminobacter amylophilus* (Stackebrandt and Hippe, 1986). The generic name "*Ruminobacter*" was originally used to describe Gram-negative, nonmotile, anaerobic chemoheterotrophic cellulose fermenting bacteria (Kaars Sijpesteijn, 1949). Prévot (1966) considered this genus as *incertae sedis*, but at the same time transferred the *Bacteroides* species "*Bacteroides amylogenes*", *B. ruminicola*, *B. succinogenes*, and the noncellulolytic species *B. amylophilus* into this genus. The taxon *Ruminobacter* was not included in the Approved Lists of Bacterial Names (Skerman et al., 1980); hence, it had no standing in nomenclature and could be proposed to accommodate the former *B. amylophilus*.

Analysis of 16S rDNA sequence is to date the most reliable method by which isolates can be affiliated to *Ruminobacter amylophilus*. Without sequence information and placement, this species can be misclassified as *Prevotella ruminicola* (Shah and Collins, 1990) or *Fibrobacter succinogenes* (Montgomery et al., 1988), organisms that were previously classified as nonmotile members of *Bacteroides*. *Ruminobacter amylophilus* differs considerably from these organisms in a combination of properties, e.g., the pattern of carbohydrates utilized, fermentation products, mol% G + C of their DNA, and fatty acid composition. *Ruminobacter* differs phenotypically from other genera previously grouped in the family *Bacteroidaceae* as defined by Holdeman et al. (1984a). *Ruminobacter* can be distinguished from *Fusobacterium* and *Leptotrichia* by the end-products of carbohydrate fermentation (Holdeman et al., 1984a). On the basis of metabolic properties *Ruminobacter* can also be distinguished unambiguously from several other genera containing Gram-negative obligate-anaerobes that have been described recently, viz. *Acidaminobacter* (Stams and Hansen, 1984), *Ilyobacter* (Stieb and Schink, 1985), *Pelobacter* (Schink and Pfennig, 1982a), *Propionigenium* (Schink and Pfennig, 1982b), and *Propionispira* (Schink et al., 1982), all of which are phylogenetically related to members of the *Clostridium-Bacillus* line of descent.

FURTHER READING

Blackburn, T.H. and P.N. Hobson. 1962. Further studies on the isolation of proteolytic bacteria from sheep rumen. J. Gen. Microbiol. *29*: 69–81.

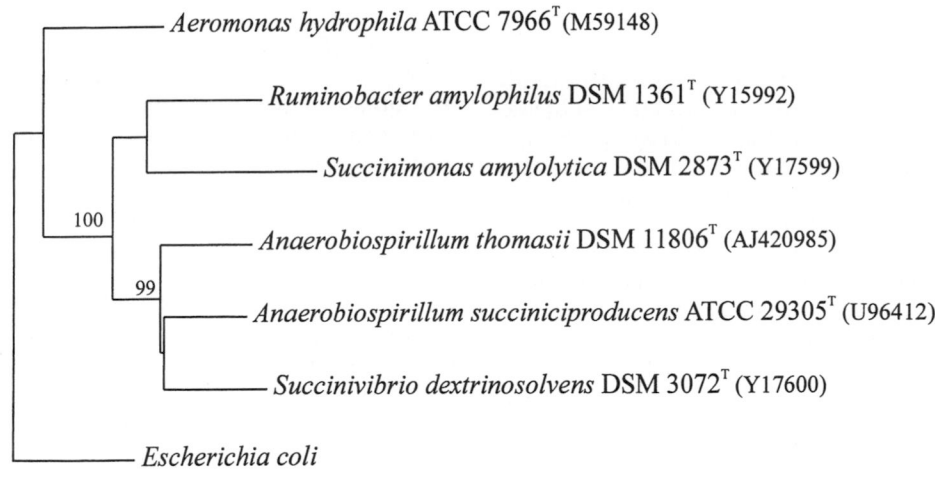

1%

FIGURE BXII.γ.188. Dendrogram showing the phylogenetic position of *Ruminobacter amylophilus* next to its phylogenetic neighbors within the *Gammaproteobacteria*. *R. amylophilus* represents the family *Succinivibrionaceae*, which includes the genera *Ruminococcus, Anaerobiospirillum, Succinivibrio,* and *Succinimonas* (Hippe et al., 1999). The tree was constructed by the neighbor-joining method (Saitou and Nei, 1987), using corrected distance values (Jukes and Cantor, 1969). Bootstrap values (expressed as percentages of 500 replications) of 70% or more are indicated at the branch points. Bar = 1% sequence divergence.

Cato, E.P., W.E.C. Moore and M.P. Bryant. 1978. Designation of neotype strains for *Bacteroides amylophilus* Hamlina and Hungate 1956 and *Bacteroides succinogenes* Hungate 1950. Int. J.Syst. Bacteriol. *28:* 491–495.

Miura, H., M. Horiguchi and T. Matsumoto. 1980. Nutritional interdependence among rumen bacteria, *Bacteroides amylophilus, Megasphaera*

elsdenii and *Ruminococcus albus.* Appl. Environ. Microbiol. *40:* 294–300.

Stackebrandt, E. and H. Hippe. 1986. Transfer of *Bacteroides amylophilus* to a new genus *Ruminobacter* gen. nov., nom. rev. as *Ruminobacter amylophilus* comb. nov. Syst. Appl. Microbiol. *8:* 204–207.

List of species of the genus Ruminobacter

1. **Ruminobacter amylophilus** (Hamlin and Hungate 1956) Stackebrandt and Hippe 1987, 179[VP] (Effective publication: Stackebrandt and Hippe 1986, 205) (*Bacteroides amylophilus* Hamlin and Hungate 1956, 552.)

am.y.lo'phi.lus. Gr. n. *amylo* starch; Gr. part. *philo* loving; M.L. adj. *amylophilus* starch loving.

The characteristics are described for the genus.

Occurs sporadically in the rumen of the bovine and sheep.

The mol% G + C of the DNA is: 40-42 (Bd).

Type strain: H 18, ATCC 29744, DSM 1361, VIP 2502B.

GenBank accession number (16S rRNA): Y15992.

Additional Remarks: GenBank accession number for 23S rRNA gene is X06765.

Genus IV. **Succinimonas** *Bryant, Small, Bouma and Chu 1958, 21*[AL]*

Marvin P. Bryant

Suc.ci.ni.mo'nas. M.L. n. *acidum succinicum* succinic acid; Gr. n. *monas* a unit, monad; M.L. fem. n. *Succinimonas* succinic acid monad.

Short, straight rods to coccobacilli, 1.0–1.5 μm × 1.0–3.0 μm, with rounded ends. No spores or resting stages are produced. Gram negative. **Motile by a single polar flagellum.**

Strictly anaerobic. Chemoorganotrophic, having a fermentative type of metabolism. **Glucose, maltose, dextrin, or starch can serve as the energy source. Succinate and acetate are the main products of glucose fermentation.** No butyrate or gas is formed. Catalase negative. Occur in the bovine rumen.

The mol% G + C of the DNA is: unknown.

Type species: **Succinimonas amylolytica** Bryant, Small, Bouma and Chu 1958, 21.

FURTHER DESCRIPTIVE INFORMATION

In the water of syneresis of RGCA slants[1] incubated overnight at 37°C, the cells are arranged as singles, pairs, and clumps. No capsules are evident. Motility is relatively slow and progressive and is rapidly lost upon exposure to air.

Surface colonies on RGCA agar in roll tubes are smooth, convex, translucent, light tan, and 0.7–1.5 mm in diameter after 3 d incubation at 37°C. Deep colonies are lenticular and 0.7–1.0

---

*Editorial Note: The chapter on *Succinimonas* is reprinted from the 1st edition of *Bergey's Manual of Systematic Bacteriology.*

1. RGCA slants (g/l): K$_2$HPO$_4$, 0.23 g; KH$_2$PO$_4$, 0.45; (NH$_4$)$_2$SO$_4$, 0.45; NaCl, 0.45; MgSO$_4$, 0.023; CaCl$_2$, 0.023; resazurin, 0.001; glucose, 0.5; cellobiose, 0.5; soluble starch, 0.5; rumen fluid (centrifuged), 400; cysteine·HCl, 0.25; Na$_2$S·9H$_2$O, 0.25; Na$_2$CO$_3$, 4.0; agar (Difco), 10.0; pH 6.7; gas phase, 100% CO$_2$. Prepared and used with the Hungate technique (Bryant, 1972).

mm in diameter. In liquid media with glucose as the energy source growth occurs as light, uniform turbidity.

The type strain of *S. amylolytica* grows well in a glucose medium with rumen fluid replaced by trypticase and yeast extract. It also grows in a chemically defined medium containing glucose, CO_2/HCO_3 buffer, minerals, B vitamins, and acetate. Ammonium ions serve as the nitrogen source and cannot be replaced by amino acids or peptides. Sulfide serves as both the reducing agent and the sulfur source. Acetate (30 mM) is a highly stimulatory supplement even though the organisms produce acetate from glucose (Bryant and Robinson, 1962; Roberton and Bryant, unpublished data). During fermentation of glucose to succinate and acetate, a net uptake of CO_2 occurs. Of the substrates tested, only glucose, maltose, dextrin, and starch are fermented. Arabinose, xylose, fructose, cellobiose, lactose, sucrose, cellulose, inulin, xylan, glycerol, esculin, mannitol, lactate, amino acids, and peptides are not fermented. H_2S and indole are not produced. Gelatin is not liquefied. The Voges–Proskauer test is variable. Nitrate is not reduced.

Growth occurs at 30° and 37°C but not at 22° or 45°C. The final pH in poorly buffered, liquid glucose medium is 5.2–5.8. Good growth occurs at pH 6.5–7.0. The upper pH limit has not been determined.

S. amylolytica appears to be nonpathogenic for humans or animals. It occurs in the rumen of cattle fed diets containing roughage and some grain, where it is involved in fermentation of starch and its hydrolytic products. It is usually present as only a small proportion of the total viable bacteria in the bovine rumen (less than 6% of total). Whether *S. amylolytica* occurs in the rumen of ruminants other than cattle or in nonruminal ecosystems is not known.

ENRICHMENT AND ISOLATION PROCEDURES

Succinimonas is isolated nonselectively in anaerobic roll tubes of RGCA medium from the rumen of cattle fed hay-grain diets. It constitutes only a small proportion of the colonies that develop after 3 d or longer of incubation at 37°C.

MAINTENANCE PROCEDURES

S. amylolytica strains can be maintained on stab-inoculated RGCA slants at −70°C for a year or more. They can also be preserved indefinitely by lyophilization.

FURTHER READING

Bryant, M.P., N. Small, C. Bouma and H. Chu. 1958. *Bacteroides ruminicola*, sp. nov. and *Succinimonas amylolytica*, gen. nov., species of succinic acid-producing anaerobic bacteria of the bovine rume. J. Bacteriol. *76*: 15–23.

Bryant, M.P. and I.M. Robinson. 1962. Some nutritional characteristics of predominant culturable ruminal bacteria. J. Bacteriol. *84*: 605–614.

List of species of the genus Succinimonas

1. **Succinimonas amylolytica** Bryant, Small, Bouma and Chu 1958, 21.[AL]

   am.y.lo.1y' ti.ca. Gr. n. *amylum* fine meal, starch; Gr. adj. *lyticus* loosening, dissolving; M.L. fem. adj. *amylolytica* starch dissolving.

The characteristics are as described for the genus. Occur in the bovine rumen.

The mol% G + C of the DNA is: unknown.

Type strain: ATCC 19206, DSM 2873, VPI 13846.

GenBank accession number (16S rRNA): Y17599.

Order XIII. "Enterobacteriales"

En.te.ro.bac.te.ri.a' les. M.L. n. *enterobacterium* an intestinal bacterium; *-ales* ending to denote order; M.L. fem. n. *Enterobacteriales* the *enterobacterium* order.

Description is the same as for the family *Enterobacteriaceae*.

Type genus: **Escherichia** Castellani and Chalmers 1919, 941.

Family I. **Enterobacteriaceae** Rahn 1937, Nom. Fam. Cons. Opin. 15, Jud. Comm. 1958a, 73; Ewing, Farmer, and Brenner 1980, 674; Judicial Commission 1981, 104

DON J. BRENNER AND J.J. FARMER III

En.te.ro.bac.te.ri.a' ce.ae. M.L. n. *enterobacterium* an intestinal bacterium; *-aceae* ending to denote a family; M.L. fem. pl. n. *Enterobacteriaceae* the family of the enterobacteria. Rahn's original derivation of the name *Enterobacteriaceae* is not certain. It may have come from his genus *Enterobacter*, or may have come from the root enterobacterium.

Gram-negative straight rods, 0.3–1.0 × 1.0–6.0 μm, except for *Arsenophonus*, which is 7–10 μm in length. Motile by peritrichous flagella, except for *Tatumella*, or nonmotile. Do not form endospores or microcysts; not acid-fast. Grow in the presence and absence of oxygen. Grow well on peptone, meat extract, and usually MacConkey's medium, except for *Calymmatobacterium* and insect symbionts, which have not been cultivated. Most grow well at 22–35°C; optimal growth and maximal biochemical capacity of a number of genera (*Yersinia*, *Hafnia*, *Xenorhabdus*, *Photorhabdus*, and many erwiniae) occurs at 25–28°C. Some grow on D-glucose as the sole source of carbon; some require vitamins and/or amino acids. **Chemoorganotrophic; having both a respira-**

tory and a fermentative metabolism. Acid and visible gas are often produced during fermentation of D-glucose, other carbohydrates and polyhydroxyl alcohols. Not halophilic. **Most are catalase-positive,** except for *Shigella dysenteriae* O group 1 and *Xenorhabdus.* **Most are oxidase negative,** except for *Plesiomonas.* **Most reduce nitrate to nitrite,** except *Saccharobacter fermentatus* (Yaping et al., 1990) and some strains of *Erwinia* and *Yersinia.*

The mol% G + C of the DNA is: 38–60 (63.5 for *S. fermentatus* (Yaping et al., 1990)).

Type genus: **Escherichia** Castellani and Chalmers 1919, 941. Designated type genus Nom. Fam. Cons. Opin. 15, Jud. Comm. 1958a, 73.

FURTHER DESCRIPTIVE INFORMATION

The definition of the family circumscribes a large biochemically and genetically related group that shows substantial heterogeneity in its ecology, host range, and pathogenic potential for man, animals, insects and plants. The phylogenetic position of the family *Enterobacteriaceae* lies within the *Gammaproteobacteria.* Its nearest neighbors are the families *Alteromonadaceae*, *Vibrionaceae*, *Aeromonadaceae*, and *Pasteurellaceae.* Phylogenetic relationships of genera within the *Enterobacteriaceae* based on analysis of the 16S rRNA sequences of the type strains of the type species are shown in Fig. BXII.γ.189. Phenotypic characteristics useful in differentiating *Enterobacteriaceae* from its nearest phenotypic neighbors are shown in Table BXII.γ.191. The genera *Vibrio*, *Photobacterium*, and *Aeromonas* are oxidase positive, have polar flagella when grown in liquid media, and do not contain the enterobacterial common antigen—characteristics which distinguish them from *Enterobacteriaceae.* However, at least two *Vibrio* species (*V. metschnikovii* and *V. gazogenes*) are oxidase negative; strains of other species are oxidase negative or weakly positive; and, under certain conditions (often on solid media), members of these genera produce peritrichous flagella. Some *Aeromonas* species show higher DNA relatedness to *E. coli*, the type species of the type genus of *Enterobacteriaceae*, than that seen between several genera within the family. 16S rRNA gene sequence comparisons do not reveal any overlap between these families.

Biochemical variability and fastidiousness of a growing number of new species added to *Enterobacteriaceae*, the proposed inclusion in the family of the genus *Plesiomonas*, and the inclusion of *Calymmatobacterium* and a number of genera that are endosymbionts of insects has made a literal description of this family difficult. *Calymmatobacterium* is very fastidious, and cultivation is rarely successful—no strains are available in culture collections. The insect symbionts in the genera *Arsenophonus* (Gherna et al., 1991; Hypsa and Dale, 1997), *Buchnera* (Munson et al., 1991b; Clark et al., 1992), and *Wigglesworthia* (Aksoy, 1995b) have not been cultivated on bacteriological media, and therefore no biochemical profiles are available. These species can only be identified by their 16S rRNA gene sequences and perhaps by their host specificity. The description of the family therefore includes many traits for which there are exceptions. For example, *Plesiomonas*, a biochemically well-defined genus with a single species, is oxidase positive genus in an extremely large, oxidase negative family. Similarly, all cultivated species except *Erwinia chrysanthemi* contain enterobacterial common antigen, a trait that is specific for the family *Enterobacteriaceae*; two members of the family *Pasteurellaceae*, *Actinobacillus equuli* and *Actinobacillus suis*, also possess this antigen (Le Minor et al., 1972; Ramia et al., 1982; Böttger et al., 1987).

The relationships of many members of *Enterobacteriaceae* have

been defined based on DNA relatedness (Wayne et al., 1987; Fox et al., 1992; Stackebrandt and Goebel, 1994). DNAs from species within most genera are at least 20% related to one another and to *Escherichia coli*, the type species of the type genus of the family. Notable exceptions are species of *Plesiomonas*, *Yersinia*, *Proteus*, *Providencia*, *Hafnia*, *Edwardsiella*, *Xenorhabdus*, and *Photorhabdus*, whose DNAs are usually 5–20% related to those of species from other genera. Most species have been analyzed by their 16S rDNA sequence. 16S rDNA sequence analysis is the method of choice for determining genera and higher taxa (Stackebrandt and Goebel, 1994); however, DNA relatedness can also be used to approximate evolutionary divergence within genera (Fig. BXII.γ.190).

The numbers of genera and species in the family have markedly increased during the past 25 years. There were 12 genera and 36 species in 1974 when the 8th edition of *Bergey's Manual of Determinative Bacteriology* was published (Buchanan and Gibbons, 1974). By 1984, when the 1st edition of *Bergey's Manual of Systematic Bacteriology* was published, the family contained 20 genera and 76 species (Krieg and Holt, 1984). The total rose to 30 genera and 107 species by the 9th edition of *Bergey's Manual of Determinative Bacteriology* in 1994 (Holt et al., 1994). In the classification used in this chapter, the family contains 44 genera and 176 named species. Unnamed published and unpublished genomospecies and presently undescribed groups undoubtedly include a number of new genera and species. A prediction of 225 or more species contained in 60 or more genera by the next edition of this *Manual* may be conservative.

Compared with the 1st edition of *Bergey's Manual of Systematic Bacteriology* and the 9th edition of *Bergey's Manual of Determinative Bacteriology*, the present volume contains proposed changes in classification, new genera, and new species. Taxonomic proposals since the last edition of the *Manual* are shown in Table BXII.γ.192 and discussed below and in the chapters describing individual genera. If there is any certainty with respect to the *Enterobacteriaceae*, it is that the family will continue to change and to pose a challenge to microbiologists in all specialties.

Plant, animal, and human disease *Enterobacteriaceae* are distributed worldwide. They are found in soil, water, fruits, meats, eggs, vegetables, grains, flowering plants and trees, and in animals from insects to man. Their pathogenicity for man and animals and economic importance, as well as their rapid generation time, ability to grow on defined media, and ease of genetic manipulation have made them the objects of intense laboratory study.

Many species are of considerable economic importance. Erwiniae and pectobacteria cause blight, wilt and soft-rot disease in corn, potatoes, apples, sugar cane, pineapples, and many other crops, often destroying substantial amounts of these crops (Starr and Chatterjee, 1972). For example, it has been estimated that plant pathogenic species causing soft rot diseases have caused at least $50 million of damage annually (Pérombelon and Kelman, 1980). The commercial and tropical fish industries are severely affected by the diseases caused by *Yersinia ruckeri* and species of *Edwardsiella* (Ewing et al., 1978; Shotts and Snieskzo, 1976).

Salmonellosis in poultry and in eggs is a worldwide problem, both for poultry farmers and as a vehicle for human disease (Williams, 1965; Von Rockel, 1965; Hall, 1965; Mishu et al., 1994). Salmonellosis is also common in pigs, cows, horses, dogs, and cats (Barnes and Sorensen, 1975; Ewing, 1969). Stillbirths and wool damage in sheep are usually caused by salmonellae (Jensen,

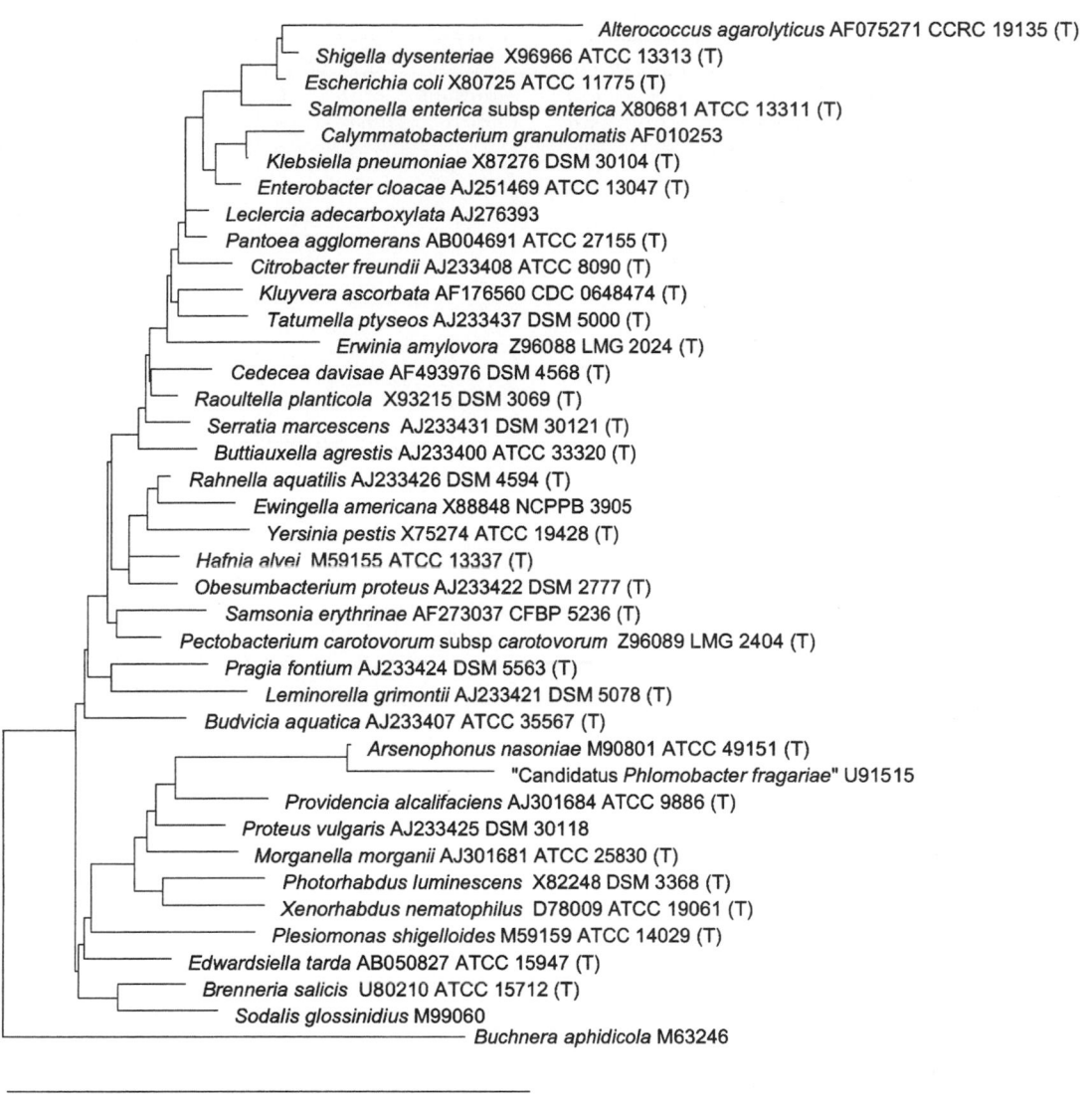

0.05

FIGURE BXII.γ.189. Phylogenetic relationships of the genera of the family *Enterobacteriaceae*. The distances in the tree were calculated using 1101 positions (the least-squares method, Jukes-Cantor model). (Courtesy T. Lilburn of the Ribosomal Database Project.)

1974). Enterotoxigenic *E. coli* strains are primarily responsible for diarrhea in lambs, piglets, and calves (Bruner and Gillespie, 1973). Klebsiellae and *Citrobacter freundii* are causes of bovine mastitis. Species of *Enterobacteriaceae* are responsible for numerous other animal infections. Examples include sexually transmitted uterine infections in horses caused by *Klebsiella pneumoniae*; infections in snakes, turtles and lizards caused by salmonellae; diarrheal and septicemic infection in rabbits, mink, and other rodents caused by yersiniae; and shigellosis in monkeys.

Salmonella serotype Typhi (*Salmonella typhi*) is the cause of typhoid fever, and *Yersinia pestis* causes bubonic and pneumonic plague. Many other *Enterobacteriaceae* are pathogenic for humans, causing a wide variety of diseases. These include diarrheal disease, usually transmitted by contaminated food or water; septicemia; respiratory disease; wound and burn infections; urinary tract infections; and meningitis. The causative agents of these diseases can be loosely divided into species that are normally pathogenic (*Salmonella*, *Shigella*, *K. pneumoniae*, *Y. pestis*, various serotypes of *E. coli*, and *Y. enterocolitica*) and species that cause

disease under certain circumstances. Members of this second group are often referred to as opportunistic pathogens. The compromised host (for example, the malnourished, diabetic, immunosuppressed, catheterized, burn, cancer, respiratory, AIDS, or elderly patient) is vulnerable to nosocomial infections caused by opportunistic pathogens. *Enterobacteriaceae* have long been responsible for a substantial percentage of nosocomial infections in the United States (Jarvis et al., 1984), including urinary tract infections, surgical wound infections, lower respiratory infections, and bacteremias.

Salmonellae, *E. coli*, and shigellae are frequent causes of foodborne disease outbreaks (Beane et al., 1990), and *Yersinia enterocolitica* also causes foodborne disease. More recently, *E. coli* O157:H7, *Salmonella* serotype Typhimurium phage type 104 (definitive type 104, resistant to at least 5 antimicrobials), and *Salmonella* serotype Enteritidis phage type 4, have emerged as major foodborne pathogens (Boyce et al., 1995; Altekruse et al., 1997). *Salmonella* serotypes Typhimurium and Enteritidis are the two most prevalent causes of human salmonellosis in the United

TABLE BXII.γ.191. Some differential characteristics of the family *Enterobacteriaceae* and their nearest relatives[a]

Characteristic	Enterobacteriaceae	Aeromonadaceae	Pasteurellaceae	Vibrionaceae
Cell diameter, microns	0.3–1.5	0.3–1.0	0.2–0.4	0.3–1.3
Straight rods	+	D	+	D
Curved rods	−	D	−	D
Motility	D	+[b]	−	+[c]
Flagellar arrangement (liquid medium):				
Polar	−[d]	+		+
Lateral	+[d]	−		−
Oxidase test	−[e]	+[f]	+	+[f]
Sodium required or stimulatory for growth	−	−	−	D
Contain enterobacterial common antigen	+[g]	−	−[g]	−
Cells contain menaquinones[h]	D	D	−	D
Parasitic on mammals and birds	D	−[c]	+	−[c]
Heme and/or nicotinamide adenine dinucleotide required for growth	−	−	D	−
Plant pathogenicity	D	−	−	−
Organic nitrogen sources required	−[c]	−	+	−[c]

[a]Symbols: see standard definitions.

[b]Except *Ruminobacter*, *Tolumonas*, and certain biogroups of *Aeromonas salmonicida*.

[c]A few exceptions may occur.

[d]Except *Plesiomonas*, which has lateral polar flagella, and *Tatumella*, which may have polar, subpolar or lateral flagella.

[e]Except *Plesiomonas*.

[f]Except *Vibrio metschnikovii* and *Vibrio gazogenes* in *Vibrionaceae* and the genus *Tolumonas* in *Aeromonadaceae*.

[g]*Erwinia chrysanthemi* does not contain the antigen; *Actinobacillus equuli* and *Actinobacillus suis* contain the antigen.

[h]*Pasteurellaceae* do contain demethylmenaquinones but not menaquinones; ubiquinones may or may not be produced. *Enterobacteriaceae*, *Vibrionaceae*, and *Aeromonadaceae* may contain menaquinones, demethylmenaquinones, and ubiquinones.

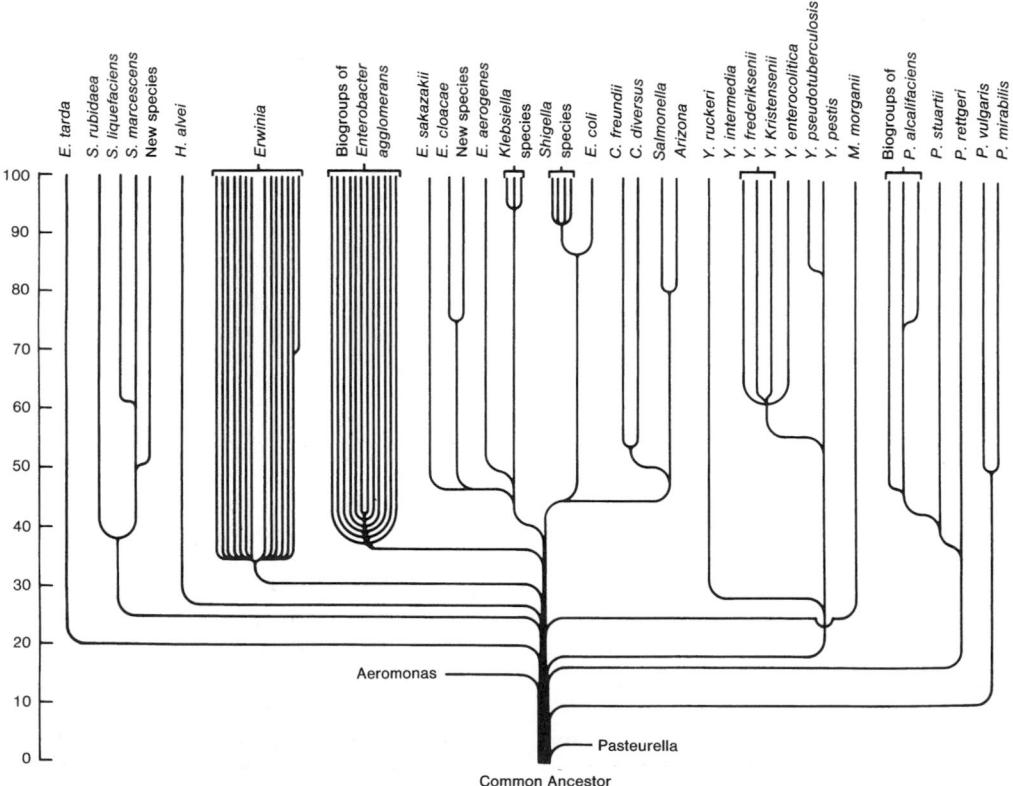

FIGURE BXII.γ.190. Divergence of *Enterobacteriaceae*. The ordinate is percentage of relatedness. This figure is a simplified attempt to depict relatedness of species of enterobacteria to other species. It assumes a common ancestor from which all of the organisms have diverged. The horizontal branches depict the degree of relatedness of the group of organisms to all organisms that have not yet branched. For example, *E. tarda* is 20% related to all organisms except *Aeromonas*, *Proteus*, *Providencia* and *Pasteurella*; *Citrobacter* species are 45% related to all species that branch above them and *C. diversus* and *C. freundii* were speciated at a point in time such that they are now 50% related.

TABLE BXII.γ.192. Comparison of the classification with that in the *Bergey's Manual of Systematic Bacteriology*, First Edition, and *Bergey's Manual of Determinative Bacteriology*, Ninth Edition[a]

Classification	Synonyms	Classification in *Bergey's Manual of Systematic Bacteriology*, First Edition 1, and *Bergey's Manual of Determinative Bacteriology*, Ninth Edition
Alterococcus agarolyticus		NL
Arsenophonus nasoniae		*A. nasoniae*
Candidatus Arsenophonus triatominarum		NL
Brenneria alni	*Erwinia alni*	NL
Brenneria nigrifluens	*Erwinia nigrifluens*	*E. nigrifluens*
Brenneria paradisiaca	*Erwinia paradisiaca*	NL
Brenneria quercina	*Erwinia quercina*	*E. quercina*
Brenneria rubrifaciens	*Erwinia rubrifaciens*	*E. rubrifaciens*
Brenneria salicis	*Erwinia salicis*	*E. salicis*
Buchnera aphidicola		NL
Budvicia aquatica		*B. aquatica*
Buttiauxella agrestis		*B. agrestis*
Buttiauxella brennerae		NL
Buttiauxella ferragutiae	Enteric group 63	Enteric group 63
Buttiauxella gaviniae	Enteric group 64	Enteric group 64
Buttiauxella izardii		NL
Buttiauxella noackiae		NL
Buttiauxella warmboldiae		NL
Calymmatobacterium granulomatis	*Klebsiella granulomatis*	*C. granulomatis*
Cedecea davisae		*C. davisae*
Cedecea lapagei		*C. lapagei*
Cedecea neteri		*C. neteri*
Cedecea species 3		Cedecea species 3
Cedecea species 5		Cedecea species 5
Citrobacter amalonaticus	*Levinea amalonatica*	*C. amalonaticus*
Citrobacter braakii		NL
Citrobacter farmeri	*C. amalonaticus* biogroup 1	*C. amalonaticus* biogroup 1
Citrobacter freundii		*C. freundii*
Citrobacter gillenii	*Citrobacter* unnamed species 10	NL
Citrobacter koseri	*L. malonatica, C. diversus*	*C. diversus*
Citrobacter murliniae	*Citrobacter* unnamed species 11	NL
Citrobacter rodentium	*Citrobacter* unnamed species 9	NL
Citrobacter sedlakii		NL
Citrobacter werkmanii		NL
Citrobacter youngae		NL
Edwardsiella hoshinae		*E. hoshinae*
Edwardsiella ictaluri		*E. ictaluri*
Edwardsiella tarda	*E. anguillimortifera*	*E. tarda*
Enterobacter amnigenus		*E. amnigenus*
Enterobacter asburiae		*E. asburiae*
Enterobacter cancerogenus	*Enterobacter taylorae, Erwinia cancerogena*	*E. taylorae*
Enterobacter cloacae		*E. cloacae*
Enterobacter dissolvens	*Erwinia dissolvens*	*Enterobacter dissolvens*
Enterobacter gergoviae		*E. gergoviae*
Enterobacter hormaechei		*E. hormaechei*
Enterobacter kobei		NL
Enterobacter nimipressuralis	*Erwinia nimipressuralis*	*Enterobacter nimipressuralis*
Enterobacter pyrinus		NL
Enterobacter sakazakii	yellow-pigmented *Enterobacter cloacae*	*E. sakazakii*
Erwinia amylovora		*E. amylovora*
Erwinia aphidicola		NL
Erwinia billingiae		NL
Erwinia mallotivora		*E. mallotivora*
Erwinia persicina	"*E. nulandii*"	*E. persicinus*
Erwinia psidii		*E. psidii*
Erwinia pyrifoliae		NL
Erwinia rhapontici	*Pectobacterium rhapontici*	*E. rhapontici*
Erwinia tracheiphila		*E. tracheiphila*
Escherichia blattae		*E. blattae*
Escherichia coli		*E. coli*
Escherichia fergusonii		*E. fergusonii*
Escherichia hermannii		*E. hermannii*
Escherichia vulneris		*E. vulneris*
Ewingella americana		*E. americana*
Hafnia alvei		*H. alvei*
Klebsiella mobilis	*Enterobacter aerogenes*	*Enterobacter aerogenes*
Klebsiella oxytoca		*K. oxytoca*
Klebsiella ornithinolytica	*Raoultella ornithinolytica*	NL
Klebsiella planticola	*Raoultella planticola*	*K. planticola*
Klebsiella pneumoniae subsp. *ozaenae*	*K. ozaenae*	*Klebsiella pneumoniae* subsp. *ozaenae*
Klebsiella pneumoniae subsp. *pneumoniae*		*Klebsiella pneumoniae* subsp. *pneumoniae*

(*continued*)

TABLE BXII.γ.192. *(cont.)*

Current classification	Synonyms	Classification in *Bergey's Manual of Systematic Bacteriology*, First Edition 1, and *Bergey's Manual of Determinative Bacteriology*, Ninth Edition
Klebsiella pneumoniae subsp. *rhinoscleromatis*	*K. rhinoscleromatis*	*Klebsiella pneumoniae* subsp. *rhinoscleromatis*
Klebsiella terrigena	*Raoultella terrigena*	*K. terrigena*
Kluyvera ascorbata		*K. ascorbata*
Kluyvera cochleae	*Enterobacter intermedius*[b]	*Enterobacter intermedium*
Kluyvera cryocrescens		*K. cryocrescens*
Kluyvera georgiana	*Kluyvera* species 3	*Kluyvera* species 3
Leclercia adecarboxylata	*Escherichia adecarboxylata, Enterobacter agglomerans* DNA group XI	*L. adecarboxylata*
Leminorella grimontii		*L. grimontii*
Leminorella richardii		*L. richardii*
Leminorella species 3		
Moellerella wisconsensis		*M. wisconsensis*
Morganella morganii subsp. *morganii*	*Proteus morganii*	*M. morganii*
Morganella morganii subsp. *sibonii*	*Proteus morganii*	*M. morganii*
Obesumbacterium proteus	*Hafnia protea*	*O. proteus*
Pantoea agglomerans	*Enterobacter agglomerans* DNA group XIII, *Erwinia herbicola, Erwinia milletiae*	*Enterobacter agglomerans*
Pantoea ananatis	*Pantoea ananas, Erwinia ananas, Enterobacter agglomerans* DNA group VI, *Erwinia uredovora*	*Erwinia ananas*
Pantoea citrea		NL
Pantoea dispersa	*E. agglomerans* DNA group III, phenon 8	*P. dispersa*
Pantoea punctata		NL
Pantoea stewartii subsp. *indologenes*		NL
Pantoea stewartii subsp. *stewartii*		*Erwinia stewartii*
Pantoea terrea		NL
Pectobacterium cacticida	*Erwinia cacticida*	*Erwinia cacticida*
Pectobacterium carotovorum subsp. *atrosepticum*	*Erwinia atroseptica, E. carotovora* subsp. *atroseptica*	*Erwinia carotovora* subsp. *atroseptica*
Pectobacterium carotovorum subsp. *betavasculorum*	*Erwinia carotovora* subsp. *betavasculorum*	*Erwinia carotovora* subsp. *betavasculorum*
Pectobacterium carotovorum subsp. *carotovorum*	*Erwinia carotovora* subsp. *carotovora*	*Erwinia carotovora* subsp. *carotovora*
Pectobacterium carotovorum subsp. *odoriferum*	*Erwinia carotovora* subsp. *odorifera*	NL
Pectobacterium carotovorum subsp. *wasabiae*	*Erwinia carotovora* subsp. *wasabiae*	*Erwinia carotovora* subsp. *wasabiae*
Pectobacterium chrysanthemi	*Erwinia chrysanthemi*	*Erwinia chrysanthemi*
Pectobacterium cypripedii	*Erwinia cypripedii*	*Erwinia cypripedii*
Candidatus Phlomobacter fragariae		NL
Photorhabdus luminescens subsp. *luminescens*	*Xenorhabdus luminescens*	*X. luminescens*
Photorhabdus luminescens subsp. *akhurstii*		NL
Photorhabdus asymbiotica		NL
Photorhabdus temperata		NL
Plesiomonas shigelloides[b]		*P. shigelloides*[b]
Pragia fontium		*P. fontium*
Proteus mirabilis		*P. mirabilis*
Proteus myxofaciens		*P. myxofaciens*
Proteus penneri		*P. penneri*
Proteus vulgaris		*P. vulgaris*
Providencia alcalifaciens		*P. alcalifaciens*
Providencia heimbachae		*P. heimbachae*
Providencia rettgeri		*P. rettgeri*
Providencia rustigianii		*P. rustigianii*
Providencia stuartii		*P. stuartii*
Rahnella aquatilis		*R. aquatilis*
Rahnella genomospecies 2		NL
Rahnella genomospecies 3		NL
Saccharobacter fermentatus		NL
Salmonella bongori	*Salmonella* subsp. *bongori, Salmonella* subsp. V	"*S. bongori*"[c], *Salmonella bongori*[d]
Salmonella enterica subsp. *arizonae*	*S. arizonae* (monophasic), *Salmonella* subsp. IIIa	*Salmonella arizonae*[e], *S. choleraesuis* subsp. *arizonae*[d]
Salmonella enterica subsp. *diarizonae*	*S. arizonae* (diphasic), *Salmonella* subsp. IIIb	*S. choleraesuis* subsp. *diarizonae*[d]
Salmonella enterica subsp. *enterica* Choleraesuis	*Salmonella* subsp. I	*Salmonella choleraesuis*[c], *Salmonella choleraesuis* subsp. *choleraesuis* Choleraesuis[d]
Salmonella enterica subsp. *enterica* Enteritidis		*Salmonella enteritidis*[c]
Salmonella enterica subsp. *enterica* Gallinarum		"*Salmonella gallinarum*"[c], *Salmonella choleraesuis* subsp. *choleraesuis* Gallinarum[d]
Salmonella enterica subsp. *enterica* Paratyphi A		"*Salmonella paratyphi-A*"[c], *Salmonella choleraesuis* subsp. *choleraesuis* Paratyphi A[d]
Salmonella enterica subsp. *enterica* Paratyphi B		NL

(continued)

TABLE BXII.γ.192. *(cont.)*

Current classification	Synonyms	Classification in *Bergey's Manual of Systematic Bacteriology*, First Edition 1, and *Bergey's Manual of Determinative Bacteriology*, Ninth Edition
Salmonella enterica subsp. *enterica* Paratyphi C		NL
Salmonella enterica subsp. *enterica* Typhi		*Salmonella typhi*[c], *Salmonella choleraesuis* subsp. *choleraesuis* Typhi[d]
Salmonella enterica subsp. *enterica* Typhimurium		*Salmonella typhimurium*[c]
Salmonella enterica subsp. *houtenae*	*Salmonella* subsp. IV	"*Salmonella houtenae*"[c] *S. choleraesuis* subsp. *houtenae*[d]
Salmonella enterica subsp. *indica*	*Salmonella* subsp. V	*S. choleraesuis* subsp. *indica*[d]
Salmonella enterica subsp. *salamae*	*Salmonella* subsp. II	"*Salmonella salamae*"[c], *S. choleraesuis* subsp. *salamae*[d]
Serratia entomophila		*S. entomophila*
Serratia ficaria		*Serratia ficaria*
Serratia fonticola		*S. fonticola*
Serratia grimesii		*S. grimesii*
Serratia liquefaciens		*S. liquefaciens*
Serratia marcescens		*S. marcescens*
Serratia odorifera		*S. odorifera*
Serratia plymuthica		*S. plymuthica*
Serratia proteamaculans subsp. *proteamaculans*		*S. proteamaculans* subsp. *proteamaculans*
Serratia proteamaculans subsp. *quinovora*		*S. proteamaculans* subsp. *quinovora*
Serratia rubidaea		*S. rubidaea*
Shigella boydii		*S. boydii*
Shigella dysenteriae		*S. dysenteriae*
Shigella flexneri		*S. flexneri*
Shigella sonnei		*S. sonnei*
Sodalis glossinidius		NL
Tatumella ptyseos		*T. ptyseos*
Trabulsiella guamensis		NL
Wigglesworthia glossinidia		NL
Xenorhabdus beddingii	*X. nematophilus* subsp. *beddingii*	*X. beddingii*
Xenorhabdus bovienii	*X. nematophilus* subsp. *bovienii*	*X. bovienii*
Xenorhabdus japonica		NL
Xenorhabdus nematophila		*X. nematophilus*
Xenorhabdus poinarii	*X. nematophilus* subsp. *poinarii*	*X. poinarii*
Yersinia aldovae		*Y. aldovae*
Yersinia bercovieri		*Y. bercovieri*
Yersinia enterocolitica		*Y. enterocolitica*
Yersinia frederiksenii		*Y. frederiksenii*
Yersinia intermedia		*Yersinia intermedia*
Yersinia kristensenii		*Y. kristensenii*
Yersinia mollaretii		*Y. mollaretii*
Yersinia pestis		*Y. pestis*
Yersinia pseudotuberculosis		*Y. pseudotuberculosis*
Yersinia rohdei		*Y. rohdei*
Yersinia ruckeri		*Y. ruckeri*
Yokenella regensburgei	*Koserella trabulsii*, *Hafnia* hybridization group 3, Enteric group 45, Enteric group 45, NIH (Japan) biogoup 9	*Y. regensburgei*

[a]NL, not listed.

[b]*Enterobacter intermedius* and *Kluyvera cochleae* have identical 16S rRNA sequences.

[c]Name used in *Bergey's Manual of Systematic Bacteriology*, First Edition, 1984.

[d]Name used in *Bergey's Manual of Determinative Bacteriology*, Ninth Edition, 1994.

States and in the United Kingdom. Outbreaks due to *E. coli* O157:H7 have occurred in the United States, Europe, Asia, and Africa. In the United States, this organism is the leading cause of hemolytic uremic syndrome, the main cause of acute kidney failure in children (Boyce et al., 1995). *Enterobacteriaceae* are also significant causes of waterborne disease outbreaks.

Maintenance Procedures The following procedures have been used for many years in the Enteric Reference Laboratory, CDC, and seem to work well for all the genera and species of *Enterobacteriaceae*. Many other procedures have been described in the literature, but there have been few, if any, good comparative studies.

Long term preservation of cultures. Compared to many other organisms, strains of *Enterobacteriaceae* are simple to grow and maintain. However, strains will die if they are allowed to dry on plating media, agar slants or in "deeps." This die off is mainly due to drying. Freezing and storage at low temperature is the method of choice for long term preservation and maintenance. An alternative method is lyophilization and storage at 4°C, which is used routinely, with good results, by the American Type Culture Collection.

Preparation of a permanent frozen stock of the "whole culture." All important cultures of *Enterobacteriaceae* should be frozen and stored in liquid nitrogen. If liquid nitrogen storage is not available, they can be stored in a laboratory freezer at −70°C, or even in a "home freezer" at −10 to −20°C. The lowest temperature available should be used for long term storage. The culture to be maintained is first inoculated heavily and grown on a blood agar plate for 24 h (or longer if it grows slowly). Growth from the heavy area of growth (designated as the "whole culture," as opposed to a single colony) is removed with a cotton swab, and a very heavy suspension is made in 10% skim milk (a variety of other suspending media are also in general use). One ml of this suspension is then removed to a sterile plastic freezer vial. A permanent marking pen is used to write the genus and species names and the strain number. The vial is then frozen in an alcohol bath kept in the freezer. Vials are then transferred to a 100-compartment storage box and stored at the lowest temperature available.

Preparation of a "working stock" culture of the single colony. This procedure is recommended for cultures that are used frequently, since it avoids the inconvenience of repeatedly going to a freezer. Growth from the top of a single, isolated colony is touched and stabbed to the bottom of a 13 × 100 mm screw-cap tube of working stock medium. The culture is grown overnight and checked to insure that the tube has good growth. The tube is sealed by closing the cap tightly. These working stocks are convenient for daily use. A small amount of growth is removed with a loop and transferred to a plate or tube of a good growth medium. After incubation, the culture is ready to use. The sealing method described above does not give an air-tight fit, thus water from the semi-solid medium will begin to evaporate at a rate dependent on the tightness of the fit. Drying is not a problem for many months, but eventually the viable count of the culture may begin to drop. To avoid this problem, two alternative sealing methods are used. A small volume (about 0.3 ml) of sterile mineral oil can be added to cover the surface (about 3 mm above the interface), which prevents evaporation of the medium. This works extremely well in preventing evaporation, but the mineral oil is messy and must be considered when the culture is transferred. A second method to prevent drying is to insert a number "000" white rubber stopper into the neck of the screw-cap tube to seal it. Cultures are conveniently stored at room temperature in the dark in 40-compartment divided boxes. Many species of *Enterobacteriaceae* survive for decades with this method, but some species die more quickly than others. To avoid the death of cultures, they should be frozen in addition to being stored at room temperature.

Differentiation of genera and species of *Enterobacteriaceae* Genera and species of the family *Enterobacteriaceae* have traditionally been differentiated based on biochemical tests. Unfortunately, the media and tests are not completely standardized, and few laboratories use exactly the same formulations or procedures. Even with these variables, this approach usually results in correct identifications of the common species of *Enterobacteriaceae*. Biochemical reactions for *Enterobacteriaceae* are presented in Table BXII.γ.193. All species were studied in a single laboratory (the Enteric Reference Laboratories at the Centers for Disease Control and Prevention, Atlanta, GA, USA) by a single set of methods (Edwards and Ewing, 1972; Farmer et al., 1980a; Hickman and Farmer, 1978). Because different methods and tests are often used in different laboratories, the percentage of positive reactions obtained may differ somewhat from those presented in the chapters on specific genera. The purpose of Table BXII.γ.193 is not to advocate any given set of tests or to put undue emphasis on the percentages obtained, but to present a comprehensive comparison derived from a single set of data ob-

tained by tests commonly done in a diagnostic laboratory. Commercial identification systems and computer-aided identification are also widely used. These methods can be supplemented by identification based on genus- and/or species-specific tests, which can be both sensitive and specific. (An example is bacteriophage O1, which is both sensitive and specific for the most common serotypes of *Salmonella*.) Several of these tests are listed in Table BXII.γ.194. Many genera and species of *Enterobacteriaceae* also have typical patterns of resistance and susceptibility to antibiotics; thus, the antibiogram of an isolate can also be used as an aid to identification (Table BXII.γ.195). Detailed discussions of practices in clinical microbiology laboratories for isolation, maintenance, storage, computer-aided identification, use of commercial test "kits", antibiotic susceptibility testing, and difficulties presented by "problem strains" of *Enterobacteriaceae* can be found in Brenner (1992b) and Farmer (1999).

From 50 to more than 200 biochemical tests have been used in phenetic or numerical taxonomic studies of *Enterobacteriaceae* (Bascomb et al., 1971; Johnson et al., 1975; Véron, 1975; Véron and Le Minor, 1975a, b). These include tests for the fermentation of a large number of carbohydrates and polyhydroxyl alcohols, tests for the ability to use a wide variety of organic substrates as the sole source of carbon and energy, and tests for the presence of specific enzymes. A number of these tests are useful for the differentiation of species, subspecies, or biogroups within *Enterobacteriaceae* (Véron and Le Minor, 1975a, b). Some tests of particular diagnostic value are for nitrate reductase type A or type B (Pichinoty and Piéchaud, 1968; Pichinoty et al., 1969), tetrathionate reductase (Richard, 1977), fermentation or growth on D-galacturonate (Le Minor et al., 1979), presence of α-glutamyl transferase (Giammanco et al., 1980), and fermentation or growth on 2-ketogluconate (Buissière et al., 1981). A summary of data obtained at the Institut Pasteur for these tests is given in Table BXII.γ.196. Other nonroutine tests of value in differentiating between phenotypically similar organisms are ascorbate and D-arabitol for *Buttiauxella* and *Kluyvera*; L-fucose, 5-ketogluconate, D-xylose, and D-sorbose to differentiate *Klebsiella* from *Enterobacter*; susceptibility to *Hafnia*-specific bacteriophage to differentiate *Hafnia* from other genera; pectinase to differentiate erwiniae from other genera; ascorbate, D-glucose fermentation at 5°C within 21 d, and irgasan susceptibility to differentiate *Kluyvera* species; tyrosine clearing to differentiate *Leminorella* species from biochemically similar species; growth in *trans*-aconitate, adonitol, benzoate, *m*-erythritol, gentisate, D-malate, L-rhamnose, and *m*-tartrate to differentiate *S. liquefaciens*, *S. proteamaculans*, and *S. grimesii*; growth in histamine, D-melizitose, and D-tartrate to differentiate *S. rubidaea* subspecies; and bioluminescence, pigment on Loeffler's blood serum, colony color on MacConkey agar, and adsorption of bromthymol blue to differentiate *Photorhabdus* and *Xenorhabdus* species. Reactions for species in these specialized tests are given in the following genus chapters.

Molecular methods can be used to identify bacteria at taxonomic levels from the family down to the strain; furthermore, molecular tests based on virulence and pathogenicity genes can be used to distinguish pathogenic and nonpathogenic isolates. Many of these methods have been described for *Enterobacteriaceae*. For example, detection of the *phoE* gene by PCR amplification provides a sensitive and specific test for *Escherichia* and *Shigella*. A growing number of genus or species-specific probes have been reported in the literature. A number of systems for identification by either complete or partial 16S rRNA gene sequencing are now under development.

TABLE BXII.γ.193. Biochemical reactions of the named species and unnamed Enteric Groups of the family *Enterobacteriaceae*[a, b]

Characteristic	*Escherichia coli*	*Escherichia coli*, inactive	*Escherichia blattae*	*Escherichia fergusonii*	*Escherichia hermannii*	*Escherichia vulneris*	*Budvicia aquatica*	*Buttiauxella agrestis*	*Buttiauxella brennerae*	*Buttiauxella ferragutiae*	*Buttiauxella gaviniae*	*Buttiauxella izardii*	*Buttiauxella noackiae*	*Buttiauxella warmboldiae*	*Cedecea davisae*	*Cedecea lapagei*
Indole production	98	80	0	98	99	0	0	0	0	0	0	0	33	0	0	0
Methyl red	99	95	100	100	100	100	93	100	100	100	100	100	100	100	100	40
Voges–Proskauer	0	0	0	0	0	0	0	0	0	0	0	0	0	0	50	80
Citrate (Simmons)	1	1	50	17	1	0	0	100	0	0	20	0	33	33	95	99
Hydrogen sulfide (TSI)	1	1	0	0	0	0	80	0	0	0	0	0	0	0	0	0
Urea hydrolysis	1	1	0	0	0	0	33	0	0	0	0	0	0	0	0	0
Phenylalanine deaminase	0	0	0	0	0	0	0	0	0	0	0	0	100	100	0	0
Lysine decarboxylase	90	40	100	95	6	85	0	0	0	100	0	0	0	0	0	0
Arginine dihydrolase	17	3	0	5	0	30	0	0	0	0	20	0	67	0	50	80
Ornithine decarboxylase	65	20	100	100	100	0	0	100	33	80	0	100	0	0	95	0
Motility	95	5	0	93	99	100	27	100	100	60	80	100	100	100	95	80
Gelatin hydrolysis (22°C)	0	0	0	0	0	0	0	0	0	0	0	0	0	0	0	0
Growth in KCN	3	1	0	0	94	15	0	80	100	40	60	67	100	33	85	100
Malonate utilization	0	0	100	35	0	85	0	60	100	0	100	100	100	100	91	99
Esculin hydrolysis	35	5	0	46	40	20	0	100	100	100	100	100	100	100	45	100
Tartrate, Jordan's	95	85	50	96	35	2	27	60	0	0	40	67	100	0	0	0
Acetate utilization	90	40	0	96	78	30	0	0	0	0	0	0	0	0	0	60
Lipase (corn oil)	0	0	0	0	0	0	0	0	0	0	0	0	0	0	91	100
DNase (25°C)	0	0	0	0	0	0	0	0	0	0	0	0	0	0	0	0
Nitrate oxidized to nitrite	100	98	100	100	100	100	100	100	100	100	100	100	100	100	100	100
Oxidase, Kovacs	0	0	0	0	0	0	0	0	0	0	0	0	0	0	0	0
ONPG test	95	45	0	83	98	100	93	100	100	100	100	100	100	100	90	99
Yellow pigment	0	0	0	0	98	50	0	0	0	0	0	0	0	0	0	0
D-Glucose, acid	100	100	100	100	100	100	100	100	100	100	100	100	100	100	100	100
D-Glucose, gas	95	5	100	95	97	97	53	100	100	100	40	100	100	100	70	100
Fermentation of:																
Adonitol	5	3	0	98	0	0	0	0	67	0	100	0	0	0	0	0
L-Arabinose	99	85	100	98	100	100	80	100	100	100	100	100	100	100	0	0
D-Arabitol	5	5	0	100	8	0	27	0	67	0	80	0	0	0	100	100
Cellobiose	2	2	0	96	97	100	0	100	100	100	100	100	100	100	100	100
Dulcitol	60	40	0	60	19	0	0	0	0	0	0	0	0	0	0	0
Erythritol	0	0	0	0	0	0	0	0	0	0	0	0	0	0	0	0
Glycerol	75	65	100	20	3	25	0	60	67	0	0	33	0	0	0	0
myo-Inositol	1	1	0	0	0	0	0	0	0	0	0	0	0	67	0	0
Lactose	95	25	0	0	45	15	87	100	0	0	60	100	0	0	19	60
Maltose	95	80	100	96	100	100	0	100	100	100	60	100	100	100	100	100
D-Mannitol	98	93	0	98	100	100	60	100	100	100	100	100	100	100	100	100
D-Mannose	98	97	100	100	100	100	0	100	100	100	100	100	100	100	100	100
Melibiose	75	40	0	0	0	100	0	100	100	0	0	67	0	0	0	0
α-Methyl-D-glucoside	0	0	0	0	0	25	0	0	0	40	0	0	33	0	5	0
Mucate	95	30	50	0	97	78	20	100	67	60	80	100	100	0	0	0
Raffinose	50	15	0	0	40	99	0	100	100	0	0	33	0	0	10	0
L-Rhamnose	80	65	100	92	97	93	100	100	33	100	100	100	100	100	0	0
Salicin	40	10	0	65	40	30	0	100	100	100	100	100	100	100	99	100
D-Sorbitol	94	75	0	0	0	1	0	0	0	0	100	0	0	0	0	0
Sucrose	50	15	0	0	45	8	0	0	0	0	0	0	0	0	100	100
Trehalose	98	90	75	96	100	100	0	100	100	100	100	100	100	100	100	100
D-Xylose	95	70	100	96	100	100	93	100	100	100	100	100	100	100	100	0

[a]Each number is the percentage of positive reactions after 2 d at 36°C unless otherwise indicated (gelatin liquefaction and deoxyribonuclease; *Photorhabdus luminescens*).

[b]Table taken from Farmer, 1999.

[c]In a Request for an Opinion published in 1987, Le Minor and Popoff proposed replacement of the type species of *Salmonella* (*Salmonella choleraesuis* subsp. *choleraesuis*) with *Salmonella enterica* as the former was considered to be a source of confusion. Although the Request was denied by the Judicial Commission, their proposal resulted in an alternative naming convention which has found widespread endorsement in the public health community. This matter was revisited in July 2002 by the Judicial Commission during the IUMS Congress in response to several new Requests for an Opinion and will likely result in a decision to replace the type strain *Salmonella choleraesuis* subsp. *choleraesuis* with *Salmonella enterica* subsp. *enterica* LT2, while preserving the former rather than placing it on the list of rejected names. We view the six subspecies of *Salmonella choleraesuis* as deprecated. Readers are also advised that the names *Salmonella enteritidis*, *Salmonella paratyphi*, *Salmonella typhi*, and *Salmonella typhimurium* are synonyms of *Salmonella enterica* subsp. *enterica* and refer to specific serovars. These names have not been deprecated at this time as they remain in use by some public health reporting agencies.

(continued)

TABLE BXII.γ.193. *(cont.)*

Characteristic	Cedecea neteri	Cedecea species 3	Cedecea species 5	Citrobacter amalonaticus	Citrobacter freundii	Citrobacter braakii	Citrobacter farmeri	Citrobacter gillenii	Citrobacter koseri (C. diversus)	Citrobacter murliniae	Citrobacter rodentium	Citrobacter sedlakii	Citrobacter werkmanii	Citrobacter youngae	Edwardsiella tarda	Edwardsiella tarda biogroup 1
Indole production	0	0	0	100	33	33	100	0	99	100	0	83	0	15	99	100
Methyl red	100	100	100	100	100	100	100	100	100	100	100	100	100	100	100	100
Voges–Proskauer	50	50	50	0	0	0	0	0	0	0	0	0	0	0	0	0
Citrate (Simmons)	100	100	100	95	78	87	10	33	99	100	0	83	100	75	1	0
Hydrogen sulfide (TSI)	0	0	0	5	78	60	0	67	0	67	0	0	100	65	100	0
Urea hydrolysis	0	0	0	85	44	47	59	0	75	67	100	100	100	80	0	0
Phenylalanine deaminase	0	0	0	0	0	0	0	0	0	0	0	0	0	0	0	0
Lysine decarboxylase	0	0	0	0	0	0	0	0	0	0	0	0	0	0	100	100
Arginine dihydrolase	100	100	50	85	67	67	85	33	80	67	0	100	0	50	0	0
Ornithine decarboxylase	0	0	50	95	0	93	100	0	99	0	100	100	0	5	100	100
Motility	100	100	100	95	89	87	97	67	95	100	0	100	100	95	98	100
Gelatin hydrolysis (22°C)	0	0	0	0	0	0	0	0	0	0	0	0	0	0	0	0
Growth in KCN	65	100	100	99	89	100	93	100	0	100	0	100	100	95	0	0
Malonate utilization	100	0	0	1	11	0	0	100	95	0	100	100	100	5	0	0
Esculin hydrolysis	100	100	100	5	0	0	0	0	1	0	0	17	0	5	0	0
Tartrate, Jordan's	0	0	0	96	100	93	93	100	90	100	100	100	100	100	25	0
Acetate utilization	0	50	50	86	44	53	80	0	75	33	0	83	100	65	0	0
Lipase (corn oil)	100	100	50	0	0	0	0	0	0	0	0	0	0	0	0	0
DNase (25°C)	0	0	0	0	0	0	0	0	0	0	0	0	0	0	0	0
Nitrate oxidized to nitrite	100	100	100	99	100	100	100	100	100	100	100	100	100	85	100	100
Oxidase, Kovacs	0	0	0	0	0	0	0	0	0	0	0	0	0	0	0	0
ONPG test	100	100	100	97	89	80	100	67	99	100	100	100	100	90	0	0
Yellow pigment	0	0	0	0	0	0	0	0	0	0	0	0	0	0	0	0
D-Glucose, acid	100	100	100	100	100	100	100	100	100	100	100	100	100	100	100	100
D-Glucose, gas	100	100	100	97	89	93	96	100	98	100	100	100	100	75	100	50
Fermentation of:																
Adonitol	0	0	0	0	0	0	0	0	99	0	0	0	0	0	0	0
L-Arabinose	0	0	0	99	100	100	100	0	99	100	100	100	100	100	9	100
D-Arabitol	100	100	100	0	0	0	0	0	98	0	0	0	0	5	0	0
Cellobiose	100	100	100	100	44	73	100	67	99	100	100	100	0	45	0	0
Dulcitol	0	0	0	1	11	33	2	0	40	100	0	100	0	85	0	0
Erythritol	0	0	0	0	0	0	0	0	0	0	0	0	0	0	0	0
Glycerol	0	0	0	60	100	87	65	67	99	100	0	83	100	90	30	0
myo-Inositol	0	0	0	0	0	0	0	0	0	0	0	0	0	5	0	0
Lactose	35	0	0	35	78	80	15	67	50	67	100	100	17	25	0	0
Maltose	100	100	100	99	100	100	100	100	100	100	100	100	100	95	100	100
D-Mannitol	100	100	100	100	100	100	100	100	99	100	100	100	100	100	0	100
D-Mannose	100	100	100	100	100	100	100	100	100	100	100	100	100	100	100	100
Melibiose	0	100	100	0	100	80	100	67	0	33	0	100	0	10	0	0
α-Methyl-D-glucoside	0	50	0	2	11	33	75	0	40	0	0	0	0	0	0	0
Mucate	0	0	0	96	100	100	100	67	95	100	100	100	100	100	0	0
Raffinose	0	100	100	5	44	7	100	0	0	33	0	0	0	10	0	0
L-Rhamnose	0	0	0	100	100	100	100	0	99	100	100	100	100	100	0	0
Salicin	100	100	100	30	0	0	9	0	15	33	0	17	0	10	0	0
D-Sorbitol	100	0	100	99	100	100	98	100	99	100	100	100	100	100	0	0
Sucrose	100	50	100	9	89	7	100	33	40	33	0	0	0	20	0	100
Trehalose	100	100	100	100	100	100	100	100	100	100	100	100	100	100	0	0
D-Xylose	100	100	100	99	89	100	100	100	100	100	100	100	100	100	0	0

[a]Each number is the percentage of positive reactions after 2 d at 36°C unless otherwise indicated (gelatin liquefaction and deoxyribonuclease; *Photorhabdus luminescens*).

[b]Table taken from Farmer, 1999.

[c]In a Request for an Opinion published in 1987, Le Minor and Popoff proposed replacement of the type species of *Salmonella* (*Salmonella choleraesuis* subsp. *choleraesuis*) with *Salmonella enterica* as the former was considered to be a source of confusion. Although the Request was denied by the Judicial Commission, their proposal resulted in an alternative naming convention which has found widespread endorsement in the public health community. This matter was revisited in July 2002 by the Judicial Commission during the IUMS Congress in response to several new Requests for an Opinion and will likely result in a decision to replace the type strain *Salmonella choleraesuis* subsp. *choleraesuis* with *Salmonella enterica* subsp. *enterica* LT2, while preserving the former rather than placing it on the list of rejected names. We view the six subspecies of *Salmonella choleraesuis* as deprecated. Readers are also advised that the names *Salmonella enteritidis*, *Salmonella paratyphi*, *Salmonella typhi*, and *Salmonella typhimurium* are synonyms of *Salmonella enterica* subsp. *enterica* and refer to specific serovars. These names have not been deprecated at this time as they remain in use by some public health reporting agencies.

(continued)

TABLE BXII.γ.193. *(cont.)*

Characteristic	Edwardsiella hoshinae	Edwardsiella ictaluri	Enterobacter cloacae	Enterobacter aerogenes	Enterobacter amnigenus biogroup 1	Enterobacter amnigenus biogroup 2	Enterobacter asburiae	Enterobacter cancerogenus (E. taylorae)	Enterobacter dissolvens	Enterobacter gergoviae	Enterobacter hormaechei	Enterobacter intermedius	Enterobacter nimipressuralis	Enterobacter pyrinus	Enterobacter sakazakii	Ewingella americana
Indole production	50	0	0	0	0	0	0	0	0	0	0	0	0	0	11	0
Methyl red	100	0	5	5	7	65	100	5	0	5	57	100	100	29	5	84
Voges–Proskauer	0	0	100	98	100	100	2	100	100	100	100	100	100	86	100	95
Citrate (Simmons)	0	0	100	95	70	100	100	100	100	99	96	65	0	0	99	95
Hydrogen sulfide (TSI)	0	0	0	0	0	0	0	0	0	0	0	0	0	0	0	0
Urea hydrolysis	0	0	65	2	0	0	60	1	100	93	87	0	0	86	1	0
Phenylalanine deaminase	0	0	0	0	0	0	0	0	0	0	4	0	0	0	50	0
Lysine decarboxylase	100	100	0	98	0	0	0	0	0	90	0	0	0	100	0	0
Arginine dihydrolase	0	0	97	0	9	35	21	94	100	0	78	0	0	0	99	0
Ornithine decarboxylase	95	65	96	98	55	100	95	99	100	100	91	89	100	100	91	0
Motility	100	0	95	97	92	100	0	99	0	90	52	89	0	43	96	60
Gelatin hydrolysis (22°C)	0	0	0	0	0	0	0	0	0	0	0	0	0	0	0	0
Growth in KCN	0	0	98	98	100	100	97	98	100	0	100	65	100	0	99	5
Malonate utilization	100	0	75	95	91	100	3	100	100	96	100	100	100	86	18	0
Esculin hydrolysis	0	0	30	98	91	100	95	90	100	97	0	100	100	100	100	50
Tartrate, Jordan's	0	0	30	95	9	0	30	0	0	97	13	100	0	0	1	35
Acetate utilization	0	0	75	50	0	0	87	35	100	93	74	0	0	0	96	10
Lipase (corn oil)	0	0	0	0	0	0	0	0	0	0	0	0	0	0	0	0
DNase (25°C)	0	0	0	0	0	0	0	0	0	0	0	0	0	0	0	0
Nitrate oxidized to nitrite	100	100	99	100	100	100	100	100	100	99	100	100	100	100	99	97
Oxidase, Kovacs	0	0	0	0	0	0	0	0	0	0	0	0	0	0	0	0
ONPG test	0	0	99	98	91	100	100	100	100	97	95	100	100	100	100	85
Yellow pigment	0	0	0	0	0	0	0	0	0	0	0	0	0	0	98	0
D-Glucose, acid	100	100	100	100	100	100	100	100	100	100	100	100	100	100	100	100
D-Glucose, gas	35	50	100	100	100	100	95	100	100	98	83	100	100	100	98	0
Fermentation of:																
Adonitol	0	0	25	98	100	0	0	0	0	0	0	0	0	0	0	0
L-Arabinose	13	0	100	100	100	100	100	100	100	99	100	100	100	100	100	0
D-Arabitol	0	0	15	100	0	0	0	0	0	97	0	0	0	0	0	99
Cellobiose	0	0	99	100	100	100	100	100	100	99	100	100	100	100	100	10
Dulcitol	0	0	15	5	0	0	0	0	0	0	87	100	0	0	5	0
Erythritol	0	0	0	0	0	0	0	0	0	0	0	0	0	0	0	0
Glycerol	65	0	40	98	0	0	11	1	0	100	4	100	0	0	15	24
myo-Inositol	0	0	15	95	100	0	0	0	0	0	0	0	0	100	75	0
Lactose	0	0	93	95	70	35	75	10	0	55	9	100	0	14	99	70
Maltose	100	100	100	99	100	100	100	99	100	100	100	100	100	100	100	16
D-Mannitol	100	0	100	100	100	100	100	100	100	99	100	100	100	100	100	100
D-Mannose	100	100	100	95	100	100	100	100	100	100	100	100	100	100	100	99
Melibiose	0	0	90	99	100	100	0	0	100	97	0	100	100	0	100	0
α-Methyl-D-glucoside	0	0	85	95	55	100	95	1	100	2	83	100	100	0	96	0
Mucate	0	0	75	90	35	100	21	75	100	2	96	100	100	0	1	0
Raffinose	0	0	97	96	100	0	70	0	100	97	0	100	0	0	99	0
L-Rhamnose	0	0	92	99	100	100	5	100	100	99	100	100	100	100	100	23
Salicin	50	0	75	100	91	100	100	92	100	99	44	100	100	100	99	80
D-Sorbitol	0	0	95	100	9	100	100	1	100	0	0	100	100	0	0	0
Sucrose	100	0	97	100	100	0	100	0	100	98	100	65	0	100	100	0
Trehalose	100	0	100	100	100	100	100	100	100	100	100	100	100	100	100	99
D-Xylose	0	0	99	100	100	100	97	100	100	99	96	100	100	0	100	13

[a]Each number is the percentage of positive reactions after 2 d at 36°C unless otherwise indicated (gelatin liquefaction and deoxyribonuclease; *Photorhabdus luminescens*).

[b]Table taken from Farmer, 1999.

[c]In a Request for an Opinion published in 1987, Le Minor and Popoff proposed replacement of the type species of *Salmonella* (*Salmonella choleraesuis* subsp. *choleraesuis*) with *Salmonella enterica* as the former was considered to be a source of confusion. Although the Request was denied by the Judicial Commission, their proposal resulted in an alternative naming convention which has found widespread endorsement in the public health community. This matter was revisited in July 2002 by the Judicial Commission during the IUMS Congress in response to several new Requests for an Opinion and will likely result in a decision to replace the type strain *Salmonella choleraesuis* subsp. *choleraesuis* with *Salmonella enterica* subsp. *enterica* LT2, while preserving the former rather than placing it on the list of rejected names. We view the six subspecies of *Salmonella choleraesuis* as deprecated. Readers are also advised that the names *Salmonella enteritidis*, *Salmonella paratyphi*, *Salmonella typhi*, and *Salmonella typhimurium* are synonyms of *Salmonella enterica* subsp. *enterica* and refer to specific serovars. These names have not been deprecated at this time as they remain in use by some public health reporting agencies.

(continued)

TABLE BXII.γ.193. *(cont.)*

Characteristic	*Hafnia alvei*	*Hafnia alvei* biogroup 1	*Klebsiella pneumoniae* subsp. *ozaenae*	*Klebsiella pneumoniae* subsp. *pneumoniae*	*Klebsiella pneumoniae* subsp. *rhinoscleromatis*	*Klebsiella oxytoca*	*Klebsiella oxytoca,* ornithine positive	*Klebsiella planticola*	*Klebsiella terrigena*	*Kluyvera ascorbata*	*Kluyvera cryocrescens*	*Kluyvera georgiana*	*Leclercia adecarboxylata*	*Leminorella grimontii*	*Leminorella richardii*	*Moellerella wisconsensis*
Indole production	0	0	0	0	0	99	100	20	0	92	90	100	100	0	0	0
Methyl red	40	85	98	10	100	20	96	100	60	100	100	100	100	100	0	100
Voges–Proskauer	85	70	0	98	0	95	70	98	100	0	0	0	0	0	0	0
Citrate (Simmons)	10	0	30	98	0	95	100	100	40	96	80	100	0	100	0	80
Hydrogen sulfide (TSI)	0	0	0	0	0	0	0	0	0	0	0	0	0	100	100	0
Urea hydrolysis	4	0	10	95	0	90	100	98	0	0	0	0	48	0	0	0
Phenylalanine deaminase	0	0	0	0	0	1	0	0	0	0	0	0	0	0	0	0
Lysine decarboxylase	100	100	40	98	0	99	100	100	100	97	23	100	0	0	0	0
Arginine dihydrolase	6	0	6	0	0	0	0	0	0	0	0	0	0	0	0	0
Ornithine decarboxylase	98	45	3	0	0	0	100	0	20	100	100	100	0	0	0	0
Motility	85	0	0	0	0	0	0	0	0	98	90	100	79	0	0	0
Gelatin hydrolysis (22°C)	0	0	0	0	0	0	0	0	0	0	0	0	0	0	0	0
Growth in KCN	95	0	88	98	80	97	100	100	100	92	86	83	97	0	0	70
Malonate utilization	50	45	3	93	95	98	100	100	100	96	86	50	93	0	0	0
Esculin hydrolysis	7	0	80	99	30	100	100	100	100	99	100	100	100	0	0	0
Tartrate, Jordan's	70	30	50	95	50	98	100	100	100	35	19	50	83	100	100	30
Acetate utilization	15	0	2	75	0	90	95	62	20	50	86	83	28	0	0	10
Lipase (corn oil)	0	0	0	0	0	0	0	0	0	0	0	0	0	0	0	0
DNase (25°C)	0	0	0	0	0	0	0	0	0	0	0	0	0	0	0	0
Nitrate oxidized to nitrite	100	100	80	99	100	100	100	100	100	100	100	100	100	100	100	90
Oxidase, Kovacs	0	0	0	0	0	0	0	0	0	0	0	0	0	0	0	0
ONPG test	90	30	80	99	0	100	100	100	100	100	100	100	100	0	0	90
Yellow pigment	0	0	0	0	0	1	0	1	0	0	0	0	37	0	0	0
D-Glucose, acid	100	100	100	100	100	100	100	100	100	100	100	100	100	100	100	100
D-Glucose, gas	98	0	50	97	0	97	100	100	80	93	95	17	97	33	0	0
Fermentation of:																
Adonitol	0	0	97	90	100	99	100	100	100	0	0	0	93	0	0	100
L-Arabinose	95	0	98	99	100	98	100	100	100	100	100	100	100	100	100	0
D-Arabitol	0	0	95	98	100	98	100	100	100	0	0	0	96	0	0	75
Cellobiose	15	0	92	98	100	100	100	100	100	100	100	100	100	0	0	0
Dulcitol	0	0	2	30	0	55	10	15	20	25	0	33	86	83	0	0
Erythritol	0	0	0	0	0	2	0	0	0	0	0	0	0	0	0	0
Glycerol	95	0	65	97	50	99	100	100	100	40	5	33	3	17	0	10
myo-Inositol	0	0	55	95	95	98	95	100	80	0	0	0	0	0	0	0
Lactose	5	0	30	98	0	100	100	100	100	98	95	83	93	0	0	100
Maltose	100	0	95	98	100	100	100	100	100	100	100	100	100	0	0	30
D-Mannitol	99	55	100	99	100	99	100	100	100	100	95	100	100	0	0	60
D-Mannose	100	100	100	99	100	100	100	100	100	100	100	100	100	0	0	100
Melibiose	0	0	97	99	100	99	100	100	100	99	100	100	100	0	0	100
α-Methyl-D-glucoside	0	0	70	90	0	98	100	100	100	98	95	100	0	0	0	0
Mucate	0	0	25	90	0	93	96	100	100	90	81	83	93	100	50	0
Raffinose	2	0	90	99	90	100	100	100	100	98	100	100	66	0	0	100
L-Rhamnose	97	0	55	99	96	100	100	100	100	100	100	83	100	0	0	0
Salicin	13	55	97	99	98	100	100	100	100	100	100	100	100	0	0	0
D-Sorbitol	0	0	65	99	100	99	100	92	100	40	45	0	0	0	0	0
Sucrose	10	0	20	99	75	100	100	100	100	98	81	100	66	0	0	100
Trehalose	95	70	98	99	100	100	100	100	100	100	100	100	100	0	0	0
D-Xylose	98	0	95	99	100	100	100	100	100	99	91	100	100	83	100	0

[a]Each number is the percentage of positive reactions after 2 d at 36°C unless otherwise indicated (gelatin liquefaction and deoxyribonuclease; *Photorhabdus luminescens*).

[b]Table taken from Farmer, 1999.

[c]In a Request for an Opinion published in 1987, Le Minor and Popoff proposed replacement of the type species of *Salmonella* (*Salmonella choleraesuis* subsp. *choleraesuis*) with *Salmonella enterica* as the former was considered to be a source of confusion. Although the Request was denied by the Judicial Commission, their proposal resulted in an alternative naming convention which has found widespread endorsement in the public health community. This matter was revisited in July 2002 by the Judicial Commission during the IUMS Congress in response to several new Requests for an Opinion and will likely result in a decision to replace the type strain *Salmonella choleraesuis* subsp. *choleraesuis* with *Salmonella enterica* subsp. *enterica* LT2, while preserving the former rather than placing it on the list of rejected names. We view the six subspecies of *Salmonella choleraesuis* as deprecated. Readers are also advised that the names *Salmonella enteritidis*, *Salmonella paratyphi*, *Salmonella typhi*, and *Salmonella typhimurium* are synonyms of *Salmonella enterica* subsp. *enterica* and refer to specific serovars. These names have not been deprecated at this time as they remain in use by some public health reporting agencies.

(continued)

TABLE BXII.γ.193. *(cont.)*

Characteristic	Morganella morganii subsp. morganii	Morganella morganii subsp. sibonii	Morganella morganii biogroup 1	Obesumbacterium proteus biogroup 2	Pantoea agglomerans	Pantoea dispersa	Photorhabdus luminescens (all tests at 25°C)	Photorhabdus DNA hybridization group 5	Pragia fontium	Proteus vulgaris	Proteus mirabilis	Proteus myxofaciens	Proteus penneri	Providencia alcalifaciens	Providencia heimbachae	Providencia rettgeri
Indole production	95	50	100	0	20	0	50	0	0	98	2	0	0	99	0	99
Methyl red	95	86	95	15	50	82	0	0	100	95	97	100	100	99	85	93
Voges–Proskauer	0	0	0	0	70	64	0	0	0	0	50	100	0	0	0	0
Citrate (Simmons)	0	0	0	0	50	100	50	20	89	15	65	50	0	98	0	95
Hydrogen sulfide (TSI)	20	7	15	0	0	0	0	0	89	95	98	0	30	0	0	0
Urea hydrolysis	95	100	100	0	20	0	25	60	0	95	98	100	100	0	0	98
Phenylalanine deaminase	95	93	100	0	20	9	0	0	22	99	98	100	99	98	100	98
Lysine decarboxylase	1	29	100	100	0	0	0	0	0	0	0	0	0	0	0	0
Arginine dihydrolase	0	0	0	0	0	0	0	0	0	0	0	0	0	0	0	0
Ornithine decarboxylase	95	64	80	100	0	0	0	0	0	0	99	0	0	1	0	0
Motility	95	79	0	0	85	100	100	100	100	95	95	100	85	96	46	94
Gelatin hydrolysis (22°C)	0	0	0	0	2	0	50	80	0	91	90	100	50	0	0	0
Growth in KCN	98	79	90	0	35	82	0	20	0	99	98	100	99	100	8	97
Malonate utilization	1	0	5	0	65	9	0	0	0	0	2	0	0	0	0	0
Esculin hydrolysis	0	0	0	0	60	0	0	0	78	50	0	0	0	0	0	35
Tartrate, Jordan's	95	100	100	15	25	9	50	60	0	80	87	100	85	90	69	95
Acetate utilization	0	0	0	0	30	100	0	20	0	25	20	0	5	40	0	60
Lipase (corn oil)	0	0	0	0	0	0	0	0	0	80	92	100	45	0	0	0
DNase (25°C)	0	0	0	0	0	0	0	0	0	80	50	50	40	0	0	0
Nitrate oxidized to nitrite	90	100	90	100	85	91	0	0	100	98	95	100	90	100	100	100
Oxidase, Kovacs	0	0	0	0	0	0	0	0	0	0	0	0	0	0	0	0
ONPG test	10	0	20	0	90	91	0	0	0	1	0	0	1	1	0	5
Yellow pigment	0	0	0	0	75	27	50	60	0	0	0	0	0	0	0	0
D-Glucose, acid	99	100	100	100	100	100	75	100	100	100	100	100	100	100	100	100
D-Glucose, gas	90	86	93	0	20	0	0	0	0	85	96	100	45	85	0	10
Fermentation of:																
Adonitol	0	0	0	0	7	0	0	0	0	0	0	0	0	98	92	100
L-Arabinose	0	0	0	0	95	100	0	0	0	0	0	0	0	1	0	0
D-Arabitol	0	0	0	0	50	100	0	0	0	0	0	0	0	0	92	100
Cellobiose	0	0	0	0	55	55	0	0	0	0	1	0	0	0	0	3
Dulcitol	0	0	0	0	15	0	0	0	0	0	0	0	0	0	0	0
Erythritol	0	0	0	0	0	0	0	0	0	1	0	0	0	0	0	75
Glycerol	5	7	100	0	30	27	0	0	0	60	70	100	55	15	0	60
myo-Inositol	0	0	0	0	15	0	0	0	0	0	0	0	0	1	46	90
Lactose	1	0	0	0	40	0	0	0	0	2	2	0	1	0	0	5
Maltose	0	0	0	50	89	82	25	0	0	97	0	100	100	1	54	2
D-Mannitol	0	0	0	0	100	100	0	0	0	0	0	0	0	2	0	100
D-Mannose	98	100	100	85	98	100	100	100	0	0	0	0	0	100	100	100
Melibiose	0	0	0	0	50	0	0	0	0	0	0	0	0	0	0	5
α-Methyl-D-glucoside	0	0	0	0	7	0	0	0	0	60	0	100	80	0	0	2
Mucate	0	7	0	0	40	0	0	0	0	0	0	0	0	0	0	0
Raffinose	0	0	0	0	30	0	0	0	0	1	1	0	1	1	0	5
L-Rhamnose	0	0	0	15	85	91	0	0	0	5	1	0	0	0	100	70
Salicin	0	0	0	0	65	0	0	0	78	50	0	0	0	1	0	50
D-Sorbitol	0	0	0	0	30	0	0	0	0	0	0	0	0	1	0	1
Sucrose	0	7	0	0	75	1	0	0	0	97	15	100	100	15	0	15
Trehalose	0	100	0	85	97	100	0	0	0	30	98	100	55	2	0	0
D-Xylose	0	0	0	15	93	100	0	0	0	95	98	0	100	1	8	10

[a]Each number is the percentage of positive reactions after 2 d at 36°C unless otherwise indicated (gelatin liquefaction and deoxyribonuclease; *Photorhabdus luminescens*).

[b]Table taken from Farmer, 1999.

[c]In a Request for an Opinion published in 1987, Le Minor and Popoff proposed replacement of the type species of *Salmonella* (*Salmonella choleraesuis* subsp. *choleraesuis*) with *Salmonella enterica* as the former was considered to be a source of confusion. Although the Request was denied by the Judicial Commission, their proposal resulted in an alternative naming convention which has found widespread endorsement in the public health community. This matter was revisited in July 2002 by the Judicial Commission during the IUMS Congress in response to several new Requests for an Opinion and will likely result in a decision to replace the type strain *Salmonella choleraesuis* subsp. *choleraesuis* with *Salmonella enterica* subsp. *enterica* LT2, while preserving the former rather than placing it on the list of rejected names. We view the six subspecies of *Salmonella choleraesuis* as deprecated. Readers are also advised that the names *Salmonella enteritidis*, *Salmonella paratyphi*, *Salmonella typhi*, and *Salmonella typhimurium* are synonyms of *Salmonella enterica* subsp. *enterica* and refer to specific serovars. These names have not been deprecated at this time as they remain in use by some public health reporting agencies.

(continued)

TABLE BXII.γ.193. *(cont.)*

Characteristic	*Providencia rustigianii*	*Providencia stuartii*	*Rahnella aquatilis*	*Salmonella bongori*	*Salmonella enterica* subsp. *enterica* [c]	*Salmonella enterica* subsp. *arizonae* [c]	*Salmonella enterica* subsp. *diarizonae* [c]	*Salmonella enterica* subsp. *houtenae* [c]	*Salmonella enterica* subsp. *indica* [c]	*Salmonella enterica* subsp. *salamae* [c]	*Salmonella* serovar *Choleraesuis* [c]	*Salmonella* serovar *Gallinarum* [c]	*Salmonella* serovar *Paratyphi* A [c]	*Salmonella* serovar *Pullorum* [c]	*Salmonella* serovar *Typhi* [c]	*Serratia marcescens*
Indole production	98	98	0	0	1	1	2	0	0	2	0	0	0	0	0	1
Methyl red	65	100	88	100	100	100	100	100	100	100	100	100	100	90	100	20
Voges–Proskauer	0	0	100	0	0	0	0	0	0	0	0	0	0	0	0	98
Citrate (Simmons)	15	93	94	94	95	99	98	98	89	100	25	0	0	0	0	98
Hydrogen sulfide (TSI)	0	0	0	100	95	99	99	100	100	100	50	100	10	90	97	0
Urea hydrolysis	0	30	0	0	1	0	0	2	0	0	0	0	0	0	0	15
Phenylalanine deaminase	100	95	95	0	0	0	0	0	0	0	0	0	0	0	0	0
Lysine decarboxylase	0	0	0	100	98	99	99	100	100	100	95	90	0	100	98	99
Arginine dihydrolase	0	0	0	94	70	70	70	70	67	90	55	10	15	10	3	0
Ornithine decarboxylase	0	0	0	100	97	99	99	100	100	100	100	1	95	95	0	99
Motility	30	85	6	100	95	99	99	98	100	98	95	0	95	0	97	97
Gelatin hydrolysis (22°C)	0	0	0	0	0	0	0	0	0	2	0	0	0	0	0	90
Growth in KCN	100	100	0	100	0	1	1	95	0	0	0	0	0	0	0	95
Malonate utilization	0	0	100	0	0	95	95	0	0	95	0	0	0	0	0	3
Esculin hydrolysis	0	0	100	0	5	1	1	0	0	15	0	0	0	0	0	95
Tartrate, Jordan's	50	90	6	0	90	5	20	65	100	50	85	100	0	0	100	75
Acetate utilization	25	75	6	100	90	90	75	70	89	95	1	0	0	0	0	50
Lipase (corn oil)	0	0	0	0	0	0	0	0	0	0	0	0	0	0	0	98
DNase (25°C)	0	10	0	0	2	2	2	0	0	0	0	10	0	0	0	98
Nitrate oxidized to nitrite	100	100	100	100	100	100	100	100	100	100	98	100	100	100	100	98
Oxidase, Kovacs	0	0	0	0	0	0	0	0	0	0	0	0	0	0	0	0
ONPG test	0	10	100	94	2	100	92	0	44	15	0	0	0	0	0	95
Yellow pigment	0	0	0	0	0	0	0	0	0	0	0	0	0	0	0	0
D-Glucose, acid	100	100	100	100	100	100	100	100	100	100	100	100	100	100	100	100
D-Glucose, gas	35	0	98	94	96	99	99	100	100	100	95	0	99	90	0	55
Fermentation of:																
Adonitol	0	5	0	0	0	0	0	5	0	0	0	0	0	0	0	40
L-Arabinose	0	1	100	94	99	99	99	100	100	100	0	80	100	100	2	0
D-Arabitol	0	0	0	0	0	1	1	5	0	0	1	0	0	0	0	0
Cellobiose	0	5	100	0	5	1	1	50	0	0	0	10	5	5	0	5
Dulcitol	0	0	88	94	96	0	1	0	67	90	5	90	90	0	0	0
Erythritol	0	0	0	0	0	0	0	0	0	0	1	1	0	0	0	1
Glycerol	5	50	13	0	5	10	10	0	33	25	0	0	10	0	20	95
myo-Inositol	0	95	0	0	35	0	0	0	0	5	0	0	0	0	0	75
Lactose	0	2	100	0	1	15	85	0	22	1	0	0	0	0	1	2
Maltose	0	1	94	100	97	98	98	100	100	100	95	90	95	5	97	96
D-Mannitol	0	10	100	100	100	100	100	98	100	100	98	100	100	100	100	99
D-Mannose	100	100	100	100	100	100	100	100	100	95	95	100	100	100	100	99
Melibiose	0	0	100	94	95	95	95	100	89	8	45	0	95	0	100	0
α-Methyl-D-glucoside	0	0	0	0	2	1	1	0	0	8	0	0	0	0	0	0
Mucate	0	0	30	88	90	90	30	0	89	96	0	50	0	0	0	0
Raffinose	0	7	94	0	2	1	1	0	0	0	1	10	0	1	0	2
L-Rhamnose	0	0	94	88	95	99	99	98	100	100	100	10	100	100	0	0
Salicin	0	2	100	0	0	0	0	60	0	5	0	0	0	0	0	95
D-Sorbitol	0	1	94	100	95	99	99	100	0	100	90	1	95	10	99	99
Sucrose	35	50	100	0	1	1	5	0	0	1	0	0	0	0	0	99
Trehalose	0	98	100	100	99	99	99	100	100	100	0	50	100	90	100	99
D-Xylose	0	7	94	100	97	100	100	100	100	100	98	70	0	90	82	7

[a]Each number is the percentage of positive reactions after 2 d at 36°C unless otherwise indicated (gelatin liquefaction and deoxyribonuclease; *Photorhabdus luminescens*).

[b]Table taken from Farmer, 1999.

[c]In a Request for an Opinion published in 1987, Le Minor and Popoff proposed replacement of the type species of *Salmonella* (*Salmonella choleraesuis* subsp. *choleraesuis*) with *Salmonella enterica* as the former was considered to be a source of confusion. Although the Request was denied by the Judicial Commission, their proposal resulted in an alternative naming convention which has found widespread endorsement in the public health community. This matter was revisited in July 2002 by the Judicial Commission during the IUMS Congress in response to several new Requests for an Opinion and will likely result in a decision to replace the type strain *Salmonella choleraesuis* subsp. *choleraesuis* with *Salmonella enterica* subsp. *enterica* LT2, while preserving the former rather than placing it on the list of rejected names. We view the six subspecies of *Salmonella choleraesuis* as deprecated. Readers are also advised that the names *Salmonella enteritidis*, *Salmonella paratyphi*, *Salmonella typhi*, and *Salmonella typhimurium* are synonyms of *Salmonella enterica* subsp. *enterica* and refer to specific serovars. These names have not been deprecated at this time as they remain in use by some public health reporting agencies.

(continued)

TABLE BXII.γ.193. *(cont.)*

Characteristic	Serratia entomophila	Serratia ficaria	Serratia fonticola	Serratia liquefaciens	Serratia marcescens biogroup 1	Serratia odorifera biogroup 1	Serratia odorifera biogroup 2	Serratia plymuthica	Serratia rubidaea	Shigella dysenteriae	Shigella boydii	Shigella flexneri	Shigella sonnei	Tatumella ptyseos	Trabulsiella guamensis	Xenorhabdus nematophilus
Indole production	0	0	0	1	0	60	50	0	0	45	25	50	0	0	40	40
Methyl red	20	75	100	93	100	100	60	94	20	99	100	100	100	0	100	0
Voges–Proskauer	100	75	9	93	60	50	100	80	100	0	0	0	0	5	0	0
Citrate (Simmons)	100	100	91	90	30	100	97	75	95	0	0	0	0	2	88	0
Hydrogen sulfide (TSI)	0	0	0	0	0	0	0	0	0	0	0	0	0	0	100	0
Urea hydrolysis	0	0	13	3	0	5	0	0	2	0	0	0	0	0	0	0
Phenylalanine deaminase	0	0	0	0	0	0	0	0	0	0	0	0	0	90	0	0
Lysine decarboxylase	0	0	100	95	55	100	94	0	55	0	0	0	0	0	100	0
Arginine dihydrolase	0	0	0	0	4	0	0	0	0	2	18	5	2	0	50	0
Ornithine decarboxylase	0	0	97	95	65	100	0	0	0	0	2	0	98	0	100	0
Motility	100	100	91	95	17	100	100	50	85	0	0	0	0	0	100	100
Gelatin hydrolysis (22°C)	100	100	0	90	30	95	94	60	90	0	0	0	0	0	0	80
Growth in KCN	100	55	70	90	70	60	19	30	25	0	0	100	0	0	100	0
Malonate utilization	0	0	88	2	0	0	0	0	94	100	0	0	0	0	0	0
Esculin hydrolysis	100	100	100	97	96	95	40	81	94	0	0	0	0	0	40	0
Tartrate, Jordan's	100	17	58	75	50	100	100	100	70	75	50	30	90	0	50	60
Acetate utilization	80	40	15	40	4	60	65	55	80	0	0	8	0	0	88	0
Lipase (corn oil)	20	77	0	85	75	35	65	70	99	0	0	0	0	0	0	0
DNase (25°C)	100	100	0	85	82	100	100	100	99	0	0	0	0	0	0	20
Nitrate oxidized to nitrite	100	92	100	100	83	100	100	100	100	99	100	99	100	98	100	20
Oxidase, Kovacs	0	8	0	0	0	0	0	0	0	0	0	0	0	0	0	0
ONPG test	100	100	100	93	75	100	100	70	100	30	10	1	90	0	100	0
Yellow pigment	0	0	0	0	0	0	0	0	0	0	0	0	0	0	0	60
D-Glucose, acid	100	100	100	100	100	100	100	100	100	0	100	100	100	100	100	80
D-Glucose, gas	0	0	79	75	0	0	13	40	30	0	0	3	0	0	100	0
Fermentation of:																
Adonitol	0	0	100	5	30	50	55	0	99	0	0	0	0	0	0	0
L-Arabinose	0	100	100	98	0	100	100	100	100	45	94	60	95	0	100	0
D-Arabitol	60	100	100	0	0	0	0	0	85	0	0	1	0	0	0	0
Cellobiose	0	100	6	5	4	100	100	88	94	0	0	0	5	0	100	0
Dulcitol	0	0	91	0	0	0	0	0	0	0	5	1	0	0	0	0
Erythritol	0	0	0	0	0	0	7	0	0	0	0	0	0	0	0	0
Glycerol	0	0	88	95	92	40	50	50	20	10	50	10	15	7	0	0
myo-Inositol	0	55	30	60	30	100	100	50	20	0	0	0	0	0	0	0
Lactose	0	15	97	10	4	70	97	80	100	0	1	1	2	0	0	0
Maltose	100	100	97	98	70	100	100	94	99	15	20	30	90	0	100	0
D-Mannitol	100	100	100	100	96	100	97	100	100	100	97	95	99	0	100	0
D-Mannose	100	100	100	100	100	100	100	100	100	100	100	100	100	100	100	80
Melibiose	0	40	98	75	0	100	96	93	99	0	15	55	25	25	0	0
α-Methyl-D-glucoside	0	8	91	5	0	0	0	70	1	0	0	0	0	0	0	0
Mucate	0	0	0	0	0	5	0	0	0	0	0	0	10	0	100	0
Raffinose	0	70	100	85	0	100	7	94	99	0	0	40	3	11	0	0
L-Rhamnose	0	35	76	15	0	95	94	0	1	30	1	5	75	0	100	0
Salicin	100	100	100	97	92	98	45	94	99	0	0	0	0	55	13	0
D-Sorbitol	0	100	100	95	92	100	100	65	1	30	43	29	2	0	100	0
Sucrose	100	100	21	98	100	100	0	100	99	1	0	1	1	98	0	0
Trehalose	100	100	100	100	100	100	100	100	100	90	85	65	100	93	100	0
D-Xylose	40	100	85	100	0	100	100	94	99	4	11	2	2	9	100	0

[a]Each number is the percentage of positive reactions after 2 d at 36°C unless otherwise indicated (gelatin liquefaction and deoxyribonuclease; *Photorhabdus luminescens*).

[b]Table taken from Farmer, 1999.

[c]In a Request for an Opinion published in 1987, Le Minor and Popoff proposed replacement of the type species of *Salmonella* (*Salmonella choleraesuis* subsp. *choleraesuis*) with *Salmonella enterica* as the former was considered to be a source of confusion. Although the Request was denied by the Judicial Commission, their proposal resulted in an alternative naming convention which has found widespread endorsement in the public health community. This matter was revisited in July 2002 by the Judicial Commission during the IUMS Congress in response to several new Requests for an Opinion and will likely result in a decision to replace the type strain *Salmonella choleraesuis* subsp. *choleraesuis* with *Salmonella enterica* subsp. *enterica* LT2, while preserving the former rather than placing it on the list of rejected names. We view the six subspecies of *Salmonella choleraesuis* as deprecated. Readers are also advised that the names *Salmonella enteritidis*, *Salmonella paratyphi*, *Salmonella typhi*, and *Salmonella typhimurium* are synonyms of *Salmonella enterica* subsp. *enterica* and refer to specific serovars. These names have not been deprecated at this time as they remain in use by some public health reporting agencies.

(continued)

TABLE BXII.γ.193. *(cont.)*

Characteristic	*Yersinia pestis*	*Yersinia aldovae*	*Yersinia bercovieri*	*Yersinia enterocolitica*	*Yersinia frederiksenii*	*Yersinia intermedia*	*Yersinia kristensenii*	*Yersinia mollaretii*	*Yersinia pseudotuberculosis*	*Yersinia rohdei*	*Yersinia ruckeri*	*Yokenella regensburgei (Koserella trabulsii)*	Enteric Group 58	Enteric Group 60	Enteric Group 68	Enteric Group 69
Indole production	0	0	0	50	100	100	30	0	0	0	0	0	0	0	0	0
Methyl red	80	80	100	97	100	100	92	100	100	62	97	100	100	100	100	0
Voges–Proskauer	0	0	0	2	0	5	0	0	0	0	10	0	0	0	50	100
Citrate (Simmons)	0	0	0	0	15	5	0	0	0	0	0	92	85	0	0	100
Hydrogen sulfide (TSI)	0	0	0	0	0	0	0	0	0	0	0	0	0	0	0	0
Urea hydrolysis	5	60	60	75	70	80	77	20	95	62	0	0	70	50	0	0
Phenylalanine deaminase	0	0	0	0	0	0	0	0	0	0	0	0	0	0	0	0
Lysine decarboxylase	0	0	0	0	0	0	0	0	0	0	50	100	100	0	0	0
Arginine dihydrolase	0	0	0	0	0	0	0	0	0	0	5	8	0	0	0	100
Ornithine decarboxylase	0	40	80	95	95	100	92	80	0	25	100	100	85	100	0	100
Motility	0	0	0	2	5	5	5	0	0	0	0	100	100	75	0	100
Gelatin hydrolysis (22°C)	0	0	0	0	0	0	0	0	0	0	0	30	0	0	0	0
Growth in KCN	0	0	0	2	0	10	0	0	0	0	15	92	100	0	100	100
Malonate utilization	0	0	0	0	0	5	0	0	0	0	0	0	85	100	0	100
Esculin hydrolysis	50	0	20	25	85	100	0	0	95	0	0	67	0	0	0	100
Tartrate, Jordan's	0	100	100	85	55	88	40	100	50	100	30	0	60	75	0	0
Acetate utilization	0	0	0	15	15	18	8	0	0	0	0	25	45	0	0	25
Lipase (corn oil)	0	0	0	55	55	12	0	0	0	0	0	30	0	0	0	0
DNase (25°C)	0	0	0	5	0	0	0	0	0	0	0	0	0	0	100	0
Nitrate oxidized to nitrite	85	100	100	98	100	94	100	100	95	88	75	100	100	100	100	100
Oxidase, Kovacs	0	0	0	0	0	0	0	0	0	0	0	0	0	0	0	0
ONPG test	50	0	80	95	100	90	70	20	70	50	50	100	100	100	0	100
Yellow pigment	0	0	0	0	0	0	0	0	0	0	0	0	0	0	0	0
D-Glucose, acid	100	100	100	100	100	100	100	100	100	100	100	100	100	100	100	100
D-Glucose, gas	0	0	0	5	40	18	23	0	0	0	5	100	85	100	0	100
Fermentation of:																
Adonitol	0	0	0	0	0	0	0	0	0	0	0	0	0	0	0	0
L-Arabinose	100	60	100	98	100	100	77	100	50	100	5	100	100	25	0	100
D-Arabitol	0	0	0	40	100	45	45	0	0	0	0	0	0	0	0	0
Cellobiose	0	0	100	75	100	96	100	100	0	25	5	100	100	0	0	100
Dulcitol	0	0	0	0	0	0	0	0	0	0	0	0	85	0	0	100
Erythritol	0	0	0	0	0	0	0	0	0	0	0	0	0	0	0	0
Glycerol	50	0	0	90	85	60	70	20	50	38	30	0	30	75	50	0
myo-Inositol	0	0	0	30	20	15	15	0	0	0	0	0	0	0	0	0
Lactose	0	0	20	5	40	35	8	40	0	0	0	0	30	0	0	100
Maltose	80	0	100	75	100	100	100	60	95	0	95	100	100	0	50	100
D-Mannitol	97	80	100	98	100	100	100	100	100	100	100	100	100	50	100	100
D-Mannose	100	100	100	100	100	100	100	100	100	100	100	100	100	100	100	100
Melibiose	20	0	0	1	0	80	0	0	70	50	0	92	0	0	0	100
α-Methyl-D-glucoside	0	0	0	0	0	77	0	0	0	0	0	0	55	0	0	100
Mucate	0	0	0	0	5	6	0	0	0	0	0	0	0	0	0	100
Raffinose	0	0	0	5	30	45	0	0	15	62	5	25	0	0	0	100
L-Rhamnose	1	0	0	1	99	100	0	0	70	0	0	100	100	75	0	100
Salicin	70	0	20	20	92	100	15	20	25	0	0	8	100	0	50	100
D-Sorbitol	50	60	100	99	100	100	100	100	0	100	50	0	100	0	0	100
Sucrose	0	20	100	95	100	100	0	100	0	100	0	0	0	0	100	25
Trehalose	100	80	100	98	100	100	100	100	100	100	95	100	100	100	100	100
D-Xylose	90	40	100	70	100	100	85	60	100	38	0	100	100	0	0	100

[a]Each number is the percentage of positive reactions after 2 d at 36°C unless otherwise indicated (gelatin liquefaction and deoxyribonuclease; *Photorhabdus luminescens*).

[b]Table taken from Farmer, 1999.

[c]In a Request for an Opinion published in 1987, Le Minor and Popoff proposed replacement of the type species of *Salmonella* (*Salmonella choleraesuis* subsp. *choleraesuis*) with *Salmonella enterica* as the former was considered to be a source of confusion. Although the Request was denied by the Judicial Commission, their proposal resulted in an alternative naming convention which has found widespread endorsement in the public health community. This matter was revisited in July 2002 by the Judicial Commission during the IUMS Congress in response to several new Requests for an Opinion and will likely result in a decision to replace the type strain *Salmonella choleraesuis* subsp. *choleraesuis* with *Salmonella enterica* subsp. *enterica* LT2, while preserving the former rather than placing it on the list of rejected names. We view the six subspecies of *Salmonella choleraesuis* as deprecated. Readers are also advised that the names *Salmonella enteritidis*, *Salmonella paratyphi*, *Salmonella typhi*, and *Salmonella typhimurium* are synonyms of *Salmonella enterica* subsp. *enterica* and refer to specific serovars. These names have not been deprecated at this time as they remain in use by some public health reporting agencies.

TABLE BXII.γ.194. Screening tests for genera and species of *Enterobacteriaceae* often isolated from human clinical specimens[a]

Organism (genus, species, or serovar)	Test or property[b, c, d]
Salmonella	Lactose[-], sucrose[-], H_2S^+, 01 phage[+e], MUCAP[+], agglutinates in polyvalent serum[d], typical colonies on media selective/differential for *Salmonella* (brilliant green agar, SS agar, Rambach agar, etc.), lysed by the *Salmonella* specific bacteriophage[e], often antibiotic resistant
Salmonella Typhi	H_2S^+ (trace amount only), agglutinates in group D serum
Shigella	Nonmotile, lysine[-], gas[-], agglutinates in polyvalent serum, biochemically inactive, often antibiotic resistant, molecular test: *phoE*[+]
Shigella dysenteriae	Agglutinates in group A serum, D-mannitol[-]
Shigella dysenteriae O1	Catalase[-], agglutinates in O1 serum, Shiga toxin[+]
Shigella flexneri	Agglutinates in group B serum, D-mannitol[+]
Shigella boydii	Agglutinates in group C serum, D-mannitol[+]
Shigella sonnei	Agglutinates in group D serum, D-mannitol[+]
Escherichia coli	Extremely variable biochemically, indole[+], MUG[+], grows at 44.5°C, sometimes antibiotic resistant, molecular test: *phoE*[+]
Escherichia coli O157:H7	Colorless colonies on sorbitol-MacConkey agar, MUG[-], D-sorbitol[-] (or delayed), agglutinates in O157 serum and H7 serum
Yersinia	Grows on CIN agar; often more active biochemically at 25°C than 36°C; motile at 25°C, nonmotile at 36°C; urea[+]
Yersinia enterocolitica (pathogenic serogroups)	CR-MOX[+], pyrazinamidase[-], salicin[-], esculin[-], agglutinates in O sera: 3; 4, 32; 5, 27; 8; 9; 13a, 13b; 18; 20; or 21
Yersinia enterocolitica O3 (a pathogenic serogroup)	D-Xylose[-], agglutinates in O3 serum, tiny colonies at 24 h on plating media
Yersinia enterocolitica nonpathogenic serogroups	CR-MOX[-], pyrazinamidase[+], salicin[+], esculin[+], no agglutination in O sera: 3; 4, 32; 5, 27; 8; 9;13a, 13b; 18; 20 or 21
Citrobacter	Citrate[+], lysine decarboxylase[-], often grows on CIN agar, strong characteristic odor
Hafnia	Lysed by *Hafnia*-specific bacteriophage[e], often more active biochemically at 25°C than 36°C
Klebsiella	Mucoid colonies, encapsulated cells, nonmotile, lysine[+], very active biochemically, ferments most sugars, VP[+], malonate[+], resistant to carbenicillin and ampicillin
Enterobacter	Variable biochemically, citrate[+], VP[+], resistant to cephalothin
Serratia	DNase[+], gelatinase[+], lipase[+], resistant to colistin and cephalothin
Serratia marcescens	L-arabinose[-]
Serratia, other species	L-arabinose[+]
Proteus-Providencia-Morganella	Phenylalanine[+], tyrosine hydrolysis[+], often urea[+], resistant to colistin
Proteus	Swarms on blood agar, pungent odor, H_2S^+, gelatin[+], lipase[+]
Proteus mirabilis	Urea[+], indole[-], ornithine decarboxylase[+], maltose[-]
Proteus vulgaris	Urea[+], indole[+], ornithine decarboxylase[-], maltose[+]
Providencia	No swarming, H_2S^-, ornithine decarboxylase[-], lipase[-]
Morganella	Very inactive biochemically, no swarming, citrate[-], H_2S^-, ornithine decarboxylase[+], gelatin[-], lipase[-], urea[+]

[a]Table taken from Farmer, 1999.

[b]Abbreviations: CIN, cefsulodin-irgasan-novobiocin agar (a plating medium selective for *Yersinia*); CR-MOX, Congo red, magnesium oxalate agar (a differential medium useful for distinguishing pathogenic from nonpathogenic strains of *Yersinia*); MUCAP, 4-methyl-umbelliferyl caprylate (a genus-specific test for *Salmonella*); MUG, 4-methyl-umbelliferyl-β-D-glucuronidase; ONPG, *o*-nitrophenyl-β-D-galactopyranoside; *phoE*, a test done by PCR that is sensitive and specific for *E. coli/Shigella*; VP, Voges–Proskauer.

[c]This table gives only the general properties of the genera, species, and serogroups; exceptions occur. See Table BXII.γ.193 for full biochemical reactions. The properties listed for a genus or group of genera generally apply for each of its species, and the properties listed for a species generally apply for each of its serogroups.

[d]Biochemical test results are given as percentages in Table BXII.γ.193. The serological tests refer to slide agglutination in group or individual antisera (O1, O3, etc.) for *Salmonella*, *Shigella*, *E. coli*, or *Yersinia*, respectively.

[e]These are bacteriophage tests useful for identification.

Taxonomy and Nomenclature Compared with the 1st edition of *Bergey's Manual of Systematic Bacteriology* and the 9th edition of *Bergey's Manual of Determinative Bacteriology*, the present volume contains nomenclatural changes, new genera, and new species. In Table BXII.γ.192, the authors' proposed classification is compared with that in the previous two *Manuals*. Nomenclatural synonyms exist for a number of species, as shown in Table BXII.γ.192.

Arsenophonus A second species has provisionally been added to the genus as *Arsenophonus triatominarum*. The *candidatus* status was used because this species has only been recognized on the basis of its 16S rDNA sequence, and has not been isolated or characterized biochemically (Hypsa and Dale, 1997).

Candidatus **Phlomobacter fragariae** The designation "Candidatus" is given to presumed new species that have not been cultivated. One such organism has been described whose 16S rDNA sequences place it within the family *Enterobacteriaceae* (Zreik

et al., 1998). It has been named, *Candidatus* Phlomobacter fragariae. It is restricted to the phloem of strawberries and is associated with marginal chlorosis of strawberry (Zreik et al., 1998).

Citrobacter *Citrobacter koseri*, *Citrobacter diversus*, and *Levinea malonatica* are all synonyms for the same species. In response to a request for an opinion (Frederiksen, 1990), the Judicial Commission issued the opinion that *C. diversus* was a *nomina rejicienda* (rejected name). Therefore, the correct name is *C. koseri*. Eight new *Citrobacter* species were recently described (Brenner et al., 1993; Schauer et al., 1995). *Citrobacter farmeri* was formerly known as a biogroup of *C. amalonaticus*. The remaining 6 named and 2 unnamed species were previously included in *C. freundii* or had been previously identified as atypical citrobacteria.

Enterobacter *Enterobacter cancerogenus* (formerly *Erwinia cancerogena*) was shown to be a senior subjective synonym of *Enterobacter taylorae* and is therefore the correct name for this species (Grimont and Ageron, 1989). In several recently proposed classifi-

TABLE BXII.γ.195. Intrinsic antimicrobial resistance in some of the common *Enterobacteriaceae*[a]

Genus/Species	Most strains are resistant to
Buttiauxella species	Cephalothin
Cedecea species	Polymyxins, ampicillin, cephalothin
Citrobacter amalonaticus	Ampicillin
Citrobacter freundii	Cephalothin
Citrobacter diversus	Cephalothin, carbenicillin
Edwardsiella tarda	Colistin
Enterobacter cloacae	Cephalothin
Enterobacter aerogenes	Cephalothin
Many other *Enterobacter* species	Cephalothin
Escherichia hermannii	Ampicillin, carbenicillin
Ewingella americana	Cephalothin
Hafnia alvei	Cephalothin
Klebsiella pneumoniae	Ampicillin, carbenicillin
Kluyvera ascorbata	Ampicillin
Kluyvera cryocrescens	Ampicillin
Proteus mirabilis	Polymyxins, tetracycline, nitrofurantoin
Proteus vulgaris	Polymyxins, ampicillin, nitrofurantoin, tetracycline
Morganella morganii	Polymyxins, ampicillin, cephalothin
Providencia rettgeri	Polymyxins, cephalothin, nitrofurantoin, tetracycline
Other *Providencia* species[b]	Polymyxins, nitrofurantoin
Serratia marcescens[c]	Polymyxins, cephalothin, nitrofurantoin
Serratia fonticola	Ampicillin, carbenicillin, cephalothin
Other *Serratia* species	Polymyxins[d], cephalothin

[a]Table taken from Farmer, 1999.

[b]Most strains of *Providencia stuartii* are also resistant to cephalothin and tetracycline.

[c]*Serratia marcescens* can also be resistant to ampicillin, carbenicillin, streptomycin, and tetracycline.

[d]Resistance to polymyxins is common in *Serratia* species, but some strains have zones of 10–12 mm or larger.

cations, *Enterobacter agglomerans* and other species in the "*E. agglomerans* group" have been classified in the genus *Pantoea*. The *E. agglomerans* group was extremely heterogenous and a number of species have been proposed for strains previously included in this species (see discussions of *Erwinia*, *Leclercia*, and *Pantoea* below and Table BXII.γ.192). *E. aerogenes* is genetically and phenotypically closer to *K. pneumoniae* than to *Enterobacter cloacae* (Brenner et al., 1972c; Steigerwalt et al., 1976; Bascomb et al., 1971; Grimont and Grimont, 1992). The names "*Klebsiella aerogenes*" and *Klebsiella mobilis* have been used for *E. aerogenes*. The type strain of *E. cloacae* is not representative of the species. If it is maintained, typical strains identified in clinical laboratories might not be called *E. cloacae* (Grimont and Grimont, 1992). This problem is dealt with in detail in the chapter on *Enterobacter*. Recently obtained biochemical test data and DNA–DNA hybridization data indicate that *Enterobacter intermedius* is a senior subjective synonym for the recently described species *Kluyvera cochleae* (A.G. Steigerwalt and J.J. Farmer, unpublished data).

Erwinia Plant isolates of erwiniae have been mainly studied by phytomicrobiologists and phytopathologists. The media, biochemical and other phenotypic tests used for their isolation, enrichment, cultivation, and identification are quite different from those used for other *Enterobacteriaceae*. The 35–37°C incubation temperature used for most other *Enterobacteriaceae* is near, at, or above the maximum growth temperatures of erwiniae. The isolation of erwiniae from humans or animals is rarely reported. It

is not known, however, if they are actually rarely occurring, whether their seeming lack of occurrence reflects improper isolation and enrichment procedures, or, if they are isolated, their identification fails. A study using optimal isolation procedures and an optimum incubation temperature would help to resolve this problem. Also needed is a study to characterize all erwiniae by tests and methods used for other *Enterobacteriaceae*. Erwiniae have been shown to be quite diverse on the basis of DNA relatedness studies (Gardner and Kado, 1972; Brenner et al., 1973b; Brenner et al., 1973b, and now 16S rRNA sequence studies (Kwon et al., 1997). Several recently proposed classifications have expanded the genus *Pantoea* to include the Herbicola or Herbicola-Lathyri group of the genus *Erwinia*. Similarly, there have been proposals to change the circumscription of the genus *Pectobacterium* to include a number of species formerly contained in the genus *Erwinia*, and the new genus *Brenneria* was recently described for other species formerly in *Erwinia* (see Table BXII.γ.192 and chapters on *Erwinia*, *Pantoea*, *Pectobacterium*, and *Brenneria*).

Escherichia* and *Shigella The four nomenspecies of *Shigella* and *E. coli* are a single genomospecies on the basis of DNA relatedness (Brenner et al., 1972a, 1973a). *Shigella* and *E. coli* strains are often extremely difficult to separate biochemically since there are aerogenic (gas-producing) shigellae and lactose-negative, anaerogenic, nonmotile *E. coli*. Such strains can cause a dysentery-like diarrhea, so pathogenicity does not provide definitive separation. Shigellae can be considered metabolically inactive biogroups of *E. coli*. Based mainly on molecular data, it has been argued that these organisms do not represent distinct species, but a continuum of closely related lines of descent from a common ancestor. In this *Manual*, they are classified as distinct species because of the ease of communication these names provide in medical microbiology and because of the resistance and confusion that would be caused by reclassification. However, the original usage implying that shigellae were pathogenic and *E. coli* was not is certainly not true. Tamura et al. (1986) proposed an alternative classification. They defined a new genus *Leclercia*, with the single species *L. adecarboxylata*, which was originally named *Escherichia adecarboxylata*.

Hafnia *Hafnia alvei* remains the only named species in this genus. DNA studies revealed two separate DNA relatedness groups within *H. alvei*, with approximately one-half of the strains in each relatedness group (Steigerwalt et al., 1976). A second species was not designated since no single biochemical test or series of tests served to unequivocally separate the two relatedness groups (F.W. Brenner and J.J. Farmer III, personal communication). "*Hafnia protea*" is an illegitimate name for *Obesumbacterium proteus*. There were two biogroups within *O. proteus*. Biogroup 2 is *O. proteus*, whereas biogroup 1 is a metabolically inactive biogroup of *H. alvei* that has become adapted to the brewery environment.

Klebsiella In 1984, Orskov proposed a classification in which *Klebsiella ozaenae* and *Klebsiella rhinoscleromatis* would be classified as subspecies of *K. pneumoniae* (Orskov, 1984a). While there is no doubt that these 3 nomenspecies are the same genomospecies on the basis of DNA relatedness studies (Brenner, Steigerwalt, and Fanning, 1972), many laboratories still accept their original classification and report them as separate species. *Klebsiella ornithinolytica* was described by Sakazaki et al. (1989a) for strains previously determined to be a biogroup of *Klebsiella planticola* by Farmer et al., 1985a. These strains have been variously referred to as ornithine-positive *K. oxytoca*, NIH (Japan) Group 12, CDC

TABLE BXII.γ.196. Additional biochemical reactions of *Enterobacteriaceae*[a, b, c]

Species	Nitrate reductase	Tetrathionate reductase	D-Galacturonate	2-keto-Gluconate	γ-Glutamyl transferase
Citrobacter amalonaticus		+		+	+
Citrobacter freundii	+, A	+	+	+	+
Citrobacter koseri		−		+	+
Edwardsiella hoshinae				−	
Edwardsiella tarda	+, B	+		−	−
Erwinia carotovora	+, A	−	+	−	+
Enterobacter aerogenes	+, A	−	+	+	+
Enterobacter cloacae	+, A	−	d	+	+
Enterobacter gergoviae				+	+
Enterobacter sakazakii				+	
Escherichia coli	+, A	−		−	+
Hafnia alvei	d, A or B	d	+	+	+
Klebsiella oxytoca	d, A	d		+	+
Klebsiella pneumoniae subsp. *ozaenae*		−		+	+
Klebsiella pneumoniae subsp. *pneumoniae*	+, A	−		+	+
Klebsiella pneumoniae subsp. *rhinoscleromatis*		−		d	−
Morganella morganii	d, A	+		−	+
Pantoea agglomerans	+, A	−	+	+	+
Proteus mirabilis	+, A	+		−	+
Proteus vulgaris	+, A	+		−	+
Providencia alcalifaciens	+, B	+		−	+
Providencia rettgeri	+, A	+		d	+
Providencia stuartii	+, A	+		−	+
Salmonella enterica subsp. *arizonae*	+, A	+	−	−	−
Salmonella enterica subsp. *diarizonae*	+, A	+	+	−	+
Salmonella enterica subsp. *enterica*	+, A	+	-	−	+
Salmonella enterica subsp. *houtenae*	+, A	+	+	−	+
Salmonella enterica subsp. *salamae*	+, A	+	+	−	+
Serratia ficaria				+	+
Serratia liquefaciens	+, A	+	+	+	+
Serratia marcescens	+, A	d	+	+	+
Serratia odorifera	+, A	−		+	+
Serratia plymuthica	+, A	−	+	+	+
Serratia rubidaea	+, A	−	+	+	+
Shigella boydii	+, A	−		−	d
Shigella dysenteriae		−		−	d
Shigella flexneri	+, A	−		−	d
Shigella sonnei	+, A	−		−	−
Yersinia enterocolitica	+, B	d		d	+
Yersinia pestis	d, B	d		-	−
Yersinia pseudotuberculosis	+, B	−		−	+
Yersinia frederiksenii	+, B	+			
Yersinia intermedia	+, A or B	+			
Yersinia kristensenii	+, B	d			

[a]Symbols: +, 90% or more of strains are positive; d, 10.0–89.9% positive; −, 0–9.9% positive; blank space, not available; A, type A nitrate reductase; B, type B nitrate reductase.

[b]The nitrate reductase test was incubated at 32°C; all other tests were incubated at 35–37°C. The γ-glutamyl transferase test was read at 24 h; all other test results were read at 48 h.

[c]Data compiled from: Pichinoty et al. (1969), Grimont (1977b), Richard (1977), Grimont et al. (1978a), Le Minor et al. (1979), Bercovier et al. (1980a), Bercovier et al. (1980b), Brenner et al. (1980e), Giammanco et al. (1980), Ursing et al. (1980a), and Buissière et al. (1981).

Klebsiella Group 47 (indole-positive, ornithine-positive); and indole-positive, ornithine-positive biogroup of *Klebsiella planticola*. Laboratories agreed on the phenotypic properties of this organism, but they disagreed on whether DNA relatedness data indicated that it should be retained as a biogroup of *K. planticola* or that it should be a separate species.

Kluyvera The recently described species *K. cochleae* (Müller et al., 1996) appears to be a junior subjective synonym for *Enterobacter intermedius* (A.G. Steigerwalt and J.J. Farmer, unpublished data). In this *Manual*, this species is not included as one of the species of *Kluyvera*, but see *Enterobacter intermedius*.

Obesumbacterium proteus. See *Hafnia* above.

Pantoea In this *Manual*, the genus *Pantoea* includes species previously contained in the *Enterobacter agglomerans* complex (including *Erwinia herbicola*, *Erwinia lathyri*, *Erwinia ananas*, *Erwinia*

uredovora, *Erwinia milletiae*, and *Erwinia stewartii*). Future proposals may expand its definition to include additional species to be defined from some of the remaining *Enterobacter agglomerans* DNA hybridization groups that have not yet been classified (Brenner et al., 1984a; Grimont and Grimont, 1992). See sections on *Enterobacter* and *Erwinia* above.

Photorhabdus* and *Xenorhabdus Substantial differences in biochemical reactions have been reported by different laboratories (Grimont et al., 1984b; Akhurst and Boemare, 1988; Farmer et al., 1989). Many of the differences are probably due to slow and weak reactions and to differences in media. In the proposed classification of Akhurst and Boemare, *Xenorhabdus beddingii* and *Xenorhabdus bovienii*, are classified as species (Akhurst and Boemare, 1988). They were originally described as subspecies of *Xenorhabdus nematophilus*. Four additional genomospecies within

Photorhabdus luminescens (*Xenorhabdus luminescens*) were not named because they are not phenotypically separable from *P. luminescens* (Farmer et al., 1989).

Proteus   The type strain of *Proteus vulgaris* is in a DNA relatedness group that includes only one other strain out of 36 tested. The remaining strains are in four new genomospecies, which include essentially all isolates currently identified as *P. vulgaris* (Brenner et al., 1995). A request for an opinion to replace the type strain was submitted (Brenner et al., 1995), and was recently granted (H. Trüper, personal communication). MacDonell and Colwell (1985) recommended that *Plesiomonas shigelloides* be classified in the genus *Proteus* because its 5S rRNA sequence is closely related to that of *P. mirabilis*. They did not formally describe the new combination ("*Proteus shigelloides*"). Such a change is not supported by phenotypic characteristics or by DNA relatedness and the new combination has not been used.

Salmonella   The classification and nomenclature of salmonellae has long been a source of confusion, even to specialists. Much of the confusion was due to the practice of naming serotypes (serovars) and equating them with species. All phylogenetic data indicate that the evolution of the genus *Salmonella* is a continuum, but several recent classifications recognize two distinct species of *Salmonella*, one of which contains six subspecies (Reeves et al., 1989a; Brenner and McWhorter-Murlin, 1998). The type species of the genus is *Salmonella choleraesuis*. Unfortunately, there is also a serotype Choleraesuis that, until recently was treated as a species. For this reason Le Minor and Popoff (1987) requested an opinion to replace the name *S. choleraesuis* with "*Salmonella enterica*". This request was not approved by the Judicial Commission (Wayne, 1994; Judicial Commission, 1991), although the wording of the refusal seemed to leave the door open for another request to remedy the confusion caused by the name *S. choleraesuis*. This request has now been resubmitted by Euzéby (1999), who also requested that "*Salmonella typhi*" be designated as a species. In the interim, the WHO International Collaborating Centre for *Salmonella* located at the Pasteur Institute in Paris, the National *Salmonella* Center located at the Centers for Disease Control and Prevention in Atlanta, and many other reference laboratories and their constituents have begun to use *S. enterica*, so both names are currently in use. In one proposed classification there are 6 subspecies of *S. choleraesuis*: subsp. *enterica* (or subspecies I or subspecies 1), for those biochemically typical serotypes previously placed in subgenus I; subsp. *salamae* (or subspecies II or subspecies 2), for biochemically atypical serotypes previously in subgenus II; subsp. *arizonae* (or subspecies IIIa or subspecies 3a) for the monophasic strains previously in "*Arizona hinshawii*" or subgenus III; subspecies subsp. *diarizonae* (or subspecies IIIb or subspecies 3b) for the diphasic strains previously in "*A. hinshawii*" or subgenus III; subsp. *houtenae* (or subspecies IV or subspecies 4) for the biochemically atypical strains previously in subgenus IV; and subsp. *indica* (or subspecies VI or subspecies 6) for more recently defined atypical strains that differ from the biochemical patterns of subspecies I-IV. The second species is *Salmonella bongori*, a small number of biochemically unique serotypes that have been classified as subsp. *bongori* (or subspecies V or subspecies 5) but are considered to be separate species in the proposed classification of Reeves et al., (1989a) and in this *Manual*. The practice in medical bacteriology is to use names for the serotypes in subsp. *enterica*. For example, the serotypes formerly reported as *S. typhi* and *S. typhimurium* would now be reported as *S. choleraesuis* (or "*S. enterica*") subsp. *enterica* (or I or 1) serotype Typhi or Typhimurium. Note that the serotype name is capitalized and nonitalicized. Alternatively these serotypes could be reported simply as *Salmonella* serotype Typhi and *Salmonella* serotype Typhimurium. Serotypes in the other subspecies are reported similarly. Some laboratories choose to use names (where the serotypes are named) and some choose to use the antigenic formulae. Serotypes in subsp. *arizonae* and subsp. *diarizonae* are not named. Therefore a typical report, would be *Salmonella* serotype 60:k:z (= *Salmonella enterica* subsp. *diarizonae* serotype 60:k:z). *Salmonella enterica* subsp. *indica* serotype Brookfield has the antigenic formula 66:z41:-. Any of the following reports would be accurate: *Salmonella* serotype Brookfield, *Salmonella* subsp. *indica* serotype Brookfield, *Salmonella* subsp. V Brookfield, as would any of the above designations with the antigenic formula used instead of the name Brookfield. Much of the literature still uses the "serotype as species" classification and uses the Linnean system of writing the genus and species in italics. Examples include *Salmonella typhi*, *Salmonella typhimurium*, and *Salmonella enteritidis*.

Yokenella   *Yokenella*, with its single species *Y. regensburgei*, was described by Kosako et al. in 1984. The genus *Koserella* with its single species *K. trabulsii*, was described by Hickman-Brenner et al. in 1985a *Y. regensburgei* and *K. trabulsii* were shown to be subjective synonyms by Kosako et al. in 1987 Other synonyms for this organism are *Hafnia* hybridization group 3, Enteric Group 45, NIH (Japan) biogroup 9. Even though *K. trabulsii* had priority by virtue of its earlier alphabetical appearance on Validation List No. 17 in the *International Journal of Systematic Bacteriology*, both groups of investigators agreed that *Yokenella* should be the genus name.

Unnamed Enteric Groups   A number of groups belonging to the family *Enterobacteriaceae* have been characterized in the Enteric Reference Laboratory at the CDC but not given genus and species names. Four Enteric Groups are briefly described below. Their biochemical profiles are included in Table BXII.γ.193.

Enteric Group 58   Enteric Group 58 consists of strains that were first recognized in 1981 and described in 1985 (Farmer et al., 1985a). The first five isolates were from human clinical specimens; four from wounds (foot, ankle, leg, and hip), and one from feces. Since the original report in 1985, the number of isolates at CDC has grown to 21, including an isolate from a case of bacteremia. Enteric Group 58 has the general properties of the family *Enterobacteriaceae*.

Enteric Group 60   Enteric Group 60 consists of four strains first recognized in 1981 and described in 1985 (Farmer et al., 1985a). All of the strains were from human clinical specimens (three from urine and one from sputum). Enteric Group 60 is inactive biochemically and was originally thought to be most like *Morganella morganii*. However, it is sensitive to colistin and tyrosine-negative, reactions incompatible with *Morganella*. Enteric Group 60 has the general properties of the family *Enterobacteriaceae*. Its closest phylogenetic relatives are not yet known.

Enteric Group 68   Enteric Group 68 consists of a small group of strains isolated from human urine that were first recognized in 1981 and described in 1985 (Farmer et al., 1985a). The group is positive for DNase, but otherwise quite different from *Serratia*. Enteric Group 68 also has the general properties of the family *Enterobacteriaceae*. Its closest phylogenetic relatives are not yet known.

Enteric Group 69   Enteric Group 69 is the name given to a group of strains isolated from refrigerated beef carcasses (Farmer et al., 1985a). The strains were phenotypically similar to *E. sa-*

kazakii in most biochemical reactions, including production of a yellow pigment. However, they were only 43% related to *E. sakazakii* by DNA–DNA hybridization. Recent DNA relatedness data by Kosako et al. (1996) indicated that *Enterobacter kobei* is the closest relative of Enteric Group 69.

Genus I. *Escherichia* Castellani and Chalmers 1919, 941T^AL

FLEMMING SCHEUTZ AND NANCY A. STROCKBINE

Esch.er.i' chi.a. M.L. fem. n. *Escherichia* named after Theodor Escherich, who isolated the type species of the genus.

Straight cylindrical rods, 1.1–1.5 × 2.0–6.0 μm, occurring singly or in pairs. Conform to the general definition of the family *Enterobacteriaceae*. Gram negative. Motile by peritrichous flagella or nonmotile. Aerobic and facultatively anaerobic having both a respiratory and a fermentative type of metabolism, but anaerogenic biotypes occur. Oxidase negative. Chemoorganotrophic. **Both acid and gas are formed from most fermentable carbohydrates, but *i*-inositol is not utilized and D-adonitol is utilized only by *Escherichia fergusonii*. Lactose is fermented by most strains of *Escherichia coli*,** but fermentation may be delayed or absent in *Escherichia blattae*, *Escherichia hermannii*, *Escherichia fergusonii*, and *Escherichia vulneris*. **Do not grow in KCN (with the exception of *E. hermannii* and a small proportion of *E. vulneris*). Usually do not produce H$_2$S.** *E. coli* occur naturally in the lower part of the intestine of warm-blooded animals, *E. blattae* in the hind-gut of cockroaches, and *E. fergusonii*, *E. hermannii*, and *E. vulneris* are found in the intestine, as well as extraintestinal sites of warm-blooded animals. Seven copies of the *rrn* operon with genes coding for 16S, 23S, and 5S rRNA are present on the chromosome of *E. coli*. Comparative sequence analysis between the genes for 16S rRNA of *E. coli*, *E. vulneris*, and *E. hermannii* and homologous genes from all eubacteria places *E. coli* and *E. vulneris* together in a tightly related cluster with shigellae, and *E. hermannii* between *Salmonella* spp. and *Citrobacter freundii* (Cilia et al., 1996). Based on 16S rRNA sequencing, escherichiae belong in the *Gammaproteobacteria*.

The mol% G + C of the DNA is: 48–59.

Type species: **Escherichia coli** (Migula 1895) Castellani and Chalmers 1919, 941 (*Bacillus coli* Migula 1895, 27.)

FURTHER DESCRIPTIVE INFORMATION

Cell morphology Escherichiae are straight, cylindrical, Gram-negative rods with rounded ends that are 1.1–1.5 μm in diameter and 2.0–6.0 μm in length. They occur singly or in pairs and can be motile by peritrichous flagella or nonmotile. Fig. BXII.γ.191 shows negatively stained preparations of each of the *Escherichia* species. See Nanninga (1985) for a comprehensive treatment of the ultrastructure of *E. coli*.

Phylogenetic and systematic treatment The genus consists of five species: *E. coli*, *E. hermannii*, *E. fergusonii*, *E. vulneris*, and *E. blattae*. Biochemical reactions that will help in differentiating between the species of *Escherichia* are listed in Table BXII.γ.197.

Biochemical reactions Escherichiae produce strong acids and usually gas from the fermentation of D-glucose (positive in the Methyl Red test) and do not produce acetyl-methyl carbinol (acetoin) (negative in the Voges–Proskauer test). Sodium acetate is frequently used as a sole carbon source, except by *E. blattae* and a majority of *E. vulneris* strains. Citrate (Simmons' citrate agar) cannot be used by *E. coli* and *E. vulneris*, whereas a smaller proportion of *E. fergusonii* and *E. hermannii* exhibit immediate or delayed use of this substrate. Growth on Simmons' citrate agar by *E. blattae* is variable and probably strain specific.

Lysine is decarboxylated by the majority of strains. Exceptions include "metabolically inactive" *E. coli* strains, the majority of enteroinvasive *E. coli* strains (EIEC), and *E. hermannii*. Ornithine is decarboxylated by all species except by *E. vulneris* and a little less than half of *E. coli* strains. Indole is produced by all species except *E. blattae* and *E. vulneris*.

Other tests that help in differentiating between the species of *Escherichia* include growth in potassium cyanide, malonate utilization, and acid production from D-adonitol, D-arabitol, cellobiose, dulcitol, lactose, D-mannitol, melibiose, D-sorbitol, and mucate. See Table BXII.γ.197.

Molecular data Findings from the comparison of 16S rDNA sequences performed with strains of *E. coli*, *Salmonella* spp., and *C. freundii* have shown a close phylogenetic relatedness between these bacteria (Ahmad et al., 1990; Cilia et al., 1996; Chang et al., 1997). Analysis of 16S rDNA sequences separates the two *Salmonella* species (*S. enterica* and *S. bongori*) from the complex of *E. coli* and *Shigella*, and shows that *E. hermannii* is more closely related to *Salmonella enterica* and *C. freundii* (Christensen et al., 1998). This analysis is unable to separate inter-operon variation of *E. coli* from strains of *E. coli* and from *Shigella*. Patterns of sequence heterogeneity in strains from the *E. coli* Collection of Reference (ECOR) (Ochman and Selander, 1984) have been located at regions V1 and V6 of cloned 16S rRNA genes (Martinez-Murcia et al., 1999).

Average DNA relatedness assessed by DNA–DNA hybridization among *Escherichia* species ranges from 29% to 94% (Table BXII.γ.198). DNAs from different strains of *E. coli* are closely related (average, 84%; Table BXII.γ.198). With the exception of *S. boydii* serotype 13, the DNAs of *E. coli* and the four *Shigella* species show such a high degree of relatedness (average 80–87%, except *S. boydii* type 13, which is about 65%) that these species should be considered as a single species (Brenner et al., 1973a). The distinction between these bacteria prevails, however, for reasons of historical/medical precedent and to avoid confusion in the literature and with existing surveillance systems.

Based on complete sequencing of the K-12 strain MG1655 (Blattner et al., 1997), it is estimated that the *E. coli* lineage diverged from the *Salmonella* lineage some 100 million years ago (Lawrence and Ochman, 1998). This is a little less than the 120–160 million years estimated by calibration of the rate of 16S rRNA evolution in bacteria (Ochman and Wilson, 1987). The chromosome of *E. coli* strain MG1655 is 4,639,221 bp and contains 4,288 open reading frames (ORFs). Approximately 18% of these ORFs represent genes that have been acquired and have persisted

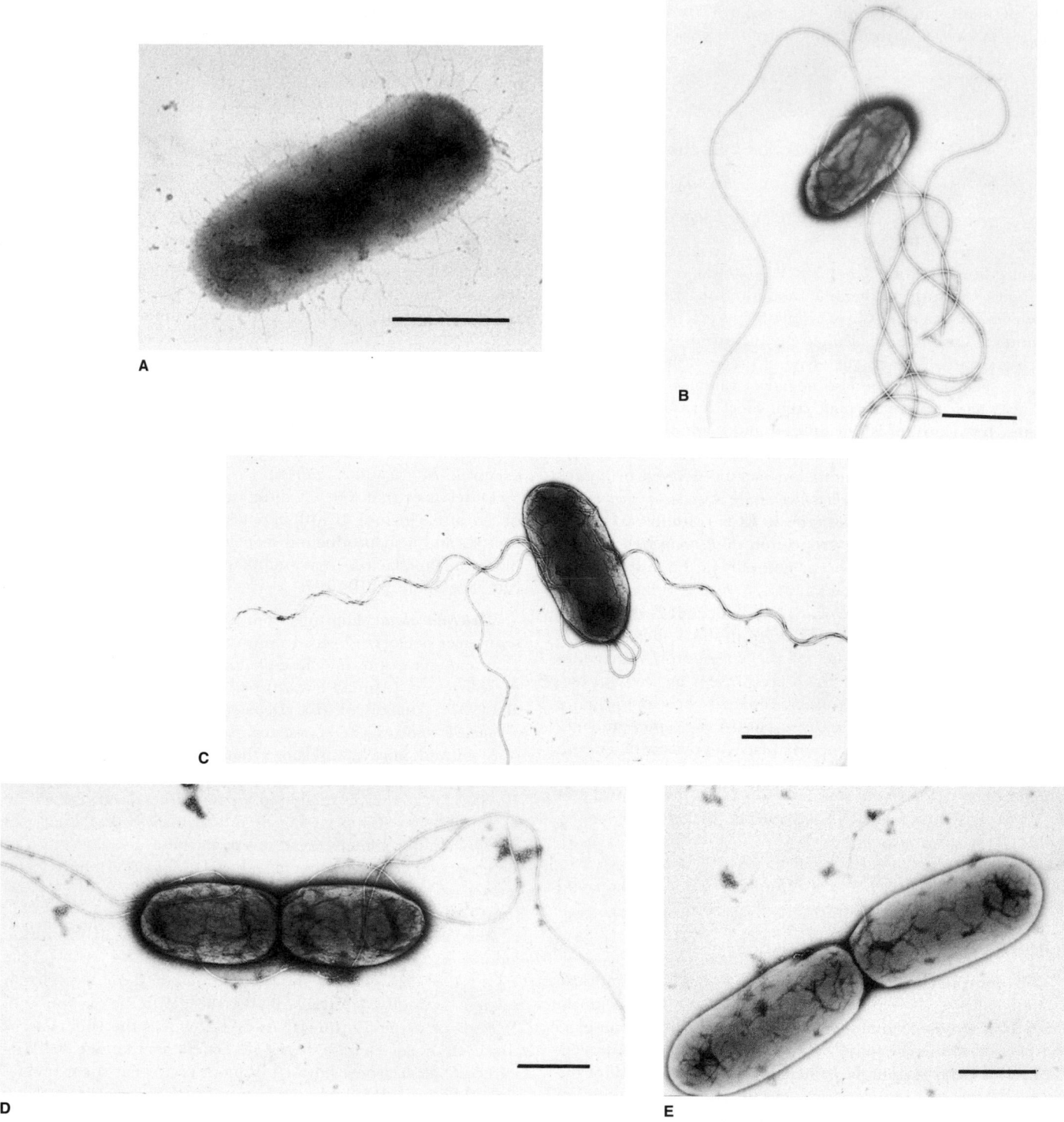

FIGURE BXII.γ.191. Electron micrographs of *E. blattae* strain 2928-78 (*A*), *E. coli* O157:H7 strain EDL933 (*B*), *E. fergusonii* strain 2460-89 (*C*), *E. hermannii* strain 2456-88 (*D*), and *E. vulneris* strain 2485-88 (*E*) prepared by negatively staining in 0.5% (w/v) uranyl acetate. Bar = 1000 nm. (Micrograph courtesy of Charles D. Humphrey, CDC.)

since divergence. This is similar to estimates based on analysis of codon usage, indicating that 16% of sequenced genes arose through horizontal transfer (Médigue et al., 1991). The ability of individual strains and lineages to acquire foreign DNA results in great heterogeneity in both individual genes and in the size of the *E. coli* chromosome. This is exemplified by the unusual

serotype O157:H7, the genome of which is 5.5 Mb in size, 859 Kb larger than that of the laboratory strain K-12 (Hayashi et al., 2001; Perna et al., 2001). The genome sizes of 14 *E. coli* strains from the five major subgroups in the ECOR (Ochman and Selander, 1984) range from 4.66–5.30 Mb (Bergthorsson and Ochman, 1995), and the uropathogenic *E. coli* strain J96 genome is

TABLE BXII.γ.197. Differentiation of the five species of *Escherichia*[a,b]

Test	*E. coli*	*E. coli* (metabolically inactive strains)	*E. blattae*	*E. fergusonii*	*E. hermannii*	*E. vulneris*
Indole	+	[+]	−	+	+	−
Citrate, Simmons	−	−	d	[−]	−[c]	−
Lysine decarboxylase	+	d	+	+	−[d]	[+]
Ornithine decarboxylase	d	[−]	+	+	+	−
Motility	+	−	−[e]	+	+	+
KCN, growth	−	−	−	−	+	[−]
Malonate utilization	−	−	+	d	−	[+]
D-Glucose, gas	+	−	+	+	+	+
Acid production from:						
D-Adonitol	−	−	−	+	−	−
D-Arabitol	−	−	−	+	−	−
Cellobiose	−	−	−	+	+	+
Dulcitol	d	d	−	d	[−]	−
Lactose	+	[−]	−	−[f]	d	[−][f]
D-Mannitol	+	+	−	+	+	+
Melibiose	[+]	d	−	−	−	+
D-Sorbitol	+	d	−	−	−	−
Mucate	+	d	d	−	+	[+]
Acetate utilization	+	d	−	+	[+]	d
Yellow pigmentation	−	−	−	−	+	d

[a]Data compiled from references Farmer (1999), Cowan et al. (1995), Holt et al. (1994), and Richard (1989). Reactions for indole for *E. fergusonii* and melibiose for *E. coli* differ slightly in these references. The reactions listed in this table are supported by our own unpublished data.

[b]Symbols: −, 0–10% positive; [−], 11–25% positive; d, 26–75% positive; [+], 76–89% positive; +, 90–100% positive. Results are for 48 h incubation at 36° ± 1°C.

[c]Delayed positive in approximately a fifth of *E. hermannii* strains.

[d]Delayed positive in a third of *E. hermannii* strains.

[e]75% of *E. blattae* strains will become motile after incubation of more than 2 d.

[f]Delayed positive in approximately two thirds of *E. fergusonii* and *E. vulneris* strains.

TABLE BXII.γ.198. DNA relatedness among escherichiae[a]

Labeled DNA from	Average percent relatedness				
	E. coli	*E. blattae*	*E. fergusonii*	*E. hermannii*	*E. vulneris*
E. coli	84	42	64	38	43
E. blattae		90		39	29
E. fergusonii	57		94	59	
E. hermannii	43	32		89	33
E. vulneris	39	29	33	36	78

[a]Data compiled from Ewing, 1986b.

reported to be 5.12 Mb (Melkerson-Watson et al., 2000) as estimated by pulsed field gel electrophoresis (PFGE).

Of the five species in the genus *Escherichia*, *E. coli* is the most studied. Because the amount of knowledge about the other four species is limited, the descriptive information to follow applies to *E. coli*, unless indicated otherwise.

Cell wall composition The chemical composition and molecular structure of the cell wall of *E. coli* has been extensively studied and is described in detail by Neidhardt and Umbarger (1996) and Park (1996). The structural rigidity of the cell wall is provided by the murein sacculus, which consists mainly of a single monomolecular sheet of murein, a complex polymer composed of roughly equal amounts of polysaccharides (N-acetylglucosamine [GlcNAc] and N-acetylmuramic acid [MurNAc]) and peptides (L-alanine, D-glutamic acid, L-*meso*-diaminopimelic acid (DAP), and D-alanine). Linear chains of alternating units of GlcNAc and MurNAc are linked together by β-1→4 glycosidic bonds, and short chains of the above peptides in alternating D and L optical isomers are attached to the sugars through amide linkages to the carboxyl groups of each muramic acid. Adjacent glycan strands are cross-linked to each other through the peptide side chains to create one giant molecule that provides structural support to the cell. Notable features of the murein of *E. coli* include the presence of a small percentage of peptide chains that either lack the D-alanine or terminate in an additional D-alanine and the absence of amidation involving the carboxyl groups of glutamic acid and DAP. A molecule of lipoprotein is also attached to about every tenth muropeptide.

Outer membrane The outer membrane of an average *E. coli* cell contains over a million molecules of lipopolysaccharide (LPS), which consist of three covalently linked domains: (1) lipid A (endotoxin), (2) the core region of phosphorylated nonrepeating oligosaccharides, and (3) the O antigen polymer of immunogenic repeating oligosaccharides (1–40 units).

Fine structure

Flagella Motile organisms of the genus typically possess 5–10 flagella per cell, which are randomly situated around the cell surface (peritrichous flagellation). The flagellar filament is about 20 nm in diameter and may be up to 20 μm long. It consists of subunits of a single protein, flagellin, which is encoded by the *fliC* gene. Fifty-three antigenically distinct types of flagellin have been described (Ørskov and Ørskov, 1984a). Electron microscopic studies of these antigenically distinct types of flagellar filaments revealed differences in surface structure; six unique flagellar morphotypes have been described (Lawn et al., 1977). Unlike *Salmonella*, most *E. coli* strains have only one flagellin gene and do not undergo phase variation. Exceptions have been described (Ratiner, 1967, 1982, 1999). See Macnab (1996) for a complete description of the structure and genes involved in motility.

Fimbriae In addition to the proteinaceous flagella, most strains have fimbriae (pili) or fibrillar proteins often extending in great numbers from the bacterial surface and far out into the surrounding medium. A typical *E. coli* K-12 cell contains 100–500 type 1 fimbriae arranged peritrichously, each with a diameter of approximately 7 nm and a length of 0.2–2.0 μm. More than 30 different fimbriae have been described in *E. coli*, which commonly expresses more than one type at a time. The characteristics, biogenesis, and classification schemes of fimbriae are reviewed by Low et al. (1996). Fimbriae have historically been classified by phenotypic properties. One widely used scheme classifies fimbriae according to their adhesive properties for red blood cells from different host species in the presence of mannosides. By this method, two main types of fimbriae are recognized: mannose-sensitive (MS) fimbriae, which are unable to agglutinate red blood cells in the presence of α-D-mannose, and mannose-resistant (MR) fimbriae, which are able to agglutinate red blood cells in the presence of this sugar. MS fimbriae, which include the so-called type 1 fimbriae (pili), are found in the majority of *E. coli* strains and comprise a group of more or less serologically related antigens. Because they are expressed by pathogens as well as commensal organisms, their role in virulence has been difficult to establish. Evidence for the role of type 1 fimbriae in virulence is reviewed by Abraham and Jaiswal (1997). Type 1 fimbriae mediate avid bacterial attachment to mucosal surfaces, to noncellular host constituents, and to various inflammatory cells. They bind to certain oligomannoside-containing glycoproteins present on mucosal surfaces, including the Tamm–Horsfall glycoprotein, which is synthesized in the kidney and present in urinary slime (Ørskov et al., 1980a); fibronectin, a glycoprotein that is a member of a family of proteins found in the extracellular matrix (ECM), plasma, and other body fluids; and laminin, a glycoprotein present in basement membranes (Kukkonen et al., 1993). The genes involved in the synthesis and regulation of type 1 fimbriae are located on the chromosome. Expression of type 1 fimbriae is subject to being turned on or off (phase variation) as a result of the inversion of a 314 base-pair fragment of DNA containing the promoter region of the gene encoding the major fimbrial subunit (*fimA*). Expression of type 1 fimbriae is influenced by environmental and growth conditions and is controlled by global regulatory factors such as leucine-responsive regulatory protein (Lrp), integration host factor (IHF) and histone-like protein H-NS. See Abraham and Jaiswal (1997) and Low et al. (1996) for a review of the environmental factors and genes involved in synthesis and regulation of type 1 fimbriae.

MR fimbriae are serologically diverse (Ørskov et al., 1980b, 1982; Ørskov and Ørskov, 1990) and often function as virulence factors to mediate adherence that is species- and organ-specific. The genes for these proteins may be located on plasmids or on the chromosome. When located chromosomally, they often cluster together with other virulence genes in regions of the chromosome referred to as pathogenicity islands (PAIs). More than any other virulence factor, the MR pili of *E. coli* illustrate the

species' capacity to adapt to the receptor-specific epithelial cells of certain hosts primarily through horizontal acquisition of gene cassettes on plasmids, phages, or other mobile DNA elements that will allow for colonization. Evidence for the horizontal transfer of fimbrial genes in *E. coli* comes from the remarkable similarity between the genetic organization of its fimbrial operons and those of other members of the family *Enterobacteriaceae* and from the strikingly low C + G content and different codon usage pattern among the fimbrial genes compared to those observed overall among other genes on the *E. coli* chromosome.

Fimbriae may also be classified based on their morphology. One group consists of thick, rod-shaped fimbriae with a diameter of 7 nm (range 3.4–8 nm), a length of 0.5–2 μm, and an axial hole diameter of 2.0–2.5 nm. Fimbriae with these dimensions are represented by the rigid pyelonephritis (P), sialic acid (S), type 1, F6 (987P), colonization factor antigen I (CFA/I), coli surface antigen 1 (CS1) and CS2 pili, and by the bundle-forming CS8 (CFA/III) and CS21 (longus), the latter with homology to the type 4 fimbrial family. Enteropathogenic *E. coli* (EPEC) is known to produce a type 4 fimbria called the bundle-forming pilus (BFP). Another flexible, bundle-forming fimbrial structure of 2–3 nm diameter, designated aggregative adherence fimbriae I (AAF/I), shows no homology to the type 4 class of fimbriae. Together with another recently described AAF/II fimbria, it exhibits sequence and organizational resemblance to the Dr family. Another group consisting of thinner, more flexible fibrillae with a width of 2–5 nm and a length of 0.5–2.0 μm is represented by F4 (K88), F5 (K99), F41, and CS3. Helical fibrillae, where two fibrillae are arranged in a helix, are represented by CS5 and CS7.

Nonfimbrial and related adhesins Bacterial adherence may also be mediated by adhesions that are afimbrial (AFA) or nonfimbrial (NFA). Some of these proteins form larger multimers that aggregate around the bacterial cell as an amorphous structure reminiscent of capsular K antigens. Afimbrial adhesins are found in both uropathogenic *E. coli* (UPEC) and in diffusely adhering *E. coli* (DAEC) and represented by the Afa/Dr family consisting of Dr (previously referred to as O75X) and Dr-II (*drb*), the F1845 pilus (*daa*), and AFAI-IV (*afa*). Dr fimbriae and related adhesins recognize different epitopes of the Dr blood group antigen (Nowicki et al., 1990) and bind to a complement-regulatory protein, the common receptor decay accelerating factor (DAF). *E. coli* strains expressing Dr fimbriae are able to enter epithelial cells by interacting with DAF (Goluszko et al., 1997). Other adhesins such as the M agglutinin and AIDA-I adhesin, a plasmid-encoded outer membrane protein involved in diffuse adherence of certain types of *E. coli* (Benz and Schmidt, 1989), are commonly present.

Colonial and cultural characteristics; life cycles Depending on the degree of polymerization of the O antigen polysaccharide, the phenotypes of strains growing on agar media are described as smooth (S) or rough (R). S forms, which usually grow on nutrient agar as convex, glistening, moist, gray colonies (2–3 mm diameter) with a defined edge or in fluid medium as turbid growth, have developed polysaccharide side chains, while R forms, which usually grow as flat, dry, dull, wrinkled colonies (1–5 mm diameter) with a blurred edge on agar and agglutinate spontaneously in fluid media, have lost their polysaccharide side chains by mutation (Lüderitz et al., 1966). There are intermediate forms between these extremes. Mucoid and slime-producing forms occur. *E. hermannii* is yellow-pigmented, as are half of the described *E. vulneris* strains. See Raetz (1996) and Hull

(1997) for a discussion of the chemical structure, biosynthesis, and biological/virulence properties of LPS.

Nutrition and growth conditions Of the range of temperatures, pH values, water activities, and pressures over which bacterial growth can occur, *E. coli* strains survive and grow over the mid-range (15–45°C) of these environmental conditions. Most strains can grow over a temperature range of approximately 40°C. The normal temperature range for balanced growth extends from 21° to 37°C; however, strains that can grow at temperatures as low as 7.5–7.8°C (Shaw et al., 1971) and as high as 49°C (Herendeen et al., 1979) have been described. A minimum growth temperature for *E. vulneris* of 1.6°C (0.8–2.6°C) was reported for a strain isolated from refrigerated meat (Ridell and Korkeala, 1997). *E. coli* is neutrophilic and will grow over the mid-range of pH, from about pH 5.0 to 9.0 (Ingraham and Marr, 1996).

Metabolism and metabolic pathways Glucose and other carbohydrates are fermented with the production of pyruvate, which is further converted into lactic, acetic, and formic acids. Part of the formic acid is split by a complex hydrogenlyase system into equal amounts of CO_2 and H_2.

Phylogeny The establishment of the *Escherichia coli* Collection of Reference (ECOR) in 1984 (Ochman and Selander, 1984) and subsequent studies of the strains in ECOR and comparisons with other *E. coli* strains have contributed substantially to our understanding of the evolution and population structure of *E. coli*. ECOR is a collection of 72 strains from humans and 16 other mammalian species from various geographical areas that have been grouped into five main groups—A, B (comprising subgroups B1 and B2), C, D, and E—according to their electrophoretic types and enzyme allele (allozyme) profiles, based on the results of multilocus enzyme electrophoresis (MLEE) (Selander et al., 1986). The original data for 35 enzymes (Selander et al., 1987) have been expanded to include allozymes of four esterase loci (Goullet and Picard, 1989), and, based on allelic variation at 38 enzyme-encoding loci, multilocus genotypes have been used to construct a dendrogram based on the neighbor joining algorithm (Herzer et al., 1990), demonstrating the clonal structure of the species. The phylogenetic groups are distinguishable but not identical by random amplified polymorphic DNA (RAPD), restriction fragment length polymorphism (RFLP) of *rrn* genes (ribotyping) (Desjardins et al., 1995), and to a lesser extent by repetitive-element PCR (rep-PCR) fingerprinting using ERIC2 and BOXA1R primers (Johnson and O'Bryan, 2000). Generating a phylogenetic tree of the ECOR and 15 O157 strains by fluorescent amplified-fragment length polymorphism (FAFLP) demonstrated close correlation with the MLEE groups of ECOR and placed the STEC/VTEC O157 strains on an outlier branch (Arnold et al., 1999). Thus, there is sufficient evidence that the ECOR strains broadly represent genotypic variation in clonal groups of *E. coli* in spite of the fact that many isolates are commensal forms from healthy carriers and that only 11 out of the 72 strains are from human disease: 1 strain from a case of asymptomatic bacteriuria, 4 from acute cystitis, and 6 from acute pyelonephritis. Other MLEE studies have showed clonal relationships of diarrheagenic *E. coli*, and a collection of 78 diarrheagenic *E. coli* (DEC) representing 15 clonal groups has been established (Whittam et al., 1993). A similar collection of STEC/VTEC strains has been collected by the STEC Center based at the National Food Safety and Toxicology Center at Michigan State University and is designed to facilitate research on the Shiga/Verocytotoxin producing *E. coli* by providing a standard reference collection of well-characterized strains and a central, on-line accessible database.*

Numerous studies have indicated that *E. coli* and the four named species of *Shigella* should be regarded as being one species (Brenner et al., 1972a, b, 1973a; Goullet, 1980; Ochman et al., 1983; Whittam et al., 1983; Hartl and Dykhuizen, 1984; Karaolis et al., 1994; Stevenson et al., 1994; Whittam, 1996; Pupo et al., 1997). Findings from MLEE studies combined with those from *mdh* (malate dehydrogenase) housekeeping gene sequence studies (Pupo et al., 1997) and ribotyping (Rolland et al., 1998) confirm that the genus *Shigella* comprises a group of closely related pathogenic *E. coli* strains and indicate that *Shigella*, EIEC, and other diarrheagenic *E. coli* strains do not have a single evolutionary origin, but are derived from different ancestral strains many times. Furthermore, pathogenic strains belonging to pathogenic groups of EIEC, EPEC, and ETEC were found to be closely related to ECOR group A strains by MLEE and have *mdh* sequences identical to five ECOR strains from group A, which is thought to represent commensal strains. It has been suggested that any *E. coli* strain may acquire virulence factors from numerous sources, including plasmids, bacteriophages, and other mobile DNA elements from a large pool of strain-specific genes whose origin could be outside the species boundaries, and that this is how a commensal form is turned into a pathogenic form (Pupo et al., 1997; Hurtado and Rodriguez-Valera, 1999; Donnenberg and Whittam, 2001). In their analysis, Lawrence and Ochman (1998) surmised that about 10% of the *E. coli* K-12 genome consists of genes that were acquired in over 200 events of lateral gene transfer, which occurred subsequent to the divergence of *E. coli* and *Salmonella* some 100 million years ago. While mutations have contributed substantially to the heterogeneity of *E. coli*, the importance of these recombinational events should not be underestimated, and transfer of smaller or bigger segments of DNA between different clonal lineages and other species has probably contributed more to the evolution of *E. coli* than anyone could have imagined. Ongoing and future studies will be directed toward increasing our understanding of the principles that govern gene patterns and the relations between individual traits in terms of their significance for virulence and host adaptation. The combined efforts should contribute to a general IDEA (Index of Diversity in Evolution and Adaptation) that will bridge the phylogenetic approach based on the clonal concept on one side and the horizontal transfer of gene cassettes on the other side.

Mutants, plasmids, phages, and bacteriocins The fact that *E. coli* mutants can easily be produced in the laboratory has substantially contributed to our understanding of many genetic mechanisms, ranging from the characterization and function of an individual gene to the description of complex operons. Plasmids carrying resistance genes (R plasmids) have been used as vectors and introduced into both laboratory strains and wild-type strains, and phages and other mobile elements such as transposons have been used widely in both research and applied biotechnology. Virulence genes are often found on plasmids (see pathogenicity section below). *E. coli* strains produce a variety of secreted antibiotically active polypeptides, bacteriocins, and microcins, which have the ability to kill or inhibit competing bacterial strains. Colicins, encoded by plasmids of *E. coli*, act on other

---

*Editorial Note: At the time of publication, this information could be obtained at http://www.shigatox.net.

E. coli or closely related bacteria that do not carry that particular Col plasmid. Small lipoproteins are important components of the secretory apparatus, which facilitates release of colicin and plasmid- and phage-encoded proteins across the outer membrane. For additional information on *E. coli* mutants, plasmids, and phages see Campbell (1996), Bachmann (1996), Helinski et al. (1996), and Harwood (1993).

Antigenic structure

O antigens The main aspect of this analysis is the O antigen determination based on antigenicity of the LPS; O group designations run from O1 to O173 but O groups O31, O47, O67, O72, O93, O94, and O122 have been removed. Also included are the provisional O groups OX3 and OX7 listed by Ewing (1986b), which will receive the designations O174 and O175, respectively. An additional six new O groups representing STEC/VTEC strains are currently being investigated and will receive O designations O176 through O181 (Scheutz, unpublished data). Subtypes exist within most O groups and are designated ab, ac, etc., e.g., O128ab and O128ac. Many of these O antigens cross-react with other O antigens and to some extent to K antigens within *E. coli*, with other members of the genus and with other enterobacteria. Recent molecular typing using primers just outside the O antigen gene cluster (*rfb*) of 148 representative O groups observed unique amplified fragments for each O group with sizes ranging from 1.7 to 20 kb (Coimbra et al., 2000). Subsequent *Mbo*II digestion of PCR-amplified products resulted in clearly identifiable and reproducible O patterns for the great majority of O groups with a variation of band numbers for each pattern ranging from 5 to 25. Computer analysis identified a total of 147 O patterns and allowed subdivision of 13 O groups. However, two or more O groups shared a pattern among 13 other O patterns. The restriction method (*rfb*-RFLP) is more rapid and may prove to be more sensitive than conventional serotyping since 100% of strains are typeable, particularly those that are O rough or nonagglutinating. Additionally, it should facilitate the typing of strains outside the existing O antigen scheme, which is restricted to include only O groups of clinical, epidemiological, or scientific relevance. The success and general application of such a typing scheme will require international collaboration to develop standardized methods for generating, comparing, and maintaining a database of O patterns.

K antigens The K antigens are the acidic capsular polysaccharide (CPS) antigens. K antigens may be separated into two distinct groups designated group I and group II. Group I antigens, which are composed of high-molecular mass (>100 kDa) CPS, are only found in strains with O groups O8, O9, O20, and O101, and are expressed at both 18° and 37°C. Group I antigens are subdivided according to the absence (IA) or presence (IB) of amino sugars on their CPS. The CPSs of group IA antigens share structural identity or resemblance to those from *Klebsiella* spp., whereas the CPSs of group IB antigens share no structural resemblance to those from other bacteria. Representative strains expressing group IA antigens do not contain the *rol* (*cld*) gene encoding the regulator of lipopolysaccharide O-chain length, whereas a similar subset of strains expressing group IB antigens contains the *rol* gene (Dodgson et al., 1996). Portions of the CPSs of some group I antigens are attached to the lipid A-core in a form that has been designated K_{LPS}, which will behave similarly to the traditional O antigens. A good example of a group I K_{LPS}

is K84, which may be operationally defined as an O antigen and was originally designated as O93.

Group II antigens, which are composed of low molecular mass (<50 kDa) CPS, are found primarily in strains with O groups that are associated with extraintestinal disease. The CPSs of many group II antigens have structural resemblance or near identity to those from Gram-positive bacteria. These antigens differ widely in composition and structural features and may be divided into subgroups based on their acidic components. Twenty to fifty percent of the CPS chains are bound to phospholipids. They were originally thought to be temperature dependent, i. e., only expressed at 37°C. However, K2, K3, K10, K11, K19, K54/K96, and K98, which are tentatively classified as group I/II antigens (Finke et al., 1990), show no temperature regulation of their capsules and, like group I antigens, do not depend on an elevated CMP-KDO concentration for capsule expression. Based on genetic data, a subset of the group I/II antigens (K3, K10, and K54/K96) has been designated group III antigens (Pearce and Roberts, 1995). A number of K antigens are closely related (indicated below by "~") or identical (indicated below by "="). The CPSs of some group IB antigens are structurally identical to the side chains of O antigens and are only considered as K antigens when co-expressed with another authentic O antigen. The following 60 different K antigens are recognized: K1, K2a/ac, K3, K4, K5, K6, K7 (=K56), K8, K9 (=O104), K10, K11, K12 (=K82), K13 (~ K20 and ~ K23), K14, K15, K16, K18a, K18ab (=K22), K19, K24, K26, K27, K28, K29, K30, K31, K34, K37, K39, K40, K41, K42, K43, K44, K45, K46, K47, K49 (=O46), K50, K51, K52, K53, K54 (~ K96), K55, K74, K84, K85ab/ac (=O141), K87 (=O32), K92, K93, K95, K97, K98, K100, K101, K102, K103, KX104, KX105, and KX106. The inclusion of an "X" before the number represents a temporary K antigen designation. A description of the serology, chemistry, and genetics of *E. coli* O and K antigens is given by Ørskov et al. (1977).

H antigens Flagellar or H antigens make up the third main group of serotyping antigens. A total of 56 H antigens have been described, but two, H13 and H22, have been removed as being *C. freundii*, and H50 has been withdrawn because it is identical to H10. Cross-reactions are also seen between the H antigens.

Fields et al. (1997) described a tentative molecular method for the differentiation of flagellar antigen groups in *E. coli* based on restriction fragment length polymorphisms (RFLP) in *fliC* using the restriction enzyme *Rsa*I. A wide variety of *fliC* restriction fragment patterns was reported among isolates of 53 different flagellar antigen groups; the majority of the RFLP patterns observed corresponded to a unique H antigen group. Limited numbers of RFLP patterns were observed among members of some H groups, suggesting that the sequence of *fliC* within certain H groups is fairly well conserved. Four patterns were observed among strains expressing the H7 antigen. Interestingly, the same pattern was detected among all *E. coli* strains of serotype O157:H7 and 16 of 18 of serotype O55:H7, reflecting the common lineage of these strains observed by multilocus enzyme typing (Whittam et al., 1993). Nevertheless, sequencing of a total of 20 H7 *fliC* genes, representing 10 different serotypes (Reid et al., 1999; Wang et al., 2000), revealed a notable polymorphism in the *fliC* gene and identified 10 sequences with differences ranging from 0.06% to 3.12%. Recently, a collection of reference strains representing 48 H types was resolved into 62 patterns (F types) using *Hha*I restriction of the *fliC* gene (Machado et al., 2000). A single F type was associated with each of 39 H types and more than one

F type was associated with the other nine H types. Antigenically related H12 and H45 gave a single F type. The determination of *Hha*I-*fliC* F types could allow deduction of all H types and subdivision of some of these. The two above-mentioned molecular typing methods hold promise of a rapid, more refined and specific typing scheme for the H antigens, which may also be helpful in determining phylogenetic relatedness between different clones of *E. coli*. Furthermore, molecular typing has the advantage of allowing typing of nonmotile strains or of strains that do not (sufficiently) express the immunoreactive H antigen, and is likely to expand the present number of significantly different H types. The observed polymorphism in a single determinant, such as the H7 *fliC* gene, stresses the importance of solid and extensive validation of molecular typing with reference to the existing serotyping scheme, and calls for caution in the interpretation of patterns obtained by DNA fingerprinting methods.

Serotyping Subdivision of *E. coli* can be carried out in many ways, but serotyping remains one of the most useful ways to subdivide the species on a global basis. This typing method is based on the many antigenic differences found in structures on the bacterial surface. A serotype is recorded in the following way: O18ac:K1:H7 or O111:H2 (the latter antigenic formula indicates that K antigens are not present in the strain). MR fimbriae, which are present only in some, often pathogenic, serotypes, can also be used for the serological characterization (Ørskov et al., 1977, 1980b; Ørskov and Ørskov, 1990) in which case the complete serotype is recorded as O4:K3:H5; F13 or O147:H19; F4ac. Serotyping procedures are described in Gross and Rowe (1985) and Ørskov and Ørskov (1984a).

Even though complete serotyping involving the many known O, K, H, and F antigens has been carried out in only a very few laboratories, it is well known that the existing number of serotypes is very high.

Antibiotic or drug sensitivity Like other Gram-negative bacteria, *E. coli* is intrinsically resistant to hydrophobic antibiotics, such as macrolides, novobiocins, rifamycins, actinomycin D, and fusidic acid (Nikaido, 1996). The structure of the outer membrane of *E. coli* and its role in mediating intrinsic resistance to these molecules was reviewed by Nikaido (1996). Resistance to these compounds is attributed, in part, to the low permeability of the outer membrane bilayer to lipophilic solutes; however, active efflux mechanisms may have a synergistic effect on resistance in certain cases (Nikaido, 1996).

Acquired resistance to aminoglycosides, beta-lactams, chloramphenicol, macrolides, sulfonamides, tetracycline, and trimethoprim has been described for *E. coli* strains (reviewed by Quintiliani and Courvalin, 1995). Acquired resistance can develop by four distinct mechanisms: alteration of the target site, enzymatic detoxification of the antibiotic, decreased drug accumulation, and bypass of an antibiotic-sensitive step. The first three mechanisms may be mediated by chromosomal mutations or the acquisition of plasmids carrying resistance genes. The fourth mechanism is primarily attributable to the horizontal transfer of antibiotic resistance genes on a plasmid or transposon. The biochemical mechanisms and genetic basis of acquired resistance to antimicrobial agents of clinical importance was discussed by Quintiliani and Courvalin (1995). Genetic methods for the detection of antibacterial resistance genes were reviewed by Tenover et al. (1995).

Pathogenicity *E. coli* is a natural and essential part of the bacterial flora in the gut of humans and animals. Most *E. coli* strains are nonpathogenic and reside harmlessly in the colon; however, certain serotypes or clones play an important role in both intestinal and extraintestinal diseases. The diverse pathogenesis of this bacterium in apparently healthy individuals is largely attributable to its possession of a variety of specific virulence factors. In hosts with compromised defenses, *E. coli* can also be an excellent opportunistic pathogen.

E. COLI IN HUMAN INTESTINAL DISEASES *E. coli* strains isolated from intestinal diseases have been grouped into at least six different main categories based on epidemiological evidence, phenotypic traits, clinical features of the disease they produce, and specific virulence factors. The currently recognized categories of diarrheagenic *E. coli* include enteropathogenic *E. coli* (EPEC) (actually a subgroup of attaching and effacing *E. coli* (A/EEC) defined as *eae* positive *E. coli* belonging to both the classical EPEC serotypes and nonclassical EPEC serotypes), enterotoxigenic *E. coli* (ETEC), enteroinvasive *E. coli* (EIEC), enteroaggregative *E. coli* (EAggEC), diffusely adherent *E. coli* (DAEC), and Shiga toxin-producing *E. coli* (STEC), which are also referred to as Vero cytotoxin-producing *E. coli* (VTEC). These categories are reviewed below with emphasis on their virulence factors.

ENTEROPATHOGENIC *E. COLI* (EPEC) Enteropathogenic *E. coli* was the first category of diarrheagenic *E. coli* to be recognized. The term *enteropathogenic E. coli* (EPEC) was originally used to refer to strains belonging to a limited number of O groups that were epidemiologically associated with infantile diarrhea (Neter et al., 1955). This rather imprecise definition, which allowed for the inclusion of a heterogeneous group of pathogens, was used for decades and became increasingly problematic as groups of *E. coli* that could produce diarrheal disease by the production of enterotoxins or invasion of intestinal epithelial cells were recognized. The confusion generated by the discovery of new pathogenic groups of *E. coli* and the findings that EPEC strains, which lacked the virulence properties of these newly recognized groups, caused disease in adult volunteers (Levine et al., 1978) prompted researchers in 1982 to define EPEC as "diarrheagenic *E. coli* belonging to serogroups epidemiologically incriminated as pathogens but whose pathogenic mechanisms have not yet been proven to be related to either heat-labile enterotoxins or heat-stable enterotoxins or *Shigella*-like invasiveness" (Edelman and Levine, 1983). As more was learned about the strains associated with infant diarrhea, the definition was refined to include only certain O:H serotypes associated with illness. Table BXII.γ.199 lists some of the O:H serotypes that have been regarded for many years as EPEC. Since 1982, advances in our understanding of the molecular aspects of EPEC pathogenesis have allowed researchers to move beyond the serologic markers that correlate with disease to develop a definition based on pathogenic characteristics. A definition adopted in 1995 identified the most important characteristics of EPEC as its ability to cause attaching and effacing (A/E) histopathology and its inability to produce Shiga toxins (Kaper, 1996). The pathogenesis of EPEC is highlighted below and has been reviewed in detail by Nataro and Kaper (1998) and Williams et al. (1997).

The first advance in understanding the pathogenesis of EPEC infection was the discovery that EPEC strains adhere to HEp-2 cells in cell culture (Cravioto et al., 1979) in a distinctive pattern termed localized adherence (LA) (Scaletsky et al., 1984). Expression of the LA phenotype in EPEC requires a plasmid re-

TABLE BXII.γ.199. O:H serotypes regarded as classical and newly recognized EPEC O:H serotypes[a,b]

O group	H antigen[c]	Comments
O26	H⁻; H11	O26:H⁻ and O26:H11 may also be STEC/VTEC[d] (Levine et al., 1987; Scotland et al., 1990; Bitzan et al., 1991)
O55	H⁻; H6; H7	O55:H7, H10 and H⁻ may also be STEC/VTEC (Dorn et al., 1989)
O86	H⁻; H8; H34	O86:H⁻ may also be EAggEC (Albert et al., 1993b; Tsukamoto and Takeda, 1993; Schmidt et al., 1995b; Smith et al., 1997b)
		O86:H8 is a new *eae-* and *bfpA*-positive type isolated in Denmark (Scheutz, unpublished data)
O88	H⁻; H25	(Tsukamoto et al., 1992)
O103	H2	New EPEC type
O111	H⁻; H2; H7; H12	O111:H⁻ may also be STEC/VTEC (Dorn et al., 1989; Bitzan et al., 1991; Caprioli et al., 1994; Cameron et al., 1995; Allerberger et al., 1996) or EAggEC (Scotland et al., 1991, 1994; Tsukamoto and Takeda, 1993; Chan et al., 1994; Schmidt et al., 1995b; Monteiro-Neto et al., 1997; Morabito et al., 1998)
O114	H⁻; H2	
O119	H⁻; H2; H6	
O125ac	H⁻; H6	O125 may also be EAggEC (Tsukamoto and Takeda, 1993; do Valle et al., 1997; Smith et al., 1997b)
O126	H⁻; H2; H21; H27	
O127	H⁻; H6; H21; H40	
O128ab	H⁻; H2; H7; H12	O128:H2 may also be STEC/VTEC (Beutin et al., 1993a)
O142	H⁻; H6; H34	
O145	H⁻; H45	New EPEC type
O157	H⁻; H8; H16; H45	New EPEC types
O158	H⁻; H23	

[a]Data from Cravioto et al. (1979), Levine and Edelman (1984), Levine et al. (1985), Scaletsky et al., 1985, Robins-Browne (1987), Gomes et al. (1989b), Knutton et al. (1989, 1991), Scotland et al. (1989, 1992, 1996), Ørskov and Ørskov (1992), Donnenberg (1995).

[b]O18:H⁻, H7, H14; O26:H34 and O44:H34 have also been listed but only in Knutton et al. (1991). O18 strains are probably not EPEC (Knutton et al., 1989; Ørskov and Ørskov, 1985). O44:H18 is now considered to belong to the group of enteroaggregative *E. coli* (Smith et al., 1994).

[c]Nonmotile strains of *E. coli* are regarded as descendants of motile strains that have lost their motility by mutation(s). Their original H antigen was often deduced from comparison of biochemical reactions (Kauffmann and Dupont, 1950; Staley et al., 1969).

[d]Abbreviations: EPEC, enteropathogenic *E. coli*; STEC/VTEC, Shiga toxin-producing *E. coli*/Vero cytotoxin-producing *E. coli*; EAggEC, enteroaggregative *E. coli*.

ferred to as the EPEC adherence factor (EAF) plasmid (Baldini et al., 1983). A 1-kb DNA probe originally thought to encode the EPEC adherence factor (EAF) necessary for LA was cloned from this plasmid (Nataro et al., 1985a) and has been used extensively as a marker to study the prevalence of EPEC infections (Nataro et al., 1985a; Echeverria et al., 1987, 1991; Gomes et al., 1989a, b; Moyenuddin et al., 1989; Senerwa et al., 1989; Cravioto et al., 1991; Kain et al., 1991; Strockbine et al., 1992; Begaud et al., 1993). EAF plasmids have been found in many EPEC serotypes and range in size from 26 to 76 MDa (Scotland et al., 1989) but typically are 50–70 MDa. Evidence supporting a role for the EAF plasmid in pathogenesis was provided by feeding studies showing that volunteers ingesting a plasmid-cured EPEC strain developed less diarrhea than those ingesting the plasmid-containing parental strain (Levine et al., 1985). Genes at two loci are necessary for expression of the LA phenotype: a cluster of 14 genes on the EAF plasmid involved in biogenesis of the bundle forming pilus (BFP), a type-IV pilus, which includes genes encoding bundlin (*bfpA*), the major structural subunit of the type-IV pilus, a prepilin peptidase, which processes pre-bundlin to its mature form, and 12 other proteins, and *dsbA* on the chromosome (Donnenberg et al., 1997).

A hallmark of the histopathology of EPEC infections is the presence of attaching and effacing (A/E) lesions in the intestinal tract. On electron micrographs of jejunal biopsies from children infected with EPEC, the bacteria are seen intimately attached to the epithelial cells on cup-like pedestals composed of depolymerized cytoskeletal proteins (Knutton et al., 1987). Microvilli

are disrupted as a result of the cytoskeletal rearrangements and effaced by vesiculation (Fig. BXII.γ.192). The intimate attachment of EPEC is mediated by a protein known as intimin, which is a 94-kDa outer membrane protein encoded by the *eae* gene (*E. coli* attaching and effacing) (Jerse and Kaper, 1991). A 1-kb fragment of the *eae* gene referred to as CVD434 has been cloned (Jerse et al., 1990) and used to screen for attaching and effacing *E. coli* A/EEC (Bokete et al., 1997), and to characterize enteropathogenic *E. coli* (Scotland et al., 1996). Intimins belong to a growing family of proteins. In human EPEC strains, intimins α, β, γ, λ (δ not now used, although it is in the literature because it is a variant of the ε variant), and ε derivatives have been found to be serotype-specific and exhibit specific different binding affinities (Agin and Wolf, 1997). Intimins are also found in rabbit EPEC RDEC-1 strains, A/EEC strains from dogs (Beaudry et al., 1996), pigs (Zhu et al., 1995), and in *C. freundii* (Int_CF) and *Citrobacter rodentium* (Int_CR). Five strains initially identified as *Hafnia alvei* were reported to contain the *eae* gene (Albert et al., 1992), and a protein referred to as Int_HA has been characterized and compared to the above intimins. However, the strains are actually unusual biotypes belonging to the genus *Escherichia* (Janda et al., 1999), most likely *E. coli*. The *eae* gene is only one of many genes located on a pathogenicity island (PAI) known as the locus of enterocyte effacement (LEE) (McDaniel et al., 1995).

EPEC strains secrete at least four proteins, Esps for EPEC secreted proteins, encoded by LEE. EspA, B, D, and Tir proteins are secreted via the type III apparatus and are required for attaching and effacing activity. Protein secretion, the transcription

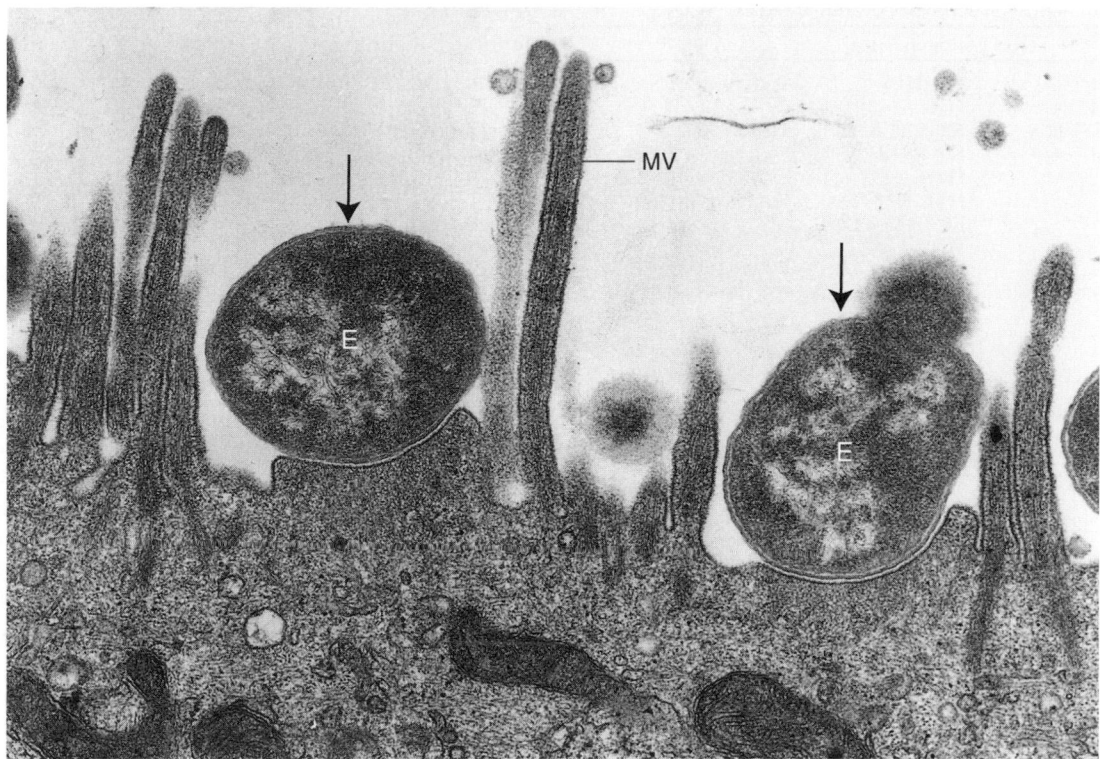

FIGURE BXII.γ.192. EM of cultured human intestinal mucosa infected with EPEC strain E2348 (*E*) for 8 h. Bacteria are seen intimately attached to the epithelial cells on cup-like pedestals composed of depolymerized short filament cytoskeletal proteins (*arrows*). Brush border microvilli (*MV*) are disrupted as a result of the cytoskeletal rearrangements and effaced by vesiculation. (Reproduced with permission from S. Knutton et al., Infection and Immunity *55*: 69–77, 1987, ©American Society for Microbiology.)

of the intimin gene (*eae*), and the synthesis of the bundle-forming pilus are all affected by growth conditions (Haigh et al., 1995; Jarvis et al., 1995; Kenny and Finlay, 1995) and a plasmid-encoded regulatory region. The plasmid locus is either referred to as the *per* (plasmid-encoded regulator) genes (*perABCD*) (Gómez-Duarte and Kaper, 1995) or as an integrated part of the bundle-forming pilus (*bfp*) operon, genes *bfpT, V*, and *W* (Tobe et al., 1996).

EPEC interferes with normal pathways of signal transduction in epithelial cells. EPEC will induce tyrosine phosphorylation; phosphorylation of myosin light chain, vinculin, and α-actinin; *in vitro* release of intracellular calcium; phospholipase C activity, resulting in elevated levels of inositol phosphates and diacylglycerol, in turn activating protein kinase C; reactions that induce host cell proteins to initiate cytoskeletal rearrangement; and bacterial uptake (Rosenshine et al., 1992; Manjarrez-Hernandez et al., 1996).

EPEC is not invasive in the same way as *Shigella, Yersinia,* and *Salmonella* (Geyid et al., 1996), although some EPEC strains may exhibit *in vitro* invasion of epithelial cells at levels comparable to those seen in EIEC (Donnenberg et al., 1989; Pelayo et al., 1999). A 4.5-kb fragment from a large so-called EPEC plasmid of an O111:H⁻ strain negative for both EAF and *eae* will confer epithelial cell invasivity and the attaching and effacing ability on a noninvasive laboratory strain (Fletcher et al., 1990, 1992). Only a smaller proportion of EPEC strains hybridize with the cloned fragment.

Of the EPEC O:H serotypes shown in Table BXII.γ.199, only 16 of the possible 175 recognized O groups are represented and only a few H antigens, e.g., H2, H6, H7, H12, H21, and H34, occur among several of these 16 O groups. The more recently described serotypes possessing EPEC-associated virulence markers have been added to the list. Some of these are O86:H8, O88:H25, O103:H2, O127:H40, O142:H34, O145:H45, O157:H8, O157:H16, and O157:H45.

ENTEROTOXIGENIC *E. COLI* (ETEC) Enterotoxigenic *E. coli* strains are important causes of diarrhea in both humans and domestic animals. ETEC strains do not invade epithelial cells but produce one or more enterotoxins that are either heat-labile (LT), which is closely related to cholera toxin, or heat-stable (ST) (Cohen and Giannella, 1995). LT-I and ST are plasmid-encoded (LT-II is chromosomally encoded) and found alone or together, and are often associated with a limited number of O:K:H serotypes and O groups. Table BXII.γ.200 shows some of the most common serotypes from humans. These enterotoxins cause intestinal secretion either by activation of guanylate (ST) or adenylate (LT) cyclase and are subdivided based on their biological activities, receptors, and chemical and antigenic properties. STh and STp are used to indicate strains of human or porcine origin. The gene encoding another heat-stable enterotoxin called enteroaggregative heat-stable toxin 1 (EAST1), originally thought to be produced by EAggEC only, may also be present in addition to the STa gene. The different variants of enterotoxins are summarized in Table BXII.γ.201.

There is host specificity among ETEC strains causing diarrhea in humans or different species of domestic animals. This is mainly due to the specific recognition between bacterial colonization factors and the epithelial receptors during host-parasite inter-

TABLE BXII.γ.200. O:(K):H serotypes of human ETEC

O group		(K):H antigen
O6	H⁻	K15:H16
O7	H⁻	H18
O8	K47:H⁻	K25:H9; K40:H9; H10; K87:H19
O9	H⁻	K9; K84:H2
O11		H27
O15		H11; H15; H45
O17		K23:H45; H18
O20	H⁻	H30
O21		H21
O25	H⁻	K7:H42; H16
O27		H7; H20; H27
O29		H?
O48		H26
O55		H7
O56	H⁻	
O63		H12; H30
O64	H⁻	
O65		H12
O71		H36
O73		H45
O77		H45
O78	H⁻	K2; H11; H12
O85		H7
O86		H2
O88		H25
O105		H?
O114	H⁻	H21
O115	H⁻	H21; H40; H51
O119		H6
O126	H⁻	H9; H12
O128ac		H7; H12; H19; H21
O133		H16
O138		K81
O139		H28
O141	H⁻	H4
O147		H?
O148		H28
O149		H4; H10; H19
O153		H10
O159	H⁻	H4; H5; H12; H20; H21; H27; H34; H37
O166		H27
O167		H5
O?		H2; H10; H28; K39:H32

TABLE BXII.γ.201. Variants of enterotoxins found in ETEC strains[a]

Toxin type	Subtypes		Comments
LT: Heat-labile enterotoxins			
LT-I	LTp (LTp-I)	LTh (LTh-I)	Associated with disease in both humans and animals
LT-II	LT-IIa	LT-IIb	No specific association with disease. Rare in human isolates
ST: Heat-stable enterotoxins			
STa (STI)	STp (STIa)	STh (STIb)	Produced by ETEC and several other Gram-negative bacteria. In ETEC, ST is often found together with the genes for EAST1.
STb (STII)			Induces histological damage in the intestinal epithelium. Most often found in porcine ETEC.

[a]Suffixes commonly used are: h, human variant; p, porcine variant.

action. Most of the fimbriae found in ETEC are typically MR fimbriae. In human ETEC strains at least 21 different surface structures called CS, for coli surface antigens, or CFA, for colonization factor antigens, usually plasmid encoded, have been described (Gaastra and Svennerholm, 1996). In animals, the most common fimbriae are F4 (which was originally described as K88), F5 (originally named K99), F6 (originally named 987P), F17, F41, F42, CS31A, F141, and F165. These fimbriae are found in enterotoxigenic strains from newborn piglets, pigs, lambs, and newborn calves. A close association between the EAST1 toxin and CS31A, which is related to F4, among pathogenic bovine *E. coli* has been suggested (Bertin et al., 1998).

More recently, the fimbrial F18 antigen (provisionally designated F107, 2134P, or 8813) has been found in both porcine ETEC and STEC/VTEC strains. It is characteristic that the fimbriae usually are found in a limited number of serotypes. The antigenic variants of F18 fimbriae (F18ab and F18ac) are biologically distinct. F18ab fimbriae are expressed poorly both *in vitro* and *in vivo* and are frequently linked with the production of Stx2e/VT2e and O group O139, while F18ac are more efficiently expressed *in vitro* and *in vivo* and most often are linked with enterotoxin (STa, STb) production, and O groups O141 and O157.

ENTEROINVASIVE *E. COLI* (EIEC)   Enteroinvasive *E. coli* are very similar to *Shigella*. Like *Shigella*, they are capable of invading and multiplying in the intestinal epithelial cells of the distal large bowel in humans. Approximately two thirds of EIEC strains are lactose negative and virtually all are lysine negative. Multiple chromosomal and plasmid genes are associated with virulence (Acheson and Keusch, 1995). EIEC are restricted to a very limited number of serotypes, most of which are nonmotile (Table BXII.γ.202).

ENTEROAGGREGATIVE *E. COLI* (EAggEC)   Enteroaggregative *E. coli* are characterized by a distinct aggregative adherence (AA) pattern to HEp-2 cells *in vitro* first described by Nataro et al. (1987). This pattern is distinguished by the prominent autoagglutination of bacterial cells to each other, to the surface of the HEp-2 cells, as well as to the glass cover slip in a characteristic layering best described as a "stacked brick" configuration. The AA pattern is plasmid-mediated (Nataro et al., 1985b) and was suspected to be a putative agent of diarrheal disease as early as 1988 (Vial et al., 1988). Volunteer studies with EAggEC have indicated that certain types will cause diarrhea and other enteric symptoms including borborygmia and cramps (summarized in

TABLE BXII.γ.202. O:H serotypes of EIEC

O group	H antigen
O28ac	H⁻
O29	H⁻
O112ac	H⁻
O115	H⁻
O121	H⁻
O124	H⁻; H7; H30; H32
O135	H⁻
O136	H⁻
O143	H⁻
O144	H⁻; H25
O152	H⁻
O159	H2
O164	H⁻
O167	H⁻; H4; H5
O173	H⁻

Nataro, 1995), and epidemiological studies have implicated EAggEC as a cause of travelers' diarrhea (Brook et al., 1994; Scotland et al., 1994). Only about half of the case-control studies that have been carried out find significantly higher isolation rates in cases than in controls (summarized by Law and Chart, 1998). Most recently, EAggEC have been associated with acute and chronic diarrhea and abdominal colic in young children in Germany (Huppertz et al., 1997), and with four outbreaks of gastroenteritis in the UK (Smith et al., 1997b). A DNA probe, referred to as CVD432, has been cloned from the plasmid of strain O42, serotype O44:H18 (formerly an EPEC O:H serotype) and used to identify EAggEC (Baudry et al., 1990). A plasmid encoded flexible, bundle-forming fimbrial structure, designated aggregative adherence fimbria I (AAF/I), is encoded by two regions (Nataro et al., 1992, 1993). Region 2 encodes a transcriptional activator of the AraC family of DNA-binding proteins (Nataro et al., 1993). At least one toxin similar to the heat-stable toxin of ETEC, ST designated EAST1, has been identified in an EAggEC strain (Savarino et al., 1991, 1993). The gene encoding EAST1 has been shown to be broadly distributed among diarrheagenic *E. coli*: 100% of 75 O157:H7 STEC/VTEC, 41% of 227 EAggEC, 41% of 149 ETEC, 22% of 65 EPEC, and 38% of 47 *E. coli* from asymptomatic children hybridized with the EAST1 DNA probe (SS126) (Savarino et al., 1996). Despite the common occurrence of the EAST1 gene in many diarrheagenic *E. coli* groups, outbreaks had not been attributed to EAST1-only-producing *E. coli* until recently, when an outbreak in Japan was associated with an O166:H? strain positive for the EAST1 gene (Nishikawa et al., 1999). In Spain, a case-control study suggested that EAST-1 positive *E. coli* strains are associated with diarrheal diseases in Spanish children, whereas EAggEC strains are not (Vila et al., 1998). The significance of these findings remains to be established. Many serotypes have been observed and some of the most commonly reported serotypes are listed in Table BXII.γ.203.

DIFFUSELY ADHERENT *E. COLI* (DAEC) Diffusely adherent *E. coli* are defined by the presence of the diffuse adherence (DA) pattern of *E. coli* strains to HEp-2 cells (Scaletsky et al., 1984; Nataro et al., 1985b). A surface fimbriae designated F1845 confers the DA phenotype, and a DNA probe has been cloned (Bilge et al., 1989). Another adhesin (designated AIDA-I) has also been associated with DA of *E. coli* of serotype O126:H27 (Benz and Schmidt, 1989). The role of DAEC in diarrhea is unclear.

SHIGA TOXIN-PRODUCING *E. COLI* (STEC) OR VERO CYTOTOXIN-PRODUCING *E. COLI* (VTEC) Shiga toxin-producing *E. coli* or Vero cytotoxin-producing *E. coli* strains are characterized by their ability to produce either one or both of at least two antigenically distinct, usually bacteriophage-mediated cytotoxins referred to as Stx1 or VT1 (first described as Shiga-like toxin I, SLTI) and Stx2 or VT2 (first described as Shiga-like toxin II, SLTII). Whereas STEC/VTEC refers to all *E. coli* strains that produce Stx/VT in culture supernatants (Konowalchuk et al., 1977, 1978), the term *enterohemorrhagic E. coli* (EHEC) has been used to refer to strains that have the same clinical and pathogenic features associated with the prototype organism *E. coli* O157:H7 (Levine, 1987). In practice, EHEC is used to describe a subgroup of STEC/VTEC that causes hemorrhagic colitis.

Shiga toxins or Vero cytotoxins belong to the Shiga toxin family, comprises the following members: Shiga toxin, which is produced by *Shigella dysenteriae* type 1, Stx1/VT1, STx2/VT2, and Stx2/VT2 variants (Stxv/VT2v). Stx/VT genes, products, and synonyms are summarized in Table BXII.γ.204. Stx2e/VT2e, one of the variants, is produced by STEC/VTEC strains causing edema disease (Marques et al., 1987), a usually fatal disease, in weanling pigs and referred to as Stx2e/VT2e. Unlike all other Stxs/VTs, this is not cytotoxic to HeLa cells and binds to a different receptor, Gb4. The functional receptor for human Stx/VT is the glycolipid globotriosyl ceramide Gb3 (galactose-α-(1-4)-galactose-β-(1-4)-glucose ceramide) (Lingwood et al., 1987; Waddell et al., 1988) found in human renal endothelial cells (Obrig et al., 1993).

Stxs/VTs inhibit protein synthesis by depurination of adenine in 28S rRNA (*N*-glycosidases), thus inhibiting the elongation factor 1 (EF-1)-dependent aminoacyl-tRNA binding to 60S ribosomal subunits (Endo et al., 1988). In Vero cells the result is cell death by apoptosis (Inward et al., 1995).

NOMENCLATURE OF SHIGA TOXINS/VEROCYTOTOXINS The recognition and investigation of cytotoxin-producing *E. coli* infections by several groups has resulted in the use of differing systems of nomenclatures for the toxins produced by these bacteria. In 1994, O'Brien et al. (1994) developed a proposal for rationalizing the nomenclature of the *E. coli* cytotoxins. In their proposal, they recommended guidelines for classifying and designating members of the toxin family and stated the toxins shall be referred to by two alternate but interchangeable names, Shiga-like toxins (SLT) and Vero cytotoxins (VT). Two years later, a proposal to simplify the Shiga-like toxin nomenclature was published by Calderwood et al. (1996). In this proposal, the word "like" was omitted and the toxins and genes were renamed to reflect their relationship to Shiga toxin, the prototype toxin for the family. To avoid confusion in the literature, it was suggested that cross-reference to existing VT nomenclature could be used. While the omission of the word "like" has received general acceptance in the scientific community, strong arguments for maintaining the existing phenotype nomenclatures for *E. coli* cytotoxins were immediately put forward (Karmali et al., 1996), and the two systems of nomenclature are still being widely used. Toxins included in the Shiga toxin/Verocytotoxin family share the following properties: DNA sequence homology and operon structure (A subunit gene immediately upstream of the B subunit gene); polypeptide subunit structure (five B subunits to one A subunit in the mature holotoxin); enzymatic activity (*N*-glycosidases); binding to specific glycolipid receptors and biological properties, including enterotoxicity in ligated rabbit ileal loops;

TABLE BXII.γ.203. O:H serotypes of the most frequently reported enteroaggregative *E. coli* (EAggEC)[a]

O group	H antigen	References listing these serotypes as EAggEC
O3	H2	Albert et al. (1993b)
O15	H18	Vial et al., 1988, Albert et al. (1993b), Tsukamoto and Takeda (1993), Scotland et al. (1994)
O44	H⁻; H18	Scotland et al. (1991, 1994), Tsukamoto and Takeda (1993), Smith et al. (1994), Schmidt et al. (1995b)
O86	H⁻	Albert et al. (1993b), Tsukamoto and Takeda (1993), Schmidt et al. (1995b), Smith et al. (1997b)
O111	H12; H21	Scotland et al. (1991, 1994), Tsukamoto and Takeda (1993), Chan et al. (1994), Schmidt et al. (1995b), Monteiro-Neto et al. (1997), Morabito et al. (1998)
O125	H9; H21	Tsukamoto and Takeda (1993), do Valle et al. (1997), Smith et al. (1997b)

[a]Many other serotypes of EAggEC have been published by several authors.

TABLE BXII.γ.204. Designation of *stx/vtx* genes, their products (Stx/VT)[a], and their previous designations

Toxin type[b]	Toxin gene	Toxin	Prototype organisms (reference)	Serotype of prototype organism	Previous designation or synonym for toxin gene	Previous designation or synonym for toxin
1	$stx_{1}/vtx1$	Stx1/VT1	EDL933 (O'Brien et al., 1984)	O157:H7	slt-I	SLT-I
			H19 (Konowalchuk et al., 1977)	O26:H11		
			H30 (Konowalchuk et al., 1977)	O26:H11		
	$stx_{-O103}/vtx1$-O103	Stx1-O103/VT1-O103	PMK1 (Mariani-Kurkdjian et al., 1993)	O103:H2		
	$stx_{-O111\text{-}PH}/vtx1$-O111-PH	Stx1-O111-PH /VT1-O111-PH	PH (Paton et al., 1993a)	O111:H⁻	$sltI$/PH	Slt-I/PH
	$stx_{-OX3}/vtx1$-OX3	Stx1-OX3/VT1-OX3	131/3 (Paton et al., 1995a)	OX3:H8	$sltI$/OX3	Slt-I/OX3
	$stx_{-O48}/vtx1$-O48	Stx1-O48/VT1-O48	94C (Paton et al., 1995a)	O48:H21	$sltI$/O48	Slt-I/O48
	$stx_{-O111\text{-}CB168}/vtx1$-O111-CB168	Stx1-O111-CB168/VT1-O111-CB168	CB168 (Paton et al., 1995a)	O111:H⁻	$sltI$/CB	Slt-I/CB
2	$stx_{2}/vtx2$	Stx2/VT2	EDL933 (O'Brien et al., 1984)	O157:H7	slt-II	SLT-II
2c	$stx_{2c}/vtx2c$	Stx2c/VT2c	E32511 (Schmitt et al., 1991)	O157:H⁻	slt-IIc	Stx2v/VT2v
						SLT-IIc
	$stx_{-O22}/vtx2$-O22	Stx2-O22/VT2-O22	KY-O19 (Lin et al., 1993a)	O22:H⁻		VT2e(pKTN1054)
	$stx_{-O157\text{-}TK\text{-}51}/vtx2$-O157-TK-51	Stx2-O157-TK-51/VT2-O157-TK-51	TK-51 (Lin et al., 1993a)	O157:H7		VT2e(pKTN1050)
	$stx_{-OX3/a\text{-}031}/vtx2$-OX3/a-031	Stx2-OX3/a-031/VT2-OX3/a-031	031 (Paton et al., 1992)	OX3:H21	$stx2_{,OX392}$	SLT-II/OX3/VT2d-OX3³ᵈ, SLT-II/OX3a
	$stx_{-OX3/b\text{-}031}/vtx2$-OX3/b-031	Stx2-OX3/b-031/VT2-OX3/b-031	031 (Paton et al., 1993b)	OX3:H21	$stx2_{,OX393}$	SLT-II/OX3/2/VT2d-OX3/2°; SLT-II/OX3b
	$stx_{-O48}/vtx2$-O48	Stx2-O48/VT2-O48	94C (Paton et al., 1995b)	O48:H21		SLT-II/O48
	$stx_{-O111\text{-}PH}/vtx2$-O111-PH	Stx2-O111-PH /VT2-O111-PH	CB168 (Paton et al., 1995b)	O111:H⁻	$stx2_{,O111}$	SLT-II/O111/VT2d-O111°
	$stx_{-O118}/vtx2$-O118	Stx2-O118/VT2-O118ᵈ	EH250 (Piérard et al., 1998)	O118:H12		Stx2d/VT2d-Ount
	$stx_{-O6}/vtx2$-O6	Stx2-O6/VT2-O6	NV206 (Bertin et al., 2001)	O6:H10	$stx2_{-NV206}$	Stx2-NV206
2dᵈ	$stx_{2d}/vtx2d$	Stx2d/VT2d	B2F1 (Ito et al., 1990)	O91:H21	$stx2vha/vtx2vha$	SLT-IIvh
					$stx2ha$	Stx2vha/VT2vha
	$stx_{2d1}/vtx2d1$	Stx2d1/VT2d1	B2F1 (Ito et al., 1990)	O91:H21		Stx2vh-a/VT2vh-a
						VT2v-a
	$stx_{2d2}/vtx2d2$	Stx2d2/VT2d2	B2F1 (Ito et al., 1990)	O91:H21	$stx2vhb/vtx2vhb$ $stx2hb$	SLT-IIvha
						Stx2vhb/VT2vhb
						Stx2vh-b/VT2vh-b VT2v-b
						SLT-IIvhb
2e	$stx_{2e}/vtx2e$	Stx2e/VT2e	412 (Gyles et al., 1988)	O139:K12:H1	slt-IIv	SLT-IIv
			S1191 (Weinstein et al., 1988)	O139:H1	slt-IIva	SLT-IIvaᵉ
					slt-IIe	SLTIIe/VTe
						VT2vp
						VT2vp1
2f	$stx_{2f}/vtx2f$	Stx2f/VT2f	H.I.8 (Gannon et al., 1990)	O128:H2	$stx2_{ev}/vtx2ev$	Stx2ev/VT2ev
			T4/97 (Schmidt et al., 2000)	O128:H2		Stx2vp2/VT2vp2
						VTev
					slt-IIvhc	SLTIIvhc
					slt-IIdᵈ	SLT-IId/VT2dᵈ

[a]Data compiled from references Bastian et al. (1998), Bertin et al. (2001), Calderwood et al. (1996), Gannon et al. (1990), Ito et al. (1990), W.M. Johnson et al. (1991), Karmali et al. (1996), Kokai-Kun et al. (2000), Lin et al. (1993a, b), Takeda et al. (1993), and Tyler et al. (1991).

[b]Toxin types are defined according to antigenic variability, differences in toxicity for tissue culture cells and/or animals, their capacity to be activated by mouse elastase, and differences in DNA or amino acid sequences.

[c]There are several toxins suffixed by d in the literature: The Stx2d/VT2d toxins of O91:H21 (Melton-Celsa and O'Brien, 1998), the VT2d (= stx2-O group/ strain designation and/or year) variants by Paton et al. (1992, 1993a, 1995b), Piérard et al. (1998), and the SLT-IId/VT2d (= Stx2f/VT2f) toxin produced by strain H.I.8 (serotype O128:H2) as proposed by Gyles (1994). We support the use of the d designation for activatable Stx2/VT2 toxins as proposed by Melton-Celsa and O'Brien, (1998).

[d]Stx2-O118/VT2-O118 (formerly known as VT2d-Ount) is expected to be nonactivatable based on analysis of the nucleotide sequence (Denis Piérard and Angela Melton-Celsa; unpublished); the original strain has been retyped as O118:H12 (Lothar Beutin and Flemming Scheutz, unpublished).

[e]This designation has also been referred to for the Stx2f/VT2f toxin produced by strain H.I.8 (serotype O128:H2).

[f]The nucleotide sequence of the former $stx2_{ev}/vtx2ev$ of strain H.I.8 (serotype O128:H2) is nearly identical to the recently published $stx2_{f}/vtx2f$ found in strain T4/97 (serotype O128:H2) from feral pigeons (Schmidt et al., 2000). As its nucleotide sequence is distinctly different from those of the other Stx2/VT2 toxins and variants as well as Stx1/VT1, we support the proposal of renaming $stx2_{ev}/vtx2ev$ as $stx2_{f}/vtx2f$.

neurotoxicity in mice; and cytotoxicity to receptor-expressing tissue culture cell lines, such as Vero and HeLa cell lines. Classification of the toxins into major toxin types, designated with Arabic numbers, is based on differences that result in no cross-neutralization by homologous polyclonal antisera and no DNA–DNA cross-hybridization of their genes under conditions of high stringency. Toxin subtypes, designated with letters added to the type name, share cross-hybridization of their genes under high stringency but show significant differences in biologic activity, including the capacity to be activated; serologic reactivity; or receptor binding. Table BXII.γ.204 summarizes the currently reported Shiga toxin/Verocytotoxin genes and toxins, prototype organisms, and their previous designations in the literature.

The distinction between toxin types 2c and 2d in Table BXII.γ.204 is based on differences in the A subunit determining whether the toxin is activatable (Stx2d/VT2d) or nonactivatable (Stx2c/VT2c). The other Stx2/VT2 subtypes described in the literature, which have not been tested for all properties, have been tentatively placed together with toxin type 2c, primarily based on similarities in nucleotide sequences that place them in a phylogenetically related Stx2/VT2 cluster including all variants. Based on their degree of overall nucleotide sequence similarity, Stx2/VT2 toxins fall into two phylogenetically distinct groups (Bastian et al., 1998; Piérard et al., 1998). Toxins in group 1, which include Stx2/VT2, Stx2c/VT2c (including variants Stx2-O22/VT2-O22, Stx2-O157-TK-51/VT2-O157-TK-51, Stx2-OX3/b-031/VT2-OX3/b-031, Stx2-O48/VT2-O48), and Stx2d/VT2d, share 99.1–99.2% and 95.9–98.5% nucleotide sequence similarity in their A and B subunits, respectively; while toxins in group 2, which include Stx2-OX3/a-031/VT2-OX3/a-031, Stx2-O111-PH/VT2-O111-PH, and Stx2-O118/VT2-O118, share 96.9–99.9 % and 99.6–100% nucleotide sequence similarity in their A and B subunits, respectively. The similarities of individual sequences between group 1 and group 2 are 93.4–96.0% for the A subunits and 86.2–89.3% for the B subunits. The extent of, and differences between, toxicity for tissue culture cells and/or animals and their capacity to be activated are not fully established for all the types. Further analysis against all the criteria necessary to allow definitive placement with appropriate other toxin types is required. In the absence of this information, suffixes have been added after the O group of the source organisms, and, when necessary, the original strain designation. Stx2/VT2 subtype toxins found in the same original strain are suffixed "/a", "/b" etc. as in Stx2-OX3/a-031/VT2-OX3/a-031 and Stx2-OX3/b-031/VT2-OX3/b-031.

For a brief presentation and practical use of the subtyping of Stx/VT genes with 4 Stx1/VT1 and 9 Stx2/VT2 oligonucleotide DNA probes, and 3 Stx1/VT1 and 8 Stx2/VT2 primer pairs, see Smith et al. (1993), Yamasaki et al. (1996), Piérard et al. (1997), and Bastian et al. (1998).

Like EPEC, some STEC/VTEC strains have been shown to cause attaching and effacing lesions *in vivo* (Hall et al., 1990), in animal models (Francis et al., 1986; Tzipori et al., 1986; Sherman et al., 1988), and *in vitro* (Knutton et al., 1989). Two separate groups have cloned, sequenced, and characterized the *eae* homologue from VTEC O157:H7 (Beebakhee et al., 1992; Yu and Kaper, 1992). Homology between EPEC and STEC/VTEC sequences was 86% and 83% at the nucleotide and amino acid levels, respectively (Yu and Kaper, 1992), and the STEC/VTEC *eae* sequence was 97% homologous to the EPEC *eae* gene for the first 2200 bp and 59% homologous over the last 800 bp (Beebakhee et al., 1992). Both *eae* sequences show 50% homology to

the central region of the *Yersinia pseudotuberculosis inv* gene (Jerse et al., 1990; Beebakhee et al., 1992), and the predicted amino acid sequence of the STEC/VTEC *eae* gene share 31% identity and 51% similarity with the invasin molecule of *Yersinia pseudotuberculosis* (Yu and Kaper, 1992).

Serotype specific heterogeneity of the *eae* gene in STEC/VTEC strains O55:H7 or H$^-$, O111:H8 and O157:H7 or H$^-$, and in O groups O26, O103, and O157 has been demonstrated (Gannon et al., 1993; Louie et al., 1994).

Almost all STEC/VTEC O157:H7 strains harbor a large 60-65 MDa plasmid (Johnson et al., 1983), designated pO157, that plays a role in virulence (Karch et al., 1987) and a small plasmid of 6.6 kb found in O157:H7 STEC/VTEC strains appear to synthesize colicin D (Bradley et al., 1991). O26:H11 strains also possess at least one plasmid in the range of 55–70 MDa. Restriction enzyme patterns of plasmids from other O:H serotypes (including O5, O91, O103, O111, O121, and O127) show a notable similarity with the large plasmids in O157 and O26 strains (Levine et al., 1987). A 3.4-kb fragment from a large plasmid of O157:H7 (prototype EDL 933) has been cloned and used as a DNA probe (referred to as CVD419) to identify EHEC plasmids (Levine et al., 1987); i.e., large plasmids found in Verocytotoxin producing *E. coli* strains. DNA probing with gene probes defining the incompatibility group of plasmids indicates that the STEC/VTEC plasmids share an approximately 23-kb fragment with EPEC plasmids and that the large plasmids of both EPEC and STEC/VTEC constitute a family of transfer-deficient Inc F-IIA plasmids (Hales et al., 1992), while sequencing of pO157 reveals high homology to the orf1 of the RepFIB replicon (Schmidt et al., 1996a).

The large plasmid of O157 encodes the EHEC-hemolysin (Ehx), which is homologous to the *E. coli* α hemolysin (Schmidt et al., 1994, 1995a), and a novel catalase-peroxidase, KatP (Brunder et al., 1996). In contrast to α hemolysin, Ehx can be detected only on blood agar plates containing washed sheep erythrocytes. The zones of hemolysis on these plates are smaller and more turbid than those caused by α hemolysin and require overnight incubation before they become visible (Beutin, 1991; Beutin et al., 1988, 1989). A role for Ehx in the pathogenesis of diarrheal disease has not been demonstrated. Because α hemolysin-producing strains are uncommon in feces, these hemolysins serve as useful phenotypic markers for the detection of the majority of STEC/VTEC organisms.

The genes encoding the Ehx constitute a typical RTX (repeats in toxin) determinant, the Ehx-operon, with the gene order CABD (Schmidt et al., 1996a). The *ehxA* gene encodes the active protein, and *ehxB* and *ehxD* share high sequence homology with other RTX transport proteins (Schmidt et al., 1995a, 1996a). Like α hemolysin, the Ehx is a highly active cytolysin of the RTX family with a similar but not identical pore-forming capacity (Schmidt et al., 1996b). The Ehx plasmid DNA probe (CVD419) covers the *ehxA* and part of the *ehxB* gene (Schmidt et al., 1995a).

Two other enterohemolysins Ehly1 and Ehly2 have been described (Beutin et al., 1993b; Stroeher et al., 1993). Ehly1 is a 33-kDa cell-associated protein encoded by a bacteriophage, ΦC3888, found in O26:H11 STEC/VTEC. Ehly1 has no known sequence homology to any other DNA or protein sequence. The Ehly2 enterohemolysin is also encoded by a bacteriophage, ΦC3208, found in O26:H11. It is in part homologous to DNA of bacteriophage λ; but completely unrelated to Ehly1 (Beutin et al., 1993b).

Most information on the source and transmission of STEC/VTEC has been learned from outbreak investigations. Findings from these investigations showed that most outbreaks are related to carriage of the organism in ruminants, especially cattle, which show no symptoms of disease. During the period from 1982 to 1993, at least 20 outbreaks of O157:H7 have been reported in the USA (summarized in Anonymous, 1994). These outbreaks have affected 1509 patients, resulting in the hospitalization of 346 patients, 86 cases of HUS, and 19 deaths. This increased dramatically in the following years with 13 outbreaks in 1993 and 30 outbreaks in 1994 (Armstrong et al., 1996). The largest multistate outbreak in the USA occurred in early 1993 with more than 700 illnesses and 4 deaths (Bell et al., 1994; Davis, 1994). It has been estimated that *E. coli* O157:H7 causes 73,000 illnesses annually in the United States and non-O157 STEC/VTEC, 37,000 illnesses; and that 91 deaths occur each year in the USA (Mead et al., 1999). In Canada, 15 outbreaks were reported in 1982–1987 with 242 cases, 24 cases of HUS, and 15 deaths (Karmali, 1989). The first recognized community outbreak of O157:H7 in Europe occurred in the UK in the summer of 1985 affecting at least 24 persons. Eleven patients were hospitalized and one died (Morgan et al., 1988). In England and Wales, O157:H7 was isolated from 39% of sporadic cases of hemorrhagic colitis (Smith et al., 1987) and 33% of sporadic HUS cases (Scotland et al., 1988). In an outbreak of HUS in the West Midlands, O157:H7 was isolated from 33% of cases (Taylor et al., 1986a; Willshaw et al., 2001). Subsequent outbreaks and sporadic cases in the UK have been reported (Salmon et al., 1989). Scotland has one of the highest rates of infection with O157 increasing from 1.37/100,000 of the population in 1989 (Thomas et al., 1996a) to 32.3/100,000 in 1996 (Reilly and Carter, 1997). The worst food poisoning outbreak with O157 STEC/VTEC in Scotland occurred in 1996 with 501 cases; 151 were hospitalized and 20 elderly died (Ahmed and Donaghy, 1998).

O157 STEC/VTEC has been isolated from outbreaks and from sporadic cases of diarrhea and HUS in many parts of the world: Canada, UK, Argentina, Germany, Central Europe, Chile, and Italy (summarized by Griffin, 1995; see also Chapters 2–9 in Kaper and O'Brien, 1998). Sakai City in Japan experienced the largest outbreak of O157 STEC/VTEC ever recorded in July 1996, which was part of several outbreaks that summer with an estimated number of a little less than 8,000 cases and 6 deaths (Infectious Disease Surveillance Center, Japan, 1997).

Other STEC/VTEC O:H serotypes have caused outbreaks of diarrhea and HUS: O111:H⁻, O145:H⁻, and O?:H19 in Japan (Kudoh et al., 1994); O26:H11 in the Czech Republic (Bielaszewská et al., 1990), USA (Brown et al., 1998), and Ireland (McMaster et al., 2001); O103:H2 in France (Mariani-Kurkdjian et al., 1993); O104:H21 in the USA (Centers for Disease Control, 1995); O111:H⁻ in Australia (Cameron et al., 1995), Italy (Caprioli et al., 1994), USA (Banatvala et al., 1996; Centers for Disease Control, 2000), Spain (Blanco et al., 1996), and France (Boudailliez et al., 1997; Mariani-Kurkdjian et al., 1997); O113:H21 in Australia (Paton et al., 1999); and O119 in France (Deschenes et al., 1996). A clone of sorbitol fermenting O157:H7 STEC/VTEC has been isolated from patients with diarrhea and HUS (Gunzer et al., 1992; Karch et al., 1993) and caused an outbreak in Germany (Ammon et al., 1999). Studies in Europe indicate that non-O157 STEC/VTEC strains are increasing in frequency as a cause of hemolytic-uremic syndrome (HUS) and found much more commonly in children with diarrhea (Verweyen et al., 1999; Scheutz et al., 2001; Tozzi et al., 2001).

Among the over 400 STEC/VTEC serotypes, and apart from O157:H⁻ and O157:H7, those in O groups O26, O103, O111,

and O145 are most commonly isolated from humans worldwide. These, along with strains that have caused outbreaks, are clearly recognized as pathogens. Table BXII.γ.205 shows the non-O157 STEC/VTEC serotypes that have been isolated from humans.

E. COLI IN HUMAN EXTRAINTESTINAL INFECTIONS

EXTRAINTESTINAL PATHOGENIC E. COLI (ExPEC)   Extraintestinal pathogenic *E. coli* (ExPEC) are *E. coli* strains that possess currently recognized extraintestinal virulence factors or have been demonstrated to possess enhanced virulence in an appropriate animal model (Russo and Johnson, 2000). ExPEC strains primarily belong to pathogenic clones of a limited number of O:K:H serotypes (Ørskov and Ørskov, 1975), usually with MR fimbriae (P and S fimbriae), siderophores (e.g., aerobactin), host defense-avoidance mechanisms such as capsules (Ørskov and Ørskov, 1977; Ørskov et al., 1982), O antigens, and serum resistance and toxins (often α-hemolysin). ExPEC is common in all age groups and may occur at almost any extraintestinal site. The most common infections include urinary tract infections (UTIs) ranging from uncomplicated to febrile to invasive, pyelonephritis, neonatal, and postneurosurgical meningitis and septicemia. This group is epidemiologically and phylogenetically distinct from commensal and intestinal strains of *E. coli* (Picard et al., 1999). Virulence genes are often located on pathogenicity islands (PAIs), which have the tendency to delete with high frequencies or may undergo duplications and amplifications. They are often associated with tRNA loci, which may represent target sites for the chromosomal integration of these elements (Hacker et al., 1997). Many produce toxins that can lyse erythrocytes of different mammalian species. The best characterized of these is α-hemolysin, which is often produced by strains causing UTIs (J.R. Johnson, 1991) and is believed to play an important role. α-Hemolysin is secreted and can be demonstrated in culture fluid filtrates (Beutin, 1991). Hemolytic colonies can also be identified by the clear zone of hemolysis produced on blood agar plates after 3–4 h of incubation.

UROPATHOGENIC E. COLI (UPEC)   This somewhat misleading acronym (clones or virulence factors are not syndrome-specific) has been used to refer to the majority of specific clonal groups of uropathogenic *E. coli* isolated from UTIs including pyelonephritis. They are characterized by a number of virulence factors that together play a role in their pathogenesis. First, UPEC is dominated by a limited number of O groups with O groups O1, O2, O4, O6, O7, O18ac, O75, O16, and O15 as the most commonly isolated (Ørskov and Ørskov, 1985). These strains are also represented by a limited number of K antigens: K1, K2, K3, K5, K12, and K13. Common serotypes include O1:K1:H7, O2:K1:H4, O4:K12:H1, O4:K12:H5, O6:K2:H1, O6:K5:H1, O6:K13:H1 (cystitis), O16:K1:H6, and O18ac:K5:H7. Furthermore, the majority of UPECs express P fimbriae of F types F7 through F16 and /or S fimbriae.

NEONATAL MENINGITIS E. COLI (NMEC)   Neonatal meningitis *E. coli* is frequently associated with O groups O7, O18ac, O1, and O6 that have the K1 antigen identical to the capsule of *N. meningitidis* type B (Sarff et al., 1975). O83:K1 strains are also common but apparently only in Europe (Ørskov and Ørskov, 1985). One of the most commonly isolated types is an S fimbriated clone of serotype O18ac:K1:H7.

E. COLI IN ANIMAL INFECTIONS   As is the case in humans, certain strains of *E. coli* can cause disease in animals. In farm animals, *E. coli* strains are associated with a variety of pathological

TABLE BXII.γ.205. Serotypes of non-O157 STEC/VTEC isolated from humans[a,b,c]

Serotype	Serotype	Serotype	Serotype	Serotype	Serotype	Serotype	Serotype	Serotype	Serotype
O1:H⁻	**O8:H21**	**O25:K2:H2**	O52:H23	**O83:H1**	O103:H18	O114:H4	O126:H20	O146:H11	O169:H⁻
O1:H1	O8:H25	O25:H14	O52:H25	**O84:H⁻**	O103:H21	O114:H48	O126:H21	O146:H14	**O171:H⁻**
O1:H2	**O9ab:H⁻**	**O26:H⁻**	O54:H21	O84:H2	O103:H25	**O114:H?**	**O126:H27**	O146:H21	O171:H2
O1:H7	O9:H7	O26:H2	**O55:H⁻**	O84:H20	O103:HNT	**O115:H10**	O127	O146:H28	**O172:H⁻**
O1:H20	O9:H21	O26:H8	**O55:H6**	**O85:H⁻**	**O104:H⁻**	O115:H18	**O128:H⁻**	O148:H28	**O172:H?**
O1:HNT	**O11:H⁻**	**O26:H11**	**O55:H7**	**O85:H10**	O104:H2	O116:H⁻	**O128ab:H2**	O150:H⁻	**O173:H2**
O2:H⁻	**O11:H2**	O26:H12	O55:H9	**O85:H23**	O104:H7	O116:H4	**O128:H7**	O150:H8	**O174:H⁻** [d]
O2:H1	O11:H8	O26:H32	**O55:H10**	**O86:H⁻**	O104:H16	O116:H10	O128:H8	O150:H10	**O174:H2**[d]
O2:K1:H2	O11:H49	O26:H46	O55:H19	O86:H10	**O104:H21**	O116:H19	O128:H10	O152:H4	O174:H8[d]
O2:H5	O12:H⁻	O27:H⁻	**O55:H?**	O86:H40	**O105ac:H18**	O117:H⁻	O128:H12	O153:H2	O174:H21[d]
O2:H6	**O14:H⁻**	O27:H30	O60:H⁻	O87:H16	O105:H19	**O117:H4**	**O128:H25**	O153:H11	O175:H16[e]
O2:H7	**O15:H⁻**	O28ab:H⁻	O64:H25	O88:H⁻	O105:H20	O117:H7	O128:H31	O153:H12	**OX176:H⁻** [f]
O2:H11	**O15:H2**	O28:H25	O65:H16	**O88:H25**	O106	O117:H8	**O128:H45**	O153:H21	**OX177:H⁻** [f]
O2:H27	O15:H8	O28:H35	O68:H⁻	O89:H⁻	O107:H27	O117:K1:H7	O129:H⁻	O153:H25	OX177:H11[f]
O2:H29	O15:H27	O30:H2	**O69:H⁻**	O90:H⁻	O109:H2	O117:H19	**O130:H11**	O153:H30	OX178:H7[f]
O2:H44	**O16:H⁻**	O30:H21	O69:H11	**O91:H⁻**	O109:H16	O117:H28	O131:H4	O153:H33	OX179:H8[f]
O3:H10	O16:H6	O30:H23	O70:H11	O91:H4	O110:H⁻	O118:H⁻	O132:H⁻	**O154:H⁻**	OX181:H15[f]
O4:H⁻	O16:H21	O37:H41	O71:H⁻	**O91:H10**	O110:H19	O118:H2	O133:H⁻	**O154:H4**	OX181:H49[e]
O4:H5	O17:H18	O38:H21	O73:H34	O91:H14	O110:H28	**O118:H12**	O133:H53	**O154:H19/20**	ONT:H⁻
O4:H10	O17:H41	O38:H26	O74	O91:H15	**O111:H⁻**	**O118:H16**	**O134:H25**	O156:H⁻	**ONT:H2**
O4:H40	**O18:H⁻**	**O39:H4**	**O75:H⁻**	**O91:H21**	**O111:H2**	**O118:H30**	**O137:H6**	O156:H4	ONT:II8
O5:H⁻	O18:H7	O39:H8	O75:H1	O91:H40	**O111:H7**	O119:H⁻	**O137:H41**	O156:H7	ONT:H18
O5:H16	O18:H12	O39:H28	**O75:H5**	O91:HNT	**O111:H8**	**O119:H5**	O138:H2	O156:H25	ONT:H19
O6:H⁻	O18:H15	O40:H2	O75:H8	O92:H3	O111:H11	**O119:H6**	O141:H⁻	O156:H27	ONT:H21
O6:H1	**O18:H?**	O40:H8	O76:H7	O92:H11	O111:H21	O119:H25	O141:H2	O156:HNT	**ONT:H25**
O6:H2	O20:H⁻	O41:H2	O76:H19	O95:H⁻	O111:H30	**O120:H19**	O141:H8	**O160:H?**	ONT:H41
O6:H4	O20:H7	O41:H26	O77:H⁻	O96:H10	O111:H34	**O121:H⁻**	O142	**O161:H⁻**	ONT:H47
O6:H12	**O20:H19**	O44	**O77:H⁻**	**O98:H⁻**	O111:H40	O121:H8	O143:H⁻	O162:H4	**ONT:K39:H48**
O6:H28	**O21:H5**	O45:H⁻	O77:H4	O98:H8	O111:H49	O121:H11	O144:H⁻	O163:H⁻	Orough:H⁻
O6:H29	**O21:H8**	O45:H2	O77:H7	**O100:H25**	**O111:H?**	**O121:H19**	**O145:H⁻**	O163:H19	Orough:H2
O6:H31	**O21:H?**	O45:H7	O77:H18	O100:H32	**O112ab:H2**	O123:H19	O145:H4	O163:H25	**Orough:H5**
O6:H34	O22:H⁻	O46:H2	O77:H41	**O101:H⁻**	O112:H19	O123:H49	**O145:H8**	**O165:H⁻**	**Orough:K1:H6**
O6:H49	**O22:H1**	**O46:H31**	O78:H⁻	O101:H9	O112:H21	O124:H⁻	O145:H16	O165:H10	Orough:K1:H7
O7:H4	**O22:H5**	O46:H38	**O79:H7**	O102:H6	O113:H2	O125:H⁻	**O145:H25**	**O165:H19**	**Orough:H11**
O7:H8	**O22:H8**	**O48:H21**	O79:H14	**O103:H⁻**	**O113:H4**	O125:H8	O145:H26	O165:H21	**Orough:H16**
O8:H⁻	O22:H16	**O49:H⁻**	O79:H23	O103:H2	O113:H5	**O125:H?**	**O145:H28**	**O165:H25**	Orough:H18
O8:H2	O22:H40	**O49:H10**	O80:H⁻	O103:H4	O113:H7	O126:H⁻	O145:H46	O166:H12	Orough:H20
O8:H9	O23:H7	O50:H⁻	O81:H?	O103:H6	**O113:H21**	O126:H2	O145:HNT	O166:H15	Orough:H21
O8:H11	O23:H16	**O50:H7**	O82:H⁻	O103:H7	O113:H32	O126:H8	O146:H⁻	O166:H28	Orough:H28
O8:H14	O23:H21	O51:H49	O82:H5	O103:H11	O113:H53	O126:H11	L		Orough:H46
O8:H19	O25:H⁻	O52:H19	O82:H8						
			O83:H⁻						

[a]Data from Scheutz et al. (2001 and unpublished results); Blanco et al. (2001), WHO (1999).

[b]Serotypes in bold represent strains isolated from patients with HUS.

[c]An updated list of STEC, with literature references, can be found at http://www.microbionet.com.au/frames/feature/vtec/brief01.html

[d]Formerly known as OX3.

[e]Formerly known as OX7.

[f]Provisional designation for new O antigens.

conditions, which include colibacillary diarrhea, colibacillary toxemia in pigs, systemic colibacillosis, coliform mastitis, and UTIs. Colibacillary diarrhea is an acute diarrheal disease due to ETEC infection, which occurs primarily in 1–3-d-old calves, lambs, and piglets. A limited number of O groups are represented among these ETEC strains. In England and Wales, the most common O groups of *E. coli* isolates from pigs with diarrhea are O149, O8, O158, O147, and O157 (Wray et al., 1993). Colibacillary toxemia in pigs can take several forms: shock in weaner syndrome, hemorrhagic enteritis, and edema disease. These disease syndromes are also attributable to *E. coli* belonging to a small number of O groups (O8, O45, O138, O139, O141, and O149). Rapid absorption of endotoxin from the bowel is hypothesized to play a role in the pathogenesis of the shock in weaner syndrome and hemorrhagic enteritis, while the toxin Stx2e/VT2e, which is produced by many of the strains having the above specified O groups, has been shown to play a role in the pathogenesis of edema disease (Macleod et al., 1991). Systemic colibacillosis occurs when septicemic strains of *E. coli* pass through the intestinal or respiratory mucosa into the bloodstream of calves, lambs, and poultry. Once they enter the bloodstream, they can cause either a generalized infection or a localized infection, such as meningitis and/or arthritis in calves and lambs or air sacculitis and pericarditis in poultry. *E. coli* strains are also an important cause of mastitis in cows. Endotoxin is believed to play a role in the inflammatory response observed during this disease.

The roles of various adherence mechanisms and toxins in the pathogenesis of the infections were reviewed by Wray and Woodward (1997). Fimbrial antigens and putative colonization factors associated with strains of *E. coli* causing disease in animals cited include F1, F4 (K88), F5 (K99), F41, F6 (987P), F17, F18, CS31A, F165, M326, C1213, F42, F11, curli, type IV pilins, and Nfa. Toxins produced by *E. coli* causing disease in animals include the heat-labile enterotoxins LTI and LTII, heat-stable enterotoxins STa and STb, cytotoxic necrotizing factors 1 and 2 (CNF1 and CNF2), and Shiga/Verocytotoxins.

Other typing methods For a description of other methods for subdivision of *E. coli*, i.e., phage typing, colicin typing, biotyping, typing by outer membrane protein (OMP) pattern, typing by antibiotic resistance patterns, and typing by direct hemagglutination, see Ørskov and Ørskov (1984a) and Sussman (1985). Phage typing is very useful for certain antigens because antisera are difficult to produce. This is particularly true for K1 (Gross et al., 1977), K3, K5, K7, K12, K13 (Nimmich et al., 1992), and K95 (Nimmich, 1994). A phage-typing scheme for STEC/VTEC O157:H7 established in 1987 (Ahmed et al., 1987) and extended in 1990 (Khakhria et al., 1990) has proven very useful in the epidemiological surveillance of STEC/VTEC O157 (Frost et al., 1993; Saari et al., 2001) infections and is applicable even in low-technology laboratories. In general, phage typing of O157 should be supplemented with one of the molecular typing methods mentioned below. Because of their discriminatory power, speed, use of commercially available reagents and equipment, and amenability to automation and electronic networking, molecular subtyping methods have become very popular for subtyping of *E. coli*, particularly strains involved in causing outbreaks of foodborne disease. Molecular subtyping methods for *E. coli* O157:H7 and other foodborne bacterial pathogens were reviewed by Barrett (1997). Some recently described methods include macro-restriction endonuclease analysis with PFGE (Preston et al., 2000; Zhang et al., 2000; Swaminathan et al., 2001); detection of insertion sequences and characterization of virulence genes by DNA probes or PCR (Thompson et al., 1998; Zhang et al., 2000); detection of amplified fragment length polymorphisms (Iyoda et al., 1999); computer identification by rRNA gene restriction patterns (Machado et al., 1998); and the analysis of randomly amplified polymorphic DNA (Hopkins and Hilton, 2001). In 1996, a molecular subtyping network in the United States, designated PulseNet, for the electronic comparison of DNA fingerprints generated by macrorestriction endonuclease analysis with PFGE was developed to subtype *E. coli* O157:H7. PulseNet has proved an exceptionally valuable tool for detecting outbreaks of *E. coli* O157:H7 infection. The development of standardized laboratory and data analysis protocols and their successful use in providing surveillance for *E. coli* O157:H7 and other foodborne bacterial pathogens was reviewed by Swaminathan et al. (2001).

ENRICHMENT AND ISOLATION PROCEDURES

Many simple agar media can be used for isolation. Media used for selective isolation from feces usually contain substances that partly or completely inhibit growth of bacteria other than *Enterobacteriaceae* (tetrathionate, deoxycholate, bile salts, etc.). The addition of Maranil (dodecylbenzolsulfonate) at a concentration of 0.005% will inhibit swarming of *Proteus* organisms. For details, see Edwards and Ewing (1972) or Kauffmann (1966) or any catalogue from one of the medium-producing companies. At Statens Serum Institut, Copenhagen, a medium developed in the media department of the institute, bromothymol blue (BTB) agar, is used.[1]

MAINTENANCE PROCEDURES

E. coli strains can be kept viable for many years in beef extract agar stabs (tightly closed, e.g., by corks soaked in melted paraffin wax) or on Dorset egg medium. Cultures are initially incubated at 37°C followed by storage in the dark at room temperature (20–22°C). After a few weeks or months, such cultures often contain many mutational forms such as R forms and acapsular forms; consequently, we prefer to store important cultures in beef broth containing 10% glycerol at −80°C. Screw-capped vials are used for easy access.

PROCEDURES FOR TESTING SPECIAL CHARACTERS

Kilian and Bülow (1976) found that a very high percentage (97%) of *Escherichia coli* and the majority (57%) of *Shigella* strains, exclusively among the *Enterobacteriaceae*, produce β-glucuronidase (GUD). Prolonged incubation of 28 h increases positivity to 99.5% (Rice et al., 1990), which is in accordance with the presence of the *uidA* (GUD) gene in all *E. coli* strains (McDaniels et al., 1996). This test (referred to as the GUD-, PGUA-, MUG-, GUR-, or GLUase-HR test) therefore is very suitable as a screening test for *E. coli*, with the unfortunate exception of most STEC/VTEC O157 strains, which are phenotypically negative. None of the other four species of the genus *Escherichia* are positive for this enzyme (Rice et al., 1991). Both genotypic and phenotypic assays for glutamate decarboxylase (GAD) used by environmental scientists have been described as highly specific for *E. coli*. Unfortunately, these tests have only recently been shown to exhibit the same specificity on a smaller collection of pathogenic isolates of *E. coli* (Grant et al., 2001).

DIFFERENTIATION OF THE GENUS *ESCHERICHIA* FROM OTHER GENERA

See Table BXII.γ.193 of the family *Enterobacteriaceae* for characteristics that can be used to differentiate this genus from other genera of the family.

TAXONOMIC COMMENTS

The identification of *Escherichia* strains seldom causes problems; however, many studies have shown that "*Escherichia* is a genus (or species) made up of phenotypically variable strains" (Farmer and Brenner, 1977). DNA–DNA hybridization studies have been an invaluable tool for solving problems in this field. The genus *Shigella* is closely related to *Escherichia*, and only historical reasons make it acceptable that these two genera are not united. Several typical EIEC types have been found that have pathogenic traits that are similar to those of *Shigella*. The Sereny test (Sérèny, 1957), which demonstrates the capacity to cause keratoconjunctivitis in the guinea pig, typical of *Shigella* strains, is also found in these special *Escherichia* strains. Day et al. (1981) described a tissue culture technique that can be used as a substitute for the Sereny test. Typically, such dysentery-associated *E. coli* strains have O antigens that are closely related or identical to *Shigella* O antigens. Brenner et al. (1972a), by DNA reassociation studies, found species-level relatedness between *Shigella* strains and these special *Escherichia* strains, as well as nonpathogenic *E. coli* strains.

---

1. Bromothymol blue agar (selective for *Enterobacteriaceae*). Combine the following ingredients: peptone (Orthana Ltd., Copenhagen), 10.0 g; NaCl, 5.0 g; yeast extract (Oxoid), 5.0 g; and distilled water, 1000 ml. The pH is adjusted to 8.0, agar powder is added, and the preparation is autoclaved at 120°C for 20 min. The following components are then added aseptically from sterile stock solutions: Maranil solution [Paste A75 (dodecylben-zolsulfonate), Henkel, Germany], 1.0 ml; sodium thiosul-

fate (50% solution), 2.0 ml; bromothymol blue (Riedel de Haen, Germany; 1.0% solution), 10.0 ml; lactose (33% solution), 27 ml; and glucose (33% solution), 1.2 ml. The pH is adjusted to 7.7–7.8. To obtain optimum results, the amount of glucose must be adjusted for every new batch of yeast extract, peptone, and agar. This medium is very useful for differentiation of lactose-fermenting colonies based on their color.

Not unexpectedly, many strains are phenotypically intermediate between *Escherichia* and *Shigella*, but for obvious reasons a special taxonomic status for such strains is not warranted. In the older literature the name Alkalescens-Dispar is used, but, as stated by Brenner (1978), this group is virtually indistinguishable from *E. coli* strains and is, in fact, a biogroup of *E. coli* that is anaerogenic, lactose-negative (or delayed), and nonmotile.

While most or all characters that classically have been used for definition of the genus *Escherichia* are chromosomally determined, several traits that are not characteristic of *Escherichia* have been found in otherwise typical *Escherichia* strains. Lautrop et al. (1971) and Layne et al. (1971) described H_2S-positive strains of *Escherichia*, and this character was plasmid-determined. It is not known which selective forces account for the simultaneous isolation of H_2S-positive *Escherichia* strains in different parts of the world.

Other "forbidden" phenotypic traits have similarly been described in *Escherichia*, many of which are plasmid-determined. Ørskov et al. (1961) found many urease-producing strains among typical serotypes from piglet diarrhea. Wachsmuth et al. (1979) demonstrated the plasmid-determined nature of a similar urease-positive phenotype in human *E. coli* strains. Citrate-utilizing *E. coli* strains were described by Washington and Timm (1976) and were found to be plasmid determined in similar strains by Sato et al. (1978). Carbon dioxide-dependent cultures can be found (Eykyn and Phillips, 1978). A citrate-positive, malonate-positive biogroup and a biogroup negative in these reactions were described (Burgess et al., 1973).

FURTHER READING

Abraham, S.N. and S. Jaiswal. 1997. Type-1 fimbriae of *Escherichia coli*. *In* Sussman (Editor), *Escherichia coli*: Mechanisms of Virulence, Cambridge University Press, Cambridge. pp. 169–192.

Cowan, S.T., K.J. Steel, G.I. Barrow and R.K.A. Feltham (Editors). 1995. Cowan and Steel's Manual for the Identification of Medical Bacteria, University of Cambridge, Cambridge.

de Graaf, F.K. and W. Gaastra. 1997. Fimbriae of enterotoxigenic *Escherichia coli*. *In* Sussman (Editor), *Escherichia coli*: Mechanisms of Virulence, Cambridge University Press, Cambridge. pp. 193–211.

Dodson, K.W., F. Jacob-Dubuisson, R.T. Striker and S.J. Hultgren. 1997. Assembly of adhesive virulence-associated pili in Gram-negative bacteria. *In* Sussman (Editor), *Escherichia coli*: Mechanisms of Virulence, Cambridge University Press, Cambridge. pp. 213–236.

Duffy, G., P. Garvey and D.A. McDowell (Editors). 2001. Verocytotoxigenic *E. coli*, Food & Nutrition Press, Inc., Conneticut. pp. 1–457.

Ewing, W.H. 1986. The Genus *Escherichia*. *In* Edwards and Ewing (Editors), Edwards and Ewing's Identification of *Enterobacteriaceae*, 4th Ed., Elsevier Science Publishing Co., New York. pp. 93–122.

Gyles, C.L. (Editor). 1994. *Escherichia coli* in Domestic Animals and Humans, CAB International, Wallingford.

Jann, K. and B. Jann (Editors). 1990. Bacterial Adhesins. Current Topics in Microbiology and Immunology, Springer-Verlag, Berlin. 151 pp.

Kaper, J.B. and A.D. O'Brien (Editors). 1998. *Escherichia coli* O157:H7 and other Shiga Toxin-Producing Strains, ASM Press, Washington, D.C. 465 pp.

Klemm, P. (Editor). 1994. Fimbriae: Adhesins, Genetics, Biogenesis, and Vaccines, CRC Press, Boca Raton.

Low, D., B. Braaten and M. van der Woude. 1996. Fimbriae. *In* Neidhardt, Curtiss, Ingraham, Lin, Low, Magasanik, Reznikoff, Riley, Schaechter and Umbarger (Editors), *Escherichia coli* and *Salmonella*: Cellular and Molecular Biology, 2nd Ed., ASM Press, Washington, D.C. pp. 146–157.

Nataro, J.P. and J.B. Kaper. 1998. Diarrheagenic *Escherichia coli*. Clin. Microbiol. Rev. *11*: 142–201.

Neidhardt, F.C., R. Curtiss, III, J.L. Ingraham, E.C.C. Lin, K.B. Low, B. Magasanik, W.S. Reznikoff, M. Riley, M. Schaechter and H.E. Umbarger (Editors), 1996. *Escherichia coli* and *Salmonella*: Cellular and Molecular Biology, 2nd Ed., ASM Press, Washington, D.C.

Sussman, M. (Editor). 1997. *Escherichia coli*: Mechanisms of Virulence, Press syndicate of the University of Cambridge, Cambridge. 639 pp.

DIFFERENTIATION OF THE SPECIES OF THE GENUS *ESCHERICHIA*

Characteristics useful in distinguishing the five species of *Escherichia* are given in Table BXII.γ.193 of the family *Enterobacteriaceae* and in Table BXII.γ.197 of the genus *Escherichia*.

List of species of the genus Escherichia

1. **Escherichia coli** (Migula 1895) Castellani and Chalmers 1919, 941^AL^ (*Bacillus coli* Migula 1895, 27.)

   co′li. Gr. n. *colon* large intestine, colon; M.L. gen. n. *coli* of the colon.

   The characteristics are as described for the genus and as listed in Table BXII.γ.193 of the family *Enterobacteriaceae*. Occurs naturally in the lower part of the intestine of warm-blooded animals, and as intestinal (some foodborne) and extraintestinal pathogens of humans and animals.

   The mol% G + C of the DNA is: 48.5–52.1 (T_m).

   Type strain: ATCC 11775, CCM 5172, CIP 54.8, DSM 30083, IAM 12119, NCDO 1989, NCTC 9001. Serotype O1:K1(L1):H7.

   GenBank accession number (16S rRNA): X80725.

   Additional Remarks: Other sequences are listed in Table BXII.γ.206.

2. **Escherichia blattae** Burgess, McDermott and Whiting 1973, 4^AL^

   blat′tae. L. fem. n. *blatta* cockroach; L. gen. n. *blattae* of the cockroach.

   The characteristics are as described for the genus and as listed in Table BXII.γ.193 of the family *Enterobacteriaceae*. *E. blattae* was isolated from the hindgut of healthy cockroaches, *Blatta orientalis*, in England (Burgess et al., 1973) and on Easter Island (Nogrady and Aubert, personal com-

TABLE BXII.γ.206. *rrn* operon sequences of *Escherichia* strains

Source and strain[a, b]	EMBL[c]	Method[d]
Escherichia coli:		
	J01859	rRNA
	J01695	*rrn*B
	V00348	*rrn*B
(PK3)	X80731	PCR
(MC4100)	X80732	PCR
CIP (ATCC 11775[T])	X80725	PCR
ATCC 25922	X80724	PCR
(K-12)	M87049	*rrn*A
(K-12)	U00006	*rrn*B
(K-12)	L10328	*rrn*C
(K-12)	U18997	*rrn*D
(K-12)	U00006	*rrn*E
	M29364	*rrn*G
(K–12)	D15061	*rrn*H
BioM	X80733	PCR
(PK3)	X80721	*rrn*A
(PK3)	X80722	*rrn*B
(PK3)	X80723	*rrn*C
(PK3)	X80727	*rrn*D
(PK3)	X80728	*rrn*E
(PK3)	X80729	*rrn*G
(PK3)	X80730	*rrn*H
Escherichia fergusonii:		
ATCC 35469	AF530475	NA[e]
Escherichia hermannii:		
BioM	X80675	rRNA
Escherichia vulneris:		
CIP (ATCC 33821[T])	X80734	PCR

[a]Some strain numbers have been lost; sequences were most probably obtained using *E. coli* K-12.

[b]Bacterial collection from which each strain is deposited: ATCC (American Type Culture Collection); BioM (BioMérieux, Marcy l'Étoile, France), CIP (Collection de l' Institut Pasteur).

[c]Accession numbers under which sequence is available.

[d]Method by which each sequence has been obtained: rRNA (total rRNA sequenced using reverse transcriptase), PCR (total PCR products sequenced using T7-DNA polymerase); *rrn*X (sequence of a single operon, X).

[e]Not available.

munication). It appears as two biotypes, one of which is citrate and malonate positive, the other negative, and it is the only species within *Escherichia* that is gluconate positive (Burgess et al., 1973). *E. blattae* has not been associated with disease either in humans or in cockroaches.

The mol% G + C of the DNA is: not determined.

Type strain: ATCC 29907, CDC 9005-74, DSM 4481, NCTC 12127.

GenBank accession number (16S rRNA): X87025.

3. **Escherichia fergusonii** Farmer, Fanning, Davis, O'Hara, Riddle, Hickman-Brenner, Asbury, Lowery and Brenner 1985c, 223[VP] (Effective publication: Farmer, Fanning, Davis, O'Hara, Riddle, Hickman-Brenner, Asbury, Lowery and Brenner 1985b, 77.)*

fer.gu.so'ni.i. M.L. masc. (substantive) *fergusonii* coined to honor the American microbiologist William W. Ferguson, who made many contributions to enteric bacteriology and was one of the first to show the role of certain strains of *E. coli* in infantile diarrhea (Farmer et al., 1985b).

The characteristics are as described for the genus and as listed in Table BXII.γ.193 of the family *Enterobacteriaceae*.

Has been isolated from human clinical specimens (stool, urine, blood, and an abdominal wound), the feces of captive raptors belonging to the order *Falconiformes* or *Strigiformes* (Bangert et al., 1988), and from unspecified sites for other warm-blooded animals.

The mol% G + C of the DNA is: not determined.

Type strain: ATCC 35469, CDC 0568-73.

Additional Remarks: Other sequences are listed in Table BXII.γ.206.

4. **Escherichia hermannii** Brenner, Davis, Steigerwalt, Riddle, McWhorter, Allen, Farmer, Saitoh and Fanning 1983a, 438[VP] (Effective publication: Brenner, Davis, Steigerwalt, Riddle, McWhorter, Allen, Farmer, Saitoh and Fanning 1982a, 705.)*

her.man'ni.i. M.L. *hermannii* of Hermann, named in honor of George J. Hermann, former chief of the Enteric Section at the CDC, for his many contributions to enteric bacteriology, and Lloyd G. Herman, formerly of the Environmental Services Branch, National Institutes of Health, Bethesda, MD, for his contributions to the study of yellow-pigmented bacteria (Brenner et al., 1982a).

The characteristics are as described for the genus and as listed in Table BXII.γ.193 of the family *Enterobacteriaceae*. Those that together distinguish it from most other members of this family include growth in the presence of KCN, fermentation of cellobiose, and production of yellow pigment. Has been isolated from human clinical specimens (wounds, sputum, lung, stool, blood, and spinal fluid) and recently from the sludge of an industrial wastewater treatment plant (Kiernicka et al., 1999). The sludge isolate shows promise for bioremediation; it grows in and degrades high concentrations of chlorobenzene.

The mol% G + C of the DNA is: 53–58 (T_m).

Type strain: ATCC 33650, CDC 980-72, DSM 4560.

Additional Remarks: Other sequences are listed in Table BXII.γ.206.

5. **Escherichia vulneris** Brenner, McWhorter, Leete Knutson and Steigerwalt 1983d, 438[VP] (Effective publication: Brenner, McWhorter, Leete Knutson and Steigerwalt 1982b, 1137.)*

vul.ner'is. L. n. *vulnus* a wound; L. gen. n. *vulneris* of a wound; *Escherichia vulneris* the *Escherichia* of a wound.

The characteristics are as described for the genus and as listed in Table BXII.γ.193 of the family *Enterobacteriaceae* (Brenner et al., 1982b). Has been isolated from human clinical specimens, primarily wounds, the majority of which occurred on the arms or legs, but also blood, throat, sputum, vagina, urine, and stool, and other warm-blooded animals. The type species was isolated from the intestine of a cowbird in Michigan, USA.

The mol% G + C of the DNA is: 58.5–58.7 (T_m).

Type strain: ATCC 33821, CDC 875-72, DSM 4564, NIH 580.

GenBank accession number (16S rRNA): X80734.

Additional Remarks: Other sequences are listed in Table BXII.γ.206.

Editorial Note: This species was formerly known as Enteric Group 10.

Editorial Note: This species was formerly known as Enteric Group 11.

Editorial Note: This species was formerly known as Enteric Group 1.

Genus II. Alterococcus Shieh and Jean 1999, 341[VP] (Effective publication: Shieh and Jean 1998, 644)

THE EDITORIAL BOARD

Al.te.ro.coc' cus. L. *alter* another; Gr. n. *coccus* a grain or berry; M.L. masc. n. *Alterococcus* another coccus.

Spherical cells 0.8–0.9 μm in diameter, occurring singly or in pairs. **Motile by means of a single flagellum.** Gram negative. **Facultatively anaerobic**, capable of aerobic and anaerobic fermentation. Chemoheterotrophic. **Oxidase and catalase positive.** Optimum temperature for growth, 45°C. No growth occurs at 30°C or 60°C. **Halophilic**; growth occurs in the presence of 1–3% NaCl (optimum, 2.0%), but not 0% or 5% NaCl. The major cellular fatty acid is *anteiso*-15-carbon acid ($C_{15:0 \text{ anteiso}}$). Cause agar liquefaction. Habitat: coastal hot springs.

The mol% G + C of the DNA is: 65.5–67.0.

Type species: **Alterococcus agarolyticus** Shieh and Jean 1999, 341 (Effective publication: Shieh and Jean 1998, 644)

FURTHER DESCRIPTIVE INFORMATION

Butyrate and propionate are formed during both aerobic and anaerobic growth in PY semisolid medium[1]. Butyrate with propionate and/or formate is produced during both aerobic and anaerobic growth in PYG broth[2].

ENRICHMENT AND ISOLATION PROCEDURES

Water samples are decimally diluted with sterile NaCl-MOPSO buffer[3] and spread on PY agar. The plates are incubated aerobically at 50°C in the dark for 3–7 d. Colonies that cause agar liquefaction on the incubated plates are purified by successive streaking.

MAINTENANCE PROCEDURES

Early stationary-phase cultures grown in PY broth are inoculated into sterile 60% seawater at a ratio of 0.25 ml in 5 ml (1:20) and stored for two or three months at 45°C.

DIFFERENTIATION OF THE GENUS ALTEROCOCCUS FROM OTHER GENERA

The genus is the only one known to contain Gram-negative, halophilic, thermophilic bacteria that degrade agar and grow both aerobically and anaerobically. The only other Gram-negative genera that contain heterotrophic, halophilic, thermophilic bacteria are *Rhodothermus, Thermotoga, Thermosipho,* and *Spirochaeta. Rhodothermus* contains obligately aerobic rods, whereas *Alterococcus* contains cocci. The genera *Thermotoga* and *Thermosipho* contain strictly anaerobic rods, whereas *Alterococcus* is facultatively anaerobic. *Spirochaeta* contains facultatively anaerobic species, but the cells are spiral, not coccoid.

TAXONOMIC COMMENTS

16S rDNA-based phylogenetic analysis of the type strain of *Alterococcus agarolyticus* has indicated that the species is most closely related to members of the family *Enterobacteriaceae.* In this edition of the *Manual,* the genus is placed in this family, which belongs to the order *Enterobacteriales* in the class *Gammaproteobacteria.*

List of species of the genus Alterococcus

1. **Alterococcus agarolyticus** Shieh and Jean 1999, 341[VP] (Effective publication: Shieh and Jean 1998, 644.)

   a.gar.o.ly' ti.cus. Malayan n. *agar* agar, a complex gelling polysaccharide from marine red algae; Gr. adj. *lyticum* dissolving; M.L. adj. *agarolyticus* agar-dissolving.

   The characteristics are as described for the genus and as listed in Table BXII.γ.207. Young colonies on agar media are white, circular, and opaque. Glucose, cellobiose, galactose, lactose, sucrose, trehalose, and xylose are fermented. Sensitive to ampicillin, chloramphenicol, erythromycin, penicillin G, and tetracycline.

   Isolated from two hot springs in the intertidal zone of Lutao Island, Taiwan.

   The mol% G + C of the DNA is: 65.8.

   Type strain: ADT3, CCRC 19135.

   GenBank accession number (16S rRNA): AF075271.

---

1. PY broth is composed of (g/l deionized water): Bacto Peptone (Difco), 4.0; Bacto yeast extract (Difco), 2.0; NaCl, 20.0; MgSO₄·7H₂O, 0.5; CaCl₂, 0.01; and 3-(*N*-morpholino)-2-hydroxypropanesulfonic acid (MOPSO; Sigma), 4.5. This broth medium is adjusted to pH 7.0. Bacto agar (Difco) is added to this medium at 5 and 15 g/l for the preparation of PY semisolid and plating media, respectively.

2. PYG medium is composed of two parts. Part 1 consists of the following ingredients dissolved in 900 ml of distilled water: Bacto Peptone (Difco), 4.0 g; Bacto yeast extract (Difco), 2.0 g; NaCl, 20.0 g; MgSO₄·7H₂O, 0.5 g; CaCl₂, 0.01g; and Tris buffer (Sigma), 6.0 g. The pH is adjusted to 7.8. Part 2 consists of glucose (5.0 g) dissolved in 100 ml of distilled water. Parts 1 and 2 are autoclaved separately and combined aseptically after cooling to room temperature.

3. NaCl-MOPSO buffer consists of (g/l deionized water): NaCl, 25.0 and MOPSO (*N*-morpholino)-2-hydroxypropanesulfonic acid), 0.45; pH 7.0.

TABLE BXII.γ.207. Phenotypic characteristics of *Alterococcus agarolyticus*[a,b]

Characteristic	Reaction
Fermentation of:	
Adonitol, D-arabinose, dulcitol, *myo*-inositol, mannitol, melibiose, sorbitol	−
Cellobiose, galactose, glucose, lactose, sucrose, trehalose, xylose	+
Mannose	W
Motility	+
Swarming	−
Luminescence	−
Agarase activity	+
Hydrolysis of starch	W
Hydrolysis of casein, DNA, gelatin, fats	−
Catalase, oxidase	+
Arginine dihydrolase, lysine decarboxylase, ornithine decarboxylase	−
Growth at:	
30°C	−
38°C	W
40–56°C	+
58°C	W
60°C	−
Growth in presence of NaCl levels of:	
0%	−
0.50%	W
1.0–3.5%	+
4.00%	W
5.00%	−

[a]Symbols: +, positive; −, negative; W, weakly positive.
[b]Data from Shieh and Jean (1998).

Genus III. **Arsenophonus** *Gherna, Werren, Weisburg, Cote, Woese, Mandelco and Brenner 1991, 564^VP*

JOHN H. WERREN

Ar.se.no.pho' nus. Gr. n. *arsen* a male; Gr. suff. *phonus* slayer; N.L. masc. n. *Arsenophonus* male-killer.

Cells are Gram-negative, nonmotile, nonsporeforming, nonflagellated, long to highly filamentous rods, which divide by septation. **The bacteria infect insect tissues.** The type species can be cultivated on cell-free media. Colonies are mucoid, gray-white, round, and convex with entire edges. Enzymatically digested proteins best serve as nitrogen sources. Does not utilize KNO_3, $(NH_4)_2SO_4$, complete defined amino acid mixtures, or acid-hydrolyzed peptones as nitrogen sources. Sugars sucrose, fructose, and D-glucose are utilized as primary carbon sources. Weak growth occurs with maltose, trehalose, cellobiose, and D-xylose. Acid is produced with D-glucose, fructose, and sucrose. Growth is negative with L-arabinose, glycerol, dulcitol, lactose, D-mannitol, raffinose, and *i*-inositol. Positive for catalase and gelatin liquefaction; negative for Voges–Proskauer, methyl red, nitrate reduction, indole, oxidase, hydrogen sulfide, o-nitrophenyl-β-D-galactopyranoside, lysine and ornithine decarboxylase, urease, and arginine dehydrolase. Cells grow at pH 6.2–8.7 (optimum pH range 7.4–8.0), and a temperature range of 15–35°C (30°C optimum).

There is currently only one validly named species in the genus (and an additional *Candidatus* species). However, **the genus may be widespread in insects**, based upon recent evidence. Therefore, the genus description may well change as additional information becomes available.

The mol% G + C of the DNA is: 39.5.

Type species: **Arsenophonus nasoniae** Gherna Werren, Weisburg, Cote, Woese, Mandelco and Brenner 1991, 564.

FURTHER DESCRIPTIVE INFORMATION

General biology The genus *Arsenophonus* was originally described from a single bacterial species that is the causative agent of the "son-killer" trait in the parasitic wasp *Nasonia vitripennis* (insect order *Hymenoptera*). Subsequently, a second species has been proposed for the genus (*Candidatus* Arsenophonus triatominarum), isolated from a reduviid bug (insect order *Heteroptera*). Despite being found in divergent insects, the two are similar in morphology, patterns of infection of host insect tissues, and 16S rDNA sequence (97.8% similarity). Biochemical characterization of "*Candidatus* Arsenophonus triatominarum" has not yet

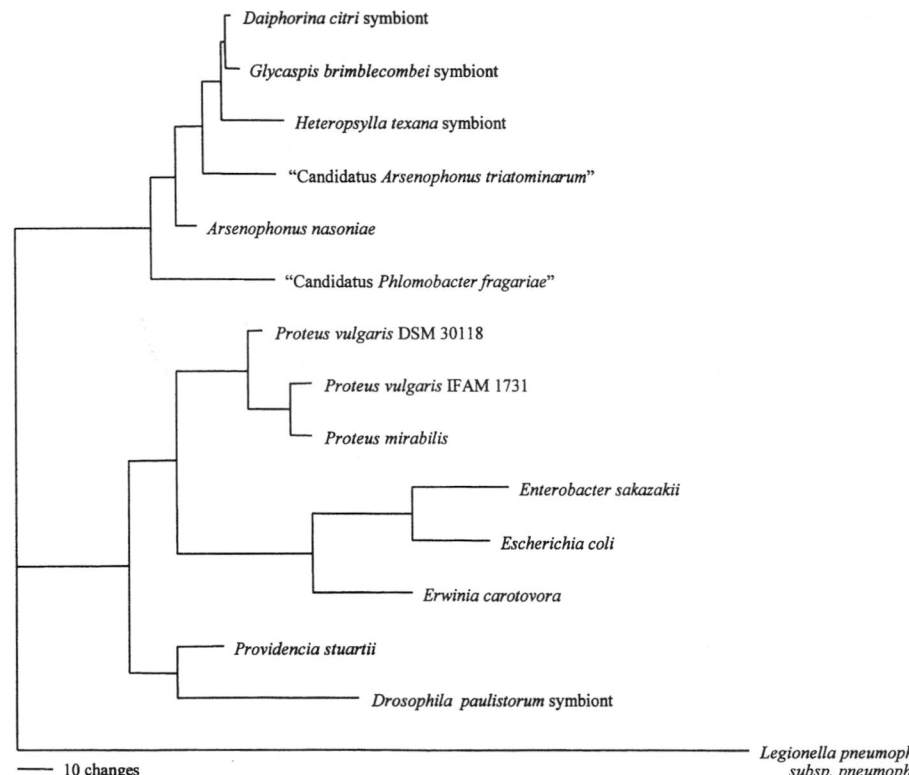

FIGURE BXII.γ.193. A phylogenetic tree based on 16S rDNA sequences is presented based on a parsimony analysis using a heuristic search and stepwise addition (D.L. Swofford, 1996, PAUP 4.0 Sinauer, Sunderland, MA) with *Legionella pneumophila* defined as the outgroup. Partial 16S rDNA nucleotide sequences (1476 sites) of *Arsenophonus nasoniae, Candidatus* A. triatominarum, other endosymbiotic and free-living bacteria were analyzed. Names of the insect host species are shown for the psyllid secondary symbionts and the unidentified bacterium from *D. paulistorum*. In all other cases the bacterial genus and species are shown. Numbers above nodes are bootstrap percentages from 1000 replicate searches. The sequence data for *Diaphorina citri* secondary symbiont was provided by T. Fukatsu (Subandiyah et al., 2000). Other sequences were obtained from the NCBI database (accession numbers AF263561, AF263562, U91786, M90801, U91515, X07652, AJ233425, X07652, AF008582, AB004746, AB035924, AJ233411, AF008581, U20273, and X73402).

been performed; nor are its effects on the host insect known. *A. nasoniae* can be grown on cell-free media; "*Candidatus* A. triatominarum" has not yet been grown in cell-free culture, but can be maintained on insect tissue culture.

Phylogenetic analysis using 16S rDNA sequences (see below) indicates that closely related bacteria are secondary symbionts in several species of psyllid insects (insect order *Homoptera*), although the taxonomic status of these bacteria has not yet been resolved. Taken together, the results suggest that *Arsenophonus* (and related bacteria) may have a wide distribution as infectious agents of insects. Another closely related bacterium, *Candidatus* Phlomobacter fragariae, causes disease in strawberry (Zreik et al., 1998).

Phylogenetic analyses Based on 16S rDNA phylogeny *Arsenophonus* is a lineage in the *Gammaproteobacteria* (Gherna et al., 1991; Hypsa and Dale, 1997), within the family *Enterobacteriaceae*. The phylogenetic analysis places *A. nasoniae* and the *Candidatus* A. triatominarum as close relatives (97.8% similarity) (Fig. BXII.γ.193). Other closely related organisms (Fig. BXII.γ.193) are secondary symbionts found in psyllid insects (98.6–97.6% similarity). Although they are closely related, the specific phylogenetic relationships between *A. nasoniae*, *Candidatus* A. tria-tominarum, and the psyllid secondary symbionts are not clearly resolved. These may eventually be placed in the genus *Arsenophonus* when further characterization is completed. *A. nasoniae*, *Candidatus* A. triatominarum, and the three psyllid secondary endosymbionts form a monophyletic clade with *Candidatus* Phlomobacter fragariae (97.0% similarity to *A. nasoniae*). This monophyletic group is strongly supported by various phylogenetic analyses. *Candidatus* Phlomobacter fragariae is found within the phloem of strawberry and is associated with strawberry marginal chlorosis disease (Zreik et al., 1998). Plant phloem associated bacteria can be vectored by phloem feeding insects such as psyllids and aphids, which may suggest a biological basis for its phylogenetic similarity to psyllid secondary endosymbionts. The apparent monophyly of this bacterium with *Arsenophonus* and the psyllid secondary endosymbionts (Subandiyah et al., 2000; Thao et al., 2000b) suggests that separate generic status for this bacterium may not be warranted. However, additional sequences and biological characterization are needed to resolve the issue. The closest known bacteria outside of the "*Arsenophonus* group" include *Proteus vulgaris*, *Proteus mirabilis*, and *Providencia stuartii* (94.2–94.3% similarity).

List of species of the genus *Arsenophonus*

1. **Arsenophonus nasoniae** Gherna, Werren, Weisburg, Cote, Woese, Mandelco and Brenner 1991, 564[VP]

   na.so'ni.ae. N.L. n. *Nasonia* a genus of parasitoid wasps; *nasoniae* of the genus *Nasonia*.

   The cells are nonmotile nonsporeforming, long rods to highly filamentous (0.40–0.57 × 6.9–10.0 μm) in culture. Filamentous cells can also be found in insect tissues. Biochemical features are presented in the genus description. *A. nasoniae* is the causative agent of the "son-killer" trait in the parasitic wasp *N. vitripennis*. *Nasonia* wasps parasitize the pupae of various dipterans (flies). They sting the host pupa and lay their eggs underneath the puparial wall. The insects hatch and feed upon the host fly to complete development. In female wasps infected with the bacterium, male eggs fail to hatch, whereas female eggs develop normally (Werren et al., 1986). Genetic experiments indicate that the bacterium acts by blocking the development of unfertilized eggs, which normally develop into males in this haplodiploid species. The son-killer bacterium is found in *N. vitripennis* populations throughout North America at frequencies around 10%, and has also been detected in the sibling species *N. longicornis* (Balas et al., 1996).

   The bacterium is introduced into the dipteran host pupa during stinging by the infected female, where it replicates and is ingested by the feeding wasp larvae (Huger et al., 1985). As a result, the bacteria can be transmitted maternally from an infected female to her developing larvae, and horizontally when more than one female parasitizes the same host. The bacteria initially invade the midgut epithelium and subsequently spread to various tissues, including brain, fat body, muscles, eyes, and hemocytes. Massive infections of tissues can be observed in adults without major negative effects on insect viability. In insect tissues, the bacteria are pleomorphic, with rods being the most common morphology. Both male and female larvae can become infected by host feeding during larval development. Female reproductive tracts are often infected with the bacteria, whereas infections are not observed in male reproductive tissues. In adult wasps, drop-like bacterial masses can be discharged from the midgut epithelium into the gut lumen, suggesting a possible additional fecal route for infectious transmission. The mechanism by which *A. nasoniae* causes lethality of male eggs is unknown. Experiments indicate that unfertilized eggs, which normally develop into males in the haplodiploid insect host, fail to develop when the female is infected. Levels of male embryo mortality can vary in different infected females (Balas et al., 1996).

   A. nasoniae can be isolated from infected insects by surface sterilization of the insect followed by maceration of the insect onto GC media with Kellogg's supplement, and serial streaking of the macerated material. Characteristic gray-white mucoid colonies will develop in 2–4 d at 26°C. Bacteria can also be isolated by antiseptic removal of hemolymph from fly pupae parasitized by infected *Nasonia* females, and inoculation of the hemolymph onto GC media with Kellogg's supplement.

   The type strain grows poorly or does not grow on many standard media. It can be grown on brain–heart infusion broth (Difco no. 0037) or solid media. However, improved growth was found on a GC media base (Difco no. 0289) supplemented with Kellogg's additive (Kellogg et al., 1963). Even on this medium, bacterial growth is slow relative to many culturable bacteria, and visible colonies may take 2–3 d at 26°C. Growth is best on solid media. Broth cultures show elevated frequencies of filamentous bacteria relative to agar base solid media. The type strain was isolated from an *N. vitripennis* son-killer strain collected near Salt Lake City, Utah.

   The mol% G + C of the DNA is: 39.5 (T_m).

   Type strain: SKI4, ATCC 49151.

   GenBank accession number (16S rRNA): M90801.

2. *Candidatus* Arsenophonus triatominarum Hypsa and Dale 1997, 1143.

   tri'a.to'min.a.rum. N.L. n. *Triatominae* subfamily of reduviid bugs; *triatominarum* of the *Triatominae*.

   Cells are nonmotile, nonsporeforming, nonflagellated highly filamentous rods (1–1.5 × >15 μm), and divide by septation. Bacteria grow within the cytoplasm of cells of the reduviid bug *Triatoma infestans*. Phylogenetic analysis of 16S ribosomal gene sequence indicates a close phylogenetic relationship to *A. nasoniae* (97.8% similarity), as do general morphological characteristics and the pattern of infection within tissues of their respective hosts. The bacterium cannot be grown on cell-free minimal medium, but can be maintained in *A. albopictus* cell culture. Biochemical characteristics of the bacterium have not yet been determined.

"*Candidatus* A. triatominarum" was isolated from the hemolymph of *Triatoma infestans*, a reduviid bug also known to be a vector of *Trypanosoma cruzi*, the causative agent of Chagas disease in humans. Within infected *T. infestans*, heavy infections can be found in neural ganglia, visceral muscles, salivary glands, nephrocytes, testes, ovaries, and dorsal vessels. In contrast to *A. nasoniae*, the bacterium could not be grown of a variety of tested cell-free media, including media upon which *A. nasoniae* can be cultured (Hypsa and Dale, 1997). Phenotypic effects of the bacterium on *T. infestans* are unknown. Strain TI1 was isolated from hemolymph of *T. infestans* individuals from a laboratory colony, and cultured in *Aedes albopictus* cell line C6/36.

The mol% G + C of the DNA is: unknown.

Deposited strain: TI1.

GenBank accession number (16S rRNA): U91786.

Genus IV. **Brenneria** Hauben, Moore, Vauterin, Steenackers, Mergaert, Verdonck and Swings 1999a, 1^VP (Effective publication: Hauben, Moore, Vauterin, Steenackers, Mergaert, Verdonck and Swings 1998a, 394.)

LYSIANE HAUBEN AND JEAN SWINGS

Bren.ne′ri.a. M.L. fem. n. *Brenneria* named after Don J. Brenner.

Cells are $0.5–1.0 \times 1.3–3.0$ µm; have rounded ends; occur singly or rarely in pairs. Gram negative. Motile by peritrichous flagella. Facultatively anaerobic, but anaerobic growth by some species is weak. Optimum temperature, 27–30°C; maximum temperature for growth is 40°C. Oxidase negative. Catalase positive. Acid is produced from fructose, D-galactose, D-glucose, D-mannose, salicin, and sucrose but not from adonitol or dulcitol. Do not possess arginine decarboxylases, arginine dihydrolase, lysine decarboxylases, ornithine decarboxylases, or starch hydrolase.

The species of the genus *Brenneria* comprise a distinct phylogenetic group, as determined by 16S rRNA gene sequence comparisons, and have 12 characteristic signature nucleotides (Table BXII.γ.223 of the genus *Erwinia*).

Brenneria species cause diseases on trees (e.g., deciduous trees and walnut trees), which include blights, cankers, wilts, necrosis, and rots. Ingress by the pathogen generally occurs through natural openings and wounds. More details on disease symptoms are described for each species.

The mol% G + C of the DNA is: 50.1–56.1 (T_m, Bd).

Type species: **Brenneria salicis** (Day 1924) Hauben, Moore, Vauterin, Steenackers, Mergaert, Verdonck and Swings 1999a, 1 (Effective publication: Hauben, Moore, Vauterin, Steenackers, Mergaert, Verdonck and Swings 1998a, 394) (*Erwinia salicis* (Day 1924) Chester 1939, 406; "*Bacterium salicis*" Day 1924, 14.)

FURTHER DESCRIPTIVE INFORMATION

Metabolic features are the same as for the genus *Erwinia*.

Pectate lyases are produced by strains of *B. rubrifaciens* (Gardner and Kado, 1976).

Fermentation end products from D-glucose are CO_2 and different combinations of succinate, lactate, formate, and acetate; some form 2,3-butanediol and some ethanol (White and Starr, 1971). Starch is not hydrolyzed beyond dextrins.

16S rDNA sequence analyses of the species of the genus *Brenneria* by Kwon et al. (1997) and Hauben et al. (1998a) are in good agreement except for the type strain of *B. salicis*, and place the genus *Brenneria* within the Enterobacteriaceae, closely related to the genera *Pectobacterium*, *Erwinia*, *Pantoea*, and *Enterobacter* (Fig. BXII.γ.201 and Table BXII.γ.223 of the genus *Erwinia*). The gene sequences of the type strains of *B. salicis* sequenced by Kwon et al. (1997) (ATCC 15712) and Hauben et al. (1998a) (LMG 2698) differ in 73 nucleotides. A genomic fingerprinting study by AFLP of 77 *B. salicis* strains revealed the profile of strain LMG 2698 to be highly similar (>80% similarity) to the ones of other authentic *B. salicis* strains, indicating that strain LMG 2698 is an authentic *B. salicis* (Hauben et al., 1998b).

Virulent or temperate phages have been isolated, characterized, and reported to be active against strains of *B. nigrifluens* and *B. rubrifaciens* (Zeitoun and Wilson, 1969).

Antisera prepared against live or heat-killed cells, nonpurified or purified immunogens, have been used for the differentiation or identification of all *Brenneria* species (De Kam, 1976; Schaad, 1979).

ENRICHMENT AND ISOLATION PROCEDURES

The isolation procedure is the same as for the genus *Erwinia*. The affected plant material is washed in tap water, then in sterile water, and dried by paper toweling. Surface sterilization (3 min in 1:10 dilution of 5.25% active sodium hypochlorite) is sometimes detrimental for isolation. Affected tissue is removed from a young lesion or the edge of older necrotic areas by a sterile scalpel; the tissue is crushed in sterile water, saline, or buffer solution, and is streaked onto a solid medium.

The isolation of some *Brenneria* species can be facilitated by use of selective-differential media, but such media are usually not necessary. *B. nigrifluens*, *B. quercina*, and *B. rubrifaciens* will grow on MS medium (Miller and Schroth, 1972) and produce characteristic colonies. A soluble pink pigment is produced by *B. rubrifaciens*.

MAINTENANCE PROCEDURES

Stock cultures of *Brenneria* species should be grown on standard media of choice at 25–30°C until good growth occurs. The cultures can be maintained for short-term storage in a refrigerator (4–5°C).

Long-term preservation is the same as for the genus *Erwinia*. The bacteria can be successfully stored as lyophilized cultures, usually suspended in a filter-sterilized mixture of 200 ml of horse serum (Oxoid SR035C) to which 1675 mg of nutrient broth (Oxoid) and 20 g of glucose in 67 ml of distilled water is added. Strains have also been stored in liquid nitrogen and in glycerol at −80°C (broth + 15% glycerol).

DIFFERENTIATION OF THE GENUS *BRENNERIA* FROM OTHER GENERA

The differential characteristics of the species of *Brenneria* are given in Tables BXII.γ.208 and BXII.γ.209.

TABLE BXII.γ.208. Diagnostic reactions for *Brenneria* species[a]

Characteristic	1. *B. salicis*	2. *B. alni*	3. *B. nigrifluens*	4. *B. paradisiaca*	5. *B. quercina*	6. *B. rubrifaciens*
Indole	−	−	−	+	−	−
Methyl red	nd	+	+	nd	−	nd
Growth factors required	nd	−	−	nd	+	nd
Endoglucanase activity	nd	+	−	nd	+	nd
Esculin hydrolase	nd	−	+	nd	−	nd
β-Galactosidase	+	−	v	+	v	+
Pectinase	nd	−	−	+	−	nd
Urease	−	+	v	−	−	−
Acid production from:						
Arabinose	−	+	+	+	−	+
Esculin	−	nd	+	nd	−	−
Maltose	nd	+	−	−	−	+
Melibiose	−	−	+	+	−	nd
Raffinose	+	−	+	+	−	−
Sorbitol	nd	−	+	−	v	−
Trehalose	nd	+	+	nd	−	nd
Xylose	−	+	+	+	−	−
Utilization of carbon sources:						
L-Arabinose	nd	nd	+	+	−	+
Citrate	+	−	v	+	+	+
Galacturonic acid	−	nd	−	+	−	−
Maltose	nd	nd	−	nd	+	−
Raffinose	+	nd	+	nd	−	−
L-Rhamnose	nd	nd	nd	+	nd	−
Sorbitol	−	nd	+	nd	−	−
Trehalose	nd	nd	+	+	−	−
Xylose	−	nd	−	+	−	−
Utilization of nitrogen sources:						
Acetamide	nd	nd	−	nd	−	+
Anthranilic acid	+	nd	−	nd	−	nd
Glycocyamine	nd	nd	+	nd	−	+
L-Serine	+	nd	+	+	−	+
Threonine	+	nd	−	+	+	+
Tryptamine	−	nd	−	+	−	−
Xanthin	+	nd	+	−	nd	+
Sensitivity toward:						
Streptomycin	nd	nd	−	+	nd	nd

[a]For symbols see standard definitions; nd, not determined.

TAXONOMIC COMMENTS

The species of the genus *Brenneria* were formerly classified in the genus *Erwinia*. We refer to the taxonomic comments of the genus *Erwinia* for a discussion of this issue.

Characteristics that differentiate *Brenneria* from the genera *Pectobacterium, Pantoea,* and *Erwinia* are given in Table BXII.γ.223 of the *Erwinia* chapter. As it is very difficult to differentiate them phenotypically, genomic methods are recommended for differentiation.

List of species of the genus Brenneria

1. **Brenneria salicis** (Day 1924) Hauben, Moore, Vauterin, Steenackers, Mergaert, Verdonck and Swings 1999a, 1[VP] (Effective publication: Hauben, Moore, Vauterin, Steenackers, Mergaert, Verdonck and Swings 1998a, 394) (*Erwinia salicis* (Day 1924) Chester 1939, 406; *"Bacterium salicis"* Day 1924, 14.)

   sa'li.cis. L. n. *salix* the willow; L. gen. n. *salicis* of the willow.

   The characteristics are as given for the genus and as listed in Tables BXII.γ.208 and BXII.γ.209.

   Grows poorly on nutrient agar but moderately well on YDC agar (0.5% yeast extract, 1% dextrose, 3% CaCO$_3$) or on glucose nutrient agar. Colonies on 0.5% starch potato agar (pH 6.5) are yellowish in 2–3 d. A bright yellow pigment is produced on autoclaved potato tissue. Craters form around colonies on the pectate gel of Paton (1959).

   A PCR based identification and detection method was developed with primers Es1a (5′-GCGGCGGACGGGTGA-GTAAA-3′) and Es4b (5′-CTAGCCTGTCAGTTTTGAATG-CT-3′), annealing at 64°C, derived from the 16S rDNA sequence of *B. salicis* (Hauben et al., 1998b).

   Antisera have been used for the identification of *B. salicis* (De Kam, 1976; 1986) and soluble antigens were detected in the leaves of *Salix alba* (De Kam, 1982).

   Causes a vascular wilt (watermark disease) of *Salix* species. The pathogen occurs mainly in the xylem vessels of the host plant. Infected willows show wilted, dried, brown-colored leaves and a watery, transparent color of the wood. Occasionally the whole tree is killed.

   The mol% G + C of the DNA is: 51.3–51.5 (T_m, Bd.).

   Type strain: ATCC 15712, ICMP 1587, LMG 2698, NCPPB 447.

   GenBank accession number (16S rRNA): U80210 and Z96097.

2. **Brenneria alni** (Surico, Mugnai, Pastorelli, Giovannetti and Stead 1996) Hauben, Moore, Vauterin, Steenackers, Mergaert, Verdonck and Swings 1999a, 1[VP] (Effective publication: Hauben, Moore, Vauterin, Steenackers, Mergaert, Verdonck and Swings 1998a, 395) (*Erwinia alni* Surico, Mugnai, Pastorelli, Giovannetti and Stead 1996, 725.)

TABLE BXII.γ.209. Additional reactions for *Brenneria* species[a]

Characteristic	1. *B. salicis*	2. *B. alni*	3. *B. nigrifluens*	4. *B. paradisiaca*	5. *B. quercina*	6. *B. rubrifaciens*
Growth in 5% NaCl	nd	+	+	nd	+	nd
Growth at 37°C	nd	+	+	nd	+	nd
$NO_3^- \rightarrow NO_2^-$	nd	−	−	nd	−	nd
H₂S from cysteine	nd	+	+	nd	+	nd
Acetoin	nd	+	+	nd	+	nd
Oxidase	−	−	−	−	−	−
Catalase	+	+	+	+	+	+
Arginine dihydrolase	−	−	−	−	−	−
Caseinase	−	−	−	−	−	−
Arginine decarboxylase	−	−	−	−	−	−
Aspartase	+	nd	+	nd	nd	nd
Gelatinase	nd	−	−	nd	−	nd
Lysine decarboxylase	−	−	−	−	−	−
Ornithine decarboxylase	−	−	−	−	−	−
Phenylalanine deaminase	−	−	−	nd	−	−
Starch hydrolase	−	−	−	−	−	−
Acid production from:						
N-Acetylglucosamine	+	nd	+	+	+	+
Adonitol	−		−	−	−	−
Amygdalin	+		+	nd	+	+
Arabitol	−	nd	−	−	−	−
Arbutine	+	nd	+	+	+	+
Cellobiose	nd	nd	nd	nd	−	−
Dextrin	−	nd	−	nd	−	−
Dulcitol	−		−	−	−	−
Erythritol	−	nd	−	−	−	−
Erythrose	−		nd	−	nd	nd
Fructose	+	+	+	+	+	+
DL-Fucose	−	nd	−	−	−	−
D-Galactose	+	+	+	+	+	+
Gentiobiose	nd	nd	nd	+	nd	nd
D-Glucose	+	+	+	+	+	+
Gluconate	nd	nd	nd	+	nd	nd
Glycerol	nd	+	+	nd	+	nd
Glycogen	−	nd	−	−	−	−
Inulin	−	nd	−	nd	−	−
Lactose	−		nd	−	−	−
D-Lyxose	nd	nd	nd	−	nd	nd
D-Mannitol	+	+	+	nd	+	+
D-Mannose	+	+	+	+	+	+
α-Methylmannoside	−	nd	−	−	−	−
α-D-Melezitose	−		nd	−	−	−
β-Methylglucoside	+	nd	+	+	+	+
β-Methylxyloside	−	nd	−	−	−	−
Rhamnose	+	nd	+	+	+	+
Ribose	+	nd	+	+	+	+
Salicin	+	+	+	+	+	+
Sorbose	−	nd	−	−	−	−
Starch	−	nd	−	−	−	−
Sucrose	+	+	+	+	+	+
Tagatose	−	nd	−	−	−	−
Xylitol	nd	nd	nd	−	nd	nd
Utilization of carbon sources:						
Adipate	−	nd	−	−	−	−
Adonitol	−	nd	−	nd	v	−
Amygdalin	nd	nd	−	nd	nd	nd
D-Arabinose	−	nd	nd	nd	nd	nd
D-Arabitol	−	nd	−	−	−	−
Arbutin	+	nd	+	+	+	+
Benzoate	−	nd	−	−	−	−
Betaine	−	nd	−	−	−	−
Butanate	nd	nd	−	nd	nd	nd
Butanol	−	nd	nd	−	−	−
Cellobiose	nd	nd	−	nd	−	nd
Dextrin	−	nd	−	nd	−	−
Dulcitol	−	nd	−	nd	−	−
meso-Erythritol	−	nd	−	nd	−	−
Esculin	+	nd	+	nd	+	+
Ethylene glycol	−	nd	−	nd	−	−
Fructose	+	nd	+	+	+	+
Fumarate	+	nd	+	+	+	+
D-Galactose	+	nd	+	+	+	+
Gallate	−	nd	−	−	−	−

(*continued*)

TABLE BXII.γ.209. *(cont.)*

Characteristic	1. *B. salicis*	2. *B. alni*	3. *B. nigrifluens*	4. *B. paradisiaca*	5. *B. quercina*	6. *B. rubrifaciens*
Gluconate	+	nd	+	+	+	+
Glucose	+	nd	+	+	+	+
Glucuronic acid	nd	nd	+	nd	nd	nd
L-Glutaminic acid	−	nd	nd	nd	nd	−
Glycerol	+	nd	+	+	+	+
Glycogen	−		−	nd	−	−
Glycol	−	nd	nd	nd	nd	nd
α-Ketoglutarate	nd	nd	nd	+	nd	nd
Lactose	−	nd	−	nd	−	−
Lactulose	−	nd	−	nd	nd	−
Lyxose	−	nd	−	nd	−	−
Malate	+	nd	+	+	+	+
Malonate	nd		−	−	nd	−
Malonic acid	−	nd	nd	nd	−	−
D-Mannitol	+	nd	+	+	+	+
Mannose	+	nd	+	+	+	+
Melibiose	+	nd	+	+	nd	nd
α-D-Melezitose	−	nd	−	nd	−	−
Methanol	−	nd	−	−	−	−
α-Methylglucoside	nd	nd	nd	−	nd	nd
β-Methylglucoside	+	nd	│	+	+	+
Mucate	nd	nd	nd	+	nd	nd
Naphthalene	−		−	nd	−	−
Oxalate	−	nd	−	−	−	−
Pectinic acid	−	nd	nd	nd	nd	−
Propanol	−	nd	−	nd	−	−
Propionate	−	nd	−	−	−	−
Ribose	+	nd	+	+	+	+
Salicin	+	nd	+	+	+	+
Sorbinic acid	−	nd	−	nd	−	−
Sorbose	−	nd	−	−	−	−
Starch	−	nd	−	nd	−	−
Succinate	+	nd	+	+	+	+
Sucrose	+	nd	+	nd	+	+
Tartrate	nd	nd	nd	−	nd	nd
Triacetine	nd	nd	nd	−	nd	nd
Xylitol	−	nd	−	−	−	−
Utilization of nitrogen sources:						
Adenine	nd	nd	nd	+	nd	nd
Alanine	+	nd	+	+	+	+
Allantoine	+	nd	+	+	+	+
Ammonium chloride	+	nd	+	+	+	+
Arginine	+	nd	+	+	+	+
Asparagine	+	nd	nd	nd	+	+
Asparaginic acid	+	nd	+	+	+	+
Betaine	−	nd	−	−	−	−
Carnosine	nd	nd	nd	+	nd	nd
Choline	−	nd	−	−	−	−
Citrulline	+	nd	+	+	+	+
Cysteamine	−	nd	−	−	−	−
Creatine	nd	nd	nd	+	nd	nd
Ethanolamine	nd	nd	+	nd	+	+
Glucosamine	+	nd	+	+	+	+
L-Glutaminic acid	+	nd	+	+	+	+
Glutathion	+	nd	+	+	+	+
Glycine	+	nd	+	+	+	+
Glycylglycine	+	nd	+	+	+	+
Guanine	nd	nd	nd	−	nd	nd
Histidine	+	nd	+	+	+	+
Hydroxyproline	−	nd	−	−	−	−
Isoleucine	+	nd	nd	nd	+	+
Kynureninic acid	−	nd	−	−	−	−
Leucine	+	nd	+	+	+	+
L-Methionine	+	nd	+	+	+	+
Nicotinic acid	nd	nd	nd	−	nd	nd
Octopine	nd	nd	nd	+	nd	nd
Ornithine	nd	nd	nd	+	nd	nd
Phenylalanine	+	nd	+	+	+	+
Quinolinic acid	−	nd	−	−	−	−
Sarcosine	−	nd	−	−	−	−
Spermidine	−	nd	−	−	−	−
Spermine	−	nd	−	−	−	−
Taurine	nd	nd	nd	−	nd	nd

(continued)

TABLE BXII.γ.209. *(cont.)*

Characteristic	1. *B. salicis*	2. *B. alni*	3. *B. nigrifluens*	4. *B. paradisiaca*	5. *B. quercina*	6. *B. rubrifaciens*
Thymine	−	nd	−	nd	−	−
Trigonelline	−	nd	−	−	−	−
L-Tryptophan	+	nd	nd	+	+	+
Tyrosine	+	nd	+	+	+	+
Ureum	+	nd	+	+	+	+
Valine	+	nd	+	+	+	+
Sensitivity toward:						
Amoxycillin	nd	nd	+	+	+	+
Ampicillin	nd	nd	+	+	+	+
Bacitracin	−	nd	−	nd	−	−
Carbenicillin	+	nd	+	+	+	+
Cephalexin	+	nd	+	+	+	+
Cephaloridine	+	nd	+	+	+	+
Cephalotine	+	nd	+	+	nd	+
Chloramphenicol	+	nd	+	+	+	+
Cloxacyclin	−	nd	−	nd	−	−
Colistine sulfate	−	nd	−	nd	−	−
Doxycylin	+	nd	+	nd	+	+
Erythromycin	−	nd	−	nd	−	−
Framycetine	+	nd	+	+	nd	+
Furazolidone	+	nd	+	+	+	+
Fusidinic acid	−	nd	−	−	−	−
Gentamicin	−	nd	−	nd	−	−
Kanamycin	+	nd	nd	+	v	nd
Lincomycin	−	nd	−	nd	nd	−
Methicillin	−	nd	−	−	−	−
Nalidixinic acid	+	nd	+	+	+	+
Neomycin	nd	nd	−	nd	−	nd
Nitrofurantoin	+	nd	+	nd	+	+
Oxytetracycline	+	nd	+	+	+	+
Penicillin	nd	nd	+	nd	+	+
Polymyxin	nd	nd	−	nd	nd	−
Spectinomycin	−	nd	−	nd	nd	nd
Sulfafurazol	−	nd	−	−	−	−
Novobiocine	nd	nd	+	nd	nd	nd
Tetracycline	+	nd	+	+	+	+

^aFor symbols see standard definitions; nd, not determined.

al' ni. L. fem. n. *alnus* alder; L. gen. n. *alni* of alder, referring to the plant from which the organism was first isolated.

The characteristics are as given for the genus and as listed in Tables BXII.γ.208 and BXII.γ.209.

Abundant growth occurs in nutrient broth and on nutrient agar, YDC (0.5% yeast extract, 1% dextrose, 3% $CaCO_3$), Miller–Schroth medium. Colonies on NSA (nutrient agar containing 5% sucrose) are umbonate with translucent margins, whitish, and 2–2.5 mm in diameter after 5 days at 27°C.

Causes bark canker disease of alder (*Alnus* sp.). At first, diseased alder plants have small necrotic cankers in the bark of the trunk and in the bark of branches, twigs, and suckers. These cankers are slightly sunken, dark brown, and irregularly circular and appear to be water soaked. They usually develop at lenticels or are localized at nodes with dead twigs or on leaf scars. As the infection progresses, the necrosis spreads laterally, becoming deeper and reaching the cambium and sometimes the first layers of wood, which then turn brown. Cankers may eventually girdle and kill a branch or tree.

The mol% G + C of the DNA is: 50.1–50.7 (T_m).

Type strain: ICMP 12481, NCPPB 3934.

GenBank accession number (16S rRNA): AJ223468.

3. **Brenneria nigrifluens** (Wilson, Starr and Berger 1957) Hauben, Moore, Vauterin, Steenackers, Mergaert, Verdonck and Swings 1999a, 1^{VP} (Effective publication: Hauben,

Moore, Vauterin, Steenackers, Mergaert, Verdonck and Swings 1998a, 395) (*Erwinia nigrifluens* Wilson, Starr and Berger 1957, 673.)

ni.gri.flu' ens. L. adj. *niger* nigra black; L. v. *fluo* flow; M.L. part. adj. *nigrifluens* black flowing.

The characteristics are as described for the genus and as listed in Tables BXII.γ.208 and BXII.γ.209.

Colonies on Bacto-EMB (Difco) agar are dark violet with a green metallic sheen. Craters form around colonies on the polypectate medium (1 liter nearly boiling distilled water, 1 ml of bromothymol blue, 6 ml of 10% $CaCl_2 \cdot H_2O$, 2 g of sodium polypectate) of Hildebrand (1971). Growth media should contain yeast extract and should be at pH 7–8.

Causes a bark necrosis of the Persian walnut (*Juglans regia*). The disease is visible as dark brown necrotic areas on the trunks or large branches. The lesions extend well beyond the discolored bark. Exudates may be produced. The lesions are superficial, but may penetrate enough to kill the phloem.

The mol% G + C of the DNA is: 56.1 (T_m, Bd).

Type strain: ATCC 13028, ICMP 1578, LMG 2694.

GenBank accession number (16S rRNA): U80203, Z96095.

4. **Brenneria paradisiaca** (Fernández-Borrero and López-Duque 1970) Hauben, Moore, Vauterin, Steenackers, Mergaert, Verdonck and Swings 1999a, 1^{VP} (Effective publica-

tion: Hauben, Moore, Vauterin, Steenackers, Mergaert, Verdonck and Swings 1998a, 396) (*Erwinia paradisiaca* Fernández-Borrero and López-Duque 1970, 22.)

pa.ra.di.si.a' ca. M.L. n. generic name *Musa paradisiaca* banana.

The characteristics are as given for the genus and as listed in Tables BXII.γ.208 and BXII.γ.209.

Causes rhizome rot of roots and soft rot of green fruit of *Musa paradisiaca.*

The mol% G + C of the DNA is: 54.7 (T_m, Bd).

Type strain: ATCC 33242, LMG 2542, NCPPB 2511.

GenBank accession number (16S rRNA): Z96096.

5. **Brenneria quercina** (Hildebrand and Schroth 1967) Hauben, Moore, Vauterin, Steenackers, Mergaert, Verdonck and Swings 1999a, 1[VP] (Effective publication: Hauben, Moore, Vauterin, Steenackers, Mergaert, Verdonck and Swings 1998a, 395) (*Erwinia quercina* Hildebrand and Schroth 1967, 253.)

quer.ci' na. L. n. *quercus* oak; L. suff. *ina* belonging to; M.L. part. adj. *quercina* oak-belonging.

The characteristics are as given for the genus and as listed in Tables BXII.γ.208 and BXII.γ.209.

Growth on PGPC agar (60 g homogenized potato, 10 g glucose, 10 g peptone, 1 g CaCO$_3$, 1 liter distilled water) is luxuriant, and, after 24 h, colonies are white, circular, and raised with entire margins. Craters form around colonies on polypectate gel of Hildebrand (1971), but no pectolytic activity was observed in potato or carrot tissue.

Small amounts of gas are produced (possibly from peptone) in a glucose peptone medium and in PGPC.

Causes copious oozing of sap from acorns ("drippy nut") and shoot blight of oak: *Quercus agrifolia* and *Q. wislizeni.* Acorns may rot somewhat, and cups may ooze after nutfall. Oozing takes place in late summer when temperatures average about 29°C. Superficially rots onion (but not potato) slices and induces profuse lateral root development in 3–4 days on slices of carrot, turnip, or beet.

The mol% G + C of the DNA is: from 54.6–55.1 (T_m, Bd).

Type strain: ATCC 29281, ICMP 1845, LMG 2724, NCPPB 1852.

GenBank accession number (16S rRNA): AJ223469.

6. **Brenneria rubrifaciens** (Wilson, Zeitoun and Fredrickson 1967) Hauben, Moore, Vauterin, Steenackers, Mergaert, Verdonck and Swings 1999a, 1[VP] (Effective publication: Hauben, Moore, Vauterin, Steenackers, Mergaert, Verdonck and Swings 1998a, 395) (*Erwinia rubrifaciens* Wilson, Zeitoun and Fredrickson 1967, 621.)

rub.ri.fac' i.ens. L. adj. *ruber* red; L. v. *facio* make; M.L. part. adj. *rubrifaciens* red-producing.

The characteristics are as given for the genus and as listed in Tables BXII.γ.208 and BXII.γ.209.

Produces a pink soluble pigment when grown on yeast extract glucose chalk agar (YDC, 0.5% yeast extract, 1% dextrose, 3% calcium carbonate (Dye, 1968). The structure of a 3-(3-carboxymethyl-4,5-dihydro-4,5-dioxo-2-pyrrolyl)-4,5,6-trihydroxy-pyridine-2-acetic acid (rubrifacine) is suggested (Feistner et al., 1984) as the main component of the pigment.

Grows poorly on nutrient agar, but well on YDC on which colonies are cream to yellow, low convex, smooth, shining with entire margins. Craters form around colonies on the polypectate gel B and C of Hildebrand (1971), but no pectolytic activity was observed on vegetable tissue.

Causes a phloem necrosis of Persian walnut trees (*Juglans regia*). Dark brown to black necrotic streaks from along the inner bark and outer sapwood. In places, the necrosis develops outward to the periderm, which breaks, and a dark exudate emerges.

The mol% G + C of the DNA is: 52.0–52.6 (T_m, Bd).

Type strain: ATCC 29291, ICMP 1915, LMG 2709, NCPPB 2020.

GenBank accession number (16S rRNA): U80207 and Z96098.

Genus V. *Buchnera* Munson, Baumann and Kinsey 1991b, 566[VP]

PAUL BAUMANN, LINDA BAUMANN AND NANCY A. MORAN

Buch.ne' ra. M.L. fem. n. *Buchnera* named for Paul Buchner, German biologist who made extensive contributions to the study of endosymbiosis.

Round or slightly oval cells, 2–5 μm in diameter. **Gram-negative** cell wall, lack flagella. Cells divide by binary fission. Do not form resting stages or endospores. **Found in bacteriocytes** of aphids in vesicles derived from the host membranes. Capable of respiration. **Cannot be cultivated** outside the aphid host. Essential for the survival of the aphid.

The mol% G + C of the DNA is: 26–27.

Type species: **Buchnera aphidicola** Munson, Baumann and Kinsey 1991b, 567.

FURTHER DESCRIPTIVE INFORMATION

Cell structure and life cycle During their reproductive phase most aphids contain within their body cavity a bilobed structure called a bacteriome consisting of 60–90 polyploid cells called bacteriocytes (Douglas and Dixon, 1987). These cells are filled with host-derived vesicles (symbiosomes) containing *Buchnera*

(Fig. BXII.γ.194) (Hinde 1971b; Griffiths and Beck, 1973; McLean and Houk, 1973; Akhtar and van Emden, 1994). This organism is spherical or oval in shape having a diameter of 2–5 μm and a cell wall resembling that of Gram-negative bacteria (Fig. BXII.γ.195). A thin line indicative of the peptidoglycan layer has been detected between the two unit membranes (Houk et al., 1977). The presence of peptidoglycan is also indicated by chemical analysis and by the alteration of cell wall structure observed upon addition of penicillin to the aphid diet (Griffiths and Beck, 1974; Houk et al., 1977).

In their most active reproductive stage, aphids are females that reproduce by parthenogenesis. The embryos develop within the mother, which gives birth to live young (Blackman, 1987). A representative aphid, *Schizaphis graminum*, when born weighs 24 mg and contains 2×10^5 cells of *Buchnera*. In 9–10 d, it reaches its maximum weight of 540 mg and contains 5.6×10^6 endo-

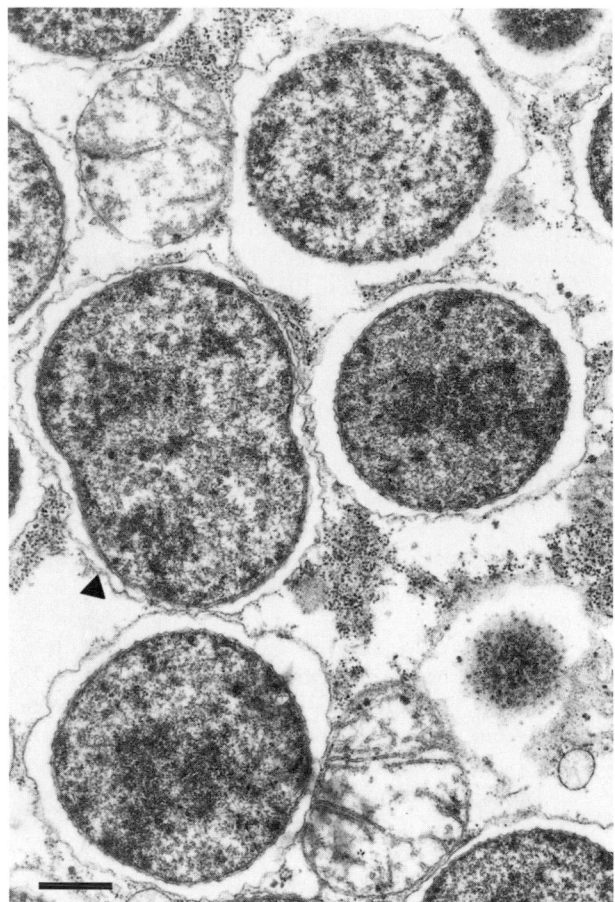

FIGURE BXII.γ.194. Electron micrograph of *Buchnera* within aphid bacteriocytes. *Arrow* indicates host-derived vesicle membrane. Bar = 1 μm. (Printed with permission of M. Kinsey and D.L. McLean.)

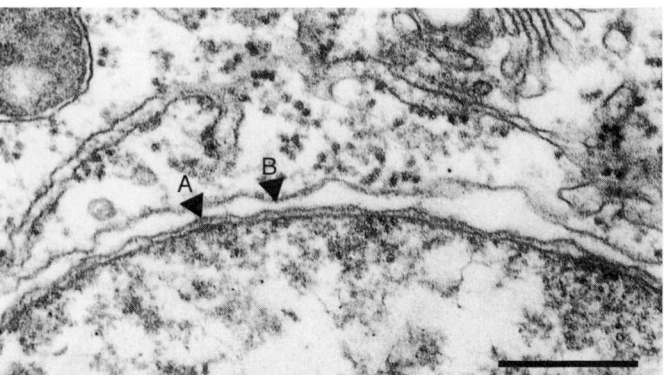

FIGURE BXII.γ.195. Electron micrograph of *Buchnera* showing the Gram-negative cell wall (*A*) and the vesicle membrane (*B*). Bar = 0.5 μm. (Printed with permission of M. Kinsey and D.L. McLean.)

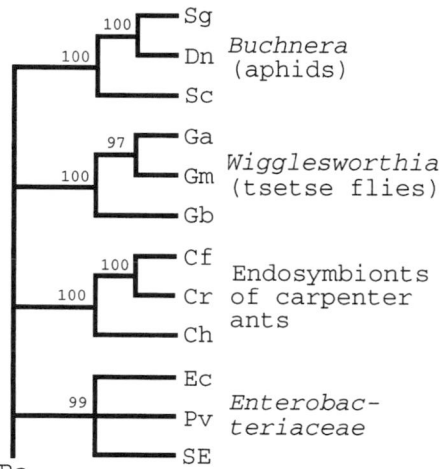

FIGURE BXII.γ.196. Phylogenetic trees resulting from parsimony analyses (D.L. Swofford, 1993, PAUP 3.1.1, Champaign, IL) using 16S rDNA nucleotide sequences (1189 sites) of *Buchnera* and other closely related insect endosymbionts as well as free-living bacteria. Numbers above nodes are bootstrap percentages from parsimony searches (1000 replicates). Data for *Wigglesworthia glossinidia* from Aksoy (1995b); the following abbreviations designate the insect hosts (in parentheses): *Ga* (*Glossina austeni*), *Gm* (*G. morsitans*), *Gb* (*G. brevipalpis*). Data for carpenter ant endosymbionts from Schröder et al. (1996); hosts: *Cf* (*Camponotus floridanus*), *Cr* (*C. rufipes*), *Ch* (*C. herculeanus*); *SE* (secondary endosymbiont of the aphid *Acyrthosiphon pisum*); *Ra* (*Ruminobacter amylophilus*).

symbionts (Baumann and Baumann, 1994). The increase in the total number of *Buchnera* parallels the increase in the weight of the aphid. The endosymbionts are partitioned between the maternal and the embryonic bacteriocytes. In mature aphids, most of the endosymbionts are found in the embryos (Humphreys and Douglas, 1997). Halfway through the reproductive period the number of maternal bacteriocytes undergoes a sharp decrease (Douglas and Dixon, 1987) and there is degradation of the endosymbionts (Hinde, 1971a; Griffiths and Beck, 1973).

Aphids may also form sexual forms; the females deposit eggs that overwinter and hatch in the spring. *Buchnera* is transmitted maternally to the embryos and the eggs. The mechanism of transmission is complex and has not been extensively studied (Hinde, 1971a; Blackman, 1987; Brough and Dixon, 1990).

Phylogenetic analyses Based on 16S rDNA phylogeny, *Buchnera* is a lineage within the *Gammaproteobacteria* (Unterman et al., 1989; Munson et al., 1991a; Moran et al., 1993; van Ham et al., 1997). The closest known organisms (Fig. BXII.γ.196) are the nonculturable *Wigglesworthia glossinidia* (endosymbionts of tsetse flies) and endosymbionts of carpenter ants as well as members of the *Enterobacteriaceae* (Aksoy, 1995b; Aksoy et al., 1995; Schröder et al., 1996). These organisms constitute four separate lineages and their relationship to each other is not clearly resolved based on 16S rDNA.

Fig. BXII.γ.197 presents the results of a phylogenetic analysis

primarily restricted to *Buchnera*. Most of the characterized endosymbionts are from aphids belonging to the family *Aphididae*. Based on 16S rDNA gene sequences (Fig. BXII.γ.197a), *Buchnera* forms one cluster within which a subcluster of endosymbionts from aphids of *Aphididae* is clearly differentiated. An additional cluster consists of endosymbionts from the aphids *Melaphis rhois* (Mr) and *Schlechtendalia chinensis* (Sc), both of which are members of the tribe Melaphidini in the family *Pemphigidae*. Relationships between the remaining endosymbionts from this family as well as from two other aphid families are not clearly resolved using 16S rDNA gene sequences. Using a portion of the chromosomal *trpB*, relationships among endosymbionts of some members of the *Aphididae* and especially within the aphid genus *Uro-*

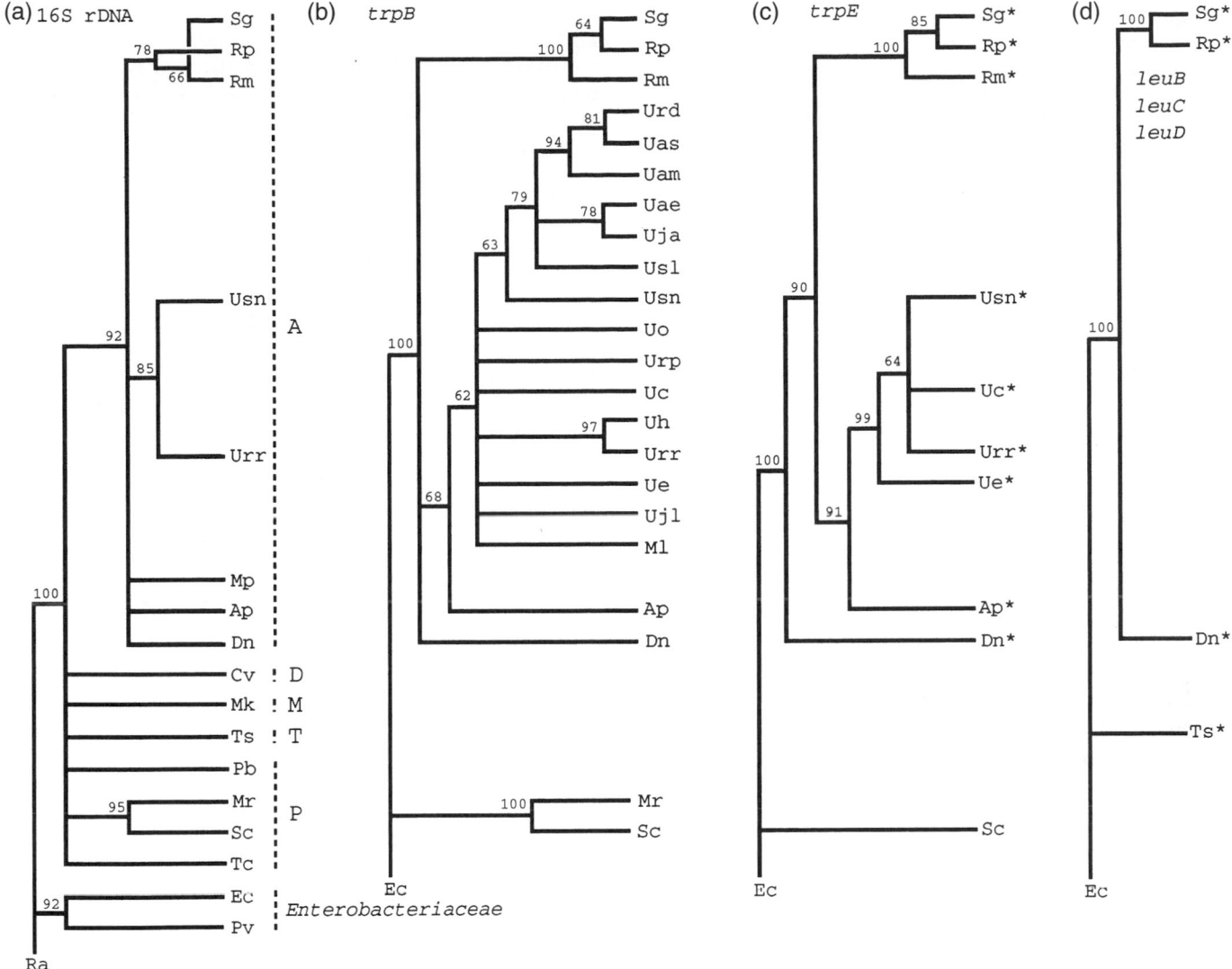

FIGURE BXII.γ.197. Phylogenetic trees resulting from parsimony analyses (D.L. Swofford, 1993, PAUP 3.1.1, Champaign, IL) using *Buchnera* (*a*) 16S rDNA (1475 sites), (*b*) *trpB* (679 sites), (*c*) *trpE* (1623 sites), and (*d*) *leuB, leuC*, and *leuD* (total 3140 sites). *, plasmid-associated genes; numbers above nodes are bootstrap percentages from parsimony searches (1000 replicates). Abbreviations designate insect hosts in the order from top to bottom (given in parentheses): *Sg* (*Schizaphis graminum*), *Rp* (*Rhopalosiphum padi*), *Rm* (*R. maidis*), *Urd* (*Uroleucon rudbeckiae*), *Uas* (*U. astronomus*), *Uam* (*U. ambrosiae*), *Uae* (*U. aeneum*), *Uja* (*U. jaceiae*), *Usl* (*U. solidaginis*), *Usn* (*U. sonchi*), *Uo* (*U. obscurum*), *Urp* (*U. rapunculoidis*), *Uc* (*U. caligatum*), *Uh* (*U. helianthicola*), *Urr* (*U. rurale*), *Ue* (*U. erigeronense*), *Ujl* (*U. jaceicola*), *Ml* (*Macrosiphoniella ludovicianae*), *Mp* (*Myzus persicae*), *Ap* (*Acyrthosiphon pisum*), *Dn* (*Diuraphis noxia*), *Cv* (*Chaitophorus viminalis*), *Mk* (*Mindarus kinseyi*), *Ts* (*Thelaxes suberi*), *Pb* (*Pemphigus betae*), *Mr* (*Melaphis rhois*), *Sc* (*Schlechtendalia chinensis*), *Tc* (*Tetraneura caerulescens*). Species of the *Enterobacteriaceae*: *Ec* (*E. coli*), *Pv* (*P. vulgaris*). Dashed lines in 4a designate aphid species within one family: *A, Aphididae; D, Drepanosiphidae; M, Mindaridae; T, Thelaxidae; P, Pemphigidae*. Data for *Ts* and *Tc* in 4a and 4d from van Ham et al. (1997); for *Rp* in 4d from Bracho et al. (1995); for other sources of data for *Buchnera* see text.

leucon could be resolved (Fig. BXII.γ.197b). These relationships are in broad agreement with more limited studies involving *Buchnera* plasmid-associated *trpE* and plasmid-associated *leuBCD* (Fig. BXII.γ.197c and d) (Bracho et al., 1995; Rouhbakhsh et al., 1996, 1997; Baumann et al., 1997, 1999; van Ham et al., 1997). In addition, the phylogenetic relationships based on *trpB* agree with the phylogeny reconstructed for the same species of *Uroleucon* using nuclear and mitochondrial sequences from the aphids (Moran et al., 1999). This pattern of congruence of phylogenies derived from chromosomal genes, plasmid genes, and host genes is strong support for a completely vertical mode of evolution, with no transfer of either bacteria or plasmids among lineages of hosts (Moran and Baumann, 1994). An implication of the

congruence is that bacterial and host ancestors representing corresponding nodes on the phylogenies occurred contemporaneously, allowing dates inferred from fossil aphids to be extended to ancestral *Buchnera* (Moran et al., 1993). Phylogenetic analysis and the rate of 16S rDNA change inferred from fossil aphids has suggested that the *Buchnera*–aphid association is the result of a single infection of an aphid ancestor that occurred 150–250 million years ago. In *Buchnera* the rate of 16S rDNA evolutionary change is about twice that of free-living bacteria and about 36 times greater than in homologous regions of the host 18S rDNA (Moran, 1996; Moran et al., 1995). The increased rate of change in endosymbionts compared to free-living bacteria extends to genes other than 16S rDNA (Moran, 1996).

Nutrition and metabolism   Insects, like other animals, are thought to require preformed ten essential amino acids (Dadd, 1985; Douglas, 1998). Plant sap (phloem), the diet of aphids, is rich in carbohydrates but is deficient in nitrogenous compounds (Dadd, 1985; Sandström and Pettersson, 1994). It has been thought that one of the functions of *Buchnera* is the synthesis of the essential amino acids and their provision to the aphid host (Dadd, 1985; Douglas, 1998). Some species of aphids are able to grow on synthetic diets even in the absence of some of the essential amino acids. The incorporation of antibiotics into such diets results in the elimination of the endosymbionts and the failure of the aphid to grow and reproduce. There is some sparing effect when essential amino acids are included in the antibiotic-containing diet. These results have been interpreted as being consistent with the synthesis of essential amino acids by the endosymbionts (Douglas, 1998). The most detailed experiments involve the essential amino acid tryptophan. Douglas and Prosser (1992) have shown there is a sparing effect of tryptophan on aphid growth in chlortetracycline-containing synthetic diet. In addition, these investigators detected the enzyme tryptophan synthase in *Buchnera*. Activity was absent in aphids treated with chlortetracycline that lacked the endosymbionts.

Using synthetic diets containing radiolabeled amino acids, it has been shown that the synthesis of the essential amino acids arginine, threonine, isoleucine, and lysine was eliminated by the inclusion of rifampicin in the diet (Liadouze et al., 1996). Similarly, experiments using radiolabeled sulfate showed that endosymbionts reduce sulfur and incorporate it into methionine and cysteine, which are made available to aphid tissue (Douglas, 1988). Sasaki et al. (1991) showed that treatment of aphids with rifampicin eliminated the incorporation of dietary [^{15}N]-glutamine into the essential amino acids arginine, histidine, isoleucine and/or leucine, phenylalanine, threonine, and valine.

Glutamine is one of the predominant amino acids in phloem (Sasaki et al., 1990; Sandström and Pettersson, 1994). Sasaki and Ishikawa (1995) have found that it is the predominant amino acid in aphid hemolymph. These investigators have shown that bacteriocytes take up glutamine and convert it to glutamate. Subsequently glutamate is taken up by *Buchnera*. The isolated endosymbionts incorporate the nitrogen from this compound into the essential amino acids isoleucine, leucine, valine, and phenylalanine as well as into a number of other nonessential amino acids.

Whitehead and Douglas (1993) isolated vesicles containing *Buchnera* (symbiosomes). The symbiosomes readily took up acetic, glutamic, and aspartic acid as well as tricarboxylic acid cycle intermediates and oxidized them to CO_2. Oxygen consumption was also detected and was greatly reduced by KCN. These results indicate a respiratory metabolism.

Genetics   *Buchnera* contains a single copy of the genes encoding for rRNAs that are organized into two transcription units consisting of 1) 16S rRNA (GenBank accession number M63246) and 2) tRNA ^{Glu}–23S rRNA–5S rRNA (GenBank accession number U09230) (Baumann et al., 1995). A single copy of these genes is consistent with the slow growth rate of the endosymbiont (Baumann et al., 1995). Upstream of the rRNA genes are sequences that are readily recognized as −35 and −10 regions of a putative promoter.

The genome size of *Buchnera* from the aphid *Acyrthosiphon pisum* has been found to be 657 kb (Charles and Ishikawa, 1999). Approximately 113 kb of the genome of *Buchnera* from the aphid *S. graminum*, 12 kb from the endosymbiont of *Diuraphis noxia*, 11

kb from the endosymbiont of *S. chinensis*, and 6 kb from the endosymbiont of *A. pisum* have been sequenced (Ohtaka et al., 1992; Baumann et al., 1995; Sato and Ishikawa, 1997b; Baumann and Baumann, 1998; Clark et al., 1998a, 1998b; Thao and Baumann, 1998). The gene content obtained from the nucleotide sequence is an indication of the metabolic capacity of an organism. In *Buchnera* from *S. graminum* 109 open reading frames were detected. Of these, 89 corresponded to known genes and the remaining 20 had similarities to open reading frames on the *E. coli* genome, which have no known function (Blattner et al., 1997). *Buchnera* contains *dnaA*, encoding a protein that initiates bidirectional chromosome replication, and *ftsZ*, encoding a protein involved in septum formation during cell division (Lai and Baumann, 1992; Baumann and Baumann, 1998). Among other genes that were found are those encoding proteins for peptidoglycan synthesis, cell division, DNA replication, DNA transcription, ribosomal proteins, amino acid tRNA synthases, ATP synthase, electron transport, protein secretion, and glycolysis. In addition, genes for three tRNAs were detected. Genes encoding homologs of proteins involved in the *E. coli* heat shock response (*groEL, groES, htrA, dnaK, dnaJ*) and the cold shock response (*hscA, hscB*) were also detected (Ohtaka et al., 1992; Hassan et al., 1996; Sato and Ishikawa, 1997b; Clark et al., 1998b). In addition, some of the genes encoding enzymes for the biosynthesis of aromatic amino acids (shikimate pathway, tryptophan branch), branched chain amino acids (isoleucine, valine, leucine), lysine, cysteine, and serine, as well as genes for the complete pathway of histidine biosynthesis, were also found. The presence of genes for enzymes of amino acid biosynthesis is in marked contrast to such fastidious organisms as *Mycoplasma genitalium* and *Borrelia burgdorferi* in which the genes encoding enzymes of amino acid biosynthesis were absent (Fraser et al., 1995, 1997). Retention of amino acid biosynthetic genes may reflect the role of these pathways in the mutualistic association with the host aphids.

The overproduction of essential amino acids for the aphid host may involve modifications of some mechanisms that regulate the levels of biosynthetic enzymes (Baumann et al., 1995). One possible modification is an increase in the level of a key regulated enzyme by gene amplification. This appears to be the case in the tryptophan biosynthetic pathway of rapidly growing aphids in which the genes for the enzyme anthranilate synthase (*trpEG*) are present as tandem repeats on plasmids. In the case of *Buchnera* from *S. graminum*, *trpEG* is present as four tandem repeats that constitute a plasmid of 14.3 kb (Lai et al., 1994). The remaining genes of the tryptophan biosynthetic pathway (*trpDC[F]BA*) are chromosomal (Munson and Baumann, 1993). There are three to four copies of the *trpEG*-containing plasmid for each of the chromosomal genes. Amplification of *trpEG* is widespread in *Buchnera* of the family *Aphididae*, which is composed of rapidly growing aphids (Fig. BXII.γ.197c) (Lai et al., 1994; Rouhbakhsh et al., 1996, 1997). In each case, the region upstream of *trpE* contains two or more copies of a nine-nucleotide sequence known as a DnaA box. Since the DnaA protein binds to this sequence and initiates DNA replication, this region may be a putative origin of plasmid replication. In the slow-growing aphid *Schlechtendalia chinensis*, *trpEG* is present as one copy on the *Buchnera* chromosome (Lai et al., 1995). Curiously, the *trpEG*-containing plasmids of *Buchnera* from *D. noxia* and from *U. sonchi* contain an intact copy of *trpEG* as well as numerous copies of *trpEG* pseudogenes (Lai et al., 1996; Baumann et al., 1997). There is currently no satisfactory explanation for this observation.

An independent case in which genes for amino acid biosynthesis are plasmid-borne involves the leucine biosynthetic pathway. In *Buchnera* from several aphids, all of the enzymes of leucine biosynthesis (*leuABCD*) are on plasmids (pLeu) of about 8 kb (Bracho et al., 1995; van Ham et al., 1997; Baumann et al., 1999). In *Buchnera* from the aphid *S. graminum* there are about 23 copies of pLeu per endosymbiont chromosomal gene (Thao et al., 1998). Aphids from which pLeu has been sequenced are given in Fig. BXII.γ.197d. These plasmids have no similarity to the *trpEG*-containing plasmids and lack DnaA boxes. Instead they contain open reading frames (*repA1*, *repA2*) that correspond to putative proteins related to RepA of the IncFII incompatibility group plasmids, which is involved in plasmid DNA replication (Bracho et al., 1995). Also encoded on pLeu is a conserved open reading frame of unknown function. A most remarkable finding is the presence of a 1.7-kb plasmid in *Buchnera* from the aphid *Tetraneura caerulescens*, which contains only *repA1* and an unidentified open reading frame (van Ham et al., 1997). It is possible that this plasmid represents a minimal-size replicon that under the right selection can be used for the amplification of other chromosomal genes.

GroEL overproduction In *Buchnera*, the chaperonin GroEL constitutes a major fraction of the total protein (Sato and Ishikawa, 1997b). In addition, GroEL is detected in the aphid hemolymph (van den Heuvel et al., 1994). Overproduction of GroEL is a characteristic of endosymbionts and pathogens in the intracellular environment (Baumann et al., 1995; Hogenhout et al., 1998). GroEL has been localized in maternal and embryonic endosymbionts by immunohistochemistry (Fukatsu and Ishikawa, 1992b). Electron micrographs indicate that the purified *Buchnera* GroEL has the characteristic double-ring appearance observed with the *E. coli* protein (Hara and Ishikawa, 1990; Filichkin et al., 1997). In addition, it has ATPase activity and in the presence of *E. coli* GroES and ATP could reconstitute denatured *Rhodospirillum rubrum* ribulose-1,5-biphosphate carboxylase (Kakeda and Ishikawa, 1991). *Buchnera* GroEL is able to complement *E. coli* mutants (Ohtaka et al., 1992).

The *Buchnera* groESL operon organization resembles that of *E. coli* (Ohtaka et al., 1992; Hassan et al., 1996; Hogenhout et al., 1998). Upstream of *groES* are nucleotide sequences characteristic of the -35 and -10 regions of σ^{32} promoters. A message of 2.1 kb (containing both *groES* and *groEL*) is made by the endosymbiont using only this promoter (Sato and Ishikawa, 1997a). It is not understood why GroES is barely detectable in the endosymbiont, in contrast to the high quantities of GroEL (Kakeda and Ishikawa, 1991; Fukatsu and Ishikawa, 1993). The genes for σ^{32} (*rpoH*) as well as *dnaKJ* have been cloned and sequenced (Sato and Ishikawa, 1997a, b). The latter are also transcribed solely from a σ^{32} promoter. In *E. coli* as well as other organisms, transcription of the *groESL* operon and the *dnaKJ* operon are in part of the σ^{32} regulon and their synthesis is increased by heat shock (Gross, 1996). It would appear that this mode of regulation is modified in *Buchnera* (Sato and Ishikawa, 1997a, b). Synthesis of *groESL* and *dnaKJ* messenger RNA is constitutive and is not increased by heat shock. This conclusion is supported by the observation that there is no increase in the level of total GroEL in aphids shifted from a temperature of 23°C to 33°C for a period of 1 d (Baumann et al., 1996).

Pathogenicity Aphids are major pests of agricultural plants (Blackman and Eastop, 1984). Although large populations of aphids can cause plant debilitation due to nutrient consumption, perhaps the major economic effect of aphids on agriculture is

due to their transmission of plant viruses. Recently, *Buchnera*-derived GroEL has been implicated in the survival of luteoviruses in the hemolymph (van den Heuvel et al., 1994; Filichkin et al., 1997; Hogenhout et al., 1998). These viruses replicate in the plant and are ingested by aphids when they feed on phloem sap. Subsequently they are transported from the digestive tract into the hemolymph and from there into the salivary gland for transmission to plants via salivary secretions. The viruses are retained in an infective form (without replication) in the hemolymph throughout the life span of the aphid. There is evidence that the GroEL that is found in the hemolymph coats the virus particles and protects them from host defenses. A region in *Buchnera* GroEL has been identified that is essential for binding to the virus (Hogenhout et al., 1998), and similarly a portion of a viral capsid protein has been identified as the region to which the endosymbiont GroEL binds (van den Heuvel et al., 1997).

Antibiotic and drug sensitivity A variety of antibiotics have been used to eliminate endosymbionts, resulting in a reduction of weight gain of the aphid and the loss of reproductive potential. These antibiotics include chloramphenicol, chlortetracyline, neomycin, penicillin, rifampicin, and streptomycin.

Additional symbionts Many species of aphids also contain rod-shaped procaryotic endosymbionts designated as the secondary (S)-endosymbionts (Buchner, 1965; Houk and Griffiths, 1980; Fukatsu and Ishikawa, 1993). These organisms are usually absent from bacteriocytes and are found in vesicles within the sheath cells that surround the bacteriome. The S-endosymbiont from the aphid *A. pisum* is a member of the *Enterobacteriaceae* (Unterman et al., 1989). In this organism, the 16S rDNA is upstream of 23S rDNA (Unterman et al., 1989; Unterman and Baumann, 1990). The S-endosymbiont of *A. pisum* is dispensable (Chen and Purcell, 1997) since strains lacking this endosymbiont have been found in nature. The S-endosymbiont can be introduced into such strains by injection of hemolymph from strains of *A. pisum* that contain the S-endosymbiont. Some aphids may also contain a rod-shaped rickettsia closely related to *Rickettsia bellii* (Chen et al., 1996).

Although most species of aphids appear to contain bacteriocytes with *Buchnera*, some species of the tribe Cerataphidini lack both (Fukatsu and Ishikawa, 1992a; Fukatsu et al., 1994). These aphids instead contain, in their body cavity, a yeast-like symbiont that based on its 18S rDNA was found to belong to the subphylum Ascomycotine of the class Pyrenomycetes (Fukatsu and Ishikawa, 1996). These symbionts were related to those of planthoppers.

ENRICHMENT AND ISOLATION PROCEDURES

Endosymbiont-enriched preparations have been obtained using methods previously described (Ishikawa, 1982; Harrison et al., 1989; Sasaki and Ishikawa, 1995). The endosymbionts require high osmolarity for the retention of their structure. The best criteria of purity have involved examination of the purified preparations by electron microscopy, which also allows determination of whether the endosymbionts are still within host-derived vesicles.

Both the aphids and the endosymbionts have a similar mol% G + C DNA content (Ishikawa, 1987; Unterman and Baumann, 1990), and consequently endosymbiont DNA cannot be separated from host DNA by CsCl density gradient centrifugation. For the routine purification of *Buchnera* from *S. graminum*, to obtain starting material for the preparation of endosymbiont-enriched DNA, the method of Sasaki and Ishikawa (1995) has been used. All of the reagents and equipment are kept on ice and the procedures are performed as rapidly as possible. Approximately 2–3 g (wet weight) aphids are transferred to a 1.5 cm diameter tissue grinder. Ten milliliters of buffer A[1] is added and the aphids are ground with a loose-fitting plunger for 5 min and the preparation passed through a double layer of a nylon mesh to remove large particulate material. The filtrate is brought to a volume of about 100 ml with buffer A and then quickly passed through a 100-μm nylon filter (Spectrum Medical Industries, Inc., Houston, Texas, USA), followed by filtration through 20-μm and 10-μm nylon filters. Only slight vacuum pressure is applied during the last two filtration steps. The volume is made up to approximately 120 ml with buffer A and 30-ml aliquots are centrifuged in a swinging bucket rotor for 6 min at 1500 × g. The pellets are gently resuspended in 1.25 ml of buffer A and used for DNA purification (Unterman et al., 1989; Munson et al., 1991a). The DNA obtained from such preparations is endosymbiont-enriched 13- to 15-fold relative to the DNA obtained from the whole aphid (Baumann et al., 1997). Further purification of the endosymbiont preparation can be obtained by centrifugation through 10–90% Percoll gradients (Pharmacia Biotech, Uppsala, Sweden) as described by Ishikawa (1982).

MAINTENANCE PROCEDURES

Established lines of *S. graminum*, the aphid that contains the type strain of *B. aphidicola*, can be readily grown on wheat or barley at 23–24°C with 1500 foot candles illumination and a photoperiod of 16 h light and 8 h dark. Seeds are planted in rich soil, and, after 5–6 d, the seedlings are inoculated with aphids. The inoculum is obtained by cutting aphid-infested plants close to the roots and shaking them over the seedlings to dislodge the aphids. New seedlings are inoculated every 7–10 d.

PROCEDURES FOR TESTING SPECIAL CHARACTERS

Buchnera is somewhat unusual in that its rRNA genes are organized into two transcription units (Baumann et al., 1995). Consequently, 16S rDNA is not upstream of 23S rDNA. In eight strains of *Buchnera*, encompassing the diversity presented in Fig. BXII.γ.197a, *aroE* has been found to be upstream of 23S rDNA (Rouhbakhsh and Baumann, 1995). This property has been used for the identification of *Buchnera* using oligonucleotide primers complementary to *aroE* and 16S rDNA and PCR (Rouhbakhsh et al., 1994). This and other PCR-based identification methods are discussed by Rouhbakhsh et al. (1994). It should be noted that the S-endosymbiont found in some aphids has 16S rDNA upstream of the 23S rDNA (Unterman and Baumann, 1990).

DIFFERENTIATION OF THE GENUS *BUCHNERA* FROM OTHER GENERA

Since different evolutionary lineages of nonculturable endosymbionts are found within aphids, carpenter ants, and tsetse flies,

their initial putative identification is at the level of the insect host (Fig. BXII.γ.196). In *Wigglesworthia* (endosymbionts of tsetse flies), 16S rDNA is upstream of 23S rRNA and this is a useful differential character (Aksoy, 1995a, b). In the carpenter ant endosymbionts the 16S rDNA is not upstream of 23S rDNA, and it appears that the rRNA genes are organized into at least two transcription units (C. Elsishans and R. Gross, personal communication). It is not known if in carpenter ant endosymbionts *aroE* is upstream of the 23S rDNA, as is the case in *Buchnera*. The members of the *Enterobacteriaceae* are mostly free-living, readily identifiable bacteria with their rRNA genes arranged in the order 16S–23S–5S.

TAXONOMIC COMMENTS

There are over 4000 species of aphids (Blackman and Eastop, 1984), of which only 28 have been characterized by molecular methods. Consequently, our conclusions concerning *Buchnera* are based on a very small sample of aphid species. 16S rDNA has been useful for showing the monophyletic origin of the aphid symbiosis. It appears also to be useful for the differentiation of major aphid subgroups, as for example the *Aphididae* and possibly other families (Fig. BXII.γ.197a). 16S rDNA, however, is far too conserved for the elucidation of relationships between endosymbionts of closely related aphids. We have had some success with the use of a portion of *trpB* for this purpose (Fig. BXII.γ.197b). However, other informational molecules should also be tried.

The name *Buchnera aphidicola* as currently used designates the whole aphid endosymbiont lineage (Fig. BXII.γ.197a). The 16S rDNA sequence difference between *Buchnera* from *S. graminum* and *Buchnera* from *S. chinensis* (Sg and Sc in Fig. BXII.γ.197a) is approximately the same as that between *E. coli* and *Proteus vulgaris*. This clearly indicates that in subsequent studies *Buchnera* should be subdivided into new species. For this purpose, it is necessary to study a molecule that is less conserved than the 16S rDNA, and it is also necessary to know the range of variation within the endosymbionts of a single aphid species. In addition, since there is co-speciation between the host and the endosymbiont, the phylogeny of the host should also be elucidated to serve as an aid in the recognition of new endosymbiont species.

FURTHER READING

Baumann, P., L. Baumann, C.-Y. Lai, D. Rouhbakhsh, N.A. Moran and M.A. Clark. 1995. Genetics, physiology, and evolutionary relationships of the genus *Buchnera*: intracellular symbionts of aphids. Annu. Rev. Microbiol. *49*: 55–94.

Douglas, A.E. 1998. Nutritional interactions in insect-microbial symbioses: aphids and their symbiotic bacteria *Buchnera*. Annu. Rev. Entomol. *43*: 17–37.

Ishikawa, H. 1989. Biochemical and molecular aspects of endosymbiosis in insects. Int. Rev. Cytol. *116*: 1–45.

Moran, N.A. and A. Telang. 1998. Bacteriocyte-associated symbionts of insects - A variety of insect groups harbor ancient prokaryotic endosymbionts. Bioscience *48*: 295–304.

List of species of the genus Buchnera

1. **Buchnera aphidicola** Munson, Baumann and Kinsey 1991b, 567[VP]

a.ph′ di.co.la. M.L. fem. n. *aphidicola* aphid dweller.

The characteristics are those described for the genus. Morphological features are depicted in Figs. BXII.γ.194 and BXII.γ.195.

The mol% G + C of the DNA is: 27 (determined from the nucleotide sequence of over 100 kb of DNA).

Type strain: Endosymbiont of the aphid *Schizaphis graminum*.

GenBank accession number (16S rRNA): M63246.

---

1. Buffer A contains 0.25 M sucrose, 35 mM Tris-HCl (pH 7.6), 25 mM KCl, 10 mM MgCl$_2$, and 1 mM dithiothreitol (Ishikawa, 1982).

Genus VI. Budvicia *Aldová, Hausner, and Gabrhelová in Bouvet, Grimont, Richard, Aldová, Hausner and Gabrhelová 1985, 63^VP*

J.J. FARMER III

Bud.vi' ci.a. L. fem. n. *Budvicia* derived from Budvicium, the Latin name of the city České Budějovice where the bacterium was first isolated.

Small, straight, rod-shaped cells, nonsporeforming, nonencapsulated, and Gram negative, conforming to the general definition of the family *Enterobacteriaceae*. Contains the enterobacterial common antigen. **Motile, producing peritrichous flagella when grown at 22°C; less motile at 36°C.** Grows on laboratory media including MacConkey agar. Grows slowly on nutrient agar, forming colonies only about 0.1 mm at 24 h, 36°C; colonies are five times larger when grown at 30°C for 24 h. More active biochemically at 25°C than 36°C. Oxidase negative, reduce nitrate to nitrite. **Positive for: H_2S production, urea hydrolysis, and fermentation of glucose, L-arabinose, L-rhamnose, D-xylose, and D-galactose.**

Negative for indole production, Voges–Proskauer, citrate utilization (Simmons), phenylalanine deaminase, lysine decarboxylase, arginine dihydrolase, ornithine decarboxylase, growth in the presence of cyanide (KCN test), malonate utilization, esculin hydrolysis, gelatin hydrolysis (22°C), lipase (corn oil), DNase, and the fermentation of sucrose, dulcitol, salicin, adonitol, *myo*-inositol, D-sorbitol, raffinose, maltose, trehalose, cellobiose, α-methyl-D-glucoside, erythritol, melibiose, glycerol, and D-mannose.

Utilize 29 of 83 carbon sources tested. **Grow at 4°, 10°, 22°, 32°, and 37°C, but not at 42°C.** Complex growth factor requirement: **absolute requirement for nicotinic acid**; other vitamins and amino acids stimulate growth. Susceptible to colistin, nalidixic acid, sulfadiazine, gentamicin, streptomycin, kanamycin, chloramphenicol, and carbenicillin (disk diffusion method on Mueller-Hinton agar); **resistant to penicillin, ampicillin, and cephalothin. Isolated from drinking and environmental water** in Europe, and from fecal specimens in United States. **No evidence that it is a pathogen for plants or animals.**

The mol% G + C of the DNA is: 46 ± 1 (Bouvet et al., 1985); 51–54 (Aldová et al., 1983).

Type species: **Budvicia aquatica** Aldová, Hausner, and Gabrhelová *in* Bouvet, Grimont, Richard, Aldová, Hausner and Gabrhelová 1985, 63.

FURTHER DESCRIPTIVE INFORMATION

Guide to the literature Since the genus name was first published in 1983, there have been a few reports in the literature. "HG group" should be included in computer literature searches to retrieve early references. A search from 1983 to May 1999 yielded only seven citations in MEDLINE and 12 citations in BIOSIS. Most are about the organism (Aldová et al., 1983, 1984; Bouvet et al., 1985), comparative taxonomic studies (Bouvet and Grimont, 1987a), or water bacteriology (Hausner et al., 1986; Schubert and Groeger-Söhn, 1998).

Sources Most isolates of *Budvicia* have been from various types of water: wells, rivers and streams, swimming pools, conduits, environmental, and sewage. Hausner et al. (1986) isolated 170 strains of *B. aquatica* from 21,300 water samples, mainly from wells and water mains. One isolate was from the digestive tract of a shrew in Spain (Bouvet et al., 1985).

B. aquatica has also been isolated from human feces in United States (Enteric Reference Laboratory, CDC, unpublished). However, there is no evidence that *B. aquatica* causes diarrhea or infections of the intestinal tract. Apparently, it has not been isolated from extraintestinal human clinical specimens.

Bacteriocin-like agents Strains of *Budvicia* produce bacteriocin-like agents that Šmarda (1987) termed aquaticins.

Isolation, identification, culture preservation, phenotypic characterization Strains of *Budvicia* are not difficult to grow, and are typical *Enterobacteriaceae* in most respects. Strains grow slowly on nutrient agar, forming colonies only about 0.1 mm at 24 h at 36°C, but about five times larger when grown at 30°C for 24 h (Bouvet et al., 1985). Similarly, the biochemical reactions may be different at the two temperatures; and strains may be more active at 25°C. Schubert and Groeger-Söhn (1998) described a method for its detection and quantification in water and sewage samples. See the chapter on the family *Enterobacteriaceae* in this *Manual* for general information on growth, plating media, working and frozen stock cultures, media and methods for biochemical testing, identification, computer programs, and antibiotic susceptibility.

Biochemical reactions and differentiation from other *Enterobacteriaceae* Table BXII.γ.193 in the chapter on the family *Enterobacteriaceae* gives the results at 36°C for *Budvicia* strains in 47 biochemical tests normally used for identification (Farmer, 2003). There are no genus- or species-specific tests or sequences for the identification of *Budvicia*. The best approach is to do a complete set of phenotypic tests (growth on plating media, biochemical tests, antibiogram, etc.) and compare the results with all the other named organisms in the family *Enterobacteriaceae*. This approach is described in more detail in the chapter on the family *Enterobacteriaceae*. The biochemical results of a test strain can be compared to all the organisms listed in Table BXII.γ.193 of the chapter on the family *Enterobacteriaceae*. Several computer programs greatly facilitate analyzing the results. Bouvet et al. (1985) point out that strains of *B. aquatica* could be misidentified as H_2S-positive strains of *Yersinia*, *Citrobacter*, or *Salmonella*.

TAXONOMIC COMMENTS

Discovery, history, nomenclature, and type strains In 1976 the first strain of an unusual H_2S-producing strain of *Enterobacteriaceae* was isolated at the Public Health Laboratory in České Budějovice, Czechoslovakia (Aldová et al., 1983). From 1976 to 1983, 27 additional strains were isolated from water samples in three different areas of Bavaria. In 1983 Aldová et al. described this collection of strains, which had been given the vernacular name "HG group". All the data indicated that this was a new group that was distinct from all of the described genera and species of *Enterobacteriaceae*. Throughout the paper Aldová et al. (1983) used the term "HG group" for the new organism; however, at the end of the paper there was a separate section with the heading "Note", which stated that after the manuscript had been submitted DNA–DNA hybridization data became available from the laboratory of P.A.D. Grimont of the Institute Pasteur confirming that the group was a new genus of *Enterobacteriaceae*. In this note they stated: "The authors accordingly propose for this genus the name *Budvicia* (from the ancient Latin name 'Budvicium' for the city of České Budějovice where the microorganism was first isolated) and for its only species the name *Budvicia aquatica*. The epithet, as is obvious, is derived from the Latin adjective 'aquaticus'."

It is clear from this paragraph that a formal nomenclatural proposal was actually made. However, since no type strain was designated for *B. aquatica*, the names *Budvicia* and *Budvicia aquatica* were not "validly published" (see the *Bacteriological Code*, pages 28–29), and thus are illegitimate names. This situation was apparently rectified when the names *Budvicia* Aldová, Hausner, Gabrhelová, Schindler, Petráš and Braná 1985a and *Budvicia aquatica* Aldová, Hausner, Gabrhelová, Schindler, Petráš and Braná 1985a gained standing in nomenclature and were validly published in Validation List 17 in the April 1985 issue of the IJSB (see page 223). Footnote f of Validation List 17 states that the type strain of *Budvicia aquatica* is 20186 (CNCTC 350); "Personal communication to the editor of the IJSB". In 1984, Aldová et al. had published a second paper *that* specifically defined this as the type strain. A sentence on page 234 of Aldová et al. (1984) stated: "As the type strain, we propose strain No. 20186 of the National Dysentery Reference Laboratory, which has been deposited at the Czechoslovak National Collection of Type Cultures, Prague under No. M 350."

However, the 1984 paper of Aldová et al. was not listed in the Literature Cited section of Validation List 17. Validation List 17 would have ended a slightly confusing matter; however, in January, 1985 there was a new set of proposals for *Budvicia* and *Budvicia aquatica*. Bouvet et al. published a paper in the *International Journal of Systematic Bacteriology* proposing *Budvicia* and *Budvicia aquatica* as a new genus and species, and their species appeared to be based on a different type strain. This latter paper made a confusing situation even more confusing. In addition, they compounded the confusion by proposing an unconventional name citation (see Nomenclatural problem 3, below). These four conflicting proposals (Aldová et al., 1983, 1984, 1985a; Bouvet et al., 1985) create four nomenclatural problems that require analysis and discussion.

Nomenclatural problem 1   Priority of *Budvicia* Bouvet, Grimont, Richard, Aldová, Hausner and Gabrhelová 1985 over *Budvicia* Aldová, Hausner, Gabrhelová, Schindler, Petráš and Braná 1985a. The first name clearly has priority because it was validly published first (as an article in the *International Journal of Systematic Bacteriology*, January 1985, pages 60–64). The second name was published in the next issue (April 1985) in Validation List 17 (page 223).

Nomenclatural problem 2   Priority of *Budvicia aquatica* Bouvet, Grimont, Richard, Aldová, Hausner and Gabrhelová 1985 whose type strain is 20186HG01 (ATCC 35567) over *Budvicia aquatica* Aldová, Hausner, Gabrhelová, Schindler, Petráš and Braná 1985a whose type strain is 20186 (CNCTC 350). The first name has priority for the same reason given above.

Nomenclatural problem 3   Are *Budvicia aquatica* Bouvet, Grimont, Richard, Aldová, Hausner and Gabrhelová 1985 and *Budvicia aquatica* Aldová, Hausner, Gabrhelová, Schindler, Petráš and Braná 1985a objective or subjective synonyms? The two species would be objective synonyms if they have the same type strain, but would be subjective synonyms if they have different type strains. The type strains were published with different, but similar, strain numbers (20186HG01 vs 20186). Whether these strain numbers are the same or different is not clear from any of the published papers.

Nomenclatural problem 4   Author citations for *Budvicia* and *Budvicia aquatica*. In their original paper in the *International Journal of Systematic Bacteriology* Bouvet et al. (1985) stated: "To recognize the priority of three authors in the delineation of the new genus and species, we propose the following citation of names: *Budvicia* Aldová, Hausner, and Gabrhelová *in* Bouvet, Grimont, Richard, Aldová, Hausner and Gabrhelová 1985, and *Budvicia aquatica* Aldová, Hausner, and Gabrhclová *in* Bouvet, Grimont, Richard, Aldová, Hausner and Gabrhelová 1985."

It seems confusing and a danger to "stability in nomenclature" to have a name citation that disagrees with the authorship of the original paper. The conventional style for author citations as specified in the Bacteriological Code has been followed here, and the reader is referred to the irregular citation proposed by Bouvet et al. (1985). However, in the chapter on *Tatumella*, an

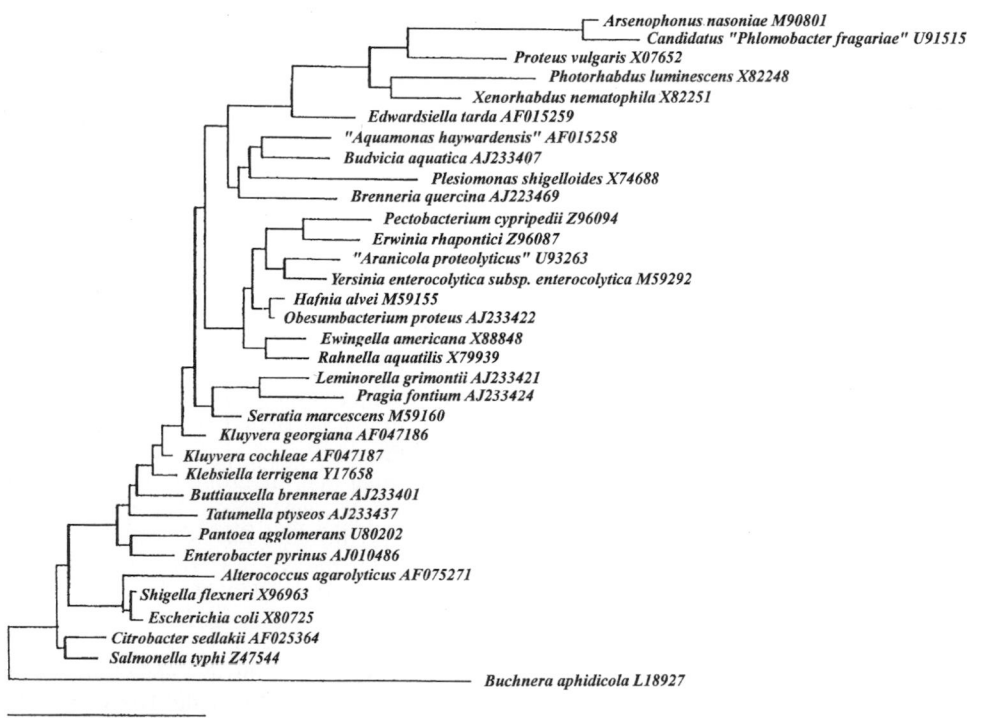

FIGURE BXII.γ.198.   Relationship of *Budvicia* and several other new *Enterobacteriaceae* to other organisms in the family based on 16S rDNA sequencing data. Also, see Figure BXII.γ.189 in the family *Enterobacteriaceae*.

irregular name citation was also used. For consistency, the irregular author citation given by Bouvet et al. (1985) was used in this chapter on *Budvicia*.

Interestingly, this is a point that can cause instability in nomenclature. Because an "authors' proposal for an unorthodox name citation" is not specifically covered in the Bacteriological Code, remedial action in the form of a rule modification or Opinion of the Judicial Commission will be needed to guide future writers.

Phylogeny based on DNA–DNA hybridization and 16S rRNA sequencing Based on the DNA–DNA hybridization data *Budvicia* is a distinct and well-defined genus of *Enterobacteriaceae* with its one species *Budvicia aquatica* (Bouvet et al., 1985). None of the other genera of *Enterobacteriaceae* are closely related. Fig.BXII.γ.198 is a tree based on 16S rDNA sequencing that confirms this relationship. The organism on the same branch as *B. aquatica* is a new H₂S producing *Enterobacteriaceae* that will be named *Aquimonas* (originally listed as *Aquamonas* in the GenBank database of sequences). This tree should be compared with the tree published by Spröer et al. (1999), which also shows the distinctness of *Budvicia*. However, this latter tree has *Pragia* and *Leminorella*, two other H₂S-producing genera of *Enterobacteriaceae*, on the same branch.

Schindler et al. (1992) agreed that *Budvicia* strains are distinct from other H₂S-producing *Enterobacteriaceae* based on SDS-PAGE protein patterns.

FURTHER READING

Aldová, E., O. Hausner, M. Gabrhelová, J. Schindler, P. Petráš and H. Braná. 1983. A hydrogen sulfide-producing gram-negative rod from water. Zentbl. Bakteriol. Mikrobiol. Hyg. 1 Abt. Orig. A. *254*: 95–108.

Bouvet, O.M.M., P.A.D. Grimont, C. Richard, E. Aldová, O. Hausner and M. Gabrhelová. 1985. *Budvicia aquatica*, gen. nov., sp. nov.: a hydrogen sulfide-producing member of the *Enterobacteriaceae*. Int. J. Syst. Bacteriol. *35*: 60–64.

List of species of the genus Budvicia

1. **Budvicia aquatica** Aldová, Hausner, and Gabrhelová *in* Bouvet, Grimont, Richard, Aldová, Hausner and Gabrhelová 1985, 63^VP

a.qua′ ti.ca. L. fem. adj. *aquatica* living in water; named to show the aquatic habitat of the organism, since all but one of the original strains were isolated from water.

The characteristics are as previously described for the genus. Biochemical characteristics of the species are given in Table BXII.γ.193 in the chapter on the family *Enterobacteriaceae*. Isolated from well water, rivers and streams, swimming pools, water conduits, environmental water, sewage, the digestive tract of a shrew, and from human fecal specimens in United States (Enteric Reference Laboratory, CDC, unpublished). There is no evidence that *B. aquatica* causes diarrhea, infections of the intestinal tract, or extraintestinal infections.

The mol% G + C of the DNA is: 46 ± 1 (T_m) (Bouvet et al., 1985); 51–54 (T_m and UV spectrophotometry) (Aldová et al., 1983).

Type strain: 20186HG01, ATCC 35567, Eb 13/82, CDC 0440-84, CNCTC 20186, DSM 5075.

GenBank accession number (16S rRNA): AJ233407.

Additional Remarks: The ATCC also lists two other strains ATCC 35566 (Eb 4/81, CNCTC 21930, M 325) and ATCC 51341 (85-01-010).

Genus VII. **Buttiauxella** *Ferragut, Izard, Gavini, Lefebvre and Leclerc 1982, 266^VP (Effective publication: Ferragut, Izard, Gavini, Lefebvre and Leclerc 1981, 40)*

PETER KÄMPFER

But.ti′ aux′ el.la. M.L. dim. ending *-ella*; M.L. fem. n. *Buttiauxella* named after René Buttiaux, a French microbiologist for his numerous contributions to the taxonomy of *Enterobacteriaceae*.

Straight rods, 0.5–0.7 × 2–3 μm, conforming to the general definition of the family *Enterobacteriaceae*. **Gram negative, motile by peritrichous flagella.** Facultatively anaerobic. Chemoorganotrophic. Psychrotolerant, growing at 4°C. D-Glucose is fermented with the production of acid and gas. Nitrate is reduced to nitrite. Oxidase negative, catalase positive. **Most strains are methyl red positive. The majority of strains are indole negative. Acids are produced from various carbohydrates, including L-arabinose, cellobiose, maltose, D-mannose, L-rhamnose, D-xylose, and salicin.** 16S rDNA sequence analysis of the type strain of *Buttiauxella agrestis* clearly places the genus in the family *Enterobacteriaceae* within the *Gammaproteobacteria*. Often isolated from fresh water, but also found in soil, and especially in the intestines of snails and slugs from various regions in the world.

The mol% G + C of the DNA is: 47–51.

Type species: **Buttiauxella agrestis** Ferragut, Izard, Gavini, Lefebvre and Leclerc 1982, 266, emend. Müller, Brenner, Fanning, Grimont and Kämpfer 1996, 60 (Effective publication: Ferragut, Izard, Gavini, Lefebvre and Leclerc 1981, 40).

FURTHER DESCRIPTIVE INFORMATION

Based on 16S rDNA sequence analysis of the type strain, *B. agrestis* was most similar to the type strain of *Klebsiella planticola* (97.9%).

Analysis of 16S rDNA of the type strains of the seven species revealed a very high degree of phylogenetic relatedness (99.1–99.7% similarity) (Spröer et al., 1999). The phenotypic description of the genus *Buttiauxella* is largely based on the studies of Gavini et al. (1976b), Ferragut et al. (1981), Farmer (1984b), Farmer et al. (1985a), Brenner (1992a), and Müller et al. (1996). Based on DNA–DNA hybridization it was obvious that the *Buttiauxella* strains comprised seven closely related species, which were very difficult to differentiate phenotypically. The subdivision of the genus *Buttiauxella* is described below (seeTaxonomic Comments).

Cells are straight rods, 0.5–0.7 × 2–3 μm, and can occur singly or in pairs. They are motile at 36°C by peritrichous flagella. *Buttiauxella* strains grow readily on many common media; no growth requirements have been described. A nutrient-rich medium (e.g., Tryptone–Soy agar, sheep-blood agar, or nutrient agar) gives best results. Optimal growth occurs at temperatures ranging from 25°C to 35°C, often yielding better growth than incubation at 37°C. The majority of strains are psychrotolerant, growing (slowly) at 4°C, but not at 41°C.

All members of the genus *Buttiauxella* produce a positive β-galactosidase reaction (ONPG test), although not all produce acid from lactose. The β-glucuronidase test is negative for all

TABLE BXII.γ.210. Characteristics of species of the genus *Buttiauxella*[a,b,c]

Test	Method (s)[d]	B. agrestis	B. brennerae	B. ferragutiae	B. gaviniae	B. izardii	B. noackiae	B. warmboldiae
Hybridization group		1	4	2	3	5	6	7
Acetate	Ut.	[+]	d	+	[+]	+	+	+
N-Acetyl-D-galactosamine	Ut.	+	+	+	d	+	+	+
N-Acetyl-L-glutamate	Ut.	+	+	+	+	+	+	+
N-Acetyl-L-glutamine	Ut.	−	−	−	[−]	−	[+]	−
N-Acetylglycine	Ut.	−	−	−	−	−	−	−
Adonitol	Acid., Ut.	−	d	−	d	−	−	−
D-Arabinose	Ut.	d	−	−	d	+	d	+
D-Arabitol	Acid., Ut.	−	d	−	d	−	−	+
L-Arginine dihydrolase	Alcal.	−	d	−	+	[−]	+	+
L-Arginine	Ut.	−	−	−	−	−	[−]	−
Citrate utilization (Simmons)	Ut.,	d	d	−	d	+	+	−
Dulcitol	Acid., Ut.	−	−	−	[−]	−	−	−
L-Fucose	Acid., Ut.	[+]	−	−	d	[+]	−	+
Glycerol	Acid.	d	−	−	−	−	−	−
3-Hydroxybenzoate	Ut.	−	−	−	−	−	−	−
3-Hydroxyphenylacetate	Ut.	−	−	[+]	−	−	−	−
Indole production	Kovacs	−	−	−	d	−	d	−
myo-Inositol	Acid.	−	−	−	−	−	−	+
KCN	Growth	+	+	+	[+]	+	+	
5-Ketogluconate	Ut.	−	+	−	+	d	+	
Lactose	Acid.	+	+	−	−	+	−	
Lactulose	Acid., Ut.	−	−	−	−	−	−	
L-Lysine decarboxylase	Alcal.	−	−	+	−	−	−	
Malonate utilization	Ut.	+	+	−	+	+	+	+
Maltitol	Acid., Ut.	[+]	+	+	[+]	d	[+]	−
Melibiose	Acid.	d	+	[+]	−	d	−	
α-Methyl-D-glucopyranoside	Acid., Ut.	−	d	[−]	−	−	−	
3-Methyl-D-glucopyranoside	Ut.	+	d	[+]	+	[+]	[+]	−
Methyl red	Acid	+	+	+	+	+	+	+
Mucate	Acid.	d	+	[−]	d	d	+	−
L-Ornithine decarboxylase	Alcal.	+	d	+	−	+	−	
Palatinose	Acid., Ut.	+	+	+	+	d	+	
Phenylacetate	Ut.	d	−	d	[−]	d	d	
L-Phenylalanine deaminase	FeCl₃ test	−	−	−	−	−	d	
3-Phenylpropionate	Ut.	−	−	−	−	−	−	
L-Proline	Ut.	[+]	+	[+]	+	+	[+]	+
Raffinose	Acid., Ut.	d	+	[−]	−	−	−	
D-Sorbitol	Acid., Ut.	d	−	+	[−]	−	−	
Sucrose	Acid.	−	−	−	[−]	−	−	
D-Tagatose	Acid., Ut.	−	−	−	[−]	−	−	
Tartrate (Jordan's)	Acid,	d	[−]	−	d	[−]	+	−
Voges–Proskauer	Acetoin	−	−	−	−	−	−	−

[a]Data adapted from Müller et al. (1996).

[b]Symbols: +, positive for 90–100% of strains; [+], positive for 75–89% of strains; d, positive for 25–74% of strains; [−], positive for 11–24% of strains; −, positive for 0–10% of strains. All reactions, unless otherwise stated, were done at 36° ± 1°C and read after 48 h.

[c]All of the strains studied were Gram-negative, oxidase-negative, catalase-positive, D-glucose-fermenting, nitrate-reducing, rod-shaped organisms. All strains grew on MacConkey agar and on Endo agar at 30°C. With few exceptions, all of the strains were positive in standard tests for motility, acid and gas production from D-glucose, fermentation of D-mannitol, salicin, L-arabinose, L-rhamnose, maltose, D-xylose, trehalose, cellobiose, D-mannose, galactose, gentiobiose, D-ribose, and arbutin, esculin hydrolysis, and o-nitrophenyl-β-D-galactopyranoside (β-galactosidase) test. They were all negative for production of hydrogen sulfide on triple sugar iron, Christensen urease activity, gelatin liquefaction (at 22°C), lipase activity (Tween 80 and corn oil), DNase activity, and production of a yellow pigment, and did not ferment 2-deoxy-D-glucose, 2-deoxy-D-ribose, erythritol, α-D-fucose, D-lyxose, and xylitol.

Almost all of the strains utilized the following compounds as sole carbon and energy sources: N-acetyl-D-glucosamine, D-alanine, L-alanine, L-arabinose, arbutin, DL-asparagine, D-cellobiose, D-fructose, D-galactose, D-galacturonate, gentiobiose, D-gluconate, D-glucosamine, D-glucuronate, L-glutamine, L-glutamate, DL-glycerate, glycerol, 4-hydroxybenzoate, 2-ketogluconate, D-lactate, DL-lactate, maltose, maltotriose, mannitol, D-mannose, methyl-α-galactoside, methyl-β-galactoside, methyl-β-D-glucoside, mucate, oxaloacetate, palatinose, protocatechuate, pyruvate, quinate, L-rhamnose, D-ribose, D-saccharate, salicin, L-serine, starch, D-trehalose, and D-xylose.

Almost all of the strains were unable to utilize the following compounds as sole carbon and energy sources: acetamide, acetamidocaprate, N-acetyl-DL-methionine, N-acetyl-L-proline, trans-aconitate, adipate, β-alanine, allantoin, altrose, DL-2-aminoadipate, 2-aminobenzoate, 3-aminobenzoate, 4-aminobenzoate, DL-2-aminobutyrate, DL-3-aminobutyrate, 4-aminobutyrate, DL-2-aminoisobutyrate, DL-3-aminoisobutyrate, 5-aminovalerate, aminoxyacetate, amygdalin, anthranilate, L-arabitol, arabonate, D-arginine, D-asparagine, azelate, benzoate, betaine, 1-butanol, 2-butanol, n-butyrate, cadaverine, caprate, caprylate, carni-tine, carnosine, L-citrulline, citraconate, L-cysteinate, dextran, DL-2,4-diaminobutyrate, diaminopimelate, 2,3-diaminopropionate, dimethylglycine, m-erythritol, ethanol, ethanolamine, ethylamine, D-fucose, D-glutamate, glutarate, glycinamide, glycogen, glycolate, glycyrrhizinate, 1-hexanol, 1,6-hexandiol, hexylamine, hippurate, histamine, D-histidine, L-histidine, L-homoserine, DL-3-hydroxybutyrate, 4-hydroxybutyrate, DL-2-hydroxycaprate, DL-2-hydroxyisobutyrate, DL-2-hydroxyisocaprate, 2-hydroxyisovalerate, DL-δ-hydroxylysine, 2-hydroxyphenyla-cetate, 4-hydroxyphenylglycine, DL-hydroxyproline, 2-hydroxyvalerate, HQ-β-glucuronide, heptanoate, indole-3-acetate, isobutyrate, isophthalate, itaconate, 2-ketoglutarate, 2-ketoisocaprate, D-leucine, L-leucine, levulinate, D-lysine, D-malate, maleate, D-mandelate, L-mandelate, D-mannoheptulose, D-melezitose, mesaconate, mesoxalate, D-methionine, L-methionine, DL-methioninesulfone, methyl-α-D-glucoside, α-methyl-D-mannoside, β-methyl-D-xyloside, L-norleucine, D-norvaline, L-norvaline, 1,8-octandiol, D-ornithine, phenoxyacetate, D-phenylalanine, phenylglycine, phenyllactate, phosphoenolpyruvate, phthalate, pimelate, poly-D-galactomannan, D-proline, propionate, protocatechuate, putrescine, salicylamide, salicylate, sarcosine, sorbate, L-sorbose, spermine, suberate, D-tartrate, meso-tartrate, tartronate, taurine, thiamine, tricarballylate, trigonelline, tropate, tryptamine, D-tryptophan, L-tryptophan, tyramine, L-tyrosine, ureidosuccinate, n-valerate, isovalerate, L-valine, xylitol, and L-xylose.

The strains varied in their ability to utilize the following compounds as sole sources of carbon and energy, and these characteristics could not be used to differentiate hybridization groups: D-alanine, L-arginine, L-aspartate, benzoate, fumarate, D-glucarate, D-glucosaminate, glycerate, glycerophosphate, glycine, glyoxylate, 4-hydroxy-benzoate, 4-hydroxyphenylacetate, inulin, DL-isocitrate, DL-isoleucine, 2-ketoglutarate, L-lactate, D-lyxose, L-lyxose, L-malate, L-mannose, phenylpyruvate, D-serine, spermidine, succinate, L-tartrate, L-threonine, and D-turanose.

B. agrestis and *B. izardii* could not be distinguished on the basis of the results of any single test. These groups had to be distinguished on the basis of their overall biochemical profiles. *B. gaviniae* and *B. noackiae* could also not be separated on the basis of the results of a single test, but could be differentiated on the basis of their N-acetyl-L-glutamine, adonitol, D-arabitol, L-fucose, mucate, and tartrate reactions. As stated in the text, these two pairs of hybridization groups exhibited the highest levels of interspecies DNA relatedness.

[d]Ut, utilization test; Acid, acidification test; Alcal, alkalinization test.

strains. Hydrogen sulfide formation, Voges–Proskauer, urease, and DNase tests are negative for all strains.

The reduction of nitrous oxide to dinitrogen has been reported in some strains of *B. agrestis* (Kaldorf et al., 1993). Various carbon sources, including carbohydrates, organic acids, and amino acids are utilized as sole sources of carbon (Gavini et al., 1976b; Ferragut et al., 1981; Müller et al., 1996; Table BXII.γ.210).

Isoprenoid quinone Q-8 is the predominant quinone type and a small amount of the menaquinone MK-8 is present in *B. agrestis* ATCC 33320^T (P. Kämpfer, unpublished results). The fatty acid composition as determined by gas chromatographic analysis of 66 strains representing all species (Kämpfer et al., 1997) differed only slightly. All strains contained the fatty acids $C_{14:0}$, $C_{16:0}$, $C_{17:0\ cyclo}$, summed feature ($C_{16:1\ isoI}$ and/or $C_{14:0\ 3OH}$), summed feature ($C_{16:1\ ω7c}$ and/or $C_{15:0\ iso\ 2OH}$), and summed feature ($C_{18:1\ ω7c}$, $C_{18:1\ ω9t}$, and/or $C_{18:1\ ω12t}$) (Kämpfer et al., 1997), a fatty acid profile typical for the family *Enterobacteriaceae*.

Antimicrobial susceptibility data are summarized by Freney et al. (1988). Members of the genus *Buttiauxella* (13 strains) were usually sensitive to aminoglycosides (gentamicin, tobramicin, and amikacin), doxocycline, and trimethoprim, but resistant to chloramphenicol. Susceptibility to ampicillin, aminoxycillin-clavulanic acid, and tiracillin was observed. Cephalothin and cefoxitin were inactive. *Buttiauxella* produces enterobacterial common antigen (Böttger et al., 1987).

Pathogenicity of *Buttiauxella* for humans and animals has not been documented. Only one *B. agrestis* strain in the study of Müller et al. (1996) was from a human clinical specimen. Farmer et al. (1985a) described seven strains of Enteric Group 59 (now *Buttiauxella noackiae*) originating from sputum and a foot wound. *Buttiauxella* is widely distributed in nature, may be isolated from foods and is occasionally isolated from human sources. Although the natural habitat of *Buttiauxella* was originally thought to be water, the majority of strains have been isolated from the intestines of snails and slugs (Müller et al., 1996).

ENRICHMENT AND ISOLATION PROCEDURES

The cultivation media used for the isolation of *Buttiauxella* are those regarded useful for other members of the family *Enterobacteriaceae*. No specific selective medium has been reported for *Buttiauxella* species. Because of the relatively low clinical significance and the presence of *Buttiauxella* in various habitats, the selective isolation of *Buttiauxella* is rarely required. In most cases, organisms belonging to the genus *Buttiauxella* are isolated with differential media not inhibitory for *Enterobacteriaceae*, such as MacConkey agar, bromothymol blue lactose agar, or phenol red lactose agar. *Buttiauxella* strains grow readily on ordinary nutrient-rich media, and no growth requirements have been described.

MAINTENANCE PROCEDURES

Buttiauxella strains can be maintained in tryptone soy agar stabs or on nutrient agar when kept at room temperature in the dark. They can be preserved by storage in broth containing 10% glycerol or in calf or bovine serum at −80°C. Lyophilization seems to be the best procedure for preservation.

DIFFERENTIATION OF THE GENUS *BUTTIAUXELLA* FROM OTHER GENERA

No single distinguishing feature is useful for the differentiation of *Buttiauxella* from the other genera of *Enterobacteriaceae* (Table BXII.γ.193 in *Enterobacteriaceae*). Biochemically and based on DNA–DNA hybridization data, *Buttiauxella* are most similar to

the genus *Kluyvera*. (Gavini et al., 1983a; Farmer et al., 1985a; Müller et al., 1996); however, the levels of DNA similarity between members of the genera *Buttiauxella* and *Kluyvera* and more than 50 other species belonging to the family *Enterobacteriaceae* were between 15% and 36% (Gavini et al., 1983a; Müller et al., 1996), indicating the separate position of these two genera. In addition, the mol% G + C ratios of *Buttiauxella* and *Kluyvera* are quite different, with values ranging from 47–51% for *Buttiauxella* and 54–58% for *Kluyvera* (Ferragut et al., 1981; Gavini et al., 1983a). *Buttiauxella* and *Kluyvera* exhibit similar fatty acid patterns, but all strains of *K. ascorbata*, *K. cryocrescens*, and *K. georgiana* produced the summed feature $C_{18:1\ ω7c}$, $C_{18:1\ ω9c}$, and/or $C_{18:1\ ω12t}$ in higher amounts (>20% of all fatty acids) than any strain of *Buttiauxella* (Kämpfer et al., 1997). Although *K. georgiana* produced this feature also in amounts >20%, this species can be separated from all *Buttiauxella* species by its high amounts of 15:0 (>8%), which is not found in *Buttiauxella* (Kämpfer et al., 1997). Further physiological and biochemical characters that are helpful for differentiation of the two genera are given in Table BXII.γ.211.

TAXONOMIC COMMENTS

Based on a numerical taxonomic study of *Enterobacteriaceae*, Gavini et al. (1976b) defined a new group of strains within the family and gave it the vernacular name "group F". Originally these 17 strains isolated from water and unpolluted soils were regarded to be similar to the genus *Citrobacter* because of their negative indole and Voges–Proskauer reactions and their positive citrate and methyl red tests. Subsequently, Ferragut et al. (1981) used DNA–DNA hybridization to compare strains of group F to each other and to named species of *Enterobacteriaceae*. Based on DNA similarity of 82–96% within group F and its low level of DNA similarity to other *Enterobacteriaceae*, they proposed the new genus *Buttiauxella* with one species *Buttiauxella agrestis*. The names *Buttiauxella* and *B. agrestis* were effectively published, but were not validated in the *International Journal of Systematic Bacteriology* before January 1, 1980. They did not appear on the Approved Lists of Bacterial Names (Skerman et al., 1980), but both names have now been validly published (Ferragut et al., 1982) and have standing in nomenclature. Four additional "group F" strains were only 62–66% related to *B. agrestis*. Based on phenotypic differences, three of these strains were placed in a group that was given the name Enteric Group 63, and the remaining strain was designated as Enteric Group 64 (Farmer et al., 1985a). In 1981, another group of strains, phenotypically similar to *Pantoea agglomerans* (formerly *Enterobacter agglomerans*), except for a positive arginine dihydrolase reaction, was reported and named Enteric Group 59 (Farmer et al., 1985a). Eight isolates, originating from sputum (6), a foot wound (1), and ham (1), were studied.

Between 1984 and 1988, Müller et al. (1995a, b, 1996)

TABLE BXII.γ.211. Differentiation of *Buttiauxella* and *Kluyvera*[a,b]

Test	*Buttiauxella*	*Kluyvera*
Mol% G + C content	47–51	54–58
Indole production	−	+[c]
Citrate (Simmons)	d	[+]
Sucrose fermentation	−	[+]
Raffinose fermentation	[−]	+
α-Methyl-D-glucose fermentation	−	+
Melibiose fermentation	d	+
N-Acetyl-L-glutamate utilization	+	−

[a]Data adapted from Ferragut et al. (1981), Gavini et al. (1983a), Müller et al. (1996)

[b]Symbols: +, positive for 90–100% of strains; [+], positive for 75–89% of strains; d, positive for 25–74% of strains; [−], positive for 11–24% of strains; − positive for 0–10% of strains.

[c]Except for *K. cochleae*.

screened the intestinal contents of snails and slugs for the presence of *Enterobacteriaceae* and, in addition to *Rahnella* isolates and the infrequently seen species *Ewingella americana* and *Serratia fonticola*, found many strains belonging to the genera *Buttiauxella* and *Kluyvera*.

Two hundred and nineteen strains belonging to the genera *Buttiauxella* and *Kluyvera* were subjected to an extensive DNA similarity study. The results indicated that the strains belonging to the genus *Buttiauxella* comprised seven closely related hybridization groups. One of these (hybridization group 1) corresponded to *B. agrestis*. Hybridization groups 2 through 7 were described as the new species: *B. ferragutiae* (formerly Enteric Group 63), *B. gaviniae* (formerly Enteric Group 64), *B. brennerae*, *B. izardii*, *B. noackiae* (formerly Enteric Group 59), and *B. warmboldiae* (Müller et al., 1996). The levels of similarity obtained for members of the seven species were generally between 45% and 65%, a good indication that all of these groups belong to a single genus. The hybridization groups that exhibited the highest levels of intergeneric similarity were hybridization groups 1 (*B. agrestis*) and 5 (*B. izardii*) and hybridization groups 3 (*B. gaviniae*) and 6 (*B. noackiae*). The relative binding ratios obtained from DNA–DNA hybridization studies between the *Buttiauxella* species are given in Table BXII.γ.212. The levels of DNA similarity between members of the *Buttiauxella* and *Kluyvera* species were generally between 15% and 30% (Müller et al., 1996).

ACKNOWLEDGMENTS

I thank Don J. Brenner for his helpful comments and critical reading of this chapter, Hans E. Müller for interesting discussions and for furnishing strains, and W. Ludwig for critical comments and his experience in 16S rDNA sequence analyses.

FURTHER READING

Brenner, D.J. 1992. Additional genera of *Enterobacteriaceae*. *In* Balows, Trüper, Dworkin, Harder and Schleifer (Editors), The Prokaryotes, 2nd Ed., Vol. 3, Springer-Verlag, New York. pp. 2922–2937.

Farmer, J.J. III 1984. Other genera of the family *Enterobacteriaceae*. *In* Krieg and Holt (Editors), Bergey's Manual of Systematic Bacteriology, 1st Ed., Vol. 1, The Williams & Wilkins Co., Baltimore. pp. 506–516.

Ferragut, C., D. Izard, F. Gavini, B. Lefebvre and H. Leclerc. 1981. *Buttiauxella*, a new genus of the family *Enterobacteriaceae*. Zentbl. Bakteriol. Mikrobiol. Hyg. 1 Abt. Orig. C 2: 33–44.

Gavini, F., B. Lefebvre and H. Leclerc. 1976. Positions taxonomiques d'entérobactéries H$_2$S-par rapport au genre *Citrobacter*. Ann. Microbiol. (Paris) *127a*: 275–295.

Müller, H.E., D.J. Brenner, G.R. Fanning, P.A.D. Grimont and P. Kämpfer. 1996. Emended description of *Buttiauxella agrestis* with recognition of six new species of *Buttiauxella* and two new species of *Kluyvera*: *Buttiauxella ferragutiae* sp. nov., *Buttiauxella gaviniae* sp. nov., *Buttiauxella brennerae* sp. nov., *Buttiauxella izardii* sp. nov., *Buttiauxella noackiae* sp. nov., *Buttiauxella warmboldiae* sp. nov., *Kluyvera cochleae* sp. nov., and *Kluyvera georgiana* sp. nov. Int. J. Syst. Bacteriol. *46*: 50–63.

List of species of the genus Buttiauxella

1. **Buttiauxella agrestis** Ferragut, Izard, Gavini, Lefebvre and Leclerc 1982, 266[VP], emend. Müller, Brenner, Fanning, Grimont and Kämpfer 1996, 60 (Effective publication: Ferragut, Izard, Gavini, Lefebvre and Leclerc 1981, 40.)

   a.gres'tis. L. masc. n. *ager, agri*; adj. *agrestis* living in the fields, so named because all original strains were isolated from unpolluted soils and water.

   B. agrestis was called DNA hybridization group 1 in the study of Müller et al. (1996). All 30 strains studied by Müller et al. (1996) were ≥73% DNA-related in 60°C reactions (range 73–100%). In the original description (Ferragut et al., 1981), which was based on the 17 strains originally described by Gavini et al. (1976b), it was observed that 60% of *B. agrestis* strains were malonate positive. In the study of Müller et al. (1996), 96% of the strains were malonate positive. Biochemical tests that are useful for differentiating *B. agrestis* from other *Buttiauxella* species are given in Table BXII.γ.210. The arginine dihydrolase, fucose, glycerol, lactose, melibiose, ornithine decarboxylase, palatinose, and D-sorbitol tests are helpful. Clinical significance, if any, is unknown. Isolated from mollusks, as well as water, soil, and human materials.

   The mol% G + C of the DNA is: 47–50 (T_m).

   Type strain: ATCC 33320, CDC 1176-81, CUETM 77-167, DSM 4586.

   GenBank accession number (16S rRNA): AJ233400.

2. **Buttiauxella brennerae** Müller, Brenner, Fanning, Grimont and Kämpfer 1996, 62[VP]

   bren'ner.ae. M.L. fem. gen. n. *brennerae* of Brenner, in honor of Frances W. Hickman-Brenner, an American microbiologist, for her contributions to the study of many genera of the *Enterobacteriaceae*.

   Biochemical characteristics are shown in Table BXII.γ.210. Tests that can be used to differentiate *B. brennerae* from other *Buttiauxella* species are the L-arabinose, arginine dihydrolase, fucose, *myo*-inositol, 5-ketogluconate, lysine decarboxylase, and malonate. Clinical significance, if any, is unknown. Isolated from mollusks.

   The mol% G + C of the DNA is: ~50 (T_m).

   Type strain: ATCC 51605, DSM 9396.

   GenBank accession number (16S rRNA): AJ233401.

3. **Buttiauxella ferragutiae** Müller, Brenner, Fanning, Grimont and Kämpfer 1996, 60[VP]

   fer'ra.gut.i.ae. M.L. fem. gen. n. *ferragutiae* of Ferragut, in honor of Carmen Ferragut, a French microbiologist, for

TABLE BXII.γ.212. Relative binding ratios of DNAs at 60°C between the species of the genus *Buttiauxella* [a,b]

Test	B. agrestis	B. brennerae	B. ferragutiae	B. gaviniae	B. izardii	B. noackiae	B. warmboldiae
B. agrestis	81 (73–100)	53 (34–58)	51 (46–54)	56 (51–64)	63 (57–71)	58 (52–64)	56
B. brennerae		85 (68–100)	50 (34–57)	59 (39–68)	53 (44–60)	55 (38–66)	50 (44–55)
B. ferragutiae			100	50 (44–56)	48 (43–56)	55 (52–58)	49
B. gaviniae				84 (72–100)	52 (49–59)	71 (67–77)	49 (42–54)
B. izardii					88 (66–100)	50 (43–62)	48 (45–58)
B. noackiae						88 (81–100)	48 (46–51)
B. warmboldiae							99 (98–100)

[a]Data adapted from Müller et al. (1996).

[b]Results are expressed as percentages, figures in parentheses indicate the range of similarity.

her contribution to the study of the genus *Buttiauxella*, previously called Enteric Group 63.

Biochemical characteristics are shown in Table BXII.γ.210. Positive lysine decarboxylase and D-sorbitol tests and negative ketogluconate and malonate tests can clearly differentiate this species from all other *Buttiauxella* species. Clinical significance, if any, is unknown.

Isolated from water and soil.

The mol% G + C of the DNA is: 48–50 (T_m).

Type strain: ATCC 51602, CDC 1180-81, CUETM 78-31, DSM 9390.

GenBank accession number (16S rRNA): AJ233402.

4. **Buttiauxella gaviniae** Müller, Brenner, Fanning, Grimont and Kämpfer 1996, 62[VP]

ga.vin'i.ae. M.L. fem. gen. n. *gaviniae* of Gavini, in honor of Françoise Gavini, a French microbiologist, for her contributions to the study of the genus *Buttiauxella*, previously called Enteric Group 64.

Biochemical characteristics are shown in Table BXII.γ.210. Positive arginine dihydrolase, 5-ketogluconate, and palatinose reactions and negative ornithine decarboxylase and raffinose reactions are useful in differentiating *B. gaviniae* from other *Buttiauxella* species. A combination of reaction results is necessary to differentiate this species from *Buttiauxella noackiae*. Clinical significance, if any, is unknown. All but one strain were isolated from mollusks.

The mol% G + C of the DNA is: ~51 (T_m).

Type strain: ATCC 51604, DSM 9393.

5. **Buttiauxella izardii** Müller, Brenner, Fanning, Grimont and Kämpfer 1996, 62[VP]

iz.ard'i.i. M.L. masc. gen. n. *izardii* of Izard, in honor of Daniel Izard, a French microbiologist, for his contribution to the study of the genus *Buttiauxella*.

Biochemical characteristics are shown in Table BXII.γ.210. No single biochemical characteristic differentiates *B. izardii* from *B. agrestis*, but a combination of several characteristics differentiates these taxa. Useful differential biochemical tests are the L-arabinose, citrate, fucose, *myo*-inositol, ornithine decarboxylase, and raffinose. Clinical significance, if any, is unknown. Isolated from mollusks.

The mol% G + C of the DNA is: ~50 (T_m).

Type strain: ATCC 51606, DSM 9397.

GenBank accession number (16S rRNA): AJ233404.

6. **Buttiauxella noackiae** Müller, Brenner, Fanning, Grimont and Kämpfer 1996, 62[VP]

no.ack'i.ae. M.L. fem. gen. n. *noackiae* of Noack, in honor of Katrin Noack, who phenotypically characterized the *Buttiauxella* strains, previously called Enteric Group 59.

Biochemical characteristics of 15 strains isolated from snails are shown in Table BXII.γ.210. These characteristics are almost identical to those described for Enteric Group 59; the only exception is the result of the lactose test. All snail isolates lack a yellow pigment and are lactose negative, but seven of the eight Enteric Group 59 strains isolated from humans and food were shown to be lactose positive (Farmer et al., 1985a). It is possible that there is a correlation between the ability to split and metabolize lactose and the ability to survive in humans. Tests that are useful in differentiating *B. noackiae* from other *Buttiauxella* species are the N-acetyl-L-glutamine, L-arginine dihydrolase, melibiose, and L-ornithine decarboxylase. A combination of several characteristics is necessary to differentiate *B. noackiae* from other *Buttiauxella* species.

Isolated from mollusks, human sputum, human wounds, and food. Clinical significance, if any, is unknown.

The mol% G + C of the DNA is: ~50 (T_m).

Type strain: ATCC 51607, DSM 9401.

GenBank accession number (16S rRNA): AJ233405.

7. **Buttiauxella warmboldiae** Müller, Brenner, Fanning, Grimont and Kämpfer 1996, 62[VP]

warm'bold.i.ae. M.L. fem. gen. *warmboldiae* of Warmbold, in honor of Sabine Warmbold, who isolated most strains of the new *Buttiauxella* species at the Staatliches Medizinaluntersuchungsamt Braunschweig.

Biochemical characteristics are shown in Table BXII.γ.210. *Myo*-inositol is utilized and acid is produced by freshly isolated strains for many months, in contrast to all other species belonging to the genus *Buttiauxella*. However, these characteristics are lost after some years of storage; therefore, they are not constitutive. Other biochemical tests that are useful in differentiating *B. warmboldiae* from other *Buttiauxella* species are the L-arabinose, arginine dihydrolase, citrate, fucose, KCN, malonate, maltitol, ornithine decarboxylase, and palatinose. This species is the species that is the most distant from all other *Buttiauxella* species and has the lowest level of DNA–DNA similarity to other *Buttiauxella* species. Clinical significance, if any, is unknown. Isolated from snails.

The mol% G + C of the DNA is: ~52 (T_m).

Type strain: ATCC 51608, DSM 9404.

GenBank accession number (16S rRNA): AJ233406.

Genus VIII. **Calymmatobacterium*** Aragão and Vianna 1913, 221[AL]

GEORGE H. BROWNELL

Ca.lym.ma.to.bac.te'ri.um. Gr. n. *calymma* mantle, sheath; Gr. dim. neut. n. *bakterion* a small rod; M.L. neut. n. *Calymmatobacterium* the sheathed rodlet.

Pleomorphic rods, 0.5–1.5 × 1.0–2.0 μm, with rounded ends. Occur singly or in clusters. **The cells exhibit single or bipolar condensation of chromatin. Capsules are present.** Gram negative. Nonmotile. The exudate from infected tissues, when stained by

Wright's stain or by Giemsa stain, demonstrates **characteristic intracellular organisms in the cytoplasm of large mononuclear phagocytes** ("Donovan bodies"). Can be cultivated *in vivo* in the yolk sac of embryonated chicken eggs, peripheral blood mononuclear cells, and human epithelial (HEp-2) monolayers, or *in vitro* on special egg yolk-containing media; has not been reproducibly cultivated in bacteriologic media. Optimum growth temperature, 37°C. Pathogenic for humans, causing **Donovanosis (granuloma inguinale).**

Editorial Note: The type species of the genus *Calymmatobacterium* has been transferred to *Klebsiella* as *Klebsiella granulomatis* (Aragão and Vianna 1913) Carter, Bowden, Bastian, Myers, Sriprakash and Kemp 1999.

The mol% G + C of the DNA is: not known.

Type species: **Calymmatobacterium granulomatis** Aragão and Vianna 1913, 221.

FURTHER DESCRIPTIVE INFORMATION

In diseased tissue smears stained by Wright's method, *C. granulomatis* occurs within the cytoplasm of large mononuclear monocytes as blue to purple pleomorphic rods surrounded by pink capsules (Fig. BXII.γ.199). The organism may occasionally be observed free in extracellular spaces. The single or bipolar condensation of chromatin gives rise to characteristic "safety-pin" forms. The ultrastructure of the intracellular organisms has been described by Davis and Collins (1969), Dodson et al. (1973), Kuberski et al. (1980), Chandra et al. (1989), Chandra and Jain (1991), and Kharsany et al. (1997). Electron micrographs reveal encapsulated bacilliforms with characteristic Gram-negative cell walls. The presence of fimbriae-like structures have been reported by some investigators: Kuberski et al. (1980), Chandra et al. (1989), and Chandra and Jain (1991).

Calymmatobacterium granulomatis has been clinically proven to be the causal agent of granuloma inguinale (Dienst et al., 1938). The disease is often referred to as donovanosis because initial lesions have been diagnosed in skin areas other than the genital region. Most researchers support the contention that donovanosis is not necessarily a venereal disease but is an infection resulting from intimate contamination and poor hygiene. Infection

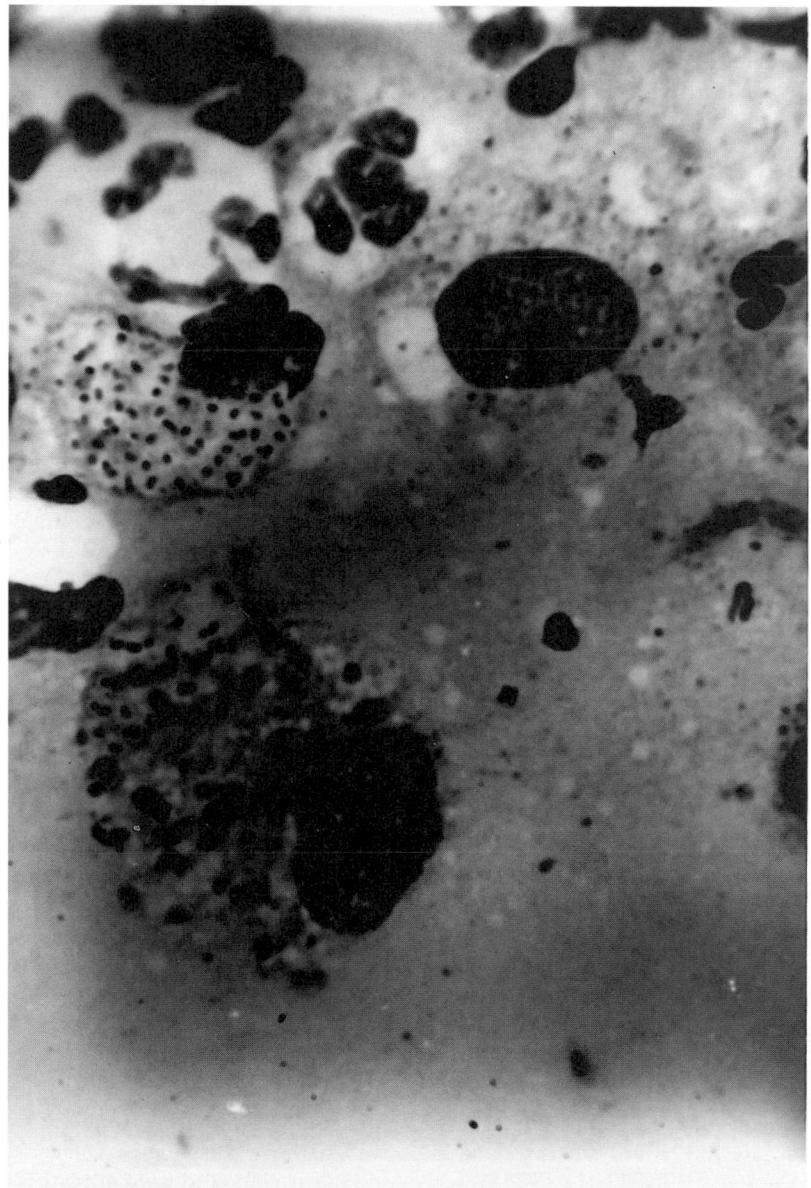

FIGURE BXII.γ.199. Large mononuclear phagocytes filled with *C. granulomatis*. Wright's stain (× 675). (Reproduced with permission from R.B. Dienst and G.H. Brownell in M.P. Starr et al. (Eds.) *The Prokaryotes: a Handbook on Habitats, Isolation and Identification of Bacteria.* p. 1410, 1981, ©Springer-Verlag, New York.)

is usually seen as a chronic, granulomatous, genital ulcerative disease. Donovanosis is an uncommon disease found in specific geographical regions including northern and central Australia, New Guinea, parts of central and south Africa, southeast India, the Caribbean, and parts of South America. The organism is pathogenic only for humans and infection cannot be produced in laboratory animals.

There are no protective antibodies produced by a patient infected with *C. granulomatis*. Once the infection occurs, the disease persists chronically and may spread through the lymphatics to all tissue unless treated with antibiotics. The patient does produce specific sensitizing antibodies as shown by skin testing (Chen et al., 1949). Antibodies can be detected in the serum of patients by complement-fixation procedures (Anderson et al., 1945b; Dulaney and Packer, 1947; Goldberg et al., 1953). The test antigens used in these techniques have included pus from granulomatous lesions, whole or ruptured *C. granulomatis*, or boiled or extracted egg yolk medium following growth of the organism. Immunological studies first noted a relationship between strains of *C. granulomatis* and *Klebsiella pneumoniae* (Packer and Goldberg, 1950). An indirect immunofluorescence test for granuloma inguinale was reported by Freinkel et al. (1992). The antigen consisted of paraffin-embedded tissue sections from lesions containing Donovan bodies. Unabsorbed sera at dilutions of 1:160 were reported to be 100% sensitive and 98% specific. After absorption with *K. pneumoniae* antigen, sera from proven granuloma inguinale patients remained reactive but at lower titers. Antigens obtained from newly reported culture methods well undoubtedly lead to more effective serological tests. (Carter et al., 1997; Kharsany et al., 1997).

ENRICHMENT AND ISOLATION PROCEDURES

At present, the only sources for *C. granulomatis* are the lesions of donovanosis, although new culture methods may soon provide reference strains. An organism was reportedly isolated from human feces (Goldberg, 1962) that had antigenic similarities to *C. granulomatis*. Early reports of the cultivation of *C. granulomatis* in cell-free media include the yolk of embryonated eggs (Anderson, 1944), fresh yolk medium (Dienst et al., 1948), Locke-yolk Dulaney slants (Dulaney et al., 1948), and egg yolk replaced by lactalbumin hydrolysate (Goldberg, 1959). Reports of the successful culturing of *C. granulomatis* in peripheral blood mononuclear cells (Kharsany et al., 1997) or, more conveniently, human epithelial (HEp-2) cells (Carter et al., 1997), should provide the materials that can provide fundamental information about this species that to date is unknown.

The following procedure is given by Morse (1980). The ulcerative lesions are cleansed with sterile, saline-soaked gauze before obtaining samples, in order to decrease contamination and remove tissue debris. Samples of tissue are removed by scraping or by means of a biopsy punch from beneath the border of the lesion, and small cleansed pieces of tissue are minced into small particles. Inoculation is made into the yolk sacs of 5-day-old embryonated eggs. After incubation for 72 h, the organisms can be detected in the yolk sac fluid.

C. granulomatis can also be isolated and grown *in vitro*. For example, a pure culture was isolated by Dienst et al. (1948) by inoculating fresh egg medium exudate aspirated from a pseudobubo of a patient. This isolate was reported to be subcultured and maintained in the same culture medium, and examination of the subcultures revealed large numbers of encapsulated organisms consistent with the morphology of *C. granulomatis*. Dienst

et al. (1948) indicated that several factors were important for isolation and cultivation: (a) maintenance of a low oxidation-reduction potential, (b) the requirement for a growth factor found in egg yolk, and (c) use of semisolid media containing 0.12% agar.

Dulaney slants have also been used for isolation and cultivation of *C. granulomatis* (Dulaney et al., 1948). After inoculation of lesion material onto a Dulaney slant, Locke's fluid is added to cover three-quarters of the slant, and the tubes are then incubated in a vertical position for 48–72 h (Morse, 1980). A semisynthetic medium has been devised by Goldberg (1959) for cultivation of laboratory strains of *C. granulomatis*. In this medium, the requirement for egg yolk is replaced by lactalbumin hydrolysate or by papaic digest of soy meal USP.

Recent efforts to culture *C. granulomatis* from biopsy material using these cell-free growth media suggest that they are unreliable. The recommended procedure is therefore the use of fresh monocytes (Kharsany et al., 1996, 1997) or the technically less demanding HEp-2 cell line. The latter employs standard *Chlamydia* culturing procedure (Carter et al., 1997).

Because diagnosis is still based on the observations of *C. granulomatis* in stained biopsy specimens and because of unreliable culture methods, this organism is a prime candidate for PCR primer targeting identification. Using PCR primers designed to target the *pho*E gene encoding for porin protein among klebsiellae and other enterobacteria, Bastian and Bowden (1996) reported amplification of *pho*E gene fragments from biopsy material showing Donovan bodies. The amplified sequences showed close correlation to those from *K. pneumoniae*, *K. rhinoscleromatis*, and *K. ozaenae*, and considerable divergence from corresponding sequences of other enterobacteria analyzed. Using the *pho*E gene primers in combination with primers for *Klebsiella scr*A (sucrose transport gene), reportedly absent in *C. granulomatis*, Carter et al. (1997) obtained *pho*E-positive and *scr*A-negative amplification from original clinical swab material as well as HEp-2 cell cultured isolates. Specific *C. granulomatis* primers have not yet been reported.

DIFFERENTIATION OF THE GENUS *CALYMMATOBACTERIUM* FROM OTHER GENERA

As of this printing, the examination of diseased tissue smears stained by Wright's blood stain or Giemsa stain remains the simplest procedure for identification for *C. granulomatis*. The characteristic appearance of the intracellular organisms (Fig. BXII.γ.199) is specific for the diagnosis of donovanosis. Species-specific PCR primers as well as immunofluorescence tests should soon be available.

TAXONOMIC COMMENTS

The coccobacillary microorganisms first observed by Donovan (1905) were frequently referred to as "Donovan bodies" when seen in tissue smears from patients with granulomatis lesions in the inguinal region. The Donovan bodies were later called *Calymmatobacterium granulomatis* by Aragão and Vianna (1913). When Anderson et al. (1945a) first isolated the organisms by yolk sac inoculation, they termed the etiologic agent of granuloma inguinale "*Donovania granulomatis*"; however, the name "*Donovania*" did not have priority over *Calymmatobacterium*.

In the first edition of *Bergey's Manual of Systematic Bacteriology*, the genus was not assigned to any family. It has been suggested that it should be placed in the family *Enterobacteriaceae* (Rake, 1948), and this relationship is supported (based on very limited

PCR-generated fragment analysis with "*Klebsiella*-like" sequences) by Bastian and Bowden (1996). Kharsany et al. (1997) observed no cross-reactivity with sera from patients showing Dovonan bodies and laboratory reference strains of *K. pneumoniae*, *K. oxytoca*, or *Enterobacter aerogenes*. Thus, the taxonomic relationships of *Ca-lymmatobacterium* to other bacterial genera are not yet understood, and in the present edition of the *Manual* it seems desirable not to ally the genus with any established family, although it must be noted that the type strain was recently transferred to the genus *Klebsiella* as *Klebsiella granulomatis* (Carter et al., 1999).

List of species of the genus Calymmatobacterium

1. **Calymmatobacterium granulomatis** Aragão and Vianna 1913, 221[AL]

   gran.u.lo′ma.tis. L. dim. n. *granulum* a small grain; Gr. suff. *-oma* a swelling or tumor; M.L. n. *granuloma* a granuloma; M.L. gen. n. *granulomatis* of a granuloma.

   The characteristics are as described for the genus and as depicted in Fig. BXII.γ.199.

   The mol% G + C of the DNA is: not known.

   Type strain: no strain extant.

Genus IX. **Cedecea** Grimont, Grimont, Farmer and Asbury 1981a, 325[VP] (Enteric Group 15 Farmer, Grimont, Grimont and Asbury 1980b, 295)

J.J. Farmer III

Ce.de′ce.a. M.L. fem. n. *Cedecea* formed from the abbreviation CDC. The named was coined by P.A.D. Grimont and F. Grimont for the Centers for Disease Control, Atlanta, Georgia, where the organisms were originally recognized as a new group and named Enteric Group 15.

Rod-shaped cells 0.6–0.7 × 1.3–1.9 μm, conforming to the general definition of the family *Enterobacteriaceae*. Gram negative. Motile, with five to nine peritrichous flagella. Grow at 15°, 20°, and 37°C. Facultatively anaerobic. **Catalase positive, strong and rapid.** Oxidase negative. Nonpigmented. Reduce nitrate to nitrite.

Positive for methyl red, Voges–Proskauer, citrate utilization (Simmons), motility at 36°C, growth in the presence of cyanide (KCN test), malonate utilization, D-glucose fermentation, and the fermentation of D-mannitol, salicin, maltose, trehalose, cellobiose, melibiose, D-arabitol, D-mannose, and D-galactose. Many strains produce visible gas during fermentation. Lipase (corn oil, Tween 40, Tween 60, Tween 80, and tributyrin) is positive, but gelatin hydrolysis, DNase, chitinase, polygalacturonase, and amylase are negative. Utilize 24 of 97 carbon sources.

Negative for indole production, H₂S production (TSI), urea hydrolysis, phenylalanine deaminase, lysine decarboxylase and the fermentation of dulcitol, adonitol, *myo*-inositol, D-sorbitol, L-arabinose, L-rhamnose, α-methyl-D-glucoside, erythritol, melibiose, glycerol, and mucate.

Susceptible to nalidixic acid, sulfadiazine, trimethoprim, gentamicin, streptomycin, kanamycin, tobramycin, amikacin, tetracycline, minocycline, chloramphenicol, carbenicillin, and furantoin (disk diffusion method on Mueller-Hinton agar); **resistant to colistin, polymyxin, penicillin, ampicillin, and cephalothin**.

Isolated from human clinical specimens that are normally sterile such as blood, urine, gallbladder, and lung tissue; also isolated from other clinical specimens such as throat, sputum, ulcers, and wounds. **An opportunistic pathogen that occasionally causes extraintestinal human infections**, or colonizes body surfaces. Isolated from water, ticks, and insects. A **rarely isolated genus of a *Enterobacteriaceae***. No 16S rRNA sequences of *Cedecea* strains have been reported.

The mol% G + C of the DNA is: 48–52.

Type species: **Cedecea davisae** Grimont, Grimont, Farmer and Asbury 1981a, 325.

Further descriptive information

Literature Since the genus was described in 1981 there have been a few reports in the literature; 11 reports cataloged in MEDLINE and 27 cataloged in BIOSIS. These have described the genus in human clinical specimens and infections, its isolation from natural sources, and its basic physiology-metabolism.

Infections in humans Most of the original isolates of *Cedecea* in the CDC collection were from human respiratory tract specimens, so it was difficult to assess clinical significance (Farmer et al., 1980b). However, a few isolates were from body sites that are normally sterile. Since the original description in 1981, there have been several case reports that suggest clinical significance. The first case of bacteremia due to a strain of *Cedecea* was caused by *Cedecea neteri* (Farmer et al., 1982) and led to the naming of this organism, which was previously without a scientific name ("*Cedecea* species 4"). Other cases of bacteremia have now been described (Perkins, et al., 1986; Aguilera et al., 1995). Several other cases have been described in which a *Cedecea* strain was isolated from a body site that is normally sterile, or thought to be clinically significant (Bae et al., 1975; Hansen and Glupczynski, 1984; Coudron and Markowitz, 1987; Anon et al., 1993). Infections due to *Cedecea* have usually been in elderly hospital patients with debilitating conditions such as heart disease, diabetes, alcoholism, and renal insufficiency. A few isolates have been from human feces, but there is no evidence that any *Cedecea* species causes diarrhea or intestinal infections.

Possible origin of the *Cedecea* strains in human clinical specimens Originally there was little information on this point. Strains of *Cedecea* have now been isolated from well water, ticks, and insects (Jang and Nishijimi, 1990; Kaaya and Okech, 1990; Pellegrini et al., 1992), which may be natural reservoirs and lead to exposure of humans. Berkowitz and Metchock (1995) described bacteria in the feces of hospitalized children that are resistant to third-generation cephalosporins. Antibiotic usage is common in hospitals and selects for strains that have intrinsic or acquired resistance. Since *Cedecea* strains have high intrinsic resistance to penicillin, ampicillin, and the cephalosporin antibiotic cephalothin, they probably have a selective advantage in the feces of hospitalized patients. The reduction of the normal host flora by antibiotic usage could favor the growth and selection of *Cedecea* strains, which might then colonize certain body sites, and under the right conditions cause infection.

Additional studies and more case reports are needed to better define the pathogenic potential of the five species in the genus. The clinical significance of a *Cedecea* is probably as an infrequent colonizer or infrequent opportunistic pathogen, particularly in older people who are debilitated. More information is needed on ecology, epidemiology, and how humans are exposed to and acquire the organism.

Bacteriology Strains of *Cedecea* grow well on media normally used in enteric bacteriology. Colonies on nutrient agar are about 1.5 mm (24 h, 37°C). They are typical *Enterobacteriaceae* in most of their properties. See the chapter on the family *Enterobacteriaceae* for general information on growth, plating media, working and frozen stock cultures, media, and methods for biochemical testing, identification, and antibiotic susceptibility. Also see Table BXII.γ.193 in the chapter on the family *Enterobacteriaceae*, which gives the percentage positive for the five *Cedecea* species on 47 biochemical tests done at the Enteric Reference Laboratories at the Centers for Disease Control (CDC) with standard media and methods (Farmer, 1995). These tests and computer analysis have proved useful for identification and for differentiating strains of *Cedecea* from other species in the family. Table BXII.γ.213 lists phenotypic tests that are useful for differentiating and identifying the five species of *Cedecea*. Additional descriptive material, including the results for 97 carbon source utilization tests, can be found in the original description of the genus (Grimont et al., 1981a). Chester and Moskowitz (1987) found that *Cedecea* strains produce catalase, and that the catalase reaction was very strong and rapid and could be useful as a screening test.

TAXONOMIC COMMENTS

History and discovery *Cedecea* was proposed as a new genus in *Enterobacteriaceae* in 1980 (Farmer et al., 1980b; Grimont et al., 1981a). In 1977 it was first recognized and named "Enteric Group 15" as a diagnostic culture was being studied at the CDC. The laboratory's STRAIN MATCHER computer program indicated a group of over a dozen other cultures that were very similar to

the diagnostic culture, and all had been reported "unidentified". The strains were lipase-positive (corn oil) and resistant to the antibiotics colistin and cephalothin. Among *Enterobacteriaceae*, these properties are unique to the genus *Serratia*, but the new group differed from *Serratia* because it was negative for DNase and gelatin hydrolysis. Originally, Enteric Group 15 was thought to be a uniform group of strains that would probably be a new species of *Serratia*, intermediate between the "typical" *Serratia* (DNase positive, lipase positive, gelatinase positive) and *Serratia fonticola*, which is negative for all three of these tests (Farmer et al., 1980b). Strains of Enteric Group 15 were sent to the Grimonts in France (Farmer et al., 1980b; Grimont et al., 1981a), who used phenotypic characterization and DNA–DNA hybridization (SI nuclease method) to characterize them. Fifteen strains of Enteric Group 15 were more closely related (Fig. BXII.γ.200) to each other (32–100%) than to strains of the six named *Serratia* species (6–10%) or to other *Enterobacteriaceae* (1–23%). Originally, Enteric Group 15 was defined as a single group of 17 strains, but further study indicated five different DNA hybridization groups (Fig. BXII.γ.200) that were also phenotypically distinct (Table BXII.γ.213). In the original paper proposing the genus *Cedecea*, the two largest hybridization groups were named *C. davisae* and *C. lapagei*. The name *Cedecea neteri* was later given to DNA hybridization group 4 (Farmer et al., 1982). The original vernacular names "*Cedecea* species 3" and "*Cedecea* species 5" are still being used for the two other DNA hybridization groups until more strains are available, or there is a compelling reason to name them (see Other Organisms, below).

FURTHER READING

Farmer, J.J., III, N.K. Sheth, J.A. Hudzinski, H.D. Rose and M.F. Asbury. 1982. Bacteremia due to *Cedecea neteri* sp. nov. J. Clin. Microbiol. *16*: 775–778.

Grimont, P.A.D., F. Grimont, J.J. Farmer, III and M.A. Asbury. 1981. *Cedecea davisae*, gen. nov., sp. nov. and *Cedecea lapagei*, sp. nov., new *Enterobacteriaceae* from clinical specimens. Int. J. Syst. Bacteriol. *31*: 317–326.

Perkins, S.R., T.A. Beckett and C.M. Bump. 1986. *Cedecea davisae* bacteremia. J. Clin. Microbiol. *24*: 675–676.

List of species of the genus Cedecea

1. **Cedecea davisae** Grimont, Grimont, Farmer and Asbury 1981a, 325[VP]

   da'vi.sae. M.L. gen. n. *davisae* named to honor Betty Davis, the American bacteriologist of the Enteric Bacteriology Lab-

   oratories, Centers for Disease Control and Prevention, Atlanta, Georgia, who made many contributions to the biochemical and serological identification of *Enterobacteriaceae* and *Vibrionaceae*.

TABLE BXII.γ.213. Differentiation of the three named and two unnamed *Cedecea* species

Test or property	Cedecea davisae	Cedecea lapagei	Cedecea neteri	Cedecea species 3	Cedecea species 5
Ornithine decarboxylase	95[a]	0	0	0	50
Sucrose fermentation	100	0	100	50	100
D-Sorbitol fermentation	0	0	100	0	100
Raffinose fermentation	10	0	0	100	100
D-Xylose fermentation	100	0	100	100	100
Melibiose fermentation	0	0	0	100	100
Malonate utilization	91	100	100	0	0
Thiamin required for growth[b]	100	0	0	0	0
Growth at 5°C[b]	100	0	0	0	0

[a]For the first seven tests each number gives the percentage positive after 2 d incubation at 36°C and is based on the data summarized by Farmer (1999). The vast majority of these positive reactions occur within 24 h. Reactions that become positive after 2 d are not considered.

[b]The results for thiamin requirement and growth at 5°C are from Grimont et al. (1981a).

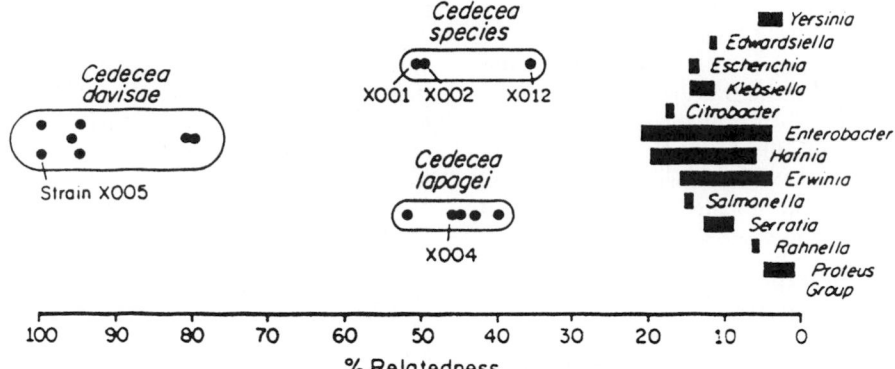

FIGURE BXII.γ.200. Relatedness by DNA–DNA hybridization of the *Cedecea davisae* type stain (X005) to 14 other *Cedecea* strains and to other genera of *Enterobacteriaceae*. Strain X004 is the type strain of *Cedecea lapagei*; X002 is the type strain of *Cedecea neteri*; X001 is the reference strain of "*Cedecea* species 3"; and X012 is the reference strain of "*Cedecea* species 5". (Redrawn with permission from J.J. Farmer III, P.A.D. Grimont, F. Grimont, and M.A. Asbury.)

The characteristics are as given for the genus. All strains require thiamin (0.01 μg/ml) for growth in minimal media and one strain also requires *p*-aminobenzoic acid (Grimont et al., 1981a). Most strains grow at 5°C, unlike the other *Cedecea* species. See Table BXII.γ.193 in the chapter on the family *Enterobacteriaceae* for a more complete phenotypic description based on the results of 47 biochemical tests. One strain produces a strong distinct aroma, the "odor of *Serratia odorifera*", which is apparently due to the pyrazine compound 3-sec-butyl-2-methoxypyrazine (Gallois and Grimont, 1985).

Isolated from human clinical specimens including blood, gallbladder, eye, urine, sputum, throat, wound, feces; probably a rare opportunistic pathogen or colonizer. Perkins et al. (1986) described a case of bacteremia in a 70-year-old woman with a history of heart disease and a diagnosis of bronchitis and chronic obstructive pulmonary disease. Other case reports have described pneumonia in an elderly diabetic with a heart condition (Bae et al., 1975) and a scrotal abscess in a patient with alcoholic liver disease (Bae and Sureka, 1983). Isolated from the cockroaches *Blatta germania* and *Blatta orientalis* (Pellegrini et al., 1992). The 16S rRNA sequence has not been determined.

The mol% G + C of the DNA is: 49–50 (T_m).

Type strain: ATCC 33431, CDC 3278-77, CIP 80.34.

Additional Remarks: The American Type Culture Collection includes four other human strains; two from sputum and one each from gallbladder and wound.

2. **Cedecea lapagei** Grimont, Grimont, Farmer and Asbury 1981a, 325[VP]

la.pa'ge.i. M.L. gen. n. *lapagei* named to honor the late Stephen Lapage, a British bacteriologist who made many contributions to the *Enterobacteriaceae*, bacterial systematics, and particularly as an editor of the Bacteriological Code.

The characteristics are as given for the genus. See Table BXII.γ.193 in the chapter on the family *Enterobacteriaceae* for a more complete phenotypic description based on the results of 47 biochemical tests. All five strains of *C. lapagei* are highly related by DNA–DNA hybridization, but strain 1554-

75 had higher divergence values and differed from the other strains in phenotype (Grimont et al., 1981a).

Isolated from human respiratory tract (throat and sputum) and from ticks. Coudron and Markowitz (1987) described an isolate from lung tissue in a case report, but concluded that its etiological role was not proven. Appears to be a tick pathogen. Brun and Texeira (1992a, b) described genital tract infections of the tick species *Boophilus microplus* that resulted in "engorged females". The 16S rRNA sequence has not been determined.

The mol% G + C of the DNA is: 48–52 (T_m).

Type strain: ATCC 33432, CDC 0485-76, CIP 80.35.

Additional Remarks: The American Type Culture Collection includes four other strains, all from human sputum.

3. **Cedecea neteri** Farmer, Sheth, Hudzinski, Rose and Asbury 1983, 438[VP] (Effective publication: Farmer, Sheth, Hudzinski, Rose and Asbury 1982, 777.)

ne'te.ri. M.L. gen. n. *neteri* named to honor Erwin Neter, an American physician and microbiologist who made many contributions to our knowledge of the family *Enterobacteriaceae*, particularly the role of this family in human disease.

The characteristics are as given for the genus. Also see Table BXII.γ.193 in the chapter on the family *Enterobacteriaceae* for a more complete phenotypic description based on the results of 47 biochemical tests.

Isolated from human blood, sputum, and wound. Farmer et al. (1982) described a case of bacteremia with possible endocarditis in a 62-year-old man with vascular heart disease and recurrent fever and chills of 4 d duration. Anon et al. (1993) described a case of peritonitis after vigorous abdominal surgery in which *Cedecea neteri* was isolated along with *Escherichia vulneris*. Aguilera et al. (1995) described a case of bacteremia in a patient with systemic lupus erythematosus.

The mol% G + C of the DNA is: not reported.

Type strain: ATCC 33855, CDC 0621-75.

Additional Remarks: The American Type Culture Collection includes two other strains, one from human blood (the case report of Farmer et al., 1982) and one from sputum.

Other Organisms

1. "Cedecea species 3"

    The CDC collection has three strains of this organism, from blood, gallbladder, and sputum.

    The mol% G + C of the DNA is: not reported.

    Deposited strain: X001, CDC 4853-73.

2. "Cedecea species 5"

    The CDC collection has only one strain of this organism, from a human toe.

    The mol% G + C of the DNA is: not reported.

    Deposited strain: X012, CDC 3699-73.

The CDC collection has at least 11 other cultures from a variety of human clinical specimens that were reported "*Cedecea* species" because they had high computer identification scores as "genus *Cedecea*" and had characteristics typical for the genus. However, the phenotypic properties of these strains differed from the five *Cedecea* species described above. The data on the two unnamed *Cedecea* species and these 11 "*Cedecea* species" clearly indicate that there are additional species of *Cedecea*. Future studies are needed to better define this collection. There have been two case reports describing *Cedecea* strains that are phenotypically different from the five named species. Hansen and Glupczynski (1984) described a *Cedecea* isolate from a cutaneous ulcer with a purulent discharge in a 79-year-old diabetic with arthritis of both lower limbs. Cultures were positive on days 2, 5, and 43. The isolate was most like *Cedecea davisae* except that it was positive for acetate utilization and D-sorbitol fermentation. The isolate was lost before it could be confirmed by a reference laboratory. Magnum and Radisch (1982) described a *Cedecea* isolate from the postmortem heart blood of a 76-year-old alcoholic man with decubiti, anemia, metabolic acidosis, renal insufficiency, and urinary incontinence. This strain was most like *Cedecea* species 3 except it was Voges–Proskauer positive, fermented sucrose, and grew on acetate agar.

Two additional *Cedecea* species have been described, but they were given the vernacular names of "*Cedecea* species 3" and "*Cedecea* species 5" rather than scientific names (Farmer et al., 1980b; Grimont et al., 1981a). Thus, the two organisms do not have standing in nomenclature, but appear to be unnamed species of *Cedecea* based on DNA–DNA hybridization and phenotypic differences (Table BXII.γ.213). See Table BXII.γ.193 in the chapter on the family *Enterobacteriaceae* for a more complete phenotypic description of this organism based on the results of 47 biochemical tests.

Genus X. **Citrobacter** *Werkman and Gillen 1932, 173*[AL]

WILHELM FREDERIKSEN

Cit.ro.bac′ter. L. n. *citrus* lemon; M.L. n. *bacter* a small rod; M.L. masc. n. *Citrobacter* a citrate-utilizing rod.

Straight rods, ~1.0 µm × 2.0–6.0 µm. Occur singly and in pairs. Conform to the general definition of the family *Enterobacteriaceae*. Usually not encapsulated. Gram negative. Usually **motile by peritrichous flagella**. Facultatively anaerobic, having both a respiratory and a fermentative type of metabolism. Grow readily on ordinary media. Colonies on nutrient agar are generally 2–4 mm in diameter, smooth, low convex, moist, translucent or opaque, and gray with a shiny surface and entire edge. Mucoid or rough forms may occur occasionally. Oxidase negative. Catalase positive. Chemoorganotrophic. **Citrate can be utilized as a sole carbon source by most strains. Lysine is not decarboxylated.** Alginate and pectate are not decomposed. D-glucose is fermented with the production of acid and gas. The methyl red test is positive; the Voges–Proskauer test is negative. Occur in the feces of humans and some animals; probably normal intestinal inhabitants. Sometimes pathogenic and often isolated from clinical specimens as opportunistic pathogens. Can also be found in soil, water, sewage, and food.

The mol% G + C of the DNA is: 50–52 (T_m).

Type species: **Citrobacter freundii** (Braak 1928) Werkman and Gillen 1932, 173 (*Bacterium freundii* Braak 1928, 140.)

FURTHER DESCRIPTIVE INFORMATION

Members of *Citrobacter* may or may not ferment lactose promptly but nearly always produce β-galactosidase. L-arabinose, maltose, L-rhamnose, trehalose, D-xylose, D-mannitol, and D-sorbitol are fermented rapidly by the majority of strains. Erythritol and *myo*-inositol are rarely attacked.

Ornithine is decarboxylated by almost all strains of *C. koseri*, *C. amalonaticus*, *C. farmeri*, *C. braakii*, *C. sedlakii*, and *C. rodentium*, but not by *C. freundii*, *C. youngae*, and *C. werkmanii*. *C. koseri* and *C. rodentium* do not grow in media containing potassium cyanide (KCN) in contrast to the other species.

Strains of *C. koseri* ferment D-adonitol and D-arabitol, but strains of the other species do not. Malonate is utilized as a sole carbon source by most strains of *C. koseri*, *C. werkmanii*, *C. sedlakii*, and *C. rodentium*, but can be used by less than 15% of the strains of the other species.

The majority of strains of *C. freundii*, *C. youngae*, *C. braakii*, and *C. werkmanii* produce abundant H_2S in the butt of Kligler iron agar and triple-sugar iron agar.

Indole is produced by *C. koseri*, *C. amalonaticus*, *C. farmeri*, and *C. sedlakii* with few exceptions, but few strains of the other species give a positive indole test.

Based on 16S rRNA gene sequence, the genus *Citrobacter* clusters in the family *Enterobacteriaceae* within the *Gammaproteobacteria*. Its nearest relatives are members of the genera *Salmonella* and *Pantoea*.

DNA–DNA hybridization experiments (Brenner et al., 1993) using the hydroxyapatite method showed intraspecific relatedness almost always above 70%, with 5% or less divergence within related sequences, and interspecific relatedness from ~30% to ~70%. *C. farmeri* was closely related to *C. amalonaticus*; *C. youngae*, *C. braakii*, and *C. werkmanii* were related to each other and to *C. freundii* at the 50–70% level. The other interspecies relations were on a lower level (~30% to ~55%).

Some strains of *C. koseri* produce bacteriocins, and many can be induced to produce bacteriophages (Hamon et al., 1974; Markel et al., 1975).

The antigenic structure was developed around the so-called Bethesda-Ballerup group that was considered close to *Salmonella* (Kauffmann and Møller, 1940; West and Edwards, 1954). After its inclusion in *Citrobacter freundii*, the group formed the basis for the elucidation of the O- and H-antigen structure of this species, and its relation to other taxa, especially within the genus *Salmonella*; see Ewing (1986a) for a detailed description. The Vi antigen of *Salmonella* serovar Typhi (*S. typhi* or *S. enterica*, subsp. *enterica*, serovar Typhi) can be found in certain strains of *C. freundii*.

The antigenic structure of *C. koseri* was studied by several groups (Gross and Rowe, 1974, 1983; Popoff and Richard, 1975; Sourek and Aldová, 1976) resulting in three different O- and H-antigen serotyping systems. Sourek and Aldová (1976) also established an O-antigen scheme for *C. amalonaticus*, which they extended in 1988 (Sourek and Aldová, 1988). Van Oye et al. (1975) found that many strains of *C. amalonaticus* reacted in different *Shigella* sera.

Miki et al. (1996) reexamined the 90 reference strains listed in the scheme for *C. freundii* by West and Edwards (1954). They found that 40 of these strains belonged to *C. youngae* as defined by Brenner et al. (1993), 25 to *C. braakii*, 13 to *C. werkmanii*, and only three to the redefined *C. freundii*. Nine O-group-29 strains formerly allocated to the Ballerup group were all identified as *C. braakii*.

With modern typing systems there seems to be less place for serotyping of citrobacters for epidemiological purposes. Both multilocus enzyme electrophoresis (Woods et al., 1992), ribotyping (Papasian et al., 1996; El Harrif-Heraud et al., 1997), PCR methods (Woods et al., 1992; El Harrif-Heraud et al., 1997), and pulsed field gel electrophoresis (Papasian et al., 1996) have been applied, often to elucidate possible outbreaks of *C. koseri* meningitis/cerebral abscesses.

A 32-kDa outer membrane protein (OMP) was found to be a marker of virulence for *C. koseri* (Kline et al.,1988), and possibly a factor of virulence (Li et al., 1990).

Antibiotic sensitivity Strains of *Citrobacter* are naturally susceptible to sulfonamides, trimethoprim, aminoglycosides, chloramphenicol, tetracycline, nalidixic acid, fluoroquinolones, nitrofurantoin, polymyxins, and fosfomycin. Like other enterobacteria they are resistant to erythromycin and other macrolides, lincosamides, fusidic acid, and vancomycin. Resistance to one or more of the former group of antibiotics may be acquired, depending on circumstances, especially antibiotic usage policy in a broad sense.

Citrobacter strains are usually extremely susceptible to the fluoroquinolones. However, like many other enterobacteria they are prone to develop resistance, although this seems to occur rarely. The majority of strains are susceptible to the aminoglycosides, although less so to streptomycin than to newer aminoglycosides.

Almost all strains of *Citrobacter* produce a β-lactamase. *C. freundii* and *C. koseri* differ in the type of β-lactamase produced, *C. freundii* being resistant to cephalosporins and susceptible to carboxypenicillins, while *C. koseri* behave in the reverse way. Resistance to a broad range of newer β-lactam antibiotics may appear in both species; however, imipenem seems to remain active. See Frederiksen and Søgaard (1992) for a broader review of antibiotics and *Citrobacter*. El Harrif-Heraud et al. (1997) described a nosocomial outbreak where six identical strains of *C. koseri* produced an extended spectrum β-lactamase (ESBL), leaving susceptibility only to imipenem, latamoxef, and some combinations with clavulanic acid. The strains harbored a plasmid mediating a SHV-4 type lactamase.

Members of the genus *Citrobacter* occur not only in feces of humans and animals with no disorder but also in water, sewage, soil, and food. Isolates of *C. freundii* recovered from fish were compared with strains from various other sources by Toranzo et al. (1994), and a possible role as a fish pathogen discussed. Citrobacters are found in clinical bacteriology not only in stools but also in urine, sputum, and specimens from bacteremia, meningitis, otitis media, wounds, abscesses, the throat, and autopsies. Their role often seems to be that of an opportunistic pathogen. Cases of neonatal meningitis caused by *C. koseri* have been reported (Gwynn and George, 1973; Kline, 1988). Khashe and Janda (1996) found that an iron-scavenging mechanism such as induction of high molecular mass proteins (72–83 kDa) could be a virulence factor for *C. koseri*.

Although *C. freundii* was once considered an enteropathogen, it seems rather to be a normal inhabitant of the intestine (Sakazaki et al., 1960), but the role in diarrhea is dubious (Sedlák, 1973, Lipsky et al., 1980). In most instances the organism is probably a normal intestinal inhabitant, but some strains produce enterotoxins like many other enterobacteria, and may then act as intestinal pathogens (Guarino et al., 1987, 1989). Septicemia and other disseminated infections, e.g., osteomyelitis, occur with low frequency (Frederiksen and Søgaard, 1992).

C. rodentium has been found only in rodents; it was shown to be the cause of transmissible murine colonic hyperplasia, a disease of laboratory mice (Schauer and Falkow, 1993; Schauer et al., 1995).

Ecology As intestinal inhabitants of humans and animals, citrobacters are excreted into the environment and found in sewage, water, and soil. Apart from this, nothing is known about the ecological role of *Citrobacter* species in the environment.

Phylogenetic position On the basis of their 16S rDNA sequences, citrobacteria group within the *Gammaproteobacteria*, with *Escherichia*, *Erwinia*, *Salmonella*, and *Serratia* species as their nearest relatives.

ENRICHMENT AND ISOLATION PROCEDURES

The majority of *Citrobacter* strains can grow in liquid enrichment media such as selenite broth and tetrathionate broth and on selective isolation media such as *Salmonella–Shigella* agar, deoxycholate–citrate agar, brilliant green agar, and bismuth sulfite agar. Colonies that ferment lactose slowly often resemble *Salmonella* colonies.

MAINTENANCE PROCEDURES

Stock cultures of *Citrobacter* strains may be maintained at room temperature in a semisolid medium containing 1.0% Bacto-casitone (Difco), 0.3% yeast extract, 0.5% NaCl, and 0.3% agar, pH 7.0. The cultures remain viable for up to a year without subculturing if they are sealed with a rubber stopper or a cork that has been soaked in hot paraffin wax. Strains may be preserved indefinitely by lyophilization. They can also be maintained for years as stab cultures in sealed tubes of meat extract agar.

DIFFERENTIATION OF THE GENUS *CITROBACTER* FROM OTHER GENERA

The differentiation of *Salmonella* from lactose-negative *Citrobacter* proved difficult ever since the "Bethesda–Ballerup" group was recognized as belonging in the genus *Citrobacter*. The most useful tests to distinguish among these lactose-negative strains are lysine decarboxylase and growth in KCN medium. The greatest difficulties arise with the rare lactose/o-nitrophenyl-β-galactosidase (ONPG) positive *Salmonella* strains, the ONPG-negative *Citrobacter* strains (also rare), and, in addition, when H_2S positive *E. coli* are encountered. Table BXII.γ.214 gives some tests useful for differentiation and correct identification of such strains.

TAXONOMIC COMMENTS

The genus *Citrobacter* was proposed by Werkman and Gillen (1932) for the citrate-utilizing "coli-aerogenes intermediates". The organisms were since described under several names, and *C. freundii* was called "*Escherichia freundii*" by Yale (1939a). The role of citrobacters as possible pathogens was first noticed by Kauffmann and Møller (1940), who described an organism called "*Salmonella ballerup*" that is presently classified in *C. freundii*. Later this biogroup of organisms was removed from the genus *Salmonella* and was called the Ballerup group (Bruner et al., 1949; Harhoff, 1949). Independently of the Ballerup group of organisms, Edwards et al. (1948) and Moran and Bruner (1949) studied a group of bacteria characterized by Barnes and Cherry (1946) and referred to it as the Bethesda group of bacteria. West and Edwards (1954) found that organisms of both the Bethesda and Ballerup groups were biochemically and serologically indistinguishable and combined the two groups into the Bethesda-Ballerup group. Moreover, West and Edwards (1954) and Møller (1954) called attention to the close biochemical relationship between members of the Bethesda-Ballerup group and strains of "*E. freundii*". Accordingly, Kauffmann (1954) reclassified the Bethesda-Ballerup group into "*E. freundii*", and later revived the genus *Citrobacter* for "*E. freundii*" (Kauffmann, 1956).

Frederiksen (1970) described a new species that he named *Citrobacter koseri*. Young et al. (1971) described a new genus, *Levinea*, which contained two species, *L. malonatica* and *L. amalonatica*. Ewing and Davis (1972a) published a paper on *Citrobacter diversus*, using the specific epithet "*diversum*" of Werkman and Gillen (1932) for a group of organisms that they considered to belong in *Citrobacter*. It became apparent that *C. koseri*, *L. malonatica*, and *C. diversus* were probably different names for the same species. DNA–DNA hybridization studies by Crosa et al. (1974) showed that *C. diversus* and *L. malonatica* belonged to one hybridization group and that this group and *C. freundii* and *L. amalonatica* were all related with binding ratios around 50–60%.

Thus, they all could be considered to belong in the genus *Citrobacter*, with no need for the genus *Levinea*.

Crosa et al. (1974) suggested moving *L. amalonatica* to *Citrobacter*, and Sakazaki et al. (1976), as a result of numerical taxonomy, suggested placing *L. amalonatica* in *Citrobacter* as a species separate from *C. freundii*. The name *C. amalonaticus* was formally proposed by Brenner and Farmer (1981, 1982). Macierevicz (1966) studied a group of organisms that were H_2S-negative and ornithine decarboxylase-positive and proposed the name "*Padlewskia*", without designating a specific epithet for the organisms of this genus. From the biochemical characteristics described it was probable that "*Padlewskia*" organisms and *C. amalonaticus* were identical. This was confirmed when strains from Macierevicz and from Young were compared by the author (unpublished observations).

Werkman and Gillen (1932) also proposed the species name "*Citrobacter intermedium*" (sic) for four of their strains. Frederiksen (1970) showed that the only extant strain, ATCC 6750, was a typical *C. freundii*. The name "*Citrobacter intermedius*", therefore, was not included on the Approved Lists of Bacterial Names in 1980, and has no nomenclatural standing.

For a time there were three names—*C. koseri*, *L. malonatica*, and *C. diversus*—for the same species, all of which were on the Lists of Approved names. Frederiksen (1990) requested that the name *C. diversus* (Werkman and Gillen, 1932) be placed on the list of rejected names, because it was incorrectly used by Ewing and Davis (1972a), as the organism they described differed in at least eight characteristics from the organism described by Werkman and Gillen as "*Citrobacter diversum*" (sic), and is thus a nomen dubium. This request was granted by the Judicial Commission (1993). The epithet *malonatica* is a junior synonym and should not be used.

The concept of what should be called *C. freundii* was changed when Brenner et al. (1993) published a study showing that a number of atypical *Citrobacter* strains (considered *C. freundii* or *Citrobacter* species) could be arranged in nine genomospecies apart from *C. koseri* and *C. amalonaticus*. They named five new species: *C. farmeri* for *C. amalonaticus* biovar 1, and *C. youngae*, *C. braakii*, *C. werkmanii*, and *C. sedlakii* for strains that had been considered to be *C. freundii* or atypical *C. freundii*. Only 9 of 66 strains were shown genotypically to belong in *C. freundii*. The remaining 57 strains were allocated to seven genomospecies that could all be differentiated phenotypically. Four of them were named (*C. youngae*, *C. braakii*, *C. werkmanii*, and *C. sedlakii*), whereas three (number 9, 10, and 11) were not named, as they contained only three strains each.

Genomospecies 9, containing strains isolated from rodents, was subsequently named *C. rodentium* by Schauer et al. (1995), who studied three additional strains isolated from mouse intestine that were previously called *C. freundii* biotype 4280 (Barthold et al., 1976).

Janda et al. (1994) examined 235 *Citrobacter* strains. Within what was called a *C. freundii* complex, 37% were found to be *C. freundii*, 24% to be *C. youngae*, 13% *C. braakii*, and 6% *C. werkmanii*.

The recent study by Miki et al. (1996) showed that the test strains from the West and Edwards (1954) scheme for the Bethesda-Ballerup group mainly belonged in *C. youngae* (40 strains), *C. braakii* (25 strains), *C. werkmanii* (13 strains), and genomospecies 10 (six strains). Only three strains could be allocated to *C. freundii* in the new sense. Nine strains of the O-29 group ("Ballerup", in which the Vi antigen may occur) were found to belong in *C. braakii*.

TABLE BXII.γ.214. The differentiation of H_2S positive *Citrobacter* from ONPG positive *Salmonella* and H_2S positive *Escherichia coli*[a]

Characteristic	Citrobacter	Salmonella	E. coli
Indole	d	−	+
Citrate	+	+	−
KCN (growth)	+	−	−
Lysine	−	+	+
Ornithine	d	+	d
Sucrose	d	−	d
Cellobiose	d	−	−
ONPG	+	+	+
H_2S (iron agar)	+	+	+

[a]For symbols, see standard definitions.

The definition of *C. freundii* has thus changed; the new species and *C. freundii* can be differentiated phenotypically and genotypically, but there are no good "key" tests to characterize them; H$_2$S-positive strains are found in several species, and so are indole-positive strains. The relative distribution within the *C. freundii* complex evidently depends on how the strains included had been selected: strains sent to reference laboratories, or "nonselected" strains from clinical microbiology laboratories (some of the strains used by Janda et al., 1994). There is evidently a need for a thorough characterization of a large unselected collection of *Citrobacter* strains, both phenotypically and genotypically, in order to evaluate the relative importance of the species of the genus as now defined by Brenner and his co-workers.

DIFFERENTIATION OF THE SPECIES OF THE GENUS *CITROBACTER*

The differential characteristics of the species of *Citrobacter* are indicated in Table BXII.γ.215. Table BXII.γ.216 lists other characteristics of the species. See also Tables BXII.γ.193, BXII.γ.194, BXII.γ.195, and BXII.γ.196 in the chapter on *Enterobacteriaceae*.

List of species of the genus Citrobacter

1. **Citrobacter freundii** (Braak 1928) Werkman and Gillen 1932, 173[AL] (*Bacterium freundii* Braak 1928, 140.)
freun'di.i. M.L. gen. n. *freundii* of Freund, named after A. Freund, the bacteriologist who first observed that trimethylene glycol was a product of fermentation.

The morphology is as given for the genus. Usually motile. Usually not encapsulated, although encapsulated strains may occur.

The colony morphology is similar to that of *Escherichia coli.*

Physiological and biochemical characteristics are presented in Tables BXII.γ.215 and BXII.γ.216. Ewing (1986a) listed only 2.1% of strains as indole positive, but among the nine strains included in the study of Brenner et al. (1993), 38% were positive. Janda et al. (1994) found two indole-positive strains among 60 strains allocated to *C. freundii* sensu Brenner et al. (1993). The urease test is positive for most strains if Christensen's method is used, but usually negative if tested for the preformed enzyme.

Found in humans and animals including mammals, birds, reptiles, amphibians, and fish. Also found in soil, water, sewage, and food. Often found in clinical specimens such as urine, throat, sputum, blood, and wound swabs as an opportunistic or secondary pathogen.

The mol% G + C of the DNA is: 50–51 (T_m).

Type strain: ATCC 8090, DSM 30039, IFO 12681, NCTC 9750.

GenBank accession number (16S rRNA): AJ233408.

2. **Citrobacter amalonaticus** (Young, Kenton, Hobbs and Moody 1971) Brenner and Farmer 1982, 266[VP] (Effective publication: Brenner and Farmer 1981, 1140) (*Levinea amalonatica* Young, Kenton, Hobbs and Moody 1971, 58.)
a.ma.lo.na'ti.cus. Gr. prefix *a* not; M.L. adj. *malonaticus* pertaining to malonate; M.L. adj. *amalonaticus* not pertaining to malonate (i.e., not able to utilize malonate).

The morphology is as given for the genus. Motile. Not encapsulated.

Colonies on nutrient agar are translucent to opaque, resembling those of *E. coli.*

Physiological and biochemical characteristics are indicated in Tables BXII.γ.215 and BXII.γ.216. Some strains are late gelatin liquefiers.

Found in the feces of humans and animals and in soil,

TABLE BXII.γ.215. Conventional tests useful in differentiating *Citrobacter* species (adapted from Brenner et al., 1993)[a]

Characteristic	1. *C. freundii*	2. *C. amalonaticus*	3. *C. braakii*	4. *C. farmeri*	5. *C. gillenii*	6. *C. koseri*	7. *C. murliniae*	8. *C. rodentium*	9. *C. sedlakii*	10. *C. werkmanii*	11. *C. youngae*
Indole	d[b]	+	d	+	−	+	+	−	+	−	d
Citrate—Simmons	d	+	d[c]	d[c]	d[c]	+	+	−[c]	d[c]	+	d[c]
H$_2$S in iron agar	d[b]	d[d]	d	−	d	−	d	−	−	+	d
Ornithine	−	+	+	+	−	+	−	+	+	−	−
KCN growth	d[c]	+	+	+	+	−	+	−	+	+	+
Malonate	d	d[d]	−	−	+	+	−	+	+	+	−
Acid from:											
Sucrose	+	d	−	+	d	d	d	−	−	−	d
Melibiose	+	−	d[c]	+	d	−	d	−	+	−	−
Raffinose	d[c]	−	−	+	−	−	d	−	−	−	−
Dulcitol	d	−	d	−	−	d	+	−	+	−	d
Adonitol	−	−	−	−	−	+	−	−	−	−	−
D-Arabitol	−	−	−	−	−	+	−	−	−	−	−

[a]For symbols see standard definitons.

[b]Previous reports (e.g., Ewing, 1986a), on many strains had indole, − (2.1%); and H$_2$S, + (93.1%).

[c]Late positive reactions may occur in some tests in addition, that may change − to d and d to +.

[d]Previous reports (e.g., Young et al., 1971) had H$_2$S, −; malonate, −.

TABLE BXII.γ.216. Carbon source utilization reactions of *Citrobacter* species (% positive reactions at 48 h) (adapted from Brenner et al., 1993)[a]

Carbon source	1. C. freundii	2. C. amalonaticus	3. C. braakii	4. C. farmeri	5. C. gillenii	6. C. koseri	7. C. murliniae	8. C. rodentium	9. C. sedlakii	10. C. werkmanii	11. C. youngae
cis-Aconitate	88	93	89	57	33	100	67	0	100	100	78
trans-Aconitate	13	0	0	57	0	25	0	0	50	83	0
Adonitol	0	0	0	0	0	100	0	0	0	0	0
D-Alanine	100	100	100	71	67	100	100	100	100	100	100
4-Aminobutyrate	25	0	17	7	0	0	33	0	50	0	22
5-Aminovalerate	38	33	50	7	0	6	67	0	33	100	43
D-Arabitol	0	0	0	0	0	100	0	0	0	0	0
Benzoate	0	73	0	93	0	0	0	0	100	0	0
Caprate	0	0	0	0	0	0	0	0	0	0	0
D-Cellobiose	88	100	94	100	67	94	100	100	100	67	78
m-Coumarate	88	0	100	0	33	0	0	0	67	100	96
Dulcitol	13	0	33	0	0	44	100	0	100	0	87
Esculin	0	0	6	0	0	0	0	0	17	0	0
Ethanolamine	0	33	0	14	0	19	0	0	17	17	9
L-Fucose	100	100	94	100	67	100	100	33	100	100	91
Gentiobiose	88	100	89	93	67	88	100	0	100	67	52
Gentisate	100	100	94	93	0	100	67	67	100	100	0
L-Glutamate	75	100	94	86	0	100	100	67	83	83	83
Glycerol	100	100	100	100	100	100	100	0	100	100	100
3-Hydroxybenzoate	100	100	100	100	0	100	67	100	100	100	0
4-Hydroxybenzoate	0	100	0	100	0	0	0	0	100	0	0
3-Hydroxybutyrate	100	20	44	7	0	6	33	0	100	67	74
myo-Inositol	100	0	6	0	67	100	0	0	100	0	0
2-Ketogluconate	100	100	100	100	100	100	100	0	100	100	100
5-Ketogluconate	100	100	100	100	100	100	100	0	0	50	96
2-Ketoglutarate	13	7	22	7	33	6	0	0	33	33	0
DL-Lactate	100	100	100	100	100	100	100	33	100	100	100
Lactose	88	20	78	50	67	56	33	100	100	17	22
Lactulose	88	0	67	7	67	0	33	0	100	17	0
D-Lyxose	63	0	56	0	0	100	100	0	0	83	9
Malonate	0	0	0	0	0	81	0	67	67	50	0
Maltitol	25	7	44	93	0	100	0	0	0	0	0
D-Melibiose	88	7	94	100	100	0	33	0	100	0	0
1-O-CH₃-α-galactoside	88	0	100	100	67	0	0	0	100	0	0
1-O-CH₃-β-galactoside	100	7	83	14	67	100	100	100	100	100	65
3-O-CH₃-D-glucose	63	87	94	86	0	13	67	0	100	100	0
1-O-CH₃-α-D-glucoside	25	13	39	86	0	94	0	0	0	0	0
1-O-CH₃-β-D-glucoside	100	100	100	100	100	100	100	0	100	100	100
Palatinose	25	13	67	100	0	100	67	0	0	0	4
Phenylacetate	25	0	0	0	0	0	0	0	0	0	0
3-Phenylpropionate	75	0	83	0	0	0	0	0	0	100	96
L-Proline	100	87	100	57	33	100	100	100	100	100	87
Propionate	88	80	72	86	0	88	100	67	83	100	78
Protocatechuate	0	100	0	100	0	0	0	0	100	0	0
Putrescine	50	0	39	0	0	0	0	100	0	67	0
D-Raffinose	75	0	6	100	67	0	33	0	0	0	0
L-Sorbose	100	87	6	100	0	100	100	0	0	83	100
Sucrose	100	0	6	100	33	44	33	0	0	0	9
D-Tagatose	13	0	0	36	0	0	0	0	0	0	4
D-Tartrate	13	0	6	0	0	0	0	0	0	100	0
L-Tartrate	0	33	17	36	0	19	33	100	17	67	22
meso-Tartrate	50	40	72	7	0	0	67	100	67	100	78
Tricarballylate	100	93	89	100	0	0	100	100	83	100	4
D-Turanose	0	0	11	36	0	6	0	0	0	0	0
L-Tyrosine	75	0	72	0	0	88	100	0	0	67	74
Xylitol	0	0	0	0	0	19	0	0	0	0	0

[a]All strains (with few single exceptions) utilized the following carbon sources: N-acetyl-D-glucosamine, L-alanine, L-arabinose, L-aspartate, citrate, D-fructose, fumarate, D-galactose, D-galacturonate, D-gluconate, D-glucosamine, D-glucose, D-glucuronate, DL-glycerate, D-malate, L-malate, maltose, maltotriose, D-mannitol, D-mannose, mucate, L-rhamnose, D-ribose, D-saccharate, L-serine, D-sorbitol, succinate D-trehalose, and D-xylose. All strains (with two single exceptions) failed to utilize the following carbon sources within 96 h: L-arabitol, betaine, caprate, caprylate, i-erythritol, glutarate, histamine, L-histidine, HQ-beta-glucuronide, itaconate, D-melezitose, quinate, trigonelline, tryptamine, and tryptophan.

water, and sewage. Also found in a variety of human clinical specimens as an opportunistic pathogen.

The mol% G + C of the DNA is: 51–52 (T_m).

Type strain: ATCC 25405, NCTC 10805.

3. **Citrobacter braakii** Brenner, Grimont, Steigerwalt, Fanning, Ageron, and Riddle 1993, 657[VP]

braak′ i.i. M.L. gen. n. braakii of Braak; named after Hendrik R. Braak, a Dutch microbiologist.

Morphology and cultural characteristics as given for the genus.

Physiological and biochemical characteristics are presented in Tables BXII.γ.215 and BXII.γ.216.

Found in human stools and isolated from animals.

The mol% G + C of the DNA is: unknown.
Type strain: ATCC 51113, CDC 80-58.
GenBank accession number (16S rRNA): AF025368.

4. **Citrobacter farmeri** Brenner, Grimont, Steigerwalt, Fanning, Ageron, and Riddle 1993, 654[VP]
far' mer.i. M.L. gen. n. *farmeri* of Farmer; named after J.J. Farmer III, an American bacteriologist.

Morphology and cultural characteristics as given for the genus.

Physiological and biochemical characteristics are presented in Tables BXII.γ.215 and BXII.γ.216.

Found in human stools, urine, wounds, and blood.

The mol% G + C of the DNA is: unknown.

Type strain: ATCC 51112, CDC 2991-81.

GenBank accession number (16S rRNA): AF025371.

5. **Citrobacter gillenii** Brenner, O'Hara, Grimont, Janda, Falsen, Aldová, Ageron, Schindler, Abbott and Steigerwalt 2000, 423[VP] (Effective publication: Brenner, O'Hara, Grimont, Janda, Falsen, Aldová, Ageron, Schindler, Abbott and Steigerwalt 1999, 2623.)
gil.len'.ii. N.L. gen. n. *gillenii* of Gillen; named after George Francis Gillen, an American microbiologist.

Morphology and cultural characteristics as given for the genus. Physiological and biochemical characteristics are present in Tables BXII.γ.215 and BXII.γ.216. Found in human stool, human urine, human blood, and environment.

The mol% G + C of the DNA is: unknown.

Type strain: ATCC 51117, CCUG 30796, CDC 4693–86, CIC 106783.

6. **Citrobacter koseri** Frederiksen 1970, 93[AL]
ko.ser.i. M.L. gen. n. *koseri* of Koser; named after Stewart A. Koser, an American bacteriologist.

The morphology is as given for the genus. Motile. Not encapsulated.

Colonies on nutrient agar are translucent to opaque, resembling those of *E. coli*. Cultures often have a distinct fecal odor.

Physiological and biochemical characteristics are presented in Tables BXII.γ.215 and BXII.γ.216.

Found in the feces of humans and animals and in soil, water, sewage, and food. Also isolated from human clinical specimens such as urine, throat, nose, and sputum and wound swabs. Causes neonatal meningitis, often complicated with cerebral abscesses, and often fatal. Some cases occur in small outbreaks.

The mol% G + C of the DNA is: 51–52 (T_m).

Type strain: ATCC 27028.

7. **Citrobacter murliniae** Brenner, O'Hara, Grimont, Janda, Falsen, Aldová, Ageron, Schindler, Abbott and Steigerwalt 2000, 423[VP] (Effective publication: Brenner, O'Hara, Grimont, Janda, Falsen, Aldová, Ageron, Schindler, Abbott and Steigerwalt 1999, 2623.)
mur.lin'.i.ae. N.L. gen. n. *murliniae* of Murlin; named after Alma C. McWhorter-Murlin, an American microbiologist.

Morphology and cultural characteristics as given for the genus. Physiological and biochemical characteristics are presented in Tables BXII.γ.215 and BXII.γ.216. Found in human stool, human wound, human blood, human urine, and food.

The mol% G + C of the DNA is: unknown.

Type strain: ATCC 51118, CCUG 30797, CDC 2970–59, CIC 104556.

8. **Citrobacter rodentium** Schauer, Zabel, Pedraza, O'Hara, Steigerwalt, and Brenner 1996, 362[VP] (Effective publication: Schauer, Zabel, Pedraza, O'Hara, Steigerwalt, and Brenner 1995, 2067.)
ro.den' ti.um. L. part. adj. used as a gen. n. *rodentium* of rodents (gnawing animals).

Morphology and cultural characteristics as given for the genus; however, four strains were nonmotile, and two showed motility only after 4 days.

Physiological and biochemical characteristics are presented in Tables BXII.γ.215 and BXII.γ.216. Citrate utilization was late or absent.

Found in rodents only; includes strains known to produce transmissible murine colonic hyperplasia, associated with the presence of an *eaeA* gene in the isolate.

The mol% G + C of the DNA is: unknown.

Type strain: ATCC 51116, CDC 1843-73.

GenBank accession number (16S rRNA): AF025363.

9. **Citrobacter sedlakii** Brenner, Grimont, Steigerwalt, Fanning, Ageron, and Riddle 1993, 657[VP]
sed.lak' i.i. M.L. gen. n. *sedlakii* of Sedlak; named after Jiri Sedlák, a Czechoslovakian bacteriologist.

Morphology and cultural characteristics as given for the genus.

Physiological and biochemical characteristics are presented in Tables BXII.γ.215 and BXII.γ.216.

Found in human stools, blood, and wounds.

The mol% G + C of the DNA is: unknown.

Type strain: ATCC 51115, CDC 4696-86.

GenBank accession number (16S rRNA): AF025364.

10. **Citrobacter werkmanii** Brenner, Grimont, Steigerwalt, Fanning, Ageron, and Riddle 1993, 657[VP]
werk' mani.i. M.L. gen. n. *werkmanii* of Werkman; named after Chester H. Werkman, an American bacteriologist.

Morphology and cultural characteristics as given for the genus.

Physiological and biochemical characteristics are presented in Tables BXII.γ.215 and BXII.γ.216.

Found in human stools and blood, and in soil.

The mol% G + C of the DNA is: unknown.

Type strain: ATCC 51114, CDC 876-58.

GenBank accession number (16S rRNA): AF025373.

11. **Citrobacter youngae** Brenner, Grimont, Steigerwalt, Fanning, Ageron, and Riddle 1993, 654[VP]
young' ae. M.L. gen. n. *youngae* of Young; named after Viola M. Young, an American bacteriologist.

Morphology and cultural characteristics as given for the genus.

Physiological and biochemical characteristics are presented in Tables BXII.γ.215 and BXII.γ.216.

Found in human stools, urine, and wounds, and isolated from animals and food.

The mol% G + C of the DNA is: unknown.

Type strain: ATCC 29935, CDC 460-61.

Genus XI. **Edwardsiella** *Ewing and McWhorter 1965, 37*[AL]

RIICHI SAKAZAKI

Ed.ward.si.el'la. M.L. dim. ending *-ella*; M.L. fem. n. *Edwardsiella* named after the American bacteriologist P.R. Edwards (1901–1966).

Straight rods, ~1.0 × 2.0–3.0 μm, conforming to the general definition of the family *Enterobacteriaceae*. Gram negative. **Motile by peritrichous flagella**, but nonmotile strains may occur. Facultatively anaerobic. Growth occurs on ordinary media with small colonies (0.5–1.0 mm in diameter) after 24 h incubation. Optimum temperature, 37°C, except for *E. ictaluri*, which prefers a lower temperature. **Nicotinamide and amino acids are required for growth.** Reduce nitrate to nitrite. Ferment D-glucose with the production of acid and often gas. Also **ferment a few other carbohydrates but are inactive compared to many taxa in the family *Enterobacteriaceae*. Usually resistant to colistin but susceptible to most other antibiotics, including penicillin.** Frequently isolated from river fish and cold-blooded animals and their environment, particularly fresh water. Pathogenic for eels, catfish, and other animals, sometimes causing economic losses; also opportunistic pathogen and possibly rare cause of gastroenteritis for humans. A member of the *Gammaproteobacteria*.

The mol% G + C of the DNA is: 53–59.

Type species: **Edwardsiella tarda** Ewing and McWhorter *in* Ewing, McWhorter, Escobar and Lubin 1965, 37.

FURTHER DESCRIPTIVE INFORMATION

Most of studies on *Edwardsiella* concentrated on *E. tarda*; little information is available on other species. Encapsulated strains possessing significant glycocalixes have not been found in members of *Edwardsiella*, but some strains may produce slime substance (Wong et al., 1989). They are afimbriate. Wong et al. (1989) and Janda et al. (1991b) reported that approximately half of strains of *E. tarda*, but not *E. hoshinae* and *E. ictaluri*, produce afimbrial mannose-resistant hemagglutination (MRHA) against guinea pig erythrocytes. Aoki and Holland (1985), who analyzed outer membrane composition of *E. tarda*, found a large number of prominent protein bands in all strains, with major bands present at 27, 35, and 46 kDa, and a cluster of three or four outer membrane proteins located in each strain at 52–56 kDa. Studies of the lipopolysaccharide of three species of *Edwardsiella* indicated that each species may be divided into different chemoforms (Nomura and Aoki, 1985). Since strains of *E. tarda* require nicotinamide and amino acids including cysteine and methionine (d'Empaire, 1969) for their growth and other species of the genus probably have similar requirements, strains of *Edwardsiella* grow less luxuriantly than other members of the *Enterobacteriaceae* and form smaller colonies on ordinary agar plates within 24 h of incubation at 37°C. In contrast with other two *Edwardsiella* species, the optimum growth temperature of *E. ictaluri* is between 25 and 30°C and growth is very slow on plating media, often requiring 2–3 d incubation at 30°C to form typical colonies 1 mm in diameter. Biochemically, *E. ictaluri* is also the least active species of *Edwardsiella*.

Plasmids ranging from 2–120 MDa are harbored in many strains of *Edwardsiella* (Lobb and Rhoades, 1987; Janda et al., 1991a). The presence of R plasmids that mediate antibiotic resistance was reported in *E. tarda* by Aoki et al. (1977). Lobb and Rhoades (1987) recognized that catfish isolates of *E. ictaluri* always harbored two plasmids of 5.7 and 4.9 kb (pCL1 and pCL2, respectively). They suggested that plasmid function may be met-

abolically important to the host bacterium, or that these plasmids may code for factors important to the virulence, or both. Moreover, Lobb et al. (1993) found that green knife fish isolates of the organisms were serologically distinct from those of catfish and harbored four plasmids with relative mobilities of 6.0, 5.7, 4.1, and 3.1 kb, of which 5.7- and 4.1-kb plasmids strongly hybridized to probes specific for pCL1 and pCL2. Hamon et al. (1969) demonstrated bacteriocin production and sensitivity within *E. tarda*.

For *E. tarda*, two independent serotyping schemes were reported by Sakazaki (1967) and Edwards and Ewing (1972). Currently, an international serotyping scheme combining the two schemes mentioned above and comprising 61 O groups and 45 H antigens has been established by Tamura et al. (1988). Lobb et al. (1993) suggested the possibility of serotyping of *E. ictaluri* as an epidemiological tool. Other typing schemes such as bacteriocin typing, bacteriophage typing, and biotyping have not been studied for *Edwardsiella*.

Strains of *Edwardsiella* are primarily resistant to colistin and polymyxin B, but are usually susceptible to other antimicrobial agents including penicillin G, ampicillin, carbenicillin, streptomycin, kanamycin, gentamicin, tetracycline, chloramphenicol, cephalosporins, sulfadiazine, and, with few exceptions, nalidixic acid (Muyembe et al., 1973; Reinhardt et al., 1985). However, Waltman et al. (1986) reported more resistance to sulfa compounds, streptomycin, and methicillin by disk diffusion testing. These differences may be linked to the source of strains tested. All three species may show large zones around penicillin-impregnated disks. This is an unusual finding for members of the *Enterobacteriaceae*. Clark et al. (1991b) found that strains of *E. tarda*, most originating from human sources, have uniform susceptibility to 22 antibiotics. Although they produced β-lactamase, all strains showed susceptibility to β-lactamase inhibitors. Members of *Edwardsiella* are apparently more susceptible to 2,4-diamino-6,7-diisopropyl pteridine (vibriostatic agent O/129, Sigma) than other *Enterobacteriaceae* (Chatelain et al., 1979; Grimont et al., 1980).

There have been no definitive studies indicating *E. tarda* as a potential cause of human intestinal diseases, but a number of papers from tropical countries such as Madagascar (Fourquet et al., 1975), Zaire (Makulu et al., 1973), Tahiti (Fourquet et al., 1975), Dominica (de Inchaustegui et al., 1976), Panama (Kourany et al., 1977), Cuba (Rakovsky and Aldová, 1965), Australia (Iveson, 1973), India (Bhat et al., 1967; Sakazaki et al., 1971), Thailand (Bockemühl et al., 1971; Ovartlarnporn et al., 1986), Viet-Nam (Nguyen-Van-Ai et al., 1975), Philippines (Tocal and Mezez, 1968), Malaysia (Gilman et al., 1971), and Singapore (Tan et al., 1977) have reported an etiologic relationship of this species to diarrheal diseases. On the other hand, *E. tarda* is also found in the feces of healthy people, although the rate of the isolation is extremely low. Onogawa et al. (1976) found only one positive culture from 97,704 food handlers and 25 positive cultures from 255,896 schoolchildren. Makulu et al. (1973) found no *E. tarda* cultures among 841 healthy subjects in Zaire. Iveson (1973) suggested that the number of isolations of *E. tarda* depends upon culture methods for stool specimens, the geographic area of the study, and the season in which the survey is performed. In those people from whose stool specimens of *E. tarda* have been isolated, the percentage of diarrheal patients has varied from 25% (Kourany et al., 1977) to more than 75% (Bockemühl et al., 1971; Makulu et al., 1973). A higher isolation rate has invariably been recognized among patients with diarrhea than among asymptomatic people (Bhat et al., 1967; Gilman et al., 1971; Makulu et al., 1973; Nguyen-Van-Ai et al., 1975). In tropical countries there is a tendency for *E. tarda* to be isolated from diarrheal stools together with well-established enteropathogens such as *Salmonella*, *Shigella*, *Vibrio cholerae*, *Vibrio parahaemolyticus*, and intestinal parasites (Bhat et al., 1967; Bockemühl et al., 1971; Sakazaki et al., 1971; Makulu et al., 1973; Kourany et al., 1977). The incidence is rare in industrialized countries, but intestinal infection with *E. tarda* was also reported by King and Adler (1964), Chatty and Gavan (1968), and Desenclos et al. (1990) in the United States, and by Vandepitte et al. (1983) in Belgium.

Although *E. tarda* may be able to cause diarrhea, the organism should not be considered as an "inherent" pathogen such as *Salmonella* and *Shigella*. Marques et al. (1984) demonstrated the invasiveness of *E. tarda* in HeLa cells. Janda et al. (1991a) confirmed this finding with HEp-2 cells. However, invasive strains of *E. tarda* give constantly negative results in the Sereny test, suggesting that they have no KepA-like locus as is found in *Shigella* (Ullah and Arai, 1983b; Marques et al., 1984). Janda and Abbott, (1993a) suggested that HEp-2 invasion by *E. tarda* was a microfilament-dependent process. Furthermore, they suggested the cell-associated hemolysin, which was originally demonstrated by Watson and White (1979), as a second major virulence marker. The hemolysin is not released in sufficient amounts under usual growth conditions. Activity of the hemolysin is enhanced by Ca^{2+} and Mg^{2+} and may play a role in infection through epithelial cell destruction leading to an inflammatory infiltrate in the intestinal mucosa, or it could destroy villus cells or disturb intestinal absorption function resulting in diarrhea. In addition to these, Ullah and Arai (1983a) and Janda et al. (1991b) suggested that mannose-resistant hemagglutinins, serum resistance, a dermatotoxin, and chondroitinase activity may be related to human pathogenicity of *E. tarda*. These associations are independent of serovars or source of strains. Whether intestinal infection with *E. tarda* in humans is linked to strains possessing these virulence factors remains to be determined. Bockemühl et al. (1983) reported that some *E. tarda* strains produced a heat-stable enterotoxin (ST) similar to that produced by enterotoxigenic *Escherichia coli*. However, the enterotoxigenic activity was recognized only in culture supernatants concentrated by ultrafiltration. Janda et al. (1991a) also tested *E. tarda* for ST-like activity, but all strains were negative.

Extraintestinal infections with *E. tarda* are well documented, although these occur only rarely in most industrialized countries. It causes meningitis, endocarditis, bacteremia, osteomyelitis, urinary tract infections, or wound infection (Gonzalez and Ruffolo, 1966; Field et al., 1967; Chatty and Gavan, 1968; Okubadejo and Alausa, 1968; Sonnenwirth and Kallus, 1968; Jordan and Hadley, 1969; Pankey and Seshul, 1969; Bockemühl et al., 1971; Sachs et al., 1974; Koshi and Lalitha, 1976; Le Frock et al., 1976; Clarridge et al., 1980; Rao et al., 1981; Sechter et al., 1983; Maserati et al., 1985; Martinez, 1987; Vohora and Torrijos, 1988; Wilson et al., 1989a; Vartian and Septimus, 1990; Hargreaves and Lucey, 1990; Coutlee et al., 1992; Walton et al., 1993; Zighelboim et al., 1992). In most cases of extraintestinal infection, *E. tarda* has been an opportunist in persons with underlying diseases or in conditions

with predisposing factors. Infants and the aged tend to be most susceptible to infections with *E. tarda*. *E. tarda* appears to cause serious infection in patients with preexisting liver disease or in those with altered levels of available iron. In cases of wound infection, patients who tend to develop severe infection often have a recent history of fishing, swimming, or injury caused by marine-related items. In extraintestinal infections with *E. tarda*, cell-associated hemolysin is also suggested as a virulence factor, in which a possible function of this hemolysin could be the acquisition of host iron through lysis of erythrocyte to release hemoglobin stores.

E. tarda has been infrequently isolated from warm-blooded animals such as dogs, pigs, cattle, monkeys, rats, panthers, skunks, seals, sea lions, and birds (Ewing et al., 1965; Wallace et al., 1966; Arambulo et al., 1967; Chamoiseau, 1967; Sakazaki, 1967; Tocal and Mezez, 1968; d'Empaire, 1969; Otis and Behler, 1973; White et al., 1973; Owens et al., 1974; Nguyen-Van-Ai et al., 1975; Kourany et al., 1977; Coles et al., 1978). *E. tarda* was associated with diseases in some of those animals, but its pathogenic role in warm-blooded animals is unknown.

Two ecological groups of hosts, cold-blooded animals and fish, could be considered as reservoirs of *E. tarda*. Since Sakazaki and Murata (1962) first suggested *E. tarda* as a normal intestinal inhabitant of snakes, it has been recognized that a wide range of reptiles and amphibians including snakes, crocodiles, alligators, toads, lizards, frogs, and turtles, are possible natural reservoirs for *E. tarda* (Wallace et al., 1966; d'Empaire, 1969; Jackson et al., 1969; White et al., 1969, 1973; Iveson, 1971; Makulu et al., 1973; Meyer and Bullock, 1973; Otis and Behler, 1973; Sharma et al., 1974; Van Der Waaiji et al., 1974; de Inchaustegui et al., 1976; Roggendorf and Müller, 1976; Bartlett et al., 1977; Kourany et al., 1977; Wyatt et al., 1979; Tan et al., 1978). On the other hand, Van Damme and Vandepitte (1980) reported the isolation of *E. tarda* from various kinds of river fish in Zaire. It is considered that river fish and their environment seem to constitute the natural habitat of *E. tarda* and to be the most possible source of human infection, at least in tropical countries. Vandepitte et al. (1983) described protracted diarrhea in an infant associated with *E. tarda*, which was possibly introduced from a tropical aquarium fish in the home of the patient. Trust and Bartlett (1974) reported the presence of *E. tarda* in water samples from aquariums containing goldfish or tropical ornamental fish. Wyatt et al. (1979) recognized *E. tarda* in freshwater catfish and in their environment. Human wound infection with this organism resulting from swimming or diving accidents was reported by Chatty and Gavan (1968) and Clarridge et al. (1980).

E. tarda can cause outbreaks of "red disease" in pond-cultured eels (Hoshina, 1962; Wakabayashi and Egusa, 1973) or "emphysematous putrifactive disease" in channel catfish (Meyer and Bullock, 1973). Kusuda et al. (1977) reported *E. tarda* as causing an outbreak in cultured sea beam. Amandi et al. (1982) showed that *E. tarda* is often isolated from salmonid fish and other fish species in the Pacific Northwest of the United States, and that it is pathogenic for chinook salmon, steelhead and rainbow trout, as well as channel catfish.

E. hoshinae is associated with animals but only eight isolates were originally reported (Grimont et al., 1980). From their paper, the presently known habitats of *E. hoshinae* include reptiles, birds, and water. Although two human isolates of this species were included in their report, those came from feces of individuals without diarrhea; thus there is no evidence that *E. hoshinae* can cause human disease. Another species of *Edwardsiella*, *E. ic-*

taluri, is the causative agent responsible for acute septicemia of catfish. Since the first publication by Hawke et al. (1981), this disease has become a significant problem in the aquaculture of channel catfish in the southeastern United States (Waltman et al., 1986). Shotts et al. (1986) suggested that pathogenesis of *E. ictaluri* may proceed through gastrointestinal tract infection. Enteric septicemia caused by *E. ictaluri* occurs when the temperature ranges from 22–28°C and progresses rapidly, resulting in significant mortality. The isolation of *E. ictaluri* from nonictalurid fish, such as the tropical danio or green knifefish, was also documented (Waltman et al., 1985). Janda et al. (1991b) reported that *E. ictaluri* lacked the invasiveness and cell-associated hemolysin recognized in *E. tarda*, and the lower frequency of these two activities in *E. hoshinae*.

ENRICHMENT AND ISOLATION PROCEDURES

E. tarda and *E. hoshinae* can grow on plating agar media used in enterobacteriology, such as MacConkey, xylose-lysine-deoxycholate (XLD), Hektoen enteric, deoxycholate-citrate, and Salmonella–Shigella agars. Like *Salmonella*, strains of *E. tarda* produce distinguishable black colonies on XLD and Hektoen enteric agars because of H_2S production. As they require some growth factors, their growth may be slower on those media than other enteric bacteria.

Makulu et al. (1973) and Wyatt et al. (1979) reported that tetrathionate or selenite broths as enrichment media may be successful for the isolation of *E. tarda*. Iveson (1971, 1973), who carried out a comparative study to evaluate enrichment media to isolate *E. tarda* from the cloacal contents of tiger snakes and human feces, showed that strontium chloride B broth[1] was the most satisfactory for isolating *E. tarda*, as well as *Salmonella*. After 24 h of incubation at 37°C or 43°C, plates of deoxycholate citrate agar are streaked from the enrichment culture.

On the other hand, Wyatt et al. (1979) evaluated several media for the isolation of *E. tarda* from catfish and the environmental water and found that the most effective isolation of *E. tarda* was obtained with selective enrichment in double-strength SS broth[2] followed by plating on single-strength SS agar. They pointed out that salmonellae were not isolated with this method. A 1-ml amount of lactose broth pre-enrichment is transferred to the double-strength SS broth. After incubation at 35°C for 24 h, cultures are streaked on an SS agar plate.

Muyembe et al. (1973) pointed out that resistance of *E. tarda* to 10 μg/ml colistin can be used in isolation. On the isolation agar plate containing 10 μg/ml colistin, most enteric bacteria are inhibited, with the exception of *Serratia*, *Proteus*, *Providencia*, *Morganella*, *Cedecea*, and some strains of *Yersinia* that are colistin-resistant. Little information is available on the isolation of *E. ictaluri*. Agar medium containing colistin is also effective for the isolation of *E. ictaluri*. The optimum growth temperature of *E. ictaluri* is between 25 and 30°C and growth is slow, often requiring 2–3 days incubation to form colonies 1 mm in diameter.

---

1. Strontium chloride B broth (g/l): Bacto-tryptone (Difco), 5.0; NaCl, 8.0; K_2HPO_4, 1.0; and $SrCl_2$, 34.0. Sterilization is done by heating at 100°C for 30 min (final pH, 5.0–5.5).

2. Double-strength SS broth (g/l): SS agar, dehydrated, 120, and glucose 5. The mixture is stirred without heat to dissolve the ingredients except agar. The agar is removed by filtration through a Whatman no. 1 filter. The broth is heated at 100°C for 5 min and dispensed to 10-ml quantities into a tube (18 × 180 mm). The tubes are steamed for 30 min.

MAINTENANCE PROCEDURES

Strains of *Edwardsiella* usually survive well at least for 5 years without subculture when the culture is stabbed into a tube with semisolid medium. The tube is tightly sealed with a rubber stopper or with paraffin-coated corks and kept at room temperature in the dark without transfer. A semisolid agar containing 1% Bacto-casitone (Difco), 0.3% yeast extract, 0.5% NaCl, and 0.3% agar (pH 7.2) is an excellent choice for this purpose. However, important cultures should also be preserved at −70°C. Freeze-drying can also be used for long-term preservation.

PROCEDURES FOR TESTING SPECIAL CHARACTERS

L-Pyrrolidonyl aminopeptidase test Dissolve 10 mg of L-pyrrolidonyl-β-naphthylamide (Sigma) in 10 ml of 95% ethanol. Dip sterile cotton swabs into the solution and rotate each swab several times with firm pressure on the inside wall of the tube to remove excess fluid. Dry the swabs overnight at 37°C and store at −20°C. For testing, make a heavy suspension (MacFarland no. 1) of the test culture into 0.2–0.3 ml of 0.1 M phosphate buffer (pH 7.4), then dip the reagent swab into the suspension and incubate at 37°C for 10 min. Finally, add a drop of cinnamaldehyde solution (*p*-dimethylaminocinnamaldehyde [Sigma] 0.5 g; glacial acetic acid 3.5 ml; ethylene glycol-1-methylester 6.0 ml; sodium lauryl sulfate 3.5 g; and distilled water 90 ml; store in a screw-capped brown container in a refrigerator). A positive result is shown by a pink or red color.

β-Galactosidase and β-glucuronidase tests Dissolve 20 mg each of 4-methylumbelliferyl-β-D-galactopyranoside and 4-methylumbelliferyl-β-D-glucuronide (Sigma) in a small volume of dimethyl sulfoxide, and bring the volume to 10 ml with 0.1 M phosphate buffer (pH 7.4). Dip sterile cotton swabs into the solution and rotate each swab several times with firm pressure on the inside wall of the tube to remove excess fluid. Dry the swabs overnight at 37°C and store at −20°C. For testing, make a heavy suspension (MacFarland no. 1) of the test culture in 0.2–0.3 ml of 0.1 M phosphate buffer (pH 7.4). Dip the reagent cotton swab in the suspension, and incubate at 37°C for 30 min. A positive reaction is indicated by development of blue fluorescence under long-wave (360 μm) ultraviolet lamp.

DIFFERENTIATION OF THE GENUS *EDWARDSIELLA* FROM OTHER GENERA

Edwardsiella is biochemically somewhat similar to *E. coli*, *Salmonella*, *Citrobacter* spp., *Leminorella* spp., *Proteus*, and *Providencia*, but is easily differentiated based on activities of L-pyrrolidonyl aminopeptidase, β-galactosidase, β-glucuronidase, and phenylalanine (or tryptophan) deaminase, indole production, and maltose and D-xylose fermentation. Table BXII.γ.217 presents differential characteristics of *Edwardsiella* and other similar members of the *Enterobacteriaceae*. In addition, susceptibilities to penicillin and colistin and to the vibriostatic agent O/129 are useful for differentiation of *Edwardsiella*. Strains of *Edwardsiella* may have zones of inhibition around a penicillin-G disk and no zone around a colistin disk, whereas many groups of the *Enterobacteriaceae* usually have the opposite patterns. *Edwardsiella* may show zones around a 10-μg disk of O/129.

TAXONOMIC COMMENTS

Edwardsiella was independently reported by several investigators early in the 1960s. Sakazaki and Murata (1962) and Sakazaki (1967) reported a new group of organisms in the family *Ente-*

TABLE BXII.γ.217. Differential characteristics of the genus *Edwardsiella* and biochemically related genera[a]

Characteristic	*Edwardsiella*	*Citrobacter*	*Escherichia coli*	*Leminorella*	*Proteus* and *Providencia*	*Salmonella*
H₂S production	D	D	−	+	D	+
Indole production	D	D	+	−	D	−
L-Pyrrolidonyl aminopeptidase	−	+	−	−	−	−
β-Galactosidase	−	+	+	−	−	−
β-Glucuronidase	−	−	+	−	−	D
Phenylalanine deaminase	−	−	−	−	+	−
Maltose fermentation	+	+	+	−	D	+
D-Xylose fermentation	−	+	+	+	D	+
Mol% G + C of DNA	53–59	50–52	48–52	52–53	38–41	53–58

[a]Symbols: +, 90–100% of strains are positive; −, 90–100% of strains are negative; D, different reaction given by different species or biogroups.

robacteriaceae that were isolated from snakes in 1959 and suggested a vernacular name "Asakusa group" for this group. King and Adler (1964) described the isolation of a culture of the organism and gave it the name "Bartholomew group". Later, Ewing et al. (1965) suggested the scientific name *Edwardsiella tarda* for a group of organisms that had been known as biotype 1483-59 in their laboratory since 1959, and indicated that these organisms were similar to those of the Asakusa and Bartholomew groups. Although the genus *Edwardsiella* with the single species *E. tarda* was proposed by Ewing et al. (1965), there was some question about this designation based upon only phenotypic characterization. Brenner et al. (1974b) demonstrated that strains of *E. tarda* were highly related by DNA hybridization, regardless of their source. They also reported that *Edwardsiella tarda* was only 8–29% related to other genera in the family *Enterobacteriaceae* and was distinct from other members of the family.

Grimont et al. (1980) described a new biogroup designated biogroup 1 of *E. tarda* that was distinguished from biochemically typical strains by its acid production from sucrose, D-mannitol, and L-arabinose, but was closely related to the latter by DNA hybridization. Walton et al. (1993) reported the isolation of a single strain of sucrose-positive, but D-mannitol- and L-arabinose-negative *E. tarda* mimicking biogroup 1 from a patient with cholelithiasis.

The second species of the genus was recognized by Grimont et al. (1980) among the organisms formerly identified as atypical *E. tarda*, and the name *Edwardsiella hoshinae* was proposed for this species. Hawke et al. (1981) reported the third species, *Edwardsiella ictaluri*, associated with septicemia and death in catfish.

Hoshina (1962) recognized an organism as an etiologic agent of "red disease" in eels and named it *"Paracolobactrum anguillimortiferum"*. Based on the International Code of Bacterial No-

menclature 1953 and 1962, *"Paracolobactrum anguillimortiferum"* of Hoshina was a legitimate name for this species. Sakazaki and Tamura (1975) suggested a new combination *"Edwardsiella anguillimortifera"* instead of *Edwardsiella tarda*, because they considered the name *E. tarda* was a junior synonym of *"Paracolobactrum anguillimortiferum"*. Farmer et al. (1976) pointed out differences of some biochemical reactions in descriptions between Hoshina (1962) and Ewing et al. (1965), and emphasized the legitimacy and validity of *Edwardsiella tarda*. Unfortunately, no type strain of *"Paracolobactrum anguillimortiferum"* was designated by Hoshina, and strains studied by him were no longer available. In this status, *"P. anguillimortiferum"* should be considered as a doubtful name, and *E. tarda* is the name that should be used to avoid unnecessary confusion in the literature.

Acknowledgments

Riichi Sakazaki, who died in 2002, spent more than forty years at the Nippon Institute of Biological Sciences. His illustrious career focused on the classification and epidemiology of human and fish pathogens in Japan, particularly pathogenic bacteria in the families *Vibrionaceae* and *Enterobacteriaceae*. His contributions to Japanese bacteriology paralleled those of Drs. Edwards and Ewing in the U.S. He discovered and described many species in the genera *Enterobacter*, *Edwardsiella*, and *Vibrio*, and he developed serotyping schemes for *V. cholerae* and *V. parahaemolyticus*, as well as for several important species of *Enterobacteriaceae*. His research was always careful and comprehensive. He was one of the first Japanese scientists to publish in English language journals, thereby making his impressive accomplishments available to the world. He was recognized as one of the world's foremost experts on the genus *Vibrio* and on many genera in *Enterobacteriaceae*. He was a member of the WHO Subcommittees on the Taxonomy of *Vibrionaceae* and on the Taxonomy of *Enterobacteriaceae* for more than thirty years. He authored chapters on these organisms in three editions of *Bergey's Manual*. He was a good friend to the Bergey's Trust, as well as to all who knew him.

DIFFERENTIATION OF THE SPECIES OF THE GENUS *EDWARDSIELLA*

Table BXII.γ.218 presents the differential characteristics of the species and biogroups of *Edwardsiella*. Table BXII.γ.219 shows additional biochemical features of those organisms. In the H₂S production test, *E. tarda* biogroup 1 and *E. hoshinae* give a weakly positive reaction in the butt of Kligler iron agar but are usually negative in the butt portion of triple sugar iron agar in which acid produced by sucrose fermentation interferes to combine hydrogen sulfide with Fe²⁺.

List of species of the genus Edwardsiella

1. **Edwardsiella tarda** Ewing and McWhorter *in* Ewing, McWhorter, Escobar and Lubin 1965, 37[AL]

   tar'da. L. fem. adj. *tarda* slow (intended meaning was "inactive," referring to the fermentation on only a few carbohydrates compared to many other *Enterobacteriaceae*).

   The characteristics are as given for the genus and listed in Tables BXII.γ.218 and BXII.γ.219.

   Occurs in a wide variety of animals, rarely in the feces

of healthy people. It is an opportunistic human pathogen, which may cause wound infection and probably some cases of diarrheal diseases.

   The mol% G + C of the DNA is: 55–58 (T_m).

   Type strain: ATCC 15947, DSM 30052.

   GenBank accession number (16S rRNA): AF053975 and AF015259.

TABLE BXII.γ.218. Differentiation of species of the genus *Edwardsiella*[a]

Characteristic	E. tarda[b]	E. tarda[c]	E. hoshinae	E. ictaluri
Indole production	+	+	d	−
H₂S production (Kligler)	+	(+)[d]	(+)[d]	−
Motility	+	+	+	−
Malonate utilization	−	−	+	−
Fermentation of:				
L-Arabinose	−	+	d	−
Sucrose	−	+	+	−
Trehalose	−	−	+	−
D-Mannitol	−	+[e]	+	−

[a]For symbols, see standard definitions.

[b]Results for most strains of *E. tarda*.

[c]Results for *E. tarda* biogroup 1.

[d]Weakly positive but may be negative in the butt of TSI agar.

[e]Some strains may be negative.

Additional Remarks: These sequences are not from the type strain.

2. **Edwardsiella hoshinae** Grimont, Grimont, Richard and Sakazaki 1981b, 216[VP] (Effective publication: Grimont, Grimont, Richard and Sakazaki 1980, 349.)

ho.shi' nae. M.L. gen. n. *hoshinae* of Hoshina; named after the late Toshikazu Hoshina, the Japanese bacteriologist who was one of the first to describe an organism that was probably an *Edwardsiella*.

The characteristics are as described for the genus and indicated in Tables BXII.γ.218 and BXII.γ.219. Hydrogen sulfide is weakly produced in Kligler iron agar but may be negative in TSI agar.

Most isolates are from animals. There is no evidence that this species causes diarrhea.

The mol% G + C of the DNA is: 56–57 (T_m).

Type strain: 2-78, ATCC 33379, CIP 78-56.

3. **Edwardsiella ictaluri** Hawke, McWhorter, Steigerwalt and Brenner 1981, 400[VP]

ic.ta.lu' ri. Ictalurus the genus name for catfish; M.L. fem. adj. *ictaluri* pertaining to catfish.

The characteristics are as described for the genus and as indicated in Tables BXII.γ.218 and BXII.γ.219.

E. ictaluri is the most fastidious of the three *Edwardsiella* species. Growth is very slow on plating media. It seems to prefer a lower temperature, although characteristic biochemical reactions are apparent at 36°C (Tables BXII.γ.218

TABLE BXII.γ.219. Biochemical characteristics of the species and biogroups of the genus *Edwardsiella* [a]

Characteristic	E. tarda[b]	E. tarda[c]	E. hoshinae	E. ictaluri
Indole production	+	+	+	−
Voges–Proskauer	−	−	−	−
Citrate utilization (Simmons)	−	−	−	−
H₂S production (Kligler)	+	(+)[d]	(+)[d]	−
Lysine decarboxylase	+	+	+	+
Arginine dihydrolase	−	−	−	−
Ornithine decarboxylase	+	+	+	d
Phenylalanine deaminase	−	−	−	−
Gelatinase	−	−	−	−
Urease (Christensen)	−	−	−	−
β-Galactosidase	−	−	−	−
β-Glucuronidase	−	−	−	−
L-Pyrrolidonyl aminopeptidase	−	−	−	−
Lipase (Tween 80)	−	−	−	−
Deoxyribonuclease	−	−	−	−
Malonate utilization	−	−	+	−
Esculin hydrolysis	−	−	−	−
Growth in KCN medium	−	−	−	−
Gas from glucose	+	d	d	d
Acid from carbohydrate:				
Maltose	+	+	+	+
L-Arabinose	−	+	d	−
Sucrose	−	+	+	−
D-Mannitol	−	+[e]	+	−
Trehalose	−	−	+	−
Salicin	−	−	d	−
D-Cellobiose, lactose, melibiose, raffinose, L-rhamnose, D-xylose, adonitol, D-arabitol, dulcitol, *i*-inositol, D-sorbitol, α-methyl-D-glucoside	−	−	−	−

[a]For symbols, see standard definitions.

[b]Results for most strains of *E. tarda*.

[c]Results for *E. tarda* biogroup 1.

[d]Weakly positive but may be negative in the butt of TSI agar.

[e]Some strains may be negative.

and BXII.γ.219). Biochemically, it is also the least active of the three *Edwardsiella* species.

The antibiotic susceptibility by disk diffusion is difficult to determine because the strains grow so poorly on Mueller-Hinton agar at 37°C. They must be incubated at 25°C instead.

Occurs as a pathogen of catfish.

The mol% G + C of the DNA is: 53 (Bd).

Type strain: SECFDL, GA 7752, ATCC 33202, CDC 1976-78.

Genus XII. **Enterobacter** Hormaeche and Edwards 1960b, 72[AL] Nom. Cons. Opin. 28, Jud. Comm. 1963, 38

PATRICK A.D. GRIMONT AND FRANCINE GRIMONT

En.te.ro.bac' ter. Gr. neut. n. *enteron* intestine; M.L. masc. n. *bacter* equivalent of bacterium, a small rod; M.L. masc. n. *Enterobacter* intestinal small rod.

Straight rods, 0.6–1.0 × 1.2–3.0 μm, conforming to the general definition of the family *Enterobacteriaceae*. **Motile by peritrichous flagella** (generally 4–6). Gram negative. **Facultatively anaerobic.** Growth occurs readily on ordinary media. Glucose is fermented with production of acid and gas (generally $CO_2:H_2 = 2:1$). Gas is not produced from glucose at 44.5°C. **Most strains give a pos-**

itive Voges–Proskauer reaction and a negative methyl red test. An alkaline reaction occurs in Simmons citrate and malonate broth. Nitrate is reduced to nitrite. H_2S is not produced from thiosulfate. Tetrathionate is not reduced. **Corn oil and tributyrin are not hydrolyzed. Gelatin, DNA, and Tween 80 are either not, or very slowly, hydrolyzed.** L-Arabinose, D-cellobiose, D-fructose,

D-galactose, D-galacturonate, gentiobiose, D-gluconate, D-glucosamine, D-glucose, D-glucuronate, 2-ketogluconate, L-malate, D-mannitol, D-mannose, D-trehalose, and D-xylose utilized by all or almost all strains, as sole source of carbon and energy. L-rhamnose utilized by all strains except *Enterobacter asburiae*. L-Arabitol, ethanolamine, itaconate, 3-phenylpropionate, L-sorbose, D-tartrate, tryptamine, and xylitol are not utilized. *meso*-Erythritol, gentisate, glutarate, and tricarballylate not utilized except by some strains of *Enterobacter gergoviae*. D-melezitose not utilized except by some strains of *Enterobacter sakazakii*. Optimum temperature for growth is 30°C. Most clinical strains grow at 37°C; some environmental strains give erratic biochemical reactions at 37°C. Widely distributed in nature; common in man and animals.

The mol% G + C of the DNA is: 52–60 (Bd).

Type species: **Enterobacter cloacae** (Jordan 1890) Hormaeche and Edwards 1960b72. Nom. Cons. Opin. 28, Jud. Comm. 1963, 38 (*Bacillus cloacae* Jordan 1890, 836.)

FURTHER DESCRIPTIVE INFORMATION

Phylogenetic treatment The phylogenetic approach using sequences of the *rrs* gene (encoding 16S rRNA) does not provide sufficient resolution when closely related enterobacterial species are studied, and branching is often unreliable. However, some sequence comparisons have corroborated DNA–DNA hybridization studies. *Enterobacter aerogenes* is more closely related to *Klebsiella* species than to *E. cloacae*, the type species of the genus *Enterobacter*. Therefore, in this volume, *E. aerogenes* has been treated with the genus *Klebsiella* under the name *K. mobilis*. *Enterobacter agglomerans* and several species of the *E. agglomerans* complex cluster together and away from *E. cloacae*. *E. agglomerans* has been transferred to a new genus, *Pantoea*, and in this volume, the *E. agglomerans* complex is treated with the genus *Pantoea*. *Enterobacter intermedius* (originally named *E. intermedium*; corrected by von Graevenitz (1990)) has a *rrs* sequence identical with that of *Kluyvera cochleae*. Furthermore, DNA–DNA hybridization has shown the type strains of *E. intermedius* and *K. cochleae* to be 99% related with a thermal instability of hybridized molecules of 1.5°C (D.J. Brenner, personal communication). Therefore, *E. intermedius* is not treated in this chapter. (See the section on Other organisms in the chapter on the genus *Kluyvera*.)

The *Enterobacter* species treated in this chapter do not cluster together in an *rrs* sequence comparison. However, after *rpoB* sequence comparison, all species and subgroups in the *E. cloacae* complex composed of *E. cloacae*, *Enterobacter asburiae*, *Enterobacter hormaechei*, and the type strain of *Enterobacter dissolvens*) form a cluster (P.A.D. Grimont, unpublished data). *Enterobacter cancerogenus/Enterobacter taylorae* and *E. agglomerans* DNA group VII (not a member of *Pantoea*) form a cluster near *Leclercia adecarboxylata*. *Enterobacter amnigenus*, *Enterobacter nimipressuralis*, and *Enterobacter kobei* constitute another cluster. *Enterobacter cowanii* clusters with *E. agglomerans* DNA group IX (not a member of *Pantoea*) (unpublished data). More work is needed to establish criteria for delineating genera and ascertaining the phylogenetic position of all *Enterobacter* species.

Nutrition and growth conditions Best results are obtained when *Enterobacter* cultures are incubated at 30°C. At 37°C, yellow pigmentation of *E. sakazakii* may be weak. *E. sakazakii* is able to grow at 44°C.

Media used in the isolation of *Enterobacter* species are similar to those used for other *Enterobacteriaceae*. On nutrient agar, *E. cloacae*, *E. gergoviae*, *E. amnigenus*, and *E. nimipressuralis* form col-

onies that are round, 2–3 mm in diameter, and slightly iridescent or flat with irregular edges (Richard, 1984). *E. sakazakii* strains forms bright yellow colonies at 25°C or pale yellow colonies at 37°C, 1–3 mm in diameter. These colonies are smooth, mucoid, or dry (Farmer et al., 1980a).

Differential media that are not inhibitory for the *Enterobacteriaceae* are often used. These are bromothymol blue-lactose agar, phenol red-lactose agar, Drigalski lactose agar, eosin methylene blue agar, or MacConkey agar. It should be remembered that some strains in a species known to produce acid from lactose might fail to do so and thus give lactose-negative colonies on differential media.

In Biotype-100 strips (BioMerieux, Craponne, France), or in a minimal medium containing ammonium sulfate as the nitrogen source, the following compounds serve as sole carbon sources for most *Enterobacter* strains: *N*-acetyl-D-glucosamine, D-alanine, L-alanine, L-arabinose, L-aspartate, D-cellobiose, D-fructose, fumarate, D-galactose, D-galacturonate, gentiobiose, D-gluconate, D-glucosamine, D-glucose, D-glucuronate, L-glutamate, glycerol, 2-ketogluconate, DL-lactate, L-malate, D-mannitol, D-mannose, L-proline, D-ribose, L-serine, succinate, D-trehalose, and D-xylose. L-rhamnose is utilized by all strains except for most *E. asburiae* strains.

Most strains cannot utilize the following substrates as sole carbon and energy sources: L-arabitol, ethanolamine, itaconate, 3-phenylpropionate, and L-sorbose. The following substrates are not utilized except by some strains of *E. gergoviae*: *meso*-erythritol, gentisate, glutarate, histamine, 3-hydroxybenzoate, and tricarballylate. D-melezitose is not utilized, except by some strains of *E. sakazakii*.

Metabolism When pyrroloquinoline quinone (PQQ) is added, gluconate is produced from D-glucose in the presence of iodoacetate under aerobic conditions by all *Enterobacter* species by means of a glucose dehydrogenase. Only *E. gergoviae*, *E. sakazakii*, *E. cowanii*, and *E. hormaechei* genomic group (*E. cloacae* group III) can produce gluconate without added PQQ (Bouvet et al., 1989). Data are not available for *E. kobei* and *E. pyrinus*. A reducing compound, 2-ketogluconate, is also produced from gluconate by *E. gergoviae* by means of a gluconate dehydrogenase (Bouvet et al., 1989). 2,5-Diketogluconate is not produced from 2-ketogluconate (Bouvet et al., 1989).

Most *Enterobacter* species and biogroups give a positive Voges–Proskauer reaction (except *E. kobei*) and Simmons' citrate tests, motility (except some strains of *E. asburiae*); acid production from D-glucose, D-mannitol, salicin, L-arabinose, L-rhamnose (except most strains of *E. asburiae*), D-xylose, trehalose, D-cellobiose, and maltose; hydrolysis of *o*-nitrophenyl-β-D-galactoside (ONPG), and nitrate reduction. All *Enterobacter* species produce gas from D-glucose.

Most *Enterobacter* species or biogroups are negative for the following tests: H₂S production from thiosulfate, phenylalanine deaminase, tetrathionate reduction, tributyrin and corn oil hydrolysis, and β-glucuronidase.

Bacteriophage and bacteriocin typing Phage typing has been used for *E. cloacae*. Gaston (1987b) isolated 76 phages active against *E. cloacae* from sewage. Of these, 26 phages were selected after numerical taxonomic analysis of their reaction patterns on 92 *E. cloacae* strains. This system has been tried on 384 isolates, 94% of which were susceptible to at least one phage. 325 phage susceptibility patterns were observed. Reproducibility of patterns was 100% when duplicate testing was on the same day, but the reproducibility was only 40% when duplicate testing was done after 18 months (Gaston, 1987a).

Three different schemes have been developed for typing *E. cloacae* strains by susceptibility to bacteriocins. Freitag and Friedrich (1981) were able to type 51 of 65 strains (78%) which were assigned to 23 bacteriocin types designated A through X (type Y contained non-typable isolates). Traub et al. (1982) found 9% of 256 *E. cloacae* isolates to be bacteriocinogenic. Bacteriocins were produced by 16 strains after induction by mitomycin. A total of 308 isolates from various clinical sources were studied and 79% fell into 52 bacteriocin types leaving 21.4% of the isolates untypable. Bauernfeind and Petermüller (1984) found 132 of 149 isolates to produce bacteriocins. With a set of eight bacteriocin-producing strains, typability of 134 clinical isolates was 96.3%. A total of 44 different bacteriocin types could be distinguished. Only 11 (8. 2%) of the isolates fell into the largest bacteriocin type.

Antigenic structure *E. cloacae* serotyping schemes have been devised. Sakazaki and Namioka (1960) distinguished 53 O- and 56 H-antigens in agglutination tests ; 170 isolates were distributed among 79 serotypes. Bacterial suspensions had to be boiled in order to be agglutinated by the O-antisera. Unfortunately, epidemiological investigations using this serotyping scheme have not been reported.

Gaston et al. (1983) devised a serotyping scheme based on heat-stable somatic (O) antigens. 28 antisera including 11 absorbed sera were used, thus defining 28 O-serogroups. Of 300 clinical isolates from 66 hospitals, 78% were typable, 11% were not agglutinated by any of the sera, and 11% were autoagglutinable in saline.

Pathogenicity *Enterobacter* species are found in the natural environment including water, sewage, vegetables, and soil. The increased prevalence of *Enterobacter* spp. as nosocomial pathogens may be due to a greater resistance to disinfectants and antimicrobial agents than that of other members of the *Enterobacteriaceae*. Contaminated medicinal agents can be sources of outbreaks. Maki and Martin (1975) showed that *E. cloacae*, *E. agglomerans* and *Serratia marcescens* multiplied better in 5% dextrose solution at 25°C than did other members of the *Enterobacteriaceae*.

The pathogenicity of *Enterobacter* spp. has been reviewed by Sanders and Sanders (1997). Most *Enterobacter* bacteremias are acquired in the hospital. *E. cloacae* predominates, followed by *E. agglomerans*, *E. sakazakii*, and others. 14–53% of bacteremias that involve *Enterobacter* spp. are polymicrobial. *Enterobacter* bacteremias occur in patients with severe underlying diseases. Cases involving previously healthy individuals can be caused by administration of a contaminated medical product. The mortality rate is similar to that due to other enteric bacilli. A study of patients undergoing cardiac surgery suggested that *E. cloacae* might be more virulent than *E. aerogenes* (*Klebsiella mobilis*). Among patients colonized by *E. cloacae*, 26% developed an infection, whereas only 7% of those colonized by *K. mobilis* developed an infection.

In the lower respiratory tract, *Enterobacter* spp. have been involved in asymptomatic colonization of respiratory secretions, purulent bronchitis, lung abscess, pneumonia, and empyema. Pneumonia occurs in lung transplant recipients, patients with severe underlying diseases, and in elderly persons who are institutionalized.

Enterobacter spp. are a common cause of nosocomial infections of surgical wounds and burns. Other infections involving the skin and soft tissues are cellulitis, fasciitis, abscesses, emphysema, and myositis. Endocarditis due to *Enterobacter* spp. occurs in intravenous drug abusers and individuals with prosthetic valves.

Enterobacter spp. have been involved in biliary sepsis. Bacteremias after hepatic transplantation or endoscopic retrograde cholangiopancreatography, hepatic gas gangrene, fulminant emphysematous cholecystitis, acute suppurative cholangitis, and peritonitis following small intestine obstruction have been reported.

Urinary tract infections due to *Enterobacter* spp. range from asymptomatic bacteriuria to pyelonephritis and urosepsis. *Enterobacter* meningitis, ventriculitis, brain abscess, and infections near foreign bodies have been reported. *Enterobacter* spp. are infrequently the cause of postoperative endophthalmitis, but consequences are devastating (loss of vision or eye). Occasionally, severe septic arthritis, osteomyelitis, infections of multiple bones and joints in infants and children, vertebral osteomyelitis, bilateral hip infections and prosthetic hip infections due to *Enterobacter* spp. have been reported (reviewed by Sanders and Sanders, 1997).

Other species in the *E. cloacae* complex (*E. dissolvens*, *E. asburiae*, *E. hormaechei*) may have been reported as *E. cloacae* due to taxonomic uncertainties. *E. hormaechei* has been isolated from wounds, sputum, and blood (O'Hara et al., 1989). An outbreak of *E. hormaechei* infection and colonization among vulnerable, low-birth-weight premature infants has been described (Wenger et al., 1997). In our opinion, the *E. hormaechei* phenotype as described by O'Hara et al. (1989) is a biogroup of a larger genomic species. Thus, the strains that caused an outbreak in Marseille hospitals were identified as belonging to the *E. hormaechei* genomic species, although they resembled *E. cloacae* in classical biochemical tests (Davin-Regli et al., 1997). Our observation is that most *E. cloacae* that cause nosocomial infections belong to the *E. cloacae* group III/ *E. hormaechei* genomic species (unpublished data).

E. asburiae strains were isolated from clinical specimens, mostly urine, respiratory tract, feces, wounds, and blood (Brenner et al., 1986). The clinical significance of this organism is not known.

Enterobacter dissolvens has not yet been reported in clinical specimens and has only been recovered from environmental sources. *E. dissolvens* was first isolated by Rosen (1922) from diseased corn and is found in rotting cornstalks.

The role of *E. sakazakii* in pathology and as a contaminant has been reviewed by Nazarowec-White and Farber (1997). Little is known about the presence of *E. sakazakii* in the environment. *E. sakazakii* has frequently been found in powdered substitutes for breast milk in many countries together with other enteric bacteria. It has also been recovered from liquid formula and milk subjected to ultra high temperature (UHT) processing. Numerous reports of *E. sakazakii* meningitis and generalized sepsis have been reported in neonates. Powdered infant formula was implicated in several cases. Even when death did not occur, severe complications or sequelae were observed (brain abscesses, necrotizing enterocolitis, hydrocephalus). There have been only three reports of *E. sakazakii* infections in adults (an ulcer in the foot of a diabetic patient, urosepsis in an elderly patient, and bacteremia) (Nazarowec-White and Farber, 1997).

E. cancerogenus (*E. taylorae*) has only rarely been associated with human infections. Septicemia and urinary tract infections in patients with underlying diseases have been observed. Wound infections after severe trauma or crush injuries have been described. Some of them were chronic or unresolving infections. An environmental source was suspected (Abbott and Janda, 1997).

E. nimipressuralis has not yet been reported in clinical specimens and has only been recovered from environmental sources.

E. amnigenus is mainly found in water, but some strains were isolated from clinical specimens such as respiratory tract, wound, or feces (Farmer et al., 1985a).

E. gergoviae occurs in a variety of environmental sources such as water, cosmetics and clinical sources in France, Africa, and the United States (Richard et al., 1976; Brenner et al., 1980b). Multiply drug resistant strains have been found in urine samples during an infection outbreak (Richard et al., 1976).

Pathogenesis Adhesive properties may be important in the establishment or maintenance of bacterial infections. Adhesins are often also hemagglutinins (HA) and may or may not be located on fimbriae. Most strains of *Enterobacter amnigenus*, *E. cloacae*, and *E. sakazakii* produce a mannose-sensitive hemagglutinin (MS-HA) associated with type-1 fimbriae, i.e., thick, channeled fimbriae having an external diameter of 7–8 nm. These fimbriae can be coated by type-1 fimbrial antiserum against *E. cloacae* 035 but not by type-1 fimbrial antiserum against *Klebsiella pneumoniae* K55/1 (Adegbola and Old, 1983b). No other hemagglutinin and fimbrial type has been observed in these species.

Seven of eight *E. gergoviae* strains produce a mannose-resistant *Klebsiella*-like hemagglutinin (MR/K-HA) that agglutinates tanned ox erythrocytes and is associated with type-3 fimbriae, i.e., thin, non-channelled fimbriae having an external diameter 4–5 nm. These fimbriae can be coated with type-3 fimbrial antiserum against *K. oxytoca* K70/1 (Adegbola and Old, 1983b). No other hemagglutinin and fimbrial type has been observed in this species.

All *E. cloacae* strains tested could adhere to and invade HEp-2 cells (Keller et al., 1998). Only two of 14 *E. cloacae* isolates adhered to the human cell line Intestine-407 (Livrelli et al., 1996).

The outer membrane protein OmpX plays a role in the invasion of rabbit ileal tissue by *E. cloacae*. Invasiveness varies directly with the level of OmpX in the outer membrane (De Kort et al., 1994).

Iron is essential for bacterial growth. In the human body, iron is complexed to carrier molecules such as transferrin (in the serum) or lactoferrin (in milk and other secretions), or sequestered within cells (in heme proteins). When growing under iron-limiting conditions, potentially pathogenic *Enterobacteriaceae* produce high affinity systems to solubilize and import the required iron. These iron-chelating compound produced are mostly of two sorts, phenolate (e.g., enterochelin) and hydroxamate (aerobactin) siderophores (Payne, 1988). All strains of *E. cloacae*, *E. gergoviae*, and *E. sakazakii* tested by Reissbrodt and Rabsch (1988) produced enterochelin. Some strains of all these species produced aerobactin. Aerobactin was first isolated from a strain of *K. mobilis* (then "*Aerobacter aerogenes*") (Gibson and Magrath, 1969). Aerobactin and cloacin DF13 bind to the same receptor sites located in the outer membrane (Van Tiel-Menkveld et al., 1982). Aerobactin is encoded by a relatively simple genetic system that has been extensively characterized. Only four genes are required for synthesis of aerobactin (Carbonetti and Williams, 1984). *E. cloacae* harbor a relatively large conjugative plasmid encoding susceptibility to cloacine DF13 as well as production and uptake of acrobactin (Krone et al., 1985).

Most clinical strains of *E. cloacae* examined have been resistant to serum bactericidal activity and could produce aerobactin and mannose-sensitive hemagglutinin (Keller et al., 1998).

Antibiotic sensitivity The antibiotic sensitivity of *Enterobacter* spp. has been reviewed by Sanders and Sanders (1997). *E. cloacae* is naturally resistant to ampicillin, cephalothin and other older cephalosporins, and cefoxitin. Most strains are sensitive to ureidopenicillins and carboxypenicillins. Strains are not very sensitive to cefamandole and cefuroxime but are more sensitive to expanded-spectrum cephalosporins and aztreonam. The highest sensitivity is to carbapenems (e.g., imipenem). Most strains are sensitive to aminoglycosides and ciprofloxacin. The sensitivity to trimethoprim-sulfamethoxazole is variable.

All species of *Enterobacter* examined to date produce a chromosomally encoded group 1 β-lactamase. This enzyme is produced only at low levels (noninducible) by *E. gergoviae* and some strains of *E. sakazakii*, which explains the greater sensitivity of these strains to ampicillin, older cephalosporins, and cefoxitin. The enzyme is inducible in *E. cloacae*, *E. cancerogenus*, *E. asburiae*, and most strains of *E. sakazakii*, and these strains are resistant to ampicillin, older cephalosporins, and cefoxitin (Sanders and Sanders, 1997).

The *ampD* mutation in *E. cloacae* stably derepresses the chromosomal β-lactamase, which is then produced at high levels, causing the strains to be resistant to extended-spectrum cephalosporins, broad-spectrum penicillin, and aztreonam. These mutants remain sensitive to carbapenems and cefepime. However, additional mutations involving permeability through the outer envelope can lead to resistance to these agents. A chromosomally encoded carbapenemase has been reported (Sanders and Sanders, 1997).

Strains can acquire plasmid-borne resistance determinants encoding group 2b (TEM-1, TEM-2, or SHV-1) or group 2d (OXA-1) β-lactamases. Strains carrying such plasmids are susceptible to extended-spectrum cephalosporins and β-lactamase inhibitor-β-lactam drug combinations. Resistance to extended-spectrum cephalosporins can also be acquired with plasmids encoding group 2be β-lactamase (extended-spectrum β-lactamases).

Resistance to aminoglycosides is due to the production of one or more aminoglycoside-inactivating enzymes. The most frequently occurring enzymes are acetylating enzymes AAC(3)-II, AAC(6′), AAC(3)-III, AAC(3)-I, and AAC(3)-V. A nucleotidylating enzyme, ANT(2″), may also occur (Sanders and Sanders, 1997).

Ecology Nitrogen-fixing strains of *E. cloacae* have been isolated from the roots of dryland and wetland rices (Ladha et al., 1983). The nitrogen-fixing *E. cloacae* strains isolated by Bally et al. (1983) belong to *E. cloacae* genomic group 5 (P.A.D. Grimont, unpublished observations). A strain identified as *E. cloacae* suppresses sporangium germination in *Pythium ultimum* (the cause of seed and root diseases of a wide variety of plants). This is achieved by a competitive utilization of fatty acids present in the exudate of germinating seeds (Van Dijk and Nelson, 2000). Some strains identified as *E. cloacae* are endophytic symbionts of corn (Hinton and Bacon, 1995).

A strain identified as *E. asburiae* is a cotton endophyte that is able to colonize internal tissues of different plant species (Quadt-Hallmann and Kloepper, 1996).

E. nimipressuralis has been described as being the causal agent of wetwood in elm trees (Carter, 1945). *Enterobacter cancerogenus* (as *Erwinia cancerogena*) has been described as the causal agent of a canker disease of poplar (*Populus* species). *Enterobacter pyrinus* causes a pear brown leaf spot disease in Korea (Chung and Cho, 1993).

ENRICHMENT AND ISOLATION PROCEDURES

All media designed for the isolation of *Enterobacteriaceae* can be used for the isolation of *Enterobacter* species: MacConkey agar, Drigalski lactose agar, Hektoen agar, deoxycholate lactose citrate agar, etc. *Enterobacter* can also grow on media for general use, such as blood agar, nutrient agar, tryptic soy agar, bromocresol purple lactose agar, etc.

There are no selective media for *Enterobacter* species. The requirements for devising a selective medium are (i) a precise and stable delineation of the genus (or species, if the medium is to be species specific), (ii) known common properties within the genus (or species) which are uncommon outside the genus (or species), and (iii) a clinical, public health or special need for such selective medium. Due to pending changes in the delineation of the genus *Enterobacter* and the species *E. cloacae*, conditions (i) and (ii) are not met. Except for some epidemiological studies, the selective isolation of *Enterobacter* spp. is rarely necessary, since the presence of *Enterobacter* spp. in pluribacterial habitats (feces, throat, skin) is clinically meaningless. The public health significance of *Enterobacter* spp. in water or foods is uncertain. Isolation of *Enterobacter* spp. from clinical specimens is done either by direct plating on blood agar, tryptic soy agar or nutrient agar (e.g., pus, urine) or by plating after prior growth in tryptic soy broth or nutrient broth (blood, pus, cerebrospinal fluid).

MAINTENANCE PROCEDURES

Strains are initially grown on tryptic soy agar at their optimum temperature and then stabbed into a medium[1] designed for maintenance of *Enterobacteriaceae* and related organisms. The cultures are stored at room temperature in a dark, dry place. Cultures may be also preserved by freeze-drying. Freeze-drying is the best procedure for preservation of pigmented strains.

PROCEDURES FOR TESTING SPECIAL CHARACTERS

Procedures for carbon source utilization (using Biotype-100 strips), glucose oxidation, gluconate- and 2-ketogluconate dehydrogenase, Voges–Proskauer (Richard's modification), and β-xylosidase tests are given in the chapter on *Serratia* in this volume.

DIFFERENTIATION OF THE GENUS *ENTEROBACTER* FROM OTHER GENERA

Because the genus is not monophyletic and may be split further, differentiation of the genus should be at the species level.

TAXONOMIC COMMENTS

The history of some species of the genus *Enterobacter* can be confusedly traced to the end of the nineteenth century. *"Bacillus lactis aerogenes"* was isolated by Escherich (1885) from milk and renamed *"Bacillus aerogenes"* by Kruse (1896) and *"Aerobacter aerogenes"* by Beijerinck (1900b). Differentiation of this organism from Friedländer's bacillus (now *Klebsiella pneumoniae*) was not clear before 1955, and most authors considered *"B. lactis aerogenes"* or *"Aerobacter aerogenes"* as nonmotile, or as containing motile and nonmotile strains (Grimbert and Legros, 1900; Edwards and Fife, 1955). This led Edwards and Fife (1955) to state that *"A. aerogenes"* strains were in fact *Klebsiella* strains.

"Bacterium cloacae" was described by Jordan (1890) and transferred to a new genus *"Cloaca"* as *"Cloaca cloacae"* by Castellani and Chalmers (1920). *Bergey's Manual* (Bergey et al., 1923) transferred this species to the genus *"Aerobacter"* as *"A. cloacae"*. Because *"Aerobacter aerogenes"* was indistinguishable (at that time) from *Klebsiella pneumoniae*, it was proposed that the species *"A. aerogenes"* disappear (Edwards and Fife, 1955) although disappearance of the type species (*"A. aerogenes"*) implied disappearance of the genus (*"Aerobacter"*).

A significant step forward occurred when Møller (1955) devised simple methods for testing amino-acid decarboxylases. The *"Cloaca"* group, being arginine-positive, could now easily be distinguished from the *Klebsiella* group (arginine-negative) by biochemical tests. This lead to the finding of motile strains of the *"Cloaca"* group which were arginine-negative and produced gas from inositol and glycerol (Hormaeche and Munilla, 1957). These were called *"Cloaca B"* (arginine-positive strains forming the *"Cloaca A group"*). After reexamination of many cultures with decarboxylase tests, Hormaeche and Edwards (1958) redefined the genus *"Aerobacter"* to include two species, *"A. aerogenes"* (*"Cloaca B"*) and *"A. cloacae"* (*"Cloaca A"*). The type species was reaffirmed to be *"A. aerogenes"*.

In an attempt to avoid confusion resulting from the reclassification in the genus *Klebsiella* of many nonmotile strains previously labeled *"A. aerogenes"*, Hormaeche and Edwards (1960a, b) proposed a new genus, *Enterobacter*, as a substitute for *"Aerobacter"*. This genus was then composed of *E. cloacae* (type species) and *E. aerogenes*. The Judicial Commission of the International Committee on Nomenclature of Bacteria placed the name *Enterobacter* on the list of conserved names (Judicial Commission, 1963).

The genus *Erwinia* has long been a depository for plant-associated members of the *Enterobacteriaceae*. Several species of this genus were found phenotypically similar to *Enterobacter* species (*Erwinia herbicola*, *Erwinia ananas**, *Erwinia uredovora*, *Erwinia milletiae*, *Erwinia dissolvens*, *Erwinia nimipressuralis*) (Lelliott, 1974; Lelliott and Dickey, 1984) or highly related to *Enterobacter* species by DNA hybridization (*Erwinia dissolvens*, *Erwinia nimipressuralis*) (Steigerwalt et al., 1976).

The other *Enterobacter* species have been described after delineation by DNA–DNA hybridization. DNA–DNA hybridization has allowed bacteriologists to modify the circumscription of some species and to find synonymies.

Enterobacter taylorae was described based on clinical strains closely related to *E. cloacae* (Farmer et al., 1985b). The species was later found synonymous to *Enterobacter cancerogenus* by both DNA relatedness and phenotypic characters (Grimont and Ageron, 1989).

Early DNA relatedness studies on *E. cloacae* (Steigerwalt et al., 1976) showed the genomic heterogeneity of this nomenspecies. In work done in our laboratory (Grimont and Grimont, 1992; and unpublished) on 49 strains previously identified as *E. cloacae*, 45 strains fell into five DNA relatedness groups (1 to 5). Group 1 contained only three strains, including the type strains of *E. cloacae* and *E. dissolvens*. DNA from these type strains showed some divergence (ΔT_m values of 4.5°C). In spite of a search for similar strains in our large laboratory collection, no other strain of group 1 could be found. Group 2 contained seven strains, including

---

1. Maintenance medium (g/liter): Bacto-peptone (Difco), 10.0; NaCl, 5.0: Bacto-agar (Difco), 10.0; pH 7.4. The medium should be dispensed into small (9.5–10 × 90 mm) screw-capped tubes.

Editorial Note: Mergaert et al. (1993) transferred *Erwinia ananas* to the genus *Pantoea* as *P. ananas*; the name was corrected to *Pantoea ananatis* by Trüper and De' Clari (1997).

reference strain CDC 1347-71 (Steigerwalt et al., 1976). Group 3 contained 15 strains, including the type strain of *Enterobacter hormaechei*. This group was slightly heterogeneous with $\Delta\,T_m$ values ranging from 0.0–4.0°C. Group 4 contained 17 strains and could be split into three subgroups based on ΔT_m values (0.0–2.5°C within subgroup and 4.0–6.5°C between subgroups). Subgroup 4a contained the type strain of *Enterobacter asburiae*. Group 5 (earlier referred to as group 6 (Bouvet et al., 1989)) contained three nitrogen-fixing strains.

Three problems are raised by this study. The first problem involves *E. hormaechei*. A group of similar strains (Enteric group 75), which differ phenotypically from *E. cloacae*, was genomically distinct from the type strain of *E. cloacae* and thus described as a new species, *E. hormaechei* (O'Hara et al., 1989). In our group 3, we found only one strain phenotypically similar to *E. hormaechei*. All other strains of group 3 fit the classical definition of *E. cloacae* (Richard, 1984). Group 3 strains (including *E. hormaechei*) produce an active glucose dehydrogenase, and this property differentiates group 3 from the other groups in the *E. cloacae* complex (Bouvet et al., 1989; P.A.D. Grimont, unpublished observations). Group 3 should be named *E. hormaechei*, and the species definition should be emended.

The second problem involves *E. asburiae*. Enteric group 17 differed from *E. cloacae* by some phenotypic features and DNA hybridization showed Enteric Group 17 (*E. asburiae*) to be distinct from the type strain of *E. cloacae*. Our subgroup 4a is genomically and phenotypically identical with *E. asburiae*. However, subgroups 4b and 4c are less differentiable from *E. cloacae* by phenotypic characteristics. Group 4 (including subgroups 4a, 4b, and 4c) should be called *E. asburiae*, and the species definition should be amended.

The third problem is in fact the cause of the other two. The choice of the neotype strain for *E. cloacae* was unfortunate. This strain was isolated from cerebrospinal fluid, which is not the usual habitat of what is currently known as *E. cloacae*. There are two possible solutions: either the nomenclatural rules are followed, in which case *E. cloacae* (our group 1) will disappear from routine clinical laboratory work (because most strains presently labeled *E. cloacae* are in fact *E. hormaechei* or *E. asburiae*), or the type strain can be changed and a new one designated from our group 2 (which can be encountered in clinical microbiology, although less frequently than *E. hormaechei* and *E. asburiae* genomic species).

Some species that were included in the genus *Enterobacter* are now excluded; these are discussed below.

The group of strongly proteolytic strains named *"Aerobacter liquefaciens"* by Grimes and Hennerty (1931) was included in the genus *Enterobacter* by Edwards and Ewing (1972). The species was transferred to the genus *Serratia* as *S. liquefaciens* by Bascomb et al. (1971).

A group of strains named *Hafnia* by Møller (1954) was included in the genus *Enterobacter* as *E. hafniae* (Ewing and Fife, 1968). This species appeared in the 8th edition of the *Bergey's Manual* (Buchanan and Gibbons, 1974) as *Hafnia alvei*. Steigerwalt et al. (1976) favored the separation of *Hafnia alvei* from the genus *Enterobacter* based on DNA relatedness studies.

E. aerogenes is closer to *Klebsiella pneumoniae* (about 55% DNA relatedness) than to *E. cloacae* (about 45% relatedness) (Brenner et al., 1972c; Steigerwalt et al., 1976). In a numerical taxonomy study (Bascomb et al., 1971), *E. aerogenes* was so similar to the genus *Klebsiella* that transfer of this species to the genus *Klebsiella* was proposed. Since the name *"K. aerogenes"* had been used for bacteria that are indistinguishable from *K. pneumoniae* by DNA relatedness, the name *K. mobilis* was proposed for *E. aerogenes* (Bascomb et al., 1971). The species is treated in the chapter on the genus *Klebsiella* as *K. mobilis*.

E. agglomerans is a very complex group of environmental bacteria that may cause opportunistic infections. The name covers many (20–40) genomic groups (Brenner et al., 1988b) or phenons (Gavini et al., 1983b; Verdonck et al., 1987). In addition to this diversity, strains of the *E. agglomerans* complex are not closely related to *E. cloacae* (the type species of the genus *Enterobacter*) by DNA relatedness. Some groups in this complex have been placed in new genera (*Rahnella aquatilis*, *Ewingella americana*, *Leclercia adecarboxylata*). A new genus, *Pantoea*, has been proposed for several groups of the *E. agglomerans* complex. These are treated in the chapter on the genus *Pantoea*.

Enterobacter intermedius (originally named *E. intermedium*; corrected by von Graevenitz (1990)) was a Voges–Proskauer positive species found in water and unpolluted soil (Izard et al., 1980). However, it has an *rrs* sequence identical to that of *Kluyvera cochleae*. Furthermore, DNA–DNA hybridization has shown the type strains of *E. intermedius* and *K. cochleae* to be 99% related with a thermal instability of hybridized molecules of 1.5°C (D.J. Brenner, personnal communication).

FURTHER READING

Janda, J.M. and S.L. Abbott. 1998. The Enterobacteria, Lippincott-Raven, Philadelphia. pp. 387.

DIFFERENTIATION OF THE SPECIES OF THE GENUS *ENTEROBACTER*

Tables BXII.γ.220 and BXII.γ.221 give the characteristics differentiating the species and unnamed genomic groups in the genus *Enterobacter*.

List of species of the genus Enterobacter

1. **Enterobacter cloacae** (Jordan 1890) Hormaeche and Edwards 1960b, 72[AL]. Nom. Cons. Opin. 28, Jud. Comm. 1963, 38 (*Bacillus cloacae* Jordan 1890, 836.)

   clo.a' cae. L. n. *cloaca* a sewer; L. gen. n. *cloacae* of sewer.

   The cell and colonial morphology are as given for the genus. Physiological and nutritional characteristics are presented in Tables BXII.γ.220, BXII.γ.221 and BXII.γ.222. Several DNA hybridization groups compose this species. Some of these DNA groups can be identified to named species (*E. dissolvens*, *E. asburiae*, or *E. hormaechei*) with different circumscriptions than originally published. Occurs in water, sewage, soil, meat, and hospital environments and on the skin and in the intestinal tracts of man and animals as a commensal. May cause nosocomial infections

   The mol% G + C of the DNA is: 52–54 (T_m).

   Type strain: ATCC 13047, CIP 60.85, DSM 30054, JCM 1232, LMG 2783, NCTC 10005.

   GenBank accession number (16S rRNA): AJ417484.

   This strain is very close by DNA–DNA hybridization to the type strain of *E. dissolvens* and may not properly represent the strains routinely identified as *E. cloacae*.

TABLE BXII.γ.220. Characteristics of the species of the genus *Enterobacter*[a]

Characteristic	E. cloacae complex[b]	E. amnigenus	E. cancerogenus	E. cowanii	E. gergoviae	E. kobei	E. nimipressuralis	E. pyrinus	E. sakazakii
Motility (36°C)	d	+	+	+	+	+	+	+	+
Yellow pigment	−	−	−	d	−	−	−	−	+
Urea hydrolyzed	−	−	−	−	+	d	−	+	−
Indole production	−	−	−	−	−	−	−	−	d
β-Xylosidase test	+	+	−	+	−	+	+	ND	+
Methyl red	d	−	−	+	−	ND	+	ND	−
Voges–Prokauer	d	+	+	+	+	−	+	+	+
Growth in KCN	+	+	+	+	−	+	+	+	+
Gelatin hydrolysis at 22°C	(d)	−	−	−	−	−	−	−	−
Deoxyribonuclease (25°)	−	−	−	−	−	−	−	−	(+)
Lysine decarboxylase	−	−	−	−	+	−	−	d	−
Arginine dihydrolase	+	+	+	−	−	+	+	−	+
Ornithine decarboxylase	+	+	+	−	+	d	+	+	+
Phenylalanine deaminase	−	−	−	−	−	−	−	−	d
Glucose dehydrogenase	D	−	−	+	+	ND	−	ND	+
Gluconate dehydrogenase	−	−	−	−	+	ND	−	+	−
Growth at 41°C	+	−	ND	+	+	+	−	ND	+
Esculin hydrolysis	d	+	−	+	+	d	+	+	+
Acetate	d	−	−	+	+	d	ND	ND	+
Acid from:									
Adonitol	d	−	−	−	−	−	−	−	−
L-Arabinose	+	+	+	+	+	+	+	+	+
D-Arabitol	d	−	−	−	+	−	ND	ND	−
Cellobiose	+	+	+	+	+	+	+	+	+
Dulcitol	D	−	−	+	−	d	−	−	−
meso-Erythritol	−	−	−	−	−	ND	−	ND	−
Glycerol	d	−	d	+	+	d	ND	+	−
myo-Inositol	d	−	−	−	−	d	d	+	(+)
Lactose	d	d	−	+	d	+	+	−	+
Maltose	+	+	+	+	+	+	+	+	+
D-Mannitol	+	+	+	+	+	+	+	+	+
Melibiose	d	+	−	+	+	+	+	+	+
α-Methylglucoside	+	d	d	−	−	+	+	ND	+
Mucate	d	+	d	d	−	d	+	ND	−
Raffinose	+	+	−	+	+	+	d	−	+
L-Rhamnose	d	+	+	+	+	+	+	+	+
Salicin	D	+	+	+	+	+	+	+	+
D-Sorbitol	+	−	−	+	−	+	+	−	−
Sucrose	+	+	−	+	+	+	−	+	+
Trehalose	+	+	+	+	+	+	+	+	+
D-Xylose	+	+	+	+	+	+	+	+	+
Utilization of:									
cis-Aconitate	d	−	+	+	d	−	−	(d)	+
trans-Aconitate	d	−	+	+	−	−	−	(d)	d
Adonitol	D	−	−	−	−	−	−	−	−
4-Aminobutyrate	−	−	−	−	+	−	−	−	+
5-Aminovalerate	−	−	−	−	+	−	−	−	−
D-Arabitol	D	−	−	−	+	−	−	+	−
Benzoate	−	−	−	−	d	−	−	−	−
Citrate	+	+	+	+	+	+	+	−	+
m-Coumarate	−	−	−	−	+	−	−	−	−
Dulcitol	D	−	−	+	−	+	−	−	d
L-Fucose	D	−	+	−	d	−	−	−	−
Gentisate	−	−	−	(+)	d	−	−	−	−
Histamine	−	−	−	−	+	−	−	−	−
3-Hydroxybenzoate	−	−	−	−	d	−	−	−	−
4-Hydroxybenzoate	−	−	−	−	+	−	−	−	−
3-Hydroxybutyrate	d	−	−	−	d	−	−	−	−
myo-Inositol	D	−	−	−	d	−	−	−	d
5-Ketogluconate	−	−	−	−	+	−	−	+	−
2-Ketoglutarate	d	−	d	−	d	−	−	d	−
Lactose	(+)	+	(d)	(+)	(+)	+	+	−	+
Lactulose	d	d	−	(+)	d	+	−	−	+
D-Lyxose	D	+	−	+	−	−	+	−	−
D-Malate	d	−	d	(+)	d	+	d	−	(d)
Malonate	d	−	(d)	−	d	d	−	−	(d)
Maltitol	+	+	−	−	−	+	+	−	+
D-Melibiose	D	+	−	+	+	+	+	−	+
1-*O*-Methyl-α-galactoside	D	+	−	+	+	+	+	−	+
3-*O*-Methyl-D-glucose	D	−	+	−	−	v	−	−	−
1-*O*-Methyl-α-D-glucoside	+	d	−	−	−	+	+	−	+

(*continued*)

TABLE BXII.γ.220. *(cont.)*

Characteristic	E. cloacae complex[b]	E. amnigenus	E. cancerogenus	E. cowanii	E. gergoviae	E. kobei	E. nimipressuralis	E. pyrinus	E. sakazakii
Mucate	d	+	+	+	−	+	+	−	−
Palatinose	+	+	−	−	−	+	+	−	+
Phenylacetate	D	−	+	−	+	−	+	−	−
L-Proline	+	d	+	+	+	+	+	−	+
Protocatechuate	−	−	−	−	+	−	−	+	−
Putrescine	D	−	+	−	d	−	−	+	+
Quinate	−	−	−	−	+	−	−	−	+
D-Raffinose	d	+	−	+	+	+	d	−	+
L-Rhamnose	D	+	+	+	+	+	+	+	+
D-Saccharate	+	+	+	+	−	+	+	−	−
D-Sorbitol	+	−	−	+	−	+	+	−	−
Sucrose	+	+	−	+	+	+	−	−	+
D-Tagatose	d	−	−	d	−	−	d	+	−
meso-Tartrate	−	−	−	−	d	−	−	−	−
Tricarballylate	−	−	−	−	d	−	−	−	−
Tryptamine	−	−	−	−	−	−	−	−	−
D-Turanose	d	(d)	−	−	−	+	d	−	−
L-Tyrosine	d	−	−	−	d	−	−	−	d

[a] +, 90–100% strains positive in 1–2 days; (+), 90–100% strains positive in 1–4 days; −, 90–100% strains negative in 4 days; d, positive or negative in 3–4 days; D, test used to differentiate species within a complex; (d), positive or negative in 3–4 days; ND, no data.

[b] The *E. cloacae* complex includes *E. cloacae*, *E. dissolvens*, *E. hormaechei*, and *E. asburiae*.

TABLE BXII.γ.221. Phenotypic properties of genomic groups within the *Enterobacter cloacae* complex[a]

Characteristic	Genomic group or subgroup[b]						
	1	2	3	4a	4b	4c	5
Glucose dehydrogenase	−	−	+	−	−	−	−
Motility	+	+	+	−	d	+	+
Malonate test	+	+	+	−	−	d	+
Esculin hydrolyzed	d	(d)	(d)	+	+	+	+
Utilization of:							
Adonitol	−	−	d	−	−	−	−
D-Arabitol	−	−	d	−	−	−	−
Dulcitol	−	d	d	−	−	d	+
Fucose	−	−	d	−	−	−	−
D-Galacturonate	+	d	+	+	+	+	+
myo-Inositol	+	+	d	+	+	+	+
Lyxose	d	−	+	+	+	d	+
D-Melibiose	+	+	d	−	+	+	+
3-Methylglucose	−	−	d	−	−	−	−
Phenylacetate	d	+	+	−	+	+	+
Putrescine	d	+	−	+	−	+	−
D-Raffinose	+	+	d	d	d	+	+
L-Rhamnose	+	+	+	−	d	−	+
D-Sorbitol	+	+	d	+	+	+	+
Xylitol	−	−	(d)	−	−	−	−

[a] +, 90–100% strains positive in 1–2 days (utilization tests) or in 1 day (other tests); (+), 90–100% strains positive in 1–4 days; −, 90–100% strains negative in 4 days; d, positve or negative in 1–4 days; (d), positive or negative in 3–4 days.

[b] The type strain of *E. dissolvens* and the present type strain of *E. cloacae* are in genomic group 1; the type strain of *E. hormaechei* is in genomic group 3, and the type strain of *E. asburiae* is in genomic group 4, subgroup 4a.

2. **Enterobacter amnigenus** Izard, Gavini, Trinel and Leclerc 1981b, 37[VP]

am.ni′ ge.nus. L. adj. *amnigenus* coming from water.

The cell and colonial morphology are as given for the genus. Physiological and nutritional characteristics are presented in Table BXII.γ.220. Isolated from water and rarely from clinical samples.

The mol% G + C of the DNA is: 60 (Bd).

Type strain: ATCC 33072, CCUG 14182, CIP 103169, DSM 4486, JCM 1237, LMG 2784, NCTC 12124..

GenBank accession number (16S rRNA): AB004749.

TABLE BXII.γ.222. Utilization of substrates by biogroups of genomic group 3 of the *E. cloacae* complex[a]

Utilization of:	Biogroup[b]						
	3a	3b	3c	3d	3e	3f	3g
Adonitol	−	−	−	−	−	+	+
D-Arabitol	−	−	−	−	−	+	+
Fucose	d	+	+	−	+	+	+
Methyl-α-D-galactoside	−	+	+	+	−	+	+
3-Methylglucose	+	−	+	−	−	+	−
D-Melibiose	−[b]	+	+	+	−	+	+
D-Raffinose	−	+	+	+	+	+	+
D-Sorbitol	−	+	+	+	+	+	+

[a] +, all strains positive in 1–2 days; −, strains negative in 4 days; d, positive or negative in 1–4 days.

[b] Five strains representing *E. hormaechei* (including the type strain) and received from the Centers for Disease Control corresponded to biogroup 3a.

3. **Enterobacter asburiae** Brenner, McWhorter, Kai, Steigerwalt and Farmer III 1988b, 220[VP] (Effective publication: Brenner, McWhorter, Kai, Steigerwalt and Farmer III 1986, 1117.)

as.bur′ i.ae. N.L. gen. n. *asburiae* named in honor of Marie Alyce Fife-Asbury, an American bacteriologist.

The cell and colonial morphology are as given for the genus. Physiological and nutritional characteristics are included in Tables BXII.γ.220 and BXII.γ.221.

The type strain belongs in DNA group 4a in the *E. cloacae* complex.

The mol% G + C of the DNA is: 55 (T_m).

Type strain: ATCC 35953, CIP 103358, JCM 6051, NCTC 12123.

GenBank accession number (16S rRNA): AB004744.

4. **Enterobacter cancerogenus** (Urosevic 1966) Dickey and Zumoff 1988, 373[VP] *Erwinia cancerogena* Urosevic 1966, 500.)

can.cer.o′ ge.nus. L. *cancer* crab, the disease cancer; L. v. *gigno* to produce; L. masc. adj. *cancerogenus* cancer-inducing.

The cell and colonial morphology are as given for the genus. Physiological and nutritional characteristics are presented in Table BXII.γ.220.

Isolated from cankers in poplars (*Populus* species) in

Czechoslovakia and in clinical specimens. *Enterobacter taylorae* (Farmer et al. 1985b; type strain ATCC 35317 = CDC 2126-81) is a junior synonym of *E. cancerogenus* (Grimont and Ageron, 1989).

The mol% G + C of the DNA is: not determined.

Type strain: ATCC 33241, CCUG 25231, CFBP 4167, CIP 103787, ICMP 5706, LMG 2693, NCBBP 2176.

GenBank accession number (16S rRNA): Z96078.

5. **Enterobacter cowanii** Inoue, Sugiyama, Kosako, Sakazaki and Yamai 2001, 1619^{VP} (Effective publication: Inoue, Sugiyama, Kosako, Sakazaki and Yamai 2000, 419.)

co.wa' ni.i. M.L. gen. n. *cowani* of Cowan, named after S.T. Cowan, British bacteriologist.

The cell morphology is as given for the genus. Colonies on nutrient agar are often yellow pigmented. Physiological and nutritional characteristics are presented in Table BXII.γ.220. Isolated from diverse clinical specimens but clinical significance is unknown. May be found in foods.

The mol% G + C of the DNA is: 53 (HPLC).

Type strain: 88-76, CIP 107300, JCM 10956.

6. **Enterobacter dissolvens** (Rosen 1922) Brenner, McWhorter, Kai, Steigerwalt and Farmer III 1988b, 220^{VP} (Effective publication: Brenner, McWhorter, Kai, Steigerwalt and Farmer III 1986, 1119 (*"Phytomonas dissolvens"* Rosen 1922, 497; *Erwinia dissolvens* (Rosen 1922) Burkholder 1948a, 472.)

dis.sol' vens. L. part. adj. *dissolvens* dissolving.

The cell and colonial morphology are as given for the genus. Physiological and nutritional characteristics are included in Tables BXII.γ.220 and BXII.γ.221. Presently, it is not possible to phenotypically differentiate *E. dissolvens* from *E. cloacae*.

The type strain belongs in DNA group 1 in the *E. cloacae* complex. Originally isolated from diseased corn.

The mol% G + C of the DNA is: 54 (Bd).

Type strain: ATCC 23373, CIP 105586, ICMP 1570, JCM 6049, LMG 2683, NCBBP 1850.

GenBank accession number (16S rRNA): AJ417485, Z96079.

The type strain belongs in the *Enterobacter cloacae* complex.

7. **Enterobacter gergoviae** Brenner, Richard, Steigerwalt, Asbury and Mandel 1980b, 1^{VP}

ger.go' vi.ae. M.L. gen. n. *gergoviae* of Gergovie Highland, France.

The cell and colonial morphology are as given for the genus. Physiological and nutritional characteristics are presented in Table BXII.γ.220. Occurs in various environmental sources such as cosmetics or water. Has also been recovered from clinical specimens.

The mol% G + C of the DNA is: 60 (Bd).

Type strain: ATCC 33028, CIP 76.1, DSM 9245, JCM 1234, LMG 5739, NCTC 11434.

GenBank accession number (16S rRNA): AB004748.

8. **Enterobacter hormaechei** O'Hara, Steigerwalt, Hill, Farmer, Fanning and Brenner 1990, 105^{VP} (Effective publication: O'Hara, Steigerwalt, Hill, Farmer, Fanning and Brenner 1989, 2048.)

hor.ma.e' che.i. M.L. gen. n. *hormaechei* of Hormaeche, named after Estenio Hormaeche, Uruguayan microbiologist.

The cell and colonial morphology are as given for the genus. Physiological and nutritional characteristics are presented in Tables BXII.γ.220 and BXII.γ.221. Oxidation of glucose into gluconate in the absence of PQQ is a characteristic of the *E. hormaechei* genomic group. Isolated from clinical specimens.

The mol% G + C of the DNA is: not determined.

Type strain: ATCC 49162, CCUG 27126, CIP 103441.

GenBank accession number (16S rRNA): AJ417450.

9. **Enterobacter kobei** Kosako, Tamura, Sakazaki and Miki 1997, 915^{VP} (Effective publication: Kosako, Tamura, Sakazaki and Miki 1996, 264.)

kobe.i. M.L. gen. n. *kobei* pertaining to the city of Kobe, Japan.

The cell and colonial morphology are as given for the genus. Physiological and nutritional characteristics are presented in Table BXII.γ.220. Isolated from diverse clinical specimens but clinical significance is unknown. May be found in foods.

The mol% G + C of the DNA is: 53 (T_m).

Type strain: ATCC BAA-260, CIP 105566, DSM 13645, JCM 8580.

10. **Enterobacter nimipressuralis** (Carter 1945) Brenner, McWhorter, Kai, Steigerwalt and Farmer 1988b, 220^{VP} (Effective publication: Brenner, McWhorter, Kai, Steigerwalt and Farmer III 1986, 1119 (*"Erwinia nimipressuralis"*) Carter 1945, 423; Dye 1969a, 83.)

ni.mi.pres.su.ra' lis. L. adv. *nimis* overmuch; L. n. *pressura* pressure; M.L. adj. *nimipressuralis* with excessive pressure.

The cell and colonial morphology are as given for the genus. Physiological and nutritional characteristics are presented in Table BXII.γ.220. Reported as the causative agent of "wetwood" disease in elm trees (Carter, 1945).

The mol% G + C of the DNA is: 55 (Bd).

Type strain: ATCC 9912, CIP 104980, ICMP 1577, JCM 6050, NCPPB 2045.

GenBank accession number (16S rRNA): Z96077.

11. **Enterobacter pyrinus** Chung, Brenner, Steigerwalt, Kim, Kim and Cho 1993b, 161^{VP}

pyr' i.nus. L. n. *pyrus* pear; L. suff. *-inus* belonging to; L. adj. *pyrinus* from pears.

The cell and colonial morphology are as given for the genus. Physiological and nutritional characteristics are presented in Table BXII.γ.220. Isolated from brown leaf spot lesions on pear trees in Korea.

The mol% G + C of the DNA is: 57–61 (T_m).

Type strain: ATCC 49851, CFBP 4168, CIP 104019, DSM 12410, ICMP 12530.

12. **Enterobacter sakazakii** Farmer, Asbury, Hickman and Brenner 1980a, 575^{VP}

sa.ka.za' ki.i. M.L. gen. n. *sakazakii* of Sakazaki; named after the Japanese bacteriologist, Riichi Sakazaki.

The cell and colonial morphology are as given for the genus. Physiological and nutritional characteristics are presented in Table BXII.γ.220. Occurs in the environment and in foods. May contaminate milk powder and subsequent nutritional preparations and then cause generalized infection (septicemia, meningitis) in newborns.

The mol% G + C of the DNA is: 57 (T_m).

Type strain: ATCC 29544, CCUG 14558, CIP 103183, DSM 4485, JCM 1233, LMG 5740, NCTC 11467.

GenBank accession number (16S rRNA): AB004746.

Genus XIII. **Erwinia** Winslow, Broadhurst, Buchanan, Krumwiede, Rogers and Smith 1920, 209[AL] emend. Hauben, Moore, Vauterin, Steenackers, Mergaert, Verdonck and Swings 1999a, 1

LYSIANE HAUBEN AND JEAN SWINGS

Er.wi'ni.a. M.L. fem. n. *Erwinia* named after Erwin F. Smith.

Straight rods, 0.5–1.0 × 1.0–3.0 µm; occur singly or in pairs. Gram negative. Motile by peritrichous flagella. Facultatively anaerobic, but anaerobic growth by some species is weak. Optimum temperature, 27–30°C; maximum temperature for growth is 40°C. Oxidase negative. Catalase positive. **Pectinase negative.** Acid is produced from fructose, D-galactose, D-glucose, and sucrose but not from adonitol, arabitol, dextrin, dulcitol, inulin, maltose, starch, and tagatose. Utilize fumarate, D-galactose, gluconate, D-glucose, glycerol, β-methylglucoside, malate, and succinate, but not L-arabitol, benzoate, butanol, methanol, oxalate, propionate, or sorbose as carbon- and energy-yielding sources. Utilize L-alanine, L-glutaminic acid, glycylglycine, and L-serine, but not kynureninic acid and trigonelline, as nitrogen sources. Sensitive to chloramphenicol, furazolidone, nalidixinic acid, oxytetracyline, and tetracycline. Do not possess arginine dihydrolase, caseinase, pectinase, phenylalanine deaminase, or urease. Associated with plants as pathogens, saprophytes, or constituents of the epiphytic flora.

The species of the genus *Erwinia* comprise a distinct phylogenetic group, as determined by 16S rRNA gene sequence comparisons, and are characterized by 14 generic signature nucleotides (Table BXII.γ.223).

Erwinia species cause plant diseases that include mainly blights and wilts. The pathogen usually starts to cause damage in the vascular tissue and then spreads throughout the plant. Ingress by the pathogen generally occurs through natural openings and wounds.

The mol% G + C of the DNA is: 49.8–54.1.

Type species: **Erwinia amylovora** (Burrill 1882) Winslow, Broadhurst, Buchanan, Krumwiede, Rogers and Smith 1920, 209 emend. Hauben, Moore, Vauterin, Steenackers, Mergaert, Verdonck and Swings 1999a, 1 (*Micrococcus amylovorus* Burrill 1882, 134.)

FURTHER DESCRIPTIVE INFORMATION

Gas production is comparatively weak or absent. Decarboxylases for arginine, lysine, or ornithine cannot be detected by Møller's method (Møller, 1955). Formation of putrescine occurs when the amino acids are decarboxylated under aerobic conditions (Zherebilo and Gvozdyak, 1976). Glutamic acid is not decarboxylated. Lipases are rarely produced. Additional characteristics of the species of the genus are given in Tables BXII.γ.223, BXII.γ.224, and BXII.γ.225.

Fermentation end products from glucose are CO_2 and different combinations of succinate, lactate, formate, and acetate; some strains form 2,3-butanediol and some ethanol (White and Starr, 1971). Starch is not hydrolyzed beyond dextrins.

Sequence analyses on 16S rDNA of the species of the genus *Erwinia* by Kwon et al. (1997) and Hauben et al. (1998a) are in good agreement and place the genus *Erwinia* within the *Enterobacteriaceae*, closely related to the genera *Pantoea*, *Pectobacterium*, *Brenneria*, and *Enterobacter* (Fig. BXII.γ.201; Table BXII.γ.223).

Burrill (1882) was the first to attribute the cause of a plant disease to a bacterium, which he named "*Micrococcus amylovorus*". In successive taxonomic studies, the bacterium was renamed "*Ba-

cillus amylovorus" (Trevisan, 1889a), "*Bacterium amylovorus*" (Chester, 1897), "*Bacterium amylovorum*" (Serbinov, 1915), and finally *Erwinia amylovora* (Winslow et al., 1920). *Erwinia* species cause plant diseases that include blights, die back, leaf spots, wilts, and crown rot. *Erwinia tracheiphila* overwinters in the bodies of cucumber beetles (*Diabrotica vittata* Fabr. and *D. duodecimpunctata* Oliv.) (Leach, 1964).

Phytopathogenicity of *Erwinia rhapontici* and *Erwinia persicina** may be due to the release of proferrorosamines (Feistner et al., 1997).

ENRICHMENT AND ISOLATION PROCEDURES

In general, *Erwinia* can be easily isolated. The affected plant material is washed in tap water, followed by sterile water, and dried by paper toweling. Surface sterilization (3 min in 1:10 dilution of 5.25% active sodium hypochlorite) is sometimes detrimental for isolation. Affected tissue is removed from a young lesion or the edge of older necrotic areas by a sterile scalpel; the tissue is crushed in sterile water, saline, or buffer solution and is streaked onto a solid medium, such as nutrient agar or yeast dextrose chalk agar (YDC) (0.5% yeast extract, 1% dextrose, 3% calcium carbonate) (Dye, 1968). The isolation of *E. tracheiphila* is more easily accomplished by aseptically cutting the affected stem, placing the two cut stem surfaces together, and then gently pulling apart, removing a portion of the threads of bacteria and placing them in nutrient broth or onto a solid medium (Burkholder, 1960b). The delicate growth of *E. tracheiphila* will appear in 3–4 d; frequent transfer is necessary, but virulence may be reduced or lost after repeated transfers.

The isolation of some *Erwinia* species can be facilitated by use of selective-differential media, but such media are usually not necessary. *E. amylovora* will grow on MS medium (Miller and Schroth, 1972) and produce characteristic colonies. Later, Schroth and Hildebrand (1980) substituted D-mannitol by D-sorbitol in the MS medium for the isolation of *E. amylovora*. The medium of Crosse and Goodman (1973)[1] or CCT medium[2] can also be used for *E. amylovora*. A soluble pink pigment is produced by *E. persicina* and *E. rhapontici* (Hao et al., 1990) on medium containing 1% casitone, 0.3% yeast extract, 0.5% NaCl, and 0.7% agar.

MAINTENANCE PROCEDURES

Stock cultures of *Erwinia* species should be grown on standard media of choice at 25–30°C until good growth occurs. The cul-

---

Editorial Note: Erwinia persicina was originally known as *Erwinia persicinus*; the name was corrected by Euzéby (1998).

1. Crosse and Goodman's medium consists of sucrose, 160 g; nutrient agar, 12 g; 0.1% crystal violet in absolute ethanol, 0.8 ml; 0.1% cycloheximide, 20 ml; 380 ml distilled water.

2. CCT medium consists of sucrose, 100 g; D-sorbitol, 10 g; 1% tergitol anionic 7, 30 ml; 0.1% crystal violet in absolute ethanol, 2 ml; nutrient agar (Difco), 23 g; 970 ml distilled water; after autoclaving add 2 ml 1% (w/v) thallium nitrate and 50 mg cycloheximide (Ishimaru and Klos, 1984).

TABLE BXII.γ.223. Diagnostic characteristics of the genera *Erwinia, Brenneria, Enterobacter, Pantoea,* and *Pectobacterium*[a]

Characteristic	Erwinia	Brenneria	Enterobacter	Pantoea	Pectobacterium
Biochemical characteristics:					
Citrate (Simmons)	nd	nd	+	nd	D
Production of acetoin	d	d	D	+	+
Arginine dihydrolase	−	−	nd	nd	D
Arginine decarboxylase	nd	−	nd	nd	−
Lysine decarboxylase	nd	−	D	nd	−
Ornithine decarboxylase	nd	−	+	nd	−
Esculine hydrolase	D	D	nd	nd	+
Caseinase	−	−	nd	nd	D
Pectinase	−	D	nd	nd	D
Phenylalanine deaminase	−	−	−	nd	D
Urease	−	D	nd	nd	−
mol% G + C of DNA	49–54	50–56	52–60	55–57	50–56
Acid production from:					
Adonitol	−	−	D	nd	−
N-acetylglucosamine	D	d	nd	nd	+
Inulin	−	nd	nd	nd	D
D-Lyxose	D	nd	nd	nd	−
Maltose	−	D	nd	+	D
D-Mannitol	D	+	nd	+	
D-Mannose	D	+	nd	+	+
L-Rhamnose	D	+	nd	nd	+
D-Ribose	D	+	nd	+	+
Salicin	D	+	nd	nd	+
Trehalose	D	D	nd	+	D
D-Xylose	D	D	nd	+	
Utilization of carbon sources:					
Acetate	D	nd	nd	nd	+
4-Aminobutyrate	nd	nd	nd	+	−
Arbutin	D	+	nd	nd	+
Citrate	+	D	+	nd	+
Maltose	nd	D	nd	nd	D
Mannose	D	+	nd	nd	+
L-Proline	D	nd	nd	nd	nd
Sucrose	nd	d	nd	nd	+
Trehalose	nd	D	nd	nd	d
16S rDNA signature nucleotides:					
E. coli numbering position:					
379	C	G	C	C	C
384	G	C	S	G	G
408	A	R	A	A	G
434	Y	Y	T	T	C
593	Y	T	C	T	Y
598	C	Y	C	C	T
599	R	R	G	A	G
638	Y	Y	C	T	C
639	G	R	G	G	A
646	G	R	G	G	A
839	C	G	C	C	C
847	G	C	G	G	G
848	H	Y	M	A	C
987	G	A	R	A	R
988	G	C	S	C	C
989	C	T	T	T	T
1216	G	A	A	A	A
1217	C	R	S	G	G
1218	C	T	Y	T	T
1308	C	T	T	C	T
1329	G	A	A	G	A

[a]For symbols see standard definitions; nd, not determined.

tures can be maintained for a maximum of 3 weeks in a refrigerator (4–5°C).

For long-term preservation, erwiniae can be successfully stored as lyophilized cultures usually suspended in a filter sterilized mixture of 200 ml horse serum (Oxoid SR035C) to which 1675 mg nutrient broth (Oxoid) and 20 g glucose in 67 ml distilled water is added. Strains have also been stored in liquid nitrogen and in glycerol at −80°C (broth + 15% glycerol). Regular viability controls are recommended.

TABLE BXII.γ.224. Diagnostic characteristics of *Erwinia* species[a]

Characteristic	1. *E. amylovora*	2. *E. billingiae*	3. *E. mallotivora*	4. *E. persicina*	5. *E. psidii*	6. *E. rhapontici*	7. *E. tracheiphila*
Pink diffusible pigment	−	nd	−	+	−	+	−
Mucoid growth	+	nd	+	nd	nd	+	−
Growth in KCN broth	−	nd	−	nd	nd	+	−
Growth in 5% NaCl	nd	nd	−	nd	nd	+	−
Growth factors required	+	nd	+	nd	nd	−	+
NO$_3^-$ → NO$_2^-$	−	+	−	+	−	+	−
Levan	+	nd	+	nd	+	nd	+
H$_2$S from cysteine	−	nd	v	nd	+	+	+
Production of acetoin	+	+	+	+	−	v	d
Esculin hydrolase	+	nd	−	nd	nd	nd	nd
Gelatinase	+	−	v	nd	v	−	−
Acid production from:							
N-Acetylglucosamine	nd	+	nd	−	nd	nd	+
Amygdalin	nd	−	nd	−	nd	nd	+
L-Arabinose	d	+	−	−	+	+	v
D-Cellobiose	−	−	−	+	−	+	−
Dulcitol	−	nd	v	−	+	d	−
Esculin	−	nd	−	nd	nd	+	−
Gentiobiose	+	+	−	−	nd	nd	−
D-Gluconate	nd	+	nd	−	nd	nd	nd
α-Methylglucoside	−	−	−	−	−	d	v
β-Methylglucoside	+	nd	+	+	+	+	−
Glycerol	−	−	−	−	−	+	v
Myo-inositol	−	nd	v	−	+	+	−
Inulin	−	nd	−	−	nd	+	−
Lactose	−	−	−	+	−	+	−
Lyxose	nd	−	nd	−	nd	+	nd
Maltose	−	nd	−	nd	nd	+	−
D-Mannose	−	+	+	nd	+	+	v
Melezitose	−	−	−	−	nd	+	−
Melibiose	−	−	−	+	nd	+	−
Raffinose	−	−	v	+	−	+	−
L-Rhamnose	−	+	−	+	+	+	v
D-Ribose	+	nd	+	+	+	+	−
Salicin	−	nd	v	nd	+	+	−
Starch	−	nd	−	−	nd	+	−
Trehalose	+	+	v	+	−	+	−
Xylitol	nd	−	nd	−	nd	+	nd
D-Xylose	−	+	v	−	−	d	−
Utilization of carbon sources:							
Acetate	+	−	+	+	−	nd	+
L-Arabinose	−	nd	−	+	+	nd	v
D-Arabitol	−	+	−	−	−	−	v
Arbutin	nd	+	−	+	nd	nd	+
D-Cellobiose	nd	nd	−	+	nd	nd	−
Citrate	+	−	+	+	nd	+	+
Formate	+	nd	−	nd	nd	+	d
Galacturonic acid	−	nd	−	nd	nd	d	−
L-Glutamate	−	nd	−	+	+	nd	−
Glycerol	nd	nd	nd	+	nd	nd	−
L-Histidine	−	nd	−	+	−	nd	−
DL-Lactate	+	nd	−	+	nd	+	−
Lactose	nd	nd	nd	+	nd	nd	−
Malonate	−	−	−	nd	−	+	−
D-Mannose	−	nd	+	+	+	nd	v
Melibiose	+	nd	−	+	−	nd	−
Proline	−	nd	−	+	−	nd	−
Raffinose	−	nd	−	+	−	nd	−
Tartrate	−	nd	−	nd	−	d	v
D-Xylose	−	nd	+	nd	−	nd	−
Utilization of nitrogen sources:							
L-Isoleucine	+	nd	−	nd	nd	nd	nd
Threonine	+	nd	−	nd	nd	nd	nd
Tryptamine	nd	nd	−	nd	nd	+	−
Xanthin	nd	nd	+	nd	nd	nd	−
Sensitivity toward:							
Furazolidone	+	nd	−	nd	nd	nd	nd

[a]For symbols see standard definitions; nd, not determined.

TABLE BXII.γ.225. Additional characteristics of *Erwinia* species[a]

Characteristic	1. *E. amylovora*	2. *E. billingiae*	3. *E. mallotivora*	4. *E. persicina*	5. *E. psidii*	6. *E. rhapontici*	7. *E. tracheiphila*
Levan	+	nd	+	nd	+	nd	+
Production of indole	−	−	−	nd	−	−	−
Oxidase	−	−	−	−	−	−	−
Catalase	+	+	+	+	+	+	+
Alkaline phosphatase	nd	nd	nd	+	nd	nd	nd
Arginine dihydrolase	−	−	−	−	−	−	−
Caseinase	−	nd	−	−	−	−	−
Arylsulfatase	nd	nd	nd	−	nd	nd	nd
Esterase	nd	nd	nd	−	nd	nd	nd
β-Galactosidase	nd	+	nd	nd	nd	nd	+
β-Glucosidase	nd	nd	nd	+	nd	nd	nd
β-Glucuronidase	nd	nd	nd	−	nd	nd	nd
Lysine decarboxylase	nd	−	nd	nd	nd	nd	−
Nucleoside phosphotransferase	nd	nd	nd	−	nd	nd	nd
Pectinase	−	−	−	−	−	−	−
Phenylalanine deaminase	−	nd	−	−	−	−	−
Starch hydrolase	−	nd	−	nd	−	nd	−
Tetrathionate reductase	nd	−	nd	−	nd	nd	nd
Tween 80 hydrolase	nd	−	nd	−	nd	nd	nd
Urease	−	−	−	−	−	−	−
Acid production from:							
Adonitol	−	−	−	−	−	−	−
Dextrin	−	nd	−	−	−	−	−
Dulcitol	−	−	−	−	−	−	−
Erythritol	nd	−	nd	nd	nd	nd	−
Fructose	+	+	+	+	+	+	+
Fucose	nd	−	nd	−	nd	nd	−
D-Galactose	+	+	+	+	+	+	+
D-Glucose	+	+	+	+	+	+	+
Inulin	−	−	−	−	−	−	−
2-Ketogluconate	nd	nd	nd	−	nd	nd	nd
5-Ketogluconate	nd	nd	nd	−	nd	nd	nd
Maltose	−	nd	−	−	−	−	−
Mannitol	v	+	+	nd	+	+	v
α-Methylmannoside	nd	−	nd	−	nd	nd	−
β-Methylxyloside	nd	−	nd	nd	nd	nd	−
Sorbitol	d	nd	v	+	+	+	v
Sorbose	nd	−	nd	−	nd	nd	−
Starch	−	−	−	−	−	−	−
Sucrose	+	nd	+	+	+	+	+
Tagatose	−	−	−	−	−	−	−
D-Turanose	nd	−	nd	−	nd	nd	nd
Utilization of carbon sources:							
N-Acetylglucosamine	nd	nd	nd	+	nd	nd	nd
Aconitate	nd	nd	nd	+	nd	nd	nd
Adipate	nd	nd	nd	−	nd	nd	−
Adipinic acid	nd	nd	nd	nd	nd	nd	−
Adonitol	nd	nd	nd	nd	nd	nd	−
DL-α-Alanine	nd	nd	nd	+	nd	nd	nd
Amygdalin	nd	nd	nd	+	nd	nd	nd
D-Arabinose	nd	nd	nd	−	nd	nd	nd
L-Arabitol	−	nd	−	−	−	−	−
L-Aspartate	nd	+	nd	+	nd	nd	nd
Benzoate	−	nd	−	−	−	−	−
Betaine	nd	nd	nd	nd	nd	nd	−
Butanol	−	nd	−	−	−	−	−
Butyrate	nd	nd	nd	−	nd	nd	nd
Caprylate	nd	nd	nd	−	nd	nd	nd
Cysteine	nd	nd	nd	−	nd	nd	nd
Dextrin	nd	nd	nd	nd	nd	nd	−
Dulcitol	nd	nd	nd	−	nd	nd	nd
Erythritol	nd	nd	nd	+	nd	nd	nd
Esculin	nd	nd	nd	+	nd	nd	nd
Ethanol	nd	nd	nd	nd	nd	nd	−
Ethylene glycol	nd	nd	nd	nd	nd	nd	−
Fructose	nd	nd	nd	+	nd	nd	nd
Fucose	nd	nd	nd	−	nd	nd	nd
Fumarate	+	nd	+	+	+	+	+
Galactose	+	nd	+	+	+	+	+
Gluconate	+	nd	+	+	+	+	+
Glucose	+	nd	+	+	+	+	+
DL-Glycerate	nd	nd	nd	+	nd	nd	nd

(continued)

TABLE BXII.γ.225. *(cont.)*

Characteristic	1. *E. amylovora*	2. *E. billingiae*	3. *E. mallotivora*	4. *E. persicina*	5. *E. psidii*	6. *E. rhapontici*	7. *E. tracheiphila*
Glycerol	+	nd	+	+	+	+	+
L-Glycine	nd	nd	nd	−	nd	nd	nd
L-Histamine	nd	nd	nd	−	nd	nd	nd
myo-Inositol	nd	nd	nd	+	nd	nd	nd
Inulin	nd	nd	nd	−	nd	nd	nd
2-Ketogluconate	nd	nd	nd	+	nd	nd	nd
5-Ketogluconate	nd	nd	nd	+	nd	nd	nd
Lactulose	nd	nd	−	nd	nd	nd	−
L-Leucine	nd	nd	nd	−	nd	nd	nd
L-Lysine	nd	nd	nd	−	nd	nd	nd
Lyxose	nd	nd	nd	−	nd	nd	−
Malate	+	+	+	+	+	+	+
Malonic acid	nd	nd	nd	nd	nd	nd	−
Maltose	nd	nd	+	+	nd	nd	+
D-Mannitol	nd	nd	nd	+	nd	nd	nd
Melezitose	nd	nd	nd	−	nd	nd	nd
Methanol	−	nd	−	−	−	−	−
β-Methylglucoside	+	nd	+	+	+	+	+
Naphthalene	nd	nd	nd	nd	nd	nd	−
L-Ornithine	nd	nd	nd	−	nd	nd	nd
Oxalate	−	nd	−	−	−	−	−
Pectinic acid	nd	nd	nd	nd	nd	nd	−
Propanol	nd	nd	nd	nd	nd	nd	−
Propionate	−	nd	−	−	−	−	−
Propionic acid	nd	nd	nd	nd	nd	nd	−
L-Rhamnose	nd	nd	nd	+	nd	nd	nd
Ribose	nd	nd	nd	+	nd	nd	+
Salicin	nd	nd	nd	+	nd	nd	+
L-Serine	nd	nd	nd	+	nd	nd	nd
Sorbinic acid	nd	nd	nd	nd	nd	nd	−
Sorbitol	+	nd	nd	+	nd	nd	+
Sorbose	−	nd	−	−	−	−	−
Spermine	nd	nd	nd	−	nd	nd	nd
Starch	nd	nd	nd	−	nd	nd	−
Succinate	+	nd	+	+	+	+	+
Sucrose	nd	nd	nd	+	nd	nd	+
D-Tagatose	nd	nd	nd	−	nd	nd	nd
L-Threonine	nd	nd	nd	−	nd	nd	nd
Trehalose	nd	nd	nd	+	nd	nd	nd
Tryptamine	nd	nd	nd	−	nd	nd	nd
D-Turanose	nd	nd	nd	−	nd	nd	nd
Urea	nd	nd	nd	−	nd	nd	nd
L-Valine	nd	nd	nd	−	nd	nd	nd
Xylitol	nd	nd	nd	nd	nd	nd	−
Utilization of nitrogen sources:							
L-Alanine	+	nd	+	+	+	+	+
Allantoin	nd	nd	nd	nd	nd	nd	+
Ammonium chloride	nd	nd	nd	nd	nd	nd	+
Anthranilic acid	nd	nd	−	nd	nd	nd	nd
Arginine	nd	nd	nd	nd	nd	nd	+
Betaine	nd	nd	nd	nd	nd	nd	−
Choline	nd	nd	nd	nd	nd	nd	−
Citrulline	nd	nd	nd	nd	nd	nd	+
Cysteamine	nd	nd	nd	nd	nd	nd	−
Glucosamine	nd	nd	nd	nd	nd	nd	+
L-Glutaminic acid	+	nd	+	+	+	+	+
Glutathion	nd	nd	nd	nd	nd	nd	+
Glycylglycine	+	nd	+	+	+	+	+
Hydroxyproline	nd	nd	nd	nd	nd	nd	−
Kynureninic acid	−	nd	−	−	−	−	−
Leucine	nd	nd	nd	nd	nd	nd	+
L-Methionine	+	nd	nd	nd	nd	nd	nd
Phenylalanine	nd	nd	nd	nd	nd	nd	+
Quinolinic acid	nd	nd	nd	nd	nd	nd	−
Sarcosine	nd	nd	nd	nd	nd	nd	−
L-Serine	+	nd	+	+	+	+	+
Spermidine	nd	nd	nd	nd	nd	nd	−
Spermine	nd	nd	nd	nd	nd	nd	−
Thymine	nd	nd	−	nd	nd	nd	−
Trigonelline	−	nd	−	−	−	−	−
L-Tryptophan	nd	nd	+	nd	nd	nd	nd
Tyramine	nd	nd	nd	nd	nd	+	nd

(continued)

TABLE BXII.γ.225. *(cont.)*

Characteristic	1. *E. amylovora*	2. *E. billingiae*	3. *E. mallotivora*	4. *E. persicina*	5. *E. psidii*	6. *E. rhapontici*	7. *E. tracheiphila*
Tyrosine	nd	nd	nd	nd	nd	nd	+
Valine	nd	nd	−	nd	nd	nd	nd
Sensitivity to antibiotics:							
Amikacin	nd	nd	nd	+	nd	nd	nd
Amoxycillin	nd	nd	nd	nd	nd	+	+
Ampicillin	nd	nd	+	nd	nd	+	+
Carbenicillin	nd	nd	+	nd	nd	nd	+
Cephalexin	nd	nd	+	nd	nd	nd	+
Cephaloridine	nd	nd	+	nd	nd	nd	+
Cephalotine	nd	nd	+	nd	nd	nd	+
Chloramphenicol	+	nd	+	+	+	+	+
Clindamycin	nd	nd	nd	+	nd	nd	nd
Erythromycin	nd	nd	nd	+	nd	+	nd
Framycetine	nd	nd	nd	nd	nd	nd	+
Furazolidone	+	nd	+	+	+	+	+
Fusidinic acid	nd	nd	−	nd	nd	nd	−
Gentamicin	nd	nd	nd	+	nd	nd	nd
Kanamycin	nd	nd	+	nd	nd	nd	+
Lincomycin	nd	nd	nd	+	nd	nd	nd
Methicillin	nd	nd	nd	nd	nd	nd	−
Minocyclin	nd	nd	nd	+	nd	nd	nd
Nalidixinic acid	+	nd	+	+	+	+	+
Nitrofurantoin	nd	nd	nd	+	nd	nd	nd
Novobiocin	nd	nd	−	nd	nd	nd	nd
Oxytetracycline	+	nd	+	+	+	+	+
Penicillin G	nd	nd	nd	−	nd	nd	nd
Polymyxin B	nd	nd	nd	+	nd	nd	nd
Spectinomycin	nd	nd	+	nd	nd	nd	nd
Streptomycin	v	nd	+	+	nd	nd	nd
Sulfafurazol	nd	nd	−	nd	nd	nd	−
Tetracycline	+	nd	+	+	+	+	+

^aFor symbols see standard definitions; nd, not determined.

DIFFERENTIATION OF THE GENUS *ERWINIA* FROM OTHER GENERA

Characteristics that differentiate *Erwinia* from the genera *Pectobacterium*, *Pantoea*, and *Brenneria* are given in Table BXII.γ.223. As it is very difficult to differentiate them phenotypically; genomic methods are recommended for differentiation.

TAXONOMIC COMMENTS

The genus *Erwinia*, named after the phytobacteriologist Erwin F. Smith, was created in 1920 to unite all Gram-negative, fermentative, nonsporulating, peritrichously flagellated plant-pathogenic bacteria (Winslow et al., 1920). The taxonomy of the genus *Erwinia* and designation of species in the genus has been complicated by the heterogeneity of the strains included in the taxon. It has been suggested in the past that members of the genus be placed into new groupings with other members of the *Enterobacteriaceae* (Starr and Mandel, 1969; White and Starr, 1971). This concept also was supported by studies of DNA–DNA relatedness (Gardner and Kado, 1972), DNA relatedness (Brenner et al., 1974a), and DNA–DNA segmental relatedness (Murata and Starr, 1974).

Waldee (1945) suggested that *Erwinia* should be limited to pathogens (*E. amylovora*, *Erwinia salicis*, and *E. tracheiphila*) that cause necrotic or wilt diseases, utilize a restricted range of carbon compounds, and usually require organic nitrogen compounds for growth, and that the biochemically more active soft rotting pathogens (*Erwinia carotovora* and *Erwinia chrysanthemi*) should be placed in a separate genus *Pectobacterium*. Although some work-

ers have supported this suggestion (Brenner et al., 1973b, 1974a), it was not generally accepted because some species are taxonomically intermediate between these two groups, i.e., they resemble *E. carotovora* in most of their characteristics but do not cause rots.

Taxonomic changes since the previous edition of *Bergey's Manual of Systematic Bacteriology* The species *Erwinia herbicola*, *Erwinia milletiae* (Gavini et al., 1989b), *Erwinia ananas*, *Erwinia uredovora*, and *Erwinia stewartii* (Mergaert et al., 1993) have been transferred to the genus *Pantoea*. The species incertae sedis *Erwinia dissolvens*, *Erwinia nimipressuralis* (Brenner et al., 1986), and *E. cancerogena* (Dickey and Zumoff, 1988) have been transferred to the genus *Enterobacter*. Five new *Erwinia* species and two new subspecies were created: *Erwinia alni* (Surico et al., 1996), *Erwinia billingiae* (Mergaert et al., 1999), *Erwinia cacticida* (Alcorn et al., 1991), *Erwinia persicina* (Hao et al., 1990), *Erwinia psidii* (Rodrigues-Neto et al., 1987), *E. carotovora* subsp. *odorifera* (Gallois et al., 1992), and *E. carotovora* subsp. *wasabiae* (Goto and Matsumoto, 1987).

Confusion exists about the organisms named *Erwinia carnegieana*. None of the original isolates are available and no strains have been isolated corresponding to the original description (Alcorn and Orum, 1988). The type strain of *E. carnegieana* NCPPB 439 is nonpectinolytic and non-plant pathogenic and was identified as a *Klebsiella pneumoniae* (Graham, 1964; Edwards and Ewing, 1972; Alcorn and Orum, 1988).

Recently obtained sequence data allowed estimations of phylogenetic relationships, shed a new light on these earlier viewpoints, and raised new perspectives for the taxonomy of the genus

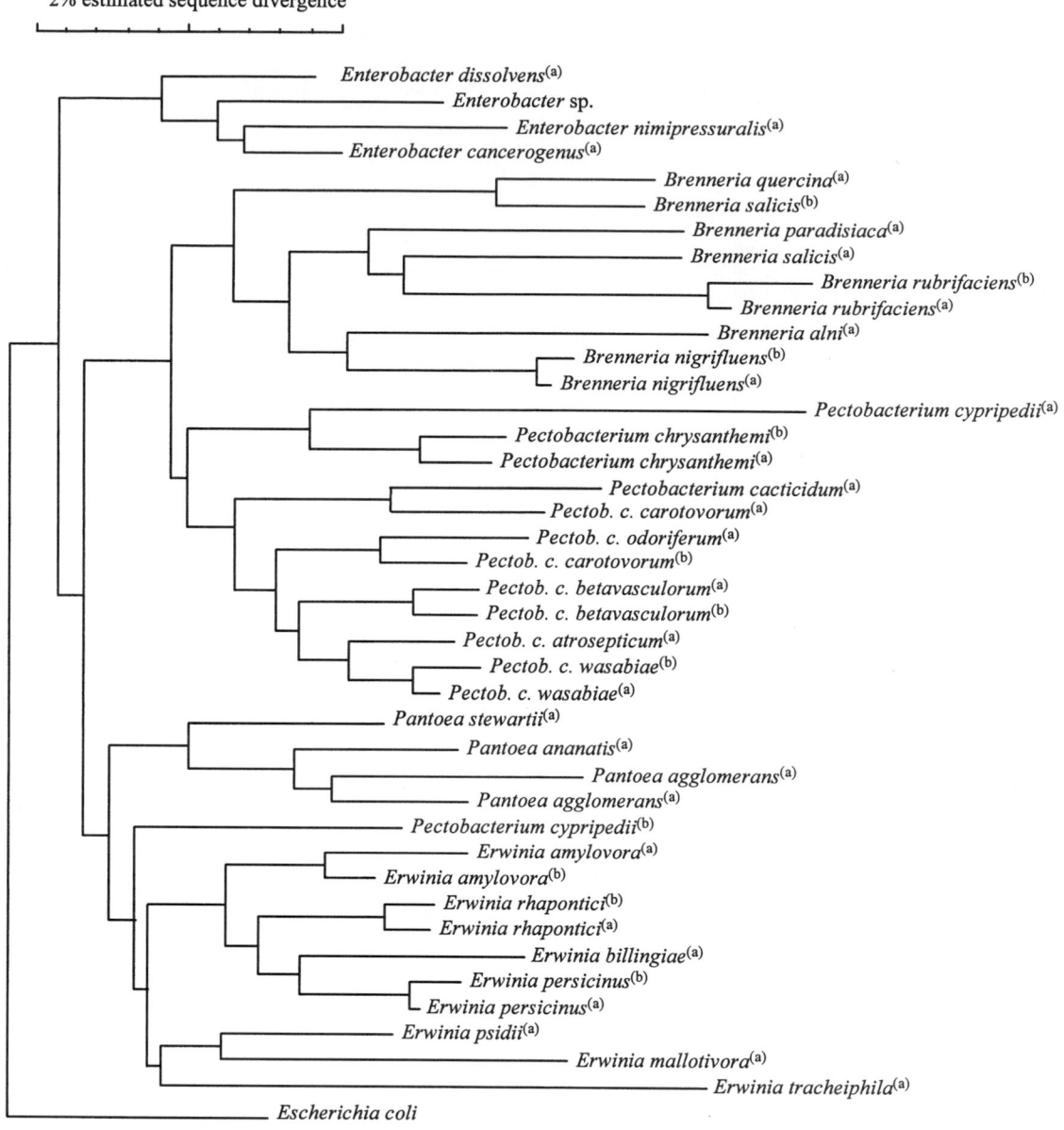

2% estimated sequence divergence

FIGURE BXII.γ.201. Neighbor-joining dendrogram depicting the estimated phylogenetic relationships among the genera *Erwinia*, *Pectobacterium*, *Brenneria*, *Pantoea*, and *Enterobacter*, based on pairwise comparisons of nearly complete 16S rDNA sequences calculated with the software package GeneCompar (Applied Maths, Kortrijk, Belgium), after removal of all unknown bases and gaps. The distance between two species is obtained by summing the lengths of the connecting horizontal branches using the scale on the top. *Pectob. c.* = *Pectobacterium carotovorum* subspecies; *(a)*: 16S rDNA sequence obtained from Hauben et al. (1998a); *(b)*: 16S rDNA sequence obtained from Kwon et al. (1997).

Erwinia. After comparison of the 16S rDNA sequences of representatives of the genus *Erwinia* with other members of the *Enterobacteriaceae*, the phylogenetic positions of the *Erwinia* species were determined, and led to the emended description of the genera *Erwinia* and *Pectobacterium* and to the description of the new genus *Brenneria* (Hauben et al., 1998a).

DIFFERENTIATION OF THE SPECIES OF THE GENUS *ERWINIA*

The characteristics of the species of *Erwinia* are given in Tables BXII.γ.223, BXII.γ.224, and BXII.γ.225. Only four strains of *E. tracheiphila* were studied, and data for this species should be treated with caution.

List of species of the genus Erwinia

1. **Erwinia amylovora** (Burrill 1882) Winslow, Broadhurst, Buchanan, Krumwiede, Rogers and Smith 1920, 209[AL] emend. Hauben, Moore, Vauterin, Steenackers, Mergaert, Verdonck and Swings 1999a, 1 (*Micrococcus amylovorus* Burrill 1882, 134.)

a.my.lo'vo.ra. Gr. n. *amylum* starch; L. v. *voro* to devour; M.L. fem. adj. *amylovora* starch-destroying.

The characteristics are as given for the genus and as listed in Tables BXII.γ.224 and BXII.γ.225.

The virulence of *E. amylovora* is often correlated with its capacity to produce extracellular polysaccharides (or EPS, e.g., amylovoran), lipopolysaccharide, a high-molecular-mass levan, a low-molecular-mass glucan (Roberts and Coleman, 1991), and levansucrase (Gross et al., 1992). Levansucrase-negative mutants show retarded development of necrotic symptoms on inoculated pear seedlings (Geier and Geider, 1993). RcsA and RcsB are two regulator proteins that interact with the promoter of the operon for amylovoran biosynthesis (Bereswill and Geider, 1997; Kelm et al., 1997). The genes *amsA* to *amsI* are required for exopolysaccharide synthesis (Bugert and Geider, 1995, 1997).

After 2–3 days at 27°C, colonies on 5% sucrose nutrient agar are typically white, domed, shining, mucoid (levan type) with radial striations, and a dense flocculent center or central ring. Nonlevan forms are isolated rarely.

Several regions of the bacterial DNA of *E. amylovora* are species specific and can be used for identification by (1) a PCR assay with primers AMSbL (5′-GCTACCAGCAGGGT-GAG-3′) and AMSbR (5′-TCATCACGATGGTGTAG-3′) derived from the ams region, with an annealing temperature of 49°C (Bereswill et al., 1995); (2) a typical banding pattern derived from arbitrarily primed PCR with oligonucleotide APT1 (5′-CAGGACGCTACTTGTGT-3′), annealing temperature 32°C, from the sequence of transposon Tn5 (Bereswill et al., 1995); (3) a PCR assay with primers, primer A (5′-CGGTTTTTAACGCTGGG-3′) and primer B (5′-GGGCAAATACTCGGATT-3′), annealing at 52°C, derived from a 29-kb plasmid (pEA29) that modulates the development of fire blight (Falkenstein et al., 1989; Bereswill et al., 1992); (4) nested PCR with primers A (5′-CGGTTT-TTAACGCTGGG-3′), B (5′-GGGCAAATACTCGGATT-3′), AJ75 (5′-CGTATTCACGGCTTCGCAGAT-3′), and AJ76 (5′-ACCCGCCAGGATAGTCGCATA-3′) with an annealing temperature of 52°C, and PCR-dot-blot hybridizations derived from pEA29 (McManus and Jones, 1995); and (5) a PCR assay with primers EaF (5′-GCGCAGTAAAGGGTGA-CAGCCCCGTACACAAAAAGGCACT-3′) and EaR (5′-CCC-TAGCCGAAACAGTGCTCTACCCCCGG-3′) derived from the 23S rDNA sequence, with an annealing temperature of 72°C (Maes et al., 1996).

Repetitive element PCR (rep-PCR) (McManus and Jones, 1994) and RAPD fingerprinting (Momol et al., 1997) can distinguish *E. amylovora* strains isolated from Pomaceae from strains isolated from *Rubus* (subfamily Rosaceae). Pulsed field gel electrophoresis (PFGE) can distinguish *E. amylovora* strains of different geographical origin (Zhang and Geider, 1997).

E. amylovora produces harpins, bacterial protein elicitors that induce hypersensitive response-like necrosis when infiltrated into nonhost plants such as tobacco (Wei et al., 1992). Harpin activates a myelin basic protein kinase in tobacco leaves (Popham et al., 1995; Adam et al., 1997) and elicits active oxygen production in suspension cells (Baker et al., 1993). The secretion of harpin occurs via a type III pathway, similar to the export system of pathogenic proteins of *Yersinia* spp. (Bogdanove et al., 1996). These proteins are encoded by the *hrp* gene cluster, consisting of seven transcriptional units (Wei and Beer, 1995; Kim et al., 1997b). The *dsp* genes also affect pathogenicity but not the eliciting of a hypersensitive response (Roberts and Coleman, 1991). The expression of the *hrp* and *dsp* genes is temperature-sensitive.

Plasmid pULB113 was found in *E. amylovora*, mediating chromosomal mobilization and R-prime formation (Chatterjee et al., 1985). The self-transmissible 56-kb plasmid Ea322 is not involved in pathogenicity (Steinberger et al., 1990). pEa8.7, an 8.7-kb plasmid was found in streptomycin- and sulfonamide-resistant strains (Palmer et al., 1997). The genes *strA* and *strB* were found in streptomycin-resistant strains (Chiou and Jones, 1991, 1995; McManus and Jones, 1994).

Cells infected with bacteriophage ERA103 produce an enzyme that degrades the extracellular polysaccharide of noninfected cells (Vandenburgh et al., 1985). *E. amylovora* cells are sensitive to bacteriophage Mu, which can be used as a genetic tool (Vanneste et al., 1990).

Several monoclonal antibodies have been produced against *E. amylovora* strains (Schaad, 1979) that can be used for the detection of the bacterium in suspensions as well as in contaminated twigs, plant fluids, and fruit tissues. The following methods can be applied: (1) Slide agglutination tests: agglutination with *E. amylovora* antiserum is a common and accurate method of determination (Lelliott, 1968; Zielke et al., 1993; Arsenijevic et al., 1994); the species is serologically homogeneous and has few agglutinogens in common with related species or with the saprophytes found in diseased material. (2) Immunofluorescence techniques (IFT) (Lin et al., 1987; Zielke et al., 1993). (3) Enzyme-linked immunosorbent assays (double-antibody sandwich [DAS]-ELISA) (Laroche and Verhoyen, 1984; Lin et al., 1987; Zielke et al., 1993). The sensitivity of the three methods is approximately the same, i.e., 10^5 cells/ml.

Pantoea agglomerans and *Pseudomonas fluorescens* are antagonistic toward *E. amylovora* (Wilson et al., 1992; Wilson and Lindow, 1993; Wodzinski et al., 1994; Kearns and Hale, 1995, 1996; Mercier and Lindow, 1996).

E. amylovora causes a necrotic disease (fire blight) of most species of the Pomaceae and of some species in other subfamilies of the Rosaceae. Early in the season blossom and tips of twigs die back. Leaves shrivel and blacken and remain attached to the blackening stem that frequently crooks characteristically at the tip. The disease may progress into larger branches or even the trunk and form a canker. A *forma specialis* has been described from raspberry (*Rubus idaeus*) by Starr and Folsom (1951).

The mol% G + C of the DNA is: 53.6–54.1 (Bd).

Type strain: ATCC 15580, ICMP 1540, LMG 2024, NCPPB 683.

GenBank accession number (16S rRNA): Z96088, U80195.

2. **Erwinia aphidicola** Harada, Oyaizu, Kosako and Ishikara 1998, 1083[VP] Harada, Oyaizu, Kosako and Ishikara 1997, 354

a.phi' di.co.la. L. *aphid* the aphid; L. suff. *-cola* dweller; M.L. n. *aphidicola* aphid dweller.

Cells are Gram-negative, oxidase-negative, catalase-positive, fermentative rods (0.5–0.6 × 1.6–2.0 μm), and motile by means of peritrichous flagella. All strains grow well on nutrient agar and common laboratory media. Colonies on peptone-yeast extract agar are circular and smooth. These organisms grow well at 25, 30, and 35°C, but not at 42°C. Cells are motile. Voges–Proskauer test is positive, and cells can reduce nitrate. Cells can utilize citrate and malonate. Pigment is not produced. Indole is not produced. No gelatin liquefaction or tryptophan deaminase activity. H_2S production from sodium thiosulfate. No arginine hydrolase activity, no lysine decarboxylase activity, no ornithine decarboxylase activity, no deoxyribonuclease activity, and no urease activity. Acid is produced from the following carbohydrates: glycerol, erythritol, L-arabinose, D-ribose, D-xylose, adonitol, D-galactose, D-glucose, D-fructose, D-mannose, L-rhamnose, inositol, D-mannitol, N-acetyl-D-glucosamine, amygdalin, arbutin, esculin, salicin, cellobiose, maltose, melibiose, sucrose, trehalose, raffinose, D-xylitol, β-gentiobiose, D-lyxose, D-fucose, D-gluconate, 2-keto-D-gluconate, and 5-keto-D-gluconate.

The mol% G + C of the DNA is: unknown.

Type strain: IAM 14479.

3. **Erwinia billingiae** Mergaert, Hauben, Cnockaert and Swings 1999, 382[VP]

bil.ling' i.ae. L. gen. n. *billingiae* of Billing, named after Eve Billing, who first isolated these organisms.

The characteristics are as given for the genus and are further listed in Tables BXII.γ.224 and BXII.γ.225.

Antisera prepared against live or heat-killed cells, non-purified or purified immunogens have been used for the differentiation or identification of *E. tracheiphila* (Elrod, 1946).

Strains were isolated from stem cankers, diseased blossoms, and immature fruits mainly of rosaceous trees, often in association with plant pathogens, and are considered as secondary invaders rather than primary pathogens.

The mol% G + C of the DNA is: 54.1–55.1 (T_m, Bd)

Type strain: LMG 2613, NCPPB 661.

GenBank accession number (16S rRNA): Y13249.

4. **Erwinia mallotivora** Goto 1976, 472[AL] emend. Hauben, Moore, Vauterin, Steenackers, Mergaert, Verdonck and Swings 1999a, 1.

mal.lo.ti' vo.ra. M.L. n. *Mallotus* a genus of trees; L. v. *voro* to devour; M.L. adj. *mallotivora Mallotus*-destroying.

The characteristics are as described for the genus and are further listed in Tables BXII.γ.224 and BXII.γ.225.

Colonies on nutrient agar without sucrose are 1.5 mm in diameter, white, raised, transparent, and circular with smooth surfaces and entire margins after 2 d. Colonies on nutrient agar with 5% sucrose are flat, white, circular with entire margins and smooth surfaces, butyrous, and transparent after 1 d; after 4 d colonies are domed, circular, white, mucoid, and translucent, and sometimes possess radial striations.

Causes a leaf spot of Akamegashiwa, Mongolian oak (*Mallotus japonicus*). Initially, dark green spots appear on the leaves next to the veins. As the disease evolves, these spots grow and turn dark brown with a yellow border. Eventually, the leaf dies and twigs wilt.

The mol% G + C of the DNA is: 49.8–51.0 (Bd).

Type strain: ATCC 29573, ICMP 5705, LMG 2708, NCPPB 2851.

GenBank accession number (16S rRNA): Z96084.

5. **Erwinia persicina** Hao, Brenner, Steigerwalt, Kosaka and Komagata 1990, 382[VP] (*"Erwinia nulandii"*)

per.si' cin.a. M.L. adj. *persicina* peach color, because of the pinkish or peach-colored pigment of this organism.

The characteristics are as described for the genus and are further listed in Tables BXII.γ.224 and BXII.γ.225.

Strains vary in their ability to utilize adonitol, L-arginine, β-gentiobiose, and glucosamine. Resistance to ampicillin, carbenicillin, cefazolin, cephalothin, colistin, kanamycin, mydecamycin, nalidixic acid, sufisoxazole, and tobramycin varies.

Strains were isolated from, but are not pathogenic for, tomato, banana, and cucumber. Strains are pathogenic to bean pods and seeds. One *E. persicina* strain was recently isolated from a human (O'Hara et al., 1998).

The mol% G + C of the DNA is: 52–54 (T_m).

Type strain: ATCC 35998, CDC 9108-82, LMG 11254.

GenBank accession number (16S rRNA): Z96086, U80205.

Additional Remarks: The type strain was isolated from a tomato by Komagata and Okada in 1962.

6. **Erwinia psidii** Rodrigues-Neto, Robbs and Yamoshiro 1988, 328[VP] (Effective publication: Rodrigues-Neto, Robbs and Yamoshiro 1987, 348.)

psi' di.i. L. n. *Psidium* generic name.

The characteristics are as described for the genus and are further listed in Tables BXII.γ.224 and BXII.γ.225.

Pathogenic to guava (*Psidium guajava* L.) causing a dieback of twigs and branches, characterized by a collapse of the vascular tissue. In artificial inoculations, the bacterium was also pathogenic to strawberry guava (*Psidium cattleianum* Lam.), jambolana (*Eugenia jambolana* Lam.), and "melaleuca" (*Melaleuca viridiflora* Brogn. and Gris.), and to all members of the Myrtaceae family.

The mol% G + C of the DNA is: is 51.7–52.2 (T_m).

Type strain: ATCC 49406, ICMP 8426.

7. **Erwinia pyrifoliae** Kim, Gardan, Rhim and Geider 1999, 905[VP]

py.ri.fo' liae. L. gen. fem. n. *pyrifoliae* of *pyrifolia*, the species name of the host plant, the Nashi pear, *Pyrus pyrifolia*.

Cells are Gram-negative, nonsporeforming, peritrichous, straight rods. The strains are facultatively anaerobic; oxidase is not produced. Nitrates are not reduced. This species conforms to the definition of the family *Enterobacteriaceae*. Strains grow on YPDA medium (1% yeast extract, 2% polypeptone, 2% glucose, and 20 μg/ml adenine hemisulfate), producing colonies that are 2 mm after 48 h at 28°C. Colonies are circular, white, well-domed, and opaque. Glucose (dextrose) is fermented without gas production. Voges–Proskauer test is (weakly) positive. Polypectate gel is not acidified or liquefied. Arginine dihydrolase, lysine decarboxylase and ornithine decarboxylase are not present. Indole is not produced from tryptophan. Reducing compounds are produced from sucrose. Acid is not produced from inulin, lactose, methyl-α-glucoside, melibiose, D-arabitol, D-arabinose, or raffinose. Acid is produced from mannitol, sorbitol, and saccharose. There is no alkalinization of malonate, citrate, or D-tartrate. β-Galactosidase is not produced. The following substrates are utilized as sole sources of carbon and energy: D-fructose, D-galactose, D-trehalose, saccharose, methyl-D-glucopyranoside, D-ribose, L-arabinose, glycerol, *myo*-inositol, D-mannitol, D-sorbitol, L-malate, *N*-acetyl-D-glucosamine, D-gluconate, succinate, fumarate, L-glutamate, and L-proline. The following substances are not utilized: D-mannose, D-sorbose, α-D-melibiose, D-raffinose, maltotriose, maltose, α-lactose, lactulose, methyl-β-galactopyranoside, D-cellobiose, β-gentiobiose, esculin, D-xylose, palatinose, α-L-rhamnose, α-L-fructose, D-arabitol, L-arabitol, xylitol, dulcitol, D-tagatose, maltitol, D-turanose, adonitol, hydroxyquinoline β-glucoside, 3-methyl-D-glucopyranoside, D-saccharate, mucate, L-tartrate, D-tartrate, *meso*-tartrate, D-malate, *trans*-aconitate, tricarballylate, D-glucuronate, D-galacturonate, keto-D-gluconate, L-tryptophan, phenylacetate, protocatechuate, *p*-hydroxybenzoate, quinate, *m*-hydroxybenzoate, benzoate, phenylpropionate, *m*-coumarate, trigonelline, betaine, putrescine, aminobutyrate, histamine, DL-lactate, caprate, caprylate, L-histidine, glutarate, DL-glycerate, aminovalerate, ethanolamine, tryptamine, itaconate, hydroxybutyrate, D-alanine, malonate, propionate, L-tyrosine, and ketoglutarate.

The pathogen was isolated from necrotic symptoms of leaves and branches of *Pyrus pyrifolia* cv. "Shingo". It is closely related to the fire blight pathogen *Erwinia amylovora* by the DNA sequence of the 16S rDNA, but distantly by its intergenic transcribed spacer region (ITS). It does not contain plasmid pEA29 (Rhim et al., 1999).

The mol% G + C of the DNA is: 52 (T_m).

Type strain: Ep16/96, CFBP 4172, CIP 106111, DSM 12163.

GenBank accession number (16S rRNA): AJ009930.

8. **Erwinia rhapontici** (Millard 1924) Burkholder 1948a, 475[AL] emend. Hauben, Moore, Vauterin, Steenackers, Mergaert, Verdonck and Swings 1999a, 1 (*Phytomonas rhapontica* (sic) Millard 1924, 11; *Pectobacterium rhapontici* (Millard 1924) Patel and Kulkarni 1951b, 80; *Erwinia carotovora* biovar rhapontici (Millard 1924) Dye 1969a, 93.)

rha.pon' ti.ci. M.L. n. *rhaponticum* specific epithet of *Rheum rhaponticum*, rhubarb; M.L. gen. n. *rhapontici* of rhubarb.

The characteristics are as described for the genus and are further listed in Tables BXII.γ.224 and BXII.γ.225. Produces proferrorosamine A.

Causes a crown rot of rhubarb (*Rheum rhaponticum*) that extends into the center of the root, pink grain of wheat that will show a cavity below the hilum (Roberts, 1974) and internal browning of hyacinth bulbs, and occurs epiphytically and saprophytically in lesions caused by other bacteria (Sellwood and Lelliott, 1978; Kokoskova, 1992). Rots potato, onion, and cucumber slices slowly, weakly, and erratically (Sellwood and Lelliott, 1978).

The mol% G + C of the DNA is: 51.0–53.1 (Bd).

Type strain: ATCC 29283, ICMP 1582, LMG 2688, NCPPB 1578.

GenBank accession number (16S rRNA): Z96087, U80206.

9. **Erwinia tracheiphila** (Smith 1895) Bergey, Harrison, Breed, Hammer and Huntoon 1923, 173[AL] emend. Hauben, Moore, Vauterin, Steenackers, Mergaert, Verdonck and Swings 1999a, 1 (*Bacillus tracheiphilus* Smith 1895, 364.) *tra.che.i' phi.la.* L. n. *trachia* the windpipe; Gr. adj. *philus* loving; M.L. adj. *tracheiphila* trachea-loving, i.e., growing in the tracheiphila of the vascular bundles.

The characteristics are as given for the genus and are further listed in Tables BXII.γ.224 and BXII.γ.225.

Antisera prepared against live or heat-killed cells, non-purified or purified immunogens have been used for the differentiation or identification of *E. tracheiphila* (Elrod, 1946).

Grows very poorly on nutrient agar but moderately well on yeast extract glucose chalk agar (YCC) (10 g glucose, 5 g yeast extract, 30 g $CaCO_3$, 1.5% agar, 1 l distilled water) or glucose nutrient agar.

Causes a vascular wilt of *Cucurbita* species, *Cirullus lanatus*, and *Cucumis melo*. At first dull wilted areas appear on leaves. Later whole leaves and stems wilt, shrivel, and die. Infection is systemic and bacterial ooze is usually obvious at the cut ends of vascular tissues.

The mol% G + C of the DNA is: 50–52 (T_m, Bd).

Type strain: ATCC 33245, LMG 2906, NCPPB 2452.

GenBank accession number (16S rRNA): Y13250.

Genus XIV. **Ewingella** *Grimont, Farmer, Grimont, Asbury, Brenner and Deval 1984a, 91*[VP]
(Effective publication: Grimont, Farmer, Grimont, Asbury, Brenner and Deval 1983a, 41)

CAROLINE M. O'HARA AND J.J. FARMER III

Ewing.el' la. M.L. dim. ending -*ella*; M.L. fem. n. *Ewingella* named to honor William H. Ewing an American bacteriologist who made many contributions to the nomenclature and classification of the families *Enterobacteriaceae* and *Vibrionaceae*.

Small rod-shaped cells, 0.6–0.7 × 1–1.8 μm, conforming to the general definition of the family *Enterobacteriaceae*. Contains the enterobacterial common antigen. Motile by 3–10 peritrichous flagella. Facultatively anaerobic. Gram negative, oxidase negative, catalase positive. **Some strains grow faster and are more active biochemically at 25°C than 36°C. Positive reactions for methyl red, Voges–Proskauer, citrate utilization (Simmons), ONPG test, nitrate reduction to nitrite, and acid production from D-glucose, D-mannitol, salicin, trehalose, D-arabitol, D-mannose, and D-ga-**

lactose. Utilize 35 of 156 carbon sources. Negative reactions for indole production, H_2S production (TSI), urea hydrolysis, phenylalanine deaminase, lysine decarboxylase, arginine dihydrolase, ornithine decarboxylase, gelatin hydrolysis (22°C), growth in KCN, malonate utilization, acetate utilization, lipase (corn oil), DNase production, pigment production, and fermentation of sucrose, dulcitol, adonitol, *myo*-inositol, D-sorbitol, L-arabinose, raffinose, maltose, D-xylose, cellobiose, α-methyl-D-glucoside, erythritol, melibiose, and mucate. No visible gas is produced during fermentation. Susceptible to colistin, nalidixic acid, sulfadiazine (most strains), gentamicin, streptomycin, kanamycin, tetracycline, chloramphenicol, and carbenicillin (disk diffusion method on Mueller-Hinton agar); **resistant to penicillin and cephalothin**; variable susceptibility to ampicillin.

Isolated from human clinical specimens that are normally sterile such as blood and urine. Also isolated from other clinical specimens such as throat, sputum, and wounds. **Isolated from outbreaks of bacteremia, and from outbreaks of pseudobacteremia traced to contaminated collection tubes.** Probably an **opportunistic pathogen** with a low capacity to cause extraintestinal infections in humans. Occurs in water, food, and mollusks. Implicated as the cause of internal stipe necrosis of cultivated mushrooms. **A rarely isolated genus of** *Enterobacteriaceae.*

The mol% G + C of the DNA is: 53.6–55.2.

Type species: **Ewingella americana** Grimont, Farmer, Grimont, Asbury, Brenner and Deval 1984a, 91 (Effective publication: Grimont, Farmer, Grimont, Asbury, Brenner and Deval 1983a, 41.)

FURTHER DESCRIPTIVE INFORMATION

Ewingella has been isolated from a variety of clinical specimens. In CDC's culture collection, blood and wound isolates account for 22 of 60 isolates. The blood isolates were usually from pneumonia or surgery patients. Five were from an outbreak of bacteremia in the intensive care unit of a community hospital (Pien and Bruce, 1986). All patients had undergone either cardiovascular or peripheral vascular surgery prior to infection. The probable source was a contaminated ice bath used to cool syringes for cardiac output determinations. *E. americana* was cultured from the bath. Twenty "blood isolates" and 15 environmental isolates were related to an outbreak of pseudobacteremia in a pediatric hospital that was traced to cross contamination from nonsterile citrated blood collection tubes (McNeil et al., 1987a). Additional case reports of bacteremia have been described by Pien et al. (1983) and DeVreese et al. (1992). In these reports, blood isolates were from immunocompetent patients who had undergone coronary bypass surgery and cholecystectomy, respectively.

Wounds of thumb, toe, hand, finger, and leg accounted for nine of 60 isolates. One isolate was thought to be a colonizer (Bear et al., 1986) of a leg wound, but there was insufficient information to evaluate the clinical significance of the other strains. Heizmann and Michel (1991) described a patient with conjunctivitis, and *E. americana* was isolated from both eyes. Strains of *E. americana* have also been isolated from respiratory specimens, including sputum, throat, and tracheal aspirate, and from urine and ear.

Clinical significance The clinical significance of *E. americana* and its ability to actually cause infections is still being evaluated. Systematic study and more good case reports are needed. The blood cultures and case reports suggest that *Ewingella* is an opportunistic pathogen. Only one isolate has come from feces, and there is no evidence that *E. americana* can actually cause diarrhea or intestinal infections.

Other isolates Environmental isolates have been from aqueous solutions in the hospital environment, water, mollusks (Müller et al., 1995a), and mushrooms (Inglis and Peberdy, 1996; Inglis et al., 1996).

Internal stipe necrosis of mushrooms In 1996 Inglis and Peberdy found that *E. americana* was often present on the commercially cultivated mushroom *Agaricus bisporus.* In a second paper Inglis et al. (1996) presented evidence that *E. americana* caused this browning reaction known as internal stipe necrosis. This is a potentially serious disease to the mushroom industry in the United Kingdom, which has sales worth over 400 million dollars.

Isolation, identification, culture preservation, phenotypic characterization Strains of *E. americana* are not difficult to grow, and are typical *Enterobacteriaceae* in most respects. See the chapter on the family *Enterobacteriaceae* for general information on growth, plating media, working and frozen stock cultures, media and methods for biochemical testing, identification, computer programs, and antibiotic susceptibility. Some strains of *E. americana* grow better, and are more active biochemically at 25°C than at 35°C.

Biochemical reactions and differentiation from other *Enterobacteriaceae* Table BXII.γ.193 in the introductory chapter on *Enterobacteriaceae* gives the results for *E. americana* in 47 biochemical tests normally used for identification (Farmer et al., 1985a; Farmer, 1999). There are no genus- or species-specific tests or sequences for identification. The best approach is to do a complete set of phenotypic tests (growth on plating media, biochemical tests, antibiogram, etc.) and compare the results with all the other named organisms in the family *Enterobacteriaceae.* This approach is described in more detail in the introductory chapter on *Enterobacteriaceae.* The biochemical results of a test strain can be compared to all the organisms listed in Table BXII.γ.193 in the chapter on the family *Enterobacteriaceae.* Several computer programs greatly facilitate analyzing the results. Table BXII.γ.226 may be helpful in differentiation.

TAXONOMIC COMMENTS

Discovery and DNA–DNA hybridization After the work had been completed on the characterization and description of the genus *Cedecea* (Grimont et al., 1981a), there remained a group of 10 strains that were phenotypically similar to, but distinct from, *Cedecea.* The group was named "Enteric Group 40" (Farmer, unpublished data; mentioned in Grimont et al., 1983a), and it was differentiated from *Cedecea* strains by a negative lipase (corn oil) reaction and several other tests. DNA–DNA hybridization indicated that Enteric Group 40 was less than 21% related to the other *Enterobacteriaceae.* Based on these data, Grimont et al. (1983a) proposed that Enteric Group 40 be reclassified as a new genus *Ewingella,* with a single species *Ewingella americana.*

Another genus named *Ewingella* In literature searches a source of possible confusion is the genus of radiolarians also named *Ewingella* (Pessagno, 1969).

Phylogeny based on 16S rDNA sequencing An unpublished sequence of 1398 bases has been deposited by P.W. Inglis for *E. americana* strain NCPPB 3905 (PI 98) described by Inglis and Peberdy (1996) and Inglis et al. (1996), which is a mushroom strain, not the type strain. A 16S rRNA tree that includes *E. americana* can be found in the section on *Budvicia* in this *Manual.* These 16S rDNA sequencing data agree with data from DNA–DNA hybridization, and both lead to the conclusion that *Ewingella* is distinct from other genera of *Enterobacteriaceae.* In the tree, *Rahnella* is the closest relative, in agreement with the tree published by Spröer et al. (1999), which shows these two genera on a distinct branch. However, the two trees differ in the placement of other organisms.

TABLE BXII.γ.226. Differentiation of *Ewingella americana* from several other *Enterobacteriaceae*[a]

Test	*Ewingella americana*	*Cedecea davisae*	"*Enterobacter agglomerans* complex"[b]	*Rahnella aquatilis*	*Serratia marcescens*
Lipase (corn oil)	0[b]	91	0	0	98
DNase at 25°C	0	0	0	0	98
Malonate utilization	0	91	65	73	3
Gas from D-glucose	0	70	20	98	55
Gelatin liquefaction at 22°C	0	0	0	0	90
Fermentation of:					
L-Arabinose	0	0	95	100	0
D-Arabitol	87	100	50	6	0
D-Sorbitol	0	0	30	96	99
Yellow pigment production at 25°C	0	0	75	0	0

[a]Each number gives the percentage positive after 2 d incubation at 36°C (unless a different temperature is indicated) and is based on the data summarized by Farmer (2003). Most positive reactions occur within 24 h.

[b]See Farmer (2003) for a more detailed description of this "vernacular name". It is a term defined for practical identification in the clinical microbiology laboratory. It includes over a dozen DNA–DNA hybridization groups that were originally included in the species *Enterobacter agglomerans*, which is now known to be a heterogeneous species.

FURTHER READING

Grimont, P.A., J.J. Farmer III, F. Grimont, M.A. Asbury, D.J. Brenner and C. Deval. 1983. *Ewingella americana* gen. nov., sp. nov., a new *Enterobacteriaceae* isolated from clinical specimens. Ann. Microbiol. Inst. Pasteur (Paris) *134A*: 39–52.

McNeil, M.M., B.J. Davis, S.L. Solomon, R.L. Anderson, S.T. Shulman, S. Gardner, K. Kabat and W.J. Martone. 1987. *Ewingella americana*: recurrent pseudobacteremia from a persistent environmental reservoir. J. Clin. Microbiol. *25*: 498–500.

Pien, F.D. and A.E. Bruce. 1986. Nosocomial *Ewingella americana* bacteremia in an intensive care unit. Arch. Intern. Med. *146*: 111–112.

List of species of the genus *Ewingella*

1. **Ewingella americana** Grimont, Farmer, Grimont, Asbury, Brenner and Deval 1984a, 91[VP] (Effective publication: Grimont, Farmer, Grimont, Asbury, Brenner and Deval 1983a, 41.)

a.mer.i.can' a. M.L. adj. *americana* to denote that the original 10 strains were isolated in the United States of America.

The characteristics are as described for the genus and as listed in Table BXII.γ.193 in the chapter on the family *Enterobacteriaceae*. Occurs in human clinical specimens and the environment. May cause internal stipe necrosis of the mushroom species *Agaricus bisporus*, which is a potentially serious disease to the mushroom industry in the United Kingdom.

In the original study, Grimont et al. (1983a) noted that all 10 strains of *E. americana* were highly related by DNA–DNA hybridization. The relative binding percentages at both 60°C and 75°C were greater than 73%. For this reason a single species was defined and named. However, strain 0679-79 had a ΔT_m value of 5.5 at 60°C, which was much higher than all the other strains, which had values of 1.2–2.5.

Grimont et al. (1983a) defined two biogroups of *E. amer-*

icana based on three phenotypic differences. Biogroup 1 was defined to include nine strains, and does not ferment L-rhamnose and D-xylose, and is β-xylosidase negative. Biogroup 2 included strain 0679-79 described above, and was positive for these three tests. Phenotypically, this strain differed in other ways (Table II of Grimont et al., 1983a).

The original 10 strains (Grimont et al., 1983a) were very active biochemically at 36°C. However, other strains studied since 1983 have been biochemically more active at 25°C than at 36°C. These strains should be studied at both temperatures.

Additional studies are needed to determine if the unusual strains described above and the newly described strains from different habitats are genetically different from the type strain of *E. americana*.

The mol% G + C of the DNA is: 53.6–55.2 (T_m).

Type strain: ATCC 33852, CCUG 14506, CDC 1468-78, CIP 8194, DSM 4580, NCTC 12157.

Additional Remarks: The American Type Culture Collection includes four other strains from human clinical specimens: ATCC 33850, blood; ATCC 338514, sputum; ATCC 33853, gallbladder; ATCC 33854, thumb wound.

Genus XV. **Hafnia** Møller 1954, 272[AL]

RIICHI SAKAZAKI

Haf' ni.a. L. fem. n. *Hafnia* the old name for Copenhagen.

Straight rods, ~1.0 × 2.0–5.0 μm, conforming to the general definition of the family *Enterobacteriaceae*. Not encapsulated. Gram negative. **Motile by peritrichous flagella at 30°C**, but nonmotile strains may occur. Facultatively anaerobic. Oxidase negative. Catalase positive. **The majority of strains utilize citrate as a sole carbon source at 30°C** after 3–4 d of incubation. Nitrate is reduced to nitrite. H₂S is not produced in the butt of Kligler iron

agar. Gelatinase, lipase, and deoxyribonuclease are not produced. Alginate is not utilized. Pectate is not decomposed. Phenylalanine deaminase is not produced. **Lysine and ornithine decarboxylase tests are positive**, but the arginine dihydrolase test is negative. D-Glucose is fermented with the production of acid and gas. **Acid is not produced from D-sorbitol, raffinose, melibiose, D-adonitol, D-arabitol, and *myo*-inositol.** The methyl red test is usually positive at 35°C and negative at 22°C. Acetylmethylcarbinol is usually produced from D-glucose at 22–28°C but may not be produced at 35°C. Occur in the feces of humans and a wide range of animals including birds; also occur in sewage, soil, water, and dairy products. A member of the *Gammaproteobacteria*.

The mol% G + C of the DNA is: 48–49.

Type species: **Hafnia alvei** Møller 1954, 272.

FURTHER DESCRIPTIVE INFORMATION

Members of *Hafnia* are able to grow at 35°C, but many of their physiological and biochemical activities at this temperature are irregular. The majority of strains are motile at 25–30°C, but may be nonmotile at 35°C. Although most strains fail to produce acetylmethylcarbinol from D-glucose at 35°C, they give a positive Voges–Proskauer reaction when incubated at 22–28°C. They produce gas from D-glucose and about 60% of them grow on Simmons citrate agar after incubation for 3–4 d at 25°C, but in many strains these reactions may be negative at 35°C. Lactose is not fermented, but plasmid-mediated lactose-positive strains may occur (Le Minor and Coynault, 1976).

Members of *Hafnia* are defined as H₂S-negative organisms. Møller (1954) and Kauffmann (1954) defined the *Hafnia* group of *Bacteria* as H₂S-producing organisms, since the majority of strains of *Hafnia alvei* slightly darken ferric chloride-gelatin medium (Kauffmann, 1954) and SIM medium (Difco). However, they fail to blacken the butt of Kligler iron agar and of triple sugar iron agar. Ewing (1960) suggested, for the H₂S test of the family *Enterobacteriaceae*, that sensitivity of the test should be placed at a certain level, and that either Kligler iron agar or triple sugar iron agar serves this purpose because each easily permits group differentiation within the family.

The serology of *Hafnia* was first studied by Stuart and Rustigian (1943), who divided their cultures of biotype 32011, the majority of which are now classified in *Hafnia* (Sakazaki and Namioka, 1957), into eight serovars. Eveland and Faber (1953) studied serologically 58 strains of biotype 32011 and reported 21 somatic and 22 flagellar antigens. Deacon (1952) also carried out a serological study on 17 cultures of *"Aerobacter cloacae"* including biotype 32011 and recognized 12 somatic and 6 flagellar antigens among the cultures. Serology of cultures that were biochemically defined as members of the genus *Hafnia* was studied by Sakazaki (1961) and Matsumoto (1963, 1964). They established an antigenic scheme of the *Hafnia* group consisting of 68 O groups and 34 H antigens. However, the scheme is no longer available, because many test strains for O and H antigens have been lost. Independent of the scheme of previous workers, Baturo and Raginskaya (1978) have published another antigenic scheme including 39 O and 35 H antigens of *H. alvei*. Intergeneric relationships of O antigens were recognized between *Hafnia* cultures and other members of the family *Enterobacteriaceae* (Stamp and Stone, 1944; Eveland and Faber, 1953; Harada et al., 1957; Sakazaki, 1961; Matsumoto, 1963, 1964; Sedlák and Slajsova, 1966). Deacon (1952) reported the diphasic variation in the H antigens of the *Hafnia* strains he studied, but Sakazaki (1961) and Matsumoto (1963) failed to confirm such variation. Some *Hafnia* strains may have K antigen and alpha antigen of Stamp and Stone (1944) (Deacon, 1952; Emslie-Smith, 1961). Sakazaki (1961) suggested that the K antigen that inhibits the O agglutination was a slime antigen.

The majority of strains of *H. alvei* are susceptible to carbenicillin, streptomycin, gentamicin, kanamycin, tetracycline, polymyxin B, and nalidixic acid, but resistant to cephalosporins and ampicillin. Washington et al. (1971) noted a definite difference between *H. alvei* and *Serratia liquefaciens* with respect to susceptibility testing of ampicillin and polymyxin B. They reported that most of the isolates of *Serratia liquefaciens* were susceptible to ampicillin at 20 µg/ml, while none of the *Hafnia* strains were. On the other hand, all of the *Hafnia* strains tested were susceptible to polymyxin B at 10 µg/ml, whereas only 6% strains of *Serratia liquefaciens* were.

A *Hafnia*-specific bacteriophage 1672 that provides a reliable tool for the identification of *Hafnia* strains was described by Guinée and Valkenburg (1968). They reported that the phage 1672 lysed all 100 strains of *H. alvei* tested, whereas it did not lyse strains of *Enterobacter, Klebsiella, Citrobacter, Serratia,* and *Salmonella.*

H. alvei occurs in humans and animals, including birds, and in natural environments such as soil, sewage, and water. McClure et al. (1957) described occurrence of the organisms in wild and caged birds, none of which appeared sick. Occasionally, the organisms cause illness. Kume (1962) described a case of equine abortion in which *H. alvei* was isolated from a fetus and lochia in pure culture. Gelev et al. (1990) reported epizootic hemorrhagic septicemia in rainbow trout associated with *H. alvei.*

Although it is difficult to assign a clear-cut clinical significance, *H. alvei* has been reported to cause septicemia (Englund, 1969; Mobley, 1971), respiratory tract infections (Washington et al., 1971; Klapholz et al., 1994; Fazal et al., 1997), meningitis (Mojtabaee and Siadati, 1978), abscesses (Washington et al., 1971; Agustin and Cunha, 1995), urinary tract infections (Whitby and Muir, 1961), and wound infections (Washington et al., 1971; Berger et al., 1977). In most cases, however, it has been found in mixed culture and seems to be an opportunistic pathogen that produces infections in patients with some underlying illness or predisposing factors such as diabetes, chronic renal failure, chronic obstructive pulmonary disease, malignancy, and HIV infection. Washington et al. (1971), who reviewed the epidemiology of *H. alvei* isolated in their laboratory, concluded that the majority of the isolates originated in the respiratory tract and were considered to be commensals, while a few were secondary invaders. They also reported that previous administration of ampicillin or cephalosporins was a common feature of patients who acquired the organisms nosocomially.

There has been a controversy on whether *H. alvei* is enteropathogenic. Kauffmann (1954) suggested that the organisms of the *Hafnia* group are probably nonpathogenic for humans. Matsumoto (1963) reported the isolation of this organism from 13% of stool specimens from apparently healthy individuals. Ewing (1986a) reported that "members of this species are not known to be incitants of gastroenteritis." On the other hand, earlier investigators (Stuart and Rustigian, 1943; Stuart et al., 1943; Deacon, 1952; Eveland and Faber, 1953) incriminated biotype 32011 of paracolon bacteria or non-lactose-fermenting *"Aerobacter cloacae"*, which are now classified in *Hafnia alvei*, as a causative agent of intestinal disorder. Harada et al. (1957) reported the isolation of *H. alvei* from sporadic cases of diarrheal diseases. Emslie-Smith (1961) suggested a possible role for *H. alvei* in gastroenteritis. In addition, Ratnam et al. (1979) reported the incrimination of *H. alvei* in a nosocomial outbreak of gastroenteritis. However, no conclusive evidence on enteropathogenicity of the *Hafnia* has been obtained. Albert et al. (1991, 1992) reported that strains of *H. alvei* isolated from diarrheal stools produced the attaching-effacing (AE) lesion in rabbits and HEp-2 cells resembling enteropathogenic *Escherichia coli* (EPEC), and that the virulence-associated gene *eaeA* was shared between EPEC and *H. alvei*. In addition, Ridell et al. (1995) found that *eaeA*-positive strains of

H. alvei have some characteristic properties in which they give negative reactions in the 2-ketogluconate and histidine assimilation tests and a positive reaction in the 3-hydroxybenzoate assimilation test in contrast with *eaeA*-negative strains. On the other hand, Ismaili et al. (1996) examined 11 strains of *H. alvei* isolated from children with diarrhea in Canada, and a Bangladesh *H. alvei* strain for AE lesion formation, presence of *eaeA* gene, profile of outer membrane protein extracts, chromosomal macrorestriction fragments, and plasmids. They found that, in contrast with the Bangladesh strain, which possessed the *eaeA* gene and forms the AE lesion, none of the Canadian strains had the *eaeA* gene, nor did they form the AE lesion. The outer membrane protein profile of all of the Canadian strains were identical to each other but differed from that of the Bangladesh strain. In addition, pulsed-field gel electrophoresis and plasmid profile analyses of the Canadian strains differed substantially from those of the Bangladesh strain. Their results indicate that there is heterogeneity among *H. alvei* strain factors associated with enteropathogenicity. Ridell et al. (1994) found an epidemiological association of *H. alvei* with diarrhea. *H. alvei* was isolated from 16% of adult Finnish tourists who visited Morocco and contracted diarrhea, but from 0% of tourists without diarrhea. *H. alvei* was isolated only from 2% of another group of adult Finnish patients with diarrhea. However, the authors found that those Finnish strains from diarrheal patients were negative for AE lesion formation and did not have the *eaeA* gene of EPEC. These recent papers suggest that some strains of *H. alvei* have the potential to cause diarrhea and that, although AE lesion formation is a virulence factor, mechanisms other than AE lesion formation may also be involved in the association of *H. alvei* with diarrhea.

ENRICHMENT AND ISOLATION PROCEDURES

H. alvei can grow on less selective isolation media for enterobacteria such as eosin-methylene blue, MacConkey, xylose–lysine–deoxycholate, and Hektoen enteric agars. On highly selective isolation media, such as Salmonella–Shigella and deoxycholate-citrate agars, many of the *Hafnia* cultures would be inhibited. The *Hafnia* organisms may not grow on bismuth sulfite agar. Colonies of *H. alvei* on less inhibitory plating agar media are relatively large, translucent, circular, low convex, and colorless, with a smooth surface and entire edge and resemble those of *Salmonella* (*Hafnia* strains are sometimes misidentified as H_2S-negative *Salmonella*). Some strains may produce pink colonies on media containing sucrose. Rarely does a strain produce mucoid colonies. For the differential isolation of *H. alvei*, MacConkey agar containing 1% sorbitol may be useful, on which colorless colonies of *H. alvei* can be distinguished from other intestinal inhabitants, because the former is lactose- and sorbitol-negative but the latter are usually positive for these sugars.

There are no selective enrichment broth media for the isolation of *H. alvei*. Many strains of *H. alvei* may grow in selenite and tetrathionate broth media, but some strains fail to grow in those broths. Ability of *H. alvei* to grow at lower temperature may be applied for cold enrichment, which is a useful procedure for the isolation of *Yersinia* strains.

MAINTENANCE PROCEDURES

Stock cultures may be maintained at room temperature in a semisolid medium consisting of 1.0% Bacto-casitone (Difco), 0.3% yeast extract, 0.5% NaCl, and 0.3% agar (pH 7.0). *Hafnia* strains remain viable for up to several years without subculture if the culture is sealed with a rubber stopper or a cork soaked in hot paraffin wax. Strains may also be preserved by lyophilization.

PROCEDURES FOR TESTING SPECIAL CHARACTERS

The specific bacteriophage 1672 was isolated from surface water with *H. alvei* 1672 as the propagating strain using the following method (Guinée and Valkenburg, 1968). A well-dried nutrient agar plate is surface-inoculated with a fresh broth culture of the strain to be tested. After decantation of the plate at room temperature for 15 min, a drop of the undiluted phage 1672 is spotted on the plate with a Pasteur pipette and the plate is again allowed to dry. Readings are made after 16–20 h of incubation at 35°C. Clear plaques with a diameter of 1–5 mm may be produced. The phage preparation is obtained after the usual purification and will contain around 10^9 plaque-forming units/ml. The phage is not inactivated by heating at 60°C for 30 min.

The L-pyrrolidonyl aminopeptidase test is performed by dissolving 10 mg of L-pyrrolidonyl-β-naphthylamide (Sigma) in 10 ml of 95% ethanol. Dip sterile cotton swabs into the solution and rotate each swab several times with firm pressure on the inside wall of the tube to remove excess fluid. Dry the swabs overnight at 37°C and store at −20°C. Make a heavy suspension (MacFarland no. 1) of the test culture into 0.2–0.3 ml of 0.1 M phosphate buffer (pH 7.4), then dip the reagent swab into the suspension and incubate at 37°C for 10 min. Finally, add a drop of cinnamaldehyde solution.[1] A positive result is shown by a pink or red color.

The β-glucuronidase test is performed by dissolving 20 mg of 4-methylumbelliferyl-β-D-glucuronide (Sigma) in a small volume of dimethyl sulfoxide and bringing the volume to 10 ml with 0.1 M phosphate buffer (pH 7.4). Dip sterile cotton swabs into the solution and rotate each swab several times with firm pressure on the inside wall of the tube to remove excess fluid. Dry the swabs overnight at 37°C and store at −20°C. For testing, make a heavy suspension (MacFarland no. 1) of the test culture in 0.2–0.3 ml of 0.1 M phosphate buffer (pH 7.4). Dip the reagent cotton swab in the suspension and incubate at 37°C for 30 min. A positive reaction is indicated by development of blue fluorescence under long wave (360 μm) ultraviolet lamp.

DIFFERENTIATION OF THE GENUS *HAFNIA* FROM OTHER GENERA

Hafnia are often misidentified as members of *Enterobacter* or *Serratia*. Members of the genus *Hafnia* give a positive reaction in lysine and ornithine decarboxylase tests but are negative in arginine dihydrolase test. In differentiating *Hafnia* from *Enterobacter* and *Serratia*, in which some species show the same reactions in those tests, the failure of *Hafnia* to ferment raffinose, D-sorbitol, adonitol, and *i*-inositol may be valuable. Moreover, members of *Hafnia* have no activity of pyrrolodonyl aminopeptidase, although *Enterobacter* and *Serratia* possess this enzyme. Stuart et al. (1943) and Eveland and Faber (1953) described a close similarity between paracolon biotype 32011, which is now classified in the *Hafnia*, and *Salmonella* in biochemical reactions. Like *Salmonella*, *Hafnia* cultures are positive in tests of lysine and ornithine decarboxylases and negative in tests of indole, urease, and fermentation of lactose and sucrose, as well as in the Voges–Proskauer

---

1. Cinnamaldehyde solution contains *p*-dimethlaminocinnamaldehyde (Sigma), 0.5 g; glacial acetic acid, 3.5 ml; ethylene glycol-1-methylester, 6.0 ml; sodium lauryl sulfate, 3.5 g; and distilled water, 90 ml. Store in a screw-capped brown container in a refrigerator.

reaction at 35°C, although *Hafnia* cultures fail to produce H$_2$S. In addition, Eveland and Faber (1953) and Harada et al. (1957) reported that the *Hafnia* cultures are often agglutinated by *Salmonella* O antisera. *Hafnia* cultures may also be misidentified with some non-lactose-fermenting *Escherichia coli* strains, because the former gives a negative reaction in tests of Voges–Proskauer and citrate utilization at 35°C. However, *E. coli* produces indole and β-glucuronidase, while *Hafnia* is negative in these two tests. On the other hand, *Hafnia* is occasionally misidentified with Shiga toxin-producing *E. coli* O157:H7, because both organisms fail to ferment sorbitol and to produce β-glucuronidase. In addition, some strains of *H. alvei* are agglutinated by *E. coli* O157 antiserum (Aleksič et al., 1992). *Yokenella regensburgei*, which was first considered to be *Hafnia*, is another organism to be differentiated from the *Hafnia*. *Y. regensburgei* is different from *Hafnia* in its ability to utilize citrate, and to produce acid from cellobiose, melibiose, and D-arabitol. *Y. regensburgei* cannot be lysed by the *Hafnia* specific bacteriophage. Table BXII.γ.227 indicates the differential characteristics between *Hafnia* and biochemically similar taxa.

TAXONOMIC COMMENTS

The bacteria of the genus *Hafnia* have been described under several names: "*Bacillus asiaticus*" and "*Bacillus asiaticus*" biovar "mobilis" (Castellani, 1912, cited by Ewing and Fife, 1968), "*Bacterium cadaveria*" (Gale and Epps, 1943), biotype 32011 of the genus "*Aerobacter*" (Stuart et al., 1943), "*Enterobacter alvei*" (Sakazaki, 1961), *Enterobacter aerogenes* subsp. "*hafniae*" (Ewing, 1963), and "*Enterobacter hafniae*" (Ewing and Fife, 1968). However, *Hafnia alvei* (Møller, 1954) is the only correct name for these organisms. Møller (1954) found a new group of organisms, in which a supposedly authentic strain of "*Bacillus paratyphi-alvei*" of Bahr (1919) was included. He proposed the name *Hafnia alvei* for this group of bacteria because he considered that Bahr's strain ought to be regarded as the type strain of this group. Sakazaki (1961) suggested the new combination "*Enterobacter alvei*" for *H. alvei*, because of its biochemical similarity to *Enterobacter*. Ewing and Fife (1968) pointed out that Bahr's strain, which had been designated as the type strain of *H. alvei* by Møller (1954), was not an authentic strain of this species, since biochemical reactions of the strain were not the same as those described by Bahr (1919). They considered therefore that the specific epithet "*alvei*" was illegitimate, and proposed the name "*Enterobacter hafniae*" for *H. alvei*. However, there is no doubt that Bahr's strain studied by Møller (1954) was a new bacterium at that time. In addition, numerical taxonomy studies by Johnson

et al. (1975) and Gavini et al. (1976a) indicated that *Hafnia* strains retained a position separate from *Enterobacter*. In DNA–DNA hybridization studies, Steigerwalt et al. (1976) reported 11–26% relatedness of *H. alvei* with *Enterobacter* and *Klebsiella*. These data appear to justify the status of the *Hafnia* as a separate genus rather than including it in the genus *Enterobacter*.

Although only a single species, *Hafnia alvei*, has been designated, Steigerwalt et al. (1976) indicated in DNA reassociation studies that there were two genomospecies that were about 50% interrelated within the genus *Hafnia*. Overall phenotypic profiles may serve to separate the second subspecies from the named species *H. alvei*; however, it has not been named. Farmer et al. (1985a) referred to this second genomospecies as *Hafnia alvei* biogroup 1.

Priest et al. (1973) proposed that *Obesumbacterium proteus* Shimwell 1964, a common brewery contaminant, should be placed in the genus *Hafnia* as "*H. protea*". They described two groups in this species by numerical analysis of phenotypic characteristics. Brenner (1981) determined DNA relatedness in both groups and found that one group appears to be a biovar of *H. alvei*, whereas the other is a new species that does not belong to the genus *Hafnia*.

Another group resembling *H. alvei*, *Yokenella regensburgei*, was first considered as a species of the genus *Hafnia*. Kosako et al. (1984), who studied the phenotypic and genotypic characteristics of both groups, confirmed that the former was distinct from members of the *Hafnia* and suggested the name *Yokenella regensburgei* for this group.

ACKNOWLEDGMENTS

Riichi Sakazaki, who died in 2002, spent more than forty years at the Nippon Institute of Biological Sciences. His illustrious career focused on the classification and epidemiology of human and fish pathogens in Japan, particularly pathogenic bacteria in the families *Vibrionaceae* and *Enterobacteriaceae*. His contributions to Japanese bacteriology paralleled those of Drs. Edwards and Ewing in the U.S. He discovered and described many species in the genera *Enterobacter*, *Edwardsiella*, and *Vibrio*, and he developed serotyping schemes for *V. cholerae* and *V. parahaemolyticus*, as well as for several important species of *Enterobacteriaceae*. His research was always careful and comprehensive. He was one of the first Japanese scientists to publish in English language journals, thereby making his impressive accomplishments available to the world. He was recognized as one of the world's foremost experts on the genus *Vibrio* and on many genera in *Enterobacteriaceae*. He was a member of the WHO Subcommittees on the Taxonomy of *Vibrionaceae* and on the Taxonomy of *Enterobacteriaceae* for more than thirty years. He authored chapters on these organisms in three editions of *Bergey's Manual*. He was a good friend to the Bergey's Trust, as well as to all who knew him.

TABLE BXII.γ.227. Differential characteristics of the genus *Hafnia* and biochemically related genera[a]

Characteristic	*Hafnia*	*Enterobacter*	*Escherichia coli*	*Escherichia* strain STEC 0157[b]	*Salmonella*	*Serratia*	*Yokenella*
H$_2$S production	−	−	−	−	+	−	−
Indole production	−	−	+	+	−	D	−
β-Galactosidase	D	+	+	+	D	+	+
β-Glucuronidase	−	−	+	−	D	−	−
L-Pyrrolidonyl aminopeptidase	−	+	−	−	−	+	+
Sorbitol fermentation	−	D	+	−	+	D	−
Hafnia specific bacteriophage lysis[c]	+	−	−	−	−	−	−
Mol% G + C of DNA	48–49	52–60	48–52		50–53	52–60	58–59

[a]Symbols: +, 90–100% of strains are positive; −, 90–100% of strains are negative; D, different reactions given by different species or biogroups.

[b]Shiga toxin-producing *Escherichia coli* O157:H7.

[c]Data from Guinée and Valkenburg (1968).

DIFFERENTIATION OF THE SPECIES OF THE GENUS *HAFNIA*

Table BXII.γ.228 presents the biochemical characteristics of *Hafnia* species, which are also useful for differentiation between *H. alvei* and unnamed genospecies (genomospecies 2).

List of species of the genus Hafnia

1. **Hafnia alvei** Møller 1954, 272[AL]

   al' ve.i. L. n *alveus* a beehive; L. gen. n. *alvei* of a beehive.

   The morphology is as given for the genus. Motility is most pronounced at 30°C. Nonmotile strains may be encountered occasionally.

   Grows readily on ordinary media. Colonies on nutrient agar are generally 2–4 mm in diameter, smooth, moist, translucent, and gray, with a shiny surface and entire edge. Rare strains may produce mucoid colonies. Physiological and biochemical characteristics are presented in Table BXII.γ.228.

   Found in the feces of humans and other animals, including birds. Also found in sewage, soil, water, and dairy products. It may be isolated in association with some pathological processes in patients with underlying illness. Some strains may cause diarrheal diseases.

   The mol% G + C of the DNA is: 48.0–48.7 (T_m).

   Type strain: ATCC 13337, DSM 30163, NCTC 8106.

   GenBank accession number (16S rRNA): M59155.

TABLE BXII.γ.228. Characteristics of *Hafnia alvei* and genomospecies 2[a]

Characteristic	*H. alvei*	Genomospecies 2
Indole production	−	−
Voges–Proskauer test (25°C)	+	+
Voges–Proskauer test (35°C)	d	d

(continued)

TABLE BXII.γ.228. *(cont.)*

Characteristic	*H. alvei*	Genomospecies 2
Citrate (Simmons) (25°C)	d	−
Citrate (Simmons) (35°C)	−	−
Lysine decarboxylase	+	+
Arginine dihydrolase	−	−
Ornithine decarboxylase	+	d
H₂S production (Kligler)	−	−
Phenylalanine deaminase	−	−
Urease (Christensen)	−	−
β-Galactosidase	+	d
β-Glucuronidase	−	−
Malonate utilization	d	d
L-Pyrrolidonyl aminopeptidase	−	−
Gelatinase	−	−
Lipase (Tween 80)	−	−
Deoxyribonuclease	−	−
Growth in KCN medium	+	−
Esculin hydrolysis	−	−
Gas from D-glucose	+	−
Acid from carbohydrates:		
L-Arabinose, maltose, L-rhamnose, D-xylose	+	−
Lactose, melibiose, raffinose, sucrose[b], adonitol, D-arabitol, dulcitol, *myo*-inositol, D-sorbitol, α-methyl-D-glucoside	−	−
Cellobiose	d	−
D-Mannitol	+	d
Trehalose	+	d
Salicin	d	d

[a]For symbols see standard definitions.

[b]Late positive reactions are given by approximately 50% of the strains of *H. alvei*.

Genus XVI. **Klebsiella** *Trevisan 1885, 105*[AL] *emend. Drancourt, Bollet, Carta and Roussselier 2001, 930*

PATRICK A.D. GRIMONT AND FRANCINE GRIMONT

Kleb.si.el' la. M.L. dim. ending *-ella;* M.L. fem. n. *Klebsiella* named after Edwin Klebs (1834–1913), a German bacteriologist.

Straight rods, 0.3–1.0 × 0.6–6.0 μm, arranged singly, in pairs or short chains; often surrounded by a **capsule.** Conform to the general definition of the family *Enterobacteriaceae.* Gram negative. **Nonmotile** (except *K. mobilis*). Facultatively anaerobic, having both a respiratory and a fermentative type of metabolism. Grow on meat extract media (except *K. granulomatis*, which has not been cultured), producing more or less dome-shaped, glistening colonies of varying degrees of stickiness depending on the strain and the composition of the medium. **Oxidase negative.** Glucose is fermented with the production of acid and gas (more CO₂ is produced than H₂), but anaerogenic strains occur. Most strains produce 2,3-butanediol as a major end product of glucose fermentation. The **Voges–Proskauer test is usually positive.** Lactic, acetic, and formic acids are formed in smaller amounts and ethanol in larger amounts than in a mixed acid fermentation. All strains utilize L-arabinose, D-arabitol, D-cellobiose, citrate, D-fructose, D-galactose, D-glucose, 2-ketogluconate, maltose, D-mannitol, D-melibiose, D-raffinose, D-trehalose, and D-xylose as sole carbon sources. With the exception of some *K. pneumoniae* subsp. *ozaenae* strains, all *Klebsiella* strains **utilize *myo*-inositol, L-rhamnose, and sucrose as sole carbon sources.** With the exception of some *K. pneumoniae* subsp. *ozaenae* and *K. pneumoniae* subsp. *rhinoscleromatis* strains, all strains utilize lactose and D-sorbitol as sole carbon sources. No strain utilizes betaine, caprate, caprylate, glutarate, itaconate, 3-phenylpropionate, and propionate. H₂S is not produced, β-glucuronides are not hydrolyzed, and L-tryptophan and L-histidine are not deaminated. Ornithine is not decarboxylated by klebsiellae strains except *K. mobilis*, *K. ornithinolytica*, and rare strains of *K. pneumoniae*. Most strains hydrolyze urea and β-galactosides. Some strains fix nitrogen. Occur in intestinal contents, clinical specimens from humans and animals (e.g., horses, swine, monkeys), soil, water, or on plants.

The mol% G + C of the DNA is: 53–58.

Type species: **Klebsiella pneumoniae** (Schroeter 1886) Trevisan 1887, 94 (*Hyalococcus pneumoniae* Schroeter 1886, 1952.)

FURTHER DESCRIPTIVE INFORMATION

Phylogenetic treatment The agent of donovanosis, *Calymmatobacterium granulomatis*, was found to be so closely related to *Klebsiella pneumoniae* by *rrs* (16S rRNA gene) sequence comparison that the species was transferred to the genus *Klebsiella* as *Klebsiella granulomatis* (Carter et al., 1999; Kharsany et al., 1999). However, it is not known whether *K. granulomatis* is a species distinct from *K. pneumoniae*.

The phylogenetic position of klebsiellae has been addressed by *rrs*, *rpoB* (Drancourt et al., 2001), *gyrA*, and *parC* (Brisse and Verhoef, 2001) sequence comparison.

The genus *Klebsiella* was found to be polyphyletic by Drancourt et al. (2001). Cluster I contained *Klebsiella pneumoniae* subsp. *pneumoniae*, *K. pneumoniae* subsp. *ozaenae*, *K. pneumoniae* subsp. *rhinoscleromatis*, and *K. granulomatis*. Cluster II contained *K. planticola*, *K. ornithinolytica*, and *K. terrigena*. Cluster III contained *K. oxytoca*. As a consequence, cluster II was proposed to constitute a new genus, *Raoultella* (Drancourt et al., 2001). The position of *Klebsiella mobilis* (*Enterobacter aerogenes*) was very close to cluster II, although the proposal by Drancourt et al. (2001) did not include that species.

A somewhat different structure was found by Brisse and Verhoef (2001), who uncovered two groups. The first group contained *K. pneumoniae* with its three subspecies, and the second group contained *K. oxytoca*, *K. planticola*, *K. ornithinolytica*, *K. terrigena*, and *K. mobilis*. Furthermore, three clusters (KpI to KpIII) were evidenced in *K. pneumoniae*, which did not correlate with the named subspecies. These clusters may have different habitats and different physiological properties (D-adonitol fermentation, for example). *K. oxytoca* was composed of two clusters, the significance of which is unknown.

Cell morphology and cell wall composition The outermost layer of *Klebsiella* bacteria consists of a large polysaccharide capsule, a characteristic that distinguishes members of this genus from most other bacteria in the family (*Escherichia coli* strains with a heat-stable K antigen may form similar capsules). The cell wall is structured similarly to other *Enterobacteriaceae*. Above the cytoplasmic membrane is the peptidoglycan layer and the outer membrane containing lipopolysaccharide (LPS). In addition, *Klebsiella* strains may possess fimbriae (pili) (Duguid, 1959).

The production of the large capsules gives rise to large mucoid colonies of a viscid consistency. The capsular material also diffuses freely into the surrounding liquid medium as extracellular capsular material.

Nutrition and growth conditions Best results with biochemical tests are obtained when *Klebsiella* cultures are incubated at 30–35°C. *Klebsiella* strains (except *K. granulomatis*) grow readily on all kinds of media commonly used to isolate *Enterobacteriaceae*, such as nutrient agar, tryptic casein soy agar, bromocresol purple lactose agar, Drigalski agar, MacConkey agar, eosin-methylene blue (EBM) agar, and bromothymol blue (BTB) agar. *K. pneumoniae* and *K. oxytoca* colonies are lactose positive, more or less dome-shaped, 3–4 mm in diameter after overnight incubation at 37°C or 30°C, with a mucoid aspect and sometimes stickiness depending on the strain and the composition of the medium.

K. planticola, *K. terrigena*, *K. ornithinolytica*, and *K. mobilis* (*Enterobacter aerogenes*) colonies are also lactose positive, 1.5–2.5 mm in diameter, dome-shaped, with a weakly mucoid aspect. *K. pneumoniae* subsp. *ozaenae*, *K. pneumoniae* subsp. *rhinoscleromatis*, and occasionally *K. pneumoniae* K1 strains grow more slowly on the same media, yielding voluminous, rounded, very mucoid, translucent, and confluent colonies in 48 h at 30°C or 37°C (Ørskov, 1981; Richard, 1982). Similar colonies, indistinguishable from those of *Klebsiella*, may be formed by other genera of the *Enterobacteriaceae*, particularly *E. coli* mucoid varieties with capsular K antigens.

The use of a carbohydrate-rich medium, such as bromothymol blue lactose agar or Hajna or Worfel–Ferguson medium, produces a better formed capsule than a carbohydrate-poor medium.

Almost all *Klebsiella* strains grow in minimal medium with ammonium ions or nitrate as sole nitrogen source and a carbon source, without growth factor requirement. Some *K. pneumoniae* K1 isolates require arginine or adenine or both as growth factors. *K. pneumoniae* subsp. *rhinoscleromatis* requires arginine and uracil but growth factor requirements of *K. pneumoniae* subsp. *ozaenae* are not fully determined (leucine, cysteine, and methionine are stimulatory). Ornithine can replace arginine for these requirements (Grimont et al., 1991).

K. granulomatis has not been grown axenically in artificial media. Cultures have been achieved *in vivo* in the yolk sac of developing chick embryo and in the developing chick embryo brain. Growth can be obtained in cell cultures, fresh mononuclear cells (Kharsany et al., 1997) or in Hep-2 human epithelial cell line (Carter et al., 1997).

Metabolism The characteristics useful for the identification of *Klebsiella* species are given in Table BXII.γ.229.

The Voges–Proskauer (VP) test is usually positive in *Klebsiella*, meaning that acetoin and 2,3-butanediol are produced during glucose fermentation. Some strains of *K. pneumoniae* subsp. *rhinoscleromatis*, do not form acetoin and 2,3-butanediol. In some strains, 2,3-butanediol is utilized and acetoin will disappear before the VP reaction is tested. Richard's modification of the VP test (Le Minor and Richard, 1993) most often gives positive results with *Klebsiella* strains (except *K. pneumoniae* subsp. *ozaenae* and subsp. *rhinoscleromatis*).

Carbon source utilization tests are conveniently obtained with Biotype-100 strips (BioMérieux, La Balme-les-Grottes, France) used with a minimal medium (Biotype Medium 1) containing 16 growth factors. *K. pneumoniae* subsp. *rhinoscleromatis* and some strains of *K. pneumoniae* subsp. *ozaenae* may need Biotype medium 2, which contains all known growth factors for enteric bacteria. The strips are examined for visual growth after 2 and 4 d (Grimont et al., 1991). Almost all *Klebsiella* strains utilize the following compounds as sole carbon and energy sources (provided growth factor requirements of some strains are met): N-acetyl-D-glucosamine, L-alanine, L-arabinose, D-arabitol, L-aspartate, D-cellobiose, citrate, D-fructose, L-fucose, fumarate, D-galactose, gentiobiose, D-gluconate, D-glucosamine, D-glucose, DL-glycerate, glycerol, *myo*-inositol, 2-ketogluconate, DL-lactate, lactose, D-malate, L-malate, maltose, maltotriose, D-mannitol, D-mannose, D-melibiose, 1-*O*-methyl-β-glucoside, L-proline, D-raffinose, D-ribose, L-serine, D-trehalose, and D-xylose. Esculin is hydrolyzed. H$_2$S, β-glucuronidase, phenylalanine deaminase, and tryptophan deaminase are not produced; DNA and tributyrin are not hydrolyzed. Almost no strain of *Klebsiella* can utilize the following com-

TABLE BXII.γ.229. Differentiation characteristics of the species and subspecies of the genus *Klebsiella*[a]

Characteristics	1a. K. pneumoniae subsp. pneumoniae	1b. K. pneumoniae subsp. ozaenae	1c. K. pneumoniae subsp. rhinoscleromatis	3. K. mobilis	4. K. oxytoca	5. K. planticola	6. K. terrigena	Species incertae sedis K. ornithinolytica
Motility	−	−	−	+	−	−	−	−
Growth at:								
5°C	−	−	−	+	−	+	+	+
41°C	+	+	+	+	+	d	−	+
44.5°C	+	nd	nd	nd	d	−	−	nd
Urea hydrolyzed	+	d	−	−	+	+	+	+
Pectate hydrolyzed	−	−	−	−	+	−	−	−
ONPG test	+	+	−	+	+	+	+	+
Indol produced	−	−	−	−	+	d	−	+
Voges–Proskauer test	+	−	−	+	+	+	+	+
Malonate test	+	−	+	+	+	+	+	+
Lysine decarboxylated	+	d	−	+	+	+	+	+
Ornithine decarboxylated	−	−	−	+	−	−	−	+
Glucose dehydrogenase:								
Without added pyrroloquinoline quinone	+	−	−	+	−	−	−	nd
With added pyrroloquinoline quinone	+	−	−	+	+	+	+	nd
Gluconate dehydrogenase	+	−	−	+	−	+	+	nd
Utilization of:								
Adonitol	d	+	(+)	+	+	+	+	+
D-Alanine	+	+	−	+	+	+	+	+
L-Arabitol	−	−	−	−	d	−	−	−
Benzoate	d	−	−	+	d	+	d	+
m-Coumarate	d	d	−	−	+	+	−	+
Dulcitol	d	−	−	d	d	d	−	d
i-Erythritol	−	−	−	−	−	(d)	−	d
D-Galacturonate	+	+	−	+	+	+	+	+
Gentisate	−	−	−	+	+	−	d	−
D-Glucuronate	+	+	−	+	+	+	+	+
Histamin	−	−	−	(+)	−	+	d	d
3-Hydroxybenzoate	−	−	−	+	+	−	(+)	−
4-Hydroxybenzoate	+	−	−	d	+	+	+	+
5-Ketogluconate	d	−	−	d	+	+	+	+
Lactulose	+	d	−	(+)	+	+	(+)	+
Malonate	d	−	d	d	(d)	(d)	−	d
Maltitol	+	+	−	+	+	+	+	+
D-Melezitose	−	−	−	−	d	−	+	−
1-*O*-Methyl-β-galactoside	+	+	−	+	+	+	+	+
3-*O*-Methyl-D-glucose	−	−	−	−	−	+	+	+
1-*O*-Methyl-α-D-glucoside	d	+	−	+	+	+	+	+
Mucate	+	d	−	+	+	+	+	(+)
Palatinose	+	+	−	+	+	+	+	+
Phenylacetate	d	d	−	+	+	+	+	+
Protocatechuate	+	d	−	+	+	+	+	+
Putrescine	d	d	−	d	(+)	+	(+)	+
Quinate	+	d	−	+	+	+	+	+
L-Rhamnose	+	d	+	+	+	+	+	+
D-Saccharate	+	d	d	+	+	+	+	(+)
D-Sorbitol	+	d	d	+	+	+	+	+
L-Sorbose	d	d	−	−	+	+	+	+
Sucrose	+	d	+	+	+	+	+	+
D-Tagatose	d	−	−	d	d	d	−	d
D-Tartrate	d	−	−	−	−	(d)	−	d
Tricarballylate	d	−	−	+	+	−	+	−
Trigonelline	−	−	−	d	−	−	−	−
Tryptamine	−	−	−	d	−	−	−	−
D-Turanose	d	d	−	d	d	(d)	−	d
L-Tyrosine	−	−	−	d	−	−	−	−
D-Xylitol	d	−	−	d	(d)	d	−	d

[a]Symbols: +, 95–100% strains positive in 1–2 d (utilization tests) or in 1 d (other tests); (+), 95–100% strains positive in 1–4 d; −, 95–100% strains negative in 4 d; d, different reactions; nd, not determined.

pounds as sole carbon and energy sources after 4-d incubation: betaine, caprate, caprylate, glutarate, itaconate, 3-phenylpropionate, and propionate. It is obvious from Table BXII.γ.229 that *K. pneumoniae* subsp. *ozaenae* and *K. pneumoniae* subsp. *rhinoscle-* romatis are biotypes of *K. pneumoniae* with less nutritional versatility. There is no substrate utilized by subsp. *ozaenae* and subsp. *rhinoscleromatis* that is not utilized by subspecies subsp. *pneumoniae*. For this reason, it is necessary to check the identification of these less active subspecies by capsular typing. Carbon source utilization tests are essential for the precise identification of *Klebsiella* species (Grimont et al., 1991). In Biotype-100 strips, *K. oxytoca* often produces a soluble yellow compound in all cupules showing some growth, and this may be analogous to the pigment produced by this species on ferric gluconate (Ørskov, 1984b). Growth on gentisate and 3-hydroxybenzoate never occurs with *K. pneumoniae* and *K. planticola*. Growth on histamine never occurs with *K. pneumoniae* and *K. oxytoca*. Growth on D-sorbose is always negative with *K. mobilis* (*E. aerogenes*) and growth on tricarballylate is always negative with *K. planticola*. Only *K. terrigena* and some strains of *K. oxytoca* can grow on D-melezitose, and only *K. planticola* and *K. terrigena* can grow on 3-*O*-methyl-D-glucose. These carbon sources are often sufficient for the identification of all *Klebsiella* species (Grimont et al., 1991).

Oxidation of glucose to gluconate (mediated by glucose dehydrogenase) in the absence of added pyrroloquinoline quinone is a distinctive property of *K. pneumoniae* subsp. *pneumoniae* and *K. mobilis* (Bouvet et al., 1989). The other species require the addition of cofactor pyrroloquinoline quinone to express glucose dehydrogenase activity. Gluconate is oxidized into 2-ketogluconate (by a gluconate dehydrogenase) by *K. pneumoniae* subsp. *pneumoniae*, *K. planticola*, *K. terrigena*, and *K. mobilis*, but not by *K. oxytoca*, *K. pneumoniae* subsp. *ozaenae*, and *K. pneumoniae* subsp. *rhinoscleromatis*. *Klebsiella* species do not oxidize 2-ketogluconate to 2,5-ketogluconate (Bouvet et al., 1989).

K. pneumoniae subsp. *pneumoniae* can grow fermentatively on glycerol thanks to a glycerol dehydrogenase type I (induced by glycerol and dihydroxyacetone) and 1,3-propanediol dehydrogenase, which are typical enzymes of the anaerobic glycerol dissimilation pathway (Bouvet et al., 1995a). Other species of *Klebsiella* cannot grow fermentatively on glycerol and possess a glycerol dehydrogenase, but no 1,3-propanediol dehydrogenase (Bouvet et al., 1995a).

Some *K. oxytoca* and *K. planticola* strains are capable of fixing nitrogen and are classified as associative nitrogen fixers (Ladha et al., 1983). Nitrogen fixation is catalyzed by the enzyme nitrogenase in the absence of combined nitrogen, and under anaerobic conditions. Nitrogen fixation in the facultative anaerobe *K. pneumoniae* occurs only under anaerobic or microaerobic conditions. Low O_2 concentrations enhance nitrogenase synthesis; higher O_2 concentrations inhibit both synthesis and activity (Hill et al., 1984).

Genetic recombination has been reported in *Klebsiella* (Matsumoto and Tazaki, 1970), and *K. pneumoniae* has been used by several workers for detailed genetic analysis of the genes involved in N_2 fixation (*nif* genes). These genes are clustered near the *his* region on the chromosome but can be mobilized and transferred to other organisms. The genetics and regulation of nitrogen fixation in *Klebsiella* have been reviewed by Gussin et al. (1986).

Klebsiella strains may harbor a *lac* plasmid giving a stronger lactose positive phenotype (Reeve and Braithwaite, 1973).

Klebsiella strains may be lysogenic, but phages used by some workers for phage typing have been isolated from stools or sewage (Slopek et al., 1967; Slopek, 1978). Many *Klebsiella* strains produce bacteriocin (klebecin) and typing sets of such producers can be selected (Slopek and Maresz-Babczyszyn, 1967; Edmondson and Cooke, 1979).

Antigenic structure   The outer membrane contains the lipopolysaccharide (LPS) that forms the O-antigen. The oligosaccharide repeating units of the O-antigen consists of only a few carbohydrate components. Some of the O-antigens are homopolysaccharides such as galactans or mannans. There are only 12 different O-antigens. O-antigen determination is difficult because it is hampered by the stable K-antigens (Ørskov and Ørskov, 1984b).

Klebsiellae are enveloped by a polysaccharide capsule of considerable thickness. This constitutes the capsular or K-antigen. The capsular polysaccharide also diffuses freely into the liquid medium as extracellular capsular material. The *Klebsiella* capsular polysaccharides are acidic and composed of repeating units. Only a few carbohydrate constituents are found. Most K-antigens contain charged monosaccharide constituents such as glucuronic and hexoses, occasionally 6-deoxyhexoses. Noncarbohydrate constituents such as formyl, acetyl, or pyruvate groups are found. Following the pioneer work of Julianelle (1926), who determined the first three capsular types, a total of 82 K-antigens (K:1 to K:82) were described, although K:73, K:75, K:76, and K:78 were invalidated. There are thus only 78 available types. Cross-reactions among established K-antigens are numerous. K-antigen determination is traditionally by the capsular swelling reaction. When bacteria and the corresponding serum are mixed on a glass slide and observed under the microscope, the capsule becomes highly refractile and easily visible. Other methods include slide or tube agglutination, indirect immunofluorescence, and countercurrent immunoelectrophoresis (Ørskov and Ørskov, 1984b).

Some K-antigens are correlated with pathogenicity. *K. pneumoniae* subsp. *rhinoscleromatis* strains have capsular antigen K:3, whereas *K. pneumoniae* subsp. *ozaenae* strains are mostly K:4 and occasionally K:3 or K:5. *K. pneumoniae* subsp. *pneumoniae* strains recovered from respiratory tract infections have generally antigens K:1 to K:6, whereas strains form other nosocomial infections (e.g., urinary tract) have often antigens K:2, K:21, K:55, K:10, or K:24. Other K-types (17, 25, 22, 43, 1, 3, 33) can also be encountered (Le Minor and Richard, 1993). Strains found in the sperm of stallions and that cause metritis and abortion in mares contaminated with such sperm are often K:1, K:2, K:7, or K:30. Nosocomial *K. oxytoca* strains have been found with K:26, K:68, K:43, and less frequently K:26, K:21, K:18, or K:47. *K. mobilis* (*E. aerogenes*) strains most of the time can show capsular swelling with *Klebsiella* antisera. K-types associated with this species are often K:68 or K:26 and occasionally K:42, K:59, K:11, or K:4 (Le Minor and Richard, 1993).

Pathogenicity   The genus *Klebsiella* is ubiquitous. Strains of this genus are found in surface water, sewage, soil, and on plants. They also colonize the mucosal surfaces of humans, horses, and swine (Podschun and Ullmann, 1998). In humans, *K. pneumoniae* is present as a commensal in the nasopharynx (detection rate, 1–6%) and in the intestinal tract (detection rate, 5–38%). In hospitals, colonization rates increase with the length of stay (detection rates, 77% in the stool, 19% in the pharynx, and 42% on the hands of patients) and hospital personnel can carry *K. pneumoniae* (Podschun and Ullmann, 1998). The high rate of colonization in patients is associated with the use of antibiotics.

The genus *Klebsiella* can be associated with different sorts of infections. *Klebsiella pneumoniae* is a cause of community-acquired bacterial pneumonia (Friedländer's pneumonia), occurring particularly in chronic alcoholics, and showing characteristic radiographic abnormalities. The fatality rate is high if untreated (Podschun and Ullmann, 1998).

K. pneumoniae subsp. *rhinoscleromatis* is the causative agent of rhinoscleroma, a chronic infection that can involve the nasal cavity (most often) or the upper airways (pharynx, larynx, trachea). Typically, a granulomatous destructive and disfiguring process results in airway obstruction. Ultimately, extensive fibrosis and scarring occur. Histological features include mixed diffuse infiltrate of the upper and lower dermis with numerous plasmocytes, lymphocytes, neutrophils, and characteristic large macrophages with foamy cytoplasm (Mikulicz cells) and that contain *Klebsiella*.

K. pneumoniae subsp. *ozaenae* has been implicated in ozena, a chronic atrophic rhinitis giving off a very bad smell. However, the pathogenic role of the bacterium in this syndrome is less clear.

Klebsiella granulomatis (formerly *Calymmatobacterium granulomatis*) is the presumed causative agent of granuloma inguinale (donovanosis), a genital ulceration that is sexually transmissible. Histological features include dense dermal infiltrate of plasmocytes, neutrophils, and large macrophages with vacuolated cytoplasms that contain intracellular bacilli (Donovan bodies). The similarity of donovanosis and rhinoscleroma histological lesions has been pointed out (Carter et al., 1999).

Klebsiella spp. (with the exception of *K. pneumoniae* subsp. *rhinoscleromatis*, *K. pneumoniae* subsp. *ozaenae*, and *K. granulomatis*) are mostly considered nosocomial pathogens. *Klebsiella pneumoniae* and *K. mobilis* (*Enterobacter aerogenes*) are most frequently involved, although *K. oxytoca* and *K. planticola*, and rarely *K. terrigena*, can be found.

The urinary tract is the most common site of infection, and *K. pneumoniae* accounts for 6–17% of all nosocomial urinary tract infections (Podschun and Ullmann, 1998). *K. pneumoniae* is also a frequent cause of bacteremia. In premature infants, *K. pneumoniae* is often involved in neonatal sepsis.

The principal reservoir of *K. pneumoniae* in the hospital is the gastrointestinal tract of patients. The principal vectors are the hands of personnel. Occasional sources are contaminated medical equipment and blood products (Podschun and Ullmann, 1998).

Renewed interest in *Klebsiella* infections stemmed from the spread of multiresistant strains. *K. pneumoniae* is naturally susceptible to aminoglycosides. However, plasmids mediating aminoside-modifying enzymes have spread in the 1970s. In a French survey, 78% of *Klebsiella* isolates were susceptible to gentamicin, tobramycin, netilmycin, and amikacin. Different phenotypes of aminoside resistance were observed, involving resistance to gentamicin, tobramycin, netilmycin, and/or amikacin (Lemozy et al., 1985). The most frequent aminoside inactivating enzymes found produced by *Klebsiella* isolates were aminoglycoside-3′-*N*-acetyltransferases and aminoglycoside -2′′-*O*-nucleotidyltransferase (Lemozy et al., 1985).

Strains of *K. pneumoniae* and *K. oxytoca* that have not acquired any resistance determinant are naturally resistant to aminopenicillins (ampicillin) and carboxypenicillins (carbenicillin) and susceptible to other β-lactam antibiotics. This is due to the production of a chromosomal penicillinase, which is inhibited by clavulanic acid. A small zone of inhibition around 100–mg carbenicillin disks is typical of this phenotype (Jarlier, 1985). Acquired resistance arose from the production of a plasmid-determined penicillinase. The strains showed a higher resistance to carbenicillin (no zone around carbenicillin disks), and resistance

to ureidopenicillins, cefalothin, cefamandole, and cefuroxime. Inactivation of amoxycillin was not blocked by clavulanic acid. However, these strains remained susceptible to third-generation cephalosporins (e.g., cefotaxime) (Jarlier, 1985). Since 1982, strains producing extended spectrum β-lactamases (ESBL) encoded by a plasmid have spread. *K. pneumoniae* strains resistant to ceftazidime were commonly producing a SHV-5 type β-lactamase in Europe and a TEM-10 or TEM-12 type in the United States. These strains account for 5% (USA) to 16% (France) of the *K. pneumoniae* tested (Podschun and Ullmann, 1998). These ESBL-producing strains have been susceptible to carbapenems (imipenem, meropenem). However, ESBL-producing *K. pneumoniae* resistant to imipenem have been isolated. They produce a β-lactamase of the AmpC type (Podschun and Ullmann, 1998). *Klebsiella mobilis* (*Enterobacter aerogenes*) was rarely encountered as a nosocomial pathogen until strains carried a plasmid encoding an ESBL. In France and Belgium, β-lactamases of type SHV-4, TEM-24, and TEM-3 were characterized in *K. mobilis*. It was shown that TEM-24 producing isolates found in several epidemics in different countries constituted mostly a single clone (Galdbart et al., 2000; De Gheldre et al., 2001; Mammeri et al., 2001).

Pathogenesis *Klebsiella* species are surrounded by a hydrophilic polysaccharide capsule, which is the first virulence factor described in klebsiellae (Toenniessen, 1914; Cryz et al., 1984). Simoons-Smit et al. (1984) showed that strains with K1, K2, K4, and K5 capsular antigens were more virulent for mice (skin model) than strains with K6 and K above 6. The loss of K antigen by K$^+$ strains resulted in the reduced virulence of their K$^-$ variants. This reduced virulence may be explained by a higher degree of phagocytosis as measured by chemiluminescence response of human polymorphonuclear leukocytes (PMNLs) and by enhanced killing by either human PMNLs or human serum or both (Simoons-Smit et al., 1986). Other studies (mouse peritonitis model) found strains with antigens K1 and K2 more virulent than strains with other K antigens (Podschun and Ullmann, 1998). The K antigen plays a crucial role in protecting the bacterium from opsonophagocytosis in the absence of specific antibodies (Williams et al., 1983, 1986b; Simoons-Smit et al., 1986). The antiphagocytic function consists of inhibiting the activation or uptake of complement components, especially C3b. In addition, the K antigen inhibits the differentiation and functional capacity of macrophages *in vitro*. The degree of virulence conferred by a particular K antigen might be connected to the mannose content of the capsular polysaccharide (Podschun and Ullmann, 1998). K2 serotype is among the most common K-type isolated from patients with urinary tract infection, pneumonia, or bacteremia. K2 is the predominant serotype of human isolates worldwide, whereas K2 strains are very rarely encountered in the environment (Podschun and Ullmann, 1998).

Strains of *K. pneumoniae*, *K. oxytoca*, *K. planticola*, and *K. terrigena* may produce thick, channeled (type-1) fimbriae associated with mannose-sensitive hemagglutination (MS-HA). *K. mobilis* (*E. aerogenes*) produces antigenically similar fimbriae. Type 1 fimbriae mediate the attachment of *K. pneumoniae* to uroepithelial cells (Williams and Tomas, 1990) and to ciliated tracheal cells *in vitro* (Fader et al., 1988).

Strains of *K. pneumoniae*, *K. oxytoca*, *K. planticola*, and *K. terrigena* may also produce thin, nonchanneled (type-3) fimbriae associated with mannose-resistant hemagglutinins of the *Klebsiella* type (MR/K-HA). These fimbriae are efficient in promoting adherence to the roots of various grasses and cereals (Haahtela and

Korhonen, 1985). They are capable of binding to endothelial cells, respiratory tract epithelial cells, and uroepithelial cells. In the kidney, they mediate bacterial adhesion to tubular basement membranes, Bowman's capsules, and renal vessels (Podschun and Ullmann, 1998). Three new putative colonization factors have also been described (Podschun and Ullmann, 1998).

The most important role of the O antigen is to protect *K. pneumoniae* from complement-mediated killing (Williams and Tomas, 1990). For this protection, O antigen chain length seems to be important (McCallum et al., 1989).

Iron is essential for bacterial growth. Brewer et al. (1982) observed that virulence was enhanced by hyperferremi. Since in the human body iron is complexed to carrier molecules such as transferrin (in the serum) or lactoferrin (in milk and other secretions), or sequestered within cells (in heme proteins), potentially pathogenic *Enterobacteriaceae* produce high-affinity systems (siderophores) to solubilize and import the required iron. The iron-chelating compounds produced are mostly of two sorts, phenolates (e.g., enterochelin) and hydroxamates (aerobactin) (Payne, 1988). Almost all strains of *Klebsiella* produce enterochelin whereas only a few produce aerobactin (Williams et al., 1987). Nassif and Sansonetti (1986) correlated the virulence of *K. pneumoniae* serotypes K1 or K2 with the presence of a 180-kb plasmid encoding the hydroxamate siderophore aerobactin. Aerobactin is an essential factor of pathogenicity and the 180-kb plasmid carries additional genes encoding other virulence factors (Nassif and Sansonetti, 1986). Some strains of *K. pneumoniae* express a ferric aerobactin uptake system without making the chelator itself. This may confer a selective advantage in mixed infections in competition with other aerobactin-producing bacteria (Williams and Tomas, 1990).

The production of cytotoxins, enterotoxins, and hemolysins have been sporadically described (Podschun and Ullmann, 1998).

Ecology Klebsiellae have been recovered from aquatic environments receiving industrial wastewaters, plant products, fresh vegetables, food with a high content of sugars and acids, frozen orange juice concentrate, sugar cane wastes, living trees, plants, and plant by-products. They are commonly associated with wood, saw dust, and waters receiving industrial effluents from pulp and paper mills and textile finishing plants. Isolates have been described in forest environments, degrading kraft-lignin, tannic acid, pine bark, and condensed tannin, or from living or decaying wood and bark or composted wood.

Klebsiella can frequently be isolated from the root surfaces of various plants. *K. pneumoniae*, *K. oxytoca*, or *K. planticola* are capable of fixing nitrogen and are classified as associative nitrogen fixers.

Strains from plants certainly need to be reidentified in the light of present taxonomic schemes. Strains of *K. pneumoniae* sensu stricto that are associated with plants differ from those associated with serious human infections. These environmental *K. pneumoniae* strains are most often able to utilize 5-ketogluconate as sole carbon source and never have capsular types K1 to K6. Strains involved in serious infection do not utilize 5-ketogluconate and may have capsular types K1 to K6 as well as other capsular types (Grimont et al., 1991).

Capsular types K1, K2, and K5 were major causes of epidemic metritis in mares in England, whereas type K7 was associated with sporadic, opportunistic genital infection. Outbreaks of metritis in mare were due to type K2 in the United States and in France.

The stallion plays an important role in the transmission of *K. pneumoniae*. Type K7 was found on the preputial skin of stallions and may be part of the normal bacterial flora in this location. Thus it is important to type *Klebsiella* isolated in the genital tract of horses to detect a stallion carrying an epidemic strain among other stallions carrying less pathogenic *K. pneumoniae* (Grimont et al., 1991).

Klebsiella have been frequently associated with bovine mastitis, and causes serious infections in other animals including rhesus monkeys, guinea pigs, or muskrats. Epidemics of fatal generalized infections among captive squirrel monkeys (*Saimiri sciureus*) in French Guyana and lemurs in a French zoo were due to *K. pneumoniae* K5 and K2, respectively. Immunization of the monkeys with the corresponding capsular polysaccharide is efficient in stopping the epidemic (Grimont et al., 1991).

ENRICHMENT AND ISOLATION PROCEDURES

The detection, isolation, and enumeration in clinical, industrial, and natural environments can be facilitated by using a selective medium.

On agar plates, although colonies develop overnight, the characteristic elevated and mucoid appearance is observed after incubation for 48 h.

The ability to utilize citrate (Cooke et al., 1979) or *myo*-inositol (Legakis et al., 1976) has been applied to the formulation of selective media. Resistance of *Klebsiella* spp. to methyl violet (Campbell and Roth, 1975), double violet (Campbell et al., 1976), potassium tellurite (Tomas et al., 1986), and carbenicillin (Thom, 1970) has been used in selective media.

Thom (1970) developed a medium based on the MacConkey agar in which lactose is replaced by inositol (1% w/v), with the addition of 100 µg of carbenicillin per ml. Bagley and Seidler (1978) devised a similar medium with only 50 µg/ml carbenicillin. On this medium, about 95% of pink-to-red colonies were verified to be *Klebsiella* spp., whereas only 1% of yellow background colonies were *Klebsiella*.

Since about 10% of *Klebsiella* strains are susceptible to 50 µg/ml of carbenicillin, the antibiotic was replaced in the above formula by tellurite (K_2TeO_3, 3 µg/ml) (Tomas et al., 1986), which is a strong inhibitor of phosphate transport in *E. coli*. Minimal inhibitory concentrations of K_2TeO_3 were 100 or 200 µg/ml for *K. pneumoniae* subsp. *pneumoniae*, *K. oxytoca*, *K. planticola*, and *K. terrigena*, 10 µg/ml for *K. pneumoniae* subsp. *ozaenae*, and 1–3 µg/ml for other *Enterobacteriaceae* (Tomas et al., 1986). In a field test, 77% of pink-to-red colonies on MacConkey–inositol–potassium tellurite agar were confirmed as *Klebsiella* spp.; however, the efficiency of plating was about 1% (Dutka et al., 1987).

Bruce et al. (1981) devised an agar medium combining Koser citrate and raffinose (carbon sources) and ornithine and low pH (for ornithine decarboxylase). On acidic Koser citrate agar with ornithine and raffinose, *Klebsiella* strains grow as yellow mucoid colonies. Other *Enterobacteriaceae* either do not grow, or produce small colorless, pink, red, or orange colonies.

For the isolation of *K. pneumoniae* and *K. oxytoca* from human feces, Van Kregten et al. (1984) have developed a medium based on the presence of two carbon sources, citrate and inositol, without inhibitor. The medium consists of Simmons citrate agar with 1% inositol. *Klebsiella* spp. appear as yellow, dome-shaped, often mucoid colonies, whereas *E. coli* appears as tiny, watery colonies. Apart from some *Enterobacter* strains, no other of bacteria grow on the medium.

Wong et al. (1985) devised a minimal medium in which the carbon source is lactose and the nitrogen source is potassium nitrate. Inhibitory compounds were deoxycholate, neutral red, and crystal violet. On this medium, *K. pneumoniae* and *K. oxytoca* grow as convex, and 1–2 mm in diameter, rather mucoid pink-to-red colonies, or larger, more watery pale red colonies with a dark red center. Non-klebsiellae either fail to grow or form colorless colonies.

MAINTENANCE PROCEDURES

Klebsiella strains can be easily maintained at room temperature in meat extract semisolid agar, at $-80°C$ in a broth medium with 10–50% (v/v) glycerol, or freeze-dried. *K. granulomatis* will not remain viable when stored at 5° or 37°C. At 25°C, viability of egg yolk cultures is maintained for 8–10 d.

DIFFERENTIATION OF THE GENUS *KLEBSIELLA* FROM OTHER GENERA

See Table BXII.γ.193 of the family *Enterobacteriaceae* for characteristics that can be used to distinguish *Klebsiella* from other genera of the family. No strain of *Klebsiella* other than *K. mobilis* is motile.

Classically, a Gram-negative capsulated isolate that is nonmotile, aerogenic, Voges–Proskauer positive, lysine decarboxylase positive, ornithine decarboxylase negative, and arginine dihydrolase negative belongs in the genus *Klebsiella*. However, inclusion of *K. mobilis* and *K. ornithinolytica* makes this classical identification more difficult at the genus level. In the following discussion, the genus *Enterobacter* will be considered without *E. aerogenes*.

Carbon source utilization tests are essential. All *Klebsiella* strains (with the exception of *K. pneumoniae* subsp. *ozaenae* and subsp. *rhinoscleromatis*) utilize the following six substrates: D-arabitol, *myo*-inositol, palatinose, quinate, D-sorbitol, and sucrose. No *Enterobacter*, *Pantoea*, or *Erwinia* species utilizes all six of these substrates. In the genus *Serratia*, only *S. ficaria* would utilize all six substrates and also L-arabitol and *i*-erythritol, which no *Klebsiella* has the simultaneous ability to utilize. Furthermore, D-adonitol is utilized by all *Klebsiella* strains except environmental strains of *K. pneumoniae*. In the genus *Enterobacter*, D-adonitol is utilized only by some strains of *Enterobacter cloacae*. Some strains of *Pantoea* also utilize D-adonitol.

TAXONOMIC COMMENTS

The first organism of the genus *Klebsiella* described was a capsulated bacillus isolated from patients with rhinoscleroma (von Frisch, 1882). Then Friedländer (1882) described a bacterium from lungs of a patient who had died of pneumoniae. The organism was subsequently named *"Hyalococcus pneumoniae"* (Schroeter, 1889). The genus *Klebsiella* was coined by Trevisan (1885) to honor the German microbiologist Edwin Klebs (1834–1913). In 1887 (Trevisan, 1887), the genus contained *Klebsiella pneumoniae* (Friedländer bacillus) and *K. rhinoscleromatis* (von Frisch's bacillus). *"Bacillus mucosus ozaenae"*, which Abel (1893) observed in nasal secretion of patients with ozena, was later included in the genus as *K. ozaenae* (Bergey et al., 1925).

The genus *Klebsiella* with three species lasted up to the eighth edition of the *Bergey's Manual of Determinative Bacteriology*.

Flügge (1886) described *"Bacillus oxytocus perniciosus"* from old milk. This organism was named *"Aerobacter oxytocum"* (Bergey et al., 1923) and *Klebsiella oxytoca* (Lautrop, 1956). For many years, the existence of this indole-positive species was questioned and

it was often considered a biogroup of *K. pneumoniae* (Edwards and Ewing, 1972; Ørskov, 1974). DNA relatedness studies showed *K. oxytoca* to be clearly distinct from *K. pneumoniae* (Jain et al., 1974; Brenner et al., 1977).

K. planticola (Bagley et al., 1981) was proposed for strains isolated primarily from botanical and soil environments (Bagley et al., 1981). *K. planticola* is distinct from other *Klebsiella* species based on DNA relatedness (Woodward et al., 1979; Izard et al., 1981a).

K. terrigena (Izard et al., 1981a), was proposed for strains isolated mainly from aquatic and soil environments (Izard et al., 1981a). It is a distinct species according to DNA relatedness studies (Woodward et al., 1979; Izard et al., 1981a).

The species described as *K. trevisanii* (Ferragut et al., 1983) was shown to be synonymous with *K. planticola* by DNA–DNA hybridization (Gavini et al., 1986).

In the first edition of the *Bergey's Manual of Systematic Bacteriology*, four species were recognized: *K. pneumoniae*, *K. oxytoca*, *K. planticola*, and *K. terrigena*.

K. ozaenae and *K. rhinoscleromatis* cannot be separated from *K. pneumoniae* by DNA relatedness (Brenner et al., 1972c). For this reason, *K. ozaenae* and *K. rhinoscleromatis* were treated as subspecies of *K. pneumoniae* in the first edition of the *Bergey's Manual of Systematic Bacteriology* (Ørskov, 1974).

For many years, *K. pneumoniae* could not be objectively separated from an organism observed by Escherich (1885), named *"Bacterium lactis aerogenes"* (Kruse, 1896), transferred to genus *"Aerobacter"* as *"A. aerogenes"*, and finally as *Enterobacter aerogenes* (Hormaeche and Edwards, 1960b). Confusion occurred since both *K. pneumoniae* and *"Aerobacter aerogenes"* fermented many carbohydrates (often with gas production), gave a positive Voges–Proskauer reaction, and reacted with *Klebsiella* capsular antisera. When Møller (1955) introduced the decarboxylase tests, *K. pneumoniae* was defined as nonmotile and ornithine decarboxylase negative, whereas *"A. aerogenes"* was defined as motile or nonmotile and ornithine decarboxylase positive (Hormaeche and Edwards, 1958). Later, a new genus *Enterobacter* was formed (Hormaeche and Edwards, 1960b) to which *"A. aerogenes"* was transferred as *E. aerogenes*.

Nomenclatural confusion occurred when a biogroup of *K. pneumoniae* was named *"Klebsiella aerogenes"* (Taylor et al., 1956). Cowan et al. (1960) subdivided *K. pneumoniae sensu lato* into *K. pneumoniae* (*sensu stricto*), *"K. aerogenes"*, *"K. edwardsii* subsp. *edwardsii"*, and *"K. edwardsii* subsp. *atlantae"*. Authentic or typical strains of *K. pneumoniae*, *K. edwardsii*, and *"K. aerogenes"* cannot be differentiated by DNA relatedness or protein electrophoresis, and thus belong to a single species, *K. pneumoniae* (Brenner et al., 1972c; Jain et al., 1974; Ferragut et al., 1989).

These problems around the epithet *aerogenes* were due to a lack of authentic cultures. Neotypes have since been designated when no original type strain was available.

Enterobacter aerogenes is much closer to *Klebsiella* species than to *Enterobacter cloacae* (the type species of that genus) based on phenotypic traits and DNA relatedness (Bascomb et al., 1971; Brenner et al., 1972c; Steigerwalt et al., 1976; Izard et al., 1980). The transfer of *E. aerogenes* to the genus *Klebsiella* has been proposed (Bascomb et al., 1971). However, since the name *"Klebsiella aerogenes"* had been used for another organism, a new name, *Klebsiella mobilis*, was coined (Bascomb et al., 1971).

Nonmotile, encapsulated, ornithine decarboxylase positive strains were described under the name *Klebsiella ornithinolytica*

(Sakazaki et al., 1989a). Apart from ornithine decarboxylase test, these strains are biochemically undistinguishable from *K. planticola*, and are identical to *K. planticola* by DNA hybridization (Farmer et al., 1985a; D.J. Brenner, personal communication). Farmer et al. (1985a) questioned the validity of *K. ornithinolytica* as a species distinct from *K. planticola*. Some ornithine decarboxylase positive strains have been shown to belong in *K. pneumoniae* by DNA relatedness criteria (Lindh and Frederiksen, 1990).

Finally, *Calymmatobacterium granulomatis*, which has not been cultured on bacteriological media, has been transferred to the genus *Klebsiella* as *Klebsiella granulomatis*, mostly based on *rrs* sequences (Carter et al., 1999). Partial *rpoB* sequencing suggested that *K. granulomatis* is very close to *K. pneumoniae* (Drancourt et al., 2001) and *phoE* sequencing showed *K. granulomatis* close to *K. pneumoniae* subsp. *rhinoscleromatis* (Carter et al., 1999). Thus, there is presently no proof that *K. granulomatis* is a species distinct from *K. pneumoniae*, and instead there are arguments to make it part of *K. pneumoniae*.

Sequence data are increasingly used to suggest or support nomenclatural changes and some of these changes can be anticipated. The taxonomic significance of the three sequence clusters observed in *K. pneumoniae* or the two sequence clusters in *K. oxytoca* (Brisse and Verhoef, 2001) is still unknown, and DNA–DNA hybridization data are needed before a proposal can be made to design new species or subspecies. Since the phylogenetic tree obtained by *rrs* gene (encoding 16S rRNA) sequence comparison showed the genus *Klebsiella* to be split in at least two groups, a new genus, *Raoultella*, was proposed to contain *R. planticola* (type species), *R. ornithinolytica*, and *R. terrigena* (Drancourt et al., 2001). More sequencing work needs to be done to evaluate this proposal since partial *rpoB* sequence failed to group *Raoultella* species in a single cluster and these sequences were only 512 nucleotides long.

DIFFERENTIATION OF THE SPECIES OF THE GENUS *KLEBSIELLA*

Table BXII.γ.229 presents the characteristics differentiating the species and subspecies of *Klebsiella*.

List of species of the genus Klebsiella

1. **Klebsiella pneumoniae** (Schroeter 1886) Trevisan 1887, 94[AL] (*Hyalococcus pneumoniae* Schroeter 1886, 1952.)
   pneu.mo'ni.ae. Gr. n. *pneumonia* pneumonia, inflammation of the lungs; M.L. gen. n. *pneumoniae* of pneumonia.

   The characteristics are as described for the genus and as listed in Table BXII.γ.229. *K. pneumoniae* can be divided into three sequence clusters (Brisse and Verhoef, 2001) unrelated to the present subdivision in subspecies. *K. pneumoniae* is normally found in the intestinal tract of humans and animals. It may be isolated in association with several pathological processes in humans, e.g., community-acquired pneumonia or nosocomial urinary tract infection. Serotype K2 is the most common K-type isolated from patients with urinary tract infection, pneumonia, or bacteremia. In animals, *K. pneumoniae* may be isolated from metritis in mares, bovine mastitis, or generalized infections in captive monkeys. Environmental strains generally utilize more carbon sources than clinical strains, but often fail to utilize D-adonitol. Intraspecies DNA relative reassociation values among strains is ~80–90% (Brenner et al., 1972c) or 73–100% (Woodward et al., 1979).

   The mol% G + C of the DNA is: 56–58 (T_m) (Seidler et al., 1975).

   Type strain: ATCC 13883, CIP 82.9, DSM 30104, JCM 1662.

   GenBank accession number (16S rRNA): X87276, Y17656, AB004753, AF130981.

   a. **Klebsiella pneumoniae** subsp. **pneumoniae** (Schroeter 1886) Trevisan 1887, 94[AL] (*Hyalococcus pneumoniae* Schroeter 1886, 1952.)
      Distinguished from the subspecies subsp. *ozaenae* and subsp. *rhinoscleromatis* by the characteristics listed in Table BXII.γ.229. Clinical strains may produce an extended-spectrum β-lactamase.

      The mol% G + C of the DNA is: 56–58 (T_m) (Seidler et al., 1975).

      Type strain: ATCC 13883, CIP 82.9, DSM 30104, JCM 1662.

      GenBank accession number (16S rRNA): X87276, Y17656, AB004753, AF130981.

   b. **Klebsiella pneumoniae** subsp. **ozaenae** (Abel 1893) Ørskov 1984c, 355[VP] (Effective publication: Ørskov 1984b, 463) (*Klebsiella ozaenae* (Abel 1893) Bergey, Harrison, Breed, Hammer and Huntoon 1925, 266; *Bacillus mucosus ozaenae* Abel 1893, 167; *Bacillus ozaenae* (Abel 1893) Lehmann and Neumann 1896, 204.)
      o.zae'nae. L. fem. n. *ozaena* ozena; L. gen. n. *ozaenae* of ozena.

      Distinguished from the subspecies subsp. *pneumoniae* and subsp. *ozaenae* by the characteristics listed in Table BXII.γ.229. Occurs in ozena and other chronic diseases of the respiratory tract.

      The mol% G + C of the DNA is: not available.

      Type strain: ATCC 11296, CIP52.211, JCM 1663, LMG 3113.

      GenBank accession number (16S rRNA): Y17654, AF130982.

   c. **Klebsiella pneumoniae** subsp. **rhinoscleromatis** (Trevisan 1887) Ørskov 1984c, 355[VP] (Effective publication: Ørskov 1984b, 464) (*Klebsiella rhinoscleromatis* Trevisan 1887, 95; *Bacterium rhinoscleromatis* (Trevisan 1887) Migula 1900, 352.)

rhi.no.scle.ro' ma.tis. M.L. adj. *rhinoscleromatis* pertaining to rhinoscleroma.

Distinguished from the subspecies subsp. *pneumoniae* and subsp. *ozaenae* by the characteristics listed in Table BXII.γ.229. Found in patients with rhinoscleroma.

The mol% G + C of the DNA is: not available.

Type strain: ATCC 13884, CIP 52.210, JCM 1664, LMG 3184.

GenBank accession number (16S rRNA): Y17657, AF130983.

2. **Klebsiella granulomatis** (Aragão and Vianna 1913) Carter, Bowden, Bastian, Myers, Sriprakash and Kemp 1999, 1698[VP] (*Calymmatobacterium granulomatis* Aragão and Vianna 1913, 221.)

gran.u.lo' ma.tis. L. dim. n. *granulum* a small grain; Gr. suff. *-oma* a swelling or tumor; M.L. n. *granuloma* a granuloma; M.L. gen. n. *granulomatis* of a granuloma.

Occurs in granuloma inguinale (donovanosis) lesions. The bacilli (Donovan bodies) are seen in large vacuolated macrophages found in dermal infiltrate of genital ulcerations. Grow in the yolk sac of developing embryo or in cell cultures (fresh mononuclear cells or Hep-2 cell line). No growth in bacteriological media. Type strain: No type culture is currently available.

The mol% G + C of the DNA is: not available.

GenBank accession number (16S rRNA): AF009171, AF010251, AF010252, AF010253.

3. **Klebsiella mobilis** Bascomb, Lapage, Willcox and Curtis 1971, 279[AL]*

mo' bi.lis. L. adj. *mobilis* movable, motile.

The characteristics are described in Table BXII.γ.229. This is the only motile species in the genus. Strains producing extended-spectrum β-lactamase occur as nosocomial pathogens. Occur in water, sewage, soil, dairy products, and the feces of humans and animals. Also an opportunistic pathogen.

The mol% G + C of the DNA is: 53–54 (Bd).

Type strain: ATCC 13048, CIP 60.86, DSM 30053, JCM 1235, LMG 2094.

GenBank accession number (16S rRNA): AB004748.

4. **Klebsiella oxytoca** (Flügge 1886) Lautrop 1956, 375[AL] (*Bacillus oxytocus perniciosus* Flügge 1886, 268.)

oxy.to' ca. Gr. *oxys* sour, acid; Gr. suff. *-tokos* bearer, producer; M.L. n. *oxytocus* acid-producer, spurious; M.L. adj. *oxytoca* (sic) acid-producing.

The characteristics are as described for the genus and as listed in Table BXII.γ.229. Present in the intestinal tract of humans and animals. Can be isolated from various pathological processes and from botanical and aquatic environments. *K. oxytoca* can be divided into two sequence clusters (Brisse and Verhoef, 2001). Encapsulated; typable with *Klebsiella* K antisera. The intraspecies DNA relative reassociation values were 75% in the study by Brenner et al. (1972c) and 95% (average value) in the study by Woodward et al. (1979).

The mol% G + C of the DNA is: 55–58 (T_m).

Type strain: ATCC 13182, CIP 103434, JCM 1665, LMG 3055.

GenBank accession number (16S rRNA): Y17655, AB004754, AF129440.

5. **Klebsiella planticola** Bagley, Seidler and Brenner 1982, 266[VP] (Effective publication: Bagley, Seidler and Brenner 1981, 109.)*

plan.ti' co.la. L. fem. n. *planta* a plant; L. suff. *-cola* dweller; M.L. fem. n. *planticola* plant-dweller.

The characteristics are as described for the genus and as listed in Table BXII.γ.229. Isolated mainly from botanical, aquatic, and soil environments. Three biovars have been described (Naemura et al., 1979). Encapsulated; typable with *Klebsiella* K antisera. The average intraspecies DNA relative reassociation value is above 75% (Woodward et al., 1979).

The mol% G + C of the DNA is: 53.9–55.4 (T_m) (Seidler et al., 1975).

Type strain: ATCC 33531, ATCC 33558, CIP 100751, DSM 3069, IFO 14939.

GenBank accession number (16S rRNA): X93215, Y17659, AB004755, AF129443.

6. **Klebsiella terrigena** Izard, Ferragut, Gavini, Kersters, De Ley and Leclerc 1981a, 125[VP]

ter.ri.ge' na. L. n. *terra* soil; L. suff. *gena* origin; M.L. n. *terrigena* from soil.

The characteristics are as described for the genus and as listed in Table BXII.γ.229. Isolated mainly from aquatic and soil environments. The average intraspecies DNA relative reassociation value is above 86% (Izard et al., 1981a).

The mol% G + C of the DNA is: 56.7 (T_m) (Izard et al., 1981a).

Type strain: ATCC 33257, CIP 80-07, DSM 2687, JCM 1687, LMG 3222.

GenBank accession number (16S rRNA): Y17658, AF129442.

Species Incertae Sedis

1. **Klebsiella ornithinolytica** Sakazaki, Tamura, Kosako and Yoshizaki 1989b, 495[VP] (Effective publication: Sakazaki, Tamura, Kosako and Yoshizaki 1989a, 205.)

or.ni.thi.no.ly' ti.ca. M.L. *ornithinum* ornithine, an amino acid; Gr. adj. *lyticus* dissolving; M.L. fem. adj. *ornithinolytica* ornithine dissolving.

The CDC group called Klebsiella Group 47 corresponds to this species (Sakazaki et al., 1989a). Physiological and nutritional characteristics are presented in Table BXII.γ.229. Isolated from clinical materials, including urine, sputa, stool, and pus. Can be found in food. Clinical significance unknown.

The mol% G + C of the DNA is: 57–58 (T_m) (Sakazaki et al., 1989a).

Type strain: ATCC 31898, CIP 103576, DSM 7464, JCM 6096.

GenBank accession number (16S rRNA): AF129441, AJ251467.

Editorial Note: Klebsiella mobilis Bascomb et al. 1971 and *Enterobacter aerogenes* Hormaeche and Edwards 1960b have the same type strain and therefore are homotypic synonyms.

Editorial Note: Klebsiella trevisanii Ferragut et al. 1983 (phenon K of Gavini et al., 1977) is a junior subjective synonym of *K. planticola* (Gavini et al., 1986).

Genus XVII. **Kluyvera** Farmer, Fanning, Huntley-Carter, Holmes, Hickman, Richard and Brenner 1981b, 382[VP] (Effective publication: Farmer, Fanning, Huntley-Carter, Holmes, Hickman, Richard and Brenner 1981a, 927)

J.J. FARMER III

Kluy' ve.ra. M.L. fem. n. *Kluyvera* named by Asai et al. (1956) to honor the Dutch microbiologist A.J. Kluyver, who made many contributions to microbial physiology and taxonomy.

Small rod-shaped cells, 0.5–0.7 × 2–3 μm, conforming to the general definition of the family *Enterobacteriaceae.* Gram negative. Motile, with peritrichous flagella. Contains the enterobacterial common antigen. Facultatively anaerobic. Catalase positive (weak). Oxidase negative. Nonpigmented. Ferment, rather than oxidize, D-glucose and other carbohydrates. Reduce nitrate to nitrite. **Positive for indole production, methyl red, citrate utilization (Simmons), ornithine decarboxylase, motility at 36°C, growth in the presence of cyanide (KCN test), malonate utilization, and esculin hydrolysis. Ferment D-glucose, lactose, sucrose, D-mannitol, salicin, L-arabinose, raffinose, L-rhamnose, maltose, D-xylose, trehalose, cellobiose, α-methyl-D-glucoside, melibiose, mucate, D-mannose, and D-galactose. Produce visible gas during fermentation. Produce large amounts of α-ketoglutaric acid during the fermentation of D-glucose.** Negative for Voges–Proskauer, H_2S production (TSI), urea hydrolysis, phenylalanine deaminase, arginine dihydrolase, gelatin hydrolysis (22°C), lipase (corn oil), DNase, and the fermentation of dulcitol, adonitol, *myo*-inositol, erythritol, and D-arabitol. Susceptible to colistin, sulfadiazine, gentamicin, kanamycin, tetracycline, and chloramphenicol (disk diffusion method on Mueller–Hinton agar); **resistant to carbenicillin; penicillin, ampicillin, and cephalothin**; variable susceptibility to nalidixic acid and streptomycin.

Some strains produce kluyveramycin, unusual reddish blue crystals of unknown composition. **Occur in human clinical specimens**, food, soil and sewage. Probably an occasional **opportunistic pathogen and a cause of extraintestinal infections of humans**.

The genus *Kluyvera* includes *Kluyvera ascorbata, K. cryocrescens*, and *K. georgiana.* There is also a fourth species with standing in nomenclature, *Kluyvera cochleae*, but it appears to be a synonym of *Enterobacter intermedius* (originally named *E. intermedium*; corrected by von Graevenitz, 1990).

The mol% G + C of the DNA is: 55.1–56.6.

Type species: **Kluyvera ascorbata** Farmer, Fanning, Huntley-Carter, Holmes, Hickman, Richard and Brenner 1981b, 382 (Effective publication: Farmer, Fanning, Huntley-Carter, Holmes, Hickman, Richard and Brenner 1981a, 927.)

FURTHER DESCRIPTIVE INFORMATION

Literature   Since *Kluyvera* was described in 1956, there have been many reports in the literature; over 40 have been cataloged in MEDLINE and many others in BIOSIS. In addition to the genus and species names, "Enteric Group 8", "*K. citrophila*", and "*K. noncitrophila*" should also be included as a search terms, although they are rarely used today.

The National Library of Medicine's Internet site Entrez (http://www.ncbi.nlm.nih.gov), which covers most of the sequence databases, includes full sequences of the 16S rRNA gene for *K. ascorbata, K. georgiana*, and *K. cochleae* (*Enterobacter intermedius*), and a partial sequences for *K. cryocrescens.* A search for the term *Kluyvera* yielded 29 sequences.

Problems in evaluating the literature   Computer literature searches often yield articles such as "First case of bacteremia due

to xx" or "Wound infections due to yy"; where "xx" and "yy" are newly described bacterial species. The reader should evaluate these reports very critically, with particular attention to the way that cultures were identified. Over half the cultures of new or unusual *Enterobacteriaceae* in these reports have been misidentified (Farmer unpublished). *Kluyvera* strains from two published studies were subsequently identified as *Serratia fonticola* (Enteric Reference Laboratory at CDC, unpublished).

Sources, strains, collections   The original collection described by Farmer et al. (1981a) included isolates from body sites that are normally sterile such as blood (2 isolates) and urine (13). Strains were also from nonsterile body sites: intestine or feces (8), throat or sputum (40), and other or unspecified (10). Other isolates were from nonhuman sources or culture collections: cow (1), water (1), sewage (4), food (1), milk (1), soil (1), hospital sink (1), and unspecified or unknown (18). These other sources suggest ways that humans and animals might come in contact with the organism. The description of the genus was based on an unweighted tabulation of 193 *Kluyvera* strains from the Enteric Reference Laboratory's collection; 149 of *K. ascorbata*, 38 of *K. cryocrescens*, and six of *Kluyvera* species 3.

Occurrence in human feces   The original report of Farmer et al. (1981a) listed eight isolates from human feces. All these presumably came from cases of diarrhea or intestinal infections, and were identified as *Kluyvera ascorbata.* Although clinical information that accompanied the cultures was limited, there was no suggestion of an etiological role for *Kluyvera.* Fainstein et al. (1982a) isolated strains of *Kluyvera* from patients with and without diarrhea, and in the title of their paper suggested that *Kluyvera* strains might have had a role in some of the diarrhea cases. Kay et al. (1990) isolated 28 *Kluyvera* strains during a 2-yr prospective case-control study of acute gastroenteritis in children younger that 2 years old, and concluded that *Kluyvera* did not appear to be a diarrheal pathogen. In summary, there is no strong evidence that *Kluyvera* strains can actually cause diarrhea or intestinal infections, but this issue warrants further study. The presence of *Kluyvera* in food and water is a possible source of these intestinal isolates.

Occurrence in extraintestinal human specimens: clinical significance   Most of the *Kluyvera* strains have been from the respiratory tract. Forty strains from the original collection (Farmer et al., 1981a) were from throat or sputum. Clinical significance and the ability to actually cause infections are still being evaluated. Systematic study and additional case reports are needed. Schwach (1979) reported three isolates of Enteric Group 8 from upper respiratory tract specimens. The strains were in mixed culture and not detected in subsequent specimens; thus they were of doubtful clinical significance. Braunstein et al. (1980) reported two cases that yielded cultures identified as Enteric Group 8 (subsequently, both were identified as *K. ascorbata*). One of these was from the sputum of a 6-year-old boy with pulmonary tuberculosis, and was not considered clinically significant. A second isolate was from gallbladder drainage fluid of a 63-year-old woman with acute pancreatitis. Based on chart review, this isolate was considered clinically significant. Of CDC's current collection of 144 *Kluyvera* strains, none has been isolated from spinal fluid, but five strains have been isolated from blood: three strains (two from France) of *K. ascorbata*, one strain of *K. cryocrescens* (a 3-month-old, at autopsy), and one of *Kluyvera* species group 3. Most of these were submitted with little or no patient information, so it was impossible to evaluate clinical significance. The five blood isolates and the growing number of literature reports (Aldová et al., 1985b; Wong, 1987; Luttrell et al., 1988; Thaller et al., 1988; Tristram and Forbes, 1988; Dollberg et al., 1990; Sierra-Madero et al., 1990; Yogev and Koszlowski, 1990; Sezer et al., 1996; Padilla et al., 1997) suggest that *Kluyvera* is more than a benign saprophyte. Most new species of *Enterobacteriaceae* have at least attained the status of "infrequent opportunistic pathogen". Based on present knowledge, this status also seems appropriate for *Kluyvera*. The respiratory tract has been the most common source for *Kluyvera*, but there is no strong evidence that it is clinically significant at this site (however, one isolate of *K. ascorbata* was from a lung at autopsy). The respiratory tract (particularly sputum) is notoriously difficult to evaluate for clinical significance. The urinary tract has been the next most common source, but it has also been difficult to document clinical significance (Tristram and Forbes, 1988).

Occurrence in other animals Bangert et al. (1988) identified strains as *Kluyvera* species along with other *Enterobacteriaceae* in a fecal culture survey of 47 captive raptors (the bird orders *Falconiformes* and *Strigiformes*).

Reporting of *Kluyvera* species Originally, Farmer et al. (1981a) were pessimistic that most clinical and public health laboratories could identify *Kluyvera* strains to species. They suggested that reference laboratories could attempt complete identification, but that clinical laboratories should not attempt the additional phenotypic testing with the addition of special tests. Normal identification methods should allow an identification to the genus level, with a report of "*Kluyvera* species". The addition of one new *Kluyvera* species, and seven new *Buttiauxella* species has made it even more difficult to identify cultures of *Kluyvera*–*Buttiauxella* to species. Further work is needed to find better tests for differentiation. In the meantime, a conservative approach would be to use terms such as "*Kluyvera* species", "*Buttiauxella* species", and "*Kluyvera*–*Buttiauxella* group" for cultures that cannot be definitively assigned a species name.

Accumulation of α-ketoglutaric acid Asai et al. (1956) assigned considerable importance to the fact that *Kluyvera* strains accumulated extremely large amounts of α-ketoglutaric acid during D-glucose fermentation. However, few other *Enterobacteriaceae* have been investigated for this property.

ENRICHMENT AND ISOLATION PROCEDURES

Strain of *Kluyvera* are not difficult to grow and are typical *Enterobacteriaceae* in most respects. See the chapter on the family *Enterobacteriaceae* for general information on growth, plating media, working and frozen stock cultures, media and methods for biochemical testing, identification, computer programs, and antibiotic susceptibility.

DIFFERENTIATION OF THE GENUS *KLUYVERA* FROM OTHER GENERA

Table BXII.γ.193 in the chapter on the family *Enterobacteriaceae* gives the results for *K. ascorbata*, *K. cryocrescens*, and *K. georgiana*

in 47 biochemical tests normally used for identification (Farmer, 1999). There are no genus- or species-specific tests or sequences for the identification of *Kluyvera*. The best approach is to do a complete set of phenotypic tests (growth on plating media, biochemical tests, antibiogram, etc.) and compare the results with all the other named organisms in the family *Enterobacteriaceae*. This approach is described in more detail in the section on the family *Enterobacteriaceae*. The biochemical results of a test strain can be compared to all the organisms listed in Table BXII.γ.193 of that section. Several computer programs greatly facilitate analyzing the results.

TAXONOMIC COMMENTS

Kluyvera is a genus with a turbulent history. Although the name was used in the literature from 1956 to 1980, the genus name, and the two original species names ("*K. citrophila*" and "*K. noncitrophila*") did not appear on the 1980 Approved Lists of Bacterial Names. The reasons for these omissions were never stated. Thus, *Kluyvera*, "*K. citrophila*", and "*K. noncitrophila*" lost standing in nomenclature. To resolve this nomenclatural problem, Farmer et al. (1981a) proposed a redefined genus *Kluyvera*.

The history of *Kluyvera* Asai et al., 1956 In 1956 and 1957, Asai and co-workers in Japan proposed the genus name *Kluyvera* for a group of "polarly flagellated" bacteria that produced large amounts of α-ketoglutaric acid during the fermentation of D-glucose. Five strains were originally studied, one from soil and four from sewage (Table BXII.γ.230). Two species names were proposed, "*K. citrophila*" and "*K. noncitrophila*", which were based on the difference among the strains in utilizing citrate as the sole source of carbon and energy (Asai et al., 1956). Asai and co-workers state (based on their interpretation of Kluyver and van Niel, 1936) that the genus was named to honor Professor A.J. Kluyver, who, with C.B. van Niel, in 1936 postulated that there may be a group of polarly flagellated organisms in the tribe *Pseudomonadeae* that have a mixed acid type of fermentation similar to *Escherichia* (called *Bacterium* in the paper). If such a group were to be discovered, it could be a separate genus, which would differentiate it from the genus *Aeromonas*, which has a butylene glycol fermentative pathway rather than a mixed acid pathway. Asai and co-workers thought they had discovered this postulated group of polarly flagellated organisms and named the group *Kluyvera* in honor of A.J. Kluyver for his many contributions to microbial metabolism and physiology. *Kluyvera* was classified in the tribe *Pseudomonadeae*, which at that time included nonfermentative genera, but also included the fermenters of the genus *Aeromonas*. Today, the family *Pseudomonadaceae* is restricted to bacteria that do not ferment glucose.

***Kluyvera* Asai et al., 1956 was "abolished" in 1962** In 1962, Asai and co-workers confirmed the observations of J.M. Shewan and Rudolph Hugh that all five of their *Kluyvera* strains actually had peritrichous rather than polar flagella (Asai et al., 1962). Thus, they proposed an alternative classification; the two species "*K. citrophila*" and "*K. noncitrophila*" were "transferred" to the genus *Escherichia* in the family *Enterobacteriaceae*. Thus, this alternative classification would result in the creation of two "new combinations"—"*Escherichia citrophila*" and "*Escherichia noncitrophila*"; however, these names were rarely used and did not appear on the 1980 Approved Lists of Bacterial Names.

History of the redefined genus *Kluyvera*: *Kluyvera* Farmer et al., 1981a The Enteric Reference Laboratory's interest in this group of organisms began in 1977 when the laboratory's

TABLE BXII.γ.230. Current nomenclature and classification (Farmer et al., 1981a) for strains originally proposed as *"Kluyvera citrophila"* and *"K. noncitrophila"*

Original name given by Asai et al. (1962)	Current (proposed) classification	ATCC number	Original strain designation	Source
"K. citrophila"	*K. ascorbata*	14236	6,β	Sewage, Japan
"K. citrophila"	*K. cryocrescens*	14237	84C,α	Soil, Tokyo, Japan
"K. citrophila"	*K. cryocrescens*	14238	11,γ	Sewage, Keihin District, Japan
"K. noncitrophila"	*K. cryocrescens*	14239	4	Sewage, Keihin District, Japan
"K. noncitrophila"	*K. cryocrescens*	14240	10	Sewage, Keihin District, Japan

STRAIN MATCHER computer programs listed over a dozen strains with almost identical biochemical reactions that had been reported as "unidentified". During this period, groups of unidentified *Enterobacteriaceae* were being defined, and this was the eighth group. Thus, the vernacular name "Enteric Group 8" was given to the group, and strains were then reported with this designation and included a comment that additional strains would be welcomed so the group could be further studied. Soon after Enteric Group 8 was defined, Holmes of the Computer Identification Laboratory, National Collection of Type Cultures (NCTC), England, reported that the NCTC collection had strains that were almost identical to our Enteric Group 8, and they had been cataloged under the name *Kluyvera* (unpublished). The NCTC stains of *Kluyvera* were then studied at CDC and the conclusions of Holmes were confirmed. Since then, Enteric Group 8 was thought of as a synonym of *Kluyvera*. The DNA–DNA hybridization studies of Fanning et al. (1979) defined two large species groups in *Kluyvera* and defined a third group of strains that became "*Kluyvera* species group 3", and was thought to represent one or more additional *Kluyvera* species. Although *Kluyvera* was "abolished" in 1962 (Asai et al., 1962), reports in the literature continued to use the name *Kluyvera*, probably because strains had been deposited in the American Type Culture Collection, NCTC, and other culture collections. The name *Kluyvera* was proposed as a redefined genus by Farmer et al. (1981a) because this was considered a better alternative than proposing a new genus. The main consideration was that the name *Kluyvera* was still being used in the literature, and that the genus *Kluyvera* was well represented in culture collections by established strains of long standing.

Phylogenetic position Based on both phenotype and DNA–DNA hybridization, Farmer et al. (1981a) thought that *Kluyvera* was a tight, well-defined genus of *Enterobacteriaceae*. Two 16S rDNA trees (Figure BXII.γ.198 in the section on *Budvicia* and Figure BXII.γ.189 in the chapter on the family *Enterobacteriaceae*) include *Kluyvera*. Unfortunately, the species of *Kluyvera* and *Buttiauxella* were not included in the 16S rDNA tree published by Spröer et al. (1999).

Additional species of *Kluyvera* There are probably several additional *Kluyvera* species. The Enteric Reference Laboratory's collection has 21 strains that were reported as "*Kluyvera* species", three as "possible-probable" *Kluyvera ascorbata*, three as *Kluyvera cryocrescens*, and two as "*Kluyvera–Buttiauxella* Group" (Enteric Reference Laboratory, 2000, unpublished). These 29 and the following strains need to be studied by techniques that measure evolutionary relatedness: the four additional strains included in *Kluyvera georgiana* by Müller et al. (1996); the three strains of *K. ascorbata* and five strains of *K. cryocrescens* that had divergence values in DNA–DNA hybridization experiments in the range 5–9 (see Table 4 of Farmer et al., 1981a); and all other "atypical strains" of *Kluyvera* and *Buttiauxella*. Detailed study of this collection would probably result in the definition of many new species.

Further studies needed New phenotypic and molecular tests are needed for the routine identification and differentiation of *Kluyvera*, *Buttiauxella*, and related organisms. The 16S rDNA sequences of all *Kluyvera* and *Buttiauxella* species need to be completed or repeated and new trees drawn. This could provide a useful way to identify strains and clarify relationships of the genera and species.

FURTHER READING

Asai, T., H. Iizuka and K. Komagata. 1964. The flagellation and taxonomy of the genera *Gluconobacter* and *Acetobacter* with reference to the existence of intermediate strains. J. Gen. Appl. Microbiol. *10*: 95–126.

Asai, T., S. Okumura and T. Tsunoda . 1956. On a new genus, *Kluyvera*. Proc. Japan Acad. *32*: 488-493.

Fainstein, V., R.L. Hopfer, K. Mills and G.P. Bodey. 1982. Colonization by or diarrhea due to *Kluyvera* species. J. Infect. Dis. *145*: 127.

Farmer, J.J. III, G.R. Fanning, G.P. Huntley-Carter, B. Holmes, F.W. Hickman, C. Richard and D.J. Brenner. 1981. *Kluyvera*: a new (redefined) genus in the family *Enterobacteriaceae*. Identification of *Kluyvera ascorbata* sp. nov. and *Kluyvera cryocrescens* sp. nov. in clinical specimens. J. Clin. Microbiol. *13*: 919–933.

Luttrell, R.E., G.A. Rannick, J.L. Soto-Hernandez and A. Verghese. 1988. *Kluyvera* species soft tissue infection: case report and review. J. Clin. Microbiol. *26*: 2650–3651.

DIFFERENTIATION OF THE SPECIES OF THE GENUS *KLUYVERA*

The original two named species of *Kluyvera*, *K. ascorbata* and *K. cryocrescens*, had almost identical results in the 47 tests commonly used to identify cultures of *Enterobacteriaceae* (Farmer et al., 1981a; see Table BXII.γ.193 in the chapter on the family *Enterobacteriaceae*). However, these species can be differentiated by several simple phenotypic tests (Table BXII.γ.231). *K. georgiana* is also almost identical to *K. ascorbata* and *K. cryocrescens* in the 47 tests commonly used to identify cultures of *Enterobacteriaceae* (Farmer et al., 1981a), which is one of the reasons it was not originally named. At present there are no simple ways to differentiate it.

TABLE BXII.γ.231. Tests useful for differentiating the three *Kluyvera* species; and between *Kluyvera* and *Buttiauxella*[a]

Test or property	K. ascorbata	K. cryocrescens	K. georgiana	Buttiauxella
Ascorbate test	97	0	nd	nd
Growth and D-glucose fermentation at 5°C within 21 d	3	100	nd	nd
Zone of inhibition around carbenicillin and cephalothin (see Figure 4 of Farmer et al., 1981a)	Small	Large	nd	nd
Growth on cefsulodin-irgasan-novobiocin agar[b]	100	0	nd	nd
Minimum inhibitory concentration (MIC) for irgasan (μg/ml)[b]	>128	≤0.25	nd	nd
Lysine decarboxylase (Møller)	97	23	100	−
Gas production during the fermentation of D-glucose	93	95	17	+
Dulcitol fermentation	25	0	33	0
Indole production	92	90	100	−
Sucrose fermentation	98	81	100	0
Citrate utilization	96	80	100	d
Raffinose fermentation	98	100	100	d
α-Methyl-D-glucoside fermentation	98	95	100	−
Melibiose fermentation	99	100	100	d
Acetate utilization	50	86	83	0

[a]Symbols: +, most strains are positive; −, most strains are negative; d, species to species variation, but many strains are negative. Each number is the percentage positive after 2 d incubation at 36°C.

[b]The data for growth on CIN agar and the irgasan MIC are from Altwegg et al. (1986); the other data are from Farmer et al. (1981a) and Farmer (1999).

List of species of the genus Kluyvera

1. **Kluyvera ascorbata** Farmer, Fanning, Huntley-Carter, Holmes, Hickman, Richard and Brenner 1981b, 382[VP] (Effective publication: Farmer, Fanning, Huntley-Carter, Holmes, Hickman, Richard and Brenner 1981a, 927.)

a.scor.ba' ta. ascorbate, a modern chemical term, a salt of ascorbic acid; M.L. fem. adj. *ascorbata* pertaining to ascorbate.

The characteristics are as described for the genus. See Table BXII.γ.193 in the chapter on the family *Enterobacteriaceae* for the results of 47 biochemical tests normally used for identification (Farmer, 1999). Occurs in human clinical specimens, water, sewage, and food. Probably an opportunistic pathogen of humans that is occasionally encountered in clinical microbiology laboratories. The type strain was isolated from human sputum.

The mol% G + C of the DNA is: 56.1–56.6 (Bd).

Type strain: ATCC 33433, CDC 0648-74.(holotype).

GenBank accession number (16S rRNA): AF176560.

Additional Remarks: The ATCC includes two other strains; ATCC 33434 and ATCC 14236 (Table BXII.γ.230).

2. **Kluyvera cryocrescens** Farmer, Fanning, Huntley-Carter, Holmes, Hickman, Richard and Brenner 1981b, 382[VP] (Effective publication: Farmer, Fanning, Huntley-Carter, Holmes, Hickman, Richard and Brenner 1981a, 927.)

cry.o.cres' cens. Gr. n. kryos cold; L. fem. pres. part. *crescens* growing; M.L. fem. adj. *cryocrescens* growing in the cold; referring to the fact the cultures grow at 4–5°C.

The characteristics are as described for the genus. See Table BXII.γ.193 in the chapter on the family *Enterobacteriaceae* for the results of 47 biochemical tests normally used

for identification (Farmer, 1999). Occurs in human clinical specimens, soil, water, sewage, and the hospital environment. Probably an opportunistic pathogen of humans that is rarely encountered in clinical microbiology laboratories.

The mol% G + C of the DNA is: 55.1 (Bd).

Type strain: ATCC 33435, CDC 2065-78 (holotype).

Additional Remarks: The ATCC includes six other strains, four of which are described in Table BXII.γ.230.

3. **Kluyvera georgiana** Müller, Brenner, Fanning, Grimont and Kämpfer 1996, 63[VP]

geor.gi.a' na. M.L. fem. adj. *georgiana* pertaining to Georgia, U.S.A., where the characterization and redefinition of *Kluyvera* was done.

Synonym: *K. georgiana* and *Kluyvera* species group 3 of Farmer et al. (1981a) are similar in several ways, but are different in circumscription as described below.

The characteristics are as described for the genus. See Table BXII.γ.193 in the chapter on the family *Enterobacteriaceae* for the results of 47 biochemical tests normally used for identification (Farmer, 1999). Currently, it is not possible to completely differentiate *K. georgiana* from *K. ascorbata* and *K. cryocrescens* with simple tests (Table BXII.γ.231). The type strain was isolated from human sputum.

Phenotypic properties as originally given by Müller et al. (1996). The species includes the type strain but no other strains. Other strains named *K. georgiana* include CDC 2774-70 (ATCC 51703), CDC 2065-76 (ATCC 51702), CDC 4246-74, and CDC 3108-76 (Müller et al., 1996). However, no data are presented to show that these four strains are highly related to the type strain by either DNA–DNA hybridization or phenotype (see the discussion below of *Kluyvera* species 3). Thus, there is a change in circumscription between the two species. In the future, additional strains that are highly related to the type strain should be included.

The mol% G + C of the DNA is: not determined.

Type strain: 189, ATCC 51603, CDC 2891-76, DSM 9409 (holotype).

Other Organisms

1. *Kluyvera cochleae* Müller, Brenner, Fanning, Grimont and Kämpfer 1996, 63[VP]

    coch' le.ae. L. fem. gen. n. *cochleae* of a snail.

    Müller et al. (1996) defined a fourth *Kluyvera* species, *Kluyvera cochleae*, which now appears to be a junior subjective synonym of *Enterobacter intermedius*. *K. cochleae* included two strains isolated from snails in Germany and one strain from a slug in Great Britain. In DNA–DNA hybridization experiments (Müller et al., 1996), the type strain of *K. cochleae* S3/1-49[T] was labeled and was highly related (86–92% at 60°C with divergence values of 0.5–1.0) to the two other strains. It was 19–24% related to strains of *Buttiauxella*, but no relatedness values were given for its relationship to the type strains of the three *Kluyvera* species or to other *Enterobacteriaceae*.

    Farmer (unpublished, see Farmer, 2003) studied the three strains of *K. cochleae* and found that they apparently metabolize D-glucose via the butanediol pathway (i.e., they are Voges–Proskauer positive). This is a very unlikely pathway to occur in *Kluyvera*, which was defined to be methyl red positive (well documented to accumulate large amounts of α-ketoglutaric acid) and Voges–Proskauer negative. Computer analysis of the phenotypic data indicated that the strains of *K. cochleae* were essentially identical to 10 reference strains of *Enterobacter intermedius* (Farmer, unpublished). *Enterobacter* is a genus that is well known to have the butanediol pathway of D-glucose metabolism (Voges–Proskauer positive). Later it was shown that the type strains of these two species have almost identical published 16S rRNA sequences. These observations led to further laboratory studies (Brenner and Steigerwalt, unpublished; Farmer, unpublished). By DNA–DNA hybridization, the type strains of the two species are very highly related (D.J. Brenner, personal communication).

    To summarize, all current data indicate that *Kluyvera cochleae* and *Enterobacter intermedius* are different names for the same organism. The two species have different type strains, so they should be considered as subjective synonyms, with *Enterobacter intermedius* being the senior subjective synonym, since it was validly published first.

2. *"Kluyvera fluvialis"*

    In an abstract, Gadaleta et al. (1996) from Buenos Aires, Argentina, described their work with "strain 21g" isolated from a polluted river. They determined its 16S rDNA sequence, compared it to the sequences of *K. ascorbata* and *K. cryocrescens*, and concluded " . . . the 21g strain is a member of an until now undescribed species of *Kluyvera* that we propose to be named *K. fluvialis*". Since the name *"Kluyvera fluvialis"* has not been effectively published or validly published, it lacks standing in nomenclature.

    The vernacular name *Kluyvera* species group 3 was originally given by Farmer et al. (1981a) to five strains (2774-70, 2065-76, 2891-76, 4246-74, and 3108-76) that were 60–68% related (with divergence values of 9–12%) by DNA–DNA hybridization to strains of *K. ascorbata* and *K. cryocrescens* (see Table 4 in Farmer et al., 1981a). None of these strains were labeled and tested against the others to determine if they all belonged to the same species. However, the authors concluded that the five strains represented one or more additional species of *Kluyvera*. Because of its uncertain circumscription and the lack of simple tests to differentiate it from the two named *Kluyvera* species, they did not give it a scientific name, designate a type strain, or deposit strains in the American Type Culture Collection.

    One of the strains (2891-76) was later designated as the type strain of *Kluyvera georgiana* by Müller et al. (1996) as described above. *K. georgiana* of Müller et al. (1996) also include the other four strains that Farmer et al. (1981a) included in *Kluyvera* species 3. However, no information is included in the paper of Müller et al. (1996) to show that these latter four strains are highly related to the type strain of *K. georgiana* by DNA–DNA hybridization or phenotype. The assignment of these four strains to *K. georgiana* or a new *Kluyvera* species awaits further study.

Genus XVIII. **Leclercia** *Tamura, Sakazaki, Kosako, and Yoshizaki 1987, 179[VP] (Effective publication: Tamura, Sakazaki, Kosako, and Yoshizaki 1986, 183)*

CAROLINE M. O'HARA AND J.J. FARMER III

Le.clerc' i.a. M.L. fem. n. *Leclercia* named to honor H. Leclerc, a French bacteriologist, who first described and named this organism *Escherichia adecarboxylata* in 1962, and who made many other contributions to enteric bacteriology.

Small rod-shaped cells, conforming to the general definition of the family *Enterobacteriaceae*. Gram negative. Motile at 36°C and 25°C with peritrichous flagella. Facultatively anaerobic. Catalase positive. Oxidase negative. **Many strains produce a nondiffusible yellow pigment that may be weak and may be lost on storage and subculture.** Ferment, rather than oxidize, D-glucose and other carbohydrates. Reduce nitrate to nitrite. **Positive for indole production, methyl red, growth in the presence of cyanide, malonate utilization, esculin hydrolysis, ONPG, and the fermentation of** lactose, D-mannitol, dulcitol, salicin, adonitol, L-arabinose, L-rhamnose, maltose, D-xylose, trehalose, cellobiose, erythritol, esculin, melibiose, D-arabitol, mucate, D-mannose, and D-galactose. **Produce visible gas during fermentation.** Negative for Voges–Proskauer, citrate utilization (Simmons), H₂S production (TSI), phenylalanine deaminase, **lysine decarboxylase, arginine dihydrolase, ornithine decarboxylase**, gelatin hydrolysis (22°C), lipase (corn oil) production, DNase production, and the fermentation of *myo*-inositol, D-sorbitol, α-methyl-D-glucoside, and erythritol.

Most strains are susceptible to colistin, nalidixic acid, sulfadiazine, gentamicin, streptomycin, kanamycin, tetracycline, chloramphenicol, carbenicillin, cephaloridine, and ampicillin (disk diffusion method on Mueller–Hinton agar); **resistant to penicillin**.

Isolated from human clinical specimens that are normally sterile such as blood and urine. Also isolated from other clinical specimens such as sputum and wounds, usually in mixed culture. Probably an **opportunistic pathogen that occasionally causes extraintestinal infections** in humans. Occasionally occurs in human feces, but there is **no evidence that it causes diarrhea or intestinal infections**. Natural reservoirs and ecological niches are not completely defined, but has been isolated from water, food, milk, and other environmental samples. Appears to be important in the processing of "dry-cured" hams. A rarely isolated genus of *Enterobacteriaceae*.

The mol% G + C of the DNA is: 52–55.

Type species: **Leclercia adecarboxylata** (Leclerc 1962) Tamura, Sakazaki, Kosako, and Yoshizaki 1987, 179 (Effective publication: Tamura, Sakazaki, Kosako, and Yoshizaki 1986, 183) (*Escherichia adecarboxylata* Leclerc 1962, 737.)

FURTHER DESCRIPTIVE INFORMATION

Literature Since the genus was described in 1986, there have been a few reports in the literature; seven cataloged in MEDLINE and 18 cataloged in BIOSIS. These reports include reviews of new *Enterobacteriaceae*; human clinical specimens and infections; occurrence in animals, plants, food, water, and the environment; taxonomic studies; and studies on basic biology such as physiology-metabolism, biochemistry, and genetics.

Leclercia **strains in reference laboratory collections** The original report by Tamura et al. (1986) described 58 isolates from human clinical specimens including blood, urine, sputum, and feces, and 27 isolates from nonhuman sources, including food, water, and the environment. However, they did not specify the number of isolates from each of these sources. Farmer et al. (1985a) described their studies with the type strain (isolated from water), and concluded that their collection also contained 6–10 isolates that were very similar. The CDC collection now includes 18 human isolates: blood (5), urine (1), wound (1), sputum (1), hand wound (1), leg wound (2), finger (1), ear (1), feces (3), and unspecified (2). It also includes 11 isolates from nonhuman sources or culture collections: intravenous fluid bottle (1), milk (1), environmental (1), and unspecified or unknown (8).

Occurrence in human clinical specimens Strains of *L. adecarboxylata* have been from a wide variety of human clinical specimens, but there have been few systematic clinical studies or case reports. Temesgen et al. (1997) found only five strains during 1984–1995 at the Mayo Clinic in Rochester, Minnesota, USA. In three cases it was isolated from wounds of the lower extremities, all in mixed culture. Another case was a pneumonia patient whose sputum yielded *L. adecarboxylata* along with other organisms. One patient had a culture of *Leclercia adecarboxylata* that was clinically significant. A 35-year-old woman with acute nonlymphocytic leukemia had undergone bone marrow transplantation. She developed bacteremia and fever while she was neutropenic, and *L. adecarboxylata* was isolated in pure culture from one of two sets of blood cultures.

Other cases of bacteremia have also been described. Daza et al. (1993) reported the case of a 45-year-old alcoholic who had severe abdominal pain, hypotension, tachycardia, and diapho-

resis 6 hours after undergoing paracentesis for the removal of ascitic fluid. Four days after undergoing a laparotomy, he developed fever and had three blood cultures positive for *L. adecarboxylata*. Otani and Bruckner (1991) reported a case in an 8½-month-old with a history of congenital gastroschisis and intestinal atresia. He came to the emergency department with shaking chills and fever about 1 hour after total parenteral nutrition through a central line catheter. The blood culture from the central line was positive for *L. adecarboxylata*. Dudkiewicz and Szewczyk (1993) described one isolate of *L. adecarboxylata* in their series of 72 cases of bacterial endocarditis.

When *L. adecarboxylata* is isolated in mixed culture from specimens such as sputum and wounds, it is difficult to determine clinical significance. Martinez et al. (1998) reported its isolation from an ulcer exudate. However, the isolation of this organism from blood cultures suggests that it is at least an opportunistic pathogen. Additional studies and case reports are needed to better define its role as a human pathogen.

Occurrence in human feces Several collections have included isolates from feces (Tamura et al., 1986), which usually are from patients with diarrhea. Cai et al. (1992) described three patients in China with diarrhea whose fecal cultures were positive for *L. adecarboxylata*, but the authors could not demonstrate an etiological role. Although it has occasionally been isolated from feces, there is no evidence that *L. adecarboxylata* can actually cause diarrhea or infections of the intestinal tract.

Occurrence in animals, plants, food, water, and the environment In addition to sources previously mentioned, *L. adecarboxylata* (referred to as Enteric Group 41 by Bangert et al., 1988) also has been isolated from captive raptors (the bird order *Falconiformes*; the falcons) (Bangert et al., 1988). *L. adecarboxylata* appears to be important in the processing of "dry-cured" hams. It was the only species present at the end of the fast-curing process, and predominated along with some other *Enterobacteriaceae* at the end of the slow-curing process (Marin et al., 1996). These environmental reservoirs suggest additional ways that hospital and other patients can come in contact with *L. adecarboxylata*. If a blood culture from a patient on total parenteral nutrition is positive for *L. adecarboxylata*, all the liquids should be cultured as possible reservoirs.

Isolation, identification, culture preservation, phenotypic characterization Strains of *Leclercia adecarboxylata* grow well on media normally used in enteric bacteriology, and are typical *Enterobacteriaceae* in most respects. The typical yellow pigment is probably present in most cultures on initial isolation, but is easily lost. See the chapter on the family *Enterobacteriaceae* for general information on growth, plating media, working and frozen stock cultures, media and methods for biochemical testing, identification, and antibiotic susceptibility. Table BXII.γ.232 below summarizes the phenotypic characteristics of *L. adecarboxylata*. Also see Table BXII.γ.193 of the chapter on the family *Enterobacteriaceae* that has the tabulated biochemical reactions of *L. adecarboxylata* and other *Enterobacteriaceae*.

TAXONOMIC COMMENTS

History Leclerc (1962) recognized a group that he referred to as "yellow-pigmented coliform bacteria" that was similar to *E. coli* in its IMViC reactions, but differed by producing a yellow pigment and fermenting many sugars. He proposed that it be recognized as a separate species in the genus *Escherichia*, *Escher-*

TABLE BXII.γ.232. Phenotypic characteristics of *Leclercia adecarboxylata* based on data from three different laboratories

Test or property	Study 1[a]	Study 2	Study 3
Number of strains studied	20	85	33
Incubation temperature[b]	30°C	35°C	36°C
Final reading for the test[b]	2 d	2 d	2 d
Yellow pigment production, 2 d	60[c] (30°C)	nd	37° (25°C)
Yellow pigment production, 7 d	nd	14 (25°C)	63 (25°C)
Indole production	90	100	100
Methyl red	nd	100	100
Voges–Proskauer	0	0	0
Citrate utilization (Simmons)	0	0	0
H₂S production	0	0 (KIA)	0 (TSI)
Urea hydrolysis	70	10	48
Phenylalanine deaminase	0	0	0
Lysine decarboxylase	0	0	0
Arginine dihydrolase	0	0	0
Ornithine decarboxylase	0	0	0
Motility, 35–37°C	nd	99	79
Motility, 22–25°C	nd	nd	60
Motility, 30°C	100	nd	nd
Gelatin hydrolysis	0	0	0 (22°C)
KCN test (% resistant)	100	100	97
Malonate utilization	100	100	93
D-Glucose, acid production	100	100	100
D-Glucose, gas production	100	99	97
Lactose fermentation	85	100	93
Sucrose fermentation	20	52	66
D-Mannitol fermentation	100	100	100
Dulcitol fermentation	70	87	86
Salicin fermentation	100	100	100
Adonitol fermentation	55	98	93
myo-Inositol fermentation	nd	0	0
D-Sorbitol fermentation	8	9	0
D-Arabinose fermentation	100	100	100
Raffinose fermentation	30	59	66
L-Rhamnose fermentation	100	100	100
Maltose fermentation	100	100	100
D-Xylose fermentation	100	100	100
Trehalose fermentation	100	100	100
Cellobiose fermentation	100	100	100
α-Methyl-D-glucoside fermentation	0	0	0
Erythritol fermentation	0	0	0
Esculin hydrolysis	100	100	100
Melibiose fermentation	100	100	100
D-Arabitol fermentation	nd	96	96
Glycerol fermentation	100 (30 d)		3
Mucate fermentation	90	65	93
Tartrate fermentation (Jordan)	70	nd	83
Acetate utilization	nd	0	28
Lipase (corn oil)	nd		0
DNase production	0	0 (25°C)	0 (25°C)
Nitrate reduction to nitrite	100	99	100
Oxidase	nd	0	0
ONPG test	100	100	100
D-Mannose fermentation	100	100	100
Tyrosine hydrolysis	nd	nd	0
D-Galactose fermentation	100	100	100
MacConkey agar, growth	nd	nd	100
Catalase	nd	nd	+
Citrate, Christensen	nd	0	0
H₂S on peptone iron agar	nd	nd	0
Kauffmann–Peterson tests:			
Citrate	nd	0	nd
D-Tartrate	0	0	nd
L-Tartrate	30	nd	nd
meso-Tartrate	0	nd	nd
Growth at 4–5°C	15 (30 d)	49 (14 d)	nd
Growth at 35–37°C	nd	nd	+

(*continued*)

TABLE BXII.γ.232. (*cont.*)

Test or property	Study 1[a]	Study 2	Study 3
Growth at 41°C	85	nd	nd
Growth at 42°C	nd	0	nd
Growth at 50°C	0	nd	nd
β-Glucuronidase	0	0	nd
Pectinase	nd	0	nd
Tetrathionate reductase	25	54	nd
β-Xylosidase	95	100	nd
Carbon source utilization:			
Number of compounds tested	123	122	nd
Number (%) utilized	26 (21.1)	31 (25.4)	nd
Amygdalin fermentation	nd	0	nd
D-Arabinose fermentation	nd	0	nd
L-Arabitol fermentation	nd	0	nd
Arbutin fermentation	nd	100	nd
D-Fructose fermentation	100	nd	nd
D-Fucose fermentation	nd	0	nd
L-Fucose fermentation	nd	0	nd
Gentiobiose fermentation	nd	100	nd
D-Gluconate fermentation	nd	100	nd
2-Ketogluconate fermentation	nd	100	nd
5-Ketogluconate fermentation	nd	0	nd
Glycogen fermentation	5	nd	nd
Inulin fermentation	nd	0	nd
D-Levulose fermentation	nd	100	nd
D-Lyxose fermentation	nd	0	nd
L-Lyxose fermentation	nd	0	nd
Melizitose fermentation	0	0	nd
Methylmannoside fermentation	nd	0	nd
D-Ribose fermentation	100	100	nd
L-Sorbose fermentation	0	0	nd
D-Tagatose fermentation	nd	0	nd
D-Turanose fermentation	nd	0	nd
Xylitol fermentation	nd	0	nd
Tween 80 hydrolysis	nd	0	nd

[a]Study 1, Izard et al. (1985); study 2, Tamura et al. (1986); study 3, Farmer (2003).

[b]The standard conditions used in each study; a different temperature or time for the final reading is indicated in parentheses. Abbreviations: TSI, triple sugar iron agar; KIA, Kligler iron agar; ONPG, o-nitrophenyl-β-D-galactopyranoside.

[c]Each number is the percentage of strains that are positive. Most positive reactions occur during the first 24 h. For a few tests in study there were not enough quantitative data, so these results are represented by symbols with the usual meaning stated in this *Manual*; nd, not determined.

ichia adecarboxylata. Ewing and Fife (1972) studied the type strain of *Escherichia adecarboxylata* and proposed a different classification. They classified it in their redefined species (which is now known to be heterogeneous) *Enterobacter agglomerans* as "a typical strain of biogroup G3".

Escherichia adecarboxylata* as a distinct group of *Enterobacteriaceae*; proposal of *Leclercia   Several other investigators considered the taxonomic position of *E. adecarboxylata* and concluded that it is different from *Escherichia*, the *Enterobacter agglomerans–Erwinia* complex, and from other *Enterobacteriaceae* (see Table BXII.γ.233) (Gavini et al., 1983b; Sakazaki et al., 1983; Izard et al., 1985; Verdonck et al., 1987). The situation was clarified when Tamura et al. (1986) used DNA–DNA hybridization to study their collection of strains. They showed that *E. adecarboxylata* was different from other taxa of *Enterobacteriaceae*. The type strain of *E. adecarboxylata* was more than 70% related to eight other strains and was 64–66% related to two additional strains. It was only 32% related to the type strain of *Enterobacter agglomerans* and only 26% related to the type strain of *Escherichia coli*. Because it was distinct by both DNA–DNA hybridization and phenotype, they proposed a new genus *Leclercia* with one species *Leclercia adecarboxylata* to include strains formerly classified as *Escherichia adecarboxylata*. Their proposal is an alternative classification that has been well accepted by the scientific community.

Phylogeny based on 16S rDNA sequencing Recently a complete 16S rDNA sequence of "strain LBV 449" (which is not the type strain) was deposited (GenBank accession number AJ276393) (De Baere et al., 2001). A 16S rRNA tree that includes *L. adecarboxylata* can be found in the chapter on the family *Enterobacteriaceae*. The 16S rDNA sequencing data agree with data from DNA–DNA hybridization experiments that *Leclercia* is distinct from other genera of *Enterobacteriaceae* (Tang et al., 1998). Unfortunately, *L. adecarboxylata* was not included in the tree published by Spröer et al. (1999).

Several sequences related to mercury resistance have also been deposited.

FURTHER READING

Leclerc, H. 1962. Étude biochemique d' *Enterobacteriaceae* pigmentées. Ann. Inst. Pasteur (Paris) *102*: 726–741.

Izard, D., J. Mergaert, F. Gavini, A. Beji, K. Kersters, J. De Ley and H. Leclerc. 1985. Separation of *Escherichia adecarboxylata* from the *Erwinia herbicola-Enterobacter agglomerans* complex and from the other enterobacteriaceae by nucleic acid and protein electrophoretic techniques. Ann. Inst. Pasteur Microbiol. *136B*: 151–168.

Tamura, K., R. Sakazaki, Y. Kosako and E. Yoshizaki. 1986. *Leclercia ade-*

TABLE BXII.γ.233. Differentiation of *Leclercia adecarboxylata* from its closest phenotypic relatives in *Enterobacteriaceae*

Test	*Leclercia adecarboxylata*	*Escherichia coli*	"*Enterobacter agglomerans* complex"[a]
Indole production	+	+	d
Methyl red	+	+	d
Lysine decarboxylase	−	+	−
Arginine dihydrolase	−	d	−
Ornithine decarboxylase	−	d	−
Yellow pigment	d	−	d
KCN	+	−	d
D-Sorbitol fermentation	−	+	d
Cellobiose fermentation	+	−	d
D-Glucose, gas	+	+	d
Dulcitol fermentation	+	d	d
D-Arabitol fermentation	+	−	−

[a]See Farmer (2003) for a more detailed description of this "vernacular name". It is a term defined for practical identification in the clinical microbiology laboratory. It includes over a dozen DNA—DNA hybridization groups that were originally included in the species *Enterobacter agglomerans*, which is now known to be a heterogeneous species.

carboxylata, gen. nov., comb. nov., formerly known as *Escherichia adecarboxylata*. Curr. Microbiol. *13*: 179–184.

Temesgen, Z., D.R. Toal and F.R. Cockerill, III. 1997. *Leclercia adecarboxylata* infections: case report and review. Clin. Infect. Dis. *25*: 79–81.

List of species of the genus *Leclercia*

1. **Leclercia adecarboxylata** (Leclerc 1962) Tamura, Sakazaki, Kosako, and Yoshizaki 1987, 179[VP] (Effective publication: Tamura, Sakazaki, Kosako, and Yoshizaki 1986, 183) (*Escherichia adecarboxylata* Leclerc 1962, 737; *Enterobacter agglomerans* biogroup G3 Ewing and Fife 1972, 10.)

   a.de.car.box.y.la′ ta. Gr. adj. *a* without; M. Fr. n. *decarboxyl* removal of a molecule of carbon dioxide from an organic compound; *adecarboxylata* without decarboxylase activity; because it has negative reactions in lysine decarboxylase, ornithine decarboxylase and arginine dihydrolase; i.e., "triple decarboxylase negative".

   The characteristics are as described for the genus; a more complete description is given in Table BXII.γ.232. Isolated from human clinical specimens, environmental samples, food, and water. Its clinical significance is not fully documented but its potential role as a pathogen is suggested by isolates from blood and similar specimens that are normally sterile. However, it may be colonizing rather than infecting nonsterile body sites. The isolates from food, drinking water, feces, and an intravenous fluid bottle suggest ways that humans come in contact with it. There is no evidence that it can cause diarrhea or intestinal infections. It should be considered a rarely isolated species of *Enterobacteriaceae*, and a possible opportunistic pathogen (extraintestinal infections only) for humans. The type strain was isolated from drinking water by Leclerc, 1962.

   The mol% G + C of the DNA is: 52–55 (T_m).

   Type strain: 1783, ATCC 23216, CIP 82.92, DSM 5077, HAMBI 1696, JCM 1667, LMG 2803.

   Additional Remarks: The American Type Culture Collection includes 7 other strains of *L. adecarboxylata*, including four strains from "apple fruit".

Other Organisms

Currently *Leclercia* has only one species, but it will be interesting to see if future studies will change or clarify the current concept or circumscription of the genus. Tamura et al. (1986) included only one species in *Leclercia*; however, Table 2 of their paper includes strains 1523 and 363, which were only 66.2% and 64.2% related to the type strain by DNA–DNA hybridization; 70% is the usual cutoff point for the level of species. They did not include divergence values (ΔT_m), so it is more difficult to determine if these two strains are as related to the type strain as the other eight strains. The CDC collection contains five strains that were identified as "possible-probable" *Leclercia adecarboxylata* because their phenotypic characteristics were similar to, but different from, strains now included in the species. These five strains, the two of Tamura et al. (1986), and similar strains should be investigated as possible additional species of *Leclercia*.

Genus XIX. Leminorella Hickman-Brenner, Vohra, Huntley-Carter, Fanning, Lowery, Brenner and Farmer 1985c, 375^VP (Effective publication: Hickman-Brenner, Vohra, Huntley-Carter, Fanning, Lowery, Brenner and Farmer 1985b, 235)

J.J. FARMER III AND FRANCES W. BRENNER

Le.mi.no.rel'la. M.L. dim. -*ella* ending; M.L. fem. n. *Leminorella* named to honor Leon Le Minor, a French microbiologist, for his many contributions to enteric bacteriology including the nomenclature, classification, and serotyping of *Salmonella*; lysogeny; metabolic plasmids; and new and rapid biochemical tests. The name also honors Simone Le Minor, who also made many contributions to enteric bacteriology as head of the National *Salmonella* Centre of France and for her research on *Serratia* serotyping.

Small rod-shaped cells, conforming to the general definition of the family *Enterobacteriaceae*. Gram-negative. **Nonmotile at 36°C and 25°C.** Contain the enterobacterial common antigen. Facultatively anaerobic. Catalase positive (strong and rapid). Oxidase negative. Nonpigmented. Ferment, rather than oxidize, D-glucose. Reduce nitrate to nitrite. **Inactive biochemically. Positive for H₂S production (TSI and PIA), tyrosine hydrolysis, and fermentation of L-arabinose, D-xylose, and L-tartrate.** Negative for indole production, Voges–Proskauer, urea hydrolysis, phenylalanine deaminase, lysine decarboxylase, arginine dihydrolase, ornithine decarboxylase, growth in the presence of cyanide (KCN test), malonate utilization, esculin hydrolysis, gelatin hydrolysis (22°C), lipase (corn oil), DNase, and the fermentation of lactose, sucrose, D-mannitol, salicin, adonitol, *myo*-inositol, D-sorbitol, raffinose, L-rhamnose, maltose, trehalose, cellobiose, α-methyl-D-glucoside, erythritol, melibiose, D-arabitol, glycerol, D-mannose, and D-galactose. Susceptible to colistin, nalidixic acid, sulfadiazine, gentamicin, kanamycin, tetracycline, chloramphenicol; **resistant to streptomycin, penicillin, ampicillin, carbenicillin, and cephalothin** (disk diffusion method on Mueller-Hinton agar).

Usually isolated from feces, but there is no evidence that strains cause diarrhea or intestinal infections. Some strains agglutinate in *Salmonella* diagnostic antisera, and **can be misidentified as being *Salmonella*. Rarely isolated from other clinical specimens.** Ecological niche may be the human intestinal tract since there are only a few isolates from other clinical specimens, and no reported isolates from animals, food, water or the environment. **A very rarely isolated genus of *Enterobacteriaceae*.**

The mol% G + C of the DNA is: not reported.

Type species: **Leminorella grimontii** Hickman-Brenner, Vohra, Huntley-Carter, Fanning, Lowery, Brenner and Farmer 1985c, 375 (Effective publication: Hickman-Brenner, Vohra, Huntley-Carter, Fanning, Lowery, Brenner and Farmer 1985b, 235.)

FURTHER DESCRIPTIVE INFORMATION

Literature Since the genus was described in 1985, there have been only a handful of reports in the literature, six cataloged in MEDLINE and 11 cataloged in BIOSIS. Except for the original report (Hickman-Brenner et al., 1985b), none of these has dealt specifically with *Leminorella*. The literature reports that have mentioned *Leminorella* include reviews and taxonomic studies of new *Enterobacteriaceae* (Schindler, et al., 1992; Gilchrist, 1995; Aleksic and Bockemühl, 1999); evaluation of commercial identification products or "kits" for *Enterobacteriaceae* (Kitch et al., 1994); and studies on physiology, metabolism, or biochemistry that included *Leminorella* strains (Bouvet and Grimont, 1987a; Satta et al., 1988; Bouvet et al., 1989; Grimont and Bouvet, 1989; Hodinka et al., 1991; Thaller et al., 1995; Hamana, 1996). In addition to the genus and species names, "Enteric Group 57" should be included as a search term in computerized literature searches, although it is rarely used today.

Sources and clinical significance Nine of the original *Leminorella* isolates were from feces, and two were from urine. Since the original paper in 1985, the CDC Enteric Reference Laboratory has received and identified nine additional human isolates, seven from feces and one each from urine and a decubitus wound. The clinical significance of *Leminorella* strains in these extraintestinal specimens is unknown, and further study and case reports are needed. Although strains of *Leminorella* have occasionally been isolated from feces of people with diarrhea, there is no evidence that it can actually cause diarrhea or intestinal infections. Its interest for clinical microbiology and public health appears to be more as a nuisance because *Leminorella* strains might be misidentified as *Salmonella* unless more complete biochemical and serological testing is done.

Original misidentification as *Salmonella* Several of the isolates from feces had been referred to the CDC's *Salmonella* reference laboratory because the strains agglutinated one or more *Salmonella* antisera. Presumably these tests were done with commercial antisera and with "live", rather than alcohol-treated, cultures, and the referring laboratory could not completely serotype the strain or confirm an identification as *Salmonella*. Four *L. grimontii* strains and one *L. richardii* strain have been reported to cross-react when tested at CDC with alcohol-treated antigens and CDC reference antisera. One of the strains agglutinated very weakly in O group B serum and in single factor O27 serum. The other strains did not agglutinate.

Isolation, identification, culture preservation, phenotypic characterization Strains of *Leminorella* grow on media normally used in enteric bacteriology, but they grow more slowly and are slower or more inactive biochemically than typical *Enterobacteriaceae*. One strain of *L. grimontii* and all four strains of *L. richardii* were negative for D-glucose fermentation after incubation for 24 h, but all became positive at 48 h (Hickman-Brenner et al., 1985b). On MacConkey agar at 24 h, strains appear as small colonies that are colorless (lactose negative). See the chapter on the family *Enterobacteriaceae* for general information on growth, plating media, working and frozen stock cultures, media and methods for biochemical testing, identification, computer programs, and antibiotic susceptibility.

Biochemical reactions Table BXII.γ.234 summarizes the biochemical reactions of the two *Leminorella* species. The tests useful for differentiation of *L. grimontii* and *L. richardii* are given in Table BXII.γ.235. Compared to other species of *Enterobacteriaceae*, strains of *Leminorella* are weak and inactive biochemically. One important characteristic is hydrogen sulfide production. Within 48 h on triple sugar iron agar, the species give an alkaline slant and a weak acid reaction in the butt with H₂S production.

Antibiotic susceptibility In their original description, Hickman-Brenner et al. (1985b) gave the zone sizes for each of the

TABLE BXII.γ.234. Biochemical reactions of *Leminorella grimontii* and *L. richardii* based on strains studied by DNA–DNA hybridization

Characteristic	*Leminorella grimontii*	*Leminorella richardii*
Number of strains	6[a]	4[a]
Indole production	0[b]	0
Methyl red	100	0
Voges–Proskauer	0	0
Citrate utilization (Simmons)	100	0
H₂S production (TSI[c])	100	100
H₂S production (PIA)	100	100
Urea hydrolysis	0	0
Phenylalanine deaminase	0	0
Lysine decarboxylase	0	0
Arginine dihydrolase	0	0
Ornithine decarboxylase	0	0
Motility at 36°C	0	0
Gelatin hydrolysis (22°C)	0	0
Growth in KCN	0	0
Malonate utilization	0	0
D-Glucose, acid production at 24 h	83	0
D-Glucose, acid production at 48 h	100	100
D-Glucose, gas production	33 (100)	0
Lactose fermentation	0	0
Sucrose fermentation	0	0
D-Mannitol fermentation	0	0
Dulcitol fermentation	83	0
Salicin fermentation	0	0
Adonitol fermentation	0	0
myo-Inositol fermentation	0	0
D-Sorbitol fermentation	0	0
L-Arabinose fermentation	100	100
Raffinose fermentation	0	0
L-Rhamnose fermentation	0	0
Maltose fermentation	0	0
D-Xylose fermentation	83	100
Trehalose fermentation	0	0
Cellobiose fermentation	0	0
α-Methyl-D-glucoside fermentation	0	0
Erythritol fermentation	0	0
Esculin hydrolysis	0	0
Melibiose fermentation	0	0
D-Arabitol fermentation	0	0
Glycerol fermentation	17 (33)	0
Mucate fermentation	100	50 (75)
Tartrate fermentation (Jordan's)	100	100
Acetate utilization	0	0
Lipase (corn oil)	0	0
DNase production (25°C)	0	0
Nitrate reduction to nitrite	100	100
Oxidase	0	0
ONPG test	0	0
Yellow pigment (25°C)	0	0
D-Mannose fermentation	0	0
Tyrosine hydrolysis	83 (100)	75 (100)
D-Galactose fermentation	0	0

[a]These are the original strains that were studied by DNA–DNA hybridization and used by Hickman-Brenner et al. (1985b) to define the species.

[b]Each number gives the percentage positive after 2 d incubation at 36°C (unless a different time or temperature is indicated; phenylalanine, oxidase, and nitrate reduction are done only at 1 d). The cumulative % positive between 3 and 7 d is given for some tests in parentheses. Some of the percentages are slightly different from those given in the original descriptions (Hickman-Brenner et al., 1985b) and are based on tests results that were repeated after the original publication.

[c]TSI, triple sugar iron agar; PIA, peptone iron agar; ONPG, o-nitrophenyl-β-D-galactopynanoside.

TABLE BXII.γ.235. Differentiation of *Leminorella grimontii* and *L. richardii*[a]

Test	Incubation time, days	*L. grimontii*	*L. richardii*
Methyl red	2	100	0
Citrate utilization (Simmons)	2	100	0
D-Glucose, acid production	1	83	0
D-Glucose, gas production	7	100	0
Dulcitol fermentation	2	83	0
Glucosaminidase activity[b]		100	0

[a]These data are based on results in Table BXII.γ.234 which included only strains that were documented to belong to these two species by DNA–DNA hybridization (Hickman-Brenner et al., 1985b). Each number gives the percentage positive at the specified indicated period; all the biochemical tests were done at 36°C.

[b]Data from Hodinka et al. (1991).

is not a unique pattern, the antibiogram can be a useful additional way of differentiating *Leminorella* from other genera of *Enterobacteriaceae*.

DNA, RNA, and protein sequences There is a full sequence (bases 1 to 1482) of the 16S rRNA gene of *L. grimontii* listed in GenBank (Spröer et al., 1999).

A 16S rDNA tree that includes *Leminorella grimontii* can be found in the chapter on the family *Enterobacteriaceae* (Fig. BXII.γ.189). The 16S rDNA sequencing data agree with data from DNA–DNA hybridization that *Leminorella* is distinct from other genera of *Enterobacteriaceae*. In the tree, *Pragia fontium* is the closest relative, in agreement with the tree published by Spröer et al. (1999) with respect to these two genera being on a distinct branch. However, the two trees differ in the placement of other organisms.

DIFFERENTIATION OF THE GENUS *LEMINORELLA* FROM OTHER GENERA

Strains of *Leminorella* can be distinguished because they do not ferment D-mannose, but hydrolyze tyrosine, much like strains of *Proteus*; however, they are negative for urea hydrolysis and phenylalanine deaminase. They differ from *Salmonella* because they are lysine and ornithine decarboxylase negative and do not ferment D-sorbitol. Also see Table BXII.γ.193 of the chapter on the family *Enterobacteriaceae*), which gives the percentage positive for 47 biochemical tests done at the Enteric Reference Laboratories at CDC with standard media and methods. These tests have proved useful for identification and for differentiating *Leminorella* from other species in the family.

TAXONOMIC COMMENTS

History and discovery In 1980 one of us (JJF, unpublished) noticed two H₂S+ strains that seemed to be distinct from other *Enterobacteriaceae*, and coded these as a "possible new group of *Enterobacteriaceae*" in the laboratory's master computer database. In 1981 diagnostic culture 1944-81 was received and studied in more detail (Hickman-Brenner et al., 1985b). It was from a stool culture of a 7-month-old patient in Hawaii with gastroenteritis. The culture had a low identification score for other *Enterobacteriaceae* in the computer program GEORGE (Farmer et al., 1985a). However, the computer program STRAIN MATCHER listed nine additional strains that had been reported as "unidentified" that were very similar biochemically. Several of these strains had been sent to CDC because they reacted weakly in *Salmonella* antisera. Five strains had been sent as *Salmonella*, or with a diagnosis of suspected salmonellosis or diarrhea; two were sent as "possible

11 strains with the 12 different antibiotics that have been used for many years in their reference laboratory as a aid in identification. There was little variation among the three *Leminorella* species, and all 11 strains were resistant to streptomycin, penicillin, ampicillin, carbenicillin, and cephalothin. Although this

Citrobacter"; and one was a "suspect H_2S^+ *Shigella*" (Hickman-Brenner et al. 1985b). The vernacular name Enteric Group 57 was coined in 1981 (Hickman-Brenner and Farmer, unpublished; see Farmer et al., 1985a) for the original group of 10 strains, and the taxonomic position of the group was unknown. Enteric Group 57 was simply thought of as being a unique group of H_2S-producing strains that was being confused with *Salmonella* in some reference laboratories. An 11th strain was received in 1982.

Proposal of the genus *Leminorella* Hickman-Brenner et al. (1985b) used DNA–DNA hybridization (hydroxyapatite, $^{32}PO_4$) to further characterize Enteric Group 58. Strain 1944-81^T was labeled and was highly related (77–97% at 60°C with divergence values of 0–0.5%) to five other Enteric Group 58 strains. Strain 1944-81 was only 3–16% related to other species of *Enterobacteriaceae*. Because it was distinct from all *Enterobacteriaceae* by both DNA–DNA hybridization and phenotype, it was given a scientific name. The group of six strains was named *Leminorella grimontii*, which was designated as the type species for the genus *Leminorella*. Five other strains were 32–60% related to 1944-81, so strain 0978-82 was chosen from this group and tested by DNA hybridization against all 11 strains. Three strains were 93–94% related to strain 0978-82 with divergence values of 0.5–1.0. This group of four strains was named *Leminorella richardii*. The remaining strain 3346-72 was 60% and 40% related to strains 1944-81 and 0978-82, respectively. It was considered to be a third *Leminorella* species, that was closer to *L. grimontii*, and was given the vernacular name "*Leminorella* species 3". Since there was only one strain, it was not given a scientific name. Although *Leminorella* species 3 was distinct by DNA hybridization, it could not be distinguished from *L. grimontii* by simple tests (Hickman-Brenner et al., 1985b). Since the original report in 1985, additional strains of *Leminorella* have been studied biochemically, but not by DNA–DNA hybridization. *Leminorella* and its two named species gained standing in nomenclature in July 1985, when the names appeared on Validation List 18 (Hickman-Brenner et al., 1985c).

Problems in routine identification In the original publication DNA hybridization was used to divide the strains into two named species, and a description was written based on this clear separation (Tables BXII.γ.234 and BXII.γ.235). Unfortunately, *Leminorella* species 3 cannot be distinguished from *L. grimontii* with simple tests (Hickman-Brenner et al., 1985b). This strain illustrates the difficulty in identifying *Leminorella* cultures to the species level without the benefit of DNA–DNA hybridization. Since the original report in 1985, the Enteric Reference Laboratory has received 10 additional strains of *Leminorella*. Based on the differential reactions listed in Table BXII.γ.235, five of these were identified as *L. grimontii*; one as a "possible-probable" *L. grimontii* and four as *L. richardii*. Five of the strains identified as *L. grimontii* had at least one test that was in disagreement with the composite results in Table BXII.γ.234. The strains identified as *L. richardii* were more typical, and agreed with the definition given in Table BXII.γ.235, except that one strain had one atypical reaction. Thus, the identification of new strains of *Leminorella* to the genus level is probably correct, but the identification to the species level is more tentative and needs confirmation by DNA–DNA hybridization or other methods. Enzyme profiles (Hodinka et al., 1991) and whole-cell protein patterns (Schindler et al., 1992) appear promising for differentiation. Further study and better and simpler methods are needed to assist routine identification.

FURTHER READING

Hickman-Brenner, F.W., M.P. Vohra, G.P. Huntley-Carter, G.R. Fanning, V.A.I. Lowery, D.J. Brenner and J.J. Farmer III. 1985. *Leminorella*, a new genus of *Enterobacteriaceae*: identification of *Leminorella grimontii* sp. nov. and *Leminorella richardii* sp. nov. found in clinical specimens. J. Clin. Microbiol. *21*: 234–239.

List of species of the genus Leminorella

1. **Leminorella grimontii** Hickman-Brenner, Vohra, Huntley-Carter, Fanning, Lowery, Brenner and Farmer 1985c, 375^{VP} (Effective publication: Hickman-Brenner, Vohra, Huntley-Carter, Fanning, Lowery, Brenner and Farmer 1985b, 235) *gri.mon' ti.i.* M.L. gen. n. *grimontii* named to honor Patrick Grimont and Francine Grimont, French microbiologists at the Pasteur Institute for their many contributions to enteric bacteriology.

   The characteristics are as given for the genus. The biochemical reactions are given in more detail in Table BXII.γ.234, and tests for the differentiation of *L. grimontii* and *L. richardii* are given in Table BXII.γ.235. Isolated from human feces, but there is no evidence that it can cause diarrhea or intestinal infections. Rarely isolated from other clinical specimens. Its clinical significance in these extra-intestinal specimens needs further study. One isolate was from mouse feces. The type strain was isolated from the feces of a 7-month-old child with gastroenteritis.

   The mol% G + C of the DNA is: not reported.

   Type strain: ATCC 33999, CDC 1944-8, DSM 5078.

   GenBank accession number (16S rRNA): AJ233421.

   Additional Remarks: The American Type Culture Collection includes four other strains documented to be *L. grimontii* by DNA–DNA hybridization: ATCC 43006 (CDC 3257-77), from human feces, California; ATCC 43005 (CDC 3244-76), from human urine of an 80-year-old woman with a urinary tract infection, Hawaii; ATCC 43007 (CDC 3595-77), from urine of a 9-year-old girl, Pennsylvania; and ATCC 43008 (CDC 0301-79), from mouse feces, Maryland.

2. **Leminorella richardii** Hickman-Brenner, Vohra, Huntley-Carter, Fanning, Lowery, Brenner and Farmer 1985c, 375^{VP} (Effective publication: Hickman-Brenner, Vohra, Huntley-Carter, Fanning, Lowery, Brenner and Farmer 1985b, 235.) *ri.char' di.i.* M.L. gen. n. *richardii* named to honor Claude Richard, a French microbiologist at the Pasteur Institute for his many contributions to enteric bacteriology.

   The characteristics are as given for the genus. The biochemical reactions are given in more detail in Table BXII.γ.234, and the differentiation of *L. richardii* from *L. grimontii* is given in Table BXII.γ.235. Isolated from human feces, but there is no evidence that it can cause diarrhea or intestinal infections. Nonhuman strains have not been reported. The type strain was isolated from the feces of a patient in Texas who had diarrhea. She also had systemic lupus erythematosus, and was on corticosteroid therapy. No other enteric pathogens were present.

The mol% G + C of the DNA is: not reported.

Type strain: ATCC 33998, CDC 0978-82.

Additional Remarks: The American Type Culture Collection includes three other strains documented to be *L. ri-* *chardii* by DNA–DNA hybridization; all were from human feces in the USA: ATCC 43009 (CDC 598-78), Indiana; ATCC 43010 (CDC 2209-80), Pennsylvania; and ATCC 43011 (CDC 2502-80), Texas.

Other Organisms

1. "Leminorella species 3"

In addition to the two named species of *Leminorella*, Hickman-Brenner et al. (1985b) also described a third species. Since there was only one strain, and because it could not be differentiated from *L. grimontii* by phenotypic tests, they gave it the vernacular name " *Leminorella* species 3".

The reference strain, ATCC 43012, was isolated from the human feces, Georgia, USA. This is the only strain of species 3, and we suggest that it be designated the type strain if species 3 is given a scientific name.

Deposited strain: ATCC 43012, CDC 3346-72.

Strains studied since 1985 have been reported without the benefit of DNA hybridization. One strain (1201-84) was reported as a "possible-probable" *Leminorella grimontii*. Five were reported as *L. grimontii* because they were more active biochemically. However, four of these had one or more atypical test results for this species. All new strains of *Leminorella* should be studied by DNA hybridization or other methods to determine their correct taxonomic position.

Genus XX. **Moellerella** *Hickman-Brenner, Huntley-Carter, Saitoh, Steigerwalt, Farmer and Brenner 1984c, 355^VP (Effective publication: Hickman-Brenner, Huntley-Carter, Saitoh, Steigerwalt, Farmer and Brenner 1984a, 462)*

J.J. Farmer III and Frances W. Brenner

Moe.ller.el′la. M.L. dim. ending -ella; M.L. fem. n. Moellerella named to honor Vagn Møller for his contributions to enteric bacteriology, especially for Moeller media for the determination of lysine decarboxylase, ornithine decarboxylase, and arginine dihydrolase, that are widely used for the identification of Enterobacteriaceae.

Small rod-shaped cells, conforming to the general definition of the family *Enterobacteriaceae*. Contain the enterobacterial common antigen. Gram negative. Nonmotile. Facultatively anaerobic. Catalase positive. Oxidase negative. Nonpigmented. Ferment, rather than oxidize, D-glucose and other carbohydrates. Reduce nitrate to nitrite. **Positive for methyl red, citrate utilization (Simmons), ONPG, and the fermentation of lactose, sucrose, adonitol, raffinose, melibiose, D-arabitol, D-mannose, and D-galactose.** Negative for indole production, Voges–Proskauer, H₂S production (TSI), urea hydrolysis, phenylalanine deaminase, lysine decarboxylase, arginine dihydrolase, ornithine decarboxylase, motility at 36°C, malonate utilization, esculin hydrolysis, gelatin hydrolysis (22°C), lipase (corn oil), and DNase, and the fermentation of dulcitol, salicin, *myo*-inositol, D-sorbitol, L-arabinose, L-rhamnose, D-xylose, trehalose, cellobiose, α-methyl-D-glucoside, erythritol, glycerol, and mucate. No visible gas produced during fermentation.

Most strains are susceptible to nalidixic acid, gentamicin, streptomycin, kanamycin, chloramphenicol, and cephalothin (disk diffusion method on Mueller-Hinton agar); **resistant to colistin, penicillin, ampicillin, and carbenicillin**. Variable susceptibility to sulfadiazine and tetracycline. Rarely isolated from human clinical specimens that are normally sterile. Most isolates have been from human feces, but there is **no evidence that it actually causes diarrhea or intestinal infections**. Natural reservoirs and ecological niches are not known. Can occur in water and food. **A rarely isolated genus of *Enterobacteriaceae* that is probably an opportunistic pathogen.**

The mol% G + C of the DNA is: not determined.

Type species: **Moellerella wisconsensis** Hickman-Brenner, Huntley-Carter, Saitoh, Steigerwalt, Farmer and Brenner 1984c, 355 (Effective publication: Hickman-Brenner, Huntley-Carter, Saitoh, Steigerwalt, Farmer and Brenner 1984a, 462.)

Further descriptive information

Literature Since *Moellerella* was described in 1984, there have been only a few reports in the literature, 11 cataloged in MEDLINE and 12 cataloged in BIOSIS. In addition to the genus and species names, "Enteric Group 46" should also be included as a search term in computerized literature searches, although it is rarely used today. The literature reports with *M. wisconsensis* include reviews and taxonomic studies of new *Enterobacteriaceae* (Farmer et al., 1985a; Richard, 1989; Gilchrist, 1995; Pokhil, 1996; Aleksic and Bockemühl, 1999); evaluation of commercial identification products or "kits" for *Enterobacteriaceae* (Gonzalez et al., 1986; Kitch et al., 1994); and surveys or comparisons of the family *Enterobacteriaceae* for metabolic pathways (Bouvet et al., 1989), acid phosphatases (Thaller et al., 1995), siderophores (Rabsch and Winkelmann, 1991), and polyamines (Hamana, 1996).

Occurrence in human feces Strains of *M. wisconsensis* were first recognized because of six strains submitted to CDC from Wisconsin; all were from human feces. Eight of the nine original cultures described by Hickman-Brenner et al. (1984b) were from feces and the other one was from water. Most isolates of *M. wisconsensis* have been from people with diarrhea, but there is no evidence that the strains are actually causing diarrhea or intestinal infections (Hickman-Brenner et al., 1984a). Marshall et al. (1986) reported the isolation of three strains of *M. wisconsensis* from 400 stool specimens screened for this organism with the aid of a new selective medium, developed in their laboratory. There was one report of *M. wisconsensis* isolation in the U.K., but its etiological role was not established in any of the diarrhea cases.

The interest of *M. wisconsensis* to clinical microbiology and public health laboratories may be more as a possible nuisance because strains have been picked as "suspect *Salmonella*" or "suspect *Yersinia*". More complete biochemical and serological testing should easily rule out these incorrect identifications.

Occurrence in extraintestinal human specimens Wallet et al. (1994) reported the isolation of *M. wisconsensis* from a bronchial aspirate of a patient with inhalation pneumonia. It was from an autopsy culture of a patient admitted in a deep coma following cardiac arrest. They concluded that the origin of the organism may have been from digestive secretions that inoculated the lower respiratory tract. This was the first isolate of *M. wisconsensis* from a human source other than stool or gallbladder. Wittke et al. (1985) reported the isolation of *M. wisconsensis* from the infected gallbladder of a 71-year-old man with typical signs of acute cholecystitis. Equal numbers of enterococci were also isolated. Serum from the patient taken on the 24th day after cholecystectomy did not contain agglutinating antibodies to heated or unheated suspensions of *M. wisconsensis*, providing negative evidence for its pathogenic role. Ohanessian et al. (1987) reported the isolation of *M. wisconsensis* along with *Hafnia alvei* from the infected gallbladder of a 77-year-old woman with coronary cardiac failure and gallstones. In 1985 the CDC's Enteric Reference Laboratory received the first isolate from human blood.

Clinical significance of human isolates The clinical significance and the ability of *M. wisconsensis* to actually cause infections should be carefully evaluated in cases that yield this new organism. Systematic study and good case reports are needed. The isolates described above suggest that it is probably an opportunistic pathogen with only a slight capacity to cause extraintestinal infections in humans.

Occurrence in other animals Bangert et al. (1988) isolated *M. wisconsensis* and other *Enterobacteriaceae* in a fecal culture survey of 47 captive raptors (the bird orders *Falconiformes* and *Strigiformes*). Giordano-Dias et al. (1997) reported isolates from the sandfly *Lutzomyia longipalpis* (*Diptera: Psychodidae*) maintained in the laboratory.

Occurrence in food and water One of the original nine strains reported by Hickman-Brenner et al. (1984a) was from a routine drinking water sample from a South Dakota (USA) town that did not chlorinate its water. Cabadajova and Kudrna (1988) of the Czech Republic isolated five strains of *M. wisconsensis* from food, but could not show an etiological role in relation to the corresponding human diarrhea cases.

Isolation, identification, culture preservation, phenotypic characterization Strains of *Moellerella* are not difficult to grow,

and are typical *Enterobacteriaceae* in most respects. See the chapter on the family *Enterobacteriaceae* for general information on growth, plating media, working and frozen stock cultures, media and methods for biochemical testing, identification, computer programs, and antibiotic susceptibility. Strains grow on media normally used in enteric bacteriology. On MacConkey agar, colonies of *M. wisconsensis* are bright red with precipitated bile around them, and thus are indistinguishable from *Escherichia coli* colonies (Hickman-Brenner et al., 1984a). Marshall et al. (1986) reported the enhanced isolation of *M. wisconsensis* from stool specimens with a selective medium developed in their laboratory that contained bacitracin and polymyxin.

Biochemical reactions and differentiation from other *Enterobacteriaceae* Table BXII.γ.193 in the chapter on the family *Enterobacteriaceae* gives the results for *Moellerella* in 47 biochemical tests normally used for identification (Farmer, 1999). There are no genus- or species-specific tests or sequences for the identification of *Moellerella*. The best approach is to do a complete set of phenotypic tests (growth on plating media, biochemical tests, antibiogram, etc.) and compare the results with all the other named organisms in the family *Enterobacteriaceae*. This approach is described in more detail in the section on the family *Enterobacteriaceae*. The biochemical results of a test strain can be compared to all the organisms listed in Table BXII.γ.193 in that chapter. Several computer programs greatly facilitate analyzing the results.

Antibiotic susceptibility In their original description, Hickman-Brenner et al. (1984a) gave the zone sizes for each of the nine strains around 12 different antibiotics (the standard "antibiogram" that has been used for many years in their reference laboratory as an aid in identification). There was some variation among the strains, but all were resistant to colistin and penicillin. Although this is not a unique pattern, the antibiogram can be a useful additional way of differentiating *M. wisconsensis* from other species of *Enterobacteriaceae*.

TAXONOMIC COMMENTS

Discovery and DNA–DNA hybridization In 1980 the vernacular name Enteric Group 46 was applied (Hickman-Brenner et al., 1984b) to a group of strains that had been studied at the Centers for Disease Control, and most had originally been sent from the Wisconsin State Laboratory of Hygiene. The strains were characterized biochemically and were phenotypically distinct from all of the described organisms in the family *Enterobacteriaceae*. By DNA–DNA hybridization the strains were 78–97% related to the type strain, with divergence values (ΔT_m) of 0–1.5. Other *Enterobacteriaceae* were only 2–32% related. Because Enteric Group 46 was distinct by both DNA–DNA hybridization and phenotype, a new genus *Moellerella* with a single species, *Moellerella wisconsensis*, was proposed (Hickman-Brenner et al., 1984a).

FURTHER READING

Hickman-Brenner, F.W. , G.P. Huntley-Carter, Y. Saitoh, A.G. Steigerwalt, J.J. Farmer, III and D.J. Brenner. 1984. *Moellerella wisconsensis* a new genus and species of *Enterobacteriaceae* found in human stool specimens. J. Clin. Microbiol. *19*: 460–463.

Marshall, A.R., I.J. Al Jumaili and A.J. Bint. 1986. The isolation of *Moellerella wisconsensis* from stool samples in the U.K. J. Infect. *12*: 31–33.

Wittke, J.W., S. Aleksic and H.H. Wuthe. 1985. Isolation of *Moellerella wisconsensis* from an infected human gallbladder. Eur. J. Clin. Microbiol. *4*: 351–352.

List of species of the genus Moellerella

1. **Moellerella wisconsensis** Hickman-Brenner, Huntley-Carter, Saitoh, Steigerwalt, Farmer and Brenner 1984c, 355[VP] (Effective publication: Hickman-Brenner, Huntley-Carter, Saitoh, Steigerwalt, Farmer and Brenner 1984a, 462.)

wis.con.sen' sis. M.L. fem. adj. *wisconsensis* pertaining to the state of Wisconsin, U.S.A., where most of the original strains were isolated.

The characteristics are as given for the genus and are summarized in Table BXII.γ.193 in the chapter on the family *Enterobacteriaceae.* Isolated from human feces but rarely from extraintestinal specimens, also isolated from animals, water, and food. The type strain was isolated from the stool culture of a 16-year-old girl in Wisconsin.

The mol% G + C of the DNA is: not determined.
Type strain: ATCC 35017, CDC 2896-78, DSM 5076.
Additional Remarks: The American Type Culture Collection includes four other strains, all from human feces.

Genus XXI. **Morganella** *Fulton 1943, 81*[AL]

J. Michael Janda and Sharon L. Abbott

Mor.ga.nel' la. M.L. dim. ending *-ella*; M.L. fem. n. *Morganella* named after H. de R. Morgan, who first studied the organism.

Straight rods, 0.6–0.7 × 1.0–1.7 μm. Gram negative. Motile by means of peritrichous flagella. Facultatively anaerobic. **Oxidative deamination of various amino acids including L-phenylalanine and L-tryptophan** (Singer and Volcani, 1955). **Urease positive. Indole positive** (Penner, 1984). Relatively few carbohydrates fermented. **Ornithine decarboxylase positive. Acid produced from D-mannose but not D-xylose. Gelatin not degraded.** The genus *Morganella* presently resides in the class *Gammaproteobacteria* in the family *Enterobacteriaceae,* based upon DNA hybridization data and not rDNA sequencing (Stackebrandt et al., 1988). 16S rDNA sequence data are presently unavailable.

The mol% G + C of the DNA is: 50 (Falkow et al., 1962).

Type species: **Morganella morganii** (Winslow, Kliger and Rothberg 1919) Fulton 1943, 81 (*Bacillus morgani* (sic) Winslow, Kliger and Rothberg 1919, 481; *Proteus morganii* (Winslow, Kliger and Rothberg 1919) Yale 1939b, 435.)

Further descriptive information

M. morganii displays morphologic characteristics typical of members of the *Enterobacteriaceae* appearing as short, straight rods. Capsules not produced (McKell and Jones, 1976). Peritrichous flagella with normal and curly curvature have been described (Leifson et al., 1955). Normal flagella have a wavelength of 2.09–2.36 μm with an amplitude of 0.51 μm. Curly flagella possess a shorter periodicity (1.14–1.18 μm) and amplitude (0.37–0.40 μm). Strains often express multiple fimbriae consisting of morphologically thin (outside diameter [o.d.] 4–5 nm) and thick forms (o.d. 7–8 nm) on the same cell (Old and Adegbola, 1982). The ultrastructure of *M. morganii* has not been investigated to any extent.

The peptidoglycan of *M. morganii* is O-acetylated at the C-6 hydroxyl group of N-acetylmuramyl residues (Clarke, 1993). The degree of O-acetylation ranges from 43.0–49.6% and appears responsible for resistance to muramidase (lysozyme) activity. The major fatty acids present in *M. morganii* lipopolysaccharide are 3-hydroxytetradecanoic ($C_{14:0\ 3OH}$), tetradecanoic ($C_{14:0}$), hexa-decanoic ($C_{16:0}$), and dodecanoic ($C_{12:0}$) acids (Vasyurenko and Chernyavskaya, 1990). Cellular fatty acid analysis of morganellae reveals hexadecanoic acid as the dominant peak with octadecenoic acid ($C_{18:1}$) and methylenehexadecanoic acid ($C_{16:0\ ante}$) as additional major peaks (Vasyurenko and Chernyavskaya, 1990). The presence of detectable amounts of dodecanoic acid separates morganellae from *Proteus* and *Providencia.* Histamine, cadaverine, and diaminopropane are the predominant polyamines synthesized under defined conditions (Hamana, 1996).

Half of all morganellae grown in broth culture show uniform turbidity with ring or pellicle formation (McKell and Jones, 1976). Growth occurs between 4 and 45°C. On nutrient agar, *M. morganii* produces smooth transparent colonies with an entire edge. Pigmentation not observed (McKell and Jones, 1976). On nonselective media, most morganellae appear as nonhemolytic, buff-colored, convex colonies, 2–3 mm in diameter after overnight incubation at 35–37°C (Janda and Abbott, 1998a). Hemolysis on blood agar may be detected by prolonged incubation (48–72 h). On media containing an aromatic amino acid such as phenylalanine agar an almond-like odor may be emitted (Müller, 1986a). Because most strains are sucrose- and lactose-negative they appear as colorless colonies on selective media such as MacConkey's, xylose-lysine-desoxycholate, eosin-methylene blue, and Salmonella–Shigella agar. Some strains are highly pleomorphic producing multiple morphovars on media such as Salmonella–Shigella agar (Janda and Abbott, 1998a).

Morganellae possess a number of metabolic features almost exclusively associated with members of the tribe *Proteeae* (*Morganella, Proteus,* and *Providencia*). Oxidative deamination of certain amino acids (e.g., phenylalanine deaminase) to form keto acids that then react with ferric compounds yielding chromogenic products is a unique characteristic of this tribe (Singer and Volcani, 1955; Ewing, 1986a). On DL-tryptophan agar, *Morganella* produces a reddish-brown melanin-like compound with a molecular weight ≤12,000 (Polster and Svobodová, 1964; Müller, 1986a). *M. morganii* also degrades L-tyrosine crystals incorporated into solid media within 24 h, presumably mediated via a tyrosine phenol-lyase (Sheth and Kurup, 1975). All proteeae, including *M. morganii*, produce extracellular bacteriolytic enzymes capable of degrading cell wall components of *Escherichia coli* and *Pseudomonas aeruginosa*. This reaction presumably occurs through extracellular secretion of a peptidoglycan-hydrolase (Branca et al., 1996). Like *Proteus*, morganellae elaborate a type II glycerol dehydrogenase that distinguishes these taxa from *Providencia* (type III); 1,3-propanediol dehydrogenase is not produced (Bouvet et al., 1995a).

The genus is biochemically homogeneous. Indole formed. Nitrates reduced. The methyl red test is positive. The Voges–Proskauer test is negative. Lysine decarboxylase activity is variable. Growth in KCN broth. Citrate, acetate, malonate, and mucate not utilized. Hydrolysis of urea. No H_2S produced on triple sugar iron (TSI) agar slants, although upon prolonged incubation (>24 h) a slight blackening at the junction may occur due to small amounts of a reddish-brown pigment being produced (Janda and Abbott, 1998a).

Acid with gas from D-glucose fermentation. Anaerobic fermentation of D-mannose. Fermentation of glycerol and trehalose is variable (Siboni, 1976). No fermentation of adonitol, L-arabinose, cellobiose, dulcitol, α-methyl-D-glucoside, *m*-inositol, lactose, maltose, D-mannitol, melibiose, raffinose, L-rhamnose, salicin, sucrose, and D-xylose.

Deoxyribonuclease, ribonuclease, alkysulfatase, arylsulfatase, lipase, lecithinase, hyaluronidase, and protease activities not produced. Chitin, elastin, pectin, mucin, and fibrin are not degraded (Janda et al., 1996b). Alkaline and acid phosphatase, leucine arylamidase, and naphthol-AS-BI-phosphohydrolase positive.

Morganella exhibits 20% relatedness to core members of the *Enterobacteriaceae* in DNA hybridization studies (Brenner et al., 1978). All *M. morganii* studied constituted a single DNA relatedness group. The average relatedness of 18 strains was 90% in 60°C reactions and 85% in 75°C reactions with <1% divergence (Brenner et al., 1978). Biochemically aberrant morganellae, exhibiting either the lysine decarboxylase-positive or ornithine decarboxylase-negative phenotype, were subsequently found to be highly related to classic *M. morganii* using the lysine decarboxylase-positive strain, CDC 1274-75, as the source of labeled DNA (Hickman et al., 1980). Hybridization studies indicated 73–92% relatedness in 60°C reactions and 79–96% relatedness in 75°C reactions.

Morganella strains harbor between one and four extrachromosomal elements ranging from 35–60 MDa (Cornelis et al., 1981; Janda et al., 1996b). Plasmids in *Morganella* that confer antibiotic resistance belong to compatibility groups N, FI, and FII (Hedges et al., 1973). Additionally, lactose fermentation and lysine decarboxylation are attributable to plasmid carriage (Le Minor and Coynault, 1976; Cornelis et al., 1981). A bacteriophage typing scheme using seven phages defined 14 lytic patterns, but plaques produced are poorly developed (Schmidt and Jeffries, 1974). Phages are lytic only for *Morganella* and have not been found to attack *Proteus* or *Providencia* (Coetzee, 1963). A bacteriocin (morganocin) system detecting both morganocin production and sensitivity revealed 33 types in 45 serologically distinct strains of *Morganella* (Senior and Vörös, 1989).

The antigenic schema of *Morganella* consists of 77 serotypes with 44 somatic (O), 4 capsular (K), and 38 flagellar (H) antigens (Vörös and Senior, 1990). In 1990, 11 additional O types were added, extending the number of somatic antigens to 55 (Vörös

and Senior, 1990). Passive hemagglutination can be used to determine the O antigen (Penner and Hennessy, 1979b).

Morganella is generally resistant to ampicillin, extended-spectrum penicillins, first-generation cephalosporins, and cefoxitin. Antimicrobial agents to which strains are susceptible include third-generation cephalosporins, aztreonam, quinolones, tobramycin, and chloramphenicol. Variable resistance has been demonstrated for gentamicin, amikacin, imipenem, and tetracycline. Disk diffusion testing may be unreliable for some cephems (Biedenbach et al., 1993).

Although isolated more often from urine than other sources, *Morganella* is an uncommon cause of urinary tract infection (UTI) (Janda and Abbott, 1998a). It is often isolated as a colonizer in cases of bacteriuria in patients undergoing long-term catheterization (Mobley and Warren, 1987). Bacteremia is rare, usually occurring in immunocompromised patients; 70% of cases are acquired nosocomially (Janda and Abbott, 1998a). Surgical patients are affected most often, with the focus of infection being wounds. Urinary tract infections are not a common source for bacteremias (McDermott and Mylotte, 1984).

M. morganii produce mannose-sensitive hemagglutinins (MSHA) (Coetzee et al., 1962); electron micrographs of MSHA strains show perifimbriation but with denser polar aggregations. Catecholate- or hydroxymate-type siderophores used for iron acquisition are not produced (Drechsel et al., 1993). However, α-keto acids generated by L-amino acid deaminases produced by *M. morganii* can form ferric complexes that are sufficiently stable to transport iron (Drechsel et al., 1993). *M. morganii* can also use exogenous siderophores (ferrichromes, rhizoferrin, and citrate) for iron acquisition. *M. morganii* strains producing heat labile enterotoxin have been reported in a group of Swedish travelers with diarrhea (Jertborn and Svennerholm, 1991). Approximately 50% of morganellae produce a cell-free hemolysin related to the α-hemolysin of *E. coli* (Koronakis et al., 1987). The cytoplasmic location of urease and its low activity optimum of pH 5.5 in *M. morganii* allows the organism to survive under acidic conditions when other urease-positive, Gram-negative rods die (Young et al., 1996a). However, *M. morganii* grows more slowly and is less efficient in producing alkaline conditions in the urine than *P. mirabilis*, which is a more frequent cause of UTIs (Senior, 1983).

ENRICHMENT AND ISOLATION PROCEDURES

Morganellae may be recovered from specimens from sterile body sites using routine enteric isolation media. Both smooth and rough morphovars may be present. For fecal specimens, MacConkey agar with methyl blue and phenolphthalein diphosphate may be helpful (Janda and Abbott, 1998a). When testing for human intestinal carriage, Rustigian and Stuart (1945) reported increases in *Morganella* recovery of 1.8–10% using tetrathionate or selenite enrichment broth prior to plating on media.

MAINTENANCE PROCEDURES

Morganella strains may be maintained at room temperature in agar deeps, especially motility agar, for several months. For long-term storage cultures should be frozen and maintained at −70°C.

DIFFERENTIATION OF THE GENUS *MORGANELLA* FROM OTHER GENERA

See the genus *Proteus*, Table BXII.γ.254, for characteristics that can be used to differentiate *Morganella* from other related genera of *Enterobacteriaceae*.

List of species of the genus Morganella

1. **Morganella morganii** (Winslow, Kliger and Rothberg 1919) Fulton 1943, 81[AL] (*Bacillus morgani* (sic) Winslow, Kliger and Rothberg 1919, 481; *Proteus morganii* (Winslow, Kliger and Rothberg 1919) Yale 1939b, 435.)

mor.ga' ni.i. M.L. gen. n. *morganii* of Morgan; named after H. de R. Morgan, a British bacteriologist who first studied the organism.

The description is the same as that for the genus. Trehalose negative, susceptible to tetracycline See Tables BXII.γ.236 and BXII.γ.237 for other characteristics.

Occurs in the feces of humans, dogs, other mammals, and reptiles. Opportunistic human pathogens.

The mol% G + C of the DNA is: 50 (T_m).

Type strain: ATCC 25830, DSM 30164, IFO 3848, NCIB 235.

a. **Morganella morganii** *subsp.* **morganii** *subsp. nov.* (Winslow, Kliger and Rothberg 1919) Fulton 1943, 81[AL] (*Bacillus morgani* [sic] Winslow, Kliger and Rothberg 1919, 481; *Proteus morganii* (Winslow, Kliger and Rothberg 1919) Yale 1939b, 435.)

Distinguished from subsp. *sibonii* by positive trehalose reaction and resistance to tetracycline. See Table BXII.γ.236 for other characteristics.

The mol% G + C of the DNA is: 50 (T_m).

Type strain: ATCC 25830, DSM 30164, IFO 3848, NCIB 235.

b. **Morganella morganii** *subsp.* **sibonii** *subsp. nov.* Jensen, Frederiksen, Hickman-Brenner, Steigerwalt, Riddle and Brenner 1992, 619[VP]

si.bo' ni.i. L. gen. n. *sibonii* of Siboni, named after Knud Siboni, a Danish microbiologist who first recognized trehalose-fermenting *Morganella morganii*.

Distinguished from subsp. *morganii* by characteristics in Table BXII.γ.236. The above subspecies designations were created in keeping with the recommendation that "genetically close organisms that diverge in phenotype" be given subspecies status; DNA relatedness data for these strains indicated the creation of subspecies as opposed to species.

The mol% G + C of the DNA is: 50.

Type strain: 8103-85, ATCC 49948.

TABLE BXII.γ.236. Characteristics of *Morganella morganii*[a,b]

Characteristic	Morganella morganii subsp. *morganii*	Morganella morganii subsp. *sibonii*
Indole	+	d
Voges–Proskauer	−	−
Utilization of Simmons citrate, malonate, mucate, acetate	−	−
Urease	+	+
H$_2$S production (triple sugar iron agar)	d	−
Phenylalanine deaminase	+	+
o-Nitrophenyl-β-galactopyranoside (ONPG) hydrolysis	D	−
Møeller amino acid decarboxylases:		
Lysine decarboxylase	d	d
Arginine dihydrolase	−	−
Ornithine decarboxylase	d	d
Motility	+	d
Gelatin liquefication 22°C	−	−
Growth in potassium cyanide (KCN)	+	d
NO$_3^-$ reduction to NO$_2^-$	+	+
Corn oil lipase	−	−
Deoxyribonuclease	−	−
Acid production from:		
Glucose, D-mannose	+	+
Trehalose, adonitol, D-arabitol	−	+
L-Arabinose, cellobiose, dulcitol, erythritol, *myo*-inositol, lactose, maltose, mannitol, α-methylglucoside, melibiose, raffinose, rhamnose, salicin, D-sorbitol, sucrose, D-xylose	−	−
Tetracycline susceptibility	+	−

[a]For symbols see standard definitions; temperature of reactions, 36 ± 1°C. All reactions are for 48 h.

[b]Adapted from Farmer (1995) and Jensen et al. (1992).

TABLE BXII.γ.237. Identification of *Morganella morganii* biogroups[a,b]

Characteristic	Biogroup						
	A	B	C	D	E	F	G
Acid production from:							
Trehalose	−	−	−	−	+	+	+
Glycerol	+[c]	d	+[c]	−	−	d[c]	d[c]
Møeller amino acid decarboxylase:							
Lysine decarboxylase	−	+	−	+	+	d	−
Ornithine decarboxylase	+	+	−	−	+	−	+
Motility	+[c]	d	d	−	+	d	+
Tetracycline susceptibility	+	+	d	+	−	−	d

[a]For symbols see standard definitions. Temperature of reactions 36 ± 1°C. All reactions are for 48 h.

[b]Adapted from Jensen et al., 1992.

[c]Positive at 3–7 d.

Genus XXII. **Obesumbacterium** *Shimwell 1963, 759*[AL]

J.J. FARMER III AND DON J. BRENNER

O.be' sum.bac.te' ri.um. L. neut. adj. *obesum* fat; L. neut. n. *bacterium* rod; M.L. neut. n. *Obesumbacterium* a fat, rod-shaped bacterium.

Pleomorphic rods 0.8–2.0 × 1.5–100 µm (short, "fat" rods predominate when grown in beer wort with live yeasts; long pleomorphic rods usually predominate when grown in most bacteriological media; some strains have been reported to display a branching cell morphology), conforming to the general definition of the family *Enterobacteriaceae*. Gram negative. **Nonmotile.** Facultatively anaerobic. Slow growing, forming colonies <0.5 mm in diameter on ordinary plating media at 24 h. Optimal growth temperature is 25–32°C; **growth at 37°C is comparatively poor. Acid formed from D-glucose and D-mannose; very few other carbohydrates are fermented.** Gas formation during fermentation appears to be variable (original description says gas is produced, but none of the strains studied produced gas). **Lysine decarboxylase is positive.** Nitrate is reduced to nitrite. **Many biochemical tests normally used for differentiation of *Enterobacteriaceae* are negative or delayed positive** (3–7 d at 36°C). **One of only three genera** (along with *Hafnia* and *Pragia*) **in *Enterobacteriaceae* to contain only gluconate dehydrogenase** in its D-glucose oxidation pathway (Bouvet et al., 1989). **Occurs as a brewery contaminant** that can survive and grow in the presence of live yeasts during beer production. Not isolated from human clinical specimens; no evidence of pathogenicity for humans or animals. The genus has a single species, *O. proteus*, with two defined biogroups (1 and 2). The description of the genus above is a "composite description" based on data for both biogroups. However, the two biogroups are actually distinct species that are phenotypically different and only distantly related by DNA–DNA hybridization (Brenner, 1981). *O. proteus* biogroup 1 is actually a biogroup of *Hafnia alvei* (Brenner, 1981), and will be referred to as *H. alvei* biogroup 1 (Farmer et al., 1985a) in this chapter.

The mol% G + C of the DNA is: 48–49.

Type species: **Obesumbacterium proteus** (Shimwell and Grimes 1936) Shimwell 1963, 759 (*Flavobacterium proteum* (sic) Shimwell and Grimes 1936, 348.)

FURTHER DESCRIPTIVE INFORMATION

Obesumbacterium, as a member of *Enterobacteriaceae*, belongs to the *Gammaproteobacteria*. Shimwell's original description of an organism he called "*Flavobacterium proteum*" centered on its cellular morphology (Shimwell, 1936; Shimwell and Grimes, 1936). He noted that it appeared as plump rods 0.8–1.2 × 1.5–4 µm when grown in wort media or when taken directly from breweries during fermentation. This morphology had no doubt led to the term "short fat rod of pitching yeasts", which had been used in breweries for many years. Shimwell (1936) also noted much pleomorphism when the organism was grown in laboratory media that were alkaline or neutral. Chains of up to 100 µm were observed under these conditions.

SDS-PAGE protein fingerprints differentiate *H. alvei* biogroup 1 (*O. proteus* biogroup 1) from *O. proteus* biogroup 2 (Fernandez et al., 1993), as do plasmid profiles, ribotyping, enteric repetitive intergenic consensus (ERIC)-PCR (Prest et al., 1994), and random amplified polymorphic DNA (RAPD) profiles (Savard et al., 1994). Plasmid profiling, ERIC-PCR, and RAPD profiles successfully differentiated between isolates of *O. proteus* biogroup 2 (Prest et al., 1994; Savard et al., 1994).

A number of reports discuss the ecology of *Obesumbacterium* in breweries (Shimwell, 1936, 1948, 1963, 1964; Shimwell and Grimes, 1936; Strandskov et al., 1953; Case, 1965). Unfortunately, no distinction has been made between biogroups 1 and 2, so it is usually impossible to determine whether these two biogroups (which are really distinct species, one of which does not belong *Obesumbacterium*; see Taxonomic Comments, below), are different in their ecology, distribution, and other factors. Future studies in breweries should resolve this problem, but only if the two distinct biogroups are differentiated. This approach was used by van Vuuren (1978) in South African breweries.

Changes in brewery practices have eliminated many bacterial contaminants that were once carried along with pitching yeasts; thus *Obesumbacterium* is isolated much less frequently than in the past (F.G. Priest, personal communication). This fact may hamper future studies on the organism.

ENRICHMENT AND ISOLATION PROCEDURES

Quantitative recovery of *O. proteus* from samples of ale yeast was best after 3 d incubation at 25°C on universal beer agar and Wallerstein Laboratories' differential medium, containing cycloheximide to inhibit yeast growth, intermediate on wort agar and yeast mannitol (YM) (Difco) agar, and unsatisfactory on MacConkey agar and membrane lauryl sulfate agar (Fernandez et al., 1993).

MAINTENANCE PROCEDURES

Long-term storage can be accomplished by quick freezing at −80°C in yeast extract peptone dextrose broth supplemented with 15% glycerol (Fernandez et al., 1993), or in 10% skim milk.

TAXONOMIC COMMENTS

Obesumbacterium was first proposed as a genus by Shimwell in 1963 (Shimwell, 1963, 1964) to accommodate the organism known as "*Flavobacterium proteus*" (Shimwell and Grimes, 1936), which was called "*Flavobacterium proteum*" in the original proposal (Shimwell and Grimes, 1936). The specific epithet "*proteus*" was chosen because the organism has a very pleomorphic cell morphology depending upon the particular growth conditions (Shimwell, 1936, 1948, 1964). The organism was first named in 1936 when Shimwell was doing studies on the "short fat rods of pitching yeasts". He gave an adequate description (based on the available techniques) of the organism, but much of his description is not helpful in identifying it. Shimwell deposited a pure culture of "*F. proteus*" (isolated from yeast of Beamish and Crawford's Brewery, Cork, Ireland) in the National Collection of Type Cultures (NCTC), England. However, when Shimwell's own culture was "lost", he wrote the NCTC and found they had also "lost" this culture (unpublished letter of 27 February 1964 from J.L. Shimwell to E.F. Lessel of the ATCC). Thus it appears that no culture has survived from those originally studied by Shimwell in writing his description of "*F. proteus*". The strain most studied is apparently ATCC 12841 (NCIB 8771; strain 42 of Strandskov and Bockelmann), which, until 1980 was only a reference strain (*Bergey's Manual of Determinative Bacteriology*, 8th ed.). This strain was isolated from lager and ale yeasts and deposited by the Shaefer

Brewing Co. (Strandskov and Bockelmann, 1955). This strain was given status, without comment, as the type strain (neotype) of *O. proteus* when the Approved Lists of Bacterial Names were issued in 1980.

There is confusion whether the current type strain of *O. proteus* is the same organism that Shimwell studied and named "*F. proteus*". In 1956, the NCTC sent three cultures (numbers 42, 2, and 41), isolated and described by Strandskov and Bockelmann (1955), to Shimwell to examine and determine whether he thought they were "*F. proteus*". For no. 42 (the current type strain of *O. proteus*) he concluded: "This is almost certainly an authentic strain. Its morphology is almost exactly that of my original isolation, namely thick (up to 2 or more μm), long (up to 100 μm or more) filaments etc. together with the usual short fat rods in fair numbers". The properties of these strains include indole, acetylmethylcarbinol, H₂S and starch, all negative; nitrite from nitrate, positive. However, probably hundreds of Gram-negative bacteria would answer to this description. Indeed, the species is very poorly characterized biochemically, its main characteristic being its extremely large cell size, and its almost incredible pleomorphism, by means of which (taken in conjunction with its presence in a brewery fermentation) it can readily be identified. In cataloguing any of the strains I suggest that no. 42 could be safely named *F. proteus*; no. 2 a little doubtfully, and no. 41 very doubtful indeed" (unpublished letter of 7 June 1956 from J.H. Shimwell to W.S. Greaves of the NCTC). Thus the current concept of *O. proteus* based on its type strain ATCC 12841 (no. 42) is not incompatible with "*F. proteus*" as defined by Shimwell based on strains that no longer exist. Unfortunately, Shimwell's original description of "*F. proteum*" fits both *H. alvei* biogroup 1 (*O. proteus* biogroup 1) and *O. proteus* biogroup 2, so it is uncertain which of these latter two organisms (or perhaps both) was originally studied by Shimwell (1936). It is even possible that it was neither since Shimwell described "*F. proteum*" as a producer of gas during fermentation of carbohydrates, but neither *H. alvei* biogroup 1 nor *O. proteus* biogroup 2 produce gas (Priest et al., 1973; see also Table BXII.γ.193 in the chapter on the family *Enterobacteriaceae*.

"*F. proteus*" had been known for many years as a brewery contaminant that can survive and grow in the presence of live yeasts during beer production. Because it fermented D-glucose and other carbohydrates, it was incompatible with a redefined genus *Flavobacterium* that was limited to oxidative rather than fermentative bacteria. Its removal from *Flavobacterium* was subsequently confirmed by Bauwens and De Ley (1981), who used DNA–rRNA hybridization to show that "*F. proteus*" was not closely related to other *Flavobacterium* species. Shimwell (1963, 1964) formed the new genus *Obesumbacterium* for "*F. proteus*" because its phenotypic properties and ecological niche differed from those of other genera. He did not assign *Obesumbacterium* to a family, but its properties (see also Table BXII.γ.193 in the chapter on the family *Enterobacteriaceae*) are compatible with those of the family *Enterobacteriaceae*.

Priest et al. (1973) determined the phenotypic properties and did DNA–DNA relatedness studies on 19 strains of *O. proteus*, including 16 brewery isolates. They defined two biogroups, which had the same mol% G + C content of DNA, 48.0–48.5. They also proposed that *O. proteus* be reclassified in the genus *Hafnia* where its citation would be "*H. protea*" (Shimwell and Grimes, 1936) Priest, Somerville, Cole and Hough 1973. The new combination could have been proposed as "*Hafnia proteus*", since

"*proteus*" is a substantive that need not agree in gender with its genus. This change in classification was accepted to some extent, but in reality both *O. proteus* and "*H. protea*" have, until recently, been rarely used in the literature. Most of the existing citations have been in journals related to brewing. The name "*Hafnia protea*" lost standing in nomenclature on 1 January 1980, because it did not appear on the Approved Lists of Bacterial Names; however, *Obesumbacterium* and *O. proteus*, with its type strain ATCC 12841, have standing in nomenclature since they did appear on the lists.

The classification of *Obesumbacterium* was clarified by Brenner and co-workers (1981), who used DNA–DNA hybridization to determine the relatedness of *O. proteus* biogroups 1 and 2 to each other and to other *Enterobacteriaceae*. *O. proteus* biogroup 1 was very highly related to *Hafnia alvei*, and it was concluded that this biogroup is a synonym of *H. alvei*. It can best be thought of as the pleomorphic, KCN-negative, nonmotile, non-gas-producing, salicin-positive, L-arabinose-negative, L-rhamnose-negative, maltose-negative, D-xylose-negative, β-galactosidase-negative biogroup of *H. alvei* that has adapted to the brewery environment. This adaptation to the brewery environment was noted by Shimwell and Grimes (1936) in the original description of "*F. proteus*": "The organism sometimes failed to grow in dilute media, probably owing to its having become accustomed to the more concentrated nature of beer-wort in the brewery." This adaptation has presumably made the organism very "sluggish" in its metabolic activities, as is reflected in its slow growth rate and diminished activity in the tests normally done for identification of *Enterobacteriaceae*. The classification of *O. proteus* biogroup 1 as *H. alvei* is further strengthened by the fact that strains of *O. proteus* biogroup 1 (but not biogroup 2) are lysed (Farmer, 1984a; Table BXII.γ.238) by the *Hafnia*-specific bacteriophage 1672 described by Guinée and Valkenburg (1968).

Van Vuuren et al. (1981) studied 10 cultures of "*Hafnia alvei-Obesumbacterium proteus*" isolated from South African lager-beer breweries. One of their five distinct biogroups (based on API 20E biochemical profiles) was very active biochemically and more like typical cultures of *H. alvei*. The other four biogroups were progressively less biochemically active with the least active group more like *O. proteus*. Based on phenotypic and DNA relatedness data (Brenner, 1981), Farmer et al. (1985a) proposed that *O. proteus* biogroup 1 and biochemically inactive strains of *H. alvei* isolated from breweries be classified in a single taxon. They proposed the names "*Hafnia alvei* biogroup 1" and "*Hafnia alvei* brewery biogroups" to aid in the recognition and identification of these strains adapted to the brewery environment. *H. alvei*, *H. alvei* biogroup 1, and *O. proteus* biogroup 2 have been listed as separate taxa in the CDC's master biochemical chart of *Enterobacteriaceae* since 1985. Based on these data, it appears that biogroups intermediate between typical *H. alvei* and *O. proteus* biogroup 1 may occur in breweries. A complete set of biochemical tests (Table BXII.γ.238, also see Table BXII.γ.193 in the chapter on the family *Enterobacteriaceae*) and lysis by the *Hafnia*-specific bacteriophage are needed to characterize isolates from breweries that resemble *Hafnia/Obesumbacterium*.

O. proteus biogroup 2 is different biochemically from *H. alvei* biogroup 1 and from *H. alvei* (Table BXII.γ.238). In DNA–DNA relatedness studies *O. proteus* biogroup 2 was only 25–30% related to *H. alvei* biogroup 1 (Brenner, 1981). In fact, its closest relative is *Escherichia blattae* to which it was 60–65% related. Thus *O. proteus*

TABLE BXII.γ.238. Differentiation of *Obesumbacterium proteus* biogroup (BG) 2, *Hafnia alvei*, and *Hafnia alvei* biogroup 1 (*Obesumbacterium proteus* biogroup 1)[a]

Characteristic	Incubation time (d)	*O. proteus* BG 2	*Hafnia alvei* wildtype	*Hafnia alvei* BG 1 [b]
Lysis by the *Hafnia*-specific bacteriophage of Guinée and Valkenburg (1968)	1	−	+	+
Strong rapid catalase[c]		−	+	+
Voges–Proskauer (22°C)	4	−	+	+
Acid production from:				
D-Mannitol	10	−	+	+
Salicin	7	−	[−]	+
D-Xylose	7	+	+	−
Esculin hydrolysis	7	−	[−]	+

[a]Symbols: +, positive for 90–100% of strains; [−], positive for 11–25% of strains; −, positive for 0–10% of strains.

[b]Formerly known as *O. proteus* biogroup 1.

[c]Strains of *O. proteus* biogroup 2 are catalase positive but the reaction is weak and takes 10–30 seconds. Strains of wild-type *H. alvei* and of *H. alvei* biogroup 1 give a very strong and rapid (1 second) catalase reaction. This is a very simple way to differentiate between strains of *H. alvei* and *O. proteus* biogroup 2.

biogroup 2 should not be included in the same species or in the same genus as *H. alvei* biogroup 1.

The type strain of *O. proteus* is ATCC 12841 (NCIB 8771). It was listed as *"H. protea"* in the 14th edition of the ATCC catalog, and has been listed as *O. proteus* in the 15th to 19th editions. This culture is lysed by the *Hafnia*-specific bacteriophage 1672, is methyl red negative, D-mannitol negative, salicin positive (6 d), maltose negative, D-xylose negative, esculin positive (7 d), and ONPG negative. Thus from Table BXII.γ.238 it clearly belongs to *H. alvei* biogroup 1, although it is somewhat more inactive or slower in its biochemical reactions than other strains. *H. alvei* biogroup 1 can be logically considered as a series of inactive biogroups of *H. alvei*. However, if it were classified as a named subspecies it could cause serious nomenclatural problems in *Hafnia* because it has priority over *H. alvei*. Since the name *Obesumbacterium* has traditionally been well known in the brewing industry, which seems to be the ecological niche for this organism, the most practical solution at present is to continue use of the names *H. alvei* biogroup 1 and *O. proteus* biogroup 2 until all the taxonomic problems are resolved and a "final" nomenclature and classification can be proposed after careful analysis.

Since *O. proteus* is now known to be a heterogeneous species, it is essential to append "biogroup 2" to the name that will correspond to the second distinct species shown by DNA–DNA hybridization and biochemical tests. Unless this is done, the intended meaning of *O. proteus* is unclear. We recommend that the term *O. proteus* biogroup 1 be discontinued, and replaced by *H. alvei* biogroup 1.

O. proteus biogroup 2 is a unique genomospecies (DNA hybridization group); however, since its type strain is in a different hybridization group than that of *H. alvei*, the species and the genus (since it is the type species of the genus) would be invalid because it is a junior synonym of *H. alvei*. To rectify this, a new type strain would have to be designated. A Request for an Opinion to do this would have to be made to the Judicial Commission. If the request were granted, the genus *Obesumbacterium* and the species *O. proteus* would again be valid under the rules of nomenclature of the Bacteriologic Code. Other alternatives remain since, based on its DNA–DNA relatedness to *Escherichia blattae*, *O. proteus* could be classified in the genus *Escherichia*. Placing it in this genus would be a simple matter—designating a new type strain and describing a new combination: *"Escherichia proteus"*. An alternative would be to classify *E. blattae* and *O. proteus* biogroup 2 together, either in a redefined genus *Obesumbacterium* or as a new genus. The results from 16S rDNA sequencing (not currently available), should be helpful in deciding on the most logical classification.

FURTHER READING

Case, A.C. 1965. Conditions controlling *Flavobacterium proteus* in brewery fermentations. J. Inst. Brew. *71*: 250–256.

Priest, F.G., H.J. Somerville, J.A. Cole and J.S. Hough. 1973. The taxonomic position of *Obesumbacterium proteus*, a common brewery contaminant. J. Gen. Microbiol. *75*: 295–307.

Shimwell, J.L. 1936. A study of the common rod bacteria of brewers' yeast. J. Inst. Brew. *42*: 119–127.

List of species of the genus Obesumbacterium

1. **Obesumbacterium proteus** (Shimwell and Grimes 1936) Shimwell 1963, 759[AL] (*Flavobacterium proteum* (sic) Shimwell and Grimes 1936, 348.)

   pro′te.us. Gr. masc. n. *proteus* the ancient Greek sea-god noted for being able to change his form at will; Gr. masc. n. *proteus* pleomorphic.

   Obesumbacterium proteus is really two different species, which can be differentiated by phenotypic tests (Table BXII.γ.238) and DNA–DNA relatedness tests. The type strain has the properties of *H. alvei* (*O. proteus*) biogroup 1 (see Taxonomic Comments, above). The description of *O. proteus* biogroup 2 is as given for the genus and as listed in Table BXII.γ.193 in the chapter on the family *Enterobacteriaceae*. Occurs in breweries where it grows in beer-wort along with yeasts early in the fermentation. There is no evidence that it is pathogenic for humans or animals.

The mol% G + C of the DNA is: 48–49 (Bd).

Type strain: 42 of Strandskov and Bockelmann (1955), ATCC 12841, DSM 2777, NCIB 8771.

GenBank accession number (16S rRNA): AJ233422.

Additional Remarks: Strain No. 42 belongs to *H. alvei* (*O. proteus*) biogroup 1. Therefore, at present there is no valid type strain for *O. proteus* biogroup 2—the true representative of the species. Strains 520, 531, and 580 of Priest et al. (1973) belong to *O. proteus* biogroup 2 by DNA–DNA hybridization and are phenotypically typical (Table BXII.γ.238). Thus, they should be considered as the candidates for the proposed type strain.

Genus XXIII. **Pantoea** Gavini, Mergaert, Beji, Mielcarek, Izard, Kersters and De Ley 1989b, 343^VP emend. Mergaert, Verdonck and Kersters 1993, 171

PATRICK A.D. GRIMONT AND FRANCINE GRIMONT

Pan.toe' a. Gr. adj. *pantoios* of all sorts and sources; M.L. fem. n. *Pantoea* [bacteria] from diverse [geographical and ecological] sources.

Straight rods, 0.5–1.3 × 1.0–3.0 μm. Nonencapsulated. Nonsporeforming. Some strains form symplasmata. **Most strains are motile and are peritrichously flagellated.** Gram negative. Colonies on nutrient agar are smooth, translucent, and more or less convex with entire margins or heterogenous in consistency and adhering to the agar. **Colonies are yellow, pale beige to pale reddish yellow, or nonpigmented. Facultatively anaerobic. Oxidase negative. Glucose dehydrogenase and gluconate dehydrogenase are produced** and are active without an added cofactor. Acid is produced from the fermentation of D-fructose, D-galactose, trehalose, and D-ribose. Most strains are **Voges–Proskauer positive. Lysine and ornithine are not decarboxylated.** Urease negative. Pectate is not degraded. H$_2$S is not produced from thiosulfate. Optimum temperature 30°C. *N*-acetyl-D-glucosamine, L-aspartate, D-fructose, D-galactose, D-gluconate, D-glucosamine, D-glucose, L-glutamate, glycerol, D-mannose, D-ribose, and D-trehalose are utilized as sole sources of carbon and energy. 5-Aminovalerate, benzoate, caprate, caprylate, *m*-coumarate, ethanolamine, gentisate, glutarate, histamine, 3-hydroxybenzoate, 4-hydroxybenzoate, 3-hydroxybutyrate, itaconate, maltitol, D-melezitose, 1-*O*-methyl-α-D-glucoside, palatinose, 3-phenylpropionate, propionate, L-sorbose, tricarballylate, tryptamine, D-turanose, and L-tyrosine are not utilized as sole sources of carbon and energy. *Pantoea* spp. are isolated from plants, seeds, fruits, soils, water, and from humans (urine, blood, wounds, internal organs), and animals. Some strains are (or have been thought to be) phytopathogenic.

The mol% G + C of the DNA is: 49.7–60.6.

Type species: **Pantoea agglomerans** (Ewing and Fife 1972) Gavini Mergaert, Beji, Mielcarek, Izard, Kersters and De Ley 1989b, 343 (*Enterobacter agglomerans* (Beijerinck 1888) Ewing and Fife 1972, 10.)

FURTHER DESCRIPTIVE INFORMATION

Much of our knowledge about the characteristics of the genus *Pantoea* was acquired when all *Pantoea*, *Leclercia* and related bacteria were called *Enterobacter agglomerans*. Therefore, in this chapter, we will refer to *Pantoea agglomerans* when this precise species is meant and to the *Enterobacter agglomerans* complex when there is insufficient taxonomic information. Older data refer to *Erwinia herbicola* or the "herbicola group". Strains in the *Enterobacter agglomerans* complex (or *Erwinia herbicola* or the "herbicola group") may or may not belong to the genus *Pantoea*.

Phylogenetic treatment Of the 124 strains of the *Enterobacter agglomerans* complex analyzed by DNA–DNA hybridization by Brenner et al. (1984a), 90 formed 13 distinct DNA groups. The remaining 34 strains did not fit any of these groups. The synonymy between these groups and named species is given in Table BXII.γ.239.

Comparison of *rrs* gene (encoding 16S rRNA) sequences showed the genus *Pantoea* (represented by *Pantoea agglomerans*, *Pantoea ananatis*, and *Pantoea stewartii*) to constitute a monophyletic cluster distinct from clusters corresponding to the genera *Erwinia*, *Pectobacterium*, *Brenneria*, and other genera of the *Enterobacteriaceae* (Hauben et al., 1998a). When all DNA groups from Brenner et al. (1984a) and named *Pantoea* species are studied

TABLE BXII.γ.239. Correspondence between DNA relatedness groups, phenons, and nomenspecies in the *Enterobacter agglomerans* complex.

Gavini et al., 1983b	Brenner et al., 1984a	Verdonck et al., 1987	Nomenspecies (older synonyms)
phenon	DNA group	phenon	
–	I	–	*Pantoea* species
–	IV	–	*Pantoea* species
(B4)[a]	V	7B	*Pantoea* species
	XIII	8	*P. agglomerans* (*Enterobacter agglomerans, Erwinia herbicola, Erwinia milletiae*)
(B4)			
(B5)	II	9	*Pantoea* species
(B5)	III	10	*Pantoea dispersa*
(B9)	–	11	–
	X	ungrouped	Close to the genus *Buttiauxella*
	VI	12	*Pantoea ananatis* (*Erwinia ananas, Erwinia uredovora*)
B8			
	XII	16	Close to the *Enterobacter cloacae* complex
ungrouped			
C	–	(17)	*Rahnella aquatilis*
D1	–	(17)	Close to *Rahnella aquatilis*
ungrouped	IX	18	*Enterobacter cowanii*
	VIII	23	Close to *Enterobacter persicina* and *Erwinia rhapontici*
(B9)			
E2, E3, E5	XI	(26)	*Leclercia adecarboxylata*
	VII	(26)	Close to *Leclercia adecarboxylata* and *Enterobacter cloacae*
E4			

[a]( ) : parentheses indicate partial correspondence.

by *rrs* and *rpoB* sequence comparisons, the following congruent clusterings are obtained:

a. *P. agglomerans* (DNA group XIII), *P. ananatis* (DNA group VI), *P. stewartii*, and DNA group V cluster together.
b. DNA groups I and II and a cluster containing *Pantoea dispersa* (DNA group III) and DNA group IV branch with the above-mentioned cluster to constitute a monophyletic *Pantoea* cluster.
c. *Pantoea citrea*, *Pantoea terrea*, and *Pantoea punctata* constitute a discrete cluster that joins the *Pantoea* cluster at a lower level.
d. *E. agglomerans* DNA groups VII, VIII, IX, X, XI, and XII branch away from the *Pantoea* cluster (unpublished data).

More taxonomic work is needed to justify the assignment of *P. citrea*, *P. terrea*, and *P. punctata* to the genus *Pantoea*.

Cell morphology Symplasmata are sausage-shaped zoogloeal masses of bacteria observed by phase contrast microscopy in hanging drop preparations from the water of syneresis in glucose nutrient agar slopes after 24 h incubation (Graham and Hodgkiss, 1967). Symplasmata have been observed in bacteria now reclassified in the genus *Pantoea* ("*Bacterium herbicola*", *Erwinia lathyri*, *Erwinia uredovora*, and *Erwinia milletiae*) (Graham and Hodgkiss, 1967). The epithet "*agglomerans*" probably refers to symplasmata (Beijerinck, 1888).

Nutrition and growth conditions Best results are obtained when *Pantoea* cultures are incubated at 30°C. Most strains grow at 37°C on plating media, but some cultures fail to grow or only produce microcolonies. A few isolates can grow on brilliant green or bismuth sulfite agar media and can produce light growth of colonies that range from very small to 2.0 mm in diameter. Many strains produce mucoid colonies at 37°C, and most cultures do so when the media are incubated at room temperature, about 25°C.

On nutrient agar, anaerogenic strains of the *E. agglomerans* complex (*Pantoea* spp.) form mucoid colonies, smooth and irregularly round colonies, or rough and wrinkled colonies that are difficult to remove with a platinum wire. This latter aspect is what led to the description of "biconvex bodies", which can be observed with a low power stereoscopic binocular microscope in 2 or 3 day-old colonies. These bodies appear to be granular structures analogous to symplasmata or down-growths of the colonies into the medium (Graham and Hodgkiss, 1967).

P. agglomerans, *P. stewartii*, and *P. ananatis* colonies are usually yellow pigmented; *P. dispersa* is often pigmented; and *P. citrea*, *P. punctata*, and *P. terrea* are never pigmented. Yellow pigmentation of *P. agglomerans* may be weak at 37°C.

In Biotype-100 strips (BioMerieux, Craponne, France) or in a minimal medium containing ammonium sulfate as the nitrogen source, the following compounds serve as sole carbon sources for most strains: *N*-acetyl-D-glucosamine, L-aspartate, D-fructose, D-galactose, D-gluconate, D-glucosamine, D-glucose, L-glutamate, glycerol, D-mannose, D-ribose, and D-trehalose.

Most strains cannot utilize the following substrates as sole carbon and energy sources: 5-aminovalerate, benzoate, caprate, caprylate, *m*-coumarate, ethanolamine, gentisate, glutarate, histamine, 3-hydroxybenzoate, 4-hydroxybenzoate, 3-hydroxybutyrate, itaconate, maltitol, D-melezitose, 1-*O*-methyl-α-D-glucoside, palatinose, 3-phenylpropionate, propionate, L-sorbose, tricarballylate, tryptamine, D-turanose, and L-tyrosine.

Metabolism Most cultures produce acid from sucrose and therefore yield acid throughout in tubes of triple sugar iron agar.

P. agglomerans gives negative reactions in all three decarboxylase tests.

Under aerobic conditions, all *Pantoea* species produce gluconate from D-glucose in the presence of iodoacetate due to a glucose dehydrogenase, without added pyrroloquinoline quinone (PQQ) (Bouvet et al., 1989). A reducing compound, 2-ketogluconate, is also produced from gluconate due to a gluconate dehydrogenase (Bouvet et al., 1989). 2,5-Diketogluconate is produced from 2-ketogluconate by *Pantoea citrea*, *P. punctata*, and *P. terrea* (Kageyama et al., 1992).

Pantoea species are negative for the following tests: H_2S production from thiosulfate, tetrathionate reduction, tributyrin and corn oil hydrolysis, and β-glucuronidase.

Mutants, plasmids, phages and phage typing, bacteriocins *Pantoea agglomerans* strain Eh318 produces two antibiotics (pantocin A and B) that inhibit *Erwinia amylovora in vitro* (Wright et al., 2001). *Erwinia herbicola* produces two lipopeptide antibiotics, herbicolicin A and B, which are active against sterol-containing fungi (Greiner and Winkelmann, 1991). Four antibiotics, agglomerins A, B, C, and D, were isolated from the culture broth of a strain identified as *Enterobacter agglomerans*; these antibiotics are active against a wide variety of anaerobic bacteria *in vitro* (Shoji et al., 1989).

The eight bacteriocin-producing *Enterobacter cloacae* strains used to type *E. cloacae* isolates by bacteriocin susceptibility are useful in typing isolates identified as *Enterobacter agglomerans* (Bauernfeind and Petermüller, 1984).

Antigenic structure The principal antigens in most strains studied under the name *E. agglomerans* (*Erwinia herbicola*) are uncharged capsular polysaccharides. These antigens are not removed from the cells by heating at 100°C for 30 min (Slade and Tiffin, 1984). Most members of the "herbicola group" are motile, but there is no report on the antigenic structure of the flagella (Slade and Tiffin, 1984). Most strains produce antigenically similar high molecular weight acidic polysaccharides.

Two schemes of serotyping have been proposed for *Erwinia herbicola*. Muraschi et al. (1965) used immunodiffusion and antigens extracted by aqueous ether and classified 55 isolates into 7 serotypes, although "some cultures were mixtures of more than one serotype". A second scheme was established by Slade (cited by Slade and Tiffin, 1984), using laboratory strains. This scheme has not been used to type new isolates.

Pathogenicity Strains of the *E. agglomerans* complex may occur in clinical samples (blood, wounds, sputum, urine), often with dubious clinical significance (von Graevenitz, 1970; Gilardi and Bottone, 1971; Pien et al., 1972). In some cases (Cooper-Smith and von Graevenitz, 1978; von Graevenitz and Palermo, 1980), clinical significance has been demonstrated. In an Ohio hospital, *E. agglomerans* accounted for four of 58 episodes of *Enterobacter* bacteremia and one of 42 cases of nosocomial bacteremia (Watanakunakorn and Weber, 1989). In 1970, *E. agglomerans* was implicated in a United States-wide and a Canadian outbreak of septicemia caused by contaminated closures on bottles of infusion fluids. Twenty-five hospitals were involved with 378 cases and 40 deaths (Maki et al., 1976).

Strains identified as *Pantoea agglomerans* have been isolated from joint fluid of patients with arthritis following injuries with plant thorns, wood slivers, or wooden splinters (Flatauer and Khan, 1978; Olenginski et al., 1991; de Champs et al., 2000).

A case of metastatic endophthalmitis caused by a strain identified as *Enterobacter agglomerans* was reported. One day following internal hemorrhoidal ligation, the patient developed anterior uveitis, followed by panophthalmitis and loss of vision (Zeiter et al., 1989).

Enterobacter agglomerans was isolated from the blood of a patient with cotton fever, a benign febrile leukocytic syndrome seen in intravenous narcotic abusers. The strain was also isolated from cotton that the patient had used to filter heroin (Ferguson et al., 1993). *Enterobacter agglomerans* is known to heavily colonize cotton and cotton plants.

Antibiotic sensitivity Most *E. agglomerans* strains are naturally resistant to ampicillin and cephalothin and susceptible to many antibiotics including aminoglycosides, carbenicillin, cefamandole, cefuroxime, and cefoxitine. Resistance to carbenicillin may occur.

Pathogenesis Contaminated medicinal agents can be sources of outbreaks. Maki and Martin (1975) showed that strains identified as *E. agglomerans* (as well as strains of *E. cloacae* and *Serratia marcescens*) multiplied in 5% dextrose solution at 25°C better than did other members of the *Enterobacteriaceae*. A single tested strain of *E. agglomerans* was devoid of hemagglutinin (Adegbola and Old, 1983a). *E. agglomerans* strains produce aerobactin. Two of nine strains of *E. agglomerans* produced a hydroxamate compound other than aerobactin (Reissbrodt and Rabsch, 1988). In another study, strains identified as *Enterobacter agglomerans* produced hydroxamate siderophores identified as ferrioxamine E. The strains had also multiple siderophore receptors in the outer membrane (Berner et al., 1988).

The lipopolysaccharide from strains identified as *E. agglomerans* (commonly found in cotton dust) can bind to the pulmonary lipid-proteinaceous lining material (surfactant) and alter its surface tension properties (DeLucca et al., 1988). This binding in the lung may change the physiological properties of surfactant and be a possible mechanism for the pathogenesis of byssinosis, an occupational respiratory disorder caused by the inhalation of cotton dust (DeLucca et al., 1988).

Ecology The *Enterobacter agglomerans* complex is ubiquitous in the environment. This complex predominates on the leaf and bract of pre- and postsenescent cotton plants (DeLucca and Palmgren, 1986). The main species of Gram-negative bacteria in cotton dusts found in mills is *E. agglomerans* (Haglind et al., 1981). The following occurrences have been reviewed by Slade and Tiffin (1984). Strains of the *Enterobacter agglomerans* complex are found on the aerial surfaces of plants and within healthy plant tissues and seeds. Nitrogen fixing strains have been found in the rhizosphere of wheat and sorghum. In fact, these bacteria are typical of the innermost part of the rhizosphere of wheat (Kleeberger et al., 1983). Nitrogen-fixing strains identified as *Enterobacter agglomerans* have also been isolated from gut of the wood-eating termite *Coptotermes formosanus* (Potrikus and Breznak, 1977). Strains of this complex have been isolated from water, paper mill process water, soil and decaying wood. They are frequently isolated from damaged plant tissues and lesions, although they are rarely considered pathogenic. A pathogenic role has been shown in some instances. Strains named *Erwinia milletiae* produce β-indolyl acetic acid (a plant hormone) causing galls on some plants in Japan (reviewed by Slade and Tiffin, 1984). Some strains initiate freezing of buffer solutions (which would normally freeze at temperatures below −10°C) at about −4°C. This ice-nucleating property plays a critical role in causing frost damage on plants (Kozloff et al., 1983; Lindow et al., 1978).

Pink disease of pineapple, caused by *Pantoea citrea*, is characterized by a dark coloration on fruit slices after autoclaving (Cho et al., 1980). This coloration is initiated by the oxidation of glucose to gluconate, then the oxidation of gluconate to 2-ketogluconate, and finally the production of 2,5-diketogluconate. The latter appears to be responsible for the dark color characteristic of the pink disease of pineapple (Pujol and Kado, 2000).

Some strains of the *Enterobacter agglomerans* complex are associated with stalk and leaf necrosis of onion (Hattingh and Walters, 1981). Strains originally named *Erwinia uredovora* (now *Pantoea ananatis*) attack the uredia of rust (*Puccinia* spp.) on wheat, oats and rye (Pon et al., 1954; Dye, 1969b).

Stewart's bacterial wilt is a disease of corn (*Zea mays*) caused by *Pantoea stewartii* subsp. *stewartii*. On corn seedlings at the three- to five-leaf stage, the disease is characterized by water-soaked lesions on leaves, leading to stunted plants with severe yield reductions in susceptible corn hybrids (Dillard and Kline, 1989; Wilson et al., 1994). The corn flea beetle, *Chaetocnema pulicaria*, is the overwintering host and vector of *P. stewartii*, and the abundance of the primary inoculum is related to corn flea beetle populations. Many countries ban import of seed corn unless it is certified free of *P. stewartii*. A ligase chain reaction assay has been proposed for the detection of *P. stewartii* in infected plant and vector material (Wilson et al., 1994).

P. stewartii subsp. *stewartii* produces an extracellular heteropolysaccharide capsule, which plays several roles in the development of Stewart's wilt on sweet corn. The capsule provides a barrier that protects the bacterium against plant host defense factors (Braun, 1982; Beck von Bodman and Farrand, 1995). It also partially contributes to the induction of water-soaking symptoms early in the development of Stewart's wilt, and it obstructs the free flow of water in the host vascular system, causing necrosis and wilting during the systemic phase of the infection (Braun, 1982; Beck von Bodman and Farrand, 1995). *P. stewartii* subsp. *stewartii* synthesizes *N*-(-3-oxohexanoyl)-L-homoserine lactone, which is an autoinducer for capsular polysaccharide biosynthesis and *P. stewartii* pathogenicity (Beck von Bodman and Farrand, 1995).

A strain identified as *Pantoea agglomerans* was isolated from necrotic spots in the leaves of a beach pea (*Lathyrus maritimus*) that grew on the shorelines of Newfoundland, Canada. The bacterium produced cellulase and amylase and was a plant wound parasite (Khetmalas et al., 1996).

Some strains of the *E. agglomerans* complex have been used for biological control of plant pathogens (such as *Erwinia amylovora* or *Xanthomonas oryzae*), either by competition for nutrients, acid production, bacteriocin production or phage production (reviewed by Slade and Tiffin, 1984). *Pantoea agglomerans* CFA-2 is effective for the biological control of postharvest pear diseases due to *Botrytis cinerea*, *Penicillium expansum*, and *Rhizopus stolonifer* (Nunes et al., 2001).

Pantoea agglomerans in fecal pellets of locusts produces large amounts of guaiacol and small amounts of phenol, both of which are components of the locust cohesion pheromone (Dillon et al., 2002).

ENRICHMENT AND ISOLATION PROCEDURES

All media designed for the isolation of *Enterobacteriaceae* can be used for the isolation of *Pantoea* species: MacConkey agar, Drigalski lactose agar, Hektoen agar, deoxycholate lactose citrate agar, etc. *Pantoea* strains can also grow on media for general use, such as blood agar, nutrient agar, tryptic soy agar, bromocresol purple lactose agar, etc. Media specifically selective for *Pantoea* strains are not available.

A differential medium, lysine-ornithine-mannitol agar containing vancomycin, was proposed for the isolation of *Pantoea agglomerans*. It yields colorless colonies from mannitol-negative strains, yellow colonies from mannitol-positive, ornithine- and lysine-decarboxylase negative strains, and greenish-blue colonies from mannitol-positive strains that produce one or both decarboxylases (Bucher and von Graevenitz, 1982).

MAINTENANCE PROCEDURES

Strains are initially grown on tryptic soy agar at their optimum temperature. They are then inoculated by stabbing a maintenance medium[1] designed for maintenance of *Enterobacteriaceae* and related organisms. The cultures are then stored at room temperature in a dark, dry place.

Cultures may be also preserved by freeze-drying. Freeze-drying is the best procedure for preservation of pigmented strains.

DIFFERENTIATION OF THE GENUS *PANTOEA* FROM OTHER GENERA

Enterobacter strains that are negative for lysine and ornithine decarboxylases and arginine dihydrolase and that may or may not produce a yellow pigment were formerly identified as *Enterobacter agglomerans*. The genus *Pantoea* is a subset of the *Enterobacter agglomerans* complex. Species of *Pantoea* are able to oxidize D-glucose to D-gluconate (glucose dehydrogenase activity) without added pyrroloquinoline quinone, and D-gluconate to 2-ketogluconate (gluconate dehydrogenase activity). In addition, they cannot utilize phenylacetate, and most strains can utilize *myo*-inositol.

TAXONOMIC COMMENTS

The genus *Erwinia* has long been a depository for plant-associated members of the family *Enterobacteriaceae*. Several species of this genus were found to be phenotypically similar to *Enterobacter* species (*Erwinia herbicola, Erwinia ananatis, Erwinia uredovora, Erwinia milletiae, Erwinia dissolvens,* and *Erwinia nimipressuralis*) (Lelliott, 1974; Lelliott and Dickey, 1984), or highly related to *Enterobacter* species by DNA hybridization (*Erwinia dissolvens* and *Erwinia nimipressuralis*) (Steigerwalt et al., 1976).

The (invalid) name *"Bacterium herbicola aureum"* was introduced by Düggeli (1904) and *"Bacterium herbicola"* is attributed to Löhnis (1911). The species was transferred to the genus *Erwinia* as *E. herbicola* by Dye (1964). It has always been considered a saprophyte associated with plants (epiphyte).

The earliest synonym of *Erwinia herbicola*, represented by an extant culture, is *"Pseudomonas trifolii"* (Hüss 1907) shown by Dye (1964) to be identical with *Erwinia herbicola*. Unfortunately, the epithet *"trifolii"* did not appear in the Approved Lists, although it had priority over the epithet *herbicola*.

Manns and Taubenhaus (1913) described *"Bacillus lathyr"* as the cause of streak disease of sweet peas (now attributed to a virus). The species was transferred to the genus *Erwinia* as *E. lathyri* by Holland (1920).

"Bacillus milletiae" was described by Kawakama and Yoshida (1920) as causing galls on millet. The species was transferred to the genus *Erwinia* as *E. milletiae* by Magrou (1937).

"Bacillus ananas" was isolated from bacterial fruitlet brown-rot of pineapple in the Philippines by Serrano (1928). It is given as *Erwinia ananas* Serrano 1928 on the Approved Lists of Bacterial Names (Skerman et al., 1989).

"Xanthomonas uredovorus" was isolated from uredia of cereal rusts (Pon et al., 1954) and reclassified in the genus *Erwinia* by Dye (1963c).

Graham (1958) reported that the pathogenic properties of *E. ananas* and *E. milletiae* are doubtful and that culturally and biochemically these species were indistinguishable from *E. lathyri*.

Dye (1969b) proposed a classification of the "herbicola group" containing *Erwinia herbicola* subsp. *herbicola* (18 species names were listed as synonyms including *E. herbicola, E. lathyri,* and *E. milletiae*), *Erwinia herbicola* subsp. *ananas, E. uredovora* and *E. stewartii*.

"Pseudomonas stewartii" was isolated by Smith (1898) from Stewart's sweet corn wilt and was transferred to *Erwinia* as *E. stewartii* by Dye (1963b).

After studying many cultures of the "herbicola-lathyri" group, including clinical isolates, Ewing and Fife (1972) exhumed an old work by Beijerinck (1888) describing *"Bacillus agglomerans"*, a bacterium that formed what might have been symplasmata. Although no culture was available, a new combination, *Enterobacter agglomerans* (Beijerinck) Ewing and Fife 1972, was proposed for the "herbicola-lathyri" group on the argument of priority. However, the Approved Lists gave the name as *Enterobacter agglomerans* Ewing and Fife 1972, thus losing the connection with Beijerinck (1888). The other synonyms found in the Approved Lists are *Erwinia herbicola* (Löhnis 1911) Dye 1964 and *Erwinia milletiae* (Kawakama and Yoshida 1920) Magrou 1937.

E. agglomerans is a very complex group of bacteria, which may cause opportunistic infections. The name covers many (20 to 40) genomic groups (Brenner et al., 1984a) or phena (Gavini et al., 1983b; Verdonck et al., 1987). In addition to this diversity, strains of the *E. agglomerans* complex are not closely related to *E. cloacae* (the type species of the genus *Enterobacter*) by DNA relatedness. Some groups in this complex have been designated as new genera (*Rahnella aquatilis, Ewingella americana, Leclercia adecarboxylata*). There is an apparent confusion in the literature, and a close examination of published papers for the presence of commonly studied strains is needed to extract convergent pieces of information. A table relating phena and DNA groups has been created after such a literature analysis (Table BXII.γ.239).

The numerical study of Gavini et al. (1983b), based on 169 strains, yielded five phena (A to E). Phenon A corresponded to *Erwinia carotovora*. Phenon B, which included strains of the *Enterobacter agglomerans* complex, was split into nine smaller phena (B1 to B9). Phenon B4 contained the type strains of *Erwinia herbicola* and *Erwinia milletiae*. Phenon B8 contained the type strain of *Erwinia ananas* and a reference strain of *Erwinia uredovora*. Phenon C corresponded to *Rahnella aquatilis*. Phenon D was split into three smaller phena (D1 to D3), with phenon D2 corresponding to *Enterobacter sakazakii*. Phenon E was split into five smaller phena (E1 to E5), with phenon E5 containing strains previously identified as *Escherichia adecarboxylata* (now *Leclercia adecarboxylata*).

A larger numerical study (Verdonck et al., 1987), based on 529 strains and including many type and reference strains, distributed 66 strains of the *Enterobacter agglomerans* complex into 21 phena. The correspondence between phena from both studies

---

1. Maintenance medium (g/liter): Bacto-peptone (Difco), 10.0, NaCl, 5.0: Bacto-agar (Difco), 10.0; pH 7.4. The medium should be dispensed into small (9.5–10 × 90 mm) screw-capped tubes.

(Gavini et al., 1983b; Verdonck et al., 1987) is given in Table BXII.γ.239.

The DNA relatedness work of Brenner et al. (1984a) revealed the extreme genomic diversity of the *Enterobacter agglomerans* complex: of 124 strains studied, 90 fell into 13 DNA groups (I to XIII) and 34 strains did not fit into any group. Furthermore, four groups (V, XI, XII, and XIII) were heterogeneous with respect to $\Delta(T_m)$ values. An interesting finding was that aerogenic strains and anaerogenic strains were not found in the same hybridization group.

Lind and Ursing (1986) identified 52 of 86 clinical isolates with *Enterobacter agglomerans sensu strictu* by DNA hybridization. In the same study, they demonstrated the synonymy of *Enterobacter agglomerans*, *Erwinia herbicola*, and *Erwinia milletiae*. This synonymy was confirmed by Beji et al. (1988) who, in addition, identified DNA group XIII (Brenner et al., 1984a) with *Enterobacter agglomerans sensu strictu*.

To separate *Enterobacter agglomerans/Erwinia herbicola* from the genera *Enterobacter* and *Erwinia*, a new genus, *Pantoea*, was proposed with *P. agglomerans* as the type species (Gavini et al., 1989b). The epithet *agglomerans* was retained because it was believed to have priority over epithet *herbicola*. Strains of DNA group III (Brenner et al., 1984a) or phenon 8 (Verdonck et al., 1987) were proposed as a new species, *Pantoea dispersa* (Gavini et al., 1989b). The DNA groups closest to *Pantoea agglomerans* are DNA groups II, III, IV, V, and VI (Lind and Ursing, 1986). It is interesting that strains of group II to VI and XIII are characterized by the presence of a glucose oxidation pathway that produces 2-ketogluconate from glucose (Bouvet et al., 1989). This finding has been extended to DNA group I (P.A.D. Grimont, unpublished observations). Thus, the genus *Pantoea* can be envisioned to include DNA groups I, II, IV, V, and VI in addition to groups XIII (*P. agglomerans*) and III (*P. dispersa*).

Erwinia ananas (Serrano 1928) corresponds to group VI of Brenner et al. (1984a) and was transferred to the genus *Pantoea*

as *P. ananas* (Mergaert et al., 1993). The name was corrected to *P. ananatis* by Trüper and De' Clari, 1997). *Erwinia uredovora* belongs to that species.

Erwinia stewartii was included in the genus *Pantoea* as *P. stewartii* subsp. *stewartii* (Smith 1898) Mergaert et al. 1993. *P. stewartii* subsp. *indologenes* was a new subspecies composed of strains phenotypically different from *Erwinia stewartii* (and *P. stewartii* subsp. *stewartii*) and resembling *P. ananatis*. However, DNA hybridization and sequence studies indicate that both subspecies should belong to the same species.

The three species (*Pantoea punctata*, *P. citrea*, and *P. terrea*) described by Kageyama et al. (1992) have the ability to oxidize 2-ketogluconate to 2,5-diketogluconate. These species differ from the other *Pantoea* species in several biochemical or nutritional characteristics. Their phylogenetic position is as a branch on the border of the *Pantoea* cluster.

DNA group XI (Brenner et al., 1984a) was identified with *Escherichia adecarboxylata* by DNA relatedness (Izard et al., 1985). *Escherichia adecarboxylata* was transferred to a new genus *Leclercia*, as *L. adecarboxylata* (Tamura et al., 1986). DNA group VII (Brenner et al., 1984a) was close to (but distinct from) *Escherichia adecarboxylata* (*Leclercia adecarboxylata*) (Izard et al., 1985). In our laboratory, strains of DNA group VII were indistinguishable from strains of phenon E4 (Gavini et al., 1983b) by carbon source utilization tests (P.A.D. Grimont, unpublished observations). Thus, DNA group VII is a good candidate as a new species of *Leclercia*.

DNA group VIII (Brenner et al., 1984a) was 64% related to *Enterobacter persicina* (a close relative of *Erwinia rhapontici*).

DNA group IX (Brenner et al., 1984a) is identical with *Enterobacter cowanii* based on nutritional tests and *rpoB* sequence analysis (unpublished data).

More work is needed to characterize unclassified strains of the *E. agglomerans* complex.

DIFFERENTIATION OF THE SPECIES OF THE GENUS *PANTOEA*

The characteristics of named or unnamed species of *Pantoea* are given in Table BXII.γ.240 . The results of carbon source utilization tests given in Table BXII.γ.240 are mostly unpublished observations (P.A.D. Grimont and E. Ageron).

The genus *Pantoea* can be divided into two groups of species: (1) the *Pantoea* core group with *P. agglomerans*, *P. dispersa*, *P. ananatis* , and *P. stewartii* and (2) the "Japanese" group with *P. citrea*, *P. punctata*, and *P. terrea*. Species of the core group utilize the following substrates as sole carbon sources which the "Japanese"

group cannot utilize: D-alanine, L-alanine, *myo*-inositol, DL-lactate, 1-*O*-methyl-β-D-glucoside, and L-serine. Furthermore, all species of the core group utilize L-arabinose and D-mannitol, whereas two species of the "Japanese" group fail to utilize these substrates. Species of the core group are negative for arginine dihydrolase and 2-ketogluconate dehydrogenase, whereas two species of the "Japanese" group are positive for arginine dihydrolase and all three species are positive for 2-ketogluconate dehydrogenase.

List of species of the genus Pantoea

1. **Pantoea agglomerans** (Ewing and Fife 1972) Gavini Mergaert, Beji, Mielcarek, Izard, Kersters and De Ley 1989b, 343[VP] (*Enterobacter agglomerans* (Beijerinck 1888) Ewing and Fife 1972, 10.)

   ag.glo' mer.ans. L. v. *agglomerare* to form into a ball; L. part. adj. *agglomerans* forming into a ball (referring to the occurrence of symplasmata bacteria in aggregates surrounded by a translucent sheath in anaerogenic strains).

   The species has all the characteristics given for the genus. Cells may form symplasmata. Strains grow well on nutrient agar at 30°C but not at 44°C. Some strains grow slowly at 4 or 41°C. The biochemical and nutritional characteristics are shown in Table BXII.γ.240. Key characteristics are: alkaline reaction in malonate broth; utilization of D-glucuronate and D-tartrate as sole carbon sources; inability to utilize *meso*-erythritol, gentiobiose, 5-ketogluconate, D-me-

TABLE BXII.γ.240. Characteristics of the named and unnamed species of the genus *Pantoea*.[a]

Characteristic	*Pantoea agglomerans*	*Pantoea ananatis*	*Pantoea citrea*	*Pantoea dispersa*	*Pantoea punctata*	*Pantoea stewartii* subsp. *stewartii*	*Pantoea stewartii* subsp. *indologenes*	*Pantoea terrea*	DNA group I	DNA group II	DNA group IV	DNA group V
Motility (36°C)	+	+	−	+	−	−	d	+	ND	d	d	d
Yellow pigment	d	+	−	d	−	+	+	−	ND	d	d	d
Indole production	−	+	−	−	−	−	+	−	−	−	−	−
Malonate (Leifson)	+	−	ND	−	ND	−	−	ND	−	d	−	d
β-Xylosidase test	d	−	ND	−	ND	ND	ND	ND	+	−	−	+
Voges–Prokauer	+	+	+	d	+	d	d	+	ND	d	d	d
Gelatin hydrolysis at 22°C	(+)	+	−	(d)	−	−	−	−	ND	d	d	+
Arginine dihydrolase	−	−	+	−	+	−	−	−	−	−	−	−
Phenylalanine deaminase	d	−	−	−	−	−	−	−	ND	+	d	d
Glucose dehydrogenase	+	+	+	+	+	ND	ND	+	+	+	+	+
Gluconate dehydrogenase	+	+	+	+	+	ND	ND	+	+	+	+	+
2-ketogluconate dehydrogenase	−	−	+	−	+	ND	ND	+	ND	−	−	−
Esculin hydrolysis	+	d	−	d	d	−	+	+	+	d	d	d
Nitrate reduced	+	d	+	d	+	−	−	+	ND	+	d	d
ONPG hydrolyzed	+	+	+	+	−	+	+	−	ND	ND	ND	ND
Acid from:												
L-Arabinose	+	+	ND	d	ND	+	+	ND	ND	+	+	+
D-Arabitol	+	+	d	+	−	−	+	−	ND	ND	ND	ND
Cellobiose	d	+	−	d	−	−	+	−	ND	d	d	d
Dulcitol	−	−	−	−	−	−	−	−	ND	−	−	d
meso-Erythritol	−	−	d	−	−	−	−	−	ND	ND	ND	ND
Glycerol	(d)	+	ND	(d)	ND	−	d	ND	ND	d	−	d
myo-Inositol	(d)	+	ND	d	ND	−	+	ND	ND	d	d	d
Lactose	d	+	+	d	−	−	+	−	+	d	d	d
Maltose	+	+	ND	+	ND	−	+	ND	ND	ND	ND	ND
D-Mannitol	+	+	+	+	−	+	+	−	ND	+	+	+
Melibiose	−	d	+	d	+	+	+	+	ND	ND	ND	ND
Raffinose	(d)	+	ND	−	ND	+	+	ND	ND	−	−	d
L-Rhamnose	+	d	−	+	−	−	d	−	ND	+	d	d
Salicin	+	+	d	(d)	−	−	+	+	ND	d	d	d
D-Sorbitol	−	+	d	−	−	−	−	d	ND	−	−	d
Sucrose	+	+	ND	+	ND	+	+	ND	ND	d	+	d
Trehalose	+	+	ND	+	ND	+	+	ND	ND	ND	ND	ND
D-Xylose	+	+	+	+	−	+	+	+	ND	+	d	+
Utilization of:												
trans-Aconitate	+	+	+	+	−	−	+	−	+	d	+	+
Adonitol	−	−	−	−	−	−	−	−	−	−	+	−
L-Arabinose	+	+	+	+	−	+	+	−	+	+	+	+
D-Arabitol	+	+	−	+	−	−	+	−	−	d	+	d
L-Arabitol	−	−	−	−	−	−	−	−	−	−	+	-
Betaine	−	−	−	−	−	−	+	−	−	−	−	−
Cellobiose	(d)	+	−	+	(+)	d	+	−	+	d	+	+
Citrate	d	+	+	+	(+)	−	+	+	+	d	+	+
Dulcitol	−	−	−	(d)	−	−	−	−	−	(d)	−	−
meso-Erythritol	−	−	+	+	−	−	−	−	−	−	d	−
L-Fucose	−	−	−	−	−	−	−	−	−	d	−	d
D-Galacturonate	(+)	+	−	+	−	d	+	−	+	+	d	+
Gentiobiose	−	+	−	+	+	−	d	−	+	d	d	−
D-Glucuronate	(+)	+	−	+	−	−	+	−	+	+	+	+
myo-Inositol	+	+	−	+	−	+	+	−	+	d	+	+
5-Ketogluconate	−	(+)	+	+	(+)	−	+	+	+	d	+	(d)
Lactose	−	+	(+)	−	−	d	+	−	+	d	−	(d)
Lactulose	−	+	+	−	−	d	+	−	+	d	−	d
D-Malate	(+)	d	+	(d)	−	+	+	−	(+)	(+)	−	(+)
Maltose	(+)	+	+	+	−	+	+	−	−	+	d	+
Maltotriose	(+)	+	+	+	−	+	+	−	−	+	d	+
D-Melibiose	−	+	−	−	−	+	+	−	−	d	−	−
1-*O*-Methyl-α-galactoside	−	(+)	−	−	−	+	+	−	−	d	−	−
3-*O*-Methyl-D-glucose	−	(d)	−	−	−	−	+	−	−	−	−	−

(continued)

TABLE BXII.γ.240. *(cont.)*

Characteristic	*Pantoea agglomerans*	*Pantoea ananatis*	*Pantoea citrea*	*Pantoea dispersa*	*Pantoea punctata*	*Pantoea stewartii* subsp. *stewartii*	*Pantoea stewartii* subsp. *indologenes*	*Pantoea terrea*	DNA group I	DNA group II	DNA group IV	DNA group V
1-*O*-Methyl-β-D-glucoside	(+)	+	−	+	−	+	+	−	+	+	+	+
Protocatechuate	−	+	−	−	−	−	+	−	−	−	−	−
Quinate	−	(+)	−	−	−	−	+	−	−	−	−	−
D-Raffinose	−	+	−	−	−	+	+	−	−	d	−	d
L-Rhamnose	+	d	−	+	−	d	d	−	+	+	d	+
D-Saccharate	(+)	(+)	−	+	−	−	+	−	+	+	+	+
D-Sorbitol	−	d	−	−	−	−	−	−	+	(d)	−	−
Sucrose	(+)	+	−	+	(+)	(+)	+	−	−	d	+	+
D-Tagatose	−	−	+	−	−	−	−	−	−	−	−	−
D-Tartrate	(+)	−	−	−	−	−	−	−	−	−	−	−
L-Tartrate	−	−	−	d	−	−	−	−	−	d	−	d
meso-Tartrate	(+)	d	−	d	−	+	+	−	−	d	−	(+)
Trigonelline	−	−	−	−	(d)	−	−	−	−	d	+	−
Xylitol	−	−	−	−	−	−	−	−	−	d	+	−
D-Xylose	+	+	−	+	−	+	+	−	+	+	−	+

[a]Symbols: +, 90–100% of strains positive in 1–2 days; (+), 90–100% of strains positive in 1–4 days; −, 90–100% of strains negative in 4 days; d, positive in 1–4 days; (d), positive in 3–4 days; ND, no data.

libiose and D-raffinose as sole carbon sources; and no production of indole and 2-ketogluconate dehydrogenase.

Isolated from plants, flowers, seeds, vegetables, water, soil and foodstuffs. Some strains are of human (wounds, blood, urine, internal organs) and animal origin. Some strains (synonym, *Erwinia milletiae*) have been reported to cause galls on *Wisteria floribunda* and *Wisteria japonica*; some strains cause galls on *Gypsophyla paniculata*, and some strains cause stalk and leaf necrosis on onion plants.

The mol% G + C of the DNA is: 55.1–56.8 (T_m).

Type strain: ATCC 27155, CCUG 539, CFBP 3845, CIP 57.51, DSM 3493, ICMP 12534, JCM 1236, LMG 1286, NCTC 9381.

GenBank accession number (16S rRNA): AB004691, AJ233423.

2. **Pantoea ananatis** corrig. (Serrano 1928) Mergaert, Verdonck and Kersters 1993, 170[VP] (*"Bacillus ananas"* Serrano 1928, 271; *Erwinia ananas* (sic) Serrano 1928; *Pantoea ananas* (sic) Mergaert, Verdonck and Kersters 1993, 170.)

a'na.na.tis. M.L. n. *ananas* generic name of the pineapple.

The species has all the characteristics given for the genus. The biochemical and nutritional characteristics are given in Table BXII.γ.240. Key characteristics are: indole production; no alkaline reaction in malonate broth; utilization of gentiobiose, D-glucuronate, 5-ketogluconate, lactose, lactulose, D-melibiose, 1-*O*-methyl-α-galactoside, protocatechuate, quinate and D-raffinose as sole carbon sources; utilization of D-sorbitol by most strains; and no utilization of *meso*-erythritol and D-tartrate.

Strains formerly classified as *Erwinia uredovora* show a stronger proteinase activity and produce a DNase.

The mol% G + C of the DNA is: 53.6–56.4 (T_m).

Type strain: ATCC 33244, CFBP 3612, CIP 105207, LMG 2665, NCPPB 1846.

GenBank accession number (16S rRNA): U80196.

3. **Pantoea citrea** Kageyama, Nakae, Yagi and Sonoyama 1992, 209[VP]

ci' tre.a. M.L. adj. *citrea* of citrus.

The cell morphology and colonial morphology are as given for the genus. The cells are nonmotile. Either nicotinic acid or nicotinamide is required for growth. Good growth occurs at 20–34°C; no growth occurs at 41°C. Colonies grown on nutrient agar at 30°C for 2 d are pale beige to pale reddish yellow. Physiological and nutritional characteristics are presented in Table BXII.γ.240. Key characteristics are: production of 2-ketogluconate dehydrogenase; positive arginine dihydrolase reaction; hydrolysis of ONPG but not esculin; utilization of L-arabinose, *meso*-erythritol, maltose, and D-tagatose as sole carbon and energy sources; and no utilization of gentiobiose. Isolated from mandarin oranges.

The mol% G + C of the DNA is: 49.7 (HPLC).

Type strain: SHS 2003, ATCC 31623, CIP 105599, CCUG 30156, DSM 13699.

4. **Pantoea dispersa** Gavini, Mergaert, Beji, Mielcarek, Izard, Kersters and De Ley 1989b, 344[VP]

dis.per' sa. L. v. *dispergere* to spread, to scatter; L. fem. part. adj. *dispersa* spread, scattered.

The species has all the characteristics of the genus. Strains grow well on nutrient agar at 30 and 41°C but not at 44°C or 4°C. Key characteristics are: no alkaline reaction in malonate broth; utilization of *meso*-erythritol, gentiobiose, D-glucuronate and 5-ketogluconate as sole carbon sources; no utilization of D-melibiose, D-raffinose, and D-tartrate as sole carbon sources; and no production of indole and 2-ketogluconate dehydrogenase. Isolated from plant surfaces, seeds, humans, and the environment.

The mol% G + C of the DNA is: 56.5–60.6 (T_m).

Type strain: ATCC 14589, CCUG 25232, CIP 103338, DSM 30073, LMG 2603.

5. **Pantoea punctata** Kageyama, Nakae, Yagi and Sonoyama 1992, 209^{VP}

punc.ta' ta. L. n. *punctum* a point; M.L. adj. *punctata* full of points.

The cell morphology and colonial morphology are as given for the genus. Cells are nonmotile. Either nicotinic acid or nicotinamide is required for growth. Good growth occurs at 20–34°C; no growth occurs at 41°C. Colonies grown on nutrient agar at 30°C for 2 d are pale beige to pale reddish yellow. Yellow pigment is not produced on nutrient agar. Physiological and nutritional characteristics are presented in Table BXII.γ.240. Key characteristics are: production of 2-ketogluconate dehydrogenase; positive arginine dihydrolase reaction; hydrolysis of esculin by some strains; no hydrolysis of ONPG; utilization of gentiobiose as a sole source of carbon and energy; and no utilization of L-arabinose, *meso*-erythritol, maltose, and D-tagatose. Isolated from mandarin oranges.

The mol% G + C of the DNA is: 50.0–50.3 (HPLC).

Type strain: SHS 2006, ATCC 31626, CIP 105598, DSM 13700.

6. **Pantoea stewartii** (Smith 1898) Mergaert, Verdonck and Kersters 1993, 170^{VP} (*"Pseudomonas stewartii"* Smith 1898, 422; *Erwinia stewartii* (Smith 1898) Dye 1963b, 504.)

stew.art' i.i. M.L. en. n. *stewartii* of Stewart; named after F.C. Stewart.

The species has all the characteristics given for the genus, except that some strains produce a capsular polysaccharide. The species is divided into two subspecies that differ strongly in vigor. Previous information about the characteristics of *Erwinia stewartii* applies only to *P. stewartii* subsp. *stewartii*, not to *P. stewartii* subsp. *indologenes*, which can hardly be differentiated from *P. ananatis* by biochemical and nutritional properties. Common characteristics of the species are listed in Table BXII.γ.240. Key characteristics are: production of a yellow pigment; no alkaline reaction in malonate broth; no reduction of nitrate to nitrite; acid production from L-arabinose, sucrose, and D-mannitol but not from D-sorbitol; utilization of L-alanine, L-arabinose, *myo*-inositol, DL-lactate, D-mannitol, D-melibiose, 1-*O*-methyl-α-galactoside, 1-*O*-methyl-β-galactoside, 1-*O*-methyl-β-D-glucoside, D-raffinose, L-serine, sucrose, *meso*-tartrate, and D-xylose as sole sources of carbon and energy; and no utilization of *meso*-erythritol and D-sorbitol. Isolated from *Zea mays*, several other grasses, *Ananas comosus*, and beetles.

The mol% G + C of the DNA is: 53.6–56.4 (T_m).

Type strain: ATCC 8199, CIP 104005, DSM 30176, LMG 2715, NCPPB 2295.

a. **Pantoea stewartii** *subsp.* **stewartii** (Smith 1898) Mergaert, Verdonck and Kersters 1993, 170^{VP} (*"Pseudomonas stewartii"* Smith 1898, 422; *Erwinia stewartii* (Smith 1898) Dye 1963b, 504.)

The biochemical and nutritional characteristics are given in Table BXII.γ.240. A few strains grow at 4°C. Some strains grow at 37°C, but no strains grow at 41°C. Key characteristics are: lack of motility; no production of indole; no hydrolysis of esculin; no acid production from D-arabitol, cellobiose, *myo*-inositol, lactose, maltose, L-rhamnose and salicin; and no utilization of D-arabitol, betaine, citrate, D-glucuronate, 5-ketogluconate, maltose, 3-*O*-methyl-D-glucose, protocatechuate, quinate, and D-saccharate as sole carbon and energy sources.

Causative agent of Stewart's bacterial wilt of corn, a vascular disease of *Zea mays*. Also isolated from the insect vector, the corn flea beetle *Chaetocnema pulicaria*.

The mol% G + C of the DNA is: 54.6–55.1 (T_m).

Type strain: ATCC 8199, CIP 104005, DSM 30176, LMG 2715, NCPPB 2295.

GenBank accession number (16S rRNA): Z96080.

b. **Pantoea stewartii** *subsp.* **indologenes** Mergaert, Verdonck and Kersters 1993, 171^{VP}

in.do.lo' gen.es. M.L. n. *indolum* indole; Gr. n. *gennao* to produce; M.L. adj. *indologenes* indole producing.

The biochemical and nutritional characteristics are given in Table BXII.γ.240. Many strains grow at 4 and 37°C and some strains grow at 41°C, but few strains grow at 44°C. Key characteristics are: motility; production of indole; hydrolysis of esculin; acid production from D-arabitol, cellobiose, *myo*-inositol, lactose, maltose, D-rhamnose and salicin; and utilization of L-arabitol, betaine, citrate, D-glucuronate, 5-ketogluconate, maltose, 3-*O*-methyl-D-glucose, protocatechuate, quinate and D-saccharate as sole carbon and energy sources.

Thought to cause leaf spot on foxtail millet (*Setaria italica*) and pearl millet (*Pennisetum americanum*) and rot of *Ananas comosus*. Also isolated from cluster bean (*Cyamopsis tetragonolobus*).

The mol% G + C of the DNA is: 56.4 (T_m).

Type strain: ATCC 51785, CIP 104006, LMG 2632.

GenBank accession number (16S rRNA): Y13251.

7. **Pantoea terrea** Kageyama, Nakae, Yagi and Sonoyama 1992, 210^{VP}

ter' re.a. L. n. *terra* soil; L. adj. *terrea* of soil.

The cell morphology and colonial morphology are as given for the genus. The cells are motile by means of one or two lateral flagella. Colonies grown on nutrient agar at 30°C for 2 d are pale beige to pale reddish yellow. Yellow pigment is not produced on nutrient agar. Physiological and nutritional characteristics are presented in Table BXII.γ.240. Key characteristics are: production of 2-ketogluconate dehydrogenase; negative arginine dihydrolase reaction; hydrolysis of esculin but not ONPG; and no utilization of L-arabinose, *meso*-erythritol, gentiobiose, maltose, and D-tagatose as sole carbon and energy sources. Isolated from soil in Japan

The mol% G + C of the DNA is: 51.0–51.9 (HPLC).

Type strain: SHS 2008, ATCC 31628, CCUG 30161, CIP 105600, DSM 13701.

Genus XXIV. **Pectobacterium** Waldee 1945, 469[AL] emend. Hauben, Moore, Vauterin, Steenakcers, Mergaert, Verdonck and Swings 1999a, 1

LYSIANE HAUBEN, FREDERIQUE VAN GIJSEGEM AND JEAN SWINGS

Pec.to.bac.te'ri.um. Gr. dim. neut. n. bakterion a small rod; M.L. neut. n. Pectobacterium a pectolytic bacterium.

Straight rods, 0.5–1.0 × 1.0–3.0 μm, rounded ends; occur singly or in pairs. Gram-negative. Motile by peritrichous flagella. Fermentative. Facultatively anaerobic. Optimum growth temperature, 27–30°C; maximum temperature for growth is 40°C. Oxidase negative. Catalase positive. Acid is produced from **N-acetylglucosamine**, fructose, D-galactose, D-glucose, D-mannose, L-rhamnose, D-ribose, salicin, and sucrose but not from adonitol, L-arabitol, D-lyxose, α-methylmannoside, L-sorbose, starch, or D-tagatose. Utilize acetate, arbutin, fructose, fumarate, D-galactose, D-glucose, glycerol, malate, D-mannitol, mannose, β-methylglucoside, ribose, salicin, succinate, and sucrose but not adipate, benzoate, betaine, butanol, gallate, methanol, oxalate, propionate, or sorbose as carbon and energy-yielding sources. Utilize L-alanine, allantoin, γ-aminobutanic acid, ammonium chloride, arginine, asparagine, asparaginic acid, citrulline, glucosamine, glutamine, L-glutaminic acid, glutathione, glycine, glycylglycine, histidine, leucine, L-methionine, phenylalanine, L-serine, L-tryptophan, tyrosine, and urea, but not anthranilic acid, betaine, choline, cysteamine, hydroxyproline, kynureninic acid, quinolinic acid, sarcosine, spermidine, spermine, trigonelline, or trimethylammonium as nitrogen sources. No decarboxylases are formed for arginine, lysine or ornithine. Do not possess tryptophan deaminase or urease. **Hydrolyze esculin but not starch.**

The species of the genus *Pectobacterium* comprise a distinct phylogenetic group, as determined by 16S rRNA gene sequence comparisons, and are characterized by 17 signature nucleotides (Table BXII.γ.223 in Genus *Erwinia*).

Pectobacterium species cause plant diseases that include blights, cankers, die back, leaf spots, wilts, discoloration of plant tissues, and especially soft rots variously described as stalk rot, crown rot, stem rot, or fruit collapse. Ingress by the pathogen generally occurs through natural openings and wounds. Soft rot is characterized by the breakdown of the plant cell wall, leading to maceration of the parenchyma, loss of electrolytes, and cell death. Pectinases and pectate lyases are mainly responsible for these symptoms.

The mol% G + C of the DNA is: 50.5–56.1.

Type species: **Pectobacterium carotovorum** (Jones 1901) Waldee 1945, 469 emend. Hauben, Moore, Vauterin, Steenackers, Mergaert, Verdonck and Swings 1999a, 1 (*Erwinia carotovora* (Jones 1901) Bergey, Harrison, Breed, Hammer and Huntoon 1923, 171; *Bacillus carotovorus* Jones 1901, 12.)

FURTHER DESCRIPTIVE INFORMATION

Metabolic features are the same as for the genus *Erwinia*. Decarboxylases for arginine, lysine, or ornithine are not present except in a few (usually 5% or less) strains of *Pectobacterium carotovorum* and *Pectobacterium chrysanthemi*.

Pectate lyases are produced by all species except by *Pectobacterium cypripedii*. Cellulases (Cx) are produced by strains of *P. carotovorum*, *P. carotovorum* subsp. *atrosepticum*, and *P. chrysanthemi* (El-Helaly et al., 1979). *P. carotovorum* and *P. chrysanthemi* produce pectinases, cellulases, hemicellulases, and proteases (Garibaldi and Bateman, 1973; Bertheau et al., 1984; Collmer and Keen, 1986; Ried and Collmer, 1986; Wandersman et al., 1986; Willis

et al., 1987). *P. carotovorum* produces an endopolygalacturonase (Lei et al., 1985; Willis et al., 1987) as well. The genes encoding these enzymes that have been mapped are located on the chromosome (Chatterjee et al., 1981; Hugouvieux-Cotte-Pattat et al., 1996). In most *Pectobacterium* species, production of the plant cell wall degrading enzyme pectin lyase (Pnl) is activated by DNA-damaging agents such as mitomycin C (MC), nalidixic acid, and UV light (Liu et al., 1994b).

Naturally occurring plasmids have been detected in strains of *P. carotovorum* and *P. chrysanthemi*, and plasmids from bacteria other than *Pectobacterium* have been introduced into strains of the foregoing *Pectobacterium* species (Lacy and Leary, 1979; Chatterjee and Starr, 1980). Plasmid-mediated transfer of chromosomal genes by conjugation also has been reported for strains of *P. carotovorum* and *P. chrysanthemi*.

Plasmid pULB113 (Van Gijsegem and Toussaint, 1982) mediates chromosomal mobilization and R-prime formation in *P. carotovorum* and *P. chrysanthemi* (Chatterjee et al., 1985).

Virulent or temperate phages have reported to be active against strains of *P. carotovorum* (Chapman et al., 1951; Faltus and Kishko, 1980; Pirhonen and Palva, 1988; Gross et al., 1991; Toth et al., 1993) and *P. chrysanthemi* (Paulin and Nassan, 1978). Bacteriocinogeny or production of bacteriocin-like substances has been noted for strains of *P. carotovorum* (Itoh et al., 1978), *P. chrysanthemi* (Echandi and Moyer, 1979), and *Pectobacterium* species from sugar beet (Stanghellini et al., 1977). Bacteriocin-resistant mutants of *P. chrysanthemi* have been isolated (Expert and Toussaint, 1985).

Lipopolysaccharide defective mutants of *P. carotovorum* (Pirhonen and Palva, 1988) and genetically engineered kanamycin resistant strains of *P. carotovorum* (Scanferlato et al., 1989; Orvos et al., 1990) have been isolated.

Antisera prepared against live or heat-killed cells, nonpurified or purified immunogens have been used for the differentiation or identification of all *Pectobacterium* species except *P. cypripedii* (Schaad, 1979). Serogroups have been determined for *P. carotovorum* (De Boer et al., 1979) and *P. chrysanthemi* (Samson and Nassan-Agha, 1978; Yakrus and Schaad, 1979).

More recently, a conductimetric assay was developed for automated detection of *P. carotovorum* subsp. *atrosepticum*, *P. carotovorum* subsp. *carotovorum*, and *P. chrysanthemi* (Fraaije et al., 1996a, 1997), and a luminescence-based assay is available for the detection of *P. carotovorum* (Grant et al., 1992; McLennan et al., 1992).

A strain of *P. carotovorum* subsp. *carotovorum* was found to produce the antibiotic 1-carbapen-2-em-3-carboxylic acid (Bainton et al., 1992b).

16S rDNA sequence analyses of the species of the genus *Pectobacterium* by Kwon et al. (1997) and Hauben et al. (1998a) agree very well, except for the type strain of *P. cypripedii*, and situate the genus *Pectobacterium* within the *Enterobacteriaceae*, closely related to the genera *Pantoea*, *Erwinia*, *Brenneria*, and *Enterobacter* (Fig. BXII.γ.201 and Table BXII.γ.223 of the genus *Erwinia*). The sequence of the type strain of *P. cypripedii*, sequenced by Kwon et al. (1997) (ATCC 29267) and Hauben et al. (1998a) (LMG 2657), differ in 53 nucleotides. The sequence reported by Hauben et al. (1998a) places *P. cypripedii* in the *Pectobacterium* cluster.

ENRICHMENT AND ISOLATION PROCEDURES

The isolation procedure is the same as for the genus *Erwinia*.

The isolation of some *Pectobacterium* species can be facilitated by use of selective-differential media, but such media are usually not necessary. Selective media have been developed for the isolation of pectolytic bacteria (Kelman and Dickey, 1980). The crystal violet pectate (CVP)[1] medium is commonly used. Miller-Schroth (Miller and Schroth, 1972) medium modified by replacement of most of the agar by sodium polypectate selectively allows the growth of pectolytic species. NaOH (105%) and MOPS (3-(-N-morpholino)propanesulfonic acid; 0.4%) are added to raise the pH and buffer the medium (Pierce and McCain, 1992). *P. chrysanthemi* strains isolated from plants grow well on LB medium[2].

MAINTENANCE PROCEDURES

Stock cultures of *Pectobacterium* species should be grown on standard media of choice at 25–30°C. The cultures can be maintained for short-term storage in a refrigerator (4–5°C); some strains of *P. chrysanthemi* are nonviable after 3–4 weeks at 4°C, but remain viable for longer periods when stored at 12°C. Long-term preservation is the same as for the genus *Erwinia*.

DIFFERENTIATION OF THE GENUS *PECTOBACTERIUM* FROM OTHER GENERA

Characteristics that differentiate *Pectobacterium* from the genera *Erwinia*, *Pantoea*, and *Brenneria* are given in Table BXII.γ.223 of the genus *Erwinia*. Apart from their pectolytic nature, it is very difficult to differentiate them phenotypically; genomic methods are recommended for differentiation.

TAXONOMIC COMMENTS

The species of the genus *Pectobacterium* were formerly classified under *Erwinia*. We refer the reader to the taxonomic comments section of the genus *Erwinia* for a discussion on this issue.

Strains of *P. chrysanthemi* have been isolated from numerous plant species and cultivars (Dickey, 1981). Six pathovars (pathovar *chrysanthemi*, pathovar *dianthicola*, pathovar *dieffenbachiae*, pathovar *paradisiaca*, pathovar *parthenii*, and pathovar *zeae*) have been designated for *P. chrysanthemi* (Dye et al., 1980). The relationship between pathogenicity, phenotypic properties and serological reactions of strains of the pathovars is not clear (Samson and Nassan-Agha, 1978; Yakrus and Schaad, 1979; Dickey, 1981; Thomson et al., 1981).

DIFFERENTIATION OF THE SPECIES OF THE GENUS *PECTOBACTERIUM*

The differential characteristics of the species of *Pectobacterium* are given in Tables BXII.γ.241, BXII.γ.242, BXII.γ.243, and BXII.γ.244. Only a small number (eight) of strains of *P. cypripedii* have been studied.

List of species of the genus Pectobacterium

1. **Pectobacterium carotovorum** (Jones 1901) Waldee 1945, 469[AL] emend. Hauben, Moore, Vauterin, Steenackers, Mergaert, Verdonck and Swings 1999a, 1 (*Erwinia carotovora* (Jones 1901) Bergey, Harrison, Breed, Hammer and Huntoon 1923, 171; *Bacillus carotovorus* Jones 1901, 12.)
ca.ro.to'vo.rum. L. n. *carota* carrot; L. v. *voro* to devour; M.L. adj. *carotovorum* carrot-devouring.

The characteristics are as described for the genus and as listed in Tables BXII.γ.241, BXII.γ.242, BXII.γ.243, and BXII.γ.244. Causes rotting, particularly of storage tissues, of a wide variety of plants and causes a vascular and parenchymal disease (blackleg) of potato plants. The small diffusible autoinducer signal molecule *N*-(β-ketocaproyl) homoserine lactone (KHL) acts as a molecular control signal for the expression of genes controlling carbapenem antibiotic biosynthesis (Bainton et al., 1992a). The *pnl* gene, encoding pectin lyase (Ohnishi et al., 1991), and the *peh* gene, encoding polygalacturonase (Peh) (Lei et al., 1992) of *P. carotovorum*, have been located and characterized.

Subspecies can be differentiated by (1) PCR with primers Y1 (5'-TTACCGGACGCCGAGCTGTGGCGT-3') and Y2 (5'-CAGGAAGATGTCGTTATCGCGAGT-3'), annealing at 65°C, and (2) restriction fragment length polymorphism (RFLP) of a *pel* gene identifying *P. carotovorum* subsp. *atrosepticum* and *P. carotovorum* subsp. *wasabiae* strains (Darrasse et al., 1994b). The pectate lyase (*pel*) genes I and III were characterized by Nikaido et al. (1985) and Yoshida et al. (1991), respectively. RAPD-PCR can differentiate *P. carotovorum* subsp. *carotovorum* and *P. carotovorum* subsp. *atrosepticum* (Maki-Valkama and Karjalainen, 1994; Parent et al., 1996) and a DNA probe, isolated from a genomic library, can differentiate the subspecies *P. carotovorum* subsp. *carotovorum*, *P. carotovorum* subsp. *atrosepticum*, and *P. carotovorum* subsp. *betavasculorum* (Ward and De Boer, 1990).

The bacteriocin carotovoricin was identified (Itoh et al., 1982) and the gene *recA*, which is required for its induction (Zink et al., 1985, Zhao and McEntee, 1990). Monoclonal antibodies were developed for detection of *P. carotovorum* subsp. *carotovorum*, *P. carotovorum* subsp. *atrosepticum*, and *P. carotovorum* subsp. *betavasculorum* (Ward and De Boer, 1989; Murray et al., 1990a; Vernon-Shirley and Burns, 1992). *Drosophila melanogaster* Meigen and *Drosophila rucksii* Coquillett can function as vectors for *P. carotovorum* subsp. *carotovorum* and *P. carotovorum* subsp. *atrosepticum* (Brewer et al., 1981).

The mol% G + C of the DNA is: 50.5–53.1 (*T_m*, Bd).

Type strain: ATCC 15713, NCPPB 312, ICMP 5702.

a. **Pectobacterium carotovorum** *subsp.* **carotovorum** (Jones 1901) Waldee 1945, 469[AL] (*Erwinia carotovora* subsp. *carotovora* Bergey, Harrison, Breed, Hammer and Huntoon 1923, 171.)

Characteristics distinguishing this subspecies from the other *Pectobacterium carotovorum* subspecies are indicated in Table BXII.γ.243.

The production of extracellular enzymes such as pectate lyase (Pel), polygalacturonase (Peh), cellulase (Cel), and protease (Prt) is activated by the cell density (quorum)-sensing signal, *N*-(3-oxohexanoyl)-L-homoserine lactone (HSL), and/or CarR (Barras et al., 1994). *RmsA*, a global repressor gene, controls the production of these

---

1. CVP medium consists of 1N NaOH, 4.5 ml; 10% CaCl₂·H₂O, 3 ml ; Bacto-agar, 1.5 g; NaNO₃, 1 g; Bacto-yeast extract, 0.05 g; blended with 300 ml boiling distilled water for 15 s, to which 15 g sodium polypectate and 200 ml boiling distilled water is added and blended (Cuppels and Kelman, 1974; Woodward and Robinson, 1990).

2. LB medium consists of (per 1 distilled water) tryptone, 10 g; yeast extract, 5 g; NaCl, 10 g.

TABLE BXII.γ.241. Diagnostic characteristic of the *Pectobacterium* species[a]

Characteristic	1. *P. carotovorum*	2. *P. cacticida*	3. *P. chrysanthemi*	4. *P. cypripedii*
Pectinase	+	+	+	−
Reducing sugars produced from sucrose	v	−	+	nd
Indole	−	−	+	−
Growth at 37°C	v	+	+	nd
Growth at 40°C	−	+	nd	nd
Production of acetoin	v	+	+	nd
Arginine dihydrolase	−	nd	d	nd
Caseinase	v	nd	+	−
Gelatinase	v	−	+	nd
Lecithinase	−	nd	+	−
Phenylalanine deaminase	−	nd	−	+
Phosphatase	−	d	+	nd
Acid production from:				
Lactose	+	nd	+	−
Malonate	−	nd	d	nd
Maltose	v	nd	−	+
Raffinose	v	nd	+	−
Trehalose	+	nd	−	+
Utilization of carbon sources:				
D-Arabinose	−	−	d	nd
L-Arabinose	+	−	+	nd
Cellobiose	+	−	+	nd
Gentiobiose	+	d	−	nd
α-Methylglucoside	d	nd	−	nd
myo-Inositol	+	−	nd	nd
Malonate	−	+	+	nd
Melibiose	v	−	+	nd
Raffinose	+	−	+	nd
Tartrate	−	−		+
Sensitivity toward:				
Erythromycin	−	−	+	+

[a]For symbols see standard definitions: nd, not determined.

enzymes and tissue macerating ability (Bainton et al., 1992b; Cui et al., 1995; McGowan et al., 1995; Mukherjee et al., 1996). The gene *aepA* (activator of extracellular protein production) controls the production of pectolytic enzymes Pel, Peh, Cel, and Prt (Liu et al., 1993c; Murata et al., 1994), and the *out* gene cluster encoding the proteins of the type II or general secretory pathway (GSP) apparatus is required for secretion of pectinases and cellulases (Reeves et al., 1993; Thomas et al., 1997). The *pnlA* gene, encodes DNA damage-inducible pectin lyase (Chatterjee et al., 1991) and *rdgB* encodes a transcriptional factor that specifically interacts with the pectin lyase structural gene *pnlA* promoter/regulatory region (Liu et al., 1997e). The *pnlA* gene is activated by DNA-damaging agents like RecA, RdgA, and RdgB. RdgA and RdgB also control bacteriocin production, phage release, and cell lysis (Barras et al., 1994).

A pleiotrophic reduced virulence (Rvi-negative) mutant was found to be defective in flagella assembly proteins (Mulholland et al., 1993). Several monoclonal antibodies have been produced against *P. carotovorum* subsp. *carotovorum* strains that can be used for identification and detection by Ouchterlony double diffusion (ODD), indirect immunofluorescence (IIF), and enzyme-linked immunosorbent assay (ELISA) (Alarcon et al., 1995), and for detection by reverse passive hemagglutination (RPH) (Koehm and Eggers-Schumacher, 1995).

Pseudomonas fluorescens strains are antagonistic (biological control agent) to *P. carotovorum* subsp. *carotovorum* (El Hendawy et al., 1998). Causes rotting, particularly of storage tissues, of a wide variety of plants, e.g., mushroom (*Agaricus* sp.), century plant (*Agave* sp.), mustard (*Brassica campestris*, *B. nigra*, and *B. juncea*), Chinese cabbage (*Brassica chinensis*), cauliflower (*Brassica oleracea* var. *botrytis*), cabbage (*Brassica oleracea* var. *capitata*), Brussels sprouts (*Brassica oleracea* var. *gemmifera*), broccoli (*Brassica oleracea* var. *italica*), turnip (*Brassica rapa* var. *rapa*), and ornamental plants. Larger fleshy organs are particularly susceptible and, once infected, they usually become softened to a pulp very quickly.

The mol% G + C of the DNA is: 50.5–53.1 (T_m, Bd).

Type strain: ATCC 15713, DSM 30168, LMG 2404.

GenBank accession number (16S rRNA): Z96089, U80197.

b. **Pectobacterium carotovorum** *subsp.* **atrosepticum** (van Hall 1902) Hauben, Moore, Vauterin, Steenackers, Mergaert, Verdonck and Swings 1999a, 1[VP] (Effective publication: Hauben, Moore, Vauterin, Steenackers, Mergaert, Verdonck and Swings 1998a, 393) (*Erwinia carotovora* subsp. *atroseptica* (van Hall 1902) Dye 1969a, 81; *Bacillus atrosepticus* van Hall 1902, 134.)
at.ro.sep' ti.cum. L. adj. *ater* black; Gr. adj. *septicus* producing a putrefaction; M.L. adj. *atrosepticum* producing a black rot.

Characteristics distinguishing this subspecies from the other *P. carotovorum* subspecies are indicated in Table BXII.γ.243. Several regions of the bacterial DNA of *P. carotovorum* subsp. *atrosepticum* seem to be subspecies-specific and can be used for identification by (1) genomic subtraction of DNA probes (Darrasse et al., 1994a);

TABLE BXII.γ.242. Additional characteristics of the *Pectobacterium* species[a]

Characteristic	1. *P. carotovorum*	2. *P. cacticida*	3. *P. chrysanthemi*	4. *P. cypripedii*
$NO_3^- \rightarrow NO_2^-$	+	+	+	nd
Production of aminopeptidase	nd	+	+	nd
Growth in 5%NaCl	v	nd	−	nd
Oxidase	−	−	−	−
Catalase	+	+	+	+
Citrate (Simmons)	v	d	+	nd
H_2S from cysteine	v	nd	v	−
β-Galactosidase	v	nd	+	nd
Arginine decarboxylase	−	−	−	−
Lysine decarboxylase	−	−	−	−
Ornithine decarboxylase	−	−	−	−
Esculin hydrolase	+	+	+	+
Starch hydrolase	−	−	−	−
Tryptophan deaminase	−	−	−	−
Urease	−	−	−	−
Acid production from:				
Adonitol	−	−	−	−
Amygdalin	nd	+	+	nd
D-Arabitol	v	nd	−	nd
L-Arabitol	−	−	−	−
L-Arabinose	nd	+	+	nd
Arbutin	nd	+	+	nd
Cellobiose	nd	+	+	nd
Citrate	v	nd	+	nd
Dulcitol	nd	−	−	−
Erythritol	nd	−	−	nd
Esculin	nd	+	+	nd
Fructose	+	+	+	+
DL-Fucose	nd	−	−	nd
Galactose	+	+	+	+
Gentiobiose	+	+	nd	nd
D-Glucose	+	+	+	+
N-acetylglucosamine	+	+	+	+
Glycerol	nd	+	+	+
Glycogen	nd	−	−	nd
5-Ketogluconate	−	−	nd	nd
Inulin	v	nd	d	−
D-Lyxose	−	−	−	−
D-Mannose	+	+	+	+
α-Methylmannoside	−	−	−	−
Melezitose	−	−	−	nd
Melibiose	nd	v	+	nd
α-Methylglucoside	nd	v	−	−
β-Methylxyloside	nd	−	−	nd
Palatinose	v	nd	−	nd
L-Rhamnose	+	+	+	+
D-Ribose	+	+	+	+
Salicin	+	+	+	+
Sorbitol	v	nd	−	nd
L-Sorbose	−	−	−	−
Starch	−	−	−	−
Sucrose	+	+	+	+
D-Tagatose	−	−	−	−
Xylitol	nd	−	−	nd
D-Xylose	nd	+	+	nd
L-Xylose	nd	−	−	nd
Utilization of carbon sources:				
Acetate	+	+	+	+
Adipate	−	−	−	−
Amygdalin	d	d	−	nd
L-Arabinose	nd	+	+	+
D-Arabitol	v	−	−	nd
Arbutin	+	+	+	+
Benzoate	−	−	−	−
Betaine	−	−	−	−
Butanol	−	−	−	−
Citrate	+	+	+	nd
Dulcitol	nd	−	−	−
Erythritol	nd	−	−	nd
Esculin	nd	+	+	nd
Fructose	+	+	+	+

(continued)

TABLE BXII.γ.242. *(cont.)*

Characteristic	1. *P. carotovorum*	2. *P. cacticida*	3. *P. chrysanthemi*	4. *P. cypripedii*
DL-Fucose	nd	−	−	nd
Fumarate	+	+	+	+
Galactose	+	+	+	+
Gallate	−	−	−	−
Gluconate	nd	+	+	+
Glucose	+	+	+	+
DL-Glycerate	nd	nd	nd	nd
Glycerol	d	+	+	+
Glycogen	−	nd	nd	nd
α-Ketoglutarate	nd	+	nd	nd
Inulin	v	−	d	nd
Lactose	d	+	d	nd
Lyxose	−	nd	nd	nd
Mallate	+	+	+	+
Maltose	d	d	−	nd
Mannitol	+	+	+	+
Mannose	+	+	+	+
Melezitose	−	nd	nd	nd
Methanol	−	−	−	−
β-Methylglucoside	+	+	+	+
Mucate	nd	nd	+	nd
Oxalate	−	−	−	−
Pectinic acid	+	+	+	nd
Propionate	−	−	−	−
L-Rhamnose	+	+	+	nd
Ribose	+	+	+	+
Saccharose	+	nd	nd	nd
Salicin	+	+	+	+
Sorbitol	v	−	−	nd
Sorbose	−	−	−	−
Starch	−	nd	nd	nd
Succinate	+	+	+	+
Sucrose	+	+	+	+
D-Tagatose	−	nd	nd	nd
meso-Tartrate	−	−	nd	nd
Trehalose	d	+	v	nd
Triacetine	ND	−	nd	nd
D-Turanose	v	−	−	nd
Xylitol	−	−	nd	nd
D-Xylose	d	ND	+	nd
Utilization of nitrogen sources:				
Adenine	nd	nd	+	nd
Alanine	+	+	+	+
Allantoin	+	+	+	+
Ammonium chloride	+	+	+	+
Anthranilic acid	−	−	−	−
Arginine	+	+	+	+
Asparagine	+	+	+	+
Asparginic acid	+	+	+	+
Betaine	−	−	−	−
γ-Aminobutanic acid	+	+	+	+
Carnosine	nd	nd	+	nd
Choline	−	−	−	−
Citrulline	+	+	+	+
Creatine	−	−	nd	nd
Cysteamine	−	−	−	−
Glucosamine	+	+	+	+
Glutamine	+	+	+	+
Glutaminic acid	+	+	+	+
Glutathion	+	+	+	+
Glycine	+	+	+	+
Glycylglycine	+	+	+	+
Guanine	−	−	nd	nd
Histidine	+	+	+	+
Hydroxypyroline	−	−	−	−
Kynureninic acid	−	−	−	−
Leucine	+	+	+	+
Methionine	+	+	+	+
Nicotinic acid	−	−	nd	nd
Octopine	nd	nd	+	nd
Ornithine	nd	nd	+	nd

(continued)

TABLE BXII.γ.242.   *(cont.)*

Characteristic	1. *P. carotovorum*	2. *P. cacticida*	3. *P. chrysanthemi*	4. *P. cypripedii*
Phenylalanine	+	+	+	+
Quinolinic acid	−	−	−	−
Sarcosine	−	−	−	−
Serine	+	+	+	+
Spermidine	−	−	−	−
Spermine	−	−	−	−
Taurine	nd	nd	−	nd
Threonine	nd	nd	+	nd
Trigonelline	−	−	−	−
Trimethylammonium	−	−	−	−
Tryptamine	−	−	nd	nd
Tryptophane	+	+	+	+
Tyrosine	+	+	+	+
Ureum	+	+	+	+
Valine	nd	nd	+	nd
Xanthin	−	−	nd	nd
Sensitivity toward:				
Amoxycillin	+	+	nd	nd
Cephalotine	+	+	nd	nd

[a]For symbols see standard definitions; nd, not determined.

TABLE BXII.γ.243.   Differentiation of *Pectobacterium carotovorum* subspecies[a]

Characteristic	*P. carotovorum* subsp. *carotovorum*	*P. carotovorum* subsp. *atrosepticum*	*P. carotovorum* subsp. *betavasculorum*	*P. carotovorum* subsp. *odoriferum*	*P. carotovorum* subsp. *wasabiae*
Reducing sugars produced from sucrose	−	+	+	+	−
Methyl red	+	+	−	nd	+
Growth in 5% NaCl	+	+	+	nd	−
Growth at 37°C	+	−	+	+	nd
Utilization of citrate	+	+	−	nd	+
H₂S from cysteine	v	v	v	−	+
Acetoin production	+	v	+	+	−
Citrate (Simmons)	+	+	−	+	
Caseinase	+	−	−	nd	+
β-Galactosidase	+	+	+	+	−
Gelatinase	+	−	v	+	+
Acid production from:					
d-Arabitol	−	−	−	+	nd
Citrate	+	+	−	nd	+
α-Methylglucoside	−	+	v	+	−
Gluconate	−	+	−	nd	nd
myo-Inositol	−	−	+	−	−
Inulin	−	−	+	−	−
Maltose	−	+	+	nd	−
Melibiose	+	+	−	+	−
Palatinose	−	+	+	+	nd
Raffinose	+	+	v	+	−
Sorbitol	−	−	−	+	−
Utilization of carbon sources:					
D-Arabitol	−	−	−	+	nd
Citrate	+	+	−	+	nd
Gluconate	−	+	+	v	nd
Inulin	−	−	+	−	nd
Melibiose	+	+	−	+	nd
Palatinose	−	+	+	+	nd
Sorbitol	−	−	−	+	−
D-Turanose	−	d	+	d	nd

[a]For symbols see standard definitions; nd, not determined.

(2) PCR detection with primers ECA1f (5′-CGGCAT-CATAAAAACACG-3′) and ECA2r (5′-GCACACTTCA-TCCAGCGA-3′), annealing at 62°C, amplifying a 690-bp DNA fragment (De Boer and Ward, 1995); (3) hybridization with a nonradioactive labeled DNA probe (Helander and Persson, 1996); (4) PCR detection with primers derived from sequences of metalloprotease-coding genes: ERWFOR (5′-ACGCATGAAATCGGCCATGC-3′) T_m 62°C, ATROREV (5′-ATCGATAATTTGATTGTCCT-3′) T_m 52°C, and CHRREV (5′-AGTGCTGCCGTACA-GCACGT-3′) T_m 64°C (Smid et al., 1995); (5) detection with a digoxigenin-labeled DNA probe, selected from an

TABLE BXII.γ.244. Additional characteristics of *Pectobacterium carotovorum* subspecies[a]

Characteristic	*P. carotovorum* subsp. carotovorum	*P. carotovorum* . subsp. atrosepticum	*P. carotovorum* subsp. betavasculorum	*P. carotovorum* subsp. odoriferum	*P. carotovorum* subsp. wasabiae
Indole production	−	−	−	−	−
Growth in 7% NaCl	+	nd	nd	nd	nd
Growth at 40°C	−	−	−	nd	−
Production of aminopeptidase	+	+	+	+	nd
Arginine dihydrolase	−	−	−	−	nd
β-Glucuronidase	nd	nd	nd	−	nd
Lecithinase	−	−	−	−	−
Production of pectolytic enzymes	+	+	+	+	+
Phenylalanine deaminase	−	−	−	nd	−
Phosphatase	−	−	−	−	−
Acid production from:					
Amygdalin	+	+	+	+	nd
L-Arabinose	+	+	+	+	nd
Arbutine	+	+	+	+	nd
Cellobiose	+	+	v	+	nd
Dulcitol	−	−	−	−	nd
Erythritol	−	−	−	−	nd
Esculin	+	+	+	+	nd
DL-Fucose	−	−	−	−	nd
Glycerol	+	+	+	+	nd
Glycogen	−	−	−	−	nd
Lactose	+	+	+	+	+
Malonate	−	−	−	nd	−
Mannitol	+	+	+	+	+
L-Rhamnose	+	+	+	+	nd
D-Ribose	+	+	+	+	nd
Salicin	+	+	+	+	nd
L-Sorbose	−	−	−	−	nd
Starch	−	−	−	−	nd
Sucrose	+	+	+	+	nd
D-Tagatose	−	−	−	−	nd
Trehalose	+	+	+	+	+
Xylitol	−	−	−	−	nd
D-Xylose	+	⊢	+	+	nd
L-Xylose	−	−	−	−	nd
β-Methylxyloside	−	−	−	−	nd
Utilization of carbon sources:					
Aconitate	nd	nd	nd	−	nd
Adonitol	nd	nd	nd	−	nd
Amygdalin	d	d	+	d	nd
D-Arabinose	−	−	−	−	nd
L-Aspartate	nd	nd	nd	+	nd
Cellobiose	+	+	v	+	+
Dulcitol	nd	nd	nd	−	nd
Erythritol	nd	nd	nd	−	nd
Fucose	nd	nd	nd	−	nd
Methyl-α-galactoside	nd	nd	nd	d	nd
Galacturonic acid	nd	nd	−	+	nd
Gentiobiose	+	+	+	+	nd
α-Methylglucoside	v	+	+	+	nd
N-Acetyl-D-glucosamine	nd	nd	nd	+	nd
L-Glutamate	nd	nd	nd	+	nd
D-Glucosamine	nd	nd	nd	+	nd
Ketogluconate	nd	nd	nd	d	nd
Glucuronate	nd	nd	nd	d	nd
Glutarate	nd	nd	nd	−	nd
DL-Glycerate	nd	nd	nd	+	nd
L-Histamine	nd	nd	nd	−	nd
Hypoxanthin	nd	−	nd	nd	nd
myo-Inositol	+	+	+	+	+
α-Ketoglutarate	+	+	+	+	+
DL-Lactate	nd	nd	nd	−	nd
Lactose	+	+	+	+	nd
Lactulose	nd	nd	nd	+	nd

(continued)

TABLE BXII.γ.244. *(cont.)*

Characteristic	*P. carotovorum* subsp. *carotovorum*	*P. carotovorum* subsp. *atrosepticum*	*P. carotovorum* subsp. *betavasculorum*	*P. carotovorum* subsp. *odoriferum*	*P. carotovorum* subsp. *wasabiae*
Malonate	−	−	−	−	−
Maltitol	nd	nd	nd	d	nd
Maltose	v	d	+	d	nd
Maltotriose	nd	nd	nd	−	nd
Mucate	nd	nd	nd	+	nd
Proline	nd	nd	nd	d	nd
Raffinose	+	+	v	+	+
Saccharate	nd	nd	nd	+	nd
L-Serine	nd	nd	nd	+	nd
D-Tagatose	nd	nd	nd	−	nd
Tartrate	−	−	−	−	−
Trehalose	+	+	+	+	nd
Triacetine	−	−	−	−	−
D-Xylose	nd	nd	nd	+	nd
Utilization of nitrogen sources:					
Hypoxanthin	nd	nd	+	nd	nd
Pyrazinamide	nd	nd	+	nd	nd
Sensitivity toward:					
Erythromycin	−	−	−	nd	−
Hypoxanthin	−	nd	nd	nd	nd

[a]For symbols see standard definitions; nd, not determined.

EcoRI digest of a cloned library (Ward and De Boer, 1994).

Several monoclonal antibodies have been produced against *P. carotovorum* subsp. *atrosepticum* strains that can be used for identification and detection in cell suspensions as well as in plant material by (1) ODD (Alarcon et al., 1995); (2) IIF with a sensitivity of up to 240 cells/ml (De Boer and McNaughton, 1987; Gorris et al., 1994; Alarcon et al., 1995); (3) ELISA with a sensitivity of up to 10^3 cells/ml tuber peel extract (Jones et al., 1993; Gorris et al., 1994; Alarcon et al., 1995; Basalp et al., 1995; Hyman et al., 1995; Pérombélon and Hyman, 1995); (4) immunofluorescence (IF) cell staining with a sensitivity of up to 10^5 cells/ml tuber peel extract (Fraaije et al., 1996b); (5) immunofluorescence colony staining (Gorris et al., 1994; Pérombélon and Hyman, 1995; Schober and Van Vuurde, 1997); and (6) slide agglutination test (McLeod and Pérombélon, 1992).

Causes a vascular and parenchymal disease (blackleg) of potato (*Solanum tuberosum*), cauliflower (*Brassica oleraceae* var. *botrytis*), tomato (*Lycopersicon esculentum*), and cabbage (*Brassica oleracea* var. *capitata*), and a storage rot of potato tubers. Bases of stems are blackened, rotted, and slimy. Plants wilt and are stunted or die. The soft rot may affect various parts of the plant, particularly the tubers in storage.

The mol% G + C of the DNA is: 51.3–53.1 (T_m, Bd).
Type strain: ATCC 33260, NCPPB 549, LMG 2386.
GenBank accession number (16S rRNA): Z96090.

c. **Pectobacterium carotovorum** *subsp.* **betavasculorum** (Thomson, Hildebrand and Schroth 1981) Hauben, Moore, Vauterin, Steenackers, Mergaert, Verdonck and Swings 1999a, 1^{VP} (Effective publication: Hauben, Moore, Vauterin, Steenackers, Mergaert, Verdonck and Swings 1998a, 393) (*Erwinia carotovora* subsp. *betavasculorum* Thomson, Hildebrand and Schroth 1981, 1040.)
be.ta.vas.cu.lo′ rum. L. n. *beta* beet; L. n. *vasculum* vascular tissue; M.L. pl. gen. n. *betavasculorum* of the beet's vascular tissues.

Characteristics distinguishing this subspecies from the other *Pectobacterium carotovorum* subspecies are indicated in Table BXII.γ.243. Strain Ecb168 was found to produce an antibiotic that suppresses growth of *P. carotovorum* subsp. *carotovorum* (Costa and Loper, 1994). Causes soft rot and vascular necrosis of sugar beet (*Beta vulgaris*) and, after artificial inoculation, of potato stems.

The mol% G + C of the DNA is: 54.4–54.7 (T_m, Bd).
Type strain: ATCC 43762, NCPPB 2795, LMG 2466.
GenBank accession number (16S rRNA): U80198, Z96091.

d. **Pectobacterium carotovorum** *subsp.* **odoriferum** (Gallois, Samson, Ageron and Grimont 1992) Hauben, Moore, Vauterin, Steenackers, Mergaert, Verdonck and Swings 1999a, 1^{VP} (Effective publication: Hauben, Moore, Vauterin, Steenackers, Mergaert, Verdonck and Swings 1998a, 394) (*Erwinia carotovora* subsp. *odorifera* Gallois, Samson, Ageron and Grimont 1992, 586.)
o.do.r.i′ fe.rum. L. masc. adj. *odoriferum* bringing odors, fragrant.

Characteristics distinguishing this subspecies from the other *Pectobacterium carotovorum* subspecies are indicated in Table BXII.γ.243. Grow on nutrient agar, producing colonies that are about 2.5 mm in diameter after 48 h at 25°C and also on yeast peptone dextrose (YPDA) medium (3 g/l yeast extract, 5 g/l peptone, 5 g/l dextrose, 15% agar). Colonies are circular, grayish (to yellowish on media containing sucrose or gelatin), slightly domed, and semitranslucent. Causes odorous soft rot of chicory (*Cichorium intybus* L.). The disease typically results in water-soaked, translucent, wet rot of the etiolated leaves. The development of slimy rot in witloof chicories is always associated with the production of a sweet, ripe, banana-like odor.

The mol% G + C of the DNA is: unknown.
Type strain: CFBP 1878, LMG 17566.
GenBank accession number (16S rRNA): AJ223407.

e. **Pectobacterium carotovorum** *subsp.* **wasabiae** (Goto and Matsumoto 1987) Hauben, Moore, Vauterin, Steenackers, Mergaert, Verdonck and Swings 1999a, 1[VP] (Effective publication: Hauben, Moore, Vauterin, Steenackers, Mergaert, Verdonck and Swings 1998a, 393) (*Erwinia carotovora* subsp. *wasabiae* Goto and Matsumoto 1987, 132.)

wa.sa' bi.a.e. L. gen. n. *wasabiae* of wasabi (*Eutrema wasabi*), name of host plant.

Characteristics distinguishing this subspecies from the other *Pectobacterium carotovorum* subspecies are indicated in Table BXII.γ.243. Colonies on yeast extract-peptone agar plates at 28°C are white, transparent, circular with entire margins, convex, and 1.0–2.0 mm in diameter after 24 h. Growth on yeast extract-peptone agar slants is moderate, white, and butyrous, with high viability. Abundant growth is obtained on yeast extract-peptone agar slants supplemented with glucose or on potato-glucose agar slants, although viability on these media is poor. Causes soft rot and an internal black discoloration on slices of wasabi rhizomes and, after artificial inoculation, also on potato tubers, carrot and radish roots, midribs of chinese cabbage and intact wasabi, and tomato and tobacco plants.

The mol% G + C of the DNA is: 51.4–51.7 (T_m).

Type strain: ATCC 43316, ICMP 9121, LMG 8444.

GenBank accession number (16S rRNA): U80199, AJ223408.

2. **Pectobacterium cacticida** (Alcorn, Orum, Steigerwalt, Foster, Fogleman, and Brenner 1991) Hauben, Moore, Vauterin, Steenackers, Mergaert, Verdonck and Swings 1999a, 1[VP] (Effective publication: Hauben, Moore, Vauterin, Steenackers, Mergaert, Verdonck and Swings 1998a, 394) (*Erwinia cacticida* (Hori 1911) Dye 1969a, 93; *Erwinia cypripedii* Alcorn, Orum, Steigerwalt, Foster, Fogleman, and Brenner 1991, 210.)

cac.ti.ci' da. Gr. n. *kaktos* prickly plant; Gr. v. *cid* to kill; L. adj. *cacticida* cactus killing.

The characteristics are as described for the genus and as listed in Tables BXII.γ.241 and BXII.γ.242. Colonies grown on PDP (40 g potato dextrose agar (Difco), 5 g agar, 20 g peptone, 1 l distilled water, pH 7.2) for 24 h at 30°C are small, smooth, glistening, circular, entire, slightly convex, ivory, and translucent, and frequently have striations. They produce a distinct odor and become opaque, off-white, and butyrous with age. Causes soft rot of *Opuntia* (cactus) fruits and pads. After artificial inoculation, also causes soft rot of saguaro, organ pipe, senita cacti, tomato fruits, potato slices, and slices of carrot roots.

The mol% G + C of the DNA is: 50.8–51.7 (T_m).

Type strain: ATCC 49481, LMG 17936, ICPB EC186.

GenBank accession number (16S rRNA): AJ223409.

3. **Pectobacterium chrysanthemi** (Burkholder, McFadden and Dimock 1953) Brenner, Steigerwalt, Miklos and Fanning 1973b, 205[AL] emend. Hauben, Moore, Vauterin, Steenackers, Mergaert, Verdonck and Swings 1999a, 1 (*Erwinia chrysanthemi* Burkholder, McFadden and Dimock 1953, 526.)

chrys.an' the.mi. M.L. n. *Chrysanthemum* generic name; M.L. gen. n. *chrysanthemi* of chrysanthemums.

The characteristics are as described for the genus and as listed in Tables BXII.γ.241 and BXII.γ.242.

Colonies on potato–glucose agar (pH 6.5) (ATCC culture medium 97) are characteristically umbonate with undulate to coralloid margins ("fried egg") at 3–6 d.

The chrysobactin-mediated iron uptake is essential in enabling the bacterium to systematically attack plants (Expert et al., 1996). Chrysobactin (N-(N^2-(2,3-dihydroxybenzoyl)-D-lysyl)-L-serine) is a compound with siderophore activity (Persmark et al., 1989). The iron status is involved in *pel* regulation. Iron acquisition involves an inductive process resulting in differential expression of two siderophore-mediated pathways in relation to external accessibility (Mahe et al., 1995).

The general protein secretory pathway was studied by mutagenesis of cellulase EGZ (Py et al., 1993). *P. chrysanthemi* produces four proteases encoded by *prtA*, *prtB*, *prtC*, and *prtG*, and they are secreted by a type I secretion system encoded by *prtD*, *prtE* (inner membrane), and *prtF* (outer membrane) (Barras et al., 1994). PrtG is an extracellular metalloprotease secreted through a signal peptide-independent secretion pathway. The COOH-terminal exposition of the last four amino acids in the secretion of PrtG plays a key role in the protease secretion pathway (Ghigo and Wandersman, 1994). These metalloproteases are secreted independently of the general export pathway encoded by the *sec* genes. They are secreted via a C-terminal secretion signal and by a secretion apparatus composed of the two inner membrane proteins, PrtD and PrtE, and the outer membrane protein PrtF. PrtD exhibits a secretion signal-regulated ATPase activity (Delepelaire, 1994). PrtD is the ATP-binding cassette (ABC) integral membrane component from the metalloprotease secretion system. PrtA and prtB are two tandem metalloprotease-encoding structural genes (Boyd and Keen, 1993). Further, *P. chrysanthemi* produces two cellulases, encoded by *celZ* (the majority) and *celY* (5% of the total activity), and four types of pectinases. The importance of these various isoenzymes in pathogenicity depends on the plant species infected (Barras et al., 1994). The pectinases are subjected to multiple regulations— KdgR, PecT, PecS, PecM—and products of pectic catabolism encoded by *ogl*, *kduI*, *kduD*, *kdgK*, and *kdgA*. Mutants were isolated, containing genetic fusions of the *kdgK*, encoding 2-keto-3-deoxygluconate (KDG) kinase (Hugouvieux-Cotte-Pattat et al., 1994) or *kdgA* genes to the *lacZ* gene of *Escherichia coli* by infection of a *lacZ* mutant of *E. chrysanthemi* with the phage Mu d(Ap lac) (Hugouvieux-Cotte-Pattat and Robert-Baudouy, 1985). A second set of secondary pectate lyases are preferentially expressed in plants (Barras et al., 1994). The *out* genes are required for the translocation across the outer membrane during secretion of pectate lyases, pectin methyl esterases, exopolygalacturonases, and CelZ. An open reading frame *outT*, located between *outB* and *outC*, has no homology with the *pul* cluster but is involved in secretion. *OutC*, *outD*, and *outE* form an operon, while *outS*, *outB*, and *outT* constitute independent transcription units. *OutT* and the *outCDE* operon are regulated by *kdgRa*, encoding a repressor that negatively regulates the expression of genes involved in pectinolysis and in pectinase secretion. *OutB* and *outS* seem to be expressed constitutively (Ji et al., 1989; Condemine et al., 1992; Nasser et al., 1994).

Mutants defective in secretion of pectinase and cellulase were obtained by chemical and insertion mutagenesis (Andro et al., 1984).

The regulation of pectinolysis was reviewed by Hugouvieux-Cotte-Pattat et al. (1996). Four types of pectinases have been identified: two pectin methyl esterases (PemA, PemB), a polygalacturonase (PehX), nine pectate lyases (PelA, PelB, PelC, PelD, PelE, PelI, PelL, PelZ, PelX), and a pectin acetyl esterase (PaeY) (Shevchik et al., 1997). PecT, encoded by *pecT*, represses the expression of pectate lyase genes *pelC*, *pelD*, *pelE*, *pelL*, and *kdgC*, activates *pelB*, and has no effect on the expression of *pelA* or the pectin methylesterase genes *pemA* and *pemB*. PecT activates its own expression (Surgey et al., 1996). PecS encodes a repressor that negatively regulates the expression of virulence factors such as pectinases or cellulases (Reverchon et al., 1994; Praillet et al., 1996, 1997). The oligogalacturonate lyase (*ogl*) gene is involved in oligogalacturonides degradation (Reverchon and Robert-Baudouy, 1987) and the *pem* gene encodes the pectin methylesterase (PME) (Laurent et al., 1993). *kduD* mutants altered in pectin degradation were isolated by chemical and Mu d(Ap lac) insertion mutagenesis (Condemine et al., 1986).

Harpins are synthesized in a nutrient-deprived medium and induce plant response, including alkalinization of the apoplastic fluid. Harpin from *P. chrysanthemi* contributes to soft-rot pathogenesis (Bauer et al., 1995). Harpins are encoded by the *hrp* genes, which have at least three biochemical functions: gene regulation, protein secretion, and production of HR elicitor proteins.

P. chrysanthemi strains can be genotypically characterized by RFLP of PCR-amplified fragments of *pel* genes (Nassar et al., 1996a) and ribotyping, which correlates well with the established pathovars (Nassar et al., 1994).

A transducing phage Phi EC2 was described by Resibois et al. (1984) and Phi EC2-resistant mutants, lipopolysaccharide-defective, were isolated by Schoonejans et al. (1987).

Several monoclonal antibodies have been produced against *P. chrysanthemi* strains that can be used for detection or identification by double–antibody sandwich (DAS)-ELISA with a sensitivity of 5×10^3 CFU/ml in carnation stem samples (Nassar et al., 1996b), time-resolved fluoroimmunoassay (TR-FIA) in potato peel extracts with a sensitivity of 10^5 cells/ml (Van De Wolf, 1993).

L-Asparaginase from *P. chrysanthemi* provides an alternative to the enzyme from *E. coli* for the effective treatment of acute lymphoblastic leukemia (Goward et al., 1992). It alters the coagulation system less severely than does *E. coli* asparaginase (Carlsson et al., 1995).

Causes vascular wilts, stunting, soft rots, spotting of leaves or parenchymal necroses of a wide range of plant species and cultivars like *Allium cepa*, *Capsicum anuum*, *Colocasia esculanta*, *Kalanchoe blossfeldiana*, *Nicotiana tabacum*, *Nopalea* sp., *Pelargonium zonale*, and *Primula* sp. The disease symptoms are usually systemic.

The mol% G + C of the DNA is: 55.1–57.1 (T_m, Bd).

Type strain: ATCC 11663, DSM 4610, LMG 2804, ICMP 5703, NCPPB 402.

GenBank accession number (16S rRNA): U80200, Z96093.

4. **Pectobacterium cypripedii** (Hori 1911) Brenner, Steigerwalt, Miklos and Fanning 1973b, 205[AL] emend. Hauben, Moore, Vauterin, Steenackers, Mergaert, Verdonck and Swings 1999a, 1 (*Erwinia carotovora* biovar cypripedii (Hori 1911) Dye 1969a, 93; *Erwinia cypripedii* (Hori 1911) Bergey, Harrison, Breed, Hammer and Huntoon 1923, 171; *Bacillus cypripedii* Hori 1911, 91.)

cyp.ri.ped'i.i. M.L. n. *Cypripedium* generic name; M.L. gen. n. *cypripedii* of cypripedium orchids.

The characteristics are as described for the genus and as listed in Tables BXII.γ.241 and BXII.γ.242. The gene cluster encoding three subunits of membrane-bound gluconate dehydrogenase (GADH) was studied by Yum et al. (1997). Causes a brown rot of cypripedium orchids (*Cypripedium* spp.). The disease usually attacks orchids with fleshy leaves. It starts with small water-soaked lesions that enlarge and become slightly sunken, brownish greasy-looking areas. It may spread down the stem and involve the growing point.

The mol% G + C of the DNA is: 54.1–54.6 (T_m, Bd).

Type strain: ATCC 29267, DSM 3873, ICMP 1591, LMG 2657.

GenBank accession number (16S rRNA): U80201, Z96094.

Genus XXV. **Candidatus** Phlomobacter *Zreik, Bové and Garnier 1998, 260*

MONIQUE GARNIER

Phlo.mo.bac.ter. Gr. n. *phlomos* bark; Gr. n. *bakterion* a small rod; M.L. neut. *Phlomobacter* rod in the bark.

Filamentous bacteria occurring in the sieve tubes of phloem tissue of plants (Figs. BXII.γ.202 and BXII.γ.203). The original description is from infected strawberry plants (*Fragaria* x *ananassa*) showing leaf marginal chlorosis and stunting in France (Nourrisseau et al., 1993). Like most other phloem-restricted bacteria, *Candidatus* Phlomobacter fragariae has resisted *in vitro* cultivation.

The mol% G + C of the DNA is: not determined.

Type species: *Candidatus* Phlomobacter fragariae Zreik, Bové and Garnier 1998, 260.

FURTHER DESCRIPTIVE INFORMATION

Phylogenetic treatment Phylogenetic analysis of the 16S rRNA gene indicates that *Candidatus* Phlomobacter is a member of the *Enterobacteriaceae* in the *Gammaproteobacteria*. Its closest cultured relative is *Arsenophonus nasoniae*, the recently characterized agent of the son-killer trait in the parasitic wasp *Nasonia vitripennis* (92% 16S rDNA identity) (Fig. BXII.γ.204). The complete sequence of *Candidatus* Phlomobacter 16S rRNA gene is deposited in GenBank under the accession number U91515. The oligo-

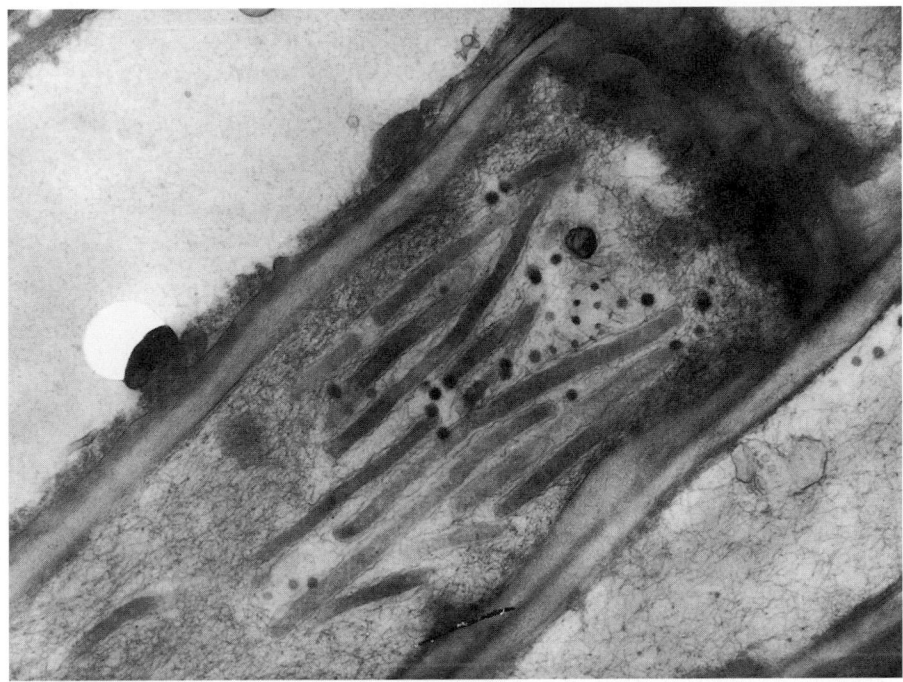

FIGURE BXII.γ.202. *Candidatus* Phlomobacter fragariae cells in a phloem sieve tube of strawberry leaves showing marginal chlorosis (× 20,000).

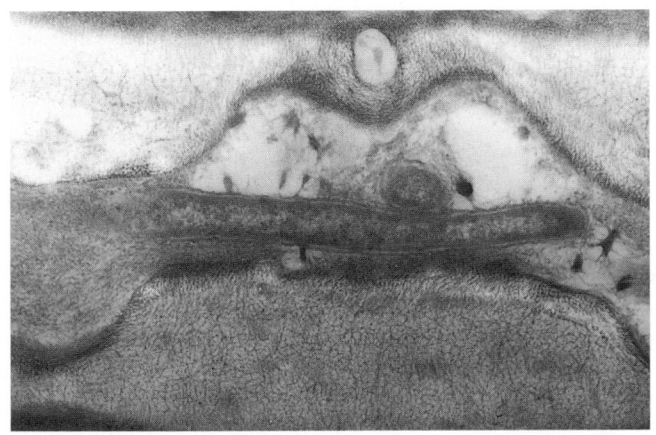

FIGURE BXII.γ.203. *Candidatus* Phlomobacter fragariae cell in a sieve tube with thickened walls (× 29,000).

nucleotide sequence complementary to unique region of the 16S rRNA is 5′-AGCAATTGACATTAGCGA-3′ (Zreik et al., 1998).

Cultivation To date, all attempts to culture *Candidatus* Phlomobacter have failed.

Strain morphology Electron microscopy measurements on thin sections show that *Candidatus* Phlomobacter fragariae cells are 0.2–0.27 μm in diameter and up to 4 μm in length. The cell envelope is 250 nm thick. There is no evidence for flagella or pili (Figs. BXII.γ.202 and BXII.γ.203).

Ecological data, host range *Candidatus* Phlomobacter are phytopathogenic bacteria affecting strawberry (*Fragaria* x *ananassa*) cultivars in which they induce marginal chlorosis of leaf

blades, stunting, and fruit malformations. At the ultrastructural level, they induce thickening of the sieve tube walls (Fig. BXII.γ.203). To date, they have not been described in other hosts, and trials to transmit the bacterium via dodder (*Cuscuta campestris*) to periwinkle (*Catharantus roseus*), the experimental plant for phloem-restricted bacteria, have failed. Vector transmission of the bacterium occurs in nature. The insect vector has been identified as *Cixius wagneri* (China). The disease was described in France, and its incidence is highest in strawberry production fields in Southern France. Strawberry nurseries, located in northern France, are not affected. Whether *Candidatus* Phlomobacter fragariae is present in countries other than France has not been investigated.

PROCEDURES FOR TESTING SPECIAL CHARACTERS

Specific identification *Candidatus* Phlomobacter can be identified by amplification and sequencing of the 16S rDNA. By sequence comparisons, two primers specific for phlomobacteria have been selected on the 16S rDNA sequence. They are efficient and specific for *Candidatus* Phlomobacter detection in plants. When used for DNA amplification in insects, these primers are not specific, as *Candidatus* Phlomobacter shares strong homologies with insect bacterial symbionts and parasites as well as with enterobacteria. Thus, specific identification from hosts other than plants requires sequencing of the 16S rDNA. Genes other than 16S rDNA have been isolated and are under study.

ACKNOWLEDGMENTS

Monique Garnier-Semancik died suddenly in May, 2003. She was Director of Research at INRA Bordeaux and served as director of a "Joined Research Unit" in which INRA researchers and University teachers work together. Most of her scientific contributions were to the field of phloem- and xylem-restricted plant pathogenic bacteria, which won her respect both in France and internationally.

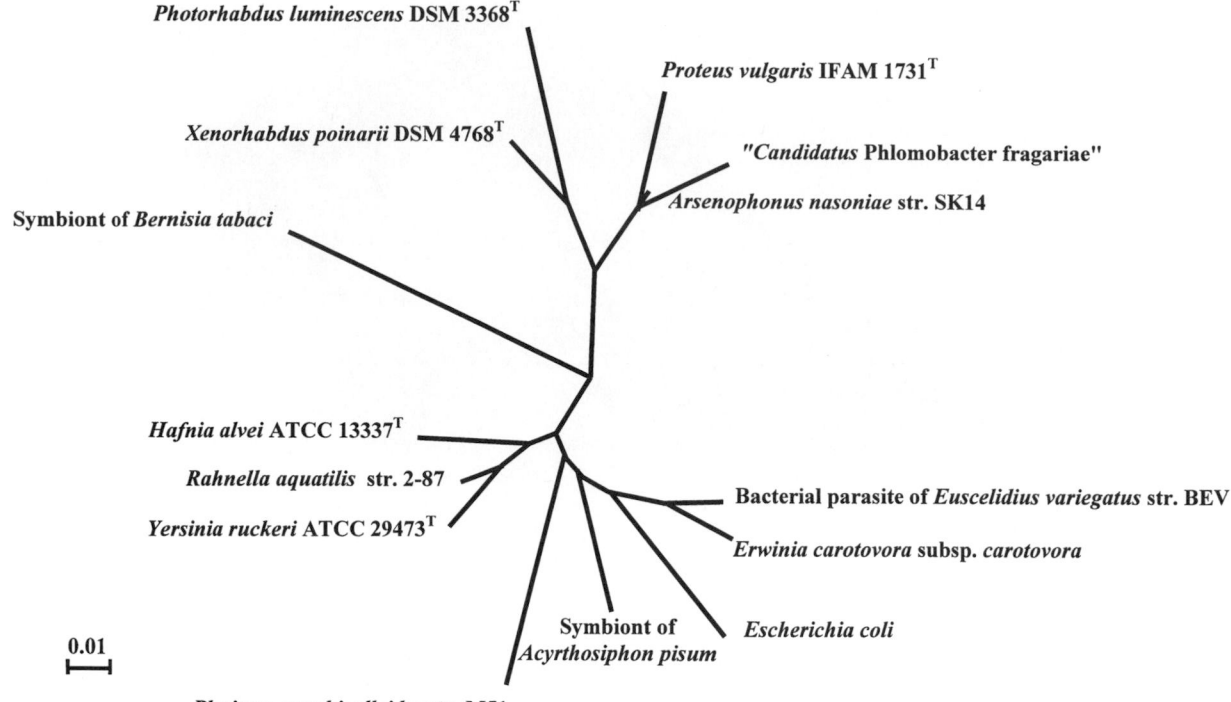

FIGURE BXII.γ.204. Phylogenetic tree constructed with the following 16S rDNA sequences from GenBank using the Suggest Tree Tool of the Ribosomal Database Project (Olsen et al., 1991): *Arsenophonus nasoniae* SK14 (M90801), *Proteus vulgaris* 1731 (X07653), *Hafnia alvei* 13337 (M59155), *Rahnella aquatilis* 2–87 (X79937), *Yersinia ruckeri* 29473 (X75275), *Plesiomonas shigelloides* M61 (M59159), *Acyrthosiphon pisum* symbiont S (M27040), *Escherichia coli* (J01859), parasite of *Euscelidius variegatus* (Z14096), *Erwinia carotovora* (M59149), *Photorhabdus luminescens* 3368 (X82248), *Xenorhabdus poinarii* 4768 (X82253), and symbiont of *Bemisia tabaci* (Z11925). The scale corresponds to 1% substitution. (Reproduced with permission from L. Zreik et al., International Journal of Systematic Bacteriology *48:* 257–261, 1998, ©International Union of Microbiological Societies.)

List of species of the genus Candidatus Phlomobacter

1. *Candidatus* Phlomobacter fragariae Zreik, Bové and Garnier 1998, 260.

   fra.ga' riae. L. gen. n. *fragaria, fragariae* of *fragaria*, strawberry.

   The description of this organism is identical to that of the genus *Candidatus* Phlomobacter.

The mol% G + C of the DNA is: not determined.
GenBank accession number (16S rRNA): U91515.

Genus XXVI. **Photorhabdus** *Boemare, Akhurst and Mourant 1993, 253*[VP]

Noël E. Boemare and Raymond J. Akhurst

Pho.to.rhab' dus. Gr. n. *photo* light; Gr. fem. n. *rhabdus* rod; M.L. masc. n. *Photorhabdus* bioluminescent rod-shaped bacterium.

Asporogenous rod-shaped cells 0.5–2 × 1–10 μm. Cell size is highly variable within and between cultures with occasional filaments up to 30 μm long. In the last stage of exponential growth and during stationary growth period, **spheroplasts may occur with an average of 2.6 μm in diameter** (10–20% of cell population), resulting from the partial disintegration of the cell wall. Proteinaceous **protoplasmic inclusions** are synthesized inside a high proportion of cells (50–80%) during the stationary period. Gram negative, motile by means of peritrichous flagella. Facultatively anaerobic, having both a respiratory and a fermentative type of metabolism. Optimum growth temperature usually ~28°C; some strains grow at 37–38°C. Most strains produce pink, red, orange, yellow, or green **pigmented colonies on nutrient agar, and especially on rich media** (tryptic soy agar, egg yolk agar). Bioluminescent, usually detectable by the dark-adapted eye; intensity varies within and between isolates and may only be detectable by photometer or scintillation counter in some isolates; only very few nonluminescent isolates have been reported. **Spontaneous phase shift occurs in subcultures inducing the appearance of phase II clones** characterized by the loss of neutral red adsorption on MacConkey agar, of production of antibiotics, and of some other properties usually exhibited by wild clones freshly isolated from the natural environment and named phase I variants. Catalase positive. **Do not reduce nitrate.** Negative for oxidase, o-nitrophenyl-β-D-galactopyranoside (ONPG), Voges–Proskauer, arginine dihydrolase, lysine and ornithine decarboxylase tests. Proteolytic for gelatin. Most strains hemolytic for sheep and/or horse blood, some producing an unusual annular hemolysis on sheep blood at 25°C (Fig. BXII.γ.205). Lipolytic activity on Tween 20; many strains lipolytic for Tweens 40, 60, 80, and/or 85. Acid production from glucose without gas. Acid produced from fructose, D-mannose, maltose, ribose, and N-acetylglucosamine; weak for acid production from glycerol. Fumarate, glucosamine, L-glutamate, L-malate, L-proline, succinate, and L-tyrosine are utilized as sole carbon and energy sources. Biochemical identification of *Photorhabdus* within the family *Enterobacteriaceae* is summarized in Table BXII.γ.245. Sequence analyses of 16S rDNA show that all *Photorhabdus* strains branch deeply within the radiation of the family *Enterobacteriaceae*, and have **a specific TGAAAG sequence at positions 208–213** (*E. coli* numbering). **The natural habitat for most strains is the intestinal lumen of entomopathogenic nematodes of the genus** *Heterorhabditis* and insects infected by these nematodes. However, some nonsymbiotic strains have been identified as opportunistic pathogens for humans, not nematodes.

The mol% G + C of the DNA is: 43–45.

Type species: **Photorhabdus luminescens** (Thomas and Poinar

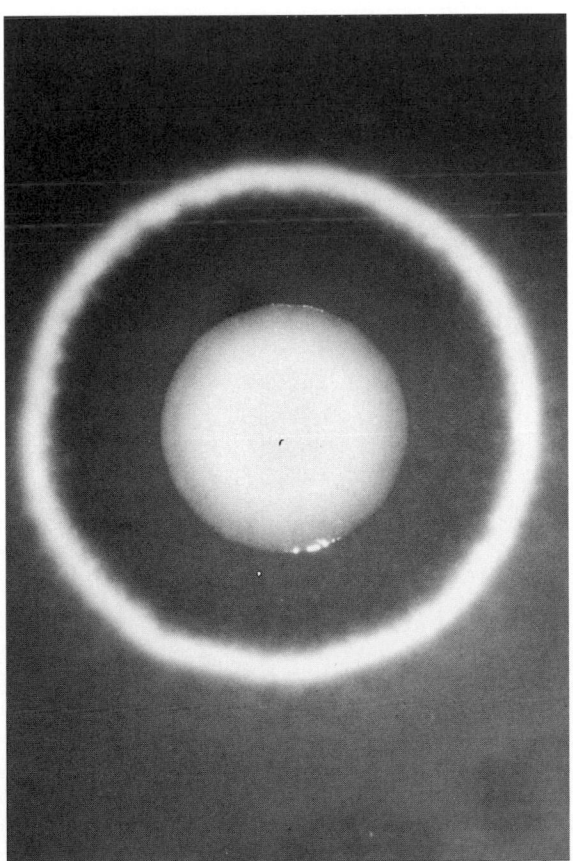

FIGURE BXII.γ.205. Annular hemolysis on sheep blood (10%) agar produced at 25°C by many *Photorhabdus* strains.

1979) Boemare, Akhurst and Mourant 1993, 254 *Xenorhabdus luminescens* Thomas and Poinar 1979, 354.)

FURTHER DESCRIPTIVE INFORMATION

The strains formerly considered as belonging to *Xenorhabdus luminescens* clearly form a DNA relatedness group that is distinct from all the other *Xenorhabdus* strains (Boemare et al., 1993; Akhurst et al., 1996). These DNA data, together with the significant differences in phenotypic characters between "*X. luminescens*" and the other *Xenorhabdus* species (Akhurst and Boemare, 1988; Boemare and Akhurst, 1988), fatty acids of the whole cellular composition (Janse and Smits, 1990), and chemotaxonomic data (Suzuki et al., 1990), led to the transfer of *X. luminescens* into a new genus, *Photorhabdus*, as *Photorhabdus luminescens* comb. nov. (Boemare et al., 1993). Table BXII.γ.245 lists the characteristics of *Photorhabdus* genus for the three species and four subspecies recognized today (Fischer-Le Saux et al., 1999).

Nevertheless, comparison of 16S rDNA sequences of the type strains of *Photorhabdus* and *Xenorhabdus* species indicate the close phylogenetic relationship of these two genera (Rainey et al., 1995). However, all *Xenorhabdus* strains could be clearly distinguished from strains of *Photorhabdus* by the occurrence of a TTCG sequence at positions 208–211 (*E. coli* numbering) of the 16S rDNA, while *Photorhabdus* have a TGAAAG sequence (Szállás et al., 1997). The nearest phylogenetic neighbor is *Proteus vulgaris* as demonstrated by PCR-RFLP (Brunel et al., 1997) and sequencing of 16S rDNA (Rainey et al., 1995; Suzuki et al., 1996; Szállás et al., 1997; Fischer-Le Saux et al., 1999).

The major cellular fatty acids of *Photorhabdus* are $C_{16:0}$ and $C_{18:1}$; $C_{15:0 \, iso}$, $C_{17:0 \, iso}$, and $C_{16:1}$ are major components in some strains (Janse and Smits, 1990; Suzuki et al., 1990). Ubiquinone-8 is the respiratory quinone in all strains (Suzuki et al., 1990).

The main difficulty when subculturing *Photorhabdus* strains is the occurrence of a colonial dimorphism that appears suddenly on agar plates. It can be easily detected by two major properties: dye adsorption and antibiotic production (Akhurst, 1980). Each strain occurs as two phase variants that can be described as morphovar, chemovar, and/or biovar. The bacterium isolated from the infective stage (*dauer* larvae) of *Heterorhabditis* was named the phase I variant (Boemare and Akhurst, 1988). Phase I colonies are mucoid and stick to the loop when streaked on plates, produce antibiotic molecules (Akhurst, 1982a), adsorb dyes when incorporated into agar (e.g., the neutral red in MacConkey agar), and are differently pigmented from phase II variants (e.g., red in phase I and yellow in phase II). Among the *Enterobacteriaceae*, cells of some strains of *Photorhabdus* may be the largest known (0.5–2 × 1–10 μm). Phase I cells are larger than phase II cells; they are pleomorphic, comprising rods (80–90%) and spheroplasts (10–20%), and harbor protoplasmic inclusions (Boemare et al., 1983) (Fig. BXII.γ.206). Phase II appears spontaneously during stationary growth period from *in vitro* culture and during nematode rearing on artificial diets. Phase II colonies are not mucoid, do not adsorb dye, and do not produce antibiotics. Several intermediate colony forms, possessing at least some phase I properties, have been recorded (Gerritsen et al., 1992); however, it is not certain that these are not mixtures of phase I and phase II. In general, variation in phase-related characters has been reported qualitatively ("+" and "−"). However, for every character that can be quantified (e.g., luminescence, antibiotic production), it is clear that the difference between phases is a matter of magnitude, not presence/absence. It is highly probable that this holds true for all phase-related characters. Table BXII.γ.246

TABLE BXII.γ.245. Main characteristics of *Photorhabdus* and characteristics differentiating species and subspecies[a]

Characteristic	*Photorhabdus luminescens*[b]	*P. luminescens* subsp. *luminescens*[b]	*P. luminescens* subsp. *laumondii*[c]	*P. luminescens* subsp. *akhurstii*[d]	*Photorhabdus asymbiotica*[e]	*Photorhabdus temperata*[f]
DNA–DNA hybridization with *E. coli*	4%	4%	4%	4%	4%	4%
Mol% G + C of DNA	43–45	43–45	43–45	43–45	43–45	43–45
Maximum growth temperature, °C	35–39	38–39	35–36	38–39	37–38	33–35
Pathogenicity for insects	+	+	+	+	+	+
Motility	+	+	+	+	+	+
Peritrichous flagella	+	+	+	+	+	+
Protoplasmic inclusions	+	+	+	+	−	+
Bioluminescence	+	+	+	+	+	+
Pigmentation	+	+	+	+	+	+
Dye	+	+	+	+	+	+
Antimicrobial production	+	+	+	+	−	+
Oxidation-fermentation	F	F	F	F	F	F
Catalase	+	+	+	+	+	+
Nitrate reduced to nitrite	−	−	−	−	−	−
Oxidase (Kovac's)	−	−	−	−	−	−
Growth in KCN	−	−	−	−	−	−
Indole production	+	+	+	d	−	[−]
Methyl red	−	−	−	−	−	−
Voges–Proskauer	−	−	−	−	−	−
Simmons citrate	d	+	d	d	+ w	d
Hydrogen sulfide production	−	−	−	−	−	−
ONPG (β-galactosidase)	−	−	−	−	−	−
Esculin hydrolysis	+	+	+	+	+	[+]
Urease (Christensen's)	d	−	[+]	d	+	d
Phenylalanine deaminase	[−]	−	d	−	−	[+]
Tryptophan deaminase	[−]	−	d	−	−	[−] w
Amino acid decarboxylases (Moeller's)						
Lysine decarboxylase	−	−	−	−	−	−
Ornithine decarboxylase	−	−	−	−	−	−
Arginine dihydrolase	−	−	−	−	−	−
D-Glucose, acid production	+	+	+	+	+	+
D-Glucose, gas production	−	−	−	−	−	−
Acid production from:						
D-Adonitol	−	−	−	−	−	−
L-Arabinose	−	−	−	−	−	−
Cellobiose	−	−	−	−	−	−
Dulcitol	−	−	−	−	−	−
Fructose	+	+	+	+	+	+
Glycerol	+ w	+ w	+ w	+ w	+ w	+ w
N-acetylglucosamine	+	+	+	+	+	+
myo-Inositol	d	+	[+]	[+]	d w	d w
Lactose	−	−	−	−	−	−
Maltose	+	+	+	+	+	+ w
D-Mannitol	d	d w	−	+	−	[−]
D-Mannose	+	+	+	+	+	+
Melibiose	−	−	−	−	−	−
α-Methyl-D-glucoside	−	−	−	−	−	−
Raffinose	−	−	−	−	−	−
L-Rhamnose	−	−	−	−	−	−
Ribose	+	+	+	+	+	+
Salicin	−	−	−	−	−	−
D-Sorbitol	−	−	−	−	−	−
Sucrose	−	−	−	−	−	−
Trehalose	[+] w	+ w	[+] w	[+] w	[+]	[+]
D-Xylose	−	−	−	−	−	−
Utilization of:						
L-Fucose	d	d	−	[+]	−	d
DL-Glycerate	[−]	d	−	−	d	+
L-Histidine	d	+	[+] w	d	d	[+]

(continued)

TABLE BXII.γ.245. *(cont.)*

Characteristic	Photorhabdus luminescens[b]	P. luminescens subsp. luminescens[b]	P. luminescens subsp. laumondii[c]	P. luminescens subsp. akhurstii[d]	Photorhabdus asymbiotica[e]	Photorhabdus temperata[f]
myo-Inositol	+	+	+	+	d	[+]
DL-Lactate	[−]	−	−	d w	−	−
D-Mannitol	d	+	−	+	−	[-]
Ribose	+	+	+	+	+	+
L-Tyrosine	+	+	+	+	d	+
Gelatin hydrolysis (Kohn's)	+	+	+	+	+	+
Lecithinase (egg yolk agar)	+	+	+	+	−	+
Lipase (Tween 20)	+	+	+	+	+	+
Lipase (Tween 80)	+	+	+	+	[+]	+
DNase	[−]	−	+	−	−	+
Annular hemolysis at 25°C on:						
Sheep blood agar	d	+	[−]	+	+	+
Horse blood agar	d	+	−	d	+	+

[a]Symbols: +, 90–100% of strains are positive; [+], 76–89% are positive; d, 26–75% are positive; [−], 11–25% are positive; −, 0–10% are positive; F, fermentative. The letter w indicates a weak reaction. Data from Akhurst et al. (1996) and Fischer-Le Saux et al. (1999). All tests were done at 28° ± 1°C except annular hemolysis.

[b]Type strain ATCC 29999 isolated from *Heterorhabditis bacteriophora* group Brecon.

[c]Type strain CIP 105565 isolated from *Heterorhabditis bacteriophora* group HP88.

[d]Type strain CIP 105564 isolated from *Heterorhabditis indica*.

[e]Type strain ATCC 43950 isolated from human blood and/or wounds.

[f]Type strain CIP 105563 isolated from *Heterorhabditis megidis* Palaearctic group.

summarizes the characters affected by phase variation. However, both phases show a similar entomopathogenic effect and share all the other bacteriological properties of members of the genus.

In natural conditions, cells of *Photorhabdus* are carried in the intestinal lumen of the free-living stage (infective juvenile L3 or *dauer* larva) of *Heterorhabditis*. The bacterial cells are stored and do not multiply in the gut of the *dauer* host, which is a nonfeeding stage. When infective juveniles infect an insect, they release their symbionts into the body cavity of the insect prey. *Photorhabdus* cells multiply inducing toxemia and septicemia and the insect dies. The carcass is a sort of monoxenic microcosm where the symbionts eliminate competitive microorganisms by using several antimicrobial barriers, such as antibiotics possessing a wide spectrum of activity (Akhurst, 1982a) and bacteriocins acting against closely related species (Boemare et al., 1992; Baghdiguian et al., 1993). Nematodes reproduce in the insect carcass, feeding on the insect remains metabolized by the symbiotic bacteria and on the bacterial biomass. When the *dauers* escape the insect cadaver to search for new prey, they carry the symbiont in their gut, ensuring the vertical transmission of the mutualistic association. Although the nematode hosts are the natural vectors of their propagation in the insects, in nutritional terms *Photorhabdus* might be considered cntomophilic rather than nematophilic microorganisms.

The life cycle of the clinical strains of *Photorhabdus* is much less certain. *Photorhabdus* has been isolated from five clinical sources in the U.S.A. (Farmer et al., 1989) and six recently in Australia (Peel et al., 1999; Gerrard et al., 2003). Isolations were variously made from tissue, blood, and sputum samples; no definite route of infection has been established. Although some patients may have been immunocompromised, this was definitcly not the case for at least two of the Australian patients. These clinical isolates were all easily cultured on standard media at 37°C. They all exhibited the annular hemolysis on sheep blood agar at 25°C. By the time they were identified, all were phase II cultures; it is not known if that is the form in which they were originally isolated.

In vitro subcultures from fresh isolates can be obtained without major difficulties. *Photorhabdus* are mesophilic bacteria able to grow between 15 and 35°C, and some strains at 37–38°C. Subculturing and all biochemical tests should be undertaken around the optimal temperature at 28°C. Usually, nutrient agar or Luria-Bertani agar are sufficient for growth. On minimal media, nicotinic acid, *para*-aminobenzoic acid, proline, tyrosine, and serine are required as growth factors, the mix of growth factors varying between strains (Grimont et al., 1984b). Minimal medium II (BioMérieux) contains all the necessary requirements to test utilization of organic compounds.

Although the shift from phase I to phase II is spontaneous, it is remarkable that the wild *dauer Heterorhabditis* almost exclusively harbor phase I *Photorhabdus* (Akhurst and Boemare, 1990). The role of phase II variants is not clarified, and today we have no convincing data to support a good explanation of their occurrence. Phase change occurs during the *in vitro* stationary period in a highly unpredictable manner (Akhurst and Boemare, 1990). The two phases of *Photorhabdus* differ significantly in respiratory activity (Smigielski et al., 1994). After periods of starvation, phase II cells recommenced growth within 2–4 h after the addition of nutrients, compared with 14 h for phase I cells, indicating a more efficient nutrient uptake ability in the former.

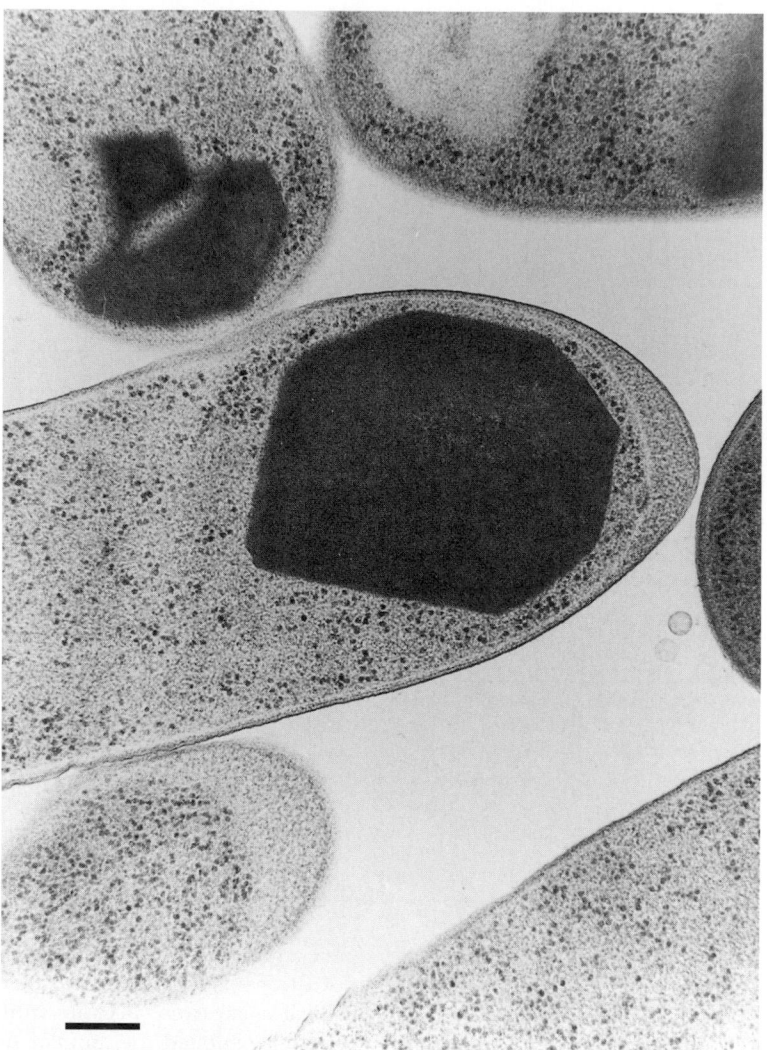

FIGURE BXII.γ.206. Electron micrograph of a fast-freeze fixation and freeze substitution of a culture on nutrient agar in stationary growth of the type strain ATCC 29304 of *Photorhabdus* showing typical inclusions in the protoplasm of the cells. Sections were contrasted with uranyl acetate and lead citrate. Bar = 0.5 μm.

TABLE BXII.γ.246. Differential characteristics of phase variants in *Photorhabdus*[a,b]

Characteristic	Phase I	Phase II
Colonies	mucoid	smooth
Colonial adhesiveness	+	−
Dye adsorption	+	−
Pigmentation	+	+[c]
Protoplasmic inclusions	+	[−] w
Bioluminescence	+	+ w
Antibiotic production	+	−
Lecithinase (egg yolk agar)	+	−

[a]Symbols: +, 90–100% of strains are positive; [−], 11–25% are positive; −, 0–10% are positive. The letter w indicates a weak reaction. All tests were done at 28° ± 1°C.

[b]Characteristics scored in this table are those that always differ between phases, but there are others that differ for the two phases of an individual strain. For those characteristics that can be quantified (e.g., bioluminescence, pigmentation, and protoplasmic inclusions), the difference between phases is a matter of magnitude, not presence/absence. It is highly probable that this holds true for all phase-related characteristics.

[c]Phase II variants of *Photorhabdus* are differently pigmented from phase I variants (e.g., colonies of strain K80 are red in phase I, and yellow in phase II).

Phase II variants may grow a little on complex media previously utilized by phase I variants (Akhurst and Boemare, 1990). However, although reciprocal phase change occurs in *Xenorhabdus*, conversion from phase II *Photorhabdus* to phase I has not been unambiguously demonstrated.

A DNA relatedness study demonstrated unequivocally that phase variation of *Photorhabdus* is a real phenomenon, and not merely an artifact due to contamination, by showing that the two phases of any strain have 100% DNA relatedness. Furthermore, restriction digests and Southern cross analysis of phase I and phase II DNA indicate that the organization of the genome is the same in the two phases (Akhurst et al., 1992).

There are several groups and subgroups within the *Photorhabdus* strains associated with nematodes (*Heterorhabditis* spp.) and human clinical specimens. By using the hydroxyapatite (HA) method to analyze DNA–DNA *Photorhabdus* heteroduplexes, two DNA relatedness groups associated with nematode (*Heterorhabditis* spp.) and another that contains the American human clinical specimens (Farmer et al., 1989) were recognized (Akhurst et al., 1996). More recently, restriction patterns obtained after amplification of the 16S rDNA allowed identification of 12 genotypes among *Photorhabdus* strains (Fischer-Le Saux et al., 1998). By hybridizing the DNA of some representative strains of each ribosomal genotype, and by using the S1 nuclease method, the previous three genomic groups were confirmed (Fischer-Le Saux et al., 1999). They exhibited between them DNA–DNA hybridization values lower than 42% with ΔT_m higher than 8.7°C. Moreover, the phylogenetic trees inferred from the complete 16S rDNA sequence analysis (neighbor joining, parsimony, and maximum likelihood) delineate the same clusters as both DNA–DNA hybridization methods. Consequently, species and subspecies among *Photorhabdus* were delineated by applying a polyphasic approach, combining 16S rDNA, DNA–DNA hybridization, and phenotypic data (Fischer-Le Saux et al., 1999) (Table BXII.γ.245). Based on DNA–DNA hybridization, 16S rDNA and *gyrB* sequencing, and phenotypic data, the new Australian clinical strains (Peel et al., 1999) will probably constitute a subspecies of *P. asymbiotica* (Akhurst et al., unpublished data). A nonluminescent strain (Akhurst and Boemare, 1986) probably constitutes yet another species (Akhurst et al., 1996).

Within *P. luminescens*, subgroups of strains, which shared very high DNA–DNA hybridization values and ΔT_m lower than 1.5°C and were separated by stable 16S rDNA branching, were identified (Fischer-Le Saux et al., 1999). Consequently, *P. luminescens* was divided into three subspecies (Table BXII.γ.245).

The molecular mechanism of phase variation is uncertain. Lipase and protease are regulated at a posttranslational level in the Irish K122 strain of *P. temperata* (Wang and Dowds, 1993), whereas the *lux* genes are posttranscriptionally regulated in the Hm strain (Hosseini and Nealson, 1995). This contrasts with *X. nematophila* in which the flagellin genes are not transcribed in phase II (Givaudan et al., 1996). These preliminary studies seem to indicate that phase variation in *Photorhabdus* is regulated at a different genetic level than in *Xenorhabdus*.

Several strains of *Photorhabdus* harbor plasmids for which no role or gene has been identified. Heating cultures at 45°C (20 min) and subculturing at 28°C, or mitomycin C treatment, may lead to complete lysis of the cultures, suggesting the occurrence of lysogenic strains (Boemare et al., 1992). Phage tail-like particles, different from those of *Xenorhabdus*, are associated with bacteriocin production in *Photorhabdus* (Baghdiguian et al., 1993).

Antibiograms (Bauer et al., 1966) must be done at 28°C and incubated for 3 days to observe clear zones. *Photorhabdus* have large zones of inhibition around disks impregnated with nalidixic acid, gentamicin, streptomycin, kanamycin, tetracycline, and chloramphenicol, but none around penicillin. Resistance is variable from strain to strain with colistin, ampicillin, carbenicillin, and cephalothin (Farmer, 1984b). It is interesting to note that one patient infected with an isolate sensitive to gentamicin *in vitro* did not respond to gentamicin treatment (Peel et al., 1999).

All strains of *Photorhabdus* have been reported to be entomopathogenic, the LD_{50} usually being <100 cells when injected into hemocoel of the insect *Galleria mellonella* (Farmer et al., 1989; Akhurst and Boemare, 1990; Akhurst and Dunphy, 1993). *Photorhabdus* has also been isolated from human wounds and blood in the U.S.A. (Farmer et al., 1989) and more recently in Australia (Peel et al., 1999). None of the infections was lethal, but some required weeks of treatment.

Photorhabdus strains are the only terrestrial bioluminescent bacteria known today. Their *lux* genes share a strong identity with the *lux* genes of the marine luminescent *Photobacterium*, and

Vibrio spp. (Frackman et al., 1990). No satisfactory demonstration has yet been provided to explain the role of luminescence in this genus.

Two ecological niches have been identified for *Photorhabdus*: one as a metabolically active form in the insect host, and the other as a quiescent form in the gut of the nonfeeding dauer nematode. The occurrence of nonsymbiotic clinical strains possessing the phase II properties of the nematophilic symbionts and the remarkable differences in respiratory activity between the two phases of the symbiotic strains suggest that there may be a third niche in the soil. However, experiments to test the symbiotic strains ability to grow and survive in external environments indicate that they disappear within a few days (Morgan et al., 1997). *Photorhabdus* strains may enter into a nonculturable but viable survival strategy, as do *Aeromonas*, *Vibrio*, *E. coli*, and *Salmonella* spp.

The specific interaction with the nematode host has been tested by gnotobiological experiments. These assays demonstrate that *Photorhabdus* isolates do not support culture of any *Steinernema* species *in vitro* (Akhurst, 1983b) but in some combinations support culture of nonhost *Heterorhabditis* spp. (Akhurst and Boemare, 1990; Han et al., 1990). Similarly, *Xenorhabdus* spp. do not support culture of *Heterorhabditis* (Akhurst, 1983b). In all the microbial ecological surveys undertaken, a *Xenorhabdus* isolate has never been recovered from *Heterorhabditis* or a *Photorhabdus* isolate from *Steinernema*. Despite the intestinal location of these bacteria, which would allow environmental contamination, the specificity of *Photorhabdus* for *Heterorhabditis* is a remarkable feature of this symbiosis (Boemare et al., 1997a; Forst et al., 1997). When accurate microbial ecology studies are undertaken, using a simple and fast method of PCR-RFLP of 16S RNA (Brunel et al., 1997), a clear correspondence between *Photorhabdus* isolates and nematode species can be seen. From a total of 75 isolates identified in the Caribbean region (Fischer-Le Saux et al., 1998), two genotypes were associated only with *Heterorhabditis bacteriophora* and another two only with *Heterorhabditis indica*.

Photorhabdus–Heterorhabditis and *Xenorhabdus–Steinernema* symbioses are widely divergent. Similar patterns of infectivity, life cycle, and mutualism with nematodes have to be considered to be the result of evolutionary convergence. The symbiotic, pathogenic, and phase variation properties that are the conditions for such associations (Boemare and Akhurst, 1994) do not necessarily imply the same physiological mechanisms. As more genes are sequenced, it will be possible to formulate a clearer picture to explain the convergent evolution of *Photorhabdus–Heterorhabditis* and *Xenorhabdus–Steinernema* symbioses (Boemare and Akhurst, 1994).

ENRICHMENT AND ISOLATION PROCEDURES

Three methods have been used for isolating *Photorhabdus* from nematodes. The "hanging drop" uses a sterile drop of insect hemolymph to which surface-disinfected dauer stage *Heterorhabditis* are added (Poinar and Thomas, 1966). The nematodes exsheath and commence development, releasing their symbiont, which can be subcultured after about 24 h. A second method is to collect under sterile conditions a drop of insect hemolymph from an insect 24 h after infection by *Heterorhabditis*, and to streak it onto nutrient agar. The third method is to crush about 100 surface-disinfected dauer *Heterorhabditis* that have lost their second stage cuticle and to streak the macerate onto nutrient agar (Akhurst, 1980). This latter method is the most rigorous method for assessing the microflora of the intestine of entomopathogenic nematodes, provided that a suitable control on the effectiveness of the surface disinfection is employed. It reveals the occurrence of *Photorhabdus* in every *Heterorhabditis* sp.

All isolations from nematodes or infected insects should be conducted at or below 28°C.

Human clinical isolates of *Photorhabdus* have been variously obtained from open wounds, tissues aspirated from unerupted lumps, blood, and sputum. These clinical isolates and those from two subspecies of *P. luminescens* can be cultured at 37–38°C and 38–39°C, respectively (Table BXII.γ.245).

MAINTENANCE PROCEDURES

The standard methods of freeze-drying and low temperature storage (liquid nitrogen or at −80°C) used for *Enterobacteriaceae* are also useful for long-term storage of *Photorhabdus* strains; −20°C is unsatisfactory. Cultures do not survive more than a few months in broth or on agar plates at room temperature, and phase variation is likely to occur in this time. Cultures can be routinely maintained for 1 month at 15°C, but storage at 4°C is unsuitable; to prevent phase variation, phase I clones have to be subcultured every week from the neutral red dye-adsorbing clones on MacConkey agar.

PROCEDURES FOR TESTING SPECIAL CHARACTERS

With the exception of the test for annular hemolysis, which should be conducted at 25°C, all biochemical tests used for phenotypic characterization of *Photorhabdus* should be conducted at 28°C, its optimal growth temperature.

Details of the techniques for identifying the phase variants have been summarized (Boemare et al., 1997b). Adsorption of dyes as described by Akhurst (1980) is the most convenient test to characterize the phase variants. MacConkey agar, or better still MacConkey agar without the bile salts, is a good medium for distinguishing phase I variants (red colonies) from phase II variants (off-white or yellow) (Boemare and Akhurst, 1988). On the NBTA medium [1] described by Akhurst (1980), the adsorption of bromothymol blue by *Photorhabdus* may be confused by the pigmentation of strains and the resulting color of the clones can be difficult to distinguish. As most *Photorhabdus* are pigmented, growth on nutrient agar is often sufficient to differentiate clones of the variants, which differ significantly in not only pigmentation but also colony morphology, with phase I being mucoid and convex whereas phase II is non-mucoid and flattened.

To test antibiotic production by *Photorhabdus*, clones of both variants are spot-inoculated on nutrient agar plates. After growth (generally 48 h), cultures are killed by chloroform vapor (30 min) and covered by fresh nutrient semisolid agar (0.6%) inoculated with a bacterial indicator such as *Micrococcus luteus* (Akhurst, 1982a). The inhibition halos of the indicator culture indicate the phase I variants.

Bioluminescence for most of phase I variants can be checked in a darkroom after 10 min for dark-adaptation of the eyes. To assess absence or weakness of such a light production in the phase II variants, by comparison with the phase I variants' luminescence, a scintillation counter, a fluorimeter, or a photomultiplier have to be used. A loopful of an agar culture of each phase variant culture is suspended in 10 ml of distilled water in a scintillation vial for immediate counting with a fully opened window setting (Grimont et al., 1984b).

DIFFERENTIATION OF THE GENUS *PHOTORHABDUS* FROM OTHER GENERA

Photorhabdus strains are easily distinguished phenotypically from all the *Xenorhabdus* spp. Bioluminescence and catalase, both physiologically very significant characters, are positive for *Photorhabdus* and negative for *Xenorhabdus*. In most strains, urease is positive and assimilation of DL-lactate is negative for *Photorhabdus* (Table BXII.γ.245); annular hemolysis on sheep blood agar was only observed with *Photorhabdus* strains (Tables BXII.γ.245 and BXII.γ.247). Bioluminescence does not cause confusion with the light-emitting marine bacteria of different families. *Photorhabdus* is differentiated from *Vibrio*, *Alteromonas*, and *Photobacterium* by having peritrichous, not polar, flagella and in not requiring sodium ions for growth.

TAXONOMIC COMMENTS

By DNA–DNA hybridization *Photorhabdus* is only 4% related to *Escherichia coli*, the type species of the type genus for the family (Table BXII.γ.245). However, *Photorhabdus* possesses the enterobacterial common antigen (Ramia et al., 1982). These data indicate that, while most biochemical tests used for differentiation of the *Enterobacteriaceae* are negative for *Photorhabdus* and although it is only distantly related to the genera that comprise the "core" of the family (Farmer, 1984b), *Photorhabdus* should be retained in the family *Enterobacteriaceae*.

It appears that the genus *Photorhabdus* evolved after the main radiation of *Xenorhabdus* species occurred (Forst et al., 1997) (see Fig. BXII.γ.213 in Genus *Xenorhabdus*). This conclusion, originally derived from the analysis of only seven strains (Rainey et al., 1995), was verified when the 16S rDNA analysis was extended to a balanced number of data sets from *Photorhabdus* and *Xenorhabdus* in comparison to other genera and families (Suzuki et al., 1996; Szállás et al., 1997; Fischer-Le Saux et al., 1999). *Photorhabdus*, *Xenorhabdus*, *Proteus*, and *Arsenophonus* can be considered sister genera because they do not branch to a common ancestor with any other group of the family *Enterobacteriaceae* (Liu et al., 1997a).

Poinar isolated a bacterium from a new nematode, *Heterorhabditis bacteriophora* (Poinar, 1975), which was apparently similar to a *Xenorhabdus* sp. but noticeably bioluminescent (Poinar et al., 1977). Poinar et al. proposed to include this luminescent species in the genus as *X. luminescens* (Thomas and Poinar, 1979). The genus *Photorhabdus* was created in 1993, according to the series of arguments described above, to accommodate this species (Boemare et al., 1993). At the present time, it includes three species, *P. luminescens*, *P. temperata*, and *P. asymbiotica*, one of which contains three subspecies, *P. luminescens* subsp. *luminescens*, *P. lumi-*

TABLE BXII.γ.247. Differential characteristics of *Photorhabdus*, *Xenorhabdus*, and *Proteus*[a]

Characteristic	Xenorhabdus	Photorhabdus	Proteus
Bioluminescence	−	+	−
Catalase	−	+	+
Annular hemolysis on sheep blood agar (25°C)	−	d	−
Urease	−	d	+
Indole	−	d	d
H₂S production	−	−	[+]
Nitrate reductase	−	−	+
Acid from D-mannose	+	+	−

[a]Symbols: +, 90–100% of strains are positive; [+], 76–89% are positive; d, 26–75% are positive; −, 0–10% are positive.

---

1. NBTA medium: nutrient agar containing 0.0025% (w/v) bromothymol blue and 0.004% (w/v) triphenyltetrazolium chloride (Akhurst, 1980).

nescens subsp. *laumondii*, and *P. luminescens* subsp. *akhurstii* (Fischer-Le Saux et al., 1999). According to the rule 65 (2) of the Code of Nomenclature, generic and subgeneric names that are modern compounds from two or more Latin or Greek words take the gender in the original language of the last component of the compound word. Consequently, *Photorhabdus* genus ending by *rhabdus* (from *rhabdos*, rod in Greek as a feminine word), becomes in modern Latin a feminine word, explaining the feminine names of species and subspecies when they are adjectives (Euzéby and Boemare, 2000). Subspecies names referring to a person take the gender of the person, such as *P. luminescens* subsp. *laumondii* and *P. luminescens* subsp. *akhurstii*.

FURTHER READING

Akhurst, R.J., R.G. Mourant, L. Baud and N.E. Boemare. 1996. Phenotypic and DNA relatedness between nematode symbionts and clinical strains of the genus *Photorhabdus* (*Enterobacteriaceae*). Int. J. Syst. Bacteriol. *46*: 1034–1041.

Boemare, N.E., R.J. Akhurst and R.G. Mourant. 1993. DNA relatedness between *Xenorhabdus* spp. (*Enterobacteriaceae*), symbiotic bacteria of entomopathogenic nematodes, and a proposal to transfer *Xenorhabdus luminescens* to a new genus, *Photorhabdus* gen. nov. Int. J. Syst. Bacteriol. *43*: 249–255.

Fischer-Le Saux, M., V. Viallard, B. Brunel, P. Normand and N.E. Boemare. 1999. Polyphasic classification of the genus *Photorhabdus* and proposal of new taxa: *P. luminescens* subsp. *luminescens* subsp. nov., *P. luminescens* subsp. *akhurstii* subsp. nov., *P. luminescens* subsp. *laumondii* subsp. nov., *P. temperata* sp. nov., *P. temperata* subsp. *temperata* subsp. nov. and *P. asymbiotica* sp. nov. Int. J. Syst. Bacteriol. *49*: 1645–1656.

Forst, S., B. Dowds, N.E. Boemare and E. Stackebrandt. 1997. *Xenorhabdus* and *Photorhabdus* spp.: bugs that kill bugs. Annu. Rev. Microbiol. *51*: 47–72.

Szallas, E., C. Koch, A. Fodor, J. Burghardt, O. Buss, A. Szentirmai, K.H. Nealson and E. Stackebrandt. 1997. Phylogenetic evidence for the taxonomic heterogeneity of *Photorhabdus luminescens*. Int. J. Syst. Bacteriol. *47*: 402–407.

List of species of the genus Photorhabdus

1. **Photorhabdus luminescens** (Thomas and Poinar 1979) Boemare, Akhurst and Mourant 1993, 254[VP] (*Xenorhabdus luminescens* Thomas and Poinar 1979, 354.)

lu.mi.nes' cens. M.L. pres. part. *luminescens* luminescing, for its luminescence.

Large rods (2 × 0.5 to 6 × 1.4 µm). Occurs as two phase variants. Luminous, with luminescence more than 100-fold greater in phase I. Maximum growth temperature in nutrient broth occurs at 35–39°C. Indole positive. Weak acid produced from fructose, N-acetyl-glucosamine, glucose, glycerol, maltose, D-mannose, ribose, and trehalose by most strains. Some strains acidify D-mannitol. Proteinaceous inclusions in the protoplasm of phase I cells; poorly produced in phase II. The natural habitat is in the intestinal lumen of entomopathogenic nematodes of *Heterorhabditis bacteriophora* (Brecon and HP88 groups), and of *Heterorhabditis indica*. Three subspecies have been described.

The mol% G + C of the DNA is: 43–45 (Bd).
Type strain: Hb, ATCC 29999, DSM 3368.
GenBank accession number (16S rRNA): X82248, D78005.

a. **Photorhabdus luminescens** *subsp.* **luminescens** (Thomas and Poinar 1979) Boemare, Akhurst and Mourant 1993, 254[VP] (*Xenorhabdus luminescens* Thomas and Poinar 1979, 354.)

Maximum growth in nutrient broth occurs at 38–39°C. Esculin hydrolysis positive and indole weakly positive. DNase, tryptophan deaminase, and urease negative. Annular hemolysis of sheep and horse blood agars. Does not use DL-lactate as sole source of carbon. D-mannitol used as sole source of carbon and energy. Symbiotically associated with nematodes from the group Brecon of *H. bacteriophora*, the type species of the genus *Heterorhabditis* (Poinar, 1975).

The mol% G + C of the DNA is: 43–45 (Bd).
Type strain: Hb, ATCC 29999, DSM 3368.
GenBank accession number (16S rRNA): X82248, D78005.

b. **Photorhabdus luminescens** *subsp.* **akhurstii** Fischer-Le Saux, Viallard, Brunel, Normand and Boemare 1999, 1654[VP]

ak.hurs' ti.i. M.L. gen. n. *akhurstii* of Akhurst, referring

to Dr. R. Akhurst, a major contributor to the bacteriological symbionts of entomopathogenic nematodes.

Maximum growth in nutrient broth occurs at 38–39°C. Esculin hydrolysis positive. Tryptophan deaminase and DNase negative. Urease and indole variable. Annular hemolysis observed on sheep blood agar and, in some strains, on horse blood agar. Utilization of DL-lactate as sole source of carbon variable; weak when positive. D-mannitol used and acidified. DL-glycerate utilization negative. Symbiotically associated with the nematode *H. indica* isolated in warm regions; the first strain (strain D1) was isolated from Australia (Darwin, Northern Territory) by Dr. R. Akhurst.

The mol% G + C of the DNA is: 43–45 (Bd).
Type strain: FRG04, CIP 105564.
GenBank accession number (16S rRNA): AJ007359.

c. **Photorhabdus luminescens** *subsp.* **laumondii** Fischer-Le Saux, Viallard, Brunel, Normand and Boemare 1999, 1654[VP]

lau.mon' di.i. M.L. gen. n. *laumondii* of Laumond, referring to Dr. C. Laumond, a major contributor to the use of entomopathogenic nematode/bacterial complexes for insect pest control.

Maximum growth in nutrient broth occurs at 35–36°C. Esculin hydrolysis, indole, and DNase positive. Tryptophan deaminase variable and urease mostly positive. Total hemolysis on sheep and horse blood agars; the *Photorhabdus* annular reaction is rare. Does not use L-fucose, DL-glycerate, DL-lactate, or D-mannitol. Symbiotically associated with nematodes of the group HP88 of *H. bacteriophora* isolated in South and North America, southern Europe, and Australia, responding to the satellite DNA probe of the nematode strain HP88 provided by the team of Dr. C. Laumond (Grenier et al., 1996).

The mol% G + C of the DNA is: 43–45 (Bd).
Type strain: TT01, CIP 105565.
GenBank accession number (16S rRNA): AJ007404.

2. **Photorhabdus asymbiotica** Fischer-Le Saux, Viallard, Brunel, Normand and Boemare 1999, 1655[VP]

a.sym.bio' ti.ca. Gr. pref. *a* not; M.L. adj. *symbioticus, -a., -um* living together; M.L. fem. adj. *asymbiotica* not symbiotic.

Rods (2 × 0.5 to 3 × 1.0 μm). Maximum growth in nutrient broth occurs at 37–38°C. Yellow or brown pigment. No phase I isolates have been detected, and isolates do not absorb dyes, sometimes weakly produce antibiotics, and are negative for lecithinase on egg yolk agar. Positive for urease, esculin hydrolysis, and for Christensen's citrate, but weakly positive on Simmons' citrate. Tryptophan deaminase negative. Indole and DNase negative. Acid produced from fructose, *N*-acetyl-glucosamine, glucose, maltose, D-mannose, and ribose; weakly produced from glycerol. Proteinaceous inclusions poorly produced. Tween 40 esterase variable. Annular hemolysis on sheep and horse blood agars. Does not use L-fucose, DL-lactate, or D-mannitol. Natural habitat uncertain. All isolates obtained from human clinical specimens.

The mol% G + C of the DNA is: 43–45 (Bd).

Type strain: 3265-86, ATCC 439500.

GenBank accession number (16S rRNA): Z76755.

3. **Photorhabdus temperata** Fischer-Le Saux, Viallard, Brunel, Normand and Boemare 1999, 1655[VP]

tem.pe.ra' ta. L. fem. part. adj. *temperata* moderate, because this species grows at moderate temperature.

Large rods (2 × 0.5 to 6 × 1.4 μm). Occurs as two phase variants. Highly luminous. Maximum growth temperature in nutrient broth occurs at 33–35°C. DNase positive. Most of the strains are indole negative. Esculin hydrolysis and phenylalanine deaminase mostly positive. Urease variable. Acid produced from fructose, *N*-acetyl-glucosamine, glucose, D-mannose, and ribose; weak acid production from glycerol and maltose. Proteinaceous inclusions in protoplasm of phase I cells; poorly produced in phase II. Annular hemolysis often occurs on sheep and horse blood agars. Uses DL-glycerate, and does not use DL-lactate as sole source of carbon. D-mannitol is not used by most strains. The natural habitat is in the intestinal lumen of entomopathogenic nematodes of *Heterorhabditis megidis*, of group NC of *H. bacteriophora*, and of *H. zealandica*.

The mol% G + C of the DNA is: 43–45 (Bd).

Type strain: XlNach, CIP 105563.

GenBank accession number (16S rRNA): AJ007405.

Genus XXVII. **Plesiomonas** *Habs and Schubert 1962, 324*[AL]

J. MICHAEL JANDA

Ple.si.o.mo' nas. Gr. masc. n. *plesios* neighbor; fem. n. *monas* unit, monad; M.L. fem. n. *Plesiomonas* neighbor monad (to *Aeromonas*).

Rod-shaped, straight cells, 0.8–1.0 × 3.0 μm (Janda, 1998). Endospores not produced. Gram negative. **Motile.** Growth on common basal agars with most strains growing on mineral media containing ammonium salts and glucose as sole nitrogen and carbon sources (Schubert, 1984). Chemoorganotrophic with both a respiratory- and fermentative-type metabolism. **Facultatively anaerobic. Acid from the catabolism of D-glucose, without gas.** Anaerogenic fermentation of a limited number of carbohydrates **including *m*-inositol. Oxidase and catalase positive. Lysine- and ornithine-decarboxylase and arginine dihydrolase positive. NaCl supplementation not required for growth. Most strains susceptible to 10 μg and 150 μg of the vibriostatic agent O/129 (2,4-diamino-6,7-diisopropylpteridine).** Few extracellular enzymes elaborated (Schubert, 1984). **Starch not hydrolyzed.** 16S rDNA and 16S rRNA analysis indicate plesiomonads belong to the *Gammaproteobacteria*, with maximum homology to members of the *Enterobacteriaceae* (Martinez-Murcia et al., 1992a; Ruimy et al., 1994). Unique genus-specific signature oligonucleotides have not been identified.

The mol% G + C of the DNA is: 51 (Sebald and Véron, 1963).

Type species: **Plesiomonas shigelloides** (Bader 1954) Habs and Schubert 1962, 324 (*Pseudomonas shigelloides* Bader 1954, 455.)

FURTHER DESCRIPTIVE INFORMATION

Contain16S rRNA signature oligonucleotides CUAAUACCG, YCACAYYG (Y = pyrimidine), CUAACUYYG, and UCACAC-CAUG at positions 170, 315, 510, and 1410 of the *Escherichia coli* numbering system indicative of *Gammaproteobacteria* (Stackebrandt et al., 1988; Martinez-Murcia et al., 1992a). Although previously classified as *Vibrionaceae*, recent phylogenetic data indicates significant evolutionary divergence between *Plesiomonas* and both the *Vibrio/Photobacterium* and *Aeromonas* groups. 5S rRNA, 16S rRNA/rDNA, and 23S rRNA sequence data demonstrate a close phylogenetic relationship between plesiomonads

and the *Enterobacteriaceae*, especially the genus *Proteus* (MacDonell et al., 1986; East et al., 1992; Martinez-Murcia et al., 1992a; Ruimy et al., 1994).

P. shigelloides typically appears as short, straight rods without observable curvature. Filamentous bacilli, two to three times the normal cell length, can occur in response to various conditions including exposure to β-lactams (Brenden et al., 1988). Capsules not detected. A heat-labile substance, termed a capsular K or masked antigen by Sakazaki (1984), is produced that inhibits agglutination of cells with homologous O antisera (Farmer et al., 1992). Two types of unsheathed flagella produced. Older cultures express lophotrichous flagellation consisting of a tuft of one to five polar flagella with a wavelength of 3.5–4.0 μm (Ewing et al., 1961; Schubert, 1984). Young cultures (18 h, 25°C) grown on peptone or nutrient based solid media produce peritrichous, curly, flagella with a shorter periodicity of 1.3–1.7 μm (Ewing et al., 1961; Inoue et al., 1991). Up to 68% of young *P. shigelloides* cultures produce lateral flagella when stained with the Flagellar Staining Solution–Shionogi (Inoue and Shimada, 1990; Inoue et al., 1991). Inclusion bodies present in log phase cells consist of polyphosphate granules containing phosphate, potassium, and magnesium (Pastian and Bromel, 1984; Ogawa and Amano, 1987). Laybourn or methylene blue stained inclusions can have bipolar ("safety pin") or central intracellular locations depending on the growth phase (Ogawa and Amano, 1987; Brenden et al., 1988). Intracellular poly-β-hydroxybutyrate granules not formed.

The enterobacterial common antigen (ECA) found exclusively in members of the *Enterobacteriaceae* is present in *Plesiomonas* (Whang et al., 1972; Ramia et al., 1982). ECA appears to be a heteropolymer consisting of *N*-acetylated amino sugars linked to the lipopolysaccharide moiety (Mäkela and Mayer, 1976). Lipid A analysis reveals a common 1,4'-bis-phosphorylated-β-1,6-linked glucosamine backbone with six fatty acid residues either ester-

or amide-linked (Basu et al., 1985). Cellular fatty acid analysis indicates an overall composition similar to other enteric bacteria and *Vibrionaceae* with hexadecanoate ($C_{16:0}$), hexadecenoate ($C_{16:1}$), and octadecanoate ($C_{18:0}$) as major peaks (Lambert et al., 1983; Yamamoto et al., 1991a). Putrescine, cadaverine, and spermidine are the predominant polyamines synthesized (Yamamoto et al., 1991a).

Ultrastructural analysis of *P. shigelloides* reveals a typical cell wall, cytoplasm, and nucleoplasm characteristic of Gram-negative *Bacteria* (Brenden et al., 1988). Electron-dense granules, 50–150 nm in diameter, can be detected as early as 4 h of incubation with maximum size (>500 nm) reached within 12–24 h (Ogawa and Amano, 1987). Electron-transparent vacuoles have also been described (Brenden et al., 1988). Thin section electron micrographs of ruthenium-stained cells demonstrate a string-like acidic mucopolysaccharide material 35 nm thick emanating from the cell surface (Fig. BXII.γ.207). Fimbriae have not been detected.

Plesiomonads produce uniform turbidity in liquid culture without pellicle formation. Autoagglutination rarely observed. Brain heart infusion broth cultures of *P. shigelloides* produce cells with a high surface charge but low hydrophobicity (Abbott et al., 1991). Colonies on nutrient and blood agars are 1.0–2.0 mm in diameter and are smooth, gray, convex, and opaque with an entire edge. Nonhemolytic. Multiple morphovars arising from individual colonies have been reported (Farmer et al., 1992). Cell-associated or diffusible pigments not elaborated; luminescence not observed. Most *Plesiomonas* produce 1.0–2.0 mm flat, colorless colonies on MacConkey, xylose–lysine–desoxycholate, Hektoen enteric, and desoxycholate agars after 24 h growth (Brenden et al., 1988). Delayed lactose fermentation on MacConkey agar by some *P. shigelloides* leads to lactose-positive variants

arising upon prolonged incubation, which suggests mixed populations. Salmonella–Shigella, eosin methylene blue, and brilliant green agars are inhibitory to some strains (Brenden et al., 1988).

Growth reported over a wide temperature range (8–45°C); optimal growth occurs at 37–38°C (Schubert, 1984; Miller and Koburger, 1986b). Most plesiomonads multiply in liquid media containing 1–4% NaCl; growth in the presence of 5% NaCl reported to be medium-dependent (Miller and Koburger, 1986b). The pH range for growth has been reported to occur between 4.0–9.0 (Schubert, 1984; Miller and Koburger, 1986b). No growth under extremely acidic (pH 3.0) or alkaline (pH 9.5) conditions.

The genus is biochemically homogeneous. Indole formed. Nitrates reduced. The methyl red test is positive. The Voges–Proskauer test (25°C, 37°C) is negative and acetoin is not produced. No growth in KCN. Citrate, gluconate, malonate, and mucate not utilized; sodium acetate is variable. Phenylalanine deaminase and urease activities absent. The string test (0.5% desoxycholate) is negative. No H_2S produced on triple sugar iron agar slants.

Anaerogenic fermentation of D-glucose, *m*-inositol, maltose, and trehalose. Acid from D-galactose, glycerol, lactose, D-mannose, melibiose, and salicin is variable. No fermentation of adonitol, L-arabinose, cellobiose, dextrin, dulcitol, esculin, α-methyl-D-glucoside, inulin, D-mannitol, melezitose, D-sorbitol, sucrose, L-rhamnose, or D-xylose.

Phosphatase and β-galactosidase (ONPG, *o*-nitrophenyl-β-D-galactopyranoside) positive. β-D-glucuronidase negative. A number of *P. shigelloides* arylamidases and esterases are detected using semiquantitative micromethods with chromogenic substrates (Manafi and Rotter, 1992). Chitinase elaborated (O'Brien and

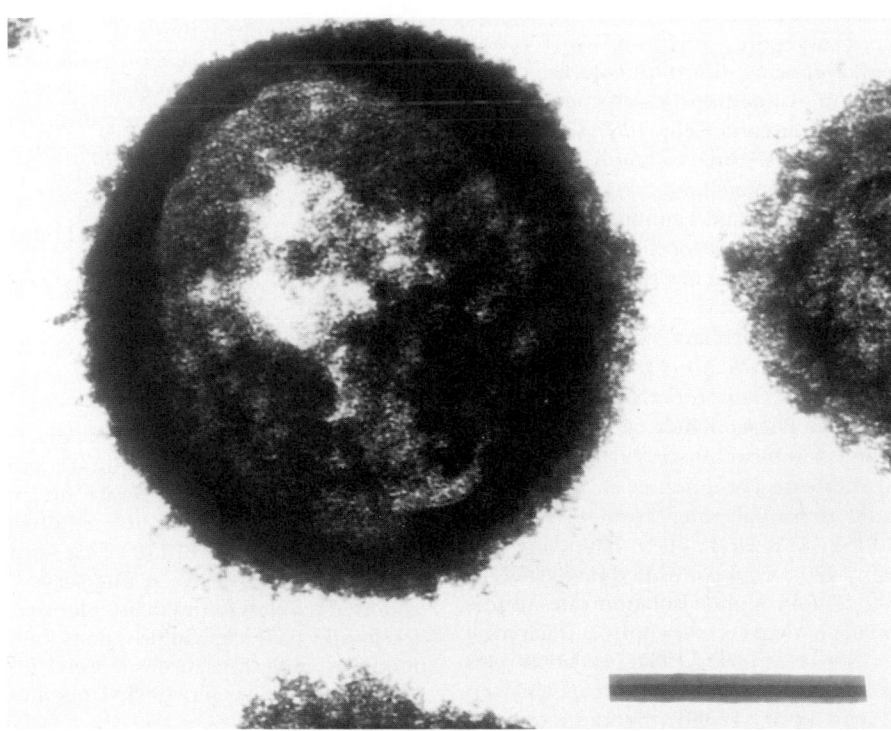

FIGURE BXII.γ.207. Ruthenium-red stained cells of *P. shigelloides* displaying external acidic mucopolysaccharide layer (× 34,000). Bar = 0.5µm. (Reproduced with permission from R.A.Brenden et al., Reviews of Infectious Disease *10*: 303–316, 1988, ©University of Chicago Press.)

Colwell, 1987). Casein, DNA, elastin, fibrinogen, gelatin, poly-pectate, starch, tyrosine, and Tween 80 are neither degraded nor hydrolyzed.

Formal DNA–DNA hybridization studies on *Plesiomonas* have not been published. Anecdotal information from one epidemiological investigation indicates that all 27 *P. shigelloides* strains constituted a single relatedness group by DNA–DNA hybridization (Holmberg et al., 1986b; Don J. Brenner, personal communication). Between 63 and 100% of *P. shigelloides* strains harbor extrachromosomal elements. Plasmids range in size from 2–8 MDa to very large extrachromosomal elements of 120–312 MDa (Herrington et al., 1987; Ølsvik et al., 1990; Abbott et al., 1991). Plasmid-cured derivatives show loss of expression of β-galactosidase activity (Herrington et al., 1987) and resistance to streptomycin (Marshall et al., 1996). *P. shigelloides* also produces a restriction endonuclease (*Psh* AI) with a novel recognition site of 5′-GACNN/NNGTC (Miyahara et al., 1990). Bacteriocin and phage typing systems have not been described.

An international antigenic scheme recognizes 96 somatic (O) and 48 flagellar (H) antigens of *P. shigelloides* (Aldová and Schubert, 1996). Some serovars show partial or complete identity to *Shigella* somatic antigens. Serovar O17, a common group found in clinical material, reacts with *Shigella* group D antisera (Abbott et al., 1991) and is identical to the O antigen of *S. sonnei* (Sakazaki, 1987). An a, b-a, c type of antigenic relationship also exists between serovar O11 and *S. dysenteriae* 8, serovar O22 and *S. dysenteriae* 7, serovar O23 and *S. boydii* 13, serovar O54 and *S. boydii* 2, and serovar O57 and *S. boydii* 9 (Sakazaki, 1987; Shimada et al., 1994a). Certain plesiomonads share type-specific antigens with *S. flexneri* 6 and common group 1 antigen of *S. dysenteriae* 1 (Albert et al., 1993a). A second unrelated scheme based upon "Schubert" antigens was developed with reference strains from pond water and water insects (Aldová and Schubert, 1996). This scheme is primarily useful in typing environmental isolates; 23 "O" and 5 "H" antigens are recognized.

P. shigelloides is uniformly susceptible *in vitro* to first-, second-, and third-generation cephalosporins, fluoroquinolones, carbapenems, monobactams, and trimethoprim-sulfamethoxazole (Reinhardt and George, 1985; Kain and Kelly, 1989; Clark et al., 1990; Clark, 1992). Widespread resistance to ampicillin, amoxicillin, carbenicillin, mezlocillin, piperacillin, and ticarcillin. Variable results described for tetracycline and aminoglycosides. Most strains produce a β-lactamase using the nitrocefin test (Reinhardt and George, 1985; Clark et al., 1990) but are susceptible to antibiotic-β-lactamase-inhibitor combinations.

The gastrointestinal tract is the primary site from which *P. shigelloides* is recovered. While definitive proof establishing *P. shigelloides* as an enteropathogen is lacking, several lines of evidence support a role in gastroenteritis. These include a low carrier rate, case-controlled investigations, well-circumscribed case reports, and outbreaks of diarrheal disease (Brenden et al., 1988). Carriage rates in asymptomatic individuals range from 0–5.5% with most reported values <0.1% (Arai et al., 1980; Holmberg and Farmer, 1984; Bozsó et al., 1986). Case-controlled investigations have generally found 3- to 20-fold higher isolation rates of plesiomonads from symptomatic patients versus controls (Pitarangsi et al., 1982; Bozsó et al., 1986; Lim et al., 1987). Isolation rates of *P. shigelloides* as high as 22% have been reported from children with bloody diarrhea (Taylor et al., 1986b). Sporadic cases of gastroenteritis have documented high numbers of plesiomonads (3×10^8 CFU/ml) in duodenal biopsies (Penn et al., 1982) and seroconversion to the lipopolysaccharide of infecting strains (van Loon et al., 1989). Two Japanese outbreaks of *P. shigelloides* diarrheal disease caused by serovars O17:H2 and O24:H5 have been associated with contaminated drinking water (Tsukamoto et al., 1978). A third less well-defined outbreak was linked to oyster consumption (Rutala et al., 1982).

P. shigelloides gastroenteritis has been reported around the world. Diarrheal disease can range from a mild enteritis to frank dysentery (Holmberg et al., 1986b; Kain and Kelly, 1989; Rautelin et al., 1995b). Cholera-like illnesses are rare (Sawle et al., 1986). All age groups are affected. Common symptoms include diarrhea with abdominal pain/cramps. Less frequent symptoms include fever, nausea, and vomiting. Risk factors associated with infection include consumption of raw shellfish, foreign travel, and exposure to reptiles and tropical fish (Davis et al., 1978; Van Damme and Vandepitte, 1980; Holmberg et al., 1986b; Tippen et al., 1989).

Extraintestinal disease principally involves septicemia. At least 20 cases have been reported in the literature to date. Most cases of bacteremia involve persons with underlying conditions including neonatality, hematologic dyscrasias, and liver disease (Clark and Janda, 1991; Delforge et al., 1995; Riley et al., 1996). Meningitis accompanying septicemia is primarily a disease of neonates (Billiet et al., 1989). Prolonged membrane rupture prior to delivery predisposes infants to infection (Clark and Janda, 1991). Infrequent extraintestinal complications of *P. shigelloides* infection include biliary disease (Claesson et al., 1984; Körner et al., 1992), pancreatic abscess (Kennedy et al., 1990), pseudoappendicitis (Fischer et al., 1988), arthritis (Gordon et al., 1983), osteomyelitis (Ingram et al., 1987), and ocular infections (Butt et al., 1997a).

Virulence factors are poorly defined (Janda, 1998). Extracellular factors elaborated by select *P. shigelloides* strains include heat-labile and heat-stable enterotoxins (Manorama et al., 1983; Gardner et al., 1987; Matthews et al., 1988) and cytotoxins (Abbott et al., 1991; Albert et al., 1993a). Most strains produce a cell-associated hemolysin that may be responsible for cytotoxic activity (Janda and Abbott, 1993b). Invasion of HeLa and HEp-2 cells has been reported (Binns et al., 1984; Albert et al., 1993a). Sereny (keratoconjunctivitis) test negative. Siderophores not produced (Daskaleros et al., 1991). Human volunteers fed 10^3–10^9 CFU of *P. shigelloides* P012 failed to develop diarrhea (Herrington et al., 1987).

Freshwater environments, including mud and sediment, are the major reservoirs for *P. shigelloides*. A Dutch study (Medema and Schets, 1993) found levels of *P. shigelloides* in surface water samples correlated with fecal pollution (*Escherichia coli*) and the trophic state (chlorophyll *a*, Secchi depth). A subsequent Brazilian investigation did not draw a similar conclusion (de Mondino et al., 1995). A survey of the Suwannee River estuary found most water and sediment samples positive for *P. shigelloides* during three sampling periods (August–February). Flora associated with the ecosystem such as eels, crabs, bream, catfish, crappie, pinfish, clams, and oysters also contained plesiomonads (Miller and Koburger, 1986a). Four of 170 (2.4%) seawater samples were found to contain *P. shigelloides* in one survey (Zakhariev, 1971). Other reported animals found to be colonized with *P. shigelloides* include freshwater fish, tropical fish, dogs, cats, sheep, cows, goats, pigs, monkeys, polecats, turtles, newts, toads, and turkey vultures (Miller and Koburger, 1985; Tippen et al., 1989).

ENRICHMENT AND ISOLATION PROCEDURES

A number of selective media have been designed to recover *Plesiomonas* from environmental samples and contaminated clinical specimens (e.g., feces). Most of these rely on brilliant green and/ or bile salts as inhibitors of competing flora and *m*-inositol, a cyclic polyhydroxyl alcohol, as the differential substrate (von Graevenitz and Bucher, 1983; Farmer et al., 1992). While virtually all *P. shigelloides* are inositol positive, only a limited number of enteric groups (*Enterobacter aerogenes*, *Klebsiella* spp., *Serratia marcescens*) are capable of fermenting this alcoholic sugar.

Inositol-brilliant green-bile salts agar (IBB)[1] has been found to be the best selective media for the recovery of plesiomonads from both clinical (von Graevenitz and Bucher, 1983) and environmental specimens (Miller and Koburger, 1986a). On IBB, *Plesiomonas* colonies appear whitish to pinkish. IBB was 90% sensitive and 100% specific for the recovery of *Plesiomonas* from artificially seeded stool specimens. Optimum *Plesiomonas*/coliform ratios were obtained using IBB (von Graevenitz and Bucher, 1983). However, thermally injured cells may not be satisfactorily recovered using IBB (Miller and Koburger, 1986a). Plesiomonas differential agar (PDA), a variation of IBB, was found superior to IBB for the recovery of *P. shigelloides* in a Bangladesh study (Huq et al., 1991). PDA incubated at 42°C produced optimal results with large colonies (3 mm). Alkaline peptone water (pH 8.6) has been found to be the best enrichment medium for plesiomonads by most investigators (von Graevenitz and Bucher, 1983; Janda, unpublished results), although some studies have found this broth to be unsatisfactory (Millership and Chattopadhyay, 1984).

MAINTENANCE PROCEDURES

Plesiomonas strains are sturdier than *Vibrio* or *Aeromonas* isolates but more fragile than common members of the *Enterobacteriaceae*. Working cultures can be maintained on motility deeps for several months. Permanent stock cultures can be prepared by lyophilization or by freezing cultures at −70°C using cryoprotective agents such as bovine serum albumin (fraction V).

PROCEDURES FOR TESTING SPECIAL CHARACTERS

For the demonstration of the properties of *Plesiomonas* strains, the same methods are used as those employed for the characterization of *Aeromonas*, *Vibrio*, and members of the *Enterobacteriaceae* (Schubert, 1984).

DIFFERENTIATION OF THE GENUS *PLESIOMONAS* FROM OTHER GENERA

Although *Plesiomonas* has been transferred to the family *Enterobacteriaceae* based upon phylogenetic relatedness, misidentifica-

tions as a *Vibrio* or *Aeromonas* can occur by virtue of the production of cytochrome oxidase. Key differential tests that separate these taxa are listed in Table BXII.γ.248.

TAXONOMIC COMMENTS

Ferguson and Henderson (1947) initially described *P. shigelloides* as a motile *Enterobacteriaceae* culture ("C27") that possessed the major antigen of *Shigella sonnei* phase I. C2 (now ATCC 14030), however, possessed a number of properties distinct from *S. sonnei* including motility, production of indole, delayed lactose fermentation, and the inability to produce acid from D-mannitol. It was therefore placed in the anaerogenic "Paracolon" group by these authors. Because of polar flagellation, Bader (1954) subsequently placed this C27 in the genus *Pseudomonas*, with the epithet *shigelloides* referring to its *Shigella*-like characteristics (Schubert, 1984). Ewing et al. (1961) proposed that this bacterium be transferred to the genus *Aeromonas* based upon a fermentative metabolism and similarities in morphology, flagella arrangement, and biochemical properties. This recommendation was soon replaced by a proposal to create a new genus for *Aeromonas shigelloides*, *Plesiomonas* (Habs and Schubert, 1962). At least two other proposals have subsequently been made to reassign *P. shigelloides* to either the genus *Fergusonia* or *Vibrio* (Vandepitte et al., 1974).

Current phylogenetic and chemotaxonomic data indicate that plesiomonads are more closely related to the *Enterobacteriaceae* than to any other group (Table BXII.γ.249). These data also include proposed 16S rRNA signature nucleotide differences between the *Enterobacteriaceae* and vibrios (Table BXII.γ.250).

Based on phylogenetic evidence a proposal has been made to reassign *P. shigelloides* to the tribe *Proteeae*, considering the 5S rRNA sequence relatedness to *Proteus mirabilis* and *Proteus vulgaris* (MacDonell and Colwell, 1985). However, this proposal seems untenable as classic defining reactions of the tribe *Proteeae* and the genus *Proteus* are entirely absent in *P. shigelloides* (Table BXII.γ.251).

---

1. Inositol-brilliant green-bile salts agar has the following composition (g/l): proteose peptone (Difco), 10.0; meat extract (Lab Lemco [Oxoid]), 5.0; NaCl, 5.0; bile salts no. 3 (Difco), 8.5; brilliant green (Merck), 0.00033; neutral red (Merck), 0.025; *meso*-inositol (Merck), 10.0; and agar (Difco), 15.0; pH adjusted to 7.2.

TABLE BXII.γ.248. Key features separating *Plesiomonas*, *Aeromonas*, and *Vibrio*

Characteristic	*Plesiomonas*	*Aeromonas*	*Vibrio*
Growth in nutrient broth/agar (Difco) containing:			
0% NaCl	+	+	D
6% NaCl	−	−	+
O/129 susceptibility:			
10 µg	d	−	D
150 µg	d	−	D
String test	−	−	+
Gas from glucose	−	D	−

TABLE BXII.γ.249. Characteristics resulting in the reassignment of *P. shigelloides*[a,b]

Characteristic	*Plesiomonas*	*Enterobacteriaceae*	*Vibrionaceae*	*Aeromonadaceae*
Possession of ECA	+	+	−	−
Sequence relatedness to **Plesiomonas**:				
5S rRNA (% homology)	100	96–97	<86	87
16S rRNA (% homology)	100	93–95	91	91
23S rRNA (% homology)	100	91	nd	88

[a]nd, not determined.

[b]Data from MacDonell et al., 1986; Martinez-Murcia et al., 1992a; East et al., 1992.

TABLE BXII.γ.250. Proposed 16S rRNA signature sequence differences[a]

	Composition in:		
Position(s)	Vibrio	E. coli	Plesiomonas
154–167	C–G	U–A	U–A
446–488	U–G	G–C	G–C
614–626	G–C	C–G	C–G
896–903	U–A	C–G	C–G
1123–1149	G–C	U–A	U–A

[a]Adapted from Dorsch et al., 1992; Martinez-Murcia et al., 1992a.

FURTHER READING

Brenden, R.A., M.A. Miller and J.M. Janda. 1988. Clinical disease spectrum and pathogenic factors associated with *Plesiomonas shigelloides* infections in humans. Rev. Infect. Dis. *10*: 303–316.

Holmberg, S.D. and J.J. Farmer III. 1984. *Aeromonas hydrophila* and *Plesiomonas shigelloides* as causes of intestinal infections. Rev. Infect. Dis. *6*: 633–639.

Miller, M.L. and J.A. Koburger. 1985. *Plesiomonas shigelloides*: an opportunistic food and waterborne pathogen. J. Food Prot. *48*: 449–457.

TABLE BXII.γ.251. Defining characteristics of the *Proteeae* and *Plesiomonas*[a]

Characteristic	Plesiomonas	Proteus	Providencia	Morganella
Phenylalanine deaminase	−	+	+	+
Swarming motility	−	+	−	−
Urea hydrolysis	−	+	D	+
Degradation of L-tyrosine crystals	−	+	+	+
Pigmentation on DL-tryptophan agar	−	+	+	+

[a]Data from Janda and Abbott (1998a).

List of species of the genus Plesiomonas

1. **Plesiomonas shigelloides** (Bader 1954) Habs and Schubert 1962, 324[AL] (*Pseudomonas shigelloides* Bader 1954, 455.) *shi.gel.loi′ des.* M.L. fem. n. *Shigella* a generic name; Gr. suffix *eides* similar; M.L. adj. *shigelloides Shigella*-like.

The description of the genus and species are given in Table BXII.γ.252.

Occurs in freshwater and marine environments and inhabitants of those ecosystems including fish, amphibia, and shellfish. Isolated from a variety of animals. Associated with gastroenteritis in healthy adults and children. Occasionally causes systemic infections primarily in persons who are immunocompromised or have underlying illnesses.

The mol% G + C of the DNA is: 51 (Ch).

Type strain: M51, ATCC 14029, CDC 3085-55, DSM 8224, NCIB 9242.

GenBank accession number (16S rRNA): M59159, X74688.

TABLE BXII.γ.252. Characteristics of *Plesiomonas shigelloides*

Test	Reaction or Result
Oxidase	+
Catalase	+
Motility	+
Amino acid decarboxylases (Møller):	
Ornithine decarboxylase	+
Lysine decarboxylase	+
Arginine dihydrolase	+
Methyl red test	+
Voges–Proskauer (25°C, 37°C)	−
Urea hydrolysis	−
ONPG hydrolysis[a]	+
NO$_3^-$ reduced to NO$_2^-$	+
Utilization of citrate (Simmons)	−
Utilization of malonate, mucate	−
Utilization of acetate	d
Acid production from D-glucose	+
Gas production from D-glucose	−
Acid production from:	
m-Inositol, maltose, trehalose	+
D-Galactose, glycerol, lactose, D-mannose, melibiose, salicin	d
Adonitol, L-arabinose, cellobiose, dextrin, dulcitol, D-mannitol, melezitose, D-sorbitol, sucrose, L-rhamnose, D-xylose	−
Gelatin liquefaction	−
Tyrosine clearing	−
Deoxyribonuclease	−

[a]ONPG, o-nitrophenyl-β-galactopyranoside.

Genus XXVIII. **Pragia** Aldová, Hausner, Brenner Kocmoud, Schindler, Potužníková, and Petráš 1988a, 187[VP]

JIRI SCHINDLER, SR.

Pra′gi.a. L. fem. n. *Pragia* of Prague, the city in which strains of the genus were identified.

Gram-negative, oxidase-negative, catalase-positive, peritrichously flagellated rods conforming to the definition of the family *Enterobacteriaceae*. Facultatively anaerobic. Colonies approximately 0.5 mm in diameter are formed on nutrient, Endo, and MacConkey agars. Colonies formed on desoxycholate-citrate agar have a black center. Nonhemolytic on sheep blood agar. A *Shigella*-like odor is produced on nutrient agar. A pigsty-like odor is produced on Endo agar overlaid with a sloping surface of agar. Growth at 4°C, **but not at 42°C**. Optimal growth is obtained at temperatures between 22 and 37°C. Fermentative, forms acid without gas from D-glucose and D-galactose. Motile, reduces nitrates to nitrite, **utilizes citrate, produces hydrogen sulfide, and oxidizes gluconate**. The majority of strains is positive for the methyl red test and delayed positive for acid production from glycerol and *myo*-inositol. Negative reactions in tests for indole production, Voges–Proskauer test, phenylalanine deaminase, lysine and ornithine decarboxylases, arginine dihydrolase, urease, deoxyribonuclease, malonate, ONPG, growth in KCN, and acid production from L-arabinose, cellobiose, dulcitol, lactose, maltose, D-mannitol, D-mannose, melibiose, raffinose, L-rhamnose, D-sorbitol, D-sorbose, starch, sucrose, and trehalose. Additional biochemical reactions are given in Table BXII.γ.193 of the chapter on *Enterobacteriaceae*.

The mol% G + C of the DNA is: 46–47.

Type species: **Pragia fontium** Aldová, Hausner, Brenner Kocmoud, Schindler, Potužníková, and Petráš 1988a, 187.

FURTHER DESCRIPTIVE INFORMATION

Pragia is a member of the *Gammaproteobacteria*. Pragia DNA is 17% or less related to members of all other genera in *Enterobacteriaceae*. Bacteriocin-like agents called fonticins were observed in five *P. fontium* strains. They were active against other *P. fontium* strains as well as against *Budvicia aquatica*, but were not active against *Escherichia coli* or *Shigella sonnei*. Similar bacteriocin-like agents, budvicins, with the same specificity and properties have been isolated from *Budvicia aquatica*. Fonticins have very narrow inhibition zones on indicator strains, are heat sensitive (inactivated at temperatures between 42 and 55°C) and are resistant to trypsin (Šmarda, 1987). They are of corpuscular nature and resemble contracted tails of T4 bacteriophage (Šmarda, 1987). Isolated from fresh water springs or fountains in the South Bohemia region of the Czech Republic (Aldová et al., 1983, 1988b). One human isolate not associated with disease. There is no evidence of pathogenicity for humans or animals.

DIFFERENTIATION OF THE GENUS *PRAGIA* FROM OTHER GENERA

Pragia is easily distinguishable from leminorellae, *B. aquatica*, *C. freundii*, and salmonellae based on whole cell protein analysis (Aldová et al., 1988a; Schindler et al., 1992). Biochemical tests useful in differentiating *Budvicia* from other H$_2$S-positive genera in *Enterobacteriaceae* are shown in Table BXII.γ.253.

List of species of the genus Pragia

1. **Pragia fontium** Aldová, Hausner, Brenner Kocmoud, Schindler, Potužníková, and Petráš 1988a, 187.[VP]

fon'ti.um. L. gen. pl. n. *fontium* from springs or fountains, the source of isolation of all but one strain.

The characteristics are as described for the genus and as listed in Table BXII.γ.193 in the chapter on *Enterobacteriaceae*. Occurs in drinking water. One isolate from the stool of a healthy human. No indication of pathogenicity for humans or animals.

The mol% G + C of the DNA is: 46–47 (T_m).

Type strain: DRL 20125, HG 16, ATCC 49100, CNTCTC Eb11/82, CCUG 18073, CDC 963-83, DSM 5563.

GenBank accession number (16S rRNA): AJ233424.

TABLE BXII.γ.253. Differentiation of *Pragia* from other H$_2$S-positive genera in *Enterobacteriaceae*[a,b]

Characteristic	Pragia	Budvicia	Edwardsiella	Salmonella	Citrobacter	Proteus
Lysine decarboxylase	−	−	+	v	−	−
Ornithine decarboxylase	−	−	+	+	v	v
Urease	−	+	−	−[c]	v	+
ONPG	−	+	−	−	+	−
Indole production	−	−	v	+	v	v
Citrate	+	−	−	+	+	v
Motility	+	v	+[d]	+	+	+
KCN, growth	−	−	−	−[e]	v	+
L-Arabinose, acid	−	v	v	+[f]	+	−
Gluconate oxidation	+	−	−	−	−	−
Growth, 42°C	−	−	+	−	−	−

[a]For symbols, see standard definitions.

[b]Reactions carried out at 37°C, unless otherwise noted.

[c]Subspecies *Salmonella choleraesuis* subsp. *arizonae* and *Salmonella choleraesuis* subsp. *diarizonae* are positive.

[d]Except for *E. ictaluri*.

[e]Subspecies *Salmonella choleraesuis* subsp. *houtenae* is positive.

[f]Except serovars choleraesuis and typhi.

Genus XXIX. **Proteus** *Hauser 1885, 12*[AL]

JOHN L. PENNER

Pro'te.us. Gr. n. *Proteus* an ocean god able to change himself into different shapes.

Straight rods, 0.4–0.8 × 1.0–3.0 μm. Gram negative. Motile by peritrichous flagella. **Most strains swarm with periodic cycles of migration producing concentric zones, or spread in a uniform film, over moist surfaces solidified with agar or gelatin.** The organisms in this genus conform to the definition of the family *Enterobacteriaceae*. **They are facultatively anaerobic, chemoorganotrophic, having both a respiratory and a fermentative type of** metabolism. Optimal growth temperature is 37°C. **Oxidase negative; catalase positive. Methyl red positive; species vary in indole production, Voges–Proskauer, and Simmons citrate tests. They oxidatively deaminate phenylalanine and tryptophan. Urea is hydrolyzed. Lysine decarboxylase negative and arginine dihydrolase negative; only *Proteus mirabilis* decarboxylates ornithine. All but *Proteus myxofaciens* decompose tyrosine to produce a clearing on**

agar media in which the insoluble amino acid is incorporated. Grow on KCN. H₂S is usually produced. Malonate is not utilized. D-glucose and a few other carbohydrates are catabolized with production of acid and usually gas. Does not produce acid from inositol or from straight chain tetra-, penta-, or hexahydroxy— alcohols, but generally produces acid from glycerol. One or more species ferment maltose, sucrose, trehalose, and D-xylose. Human pathogens, causing urinary tract infections; also are secondary invaders, causing septic lesions at other sites of the body. Occurs in the intestines of humans and a wide variety of animals; also occurs in manure, soil, and polluted waters. *P. myxofaciens* has been isolated only from gypsy moth larvae. Based on 16S rDNA sequence analysis, *Proteus* belongs to the family *Enterobacteriaceae* within the *Proteobacteria* (Woese et al., 1985; Niebel et al., 1987; Stackebrandt et al., 1988).

The mol% G + C of the DNA is: 38–41 (Falkow et al., 1962).

Type species: **Proteus vulgaris** Hauser 1885, 12 emend. Brenner, Hickman-Brenner, Holmes, Hawkey, Penner, Grimont and O' Hara 1995, 870.

FURTHER DESCRIPTIVE INFORMATION

Comparative analysis of 16S ribosomal RNA sequences (the oligonucleotide cataloging method) showed that *P. mirabilis* belongs to the *Gammaproteobacteria*, which includes the family *Enterobacteriaceae* (Woese et al., 1985; Stackebrandt et al., 1988). DNA relatedness studies have shown that members of the genus *Proteus* are only distantly related to the other species in the family *Enterobacteriaceae* and from an evolutionary standpoint they are therefore at the periphery of this family (Brenner et al., 1978). *P. mirabilis* strains are a highly related group. They are 55% related to *P. vulgaris* and to *P. penneri* strains. *P. penneri* consists of a highly related group of indole-negative strains that ferment salicin and hydrolyze esculin. They were classified as *P. vulgaris* biogroup 1 (Brenner et al., 1978) before they were recognized as a new species (Hickman et al., 1982c). At present, *P. vulgaris* consists of two biogroups. *P. vulgaris* biogroup 2 consists of strains that are indole positive, that ferment salicin and hydrolyze esculin. Strains of this biogroup conform to the description of the genus and include the vast majority of the *P. vulgaris* strains isolated from clinical sources. Computerized analysis of electrophoretic protein patterns and DNA relatedness tests showed that *P. vulgaris* biogroup 3 strains (indole positive, salicin negative, and esculin negative) are a heterogeneous group that includes four genomospecies, one of which includes the type strain (Costas et al., 1993; Brenner et al., 1995). Future studies will likely lead to the identification of new species within this biogroup.

On media solidified with gelatin or agar, the bacteria migrate from the point of inoculation to spread over the surface of the medium. The phenomenon, known as swarming, is attributed to the dimorphic nature of the bacteria. The short mononucleoid, peritrichously flagellated rods that occur singly in liquid media, referred to as swimmers, differentiate when transferred to solid media to produce long, nonseptate, multinucleoid swarmers with large increases in the number of flagella (Belas, 1992). Glutamine has been identified as the chemoattractant that signals the surface-induced differentiation (Allison et al., 1993). Differentiation into swarmer cells is associated with marked increases in synthesis of lipopolysaccharide with long O-polymers and a higher fluidity of the outer membrane (Sidorczyk and Zych, 1986). The production also of extracellular slime forming a thin cocoon covering swarms of cells (VanderMolen and Williams, 1977) or capsular polysaccharide containing galacturonic acid

and *N*-acetylgalactosamine facilitates migration of swarmer cells (Stahl et al., 1983; Gygi et al., 1995). Swarmer cells are the virulent forms of *Proteus* and during differentiation a coordinated increase occurs in the synthesis of urease, hemolysins, and proteases (Jin and Murray, 1987; Allison et al., 1992, 1994).

Three phases have been recognized in the process of swarming. In the first phase, differentiation of swimmer cells into swarmer cells takes place. In the second phase, groups of swarmer cells migrate from the point of inoculation and continue to migrate until either a change in direction occurs or the number of swarmers is reduced by a loss of some along the way. In the third phase, migration ceases, which results in consolidation. At this point, the swarmer cells form septa to become short cells that multiply for a period of time before producing another generation of swarmer cells (Williams and Schwarzhoff, 1978; Rózalski et al., 1997). Cyclic repetitions of the phases leads to concentric rings surrounding the site of inoculation. Some strains (or variants) produce a uniform film (referred to as spreading) without periodic cycles (C variant of Belyavin, 1951; Z variant of Coetzee and Sacks, 1960). Some strains neither spread nor swarm and merely form distinct colonies.

The swarming of *Proteus* makes it difficult to isolate bacteria of other species from pathological specimens plated on agar media and to prevent swarming; inhibitors such as bile salts and detergents are incorporated in the media (Kopp et al., 1966). The compound *p*-nitrophenyl glycerol, at concentrations of 0.1–0.3 mM, inhibits swarming without affecting flagellation or motility, and because of its low toxicity to *Proteus* and other bacteria, its use has been advocated (Williams, 1973). With continued incubation, however, the *Proteus* strains eventually swarm in the presence of this compound and thus isolations of *Campylobacter* species and other bacteria that require longer periods of incubation require other measures such as the incorporation of antibiotics in the isolation media (Firehammer, 1987). The incorporation of charcoal in the medium has been reported to suppress swarming without interfering with the growth of other bacterial species (Alwen and Smith, 1967). Other methods of preventing swarming include reduction of the salt concentration of the media and increasing the concentration of agar to 4% (New Zealand agar) or 7% (Japanese agar).

In the Dienes test, swarms of two strains that do not penetrate into each other but form a sharp line between them indicate that the strains are different but strains that produce swarms that merge into each other without a line of demarcation are interpreted to be the same (Dienes, 1946). This test has been used in epidemiological investigations of *Proteus* infections, mostly of *P. mirabilis*, to determine identity and nonidentity of clinical isolates (Story, 1954; Skirrow, 1969). However, strains of different biovars have been found to swarm into each other (Kippax, 1957), and, in some cases, results obtained with the test fail to correlate with bacteriophage typing results (Hickman and Farmer, 1976). The production of the demarcation line is apparently unrelated to the flagellar (H) antigens (Sourek, 1968; Skirrow, 1969) but appears to be dependent on both the bacteriocins produced by the swarming strains and the bacteriocins to which they are susceptible (Senior, 1977a). The Dienes test is useful in combination with other procedures for epidemiological typing and when its limitations are recognized.

A well-known distinguishing feature of the genus is the production of urease. It is an important taxonomic criterion that differentiates the genus from most other species of *Enterobacteriaceae* except for the genera *Morganella* and *Providencia*, some species of which also produce the enzyme. The enzyme is constitutive in most strains of *P. mirabilis* and inducible in *Proteus vulgaris*, *Proteus penneri*, and some *P. mirabilis* strains (Mobley et al., 1987; 1995; McLean et al., 1988; Mobley and Hausinger, 1989). The enzyme is a 212–280 kDa protein consisting of three subunits (Jones and Mobley, 1989; Mobley and Hausinger, 1989) and is known to have an important role in the development of human urinary tract infections (Mobley, 1996).

An important distinguishing feature of *Proteus*, *Providencia*, and *Morganella* is their ability to oxidatively deaminate a variety of amino acids, producing keto acids and ammonia (Bernheim et al., 1935; Stumpf and Green, 1944; Singer and Volcani, 1955). The carboxylic acids bind iron and have siderophore activity (Evanylo et al., 1984; Drechsel et al., 1993). Addition of ferric chloride solution to keto acids in aqueous solution produces different colors dependent upon the amino acid from which the keto acid was produced (Singer and Volcani, 1955), and the same colors are produced when ferric chloride solution is added to bacteria grown on nutrient media supplemented with the amino acids. This is the basis of the diagnostic test used in the clinical laboratory to distinguish these three genera from other *Enterobacteriaceae*. Tests for phenylalanine deaminase and tryptophan deaminase are widely used (Henriksen, 1950; Thibault and Le Minor, 1957). The gene encoding an amino acid deaminase (*aad*) in *P. mirabilis* has been identified (Massad et al., 1995). The probe for the *aad* gene does not hybridize with DNA from *Providencia* and *Morganella* strains, suggesting that the deaminases of these three genera are unrelated. As with ureases, an end product of the deaminases is ammonia, which is now known to have an important role in pathogenesis by elevating pH, but the contributions of the deaminases in the pathogenic process have yet to be determined (Rózalski et al., 1997).

The production of amines by *Proteus* species and *Morganella morganii* when grown in nutrient broth is a distinguishing biochemical property of the two genera (Proom and Woiwod, 1951). The inability of *Providencia rettgeri*, classified during that period as *Proteus rettgeri*, to produce amines under these conditions was cited as an important criterion in their proposal for removing this species from the genus *Proteus*. Simplified tests to detect amines for differentiation of these genera from other *Enterobacteriaceae* in the clinical laboratory have not been developed. However, the amines continue to be of interest, particularly with the finding that a carcinogenic nitrosamine (*N*-nitrosodimethylamine) is produced in urine by *P. mirabilis* during the course of a urinary tract infection (Brooks et al., 1972).

The antigens of the genus *Proteus* that are used for differentiating the serovars are the lipopolysaccharide (O) antigens, the flagellar (H) antigens, and the capsular (K or C) antigens. The O antigenic scheme of Kauffmann and Perch for *P. vulgaris* and *P. mirabilis* (Kauffmann, 1966) lists 49 serovars. Tube agglutination tests and cell suspensions boiled for 1 h were used to determine the O specificities. Additional serovars have been identified that are not included in the scheme of Kauffmann and Perch (Larsson and Olling, 1977; Penner and Hennessy, 1980). Seventeen serovars of the Kauffmann and Perch scheme were found to include only *P. vulgaris* strains, 27 included only *P. mirabilis* strains, and five included strains of both species. Since isolates generally agglutinate in antisera against the same species, separation of the serovars to provide individual schemes for each species facilitated routine serotyping (Penner and Hennessy, 1980). In the system developed by Penner and Hennessy for serotyping *P. vulgaris* and *P. mirabilis*, the passive (indirect) hem-

agglutination technique was employed to determine the specificities of soluble heat extracted antigens. Their system included 22 *P. vulgaris* and 32 *P. mirabilis* strains from the Kauffmann and Perch scheme, five new serovars of *P. vulgaris*, and six new serovars of *P. mirabilis*. However, from the large numbers of untypeable smooth strains isolated in epidemiological studies, it is clear that a substantial number of serovars remain undefined (Larsson and Olling, 1977; Larsson, 1980, 1984; Penner and Hennessy, 1980). Reports differ on the distributions of serotypes of *P. mirabilis*, but among those listed the most frequently occurring are isolates of O serotypes 3, 6, 10, 27, and 28 (Lányi, 1956; de Louvois, 1969; Larsson and Olling, 1977; Penner and Hennessy, 1980; Rózalski et al., 1997). In these earlier studies the strains of *P. vulgaris* were not differentiated into biogroups and thus the collections included a few biogroup 1 strains that would now be classified as *P. penneri* (Hickman et al., 1982c). A system for serotyping *P. penneri* has not been developed. However, one serotypically homogeneous group of strains collected from three countries has been identified and designated group O61 (Sidorczyk et al., 1996). The molecular structures of the lipopolysaccharides of at least 10 different strains of *P. penneri* have been determined (Zych et al., 1997). Examination of the variety of structures suggests the possibility that more serotypes will be defined in the future.

Much research has recently focused on the structures of the lipopolysaccharide (LPS) of *P. vulgaris*, *P. mirabilis*, and *P. penneri*, and progress in this area has been the subject of a major review (Rózalski et al., 1997). The structure of lipid A is comparable to the lipid A of *Escherichia coli* and salmonellae but differs in its fatty acid composition and in the presence of a 4-deoxy-arabinosyl residue that substitutes the phosphate residue of glucosamine (Sidorczyk et al., 1983). The biological activity of *Proteus* lipid A is comparable to that of *E. coli* and salmonellae (Sidorczyk et al., 1978). The core oligosaccharides of *P. vulgaris* and *P. mirabilis* consist of Kdo (3-deoxy-D-*manno*-octulosonic acid), heptoses, and hexoses, as in *E. coli* and salmonellae, but differ in that they have terminal or chain-linked galacturonic acids that contribute significantly to the immunospecificity of the oligosaccharide (Kotelko, 1986; Sidorczyk et al., 1987). The O-polysaccharide polymers consist of repeating units composed of hexoses, hexosamines, 6-deoxy sugars, uronic acids with amino acids attached to their carboxyl groups, and other noncarbohydrate acidic components (Sidorczyk et al., 1975; Kotelko, 1986; Sidorczyk et al., 1995).

P. vulgaris strains OX19 and OX2 and *P. mirabilis* strain OXK are the test strains for O antigens 1, 2, and 3, respectively, in the antigenic scheme of Kauffmann and Perch (Kauffmann, 1966). Antibodies in sera from patients with rickettsial infections agglutinate either one or two of these strains. This reaction is the basis of the Weil-Felix test, which is a presumptive test for rickettsial infections. Sera from patients with endemic or epidemic typhus agglutinate the OX19 strain, patients infected with the spotted fever group of rickettsiae agglutinate both OX19 and OX2 strains, and patients with scrub typhus agglutinate the OXK strain. The reactions have been suspected to be due to LPS and the finding that sera from patients with spotted fever react with LPS from both the rickettsial strain and the *P. vulgaris* OX2 strain indicates this to be the case (Amano et al., 1993). Since *Proteus* infections may also stimulate antibody production and *P. mirabilis* strains with the O3 antigen are among the most frequently isolated, positive Weil-Felix tests should be interpreted with caution. Furthermore, the heterophile antigens to which the cross-reactions in the Weil-Felix test are attributed also occur in *Legionella bozemanii* and *Legionella micdadei*, indicating that these antigenic specificities are more widely distributed than originally believed (Sompolinsky et al., 1986; Westfall et al., 1986).

Analysis of the flagellar (H) antigens of *P. vulgaris* and *P. mirabilis* was carried out by Perch (1948). A total of 19 were identified and included in the Kauffmann and Perch antigenic scheme (Kauffmann, 1966). Cross-reactions among the H antigens are numerous and complex. Five antigenic factors were identified in the H 1 complex and other H antigens showed even greater complexity. The use of H antigens for routine serotyping of *P. vulgaris* and *P. mirabilis* strains has not been adopted and has been limited essentially to the initial studies of Kauffmann and Perch (Perch, 1948). The flagellar antigens of *P. penneri* have not been investigated.

Capsular antigens, unlike the K antigens of other *Enterobacteriaceae*, were reported to be present in the genus *Proteus* (Namioka and Sakazaki, 1959). These antigens, designated C antigens to distinguish them from the *E. coli* L and B antigens, inhibit O agglutination, retain antibody binding capacity after 1 h at 100°C, and have antigenic specificities restricted to certain O groups. The authors suggested that numerous serological types of C antigens occur in the genus *Proteus*, but a system for identifying them has not been developed. Research has focused on the role of capsules in facilitating swarming (Stahl et al., 1983), and a polysaccharide that evidently has this effect has been identified and chemically characterized (Gygi et al., 1995). Investigations are also under way to examine the role of polysaccharide capsules or glycocalyces in the formation of biofilms during infections of the urinary tract. It has been hypothesized that the biofilm is a major virulence factor because it enhances bacterial adhesion, surrounds the bacteria to protect them from the immune response and from exposure to antibiotics, and provides a microenvironment in the kidney and bladder that is conducive to stone formation (Nickel et al., 1985; McLean et al., 1988; Clapham et al., 1990).

Since the last edition of the *Manual*, numerous investigations of the surface structures of *P. vulgaris*, *P. mirabilis*, and *P. penneri* have resulted in clarification of the nature of the well-known hemagglutinins and fimbriae and in the discovery of several new types of fimbriae. A greater understanding of the mechanisms of pathogenesis has been gained. The MR/P (mannose resistant *Proteus*-like) fimbriae, also known as type IV fimbriae, have been the most extensively studied. These are highly immunogenic proteins 7 nm in diameter. The MR/P fimbriae have been isolated and purified and the genes involved in their determination have been identified (Sareneva et al., 1990; Bahrani et al., 1991; Bahrani and Mobley, 1993, 1994). Evidence obtained in experiments using the mouse model indicates that these fimbriae are associated with colonization of the upper regions of the urinary tract (Bahrani et al., 1994).

The MR/K (mannose resistant, *Klebsiella*-like) type 3 hemagglutinins are thin (diameter of 4–5 nm) and hemagglutinate tanned erythrocytes. Two antigenic types have been found but neither type is antigenically related to *Klebsiella* fimbriae (Old and Adegbola, 1982). MR/K hemagglutinins are more characteristic of *P. penneri* than of *P. mirabilis* (Yakubu et al., 1989). *Proteus* strains with MR/K hemagglutinins have a high affinity for the materials of which catheters are made, and this feature has been cited as a factor in prolonged chronic urinary tract infections of catheterized patients (Mobley et al., 1988; Yakubu et al., 1989; Roberts et al., 1990).

Of the newly discovered fimbriae, one type designated PMF for *P. mirabilis* fimbriae, has been suggested to be important in the colonization of the bladder, but not the kidney (Bahrani et al., 1993; Massad et al., 1994b). Another type, referred to as ambient temperature fimbriae (ATF) of *P. mirabilis*, is optimally produced at 23°C and apparently does not have a role in pathogenesis (Massad et al., 1994a). A third type, also expressed by *P. mirabilis*, was referred to as nonagglutinating fimbriae (NAF) (Tolson et al., 1995). It does not have the hemagglutinating properties of MR/P or MR/K fimbriae, but adheres to uroepithelial cells, indicating possible importance as a virulence factor in establishing urinary tract infections.

The ability of *Proteus* species to produce cytotoxic hemolysins has been known since 1919 (Wenner and Rettger, 1919), but definitive studies had to await the development of molecular biological technologies. With the aid of new techniques, it has been established that *Proteus* species produce two distinct hemolysins (Welch, 1987). The HlyA hemolysin is genetically related to the α-hemolysin of *E. coli*, has a molecular size of 110 kDa, and is a Ca^{2+}-dependent, pore-forming cytotoxin (Koronakis et al., 1987, Welch, 1991). The HpmA hemolysin (166 kDa) is related to the ShlA hemolysin of *Serratia marcescens* and is calcium independent (Uphoff and Welch, 1990). A protein, HpmB (63 kDa), which is closely related to the ShlB protein of *S. marcescens* is responsible for the transport and activation of the HpmA protein (Swihart and Welch, 1990b; Uphoff and Welch, 1990). Nearly all the *P. vulgaris* and *P. mirabilis* strains produce the HpmA hemolysin, but only a few *P. vulgaris* produce the HlyA hemolysin (Swihart and Welch, 1990a). Most strains of *P. penneri* produce the HpmA hemolysin and most also produce the HlyA hemolysin (Rózalski and Kotelko, 1987; Lukomski et al., 1990; Senior, 1993). Expression of hemolysins is associated with increased cytotoxicity, cell invasiveness, and a lower LD_{50} in mice, but is not associated with increased kidney colonization or histopathological changes in murine urinary tract infections (Peerbooms et al., 1983, 1984, 1985; Mobley and Chippendale, 1990).

Proteus spp. are among the few Gram-negative species capable of producing proteolytic enzymes that degrade immunoglobulins. Strains of *P. vulgaris*, *P. mirabilis*, and *P. penneri* have been shown to produce immunoglobulin A (IgA) proteases (Senior et al., 1988; Wassif et al., 1995). Analysis of the proteolytic activity of one *P. mirabilis* strain isolated from a patient with urinary tract infection showed that it has a much wider range of activity than the usual IgA proteases. Both serum and secretory IgA1 and IgA2, both the free and IgA-bound secretory component, and IgG are susceptible targets of this enzyme but its role as a virulence factor in urinary tract infections remains uncertain (Loomes et al., 1990). A polypeptide with protease activity that is secreted by uropathogenic *P. mirabilis* has been implicated as a virulence factor in enhancing the ability of the organisms to infect the urinary tract (Zhao et al., 1999).

The antibiotic susceptibility of *P. myxofaciens* is not known but a feature common to the other three *Proteus* species is an intrinsic resistance to bacitracin and the polymyxins (Potee et al., 1954; Von Graevenitz and Nourbakhsh, 1972; Shimizu et al., 1977). Susceptibility to colistin (colymycin) is rare among *Proteus* species, and if a strain is susceptible to this antibiotic, it is most likely not to be a *Proteus* (Li and Miller, 1970). These cyclic polycationic antibiotics are believed to bind to the charged Kdo-lipid A region of the LPS to cause membrane disruption. Substitution of the phosphate ester bonds of the lipid A with L-arabinoso-4-amine in *Salmonella typhimurium* leads to resistance of the bacteria to polymyxin (Vaara et al., 1981). *Proteus* strains have lipid A that contains L-arabinoso-4-amine naturally, and this has been suggested as a contributing factor for their natural resistance to these antibiotics (Sidorczyk et al., 1987; Rózalski et al., 1997).

P. mirabilis is the species most frequently isolated from human infections and is generally also the most susceptible to antibiotics, but some very resistant isolates have been reported (Penner et al., 1982). Increasing resistance, particularly to tetracyclines, has been noted since the 1950s (Potee et al., 1954; Chiu and Hoeprich, 1961). The majority of the strains are susceptible to chloramphenicol, but many carry an inducible chloramphenicol acetylase gene (Charles et al., 1985) that could lead to failure in the treatment of brain abscesses (Hawkey, 1997). The occurrence of both nitrofurantoin-sensitive and nitrofurantoin-resistant strains in clinical specimens has been reported.

P. mirabilis is the most susceptible of the three species to the β-lactam group of antibiotics, although resistant strains occur and the necessity of performing antibiotic susceptibility tests is stressed (Barry and Hoeprich, 1973; Shafi and Datta, 1975). *P. mirabilis*, unlike the other two species, does not appear to carry a chromosomal cephalosporinase or an AmpC type penicillinase, and thus many infections can be treated with either cephalosporins or ampicillin (Livermore, 1995). Only rarely is imipenem-resistant *P. mirabilis* isolated, and it has been suggested to be due to altered penicillin-binding proteins (Neuwirth et al., 1995).

Most strains of *P. mirabilis* are susceptible to aminoglycosides, but some strains may acquire resistance factors (R factors) that cause resistance to gentamicin and kanamycin. Amikacin is effective against *P. mirabilis* strains but some have been reported to be resistant to this antibiotic as well as to gentamicin (Briedis and Robson, 1976; Drasar et al., 1976). Resistant *P. mirabilis* can be anticipated to be more frequently encountered in the future, especially in hospital outbreaks, as increased usage has been shown in the past to lead to increased resistance.

Antibiotics generally used to treat infections caused by *P. mirabilis* are trimethoprim or one of the fluoroquinolones such as ciprofloxacin. However, susceptibility tests should be performed for each isolate (Hawkey, 1997).

P. vulgaris is separable into two biogroups, but most of the data on antibiotic susceptibility do not distinguish strains on this basis, and results of antibiotic susceptibility studies in earlier periods did not distinguish between *P. vulgaris* and *P. penneri*. Assessments of the susceptibilities of *P. vulgaris* have been made based on mixtures of strains. However, it is now clear that *P. vulgaris* strains are more resistant than *P. mirabilis* strains. The high levels of *P. vulgaris* resistance to penicillins and cephalosporins is a feature distinguishing it from *P. mirabilis*. Carbenicillin is the only penicillin that has been shown to have significant activity against *P. vulgaris* strains. Most strains, however, are susceptible to aminoglycosides. A small percentage are resistant to gentamicin and kanamycin, but, in most cases, are susceptible to amikacin. Both chloramphenicol-resistant and chloramphenicol-susceptible *P. vulgaris* strains have been reported to be present in clinical specimens, but it is not known what proportion of these would now be classified as *P. penneri*, which are characteristically resistant to this antibiotic (Hickman et al., 1982c).

The high levels of resistance of *P. penneri* strains to chloramphenicol is a definitive criterion in the description of the species (Hickman et al., 1982c). Although increasing numbers of laboratories are identifying *P. penneri* strains, the numbers reported in susceptibility studies are relatively small for a general assessment of the susceptibility of the species. Reports from several laboratories, however, confirm the resistance to chloram-

phenicol as a species characteristic (Hickman et al., 1982c; Hawkey et al., 1983; Fuksa et al., 1984; Piccolomini et al., 1987). It also appears that the species is more resistant to penicillins than is *P. vulgaris*. In a study of 45 isolates, it was found that ceftizoxime, ceftazidime, moxalactam, and cefoxitin would be effective for treatment of *P. penneri* infections (Fuksa et al., 1984). In another study of 39 isolates, imipenem was found to be the most effective antibiotic against isolates of this species (Piccolomini et al., 1987). Isolates were also found to be susceptible to four aminoglycosides (gentamicin, tobramycin, netilmicin, and amikacin), and their use was recommended for treatment of systemic infections (Fuksa et al., 1984). Isolates of the three species are susceptible to the quinolones, particularly to ciprofloxacin and norfloxacin, but a small number of *P. penneri* isolates show higher levels of resistance (Hawkey and Hawkey, 1984).

P. vulgaris, *P. mirabilis*, and *P. penneri* are important human pathogens causing primary and secondary infections. Of the three species, *P. mirabilis* is most frequently isolated from clinical specimens and is one of the most important pathogens of the urinary tract. Underlying conditions such as diabetes or structural abnormalities of the urinary tract are associated with infections acquired outside the hospital (Grossberg et al., 1962; Wallace and Petersdorf, 1971). Hospital-acquired infections commonly occur in patients with predisposing conditions such as catheterization, urological instrumentation, or surgery of the urinary tract. Infectious strains are transmitted from the intestinal flora of the patient or contracted through transmission from other patients or from a common reservoir (Dutton and Ralston, 1957; Kippax, 1957; Chow et al., 1979). *Proteus* urinary tract infections may give rise to bacteremias that are difficult to treat and often fatal. *Proteus* bacteria may cause infections at other sites of the body if favorable conditions exist. They may cause infections of wounds, burns, and respiratory tracts and have been isolated from the eyes, ears, and the throat. *Proteus* strains may also infect the neonatal umbilical stump, and fatal bacteremias and meningitis may result (Becker, 1962; Librach, 1968; Shortland-Webb, 1968; Burke et al., 1971). *P. mirabilis* has also been isolated from both children and adults with osteomyelitis (Levy and Ingall, 1967; Meyers et al., 1973).

Proteus species have been implicated as causative agents of enteritis and were isolated from 46.2% of infants with diarrhea (Lányi, 1956). However, in light of the fact that approximately one-quarter of the population are intestinal carriers of *Proteus*, it is difficult to assess this finding as convincing evidence of a role for strains of this genus in the etiology of human enteritis. Furthermore, species of the genus *Campylobacter* that are now known to be major causes of human diarrhea were not recognized at the time of the earlier studies.

Recent research on the modification of peptidoglycan (murein) by bacterial enzymes of Gram-positive and Gram-negative bacteria suggest possible involvement of *P. vulgaris* and *P. mirabilis* in the development of arthritic syndromes. It has been demonstrated that one strain of *P. vulgaris* and 17 strains of *P. mirabilis* possess acetyltransferase activity for converting the muramyl residue of peptidoglycan to the 2,6-diacetylmuramyl derivative (Fleck et al., 1971; Gmeiner and Kroll, 1981; Dupont and Clarke, 1991; Clarke and Dupont, 1992). This has the effect of creating resistance of the peptidoglycan to the hydrolytic enzymes in phagocytic cells and serum, resulting in the persistence and circulation in the host of high molecular weight peptidoglycan fragments, which have been implicated in animal models as the inciting agents of rheumatoid arthritis (Chedid et al., 1978; Cromartie, 1981; Clarke and Dupont, 1992). It is also noteworthy that antibodies against *Proteus* species have been detected in patients with rheumatoid arthritis (Ebringer et al., 1985). The possibility that organisms of the genus *Proteus* could be involved in the development of rheumatoid arthritis is an interesting concept, but more research is required (Clarke and Dupont, 1992).

Epidemiological typing is performed at present only for *P. vulgaris* and *P. mirabilis*. For reference work, O-serotyping is the method widely accepted as the standard but it is carried out only in a few laboratories that have the large set of typing antisera (Perch, 1948; Kauffmann, 1966; Penner and Hennessy, 1980; Larsson, 1984; Hawkey et al., 1986b). The antisera are not commercially available, and other systems for typing have been introduced and tested with the objective of enabling clinical laboratories to conduct epidemiological studies without the high costs associated with O-serotyping. Among these alternative methods is the Dienes test (described above), which can be readily performed but lacks specificity (Kippax, 1957; Hickman and Farmer, 1976). The incompatibility of two swarms that produce a line of demarcation between them is believed to be due to both the production of and susceptibility to bacteriocins (Senior, 1977a). Typing based on sensitivity to bacteriocins (proticins) has been advocated since 1965 (Cradock-Watson, 1965), and several systems have been developed (Al-Jumaili, 1975; Senior, 1977b; Kusek and Herman, 1980). In a study using the proticine production and proticine sensitivity method, O serotyping, and the Dienes test, a highly discriminating system for identifying *P. mirabilis* and *P. vulgaris* strains, was achieved (Senior and Larsson, 1983). The results also showed that the Dienes test was useful in determining the presence of a common type among a small number of isolates, but not when applied to a large number of isolates for which testing of many combinations is required. In such cases, proticine typing or O serotyping are more appropriate. The greatest discrimination is provided by using both proticine and O serotyping.

Lytic bacteriophage may also be used to differentiate strains of *P. vulgaris* and *P. mirabilis*; several groups of investigators have described typing schemes (Vieu and Capponi, 1965; France and Markham, 1968; Izdebska-Szymona et al., 1971; Schmidt and Jeffries, 1974; Hickman and Farmer, 1976; Bergan, 1978).

A high level of discrimination among *P. mirabilis* strains may also be achieved by testing the strains for growth inhibition in a selected group of chemicals that does not include antibiotics (Kashbur et al., 1974). This method of typing, referred to as resistotyping, was found to compare favorably with serotyping, phage typing, proticine typing, and the Dienes test.

Traditional methods of typing are now being supplanted by new techniques as advances are made in molecular biology. Protein profiles of outer membranes and multilocus enzyme electrophoresis analysis have been used to differentiate strains of *P. vulgaris* and *P. mirabilis* (Kappos et al., 1992). Arbitrarily primed polymerase chain reaction (AP-PCR) and ribotyping have been exploited in a study of *P. mirabilis* isolated in a pediatric hospital (Bingen et al., 1993). With increasing interest in epidemiology, along with expected future simplifications of DNA fingerprinting methods, it is anticipated that more studies will be undertaken to advance our understanding of the epidemiology of these pathogens.

Early studies have shown that *Proteus* strains are widely distributed in nature, occurring in manure, soil, and polluted water, where they are believed to have important roles in the decomposition of organic matter (Wilson and Miles, 1975). They are also present in the intestinal contents of both wild and domesticated animals as well as in the intestines of patients and healthy humans. Incidences of 17–24% for *P. mirabilis* and 3–5% for *P. vulgaris* in healthy humans have been reported (Rustigian and Stuart, 1945; Krikler, 1953). These results were obtained without knowledge that *P. penneri* existed in the genus. Special notice of indole-negative strains was usually not taken, as these would have been considered as atypical indole-negative *P. vulgaris* or atypical ornithine decarboxylase-negative *P. mirabilis* (Müller, 1986a). Thus, their strains identified as *P. vulgaris* may or may not have included *P. penneri* strains. It is noteworthy that in a recent study of bacteria in fecal and urine contaminated bedding in calf-rearing units, *P. vulgaris* and *P. mirabilis*, but no *P. penneri*, strains were isolated (Hawkey et al., 1986b). It seems premature, therefore, to assume that the distribution of *P. penneri* in nature is similar to that of *P. vulgaris*. However, it has been found that *P. penneri* does occur in human feces. Müller (1986a) isolated 13 (0.9%) *P. penneri* from 1422 healthy humans and 20 (1.6%) from 1271 patients. The ecological importance of *Proteus* in the intestine is not known. The hydrolysis of urea is probably minor in comparison to that produced by the large population of urease-producing anaerobes (Sabbaj et al., 1970; Brown et al., 1971), but the oxidative deamination of amino acids may be significant (Drasar and Hill, 1974).

Other human sites of *Proteus* carriage include the urinary tract, the groin area, the vagina, and the prepuce of male infants. The high carriage rate in the intestine has been implicated in the colonization of the groin area (Ehrenkranz et al., 1989) and subsequent development of autoinfections and cross-infections of the urinary tract (Burke et al., 1971). Urine may be considered a habitat of *Proteus* and the source of hospital acquired infections (Hickman and Farmer, 1976; Warren, 1996). The carriage of *Proteus* in the maternal vagina is reported be the source of neonatal infections (Bingen et al., 1993). The carriage of *P. mirabilis* in the prepuce of the infant is believed to be the source of infections in male babies (Glennon et al., 1988).

P. myxofaciens has been isolated only from living and dead gypsy moth larvae (*Porthetria dispar*), but a role in the production of disease in these insects has not been established (Cosenza and Podgwaite, 1966).

ENRICHMENT AND ISOLATION PROCEDURES

Isolation media in the clinical laboratory are formulated to inhibit the swarming of *Proteus* spp. and to select the well-known pathogens such as *Salmonella* and *Shigella*. Such media are suitable for the isolation of *Proteus* spp., but special media for the primary isolation of *Proteus* spp. have been designed (Zarett and Doetsch, 1949; Malinowski, 1966; Xilinas et al., 1975; Hawkey et al., 1986a).

For enrichment of *Proteus* species in fecal samples, tetrathionate or selenite broth have been found to increase the rate of isolation of *P. mirabilis* from 8.2% to 23.6% and *P. vulgaris* from 0% to 2.7% when enrichment preceded plating (Hynes, 1942; Rustigian and Stuart, 1945).

MAINTENANCE PROCEDURES

Proteus strains may be maintained on trypticase soy agar at 4°C with monthly transfers or may be preserved indefinitely by lyophilization. At the Centers for Disease Control and Prevention, cultures in tubes of blood agar base or trypticase soy agar sealed with a cork or rubber stopper remain viable at room temperature for many years without transfer (J.J. Farmer and F.W. Hickman, personal communication).

DIFFERENTIATION OF THE GENUS *PROTEUS* FROM OTHER GENERA

Key characteristics for differentiating the genera *Proteus*, *Providencia*, and *Morganella* are shown in Table BXII.γ.254.

FURTHER READING

Belas, R. 1992. The swarming phenomenon of *Proteus mirabilis*. ASM News. *58*: 15–22.

Brenner, D.J., J.J. Farmer, III, G.R. Fanning, A.G. Steigerwalt, P. Klykken, H.G. Wathen, F.W. Hickman and W.H. Ewing. 1978. Deoxyribonucleic acid relatedness in species of *Proteus* and *Providencia*. Int. J. Syst. Bacteriol. *28*: 269–282.

Brenner, D.J., F.W. Hickman-Brenner, B. Holmes, P.M. Hawkey, J.L. Penner, P.A. Grimont and C.M. O'Hara. 1995. Replacement of NCTC 4175, the current type strain of *Proteus vulgaris*, with ATCC 29905. Request for an opinion. Int. J. Syst. Bacteriol. *45*: 870–871.

Clarke, A.J. and C. Dupont. 1992. *O*-acetylated peptidoglycan: its occurrence, pathobiological significance, and biosynthesis. Can. J. Microbiol. *38*: 85–91.

Kotelko, K. 1986. *Proteus mirabilis*: taxonomic position, peculiarities of growth, components of the cell envelope. Curr. Top. Microbiol. Immunol. *129*: 181–215.

Larsson, P. 1984. Serology of *Proteus mirabilis* and *Proteus vulgaris*. Methods Microbiol. *14*: 187–214.

Mobley, H.L. and R.P. Hausinger. 1989. Microbial ureases: significance, regulation, and molecular characterization. Microbiol. Rev. *53*: 85–108.

TABLE BXII.γ.254. Characteristics differentiating *Proteus*, *Morganella*, and *Providencia*[a,b]

Characteristic	Proteus	Morganella	Providencia
Swarming	+	−	−
H₂S production	+	−	−
Gelatin hydrolysis	+	−	−
Lipase (corn oil)	+	−	−
Utilization of citrate (Simmons)	D	−	+
Ornithine decarboxylase	D	+	−
Acid production from:			
Maltose	D	−	−
D-Mannose	−	+	+
Acid from one or more of the following polyhydric alcohols:			
i-Inositol, D-mannitol, adonitol, D-arabitol, erythritol	−	−	+[c]

[a]For symbols, see standard definitions.

[b]Temperature of reactions, 36° ± 1°C. All reactions for 48 h.

[c]One *Providencia* species (*P. rustigianii*) does not produce acid from polyhydric alcohols.

Mobley, H.L., M.D. Island and R.P. Hausinger. 1995. Molecular biology of microbial ureases. Microbiol. Rev. *59*: 451–480.

Rózalski, A., Z. Sidorczyk and K. Kotelko. 1997. Potential virulence factors of *Proteus bacilli*. Microbiol. Mol. Biol. Rev. *61*: 65–89.

DIFFERENTIATION OF THE SPECIES OF THE GENUS *PROTEUS*

The key differentiating characteristics for the four species of the genus *Proteus* are shown in Table BXII.γ.255. Additional characteristics of the species are provided in Table BXII.γ.256.

List of species of the genus Proteus

1. **Proteus vulgaris** Hauser 1885, 12[AL] emend. Brenner, Hickman-Brenner, Holmes, Hawkey, Penner, Grimont and O'Hara 1995, 870.

   vul.ga' ris. L. adj. *vulgaris* common.

   Morphological characteristics as described for the genus. Other characteristics are listed in Tables BXII.γ.254, BXII.γ.255, and BXII.γ.256. Since the last edition of the *Manual*, indole-negative strains formerly classified as *P. vulgaris* biogroup 1 have been assigned to a new species, *Proteus penneri*. *P. vulgaris* at present includes two biogroups. Biogroup 2 strains are positive for indole, salicin, and esculin and have other reactions characteristic of the vast majority of strains in the species. The present type strain and a very small heterogeneous group belong to biogroup 3, which is indole positive but negative in both the salicin and esculin reactions. Additional species will probably be identified in these biogroups.

   Some strains are hemolytic on blood agar. Generally resistant to penicillins and cephalosporins. Generally more resistant to antibiotics than *P. mirabilis*. Causes infections of the human urinary tract and other sites but not as frequently as *P. mirabilis*. Widely distributed in nature, occurring in human and animal intestines and in contaminated soil and water.

   The mol% G + C of the DNA is: 39.3 ± 1.2 (T_m) (Falkow et al., 1962).

   Type strain: ATCC 29905, DSM 13387.

2. **Proteus mirabilis** Hauser, 1885, 34.[AL]

   mi.ra' bi.lis. L. adj. *mirabilis* wonderful.

   Morphological characteristics are as described for the genus. Other characteristics are listed in Tables BXII.γ.254, BXII.γ.255, and BXII.γ.256. Characteristically indole and maltose negative and ornithine decarboxylase positive. Some strains are hemolytic on blood agar. More frequently isolated from clinical specimens than *P. vulgaris*. Most common site of infection is the human urinary tract. Generally susceptible to penicillins and cephalosporins and more susceptible than *P. vulgaris*.

   The mol% G + C of the DNA is: 39.3 ± 1.45 (T_m) (Falkow et al., 1962).

   Type strain: ATCC 29906.

   GenBank accession number (16S rRNA): AF008582.

3. **Proteus myxofaciens** Cosenza and Podgwaite 1966, 188.[AL]

   myx.o.fac' i.ens. Gr. fem. n. *myxa* slime; M.L. masc. n. *faciens* producing; *myxofaciens* slime-producing (bacteria).

   Morphological characteristics as described for the genus. Other characters are listed in Tables BXII.γ.254, BXII.γ.255, and BXII.γ.256. Only one isolate studied in detail. Thin film of growth on solid media. Produces highly viscous slime when cultured in trypticase soy broth. Hemolytic on blood agar. Isolated from living and dead gypsy moth larvae (*Porthetria dispar* L).

   The mol% G + C of the DNA is: not determined.

   Type strain: ATCC 19692.

TABLE BXII.γ.255. Differentiation of the species and biogroups of the genus *Proteus*[a,b]

Characteristic	P. vulgaris BG2	P. vulgaris BG3	P. mirabilis	P. myxofaciens[c]	P. penneri
Indole production	+	+	−	−	−
Ornithine decarboxylase	−	−	+	−	−
Acid from:					
Maltose	+	+	−	+	+
D-Xylose	+	+	+	−	+
Salicin	+	−	−	−	−
Esculin hydrolysis	+	−	−	−	−
Tyrosine clearing	+	+	+	−	+
Slime producing[d]	−	−	−	+	−

[a]For symbols, see standard definitions.

[b]Temperature of all reactions, except slime production, at 36° ± 1°C.

[c]Reactions based on study at one strain (ATCC 19692).

[d]Slime production at 25°C in trypticase soy broth.

TABLE BXII.γ.256. Other characteristics of species of genus *Proteus*[a,b]

Characteristic	P. vulgaris[c]	P. mirabilis	P. myxofaciens[d]	P. penneri[e]
Phenylalanine deaminase	+	+	+	+
Urease	+	+	+	+
NO_3^- reduced to NO_2^-	+	+	+	+
Motility	+	+	+	d
Swarming	+	+	+	d
Gelatin liquefaction (22°C)	+	+	+	d
H₂S production (TSI)	+	+	+ (3–4)[f]	−
Growth in KCN	+	+	+	+
Acid from glucose	+	+	+	+
Methyl red test	+	+		+
Voges–Proskauer test	−	d	+	−
Citrate utilization (Simmons)	d	d	+	d
Tartrate utilization	+	d	+	+
Acetate utilization	d	d	+	−
Lipase activity (corn oil)	d	d	−	d
Deoxyribonuclease (25°C)	d	d	−	d
Oxidase test	−	−	−	−
ONPG hydrolysis	−	−	−	−
Pectate liquefaction	−	−	−	−
Malonate utilization	−	−	−	−
L-Lysine	−	−	−	−
L-Arginine	−	−	−	−
L-Ornithine	−	+	−	−
Acid production from:				
Adonitol	−	−	−	−
L-Arabinose	−	−	−	−
D-Arabitol	−	−	−	−
Cellobiose	−	−	−	−
Dulcitol	−	−	−	−
Erythritol	−	−	−	−
Glycerol	d	+	+	+ (7)
i-Inositol	−	−	−	−
Lactose	−	−	−	−
D-Mannitol	−	−	−	−
D-Mannose	−	−	−	−
α-CH₃-glucoside	d	−	+	+ (7)
Raffinose	−	−	−	−
L-Rhamnose	−	−	−	−
D-Sorbitol	−	−	−	−
Sucrose	+	d	−	+
Esculin hydrolysis	d	−	−	−
Mucate	−	−	−	−
ONPG	−	−	−	−

[a]For symbols, see standard definitions.

[b]Biochemical reaction at 36° ± 1°C unless otherwise indicated.

[c]Includes strains of biogroup 2 and biogroup 3 (Brenner et al., 1978).

[d]Reactions based on study of one strain (Brenner et al., 1978).

[e]Reactions as reported by Hickman et al. (1982c).

[f]Number in parentheses indicates number of days for test.

4. **Proteus penneri** Hickman, Steigerwalt, Farmer III and Brenner 1983, 438[VP] (Effective publication: Hickman, Steigerwalt, Farmer III and Brenner 1982c, 1100.)

penn.er' i. M.L. gen. n. *penneri* of Penner; named after the Canadian microbiologist J.L. Penner.

Morphological characteristics are as described for the genus. Other characteristics are listed in Tables BXII.γ.254, BXII.γ.255, and BXII.γ.256. Formerly classified as *P. vulgaris* biogroup 1. Strains are characteristically negative in indole, maltose, salicin, and esculin reactions. Some strains are hemolytic on blood agar. A pathogen of the human urinary tract. Very resistant to chloramphenicol.

The mol% G + C of the DNA is: 38 (T_m).

Type strain: ATCC 33519, CDC 1808-73.

Genus XXX. **Providencia** Ewing 1962, 96[AL]

JOHN L. PENNER

Pro.vi.den' ci.a. M.L. fem. n. *Providencia* Providencia, named after the city of Providence, Rhode Island, U.S.A.

Straight rods, 0.6–0.8 × 1.5–2.5 μm, conforming to the general definition of the family *Enterobacteriaceae*. Gram negative. Motile by peritrichous flagella. They are **facultatively anaerobic**, chemoorganotrophic, having **both a respiratory and a fermentative type metabolism**. Optimal growth temperature is 37°C. **Oxidatively deaminate phenylalanine and tryptophan. Acid is produced from D-mannose. All except *Providencia rustigianii* produce acid from one or more of the following polyhydric alcohols**: inositol, D-mannitol, adonitol, D-arabitol, and erythritol. Except for *Providencia heimbachae* all are indole positive. Tartrate (Jordan) is utilized by all but a few strains of *P. heimbachae*. Simmons citrate is utilized by all species except *P. heimbachae* and some strains of *P. rustigianii*. Isolated from diarrhetic stools, urinary tract infections, wounds, burns, bacteremias, and contaminated environmental sources. DNA relatedness tests have shown that *Providencia* are more closely related to *Proteus vulgaris* and *Proteus mirabilis* than to other members of the family *Enterobacteriaceae*. Although 16S RNA sequence data are not available for the *Providencia* species, the close relationship of *Providencia* to *Proteus* indicates that *Providencia*, like *Proteus*, belongs to the *Gammaproteobacteria* within the family *Enterobacteriaceae*.

The mol% G + C of the DNA is: 39–42 (Falkow et al., 1962).

Type species: **Providencia alcalifaciens** (de Salles Gomes 1944) Ewing 1962, 96 (*Eberthella alcalifaciens* de Salles Gomes 1944, 183; *Proteus inconstans* (Ornstein 1921) Shaw and Clarke 1955, 155.)

FURTHER DESCRIPTIVE INFORMATION

Since the last edition of the *Manual*, the genus has been extended to include two new species, *P. rustigianii* and *P. heimbachae*. All *Providencia* species, like those of *Proteus* and *Morganella*, deaminate phenylalanine and other amino acids to produce α-keto and α-hydroxycarboxylic acids (Bernheim et al., 1935; Stumpf and Green, 1944; Singer and Volcani, 1955). The carboxylic acids that actively bind iron have siderophore activity (Drechsel et al., 1993). A probe for the amino acid deaminase (*aad*) of *Proteus mirabilis* does not hybridize with DNA from *Providencia* and *Morganella* strains, indicating unrelatedness of the deaminases of these three genera (Massad et al., 1995). In the clinical laboratory, tests for phenylalanine deaminase are commonly used to identify strains belonging to these three genera (Henriksen, 1950; Thibault and Le Minor, 1957).

Providencia, like *Proteus* and *Morganella*, decomposes tyrosine to produce a clearing on the agar medium in which the insoluble amino acid is incorporated (Sheth and Kurup, 1975) and produces a reddish-brown pigment on nutrient agar containing 5% tryptophan (Polster and Svobodová, 1964) or other aromatic amino acids (Müller et al., 1986). *Providencia* species, with the exception of *P. rustigianii*, differ from *Proteus* and *Morganella* in their ability to produce acid from inositol and straight-chain tetra-, penta-, or hexahydroxy- alcohols, and the species are differentiated based on their reactions on these substrates. *Providencia rettgeri* is the most noted for this property, as most strains of the species are capable of producing acid from several or all of these polyhydric alcohols (Penner et al., 1975). In contrast, *P. rustigianii* strains do not produce acid from any of the polyols. Yellowish orange-centered colonies are known to be produced on deoxycholate citrate agar by *P. alcalifaciens*, *P. rettgeri*, and *P. stuartii* (Cook, 1948; Buttiaux et al., 1954). *P. heimbachae* and *P. rustigianii* have not been examined for this property. The color is apparently caused by the precipitation of ferric hydroxide because of the alkalinity produced by the growth of the bacteria on the medium (Catsaras et al., 1965).

Urease is produced characteristically by strains of *P. rettgeri* and variably by strains of *P. stuartii* (Penner et al., 1976c; Farmer et al., 1977; Mobley et al., 1985). In a hospital outbreak of urinary tract infections, some patients were found to be infected with both urease-positive and urease-negative strains of *P. stuartii* (Penner et al., 1976c). The proportion of the urease-positive strains in *P. stuartii* has been estimated to be as low as 15–16% (Brenner et al., 1978; Penner et al., 1979b) and as high as 24% and 30% (Mobley et al., 1985; Mobley et al., 1986a). The urease enzyme of at least some *P. stuartii* strains was found to be transmissible (Grant et al., 1981) on an 82-kb plasmid (Mobley et al., 1986b). Its molecular weight has been estimated to be 375,000 ± 35,000 (Mobley et al., 1986b). The presence of the plasmid in endemic strains of some hospitals could be the cause of variations among hospitals in the frequencies of isolations of urease-positive *P. stuartii* (Penner et al., 1979b) and the failure to identify the species correctly with commercial identification systems (Cornaglia et al., 1988). A key test for separating *P. stuartii* from *P. rettgeri*, particularly from those aberrant *P. rettgeri* in fermenting D-arabitol or erythritol or both, is the fermentation of trehalose (Fischer et al., 1989). Of the *Providencia* species, only *P. stuartii* strains ferment this compound, and thus the test serves to separate this species from others of the genus.

The production of indole is an important test in the identification of *Providencia*. All species give positive reactions in this test except *P. heimbachae* (Müller et al., 1986). However, it should be noted that strains of *P. heimbachae* may produce positive reactions after 1 d of incubation when tryptone or peptone media with 0.1% tryptophan are used (Müller, 1986b). To avoid false-positive reactions tryptone media without added tryptophan has been recommended (Müller, 1986b).

Although the strains of *Providencia* are cited as unable to produce acid from lactose, some variants have been reported. An unusual strain of *P. rettgeri* that fermented lactose was reported to be associated with an outbreak of nosocomial urinary tract infection (Traub et al., 1971) and a group of 31 lactose positive isolates of *P. rettgeri* were obtained from hospitalized patients with urinary tract infections (Richard et al., 1974). In the latter study, it was found that the lactose-positive trait was not transferable by conjugation. Five lactose-positive strains were also identified among a collection of 699 *P. stuartii* obtained from urinary tracts of catheterized patients (Mobley et al., 1985). The lactose fermentation was transferable by a 150-kb plasmid (Mobley et al., 1985). In addition to urease and lactose fermentation, acid production from sucrose has also been shown to be plasmid-mediated in *P. stuartii*, indicating that variants in this species are more likely to be the results of plasmid-encoded traits than is generally recognized (Mobley et al., 1985).

Antigenic structure　Like other Gram-negative species, *Providencia* possesses the surface arrays of antigenic molecules that may serve as markers for differentiation of the strains. The lipopolysaccharide (O) antigens have been the most widely used for serotyping. The original serotyping scheme for *Providencia* included both *P. alcalifaciens* and *P. stuartii* and recognized 56 O antigens, 28 flagellar (H) antigens, and two capsular (K) antigens (Ewing et al., 1954). The O serotyping scheme has since been separated according to species and extended, so that currently 46 O antigens for *P. alcalifaciens* and 17 O antigens for *P. stuartii* may be identified (Penner et al., 1976b, 1979a, b). The original scheme proposed for serotyping *P. rettgeri* listed 34 O antigens and 26 H antigens (Namioka and Sakazaki, 1958). The O serotype reference strains (serostrains) of this scheme were included in a revised and extended scheme that currently recognizes 93 O specificities (Penner et al., 1974, 1976a). In the schemes for O serotyping the three *Providencia* species by Penner and colleagues, the passive (indirect) hemagglutination technique was used instead of the slide agglutination technique to identify the O antigens because it provided greater specificity, sensitivity, and reproducibility of results. This is due to the 228-kDa lipoglycoprotein receptor on the erythrocyte membrane that binds lipopolysaccharide specifically and does not bind proteins or polysaccharides (Springer et al., 1973). Thus, the possible participation in passive hemagglutination of flagellar, fimbrial, or capsular antigens is largely excluded from the antigen-antibody reactions. The O serotyping schemes have been applied in a number of epidemiological investigations and in the study that advocated the reassignment of the *P. rettgeri* biogroup 5 strains to *P. stuartii* (Penner et al., 1975; Penner and Hennessy, 1977). At present, serotyping schemes for *P. heimbachae* and *P. rustigianii* have not been developed. However, three strains of *P. rustigianii*, NCTC 9259, NCTC 9195, and NCTC 9235, originally classified as *P. alcalifaciens* biogroup 3 (Ewing et al., 1972), were included in the present serotyping scheme for *P. alcalifaciens* as the reference strains for serotypes O:10, O:14, and O:54, respectively

(Penner et al., 1976b). Strains of these serotypes have not been identified in any epidemiological study on *Providencia* (Penner et al., 1979a).

Although it is highly probable that strains of the same O serotype from the same epidemiological setting are the same and thus important for investigating outbreaks, strains from different sources that have the same O antigens may not be the same. In such cases, a second method of discriminating between isolates is necessary. Using plasmid content to differentiate strains of *P. rettgeri*, it was shown that environmental strains of serotype O:58 were not identical (Hawkey et al., 1986b).

Although flagellar (H) antigens have been defined for some of the serotype reference strains of *P. alcalifaciens* and *P. stuartii* (Ewing et al., 1954) and for some *P. rettgeri* (Namioka and Sakazaki, 1958; Penner and Hinton, 1973), practical use of H typing has been limited to a few isolates examined in the earlier studies. Capsular antigens have also not been used as biomarkers for differentiating strains of this genus. One capsular antigen of *P. rettgeri* has been examined and found to be unlike the typical K antigens of *Enterobacteriaceae* (Namioka and Sakazaki, 1959). It is different in that it retains antibody-binding capacity after heat treatment at 100°C for 1 h and the antigenic specificity is restricted to particular O groups. The authors proposed the designation of C to distinguish this capsular antigen from the *Escherichia coli* L and B antigens and suggested that there are numerous serological types of C antigens.

Numerical analysis of electrophoretic protein patterns　Serotyping based on the O antigens has its drawbacks for routine work in the clinical laboratory, because large numbers of commercially unavailable antisera are required. A promising alternative to serotyping is the application of high resolution polyacrylamide gel electrophoresis (PAGE) of proteins combined with computerized analysis of patterns This approach has been used successfully to determine the PAGE types among strains of *P. alcalifaciens*, *P. rettgeri*, *P. stuartii*, and *P. rustigianii* (Costas et al., 1987, 1989, 1990; Holmes et al., 1988). No strains tested by this procedure were serotyped, and although the two methods have not been correlated, the results obtained by determination of the PAGE types suggest the method is as discriminating as serotyping (Costas et al., 1990). Moreover, computerized analysis of electrophoretic protein patterns correlate well with DNA relatedness studies.

Other methods of differentiating strains of *Providencia* species　Results have been shown that rDNA fingerprinting also permits discrimination among strains of *P. stuartii* and their differentiation from other species of *Providencia* and *Proteus* (Owen et al., 1988).

Another alternative to serotyping is to discriminate among strains based on their sensitivity and resistance to bacteriocins. An extensive scheme employing 12 bacteriocins (provicines) to discriminate among isolates of *P. alcalifaciens* and *P. stuartii* demonstrated that 37 sensitivity patterns occurred among 300 isolates of these two species, most of which were *P. stuartii* (Al-Jumaili and Fenwick, 1978). One provicine sensitivity pattern accounted for 32 (~10%) of the isolates, seven patterns accounted for ~40%, and the remaining 29 patterns accounted for 40%. The number of patterns and the distribution of the isolates among the different patterns indicated that this system of typing would be applicable in epidemiological investigations of *Providencia* infections caused by these two species.

Fimbriae and hemagglutination In *Providencia*, as in other *Enterobacteriaceae*, fimbriae of different types are present but their distribution is more complex than in most other genera of the family (Old and Adegbola, 1982). The presence of six different types is demonstrable by electron microscopy. The *Providencia* fimbriae that have been examined in most detail are those of *P. stuartii* (Old and Scott, 1981; Old and Adegbola, 1982; Mobley et al., 1988). Three fimbrial types have been examined for hemagglutinating and adhesive properties. Cells expressing mannose-sensitive (MS) type 2 fimbriae agglutinate guinea pig erythrocytes in the absence but not in the presence of mannose. Mannose-resistant, *Klebsiella*-like (MR/K) type 3 fimbriae agglutinate only tannic acid-treated cells, and mannose-resistant, *Proteus*-like (MR/P) type 4 fimbriae agglutinate both tannic acid-treated and untreated erythrocytes. Evidence suggests that the MR/K fimbriae of this species enable binding to catheter material, an important factor in contributing to its long-term persistence in the urinary tract of infected patients with indwelling catheters (Mobley et al., 1988).

Deoxyribonucleic acid relatedness of *Providencia* species *Providencia* are most closely related to *Proteus*, but both genera are only distantly related to *E. coli* and to other species in the family *Enterobacteriaceae* (Coetzee, 1972; Brenner et al., 1978; Brenner, 1984a). In the biochemical characterization of *P. alcalifaciens* four biogroups were recognized (Ewing et al., 1972). DNA relatedness tests showed that biogroups 1 and 2 are a homogeneous group, but biogroup 3 strains were found to be only 44–49% related to biogroups 1 and 2 and 26–33% related to *P. stuartii* (Brenner et al., 1978; Hickman-Brenner et al., 1983b). The biogroup 3 strains were transferred from *P. alcalifaciens* and assigned to a new species, *P. rustigianii* (Hickman-Brenner et al., 1983b). Another group of strains isolated mostly from feces of several species of penguins resembled most closely the biogroup 3 strains and were placed in another newly described species, *P. friedericiana* (Müller, 1983). In a subsequent collaborative study between Müller and the group at the Centers for Disease Control and Prevention, DNA hybridization testing showed that strains of *P. friedericiana* are very closely related to *P. rustigianii* and it was concluded that the two groups should be included in the same species (Hickman-Brenner et al., 1986). *P. rustigianii* was described first and the species name was validated before *P. friedericiana* and therefore *P. rustigianii* has priority and is the senior of the two subjective synonyms (Hickman-Brenner et al., 1986).

A biogroup of strains from penguins that were initially included in *P. friedericiana* and one strain listed under the CDC Enteric Group 78 (strain CDC 1519-73) were found to constitute a separate hybridization group and have been assigned to *P. heimbachae*, a newly described species (Müller et al., 1986).

The *P. alcalifaciens* biogroup 4 strains, originally reported to be inositol-negative were found on reexamination to be inositol-positive, a characteristic reaction of *P. stuartii*. They are also highly related to *P. stuartii* and have low levels of relatedness to *P. alcalifaciens*, *P. rettgeri*, and to *Proteus*, and have therefore been transferred to *P. stuartii* (Brenner et al., 1978).

In the earlier classification of the *Proteus-Providencia* group, *P. rettgeri* was included in the genus *Proteus* (Ewing and Davis, 1972b). In a biochemical characterization of this species, five biogroups were recognized (Penner et al., 1975). The biogroup 5 strains are urease-positive, a characteristic reaction of strains in the genus *Proteus*, but in other respects they resemble urease-negative *P. stuartii* (Penner et al., 1975, 1976c; Penner and Hennessy, 1977). On the basis of DNA relatedness tests they were found to be inseparable from *P. stuartii* and to be only 30–35% related to typical *Proteus rettgeri* biogroups 1, 2, and 3 and were therefore transferred to *P. stuartii* (Brenner et al., 1978). DNA of the three strains of *P. rettgeri* biogroup 4 (salicin and rhamnose negative) and other strains that have reactions atypical of *P. rettgeri* biogroups 1, 2, and 3 have not been examined for relatedness.

Antimicrobial susceptibilities New insights into species differences in antimicrobial susceptibilities have been noted since the revisions in the classification of genus *Providencia* in 1978 (Brenner et al., 1978). Studies using earlier classifications included urease-positive *P. stuartii* in collections identified as *Proteus rettgeri* in examining strains for antimicrobial susceptibility. Since the general adoption of the revised classification, it has become evident that, with the exception of a few unusual strains, *P. stuartii* is the most resistant species in the genus *Providencia* and one of the most resistant species in the family *Enterobacteriaceae*. *P. alcalifaciens* is the least resistant, being susceptible to cephalosporins, aminoglycosides, and quinolones, whereas *P. rettgeri* is intermediate to *P. alcalifaciens* and *P. stuartii*. Most strains of *P. rettgeri* are susceptible to gentamicin, tobramycin, and amikacin but are markedly resistant to quinolones. Most strains of *P. stuartii* are resistant to gentamicin, tobramycin, and the quinolones (including ciprofloxacin) but are susceptible to amikacin (Penner and Preston, 1980; Penner et al., 1982; Hawkey and Hawkey, 1984). Strains of *P. stuartii* that are also resistant to carbenicillin (due to a 47-kb plasmid) have also been identified (Hawkey et al., 1985). Some of the newer β-lactams, however, may be useful alternatives to amikacin in the treatment of serious *P. stuartii* infections (Hawkey, 1984). Only 11 strains of *P. rustigianii* and 13 of *P. heimbachae* have been examined for antimicrobial susceptibilities. They generally resemble *P. alcalifaciens* in their susceptibility to cephalosporins and aminoglycosides, but a small number differ in resistance to particular antibiotics (Hickman-Brenner et al., 1983b; Müller et al., 1986). 73% of the *P. rustigianii* are resistant to colistin and 100% and 77% of the *P. heimbachae* are resistant to tetracycline and cephalothin, respectively. However, it should not be concluded that these are characteristic species differences because of the small numbers of isolates that were available for study.

Pathogenicity *P. alcalifaciens* has been isolated from human feces, particularly from those of the pediatric population ever since the organism was first described, but whether this species is an etiological agent of human gastroenteritis is still a matter of debate and speculation. Evidence in support of a pathogenic role is the finding in several studies that the organisms are more frequently isolated from patients with diarrhea than from normal individuals (Prakash et al., 1966; Bhat et al.,1971a, b; Albert et al., 1995a, 1998; Guth and Perrella, 1996). Serotyping has shown that isolates of serotype O:3 are more frequently associated with pediatric diarrheas than are other serotypes, and this serotype is not frequently found in healthy children (Carpenter, 1964; Bhat et al., 1971a, b; Penner et al., 1979a). Other serotypes identified less frequently in human cases of diarrhea are O:12, O:13, and O:22 (Stenzel, 1961; Bhat et al., 1971a). In the earlier studies, examinations were not made for the more recently recognized pathogens such as rotaviruses, campylobacteria, pathogenic *E. coli*, *Yersinia enterocolitica*, and intestinal parasites, and therefore the importance of fecal isolates of *P. alcalifaciens* has remained unclear. If the organisms of this species are not causal agents, it may be that they merely flourish in the intestinal environment created by known diarrhea-causing agents. However, in an investigation of traveler's diarrhea it was found that 10% of returning travelers had stools positive for *P. alcalifaciens*, whereas only 1.3% of stools from a control population with diarrhea were positive (Haynes and Hawkey, 1989). No well-known pathogens were found in any of the samples that were positive for *P. alcalifaciens*, and the authors cited this observation in support of a role for this species in producing human diarrhea. Moreover, strains of this species isolated from diarrheal stools have been shown to be invasive for HEp-2 cells, to be able to produce diarrhea in adult rabbits with removable ileal ties (RITARD model), and to invade the rabbit intestinal mucosa (Albert et al., 1995a, 1998). In addition, some patients infected only with *P. alcalifaciens* experienced manifestations of invasive diarrhea (Albert et al., 1998). In another study, 50% of the *P. alcalifaciens* strains isolated in pure culture or from stool specimens in which no other enteropathogen was identified, were found to be invasive for HeLa cells (Guth and Perrella, 1996). Most of the invasive strains were isolated from diarrheal stools but the invasive characteristic was not present in all *P. alcalifaciens* strains that were isolated from patients with diarrhea. The production of cytotoxins or enterotoxins could not be detected in the invasive *P. alcalifaciens* isolates and their invasiveness could therefore not be attributed to these well-known virulence factors (Albert et al., 1995a, 1998; Guth and Perrella, 1996). It should be noted that an *E. coli* strain that can colonize the small bowel mucosa is unable to produce any of the well-known toxins, but can cause fluid accumulation and diarrhea in the rabbit model (Wanke and Guerrant, 1987). The mechanisms for the production by the *E. coli* strain of the fluid and diarrhea are unknown but the possibility of similar, yet unknown mechanisms, should be considered before discounting an etiological role in the enteropathogenicity of at least some strains of *P. alcalifaciens*. Although evidence is accumulating in support of an etiological role, more direct studies are necessary to provide convincing evidence for the inclusion of this species as a true member of the human enteropathogens.

The urinary tract of the compromised or catheterized patient is the most common site of *P. rettgeri* and *P. stuartii* infections. The rise in importance of these two species is associated with their tendency to cause nosocomial infections and with their marked resistance to numerous antibiotics (Wenzel et al., 1976; Whiteley et al., 1977; Kopf and Freitag, 1979; Penner et al., 1979b, 1981; Kocka et al., 1980; McHale et al., 1981; Warren et al., 1982; Hollick et al., 1984; Warren, 1986). Of the two species, *P. stuartii* is significantly more pathogenic and more resistant (Hawkey, 1984). Earlier studies on *P. rettgeri* often included urease-positive *P. stuartii* strains, thus occluding the significant differences between the two species (Lindsey et al., 1976; Penner et al., 1979b). Since the revision in classification, *P. stuartii* have been the subject of numerous studies to determine the basis of their pathogenicity. It has been shown that the majority of the *P. stuartii* isolated from urinary tract infections adhere to uroepithelial cells *in vitro*, suggesting that this property may account for the persistence of the bacteria in the urinary tract (Mobley et al., 1986a). Further studies have yielded evidence that strains bearing the MR/K fimbriae (mannose-resistant *Klebsiella*-like hemagglutinins) have the exceptional ability of binding to catheters, a characteristic that has been hypothesized to enhance their resistance to antimicrobials and to increase the duration of infection in patients with urinary tract catheters (Mobley et al., 1986a). *P. stuartii* can cause serious

infections in surgical and burn wounds and often give rise to fatal septicemias (Curreri et al., 1973; Milstoc and Steinberg, 1973; Wenzel et al., 1976).

In elderly catheterized patients who often have high levels of urinary indoxyl sulfate and have colonization of their urinary tract with *P. stuartii*, the urinary catheter bag develops an intense purple color. This purple urine bag syndrome has been determined to be due to the presence in the organisms of indoxyl sulfatase activity, which produces indigo through the decomposition of the urinary indoxyl sulfate (Dealler et al., 1988).

Although *P. stuartii* has been isolated from human feces, there is no evidence that the species causes infections of the human gastrointestinal tract, but in an investigation of neonatal diarrhea in calves (calf scours), evidence was obtained that suggested an etiological role for *P. stuartii* (Waldhalm et al., 1969) or a contributing role by acting synergistically with a neonatal calf diarrhea virus in the production of such diarrheas (Waldhalm et al., 1974b).

P. rustigianii strains have been isolated from human feces, but only infrequently. In a collection of 891 *Providencia* isolates made by Ewing et al. (1972), only 11 (<2%) were identified as *P. rustigianii* and three of these were reported to have been isolated from stool cultures (Hickman-Brenner et al., 1983b). Of the 19 isolates of this species studied by Costas et al. (1987), 10 were also from human feces (Costas et al., 1987). These studies indicate that bacteria of this species can colonize the human intestinal tract, but the clinical significance of this is not known. Strains of this species were also isolated from penguin feces by Müller (1983), but the significance of the presence of these organisms in the intestine to the health of the animals has not been investigated.

All 13 isolates of *P. heimbachae* that have been studied are from healthy penguins except one isolate that was obtained from an aborted bovine fetus (Müller et al., 1986). The medical and veterinary importance of these organisms remains unknown.

Habitats   Early studies to determine the habitats of the *Providencia* were carried out without the advantage of the recently revised classifications, and the prevailing view was that these organisms are widespread in nature, occurring in the intestines of various animals and polluted environmental sources (Müller, 1972). Uncertainty remains on the significance of their presence in the human intestine and only recently have systematic studies been reported to address this issue. From examinations of 1108 stool cultures of pediatric patients 56 isolates of *P. alcalifaciens* were found, and in the same period of time 25 isolates were recovered in routine from three adult hospitals (Penner et al., 1979a). Clearly organisms of this species can be found in the human intestine without special enrichment techniques, but their association with the production of diarrhea remains to be demonstrated.

It has been suspected that *P. stuartii* may also be found in the human intestine, but it is rarely isolated in the clinical laboratory from stool samples. As shown in two studies using special techniques such as pre-enrichment or selective differential media, the organisms of this species occur in the human intestine more frequently than previously believed and may be an important and previously underestimated nosocomial reservoir (Hawkey et al., 1982a, b; Stickler et al., 1985). The major source of the organisms in the hospital setting was found to be the colonized skin of the groin area of patients that were fecal carriers of *P. stuartii* (Stickler et al., 1985). *P. stuartii* was not isolated from raw sewage and sewage contaminated water samples (Stickler et al., 1985). In a survey of the *Proteus*, *Providencia*, and *Morganella* in the environment of calf rearing units, 127 isolates were recovered of which 50 were *Providencia* (Hawkey et al., 1986b). One was classified as *P. alcalifaciens*, 35 as *P. rettgeri*, and 14 as *P. stuartii*, and it was suggested that a potential cause of colonization of the human intestine by these organisms is the consumption of meat (Hawkey et al., 1986b).

P. rettgeri was first isolated by Rettger from poultry (Hadley et al., 1918). Its occurrence in the intestines of reptiles, frogs, and a duck, and its presence in natural waters have been reported (Müller, 1972; Penner and Hennessy, 1979a). Of 112 isolates collected from the clinical laboratory of a large hospital, 106 were from urine specimens and six were from stool samples (Penner and Hennessy, 1979a). In a study of the environment of calf rearing units in England, *P. rettgeri* was found to be second to *Proteus vulgaris* as the most frequently occurring species (Hawkey et al., 1986b). It thus appears that the species is widely distributed in nature, but is not a common resident of the human intestinal tract.

Studies to determine the habitats of *P. rustigianii* and *P. heimbachae* have not been conducted. In the first description of *P. rustigianii* only 11 isolates were available. Three of these were isolated from stools, but the sources of the other eight isolates are unknown (Hickman-Brenner et al., 1983b). Another group of *P. rustigianii* isolates (originally classified as *P. friedericiana*) were all isolated from feces of different species of penguins (Müller, 1983). Strains of *P. heimbachae* have been isolated only from fecal specimens of penguins and from one aborted bovine fetus (Müller et al., 1986).

ENRICHMENT AND ISOLATION PROCEDURES

Media used in the clinical laboratory for isolation of *Enterobacteriaceae* may be used to isolate *Providencia*. Tetrathionate or selenite broths may be used for enrichment.

A selective differential medium for the detection of *P. alcalifaciens* has been developed by Senior (1997). After enrichment of fecal samples in tetrathionate broth, the broth cultures are plated on a solid agar medium that contains phenol red dye and three sugars (xylose, mannitol, and galactose) that *P. alcalifaciens* does not ferment. Since other enteric bacteria that are present in the tetrathionate broth ferment one or more of these sugars, they produce lemon-yellow (acid-forming) colonies. Because strains of *P. alcalifaciens* do not ferment sugars in this medium, the strains produce red (alkaline) colonies.

Enrichment in heart infusion broth (2.5%) followed by plating on MacConkey agar containing 5 mg/l gentamicin sulfate has been found useful for the isolation of *P. stuartii* (Hawkey et al., 1982b). A selective differential medium for *P. stuartii* contains colistin (as the selective agent) and inositol in a nutrient agar base with bromothymol blue (as the acid-alkali indicator). After overnight incubation at 37°C, *P. stuartii* produces yellow inositol-fermenting colonies, *Proteus mirabilis* produces green-centered colonies, and *Morganella morganii* produces blue colonies (Stickler et al., 1985).

MAINTENANCE PROCEDURES

Providencia strains may be maintained on trypticase soy agar at 4°C with monthly transfers or may be preserved indefinitely by lyophilization. The Enteric Section, Centers for Disease Control and Prevention, stores cultures at room temperature in tubes of blood agar base or trypticase soy agar. These tubes are sealed with a cork or rubber stopper, and the cultures have remained viable for many years without transfer (J.J. Farmer III and F. Hickman-Brenner, personal communication).

DIFFERENTIATION OF THE GENUS *PROVIDENCIA* FROM OTHER GENERA

Characteristics and biochemical reactions for differentiating *Providencia* from the closely related genera, *Proteus* and *Morganella*, are described in Table BXII.γ.254 in the chapter on the genus *Proteus*. An important distinguishing characteristic of the *Providencia* (except *P. rustigianii*) is the ability to produce acid from one or more polyhydric alcohols. Biochemical reactions for differentiating *Providencia* from other species in the family *Enterobacteriaceae* are provided in Tables BXII.γ.193, BXII.γ.194, BXII.γ.195, and BXII.γ.196 in the chapter on *Enterobacteriaceae*.

TAXONOMIC COMMENTS

Since the last edition of the *Manual*, two new species have been included in the genus *Providencia*. Strains formerly classified as *P. alcalifaciens* biogroup 3 (Ewing et al., 1972) were found to constitute a separate species based on DNA relatedness tests and were assigned the epithet *rustigianii* in honor of Robert Rustigian who is noted for his early work on the genus *Proteus* (Hickman-Brenner et al., 1983b). A group of well-defined strains isolated from fecal samples obtained from penguins were found to represent a new species of the genus, and the epithet *friedericiana* was proposed to recognize the technical contributions of Friederike Heimbach (Müller, 1983). Both of these species gained standing in nomenclature but studies of strains exchanged between the two laboratories led to the finding that the species were the same. Since *P. rustigianii* was validated earlier it has priority and *P. friedericiana* is therefore a (junior) subjective synonym of *P. rustigianii* (Hickman-Brenner et al., 1983b). A biogroup of *P. friedericiana* and a strain (CDC 1519-73) of the "Enteric Group 78" were found in DNA relatedness tests to constitute a separate species that has been assigned the epithet *heimbachae* in honor of Friederike Heimbach, who isolated the original 12 strains of this species (Müller et al., 1986).

FURTHER READING

Albert, M.J., M. Ansaruzzaman, N.A. Bhuiyan, P.K. Neogi and A.S. Faruque. 1995. Characteristics of invasion of HEp-2 cells by *Providencia alcalifaciens*. J. Med. Microbiol. *42*: 186–190.

Albert, M.J., A.S. Faruque and D. Mahalanabis. 1998. Association of *Providencia alcalifaciens* with diarrhea in children. J. Clin. Microbiol. *36*: 1433–1435.

Brenner, D.J., J.J. Farmer, III, G.R. Fanning, A.G. Steigerwalt, P. Klykken, H.G. Wathen, F.W. Hickman and W.H. Ewing. 1978. Deoxyribonucleic acid relatedness in species of *Proteus* and *Providencia*. Int. J. Syst. Bacteriol. *28*: 269–282.

Hawkey, P.M. 1984. *Providencia stuartii*: a review of a multiply antibiotic-resistant bacterium. J. Antimicrob. Chemother. *13*: 209–226.

Hickman-Brenner, F.W., J.J. Farmer III, A.G. Steigerwalt and D.J. Brenner. 1983. *Providencia rustigianii*: a new species in the family *Enterobacteriaceae* formerly known as *Providencia alcalifaciens* biogroup 3. J. Clin. Microbiol. *17*: 1057–1060.

Müller, H.E., C.M. O'Hara, G.R. Fanning, F.W. Hickman-Brenner, J.M. Swenson and D.J. Brenner. 1986. *Providencia heimbachae*, a new species of *Enterobacteriaceae* isolated from animals. Int. J. Syst. Bacteriol. *36*: 252–256.

Warren, J.W. 1986. *Providencia stuartii*: a common cause of antibiotic-resistant bacteriuria in patients with long-term indwelling catheters. Rev. Infect. Dis. *8*: 61–67.

List of species of the genus Providencia

1. **Providencia alcalifaciens** (de Salles Gomes 1944) Ewing 1962, 96[AL] (*Eberthella alcalifaciens* de Salles Gomes 1944, 183; *Proteus inconstans* (Ornstein 1921) Shaw and Clarke 1955, 155.)

al.cal.i.fac′i.ens. Fr. n. *alcali* alkali; L. v. *facere* to do, make; L. part. adj. *faciens* making; M.L. part. adj. *alcalifaciens* alkali-producing.

The characteristics are as described for the genus and as listed in Tables BXII.γ.257 and BXII.γ.258. Urease negative. Most strains are susceptible to penicillins, cephalosporins, and aminoglycosides. Organisms are generally isolated from diarrheic stools but also from some normal stools. The species is suspected to be a causal agent of infant diarrhea and has been implicated as a cause of traveler's diarrhea but firm evidence for an etiological role in the production of diarrhea has not been established.

The mol% G + C of the DNA is: 43 (T_m) (Owen et al., 1987).

Type strain: ATCC 9886, DSM 30120.

2. **Providencia heimbachae** Müller, O'Hara, Fanning, Hickman-Brenner, Swenson and Brenner 1986, 255.[VP]

heim.bach′ae. M.L. gen. fem. n. *heimbachae* of Heimbach, named to honor Friederike Heimbach who isolated the 12 original strains of the species.

The characteristics are as described for the genus and as listed in Tables BXII.γ.257 and BXII.γ.258. It can be differentiated from other *Providencia* species by its negative reactions for Simmons citrate, growth in the presence of KCN, acid production from trehalose, urease, and indole production, and by positive tests for acid production from adonitol, D-arabitol, D-galactose, and L-rhamnose. Generally susceptible to penicillins, cephalosporins, and aminoglycosides, but most strains are resistant to cephalothin and tetracycline. Only 13 strains have been available for study. All were isolated from healthy penguins except for one that was isolated from an aborted bovine fetus. The medical and veterinary significance of these organisms is unknown.

The mol% G + C of the DNA is: 39.6 (T_m) (Owen et al., 1987).

Type strain: ATCC 35613, DSM 3591.

3. **Providencia rettgeri** (Hadley, Elkins and Caldwell 1918) Brenner, Farmer, Steigerwalt, Klykken, Wathen, Hickman and Ewing 1978, 269[AL] (*Bacterium rettgeri* Hadley, Elkins and Caldwell 1918, 180; *Proteus rettgeri* (Hadley, Elkins and Caldwell 1918) Rustigian and Stuart 1943, 242.)

rett′ge.ri. M.L. gen. n. *rettgeri* of Rettger, named after L.F. Rettger, the American bacteriologist who first isolated the organism in 1904.

The characteristics are as described for the genus and as listed in Tables BXII.γ.257 and BXII.γ.258. Virtually all strains hydrolyze urea and are distinguished from the other *Providencia* by their ability to produce acid from a variety of polyhydric alcohols. Many strains are resistant to the penicillins and cephalosporins, but the strains are generally more susceptible to aminoglycosides than are *P. stuartii* strains. Generally isolated from urine specimens of hospitalized and catheterized patients and less frequently from other sites. May cause nosocomial infections. Rarely isolated from stool specimens.

The mol% G + C of the DNA is: 40.5 (T_m) (Owen et al., 1987).

Type strain: ATCC 29944, DSM 4542.

4. **Providencia rustigianii** Hickman-Brenner, Farmer, Steigerwalt and Brenner 1983a, 673[VP] (Effective publication: Hickman-Brenner, Farmer, Steigerwalt and Brenner 1983b, 1060.)

rus.tig.i.an′i.i. M.L. gen. n. *rustigianii* of Rustigian, named in honor of Robert Rustigian who did early studies on the *Proteus* group.

The characteristics are as described for the genus and as listed in Tables BXII.γ.257 and BXII.γ.258. Strains are differentiated from other *Providencia* species by their production of acid from D-galactose but not from adonitol, *i*-inositol, and trehalose. Only 11 strains have been examined for antibiotic susceptibly. Most are susceptible to cephalosporins and synthetic penicillins but resistant to natural penicillin and colistin. They may colonize the human intestinal tract but the medical significance of this is not known.

The mol% G + C of the DNA is: 41.8 (T_m) (Owen et al., 1987).

Type strain: ATCC 33673, DSM 4541.

5. **Providencia stuartii** (Buttiaux, Osteux, Fresnoy and Mor-

TABLE BXII.γ.257. Differential characteristics of the genus *Providencia*

Characteristic[a]	1. *P. alcalifaciens*	2. *P. heimbachae*	3. *P. rettgeri*	4. *P. rustigianii*	5. *P. stuartii*
Indole production	+	−	+	+	+
Citrate (Simmons)	+	−	+	d	+
Urea	−	−	+	−	d
Motility	+	d	+	d	d
Growth in KCN	+	−	+	+	+
Gas from D-Glucose	d	−	−	d[b]	−
Acid production from:					
Adonitol	+	+	+	−	−
D-Arabitol	−	+	+	−	−
D-Galactose	−	+	+	+	+
i-Inositol	−	d	+	−	+
D-Mannitol	−	−	+	−	−
L-Rhamnose	−	+	d	−	−
Trehalose	−	−	−	−	+

[a]All tests were performed at 36° ± 1°C.

[b]Minimal gas production.

TABLE BXII.γ.258. Other characteristics of the species of the genus *Providencia*

Characteristic[a]	1. *P. alcalifaciens*[b]	2. *P. heimbachae*[c]	3. *P. rettgeri*[b]	4. *P. rustigianii*[d]	5. *P. stuartii*[b]
Methyl red	+	d	+	d	+
Voges–Proskauer	−	−	−	−	−
H₂S on TSI	−	−	−	−	−
Phenylalanine	+	+	+	+	+
L-Lysine (Møller's)	−	−	−	−	−
L-Arginine (Møller's)	−	−	−	−	−
L-Ornithine (Møller's)	−	−	−	−	−
Gelatin (22°C)	−	−	−	−	−
Malonate	−	−	−	−	−
D-Glucose (acid production)	+	+	+	+	+
D-Glucose (gas production)	d	d (7)[e]	d	d	−
Acid from:					
L-Arabinose	−	−	−	−	−
Dulcitol	−	−	−	−	−
Erythritol	−	−	d	−	−
Glycerol	d	d (7)	d	−	d
Lactose	−	−	−	−	−
Maltose	−	+ (7)	−	−	−
D-Mannose	+	+	+	+	+
Melibiose	−	−	−	−	−
α-CH₃-glucoside	−	−	−	−	−
Raffinose	−	−	−	−	−
Salicin	−	−	d	−	−
D-Sorbitol	−	−	−	−	−
Sucrose	d	−	d	+ (7)	d
D-Xylose	−	d (7)	d	−	−
Esculin hydrolysis	−	−	d	−	−
Acid from mucate	−	−	−	−	−
Tartrate (Jordan's)	+	d	+	+	+
Acetate utilization	d	−	d	−	d
Lipase (corn oil)	−	−	−	−	−
DNase 25°C	−	−	−	−	−
NO₃→NO₂	+	+	+	+	+
Oxidase production	−	−	−	−	−
ONPG production	−	−	−	−	−
Tyrosine clearing	+	+	+	+	+

[a]Biochemistry reactions at 36° ± 1°C unless otherwise stated.

[b]Data from Brenner et al. (1978).

[c]Data from Müller et al. (1986).

[d]Data from Hickman-Brenner et al. (1983b).

[e]Numbers in parentheses indicate number of days for test.

iamez 1954) Ewing 1962, 96[AL] (*Proteus stuartii* Buttiaux, Osteux, Fresnoy and Moriamez 1954, 385; *Proteus inconstans* (Ornstein 1921) Shaw and Clarke 1955, 155.)

stu.ar'ti.i. M.L. gen. n. *stuartii* of Stuart, named after C.A. Stuart, American bacteriologist who did much of the early work on *Providencia*.

The characteristics are as described for the genus and as listed in Tables BXII.γ.257 and BXII.γ.258. Highly resistant to antibiotics. Some strains are among the most resistant of the *Enterobacteriaceae*. Strains may be urease positive or urease negative. Isolated most often from urine specimens of hospitalized and catheterized patients. Some strains have the ability to adhere to catheters and thus enhance resistance to treatment and increase the duration of the infection. Less frequently isolated from wounds, burns, and bacteremias. They occur in small numbers in the intestines of some individuals and are rarely isolated routinely from stool cultures in the clinical laboratory but serve as the source for the colonization of the skin of the groin area, which may lead to autoinfections and cross-infections.

The mol% G + C of the DNA is: 40.7 (T_m) (Owen et al., 1987).

Type strain: ATCC 29914, DSM 4539.

Genus XXXI. Rahnella Izard, Gavini, Trinel and Leclerc 1981c, 382[VP] (Effective publication: Izard, Gavini, Trinel and Leclerc 1979, 174)

PETER KÄMPFER

Rah'nel.la. M.L. dim. ending -*ella*; M.L. fem. n. *Rahnella* named after Otto Rahn, the German-American microbiologist and who proposed the name *Enterobacteriaceae* in 1937.

Straight rods 0.5–0.7 × 2–3 μm, conforming to the general definition of the family *Enterobacteriaceae*. **Gram negative, motile by peritrichous flagella when grown at 25°C.** Facultatively anaerobic. Chemoorganotrophic. **Psychrotolerant,** growing at 4°C. D-Glucose is fermented with the production of acid and, for the majority of strains, gas. Nitrate is reduced to nitrite. Oxidase negative, catalase positive. **Negative for lysine and ornithine decarboxylases and for arginine dihydrolase. Most strains are (weakly) positive for phenylalanine deaminase (after 48 h), methyl red, and Voges–Proskauer reaction. Acids are produced from various carbohydrates, including L-arabinose, cellobiose, lactose, maltose, mannose, L-rhamnose, raffinose, D-xylose, and salicin.** Sequence analyses of the 16S rDNA of four *Rahnella* strains clearly placed the genus within the *Gammaproteobacteria* in the family *Enterobacteriaceae*. Highest 16S rDNA sequence similarity to *Yersinia enterocolitica* ATCC 9610^T (97.7%). By omitting the hypervariable regions V1 and V5 (corresponding to *E. coli* base positions 147–1490; Brosius et al., 1978), *Hafnia alvei* ATCC 13337^T showed the highest sequence similarity (96.63%) to *R. aquatilis* ATCC 33071^T. Most often isolated from fresh water, but also found in the intestine of snails and various other environmental habitats, including soils and the rhizosphere. Can occasionally be isolated from foods or human clinical specimens, including wound infections, bacteremias, feces from patients with acute gastroenteritis, and septicemia, especially from immunocompromised patients. The genus currently includes three genomospecies (*R. aquatilis* [*Rahnella* genomospecies 1], *Rahnella* genomospecies 2, and *Rahnella* genomospecies 3), which cannot be phenotypically differentiated.

The mol% G + C of the DNA is: 51–56.

Type species: **Rahnella aquatilis** Izard, Gavini, Trinel and Leclerc 1981c, 382 (Effective publication: Izard, Gavini, Trinel and Leclerc 1979, 174.)

FURTHER DESCRIPTIVE INFORMATION

A comparison of the total 16S rDNA sequence of the type strains of *R. aquatilis* with other representatives of the family *Enterobacteriaceae* showed highest similarity (97.7%) with the type strain of *Yersinia enterocolitica*. Eight representative strains representing all three *Rahnella* genomospecies were used for 16S rRNA gene amplification and sequencing (Brenner et al., 1998). The hypervariable regions V1 and V5 were omitted for sequence comparison resulting in a sequence that corresponded to *E. coli* base positions 147–1490 (Brosius et al., 1978). The eight strains showed sequence similarities ranging from 97.7–100% and fell in three distinct groups (Brenner et al., 1998). Three additional strains sequences already deposited in GenBank also grouped with *R. aquatilis* type strain, exhibiting 99.7–100% sequence similarity. The second group contained five strains used in the DNA–DNA relatedness experiments (Brenner et al., 1998), four of which were in genomospecies 2 and one in genomospecies 1 (*Rahnella aquatilis sensu stricto*). Two further strains, for which sequences were deposited in GenBank, were also found in this group. This observation supports the advisability of performing DNA–DNA hybridization experiments on strains showing 97% or greater sequence similarity, before considering them to belong to the same species (Brenner et al., 1998). The third group contained only strain DSM 30078, representing *Rahnella* genomospecies 3.

The phenotypic description of the genus *Rahnella* is largely based on the studies of Gavini et al. (1976a), Izard et al. (1979), Farmer (1984b), Farmer et al. (1985a), Brenner (1992a), and Brenner et al. (1998). In the study of Brenner et al. (1998), a total of 51 *Rahnella* strains were included, and the results indicated that the strains comprise three closely related species (genomospecies), which cannot be differentiated phenotypically. Their subdivision is described below (see Taxonomic Comments).

Rahnella strains grow readily on all kinds of media and no growth requirements have been described. A nutrient-rich medium (e.g., tryptone–soy agar, sheep-blood agar, or nutrient agar) gives best results. Incubation at temperatures ranging from 25–35°C results often in better growth that incubation at 37°C. The majority of strains are psychrotolerant, growing (slowly) at temperatures of 4°C. Growth was reported even at temperatures of 1°C ± 1°C (Davis and Eyles, 1992).

Members of the genus *Rahnella* ferment lactose and produce β-galactosidase (ONPG test). Acid production from D-xylose is observed in parallel with strong β-xylosidase activity. β-glucuronidase is negative in all strains. Indole production, hydrogen sulfide formation, tyrosine deaminase, urease, and DNase tests are negative. Nitrogen fixation (Berge et al., 1991; Kim et al., 1998) and mineral phosphate solubilization (Kim et al., 1998) has been reported in some strains of *R. aquatilis*. Various carbon sources, including carbohydrates, organic acids, and amino acids, are utilized as sole sources of carbon (Brenner et al., 1998).

Isoprenoid quinone Q-8 is the predominant quinone and small amounts of the menaquinone MK-8 are found in *R. aquatilis* ATCC 33071^T (P. Kämpfer, unpublished results). The fatty acid composition as determined by gas chromatographic analysis of all 51 strains used in the study of Brenner et al. (1998) differed only slightly (P. Kämpfer, unpublished results). All strains contained the fatty acids $C_{12:0}$, $C_{14:0}$, $C_{16:0}$, $C_{17:0\ cyclo}$, $C_{19:0\ cyclo}$, summed feature ($C_{16:1\ iso\ I}$ and/or $C_{14:0\ 3OH}$), summed feature ($C_{16:1\ \omega7c}$ and/or $C_{15:0\ iso\ 2OH}$), and summed feature ($C_{18:1\ \omega7c}$, $C_{18:1\ \omega9t}$, and/or $C_{18:1\ \omega12t}$), a fatty acid type typical of a member of the family *Enterobacteriaceae*. The dominant polyamines, as determined by high-performance liquid chromatography, were putrescine and cadaverine (Hamana, 1996). Minor amounts of diaminopropane were found in both strains studied; one strain also contained agmatine, whereas the other contained spermidine.

Members of the genus *Rahnella* are usually sensitive to aminoglycosides, cefotaxime, doxocyline, trimethoprim, gentamicin, tetracycline, and tobramycin, but resistant to chloramphenicol. Susceptibility to β-lactams differs among strains. Antimicrobial susceptibility data are summarized by Funke and Rosner (1995). They state that aminopenicillins and first-generation cephalosporins have only limited activity against *R. aquatilis*, whereas ureidopenicillins and carbapenems are more active. *R. aquatilis* is usually susceptible to aminoglycosides and quinolones. *Rahnella* produces enterobacterial common antigen (Böttger et al., 1987).

The pathogenicity of *Rahnella* in humans and animals is not clearly established. Strains isolated from humans appear to be opportunistic rather than true pathogens.

Rahnella are widely distributed in nature. They may be isolated from foods and occasionally from a wide variety of human sources. The natural habitat of *Rahnella* seems to be water, and all of the isolates of Gavini et al. (1976a) were from waters in France. Several American water isolates have also been identified as *R. aquatilis*. The study of the intestinal content of snails for the presence of *Enterobacteriaceae* (Brenner et al., 1998) resulted in the isolation of infrequently seen enterobacterial species, among them *Rahnella* spp. Rhodes et al. (1998) found *Rahnella* spp. in the intestinal content associated with 12,000-year-old mastodon remains. The study of mountain soils for psychrotrophic *Enterobacteriaceae* resulted in the isolation of *Rahnella* (Horie et al., 1985), and nitrogen-fixing strains were isolated from the rhizosphere of wheat and maize (Berge et al., 1991; Heulin et al., 1994). Heron et al. (1993) found *Rahnella* isolated among enterobacteria associated with grass and silages. The microbiological study of prepared salad vegetables and sprouts resulted in the isolation of *Rahnella* (Geiges et al., 1990). In minced meat, fish, and milk, *Rahnella* were frequently encountered by Lindberg et al. (1998), and one fish isolate was found to contain the gene of *E. coli* heat-labile toxin (lt). Hamze et al. (1991) found *R. aquatilis* as a potential contaminant in lager beer breweries. Two strains of *Rahnella* were isolated from buckwheat seeds (Iimura and Hosono, 1996).

R. aquatilis* may also occasionally occur in human clinical specimens. The first case reported was a strain from a burn wound (Farmer et al., 1985a). Maraki et al. (1994) reported a surgical wound infection with *Rahnella*. Two *Rahnella* isolates were recovered from the feces of two patients with acute gastroenteritis, one of whom was an AIDS patient (Reina and Lopez, 1996a). The strains were resistant to ampicillin, cephalothin, and cefoxitin. Funke and Rosner (1995) reported *R. aquatilis* bacteremia in an HIV-infected intravenous drug-abuser. Caraccio et al. (1994) isolated *Rahnella* from a bacteremic patient with chronic renal failure. Matsukura et al. (1996) isolated *Rahnella* aquatilis from a blood culture obtained from a case of endocarditis. *R. aquatilis* was found in the blood of a diabetic patient and a patient with laryngeal carcinoma (Oh and Tay, 1995). Both patients recovered from the infection after treatment with parenteral antibiotics. Additional cases were described by Goubau et al. (1988), Harrell et al. (1989), and Hoppe et al. (1993). Literature on *R. aquatilis* infections in humans are reviewed by Alballaa et al. (1992), Maraki et al. (1994), and Funke and Rosner (1995).

ENRICHMENT AND ISOLATION PROCEDURES

Media used for the isolation of *Rahnella* are similar to those used for other members of the family *Enterobacteriaceae*. No specific selective medium is available for *Rahnella* species. As pointed out by Grimont and Grimont (1992), the requirements for a selective medium can be summarized as follows: (1) a precise and stable delincation of the taxon (genus or species); (2) known and stable common nutritional properties uncommon outside the group (genus or species); and (3) a special, in most cases clinical, or public health requirement for such a medium. Because of the low clinical significance and the presence of *Rahnella* in various habitats, the selective isolation of *Rahnella* is rarely required, except for epidemiological studies. In most cases, organisms belonging to the genus *Rahnella* are isolated with differential media not inhibitory for *Enterobacteriaceae*, such as MacConkey agar, bromothymol blue lactose agar, or phenol red lactose agar. *Rahnella* strains grow readily on nutrient-rich media, and no particular growth requirements have been described.

MAINTENANCE PROCEDURES

Rahnella strains can be maintained in tryptone soy agar stabs or on nutrient agar when kept at room temperature in the dark. They can be preserved by storage in broth containing 10% glycerol or in calf or bovine serum at −80°C. Lyophilization seems to be the best procedure for preservation.

DIFFERENTIATION OF THE GENUS *RAHNELLA* FROM OTHER GENERA

The genus *Rahnella* has no single distinguishing feature useful for differentiation from the other genera of *Enterobacteriaceae* (Table BXII.γ.193 of the *Enterobacteriaceae*). *Rahnella* does not produce a yellow pigment, is negative for lysine and ornithine decarboxylases, and is weakly positive for phenylalanine deaminase. These characters may be helpful in differentiation of *Rahnella* from the genera *Pantoea* and *Erwinia*.

TAXONOMIC COMMENTS

Based on a numerical taxonomic study of *Enterobacteriaceae*, Gavini et al. (1976a) defined a new group within the family and gave it the vernacular name "group H2". Subsequently, Izard et al. (1979) used DNA–DNA hybridization to compare strains of group H2 to each other and to named species of *Enterobacteriaceae*. Based on the close relatedness within group H2 and the low relatedness to other *Enterobacteriaceae*, they proposed the new genus *Rahnella* with one species *Rahnella aquatilis*, and designated a holotype strain. The name *Rahnella* was chosen to honor Otto Rahn, the German-American microbiologist for his many contributions to systematic bacteriology and for his proposal of the family name *Enterobacteriaceae* in 1937. Because all of the isolates investigated by Izard et al. (1979) were from water, the species name *aquatilis* was chosen. The names *Rahnella* and *R. aquatilis* were effectively published (Izard et al., 1979) but were not validated in the *International Journal of Systematic Bacteriology* before 1 January 1980. They did not appear on the Approved Lists of Bacterial Names (Skerman et al., 1980), but both names have now been validly published (Izard et al., 1981c) and have standing in nomenclature. From 1979 to now, *Rahnella* was isolated from various other environmental habitats, including soils and the rhizosphere, and from food, indicating a wide distribution in nature. Furthermore, several clinical isolates were reported, in most cases from immunocompromised patients. Between 1984 and 1988, Müller et al. (1995a, b, 1996) screened the intestinal contents of snails and slugs for the presence of *Enterobacteriaceae* and found several *Rahnella* isolates in addition to the genera *Buttiauxella*, *Kluyvera*, and others. In a subsequent study 51 *Rahnella* and *Rahnella*-like organisms were subjected to an extensive DNA–DNA hybridization study. The results indicated that the strains comprised three closely related (genomo-)species within the genus, one corresponding to *R. aquatilis*. The relative binding ratios obtained from DNA–DNA hybridization studies between the *Rahnella* genomospecies are given in Table BXII.γ.259. For the two new species the vernacular names *Rahnella* genomospecies 2 and *Rahnella* genomospecies 3 were proposed (Brenner et al., 1998). It was not possible to differentiate these genomospecies on the basis of physiological and biochemical tests.

ACKNOWLEDGMENTS

I thank Don J. Brenner for his helpful comments and critical reading of this chapter, Hans E. Müller for interesting discussions and furnishing

TABLE BXII.γ.259. Relative binding ratios of DNAs at 60°C (given as percentages/ranges of percentages) between the genomospecies of the genus *Rahnella*[a]

Test	R. aquatilis (Rahnella genomospecies 1)	Rahnella genomospecies 2	Rahnella genomospecies 3
R. aquatilis	88 (71–100)	64 (42–81)	54 (45–64)
Rahnella genomospecies 2		93 (70–100)	54 (49–61)
Rahnella genomospecies 3			100

[a]Data adapted from Brenner et al. (1998).

TABLE BXII.γ.260. Characteristics of (genomo)species of the genus *Rahnella*[a,b]

Biochemical reaction	*Rahnella aquatilis* (Genomospecies 1)	*Rahnella* (Genomospecies 2)	*Rahnella* (Genomospecies 3)
Methyl red	[+]	d	+
Voges–Proskauer	+	d	+
Citrate utilization	[+] (+)	d ([+])	−
Phenylalanine deaminase	+	d	+
Arginine dihydrolase	−	−	−
Lysine decarboxylase	−	−	−
Ornithine decarboxylase	− ([−])	−	−
Motility	d	d	−
Gelatin liquefaction (22°C)	−	− (d)	−
KCN, growth	− ([−])	−	−
Malonate	[+]	d ([+])	+
Gas from D-glucose	[+]	[+]	−
Acid from:			
D-Arabitol	−	−	+
Cellobiose	+	+	+
Dulcitol	[+] (+)	[+] (+)	−
Maltose	+	+	+
D-Mannose	+	+	+
Melibiose	+	+	+
Methyl-α-D-glucoside	−	−	−
Raffinose	+	+	+
L-Rhamnose	+	+	−
D-Sorbitol	+	+	+
Sucrose	+	+	+
D-Xylose	+	+	(+)
Glycerol	− (d)	− (d)	
Esculin hydrolysis	+	+	+
Mucate	d	−	
Tartrate	d	[+]	+
Acetate	[−]	−	−
Lipase (corn oil)	−	−	−
Lipase (Tween 80)	[+]	d	−
Utilization of:			
N-Acetyl-D-galactosamine	−	−	−
D-Fructose	[+]	d	+
Sucrose	+	+	+
Acetate	−	d	−
cis-Aconitate	d	[−]	−
trans-Aconitate	−	−	−
4-Aminobutyrate	−	−	−
DL-Lactate	+	+	−
2-Oxoglutarate	d	−	−
L-Alanine	+	d	+
L-Aspartate	+	+	+
L-Histidine	d	−	+
L-Ornithine	d	−	−
L-Phenylalanine	−	−	−
L-Proline	d	−	−
Phenylacetate	−	−	−
Hydrolysis of:			
pNP-β-D-Galactopyranoside	−	−	−
pNP-α-D-Glucopyranoside	d	−	−
pNP-β-D-Glucopyranoside	+	[−]	+
2-Deoxythymidine-5′-pNP-phosphate	−	[−]	−
L-Glutamate-γ-3-carboxy-pNA	+	[+]	−
L-Proline-pNA	[+]	−	+

[a]Symbols: +, positive for 90–100% of strains; [+], positive for 75–89% of strains; d, positive for 26–74% of strains; [−], positive for 11–25% of strains; −, positive for 0–10% of strains.

[b]Data adapted from Brenner et al. (1998). All reactions, unless otherwise stated were done at 36° ± 1°C and read after 48 h. Numbers in parentheses indicate the percentage of strains giving a positive reaction within 7 d. All strains grew at 4°C, 25°C, and 37°C, and gave positive reactions within 48 h in tests for fermentation of L-arabinose, D-galactose, D-glucose, lactose, D-mannitol, salicin, and trehalose, and in the tests for nitrate reduction to nitrite, catalase, motility at 25°C, and gelatin liquefaction at 22°C. All strains were oxidase-negative and gave negative reactions after 7 d in tests for fermentation of adonitol, erythritol, and *i*-inositol, and in tests for DNase, hydrogen sulfide, indole production, oxidase, tyrosine deaminase, urease, and yellow pigment. All strains gave positive reactions in tests for utilization of N-acetyl-D-glucosamine, L-arabinose, p-arbutin, D-cellobiose, D-galactose gluconate, D-glucose, D-maltose, D-mannose, α-D-melibiose, L-rhamnose, D-ribose, salicin, D-trehalose, D-xylose, D-mannitol, D-sorbitol, citrate, fumarate, L-malate, pyruvate, and L-serine, and positive reactions in tests for hydrolysis of pNP-β-D-galactopyranoside, pNP-β-D-xylopyranoside, bis-pNP-phosphate, pNP-phenyl-phophonate, pNP-phosphoryl choline, and L-alanine-pNA (pNP = para-nitrophenyl-; pNA = para-nitroanilide). All strains gave negative reactions in tests for utilization of adonitol, *i*-inositol, putrescine, propionate, adipate, azelate, glutarate, DL-3-hydroxybutyrate, mesaconate, suberate, β-alanine, L-leucine, L-tryptophan, 3-hydroxybenzoate, and 4-hydroxybenzoate.

with strains, and Wolfgang Ludwig for critical comments and his experience in 16S rRNA sequence analyses.

FURTHER READING

Brenner, D.J. 1992. Additional genera of *Enterobacteriaceae. In* Balows, Trüper, Dworkin, Harder and Schleifer (Editors), The Prokaryotes, 2nd Ed., Vol. 3, Springer-Verlag, New York. pp. 2922–2937.

Brenner, D.J., H.E. Müller, A.G. Steigerwalt, A.M. Whitney, C.M. O'Hara and P. Kämpfer. 1998. Two new *Rahnella* genomospecies that cannot be phenotypically differentiated from *Rahnella aquatilis.* Int. J. Syst. Bacteriol. *48*: 141–149.

Farmer, J.J. III 1984. Other genera of the family *Enterobacteriaceae. In* Krieg and Holt (Editors), Bergey's Manual of Systematic Bacteriology, 1st Ed., Vol. 1, The Williams & Wilkins Co, Baltimore. pp. 506–516.

Gavini, F., C. Ferragut, B. Lefebre and H. Leclerc. 1976. Étude taxonomique d'entérobactéries appartenant ou apparentées au genre *Enterobacter.* Ann. Inst. Pasteur Microbiol. *127B*: 317–335.

Izard, D., F. Gavini, P.A. Trinel and H. Leclerc. 1979. *Rahnella aquatilis,* nouveau membre de la famille des *Enterobacteriaceae.* Ann. Inst. Pasteur Microbiol. *130A*: 163–177.

List of species of the genus Rahnella

1. **Rahnella aquatilis** Izard, Gavini, Trinel and Leclerc 1981c, 382[VP] (Effective publication: Izard, Gavini, Trinel and Leclerc 1979, 174.)

   a.qua' ti.lis. L. adj. *aquatilis* living in water.

   Represents *Rahnella* genomospecies 1 of the study of Brenner et al. (1998). The characteristics are as described for the genus. Detailed characteristics are given in Table BXII.γ.260.

   Occurs in freshwater, but also in environmental habitats like soil. May occasionally be isolated from human clinical specimens; in most cases from immunocompromised hosts. Clinical significance is unknown. Mean DNA–DNA relatedness among the 21 strains studied by Brenner et al. (1998), including 9 of the 10 original water isolates described by Izard et al. (1979) was 88% in 60°C reactions (range 72–100%). Mean relatedness in 75°C reactions was 81% (range 61–100%).

   The mol% G + C of the DNA is: 51–56 (T_m).

   Type strain: 133, ATCC 33071, CIP 78-65, DSM 4594.

   GenBank accession number (16S rRNA): AJ233426.

Other Organisms

1. *Rahnella* genomospecies 2.

   This species was proposed on the basis on DNA–DNA hybridization experiments and comprises 30 strains (Brenner et al., 1998). Phenotypically it is indistinguishable from *Rahnella aquatilis.* A combination of the utilization tests with L-histidine, L-ornithine, and L-proline (Table BXII.γ.260; Brenner et al., 1998) may be helpful in differentiation from *Rahnella aquatilis,* but does not allow a clear (genomo)-species identification. All 30 strains studied by Brenner et al. (1998) were 83% or more related in 60°C reactions (range 70–100%). In 75°C reactions the strains were 74% or more related, with one exception showing only 68% relatedness to the labeled reference organism.

   Strains were isolated from water, intestinal contents of snails, and clinical material, including stored donated human blood.

   Deposited strain: SM S7/1-576.

2. *Rahnella* genomospecies 3.

   This species was also proposed based on DNA–DNA hybridization experiments (Brenner et al., 1998) and is represented by a single strain: DSM 30078 (isolated from minced meat). Phenotypically it can be separated from *Rahnella aquatilis* and *Rahnella* genomospecies 2 (Table BXII.γ.260), but without results from different strains one cannot determine whether the results are strain- or species-specific.

   Deposited strain: DSM 30078.

Genus XXXII. **Saccharobacter** Yaping, Xiaoyang and Jiaqu 1990, 412[VP]

DON J. BRENNER

Sac.cha.ro.bac' ter. L. n. *saccharum* sugar; M.L. n. *bacter* a rod; M.L. masc. n. *Saccharobacter* a sugar rod.

Small straight rods, 0.5–0.9 × 1.0–1.0 µm. **Gram negative, motile by peritrichous flagella.** Facultatively anaerobic. Chemoorganotrophic. Colonies on glucose-yeast extract agar are opaque and milky white, smooth, and low convex with entire margins. Oxidase negative, catalase positive. Nonpigmented. Nonsporeforming. No lipid granules present. Optimal growth temperature range is from 30–46°C. **D-glucose is fermented with the production of ethanol, CO_2, and small amounts of acids, but no hydrogen. Indole negative, methyl red positive, Voges–Proskauer test positive, citrate positive.** H_2S production, gelatin hydrolysis, urease, nitrogen reduction, lysine decarboxylase, and ornithine decarboxylase are negative; arginine dihydrolase is delayed positive (4 d). Acid is produced from D-glucose, L-arabinose, esculin, D-fructose, D-galactose, maltose, D-mannitol, melibiose, L-rhamnose, L-sorbose, starch, sucrose, trehalose, D-xylose. Lactose fermentation is delayed positive after 9 d. Dulcitol, gluconate, *myo*-inositol, and raffinose are not fermented. Malonate is used as a sole carbon source; ammonium sulfate, yeast extract, urea, phenylalanine, glutamine, and tryptone serve as sole nitrogen sources, but sodium glutamate does not. Produces β-galactosidase and grows in the presence of KCN. **Grows in the presence of 35% D-glucose and 6% NaCl, but not 40% D-glucose or 8% NaCl.** 16S rDNA sequence has not been determined.

The mol% G + C of the DNA is: 63.5.

Type species: **Saccharobacter fermentatus** Yaping, Xiaoyang, and Jiaqu 1990, 412.

FURTHER DESCRIPTIVE INFORMATION

Carbohydrates are presumably degraded by the Embden-Meyerhof-Parnas pathway, as indicated by the absence of 6-phosphogluconate dehydratase and 2-keto-3-deoxygluconate-6-phosphate aldolase and the presence of fructose diphosphate aldolase. Degrades 1 mol of glucose to approximately 2 mol of ethanol and 2 mol of CO_2.

MAINTENANCE PROCEDURES

Cultures are grown at 30°C and maintained on glucose-peptone-yeast extract agar. Lyophilization is used for long-term storage.

DIFFERENTIATION OF THE GENUS SACCHAROBACTER FROM OTHER GENERA

Saccharobacter is easily differentiated from *Zymomonas*, another ethanol-producing genus. *Saccharobacter* is peritrichously flagellated, whereas *Zymomonas* has polar flagella. *Zymomonas* ferments only D-glucose, fructose, and sucrose, whereas *Saccharobacter* ferments a number of other carbohydrates (see above). *Zymomonas* degrades sugars by the Entner-Duoderoff pathway, while *Saccharobacter* ferments sugars by the Embden-Meyerhof-Parnas pathway. *Saccharobacter* can easily be differentiated from various species of *Enterobacteriaceae* based on its indole, methyl red, Voges–Proskauer, and citrate (IMViC) reactions, together with its negative nitrate reductase reaction, its positive phenylalanine de-

aminase reaction, and its sugar fermentation pattern (see chapter on *Enterobacteriaceae*).

TAXONOMIC COMMENTS

Since 16S rDNA sequencing has not been done on the type species, it is impossible to accurately place the genus *Saccharobacter* phylogenetically. *Saccharobacter* appears to be phenotypically similar to members of the family *Enterobacteriaceae* based on a number of key characteristics. These include negative oxidase and positive catalase reactions, peritrichous flagella, facultatively anaerobic metabolism, and fermentation of D-glucose and other carbohydrates. It differs from *Enterobacteriaceae* in the end products of fermentation, which in *Enterobacteriaceae* are mixed acids or 2,3-butanediol, with no more then 0.5 mol of ethanol from 1 mol of D-glucose. *Saccharobacter* also differs from most, if not all, members of *Enterobacteriaceae* by the high mol% G + C content.

FURTHER READING

Yaping, J., L. Xiaoyang and Y. Jiaqi. 1990. *Saccharobacter fermentatus* gen. nov., sp. nov., a new ethanol-producing bacterium. Int. J. Syst. Bacteriol. *40*: 412–414.

List of species of the genus Saccharobacter

1. **Saccharobacter fermentatus** Yaping, Xiaoyang and Jiaqu 1990, 412[VP]

    fer.men.ta' tus. M.L. adj. *fermentatus* fermentative.

    The characteristics are those given for the genus. All

extant strains were isolated from squeezed leaf juice of agave. Isolated from squeezed leaf juice of agave in Wuhan, People's Republic of China.

   The mol% G + C of the DNA is: 63.5 (T_m).
   Type strain: WVB8512.

Genus XXXIII. **Salmonella** Lignières 1900, 389[AL]

MICHEL Y. POPOFF AND LÉON E. LE MINOR

Sal.mon.el' la. M.L. -*ella* dim. ending; M.L. fem. n. *Salmonella* named after D.E. Salmon, an American bacteriologist.

Straight rods, 0.7–1.5 × 2.0–5.0 μm, conforming to the general definition of the family *Enterobacteriaceae*. Gram negative. **Usually motile** (peritrichous flagella). Facultatively anaerobic. Colonies are generally 2–4 mm in diameter. Nitrates are reduced to nitrites. **Gas is usually produced from D-glucose.** Hydrogen sulfide is usually produced on triple-sugar iron agar. Indole negative. **Citrate is usually utilized as a sole carbon source.** Lysine and ornithine decarboxylase (Møller's) reactions are usually positive. Urease negative. Phenylalanine and tryptophan are not oxidatively deaminated. Sucrose, salicin, inositol, and amygdalin are usually not fermented. Lipase and deoxyribonuclease are not produced. Pathogenic for humans, causing enteric fevers, gastroenteritis, and septicemia; may also infect many animal species besides humans. Some serovars are strictly host-adapted.

   The mol% G + C of the DNA is: 50–53.
   Type species: **Salmonella choleraesuis** (Smith 1894) Weldin 1927, 155 (*Bacillus cholerae suis* Smith 1894, 9.)

FURTHER DESCRIPTIVE INFORMATION

Although most salmonellae are motile, the serovar Gallinarum (or serovar Pullorum) is always nonmotile.

   Certain *Salmonella* serovars, such as serovar Abortusovis, may form unusually small colonies (~1 mm diameter), whereas most types form larger colonies (2–4 mm).

Most salmonellae are aerogenic. However, serovar Typhi, an important exception, never produces gas. Anaerogenic variants of normally gas-producing *Salmonella* serovars may occur; this is particularly common with serovar Dublin.

   Hydrogen sulfide is produced by most salmonellae, but a few serovars do not produce it (e.g., most strains of serovar Paratyphi A and some strains of serovar Choleraesuis).

   Citrate is generally utilized by salmonellae, but some serovars do not use it (particularly host-adapted serovars such as serovar Typhi and serovar Paratyphi A).

   The lysine decarboxylase reaction (Møller's) is positive for most salmonellae, but an important exception is serovar Paratyphi A. Most salmonellae are also positive for ornithine decarboxylase reaction (Møller's), but serovar Typhi is negative.

   Other biochemical characteristics of the genus are indicated in Tables BXII.γ.193, BXII.γ.194, and BXII.γ.196 in the chapter on the family *Enterobacteriaceae*. Subdivision of the genus *Salmonella* into species and subspecies (Le Minor and Popoff, 1987; Reeves et al., 1989a) based on biochemical characteristics is shown in Table BXII.γ.261. Kauffmann (1960, 1963a, b, 1964) subdivided the genus *Salmonella* into "subgenera" (see Taxonomic Comments). These subdivisions correspond more closely to species or subspecies in other groups of bacteria, but whatever rank is assigned to them, the worthiness of these subdivisions was confirmed by Rohde (1965, 1966, 1967).

TABLE BXII.γ.261. Differential characteristics of the *Salmonella* species and subspecies[a]

Characteristic	*S. enterica* subsp. *enterica*	*S. enterica* subsp. *arizonae*	*S. enterica* subsp. *diarizonae*	*S. enterica* subsp. *houtenae*	*S. enterica* subsp. *indica*	*S. enterica* subsp. *salamae*	*S. bongori*
Dulcitol	+	−	−	−	d	+	+
ONPG (2 h)	−	+	+	−	d	−	+
Malonate	−	+	+	−	−	+	−
Gelatinase	−	+	+	+	+	+	−
Sorbitol	+	+	+	+	−	+	+
Culture with KCN	−	−	−	+	−	−	+
L(+)-Tartrate[b]	+	−	−	−	−	−	−
Galacturonate	−	−	+	+	+	+	+
γ-Glutamyltransferase	+[c]	−	+	+	+	+	+
β-Glucuronidase	d	−	+	−	d	d	−
Mucate	+	+	− (70%)	−	+	+	+
Salicine	−	−	−	+	−	−	−
Lactose	−	− (75%)	+ (75%)	−	d	−	−
Lysis by phage O1	+	−	+	−	+	+	d
Habitat:							
Warm-blooded animals	+						
Cold-blooded animals		+	+	+	+	+	+

[a]Symbols: + , positive for 90% or more of strains in 1–2 days; d, positive for 11–89% of strains in 1–2 days; − , positive for 0–10% of strains in 1–2 days, unless otherwise indicated in the table. The temperature for all reactions is 37°C.

[b]D-tartrate.

[c] Typhimurium d, Dublin − .

Division into serovars The antigenic formulae of *Salmonella* serovars are listed in the Kauffmann–White scheme (Table BXII.γ.262). They are composed of numbers and letters given to the different O (somatic), Vi (capsular), and H (flagellar) antigens. Only those antigens of primary diagnostic importance are indicated in the Kauffmann–White scheme. Antigenic formulae (for example 6,7,[Vi]:c:1,5) represent the O antigenic factors, the Vi capsular antigen when present, the first phase of the H antigen, and the second phase of the H antigen, respectively. Those formulae with major O antigenic factors in common are collected into an O group and arranged alphabetically by the first phase of the H antigen within the group.

Lysogenization by certain converting phages may produce changes in the O antigenic formulae of salmonellae. In groups O:2, O:4, and O:9, presence of O:1 antigenic factor is associated with lysogenization (Iseki and Kashiwagi, 1955, 1957; Zinder, 1957; Stocker, 1958), but presence or absence of this factor in strains of these groups does not change the name of the serovar (for example, serovar Typhimurium applies to both O:1-positive and O:1-negative strains). Factors associated with phage conversion are underlined in the Kauffmann–White scheme. Converting phages of *Salmonella* are identical in morphology (Vieu et al., 1965), but their action is limited to certain O groups and they are serologically different from one another (Le Minor, 1968).

The specificities of O factors in *Salmonella* are determined by the composition and structure of the polysaccharides. Specificity is modified during smooth to rough mutation and by bacteriophage conversions (see reviews by Lüderitz et al., 1971; Stocker and Mäkelä, 1971, 1978). The only difference between the 4,12 and the 9,12 O-specific repeating units is in the di-deoxyhexose branch unit attached to the mannose, which is abequose in O:4,12 and tyvelose in O:9,12. In the conversion of O:3,10 to O:3,15, the terminal acetyl radical of the chain is suppressed and the α-lineage between galactose and mannose is transformed into a β-linkage. Other modifications of the specificity of O antigens may occur after mutation(s), resulting in new specificities called

T1 and T2 by Kauffmann (1956) or in different R types (reviewed by Lüderitz et al., 1971; Stocker and Mäkelä, 1971, 1978).

Surface antigens, commonly observed in other members of the *Enterobacteriaceae* family (e.g., *Escherichia coli* and *Klebsiella*), may be found in some *Salmonella* serovars. Surface antigens in *Salmonella* may mask O antigens, and the culture will not agglutinate with O antisera. The Vi antigen is a surface antigen found mainly in serovar Typhi and serovar Paratyphi C, as well as in a few strains of serovar Dublin. Heating at 100°C generally solubilizes Vi antigen, and heat-treated bacteria agglutinate with proper O antisera. Biogenesis of the Vi polysaccharide is governed by a set of genes located at the *viaB* locus, which is specific to Vi-expressing strains. Expression of Vi antigen is controlled by two two-component regulatory systems, OmpR-EnvZ and RcsB-RcsC (reviewed by Virlogeux-Payant and Popoff, 1996). One signal for this complex regulatory control was shown to be osmolarity (Pickard et al., 1994).

Subdivision of serovars Biovars are different sugar fermentation patterns shown by strains of the same serovar. They are determined by the presence or absence of enzymes and hence are genetically determined. Biovars may serve as markers and be of interest epidemiologically (for example, the xylose-positive and xylose-negative character of serovar Typhi).

Phagovars are determined by the sensitivity of cultures to a series of bacteriophages at appropriate dilutions. Phage typing of serovar Typhi and other salmonellae that possess the Vi antigen (serovar Paratyphi C and a few strains of serovar Dublin) is based on a series of adapted phages from phage Vi II of Craigie and Yen (1938). Phage typing of serovar Paratyphi B (Felix and Callow, 1943) and serovar Typhimurium (Anderson, 1964) uses a different series of phages. Analogous methods have been proposed for other serovars of *Salmonella*, some of them making use of the lysogenicity of the strains.

Other subdivisions of serovars may be based on production of, or on sensitivity to, bacteriocins, and on resistance to antibiotics.

TABLE BXII.γ.262. Antigenic formulae of the serovars of the genus *Salmonella*: the Kauffmann–White scheme[a]

Type[c]	Somatic (O) antigen	Flagellar (H) antigen Phase 1	Phase 2
Group O:2 (A)			
Paratyphi A	1,2,12	a	[1,5]
Nitra	2,12	g,m	
Kiel	1,2,12	g,p	
Koessen	2,12	l,v	1,5
Group O:4(B)			
Kisangani	1,4,[5],12	a	1,2
Hessarek	4,12,27	a	1,5
Fulica	4,[5],12	a	
Arechavaleta	4,[5],12	a	1,7
Bispebjerg	1,4,[5],12	a	e,n,x
Tinda	1,4,12, 27	a	e,n,z_{15}
II	1,4,[5],12, 27	a	e,n,x
Huettwillen	1,4,12	a	l,w
Nakuru	1,4,12, 27	a	z_6
II	1,4,12, 27	a	z_{39}
Paratyphi B	1,4,[5],12	b	1,2
Limete	1,4,12, 27	b	1,5
II	4,12	b	1,5
Canada	4,12,27	b	1,6
Uppsala	1,4,12, 27	b	1,7
Abony	1,4,[5],12, 27	b	e,n,x
II	1,4,12, 27	b	[e,n,x]
Wagenia	1,4,12, 27	b	e,n,z_{15}
Wien	1,4,12, 27	b	l,w
Tripoli	1,4,12, 27	b	z_6
Schleissheim	4,12,27	b	
Legon	1,4,12, 27	c	1,5
Abortusovis	4,12	c	1,6
Altendorf	4,12,27	c	1,7
Bissau	4,12	c	e,n,x
Jericho	1,4,12, 27	c	e,n,z_{15}
Hallfold	1,4,12, 27	c	l,w
Bury	4,12,27	c	z_6
Stanley	1,4,[5],12, 27	d	1,2
Eppendorf	1,4,12, 27	d	1,5
Brezany	1,4,12, 27	d	1,6
Schwarzengrund	1,4,12, 27	d	1,7
II	4,12	d	e,n,x
Sarajane	1,4,[5],12, 27	d	e,n,x
Duisburg	1,4,12, 27	d	e,n,z_{15}
Mons	1,4,12, 27	d	l,w
Ayinde	1,4,12, 27	d	z_6
Saintpaul	1,4,[5],12	e,h	1,2
Reading	1,4,[5],12	e,h	1,5
Eko	4,12	e,h	1,6
Kaapstad	4,12	e,h	1,7
Chester	1,4,[5],12	e,h	e,n,x
Sandiego	4,[5],12	e,h	e,n,z_{15}
II	4,12	e,n,x	1,2,7
II	1,4,12, 27	e,n,x	1,[5],7
Derby	1,4,[5],12	f,g	[1,2]
Agona	1,4,12	f,g,s	[1,2]
II	1,4,[5],12	f,g,t	z_6:z_{42}
Essen	4,12	g,m	
Hato	1,4,[5],12	g,m,s	
II	1,4,12, 27	g,[m],[s],t	e,n,x
II	1,4,12, 27	g,[m],t	[1,5]
II	4,12	g,m,t	z_{39}
California	4,12	g,m,t	[z_{67}]
Kingston	1,4,[5],12, 27	g,s,t	[1,2]
Budapest	1,4,12, 27	g,t	
Travis	4,[5],12	g,z_{51}	1,7
Tennyson	4,5,12	g,z_{51}	e,n,z_1
II	4,12	g,z_{62}	
Banana	1,4,[5],12	m,t	[1,5]
Madras	4,[5],12	m,t	e,n,z_{15}

TABLE BXII.γ.262. *(cont.)*

Type[c]	Somatic (O) antigen	Flagellar (H) antigen Phase 1	Phase 2
Typhimurium	1,4,[5],12	i	1,2
Lagos	1,4,[5],12	i	1,5
Agama	4,12	i	1,6
Farsta	4,12	i	e,n,x
Tsevie	4,12	i	e,n,z_{15}
Gloucester	1,4,12, 27	i	l,w
Tumodi	1,4,12	i	z_6
II	4,12,27	i	z_{35}
Massenya	1,4,12, 27	k	1,5
Neumuenster	1,4,12, 27	k	1,6
II	1,4,12, 27	k	1,6
Ljubljana	4,12,27	k	e,n,x
Texas	4,[5],12	k	e,n,z_{15}
Fyris	4,[5],12	l,v	1,2
Azteca	4,[5],12,27	l,v	1,5
Clackamas	4,12	l,v	1,6
Bredeney	1,4,12, 27	l,v	1,7
Kimuenza	1,4,12, 27	l,v	e,n,x
II	1,4,12, 27	l,v	e,n,x
Brandenburg	1,4,[5],12, 27	l,v	e,n,z_{15}
II	1,4,12, 27	l,v	z_{39}
Mono	4,12	l,w	1,5
Togo	4,12	l,w	1,6
II	4,12	l,w	e,n,x
Ayton	1,4,12, 27	l,w	z_6
Kunduchi	1,4,[5],12, 27	l,[z_{13}],[z_{28}]	1,2
Tyresoe	1,4,12, 27	l,[z_{13}],z_{28}	1,5
Haduna	4,12	l,z_{13},[z_{28}]	1,6
Kubacha	1,4,12, 27	l,z_{13},z_{28}	1,7
Kano	1,4,12, 27	l,z_{13},z_{28}	e,n,x
Vom	1,4,12, 27	l,z_{13},z_{28}	e,n,z_{15}
Reinickendorf	4,12	l,z_{28}	e,n,x
II	4,12	l,z_{28}	
Heidelberg	1,4,[5],12	r	1,2
Bradford	4,12,27	r	1,5
Winneba	4,12	r	1,6
Remo	1,4,12, 27	r	1,7
Bochum	4,[5],12	r	l,w
Southampton	1,4,12, 27	r	z_6
Drogana	1,4,12, 27	r,i	e,n,z_{15}
Africana	4,12	r,i	l,w
Coeln	1,4,[5],12	y	1,2
Trachau	4,12,27	y	1,5
Finaghy	4,12	y	1,6
Teddington	1,4,12, 27	y	1,7
Ball	1,4,12, 27	y	e,n,x
Jos	1,4,12, 27	y	e,n,z_{15}
Kamoru	4,12,27	y	z_6
Shubra	4,[5],12	z	1,2
Kiambu	1,4,12	z	1,5
II	1,4,12, 27	z	1,5
Loubomo	4,12	z	1,6
Indiana	1,4,12	z	1,7
II	4,12	z	1,7
Neftenbach	4,12	z	e,n,x
II	1,4,12, 27	z	e,n,x
Koenigstuhl	1,4,[5],12	z	e,n,z_{15}
Preston	1,4,12	z	l,w
Entebbe	1,4,12, 27	z	z_6
II	4,12	z	z_{39}
Stanleyville	1,4,[5],12, 27	z_4,z_{23}	[1,2]
Vuadens	4,12,27	z_4,z_{23}	z_6
Kalamu	4,[5],12	z_4,z_{24}	[1,5]
Haifa	1,4,[5],12	z_{10}	1,2
Ituri	1,4,12	z_{10}	1,5
Tudu	4,12	z_{10}	1,6
Albert	4,12	z_{10}	e,n,x
Tokoin	4,12	z_{10}	e,n,z_{15}

(continued)

TABLE BXII.γ.262. *(cont.)*

Type[c]	Somatic (O) antigen	Phase 1	Phase 2
		\multicolumn{2}{c}{Flagellar (H) antigen}	

Type[c]	Somatic (O) antigen	Phase 1	Phase 2
Mura	1,4,12	z_{10}	l,w
Fortune	1,4,12, 27	z_{10}	z_6
Vellore	1,4,12, 27	z_{10}	z_{35}
Brancaster	1,4,12, 27	z_{29}	
II	1,4,12	z_{29}	e,n,x
Pasing	4,12	z_{35}	1,5
Tafo	1,4,12, 27	z_{35}	1,7
Sloterdijk	1,4,12, 27	z_{35}	z_6
Yaounde	1,4,12, 27	z_{35}	e,n,z_{15}
Tejas	4,12	z_{36}	
Wilhelmsburg	1,4,[5],12, 27	z_{38}	[e,n,z_{15}]
II	1,4,12, 27	z_{39}	1,[5],7
Thayngen	1,4,12, 27	z_{41}	1,(2),5
Maska	1,4,12, 27	z_{41}	e,n,z_{15}
Abortusequi	4,12		e,n,x
Group O:7 (C1)[d]			
Sanjuan	6,7	a	1,5
II	6,7,14	a	1,5
Umhlali	6,7	a	1,6
Austin	6,7	a	1,7
Oslo	6,7,14	a	e,n,x
Denver	6,7	a	e,n,z_{15}
Coleypark	6,7,14	a	l,w
Damman	6,7	a	z_6
II	6,7	a	z_6
II	6,7	a	z_{42}
Brazzaville	6,7	b	1,2
Edinburg	6,7,14	b	1,5
Adime	6,7	b	1,6
Koumra	6,7	b	1,7
Lockleaze	6,7,14	b	e,n,x
Georgia	6,7	b	e,n,z_{15}
II	6,7	b	[e,n,x]:z_{42}
Ohio	6,7,14	b	l,w
Leopoldville	6,7	b	z_6
Kotte	6,7	b	z_{35}
II	6,7	b	z_{39}
Hissar	6,7,14	c	1,2
Paratyphi C	6,7,[Vi]	c	1,5
Choleraesuis	6,7	c	1,5
Typhisuis	6,7	c	1,5
Birkenhead	6,7	c	1,6
Schwabach	6,7	c	1,7
Namibia	6,7	c	e,n,x
Kaduna	6,7,14	c	e,n,z_{15}
Kisii	6,7	d	1,2
Isangi	6,7,14	d	1,5
Kivu	6,7	d	1,6
Kambole	6,7	d	1,[2],7
Amersfoort	6,7,14	d	e,n,x
Gombe	6,7,14	d	e,n,z_{15}
Livingstone	6,7,14	d	l,w
Wil	6,7	d	l,z_{13},z_{28}
Nieukerk	6,7,14	d	z_6
II	6,7	d	z_{42}
Larochelle	6,7	e,h	1,2
Lomita	6,7	e,h	1,5
Norwich	6,7	e,h	1,6
Nola	6,7	e,h	1,7
Braenderup	6,7,14	e,h	e,n,z_{15}
II	6,7	e,n,x	1,6:z_{42}
Rissen	6,7,14	f,g	
Eingedi	6,7	f,g,t	1,2,7
Afula	6,7	f,g,t	e,n,x
Montevideo	6,7,14	g,m,[p],s	[1,2,7]
II	6,7	g,m,[s],t	e,n,x
II	6,7	(g),m,[s],t	1,5
II	6,7	g,[m],s,t	[z_{42}]

(continued)

TABLE BXII.γ.262. *(cont.)*

Type[c]	Somatic (O) antigen	Phase 1	Phase 2
Othmarschen	6,7,14	g,m,[t]	
Menston	6,7	g,s,[t]	[1,6]
II	6,7	g,t	e,n,x:z_{42}
Riggil	6,7	g,(t)	
Alamo	6,7	g,z_{51}	1,5
IV	6,7	g,z_{51}	
Haelsingborg	6,7	m,p,t,[u]	
Winston	6,7	m,t	1,6
Oakey	6,7	m,t	z_{64}
II	6,7	m,t	
Oranienburg	6,7,14	m,t	[z_{57}]
Augustenborg	6,7,14	i	1,2
Oritamerin	6,7	i	1,5
Garoli	6,7	i	1,6
Lika	6,7	i	1,7
Athinai	6,7	i	e,n,z_{15}
Norton	6,7	i	l,w
Stuttgart	6,7,14	i	z_6
Galiema	6,7,14	k	1,2
Thompson	6,7,14	k	1,5
Daytona	6,7	k	1,6
Baiboukoum	6,7	k	1,7
Singapore	6,7	k	e,n,x
Escanaba	6,7	k	e,n,z_{15}
IIIb	6,7	(k)	z:[z_{54}]
II	6,7	k	[z_6]
Concord	6,7	l,v	1,2
Irumu	6,7	l,v	1,5
Mkamba	6,7	l,v	1,6
Kortrijk	6,7	l,v	1,7
Bonn	6,7	l,v	e,n,x
Potsdam	6,7,14	l,v	e,n,z_{15}
Gdansk	6,7,14	l,v	z_6
Coromandel	6,7	l,v	z_{35}
IIIb	6,7	l,v	z_{53}
Gabon	6,7	l,w	1,2
Colorado	6,7	l,w	1,5
II	6,7	l,w	1,5,7
II	6,7	l,w	z_{42}
Nessziona	6,7	l,z_{13}	1,5
Kenya	6,7	l,z_{13}	e,n,x
Neukoelln	6,7	l,z_{13},[z_{28}]	e,n,z_{15}
Makiso	6,7	l,z_{13},z_{28}	z_6
Strathcona	6,7	l,z_{13},z_{28}	1,7
II	6,7	l,z_{28}	1,5:[z_{42}]
II	6,7	l,z_{28}	e,n,x
II	6,7	l,z_{28}	z_6
Virchow	6,7	r	1,2
Infantis	6,7,14	r	1,5
Nigeria	6,7	r	1,6
Colindale	6,7	r	1,7
Papuana	6,7	r	e,n,z_{15}
Grampian	6,7	r	l,w
Richmond	6,7	y	1,2
Bareilly	6,7,14	y	1,5
Oyonnax	6,7	y	1,6
Gatow	6,7	y	1,7
Hartford	6,7	y	e,n,x:[z_{67}]
Mikawasima	6,7,14	y	e,n,z_{15}
Chile	6,7	z	1,2
Poitiers	6,7	z	1,5
II	6,7	z	1,5
Oakland	6,7	z	1,6,[7]
Cayar	6,7	z	e,n,x
II	6,7	z	e,n,x
Businga	6,7	z	e,n,z_{15}
Bruck	6,7	z	l,w
II	6,7	z	z_6

(continued)

TABLE BXII.γ.262. *(cont.)*

Type[c]	Somatic (O) antigen	Flagellar (H) antigen Phase 1	Flagellar (H) antigen Phase 2
II	6,7	z	z_{39}
II	6,7	z	z_{42}
Obogu	6,7	z_4,z_{23}	1,5
Planckendael	6,7	z_4,z_{23}	1,6
Aequatoria	6,7	z_4,z_{23}	e,n,z_{15}
Goma	6,7	z_4,z_{23}	z_6
IV	6,7	z_4,z_{23}	
II	6,7	z_4,z_{24}	z_{42}
Somone	6,7	z_4,z_{24}	
IV	6,7	z_4,z_{24}	
II	6,7	z_6	1,7
Menden	6,7	z_{10}	1,2
Inganda	6,7	z_{10}	1,5
Eschweiler	6,7	z_{10}	1,6
Ngili	6,7	z_{10}	1,7
Djugu	6,7	z_{10}	e,n,x
Mbandaka	6,7,<u>14</u>	z_{10}	e,n,z_{15}
Jerusalem	6,7,<u>14</u>	z_{10}	l,w
Redba	6,7	z_{10}	z_6
Omuna	6,7	z_{10}	z_{35}
Tennessee	6,7,<u>14</u>	z_{29}	[1,2,7]
II	6,7	z_{29}	[z_{42}]
Tienba	6,7	z_{35}	1,6
Palime	6,7	z_{35}	e,n,z_{15}
Tampico	6,7	z_{36}	e,n,z_{15}
II	6,7	z_{36}	z_{42}
IV	6,7	z_{36}	
Rumford	6,7	z_{38}	1,2
Lille	6,7,<u>14</u>	z_{38}	
IIIb	6,7,<u>14</u>	z_{39}	1,2
II	6,7	z_{39}	1,5,7
VI	6,7	z_{41}	1,7
Hillsborough	6,7	z_{41}	l,w
Tamilnadu	6,7	z_{41}	z_{35}
II	6,7	z_{42}	1,7
Bulovka	6,7	z_{44}	
II	6,7		1,6
Group O:8 (C2–C3)[e]			
Be	8,<u>20</u>	a	[z_6]
Valdosta	6,8	a	1,2
Doncaster	6,8	a	1,5
Curacao	6,8	a	1,6
Nordufer	6,8	a	1,7
Narashino	6,8	a	e,n,x
II	6,8	a	e,n,x
Leith	6,8	a	e,n,z_{15}
II	6,8	a	z_{39}
II	6,8	a	z_{52}
Djelfa	8	b	1,2
Skansen	6,8	b	1,2
Korbol	8,<u>20</u>	b	1,5
Nagoya	6,8	b	1,5
II	6,8	b	1,5
Stourbridge	6,8	b	1,6
Sanga	8	b	1,7
Eboko	6,8	b	1,7
Konstanz	8	b	e,n,x
Gatuni	6,8	b	e,n,x
Shipley	8,<u>20</u>	b	e,n,z_{15}
Presov	6,8	b	e,n,z_{15}
Bukuru	6,8	b	l,w
Tounouma	8,<u>20</u>	b	z_6
Banalia	6,8	b	z_6
Wingrove	6,8	c	1,2
Utah	6,8	c	1,5
Bronx	6,8	c	1,6
Belfast	6,8	c	1,7
Alexanderpolder	8	c	l,w

(continued)

TABLE BXII.γ.262. *(cont.)*

Type[c]	Somatic (O) antigen	Flagellar (H) antigen Phase 1	Flagellar (H) antigen Phase 2
Santiago	8,<u>20</u>	c	e,n,x
Belem	6,8	c	e,n,x
Quiniela	6,8	c	e,n,z_{15}
Tado	8,<u>20</u>	c	z_6
Virginia	8	d	1,2
Muenchen	6,8	d	1,2:[z_{67}]
Yovokome	8,<u>20</u>	d	1,5
Manhattan	6,8	d	1,5
Portanigra	8,<u>20</u>	d	1,7
Dunkwa	6,8	d	1,7
Sterrenbos	6,8	d	e,n,x
Herston	6,8	d	e,n,z_{15}
Labadi	8,<u>20</u>	d	z_6
II	6,8	d	z_6:z_{42}
Bardo	8	e,h	1,2
Newport	6,8,<u>20</u>	e,h	1,2:[z_{67}]
Ferruch	8	e,h	1,5
Kottbus	6,8	e,h	1,5
Cremieu	6,8	e,h	1,6
Atakpame	8,<u>20</u>	e,h	1,7
Tshiongwe	6,8	e,h	e,n,z_{15}
Rechovot	8,<u>20</u>	e,h	z_6
Sandow	6,8	f,g	e,n,z_{15}
II	6,8	f,g,m,t	[e,n,x]
Emek	8,<u>20</u>	g,m,s	
Chincol	6,8	g,m,[s]	[e,n,x]
II	6,8	g,m,t	1,7
Reubeuss	8,<u>20</u>	g,m,t	
Alminko	8,<u>20</u>	g,s,t	
Nanergou	6,8	g,s,t	
Yokoe	8,<u>20</u>	m,t	
II	6,8	m,t	1,5
II	6,8	m,t	e,n,x
Bassa	6,8	m,t	
Lindenburg	6,8	i	1,2
Bargny	8,<u>20</u>	i	1,5
Takoradi	6,8	i	1,5
Warnow	6,8	i	1,6
Malmoe	6,8	i	1,7
Bonariensis	6,8	i	e,n,x
Aba	6,8	i	e,n,z_{15}
Magherafelt	8,<u>20</u>	i	l,w
Cyprus	6,8	i	l,w
Kentucky	8,<u>20</u>	i	z_6
Kallo	6,8	k	1,2
Haardt	8	k	1,5
Blockley	6,8	k	1,5
Schwerin	6,8	k	e,n,x
Charlottenburg	6,8	k	e,n,z_{15}
Pakistan	8	l,v	1,2
Litchfield	6,8	l,v	1,2
Loanda	6,8	l,v	1,5
Amherstiana	8	l,v	1,6
Manchester	6,8	l,v	1,7
Holcomb	6,8	l,v	e,n,x
II	6,8	l,v	e,n,x
Edmonton	6,8	l,v	e,n,z_{15}
Fayed	6,8	l,w	1,2
II	6,8	l,w	z_6:z_{42}
Hiduddify	6,8	l,z_{13},z_{28}	1,5
Breukelen	6,8	$l,z_{13},[z_{28}]$	e,n,z_{15}
II	6,8	l,z_{28}	e,n,x
Bsilla	6,8	r	1,2
Hindmarsh	8,<u>20</u>	r	1,5
Bovismorbificans	6,8,<u>20</u>	r,[i]	1,5
Noya	8	r	1,7
Akanji	6,8	r	1,7
Cocody	8,<u>20</u>	r,i	e,n,z_{15}

(continued)

TABLE BXII.γ.262. (*cont.*)

Type[c]	Somatic (O) antigen	Flagellar (H) antigen Phase 1	Phase 2
Hidalgo	6,8	r,[i]	e,n,z_{15}
Brikama	8,20	r,[i]	l,w
Goldcoast	6,8	r	l,w
Altona	8,20	r,[i]	z_6
Giza	8,20	y	1,2
Brunei	8,20	y	1,5
Tananarive	6,8	y	1,5
Bulgaria	6,8	y	1,6
II	6,8	y	1,6:z_{42}
Alagbon	8	y	1,7
Inchpark	6,8	y	1,7
Sunnycove	8	y	e,n,x
Daarle	6,8	y	e,n,x
Praha	6,8	y	e,n,z_{15}
Kraligen	8,20	y	z_6
Benue	6,8	y	l,w
Sindelfingen	8,20	y	l,w
Mowanjum	6,8	z	1,5
II	6,8	z	1,5
Phaliron	8	z	e,n,z_{15}
Kalumburu	6,8	z	e,n,z_{15}
Kuru	6,8	z	l,w
Daula	8,20	z	z_6
Bellevue	8	z_4,z_{23}	1,7
Lezennes	6,8	z_4,z_{23}	1,7
Breda	6,8	z_4,z_{23}	e,n,x
Chailey	6,8	z_4,z_{23}	e,n,z_{15}
Dabou	8,20	z_4,z_{23}	l,w
Corvallis	8,20	z_4,z_{23}	[z_6]
Albany	8,20	z_4,z_{24}	
Duesseldorf	6,8	z_4,z_{24}	
Tallahassee	6,8	z_4,z_{32}	
Bazenheid	8,20	z_{10}	1,2
Zerifin	6,8	z_{10}	1,2
Paris	8,20	z_{10}	1,5
Mapo	6,8	z_{10}	1,5
Cleveland	6,8	z_{10}	1,7
Istanbul	8	z_{10}	e,n,x
Hadar	6,8	z_{10}	e,n,x
Chomedey	8,20	z_{10}	e,n,z_{15}
Glostrup	6,8	z_{10}	e,n,z_{15}
Remiremont	8,20	z_{10}	l,w
Molade	8,20	z_{10}	z_6
Wippra	6,8	z_{10}	z_6
II	6,8	z_{29}	1,5
II	8	z_{29}	e,n,x:z_{42}
Tamale	8,20	z_{29}	[e,n,z_{15}]
Uno	6,8	z_{29}	[e,n,z_{15}]
II	6,8	z_{29}	e,n,x
Kolda	8,20	z_{35}	1,2
Yarm	6,8	z_{35}	1,2
Angers	8,20	z_{35}	z_6
Apeyeme	8,20	z_{38}	
Diogoye	8,20	z_{41}	z_6
Aesch	6,8	z_{60}	1,2
Group O:9 (D1)			
Sendai	1,9,12	a	1,5
Miami	1,9,12	a	1,5
II	9,12	a	1,5
Os	9,12	a	1,6
Saarbruecken	1,9,12	a	1,7
Lomalinda	1,9,12	a	e,n,x
II	1,9,12	a	e,n,x
Durban	9,12	a	e,n,z_{15}
II	9,12	a	z_{39}
II	1,9,12	a	z_{42}
Onarimon	1,9,12	b	1,2
Frintrop	1,9,12	b	1,5

(*continued*)

TABLE BXII.γ.262. (*cont.*)

Type[c]	Somatic (O) antigen	Flagellar (H) antigen Phase 1	Phase 2
II	1,9,12	b	e,n,x
II	1,9,12	b	z_6
II	1,9,12	b	z_{39}
Goeteborg	9,12	c	1,5
Ipeko	9,12	c	1,6
Elokate	9,12	c	1,7
Alabama	9,12	c	e,n,z_{15}
Ridge	9,12	c	z_6
Ndolo	1,9,12	d	1,5
Tarshyne	9,12	d	1,6
Eschberg	9,12	d	1,7
II	9,12	d	e,n,x
Bangui	9,12	d	e,n,z_{15}
Zega	9,12	d	z_6
Jaffna	1,9,12	d	z_{35}
II	9,12	d	z_{39}
Typhi	9,12[Vi]	d	
Bournemouth	9,12	e,h	1,2
Eastbourne	1,9,12	e,h	1,5
Westafrica	9,12	e,h	1,7
Israel	9,12	e,h	e,n,z_{15}
II	9,12	e,n,x	1,[5],7
II	9,12	e,n,x	1,6
Berta	1,9,12	[f],g,[t]	
Enteritidis	1,9,12	g,m	
Blegdam	9,12	g,m,q	
II	1,9,12	g,m,[s],t	[1,5,7]:[z_{42}]
II	1,9,12	g,m,s,t	e,n,x
Dublin	1,9,12[Vi]	g,p	
Naestved	1,9,12	g,p,s	
Rostock	1,9,12	g,p,u	
Moscow	9,12	g,q	
II	9,12	g,s,t	e,n,x
Newmexico	9,12	g,z_{51}	1,5
II	1,9,12	g,z_{62}	[e,n,x]
Antarctica	9,12	g,z_{63}	
II	9,12	m,t	e,n,x
Pensacola	1,9,12	m,t	[1,2]
II	1,9,12	m,t	1,5
II	1,9,12	m,t	z_{39}
Seremban	9,12	i	1,5
Claibornei	1,9,12	k	1,5
Goverdhan	9,12	k	1,6
Mendoza	9,12	l,v	1,2
Panama	1,9,12	l,v	1,5
Kapemba	9,12	l,v	1,7
Zaiman	9,12	l,v	e,n,x
II	9,12	l,v	e,n,x
Goettingen	9,12	l,v	e,n,z_{15}
II	9,12	l,v	z_{39}
Victoria	1,9,12	l,w	1,5
II	1,9,12	l,w	e,n,x
Itami	9,12	l,z_{13}	1,5
Miyazaki	9,12	l,z_{13}	1,7
Napoli	1,9,12	l,z_{13}	e,n,x
Javiana	1,9,12	l,z_{28}	1,5
Kotu	9,12	l,z_{28}	1,6
II	9,12	l,z_{28}	1,5:[z_{42}]
II	9,12	l,z_{28}	e,n,x
Jamaica	9,12	r	1,5
Camberwell	9,12	r	1,7
Campinense	9,12	r	e,n,z_{15}
Lome	9,12	r	z_6
Powell	9,12	y	1,7
II	1,9,12	y	z_{39}
Mulhouse	1,9,12	z	1,2
Lawndale	1,9,12	z	1,5
Kimpese	9,12	z	1,6

(*continued*)

TABLE BXII.γ.262. *(cont.)*

Type[c]	Somatic (O) antigen	Phase 1	Phase 2
		Flagellar (H) antigen	
II	1,9,12	z	1,7
II	1,9,12	z	z_6
II	9,12	z	z_{39}
Wangata	1,9,12	z_4,z_{23}	[1,7]
Natal	9,12	z_4,z_{24}	
Franken	9,12	z_6	z_{67}
Portland	9,12	z_{10}	1,5
Treguier	9,12	z_{10}	z_6
Ruanda	9,12	z_{10}	e,n,z_{15}
II	9,12	z_{29}	1,5
II	1,9,12	z_{29}	e,n,x
Penarth	9,12	z_{35}	z_6
Elomrane	1,9,12	z_{38}	
II	1,9,12	z_{39}	1,7
Ottawa	1,9,12	z_{41}	1,5
II	1,9,12	z_{42}	1,[5],7
Gallinarum	1,9,12		
Group O:9,46 (D2)			
Baildon	9,46	a	e,n,x
Doba	9,46	a	e,n,z_{15}
Cheltenham	9,46	b	1,5
Zadar	9,46	b	1,6
Worb	9,46	b	e,n,x
II	9,46	b	e,n,x
Bamboye	9,46	b	l,w
Linguere	9,46	b	z_6
Kolar	9,46	b	z_{35}
Itutaba	9,46	c	z_6
Ontario	9,46	d	1,5
Quentin	9,46	d	1,6
Strasbourg	9,46	d	1,7
Olten	9,46	d	e,n,z_{15}
Plymouth	9,46	d	z_6
Bergedorf	9,46	e,h	1,2
Waedenswil	9,46	e,h	1,5
Guerin	9,46	e,h	z_6
II	9,46	e,n,x	1,5,7
Wernigerode	9,46	f,g	
Hillingdon	9,46	g,m	
Macclesfield	9,46	g,m,s	1,2,7
II	9,46	g,[m],[s],t	[e,n,x]
Gateshead	9,46	g,s,t	
II	9,46	g,z_{62}	
II	9,46	m,t	e,n,x
Sangalkam	9,46	m,t	
Mathura	9,46	i	e,n,z_{15}
Potto	9,46	i	z_6
Marylebone	9,46	k	1,2
Cochin	9,46	k	1,5
Ceyco	9,46	k	z_{35}
India	9,46	l,v	1,5
Geraldton	9,46	l,v	1,6
Toronto	9,46	l,v	e,n,x
Ackwepe	9,46	l,w	
Nordrhein	9,46	l,z_{13},z_{28}	e,n,z_{15}
Deckstein	9,46	r	1,7
Shoreditch	9,46	r	e,n,z_{15}
Sokode	9,46	r	z_6
Benin	9,46	y	1,7
Irchel	9,46	y	e,n,x
Nantes	9,46	y	l,w
Mayday	9,46	y	z_6
II	9,46	z	1,5
II	9,46	z	e,n,x
Bambylor	9,46	z	e,n,z_{15}
Ekotedo	9,46	z_4,z_{23}	
II	9,46	z_4,z_{24}	$z_{39}:z_{42}$
Ngaparou	9,46	z_4,z_{24}	

(continued)

TABLE BXII.γ.262. *(cont.)*

Type[c]	Somatic (O) antigen	Phase 1	Phase 2
		Flagellar (H) antigen	
Lishabi	9,46	z_{10}	1,7
Inglis	9,46	z_{10}	e,n,x
Mahina	9,46	z_{10}	e,n,z_{15}
Louisiana	9,46	z_{10}	z_6
II	9,46	z_{10}	z_6
II	9,46	z_{10}	z_{39}
Ouakam	9,46	z_{29}	
Hillegersberg	9,46	z_{35}	1,5
Basingstoke	9,46	z_{35}	e,n,z_{15}
Trimdon	9,46	z_{35}	z_6
Fresno	9,46	z_{38}	
II	9,46	z_{39}	1,7
Wuppertal	9,46	z_{41}	
Group O:9,46,27 (D3)			
II	1,9,12,46,27	a	z_6
II	1,9,12,46,27	c	z_{39}
II	9,12,46,27	g,t	e,n,x
II	1,9,12,46,27	l,z_{13},z_{28}	z_{39}
II	1,9,12,46,27	y	z_{39}
II	1,9,12,46,27	z_4,z_{24}	1,5
II	1,9,12,46,27	z_{10}	1,5
II	1,9,12,46,27	z_{10}	e,n,x
II	1,9,12,46,27	z_{10}	z_{39}
Group O:3,10 (E1)[f]			
Aminatu	3,10	a	1,2
Goelzau	3,10[15]	a	1,5
Oxford	3,10[15][15,34]	a	1,7
Masembe	3,10	a	e,n,x
II	3,10	a	e,n,x
Galil	3,10	a	e,n,z_{15}
II	3,10	a	l,v
II	3,10	a	z_{39}
Kalina	3,10	b	1,2
Butantan	3,10[15][15,34]	b	1,5
Allerton	3,10	b	1,6
Huvudsta	3,10	b	1,7
Benfica	3,10	b	e,n,x
II	3,10	b	e,n,x
Yaba	3,10[15]	b	e,n,z_{15}
Epicrates	3,10	b	l,w
Wilmington	3,10	b	z_6
Westminster	3,10[15]	b	z_{35}
II	3,10	b	z_{39}
Asylanta	3,10	c	1,2
Gbadago	3,10[15]	c	1,5
Ikayi	3,10[15]	c	1,6
Pramiso	3,10	c	1,7
Agege	3,10	c	e,n,z_{15}
Anderlecht	3,10	c	l,w
Okefoko	3,10	c	z_6
Stormont	3,10	d	1,2
Shangani	3,10[15]	d	1,5
Lekke	3,10	d	1,6
Onireke	3,10	d	1,7
Souza	3,10[15]	d	e,n,x
II	3,10	d	e,n,x
Madjorio	3,10	d	e,n,z_{15}
Birmingham	3,10[15]	d	l,w
Weybridge	3,10	d	z_6
Maron	3,10	d	z_{35}
Vejle	3,10[15]	e,h	1,2
Muenster	3,10[15][15,34]	e,h	1,5
Anatum	3,10[15][15,34]	e,h	1,6
Nyborg	3,10[15]	e,h	1,7
Newlands	3,10[15,34]	e,h	e,n,x
Lamberhurst	3,10	e,h	e,n,z_{15}
Meleagridis	3,10[15][15,34]	e,h	l,w
Sekondi	3,10	e,h	z_6

(continued)

TABLE BXII.γ.262. *(cont.)*

Type[c]	Somatic (O) antigen	Flagellar (H) antigen Phase 1	Phase 2
II	3,10	e,n,x	1,7
Regent	3,10	f,g,[s]	[1,6]
Alfort	3,10	f,g	e,n,x
Suberu	3,10	g,m	
Amsterdam	3,10[15][15,34]	g,m,s	
II	3,10[15]	g,m,s,t	[1,5]
Westhampton	3,10[15][15,34]	g,s,t	
Bloomsbury	3,10	g,t	1,5
II	3,10	g,t	
II	3,10	m,t	1,5
Southbank	3,10[15][15,34]	m,t	[1,6]
II	3,10	m,t	e,n,x
Cuckmere	3,10	i	1,2
Amounderness	3,10	i	1,5
Tibati	3,10	i	1,6
Truro	3,10	i	1,7
Bessi	3,10	i	e,n,x
Falkensee	3,10[15]	i	e,n,z_{15}
Hoboken	3,10	i	l,w
Yeerongpilly	3,10	i	z_6
Wimborne	3,10	k	1,2
Zanzibar	3,10[15]	k	1,5
Serrekunda	3,10	k	1,7
Yundum	3,10	k	e,n,x
Marienthal	3,10	k	e,n,z_{15}
Newrochelle	3,10	k	l,w
Nchanga	3,10[15]	l,v	1,2
Sinstorf	3,10	l,v	1,5
London	3,10[15]	l,v	1,6
Give	3,10[15][15,34]	[d],l,v	1,7
II	3,10	l,v	e,n,x
Ruzizi	3,10	l,v	e,n,z-15
II	3,10	l,v	z_6
Sinchew	3,10	l,v	z_{35}
Assinie	3,10	l,w	z_6
Freiburg	3,10	l,z_{13}	1,2
Uganda	3,10[15]	l,z_{13}	1,5
Fallowfield	3,10	l,z_{13},z_{28}	e,n,z_{15}
Hoghton	3,10	l,z_{13},z_{28}	z_6
II	3,10	l,z_{28}	1,5
Joal	3,10	l,z_{28}	1,7
Lamin	3,10	l,z_{28}	e,n,x
II	3,10	l,z_{28}	e,n,x
II	3,10	l,z_{28}	z_{39}
Ughelli	3,10	r	1,5
Elisabethville	3,10[15]	r	1,7
Simi	3,10	r	e,n,z_{15}
Weltevreden	3,10[15]	r	z_6
Seegefeld	3,10	r,i	1,2
Dumfries	3,10	r,i	1,6
Amager	3,10[15]	y	1,2
Orion	3,10[15][15,34]	y	1,5
Mokola	3,10	y	1,7
Ohlstedt	3,10[15]	y	e,n,x
Bolton	3,10	y	e,n,z_{15}
Langensalza	3,10	y	l,w
Stockholm	3,10[15]	y	z_6
Fufu	3,10	z	1,5
II	3,10	z	1,5
Harleystreet	3,10	z	1,6
Huddinge	3,10	z	1,7
II	3,10	z	e,n,x
Clerkenwell	3,10	z	l,w
Landwasser	3,10	z	z_6
II	3,10	z	z_{39}
Adabraka	3,10	z_4,z_{23}	[1,7]
Wagadugu	3,10	z_4,z_{23}	z_6
Florian	3,10[15]	z_4,z_{24}	

(continued)

TABLE BXII.γ.262. *(cont.)*

Type[c]	Somatic (O) antigen	Flagellar (H) antigen Phase 1	Phase 2
II	3,10	z_4,z_{24}	
Okerara	3,10	z_{10}	1,2
Lexington	3,10[15][15,34]	z_{10}	1,5
Harrisonburg	3,10[15][15,34]	z_{10}	1,6
Coquilhatville	3,10	z_{10}	1,7
Kristianstad	3,10	z_{10}	e,n,z_{15}
Biafra	3,10	z_{10}	z_6
Everleigh	3,10	z_{29}	e,n,x
II	3,10	z_{29}	[e,n,x]
Jedburgh	3,10[15]	z_{29}	
Zongo	3,10	z_{35}	1,7
Shannon	3,10	z_{35}	l,w
Cairina	3,10	z_{35}	z_6
Macallen	3,10	z_{36}	
Bolombo	3,10	z_{38}	[z_6]
II	3,10	z_{38}	z_{42}
II	3,10	z_{39}	1,[5],7
Pietersburg	3,10[15,34]	z_{69}	1,7
Group O:1,3,19 (E4)			
Niumi	1,3,19	a	1,5
Juba	1,3,19	a	1,7
Gwoza	1,3,19	a	e,n,z_{15}
Alkmaar	1,3,19	a	l,w
Gnesta	1,3,19	b	1,5
Visby	1,3,19	b	1,6
Tambacounda	1,3,19	b	e,n,x
Kande	1,3,19	b	e,n,z_{15}
Broughton	1,3,19	b	l,w
Accra	1,3,19	b	z_6
Eastglam	1,3,19	c	1,5
Bida	1,3,19	c	1,6
Madiago	1,3,19	c	1,7
Ahmadi	1,3,19	d	1,5
Liverpool	1,3,19	d	e,n,z_{15}
Tilburg	1,3,19	d	l,w
Niloese	1,3,19	d	z_6
Vilvoorde	1,3,19	e,h	1,5
Hayindogo	1,3,19	e,h	1,6
Sanktmarx	1,3,19	e,h	1,7
Sao	1,3,19	e,h	e,n,z_{15}
Calabar	1,3,19	e,h	l,w
Rideau	1,3,19	f,g	
Petahtikve	1,3,19	f,g,t	1,7
Maiduguri	1,3,19	f,g,t	e,n,z_{15}
Kouka	1,3,19	g,m,[t]	
Senftenberg	1,3,19	g,[s],t	
Cannstatt	1,3,19	m,t	
Stratford	1,3,19	i	1,2
Chichester	1,3,19	i	1,6
Machaga	1,3,19	i	e,n,x
Avonmouth	1,3,19	i	e,n,z_{15}
Zuilen	1,3,19	i	l,w
Taksony	1,3,19	i	z_6
Oersterbro	1,3,19	k	1,5
Bethune	1,3,19	k	1,7
Ngor	1,3,19	l,v	1,5
Parkroyal	1,3,19	l,v	1,7
Svedvi	1,3,19	l,v	e,n,z_{15}
Fulda	1,3,19	l,w	1,5
Westerstede	1,3,19	l,z_{13}	1,2
Winterthur	1,3,19	l,z_{13}	1,6
Lokstedt	1,3,19	l,z_{13},z_{28}	1,2
Stuivenberg	1,3,19	l,[z_{13}]z_{28}	1,5
Bedford	1,3,19	l,z_{13},z_{28}	e,n,z_{15}
Tomelilla	1,3,19	l,z_{28}	1,7
Kindia	1,3,19	l,z_{28}	e,n,x
Yalding	1,3,19	r	e,n,z_{15}
Fareham	1,3,19	r,i	l,w

(continued)

TABLE BXII.γ.262. *(cont.)*

Type[c]	Somatic (O) antigen	Phase 1	Phase 2
Gatineau	1,3,19	y	1,5
Thies	1,3,19	y	1,7
Slade	1,3,19	y	e,n,z₁₅
Kinson	1,3,19	y	e,n,x
Krefeld	1,3,19	y	l,w
Korlebu	1,3,19	z	1,5
Kainji	1,3,19	z	1,6
Lerum	1,3,19	z	1,7
Schoeneberg	1,3,19	z	e,n,z₁₅
Carno	1,3,19	z	l,w
Hongkong	1,3,19	z	z₆
Sambre	1,3,19	z₄,z₂₄	
Dallgow	1,3,19	z₁₀	e,n,z₁₅
Llandoff	1,3,19	z₂₉	[z₆]
Ochiogu	1,3,19	z₃₈	[e,n,z₁₅]
Chittagong	1,3,10,19	b	z₃₅
Bilu	1,3,10,19	f,g,t	1,(2),7
Ilugun	1,3,10,19	z₄,z₂₃	z₆
Dessau	1,3,15,19	g,s,t	
Cannonhill	1,3,15,19	y	e,n,x
Group O:11 (F)			
II	11	a	d:e,n,z₁₅
Gallen	11	a	1,2
Marseille	11	a	1,5
VI	11	a	1,5
Toowong	11	a	1,7
Luciana	11	a	e,n,z₁₅
Epinay	11	a	l,z₁₃,z₂₈
II	11	a	z₆:z₄₂
Atento	11	b	1,2
Leeuwarden	11	b	1,5
Wohlen	11	b	1,6
VI	11	b	1,7
VI	11	b	e,n,x
Pharr	11	b	e,n,z₁₅
Chiredzi	11	c	1,5
Brindisi	11	c	1,6
II	11	c	e,n,z₁₅
Woodinville	11	c	e,n,x
Ati	11	d	1,2
Gustavia	11	d	1,5
Chandans	11	d	[e,n,x]:[r]
Findorff	11	d	z₆
Chingola	11	e,h	1,2
Adamstua	11	e,h	1,6
Redhill	11	e,h	l,z₁₃,z₂₈
Abuja	11	g,m	1,5
Missouri	11	g,s,t	
II	11	g,[m],s,t	z₃₉
IV	11	g,z₅₁	
Moers	11	m,t	
II	11	m,t	e,n,x
Aberdeen	11	i	1,2
Brijbhumi	11	i	1,5
Heerlen	11	i	1,6
Veneziana	11	i	e,n,x
Pretoria	11	k	1,2
Abaetetuba	11	k	1,5
Sharon	11	k	1,6
Colobane	11	k	1,7
Kisarawe	11	k	e,n,x,[z₁₅]
Mannheim	11	k	l,w
Amba	11	k	l,z₁₃,z₂₈
IIIb	11	k	z₅₃
Stendal	11	l,v	1,2
Maracaibo	11	l,v	1,5
Fann	11	l,v	e,n,x
Bullbay	11	l,v	e,n,z₁₅

Type[c]	Somatic (O) antigen	Phase 1	Phase 2
IIIb	11	l,v	z
IIIb	11	l,v	z₅₃
Glidji	11	l,w	1,5
Tours	11	l,z₁₃	1,2
Connecticut	11	l,z₁₃,z₂₈	1,5
Osnabrueck	11	l,z₁₃,z₂₈	e,n,x
II	11	l,z₂₈	e,n,x
Senegal	11	r	1,5
Rubislaw	11	r	e,n,x
Clanvillian	11	r	e,n,z₁₅
Euston	11	r,i	e,n,x,z₁₅
Volta	11	r	l,z₁₃,z₂₈
Solt	11	y	1,5
Jalisco	11	y	1,7
Herzliya	11	y	e,n,x
Crewe	11	z	1,5
Maroua	11	z	1,7
II	11	z	e,n,x
Nyanza	11	z	z₆:[z₈₃]
II	11	z	z₃₉
Remete	11	z₄,z₂₃	1,6
Etterbeek	11	z₄,z₂₃	e,n,z₁₅
IIIa	11	z₄,z₂₃	
IV	11	z₄,z₂₃	
Yehuda	11	z₄,z₂₄	
IV	11	z₄,z₃₂	
Wentworth	11	z₁₀	1,2
Straengnaes	11	z₁₀	1,5
Telhashomer	11	z₁₀	e,n,x
Lene	11	z₃₈	
Maastricht	11	z₄₁	1,2
II	11		1,5
Group O:13 (G)[g]			
Chagoua	1,13,23	a	1,5
II	1,13,23	a	1,5
Mim	13,22	a	1,6
II	13,22	a	e,n,x
Wyldegreen	1,13,23	a	l,w
Marshall	13,22	a	l,z₁₃,z₂₈
II	1,13,23	a	z₄₂
Ibadan	13,22	b	1,5
Mississippi	1,13,23	b	1,5
Oudwijk	13,22	b	1,6
II	1,13,23	b	[1,5]:z₄₂
Bracknell	13,23	b	1,6
Rottnest	1,13,22	b	1,7
Vaertan	13,22	b	e,n,x
Ullevi	1,13,23	b	e,n,x
Bahati	13,22	b	e,n,z₁₅
Durham	13,23	b	e,n,z₁₅
Sanktjohann	13,23	b	l,w
II	1,13,22	b	z₄₂
Haouaria	13,22	c	e,n,x,z₁₅
Handen	1,13,23	d	1,2
Mishmarhaemek	1,13,23	d	1,5
Friedenau	13,22	d	1,6
Wichita	1,13,23	d	1,6
Grumpensis	1,13,23	d	1,7
II	13,23	d	e,n,x
Diguel	1,13,22	d	e,n,z₁₅
Telelkebir	13,23	d	e,n,z₁₅
Putten	13,23	d	l,w
Isuge	13,23	d	z₆
Tschangu	1,13,23	e,h	1,5
Willemstad	1,13,22	e,h	1,6
Vridi	1,13,23	e,h	l,w
II	1,13,23	c,n,x	1,[5],7
Raus	13,22	f,g	e,n,x

(continued)

(continued)

TABLE BXII.γ.262. *(cont.)*

Type[c]	Somatic (O) antigen	Phase 1	Phase 2
		Flagellar (H) antigen	
Havana	1,13,23	f,g,[s]	
Bron	13,22	g,m	[e,n,z_{15}]
Agbeni	1,13,23	g,m,[s],[t]	
II	1,13,22	g,m,t	[1,5]
II	1,13,23	g,m,s,t	1,5
II	1,13,23	g,m,[s],t	[e,n,x]
II	1,13,23	g,m,s,t	z_{42}
Congo	13,23	g,m,s,t	
Newyork	13,22	g,s,t	
Okatie	13,23	g,[s],t	
II	1,13,22	g,t	1,5
II	1,13,23	g,t	1,5
II	1,13,23	g,[s],t	z_{42}
IIIa	1,13,23	g,z_{51}	
Washington	13,22	m,t	
II	1,13,23	m,t	1,5
II	1,13,23	m,t	e,n,x
II	13,22	m,t	z_{42}:z_{39}
II	1,13,23	m,t	z_{42}
Kintambo	13,23	m,t	
V	1,13,22	i	
Idikan	1,13,23	i	1,5
Jukestown	13,23	i	e,n,z_{15}
Kedougou	1,13,23	i	l,w
II	13,22	k	1,5:z_{42}
Marburg	13,23	k	
II	13,23	k	z_{41}
Lovelace	13,22	l,v	1,5
IIIb	13,22	l,v	1,5,7
Borbeck	13,22	l,v	1,6
Nanga	1,13,23	l,v	e,n,z_{15}
II	13,23	l,w	e,n,x
Taiping	13,22	l,z_{13}	e,n,z_{15}
II	13,22	l,z_{28}	1,5
II	13,23	l,z_{28}	1,5
II	13,23	l,z_{28}	z_6
II	1,13,23	l,z_{28}	z_{42}
V	13,22	r	
Adjame	13,23	r	1,6
Linton	13,23	r	e,n,z_{15}
Tanger	1,13,22	y	1,6
Yarrabah	13,23	y	1,7
Ordonez	1,13,23	y	l,w
Tunis	1,13,23	y	z_6
II	1,13,23	z	1,5
Poona	1,13,22	z	1,6
Farmsen	13,23	z	1,6
Bristol	13,22	z	1,7
Tanzania	1,13,22	z	e,n,z_{15}
Worthington	1,13,23	z	l,w
II	1,13,23	z	z_{42}
II	13,22	z	
Ried	1,13,22	z_4,z_{23}	[e,n,z_{15}]
IIIa	13,22	z_4,z_{23}	
Ajiobo	13,23	z_4,z_{23}	
IIIa	13,23	z_4,z_{23},[z_{32}]	
Romanby	1,13,23	z_4,z_{24}	
IIIa	1,13,23	z_4,z_{24}	
Roodepoort	1,13,22	z_{10}	1,5
II	1,13,22	z_{10}	z_6
Sapele	13,23	z_{10}	e,n,z_{15}
Demerara	13,23	z_{10}	l,w
II	13,22	z_{29}	1,5
II	13,22	z_{29}	e,n,x
II	1,13,23	z_{29}	e,n,x
Agoueve	13,22	z_{29}	
Cubana	1,13,23	z_{29}	
Mampong	13,22	z_{35}	1,6

(continued)

TABLE BXII.γ.262. *(cont.)*

Type[c]	Somatic (O) antigen	Phase 1	Phase 2
		Flagellar (H) antigen	
Nimes	13,22	z_{35}	e,n,z_{15}
Anna	13,23	z_{35}	e,n,z_{15}
Leiden	13,22	z_{38}	
Fanti	13,23	z_{38}	
II	13,22	z_{39}	1,7
II	1,13,23	z_{39}	1,5,7
II	1,13,23	[z_{42}]	1,[5],7
II	13,23		1,6
Group O:6,14 (H)			
Garba	1,6,14,25	a	1,5
VI	[1],6,14	a	1,5
VI	1,6,14,25	a	e,n,x
Banjul	1,6,14,25	a	e,n,z_{15}
Ndjamena	1,6,14,25	b	1,2
Kuntair	1,6,14,25	b	1,5
Tucson	[1],6,14,[25]	b	1,7
IIIb	(6),14	b	e,n,x
Blijdorp	1,6,14,25	c	1,5
Kassberg	1,6,14,25	c	1,6
Runby	1,6,14,25	c	e,n,x
Minna	1,6,14,25	c	l,w
Finkenwerder	[1],6,14,[25]	d	1,5
Woodhull	1,6,14,25	d	1,6
Midway	6,14,24	d	1,7
Florida	[1],6,14,[25]	d	1,7
Lindern	6,14[24]	d	e,n,x
Charity	[1],6,14,[25]	d	e,n,x
Teko	1,6,14,25	d	e,n,z_{15}
Encino	1,6,14,25	d	l,z_{13},z_{28}
Albuquerque	1,6,14,24	d	z_6
Bahrenfeld	6,14,24	e,h	1,5
Onderstepoort	1,6,14,[25]	e,h	1,5
Magumeri	1,6,14,25	e,h	1,6
Beaudesert	[1],6,14,[25]	e,h	1,7
Warragul	[1],6,14,[25]	g,m	
Caracas	[1],6,14,[25]	g,m,s	
Sylvania	[1],6,14,[25]	g,p	
Catanzaro	6,14	g,s,t	
II	1,6,14	m,t	1,5
II	6,14	m,t	e,n,x
Kaitaan	1,6,14,25	m,t	
Mampeza	1,6,14,25	i	1,5
Buzu	[1],6,14,[25]	i	1,7
Schalkwijk	6,14	i	e,n,z_{15}
Moussoro	1,6,14,25	i	e,n,z_{15}
Harburg	[1],6,14,[25]	k	1,5
II	6,14,[24]	k	1,6
II	6,14	k	e,n,x
IIIb	(6),14	k	z
II	1,6,14	k	z_6:z_{42}
IIIb	(6),14	k	z_{53}
Boecker	[1],6,14,[25]	l,v	1,7
Horsham	1,6,14,[25]	l,v	e,n,x
IIIb	(6),14	l,v	z
IIIb	(6),14	l,v	z_{35}
IIIb	(6),14	l,v	z_{53}
Aflao	1,6,14,25	l,z_{28}	e,n,x
Istoria	1,6,14,25	r,i	1,5
IIIb	(6),14	r	z
Surat	[1],6,14,[25]	r,[i]	e,n,z_{15}
Carrau	6,14,[24]	y	1,7
Madelia	1,6,14,25	y	1,7
Fischerkietz	1,6,14,25	y	e,n,x
Mornington	1,6,14,25	y	e,n,z_{15}
Homosassa	1,6,14,25	z	1,5
Kanifing	1,6,14,25	z	1,6
Soahanina	6,14,24	z	e,n,x
Sundsvall	[1],6,14,[25]	z	e,n,x

(continued)

TABLE BXII.γ.262. *(cont.)*

Type[c]	Somatic (O) antigen	Flagellar (H) antigen Phase 1	Phase 2
Royan	1,6,14,25	z	e,n,z_{15}
Poano	1,6,14,25	z	l,z_{13},z_{28}
Arapahoe	6,14	z_4,z_{23}	1,5
Bousso	1,6,14,25	z_4,z_{23}	e,n,z_{15}
IV	6,14	z_4,z_{23}	
Chichiri	6,14,24	z_4,z_{24}	
Uzaramo	1,6,14,25	z_4,z_{24}	
Nessa	1,6,14,25	z_{10}	1,2
VI	1,6,14,25	z_{10}	1,(2),7
II	1,6,14	z_{10}	1,5
Laredo	1,6,14,25	z_{10}	1,6
IIIb	(6),14	z_{10}	e,n,x,z_{15}
IIIb	(6),14	z_{10}	z
II	1,6,14	z_{10}	$z_6{:}z_{42}$
IIIb	6,14	z_{10}	z_{53}
Potosi	6,14	z_{36}	1,5
Sara	1,6,14,25	z_{38}	e,n,x
II	1,6,14	z_{42}	1,6
IIIb	6,14	z_{52}	e,n,x,z_{15}
IIIb	1,6,14,25	z_{52}	z_{35}
Group 0:16 (I)			
Hannover	16	a	1,2
Brazil	16	a	1,5
Amunigun	16	a	1,6
Nyeko	16	a	1,7
Togba	16	a	e,n,x
Fischerhuette	16	a	e,n,z_{15}
Heron	16	a	z_6
Hull	16	b	1,2
Wa	16	b	1,5
Glasgow	16	b	1,6
Hvittingfoss	16	b	e,n,x
II	16	b	e,n,x
Sangera	16	b	e,n,z_{15}
Vegesack	16	b	l,w
Malstatt	16	b	z_6
II	16	b	z_{39}
II	16	b	z_{42}
Vancouver	16	c	1,5
Gafsa	16	c	1,6
Shamba	16	c	e,n,x
Hithergreen	16	c	e,n,z_{15}
Yoruba	16	c	l,w
Oldenburg	16	d	1,2
Sculcoates	16	d	1,5
II	16	d	1,5
Sherbrooke	16	d	1,6
Gaminara	16	d	1,7
Barranquilla	16	d	e,n,x
Nottingham	16	d	e,n,z_{15}
Caen	16	d	l,w
Barmbek	16	d	z_6
Malakal	16	e,h	1,2
Saboya	16	e,h	1,5
Rhydyfelin	16	e,h	e,n,x
Weston	16	e,h	z_6
II	16	e,n,x	1,(5),7
II	16	e,n,x	$1,6{:}z_{42}$
Tees	16	f,g	
Adeoyo	16	g,m,[t]	
Nikolaifleet	16	g,m,s	
II	16	g,[m],[s],t	$[1,5]{:}[z_{42}]$
II	16	g,[m],[s],t	[e,n,x]
Cardoner	16	g,s,t	
II	16	m,t	e,n,x
II	16	m,t	$[z_{42}]$
Mpouto	16	m,t	
Amina	16	i	1,5

(continued)

TABLE BXII.γ.262. *(cont.)*

Type[c]	Somatic (O) antigen	Flagellar (H) antigen Phase 1	Phase 2
Agbara	16	i	1,6
Wisbech	16	i	1,7
Frankfurt	16	i	e,n,z_{15}
Pisa	16	i	l,w
Abobo	16	i	z_6
IIIb	16	i	z_{35}
Szentes	16	k	1,2
Nuatja	16	k	e,n,x
Orientalis	16	k	e,n,z_{15}
IIIb	16	k	z
IIIb	16	(k)	z_{35}
IIIb	16	k	z_{53}
IIIb	16	l,v	1,5,7
Shanghai	16	l,v	1,6
Welikade	16	l,v	1,7
Salford	16	l,v	e,n,x
Burgas	16	l,v	e,n,z_{15}
IIIb	16	l,v	$z{:}[z_{61}]$
Losangeles	16	l,v	z_6
IIIb	16	l,v	z_{35}
IIIb	16	l,v	z_{53}
Zigong	16	l,w	1,5
Westeinde	16	l,w	1,6
Brooklyn	16	l,w	e,n,x
Lomnava	16	l,w	e,n,z_{15}
II	16	l,w	z_6
Mandera	16	l,z_{13}	e,n,z_{15}
Enugu	16	$l,[z_{13}],z_{28}$	[1,5]
Battle	16	l,z_{13},z_{28}	1,6
Ablogame	16	l,z_{13},z_{28}	z_6
II	16	l,z_{28}	z_{42}
Rovaniemi	16	r,i	1,5
Ivory	16	r	1,6
Brunflo	16	r	1,7
Annedal	16	r,i	e,n,x
Zwickau	16	r,i	e,n,z_{15}
Saphra	16	y	1,5
Akuafo	16	y	1,6
Kikoma	16	y	e,n,x
Avignon	16	y	e,n,z_{15}
Gerland	16	z	1,5
Fortlamy	16	z	1,6
Lingwala	16	z	1,7
II	16	z	e,n,x
Brevik	16	z	$e,n,[x],z_{15}$
Bouake	16	z	z_6
II	16	z	z_{42}
Kibi	16	z_4,z_{23}	[1,6]
II	16	z_4,z_{23}	
IV	16	z_4,z_{23}	
II	16	z_4,z_{24}	
IV	16	z_4,z_{32}	
II	16	z_6	1,6
Badagry	16	z_{10}	1,5
IIIb	16	z_{10}	1,5,7
Lisboa	16	z_{10}	1,6
IIIb	16	z_{10}	e,n,x,z_{15}
Redlands	16	z_{10}	e,n,z_{15}
Angouleme	16	z_{10}	z_6
Saloniki	16	z_{29}	
II	16	z_{29}	1,5
II	16	z_{29}	e,n,x
Trier	16	z_{35}	1,6
Dakota	16	z_{35}	e,n,z_{15}
II	16	z_{35}	e,n,x
IV	16	z_{36}	
II	16	z_{36}	e,n,z_{15}
Naware	16	z_{38}	

(continued)

TABLE BXII.γ.262. *(cont.)*

Type[c]	Somatic (O) antigen	Phase 1	Phase 2
Grancanaria	16	z_{39}	[1,6]
II	16	z_{42}	1,(5),7
IIIb	16	z_{52}	z_{35}
Group O:17 (J)			
Bonames	17	a	1,2
Jangwani	17	a	1,5
Kinondoni	17	a	e,n,x
Kirkee	17	b	1,2
Dahra	17	b	1,5
II	17	b	e,n,x,z_{15}
Bignona	17	b	e,n,z_{15}
II	17	b	z_6
Luedinghausen	17	c	1,5
Victoriaborg	17	c	1,6
II	17	c	z_{39}
Berlin	17	d	1,5
Karlshamn	17	d	e,n,z_{15}
Niamey	17	d	l,w
Jubilee	17	e,h	1,2
II	17	e,n,x,z_{15}	1,6
II	17	e,n,x,z_{15}	1,[5],7
II	17	g,m,s,t	
Lowestoft	17	g,s,t	
II	17	g,t	$[e,n,x,z_{15}]$
II	17	g,t	z_{39}
Bama	17	m,t	
II	17	m,t	
Ahanou	17	i	1,7
IIIb	17	i	z_{35}
Irenea	17	k	1,5
Warri	17	k	1,7
Matadi	17	k	e,n,x
Zaria	17	k	e,n,z_{15}
IIIb	17	k	z
II	17	k	
Morotai	17	l,v	1,2
Michigan	17	l,v	1,5
Lancaster	17	l,v	1,7
Carmel	17	l,v	e,n,x
IIIb	17	l,v	e,n,x,z_{15}
IIIb	17	l,v	z_{35}
Granlo	17	l,z_{28}	e,n,x
Lode	17	r	1,2
IIIb	17	r	z
II	17	y	
Tendeba	17	y	e,n,x
Hadejia	17	y	e,n,z_{15}
Gori	17	z	1,2
Warengo	17	z	1,5
II	17	z	1,7
Tchamba	17	z	e,n,z_{15}
II	17	z	$l,w:z_{42}$
IIIa	17	z_4,z_{23}	
IIIa	17	z_4,z_{23},z_{32}	
IIIa	17	z_4,z_{24}	
IIIa	17	z_4,z_{32}	
Djibouti	17	z_{10}	e,n,x
IIIb	17	z_{10}	e,n,x,z_{15}
IIIb	17	z_{10}	z
II	17	z_{10}	
Kandla	17	z_{29}	
IIIa	17	z_{29}	
IV	17	z_{29}	
Aachen	17	z_{35}	1,6
IIIa	17	z_{36}	
IV	17	z_{36}	
Group O:18 (K)			
Brazos	6, 14,18	a	e,n,z_{15}

TABLE BXII.γ.262. *(cont.)*

Type[c]	Somatic (O) antigen	Phase 1	Phase 2
Fluntern	6, 14,18	b	1,5
Rawash	6, 14,18	c	e,n,x
Groenekan	18	d	1,5
Usumbura	18	d	1,7
Pontypridd	18	g,m	
IIIa	18	g,z_{51}	
II	18	m,t	1,5
Langenhorn	18	m,t	
Memphis	18	k	1,5
IIIb	18	(k)	z_{53}
IIIb	18	(k)	z_{54}
IIIb	18	l,v	e,n,x,z_{15}
Orlando	18	l,v	e,n,z_{15}
IIIb	18	l,v	z
IIIb	18	l,v	z_{53}
Toulon	18	l,w	e,n,z_{15}
Tennenlohe	18	r	1,5
IIIb	18	r	z
II	18	y	e,n,x,z_{15}
Potengi	18	z	
Cerro	6, 14,18	z_4,z_{23}	[1,5]
Aarhus	18	z_4,z_{23}	z_{64}
II	18	z_4,z_{23}	
IIIa	18	z_4,z_{23}	
Blukwa	6, 14,18	z_4,z_{24}	
II	18	z_4,z_{24}	
IIIa	18	z_4,z_{32}	
IIIb	18	z_{10}	e,n,x,z_{15}
Leer	18	z_{10}	1,5
Carnac	18	z_{10}	z_6
II	18	z_{10}	z_6
II	18	z_{36}	
IV	18	z_{36},z_{38}	
Sinthia	18	z_{38}	
Delmenhorst	18	z_{71}	
Cotia	18		1,6
Group O:21 (L)			
Assen	21	a	[1,5]
II	21	b	1,5
Ghana	21	b	1,6
Minnesota	21	b	e,n,x
Hydra	21	c	1,6
Rhone	21	c	e,n,x
II	21	c	e,n,x
IIIb	21	c	e,n,x,z_{15}
Spartel	21	d	1,5
Magwa	21	d	e,n,x
Madison	21	d	z_6
Good	21	f,g	e,n,x
II	21	g,[m],[s],t	
IIIa	21	g,z_{51}	
IV	21	g,z_{51}	
II	21	m,t	
Diourbel	21	i	1,2
IIIb	21	i	1,5,7
IIIb	21	i	e,n,x,z_{15}
IIIb	21	k	e,n,x,z_{15}
IIIb	21	k	z
Surrey	21	k	1,2,5
IIIb	21	l,v	z
IIIb	21	l,v	z_{57}
Keve	21	l,w	
Jambur	21	l,z_{28}	e,n,z_{15}
Mountmagnet	21	r	
IIIb	21	r	z
Ibaragi	21	y	1,2
Ruiru	21	y	e,n,x
II	21	z	

(continued)

TABLE BXII.γ.262. *(cont.)*

Type[c]	Somatic (O) antigen	Flagellar (H) antigen Phase 1	Phase 2
Baguida	21	z_4,z_{23}	
IIIa	21	z_4,z_{23}	
IV	21	z_4,z_{23}	
II	21	z_4,z_{24}	
IIIa	21	z_4,z_{24}	
IV	21	z_4,z_{32}	
IIIb	21	z_{10}	e,n,x,z_{15}
IIIb	21	z_{10}	z
II	21	z_{10}	$[z_6]$
IIIb	21	z_{10}	z_{53}
IIIa	21	z_{29}	
Gambaga	21	z_{35}	e,n,z_{15}
IV	21	z_{36}	
IIIb	21	z_{65}	e,n,x,z_{15}
Group O:28 (M)			
Solna	28	a	1,5
Dakar	28	a	1,6
Bakau	28	a	1,7
Seattle	28	a	e,n,x
II	28	a	e,n,x
Honelis	28	a	e,n,z_{15}
Dibra	28	a	z_6
Moero	28	b	1,5
Ashanti	28	b	1,6
Bokanjac	28	b	1,7
Soumbedioune	28	b	e,n,x
II	28	b	e,n,x
Langford	28	b	e,n,z_{15}
Freefalls	28	b	l,w
II	28	b	z_6
Hermannswerder	28	c	1,5
Eberswalde	28	c	1,6
Halle	28	c	1,7
Dresden	28	c	e,n,x
Wedding	28	c	e,n,z_{15}
Techimani	28	c	z_6
Amoutive	28	d	1,5
Hatfield	28	d	1,6
Mundonobo	28	d	1,7
Mocamedes	28	d	e,n,x
Patience	28	d	e,n,z_{15}
Cullingworth	28	d	l,w
Kpeme	28	e,h	1,7
Gozo	28	e,h	e,n,z_{15}
II	28	e,n,x	1,7
Friedrichsfelde	28	f,g	
Yardley	28	g,m	1,6
Abadina	28	g,m	$[e,n,z_{15}]$
II	28	g,(m),[s],t	1,5
Croft	28	g,m,s	$[e,n,z_{15}]$
II	28	g,m,t	e,n,x
II	28	g,m,t	z_{39}
II	28	g,s,t	e,n,x
Ona	28	g,s,t	
II	28	m,t	[e,n,x]
Vinohrady	28	m,t	$[e,n,z_{15}]$
Morillons	28	m,t	1,6
Doorn	28	i	1,2
Cotham	28	i	1,5
Volkmarsdorf	28	i	1,6
Dieuppeul	28	i	1,7
Warnemuende	28	i	e,n,x
Kuessel	28	i	e,n,z_{15}
Douala	28	i	l,w
Guildford	28	k	1,2
Ilala	28	k	1,5
Adamstown	28	k	1,6
Ikeja	28	k	1,7

(continued)

TABLE BXII.γ.262. *(cont.)*

Type[c]	Somatic (O) antigen	Flagellar (H) antigen Phase 1	Phase 2
Taunton	28	k	e,n,x
Ank	28	k	e,n,z_{15}
Leoben	28	l,v	1,5
Vitkin	28	l,v	e,n,x
Nashua	28	l,v	e,n,z_{15}
Ramsey	28	l,w	1,6
Catalunia	28	l,z_{13},z_{28}	1,5
Penilla	28	l,z_{13},z_{28}	e,n,z_{15}
II	28	l,z_{28}	1,5
Fajara	28	l,z_{28}	e,n,x
Bassadji	28	r	1,6
Kibusi	28	r	e,n,x
II	28	r	e,n,z_{15}
Fairfield	28	r	l,w
Chicago	28	r,[i]	1,5
Banco	28	r,i	1,7
Sanktgeorg	28	r,[i]	e,n,z_{15}
Oskarshamn	28	y	1,2
Nima	28	y	1,5
Pomona	28	y	$1,7:[z_{60}]$
Kitenge	28	y	e,n,x
Telaviv	28	y	e,n,z_{15}
Shomolu	28	y	l,w
Selby	28	y	z_6
Vanier	28	z	1,5
II	28	z	1,5
Doel	28	z	1,6
Ezra	28	z	1,7
Brisbane	28	z	e,n,z_{15}
II	28	z	z_{39}
Cannobio	28	z_4,z_{23}	1,5
Teltow	28	z_4,z_{23}	1,6
Babelsberg	28	z_4,z_{23}	$[e,n,z_{15}]$
Rogy	28	z_{10}	1,2
Farakan	28	z_{10}	1,5
Libreville	28	z_{10}	1,6
Malaysia	28	z_{10}	1,7
Umbilo	28	z_{10}	e,n,x
Luckenwalde	28	z_{10}	e,n,z_{15}
Moroto	28	z_{10}	l,w
IIIb	28	z_{10}	z
Djermaia	28	z_{29}	
II	28	z_{29}	1,5
II	28	z_{29}	e,n,x
Konolfingen	28	z_{35}	1,6
Balili	28	z_{35}	1,7
Santander	28	z_{35}	e,n,z_{15}
Aderike	28	z_{38}	e,n,z_{15}
Group O:30 (N)			
Overvecht	30	a	1,2
Zehlendorf	30	a	1,5
Guarapiranga	30	a	e,n,x
Doulassame	30	a	e,n,z_{15}
II	30	a	z_{39}
Louga	30	b	1,2
Aschersleben	30	b	1,5
Urbana	30	b	e,n,x
Neudorf	30	b	e,n,z_{15}
II	30	b	z_6
Zaire	30	c	1,7
Morningside	30	c	e,n,z_{15}
II	30	c	z_{39}
Messina	30	d	1,5
Livulu	30	e,h	1,2
Torhout	30	e,h	1,5
Godesberg	30	g,m,[t]	
II	30	g,m,s	e,n,x
Giessen	30	g,m,s	

(continued)

TABLE BXII.γ.262. *(cont.)*

Type[c]	Somatic (O) antigen	Phase 1	Phase 2
Sternschanze	30	g,s,t	
II	30	g,t	
Wayne	30	g,z_{51}	
II	30	m,t	
Landau	30	i	1,2
Morehead	30	i	1,5
Mjordan	30	i	e,n,z_{15}
Soerenga	30	i	l,w
Hilversum	30	k	1,2
Ramatgan	30	k	1,5
Aqua	30	k	1,6
Angoda	30	k	e,n,x
Odozi	30	k	e,n,[x],z_{15}
II	30	k	e,n,x,z_{15}
Ligeo	30	l,v	1,2
Donna	30	l,v	1,5
Ockenheim	30	l,z_{13},z_{28}	1,6
Morocco	30	l,z_{13},z_{28}	e,n,z_{15}
II	30	l,z_{28}	z_6
Grandhaven	30	r	1,2
Gege	30	r	1,5
Matopeni	30	y	1,2
Bietri	30	y	1,5
Steinplatz	30	y	1,6
Baguirmi	30	y	e,n,x
Nijmegen	30	y	e,n,z_{15}
Stoneferry	30	z_4,z_{23}	
Bodjonegoro	30	z_4,z_{24}	
II	30	z_6	1,6
Sada	30	z_{10}	1,2
Senneville	30	z_{10}	1,5
Kumasi	30	z_{10}	e,n,z_{15}
II	30	z_{10}	e,n,x,z_{15}
Aragua	30	z_{29}	
Kokoli	30	z_{35}	1,6
Wuiti	30	z_{35}	e,n,z_{15}
Ago	30	z_{38}	
II	30	z_{39}	1,7
Group O:35 (O)			
Umhlatazana	35	a	e,n,z_{15}
Tchad	35	b	
Gouloumbo	35	c	1,5
Yolo	35	c	[e,n,z_{15}]
II	35	d	1,5
Dembe	35	d	l,w
Gassi	35	e,h	z_6
Adelaide	35	f,g	
Ealing	35	g,m,s	
II	35	g,m,s,t	
Ebrie	35	g,m,t	
Anecho	35	g,s,t	
II	35	g,t	1,5
II	35	g,t	z_{42}
Agodi	35	g,t	
IIIa	35	g,z_{51}	
Monschaui	35	m,t	
II	35	m,t	
IIIb	35	i	e,n,x,z_{15}
Gambia	35	i	e,n,z_{15}
Bandia	35	i	l,w
IIIb	35	i	z
IIIb	35	i	z_{35}
IIIb	35	i	z_{53}
IIIb	35	k	e,n,x,z_{15}
IIIb	35	k	z
IIIb	35	(k)	z
IIIb	35	(k)	z_{35}
IIIb	35	k	z_{53}

Type[c]	Somatic (O) antigen	Phase 1	Phase 2
IIIb	35	l,v	1,5,7
IIIb	35	l,v	z_{35}:[z_{67}]
II	35	l,z_{28}	
IIIb	35	r	e,n,x,z_{15}
Massakory	35	r	l,w
IIIb	35	r	z
IIIb	35	r	z_{35}
IIIb	35	r	z_{61}
Alachua	35	z_4,z_{23}	
IIIa	35	z_4,z_{23}	
Westphalia	35	z_4,z_{24}	
IIIa	35	z_4,z_{32}	
Camberene	35	z_{10}	1,5
Enschede	35	z_{10}	l,w
Ligna	35	z_{10}	z_6
IIIb	35	z_{10}	z_{35}
II	35	z_{29}	e,n,x
Widemarsh	35	z_{29}	
IIIa	35	z_{29}	
IIIa	35	z_{36}	
Haga	35	z_{38}	
IIIb	35	z_{52}	1,5,7
IIIb	35	z_{52}	e,n,x,z_{15}
IIIb	35	z_{52}	z
IIIb	35	z_{52}	z_{35}
Group O:38 (P)			
Oran	38	a	e,n,z_{15}
II	38	b	1,2
Rittersbach	38	b	e,n,z_{15}
Sheffield	38	c	1,5
Kidderminster	38	c	1,6
II	38	d	[1,5]
Thiaroye	38	e,h	1,2
Kasenyi	38	e,h	1,5
Korovi	38	g,m,[s]	
II	38	g,t	
IIIa	38	g,z_{51}	
IV	38	g,z_{51}	
Rothenburgsort	38	m,t	
Mgulani	38	i	1,2
Lansing	38	i	1,5
IIIb	38	i	z
IIIb	38	i	z_{53}
Echa	38	k	1,2
Mango	38	k	1,5
Inverness	38	k	1,6
Njala	38	k	e,n,x
IIIb	38	k	e,n,x,z_{15}
IIIb	38	k	z
IIIb	38	k	z_{53}
IIIb	38	(k)	1,5,7
IIIb	38	(k)	z
IIIb	38	(k)	z_{35}
IIIb	38	(k)	z_{54}
IIIb	38	(k)	z_{54}
Alger	38	l,v	1,2
Kimberley	38	l,v	1,5
Taylor	38	l,v	e,n,z_{15}
Roan	38	l,v	e,n,x
IIIb	38	l,v	z
IIIb	38	l,v	z_{35}
IIIb	38	l,v	z_{53}:[z_{54}]
Lindi	38	r	1,5
IIIb	38	r	1,5,7
Emmastad	38	r	1,6
IIIb	38	r	e,n,x,z_{15}
IIIb	38	r	z:[z_{57}]
IIIb	38	r	z_{35}

(continued)

TABLE BXII.γ.262. *(cont.)*

Type[c]	Somatic (O) antigen	Phase 1	Phase 2
		Antigenic formulas[b]	
		Flagellar (H) antigen	
Freetown	38	y	1,5
Colombo	38	y	1,6
Perth	38	y	e,n,x
Stachus	38	z	
Yoff	38	z_4,z_{23}	1,2
IIIa	38	z_4,z_{23}	
IV	38	z_4,z_{23}	
Bangkok	38	z_4,z_{24}	
Neunkirchen	38	z_{10}	
IIIb	38	z_{10}	z
IIIb	38	z_{10}	z_{53}
Klouto	38	z_{38}	
IIIb	38	z_{52}	z_{35}
IIIb	38	z_{52}	z_{53}
IIIb	38	z_{53}	
IIIb	38	z_{61}	$[z_{53}]$
Group O:39 (Q)			
II	39	a	z_{39}
Wandsworth	39	b	1,2
Abidjan	39	b	l,w
II	39	c	e,n,x
Logone	39	d	1,5
Mara	39	e,h	1,5
II	39	e,n,x	1,7
II	39	[g],m,t	[e,n,x]
Hofit	39	i	1,5
Cumberland	39	i	e,n,x
Alma	39	i	e,n,z_{15}
Champaign	39	k	1,5
II	39	l,v	1,5
Kokomlemle	39	l,v	e,n,x
Oerlikon	39	l,v	e,n,z_{15}
II	39	l,z_{28}	e,n,x
II	39	l,z_{28}	z_{39}
Anfo	39	y	1,2
Windermere	39	y	1,5
Hegau	39	z_{10}	
II	39		1,7
Group O:40 (R)			
Shikmonah	40	a	1,5
Greiz	40	a	z_6
II	1,40	a	z_6
II	40	a	z_{39}
Riogrande	40	b	1,5
Saugus	40	b	1,7
Johannesburg	1,40	b	e,n,x
Duval	1,40	b	e,n,z_{15}
Benguella	40	b	z_6
II	40	b	
II	1,40	c	e,n,x,z_{15}
II	1,40	c	z_{39}
Driffield	1,40	d	1,5
II	40	d	
Tilene	1,40	e,h	1,2
II	1,40	e,n,x	1,[5],7
II	1,40	e,n,x,z_{15}	1,6
Bijlmer	1,40	g,m	
II	1,40	g,[m],[s],[t]	e,n,x
II	1,40	g,[m],[s],t	1,5
II	1,40	g,t	e,n,x,z_{15}
II	40	g,t	z_{39}
IV	1,40	g,t	
II	1,40	g,[m],[s],t	z_{42}
IIIa	40	g,z_{51}	
IIIb	40	g,z_{51}	e,n,x,z_{15}
IV	1,40	g,z_{51}	
II	40	m,t	z_{39}
II	1,40	m,t	z_{42}

(continued)

TABLE BXII.γ.262. *(cont.)*

Type[c]	Somatic (O) antigen	Phase 1	Phase 2
		Antigenic formulas[b]	
		Flagellar (H) antigen	
IV	40	m,t	
IIIb	40	i	1,5,7
Goulfey	1,40	k	1,5
Allandale	1,40	k	1,6
Hann	40	k	e,n,x
II	1,40	k	e,n,x,z_{15}
IIIb	40	k	$z:z_{57}$
II	40	k	z_6
IIIb	40	k	z_{53}
Millesi	1,40	l,v	1,2
Canary	40	l,v	1,6
II	40	l,v	e,n,x
IIIb	40	l,v	z
IIIb	40	l,v	z_{53}
Overchurch	1,40	l,w	[1,2]
Tiko	1,40	l,z_{13},z_{28}	1,2
Bukavu	1,40	l,z_{28}	1,5
II	1,40	l,z_{28}	$1,5:z_{42}$
Santhiaba	40	l,z_{28}	1,6
II	1,40	l,z_{28}	z_{39}
Odienne	40	y	1,5
II	1,40	z	1,5
Casamance	40	z	e,n,x
Nowawes	40	z	z_6
II	1,40	z	z_6
II	1,40	z	z_{39}
II	40	z	z_{42}
IIIa	40	z_4,z_{23}	
IV	1,40	z_4,z_{23}	
II	40	z_4,z_{24}	z_{39}
IIIa	40	z_4,z_{24}	
IV	40	z_4,z_{24}	
IIIa	40	z_4,z_{32}	
IV	40	z_4,z_{32}	
II	1,40	z_6	1,5
Trotha	40	z_{10}	z_6
IIIb	40	z_{10}	z_{35}
Omifisan	1,40	z_{29}	
IIIa	40	z_{29}	
II	1,40	z_{35}	e,n,x,z_{15}
Yekepa	1,40	z_{35}	e,n,z_{15}
V	1,40	z_{35}	
IIIa	40	z_{36}	
II	1,40	z_{39}	$1,5:z_{42}$
II	1,40	z_{39}	1,6
IIIb	40	z_{39}	1,6
II	40	z_{39}	1,7
Karamoja	1,40	z_{41}	1,2
II	1,40	z_{42}	1,6
II	1,40	$[z_{42}]$	1,(5),7
V	1,40	z_{81}	
Group O:41 (S)			
Burundi	41	a	
II	41	b	1,5
Vaugirard	41	b	1,6
VI	41	b	1,7
Vietnam	41	b	z_6
Sica	41	b	e,n,z_{15}
IIIb	41	c	e,n,x,z_{15}
II	41	c	z_6
Egusi	41	d	1,5
II	41	d	z_6
II	41	g,m,s,t	z_6
II	41	g,t	
IIIa	41	g,z_{51}	
Leatherhead	41	m,t	1,6
Samaru	41	i	1,5
Verona	41	i	1,6

(continued)

TABLE BXII.γ.262. *(cont.)*

Type[c]	Somatic (O) antigen	Phase 1	Phase 2
Ferlo	41	k	1,6
II	41	k	1,6
II	41	k	z_6
IIIb	41	(k)	z_{35}
II	41	l,z_{13},z_{28}	e,n,x,z_{15}
Lubumbashi	41	r	1,5
II	41	z	1,5
Bofflens	41	z_4,z_{23}	1,7
Waycross	41	z_4,z_{23}	$[e,n,z_{15}]$
IIIa	41	z_4,z_{23}	
IV	41	z_4,z_{23}	
IIIa	41	z_4,z_{23},z_{32}	
Ipswich	41	z_4,z_{24}	1,5
IIIa	41	z_4,z_{24}	
IIIa	41	z_4,z_{32}	
II	41	z_{10}	1,2
Leipzig	41	z_{10}	1,5
Landala	41	z_{10}	1,6
Inpraw	41	z_{10}	e,n,x
II	41	z_{10}	e,n,x,z_{15}
II	41	z_{10}	z_6
Lodz	41	z_{29}	
IIIa	41	z_{29}	
IV	41	z_{29}	
Ahoutoue	41	z_{35}	1,6
IIIa	41	z_{36}	
Offa	41	z_{38}	
IV	41	z_{52}	
II	41		1,6
Group O:42 (T)			
Faji	1,42	a	e,n,z_{15}
II	42	b	1,5
Orbe	42	b	1,6
II	42	b	e,n,x,z_{15}
Tomegbe	1,42	b	e,n,z_{15}
Egusitoo	1,42	b	z_6
II	42	b	z_6
Antwerpen	1,42	c	e,n,z_{15}
Kampala	1,42	c	z_6
II	42	d	z_6
II	42	e,n,x	1,6
II	42	g,t	
Maricopa	1,42	g,z_{51}	1,5
IIIa	42	g,z_{51}	
IV	1,42	g,z_{51}	
II	42	m,t	$[e,n,x,z_{15}]$
Waral	1,42	m,t	
Kaneshie	1,42	i	l,w
Borromea	42	i	1,6
Middlesbrough	1,42	i	z_6
Haferbreite	42	k	1,6
IIIb	42	k	e,n,x,z_{15}
IIIb	42	k	z
Gwale	1,42	k	z_6
IIIb	42	(k)	z_{35}
IIIb	42	l,v	1,5,7
II	42	l,v	e,n,x,z_{15}
IIIb	42	l,v	e,n,x,z_{15}
Coogee	42	l,v	e,n,z_{15}
IIIb	42	l,v	z
IIIb	42	l,v	z_{53}
II	1,42	l,w	e,n,x
II	1,42	$l,[z_{13}],z_{28}$	z_6
Sipane	1,42	r	e,n,z_{15}
Brive	1,42	r	l,w
IIIb	42	r	z
IIIb	42	r	z_{53}
II	42	r	

Type[c]	Somatic (O) antigen	Phase 1	Phase 2
IIIa	42	r	
Spalentor	1,42	y	e,n,z_{15}
Harvestehude	1,42	y	z_6
II	42	z	1,5
Ursenbach	1,42	z	1,6
II	42	z	e,n,x,z_{15}
Melbourne	42	z	e,n,z_{15}
II	42	z	z_6
Gera	1,42	z_4,z_{23}	1,6
Broc	42	z_4,z_{23}	e,n,z_{15}
IIIa	42	z_4,z_{23}	
Toricada	1,42	z_4,z_{24}	
IIIa	42	z_4,z_{24}	
IV	1,42	z_4,z_{24}	
II	42	z_6	1,6
II	42	z_{10}	1,2
II	42	z_{10}	e,n,x,z_{15}
IIIb	42	z_{10}	e,n,x,z_{15}
IIIb	42	z_{10}	z
Loenga	1,42	z_{10}	z_6
II	42	z_{10}	z_6
IIIb	42	z_{10}	z_{35}
IIIb	42	z_{10}	z_{67}
Djama	1,42	z_{29}	[1,5]
Kahla	1,42	z_{35}	1,6
Hennekamp	42	z_{35}	e,n,z_{15}
Tema	1,42	z_{35}	z_6
Weslaco	42	z_{36}	
IV	42	z_{36}	
Vogan	1,42	z_{38}	z_6
Taset	1,42	z_{41}	
IIIb	42	z_{52}	z
Group O:43 (U)			
Graz	43	a	1,2
Berkeley	43	a	1,5
II	43	a	1,5
II	43	a	z_6
Niederoderwitz	43	b	
II	43	b	z_{42}
Montreal	43	c	1,5
Orleans	43	d	1,5
II	43	d	e,n,x,z_{15}
II	43	d	z_{39}
II	43	d	z_{42}
II	43	e,n,x,z_{15}	1,(5),7
II	43	e,n,x,z_{15}	1,6
Milwaukee	43	f,g,[t]	
II	43	g,m,[s],t	$[z_{42}]$
II	43	g,t	[1,5]
IIIa	43	g,z_{51}	
IV	43	g,z_{51}	
II	43	g,z_{62}	e,n,x
Mbao	43	i	1,2
Voulte	43	i	e,n,x
Thetford	43	k	1,2
Ahuza	43	k	1,5
IIIb	43	k	z
IIIb	43	l,v	z_{53}
Sudan	43	l,z_{13}	
II	43	l,z_{13},z_{28}	1,5
IIIb	43	r	e,n,x,z_{15}
IIIb	43	r	z
IIIb	43	r	z_{53}
Farcha	43	y	1,2
Kingabwa	43	y	1,5
Ogbete	43	z	1,5
II	43	z	1,5
Arusha	43	z	e,n,z_{15}

(continued)

TABLE BXII.γ.262. *(cont.)*

Type[c]	Somatic (O) antigen	Phase 1	Phase 2
II	43	z_4,z_{23}	
IIIa	43	z_4,z_{23}	
IV	43	z_4,z_{23}	
IIIa	43	z_4,z_{24}	
IV	43	z_4,z_{24}	
IV	43	z_4,z_{32}	
Adana	43	z_{10}	1,5
II	43	z_{29}	e,n,x
II	43	z_{29}	z_{42}
Makiling	43	z_{29}	
IV	43	z_{29}	
Ahepe	43	z_{35}	1,6
IIIa	43	z_{36}	
IV	43	z_{36},z_{38}	
Irigny	43	z_{38}	
II	43	z_{42}	1,5,7
IIIb	43	z_{52}	z_{53}
Group O:44 (V)			
IV	44	a	
Niakhar	44	a	1,5
Tiergarten	44	a	e,n,x
Niarembe	44	a	l,w
Sedgwick	44	b	e,n,z_{15}
Madigan	44	c	1,5
Quebec	44	c	e,n,z_{15}
Bobo	44	d	1,5
Kermel	44	d	e,n,x
Fischerstrasse	44	d	e,n,z_{15}
Palamaner	1,44	d	z_{35}
II	1,44	e,n,x	1,6
Vleuten	44	f,g	
Gamaba	1,44	g,m,[s]	
Splott	44	g,s,t	
II	44	g,t	z_{42}
IIIb	44	g,t	$1,5:z_{42}$
Carswell	44	g,z_{51}	
IV	44	g,z_{51}	-
Muguga	44	m,t	
Maritzburg	1,44	i	e,n,z_{15}
Lawra	44	k	e,n,z_{15}
Malika	44	l,z_{28}	1,5
Brefet	44	r	e,n,z_{15}
V	44	r	
Uhlenhorst	44	z	l,w
Bolama	44	z	e,n,x
Kua	44	z_4,z_{23}	
Ploufragan	1,44	z_4,z_{23}	e,n,z_{15}
II	44	z_4,z_{23}	
IIIa	44	z_4,z_{23}	
IV	44	z_4,z_{23}	
IIIa	44	z_4,z_{23},z_{32}	
Christiansborg	44	z_4,z_{24}	
IIIa	44	z_4,z_{24}	
IV	44	z_4,z_{24}	
IIIa	44	z_4,z_{32}	
IV	1,44	z_4,z_{32}	
Guinea	1,44	z_{10}	1,7
Llobregat	44	z_{10}	e,n,x
II	44	z_{29}	$e,n,x:z_{42}$
Zinder	44	z_{29}	
IV	44	z_{29}	
IV	44	$z_{36},[z_{38}]$	
Koketime	44	z_{38}	
II	1,44	z_{39}	e,n,x,z_{15}
V	44	z_{39}	
Group O:45 (W)			
VI	45	a	e,n,x
Meekatharra	45	a	e,n,z_{15}

TABLE BXII.γ.262. *(cont.)*

Type[c]	Somatic (O) antigen	Phase 1	Phase 2
II	45	a	z_{10}
Riverside	45	b	1,5
Fomeco	45	b	e,n,z_{15}
Deversoir	45	c	e,n,x
Dugbe	45	d	1,6
Karachi	45	d	e,n,x
Warmsen	45	d	e,n,z_{15}
Suelldorf	45	f,g	
Tornow	45	g,m,[s],[t]	
II	45	g,m,s,t	1,5
II	45	g,m,s,t	e,n,x
II	45	g,m,t	e,n,x,z_{15}
Binningen	45	g,s,t	
IIIa	45	g,z_{51}	
IV	45	g,z_{51}	
II	45	m,t	1,5
Apapa	45	m,t	
Verviers	45	k	1,5
Casablanca	45	k	1,7
Cairns	45	k	e,n,z_{15}
Imo	45	l,v	$[e,n,z_{15}]$
Kofandoka	45	r	e,n,z_{15}
II	45	z	1,5
Yopougon	45	z	e,n,z_{15}
II	45	z	z_{39}
IIIa	45	z_4,z_{23}	
IV	45	z_4,z_{23}	
Transvaal	45	z_4,z_{24}	
IIIa	45	z_4,z_{24}	
IIIa	45	z_4,z_{32}	
Aprad	45	z_{10}	
Jodhpur	45	z_{29}	
II	45	z_{29}	1,5
II	45	z_{29}	e,n,x
II	45	z_{29}	z_{42}
IIIa	45	z_{29}	
Lattenkamp	45	z_{35}	$\overline{1,5}$
Balcones	45	z_{36}	
IV	45	z_{36},z_3	
Group O:47 (X)			
II	47	a	1,5
II	47	a	e,n,x,z_{15}
Wenatchee	47	b	1,2
II	47	b	1,5
II	47	b	e,n,x,z_{15}
Sya	47	b	z_6
II	47	b	z_6
IIIb	47	c	1,5,7
Kodjovi	47	c	1,6
IIIb	47	c	$e,n,x,z_{15}:[z_{57}]$
IIIb	47	c	z
IIIb	47	c	z_{35}
Stellingen	47	d	e,n,x
II	47	d	z_{39}
II	47	e,n,x,z_{15}	1,6
Sljeme	1,47	f,g	
Luke	1,47	g,m	
II	47	g,t	e,n,x
IIIa	47	g,z_{51}	
Mesbit	47	m,t	e,n,z_{15}
IIIb	47	i	e,n,x,z_{15}
Bergen	47	i	e,n,z_{15}
IIIb	47	i	z
IIIb	47	i	z_{35}
IIIb	47	i	$z_{53}:[z_{57}]$
Staoueli	47	k	1,2
Bootle	47	k	1,5
IIIb	47	k	1,5,7

(continued)

(continued)

TABLE BXII.γ.262. *(cont.)*

Type[c]	Somatic (O) antigen	Flagellar (H) antigen Phase 1	Phase 2
Dahomey	47	k	1,6
IIIb	47	k	e,n,x,z_{15}
Lyon	47	k	e,n,z_{15}
IIIb	47	k	z
IIIb	47	k	z_{35}
IIIb	47	k	z_{53}
IIIb	47	l,v	1,5,(7)
IIIb	47	l,v	e,n,x,z_{15}
IIIb	47	l,v	z
IIIb	47	l,v	z_{35}
IIIb	47	l,v	z_{53}
IIIb	47	l,v	z_{57}
IV	47	l,v	
Teshie	1,47	l,z_{13},z_{28}	e,n,z_{15}
IIIb	47	r	e,n,x,z_{15}
Dapango	47	r	1,2
IIIb	47	r	1,5,7
IIIb	47	r	z
IIIb	47	r	z_{35}
IIIb	47	r	$z_{53}:[z_{60}]$
IIIa	47	r	
Moualine	47	y	1,6
Blitta	47	y	e,n,x
Mountpleasant	47	z	1,5
Kaolack	47	z	1,6
II	47	z	e,n,x,z_{15}
II	47	z	z_6
Tabligbo	47	z_4,z_{23}	e,n,z_{15}
Binche	47	z_4,z_{23}	l,w
Bere	47	z_4,z_{23}	z_6
IIIa	47	z_4,z_{23}	
Tamberma	47	z_4,z_{24}	
II	47	z_6	1,6
IIIb	47	z_{10}	1,5,7
Namoda	47	z_{10}	e,n,z_{15}
IIIb	47	z_{10}	z
IIIb	47	z_{10}	z_{35}
II	47	z_{29}	e,n,x,z_{15}
Ekpoui	47	z_{29}	
IIIa	47	z_{29}	
Bingerville	47	z_{35}	e,n,z_{15}
IV	47	z_{36}	
Alexanderplatz	47	z_{38}	
Quinhon	47	z_{44}	
IIIb	47	z_{52}	1,5,7
IIIb	47	z_{52}	e,n,x,z_{15}
IIIb	47	z_{52}	z
IIIb	47	z_{52}	z_{35}
Group O:48 (Y)			
Hisingen	48	a	1,5,7
II	48	a	z_6
II	48	a	z_{39}
II	48	b	z_6
V	48	b	
IIIb	48	c	z
II	48	d	1,2
II	48	d	z_6
Buckeye	48	d	
Fitzroy	48	e,h	1,5
II	48	e,n,x,z_{15}	z_6
II	48	g,m,t	
IIIa	48	g,z_{51}	
IV	48	g,z_{51}	
IIIb	48	i	z
IIIb	48	i	$z_{35}:[z_{57}]$
IIIb	48	i	z_{53}
IIIb	48	i	z_{61}
V	48	i	

TABLE BXII.γ.262. *(cont.)*

Type[c]	Somatic (O) antigen	Flagellar (H) antigen Phase 1	Phase 2
IIIb	48	k	1,5,(7)
II	48	k	e,n,x,z_{15}
IIIb	48	k	e,n,x,z_{15}
Dahlem	48	k	e,n,z_{15}
IIIb	48	k	z
IIIb	48	k	z_{35}
II	48	k	z_{39}
IIIb	48	k	z_{53}
IIIb	48	(k)	z_{53}
Australia	48	l,v	1,5
IIIb	48	l,v	1,5,(7)
IIIb	48	l,v	z
IIIb	48	r	e,n,x,z_{15}
IIIb	48	r	z
Toucra	48	z	1,5
II	48	z	1,5
IIIb	48	z	1,5,7
IIIa	48	z_4,z_{23}	
IV	48	z_4,z_{23}	
IIIa	48	z_4,z_{23},z_{32}	
Djakarta	48	z_4,z_{24}	
IIIa	48	z_4,z_{24}	
IIIa	48	z_4,z_{32}	
IV	48	z_4,z_{32}	
II	48	z_{10}	[1,5]
VI	48	z_{10}	1,5
Isaszeg	48	z_{10}	e,n,x
IIIb	48	z_{10}	e,n,x,z_{15}
IIIb	48	z_{10}	z
II	48	z_{29}	
IV	48	z_{29}	
IIIb	48	z_{35}	z_{52}
V	48	z_{35}	
IIIa	48	z_{36}	
IV	48	$z_{36},[z_{38}]$	
V	48	z_{39}	
V	48	z_{41}	
IIIb	48	z_{52}	e,n,x,z_{15}
IIIb	48	z_{52}	z
V	48	z_{65}	
V	48	z_{81}	
Group O:50 (Z)			
IV	50	a	
Rochdale	50	b	e,n,x
II	50	b	z_6
IV	50	b	
Hemingford	50	d	1,5
IV	50	d	
II	50	e,n,x	1,7
II	50	g,[m],s,t	[1,5]
IV	50	g,z_{51}	
II	50	g,z_{62}	e,n,x
II	50	m,t	$z_6:z_{42}$
IIIb	50	i	1,5,7
IIIb	50	i	e,n,x,z_{15}
IIIb	50	i	z
IIIb	50	k	1,5,7
II	50	k	$e,n,x:z_{42}$
IIIb	50	k	e,n,x,z_{15}
IIIb	50	k	z
IIIb	50	(k)	z
II	50	k	z_6
IIIb	50	k	z_{35}
IIIb	50	(k)	z_{35}
IIIb	50	k	z_{53}
Fass	50	l,v	1,2
IIIb	50	l,v	e,n,x,z_{15}
IIIb	50	l,v	z

(continued)

(continued)

TABLE BXII.γ.262. *(cont.)*

Type[c]	Somatic (O) antigen	Phase 1	Phase 2
IIIb	50	l,v	z_{35}
VI	50	l,v	z_{67}
II	50	l,w	$e,n,x,z_{15}:z_{42}$
II	50	l,z_{28}	z_{42}
IIIb	50	r	1,5,(7)
IIIb	50	r	e,n,x,z_{15}
IIIb	50	r	z
IIIb	50	r	z_{35}
IIIb	50	r	z_{53}
Dougi	50	y	1,6
II	50	z	e,n,x
IIIb	50	z	z_{52}
IIIa	50	z_4,z_{23}	
IV	50	z_4,z_{23}	
IIIa	50	z_4,z_{23},z_{32}	
IIIa	50	z_4,z_{24}	
IV	50	z_4,z_{24}	
IIIa	50	z_4,z_{32}	
IV	50	z_4,z_{32}	
IIIb	50	z_{10}	z
II	50	z_{10}	$z_6:z_{42}$
IIIb	50	z_{10}	z_{53}
Ivorycoast	50	z_{29}	
IIIa	50	z_{29}	
IIIa	50	z_{36}	
II	50	z_{42}	1,7
IIIb	50	z_{52}	1,5,7
IIIb	50	z_{52}	z_{35}
IIIb	50	z_{52}	z_{53}
Group O:51			
IV	51	a	
Tione	51	a	e,n,x
Karaya	51	b	1,5
IV	51	b	
II	51	c	
Gokul	1,51	d	1,5
Meskin	51	e,h	1,2
II	51	g,s,t	e,n,x
IIIa	51	g,z_{51}	
Kabete	51	i	1,5
Dan	51	k	e,n,z_{15}
IIIb	51	k	z_{35}
Harcourt	51	l,v	1,2
Overschie	51	l,v	1,5
Dadzie	51	l,v	e,n,x
IIIb	51	l,v	z
Moundou	51	l,z_{28}	1,5
II	51	l,z_{28}	z_6
II	51	l,z_{28}	z_{39}
Lutetia	51	r,i	l,z_{13},z_{28}
Antsalova	51	z	1,5
Treforest	1,51	z	1,6
Lechler	51	z	e,n,z_{15}
IIIa	51	z_4,z_{23}	
IV	51	z_4,z_{23}	
IIIa	51	z_4,z_{24}	
IIIa	51	z_4,z_{32}	
Bergues	51	z_{10}	1,5
II	51	z_{29}	e,n,x,z_{15}
II	51		1,7
Group O:52			
Uithof	52	a	1,5
Ord	52	a	e,n,z_{15}
Molesey	52	b	1,5
Flottbek	52	b	e,n,x
II	52	c	k
Utrecht	52	d	1,5
II	52	d	e,n,x,z_{15}
II	52	d	z_{39}
Butare	52	e,h	1,6
Derkle	52	e,h	1,7
Saintemarie	52	g,t	
II	52	g,t	
Bordeaux	52	k	1,5
IIIb	52	k	z_{35}
IIIb	52	(k)	z_{35}
IIIb	52	k	z_{53}
IIIb	52	l,v	z_{53}
II	52	z	z_{39}
IIIb	52	z	z_{52}
II	52	z_{39}	1,5,7
II	52	z_{44}	1,5,7
Group O:53			
II	53	c	1,5
II	53	d	1,5
II	1,53	d	z_{39}
II	53	d	z_{42}
IIIa	53	g,z_{51}	
IV	1,53	g,z_{51}	
IIIb	53	i	z
IIIb	53	k	e,n,x,z_{15}
IIIb	53	k	z
IIIb	53	(k)	z
IIIb	53	(k)	z_{35}
IIIb	53	k	z_{53}
IIIb	53	l,v	e,n,x,z_{15}
IIIb	53	l,v	z
IIIb	53	l,v	z_{35}
II	53	l,z_{28}	e,n,x
II	53	l,z_{28}	z_6
II	53	l,z_{28}	z_{39}
IIIb	53	r	z
IIIb	53	r	z_{35}
IIIb	53	r	z_{68}
II	53	z	1,5
IIIb	53	z	1,5,(7)
II	53	z	z_6
IIIa	53	z_4,z_{23}	
IV	53	z_4,z_{23}	
IIIa	53	z_4,z_{23},z_{32}	
II	53	z_4,z_{24}	
IIIa	53	z_4,z_{24}	
IIIb	53	z_{10}	z
IIIb	53	z_{10}	z_{35}
IIIa	53	z_{29}	
IV	1,53	z_{36},z_{38}	
IIIb	53	z_{52}	z_{35}
IIIb	53	z_{52}	z_{53}
Leda	53		1,6
Group O:54			
Tonev	21,54	b	e,n,x
Winnipeg	54	e,h	1,5
Rossleben	3,54	e,h	1,6
Borreze	54	f,g,s	
Uccle	3,54	g,s,t	
Newholland	4,12,54	m,t	
Poeseldorf	8,20,54	i	z_6
Ochsenwerder	6,7,54	k	1,5
Czernyring	54	r	1,5
Steinwerder	3,15,54	y	1,5
Yerba	54	z_4,z_{23}	
Canton	54	z_{10}	e,n,x
Barry	54	z_{10}	e,n,z_{15}
Group O:55			
II	55	k	z_{39}

(continued)

TABLE BXII.γ.262. *(cont.)*

Type[c]	Somatic (O) antigen	Phase 1	Phase 2
Group O:56			
II	56	b	
II	56	d	
II	56	e,n,x	1,7
II	56	l,v	z_{39}
II	56	l,z_{28}	
II	56	z	z_6
IIIa	56	z_4,z_{23}	
IIIa	56	z_4,z_{23},z_{32}	
II	56	z_{10}	e,n,x
IIIa	56	z_{29}	
Group O:57			
Antonio	57	a	z_6
II	57	a	z_{42}
Maryland	57	b	1,7
Batonrouge	57	b	e,n,z_{15}
IIIb	57	c	e,n,x,z_{15}
IIIb	57	c	$z:[z_{60}]$
II	57	d	1,5
II	57	g,[m],s,t	z_{42}
II	57	g,t	
IIIb	57	i	e,n,x,z_{15}
IIIb	57	i	z
IIIb	57	k	e,n,x,z_{15}
IV	57	z_4,z_{23}	
IIIb	57	z_{10}	z
II	57	z_{29}	z_{42}
II	57	z_{39}	e,n,x,z_{15}
II	57	z_{42}	$1,6:z_{53}$
Group O:58			
II	58	a	z_6
II	58	b	1,5
II	58	c	z_6
II	58	d	z_6
IIIb	58	i	e,n,x,z_{15}
IIIb	58	k	z
IIIb	58	l,v	e,n,x,z_{15}
IIIb	58	l,v	z_{35}
II	58	l,z_{13},z_{28}	1,5
II	58	l,z_{13},z_{28}	z_6
IIIb	58	r	e,n,x,z_{15}
IIIb	58	r	z
IIIb	58	r	$z_{53}:[z_{57}]$
II	58	z_6	1,6
II	58	z_{10}	1,6
IIIb	58	z_{10}	e,n,x,z_{15}
II	58	z_{10}	z_6
IIIb	58	z_{10}	z_{53}
II	58	z_{39}	e,n,x,z_{15}
IIIb	58	z_{52}	z
IIIb	58	z_{52}	z_{35}
Group O:59			
IIIb	59	c	e,n,x,z_{15}
IIIb	59	i	e,n,x,z_{15}
IIIb	59	i	z
IIIb	59	i	z_{35}
IIIb	59	(k)	e,n,x,z_{15}
II	59	k	(z)
IIIb	59	(k)	z
IIIb	59	(k)	z_{35}
IIIb	59	k	z_{53}
IIIb	59	l,v	z
IIIb	59	l,v	z_{53}
IIIb	59	r	z_{35}
II	1,59	z	z_6
IIIa	59	z_4,z_{23}	
IIIb	59	z_{10}	z_{53}
IIIb	59	z_{10}	z_{57}
IIIa	59	z_{29}	
IIIa	59	z_{36}	
IIIb	59	z_{52}	z_{53}
Group O:60			
II	60	b	
II	60	g,m,t	z_6
IIIb	60	i	e,n,x,z_{15}
IIIb	60	i	z
IIIb	60	i	z_{35}
IIIb	60	k	z
IIIb	60	k	z_{35}
IIIb	60	(k)	z_{53}
IIIb	60	l,v	z
IIIb	60	r	e,n,x,z_{15}
IIIb	60	r	z
IIIb	60	r	z_{35}
IIIb	60	r	z_{53}
II	60	z	e,n,x
IIIb	60	z_{10}	z
IIIb	60	z_{10}	z_{35}
IIIb	60	z_{10}	z_{53}
II	60	z_{29}	e,n,x
V	60	z_{41}	
IIIb	60	z_{52}	1,5,[7]
IIIb	60	z_{52}	z
IIIb	60	z_{52}	z_{35}
IIIb	60	z_{52}	z_{53}
Group O:61			
IIIb	61	c	1,5,(7)
IIIb	61	c	z_{35}
IIIb	61	i	e,n,x,z_{15}
IIIb	61	i	z
IIIb	61	i	z_{35}
IIIb	61	i	z_{53}
IIIb	61	k	1,5,(7)
IIIb	61	k	z_{35}
IIIb	61	(k)	z_{53}
IIIb	61	l,v	$1,5,7:[z_{57}]$
IIIb	61	l,v	z
IIIb	61	l,v	z_{35}
IIIb	61	r	1,5,7
IIIb	61	r	z
IIIb	61	r	z_{35}
IIIb	61	r	z_{53}
IIIb	61	z_{10}	z_{35}
V	61	z_{35}	
IIIb	61	z_{52}	1,5,7
IIIb	61	z_{52}	z
IIIb	61	z_{52}	z_{35}
IIIb	61	z_{52}	z_{53}
Group O:62			
IIIa	62	g,z_{51}	
IIIa	62	z_4,z_{23}	
IIIa	62	z_4,z_{32}	
IIIa	62	z_{29}	
IIIa	62	z_{36}	
Group O:63			
IIIa	63	g,z_{51}	
IIIa	63	z_4,z_{23}	
IIIa	63	z_4,z_{32}	
IIIa	63	z_{36}	
Group O:65			
IIIb	65	c	1,5,7
IIIb	65	c	z
IIIb	65	c	z_{53}
II	65	g,t	
IIIb	65	i	e,n,x,z_{15}
IIIb	65	(k)	z

(continued)

TABLE BXII.γ.262. *(cont.)*

Type[c]	Somatic (O) antigen	Flagellar (H) antigen	
		Phase 1	Phase 2
IIIb	65	(k)	z_{35}
IIIb	65	(k)	z_{53}
IIIb	65	l,v	e,n,x,z_{15}
IIIb	65	l,v	z
IIIb	65	l,v	z_{35}
IIIb	65	l,v	z_{53}
IIIb	65	r	z_{35}
IIIb	65	z_{10}	e,n,x,z_{15}
IIIb	65	z_{10}	z
IIIb	65	z_{52}	e,n,x,z_{15}
IIIb	65	z_{52}	z
IIIb	65	z_{52}	z_{35}
IIIb	65	z_{52}	z_{53}
II	65		1,6
Group O:66			
V	66	z_{35}	
V	66	z_{39}	
V	66	z_{41}	
V	66	z_{65}	
V	66	z_{81}	
Group O:67			
Crossness	67	r	1,2

[a]Symbols: [ ], O (not underlined) or H factor that may be present or absent without relation to phage conversion, i.e., factor [5] of O:4 (B) group. When H factors are in square brackets, this means that they are exceptionally found in wild strains. For example, most strains of Paratyphi A possess the monophasic antigen phase 1 H:a. In rare cases, diphasic strains with H:1,5 as H antigen phase 2 may be isolated. For this reason, [1,5] is mentioned in square brackets in the formula of this serovar. ( ), O or H factor weakly agglutinable. The factor (k) corresponding to Arizona factor H:22 is weakly agglutinable by k standard serum, but is normally agglutinable by k polyvalent serum. Symbols for somatic factors determined by phage conversion are underlined (example 6, 14,18). They are present only if the culture is lysogenized by the corresponding converting phage. These factors are usually added to the factors present in nonconverted strain (for example 6,7 →6,7,14). In 0:3,10 group, factors 15 or 15, 34 take the place of factor 10. For this reason, these factors are underlined and quoted into square brackets in this group. These underlined factors are mentioned in the table for serovars in which they were found. It is likely that this situation may be encountered for all serovars in the same group O.

[b]The subfactors of O factors—40, 47, 48, and 50—are no longer mentioned, as their identification is unnecessary in current practice. O and H factors having the same symbol in the Kauffmann–White scheme are always related, but not always identical in different serovars. Table BXII.γ.262 of the antigenic formulae of the *Salmonella* is a scheme established for a diagnostic purpose. Details unnecessary for the diagnosis of serovars are not given in this scheme (e.g., R-phase of the H antigen are not indicated).

[c]Names of the serovars of *S. enterica* subsp. *enterica*. For other subspecies of *S. enterica*, the subspecies to which the serovars belong are indicated by the following symbols: II, serovars of *S. enterica* subsp. *salamae*; IIIa, serovars of *S. enterica* subsp. *arizonae*; IIIb, serovars of *S. enterica* subsp. *diarizonae*; IV, serovars of *S. enterica* subsp. *houtenae*; VI, serovars of *S. enterica* subsp. *indica*; V, serovars of *S. bongori* (the symbol "V" was retained to avoid confusion with serovar name of *S. enterica* subsp. *enterica*).

[d]Strains of this group maybe lysogenized by phage 14 (O:6,7, → O:6,7,14). The strains possessing O:6,7,14 had been classified in a special group, C4. They are now classified into group C1. Names formerly given to serovars of group C4 are deleted.

[e]Groups O:6,8(C2) and O:8(C3) differentiated only by presence or absence of O:6 factor, were lumped together in a single group O:8.

[f]Strains of this group maybe lysogenized by phage e15 (O:3,10→ O:3,15), then by phage e34 (O:3,15→ O:3,15,34). In these cases, factors O:15 or O:15,34 take the place of factor O:10 which is no longer agglutinable. The strains of O structure O:3,15 were formerly classified in a special group, E2, and the strains of O structure O:3,15,34 in another group, E3. They are now classified with the strains O:3,10 into group E1. Factors O:15 and O:15,34 are given in square brackets when they have been found in wild strains.

[g]Groups formerly called O:13,22(G1) and O:13,23 (G2) were lumped together in a single group O:13.

Genetics Analysis of 16S rRNA gene sequences indicates that salmonellae belong to the *Gammaproteobacteria* (Chang et al., 1997). *S. enterica* and *S. bongori* (Table BXII.γ.261) were further separated by 16S rDNA analysis and found to be closely related to the *E. coli* and *Shigella* complex by both 16S and 23S rDNA analyses (Christensen et al., 1998).

A total of 1160 genes are located on the genetic map of Typhimurium strain LT2: 1081 are on the circular chromosome, 29 on the 90-kb virulence-associated plasmid, and 50 are not yet mapped (see review by Sanderson et al., 1996). The linkage map of serovar Typhimurium was first determined by F-mediated conjugation. Hfr strains may be selected after F plasmid transfer. Conjugative chromosomal transfer may occur from *Salmonella* to *E. coli*, from *E. coli* to *Salmonella*, and from one serovar of *Salmonella* to another. Chromosomal genes responsible for O, Vi, and H antigens can be transferred from one genus to the other (Iino and Lederberg, 1964). Crosses may be used to localize the regions of the bacterial chromosome that specify avirulence for mice (Krishnapillai and Baron, 1964) or to study the role of O antigen factors in the virulence of *Salmonella* (Mäkelä et al., 1973).

The physical map of the chromosome of several *Salmonella* serovars is now available. Comparisons of these maps show that serovar Typhimurium (Liu and Sanderson, 1992; Liu et al., 1993b), serovar Enteritidis (Liu et al., 1993a), serovar Paratyphi B (Liu et al., 1994a), serovar Paratyphi A (Liu and Sanderson, 1995a), and serovar Typhi (Liu and Sanderson, 1995b) share a common basic genomic structure: genome sizes are all between 4600 and 4800 kb, the order of genes on chromosome segments is usually the same, and all have seven *rrn* operons. Relative to other *Salmonella* serovars, the genomic cleavage map of serovar Paratyphi A shows an insertion of about 100 kb between *rrnH-G* and *proB*, and an inversion of half the genome between *rrnH* and *rrnG*. The chromosome of serovar Typhi has undergone major genetic rearrangements (Liu and Sanderson, 1995b), including (1) homologous recombination between the seven *rrn* operons; (2) inversion that covers the replication terminus region; and (3) at least three insertions, one of which is up to 118 kb long and contains the *viaB* locus for Vi antigen biosynthesis.

As with other *Enterobacteriaceae*, salmonellae may harbor "foreign" replicons (such as temperate phages and plasmids) that may code for virulence determinants, for antibiotic resistance, for antigenic changes of O antigen, and for metabolic characteristics commonly used in diagnostic identification, e.g., lactose or sucrose fermentation (Le Minor et al., 1973, 1974). Thus it is unwise to exclude *Salmonella* solely based on a positive lactose or sucrose reaction. It is also more difficult to identify salmonellae when a pleiotrophic mutation occurs, such as one that simultaneously affects nitrate, tetrathionate, and thiosulfate reductase as well as hydrogenlyase (Le Minor, et al., 1969).

About 5% of *Salmonella* strains produce bacteriocins active against *E. coli*, *Shigella*, and/or *Salmonella* (Fredericq, 1948). Most of these bacteriocins adsorb to the same receptor as that for colicins B, E1, E2, or I. *Salmonella* bacteriocins differ from colicins *sensu stricto* by their activity spectra on colicin indicator strains. Some of these *Salmonella* bacteriocins are not even active against colicin indicator strains but are active against *Salmonella* strains only (Hamon and Péron, 1966).

Susceptibility to the O1 phage Most strains of the genus *Salmonella* are susceptible to the O1 phage of Felix and Callow (1943). This phage is highly specific for *Salmonella*, lysing more than 98% of the strains studied in routine *Salmonella* diagnosis (Kallings, 1967). Susceptibility of *Salmonella* species and subspecies to phage O1 is reported in Table BXII.γ.261. Mutations conferring resistance to O1 phage have been studied by Lindberg and Holme (1969), MacPhee et al. (1975), and Hudson et al. (1978).

A *Salmonella* phage that infects only flagellated bacteria was isolated by Sertic and Boulgakov (1936). Sensitivity to this phage depends on the antigenic specificity of the H antigen. For example, bacteria with antigens of the "G" complex are resistant (Meynell, 1961).

Pathogenicity *Salmonella* serovars may be strictly adapted to one particular host (these serovars are auxotrophic), may be ubiquitous (found in a large number of animal species), or may be of still unknown pathogenicity.

Serovars adapted to humans (e.g., serovar Typhi, serovar Paratyphi A, and serovar Sendai) usually cause severe diseases with septicemia-typhoidic syndrome. They are not naturally pathogenic for other animal species. Salmonellosis is transmitted from person to person, without an intermediate host, through fecal contamination of water and food. The incidence is higher in developing countries with poor hygiene. Typhoid fever, a systemic infection caused by serovar Typhi, remains a major public concern. The World Health Organization has estimated that there are more than 16.6 million typhoid cases per year worldwide, causing 600,000 deaths yearly (Ivanoff and Levine, 1997).

Other serovars are adapted to one animal species: serovar Abortusovis is adapted to sheep and is a major cause of abortion in ewes; serovar Typhisuis and serovar Gallinarum are adapted to swine and poultry, respectively.

Ubiquitous *Salmonella* serovars (e.g., serovar Typhimurium) are mainly responsible for food-borne infections. Low infective doses ($<10^3$) are sufficient to cause clinical symptoms (Blaser and Newman, 1982). Salmonellosis of newborns and infants (who are more susceptible to infections than adults) presents diverse clinical symptoms, from a grave typhoid-like illness with septicemia to a mild or asymptomatic infection. In pediatric wards, the infection is usually transmitted by the hands of personnel.

The entrance of a serovar into a food chain may be the origin of its importation into a country. For example, many countries have become infected with serovar Hadar introduced by imported turkeys, or by serovar Enteritidis introduced by imported poultry and eggs.

After recovery from a clinical case of salmonellosis, some patients, although asymptomatic, remain carriers for weeks, months, or years (i.e., continue to eliminate salmonellae in feces). Carriage contributes to the dissemination of salmonellosis, especially if the diagnosis of the carrier state is not monitored by periodic stool cultures. Antibiotics that are active in curing the disease are usually ineffective in the treatment of the carrier state.

Strains of *Salmonella* from urine are often of the R form. Bilharziosis, a parasitic infection caused by *Schistosoma*, has to be controlled in *Salmonella* carriers (Lo Verde et al., 1980). Sickle-cell anemia must be suspected in cases of osteomyelitis due to *Salmonella* in black children (Vandepitte et al., 1953).

The first step in *Salmonella* pathogenicity is the invasion of the small bowel mucosa. Electron microscopic studies showed that *Salmonella* rapidly adhered to and entered M cells of the follicle-associated epithelium and subsequently invaded absorptive enterocytes. This essential feature of *Salmonella* pathogenesis may be conceptualized as a two-step process. Bacteria may adhere to the target cell first, thereupon inducing an endocytic event that results in internalization of *Salmonella*. The intimate interaction between the bacterium and the host cell appears to trigger cytoskeletal rearrangements and membrane ruffling of the epithelial cell, which ultimately leads to bacterial uptake by macropinocytosis. Complexity of the mechanisms governing *Salmonella* entry into epithelial cells is reflected by the large number of loci involved in this process. Most of them are clustered at centisome 63 of the *Salmonella* chromosome (reviewed by Galan, 1996a, 1996b). In the second step of *Salmonella* pathogenesis, bacteria gain access to the mesenteric lymph nodes, drain through the lymphatics to the thoracic duct into the blood, and ultimately infect the liver and spleen. It is now well established that pathogenic serovars of *Salmonella* other than serovar Typhi contain a plasmid that is essential for systemic dissemination in the appropriate host. Although the role of this virulence-associated plasmid is not fully understood, it is currently thought that plasmid products enhance bacterial growth within the reticuloendothelial system of the host (reviewed by Gulig et al., 1993). In human beings, systemic dissemination of serovar Typhi in the reticuloendothelial system requires expression of the Vi antigen. It was shown that the presence of Vi antigen was associated with resistance to the bactericidal effect of serum, resistance to activation of complement by the alternative pathway, resistance to opsonization by inhibition of C3b binding to bacteria, and resistance to postphagocytic oxidative burst (reviewed by Virlogeux-Payant and Popoff, 1996).

Antibiotic and drug sensitivity Similarly to *E. coli*, *Salmonella* strains can readily acquire plasmids that contain genes conferring resistance to antibiotics. Multiple resistance is selected for when antibiotics are used extensively in hospitals or added to animal feed. The same plasmids may be found in strains of human or animal origin (Anderson et al., 1975). Since around 1990, strains of serovar Typhi have emerged that are resistant to most previously useful oral antibiotics. The antimicrobials that remain effective are relatively expensive (e.g., fluoroquinolones and ceftriaxone) and some must be administered parentally (e.g., ceftriaxone), thereby posing a quandary for developing countries (reviewed by Ivanoff and Levine, 1997). Similarly, multiple antibiotic-resistant strains of serovar Typhimurium definitive type (DT) 104 have emerged recently (Wall et al., 1994; Glynn et al., 1998; Poppe et al., 1998). They are pathogenic for humans and animals, particularly cattle. The DT104 strains were isolated in the United States, Canada, and Europe. More prudent use of antimicrobial agents in farm animals and more effective disease prevention on farms are necessary to reduce the dissemination of this multi-drug-resistant pathogen (Glynn et al., 1998).

Ecology Although some *Salmonella* serovars are strictly host adapted, the majority have a wide host range (e.g., serovar Typhimurium). Some are localized in a particular region of the globe (e.g., serovar Sendai in the Far East, serovar Berta in North America), but others are ubiquitous (e.g., serovar Typhimurium). Strains belonging to *S. enterica* subsp. *salamae*, subsp. *arizonae*, and subsp. *diarizonae* are frequently isolated from the intestinal contents of cold-blooded animals and only rarely from human beings and warm-blooded animals. Strains of subsp. *houtenae* and *S. bongori* are isolated chiefly from the environment and are rarely pathogenic for humans.

ENRICHMENT AND ISOLATION PROCEDURES

Isolation from blood is done according to the classical method for hemoculture. A biphasic culture bottle containing a vertical agar layer along one side and a broth medium at the bottom (Castaneda, 1947; Hall et al., 1979; Krieg and Gerhardt, 1981) prepared with tryptic soy agar/broth containing 2% sodium citrate is convenient. Isolated colonies grow on the agar layer. Identification is usually done by (a) diagnosis of the family *Enterobacteriaceae*, (b) diagnosis of the genus *Salmonella* (diagnosis of the subspecies for strains isolated from blood cultures is not routinely necessary, because almost all blood isolates belong to subsp. *enterica*), (c) diagnosis of the serovar, (d) determination of the antibiotic susceptibility pattern, and (e) further study of the biovar and phagovar if indicated.

Selective procedures are needed for the isolation of *Salmonella* from specimens containing mixed bacterial flora (fecal samples, autopsy samples, food, environmental samples, etc.). Enrichment (i.e., an increased ratio of *Salmonella* cells to other bacterial cells during incubation) is obtained using liquid nutrient media containing selective agents that inhibit or retard growth of bacteria other than *Salmonella*. Use of enrichment media is essential when the number of salmonellae in a sample is very low, i.e., when the probability of finding colonies by direct isolation is low. Three media may be recommended for general use: (a) the tetrathionate medium of Muller (1923); (b) Muller's medium modified by Kauffmann (1935) by addition of bile and brilliant green; and (c) selenite F broth devised by Leifson (1936). Tetrathionate and selenite broth are suitable for all *Salmonella* serovars. Tetrathionate bile brilliant-green medium is suitable for all serovars, except host-adapted serovars such as serovar Typhi. Enrichment media should be heavily inoculated, e.g., 0.5 ml of fecal suspension per 10 ml of medium. After incubation for 18 h at 37°C, a loopful of enrichment culture is streaked onto agar plating medium.

The same enrichment media may be used for detection of salmonellae in water. The simplest method is to add one volume of the water sample to an equal volume of double-strength medium. For detecting salmonellae in food, a generally suitable procedure is to inoculate 25 g of the suspected food into 225 ml of selenite F broth, incubate for 24 h, and isolate on selective agar media. In the case of a dehydrated food, nutrient broth containing the sample is incubated overnight before inoculation of enrichment media.

Agar media are used for isolation of salmonellae. Streaking a loopful of enrichment culture or a suspension of the sample (e.g., stool) should be done carefully to obtain the greatest number of perfectly isolated colonies. Because the most discriminating character is lactose fermentation, the majority of media for isolation contain lactose and a pH indicator. In addition, the media contain selective agents to inhibit the growth of non-*Salmonella* organisms and the swarming of *Proteus mirabilis* and *P. vulgaris*. Some media also contain ferrous citrate detection of H2S-producing bacteria.

Examples of media of moderate selectivity are (a) MacConkey agar, which contains lactose, neutral red, and the selective inhibitors crystal violet and bile salts; lactose-positive colonies are red, lactose-negative colonies are colorless; and (b) desoxycholate citrate agar, which contains lactose, neutral red, and the selective agent desoxycholate; ferric ammonium citrate is included as an indicator of H2S production; lactose-positive colonies are red; lactose-negative colonies are colorless; if H2S is produced, the inner part of the colony is black.

Examples of media of higher selectivity are the following: (a) Salmonella–Shigella (SS) agar, which contains lactose, neutral red, and the selective agents brilliant green and bile salts. Ferric citrate is an indicator of H2S production. The appearance of colonies is the same as on desoxycholate citrate agar. (b) Brilliant green agar, which contains lactose, phenol red, and the selective agent brilliant green. This medium is easy to prepare and is suitable for all salmonellae except host-adapted serovars. It is not suitable for shigellae. Lactose-positive colonies are green, lactose-negative colonies are pink. All of the above-mentioned media were reviewed by Kauffmann (1966) and Edwards and Ewing (1972). (c) Hektoen enteric medium (King and Metzger, 1968), which contains lactose, sucrose and salicin, a mixture of bromothymol blue and Andrade's pH indicator, ferric citrate to detect H2S production, and sodium desoxycholate as a selective inhibitor. Colonies that do not ferment any of the three sugars (e.g., *Salmonella*) are blue-green, with a black center if H2S is produced. Colonies fermenting one or more of the sugars (e.g., *E. coli, Enterobacter cloacae*) are salmon-colored. This medium is suitable for all *Salmonella* serovars and for shigellae.

A general procedure for the detection of salmonellae in feces or food is as follows. A suspension of the sample in saline is streaked onto the chosen isolation medium and also inoculated into an enrichment broth. After overnight incubation, the plating medium is examined for suspect colonies (lactose negative, H2S positive or negative); also, a loopful of the enrichment culture is streaked onto another plate of selective agar medium. After overnight incubation, this plate is also examined for suspect colonies. A quick screening of several suspect colonies is done by inoculating each into a few drops of urea medium and incubating at 37°C for 2 h. Biochemical characterization is continued only for urease-negative colonies (urease-positive colonies growing at 18 h are likely to be *Proteus*) by inoculating, e.g., triple sugar–iron (TSI) or Kligler-Hajna medium. *Salmonella* must be differentiated mainly from *Citrobacter freundii, P. mirabilis*, and, in food bacteriology, *Alteromonas putrefaciens*. To detect *Salmonella* belonging to subsp. *arizonae* and serovar Diarizonae, attention should be given to lactose-positive, H2S-positive colonies on plating media.

MAINTENANCE PROCEDURES

Salmonella cultures remain viable for many years when stored on peptone agar (meat extract, 5.0 g; peptone, 10.0 g; NaCl, 3.0 g; Na2HPO4·12H2O, 2.0 g; agar, 10.0 g; distilled water, 1000 ml; pH 7.4) and distributed into small, tightly stoppered, screw-capped tubes. This medium is stab-inoculated and kept in the dark at room temperature. Lyophilization and freezing at or below −70°C give very good results. For lyophilization, it is necessary to isolate each subculture and to select a colony with the desired serologic characteristics.

DIFFERENTIATION OF THE GENUS *SALMONELLA* FROM OTHER GENERA

Characteristics useful for differentiating the genus *Salmonella* from other *Enterobacteriaceae* are given in Tables BXII.γ.193, BXII.γ.194, and BXII.γ.196 of the chapter on the family *Enterobacteriaceae*.

TAXONOMIC COMMENTS

The species concept in the genus *Salmonella* has evolved in several overlapping phases. The typhoid bacillus was first observed by Eberth (1880) in spleen sections and mesenteric lymph nodes from a patient who died from typhoid fever. During the following decade, other bacteria were isolated from clinical cases of typhoid fever, but they were distinct from the typhoid bacillus both culturally and serologically. The genus *Salmonella* was created by Lignières (1900). Hence, *Salmonella* strains were considered different species when isolated from different conditions or different hosts. However, it was soon realized that a number of these so-called species were ubiquitous, and emphasis shifted to antigenic properties.

Serological analysis of *Salmonella* O and H antigens was initiated by White (1926). This work was extended by Kauffmann (1961), who defined species as "a group of related sero-fermentative phage types". Consequently, each *Salmonella* serovar was considered as a species. But, as most serovars could not be distinguished by biochemical tests, Kauffmann (1966) divided the genus *Salmonella* into four subgenera (I–IV) and continued to apply his concept of "one serovar–one species". Nevertheless, phenotypic studies and numerical taxonomy showed that, apart from host-adapted serovars, *Salmonella* serovars within each subgenus were biochemically indistinguishable (reviewed by Le Minor and Popoff, 1987). For this reason, Kauffmann's subgenera were considered to be species (Le Minor et al., 1970b).

Later, DNA relatedness studies demonstrated that all *Salmonella* serovars formed a single DNA hybridization group with seven subgroups delineated by studies of the thermal stability of hybrids (Crosa et al., 1973a; Stoleru et al., 1976; Le Minor et al., 1986). The seven DNA subgroups could be differentiated by using biochemical characteristics. Correlation between DNA subgroups and Kauffmann's subgenera was close except that subgenus III was split into two DNA subgroups containing monophasic and diphasic serovars, respectively.

Based on these numerical taxonomy and DNA relatedness studies, it was proposed that the genus *Salmonella* should consist of a single species that could be divided into seven subspecies (I–VII), although subspecies V might possibly represent a second species (Le Minor et al., 1982, 1986). The strict application of the Bacteriological Code (Rules Revision Committee, Judicial Commission, International Committee on Systematic Bacteriology, 1985) led Le Minor et al. (1982, 1986) to propose *S. choleraesuis* as the species name, having seven subspecies designated subsp. *choleraesuis*, subsp. *salamae*, subsp. *arizonae*, subsp. *diarizonae*, subsp. *houtenae*, subsp. *bongori*, and subsp. *indica*. Subspecies names were validated by announcement in the *International Journal of Systematic Bacteriology* (1985, 1987). However, the name *S.*

choleraesuis could be considered as an ambiguous name since this name was used for the serovar possessing the antigenic formula 6,7:c:1,5 (Table BXII.γ.262). To avoid further confusion, it was proposed that the type species of the genus *Salmonella* be *S. enterica* (Le Minor and Popoff, 1987). This name, first proposed by Kauffmann and Edwards (1952), received unanimous support from the members of the Subcommittee of *Enterobacteriaceae* of the International Committee on Systematic Bacteriology at the XIV International Congress of Microbiology (Manchester, U.K., 1986). However, the request to change the name of the type species to *S. enterica* was not approved by the Judicial Commission. A second request has been made and, in the interim, the WHO collaborating center for reference and research on *Salmonella* and national *Salmonella* reference laboratories are using *S. enterica*.

Finally, by using multilocus enzyme analysis, it was shown that *S. enterica* subsp. *bongori* had evolved considerably from other subspecies (Reeves et al., 1989a). Based on this divergence and on DNA relatedness data (Le Minor et al., 1982; 1986), Reeves et al. (1989a) proposed that *S. enterica* subsp. *bongori* be elevated to the level of species in the new combination *S. bongori*.

The practice of giving names to the serovars of *S. enterica* subsp. *enterica* should continue since the diagnostic use of the Kauffmann–White scheme (Table BXII.γ.262) is overridingly important and since these names are very familiar to microbiologists and physicians. However, these names must no longer be italicized and the first letter must be a capital. Serovars of the other subspecies of *S. enterica* and those of *S. bongori* should be designated only by their antigenic formulae.

DIFFERENTIATION OF THE SPECIES OF THE GENUS *SALMONELLA*

Table BXII.γ.261 presents the biochemical characteristics differentiating the species and subspecies of the genus *Salmonella*. The antigenic formulae of *Salmonella* serovars (i.e., the Kauffmann–

White scheme) is given in Table BXII.γ.262. An alphabetical listing of serovars indicating the O groups is presented in Table BXII.γ.263.

List of species of the genus Salmonella

1. **Salmonella enterica** (ex Kauffmann and Edwards 1952) Le Minor and Popoff 1987, 466.
   en.te.ri' ca. Gr. n. *enteron* gut; L. adj. *enterica* of the gut.

   The description of *S. enterica* is that of *S. choleraesuis sensu* (Le Minor et al., 1982). The characteristics are as described for the genus (see Tables BXII.γ.193, BXII.γ.194, and BXII.γ.196 of the chapter on the family *Enterobacteriaceae* and as listed in Table BXII.γ.261. *S. enterica* can be divided into six subspecies (Le Minor et al., 1982, 1986; Reeves et al. 1989a).

   The type strain carries a virulence-associated plasmid of about 90 kb. Antigenic formula 4,5,12:i:1,2 (serovar Typhimurium).

   The mol% G + C of the DNA is: 50–53 (hydrolysis and chromatography, Bd, T_m).

   Type strain: LT2, ATCC 43971, CIP 60.62, NCIB 11450.

   a. **Salmonella enterica** subsp. **enterica** (ex Kauffmann and Edwards 1952) Le Minor and Popoff 1987, 467.

      Distinguished from other *S. enterica* subspecies by the characteristics reported in Table BXII.γ.261. Contains at least 1443 serovars (Popoff et al., 1998). Isolated mainly from humans and warm-blooded animals.

      A list of selected serovars belonging to *S. enterica* subsp. *enterica* is given below.

      The mol% G + C of the DNA is: see species description.
      Type strain: LT2, ATCC 43971, CIP 60.62, NCIB 11450.

   i. **Salmonella enterica** subsp. **enterica** *serovar* **Choleraesuis** (Smith 1894) Weldin 1927, 155 (*Bacillus cholerae suis* Smith 1894, 9.)
      chol.er.ae.su' is. Gr. n. *cholera* cholera; L. n. *sus* hog. M.L. gen. n. *suis* of a hog; M.L. gen. n. *choleraesuis* of hog cholera.

      Antigenic formula: 6,7:c:1,5. The detailed O antigen formula is normally 6_2,7, but this may be transformed by lysogenization into 6_1,7 or 6_2,7,14. Arabinose and trehalose are not fermented; dulcitol is slowly and irregularly fermented. Those strains that produce H_2S are designated as serovar Choleraesuis biovar Kunzendorf. Pathogenic for humans and animals.

      The mol% G + C of the DNA is: see species description.

      Deposited strain: ATCC 13312, NCTC 5735.

   ii. **Salmonella enterica** subsp. **enterica** *serovar* **Enteritidis** (Gaertner 1888) Castellani and Chalmers 1919, 939 (*Bacterium enteritidis* Gaertner 1888, 573.)
      en.te.ri'ti.dis. Gr. n. *enteron* gut, intestine; M.L. n. *enteritis* enteritis, inflammation of the intestine; M.L. gen. n. *enteritidis* of enteritis.

      Antigenic formula: 1,9,12:g,m: –. The presence of factor O1 is connected with lysogenization. Ubiquitous and frequently the cause of infections in humans and

TABLE BXII.γ.263. Alphabetical list of names of *Salmonella* serovars, indicating the O groups

Serovar	O Group
Aachen	J
Aarhus	K
Aba	C2–C3
Abadina	M
Abaetetuba	F
Aberdeen	F
Abidjan	Q
Ablogame	I
Abobo	I
Abony	B
Abortusequi	B
Abortusovis	B
Abuja	F
Accra	E4
Ackwepe	D2
Adabraka	E1
Adamstown	M
Adamstua	F
Adana	U
Adelaide	O
Adeoyo	I
Aderike	M
Adime	C1
Adjame	G
Aequatoria	C1
Aesch	C2–C3
Aflao	H
Africana	B
Afula	C1
Agama	B
Agbara	I
Agbeni	G
Agege	E1
Ago	N
Agodi	O
Agona	B
Agoueve	G
Ahanou	J
Ahepe	U
Ahmadi	E4
Ahoutoue	S
Ahuza	U
Ajiobo	G
Akanji	C2–C3
Akuafo	I
Alabama	D1
Alachua	O
Alagbon	C2–C3
Alamo	C1
Albany	C2–C3
Albert	B
Albuquerque	H
Alexanderplatz	X
Alexanderpolder	C2–C3
Alfort	E1
Alger	P
Alkmaar	E4
Allandale	R
Allerton	E1
Alma	Q
Alminko	C2–C3
Altendorf	B
Altona	C2–C3
Amager	E1
Amba	F
Amersfoort	C1
Amherstiana	C2–C3
Amina	I
Aminatu	E1
Amounderness	E1

(continued)

TABLE BXII.γ.263. *(cont.)*

Serovar	O Group
Amoutive	M
Amsterdam	E1
Amunigun	I
Anatum	E1
Anderlecht	E1
Anecho	O
Anfo	Q
Angers	C2–C3
Angoda	N
Angouleme	I
Ank	M
Anna	G
Annedal	I
Antarctica	D1
Antonio	57
Antsalova	51
Antwerpen	T
Apapa	W
Apeyeme	C2–C3
Aprad	W
Aqua	N
Aragua	N
Arapahoe	H
Arechavaleta	B
Arusha	U
Aschersleben	N
Ashanti	M
Assen	L
Assinie	E1
Asylanta	E1
Atakpame	C2–C3
Atento	F
Athinai	C1
Ati	F
Augustenborg	C1
Austin	C1
Australia	Y
Avignon	I
Avonmouth	E4
Ayinde	B
Ayton	B
Azteca	B
Babelsberg	M
Badagry	I
Baguida	L
Baguirmi	N
Bahati	G
Bahrenfeld	H
Baiboukoum	C1
Baildon	D2
Bakau	M
Balcones	W
Balili	M
Ball	B
Bama	J
Bamboye	D2
Bambylor	D2
Banalia	C2–C3
Banana	B
Banco	M
Bandia	O
Bangkok	P
Bangui	D1
Banjul	H
Bardo	C2–C3
Bareilly	C1
Bargny	C2–C3
Barmbek	I
Barranquilla	I
Barry	54
Basingstoke	D2

(continued)

TABLE BXII.γ.263. *(cont.)*

Serovar	O Group
Bassa	C2–C3
Bassadji	M
Batonrouge	57
Battle	I
Bazenheid	C2–C3
Be	C2–C3
Beaudesert	H
Bedford	E4
Belem	C2–C3
Belfast	C2–C3
Bellevue	C2–C3
Benfica	E1
Benguella	R
Benin	D2
Benue	C2–C3
Bere	X
Bergedorf	D2
Bergen	X
Bergues	51
Berkeley	U
Berlin	J
Berta	D1
Bessi	E1
Bethune	E4
Biafra	E1
Bida	E4
Bietri	N
Bignona	J
Bijlmer	R
Bilu	E4
Binche	X
Bingerville	X
Binningen	W
Birkenhead	C1
Birmingham	E1
Bispebjerg	B
Bissau	B
Blegdam	D1
Blijdorp	H
Blitta	X
Blockley	C2–C3
Bloomsbury	E1
Blukwa	K
Bobo	V
Bochum	B
Bodjonegoro	N
Boecker	H
Bofflens	S
Bokanjac	M
Bolama	V
Bolombo	E1
Bolton	E1
Bonames	J
Bonariensis	C2–C3
Bonn	C1
Bootle	X
Borbeck	G
Bordeaux	52
Borreze	54
Borromea	T
Bouake	I
Bournemouth	D1
Bousso	H
Bovismorbificans	C2–C3
Bracknell	G
Bradford	B
Braenderup	C1
Brancaster	B
Brandenburg	B
Brazil	I
Brazos	K

(continued)

TABLE BXII.γ.263. *(cont.)*

Serovar	O Group
Brazzaville	C1
Breda	C2–C3
Bredeney	B
Brefet	V
Breukelen	C2–C3
Brevik	I
Brezany	B
Brijbhumi	F
Brikama	C2–C3
Brindisi	F
Brisbane	M
Bristol	G
Brive	T
Broc	T
Bron	G
Bronx	C2–C3
Brooklyn	I
Broughton	E4
Bruck	C1
Brunei	C2–C3
Brunflo	I
Bsilla	C2–C3
Buckeye	Y
Budapest	B
Bukavu	R
Bukuru	C2–C3
Bulgaria	C2–C3
Bullbay	F
Bulovka	C1
Burgas	I
Burundi	S
Bury	B
Businga	C1
Butantan	E1
Butare	52
Buzu	H
Caen	I
Cairina	E1
Cairns	W
Calabar	E4
California	B
Camberene	O
Camberwell	D1
Campinense	D1
Canada	B
Canary	R
Cannobio	M
Cannonhill	E4
Cannstatt	E4
Canton	54
Caracas	H
Cardoner	I
Carmel	J
Carnac	K
Carno	E4
Carrau	H
Carswell	V
Casablanca	W
Casamance	R
Catalunia	M
Catanzaro	H
Cayar	C1
Cerro	K
Ceyco	D2
Chagoua	G
Chailey	C2–C3
Champaign	Q
Chandans	F
Charity	H
Charlottenburg	C2–C3
Cheltenham	D2

(continued)

TABLE BXII.γ.263. *(cont.)*

Serovar	O Group
Chester	B
Chicago	M
Chichester	E4
Chichiri	H
Chile	C1
Chincol	C2–C3
Chingola	F
Chiredzi	F
Chittagong	E4
Choleraesuis	C1
Chomedey	C2–C3
Christiansborg	V
Clackamas	B
Claibornei	D1
Clanvillian	F
Clerkenwell	E1
Cleveland	C2–C3
Cochin	D2
Cocody	C2–C3
Coeln	B
Coleypark	C1
Colindale	C1
Colobane	F
Colombo	P
Colorado	C1
Concord	C1
Congo	G
Connecticut	F
Coogee	T
Coquilhatville	E1
Coromandel	C1
Corvallis	C2–C3
Cotham	M
Cotia	K
Cremieu	C2–C3
Crewe	F
Croft	M
Crossness	67
Cubana	G
Cuckmere	E1
Cullingworth	M
Cumberland	Q
Curacao	C2–C3
Cyprus	C2–C3
Czernyring	54
Daarle	C2–C3
Dabou	C2–C3
Dadzie	51
Dahlem	Y
Dahomey	X
Dahra	J
Dakar	M
Dakota	I
Dallgow	E4
Damman	C1
Dan	51
Dapango	X
Daula	C2–C3
Daytona	C1
Deckstein	D2
Delmenhorst	K
Dembe	O
Demerara	G
Denver	C1
Derby	B
Derkle	52
Dessau	E4
Deversoir	W
Dibra	M
Dieuppeul	M
Diguel	G

TABLE BXII.γ.263. *(cont.)*

Serovar	O Group
Diogoye	C2–C3
Diourbel	L
Djakarta	Y
Djama	T
Djelfa	C2–C3
Djermaia	M
Djibouti	J
Djugu	C1
Doba	D2
Doel	M
Doncaster	C2–C3
Donna	N
Doorn	M
Douala	M
Dougi	Z
Doulassame	N
Dresden	M
Driffield	R
Drogana	B
Dublin	D1
Duesseldorf	C2–C3
Dugbe	W
Duisburg	B
Dumfries	E1
Dunkwa	C2–C3
Durban	D1
Durham	G
Duval	R
Ealing	O
Eastbourne	D1
Eastglam	E4
Eberswalde	M
Eboko	C2–C3
Ebrie	O
Echa	P
Edinburg	C1
Edmonton	C2–C3
Egusi	S
Egusitoo	T
Eingedi	C1
Eko	B
Ekotedo	D2
Ekpoui	X
Elisabethville	E1
Elokate	D1
Elomrane	D1
Emek	C2–C3
Emmastad	P
Encino	H
Enschede	O
Entebbe	B
Enteritidis	D1
Enugu	I
Epicrates	E1
Epinay	F
Eppendorf	B
Escanaba	C1
Eschberg	D1
Eschweiler	C1
Essen	B
Etterbeek	F
Euston	F
Everleigh	E1
Ezra	M
Fairfield	M
Fajara	M
Faji	T
Falkensee	E1
Fallowfield	E1
Fann	F
Fanti	G

(continued)

(continued)

TABLE BXII.γ.263. *(cont.)*

Serovar	O Group
Farakan	M
Farcha	U
Fareham	E4
Farmsen	G
Farsta	B
Fass	Z
Fayed	C2–C3
Ferlo	S
Ferruch	C2–C3
Finaghy	B
Findorff	F
Finkenwerder	H
Fischerhuette	I
Fischerkietz	H
Fischerstrasse	V
Fitzroy	Y
Florian	E1
Florida	H
Flottbek	52
Fluntern	K
Fomeco	W
Fortlamy	I
Fortune	B
Franken	D1
Frankfurt	I
Freefalls	M
Freetown	P
Freiburg	E1
Fresno	D2
Friedenau	G
Friedrichsfelde	M
Frintrop	D1
Fufu	E1
Fulda	E4
Fulica	B
Fyris	B
Gabon	C1
Gafsa	I
Galiema	C1
Galil	E1
Gallen	F
Gallinarum	D1
Gamaba	V
Gambaga	L
Gambia	O
Gaminara	I
Garba	H
Garoli	C1
Gassi	O
Gateshead	D2
Gatineau	E4
Gatow	C1
Gatuni	C2–C3
Gbadago	E1
Gdansk	C1
Gege	N
Georgia	C1
Gera	T
Geraldton	D2
Gerland	I
Ghana	L
Giessen	N
Give	E1
Giza	C2–C3
Glasgow	I
Glidji	F
Glostrup	C2–C3
Gloucester	B
Gnesta	E4
Godesberg	N
Goelzau	E1

TABLE BXII.γ.263. *(cont.)*

Serovar	O Group
Goeteborg	D1
Goettingen	D1
Gokul	51
Goldcoast	C2–C3
Goma	C1
Gombe	C1
Good	L
Gori	J
Goulfey	R
Gouloumbo	O
Goverdhan	D1
Gozo	M
Grampian	C1
Grancanaria	I
Grandhaven	N
Granlo	J
Graz	U
Greiz	R
Groenekan	K
Grumpensis	G
Guarapiranga	N
Guerin	D2
Guildford	M
Guinea	V
Gustavia	F
Gwale	T
Gwoza	E4
Haardt	C2–C3
Hadar	C2–C3
Hadejia	J
Haduna	B
Haelsingborg	C1
Haferbreite	T
Haga	O
Haifa	B
Halle	M
Hallfold	B
Handen	G
Hann	R
Hannover	I
Haouaria	G
Harburg	H
Harcourt	51
Harleystreet	E1
Harrisonburg	E1
Hartford	C1
Harvestehude	T
Hatfield	M
Hato	B
Havana	G
Hayindogo	E4
Heerlen	F
Hegau	Q
Heidelberg	B
Hemingford	Z
Hennekamp	T
Hermannswerder	M
Heron	I
Herston	C2–C3
Herzliya	F
Hessarek	B
Hidalgo	C2–C3
Hiduddify	C2–C3
Hillegersberg	D2
Hillingdon	D2
Hillsborough	C1
Hilversum	N
Hindmarsh	C2–C3
Hisingen	Y
Hissar	C1
Hithergreen	I

(continued)

(continued)

TABLE BXII.γ.263. *(cont.)*

Serovar	O Group
Hoboken	E1
Hofit	Q
Hoghton	E1
Holcomb	C2–C3
Homosassa	H
Honelis	M
Hongkong	E4
Horsham	H
Huddinge	E1
Huettwillen	B
Hull	I
Huvudsta	E1
Hvittingfoss	I
Hydra	L
Ibadan	G
Ibaragi	L
Idikan	G
Ikayi	E1
Ikeja	M
Ilala	M
Ilugun	E4
Imo	W
Inchpark	C2–C3
India	D2
Indiana	B
Infantis	C1
Inganda	C1
Inglis	D2
Inpraw	S
Inverness	P
Ipeko	D1
Ipswich	S
Irchel	D2
Irenea	J
Irigny	U
Irumu	C1
Isangi	C1
Isaszeg	Y
Israel	D1
Istanbul	C2–C3
Istoria	H
Isuge	G
Itami	D1
Ituri	B
Itutaba	D2
Ivory	I
Ivorycoast	Z
Jaffna	D1
Jalisco	F
Jamaica	D1
Jambur	L
Jangwani	J
Javiana	D1
Jedburgh	E1
Jericho	B
Jerusalem	C1
Joal	E1
Jodhpur	W
Johannesburg	R
Jos	B
Juba	E4
Jubilee	J
Jukestown	G
Kaapstad	B
Kabete	51
Kaduna	C1
Kahla	T
Kainji	E4
Kaitaan	H
Kalamu	B
Kalina	E1

TABLE BXII.γ.263. *(cont.)*

Serovar	O Group
Kallo	C2–C3
Kalumburu	C2–C3
Kambole	C1
Kamoru	B
Kampala	T
Kande	E4
Kandla	J
Kaneshie	T
Kanifing	H
Kano	B
Kaolack	X
Kapemba	D1
Karachi	W
Karamoja	R
Karaya	51
Karlshamn	J
Kasenyi	P
Kassberg	H
Kedougou	G
Kentucky	C2–C3
Kenya	C1
Kermel	V
Keve	L
Kiambu	B
Kibi	I
Kibusi	M
Kidderminster	P
Kiel	A
Kikoma	I
Kimberley	P
Kimpese	D1
Kimuenza	B
Kindia	E4
Kingabwa	U
Kingston	B
Kinondoni	J
Kinson	E4
Kintambo	G
Kirkee	J
Kisangani	B
Kisarawe	F
Kisii	C1
Kitenge	M
Kivu	C1
Klouto	P
Kodjovi	X
Koenigstuhl	B
Koessen	A
Kofandoka	W
Koketime	V
Kokoli	N
Kokomlemle	Q
Kolar	D2
Kolda	C2–C3
Konolfingen	M
Konstanz	C2–C3
Korbol	C2–C3
Korlebu	E4
Korovi	P
Kortrijk	C1
Kottbus	C2–C3
Kotte	C1
Kotu	D1
Kouka	E4
Koumra	C1
Kpeme	M
Kraligen	C2–C3
Krefeld	E4
Kristianstad	E1
Kua	V
Kubacha	B

(continued)

(continued)

TABLE BXII.γ.263. *(cont.)*

Serovar	O Group
Kuessel	M
Kumasi	N
Kunduchi	B
Kuntair	H
Kuru	C2–C3
Labadi	C2–C3
Lagos	B
Lamberhurst	E1
Lamin	E1
Lancaster	J
Landala	S
Landau	N
Landwasser	E1
Langenhorn	K
Langensalza	E1
Langford	M
Lansing	P
Laredo	H
Larochelle	C1
Lattenkamp	W
Lawndale	D1
Lawra	V
Leatherhead	S
Lechler	51
Leda	53
Leer	K
Leeuwarden	F
Legon	B
Leiden	G
Leipzig	S
Leith	C2–C3
Lekke	E1
Lene	F
Leoben	M
Leopoldville	C1
Lerum	E4
Lexington	E1
Lezennes	C2–C3
Libreville	M
Ligeo	N
Ligna	O
Lika	C1
Lille	C1
Limete	B
Lindenburg	C2–C3
Lindern	H
Lindi	P
Linguere	D2
Lingwala	I
Linton	G
Lisboa	I
Lishabi	D2
Litchfield	C2–C3
Liverpool	E4
Livingstone	C1
Livulu	N
Ljubljana	B
Llandoff	E4
Llobregat	V
Loanda	C2–C3
Lockleaze	C1
Lode	J
Lodz	S
Loenga	T
Logone	Q
Lokstedt	E4
Lomalinda	D1
Lome	D1
Lomita	C1
Lomnava	I
London	E1

Serovar	O Group
Losangeles	I
Loubomo	B
Louga	N
Louisiana	D2
Lovelace	G
Lowestoft	J
Lubumbashi	S
Luciana	F
Luckenwalde	M
Luedinghausen	J
Luke	X
Lutetia	51
Lyon	X
Maastricht	F
Macallen	E1
Macclesfield	D2
Machaga	E4
Madelia	H
Madiago	E4
Madigan	V
Madison	L
Madjorio	E1
Madras	B
Magherafelt	C2–C3
Magumeri	H
Magwa	L
Mahina	D2
Maiduguri	E4
Makiling	U
Makiso	C1
Malakal	I
Malaysia	M
Malika	V
Malmoe	C2–C3
Malstatt	I
Mampeza	H
Mampong	G
Manchester	C2–C3
Mandera	I
Mango	P
Manhattan	C2–C3
Mannheim	F
Mapo	C2–C3
Mara	Q
Maracaibo	F
Marburg	G
Maricopa	T
Marienthal	E1
Maritzburg	V
Maron	E1
Maroua	F
Marseille	F
Marshall	G
Maryland	57
Marylebone	D2
Masembe	E1
Maska	B
Massakory	O
Massenya	B
Matadi	J
Mathura	D2
Matopeni	N
Mayday	D2
Mbandaka	C1
Mbao	U
Meekatharra	W
Melbourne	T
Meleagridis	E1
Memphis	K
Menden	C1
Mendoza	D1

(continued)

TABLE BXII.γ.263. *(cont.)*

Serovar	O Group
Menston	C1
Mesbit	X
Meskin	51
Messina	N
Mgulani	P
Miami	D1
Michigan	J
Middlesbrough	T
Midway	H
Mikawasima	C1
Millesi	R
Milwaukee	U
Mim	G
Minna	H
Minnesota	L
Mishmarhaemek	G
Mississippi	G
Missouri	F
Miyazaki	D1
Mjordan	N
Mkamba	C1
Mocamedes	M
Moero	M
Moers	F
Mokola	E1
Molade	C2–C3
Molesey	52
Mono	B
Mons	B
Monschaui	O
Montevideo	C1
Montreal	U
Morehead	N
Morillons	M
Morningside	N
Mornington	H
Morocco	N
Morotai	J
Moroto	M
Moscow	D1
Moualine	X
Moundou	51
Mountmagnet	L
Mountpleasant	X
Moussoro	H
Mowanjum	C2–C3
Mpouto	I
Muenchen	C2–C3
Muenster	E1
Muguga	V
Mulhouse	D1
Mundonobo	M
Mura	B
Naestved	D1
Nagoya	C2–C3
Nakuru	B
Namibia	C1
Namoda	X
Nanergou	C2–C3
Nanga	G
Nantes	D2
Napoli	D1
Narashino	C2–C3
Nashua	M
Natal	D1
Naware	I
Nchanga	E1
Ndjamena	H
Ndolo	D1
Neftenbach	B
Nessa	H

Serovar	O Group
Nessziona	C1
Neudorf	N
Neukoelln	C1
Neumuenster	B
Neunkirchen	P
Newholland	54
Newlands	E1
Newmexico	D1
Newport	C2–C3
Newrochelle	E1
Newyork	G
Ngaparou	D2
Ngili	C1
Ngor	E4
Niakhar	V
Niamey	J
Niarembe	V
Niederoderwitz	U
Nieukerk	C1
Nigeria	C1
Nijmegen	N
Nikolaifleet	I
Niloese	E4
Nima	M
Nimes	G
Nitra	A
Niumi	E4
Njala	P
Nola	C1
Nordrhein	D2
Nordufer	C2–C3
Norton	C1
Norwich	C1
Nottingham	I
Nowawes	R
Noya	C2–C3
Nuatja	I
Nyanza	F
Nyborg	E1
Nyeko	I
Oakey	C1
Oakland	C1
Obogu	C1
Ochiogu	E4
Ochsenwerder	54
Ockenheim	N
Odienne	R
Odozi	N
Oerlikon	Q
Oersterbro	E4
Offa	S
Ogbete	U
Ohio	C1
Ohlstedt	E1
Okatie	G
Okefoko	E1
Okerara	E1
Oldenburg	I
Olten	D2
Omifisan	R
Omuna	C1
Ona	M
Onarimon	D1
Onderstepoort	H
Onireke	E1
Ontario	D2
Oran	P
Oranienburg	C1
Orbe	T
Ord	52
Ordonez	G

(continued)

Serovar	O Group
Orientalis	I
Orion	E1
Oritamerin	C1
Orlando	K
Orleans	U
Os	D1
Oskarshamn	M
Oslo	C1
Osnabrueck	F
Othmarschen	C1
Ottawa	D1
Ouakam	D2
Oudwijk	G
Overchurch	R
Overschie	51
Overvecht	N
Oxford	E1
Oyonnax	C1
Pakistan	C2–C3
Palamaner	V
Palime	C1
Panama	D1
Papuana	C1
Paratyphi A	A
Paratyphi B	B
Paratyphi C	C1
Paris	C2–C3
Parkroyal	E4
Pasing	B
Patience	M
Penarth	D1
Penilla	M
Pensacola	D1
Perth	P
Petahtikve	E4
Phaliron	C2–C3
Pharr	F
Pietersburg	E1
Pisa	I
Planckendael	C1
Ploufragan	V
Plymouth	D2
Poano	H
Poeseldorf	54
Poitiers	C1
Pomona	M
Pontypridd	K
Poona	G
Portanigra	C2–C3
Portland	D1
Potengi	K
Potosi	H
Potsdam	C1
Potto	D2
Powell	D1
Praha	C2–C3
Pramiso	E1
Presov	C2–C3
Preston	B
Pretoria	F
Putten	G
Quebec	V
Quentin	D2
Quinhon	X
Quiniela	C2–C3
Ramatgan	N
Ramsey	M
Raus	G
Rawash	K
Reading	B
Rechovot	C2–C3

Serovar	O Group
Redba	C1
Redhill	F
Redlands	I
Regent	E1
Reinickendorf	B
Remete	F
Remiremont	C2–C3
Remo	B
Reubeuss	C2–C3
Rhone	L
Rhydyfelin	I
Richmond	C1
Rideau	E4
Ridge	D1
Ried	G
Riggil	C1
Riogrande	R
Rissen	C1
Rittersbach	P
Riverside	W
Roan	P
Rochdale	Z
Rogy	M
Romanby	G
Roodepoort	G
Rossleben	54
Rostock	D1
Rothenburgsort	P
Rottnest	G
Rovaniemi	I
Royan	H
Ruanda	D1
Rubislaw	F
Ruiru	L
Rumford	C1
Runby	H
Ruzizi	E1
Saarbruecken	D1
Saboya	I
Sada	N
Saintemarie	52
Saintpaul	B
Salford	I
Saloniki	I
Samaru	S
Sambre	E4
Sandiego	B
Sandow	C2–C3
Sanga	C2–C3
Sangalkam	D2
Sangera	I
Sanjuan	C1
Sanktgeorg	M
Sanktjohann	G
Sanktmarx	E4
Santander	M
Santhiaba	R
Santiago	C2–C3
Sao	E4
Sapele	G
Saphra	I
Sara	H
Sarajane	B
Saugus	R
Schalkwijk	H
Schleissheim	B
Schoeneberg	E4
Schwabach	C1
Schwarzengrund	B
Schwerin	C2–C3
Sculcoates	I

(continued)

(continued)

TABLE BXII.γ.263. *(cont.)*

Serovar	O Group
Seattle	M
Sedgwick	V
Seegefeld	E1
Sekondi	E1
Selby	M
Sendai	D1
Senegal	F
Senftenberg	E4
Senneville	N
Seremban	D1
Serrekunda	E1
Shamba	I
Shangani	E1
Shanghai	I
Shannon	E1
Sharon	F
Sheffield	P
Sherbrooke	I
Shikmonah	R
Shipley	C2–C3
Shomolu	M
Shoreditch	D2
Shubra	B
Sica	S
Simi	E1
Sinchew	E1
Sindelfingen	C2–C3
Singapore	C1
Sinstorf	E1
Sinthia	K
Sipane	T
Skansen	C2–C3
Slade	E4
Sljeme	X
Sloterdijk	B
Soahanina	H
Soerenga	N
Sokode	D2
Solna	M
Solt	F
Somone	C1
Soumbedioune	M
Southampton	B
Southbank	E1
Souza	E1
Spalentor	T
Spartel	L
Splott	V
Stachus	P
Stanley	B
Stanleyville	B
Staoueli	X
Steinplatz	N
Steinwerder	54
Stellingen	X
Stendal	F
Sternschanze	N
Sterrenbos	C2–C3
Stockholm	E1
Stoneferry	N
Stormont	E1
Stourbridge	C2–C3
Straengnaes	F
Strasbourg	D2
Stratford	E4
Strathcona	C1
Stuivenberg	E4
Stuttgart	C1
Suberu	E1
Sudan	U
Suelldorf	W

(continued)

TABLE BXII.γ.263. *(cont.)*

Serovar	O Group
Sundsvall	H
Sunnycove	C2–C3
Surat	H
Surrey	L
Svedvi	E4
Sya	X
Sylvania	H
Szentes	I
Tabligbo	X
Tado	C2–C3
Tafo	B
Taiping	G
Takoradi	C2–C3
Taksony	E4
Tallahassee	C2–C3
Tamale	C2–C3
Tambacounda	E4
Tamberma	X
Tamilnadu	C1
Tampico	C1
Tananarive	C2–C3
Tanger	G
Tanzania	G
Tarshyne	D1
Taset	T
Taunton	M
Taylor	P
Tchad	O
Tchamba	J
Techimani	M
Teddington	B
Tees	I
Tejas	B
Teko	H
Telaviv	M
Telelkebir	G
Telhashomer	F
Teltow	M
Tema	T
Tendeba	J
Tennenlohe	K
Tennessee	C1
Tennyson	B
Teshie	X
Texas	B
Thayngen	B
Thetford	U
Thiaroye	P
Thies	E4
Thompson	C1
Tibati	E1
Tienba	C1
Tiergarten	V
Tiko	R
Tilburg	E4
Tilene	R
Tinda	B
Tione	51
Togba	I
Togo	B
Tokoin	B
Tomegbe	T
Tomelilla	E4
Tonev	54
Toowong	F
Torhout	N
Toricada	T
Tornow	W
Toronto	D2
Toucra	Y
Toulon	K

(continued)

Serovar	O Group
Tounouma	C2–C3
Tours	F
Trachau	B
Transvaal	W
Travis	B
Treforest	51
Treguier	D1
Trier	I
Trimdon	D2
Tripoli	B
Trotha	R
Truro	E1
Tschangu	G
Tsevie	B
Tshiongwe	C2–C3
Tucson	H
Tudu	B
Tumodi	B
Tunis	G
Typeb	O
Typhi	D1
Typhimurium	B
Typhisuis	C1
Tyresoe	B
Uccle	54
Uganda	E1
Ughelli	E1
Uhlenhorst	V
Uithof	52
Ullevi	G
Umbilo	M
Umhlali	C1
Umhlatazana	O
Uno	C2–C3
Uppsala	B
Urbana	N
Ursenbach	T
Usumbura	K
Utah	C2–C3
Utrecht	52
Uzaramo	H
Vaertan	G
Valdosta	C2–C3
Vancouver	I
Vanier	M
Vaugirard	S
Vegesack	I
Vejle	E1
Vellore	B
Veneziana	F
Verona	S
Verviers	W
Victoria	D1
Victoriaborg	J
Vietnam	S
Vilvoorde	E4
Vinohrady	M
Virchow	C1
Virginia	C2–C3
Visby	E4
Vitkin	M
Vleuten	V
Vogan	T
Volkmarsdorf	M
Volta	F
Vom	B
Voulte	U
Vridi	G
Vuadens	B
Wa	I
Waedenswil	D2

(continued)

Serovar	O Group
Wagadugu	E1
Wagenia	B
Wandsworth	Q
Wangata	D1
Waral	T
Warengo	J
Warmsen	W
Warnemuende	M
Warnow	C2–C3
Warragul	H
Warri	J
Washington	G
Waycross	S
Wayne	N
Wedding	M
Welikade	I
Weltevreden	E1
Wenatchee	X
Wentworth	F
Wernigerode	D2
Weslaco	T
Westafrica	D1
Westeinde	I
Westerstede	E4
Westhampton	E1
Westminster	E1
Weston	I
Westphalia	O
Weybridge	E1
Wichita	G
Widemarsh	O
Wien	B
Wil	C1
Wilhelmsburg	B
Willemstad	G
Wilmington	E1
Wimborne	E1
Windermere	Q
Wingrove	C2–C3
Winneba	B
Winnipeg	54
Winston	C1
Winterthur	E4
Wippra	C2–C3
Wisbech	I
Wohlen	F
Woodhull	H
Woodinville	F
Worb	D2
Worthington	G
Wuiti	N
Wuppertal	D2
Wyldegreen	G
Yaba	E1
Yalding	E4
Yaounde	B
Yardley	M
Yarm	C2–C3
Yarrabah	G
Yeerongpilly	E1
Yehuda	F
Yekepa	R
Yerba	54
Yoff	P
Yokoe	C2–C3
Yolo	O
Yopougon	W
Yoruba	I
Yovokome	C2–C3
Yundum	E1
Zadar	D2

(continued)

TABLE BXII.γ.263. *(cont.)*

Serovar	O Group
Zaiman	D1
Zaire	N
Zanzibar	E1
Zaria	J
Zega	D1
Zehlendorf	N
Zerifin	C2–C3
Zigong	I
Zinder	V
Zongo	E1
Zuilen	E4
Zwickau	I

animals; since the last decade, very frequent agent of *Salmonella* gastroenteritis in humans.

The mol% G + C of the DNA is: see species description.

Deposited strain: ATCC 13076.

iii. **Salmonella enterica** *subsp.* **enterica** *serovar* **Gallinarum** (Klein 1889) Bergey, Harrison, Breed, Hammer and Huntoon 1925, 236 (*Bacillus gallinarum* Klein 1889, 689; *Bacterium pullorum* Rettger 1909, 123; *Salmonella gallinarum-pullorum* Taylor, Bensted, Boyd, Carpenter, Dowson, Lovell, Taylor, Thornton, Wilson and Shaw 1952, 140.)
gal.li.na′rum. L. n. *gallina* hen; L. gen. pl. n. *gallinarum* of hens.

Antigenic formula: 1,9,12: − : −. The presence of factor O1 is connected with lysogenization. Always nonmotile. May be subdivided into biovars based on fermentation characteristics, production of gas, and production of H₂S. Does not grow on a minimal defined medium. Isolated chiefly from chickens and other birds. Causative agent of fowl typhoid.

The mol% G + C of the DNA is: see species description.

iv. **Salmonella enterica** *subsp.* **enterica** *serovar* **Paratyphi A** (Brion and Kayser 1902) Castellani and Chalmers 1919, 939 (*Bacterium paratyphi* Kayser 1902, 426; *Bacterium paratyphi* typhus A Brion and Kayser 1902, 613.)
pa.ra.ty′phi. Gr. prep. *para* alongside of; Gr. n. *typhus* a stupor; M.L. gen. n. *paratyphi A* of type A typhoid-like infection.

Antigenic formula: 1,2,12:a: −. The presence of factor O1 is connected with lysogenization. Aerogenic. Ferments arabinose but no xylose. The majority of strains do not produce H₂S, and in this respect serovar Paratyphi A is unlike most other salmonellae. Lysine decarboxylase reaction is weak or negative. Pathogenic only for humans.

The mol% G + C of the DNA is: see species description.

v. **Salmonella enterica** *subsp.* **enterica** *serovar* **Paratyphi B** (Brion and Kayser 1902) Bergey, Harrison, Breed, Hammer and Huntoon 1923, 213 (*Bacterium paratyphi* typhus B Brion and Kayser 1902, 613; *Bacillus schottmuelleri* Winslow, Kligler and Rothberg 1919, 479.)
pa.ra.ty′phi. Gr. prep. *para* alongside of; Gr. n. *typhus* a stupor; M.L. gen. n. *paratyphi B* of type B typhoid-like infection.

Antigenic formula; 1,4,[5],12:b:1,2. The presence of

factor O1 is connected with lysogenization. Produces a slime layer when grown on a medium containing 0.5% glucose and 0.2 M sodium phosphate, pH 7. Negative for D-tartrate. Causes enteric fever in humans and very rarely infects animals. A variant known as biovar Java is positive for D-tartrate, fails to produce a slime layer, and usually causes enteritis in humans and not uncommonly in animals as well (Kauffmann, 1941). Some strains are intermediate between these two extremes.

The mol% G + C of the DNA is: see species description.

vi. **Salmonella enterica** *subsp.* **enterica** *serovar* **Paratyphi C** Hirschfeld 1919, 296 (Paratyphoid C bacillus Hirschfeld 1919, 296; *Salmonella hirschfeldii* Weldin 1927, 161; *Salmonella* paratyphi-C International Salmonella Subcommittee 1934.)
pa.ra.ty′phi. Gr. prep. *para* alongside of; Gr. n. *typhus* a stupor; M.L. gen. n. *paratyphi C* of type C typhoid-like infection.

Antigenic formula; 6,7,[Vi]:c:1,5. Ferments dulcitol and trehalose; produces H₂S. Arabinose fermentation is variable.

The mol% G + C of the DNA is: see species description.

vii. **Salmonella enterica** *subsp.* **enterica** *serovar* **Typhi** (Schroeter 1886) Warren and Scott 1930, 416 (*Bacillus typhi* Schroeter 1886, 165.)
ty′phi. Gr. n. *typhus* a stupor; M.L. gen. n. *typhi* of typhoid.

Antigenic formula: 9,12,[Vi]:d: −. Wild strains may possess H antigen z66 instead of H antigen d (Guinée et al., 1981). Does not grow on Simmons' citrate medium or on a minimal defined medium; requires at least tryptophan as a growth factor. Does not produce gas from glucose or other sugars. Fermentation of xylose is variable. Many strains are agglutinated by anti-Vi serum and are inagglutinable by anti-09 serum; their colonies (V colonies) are opaque and have an iridescent appearance when examined by transmitted light. Colonies of intermediate appearance agglutinable by both Vi and O antisera may occur (VW colonies). Pathogenic only for humans, causing typhoid (enteric) fever; transmitted by water or food contaminated by human excreta.

The mol% G + C of the DNA is: see species description.

Deposited strain: ATCC 19430.

viii. **Salmonella enterica** *subsp.* **enterica** *serovar* **Typhimurium** (Loeffler 1892) Castellani and Chalmers 1919, 939 (*Bacterium typhimurium* Loeffler 1892, 134.)
ty.phi.mu′ri.um. Gr. n. *typhus* a stupor; L. n. *mus* mouse; L. gen. pl. n. *murium* of mice; M.L. gen. pl. n. *typhimurium* typhoid of mice.

Antigenic formula: 1,4,[5],12:i:1,2. The presence of factor O1 follows lysogenization by a converting phage named *iota* or PLT22. Ubiquitous and frequently the cause of infections in humans and animals; very frequent agent of *Salmonella* gastroenteritis in man.

The mol% G + C of the DNA is: see species description.

Deposited strain: ATCC 13311.

b. **Salmonella enterica** *subsp.* **arizonae** (Borman 1957) Le Minor and Popoff 1987, 467 (*Salmonella choleraesuis* subsp. *arizonae* (Borman 1957) Le Minor, Véron and Popoff 1985, 375; *Salmonella arizonae* (Borman 1957) Kauffmann *in* van Oye 1964; *Paracolobactrum arizonae* Borman 1957, 347.)

a.ri.zo' nae. M.L. gen. n. *arizonae* of Arizona, a state in the United States.

Distinguished from other *S. enterica* subspecies by the characteristics reported in Table BXII.γ.261. Contains at least 94 serovars (Popoff et al., 1998). All serovars belonging to this subspecies are monophasic for the H antigen. Isolated mainly from cold-blooded animals and environment.

Antigenic formula: $51:z_4,z_{23}: -$.

The mol% G + C of the DNA is: see species description.

Type strain: ATCC 13314, CIP 82.30, NCTC 8297.

c. **Salmonella enterica** *subsp.* **diarizonae** (Le Minor, Véron and Popoff 1985) Le Minor and Popoff 1987, 467 (*Salmonella choleraesuis* subsp. *diarizonae* Le Minor, Véron and Popoff 1985, 375.)

di.a.ri.zo' nae. Gr. adj. *dis* twice, two; M.L. gen. n. *arizonae* of Arizona, a state in the United States.

Distinguished from other *S. enterica* subspecies by the characteristics reported in Table BXII.γ.261. Contains at least 323 serovars (Popoff et al., 1998). All serovars belonging to this subspecies are diphasic for the H antigen. Isolated mainly from cold-blooded animals and environment.

Antigenic formula: $6,7:l,v:z_{53}$.

The mol% G + C of the DNA is: see species description.

Type strain: ATCC 43973, CIP 82.31, NCTC 10060.

d. **Salmonella enterica** *subsp.* **houtenae** (Kauffmann 1962) Le Minor and Popoff 1987, 467 (*Salmonella choleraesuis* subsp. *houtenae* (Kauffmann 1962) Le Minor, Rohde and Taylor 1970b, 209; *Salmonella houtenae* Kauffmann 1962, 353.)

hou' te.nae. M.L. gen. n. *houtenae* of Houten, a town in Holland.

Distinguished from other *S. enterica* subspecies by the characteristics reported in Table BXII.γ.261. Contains at least 70 serovars (Popoff et al., 1998). All serovars belonging to this subspecies are monophasic for the H antigen. Isolated mainly from cold-blooded animals and environment.

Antigenic formula: $41:g,z_{51}: -$.

The mol% G + C of the DNA is: see species description.

Type strain: ATCC 43974, CIP 82.32, NCTC 10401.

e. **Salmonella enterica** *subsp.* **indica** (Le Minor, Popoff, Laurent and Hermant 1987) Le Minor and Popoff 1987, 467 (*Salmonella choleraesuis* subsp. *indica* Le Minor, Popoff, Laurent and Hermant 1987, 179.)

in.di' ca. L. adj. *indica* of India.

Distinguished from other *S. enterica* subspecies by the characteristics reported in Table BXII.γ.261. Contains at least 11 serovars (Popoff et al., 1998). Isolated mainly from cold-blooded animals and environment.

Antigenic formula: 1,6,14,25:a:e,n,x.

The mol% G + C of the DNA is: see species description.

Type strain: ATCC 43976, CIP 102501.

f. **Salmonella enterica** *subsp.* **salamae** (Le Minor, Rohde and Taylor 1970b) Le Minor and Popoff 1987, 467 (*Salmonella choleraesuis* subsp. *salamae* (Le Minor, Rohde and Taylor 1970b) Le Minor, Véron and Popoff 1985, 375) ("*Salmonella salamae*" Le Minor, Rohde and Taylor 1970b, 209.)

sa.la' mae. M.L. gen. n. *salamae* of (Dare-es) salaam.

Distinguished from other *S. enterica* subspecies by the characteristics reported in Table BXII.γ.261. Contains at least 488 serovars (Popoff et al., 1998). Isolated mainly from cold-blooded animals and environment.

Antigenic formula: 1,9,12:l,w:e,n,x.

The mol% G + C of the DNA is: see species description.

Type strain: ATCC 43972, CIP 82.29, NCTC 5773.

2. **Salmonella bongori** (Le Minor, Véron and Popoff 1985); Reeves, Evins, Heiba, Plikaytis and Farmer 1989b, 371[VP] (Effective publication: Reeves, Evins, Heiba, Plikaytis and Farmer 1989a, 319) (*Salmonella choleraesuis* subsp. *bongori* Le Minor, Véron and Popoff 1985, 375.)

bon' gori. M.L. gen. n. *bongori* of Bongor, a town in Chad.

The characteristics are as described for the genus (see also Table BXII.γ.193 in the chapter on the family *Enterobacteriaceae*) and as listed in Table BXII.γ.261. All serovars belonging to *S. bongori* are monophasic for the H antigen. Contains at least 20 serovars (Popoff et al., 1998). Isolated mainly from cold-blooded animals and environment.

Antigenic formula: $66:z_{41}: -$.

The mol% G + C of the DNA is: 51.8 ± 0.6 (T_m).

Type strain: ATCC 43975, CIP 82.33.

GenBank accession number (16S rRNA): AF029227.

Genus XXXIV. *Serratia* Bizio 1823, 288[AL]

FRANCINE GRIMONT AND PATRICK A.D. GRIMONT

Ser.ra' ti.a. M.L. fem. n. *Serratia* named after Serafino Serrati, an Italian physicist.

Straight rods, $0.5–0.8 \times 0.9–2.0$ μm in length, with rounded ends. Conform to the general definition of the family *Enterobacteriaceae*. Gram negative, generally **motile**, by means of peritrichous flagella. **Facultatively anaerobic. Nitrate and chlorate are reduced anaerobically. Growth factors are generally not required.** Colonies on nutrient agar are most often opaque, somewhat iridescent, and either **white, pink, or red in color.** Almost all strains can grow at temperatures between 10 and 36°C, at pH 5–9, and in the presence of 0–4% (w/v) NaCl. The catalase reaction is strongly positive. **D-Glucose is fermented** through the Embden–Meyerhof pathway. The major glucose entry route involves a phosphoenolpyruvate-dependent phosphotransferase system with both enzyme II[Glc] (glucose permease) and enzyme II[Man] (mannose permease). Glucose is also oxidized to gluconate in the presence of pyrroloquinoline quinone. **Gluconate is oxidized to 2-ketogluconate. Acetoin is produced from pyruvate** by all species except *S. fonticola*. Fructose, D-galactose, maltose, D-mannitol, D-mannose, ribose, and trehalose are fermented and utilized as sole carbon sources. L-fucose is fermented and utilized as sole carbon source by all species except *S. fonticola*. L-sorbose is not fermented or utilized as sole carbon source. All species but *S. fonticola* fail to ferment or utilize dulcitol and tagatose. *N*-acetylglucosamine, D-alanine, L-alanine, citrate, D-galacturonate, D-glucosamine, D-glucuronate, 2-ketogluconate, L-proline, putrescine, L-serine are utilized as sole carbon sources by most strains. Caprate, caproate, **caprylate**, and tyrosine are **utilized** as sole carbon sources by all species except *S. fonticola*. 5-Aminovalerate, butyrate, *m*-coumarate, ethanolamine and tryptamine are not utilized as sole carbon sources. All species except *S. fonticola* fail to utilize 3-phenylpropionate. All species except *S. entomophila* fail to utilize itaconate. Phenylalanine, histidine, and tryptophan deaminases and thiosulfate reductase (H_2S from thiosulfate) are not produced. *o*-**Nitrophenyl-β-D-galactopyranoside** (ONPG) is **hydrolyzed by most strains**. Esculin is hydrolyzed by most strains except *S. proteamaculans* subsp. *quinovora*. **Extracellular enzymes** of all species except *S. fonticola* **hydrolyze DNA, lipids** (tributyrin, corn oil) **and proteins** (gelatin, casein), but not starch (in four days), polygalacturonic acid, or pectin. Tween-80 is hydrolyzed by all species except *S. odorifera*. The organisms **occur in the natural environment** (soil, water, plant surfaces) or **as opportunistic human pathogens**.

The mol% G + C of the DNA is: 52–60.

Type species: **Serratia marcescens** Bizio 1823, 288.

FURTHER DESCRIPTIVE INFORMATION

Phylogenetic affinities among *Serratia* species have been studied (Fig. BXII.γ.208) (Dauga et al., 1990; Spröer et al., 1999). Three clusters can be observed in all studies: cluster I includes *S. marcescens* and *S. rubidaea*, cluster II includes psychrotrophic species (*S. proteamaculans*, *S. grimesii*, *S. liquefaciens*, *S. plymuthica*, and *S. fonticola*), and cluster III includes *S. ficaria* and *S. entomophila*. Depending on the analysis, *S. odorifera* may branch with cluster I or after clusters II and III have merged. Sequence comparisons leave no doubt as to the affiliation of *S. fonticola* to the genus *Serratia* and especially to the psychrophilic group of species.

Cells of *Serratia* rarely show a visible capsule in India ink mounts, although mucoid colonies can be observed in *S. plymuthica* and occasionally in other *Serratia* species; however, cells of *S. odorifera* possess a microcapsule which can be evidenced by the quellung reaction (capsular swelling) using *Klebsiella* anticapsule K4 or K68 sera (Richard, 1979). Polysaccharides excreted by cells of *S. marcescens* can be extracted from the cell surface layer or from the culture medium. These polysaccharides contain chiefly D-glucose and glucuronic acid and lower proportions of D-mannose, heptose, L-fucose and L-rhamnose (Adams and Martin, 1964; Adams and Young, 1965).

The major fatty acid components found in whole-cell methanolysates are $C_{14:0\ 3OH}$, $C_{16:0}$, $C_{16:1}$, $C_{18:1}$, and $C_{18:2}$, contributing 50–80% of the components in each strain. $C_{14:0}$ contributes 3.7–9.4% whereas other components contribute less than 3% each (Bergan et al., 1983).

Colony diameters are ~1.5–2.0 mm after overnight growth on nutrient agar. Swarming does not occur.

Two different pigments can be produced by various *Serratia* strains: prodigiosin and pyrimine (Williams and Qadri, 1980). Prodigiosin, a nondiffusible, water-insoluble, red pigment bound to the cell envelope, is produced by two biogroups (A1 and A2) of *S. marcescens* and by most strains of *S. plymuthica* and *S. rubidaea*. Prodigiosin-producing colonies are totally red or show either a red center, a red margin, or red sectors. The exact color given by the pigment depends upon cultural conditions (e.g., amino acids, carbohydrates, pH, inorganic ions, temperature) and may include orange, pink, red, or magenta. Prodigiosin is best produced on peptone-glycerol agar at 20–35°C. The temperature range for pigment production is 12–36°C. Prodigiosin is not produced anaerobically. Chemically, prodigiosin is a tripyrrole derivative, 2-methyl-3-amyl-6-methoxyprodigiosene (prodigiosene is 5-(2-pyrryl)-2,2′-dipyrrylmethene). In the cell, prodigiosin is formed by condensation of a volatile 2-methyl-3-amylpyrrol (MAP) and a nonvolatile 4-methoxy-2-2′-bipyrrole-5-carboxaldehyde (MBC). Several classes of nonpigmented mutants that are blocked on either the MAP pathway or the MBC pathway have been isolated. Syntrophic pigmentation may occur when two different class mutants are grown side by side (Williams and Qadri, 1980).

Pyrimine, a water-soluble, diffusible pink pigment (Williams and Qadri, 1980), is produced by some strains of *S. marcescens* biogroup A4. Ferrous iron is required for the production of pyrimine. Pyrimine is L-2(2-pyridyl)-D′-pyrroline-5-carboxylic acid. When pyrimine is produced, the agar medium turns pink while the colonies are white to pinkish.

A yellow diffusible pigment, 2-hydroxy-5-carboxymethylmuconic acid semialdehyde, is produced from the meta cleavage of 3,4-dihydroxyphenylacetic acid (3,4-DHP) by the enzyme 3,4-DHP 2,3-dioxygenase (Trias et al., 1988), induced by tyrosine in all *S. marcescens* strains. At present, only *S. marcescens* strains of biotype A8a which have lost the ability to grow on aromatic compounds can produce the yellow pigment. A new reddish violet pigment, a peptide-ferropyrimine complex, produced by a *S. marcescens* O5:H1 was described by Suzuki et al. (1993b).

Cultures can produce two kinds of odors, a fishy to urinary odor attributed to trimethylamine (mixed with some NH_3), or a musty, potato-like odor attributed to alkyl-methoxypyrazine. The musty odor produced by *S. odorifera*, *S. ficaria*, and a few strains of *S. rubidaea* is due to 3-isopropyl-2-methoxy-5-methylpyrasine (Gallois and Grimont, 1985). All other strains and species produce the fishy-urinary odor.

All species except *S. marcescens* and *S. rubidaea* can grow readily at 4–5°C and several grow at 40°C (*S. marcescens* and several strains of *S. rubidaea* and *S. odorifera*); however, the temperature of 37°C is not favorable for the isolation of *S. plymuthica*. When *S. liquefaciens* and *S. plymuthica* are studied, many tests that are positive at 28–35°C give negative results at 37°C (e.g., Voges–Proskauer, decarboxylases, tetrathionate reductase tests).

A strong catalase activity, which can be evidenced with 3% (or

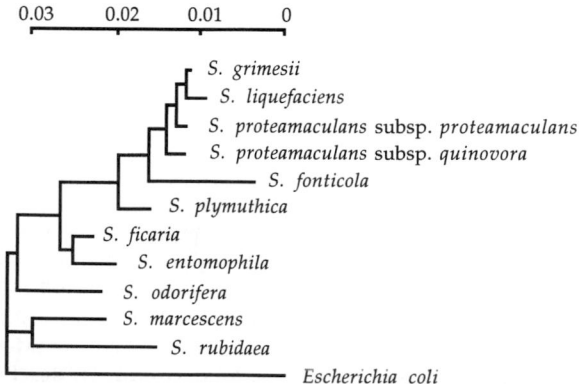

FIGURE BXII.γ.208. Phylogenetic tree of the genus *Serratia*. The tree was built by the neighbor-joining method (Saitou and Nei, 1987). Scale is in K_{nuc} according to Jukes and Cantor (1969).

less) H_2O_2, is produced by *Serratia* strains (Taylor and Achanzar, 1972).

There is no sodium ion requirement for growth in the genus *Serratia*; however, the optimum concentration of NaCl for growth is ~0.5% (w/v) for *S. marcescens* or 1% (w/v) for *S. rubidaea* (Grimont and Grimont, unpublished results). Tolerance to NaCl ranges from 5 to 6% (w/v) for *S. plymuthica*, to 10% (w/v) for *S. rubidaea*.

In Biotype-100 strips (BioMerieux, Craponne, France) or a minimal medium containing ammonium sulfate as the nitrogen source, the following compounds serve universally as sole carbon sources for all *Serratia* strains: *N*-acetylglucosamine, *cis*-aconitate, citrate, D-fructose, D-galactose, D-galacturonate, D-gluconate, D-glucosamine, D-glucose, *myo*-inositol, 2-ketogluconate, maltose, maltotriose, D-mannitol, D-mannose, D-ribose, and D-trehalose. Most strains of all species can utilize D-alanine, L-alanine, D-glucuronate, glycerate, glycerol, lactate, putrescine, L-serine, and L-tryptophan. All species except some or all strains of *S. fonticola* can utilize L-aspartate, L-fucose, fumarate, L-glutamate, L-malate, L-proline, and succinate. Utilization of caprate, caproate, and caprylate by all species except *S. fonticola* is better observed in a minimal agar containing ammonium sulfate as the nitrogen source, rather than in Biotype-100 strips. The following compounds are never utilized as sole carbon sources (Biotype-100 strips): DL-5-aminovalerate, *m*-coumarate, ethanolamine, glutarate, sorbose, and tryptamine. The following carbon sources are never utilized by *Serratia* species except some or all *S. fonticola* strains: dulcitol, 3-phenylpropionate, and tagatose (Grimont et al., 1977b, 1978b, 1979b; Grimont and Grimont, 1978b).

Characteristic extracellular enzymes are produced by most species. All species except *S. fonticola* can hydrolyze DNA, gelatin, soluble casein, tributyrin, and corn oil. Only rare strains fail to produce one or more of these extracellular enzymes. All species, except *S. odorifera*, can hydrolyze Tween 80. Chitin is hydrolyzed by all species except *S. rubidaea*, *S. odorifera*, and *S. fonticola*. Lecithin is also hydrolyzed by many strains. Spot-inoculated starch agar (Starch agar: nutrient agar containing 0.5% [w/v] soluble starch), incubated for four days and then flooded with Lugol's iodine, shows no zone of clearing (Grimont et al., 1977b); however, longer incubation (6–14 days) may allow detection of some amylase-producing strains (M. Popoff, personal communication).

A red-pigmented *S. marcescens* has been found to produce a carboxymethyl cellulase (Thayer, 1978). Depolymerization of a carboxymethylcellulose gel is faster with *S. marcescens*, *S. rubidaea*, and *S. liquefaciens* than with *S. odorifera*, *S. ficaria*, and *S. plymuthica* (unpublished results).

Up to 11 proteinases have been revealed by agar gel electrophoresis. Each strain produces one to four different proteinases. Different species have different proteinase patterns (Grimont et al., 1977a). These have been used to type strains (Grimont and Grimont, 1978c). Isoelectric points of the 11 proteinases are between pH 3.6 and pH 6.0 (Grimont and Grimont, 1978b).

Fructose, maltose, D-mannitol, D-mannose, ribose, and trehalose are fermented by all strains. Most strains ferment glycerol and *myo*-inositol. Fermentation of D-glucose is not prevented by 0.001 M iodoacetate (Grimont et al., 1977b, 1978b, 1979b; Grimont and Grimont, 1978b), an inhibitor of the Embden–Meyerhof–Parnas glycolytic pathway and other enzymic reactions. *Serratia* species can produce gluconate-6-phosphate dehydrase and 2-keto-3-deoxygluconate-6-phosphate aldolase (Kersters and De Ley, 1968), which are the characteristic enzymes of the Entner–Doudoroff pathway.

Under aerobic conditions, gluconate is produced from D-glucose in the presence of iodoacetate by all *Serratia* species due to a glucose dehydrogenase, when pyrroloquinoline quinone (PQQ) is added. Only *S. marcescens*, *S. odorifera*, *S. rubidaea*, *S. entomophila*, *S. ficaria*, and some strains of *S. plymuthica* can produce gluconate without added PQQ (Bouvet et al., 1989).

A reducing compound, 2-ketogluconate, is also produced from gluconate by all species due to a gluconate dehydrogenase (Grimont et al., 1977b, 1978a, 1979b; Bouvet et al., 1989).

2,5-Diketogluconate is produced from 2-ketogluconate at 20°C (not at 30°C) due to a 2-ketogluconate dehydrogenase by *S. marcescens*, *S. liquefaciens*, and *S. grimesii* (Bouvet et al., 1989).

In anaerobic glycerol dissimilation, a glycerol dehydrogenase induced by glycerol and not by hydroxyacetone is present in *S. marcescens*, *S. proteamaculans*, *S. liquefaciens*, *S. grimesii*, and *S. fonticola*, but not in *S. rubidaea*, *S. ficaria*, *S. odorifera*, *S. entomophila*, and *S. plymuthica* (Bouvet et al., 1995a). The Voges–Proskauer (VP) test, when done on a 3-day-old culture in Clark–Lubs medium, is negative for 40% of strains of *S. plymuthica*, although acetoin can be detected after incubation for 18 h by use of a sensitive method (Richard, 1972). These strains, which are VP-negative after 3 days of incubation, can utilize 2,3-butanediol as a sole carbon source (Grimont et al., 1978b). *Serratia* strains, other than *S. fonticola*, that cannot produce acetoin from pyruvate (under any experimental conditions) are very rare. *S. fonticola* is always negative for acetoin. A tiny gas bubble is commonly produced by *S. marcescens* in a peptone–water–glucose medium with Durham tube. *S. plymuthica* and *S. liquefaciens* produce a large amount of gas. The end products of glucose fermentation by *S. marcescens* are 2,3-butanediol, ethanol, formate, lactate, succinate, and CO_2 with small amounts of acetate, acetoin, and glycerol and very little or no H_2 (Neish et al., 1948; White and Starr, 1971). The end products yielded by *S. plymuthica* are 2,3-butanediol, ethanol, lactate, succinate, CO_2, H_2, and small amounts of formate, acetate, acetoin, and glycerol (Neish et al., 1948). The 2,3-butanediol produced by *S. marcescens* is mostly a *meso*-isomer, whereas *S. plymuthica* is unique in producing a levo-rotatory 2,3-butanediol (Neish et al., 1948).

Transduction systems in *S. marcescens* have been described (Kaplan and Brendel, 1969; Matsumoto et al., 1973). The earliest genetic transfer described in *S. marcescens* (Belser and Bunting, 1956) is also suggestive of a transduction mechanism. *S. marcescens* was transformed with plasmid pBR322 by Reid et al. (1982). Transformants were selected based on resistance to high levels of ampicillin.

All species of the genus *Serratia* have been delineated by DNA–DNA hybridization. All species of the genus except *S. fonticola* share a number of phenotypic properties. *S. fonticola* was included in the genus *Serratia* because of significant DNA relatedness with other *Serratia* species (Steigerwalt et al., 1976; Gavini et al., 1979); this inclusion was confirmed by comparison of 16S rDNA gene sequences (Dauga et al., 1990). A summary of DNA relatedness within the genus *Serratia* is given in Table BXII.γ.264.

Lactose plasmids have been demonstrated in *S. liquefaciens* (Le Minor et al., 1974) and in *S. marcescens* (C. Coynault, personal communication). Antibiotic resistance plasmids of incompatibility groups S (= H_2), C, L/M, P, W, and FII have been identified in *S. marcescens*. Plasmids of groups M and N have been found in *S. liquefaciens* (Hedges, 1980). Replicon typing, using cloned DNA probes, identified plasmid groups FIB, FIC, FIIA, H12, L/M, N, B/O, P, W, Y, and Com9 in multiresistant *S. marcescens* strains (Llanes et al., 1994).

TABLE BXII.γ.264. DNA relatedness among *Serratia* species[a,b]

Species	Source of labeled DNA										
	S. marcescens[c]	*S. entomophila*	*S. fonticola*	*S. ficaria*[d]	*S. grimesii*	*S. liquefaciens*	*S. odorifera*[c]	*S. plymuthica*	*S. proteamaculans* subsp. *proteamaculans*	*S. proteamaculans* subsp. *quinovora*	*S. rubidaea*
S. marcescens[c]	**92 ± 5.5** **d = 0–3.2**	42 ± 6.3 d = 7.5–8.0	50 ± 5.1	46 ± 4.7 d = 8.5–9.5	36	29	31 ± 2.8 d = 10.5–11.5	29	34	33	26 ± 2.1
S. entomophila	ND	**84 ± 8.2** **d = 0–1.0**	ND	62 ± 5.4 d = 6.0–9.0	ND	ND	ND	ND	ND	ND	ND
S. fonticola	57 ± 4.9 d = 12.0–14.0	28	**88 ± 5.5**	40 d = 10.0	22	23	38 ± 2.8 d = 13.0	21	29	22	16
S. ficaria	ND	47 ± 6.4 d = 5.5–8.5	ND	**91 ± 4.2** **d = 0–2.0**	44	38	ND	44	41	47 d = 10.0	27 ± 3.4
S. grimesii	53 ± 1.4 d = 12.0–12.5	34 ± 6.0	ND	29	**97 ± 6.4** **d = 0–4.0**	36 ± 3.8 d = 9.5–14.0	32 ± 1.1 d = 13.5	36 ± 4.6 d = 11.0–15.5	42 ± 5.5 d = 9.5–10.5	44 ± 3.4 d = 8.5–15.0	18 ± 2.6
S. liquefaciens	62 ± 1.1 d = 10.5–12.5	48	41 ± 2.1	33 ± 1.7	47 ± 7.0 d = 10.0–11.5	**86 ± 7.5** **d = 0–3.0**	39 ± 3.5 d = 11.5	44 ± 5.1 d = 8.5–12.5	45 ± 7.1 d = 7.5–9.5	48 ± 5.7 d = 8.0–9.5	19 ± 5.0
S. odorifera	ND	40	ND	33 ± 2.8 d = 16.5	29	26	**78 ± 13.9** **d = 0–4.0**	28	28	25	21 ± 4.2
S. plymuthica	62 d = 10.0	42	44 ± 5.7	43 ± 4.6 d = 13.0	43 ± 4.1 d = 9.5–12.0	39 ± 5.2 d = 8.5–10.0	38 ± 2.0 d = 11.0	**85 ± 14.6** **d = 0–4.5**	52 ± 2.3 d = 8.5–9.5	44 ± 4.7 d = 9.0–12.0	20 ± 2.5
S. proteamaculans subsp. *proteamaculans*	ND	40	ND	41 ± 5.6	52 ± 5.6 d = 9.0–13.0	42 ± 5.4 d = 9.0–13.0	ND	45 ± 5.7 d = 9.0–12.0	**81 ± 11.0** **d = 0–6.5**	64 ± 5.3 d = 5.5–12.0	21 ± 4.8
S. proteamaculans subsp. *quinovora*	ND	ND	ND	ND	52 ± 3.5 d = 8.5–10.0	44 ± 2.5 d = 8.5–10.5	ND	48 ± 4.4 d = 9.0–12.0	58 ± 7.4 d = 5.0–6.0	**88 ± 6.7** **d = 0–4.5**	ND
S. rubidaea	49 ± 1.7 d = 10.0–11.0	34	38 ± 18.7	26 ± 3.0 d = 12.0	24	24	32 ± 3.2 d = 10.0	26	26	30	**88 ± 20.6** **d = 0–6.0**

[a]All DNA relatedness values were obtained with the S1 nuclease method except otherwise stated. Divergence (d) given rounded to the nearest 0.5.

[b]Data from Steigerwalt et al. (1976), Grimont et al. (1978a, 1979b, 1982b, 1988), and Gavini et al. (1979). Data in the *S. rubidaea* column are unpublished.

[c]DNA relatedness values obtained with the hydroxyapatite method.

[d]DNA relatedness values obtained with the filter method.

Bacteriophages active on *Serratia* are easily found in river water or sewage. Phages that are active on one species of *Serratia* are usually active on strains of other species of that genus, but rarely on strains of other genera (Grimont and Grimont, 1978b). Lysogeny is very common in *Serratia* species (Prinsloo, 1966). Several phage typing systems have been studied (Pillich et al., 1964; Hamilton and Brown, 1972; Farmer, 1974; Grimont, 1977a).

Bacteriocins produced by *Serratia* are of two kinds: (a) a trypsin-resistant, acid-sensitive (pH 2) structure (Hamon and Péron, 1961) called "group A bacteriocin" by Prinsloo (1966) and later found by electron microscopy to resemble phage tails (Traub, 1972); and (b) a trypsin-sensitive, acid-resistant protein (Hamon and Péron, 1961) called "group B bacteriocin" by Prinsloo (1966). Bacteriocins produced by one species of *Serratia* frequently cross-react with other species of this same genus. *Serratia* bacteriocins are also frequently active on *Escherichia coli* K12. *S. marcescens* strains produce group A and/or group B bacteriocins. *S. rubidaea* strains produce only group A bacteriocins. *S. liquefaciens* and *S. ficaria* produce only group B bacteriocins. *S. odorifera* produces neither group A nor group B bacteriocins (Hamon and Péron, 1979; Y. Hamon, personal communication). Bacteriocin typing can be used for epidemiological purposes (Traub, 1980).

The antigenic structure of *S. marcescens* has been described. The present scheme consists of 28 somatic antigens (O1 to O28) and 25 flagellar antigens (H1 to H25) (Edwards and Ewing, 1972; Le Minor and Pigache, 1978; Traub and Fukushima, 1979a, b; Le Minor and Sauvageot-Pigache, 1981; Traub, 1991; Aucken et al., 1996). Subdivision of antigens O5 (into O5a, O5b, O5c), O10 (into O10a, O10b), and O16 (into O16a, O16b, O16c, O16d) has been proposed (Le Minor and Sauvageot-Pigache, 1981). Cross-reactions between factors O6 and O14 are very extensive and the distinction between these two factors does not seem worthwhile. Serovar O27 cross-reacts with the O4 serovar strain, and serovar O28 with the O5 serovar strain. The O-typing scheme was improved by separating capsular material from O-antigen (Gaston and Pitt, 1989a, b). H antigens are monophasic in *S. marcescens* (Aucken et al., 1996).

Four serovars (O1:H1, O2:H1, O3:H1, and O4:H1), all sharing a common H antigen, were described among *S. ficaria* strains (Grimont and Deval, 1982).

Resistance to cephalothin, colistin, and polymyxin (with respect to achievable serum levels of antibiotics) is very frequent in the genus and almost constant in *S. marcescens*. With the antibiotic disk method, a zone phenomenon develops around disks impregnated with colistin and polymyxin: the inhibition zone contains colonies close to the disk. However, this zone phenomenon is not restricted only to *Serratia*. Resistance to tetracycline and ampicillin is very frequent in *S. marcescens* and rare in other *Serratia* species. Plasmid-determined resistance to aminoglycoside antibiotics, carbenicillin, chloramphenicol, trimethoprim, sulfonamides, and mercury ions can be found in clinical strains of *S. marcescens*. "Third generation" cephalosporins are still active on *S. marcescens* and in a multicenter survey in the USA, fewer than 8% of *S. marcescens* isolates were resistant to piperacillin, 3–4% were resistant to ceftazidime, ceftriaxone, and/or cefotaxime, and 0.3% were resistant to imipenem (Jones, 1998). Resistance to cetyl trimethylammonium chloride (1.5 mg/ml) and thallus acetate (0.8 mg/ml) is very frequent (Grimont et al., 1977b). Of all the *Serratia* species, *S. marcescens* is the most resistant to antibiotics, antiseptics, and metal ions; *S. plymuthica* is the least resistant to these antimicrobials.

Healthy humans are not often infected by *Serratia*. *S. marcescens* is a prominent opportunistic pathogen for hospitalized human patients. At present, *S. marcescens* is the only known nosocomial species of *Serratia*. Clinically, *Serratia* infections do not differ from infections by other opportunistic pathogens (von Graevenitz, 1980). Other *Serratia* species can be involved in respiratory tract infection or colonization and septicemia, especially when these bacteria are accidentally injected into the body (e.g., contaminated perfusion or irrigation liquid) (Grimont and Grimont, 1978b).

S. marcescens and *S. liquefaciens* are known to infect and cause mortality in a variety of insects which can be serious pests of crops, ornamentals, and turf throughout the world (Klein and Kaya, 1995). Commercial utilization of *S. entomophila* against the grass grub *Costelytra zealandica* in New Zealand pastures has been achieved (Klein and Jackson, 1992). This bacterium turns the grubs a honey or amber color. Pathogenic strains of *S. entomophila* colonize the larva gut, adhere to the crop, and induce starvation, which causes depletion of the fat bodies. Pathogenicity is correlated with the production of lecithinase, proteinase, and chitinase (Kaska, 1976; Lysenko, 1976).

Mastitis in cows and other animal infections have been associated with *Serratia* species (Grimont and Grimont, 1978b). Pathogenicity in experimental animals is of the type expected of a Gram-negative bacterium. Experimental depression of phagocytic cell number or function in animals enhances susceptibility to *Serratia* infections (Simberkoff, 1980).

A typical hypersensitivity reaction is produced by inoculation of plants such as tobacco and king protea with *Serratia* (Lakso and Starr, 1970; Grimont et al. 1978b). *S. proteamaculans* was isolated from a leaf spot disease of *Protea cynaroïdes* (Paine and Stanfield, 1919) and *S. marcescens* (under the name *Erwinia amylovora* var. *alfalfae*) was isolated from a root disease of alfalfa (Shinde and Lukezic, 1974a).

Serratia species occur on plants, in the digestive tract of rodents (unpublished data), and in soil and water. *S. ficaria* is especially associated with the fig/fig-wasp ecosystem (Grimont et al., 1979b).

ENRICHMENT AND ISOLATION PROCEDURES

Fecal samples (diluted with distilled water) or plant material washings are inoculated onto caprylate–thallus (CT) agar (Starr et al., 1976). After 2–5 d, the growth is removed by scraping and tested for deoxyribonuclease (DNase) activity. DNase-positive cultures are then purified by streaking on a nonselective medium (e.g., tryptic soy agar). Different colonial types are tested for DNase, and DNase-positive isolates are then thoroughly characterized and identified This procedure allows isolation of all *Serratia* species as defined in this chapter. *Providencia*, *Acinetobacter*, and fluorescent *Pseudomonas* strains can grow on CT agar when samples contain large numbers of these organisms. Other selective media based on DNase production and antibiotic resistance have been proposed (Cate, 1972; Berkowitz and Lee, 1973; Farmer et al., 1973). These antibiotic-containing media are efficient for the isolation of *S. marcescens* but may not be as reliable for more sensitive species (e.g., *S. plymuthica*).

MAINTENANCE PROCEDURES

For short-term preservation (several months), heavy suspensions of bacteria in sterile distilled water are made from bacterial growth scraped with a platinum loop from a nutrient agar slant. The suspensions are stored at room temperature. For longer preservation (several years), screw-capped tubes containing semisolid nutrient agar are stab-inoculated. After overnight growth at 30°C, the tubes are tightly closed and kept at room temperature in the dark. Maintenance failure may occur if the tube is not protected from desiccation by a rubber seal in the screw cap. Rubber corks dipped in melted paraffin wax may be preferred in place of screw caps.

Bacterial suspensions in brain–heart infusion supplemented with 50% glycerol and glass beads can be frozen at −80°C (cryoconservation). When needed, a glass bead can be taken without thawing the cryoconservation and transferred to a sterile broth for subculture. For long-term preservation (over 5 years), freeze-drying is preferred.

PROCEDURES FOR TESTING SPECIAL CHARACTERS

Carbon source utilization test This test is done by using Biotype-100 strips (bioMérieux, Craponne, France) that contains 99 pure carbon sources. Bottles of Biotype medium 1 are inoculated with a calibrated bacterial suspension (Grimont et al., 1996). The strips are incubated at 30°C for 4 d. Growth is scored visually after 2 and 4 days by comparison with the control cupule (without carbon source). The incubation day when growth is observed (1–4) is recorded for each cupule. Program Recognizer (Taxotron package, Institut Pasteur, Paris, France) can be used to enter the Biotype-100 data in an Apple Macintosh for automatic identification (Grimont et al., 1996).

Glucose oxidation test A 1-liter portion of glucose oxidation medium is composed of basal medium containing 8 g of nutrient broth (Difco Laboratories, Detroit, Mich.), 0.02 g of bromcresol purple, and 0.62 g of $MgSO_4 \cdot 7H_2O$ (2.5 mM). To 9.2 ml of autoclaved (121°C, 15 min) basal medium is added 0.4 ml of a filter-sterilized 1 M D-glucose solution (final concentration, 40 mM). This medium is supplemented with 0.4 ml of a fresh, sterile 25 mM iodoacetate solution (final concentration, 1 mM). The glucose oxidation medium is distributed (0.5–ml portions) into glass tubes (11 by 75 mm), which are plugged with sterile cotton wool. Bacteria grown overnight at 20 or 30°C on tryptocasein soy agar (Biorad, Marnes-la Coquette, France) supplemented with 0.2% (w/v) D-glucose are collected with a platinum loop, suspended in a sterile 2.5 mM $MgSO_4$ solution and adjusted to an absorbance at 600 nm of about 4. Glucose oxidation medium is then inoculated with 50 µl of bacterial suspension and vigorously shaken at 270 strokes per minute overnight at 20 or 30°C. The test is positive when a yellow color develops (acid production). In a negative test, the medium remains purple. When negative, the glucose oxidation test is repeated in the presence of 10 µM pyrroloquinoline quinone (PQQ). The control strains used are the type strains of *S. marcescens* (positive without requirement for PQQ), *S. liquefaciens* (positive only when PQQ is supplied), and *Hafnia alvei* (negative) (Bouvet et al., 1989).

Gluconate- and 2-ketogluconate dehydrogenase tests The gluconate dehydrogenase test is done as follows (Bouvet et al., 1989). Bacteria are grown overnight at 20°C on tryptocasein soy agar supplemented with 0.2% D-gluconate, then collected with a platinum loop and suspended in sterile distilled water to an absorbance (at 600 nm) of about 4. The reaction medium contains 0.2 M acetate buffer (pH 5), 1% (w/v) Triton X-100, 2.5 mM $MgSO_4$, 75 mM gluconate and (added immediately before use) 1 mM iodoacetate. A control medium contains the same ingredients except gluconate. The reaction medium and the control medium are dispensed (100-µl portions) into 96-well microtiter plates (Dynatech AG, Denkendorf, FRG). Bacterial suspen-

sions (10 µl) are added to the reaction and control media and the microtiter plates are incubated at 20°C for 20 min. Then, 10-µl portions of a 0.1 M potassium ferricyanide solution (kept in the dark at room temperature for no longer than 1 week) are added to the wells and the plates are gently shaken and incubated at 20°C for 40 min. Then 50-µl portions of a reagent ($Fe_2(SO_4)_3$, 0.6 g; SDS, 0.36 g; 85% phosphoric acid, 11.4 ml; distilled water to 100 ml) are added to the wells. The plates are examined for the development of a green to blue color (due to Prussian blue) within 15 min at room temperature. The color in the uninoculated control medium remains yellow. The suggested control strains used are *Escherichia coli* K12 (negative) and the type strain of *Serratia marcescens* (positive).

The 2-ketogluconate dehydrogenase test is the same as above, except that the control and reaction media are adjusted to pH 4.0 and 2-ketogluconate is used in place of gluconate in the reaction medium. Suggested control strains are *Escherichia coli* K12 (negative) and the type strain of *Serratia marcescens* (positive at 20°C, not at 30°C) (Bouvet et al., 1989).

Glycerol dehydrogenase test (Bouvet et al., 1995a) For the detection of glycerol dehydrogenase, bacteria are grown overnight at 30°C on tryptocasein soy agar plates (TCS, BioRad, Marnes-la-Coquette, France) supplemented with 1% glycerol or 70 mM hydroxyacetone. Bacterial growth is collected, suspended in reaction buffer [0.1 M K_2CO_3 and 30 mM $(NH_4)_2SO_4$, adjusted to pH 9.0] to an absorbance of 0.6, and bacterial suspensions are dispensed into 96-well microtiter plates. Then, 30-µl aliquots of a reagent (NAD^+, 210 mg; glycerol, 600 ml; nitro-blue tetrazolium, 42 mg; phenazine methosulfate, 2 mg; distilled water to 10 ml) are added to the wells and incubated in the dark with shaking. The plate is examined for a purple color developed within 15 to 30 min (presence of a glycerol dehydrogenase). The color in the uninoculated control medium remains yellow. Suggested control strains are the type strain of *Serratia liquefaciens* (glycerol dehydrogenase induced by glycerol, not by hydroxyacetone) and the type strain of *Enterobacter asburiae* (dehydrogenase induced by hydroxyacetone, not by glycerol) (Bouvet et al., 1995a).

Voges–Proskauer test (Richard's modification) Clark and Lubs medium (BBL) is dispensed in large 22 × 215 mm tubes (0.5 ml per tube) and inoculated with 0.05 ml of a heavy bacterial suspension in distilled water. After incubation at 30°C for 18 h, 0.5 ml of α-naphthol solution (6% w/v alcoholic solution) and 0.5 ml of 4 M NaOH are added. The tubes are shaken, heated a few seconds in a Bunsen flame, and examined for a red color (Richard, 1972).

Tetrathionate reduction The medium of Le Minor et al. (1970a) contains: peptone (Difco), 10.0 g, NaCl, 5.0 g; $K_2S_4O_6$, 5.0 g; bromothymol blue (0.2% aqueous solution), 25 ml; and distilled water to 1 liter. Adjust the pH to 7.4, sterilize by filtration, and dispense into 12 × 120 mm tubes (4 ml per tube). The size of the tubes (for a rather limited aeration) is critical. Inoculated tubes are incubated at 30°C for 24 h and examined for a yellow color (tetrathionate reduction).

β-Xylosidase Paper disks (0.5 cm) are loaded with 0.1 ml of a 2% (w/v) aqueous solution of *p*-nitrophenyl-β-D-xylopyranoside and kept dry in a tightly-capped flask at 4°C. The test is performed exactly like the β-galactosidase test, but with *p*-nitrophenyl-β-D-xylopyranoside disks in place of ONPG disks (Brisou et al., 1972).

H-Immobilization test The motility of each isolate to be typed must be enhanced by passage through a 0.3% semisolid agar U-tube.

The following autoclaved semisolid medium is dispensed in 2.0-ml volumes into small (92 × 13 mm) screw-capped tubes: tryptic peptone, 20.0 g; D-mannitol, 2.0 g; KNO$_3$, 1.5 g; phenol red solution (1%), 4 ml; agar, 4.5 g. distilled water, 1000 ml; pH 7.4. The tubes of semisolid medium are melted (boiling water bath), cooled to 50°C in a water bath, supplemented with 0.05 ml of each serum dilution under sterile conditions, and allowed to gel.

Tubes with serum dilutions (and control tubes without serum) are stab-inoculated with a highly motile culture. After overnight incubation, tubes are examined for immobilization. This H-immobilization test is very specific and much easier to perform than the classical H-agglutination (Le Minor and Pigache, 1977).

DIFFERENTIATION OF THE GENUS *SERRATIA* FROM OTHER GENERA

Table BXII.γ.265 provides the primary characteristics that can be used to differentiate the genus *Serratia* from biochemically similar taxa.

TAXONOMIC COMMENTS

A number of changes have been made since the eighth edition of the *Manual of Determinative Bacteriology*, in which it was indicated that the genus *Serratia* was composed of only one species, *S. marcescens* (the type species). Objective approaches such as numerical taxonomy and DNA relatedness applied to strains recovered from diverse habitats delineated an increasing number of species in the genus. Seven species were mentioned in the first edition of the *Bergey's Manual of Systematic Bacteriology* (Grimont and Grimont, 1984) and 10 species were mentioned in the second edition of *The Prokaryotes* (Grimont and Grimont, 1995). Ten species are currently known to belong in the genus *Serratia*.

Transfer of *Enterobacter liquefaciens* to the genus *Serratia* was first proposed by Barbe (1969) and supported by studies on bacteriocin cross-reactions between *S. marcescens* and *E. liquefaciens*

(Hamon et al., 1970). Valid publication of the new combination *S. liquefaciens* followed a numerical taxonomy study (Bascomb et al., 1971).

A phenon named "biovar 2" (Bascomb et al., 1971) and "phenon B" (Grimont and Dulong de Rosnay, 1972) was thought identical to *"Bacterium rubidaeum"* Stapp 1940 and named *S. rubidaea* (Ewing et al., 1973). The same phenon was also identified as *S. marinorubra* Zobell and Upham 1944 (Grimont et al., 1977b). *S. rubidaea* and *S. marinorubra* were based on different type strains (ATCC 27593 and ATCC 27614, respectively). The *Approved Lists of Bacterial Names*, however, gives both names (*S. rubidaea* and *S. marinorubra*) with the same type strain (viz. ATCC 27614, the type strain of *S. rubidaea*). Hence, both names, which were subjective synonyms, are now objective synonyms and redundant. To avoid further confusion, the name *S. rubidaea* (Stapp) Ewing (1986a) should now be used exclusively to designate the same (*S. rubidaea–S. marinorubra*) taxon.

The ancient species *S. plymuthica* (Lehmann and Neumann 1896) Breed, Murray, and Hitchens 1948 was shown to be a valid species by numerical taxonomy (Grimont et al., 1977b) and by DNA–DNA hybridization (Grimont et al., 1978a).

Three species, *S. odorifera* (Grimont et al., 1978a), *S. ficaria* (Grimont et al., 1979b), and *S. entomophila* (Grimont et al., 1988), were defined by DNA relatedness, carbon source utilization tests, and by standard biochemical tests. DNA relatedness studies have shown that *S. marcescens*, *S. plymuthica*, *S. rubidaea*, *S. odorifera*, and *S. ficaria* are homogeneous and discrete genomospecies (Steigerwalt et al., 1976; Grimont et al., 1978a, 1979b, 1988).

S. liquefaciens sensu lato was shown to be heterogeneous (Steigerwalt et al., 1976) and later found to be composed of several genomospecies. One biovar (Clc) of *S. liquefaciens* was identified as *Erwinia proteamaculans* (Paine and Stanfield 1919) Dye 1966a and renamed *S. proteamaculans* (Grimont et al., 1978b). When DNA binding ratios were examined without studies on the thermal stability of hybridized molecules, *S. proteamaculans* was thought to be a subjective synonym of *S. liquefaciens* (Grimont et al., 1978b). Reexamination of DNA relatedness (including thermal stability studies) in *S. liquefaciens sensu lato* disclosed at least

TABLE BXII.γ.265. Differential characteristics of the genus *Serratia* and other biochemically similar taxa[a]

Characteristics	Serratia spp.[b]	Serratia fonticola	Pantoea	Enterobacter cloacae	Pectobacterium[c]	Klebsiella[d]
Carbon source utilization:[e]						
4-Aminobutyrate (M)	+	−	+	d	−	d
4-Aminovalerate (B,M)	−	−	−	−	−	d
Arginine (M)	−	−	−	−	−	+
Caprate (M)	+	−	−	−	−	−
Caproate (M)	+	−	−	−	−	−
Caprylate (M)	+	−	−	−	−	−
D-Dulcitol (B,M)	−	+	−	−	−	−
L-Fucose (B,M)	+	−	−	−	−	+
3-Phenylpropionate (B,M)	−	+	−	−	−	−
Tagatose (B,M)	−	+	−	−	−	d
Tyrosine (M)	+	-	−	−	−	d
Voges–Proskauer test	+	−	+	+	+	+
Gelatin hydrolyzed	+	−	d	d	d	d
Tributyrin hydrolyzed	+	−	−	−	−	−
Deoxyribonuclease	+	−	−	−	d	−
Gluconate dehydrogenase	+	+	+	−	−	d
Mol% G+C of DNA	52–60	49–52	53–56	53	51–54	54–57

[a]Symbols: see standard definitions.

[b]Except *S. fonticola*.

[c]According to Bouvet et al. (1989).

[d]Including *Klebsiella pneumoniae* and *K. mobilis* (*Enterobacter aerogenes*).

[e]Determined in Biotype-100 (B) or minimal agar (M).

three genomospecies: *S. liquefaciens sensu stricto, S. proteamaculans* (Grimont et al., 1982b), and a third group containing strain ATCC 14460 and named *S. grimesii* (Grimont et al., 1982a, b).

A group of strains called "*Citrobacter* lysine + " or "*Citrobacter*-like" was found to be significantly related to the genus *Serratia* in DNA–DNA hybridization studies (Crosa et al., 1974). This genomospecies has been named *Serratia fonticola* Gavini et al. 1979; however, a difficulty is that *S. fonticola* does not have the key characteristics of the genus *Serratia*. Furthermore, *Serratia* phages that are active on strains of any *Serratia* species (as defined herein) have been found to be inactive on all *S. fonticola* strains tested (unpublished data). Bacteriocins from *Serratia* are also inactive on *S. fonticola* (Hamon, personal communication). However, the 16S rRNA gene sequence of *S. fonticola* branches within the psychrotolerant *Serratia* cluster (*S. liquefaciens, S. proteama-* *culans, S. grimesii,* and *S. plymuthica*) and this justifies the inclusion of *S. fonticola* in the genus *Serratia* (Dauga et al., 1990).

FURTHER READING

Ewing, W.H. 1986. Edwards and Ewing's identification of *Enterobacteriaceae*, 4th Ed., Elsevier, New York.

Grimont, P.A.D. and F. Grimont. 1978. The genus *Serratia*. Annu. Rev. Microbiol. *32*: 221–248.

Grimont, F. and P. Grimont. 1995. The genus *Serratia. In* Balows, Trüper, Dworkin, Harder and Schleifer (Editors), The Prokaryotes. A Handbook on the Biology of Bacteria: Ecophysiology, Isolation, Identification, Applications, Springer-Verlag, New York. pp. 2822–2848.

Grimont, P.A.D., F. Grimont, H.L.C. Dulong de Rosnay and P.H.A. Sneath. 1977. Taxonomy of the genus *Serratia*. J. Gen. Microbiol. *98*: 39–66.

von Graevenitz, A. and S.J. Rubin (Editors). 1980. The genus *Serratia*, CRC Press, Boca Ration, Florida.

DIFFERENTIATION OF THE SPECIES OF THE GENUS *SERRATIA*

The differential characteristics of the species of *Serratia* are indicated in Table BXII.γ.266. Other characteristics of the species are listed in Table BXII.γ.267.

List of species of the genus Serratia

1. **Serratia marcescens** Bizio 1823, 288[AL]

   mar.ces' cens. M.L. v. *marcesco* to fade; L. part. adj. *marcescens* fading away.

   The cell morphology and colonial morphology are as given for the genus. Prodigiosin or pyrimine can be produced.

   Physiological and nutritional characteristics are presented in Tables BXII.γ.266 and BXII.γ.267.

   A biotyping system based on pigment production, tetrathionate reduction and utilization of *meso*-erythritol, trigonelline, quinate, benzoate, 3-hydroxybenzoate, 4-hydroxybenzoate, and DL-carnitine as sole carbon sources, has been described (Grimont and Grimont, 1978a). DL-carnitine is not in Biotype-100 strips and can be replaced by D-malate and *meso*-tartrate. Groups of biovars (called biogroups) (Table BXII.γ.268) correspond to definite, nonoverlapping sets of serovars (Table BXII.γ.269) (Grimont et al., 1979a).

   Nonpigmented biogroups A3 and A4 are ubiquitous. Nonpigmented biogroups A5/8 and TCT are almost confined to hospitalized patients. Pigmented biogroups A1 and A2/6 are found in the natural environment and occasionally in human patients.

   The mol% G + C of the DNA is: 57.5–60 (T_m, Bd).

   Type strain: ATCC 13880, CIP 103235, DSM 30121, DSM 47, JCM 1239, NCDC 813-60, NCIB 9155, NCTC 10211.

   GenBank accession number (16S rRNA): AJ233431, M59160.

2. **Serratia entomophila** Grimont, Jackson, Ageron and Noonan 1988, 5[VP]

   en.to.mo' phi.la. Gr. n. *entomon* insect; Gr. v. *phylein* love; L. fem. adj. *entomophila* insect loving.

   The cell morphology and colonial morphology are as given for the genus. Prodigiosin is not produced.

   Physiological and nutritional characteristics are presented in Tables BXII.γ.266 and BXII.γ.267.

   Two biotypes can be delineated (Table BXII.γ.270).

   Isolated from larvae of *Costelytra zealandica* (grass grub) with amber disease, and from the environment. No strain has been identified as being involved in a human, animal (other than insect), or plant disease.

   The mol% G + C of the DNA is: 58 (T_m).

   Type strain: A1, ATCC 43705, CIP 102919, DSM 12358.

   GenBank accession number (16S rRNA): AJ233427.

3. **Serratia ficaria** Grimont, Grimont and Starr 1981d, 216[VP] (Effective publication: Grimont, Grimont and Starr 1979b, 282.)

   fi.ca' ri.a. M.L. fem. adj. *ficaria* of figs.

   The cell morphology and colonial morphology are as given for the genus. Prodigiosin is not produced. Cultures give off a musty, potato-like odor.

   Physiological and nutritional characteristics are presented in Tables BXII.γ.266 and BXII.γ.267.

   Associated with the fig/fig-wasp biological cycle. Occasionally found on plants other than fig trees.

   The mol% G + C of the DNA is: 59.6 (T_m).

   Type strain: ATCC 33105, CIP 79-23, DSM 4569, ICPB 4050, JCM 1241.

   GenBank accession number (16S rRNA): AJ233428, AB004745.

4. **Serratia fonticola** Gavini, Ferragut, Izard, Trinel, Leclerc, Lefebvre and Mossel 1979, 98[AL]

   fon.ti' co.la. M.L. n. *fons, fontis* spring, fountain; L. suff. *-cola* dweller; M.L. n. *fonticola* spring-dweller.

   The cell morphology and colonial morphology are as given for the genus. Prodigiosin is not produced.

   Does not share the key characteristics of the genus *Serratia*.

   Physiological and nutritional characteristics are presented in Tables BXII.γ.266 and BXII.γ.267.

   Occurs in freshwater.

   The mol% G + C of the DNA is: 48.8–52.5 (T_m).

   Type strain: ATCC 29844, CIP 78.64, CCUG 37824, DSM 4576, JCM 1242, LMG 7882.

   GenBank accession number (16S rRNA): AJ233429.

TABLE BXII.γ.266. Characteristics differentiating the species of the genus *Serratia*[a]

Characteristics	1. S. marcescens	2. S. entomophila	3. S. ficaria	4. S. fonticola	5. S. grimesii	6. S. liquefaciens	7. S. odorifera	8. S. plymuthica	9. S. proteamaculans	10. S. rubidaea
Prodigiosin production	d	−	−	−	−	−	−	d	−	+
Potato-like odor	−	−	+	−	−	−	+	−	−	d
Indole production	−	−	−	−	−	−	+	−	−	−
Lysine decarboxylase	+	−	−	+	+	+	+	−	+	d
Ornithine decarboxylase (Møller)	+	−	−	+	+	+	d	−	+	−
Arginine decarboxylase (Møller)	−	−	−	−	+	−	−	−	−	−
Tween 80 hydrolysis	+	+	+	+	+	+	−	+	+	+
Malonate test	−	−	−	+	−	−	−	−	−	d
Carbon source utilization:										
Adonitol	+	+	+	+	−	−	+	−	d	+
L-Arabinose	−	−	+	+	+	+	+	+	+	+
D-Arabitol	−	d	+	+	−	−	−	−	−	+
L-Arabitol	+	d	+	+	−	−	+	−	−	−
Betaine	−	−	−	−	−	−	−	d	−	+
D-Cellobiose	−	+	+	d	d	+	+	+	d	+
Dulcitol	−	−	−	+	−	−	−	−	−	−
i-Erythritol	d	−	+	+	−	−	d	−	d	+
L-Fucose	+	+	+	−	+	+	+	+	+	+
Gentiobiose	−	+	+	+	d	+	d	+	+	+
Itaconate	−	+	−	−	−	−	−	−	−	−
5-Ketogluconate	+	+	+	d	+	+	d	d	+	−
Malitol	−	−	+	d	+	d	−	+	+	+
D-Melezitose	−	−	+	−	d	d	−	+	+	d
D-Melibiose	−	−	+	+	+	+	+	+	+	+
Mucate	−	−	+	d	−	−	+	d	−	+
Palatinose	−	−	+	+	−	d	−	+	+	+
3-Phenylpropionate	−	−	−	d	−	−	−	−	−	−
Quinate	d	d	+	−	−	−	−	+	−	+
D-Raffinose	−	−	+	+	+	+	d	+	+	+
L-Rhamnose	−	−	+	+	−	−	+	−	d	−
D-Saccharate	−	−	+	+	−	−	+	d	−	+
D-Sorbitol	+	−	+	+	+	+	+	d	d	−
D-Tagatose	−	−	−	+	−	−	−	−	−	−
meso-Tartrate	d	−	−	−	−	+	+	−	−	+
Trigonelline	d	−	d	−	−	−	+	−	−	+
D-Turanose	−	−	+	d	+	d	−	+	+	+
Xylitol	+	−	+	d	−	−	d	−	−	d
D-Xylose	−	d	+	d	+	+	+	+	+	+

[a]Data from Grimont and Grimont (1995) and Gavini et al. (1979). For symbols see standard definitions.

5. **Serratia grimesii** Grimont, Grimont and Irino 1983b, 438[VP] (Effective publication: Grimont, Grimont and Irino 1982a, 73.)

gri.me' sii. M.L. gen. masculine form of Grimes.

S. grimesii has been isolated from the natural environment and from human clinical specimens.

Physiological and nutritional characteristics are presented in Tables BXII.γ.266 and BXII.γ.267.

Biogroup ADC might represent a subspecies of *S. grimesii.*

The mol% G + C of the DNA is: not determined.

Type strain: ATCC 14460, CIP 103361, DSM 30063, IFO 13537, JCM 5910.

GenBank accession number (16S rRNA): AJ233430.

6. **Serratia liquefaciens** (Grimes and Hennerty 1931) Bascomb, Lapage, Willcox and Curtis 1971, 293[AL] ("*Aerobacter liquefaciens*" Grimes and Hennerty 1931, 93.)

li.que.fa' ciens. M.L. part. adj. *liquefaciens* dissolving.

The cell morphology and colonial morphology are as given for the genus. Prodigiosin is not produced.

Physiological and nutritional characteristics are presented in Tables BXII.γ.266 and BXII.γ.267.

S. liquefaciens sensu lato was composed of several biovars and some of these were found to constitute genomospecies which were subsequently given species status: biovar C1ab (including the type strain of *S. liquefaciens*) corresponded to *S. liquefaciens sensu stricto*; biovars C1c (including the type strain of *S. proteamaculans*), EB, RB, and RQ corresponded to *S. proteamaculans*; and biovars C1d and ADC corresponded to *S. grimesii* (Table BXII.γ.271) (Grimont et al., 1977b, 1982a, b).

S. liquefaciens is the most prevalent *Serratia* species in the natural environment (plants, digestive tract of rodents). Occasionally encountered as an opportunistic pathogen.

The mol% G + C of the DNA is: 53–54 (T_m, Bd).

Type strain: ATCC 27592, CIP 103238, DSM 4487, JCM 1245, LMG 7884, NCTC 12962.

GenBank accession number (16S rRNA): AJ306725, AB004752.

It should be mentioned that the strain formerly considered to be the type strain of "*Aerobacter liquefaciens*" or "*Aerobacter lipolyticus*" by Grimes (1961) was ATCC 14460 (now the type strain of *S. grimesii*). Since this strain was considered atypical compared to other strains labeled as *S. liquefaciens*, another strain (ATCC 27592) was given as type strain of *S. liquefaciens* in the Approved Lists.

TABLE BXII.γ.267. Other characteristics of the species of the genus *Serratia*[a]

Characteristics	1. S. marcescens	2. S. entomophila	3. S. ficaria	4. S. fonticola	5. S. grimesii	6. S. liquefaciens	7. S. odorifera	8. S. plymuthica	9. S. proteamaculans	10. S. rubidaea
Carbon source utilization:										
N-Acetyl-D-glucosamine	+	+	+	+	+	+	+	+	+	+
cis-Aconitate	+	+	+	+	+	+	+	+	+	+
trans-Aconitate	+	+	+	d	−	−	+	d	d	+
D-Alanine	+	+	+	d	+	+	+	+	+	+
L-Alanine	+	+	+	d	d	+	+	+	+	+
4-Aminobutyrate	d	d	+	−	+	+	+	d	+	+
5-Aminovalerate	−	−	−	−	−	−	−	−	−	−
L-Aspartate	+	+	+	d	+	+	+	+	+	+
Benzoate	d	d	d	−	d	−	−	d	d	d
Caprate	+	d	+	−	d	+	d	−	d	d
Caprylate	d	d	+	−	d	d	d	−	d	d
Citrate	+	+	+	+	+	+	+	+	+	+
m-Coumarate	−	−	−	−	−	−	−	−	−	−
Ethanolamine	−	−	−	−	−	−	−	−	−	−
D-Fructose	+	+	+	+	+	+	+	+	+	+
Fumarate	+	+	+	d	+	+	+	+	+	+
D-Galactose	+	+	+	+	+	+	+	+	+	+
D-Galacturonate	+	+	+	+	+	+	+	+	+	+
Gentisate	d	−	−	d	−	−	−	−	d	−
D-Gluconate	+	+	+	+	+	+	+	+	+	+
D-Glucosamine	+	+	+	+	+	+	+	+	+	+
D-Glucose	+	+	+	+	+	+	+	+	+	+
D-Glucuronate	+	+	+	+	+	+	+	+	d	+
L-Glutamate	+	+	+	d	+	+	+	+	+	+
Glutarate	−	−	−	−	−	−	−	−	−	−
DL-Glycerate	+	+	+	+	+	+	+	d	+	+
Glycerol	+	+	+	+	+	+	+	+	+	+
Histamine	−	−	−	−	−	−	−	−	−	d
3-Hydroxybenzoate	d	−	−	d	−	−	−	−	−	−
4-Hydroxybenzoate	d	−	d	−	−	−	−	−	−	−
3-Hydroxybutyrate	d	−	d	−	−	−	d	−	−	−
myo-Inositol	+	+	+	+	+	+	+	+	+	+
2-Ketogluconate	+	+	+	+	+	+	+	+	+	+
2-Ketoglutarate	d	d	d	d	+	d	+	d	d	d
L-Lactate	+	+	d	d	+	+	+	d	+	+
Lactose	−	−	d	d	+	d	+	+	d	+
Lactulose	−	−	−	+	d	−	d	d	−	d
D-Lyxose	+	d	d	d	+	+	−	−	+	−
D-Malate	d	d	−	d	d	+	+	d	−	d
L-Malate	+	+	+	d	+	+	+	+	+	+
Malonate	−	−	−	d	−	−	−	−	−	d
Maltose	+	+	+	+	+	+	+	+	+	+
Maltotriose	+	+	+	+	+	+	+	+	+	+
D-Mannitol	+	+	+	+	+	+	+	+	+	+
D-Mannose	+	+	+	+	+	+	+	+	+	+
1-*O*-Methyl-α-galactoside	−	−	+	+	+	+	+	+	−	+
1-*O*-Methyl-β-galactoside	d	−	d	+	d	d	+	+	−	+
3-*O*-Methyl-D-glucose	−	−	−	d	d	d	d	d	+	−
1-*O*-Methyl-α-D-glucoside	−	−	d	d	−	d	−	d	d	+
1-*O*-Methyl-β-D-glucoside	+	+	+	+	+	+	+	+	+	+
Phenylacetate	d	d	d	d	d	+	d	d	d	+
L-Proline	+	+	+	d	d	+	+	+	+	+
Protocatechuate	d	−	d	−	d	d	−	+	d	d
Propionate	d	d	d	−	−	−	−	d	−	−
Putrescine	+	+	+	+	+	+	+	d	+	+
D-Ribose	+	+	+	+	+	+	+	+	+	+
L-Serine	+	+	+	d	+	+	+	+	+	+
L-Sorbose	−	−	−	−	−	−	−	−	−	−
Succinate	+	+	+	d	+	+	+	+	+	+
Sucrose	+	+	+	d	+	+	d	+	+	+
D-Tartrate	−	−	−	d	−	−	d	−	−	d
L-Tartrate	−	−	−	−	−	−	d	−	−	d
D-Trehalose	+	+	+	+	+	+	+	+	+	+
Tricarballylate	−	−	−	d	−	−	−	−	−	d
Tryptamine	−	−	−	−	−	−	−	−	−	−
L-Tyrosine	d	d	+	−	d	d	+	d	d	+

(continued)

TABLE BXII.γ.267. *(cont.)*

Characteristics	1. S. marcescens	2. S. entomophila	3. S. ficaria	4. S. fonticola	5. S. grimesii	6. S. liquefaciens	7. S. odorifera	8. S. plymuthica	9. S. proteamaculans	10. S. rubidaea
Oxidation of:										
Glucose to gluconate (without cofactor)	+	+	+	−	−	−	+	d	−	+
2-Ketogluconate to 2,5-diketogluconate	+	−	−	−	+	+	−	−	−	−
Glycerol dehydrogenase induced by glycerol, not hydroxyacetone	+	−	−	+	+	+	−	−	+	−
Acid produced from:										
Adonitol	d	+	+	+	−	−	v	−	−	+
L-Arabinose	−	−	+	+	+	+	+	+	+	+
D-Melibiose	−	−	+	+	+	+	+	+	+	+
myo-Inositol	d	−	+	+	+	+	+	d	+	d
D-Raffinose	−	−	+	+	+	+	d	+	+	+
L-Rhamnose	−	−	+	v	−	−	+	−	−	−
D-Sorbitol	+	−	+	+	+	+	+	d	+	−
Sucrose	+	+	+	−	+	+	d	+	+	+
D-Xylose	−	d	+	d	+	+	+	+	+	+
Growth at 5°C	−	+	+	+	+	+	+	+	+	−
Growth at 37°C	+	+	+	+	+	+	+	d	+	+
Growth at 40°C	+	+	−	+	−	−	+	−	−	d
Growth in NaCl:										
7% (w/v)	+	+	+	+	d	d	+	d	d	+
8.5% (w/v)	d	−	−	−	−	−	+	−	−	+
10% (w/v)	−	−	−	−	−	−	−	−	−	d
Tetrathionate reduced	d	−	−	+	+	+	−	−	+	−
Gas from glucose	−	−	−	d	+	+	−	d	+	−
β-Xylosidase	−	−	d	+	−	−	+	d	−	+
Tween 40 hydrolysis	+	+	+	+	+	+	+	+	+	+
Tween 60 hydrolysis	+	+	+	+	+	+	d	+	+	+
Chitin hydrolysis	+	d	+	−	d	d	−	+	d	+
Methyl red test	−	−	−	+	d	d	+	d	d	−
L-Histidine deaminase	−	−	−	−	−	−	−	−	−	−
Tryptophan deaminase	−	−	−	−	−	−	−	−	−	−
β-glucuronidase	−	−	−	−	−	−	−	−	−	−
Esculin hydrolysis	+	+	+	+	+	+	+	+	d	+

[a]Data from Grimont and Grimont (1995) and Gavini et al. (1979). For symbols see standard definitions.

TABLE BXII.γ.268. Identification of *S. marcescens* biogroups and biovars[a]

	Biogroups																	
	A1		A2		A6	A3				A4		A5	A8					
Characteristics	a	b	a	b	a	a	b	c	d	a	b		a	b	c	TCT	TC	TT
Prodigiosin production	+	+	+	+	+	−	−	−	−	−	−	−	−	−	−	−	−	−
Growth on:[b]																		
meso-Erythritol	+	+	+	+	+	+	+	+	+	+	+	+	+	−	−	−	−	−
Benzoate	+	+	−	−	−	−	−	−	−	−	−	−	−	−	−	−	−	−
Quinate and/or 4-hydroxybenzoate	−	−	−	−	+	+	−	−	−	+	−	+	+	+	+	−	−	−
3-Hydroxybenzoate	−	−	−	−	−	+	+	−	−	−	−	−	+	−	−	−	−	−
Trigonelline	−	−	−	−	−	−	+	−	+	−	+	+	+	+	−	−	−	+
D-Malate/*m*-tartrate	+	−	−	+	d	−	−	d	−	+	+	+	d	−	−	+	+	−
Gentisate	−	−	+	+	+	+	+	d	−	+	+	−	d	+	d	−	−	−
Tetrathionate reduction	+	+	+	+	+	+	+	+	+	−	−	+	+	+	+	+	+	+

[a]For symbols see standard definitions.

[b]Carbon source utilization test.

7. **Serratia odorifera** Grimont, Grimont, Richard, Davis, Steigerwalt and Brenner 1978a, 461[AL]

o.do.ri.fe′ra. M.L. fem. adj. *odorifera* bringing odors, fragrant.

The cell morphology and colonial morphology are as given for the genus. Prodigiosin is not produced. Cultures give off a musty, potato-like odor.

Physiological and nutritional characteristics are presented in Tables BXII.γ.266 and BXII.γ.267. Two biovars can be recognized (Table BXII.γ.272).

The capsular antigen reacts with *Klebsiella* antisera K4 or K68.

Rare opportunistic pathogen. Occasionally isolated from plants or food.

The mol% G + C of the DNA is: 54.6 (T_m).

TABLE BXII.γ.269. Correspondence between serovars and biogroups in *Serratia marcescens*

Biogroup[a]	O:H serovars[b]
A1	5:2; 5:3; 5:13; 5:23; 10:6; 10:13; 28:2
A2/6	5:23; 6,14:2; 6,14:3; 6,14:8; 6,14:9; 6,14:10; 6,14:13; 8:3; 13:5
A3	3:5; 3:11; 4:5; 4:18; 5:6; 5:15; 6,14:5; 6,14:6; *6,14:16*; 6,14:20; *9:9*; *9:11*; *9:15*; 9:17; 12:5; 12:9; *12:10*; 12:11; *12:15*; *12:16*; 12A:17; *12:18*; 12:20; *12:26*; 13:11; *13:17*; 15:3; 15:5; 15:8; 15:9; 17:4; 18:21; *18:26*; 22:11; *23:19*, 26:20
A4	1:1; 1:4; 2:1; 2:8; 3:1; 4:1; 4:4; 5:1; 5:6; 5:8; 5:24; 9:1; 13:1; 13:11; 13:13
A5/8	*2:4*; 3:12; 3,21:12; 4:12; 5:4; 6,14:4; 6,14:12; *8:4*; 8:12; 15:12; 21:12, 25:12
TCT	*1:7*; 2:7; *4:7*; 5:7; 5:19; *7:7*; 7:23; 10:9; 11:4; 13:7; 13:12; 16:19; *18:9*; 18:16; *18:19*; 19:14; *19:19*; 24:6; 27:–
TC	10:8; 20:12

[a]Biogroup A1 is composed of biotypes A1a and A1b; A2/6 of A2a, A2b, A6; A3 of A3a, A3b, A3c, A3d; A4 of A4a, A4b; A5/8 of A5, A8a, A8b, and A8.

[b]Serovars for which exceptions to the correspondence occur are in italics.

TABLE BXII.γ.270. Identification of *Serratia entomophila* biotypes[a]

Characteristics	Biotype 1b	Biotype 2
Growth on:[b]		
D-Arabitol	+	–
L-Arabitol	–	+
D-Malate	–	d
D-Sorbitol	–	d
Quinate	+	d
D-Xylose	–	+

[a]For symbols see standard definitions.

[b]The type strain corresponds to biotype 1.

Type strain: ATCC 33077, CIP 79-1, DSM 4582, ICPB 3995, NCTC 11214.

GenBank accession number (16S rRNA): AJ233432.

8. **Serratia plymuthica** (Lehmann and Neumann 1896) Breed, Murray and Hitchens 1948, 481[AL] (*Bacterium plymuthicum* (sic) Lehmann and Neumann 1896, 264.)

ply.mu' thi.ca. M.L. adj. *plymuthica* pertaining to Plymouth, UK.

The cell morphology and colonial morphology are as given for the genus. Prodigiosin is produced by most strains.

Physiological and nutritional characteristics are presented in Tables BXII.γ.266 and BXII.γ.267.

Most *S. plymuthica* strains studied were isolated from freshwater. Very rarely found in human sputum. No human infection reported.

The mol% G + C of the DNA is: 53.5–56.5 (T_m).

Type strain: ATCC 183, CIP 103239, DSM 4540, JCM 1244.

GenBank accession number (16S rRNA): AJ233433.

S. plymuthica is cited on the Approved Lists of Bacterial Names (Skerman et al., 1980) as *Serratia plymuthica* (Dyar 1895) Bergey, Harrison, Breed, Hammer and Huntoon 1923, 88. This is incorrect, for reasons discussed by Grimont et al. (1977b).

9. **Serratia proteamaculans** (Paine and Stansfield 1919) Grimont, Grimont and Starr 1978b, 503[AL] (*Pseudomonas proteamaculans* Paine and Stansfield 1919, 38.)

pro.te.a.ma.cu' lans. M.L. n. *Protea* a plant generic name; L. v. *maculo* to spot; M.L. part. adj. *proteamaculans* spotting *Protea*.

S. proteamaculans and *S. liquefaciens* were thought to be synonymous based on DNA relatedness (Grimont et al., 1978b). However, subsequent observation of significant thermal instability of reassociated DNA fragments supported the separation of both species (Grimont et al., 1982b).

The cell morphology and colonial morphology are as given for the genus. Prodigiosin is not produced.

Physiological and nutritional characteristics are presented in Tables BXII.γ.266 and BXII.γ.267.

S. proteamaculans is found in the natural environment (plants, wild rodents, insects, and water) but exceptionally from human clinical specimens).

The mol% G + C of the DNA is: not determined.

Type strain: ATCC 19323, CIP 103236, DSM 4543, NCPPB 245.

According to J-P. Euzeby (http://www.bacterio.cict.fr/corrections2.html), the spelling should have been *proteimaculans*. However, since the name appeared in the Approved Lists, no correction is allowed.

a. **Serratia proteamaculans** subsp. **proteamaculans** (Paine and Stansfield 1919) Grimont, Grimont and Starr 1978b, 503[AL] (*Pseudomonas proteamaculans* Paine and Stansfield 1919, 38.)

The subspecies includes biovars C1c, EB, and RB. The type strain belongs to biotype C1c. Differential characteristics are given in Table BXII.γ.271.

The mol% G + C of the DNA is: not determined.

Type strain: ATCC 19323, CIP 103236, DSM 4543, NCPPB 245

GenBank accession number (16S rRNA): AJ233434.

b. **Serratia proteamaculans** subsp. **quinovora** Grimont, Grimont and Irino 1983b, 438[VP] (Effective publication: Grimont, Grimont and Irino 1982a, 71.)

qui.no' vo.ra. M.L. *quinate*, from Spanish *quina* quinine; and L. v. *voro* to devour; M.L. fem. adj. *quinovora* quinate devouring.

This subspecies corresponds to biotype RQ. Quinate utilization was observed for all strains of this subspecies in a minimal agar when an unwashed inoculum was used. In Biotype-100 strips, no growth is observed on quinate when Biotype Medium 1 is used, some strains grow from quinate when Biotype Medium 2 (containing more growth factors) is used.

The mol% G + C of the DNA is: not determined.

Type strain: ATCC 33765, CIP 81-95, DSM 4597.

GenBank accession number (16S rRNA): AJ233435.

10. **Serratia rubidaea** (Stapp 1940) Ewing, Davis, Fife and Lessel 1973, 224[AL] (*"Bacterium rubidaeum"* Stapp 1940, 259; *Serratia marinorubra* Zobell and Upham 1944, 255.)

ru.bi' dae.a. L. *Rubus idaeus* raspberry, contracted and made to agree in gender with *Serratia*.

The cell morphology and colonial morphology are as given for the genus. Prodigiosin is produced by most strains.

Physiological and nutritional characteristics are presented in Tables BXII.γ.266 and BXII.γ.267.

Three biotypes (B1, B2, B3) correspond to subspecies "*S. rubidaea* subsp. *burdigalensis*", "*S. rubidaea* subsp. *rubidaea*", and "*S. rubidaea* subsp. *colindalensis*" (Grimont et al., manuscript in preparation) (Table BXII.γ.273).

S. rubidaea strains are rarely isolated, either in the natural environment or in human patients. May be found in ripe coconuts (Grimont et al., 1981c).

The mol% G + C of the DNA is: 53.5–58.5 (T_m).

Type strain: Ewing 2199-72, ATCC 27593, CIP 103234, DSM 4480, JCM 1240.

GenBank accession number (16S rRNA): AB004751, AJ233436.

TABLE BXII.γ.271. Identification of species in the *Serratia liquefaciens* complex[a]

Characteristics	*S. liquefaciens* C1ab	*S. proteamaculans* subsp. *proteamaculans* C1c[b]	EB	RB	*S. proteamaculans* subsp. *quinovora* RQ	*S. grimesii* C1d[c]	ADC
Carbon source utilization test:[d]							
trans-Aconitate	−	+	+	d	+	−	+
Benzoate	−	−	d	d	−	+	−
m-Erythritol	−	−	+	−	−	−	−
Gentisate	−	−	−	+	−	−	−
D-Malate	+	−	−	−	d	v	v
L-Rhamnose	−	−	−	+	d	−	−
m-Tartrate	+	−	−	−	d	−	−
Arginine decarboxylase	−	−	−	−	−	+	+
Esculin hydrolysed	+	+	+	+	−	+	+

[a]For symbols see standard definitions.

[b]The type strain of *S. proteamaculans* (ATCC 19323) corresponds to biotype C1c.

[c]The type strain of *S. grimesii* (ATCC 14460) corresponds to biotype C1d.

[d]Biotype-100 strips.

TABLE BXII.γ.272. Identification of *Serratia odorifera* biotypes[a]

Characteristics	Biotype 1[b]	Biotype 2
Growth on:		
m-Erythritol	−	+
L-Fucose	d	+
D-Raffinose	+	−
Sucrose	+	−
D-Tartrate	+	−
Ornithine decarboxylase	+	−
Acid from sucrose	+	−
Acid from raffinose	+	−[c]

[a]For symbols see standard definitions.

[b]The type strain corresponds to biotype 1.

[c]Some strains positive in 3–7 days.

TABLE BXII.γ.273. Identification of *Serratia rubidaea* biotypes[a]

Characteristics	Biotype[b] B1	B2	B3
Growth on:			
Histamine	d	−	d
D-Melezitose	−	+	+
D-Tartrate	+	−	d
Tricarballylate	−	d	−
Voges–Proskauer (O'Meara)	+	−	d
Lysine decarboxylase	+	+	−
Malonate (Leifson)	+	+	−

[a]For symbols see standard definitions.

[b]The biotypes were shown to correspond to subspecies designated as "*S. rubidaea* subsp. *burdigalensis*" (B1), "*S. rubidaea* subsp. *rubidaea*" (B2), and "*S. rubidaea* subsp. *colindalensis*" (B3).

Genus XXXV. **Shigella** *Castellani and Chalmers 1919, 936*[AL]

NANCY A. STROCKBINE AND ANTHONY T. MAURELLI

Shi.gel′la. M.L. dim. *-ella* ending; M.L. fem. n. *Shigella* named after K. Shiga, the Japanese bacteriologist who first discovered the dysentery bacillus.

Straight rods, 1–3 × 0.7–1.0 μm, that conform to the general definition of the family *Enterobacteriaceae* and contain the enterobacterial common antigen. Gram negative. Nonmotile. Nonpigmented. Facultatively anaerobic, having both a respiratory and a fermentative type of metabolism. Catalase positive (with exceptions in *Shigella dysenteriae*). Oxidase negative. Chemoorganotrophic. **Ferment sugars without gas production** (a few exceptions produce gas). **Salicin, adonitol, and *myo*-inositol are not fermented**. Strains of *Shigella sonnei* ferment lactose and sucrose upon extended incubation; however, other species do not utilize these substances in conventional medium. **Do not utilize citrate, malonate, or sodium acetate** (with exceptions in *Shigella flexneri* for sodium acetate) as a sole carbon source. **Do not grow in KCN or produce H$_2$S. Do not decarboxylate lysine.** Reduce nitrates to nitrites. Intestinal pathogens of humans and other primates, causing bacillary dysentery. Based on 16S rDNA sequencing, shigellae belong in the *Gammaproteobacteria*.

The mol% G + C of the DNA is: 49–53 (Laskin and Lechevalier, 1981).

Type species: **Shigella dysenteriae** (Shiga 1898) Castellani and Chalmers 1919, 935, Epit. Spec. Cons. Opinion 11 of the Jud. Comm. 1954a, 149 (*Bacillus dysenteriae* Shiga 1898, 817.)

FURTHER DESCRIPTIVE INFORMATION

Phylogenetic and systematic treatment Scientific evidence accumulated to date strongly supports the view that *Shigella* species are biotypes/pathotypes or clones of *Escherichia coli*. As early as 1957, Luria and Burrous (1957) showed that *E. coli* and the majority of *Shigella* species are in the same fertility system and suggested that genetic recombination between these organisms could play a role in the evolution of some *Shigella* serotypes and, more generally, in the evolution of natural populations of these bacteria. In the early 1970s, Brenner et al. (1972a, 1973a) published findings from DNA–DNA reassociation studies that provided the first direct insights into the evolutionary relationships between shigellae and *E. coli*. These authors found that, with the exception of *S. boydii* 13 (see discussion under Taxonomic comments), *Shigella* species were as related to *E. coli* (>75% nucleotide similarity) as they were to each other.

Given the variability that occurs within the defined *E. coli* species, the question then is whether the four species of *Shigella* represent the evolution of new species from a common ancestor or whether these organisms are clones of *E. coli*. Findings from multilocus enzyme electrophoresis (MLEE) analyses strongly support the conclusion that shigellae are more appropriately regarded as pathotypes or clones of *E. coli*. Principal component analysis of 302 electrophoretic types (ETs) from more than 1600 human and animal isolates of *E. coli* and 123 strains of the four species of *Shigella* showed that the *Shigella* strains clustered within two of the three overlapping groups of *E. coli* strains detected, rather than forming a group apart from those comprised by *E. coli*, as would be expected of a distinct genus (Ochman et al., 1983). Furthermore, with the exception of *S. boydii* 13, strains representing all four species of *Shigella* did not cluster in groups corresponding to each species. A dendrogram of the 23 ETs from *Shigella* constructed by average linkage clustering revealed two major clusters each containing strains of three or more species; one group consisted of ETs representing *S. flexneri*, *S. dysenteriae*, and *S. boydii* strains and the other consisted of ETs representing strains from all four species, with *S. sonnei* being most distantly related (Ochman et al., 1983).

MLEE findings from the study by Pupo et al. (1997) also showed that strains from the four species of *Shigella* clustered in two separate groups, with one group containing strains from the majority of *Shigella* species and the other group containing only *S. flexneri* strains. With the exception of one *S. flexneri* serotype, the *Shigella* species appeared as clones within the *E. coli* species. Nucleotide sequence analysis of the housekeeping gene maltose dehydrogenase (*mdh*) (Pupo et al., 1997) and analysis of restriction fragment length polymorphisms of rDNA (ribotyping) (Rolland et al., 1998) also yielded largely similar clustering of shigellae, but showed somewhat different distributions of the *Shigella* spp. among the *E. coli* phylogenetic groups.

A careful analysis of 16S ribosomal DNA sequences by Cilia et al. (1996) also placed shigellae and *E. coli* in the same phylogenetic group, supporting the concept that *Shigella* species and *E. coli* are five nomenspecies that represent a single genomospecies. In fact, the variation of the 16S ribosomal RNA genes in the seven *rrn* operons of a single strain of *E. coli* shared a clade

with *S. dysenteriae*, *S. flexneri*, and *S. sonnei* sequences. There were informative base alleles that were shared by both *E. coli* and *Shigella* species.

Recent findings reported by Pupo et al. (2000) from the examination of nucleotide sequence similarity at eight housekeeping genes in four regions around the chromosome of *Shigella* and selected nonpathogenic *E. coli* strains (ECOR set) revealed three clusters of closely related strains, each including exclusively *Shigella* strains but from more than one species. *S. sonnei* and three *S. dysenteriae* strains from *S. dysenteriae* serotypes 1, 8, and 10 fell outside the three main clusters but were well within the population structure of *E. coli*. The presence of these seven different groups (clusters 1–3, *S. sonnei*, *S. dysenteriae* serotypes 1, 8, and 10) within *E. coli* suggests that the *Shigella* phenotype has arisen as many as seven times from ancestors that were *E. coli*. Pupo et al. (2000) propose that the many independent origins of the *Shigella* phenotype are an example of convergent evolution. Phenotypes that may have arisen by convergent evolution include the loss of catabolic pathways involving lactose fermentation, mucate utilization, lysine decarboxylation, the loss of motility, and the acquisition of the invasion plasmid (Pupo et al., 2000).

It has been suggested that *E. coli* arose from a common ancestor with *Salmonella* and diverged largely as the result of the acquisition of multiple genomic fragments (Lawrence and Ochman, 1998). It appears that *Shigella* differs from *E. coli* largely by the loss of genomic fragments. Maurelli et al. (1998) showed that shigellae lack a large fragment of the genome relative to *E. coli* K-12. It is interesting to note that the missing fragment included the *cadA* gene discussed in more detail below. The same pattern has been noted with another genomic fragment that derives from a cryptic prophage in *E. coli* K-12 and includes the *ompT* gene (Nakata et al., 1993). This gene encodes a surface protease that when introduced into *Shigella* inhibited the ability of the organism to spread among host cells. Another fascinating deletion occurs in *S. sonnei*. A large portion of the O-antigen gene cluster has been excised, leaving remnants of genes normally found at the ends of the cluster fused together (Lai et al., 1998). The clone carries a plasmid that has O-antigen genes that apparently arose in *Plesiomonas shigelloides* (Houng and Venkatesan, 1998; Chida et al., 2000). It has been postulated that the acquisition of the new O-antigen was a prerequisite for *S. sonnei* emerging in its modern niche and that this might have occurred about 10,000 years ago (Lai et al., 1998).

In contrast to the concept of species being defined as a group having a monophyletic origin, a species may be defined as a group of organisms sharing genetic drift (Hey, unpublished studies). By this latter definition, *Shigella* species might be classified independently of *E. coli*. Factors that affect the sharing of genetic drift among bacteria are natural selection and recombinant exchange of DNA. Shigellae are considered to be host-adapted to primates and thus have a narrower niche than *E. coli*. It is difficult to conclusively define species of bacteria based on recombination-exchange partners. Islands of DNA that clearly came from other species can frequently be found in bacterial genomes. However, it does appear that the frequency of recombination is proportional to the closeness of the species. Matsutani and Ohtsubo (1993) surveyed the distribution of five insertion elements (IS1, IS600, IS629, IS630, and IS640) originally found in *S. sonnei*. In general, the elements were found in high copy numbers in *S. sonnei*, *S. dysenteriae*, *S. flexneri*, and *S. boydii* (often more than 20 copies per genome), in moderate numbers in *E. coli*, and occasionally, albeit rarely, in other *Enterobacteriaceae*. Thus, it may be that *Shigella* species are in the process of drifting apart from *E. coli*. The alternative hypothesis is that the insertion elements are more readily duplicated within clones that they exist in rather than across clones. However, the conclusion is the same since it would mean that the elements are more ancient in shigellae and this is an example of genetic drift that is not shared with *E. coli*.

If history were to be rewritten with the privilege of having our current understanding of the pathogenesis and diversity extent within *E. coli*, a very different nomenclature for *Shigella* would undoubtedly be adopted. Although phylogenetically it would be better to treat shigellae as pathotypes of *E. coli*, the members of the genus *Shigella* continue to be divided for historical and medical purposes into four species or subgroups: *S. dysenteriae* (subgroup A), *S. flexneri* (subgroup B), *S. boydii* (subgroup C), and *S. sonnei* (subgroup D). The useful information communicated about the disease caused by these organisms (shigellosis) through the genus epithet *Shigella* has preserved and will likely continue to preserve the current nomenclature until such time when there is a greater need to communicate the natural relationships that exist between these strains.

Biochemical characteristics of the genus are listed in Table BXII.γ.274. The biochemical reactions that are useful for differentiating the four species of *Shigella* are listed in Table BXII.γ.275. The ability to ferment D-mannitol is a particularly important differential biochemical trait for classifying the shigellae. Although exceptional strains occur, members of subgroup A are unable to ferment D-mannitol, while members of the other subgroups (B, C, and D) are able to ferment this sugar. Members of the genus *Shigella* have distinctive antigenic structures by which they are divided into a variety of serotypes and subserotypes. It is the combination of biochemical traits, particularly the D-mannitol phenotype, and antigenic properties that have guided the classification of members of this genus. The current schema for classifying *Shigella*, which recognizes 49 serotypes and subserotypes, is shown in Table BXII.γ.276.

Cell morphology Shigellae are nonmotile, Gram-negative rods with rounded ends. They are typically 1–3 μm in length and 0.7–1 μm in diameter. Fig. BXII.γ.209 shows negatively stained preparations of *S. dysenteriae* 1 and *S. flexneri* 2a. The bacterial cells in this figure are nonflagellated, which is typical of the members belonging to this genus. Under certain culture conditions, however, some cells in a culture have been observed to express flagella. See the section below on fine structure for a discussion of exceptions to the presence of flagella.

Cell wall composition The cell walls of shigellae resemble those of other Gram-negative bacteria in structure, composition, and endotoxic activity. When serologically characterizing living (unheated) isolates that are biochemically consistent with *Shigella*, it is important to be aware that living bacterial cells of certain strains of *Shigella* can produce a substance that will block their agglutination with O antisera. The masking effects of this substance, originally referred to as a K antigen of the B variety or B antigen, can be inactivated by heat. Treatment of nonagglutinable cultures at 100°C for 1 h restores their ability to be agglutinated by O antisera. Blocking substances also occur in certain *E. coli* strains and were shown by Ørskov et al. (1977) to be chemically and antigenically indistinguishable from the O antigens present in the cell-wall lipopolysaccharides of these bacteria. Blocking substances (B antigens) are no longer recognized as distinct antigens; however, their presence in unheated cells remains important from a practical standpoint.

TABLE BXII.γ.274. Characteristics of the genus *Shigella*[a]

Test or Substrate	Result
H₂S (triple sugar iron agar)	−
Urea hydrolysis	−
Indole production	d[b]
Methyl red	+
Voges–Proskauer	−
Citrate (Simmons')	−
Citrate (Christenson's)	−
Growth in KCN	−
Motility at 37°C	−
Gelatin hydrolysis at 22°C	−
Lysine decarboxylase	−
Arginine dihydrolase	d
Ornithine decarboxylase	d[c]
Phenylalanine deaminase	−
D-Glucose:	
Acid	+
Gas	−[d]
Acid from:	
Adonitol	−
α-Methyl-D-glucoside	−
L-Arabinose	d
Cellobiose	−
Dulcitol	d
Erythritol	−
Glycerol	d
myo-Inositol	−
Lactose	d[e]
Maltose	d
D-Mannitol	d[f]
Raffinose	d
L-Rhamnose	d
Salicin	−
D-Sorbitol	d
Sucrose	d[g]
Trehalose	d
D-Xylose	d
Malonate (sodium) utilization	−
Mucate (sodium) utilization	−
Acetate (sodium) utilization	−
Esculin hydrolysis	−
β-Galactosidase (ONPG)[h]	d[i]
Nitrate to nitrite	+
Oxidase	−

[a]For symbols see standard definitions. Adapted from findings reported by Ewing (1971) based on the examination of 5166 cultures representative of all four species. Unless specified otherwise, results are for 48 h incubation at 36° ± 1°C.

[b]Some strains of some serotypes of *S. dysenteriae*, *S. flexneri*, and *S. boydii* produce indole while strains of other serotypes are always negative. *S. sonnei* is always negative.

[c]Strains of *S. boydii* 13 and *S. sonnei* are positive.

[d]Some biotypes of *S. flexneri* 6 are positive; positive strains of *S. boydii* 13 and 14 have been described.

[e]Strains of *S. sonnei* are usually positive after several days of incubation; positive strains of *S. flexneri* 2a, *S. boydii* 9, and *S. boydii* 15 have been described. Some strains of *S. dysenteriae* 1 ferment lactose slowly; however, all are positive for β-galactosidase.

[f]Strains of *S. dysenteriae* are negative; negative biotypes of *S. flexneri* 4a ("*S. rabaulensis*", "*S. rio*") and *S. flexneri* 6 (Newcastle biotype) occur; negative biotypes of *S. sonnei* occur rarely.

[g]Strains of *S. sonnei* are usually positive after several days of incubation.

[h]ONPG, *o*-nitrophenyl-β-D-galactopyranoside.

[i]Strains of *S. dysenteriae* 1 and *S. sonnei* are positive; some positive strains of *S. flexneri* 2a and *S. boydii* 9 have been described.

Fine structure

Flagella By standard methods (radial spread from the site of inoculation in medium containing 0.3–0.4% agar), shigellae are nonmotile. Over the past decade, various investigators have shown that *Shigella* strains have the genes involved in producing flagella and, under certain conditions, can express these genes.

TABLE BXII.γ.275. Distinguishing biochemical reactions of the four species of *Shigella*[a]

Test	*S. dysenteriae*	*S. boydii*	*S. flexneri*	*S. sonnei*
Indole production	d	d	d	−
Arginine dihydrolase	−	d	−	−
Ornithine decarboxylase	−	−	−	+
Acid production from:				
Lactose	−	−	−	(+)[b]
Sucrose	−	−	−	(+)
D-Mannitol	−	+	+	+
Dulcitol	d	d	−	−
D-Sorbitol	d	d	d	−
Raffinose	−	−	d	(+)
D-Xylose	−	d	−	−
Melibiose	−	d	d	−
β-Galactosidase (ONPG)[c]	d	d	−	+

[a]For symbols see standard definitions. Results are for 48 h incubation at 36° ± 1°C under conditions described by Ewing (1971).

[b](+) = 75% or more of strains gave a positive reaction after 3 or more days.

[c]ONPG, *o*-nitrophenyl-β-D-galactopyranoside.

Direct evidence for the existence of flagellum-related sequences in *Shigella* was reported by Tominaga et al. (1994), who identified and characterized cryptic genes for flagellin (*fliC*) in *S. flexneri* and *S. sonnei*. The cloned *fliC* gene from *S. flexneri* produced normal-type flagella when introduced into a strain of *E. coli* deleted for *fliC*, while the cloned *fliC* gene from *S. sonnei* produced curly-type flagella in this host strain. Findings from comparisons of nucleotide sequences of the *fliC* genes from *S. flexneri* and *S. sonnei* with those from *E. coli* and *Salmonella* serotype Typhimurium revealed various regions of high similarity between these bacteria. Similarities at and near the 5′ and 3′ constant regions of the *S. flexneri* and *E. coli fliC* genes suggested the *fliC* gene from *S. flexneri* is an *E. coli* type gene, while similarity observed between the *fliC* gene of *S. sonnei* at the operator and 3′ constant region of *Salmonella* serotype Typhimurium and the downstream sequence of the *E. coli fliC* gene suggest the *S. sonnei* gene has undergone horizontal transfer and recombination (Tominaga et al., 1994). Coimbra et al. (2001) recently examined restriction fragment polymorphisms in the *fliC* genes of representative strains from each of the serotypes within the four species of *Shigella*. Seventeen F types, defined by *fliC* patterns sharing most of their bands, were observed among the 120 strains tested. Although O antigenic relationships between certain *E. coli* strains and shigellae are well documented, none of the F types observed among the tested *Shigella* strains matched the 62 F types previously described for *E. coli* (Machado et al., 2000). This method is a promising tool for the confirmation of atypical isolates as shigellae.

Prompted by reports that shigellae possessed the genes for flagella, Girón (1995) reexamined them for flagella by electron microscopy. He observed that strains of all four *Shigella* species, when cultivated under certain conditions, produced flagella. Motility was observed only by using low-concentration motility agar (0.175–0.2% agar), and genetic and environmental factors involved in the regulation of flagella are not yet identified. Expression of flagellated cells within a culture ranged from 1:300 to 1:1000, with flagellated cells typically producing one polar flagellum or occasionally 2–3 flagella in a semipolar topology. When expressed, the flagellum of *Shigella* is approximately 10 μm long and 12–14 nm in diameter and typically emanates from one pole of the bacterium. Putative flagellins of 33–38 kDa, which share immunologic similarities with *E. coli*, *Salmonella* spp., and

Proteus mirabilis flagellins, have also been identified (Girón, 1995).

It should be emphasized, however, that shigellae are nonmotile under standard assay conditions. Al Mamun et al. (1997) showed that members of the four *Shigella* species became nonmotile by undergoing various kinds of stable mutations in different flagellar genes. Findings from these investigators suggest that the loss of flagella in *S. boydii* is attributable to a defect or deletion in the *fliF* operon, while the primary cause for nonmotility in *S. sonnei* is a deletion in the *fliD* gene. In *S. dysenteriae* and *S. flexneri*, IS1 insertions in the *flhD* master operon may be the primary cause of motility loss (Al Mamun et al., 1996).

Fimbrial adhesins In initial studies, *S. flexneri* was the only species of *Shigella* that was found to produce fimbriae (Duguid and Gillies, 1957). The fimbriae of *S. flexneri* were similar to those produced by *E. coli*; their width was uniformly about 0.01 μm and their length was between 0.3 and 2.0 μm. In these early studies, *S. flexneri* strains were found to readily undergo a reversible mutation between a fimbriate, hemagglutinating phase and a nonfimbriate, nonhemagglutinating phase. Findings by Snellings et al. (1997) confirmed the presence of fimbriae (type 1) on strains of *S. flexneri* and demonstrated their presence on recent clinical isolates of *S. boydii* and *S. dysenteriae*. Analysis of sequences upstream of the type 1 fimbriae subunit gene, *fimA*, showed that random phase variation between fimbriated and afimbriated states in *Shigella* was accompanied by the genomic rearrangement associated with phase variation in *E. coli* (Snellings et al., 1997). *Shigella* appears to switch between phases at a much lower frequency than *E. coli*, and this in combination with inappropriate growth conditions could partially explain the difficulty researchers have had demonstrating fimbriae in *Shigella*. The role fimbriae play in the pathogenesis of shigellae is not known. Some investigators have speculated that these structures may facilitate the formation of pellicles at the water–air interface in aquatic environments and thereby promote the survival of bacteria between outbreaks of disease (Snellings et al., 1997). Studies have also demonstrated the presence of type 3 and type 4 fimbrial adhesions and of afimbrial adhesions in *Shigella* strains (Qadri et al., 1989; Utsunomiya et al., 1992).

Another surface structure of note with respect to *Shigella* is the thin aggregative fimbria called curli. Curli have been demonstrated in a wide variety of *E. coli* and *Salmonella* spp. and are believed to mediate formation of biofilms or bacterial attachment to host intestinal cells. Rather than their presence, it is their complete absence from all four species of *Shigella*, as well as from enteroinvasive *E. coli* strains, that is notable here (Sakellaris et al., 2000). Shigellae are unable to produce curli because of insertions or deletions in the curli locus (*csg*); enteroinvasive *E. coli* strains are similarly affected or have another uncharacterized lesion(s) resulting in the lack of curli expression (Sakellaris et al., 2000). Since expression of curli occurs in a variety of *E. coli* strains, the widespread loss of curli from *Shigella* and enteroinvasive *E. coli* strains, which have a very similar mechanism of pathogenesis to *Shigella*, may represent a pathoadaptive mutation (Sakellaris et al., 2000). It is possible that during the divergence of *Shigella* from *E. coli*, the expression of curli in the new virulence niche was a selective disadvantage for *Shigella*.

Colonial or cultural characteristics On agar medium, colonies of *Shigella* strains can appear smooth and glistening or rough and dry. The degree of smoothness displayed depends in large part on the extent of polymerization of the O-antigen on the

TABLE BXII.γ.276. Serologic schema, antigenic formulae, and earlier designations of *Shigella* species

Subgroup and Species	Serotype/Subserotype	Antigenic Formula[a]	Main earlier designations or synonyms
Subgroup A			
S. dysenteriae	1		"*Bacterium shigae*", "*S. shigae*" (Shiga, 1898)
	2		"*S. ambigua*", "*S. schmitzii*" (Schmitz, 1917)
	3		"*S. largei*" Q771 (Large and Sankaran, 1934), *S. arabinotarda* A
	4		"*S. largei*" Q1167 (Large and Sankaran, 1934), *S. arabinotarda* B
	5		"*S. largei*" Q1030 (Large and Sankaran, 1934)
	6		"*S. largei*" Q454 (Large and Sankaran, 1934)
	7		"*S. largei*" Q902 (Large and Sankaran, 1934)
	8		Serotype 599-52 (Ewing et al., 1952b)
	9		Serotype 58 (Cox and Wallace, 1948)
	10		Serotype 2050-52 (Ewing, 1953)
	11		Serotype 3873-50 (Ewing and Hucks, 1952)
	12		Serotype 3341-55 (Ewing et al., 1958)
	13		Serotype I9809-73 (Shmilovitz et al., 1985)
	14		Serotype E22383 (Gross et al., 1989)
	15		Serotype E23507 (Gross et al., 1989)
Subgroup B			
S. flexneri	1a	I:4	V (Andrewes and Inman, 1919)
	1b	I:(4),6	VZ (Andrewes and Inman, 1919)
	2a	II:3,4	W (Andrewes and Inman, 1919)
	2b	II:7,8	WX (Andrewes and Inman, 1919)
	3a	III:(3,4), 6,7,8	Z (Andrewes and Inman, 1919)
	3b	III:(3,4),6	
	4a	IV:(3,4)	103 (Boyd, 1931)
	4b	IV:6	103Z
	4c	IV:7,8	(Pryamukhina and Khomenko, 1988)
	5a	V:(3,4)	P119 and P119X (Boyd, 1932a, b)
	5b	V:7,8	(Petrovskaya and Khomenko, 1979)
	6	VI:4	Boyd 88 (Boyd, 1931) Manchester bacillus, Newcastle bacillus, "*S. newcastle*"
	X	−:7,8,	X (Andrewes and Inman, 1919)
	Y	−:3,4	Y (Andrewes and Inman, 1919)
Subgroup C			
S. boydii	1		170 (Boyd, 1932a, b; Ewing, 1949)
	2		P288 (Boyd, 1932a, b; Ewing, 1949)
	3		D.1 (Boyd, 1932a, b; Ewing, 1949)
	4		P274 (Boyd, 1932a, b; Ewing, 1949)
	5		P143 (Boyd, 1938; Ewing, 1949)
	6		D19 (Boyd, 1932a, b; Ewing, 1949)
	7		Type T, Lavington I, "*S. etousae*" (Ewing, 1946)
	8		Serotype 112 (Cox and Wallace, 1948)
	9		Serotype 1296/7 (Boyd, 1946; Ewing et al., 1951)
	10		Serotype 430 (Ewing and Taylor, 1951) and D15 (Szturm et al., 1950)
	11		Serotype 34 (Ewing and Taylor, 1951)
	12		Serotype 123 (and "M") (Ewing and Hucks, 1952)
	13		Serotype 425 (Ewing and Hucks, 1952)
	14		Serotype 2770-51 (Ewing and Hucks, 1952)
	15		Serotype 703 (Ewing et al., 1952a)
	16		Serotype 2710-54 (Ewing et al., 1958)
	17		Serotype 3615-53 (Ewing et al., 1958)
	18		Serotype E10163 (1344-78) (Gross et al., 1980)
	19		Serotype E16553 (Gross et al., 1982)
Subgroup D			
S. sonnei			Sonne-Duval, Sonne III, Kruse E, "*S. ceylonensis*" A (Duval, 1904; Sonne, 1915)

[a]The serologic subdivision of serotypes is done for only *S. flexneri*. The roman numerals designate the "type" or serotype-specific antigen and the Arabic numerals designate group antigens that are shared among members of the different types. Different subserotypes are defined by the combination of group antigens they possess. Group antigens shown in parentheses may be expressed in varying amounts (undetectable to strongly positive) by members of certain subserotypes. For the reactions of *S. flexneri* in typing antisera see Ewing (1986c). For *S. flexneri* serotype 3, only two subserotypes are currently recognized; strains previously designated as subserotype 3c are now identified as subserotype 3b [III:(3,4), 6] (Petrovskaya and Khomenko, 1979; Brenner, 1984b).

lipopolysaccharide (LPS) molecule. For *S. sonnei*, colonies of freshly isolated strains with a smooth appearance are termed form I and are usually virulent. Form I colonies are genetically unstable and dissociate to rough-appearing colonies termed form II, which lack 2-amino-deoxy-L-altruonic acid from their LPS O-repeating unit (Kontrohr, 1977). This form variation in *S. sonnei* is associated with the loss of the 180–220 kilobase pairs (kbp) invasion plasmid. Colonial variation, which also correlates with the presence of the invasion plasmid, has been described for colonies growing on agar containing the dye Congo red. Smooth colonies that appear red on this agar are virulent, while colorless colonies are avirulent (Maurelli et al., 1984a). The dye interacts

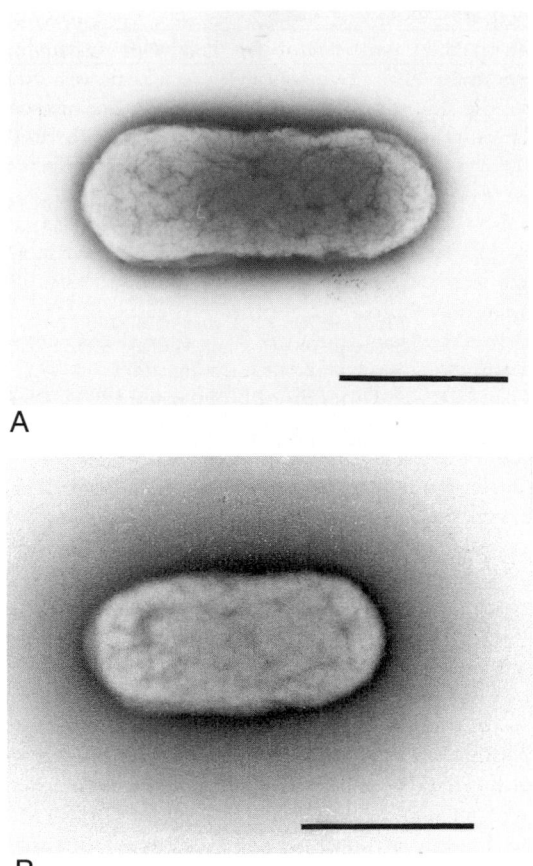

A

B

FIGURE BXII.γ.209. Electron micrograph of (*A*) *S. dysenteriae* 1 strain A5468 and (*B*) *S. flexneri* 2a strain 3342-87 prepared by negatively staining in 0.5% uranyl acetate. Bar = 1000 nm. (Micrograph courtesy of Charles D. Humphrey, CDC.)

with outer membranes and outer membrane proteins, but not with lipopolysaccharides. Only smooth colonies are scored for virulence using this dye because rough colonies, which are avirulent, can nonspecifically bind the dye, presumably by exposure of non-virulence-associated outer membrane proteins.

Nutrition and growth conditions The shigellae are aerobes and facultative anaerobes. Their optimal growth temperature is about 37°C. They generally grow less rapidly than most strains of *E. coli* and other members of the family *Enterobacteriaceae*. The average interval between cell divisions of *S. dysenteriae* in milk was reported to be 23 min compared with 12.5 min for *E. coli* (Sinclair, 1972). Ahmed et al. (1988) studied the nutritional requirements of 375 clinical isolates from the four species of *Shigella* for growth in a minimal medium containing glucose, ammonium sulfate, and inorganic salts and observed that most isolates would not grow in this medium. They found that virtually all isolates that failed to grow in the minimal medium could grow if the medium was supplemented with methionine, nicotinic acid, and tryptophan. Methionine and tryptophan appeared to be an obligatory requirement for *S. dysenteriae* 1 strains, while the combination of nicotinic acid and tryptophan was required for *S. dysenteriae* 2 strains.

In contrast to *Salmonella*, *Shigella* species and certain isolates of *E. coli* are acid-resistant (Gordon and Small, 1993; Small et al., 1994). Stationary phase cells of *S. flexneri* have the ability to sur-

vive for several hours at pH 2.5, which is hypothesized to contribute to their low infective dose by allowing them to safely pass through the stomach before colonizing the intestinal tract. The ability to survive exposure to acid is a complex phenotype, which depends on growth phase, media, and possibly a variety of proteins located at different subcellular locations. At least three genes, *hdeA*, *rpoS*, and *gadC*, play a role in mediating the acid-resistance phenotype under various conditions (Waterman and Small, 1996).

Metabolism and metabolic pathways Like *E. coli*, shigellae are chemoorganotrophic and have both a respiratory and a fermentative type of metabolism. They are generally less metabolically active than typical *E. coli* strains. They produce pyruvate from the fermentation of glucose and other sugars and convert it primarily by mixed acid (formic acid) fermentation to formic acid, acetic acid, and ethanol. In contrast to *E. coli* strains, shigellae do not produce gas from the fermentation of sugars (anaerogenic), because they lack the complex hydrogenlyase system (formic hydrogenlyase) that splits formic acid into equal amounts of CO_2 and H_2. Exceptions include some strains of *S. flexneri* serotype 6 and *S. boydii* serotypes 13 and 14. In addition to the traits described above, members of the genus *Shigella* do not hydrolyze urea and do not produce detectable amounts of acetoin (acetyl methyl carbinol) or 2,3-butyleneglycol (Voges–Proskauer test negative). They also do not produce phenylalanine deaminase and do not produce acid from α-D-methylglucoside, erythritol, or esculin. They are uniformly negative for lysine decarboxylase. The pathogenic and evolutionary significance of this negative phenotype is addressed below in the pathogenicity section.

Shigella strains vary in their ability to metabolize glycerol. A study characterizing the glycerol dehydrogenase activity among *Shigella* species showed that the type of glycerol dehydrogenase produced could be used to help differentiate *Shigella* species from *E. coli* (Bouvet et al., 1995b). In this study, no strain of *Shigella*, with the exception of the Manchester biotype of *S. flexneri* serotype 6, was found to produce an NAD$^+$-linked glycerol dehydrogenase, glyDH-II, while the majority of *E. coli* strains contained this enzyme.

Genetics Extensive genetic homology between *Shigella* and *E. coli* has permitted the use of classical genetic techniques, such as conjugal transfer and integration of chromosomal material between *E. coli* and *Shigella* (Luria and Burrous, 1957; Falkow et al., 1963; Formal et al., 1970, 1971; Sansonetti et al., 1983) and the generation of R plasmids (Timmis et al., 1985), to construct maps of the *Shigella* chromosome based on the known map for *E. coli* K-12. A *Not*I restriction map of the chromosome of *S. flexneri* 2a, with the assignment of nine virulence-associated loci identified by Tn5 insertions, was constructed by Okada et al. (1991). Genes that play a role in pathogenesis have received extensive study and were reviewed by Parsot and Sansonetti (1996), and are highlighted below in the pathogenicity section. Genes involved in O-antigen synthesis have also received considerable attention and are reviewed by Brahmbhatt et al. (1992) and discussed below in the antigenic structure section.

The work of Brenner et al. (1972a, 1973a) clearly established the genetic basis for the close biochemical and serologic relatedness between *E. coli* and *Shigella* that had been appreciated for many years. Their findings from DNA reassociation reactions followed by thermal elution chromatography on hydroxyapatite showed that, with the exception of *S. boydii* 13 strains, strains of

Shigella from all four species were as related to each other (sharing 80% or more of their nucleotide sequences) as they were to strains of *E. coli* (Brenner et al., 1973a). Estimates of genome size by I-*Ceu*I macrorestriction analysis revealed that the chromosome sizes of all four *Shigella* species were in a similar range to those observed for *E. coli* (Shu et al., 2000). By this analysis, the genome sizes for the type strains of the four species of *Shigella* were 4.415 × 10⁶ bp for *S. dysenteriae*, 4.792 × 10⁶ bp for *S. flexneri*, 4.645 × 10⁶ bp for *S. boydii*, and 4.501 × 10⁶ bp for *S. sonnei*, while those for *E. coli* were 4.653 × 10⁶ bp and 4.816 × 10⁶ bp for two *E. coli* strains (Shu et al., 2000). The genome sequences of two different isolates of *S. flexneri* 2a have been published and are deposited in GenBank under accession numbers AE014073 and AE005674 (Jin et al. (2002); Wei et al., 2003).

Published ribosomal RNA sequences for the four species of *Shigella* and of *E. coli* are nearly identical, with more than 99% homology (Shu et al., 2000). Like *E. coli*, *Shigella* strains have seven ribosomal RNA operons as determined by restriction analysis with I-*Ceu*I, an enzyme that specifically cuts a 26-bp site in the 23S rDNA sequence in *rrn* operons (Shu et al., 2000). The mapping of seven *E. coli* genes, each known to reside on a different I-*Ceu*I fragment in *E. coli* K-12, revealed that some chromosomal rearrangements involving the fragments corresponding to fragments D and E of *E. coli* K-12 took place in *S. dysenteriae* and *S. flexneri*. Available findings did not permit a determination of the mechanism (translocation versus inversion of fragments) for these rearrangements. Hybridization patterns of *S. boydii* and *S. sonnei* strains were similar to those observed for *E. coli* strains, while the patterns for the *S. dysenteriae* and *S. flexneri* strains were each distinct.

Shigellae can carry a variety of insertion sequences (IS elements) including IS1 (Nyman et al., 1981; Kharat and Mahadevan, 2000); iso-IS1 (Ohtsubo et al., 1981); IS 2 (Soldati and Piffaretti, 1991); IS5 (Schoner and Schoner, 1981); IS200 (Gibert et al., 1990); IS 630 (Houng and Venkatesan, 1998); and IS911 (Prere et al., 1990). IS elements comprise over 6% of the sequenced genome of *S. flexneri* 2a; 314 IS elements were identified in strain 301 and 284 IS elements in strain 2457T (Jin et al. (2002); Wei et al., 2003). Some of these, such as IS 1 and iso-IS1, are present in large numbers in the chromosome of *Shigella* (50 to >150 copies per chromosome, respectively) and have been associated with negative phenotypes. For example, the inability of *S. sonnei* to utilize the β-glucoside salicin has been shown to involve the insertional inactivation of the gene *bglB*, encoding phospho-β-glucosidase B, by a novel IS (Kharat and Mahadevan, 2000). In contrast to adverse effects of IS elements on some genes, the IS 630 element appears to be required for the stable expression of form I antigen in *S. sonnei* (Houng and Venkatesan, 1998).

Plasmids, phages and phage typing, bacteriocins Shigellae carry a variety of plasmids. The most notable is the large (180–220 kbp) invasion plasmid, which belongs to the RepFIIA family of replicons (Silva et al., 1988) and is present in all virulent strains of *Shigella*. This plasmid carries genes that play an essential role in these organisms' ability to cause invasive disease and appears to contain regions from different origins (mol% G + C values ranging from 30% for *virF* to 49% for *sepA*). The complete DNA sequence of the virulence plasmid of *S. flexneri* has been determined (Buchrieser et al., 2000; Venkatesan et al., 2001). The organization and origin of the genes encoded on the plasmid are discussed in these articles. In some serotypes, plasmids carry genes involved in synthesis of the O antigen (Kopecko et al., 1980; Watanabe and Timmis, 1984). Antibiotic resistance plasmids are also common among the shigellae. Because of the variety of plasmids carried by *Shigella*, plasmid profile analysis has been used widely to discriminate between strains during outbreak investigations. The use of this technique has declined in recent years in favor of molecular methods that examine differences in total cellular DNA and are less influenced by environmental or host pressures that can result in loss or acquisition of plasmids.

Shigellae have also been reported to carry Hsd (host specificity for DNA) plasmids. Strains of *S. flexneri* and *S. sonnei* carrying plasmid-encoded isoschizomers of either *Eco*RII or *Nci*I (Lee et al., 1997) and a strain of *S. boydii* serotype 13 carrying a plasmid-encoded isoschizomer of *Nru*I (Mise et al., 1986) have been described.

Shigellae are susceptible to a variety of bacteriophages, and this property has permitted the development of several phage typing systems (Thomen and Frobisher, 1945; Hammarström, 1947, 1949; Slopek et al., 1968, 1973; Lazlo et al., 1973; Pruneda and Farmer, 1977). A detailed description of the methods and the different bacteriophages used for typing *Shigella* was reported by Bergan (1979). Systems described to date, which typically use a panel of 10–20 bacteriophages to detect different patterns of lysis on a test strain, have focused on discriminating between strains within a serogroup or serotype. Discrimination between strains is generally good; for example, the method described by Lazlo et al. (1973) identified 90 types among more that 4000 strains of *S. flexneri* tested, and that of Pruneda and Farmer (1977) discriminated 87 different types among 265 strains of *S. sonnei*. Control of assay conditions, particularly the media and the smoothness of the cultures, is important for obtaining reproducible results. While the majority of strains can be characterized by phage typing, common phage types exist and it may be necessary to apply additional typing methods to achieve adequate discrimination.

Genes carried by some bacteriophages in shigellae have been identified. Among the serotypes of *S. flexneri*, bacteriophages have been demonstrated to carry genes, such as acetyltransferase, glucosyltransferase, or novel O-antigen polymerase genes, that play a role in determining the type and group determinants of the O antigens (Gemski et al., 1975; Lindberg et al., 1978; Clark et al., 1991a; Huan et al., 1997a, b; Mavris et al., 1997; Guan et al., 1999). Because many bacteriophages use O-antigenic polysaccharide chains as receptors for adsorption and infection, the modification of the host bacterial O antigens by the incoming phages is hypothesized to be a protective mechanism for excluding homologous phages from entering and may also help lysogenized cells evade preexisting host immunity.

Findings reported by McDonough and Butterton (1999) provide evidence, in the sequences surrounding the Shiga toxin gene, suggesting that *S. dysenteriae* 1 was once lysogenized by a Shiga toxin-encoding lambdoid prophage, which became defective as a result of deletions after IS element insertions and rearrangements. These authors also provided evidence that Shiga toxin-encoding phages from *E. coli* can be stably introduced into *Shigella*. Other than *S. dysenteriae* 1, Shiga toxin-producing strains of *Shigella* are extremely rare.

Antigenic structure The different serotypes are distinguished by antigenic determinants that reside in the O antigen, which is part of the LPS molecule. The LPS molecule consists of three parts: lipid A, which is made up of sugars and fatty acids and anchors the LPS molecule in the outer membrane; the core, which is made up of a single sequence of heptoses and hexoses and links the lipid A to the O antigen; and the O-antigen chain, which is composed of a repetitive sequence of hexoses and extends from the surface of the bacterium. The genetics of O antigen biosynthesis have been extensively studied and are reviewed by Schnaitman and Klena (1993) and Whitfield (1995). O antigens may contain from 10 to as many as 30 repeats of an oligosaccharide unit (O unit) each composed of three to six sugars. The diversity of O antigens is a result of varying combinations of sugars in the O unit, the type of chemical linkages between them, and the presence of nonsugar moieties such as O-acetyl residues or amino acids. Six to nineteen genes may be involved in the synthesis of the O antigen, and these are typically located together on the chromosome in a region (10 kbp or greater) called the *rfb* cluster. Depending on the sugars making up the O unit, serotypes may have completely different sets of genes. For certain shigellae, such as *S. dysenteriae* 1, *S. sonnei*, and *S. flexneri* 1–5, one or more genes for the synthesis of the O unit are carried on plasmids or lysogenic bacteriophages. Like *E. coli* and other Gram-negative bacteria, *Shigella* have a conserved 39-bp sequence, designated JUMPstart for "just upstream of many polysaccharide-associated gene starts," located in the noncoding region upstream of the *rfb* cluster (Hobbs and Reeves, 1994). This region has been used to develop a molecular method for serotyping *Shigella* isolates (Coimbra et al., 1999).

The O antigens of most serotypes of *Shigella* are identical or partially related to the O antigens of *E. coli* (Ewing, 1953, 1986c). At least 13 identical and numerous reciprocal serologic relationships exist between *Shigella* and *E. coli* (Ewing, 1986c).

With the acceptance of provisional serotypes E22383 and E23507 as new serotypes (Ansaruzzaman et al., 1995), *S. dysenteriae* contains 15 serotypes, each with a distinctive antigen by which it can be recognized; there are few cross-reactions, either within the species or with other species that present difficulties for serotyping strains. Of the different serotypes within this species, the O antigen of *S. dysenteriae* 1 has been the most extensively studied because of efforts to develop LPS-based antidysentery vaccines. Genes involved in the synthesis of its O antigen have been identified, and their structure and function were reviewed by Schnaitman and Klena (1993). Eight of these genes are located together in the chromosome and two genes (*rfp* and *rfe*) lie outside the chromosomal cluster; *rfp* is carried on a 9-kbp multicopy plasmid.

S. flexneri contains eight serotypes and nine subserotypes. The serotypes are antigenically related, but each has a qualitatively distinct major (type) antigen; the group antigens are shared by other members of the species. Because of the important intragroup relations, highly absorbed sera are needed for the detailed serotyping of *S. flexneri*. The immunochemical and genetic bases of the complex antigenic structures of the species have been summarized by Petrovskaya and Bondarenko (1977). The O antigens of all serotypes, except *S. flexneri* 6, contain group antigens 3, 4 as a main primary structure. The type-specific antigens I, II, IV, and V and the group antigens 7, 8 are all the result of phage conversion of the 3, 4 antigens resulting in the incorporation of α-glycosyl and/or glucosyl groups to the common O-repeating unit of the LPS molecule. Type-specific antigen III and group antigen 6 differ from the above antigens in that they contain acetyl groups. Nevertheless, these antigens are also formed as a result of phage conversion of the 3, 4 antigens. The genes in-

volved in synthesis of the common O-repeating unit are located on the chromosome in the *rfb* cluster.

The classification of *S. flexneri* 6 as a serotype of *S. flexneri* was questioned by Petrovskaya and Bondarenko (1977), who proposed that it be transferred to *S. boydii* because the immunochemical structure and genetics of its O antigen were more consistent with those found among serotypes of *S. boydii* than among serotypes 1-5, Y and X of *S. flexneri*. In 1984, the International Committee on Systematic Bacteriology Subcommittee on the Taxonomy of *Enterobacteriaceae* considered this proposal and upheld the 1973 recommendation of the Subcommittee Working Group on *Shigella* to reject reclassification of *S. flexneri* 6 as a serotype of *S. boydii* on the grounds that it would have no practical value and would cause confusion by altering a well-accepted classification system (Brenner, 1984b).

S. boydii contains 19 serotypes and each has a qualitatively distinct antigen; there may be some cross-reactions with antisera to other *Shigella* species, but these seldom interfere with diagnosis. Serotypes 10 and 11 share a major antigen, although each possesses a specific antigen.

S. sonnei contains only one serotype but may exist in one of two forms or "phases": form I (smooth colonial appearance) and form II (rough colonial appearance). Each form has a distinctive antigen, and antiserum containing agglutinins for both phases should be used for identification. Form I cells, which are virulent and carry the 180-kbp invasion plasmid, are typically isolated from acutely ill individuals. These cells can rapidly and irreversibly switch to form II cells, which are avirulent, when the invasion plasmid is lost. Form II cells may be isolated from ill individuals; however, they are more frequently isolated from individuals during convalescence and toward the end of an outbreak.

The O antigen unit of form I cells is structurally unique compared with those from other shigellae or *E. coli*, and it is chemically identical to that found in serotype 17 of *Plesiomonas shigelloides* (Rauss et al., 1970; Kenne et al., 1980). With the exception of two (*wzz* and *wbg*), the genes responsible for expression of the form I O antigen are located on the invasion plasmid and are nearly identical to those present in *P. shigelloides* (Shepherd et al., 2000). Factors influencing the loss of the plasmid carrying these genes are not well understood; however, Houng and Venkatesan (1998) propose that an insertion element present in the plasmid may be necessary for the stable expression of the form I antigen. These authors discovered the insertion element IS *630* within the O antigen gene cluster of all virulent strains of *S. sonnei* and observed that recombinant strains of *E. coli* carrying the cloned *S. sonnei* O antigen genes could stably express the form I antigen only when the IS *630* element was present.

While the form I O antigen expressed by *S. sonnei* is encoded by the plasmid-borne O antigen genes from *P. shigelloides* (Shepherd et al., 2000), findings by Lai et al. (1998) show that *S. sonnei* once had a typical chromosomal O antigen gene cluster. The remnant O antigen gene cluster apparently arose from a deletion due to homologous recombination between *manB* genes present in the adjacent O antigen and colanic acid gene clusters. The remnant chromosomal O antigen genes are not functional.

In addition to the recognized serotypes of shigellae, a number of provisional *Shigella* serotypes have been described. These may be added to the serotyping schema in the future, but in the meantime they remain *sub judice*, and antisera for their identification are usually available only at a very few reference laboratories. Provisional serotypes under consideration at present include serotypes E670/74 (Gross et al., 1989), 3162-96 (Kuijper et al., 1997), 93-119 (Matsushita et al., 1997), 96-204 (Matsushita et al., 1998), and 96-265 (Frank Rogers, National Laboratory for Enteric Pathogens, Health Canada, Winnipeg, Canada; personal communication), which are biochemically consistent with members of subgroup A; serotypes Y394 (Wehler and Carlin, 1988), 88-893 (Matsushita et al., 1992a), 89-141 (Matsushita et al., 1992b), which are biochemically and antigenically consistent with members of subgroup B; and serotypes 1621-54 (Ewing et al., 1958), E28938 (Gross et al., 1989), and 99-4528 (Frank Rogers, National Laboratory for Enteric Pathogens, Health Canada, Winnipeg, Canada; personal communication), which are biochemically consistent with members of subgroup C.

Pathogenicity Shigellae are pathogens of humans and other primates. Although there have been occasional reports of infections in dogs, other animals are resistant to infection. Laboratory animals such as mice, rabbits, and guinea pigs may be infected orally but only following starvation and treatment with gastric antacids and antiperistaltic agents.

Shigellosis (bacillary dysentery) is transmitted orally through contaminated food and water or by direct fecal-oral spread. A review of the literature estimated the annual number of cases throughout the world to be 164.7 million, of which 163.2 million were in developing countries (with 1.1 million deaths). A total of 69% of all episodes and 61% of all deaths attributable to shigellosis involved children less than 5 years of age (Kotloff et al., 1999). Studies of volunteers have demonstrated that ingestion of as few as 200 organisms is sufficient to cause dysentery (DuPont et al., 1989). The incubation period is from 1 to 7 days with symptoms commonly manifesting on day 3. Shigellosis often (but not always) begins with a watery diarrhea that precedes the characteristic dysentery symptoms. The diarrhea phase probably results from production of enterotoxins by the bacteria as they transit through the small intestine.

The lesions of bacillary dysentery are usually restricted to the rectum and large intestine, but in severe cases part of the terminal ileum may be affected. Shigella penetrates into the epithelial cells lining the colon, multiplies within these cells, and spreads from cell to cell through the mucosa. The foci of infected cells coalesce to form abscesses. Small volume bloody and mucoid stools contain dead cells along with mucus and large numbers of bacteria. Typically there is acute inflammation with ulceration of the epithelium, and the presence of polymorphonuclear leukocytes in the stools is consistent with the inflammatory nature of shigellosis. Common clinical signs include fever, severe abdominal pain, and cramping. The most severe forms of shigellosis are caused by strains of S. dysenteriae 1, which also produces a potent cytotoxin (Shiga toxin) that has been shown to play a role in the severity of the illness it causes (Fontaine et al., 1988). Shiga toxin-producing strains can cause hemolytic uremic syndrome. S. sonnei strains cause milder forms of the disease, while S. flexneri and S. boydii strains can cause either severe or mild illness. Bacillary dysentery is a self-limiting disease. The organisms rarely spread deeper than the lamina propria, and bloodstream involvement is uncommon.

Treatment of shigellosis with appropriate antimicrobial therapy results in decreased duration of both symptoms and excretion of the organism (Haltalin et al., 1967; Salam and Bennish, 1991); however, empirical antimicrobial therapy for this illness has become complicated by the increasing incidence of multiple drug resistance among Shigella strains. Of historical note, the first observation of multiple, transferable drug resistance in bacteria

was in Shigella in Japan (Ochiai et al., 1959). Over the years, Shigella species have become increasingly resistant to most of the widely used antibiotics (Bennish et al., 1992). Multiple antibiotic resistance in S. dysenteriae 1 is so widespread that it is uniformly susceptible only to the fluoroquinolones (Sack et al., 1997). Antimicrobial resistance surveys of Shigella species conducted during the 1990s in developed and developing countries showed that multiple drug resistance exists in strains of all species (Keusch and Bennish, 1998; Mates et al., 2000; Replogle et al., 2000). The resistance patterns varied by species and by country, highlighting the need for monitoring of local antimicrobial resistance patterns to develop appropriate empirical therapy. Because Shigella infections are typically self-limiting and because antibiotic resistance frequently develops after treatment, some have proposed that antimicrobial therapy be reserved for the most severely ill patients, immunocompromised patients, and patients for whom eliminating carriage is a public health priority (Weissman et al., 1973; Replogle et al., 2000). Others feel that, because the infection is generally transmitted from person to person and the infected person represents the major reservoir of infection, each patient with a positive stool culture or with known bacillary dysentery should be treated to prevent the spread of infection (DuPont, 2000).

The hallmarks of Shigella pathogenicity are induction of diarrhea, the ability to invade eucaryotic cells, multiplication inside these cells, and spread from cell to cell. The rabbit ileal loop assay has been used as an experimental model to demonstrate production of at least two different enterotoxins by S. flexneri, which may play a role in the diarrhea caused by the organism (Fasano et al., 1995; Nataro et al., 1995). The invasive properties of Shigella have been demonstrated in experimental infection of monkeys (LaBrec et al., 1964) as well as in the Serény test, which measures the ability of the bacteria to invade and produce keratoconjunctivitis in the guinea pig eye (Sérèny, 1957). Invasion can also be demonstrated in mammalian cells in tissue culture (LaBrec et al., 1964; Ogawa et al., 1967; Day et al., 1981). The ability of the bacteria to spread from cell to cell after invasion is measured by the plaque assay (Oaks et al., 1985). While shigellae are killed after being taken up by polymorphonuclear leukocytes (Mandic-Mulec et al., 1997), they induce apoptosis in macrophages and kill these cells after uptake (Zychlinsky et al., 1992). S. dysenteriae serotype 1 produces Shiga toxin, a potent inhibitor of eucaryotic protein synthesis (O'Brien et al., 1992). Shiga-like toxins are also produced by enterohemorrhagic strains of E. coli (O'Brien et al., 1984).

Shigella strains invade the intestinal mucosal surface via a pathogen-directed phagocytic process that actively involves elements of the host cytoskeleton (Tran Van Nhieu et al., 2000). Unlike most other intracellular bacterial pathogens, however, Shigella strains rapidly lyse the endocytic vacuole upon entry and are released free into the host cell cytoplasm where they replicate. Shigellae are actively motile during growth inside the host cell. This motility is unusual in that it is not driven by bacterial-based flagella. Rather a bacterial protein, expressed in the outer membrane at one pole of the bacterium, catalyzes the polymerization of host cell actin filaments (Makino et al., 1986; Bernardini et al., 1989). Formation of actin tracks literally propels the bacterium through the cytoplasm as actin monomers polymerize into long filaments from one pole of the bacterium. This motility is required for cell-to-cell spread of Shigella. Mutant strains of Shigella that fail to display this intracellular motility produce a greatly attenuated form of disease in the monkey model, thus confirming this phenotype as a critical hallmark of *Shigella* pathogenesis (Sansonetti et al., 1991).

Virulence in *Shigella* spp. is dependent on expression of a group of genes encoded on a large plasmid. The role of this plasmid in *Shigella* virulence was first demonstrated in *S. flexneri* and *S. sonnei* (Sansonetti et al., 1981, 1982). Subsequently, all virulent species of *Shigella* and enteroinvasive strains of *E. coli* were shown to carry a 180–220-kbp plasmid that shares substantial DNA homology with these prototype virulence plasmids (Sansonetti et al., 1985). Introduction of these plasmids into plasmid-cured strains of *Shigella* or laboratory strains of *E. coli* K-12 imparts on the bacteria the ability to invade mammalian cells in tissue culture. Thus, the *Shigella* virulence plasmid encodes all the genes required for the invasive phenotype of *Shigella*. The DNA sequence of the virulence plasmid of *S. flexneri* 5a has been determined and deposited in GenBank under accession numbers AF348706, AL391753, and AF386526 (Buchrieser et al., 2000; Venkatesan et al., 2001; Jin et al. (2002)).

A 37-kbp region of the plasmid contains the minimal sequence needed for invasion (Maurelli et al., 1985; Baudry et al., 1987). The genes in this region are roughly grouped into two clusters transcribed from opposite strands. They encode the Ipa proteins, required for inducing uptake of the bacteria by mammalian cells, and Mxi/Spa proteins of a dedicated type III secretion apparatus for transport of the Ipa proteins outside the bacteria. Type III secretion systems are specialized transport systems found in animal and plant pathogens and are responsible for mediating interactions between the bacterial pathogen and its eucaryotic host (Cornelis and Van Gijsegem, 2000). The Mxi and Spa proteins of the Shigella type III secretion system show strong homologies with the analogous proteins of the type III secretion systems of *Salmonella* Typhimurium, enteropathogenic and enterohemorrhagic *E. coli*, and *Yersinia* spp. The *Shigella ipa* genes (*ipaBCDA*) encode the secreted effector molecules that mediate the bacteria-induced phagocytic (invasion) event. The Ipa proteins (invasion plasmid antigens) are the immunodominant antigens recognized by convalescent-phase sera from shigellosis patients and challenged monkeys (Maurelli et al., 1985; Oaks et al., 1986). Unlike the genes of the virulence plasmid that are involved in invasion and postinvasion steps, the chromosomal genes associated with virulence of *Shigella* appear to mostly encode modulators of pathogenicity.

Of interest from the point of view of bacterial evolution and pathogenesis are the genes that are present in nonpathogenic strains of *E. coli* but are missing from the chromosome of *Shigella*. A cryptic prophage in the *E. coli* K-12 chromosome that is missing in shigellae encodes an outer membrane protease, OmpT. When the *ompT* gene is introduced into *Shigella* by conjugation, the strain remains invasive but loses the ability to spread from cell to cell. This phenotype is due to OmpT protease degradation of IcsA, the bacterial protein responsible for actin polymerization and motility (Nakata et al., 1993). Another example of a missing genetic locus in the *Shigella* genome is *cadA*, the gene for lysine decarboxylase. Although lysine decarboxylase activity is present in >85% of *E. coli* strains, it is absent in all strains of *Shigella* spp. and enteroinvasive *E. coli*. When *cadA* is reintroduced into *S. flexneri*, the decarboxylation of lysine generates cadaverine. The *Shigella* enterotoxins are inhibited by cadaverine (Maurelli et al., 1998). Thus, these genes can be considered antivirulence genes and the missing genetic locus, a black hole. The example of *Shigella* illustrates another way that pathogens evolve from their nonpathogenic commensal relatives by both acquiring genes (e.g., the virulence plasmid) that contribute to virulence and deleting genes that are incompatible with expression of these new virulence traits.

Growth temperature is the critical stimulus for regulation of virulence in *Shigella*. Wild-type *Shigella* strains grown at 37°C display all the attributes of virulence, whereas the same strains grown at 30°C are avirulent in the Serény test and fail to invade tissue culture cells (Maurelli et al., 1984b). The loss of virulence is reversible and the organisms regain full virulence (invasion and ability to provoke keratoconjunctivitis) after the growth temperature is shifted to 37°C. Paradoxically, growth of *Shigella* at 37°C in the laboratory leads to spontaneous deletion or loss of the virulence plasmid. Evidence suggests that expression of virulence genes *in vitro* is not well tolerated by the bacteria and leads to selection of variants that no longer express these genes either due to mutations in regulatory genes or deletion of plasmid sequences (Schuch and Maurelli, 1997). Strains grown at 30°C retain the virulence plasmid intact.

Despite many years of effort, an effective vaccine against shigellosis still has not been developed. Formulations have included killed whole cells, live attenuated strains, subunit vaccines (conjugate and proteosome), and ribosomal vaccines. The World Health Organization has placed *Shigella* vaccine development on its list of priorities for the Global Program for Vaccines and Immunization (WHO, 1997).

Ecology Shigellae are host-adapted to humans and subhuman primates, where they typically inhabit the intestinal tract. *Shigella* infections in captive primates are not uncommon; however, two groups of investigators (Takasaka et al., 1964; Carpenter and Cooke, 1965) found no evidence that wild primates living without human contact are infected. During the acute stage of the illness, shigellae are typically shed in large numbers in the feces. The number of organisms dramatically declines during recovery, and in some individuals a carrier state may develop in which organisms may persist in the feces for several weeks after the symptoms have subsided. The spread of *Shigella* infection is typically from person to person by fecal–oral transmission or from contaminated food or water. In situations without good sewage disposal and poor hygiene conditions, flies may act as vectors (Levine and Levine, 1991).

ENRICHMENT AND ISOLATION PROCEDURES

Food and water The minimum infecting dose of shigellae is small (10–100 organisms), and occurrence of the organisms in food, milk, and water may be significant even when only a small number of organisms are present. There are no reliable and effective enrichment methods, however, and the true incidence of *Shigella* contamination of foodstuffs cannot be accurately determined. The GN (Gram-negative) broth of Hajna (1955) may be useful for enrichment of *Shigella*, and it is recommended that the investigation of foodstuffs include an enrichment step using this medium. Subsequent steps in the isolation of *Shigella* from foods should follow the procedure recommended for fecal specimens.

Fecal specimens Freshly passed stools should be examined, although, if this is not possible, fecal swabs or rectal swabs showing marked fecal staining may be used. The specimens should be collected during the acute stage of the disease and before any chemotherapy is started. Specimens should be examined as soon after collection as possible. If the specimen includes blood and mucus, these should be included in the portion examined.

Although GN broth and Selenite broth are commonly used, no enrichment medium for *Shigella* consistently improves the recovery rate beyond that obtained by direct plating alone. Because some strains of *Shigella* grow poorly on inhibitory media, both a relatively noninhibitory selective medium (e.g., MacConkey) and an inhibitory medium (e.g., xylose desoxycholate, deoxycholate citrate agar, or Hektoen enteric agar) should be used for isolation. Salmonella–Shigella (SS) agar is not recommended because it inhibits the growth of some strains of *S. dysenteriae* 1. Instructions for preparation of these media are given by Ewing (1986d) and Atlas (1997). Specimens are streaked onto the chosen media, and after overnight incubation at 35–37°C, lactose- or xylose-nonfermenting colonies are selected for further examination. Even when stool specimens from acute dysentery are examined, there may be only a scanty growth of *Shigella*. Suspicious colonies should be tested by biochemical and serologic methods (direct slide agglutination or coagglutination of antibody-coated latex beads) to confirm the identification of shigellae. For *S. flexneri*, which is the most difficult to prepare typing antisera against because of its shared group antigens, a panel of monoclonal antibodies has been developed for typing (Carlin et al., 1989). In some reference laboratories, molecular methods, such as DNA hybridization (Boileau et al., 1984), PCR (Frankel et al., 1990; Sethabutr et al., 1993; Yavzori et al., 1994; Houng et al., 1997; Villalobo and Torres, 1998; Sethabutr et al., 2000), enzyme-linked immunosorbent assays (Floderus et al., 1995; Pal et al., 1997), and immunomagnetic capture assays (Islam et al., 1993), targeting O antigens or invasion-associated genes/proteins are being used to detect shigellae and enteroinvasive *E. coli*.

Laboratory surveillance for *Shigella* infections is conducted in many countries and has been valuable for investigating outbreaks of shigellosis and for guiding vaccine development efforts. Serotyping has been and continues to be the primary method for discriminating between strains. During outbreaks or for highly prevalent serotypes such as *S. sonnei*, isolates are often subjected to a variety of additional phenotypic or molecular typing methods. Virtually all the secondary typing methods are predicated upon knowledge of the isolate's serotype. Other methods commonly used for differentiating between *Shigella* strains include colicin typing and phage typing. The colicin typing scheme for *S. sonnei* described by Abbot and Shannon (1958) has been used widely and distinguishes 14 types using 15 indicator strains (see Procedures for Testing Special Characters). Phage-typing schemes have also been used by many investigators. While a number of schemes have been described for *S. flexneri* and *S. sonnei*, only a few have been reported for *S. dysenteriae* and *S. boydii* (reviewed by Bergan, 1979). A variety of molecular subtyping methods have been applied to *Shigella*, which include ribotyping (Hinojosa-Ahumada et al., 1991; Faruque et al., 1992; Nastasi et al., 1993), IS element typing (Soldati and Piffaretti, 1991), enterobacterial repetitive intergenic consensus (ERIC) sequence-based PCR (Liu et al., 1995a), randomly amplified polymorphic DNA (Bando et al., 1998), and macrorestriction endonuclease analysis with pulsed-field gel electrophoresis (PFGE) (Soldati and Piffaretti, 1991; Liu et al., 1995a; Litwin et al., 1997). Because of its discriminatory potential, PFGE is the method preferred by many investigators for molecular subtyping studies, particularly for epidemiologic investigations of *S. sonnei* infections. This method proved invaluable in the investigation of a protracted outbreak of *S. sonnei* infections among members of traditionally observant Jewish communities in North America (Sobel et al., 1998) and was critical in an investigation of geographically unrelated cases of *S. sonnei* infection, identifying parsley from a single farm as the vehicle of infection (Centers for Disease Control, 1999).

MAINTENANCE PROCEDURES

Shigella cultures are maintained best frozen in liquid nitrogen. Liquid culture medium, such as trypticase soy broth, containing 10–20% glycerol is commonly used as a freezing medium. Cultures of *Shigella* may also be lyophilized or maintained on blood agar base or Dorset egg medium (stabbed slants or deeps) at room temperature; however, rough and degraded variants frequently arise under these conditions.

PROCEDURES FOR TESTING SPECIAL CHARACTERS

For colicin typing of *S. sonnei*, the organism under investigation is inoculated heavily in a broad streak across a blood agar plate and incubated at 37°C for 24 h. The bacterial growth is then removed from the agar by scraping with a glass slide, and the organisms remaining are killed with chloroform. The 15 indicator strains are streaked onto the plate at right angles to the original line of growth. After further incubation for 8–12 h, the patterns of inhibition of growth of the indicator strains can be examined and compared with a key. It is important that controls be included in every batch of tests.

DIFFERENTIATION OF THE GENUS *SHIGELLA* FROM OTHER GENERA

The biochemical identification of *Shigella* is complicated by the similarity of some strains of other genera. In particular, strains of *Hafnia alvei*, *Providencia* spp., *Aeromonas* spp., and atypical *E. coli* can cause difficulties. Advances in our knowledge of the genetics of microorganisms have allowed the development of new approaches for identifying and detecting bacteria. Molecular strategies have been developed to facilitate microbial identification. Spierings et al. (1993) described a PCR assay targeting DNA sequences that encode the hypervariable surface-exposed regions of the outer membrane protein PhoE, which shows a high degree of species specificity and can differentiate strains of *Shigella* and *E. coli* from strains of other enteric bacteria.

Nonlactose-fermenting or anaerogenic strains of *E. coli* are commonly quite difficult to differentiate from shigellae. Of particular interest are members of the so-called Alkalescens Dispar (A-D) group, which contains nonmotile, anaerogenic biotypes of *E. coli*. These are best differentiated from *Shigella* by means of the Christensen's citrate and lysine decarboxylase tests, in which *Shigella* is always negative. The members of the A-D group were divided into eight serogroups based on their O antigens (Frantzen, 1950), although most of these are identical with or closely related to *E. coli* antigens.

TAXONOMIC COMMENTS

The genus *Shigella* consists of nonmotile organisms that conform to the definition of the family *Enterobacteriaceae* and have the biochemical and antigenic properties described above. Circumscription of the genus is problematic because of the close genetic relationship that exists between *Shigella* spp. and *E. coli*. There are no unique traits exclusively associated with *Shigella*. The negative phenotypes identified with *Shigella* spp. can occur in *E. coli*, particularly among the nonmotile, anaerogenic biotypes of *E. coli*, while positive phenotypes seen among typical *E. coli* isolates, such as lactose or sucrose fermentation and gas production from the fermentation of glucose, can occasionally occur in *Shigella*. Some *E. coli* strains (enteroinvasive *E. coli*) can also cause a dysentery-like illness, which is mediated by similar plasmid-encoded virulence genes as those possessed by *Shigella*.

Although pathogenicity (i.e., the ability to cause dysentery) was not endorsed by the *Enterobacteriaceae* Subcommittee of the International Committee on Bacteriological Nomenclature for the classification of *Enterobacteriaceae* (Carpenter, 1963), it has influenced and continues to influence the classification of *Shigella* by virtue of the fact that only strains that are capable of causing dysentery receive consideration for addition to this group. Strains that do not cause dysentery but yield biochemical reactions consistent with those produced by *Shigella* are not considered for inclusion in this genus. The preservation of this association between bacteria classified as *Shigella* and a distinct form of diarrheal illness (dysentery) has sustained a functionally useful nomenclature that might otherwise have yielded over the years to persuasive findings that reveal a sufficiently close relationship between *Shigella* and *E. coli* to consider them the same species (Brenner et al., 1972a, 1973a).

The names adopted by Pupo et al. (2000), (*Escherichia coli* clone Dysenteriae, *Escherichia coli* clone Flexneri, *Escherichia coli* clone Boydii, and *Escherichia coli* clone Sonnei), reflect the well-documented natural relationships that exist between *Shigella* and *E. coli* and have merit from a phylogenetic perspective; however, their widespread adoption will happen only when it is important to users to communicate these relationships through the bacterial names. Until that time, the medically useful link between the genus epithet *Shigella* and shigellosis, the term for the disease caused by these bacteria, will sustain the current nomenclature.

Findings from numerous studies raise questions about the inclusion of *S. boydii* 13 strains in the genus *Shigella*. The DNA–DNA reassociation findings reported by Brenner et al. (1982c) show that *S. boydii* 13 strains were related to each other, but clearly separable from other *Shigella* and *E. coli* (average 65% relatedness). Findings from MLEE (Ochman et al., 1983; Pupo et al., 1997), nucleotide sequence analysis of housekeeping genes (Pupo et al., 2000), ribotyping (Rolland et al., 1998), and esterase electrophoretic polymorphisms (Goullet and Picard, 1987) also show a distant relationship between *S. boydii* 13 strains and other shigellae.

FURTHER READING

Centers for Disease Control and Prevention. 1999. Laboratory Methods for the Diagnosis of Epidemic Dysentery and Cholera, CDC, Atlanta.

Dorman, C.J. and M.E. Porter. 1998. The *Shigella* virulence gene regulatory cascade: a paradigm of bacterial gene control mechanisms. Mol. Microbiol. *29*: 677–684.

Ewing, H.E. 1986. Edwards and Ewing's Identification of *Enterobacteriaceae*, 4th Ed., Elsevier Science Publishing Co., Inc., New York.

Keusch, G.T. and M.L. Bennish. 1998. Shigellosis. *In* Evans and Brachman (Editors), Bacterial Infections of Humans. Epidemiology and Control, 3rd Ed., Plenum Medical Book Co., New York. pp. 631–656.

Menard, R., C. Dehio and P.J. Sansonetti. 1996. Bacterial entry into epithelial cells: the paradigm of *Shigella*. Trends Microbiol. *4*: 220–226.

Sansonetti, P. and A. Phalipon. 1996. Shigellosis: from molecular pathogenesis of infection to protective immunity and vaccine development. Res. Immunol. *147*: 595–602.

DIFFERENTIATION OF THE SPECIES OF THE GENUS *SHIGELLA*

Biochemical characteristics useful for differentiating the species of *Shigella* are listed in Table BXII.γ.275.

List of species of the genus Shigella

1. **Shigella dysenteriae** (Shiga 1898) Castellani and Chalmers 1919, 935, Epit. Spec. Cons. Opinion 11 of the Jud. Comm. 1954a, 149^AL (*Bacillus dysenteriae* Shiga 1898, 817.)*
   dys.en.te′ri.ae. Gr. n. *dysenteria* dysentery; M.L. gen. n. *dysenteriae* of dysentery.

   Colonies of serotype 1 often have a pinkish tinge on Leifson's deoxycholate citrate agar. Catalase is not produced by serotype 1, but is usually produced by strains of other serotypes. D-mannitol is not fermented. Dulcitol is fermented by strains of serotype 5. Gas from the fermentation of sugars is not produced. Indole is not produced by serotype 1, but is always produced by strains of serotype 2; strains of other serotypes vary in indole production. All the serotypes have, at one time or another, been known by other designations, and these are shown in Table BXII.γ.276. Members are intestinal pathogens of humans and subhuman primates, causing bacillary dysentery. Humans are the primary reservoir. A long-term carrier state occurs in a small percentage of cases (Levine et al., 1973). Serotype 1 causes more severe disease than other serotypes and produces a potent protein exotoxin (Shiga toxin), which plays a role in the severity of the disease attributable to this serotype and in the development of hemolytic uremic syndrome in some individuals. Large epidemics in developing countries are commonly caused by serotype 1. Disease caused by other serotypes may be mild or severe.

   The mol% G + C of the DNA is: 53 (chemical analysis; Laskin and Lechevalier (1981)).
   Type strain: ATCC 13313, CIP 57.28, NCTC 4837.
   GenBank accession number (16S rRNA): X96966.
   Additional Remarks: Neotype strain ATCC 13313 was designated by Judicial Commission, 1963.

2. **Shigella boydii** Ewing 1949, 634, Epit. Spec. Cons. Opinion 11 of the Jud. Comm. 1954a, 149^AL*
   boy′di.i. M.L. gen. n. *boydii* of Boyd; named after Sir John Boyd, a British bacteriologist.

   Catalase is produced. D-mannitol is fermented. Dulcitol is usually fermented by serotypes 2, 3, 4, 6, and 10, but this may be delayed. D-xylose fermentation is variable. Indole may or may not be produced. Gas-producing biotypes of *S. boydii* serotype 13 (Rowe et al., 1975) and serotype 14 (Carpenter, 1961) have been described. Members are intestinal pathogens of humans and subhuman primates, causing bacillary dysentery. Disease may be mild or severe. Humans are the primary reservoir. A long-term carrier state occurs in a small percentage of cases (Levine et al., 1973).
   The mol% G + C of the DNA is: not determined.
   Type strain: ATCC 8700, CIP 82.50, DSM 7532, NCTC 12985.

3. **Shigella flexneri** Castellani and Chalmers 1919, 937, Epit.

Spec. Cons. Opinion 11 of the Jud. Comm. 1954a, 149[AL]* *flex'ner.i.* M.L. gen. n. *flexneri* of Flexner, named after Simon Flexner, an American bacteriologist.

Catalase is produced. D-mannitol is fermented, except by biotype Newcastle, serotype 6 and a D-mannitol negative, D-xylose positive biotype of serotype 4a (sometimes known as "*S. rabaulensis*"). Dulcitol is fermented by certain biotypes of serotype 6 (see Table BXII.γ.275), some of which produce gas from fermentable sugars. Indole is not produced by serotype 6; in other serotypes indole production is variable. Members are intestinal pathogens of humans and subhuman primates, causing bacillary dysentery. Humans are the primary reservoir. Disease may be mild or severe. A long-term carrier state occurs in a small percentage of cases (Levine et al., 1973). Infection in individuals with the HLA-B27 histocompatibility antigen can be complicated by Reiter chronic arthritis syndrome.

The mol% G + C of the DNA is: 49 (chemical analysis) (Laskin and Lechevalier, 1981) and 50.9 (nucleotide sequence analysis) (Jin et al., 2002) and (Wei et al., 2003).

Type strain: ATCC 29903, CIP 82.48, DSM 4782.

GenBank accession number (16S rRNA): X96963; complete genome: AE005673 and AE014073.

4. **Shigella sonnei** (Levine 1920) Weldin 1927, 182, Epit. Spec.

Cons. Opinion 11 of the Jud. Comm. 1954a, 149[AL] (*Bacterium sonnei* Levine 1920, 31.)*

son'ne.i. M.L. gen. n. *sonnei* of Sonne, named after Carl Sonne, a Danish bacteriologist.

On deoxycholate citrate agar colonies are at first colorless, but after a few days show bright pink papillae consisting of lactose-fermenting cells. On MacConkey agar, phase I colonies are indistinguishable from colonies of other shigellas, but phase II colonies are larger, flatter, and more translucent, and have an irregular edge. On subculture, phase I colonies produce both phase I and phase II colonies, but phase II colonies give rise to phase II colonies only. D-mannitol is fermented rapidly, lactose and sucrose more slowly. Some strains may ferment D-xylose. Catalase is produced. Indole is not produced. Ornithine is decarboxylated; arginine may be decarboxylated. Members are intestinal pathogens of humans and subhuman primates, causing bacillary dysentery. Humans are the primary reservoir. A long-term carrier state occurs in a small percentage of cases (Levine et al., 1973). Illness is typically milder than that caused by members of the other *Shigella* subgroups.

The mol% G + C of the DNA is: 51 (CsCl buoyant density).

Type strain: ATCC 29930, CIP 82.49, DSM 5570, NCTC 12984.

Other Organisms

S. boydii 13 strains were first described in 1952 by Ewing and Hucks (1952) under the provisional designation "serotype 425". Six years later, serotype 425 was added to the *Shigella* scheme as *S. boydii* 13 (Ewing et al., 1958). The initial description of the serotype included 10 strains: eight from cases and two from carriers isolated from persons from Italy, Belgian Congo, Egypt, and the United States. Except for differences in their ability to grow on ammonium salts glucose agar, the strains were biochemically indistinguishable from each other and most closely resembled shigellae of subgroup C. In 1973, Brenner et al. (1973a) reported evidence from DNA–DNA reassociation studies of *Shigella* species challenging the placement of *S. boydii* 13 strains in the genus *Shigella*. These investigators found that the *S. boydii* 13 strains they examined were highly interrelated but averaged only about 65% relatedness to other shigellae or members of the genus *Escherichia*. In a subsequent study, Brenner et al. (1982c) published findings from additional *S. boydii* 13 strains that extended and confirmed the earlier observation that *S. boydii* 13 strains,

both anaerogenic and aerogenic varieties, are separable from other shigellae and *E. coli* based on DNA relatedness. The DNA–DNA reassociation evidence clearly demonstrates that *S. boydii* 13 strains represent a new, as yet unnamed, species. More recent findings from MLEE (Ochman et al., 1983; Pupo et al., 1997), nucleotide sequence analysis of housekeeping genes (Pupo et al., 2000), ribotyping (Rolland et al., 1998), and esterase electrophoretic polymorphisms (Goullet and Picard, 1987) are consistent with a distant relationship between *S. boydii* 13 strains and the other shigellae. At present, *S. boydii* 13 strains are primarily isolated by clinical laboratories and routinely keyed out as belonging to the genus *Shigella*. Because practical and logical factors are important in considering changes to existing systems of taxonomy and nomenclature (Brenner et al., 1973a), the renaming or reclassification of these bacteria is dependent on the development of methods that are useful for their identification in clinical laboratories.

Genus XXXVI. **Sodalis** *Dale and Maudlin 1999, 273[VP]*

COLIN DALE, SERAP AKSOY, SUSAN C. WELBURN, IAN MAUDLIN AND AHARON OREN

so.da'lis. L. masc. n. *sodalis* a companion.

Cells rod-shaped. Nonmotile. Gram negative. Endospore formation not observed. **Microaerophilic.** Chemoorganotrophic. Axenic growth occurs in media with enzymatically digested proteins serving as carbon and nitrogen sources. **Certain carbohydrates are used with the production of acids.** The optimum temperature for growth is 25°C, with little or no growth occurring at temperatures >30°C. Catalase and oxidase negative. **Found as secondary intracellular symbiotic bacteria in insects.**

The mol% G + C of the DNA is: 53.5.

Type species: **Sodalis glossinidius** Dale and Maudlin 1999, 273.

Editorial Note: This species is also known as subgroup B; formerly known as "*S. paradysenteriae*" Flexner.

Editorial Note: This species is also known as subgroup D.

FURTHER DESCRIPTIVE INFORMATION

The genus *Sodalis* consists of maternally transmitted intracellular symbiotic bacteria inhabiting the hemolymph and other organs of insects. Most intracellular symbiotic bacteria of insects are fastidious, and have thus far proved refractory to conventional culture techniques. *Sodalis glossinidius*, a multitissue secondary intracellular symbiont of the tsetse fly (*Glossina morsitans*), was the first true intracellular symbiont from insects to be cultivated *in vitro*. Cultivation was first achieved through the use of a mosquito (*Aedes albopictus*) feeder cell culture system. Subsequently, the organism was isolated in pure culture on a semidefined solid medium under microaerobic conditions. Therefore, *Sodalis* is the first example of the successful cultivation of an insect secondary intracellular symbiont on an agar-based medium. Phylogenetic characterization of *Sodalis* strains from different tsetse species revealed no differences based on their 16S rDNA sequences (Chen et al., 1999b). Other insects related to the tsetse fly carry bacterial symbionts with 16S rDNA sequences very similar to the 16S rDNA of *Sodalis*, but these symbionts are still awaiting isolation.

Sodalis glossinidius, the only species of the genus described thus far, is microaerophilic and can be cultivated only on solid media under a reduced-oxygen atmosphere. An atmosphere of 5% oxygen and 95% carbon dioxide is optimal for growth. Alternatively, aerotolerance-enhancing supplements such as catalase or fresh horse blood as a source of catalase activity can be added to the plates to enable growth in air.

The biochemical capacities of *Sodalis glossinidius* are rather restricted. Among the few carbohydrates used for growth the most efficient are *N*-acetyl-D-glucosamine and raffinose, which are metabolized with massive acid production. Glucose, glycol chitosan, mannitol, and sorbitol are metabolized to a lesser extent.

Sodalis glossinidius has a chromosome of about 2 Mb and about 134 kb plasmid DNA (Akman et al., 2001). Its chromosomal DNA is subject to extensive adenine and cytosine methylation. Hybridization of *Sodalis* DNA to *Escherichia coli* macroarrays revealed the presence of about 1800 orthologs, indicating that *Sodalis* has retained many of the capabilities of free-living bacteria. Based on this analysis, it appears that *Sodalis* has retained many genes involved in transcription, translation, regulation, nucleic acid, and amino acid biosynthetic pathways. However, it is possible that *Sodalis* might have lost genes in carbon compound catabolism, central intermediary metabolism, and fatty acid and phospholipid metabolism, which could account for the organism's restricted biochemical capacities.

To enter into insect cells, *Sodalis* uses a type III secretion system invasion gene (*invC*), very similar to virulence determinants previously identified in the genera *Salmonella* and *Shigella* (Dale et al., 2001).

ENRICHMENT AND ISOLATION PROCEDURES

Sodalis glossinidius has been isolated from the hemolymph of *Glossina morsitans morsitans* in coculture with the mosquito *Aedes albopictus* cell line C6/36 (Welburn et al., 1987). Pure culture isolation was achieved through the use of solid-phase culture on serum-free Mitsuhashi-Maramorosch basal medium (Igarashi, 1978) with 1% Bacto-agar (Difco) under a microaerobic atmosphere (O_2/CO_2; 5:95). For growth under an air atmosphere, aerotolerance-enhancing supplements such as catalase or fresh horse blood cells (as source of catalase activity) should be added.

MAINTENANCE PROCEDURES

Sodalis glossinidius can be maintained by coculture in *Aedes albopictus* C6/36 cells at 25°C in liquid Mitsuhashi-Maramorosch medium (Igarashi, 1978) supplemented with 20% (v/v) heat-inactivated fetal calf serum by passaging the cells every 10 d (Welburn et al., 1987). For long-time storage bacterial cells are frozen in Mitsuhashi-Maramorosch medium supplemented with 20% (v/v) heat-inactivated fetal calf serum and 15% (v/v) glycerol.

TAXONOMIC COMMENTS

Sodalis was classified in the family *Enterobacteriaceae* based on phenotypic tests and its 16S rDNA sequence. As a microaerophilic insect symbiont, it differs sufficiently from the other described genera within the *Enterobacteriaceae* to warrant classification as a separate genus. Different species and subspecies of tsetse flies harbor intracellular symbionts having almost identical 16S rDNA sequences (Beard et al., 1993; Aksoy et al., 1997). In addition, other insects, including flour and rice weevils and certain psyllids also harbor intracellular symbionts with closely related 16S rDNA sequences (Thao et al., 2000a).

Sodalis harbors type III secretion system invasion genes that are phylogenetically related to those found previously in *Salmonella enterica* and *Shigella flexneri*.

FURTHER READING

Akman, L., R.V.M. Rio, C.B. Beard and S. Aksoy. 2001. Genome size determination and coding capacity of *Sodalis glossinidius*, an enteric symbiont of tsetse flies, as revealed by hybridization to *Escherichia coli* gene arrays. J. Bacteriol. *183*: 4517–4525.

Dale, C. and I. Maudlin. 1999. *Sodalis* gen. nov. and *Sodalis glossinidius* sp. nov., a microaerophilic secondary endosymbiont of the tsetse fly *Glossina morsitans* morsitans. Int. J. Syst. Bacteriol. *49*: 267–275.

Dale, C., S.A. Young, D.T. Haydon and S.C. Welburn. 2001. The insect endosymbiont *Sodalis glossinidius* utilizes a type III secretion system for cell invasion. Proc. Natl. Acad. Sci. U.S.A. *98*: 1883–1888.

List of species of the genus Sodalis

1. **Sodalis glossinidius** Dale and Maudlin 1999, 273.[VP]

   glos.si.ni' di.us. N.L. adj. *glossinidius* of the genus *Glossinia*.

   The characteristics are as described for the genus, with the following additional information. The cells are 1.0–1.5 × 2.0–12.0 μm. When grown intracellularly in *Aedes albopictus* cell culture, the organisms appear as pleomorphic rods. Temperature range for growth: 18–28°C. Colonies on Mitsuhashi-Maramorosch agar (Igarashi, 1978) are shiny, off-white, and concave, with entire edges.

   Negative for DNase, gelatinase, urease, nitrate reductase, indole production, hippurate hydrolysis, arginine dihydrolase, lysine decarboxylase, ornithine decarboxylase, phenylalanine decarboxylase, and starch hydrolysis. Produces α-galactosidase and β-*N*-acetylglucosaminidase. Does not produce α-fucosidase, β-galactosidase, α-glucosidase, β-glucosidase, β-glucuronidase, α-mannosidase, and β-xylosidase. Utilization of *N*-acetyl-D-glucosamine and raffinose accompanied by a high level of acid production. Utilizes glucose, glycol chitosan, mannitol, and sorbitol with accompanying weak acid production. No increase in growth is detected when any of the following carbon sources are incorporated in the media: acetic acid, adonitol, δ-aminovaleric acid, L-arabinose, *n*-butanol, citric acid, dulcitol, ethanol, fructose, fumaric acid, galactose, glycerol, glycolic acid, histamine, *p*-hydroxybenzoic acid, α-ketoglutaric acid, lactose, D,L-malic acid, maltose, mannose, melibiose, methyl-α-D-glucopyranoside, *myo*-inositol, n-propanol, pyruvic acid, quinic acid, rhamnose, ribose, saccharic acid, salicin, sarcosine, sorbose, succinic acid, sucrose, starch, trehalose, or xylose.

Contains a number of large extrachromosomal DNA elements.

Found as secondary intracellular symbionts in the midgut, fat body, and hemolymph of the tsetse fly *Glossina morsitans morsitans*.

The mol% G + C of the DNA is: 53.5 (T_m).

Type strain: M1, NCIMB 13495.

GenBank accession number (16S rRNA): M99060.

Genus XXXVII. **Tatumella** *Hollis, Hickman and Fanning 1982, 267^{VP} (Effective publication: Hollis, Hickman, Fanning, Farmer, Weaver and Brenner 1981, 86) (Group EF-9 Hollis, Hickman, Fanning, Brenner and Weaver 1980)*

J.J. FARMER III

Ta.tum.el′ la. M.L. dim. neut. -*ella* ending; M.L. fem. n. *Tatumella* named to honor Harvey Tatum, an American bacteriologist who made many contributions to our understanding of the classification and identification of fermentative and nonfermentative bacteria of medical importance.

Small rod-shaped cells 0.6–0.8 × 0.9–3 μm, conforming to the general definition of the family *Enterobacteriaceae*. Contains the enterobacterial common antigen. Gram negative. **Nonmotile at 36°C; over half the strains are motile by means of polar, subpolar, or lateral flagella when grown at 25°C.** Facultatively anaerobic. Catalase positive (weak and slow). Oxidase negative. Nonpigmented. **Stock cultures often die within a few weeks on laboratory media. Biochemically more active at 25°C than at 36°C.** Ferment, rather than oxidize, D-glucose; without the formation of visible gas. Reduce nitrate to nitrite. **Very inactive biochemically; positive tests only for Voges–Proskauer (Coblentz method), phenylalanine deaminase, and fermentation of sucrose, trehalose, and D-mannose. Negative for most tests**: indole production, methyl red, Voges–Proskauer (O'Meara method), citrate utilization (Simmons), H_2S production (TSI), urea hydrolysis, lysine decarboxylase, arginine dihydrolase, ornithine decarboxylase, growth in the presence of cyanide (KCN test), malonate utilization, esculin hydrolysis, ONPG, gelatin hydrolysis (22°C), lipase (corn oil), DNase, gas production during fermentation, and the fermentation of lactose, D-mannitol, dulcitol, adonitol, *myo*-inositol, D-sorbitol, L-arabinose, raffinose, L-rhamnose, maltose, D-xylose, cellobiose, α-methyl-D-glucoside, erythritol, D-arabitol, glycerol, and mucate. Have very large zones of inhibition around antibiotics; susceptible to colistin, nalidixic acid, sulfadiazine, gentamicin, streptomycin, kanamycin, tetracycline, chloramphenicol, carbenicillin, ampicillin, and cephalothin (disk diffusion method on Mueller–Hinton agar). **Large zone of inhibition around a penicillin G (10 U) disk, in contrast to most other** *Enterobacteriaceae*.

Isolated from human clinical specimens, mainly from the **respiratory tract** where clinical significance is questionable. Also isolated from blood cultures suggesting clinical significance. Probably **a rare opportunistic pathogen or colonizer of humans**.

The mol% G + C of the DNA is: 53–54.

Type species: **Tatumella ptyseos** Hollis, Hickman and Fanning 1982, 267 (Effective publication: Hollis, Hickman, Fanning, Farmer, Weaver and Brenner 1981, 86.)

FURTHER DESCRIPTIVE INFORMATION

Literature Since *Tatumella* was described in 1981, there have been only a few reports in the literature. In addition to the genus

and species names, "EF-9" should also be included as a search term in computerized literature searches, although it is rarely used today.

The *Tatumella* literature reports include reviews and taxonomic studies of new *Enterobacteriaceae* (Gilchrist, 1995; Aleksic and Bockemühl, 1999), and surveys or comparisons of the family *Enterobacteriaceae* for evolutionary relatedness based on 16S rRNA sequencing (Spröer et al., 1999), the enterobacterial common antigen (Ramia et al., 1982), catalase production (Chester and Moskowitz, 1987), metabolic pathways (Bouvet and Grimont, 1987a; Bouvet et al., 1989), polyamines (Hamana, 1996), and siderophores (Rabsch and Winkelmann, 1991).

Sources, strains, collections Strains of *Tatumella* have been isolated from a variety of human clinical specimens, but not from other sources. The original collection of Hollis et al. (1981) included isolates from body sites that are normally sterile such as blood (three isolates) and urine (one), but 38 were from throat, pharynx, tracheal aspirate, or sputum. One was from feces.

Clinical significance of human isolates The clinical significance and the ability of *T. ptyseos* to actually cause infections should be carefully evaluated in cases that yield this organism. Systematic study and good case reports are needed. Tan et al. (1989) reported a case of presumed sepsis in a neonate with jaundice who recovered uneventfully after treatment with antibiotics. This report and the three blood isolates described by Hollis et al. (1981) suggest that *T. ptyseos* should at least be considered as a rare opportunistic pathogen.

Isolation, identification, culture preservation, phenotypic characterization Strains of *Tatumella* are not difficult to grow, but may grow more slowly than typical *Enterobacteriaceae*. Stock cultures may die within a few weeks on agar or in semisolid stock culture media. This characteristic is very unusual, since most cultures of *Enterobacteriaceae* can be kept almost indefinitely in sealed tubes kept at room temperature. *Tatumella* cultures frozen in 5% rabbit blood and stored at −40°C to −60°C have remained viable after storage for up to 14 years. This latter method or storage in liquid nitrogen (or perhaps freeze-drying) should be suitable for long-term preservation. See the chapter on the family *Enterobacteriaceae* for general information on growth, plating media, working and frozen stock cultures, media and methods for biochemical testing, identification, computer programs, and antibiotic susceptibility. On MacConkey agar at 24 h, all strains grew and produced colorless (lactose negative) colonies.

Biochemical reactions and differentiation from other *Enterobacteriaceae* Table BXII.γ.193 in the chapter on the family *Enterobacteriaceae* gives the results for *Tatumella* in 47 biochemical tests normally used for identification (Farmer, 1999). The most striking feature of *T. ptyseos* strains is their biochemical inactivity. They are also more active biochemically at 25°C than 36°C. The flagellation of *Tatumella* is also unusual. Strains are nonmotile at 36°C, but 66% are motile at 25°C. No flagella are produced on most cells, but those seen are polar, subpolar, or lateral rather than peritrichous. In contrast to most other *Enterobacteriaceae*, cultures of *T. ptyseos* have large zones of inhibition around 10 U penicillin G disks (range of 15–36 mm, mean of 24 mm, standard deviation of 4.6 mm). There are no genus- or species-specific tests or sequences for the identification of *Tatumella*. The best approach is to do a complete set of phenotypic tests (growth on plating media, biochemical tests, antibiogram, etc.) and compare the results with all the other named organisms in the family *Enterobacteriaceae*. This approach is described in more detail in the section on *Enterobacteriaceae*. The biochemical results of a test strain can be compared to all the organisms listed in Table BXII.γ.193 in the chapter on the family *Enterobacteriaceae*. Several computer programs greatly facilitate analyzing the results.

TAXONOMIC COMMENTS

History The name *Tatumella* was coined by Hollis et al. (1981) for a group of organisms that had previously been known as "Group EF-9", a name coined and used by the Special Bacteriology Section at the Centers for Disease Control (Hollis et al., 1980). Group EF-9 had been known for many years, but its taxonomic position had not been studied.

Circumscription of *T. ptyseos* (Hollis et al., 1981) Twenty-seven "suspect" strains of Group EF-9 were studied by DNA–DNA hybridization, biochemical reactions, and antibiotic susceptibility. Twenty of the strains were very highly related to the type strain (89% or greater with ΔT_m values of 0–0.5) by DNA–DNA hybridization (60°C, hydroxyapatite method with ^{32}P). A group of six strains was less related (80–90% with ΔT_m values of 5.4–7.8), but was included in the species *T. ptyseos*. These strains may eventually be classified as one or more distinct biogroups. One strain (H34) was considerably less related (66% related at 60°C, which dropped to 44% at 75°C), and was not included in *T. ptyseos*. This probably represents a second species of *Tatumella*. By DNA hybridization, other taxa in the family *Enterobacteriaceae* were related by 7–38%, including 25–30% relatedness of *Escherichia*, the type genus of the family. Based on the DNA–DNA hybridization and phenotypic data, Group EF-9 was proposed as a new genus, *Tatumella*, with *T. ptyseos* as the only species (Hollis et al., 1981).

Phylogeny based on 16S rRNA sequencing A 16S rRNA tree that includes *T. ptyseos* can be found in Fig. BXII.γ.198 in the chapter on the genus *Budvicia* in this *Manual*. The 16S rRNA sequencing data agree with DNA–DNA hybridization and phenotypic data that *Tatumella* is a distinct genus of *Enterobacteriaceae*. This tree generally agrees with the tree published by Spröer et al. (1999), in that *Tatumella* is on a distinct branch. However, the two trees differ in the placement of the closest neighbors and other organisms (see also Fig.BXII.γ.189 in the chapter on the family *Enterobacteriaceae*).

Nomenclatural problem: the irregular author citation in the names *Tatumella* Hollis, Hickman and Fanning 1982, 267 [VP] and *Tatumella ptyseos* Hollis, Hickman and Fanning 1982, 267 [VP] The author citation for the names *Tatumella* and *Tatumella ptyseos* has traditionally been written as "Hollis, Hickman and Fanning 1982, 267 [VP]" rather than the expected citation of "Hollis, Hickman, Fanning, Farmer, Weaver and Brenner 1982, 267[VP]". The original paper (Hollis et al., 1981) had six authors rather than three, but on page 86 the authors stated: "We propose the following citations of the genus and species, which recognizes the greater contribution of three of the authors: *Tatumella* Hollis, Hickman and Fanning; *Tatumella ptyseos* Hollis, Hickman and Fanning."

This irregular author citation continued through correspondence with the editor of the *International Journal of Systematic Bacteriology* that requested validation of the genus and species name. In Validation List 8, the request for a "three-author citation" was granted and was without comment since only three authors, "Hollis, Hickman and Fanning", were listed under the column heading "Author(s)" for both names. This unusual citation appears to conflict with the rules of the Bacteriological Code (International Code of Nomenclature of Bacteria) for citing authors of new taxa. Although this is a small point, it is one that can lead to instability in nomenclature. The problem could be addressed by amending the Bacteriological Code, or through an Opinion of the Judicial Commission.

FURTHER READING

Hollis, D.G., F.W. Hickman, G.R. Fanning, J.J. Farmer, III, R.E. Weaver and D.J. Brenner. 1981. *Tatumella ptyseos* gen. nov., sp. nov., a member of the family *Enterobacteriaceae* found in clinical specimens. J. Clin. Microbiol. *14*: 79–88.

List of species of the genus Tatumella

1. **Tatumella ptyseos** Hollis, Hickman and Fanning 1982, 267[VP] (Effective publication: Hollis, Hickman, Fanning, Farmer, Weaver and Brenner 1981, 86.)

   pty' se.os. Gr. n. *ptyseos* a spitting (or less literally from sputum, the most common source of clinical isolates).

   The characteristics are as given for the genus and the results for 47 biochemical tests are given in Table BXII.γ.193 of the chapter on the family *Enterobacteriaceae*. Isolated from human clinical specimens, primarily from the respiratory tract; should probably be considered a rare opportunistic pathogen.

   The mol% G + C of the DNA is: 53–54 (Bd).
   Type strain: H36, ATCC 33301, CDC D6168, CDC 9591-78, DSM 5000.
   GenBank accession number (16S rRNA): AJ233437.
   Additional Remarks: Hollis et al. (1981) include 25 other strains in the species; however, six of these were less related to the type strain that the others (see previous discussion). One additional strain (very highly related to the type strain) was also deposited in the American Type Culture Collection (ATCC 33302 [H3 9558-78 = A7744]), from human sputum, Connecticut, USA.

Genus XXXVIII. **Trabulsiella** McWhorter, Haddock, Nocon, Steigerwalt, Brenner, Aleksic, Bockemühl and Farmer 1992, 327[VP] (Effective publication: McWhorter, Haddock, Nocon, Steigerwalt, Brenner, Aleksic, Bockemühl and Farmer 1991, 1482) (Enteric Group 90 Murlin, Brenner, Steigerwalt and Farmer 1988)

JOHN A. LINDQUIST AND J.J. FARMER III

Tra.bul.si.el'la. M.L. dim. ending *-ella;* M.L. fem. n. *Trabulsiella* named to honor L.R. Trabulsi, a Brazilian bacteriologist who did many important studies on the enteric pathogens of the family *Enterobacteriaceae*.

Rod-shaped cells, conforming to the general definition of the family *Enterobacteriaceae*. Gram negative. Motile. Facultatively anaerobic. Catalase positive. Oxidase negative. Ferment, rather than oxidize, D-glucose and other carbohydrates and produce visible gas during fermentation. Reduce nitrate to nitrite. **Positive for methyl red, citrate utilization (Simmons), H₂S production (TSI), lysine decarboxylase, ornithine decarboxylase, motility at 36°C, growth in the presence of cyanide (KCN test), esculin hydrolysis, ONPG, tyrosine hydrolysis, and the fermentation of D-mannitol, salicin, D-sorbitol, L-arabinose, L-rhamnose, maltose, D-xylose, trehalose, cellobiose, esculin, mucate, D-mannose, and D-galactose. Arginine dihydrolase is positive, but delayed.** Negative for Voges–Proskauer, urea hydrolysis, phenylalanine deaminase, malonate utilization, gelatin hydrolysis (22°C), lipase (corn oil), DNase, and the fermentation of sucrose, dulcitol, adonitol, *myo*-inositol, raffinose, α-methyl-D-glucoside, erythritol, melibiose, and D-arabitol. **Biochemically similar to, and can be confused with, *Salmonella* in routine screening tests. Do not completely type in *Salmonella* O or H antisera, and are negative in the *Salmonella*-specific test: lysis by bacteriophage O1, MUCAP (4-methylumbelliferyl caprylate), and a commercial gene probe (Gene-Trak Systems).** Susceptible to colistin, nalidixic acid, gentamicin, streptomycin, kanamycin, tetracycline, chloramphenicol, and carbenicillin; **resistant to penicillin, ampicillin, and cephalothin**; variable resistance to sulfadiazine and carbenicillin (disk diffusion method on Mueller–Hinton agar).

Ecological niches are not known, but have been isolated from environmental samples and human feces. However, there is **no evidence that *Trabulsiella* causes diarrhea or intestinal infections**. Apparently a **very rarely isolated genus of *Enterobacteriaceae*.**

The mol% G + C of the DNA is: not determined.

Type species: **Trabulsiella guamensis** McWhorter, Haddock, Nocon, Steigerwalt, Brenner, Aleksic, Bockemühl and Farmer 1992, 327 (Effective publication: McWhorter, Haddock, Nocon, Steigerwalt, Brenner, Aleksic, Bockemühl and Farmer 1991, 1483.)

FURTHER DESCRIPTIVE INFORMATION

Literature Since *Trabulsiella* was described in 1991, there have been only a few reports in the literature. In addition to the genus and species names, "Enteric Group 90" should also be included as a search term in computerized literature searches, although it is rarely used today.

Sources Six strains of *Trabulsiella* were isolated from vacuum cleaner dust on the island of Guam, three were from human stool specimens (one from New York, USA, and two from Germany), one was from wheat flour from Oregon, USA, and two were from environmental material (not further specified) in Malaysia. Many of these isolates were originally isolated and studied as "suspect *Salmonella* cultures".

Isolation, identification, culture preservation, phenotypic characterization Strains of *Trabulsiella* are not difficult to grow,

and are typical *Enterobacteriaceae* in most respects. See the chapter on the family *Enterobacteriaceae* for general information on growth, plating media, working and frozen stock cultures, media and methods for biochemical testing, identification, computer programs, and antibiotic susceptibility. Strains of *Trabulsiella* will often appear as typical *Salmonella* strains on some enteric plating and screening media; however, they do not type completely in *Salmonella* O and H antisera.

Biochemical reactions and differentiation from other *Enterobacteriaceae* Table BXII.γ.193 in the section on the family *Enterobacteriaceae* gives the results for *Trabulsiella* in 47 biochemical tests normally used for identification (Farmer, 1999). There are no genus- or species-specific tests or sequences for the identification of *Trabulsiella*. The best approach is to do a complete set of phenotypic tests (growth on plating media, biochemical tests, antibiogram, etc.) and compare the results with all the other named organisms in the family *Enterobacteriaceae*. This approach is described in more detail in the chapter on the family *Enterobacteriaceae*. The biochemical results of a test strain can be compared to all the organisms listed in Table BXII.γ.193 of that chapter. Several computer programs greatly facilitate analyzing the results. *Trabulsiella* strains are phenotypically most like *Salmonella* subgroups 4 and 5, which are rarely isolated from human clinical specimens.

TAXONOMIC COMMENTS

In 1985 the vernacular name Enteric Group 90 was coined for a group of strains that had been studied at the *Salmonella* Laboratory at the Centers for Disease Control and Prevention. From 1985 to 1988 eight strains were characterized that were phenotypically distinct from all of the described organisms in the family *Enterobacteriaceae*. They had been submitted as "biochemically *Salmonella*, but will not serotype". Since the strains were biochemically very close to *Salmonella*, particularly to *Salmonella* DNA hybridization groups 4 and 5, they were originally suspected as being a new DNA hybridization group in the genus *Salmonella*. However, by DNA–DNA hybridization the strains were very highly related to each other, but were more distantly related to other organisms in the family *Enterobacteriaceae*. Strain ATCC 49490 (CDC 370-85) that eventually became the type strain of *Trabulsiella guamensis* was 41% related by DNA–DNA hybridization to *Salmonella* serotype Typhimurium and to a strain of *Salmonella* DNA hybridization group 3b. Less relatedness was found to *Kluyvera ascorbata* (39%), *Shigella flexneri* (38%), *Klebsiella terrigena* (38%), and strain 6003-71 of the *Enterobacter agglomerans* complex (38%). Other *Enterobacteriaceae* were 6–37% related. Because strains of Enteric Group 90 were distinct by both DNA–DNA hybridization and phenotype, a new genus *Trabulsiella* with a single species *Trabulsiella guamensis* was proposed (McWhorter et al., 1991). *Trabulsiella* and *Trabulsiella guamensis* gained standing in nomenclature in 1992 when they appeared on Validation List 41 (McWhorter et al., 1992).

FURTHER READING

McWhorter, A.C., R.L. Haddock, F.A. Nocon, A.G. Steigerwalt, D.J. Brenner, S. Aleksic, J. Bockemuehl and J.J. Farmer, III. 1991. *Trabulsiella guamensis*, a new genus new species of the family *Enterobacteriaceae* that resembles *Salmonella* subgroups 4 and 5. J. Clin. Microbiol. *29*: 1480–1485.

List of species of the genus Trabulsiella

1. **Trabulsiella guamensis** McWhorter, Haddock, Nocon, Steigerwalt, Brenner, Aleksic, Bockemühl and Farmer 1992, 327[VP] (Effective publication: McWhorter, Haddock, Nocon, Steigerwalt, Brenner, Aleksic, Bockemühl and Farmer 1991, 1483.)

   guam.en'sis. M.L. adj. *guamensis* pertaining to Guam, the largest island of the Micronesian group of the Pacific Ocean, where the first strains were isolated.

   The characteristics are as given for the genus. Isolated from environmental samples, food, and human feces. However, there is no evidence that *Trabulsiella guamensis* can cause diarrhea or intestinal infections. Apparently it is a very rarely isolated species of *Enterobacteriaceae*, whose importance in microbiology may be more as a nuisance, because of its biochemical similarity to *Salmonella* and possible misidentification as *Salmonella*. The type strain was isolated from vacuum cleaner contents on the island of Guam obtained during a survey to isolate *Salmonella* from environmental reservoirs, and to look at the role of reptile feces as a source of *Salmonella* infections. Vacuum cleaner contents were a convenient sample of "indoor dirt".

   In the addendum section of their paper, McWhorter et al. (1991) described two different biochemical patterns given by *T. guamensis* strains, but they did not name them as biogroups. The eight strains from the United States and Guam were negative for indole production, gelatin hydrolysis (film method, 36°C), and esculin hydrolysis (within 2 d), but the four strains from Germany and Malaysia were positive. Strain 2421-87 (feces, Germany) was the only one of the original eight strains studied that had this latter pattern, but it was 98% related to the type strain by DNA–DNA hybridization at 60°C and the value did not drop when the temperature was raised to 75°C. Thus, the available data indicate that the two groups of strains should be considered as biogroups of *T. guamensis*.

   The mol% G + C of the DNA is: not determined.

   Type strain: ATCC 49490, CDC 0370-85.

   Additional Remarks: The American Type Culture Collection includes four other strains of *Trabulsiella guamensis* including strain ATCC 49493 (CDC 2421-87, from human feces, Germany) that is from the indole-positive biogroup described above.

   The biochemical and geographic differences of the two groups of strains may reflect evolutionary divergence that should be investigated with methods other than DNA–DNA hybridization that measure relatedness. J.J. Farmer (unpublished data) recently found a group of three strains in the collection of the Enteric Reference Laboratory, CDC that were very similar to *Trabulsiella*, but were H₂S negative. These could represent a second species of *Trabulsiella*.

Genus XXXIX. Wigglesworthia Aksoy 1995b, 849[VP]

SERAP AKSOY

Wigg.les.worth' i.a. M.L. fem. n. *Wigglesworthia* named after the parasitologist W.B. Wigglesworth.

Obligate intracellular, bacteriome-tissue associated, primary endosymbionts of tsetse flies. In the adult fly, the bacteria lie free in the cytoplasm of differentiated epithelial cells, bacteriocytes, which form the U-shaped bacteriome organ (previously referred to as mycetome) in the anterior midgut (Figs. BXII.γ.210A, B). The endosymbionts are rod shaped (2–4 × 8–10 μm) with a typical Gram-negative cell wall and inner membrane structure, lack flagella (Fig. BXII.γ.210C). The cells appear to divide by binary fission. The bacteria are obligate intracellular organisms, as their experimental elimination from tsetse by procaryotic-specific antibiotic treatment, by lysozyme, or with symbiont-specific antibodies provided in the blood meal results in retarded growth of the insect and a decrease in egg production, causing loss of the ability of these aposymbiotic hosts to reproduce (Hill et al., 1973; Nogge, 1976, 1978, 1980). In addition to *Wigglesworthia*, tsetse harbors a secondary (S)-symbiont in midgut tissue. The specific function(s) of *Wigglesworthia* and S-symbionts are not known; however, since the ability to reproduce can be partially restored when the aposymbiotic tsetse receive a blood meal that is supplemented with B-complex vitamins (thiamine, pantothenic acid, pyridoxine, folic acid, and biotin), it is suggested that the gut-tissue associated endosymbionts may play a role in metabolism that involves these compounds (Nogge, 1981). The recently sequenced genome of *Wigglesworthia* shows the presence of various biosynthetic pathways for vitamin metabolites (Akman et al., 2002). The 16S rDNA-based phylogeny of *Wigglesworthia* characterized from eight different tsetse species displays concordance with their insect host species, indicating an ancient association for this bacterium with an ancestral tsetse that has subsequently radiated along with the host insect species (Chen et al., 1999b). The bacterium is not present in the ovaries, but may be transmitted to the intrauterine larvae of tsetse via its milk-gland secretions.

Type species: **Wigglesworthia glossinidia** Aksoy 1995b, 849.

FURTHER DESCRIPTIVE INFORMATION

Various insects that rely on a single food source (blood or phloem) have established symbiotic associations with microorganisms in order to supplement their limited diets. Stuhlman found bacteria in the intestines of tsetse (Stuhlmann, 1907) and Roubaud described a mycetome (bacteriome)-like organ in midgut (Rouboud, 1919) that was confirmed by Wigglesworth and Buxton (Wigglesworth, 1929; Buxton, 1955). In ultrastructural studies, Reinhardt et al. (1972) described the large bacteroid found in tsetse mycetomes, and Pinnock and Hess (1974) distinguished these organisms from the smaller S-symbionts found in midgut tissue.

The well-defined bacteriome structure in tsetse guts can be

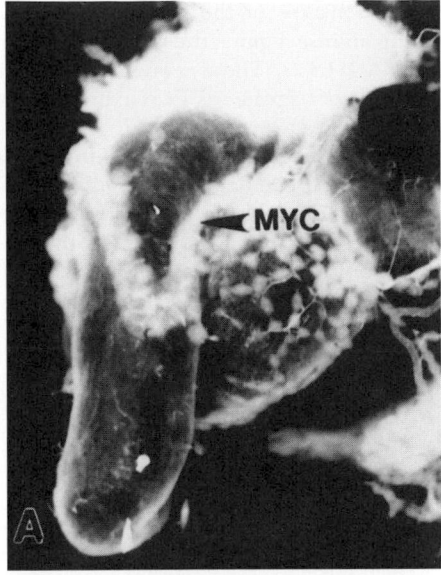

A

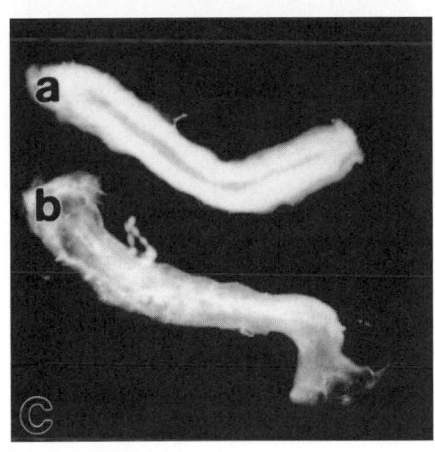

C

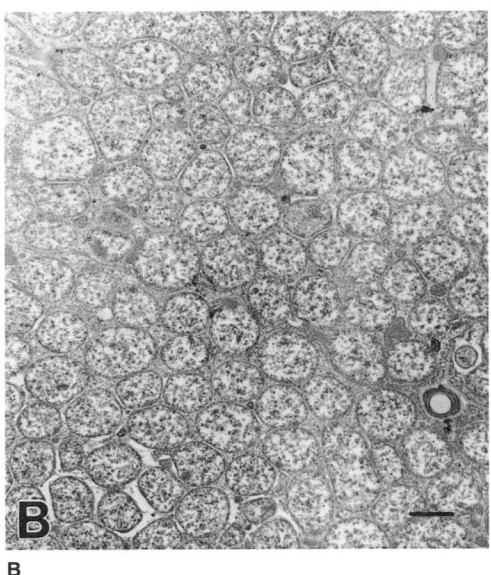

B

FIGURE BXII.γ.210. Bacteriome structure and *Wigglesworthia* in bacteriocytes. *A*, The *arrow* points at the U-shaped bacteriome (previously called mycetome, MYC) structure in the anterior midgut region of *G. m. morsitans*. (Reproduced with permission from S. Aksoy et al., Insect Molecular Biology, *4:* 15–22, 1995, ©Blackwell Science Ltd., Oxford.) *B*, Electron micrograph of *Wigglesworthia* showing the packed endosymbionts in the bacteriocytes of *G. morsitans*. The endosymbionts are surrounded by a double-membrane (× 5000; bar = 4 μm). *C*, Dissected bacteriomes: (*a*) from a female maintained on a regular blood meal, and (*h*) from female that has received an antibiotic containing blood meal for one week. The bacteriome structure in the presence of antibiotics has disintegrated. (Reproduced with permission from S. Aksoy et al., Insect Molecular Biology, *4:* 15–22, 1995, ©Blackwell Science Ltd., Oxford.)

dissected and transiently maintained in tissue culture (Wink, 1979). Gene expression studies *in vivo* and *in vitro* have shown one of the major gene products of *Wigglesworthia* to be a molecular-chaperonin, the *groEL* homolog of *Escherichia coli* (Aksoy, 1995a). The overexpression of stress-related proteins has also been reported for the bacteriome endosymbionts of aphids (Kakeda and Ishikawa, 1991) and may be due to the intracellular nature of these microorganisms. PCR-based experiments show that *Wigglesworthia* is not present in ovary tissue (O'Neill et al., 1993), thus ruling out a transovarial transmission route; however, previous microscopy studies have reported the presence of microorganisms with similar morphology in milk-gland tissue of tsetse (Ma and Denlinger, 1974). Thus, milk-gland secretions may provide a route for the maternal transmission of *Wigglesworthia* to the intrauterine larvae of tsetse.

The S-symbionts of tsetse harbored in midgut-tissue and *Wolbachia* harbored in ovary tissue are morphologically different from *Wigglesworthia*. They are short rods (1–2 μm in length), and inside host cells they are surrounded by host membranes. There is a clear lytic zone around the cells reminiscent of true *Rickettsiaceae* to which the genus *Wolbachia* is related. Since phylogenetic

analysis of tsetse S-symbionts has shown them to be enteric microorganisms, it is not clear what this lytic zone signifies (Reinhardt et al., 1972). The organization and copy number of the 16S rDNA gene in *Wigglesworthia* is different from *Buchnera* (the phylogenetically closely related P-endosymbionts of aphids), tsetse S-symbionts, and *E. coli*. The free-living bacterium *E. coli* and tsetse S-symbionts have multiple copies of rDNA operons encoding the 16S and 23S gene products. *Wigglesworthia* was shown to have a single operon encoding its 16S rDNA by Southern blot analysis (Aksoy, 1995a). The genome sequence analysis now indicates that there are two copies of the rDNA operon in *Wigglesworthia*, unlike the single operon found in *Buchnera* (Akman et al., 2002). In *Buchnera*, the 16S and 23S rDNA genes are not genetically linked, while in *Wigglesworthia* they are found to be transcriptionally linked (Aksoy, 1995a; Akman et al., 2002).

TAXONOMIC COMMENTS

While it has not been possible to maintain the bacteriome-tissue symbionts *in vitro*, recent PCR-based amplification techniques and the use of nucleic acid sequences in phylogenetic reconstructions have provided additional insight into the relationships among these intracellular bacteria. Their characterization from different species of aphids, whiteflies, carpenter ants, and tsetses have shown that they each form unique lineages within the *Gammaproteobacteria*, while the organisms from mealybugs are in the *Betaproteobacteria* and those associated with cockroaches are members of flavibacteria. Their phylogeny in each case has been found to parallel the phylogeny of their host insect species, indicating the ancient and obligatory nature of these symbiotic relationships (Baumann et al., 1998). The phylogeny of *Wigglesworthia* was analyzed from different flies representing the four species groups of genus *Glossina* (*morsitans, palpalis, machadomyia,* and *fusca*). For this analysis, the 16S rDNA sequence that is present as a single-copy gene was used (Aksoy et al., 1995; Chen et al., 1999b). The

phylogenetic relationships of the tsetse host species were also analyzed by using an insect gene, the internal transcribed spacer-2 region of rDNA (ITS-2). These sequences were then used to generate phylogenetic alignments by maximum parsimony and two representative trees are shown in Fig. BXII.γ.211: for *Wigglesworthia* (Fig. BXII.γ.211b) and for the tsetse host species (Fig. BXII.γ.211a). The alignment of *Wigglesworthia* included the tsetse midgut S-symbiont clade and the S-symbionts of the weevil, *Sitophylus zeamais; Escherichia coli* was chosen as the out group. All trees generated could distinguish the tsetse symbionts, *Wigglesworthia,* and midgut S-symbionts, as distinct lineages within the *Gammaproteobacteria.* While within the S-symbiont clade no differences were observed in their 16S rDNA sequence, *Wigglesworthia* analyzed from the fusca group (*G. brevipalpis*) was found

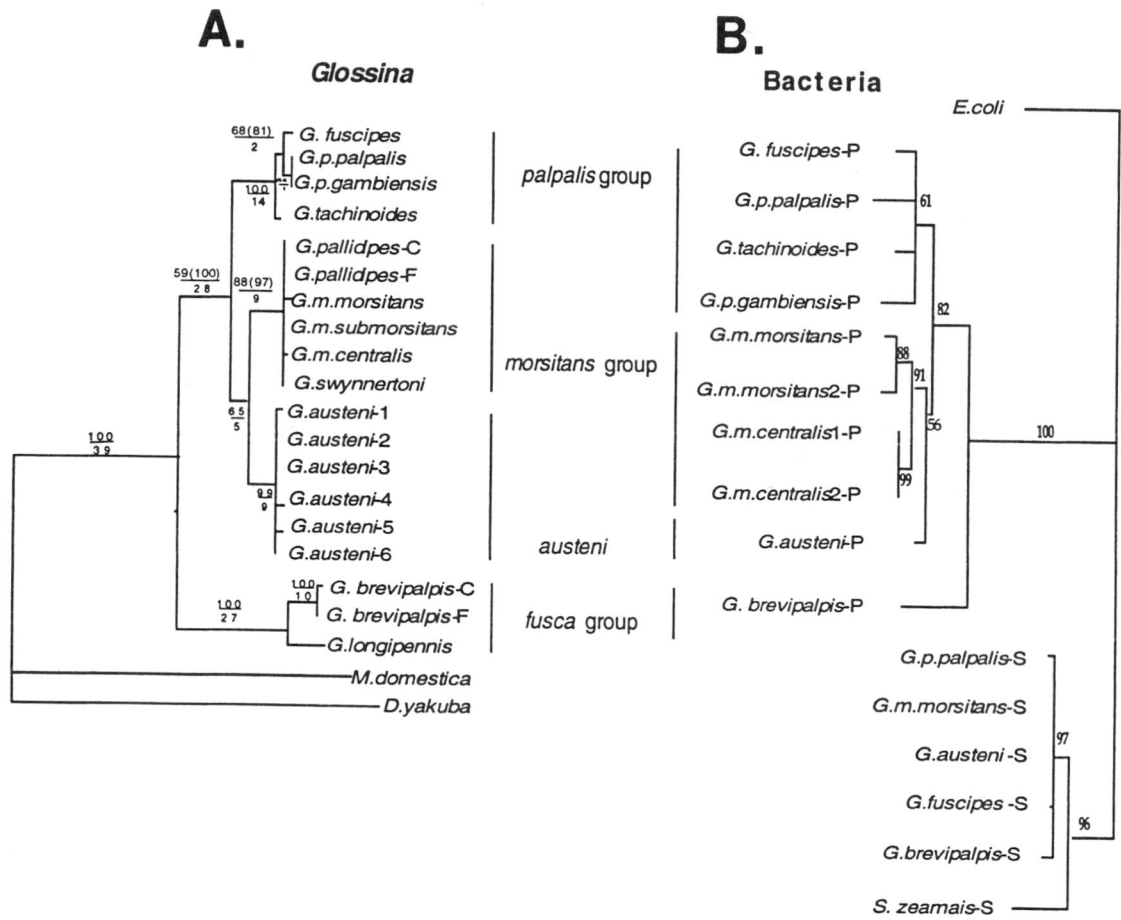

FIGURE BXII.γ.211. The phylogenetic placement of *Glossina* species and their symbionts. (Reproduced with permission from X. Chen et al., Journal of Molecular Evolution, *48:* 49–58, 1999, ©Springer-Verlag.) *A,* The phylogenetic tree constructed by maximum parsimony using the ITS-2 sequence. Tree length = 292, CI = 0.897. The bootstrap confidence values (300 replications) are presented at the nodes; the values in parentheses are bootstrap values where gaps have been included as weighted characters; numbers below denote branch length values. *D. yakuba* and *M. domestica* sequences are used as out groups. The grouping of the *palpalis, morsitans, austeni,* and *fusca* species are indicated. The origination of the *G. austeni* flies are as follows: *1,* colony maintained in Seibersdorf tsetse laboratories originating from Tanzania; *2,* field collected from Zanzibar; *3,* field collected from Shimba Hills (Kenya); *4,* colony established from Shimba Hills (Kenya); *5* and *6,* field collected from Hell's Gate Park in South Africa. *C* denotes flies analyzed from colonies. *F* denotes field isolates. *G. pallidipes*-C is from the colony at International Livestock Research Institute (ILRI) originating from Shimba Hills area of Kenya. *G. pallidipes*-F was a field collected fly from Nguruman region of Kenya. *G. brevipalpis*-C is from the colony at ILRI established from Kenya. *G. brevipalpis*-F is a field sample from South Africa. *B,* Phylogenetic analysis of *Wigglesworthia* and midgut S-symbionts based on their 16S rDNA sequence analysis. P denotes primary symbiont *Wigglesworthia,* and S denotes the secondary symbionts analyzed from the corresponding species of *Glossina.* E. coli was used as the out group. The analysis included 893 sites, of which 205 were variable and 121 were informative. One representative tree is shown (tree length 216, CI = 0.843, bootstrap values for 500 replications are shown at the nodes). *G. m. centralis*-1 was obtained from the colony maintained at ILRI in Kenya, while *G. m. centralis*-2 is from the colony maintained at Alberta and has originated from Zambia. The GenBank accession numbers for the 16S rDNA loci of *Wigglesworthia* from *G. p. gambiensis, G. m. morsitans*-2, *G. m. centralis*-1, *G. m. centralis*-2, *G. austeni,* and *G. fuscipes* are AF022875–AF022880, respectively.

to occupy the deepest branch. The *Wigglesworthia* sequences analyzed from the *morsitans* and *palpalis* group species formed sister taxa. The placement of *Wigglesworthia* from *G. austeni* (*machadomyia*), however, was less certain, since equally robust trees could be generated that differed only in the placement of *G. austeni* with respect to the other two subgenera. The low bootstrap-analysis confidence value (56) at the node where *G. austeni* symbiont forms a sister group with the *morsitans* species demonstrates this uncertainty. The ITS-2 sequence-based alignment (Fig. BXII.γ.211a) displayed similar phylogenetic relationships among the tsetse host species; the fusca flies were found to be the most divergent, while the *palpalis*, *morsitans*, and *machadomyia* subgenera were of more recent origin, and again the evolutionary relatedness of *machadomyia* with respect to the *palpalis* and *morsitans* subgenera could not be resolved with high confidence. The comparative analysis of the two phylogenies indicates their strong similarity, arguing for concordance of *Wigglesworthia* with their host tsetse species.

List of species of the genus Wigglesworthia

1. **Wigglesworthia glossinidia** Aksoy 1995b, 849[VP]

glos.si.nid′ i.a. M.L. fem. adj. *glossinidia* referring to *Glossina*, the genus of tsetse flies.

The characteristics are as described for the genus and as depicted in Fig. BXII.γ.210.

The mol% G + C of the DNA is: not known.

Type strain: endosymbiont of Glossina morsitans morsitans.

GenBank accession number (16S rRNA): L37339.

Genus XL. **Xenorhabdus** Thomas and Poinar 1979, 354[AL] emend. Thomas and Poinar 1983, 878

RAYMOND J. AKHURST AND NOËL E. BOEMARE

Xe.no.rhab′ dus. Gr. n. *xenos* unwanted guest (less literally, pathogen); Gr. fem. n. *Rhabdus* rod; M.L. masc. n. *Xenorhabdus* pathogenic rod-shaped bacterium.

Asporogenous, rod-shaped cells 0.3–2 × 2–10 μm and occasionally with filaments 15–50 μm in length. Spheroplasts, averaging 2.6 μm in diameter, appear in the last third of exponential growth. **Proteinaceous crystalline inclusions** develop in a large proportion of cells in stationary phase cultures. Gram negative. Motile by means of peritrichous flagella. Swarming may occur on 0.6–1.2% agar. Facultatively anaerobic, with both respiratory and fermentative types of metabolism. **Optimum temperature usually ~28°C or less**; a few strains grow at 40°C. Acid (no gas) production from glucose; ferment some other carbohydrates but poorly. **No catalase activity; nitrate is not reduced to nitrite.** Negative for most tests used to differentiate *Enterobacteriaceae*. Lipase detected with Tween 20 and egg yolk agar; most strains lipolytic on Tween 40, 60, 80, and/or 85. Positive for deoxyribonuclease and protease. **Phase shift occurs to varying degrees in stationary phase cultures**, giving rise to phase II, which lacks dye adsorption, antibiotic production, protein inclusions, and some other characteristics of the phase I isolated from the natural environment. **Only known from the intestinal lumen of entomopathogenic nematodes of the family** *Steinernematidae* and insects infected by these nematodes.

The mol% G + C of the DNA is: 43–50 (Bd).

Type species: **Xenorhabdus nematophila** (Poinar and Thomas 1965) Thomas and Poinar 1979, 355 (*Achromobacter nematophilus* Poinar and Thomas 1965, 249.)

FURTHER DESCRIPTIVE INFORMATION

Analyses of 16S rDNA sequences show that *Xenorhabdus* is most closely related to *Photorhabdus*, a genus composed largely of bacteria symbiotically associated with entomopathogenic nematodes of the family *Heterorhabditidae*. The next nearest phylogenetic neighbors are *Proteus vulgaris* and *Arsenophonus nasoniae* (Suzuki et al., 1996; Brunel et al., 1997; Liu et al. 1997a; Szállás et al., 1997).

Xenorhabdus species are insect pathogenic bacteria that occur naturally in the intestinal vesicle of nonfeeding infective stage entomopathogenic nematodes of the family *Steinernematidae* (Bovien, 1937; Poinar and Leutenegger, 1968; Bird and Akhurst, 1983). After invading an insect host, the nematode commences development, releasing *Xenorhabdus* into the nutrient-rich hemolymph. The bacteria proliferate, killing the insect host and producing suitable nutrient conditions for growth and reproduction of the nematodes as well as an array of antibiotics and bacteriocins to minimize competition. As the nutrient source becomes depleted, the immature nematodes develop into the infective stage that will transport *Xenorhabdus* to a new nutrient source. The association between each *Xenorhabdus* species and nematode is specific, with each nematode only naturally associated with a single *Xenorhabdus* species. However, *X. bovienii*, *X. poinarii*, and *X. beddingii* are each associated with more than one *Steinernema* species (Akhurst, 1983a, 1986b; Fischer-Le Saux et al., 1998).

A highly significant feature of bacteria of the genus *Xenorhabdus* is a phase variation that occurs in stationary phase (Akhurst, 1980). Only phase I *Xenorhabdus* have been detected in nature, but under *in vitro* conditions a variable proportion undergoes a profound change affecting colony and cell morphologies, motility, endo- and exo-enzymes, including respiratory enzymes, and secondary metabolites (Akhurst, 1980; Boemare and Akhurst, 1988; Smigielski et al., 1994; Givaudan et al. 1995; Fodor et al. 1997; Table BXII.γ.277). The timing and extent of phase change is largely unpredictable, but the rate of change from phase I to phase II is generally greater than the reverse. Phase I *Xenorhabdus* provide and protect essential nutrients for the nematodes by killing and metabolizing the insect host and producing a range of antimicrobial agents. Although phase II variants may also kill the insect host, they are less effective in providing growth conditions for the nematodes (Akhurst, 1980, 1982a), and have never been found associated with naturally occurring nematodes; the role of phase II is uncertain.

Xenorhabdus cells are rod-shaped cells that are highly variable in size, ranging from 0.3 × 2 μm to 2 × 10 μm with occasional filaments 15–50 μm in length. Varying proportions of the cells occur as spheroplasts (diameter ~2.6 μm) in stationary phase cultures. Proteinaceous inclusion bodies, commonly two morphs, are evident in many of the rod-shaped phase I cells in stationary phase (Fig. BXII.γ.212). The glycocalyx surrounding *Xenorhabdus* cells is irregular in thickness, with a mean depth of 142 nm in phase I *X. nematophila** and 49 nm in phase II (Brehélin et al.,

---

**Editorial Note: Xenorhabdus nematophila* was originally named *Xenorhabdus nematophilus*; the name was corrected by Euzéby and Boemare (2000).

TABLE BXII.γ.277. Differential characteristics for phase I and II of *Xenorhabdus* and *Photorhabdus* species[a]

Characteristic[b]	1. *X. nematophila*		2. *X. beddingii*		3. *X. bovienii*		4. *X. japonica*		5. *X. poinarii*	
	Phase I	Phase II	Phase I	Phase II	Phase I	Phase II	Phase I	Phase II	Phase I	Phase II
Colonial properties:										
Morphology[c]	g	t	g	t	g	t	g	nr	g	t
Dye adsorption	+	−	+	−	+	−	+	−	+	−
Pigmentation[d]	ow	ow	lb	ow	y	ow	yb	nr	b	ow
Ultrastrutural elements and cytological properties:										
Protoplasmic inclusions	+	−	+	−	+	−	nr	nr	+	−
Enzymatic activities:										
Antibiotics	+	+w	+	−	+	−	+	−	d	d
Lecithinase (egg yolk agar)	d	−	+	−	[+]w	−	+	−	−	−
Phospholipase (lecithin agar)	+	dw	+	−	+	−	nr	nr	[−]	−
Lipolysis (Tween 80)	[−]	+	+	+	+	+	−	−	+	+
Simmons citrate	+	+	+	+	−	+	nr	nr	d	+
Phenylalanine deaminase	−	d	−	−	−	−	+	−	−	−
Gelatin hydrolysis	+	d	+	+	+	+	+	−	+	+

[a]All tests were done at 28° ± 1°C for *Xenorhabdus* and *Photorhabdus*. Data compiled from the following references: Akhurst (1980, 1982a, 1986a), Boemare and Akhurst (1988); Brehélin et al. (1993); Nishimura et al. (1994).

[b]Symbols: +, 90–100% of strains are positive; [+], 76–89% are positive; d, 26–75% are positive; [−], 11–25% are positive; −, 0–10% are positive; w, weak reaction.

[c]g, granular; t, translucent; nr, not reported.

[d]ow, off-white; y, yellow; b, brown; lb, light brown; yb, yellowish brown.

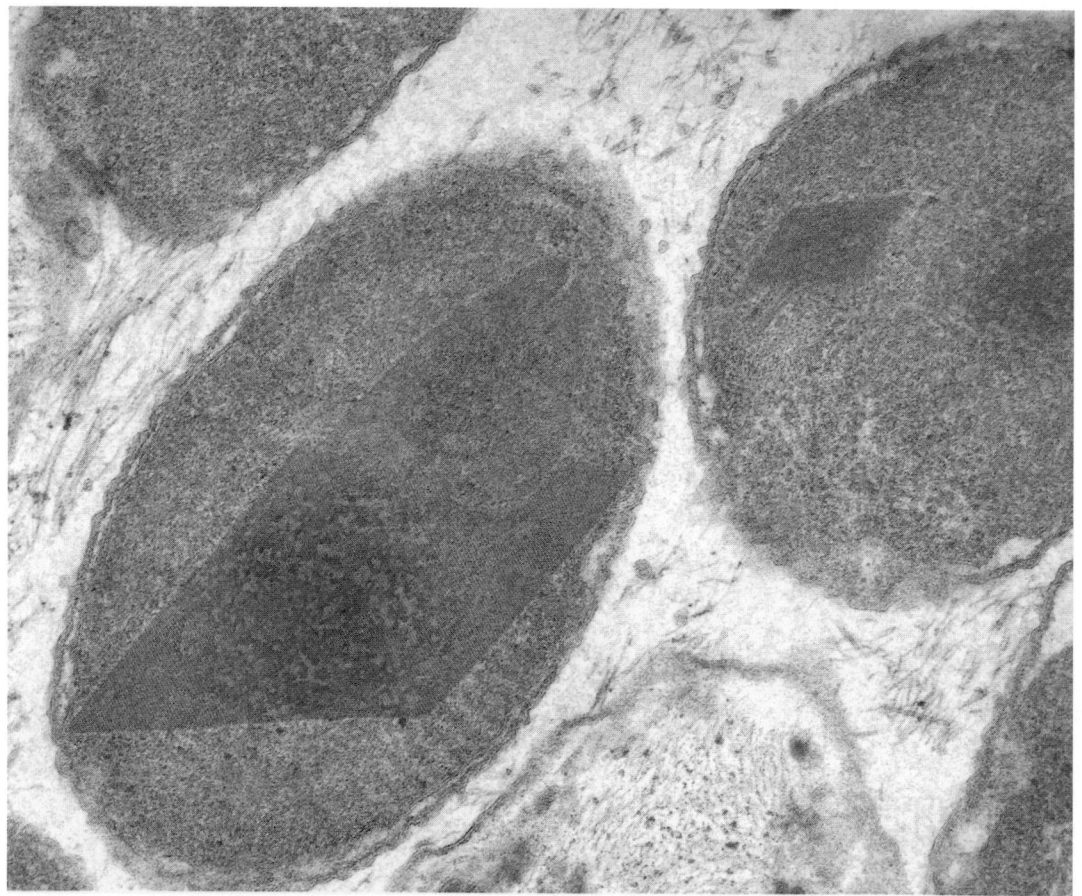

FIGURE BXII.γ.212. Protein inclusions in *X. nematophila*, showing a rhomboidal crystalline inclusion and an ovoid inclusion body.

1993). Phase I, but not phase II, cells have a peritrichous array of fimbriae with diameter of 6.4 nm, morphologically similar to the type I fimbriae of *E. coli* (Binnington and Brooks, 1993; Moureaux et al., 1995).

Phase I *Xenorhabdus* such as those isolated from the nematode associates or insects infected by the nematodes form convex, circular colonies with slightly irregular margins and a diameter of 1.5–2 mm after 4 d at 28°C; they also have a slightly granular appearance and, in some species, are pigmented (yellow, brown). These colonies adsorb dyes, such as bromothymol blue and neutral red, taking on intense coloration. Phase II variants generally form similar colonies but are flatter and wider (diameter 2.5–3.5 mm after 4 d at 28°C) and with lesser pigmentation; these colonies adsorb dyes only very weakly and no coloration is evident in 4-d colonies.

The major cellular fatty acids of *Xenorhabdus* are $C_{16:0}$, $C_{16:1}$, $C_{18:1}$, and $C_{17:0\ cyc}$, and the respiratory quinone system is ubiquinone-8 (Suzuki et al., 1990).

Xenorhabdus species are easily grown *in vitro* on a range of complex liquid and solid media and in minimal media supplemented with nicotinic acid, *p*-aminobenzoic acid, serine, tyrosine, and/or proline (Grimont et al., 1984b); nutrient agar is suitable for all strains. They are mesophilic; most grow between 15 and 30°C but strains growing at 4°C or at 40°C have been isolated.

On the basis of 16S rDNA sequence data, *Xenorhabdus* can be distinguished from its nearest phylogenetic neighbor, *Photorhabdus* (Fig. BXII.γ.213), by the sequence TTCG at positions 208–211 (*E. coli* numbering) of the 16S rDNA where *Photorhabdus* has a longer version (TGAAAG) (Szállás et al., 1997). DNA–DNA hybridization demonstrated the existence of a number of undescribed species in the genus (Fig. BXII.γ.214). Data have been recorded for some of the strains that would be assigned to new *Xenorhabdus* species (Boemare and Akhurst, 1988) but too few to warrant a decision on their taxonomic status; there are no 16S rDNA data for any of these strains.

Conjugation of plasmids from *E. coli* has been applied successfully for *X. nematophila* and *X. bovienii* (Smigielski and Akhurst, 1992; Francis et al. 1993; Forst and Nealson, 1996). In contrast, transformation of *Xenorhabdus* has not been generally successful. Although Xu et al. (1989) reported the transformation of the type strain of *X. nematophila* with a broad host-range vector, they and other workers have been unsuccessful with other strains of *X. nematophila* and other *Xenorhabdus* spp. The increase in transformation efficiency after the vector had been passed through *X. nematophila* suggested the presence of a restriction modification system. This suggestion was further supported by the demonstration that *Xenorhabdus* DNA is not easily digested by several common restriction enzymes (Akhurst et al., 1992). Endonuclease activity has subsequently been detected in a range of *Xenorhabdus* spp. (Akhurst et al., 1992).

Hypovirulent and avirulent mutants have been produced in *X. nematophila* by chemical and transposon mutagenesis (Xu et al., 1991; Dunphy, 1994). Francis et al. (1993) developed a transposon mutagenesis system for *X. bovienii* by constitutively expressing *lamB* on the surface of the bacterium, allowing them to be infected with lambda particles carrying the Tn*10* transposon. This process produced various dye-binding, lipase, protease, hemolytic, and DNase mutants.

Couche et al. (1987) demonstrated the presence of one or two small plasmids (3.6–12 kb) in some, but not all, *X. nematophila* and *X. bovienii* strains. The plasmid profiles did not differ between phases I and II. Smigielski and Akhurst (1994) reported two

megaplasmids (71.8 and 118.5 kb) as well as two additional plasmids (6.5 and 17 kb) in the A24 strain of *X. nematophila*. They also demonstrated that all strains of *X. nematophila*, *X. bovienii*, *X. beddingii*, and *X. poinarii* contained megaplasmids (48 to >680 kb) and that there were no differences in megaplasmid profile between phases I and II.

The presence of lysogenic phage in and production of bacteriocins by *X. nematophila* was demonstrated by Boemare et al. (1992). Lysis of both phases of *X. nematophila*, *X. bovienii*, and *X. beddingii* in response to mitomycin C or heat shock released complete and partial phages. In addition to inhibiting nonhost *Xenorhabdus*, the bacteriocins inhibited *P. luminescens*, *Proteus vulgaris*, and *Morganella morganii* but no other Gram-negative or Gram-positive bacteria tested, indicating that the bacteriocins act against closely related genera in contrast to the antibiotics produced by phase I variants that have a wide spectrum of activity. The *X. nematophila* bacteriocin xenorhabdicin was shown to consist of two major protein bands, corresponding to the sheath and core, and five minor bands; the phage head capsid had one major and two minor subunits (Thaler et al., 1995).

Antibiograms scored after 3 d at 28°C show that *Xenorhabdus* spp. are inhibited by streptomycin, neomycin, gentamicin, tetracycline, kanamycin, and colistin, but not penicillin. Most strains are resistant to ampicillin and cephaloridine and, to a lesser extent, furazolidone, whereas resistance to chloramphenicol is limited. Resistance to streptomycin, tetracycline, and kanamycin after selection has been demonstrated for *X. nematophila*.

Xenorhabdus spp. are insect pathogens only when delivered into the insect hemocoel, either by their nematode symbiont or by injection; they are not pathogenic when applied topically. Most are highly toxic for larvae of the greater wax moth, *Galleria mellonella*, with an LD_{50} of <20 cells (Akhurst and Dunphy, 1993). *X. poinarii* has very little pathogenicity for *G. mellonella*, (LD_{50} = 5000 cells) when injected alone, although it is highly pathogenic when co-injected with axenic *Steinernema glaseri*, its natural host (Akhurst, 1986b). *X. nematophila* produces an orally active toxin complex of the same family as those produced by *Photorhabdus*, *Serratia entomophila*, and *Yersinia pestis* (Glare and Hurst, 2002) and a smaller single unit toxin that is toxic by injection (East et al., 1998). Their toxicity varies between insects, with *X. nematophila* having an LD_{50} of about 500 for *Hyalophora cecropia* caterpillars (Götz et al., 1981) and no toxicity to maggots of the genus *Chironomus* (Götz and Boman, 1985). *Xenorhabdus* spp. do not have any demonstrable effects on vertebrates (Poinar and Thomas, 1967; Poinar et al., 1982; Obendorf et al. 1983).

Xenorhabdus species have been found only in the intestinal tract of infective-stage nematodes of the genus *Steinernema* (syn. *Neoaplectana*) and in insects killed by these nematodes. There is a high degree of specificity in these associations, with each nematode species being naturally associated with only one *Xenorhabdus* species (Akhurst, 1983a; Table BXII.γ.278). This specificity is determined by the ability of the nematode to retain the bacterium in its intestinal vesicle (Bird and Akhurst, 1983; Akhurst and Boemare, 1990). Although there have been reports of other bacteria being associated with *Steinernema* (Lysenko and Weiser, 1974; Mracek, 1977; Aguillera et al., 1993), closer examinations have always demonstrated that only *Xenorhabdus* have a specific association (Akhurst, 1982b; Bonifassi et al., 1999).

The infective stage nematodes act as vectors, transporting the bacteria into the insect via natural orifices (mouth, anus, spiracles) and then into the hemocoel. The infective (dauer) nematode, a third-stage juvenile with closed mouth and anus, recom-

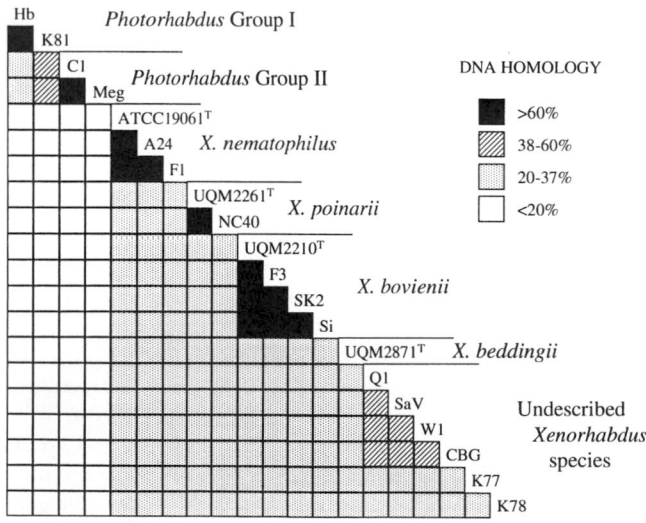

FIGURE BXII.γ.213. Phylogenetic tree showing the relationship of the genus *Xenorhabdus* to members of the *Gammaproteobacteria*. Note that since the publication of this figure, *Chromatium vinosum* has been transferred to the genus *Allochromatium* as *Allochromatium vinosum* (Imhoff et al., 1998b) and that *Vibrio marinus* has been transferred to the genus *Moritella* as *Moritella marina* (Urakawa et al., 1998, 1999b); in addition, the spelling of the species name *japonicus* has been corrected to *japonica* and that of *nematophilus* to *nematophila* (Euzéby and Boemare, 2000). (Reprinted with permission from T. Suzuki et al., Journal of Basic Microbiology, 5: 351–354, 1996 ©Wiley-VCH Verlag, Berlin.)

FIGURE BXII.γ.214. DNA relatedness of 20 strains of bacterial symbionts of entomopathogenic nematodes (modified from Boemare et al., 1993), indicating that strains representing a number of undescribed species have been isolated. In addition, the spelling of the species name *nematophilus* has since been corrected to *nematophila*.

mences development in the hemocoel, releasing its symbiotic bacterium and an inhibitor of the inducible antibacterial enzymes (Poinar and Himsworth, 1967; Götz et al., 1981). As *Xenorhabdus* multiplies it provides essential nutrients for nematode maturation and reproduction (Poinar and Thomas, 1966) and produces antibiotics (Paul et al., 1981; Akhurst, 1982a; McInerney et al. 1991a, b; Li et al., 1996).

Xenorhabdus alternates between a nutrient rich (insect) and nutrient poor (nematode) existence. Forst et al. (1997) hypothesized that phase II may be induced by the nematode gut conditions and better adapted to the nutrient poor conditions of the intestinal vesicle of the nonfeeding nematode. However, this hypothesis does not account for the fact that the bacteria isolated from field-collected infective stage juveniles are inevitably phase I. Smigielski et al. (1994) found differences in the activity of respiratory enzymes of phases I and II that indicate the greater potential of phase II to survive in soil environments than phase I. The lack of a record of isolation of *Xenorhabdus* directly from soil may be due to its slow growth, lack of a suitable selective medium, and/or the difficulty of identifying phase II, which has few positive characters in most standard tests for identifying bacteria. However, Morgan et al. (1997) found that *X. nematophila* declined very quickly in river water and soil, becoming undetectable after 2 d and 7 d, respectively. Although the phase status

TABLE BXII.γ.278. Biochemical characterization of the genus *Xenorhabdus*[a]

Characteristic[b]	1. *X. nematophila*	2. *X. beddingii*	3. *X. bovienii*	4. *X. japonica*	5. *X. poinarii*
Mol% G + C of DNA	43–48	45.5–50	44–47	45.9	42.6–49
Isolated from:					
S. carpocapsae	+	−	−	−	−
S. feltiae, S. intermedium, S. affine, S. kraussei	−	−	+	−	+
S. glaseri, S. cubanum	−	−	−	−	+
Unidentified *Steinernema* spp. (*Steinernema longicaudum?*)	−	+	−	−	−
S. kushidai	−	−	−	+	−
Pathogenicity for lepidopteran insects	+	+	+	−	−
Motility	d	d	d	d	d
Peritrichous flagella	+	+	+	+	+
Protoplasmic inclusions	d	d	d	nr	d
Bioluminescence	−	−	−	−	−
Pigmentation[c]	ow	lb	y	yb	br
Dyes adsorption	d	d	d	d	d
Antimicrobial production	d	d	d	d	d
Oxidation-fermentation	F	F	F	F	F
Catalase	−	−	−	−	−
Nitrate reduced to nitrite	−	−	[−]w	−	−
Oxidase, Kovac's	−	−	−	−	−
KCN, growth in	−	−	−	−	−
Indole production	−	−	−	−	−
Methyl red	−	−	−	−	−
Voges–Proskauer	−	−	−	−	−
Simmons citrate	+	+	+	−	+
Hydrogen sulfide production	−	−	−	−	−
ONPG (β-galactosidase)	−	−	−	−	−
Esculin hydrolysis	−	+	−	−	d
Urease, Christensen's	−	−	−	−	−
Phenylalanine deaminase	d	−	[−]	dw	[−]
Tryptophan deaminase	−	+w	+	−	−
Lysine decarboxylase	−	−	−	−	−
Ornithine decarboxylase	−	−	−	−	−
Arginine dihydrolase	−	−	−	−	−
D-Glucose, acid production	+	+	+	+	+
D-Glucose, gas production	−	−	−		−
Acid production from:					
D-Adonitol	−	−	−	−	−
L-Arabinose	−	−	−	−	−
Cellobiose	−	−	−	−	−
Dulcitol	−	−	−		
Fructose	+	+	+w	+	+
Glycerol	+w	+	+	+	+
N-Acetylglucosamine	+	+	+	+	+
myo-Inositol	+w	−	dw	−	−
Lactose	−	−	−	−	−
Maltose	+	+	+	+	+
D-Mannitol	−	−	−	−	−
D-Mannose	+	+	+	+	+
Melibiose	−	−	−	−	−
α-Methyl-D-glucoside	−	−	−		
Raffinose	−	−	−		
L-Rhamnose	−	−	−		
Ribose	−	+	+	−	−
Salicin	−	+	−		−
D-Sorbitol	−	−	−	−	−
Sucrose	−	−	−		
Trehalose	+	+	+w	+	+
D-Xylose	−	−	−	−	−
Utilization of:					
Diaminobutane	−	−	[+]	−	−
L-Fucose	−	−	−	−	−
DL-Glycerate	+	+	+	−	[−]
L-Histidine	−	+	+		[+]
myo-Inositol	+w	−	dw	−	−
DL-Lactate	+	+	+	+	+
D-Mannitol	−	−	−	−	−
Ribose	−	+	[+]	−	−
L-Tyrosine	−	+	[+]	−	−
Gelatin hydrolysis, Kohn's	+	+	+	d	+
Lecithinase (egg yolk agar)	d	d	d	d	−
Lipase (Tween 80)	d	+w	+	−	+w
DNase	+	+	+	+	+w

[a]All tests were done at 28° ± 1°C for *Xenorhabdus*. Data from Akhurst (1986a,b); Boemare and Akhurst (1988); Yamanaka et al. (1992); Brehélin et al. (1993); Nishimura et al. (1994); Fischer-Le Saux et al. (1998).

[b]Symbols: +, 90–100% of strains are positive; [+], 76–89% are positive; d, 26–75% are positive; [−], 11–25% are positive; −, 0–10% are positive; w, weak reaction; nr, not reported; F, fermentative.

[c]ow, off-white; y, yellow; br, brown; lb, light brown; yb, yellowish brown.

of *X. nematophila* used in that study was not specified, it seems likely that they used phase I. At this time, there is no satisfactory explanation of the ecological role of phase II.

ENRICHMENT AND ISOLATION PROCEDURES

Xenorhabdus spp. generally grow well at 25–28°C on nutrient or similar agar (e.g., Luria-Bertani, trypticase soy), but best growth has been achieved with yeast extract-salts (YS; Dye, 1968) and medium X (Götz et al., 1981) agars. *Xenorhabdus* can be isolated from the infective-stage nematodes by the hanging drop technique or by maceration (Poinar and Thomas, 1966; Akhurst, 1980). For both methods, the infective juveniles must first be surface sterilized; this is readily achieved by immersing a small number of live infective-stage nematodes (<100), free of debris, in 0.1% merthiolate for 1 h at room temperature and then rinsing thoroughly in several changes of sterile water. In the hanging-drop technique, individual surface-sterilized infective juveniles are transferred to a drop of aseptically collected insect hemolymph on a coverslip that is then inverted over a cavity to prevent desiccation and incubated at 25°C until the nematodes commence development (1–3 d). At this time, they void their symbiotic bacteria, which can be subcultured from the hemolymph onto an agar medium (e.g., nutrient agar) 1 d later. A more rapid method involves the maceration of 50–100 surface-sterilized infective juveniles in a nutrient broth by means of a tissue homogenizer. The macerate should be plated onto an agar medium immediately (10–100 µl aliquots) and incubated at 28°C for 3 d. The inclusion of suitable controls to confirm that the surface-sterilization procedure has been effective is essential for both methods. *Xenorhabdus* can also be isolated by the less labor-intensive method of collecting hemolymph from an insect (e.g., *G. mellonella*) larva within 24 h of its death after infection by *Steinernema*. With this last method, bacteria other than *Xenorhabdus* may also be isolated; these bacteria may be carried into the host on the exterior of the nematodes or may be picked up into the hemolymph from the insect cuticle. Contamination by other bacteria can be minimized by burying the insect in clean, damp sand, adding a small number of nematodes to the surface of the sand and incubating at 20–25°C until one to five nematodes infect the insect. This last method is better suited to reisolation of a *Xenorhabdus* strain rather than to determining the identity of the bacteria specifically associated with a nematode species.

MAINTENANCE PROCEDURES

Storage in 20% glycerol at −70°C or below is very useful for long-term maintenance of *Xenorhabdus* cultures; they do not store well at −20°C. Short-term storage (less than 1 month) is best conducted at 10–15°C because survival on agar or in broth at 4°C is very poor, and when cultures are maintained at temperatures in excess of 15°C there is a significant risk of a proportion of the culture undergoing phase change.

PROCEDURES FOR TESTING SPECIAL CHARACTERS

All tests for phenotypic characterization of *Xenorhabdus* spp. should be conducted at 28°C.

Dye adsorption in most *Xenorhabdus* species can be tested on nutrient agar containing 0.0025% (w/v) bromothymol blue and 0.004% (w/v) triphenyltetrazolium chloride (NBTA; Akhurst, 1980). Dye-adsorbing phase I colonies will appear dark blue, while nonadsorbent phase II colonies will be red. As *X. poinarii* does not adsorb bromothymol blue, dye adsorption in this species can be assessed on MacConkey agar; adsorbing colonies are dark

red (Akhurst, 1986b). Congo red (Francis et al., 1993) and some other dyes can also be used for most *Xenorhabdus* spp.

Insect pathogenicity is tested by injection of 100, 1000, and 10,000 cells (total count) from a 24-h broth culture into final instar *G. mellonella* larvae. The injected larvae should be placed on dry filter paper in Petri dishes and incubated at 25°C for 3 d. Most *Xenorhabdus* spp. will kill >50% at a dosage of 100 cells; all will kill >50% at 10,000 cells. The use of a less susceptible host (e.g., *Manduca sexta*) has enabled detection of differences in pathogenicity between the two phases of a strain (Volgyi et al., 1998).

Antibiotic production is tested by spot inoculating *Xenorhabdus* spp. onto nutrient agar and incubating at 28°C for 3 d. At this time the bacteria are killed by exposure to chloroform vapor for ~1 h. After the chloroform has evaporated from the agar, semi-solid nutrient agar (0.5%) inoculated with a suitable indicator organism (*Micrococcus luteus* or another Gram-positive species) is poured to form a thin layer. After incubation at 28°C overnight, a halo of inhibition around antibiotic-producing colonies will be evident.

DIFFERENTIATION OF THE GENUS *XENORHABDUS* FROM OTHER GENERA

Xenorhabdus spp. are easily distinguished from other *Enterobacteriaceae* by the absence of catalase in the former and, with the exception of *Photorhabdus*, by their inability to reduce nitrate. The key characteristics for differentiating *Xenorhabdus* from its most closely related genera are shown in Table BXII.γ.279.

TAXONOMIC COMMENTS

The first symbiotic bacterium isolated from entomopathogenic nematodes was described as a new species *Achromobacter nematophilus* (Poinar and Thomas, 1965). With the rejection of the genus *Achromobacter* (Hendrie et al., 1974), this new species could not be accommodated within any existing genus. Thomas and Poinar (1979) described a new genus, *Xenorhabdus*, to accommodate the bacterial symbionts of entomopathogenic nematodes as two species, *X. nematophila*, symbionts of the family *Steinernematidae*, and "*X. luminescens*", associated with the *Heterorhabditidae*. These bacteria have low DNA–DNA relatedness (4%) to the type species of the type genus of the family *Enterobacteriaceae* (Farmer, 1984b) and lack nitrate reductase, which is positive for all other genera in the family. However, they do have the enterobacterial common antigen (Ramia et al., 1982), and phylogenetic analyses based on 16S rDNA (Suzuki et al., 1996; Brunel et al., 1997; Liu et al., 1997a; Szállás et al., 1997) confirm their relatedness to the *Enterobacteriaceae*.

From a phenotypic study of bacterial symbionts of the *Steinernematidae* and *Heterorhabditidae*, five groups were recognized

TABLE BXII.γ.279. Characteristics differentiating *Xenorhabdus*, *Photorhabdus*, and *Proteus*[a]

Characteristic	Xenorhabdus	Photorhabdus	Proteus
Bioluminescence	−	+	−
Catalase	−	+	+
Annular hemolysis	−	d	−
Urea hydrolysis	−	d	+
Indole	−	d	d
H₂S production	−	−	[+]
Nitrate reductase	−	−	+
Acid from mannose	+	+	−

[a]For symbols see standard definitions.

within the genus and the subdivision of *X. nematophila* into subspecies proposed (Akhurst, 1983b). A more comprehensive phenotypic study (Boemare and Akhurst, 1988) led to the elevation of the subspecies to species status, as *X. nematophila*, *X. bovienii*, *X. poinarii*, and *X. beddingii* (Akhurst and Boemare, 1988). *Xenorhabdus japonica*[1], symbiotically associated with *Steinernema kushidai*, was described later (Nishimura et al., 1994). DNA–DNA hybridization (Suzuki et al., 1990; Akhurst et al., 1996) and 16S rDNA (Suzuki et al., 1996; Brunel et al., 1997; Liu et al., 1997a; Szállás et al., 1997) analyses validated the inclusion of these five species in, and the exclusion of *P. luminescens* from, the genus.

DNA–DNA hybridization analysis indicates that there are more than the five *Xenorhabdus* species described to date (Fig. BXII.γ.214). Too few isolates of the other "species" have been characterized phenotypically to allow description of new species.

FURTHER READING

Bedding, R., R. Akhurst and H. Kaya. 1993. Nematodes and the Biological Control of Insect Pests, CSIRO, Melbourne.

Boemare, N.E., R.J. Akhurst and R.G. Mourant. 1993. DNA relatedness between *Xenorhabdus* spp. (*Enterobacteriaceae*), symbiotic bacteria of entomopathogenic nematodes, and a proposal to transfer *Xenorhabdus luminescens* to a new genus, *Photorhabdus* gen. nov. Int. J. Syst. Bacteriol. *43*: 249–255.

Forst, S., B. Dowds, N.E. Boemare and E. Stackebrandt. 1997. *Xenorhabdus* and *Photorhabdus* spp.: bugs that kill bugs. Annu. Rev. Microbiol. *51*: 47–72.

Szallas, E., C. Koch, A. Fodor, J. Burghardt, O. Buss, A. Szentirmai, K.H. Nealson and E. Stackebrandt. 1997. Phylogenetic evidence for the taxonomic heterogeneity of *Photorhabdus luminescens*. Int. J. Syst. Bacteriol. *47*: 402–407.

List of species of the genus *Xenorhabdus*

1. **Xenorhabdus nematophila** (Poinar and Thomas 1965) Thomas and Poinar 1979, 355[AL] (*Achromobacter nematophilus* Poinar and Thomas 1965, 249.)

   ne.ma.to' phi.la. Modern entomological term n. *nematode;* Gr. adj. *philus* loving or having affinity for; M.L. adj. *nematophila* nematode-loving.

   The species characteristics are listed in Tables BXII.γ.277 and BXII.γ.278. No isolates known to grow at temperatures in excess of 34°C. Neither phase is pigmented. Most isolates sensitive to furazolidone.

   Only found associated with one species of nematode, *Steinernema carpocapsae*, but distributed globally.

   The mol% G + C of the DNA is: 43–48 (Bd).

   Type strain: ATCC19061, DSM 3370.

   GenBank accession number (16S rRNA): D78009, X82251.

2. **Xenorhabdus beddingii** (Akhurst 1986a) Akhurst and Boemare 1993, 864[VP] (Effective publication: Akhurst and Boemare 1988, 1844) (*Xenorhabdus nematophilus* subsp. *beddingii* Akhurst 1986a, 456.)

   bed.din' gi.i. M.L. gen. n. *beddingii* of Bedding, named for R.A. Bedding, who made significant contributions to the development of *Xenorhabdus/Steinernema* associations for insect pest control.

   The species characteristics are listed in Tables BXII.γ.277 and BXII.γ.278. All isolates grow at 34°C, some at 38°C. Hydrolyses esculin. Inhibited by cephaloridine and ampicillin. The brown pigmentation is not strong. Phase I species is highly unstable, producing the very stable phase II.

   Associated with two undescribed species of *Steinernema* from Australia, one of which may be *Steinernema longicaudum*, which was described from China.

   The mol% G + C of the DNA is: 45.5–50 (Bd, HPLC).

   Type strain: UQM 2871 (phase I of strain Q58), ATCC 49542, DSM 4764.

   GenBank accession number (16S rRNA): D78006, X82254.

3. **Xenorhabdus bovienii** (Akhurst 1983b) Akhurst and Boemare 1993, 864[VP] (Effective publication: Akhurst and Boemare 1988, 1843) (*Xenorhabdus nematophilus* subsp. *bovienii* Akhurst 1983b, 45.)

   bo.vi.en' i.i. M.L. gen. n. *bovienii* of Bovien, named for P. Bovien who first reported the presence of bacteria in the intestinal vesicle of a *Steinernema* species.

   The species characteristics are listed in Tables BXII.γ.277 and BXII.γ.278. No growth at 34°C; some strains will grow at 5°C. Resistant to carbenicillin.

   Associated with several species of entomopathogenic nematode (*Steinernema feltiae, Steinernema intermedium, Steinernema kraussei, Steinernema affine*) in temperate regions.

   The mol% G + C of the DNA is: 44.3 (HPLC)–46.9 (Bd).

   Type strain: UQM2210 (phase I of strain T228), ATCC 35271, DSM 4766.

   GenBank accession number (16S rRNA): X82254, D78007.

4. **Xenorhabdus japonica** Nishimura, Hagiwara, Suzuki and Yamanaka 1995, 619[VP] (Effective publication: Nishimura, Hagiwara, Suzuki and Yamanaka 1994, 209.)

   ja.po' ni.ca. M.L. adj. *japonica* of Japan.

   The species characteristics are listed in Tables BXII.γ.277 and BXII.γ.278. Does not grow at 37°C. Arginine dihydrolase activity detected in phase II. Pigmentation is yellowish brown.

   Only known to be associated with *Steinernema kushidai* in Japan.

   The mol% G + C of the DNA is: 45.9 (HPLC).

   Type strain: IAM 14265.

   GenBank accession number (16S rRNA): D78008, Z76739.

5. **Xenorhabdus poinarii** (Akhurst 1983b) Akhurst and Boemare 1993, 864[VP] (Effective publication: Akhurst and Boemare 1988, 1843 (*Xenorhabdus nematophilus* subsp. *poinarii* Akhurst 1983b, 45.)

   poi.nar' i.i. M.L. gen. n. *poinarii* , of Poinar, named for G.O. Poinar, Jr., who made major contributions to the understanding of entomopathogenic nematode/bacterial interactions.

   The species characteristics are listed in Tables BXII.γ.277 and BXII.γ.278. This is the most heat tolerant *Xenorhabdus*, with all strains growing at 36°C and some at 40°C. The intensity of pigmentation in phase I varies from light to reddish brown. Antibiotic production by phase I and phase II is variable. Associated with *Steinernema glaseri* and *Steinernema cubanum* in the USA and Caribbean. The nematode/bacterium association does not appear to be as strong for

---

1. Euzéby and Boemare (2000) noted that *-rhabdus* has a feminine root and so species names for genera ending in *-rhabdus* should also be feminine.

this species as for other *Xenorhabdus*; its symbiosis may be more primitive. It is not pathogenic for greater wax moth (*Galleria mellonella*) larvae unless associated with its nematode partner.

The mol% G + C of the DNA is: 42.6 (HPLC)–49 (Bd).
Type strain: UQM 2216 (phase I of strain G), ATCC 35272, DSM 4768.
GenBank accession number (16S rRNA): D78010, X82253

Genus XLI. **Yersinia** *Van Loghem 1944, 15*[AL]

EDWARD J. BOTTONE, HERVÉ BERCOVIER AND HENRI H. MOLLARET

Yer.si′ni.a. M.L. fem. n. *Yersinia* named for the French bacteriologist A.J.E. Yersin, who first isolated the causal organism of plague in 1894.

Straight rods to coccobacilli, 0.5–0.8 × 1–3 μm. Endospores are not formed. Capsules are not present, but an envelope occurs in *Y. pestis* strains grown at 37°C and in cells from clinical sources. Gram negative, **nonmotile at 37°C, but motile with peritrichous flagella when grown below 30°C, except for *Y. pestis*, which is always nonmotile**. Growth occurs on ordinary bacteriological media. Colonies of yersiniae are translucent to opaque, 0.1–1.0 mm in diameter after 24 h. Optimum growth temperature, 28–29°C. Facultatively anaerobic, having both a respiratory and a fermentative type of metabolism. Oxidase negative. Catalase positive. Nitrate is reduced to nitrite with a few exceptions in specific biovars. D-glucose and other carbohydrates are fermented with **acid production but little or no gas. Phenotypic characteristics are often temperature dependent, and usually more characteristics are expressed by cultures incubated at 25–29°C than at 35–37°C.** The enterobacterial common antigen is present in all species investigated. Widely distributed in nature with some species adapted to specific animal hosts and humans. Several species are pathogenic for humans and animals including *Y. pestis*, the causative agent of plague. A significant cause of food-borne and waterborne disease. Belongs to the *Gammaproteobacteria*.

The mol% G + C of the DNA is: 46–50.

Type species: **Yersinia pestis** (Lehmann and Neumann 1896) Van Loghem 1944, 15 (*Bacterium pestis* Lehmann and Neumann 1896, 194; *Yersinia pseudotuberculosis* subsp. *pestis* Bercovier, Mollaret, Alonso, Brault, Fanning, Steigerwalt and Brenner 1981a, 383.)

FURTHER DESCRIPTIVE INFORMATION

Cells of *Yersinia* species are small, coccoid-shaped Gram-negative bacilli that resemble cells of *Pasteurellaceae* rather than cells of *Enterobacteriaceae*. Pleomorphism occurs depending on the type of medium used and the temperature of incubation. Rods, coccobacilli, and small chains of four to five elements (especially in liquid media) can occur. Gram, Giemsa, or Wayson-stained smears reveal a more pronounced tendency to bipolar staining in *Y. pestis* than in the other species. Spores or specific inclusions are not formed. No definite capsule occurs, but *Y. pestis* displays a carbohydrate-protein envelope termed capsular antigen or fraction I (F-1) when cultured at 37°C (Burrows, 1963), or in direct smears taken from infected hosts (mice, guinea pigs, humans). L forms have been described for *Y. enterocolitica* (Pease, 1979).

Yersinia species are nonmotile after growth at 37°C but motile at 22–29°C, except *Y. pestis*, which is nonmotile irrespective of incubation temperatures. Fresh isolates of *Y. enterocolitica* and *Y. pseudotuberculosis* may require a few subcultures to express their motility. Motile cells have 2–15 peritrichous flagella characterized by a long wavelength (Nilehn, 1969).

Yersinias do not differ from other *Enterobacteriaceae* in their

fine structure and overall cell wall composition. Lipopolysaccharides (O antigens) have been isolated and characterized (Davies, 1958; Rische et al., 1973). The whole-cell lipid composition of all *Yersinia* species investigated exhibits a pattern shared with other *Enterobacteriaceae* (Tornabene, 1973; Jantzen and Lassen, 1980). *Y. pestis* lipopolysaccharide has core components in common with other *Enterobacteriaceae* but lacks extended O-group side chains (Perry and Fetherston, 1997).

Placement of the genus *Yersinia* within the *Enterobacteriaceae* family is supported by both biochemical and DNA–DNA relatedness studies. Further studies (Ahmad et al., 1990) comparing rDNA sequence analysis to clustering based on aromatic amino acid synthesis placed the genus *Yersinia* in one of three "enteroclusters" along with *Cedecea*, *Edwardsiella*, *Hafnia*, *Kluyvera*, *Proteus*, *Providencia*, and *Morganella*. Sequence analysis of the 16S rRNA gene performed by Ibrahim and colleagues (1993) of representatives of 10 of the 11 *Yersinia* species reveal that yersiniae form a coherent cluster within the *Gammaproteobacteria* with sequence similarities ranging from 96.9–99.8%. Within the *Gammaproteobacteria* the closest relative is *Hafnia alvei* (96.5% sequence similarity).

Intragenerically, phylogenetic analysis disclosed five sublines, with *Y. enterocolitica*, *Y. rohdei*, and *Y. ruckeri* forming separate sublines. A separate subline was formed by *Y. pestis*, *Y. pseudotuberculosis*, and *Y. kristensenii*, while *Y. mollaretii*, *Y. intermedia*, *Y. bercovieri*, *Y. aldovae*, and *Y. kristensenii* formed a fifth subline.

Yersinia species grow on nutrient agar without enrichment. A small colony diameter differentiates yersiniae from all other *Enterobacteriaceae*. After incubation for 24–30 h at 30°C or 37°C, *Y. pestis* forms minute colonies (0.1 mm) that can be discerned only with difficulty by the naked eye. After 48 h their diameter increases to 1.0–1.5 mm. The colonies are slightly opaque, butyrous, smooth, round, and have somewhat irregular edges. The use of enriched media (serum, blood, yeast extract) does not dramatically improve growth, and after 48 h the colony sizes are similar to those found on nutrient agar. All other *Yersinia* species grown on nutrient agar at 25–37°C produce visible colonies in 24 h. The colonies reach a diameter of 1.0–1.5 mm after 24–30 h, and 2.0–3.0 mm after 48 h. After 18 h they are translucent, smooth, and round with irregular edges, but after 48 h the centers become elevated and the edges become more regular, producing a "Chinese hat" shape. *Y. pestis* colonies may appear slightly mucoid. When cultured for 48 h, all *Yersinia* species dissociate into small (0.5 mm) and large (2 mm) colonies. This phenomenon appears to depend on the medium used (Bercovier et al., 1979).

Growth is moderate in liquid media: incubation of *Yersinia* for 48 h will yield the same turbidity that occurs in 18 h with other *Enterobacteriaceae*. When grown in nutrient broth, clumps of *Y.*

pestis cells adhere to the side of the tube, which slowly forms a deposit at the bottom of the tube while the supernatant remains relatively clear; this is followed by the appearance of a pellicle, which in turn disintegrates to form flocculent masses and a larger deposit. This phenomenon is attenuated in peptone water. *Y. pseudotuberculosis* occasionally grows in a manner similar to that of *Y. pestis*. All other *Yersinia* species give uniform turbidity in nutrient broth and in peptone water.

Y. pestis and *Y. pseudotuberculosis* give variable growth responses on MacConkey agar. All other species grow well on this medium, with colonies reaching a size similar to that observed on nutrient agar. On Salmonella–Shigella agar incubated at 25°C, *Y. pestis* hardly grows at all, whereas all the other species produce pinpoint colonies in 24–30 h. When incubated on this medium at 37°C, *Y. enterocolitica* is only partially inhibited, whereas all other species are severely inhibited (Nilehn, 1969; Bottone, 1977; Bercovier et al., 1979).

All *Yersinia* species except *Y. pestis* can grow at 25°C on synthetic mineral-salt media with various carbohydrates as the energy source (Burrows and Gillett, 1966; Bercovier et al., 1979). *Y. pestis* requires L-methionine, L-isoleucine, L-valine, and L-phenylalanine, and or L-threonine for growth. When incubated at 37°C on synthetic mineral-salt media all *Yersinia* species become auxotrophic, and the addition of biotin and thiamine, and pantothenate and glutamic acid for *Y. pestis*, is necessary to promote growth (Burrows and Gillett, 1966; Brubaker, 1991). The growth of *Y. pestis* on such media is further enhanced by the addition of a reducing agent, and by incubation in a CO_2-enriched atmosphere (Brubaker, 1991). Virulent strains of *Y. pestis* require Ca^{2+} or ATP for growth at 37°C but not at 25°C (Zahorchak et al., 1979). This temperature-dependent requirement for Ca^{2+}, mediated by the Lcr plasmid, has also been described for virulent strains of *Y. pseudotuberculosis* and *Y. enterocolitica*.

The growth range is 4–42°C, with an optimum temperature of 28–29°C. *Y. pestis* and *Y. pseudotuberculosis* tolerate a pH range of 5.0–9.6; other *Yersinia* species can grow in a pH range of 4.0–10.0. The optimum pH for all species is 7.2–7.4. *Yersinia* species can grow in peptone water without the addition of NaCl. *Y. pestis* and *Y. pseudotuberculosis* tolerate up to 3.5% NaCl, and the other species can tolerate up to 5% NaCl. *Y. pseudotuberculosis*, but not *Y. enterocolitica*, which lacks tellurite reductase, is the only species that grows well on media containing 0.06% tellurite (Brzin, 1968).

Yersinia species do not differ significantly from other *Enterobacteriaceae* in their general metabolism (Brubaker, 1991) (Table BXII.γ.280). They produce acid during fermentation of D-glucose. *Y. enterocolitica*, *Y. frederiksenii*, and *Y. intermedia* produce acetoin (positive Voges–Proskauer test) when incubated at 28°C, whereas this characteristic is variable for *Y. ruckeri* and is absent in *Y. pestis* and *Y. pseudotuberculosis*. No *Yersinia* species produces acetoin at 37°C (Table BXII.γ.281).

The main physiological and biochemical characteristics of the various *Yersinia* species are given in Table BXII.γ.282. *Yersinia* ferment carbohydrates usually without gas production, a characteristic that is constant for *Y. pestis* and *Y. pseudotuberculosis*, but other species may produce a few bubbles after 2–3 days at 28°C. Because the optimum growth temperature of yersiniae is 28–29°C, some biochemical activities are often temperature-dependent (cellobiose and raffinose fermentation, ornithine decarboxylase, ONPG (*o*-nitrophenyl-β-D-galactopyranoside) hydrolysis, indole production, and the Voges–Proskauer reaction) and are more constantly expressed at 28°C rather than at 37°C. All species

except *Y. intermedia* reduce nitrate to nitrite by a type B nitrate reductase; *Y. intermedia* strains have either a type A nitrate reductase, like most *Enterobacteriaceae*, or a type B reductase. ONPG activity of yersiniae does not correspond to a true β-galactosidase, but only to an ONPG-ase (Le Minor and Coynault, 1976). In addition to the characteristics given in Tables BXII.γ.281 and BXII.γ.282, *Yersinia* species are able to attack polypectate in 5–7 days and starch in 3–7 days. Yersiniae are neither hemolytic nor proteolytic, except *Y. ruckeri*, which liquefies gelatin, and some strains of *Y. pestis*, which have fibrinolytic and coagulase activity linked to the production of Pesticin 1. Plasminogen activator coagulase activity and pesticinogeny are plasmid (pPCD1) encoded. Lecithinase activity in *Y. enterocolitica* is strain-dependent. *Y. pseudotuberculosis*, *Y. enterocolitica*, and *Y. ruckeri* strains have a lipase that is active on corn oil, but only *Y. intermedia*, *Y. frederiksenii*, and *Y. enterocolitica* biovar I express a lipase-esterase that is active on Tween 80.

Transformation of auxotrophic strains of *Y. enterocolitica* by prototrophic strains using the Juni-Janik technique has been reported (Callahan and Koroma, 1979). F lac + episomes from *E. coli* have been transferred to *Y. pestis* (Martin and Jacob, 1962), to *Y. pseudotuberculosis* (Lawton et al., 1968b), and to *Y. enterocolitica* (Cornelis and Colson, 1975), but usually with a low frequency (10^{-4}–10^{-6}) . This has allowed chromosomal mapping of *Y. pseudotuberculosis* (Lawton and Stull, 1971; McMahon, 1973). Gene transfer by conjugation between *Y. pseudotuberculosis* and *Y. pestis* has also been demonstrated (Lawton et al., 1968a).

R factors have been transferred to *Y. pestis* and *Y. pseudotuberculosis* (Ginoza and Matney, 1963) and to *Y. enterocolitica* (Knapp and Lebek, 1967). Wild strains of *Yersinia* carrying R plasmids (Cornelis et al., 1973; Kanazawa and Ikemura, 1979) appear to be rare. This could be explained, at least for *Y. enterocolitica*, by the presence of a restriction-modification system (Cornelis and Colson, 1975). A self-transmissible plasmid coding for lactose fermentation has been described in *Y. enterocolitica* (Cornelis et al., 1976). This plasmid is freely transmissible between strains of *Y. enterocolitica* and *E. coli*.

Other plasmids related to various virulence tests (Ca^{2+} dependency, autoagglutination, lethality for mice and gerbils, Sereny test) have been demonstrated in *Y. pestis* (Ferber and Brubaker, 1981; Brubaker, 1991), *Y. pseudotuberculosis* (Gemski et al., 1980b), and *Y. enterocolitica* (Gemski et al., 1980a; Zink et al., 1980). These plasmids of 40–48 MDa molecular weight constitute a family of related plasmids (Ben-Gurion and Shafferman, 1981; Portnoy et al., 1981). *Y. pestis* and *Y. pseudotuberculosis* have never been found to be lysogenic, whereas of 1252 strains of *Y. enterocolitica* studied, 86.4% were lysogenic when grown at 25°C but not at 37°C (Nicolle et al., 1973). Phages active on *Y. pestis* and *Y. pseudotuberculosis* have been described (Gunnison et al., 1951; Girard, 1953), but they are not host-specific and are used only for presumptive bacteriological diagnosis. Coliphages T2, T3, and T7 are also active on *Y. pseudotuberculosis* and *Y. pestis* (Hertman, 1964). A phage typing system, useful in epidemiological investigations has been developed for *Y. enterocolitica* (Nicolle et al., 1973): strains of *Y. enterocolitica* serogroup 0:3 are associated with phagovar VIII in Europe, IXa in the Republic of South Africa, and lXb in Canada and the United States (Bottone, 1997).

Strains of *Y. pestis* produce a bacteriocin active on *Y. pseudotuberculosis* (Ben-Gurion and Hertman, 1958). This was named Pesticin I by Brubaker and Surgalla (1962) after they detected a second bacteriocin (Pesticin 11) that was produced by *Y. pestis* and *Y. pseudotuberculosis*. Pesticin I has a narrow host range, being

TABLE BXII.γ.280. Differentiation of *Yersinia* from other genera[a]

Test	Yersinia	Citrobacter	Enterobacter	Escherichia	Hafnia	Klebsiella	Proteus	Salmonella
Voges–Proskauer	−	−	+[b]	−	[+]	d	d	−
Citrate (Simmons)	−	+	+	[−]	−	d	d	+
H₂S production	−	d	−	−	−	−	d	+
Phenylalanine deaminase	−	−	−	−	−	−	+	−
Lysine decarboxylase	−[c]	−	[−][d]	[+]	+	[+]	−	+
Motility, 37°C	−	+	+[b]	[+]	[+]	−	+	+
Motility, 25°C	+[e]	+	+[b]	[+]	[+]	−	+	+
KCN, growth	−[c]	d	+[f]	−[g]	+	+	+	−[h]
Malonate utilization	−	d	+[i]	d	d	+[j]	−	d
D-Glucose, gas	− or W	+	+[k]	+	+	[+]	[+]	+
L-Arabinose, acid	+[l]	+	+	+	+	+	−	+
D-Mannitol, acid	+[m]	+	+	+[n]	+	+	−	+
Mucate, acid	−	+	d	d	−	+[o]	−	+[p]

[a]For symbols see standard definitions; W, weak.

[b]Except *E. asburiae.*

[c]Except some strains of *Y. ruckeri.*

[d]*E. aerogenes* and *E. gergoviae* are positive.

[e]Except *Y. pestis* and some strains of *Y. ruckeri.*

[f]Except *E. gergoviae* and some strains of *E. agglomerans.*

[g]Except *E. hermannii* and a few strains of *E. vulneris.*

[h]Except *S. bongori* and some strains of *S. choleraesuis* subsp. *houtenae.*

[i]Except *E. asburiae* and *E. sakazakii.*

[j]Except *K. ozaenae.*

[k]Except *E. agglomerans.*

[l]Except *Y. ruckeri* and some strains of *Y. aldovae, Y. kristensenii,* and *Y. pseudotuberculosis.*

[m]Except some strains of *Y. aldovae.*

[n]Except *E. blattae.*

[o]Except *K. ozaenae* and *K. rhinoscleromatis.*

[p]Except *S. choleraesuis* subsp. *diarizonae* and *S. choleraesuis* subsp. *houtenae.*

active against a few strains of *E. coli,* serogroup IA and IB *Y. pseudotuberculosis* (Burrows, 1963), and serogroup 0:8 *Y. enterocolitica* (Brubaker, 1991). *Y. pestis* strains that produce Pesticin I also elaborate a fibrinolytic factor and a coagulase (Brubaker, 1972). A bacteriocin-like activity associated with the presence of phage tails has been described in *Y. enterocolitica* (Nicolle et al., 1973). *Y. intermedia* produces a bacteriocin-like substance at 25°C but not at 37°C that is active on certain strains of *Y. enterocolitica, Y. intermedia, Y. frederiksenii,* and *Y. kristensenii* (Bottone et al., 1979). Cafferkey et al. (1989) have shown that bacteriocin production by "avirulent" strains of serogroup 0:3 occurs at both 25°C and at 37°C and is active against isolates of serogroups 0:3, and single strains tested of 0:8 and 0:9 irrespective of the presence of the virulence plasmid.

The antigenic structure of *Yersinia* species is complex, but antigens are shared by *Y. pestis, Y. pseudotuberculosis,* and *Y. enterocolitica.* The common enterobacterial antigen has been found in these species (Le Minor et al., 1972; Maeland and Digranes, 1975). The fraction I envelope antigen (FI) of *Y. pestis* is best produced when cultures are incubated at 37°C on protein-rich media (Fox and Higuchi, 1958). This antigen, a large gel-like capsule or envelope, is heat labile (10 min at 100°C), water soluble, and dissociates during growth into a glycoprotein (FIA) and a carbohydrate-free protein (FIB). Passive hemagglutination with FI antigen is used for serologic surveys in plague foci. The presence of this antigen has also been demonstrated in *Y. pseudotuberculosis* (Quan et al., 1965). V and W antigens expressed by virulent strains of *Y. pestis* cultivated at 37°C appear to be related to the presence of a 45 MDa plasmid (Ben-Gurion and Shafferman, 1981; Ferber and Brubaker, 1981). Production of plasmid-

mediated V and W antigens has also been described in *Y. pseudotuberculosis* (Gemski et al., 1980b) and in *Y. enterocolitica* (Gemski et al., 1980a). The somatic antigen of *Y. pestis* is rough (R antigen) and therefore no serogroups have been described in this species. This R antigen is also present in *Y. pseudotuberculosis* (Thal and Knapp, 1971). In addition, *Y. pestis* and *Y. pseudotuberculosis* share at least 11 of 18 antigens (Lawton et al., 1960). *Y. pestis* and *Y. enterocolitica* express common protein antigens (Barber and Eylan, 1976). The antigenic scheme for *Y. pseudotuberculosis* (Thal and Knapp, 1971) comprises eight main thermostable serogroups (I to VIII) with nine subtypes (IA, IB; IIA, IIB, IIC; III; IVA, IVB; VA, VB; VI; VII; VIII), and five thermolabile flagellar H antigens (a–e). Antigenic relationships have been demonstrated between *Y. pseudotuberculosis* (serogroups 11, IV, IVA, and VI) and *Salmonella* serogroups B and D, *E. coli* serogroups 017, 055, and 077, and *Enterobacter cloacae* (Knapp, 1968; Mair and Fox, 1973).

Wauters et al. (1972) initially described 34 different O antigen and 20 H antigen serogroups in *Y. enterocolitica.* This classification included some serogroups defined by strains belonging to *Y. intermedia* (O:17) and *Y. kristensenii* (O:11, 0:12, 0:28). Wauters et al. (1991) subsequently expanded the O and H antigenic schema to include approximately 60 serogroups. Serogroup expansion has also shown sharing of antigens between pathogenic and nonpathogenic *Y. enterocolitica* and *Y. bercovieri* (O:8), *Y. frederiksenii* (O:3), and *Y. mollaretii* (O:3). Nevertheless, serogrouping in conjunction with biogrouping (Table BXII.γ.283), is a useful epidemiological marker. Cross-reactions occur between *Y. enterocolitica* serogroup O:9 and *Brucella* species (Hurvell and Lindberg, 1973; Corbel, 1975).

Yersinia species are susceptible *in vitro* to the following anti-

TABLE BXII.γ.281. Reactions of *Yersinia* spp. at 25–28°C and 37°C[a]

Test	1. Y. pestis	2. Y. aldovae	3. Y. bercovieri	4. Y. enterocolitica	5. Y. frederiksenii	6. Y. intermedia	7. Y. kristensenii	8. Y. mollaretii	9. Y. pseudotuberculosis	10. Y. rohdei	11. Y. ruckeri
Voges–Proskauer:											
37°C	−	−	−	−	−	−	−	−	−	−	−
25-28°C	−	+	−	+	+	+	−	−	−	−	−
Citrate, Simmons:											
37°C	−	−	−	−	[−][b]	−	−	−	−	−	−
25-28°C	−	d	−	−	d	−	−	−	−	[+][c]	−
Urease:											
37°C	−	[+]	d	[+]	[+]	[+]	[+]	[−]	+	d	−
25-28°C	−	+	+	+	+	+	+	+	+	d	−
Ornithine decarboxylase:											
37°C	−	d	[+]	+	+	+	+	[+]	−	[−]	+
25-28°C	−	+	+	+	+	+	+	+	−	[+]	+
Motility:											
37°C	−	−	−	−	−	−	−	−	−	−	−
25-28°C	−	+	+	+	+	+	+	+	+	+	[+]
Glycerol, acid:											
37°C	d	−	−	+	[+]	d	d	[−]	d	d	d
25-28°C	d	+	[+]	+	+	+	+	+	+	[+]	d
myo-Inositol, acid:											
37°C	−	−	−	d	[−]	[−]	[−]	−	−	−	−
25-28°C	−	+	d	d	−	[+]	d	d	−	−	−
Melibiose, acid:											
37°C	[−]	−	−	−	−	[+]	−	−	d	d	−
25-28°C	d	−	−	−	−	+	−	−	+	d	−
Raffinose, acid:											
37°C	−	−	−	−	d	d	−	−	[−]	d	−
25-28°C	−	−	−	−	−	+	−	−	[−]	d	−
L-Rhamnose, acid:											
37°C	−	−	−	−	+	+	−	−	d	−	−
25-28°C	−	+	−	−	+	+	−	−	+	−	−
Salicin, acid:											
37°C	d	−	[−]	[−]	+	+	[−]	[−]	[−]	−	−
25-28°C	[+]	+	−	[−]	+	+	−	[−]	d	−	−
D-Xylose, acid:											
37°C	+	d	+	d	+	+	[+]	d	+	d	d
25-28°C	+	+	+	d	+	+	+	+	+	+	d
Mucate, acid:											
37°C	−	−	−	−	−	−	−	−	−	−	−
25-28°C	−	d	+	−	[−]	d	−	+	−	−	−
Esculin hydrolysis:											
37°C	d	−	[-]	[-]	[+]	[−]	−	−	+	−	−
25-28°C	+	+	[+]	d	+	+	−	[−]	+	−	−

[a]For symbols see standard definitions.
[b][-], 11–25% positive.
[c][+], 26–75% positive.

microbial agents: tetracycline, chloramphenicol, aminoglycosides (amikacin, streptomycin, gentamicin, kanamycin, and neomycin), sulfonamides (alone or in combination with trimethoprim), imipenem, aztreonam, and fluoroquinilones (Hoogkamp-Korstanje, 1987; Bonacorsi et al., 1994). They are variably susceptible to colistin and are resistant to erythromycin and novobiocin. *Y. pestis* and *Y. pseudotuberculosis* are usually susceptible to β-lactam antibiotics, but their susceptibility to penicillin is in the range of susceptible to intermediate. Resistance to ampicillin (Borowski and Zaremba, 1973) and to streptomycin (Kanazawa and Ikemura, 1979) has been described for *Y. pseudotuberculosis, Y. enterocolitica,* and *Y. intermedia* (Bottone, 1977). *Y. frederiksenii* and *Y. kristensenii* are resistant to penicillin and slightly susceptible or resistant to other β-lactam antibiotics (ampicillin, carbenicillin, cephalothin) (Bercovier et al., 1979). The level of resistance is strain dependent (Zaremba and Aldová, 1979) and temperature dependent (Chester and Stotzky, 1976). Antibiotic susceptibility patterns of *Y. enterocolitica* are serogroup specific. *Y. enterocolitica* strains produce both a constitutive β-lactamase (active on am-

picillin, carbenicillin, penicillin, and cephalosporins) and an inducible β-lactamase active only on cephalosporins and penicillin (Cornelis and Abraham, 1975). *Y. enterocolitica* strains that are resistant to tetracycline, chloramphenicol, streptomycin, and kanamycin have been reported (Zaremba and Aldová, 1979). Newer β-lactam antibiotics such as ceftriaxone, ceftazidime, cefotaxime, and moxalactam have excellent activity against *Y. enterocolitica,* as do imipenem and aztreonam (Hornstein et al., 1985).

Y. pestis is the causative agent of plague. Plague is primarily a disease of wild rodents. *Y. pestis* is transmitted among wild rodents by flea bites or ingestion of contaminated animal tissues (Butler, 1983). In fleas, the bacterium multiplies and blocks the esophagus and the pharynx. The fleas regurgitate the bacteria when they take their next blood meal and transmit *Y. pestis* to humans if no other hosts are available. Infective flea bites produce the typical bubonic form of plague within 2–8 d. *Y. pestis* multiplies intracellularly in macrophages and extracellularly and proceeds through the lymphatic system. The lymph nodes near the flea

TABLE BXII.γ.282. Differential characteristics of the species of the genus *Yersinia*

Test[a]	1. Y. pestis	2. Y. aldovae	3. Y. bercovieri	4. Y. enterocolitica	5. Y. frederiksenii	6. Y. intermedia	7. Y. kristensenii	8. Y. mollaretii	9. Y. pseudotuberculosis	10. Y. rohdei	11. Y. ruckeri
Pathogenic	+	(+)[b]	(+)	+	(+)	(+)	(+)	(+)	+	(+)	(+)
Mol% G + C content	46	46	50	48.5 ± 0.5	48	48.5 ± 0.5	48.5 ± 0.5	50–51	46.5	48.7–49.4	48.0 ± 0.5
Motility (22°C)	−[c]	−	−	+	−	+	−	−	+	−	
Nitrate reductase	V[d,e]	+	+	+	+	+	+	+	+	+	+
Urease	−[c]	−	d	+	+	+	+	−	+	d	−
Simmons citrate	−	d	−	−	V	+22°C or −	−	−	−	+	−
Ornithine decarboxylase	−	d	+	+	+	+	+	+	−	+	+
Acetylmethyl-carbinol (22°C)	−	+	+	+	+	+	−	+	−	+	+
B-Galactosidase	d	−	−	+	+	+	d	−	d	d	d
Indole	−	−	−	V	+	+	V	−	−	−	−
H$_2$S production	−	−	−	−	−	−	−	−	−	−	−
Aesculin hydrolysis	+	−	−	V	+	+	−	−	+	−	−
Gelatin	−	−	−	−	−	−	−	−	−	−	−
Methyl red	+[e]	+[d]	V	V	+	+	+	+	+	+	+
Glucose	+	+	+	+	+	+	+	+	+	+	+
Lactose	−	−	−	−	d	d	−	d	−	−	−
Sucrose	−	−	+	+	+	+	−	+	−	+	−
Glycerol	V[d]	−	−	+	+	d	d	−	V	d	d
α-Methylglucoside	−	−	−	−	−	+ (22°C)	−	−	−	−	−
Cellobiose	−	d	+	+	+	+	+	+	−[c]	−	−
Melibiose	V[d]	−	−	−	−	22°C[c]	−[c]	−	+	d	−
Mucate	−	d	+	−	V	V	−	+	−	+	−
Raffinose	−	−	−	−	d	d	−	−	−	d	−
L-Rhamnose	−	−	−	−	d	d	−	−	+[c]	−	−
Trehalose	+	+	+	V	V	+	−	+	+	+	+
Sorbose	−	NR[f]	NR	V	+	+	+	NR	−	NR	NR
Sorbitol	d	d	+	+	+	+	+	+	−[c]	+	d
Arabinose	+	+	+	+	+	+	+	+	+	+	−
Maltose	V[d]	−	+	+	+	+	+	d	+	−	+
D-Xylose	+	d	+	+	+	+	+	d	+	d	−
Mannitol	+	+	+	+	+	+	+	+	+	+	+
Pyrazinamidase	−	−	V	V	+	+	+	−	−	+	+
Usual habitat:											
Animals				+	+		+		+		+
Birds									+		
Dogs										+	
Fish		+			+	+					+
Fleas	+										
Mammals	+										
Rodents	+					+				+	
Humans	+	+	+	+		+		+	+	+	+
Environment				+			+				
Soil			+				+				
Sewage						+					
Water		+			+	+				+	+
Food						+			+		+
Vegetables			+						+		

[a]All *Yersinia* spp. are fermentative; negative for H$_2$S, oxidase, arginine dihydrolase, phenylalanine deaminase, malonate, dulcitol, DNase (25°C), and pigmentation; all are positive for mannitol and mannose. Motile species have peritrichous flagella, and motility is more pronounced at temperatures less than or equal to 30°C. Except for delayed reactions in some *Y. ruckeri*, all species are negative for gelatin hydrolase and lysine decarboxylase.

[b](+), Opportunistic pathogens.

[c]Key test to differentiate species.

[d]Key test to differentiate biovars.

[e]V, Strain instability or variable.

[f]NR, Not reported.

bite are the first to become inflamed, enlarged, and painful, and constitute the bubo. As the evolution of the infection is usually rapid with massive growth of *Y. pestis* in the blood, characteristic lesions are not found in the spleen or liver at autopsy. Untreated, the disease evolves in 5–10 d to profound septicemia in which *Y. pestis* may be seen in peripheral blood smears (Mann et al., 1984). Secondary pneumonia may result from hematogenous spread. From the latter, primary pneumonic plague can spread

TABLE BXII.γ.283. Differentiation of biogroups of *Yersinia enterocolitica*[a]

Test	Biogroup Reaction					
	1A	1B[b]	2	3	4	5
Lipase activity	+	+	−	−	−	−
Salicin (acid production in 24 h)	+	−	−	−	−	−
Esculin hydrolysis (24 h)	±	−	−	−	−	−
Xylose (acid production)	+	+	+	+	−	V
Trehalose (acid production)	+	+	+	+	+	−
Indole production	+	+	V	−	−	−
Ornithine decarboxylase	+	+	+	+	+	+(+)
Voges–Proskauer test	+	+	+	+	+	+(+)
Pyrazinamidase activity	+	−	−	−	−	−
Sorbose (acid production)	+	+	+	+	+	−
Inositol (acid production)	+	+	+	+	+	+
Nitrate reduction	+	+	+	+	+	−

[a]Symbols: +, positive; −, negative; (+), delayed positive; V, variable.

[b]Biogroup 1B is comprised mainly of strains isolated in the United States.

from person to person or from animals to humans by means of droplets. Both bubonic and pneumonic plague have been transmitted to humans by the domestic cat—bubonic through a scratch and bite from an infected cat (Thornton et al., 1975; Weniger et al., 1984), and pneumonic through face-to-face exposure to a cat with pneumonic plague (Werner et al., 1984; Doll et al., 1994). In this clinical form, death generally occurs in less than 4 d. Pestis minor cases, in which the bacteria remain self-limited in buboes followed by self-cures have been described in endemic plague areas (Pollitzer, 1954).

Virulence of *Y. pestis* is associated with several factors (Surgalla et al., 1968; Perry and Fetherston, 1997). Among these are synthesis of the fraction I (FI) gel-like antiphagocytic capsule or surface antigen, VW antigens (associated with Ca^{2+} dependency for growth on magnesium oxalate medium, autoagglutination in broth cultures at 37°C, pigment production (incorporation of Congo red dye or hemin into cell surfaces, which results in dark greenish-brown or red colonies at 26°C but not at 37°C), presence of siderophore (yersiniabactin) iron-acquisition system, pH6 antigen, a fibrillar adhesin surface structure that may facilitate entry into macrophages (Straley, 1993) and is synthesized in host macrophage lysosomes (Lindler and Tall, 1993), and serum resistance necessary for growth in blood and transmission between insect vector and mammalian hosts. Pesticin production (Pst plasmid) is necessary for dissemination of *Y. pestis* from peripheral sites of inoculation; Pst is a 9.5-kb plasmid that also encodes plasminogen activator, a fibrinolysin, coagulase activity, which may be limited to rabbit plasma (Brubaker, 1991), and murine toxin, a 6-adrenergic antagonist that is highly lethal for mice and rats, causing circulatory collapse. LD_{50} dose for mice inoculated with strains expressing the aforementioned virulence factors is 1–10 organisms. Avirulent strains of *Y. pestis* never produce VW antigens except in the case of the vaccine strain EV76, whose attenuated virulence has resulted from a mutation in its iron metabolism. Virulent strains and the EV76 strain harbor a 45-MDa plasmid. In contrast to the VW antigens, the lack of any of the other virulence factors does not completely abolish the virulence of *Y. pestis* strains.

Y. pseudotuberculosis is responsible for epizootics in nearly all animal species, especially in rodents and birds. Animals are usually contaminated by the oral route and, after 1–2 weeks of incubation, the bacteria are found in mesenteric lymph nodes. The main symptoms are mesenteric adenitis and chronic diarrhea. Infection evolves either in self-cure, or in fatal septicemia. *Y.*

pseudotuberculosis is an intracellular parasite and, like *Y. pestis*, reaches the lymphatic system. At autopsy, caseous lesions are found in Peyer's patches, mesenteric lymph nodes, the spleen, and the liver. Humans orally contaminated by *Y. pseudotuberculosis* develop either a mesenteric adenitis, which simulates acute appendicitis, or, in the compromised host, a severe septicemia. *Y. pestis* and *Y. pseudotuberculosis* share a common 70-kb plasmid low calcium response (LCR) that encodes for calcium dependency and four outer membrane proteins, and the V and W antigens. *Y. pseudotuberculosis* chromosomally controlled virulence factors include the outer membrane protein invasin, which promotes host cell penetration (Isberg and Falkow, 1985).

Y. enterocolitica has been recognized as pathogenic for chinchillas, hares, monkeys, and humans. The pathogenicity for animals is similar to that of *Y. pseudotuberculosis*. In humans the most common presentation is acute enteritis in children. Concurrent mesenteric lymphadenitis and terminal ileitis simulating appendicitis may also be present. In young adults, acute terminal ileitis and mesenteric lymphadenitis appear to be more common. The extent of gastrointestinal tract pathology depends on the serogroup of the invading strain and age and underlying status of the host. In adults secondary clinical forms of infection include arthritis and erythema nodosum. Septicemia with metastatic abscesses is a rarer complication and is usually associated with immunosuppression or occurs in the setting of iron overload in individuals receiving the iron chelator desferrioxamine (Robins-Browne and Pipic, 1985). Septicemia and profound shock has been associated with transfusion of *Y. enterocolitica* contaminated blood (reviewed by Bottone, 1999). Secondary immunologically mediated sequelae include arthritis and erythema nodosum. Arthritis is closely associated with the presence of the histocompatibility antigen HLA-27 (Dequeker et al., 1980), especially among Scandinavians, and *Y. enterocolitica* serogroup 0:3 biotype 4, phage type 8, and serogroup 0:9 infections. Production of a heat-stable enterotoxin (ST) resembling *E. coli* ST (Okamoto et al., 1981) has been demonstrated *in vitro* (Pai and Mors, 1978), but its role in pathogenicity is not clear: *Y. enterocolitica* strains do not produce ST when incubated *in vitro* at temperature above 30°C, and direct proof of ST production *in vivo* has not been reported.

Virulent plasmid-containing strains of *Y. pestis, Y. pseudotuberculosis*, and *Y. enterocolitica* rapidly become avirulent when grown at 37°C, which results in the loss of the virulence plasmid. Cross-immunity among these three species has been demonstrated

(Thal, 1973; Alonso et al., 1978). Human chemoprophylaxis with streptomycin (drug of choice), tetracycline, and chloramphenicol; vaccination; and the spreading of insecticides and rodenticides are the suggested measures for controlling plague. For persons exposed to *Y. pestis*, tetracycline, doxycyline, and trimethoprim-sulfamethoxazole are recommended (Poland, 1989).

The pathogenicity of *Y. intermedia, Y. kristensenii, Y. frederiksenii, Y. bercovieri*, and *Y. mollaretii* in humans and animals is not clearly established. They appear more like opportunistic pathogens than true pathogens (Bercovier et al., 1978; Wauters et al., 1988; Bottone, 1997). ST-producing strains of these three species have been described (Kapperud, 1980), but their clinical significance is still unknown. *Y. ruckeri* is a fish pathogen responsible for red mouth disease, especially in rainbow trout. An inflammation of the mouth and the throat is the main characteristic of the disease, which is enzootic (Rucker, 1966). The bacterium is usually isolated from the kidneys of fish undergoing a systemic infection.

Epidemiology The geographical distribution of *Y. pestis* is widespread, and the organism has been isolated from all continents except Australia and Antarctica. Plague is enzootic in Africa (Central, East, and South Africa), in North and South America, certain regions of Asia (Southeast Asia, Mongolia), and the former USSR, India, and southwestern United States and the Pacific coastal area (Perry and Fetherston, 1997). Between epidemics, *Y. pestis* remains localized in enzootic or maintenance hosts (Butler, 1983) and has been isolated from more than a hundred different naturally infected species of rodents, but rarely from predatory animals (carnivores and birds, the latter being resistant to the infection). The spread of plague is usually accomplished by the epizootic cycle of rodents to fleas and fleas to rodents. The reservoir for *Y. pestis* is soil contaminated by infected dead fleas and rodents in which the microorganism survives for months in deep rodent burrows.

Rodents coming from noninfected areas become infected when they dig burrows in previously contaminated areas (Mollaret et al., 1963). This cycle constitutes "sylvatic plague". When urban rodents come in contact with rural rodents, *Y. pestis* can spread between rodents and to humans through flea bites. The epidemiology of plague is thus linked to the ecology of both fleas and rodents.

Y. pseudotuberculosis is distributed worldwide. It has been found in numerous animal species, especially rodents and birds, in soil, and in humans (Wetzler, 1970). In Japan, cats and dogs have been associated with human cases (Fukushima et al., 1989). Wild animals, which are often asymptomatic carriers, are considered the reservoir of the bacteria. Humans and animals are contaminated orally either by direct contact with sick or asymptomatic animals or through food contaminated by the excretions of these animals. The incidence of *Y. pseudotuberculosis* infection varies with the season and is highest during colder months. *Yersinia* species multiply at 4°C and therefore have a selective advantage over other bacteria at low temperatures; this explains why *Y. pseudotuberculosis, Y. enterocolitica, Y. frederiksenii*, and *Y. kristensenii* are more frequently isolated from the environment during cold rather than hot seasons. Human and animal infections follow this seasonal distribution as well.

Y. enterocolitica has been isolated from a wide variety of sources (live and inanimate) in every country in which it has been sought and probably has a worldwide distribution (Mollaret et al., 1979). As shown in Table BXII.γ.284, biovar IA strains are ubiquitous, having been found in a wide range of animal and environmental sources (including foods), whereas other biovars or serogroups are frequently associated with a specific host (Bercovier et al., 1980a): biovar 5 strains have been isolated mainly from hares in Europe; biovar 4, serogroup 0:3 strains are responsible for most human gastrointestinal infections in Europe, Canada, and the Republic of South Africa, and recently in the United States as well (Bottone et al., 1987; Ostroff, 1995). Serogroup 0:8, heretofore the most frequently isolated *Y. enterocolitica* strain in the United States, has decreased in frequency in the United States, and is only sporadically reported in other parts of the world (Ostroff, 1995). Serogroup 0:9 strains are the second most common *Y. enterocolitica* isolates in Europe and Japan.

Y. intermedia and *Y. frederiksenii* have been identified in Europe, the United States, Australia and New Zealand, Israel, and Japan. These two species have been isolated mainly from fresh water and foods and only rarely from nonirrigated soil, humans or animals other than fish (Kapperud, 1977; Bercovier et al., 1978; Brenner et al., 1980a; Ursing et al., 1980a). *Y. kristensenii* has been found in Europe, the United States, Japan, and Australia. Strains of this species have been isolated mainly from soil, foods, and asymptomatic animals; isolates from other environmental sources and from human infection are rare (Bottone and Robin, 1977; Bercovier et al., 1980b). *Y. bercovieri* has been recovered from terrestrial ecosystems (Bercovier et al., 1978), while *Y. mollaretii* has been isolated from human stool specimens, most raw vegetables, soil, and drinking water (Fukushima, 1985; Kaneko and Maruyama, 1987). Evidence of human disease potential for either *Y. bercovieri* or *Y. mollaretii* are lacking.

Y. ruckeri, which is taxonomically tentative in the genus *Yersinia*, has been encountered mainly in the United States and Canada as a natural component of fresh water ecosystems. *Y. ruckeri* causes enteric red mouth disease of salmon and fish when fish are exposed to large numbers of bacteria (Ross et al., 1966). The disease is usually enzootic and occasionally epizootic in fish hatcheries especially under conditions of stress or poor environmental factors (Stevenson, 1997). Five serovars and two biotypes of *Y. ruckeri* have been identified. Serovar I strains are usually D-sorbitol positive, while serovar 2 isolates are usually D-sorbitol negative (Stevenson and Daly, 1982; Stevenson and Airdrie, 1984). A 36-MDa transferable plasmid encoding resistance to tetracycline and sulfonamides has been identified in some *Y. ruckeri* strains (DeGrandis and Stevenson, 1985).

ENRICHMENT AND ISOLATION PROCEDURES

Isolation of *Yersinia* strains from noncontaminated normally sterile samples (blood, lymph nodes) can be performed by using blood agar or nutrient agar incubated for 48 h at 28°C, or 24 h at 37°C followed by 24 h at room temperature. The isolation of *Y. pestis* from contaminated samples requires inoculation (subcutaneously or percutaneously) of animals (guinea pigs, mice, or rats). The organism can be recovered postmortem from the spleen, liver, or lymph nodes of the inoculated animals. All other *Yersinia* species can be isolated from clinical or food samples by inoculating standard or special selective media (reviewed by Bottone, 1992), of which cefsulodin-irgasan-novobiocin (CIN) agar (Schiemann, 1979), and virulent *Yersinia* enterocolitica (VYE) agar formulated by Fukushima (1987) seem the most useful. The latter medium was developed because of the growth inhibition of many *Y. bercovieri* and *Y. pseudotuberculosis* strains on CIN agar (Fukushima and Gomyoda, 1986). All these media should preferably be incubated for 48 h at 28–29°C or for 24 h at 37°C followed by 48–72 h at room temperature. Recovery of *Yersinia*

TABLE BXII.γ.284. Correlation of biogroup and serogroup with ecologic and geographic distribution

Biogroup	Associated with human infections	Serogroup(s)	Ecologic distribution
1B	+	O:8, O:4, O:13a, 13b, O:18, O:20, O:21	Environment, pigs (O:8), mainly in the U.S.A.
2	+	O:9, O:5, 27	Pigs, Europe (O:9), United States (O:5, 27), Japan (O:5, 27)
3	+	O:1, 2, 3, O:5, 27	Chinchilla, pigs (O:5, 27)
4	+	O:3	Pigs, Europe, U.S.
5	+	O:2,3	Hare
1A[a]	−	O:5, O:6, 30, O:7, 8, O:18, O:46, nontypeable	Environment, pigs, food, water animal and human feces, U.S.

[a] *Y. enterocolitica* isolates comprising biogroup 1A may be opportunistic pathogens in patients with underlying disorders.

strains from contaminated samples can be improved by various cold enrichment techniques (van Pee and Straiger, 1979; Lee et al., 1980).

MAINTENANCE PROCEDURES

Stab inoculations of *Yersinia* strains in conventional stock culture media stored in the dark at room temperature or at 4°C provide living cultures for 10 years or more, if the tubes are tightly sealed. Lyophilization and deep-freeze storage in 10% glycerol are suitable preservation techniques. To keep a strain fully virulent, it should never be subcultured at 37°C, but always at 25–28°C.

PROCEDURES FOR TESTING SPECIAL CHARACTERS

Methods to test tetrathionate reduction, tellurite reduction, and the type of nitrate reductase have been described or referenced by Bercovier et al. (1979). The Ca^{2+}-dependency of virulent *Yersinia* strains is evaluated on magnesium oxalate medium[1] (Higuchi and Smith, 1961) as follows. Inoculate 0.1 ml of a bacterial suspension (10^5 cells/ml) onto two plates: one is incubated at 37°C, the other at 26°C. Check colony numbers on the two plates after 2–3 d. Colonies growing at 26°C but not at 37°C are Ca^{2+}-dependent. A fully virulent strain should give confluent growth at 26°C, whereas only 10–100 colonies should appear at 37°C. The autoagglutination test (Laird and Cavanaugh, 1980) to detect virulent *Yersinia* strains is performed by inoculating 10 or more isolated colonies, individually into a pair of tubes (13 × 100 mm) containing 2 ml of RPMI-1640 medium containing 10% fetal calf serum and 25 ml HEPES buffer (N-2-hydroxyethylpiperazine-N′-2-ethanesulfonic acid). One tube is incubated at 37°C, the other at 26°C. After incubation at 26°C for 18 h, virulent isolates give a uniform turbidity; at 37°C a layer of agglutinated bacteria appears at the bottom of the tube and the overlying supernatant remains clear. Avirulent strains give uniform turbidity at both 26°C and 37°C, and rough strains show spontaneous agglutination at both temperatures.

DIFFERENTIATION OF THE GENUS *YERSINIA* FROM OTHER GENERA

Characteristics useful for differentiating *Yersinia* from other physiologically similar genera are listed in Table BXII.γ.280, and in Tables BXII.γ.193, BXII.γ.194, and BXII.γ.196 in the chapter on the family *Enterobacteriaceae*.

---

1. Magnesium oxalate medium consists of blood agar base (BBL, or any other manufacturer if the Ca^{2+} content of the base is low), 40.0 g; distilled water, 830 ml. Sterilize at 121°C for 15 min and cool to 45°C. From stock solution sterilized by filtration, aseptically add the following ingredients: $MgCl_2$ solution (23.8 g/l), 80 ml; sodium oxalate solution (33.5 g/l), 80 ml; and glucose solution (180.1 g/l), 10 ml.

TAXONOMIC COMMENTS

The genus *Yersinia* was proposed by van Loghem (1944) in order to separate *Y. pestis* and *Y. pseudotuberculosis* (formerly in the genus *Pasteurella*) from *Pasteurella* species *sensu stricto* (i.e., *P. multocida*, etc.), from which they differ in their negative oxidase reaction and in their DNA base composition. The genus *Yersinia* belongs to the family *Enterobacteriaceae*. *E. coli* tDNA (i.e., the genes coding for transfer RNA) and *Y. pestis* tDNA are 63% related (Brenner et al., 1976), a value similar to that found for *E. coli* tDNA and *Hafnia alvei* tDNA. All *Yersinia* species express the common enterobacterial antigen. Their physiological characteristics and their fatty acid content are similar to those of all *Enterobacteriaceae* species. The mol% G + C range of *Yersinia* species is 46–50 and is consistent with that for *Enterobacteriaceae* species.

The genus *Yersinia* currently consists of 11 different species. Based on DNA–DNA hybridization studies, all of these species are more closely related to each other than to any other *Enterobacteriaceae* species (Brenner et al., 1980e; Bercovier et al., 1984; Aleksic et al., 1987; Wauters et al., 1988). The genus *Yersinia* can be considered a very homogeneous taxon.

DNA relatedness among *Yersinia* species is 40% or higher, except for *Y. ruckeri*, which is at most 38% related to other *Yersinia* species. DNAs of *Y. ruckeri* strains have been shown to be 30% related to *Serratia* species (Ewing et al., 1978). *Y. ruckeri* was included in *Yersinia* because its mol% G + C of 48 is closer to that of *Yersinia* species than to that of *Serratia* species. Because the phenotypic characteristics of *Y. ruckeri* are very different from those of other *Yersinia* species (Tables BXII.γ.281 and BXII.γ.282), it might constitute a new genus by itself. Phylogenetic studies would be helpful in clarifying this problem.

Strains of *Y. enterocolitica* belonging to the five different biovars (Table BXII.γ.283), including the metabolically inactive biovar 5 strains, constitute a homogeneous genomospecies (Bercovier et al., 1980a). The strains originally described as *Y. enterocolitica*-like organisms or atypical *Y. enterocolitica* have been separated into three different species: *Y. intermedia* (Brenner et al., 1980a), *Y. frederiksenii* (Ursing et al., 1980a), and *Y. kristensenii* (Bercovier et al., 1980c). *Y. frederiksenii* consists of three genetic groups based on DNA–DNA hybridization (Ursing et al., 1980a). For practical reasons, because there are no phenotypic differences among the three genetic groups, only one species, *Y. frederiksenii*, has been proposed for these rhamnose-positive strains. More study of phenotypic characteristics is needed to separate the three genetic groups.

The DNAs of *Y. pestis* strains, regardless of biovar, and of *Y. pseudotuberculosis* are 90% or more interrelated. This explains the antigenic and biochemical similarities of the two species. Based on DNA data, Bercovier et al. (1980b) proposed that the two species constitute a single species, divided into two subspecies: *Y. pseudotuberculosis* subsp. *pseudotuberculosis* and *Y. pseudotuberculosis* subsp. *pestis*. Presently, however, the two species remain distinct, largely because of fear of omitting the subspecies epithet subsp. *pestis*, which may present a potential hazard for laboratory-acquired infections (Williams, 1983).

Ursing et al. (1980b) have shown, based on DNA and physiological data, that *Y. philomiragia* (Jensen et al., 1969) is not related to the genus *Yersinia*, and furthermore, that it is not a member of the family *Enterobacteriaceae*. This species has now been assigned to the genus *Francisella* as *F. philomiragia*.

FURTHER READING

Bercovier, H., J.M. Alonso, Z. Bentaiba, J. Brault and H.H. Mollaret. 1979. Contribution to the definition and taxonomy of *Yersinia enterocolitica*. Contr. Microbiol. Immunol. *5*: 12–22.

Bottone, E.J. 1977. *Yersinia enterocolitica*: a panoramic view of a charismatic microorganism. Crit. Rev. Microbiol. *5*: 211–241.

Bottone, E.J. 1997. *Yersinia enterocolitica*: the charisma continues. Clin. Microbiol. Rev. *10*: 257–276.

Bottone, E.J. 1999. *Yersinia enterocolitica*: overview and epidemiologic correlates. Microbes Infect. *1*: 323–333.

Brenner, D.J., J. Ursing, H. Bercovier, A.G. Steigerwalt, G.R. Fanning, J.M. Alonso and H.H. Mollaret. 1980. Deoxyribonucleic acid relatedness in *Yersinia enterocolitica* and *Yersinia enterocolitica*-like organisms. Curr. Microbiol. *4*: 195–200.

Brubaker, R.R. 1991. Factors promoting acute and chronic diseases caused by yersiniae. Clin. Microbiol. Rev. *4*: 309–324.

Fukushima, H. 1987. New selective agar medium for isolation of virulent *Yersinia enterocolitica*. J. Clin. Microbiol. *25*: 1068–1073.

Kaneko, S. and T. Maruyama. 1987. Pathogenicity of *Yersinia enterocolitica* serotype *O:3*, biotype 3 strains. J. Clin. Microbiol. *25*: 454–455.

Ostroff, S. 1995. *Yersinia* as an emerging infection: epidemiologic aspects of yersiniosis. Contrib. Microbiol. Immunol. *13*: 5–10.

Wauters, G., M. Janssens, A.G. Steigerwalt and D.J. Brenner. 1988. *Yersinia mollaretii* sp. nov. and *Yersinia bercovieri* sp. nov., formerly called *Yersinia enterocolitica* biogroups 3a and 3b. Int. J. Syst. Bacteriol. *38*: 424–429.

Wauters, G., K. Kandolo and M. Janssens. 1987. Revised biogrouping scheme of *Yersinia enterocolitica*. Contrib. Microbiol. Immunol. *9*: 14–21.

DIFFERENTIATION OF THE SPECIES OF THE GENUS *YERSINIA*

Characteristics useful in differentiating the various species of *Yersinia* are listed in Table BXII.γ.282.

List of species of the genus Yersinia

1. **Yersinia pestis** (Lehmann and Neumann 1896) Van Loghem 1944, 15[AL] (*Bacterium pestis* Lehmann and Neumann 1896, 194; *Yersinia pseudotuberculosis* subsp. *pestis* Bercovier, Mollaret, Alonso, Brault, Fanning, Steigerwalt and Brenner 1981a, 383.)

pes' tis. L. n. *pestis* plague, pestilence.

The characteristics are as described for the genus and as listed in Tables BXII.γ.281 and BXII.γ.282. Three biogroups have been described in relation to the geographical distribution of the organism: (a) biogroup antiqua produces acid aerobically from glycerol, reduces nitrate to nitrite, does not ferment melibiose, and is found in Central Asia and Central Africa; (b) biogroup medievalis produces acid from both glycerol and melibiose, but does not reduce nitrate to nitrite; it is found in Iran and the former USSR; and (c) biogroup orientalis (synonym: oceanic) does not produce acid from either glycerol or melibiose but reduces nitrate to nitrite and is distributed worldwide. Some rare atypical strains positive in their reactions for urease and L-rhamnose have been reported. *Y. pestis* is the causative agent of plague. The disease can be reproduced experimentally in mice, rats, guinea pigs, and monkeys.

The mol% G + C of the DNA is: 46 (T_m).

Type strain: ATCC 19428, NCTC 5923.

2. **Yersinia aldovae** Bercovier, Steigerwalt, Guiyoule, Huntley-Carter, and Brenner 1984, 171[VP]

al.do' vae. M.L. gen. n. *aldovae* in honor of Eva Aldova, the Czechoslovakian microbiologist who first isolated the bacterium.

The characteristics are as described for the genus and as listed in Tables BXII.γ.281 and BXII.γ.282. The species is composed of isolates previously referred to as *Y. enterocolitica*-like, group X2 (Bercovier et al., 1980a). Two biovars exist—one typically sucrose negative, and one sucrose positive. Differentiation of *Y. aldovae* from urease-positive *Hafnia alvei* may be achieved by production of gas by *H. alvei* from D-glucose at 36°C and susceptibility to *Hafnia* specific bacteriophage (Guinée and Valkenburg, 1968). *Y. aldovae* has been isolated from aquatic ecosystems—drinking water, river water, fish, and rarely soil. *Y. aldovae* has not been implicated in animal infections and has not been isolated from humans.

The mol% G + C of the DNA is: 48 (T_m).

Type strain: ATCC 35236, CNY 6065.

GenBank accession number (16S rRNA): X75277.

3. **Yersinia bercovieri** Wauters, Janssens, Steigerwalt, and Brenner 1988, 428[VP]

ber.co.vi.e' ri. M.L. n. *bercovieri* in honor of Herve Bercovier, who first described biogroups 3A and 3B and has made outstanding contributions to the taxonomy and ecology of yersiniae.

Formerly called *Y. enterocolitica* biogroup 3b (Bercovier et al., 1978). Later transferred to *Y. enterocolitica* biogroup 6 (Wauters et al., 1987) because of negative Voges–Proskauer reaction. *Y. bercovieri* must be distinguished from rarely occurring pathogenic strains of serogroup 0:3 that are Voges–Proskauer-negative variants of biogroup 3 (Kaneko and Maruyama, 1987). The species conforms to the definition of the family *Enterobacteriaceae* and the genus *Yersinia*. The characteristics are as described for the genus and as listed in Tables BXII.γ.281 and BXII.γ.282. *Y. bercovieri* strains have been isolated from human stool specimens, animals, raw vegetables, soil, and water. There is no evidence of pathogenicity for humans.

The mol% G + C of the DNA is: 50 (T_m).

Type strain: ATCC 43970, CDC 2475-87.

4. **Yersinia enterocolitica** (Schleifstein and Coleman 1943)

Frederiksen 1964, 104[AL] (*Bacterium enterocoliticum* Schleifstein and Coleman 1943, 56.)

en.ter.o.co.li' ti.ca. Gr. n. *enteron* intestine; Gr. n. *colon* of the colon; Gr. suff. *iticos* pertaining to; M.L. fem. adj. *enterocolitica* pertaining to the intestine and colon.

The characteristics are as described for the genus and listed in Tables BXII.γ.281 and BXII.γ.282. The Voges–Proskauer test is usually positive at 22–28°C and negative at 37°C. Biovars of *Yersinia enterocolitica* are listed in Table BXII.γ.283 and, like phagovars and serogroups, are useful epidemiological tools. Rare atypical strains that are either positive for their reactions on Simmons' citrate and for acid production from lactose and raffinose (due to a metabolic plasmid) or negative for urease activity have been reported. When incubated at 20°C, *Y. enterocolitica* strains produce a broad-spectrum mannose-resistant hemagglutinin that is lost at 37°C (MacLagan and Old, 1980). *Y. enterocolitica* causes diarrhea, terminal ileitis, mesenteric lymphadenitis, autoimmune arthritis, abscesses, and septicemia in humans. The disease can be reproduced experimentally in mice, gerbils, and monkeys. Other human infections are listed in Table BXII.γ.285. The species has been isolated from a wide variety of sources in the environment (live and inanimate) including foods and from healthy humans and animals, especially the pig.

The mol% G + C of the DNA is: 48.5 ± 1.5 (T_m, Bd).

Type strain: 161, ATCC 9610, CIP 80-27, DSM 4780.

GenBank accession number (16S rRNA): M59292.

Additional Remarks: The type strain belongs to biovar I B, serogroup 0:8, and phagovar X.

5. **Yersinia frederiksenii** Ursing, Brenner, Bercovier, Fanning, Steigerwalt, Brault and Mollaret 1981, 217[VP] (Effective publication: Ursing, Brenner, Bercovier, Fanning, Steigerwalt, Brault and Mollaret 1980a, 213.)

fred.er.ik.sen' i.i. M.L. gen. n. *frederiksenii* of Frederiksen; named after the Danish microbiologist Wilhelm Frederiksen, who made a substantial contribution to the study of the genus *Yersinia*.

The characteristics are as described for the genus and as listed in Tables BXII.γ.281 and BXII.γ.282. This species is composed of three different genomospecies. One group is positive for β-xylosidase and citrate (Simmons), and the type strain belongs to this group. The other two groups are variable or negative for these tests. More phenotypic studies are needed to differentiate the three groups. Some strains are able to ferment raffinose and lactose when they harbor a metabolic plasmid. *Y. frederiksenii* has been isolated mainly from fresh water sources, fish, foods, and occasionally from healthy or sick humans and animals.

The mol% G + C of the DNA is: 48 (T_m).

Type strain: 6175, CIP 80-29.

6. **Yersinia intermedia** Brenner, Bercovier, Ursing, Alonso, Steigerwalt, Fanning, Carter and Mollaret 1981, 217[VP] (Effective publication: Brenner, Bercovier, Ursing, Alonso, Steigerwalt, Fanning, Carter and Mollaret 1980a, 207.)

in.ter.me' di.a. L. fem. adj. *intermedia* intermediate; here it implies that biochemical reactions of this species seem midway between *Y. enterocolitica* and *Y. pseudotuberculosis*.

The characteristics are as described for the genus and as listed in Tables BXII.γ.281 and BXII.γ.282. Media with a high bile salt content (0.8%) are inhibitory, especially when incubated at 37°C. Some biochemical characteristics (citrate utilization; cellobiose, L-rhamnose, and raffinose fermentation) are always expressed at 25–28°C but are inconstant at 37°C. Either a type A or a type B nitrate reductase is present. Eight biovars have been described (Brenner et al., 1980a) based on the fermentation of melibiose, L-rhamnose, α-methyl-D-glucoside, and raffinose, and on the utilization of citrate (Simmons). Of the strains studied, 96% are positive for at least four of these five tests. *Y. intermedia* has been isolated mainly from fresh water sources, fish, foods, and occasionally from both sick and healthy humans (Bottone et al., 1974; Punsalang et al., 1987).

The mol% G + C of the DNA is: 48.5 ± 0.5 (T_m, Bd).

Type strain: 3953, Bottone 48, Chester 48, ATCC 29909, CIP 80-28.

7. **Yersinia kristensenii** Bercovier, Ursing, Brenner, Steigerwalt, Fanning, Carter and Mollaret 1981b, 217[VP] (Effective publication: Bercovier, Ursing, Brenner, Steigerwalt, Fanning, Carter and Mollaret 1980c, 219.)

kris.ten.se' ni.i. M.L. gen. n. *kristensenii* of Kristensen, named after the Danish microbiologist Martin Kristensen, who first isolated this organism.

The characteristics are as described for the genus and as listed in Tables BXII.γ.281 and BXII.γ.282. Growth is delayed (7 d) when cultures are incubated at 41°C and even at 37°C for some isolates. Some strains utilize citrate (Simmons) after 7-d incubation at 25°C. Most strains produce a "musty" or "cabbage-like" odor when grown on nutrient agar. Some strains produce an enterotoxin (ST) when incubated at 22°C and also at 37°C (Kapperud, 1980). *Y. kristensenii* strains have been isolated mainly from environmental sources such as soil, fresh water, foods, and rarely from healthy human and animal carriers. *Y. kristensenii* may occasionally be associated with human enteritis (Bottone and Robin, 1977), and *Y. kristensenii* isolates may be lethal to mice pretreated with iron dextrin (Robins-Browne et al., 1991).

The mol% G + C of the DNA is: 48.5 ± 0.5 (T_m, Bd).

Type strain: 105, ATCC 33638, CIP 80-30.

TABLE BXII.γ.285. Spectrum of *Yersinia enterocolitica* infections

Type of infection	Manifestation/population
Gastrointestinal	Enterocolitis, predominantly in young children; concomitant bacteremia may also be present in infants
	Pseudoappendicitis syndrome (children older than 5 years; adults), acute mesenteric lymphadenitis; terminal ileitis
Septicemia	Especially in immunosuppressed individuals and those in iron overload or being treated with deferrioxamine; transfusion related (usually leads to septic shock syndrome)
Metastatic	Focal abscesses: liver, kidney, spleen, lung; cutaneous manifestations: cellulitis, pyomyositis, pustules, and bullous lesions; pneumonia, cavitary pneumonia; meningitis; panophthalmitis; endocarditis, infected mycotic aneurysm; osteomyelitis
Postinfection sequelae	Arthritis (associated with HLA-B-27), myocarditis, glomerulonephritis, erythema nodosum

8. **Yersinia mollaretii** Wauters, Janssens, Steigerwalt, and Brenner 1988, 427[VP]

mol.la.re' ti.i. M.L. gen. n. *mollaretii* in honor of Henri H. Mollaret, head of the National *Yersinia* Center at the Pasteur Institute in Paris, for his years of study on the classification and epidemiology in the genus *Yersinia.*

Formerly called *Y. enterocolitica* biogroup 3A (Bercovier et al., 1978). Later transferred to *Y. enterocolitica* biogroup 6 (Wauters et al., 1987) because of negative Voges-Proskauer reaction (typical serotype 0:3 strains are Voges–Proskauer positive). *Y. mollaretii* must be distinguished from rarely occurring pathogenic strains of serogroup 0:3 that are Voges–Proskauer-negative variants of biogroup 3 (Kaneko and Maruyama, 1987). The characteristics are as described for the genus and as listed in Tables BXII.γ.281 and BXII.γ.282. The species conforms to the definition of the family *Enterobacteriaceae* and the genus *Yersinia. Y. mollaretii* strains have been isolated from human stool specimens, meat, raw vegetables, soil, and drinking water, and lack virulence markers. There is no evidence of pathogenicity for humans.

The mol% G + C of the DNA is: 50–51 (T_m).

Type strain: ATCC 43969, CDC 2465-87.

9. **Yersinia pseudotuberculosis** (Pfeiffer 1889) Smith and Thal 1965, 220[AL] (*Bacillus pseudotuberculosis* Pfeiffer 1889, 5; *Yersinia pseudotuberculosis* subsp. *pseudotuberculosis* (Pfeiffer 1889) Smith and Thal 1965, 220.)

pseu.do.tu.ber.cu.lo' sis. Gr. adj. *pseudes* false; M.L. fem. n. *tuberculosis* tuberculosis; M.L. gen. n. *pseudotuberculosis* of false tuberculosis.

The characteristics are as described for the genus and as listed in Tables BXII.γ.281 and BXII.γ.282. Some freshly isolated strains may require subculturing before expressing their motility. Strains belonging to serogroup IV are citrate positive (Simmons) and malonate positive. Up to 5% of *Y. pseudotuberculosis* strains have been reported to produce acid from adonitol. Some strains, mostly of serogroup III, produce an exotoxin that differs from the *Y. pestis* toxin. Its biological activity is not well defined. *Y. pseudotuberculosis* is a human and animal pathogen responsible for mesenteric lymphadenitis, diarrhea, and septicemia. The disease can be reproduced experimentally in guinea pigs challenged per os and in mice.

The mol% G + C of the DNA is: 46.5 (T_m).

Type strain: ATCC 29833, DSM 8992, NCTC 10275.

Additional Remarks: The type strain belongs to serogroup I.

10. **Yersinia rohdei** Aleksic, Steigerwalt. Bockemühl, Huntley-Carter, and Brenner 1987, 330[VP]

roh' de.i. M.L. gen. n. *rohdei* of Rohde; named in honor of the late Rolf Rohde, who founded the National Reference Center for *Salmonella* in Hamburg, Federal Republic of Germany, and who made many significant contributions to the diagnostic and serological identification of *Enterobacteriaceae*, especially *Salmonella.*

The characteristics are as described for the genus and listed in Tables BXII.γ.281 and BXII.γ.282. Two biovars are recognized based on positive (biovar 1) or negative (biovar 2) reactions for both raffinose and melibiose. *Y. rohdei* strains were isolated from the feces of dogs and humans and from surface water. The clinical significance of *Y. rohdei* as a diarrheal agent in humans and animals is uncertain.

The mol% G + C of the DNA is: 49.1 (T_m).

Type strain: H271-36/78, ATCC 43380.

Additional Remarks: The type strain was isolated from dog feces in Germany in 1978.

11. **Yersinia ruckeri** Ewing, Ross, Brenner and Fanning 1978, 37[AL]

ru' cker.i. M.L. gen. n. *ruckeri* of Rucker; named after R.R. Rucker, who studied the red mouth disease and its etiological agents.

The characteristics are as described for the genus and as listed in Tables BXII.γ.281 and BXII.γ.282. The cells are 1×2–3 μm. Filaments can be seen in old cultures (48 h at 22°C). Colonies on nutrient agar are smooth, circular, and slightly raised. Growth is delayed or inhibited on Salmonella–Shigella agar incubated at 37°C but not at 22°C. Corn oil is hydrolyzed when the test is performed at 22°C but not at 37°C. *Y. ruckeri* is one of the agents responsible for red mouth disease in rainbow trout. The disease can be transmitted experimentally from fish to fish. The organism was initially isolated only in North America, but reports of enteric red mouth disease caused by *Y. ruckeri* have also assumed increasing importance in fish farms in Europe (Home et al., 1984). First isolated in 1959 from a dead muskrat found in a marshy area at the Bear River Migratory Bird Refuge in northern Utah. Other strains have been isolated from water in the same area.

The mol% G + C of the DNA is: 48 ± 0.5 (Bd).

Type strain: ATCC 29473.

GenBank accession number (16S rRNA): X75275.

Genus XLII. **Yokenella** *Kosako, Sakazaki, and Yoshizaki 1985, 224[VP] (Effective publication: Kosako, Sakazaki, and Yoshizaki 1984, 124)*

J.J. FARMER III AND FRANCES W. BRENNER

Yoken.el' la. dim. ending *-ella;* Japanese abbreviation "*Yoken*" that stands for the National Institute of Health, Tokyo, Japan; M.L. fem. n. *Yokenella* pertaining to Yoken, the National Institute of Health, Tokyo, Japan where the new group of organisms was recognized and studied.

Rod-shaped cells, conforming to the general definition of the family *Enterobacteriaceae.* Contain the enterobacterial common antigen. Gram negative. Motile, with peritrichous flagella. Facultatively anaerobic. Catalase positive (weak). Oxidase negative.

Ferment, rather than oxidize, D-glucose and other carbohydrates. Reduce nitrate to nitrite.

Positive for methyl red, citrate utilization (Simmons), lysine decarboxylase, ornithine decarboxylase, motility at 36°C, and

growth in the presence of cyanide (KCN test), β-galactosidase (ONPG test), tetrathionate reductase, and the fermentation of D-mannitol, L-arabinose, L-rhamnose, maltose, D-xylose, trehalose, cellobiose, melibiose, D-mannose, D-galactose, L-fucose, and D-gluconate. Produce visible gas during fermentation.

Negative for indole production, Voges–Proskauer, H₂S production (TSI), urea hydrolysis, phenylalanine deaminase, arginine dihydrolase, malonate utilization, gelatin hydrolysis (22°C), lipase (corn oil and Tween 80), DNase, esculin hydrolysis, pectinase, chitinase, β-xylosidase, growth at 4°C, and the fermentation of lactose, sucrose, dulcitol, salicin, adonitol, *myo*-inositol, D-sorbitol, raffinose, α-methyl-D-glucoside, erythritol, D-arabitol, glycerol, and mucate.

Susceptible to nalidixic acid, sulfadiazine, gentamicin, kanamycin, tetracycline, and chloramphenicol (disk diffusion method on Mueller-Hinton agar); **usually resistant to colistin, penicillin, ampicillin, carbenicillin, and cephalothin**; variable susceptibility to streptomycin.

Ecological niches are not known, but has been **isolated from human clinical specimens**, insects, and water. **Probably an opportunistic pathogen** for humans. A **rarely isolated member of the *Enterobacteriaceae*.**

The mol% G + C of the DNA is: 58–59.3 (T_m).

Type species: **Yokenella regensburgei** Kosako, Sakazaki, and Yoshizaki 1985, 224 (Effective publication: Kosako, Sakazaki, and Yoshizaki 1984, 124.)

FURTHER DESCRIPTIVE INFORMATION

Different names for the same organism It is essential for the reader to understand that three vernacular names (*"Hafnia"* species 3, Enteric Group 45, and NIH biogroup 9) and one scientific name (*Koserella trabulsii*) have been used in the literature for the organism that is known today as *Yokenella regensburgei*. This is explained in the section Taxonomic Comments. Until recently, the names *Koserella* and *K. trabulsii* appeared in the literature, so we have inserted these names in parentheses if they were used in the original article.

Literature Since *Yokenella* and *Koserella* were described in 1984–85, there have been several reports in the literature. In addition to the two genus and two species names, *"Hafnia* species 3", "Enteric Group 45", and "NIH biogroup 9" should also be included in computerized literature searches, although these names are rarely used today.

Sources, strains, collections Strains of *Y. regensburgei* have been isolated from a variety of human clinical specimens and other sources. The original isolates of Kosako et al. (1984) were from human clinical specimens; urine (three isolates), wound (one), and abscess (one), and six isolates from insect intestine. The original collection of Hickman-Brenner et al. (1985a) included the type strain for *K. trabulsii* isolated from a wrist wound of a 28-year-old man. The other isolates were from human wounds (four), sputum (two), throat (one), stool (one), knee fluid (one), well water (one), and unknown (one). Since that original report, the CDC has received 10 additional human strains (three wound, two blood, two sputum, one urine, one lung biopsy, and one unknown). The first blood isolate was from a 35-year-old woman who had a bone marrow transplant. Group G streptococci and *Klebsiella pneumoniae* were also isolated. No further clinical information was given for the remaining strains.

Isolates from other sources Six of the original strains reported by Kosako et al. (1984) were originally isolated by F. Haas

(Germany) from the intestine of *Heteroptera* (*Pyrrhcoris apterus*). Cassel-Beraud and Richard (1988) studied stools of 88 insect eating bats (*Chaerephon pumila*), which yielded many different bacterial strains. Strains of *K. trabulsii* were identified among the 20 different species identified, most of which were *Enterobacteriaceae*.

Clinical significance Whether *Yokenella* strains can actually cause infections is still being evaluated. Systematic study and additional case reports are needed. Abbott and Janda (1994a) reported two cases that yielded *Y. regensburgei*; one was a 74-year-old man with a "septic knee", and the other was a 35-year-old immunocompromised woman without overt signs of sepsis but with a positive blood culture. Yague et al. (1989) described two cases (*K. trabulsii*) from Spain. The first case involved an isolate from the stool of a 4-year-old boy. *Campylobacter jejuni* was also isolated and was considered to be the causative agent. The second case involved an isolate from the urine of an 88-year-old man with fever and respiratory problems. The colony count was over 100,000 CFU/ml, but there was no evidence of a urinary tract infection. *K. trabulsii* was not considered significant in either case.

Other literature The literature includes reviews and taxonomic studies of new *Enterobacteriaceae* (Richard, 1989; Gilchrist, 1995; Pokhil, 1996; Aleksic and Bockemühl, 1999), and surveys or comparisons of the family *Enterobacteriaceae* for acid phosphatases (Thaller et al., 1995) and polyamines (Hamana, 1996).

Isolation, identification, culture preservation, phenotypic characterization Strain of *Y. regensburgei* are not difficult to grow, and are typical *Enterobacteriaceae* in most respects. See the chapter on the family *Enterobacteriaceae* for general information on growth, plating media, working and frozen stock cultures, media and methods for biochemical testing, identification, computer programs, and antibiotic susceptibility.

Biochemical reactions and differentiation from other Enterobacteriaceae Table BXII.γ.193 in the chapter on the family *Enterobacteriaceae* gives the results for *Y. regensburgei* in 47 biochemical tests normally used for identification (Farmer, 1999). There are no genus- or species-specific tests or sequences for the identification. The best approach is to do a complete set of phenotypic tests (growth on plating media, biochemical tests, antibiogram, etc.) and compare the results with all the other named organisms in the family *Enterobacteriaceae*. This approach is described in more detail in the chapter on the family *Enterobacteriaceae*. The biochemical results of a test strain can be compared to all the organisms listed in Table BXII.γ.193 in that section. Several computer programs greatly facilitate analyzing the results. Biochemically, *Yokenella regensburgei* is closest to *Hafnia alvei*. These two organisms can be differentiated because *Y. regensburgei* is Voges–Proskauer and glycerol negative; citrate (Simmons), cellobiose, and melibiose positive; and resistant to colistin and to the *Hafnia*-specific bacteriophage of Guinée and Valkenburg (1968). It also gives a weak, delayed, catalase reaction. These are in contrast to the reactions of *Hafnia alvei*.

TAXONOMIC COMMENTS

Today there is general agreement on the use of the names *Yokenella* and *Yokenella regensburgei*. However, the names *Koserella* and *Koserella trabulsii* also appear in the literature. Table BXII.γ.192 in the chapter on the family *Enterobacteriaceae* summarizes how these organisms were discovered, named, and how *Yokenella regensburgei* eventually displaced *Koserella trabulsii*.

Agreement to use *Yokenella* **rather than** *Koserella* Although the CDC workers did not agree with the manner in which Kosako and Sakazaki (1991) proceeded in trying to solve the nomenclatural problem of priority, they decided to use *Yokenella* and *Y. regensburgei* instead of *Koserella* and *K. trabulsii*. In McWhorter et al. (1991), one of us (JJF) wrote: "We now acknowledge that *Y. regensburgei* has priority over *K. trabulsii* because *Y. regensburgei* was published first and gained standing in nomenclature in the same issue of the *International Journal of Systematic Bacteriology* as *K. trabulsii*. In a nomenclatural sense, *K. trabulsii* should now be considered a 'junior subjective synonym' of *Y. regensburgei* and thus an illegitimate name. We will now use the name *Y. regensburgei* instead of the name *K. trabulsii*."

McWhorter et al., 1991 conceded that only quirks in the validation process had given priority to *Koserella* over *Yokenella*, and agreed that since the names *Yokenella* and *Yokenella regensburgei* were published first and should have appeared first in a validation list, they should be used. Publications, databases, computer programs, and biochemical charts of CDC have reflected this use of *Yokenella* rather than *Koserella*. For example, compare Table 1 of Farmer (1999) that used *K. trabulsii* with Table 1 of Farmer (1995) that used *Y. regensburgei*. This informal acceptance resolved the issue as far as usage; however, only an Opinion of the Judicial Commission in the future can resolve the nomenclatural problem described below.

Nomenclatural problem — do *Yokenella* **and** *Y. regensburgei* **have priority over** *Koserella* **and** *K. trabulsii?* Under the rules of the Bacteriological Code, the answer to this question is unclear. Kosako and Sakazaki (1991) argued clearly and logically for their priority, but this was clearly an opinion of Kosako and Sakazaki, which has no judicial standing under the Bacteriological Code. Unfortunately, they did not end the title of their paper with "Request for an Opinion" as previously agreed to resolve the problem of priority (Kosako et al., 1987). The rules of the Bacteriological Code are interpreted differently by different people, and their application in unique circumstances is also a matter of interpretation. Thus, the question of priority remains as only a small nomenclatural problem because the matter of usage was settled in the early 1990s.

FURTHER READING

Abbott, S.L. and J.M. Janda. 1994. Isolation of *Yokenella regensburgei* (*Koserella trabulsii*) from a patient with transient bacteremia and from a patient with a septic knee. J. Clin. Microbiol. *32*: 2854–2855.

Hickman-Brenner, F.W., G.P. Huntley-Carter, G.R. Fanning, D.J. Brenner and J.J. Farmer. 1985. *Koserella trabulsii*, a new genus and species of *Enterobacteriaceae* formerly known as enteric group-45. J. Clin. Microbiol. *21*: 39–42.

Kosako, Y., R. Sakazaki and E. Yoshizaki. 1984. *Yokenella regensburgei* gen. nov., sp. nov.: a new genus and species in the family *Enterobacteriaceae*. Japan J. Med. Sci. Biol. *37*: 117–124.

List of species of the genus Yokenella

1. **Yokenella regensburgei** Kosako, Sakazaki, and Yoshizaki 1985, 224[VP] (Effective publication: Kosako, Sakazaki, and Yoshizaki 1984, 124.)

   re.gens.bur'ge.i. M.L. gen. *regensburgei* pertaining to Regensburg, Germany, where the type strain of *Y. regensburgei* and five other strains in the original paper (Kosako et al., 1984) were isolated from insects.

   The characteristics are as given for the genus. Isolated from human clinical specimens (blood, urine, feces, throat and sputum, wounds, abscess, and others), insect intestines, and well water. Its clinical significance is not known, but if it is pathogenic for humans it is probably a very weak pathogen. However, it is also possible that it is colonizing rather than infecting body sites that are normally not sterile. There is no evidence that it can cause diarrhea or intestinal infections. It should be considered a rarely isolated species of *Enterobacteriaceae* that is probably an opportunistic pathogen for humans. The type strains was isolated by F. Haas from the intestine of an insect (*Pyrrhocoris apterus*) in Regensburg, Germany.

   The mol% G + C of the DNA is: 58–59.3 (T_m).

   Type strain: ATCC 49455, CDC 3020-86, NIH 725-83, JCM 2403.

   Additional Remarks: The American Type Culture Collection includes the five strains of *Y. regensburgei* that were originally deposited by Hickman-Brenner et al. (1985a) as *Koserella trabulsii*. Four were from human clinical specimens and one was from well water.

Order XIV. **Pasteurellales** *ord. nov.*

GEORGE M. GARRITY, JULIA A. BELL AND TIMOTHY LILBURN

Pas.teu.rel.la'les. M.L. fem. n. *Pasteurella* type genus of the order; -*ales* suffix to denote order; M.L. fem. n. *Pasteurellales* the *Pasteurella* order.

The order *Pasteurellales* was circumscribed for this volume on the basis of phylogenetic analysis of 16S rDNA sequences; the order contains the family *Pasteurellaceae*.

Description is the same as for the family *Pasteurellaceae*.

Type genus: **Pasteurella** Trevisan 1887, 94 (Nom. Cons. Opin. 13, Jud. Comm. 1954b, 153.)

Family I. **Pasteurellaceae** Pohl 1981b, 382[VP] (Effective publication: Pohl 1979, 81)

INGAR OLSEN, FLOYD E. DEWHIRST, BRUCE J. PASTER AND HANS-JÜRGEN BUSSE

Pas.teu.rel.la' ce.ae. M.L. fem. n. *Pasteurella* type genus of the family; *-aceae* suffix to denote a family; M.L. fem. pl. n. *Pasteurellaceae* the *Pasteurella* family.

Cells are straight, rigid, **coccoid to rod-shaped**, usually **0.2–0.4 × 0.4–2.0 μm**, nonsporeforming, **Gram negative**, and **nonmotile**. Pleomorphism with cell swelling and formation of filaments can be seen. Demethylmenaquinones, menaquinones, and ubiquinones may or may not be produced. Organisms are **aerobic, with varying degrees of microaerophilia, or facultatively anaerobic.** Both **respiratory and fermentative forms of metabolism** can be found among these organisms, which are **chemoorganotrophic**. Optimal growth temperature is 37°C. Acid is produced through fermentation of D-glucose, other carbohydrates, sugar alcohols, or glycosides. Conventional fermentation test media may not detect accumulation of acid fermentation products from some of the most fastidious species. Usually anaerogenic, but gas-producing species do occur. Fumarate is usually used as a terminal electron acceptor in demethylmenaquinone-mediated anaerobic respiration. Characteristically, **oxidase, catalase, and alkaline phosphatase reactions are positive**, but negative reactions can be detected in some species. **Nitrates are reduced to nitrites.** For primary isolation, complex media supplemented with yeast extract, serum, or whole blood lysate are required. Furthermore, organic nitrogen sources, amino acids, B vitamins, nicotinamide adenine nucleotides, hematin, or protoporphyrin may be needed. Members of the family are **obligate parasites in vertebrates (primarily mammals and birds)**. They inhabit mucous membranes of the alimentary, genital, or respiratory tracts and are often pathogenic in humans, other mammals, birds, and reptiles. Species are generally susceptible to benzylpenicillin and other β-lactam antibiotics, inhibitors of 70S ribosomal protein biosynthesis, as well as sulfonamides, trimethoprim, erythromycin, and colistin. The genome size is 1.2–2.2×10^9 Da.

The mol% G + C of the DNA is: 37–45.

Type genus: **Pasteurella** Trevisan 1887, 94 (Nom. Cons. Opin. 13, Jud. Comm. 1954b, 153.)

Circumscription, position and rank Sneath and Johnson (1973), who made numerical phenetic analyses on a restricted number of species, were probably the first authors to indicate that the genera *Actinobacillus*, *Haemophilus*, and *Pasteurella* could form a family. Mannheim et al. (1980) suggested from genetic studies that the genera *Actinobacillus*, *Haemophilus*, and *Pasteurella* should rank as a family, although some species were misclassified and should be removed. In 1979, Pohl proposed the name *Pasteurellaceae* for the family and the maintenance of the three genera as distinct entities (Pohl, 1979). From the results of DNA reassociation experiments, Pohl (1981a) suggested that some rearrangement of the species in the various genera was indicated. Single-linkage clustering of the DNA hybridization data revealed complex intra- and intergeneric relations at or above the 30% DNA–DNA reassociation level. Furthermore, phylogenetic studies based on 16S rDNA sequencing concluded that taxonomic division of the family into phylogenetically and phenotypically coherent genera was problematic (Dewhirst et al., 1992, 1993, 1995; Paster et al., 1995). Nevertheless, such studies contributed to clarification of the taxonomic position of several taxa within the *Pasteurellaceae* (Holmes, 1998).

The genus *Lonepinella* was effectively published (Osawa et al., 1995b) and validated (Osawa et al., 1996) as a fourth genus of the family. This genus was described to encompass isolates from the feces of koala that degrade a tannin–protein complex. So far, the genus consists of only one species, *Lonepinella koalarum*. 16S rDNA sequence analyses have clearly demonstrated that the genus *Lonepinella* clusters within the family *Pasteurellaceae*.

The genus *Mannheimia*, previously the *P. haemolytica* complex, has been validly published as a fifth genus (Angen et al., 1999a). Seven species were identified, five of which were formally named.

The genus *Phocoenobacter* has been validly published as a sixth genus in the family *Pasteurellaceae* and contains the single species *Phocoenobacter uteri* (Foster et al., 2000).

rRNA–DNA hybridization studies showed that the family *Pasteurellaceae* forms one cluster that is linked with both branches of the genus *Alteromonas*, the family *Enterobacteriaceae*, the family *Vibrionaceae*, and the family *Aeromonadaceae* (De Ley et al., 1990). All these clusters constitute superfamily I *sensu* De Ley which together with superfamily II corresponds to the *Gammaproteobacteria*.

16S rDNA sequencing of 54 strains (Dewhirst et al., 1992) and subsequently of 70 strains (Dewhirst et al., 1993) confirmed that the family *Pasteurellaceae* is located in the *Gammaproteobacteria* and demonstrated that the family *Pasteurellaceae* forms a compact phylogenetic group that is distinct from neighboring families. The sequence signatures reported previously for the family (Dewhirst et al., 1992) are still useful, particularly 667 A, 739 U, 831 G, and 855 C. However, if a full sequence is available, it is much better to place the sequence in a phylogenetic tree with appropriate outgroups than to examine only a few signature bases. The DNA probe 5′-ACCAACTACCTAATCCCACTTGGG-3′ hybridizes with DNA (at ~80%) from most of the members of the family, but not with members of other taxa, so the specificity is 100% (Dewhirst et al., 1992).

The phenotypic traits that distinguish the *Pasteurellaceae* from other families and genera are presented in Table BXII.γ.286. The dearth of clearly distinguishing phenotypic traits is a reason that phylogenetic placement at the genus and higher level must rely heavily on molecular sequence data. Nonmotile and nutritionally fastidious strains of *Enterobacteriaceae*, *Bacillus*, or aerotolerant clostridia, as well as nonfermentative organisms, can be mistaken for species of *Pasteurellaceae* (Mannheim, 1984).

The cellular fatty acid content of *Pasteurellaceae* is uniform, with only minor variations. Separation from the *Neisseriaceae* and from *Moraxella* is possible (Mutters et al., 1993). In addition, the distribution of phospholipids is uniform within *Pasteurellaceae*. Cellular carbohydrates show a common pattern within all members of the family, but characteristic profiles discriminate between groups, often to the species level (Mutters et al., 1993).

Respiratory quinones had been previously considered useful in defining the family. When selected members of the family *Pasteurellaceae* were investigated for their lipoquinone contents using high-performance liquid chromatography (Kroppenstedt and Mannheim, 1989), menaquinones (MK-7 or MK-8 or both), demethylmenaquinones (DMK-7 or DMK-8 or both) and ubiquinones (Q-7 or Q-8 or both) were detected in varying proportions. Similar results were obtained in another study concerning strains of *Pasteurella* that were grown under strictly aerobic conditions (Kainz et al., 2000). The view that *Pasteurellaceae* can be distinguished from the *Enterobacteriaceae* and other fermenting Gram-negative bacteria by lack of menaquinones can therefore not be supported. Although the situation is more complex than had previously been recognized, the distribution patterns of lipoquinone structural types and their isoprenologue analogs remain a reliable chemotaxonomic tool for these bacteria.

Genera and Taxonomic Problems There are six genera in the family *Pasteurellaceae*: *Pasteurella*, *Actinobacillus*, *Haemophilus*, *Lonepinella*, *Mannheimia* and *Phocoenobacter*. Phenotypic traits that are useful for distinguishing the genera are listed in Table BXII.γ.287. As can be seen by the distressingly large number of traits that are variable, there is little phenotypic consistency

within genera, even when the genera are restricted to monophyletic groups based on DNA–DNA hybridization or 16S rDNA sequence analysis. The phylogeny and taxonomy of the family are clearly problematic. Several approaches to elucidating taxonomy have been used, including numerical taxonomy (Sneath and Johnson, 1973; Olsen, 1993), DNA–DNA hybridization (Mannheim and Pohl, 1980; Mannheim et al., 1980; Mannheim, 1983; Escande et al., 1984), rRNA–DNA hybridization (De Ley et al., 1990), genetic transformation (Albritton et al., 1986), chemotaxonomy (Olsen, 1993), and content of cellular fatty acids (Mutters et al., 1993), quinones (Mannheim, 1981; Kroppenstedt and Mannheim, 1989); Engelhard et al., 1991, 1992; Mutters et al., 1993; Kainz et al., 2000), polyamines (Busse et al., 1997), and cellular carbohydrates (Mutters et al., 1993). While these methods have clearly advanced our understanding of the taxonomy of the *Pasteurellaceae* (Bisgaard, 1995), it is now recognized that 16S sequencing is the most useful method for determining phylogeny and unraveling taxonomy at the genus and higher levels

TABLE BXII.γ.286. Differentiation of *Pasteurellaceae* from other families[a]

Characteristic	Pasteurellaceae	Aeromonadaceae	Alteromonadaceae	Enterobacteriaceae	Vibrionaceae
Motility	−	+	+	D	[+]
Na+ required or stimulates growth	−	−	+	−	+
Parasitic in vertebrates	+	+	−	D	[−]
V-factor requirement	D	−	−	−	−
Organic nitrogen sources required	+	−	−	[−]	[−]
Mol% G + C of DNA	38–47	57–63	38–50	52–60	38–51

[a]For symbols see standard definitions; also [−], more then 70% negative; [+], more than 70% positive.

TABLE BXII.γ.287. Differential characteristics for members of the family *Pasteurellaceae*[a]

Characteristics	Haemophilus[b]	Actinobacillus[b]	Lonepinella[c]	Mannheimia[d]	Pasteurella[b]	Phocoenobacter[e]
Sticky colonies	−	d	−	−	−	−
Hemolysis on SBA	d	d	−	d	−	−
V-factor dependency	+	−[f]	−	−	−	−
X-factor dependency	+	−	−	−	−	−
Urease	d	+	−	−	−[g]	−
Ornithine decarboxylase	−[h]	−	−	d	d	−
Indole	d	−	−	d	+	−
α-Fucosidase	−	d	nd	d	−	−
Acid production from:						
D-Glucose	+	+	+	+	+	+
D-Mannitol	−	+[i]	−	+	d	−
D-Mannose	−	d	nd	−	+	−
Melibiose	−	d	+	d	−	−
i-Inositol	−	d	−	d	−	−
D-Sorbitol	−	d	−	d	d	−
Trehalose	−	d	nd	−	d	−

[a]For symbols see standard definitions; nd, not determined.
[b]*Sensu stricto* species only as shown in Fig. BXII.γ.215.
[c]Osawa (1992).
[d]Angen et al. (1999a).
[e]Foster et al. (2000).
[f]Biotype 1 strains of *A. pleuropneumoniae*, *A. indolicus*, *A. minor*, and *A. porcinus* are positive.
[g]*P. dagmatis* is positive.
[h]Some strains of *H. influenzae* are positive.
[i]*A. suis* is negative.

(Olsen and Woese, 1993), while DNA–DNA hybridization is the most useful at the species level.

Organisms of the family *Pasteurellaceae* infect most vertebrate species. In almost every vertebrate species examined carefully, multiple, often unique, *Pasteurellaceae* species have been recovered. Only a subset of these taxa has been formally named, but a large number of strains have found their way into culture collections, and many have been characterized and placed in tentative taxonomic groups. With the advent of 16S rDNA sequence analysis, it has been possible to rapidly examine strains representing these taxa and place them in an overall *Pasteurellaceae* phylogenetic tree. 16S rRNA-based phylogenies have previously been presented by a number of investigators (Dewhirst et al., 1992, 1993) but substantial new data exist, which are included in Fig. BXII.γ.215. At this time, over 230 16S rDNA sequences have been deposited in GenBank, and 114 nonredundant sequences were selected for inclusion in this tree. The neighbor-joining method (Saitou and Nei, 1987) was used for tree construction. Bootstrap values greater than 40% are shown. It can be seen from the tree that species named *Haemophilus*, *Actinobacillus*, and *Pasteurella* are intermixed throughout the tree. If the family is to be divided into phenotypically consistent monophyletic groups, then there will need to be about 20 genera in this family. The genus *Mannheimia* is probably the most distinct and well-defined cluster in the family. The delineation of almost every other cluster is somewhat subjective and in almost every case will require additional investigation before the cluster can be proposed as a genus. Because the taxonomy of the *Pasteurellaceae* is in the process of major revision, we felt it useful to present a phylogeny representing as much of the diversity of the family as possible, regardless of whether the taxa have been formally named.

Named genera

Actinobacillus sensu stricto. The genus *Actinobacillus sensu stricto* should be comprised of the following named species: *A. lignieresii*, *A. pleuropneumoniae*, *A. ureae*, *A. hominis*, *A. equuli*, and *A. suis*. *A. arthritidis* (Bisgaard Taxon 9), Bisgaard Taxon 8, and taxa represented by strains Kunstyr 570 (MCCM 00149), Bisgaard F64 (CCUG 28015), and SSI P585 (CCUG 19799) should be included. This is in general agreement with the proposal of Escande et al. (1984) and Mutters et al. (1989) that the genus *Actinobacillus* should be limited to *A. lignieresii*, *A. equuli*, *A. capsulatus*, *A. suis*, *A. hominis*, *A. pleuropneumoniae*, *A. arthritidis*, and Bisgaard's taxa 5 and 11. 16S rDNA sequencing data suggest addition of *H. parahaemolyticus* and *A. hominis* (Dewhirst et al., 1992, 1993). The type strain of *A. capsulatus* does not cluster with true actinobacilli (Dewhirst et al., 1992, 1993). See discussion under Capsulatus Cluster, below. Bisgaard Taxon 5 falls elsewhere in the tree.

Mannheimia. This genus was recently created from strains that had previously been classified as *Pasteurella haemolytica* (Angen et al., 1999a). Five species have been named: *Mannheimia haemolytica*, *M. glucosida*, *M. granulomatis*, *M. ruminalis*, and *M. varigena*. There are strains representing at least two additional *Mannheimia* species.

Lonepinella. A species isolated from koalas has been given the name *Lonepinella koalarum*.

Pasteurella sensu stricto. The genus *Pasteurella sensu stricto* should be limited to the species *P. multocida*, *P. dagmatis*, *P. canis*, and *P. stomatis*. The three *P. multocida* subspecies are nearly identical by 16S rDNA sequences. Avian species that were previously included by DNA–DNA hybridization studies appear to fall into a separate genus level cluster (see Avian Cluster, below). Bisgaard Taxon 13 strain CCUG 16497 appears to be *P. multocida*. A strain identified as *P. pneumotropica* (NCTC 10827) is not related to the type strain (Rodent Cluster) and appears to represent a novel species in the genus. *Pasteurella* species B branches substantially deeper than do other members of the genus but may warrant inclusion in the genus.

The quinone content of strains of the genus *Pasteurella*, including *P. avium*, *P. canis*, *P. dagmatis*, *P. gallinarum*, *P. langaa*, *P. multocida*, *P. stomatis*, *P. volantium*, "*Pasteurella leonis*", *Pasteurella* species A, and *Pasteurella* species B, has been reinvestigated (Kainz et al., 2000), with cells grown in liquid PPLO broth under strictly aerobic conditions. Based on the quinone composition, the species could be grouped into two clusters. Cluster I, consisting of the species *P. avium*, *P. gallinarum*, *P. volantium*, and *Pasteurella* species A, was characterized by the two predominant compounds ubiquinone Q-7 and Q-8. These species correspond to the Avian Cluster in Fig. BXII.γ.215. Cluster II, consisting of the species *P. canis*, *P. dagmatis*, *P. langaa*, *P. multocida*, *P. stomatis*, and *Pasteurella* species B, contained only ubiquinone Q-8 as the major compound and corresponds to *Pasteurella sensu stricto* of Fig. BXII.γ.215 (Mutters et al., 1993). These two clusters may be also distinguished by the presence of *sym*-norspermidine in *Pasteurella sensu stricto* (Busse et al., 1997), whereas this triamine has not been detected in representatives of the Avian Cluster in Fig. BXII.γ.215.

Haemophilus sensu stricto. In the previous edition of *Bergey's Manual*, a primary characteristic of the genus *Haemophilus* was X or V-factor requirement. It is now recognized that factor-requiring species are spread throughout the family. Based primarily on the DNA–DNA hybridization data of Burbach (1987), Mutters et al. (1989) proposed restricting the genus *Haemophilus sensu stricto* to include *H. influenzae*, *H. aegyptius*, *H. haemolyticus*, "*H. intermedius*", and *H. parainfluenzae*. The 16S rRNA tree suggests exclusion of *H. parainfluenzae*. As shown in Fig. BXII.γ.215, strains representing serotype e and f of *H. influenzae* appear distinct from the other serotypes of *H. influenzae*. The tree shows the positions of "*H. intermedius* subsp. *intermedius*", "*H. intermedius* subsp. *gazogenes*", and "*H. quentini*".

Phocoenobacter. A bacterium isolated from a porpoise has been given the name *Phocoenobacter uteri*.

Clusters that represent potential additional genera

Parainfluenzae Cluster. *H. parainfluenzae* does not belong in the genus *Haemophilus*, but rather forms its own cluster. The species is known to be heterogeneous and may represent multiple species.

Porcine Cluster. This cluster contains four taxa in two subclusters: *Actinobacillus minor*, *Haemophilus* "minor group" 202, *A. pleuropneumoniae*, and *H. paraphrohaemolyticus*.

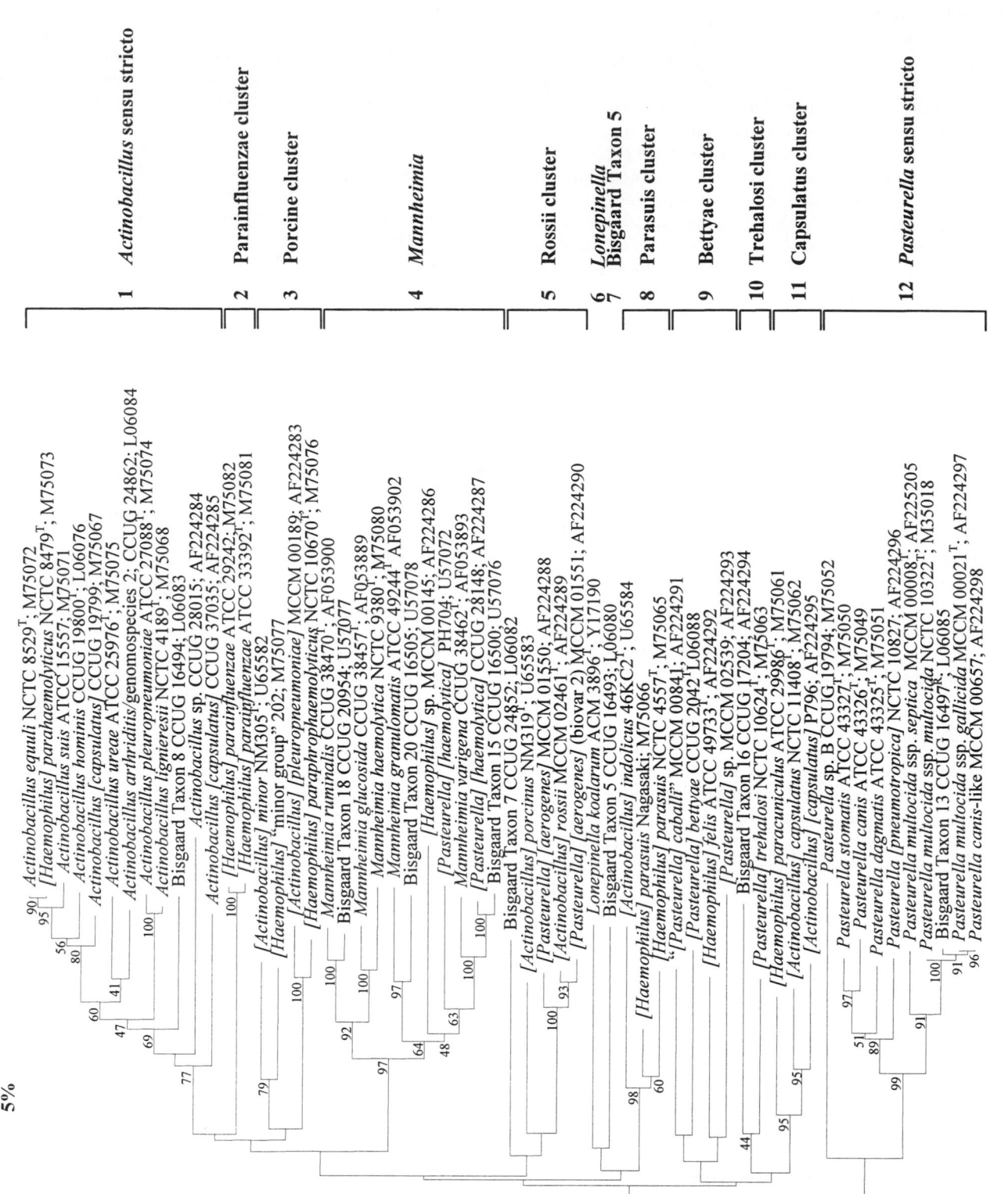

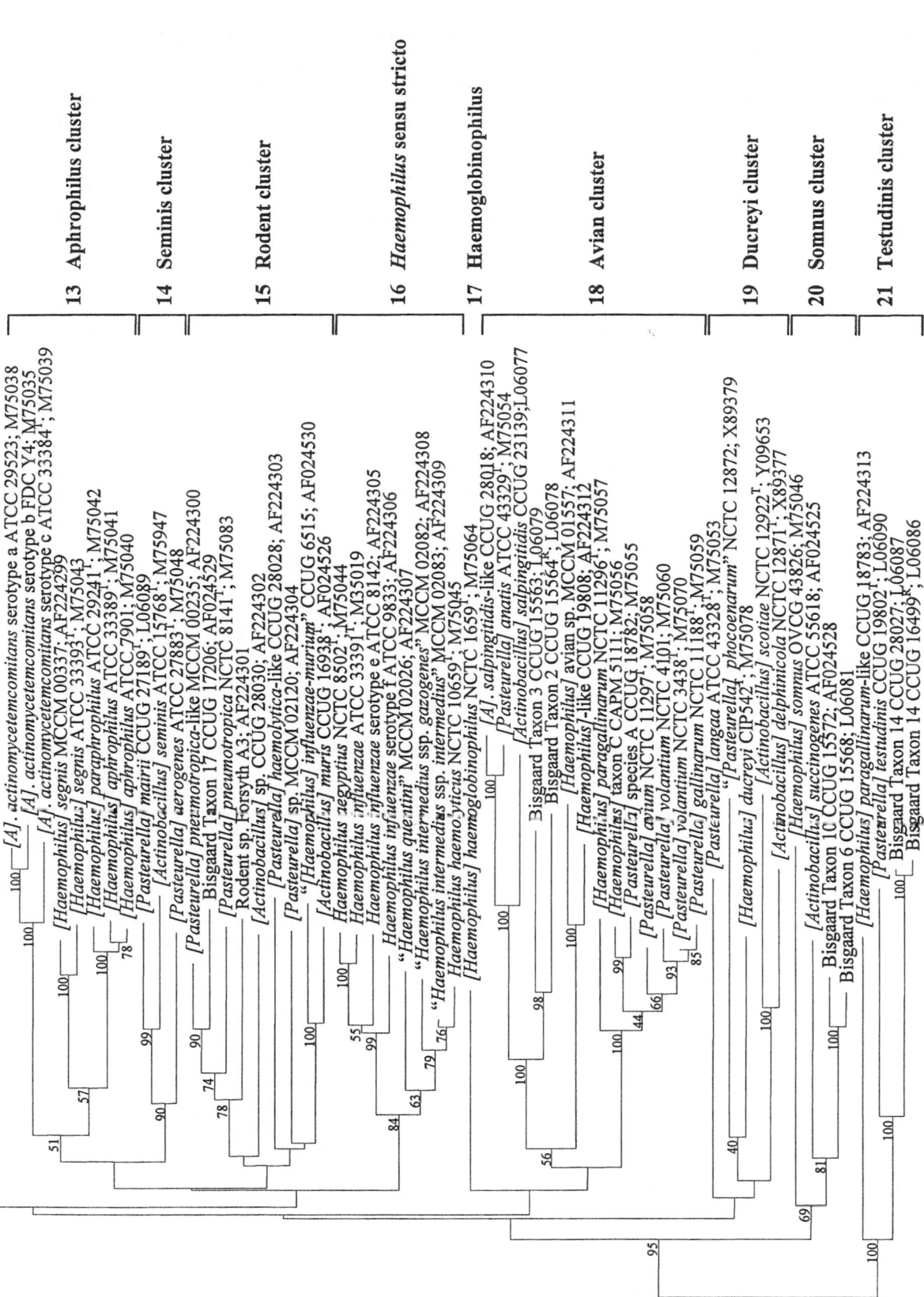

FIGURE BXII.γ.215. Phylogeny of the *Pasteurellaceae*. The family *Pasteurellaceae* forms a coherent phylogenetic subgroup within the *Gammaproteobacteria* that is divided into 21 phylogenetic clusters. Five of the clusters are represented by the named genera *Actinobacillus sensu stricto*, *Haemophilus sensu stricto*, *Pasteurella sensu stricto*, *Mannheimia*, and *Lonepinella*. "*Sensu stricto*" is defined herein as a distinct phylogenetic clade. The remaining clusters contain generically misclassified species of *Actinobacillus*, *Haemophilus*, and *Pasteurella*, as well as unnamed species of *Pasteurellaceae* from a variety of culture collections. A bracket around the genus or specific epithet denotes phylogenetic misclassification. The strain number and GenBank accession number of the 16S rRNA sequence are shown. Abbreviations for culture collections are as follows: ATCC, American Type Culture Collection, Manassas, VA; NCTC, National Collection of Type Cultures, London, England; MCCM, Medical Culture Collection of Microorganisms, Marburg, Germany; CCUG, Culture Collection, University of Göteborg, Göteborg, Sweden; CAPM, Collection of Animal Pathogenic Microorganisms, Brno, Czech Republic; Forsyth, The Forsyth Institute, Boston, MA; CIP, Collection de l'Institut Pasteur, Paris, France; ACM, Australian Collection of Microorganisms, Brisbane, Australia. Bar = 5% difference in nucleotide sequences. 200 bootstrap trees were generated, and bootstrap confidence levels (shown as percentages at the nodes) were determined.

Rossii Cluster. This cluster includes *Actinobacillus rossii*, *A. porcinus*, and strains that are considered part of the *Pasteurella aerogenes* complex, but do not fall with the type strain, which is in the Seminis Cluster. The Rossii Cluster may also include the more deeply branching Bisgaard Taxon 7, but this taxon is somewhat unstable in its branching position within the tree. See discussion for Bisgaard Taxon 5.

Bisgaard Taxon 5. The phylogenetic position of this taxon is problematic in that its position in the tree is unstable. Changes of taxa included in the analysis cause Bisgaard Taxon 5 to jump from place to place in the tree. Molecular data from other genes may be required to define its true position.

Parasuis Cluster. This cluster includes *H. parasuis*, *A. indolicus*, and a third taxon represented by *H. parasuis* strain Nagasaki.

Bettyae Cluster. This cluster includes *P. bettyae*, "*P. caballi*", *H. felis*, and guinea pig conjunctivitis strain Stenzel T 391 (HIM 1028-4).

Trehalosi Cluster. *P. trehalosi* forms a cluster with Bisgaard Taxon 16. It may be possible to join the Trehalosi and Capsulatus Clusters.

Capsulatus Cluster. This cluster includes *A. capsulatus*, *H. paracuniculus*, and a third taxon represented by *A. capsulatus* strain SSI P796. The taxonomic position of *A. capsulatus* has been controversial because DNA–DNA hybridization placed it with *Actinobacillus sensu stricto* (Escande et al., 1984; Mutters et al., 1989). However, sequences for five isolates (SSI P243^T, P244, P572, P662, and P796) all fall in the Capsulatus cluster. The horse isolate, SSI P585(CCUG 19799), falls in the middle of the *Actinobacillus sensu stricto* cluster, and if DNA from this strain was inadvertently used in place of that of the type strain (SSI P 243), it could explain the older DNA–DNA hybridization results. A hamster "*A. capsulatus*" strain, MCCM 149, branches with a duck isolate, Bisgaard F64 (CCUG 28015), to form the deepest branch of the *Actinobacillus sensu stricto* cluster. It would appear that *A. capsulatus*-like strains from species other than rabbit represent multiple unique species.

Aphrophilus Cluster. This cluster includes *H. segnis*, *H. aphrophilus*, *H. paraphrophilus*, and an *H. segnis*-like taxon represented by strain MCCM 00337. Based on rRNA–DNA hybridizations, De Ley et al. (1990) suggested that the Aphrophilus cluster be split off from the *Haemophilus sensu stricto*. We would suggest including *A. actinomycetemcomitans* in this cluster to be consistent with the DNA–DNA hybridization data of Potts et al. (1985).

Seminis Cluster. This cluster includes *A. seminis*, *P. mairii*, and *P. aerogenes*. Other strains of the *P. aerogenes* complex, such as the Rossii Cluster, fall elsewhere in the tree.

Rodent Cluster. This genus level cluster includes nine species, all isolated from rodents or guinea pigs. Named species include *P. pneumotropica*, *A. muris*, and "*H. influenzae-murium*". Also included are Bisgaard Taxon 17 (rat), strain Kunstyr 507 (MCCM 02120 [hamster]), Mannheim Michael A (CCUG 28028 [guinea pig]), Kunstyr 246 (CCUG 28030 [hamster]), Mannheim A/5a (MCCM 00235 [rat]), and Forsyth A3

(mouse). This large cluster clearly warrants further investigation for promotion to genus status.

Haemoglobinophilus Cluster. This cluster contains the single species *H. haemoglobinophilus*.

Avian Cluster. *P. volantium*, *P. gallinarum*, and *P. avium* were placed in the *Pasteurella sensu stricto* based on DNA–DNA hybridization (Pohl, 1981a), but by 16S rDNA sequence analysis, they fall into a large cluster of avian species quite distant from the *Pasteurella sensu stricto*. This avian cluster also includes "*Actinobacillus salpingitidis*", *P. anatis*, *H. paragallinarum*, Bisgaard Taxon 2, Bisgaard Taxon 3, *Haemophilus* taxon C, *Pasteurella* species A, and strains representing additional species such as Wuthe DS 1801 (MCCM 01557), "*A. salpingitidis*" CCUG 28018, and *Haemophilus* sp. CCUG 19808. Members of the subcluster consisting of *P. avium*, *P. gallinarum*, *P. volantium*, *Pasteurella* species A, and *H. paragallinarum* contain a polyamine pattern with the predominant compounds 1,3-diaminopropane and spermidine. Strains of Bisgaard Taxon 2 and Taxon 3 are characterized by polyamine patterns with the major compounds putrescine and spermidine (Taxon 2) and putrescine, cadaverine, and spermidine (Taxon 3). The third subcluster, consisting of two strains of "*A. salpingitidis*" and *P. anatis*, contain either the predominant compounds 1,3-diamonipropane and spermidine or additionally putrescine (Busse et al., 1997).

Ducreyi Cluster. This cluster includes *H. ducreyi*, *P. langaa*, and the three cetacean species *A. delphinicola*, *A. scotiae*, and "*P. phocoenarum*" (now *Phocoenobacter uteri*).

Somnus Cluster. This cluster includes "*Haemophilus somnus*", *A. succinogenes*, Bisgaard Taxon 6, and Bisgaard Taxon 10.

Testudinis Cluster. This cluster is the most deeply branching of the family *Pasteurellaceae*. It contains *P. testudinis*, Bisgaard Taxon 14, and the *H. paragallinarum*-like strain CCUG 18783. This cluster is distinct from the rest of the *Pasteurellaceae* in having the common *Proteobacteria* signature 665 A rather than the *Pasteurellaceae* signature 665 G.

The resolution of 16S rDNA sequence analysis may be inadequate to delineate the deeper phylogenetic branching in the *Pasteurellaceae*. Multi-gene sequence analysis has been used effectively to type species (Maiden et al., 1998). Similar methods can also be used to examine species- and genus-level relationships. With the complete genome of *H. influenzae* as well as extensive information for the unfinished genomes of *A. actinomycetemcomitans* and *P. multocida* available, it is reasonably easy to design primers to amplify and sequence essentially any chosen gene. Additional phylogenetic studies using genes such as *recA*, *dnaJ*, or *fur* may provide sufficient information to resolve the deeper branches in the family *Pasteurellaceae*. Phylogenetic analysis of genes other than 16S rRNA would also provide a valuable check on whether the 16S tree represents a consensus species tree, or only an evolutionary tree for one specific gene. Examination of local gene order may also be useful for defining genera. While molecular data will no doubt be the primary basis for resolving taxonomic issues, a polyphasic approach to describing the genera is essential. Promising phenotypic traits that could aid taxonomic analysis are polyamine patterns, quinone composition, and carbohydrate profiles.

Genus I. **Pasteurella** Trevisan 1887, 94[AL] Nom. Cons. Opin. 13, Jud. Comm. 1954b, 153

REINIER MUTTERS, HENRIK CHRISTENSEN AND MAGNE BISGAARD

Pas.teu.rel'la. M.L. dim. fem. n. *Pasteurella* named after Louis Pasteur.

Coccobacilli or rods, generally, 0.3–1.0 × 1.0–2.0 μm. Depending on the growth stage, cells occur singly, in pairs, or less frequently in short chains. Sometimes threads or filaments are formed resulting in marked pleomorphism. Pleomorphism occurs mainly in old cultures. **Gram negative,** although **bipolar staining often can be observed.** In tissues, *P. multocida* often shows bipolar staining with Giemsa or Wright's stain. Not acid fast. **Endospores are not formed. Nonmotile. Aerobic to microaerophilic or facultatively anaerobic.** Chemoorganotrophic with both oxidative and fermentative types of metabolism. Electron transport system is cytochrome-based with oxygen, nitrate, or fumarate as the terminal electron acceptor. **Nitrate reductase is produced. Oxidase positive, alkaline phosphatase positive**, and almost always **catalase positive. Most species are V-factor and X-factor independent, but V-factor-requiring strains do occur.** Even after specific growth factors have been provided, **complex media may be required to obtain better growth.** Optimum growth temperature 35–37°C. Arginine dihydrolase and lysine decarboxylase negative. Gelatin is not liquefied within 48 h. No growth occurs in Simmons citrate medium. Acid is produced from D-glucose, D-galactose, D-fructose, D-mannose, and sucrose. Acid is not produced from L-sorbose, L-rhamnose, *m*-inositol, and adonitol. Parasitic in vertebrates, particularly mammals and birds. Genome molecular weights range from 1.4 × 10⁹ to 1.9 × 10⁹.

The mol% G + C of the DNA is: 37.7–45.9.

Type species: **Pasteurella multocida** (Lehmann and Neumann 1899) Rosenbusch and Merchant 1939, 85 (*Bacterium multocidum multocidum* (sic) Lehmann and Neumann 1899, 196; *Pasteurella gallicida* (Burrill 1883) Buchanan 1925, 414.)

FURTHER DESCRIPTIVE INFORMATION

Phylogeny Based on 16S rDNA sequence comparison of selected strains, the genus was separated into two phylogenetic groups (Dewhirst et al., 1992, 1993). The core group around the type species included the other species from mammalian hosts: *P. canis, P. stomatis, P. dagmatis*, and *Pasteurella* sp. B. The avian species *P. avium, P. gallinarum, P. volantium*, and *Pasteurella.* sp. A form a separate cluster together with the type strain of *Haemophilus paragallinarum*. Based on 16S rRNA sequences *P. langaa* and *P. anatis* were not phylogenetically related to the other *Pasteurella* species at the genus level.

Cell wall composition The cell wall and the lipopolysaccharide (LPS) composition of *P. multocida* were comparable to most Gram-negative bacteria and contained lipid A, 2-keto-3-deoxyoctonate, L-glycero-D-mannoheptose, glucose, and glucosamine, and might contain galactose, rhamnose, D-glycero-D-mannoheptose, and galactosamine (Rimler and Rhoades, 1989). Polyamine analysis showed members of the genus *Pasteurella* to possess 1,3-diaminopropane, putrescine, cadaverine, spermidine, and spermine, and for some species *sym*-norspermidine in the cell wall. The unusual triamine *sym*-norspermidine was found only in *P. multocida, P. canis, P. dagmatis, P. stomatis*, and *Pasteurella* sp. B, but not in other *Pasteurellaceae*. This character is specific for the 16S rRNA subcluster 12 of Olsen et al. (see Fig. BXII.γ.215 of the chapter describing the family *Pasteurellaceae*, this *Manual*). Species of *Pasteurella* in 16S rRNA cluster 18 contained 1,3-diaminopropane and spermidine. In accordance with the 16S rRNA

analysis, the polyamine pattern of the remaining species of *Pasteurella* diverged from these two groups. *P. anatis* diverged from species of 16S rRNA cluster 18 by high putrescine content (Busse et al., 1997). The isoprenoid quinone composition of true members of the genus *Pasteurella* exhibits high amounts of quinones with the chain length eight: ubiquinone Q-8, menaquinone MK-8, and demethylmenaquinone DMK-8 (Mutters et al., 1993).

The capsule of *P. multocida* is composed of carbohydrates and is very hydrophilic (Rimler and Rhoades, 1989). The biosynthesis and genetic background of capsule formation has been described recently (Boyce and Adler, 2000).

Hyaluronic acid has been reported in the capsular material and the biosynthesis pathway for hyaluronan has been described for *P. multocida* (DeAngelis, 1999).

Outer membrane proteins (OMPs) OMP analysis by SDS-PAGE has been reported for *P. multocida*. Specific OMP profiles of the 16 different serotypes were not given and it was found that *in vivo* propagation led to the expression of additional OMPs (Choi et al., 1989a). The expression of OMP by *P. multocida* was found to be influenced by iron and, in particular, molecules of 76, 84, and 94 kDa were expressed under iron-restricted conditions (Choi-Kim et al., 1991; Zhao et al., 1995). A 37.5-kDa porin of *P. multocida* affected bovine neutrophils (Galdiero et al., 1998). In *P. testudinis*, iron regulated OMPs were also identified (Snipes et al., 1995). Cross-reaction was found between iron-regulated OMPs in pigs and fowl (Zhao et al., 1995). Strains of *P. trehalosi* representing the four serotypes showed only four OMP profiles by SDS-PAGE electrophoresis; however, substantial diversity was observed between OMP profiles of *M. haemolytica* and *P. trehalosi* (Davies and Quirie, 1996), confirming the reclassification of the two taxa at the genus level (see *Mannheimia* chapter).

Monoclonal antibodies raised against different OMPs of *P. multocida* were able to specifically detect strains of capsular type D of *P. multocida* or of the species *P. stomatis, P. gallinarum, P. bettyae, Pasteurella* sp. B, and *P. canis* but not other Gram-negative bacteria, suggesting variable levels of epitope specificity (Marandi and Mittal, 1995, 1996). A relationship between *P. multocida, P. gallinarum, Haemophilus paragallinarum, P. volantium*, and *P. avium* was inferred through the binding of a polyclonal antibody raised against a putative porin of *P. multocida*. The *N*-terminal amino acid sequence confirmed relationships to other Gram-negative bacteria and to members of *Pasteurellaceae* in particular (Lubke et al., 1994; Hartmann et al., 1996).

Outer membrane proteins and capsular material including hyaluronic acid of serotype A have been characterized and are probably associated with virulence by interacting with host immune system (Rimler and Rhoades, 1989; Boyce and Adler, 2000; Hunt et al., 2000).

Lipopolysaccharides (LPS) LPS of 13 Heddleston serotypes of *P. multocida* all contained glucose, 2-keto-3-deoxyoctonate, and heptose. Two isomers of heptose were found in serotypes 2 and 5. Rhamnose was identified with LPS of serotype 9 and galactose found in all serotypes except serotype 11 (Rimler et al., 1984; Conrad et al., 1996). A strain of serotype 8 contained galactose also and heptose. The amino sugars galactosamine, glucosamine, glucosamine-6-phosphate, and 3-deoxy-*d*-manno-2-octylosonic acid were found in a strain of serotype 8 (Conrad et al., 1996). SDS-PAGE analysis of the serotypes of *P. multocida* showed similarities but not identity. Capsulation did not affect the LPS profile (Rimler, 1990). Purified LPS of *P. multocida* was found to cause suppurative airsacculitis, pleuritis, and pneumonia in turkeys (Kunkle and Rimler, 1998). LPSs of *P. multocida* capsular type A were found to affect the humoral and cell-mediated immune response (Maslog et al., 1999). Similar to the analysis of OMP, high homogeneity of LPS profiles was found within *P. trehalosi* by SDS-PAGE electrophoresis. (Davies and Quirie, 1996).

Fatty acids   The dominant fatty acids in *P. multocida* serotype 8 were 3-hydroxytetradecanoate and tetradecanoate (Conrad et al., 1996) and *P. multocida* could be distinguished from other Gram-negative bacteria by the presence of $C_{14:0\ 3OH}$ (Dees et al., 1981). The fatty acids $C_{14:0}$, $C_{16:1}$, $C_{16:0}$, $C_{14:0\ 3OH}$, $C_{18:2}$, $C_{18:1}$, and $C_{18:0}$ were found in human strains of *P. multocida* subsp. *multocida*, *P. multocida* subsp. *septica*, *P. canis*, *P. stomatis*, and *P. dagmatis*; however, a differentiation of these species was not possible on the basis of fatty acid analysis. (Holst et al., 1992). The fatty acid profiles of the species of the family *Pasteurellaceae* proved to be indistinguishable (Mutters et al., 1993). The growth medium was found to affect the fatty acid profiling of members of *Pasteurella* (Boot et al., 1999).

Fine structure   Type 4 fimbriae and the corresponding genes of representative serovars of *P. multocida* have been characterized (Ruffolo et al., 1997; Doughty et al., 2000). Pili of rigid and curly types have been observed on both capsulated and noncapsulated strains of *P. multocida* (Rebers et al., 1988; Isaacson and Trigo, 1995).

Colonial and cultural characters   On standard agar plates for fastidious Gram-negative bacteria like chocolate agar or Columbia blood agar, *Pasteurella* mostly appear as regular, smooth, convex, grayish, nontransparent, circular colonies. The colonies are generally small with a diameter of 0.5–2 mm after 24 h incubations at 37°C. β-hemolysis on bovine or ovine blood agar is not seen, but a greenish discoloration may occur. Occasionally, yellowish-pigmented colonies can be seen, a phenomenon that is commonly associated with strains of *P. dagmatis* and *P. canis*. Isolates of *P. multocida* from the respiratory tract of ruminants, pigs, and rabbits may form large, watery, mucoid colonies that may collapse after 48 h incubation. Growth in broth usually causes turbidity, but granular growth may occur.

Nutrition and growth conditions   Growth requirements of *P. multocida* have been reviewed by Rimler and Rhoades (1989). Cysteine, nicotinamide, pantothenate, and thiamine were required for growth. Apart from these specific growth factors, the different species of *Pasteurella* seem to exhibit variation in nutritional needs and preferred growth conditions. Systematic investigations based on international reference strains, however, do not exist.

V-factor requirement, usually described as nicotinamide adenine dinucleotide (NAD), has been reported for *P. avium*, *P. volantium*, *Pasteurella* sp. A (Mutters et al., 1985b), and *P. multocida* subsp. *multocida* (Krause et al., 1987). Occurrence of V-factor-independent isolates of *P. avium*, *P. volantium*, and *Pasteurella* sp. A have subsequently been reported by Bisgaard (1993), Bragg et al. (1997), and Bisgaard et al. (1999). However, none of these isolates was shown to belong to the respective species based on genetic characterization. Sources of V-factor and demonstration of V-factor requirement is discussed elsewhere (see *Haemophilus*

chapter). The temperature range for growth is 22–44°C, with an optimum growth temperature of 35–37°C. Increased carbon dioxide may be required for surface growth of certain isolates. Metabolic pathways have been investigated for *P. multocida* and reviewed by Rimler and Rhoades (1989).

Genetics   The genomic sequence was determined in the avian strain Pm 70 of *P. multocida* and found to be 2,257,487 bp long with 2014 predicted coding regions (May et al., 2001). Strains of *P. multocida* have been found to possess six (Liu et al., 1999; May et al., 2001) or five (Hunt et al., 1998) ribosomal operons. Similar, but not identical, genome structure, was determined by I-*Ceu*I restriction analysis; however, a few bovine strains diverged (Liu et al., 1999). Few genes have been characterized and those exclusively in *P. multocida*. The catabolic pathways for asparagine, histidine, leucine, lysine, and phenylalanine were absent from the *P. multocida* genome (May et al., 2001).

Virulence genes   Best understood is the formation and activity of the dermonecrotic toxin found in some representatives of capsular types A and D of *P. multocida* that cause atrophic rhinitis in pigs (Hunt et al., 2000). Other virulence factors are incompletely understood in *P. multocida*, the only species studied in more detail. Two gene regions showed homology to the filamentous hemagglutinin gene of *Bordetella pertussis*, coding for proteins important in host cell binding and immunity (Fuller et al., 2000; May et al., 2001). In addition, virulence genes were identified homologous to hemolysin, hemolysin-binding protein, secretion accessory protein, Ton-dependent transport, and adherence (Fuller et al., 2000). Genes of the RTX toxin gene family have not been found except in *P. aerogenes* (Kuhnert et al., 1997). The gene cluster coding for leukotoxin has been found in all serotypes of *P. trehalosi* and the sequence of *lktA* characterized in detail (Shewen and Wilkie, 1982; Burrows et al., 1993; Davies et al., 2001). Capsulation and piliation are potential virulence factors whose genes have been characterized (Boyce and Adler, 2000; Doughty et al., 2000).

Plasmids and transposons   Plasmids of 1.3–100 kb have been found in *P. multocida* (Hunt et al., 2000). Plasmids have been found related to resistance to streptomycin, sulfonamides, tetracycline, penicillins, kanamycin, and chloramphenicol (Hunt et al., 2000). The possession of identical nonconjugative R plasmids of *P. multocida* did not follow the clonal population structure (Ikeda and Hirsh, 1990). The only transposable element identified in *P. multocida* has been Tn5706 (Hunt et al., 2000), but Tn10 was able to integrate into the chromosome of *P. multocida* and is important in generating mutants (Lee and Henk, 1996).

Population genetics   The population structure of *P. multocida* and *P. trehalosi* have been concluded to be clonal as determined by MLEE (multilocus enzyme electrophoresis), with genetic diversity of between 0.289 and 0.474 (Davies et al. 1997; Blackall et al., 1998; May et al., 2001).

Antigenic structure   *P. multocida* has been the subject of numerous antigenic and serologic studies. However, only a few serological typing systems have gained wide acceptance. Four types (I, II, III, and IV) were recognized by Roberts (1947), based on passive protection of mice by serum against live challenge organisms. The most commonly used capsule typing system used so far was developed by Carter (1955). This typing system is based on passive hemagglutination of erythrocytes sensitized by capsule antigen, and five capsule types (A, B, D, E, and F) are presently distinguished (Rimler and Rhoades, 1987, 1989). Presumptive identification of capsular types A, D, and F by capsule depoly-merizations with mucopolysaccharidases has been reported by Rimler (1994). Subsequent cloning and sequencing of the entire capsular biosynthetic loci of *P. multocida* strains X-73 (A:1) (Chung et al., 1998b) and M 1404 (B:2) (Boyce et al., 2000), and nucleotide sequence analysis of the biosynthetic region from each of the remaining three capsule types D, E, and F, identified capsule-specific regions and allowed development of a highly specific multiplex capsular PCR assay (Townsend et al., 2001).

A total of 11 different O or somatic types of *P. multocida*, based on the use of acid-treated cells and agglutinin-absorption procedures, were reported by Namioka (1978). A more commonly used system for somatic antigen typing based on gel diffusion precipitin tests was developed by Heddleston et al. (1972). Sixteen serotypes (1–16) were recognized by Heddleston et al. (1972). Serological typing of *P. multocida* has been reviewed and discussed by Rimler and Rhoades (1989). Comparative studies by Brogden and Packer (1979) indicated that a serotype determination by one method did not correlate with serotyping by other methods. The relationship between subspecies and serovars of *P. multocida* obtained by published serotyping systems remains to be elucidated. Average linkage cluster analysis of precipitate values obtained by crossed immunoelectrophoresis revealed a close antigenic relationship between species of the genus *Pasteurella* as defined by Mutters et al. (1985a, b). Avian and mammalian species clustered separately, just as *P. anatis* and *P. langaa* made up a separate cluster (Schmid et al., 1991), confirming subsequent 16S rDNA similarity studies. The immunogens of *Pasteurella* have been reviewed by Confer (1993).

Antibiotic sensitivity Minimum inhibitory concentrations (MICs) of selected antimicrobial agents have been reviewed by Rimler and Rhoades (1989). Most human isolates of *Pasteurella* are susceptible to penicillin (Holst et al., 1992), the antibiotic of choice for local wound infections. Other β-lactam antibiotics should be effective as well, although in the case of the carbapenems, meropenem was more active than imipenem (Jorgensen et al., 1991). Macrolides and, especially for urinary and respiratory infections, fluoroquinolones such as ciprofloxacin appear to be useful in human and animal infections (Gaillot et al., 1995; Hanan et al., 2000). Most strains are also susceptible to tetracyclines (Holst et al., 1992). Marbofloxacin, a fluoroquinolone exclusively for veterinary use could be useful in the treatment of diseases in dogs and cats (Spreng et al., 1995). Tilmicosin, a macrolide antibiotic for use in veterinary medicine, proved to be effective in the treatment of pasteurellosis in rabbits (McKay et al., 1996).

Ecology With the exception of certain strains of *P. multocida*, organisms classified as *Pasteurella* are usually regarded as opportunistic pathogens that may colonize and form part of the indigenous flora of the mucous membranes of the upper respiratory and lower genital tracts.

Although numerous investigations have been carried out to determine the prevalence of these organisms, lack of sensitive and specific tests for their isolation and identification have prevented insight into their real distribution (Bisgaard, 1993). Previous investigations have mainly focused on the respiratory tract. Detection of a high proportion of cloacal carriers in poultry of *P. multocida* (Muhairwa et al., 2000) emphasizes the importance of sampling mucosal membranes other than the respiratory tract.

P. multocida is distributed worldwide among terrestrial as well as aquatic species of mammals and birds, and any species of these groups should be considered as a possible host. Although the host spectrum seems very large compared to other *Pasteurellaceae*, indications also exist that certain subtypes have developed, the disease potential of which seems to be limited to only a few species (see pathogenicity).

P. canis biovar 1, *P. stomatis*, *P. dagmatis*, and *Pasteurella* sp. B are mainly associated with the oral and nasal mucosa of dogs and cats (Bisgaard, 1993; Ganiere et al., 1993; Muhairwa et al., 2001). However, isolates of *P. dagmatis* from rodents and a scarlet macaw have also been reported (Bisgaard, 1993; Bisgaard et al., 1999). These isolates, however, remain to be characterized genetically.

P. gallinarum is normally associated with different pathological lesions in poultry, and its occurrence in healthy birds remains to be investigated (Bisgaard, 1993). A single isolate has been reported from a healthy duck (Muhairwa et al., 2001). Publications on isolation from animal species other than birds should be questioned (Bisgaard and Mutters, 1986a; Boot and Bisgaard, 1995; Frederiksen and Tønning, 2001). The natural habitats of the remaining species of *Pasteurella* belonging to the 16S rRNA cluster 18 of Olsen et al. (see Fig. BXII.γ.215 of the chapter describing the family *Pasteurellaceae*, this *Manual*) seem to be associated with birds only. Isolates of *P. avium* biovar 2 associated with pneumonia in cattle should be reinvestigated, since recent investigations have shown that nearly identical 16S rDNA sequences were observed for *P. multocida* and biovars 2 of *P. canis* and *P. avium* (Christensen et al., unpublished results). *P. multocida* and biovar 2 of *P. canis* and biovar 2 of *P. avium* also expressed similar OMP profiles (Abdullahi et al., 1990).

Pathogenicity Among species classified as *Pasteurella*, *P. multocida* has been recognized as an important veterinary pathogen for more than a century. It causes a wide range of diseases in animals. Capsule type A serotypes 1, 3, and 4 are recognized as the primary cause of fowl cholera in poultry and wild birds (Rhodes and Rimler, 1989). Disease may appear as an acute septicemia characterized by disseminated intravascular coagulation, petecchial or ecchymotic hemorrhages, multifocal necroses, and fibrinous pneumonia. Chronic infections may involve a variety of local infections (Christensen and Bisgaard, 1997, 2000). *P. multocida* capsule types B and E are associated with hemorrhagic septicemia of cattle, water buffaloes, and occasionally other species, resulting in major economic losses, mainly in Southeast Asia (Carter and De Alwis, 1989; De Alwis, 1995). Respiratory diseases in cattle, including bronchopneumonia in feedlot cattle and enzootic pneumonia of calves less than 6 months old, are mainly associated with capsule type A (Frank, 1989). Outbreaks of septicemia in fallow deer have also been reported (Eriksen et al., 1999). Infections of major economic importance in pigs include atrophic rhinitis and bronchopneumonia (Chanter and Rutter, 1989; Gardner et al., 1994). These syndromes are caused by capsular types A and D. Severe cases of atrophic rhinitis are mainly associated with capsule type D. Pulmonary lesions may result in blood-borne dissemination to the kidneys in pigs (Buttenschøn and Rosendal, 1990). In addition to major diseases in production animals, *P. multocida* is recovered from a wide range of sporadic infections in many other species, including laboratory animals (Manning et al., 1989), dogs and cats (Mohan et al., 1997), and other mammals (DiGiacomo et al., 1989).

Although *P. multocida* has challenged researchers for more than a century, mechanisms behind virulence and pathogenesis have remained unclear, as demonstrated in recent literature (Christensen and Bisgaard, 1997, 2000). Successful isolation of DNA fragments containing the toxin-encoding gene from toxigenic *P. multocida* and cloning these into *E. coli* (Petersen and Foged, 1989; Kamps et al., 1990; Lax and Chanter, 1990) significantly improved our understanding of the pathogenesis of atrophic rhinitis. Sequencing of the entire genome of a common avian clone of *P. multocida* (May et al., 2001) provides a foundation for future research into virulence factors and mechanisms of pathogenesis and host specificity of this pathogen that may kill affected animals within 24 h.

P. multocida and other species of the 16S rRNA cluster 12 are considered as zoonotic pathogens (Bisgaard et al., 1994). Most human infections associated with *Pasteurella* species result from animal bites. The species usually observed in these infections are *P. multocida* subsp. *multocida* and subsp. *septica*, *P. canis*, *P. dagmatis*, and *P. stomatis* (Holst et al. 1992; Escande and Lion, 1993; Matsui et al., 1996). Like other opportunistic pathogens, *Pasteurella* species can be associated with different clinical syndromes, i.e., necrotizing fasciitis (Hamamoto et al., 1995), chronic lung abscess (Machiels et al., 1995), endocarditis (Genne et al., 1996), meningitis (Boocock and Bowley, 1995; Armstrong et al., 2000), pulmonary diseases such as pneumonia (Ory et al., 1998), peritonitis (Wallet et al., 2000), septicemia (Greif et al., 1986), periocular abscess and cellulitis (Hutcheson and Magbalon, 1999), and granulomatous hepatitis (Chateil et al., 1998). Usually human infections result from inoculation of animal secretions via bites or direct contact with animals carrying pasteurellae in their normal bacterial flora.

ENRICHMENT AND ISOLATION PROCEDURES

Species of *Pasteurella* are somewhat fastidious and V-factor-requiring species are traditionally isolated on enriched agar media supplemented with 5% serum or blood and cross-inoculated with a *Staphylococcus* or an *Acinetobacter* strain to provide the V-factor. However, in mixed cultures pasteurellae may be overgrown. Strains of *P. multocida* can be obtained from contaminated material by subcutaneous or intraperitoneal inoculation of mice (Muhairwa et al., 2001). After death, cultures are made from the spleen by ordinary methods. Several selective media have been developed for isolation of *P. multocida* (Rimler and Rhoades, 1989; Moore et al., 1994b; Lee et al., 2000a), and transport media have been investigated (Kawamoto et al., 1997). Comparative investigations, however, remain to be performed.

Short-term survival of pasteurellae for about a week is possible on complex solid media like chocolate agar stored at room temperature or preferably at 4°C in a plastic jar or plastic bag to avoid desiccation. In the case of V-factor-dependent pasteurellae, shorter intervals should be used. Cultures will remain viable for years when frozen at −70°C in liquid media such as protease peptone broth containing 20% glucose or glycerol. Recommended procedures for long-term survival also include lyophilization or storage in liquid nitrogen of cells in nutrient broth containing 20% glucose or glycerol. Both techniques preclude selection of mutants that occur during repeated subculturing.

DIFFERENTIATION OF THE GENUS *PASTEURELLA* FROM OTHER GENERA

Pasteurellae (Table BXII.γ.288) share the common features of the family *Pasteurellaceae* (see chapter *Pasteurellaceae*). They do not grow in Simmons citrate medium. Nitrate reduction is positive. No acid is produced from D-adonitol and L-sorbose. They do not contain arginine dihydrolase, but do produce alkaline phosphatase. In contrast to *Yersinia* they are nonmotile at 22°C, and do not grow in the presence of 4.5% NaCl. Differentiation

from *Kingella* and other fastidious Gram-negative rods is possible based on several features, including positive tests for oxidase and catalase and acid production from sucrose and other carbohydrates. Distinction from other genera of the family *Pasteurellaceae*, especially from *Actinobacillus*, has been difficult based on the original concepts of these genera. Despite reallocation of species and outlining of new genera as indicated by Olsen et al. (see Fig. BXII.γ.215 of the chapter describing the family *Pasteurellaceae*, this *Manual*), available data indicate that it will still be difficult to divide the family *Pasteurellaceae* into genera that are phenotypically and phylogenetically coherent.

The pattern of typical biochemical characteristics of *Pasteurellaceae*, combined with the common features of *Pasteurella* (see Table BXII.γ.288), describes true pasteurellae. Although the taxonomy of the genus is still under investigation, some phenotypic features can be defined that are useful in discriminating *Pasteurella* from both other named genera and unnamed and not formally recognized genus-like groups within the family *Pasteurellaceae*. True members of the genus never exhibit hemolysis on ordinary media. Oxidase reaction and usually catalase reaction are positive. Normally, gas is not produced from the fermentation of carbohydrates; occasionally, very small amounts of gas can be detected. The sole *Pasteurella* species exhibiting gas production is *P. dagmatis*. Differences in urease and ornithine decarboxylation can also be used for separation of the genera *Pasteurella* and *Actinobacillus*. With the exception of *P. dagmatis*, all pasteurellae are urease negative, whereas all actinobacilli are positive and, with the exception of *P. dagmatis* and *P. stomatis*, all *Pasteurella sensu stricto* (cluster 12 of Olsen et al.; see Fig. BXII.γ.215 of the chapter describing the family *Pasteurellaceae*, this *Manual*) are ornithine positive, whereas actinobacilli are negative. Differences in indole can also be used for separation of genuine pasteurellae (positive) and actinobacilli (negative), but single indole-negative isolates of *P. multocida* occur. Certain carbohydrates, like L-rhamnose and *m*-inositol, are not cleaved. On the other hand, all pasteurellae produce acid from the six-carbon carbohydrates D-glucose, D-fructose, D-mannose, and D-galactose in contrast to many *Haemophilus* and *Actinobacillus* species. Differences in D-mannose and indole separate the genera *Mannheimia* and *Pasteurella*.

TAXONOMIC COMMENTS

In the mid-1980s, the taxonomy of the genus *Pasteurella* was studied by DNA–DNA hybridization (Mutters et al., 1985a, b). More recently, the genus was investigated in the context of the whole family by 16S rDNA sequencing (see chapter *Pasteurellaceae*). The results of the DNA hybridization studies led to the reclassification of the genus *Pasteurella sensu stricto*. According to this method, *Pasteurella* contains species interrelated at or above a DNA binding level of 55% based on the optical DNA hybridization method (cit. Mutters et al., 1985a). From the six *Pasteurella* species in the last edition of *Bergey's Manual of Systematic Bacteriology* only the type species of the genus, *Pasteurella multocida* and *P. gallinarum*, are recognized as true pasteurellae. *P. multocida* was subdivided in to three subspecies— subsp. *multocida*, subsp. *septica*, and subsp. *gallicida*—according to DNA-binding data and differences in acid production from D-sorbitol and dulcitol; 1.4–2% 16S rRNA sequence variation was found among *P. multocida* subsp. *septica* and *P. multocida* subsp. *multocida* and subsp. *gallicida*, while subsp. *multocida* and subsp. *gallicida* were nearly identical (Boerlin et al., 2000; Petersen et al., 2001). Partial *atpD* DNA sequence comparison confirmed the 16S rRNA results in that the subspecies

TABLE BXII.γ.288. Phenotypic characters of *Pasteurella* species[a]

Characteristic	Species classified with the16S rRNA cluster 12 (genus *Pasteurella sensu stricto*)					Species classified with the 16S rRNA cluster 18					
	P. multocida	*P. canis*	*P. dagmatis*	*P. stomatis*	*Pasteurella* sp. B	*P. anatis*	*P. avium*	*P. gallinarum*	*P. langaa*	*P. volantium*	*Pasteurella* sp. A
Gram stain	−	−	−	−	−	−	−	−	−	−	−
Motility, 22°C and 37°C	−	−	−	−	−	−	−	−	−	−	−
Catalase	+	+	+	+	+	+	+	+	−b	+	+
Oxidase	+	+	d	+	+	d	+	+	d	+	+
Glucose (Hugh and Leifson)	F	F	F	F	F	F	F	F	F	F	F
Symbiotic growth	−	−	−	−	−	−	+	−	−	+	+
Porphyrin test	+	+	+	+	+	+	+	+	+	+	+
β-Hemolysis (bovine blood)	−	−	−	−	−	−	−	−	−	−	−
Citrate (Simmons)	−	−	−	−	−	−	−	−	−	−	−
Mucate, acid	−	−	−	−	−	−	−	−	−	−	−
Malonate, base	−	−	−	−	−	−	−	−	−	−	−
H₂S/TSI	−	−	−	−	−	−	−	−	−	−	−
KCN, growth	−	−	−	−	−	−	−	−	−	−	−
Methyl-red, 37°C	−	−	−	−	−	−b	−	−	−	−	−
Voges–Proskauer, 37°C	−	−	−	−	−	−	−	−	−	−	−
Nitrate, reduction	+	+	+	+	+	+	+	+	+	+	+
Nitrate, gas	−	−	−	−	−	−	−	−	−	−	−
Urease	−	−	+	−	−	−	−	−	−	−	−
Alanine aminopeptidase	+	+	+	+	+	+	+	+	+	+	+
Arginine dehydrolase	−	−	−	−	−	−	−	−	−	−	−
Lysine decarboxylase	−	−	−	−	−	−	−	−	−	−	−
Ornithine decarboxylase	+	+	−	−	+	−	−	−	−	d	−
Phenylalanine deaminase	−	−	−	−	−	−	−	−	−	−	−
Indole	+	+	+	+	d	−	−	−	−	−	−
Phosphatase	+	+	+	+	+	+	+	+	+	+	+
Gelatinase	−	−	−	−	−	−	−	−	−	−	−
Tween 20	−	−	−	−	−	−	−	−	−	−	−
Tween 80	−	−	−	−	−	−	−	−	−	−	−
McConkey, growth	−	−	−	−	−	d	−	−	+	−	−
Pigment	−	−	−	−	−	−	−	−	−	−	−
Glycerol	d	−	d	−	−	(+)	−	d	−	−	d
meso-Erythrol	−	−	−	−	−	−	−	−	−	−	−
Adonitol	−	−	−	−	−	−	−	−	−	−	−
D(+)Arabitol	−	−	−	−	−	−	−	d	−	−	−
Xylitol	−	−	−	−	+	−	−	−	−	−	d
L(+)Arabinose	−	−	−	−	−	−	−	−	−	−	+/(+)
D(−)Arabinose	d	−	d	−	−	−	−	d	−	d	−
D(−)Ribose	+	+	+	+	+	+	+	+	+	(+)	+
D(+)Xylose	d	−	−	−	+	+	d	d	−	d	d
L(−)Xylose	−	−	−	−	−	−	−	−	−	−	−
Dulcitol	d	−	−	−	+	−	−	−	−	−	−
meso-Inositol	−	−	−	−	−	−	−	d	−	−	−
D(−)Mannitol	+	−	−	−	+	+	−	−	+	+	d
D(−)Sorbitol	d	−	−	−	−	−	−	−	−	d	−
D(−)Fructose	+	+	+	+	+	+	+	+	+	+	+
D(+)Fucose	−	−	−	−	−	−	−	−	−	−	−
L(−)Fucose	d	−	d	−	−	−	−	d	−	d	+
D(+)Galactose	+	+	+	+	+	+	+	+	+	+	+
D(+)Glucose, acid	+	+	+	+	+	+	+	+	+	+	+
D(+)Glucose, gas	−	−	−b	−	−	−	−	−	−	−	−
D(+)Mannose	+	+	+	+	+	+	+	+	+	+	+
L(+)Rhamnose	−	−	−	−	−	−	−	−	−	−	−
L(−)Sorbose	−	−	−	−	−	−	−	−	−	−	−
Cellobiose	−	−	−	−	−	−	−	−	−	−	−
β-Glucosidase (NPG)	−	−	−	−	−	−	−	−	−	−	−
Lactose	−	−	d	−	−	d	−	d	+	d	−
ONPG	−	−	d	−	−	d	−	d	+	+	d
Maltose	−	−	+	−	+/(+)	−	−	+	−	+	d
D(+)Melibiose	−	−	−	−	−	−	−	−	−	−	−
Sucrose	+	+	+	+	+	+	+	+	+	+	+
Trehalose	d	d	d	+	+	+	+	+	−	+	+
D(+)Melizitose	−	−	−	−	−	−	−	−	−	−	−
Raffinose	−	−	d	−b	−	d	−	d	−	−	(+)
Dextrin	−	−	+/(+)	−	+/(+)	−	−	+	−	+	d
D(+)Glycogen	−	−	−	−	−	−	−	−	−	−	−
Inulin	−	−	−	−	−	−	−	−	−	−	−
Esculin	−	−	−	−	−	−	−	−	−	−	−
Amygdalin	−	−	−	−	−	−	−	−	−	−	−

(*continued*)

TABLE BXII.γ.288. *(cont.)*

Characteristic	Species classified with the16S rRNA cluster 12 (genus *Pasteurella sensu stricto*)					Species classified with the 16S rRNA cluster 18					
	P. multocida	*P. canis*	*P. dagmatis*	*P. stomatis*	*Pasteurella* sp. B	*P. anatis*	*P. avium*	*P. gallinarum*	*P. langaa*	*P. volantium*	*Pasteurella* sp. A
Arbutin	−	−	−	−	−	−	−	−	−	−	−
Gentiobiose	−	−	−	−	−	−	−	−	−	−	−
Salicin	−	−	−	−	−	−	−	−	−	−	−
D(+)Turanose	−	−	−	−	−	−	−	−	−	−	−
β-*N*-CH₃-Glucosamid	−	−	−	−	−	−	−	−	−	−	−
α-Fucosidase (ONPF)	−	−	−	−	−	−	−	−	−	−	−
α-Galactosidase	−	−	−	−	−	−	−	−	−	−	−
α-Glucosidase (PNPG)	d	d	+	+	+	+	+	+	−	+	+
β-Glucuronidase (PGUA)	−	−	−	−	−	−	−	−	−	−	−
α-Mannosidase	−	−	−	−	−	−	−	−	−	−	−
β-Xylosidase (ONPX)	−	−	−	−	−	−	−	−	−	−	−

[a]Symbols: +, 90% or more of the strains positive within 1–2 d; (+), 90% or more of the strains positive within 3–14 d; −, less than 10% of strains positive within 14 d; d, 11–89% of the strains are positive; w, weak positive; F, fermentative; ONPG, *o*-nitrophenyl-β-D-galactopyranoside. Incubation temperature 37°C.
[b]Weak positive reaction might occur.

of *P. multocida* differed, with *P. multocida* subsp. *septica* diverged the most from the others (Petersen et al., 2001).

The former biotype 6 or dog-type strains of *P. multocida* that contained D-mannitol and D-sorbitol negative strains of *P. multocida* were classified as the new species *P. canis*. Similar to *P. canis*, but ornithine negative, is *P. stomatis*, which proved to be closely related to *P. canis* and the former *Haemophilus avium*. This V-factor-dependent species of avian origin was classified as *P. avium*. Biotype Henriksen strains of *P. pneumotropica* were classified as *P. dagmatis*, *P. canis*, and *P. stomatis*. *P. dagmatis* is mostly isolated from bite wounds, mainly inflicted by dogs. In addition to the avian species, *P. gallinarum* and *P. avium*, three new species from avian hosts were described, the V-factor-dependent *P. volantium* and two V-factor-independent species, *P. langaa* and *P. anatis*. Two species remained unnamed, but were located in the genus, *Pasteurella* species A and *Pasteurella* species B. *Pasteurella* species A resembles a collection of V-factor-dependent strains formerly designated as *Haemophilus avium*-like. Based on DNA–DNA binding data, all these species were classified as *Pasteurella sensu stricto* (Mutters et al., 1985a, b). Based on 16S rRNA sequence data from selected strains, the genus was separated into two phylogenetic groups (Dewhirst et al., 1992, 1993).

The core group around the type species included the other species from mammalian hosts: *P. canis*, *P. stomatis*, *P. dagmatis*, and *Pasteurella* sp. B. The avian species *P. avium*, *P. gallinarum*, *P. volantium*, and *Pasteurella* sp. A form a separate cluster together with the type strain of *Haemophilus paragallinarum*, and reclassification of this group seems justified based on 16S rDNA phylogenetic analysis and the variations in *atpD* DNA sequences (Petersen et al., 2001).

P. langaa and *P. anatis* were not phylogenetically related to any of the present members of the genus, and should be excluded from the genus *sensu stricto*, just as reconsiderations are needed as to exclusion of the other avian taxa.

Both DNA hybridization and rRNA sequence data suggest the exclusion of species from the genus that were listed in the last edition of *Bergey's Manual of Systematic Bacteriology*. *P. ureae* was reclassified as *Actinobacillus ureae* (Mutters et al. 1986), while *P. pneumotropica* forms a large cluster with several new species requiring the rank of at least one new genus (Ryll et al., 1991). The *M. haemolytica* complex of ruminants contains distinct genetic and phenotypic groups. So far seven species have been outlined, five of which were named in the new genus *Mannheimia*. *P. aerogenes* (McAllister and Carter, 1974), *P. bettyae* (Sneath and Stevens, 1990), *P. caballi* (Schlater et al., 1989), *P. mairii* (Sneath and Stevens, 1990), *P. testudinis* (Snipes and Biberstein, 1982), and *P. trehalosi* (Sneath and Stevens, 1990) all represent misnamed pasteurellas.

Further taxa provisionally affiliated with *Pasteurella* based on phenotypic characters, such as the gas-producing SP-group, could not be integrated in the genus.

FURTHER READING

Bisgaard, M., W. Fredericksen, W. Mannheim and R. Mutters. 1994. Zoonoses caused by organisms classified with *Pasteurellaceae*. *In* Beran (Editor), Handbook of Zoonoses, Section A: Bacterial, Rickettsial, Chlamydial, and Mycotic, 2nd ed., CRC Press, London. 203–208.

Dewhirst, F.E., B.J. Paster, I. Olsen and G.J. Fraser. 1992. Phylogeny of 54 representative strains of species in the family *Pasteurellaceae* as determined by comparison of 16S rRNA sequences. J. Bacteriol. *174*: 2002–2013.

Mutters, R., P. Ihm, S. Pohl, W. Frederiksen and W. Mannheim. 1985. Reclassification of the genus *Pasteurella* Trevisan 1887 on the basis of deoxyribonucleic acid homology, with proposals for the new species *Pasteurella dagmatis*, *Pasteurella canis*, *Pasteurella stomatis*, *Pasteurella anatis* and *Pasteurella langaa*. Int. J. Syst. Bacteriol. *35*: 309–322.

Mutters, R., K. Piechulla, K.H. Hinz and W. Mannheim. 1985. *Pasteurella avium* (Hinz and Kunjara 1977) comb. nov. and *Pasteurella volantium* sp. nov. Int. J. Syst. Bacteriol. *35*: 5–9.

DIFFERENTIATION OF THE SPECIES OF THE GENUS *PASTEURELLA*

Allocation of isolates to existing species of *Pasteurella* usually does not present a problem if, as suggested by Bisgaard (1993), extended characterization and reference strains/data are used. If sequencing facilities are available, partial or full sequencing of 16S rRNA genes should be included. Correct allocation to existing taxa and of newly discovered taxa, however, still remains problematic, and cooperation with reference laboratories is recommended. The differential characteristics of species of *Pasteurella* are given in Table BXII.γ.289, while characters used for separation of subspecies of *P. multocida* are shown in Table BXII.γ.290.

TABLE BXII.γ.289. Characters used for separation of *Pasteurella* spp.[a]

Characteristic	P. multocida	P. canis	P. dagmatis	P. stomatis	Pasteurella sp. B	P. anatis	P. avium	P. gallinarum	P. langaa	P. volantium	Pasteurella sp. A
Catalase	+	+	+	+	+	+	+	+	−	+	+
Symbiotic growth	−	−	−	−	−	−	+	−	−	+	+
Urease	−	−	+	−	−	−	−	−	−	−	−
Ornithine decarboxylase	+	+	−	−	+	−	−	−	−	d	−
Xylitol	−	−	−	−	+	−	−	−	−	−	d
L(+)Arabinose	−	−	−	−	−	−	−	−	−	−	+/(+)
D(+)Xylose	d	−	−	−	+	+	d	d	−	d	d
D(−)Mannitol	+	−	−	−	+	+	−	−	+	+	d
Maltose	−	−	+	−	+/(+)	−	−	+	−	+	d
Dextrin	−	−	+/(+)	−	+/(+)	−	−	+	−	+	d
PNPG	d	d	+	+	+	+	+	+	−	+	+

[a]Symbols: +, 90% or more of the strains positive within 1–2 d; (+), 90% or more of the strains positive within 3–14 d; −, less than 10% of the strains are positive within 14 d; d, 11–89% of the strains are positive.

List of species of the genus Pasteurella

1. **Pasteurella multocida** (Lehmann and Neumann 1899) Rosenbusch and Merchant 1939, 85[AL] (*Bacterium multocidum multocidum* (sic) Lehmann and Neumann 1899, 196; *Pasteurella gallicida* (Burrill 1883) Buchanan 1925, 414.) *mul.to.ci' da.* L. adj. *multus* many; L. adj. suff. *-cidus* from; L. v. *caedere* to kill; M.L. fem. adj. *multocida* many killing, i.e., pathogenic for many species of animals.

Usually forms coccoid cells or short rods on solid media containing blood. Usually bipolar stained in Gram stain. Pleomorphic rods and short filaments can be seen in broth media and older cultures. Many strains produce capsules. Growth on solid media is best on blood containing agar like sheep blood or chocolate agar. Colonies reach a diameter of 1–2 mm with a light grayish color. Mucoid, smooth, and rough colonies are produced. Nonhemolytic. The indole-producing strains exhibit a distinct odor. The optimal growth temperature for avian strains might be as high as 42°C, compared to 36°C for mammalian strains. Growth can be observed in a wide mesophilic range from 25–42°C.

In addition to the features common to all members of the genus, the three subspecies share the following biochemical reactions. Acid usually produced from D-mannitol. Ornithine is decarboxylated. Usually no V-factor requirement, but V-factor-requiring strains may occur occasionally (Krause et al., 1987). Major differences in phenotypic characters have been reported for *P. multocida* (Heddleston, 1976; Biberstein et al., 1991; Bisgaard et al., 1991b; Fegan et al., 1995). The physiological characteristics of the species are presented in Tables BXII.γ.288, BXII.γ.289, and BXII.γ.290. Several typing systems have been developed and applied for epidemiological studies of *P. multocida* (Nielsen and Rosdahl, 1990; Wilson et al., 1993; Blackall et al., 1999b, 2000; Fussing et al., 1999; Loubinoux et al., 1999; Blackall and Miflin, 2000; Bowles et al., 2000; Gunawardana et al., 2000). Strains of the species are isolated from most mammals, including humans and birds, in which a wide range of diseases are reported due to *P. multocida*. The species belong to 16S rRNA cluster 12. The molecular weight of genomic DNA ranges from 1.5×10^9 to 1.9×10^9.

The mol% G + C of the DNA is: 40.8–43.9 (T_m).

Type strain: ATCC 43137, NCTC 10322.

GenBank accession number (16S rRNA): AF294410, M35018.

a. **Pasteurella multocida** subsp. **multocida** (Lehmann and Neumann 1899) Rosenbusch and Merchant 1939, 85[AL] (*Bacterium multocidum multocidum* Lehmann and Neumann 1899, 196; *Pasteurella gallicida* (Burrill 1883) Buchanan 1925, 414.)

Characteristics are those of the species as given above. Differentiated from the other subspecies by its production of acid from D-sorbitol, but not from dulcitol. Other reactions useful in differentiating between the subspecies are shown in Table BXII.γ.290. The species belongs to 16S rRNA cluster 12. The molecular weight of genomic DNA ranges from $1.5–1.7 \times 10^9$.

The mol% G + C of the DNA is: 40.8 to 43.9 (T_m).

Type strain: ATCC 43137, NCTC 10322.

GenBank accession number (16S rRNA): AF294410, M35018.

b. **Pasteurella multocida** subsp. **gallicida** Mutters, Ihm, Pohl, Frederiksen and Mannheim 1985a, 319[VP]

gal.li.ci' da. L. fem. n. *gallina* hen; L. adj. suff. *-cida* kill; M.L. adj. *gallicida* hen killing, referring to pathogenicity for poultry.

Characteristics are those of the species as given above. Differentiation from the other subspecies by its ability to form acid from D-sorbitol and dulcitol. Other reactions useful in differentiating between the subspecies are shown in Table BXII.γ.290. The species belongs to 16S rRNA cluster 12. The molecular weight of genomic DNA ranges from $1.5–1.8 \times 10^9$.

TABLE BXII.γ.290. Characters used for identification of subspecies of *Pasteurella multocida*[a]

Characteristic	P. multocida subsp. multocida	P. multocida subsp. gallicida	P. multocida subsp. septica
Dulcitol	−	+	−
D(−)Sorbitol	+	+	−
L(−)Fucose	d	−	d
Trehalose	d	−	+
α-Glucosidase (PNPG)	d	−	+

[a]Symbols: +, 90% or more of the strains positive; −, 90% or more of the strains negative; d, 11–89% of the strains positive.

The mol% G + C of the DNA is: 41.2–42.5 (T_m).

Type strain: ATCC 51689, NCTC 10204.

GenBank accession number (16S rRNA): AF224297, AF294412, AF326323.

    c. **Pasteurella multocida** *subsp.* **septica** Mutters, Ihm, Pohl, Frederiksen and Mannheim 1985a, 319[VP]

sep'ti.ca. M.L. fem. adj. *septica* poisoning, infecting.

Characteristics are those of the species as given above. Differentiation from the other subspecies by its negative reactions for acid production from D-sorbitol and dulcitol. Other reactions useful in differentiating between the subspecies are shown in Table BXII.γ.290. The species belongs to 16S rRNA cluster 12. The molecular weight of genomic DNA ranges from 1.5–1.6 × 10⁹.

The mol% G + C of the DNA is: 41.5–43.5 (T_m).

Type strain: ATCC 51687, NCTC 11995, CIP A125.

GenBank accession number (16S rRNA): AF225205, AF294411, AF326325.

2. **Pasteurella anatis** Mutters, Ihm, Pohl, Frederiksen and Mannheim 1985a, 320[VP]

a.na'tis. L. fem. gen. n. *anas* duck, of a duck.

Cells are coccobacillary. Colonies are circular, smooth, and grayish-white with a diameter of 1.5–2 mm. Nonhemolytic. Strains have no V-factor requirement. They do not decarboxylate ornithine, do not produce indole, and do not hydrolyze urea. Acid is produced from trehalose, D-mannitol, and D-xylose. No acid is produced from L-arabinose, maltose, D-sorbitol, or dulcitol. Members of the species have been isolated from the intestinal and respiratory tracts of ducks. The species belongs to 16S rRNA cluster 18. The molecular weight of genomic DNA ranges from 1.8–1.9 × 10⁹.

The mol% G + C of the DNA is: 39.9–42.3 (T_m).

Type strain: ATCC 43329, NCTC 11413.

GenBank accession number (16S rRNA): AF228001, M75054.

3. **Pasteurella avium** (Hinz and Kunjara 1977) Mutters, Piechulla, Hinz, and Mannheim 1985b, 8[VP] (*Haemophilus avium* Hinz and Kunjara 1977, 324.)

a'vi.um. L. fem. gen. pl. n. *avis* bird, of birds.

Coccoid to pleomorphic rods, occurring singly or in short chains. In liquid culture media, filamentous forms can be seen. Colonies on chocolate agar are usually smooth, convex, slightly yellowish or grayish-white. Nonhemolytic. The species exhibits the common biochemical characteristics of *Pasteurella sensu stricto*. Biotype 1 strains require V-factor, while biotype 2 strains, previously reported as the ornithine decarboxylase-negative type of Bisgaard Taxon 13, are V-factor independent. Acid is produced from glucose without gas formation and from D(+)galactose, D(+)mannose, D(−)fructose, and trehalose. No acid is produced from L(+)arabinose, L(+)rhamnose, raffinose, maltose, D(−)mannitol, D(−)sorbitol, or dulcitol. Indole is not produced, ornithine decarboxylase negative. Found in the hearts and infraorbital sinuses of chickens (biovar 1) and in the lungs of calves suffering from pneumonia (biovar 2). The species belongs to 16S rRNA cluster 18. The molecular weight of genomic DNA is 1.9 × 10⁹.

The mol% G + C of the DNA is: 43–45 (T_m).

Type strain: ATCC 29546, IPDH 2654.

GenBank accession number (16S rRNA): M75058.

Additional Remarks: The reference strain for biovar 2 is CCUG 16497 (Bisgaard's ornithine negative taxon 13).

4. **Pasteurella canis** Mutters, Ihm, Pohl, Frederiksen and Mannheim 1985a, 320[VP]

ca'nis. L. gen. n. *canis* of a dog.

Colonial morphology resembles that of *Pasteurella multocida*, although biotype 1 colonies of *P. canis* usually reach a diameter of only 1 mm. Nonhemolytic. Growth optimum is 36°C. Strains are V-factor independent. Ornithine decarboxylase positive. Acid is not produced from L-arabinose, raffinose, lactose, maltose, D-mannitol, D-sorbitol, or dulcitol. Urease negative. Biotype 1 strains exhibit positive reactions for indole; biotype 2 strains are indole negative. *P. canis* biotype 1 is recovered mainly from the oral cavities of dogs and is often isolated from dog-bite wounds in humans. Biotype 2 strains have been isolated from pneumoniae from calves. The species belongs to 16S rRNA cluster 12. The molecular weight of genomic DNA ranges from 1.4–1.6 × 10⁹.

The mol% G + C of the DNA is: 37.7–39.8 (T_m).

Type strain: ATCC 43326, NCTC 11621.

GenBank accession number (16S rRNA): M75049.

Additional Remarks: The reference strain for biovar 2 is CCUG 16498 (Bisgaard's ornithine positive taxon 13).

5. **Pasteurella dagmatis** Mutters, Ihm, Pohl, Frederiksen and Mannheim 1985a, 319[VP]

dag.ma'tis. Gr. fem. gen. n. *dagma* bite, from a bite.

Cells coccoid or short rods. Colonies on blood containing media are usually smooth and grayish white. Optimal growth temperature is 36°C under aerobic conditions. Nonhemolytic. Does not require V-factors. Produces small amounts of gas from D-glucose; produces acid from maltose. No acid is produced from D-xylose, L-arabinose, D-mannitol, D-sorbitol, and dulcitol. Tests for glycosides are negative. Positive reactions are obtained for indole and urease. Gelatin may be liquefied after more than 14 d of incubation. The biochemical characteristics are shown in Tables BXII.γ.288 and BXII.γ.289. *P. dagmatis* strains have been isolated from dogs and cats, as well as from human local infections resulting from animal bites or from septicemia and endocarditis (Frederiksen, 1989). The species belongs to 16S rRNA cluster 12. The molecular weight of genomic DNA ranges from 1.5–1.7 × 10⁹.

The mol% G + C of the DNA is: 38.9–41.5 (T_m).

Type strain: ATCC 43325, NCTC 11617.

GenBank accession number (16S rRNA): M75051.

6. **Pasteurella gallinarum** Hall, Heddleston, Legenhausen and Hughes 1955, 604[AL]

gal.li.na'rum. L. fem. gen. pl. n. *gallina* hen, of hens.

Cells are coccoid to rod shaped; bipolar staining occurs. Colonies on blood containing solid media are circular, smooth, and convex with a grayish appearance and reach diameters of 1.5 mm. Nonhemolytic. Not V-factor dependent. Optimal growth temperature is 36°C. Produce acid from maltose and trehalose. No acid is produced from D-mannitol, D-sorbitol, dulcitol, and L-arabinose. Tests for ornithine decarboxylase, indole formation, and urease are negative. Production of acid from *meso*-inositol may occur. *P. gallinarum* is isolated from different lesions in fowl. The species belongs to 16S rRNA cluster 18. The molecular weight of genomic DNA ranges from 1.5–1.6 × 10⁹.

The mol% G + C of the DNA is: 41.2–44.8 (T_m).
Type strain: ATCC 13361.
GenBank accession number (16S rRNA): M75059.

7. **Pasteurella langaa** Mutters, Ihm, Pohl, Frederiksen and Mannheim 1985a, 320[VP]

lan' gaa. L. fem. n. *langaa* referring to the village of Langaa in Denmark.

Cells coccobacillary. Colonies similar to but smaller than those of *P. anatis*. Nonhemolytic. No V-factor requirement. Acid is produced from D-mannitol and lactose; tests for ornithine decarboxylation, indole formation, urease activity, acid from L-arabinose, D-xylose, raffinose, trehalose, maltose, and D-sorbitol are negative. Strains of the species are found in the respiratory tracts of apparently healthy chickens. The species belongs to 16S rRNA cluster 19. The molecular weight of genomic DNA ranges from 1.7–1.9 × 10⁹.

The mol% G + C of the DNA is: 43.9–45.3 (T_m).
Type strain: ATCC 43328, NCTC 11411.
GenBank accession number (16S rRNA): M75053.

8. **Pasteurella stomatis** Mutters, Ihm, Pohl, Frederiksen and Mannheim 1985a, 320[VP]

sto.ma' tis. Gr. gen. n. *stoma* throat, of the throat.

Cells are coccoid to rod shaped with bipolar staining ends in the Gram stain. On blood containing solid media small colonies of about 1 mm in diameter with grayish or semitranslucent appearance are formed. Nonhemolytic.

Optimal growth temperature is 36°C. Strains fulfill the general criteria of the genus *Pasteurella sensu stricto*. Additionally they are V-factor independent, urease negative, and do not decarboxylate ornithine. No acid is produced from L-arabinose, D-xylose, raffinose, lactose, maltose, D-mannitol, D-sorbitol, or dulcitol. Indole positive. Occurs in the oral cavity or respiratory tract of dogs and cats. The species belongs to 16S rRNA cluster 12. The molecular weight of genomic DNA ranges from 1.5–1.6 × 10⁹.

The mol% G + C of the DNA is: 40.4–43.5 (T_m).
Type strain: ATCC 43327, NCTC 11623.
GenBank accession number (16S rRNA): M75050.

9. **Pasteurella volantium** Mutters, Piechulla, Hinz and Mannheim 1985b, 9[VP]

vo.lan' ti.um. L. fem. gen. pl. n. *volantium* from fowl.

Cells coccoid. Occur singly or in short chains. Mesophilic, facultatively aerobic, and V-factor dependent. Nonhemolytic. On chocolate agar, colonies are smooth, convex, and may produce yellowish pigments. Phenotypic features are similar to *P. avium*, but acid production from maltose and D-mannitol is positive. Some strains decarboxylate ornithine, some strains produce acid from D-sorbitol. Strains are isolated from the wattles of domestic fowl, one strain has been isolated from human tongue (Kilian, 1976). The species belongs to 16S rRNA cluster 18. The molecular weight of genomic DNA ranges from 1.5–1.9 × 10⁹.

The mol% G + C of the DNA is: 44–45 (T_m).
Type strain: ATCC 14385, NCTC 3438.
GenBank accession number (16S rRNA): M75070.

Other Organisms

1. *Pasteurella* species A (Mutters, Piechulla, Hinz, and Mannheim 1985b)

Strains of the unnamed *Pasteurella* species A resemble a collection of heterogeneous strains formerly classified as *Haemophilus avium*. All strains have a V-factor requirement and produce acid from L-arabinose, but not from D-sorbitol; the ornithine decarboxylase test is negative. Strains are isolated from poultry and wild birds. The species belongs to 16S rRNA cluster 18. The molecular weight of genomic DNA ranges from 1.7–2.1 × 10⁹.

The mol% G + C of the DNA is: 44–45.9 (T_m).
Deposited strain: IPDH 280, HIM 789-5.
GenBank accession number (16S rRNA): M75055.

2. *Pasteurella* species B (Mutters, Piechulla, Hinz and Mannheim 1985b)

Cells are coccobacillary to rod shaped. Colonies are similar to *P. canis* or *P. stomatis*. Nonhemolytic. No V-factor dependency. Indole formation and ornithine decarboxylase tests are positive. Acid is produced from trehalose, maltose, D-xylose, and dulcitol. Acid is not produced from L-arabinose, D-mannitol, or D-sorbitol. Test for urease is negative. Strains are isolated from different animal hosts, but also from human dog-bite wounds or cat scratches. The species belongs to 16S rRNA cluster 12. The molecular weight of genomic DNA is 1.9 × 10⁹.

The mol% G + C of the DNA is: 38.9–40.0 (T_m).
Deposited strain: CCUG 19794, SSI P 683.
GenBank accession number (16S rRNA): M75052.

Species Incertae Sedis

1. **Pasteurella aerogenes** McAllister and Carter 1974, 920[AL]

a.e.ro' gen.es. Gr. masc. n. *aer* air; L. v. *genere* to produce; M.L. adj. *aerogenes* gas-producing.

The species was excluded from the genus *Pasteurella* by DNA–DNA hybridization (Mutters et al., 1985a) and 16S rRNA hybridization (De Ley et al., 1990). Original isolates obtained from pigs. The species belongs to 16S rRNA cluster 14. Other isolates associated with 16S rRNA cluster 5. The molecular weight of genomic DNA is 1.6–2.0 × 10⁹.

The mol% G + C of the DNA is: 41.8 (T_m).
Type strain: ATCC 27883.
GenBank accession number (16S rRNA): M75048.

2. **Pasteurella bettyae** Sneath and Stevens 1990, 151[VP]

be' tty.ae. M.L. gen. n. *bettyae* of Betty, to commemorate Elizabeth "Betty" O. King.

The species resembles the former CDC group HB-5 and was excluded genetically as well as by DNA–DNA hybridization data from the genus *sensu stricto* as by its phenotypic features, i.e., negative reactions for oxidase, no acid from D-galactose, fructose, D-mannose, or sucrose. Strains are obtained from human Bartholin gland abscesses and human finger infections. The species belongs to 16S rRNA cluster 9.

The mol% G + C of the DNA is: not determined.
Type strain: ATCC 23273, NCTC 10535.
GenBank accession number (16S rRNA): L06088.

3. **Pasteurella caballi** Schlater, Brenner, Steigerwalt, Moss, Lambert and Packer 1990, 320[VP] (Effective publication: Schlater, Brenner, Steigerwalt, Moss, Lambert and Packer 1989, 2173.)

ca.ba' lli. Gr. n. gen. *cabilli* from a horse.

Strains of this taxon are different from true members of the genus by their aerogenic capacity and their negative catalase reaction. Some strains produce acid from *meso*-inositol and from L-rhamnose. Isolates are obtained from respiratory and genital tract infections in horses and also from humans who have had contact with horses. Cases of wound infections (Bisgaard et al., 1991a) and a wound infection after horse bite (Escande et al. 1997) have been reported. The species belongs to 16S rRNA cluster 9.

The mol% G + C of the DNA is: 41–42 (T_m).

Type strain: ATCC 49197.

GenBank accession number (16S rRNA): AF224291.

4. **Pasteurella granulomatis** Ribeiro, Carter, Frederiksen and Riet-Correa 1990, 105[VP] (Effective publication: Ribeiro, Carter, Frederiksen and Riet-Correa 1989, 1402.)

gran.nu.lo' ma.tis. L. dim. n. *granulum* grain; Gr. suff. *oma* a swelling or tumor; M.L. gen. n. *granulomatis* of a granuloma.

The nonhemolytic species was transferred to the genus *Mannheimia* as *Mannheimia granulomatis* Angen, Mutters, Caugant, Olsen and Bisgaard 1999a (see chapter on *Mannheimia*).

The mol% G + C of the DNA is: 39.2 (T_m).

Type strain: 26, ATCC 49244.

5. **Pasteurella lymphangitidis** Sneath and Stevens 1990, 151[VP]

lymph.ang.iti' d.is. M.L. gen. n. *lymphangitidis* pertaining to lymphangitis, inflammation of the lymph nodes.

DNA–DNA hybridization data exclude the former BL organisms isolated from lymphangitis in *Bos indicus* in Southern India from the genus *Pasteurella* (Mutters et al., 1985a). The biochemical characteristics of a negative oxidase reaction, lack of nitrate reduction, and no acid formation from D-galactose and D-mannose phenotypically exclude this group from the genus.

The mol% G + C of the DNA is: not determined.

Type strain: ATCC 49635, NCTC 10547.

6. **Pasteurella mairii** *corrig.* Sneath and Stevens 1990, 151[VP]

mai' ri.i. M.L. gen. n. *mairii* of Mair, after Nicholas S. Mair, who isolated the organism.

DNA hybridization data showed only low degrees of relatedness with *Pasteurella sensu stricto* as well as with other selected *Pasteurella*-named species (Mutters et al., 1985a). Strains are isolated from abortion in pigs and from septicemia in piglets. The species belongs to 16S rRNA cluster 14.

The mol% G + C of the DNA is: not determined.

Type strain: ATCC 49633, NCTC 10699.

GenBank accession number (16S rRNA): AF024532.

7. **Pasteurella pneumotropica** Jawetz 1950, 179[AL]

pneu.mo.tro' pi.ca Gr. n. *pneumon* lung; Latinized Gr. adj. *tropicus* tropic, circle, M.L. fem. adj. *pneumotropica* having an affinity for the lungs.

Based on the molecular data, the two biovars Jawetz and Heyl resemble two distinct species outside the genus *Pasteurella*. The former Henriksen biovar has been reclassified as *Pasteurella dagmatis*. Overall, the *P. pneumotropica* complex remains heterogeneous, containing several unclassified new species (Ryll et al., 1991). The species belongs to 16S rRNA cluster 15. The molecular weight of genomic DNA is 1.5–1.6 × 10[9].

The mol% G + C of the DNA is: 40.3–42.8 (T_m).

Type strain: ATCC 35149, NCTC 8141.

GenBank accession number (16S rRNA): M75083.

8. **Pasteurella testudinis** Snipes and Biberstein 1982, 209[VP]

tes.tu' di.nis. L. n. *testudo* tortoise; L. gen. n. *testudinis* of the tortoise.

The species has been recovered from healthy Californian desert tortoises (*Gopherus agassizii*) and from tortoises with upper respiratory tract disease (Snipes and Biberstein, 1982; Snipes et al., 1995). The species belongs to 16S rRNA cluster 21. The molecular weight of genomic DNA is 1.6 × 10[9].

The mol% G + C of the DNA is: 46.8–47.4 (T_m).

Type strain: ATCC 33688, UCD 90-23-79n.

GenBank accession number (16S rRNA): L06090.

9. **Pasteurella trehalosi** Sneath and Stevens 1990, 152[VP]

tre.hal.osi. M.L. gen. n. *trehalosum* pertaining to the carbohydrate trehalose.

The species includes the biovar T of the former *Pasteurella haemolytica*. Genotypic investigations have shown that the trehalose-positive biovar does not belong to the genus *Pasteurella* (Mutters et al., 1985a; De Ley et al., 1990; Dewhirst et al., 1993). Hemolysis on sheep blood agar and lack of acid formation from D(+) galactose also excludes the species from *Pasteurella*. Although these strains show no close affiliation to the genus *Pasteurella*, they were recently named *Pasteurella trehalosi* (Sneath and Stevens, 1990). The species belongs to 16S rRNA cluster 10. (The type strain has not been sequenced.) The molecular weight of genomic DNA is 1.8 × 10[9].

The mol% G + C of the DNA is: 42.6 (T_m).

Type strain: S110, ATCC 29703, NCTC 10370.

Genus II. **Actinobacillus** Brumpt 1910, 849[AL]

INGAR OLSEN AND KRISTIAN MØLLER

Ac.ti.no.ba.cil' lus. Gr. n. *actis* a ray; L. dim. masc. n. *bacillus* a small staff or rod; M.L. masc. n. *Actinobacillus* ray bacillus or rod.

Cells, measuring 0.4 ± 0.1 × 1.0 ± 0.4 µm, are spherical, oval, or rod-shaped (Phillips, 1984). **Most often bacillary but sometimes interspersed with coccal elements that may lie at the pole of a larger form, producing the characteristic "Morse-code"** form. Cell forms up to 6 µm in length may appear when grown on media containing glucose or maltose. Cells are single or arranged in pairs or, more rarely, in chains. Endospores are not formed. Gram negative, but staining is irregular. Not acid fast.

India ink may demonstrate small amounts of extracellular slime in wet preparations. **Nonmotile.** Organisms are aerobic, microaerobic, facultatively anaerobic, or chemoorganotrophic, having both respiratory and fermentative types of metabolism. After growth for 24 h on blood agar, translucent colonies, usually 1–2 mm in diameter appear. Surface colonies have low viability and may die in 2–7 days. Growth may be very sticky upon primary cultivation, making it difficult to remove colonies completely from the agar surface. Optimum growth temperature is 37°C. Temperature range for growth is 25–42°C.

Glucose is fermented, with production of acid but not gas. Inulin is not fermented, and inositol is fermented by *A. rossii* and infrequently by *A. porcinus* (Sneath and Stevens, 1985; Møller and Kilian, 1990; Møller et al., 1996a). Only *A. indolicus* is indole positive. **Nitrate reduction is positive** or variable (variable in *A. porcinus* [Møller et al., 1996a] and in *A. seminis* [van Tonder, 1979]). Actinobacilli are parasitic or commensal in humans, sheep, cattle, horses, pigs, other mammals, and birds. They are often commensals in the alimentary and genital tract and respiratory mucosa but can also occur as pathogens and cause a variety of lesions.

Genus *Actinobacillus* belongs to the family *Pasteurellaceae*. Virtually complete 16S rRNA sequencing has shown that this family is located in the *Gammaproteobacteria* (Dewhirst et al., 1992). Families closely related to the *Pasteurellaceae* are the *Enterobacteriaceae*, *Vibrionaceae*, and *Aeromonadaceae*. More distant relatives within the *Gammaproteobacteria* are the *Pseudomonas fluorescens* complex, the *Moraxellaceae*, and the *Cardiobacteriaceae*. DNA–rRNA hybridization found that the members of the *Pasteurellaceae* are most closely related to members of the *Enterobacteriaceae*, the *Vibrionaceae*, the *Aeromonadaceae*, and the genus *Alteromonas* (De Ley et al., 1990).

When 16S rRNA sequences were determined for over 70 strains of species in the family *Pasteurellaceae*, the strains were found to fall in seven clusters (Dewhirst et al., 1992, 1993). The first cluster, which is split into three subclusters, includes *Haemophilus sensu stricto*. This cluster contains the type strains of *H. influenzae*, *H. aegyptius*, *H. aphrophilus*, *H. haemolyticus*, *H. paraphrophilus*, *H. segnis*, and *A. actinomycetemcomitans*. It also contains *H. aphrophilus* NCTC 7901 and *A. actinomycetemcomitans* FDC Y4, ATCC 29522, ATCC 29523, and ATCC 29524. Cluster 2, which splits into two subclusters, contains *P. mairii* CCUG 27189, Bisgaard taxon 6 CCUG 15568, and the type strains of *A. seminis* and *P. aerogenes*. Cluster 3, including the genus *Pasteurella sensu stricto*, contains the type strains of *P. multocida*, *P. anatis*, *P. avium*, *P. canis*, *P. dagmatis*, *P. gallinarum*, *P. langaa*, *P. stomatis*, *P. volantium*, *P. trehalosi*, *H. haemoglobinophilus*, *H. parasuis*, *H. paracuniculus*, *H. paragallinarum*, "*A. salpingitidis*", and *A. capsulatus*. It also comprises Bisgaard taxa 2, 3, 7, and 13, *Pasteurella* species A CCUG 18782, *Pasteurella* species B CCUG 19974, *Haemophilus* taxon C CAPM 5111, *H. parasuis* type 5 Nagasaki, and *P. volantium* NCTC 4101. Four subclusters exist in Cluster 3. The species belonging to *Pasteurella sensu stricto* fall in three of the four disconnected clusters. Cluster 4 contains the genus *Actinobacillus sensu stricto*, which includes the type strains of *A. lignieresii*, *A. equuli*, *A. pleuropneumoniae*, *A. suis*, *A. ureae*, *A. hominis*, *H. parahaemolyticus*, *H. paraphrohaemolyticus*, *H. ducreyi*, and *M. haemolytica*. This cluster also comprises *Actinobacillus* species strain CCUG 19799 (Bisgaard Taxon 11), *A. suis* ATCC 15557, *Haemophilus* "minor group" strain 202, *P. bettyae*, and Bisgaard taxa 5, 8, and 9 (*A. arthritidis* and genomospecies 2). Cluster 4 falls in two subclusters. Each species in these subclusters branch slightly deeper than do those in the cluster above, giving no logical point at which to end the genus

Actinobacillus sensu stricto. The type strains of *P. pneumotropica* and *A. muris* form a fifth group. The sixth group is constituted by the type strain of *H. parainfluenzae* and by *H. parainfluenzae* ATCC 29242. Two strains of Bisgaard Taxon 14 and the type strain of *P. testudinis* form a seventh cluster.

The mol% G + C of the DNA is: 35.5–46.9.

Type species: **Actinobacillus lignieresii** Brumpt 1910, 849.

Further descriptive information

Cell morphology Actinobacilli show marked pleomorphism. Bacillary forms predominate, but coccobacilli and longer filaments can also be seen. The latter can break up into short bacillary forms or granules, which may give an appearance of streptobacilli or streptococcal chains.

Capsules are produced by most species. Extracellular slime can be seen especially in *A. equuli* and *A. suis*.

Cell constituents

Fatty acids Cellular fatty acid patterns of established and newly described taxa in the family *Pasteurellaceae*, as determined by capillary gas chromatography, have been found to be homogenous with minor variations (Mutters et al., 1993). In this case, discrimination of species or groups within the family could not be achieved. Separation from the *Neisseriaceae* and *Moraxella* was possible. All *Pasteurellaceae* strains investigated show proportions of >15% for unbranched tetradecanoic acid $C_{14:0}$, hexadecanoic acids $C_{16:1}$, and $C_{16:1}$ $_{cis 9}$, and moderate amounts of 3-hydroxy-dodecanoic acid $C_{12:0\ 3OH}$, tetradecanoic acid $C_{14:5}$, pentadecanoic acid $C_{15:0}$, octadecanoic acid $C_{18:0}$, eikosanoic acids $C_{20:0}$ and $C_{20:4}$ $_{cis 5,8,11}$. In contrast to this, however, Boot et al. (1999) have found that principal component analysis of fatty acid data obtained under standardized growth conditions may discriminate among *Pasteurellaceae* species.

By using a column with a nonpolar stationary phase, it is possible to analyze free fatty acids from bacterial cells directly in the gas chromatograph without derivatization (Brondz and Olsen, 1983; Brondz et al., 1983). While *A. actinomycetemcomitans* strains can be divided into three groups based on their free fatty acid content, *H. aphrophilus* is homogeneous. However, the distribution of bound cellular fatty acids from whole cells is not so specific as to allow discrimination between *A. actinomycetemcomitans* and *H. aphrophilus* (Calhoon et al., 1981; Jantzen et al., 1981; Brondz and Olsen, 1984b).

The lipid content of lipopolysaccharide (LPS) from *A. actinomycetemcomitans* constitutes 35.4%; in *H. aphrophilus* it is only 18.4% (Brondz and Olsen, 1989).

3-Hydroxytetradecanoic acid and tetradecanoic acid, which are the sole LPS acids, make up 21.1% and 14.3%, respectively, of fatty acids in *A. actinomycetemcomitans*, and 10.9% and 7.5% in *H. aphrophilus*. Variation in the lipid content of LPS acids may explain in part the difference in periodontopathogenic potential between *A. actinomycetemcomitans* and *H. aphrophilus*.

When fatty acids of isolated outer membrane vesicles in species of *Actinobacillus*, *Haemophilus*, and *Pasteurella* were analyzed by gas chromatography and the data treated with multivariate analysis, it was found that *A. actinomycetemcomitans* can be assigned to one group (class 1) and *H. influenzae* to another (class 2) (Olsen et al., 1995). *H. aphrophilus*, *H. paraphrophilus*, and *P. multocida* are distinct from these classes, with the exception of the type strain of *H. aphrophilus*, which falls in class 1 (borderline). The abundance of 3-OH fatty acids indicates that the outer membrane vesicles are rich in LPS. Multivariate analysis of fatty acid data from such vesicles can distinguish two groups of *A. actinomycetemcomitans* (Brondz et al., 1995). One group contains strains ATCC 33384^T, ATCC 29522, SUNY 366, HK 435, and Q 1247; the other group comprises ATCC 29524, ATCC 29523, FDC 2112, FDC 511, and SUNY 489.

PHOSPHOLIPIDS   The distribution of phospholipids, as determined by thin layer chromatography, is uniform within the family *Pasteurellaceae* (Mutters et al., 1993). Phosphatidylethanolamine has been observed in major amounts, phosphatidylglycerine in smaller amounts, and lysophosphatidylethanolamine in only petty amounts.

LIPOQUINES   The lipoquine content can be used to discriminate between groups within the family *Pasteurellaceae*, but it does not necessarily reflect the degree of genomic relatedness (Mutters et al., 1993). *Actinobacillus* exhibits high amounts of menaquinone-7 (MK-7) (30%) and demethylmenaquinone-7 (DMK-7) (54%). Representative strains of *Haemophilus* contains mostly DMK-6, with smaller amounts (~10% each) of MK-8 and DMK-7. A group consisting of *A. actinomycetemcomitans*, *H. aphrophilus*, *H. segnis*, and *P. bettyae* has a profile similar to that of *Haemophilus*, but does not contain MK-8. Another group comprising *M. haemolytica* and Bisgaard Taxon 3 has a lipoprotein pattern between those of *Actinobacillus* and *Haemophilus*, whereas the patterns of "*Histophilus*", "*A. salpingitidis*", and *A. rossii* are completely different.

Of 16 strains of members of the *Pasteurellaceae* studied by high-performance liquid chromatography, 12 have been found to produce menaquinones, mostly as minor components, in addition to demethylmenaquinones (Kroppenstedt and Mannheim, 1989). These organisms represent the genus *Actinobacillus*, the genus *Pasteurella sensu stricto*, the *Mannheimia haemolytica* complex, *H. paragallinarum*, *H. parasuis*, *H. aphrophilus*, *P. aerogenes*, and Bisgaard Taxon 14. *P. pneumotropica* biotype Jawetz, members of the genus *Haemophilus sensu stricto*, and *A. actinomycetemcomitans* do not contain menaquinones. The view that the family *Pasteurellaceae* can be differentiated from the *Enterobacteriaceae* and other fermenting Gram-negative bacteria by a lack of menaquinones therefore cannot be maintained.

CARBOHYDRATES   There is a common pattern of cellular sugars among all members of the family *Pasteurellaceae*, and characteristic carbohydrate profiles discriminate between groups, often to the species level (Mutters et al., 1993). The sugar patterns distinguish, for example, *P. stomatis* from Bisgaard's Taxon 16, mannose-positive *A. hominis* strains from *A. equuli*, and *A. actinomycetemcomitans* from *H. aphrophilus*, *H. segnis*, and intermediate strains. Sugar profiles, established by gas chromatography of whole cells after methanolysis and derivatization with trifluoroacetic acid anhydride have demonstrated that *H. aphrophilus* differs from *A. actinomycetemcomitans*, *A. ureae*, *H. paraphrophilus*, *H. influenzae* type b, *M. haemolytica*, and *P. multocida* by its lack of D-glycero-D-mannoheptose (Brondz and Olsen, 1984d, 1985a). Whole defatted cells, which are intermediates between whole cells and isolated cell membranes, give the same sugar profiles as whole cells, reaffirming that D-glycero-D-mannoheptose is present in *A. actinomycetemcomitans* and absent in *H. aphrophilus* (Brondz and Olsen, 1984a). Further, multivariate analysis of quantitative chemical and enzymatic characterization data has demonstrated that D-glycero-D-mannoheptose is a true chemotaxonomic marker for these organisms (Brondz et al., 1990). LPS is the primary source of D-glycero-D-mannoheptose in *A. actino-*

mycetemcomitans and *H. paraphrophilus* (Brondz and Olsen, 1984c, 1985b). Whereas *A. actinomycetemcomitans* LPS contains 7.8% D-glycero-D-mannoheptose and 11.3% L-glycero-D-mannoheptose, *H. aphrophilus* LPS contains 17.4% L-glycero-D-mannoheptose (Brondz and Olsen, 1989).

PROTEINS   One-dimensional SDS-PAGE of whole cell proteins has been used to characterize 39 strains representing all biovars established within the taxa 2 and 3 complex of Bisgaard and 2 strains belonging to the avian *M. haemolytica*–"*A. salpingitidis*" complex (Bisgaard et al., 1993). There is no correlation between results obtained from protein profiling and those previously established by DNA–DNA hybridization. Comparison of phena defined by protein profiling with species/groups established by DNA–DNA hybridization, chemotyping, and biotyping has shown that the best correlation exists between DNA–DNA hybridization and biotyping.

Both one- and two-dimensional electrophoresis of cellular proteins separate *A. actinomycetemcomitans* and *H. aphrophilus* (Calhoon et al., 1981; Jellum et al., 1984). The high resolution of the 2-D electrophoresis distinguishes strains within species, in contrast to 1-D electrophoresis. The 2-D whole-cell protein profiles of *A. actinomycetemcomitans*, *A. ureae*, *H. paraphrophilus*, *H. aphrophilus*, *H. influenzae*, *P. multocida*, and *M. haemolytica* have also been compared (Olsen et al., 1987). The protein patterns of the genera *Actinobacillus*, *Haemophilus*, and *Pasteurella* are quite distinct. *A. actinomycetemcomitans* can easily be distinguished from the other species tested, including *A. ureae*, *H. aphrophilus*, *H. paraphrophilus*, and *M. haemolytica*. The protein patterns of *P. multocida* and *M. haemolytica* also differ. However, it is difficult to distinguish strains of *H. paraphrophilus* and *H. aphrophilus*.

When outer membrane proteins of *A. actinomycetemcomitans* and *H. aphrophilus* strains were compared by means of SDS-PAGE and immunoblots with rabbit antiserum, the most prominent feature was found to be the absence of a heat-modifiable protein in *H. aphrophilus* (Bolstad et al., 1990). Another study has demonstrated a remarkable degree of homology between the outer membrane proteins of *A. actinomycetemcomitans* and *H. aphrophilus* (DiRienzo and Spieler, 1983). In addition, the principal heat-modifiable outer membrane protein of *A. actinomycetemcomitans* shows N-terminal sequence homology and immunologic cross reactivity with the OmpA proteins of other Gram-negative bacteria (Wilson, 1991).

POLYAMINES   Analysis of polyamines is a promising method for discriminating members of the family *Pasteurellaceae* (Busse et al., 1997). Members of the genus *Actinobacillus sensu stricto* contain 1,3-diaminopropane as the predominant compound. In addition, in most of the species of the genus *Haemophilus*, 1,3-diaminopropane is the major substance. In contrast, however, "*H. intermedius* subsp. *gazogenes*" and *H. parainfluenzae* contain high levels of 1,3-diaminopropane, cadaverine, and putrescine. *H. parainfluenzae* is phylogenetically only distantly related to the type species of the genus *Haemophilus* (Dewhirst et al., 1993). The phylogenetic diversity of *Pasteurella* is also to some extent reflected by different polyamine patterns.

ENZYMES   Multivariate analysis of data (API-ZYM micromethod) from 19 enzyme activities in the outer membrane vesicles/fragments from clinically important species of the *Actinobacillus*–*Haemophilus*–*Pasteurella* group has enabled distinction among *A. actinomycetemcomitans*, *H. aphrophilus*, and *H. paraphrophilus* (Myhrvold et al., 1992). *A. actinomycetemcomitans* is divided into two strain groups (classes). *A. lignieresii* falls outside or on the borderline of the *A. actinomycetemcomitans* class. *A. ureae*, *H. influenzae*, *H. parainfluenzae*, *M. haemolytica*, and *P. multocida* appear outside the *A. actinomycetemcomitans*, *H. aphrophilus*, and *H. paraphrophilus* classes. The API-ZYM method also distinguishes whole cells of *A. actinomycetemcomitans*, *A. lignieresii*, and *H. aphrophilus* (Slots, 1981).

The genetic diversity of 52 field isolates, mainly from Australia, and 15 reference strains of *A. pleuropneumoniae* has been examined with multilocus enzyme electrophoresis (MEE) (Hampson et al., 1993), and 33 electrophoretic types (ETs), with a mean genetic diversity per locus of 0.312, were found. Australian strains of serovars 1, 2, 5, and 7 belong to the same clonal lines as strains of these serovars from other countries. Distinct clones of serovars 3, 7, 11, and 12 are also detected. The type strains of serovars 1, 9, and 11 are placed in the same ET, and strains of *A. pleuropneumoniae* biovar 2 are closely related to biovar 1 strains.

MEE has also been used for clonal analysis of the *A. pleuropneumoniae* population in a geographically restricted Danish area (Møller et al., 1992). Sixty-six percent of 250 isolates from lungs of pigs with pleuropneumonia and from tonsils of apparently healthy pigs at slaughter belong to three ETs, out of a total of 37 ETs detected. While five biotype 2 isolates constitute a separate ET closely related to biotype 1 isolates, the type strain of the species constitutes its own ET, with a genetic distance of 0.30 from its nearest neighbor.

Genetic relationships among isolates assigned to *A. actinomycetemcomitans*, *H. aphrophilus*, and *H. paraphrophilus* can be examined based on electrophoretically demonstrable allelic variation in 14 structural genes coding for metabolic enzymes (Caugant et al., 1990). Among the 51 isolates that have been analyzed, 25 ETs are detected, with a genetic diversity of 0.753 per locus. Eleven ETs are seen with cluster analysis; these represent the genotypes of all 17 isolates assigned to *A. actinomycetemcomitans*. The remaining 14 ETs represent the genotypes of 34 isolates of *H. aphrophilus* and *H. paraphrophilus*. All strains of *H. aphrophilus* are closely related, except for ATCC 13252. Strains assigned to *H. paraphrophilus* include distantly related lineages, some of which are similar to *H. aphrophilus*. There is no significant overall genetic similarity between *A. actinomycetemcomitans* and the *Haemophilus* species examined.

Bacteriolysis The ability of ethylenediaminetetraacetic acid (EDTA) and hen egg white lysozyme to induce bacteriolysis has been tested in major clinical species of the *Actinobacillus–Haemophilus–Pasteurella* group (Olsen and Brondz, 1985). *A. actinomycetemcomitans* lyses more easily than does *H. aphrophilus*. The former can be divided into two strain groups according to lysis patterns, while *H. aphrophilus* is homogenous. In one group of *A. actinomycetemcomitans* (group I), EDTA has a considerable lytic effect, which is not increased by supplementation with lysozyme. In the other group of *A. actinomycetemcomitans* (group II), the lytic effect of EDTA is much less, but lysozyme has a considerable supplementary lytic effect. Maximal lysis of *A. actinomycetemcomitans* (ATCC 29522) occurs at pH 8.0 in the presence of EDTA and at pH 7.6 with EDTA–lysozyme. In *H. aphrophilus* (ATCC 33389^T) maximal lysis occurs at pH 9.0 with EDTA and at pH 9.2 with EDTA–lysozyme. When *H. influenzae*, *H. paraphrophilus*, *P. multocida*, *M. haemolytica*, and *A. ureae* were similarly tested, the group I strains of *A. actinomycetemcomitans* were found to be the most rapidly lysed by EDTA. *H. paraphrophilus* is the least sensitive of the Gram-negative organisms examined, but it is less resistant

than *Micrococcus luteus*. The latter organism is the most sensitive to lysozyme, followed by *A. ureae* and the group II strains of *A. actinomycetemcomitans*. The group I strains of *A. actinomycetemcomitans*, *H. paraphrophilus*, and *M. haemolytica* are the least sensitive.

Multivariate analysis of 41 quantitative variables, including data from lysis kinetics after exposure to both EDTA and EDTA–lysozyme, cell sugar and fatty acid contents, methylene blue reduction, and API ZYM enzymatic assessment of whole cells and outer membrane vesicles/fragments, divides *A. actinomycetemcomitans* strains into two groups (Brondz and Olsen, 1993). One group contains ATCC 33384^T, ATCC 29522, FDC 2112, and FDC 2043; the other group comprises ATCC 29524, ATCC 29523, FDC 2097, FDC 511, and FDC Y4. Both groups are distinct from members of the genera *Haemophilus* and *Pasteurella*.

Cultural and biochemical characteristics The sticky properties of actinobacilli cultures when grown on nutrient agar or blood are marked with some species, but may be lost on subcultivation. The same stickiness may be revealed when cultures are grown in liquid media, especially those containing glucose. The increased viscosity is especially seen with *A. equuli* and to a lesser extent with *A. suis*. Stickiness can be demonstrated by drawing a strand of sticky material between the surface of the broth and the inoculating loop.

A few species exhibit hemolytic activity when cultured on blood agar. Evidence has been given for hemolytic *A. lignieresii* in horses (Samitz and Biberstein, 1991). A narrow zone of hemolysis may occasionally surround *A. minor* colonies when they are tested for the Christie Atkins and Munch–Petersen (CAMP) phenomenonon. Usually, *A. pleuropneumoniae* shows complete hemolysis on ovine and bovine blood agar, but occasionally, biovar 1 isolates may demonstrate incomplete or no hemolysis. However, the CAMP-reaction is always positive (Møller and Kilian, 1990). A zone of complete hemolysis always surrounds *A. suis* when it is cultured on ovine or bovine blood agar. In contrast, only partial hemolysis is observed when *A. suis* is grown on horse blood agar. Hemolytic *A. suis*-like organisms have been isolated from the oral cavity of horses (Bisgaard et al., 1984; Samitz and Biberstein, 1991).

Pigmented colonies are not produced, except in the case of *A. suis*, which reveals a creamy yellow color when its broth-grown cultures are centrifuged and the pellet resuspended in saline (Kim et al., 1976).

Usually, it takes 24–48 h for fermentation of carbohydrates and alcohols to occur, but some species or strains may need up to 7 days. Acid, but no gas is formed. Some substrates are fermented rather slowly. *A. succinogenes* accumulates high concentrations of succinic acid (Guettler et al., 1999).

A small number of *A. equuli* strains are gelatinase positive when tested by the gelatine plate method of Frazier (1926) (Frederiksen, 1973; Vallée et al., 1974).

Genome size/pulsed-field gel electrophoresis The genome size of the type strain (serotype 1) of *A. pleuropneumoniae*, as determined by pulsed-field gel electrophoresis (PFGE) of *Asd*I and *Apa*I digested chromosomal DNA, was 2404 ± 40 kb (Chevallier et al., 1998). The chromosome sizes for the reference strains of the other serotypes ranged between 2.3 and 2.4 Mb. Different PFGE patterns demonstrate heterogeneity of the chromosomal structure among different field strains of serotype 1, 5a, and 5b, while strains of serotype 9 show homogenous *Apa*I patterns. PFGE suggests that strains of *A. pleuropneumoniae* serotype 2 obtained from Europe and North America are clonally related.

When *A. equuli* strains from 174 horses were examined by PFGE and restriction enzyme electrophoresis, a high degree of strain variability was found within each horse population (10 farms) and some variability over time was found between strains isolated from the same horse (Sternberg, 1998).

Valcarcel et al. (1997), using PFGE, found that the genome of *A. actinomycetemcomitans* contains a single chromosome of 2.3 Mb, estimated from the sum of restriction fragments generated with rare cutting endonucleases. Large plasmids with sizes ranging from 35–300 kb have been detected, and extrachromosomal elements constitute over 20% of the total genome in some strains.

16S rRNA data/phylogeny Phylogeny of the family *Pasteurellaceae* has been determined by comparison of 16S rRNA sequences (Dewhirst et al., 1992, 1993). The family has been divided into seven clusters, of which the fourth cluster contains the type strains of *A. lignieresii, A. equuli, A. pleuropneumoniae, A. suis, A. ureae, H. parahaemolyticus, H. parainfluenzae, H. paraphrohaemolyticus, H. ducreyi,* and *M. haemolytica*. This cluster also includes *Actinobacillus* sp. strain CCUG 19799 (Bisgaard Taxon 11), *A. suis* ATCC 15557, *H. ducreyi* ATCC 27722 and HD 3500, *Haemophilus* "minor group" strain 202, and *H. parainfluenzae* ATCC 29242. *A. hominis* is closely related to *Actinobacillus sensu stricto* species in Cluster 4A, while Bisgaard taxa 5, 8, 9 (*A. arthritidis* and genomospecies 2), and *P. bettyae* fall in Cluster 4B. The type strain of *A. actinomycetemcomitans* falls in the first cluster and also contains *A. actinomycetemcomitans* FDC Y4, ATCC 29522, ATCC 29523, and ATCC 29524, and *H. aphrophilus* NCTC 7901. The type strain of *A. seminis* is included in the second cluster, and the type strains of *A. capsulatus* and "*A. salpingitidis*" in the third cluster. *A. muris* is related to *P. pneumotropica,* and both are found in Cluster 5. The branching is extremely complex. For further information regarding these clusters, see genus definition at the head of this chapter.

The restriction pattern profiles of *Asc*I-, *Apa*I-, and *Nhe*I-digested chromosomes of different serotypes of *A. pleuropneumoniae* demonstrate a high degree of polymorphism (Chevallier et al., 1998). Analysis of the macrorestriction pattern polymorphism has provided a determination of phylogenetic relationships between the serotype reference strains. These relationships reflect to some extent groups of serotypes known to cross-react serologically.

Fussing et al. (1998b) has characterized and determined intraspecies relatedness of *A. pleuropneumoniae,* as well as its relationship to *A. lignieresii* and *A. suis,* by sequence analysis of the ribosomal operon. Amplification and sequence analysis of the 16S–23S rDNA ribosomal intergenic sequence (RIS) from the three species gives two RISs differing by about 100 bp. The larger RISs differ among the three species. A species-specific area has been found for *in situ* probe detection of *A. pleuropneumoniae.*

DNA–DNA hybridization DNA–DNA hybridization has been used extensively by the Mannheim group to determine the genetic relationships within and among genera of the family *Pasteurellaceae* (e.g., Mannheim and Pohl, 1980; Mannheim et al., 1980; Mannheim, 1983). DNA–DNA hybridization demonstrates the relative closeness of *Actinobacillus* and *Pasteurella* (Mannheim, 1983). Recently a micro-well format DNA hybridization method has been developed and validated on 23 strains belonging to *Actinobacillus* species, avian [*Pasteurella*] *haemolytica*-like bacteria

and *Mannheimia* species (Christensen et al., 2000). Based on DNA homology studies, Escande et al. (1984) have suggested that the genus *Actinobacillus* be limited to *A. lignieresii, A. suis, A. capsulatus,* an unnamed equine *Actinobacillus* species (strain CCM 5500), and the generically misnamed *Pasteurella ureae. A. actinomycetemcomitans,* "*A. salpingitidis*", and *A. seminis* have been suggested to not belong in the genus.

The degree of genetic relatedness of *A. pleuropneumoniae* to selected members of the family *Pasteurellaceae* has been determined by free solution DNA–DNA hybridization (Borr et al., 1991). Representative strains of all 12 serotypes of *A. pleuropneumoniae* form a homogenous group exhibiting 74–90% sequence homology with *A. pleuropneumoniae* serotype 1. All serotypes show a high degree of genetic relatedness to *A. lignieresii*. Little homology has been demonstrated between *A. pleuropneumoniae* strains and selected *Haemophilus* and *Pasteurella* species.

A. pleuropneumoniae serotype reference strains and 204 field strains representing all 12 serotypes and biovars 1 and 2 have been analyzed for the presence of a number of toxin genes by DNA–DNA hybridization with specific gene probes (Beck et al., 1994). The patterns of *apx* genes and those of the expressed Apx toxins in biovar 1 field strains are the same as the patterns in the corresponding serotype reference strain. Nearly all strains expressed their *apx* genes and produced the same Apx toxins as their serotype reference strain.

The DNA homology between serovars (1–13) of *M. haemolytica* and other species (*P. multocida* and actinobacilli) has been found to be low (Bingham et al., 1990), suggesting essentially no genetic relationship of *M. haemolytica* with the ATCC reference strains of *Pasteurella* or *Actinobacillus.*

"*H. somnus*" DNA shows 43%, 46%, and 58% homology with the DNA from *A. lignieresii, H. influenzae,* and *H. parainfluenzae,* respectively (Gonzales and Bingham, 1983).

Based on DNA relatedness Coykendall et al. (1983) have suggested that *A. actinomycetemcomitans* is distinct from *H. aphrophilus* and *H. paraphrophilus.* Furthermore, DNA–DNA hybridization has been used by Tønjum et al. (1990) to identify clinical isolates as *A. actinomycetemcomitans* and *H. aphrophilus.* The results obtained with hybridization are supported by interstrain-to-intrastrain genetic transformation ratios. Distinction between *A. actinomycetemcomitans* and *H. aphrophilus* is generally clear-cut with respect to both hybridization and genetic transformation. Distinction between these two species has also been obtained with DNA–DNA hybridization using total genomic DNA probes (Tønjum et al., 1989).

DNA–rRNA hybridization DNA–rRNA hybridization studies demonstrate at least seven rRNA branches in the family *Pasteurellaceae* (De Ley et al., 1990). The *A. actinomycetemcomitans, H. aphrophilus, H. ducreyi,* and "*Histophilus ovis*" rRNA branches are separate and quite remote from the three authentic genera of the family. DNA–rRNA hybridization with labeled probes is therefore an excellent method to decide whether an organism belongs to the family *Pasteurellaceae* or not.

When chromosomal DNA from *A. actinomycetemcomitans* is digested with the restriction endonucleases *Cla*I, *Bam*HI, *Bgl*I, or *Hind*III and hybridized to the rrnB ribosomal RNA operon of the *Escherichia coli* chromosome, isolates from the same individual and, belonging to the same serotype are found to be genetically identical, but those belonging to different serotypes are nonidentical (Saarela et al., 1993). Isolates of the same or different serotypes are genetically nonidentical in different individuals.

Typing systems

BIOCHEMICAL TYPING The possibility of using biochemical differences among strains of a given serotype as epidemiological markers has been examined in *A. pleuropneumoniae*. Sirois and Higgins (1991) applied 38 biochemical and physiological tests to distinguish 67 strains belonging to serotype 1 and 5. Three fermentation tests have allowed the classification of serotype 1 strains into six phenotypic groups and serotype 5 strains into four of these groups. Groups II and III contain exclusively serotype 1 strains, whereas the majority of strains in groups I and IV belongs to serotype 1 and 5, respectively. The latter contains almost all the serotype 5 strains studied.

SEROTYPING Phillips (1967) has conducted serological and antigenic structure typing of *A. lignieresii*. Serological characterization of *A. pleuropneumoniae* in Australian pigs from 1993–1996 demonstrates that the dominant serovars are 12, 1, and 7 (Blackall et al., 1999a). In pigs from Quebec, serotype 1 is predominant, while serotype 5 is second in prevalence (Mittal et al., 1992). Both subtypes of serotype 5 (5a and 5b) are present in Quebec. Serotypes 3, 6, 7, 8, 10, and 12 have been isolated in small numbers, whereas serotypes 4, 9, and 11 are not present. Quantitation of the type- and group-specific antigens by the coagglutination test distinguishes serotype 1 and 9 strains from those of serotype 11 (Mittal et al., 1993).

Long-chain lipopolysaccharides have been used for the detection of antibodies in animals infected with *A. pleuropneumoniae* serotypes 4 and 7 (Gottschalk et al., 1997).

Using a monoclonal antibody-based polystyrene agglutination test, Dubreuil et al. (1996) have identified serotype 5a and 5b strains of *A. pleuropneumoniae*. Mittal et al. (1988b) have found serotype 6 strains of *A. pleuropneumoniae* to be antigenically closer to serotype 8 than to serotypes 3 or 5. A combination of serological tests, such as coagglutination followed by 2-mercaptoethanol tube agglutination, and immunodiffusion tests distinguish serotype 6 strains from those of other cross-reacting serotypes. Serotype-3 strains of *A. pleuropneumoniae* possess epitopes shared with almost all other serotypes (Mittal et al., 1988a).

Serotype-specific rabbit antisera against *A. actinomycetemcomitans* have been used to serotype 177 isolates of this organism from 136 periodontally healthy or diseased subjects by indirect immunofluorescence and/or immunodiffusion assays (Asikainen et al., 1991). Serotype b is dominant in subjects with periodontal disease, and serotype c is the most common serotype in healthy subjects. Most subjects are infected with only one serotype, and *A. actinomycetemcomitans* infections are relatively serotype-stable. Nonserotypeable *A. actinomycetemcomitans* isolates may originate from serotypeable isolates, especially those of serotype c (Paju et al., 1998).

Sensitized latex particles (SLP) for detection of capsular polysaccharides have been applied to the typing and diagnosis of *A. pleuropneumoniae* (Inzana, 1995). The SLP are agglutinated by all strains of the homologous serotype, but not by heterologous serotypes or strains of *A. suis*, *P. multocida*, or *H. parasuis*.

RESTRICTION MAPPING Minor differences in the cleavage patterns of 17 serotype 7 strains of *A. pleuropneumoniae*, detectable by restriction endonuclease analysis (REA), enable grouping of these strains into seven profiles (Wards et al., 1998).

DNA fingerprinting has also been used to differentiate 12 reference strains, each representing one serotype, and 11 field strains of *A. pleuropneumoniae* belonging to serotype 9, which is

the most frequent *Actinobacillus* species in the Czech Republic (Rychlik et al., 1994). Nine of the reference strains can be distinguished from each other and from serotypes 1, 9, and 11. A close relationship exists between the latter three. Among the 12 serotype 9 strains, four DNA types can be identified. REA is also powerful for typing and discerning genetic heterogeneity and homogeneity among *A. actinomycetemcomitans* strains (Han et al., 1991).

Genomic DNA fingerprinting by restriction-fragment-end-labeling classifies 31 strains of *A. actinomycetemcomitans* into 29 distinct types based on polymorphism (van Steenbergen et al., 1995). Serotype-specific fragments can be assigned for the three serotypes investigated.

Restriction fragment-length polymorphism (RFLP) has been used to study the epidemiology and pathogenesis of *A. actinomycetemcomitans* (DiRienzo and Slots, 1990). RFLP type B of *A. actinomycetemcomitans*, corresponding to reference strain JP2, seems to be particularly virulent, as indicated by the presence of this type in three subjects who converted from a healthy periodontal state to localized juvenile periodontitis.

RIBOTYPING Ribotyping is more efficient than biochemical fingerprinting in detecting differences between isolates of *A. equuli* from healthy and diseased horses (Sternberg and Brandstrom, 1999). There is a poor correlation between the methods. In *A. pleuropneumoniae*, ribotyping has been evaluated as an epidemiological tool and applied to study infections caused by this organism in Danish pig herds (Fussing et al., 1998a). This assessment was based on 13 reference strains and 106 epidemiologically unrelated field strains representing the nine serotypes of biotype 1 and serotype 14 of biotype 2. The enzymes *Cfo*I and *Hind*II were used to provide ribotype patterns. Ribotyping of the reference strains resulted in 10 *Cfo*I types and 11 *Hind*III types. From the Danish herd strains, 17 different *Cfo*I ribotypes and 24 *Hind*III ribotypes were established. Combining *Hind*III- and *Cfo*I-ribotyping divides the herd strains into 26 different types.

The genomic relationships among 112 *A. pleuropneumoniae* serotype 2 strains collected throughout Europe and North America have been examined by Fussing (1998). *Hind*III ribotyping of the strains produced five ribotypes with high similarity. Sequence analysis of the ribosomal intergenic region of strains representing each ribotype and each country demonstrates no differences. PFGE patterns of strains representing all countries show high similarity, suggesting that strains of *A. pleuropneumoniae* serotype 2 are clonal.

Six different typing methods have been compared for *A. actinomycetemcomitans* (van Steenbergen et al., 1994). Ribotyping provides unequivocal distinction among different strains, as compared to restriction endonuclease results. Serotyping, biotyping, and SDS-PAGE patterns of outer membrane proteins are reproducible, but have a low discriminatory potential. Ribotyping shows intrafamilial similarity among *A. actinomycetemcomitans* isolates (Alaluusua et al., 1993) and supports separation of *A. actinomycetemcomitans* from *H. aphrophilus* and *H. paraphrophilus* (Sedlácek et al., 1993).

PCR-ribotyping, repetitive extragenic palindromic element (REP)-based PCR and enterobacterial repetitive intergenic consensus (ERIC)-based PCR have been used for strain differentiation in epidemiological studies of *A. seminis* (Appuhamy et al., 1998). PCR-ribotyping gives the simplest pattern and is useful in differentiating *A. seminis* strains and distinguishing this species from related species.

REA OF PCR-AMPLIFIED 16S RRNA GENES   *A. actinomycetem-comitans*, *H. aphrophilus*, and *H. paraphrophilus* can be rapidly distinguished by REA of PCR-amplified 16S rRNA genes (Riggio and Lennon, 1997).

MULTIPLEX PCR   A multiplex PCR assay has been developed for detection of *A. pleuropneumoniae* and identification of serotype 5 isolates (Lo et al., 1998). DNA sequences specific to the conserved export and serotype-specific biosynthesis regions of the capsular polysaccharide of *A. pleuropneumoniae* serotype 5 are applied as primers to amplify 0.7 and 1.1 kb DNA fragments, respectively. The multiplex PCR assay rapidly detects *A. pleuropneumoniae* and distinguishes serotype 5 strains from other serotypes.

AP-PCR   Hennessy et al. (1993) have recommended the use of M13 and T3–T7 oligodeoxynucleotide primers for diagnostic and epidemiological identification of *A. pleuropneumoniae* by arbitrarily primed (AP) PCR techniques. In *A. actinomycetemcomitans*, the isolates of a given AP-PCR genotype usually belong to the same serotype (Asikainen et al., 1995).

TOXIN TYPING   A PCR method that has been developed for toxin typing of *A. pleuropneumoniae* is useful in differentiating strains for diagnostics and epidemiology and in detecting serotypes with atypical toxin patterns (Frey et al., 1995, 1996).

Antigenic structure   Actinobacilli have a complex antigenic structure. Heat-stable somatic antigens and heat-labile surface antigens have been demonstrated in *A. lignieresii*, *A. equuli*, *A. pleuropneumoniae*, and *A. suis* (Phillips, 1967; Mráz, 1969a; Kim, 1976; Nielsen, 1990). Species may be divided into different antigenic types. Six antigenic types, 1–6, have been detected in *A. lignieresii*, with 2 subtypes, 1a and 4c, based on heat-stable antigens (Phillips, 1967). Most cattle strains in the UK belong to type 1 and most sheep strains to types 2, 3, and 4. Diverging distributions may be found in other geographic areas. In Japan, type 3 is found in cattle and the predominant type is 5 (Nakazawa et al., 1979). In *A. lignieresii* heat-labile antigens associated with extracellular slime may cause cross reaction in agglutinating tests if unheated suspensions of the different antigenic types are used. In *A. equuli*, both heat-labile and heat-stabile antigens have been demonstrated, and a number of antigenic groups have been detected. Mráz (1968) has demonstrated the presence of antigenic groups in this organism, and Kim (1976) has distinguished 28 groups based on heat-stable O antigens. There is no evidence of host specificity among groups.

In *A. pleuropneumoniae*, 14 serovars have been recognized based on soluble surface antigens composed of capsular polysaccharides and lipopolysaccharides (in addition serovars 1 and 5 can be subdivided into 1a, 1b and 5a, 5b, respectively) (Jolie et al., 1994; Nielsen et al., 1997b).

A. rossii shows serological cross-reactions with *A. lignieresii*, *A. equuli*, *A. suis*, *A. seminis*, and *M. haemolytica* (Ross et al., 1972).

No detailed study has been made of the antigens of *A. suis*, but there is a marked uniformity in the antigenic structure of 16 strains taken from Britain, Denmark, Germany, and the United States (Kim et al., 1976).

There are antigenic cross relationships between *A. suis*, *A. lignieresii*, and *A. equuli* (Haupt, 1934; Valleé et al., 1963, 1974; Wetmore et al., 1963; Bouley, 1966; Kim, 1976). Furthermore, actinobacilli and pasteurellae share antigens (Mráz, 1969a, 1977; Ross et al., 1972).

Pulverer and Ko (1972) have found 6 heat-stable agglutinating antigens in *A. actinomycetemcomitans* and have established 24 ag-

glutinating patterns among 100 strains. King and Tatum (1962) have detected 3 precipitating antigens distributed singly among different strains. This has been confirmed by Zambon et al. (1983), who named the 3 serotypes a, b, and c. Later, serotypes d and e were added (Saarela et al., 1992; Gmür et al., 1993). Zambon et al. (1983) have found that *A. actinomycetemcomitans* cross-reacts with *H. aphrophilus*.

Tube agglutination with somatic antigens reveals 5 serotypes in "*A. salpingitidis*" that differ from other actinobacilli and closely related pasteurellae (Mráz et al., 1976). Presumed existence of a thermostable envelope antigen has been found by gel precipitation, according to Ouchterlony (1968). Its low specificity demonstrates the resemblance of all species within the genus, as well as in the group of *Actinobacillus–Pasteurella* (Mráz et al., 1976).

Pathogenicity and sources   Actinobacilli have low pathogenicity in experimental animals. Efforts to produce specific lesions with *A. lignieresii* in laboratory animals have usually failed. Chick and mice embryos may be more susceptible than guinea pigs and rabbits (Mráz, 1969a, b). Mice that are injected intraperitoneally and chick embryos affected by injection into the allantoic cavity usually die (Holmes, 1998). *A. equuli* is not pathogenic in mice, and *A. equuli* and *A. suis* are not detrimental to guinea pigs and rabbits. *A. pleuropneumoniae* causes infection in mice by the intraperitoneal and intranasal route, and *A. suis* infects mice by the intraperitoneal route. *A. ureae* is slightly pathogenic for guinea pigs, mice, and rabbits (Henriksen and Jyssum, 1961).

Actinobacilli were first found associated with cattle, but have since been isolated from other animals and humans (Peel et al., 1991). In domestic animals, they occur both as pathogens and commensals. As commensals, actinobacilli are present in the alimentary, respiratory, and genital tracts of normal animals. *A. lignieresii* has a commensal role in the mouth and rumen of healthy cattle and sheep (Phillips, 1961, 1964; Bisgaard et al., 1986). Invasion and establishment within tissue usually promote the transition from commensalism to pathogenicity. This implies that actinobacilli often operate as opportunists. A number of human infections are the result of animal bites. Actinobacilli usually cause sporadic disease, but sometimes a group of animals is affected by a common trigger factor.

A. lignieresii produces actinobacillosis in cattle and sheep. Several predisposing factors exist, and there is considerable variation in pathogenicity between strains, some of which do not cause disease at all (Magnusson, 1928). Lignières and Spitz (1902) isolated an organism from actinomycotic lesions of soft tissue in cattle that they named "l'actinobacille" due to the nature of the lesion, described as actinobacillosis. In cattle, chronic granulomatous lesions are found most frequently on the tongue ("wooden tongue") and in other soft tissues of the head and upper alimentary tract. Subcutaneous injection into cattle may produce an abscess similar to that occurring naturally. Granules consisting of rods surrounded by clubs are present in the pus. Lymph nodes in the region are invariably affected. Christiansen (1917) was the first to describe the organism in sheep as "*Bacterium purifaciens*". Tunnicliff (1941) concluded that it was similar to the organism isolated from cattle. Skin, lungs, testes, and mammary glands are the sites most commonly affected in sheep.

In addition to infections in cattle and sheep, *A. lignieresii* has been isolated from an epidural abscess (Chladeck and Ruth, 1976), an intramandibular phlegmon (Zaharija et al., 1979), and an enlarged tongue (Baum et al., 1984) of horses. *A. lignieresii* infections in humans bitten by horses and sheep have been reported (Dibb et al., 1981; Clark et al., 1984; Peel et al., 1991; Benaoudia et al., 1994). *A. arthritidis* has been isolated from the cavum oris of apparently healthy horses, as well as from joints of horses with arthritis, and from horses and foals with septecemia (Christensen et al., 2002a).

A. suis is an opportunistic pathogen of swine of all ages and has been isolated from healthy pigs (Cutlip et al., 1972; van Ostaaijen et al., 1997). Generally, *A. suis* is associated with septicemia and death in suckling and newly weaned piglets. However, septicemia in mature swine has also been reported (Miniats et al., 1989; Yaeger, 1996). Localized infections, such as endocarditis, polyarthritis, subcutaneous abscesses, and pneumonia, have also been observed (Zimmermann, 1964; Mair et al., 1974). Human infection has been reported as the result of a pig bite (Escande et al., 1996). It is likely that equine strains of organisms, described as *A. suis* or hemolytic variants of *A. equuli* may constitute a separate group of organisms, provisionally designated taxon 11 (Bisgaard et al., 1984).

A. equuli causes septicemia, often in association with nephritis, arthritis, pneumonia, pleuritis, or enteritis in young foals, a condition referred to as "sleepy foal disease" or "joint ill." Adult horses may suffer from arthritis, pneumonia, endocarditis, pericarditis, nephritis, meningitis, and metritis. Abortion may occur in pregnant animals. Golland et al. (1994) have described 15 cases of peritonitis in horses. In piglets that are a few days old, *A. equuli* may cause septicemia (Pedersen, 1977), while older pigs may suffer from arthritis, endocarditis, osteomyelitis, nephritis, and pneumonia. *A. equuli* occurs as a commensal in the equine intestinal tract and mouth. It has not been isolated from normal swine. An *A. equuli*-like bacterium has been isolated from an infected horse bite wound in a human being (Peel et al., 1991).

A. hominis was originally isolated from patients with chronic lung disease, but it is also an etiologic agent of septicemia in hepatic failure (Wüst et al., 1991) and pleural empyema (Friis-Møller and Kharazmi, 1988). It has not been isolated from animals.

A. ureae has been isolated from the human respiratory tract, sputum, and cerebrospinal fluid (Henriksen and Jyssum, 1961; Jones, 1962; Verhaegen et al., 1988). In 11 cases, it has been the primary pathogen of meningitis (Kingsland and Guss, 1995). It has also been isolated from bone marrow infection (Avlami et al., 1997), chronic bronchitis, pneumonia, sepsis, and peritonitis (Kingsland and Guss, 1995). Strains vary in virulence and have no animal host (Kolyvas et al., 1978).

A. pleuropneumoniae is the etiologic agent of a highly contagious and often fatal acute or chronic pleuropneumonia in swine. As opposed to other actinobacilli, *A. pleuropneumoniae* is a primary pathogen. Pigs that survive the infection often become subclinical carriers. *A. pleuropneumoniae* has been isolated from 42% of tonsils from 303 chronically infected slaughterhouse swine (Møller and Kilian, 1990). It is rarely recovered from the nasal cavity and the lungs. Although isolates of all serovars may cause disease, significant differences in their pathogenic potential have been reported (Rosendal et al., 1985; Jacobsen et al., 1996). Several virulence factors have been described, including capsules, lipopolysaccharides, outer membrane proteins, transferrin binding proteins, and Apx toxins (Haesebrouck et al., 1997). The capsule consists of derivatized repeating oligosaccharides. The DNA region involved in the export of capsular polysaccharide by *A. pleuropneumoniae* serovar 5a has been characterized (Ward and Inzana, 1997). Among the different bioand serovars of *A. pleuropneumoniae*, three exotoxins have been

described: ApxI and ApxII, which are hemolytic, and ApxIII, which is non-hemolytic (Frey et al., 1993; Jansen, 1994). They are toxic to porcine alveolar macrophages and neutrophiles and belong to the pore-forming RTX toxins, which are widely spread among pathogenic Gram-negative bacteria, including the *Pasteurellaceae* family (Frey et al., 1994). *A. lignieresii*, *A. equuli*, and *A. suis* also possess RTX cytolysin genes, and *A. suis* produces proteins similar to ApxI and ApxII (Burrows and Lo, 1992; Kamp et al., 1994).

A. scotiae has been isolated from porpoises (*Phocoena phocoena*) stranded on the Scottish coastline (Foster et al., 1998). One strain (M2000/95/1^T) was recovered from the brain, lung, spleen, liver, mesenteric lymph node, blood, and small intestine of one animal that died of septicemia.

A. delphinicola has been isolated from harbor porpoises, a Sowerby's beaked whale, and a striped dolphin (Foster et al., 1996). The pathological significance of this organism for the sea mammal species is not clear.

A. indolicus, *A. minor*, and *A. porcinus* are isolated from the upper respiratory tracts of pigs, where they constitute part of the residing mucosal flora (Møller et al., 1996a). They are not pathogenic to pigs, as indicated by inoculation experiments (Rosendal et al., 1985; Brandreth and Smith, 1986).

A. rossii has been isolated from the vaginas of postparturient sows and aborted piglets (Ross et al., 1972; Sneath and Stevens, 1990).

A. capsulatus has been described in rabbits, where it acts as an agent of arthritis (Arseculeratne, 1962; Phillips, 1984), in hamsters (Krause et al., 1989), and in the lungs, livers, and/or spleen tissue of snowshoe hares (Zarnke and Schlater, 1988). Isolation of *A. capsulatus* in pure culture from animals with fatal infections indicates that it may act as a primary pathogen. Inoculation experiments have shown that domestic rabbits injected intravenously with *A. capsulatus* develop lung and liver lesions similar to those observed in naturally infected hares (Arseculeratne, 1962).

A. muris has been isolated from the pharyngeal flora of healthy white laboratory mice and from uteri and vaginas of apparently healthy mice (Bisgaard, 1986).

A. succinogenes has been cultured from the bovine rumen (Guettler et al., 1999). It produces succinate-pathway enzymes in much higher amounts than do other succinic-producing bacteria (van der Werf et al., 1997). Its general properties are suggestive of a microorganism adapted to a symbiotic role in the rumen.

A. seminis has been associated with epididymitis and orchitis in rams (Baynes and Simmons, 1960, 1968; van Tonder, 1973, 1979; Erasmus et al., 1982; Sponenberg et al., 1983; De Wet and Erasmus, 1984), ovine contagious epididymitis (Ajai et al., 1980; Burgess, 1982), mastitis in ewes (Alsenosy and Dennis, 1985), polyarthritis and posthitis in sheep (Watt et al., 1970) and abortion in sheep (Foster et al., 1999).

A. actinomycetemcomitans is indigenous to humans. The pathogenicity of *A. actinomycetemcomitans* in humans has been doubted, since it is detected with actinomycetes in more than 30% of actinomycotic lesions (Heinrich and Pulverer, 1959b). However, recent evidence implicates *A. actinomycetemcomitans* in the etiology of destructive periodontal diseases in man (e.g., Zambon, 1985; Zambon et al., 1988; Haffajee and Socransky, 1994). It has also been isolated from abscesses (abdominal [Garner, 1979], brain [Martin et al., 1967; Garner, 1979], facial [Page and King, 1966], hand [Mauff et al., 1983], mediastinal [Garland and Prichard, 1983], and thyroid [Burgher et al., 1973]), as well as from actinomycosis (Colebrook, 1920), endarteritis (Symbas et al., 1967), endocarditis (Mitchell and Gillespie, 1964), meningitis (Page and King, 1966), pneumonia (Meyers et al., 1971), septicemia (Page and King, 1966), urinary tract infection (Townsend and Gillenwater, 1969), and vertebral osteomyelitis (Muhle et al., 1979).

"*A. salpingitidis*" is isolated from the respiratory tract and oviduct of clinically healthy domestic fowl, from salpingo-peritonitis in hens, and from organs of dead chicks and young turkeys (Mráz et al., 1976). It is a commensal in the respiratory tract and oviduct of domestic fowl. Independent infections arise on the basis of previous weakening; the others are of mixed or secondary character (e.g., mycoplasmosis, organ tuberculosis, leukocytosis, parasitic invasion). "*A. salpingitidis*" is pathogenic for 5-day-old chicks and eventually for white mice. Guinea pigs and rabbits are not susceptible.

ENRICHMENT AND ISOLATION PROCEDURES

Enriched media (blood agar or serum agar) used for the isolation of pathogens from animal tissues can also be used for the primary isolation of actinobacilli from tissues. For the V-factor-dependent actinobacilli (*A. indolicus*, *A. minor*, *A. pleuropneumoniae* biovar 1, and *A. porcinus*), enriched media should include NAD, or a nurse-strain should be used as described in the *Haemophilus* chapter. For primary isolation of *A. pleuropneumoniae* from pleuropneumonic lungs, enriched[1] bovine blood agar plates cross-streaked with a β-hemolytic *Staphylococcus aureus* strain will support the growth and enhance the hemolytic pattern. Media suitable for serotyping and genetic characterization of *A. pleuropneumoniae* are chocolate agar (see Genus *Haemophilus*, this edition), PPLO agar (Nicolet, 1971), and brain heart infusion broth supplemented with 10 mg/l NAD (Møller et al., 1996a).

Selective media may prevent overgrowth of actinobacilli when they are present in mixed cultures. Such media have been used to recover *A. lignieresii* from the mouths of normal cattle (Till and Palmer, 1960) and from the rumens of normal cattle and sheep (Phillips, 1961, 1964), *A. indolicus*, *A. minor*, and *A. porcinus* from the upper respiratory tract (Møller and Kilian, 1990), and *A. actinomycetemcomitans* from the periodontal pocket (Slots, 1982b). Selective and indicator media for isolation of *A. pleuropneumoniae* from the pharyngeal tonsils of chronically infected swine have been described (Møller et al., 1993; Sidibé et al., 1993; Jacobsen and Nielsen, 1995).

Gram and Ahrens (1998) have developed a PCR method for specific detection of *A. pleuropneumoniae*. When used to detect *A. pleuropneumoniae* from tonsils of carrier pigs, the PCR method, combined with cultivation, shows a high sensitivity as compared to cultivation alone. The PCR amplifies the outer membrane lipoprotein (*omlA*) gene. A multiplex PCR for specific detection of *A. pleuropneumoniae* serovar 5 has been developed by Lo et al. (1998). DNA sequences specific to the conserved export and serotype-specific biosynthesis regions of the capsular polysaccharide of *A. pleuropneumoniae* serovar 5 are used as primers. Frey et al. (1995) have developed a PCR method that allows determination of the activator, structural, and secretion genes of the three toxins ApxI, ApxII, and ApxIII in *A. pleuropneumoniae*. The profile of the individual genes in the three Apx operons is con-

stant for a given serotype. Despite the high similarity between 16S rDNA from *A. pleuropneumoniae* and *A. lignieresii*—with differences in only three positions—a species-specific oligonucleotide for *in situ* detection of *A. pleuropneumoniae* has been constructed (Fussing et al., 1998b).

MAINTENANCE PROCEDURES

Most strains die on blood agar or serum agar within 4 d (Phillips, 1984). Cultures grown in Robertson's cooked medium may be stored for up to 4 weeks, and when lyophilized in sterile rabbit serum, 20% peptone solution, or skim milk, for up to 20 years. When heavy suspensions of cultures are grown on nutrient agar, blood agar, chocolate agar, or PPLO agar and washed off in the solutions described for lyophilization, they may be stored at −70°C for up to 2 years without loss of viability.

PROCEDURES FOR TESTING SPECIAL CHARACTERS

V-factor requirement V-factor requirement is demonstrated on agar medium lacking the V factor by cross-inoculation of the plate with a nurse strain (e.g., *Staphylococcus* or *Pseudomonas*) or application of a V-factor-impregnated paper disk (Genus *Haemophilus*, this edition).

Fermentation of carbohydrates Fermentation of carbohydrates for the V-factor-dependent species can be examined in phenol red broth base (Difco) supplemented with 1% of the respective carbohydrates and 10 μg/ml NAD (Genus *Haemophilus*, this edition).

Micromethods for examination of V-factor-dependent actinobacilli The micromethods do not require growth and are recommended for the detection of indole production, urease, ornithine and lysine decarboxylases, arginine dihydrolase, and glycosidase activities (Genus *Haemophilus*, this edition).

DIFFERENTIATION OF THE GENUS *ACTINOBACILLUS* FROM OTHER GENERA

Actinobacilli are distinguished from enterobacteria by their smaller genome, (e.g., *A. pleuropneumoniae*, with 2404 ± 40 kb) (Chevallier et al., 1998), obligatory parasitism, and absence of motility. They differ from the vibrios by their smaller genome, absence of flagella, and other phenotypic properties.

Complex genomic relationships have been shown by DNA–DNA hybridization to exist between members of the genera of the family *Pasteurellaceae* (Pohl, 1979). The phenotypic distinction of *Actinobacillus* and *Pasteurella* is difficult to determine with the existing composition of the genera (Sneath and Stevens, 1985). DNA–DNA hybridization studies show that the traditional genus names in *Pasteurellaceae* can be kept only if the genera are restricted to entities clustering at or above 50% DNA binding level with the type species of the genera (Mutters et al., 1985a). This is a practical choice, since there is no consensus as to which level genera should be defined within different families. rRNA–DNA hybridization identifies seven rRNA branches, in addition to numerous unclustered taxa within the family *Pasteurellaceae* (De Ley et al., 1990). 16S rRNA sequencing of 70 strains representing different taxa within *Pasteurellaceae* has been performed by Dewhirst et al. (1992, 1993). The rRNA branches (*P. multocida*, *A. lignieresii*, *H. influenzae*, *H. aphrophilus*, *A. actinomycetemcomitans*, and *H. ducreyi*) correspond to the 16S rRNA subclusters 3B, 4A, 1C, 1B, 1A, and 2B, respectively (Dewhirst et al., 1993). These findings indicate that the genera defined by DNA–DNA hybrid-

---

1. Meat blood agar: 500 g of chopped, lean meat cooked for 20 min in 1 liter of water, supplemented with 10 g/l Bacto peptone (Difco, Detroit, USA), 5 g/l NaCl, and 10 g/l bacteriological agar no. 1 (Oxoid, Unipath Ltd., Basingstoke, England), and cooked for 20 min. pH adjusted to 7.7–7.6 by NaOH. Autoclaved at 120°C for 20 min, and 5% calf's blood added at a temperature of 45°C.

ization do not represent phylogenetic groups and that further reorganization within the family is necessary (Angen, 1997).

The close similarity between actinobacilli and the genera *Haemophilus* and *Pasteurella* necessitates comparison of biochemical characters at the species level. Boot and Bisgaard (1995) have reclassified 30 *Pasteurellaceae* strains isolated from rodents. Strains previously reported as *Actinobacillus* spp. have been reclassified as *P. pneumotropica* biotype Jawetz, *P. dagmatis*, or taxon 22. These findings underline the importance of extended characterization of isolates and comparison with reference strains to avoid misclassification within the family *Pasteurellaceae*. The main features of differentiation between these genera and others are given in the *Pasteurellaceae* chapter, Table BXII.γ.287.

TAXONOMIC COMMENTS

In *Bergey's Manual of Determinative Bacteriology*, 9th Ed. (Holt et al., 1994), the genus *Actinobacillus* included *A. actinomycetemcomitans*, *A. capsulatus*, *A. equuli*, *A. hominis*, *A. lignieresii*, *A. muris*, *A. pleuropneumoniae*, *A. rossii*, *A. seminis*, *A. suis*, and *A. ureae*. The new species *A. minor*, *A. porcinus*, and *A. indolicus*, created by Møller et al. (1996a), and *A. delphinicola*, established by Foster et al. (1996), were later added to the genus. Most recently, *A. scotiae* and *A. succinogenes* have been included in the genus *Actinobacillus* (Foster et al., 1998; Guettler et al., 1999) and Bisgaard Taxon 9 has been classified as *A. arthritidis* and genomospecies 2 (Christensen et al., 2002a).

Within the family *Pasteurellaceae*, genomic relationships have indicated that several groups/complexes stand for new genera: the *H. aphrophilus* group, the "*A. salpingitidis*" group, the "*Histophilus ovis*"/"*H. somnus*" group, and the former *P. haemolytica* and *P. pneumotropica* complexes. The rRNA branches containing *H. aphrophilus*, *A. actinomycetemcomitans*, *H. ducreyi*, and *H. ovis* represent candidates for new genera (De Ley et al., 1990). Based on DNA hybridization data, Escande et al. (1984) and Mutters et al. (1989) have proposed that *Actinobacillus* be limited to *A. lignieresii*, *A. equuli*, *A. capsulatus*, *A. suis*, *A. ureae*, *A. hominis*, *A. pleuropneumoniae*, *A. arthritidis*, and Bisgaard's taxa 5 and 11. 16S rRNA sequencing (Dewhirst et al., 1992) has demonstrated true actinobacilli in cluster 4A, which contains *A. lignieresii*, *A. pleuropneumoniae*, Bisgaard's Taxon 11, *A. ureae*, *A. equuli*, *A. suis*, and *H. parahaemolyticus*. *A. hominis* has subsequently been added (Dewhirst et al., 1993). Sequencing of additional strains includes *A. arthritidis*, Bisgaard's taxa 5, 8, and *P. bettyae* in cluster 4 (Dewhirst et al., 1993). Of these, Bisgaard's taxa 8 and 9 are closely related to *Actinobacillus sensu stricto*. *A. capsulatus* does not cluster with the true actinobacilli (Dewhirst et al., 1992, 1993). This organism, along with *A. seminis*, *P. mairii*, *P. aerogenes*, and *H. paracuniculus*, may represent a new genus within *Pasteurellaceae* (Dewhirst et al., 1993).

Dewhirst et al. (1992) have pointed out that in cluster 4, each species branches slightly deeper than the one above, giving no logical point at which to end the genus *Actinobacillus sensu stricto*. In addition, Møller et al. (1996a) have emphasized the problem of defining natural genera in the family *Pasteurellaceae*, because NAD-dependent species are distributed throughout the family. Although new genera may be warranted to accommodate species such as *A. actinomycetemcomitans*, *H. aphrophilus*, *H. parasuis*, *H. segnis* and others, the natural borders for such genera may not be easy to define. Reorganization without a logical basis may cause unnecessary confusion and frustration among clinical microbiologists (Møller et al. (1996a)).

It is clear from phylogenetic studies that *A. actinomycetemco-mitans* is not a true actinobacillus, because *Actinobacillus sensu stricto* falls in cluster 4A and *A. actinomycetemcomitans* in cluster 1A (Dewhirst et al., 1992). Furthermore, the representatives of serotypes a, b, and c of *A. actinomycetemcomitans* diverge in the established phylogenetic tree at a level greater than that seen between many species. This may arrant the establishment of subspecies from these serotypes. However, not all serotypes have yet been sequenced.

According to rRNA–DNA relatedness results (De Ley et al., 1990) and 16S rRNA sequencing data (Dewhirst et al., 1992, 1993), *A. actinomycetemcomitans*, "*A. salpingitidis*", *A. muris*, and *A. seminis* should be excluded from *Actinobacillus sensu stricto*. *A. actinomycetemcomitans* and *H. aphrophilus* are related to the genus *Haemophilus sensu stricto* (Dewhirst et al., 1992). Because *A. actinomycetemcomitans* is phenotypically different from *Haemophilus* and has a long evolutionary branch, it may be placed in its own genus (Dewhirst et al., 1992). Potts et al. (1985) have suggested that *A. actinomycetemcomitans* be transferred to the genus *Haemophilus*. However, the International Committee on Taxonomy Subcommittee on *Pasteurellaceae* has rejected this proposal because *A. actinomycetemcomitans*, although similar to *H. aphrophilus*, is not convincingly related to *H. influenzae*, the type species of the genus *Haemophilus*. Previously, Kilian (1976) had pointed out that *A. actinomycetemcomitans* and *H. aphrophilus* do not require hemin and NAD for growth and therefore, at that time, did not meet the criteria for inclusion into the genus *Haemophilus*.

Most recently, the Dewhirst group has established a phylogenetic tree for the family *Pasteurellaceae* based on 114 nonredundant 16S rRNA sequences (Family *Pasteurellaceae*, Fig. BXII.γ.215). The tree contains 21 phylogenetic clusters. One of these is the *Actinobacillus sensu stricto*, which contains *A. lignieresii*, *A. suis*, *A. equuli*, *A. hominis*, *A. ureae*, and *A. pleuropneumoniae*. *H. parahaemolyticus* was also included. Some unnamed species, such as Bisgaard Taxa 8 and 9 (*A. arthritidis* and genomospecies 2), have also joined this group, as have taxa represented by strains Kunstyr 570 (MCCM 00149), Bisgaard F64 (CCUG 28015), and SSI P585 (CCUG 19799). The genus *Actinobacillus* should probably include these organisms, as proposed by Escande et al. (1984) and Mutters et al. (1989).

The remaining clusters, except for *Haemophilus sensu stricto*, *Pasteurella sensu stricto*, *Mannheimia*, and *Lonepinella*, contain generically misclassified species of *Actinobacillus*, *Haemophilus*, and *Pasteurella* and unnamed species of *Pasteurellaceae* from a variety of culture collections. Thus, *A. minor* falls with *Haemophilus* "minor group" 202 and *A. pleuropneumoniae* with *H. paraphrohaemolyticus*, each in two subclusters of the Porcine Cluster. *A. rossii* and *A. porcinus* comprises the Rossii Cluster, together with strains considered to be part of the *P. aerogenes* complex, but do not appear with the type strain of *P. aerogenes*, which is in the Seminis Cluster. In addition, the more deeply branching and unstable Bisgaard Taxon 7 may belong to the Rossii Cluster. *A. indolicus* falls in the Parasuis Cluster, together with *H. parasuis* and a taxon represented by *H. parasuis* strain Nagasaki. *A. capsulatus* is contained in the Capsulatus Cluster, together with *H. paracuniculus* and a third taxon represented by *A. capsulatus* strain SSI P796. Previously, DNA–DNA hybridization had placed *A. capsulatus* in the *Actinobacillus sensu stricto* (Escande et al., 1984; Mutters et al., 1989). However, the sequences for five isolates (SSI P243[T], P244, P572, P662, and P796) fall in the Capsulatus Cluster. This is in contrast to a horse isolate (SSI P585 [CCUG 19799]), which falls in the middle of the *Actinobacillus sensu stricto* Cluster, as well as a hamster "*A. capsulatus*" strain (MCCM 149) and a duck isolate

(Bisgaard F64 [CCUG 28015]), which form the deepest branch of the *Actinobacillus sensu stricto* Cluster. These findings suggest that *A. capsulatus*-like strains from species other than rabbit might represent unique species. *A. actinomycetemcomitans* falls in the Aphrophilus Cluster, together with *H. segnis, H. aphrophilus, H. paraphrophilus,* and an *H. segnis*-like taxon represented by strain MCCM 00337. This is consistent with rRNA–DNA studies by De Ley et al. (1990), which suggested that the Aphrophilus Cluster be separated from the *Haemophilus sensu stricto. A. seminis* is contained in the Seminis Cluster with *P. mairii* and *P. aerogenes.* The Rodent Cluster comprises 9 species, with *A. muris, P. pneumotropica,* and "*H. influenzae-murium*" as named species. This cluster also contains Bisgaard Taxon 17 (rat), strain Kunstyr 507 (MCCM

02120; hamster), Mannheim Michael A (CCUG 28028; guinea pig), Kunstyr 246 (CCUG 28030; hamster), Mannheim A/5a (MCCM 00235; rat), and Forsyth A3 (mouse). "*A. salpingitidis*" falls in the Avian Cluster, together with *P. volantium, P. gallinarum, P. avium, P. anatis, H. paragallinarum,* Bisgaard Taxon 2 and 3, *Haemophilus* taxon C, *Pasteurella* species A, and strains representing species such as Wuthe DS 1801 (MCCM 01557), "*Actinobacillus salpingitidis*" CCUG 28018, and *Haemophilus* sp. CCUG 19808. *A. delphinicola* and *A. scotiae* fall in the Ducreyi Cluster, together with *H. ducreyi, P. langaa,* and *Phocoenobacter uteri,* while *A. succinogenes* appears in the Somnus Cluster with "*H. somnus*" and Bisgaard Taxon 6 and 10.

DIFFERENTIATION OF THE SPECIES OF THE GENUS *ACTINOBACILLUS*

Differential characteristics of species of *Actinobacillus* are given in Tables BXII.γ.291 and BXII.γ.292.

In the following list of *Actinobacillus* species, *A. lignieresii, A. suis, A. equuli, A. hominis, A. ureae,* and *A. pleuropneumoniae* are

considered species belonging to *Actinobacillus sensu stricto.* The other species in the list are pending placement in other genera of the family *Pasteurellaceae.*

List of species of the genus Actinobacillus

1. **Actinobacillus lignieresii** Brumpt 1910, 849[AL]

   lig.ni.e.re' si.i. M.L. gen. n. *lignieresii* of Lignières, named for J. Lignières, one of the bacteriologists who first isolated this organism.

   Cells are small and rod-shaped, often showing shorter coccobacillary forms. In media with fermentable carbohydrates such as glucose and maltose, long bacillary or filamentous forms can be seen. Shorter bacillary or filamentous forms are more common on media containing blood or serum. Small granules can be detected scattered among the cells and often lying at the pole of the bacillary or filamentous form. This gives the characteristic "Morse-code" form (Phillips, 1960).

   The rods, which measure 1.15–1.25 × 0.4 μm, are nonmotile, nonsporing, and non-acid-fast. Slime, but not a capsule, can be demonstrated in wet India ink preparations. Cells stain easily, particularly with carbol fuchsin. They may show bipolar staining.

   Viscous colonies, 1–2 mm in diameter, are formed on primary isolation. The sticky nature of the colonies is lost with repeated subcultivation. The colony size may increase up to 4 mm upon prolonged cultivation. *A. lignieresii* grows on MacConkey agar. There is no hemolysis on sheep blood agar. The growth is poor and not visible for some days in a gelatine stab. There is no liquefaction. Broth cultures are turbid with little deposit. Growth is improved by the addition of serum.

   The best fermentation with carbohydrates is obtained in a peptone–Lemco base with 1% carbohydrate and bromothymol blue as an indicator (Holmes, 1998). Tests should not be prolonged beyond 14 d. The organism produces acid without gas promptly from maltose, mannitol, and sucrose, but with a delayed reaction in lactose (5–7 d). Cellobiose, melibiose, rhamnose, salicin, and trehalose are not fermented. Acid is also produced in litmus milk, but no clot is formed. The oxidase reaction is generally positive, while the catalase reaction varies between strains. Esculin and sodium hippurate are not hydrolysed. Produces actinobacillosis in cattle and sheep and has been isolated from an epidural abcess, an intramandibular phlegmon, and an

enlarged tongue of a horse. *A. lignieresii* infections in humans bitten by horses or sheep have occurred.

Recently Christensen et al. (2002a) have shown that equine isolates are phenotypically indistinguishable from *A. lignieresii* and that they belong to a separate species tentatively designated genomospecies 1.

The optimum temperature for growth is 37°C. Only slight growth occurs at 20°C and no growth at 44°C.

The mol% G + C of the DNA is: 41.8–42.6 (T_m) (Boháček and Mráz, 1967).

Type strain: ATCC 49236, NCTC 4189.

GenBank accession number (16S rRNA): M75068.

2. **Actinobacillus actinomycetemcomitans** (Klinger 1912) Topley and Wilson 1936, 279[AL] (*Bacterium actinomycetemcomitans* Klinger 1912, 198.)

   ac.ti.no.my.ce.tem.co' mi.tans. Gr. n. *actis* a ray; Gr. n. *myces* a fungus; M.L. n. *actinomyces* ray fungus; L. part. adj. *comitans* accompanying; M.L. part. adj. *actinomycetemcomitans* accompanying an actinomycete.

   Cells are nonmotile cocci, measuring 0.7 ± 0.1 × 1.0 ± 0.4 μm, or are rod-shaped (Phillips, 1984; Zambon, 1985). They may occur singly, in pairs, or in small clumps. The rod-shaped cells are seen more frequently in agar cultures than in broth or gelatine cultures. While cells may be distinct rod forms in culture, many more coccal forms are seen in actinomycotic lesions.

   Colonies on agar are small, <0.5 mm in diameter after 24 h, but enlarge to 2–3 mm after 24 h (Holmes, 1998). They adhere to the agar medium, are difficult to break up, and are described as starlike (Colebrook, 1920; Blix et al., 1990) or like "crossed cigars" (Heinrich and Pulverer, 1959a). Colonies may sometimes have a rough surface and pitting after several days of cultivation. They become mucoid and nonadherent after repeated subculture (Slots, 1982a).

   Cells have a microcapsule. Primary broth colonies form tiny, discrete colonies that cling to the sides of glass tubes, while subsequent cultures become less adherent and exhibit uniform turbidity (Zambon, 1985). The organism grows poorly in air but well in 5% CO_2 (capnophilic) or under

TABLE BXII.γ.291. *Differential characteristics of* Actinobacillus *species*[a]

Characteristics	1. A. lignieresii	2. A. actinomycetemcomitans	3. Actinobacillus arthritidis	4. A. capsulatus	5. A. delphinicola	6. A. equuli	7. A. hominis	8. A. indolicus	9. A. minor	10. A. muris	11. A. pleuropneumoniae	12. A. porcinus	13. A. rossii	14. A. scotiae	15. A. seminis	16. A. succinogenes	17. A. suis	18. A. ureae	19. "A. salpingitidis"
Host:																			
Bovine rumen																+			
Cattle	+																		
Cetaceans					+									+					
Domestic fowl																			+
Horse			+			+													
Lamb																			
Humans		+					+											+	
Mouse										+									
Rabbit				+															
Sheep	+														+				
Swine								+	+		+	+	+				+		
V-factor dependency	d	−	−	−	−	+	+	+	+	−	+	+	+	−	−	−	+	−	+
Catalase	+	+[b]	+	+	−	−	−	+	+	+	d	+	+	+	+	+	−	+	(−)
Oxidase	+	−	+	+	+	d	+	+	d	+	d	−	+	+	d	+	d	+	+
Hemolysis on SBA	−	−	−	−[c]	−	d	+	d	−[c]	+	+[d]	d	d	[c]	−	−	+	+	−
Urease	−	−	+	+	−	−	−	d	−	−	+	−	−	+	−	−	+	−	−
Indole	+	−	−	−	−	+	+	−	+	−	−	+	+	+	−	−	−	+	−
ONPG[e]	d	−[f]	+	+	−	d	−	+	+	−	+	d	d	+	−	+	d	−	+
Acid production from:																			
D-Mannitol	+	d	+	+	−	+	+	+	+	+	+	d	+	−	d[f]	+	+	+	+
Raffinose	d	−	+	+	−	+	+	−	−	+	d	d	d	−	−	+	+	−[g]	+
Salicin	−	−	−	+	−	+	+	+	+	+	−	−	−	−	−	+	+	+	−
D-Sorbitol	d	−	−	+	−	d	−	−	+	−	−	d	+	−	−	+	−	−	−[d]
Sucrose	+	−	+	+	−	+	+	+	+	+	+	d	−	−	−	+	+	−[g]	+
Trehalose	−	−	−	+	−	+	−	d	d	+	−	d	−	−	−	−	+	−	+
D-Xylose	+	d	+	+	−	+	−	d	−	−	+	+	+	+	+	+	+	+	
Mol% G + C content[h]	41.8–42.6	42.7	nd	42.4	nd	40.0–41.8	40.9	35.5	38.2	46.9	42.2–43.2	41.4	41.9	40.5	43.7	45	40.5	41.2–43.7	39.6–42.9 (mean 41.5)

[a]For symbols see standard definitions; nd, not determined. *A. lignieresii, A. equuli, A. hominis, A. pleuropneumoniae, A. suis,* and *A. ureae* belong to *Actinobacillus sensu stricto.* Remaining species will eventually be transferred to a number of genera in the family *Pasteurellaceae.*

[b]Different reactions are given by different authors.

[c]Slight hemolysis on SBA may be apparent if incubated longer than 48 h.

[d]Hemolysis on SBA may be very slight or absent.

[e]ONPG reaction, hydrolysis of orthonitrophenyl – β-D-galactopyranoside.

[f]Late reaction.

[g]Most strains positive.

[h]For methods of mol% G + C estimation, see text.

TABLE BXII.γ.292. Additional characteristics of *Actinobacillus* species.[a]

Characteristic	A. lignieresii	A. actinomycetemcomitans	A. arthritidis	A. capsulatus	A. delphinicola	A. equuli	A. hominis	A. indolicus	A. minor	A. muris	A. pleuropneumoniae	A. porcinus	A. rossii	A. scotiae	A. seminis	A. succinogenes	A. suis	A. ureae	"A. saphingtidis"
Growth on MacConkey agar	+	–	d	+	–	+	–	–	–	–	–		d	–	–		+	–	+
Phosphatase	+	+	+	+	+	+	+	+	+	–	+	d	+		–	+	+	–	–
Gelatinase	–	–	–	–	+	d	–	–	–	–	–	–	–	–	–	–	–	–	–
Hydrogen sulfide produced	+	–	–	–	–	d	–	+	+	–	+	d	–	–	–	–	–	–	–
Methyl red test	–	–	–	–	–	–	–	–	–	–	–	–	–	+	–	–	–	+	–
Voges-Proskauer test	d	–	–	–	+	–	–	–	–	–	–	–	–	+	–	–	–	–	–
Lysine decarboxylase	–	–	–	–	d	–	–	–	–	–	–	–	–	–	–	–	–	–	–
Ornithine decarboxylase	–	–	–	–	d	–	–	–	–	–	–	–	–	d	d	–	–	–	–
Arginine dihydrolase	–	–	–	–	d	–	–	–	–	–	–	–	–	–	–	–	–	–	–
Fermentation (acid, no gas) of:																			
Adonitol	–	–	–	–		–	–	–	–	–	–	–	–	–	–	–	–	–	–
L-Arabinose	d	–	d	–		d	–	–	–	–	–	d	+	–	–	+	+	–	d
Cellobiose	–	v	d	–	–	–	–	–	–	+	–	–	–	–	d	d	+	–	–
Dextrin	+	–	d	–	–	+	–	–	–	–	–	–	–	–	–	+	+	–	d
Dulcitol	–	–	–	–	–	–	d	–	–	–	–	–	–	–	–	–	–	–	–
Esculin	+	–	–	–	–	–	–	–	–	–	–	–	–	–	–	–	+	–	–
D-Fructose	+	+	+	+	–	+	+	+	d	+	+	–	–	–	–	+	+	+	+
D-Galactose	+	d	+	+	+	+	+	+	+	+	+	d	+	d	d	+	+	–	+
D-Glucose	+	+	+	+	+	+	–	–	–	–	+	d	+	+	+	+	+	+	+
Glycerol	d	–	d	–	–	d	–	–	–	+	–	–	–	–	–	+	+	+	d
Inositol	–	–	–	–	–	–	–	–	–	d	–	d	+	–	–	–	–	–	d
Inulin	–	–	–	–	–	–	–	–	–	–	–	–	–	–	–	–	–	–	–
Lactose	+	+	+	+	–	+	+	d	+	–	d	d	d	+	d	+	+	–	+
Maltose	+	+	+	+	+	+	+	+	+	+	+	d	d	–	–	+	+	+	d
D-Mannose	–	–	+	+	+	+	d	+	+	+	+	d	d	+	d	+	+	–	+
Melibiose	+	–	–	–	–	+	+	d	d	+	–	d	–	–	–	+	+	–	+
L-Rhamnose	–	–	–	–	–	–	–	–	d	–	–	–	–	–	–	–	–	–	–
Starch	–	d	–	–	–	d	–	d	d	–	–	d	d	–	–	–	d	–	d

[a]Symbols: see standard definitions.

anaerobic conditions (Holm, 1954; Slots, 1982a). Optimum growth at 37°C (Heinrich and Pulverer, 1959a).

Slots (1982a) has studied 135 biochemical characters in 6 reference strains of *A. actinomycetemcomitans* and 130 strains freshly isolated from the oral cavity. They all decompose hydrogen peroxide, are oxidase negative and benzidine positive, reduce nitrate, produce strong alkaline and acid phosphatases, and ferment fructose, glucose, and mannose. Variable fermentation results have been obtained with dextrin, maltose, mannitol, and xylose. Carbohydrate fermentation has been used to subgroup *A. actinomycetemcomitans* into 3–10 biochemical groups based on the variable fermentation of key sugars, including dextrin, galactose, maltose, mannitol, and xylose (King and Tatum, 1962; Pulverer and Ko, 1970; Slots et al., 1980). *A. actinomycetemcomitans* is indole negative and produces catalase.

A number of chemotaxonomic techniques have been applied to establish the relationship between *A. actinomycetemcomitans* and closely related species (Olsen, 1993; Olsen et al., 1999). Indigenous to humans where recent evidence implicates *A. actinomycetemcomitans* in the etiology of destructive periodontal diseases. It has also been isolated from abcesses and from cases of actinomycosis, endarteritis, endocarditis, meningitis, pneumonia, septicemia, urinary tract infections, and vertebral osteomyelitis.

The mol% G + C of the DNA is: 42.7 (T_m).

Type strain: ATCC 33384, NCTC 9710.

GenBank accession number (16S rRNA): M75039.

3. **Actinobacillus arthritidis** Christensen, Bisgaard, Angen and Olsen 2002a, 1244[VP]

ar.thri' ti.dis. N.L. fem. gen. n. *arthritidis* of arthritis; from Gr. n. *árthron* joint; L. suff. *-idis* used in names of inflammations.

Cells are pleomorphic without distinct morphology. On bovine blood agar the colonies are nonhemolytic, circular, raised, and regular with an entire margin. The surface of the colonies is generally smooth, shiny, grayish and nontransparent. Mucoid and watery cultures also exist. A diameter of 2–3 mm is often observed after 24 h aerobic incubation at 37°C. The species is positive in the following tests: catalase, oxidase, fermentative reaction in Hugh–Leifson medium with (+)-D-glucose, porphyrin test, nitrate reduction urease, alanine aminopeptidase, phosphatase, ONPG, α-galactosidase, and production of acid from (−)-D-ribose, (+)-D-xylose, (−)-D-mannitol, (−)-D-sorbitol, (−)-D-fructose, (+)-D-galactose, (+)-D-glucose, (+)-D-mannose, lactose, sucrose, and raffinose. Negative results are obtained in Gram staining. Motility at 22 and 37°C, symbiotic growth, β-hemolysis, Simmons citrate, mucateacid, malonate-base, H₂S/TSI, growth in the presence of KCN, methyl red and Voges–Proskauer at 37°C, production of gas from nitrate, arginine dihydrolase, lysine decarboxylase, ornithine decarboxylase, phenylalanine deaminase, indole production, gelatinase, hydrolysis of Tweens 20 and 80, pigment production, α-glucosidase (PNPG), β-glucosidase (NPG), α-fucosidase, (ONPF), α-glucuronidase (PGUA), α-mannosidase, and β-xylosidase. Negative for production of acid from *m*-erthyritol, adonitol, (+)-D-arabitol, xylitol, (−)-L-xylose, dulcitol, *m*-inositol, (+)-D-fucose, (+)-L-rhamnose, (−)-L-sorbose, cellobiose, trehalose, (+)-D-melezitose, (+)-D-glycogen, inulin, esculin, amyg-

dalin, arbutin, gentiobiose, salicin, (+)-D-turanose, and β-*N*-CH₃ glucosamide. No gas produced from (+)-D-glucose.

Strains of *A. arthritidis* have been isolated from the oral cavity of apparently healthy horses, as well as from diseased foals and horses (mainly those with septicemia and arthritis). Isolates phenotypically indistinguishable from *A. arthritidis*, except for being sorbitol negative are tentatively designated genomospecies 2 (Christensen et al., 2002a.).

The mol% G + C of the DNA is: unknown.

Type strain: Wetmore 1706, ATCC 13376, CCUG 24862.

GenBank accession number (16S rRNA): AF247712.

4. **Actinobacillus capsulatus** Arseculeratne 1962, 38[AL]

cap.su.la' tus. L. n. *capsula* a small chest, capsule; M.L. masc. adj. *capsulatus* encapsulated.

Cells are rod-shaped. Filamentous bacilli that fragment into minute, coccoid bodies are seen in old cultures (Phillips, 1984). Moniliform bodies are formed in 5-d-old cultures on Loeffler's serum. Capsules are present. Primary cultures will not grow on nutrient agar or in nutrient broth. However, subcultures grow as pinpoint colonies or as a faint turbidity. Colonies on sheep blood agar are very sticky. "Flower head" colonies are produced on rabbit blood agar (Phillips, 1984). Growth is sparse on Loeffler's serum, but profuse on Dorset egg, where colonies survive for 10 d. In serum broth, small discrete mural colonies occur.

Optimum temperature is 37°C. No growth at 22°C. Cells are killed after 10 min at 60°C. Differential biochemical characters are given in Tables BXII.γ.291and BXII.γ.292.

A. capsulatus has been isolated from rabbits, hamsters, and snowshoe hares.

The mol% G + C of the DNA is: 42.4 (T_m) (Mannheim et al., 1980).

Type strain: Frederiksen P243, ATCC 51571, NCTC 11408.

GenBank accession number (16S rRNA): M75062.

5. **Actinobacillus delphinicola** Foster, Ross, Malnick, Willems and Garcia 1996, 652[VP]

del.phi.ni' co.la. L. n. *delphinus* dolphin; L. n. *cola* dweller; M.L. n. *delphinicola* dolphin dweller.

Cells are nonmotile, facultatively anaerobic, pleomorphic rods (Foster et al., 1996). Added CO₂ is required for growth.

Colonies on CSBA (Columbia agar supplemented with 5% citrated sheep blood) incubated at 37°C in an atmosphere containing 10% added CO₂ are circular, entire, low, convex, smooth, gray, and 0.75–1 mm in diameter after 24 h. They are nonhemolytic or weakly hemolytic. Blood and serum enhance growth. There is no growth on MacConkey agar. Growth occurs at 42°C, but not at 22°C. Acid is produced from glucose and mannose, but not from adonitol, dulcitol, galactose, inositol, inulin, lactose, maltose, melibiose, raffinose, rhamnose, salicin, sorbitol, sucrose, trehalose, and xylose.

Catalase negative and oxidase positive. Nitrate is not reduced. Voges–Proskauer test is positive. Urease, H₂S, and indole tests are negative. Alkaline phosphatase, esterase–lipase, leucine arylamidase, and acid phosphatase are produced in large amounts. Naphthol-AS-BI-phosphohydrolase is generated, usually in large amounts, and esterase is produced in lesser amounts, as assessed by the API ZYM system.

Some strains may hydrolyze arginine and decarboxylate ornithine and/or lysine. Isolated from harbor porpoises, a Sowerby's beaked whale, and a striped dolphin.

The mol% G + C of the DNA is: not determined.

Type strain: M906/93, NCTC 12870.

GenBank accession number (16S rRNA): X89377.

Additional Remarks: The 16S rRNA sequence is not from the type strain.

6. **Actinobacillus equuli** (van Straaten 1918) Haupt 1934, 513[AL] (*Bacillus equuli* van Straaten 1918, 75.)
e.quu'li. L. n. *equulus* foal; L. gen. n. *equuli* of a foal.

Cells are rod-shaped, but vary markedly depending upon the growth medium. Longer, filamentous forms, similar to those of *A. lignieresii*, are seen when the medium contains glucose or maltose. On nutrient or blood agar, *A. equuli* colonies are so viscous that a string of material is formed between the medium and inoculating loop when an attempt is made to lift the colony from the medium (Holmes, 1998). Stickiness is not lost upon repeated subculturing. When first isolated from clinical material, colonies are usually rough, but they may become smooth on repeated subculture. The viscous character is also present in liquid culture, and it is not lost on repeated subculture.

Growth occurs between 20°C and 39°C. Some strains will grow at 44°C. Glycerol and mannose are usually fermented slowly. Fermentation may be delayed with dextrin, fructose, maltose, melibiose, raffinose, sucrose, trehalose, and xylose. Most strains fail to ferment arabinose, cellobiose, salicin, or sorbitol, but some strains ferment them either promptly or slowly.

Causes "sleepy foal disease". Adult horses may also be infected and abortions may occur in pregnant animals. *A. equuli* may also infect pigs.

The mol% G + C of the DNA is: 40.0–41.8 (T_m) (Boháček and Mráz, 1967).

Type strain: ATCC 19392, NCTC 8529.

GenBank accession number (16S rRNA): M75072.

7. **Actinobacillus hominis** Friis-Møller 1985, 375[VP] (Effective publication: Friis-Møller 1981, 156.)
ho'mi.nis. L. masc. n. *homo* man; L. gen. n. *hominis* of man.

Cells are small, nonmotile, polymorphic rods with small capsules (Friis-Møller, 1981). The capsule is best developed on chocolate agar. On nutrient agar, cultures grow with capsulated and noncapsulated variants. No growth is observed on MacConkey agar and 7.5% salt agar.

On 5% blood agar, colonies are 1.0 mm, grayish-white, and soft mucoid, without hemolysis after 24 h at 35°C. After 24 h, the colonies are flatter, with concentric rings. Colonies are up to 2 mm on nutrient agar, and on chocolate agar are 1.5–2.5 mm and very mucoid. There is no CO_2 or V-factor-requirement. Growth is very poor at 22°C, and no growth is seen at 4°C. In semisolid medium, colonies are facultatively anaerobic.

Strains are CAMP negative and possess very strong alkaline and acid phosphatase and esterase (C_4) activity, as shown by API ZYM tests. Nearly all strains have valine arylamidase and some α-fucosidase activity. All isolates have α-galactosidase and some α-glucosidase activity (Friis-Møller, 1981). Other biochemical characters are listed in Tables BXII.γ.291 and BXII.γ.292. Originally isolated from

patients with chronic lung disease, but also isolated from cases of septicemia in hepatic failure, and pleural empyema. Not isolated from animals. Originally isolated from patients with chronic lung disease, but also isolated from cases of septicemia in hepatic failure, and pleural empyema. Not isolated from animals.

The mol% G + C of the DNA is: 40.9 (Bd).

Type strain: P 578, CCUG 19800, ATCC 49457, P578 NCTC 11529.

GenBank accession number (16S rRNA): L06076.

8. **Actinobacillus indolicus** Møller, Fussing, Grimont, Paster, Dewhirst and Kilian 1996a, 956[VP]
in.do'li.cus. M.L. adj. *indolicus* pertaining to indole, which is formed by the organism.

Cells are small rods with lengths varying from 0.8–2.4 μm (Møller et al., 1996a). Colonies are grayish and opaque on chocolate agar and 2 mm in diameter after 48 h of incubation. Cultures have a characteristic pungent smell. Isolates show satellite growth on blood agar plates when cross-inoculated with an appropriate feeder strain.

Galactose, glucose, mannose, maltose, and sucrose are fermented. Arabinose, inulin, esculin, salicin, mannitol, sorbitol, and inositol are not fermented. Lysine and ornithine are not decarboxylated. Cells are V-factor-dependent, X-factor-independent, nonmotile, nonhemolytic, urease negative, and catalase and indole positive.

Isolated from the upper respiratory tracts of pigs.

The mol% G + C of the DNA is: 35.5 (HPLC).

Type strain: 46KC2, CIP 105316, CCUG 39029.

GenBank accession number (16S rRNA): U65584.

9. **Actinobacillus minor** Møller, Fussing, Grimont, Paster, Dewhirst and Kilian 1996a, 955[VP]
mi'nor. L. comp. adj. *minor* less, smaller, referring to the name of this taxon used previously (Møller and Kilian, 1990).

Cells appear as small rods, varying in length from 1.6 to 2.4 μm. They are V-factor-dependent but X-factor-independent. Furthermore, they are nonmotile, nonhemolytic, urease positive, and indole negative. Colonies on chocolate agar are smooth, grayish, and approximately 0.8 mm in diameter after 48 h of incubation. Isolates show satellite growth on blood agar plates when cross-inoculated with a nurse strain. Glucose, mannose, maltose, sucrose, and raffinose are fermented, but arabinose, inulin, esculin, mannitol, sorbitol, and inositol are not. Lysine and ornithine are not decarboxylated. *A. minor* has been isolated from the upper respiratory tracts of pigs.

The mol% G + C of the DNA is: 38.2 (HPLC).

Type strain: NM305, CIP 105314, CCUG 38923.

GenBank accession number (16S rRNA): U65582.

10. **Actinobacillus muris** Bisgaard 1988, 220[VP] (Effective publication: Bisgaard 1986, 1)
mu'ris. L. n. *mus* the mouse; L. gen. n. *muris* of the mouse.

The name *Actinobacillus muris* was used by Wilson and Miles (1964) as a name for the organism called *Streptobacillus moniliformis*. It was clearly a junior synonym and therefore illegitimate (Bisgaard, 1986). Bisgaard (1986) proposed the name *A. muris* for the organism listed here, since it was not on the Approved List of Bacterial Names.

Cells are nonmotile rods. The type strain is identical to

P. ureae strain Ackerman 80–443D (Ackerman and Fox, 1981). Surface cultures on bovine blood agar are circular, low convex, and regular with an entire margin (Bisgaard, 1986). Colonies have a smooth, shiny, opaque, and grayish surface and, at 37°C, reach a diameter of 1.5–2.0 mm after 24 h and 3.5–4.0 mm after 48 h. No hemolysis. Growth is butyrous and easily emulsified and cultures have no characteristic odor. Colonies are easily removed from the agar surface.

Ribose, mannitol, fructose, glucose, mannose, cellobiose, maltose, melibiose, sucrose, trehalose, melezitose, raffinose, and salicin are fermented. Catalase, oxidase, nitrate, and urease tests are positive.

The genome molecular mass is 1.4×10^9 Da (Bisgaard, 1986). *A. muris* has been isolated from the pharynx, uteri, and vaginas of mice.

The mol% G + C of the DNA is: 46.9 (T_m) (Piechulla et al., 1985a).

Type strain: Ackerman 80-443D, ATCC 49577, CCUG 16938, NCTC 12432.

GenBank accession number (16S rRNA): AF024526.

11. **Actinobacillus pleuropneumoniae** (Shope 1964) Pohl, Bertschinger, Frederiksen and Mannheim 1983, 513[VP] (*Haemophilus pleuropneumoniae* Shope 1964, 362.)

pleu.ro.pneu.mo′ niae. Gr. n. *pleura* lung, sac; Gr. n. *pneumon* the lungs; M.L. fem. gen. n. *pleuropneumoniae* of pleuropneumonia.

Matthews and Pattison (1961) were the first to describe this organism, but they called it *H. parainfluenzae* and did not propose a new species (Kilian and Biberstein, 1984). Pohl et al. (1983) transferred the organism from the genus *Haemophilus* to *Actinobacillus* on the basis of phenotypic and DNA relatedness. It contains two provisional biotypes. Biovar 1 consists of V-factor-dependent strains hitherto classified as *H. pleuropneumoniae* (Matthews and Pattison) Shope strains, as described by Kilian et al. (1978), with strain Shope 4074 (CCM 5869 ATCC 27088) as the type strain. Biovar 2 consists of V-factor-independent strains and is represented by strain Bertschinger 2008/76 Frederiksen P597 = HIM 677-314.

According to the emended description by Pohl et al. (1983), cells are nonmotile, nonsporeforming, small, and coccoid to rod-shaped, occurring singly, in pairs, or in short chains. Cells are not acid-fast and contain demethylmenaquinone and ubiquinone.

Colonies on chocolate agar are smooth and grayish white and reach 3 mm in diameter after 48 h in a candle jar at 37°C. On bovine or sheep blood agar, the majority of isolates produce a narrow zone of complete hemolysis. The hemolysin acts synergistically with *Staphylococcus aureus* β-toxin, resulting in a positive CAMP reaction. Capsulated strains produce iridescent colonies on clear agar media (e.g., Levinthal's agar and PPLO agar). Biovar 1 isolates show satellite growth on blood agar plates when cross-inoculated with a nurse strain.

Aerobic, microaerophilic in primary isolation, facultatively anaerobic. No additional CO_2 is required for growth in subculture. Mesophilic. Chemoorganotrophic, with pronounced growth response to fermentable carbohydrates in peptone media. Minimal nutritional requirements are not known.

Biovar 1 is the dominant cause of porcine pleuropneumonia around the world. A total of 14 serovars have been recognized (and biovars 1 and 5 subdivided into 1a, 1b and 5a, 5b, respectively [Nielsen, 1990; Nielsen et al., 1997b]). X factors are not required. Attention is drawn to the cultural resemblance of *A. pleuropneumoniae* biovar 2 to *A. suis* and *Mannheimia varigena* (Bisgaard's Taxon 15) (Angen et al., 1999a; Genus *Pasteurella*, this edition).

Some biochemical reactions are given in Tables BXII.γ.291 and BXII.γ.292. A urease-negative variant of *A. pleuropneumoniae* has been described (Blanchard et al., 1993).

The mol% G + C of the DNA is: 43.2 or 42.2 (T_m) (Kilian et al., 1978); 42.2 (T_m) (strain Bertschinger 2008/76).

Type strain: 4074, ATCC 27088, NCTC 12370.

GenBank accession number (16S rRNA): M75074.

Additional Remarks: GenBank accession number for 16S rRNA of MCCM 00189 is AF224283.

12. **Actinobacillus porcinus** Møller, Fussing, Grimont, Paster, Dewhirst and Kilian 1996a, 956[VP]

por.ci′ nus. M.L. adj. *porcinus* pertaining to pigs, hogs.

Cells are small rods (length 0.2–2.4 μm) (Møller et al., 1996a). Colonies on chocolate agar are smooth and translucent to grayish. Some isolates grow poorly on chocolate agar, and their colonies are only 0.2 mm in diameter. Others grow readily on chocolate agar and form colonies that are approximately 0.6 mm in diameter after 48 h of incubation. Isolates show satellite growth on blood agar plates when cross-inoculated with a nurse strain. The fermentation of most carbohydrates is variable. No fermentation of inulin, esculin, and salicin. *A. porcinus* is V-factor-dependent, X-factor-independent, nonmotile, nonhemolytic, and urease, catalase, and indole negative. Lysine and ornithine are not decarboxylated. Isolated from the upper respiratory tracts of pigs.

The mol% G + C of the DNA is: 41.4 (HPLC).

Type strain: NM319, CIP 105315, CCUG 38924.

GenBank accession number (16S rRNA): U65583.

13. **Actinobacillus rossii** Sneath and Stevens 1990, 151[VP]

ros.si′ i. M.L. gen. n. *rossii* of Ross, named after R.F. Ross, who, with his colleagues, isolated the organism (Ross et al., 1972).

Phenon 17 of Sneath and Stevens (1985); group "Ross" of Kilian and Frederiksen (1981b).

Cells are small, nonmotile bacilli, seldom coccobacilli that are often >2 mm long (Sneath and Stevens, 1990). Endospores are not formed. Growth is aerobic and facultatively anaerobic. Mesophilic. Surface colonies grown aerobically on sheep blood agar are round, grayish, semitransparent, and about 2 mm in diameter after 48 h at 37°C. Occasionally, weak hemolysis without marked greening of the erythrocytes is seen. *A. rossii* produces acid from inositol and sorbitol, but not from sucrose. The organism also reduces Janus green (Sneath and Stevens, 1985). Isolated from the vaginas of postparturient sows and aborted piglets.

The mol% G + C of the DNA is: 41.9 (T_m) (Mannheim et al., 1980).

Type strain: ATCC 27072, NCTC 10801.

14. **Actinobacillus scotiae** Foster, Ross, Patterson, Hutson and Collins 1998, 933[VP]

sco.ti.ae. L. n. *scotiae* classical Latin name of Scotland; L. gen. n. *scotiae* of Scotland, where the isolates were collected from porpoises.

Cells are pleomorphic, nonmotile, Gram-negative rods. They are facultatively anaerobic. Added CO_2 is required for growth. Colonies on Columbia agar (Difco) supplemented with 5% citrated sheep blood that are incubated at 37°C in an atmosphere of 10% added CO_2 are 0.5 mm in diameter after 24 h. They are weakly hemolytic. Blood or serum enhances growth. No requirement of X or V factors. No growth on MacConkey agar. Growth occurs at 25°C, but not at 42°C. Catalase negative and oxidase positive. Nitrate is reduced. Acid is produced from glucose, lactose, and maltose, but not from adonitol, arabinose, dulcitol, inositol, mannose, melibiose, raffinose, rhamnose, salicin, sorbitol, sucrose, trehalose, or xylose. Some strains produce acid from galactose. Urease and Voges–Proskauer positive, but indole negative. Some strains generate ornithine decarboxylase. Lysine decarboxylase and arginine dihydrolase are not produced. The type strain is positive for ornithine decarboxylase and negative for acid production from galactose. *A. scotiae* has been isolated from porpoises.

The mol% G + C of the DNA is: 40.5 (T_m).

Type strain: M2000/95/1, ATCC 27072, NCTC 12922.

GenBank accession number (16S rRNA): Y09653.

15. **Actinobacillus seminis** (ex Baynes and Simmons 1960) Sneath and Stevens 1990, 151[VP]

sem.i'nis. L. n. *semen* seed; L. gen. n. *seminis* of semen.

Phenon 14 of Sneath and Stevens (1985); "*Actinobacillus seminis*" of Kilian and Frederiksen (1981b) and Phillips (1984).

Cells are small, nonmotile bacilli and coccobacilli that are often >2 μm long (Sneath and Stevens, 1990). Gram reaction is negative. Endospores are not formed. Growth is aerobic and facultatively anaerobic, as well as mesophilic. On first isolation, 5% CO_2 is often required. On sheep blood agar, surface colonies grown aerobically are round, grayish, semitransparent, and about 1 mm in diameter after 48 h at 37°C. Colonies do not produce hemolysis, but sometimes produce greening of the erythrocytes. *A. seminis* has been associated with epididymitis and orchitis in rams; ovine contagious epididymitis; mastitis in ewes; and polyarthritis, posthitis, and abortion in sheep.

A. seminis has been associated with epididymitis and orchitis in rams; ovine contagious epididymitis; mastitis in ewes; and polyarthritis, posthitis, and abortion in sheep.

The organism produces slow and weak fermentation of carbohydrates and no acid from mannose. It also gives a negative phosphatase reaction (Sneath and Stevens, 1985).

The mol% G + C of the DNA is: 43.7 (T_m) (Mannheim et al., 1980).

Type strain: ATCC 15768, NCTC 10851.

GenBank accession number (16S rRNA): M75047.

16. **Actinobacillus succinogenes** Guettler, Rumler and Jain 1999, 214[VP]

suc.ci.no'ge.nes. M.L. n. *acidum succinicum* succinic acid; Gr. v. *gennaio* to produce; M.L. adj. *succinogenes* succinic-acid-producing.

Cells are nonmotile, pleomorphic, Gram-negative rods (0.8 × 1 μm). Filamentous cells are occasionally seen.

"Morse code" forms and chains are common in actively growing broth cultures. Cells within an extracellular matrix are common in aerobically grown cultures. Chemoorganotrophic. Yeast extract and CO_2 stimulate growth. Colonies on TSB agar are circular, entire, gray, translucent, and 1–1.5 mm in diameter after 24 h incubation at 37°C with CO_2. Growth occurs at 37–39°C, but not at 20°C or 45°C. Catalase and oxidase positive. Alkaline phosphatase positive. Acid, but not gas, is produced from glucose and fructose within 24 h. Acid is also produced from arabinose, ribose, xylose, galactose, mannose, mannitol, sorbitol, amygdalin, arbutin, esculin, salicin, cellobiose, maltose, lactose, sucrose, raffinose, β-gentiobiose, arabitol, gluconate, and 5-ketogluconate. Acid is not produced from dulcitol, inositol, inulin, glycerol, erythritol, arabinose, xylose, adonitol, methyl β-xyloside, sorbose, rhamnose, methyl α-D-mannoside, methyl α-D-glucoside, N-acetylglucosamine, melibiose, trehalose, melezitose, amidon, glycogen, xylitol, turanose, xylose, tagatose, fucose, arabitol, and 2-ketogluconate. β-Galactosidase, arginine-β-arylamidase, and leucine arylamidase are produced in large amounts. Nitrates are reduced to nitrites. Indole and urease are not produced. Rapid and extensive growth is observed when utilizing arabinose, cellobiose, fructose, galactose, glucose, lactose, maltose, mannitol, mannose, sucrose, xylose, or salicin in media with yeast extract and corn steep liquor. Marked production of succinic acid in presence of CO_2 and high substrate concentration. *A. succinogenes* has been cultured from the bovine rumen.

The mol% G + C of the DNA is: 45 (HPLC) (Mesbah et al., 1989).

Type strain: 130Z, ATCC 55618.

GenBank accession number (16S rRNA): AF024525.

17. **Actinobacillus suis** van Dorssen and Jaartsveld 1962, 456[AL]

su'is. L. n. *sus* the pig, swine; L. gen. n. *suis* of the pig.

Cells are usually rod-shaped. Considerable variability may be seen, with long rods and filaments of variable length among cells grown on media both with and without glucose. Colonies on nutrient agar or blood agar are sticky and adhere to the medium. Growth is less viscous than in *A. equuli*. Stickiness increases with prolonged incubation of up to 72 h. In older colonies, a transparent border zone is developed, giving the appearance of a fried egg. The marked adherence to the medium may be lost upon repeated subculturing. Viscous growth also occurs in nutrient broth, but is not so marked as in *A. equuli*. After centrifugation, broth-grown sedimented cells are creamy yellow. Nutrient agar and nutrient broth cultures die within 15 d at 4°C.

A. suis usually ferments arabinose, dextrin, and galactose promptly, but a few strains may show slow fermentation with these substrates. Acid is also formed from cellobiose, lactose, maltose, melibiose, salicin, and sucrose, but not from mannitol and rhamnose. Glycerol and mannose are usually slowly fermented. Most strains of *A. suis* (and *A. equuli*), but not *A. lignieresii*, ferment melibiose and trehalose (Holmes, 1998). All strains of *A. suis*, but no strains of *A. equuli* and *A. lignieresii*, hydrolyze esculin. *A. suis* is an opportunistic pathogen of swine. Human infections have been reported as the result of pig bites.

The mol% G + C of the DNA is: 40.5 (T_m) (Mráz, 1968).

Type strain: ATCC 33415, CCM 5586.

GenBank accession number (16S rRNA): AF015299.

18. **Actinobacillus ureae** (Jones 1962) Mutters, Pohl and Mannheim 1986, 343^{VP} (*Pasteurella ureae* Jones 1962, 150.)

u.re' ae. Gr. n. *urum* urine; M.L. gen. n. *ureae* of urine.

Cells are small, rod-shaped, and pleomorphic depending on the growth medium. Occasionally, bipolar staining is seen. Growth is best on media containing blood or serum. It occurs from 25–40°C, with optimal growth at 37°C. Differential biochemical characters are given in Tables BXII.γ.291 and BXII.γ.292. *A. ureae* occurs infrequently in the noses of healthy humans and occasionally causes ozaena and other infections in the respiratory tract.

The mol% G + C of the DNA is: 41.2–43.7 (T_m) (Pohl, 1979; Mutters et al., 1984).

Type strain: Henriksen 3520/59, ATCC 25976, NCTC 10219.

GenBank accession number (16S rRNA): M75075.

19. **"Actinobacillus salpingitidis"** (Kohlert 1968) Mráz, Vladík and Boháček 1976.

Cells are coccobacteria of 0.5 ± 0.1 × 1.5 ± 0.5 μm, with a tendency for bipolar staining (Mráz et al. 1976). Longer polymorphic rods and curved filaments occur. In the microscope, cells appear individually, in pairs, and in short chains. Long chains can be found in broth cultures. Nonmotile. Gram negative.

Colonies on nutrient agar are 0.1 mm in diameter. Some strains show a mixture of colony sizes (0.1–0.5 mm). They glisten in transmitted light. Older colonies of larger types reach 1 mm and acquire a grayish appearance. Colonies on nutrient agar with sheep's blood are 0.2–1 mm in diameter, exhibiting a 0.5–1.5 mm zone of expressive or total hemolysis. After 2 d of incubation, they measure 0.5–2 mm. The larger zones turn grayish and have an umbonate profile. Colonies on nutrient agar with lamb's blood, according to Smith (1962), have a simple zone of expressive or total hemolysis. MacConkey agar with crystal violet (BioQuest)

yields visible growth during 48 h of incubation. Endo agar provides no growth, except in sporadic cases. Nutrient broth yields homogeneous or fine-granulated turbidity with possible larger flakes. From the third day of incubation, a whitish-gray sediment is formed. With slight shaking the content of the test tube may be easily redispersed. Growth in litmus milk gives no changes or sporadic delayed discoloration.

Metabolism is fermentative. During 2 days of incubation, acid, but no gas, is formed from glucose, xylose, levulose, mannose, galactose, saccharose, trehalose, and mannitol; within 14 d acid is also formed from lactose (3–12 d), raffinose (2–10 d), and glycerol (2–7 d). Results are variable and, for a great part, delayed for arabinose, maltose, inulin, starch, dextrin, sorbitol, and inositol, while rhamnose, cellobiose, melibiose, adonitol, dulcitol, salicin, and esculin remain negative.

Nitrites are produced from nitrates. Urea is not hydrolysed. Indole is not produced. Hydrogen production is variable. The methyl red test is negative. Acetoin is not produced. Ammonia production is variable. Catalase is positive. Oxidase is usually positive, and cytochrome oxidase is negative. Arginine dihydrolase and lysine and ornithine decarboxylases are negative. β-Galactosidase is positive. Facultatively anaerobic. Growth at 30–40°C. Temperature optimum is 37°C. Isolated from the respiratory tract and oviduct of healthy domestic fowl, from salpingo-peritonitis in hens, and from the organs of dead chicks and young turkeys.

The mol% G + C of the DNA is: 39.6–42.9; mean value 41.5 (T_m).

Deposited strain:

GenBank accession number (16S rRNA): L06077.

Additional Remarks: Suggested neotype strain: CCM 5974 (strain 556/71 of Mráz et al., 1976) = CCUG 23139.

Genus III. **Haemophilus** Winslow, Broadhurst, Buchanan, Krumwiede, Rogers and Smith 1917, 561^{AL}

MOGENS KILIAN

Hae.mo'phi.lus. Gr. n. *haima* blood; Gr. n. *philos* lover; M.L masc. *Haemophilus* blood-lover.

Minute to medium-sized **coccobacilli or rods**, generally less than 1 μm in width and variable in length, sometimes forming threads or filaments and showing **marked pleomorphism**. Gram negative. **Nonmotile. Aerobic or facultatively anaerobic. Require preformed growth factors present in blood**, particularly X factor (protoporphyrin IX or protoheme) and/or V factor (nicotinamide adenine dinucleotide [NAD] or NAD phosphate [NADP]). Even after specific growth factors have been provided, growth is best on complex media. Optimum temperature, 35–37°C. **Nitrates are reduced to or beyond nitrites.** Oxidase and catalase reactions vary among strains. **Chemoorganotrophic. All species can attack carbohydrates fermentatively**, yielding acetic, lactic, and succinic acids as end products in glucose broth. Occur as **obligate parasites on the mucous membranes of humans and a variety of animal species.** Several 16S rRNA sequence signatures for the family *Pasteurellaceae* have been demonstrated, but none of these is specific for the genus *Haemophilus* as presently defined (Dewhirst et al., 1992).

The mol% G + C of the DNA is: 37–44.

Type species: **Haemophilus influenzae** (Lehmann and Neumann 1896) Winslow, Broadhurst, Buchanan, Krumwiede, Rogers, and Smith 1917, 561 (*Bacterium influenzae* Lehmann and Neumann 1896, 187.)

FURTHER DESCRIPTIVE INFORMATION

Phylogenetic position The genus *Haemophilus* is located in the *Gammaproteobacteria* and is a member of the family *Pasteurellaceae* (Sneath and Johnson, 1973; Pohl, 1979, 1981a; Mannheim et al., 1980; Mannheim 1981; Chuba et al., 1988; De Ley et al., 1990; Dewhirst et al., 1992, 1993).

Cell morphology Cells of *Haemophilus* species tend to occur as individual short rods or coccobacilli. Filament formation is environmentally influenced and develops as cultures age and are under less than optimum conditions of growth (Fig. BXII.γ.216). Capsules are present in some species, notably *H. influenzae, H.*

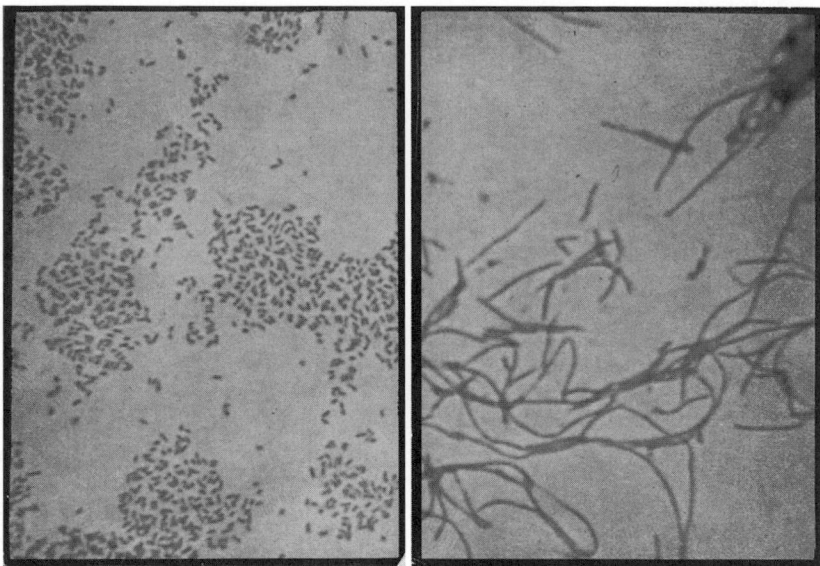

FIGURE BXII.γ.216. Cell morphology of a strain of *H. influenzae* (*left*) and a strain of *H. parainfluenzae* (*right*), illustrating the extent of pleomorphism that may be observed in bacteria belonging to this genus.

paragallinarum, and *H. parasuis* and are of particular interest in *H. influenzae* and *H. paragallinarum*, where they play a part in pathogenesis, determination of type specificity, and induction of anti-infective immunity. Occasional capsulate strains of *H. parainfluenzae* have been described (Sims, 1970).

Cytoplasmic membrane composition   The membrane phospholipids of *Haemophilus* spp. are similar to those of other Gram-negative bacteria, with phosphatidylethanolamine being the major type. Smaller quantities of phosphatidylglycerine have also been detected. It has been suggested that trace amounts of lysophosphatidylethanolamine may be a characteristic marker of *Pasteurellaceae* in general (Mutters et al., 1993).

Cell wall composition   The cell walls resemble those of other Gram-negative bacteria in structure, composition, and endotoxic activity. Unlike the lipopolysaccharide (LPS) of *Enterobacteriaceae*, *H. influenzae* LPS lacks long polysaccharide O chains and, thus, may be described more accurately as a lipooligosaccharide (LOS). It consists of a membrane-anchoring lipid A, an inner core of a singly phosphorylated 2-keto-3-deoxyoctulosonic acid (KDO) linked to three heptose molecules, and an outer core oligosaccharide consisting of a heteropolymer of glucose and galactose and a sialylated terminal lactosamine (Zamze and Moxon, 1987; Phillips et al. 1993; Roche and Moxon, 1995; Risberg et al., 1997). There is marked intra- and inter-strain heterogeneity in the composition and linkage of these saccharide units in the outer core of *H. influenzae* LPS and hence in the epitopes expressed (van Alphen et al., 1990; Roche and Moxon, 1995). The lipid A backbone is a glucosamine disaccharide substituted by two phosphate groups, as in enterobacterial lipid A, but with a simpler fatty acid composition of only tetradecanoic acid and its 3-hydroxylated derivative (Maskell et al., 1992). KDO is present in lesser amounts than in the LPS of *Enterobacteriaceae*: 0.4–1.5% of LPS in *H. influenzae*, compared to 5–8% in *Salmonella* spp. (Flesher and Insel, 1978).

Numerous genes are involved in the LPS biosynthesis by *H. influenzae*. These include the *lic* genes organized in three chromosomal loci (*lic1*, *lic2*, and *lic3*) (Weiser et al., 1990; Maskell et

al., 1992; Hood et al., 1996). The first gene (*lic1A*, *lic2A*, *lic3A*) of each of these multigene loci contains variable numbers of tandem repeats of the tetramer CAAT which, by facilitating slipped-strand mispairing, are responsible for the variable expression of LPS epitopes (Weiser et al., 1989; Maskell et al., 1992; High et al., 1996). An additional source of variation is the on–off switching in the expression of phosphorylcholine (ChoP). Choline is taken up from the environment by *H. influenzae* and incorporated into its LPS as ChoP under control of the *licA* gene in the *lic1* locus. The translational switch is based on variation in the number of CAAT repeats in the *licA* gene (Kolberg et al., 1997; Weiser et al., 1997). The frequency of spontaneous, reversible gain and loss (phase variation) of LPS oligosaccharide epitopes expressed by genes in the three chromosomal loci *lic1*, *lic2*, and *lic3* is 10^{-2}–10^{-3}/generation (Weiser et al., 1990).

As in other members of the family *Pasteurellaceae*, the number of cell wall fatty acids found in *Haemophilus* species is comparatively low. The general pattern is characterized by relatively large amounts of *n*-tetradecanoate ($C_{14:0}$), 3-hydroxy-tetradecanoate ($C_{14:0\ 3OH}$), hexadecanoate ($C_{16:1}$), and *n*-hexadecanoate ($C_{16:0}$). The three C_{18} fatty acids, octadecadienoate ($C_{18:2}$), octadecenoate ($C_{18:1}$), and *n*-octadecanoate ($C_{18:0}$) are also present, but in low concentrations. Small species-to-species variations in the relative amounts of these fatty acids have been demonstrated, but discrimination based on such differences cannot be achieved (Jantzen et al., 1981; Brondz and Olsen, 1984b; Brondz and Olsen, 1989; Mutters et al., 1993).

Analyses of polyamine patterns of whole cells of representative strains of species within the family *Pasteurellaceae* reveal unexpected diversity compared to that observed for other Gram-negative rods. At least seven different polyamine patterns are present among members of the family. All current species of *Haemophilus*, with the exception of *H. felis*, show a predominance of 1,3-diaminopropane (Table BXII.γ.293) (Busse et al., 1997). However, the extent of intra-species diversity is not yet known.

Outer membrane proteins (OMPs) have been the subject of significant interest, particularly in *H. influenzae*, as a means of typing and as potential vaccine candidates. There have been 25–

TABLE BXII.γ.293. Polyamine patterns of *Haemophilus* species and selected related taxa[a]

Polyamine content (percentage of total polyamines)[b]	*Haemophilus influenzae* Type	*Haemophilus aegyptius* Type	*Haemophilus aegyptius* MCCM 00678	*Haemophilus aphrophilus* Type	*Haemophilus ducreyi* Type	*Haemophilus felis* Type	*Haemophilus felis* MCCM 02064	*Haemophilus paracuniculus* Type	Type	*Haemophilus paragallinarum* xx36	*Haemophilus paragallinarum* xx37	*Haemophilus parahaemolyticus* Type	*Haemophilus parainfluenzae* Type	*Haemophilus segnis* Type	Type	*Actinobacillus actinomycetemcomitans* Type	"*Histophilus ovis*" MCCM 00330	"*Histophilus ovis*" MCCM 02133
DAP	96.2	91.4	62.9	98.2	79.1	2.5	33.4	71.6	84.9	73.7	70.6	86.7	40.2	65.4	99.7	97.6	88.5	78.7
PUT	1.3	8.2	22.3	0.4	0.1	70.4	1.9	0.2	0.3	2.6	3.3	0	21.3	32.4	0	0	5.9	1.1
CAD	0	0.5	14.7	0.2	0.5	5.6	0	0.2	0.6	0.5	tr	0	39.5	0.4	0	0	1.3	0.6
TYR	0.9	0	0	0	tr	0	40.0	tr	tr	0	0	0	0	0	0	0	0	0
NSPD	0	0	0	0	11.5	17.7	20.5	9.3	9.1	20.5	24.7	5.5	0	0	0	0	0	0
SPD	1.4	tr	0	0.9	0	3.7	0	0	0	0	0	0	0	1.7	0.3	1.2	4.3	10.9
HSPD	0	0	0	0	0	0	0	0.8	tr	0	0	0	0	0	0	0	0	0
SPM	0.1	tr	0	0.2	8.9	tr	4.2	17.9	5.1	2.7	1.5	7.8	0	0	0	1.2	tr	8.8

[a]Compiled from Busse et al. (1997).

[b]Abbreviations: DAP, 1,3-diaminopropane; PUT, putrescine; CAD, cadaverine; TYR, tyramine; NSPD, *sym*-norspermidine; SPD, spermidine; HSPD, *sym*-homospermidine; SPM, spermidine; tr, trace (less than 0.05 mmol/g dry weight).

35 OMPs detected in *H. influenzae* strains, with only a few accounting for 80% of the total OMPs (Loeb and Smith, 1982). The nomenclature of the major proteins has been reviewed by van Alphen (1993). The major OMPs are P1 (MW 43–50 kDa), P2 (a doublet of MW 43–50 kDa with porin activity), P4 (a lipoprotein of MW 30 kDa), P5 (2 conformers of the same protein), and P6 (a lipoprotein of MW 16.6 kDa). The protein originally designated P3 consists of both the non-heat-modified form of P1 and the heat-modified form of P5. The size polymorphism of proteins P1, P2, and P5 forms the basis for OMP typing of *H. influenzae*. While P2 is conserved in *H. influenzae* serovar b strains, it is extremely diverse and shows antigenic drift in noncapsulated strains (Duim et al., 1994; Murphy, 1994). P4 and P6 are reported to be conserved in *H. influenzae* (Granoff and Munson, 1986; Nelson et al., 1988; Barenkamp, 1992; van Alphen, 1993; Gilsdorf, 1998). OMP profile diversity has also been detected among isolates of *H. paragallinarum* (Blackall et al., 1990b).

Stable wall-deficient variants (L-forms) have been produced experimentally from *H. influenzae* (Roberts et al., 1974).

Fine structure The cell wall of *Haemophilus* species is typical of Gram-negative bacteria, having an ultrastructure composed of multiple wavy outer membranes and a poorly defined plasma membrane, with an intervening electron-transparent space (Sherwin and Wilkins, 1973; Kilian and Theilade, 1975; Doern and Buckmire, 1976; Kilian and Theilade, 1978; Holt et al., 1980). The entire cell wall, including the cell membrane, averages 20 nm in thickness. Vesicular structures ("blebs") have been demonstrated on the outer wall of several *Haemophilus* species (Holt et al., 1980). These are morphologically identical to lipopolysaccharide vesicles and are released into the surroundings. Methods for the isolation of the inner and outer membranes from *H. influenzae* serovar b cells have been devised by Loeb et al. (1981).

Fresh clinical isolates of *H. influenzae* usually express peritrichous pili with hemagglutinating properties. However, after serial subcultivations most isolates, either capsular or noncapsular, do not express such pili. Exceptions are *H. aegyptius* and the original Brazilian purpuric fever (BPF)-associated clone of *H. influenzae*, which are more stably piliated (Gilsdorf et al. 1997; Read et al., 1998). The hemagglutinating pili are 4.7–18 nm in diameter and 209–453 nm in length, with a hollow core. They are composed of a pilin protein subunit of M_r 21,000. Pilus expression is governed by a 5-gene operon (*hifA–hifE*) and oscillates between phase off and phase on at a rate of about 10^{-4}/generation. This phase variation is likely due to slip-strand mispairing in a poly(TA) dinucleotide tract that separates the divergent, bidirectional *hifA* and *hifB–E* promoters, altering the spacing upstream of the transcriptional start sites (van Ham et al., 1993). Some strains, including the previously mentioned conjunctival *Haemophilus* taxa (*H. aegyptius* and BPF-clone), possess duplicated pilus gene clusters, which may explain the higher stability of pilus expression (Read et al., 1998). In addition to the hemagglutinating pili, other types of adherence organelles that do not mediate hemagglutination have been described for *H. influenzae* (Jacques and Paradis, 1998). Some strains of *H. influenzae* biotype IV, which constitute a distinct evolutionary lineage (see below) and may be isolated from neonatal infections or the female genital tract, possess peritrichous pili that do not hemagglutinate, but mediate binding to HeLa cells (Rosenau et al., 1993). Fimbrial structures that are thinner than pili have been demonstrated in otitis media isolates of noncapsular *H. influenzae* (Bakaletz et al., 1988). In addition, different morphological types of

hemagglutinating and non-hemagglutinating pili were described by Brinton et al. (1989), but their relationship to those mentioned above has not been clarified. Fine tangled pili composed of a unique major protein subunit of M_r 24 kDa have been described in strains of *H. ducreyi* (Brentjens et al., 1996). Surface-associated thin hairlike structures of an unidentified nature have been demonstrated in *H. aphrophilus*, *H. paraphrophilus* (Holt et al., 1980), occasional strains of *H. parainfluenzae* (Kahn and Gromkova, 1981), and *H. parasuis* (Munch et al., 1992).

Colonial and cultural characters Surface colonies of *Haemophilus* species on sufficiently rich media are usually nonpigmented or slightly yellowish, flat, and convex, and they attain a diameter of 0.5–2.0 mm within 48 h at 37°C. *H. influenzae* undergoes spontaneous phase variation in colony opacity. Weiser et al. (1995) have identified a gene locus that contributes to opacity variation and is associated with the ability to colonize the nasopharynx of infant rats. Most species produce smooth colonies, but some variation is seen, particularly in *H. parainfluenzae* and *H. aphrophilus*. Some species show β hemolysis on blood agar (see Table BXII.γ.294). Growth in broth media usually shows even turbidity, but strains of *H. aphrophilus*, *H. paraphrophilus*, and *H. parainfluenzae* show granular growth with heavy deposits, due to aggregation of cells.

Nutrition and growth conditions Apart from the specific growth factor requirements, the various species of *Haemophilus* exhibit some variation in their nutritional needs and preferred growth conditions. The most universally satisfactory propagative media are chocolate agar and Levinthal media (agar and broth). The former has the virtue of relative ease of preparation, the latter that of transparency, which facilitates the recognition of colonial phases and dissociation phenomena. Colonial iridescence, a property highly correlated with encapsulation, is most readily recognized on Levinthal's agar.

The most critical ingredients of any medium for *Haemophilus* spp., whether for propagation or for characterization, are the growth factors X and/or V ("vitamin-like"). All members of the genus require one or both of these growth factors (few exceptional isolates, see below), and the exact requirement is one of the key characteristics of the individual species (see Table BXII.γ.294). X factor is usually protoporphyrin IX but, under conditions where iron cannot be obtained from another source, the iron-containing protoheme is required (Lwoff and Lwoff, 1937; White and Granick, 1963; Pidcock et al., 1988). Blood or blood derivatives, including hemin, are the traditional and adequate sources of X factor. The customary 5% blood used in blood and chocolate agar is ample. When crystalline hemin is used, required amounts vary between 0.1 and 10 µg/ml for *H. influenzae* (Gilder and Granick, 1947; Brumfitt, 1959; Biberstein and Spencer, 1962; Evans et al., 1974), while the optimum is 200 µg/ml for *H. ducreyi* (Hammond et al., 1978a). The X factor is heat stable and remains active in autoclaved media. The requirement for X factor is substantially reduced, but not eliminated, during anaerobic growth (Gilder and Granick, 1947). The X factor requirement is due to lack of some or all of the enzymes involved in the biosynthesis of protoporphyrin from δ-aminolevulinic acid (Biberstein et al., 1963; White and Granick, 1963) (Fig. BXII.γ.217). Isolates of *H. aphrophilus* often become independent of X factor upon subcultivation (King and Tatum, 1962; Sutter and Finegold, 1970; Kilian, 1976). The biochemical basis for this is not known, but it may suggest that the apparent X

TABLE BXII.γ.294. Differential characteristics of the species of the genus *Haemophilus*

Characteristics	1. H. influenzae	2. H. aegyptius	3. H. aphrophilus	4. H. avium	5. H. ducreyi	6. H. haemoglobinophilus	7. H. haemolyticus	8. H. paracuniculus	9. H. paragallinarum	10. H. parahaemolyticus	11. H. parainfluenzae	12. H. paraphrohaemolyticus	13. H. paraphrophilus	14. H. parasuis	15. H. segnis	a. "H. somnus"	b. "H. agni"
V-factor requirement	+	+	−	+	−	−	+	+	+	+	+	+	+	+	+	−	−
ALA→porphyrins	−	−	W	+	−	−	+	+	+	+	+	+	+	+	+	+	+
Indole production[a]	d	+	−	+	−	+	d	+	−	−	d	−	−	−	−	+	−
Urease[a]	d	−	−	−	−	−	+	+	−	+	d	+	−	−	−	−	d
Ornithine decarboxylase[a]	d	−	−	−	−	−	−	+	−	−	d	−	−	−	−	−	+
Arginine dihydrolase	−	−	−	−	−	−	−	+	−	−	−	−	−	−	−	−	−
Hemolysis	−	W	−	+	d	+	W	−	+	+	−[b]	+	−	−	W	+	−
D-Glucose, acid production	+	W	+	+	W	+	+	+	+	+	+	+	+	+	W	+	+
D-Glucose, gas production	−	−	+	−	−	−	d	−	−	d	d	−	+	−	−	−	−
Acid from:																	
D-Fructose	−	−	+	+	−	+	W	+	+	+	+	+	+	+	W	+	+
Sucrose	−	−	+	+	−	+	−	+	+	+	+	+	+	+	W	−	−
Lactose	−	−	+	+	−	−	−	−	−	−	−	−	+	−	−	−	−
D-Xylose	+	W	−	−	−	+	d	−	d	−	−	−	−	d	−	+	W
D-Ribose	+	−	d	−	−	d	+	−	+	−	−	−	d	−	−	−	−
D-Mannose	−	−	+	+	−	+	−	−	+	−	+	−	+	+	−	−	−
D-Mannitol	−	−	−	+	−	+	−	−	+	−	−	−	+	+	−	+	W
D-Sorbitol	−	−	−	−	−	−	−	−	d	−	−	−	−	−	−	+	−
β-Galactosidase (ONPG test)[b]	−	−	+	+	−	d	+	+	+	−	d	d	+	−	d	d	−
α-Fucosidase	+	+	−	+	−	+	+	−	−	d	d	d	−	+	−	−	+
Catalase	+	+	−	+	+	+	+	+	−	d	−	+	−	+	d	+	−
CO₂ enhances growth	−	−	+	+	−	−	−	+	+	+	+	+	+	+	d	−	−
Alkaline phosphatase	+	+	+	+	+	+	+	+	+	+	+	+	+	+	+	+	−
IgA1 protease	+	+	−	−	−	−	−	−	−	−	−	−	−	−	−	−	−

[a] For variations within *H. influenzae* and *H. parainfluenzae* see definition of biovars (Table BXII.γ.296).

[b] Some strains of *H. parainfluenzae* biovar II show weak (w) hemolysis.

PATHWAY OF PORPHYRIN BIOSYNTHESIS   PORPHYRIN TEST

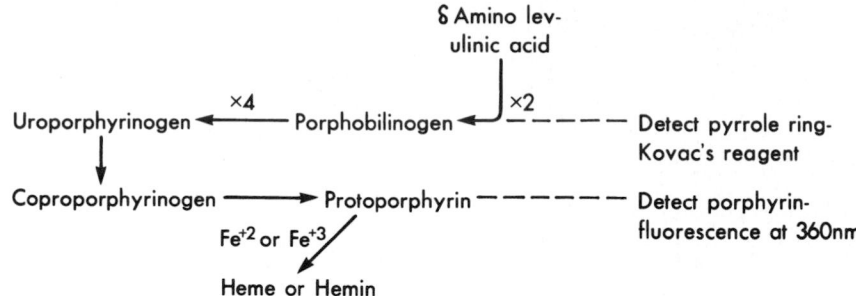

FIGURE BXII.γ.217. Steps in the biosynthesis of heme, with indications of compounds that are demonstrated in the Porphyrin test for X-factor requirement.

factor dependency in fresh isolates of *H. aphrophilus* is not due to the enzyme deficiencies in the porphyrin biosynthetic pathway described for *H. influenzae*. Indeed, White and Granick (1963) reported that one of Khairat's (1940) original isolates had all the enzymes of the biosynthetic pathway characteristic for hemin-independent species.

The V factor is minimally equivalent to nicotinamide mono-nucleotide (NMN) or nicotinamide riboside (NR), but is usually described as nicotinamide adenine dinucleotide (NAD). It is questionable if NAD phosphate (NADP) can serve as a general substitute for NAD (O'Reilly and Niven, 1986). Pyridine compounds that can serve as V factor can be characterized by possession of an intact pyridine-ribose bond in the β-configuration and a pyridine carboxamide bond at position 3. When NAD is supplied, it is cleaved by extracytoplasmic enzymes to yield NMN or NR, which is taken up and used in the biosynthesis of NAD (Wheat and Pittman, 1960; O'Reilly and Niven, 1986; Cynamon et al., 1988). It appears that none of the members of the family *Pasteurellaceae* are capable of *de novo* synthesis of NAD. V-factor dependency in some species can be explained by their inability to catalyze the formation of the nicotinamide–ribose bond, while all members of the family are incapable of amidating the carbonyl group on a nicotinate residue. Thus, the difference between V-factor dependence and independence within the family is much more subtle (the presence of nicotinamide phosphoribosyltrans-ferase) than hitherto believed (reviewed by Niven and O'Reilly, 1990). Occurrence of V-factor-independent variants of *H. parainfluenzae* from human infections and of poultry isolates of *H. paragallinarum* has been detected in South Africa (Gromkova and Koornhof, 1990; Windsor et al., 1991; Mouahid et al., 1992; Miflin et al., 1995). The property in *H. parainfluenzae* isolates is associated with a 5.25 kb plasmid that can be transferred to *H. influenzae* and confer independence of the V factor (Windsor et al., 1991). Recent studies demonstrate that this plasmid is normally present in strains of *H. ducreyi* (Martin et al., 2001).

The V factor is inactivated by autoclaving and is therefore absent in traditional agar or broth media. Although present in blood, V factor is unavailable in blood agar owing in part to its intracellular location and in part to the presence of NADase in the blood of many species (Krumwiede and Kuttner, 1938). In heated blood agar ("chocolate agar"), in which the blood is added to the basic medium at a temperature of approximately 70°C, the NAD is liberated from the blood cells and the NADase activity destroyed by the high temperature. Traditional sources

of V factor, apart from blood, have been yeast derivatives (Thjötta and Avery, 1921), which must be added after autoclaving of the medium. When met by crystalline NAD, V factor requirements of *H. influenzae* were found to range from 0.2–1 μg/ml; those of *H. parainfluenzae*, from 1–5 μg/ml, with some strains requiring as much as 25 μg/ml for optimum growth (Evans et al., 1974).

When inoculated on media deficient in V factor but containing the X factor (e.g., blood agar) *Haemophilus* colonies cluster around contaminant colonies of other bacteria which produce the critical factors in excess (Grassberger, 1897). This phenomenon is called satellitism (Fig. BXII.γ.218) and is sometimes utilized for propagation and characterization of *Haemophilus* spp. It may be demonstrated by inoculating a medium such as blood agar with a strain requiring the V factor. The inoculated area is crossed with a single streak of *Staphylococcus aureus*. Early growth of the *Haemophilus* strain is confined to the area immediately adjacent to the line of staphylococcal growth. It gradually spreads peripherally but remains heaviest in the area nearest the "feeder" streak, reflecting the diffusion gradient of the limiting growth factor. Satellitism on blood agar is strongly suggestive of a V-factor requirement. However, strains of strongly hemolytic strains, particularly *H. parahaemolyticus*, do not show satellitism, due to liberation of ample V factor from the lysed blood cells (Kilian and Poulsen, in preparation). For such strains, the V-factor requirement can be demonstrated on a blood-free agar medium cross-inoculated with *S. aureus* or supplied with a NAD-containing paper disk, or on a medium to which blood is added before autoclaving.

The optimum growth temperature is 35–37°C and the minimum temperature is 20–25°C. Most *Haemophilus* strains are killed by heating at 55°C for 30 min. *H. influenzae* grows better under aerobic than anaerobic conditions. Raised carbon dioxide tension is beneficial, and may be required for surface growth of a number of species, including *H. paragallinarum, H. aphrophilus, H. paraphrophilus,* and *H. paraphrohaemolyticus*. In the case of *H. paragallinarum* propagated in liquid media, CO_2 has been found to replace a 5% serum requirement observed for some strains under fully aerobic conditions (Rimler et al., 1976; Blackall and Reid, 1982). A serum requirement also exists for *H. parasuis* and *H. ducreyi*. The precise role of serum in these situations is not known. Detoxification of media constituents has been suggested (Page, 1962). Sodium chloride at a concentration of 1–1.5% is essential for the growth of *Haemophilus paragallinarum* (Rimler et al., 1977). Completely synthetic media have been devised re-

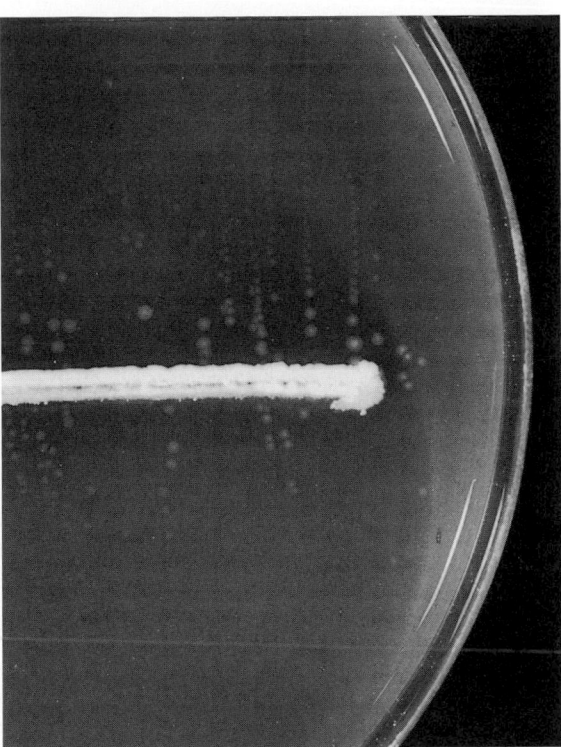

FIGURE BXII.γ.218. Satellitic growth of *H. influenzae* colonies in the vicinity of a streak of *Staphylococcus aureus* that provides the V factor.

peatedly, particularly for *H. influenzae* and *H. parainfluenzae* (Herbst and Snell, 1949; Talmadge and Herriott, 1960; Butler, 1962; Wolin, 1963; Herriott et al., 1970; Klein and Luginbuhl, 1979). In all instances, the adequacy of the medium was tested either with a very limited number of isolates, or was found to support the growth of only a portion of the strains tested. The simplest (Klein and Luginbuhl, 1979) and one of the most complex (Herriott et al., 1970), when tested extensively in parallel, performed exactly alike with regard to their growth-supporting properties.

Metabolism Enzymes of the Embden–Meyerhof–Parnas, hexose monophosphate, Entner–Doudoroff, and tricarboxylic acid cycle pathways have been demonstrated in several *Haemophilus* species (Klein, 1940; White, 1966; Holländer, 1976). Glycolytic growth in the "classical" (White and Sinclair, 1971) sense, however (i.e., in the absence of a functional electron transport chain), has been observed only in X factor-requiring members of the genus, e.g., *H. influenzae*, *H. aegyptius*, and *H. haemoglobinophilus* (White, 1963). In the *H. parainfluenzae* strain studied, glycolysis (as well as oxidative catabolism of glucose) occurred only in the presence of a functional electron transport system and a terminal electron acceptor, viz. oxygen, nitrate, or fumarate (White, 1966; White and Sinclair, 1971). The electron transport chain in the species examined proceeds from flavoproteins (dehydrogenases) via cytochromes *b*, *c*, *d*, and *a* to demethylmenaquinone (DMK) to cytochrome *o* and/or oxygen or nitrate (White and Sinclair, 1971; Holländer, 1976). Strains of *Haemophilus* spp. contain mainly DMK-6 (mean value 78%) with smaller amounts (~10%) of menaquinone-8 (MK-8) and DMK-7. Only *H. aphrophilus* and *H. segnis* lack MK-8. *H. haemoglobinophilus* is unique among the present *Haemophilus* spp. in that it contains exclusively MK-7 (Mutters et al., 1993).

A comprehensive overview of the metabolism of *H. influenzae* as compared with *Escherichia coli* deduced from the genome sequences has been presented by Tatusov et al. (1996).

Genetics *H. influenzae* is the first free-living organism whose genome has been fully sequenced. The 1.83-Mb sequence of the strain Rd (rough variant of a capsular serovar d strain) contains a total of 1743 genes, of which more than 40% have no apparent counterparts of known function in other procaryotes that have been analyzed (Fleischmann et al., 1995). Genetic maps of a strain of *H. parainfluenzae* (genome size 2.34 Mb) and *H. influenzae* serovar b strain Eagan have been constructed by the use of pulsed field gel electrophoresis (Kauc and Goodgal, 1989; Butler and Moxon, 1990). Numerous genes and operons encoding virulence and other properties of *H. influenzae* and other *Haemophilus* species have been cloned and sequenced (available from the National Center for Biotechnology Information).[1]

The genome sizes of *Haemophilus* species range between 1.8 and 2.8 Mb, with *H. influenzae* being the smallest and *H. ducreyi* the largest among species that have humans as their natural habitat (Burbach, 1987). The overall genetic diversity of the genus *Haemophilus*, as revealed by DNA–DNA hybridization studies, is considerable. The most comprehensive strain materials have been examined by Pohl (1979, 1981a) and Burbach (1987), who observed the following binding ratios to DNA of the type species *H. influenzae*: *H. aegyptius*, 70%; *H. haemolyticus* 48–52%; *H. parainfluenzae*, 15–33%; *H. aphrophilus*, 17%; *H. segnis*, 12%; *H. parahaemolyticus*, 14%; *H. paraphrohaemolyticus*, 16%; *H. parasuis* 26%; *H. paracuniculus*, 10%; and *H. haemoglobinophilus*, 22%. Relatively higher binding values, but with the same range, were reported in several other studies. Albritton et al. (1984) and Casin et al. (1986) found more than 90% homology between *H. influenzae* and *H. aegyptius*, whereas the lowest binding ratio (18%) in the study by Albritton et al. (1984) was observed between *H. influenzae* and *H. ducreyi*.

DNA–DNA hybridization studies also have revealed considerable heterogeneity within species. Burbach (1987) found that intra-species homologies among the species subjected to detailed analysis may be as low as 50% within *H. influenzae* and *H. parainfluenzae* and that certain strains of the two species show higher inter-species homology. This problem was solved by creating a new species designated "*H. intermedius*" (see discussion below) (Burbach, 1987).

Available data on 16S rRNA sequences of *Haemophilus* species are presented in Fig. BXII.γ.215. of the chapter on *Pasteurellaceae* by Olsen et al. (this *Manual*). It is notable that relative relationships revealed by DNA–DNA hybridization data are not always reflected in the relationships suggested by the 16S rRNA sequence homologies.

Although many gene sequences from *H. influenzae* reveal mosaic-like patterns and thus evidence of recombination, the population structure of at least encapsulated forms is largely clonal. Multilocus enzyme electrophoresis (MLEE) analysis of over 2000 capsulated *H. influenzae* strains from 30 countries has revealed a limited number of distinct evolutionary lineages, with the major branches corresponding closely to the six capsular serovars (Musser et al., 1990). However, two distinct branches are observed among strains expressing a serovar b capsule. Only one of these branches (division I) contains strains regularly isolated from in-

---

1. At the time this manual went to press, this information could be accessed at www.ncbi.nlm.nih.gov/htbin-post/Taxonomy/wgetorg?id 12&lvl= 3.

vasive infections, and only 8–9 clones from this branch appear to be responsible for the majority of *H. influenzae* serovar b disease worldwide (Musser et al., 1988, 1990). The population of nonencapsulated *H. influenzae*, which is distinct from that containing capsulated strains, appears to be very large and heterogeneous (Musser et al., 1986; Porras et al., 1986). Whether this significant clonal diversity among non-encapsulated *H. influenzae* is due to more extensive recombination affecting the population structure or merely reflects a larger population size has not been properly analyzed.

Studies of mainly Australian isolates of *H. parasuis* using MLEE analysis also reveal a clonal population structure with two major divisions that partly correlate with serovars (Blackall et al., 1997).

Several mechanisms of horizontal gene transfer have been detected in *Haemophilus* species. Alexander and Leidy (1951) utilized capsulation as a genetic marker to demonstrate the occurrence of transformation in *H. influenzae*. Noncapsulated R variants derived from each of the six capsular serovars a–f were examined for the ability to be transformed by DNA extracted from capsulated donor strains. Strain Rd proved to be the most dependable recipient and could be transformed to capsular types a, b, c, d, e, or f (or combinations of these), according to the capsular type of the DNA donor (Alexander et al., 1954). Catlin and Tartagni (1969) have detected serovar b antigen by immunofluorescence microscopy 40 min after adding transforming DNA to a population of strain Rd. Antibiotic resistance genes located in the chromosome, as well as on small plasmids, have been transferred between *H. influenzae* strains by genetic transformation (Stuy, 1979).

The transformation frequency for *H. influenzae* strain Rd in a rich medium increases from 10^{-7} in early exponential growth phase to about 10^{-4} in late exponential phase. The addition of cAMP to early-phase cultures increases the frequency of transformation to 10^{-4} (Wise et al., 1973; Redfield, 1991). Furthermore, under transient anaerobiosis or transfer to starvation medium, *H. influenzae* transformation frequencies are optimized to approximately 10^{-2} (Herriott et al., 1970). The increased competence under such conditions is paralleled by an altered envelope composition (Zoon and Scocca, 1975). A specific binding protein for homologous double-stranded DNA has been detected in the cell envelope of *H. influenzae* (strain Rd) (Deich and Smith, 1980). This membrane protein recognizes a specific 11-nt sequence that appears with much higher frequency in *Haemophilus* than in other genera (Sisco and Smith, 1979; Danner et al., 1980). Both intra- and interspecific transformation have been demonstrated *in vitro* in several other *Haemophilus* species, and the transformation efficiencies under experimental conditions reflect the overall genetic relationships (Leidy et al., 1956, 1959, 1965; Schaeffer, 1958; Steinhart and Herriott, 1968; Beattie and Setlow, 1970; Albritton et al., 1984).

Screening of *H. influenzae* strain Rd by various random insertional mutagenesis protocols have demonstrated that at least 5 gene loci are required for natural transformation (Larson and Goodgal, 1991; Redfield, 1991; Tomb et al., 1991; Chandler, 1992; Zulty and Barcak, 1995; Gwinn et al., 1997, 1998). Among the identified proteins are the type IV pilin-like protein and a drug-efflux transporter (Dougherty and Smith, 1999).

Plasmids of several sizes have been demonstrated in strains of *H. influenzae* and other *Haemophilus* species. Plasmids coding for resistance to a number of antibiotics have been demonstrated in *H. influenzae*, *H. parainfluenzae*, *H. parahaemolyticus*, and *H. ducreyi*. Two types of R plasmids occur in *H. influenzae*: the large 30–40

MDa conjugative plasmids and the small 2.5–4.4 MDa non-conjugative plasmids (De Graaf et al., 1976; Elwell et al., 1977; van Klingeren et al., 1977; Kaulfers et al., 1978; Laufs et al., 1979; Albritton, 1984). The large self-transferable plasmids carry the complete TnA or Tn10-like transposons in addition to antibiotic resistance genes and are often integrated into the chromosome (Stuy 1980; Levy et al., 1993). They have highly homologous core regions regardless of geographic origin and the antibiotic resistance markers carried (Elwell et al., 1977; Laufs and Kaulfers, 1977), and are closely related to plasmids coding for β-lactamase in *Enterobacteriaceae* (Laufs et al., 1978, 1979). Transfer of large plasmids coding for resistance to one to several antibiotics has been shown among strains of *H. influenzae* and between *H. influenzae* and *Escherichia coli* by a mechanism that requires cell-to-cell contact (Thorne and Farrar, 1975; van Klingeren et al., 1977; Stuy, 1979). The genetic transfer is likely mediated by a conjugation-like process. The same transfer mechanism has been demonstrated in *H. ducreyi* (Brunton et al., 1979; Deneer et al., 1982). Isolates of *H. ducreyi* may contain multiple plasmids of different sizes (1.8, 2.6, 2.8, 3.2, 5.4, 5.7, and 7.0 MDa) (Sarafian et al., 1991; Maclean et al., 1992).

The small R plasmids are highly related to gonococcal plasmids and usually carry only a fraction of the TnA sequence (de Graaf et al., 1976; Laufs et al., 1979). However, two sizes of plasmids (7.0 and 5.7 MDa) encoding β-lactamase with physically complete and functional TnA sequences have been identified in isolates of *H. ducreyi* (Brunton et al., 1982). The complete nucleotide sequence of a small (4.8 kb) yet broad-host-range plasmid encoding resistance to sulfonamides, streptomycin, and kanamycin in *H. ducreyi* has been reported (Dixon et al., 1994). Small cryptic plasmids have been detected in several *Haemophilus* species (Elwell et al., 1977; Stuy, 1979; Albritton et al., 1982b). A small 5.25 kb plasmid has been found to confer independence of V factor in occasional isolates of *H. parainfluenzae* (Windsor et al., 1991) and in *H. ducreyi* (Martin et al., 2001). Plasmids have not been detected in 75 *H. paragallinarum* isolates, including 20 antibiotic-resistant strains (Blackall, 1988), but more recently, a 6 kb plasmid was detected in one isolate (Blackall et al., 1991).

A ~24 kb cryptic plasmid (3031) has been associated with isolates of *H. influenzae* from Brazilian purpuric fever (Brenner et al., 1988a), although not all isolates possess it (Tondella et al., 1995).

Four different bacteriophages of *H. influenzae* have been reported: HP1 (and the mutants cl and c2), HP3, S2, and N3 (Boling et al., 1973; Stuy, 1978). The latter is morphologically distinct, with a longer tail and a contractile sheath. Differences in phage sensitivity and lysogeny have been demonstrated among *H. influenzae* serovars and a number of *Haemophilus* species (Stuy, 1978).

A variety of restriction endonucleases are produced by members of the genus *Haemophilus*. More than 20 different enzymes have been isolated and characterized.

Strains of *H. influenzae* serovar b have been shown to produce a bacteriocin to which other serovars and non-encapsulated *H. influenzae*, *H. parahaemolyticus*, and some strains of *H. parainfluenzae* and *H. haemolyticus* are sensitive (Venezia and Robertson, 1975; Stuy, 1978).

Antigenic structure There is a great deal of antigenic diversity in the genus and even within several species: *H. influenzae* (Pittman, 1931), *H. parasuis* (Bakos, 1955), *H. paragallinarum* (Page, 1962; Kume et al., 1980; Blackall et al., 1990a), *H. paraphrophilus*, and *H. haemoglobinophilus* (Frazer et al., 1975). In species in which isolates may possess a capsule, the capsular antigens furnish the basis for serovar specificity. *H. influenzae* isolates that lack a capsule are often referred to as "nontypeable," which is a misnomer that should·be substituted by "non-encapsulated." All the capsular antigens studied in sufficient detail have proven to be polysaccharides. They are serologically quite specific, although more than one antigenic determinant may be present in a capsule (Williamson and Zinnemann, 1951, 1954; Branefors-Helander, 1972; Lucas, 1988). Capsular structure and composition have been studied most thoroughly for *H. influenzae*, for which Pittman (1931 and unpublished results) originally described six serologically distinguishable types a–f. The structure of the serovar a–f capsular polysaccharides (and a variant of serovar e, designated e′) is shown in Fig. BXII.γ.219. All six polysaccharides, except those of serovars d and e, contain phosphate and may be characterized as teichoic acids (Crisel et al., 1975; Branefors-Helander, 1977; Branefors-Helander et al., 1979; Egan et al., 1980a, b; Branefors-Helander et al., 1980). Little is known of the physicochemical properties of the capsules of other *Haemophilus* species, apart from *H. parasuis*, in which some, but not all, capsules appear to be composed of an acidic polysaccharide (Morozumi and Nicolet, 1986).

Capsule production in *H. influenzae* depends on a cluster of genes in an 18 kb chromosomal locus termed *cap*, which consists of three major regions. *Cap* region 1 contains four genes (*bex A–D*) necessary for capsular export and, in combination with *cap* region 3, forms the *cap* chassis. *Cap* region 2, located in the middle of the chassis, is the serovar-specific locus, which contains four genes involved in capsule biosynthesis. The *cap* chassis is flanked by the insertion sequence IS1016, forming a compound transposon. In most clinical isolates of *H. influenzae* serovar b (division I isolates), the *cap* locus is composed of multiple (usually two) tandemly repeating *cap* units, and is unique in having a partial deletion of most of the IS1016-*bex A* region, which allows

amplification, as well as irreversible loss of capsule expression (Kroll et al., 1993; Roche and Moxon, 1995). As many as 20% of cells in late-exponential liquid cultures of these strains have lost the ability to produce a capsule and will grow on agar medium as small colonies with an irregular surface, containing pleomorphic, often filamentous bacteria with cytoplasmic vacuoles (Roche and Moxon, 1995). A significant proportion of clinical isolates of *H. influenzae* serovar b contain more than 2 copies of the *cap* locus and produce large amounts of capsular polysaccharide (Hoiseth et al., 1992). In contrast, serovar b isolates belonging to phylogenetic division 2 (Musser et al., 1988) have only one copy, like strains of serovars c and d. Serovar a strains have a tandem duplication of intact *cap* loci. PCR-based capsular genotyping has been shown to accurately determine the capsular states of *H. influenzae* isolates, including strains that have lost expression of capsular polysaccharide (Leaves et al., 1995). Occasional clinical isolates of serovar a from Africa have revealed the same deletion in the *cap* locus as found in serovar b isolates. This deletion may be associated with enhanced virulence of some serovar a clones (Kroll et al., 1994). While the majority of noncapsulated *H. influenzae* isolates are not the progeny of capsulate ancestors, some studies have demonstrated isolates that contain *cap*-locus-specific sequences (St. Geme et al., 1994). The reported hybridization of the genome of Brazilian purpuric fever-associated isolates of *H. influenzae* with a *cap* locus probe (Carlone et al., 1989) is, however, due to the presence of IS1016 elements rather than capsule genes (Dobson et al., 1992).

Bacterial antigens cross-reactive with the *H. influenzae* serovar b capsule have been detected across a wide and unrelated variety of bacteria, including *Staphylococcus aureus*, *Staphylococcus epidermidis*, *Streptococcus pyogenes*, *Streptococcus pneumoniae* (serovars 6, 15a, 29, 35a), *E. coli* (K 100), *Lactobacillus plantarum*, *Enterococcus faecium*, *Bacillus alvei*, and *Bacillus pumilus* (Alexander, 1958; Bradshaw et al., 1971; Argaman et al., 1974).

Serovar	Structure	
a	4)-β-D-Glc-(1→4)-D-ribitol-5-(PO₄→	
b	3)-β-D-Rib-(1→1)-D-ribitol-5-(PO₄→	
c	4)-β-D-GlcNac-(1→3)-α-D-Gal-1-(PO₄→ 3 ↑ R	R = OAc (0.8) H (0.8)
d	4)-β-D-GlcNac-(1→3)-β-D-ManANAc-(1→ 6 ↑ | R	R = L-serine (0.41) L-threonine (0.14) L-alanine (0.41)
e	3)-β-D-GlcNac-(1→4)-β-D-ManANAc-(1→	
e′	3)-β-D-GlcNac-(1→4)-β-D-ManANAc-(1→ 3 ↑ 2 β-D-fructose	
f	3)-β-D-GalNAc-(1→4)-α-D-GalNAc-1-(PO₄→ 3 ↑ OAc	

FIGURE BXII.γ.219. Structures of *H. influenzae* capsular polysaccharides. Compiled from data published by Crisel et al., 1975; Branefors-Helander, 1977; Branefors-Helander et al., 1979, 1981; Egan et al., 1980a, b; , Tsui et al., 1981a b; Zon and Robbins, 1983; Byrd et al., 1987. Ribose and fructose are in the furanose ring form. Glc, Gal, GlcNAc, and ManANAc are in the pyranose ring form.

Among somatic antigens, obtained by various extraction procedures and sonic disruption, cross-reactions often occur between serovars within species (Branefors-Helander, 1979; Blackall, 1989) and between species within the genus (Tunevall, 1953; Omland, 1964; Branefors-Helander, 1979; Schiøtz et al., 1979). Among the surface proteins expressed by *H. influenzae* strains, some (P4, P6, lipoprotein D, D15) are immunologically highly conserved, while many others (e.g., P1, P2, P5, pili) show significant antigenic diversity. Antigenic diversity has also been demonstrated for cell wall lipooligosaccharide and the secreted IgA1 protease. The genetic mechanisms creating this diversity include point mutations, phase variation (e.g., alteration of promoter structure and translational frameshift due to slipped strand mispairing), and horizontal gene transfer and homologous recombination (reviewed by Gilsdorf, 1998).

Based on heat-stable somatic antigen extracts, a total of 15 serovars (1 through 15) have been detected among isolates of *H. parasuis* (Kielstein and Rapp-Gabrielson, 1992). Subsequent studies have suggested additional antigenic diversity (Blackall et al., 1997). Several serotyping schemes for *H. paragallinarum* have been reported. The first serological typing system based on whole cell plate agglutination was developed by Page (1962), who recognized three serovars (A, B, and C). This and other similar typing schemes detect several distinct antigens, of which some are common to all members of the species (Blackall, 1989). Other typing systems for *H. paragallinarum* are based on somatic heat-stable antigens and a hemagglutinin (Hinz, 1980; Kume et al., 1983). The Kume scheme, which is based on a heat-labile, trypsin-sensitive hemagglutinin (HA-L), originally recognized three serogroups, termed I, II, and III, and a total of seven serovars, termed HA-1 to HA-7. The serovars are based on minor differences that can be detected with absorbed antisera. Two additional hemagglutin serovars have been described by Blackhall and his coworkers (Eaves et al., 1989; Blackall et al., 1990a), who also proposed a rationalized nomenclature that allows easy addition of new serovars and reconciles the schemes of Page and Kume. Under this altered scheme, the three recognized serogroups I, II, and III are renamed A, C, and B, respectively, and correspond to Page's serovars. Within each of the serogroups, the hemagglutinin serovars are numbered sequentially, allowing new serovars to be added in numerical order. The nine currently recognized serovars are termed A-1, A-2, A-3, A-4, B-1, C-1, C-2, C-3, and C-4 (Blackall et al., 1990a).

Several *H. influenzae* LPS oligosaccharide epitopes are similar to those in pathogenic *Neisseria* species and cross-react with human glycosphingolipid epitopes (Roche and Moxon, 1995). Studies with monoclonal antibodies have demonstrated species-specific epitopes in the LPS of *H. ducreyi* (Borrelli et al. 1995).

Comprehensive proteomic maps of primarily *H. influenzae*, but also including *H. parainfluenzae*, *H. haemolyticus*, and *H. parahaemolyticus*, have been reported by several research groups (Cash et al., 1997; Fountoulakis et al., 1997, 1998a, 1998b).

Antibiotic sensitivity Wild-type strains of the *Haemophilus* species are susceptible to therapeutically achievable concentrations of a wide range of antibiotics. Representative minimum inhibitory concentrations (MICs) of common antibiotics for *H. influenzae* are (µg/ml): benzyl penicillin, 1–2; ampicillin/amoxycillin, 0.5; cefuroxime, 0.5; cefotaxime, 0.06; ceftrioaxone, 0.03; cefaclor, 4; imipenem, 1; chloramphenicol, 0.5; tetracycline, 1; rifampicin, 1; erythromycin, 0.5–8; ciprofloxacin, 0.015; gentamicin, 1; sulfamethoxazole, 4; and trimethoprim, 0.5 (Butt et al., 1997b; Slack and Jordens, 1998).

The macrolide (erythromycin, oleandomycin) and lincosamide antibiotics (lincomycin, clindamycin) are generally less active, and bacitracin has so little effect on *H. influenzae* and other *Haemophilus* species that it is used, in a concentration of 5–19 U/ml, in media for the selective isolation of haemophili (Hovig and Aandahl, 1969; Ederer and Schurr, 1971; see below).

Since 1973, ampicillin resistance among *Haemophilus* species has become an increasing problem. Markedly differing rates of resistance among clinical isolated have been reported in various parts of the world, in some areas exceeding 60% (Campos et al., 1984, 1986). Ampicillin resistance is predominantly due to plasmid-encoded production of β-lactamases, which belong most often to the TEM-1-type, less often to the ROB-1-type, and occasionally to the VAT-1 type (Williams et al., 1974; Elwell et al., 1975; Bell and Plowman, 1980; Markowitz, 1980; Rubin et al., 1981; Maclean et al., 1992; Scriver et al., 1994; Vali et al., 1994; Shanahan et al., 1996). It has been suggested that the ROB-1 β-lactamase has an animal reservoir, as it has been detected in isolates of *Actinobacillus pleuropneumoniae* (Medeiros et al., 1986). Occasional isolates of *H. influenzae* show resistance to the β-lactam agents as a result of alterations in penicillin-binding proteins (Powell, 1988; Mendelman et al., 1990). Plasmid-encoded resistance to other antimicrobics has been observed. These include chloramphenicol (Manten et al., 1976; van Klingeren et al., 1977), tetracycline (Dang Van et al., 1975a; Bryan, 1978), and kanamycin (Dang Van et al., 1975b). Plasmids that simultaneously transfer ampicillin, chloramphenicol, and tetracycline resistance have been found in *H. influenzae* (Bryan, 1978). While tetracycline resistance appears to occur at rates comparable to penicillin–ampicillin resistance in some populations (Piot et al., 1977; Green et al., 1979), chloramphenicol resistance shows significant variation in prevalence (Brotherton et al., 1976; Ward et al., 1978; Butt et al., 1997b; Slack and Jordens, 1998). Resistance to trimethoprim and rifampicin has been observed (Green et al., 1979; Powell et al., 1992; Murphy et al., 1981). Chloramphenicol resistance is almost always due to the production of a plasmid-encoded chloramphenicol acetyl transferase (CAT) type II (Roberts et al., 1980). Non-CAT-mediated resistance is chromosomally encoded and due to a permeability barrier associated with the loss of an outer membrane protein (Burns et al., 1985).

H. parainfluenzae generally shows resistance levels higher than those of nonresistant *H. influenzae* to the antimicrobial agents commonly used on *Haemophilus* species (Kamme, 1969; Mayo and McCarthy, 1977). Both β-lactamase and non-β-lactamase-mediated resistance to ampicillin have been observed in *H. parainfluenzae*, usually at a higher rate than in *H. influenzae* (Green et al., 1979; Kauffman et al., 1979; Walker and Smith, 1980). Transmissible chloramphenicol resistance has been described in *H. parainfluenzae* and credited to CAT-II (Cavanagh et al., 1975; Shaw et al., 1978), while plasmid-encoded aminoglycoside resistance in the same species has been found to be mediated by a phosphotransferase (Le Goffic et al., 1977). Similar observations have been made for isolates of *H. parahaemolyticus* and *H. paraphrophilus* (Jones et al., 1976; Green et al., 1979).

The following minimum inhibitory concentrations (µg/ml) were observed on 19 isolates of *H. ducreyi* (Hammond et al., 1978b): vancomycin, 8–128; polymyxin, 32–128; penicillin G or ampicillin, 4.0 (unless β-lactamase-positive); cloxacillin, 32–64; cephalothin, 4.0–8.0; tetracycline, 0.4–32; doxycycline, 0.25–8.0; chloramphenicol, ≤4.0; rifampin, ≤4.0; sulfisoxazole, ≤8.0; and nalidixic acid, ≤8.0. Plasmid-mediated resistance to sulfonamides, aminoglycosides, tetracyclines, chloramphenicol, and β-lactam antibiotics have been demonstrated in *H. ducreyi*. It is not unusual for a single *H. ducreyi* isolate to contain multiple resistance plasmids, including more than one plasmid that confers resistance to β-lactam antibiotics (Trees and Morse, 1995).

Resistance to streptomycin is common in strains of *H. paragallinarum*, and strains resistant to tetracycline and sulfonamides also occur. There is no specific knowledge about resistance mechanisms in *H. paragallinarum*. Plasmids are not regularly associated with antibiotic resistance in this species (Blackall, 1988; Blackall et al., 1991).

Pathogenicity Among the *Haemophilus* species that colonize man, *H. influenzae* is clearly the most important from a clinical point of view. Although not responsible for epidemic influenza, as was originally believed, it is involved in a variety of numerically important and severe infections (for review see Turk and May, 1967). These infections can be divided into two groups: (a) acute, pyogenic, and usually invasive infections in which *H. influenzae* (almost exclusively serovar b) is the primary pathogen, and (b) noninvasive infections (often chronic) in which primarily non-encapsulated *H. influenzae* play an important, though in many cases probably a secondary, part.

The acute infections include meningitis in children, of which *H. influenzae* is one of the three leading causes, and other septicemic conditions with local implications, such as epiglottitis, cellulitis, arthritis, and osteomyelitis. More than 95% of *H. influenzae* isolates from these conditions produce a serovar b capsule, and belong to a limited number of clones (Musser et al., 1990), the majority of which are of biovar I (Kilian, 1976; Albritton et al., 1978; Kilian et al., 1979; Oberhofer and Bach, 1979). The introduction in many countries of a conjugated vaccine based on the *H. influenzae* serotype b capsular polysaccharide and one of several protein carriers has virtually eliminated invasive infections due to this serovar (Eskola et al., 1987; Jordens and Slack, 1995; Peltola, 2000). A relatively high prevalence of serovar a isolates have been observed from invasive infection in Papua New Guinea and in White Mountain Apache Indian children (Losonsky et al., 1984; Gratten et al., 1985). Capsulated strains of several serovars may also cause pneumonia. Noncapsulated strains of *H. influenzae* are often implicated in chronic bronchitis, sinusitis, conjunctivitis, and otitis media and have recently been recognized as a major cause of lower respiratory disease in children in the developing world (Foxwell et al., 1998). They occur frequently in the lower respiratory tract of patients with cystic fibrosis during acute exacerbations. Isolates from such conditions are genetically very diverse (Porras et al., 1986; van Alphen et al., 1997), but usually belong to the biovars II and III, the same biovars that colonize the nasopharynx of healthy individuals (Kilian, 1976; Albritton et al., 1978; Oberhofer and Bach, 1979; Granato et al.,1983; Harper and Tilse, 1991). *H. parainfluenzae* biovars identical to those that predominate in the upper respiratory tract are also a relatively frequent finding in respiratory secretions from patients with chronic bronchitis, but their pathogenic role is not clear (Taylor et al., 1992). Occasionally, *H. influenzae* and *H. parainfluenzae* are isolated from genital, pelvic, and urinary tract infections (Quentin et al., 1989; Morgan and Hamilton-Miller, 1990). *H. parahaemolyticus* is rarely isolated from healthy individuals but may be a pathogenic factor in abscesses in the oral cavity and in upper and lower respiratory tract infections (Sims, 1970; Kilian and Poulsen, unpublished studies).

H. aegyptius is a frequent cause of acute and contagious conjunctivitis, mainly in hot climates. Due to difficulties in differentiating *H. aegyptius* from *H. influenzae*, which also causes con-

junctivitis, the natural history of *H. aegyptius* infections is poorly understood. There are indications, however, that *H. aegyptius* is associated with a more acute form of conjunctivitis and that the organism, in contrast to *H. influenzae*, can colonize eyes without predisposing conditions being present (see discussion of taxonomic relationship between the two species below). Particular clones of the *H. influenzae*/ *H. aegyptius* complex are causing a disease called Brazilian purpuric fever (BPF), which is a fulminant and often fatal infection that starts as conjunctivitis and primarily affects young children. An outbreak of systemic infections due to *H. aegyptius*-like bacteria in sheep in Nigeria has been reported (Akpavie et al., 1994).

H. ducreyi is the causative agent of the venereal disease soft chancre, or chancroid, which is particularly common in Africa, Asia, and Latin America. There has been renewed interest in this pathogen and in chancroid owing to the association between genital ulcers and HIV infection (Trees and Morse, 1995).

H. parainfluenzae, *H. parahaemolyticus*, *H. paraphrohaemolyticus*, *H. segnis*, *H. aphrophilus*, and *H. paraphrophilus*, which occur in the mouth and oropharynx of healthy individuals, are opportunistic pathogens. The oral cavity is usually a likely source of the etiologic agents in infections caused by these organisms. Such infections include endocarditis, brain abscesses, dental abscesses, jaw infections, and infections following human bites or finger sucking (for review, see Frederiksen and Kilian, 1981; Frederiksen, 1993).

The clinical condition of swine in which *H. parasuis* is most frequently encountered is Glässer's disease (polyserositis), a systemic stress-related infection producing fibrinous inflammation of the membranes lining the large body cavities, joints, and meninges. While no antecedent/virus infection has been invoked as a precipitant of this condition, stressful events like weaning, weather changes, or movement to new quarters commonly, but not invariably, precede it (for review see Bisgaard, 1993).

H. paragallinarum causes an infection of the upper respiratory tract of chickens called fowl coryza. It begins in the nasal passages and sinuses and spreads to the conjunctivae, and can extend to the air sacs and lungs (Blackall, 1989).

H. haemoglobinophilus is of low pathogenicity. On rare occasions, it has been implicated in urogenital inflammatory disease of dogs. *H. felis* has been isolated from the lower respiratory tract of a cat with chronic obstructive pulmonary disease. Its pathogenic potential probably resembles that of *H. parainfluenzae* in humans (Inzana et al., 1992). *H. paracuniculus* was isolated from the small intestine of rabbits, but its pathogenic potential is unknown (Targowski and Targowski, 1979).

Ecology *Haemophilus* species are obligate parasites and form part of the indigenous flora of the mucous membranes of the human upper respiratory tract, including the oral cavity, and may occasionally be isolated from the vagina and intestinal canal. Humans are the only natural host of the following species: *H. influenzae*, *H. aegyptius*, *H. haemolyticus*, *H. ducreyi*, *H. parainfluenzae*, *H. parahaemolyticus*, *H. paraphrohaemolyticus*, *H. aphrophilus*, *H. paraphrophilus*, and *H. segnis*. *Haemophilus* species constitute approximately 10% of the total bacterial flora in the human upper respiratory tract, and *H. parainfluenzae* accounts for the largest proportion of *Haemophilus* species in the pharynx and oral cavity at all ages. While *H. parainfluenzae* is ubiquitous in the human pharynx and oral cavity, *H. influenzae* is found exclusively behind the palatinal arches (Sims 1970; Kilian and Schiøtt, 1975; Kuklinska and Kilian, 1984; Liljemark et al., 1984). Point prevalence studies of children have shown carriage rates for *H. in-*

fluenzae of 25–84% in the pharynx, although in the absence of local infection, they usually amount to less than 1% of the total microbiota. In contrast to adults, who are less frequently colonized, children often carry multiple clones that show rapid turnover (Kuklinska and Kilian, 1984; Spinola et al., 1986; Trottier et al., 1989; Faden et al., 1995; Smith-Vaughan et al., 1997). Among *H. influenzae* isolates from the upper respiratory tract, only a minority (2–7%) were found by most studies conducted in open communities to be encapsulated. The relative frequencies of the six capsular serovars a–f in such studies were 15, 40, 4, 6, 19, and 15%, respectively (reviewed by Moxon, 1986). Significantly higher nasopharyngeal carriage rates of serovar b strains have been observed among children in institutions such as orphanages and day care centers (Turk and May, 1967). The general vaccination of infants with a conjugate vaccine that includes the serovar b capsular polysaccharide has resulted in a significant reduction of oropharyneal carriage of *H. influenzae* serovar b in some countries (Takala et al., 1991).

H. parainfluenzae is carried by most individuals on the mucosal surfaces of the pharynx and oral cavity, and as members of the biofilm (dental plaque) forming on tooth surfaces. Dynamic colonization patterns similar to those observed for *H. influenzae* have been demonstrated for *H. parainfluenzae* in the pharynx of adults (Kerr et al., 1993). *H. aphrophilus*, *H. paraphrophilus*, and *H. segnis* have a particular predilection for the dental plaque of humans, and the two former species occur primarily in the bacterial deposits between teeth and in deepened gingival crevices in patients with periodontal disease (Kilian and Schiøtt, 1975; Liljemark et al., 1984). *H. haemolyticus* may also be isolated from subgingival dental plaque (Kilian, unpublished observations). The carriage rate of *Haemophilus* species in the anogenital area is approximately 10% or less (Sturm, 1986; Drouet et al., 1989; Martel et al., 1989), although *Haemophilus* spp. have been recovered from over a quarter of stool specimens (Palmer, 1981). Occurrence of *H. ducreyi* in healthy individuals has not been documented.

Haemophilus-like bacteria (V-factor dependent) are also isolated from a variety of mammalian and avian species, including laboratory rodents and rabbits (for review see Grebe and Hinz, 1975; Kilian and Frederiksen, 1981a; Rayan et al., 1987; Nicklas 1989; Busse et al., 1997), but it is questionable whether these belong in the genus *Haemophilus* as presently defined. The only species associated with animals that are still formally (though probably incorrectly) included in the genus *Haemophilus* are *H. parasuis*, *H. paragallinarum*, *H. paracuniculus*, *H. felis*, and *H. haemoglobinophilus*.

H. parasuis has its natural habitats in pigs and is part of the resident microbiota of the upper respiratory tract, mainly in the nasal cavity. The carriage rate may differ between herds (Smart and Miniats, 1989; Møller and Kilian, 1990). Using a selective culture medium, examination of 233 Danish slaughterhouse pigs representing different herds yielded growth of *H. parasuis* from the nasal cavity of 30% of animals (Møller et al., 1993).

H. paragallinarum is found in the respiratory tracts of poultry, but is also an important pathogen (Blackall, 1989p). *H. haemoglobinophilus* is a frequent commensal inhabitant of the lower genital tract of dogs. *H. felis* was isolated from the nasopharynxes of 6 of 28 apparently healthy cats (Inzana et al., 1992).

ENRICHMENT AND ISOLATION PROCEDURES

The traditional way of demonstrating *Haemophilus* organism in samples from mucosal membranes is to inoculate a blood agar plate and cross-inoculate it with a *Staphylococcus* strain to provide the V factor. Most *Haemophilus* species will grow to sizeable colonies in the vicinity of the feeder strain after overnight incubation at 35–37°C in air supplemented with 5–10% extra CO_2. However, in mixed cultures haemophili may easily be overgrown. Thus, respiratory samples often give rise to heavy growth of haemophili on selective media in cases where no apparent *Haemophilus* colonies are detectable by routine cultivation methods.

Chocolate agar supplemented with 100–300 µg/ml (6.3–18.9 U/ml) of bacitracin is a very satisfactory medium for the selective isolation of species of *Haemophilus* (Hovig and Aandahl, 1969; Ederer and Schurr, 1971). Some pharyngeal and oral *Neisseria* species grow on this medium, as do *Actinobacillus actinomycetemcomitans* and *Eikenella corrodens*. An alternative selective medium that gives equally satisfactory results is chocolate agar supplemented with bacitracin (5 U/ml) and cloxacillin (5 µg/ml) (Sims, 1970). A selective agar medium that allows differentiation of *H. influenzae* and *H. parainfluenzae* (and other V-factor dependent species) based on the detection of sucrose fermentation by the latter has been described by Roberts et al. (1987) and in modified form by Taylor et al. (1990).

For the more fastidious haemophili (e.g., *H. ducreyi* and *H. aegyptius*), higher isolation rates are obtained with chocolate agar enriched with 1% IsovitaleX (BBL) (Hammond et al., 1978a; Vastine, et al., 1974). GC-HgS agar consisting of GC agar (GIBCO Laboratories) supplemented with 3 mg of vancomycin per liter, 1% hemoglobin, 5% fetal bovine serum, 1% IsoVitaleX (BD Diagnostics) or other comparable enrichment has a high sensitivity for isolation of *H. ducreyi* from clinical specimens; so has MH-HB agar, i.e., Mueller–Hinton agar (BBL) supplemented with 5% chocolatized horse blood, 1% IsoVitaleX, and 3 mg/l of vancomycin. It is generally accepted that a combination of two media, e.g. GC-HgS and MH-HB, is optimum for the isolation of *H. ducreyi*, possibly because of differences in nutritional requirements between strains, but the sensitivity for *H. ducreyi* is never 100%. Some lots of fetal calf serum inhibit growth of *H. ducreyi*. Fetal calf serum can be replaced by either activated charcoal or bovine albumin, but not by newborn calf serum (for a review see Trees and Morse, 1995). Some strains of *H. paragallinarum* require chicken serum for growth, while other strains simply show improved growth in its presence (Hinz, 1973; Blackall and Reid, 1982). Sodium chloride at a concentration of 1.0–1.5% is essential for the growth of *H. paragallinarum* (Rimler et al., 1977).

MAINTENANCE PROCEDURES

Plate and tube cultures of haemophili usually survive no longer than 1 week without subcultivation. Survival is often better at room temperature than at 4°C. The most satisfactory way of conserving cultures is by lyophilization, e.g., in skim milk. Levinthal broth cultures remain viable for at least 2 years when frozen at −70°C in sealed glass ampules.

PROCEDURES FOR TESTING SPECIAL CHARACTERS

Test for porphyrin synthesis (X-factor requirement) Like many other bacteria, hemin-independent *Haemophilus* strains excrete porphobilinogen (PBG) and porphyrins, intermediates in the hemin biosynthetic pathway (Fig. BXII.γ.217), when supplied with δ-aminolevulinic acid (ALA). In contrast, hemin-requiring strains do not excrete these compounds because they lack the enzymes responsible for their synthesis (Biberstein et al., 1963). This is the rationale for the porphyrin test, which is the easiest and most reliable means of determining X-factor requirement in *Haemophilus* organisms (Kilian, 1974).

A substrate of the following composition is used: δ-aminolevulinic acid, (Sigma), 2 mM; and $MgSO_4$, 0.08 mM; in 0.1 M phosphate buffer, pH 6.9. The substrate is distributed in 0.5-ml quantities in small glass tubes and is inoculated with a heavy loopful of bacteria from an agar plate culture. After incubation for 4 h at 37°C, the result is read under Wood's light (360 nm), preferably in a dark room. A red fluorescence from the bacterial cells and/or the fluid is indicative of porphyrins (Fig. BXII.γ.217), meaning that the strain is independent of the X factor.

An alternative method of reading is to add 0.5 ml of Kovacs' reagent, shake vigorously, and allow the phases to separate. A red color in the lower water phase, indicative of porphobilinogen (Fig. BXII.γ.217), means that the strain is independent of X factor. With this method of reading, an inoculated "substrate" without ALA must be included to avoid false positive reactions due to indole, which also gives a red color with Kovacs' reagent (Fig. BXII.γ.217).

The traditional use of paper disks impregnated with X factor on agar media cannot be recommended for the demonstration of an X-factor requirement. Even with care, use of this method results in misidentifications approaching 20% (M. Kilian and K.R. Eriksen, unpublished results). Such misidentifications undoubtedly account for most of the reported isolates of *H. parainfluenzae* from invasive infections.

V-factor requirement The satellite phenomenon mentioned above is widely used as a means of determining V-factor requirement. This phenomenon may be demonstrated on an agar medium lacking the V factor by cross-inoculating the plate with an appropriate feeder strain (e.g., *Staphylococcus* or *Pseudomonas*). More unequivocal results may be obtained by the application of a V-factor-impregnated paper disk. Media in which all ingredients have been autoclaved are consistently free of V factor, whereas ordinary blood agar contains varying amounts of available V factor. A convincing satellite phenomenon is, therefore, sometimes difficult to achieve on ordinary blood agar plates, particularly with strains of hemolytic species. The best results are obtained on a blood agar medium in which the blood (5–10%) has been added before autoclaving. This medium is completely devoid of V factor but otherwise satisfies all growth requirements of the *Haemophilus* species, including the X factor.

Fermentation of carbohydrates Fermentation of carbohydrates can be studied in phenol red broth base (Difco) supplemented with 1% of the respective carbohydrates and 10 mg/l each of NAD and hemin, the former added as a solution sterilized by filtration (Kilian, 1976). In this medium, most strains of *H. parainfluenzae*, *H. parahaemolyticus*, *H. aphrophilus*, *H. paraphrophilus*, and *H. haemolyticus* release hydrogen and CO_2 during fermentation of glucose (Kilian, 1976). This activity may be demonstrated by inserting an inverted Durham tube in the glucose broth. *H. paracuniculus* is reported to be unable to grow in phenol red broth base supplemented with V factor. Purple broth base supplemented with NAD (10 mg/l) is recommended for the examination of fermentation reactions in this organism (Targowski and Targowski, 1979).

Hemagglutination The hemagglutinating activity found in most fresh isolates of *H. influenzae* and *H. aegyptius*, and in some isolates from fowl and swine is demonstrated as described by Davis et al. (1950). The property is easily lost upon subcultivation of isolates.

Typing of capsules Commercial antisera are available for the differentiation of the six capsular serovars of *H. influenzae*, and several techniques may be applied. Slide agglutination is the traditional method and gives reliable results when performed with care. It is wise to consider as positive only those agglutinations that occur within a few seconds. Delayed reactions should be interpreted very cautiously, as commercial antisera also contain antibodies against somatic antigens. Although more time-consuming, the use of the Quellung reaction greatly strengthens the reliability of typing. An alternative serological method that gives reliable results is coagglutination using commercial reagents (Himmelreich et al., 1985). Typing may also be achieved by demonstration of serovar-specific gene sequences. A probe (pUO38) that recognizes the cap region of all six serovars reveals 14 distinct banding patterns in *Eco*RI-digested chromosomal DNA, which can be directly translated into serovars (Kroll et al., 1991). Capsular typing by PCR is another method that accurately determines the capsular states of *H. influenzae* isolates, including b⁻ strains. It is easier to perform than probe analysis, as there is no need for DNA extractions, Southern blotting, and hybridization. The method furthermore recognizes four cap b patterns not detected by pUO38 hybridization (Leaves et al., 1995). Details about probes and primers are provided by Herbert et al. (1998).

Micromethods for Examination of *Haemophilus* Strains Micromethods that do not require growth are recommended for the detection of indole production, urease, ornithine, and lysine decarboxylases, and arginine dihydrolase and for the demonstration of glycosidase activities. All these tests use 0.5-ml quantities of the substrate and are inoculated by suspension of a heavy loopful of bacteria in the substrate. The results can usually be read after incubation for 4 h at 37°C, but the incubation may be extended to 24 h.

Indole test The substrate is 0.1% L-tryptophan in 0.067 M phosphate buffer at pH 6.8. After incubation, 0.5 ml of Kovacs' reagent is added and the mixture shaken. A red color in the upper alcohol phase indicates the presence of indole (Clark and Cowan, 1952).

Urease test The basal medium consists of (per 100 ml of distilled water): KH_2PO_4, 0.1 g; K_2HPO_4, 0.1 g; NaCl, 0.5 g; and phenol red solution (1:500, prepared by dissolving 0.2 g in 92 ml distilled water + 8 ml 1 N NaOH), 0.5 ml. pH is adjusted to 7.0 with 5 N NaOH, autoclave, and add 10.4 ml of urea solution (20% aqueous, sterilized by filtration). A red color that develops within 4 h after inoculation indicates urease activity (Lautrop and Lacy, 1960).

Amino acid decarboxylases Ornithine and lysine decarboxylases and arginine dihydrolase can be demonstrated in Møller's medium (Møller, 1955). When this is heavily inoculated with positive strains, a purple color develops within 4 h.

Glycosidases Glycosidase activities can be demonstrated in 0.1% buffered solutions (w/v) of the respective chromogenic nitrophenol derivatives (Kilian, 1978). A yellow color developing within 4 h indicates glycosidase activity.

Most of the biochemical tests mentioned above can be performed with commercial micromethod kits, such as API 10E, API 20E, HNID (Quentin et al., 1992), the Minitek System (BBL) (Oberhofer and Bach, 1979), API NH (Barbé et al., 1994), Rapid NH and RIM-*Haemophilus* (Barbé et al., 1994).

DIFFERENTIATION OF THE GENUS *HAEMOPHILUS* FROM OTHER GENERA

According to the definition of the genus employed in the previous edition of this *Manual*, the demonstrable need of X or V factor on the part of a Gram-negative rod or coccobacillus would qualify that organism as a *Haemophilus* strain. Conversely, the absence of such need would exclude a bacterium from the genus (Kilian and Biberstein, 1984). However, in the intervening period several V-factor-dependent species have been transferred from the genus *Haemophilus* to other genera within and outside the family *Pasteurellaceae* based on genetic data. "*H. pleuropneumoniae*" has been transferred to the genus *Actinobacillus* (Pohl et al., 1983). "*Haemophilus avium*", which turned out to be a heterogeneous taxon, has been transferred to the genus *Pasteurella* as *P. avium*, *P. volantium*, and an additional, yet unnamed, taxon designated *Pasteurella* species A (Mutters et al., 1985b; Piechulla et al., 1985b). Finally, "*H. equigenitalis*" has been reclassified as *Taylorella equigenitalis* (Sugimoto et al., 1983).

As a result of the reclassification of these species, V-factor requirement is now a property of some species of both *Actinobacillus* and *Pasteurella*. In addition, three V-factor-dependent taxa of porcine origin have been proposed as new species within the genus *Actinobacillus*, i.e., *A. minor* (formerly *Haemophilus* taxon "minor group" of Kilian et al., 1978), *A. porcinus*, and *A. indolicus* (Møller et al., 1996a). Conversely, it has become clear that some of the remaining *Haemophilus* species (i.e., *H. parainfluenzae* and *H. paragallinarum*) may include occasional members that are independent of the V factor, at least in some cases as a result of a transferable plasmid (see above). Similarly, as discussed above, isolates of *H. aphrophilus* often lack a demonstrable requirement for X factor, and the apparent requirement in fresh isolates is not due to a lack of enzymes in the biosynthetic pathway of heme as in *H. influenzae*. As a result, the previous definition of the genus is no longer valid.

Despite the reallocation of some species, the definition of the genus *Haemophilus* and the delineation of the individual genera in the family *Pasteurellaceae* have not become any clearer. As discussed by Olsen et al. in the chapter describing the family *Pasteurellaceae* (this *Manual*), available data suggest that it will be difficult to divide the family *Pasteurellaceae* into genera that are phenotypically and phylogenetically coherent. However, there is no doubt that several of the species currently assigned to the genus *Haemophilus* belong elsewhere. Based primarily on DNA–DNA hybridization data of Burbach (1987), Mutters et al. (1989) have proposed restricting the genus *Haemophilus* sensu stricto to *H. influenzae*, *H. aegyptius*, *H. haemolyticus*, "*H. intermedius*", and *H. parainfluenzae*. In contrast, phylogenetic analyses by 16S rRNA sequence comparisons suggest that the genus should instead be restricted to *H. influenzae*, *H. aegyptius*, *H. haemolyticus*, and "*H. intermedius*"; whereas *H. parainfluenzae* is only distantly related. *H. aphrophilus*, *H. paraphrophilus*, and *H. segnis* belong to a separate cluster related to *Actinobacillus actinomycetemcomitans* (see Fig BXII.γ.215 of Olsen et al. in the chapter describing the family *Pasteurellaceae*, this *Manual*).

Allocation of isolates to one of the existing species of *Haemophilus* usually presents no problems. This can be achieved in the clinical microbiological laboratory by examination of phenotypic traits or, if sequencing facilities are available, by partial sequencing of 16S rRNA genes. However, a decision concerning the correct allocation of newly discovered taxa to any one of the existing genera within *Pasteurellaceae* is problematic. There are at present no phenotypic traits known that correlate with the 16S rRNA sequence homology data and that would provide a workable basis for reconciling the genetic information with the practical needs for laboratory identification of genera and species. Thus, at present, the decision would be made based on genetic data, in particular, comparison of 16S rDNA sequences of an unknown isolate with available sequences for the respective type species, but even then the most natural allocation would not necessarily be clear-cut.

Of all the secondary criteria usually cited as characteristic of the genus, such as carbohydrate fermentation, alkaline phosphatase activity, and nitrate reduction, only the lattermost is without exception among the currently recognized species, and none are exclusive. Other chemotaxonomic criteria are discussed by Olsen, et al. (see the chapter on the family *Pasteurellaceae*, this *Manual*). There are no motile members of the genus, and the question of whether motility would exclude an organism meeting the other standards of the genus has never arisen.

Organisms with growth factor requirements comparable to those of *Haemophilus* spp. occur sporadically among other, unrelated taxa (Jensen and Thofern, 1953; Beljanski, 1955; Caldwell et al., 1965). The basis for excluding them from the genus *Haemophilus* is, in addition to genetic data such as significant differences from *H. influenzae* in 16S rRNA genes, based on morphological, tinctorial, and cultural characteristics identifying them with their parent taxa.

TAXONOMIC COMMENTS

In addition to the described problems related to the definition of the genus, several other taxonomic problems remain to be solved in the genus *Haemophilus*. These problems include species that seem to belong not in the genus, but rather elsewhere in the family *Pasteurellaceae*, as well as problems related to some of the "true" *Haemophilus* species. According to phylogenetic analyses based on 16S rDNA sequence comparisons, the species *H. ducreyi*, *H. parainfluenzae*, *H. parahaemolyticus*, *H. paraphrohaemolyticus*, *H. haemoglobinophilus*, *H. parasuis*, *H. paragallinarum*, *H. paracuniculus*, and *H. felis* do not cluster with the type species *H. influenzae* and seem to be misplaced in the genus *Haemophilus* (see Fig BXII.γ.215 of Olsen et al. in the chapter on the family *Pasteurellaceae*, this *Manual*).

The taxonomic position of *H. ducreyi* has been questioned for a number of years in spite of its requirement for X factor and a mol% G + C content of DNA within the accepted range for *Haemophilus* species. DNA hybridization data indicate that *H. ducreyi* is unrelated to *H. influenzae* (Casin et al., 1985). Sequence comparisons of 16S rDNA of selected isolates of *H. ducreyi*, including the type strain (CIP542), with representatives of a large number of species support its exclusion from the genus *Haemophilus* but confirm that it belongs in the family *Pasteurellaceae* (Dewhirst et al., 1992).

H. parainfluenzae is a genetically very diverse species. Although Mutters et al. (1993) included it in *Haemophilus* sensu stricto, 16S rRNA sequence data of a limited number of strains reveal that it is phylogenetically only distantly related to *H. influenzae* and belongs elsewhere in the family *Pasteurellaceae*. Further studies of a larger number of representative strains reflecting the diversity of the species are required to establish the natural affiliation of this species. Likewise, *H. parahaemolyticus* clusters with *A. lignieresii*, but the 16S rRNA sequence on which this position is based appears to be incorrect (Hedegaard et al., 2001). *H. parasuis* clusters with *Actinobacillus indolicus*; *H. paracuniculus*, with *Acti-*

nobacillus capsulatus; and *H. paragallinarum*, with a cluster of avian species including *Pasteurella gallinarum* (see discussion of the family *Pasteurellaceae* by Olsen et al. in the chapter on the family *Pasteurellaceae* in this *Manual*). Solutions to these problems must await clarification of the overall taxonomic structure of the family *Pasteurellaceae*. The fact that results of DNA reassociation and 16S rRNA sequencing are in some cases in conflict emphasizes the need for more comprehensive genetic approaches.

Numerous taxonomic questions need to be solved, even within the genus *Haemophilus sensu stricto*. Several of these problems are associated with *H. influenzae* and its close relative *H. aegyptius*. Based on cluster analyses of comprehensive DNA–DNA reassociation data, Burbach (1987) recognized two subgroupings within *H. influenzae* and proposed them as subspecies, designated "*H. influenzae* subsp. *influenzae*" and "*H. influenzae* subsp. *meningitidis*", respectively. "*H. influenzae* subsp. *influenzae*" includes the type and consists of noncapsular strains and encapsulated strains of serovars a, c, and f, all belonging to biovars II, III, and VII. "*H. influenzae* subsp. *meningitidis*" includes encapsulated strains of serovars a, b, d, and e, as well as noncapsular strains. Biovars I, II, IV, and V are represented among these strains. Excepting a single biovar II strain included in " subsp. *meningitidis*", a positive ornithine decarboxylase test can distinguish " subsp. *meningitidis*" from "*H. influenzae* subsp. *influenzae*". However, with regard to serovars and biovars, these subgroups are not consistent with those observed by cluster analysis of multilocus enzyme electrophoresis data (Musser et al., 1986, 1988).

Among strains previously identified as *H. influenzae*, Burbach (1987) furthermore observed a distinct cluster of three strains with biochemical properties as described for biovar VIII (Sottnek and Albritton, 1984). The cluster is only distantly related to *H. influenzae* but is close to *H. haemolyticus*. The designation "*H. intermedius* subsp. *gazogenes*" was applied to this taxon, as the strains produce gas in glucose broth. Three other strains previously identified as *H. parainfluenzae* biovars III and IV are closely related to "*H. intermedius* subsp. *gazogenes*", in spite of differences in X-factor requirement, and these were designated "*H. intermedius* subsp. *intermedius*". Analysis of 16S rRNA sequences of two representative strains confirm that these taxa are distinct from *H. influenzae* and *H. parainfluenzae*, respectively, and are closely related to *H. haemolyticus* (see Fig BXII.γ.215. of Olsen et al. in the chapter on the family *Pasteurellaceae*, this *Manual*). However, the species "*H. intermedius*" has not been formally validly named and, phenotypically, appears difficult to handle in practice.

Studies from North America and France have revealed a relatively high prevalence of noncapsular *H. influenzae* biovar IV among isolates from cases of neonatal bacteremia and meningitis and obstetric infections (Albritton et al., 1982a; Wallace et al., 1983; Quentin et al., 1990). Phylogenetic analyses performed by multilocus enzyme electrophoresis, DNA–DNA hybridization, and rRNA gene polymorphism strongly suggest that such strains form a separate species related to *H. influenzae*, *H. aegyptius*, and *H. haemolyticus* (Quentin et al., 1990, 1993, 1996). It is not clear if all isolates with the characteristics of biovar IV belong in this taxon. This taxon is provisionally referred to as "*Haemophilus quentini*" in Fig. BXII.γ.215 of Olsen et al. (see chapter on the family *Pasteurellaceae*, this *Manual*).

Several studies have addressed the relationship between *H. influenzae* and *H. aegyptius* and the question of whether both deserve specific recognition. Phenotypically, *H. aegyptius* is similar, but not identical to *H. influenzae* (Mazloum et al., 1982). In the three biochemical tests used to distinguish the biovars of *H.*

influenzae, *H. aegyptius* is identical to biovar III (Kilian, 1976; Mazloum et al., 1982), for which reason many researchers have identified all conjunctival isolates with those characteristics as *H. aegyptius*. This may explain some of the conflicting results that have been published on the degree of phenotypic relatedness of the two taxa. Clearly, "true" *H. influenzae* (biovars II and III) are frequently also isolated from cases of conjunctivitis in temperate climates (Kilian et al., 1976). DNA–DNA hybridization data, transformation experiments, and 16S rRNA sequence comparisons all confirm that the two species are closely related (Leidy et al., 1959; Leidy et al., 1965; Pohl, 1981a; Albritton et al., 1984; Casin et al., 1986; Burbach, 1987). While Casin et al. (1986) concluded that the two taxa are indistinguishable, Burbach (1987) found that *H. aegyptius* strains form a distinct entity. The latter conclusion is supported by phenotypic differences and by studies of outer membrane profiles, which have revealed a distinct pattern in strains of *H. aegyptius* (Carlone et al., 1985). Although these differences indicate that *H. aegyptius* is a distinct subpopulation, they do not necessarily show that both *H. influenzae* and *H. aegyptius* warrant status as separate species.

Based on their DNA reassociation data, Casin et al. (1986) argued that *H. aegyptius* be considered a junior subjective synonym of *H. influenzae*. However, this is incorrect. While *H. influenzae* is the type species, the epithet *aegyptius* Trevisan 1889b has formal priority over the epithet *influenzae* Lehmann and Neumann 1896. "To avoid confusion" Brenner and coworkers (1988a) used the informal designation "*H. influenzae* biovar aegyptius" to denote strains formerly called *H. aegyptius*. The term was also applied to isolates from the disease Brazilian purpuric fever (BPF) and subsequently has been used frequently in the literature on this pathogen. Most isolates from the original outbreak and subsequent cases of BPF in Brazil belong to at least three different clones ("BPF clones"), although minor differences in ribotype and plasmid content have been observed within clones. Isolates from occasional cases in Australia are not closely related to the Brazilian clone (Musser and Selander, 1990). The Brazilian BPF clone fails to ferment D-xylose (Brenner et al., 1988a), like strains of *H. aegyptius* and occasional strains of *H. influenzae* (Mazloum et al., 1982). Recent phylogenetic and other analyses demonstrate that BPF clone isolates and *H. aegyptius* are two recently diverged but distinct populations that are separate from *H. influenzae* (Kilian, Poulsen, and Lomholt, unpublished). Thus, the specific status and nomenclature of the two present taxa *H. influenzae* and *H. aegyptius* remain an unsolved problem.

Collectively, the observations described above emphasize that there is a need for comprehensive population genetic analyses of *H. influenzae* and *H. parainfluenzae* strains that fully reflect the genetic diversity within these species in order to clearly define them, as well as neighboring species.

The two species *H. aphrophilus* and *H. paraphrophilus* are differentiated by the requirement for V factor by the latter (Table BXII.γ.294). *H. aphrophilus* has an apparent X-factor requirement when freshly isolated, but contains all the enzymes of the heme biosynthetic pathway (White and Granick, 1963). On subculture, variant colonies consisting of X factor-independent cells arise and apparently supplant the original population (Boyce et al., 1969). Strains carried by type culture collections usually do not have the X-factor requirement. In practice, *H. aphrophilus* is identified by criteria other than the growth factor requirement. Numerous genetic and chemotaxonomic analyses have been applied to strains of the two species to clarify their taxonomic relationship. Results of DNA–DNA hybridization, genetic transformation

(Pohl., 1981a; Tanner et al., 1982; Potts et al., 1985; Tønjum et al., 1990), 16S rRNA sequencing (Dewhirst et al., 1992), ribotyping (Sedlácek et al., 1993), multilocus enzyme electrophoresis (Caugant et al., 1990), and whole cell protein mapping by two-dimensional gel electrophoresis (Olsen et al., 1987) fail to identify two separate groups. In contrast, according to Brondz and Olsen (1990) and Myhrvold et al. (1992), multivariate statistical analyses of LPS carbohydrates and enzyme data support their recognition as separate taxa. No formal action has so far been taken, and the two species are maintained as separate species in this chapter.

A proposal was made (Potts et al., 1985) to transfer the species *Actinobacillus actinomycetemcomitans*, which is associated with periodontal disease in humans, to the genus *Haemophilus*, based on its close genetic relationship to *H. aphrophilus* (Tanner et al., 1982; Potts et al., 1985). However, due to the rather distant relationship with the type species *H. influenzae*, as judged by DNA reassociation studies, this proposal has not been generally supported. Comprehensive phylogenetic analyses based on 16S rDNA sequence comparisons reveal that the species clusters neither with *H. influenzae* nor with *Actinobacillus lignieresii* but do confirm its close association with *H. aphrophilus*, *H. paraphrophilus*, and *H. segnis* (see Fig. BXII.γ.215 of Olsen et al. in the chapter on the family *Pasteurellaceae*, this *Manual*).

The previous edition of *Bergey's Manual of Systematic Bacteriology* listed three taxa as *species incertae sedis* under the genus *Haemophilus*: "*Haemophilus somnus*", "*Haemophilus agni*", and "*Haemophilus equigenitalis*" (Kilian and Biberstein, 1984). The lattermost

was subsequently transferred to a new genus *Taylorella* (Sugimoto et al., 1983). "*Haemophilus somnus*", which is associated with cattle as both a commensal and a recognized pathogen, and the phenotypically and genetically related "*Haemophilus agni*" are not genetically related to the genus *Haemophilus*, as demonstrated by DNA–DNA hybridization (Piechulla et al., 1986), DNA–rRNA hybridization (de Ley et al., 1990), and the degree of 16S rRNA sequence homology (Dewhirst et al., 1993). They may represent a new genus within the family *Pasteurellaceae* (Bisgaard, 1995). De Ley et al., (1990) suggested that both be transferred to genus "*Histophilus*", "*Histophilus ovis*" being a senior synonym for "*Haemophilus agni*". However, none of these names are valid, and no types have been designated. As the taxa are still often referred to as "*Haemophilus somnus*" and "*Haemophilus agni*", the descriptions are maintained here under Other Organisms.

ACKNOWLEDGMENTS

In the previous edition of this *Manual*, the chapter on the genus *Haemophilus* was co-authored with Dr. E.L. Biberstein. I am grateful for his many contributions and for what remains in this version.

FURTHER READING

Kilian, M., Frederiksen, E.L. and Biberstein, E.L. 1981. *Haemophilus, Pasteurella* and *Actinobacillus*, Academic Press, London.

Slack, M.P.E. and Jordens, J.Z. 1998. *Haemophilus. In* Balows and Duerden (Editors), Topley and Wilson's Microbiology and Microbial Infections, Vol. 2. Systematic Bacteriology, Arnold, London. p. 1167–1190.

Turk, D.C. and May, J.R.. 1967. *Haemophilus influenzae*. Its Clinical Importance, The English Universities Press, London.

DIFFERENTIATION OF THE SPECIES OF THE GENUS *HAEMOPHILUS*

The differential characteristics of the species of *Haemophilus* are presented in Table BXII.γ.294. Other characteristics of the species are presented in Table BXII.γ.295.

List of species of the genus Haemophilus

1. **Haemophilus influenzae** (Lehmann and Neumann 1896) Winslow, Broadhurst, Buchanan, Krumwiede, Rogers and Smith 1917, 561[AL] (*Bacterium influenzae* Lehmann and Neumann 1896, 187.)

   in.flu.en'zae. Italian n. *influenza* influenza; M.L. gen. n. *influenzae* of influenza.

   Coccobacilli or small regular rods 0.3–0.5 × 0.5–3.0 μm. Colonies on chocolate agar are smooth, low, convex, grayish, and translucent and attain a diameter of 0.5–1.0 mm in 24 h. Capsulated strains usually produce larger and more mucoid colonies (1–3 mm), which show a tendency to coalesce with no visible line of demarcation. On transparent agar media, colonies of capsulated strains show iridescence when examined under obliquely transmitted light.

   Among encapsulated isolates, six serovars (a–f) have been identified (Pittman, 1931; M. Pittman, unpublished results). The structure of the serovar a–f capsular polysaccharides is shown in Fig. BXII.γ.219. A total of 41 capsular and somatic antigenic determinants have been detected in a strain of *H. influenzae* serovar b by crossed immunoelectrophoresis. A considerable number of these determinants are related to antigenic determinants of strains of *H. haemolyticus*, *H. parainfluenzae*, and various enterobacteria (Schiøtz et al., 1979).

   H. influenzae strains may be assigned to any of seven biovars (I-VII) based on three biochemical characteristics, indole production, and urease and ornithine decarboxylase activities (Tables BXII.γ.294 and BXII.γ.296) (Kilian, 1976; Oberhofer and Bach, 1979; Gratten 1983). Strains described as *H. influenzae* biovar VIII (Sottnek and Albritton, 1984) were subsequently shown to be distinct from *H. influenzae* and were assigned to the not validly published taxon "*H. intermedius* subsp. *gazogenes*" (see above). The majority of strains with a serovar b capsule are of biovar I, an observation that first suggested that serovar b strains are distinct from non-capsulated strains, most of which belong to biovars II and III. All isolates so far examined with a serovar d and e capsule are of biovar IV (Kilian, 1976).

   H. influenzae possesses neuraminidase activity (Müller and Hinz, 1977). Virtually all strains produce an extracellular endopeptidase (IgA1 protease) capable of inducing specific cleavage of a proline–serine or a proline–threonine peptide bond in the hinge region of human immunoglobulin A1 (Kilian et al. 1996).

   H. influenzae is present in the nasopharynxes of a majority of healthy children. The carriage rate in adults is somewhat lower (Kilian and Frederiksen, 1981a). It is rarely encountered in the human oral cavity and has not been

TABLE BXII.γ.295. Further characteristics of the species of the genus *Haemophilus*

Characteristics	1. *H. influenzae*	2. *H. aegyptius*	3. *H. aphrophilus*	4. *H. avium*	5. *H. ducreyi*	6. *H. haemoglobinophilus*	7. *H. haemolyticus*	8. *H. paracuniculus*	9. *H. paragallinarum*	10. *H. parahaemolyticus*	11. *H. parainfluenzae*	12. *H. paraphrohaemolyticus*	13. *H. paraphrophilus*	14. *H. parasuis*	15. *H. segnis*	a. "*H. somnus*"	a. "*H. somnus*"
Lysine decarboxylase	d	−	−	−	−	−	−	−	−	−	d	−	−	−	−	−	−
Oxidase	+	+	−	−	+	+	+	+	−	+	+	+	+	d	−	+	+
H₂S (lead acetate)	−	−	−	+	−	d	+	−	−	+	+	+	+	−	−	−	−
Acid production from:																	
L-Arabinose	−	−	−	−	−	−	−	−	−	−	−	−	−	−	−	d	W
L-Rhamnose	−	−	−	−	−	−	−	−	−	−	−	−	−	−	−	−	−
D-Galactose	+	Wᵃ	+	+	−	+	+	−	−	d	+	d	+	+	W	d	W
Sorbose	−	−	−	d	−	−	−	−	−	−	−	−	−	−	−	−	−
Cellobiose	−	−	+	+	−	−	−	−	−	−	−	−	+	−	−	−	−
Maltose	+	W	+	+	−	+	+	+	+	+	+	+	+	+	W	+	W
Melibiose	−	−	−	−	−	−	−	−	−	−	−	−	+	−	−	−	−
Trehalose	−	−	−	−	−	−	−	−	−	−	−	−	+	−	−	+	−
Melizitose	−	−	+	−	−	−	−	−	−	−	−	−	+	−	−	−	−
Raffinose	−	−	−	−	−	−	−	−	−	−	−	−	+	+	−	−	−
Inulin	−	−	−	−	−	−	−	−	−	−	d	−	−	−	−	−	−
Dulcitol	−	−	d	−	−	−	−	−	−	−	−	−	−	−	−	−	−
Glycerol	−	−	−	−	−	−	−	−	−	−	d	−	−	d	−	d	−
meso-Erythritol	−	−	d	−	−	−	−	−	−	−	−	−	−	−	−	−	−
Inositol	−	−	−	−	−	−	−	−	−	−	−	−	−	d	−	−	−
Xylitol	−	−	−	−	−	−	−	−	−	−	−	−	−	−	−	d	−
Esculin, salicin, adonitol	−	−	−	−	−	−	−	−	−	−	−	−	−	−	−	−	−
α-Galactosidase	−	−	d	+	−	−	−	−	d	−	−	−	−	d	−	−	−
α-Glucosidase	−	−	−	−	−	−	−	−	−	−	−	−	d	d	−	−	−
β-Glucosidase	−	−	−	−	−	−	−	−	−	−	−	−	−	d	−	−	−
α-Mannosidase	−	−	−	−	−	−	−	−	−	−	−	−	−	−	−	−	−
β-Xylosidase	d	−	−	−	−	−	−	−	d	d	−	−	−	−	−	−	−
β-Glucoronidase	+	−	−	−	−	−	−	−	−	−	−	−	−	+	−	−	−
Nitrate reduction	+	+	+	+	+	+	+	+	+	+	+	+	+	+	+	+	+
Nitrite reduction	−	−	+	−	−	d	d	−	−	+	+	+	+	−	d	−	−

ᵃW, weak reaction.

TABLE BXII.γ.296. Key to the differentiation of the biovars of *H. influenzae* and *H. parainfluenzae*

Characteristic	H. influenzae biovars							H. parainfluenzae biovars							
	I	II	III	IV	V	VI	VII	I	II	III	IV	V	VI	VII	VIII
Indole	+	+	−	−	+	−	+	−	−	−	+	−	+	+	+
Urease	+	+	+	+	−	−	−	−	+	+	+	−	−	+	+
Ornithine decarboxylase	+	−	−	+	+	+	−	+	+	−	+	−	+	−	−

detected in any animal species apart from chimpanzees. Capsulated strains are harbored intermittently in the nasopharynx of a minority of healthy individuals (3–7%). Serovars b and f are the serovars most frequently encountered, whereas serovar c strains are rare (Moxon, 1986).

H. influenzae was isolated originally from cases of endemic influenza and was regarded as its causative agent at the time. It is frequently isolated from chronic infections of the upper and lower human respiratory tract, paranasal sinuses, middle ears, and conjunctivae, in which conditions *H. influenzae* plays an important, though often probably a secondary, part. Strains implicated in such conditions are usually noncapsulated and belong to the biovars II or III like the majority of isolates from healthy upper respiratory tracts (Kilian and Frederiksen, 1981a). Encapsulated strains of serovar b are among the three most common causes of bacterial meningitis in children and occasionally cause acute epiglottitis (obstructive laryngitis) cellulitis, osteomyelitis, and joint infections (Turk and May, 1967).

The mol% G + C of the DNA is: 39 (T_m).

Type strain: 680 of Pittman (noncapsulated, biovar II) ATCC 33391, NCTC 8143.

GenBank accession number (16S rRNA): M35019.

Additional Remarks: Reference strains: NCTC 8466 (serovar a, biovar I); NCTC 7279 (serovar b, biovar I); NCTC 8469 (serovar c, biovar II); NCTC 8470 (serovar d, biovar IV); NCTC 10479 (serovar e, biovar IV); NCTC 8473 (serovar f, biovar I); NCTC 4560 (noncapsulated, biovar III); NCTC 11394 (noncapsulated, biovar V).

2. **Haemophilus aegyptius** (Trevisan 1889b) Pittman and Davis 1950, 413[AL] (*Bacillus aegyptius* Trevisan 1889b, 13.)

ae.gyp′ti.us. L. adj. *aegyptius* Aegyptian.

The common name of the organism is Koch–Weeks bacillus. Long slender rods 0.2–0.3 × 2.0–3.0 μm. Growth of freshly isolated strains on chocolate agar is slow. After 48 h, colonies attain a diameter of about 0.5 mm and are smooth, low, convex, grayish, and translucent. The species does not grow on Tryptic soy agar (Difco) with X and V factors added, in contrast to *H. influenzae*. Forms comet-like colonies in semisolid agar media (Pittman and Davis, 1950).

Capsules have not been demonstrated. Bacteria of this species produce an extracellular IgAl-cleaving endopeptidase (Lomholt and Kilian, 1995; Kilian et al., 1996) and have neuraminidase activity (Müller and Hinz, 1977).

Features which may be of use in distinguishing *H. aegyptius* from *H. influenzae* include poorer growth on most media, a lack of indole production, and D-xylose fermentation, the distinct bacillary morphology, a hemagglutinating activity, the comet-like growth in semisolid media, and the susceptibility to troleandomycin. However, none of these features unequivocally differentiate the two species.

Some strains lack the enzyme ferrochelatase, which in-

serts iron into protoporphyrin IX, and thus require protoheme as X-factor (White and Granick, 1963; Kilian et al., 1976; Ruckelshausen and Holländer, 1978). The organism causes acute or subacute infectious conjunctivitis in hot climates. It has not been demonstrated in healthy individuals. Particular clones ("BPF clones") closely related to *H. aegyptius* have been associated with Brazilian purpuric fever. The clones have been referred to as *H. influenzae* biogroup *aegyptius* (see comments above). They are characterized by particular patterns of outer membrane proteins and alleles of genes encoding house-keeping enzymes, rRNA gene restriction pattern, resistance to serum bactericidal activity, a 24-MDa plasmid (not present in all isolates), a unique antigenic variant of IgA1 protease, and the lack of D-xylose fermentation (Brenner et al., 1988a; Irino et al., 1988; Porto et al., 1989; Musser and Selander, 1990; Lomholt and Kilian, 1995). The taxonomic questions associated with *H. aegyptius* and its close relative *H. influenzae* are discussed above.

The mol% G + C of the DNA is: 39 (T_m).

Type strain: 180a of Pittman, ATCC 11116.

GenBank accession number (16S rRNA): M75044.

3. **Haemophilus aphrophilus** Khairat 1940, 505[AL]

a.phro′phi.lus. Gr. n. *aphros* foam; Gr. adj. *philos* loving; M.L. adj. *aphrophilus* foam loving.

Short regular rods 0.45–0.55 × 1.5–1.7 μm with only occasional filamentous forms. Colonies on chocolate agar incubated in air supplemented with 10% extra CO_2 are high convex, granular, yellowish, and opaque and reach a diameter of 1.0–1.5 mm within 24 h. When plates are incubated without extra CO_2, the growth is characteristically stunted, with very small colonies interspersed with a few larger colonies. Grows on blood agar incubated in air plus CO_2 without feeder strain. Growth in broth media is granular, with a heavy sediment on the bottom of the tube and adhering colonies on the walls, which are difficult to remove.

H. aphrophilus possesses all the enzymes of the hemin biosynthetic pathway characteristic for hemin-independent species (White and Granick, 1963). Accordingly, it gives a positive, though usually weak, reaction for both porphobilinogen and porphyrins in the porphyrin test. The reason for an apparent need for hemin-containing media at primary isolation remains to be elucidated.

H. aphrophilus is a frequent member of the microflora of human dental plaque, particularly between the teeth and in the gingival pockets (Kraut et al., 1972; Kilian and Schiøtt, 1975). Occasionally causes endocarditis and brain abscesses in humans. Has been isolated from spinal fluid, wounds, and jaw infections (King and Tatum, 1962).

The mol% G + C of the DNA is: 42 (T_m).

Type strain: PM1, ATCC 33389, NCTC 5906.

GenBank accession number (16S rRNA): M75041.

4. **Haemophilus ducreyi** (Neveu-Lemaire 1921) Bergey, Harrison, Breed, Hammer and Huntoon 1923, 271[AL] (*Coccobacillus ducreyi* Neveu-Lemaire 1921, 20.)

du.crey'i. M.L. gen. n. *ducreyi* of Ducrey; named after Ducrey, the bacteriologist who first isolated this organism.

Slender rods in pairs or chains, measuring 0.5×1.5–2.0 μm, often seen in "school of fish"-like arrangements. Capsules have not been detected. Growth is poor on most laboratory media. Colonies on chocolate agar after 72 h are predominantly small (~0.5 mm in diameter), flat, smooth, grayish, and translucent, but often with a few interspersed larger colonies having an otherwise identical appearance. Colonies may be slid intact across the surface of agar plates. Growth on blood agar is very sparse, and there is no satellite growth around *Staphylococcus* colonies. Produces a cell-associated hemolysin with cytotoxic activity (Alfa et al., 1996) and a soluble cytotoxin (Cope et al., 1997). Either one or both of these toxins results in weak β hemolysis on blood agar plates.

Clinical isolates usually appear to be asaccharolytic. However, under favorable growth conditions, some strains show a late positive reaction for glucose fermentation. The species is inert in most of the traditional biochemical tests (Tables BXII.γ.294 and BXII.γ.295), but produces a wide array of peptidases. At least seven variants have been detected among 105 clinical isolates based on cell protein profiles (Odumeru et al., 1983). Causes the human venereal disease known as soft chancre or chancroid. A carrier state in healthy individuals has not been detected.

The mol% G + C of the DNA is: 38 (T_m).

Type strain: CIP 542, DSM 8925.

GenBank accession number (16S rRNA): M63900, M75078.

5. **Haemophilus felis** Inzana, Johnson, Shell, Møller and Kilian 1999, 341[VP] (Effective publication: Inzana, Johnson, Shell, Møller and Kilian 1992, 2111.)

fe'lis. L. n. *felis* a cat; L. gen. n. *felis* of a cat.

Coccobacilli or small regular rods, 0.45–0.55×1.5–1.7 μm, with occasional filamentous forms. Colonies on chocolate agar incubated in air plus 10% extra CO_2 are raised, smooth, and rounded and reach a diameter of 0.5–1.0 mm after 24 h. Colonies on brain–heart infusion agar supplemented with NAD are very adherent and show yellow pigmentation when collected on a loop. On blood agar, satellitic growth is observed around a feeder strain of *Staphylococcus* or an NAD-impregnated paper disk. Weak hemolysis is observed after incubation for 3 d. Requires increased concentration of CO_2 in incubation atmosphere for primary isolation. This requirement is lost upon subcultivation. Has been isolated from the lower respiratory tract of a cat with chronic obstructive pulmonary disease and from the nasopharynxes of 6 of 28 apparently normal cats.

The mol% G + C of the DNA is: 38 (T_m).

Type strain: TI189, ATCC 49733.

GenBank accession number (16S rRNA): AF224292.

6. **Haemophilus haemoglobinophilus** (Lehmann and Neumann 1907) Murray 1939, 309[AL] (*Bacterium haemoglobinophilus* (Lehmann and Neumann 1907, 270.)

hae.mo.glo.bi.no'phi.lus. M.L. n. *haemoglobinum* hemoglobin; Gr. adj. *philos* loving; M.L. adj. *haemoglobinophilus* hemoglobin-loving.

Small, slightly pleomorphic, noncapsuled rods. Colonies on chocolate agar are smooth, convex, and translucent with a small granular area on top, and reach a diameter of 1–2 mm within 24 h. In contrast to other members of the genus, they grow well on blood agar and show no satellite growth around a streak of *Staphylococcus*.

Differs from all other *Haemophilus* species in lacking alkaline phosphatase. Belongs to the normal flora of the preputial sac of dogs, and is probably of low pathogenicity.

The mol% G + C of the DNA is: 38 (T_m).

Type strain: XIII of Kristensen, NCTC 1659.

GenBank accession number (16S rRNA): M75064.

7. **Haemophilus haemolyticus** Bergey, Harrison, Breed, Hammer and Huntoon 1923, 269[AL]

hae.mo.ly'ti.cus. Gr. n. *haema* blood; Gr. adj. *lyticus* loosening, dissolving; M.L. adj. *haemolyticus* blood dissolving.

Small, regular, noncapsuled coccobacilli or short rods, with occasional filamentous forms. Colonies on chocolate agar are smooth, convex, grayish, and translucent and reach a diameter of 0.5–1.5 mm within 24 h. Produce clear hemolytic zones on bovine or sheep blood agar. However, the hemolytic activity may be lost upon subcultivation. Is found in the nasopharynx of a minority of the healthy human population and often in the bacterial deposits on human teeth below the gingival margin (subgingival dental plaque). No pathogenic potential has ever been demonstrated.

The mol% G + C of the DNA is: 39 (T_m).

Type strain: AQ/3273 of Mclves, NCTC 10659.

GenBank accession number (16S rRNA): M75045.

8. **Haemophilus paracuniculus** Targowski and Targowski 1984, 355[VP] (Effective publication: Targowski and Targowski 1979, 36.)

pa.ra.cu.ni'cu.lus. Gr. prep. *para* alongside of, resembling; L. n. *cuniculus* rabbit, or possibly the specific epithet of "*Haemophilus cuniculus*", a species that has never been described; M.L. adj. *paracuniculus* like "*Haemophilus cuniculus*".

Small coccobacilli or pleomorphic rods with occasional filamentous forms. In broth media supplemented with V factor, the organism forms chains of rods. Colonies on chocolate agar are smooth, high convex, grayish, and opaque. Capsules have not been described. Does not grow in phenol red broth base supplemented with V factor. Has been isolated from the gastrointestinal tract of rabbits with mucoid enteritis. The pathogenic significance of the organism has not been established.

The mol% G + C of the DNA is: 40 (Bd).

Type strain: RF15888, ATCC 29986.

GenBank accession number (16S rRNA): M75061.

9. **Haemophilus paragallinarum** Biberstein and White 1969, 77[AL]

pa.ra.gal.li.na'rum. Gr. prep. *para* alongside of, resembling; M.L. gen. pl. n. *gallinarum* specific epithet; M.L. adj. *paragallinarum* resembling (*Haemophilus*) *gallinarum*.

Coccobacilli to pleomorphic rods with occasional filamentous forms. Most strains are capsulated. Colonies on chocolate agar are smooth, convex, grayish, and semiopaque and attain a diameter of 0.5–1.0 mm within 48 h in air supplemented with 10% CO_2. Growth is feeble in air without extra CO_2. Young cultures (8- to 24-h-old) of capsulated strains produce iridescent colonies on transparent agar media. Growth is enhanced by serum added to cultivation media (Hinz, 1973; Rimler et al., 1976). Occasional V-factor-independent strains have been isolated in South Africa (Bisgaard Taxon 31) (Mouahid et al., 1992; Miflin et al., 1995).

Several serotyping schemes have been proposed. Three serovars (A–C) have been defined based on heat-labile surface antigens (Page, 1962). Kato and Tsubahara (1962) have used the designations I, II, and III for three serovars, of which serovar III may be a noncapsulated variant of serovar I (Kume et al., 1980). Hinz (1980) has established six serovars based on heat-stable somatic antigens. Nine serovars based on a hemagglutinin are currently recognized according to the proposal by Blackall et al. (1990a), which combines some of the former schemes in one, with the designations A-1, A-2, A-3, A-4, B-1, C-1, C-2, C-3, and C-4.

Produces neuraminidase and N-acetylneuraminate pyruvate lyase (Hinz and Müller, 1977). Present in the respiratory tract of poultry. Causes respiratory tract disease in chickens known as infectious coryza. The species includes organisms previously labeled "*Haemophilus gallinarum*".

The mol% G + C of the DNA is: 42 (T_m).

Type strain: IPDH 2403, serovar A, ATCC 29545.

GenBank accession number (16S rRNA): M75057.

10. **Haemophilus parahaemolyticus** Pittman 1953, 750[AL]

pa.ra.hae.mo.ly' ti.cus. Gr. prep. *para* alongside of, resembling; M.L. n. *haemolyticus* specific epithet; M.L. adj. *parahaemolyticus* resembling (*Haemophilus*) *haemolyticus*.

Small, pleomorphic rods, usually with long, filamentous forms. Growth on chocolate agar is similar to that of *H. parainfluenzae*. A distinct β-hemolytic zone is produced on blood agar plates, most clearly detectable on bovine or sheep blood agar. Growth on blood agar usually is not restricted to the area around a feeder strain, presumably due to the release of ample V factor from blood cells that lyse as a result of the hemolytic activity.

Has been associated with acute pharyngitis, purulent oral infections, occasional cases of endocarditis, and exacerbation of chronic lower respiratory tract infections. Strains of this species produce an extracellular IgA1 protease closely related to the *H. influenzae* endopeptidase, which is capable of cleaving human IgAl (Kilian and Poulsen, unpublished studies). Tuyau and Sims (1974) observed neuraminidase activity in members of this species. However, this cannot be confirmed with well-defined chromogenic substrates (Kilian and Poulsen, unpublished studies). Hemolytic porcine isolates that were previously assigned to this species belong in the species *Actinobacillus pleuropneumoniae*.

The mol% G + C of the DNA is: 40–41 (T_m).

Type strain: 536, NCTC 8479.

GenBank accession number (16S rRNA): AJ295746.

11. **Haemophilus parainfluenzae** Rivers 1922, 431[AL]

pa.ra.in.flu.en'zae. Gr. prep. *para* alongside of, resembling; M.L. n. *influenzae* specific epithet; M.L. gen. n. *parainfluenzae* intended to mean like the species *H. influenzae*.

Small, pleomorphic rods, usually with long, filamentous forms. Occasional strains possessing a capsule have been described (Sims, 1970). Colonies on chocolate agar are grayish white or yellowish opaque and reach a diameter of

1–2 mm after 24 h. Some strains produce flat, smooth colonies with an entire edge, others show a serrated edge, and yet others grow as very rough wrinkled colonies. The irregular forms of colonies are usually coherent in texture and can be slid intact across the surface of the agar plate. Strains growing as rough-type colonies often convert into the smooth type after some *in vitro* transfers. Some isolates show week β-hemolysis on blood agar, but the property is often lost after several subcultivations. Growth in broth media may or may not be granulated.

Eight biovars have been identified based on three biochemical reactions: indole formation, and urease and ornithine decarboxylase activities (Table BXII.γ.294) (Kilian, 1976; Oberhofer and Bach, 1979; Bruun et al., 1984a; Doern and Chapin, 1987). It is not clear if strains with the characteristics of biovar V do indeed belong in the species *H. parainfluenzae*. They are phenotypically similar to *H. segnis* and *H. paraphrophilus*. The majority of human isolates can be assigned to biovars I–III.

H. parainfluenzae is ubiquitous in the human oral cavity and pharynx and may be present in the normal vaginal flora (Sims, 1970; Tuyau and Sims, 1975; Kilian and Schiøtt, 1975; Liljemark et al., 1984). Organisms closely resembling *H. parainfluenzae* have been isolated from monkeys, swine, rabbits, and rats (Kilian and Frederiksen, 1981a). Tuyau and Sims (1974) reported that the species has neuraminidase activity. However, this cannot be confirmed with well-defined chromogenic substrates. Is of low pathogenicity, but is occasionally implicated in endocarditis and lower respiratory tract infections in humans.

The mol% G + C of the DNA is: 40–41 (T_m).

Type strain: ATCC 33392, NCTC 7857 (biovar I).

GenBank accession number (16S rRNA): M75081.

Additional Remarks: Reference strains include NCTC 10665 (biovar II), NCTC 11607 (biovar III).

12. **Haemophilus paraphrohaemolyticus** Zinnemann, Rogers, Frazer and Deveraj 1971, 143[AL]

par.aph.ro.hae.mo.ly' ti.cus. M.L adj. *paraphro* resembling (*Haemophilus*) *aphrophilus*; M.L. adj. *haemolyticus* blood dissolving; M.L. adj. *paraphrohaemolyticus* like *H. aphrophilus*, but hemolytic.

When grown in air with 10% extra CO_2, it forms short to medium-length rods, 0.75–2.5 × 0.4–0.5 μm, with occasional short filaments; in air without added CO_2, it forms short to long, coarse rods with involution forms and twisted filaments. Requires increased CO_2 tension in incubation atmosphere for satisfactory growth and good β-hemolysis on blood agar. Apart from the CO_2 requirement, the biochemical, physiological, and ecological characteristics are similar to those of β-hemolytic isolates of *H. parainfluenzae*. Has been isolated from human sore throats, ulcers of the mouth, sputum, and urethral discharge of adult males, but pathogenic significance is unknown.

The mol% G + C of the DNA is: 40–41 (T_m).

Type strain: L1, NCTC 10670.

GenBank accession number (16S rRNA): M75076.

13. **Haemophilus paraphrophilus** Zinnemann, Rogers, Frazer and Boyce 1968, 418[AL]

pa.ra.phro' phi.lus. Gr. prep. *para* alongside of, resembling; M.L. adj. *aphrophilus* specific epithet; M.L. adj. *paraphrophilus* resembling (*Haemophilus*) *aphrophilus*.

Short, regular rods with occasional filamentous forms. Involution forms occur under fully aerobic incubation. Growth characteristics are identical to those described for *H. aphrophilus*. In contrast to *H. aphrophilus*, *H. paraphrophilus* requires V factor and is not dependent on X factor on primary isolation. The two species have otherwise almost identical properties (Tables BXII.γ.294 and BXII.γ.295) and are genetically closely related (see discussion under Taxonomic comments).

Found as a member of the normal flora of the human oral cavity and pharynx (Kilian and Frederiksen, 1981a). May cause subacute endocarditis, paronychia, and brain abscesses and has been isolated from osteomyelitis of the jaw, an inflamed appendix, urine of children with congenital malformation of urogenital tract, and the vaginas of mature women.

The mol% G + C of the DNA is: 42 (T_m).

Type strain: Reece, ATCC 29241, NCTC 10557.

GenBank accession number (16S rRNA): M75042.

14. **Haemophilus parasuis** Biberstein and White 1969, 77[AL]

pa.ra.su' is. Gr. prep. *para* alongside of, resembling; M.L. n. *suis* specific epithet; M.L. adj. *parasuis* (*Haemophilus*) *suis*-like.

Thin, pleomorphic rods of varying length. Growth on chocolate agar is very feeble after 48–72 h. The colonies are smooth, grayish, and translucent and reach a diameter of about 0.5 mm. A marked enhancement of growth on chocolate agar, with colonies attaining a diameter of 1–2 mm, is seen around a streak of a *Staphylococcus* (Biberstein et al., 1977).

Some isolates form capsules, but capsule production does not appear to be associated with virulence. Some, but not all, capsules precipitate with Cetavlon and appear to be acidic polysaccharides (Morozumi and Nicolet, 1986). Four serovars (A–D) are designated, based on capsular polysaccharides (Bakos et al., 1952). Noncapsulated strains are antigenically heterogeneous. Based on heat-stable antigen extracts, a total of 15 serovars (1–15) have been described, and subsequent studies have suggested additional antigenic diversity (Kielstein and Rapp-Gabrielson, 1992; Blackall et al., 1997). Two variants (PAGE patterns I and II) have been identified based on differences in patterns of cellular proteins observed by polyacrylamide gel electrophoresis (Nicolet and Krawinkler, 1981; Morozumi and Nicolet, 1986).

Member of the normal flora of the upper respiratory tract of swine. Causes respiratory tract infections and polyserositis (Glässer's disease). Strains in culture collections labeled *"Haemophilus suis"* require only V factor and belong to *H. parasuis* according to the proposal of Biberstein and White (1969).

The mol% G + C of the DNA is: 41–42 (T_m).

Type strain: 1374 of Shope, NCTC 4557.

GenBank accession number (16S rRNA): M75065.

15. **Haemophilus segnis** Kilian 1976, 35[AL]

seg' nis. L. adj. *segnis* slow, sluggish.

Pleomorphic rods, often showing a predominance of irregular, filamentous forms. Colonies on chocolate agar are smooth or granular, convex, grayish white, and opaque and reach a diameter of about 0.5 mm after incubation for 48 h. Growth in fermentation media is slow, and reactions are negative or weakly positive. Fermentation of sucrose is usually stronger than fermentation of glucose. Is a regular member of the human oral flora, particularly in dental plaque, and can be isolated from the pharynx. Has been isolated in pure culture from a pancreatic abscess.

The mol% G + C of the DNA is: 43–44 (T_m).

Type strain: HK 316, ATCC 33393, NCTC 10977.

GenBank accession number (16S rRNA): M75043.

Other Organisms

1. *"Haemophilus somnus"** Bailie 1966, 64 (*Haemophilus*-like organism, Kennedy, Biberstein, Howarth, Frazier and Dungworth 1960, 403; *Actinobacillus actinoides*-like organism, Bailie, Anthony and Weide 1966, 165; *Actinobacillus* sp., Gossling 1966, 18; *Haemophilus somnifer* Miles, Anthony and Dennis 1972, 431.)

som' nus. L. masc. *somnus* sleep.

Gram-negative short rods, coccobacilli, or filaments. Most strains noncapsulated. Nonmotile. Not acid-fast. Colonial characteristics: 0.2–0.6 mm in diameter, raised, circular, smooth, entire colonies on beef blood agar after 24 h (Kennedy et al., 1960); 0.5–1.5 mm on PPLO–*Haemophilus* agar (Nicolet, 1971; Corboz and Nicolet, 1975); 1–2 mm after 3 d on brain–heart infusion calf blood–yeast extract plates (Garcia-Delgado et al., 1976). Prolonged incubation on blood agar produces colonies with a slightly granular appearance, papillate centers, and flattened peripheries. Most authors report no or weak hemolytic activity.

There is usually no growth of fresh isolates on infusion, tryptose, trypticase soy, serum, or hemoglobin agar under any atmospheric condition. Growth occurs on blood and chocolate agars, and on serum–yeast extract PPLO–agar under 5–20% carbon dioxide. Poor or no growth in ambient air, or anaerobically on any solid medium (Kennedy et al., 1960; Garcia-Delgado et al., 1976). Satellitic growth around *S. aureus* occurs in an atmosphere of increased CO_2, but the growth factor supplied by *S. aureus* has not been identified (Kennedy et al., 1960). Occasional strains lack the special atmospheric and nutritional requirements described, and all cultures apparently adapt gradually to fully aerobic growth. No growth response to X and V factors. Biochemical activities are shown in Tables BXII.γ.294 and BXII.γ.295.

All isolates examined have antigens in common, regardless of geographic and anatomic origin (Shigidi and Hoerlein, 1970; Corboz and Nicolet, 1975; Garcia-Delgado et al., 1976; Corboz and Wild, 1981). Cross-reactions at low titer with a number of other mostly Gram-negative bacteria have been reported (Miller et al., 1975). Significant diversity has been detected among isolates based on biochemical activities, plasmid profiles, and DNA fingerprinting (Fussing and Wegener, 1993).

Antimicrobial susceptibility: Only rough, qualitative data exist on antimicrobial susceptibility; these are based on nonstandardized disk diffusion tests. These suggest susceptibility to ampicillin, bacitracin, cephaloridine, chloropheni-

---

Editorial Note: "Haemophilus somnus" has been described as a member of the genus Haemophilus. By current standards it does not qualify for inclusion in the genus.

col, dihydrostreptomycin, erythromycin, novobiocin, penicillin G, polymyxin B, and tetracycline, and resistance to lincomycin, neomycin, and sulfonamides (Garcia-Delgado et al., 1976).

Pathogenicity: "*H. somnus*" causes septicemia and meningoencephalomyelitis in cattle (Kennedy et al., 1960) and is involved in respiratory (Brown et al., 1970) and genital infections (Waldhalm et al., 1974a), including abortions (Chladeck, 1975), in that species. Experimental infections of mice, guinea pigs, rabbits, and sheep have been reported (Kennedy et al., 1960; Panciera et al., 1968; Miles et al., 1972).

Ecology: The reservoir of "*H. somnus*" appears to be the mucous membranes of the normal bovine respiratory and genital tracts (Corstvet et al., 1973; Waldhalm et al., 1974a).

The mol% G + C of the DNA is: 37.3 ± 0.2 (T_m).

GenBank accession number (16S rRNA): M75046.

2. "*Haemophilus agni*"* Kennedy, Frazier, Theilen and Biberstein 1958, 645.

ag' ni. L. gen. n. *agri* of the lamb.

Morphology: Gram-negative, nonmotile, non-acid-fast, pleomorphic rods or coccobacilli, 0.3–0.7 × 0.5–2.0 μm. Long, thread-like forms containing spherical and thickened fusiform bodies occur. Pleomorphism diminishes on passage. Presence of capsules has been suggested.

Colonial characteristics: Colonial characteristics in 5–10% CO_2: the colonies are convex and translucent and range from <0.5–1.5 mm. Colonies become flattened peripherally and acquire more sharply contoured edges but not significantly larger size on further incubation.

Nutrition and growth conditions: Very slight growth on blood agar in air. Colonial growth on blood agar under 5–10% CO_2, as described above. Satisfactory growth occurs on hemoglobin cystine agar and chocolate agar. No growth response to X and V factors. No satellitic growth with staphylococci. No growth on MacConkey, tryptose, 10% equine serum (with and without 1% yeast hydrolysate), mycoplasma agar, coagulated blood serum, or gelatin.

Metabolism: Metabolism and biochemical activities are shown in Tables BXII.γ.294 and BXII.γ.295.

Antigenic structure: This organism cross-reacts consistently with "*H. somnus*".

Antimicrobial susceptibility: Antimicrobial susceptibility has not been reported.

Pathogenicity: This organism is associated with septicemia, meningitis, polyarthritis myositis, pneumonia, and mastitis of sheep. Of experimental animals, only suckling mice are susceptible.

Ecology: The reservoir of "*H. agni*" has not been determined.

An organism occupying a position close to both "*H. agni*" and "*H. somnus*" has been encountered in a number of inflammatory diseases of sheep in Australia (Roberts 1956; Claxton and Everett 1966; Rahaley and White 1977) and New Zealand (Kater et al., 1962) and named "*Histophilus ovis*" (Roberts, 1956). De Ley et al., (1990) consider "*Histophilus ovis*" to be a senior synonym of "*H. agni*".

The mol% G + C of the DNA is: 36.8 ± 0.3 (T_m).

Genus IV. **Lonepinella** *Osawa, Rainey, Fujisawa, Lang, Busse, Walsh and Stackebrandt 1996, 362^VP (Effective publication: Osawa, Rainey, Fujisawa, Lang, Busse, Walsh and Stackebrandt 1995b, 372)*

RO OSAWA AND ERKO S. STACKEBRANDT

Lone.pin.el' la. M.L. fem. n. *Lonepinella* named after Lone Pine Koala Sanctuary, a private zoo in Australia.

Cells coccoidal, or short to extremely filamentous and rod-shaped. Facultatively anaerobic. Nonmotile. **Neither X nor V factor is required for growth.** No growth under 15°C or over 40°C; optimum, 37°C. Indole produced. Urea not hydrolyzed. **Produces gallic acid from tannic acid (tannase positive), and decarboxylates gallic acid to pyrogallol. Oxidase and catalase reactions are negative.** Acid, but not gas, is produced from D-glucose and some other carbohydrates. **Occur in the normal intestinal flora of the koala.** Analysis of 16 S rDNA places the genus into the family *Pasteurellaceae*, in the *Gammaproteobacteria*.

The mol% G + C of the DNA is: 37.5.

Type species: **Lonepinella koalarum** Osawa, Rainey, Fujisawa, Lang, Busse, Walsh and Stackebrandt 1996, 362 (Effective publication: Osawa, Rainey, Fujisawa, Lang, Busse, Walsh and Stackebrandt 1995b, 372.)

FURTHER DESCRIPTIVE INFORMATION

Based upon 16S rDNA analysis, isolates from the arboreal marsupial *Phascolarctos* (koala) form a separate branch within the

radiation of members of the genera *Pasteurella*, *Haemophilus*, and *Actinobacillus*. Cells of *Lonepinella* strains are pleomorphic, including coccoidal forms (0.5–0.6 × 1.0–1.2 μm), short rods (0.5 × 3-5 μm), and extremely filamentous rods (0.5 × 50-100 μm), especially after prolonged incubation on plate media (Fig. BXII.γ.220). The organisms are nonmotile.

Polyamine composition of *Lonepinella* strains is characterized by the co-occurrence of putrescine and small amounts of diaminopropane, while cadaverine, tyramine, *sym*-norspermidine, spermidine, and spermine are lacking (Osawa et al., 1995b). This is in contrast to the other members of the family *Pasteurellaceae*, which have different polyamines (Busse et al., 1997). In addition, *Lonepinella* strains contain an unique, unidentified substance not reported in any other bacteria.

Lonepinella strains grow in nutrient broth supplemented with 0.1% yeast extract with uniform turbidity. On horse-blood supplemented Wilkins–Chalgren anaerobe agar or brain–heart infusion agar, smooth, opaque, more or less convex, gray to white colonies 1.0–3.0 mm in diameter without hemolysis are formed after 3 d at 37°C. No pigment is produced. Although *Lonepinella* strains are facultatively anaerobic, anaerobic conditions retard or suppress growth. Colony formation is also markedly suppressed on MacConkey agar. Optimum growth occurs at 37°C

Editorial Note: "Haemophilus agni" has been described as a member of the genus *Haemophilus*. By current standards it does not qualify for inclusion in the genus.

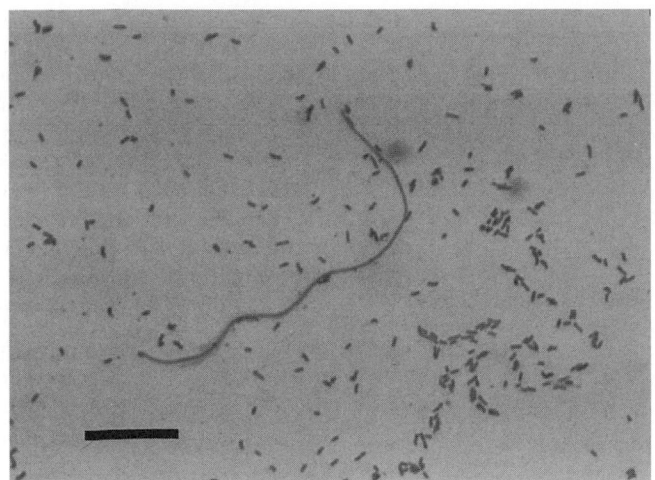

FIGURE BXII.γ.220. Gram-stained cells of *Lonepinella koalarum,* including an extremely long cell, from a colony formed on Wilkins–Chalgren anaerobe agar supplemented with 5% defibrinated horse blood after incubation at 37°C for 48 h. Bar = 20 μm.

when cultured in brain heart infusion. No growth occurs at 15°C or 40°C. Neither X nor V factor is required for growth.

After 3–4 d microaerophilic incubation at 37°C, *Lonepinella* strains form clear zones around colonies on a nonselective plate medium, the surface of which is treated with filter-sterilized tannic acid solution (Fig. BXII.γ.221). This indicates the ability to digest the tannin–protein complex formed on surface of the medium (Osawa, 1992).

Glucose, melibiose, rhamnose, sucrose, amygdalin, and L-arabinose are fermented without gas. D-mannitol, *myo*-inositol, and glucitol are not utilized. The Voges–Proskauer test and the β-galactosidase reaction are positive. The following enzyme activities are negative: oxidase, catalase, urease, arginine dihydrolase, lysine decarboxylase, ornithine decarboxylase, tryptophan deaminase, and gelatinase. H₂S and indole not produced.

Lonepinella is unique by virtue of its "tannase" (tannin acyl-hydrolase) activity, which hydrolyzes gallotannin to yield gallic acid. It was long believed that tannase could be produced only by fungal strains belonging to the genera *Aspergillus* (Pourrat et al., 1987) and *Candida* (Aoki et al., 1976), but recent studies have revealed an almost identical enzymatic activity in a bacterial species, *Streptococcus bovis* biotype I (currently proposed as a new species, "*S. gallolyticus*"; Osawa et al., 1995a]) (Osawa, 1990; Osawa et al., 1993c). *L. koalarum* is considered to be yet another bacterial species with tannase activity. This activity enables the organism to degrade the tannin–protein complex. Furthermore, the organism decarboxylates the released gallic acid to pyrogallol (Osawa et al., 1993c).

16S rDNA analysis revealed that *Lonepinella* does not belong to the family *Enterobacteriaceae,* but instead forms a phylogenetic cluster within the family *Pasteurellaceae.* Similarity values for strains of *Lonepinella* and members of the genera *Pasteurella, Haemophilus,* and *Actinobacillus* range from 93 to 94.5%. A similar range is found for the intra- and intergeneric relationships of the three genera. EMBL accession numbers are Y17189 (ACM 3666ᵀ), Y17190 (strain ACM 3896), and Y17191 (strain ACM 3496). The mol% G + C content of the DNA, as determined by the thermal melting point method (Marmur and Doty, 1962), is 37.5.

Antibiotic sensitivity testing of two *L. koalarum* strains has indicated their resistance to vancomycin, but sensitivity to the following antimicrobials: kanamycin, colistin, chloramphenicol, erythromycin, oxytetracycline, penicillin, gentamicin, rifampin, and ampicillin (Osawa et al., 1995b).

Research on the ecology of *L. koalarum* has been stimulated by frequent isolation of the organism from fecal microflora of koala, an arboreal marsupial animal feeding exclusively on eucalyptus leaves (Osawa, 1992). The organism makes up more than 60% of the facultative anaerobic fecal microflora, and colonizes

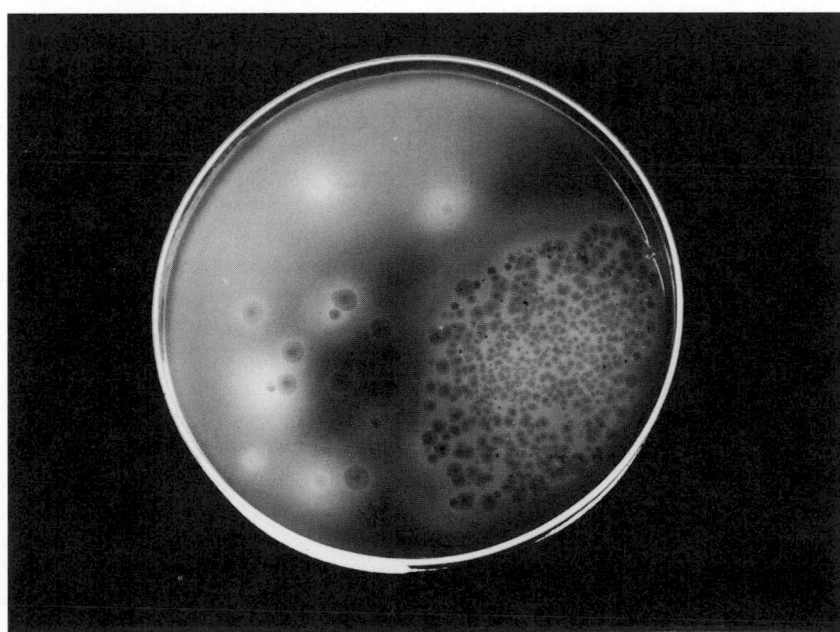

FIGURE BXII.γ.221. Growth of *Lonepinella koalarum,* with clear zones formed around colonies on vancomycin- and tannin-treated Wilkins–Chalgren anaerobe agar.

on the wall of the animal's well-developed caecum and proximal colon (Osawa et al., 1993a). Furthermore, a quantitative study on the microflora in "pap," a special maternal feces consumed by the juvenile around the time of dietary transition from maternal milk to eucalyptus leaves, revealed that "pap" has markedly higher viable counts of *L. koalarum* than the normal feces (Osawa et al., 1993b). The evidence strongly suggests that the organism plays a significant role in the digestion of a tannin-rich diet in the alimentary tract of the host animal.

ENRICHMENT AND ISOLATION PROCEDURES

The best way of isolating *Lonepinella* from fecal samples is to inoculate a plate containing a selective agar, vancomycin- and tannin-treated Wilkins–Chalgren anaerobe agar (Osawa, 1992), and incubate it microaerophilically (5–15% O_2) at 37°C for 3 d. After incubation, *L. koalarum* colonies are easily distinguished from others in mixed cultures, because they are generally about 1–2 mm in diameter with distinct clear zones around them (due to degradation of the tannin–protein complex formed on the surface of the plate). However, the plate allows growth of some enterobacterial species, including *Escherichia coli* and *Klebsiella*; thus, detection of the target colony might be difficult in fecal samples containing a large number of enterobacteria. The organism is readily cultured in conventional enrichment liquid media, such as nutrient broth and heart infusion broth, incubated at 37°C.

MAINTENANCE PROCEDURES

Plate and broth cultures of *Lonepinella* species usually survive no longer than 1 week without subcultivation. Survival is better at room temperature than at 4°C. The most satisfactory way of conserving cultures is by lyophilization, e.g., in skim milk. Wilkins–Chalgren anaerobe agar cultures remain viable for at least 1 year when frozen at − 70°C in sealed glass ampoules.

PROCEDURES FOR TESTING SPECIAL CHARACTERS

Test for tannase activity   Tannase activity may be directly detected by the clear-zone formation of colonies grown on the selective tannin-treated plate medium (Osawa, 1992). The specific clear zone is, however, sometimes difficult to recognize due to overgrowth of other bacteria in the samples. For a more definite and reliable detection of tannase activity, a visual reading method developed by Osawa and Walsh (1993) should be performed. This method is based on two phenomena: (i) the ability of tannase to hydrolyze methyl gallate, which has a molecular structure similar to gallotannin, to release free gallic acid and (ii) the green coloration of gallic acid after prolonged exposure to oxygen under alkaline conditions. Briefly, the isolate is cul-

tured on Wilkins-Chalgren anaerobe agar aerobically at 37°C for 3 d. After incubation, fresh culture on the plate is suspended in substrate medium containing NaH_3PO_4 (33 mM) and methyl gallate (20 mM) to prepare a dense suspension (equivalent to a no. 3 McFarland turbidity standard). The suspension is then incubated aerobically at 37°C for 24 h. After incubation, the suspension is alkalinized with an equal amount of saturated $NaHCO_3$ solution and left in the atmosphere at room temperature for at least 15 min. The development of any green coloration in the medium is an indication of tannase activity.

Test for gallate decarboxylase activity   In addition to tannase activity, *Lonepinella* strains are capable of decarboxylating gallic acid to pyrogallol. The most rapid and convenient method to detect this activity was developed by Osawa and Walsh (1995). Briefly, the isolate is inoculated into Wilkins–Chalgren anaerobe broth or nutrient broth supplemented with gallic acid (10 mM), and incubated anaerobically at 37°C for 3 d. After incubation, the culture is centrifuged, and the supernatant added to an equal amount of saturated $NaHCO_3$ solution. Any orange coloration of the medium that develops is an indication of positive gallate decarboxylating activity.

DIFFERENTIATION OF THE GENUS *LONEPINELLA* FROM OTHER GENERA

See Table BXII.γ.297 for major phenotypic characteristics that differentiate *Lonepinella* from *Pasteurella*, *Haemophilus*, and *Actinobacillus*.

TAXONOMIC COMMENTS

Lonepinella strains were initially placed within the family *Enterobacteriaceae*, since they are facultatively anaerobic, oxidase-negative, Gram-negative bacilli and reduce nitrates to nitrites, thus phenotypically resembling *Enterobacter agglomerans* (Osawa, 1992). However, further comparison with representatives of the *Gammaproteobacteria* has shown them to group within the radiation of organisms presently assigned to the family *Pasteurellaceae* (Osawa et al., 1995b). The phylogenetic placement of these *Phascolarctos* isolates within members of the family *Pasteurellaceae* was unexpected, as none of the species of the family known at the time, *Pasteurella*, *Haemophilus*, or *Actinobacillus*, had been reported to produce tannase and gallate decarboxylase (Mannheim, 1984). The rationale for describing the isolates as a species of a new genus, *Lonepinella*, was based on these physiological properties, their origin from feces of the koala, and the phylogenetic distinctness from the type species of the *Pasteurellaceae* genera. As demonstrated by the studies of Dewhirst et al. (1992, 1993), the phylogenetic structure of the family does not reflect the taxo-

TABLE BXII.γ.297.   Differential phenotypic characteristics of the genus *Lonepinella* and other genera[a]

Characteristics	*Lonepinella*	*Actinobacillus*	*Haemophilus*	*Pasteurella*
V-factor requirement	−	−	+	−
Growth on MacConkey agar	−	+	−	D
Production of:				
Tannase	+	−	−	−
Gallate decarboxylase	+	−	−	−
Catalase	−	D	D	D
Indole	−	−	D	D
Ornithine decarboxylase	−	−	D	D
Acetoin (VP reaction)	+	D		−

[a]For symbols, see standard definitions.

nomic affiliation of species into genera to which they presently belong (Fig. BXII.γ.222). On the other hand, the branching points of deeply rooting lineages have no high statistical significance, as is demonstrated by low (<80%) bootstrap values (Fig. BXII.γ.222). Thus, the branching point of the species *Lonepinella*

koalarum changes with the number and selection of reference species. However, the distinctness of the *Lonepinella* lineage has not been changed by either the composition of strains included in the phylogenetic trees or by the algorithm used for their generation.

List of species of the genus Lonepinella

1. **Lonepinella koalarum** Osawa, Rainey, Fujisawa, Lang, Busse, Walsh and Stackebrandt 1996, 362[VP] (Effective publication: Osawa, Rainey, Fujisawa, Lang, Busse, Walsh and Stackebrandt 1995b, 372.)

   koala' rum. Eng. n. *koala* an arboreal marsupial, *Phascolarctus cinereus*, called "koala"; L. gen. pl. *koalarum* of koalas.

   The description is as given for the genus. Cells coccoidal, 0.5–0.6 × 1.0–1.2 µm, or short to extremely filamentous and rod-shaped, 0.5–0.6 um × 3–100 µm. D-glucose, sucrose, and amygdalin are utilized at 37°C within 48 h, with the production of acid, but not gas. D-mannitol, *myo*-inositol, and glucitol not utilized. Tannase and gallate decarboxylase positive.

   Occurs in the normal intestinal flora of koalas; probably providing the animals with digests of tannin-rich eucalyptus leaves in their hind-gut.

   The mol% G + C of the DNA is: 37.5 (T_m).
   Type strain: ACM 3666, ATCC 700131, DSMZ 10053.
   GenBank accession number (16S rRNA): Y17189.

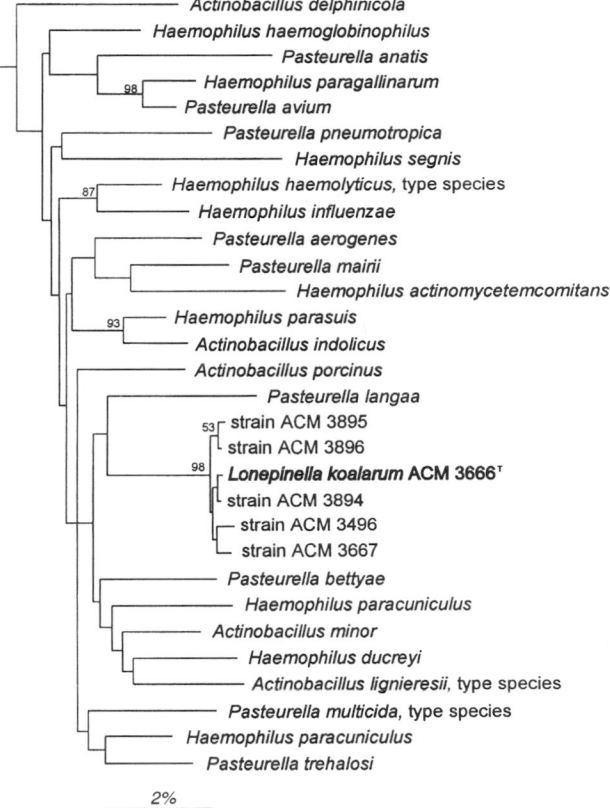

FIGURE BXII.γ.222. Dendrogram of 16S rDNA relationships of strains of the species *Lonepinella koalarum* and some selected members of the family *Pasteurellaceae* that represent individual lines of descent (Dewhirst et al., 1992, 1993; Osawa et al., 1995b). The type species of *Pasteurella*, *Haemophilus*, and *Actinobacillus* are indicated. The dendrogram was generated using the algorithm of Saitou and Nei (1987) on dissimilarity values, as calculated by Jukes and Cantor (1969). The sequences of some members of the *Enterobacteriaceae* were used as the root. Numbers refer to bootstrap values (300 trees). Only values higher than 80% are presented. Bar = 2% sequence divergence.

Genus V. **Mannheimia** Angen, Mutters, Caugant, Olsen and Bisgaard 1999a, 82[VP]

REINIER MUTTERS, ØYSTEIN ANGEN AND MAGNE BISGAARD

Mann.hei' mi.a. L. n. fem. *Mannheimia* named in tribute to the German microbiologist Walter Mannheim.

Gram-negative, nonmotile rods or coccobacilli. Cell diameters of ~0.5 × 1.2 µm are similar to those of *Pasteurella* species. **Some species exhibit bipolar staining.** Endospores are not formed. Growth is mesophilic and facultatively anaerobic or microaerophilic. Neither X nor V factor is required for growth. Glucose is fermented without gas production. **Oxidase reaction is normally** **positive**, but may be variable. **Alkaline phosphatase test is positive** and **nitrate is reduced**. Simmon's citrate, arginine dihydrolase, and Voges–Proskauer tests are negative, and there is no fermentation of D-adonitol or L-sorbose. All strains ferment D-mannitol. Urease reaction is negative. Trehalose and D-mannose are not fermented. Cultural characteristics are typical of those of the family *Pasteurellaceae*. A zone of **β-hemolysis is commonly observed**, depending on media, type of blood, and pH.

The mol% G + C of the DNA is: 39.2–43.9.

Type species: **Mannheimia haemolytica** (Newsom and Cross 1932) Angen, Mutters, Caugant, Olsen and Bisgaard 1999a, 82 (*Pasteurella haemolytica* Newsom and Cross 1932, 715.)

FURTHER DESCRIPTIVE INFORMATION

Maximum likelihood analysis of 16S rDNA sequences of *Mannheimia* strains results in five distinct but affiliated clusters (Angen et al., 1999a). The mean sequence differences within each cluster range from 0.9 to 2.5%, while the mean sequence differences between the five clusters range from 2.8 to 4.4%. Bootstrap analysis identifies clusters II–V as monophyletic in more than 80% of the repeats. Strains of cluster I, however, cluster together in only 39% of the repeats.

Cluster I includes the type strain of the genus, *M. haemolytica* (NCTC 9380, GenBank accession no. AF060699). Differences in only two nucleotide positions separate this strain from strain U57066 which has the same sequence as strains of serovars 1, 5–9, 12–14, 16, and UG1 of the former *P. haemolytica* (Davies et al., 1996). These two sequences cluster together in 71% of the bootstrap trees. Cluster I also includes *M. glucosida*, the sequences of which show considerable heterogeneity and do not appear as a monophyletic group in the phylogenetic analysis (Angen et al., 1999a). Clusters II–IV contain *M. ruminalis*, *M. granulomatis*, and *M. varigena*, respectively. Cluster V currently includes only unnamed taxa. The genus *Mannheimia* is most closely related to the genus *Actinobacillus*, as evaluated by distances to the type species of *Pasteurella*, *Actinobacillus*, and *Haemophilus*.

The polyamine patterns of the five named species of *Mannheimia* can be obtained from the publication of Busse et al. (1997). Members of the genus are characterized by the presence of 1,3-diaminopropane as the predominant compound (77.6–95.2%), followed by spermine (2.2–12.8%), spermidine (1.7–11.6%), putrescine (0.8–3.8%), and cadaverine (0.2–1.1%). The variations observed are on the same order as those observed for the type strains of the three subspecies of *P. multocida*.

Surface colonies on bovine blood agar are generally circular, slightly raised, and regular, with an entire margin. The surface of the colonies is regular, smooth, shiny, and non-transparent, with a grayish tinge. A diameter of ~1–2 mm is observed after 24 h aerobic incubation at 37°C. A narrow zone of β-hemolysis normally surrounds the colonies, which have a butyrous consistency, but weakly hemolytic strains are often isolated. The colony size of strains isolated from pigs is generally smaller. These strains, however, often produce a very distinct zone of β-hemolysis.

Organisms in Gram-stained preparations from the above cultures are rather pleomorphic without a distinctive cellular morphology. All strains grow well in commonly used broths and in conventional media used for phenotypic characterization. Growth occurs between 25°C and 40°C, with an optimum temperature of 37°C.

As with most *Pasteurella sensu stricto* and actinobacilli, the recovery of *Mannheimia* spp. from clinical specimens does not normally present a problem unless the material is contaminated. When plates are overgrown by *Proteus* spp., the specimen should be replated on blood agar containing an increased amount of agar or a detergent. Media such as trypticase soy blood agar base (BBL), Difco–tryptose blood agar base, or Oxoid–Columbia blood agar base containing 5% bovine or ovine blood are generally preferred for initial isolation. The degree of hemolysis is dependent on the medium used and on the animal species from which the blood originates. A selective medium for isolation of *M. haemolytica* from specimens containing mixed populations of bacteria has been described by Morris (1958).

DNA–DNA hybridization data are available for 102 pairs of strains (Angen et al., 1999a). DNA binding of 91% is observed between the type strain of *M. haemolytica* and the most divergent strain within the ribotype cluster containing the type strain. The different biovars of *M. glucosida* cluster above 85% DNA binding by single linkage. The group of the *M. glucosida* biovars clusters with the type strain of *M. haemolytica* at a DNA binding level of 71%. Strains classified as *M. ruminalis* cluster at 88% DNA binding, while strains identified as *M. granulomatis* cluster at a DNA binding level of 86%. A maximum level of 85% DNA binding is observed for strains identified as *M. varigena*.

Several candidates for new species of *Mannheimia* have been identified by Angen et al. (1999a). However, several taxa allocated to this genus remain to be characterized by DNA–DNA hybridization. Further DNA–DNA hybridizations are necessary before conclusions as to the taxonomic interrelationships between these taxa can be drawn.

Information as to the use of ribotyping, random amplified polymorphic DNA analysis, and pulsed-field gel electrophoresis for discrimination of the former *P. haemolytica* complex may be obtained from the publication of Kodjo et al. (1999). For review of the structure, function, and properties of cellular and extracellular components of *P. haemolytica* see Adlam (1989), while the molecular approach for isolating genes from *M. haemolytica* A1 that code for secreted and soluble antigens has been reviewed by Lo (1995). Major outer membrane proteins of *M. haemolytica* serovars 1–15 have been investigated by Morton et al. (1996).

Since previous publications based upon a few phenotypic tests and subsequent serotyping might be questioned, reinvestigations are needed to obtain solid data on the ecology, antibiotic sensitivity, and pathogenicity of species of the genus *Mannheimia*.

MAINTENANCE PROCEDURES

Proper maintenance of cultures of members of the genus *Mannheimia* includes subculturing on complex media. Cultures remain viable for no longer than a week under refrigeration. Plates should be kept in plastic bags to avoid drying. For long term preservation, lyophilization or storage at −80°C in nutrient broth containing glycerol or a mixture of glucose and serum is recommended.

List of species of the genus Mannheimia

1. **Mannheimia haemolytica** (Newsom and Cross 1932) Angen, Mutters, Caugant, Olsen and Bisgaard 1999a, 82[VP] (*Pasteurella haemolytica* Newsom and Cross 1932, 715.)

   hae.mo.ly′ti.ca. Gr. n. *haima* blood; Gr. n. *lysis* solution; adj. from *lyticos*, Latinized as M.L. fem. adj. *lytica* dissolving, referring to the hemolysis seen on blood agar.

   Corresponds to *P. haemolytica* biogroup 1 (Bisgaard and Mutters, 1986b). Cells are Gram-negative, nonmotile, small rods and coccobacilli. Colonies are regular, smooth, and grayish on blood agar and are 1–2 mm in diameter after 24 h of incubation. Most strains show a characteristic β-hemolysis on bovine blood agar. Biochemical characteristics important for characterization are listed in Tables BXII.γ.298 and BXII.γ.299 and under the description of genus *Mannheimia*. D-sorbitol, D-xylose, maltose, and dextrin are fermented. No strains ferment L-arabinose or glucosides. Strains are negative in ornithine decarboxylase and NPG (β-glucosidase) and positive in ONPF (α-fucosidase). Isolated from pneumonia in cattle and sheep and from septicemia in lambs and mastitis in ewes. Some of the serotypes are probably part of the resident microflora of the upper respiratory tract of ruminants (Frank, 1989; Gilmour and Gilmour, 1989).

A total of 17 serovars have been described within *M. haemolytica*, among which serovars T3, 4, 10, and 15 now are classified as *"Pasteurella trehalosi"*. Representatives of serovars A1, 2, 5, 6, 7, 8, 9, 12, 13, 14, and 16 are found in *Mannheimia haemolytica*. Serotype 11 has been reclassified as *Mannheimia glucosida*.

Virulence factors and pathogenesis of *M. haemolytica* in cattle have been reviewed by Confer et al. (1995). Most studies on pathogenesis and virulence factors have been done with strains classified as serotype A1, which should represent *M. haemolytica* (Bisgaard and Mutters, 1986b; Davies et al., 1996; Angen et al., 1999a). Several potential virulence factors have been described, including the polysaccharide capsule (Chae et al., 1990), LPS (Brogden, 1992), neuraminidase (Straus et al., 1993), transferrin binding proteins (Ogunnariwo and Schryvers, 1990), and sialoglycoprotease (Abdullah et al., 1991). Production of a leucotoxin has been observed in all serotypes, as well as in several untypable strains (Chang et al., 1987; Burrows et al., 1993). It has been shown to be specific for bovine macrophages, neutrophils, lymphocytes, and platelets (Clinkenbeard and Upton, 1991), and several studies have indicated that it plays a major role in the pathogenesis of pneumonic pasteurellosis (Henricks et al., 1992; Hughes et al., 1994; Tatum et al., 1998).

The type strain was isolated in 1956 from sheep in the UK. The type strain represents serovar A2, is ONPX (β-xylosidase) and ONPG negative and shows a late *meso*-inositol fermentation.

The mol% G + C of the DNA is: 43.6 (T_m).

Type strain: J.A.Watt 1266A&B, CCUG 12392, NCTC 9380.

GenBank accession number (16S rRNA): AF060699.

2. **Mannheimia glucosida** Angen, Mutters, Caugant, Olsen and Bisgaard 1999a, 83[VP]

glu.co' si.da. Gr. adj. *glykos* sweet; L. adj.-forming suff. *-id (-us)*; fem. *-ida*; M.L. adj. *glucosida* pertaining to glucosides, which are fermented by almost all biovars.

Corresponds to Frederiksen's third taxon of *M. haemolytica* (Frederiksen, 1973), classified as *P. haemolytica* biogroups 3 and 5 by Bisgaard and Mutters (1986b) and reclassified as *M. haemolytica* biogroups 3A–H by Angen et al. (1997). The species also includes NPG and *meso*-inositol positive strains of *M. haemolytica* biogroup 9 (Bisgaard and Mutters, 1986b). *M. glucosida* can be divided into 9 biovars, A–I (Angen et al., 1999a). Cells are Gram-negative, nonmotile, small rods. Colonies are regular, smooth, and grayish on blood agar and are 1–2 mm in diameter after 24 h of incubation. All strains are hemolytic on bovine blood agar. All strains ferment D-sorbitol, D-xylose, maltose, and dextrin. For the ONPG and NPG tests, all biovars are positive. The biovars are separated by differences in ornithine decarboxylase, ONPX, ONPF, and fermentation of L-arabinose and glucosides. Further biochemical characteristics are listed in Tables BXII.γ.298 and BXII.γ.299 and under the description of genus *Mannheimia*. Typeable strains within *M. glucosida* all belong to serovar A11 (Angen et al., 1999b). Most strains investigated have been isolated from sheep, although isolates of serotype 11 have been reported from cattle (Quirie et al., 1986). This group of bacteria is normally not associated with disease and probably represents part of the resident microflora in the upper respiratory tract (Biberstein and Gillis, 1962). Murray et al. (1992) have shown that the iron binding proteins in serotype A11 differed from other serotypes. Sialoglycoprotease and neuraminidase activity have not been observed within serotype A11 (Lee et al., 1994). Serotype A11 produces a leucotoxin that is serologically and functionally similar to leucotoxin from *M. haemolytica* (A1), but immunoblots have shown that strains of serotype A11 seem to produce more variants of these toxins than do the other serotypes (Burrows et al., 1993).

The type strain was originally received as B558 from Moredun Research Institute. It was isolated from ovine lung in the UK, and belongs to biovar B and serotype 11 (Angen et al., 1997).

The mol% G + C of the DNA is: 41.6 (T_m).

TABLE BXII.γ.298. Phenotypic characters separating existing species of *Mannheimia*[a]

Characteristics	1. *M. haemolytica*	2. *M. glucosida*	3. *M. granulomatis*	4. *M. ruminalis*	5. *M. varigena*
β-Hemolysis (bovine blood)	+	+	−	−	+
Ornithine decarboxylase	−	d	−	−	d
L(+)Arabinose	−	d	−	−	+
D(+)Xylose	+	+	d	−	+
meso-Inositol	d	(+)	d	−	d
D(−)Sorbitol	+	+	+	d	−
L(+)Rhamnose	−	−	−	−	d[b]
β-Glucosidase (NPG)	−	+	+	−	−[b]
Glycosides[c]	−	d	d	−	−[b]
α-Fucosidase (ONPF)	+	+	−	d	−[b]
β-Xylosidase (ONPX)	−	d	d	−	−

[a]Symbols: +, 90% or more of the strains positive within 1–2 days; (+), 90% or more of the strains positive within 3–14 days; −, 10% or less of the strains are positive within 14 days; d, 11–89% of the strains are positive; w, weak positive.

[b]Strains of Bisgaard Taxon 36 are positive.

[c]Glycosides: cellobiose, esculin, amygdalin, arbutin, gentiobiose, and salicin.

TABLE BXII.γ.299. Additional phenotypic characters of existing species of *Mannheimia*[a,b]

Characteristics	1. *M. haemolytica*	2. *M. glucosida*	3. *M. granulomatis*	4. *M. ruminalis*	5. *M. varigena*
Gram stain	−	−	−	−	−
Motility at 22°C and 37°C	−	−	−	−	−
Catalase	+	+	d	+	+
Oxidase	+	+	d	d	+
Glucose (Hugh and Leifson)	F	F	F	F	F
Symbiotic growth	−	−	−	−	−
Porphyrin test	+	+	+	+	+
Citrate (Simmons)	−	−	−	−	−
Mucate, acid	−	−	−	−	−
Malonate, base	−	−	−	−	−
H₂S/TSI	−	−	−	−	−
KCN, growth	−	−	−	−	−
Methyl-red, 37°C	−	−	−	−	−
Voges–Proskauer, 37°C	−	−	−	−	−
Nitrate, reduction	+	+	+	+	+
Nitrate, gas	−	−	−	−	−
Urease	−	−	−	−	−
Alanine aminopeptidase	+	+	+	+	+
Arginine dehydrolase	−	−	−	−	−
Lysine decarboxylase.	−	−	−	−	−
Phenylalanine deaminase	−	−	−	−	−
Indole	−	−	−	−	−
Phosphatase	+	+	+	+	+
Gelatinase	−	−	−	−	−
Tween 20	−	−	−	−	−
Tween 80	−	−	−	−	−
Growth on MacConkey agar	d	d	−	d	−
Pigment	−	−	−	−	−
Glycerol	d	(+)	(+)	d	(+)
meso-Erythritol	−	−	−	−	−
Adonitol	−	−	−	−	−
D(+)Arabitol	−	−	−	−	−
Xylitol	−	−	−	−	−
D(−)Arabinose	d	w	−	d	d
D(−)Ribose	+	+	+	+	+
L(−)Xylose	−	−	−	−	−
Dulcitol	−	−	−	−	−
D(−)Mannitol	+	+	+	+	+
D(−)Fructose	+	+	+	+	+
D(+)Fucose	−	−	−	−	−
L(−)Fucose	+	+/(+)	d	d	(+)
D(+)Galactose	+	+/(+)	+/(+)	+	+/(+)
D(+)Glucose, acid	+	+	+	+	+
D(+)Glucose, gas	−	−	−	−	−
D(+)Mannose	−	−	−	−	−
L(−)Sorbose	−	−	−	−	−
Lactose	+	+/(+)	d	(+)	d
o-Nitrophenyl-β-D-galactopyranoside (ONPG)	+	+	d	+	+
Maltose	+	+	+/(+)	d	d
D(+)Melibiose	−	−	−	−	−[c]
Sucrose	+	+	+	+	+
Trehalose	−	−	−	−	−
D(+)Melizitose	−	−	−	−	−
Raffinose	(+)	d	d	d	d
Dextrin	+	+/(+)	d	d	d
D(+)Glycogen	−	−	−	−	−
Inulin	−	−	−	−	−

(*continued*)

Type strain: P925, CCUG 38457.
GenBank accession number (16S rRNA): AF053889.

3. **Mannheimia granulomatis** (Ribeiro, Carter, Frederiksen and Riet-Correa 1990) Angen, Mutters, Caugant, Olsen and Bisgaard 1999a, 82[VP] (*Pasteurella granulomatis* Ribeiro, Carter, Frederiksen, and Riet-Correa 1990, 105.)

gran.nu.lo'ma.tis. L. dim. n. *granulum* a small grain; Gr. suff. *-oma* a swelling or tumor; M.L. n. *granuloma* a granuloma; M.L. n. *granulomatis* of a granuloma.

Cells are Gram negative, nonmotile, small rods and coccobacilli. Colonies are regular, smooth, and grayish on blood agar and are 1–2 mm in diameter after 24 h of incubation. No strains were hemolytic on bovine blood agar, but they have been reported as hemolytic using ovine blood (Ribeiro et al., 1989). All strains are negative for ornithine decarboxylase and ONPF and positive for NPG and fermentation of D-sorbitol. Strains are variable in fermentation of D-xylose, maltose, dextrin, glucosides, and *meso*-inositol.

TABLE BXII.γ.299. *(cont.)*

Characteristics	1. *M. haemolytica*	2. *M. glucosida*	3. *M. granulomatis*	4. *M. ruminalis*	5. *M. varigena*
D(+)Turanose	−	−	−	−	−
β-*N*-CH₃-Glucosamide	−	−	−	−	−
α-Galactosidase	−	−	−	−	−ᶜ
α-Glucosidase (PNPG)	−	−	−	−	−
β-Glucuronidase (PGUA)	−	−	−	d	−
α-Mannosidase	−	−	−	−	−

ᵃSymbols: +, 90% or more of the strains positive within 1–2 days; (+), 90% or more of the strains positive within 3–14 days; −, 10% or less of the strains are positive within 14 days; d, 11–89% of the strains are positive; w, weak positive; +/(+), positive or delayed positive reaction later than 48 h; F, fermentation.

ᵇIncubation temperature 37°C.

ᶜstrains of Bisgaard Taxon 36 are positive.

Further biochemical characteristics are listed in Tables BXII.γ.298 and BXII.γ.299 and under the description of genus *Mannheimia*. The species includes the following taxa: *Pasteurella granulomatis*, isolated from bovine panniculitis (Ribeiro et al., 1989; 1990; Riet-Correa et al., 1992); Bisgaard Taxon 20 biovars 1 and 2, isolated from bronchopneumonia and conjunctivitis in leprine species and deer (Devriese et al., 1991); and *M. haemolytica* biogroup 3J, isolated from deer and the oral cavity of cattle (Angen et al., 1997). No virulence factors have yet been described.

The type strain was isolated from bovine panniculitis in Brazil and was previously classified as *Pasteurella granulomatis* (Ribeiro et al., 1989). The type strain is ONPG positive and ferments maltose, dextrin, glucosides, and *meso*-inositol, but not D-xylose.

The mol% G + C of the DNA is: 39.2 (*T_m*).

Type strain: 26, ATCC 49244.

GenBank accession number (16S rRNA): AF053902.

4. **Mannheimia ruminalis** Angen, Mutters, Caugant, Olsen and Bisgaard 1999a, 83.ᵛᴾ

ru.mi.na' lis. L. n. *rumen* stomach, especially first stomach of ruminants; L. adj.-forming suff. *-alis*; M.L. adj. *ruminalis* referring to ruminants, from which the bacteria have been isolated.

Cells are Gram-negative, nonmotile, nonhemolytic, small rods or coccobacilli. Colonies are regular, smooth, and grayish on blood agar and are 1–2 mm in diameter after 24 h of incubation. All strains are ornithine decarboxylase, NPG, ONPF, and ONPX negative and do not ferment L-arabinose, *meso*-inositol, or glucosides (Table BXII.γ.298). Further biochemical characteristics are given in the description of genus *Mannheimia*. *M. ruminalis* consists of two distinct phenotypic groups. Biovar 1 consists of strains that do not ferment maltose or dextrin, show variable reactions for fermentation of D-sorbitol and D-xylose, and correspond to Bisgaard Taxon 18 biovars 1 and 3 (Bisgaard et al., 1986). Biovar 2 strains ferment maltose and dextrin and correspond to *M. haemolytica* biogroup 8D (Angen et al., 1997). All strains have been isolated from the rumen of sheep or cattle and are not associated with disease. Virulence factors have not been described. The strains have previously been called *Actinobacillus lignieresii* (Phillips, 1961, 1964, 1966).

The type strain was originally isolated as strain S1/15/6 from a sheep rumen in the UK by J.E. Phillips, and previously was classified as Bisgaard Taxon 18 biovar 1 (Bisgaard et al., 1986). The type strain belongs to biovar 1 and does not ferment D-sorbitol and D-xylose.

The mol% G + C of the DNA is: not known.

Type strain: HPA 92,CCUG 38470.

Additional Remarks: AF053900.

5. **Mannheimia varigena** Angen, Mutters, Caugant, Olsen and Bisgaard 1999a, 83.ᵛᴾ

va.ri' gena. L. fem. adj. *varia* varied; L. adj.-forming element *-gena* originating from; L. adj. *varigena*, of different origin, referring to the fact that strains of this species have been isolated both from ruminants and pigs.

Cells are Gram-negative, nonmotile, small rods or coccobacilli. Colonies are regular, smooth, and grayish on blood agar and are 1–2 mm in diameter after 24 h of incubation. Most strains are hemolytic on bovine blood agar. All strains ferment L-arabinose and D-xylose, but not D-sorbitol. Most strains do not ferment glucosides; those fermenting glucosides also ferment D-melibiose. Most strains are ornithine decarboxylase positive (biovar 1) and correspond to *M. haemolytica* biogroup 6, Bisgaard Taxon 36, and Bisgaard Taxon 15 biovar 1. Ornithine-decarboxylase-negative strains (biovar 2) were isolated from pigs and correspond to Bisgaard Taxon 15 biovar 2. Strains of *M. varigena* have been isolated from pneumonia, mastitis, and septicemia in cattle, as well as from the oral cavity, rumen, and intestines (Angen et al., 1997). Isolates from pigs have been associated with septicemia, enteritis, and pneumonia and have also been isolated from the upper respiratory tract (Bisgaard, 1984; McLaughlin et al., 1991). Some of these strains have been shown to produce a leucotoxin toxic against leucocytes from both cattle and pigs (Chang et al., 1993).

The type strain was isolated from a bovine with pneumonia in West Germany, and was previously classified as *M. haemolytica* biogroup 6 (Bisgaard and Mutters, 1986b). This strain is indole negative, ONPG and ONPF positive, and NPG negative and ferments maltose and *meso*-inositol, but not dextrin (Tables BXII.γ.298 and BXII.γ.299).

The mol% G + C of the DNA is: 41.7 (*T_m*).

Type strain: 177, CCUG 38462.

GenBank accession number (16S rRNA): AF053893.

Genus VI. *Phocoenobacter* Foster, Ross, Malnick, Willems, Hutson, Reid and Collins 2000, 138[VP]

GEOFFREY FOSTER AND MATTHEW D. COLLINS

Pho.coe.no.bac' ter. M.L. n. *phocoena* derived from Gr. n. *phokaina* porpoise; M.L. masc. n. *bacter* rod; M.L. masc. n. *phocoenobacter* a rod from a porpoise.

Cells spherical, ovoid or rod shaped, 0.2–0.5 × 0.5–1.0 μm. Occasional longer forms occur. **Some cells demonstrate bipolar staining.** Gram negative. **Nonmotile. Facultative anaerobes.** Optimum temperature for growth, 37°C. Do not require X or V factors for growth. No growth occurs on MacConkey agar. **Catalase negative, oxidase positive. Nitrates are reduced to nitrites. Voges–Proskauer positive.** Arginine dihydrolase, ornithine decarboxylase, lysine decarboxylase, urease and indole are not produced. Acid produced from glucose. Isolated from a porpoise.

The mol% G + C of the DNA is: 41.5.

Type species: **Phocoenobacter uteri** Foster, Ross, Malnick, Willems, Hutson, Reid and Collins 2000, 139.

FURTHER DESCRIPTIVE INFORMATION

Cells are pleomorphic after isolation, with coccal, coccobacillary, and bacillary forms all present. After 48 h incubation on sheep blood agar, colonies are 0.5 mm in diameter, circular, entire, low convex, smooth, and gray. Colonies are not sticky on primary isolation. Nonhemolytic on sheep blood agar initially, although alpha hemolysis is evident in older cultures. Growth on Columbia agar without serum or blood is absent but a scanty growth may be obtained on tryptic soy agar without additives. Grows at 37°C and 22°C, but not at 42°C.

Because the genus contains only one species, all the characteristics given for the genus also describe the species *Phocoenobacter uteri.*

ENRICHMENT AND ISOLATION PROCEDURES

Phocoenobacter strains can be isolated on blood-containing media. Growth is poor or absent on media without blood. Increased CO_2 does not enhance the number or size of the colonies.

MAINTENANCE PROCEDURES

Surface cultures survive up to a week on Columbia sheep blood agar and slightly longer on blood-enriched nutrient agar slopes. Both lyophilization and storage on cryobeads at −80°C are effective for long-term maintenance.

DIFFERENTIATION OF THE GENUS *PHOCOENOBACTER* FROM OTHER GENERA

Phocoenobacter can be differentiated from all other members of the *Pasteurellaceae* by the following tests. It can be separated from *Pasteurella* by its negative catalase, indole, and D-mannose reactions; from *Haemophilus* by its positive Voges–Proskauer reaction and lack of X or V requirement; from *Actinobacillus* by not producing urease; from *Mannheimia* by its negative catalase and mannitol fermentation reactions; and from *Lonepinella* by its positive oxidase reaction.

TAXONOMIC COMMENTS

The genus *Phocoenobacter* was described to accommodate a Gram-negative pleomorphic rod originating from the uterus of a porpoise. The genus is phylogenetically related to the *Pasteurellaceae.*

The family *Pasteurellaceae* is phylogenetically very complex and there is extensive intermixing of species of some of its constituent genera. The genus *Phocoenobacter* forms a distinct subline within the *Pasteurellaceae* and does not display a particularly close affinity with any described species of the family. The nearest named relative of *Phocoenobacter* in terms of 16S rDNA sequence similarity is *Haemophilus ducreyi,* but treeing analysis shows this association is statistically not significant.

List of species of the genus *Phocoenobacter*

1. **Phocoenobacter uteri** Foster, Ross, Malnick, Willems, Hutson, Reid and Collins 2000, 139[VP]

   u' te.ri. L. masc. n. *uteris* uterus.

   The characteristics are as described for the genus and as listed in Table BXII.γ.300. Acid is produced from glucose but not from adonitol, dulcitol, galactose, inositol, inulin, lactose, maltose, mannitol, mannose, melibiose, raffinose, rhamnose, salicin, sorbitol, sucrose, trehalose, or xylose. β-galactosidase positive. Alkaline phosphatase, esterase lipase, acid phosphatase, and naphthol- AS-BI-phosphohydrolase are positive. Leucine arylamidase is weakly positive.

   The mol% G + C of the DNA is: 41.5 (T_m).

   Type strain: NCTC 12872, M1063U/93.

   GenBank accession number (16S rRNA): X89379.

TABLE BXII.γ.300. Characteristics of *Phocoenobacter uteri*[a]

Characteristics	Reaction or result
Production of:	
Catalase	−
Oxidase	+
Indole	−
Arginine dihydrolase	−
Lysine decarboxylase	−
Ornithine decarboxylase	−
Acetoin	+
β-Galactosidase (ONPG)	+
Fermentation of:	
Glucose	+
Adonitol, dulcitol, galactose, inositol, inulin, lactose, maltose, mannitol, mannose, melibiose, raffinose, rhamnose, salicin, sorbitol, sucrose, trehalose, xylose	−
Growth on MacConkey agar	−
X or V factor requirement	−

[a]Symbols: +, positive; −, negative.

Bibliography

Aamand, J., T. Ahl and E. Spieck. 1996. Monoclonal antibodies recognizing nitrite oxidoreductase of *Nitrobacter hamburgensis, N. winogradskyi,* and *N. vulgaris.* Appl. Environ. Microbiol. *62:* 2352–2355.

Abbass, Z. and Y. Okon. 1993a. Physiological properties of *Azotobacter paspali* in culture and the rhizosphere. Soil Biol. Biochem. *25:* 1061–1073.

Abbass, Z. and Y. Okon. 1993b. Plant growth promotion by *Azotobacter paspali* in the rhizosphere. Soil Biol. Biochem. *25:* 1075–1083.

Abbot, J.D. and R. Shannon. 1958. A method for typing *Shigella sonnet,* using colicine production as a marker. J. Clin. Pathol. *11:* 71–77.

Abbott, S.L., W.K.W. Cheung, S. Kroske Bystrom, T. Malekzadeh and J.M. Janda. 1992. Identification of *Aeromonas* strains to the genospecies level in the clinical laboratory. J. Clin. Microbiol. *30:* 1262–1266.

Abbott, S.L. and J.M. Janda. 1994a. Isolation of *Yokenella regensburgei* (*Koserella trabulsii*) from a patient with transient bacteremia and from a patient with a septic knee. J. Clin. Microbiol. *32:* 2854–2855.

Abbott, S.L. and J.M. Janda. 1994b. Severe gastroenteritis associated with *Vibrio hollisae* infection: report of two cases and review. Clin. Infect. Dis. *18:* 310–312.

Abbott, S.L. and J.M. Janda. 1997. *Enterobacter cancerogenus* ("*Enterobacter taylorae*") — Infections associated with severe trauma or crush injuries. Am. J. Clin. Pathol. *107:* 359–361.

Abbott, S.L., W.K.W. Cheung and J.M. Janda. 2003. The genus *Aeromonas*: Biochemical characteristics, atypical reactions, and phenotypic identification schemes. J. Clin. Microbiol. *41:* 2348–2357.

Abbott, S.L., R.P. Kokka and J.M. Janda. 1991. Laboratory investigations on the low pathogenic potential of *Plesiomonas shigelloides.* J. Clin. Microbiol. *29:* 148–153.

Abbott, S.L., L.S. Seli, M.J. Catino, M.A. Hartley and J.M. Janda. 1998. Misidentification of unusual *Aeromonas* species as members of the genus *Vibrio*: a continuing problem. J. Clin. Microbiol. *36:* 1103–1104.

Abbott, S.L., H. Serve and J.M. Janda. 1994. Case of *Aeromonas veronii* (DNA Group 10) bacteremia. J. Clin. Microbiol. *32:* 3091–3092.

Abdullah, K.M., R.Y. Lo and A. Mellors. 1991. Cloning, nucleotide sequence, and expression of the *Pasteurella haemolytica* A1 glycoprotease gene. J. Bacteriol. *173:* 5597–5603.

Abdullahi, M.Z., N.J.L. Gilmour and I.R. Poxton. 1990. Outer-membrane proteins of bovine strains of *Pasteurella multocida* type-A and their doubtful role as protective antigens. J. Med. Microbiol. *32:* 55–61.

Abdurashitov, M.A., E.V. Kileva, T.V. Myakisheva, V.S. Dedkov, A.V. Shevchenko and S.K. Degtyarev. 1997. *AccBSI*: a new restriction endonuclease from *Acinetobacter calcoaceticus* BS. Priklad. Biokhim. Mikrobiol. *33:* 556–558.

Abel, R. 1893. Bakteriologische Studien über Ozaena simplex. Zentralbl. Bakteriol. Parasitenk. Infektionskr. Hyg. Abt. I Orig. *13:* 161–173.

Abeyta, C.J., C.A. Kaysner, M.M. Wekell, J.J. Sullivan and G.N. Stelma. 1986. Recovery of *Aeromonas hydrophila* from oysters implicated in an outbreak of foodborne illness. J. Food Prot. *49:* 643–646.

Abou-Akkada, A.R. and T.H. Blackburn. 1963. Some observations on the nitrogen metabolism of rumen proteolytic bacteria. J. Gen. Microbiol. *31:* 461–469.

Abraham, S.N. and S. Jaiswal. 1997. Type-1 fimbriae of *Escherichia coli. In* Sussman (Editor), *Escherichia coli*: Mechanisms of Virulence, Cambridge University Press, Cambridge. pp. 169–192.

Abu, G.O., R. Weiner and R.R. Colwell. 1994. Glucose metabolism and polysaccharide accumulation in the marine bacterium, *Shewanella colwelliana.* World J. Microbiol. Biotechnol. *10:* 543–546.

Acheson, D.W.K. and G.T. Keusch. 1995. *Shigella* and enteroinvasive *Escherichia coli. In* Blaser, Smith, Ravdin, Greenberg and Guerrant (Editors), Infections of the Gastrointesinal Tract, Raven Press, New York. pp. 763–784.

Ackerman, J.I. and J.G. Fox. 1981. Isolation of *Pasteurella ureae* from the reproductive tracts of congenic mice. J. Clin. Microbiol. *13:* 1049–1053.

Ackermann, H.W., G. Brochu and H.P.E. Konjin. 1994. Classification of *Acinetobacter* phages. Arch. Virol. *135:* 345–354.

Actis, L.A., M.E. Tolmasky, L.M. Crosa and J.H. Crosa. 1993. Effect of iron-limiting conditions on growth of clinical isolates of *Acinetobacter baumannii.* J. Clin. Microbiol. *31:* 2812–2815.

Adam, A.L., S. Pike, M.E. Hoyos, J.M. Stone, J.C. Walker and A. Novacky. 1997. Rapid and transient activation of a myelin basic protein kinase in tobacco leaves treated with harpin from *Erwinia amylovora.* Plant Physiol. *115:* 853–861.

Adamo, S.A. 1998. The specificity of behavioral fever in the cricket *Acheta domesticus.* J. Parasitol. *84:* 529–533.

Adams, C.A., B. Austin, P.G. Meaden and D. McIntosh. 1998. Molecular characterization of plasmid-mediated oxytetracycline resistance in *Aeromonas salmonicida.* Appl. Environ. Microbiol. *64:* 4194–4201.

Adams, G.A. and S.M. Martin. 1964. Extracellular polysaccharides of *Serratia marcescens.* Can. J. Biochem. *42:* 1403–1413.

Adams, G.A., C. Quadling, M. Yaguchi and T.G. Tornabene. 1970. The chemical composition of cell-wall lipopolysaccharides from *Moraxella duplex* and *Micrococcus calco-aceticus.* Can. J. Microbiol. *16:* 1–8.

Adams, G.A. and R. Young. 1965. Capsular polysaccharides of *Serratia marcescens.* Can. J. Biochem. *43:* 1499–1512.

Adams, R., J. Bygraves, M. Kogut and N.J. Russell. 1987. The role of osmotic effects in haloadaptation of *Vibrio costicola.* J. Gen. Microbiol. *133:* 1861–1870.

Adams, R.L. and N.J. Russell. 1992. Interactive effects of salt concentration and temperature on growth and lipid composition in the moderately halophilic bacterium *Vibrio costicola.* Can. J. Microbiol. *38:* 823–827.

Adegbola, R.A. and D.C. Old. 1983a. Fimbrial and non-fimbrial haemagglutinins in *Enterobacter aerogenes.* J. Med. Microbiol. *19:* 35–43.

Adegbola, R.A. and D.C. Old. 1983b. Fimbrial hemagglutinins in *Enterobacter* species. J. Gen. Microbiol. *129:* 2175–2180.

Adhikari, T.B., C.M. Vera Cruz, Q. Zhang, R.J. Nelson, D.Z. Skinner, T.W. Mew and J.E. Leach. 1995. Genetic diversity of *Xanthomonas oryzae* oryzae in Asia. Appl. Environ. Microbiol. *61:* 966–971.

Adkins, J.P., M.T. Madigan, L. Mandelco, C.R. Woese and R.S. Tanner. 1993. *Arhodomonas aquaeolei* gen. nov., sp. nov., an aerobic, halophilic bacterium isolated from a subterranean brine. Int. J. Syst. Bacteriol. *43*: 514–520.

Adlam, C. 1989. The structure, function and properties of cellular and extracellular components of *Pasteurella haemolytica*. *In* Adlam and Rutter (Editors), *Pasteurella* and Pasteurellosis, Academic Press, London. pp. 75–92.

Agin, T.S. and M.K. Wolf. 1997. Identification of a family intimins common to *Escherichia coli* causing attaching-effacing lesions in rabbits, humans, and swine. Infect. Immun. *65*: 320–326.

Aguilera, A., J. Pascual, E. Loza, J. Lopez, G. Garcia, F. Liano, C. Quereda and J. Ortuno. 1995. Bacteraemia with *Cedecea neteri* in a patient with systemic lupus erythematosus. Postgrad. Med. J. *71*: 179–180.

Aguillera, M.M., N.C. Hodge, R.E. Stall and G.C. Smart, Jr.. 1993. Bacterial symbionts of *Steinernema scapterisci*. J. Invertebr. Pathol. *62*: 68–72.

Agustin, E.T. and B.A. Cunha. 1995. Buttock abscess due to *Hafnia alvei*. Clin. Infect. Dis. *20*: 1426.

Ahlquist, G.A., C.A. Fewson, D.A. Ritchie, J. Podmore and V. Rowell. 1980. Competence for genetic transformation of *Acinetobacter calcoaceticus* NCIB 8250. FEMS Microbiol. Lett. *7*: 107–109.

Ahmad, A.A., J.P. Barry and D.C. Nelson. 1999. Phylogenetic affinity of a wide, vacuolate, nitrate-accumulating *Beggiatoa* sp. from Monterey Canyon, California, with *Thioploca* spp. Appl. Environ. Microbiol. *65*: 270–277.

Ahmad, S. and R.A. Jensen. 1987. Evolution of the biochemical pathway for aromatic amino acid biosynthesis in *Serpens flexibilis* in relationship to its phylogenetic position. Arch. Microbiol. *147*: 8–12.

Ahmad, S., W.G. Weisburg and R.A. Jensen. 1990. Evolution of aromatic amino acid biosynthesis and application to the fine-tuned phylogenetic positioning of enteric bacteria. J. Bacteriol. *172*: 1051–1061.

Ahmed, R., C. Bopp, A. Borczyk and S. Kasatiya. 1987. Phage-typing scheme for *Escherichia coli* O157:H7. J. Infect. Dis. *155*: 806–809.

Ahmed, S. and M. Donaghy. 1998. An outbreak of *Echerichia coli* O157:H7 in Central Scotland. *In* Kaper and O'Brien (Editors), *Escherichia coli* O157:H7 and Other Shiga Toxin-Producing Strains, 1st Ed., ASM Press, Washington, D.C. pp. 59–65.

Ahmed, Z.U., M.R. Sarker and D.A. Sack. 1988. Nutritional requirements of shigellae for growth in a minimal medium. Infect. Immun. *56*: 1007–1009.

Aikimbaev, M.A. 1966. Taxonomy of Genus *Francesella*. Rep. Acad. Sci. Kaz. SSR Ser. Biol. *5*: 42–44.

Ajai, C.O., J.E. Cook and S.M. Dennis. 1980. Diagnosing ovine epididymitis by immunofluorescence. Vet. Rec. *107*: 421–424.

Akagawa, M. and K. Yamasato. 1989. Synonymy of *Alcaligenes aquamarinus*, *Alcaligenes faecalis* subsp. *homari*, and *Deleya aesta*: *Deleya aquamarina* comb. nov. as the type species of the genus *Deleya*. Int. J. Syst. Bacteriol. *39*: 462–466.

Akagawa-Matsushita, M., T. Itoh, Y. Katayama, H. Kuraishi and K. Yamasato. 1992a. Isoprenoid quinone composition of some marine *Alteromonas*, *Marinomonas*, *Deleya*, *Pseudomonas* and *Shewanella* species. J. Gen. Microbiol. *138*: 2275–2281.

Akagawa-Matsushita, M., Y. Koga and K. Yamasoto. 1993. DNA relatedness among nonpigmented species of *Alteromonas* and synonomy of *Alteromonas haloplanktis* (ZoBell and Upham 1944) Reichelt and Baumann 1973 and *Alteromonas tetraodonis* Simidu et al. 1990. Int. J. Syst. Bacteriol. *43*: 500–503.

Akagawa-Matsushita, M., M. Matsuo, Y. Koga and K. Yamasato. 1992b. *Alteromonas atlantica* sp. nov. and *Alteromonas carrageenovora* sp. nov., bacteria that decompose algal polysaccharides. Int. J. Syst. Bacteriol. *42*: 621–627.

Akan, M., A. Eyigor and K.S. Diker. 1998. Motile aeromonads in the feces and carcasses of broiler chickens in Turkey. J. Food Prot. *61*: 113–115.

Akhtar, S. and H.F. Van Emden. 1994. Ultrastructure of the symbionts and mycetocytes of bird cherry aphid (*Rhopalosiphum padi*). Tissue Cell *26*: 513–522.

Akhurst, R.J. 1980. Morphological and functional dimorphism in *Xenorhabdus* spp., bacteria symbiotically associated with the insect patho-

genic nematodes *Neoaplectana* and *Heterorhabditis*. J. Gen. Microbiol. *121*: 303–310.

Akhurst, R.J. 1982a. Antibiotic activity of *Xenorhabdus* spp., bacteria symbiotically associated with insect pathogenic nematodes of the families *Heterorhabditidae* and *Steinernematidae*. J. Gen. Microbiol. *128*: 3061–3066.

Akhurst, R.J. 1982b. A *Xenorhabdus* sp. (*Eubacteriales: Enterobacteriaceae*) symbiotically associated with *Steinernema kraussei* (Nematoda: Steinernematidae). Rev. Nematol. *5*: 277–280.

Akhurst, R.J. 1983a. *Neoaplectana* species: specificity of the association between bacteria of the genus *Xenorhabdus* spp. Experimental Parasitology. *55*: 258–263.

Akhurst, R.J. 1983b. Taxonomic study of *Xenorhabdus*, a genus of bacteria symbiotically associated with insect pathogenic nematodes. Int. J. Syst. Bacteriol. *33*: 38–45.

Akhurst, R.J. 1986a. *Xenorhabdus nematophilus* subsp. *beddingii*, new subspecies (*Enterobacteriaceae*): A new subspecies of bacteria mutualistically associated with entomopathogenic nematodes. Int. J. Syst. Bacteriol. *36*: 454–457.

Akhurst, R.J. 1986b. *Xenorhabdus nematophilus* subsp. *poinarii*: its interaction with insect pathogenic nematodes. Syst. Appl. Microbiol. *8*: 142–147.

Akhurst, R.J. and N.E. Boemare. 1986. A nonluminescent strain of *Xenorhabdus luminescens* (*Enterobacteriaceae*). J. Gen. Microbiol. *132*: 1917–1922.

Akhurst, R.J. and N.E. Boemare. 1988. A numerical taxonomic study of the genus *Xenorhabdus* (*Enterobacteriaceae*) and proposed elevation of the subspecies of *Xenorhabdus nematophilus* to species. J. Gen. Microbiol. *134*: 1835–1845.

Akhurst, R.J. and N.E. Boemare. 1990. Biology and taxonomy of *Xenorhabdus*. *In* Gaugler and Kaya (Editors), Entomopathogenic Nematodes in Biological Control, CRC Press, Inc., Boca Raton. pp. 75–90.

Akhurst, R.J. and N.E. Boemare. 1993. *In* Validation of the publication of new names and new combinations previously effectively published outside the IJSB. List No. 47. Int. J. Syst. Bacteriol. *43*: 864–865.

Akhurst, R.J. and G.B. Dunphy. 1993. Tripartite interactions between symbiotically associated entomopathogenic bacteria, nematodes, and their insect hosts. *In* Beckage, Thompson and Federici (Editors), Parasites and Pathogens of Insects, Vol. 2, Academic Press, San Diego. pp. 1–23.

Akhurst, R.J., R.G. Mourant, L. Baud and N.E. Boemare. 1996. Phenotypic and DNA relatedness between nematode symbionts and clinical strains of the genus *Photohabdus* (*Enterobacteriaceae*). Int. J. Syst. Bacteriol. *46*: 1034–1041.

Akhurst, R.J., A.J. Smigielski, J. Mari, N.E. Boemare and R.G. Mourant. 1992. Restriction analysis of phase variation in *Xenorhabdus* spp. (*Enterobacteriaceae*), entomopathogenic bacteria associated with nematodes. Syst. Appl. Microbiol. *15*: 469–473.

Akman, L., R.V.M. Rio, C.B. Beard and S. Aksoy. 2001. Genome size determination and coding capacity of *Sodalis glossinidius*, an enteric symbiont of tsetse flies, as revealed by hybridization to *Escherichia coli* gene arrays. J. Bacteriol. *183*: 4517–4525.

Akman, L., A. Yamashita, H. Watanabe, K. Oshima, T. Shiba, M. Hatori and S. Aksoy. 2002. Genome sequence of the endocellular obligate symbiont of tsetse flies, *Wigglesworthia glossinidia*. Nat. Genet. *32*: 402–407.

Akpavie, S.O., A.T. Ajuwape and J.O. Ikheloa. 1994. A clinical note on *Haemophilus aegyptius* infection in sheep in Nigeria. Rev. Elev. Med. Vet. Pays Trop. *47*: 177–179.

Akporiaye, E.T., J.D. Rowatt, A.A. Aragon and O.G. Baca. 1983. Lysosomal response of a murine macrophage-like cell line persistently infected with *Coxiella burnetii*. Infect. Immun. *40*: 1155–1162.

Aksoy, S. 1995a. Molecular analysis of the endosymbionts of tsetse flies: 16S rDNA locus and over-expression of a chaperonin. Inscct Mol. Biol. *4*: 23–29.

Aksoy, S. 1995b. *Wigglesworthia* gen. nov. and *Wigglesworthia glossinidia* sp. nov., taxa consisting of the mycetocyte-associated, primary endosymbionts of tsetse flies. Int. J. Syst. Bacteriol. *45*: 848–851.

Aksoy, S., X. Chen and V. Hypsa. 1997. Phylogeny and potential transmission routes of midgut- associated endosymbionts of tsetse (*Diptera*: *Glossinidae*). Insect Mol. Biol. *6*: 183–190.

Aksoy, S., A.A. Pourhosseini and A. Chow. 1995. Mycetome endosymbionts of tsetse flies constitute a distinct lineage related to *Enterobacteriaceae*. Insect Mol. Biol. *4*: 15–22.

Al Mamun, A.A., A. Tominaga and M. Enomoto. 1996. Detection and characterization of the flagellar master operon in the four *Shigella* subgroups. J. Bacteriol. *178*: 3722–3726.

Al Mamun, A.A., A. Tominaga and M. Enomoto. 1997. Cloning and characterization of the region III flagellar operons of the four *Shigella* subgroups: genetic defects that cause loss of flagella of *Shigella boydii* and *Shigella sonnei*. J. Bacteriol. *179*: 4493–4500.

Al-Jumaili, I.J. 1975. Bacteriocin typing of *Proteus*. J. Clin. Pathol. *28*: 784–787.

Al-Jumaili, I.J. and G.A. Fenwick. 1978. Bacteriocine typing of *Providencia* isolates. Zentralbl. Bakteriol. Parasitenk. Infektionskr. Hyg. Abt. I Orig. A. *240*: 202–207.

Al-Khoja, M.S. and J.H. Darrell. 1979. The skin as the source of *Acinetobacter* and *Moraxella* species occurring in blood cultures. J. Clin. Pathol. *32*: 497–499.

Al-Mousawi, A.H. and P.E. Richardson. 1990. A comparison of fine structural features of *Xanthomonas campestris* pv. *malvacearum* growth *in vitro* and in cotton plant lines. Arab Gulf J. Sci. Res. *8*: 141–152.

Alaluusua, S., M. Saarela, H. Jousimies-Somer and S. Asikainen. 1993. Ribotyping shows intrafamilial similarity in *Actinobacillus* actinomycetemcomitans isolates. Oral Microbiol. Immunol. *8*: 225–229.

Alarcon, B., M.T. Gorris, M. Cambra and M.M. Lopez. 1995. Serological characterization of potato isolates of *Erwinia carotovora* subsp. *atroseptica* and subsp. *carotovora* using polyclonal and monoclonal antibodies. J. Appl. Bacteriol. *79*: 592–602.

Alary, M. and J.R. Joly. 1991. Risk factors for contamination of domestic hot water systems by legionellae. Appl. Environ. Microbiol. *57*: 2360–2367.

Alballaa, S.R., S.M. Qadri, O. Al-Furayh and K. Al-Qatary. 1992. Urinary tract infection due to *Rahnella aquatilis* in a renal transplant patient. J. Clin. Microbiol. *30*: 2948–2950.

Albert, M.J. 1994. *Vibrio cholerae* O139 Bengal. J. Clin. Microbiol. *32*: 2345–2349.

Albert, M.J., K. Alam, M. Islam, J. Montanaro, A.S.M.H. Rahman, K. Haider, M.A. Hossain, A.K.M.G. Kibriya and S. Tzipori. 1991. *Hafnia alvei*, a probable cause of diarrhea in humans. Infect. Immun. *59*: 1507–1513.

Albert, M.J., M. Ansaruzzaman, N.A. Bhuiyan, P.K. Neogi and A.S. Faruque. 1995a. Characteristics of invasion of HEp-2 cells by *Providencia alcalifaciens*. J. Med. Microbiol. *42*: 186–190.

Albert, M.J., M. Ansaruzzaman, F. Qadri, A. Hossain, A.K.M.G. Kibriya, K. Haider, S. Nahar, S.M. Faruque and A.N. Alam. 1993a. Characterisation of *Plesiomonas shigelloides* strains that share type-specific antigen with *Shigella flexneri* 6 and common group 1 antigen with *Shigella flexneri* spp. and *Shigella dysenteriae* 1. J. Med. Microbiol. *39*: 211–217.

Albert, M.J., M. Ansaruzzaman, T. Shimada, A. Rahman, N.A. Bhuiyan, S. Nahar, F. Qadri and M.S. Islam. 1995b. Characterization of *Aeromonas trota* strains that cross-react with *Vibrio cholerae* O139 Bengal. J. Clin. Microbiol. *33*: 3119–3123.

Albert, M.J., N.A. Bhuiyan, A. Rahman, A.N. Ghosh, K. Hultenby, A. Weintraub, S. Nahar, A.K. Kibriya, M. Ansaruzzaman and T. Shimada. 1996. Phage specific for *Vibrio cholerae* O139 Bengal. J. Clin. Microbiol. *34*: 1843–1845.

Albert, M.J., N.A. Bhuiyan, K.A. Talukder, A.S. Faruque, S. Nahar, S.M. Faruque, M. Ansaruzzaman and M. Rahman. 1997. Phenotypic and genotypic changes in *Vibrio cholerae* O139 Bengal. J. Clin. Microbiol. *35*: 2588–2592.

Albert, M.J., S.M. Faruque, M. Ansaruzzaman, M.M. Islam, K. Haider, K. Alam, I. Kabir and R. Robins-Browne. 1992. Sharing of virulence-associated properties at the phenotypic and genetic levels between enteropathogenic *Escherichia coli* and *Hafnia alvei*. J. Med. Microbiol. *37*: 310–314.

Albert, M.J., A.S. Faruque and D. Mahalanabis. 1998. Association of *Providencia alcalifaciens* with diarrhea in children. J. Clin. Microbiol. *36*: 1433–1435.

Albert, M.J., F. Qadri, A. Haque and N.A. Bhuiyan. 1993b. Bacterial clump formation at the surface of liquid culture as a rapid test for identification of enteroaggregative *Escherichia coli*. J. Clin. Microbiol. *31*: 1397–1399.

Albibi, R., J. Chen, O. Lamikanra, D. Banks, R.L. Jarret and B.J. Smith. 1998. RAPD fingerprinting *Xylella fastidiosa* Pierce's disease strains isolated from a vineyard in north Florida. FEMS Microbiol. Lett. *165*: 347–352.

Albritton, W.L. 1984. Resistance plasmids of *Haemophilus* and *Neisseria*. *In* Bryan (Editor), Antimicrobial Drug Resistance, Academic Press, Orlando. 515–527.

Albritton, W.L., J.L. Brunton, M. Meier, M.N. Bowman and L.A. Slaney. 1982a. *Haemophilus influenzae*: comparison of respiratory tract isolates with genitourinary tract isolates. J. Clin. Microbiol. *16*: 826–831.

Albritton, W.L., J.L. Brunton, L. Slaney and I.W. Maclean. 1982b. Plasmid-mediated sulfonamide resistance in *Haemophilus ducreyi*. Antimicrob. Agents Chemother. *21*: 159–165.

Albritton, W.L., S. Penner, L. Slaney and J. Brunton. 1978. Biochemical characteristics of *Haemophilus influenzae* in relationship to source of isolation and antibiotic resistance. J. Clin. Microbiol. *7*: 519–523.

Albritton, W.L., J.K. Setlow, M.L. Thomas and F.O. Sottnek. 1986. Relatedness within the family *Pasteurellaceae* as determined by genetic transformation. Int. J. Syst. Bacteriol. *36*: 103–106.

Albritton, W.L., J.K. Setlow, M. Thomas, F. Sottnek and A.G. Steigerwalt. 1984. Heterospecific transformation in the genus *Haemophilus*. Mol. Gen. Genet. *193*: 358–363.

Albus, A.M., E.C. Pesci, L.J. Runyen-Janecky, S.E.H. West and B.H. Iglewski. 1997. Vfr controls quorum sensing in *Pseudomonas aeruginosa*. J. Bacteriol. *179*: 3928–3935.

Alcorn, S.M. and T.V. Orum. 1988. Rejection of the names *Erwinia carnegieana* Standring 1942 and *Pectobacterium carnegieana* (Standring 1942) Brenner, Steigerwalt, Miklos and Fanning 1973. Request for an opinion. Int. J. Syst. Bacteriol. *38*: 132–134.

Alcorn, S.M., T.V. Orum, A.G. Steigerwalt, J.L.M. Foster, J.C. Fogleman and D.J. Brenner. 1991. Taxonomy and pathogenicity of *Erwinia cacticida*, sp. nov. Int. J. Syst. Bacteriol. *41*: 197–212.

Alday-Sanz, V., H. Rodger, T. Turnbull, A. Adams and R.H. Richards. 1994. An immunohistochemical diagnostic test for rickettsial disease. J. Fish Dis. *17*: 189–191.

Aldová, E., O. Hausner, D.J. Brenner, Z. Kocmoud, J. Schindler, B. Potužníková and P. Petráš. 1988a. *Pragia fontium*, gen. nov., sp. nov., of the family *Enterobacteriaceae*, isolated from water. Int. J. Syst. Bacteriol. *38*: 183–189.

Aldová, E., O. Hausner and M. Gabrhelová. 1984. *Budvicia*—a new genus of *Enterobacteriaceae*. Data on phenotypic characters. J. Hyg. Epidemiol. Microbiol. Immunol. (Prague) *28*: 234–237.

Aldová, E., O. Hausner, M. Gabrhelová, J. Schindler, P. Petráš and H. Braná. 1983. A hydrogen sulfide-producing gram-negative rod from water. Zentbl. Bakteriol. Mikrobiol. Hyg. 1 Abt. Orig. A. *254*: 95–108.

Aldová, E., O. Hausner, M. Gabrhelová, J. Schindler, P. Petráš and H. Braná. 1985a. *In* Validation of the publication of new names and new combinations previously effectively published outside the IJSB. List No. 17. Int. J. Syst. Bacteriol. *35*: 223–225.

Aldová, E., O. Hausner, Z. Kocmoud, J. Schindler and P. Petráš. 1988b. A new member of the family *Enterobacteriaceae*—*Pragia fontium*. J. Hyg. Epidemiol. Microbiol. Immunol. (Prague) *32*: 433–436.

Aldová, E., O. Hausner, A. Svihalkova, K. Laznickova, J. Sobotkova, J. Smolka and O. Horackova. 1985b. First strains of the genus *Kluyvera* in Czechoslovakia. Zentbl. Bakteriol. Mikrobiol. Hyg. Ser. A *260*: 8–14.

Aldová, E. and R.H.W. Schubert. 1996. Serotyping of *Plesiomonas shigelloides* - a tool for understanding ecological relationships. Med. Microbiol. Lett. *5*: 33–39.

Aldrich, J.H., A.B. Gould and F.G. Martin. 1992. Distribution of *Xylella fastidiosa* within roots of peach. Plant Dis. *76*: 885–888.

Aleksic, S. and J. Bockemühl. 1999. *Yersinia* and other *Enterobacteriaceae.* *In* Murray, Baron, Pfaller, Tenover and Yolken (Editors), Manual of Clinical Microbiology, 7th Ed., American Society for Microbiology, Washington, D.C. 483–496.

Aleksič, S., H. Karch and J. Bockemühl. 1992. A biotyping scheme for Shiga-like (Vero) toxin-producing *Escherichia coli* O157 and a list of serological cross-ractions between O157 and other gram-negative bacteria. Zentbl. Bakteriol. *276*: 221–230.

Aleksic, S., A.G. Steigerwalt, J. Bockemuhl, G.P. Huntleycarter and D.J. Brenner. 1987. *Yersinia rohdei* sp. nov. isolated from human and dog feces and surface water. Int. J. Syst. Bacteriol. *37*: 327–332.

Alexander, H.E. 1958. The *Hemophilus* group. *In* Dubos (Editor), Bacterial and Mycotic Infections of Man, 3rd ed., Lippincott Co., Philadelphia. 474.

Alexander, H.E. and G. Leidy. 1951. Induction of heritable new type in type-specific strains of *H. influenzae.* Proc. Soc. Exp. Biol. Med. *78*: 625–626.

Alexander, H.E., G. Leidy and E. Hahn. 1954. Studies on the nature of *Hemophilus influenzae* cells susceptible to heritable changes by deoxyribonucleic acids. J. Exp. Med. *99*: 505–533.

Alfa, M.J., P. DeGagne and P.A. Totten. 1996. *Haemophilus ducreyi* hemolysin acts as a contact cytotoxin and damages human foreskin fibroblasts in cell culture. Infect. Immun. *64*: 2349–2352.

Ali, A., A. Carnahan, M. Altwegg, J. Lüthy-Hottenstein and S. Joseph. 1996a. *Aeromonas bestiarum,* sp. nov. (formerly genomospecies DNA Group 2 *A. hydrophia*), a new species isolated from non-human sources. Med. Microbiol. Lett. *5*: 156–165.

Ali, A., A. Carnahan, M. Altwegg, J. Lüthy-Hottenstein and S. Joseph. 1996b. *In* Validation of the publication of new names and new combinations previously effectively published outside the IJSB. List No. 59. Int. J. Syst. Bacteriol. *46*: 1189–1190.

Allen, D.A., B. Austin and R.R. Colwell. 1983. *Aeromonas media,* new species isolated from river water. Int. J. Syst. Bacteriol. *33*: 599–604.

Allerberger, F., D. Rossboth, M.P. Dierich, S. Aleksic, H. Schmidt and H. Karch. 1996. Prevalence and clinical manifestations of Shiga toxin-producing *Escherichia coli* infections in Austrian children. Eur. J. Clin. Microbiol. Infect. Dis. *15*: 545–550.

Allison, C., N. Coleman, P.L. Jones and C. Hughes. 1992. Ability of *Proteus mirabilis* to invade human urothelial cells is coupled to motility and swarming differentiation. Infect. Immun. *60*: 4740–4746.

Allison, C., L. Emödy, N. Coleman and C. Hughes. 1994. The role of swarm cell differentiation and multicellular migration in the uropathogenicity of *Proteus mirabilis.* J. Infect. Dis. *169*: 1155–1158.

Allison, C., H.C. Lai, D. Gygi and C. Hughes. 1993. Cell differentiation of *Proteus mirabilis* is initiated by glutamine, a specific chemoattractant for swarming cells. Mol. Microbiol. *8*: 53–60.

Alm, R.A., J.P. Hallinan, A.A. Watson and J.S. Mattick. 1996. Fimbrial biogenesis genes of *Pseudomonas aeruginosa*: *pilW* and *pilX* increase the similarity of type 4 fimbriae to the GSP protein-secretion systems and *pilY1* encodes a gonococcal *PilC* homologue. Mol. Microbiol. *22*: 161–173.

Alm, R.A. and P.A. Manning. 1990. Biotype-specific probe for *Vibrio cholerae* serogroup O1. J. Clin. Microbiol. *28*: 823–824.

Alm, R.A. and J.S. Mattick. 1995. Identification of a gene, *pilV*, required for type-4 fimbrial biogenesis in *Pseudomonas aeruginosa*, whose product possesses a pre-pilin-like leader sequence. Mol. Microbiol. *16*: 485–496.

Alm, R.A. and J.S. Mattick. 1997. Genes involved in the biogenesis and function of type-4 fimbriae in *Pseudomonas aeruginosa*. Gene *192*: 89–98.

Almendras, F.E., I.C. Fuentealba, S.R.M. Jones, F. Markham and E. Spangler. 1997. Experimental infection and horizontal transmission of *Piscirickettsia salmonis* in freshwater-raised Atlantic salmon, *Salmo salar* L. J. Fish Dis. *20*: 409–418.

Alonso, A. and J.L. Martínez. 1997. Multiple antibiotic resistance in *Stenotrophomonas maltophilia.* Antimicrob. Agents Chemother. *41*: 1140–1142.

Alonso, A., P. Sanchez and J.L. Martinez. 2000. *Stenotrophomonas malto-* *philia* D457R contains a cluster of genes from gram-positive bacteria involved in antibiotic and heavy metal resistance. Antimicrob. Agents Chemother. *44*: 1778–1782.

Alonso, J.L., M.S. Botella, I. Amoros and M.A. Alonso. 1994. The occurrence of mesophilic aeromonads species in marine recreational waters of Valencia (Spain). J. Environ. Sci. Health Part A Environ. Sci. Eng. *29*: 615–628.

Alonso, J.M., A. Joseph-Francois, D. Mazigh, H. Bercovier and H.H. Mollaret. 1978. Résistance à la peste de souris experimentalement infectées par *Yersinia enterocolitica.* Ann. Microbiol. (Inst. Pasteur) *129B*: 203–207.

Alonso, J.L., I. Amoros and M.A. Alonso. 1996. Differential susceptibility of aeromonads and coliforms to cefsulodin. Appl. Environ. Microbiol. *62*: 1885–1888.

Alsenosy, A.M. and S.M. Dennis. 1985. Pathology of acute experimental *Actinobacillus seminis* mastitis in ewes. Aust. Vet. J. *62*: 234–237.

Altekruse, S.F., M.L. Cohen and D.L. Swerdlow. 1997. Emerging foodborne diseases. Emerg. Infect. Dis. *3*: 285–293.

Altorfer, R., M. Altwegg, J. Zollinger-Iten and A. Von Graevenitz. 1985. Growth of *Aeromonas* spp. on cefsulodin-irgasan-novobiocin agar selective for *Yersinia enterocolitica.* J. Clin. Microbiol. *22*: 478–480.

Altwegg, M. 1985. *Aeromonas caviae*: an enteric pathogen? Infection *13*: 228–230.

Altwegg, M. 1990. Taxonomy and epidemiology of *Aeromonas* species: the value of new typing methods, Thesis, Univeristy of Zurich.

Altwegg, M. 1993. A polyphasic approach to the classification and identification of *Aeromonas* strains. Med. Microbiol. Lett. *2*: 200–205.

Altwegg, M. 1996. Subtyping methods. *In* Austin, Altwegg, Gosling and Joseph (Editors), The Genus *Aeromonas*, 1st Ed., John Wiley & Sons, Ltd., Chichester. pp. 109–125.

Altwegg, M. 1999. *Aeromonas. In* Murray, Baron, Pfaller, Tenover and Yolken (Editors), Manual of Clinical Microbiology, 7th Ed., ASM Press, Washington, D.C. pp. 507–516.

Altwegg, M. and H.K. Geiss. 1989. *Aeromonas* as a human pathogen. Crit. Rev. Microbiol. *16*: 253–286.

Altwegg, M. and J. Lüthy-Hottenstein. 1992. Behaviour of *Aeromonas* species in Cary-Blair transport medium at various temperatures. Eur. J. Clin. Microbiol. Infect. Dis. *11*: 79–80.

Altwegg, M., M.W. Reeves, R. Altwegg-Bissig and D.J. Brenner. 1991. Multilocus enzyme analysis of the genus *Aeromonas* and its use for species identification. Zentbl. Bakteriol. *275*: 28–45.

Altwegg, M., A.G. Steigerwalt, R. Altwegg-Bissig, J. Lüthy-Hottenstein and D.J. Brenner. 1990. Biochemical identification of *Aeromonas* genospecies isolated from humans. J. Clin. Microbiol. *28*: 258–264.

Altwegg, M., I.J. Zollinger and A. von Graevenitz. 1986. Differentiation of *Kluyvera cryocrescens* from *Kluyvera ascorbata* by irgasan susceptibility testing. Ann. Inst. Pasteur (Paris) *137A*: 159–168.

Alvarez, A.M., A.A. Benedict and C.Y. Mizumoto. 1985. Identification of xanthomonads and grouping of strains of *Xanthomonas campestris* pathovar *campestris* with monoclonal antibodies. Phytopathology *75*: 722–728.

Alvarez, A.M., A.A. Benedict, C.Y. Mizumoto, L.W. Pollard and E.L. Civerolo. 1991. Analysis of *Xanthomonas campestris* pathovar *citri* and *Xanthomonas campestris* pathovar *citrumelo* with monoclonal antibodies. Phytopathology *81*: 857–865.

Alvarez, A.M., S. Schenck and A.A. Benedict. 1996. Differentiation of *Xanthomonas albilineans* strains with monoclonal antibody reaction patterns of DNA fingerprints. Plant Pathol. (Oxf.) *45*: 358–366.

Alves, M.P., F.A. Rainey, M.F. Nobre and M.S. da Costa. 2003. *Thermomonas hydrothermalis* sp. nov., a new slightly thermophilic gamma-proteobacterium isolated from a hot spring in central Portugal. System. Appl. Microbiol. *26*: 70–75.

Alwen, J. and D.G. Smith. 1967. A medium to suppress swarming of *Proteus* species. J. Appl. Bacteriol. *30*: 389–394.

Amako, K., K. Okada and S. Miake. 1984. Evidence for the presence of a capsule in *Vibrio vulnificus.* J. Gen. Microbiol. *130*: 2741–2743.

Amandi, A., S.F. Hiu, J.S. Rohovec and J.L. Fryer. 1982. Isolation and characterization of *Edwardsiella tarda* from fall chinook salmon (*Oncorhynchus tshawytscha*). Appl. Environ. Microbiol. *43*: 1380–1384.

Amann, R.I., W. Ludwig, R. Schulze, S. Spring, E. Moore and K.H. Schlei-fer. 1996. rRNA-targeted oligonucleotide probes for the identification of genuine and former pseudomonads. Syst. Appl. Microbiol. *19*: 501–509.

Amano, K., M. Fujita and T. Suto. 1993. Chemical properties of lipo-polysaccharides from spotted fever group Rickettsiae and their com-mon antigenicity with lipopolysaccharides from *Proteus* species. Infect. Immun. *61*: 4350–4355.

Amano, K. and J.C. Williams. 1984. Sensitivity of *Coxiella burnetii* pepti-doglycan to lysozyme hydrolysis and correlation of sacculus rigidity with peptidoglycan-associated proteins. J. Bacteriol. *160*: 989–993.

Amano, K., J.C. Williams, S.R. Missler and V.N. Reinhold. 1987. Structure and biological relationships of *Coxiella burnetii* lipopolysaccharides. J. Biol. Chem. *262*: 4740–4747.

Amaral, J.F., C. Teixeira and E.D. Pinheiro. 1956. O bactério causador da mancha aureolada do cafeeiro. Arg. Inst. Biol. (Saõ Paulo). *23*: 151–155.

Amaro, C., R. Aznar, E. Alcaide and M.L. Lemos. 1990. Iron-binding compounds and related outer membrane proteins in *Vibrio cholerae* non-O1 strains from aquatic environments. Appl. Environ. Microbiol. *56*: 2410–2416.

Amaro, C. and E.G. Biosca. 1996. *Vibrio vulnificus* biotype 2, pathogenic for eels, is also an opportunistic pathogen for humans. Appl. Environ. Microbiol. *62*: 1454–1457.

Amaro, C., E.G. Biosca, B. Fouz, E. Alcaide and C. Esteve. 1995. Evidence that water transmits *Vibrio vulnificus* biotype 2 infections to eels. Appl. Environ. Microbiol. *61*: 1133–1137.

Amaro, C., E.G. Biosca, B. Fouz and E. Garay. 1992. Electrophoretic analysis of heterogeneous lipopolysaccharides from various strains of *Vibrio vulnificus* biotypes 1 and 2 by silver staining and immunoblot-ting. Curr. Microbiol. *25*: 99–104.

Ammon, A., L.R. Petersen and H. Karch. 1999. A large outbreak of he-molytic uremic syndrome caused by an unusual sorbitol-fermenting strain of *Escherichia coli* O157:H⁻. J. Infect. Dis. *179*: 1274–1277.

Amuthan, G. and A. Mahadevan. 1994. Plasmid and pathogenicity in *Xanthomonas oryzae* pathovar *oryzae*, the bacterial blight pathogen of *Oryza sativa*. J. Appl. Bacteriol. *76*: 529–538.

Ancuta, P., T. Pedron, R. Girard, G. Sandström and R. Chaby. 1996. Inability of the *Francisella tularensis* lipopolysaccharide to mimic or to antagonize the induction of cell activation by endotoxins. Infection and Immunity. *64*: 2041–2046.

Anderson, A.J. and E.A. Dawes. 1990. Occurrence, metabolism, metabolic role, and industrial uses of bacterial polyhydroxyalkanoates. Micro-biol. Rev. *54*: 450–472.

Anderson, A.J., A.J. Hacking and E.A. Dawes. 1987. Alternative pathways for the biosynthesis of alginate from fructose and glucose in *Pseu-domonas mendocina* and *Azotobacter vinelandii*. J. Gen. Microbiol. *133*: 1045–1052.

Anderson, E.S. 1964. The phage typing of Salmonellae other than *S. typhi*. *In* Oye (Editor), The World Problem of Salmonellosis, Junk, The Hague. pp. 84–110.

Anderson, E.S., G.O. Humphreys and G.A. Willshaw. 1975. The molecular relatedness of *R factors* in enterobacteria of human and animal origin. J. Gen. Microbiol. *91*: 376–382.

Anderson, K. 1944. The cultivation from granuloma inguinale of micro-organisms having characteristics of Donovan bodies in yolk sac of chick embryos. Science *97*: 560–561.

Anderson, K., W. de Monbreun and E. Goodpasture. 1945a. An etiologic consideration of *Donovania granulomatis* cultivated from granuloma inguinale (three cases) in embryonic egg yolk. J. Exp. Med. *81*: 25–40.

Anderson, K., E.W. Goodpasture and W.A. de Monbreun. 1945b. Im-munologic relationship of *Donovania granulomatis* to granuloma in-guinale. J. Exp. Med. *81*: 41–50.

Anderson, M.J. and S.T. Nameth. 1990. Development of a polyclonal antibody-based serodiagnostic assay for the detection of *Xanthomonas campestris* pathovar *pelargonii* in geranium plants. Phytopathology *80*: 357–360.

Anderson, R. and A.R. Bhatti. 1986. Fatty acid distribution in the phos-pholipids of *Francisella tularensis*. Lipids. *21*: 669–671.

Ando, S., S.I. Kato and K. Komagata. 1989. Phylogenetic diversity of methanol-utilizing bacteria deduced from their 5S ribosomal RNA sequences. J. Gen. Appl. Microbiol. *35*: 351–361.

Andreev, L.V. and V.F. Gal'chenko. 1978. Fatty acid composition and identification of methanotrophic bacteria. Dokl. Akad. Nauk. SSSR *269*: 1461–1468.

Andreev, L.V. and V.F. Gal'chenko. 1983. Phospholipid composition and differentiation of methanotrophic bacteria. J. Liq. Chromatogr. *6*: 2699–2708.

Andreeva, N.B., T.A. Sorokina and I.A. Khmel. 1996. Chitinolytic activity of pigmented *Pseudomonas* and *Xanthomonas* bacteria. Microbios. *87*: 53–57.

Andrewes, A.G., S. Hertzberg, S. Liaaen-Jensen and M.P. Starr. 1973. *Xanthomonas* pigments. 2. The *Xanthomonas* "carotenoids" - non-ca-rotenoid brominated aryl-polyene esters. Acta Chem. Scand. *27*: 2383–2395.

Andrewes, A.G., C.L. Jenkins, M.P. Starr, M.P. Shepherd and H. Hope. 1976. Structure of xanthomonadin 1, a novel dibrominated aryl pol-yene pigment produced by the bacterium *Xanthomonas juglandis*. Te-trahedr. Lett. *45*: 4023.

Andrewes, F.W. and A.C. Inman. 1919. A study of the serological races of the Flexner group of dysentery bacilli, Med. Research Committee, Special Report Series, No. 42, London.

Andrews, H.J. 1986. *Acinetobacter* bacteriocin typing. J. Hosp. Infect. *7*: 169–175.

Andro, T., J.P. Chambost, A. Kotoujansky, J. Cattaneo, Y. Bertheau, F. Barras, F. Van Gijsegem and A. Coleno. 1984. Mutants of *Erwinia chrysanthemi* defective in secretion of pectinase and cellulase. J. Bac-teriol. *160*: 1199–1203.

Angeles-Ramos, R., A.K. Vidaver and P. Flynn. 1991. Characterization of epiphytic *Xanthomonas campestris* pathovar *phaseoli* and pectolytic xan-thomonads recovered from symptomless weeds in the Dominican Re-public (West Indies). Phytopathology *81*: 677–681.

Angen, Ø. 1997. Taxonomy of the ruminant, porcine and leprine [*Pas-teurella*] *haemolytica*-complex. Comparative investigation of methods used for characterization, classification and identification, Thesis, The Royal Veterinary and Agricultural University, Copenhagen.

Angen, Ø., B. Aalbaek, E. Falsen, J.E. Olsen and M. Bisgaard. 1997. Relationships among strains classified with the ruminant *Pasteurella haemolytica*-complex using quantitative evaluation of phenotypic data. Zentbl. Bakteriol. *285*: 459–479.

Angen, Ø., R. Mutters, D.A. Caugant, J.E. Olsen and M. Bisgaard. 1999a. Taxonomic relationships of the (*Pasteurella*) *haemolytica* complex as evaluated by DNA–DNA hybridizations and 16S rRNA sequencing with proposal of *Mannheimia haemolytica* gen. nov., comb. nov., *Mann-heimia granulomatis* comb. nov., *Mannheimia glucosida* sp. nov., *Mann-heimia ruminalis* sp. nov. and *Mannheimia varigena* sp. nov. Int. J. Syst. Bacteriol. *49*: 67–86.

Angen, Ø., M. Quirie, W. Donachie and M. Bisgaard. 1999b. Investiga-tions on the species specificity of *Mannheimia* (*Pasteurella*) *haemolytica* serotyping. Vet. Microbiol. *65*: 283–290.

Angst, E.C. 1929. Some new agar-digesting bacteria. Publ. Puget Sound Biol. Sta. *27*: 49–63.

Anon, M.T., L.M. Ruiz Velasco, E. Borrajo, C. Giner, M. Sendino and R. Canton. 1993. *Escherichia vulneris* infection. Report of 2 cases (Trans-lated from Spanish). Enferm. Infecc. Microbiol. Clin. *11*: 559–561.

Anonymous (USDA:APHIS) 1994. *Escherichia coli* O157:H7. Issues and Ramifications, 1–7.12.

Ansaruzzaman, M., A.K. Kibriya, A. Rahman, P.K. Neogi, A.S. Faruque, B. Rowe and M.J. Albert. 1995. Detection of provisional serovars of *Shigella dysenteriae* and designation as *S. dysenteriae* serotypes 14 and 15. J. Clin. Microbiol. *33*: 1423–1425.

Anthony, L.S.D., M. Gu, S.C. Cowley, W.W.S. Leung and F.E. Nano. 1991. Transformation and allelic replacement in *Francisella* spp. J. Gen. Mi-crobiol. *137*: 2697–2704.

Anzai, Y., H. Kim, J.Y. Park, H. Wakabayashi and H. Oyaizu. 2000. Phylogenetic affiliation of the pseudomonads based on 16S rRNA sequence. Int. J. Syst. Evol. Microbiol. *50*: 1563–1589.

Anzai, Y., Y. Kudo and H. Oyaizu. 1997. The phylogeny of the genera *Chryseomonas, Flavimonas*, and *Pseudomonas* supports synonymy of these three genera. Int. J. Syst. Bacteriol. *47*: 249–251.

Aoki, K., K. Ohtsuka, R. Shinke and H. Nishira. 1984. Microbial metabolism of aromatic amines. 4. Rapid biodegradation of aniline by *Frateuria* species ANA-18 and its aniline metabolism. Agric. Biol. Chem. *48*: 865–872.

Aoki, K., R. Shinke and H. Nishira. 1976. Purification and some properties of yeast tannase. Agric. Biol. Chem. *40*: 79–85.

Aoki, T. 1988. Drug-resistant plasmids from fish pathogens. Microbiol. Sci. *5*: 219–223.

Aoki, T. 1992. Chemotherapy and drug resistance in fish farms in Japan. *In* Shariff, Subasinghe and Arthur (Editors), Diseases in Asian Aquaculture, Fish Health Section, Asian Fisheries Society, Manila, Philippines. 519–529.

Aoki, T., T. Arai and S. Egusa. 1977. Detection of R plasmids in naturally occurring fish-pathogenic bacteria, *Edwardsiella tarda*. Microbiol. Immunol. *21*: 77–83.

Aoki, T., S. Egusa, T. Kimura and T. Watanabe. 1971. Detection of R factors in naturally occurring *Aeromonas salmonicida* strains. Appl. Microbiol. *22*: 716–717.

Aoki, T. and B.I. Holland. 1985. The outer membrane proteins of fish pathogens *Aermonas hydrophila, Aeromonas salmonicida* and *Edwardsiella tarda*. FEMS Microbiol. Lett. *27*: 299–305.

Appanna, V.D., L.G. Gazso, J. Huang and M. St. Pierre. 1996. Cesium stress and adaptation in *Pseudomonas fluorescens*. Bull. Environ. Contam. Toxicol. *56*: 833–838.

Appanna, V.D. and R. Hamel. 1996. Aluminum detoxification mechanism in *Pseudomonas fluorescens* is dependent on iron. FEMS Microbiol. Lett. *143*: 223–228.

Appanna, V.D., R.E. Mayer and M. St. Pierre. 1995. Aluminum tolerance of *Pseudomonas fluorescens* in a phosphate-deficient medium. Bull. Environ. Contam. Toxicol. *55*: 404–411.

Appanna, V.D. and M. St. Pierre. 1996. Cellular response to a multiple-metal stress in *Pseudomonas fluorescens*. J. Biotechnol. *48*: 129–136.

Appuhamy, S., J.G. Coote, J.C. Low and R. Parton. 1998. PCR methods for rapid identification and characterization of *Actinobacillus seminis* strains. J. Clin. Microbiol. *36*: 814–817.

Aragão, H. and G. Vianna. 1913. Pesquizas sobre o *Granuloma venereo*. (Untersuchungen über das *Granuloma venereum*). Mem. Inst. Oswalo Cruz *5*: 211–238.

Arahal, D.R., A.M. Castillo, W. Ludwig, K.-H. Schleifer and A. Ventosa. 2002. Proposal of *Cobetia marina* gen. nov., comb. nov., within the family *Halomonadaceae*, to include the species *Halomonas marina*. Syst. Appl. Microbiol. *25*: 207–211.

Arahal, D.R., M.T. Garcia, W. Ludwig, K.H. Schleifer and A. Ventosa. 2001a. Transfer of *Halomonas canadensis* and *Halomonas israelensis* to the genus *Chromohalobacter* as *Chromohalobacter canadensis* comb. nov and *Chromohalobacter israelensis* comb. nov. Int. J. Syst. Evol. Microbiol. *51*: 1443–1448.

Arahal, D.R., M.T. Garcia, C. Vargas, D. Cánovas, J.J. Nieto and A. Ventosa. 2001b. *Chromohalobacter salexigens* sp. nov., a moderately halophilic species that includes *Halomonas elongata* DSM 3043 and ATCC 33174. Int. J. Syst. Evol. Microbiol. *51*: 1457–1462.

Arai, T., N. Ikejima, T. Itoh, S. Sakai, T. Shimada and R. Sakazaki. 1980. A survey of *Plesiomonas shigelloides* from aquatic environments, domestic animals, pets and humans. J. Hyg. *84*: 203–211.

Arambulo, P.V., N.C. Westerlund, R.V. Sarmiento and A.S. Abaga . 1967. Isolation of *Edwardsiella tarda*: a new genus of *Enterobacteriaceae* from pig bile in the Philippines. Far East Med. J. *3*: 385–386.

Araujo, R.M., R.M. Arribas, F. Lucena and R. Pares. 1989. Relation between *Aeromonas* and fecal coliforms in fresh waters. J. Appl. Bacteriol. *67*: 213–218.

Arcos, M.L., A. de Vicente, M.A. Morinigo, P. Romero and J.J. Borrego. 1988. Evaluation of several selective media for recovery of *Aeromonas*

hydrophila from polluted waters. Appl. Environ. Microbiol. *54*: 2786–2792.

Arduino, M.J., F.W. Hickman-Brenner and J.J. Farmer, III. 1988. Phenotypic analysis of 132 *Aeromonas* strains representing 12 DNA hybridization groups. J. Diarrh. Dis. Res. *6*: 138.

Argaman, M., T.Y. Liu and J.B. Robbins. 1974. Polyribitol phosphate: an antigen of four gram-positive bacteria cross-reactive with the capsular polysaccharide of *Haemophilus influenzae*, type b. J. Immunol. *112*: 649–655.

Ark, P.A. 1939. Bacterial leaf spot of maple. Phytopathology *29*: 968–970.

Ark, P.A. 1940. Bacterial stalk rot of field corn caused by *Phytomonas lapsa* n. sp. Phytopathology *30*: 1.

Ark, P.A. and J.T. Barrett. 1946. A new bacterial leaf-spot of greenhouse-grown gardenias. Phytopathology *36*: 865–868.

Ark, P.A. and M.W. Gardner. 1936. Bacterial leaf spot of *Primula*. Phytopathology *26*: 1050–1055.

Ark, P.A. and C.M. Tompkins. 1946. Bacterial leaf blight of bird's nest fern. Phytopathology *36*: 758–761.

Arlat, M., C.L. Gough, C.E. Barber, C. Boucher and M.J. Daniels. 1991. *Xanthomonas campestris* contains a cluster of *hrp* genes related to the larger *hrp* cluster of *Pseudomonas solanacearum*. Mol. Plant-Microbe Interact. *4*: 593–601.

Armstrong, G.L., J. Hollingsworth and J.G. Morris, Jr.. 1996. Emerging foodborne pathogens: *Escherichia coli* O157:H7 as a model of entry of a new pathogen into the food supply of the developed world. Epidemiol. Rev. *18*: 29–51.

Armstrong, G.R., R.A. Sen and J. Wilkinson. 2000. *Pasteurella multocida* meningitis in an adult: case report. J. Clin. Pathol. *53*: 234–235.

Arnau, J. and M. Garriga. 1993. Black spot in cured meat-products. Fleischwirtschaft *73*: 1412–1413.

Arnau, J. and M. Garriga. 2000. The effect of certain amino acids and browning inhibitors on the 'black spot' phenomenon produced by *Carnimonas nigrificans*. J. Sci. Food Agric. *80*: 1655–1658.

Arnaud, G. 1920. Une maladie bactérienne du lierre (*Hedera helix* L.). C. R. Hebd. Séances Acad. Sci. Ser. D. *171*: 121–122.

Arnold, C., L. Metherell, G. Willshaw, A. Maggs and J. Stanley. 1999. Predictive fluorescent amplified-fragment length polymorphism analysis of *Escherichia coli*: High-resolution typing method with phylogenetic significance. J. Clin. Microbiol. *37*: 1274–1279.

Arora, S.K., B.W. Ritchings, E.C. Almira, S. Lory and R. Ramphal. 1996. Cloning and characterization of *Pseudomonas aeruginosa fliF*, necessary for flagellar assembly and bacterial adherence to mucin. Infect. Immun. *64*: 2130–2136.

Arseculeratne, S.N. 1962. Actinobacillosis in joints of rabbits. J. Comp. Pathol. *72*: 33–39.

Arsenijevic, M., S. Durisic and S. Mitrev. 1994. Serological identification of *Erwinia amylovora* bacterium, a pomaceous tree pathogen. Zast. Bilja. *45*: 273–278.

Artiguenave, F., R. Vilagines and C. Danglot. 1997. High-efficiency transposon mutagenesis by electroporation of a *Pseudomonas fluorescens* strain. FEMS Microbiol. Lett. *153*: 363–369.

Aruga, S., Y. Kamagata, T. Kohno, S. Hanada, K. Nakamura and T. Kanagawa. 2002. Characterization of filamentous Eikelboom type 021N bacteria and description of *Thiothrix disciformis* sp. nov. and *Thiothrix flexilis* sp. nov. Int. J. Syst. Evol. Microbiol. *52*: 13009–1316.

Asai, T. 1968. Acetic acid bacteria. Classification and biochemical activities, University of Tokyo Press, Tokyo.

Asai, T., H. Iizuka and K. Komagata. 1962. The flagellation of genus *Kluyvera*. J. Gen. Appl. Microbiol. *8*: 187–191.

Asai, T., S. Okumura and T. Tsunoda . 1956. On a new genus, *Kluyvera*. Proc. Japan Acad. *32*: 488–493.

Ash, C., A.J. Martinez-Murcia and M.D. Collins. 1993a. Identification of *Aeromonas schubertii* and *Aeromonas jandaei* by using a polymerase chain reaction-probe test. FEMS Microbiol. Lett. *108*: 151–155.

Ash, C., A.J. Martinez-Murcia and M.D. Collins. 1993b. Molecular identification of *Aeromonas sobria* by using a polymerase chain reaction-probe test. Med. Microbiol. Lett. *2*: 80–86.

Ashby, S.F. 1929. Gumming disease of sugar cane. Trop. Agr. Trinidad. *6*: 135–138.

Asikainen, S., C. Chen and J. Slots. 1995. *Actinobacillus actinomycetemcomitans* genotypes in relation to serotypes and periodontal status. Oral Microbiol. Immunol. *10*: 65–68.

Asikainen, S., C.H. Lai, S. Alaluusua and J. Slots. 1991. Distribution of *Actinobacillus actinomycetemcomitans* serotypes in periodontal health and disease. Oral Microbiol. Immunol. *6*: 115–118.

Asperger, O. and H.P. Kleber. 1991. Metabolism of alkanes by *Acinetobacter*. *In* Towner and Bergogne-Bérézin (Editors), The Biology of *Acinetobacter*. Taxonomy, Molecular Biology, Physiology, Industrial Relevance, Plenum Press, New York. pp. 323–350.

Assinder, S.J. and P.A. Williams. 1990. The TOL plasmids: determinants of the catabolism of toluene and the xylenes. Adv. Microb. Physiol. *31*: 1–69.

Asturias, J.A., E. Diaz and K.N. Timmis. 1995. The evolutionary relationship of biphenyl dioxygenase from Gram-positive *Rhodococcus globerulus* P6 to multicomponent dioxygenases from Gram-negative bacteria. Gene *156*: 11–18.

Atlas, R. 1997. Handbook of Microbiological Media, 2nd ed., CRC Press, Boca Raton.

Attafuah, A. and J.F. Bradbury. 1989. *Pseudomonas antimicrobica*, a new species strongly antagonistic to plant-pathogens. J. Appl. Bacteriol. *67*: 567–573.

Attafuah, A. and J.F. Bradbury. 1990. *In* Validation of the publication of new names and new combinations previously published outside the IJSB, list no. 34. Int. J. Syst. Bacteriol. *40*: 320–321.

Au, S., K.L. Roy and R.G. von Tigerstrom. 1991. Nucleotide sequence and characterization of the gene for secreted alkaline phosphatase from *Lysobacter enzymogenes*. J. Bacteriol. *173*: 4551–4557.

Aucken, H.M., S.G. Wilkinson and T.L. Pitt. 1996. Immunochemical characterization of two new O serotypes of *Serratia marcescens* O27 and O28. FEMS Microbiol. Lett. *138*: 77–82.

Audureau, A. 1940. Étude du genre *Moraxella*. Ann. Inst. Pasteur (Paris) *64*: 126–166.

Auling, G., H.-J. Busse, M. Hahn, H. Hennecke, R.-M. Kroppenstedt, A. Probst and E. Stackebrandt. 1988. Phylogenetic heterogeneity and chemotaxonomic properties of certain Gram-negative aerobic carboxydobacteria. Syst. Appl. Microbiol. *10*: 264–272.

Auling, G., H.-J. Busse, F. Pilz, L. Webb, H. Kneifel and D. Claus. 1991. Rapid differentiation, by polyamine analysis, of *Xanthomonas* strains from phytopathogenic pseudomonads and other members of the class *Proteobacteria* interacting with plants. Int. J. Syst. Bacteriol. *41*: 223–228.

Auling, G., M. Reh, C.M. Lee and H.G. Schlegel. 1978. *Pseudomonas pseudoflava* a new species of hydrogen-oxidizing bacteria: its differentiation from *Pseudomonas flava* and other yellow-pigmented, Gram-negative, hydrogen oxidizing species. Int. J. Syst. Bacteriol. *28*: 82–95.

Austen, R.A. and N.W. Dunn. 1980. Regulation of the plasmid-specified naphthalene catabolic pathway of *Pseudomonas putida*. J. Gen. Microbiol. *117*: 521–528.

Austin, B. 1982. Taxonomy of bacteria isolated from a coastal, fish rearing unit. J. Appl. Bacteriol. *53*: 253–268.

Austin, B. 1993. Recovery of atypical isolates of *Aeromonas salmonicida*, which grow at 37°C, from ulcerated non-salmonids in England. J. Fish. Dis. *16*: 165–168.

Austin, B. and C. Adams. 1996. Fish pathogens. *In* Austin, Altwegg, Gosling and Joseph (Editors), The Genus *Aeromonas*, 1st Ed., John Wiley & Sons, Ltd., Chichester. pp. 197–244.

Austin, B. and D.A. Austin. 1990. Expression of motility in strains of the nonmotile species *Aeromonas media*. FEMS Microbiol. Lett. *68*: 123–134.

Austin, B. and D.A. Austin. 1993. Bacterial Fish Pathogens, Ellis Horwood, Chichester, UK. pp. 266–279.

Austin, B., D.A. Austin, I. Dalsgaard, B.K. Gudmundstottir, S. Hoie, J.M. Thornton, J.L. Larsen, B. O'Hici and R. Powell. 1998a. Characterization of atypical *Aeromonas salmonicida* by different methods. Syst. Appl. Microbiol. *21*: 50–64.

Austin, B., M. Goodfellow and C.H. Dickinson. 1978. Numerical taxonomy of phylloplane bacteria isolated from *Lolium perenne*. J. Gen. Microbiol. *104*: 139–155.

Austin, D.A., D. McIntosh and B. Austin. 1989a. Taxonomy of fish associated *Aeromonas* spp., with the description of *Aeromonas salmonicida* subsp. *smithia*, subsp. nov., sp. nov. Syst. Appl. Microbiol. *11*: 277–290.

Austin, D.A., D. McIntosh and B. Austin. 1989b. *In* Validation of the publication of new names and new combinations previously effectively published outside the IJSB. List No. 31. Int. J. Syst. Bacteriol. *39*: 495–497.

Austin, D.A. and M.O. Moss. 1986. Numerical taxonomy of red-pigmented bacteria isolated from a lowland river, with the description of a new taxon, *Rugamonas rubra* gen. nov., sp. nov. J. Gen. Microbiol. *132*: 1899–1909.

Austin, D.A. and M.O. Moss. 1987. *In* Validation of the publication of new names and new combinations previously effectively published outside the IJSB. List No. 23. Int. J. Syst. Bacteriol. *37*: 179–180.

Austin, D.A., P.A. Robertson, D.K. Wallace, H. Daskalov and B. Austin. 1998b. Isolation of *Aeromonas salmonicida* in association with purple-pigmented bacteria in sediment from a Scottish loch. Lett. Appl. Microbiol. *27*: 349–351.

Avakyan, A.A. and V.L. Popov. 1984. *Rickettsiaceae* and *Chlamydiaceae*: comparative electron microscopic studies. Acta Virol. *28*: 159–173.

Avakyan, A.A., V.L. Popov, S.M. Chebanov, A.A. Shatkin, V.E. Sidorov and R.I. Kudelina. 1983. Comparison of the ultrastructure of small dense forms of chlamydiae and *Coxiella burnetii*. Acta Virol. *27*: 168–172.

Averhoff, B., L. Gregg-Jolly, D. Elsemore and L.N. Ornston. 1992. Genetic analysis of supraoperonic clustering by use of natural transformation in *Acinetobacter calcoaceticus*. J. Bacteriol. *174*: 200–204.

Avlami, A., C. Papalambrou, M. Tzivra, E. Dounis and T. Kordossis. 1997. Bone marrow infection caused by *Actinobacillus ureae* in a rheumatoid arthritis patient. J. Infect. *35*: 298–299.

Axenfeld, T. 1897. Weitere erfahrungen über die chronische diplocbacillen conjunctivitis. Klin. Wochenschr. *34*: 847–849.

Ayers, T.T., C.L. Lefebvre and H.W. Johnson. 1939. Bacterial wilt of lespedeza. U.S. Dept. Agr. Tech. Bull. *704*: 1–22.

Azelvandre, P. 1993. Les deferrioxamines E et D2, sidérophores de *Pseudomonas stutzeri*, Université Louis-Pasteur, Strasbourg, France.

Aznar, R., E. Alcaide and E. Garay. 1992. Numerical taxonomy of pseudomonads isolated from water, sediment and eels. Syst. Appl. Microbiol. *15*: 235–246.

Aznar, R., W. Ludwig, R.I. Amann and K.H. Schleifer. 1994. Sequence determination of rRNA genes of pathogenic *Vibrio* species and whole-cell identification of *Vibrio vulnificus* with rRNA-targeted oligonucleotide probes. Int. J. Syst. Bacteriol. *44*: 330–337.

Baas-Becking, L.G.M. 1925. Studies on the sulphur bacteria. Ann. Bot. *39*: 613–650.

Babalova, M., J. Blahova, M. Lesicka-Hupkova, V. Krcmery, Sr. and K. Kubonova. 1995. Transfer of ceftazidime and aztreonam resistance from nosocomial strains of *Xanthomonas* (*Stenotrophomonas*) *maltophilia* to a recipient strain of *Pseudomonas aeruginosa* ML-1008. Eur. J. Clin. Microbiol. Infect. Dis. *14*: 925–927.

Babenzien, H.D. 1965. Über Vorkommen und Kultur von *Nevskia ramosa*. Zentralbl. Bakteriol. Parasitenkd. Infektionskr. Abt. I. Suppl. *1*: 111–116.

Babenzien, H.D. 1966. Untersuchungen zur Biologie des Neustons. Verh. Internat. Verein. Limnol. *16*: 1503–1511.

Babenzien, H.D. 1967. Zur Biologie von *Nevskia ramosa*. Z. Allg. Mikrobiol. *7*: 89–96.

Babenzien, H.D. 1991. *Achromatium oxaliferum* and its ecological niche. Zentbl. Mikrobiol. *146*: 41–49.

Babenzien, H.D. and P. Hirsch. 1974. Genus *Nevskia*. *In* Buchanan and Gibbons (Editors), Bergey's Manual of Determinative Bacteriology, 8th Ed., The Williams & Wilkins Co., Baltimore. 161–162.

Babudieri, B. 1959. Q fever: a zoonosis. Adv. Vet. Sci. *5*: 81–182.

Baca, O.G., E.T. Akporiaye, A.S. Aragon, I.L. Martinez, M.V. Robles and N.L. Warner. 1981. Fate of phase I and phase II *Coxiella burnetii* in several macrophage-like tumor cell lines. Infect. Immun. *33*: 258–266.

Baca, O.G. and D. Paretsky. 1983. Q fever and *Coxiella burnetii*: a model for host-parasite interactions. Microbiol. Rev. *47*: 127–149.

Baca, O.G., T.O. Scott, E.T. Akporiaye, R. DeBlassie and H.A. Crissman. 1985. Cell cycle distribution patterns and generation times of L929 fibroblast cells persistently infected with *Coxiella burnetii*. Infect. Immun. *47*: 366–369.

Bachmann, B.J. 1996. Derivations and genotypes of some mutant derivatives of *Escherichia coli* K-12. *In* Neidhardt, Curtiss, Ingraham, Lin, Low, Magasanik, Reznikoff, Riley, Schaechter and Umbarger (Editors), *Escherichia coli* and *Salmonella*: Cellular and Molecular Biology, 2nd Ed., ASM Press, Washington, D.C. pp. 2460–2488.

Backhus, D.A., F.W. Picardal, S. Johnson, T. Knowles, R. Collins, A. Radue and S. Kim. 1997. Soil- and surfactant-enhanced reductive dechlorination of carbon tetrachloride in the presence of *Shewanella putrefaciens* 200. J. Contam. Hydrol. *28*: 337–361.

Bader, R.E. 1954. Uber die Herstellung eines agglutinierenden Serums gegen die Rundform von *Shigella sonnei* mit einem Stamm der Gattung *Pseudomonas*. Z. Hyg. Infektionskr. *140*: 450–456.

Badran, H., R. Sohoni, T.V. Venkatesh and H.K. Das. 1999. Construction of a *recF* deletion mutant of *Azotobacter vinelandii* and its characterization. FEMS Microbiol. Lett. *174*: 363–369.

Bae, B.H. and S.B. Sureka. 1983. *Cedecea davisae* isolated from scrotal abscess. J. Urol. *130*: 148–149.

Bae, B.H.C., S.B. Sureka and J.A. Ajamy. 1975. Enteric group 15 (*Enterobacteriaceae*) associated with pneumonia. J. Clin. Microbiol. *14*: 596–597.

Baggi, G., P. Barbieri, E. Galli and S. Tollari. 1987. Isolation of a *Pseudomonas stutzeri* strain that degrades *o*-xylene. Appl. Environ. Microbiol. *53*: 2129–2132.

Baghdiguian, S., M.H. Boyer Giglio, J.O. Thaler, G. Bonnot and N.E. Boemare. 1993. Bacteriocinogenesis in cells of *Xenorhabdus nematophilus* and *Photorhabdus luminescens*: *Enterobacteriaceae* associated with entomopathogenic nematodes. Biol. Cell *79*: 177–185.

Bagley, S.T. and R.J. Seidler. 1978. Primary *Klebsiella* identification with MacConkey–inositol–carbenicillin agar. Appl. Environ. Microbiol. *36*: 536–538.

Bagley, S.T., R.J. Seidler and D.J. Brenner. 1981. *Klebsiella planticola* sp. nov. a new species of enterobacteriaceae found primarily in nonclinical environments. Curr. Microbiol. *6*: 105–109.

Bagley, S.T., R.J. Seidler and D.J. Brenner. 1982. *In* Validation of the publication of new names and new combinations previously effectively published outside the IJSB. List No. 8. Int. J. Syst. Bacteriol. *32*: 266–268.

Bahr, H. and U. Schwartz. 1956. Untersuchungen zur Ökologie farblose fadifer Schwefelmikroben. Biol. Zeit. *75*: 451–464.

Bahr, L. 1919. Paratyfus hos Honningbien samt nogle undersøgelser verdrørende Forekomsten af Bakterierhenhorende til Coli-tyfus gruppen. I. Honningbienstarm Scand. Vet. Tidsskrift. *9*: 25–40, 45–60.

Bahrani, F.K., S. Cook, R.A. Hull, G. Massad and H.L.T. Mobley. 1993. *Proteus mirabilis* fimbriae: N-terminal amino acid sequence of a major fimbrial subunit and nucleotide sequence of the genes from two strains. Infect. Immun. *61*: 884–891.

Bahrani, F.K., D.E. Johnson, D. Robbins and H.L. Mobley. 1991. *Proteus mirabilis* flagella and MR/P fimbriae: isolation, purification, N-terminal analysis, and serum antibody response following experimental urinary tract infection. Infect. Immun. *59*: 3574–3580.

Bahrani, F.K., G. Massad, C.V. Lockatell, D.E. Johnson, R.G. Russell, J.W. Warren and H.L. Mobley. 1994. Construction of an MR/P fimbrial mutant of *Proteus mirabilis*: role in virulence in a mouse model of ascending urinary tract infection. Infect. Immun. *62*: 3363–3371.

Bahrani, F.K. and H.L. Mobley. 1993. *Proteus mirabilis* MR/P fimbriae: molecular cloning, expression and nucleotide sequence of the major fimbrial subunit gene. J. Bacteriol. *175*: 457–464.

Bahrani, F.K. and H.L. Mobley. 1994. *Proteus mirabilis* MR/P fimbrial operon: genetic organization, nucleotide sequence, and conditions for expression. J. Bacteriol. *176*: 3412–3419.

Bailie, W.E. 1966. Characterization of *Haemophilus somnus*, new species, a microorganism isolated from infectious thromboembolic meningoencephalitis of cattle, Kansas State University Manhattan, Kansas.

Bailie, W.E., H.D. Anthony and K.D. Weide. 1966. Infectious thromboembolic meningoencephalomyelitis (sleeper syndrome) in feedlot cattle. J. Am. Vet. Med. Assoc. *148*: 162–166.

Baillie, A., W. Hodgkiss and J.R. Norris. 1962. Flagellation of *Azotobacter* spp. as demonstrated by electron microscopy. J. Appl. Bacteriol. *25*: 116–119.

Baine, W.B., J.K. Rasheed, J.C. Feeley, G.W. Gorman and J. Casida, L.E.. 1978. Effect of supplemental L-tyrosine on pigment production in cultures of the Legionnaires' disease bacterium. Curr. Microbiol. *1*: 93–94.

Bainton, N.J., B.W. Bycroft, S.R. Chhabra, P. Stead, L. Gledhill, P.J. Hill, C.E.D. Rees, M.K. Winson, G.P.C. Salmond, G.S.A.B. Stewart and P. Williams. 1992a. A general role for the lux autoinducer in bacterial cell signalling: control of antibiotic biosynthesis in *Erwinia*. Gene *116*: 87–91.

Bainton, N.J., P. Stead, S.R. Chhabra, B.W. Bycroft, G.P. Salmond, G.S. Stewart and P. Williams. 1992b. N-(3-oxohexanoyl)-L-homoserine lactone regulates carbapenem antibiotic production in *Erwinia carotovora*. Biochem. J. *288*: 997–1004.

Bakaletz, L.O., B.M. Tallan, T. Hoepf, T.F. DeMaria, H.G. Birck and D.J. Lim. 1988. Frequency of fimbriation of nontypable *Haemophilus influenzae* and its ability to adhere to chinchilla and human respiratory epithelium. Infect. Immun. *56*: 331–335.

Baker, C.N., D.G. Hollis and C. Thornsberry. 1985. Antimicrobial susceptibility testing of *Francisella tularensis* with a modified Mueller-Hinton broth. J. Clin. Microbiol. *22*: 212–215.

Baker, C.J., E.W. Orlandi and N.M. Mock. 1993. Harpin, an elicitor of the hypersensitive response in tobacco caused by *Erwinia amylovora*, elicits active oxygen production in suspension cells. Plant Physiol. *102*: 1341–1344.

Baker, N.R. 1990. Adherance and the role of alginate. *In* Gacesa and Russell (Editors), *Pseudomonas* Infection and Alginates - Biochemistry, Genetics, and Pathology, Chapman and Hall, London. 95–108.

Bakken, J.S., C.C. Sanders, R.B. Clark and M. Hori. 1988. Beta-lactam resistance in *Aeromonas* spp. caused by inducible beta-lactamases active against penicillins, cephalosporins, and carbapenems. Antimicrob. Agents Chemother. *32*: 1314–1319.

Bakopoulos, V., A. Adams and R.H. Richards. 1997. Serological relationship of *Photobacterium damsela* subsp. *piscicida* isolates (the causative agent of fish pasteurellosis) determined by western blot analysis using six monoclonal antibodies. Dis. Aquat. Org. *28*: 69–72.

Bakos, K. 1955. Studien über *Haemophilus suis*, mit besonderer Berücksichtigung der serologischen Differenzierung seiner Stäamme. *In* Appelberg's Boktrykkeri AB, Uppsala.

Bakos, K., A. Nilsson and E. Thal. 1952. Untersuchungen über *Haemophilus suis*. Nord. Vet. Med. *4*: 241–255.

Balajee, S. and A. Mahadevan. 1989. Evidence for a dissimilatory plasmid in *Azotobacter chroococcum*. FEMS Microbiol. Lett. *65*: 223–227.

Balas, M.T., M.H. Lee and J.H. Werren. 1996. Distribution and fitness effects of the son-killer bacterium in *Nasonia*. Evol. Ecol. *10*: 593–607.

Balashova, V.V., I.Y. Vedenina, G.E. Markosyan and G.A. Zavarzin. 1974. The auxotrophic growth of *Leptospirillum ferroxidans*. Mikrobiologiya *43*: 581–585.

Balch, W.E., G.E. Fox, L.J. Magrum, C.R. Woese and R.S. Wolfe. 1979. Methanogens: reevaluation of a unique biological group. Microbiol. Rev. *43*: 260–296.

Baldani, J.I., B. Pot, G. Kirchhof, E. Falsen, V.L.D. Baldani, F.L. Olivares, B. Hoste, K. Kersters, A. Hartmann, M. Gillis and J. Döbereiner. 1996. Emended description of *Herbaspirillum*: inclusion of "*Pseudomonas*" *rubrisubalbicans*, a mild plant pathogen, as *Herbaspirillum rubrisubalbicans* comb. nov.; and classification of a group of clinical isolates (EF group 1) as *Herbaspirillum* species 3. Int. J. Syst. Bacteriol. *46*: 802–810.

Baldini, M.M., J.B. Kaper, M.M. Levine, D.C. Candy and H.W. Moon.

1983. Plasmid-mediated adhesion in enteropathogenic *Escherichia coli*. J. Pediatr. Gastroenterol. Nutr. *2*: 534–538.

Bali, A., G. Blanco, S. Hill and C. Kennedy. 1992. Excretion of ammonium by a *nifL* mutant of *Azotobacter vinelandii* fixing nitrogen. Appl. Environ. Microbiol. *58*: 1711–1718.

Ballard, R.W. 1970. The taxonomy of some phytopathogenic *Pseudomonas* species, University of California, Berkley.

Ballard, R.W., N.J. Palleroni, M. Doudoroff, R.Y. Stanier and M. Mandel. 1970. Taxonomy of the aerobic pseudomonads: *Pseudomonas cepacia*, *P. marginata*, *P. alliicola*, and *P. caryophylli*. J. Gen. Microbiol. *60*: 199–214.

Ballester, M., J.M. Ballester and J.P. Belaich. 1977. Isolation and characterization of a high molecular weight antibiotic produced by a marine bacterium. Microb. Ecol. *3*: 289–303.

Bally, R., D. Thomas-Bauzon, T. Heulin, J. Balandreau and C. Richard. 1983. Determination of the most frequent N₂-fixing bacteria in a rice rhizosphere. Can. J. Microbiol. *29*: 881–887.

Banatvala, N., M.M. Debeukelaer, P.M. Griffin, T.J. Barrett, K.D. Greene, J.H. Green and J.G. Wells. 1996. Shiga-like toxin-producing *Escherichia coli* O111 and associated hemolytic-uremic syndrome: A family outbreak. Pediatr. Infect. Dis. J. *15*: 1008–1011.

Bando, S.Y., G.R. do Valle, M.B. Martinez, L.R. Trabulsi and C.A. Moreira-Filho. 1998. Characterization of enteroinvasive *Escherichia coli* and *Shigella* strains by RAPD analysis. FEMS Microbiol. Lett. *165*: 159–165.

Bang, S.S., P. Baumann and K.H. Nealson. 1978. Phenotypic characterization of *Photobacterium logei* (sp. nov.), a species related to *Photobacterium fischeri*. Curr. Microbiol. *1*: 285–288.

Bang, S.S., L. Baumann, M.J. Woolkalis and P. Baumann. 1981. Evolutionary relationships in *Vibrio* and *Photobacterium* as determined by immunological studies of superoxide dismutase. Arch. Microbiol. *130*: 111–120.

Bangera, M.G. and L.S. Thomashow. 1996. Characterization of a genomic locus required for synthesis of the antibiotic 2,4-diacetylphloroglucinol by the biological control agent *Pseudomonas fluorescens* Q2-87. Mol. Plant-Microbe Interact. *9*: 83–90.

Bangert, R.L., A.C.S. Ward, E.H. Stauber, B.R. Cho and P.R. Widders. 1988. A survey of the aerobic bacteria in the feces of captive raptors. Avian Dis. *32*: 53–62.

Barbaree, J.M. 1993. Selecting a subtyping technique for use in investigations of legionellosis epidemics. *In* Barbaree, Breiman and Dufour (Editors), *Legionella*: Current Status and Emerging Perspectives, American Society for Microbiology, Washington, D.C. pp. 169–172.

Barbé, G., M. Babolat, J.M. Boeufgras, D. Monget and J. Freney. 1994. Evaluation of API NH, a new 2-hour system for identification of *Neisseria* and *Haemophilus* species and *Moraxella catarrhalis* in a routine clinical laboratory. J. Clin. Microbiol. *32*: 187–189.

Barbe, J 1969. Organization méthodique de l'étude des caractères enzymatiques des bactéries de la tribu des *Klebsielleae*. Application à la classification, Université Marseille Marseille, France. 247 pp.

Barber, C. and E. Eylan. 1976. Immunochemical relations of *Yersinia enterocolitica* with *Yersinia pestis* and their connection with other *Enterobacteriaceae*. Microbios. Lett. *3*: 25–29.

Barbeyron, T., B. Henrissat and B. Kloareg. 1994. The gene encoding the kappa-carrageenase of *Alteromonas carrageenovora* is related to beta-1,3-1,4-glucanases. Gene *139*: 105–109.

Barea, J.M. and M.E. Brown. 1974. Effects on plant growth produced by *Azotobacter paspali* related to synthesis of plant growth regulating substances. J. Appl. Bacteriol. *37*: 583–593.

Barenkamp, S.J. 1992. Outer membrane proteins and lipopolysaccharides of nontypeable *Haemophilus influenzae*. J. Infect. Dis. *165 (Suppl 1)*: S181–184.

Barja, J.L., Y. Santos, I. Huq, R.R. Colwell and A.E. Toranzo. 1990. Plasmids and factors associated with virulence in environmental isolates of *Vibrio cholerae* non-O1 in Bangladesh. J. Med. Microbiol. *33*: 107–114.

Barnes, D.M. and D.K. Sorensen. 1975. Salmonellosis. *In* Dunne and Leman (Editors), Diseases of Swine, 4th Ed., Iowa State University, Ames. pp. 554–564.

Barnes, L.A. and W.B. Cherry. 1946. A group of paracolon organisms having apparent pathogenicity. Am. J. Public Health. *36*: 481–483.

Barnett, T.C. and S.M. Kirov. 1999a. Expression and function of the cloned type IV *Aeromonas pilus* (Tap). 6th International *Aeromonas/Plesiomonas* Symposium, Chicago, Illinois. p. 25.

Barnett, T.C. and S.M. Kirov. 1999b. The type IV *Aeromonas pilus* (Tap) gene cluster is widely conserved in *Aeromonas* species. Microb. Pathog. *26*: 77–84.

Barnett, T.C., S.M. Kirov, M.S. Strom and K. Sanderson. 1997. *Aeromonas* spp. possess at least two distinct type IV pilus families. Microb. Pathog. *23*: 241–247.

Barras, F., F.v. Gijsegem and A.K. Chatterjee. 1994. Extracellular enzymes and pathogenesis of soft-rot *Erwinia*. Annu. Rev. Phytopathol. *32*: 201–234.

Barrett, E.L., R.E. Solanes, J.S. Tang and N.J. Palleroni. 1986. *Pseudomonas fluorescens* biovar V: its resolution into distinct component groups and the relationship of these groups to other *Pseudomonas fluorescens* biovars, to *Pseudomonas putida*, and to psychrotrophic pseudomonads associated with food spoilage. J. Gen. Microbiol. *132*: 2709–2721.

Barrett, T.J. 1997. Molecular fingerprinting of foodborne pathogenic bacteria: An introduction to methods, uses and problems. *In* Tortorello and Gendel (Editors), Food Microbiological Analysis: New Technologies, Marcel Dekker, New York. pp. 249–264.

Barrett Ralston, E. 1972. Some contributions to the taxonomy of the genus *Pseudomonas*, University of California, Berkley.

Barry, A.L. and P.D. Hoeprich. 1973. *In vitro* activity of cephalothin and three penicillins against *Escherichia coli* and *Proteus* species. Antimicrob. Agents Chemother. *4*: 354–360.

Barthold, S.W., G.L. Coleman, P.N. Bhatt, G.W. Osbaldiston and A.M. Jonas. 1976. The etiology of transmissible murine colonic hyperplasia. Lab. Anim. Sci. *26*: 889–894.

Bartlett, K.H., T.J. Trust and H. Lior. 1977. Small pet aquarium frogs as a source of *Salmonella*. Appl. Environ. Microbiol. *33*: 1026–1029.

Barton, H.A., Z. Johnson, C.D. Cox, A.I. Vasil and M.L. Vasil. 1996. Ferric uptake regulator mutants of *Pseudomonas aeruginosa* with distinct alterations in the iron-dependent repression of exotoxin A and siderophores in aerobic and microaerobic environments. Mol. Microbiol. *21*: 1001–1017.

Bartosch, S., I. Wolgast, E. Spieck and E. Bock. 1999. Identification of nitrite-oxidizing bacteria with monoclonal antibodies recognizing the nitrite oxidoreductase. Appl. Environ. Microbiol. *65*: 4126–4133.

Bartowsky, E.J., S.R. Attridge, C.J. Thomas, G. Mayrhofer and P.A. Manning. 1990. Role of the P plasmid in attenuation of *Vibrio cholerae* O1. Infect. Immun. *58*: 3129–3134.

Barua, D. and W.B.I. Greenough (Editors). 1992. Cholera, Plenum Publishing Co., New York and London. 372 pp.

Basalp, A., B. Cirakoglu and E. Bermek. 1995. A simple and accurate way for diagnosis of *Erwinia carotovora* subsp. *atroseptica*. Turk. J. Biol. *19*: 323–329.

Bascomb, S., S.P. Lapage, M.A. Curtis and W.R. Willcox. 1971. Numerical classification of the tribe *Klebsielleae*. J. Gen. Microbiol. *66*: 279–295.

Bascombe, S. and R.M. Jackson. 1965. *Rhizobium* culture collection.

Bashan, Y. and I. Assouline. 1983. Complementary bacterial enrichment techniques for the detection of *Pseudomonas syringae* pathovar *tomato* and *Xanthomonas campestris* pathovar *vesicatoria* in infested tomato and pepper seeds. Phytoparasitica. *11*: 187–194.

Bashan, Y., Y. Okon and Y. Henis. 1982. A note on a new defined medium for *Pseudomonas tomato*. J. Appl. Bacteriol. *52*: 297–298.

Baskerville, A., R.B. Fitzgeorge, M. Broster and P. Hambleton. 1983. Histopathology of experimental Legionnaires disease in guinea pigs, rhesus monkeys and marmosets. J. Pathol. *139*: 349–362.

Basnyat, S.R. and Y.S. Kulkarni. 1979. New bacterial leafspot of *Centella asiatica* L. Urban. Biovigyanam. *5*: 179–180.

Bast, E. 1977. Utilization of nitrogen compounds and ammonia assimilation by chromatiaceae. Arch. Microbiol. *113*: 91–94.

Bastian, I. and F.J. Bowden. 1996. Amplification of *Klebsiella*-like sequences from biopsy samples from patients with donovanosis. Clin. Infect. Dis. *23*: 1328–1330.

Bastian, S.N., I. Carle and F. Grimont. 1998. Comparison of 14 PCR systems for the detection and subtyping of *stx* genes in Shiga-toxin-producing *Escherichia coli*. Res. Microbiol. *149*: 457–472.

Basu, S., R.N. Tharanathan, T. Kontrohr and H. Mayer. 1985. Chemical structure of the lipid A component of *Plesiomonas shigelloides* and its taxonomical significance. FEMS Microbiol. Lett. *28*: 7–10.

Baturo, A.P. and V.P. Raginskaya. 1978. Antigenic schema for the hafniae. Int. J. Syst. Bacteriol. *28*: 126–127.

Baudry, B., A.T. Maurelli, P. Clerc, J.C. Sadoff and P.J. Sansonetti. 1987. Localization of plasmid loci necessary for the entry of *Shigella flexneri* into HeLa cells, and characterization of one locus encoding four immunogenic polypeptides. J. Gen. Microbiol. *133*: 3403–3413.

Baudry, B., S.J. Savarino, P. Vial, J.B. Kaper and M.M. Levine. 1990. A sensitive and specific DNA probe to identify enteroaggregative *Escherichia coli*, a recently discovered diarrheal pathogen. J. Infect. Dis. *161*: 1249–1251.

Bauer, A.W., W.M.M. Kirby, J.C. Sherris and M. Turck. 1966. Antibiotic susceptibility testing by a standardized single disk method. Am. J. Clin. Pathol. *45*: 493–496.

Bauer, C.A., G.D. Brayer, A.R. Sielecki and M.N.G. James. 1981. Active site of α-lytic protease. Eur. J. Biochem. *129*: 289–294.

Bauer, D.W., Z.M. Wei, S.V. Beer and A. Collmer. 1995. *Erwinia chrysanthemi* harpinEch: an elicitor of the hypersensitive response that contributes to soft-rot pathogenesis. Mol. Plant-Microbe Interact. *8*: 484–491.

Bauer, R.T. 1987. Stomatopod grooming behavior: functional morphology and amputation experiments in *Gonodactylus oerstedii*. J. Crustac. Biol. *7*: 414–432.

Bauernfeind, A. and C. Petermüller. 1984. Typing of *Enterobacter* spp. by bacteriocin susceptibility and its use in epidemiological analysis. J. Clin. Microbiol. *20*: 70–73.

Bauld, J., J.L. Favinger, M.T. Madigan and H. Gest. 1987. Obligately halophilic *Chromatium vinosum* from Hamelin Pool, Shark Bay, Australia. Curr. Microbiol. *14*: 335–340.

Baum, K.H., S.J. Shin, W.C. Rebhun and V.H. Patten. 1984. Isolation of *Actinobacillus lignieresii* from enlarged tongue of a horse. J. Am. Vet. Med. Assoc. *185*: 792–793.

Baumann, L., S.S. Bang and P. Baumann. 1980a. Study of relationship among species of *Vibrio*, *Photobacterium*, and terrestrial enterobacteria by an immunological comparison of glutamine synthetase and superoxide dismutase. Curr. Microbiol. *4*: 133–138.

Baumann, L. and P. Baumann. 1973. Enzymes of glucose catabolism in cell-free extracts of non-fermentative marine eubacteria. Can. J. Microbiol. *19*: 302–304.

Baumann, L. and P. Baumann. 1974. Regulation of aspartokinase activity in non-fermentative, marine eubacteria. Arch. Microbiol. *95*: 1–18.

Baumann, L. and P. Baumann. 1978. Studies of relationship among terrestrial *Pseudomonas*, *Alcaligenes*, and enterobacteria by an immunological comparison of glutamine synthetase. Arch. Microbiol. *119*: 25–30.

Baumann, L. and P. Baumann. 1994. Growth kinetics of the endosymbiont *Buchnera aphidicola* in the aphid *Schizaphis graminum*. Appl. Environ. Microbiol. *60*: 3440–3443.

Baumann, L. and P. Baumann. 1998. Characterization of *ftsZ*, the cell division gene of *Buchnera aphidicola* (endosymbiont of aphids) and detection of the product. Curr. Microbiol. *36*: 85–89.

Baumann, L., P. Baumann, M. Mandel and R.D. Allen. 1972. Taxonomy of aerobic marine eubacteria. J. Bacteriol. *110*: 402–429.

Baumann, L., P. Baumann, N.A. Moran, J. Sandström and M.L. Thao. 1999. Genetic characterization of plasmids conaining genes encoding enzymes of leucine biosynthesis in endosymbionts (*Buchnera*) of aphid species. J. Mol. Evol. *48*: 77–85.

Baumann, L., R.D. Bowditch and P. Baumann. 1983a. Description of *Deleya* gen. nov. created to accommodate the marine species *Alcaligenes aestus*, *Alcaligenes pacificus*, *Alcaligenes cupidus*, *Alcaligenes venustus*, and *Pseudomonas marina*. Int. J. Syst. Bacteriol. *33*: 793–802.

Baumann, L., M.A. Clark, D. Rouhbakhsh, P. Baumann, N.A. Moran and D.J. Voegtlin. 1997. Endosymbionts (*Buchnera*) of the aphid *Uroleucon*

sonchi contain plasmids with *trpEG* and remnants of *trpE* pseudogenes. Curr. Microbiol. *35*: 18–21.

Baumann, P. 1968. Isolation of *Acinetobacter* from soil and water. J. Bacteriol. *96*: 39–42.

Baumann, P., S.S. Bang and L. Baumann. 1978. Phenotypic characterization of *Beneckea anguillara* biotypes I and II. Curr. Microbiol. *1*: 85–88.

Baumann, P. and L. Baumann. 1977. Biology of the marine enterobacteria: genera *Beneckea* and *Photobacterium*. Annu. Rev. Microbiol. *31*: 39–61.

Baumann, P. and L. Baumann. 1981a. The marine Gram-negative eubacteria. *In* Starr, Stolp, Trüper, Balows and Schlegel (Editors), The Prokaryotes, a Handbook on Habitats, Isolation and Identification of Bacteria, 1st Ed., Springer-Verlag, New York. pp. 1352–1394.

Baumann, P. and L. Baumann. 1981b. The marine Gram-negative eubacteria: Genera *Photomicrobium*, *Benecka*, *Alteromonas*, *Pseudomonas*, and *Alcaligenes*. *In* Starr, Stolp, Trüper, Balows and Schlegel (Editors), The Prokaryotes. A Handbook on the Biology of Bacteria: Ecophysiology, Isolation, Identification, Applications, Vol. 1302-1331, Springer-Verlag, New York.

Baumann, P., L. Baumann, S.S. Bang and M.J. Woolkalis. 1980b. Re-evaluation of the taxonomy of *Vibrio*, *Beneckea*, and *Photobacterium*: abolition of the genus *Beneckea*. Curr. Microbiol. *4*: 127–132.

Baumann, P., L. Baumann, S.S. Bang and M.J. Woolkalis. 1981. *In* Validation of the publication of new names and combinations previously effectively published outside the IJSB. List No. 6. Int. J. Syst. Bacteriol. *31*: 215–218.

Baumann, P., L. Baumann, S.S. Bang and M.J. Woolkalis. 1982. *In* Validation of the publication of new names and combinations previously effectively published outside the IJSB. List No. 8. Int. J. Syst. Bacteriol. *32*: 266–268.

Baumann, P., L. Baumann, R.D. Bowditch and B. Beaman. 1984a. Taxonomy of *Alteromonas*: *Alteromonas nigrifaciens* sp. nov., nom. rev.; *Alteromonas macleodii* and *Alteromonas haloplanktis*. Int. J. Syst. Bacteriol. *34*: 145–149.

Baumann, P., L. Baumann, C.-Y. Lai, D. Rouhbakhsh, N.A. Moran and M.A. Clark. 1995. Genetics, physiology, and evolutionary relationships of the genus *Buchnera*: intracellular symbionts of aphids. Annu. Rev. Microbiol. *49*: 55–94.

Baumann, P., L. Baumann and M.A. Clark. 1996. Levels of *Buchnera aphidicola* chaperonin GroEL during growth of the aphid *Schizaphis graminum*. Curr. Microbiol. *32*: 279–285.

Baumann, P., L. Baumann, M.A. Clark and M.L. Thao. 1998. *Buchnera aphidicola*: the endosymbiont of aphids. ASM News. *64*: 203–209.

Baumann, P., L. Baumann and M. Mandel. 1971a. Taxonomy of marine bacteria: the genus *Beneckea*. J. Bacteriol. *107*: 268–294.

Baumann, P., L. Baumann, M. Mandel and R.D. Allen. 1971b. Taxonomy of marine bacteria: *Beneckea nigrapulchrituda* sp. n. J. Bacteriol. *108*: 1380–1383.

Baumann, P., R.D. Bowditch, L. Baumann and B. Beaman. 1983b. Taxonomy of marine *Pseudomonas* species *Pseudomonas stanieri* sp. nov., *Pseudomonas perfectomarina* sp. nov., nom. rev., *Pseudomonas nautica*, and *Pseudomonas doudoroffii*. Int. J. Syst. Bacteriol. *33*: 857–865.

Baumann, P., M. Doudoroff and R.Y. Stanier. 1968a. Study of the *Moraxella* group. I. Genus *Moraxella* and the *Neisseria catarrhalis* group. J. Bacteriol. *95*: 58–73.

Baumann, P., M. Doudoroff and R.Y. Stanier. 1968b. A study of the *Moraxella* group. II. Oxidative-negative species (genus *Acinetobacter*). J. Bacteriol. *95*: 1520–1541.

Baumann, P., A.L. Furniss and J.V. Lee. 1984b. Genus I. *Vibrio*. *In* Krieg and Holt (Editors), Bergey's Manual of Systematic Bacteriology, 1st Ed., Vol. 1, The Williams & Wilkins Co., Baltimore. pp. 518–538.

Baumann, P., A.L. Furniss and J.V. Lee. 1984c. *In* Validation of the publication of new names and new combinations previously effectively published outside the IJSB. List No. 15. Int. J. Syst. Bacteriol. *34*: 355–357.

Baumann, P., M.J. Gauthier and L. Baumann. 1984d. Genus *Alteromonas*. *In* Krieg and Holt (Editors), Bergey's Manual of Systematic Bacteriology, 1st Ed., Vol. 1, The Williams & Wilkins Co., Baltimore. pp. 343–352.

Baumann, P. and R.H.W. Schubert. 1984. Family II. *Vibronaceae. In* Krieg and Holt (Editors), Bergey's 's Manual of Systematic Bacteriology, 1st Ed., Vol. 1, The Williams & Wilkins Co., Baltimore. pp. 516–517.

Bauwens, M. and J. De Ley. 1981. Improvements in the taxonomy of *Flavobacterium* by DNA:rRNA hybridizations. *In* Reichenbach and Weeks (Editors), The *Flavobacterium–Cytophaga* Group, Verlag Chemie, Weinheim. 27–31.

Bavendamm, W. 1924. Die farblosen und roten Schwefelbakterien des Süss- und Salz-wassers. *In* Kolkwitz (Editor), Pflanzenforschung, Gustav Fischer Verlag, Jena. pp. 7–156.

Bayly, R.C., A. Duncan, M. Schembri, A. Smertjis, G. Vasiliadis and W.G.C. Raper. 1991. Microbiological and genetic aspects of the synthesis of polyphosphate by species of *Acinetobacter*. Water Sci. Technol. *23*: 747–754.

Baynes, I.D. and G.C. Simmons. 1960. Ovine epididymitis caused by *Actinobacillus seminis* n. sp. Aust. Vet. J. *36*: 454–459.

Baynes, I.D. and G.C. Simmons. 1968. Clinical and pathological studies of Border Leicester rams naturally infected with *Actinobacillus seminis*. Aust. Vet. J. *44*: 339–343.

Bazzi, C., E. Stefani and M. Zaccardelli. 1994a. SDS-PAGE: a tool to discriminate *Xylella fastidiosa* from other endophytic grapevine bacteria. Bull. OEPP *24*: 121–127.

Bazzi, C., M. Zaccardelli and F. Niepold. 1994b. Monospecific antiserum is suitable for the selective detection of *Xylella fastidiosa*. Microbiol. Res. *149*: 337–341.

Beane, N.H., P.M. Griffin, J.S. Goulding and C.B. Ivey. 1990. Foodborne disease outbreaks, 5-year summary, 1983–1987. CDC Surveillance Summaries. *39*: 15–57.

Bear, N., K.P. Klugman, L. Tobiansky and H.J. Koornhof. 1986. Wound colonization by *Ewingella americana*. J. Clin. Microbiol. *23*: 650–651.

Beard, C.B., S.L. O'Neill, P. Mason, L. Mandelco, C.R. Woese, R.B. Tesh, F.F. Richards and S. Aksoy. 1993. Genetic transformation and phylogeny of bacterial symbionts from tsetse. Insect Mol. Biol. *1*: 123–131.

Beattie, K.L. and J.K. Setlow. 1970. Transformation between *Haemophilus influenzae* and *Haemophilus parainfluenzae*. J. Bacteriol. *104*: 390–400.

Beaudry, M., C. Zhu, J.M. Fairbrother and J. Harel. 1996. Genotypic and phenotypic characterization of *Escherichia coli* isolates from dogs manifesting attaching and effacing lesions. J. Clin. Microbiol. *34*: 144–148.

Beaulieu, C., G.V. Minsavage, B.I. Canteros and R.E. Stall. 1991. Biochemical and genetic analysis of a pectate lyase gene from *Xanthomonas campestris* pathovar *vesicatoria*. Mol. Plant-Microbe Interact. *4*: 446–451.

Beck, M., J.F. van den Bosch, I.M. Jongenelen, P.L. Loeffen, R. Nielsen, J. Nicolet and J. Frey. 1994. RTX toxin genotypes and phenotypes in *Actinobacillus pleuropneumoniae* field strains. J. Clin. Microbiol. *32*: 2749–2754.

Beck von Bodman, S. and S.K. Farrand. 1995. Capsular polysaccharide biosynthesis and pathogenicity in *Erwinia stewartii* require induction by an *N*-acylhomoserine lactone autoinducer. J. Bacteriol. *177*: 5000–5008.

Becker, A.H. 1962. Infections due to *Proteus mirabilis* in newborn nursery. Am. J. Dis. Child. *104*: 355–359.

Becking, J.H. 1984. Genus *Beijerinckia. In* Kreig and Holt (Editors), Bergey's Manual of Systematic Bacteriology, 1st Ed., Vol. 1, The Williams &Wilkins Co., Baltimore. pp. 311–321.

Becking, J.H. 1992. The family *Azotobacteriaceae. In* Balows, Trüper, Dworkin, Harder and Schleifer (Editors), The Prokaryotes, 2nd ed., Vol. 3, Springer-Verlag, New York. 3144–3170.

Beckman, W. and T.G. Lessie. 1979. Response of *Pseudomonas cepacia* to beta-lactam antibiotics - utilization of penicillin-G as the carbon source. J. Bacteriol. *140*: 1126–1128.

Beebakhee, G., M. Louie, J. De Azavedo and J. Brunton. 1992. Cloning and nucleotide sequence of the *eae* gene homologue from entero-

hemorrhagic *Escherichia coli* serotype O157 H7. FEMS Microbiol. Lett. *91*: 63–68.

Begaud, E., P. Jourand, M. Morillon, D. Mondet and Y. Germani. 1993. Detection of diarrhoeagenic *Escherichia coli* in children less than ten years old with and without diarrhea in New Caledonia using seven acetylaminofluorene-labeled DNA probes. Am. J. Trop. Med. Hyg. *48*: 26–34.

Behki, R.M. and S.M. Lesley. 1972. Deoxyribonucleic acid degradation and the lethal effect by myxin in *Escherichia coli*. J. Bacteriol. *109*: 250–261.

Behrendt, U., A. Ulrich, P. Schumann, W. Erler, J. Burghardt and W. Seyfarth. 1999. A taxonomic study of bacteria isolated from grasses: a proposed new species *Pseudomonas graminis* sp. nov. Int. J. Syst. Bacteriol. *49*: 297–308.

Beijerinck, M.W. 1888. Cultur des *Bacillus radicola* aus den Knöllchen. Bot. Ztg. *46*: 740–750.

Beijerinck, M.W. 1889. Le *Photobacterium luminosum*. Bacterie luminosum de la Mer Nord. Arch. Neer. Sci. *23*: 401–427.

Beijerinck, M.W. 1900a. On different forms of hereditary variation in microbes. Proc. Acad. Sci. Amst. *3*: 352-365.

Beijerinck, M.W. 1900b. Schwefelwasser Stoffbildung in den Stadtgraben und Aufstellung der Gattung *Aerobacter*. Zentral. Bakteriol. Abt. 2. *6*: 193–206.

Beijerinck, M.W. 1901. Über oligonitrophile Mikroben. Zentralbl. Bakteriol. Parasitenkd. Infektionskr. Hyg. Abt. II. *7*: 561–582.

Beijerinck, M.W. 1911. Über pigmentbildung bei essigbakterien. Proc. K. Ned. Akad. Wet. *13*: 1066–1077.

Bein, S.J. 1954. A study of certain chromogenic bacteria isolated from "Red Tide" water with a description of a new species. Bull. Mar. Sci. Gulf Caribb. *4*: 110–119.

Beji, A., J. Mergaert, F. Gavini, D. Izard, K. Kersters, H. Leclerc and J. De Ley. 1988. Subjective synonymy of *Erwinia herbicola*, *Erwinia milletiae*, and *Enterobacter agglomerans* and redefinition of the taxon by genotypic and phenotypic data. Int. J. Syst. Bacteriol. *38*: 77–88.

Belas, R. 1989. Sequence analysis of the *agrA* gene encoding beta-agarase from *Pseudomonas atlantica*. J. Bacteriol. *171*: 602–605.

Belas, R. 1992. The swarming phenomenon of *Proteus mirabilis*. ASM News. *58*: 15–22.

Beljanski, M. 1955. Isolement des mutants d'*Escherichia coli* streptomycin-résistants dépourvus d'enzymes respiratoires. Action de l'hémine sur la formation de ces enzymes chez le mutant H7. CR Hebd. Acad. Sci. Paris *240*: 374–377.

Bell, B.P., M. Goldoft, P.M. Griffin, M.A. Davis, D.C. Gordon, P.I. Tarr, C.A. Bartleson, J.H. Lewis, T.J. Barrett, J.G. Wells, R. Baron and J. Kobayashi. 1994. A multistate outbreak of *Escherichia coli* O157:H7-associated bloody diarrhea and hemolytic uremic syndrome from hamburgers. JAMA (J. Am. Med. Assoc.). *272*: 1349–1353.

Bell, S.M. and D. Plowman. 1980. Mechanisms of ampicillin resistance in *Haemophilus influenzae* from respiratory tract. Lancet *8163*: 279–280.

Belser, W.L. and M.I. Bunting. 1956. Studies on a mechanism providing for genetic transfer in *Serratia marcescens*. J. Bacteriol. *72*: 582–592.

Beltrametti, F., A.M. Marconi, G. Bestetti, C. Colombo, E. Galli, M. Ruzzi and E. Zennaro. 1997. Sequencing and functional analysis of styrene catabolism genes from *Pseudomonas fluorescens* ST. Appl. Environ. Microbiol. *63*: 2232–2239.

Belyavin, G. 1951. Cultural and serological phases of *Proteus vulgaris*. J. Gen. Microbiol. *5*: 197–207.

Ben-Gurion, R. and I. Hertman. 1958. Bacteriocin-like material produced by *Pasteurella pestis*. J. Gen. Microbiol. *19*: 289–297.

Ben-Gurion, R. and A. Shafferman. 1981. Essential virulence determinants of different *Yersinia* species are carried on a common plasmid. Plasmid *5*: 183–187.

Benaoudia, F., F. Escande and M. Simonet. 1994. Infection due to *Actinobacillus lignieresii* after a horse bite. Eur. J. Clin. Microbiol. Infect. Dis. *13*: 439–440.

Bender, C.L., D.K. Malvick, K.E. Conway, S. George and P. Pratt. 1990.

Characterization of pXV10A, a copper resistance plasmid in *Xanthomonas campestris* pv. *vesicatoria*. Appl. Environ. Microbiol. *56*: 170–175.

Benedict, A.A., A.M. Alvarez, J. Berestecky, W. Imanaka, C.Y. Mizumoto, L.W. Pollard, T.W. Mew and C.F. Gonzalez. 1989. Pathovar-specific monoclonal antibodies for *Xanthomonas campestris* pv. *oryzae* and for *Xanthomonas campestris* pathovar *oryzicola*. Phytopathology *79*: 322–328.

Benedict, A.A., A.M. Alvarez, E.L. Civerolo and C.Y. Mizumoto. 1985. Delineation of *Xanthomonas campestris* pv. *citri* strains with monoclonal antibodies. Phytopathology *75*: 1352.

Benedict, A.A., A.M. Alvarez and L.W. Pollard. 1990. Pathovar-specific antigens of *Xanthomonas campestris* pv. *begoniae* and *Xanthomonas campestris* pathovar *pelargonii* detected with monoclonal antibodies. Appl. Environ. Microbiol. *56*: 572–574.

Benediktsdøttir, E., L. Verdonck, C. Sproer, S. Helgasön and J. Swings. 2000. Characterization of *Vibrio viscosus* and *Vibrio wodanis* isolated at different geographical locations: a proposal for reclassification of *Vibrio viscosus* as *Moritella viscosa* comb. nov. Int. J. Syst. Evol. Microbiol. *50*: 479–488.

Benenson, A.S., M.R. Islam and W.B.I. Greenough. 1964. Rapid identification of *Vibrio cholerae* by darkfield microscopy. Bull. Wld.Hlth. Org. *30*: 827–831.

Bennasar, A., R. Rosselló-Mora, J. Lalucat and E.R.B. Moore. 1996. 16S rRNA gene sequence analysis relative to genomovars of *Pseudomonas stutzeri* and proposal of *Pseudomonas balearica* sp. nov. Int. J. Syst. Bacteriol. *46*: 200–205.

Bennekov, T., H. Colding, B. Ojeniyi, M.W. Bentzon and N. Hoiby. 1996. Comparison of ribotyping and genome fingerprinting of *Psudeomonas aeruginosa* isolates from cystic fibrosis patients. J. Clin. Microbiol. *34*: 202–204.

Bennish, M.L., M.A. Salam, M.A. Hossain, J. Myaux, E.H. Khan, J. Chakraborty, F. Henry and C. Ronsmans. 1992. Antimicrobial resistance of *Shigella* isolates in Bangladesh, 1983–1990: increasing frequency of strains multiply resistant to ampicillin, trimethoprim–sulfamethoxazole, and nalidixic acid. Clin. Infect. Dis. *14*: 1055–1060.

Benson, R.F., W.L. Thacker, M.I. Daneshvar and D.J. Brenner. 1996. *Legionella waltersii* sp. nov. and an unnamed *Legionella* genomospecies isolated from water in Australia. Int. J. Syst. Bacteriol. *46*: 631–634.

Benson, R.F., W.L. Thacker, F.C. Fang, B. Kanter, W.R. Mayberry and D.J. Brenner. 1990a. *Legionella sainthelensi* serogroup 2 isolated from patients with pneumonia. Res. Microbiol. *141*: 453–464.

Benson, R.F., W.L. Thacker, J.A. Lanser, N. Sangster, W.R. Mayberry and D.J. Brenner. 1991a. *Legionella adelaidensis*, a new species isolated from cooling tower water. J. Clin. Microbiol. *29*: 1004–1006.

Benson, R.F., W.L. Thacker, J.A. Lanser, N. Sangster, W.R. Mayberry and D.J. Brenner. 1991b. *In* Validation of the publication of new names and new combinations previously effectively published outside the IJSB. List No. 39. Int. J. Syst. Bacteriol. *41*: 580–581.

Benson, R.F., W.L. Thacker, B.B. Plikaytis and H.W. Wilkinson. 1987. Cross-reactions in *Legionella* antisera with *Bordetella pertussis* strains. J. Clin. Microbiol. *25*: 594–596.

Benson, R.F., W.L. Thacker, R.P. Waters, P.A. Quinlivan, W.R. Mayberry, D.J. Brenner and H.W. Wilkinson. 1989. *Legionella quinlivanii*, sp. nov., isolated from water. Curr. Microbiol. *18*: 195–197.

Benson, R.F., W.L. Thacker, R.P. Waters, P.A. Quinlivan, W.R. Mayberry, D.J. Brenner and H.W. Wilkinson. 1990b. *In* Validation of the publication of new names and new combinations previously effectively published outside the IJSB. List No. 32. Int. J. Syst. Bacteriol. *40*: 105–106.

Benson, R.F., W.L. Thacker, H.W. Wilkinson, R.J. Fallon and D.J. Brenner. 1988. *Legionella pneumophila* serogroup 14 isolated from patients with fatal pneumonia. J. Clin. Microbiol. *26*: 382.

Benz, I. and M.A. Schmidt. 1989. Cloning and expression of an adhesin (AIDA-I) involved in diffuse adherence of enteropathogenic *Escherichia coli*. Infect. Immun. *57*: 1506–1511.

Berbari, E.F., F.R. Cockerill, III and J.M. Steckelberg. 1997. Infective endocarditis due to unusual or fastidious microorganisms. Mayo Clin. Proc. *72*: 532–542.

Bercovier, H., J.M. Alonso, Z. Bentaiba, J. Brault and H.H. Mollaret. 1979. Contribution to the definition and taxonomy of *Yersinia enterocolitica*. Contr. Microbiol. Immunol. *5*: 12–22.

Bercovier, H., J. Brault, N. Barre, M. Treignier, J.M. Alonso and H.H. Mollaret. 1978. Biochemical, serological and phage typing characteristics of 459 *Yersinia* strains isolated from a terrestrial ecosystem. Curr. Microbiol. *1*: 353–357.

Bercovier, H., D.J. Brenner, J. Ursing, A.G. Steigerwalt, G.R. Fanning, J.M. Alonso, G.P. Carter and H.H. Mollaret. 1980a. Characterization of *Yersinia enterocolitica* sensu stricto. Curr. Microbiol. *4*: 201–206.

Bercovier, H. and H.H. Mollaret. 1984. Genus *Yersinia*. *In* Krieg and Holt (Editors), Bergey's Manual of Systematic Bacteriology, 1 Ed., Vol. 1, The Williams & Wilkins Co., Baltimore. pp. 503–506.

Bercovier, H., H.H. Mollaret, J.M. Alonso, J. Brault, G.R. Fanning, A.G. Steigerwalt and D.J. Brenner. 1980b. Intra- and interspecies relatedness of *Yersinia pestis* by deoxyribonucleic acid hybridization and its relationship to *Yersinia pseudotuberculosis*. Curr. Microbiol. *4*: 225–230.

Bercovier, H., H.H. Mollaret, J.M. Alonso, J. Brault, G.R. Fanning, A.G. Steigerwalt and D.J. Brenner. 1981a. *In* Validation of the publication of new names and new combinations previously effectively published outside the IJSB. List No. 7. Int. J. Syst. Bacteriol. *31*: 382–383.

Bercovier, H., A.G. Steigerwalt, M. Derhi Cochin, C.W. Moss, H.W. Wilkinson, R.F. Benson and D.J. Brenner. 1986. Isolation of Legionellae from oxidation ponds and fishponds in Israel and description of *Legionella israelensis*, sp. nov. Int. J. Syst. Bacteriol. *36*: 368–371.

Bercovier, H., A.G. Steigerwalt, A. Guiyoule, G. Huntley-Carter and D.J. Brenner. 1984. *Yersinia aldovae* (formerly *Yersinia enterocolitica* like Group X2 a new species of *Enterobacteriaceae* isolated from aquatic ecosystems. Int. J. Syst. Bacteriol. *34*: 166–172.

Bercovier, H., J. Ursing, D.J. Brenner, A.G. Steigerwalt, G.R. Fanning, G.P. Carter and H.H. Mollaret. 1980c. *Yersinia kristensenfi*: a newspecies of *Enterobacteriaceae* composed of sucrose-negative strains (formerly called atypical *Yersinia enterocolitica* or *Yersinia enterocolitica*-like). Curr. Microbiol. *4*: 219–224.

Bercovier, H., J. Ursing, D.J. Brenner, A.G. Steigerwalt, G.R. Fanning, G.P. Carter and H.H. Mollaret. 1981b. *In* Validation of the publication of new names and new combinations previously effectively published outside the IJSB. List No. 6. Int. J. Syst. Bacteriol. *31*: 215–218.

Berdal, B.P. and E. Söderlund. 1977. Cultivation and isolation of *Francisella tularensis* on selective chocolate agar, as used routinely for the isolation of gonococci. Acta Pathol Microbiol. Scand. Sect. B. *85*: 108–109.

Berendes, F. 1997. Validation of the publication of new names and new combinations previously effectively published outside the IJSB. List No. 60. Int. J. Syst. Bacteriol. *47*: 242.

Berendes, F., G. Gottschalk, E. Heine-Dobbernack, E.R.B. Moore and B.J. Tindall. 1996. *Halomonas desiderata* sp. nov., a new alkaliphilic, halotolerant and denitrifying bacterium isloted from a municipal sewage works. System. Appl. Microbiol. *19*: 158–167.

Berenstein, D. 1982. Weigle reactivation in *Acinetobacter calcoaceticus*. Photochem. Photobiol. *35*: 579–582.

Berenstein, D. 1986. Prophage induction by UV light in *Acinetobacter calcoaceticus*. J. Gen. Microbiol. *132*: 2633–2636.

Bereswill, S., P. Bugert, I. Bruchmuller and K. Geider. 1995. Identification of the fire blight pathogen, *Erwinia amylovora*, by PCR assays with chromosomal DNA. Appl. Environ. Microbiol. *61*: 2636–2642.

Bereswill, S. and K. Geider. 1997. Characterization of the *rcsB* gene from *Erwinia amylovora* and its influence on exoploysaccharide synthesis and virulence of the fire blight pathogen. J. Bacteriol. *179*: 1354–1361.

Bereswill, S., A. Pahl, P. Bellemann, W. Zeller and K. Geider. 1992. Sensitive and species-specific detection of *Erwinia amylovora* by polymerase chain reaction analysis. Appl. Environ. Microbiol. *58*: 3522–3526.

Beretta, M.J.G., G.A. Barthe, T.L. Ceccardi, R.F. Lee and K.S. Derrick. 1997. A survey of strains of *Xylella fastidiosa* in citrus affected by citrus variegated chlorosis and citrus blight in Brazil. Plant Dis. *81*: 1196–1198.

Berg, G., P. Marten and G. Ballin. 1996. *Stenotrophomonas maltophilia* in the rhizosphere of oilseed rape - occurrence, characterization and interaction with phytopathogenic fungi. Microbiol. Res. *151*: 19–27.

Bergan, T. 1978. Phage typing of *Proteus. In* Bergan and Norris (Editors), Methods in Microbiology, Vol. 11, Academic Press, London. pp. 243–258.

Bergan, T. 1979. Bacteriophage typing of *Shigella. In* Bergan and Norris (Editors), Methods in Microbiology, Vol. 13, Academic Press, New York.

Bergan, T. 1981. Human- and animal-pathogenic members of the genus *Pseudomonas. In* Starr, Stolp, Trüper, Balows and Schlegel (Editors), The Prokaryotes, a handbook on habitats, isolation and identification of bacteria, Springer-Verlag, Berlin. 666–700.

Bergan, T., P.A.D. Grimont and F. Grimont. 1983. Fatty acids of *Serratia* determined by gas chromatography. Curr. Microbiol. *8*: 7–11.

Bergan, T. and A.K. Vaksvik. 1983. Taxonomic implications of quantitative transformation in *Acinetobacter calcoaceticus.* Zentbl. Bakteriol. Mikrobiol. Hyg. Ser. A *254*: 214–228.

Berge, O., T. Heulin, W. Achouak, C. Richard, R. Bally and J. Balandreau. 1991. *Rahnella aquatilis*, a nitrogen-fixing enteric bacterium associated with the rhizosphere of wheat and maize. Can. J. Microbiol. *37*: 195–203.

Berger, S.A., S.C. Edberg and R.S. Klein. 1977. *Enterobacter hafniae* infection: report of two cases and review of the literature. Am. J. Med. Sci. *273*: 101–104.

Berger, U. 1962. Über das vorkommen von neisserien bei einigen tieren. Z. Hyg. Infektionskr. *148*: 445–457.

Bergey, D.H., F.C. Harrison, R.S. Breed, B.W. Hammer and F.M. Huntoon. 1923. Bergey's Manual of Determinative Bacteriology, 1st Ed., The Williams & Wilkins Co., Baltimore. pp. 1–442.

Bergey, D.H., F.C. Harrison, R.S. Breed, B.W. Hammer and F.M. Huntoon. 1925. Bergey's Manual of Determinative Bacteriology, 2nd ed., The Williams & Wilkins Co., Baltimore. 1–462.

Bergey, D.H., F.C. Harrison, R.S. Breed, B.W. Hammer and F.M. Huntoon. 1930. Bergey's Manual of Determinative Bacteriology, 3rd edition, The Williams & Wilkins Co, Baltimore. 1–589.

Bergman, A.M. 1909. Die rote Beulenkrankheit des Aals. Ber. Bayer. Biol. Vers. Sta. *2*: 10–54.

Bergogne-Bérézin, E., M.L. Joly-Guillou and K.J. Towner (Editors). 1996. *Acinetobacter*: Microbiology, Epidemiology, Infections, Management, CRC Press, Boca Raton.

Bergogne-Bérézin, E. and K.J. Towner. 1996. *Acinetobacter* spp. as nosocomial pathogens: microbiological, clinical, and epidemiological features. Clin. Microbiol. Rev. *9*: 148–165.

Bergthorsson, U. and H. Ochman. 1995. Heterogeneity of genome sizes among natural isolates of *Escherichia coli.* J. Bacteriol. *177*: 5784–5789.

Berkowitz, D.M. and W.S. Lee. 1973. A selective medium for isolation and identification of *Serratia marcescens.* Abstr. Annu. Mtg. Amer. Soc. Microbiol, p. 105.

Berkowitz, F.E. and B. Metchock. 1995. Third generation cephalosporin-resistant gram-negative bacilli in the feces of hospitalized children. Pediatr. Infect. Dis. J. *14*: 97–100.

Bernagozzi, M., F. Bianucci, E. Scerre and R. Sacchetti. 1994. Assessment of some selective media for the recovery of *Aeromonas hydrophila* from surface waters. Zentbl. Hyg. Umweltmed. *195*: 121–134.

Bernard, C. and T. Fenchel. 1995. Mats of colourless sulphur bacteria. II. Structure, composition of biota and successional patterns. Mar. Ecol. Prog. Ser. *128*: 171–179.

Bernard, K., S. Tessier, J. Winstanley, D. Chang and A. Borczyk. 1994. Early recognition of atypical *Francisella tularensis* strains lacking a cysteine requirement. J. Clin. Microbiol. *32*: 551–553.

Bernardini, M.L., J. Mounier, H. d'Hauteville, M. Coquis-Rondon and P.J. Sansonetti. 1989. Identification of *icsA*, a plasmid locus of *Shigella flexneri* that governs bacterial intra- and intercellular spread through interaction with F-actin. Proc. Natl. Acad. Sci. U.S.A. *86*: 3867–3871.

Bernards, A.T., L. Dijkshoorn, J. Van Der Toorn, B.R. Bochner and C.P.A. Van Boven. 1995. Phenotypic characterisation of *Acinetobacter* strains of 13 DNA–DNA hybridisation groups by means of the Biolog system. J. Med. Microbiol. *42*: 113–119.

Bernards, A.T., J. Van Der Toorn, C.P.A. Van Boven and L. Dijkshoorn. 1996. Evaluation of the ability of a commercial system to identify *Acinetobacter* genomic species. Eur. J. Clin. Microbiol. Infect. Dis. *15*: 303–308.

Berner, I., S. Konetschny-Rapp, G. Jung and G. Winkelmann. 1988. Characterization of ferrioxamine E as the principal siderophore of *Erwinia herbicola* (*Enterobacter agglomerans*). Biol. Met. *1*: 51–56.

Bernheim, F., M.L.C. Bernheim and M.D. Webster. 1935. Oxidation of certain amino acids by "resting" *Bacillus proteus.* J. Biol. Chem. *110*: 165–172.

Berniac, M. 1974. Une maladie bactérienne de *Xanthosoma sagittifolium* (L.) Schott. Ann. Phytopathol. *6*: 197–202.

Bernoth, E.M. and G. Artzt. 1989. Atypical *Aeromonas salmonicida* in fish tissue may be overlooked by sole reliance on furunculosis agar. Bull. Eur. Assoc. Fish Pathol. *9*: 5–6.

Berridge, E.M. 1924. The influence of hydrogen-ion concentration on the growth of certain bacterial plant parasites and saprophytes. Ann. Appl. Biol. *11*: 73–85.

Bersa, E. 1920. Über das Vorkommen von kohlensaurem Kalk in einer Gruppe von Schwefelbakterien. Sitzungsber. Saechs Akad. Wiss. Leipzig Math-Naturwiss. Kl. Abt. I. *129*: 231–259.

Bertheau, Y., E. Madgidi-Hervan, A. Kotoujansky, C. Nguyen-The, T. Andro and A. Coleno. 1984. Detection of depolymerase isoenzymes after electrophoresis or electrofocusing, or in titration curves. Anal. Biochem. *139*: 383–389.

Berthier, Y., D. Thierry, M. Lemattre and J.L. Guesdon. 1994. Isolation of an insertion sequence (IS1051) from *Xanthomonas campestris* pathovar *dieffenbachiae* with potential use for strain identification and characterization. Appl. Environ. Microbiol. *60*: 377–384.

Bertin, Y., K. Boukhors, N. Pradel, V. Livrelli and C. Martin. 2001. Stx2 subtyping of Shiga toxin-producing *Escherichia coli* isolated from cattle in France: detection of a new Stx2 subtype and correlation with additional virulence factors. J. Clin. Microbiol. *39*: 3060–3065.

Bertin, Y., C. Martin, J.P. Girardeau, P. Pohl and M. Contrepois. 1998. Association of genes encoding P fimbriae, CS31A antigen and EAST 1 toxin among CNF1-producing *Escherichia coli* strains from cattle with scpticemia and diarrhea. FEMS Microbiol. Lett. *162*: 235–239.

Bertone, S., M. Giacomini, C. Ruggiero, C. Piccarolo and L. Calegari. 1996. Automated systems for identification of heterotrophic marine bacteria on the basis of their fatty acid composition. Appl. Environ. Microbiol. *62*: 2122–2132.

Bertoni, G., F. Bolognese, E. Galli and P. Barbieri. 1996. Cloning of the genes for and characterization of the early stages of toluene and *o*-xylene catabolism in *Pseudomonas stutzeri* OX1. Appl. Environ. Microbiol. *62*: 3704–3711.

Betz, J.L., P.R. Brown, M.J. Smyth and P.H. Clarke. 1974. Evolution in action. Nature (Lond.) *247*: 261–264.

Betz, J.L. and P.H. Clarke. 1972. Selective evolution of phenylacetamide-utilizing strains of *Pseudomonas aeruginosa.* J. Gen. Microbiol. *73*: 161–174.

Beutin, L. 1991. The different hemolysins of *Escherichia coli.* Med. Microbiol. Immunol. (Berl). *180*: 167–182.

Beutin, L., D. Geier, H. Steinrück, S. Zimmermann and F. Scheutz. 1993a. Prevalence and some properties of verotoxin (Shiga-like toxin) producing *Escherichia coli* in seven different species of healthy domestic animals. J. Clin. Microbiol. *31*: 2483–2488.

Beutin, L., M.A. Montenegro, I. Ørskov, F. Ørskov, J. Prada, S. Zimmermann and R. Stephan. 1989. Close association of verotoxin (Shiga-like toxin) production with enterohemolysin production in strains of *Escherichia coli.* J. Clin. Microbiol. *27*: 2559–2564.

Beutin, L., J. Prada, S. Zimmermann, R. Stephan, I. Ørskov and F. Ørskov. 1988. Enterohemolysin, a new type of hemolysin produced by some strains of enteropathogenic *Escherichia coli* EPEC. Zentbl. Bakteriol. Mikrobiol. Hyg. Ser. A *267*: 576–588.

Beutin, L., U.H. Stroeher and P.A. Manning. 1993b. Isolation of enterohemolysin (Ehly2)-associated sequences encoded on temperate phages of *Escherichia coli.* Gene *132*: 95–99.

Beveridge, T.J., S.A. Makin, J.L. Kadurugamuwa and Z.S. Li. 1997. Interactions between biofilms and the environment. FEMS Microbiol. Rev. *20*: 291–303.

Beveridge, W.I.B. 1941. Foot-rot in sheep: a transmissible disease due to infection with *Fusiformis nodosus* (n. sp.). Studies on its causes, epidemiology and control. Counc. Sci. Indust. Res. Aust. Bull. *140*: 1–56.

Bezrukova, L.V., N. Y.I., A.I. Nesterov, V.F. Gal'chenko and M.V. Ivanov. 1983. Comparative serological analysis of methanotrophic bacteria. Mikrobiologiya *52*: 800–805.

Bhat, P., R.M. Meyers and R.A. Feldman. 1971a. Providence group of organisms in the aetiology of juvenile diarrhoea. Indian J. Med. Res. *59*: 1010–1018.

Bhat, P., R.M. Myers and K.P. Carpenter. 1967. *Edwardsiella tarda* in a study of juvenile diarrhoea. J. Hyg. *65*: 293–298.

Bhat, P., S. Shanthakumari and R.M. Meyers. 1971b. The Providence group: subgroups including biotypes and serogroups of faecal strains isolated in Vellore. Indian J. Med. Res. *59*: 1184–1189.

Bibb, W.F., P.M. Arnow, D.L. Dellinger and S.R. Perryman. 1983. Isolation and characterization of a 7th serogroup of *Legionella pneumophila*. J. Clin. Microbiol. *17*: 346–348.

Bibb, W.F., R.J. Sorg, B.M. Thomason, M. Hicklin, A.G. Steigerwalt, D.J. Brenner and M.R. Wulf. 1981. Recognition of a second serogroup of *Legionella longbeachae*. J. Clin. Microbiol. *14*: 674–677.

Biberstein, E.L. and M.G. Gillis. 1962. The relationship of the antigenic types to the A and T types of *Pasteurella haemolytica*. J. Comp. Pathol. *72*: 316–320.

Biberstein, E.L., A. Gunnarsson and B. Hurvell. 1977. Cultural and biochemical criteria for the identification of *Haemophilus* cultures from swine. Am. J. Vet. Res. *38*: 7–11.

Biberstein, E.L., S.S. Jang, P.H. Kass and D.C. Hirsh. 1991. Distribution of indole-producing urease-negative pasteurellas in animals. J. Vet. Diagn. Invest. *3*: 319–323.

Biberstein, E.L., P.D. Mini and M.G. Gills. 1963. Action of *Haemophilus* cultures on α-aminolevulinic acid. J. Bacteriol. *86*: 814–819.

Biberstein, E.L. and P.D. Spencer. 1962. Oxidative metabolism of *Haemophilus* species grown at different levels of hemin supplementation. J. Bacteriol. *84*: 916–920.

Biberstein, E.L. and D.C. White. 1969. A proposal for the establishment of two new *Haemophilus* species. J. Med. Microbiol. *2*: 75–78.

Biebl, H. and N. Pfennig. 1978. Growth yields of green sulfur bacteria in mixed cultures with sulfur and sulfate reducing bacteria. Arch. Microbiol. *117*: 9–16.

Biedenbach, D.J., R.N. Jones and M.E. Erwin. 1993. Interpretive accuracy of the disk diffusion method for testing newer orally administered cephalosporins against *Morganella morganii*. J. Clin. Microbiol. *31*: 2828–2830.

Bielaszewská, M., L. Srámková, J. Janda, K. Bláhová and H. Ambrozová. 1990. Verotoxigenic (enterohaemorrhagic) *Escherichia coli* in infants and toddlers in Czechoslovakia. Infection *18*: 352–356.

Biggins, J. and W.E. Dietrich, Jr. 1968. Respiratory mechanisms in the *Flexibacteriaceae*. I. Studies on the terminal oxidase system of *Leucothrix mucor*. Arch. Biochem. Biophys. *128*: 40–50.

Bik, E.M., A.E. Bunschoten, R.D. Gouw and F.R. Mooi. 1995. Genesis of the novel epidemic *Vibrio cholerae* O139 strain: evidence for horizontal transfer of genes involved in polysaccharide synthesis. Embo J. *14*: 209–216.

Bilge, S.S., C.R. Clausen, W. Lau and S.L. Moseley. 1989. Molecular characterization of a fimbrial adhesin F1845 mediating diffuse adherence of diarrhea-associated *Escherichia coli* to HEp-2 cells. J. Bacteriol. *171*: 4281–4289.

Billiet, J., S. Kuypers, S. Van Lierde and J. Verhaegen. 1989. *Plesiomonas shigelloides* meningitis and septicaemia in a neonate: report of a case and review of the literature. J. Infect. *19*: 267–271.

Billing, E. 1963. The value of phage sensitivity tests for the identification of phytopathogenic *Pseudomonas* species. J. Appl. Bacteriol. *26*: 193–210.

Billing, E. 1970a. Further studies on the phage sensitivity and the deter-

mination of phytopathogenic *Pseudomonas* species. J. Appl. Bacteriol. *33*: 478–491.

Billing, E. 1970b. *Pseudomonas viridiflavca* (Burkholder 1930; Clara 1934). J. Appl. Bacteriol. *33*: 492–500.

Billing, E. and C.M.E. Garrett. 1980. Phages in the identification of plant pathogenic bacteria. *In* Goodfellow and Board (Editors), Microbial Classification and Identification. Symposium No. 8, Academic Press, London. pp. 319–338.

Billington, S.J., A.S. Huggins, P.A. Johanesen, P.K. Crellin, J.K. Cheung, M.E. Katz, C.L. Wright, V. Haring and J.I. Rood. 1999. Complete nucleotide sequence of the 27-kilobase virulence related locus (*vrl*) of *Dichelobacter nodosus*: evidence for extrachromosomal origin. Infect. Immun. *67*: 1277–1286.

Billington, S.J., J.L. Johnston and J.I. Rood. 1996a. Virulence regions and virulence factors of the ovine footrot pathogen, *Dichelobacter nodosus*. FEMS Microbiol. Lett. *145*: 147–156.

Billington, S.J., B.H. Jost and J.I. Rood. 1995. A gene region in *Dichelobacter nodosus* encoding a lipopolysaccharide epitope. Microbiology *141*: 945–957.

Billington, S.J., M. Sinistaj, B.F. Cheetham, A. Ayres, E.K. Moses, M.E. Katz and J.I. Rood. 1996b. Identification of a native *Dichelobacter nodosus* plasmid and implications for the evolution of the *vap* regions. Gene *172*: 111–116.

Bingen, E., C. Boissinot, P. Desjardins, H. Cave, N. Brahimi, N. Lambert-Zechovsky, E. Denamur, P. Blot and J. Elion. 1993. Arbitrarily primed polymerase chain reaction provides rapid differentiation of *Proteus mirabilis* isolates from a pediatric hospital. J. Clin. Microbiol. *31*: 1055–1059.

Bingham, D.P., R. Moore and A.B. Richards. 1990. Comparison of DNA:DNA homology and enzymatic activity between *Pasteurella haemolytica* and related species. Am. J. Vet. Res. *51*: 1161–1166.

Bingle, W.H., P.W. Whippey, J.L. Doran, R.G.E. Murray and W.J. Page. 1987. Structure of the *Azotobacter vinelandii* surface layer. J. Bacteriol. *169*: 802–810.

Binnington, K.C. and L. Brooks. 1993. Fimbrial attachment of *Xenorhabdus nematophilus* to the intestine of *Steinernema carpocapsae*. *In* Bedding, Akhurst and Kaya (Editors), Nematodes and the Biological Control of Insect Pests, CSIRO Publications, Melbourne. pp. 147–155.

Binns, M.M., S. Vaughan, S.C. Sanyal and K.N. Timmis. 1984. Invasive ability of *Plesiomonas shigelloides*. Zentbl. Bakteriol. Mikrobiol. Hyg. Ser. A *257*: 343–347.

Bird, A.F. and R.J. Akhurst. 1983. The nature of the intestinal vesicle in nematodes of the family *Steinernematidae*. International Journal for Parasitology. *13*: 599–606.

Bird, C.W., C.M. Lynch, F.J. Pirr, W.W. Reid, C.J.W. Brooks and B.C. Middleditch. 1971. Steroids and squalene in *Methylococcus capsulatus* grown on methane. Nature *230*: 473–474.

Birtles, R.J., N. Doshi, N.A. Saunders and T.G. Harrison. 1991. Second serogroup of *Legionella quinlivanii* isolated from two unrelated sources in the United Kingdom. J. Appl. Bacteriol. *71*: 402–406.

Birtles, R.J., T.J. Rowbotham, D. Raoult and T.G. Harrison. 1996. Phylogenetic diversity of intra-amoebal legionellae as revealed by 16S rRNA gene sequence comparison. Microbiology (Reading) *142*: 3525–3530.

Bisgaard, M. 1984. Comparative investigations of *Pasteurella haemolytica* sensu stricto and so-called *P. haemolytica* isolated from different pathological lesions in pigs. Acta Pathol. Microbiol. Scand. Sect. B Microbiol. Immunol. *92*: 201–207.

Bisgaard, M. 1986. *Actinobacillus muris* sp. nov. isolated from mice. Acta Pathol. Microbiol. Scand. Sect. B Microbiol. Immunol. *94*: 1–8.

Bisgaard, M. 1988. *In* Validation of the publication of new names and new combinations previously effectively published outside the IJSB. List No. 25. Int. J. Syst. Bacteriol. *38*: 220–222.

Bisgaard, M. 1993. Ecology and significance of *Pasteurellaceae* in animals. Zentbl. Bakteriol. *279*: 7–26.

Bisgaard, M. 1995. Taxonomy of the family *Pasteurellaceae* Pohl 1981. *In Haemophilus, Actinobacillus,* and *Pasteurella*. Proceedings of the Third International Conference on *Haemophilus, Actinobacillus,* and *Pasteu-*

rella, July 31, August 4, 1993, Plenum Press, London, Edinburgh, Scotland. 1–7.

Bisgaard, M., D.J. Brown, M. Costas and M. Ganner. 1993. Whole cell protein profiling of *Actinobacillus*-like strains classified as taxon 2 and taxon 3 according to Bisgaard. Zentralbl Bakteriol. *279*: 92–103.

Bisgaard, M., W. Fredericksen, W. Mannheim and R. Mutters. 1994. Zoonoses caused by organisms classified with *Pasteurellaceae*. *In* Beran (Editor), Handbook of Zoonoses, Section A: Bacterial, Rickettsial, Chlamydial, and Mycotic, 2nd ed., CRC Press, London. 203–208.

Bisgaard, M., O. Heltberg and W. Frederiksen. 1991a. Isolation of *Pasteurella caballi* from an infected wound on a veterinary surgeon. Apmis *99*: 291–294.

Bisgaard, M., K.H. Hinz, K.D. Petersen and J.P. Christensen. 1999. Identification of members of the *Pasteurellaceae* isolated from birds and characterization of two new taxa isolated from psittacine birds. Avian Pathol. *28*: 369–377.

Bisgaard, M., S.B. Houghton, R. Mutters and A. Stenzel. 1991b. Reclassification of German, British and Dutch isolates of so-called *Pasteurella multocida* obtained from pneumonic calf lungs. Vet. Microbiol. *26*: 115–124.

Bisgaard, M. and R. Mutters. 1986a. Characterization of some previously unclassified *Pasteurella* spp. obtained from the oral cavity of dogs and cats and description of a new species tentatively classified with the family *Pasteurellaceae* Pohl 1981 and provisionally called taxon-16. Acta Pathol. Microbiol. Immun. B. *94*: 177–184.

Bisgaard, M. and R. Mutters. 1986b. Re-investigations of selected bovine and ovine strains previously classified as *Pasteurella haemolytica* and description of some new taxa within the *Pasteurella haemolytica*-complex. Acta Pathol. Microbiol. Immunol. Scand. Sect. B Microbiol. Immunol. *94*: 185–193.

Bisgaard, M., J.E. Phillips and W. Mannheim. 1986. Characterization and identification of bovine and ovine *Pasteurellaceae* isolated from the oral cavity and rumen of apparently normal cattle and sheep. Acta Pathol. Microbiol. Scand. Sect. B Microbiol. Immunol. *94*: 9–17.

Bisgaard, M., K. Piechulla, Y.T. Ying, W. Frederiksen and W. Mannheim. 1984. Prevalence of organisms described as *Actinobacillus suis* or haemolytic *Actinobacillus equuli* in the oral cavity of horses. Comparative investigations of strains obtained and porcine strains of *A. suis* sensu stricto. Acta Pathol. Microbiol. Scand. Sect. B Microbiol. Immunol. *92*: 291–298.

Bisharat, N., V. Agmon, R. Finkelstein, R. Raz, G. Ben-Dror, L. Lerner, S. Soboh, R. Colodner, D.N. Cameron, D.L. Wykstra, D.L. Swerdlow and J.J. Farmer, III. 1999. Clinical, epidemiological, and microbiological features of *Vibrio vulnificus* biogroup 3 causing outbreaks of wound infection and bacteraemia in Israel. Israel *Vibrio* Study Group. Lancet *354*: 1421–1424.

Bisharat, N. and R. Raz. 1996. *Vibrio* infection in Israel due to changes in fish marketing. Lancet *348*: 1585–1586.

Bishop, P.E. and W.J. Brill. 1977. Genetic analysis of *Azotobacter vinelandii* mutant strains unable to fix nitrogen. J. Bacteriol. *130*: 954–956.

Bishop, P.E., D.M.L. Jarlenski and D.R. Hetherington. 1980. Evidence for an alternative nitrogen fixation system in *Azotobacter vinelandii*. Proc Natl Acad Sci U S A. *77*: 7342–7346.

Bishop, P.E. and R.D. Joerger. 1990. Genetics and molecular biology of alternative nitrogen fixation systems. Annu. Rev. Plant Physiol. Plant Mol. Biol. *41*: 109–125.

Bishop, P.E., T.M. Rizzo and K.F. Bott. 1985. Molecular cloning of *nif* DNA from *Azotobacter vinelandii*. J. Bacteriol. *162*: 21–28.

Bishop, P.E., M.A. Supiano and W.J. Brill. 1977. Technique for isolating phage for *Azotobacter vinelandii*. Appl. Environ. Microbiol. *33*: 1007–1008.

Bissett, M.L., J.O. Lee and D.S. Lindquist. 1983. New serogroup of *Legionella pneumophila*, serogroup 8. J. Clin. Microbiol. *17*: 887–891.

Bitzan, M., H. Karch, M.G. Maas, T. Meyer, H. Rüssmann, S. Aleksic and J. Bockemühl. 1991. Clinical and genetic aspects of Shiga-like toxin production in traditional enteropathogenic *Escherichia coli*. Zentbl. Bakteriol. *274*: 496–506.

Bizio, B. 1823. Lettera di Bartolomeo Bizio al chiarissimo canonico Angelo Bellani sopra il fenomeno della polenta porporina. Biblioteca Italiana o sia Giornale di Letteratura Scienze e Arti (Anno VIII). *30*: 275–295.

Blackall, L.L., A.C. Hayward and L.I. Sly. 1985. Cellulolytic and dextranolytic Gram-negative bacteria - revival of the genus *Cellvibrio*. J. Appl. Bacteriol. *59*: 81–97.

Blackall, L.L., A.C. Hayward and L.I. Sly. 1986. *In* Validation of the publication of new names and new combinations previously effectively published outside the IJSB. List No. 20. Int. J. Syst. Bacteriol. *36*: 354–356.

Blackall, P.J. 1988. Antimicrobial drug resistance and the occurrence of plasmids in *Haemophilus paragallinarum*. Avian Dis. *32*: 742–747.

Blackall, P.J. 1989. The avian *haemophili*. Clin. Microbiol. Rev. *2*: 270–277.

Blackall, P.J., R. Bowles, J.L. Pahoff and B.N. Smith. 1999a. Serological characterisation of *Actinobacillus pleuropneumoniae* isolated from pigs in 1993 to 1996. Aust. Vet. J. *77*: 39–43.

Blackall, P.J., L.E. Eaves and C.J. Morrow. 1991. Comparison of *Haemophilus paragallinarum* isolates by restriction endonuclease analysis of chromosomal DNA. Vet. Microbiol. *27*: 39–47.

Blackall, P.J., L.E. Eaves and D.G. Rogers. 1990a. Proposal of a new serovar and altered nomenclature for *Haemophilus paragallinarum* in the Kume hemagglutinin scheme. J. Clin. Microbiol. *28*: 1185–1187.

Blackall, P.J., N. Fegan, G.T.I. Chew and D.J. Hampson. 1998. Population structure and diversity of avian isolates of *Pasteurella multocida* from Australia. Microbiology (UK) *144*: 279–289.

Blackall, P.J., N. Fegan, G.T.I. Chew and D.J. Hampson. 1999b. A study of the use of multilocus enzyme electrophoresis as a typing tool in fowl cholera outbreaks. Avian Pathol. *28*: 195–198.

Blackall, P.J., N. Fegan, J.L. Pahoff, G.J. Storie, G.B. McIntosh, R.D.A. Cameron, D. O'Boyle, A.J. Frost, M.R. Bara, G. Marr and J. Holder. 2000. The molecular epidemiology of four outbreaks of porcine pasteurellosis. Vet. Microbiol. *72*: 111–120.

Blackall, P.J. and J.K. Miflin. 2000. Identification and typing of *Pasteurella multocida*: a review. Avian Pathol. *29*: 271–287.

Blackall, P.J. and G.G. Reid. 1982. Further characterization of *Haemophilus paragallinarum* and *Haemophilus avium*. Vet. Microbiol. *7*: 359–367.

Blackall, P.J., D.G. Rogers and R. Yamamoto. 1990b. Outer-membrane proteins of *Haemophilus paragallinarum*. Avian Dis. *34*: 871–877.

Blackall, P.J., D.J. Trott, V. Rapp-Gabrielson and D.J. Hampson. 1997. Analysis of *Haemophilus parasuis* by multilocus enzyme electrophoresis. Vet. Microbiol. *56*: 125–134.

Blackburn, T.H. 1968. Protease production by *Bacteroides amylophilus* strain H18. J. Gen. Microbiol. *53*: 27–36.

Blackburn, T.H. and W.A. Hallah. 1974. The cell-bound protease of *Bacteroides amylophilus* H18. Can. J. Microbiol. *20*: 435–441.

Blackburn, T.H. and P.N. Hobson. 1962. Further studies on the isolation of proteolytic bacteria from sheep rumen. J. Gen. Microbiol. *29*: 69–81.

Blackman, R.L. 1987. Reproduction, cytogenetics and development. *In* Minks and Harrewijn (Editors), Aphids, Their Biology, Natural Enemies and Control, Vol. 2A, Elsevier, Amsterdam. pp. 163–195.

Blackman, R.L. and V.F. Eastop. 1984. Aphids on the World's Crops, Wiley, Chichester.

Blake, P.A., M.H. Mersen, R.E. Weaver, D.G. Hollis and P.C. Heublein. 1979. Disease caused by a marine *Vibrio*. Clinical characteristics and epidemiology. N. Engl. J. Med. *300*: 1–5.

Blanchard, P.C., R.L. Walker and I. Gardner. 1993. Pleuropneumonia in swine associated with a urease-negative variant of *Actinobacillus pleuropneumoniae* serotype. J. Vet. Diagn. Invest. *5*: 279–282.

Blanco, G., F. Ramos, J.R. Medina and M. Tortolero. 1990. A chromosomal linkage map of *Azotobacter vinelandii*. Mol. Gen. Genet. *224*: 241–247.

Blanco, J., M. Blanco, J.E. Blanco, A. Mora, M.P. Alonso, E.A. González and M.I. Bernádez. 2001. Epidemiology of verocytotoxigenic *E. coli* (VTEC) in ruminants. *In* Duffy, Garvey and McDowell (Editors), Verocytotoxigenic *E. coli*, Food & Nutrition Press, Inc., Trumbull, CT. pp. 113–148.

Blanco, J.E., M. Blanco, M.E. Molinero, E. Peiró, A. Mora and J. Blanco. 1996. An outbreak of gastroenteritis associated with verotoxin-pro-

928　　　　　　　　　　　　　　　　　　　　　　　BIBLIOGRAPHY

ducing *Escherichia coli* (VTEC) O111:H-VT1 + *eae* +. Alimentaria. *34*: 109–113.

Bland, J.A. and T.D. Brock. 1973. The marine bacterium *Leucothrix mucor* as an algal epiphyte. Mar. Biol. *23*: 283–292.

Bland, J.A. and J.T. Staley. 1978. Observations on the biology of *Thiothrix*. Arch. Microbiol. *117*: 79–87.

Blaser, M.J. and L.S. Newman. 1982. A review of human salmonellosis: I. infective dose. Rev. Infect. Dis. *4*: 1096–1106.

Blattner, F.R., G.I. Plunkett, III, C.A. Bloch, N.T. Perna, V. Burland, M. Riley, J. Collado-Vides, J.D. Glasner, C.K. Rode, G.F. Mayhew, J. Gregor, N.W. Davis, H.A. Kirkpatrick, M.A. Goeden, D.J. Rose, B. Mau and Y. Shao. 1997. The complete genome sequence of *Escherichia coli* K-12. Science *277*: 1453–1462.

Blix, I.J., H.R. Preus and I. Olsen. 1990. Invasive growth of *Actinobacillus actinomycetemcomitans* on solid medium (TSBV). Acta Odontol Scand. *48*: 313–318.

Blouse, L. and R. Twarog. 1966. Properties of four *Herellea* phages. Can. J. Microbiol. *12*: 1023–1030.

Bock, E. and H.-P., Koops. 1992. The genus *Nitrobacter* and related genera. *In* Balows, Trüper, Dworkin, Harder and Schleifer (Editors), The Prokaryotes. A Handbook on the Biology of Bacteria: Ecophysiology, Isolation, Identification, Applications, 2nd ed., Vol. 3, Springer-Verlag, New York. pp. 2302–2309.

Bock, E., H. Sundermeyer-Klinger and E. Stackebrandt. 1983. New facultative lithoautotrophic nitrite-oxidizing bacteria. Arch. Microbiol. *136*: 281–284.

Bockemühl, J., V. Aleksic, R. Wokasch and S. Aleksic. 1983. Pathogenicity tests with strains of *Edwardsiella tarda*: detection of a heat-stable enterotoxin. Zentralbl. Bakteriol. Parasitenkd. Hyg. Infektionskr., Abt. I Orig. *A255*: 464–471.

Bockemühl, J., R. Pan Urai and F. Burkhardt. 1971. *Edwardsiella tarda* associated with human disease. Pathol. Microbiol. *37*: 393–401.

Bocklage, H., K. Heeger and B. Müller-Hill. 1991. Cloning and characterization of the *Mbo*II restriction-modification system. Nucleic Acids Res. *19*: 1007–1014.

Bodrossy, L., E.M. Holmes, A.J. Holmes, K.L. Kovacs and J.C. Murrell. 1997. Analysis of 16S rRNA and methane monooxygenase gene sequences reveals a novel group of thermotolerant and thermophilic methanotrophs, *Methylocaldum* gen. nov. Arch. Microbiol. *168*: 493–503.

Bodrossy, L., E.M. Holmes, A.J. Holmes, K.L. Kovacs and J.C. Murrell. 1998. *In* Validation of the publication of new names and new combinations previously effectively published outside the IJSB. List No. 66. Int. J. Syst. Bacteriol. *48*: 631–632.

Bodrossy, L., J.C. Murrell, H. Dalton, M. Kalman, L.G. Puskas and K.L. Kovacs. 1995. Heat-tolerant methanotrophic bacteria from the hot water effluent of a natural gas field. Appl. Environ. Microbiol. *61*: 3549–3555.

Boemare, N.E. and R.J. Akhurst. 1988. Biochemical and physiological characterization of colony form variants in *Xenorhabdus* spp. (*Enterobacteriaceae*). J. Gen. Microbiol. *134*: 751–762.

Boemare, N.E. and R.J. Akhurst. 1994. DNA homology between *Xenorhabdus* and *Photorhabdus* spp.: convergent evolution of two different genera. *In* Burnell, Ehlers and Masson (Editors), COST 812-Biotechnology: Genetics of entomopathogenic nematode-bacterium complexes, European Commission, Directorate-General XII, Science, Research and Development Environment Research Programme, Luxembourg. pp. 59–69.

Boemare, N.E., R.J. Akhurst and R.G. Mourant. 1993. DNA relatedness between *Xenorhabdus* spp. (*Enterobacteriaceae*), symbiotic bacteria of entomopathogenic nematodes, and a proposal to transfer *Xenorhabdus luminescens* to a new genus, *Photorhabdus* gen. nov. Int. J. Syst. Bacteriol. *43*: 249–255.

Boemare, N.E., A. Givaudan, M. Brehélin and C. Laumond. 1997a. Symbiosis and pathogenicity of nematode-bacterium complexes. Symbiosis *22*: 21–45.

Boemare, N., C. Louis and G. Kuhl. 1983. Étude ultrastructurale des cristaux chez *Xenorhabdus* spp. bactéries inféodées aux nématodes

entomophages Steinernematidae et Heterorhabditidae. C. R. Seances Soc. Biol. Fil. *177*: 107–115.

Boemare, N.E., M.-H. Boyer-Giglio, J.-O. Thaler, R.J. Akhurst and M. Brehélin. 1992. Lysogeny and bacteriocinogeny in *Xenorhabdus nematophilus* and other *Xenorhabdus* spp. Appl. Environ. Microbiol. *58*: 3032–3037.

Boemare, N.E., J.O. Thaler and A. Lanois. 1997b. Simple bacteriological tests for phenotypic characterization of *Xenorhabdus* and *Photorhabdus* phase variants. Symbiosis *22*: 167–175.

Boerlin, P., H.H. Siegrist, A.P. Burnens, P. Kuhnert, P. Mendez, G. Prétat, R. Lienhard and J. Nicolet. 2000. Molecular identification and epidemiological tracing of *Pasteurella multocida* meningitis in a baby. J. Clin. Microbiol. *38*: 1235–1237.

Bogdanove, A.J., Z.M. Wei, L. Zhao and S.V. Beer. 1996. *Erwinia amylovora* secretes harpin via a type III pathway and contains a homolog of yopN of *Yersinia* spp. J. Bacteriol. *178*: 1720–1730.

Bogosian, G., P.J.L. Morris and J.P. O'Neil. 1998. A mixed culture recovery method indicates that enteric bacteria do not enter the viable but nonculturable state. Appl. Environ. Microbiol. *64*: 1736–1742.

Boháček, J. and O. Mráz. 1967. Basengehalt der Deoxyribonukleinsäure bei den Arten *Pasteurella haemolytica*, *Actinobacillus lignieresii* und *Actinobacillus equuli*. Zentralbl. Bakteriol. Parasitenkd. Infektionskr. Hyg. 1 Abt. Orig. *202*: 468–478.

Boher, B., K. Kpemoua, M. Nicole, J. Luisetti and J. Geiger. 1995. Ultrastructure of interactions between cassava and *Xanthomonas campestris* pathovar *manihotis*: cytochemistry of cellulose and pectin degradation in a susceptible cultivar. Phytopathology *85*: 777–788.

Boher, B., M. Nicole, M. Potin and J.P. Geiger. 1997. Extracellular polysaccharides from *Xanthomonas axonopodis* pathovar *manihotis* interact with cassava cell walls during pathogenesis. Mol. Plant-Microbe Interact. *10*: 803–811.

Bohn, G.W. and J.C. Maloit. 1946. Bacterial spot of native golden currant (*Ribes aureum*). J. Agr. Res. *73*: 281–290.

Boileau, C.R., H.M. d'Hauteville and P.J. Sansonetti. 1984. DNA hybridization technique to detect *Shigella* species and enteroinvasive escherichia coli. J. Clin. Microbiol. *20*: 959–961.

Boisvert, H., R. Chatelain and M.J. Bassot. 1967. Etude d'un *Photobacterium* isole de l'organe luminex de poissons Leiognathidae. Ann. Inst. Pasteur (Paris) *112*: 520–524.

Bokete, T.N., T.S. Whittam, R.A. Wilson, C.R. Clausen, C.M. O'Callahan, S.L. Moseley, T.R. Fritsche and P.I. Tarr. 1997. Genetic and phenotypic analysis of *Escherichia coli* with enteropathogenic characteristics isolated from Seattle children. J. Infect. Dis. *175*: 1382–1389.

Boldur, I., A. Cohen, L.R. Tamarin and D. Sompolinsky. 1987. Isolation of *Legionella pneumophila* from calves and the prevalence of antibodies in cattle, sheep, horses, antelopes, buffaloes and rabbits. Vet. Microbiol. *13*: 313–320.

Boling, M.E., D.P. Allison and J.K. Setlow. 1973. Bacteriophage of *Haemophilus influenzae*. III. Morphology, DNA homology, and immunity properties of HPlcl, S2, and the defective bacteriophage from strain Rd. J. Virol. *11*: 585–591.

Bolstad, A.I., T. Kristoffersen, I. Olsen, H.R. Preus, H.B. Jesen, E.N. Vasstrand and V. Bakken. 1990. Outer membrane proteins of *Actinobacillus actinomycetemcomitans* and *Haemophilus aphrophilus* studied by SDS-PAGE and immunoblotting. Oral Microbiol. Immunol. *5*: 155–161.

Bonacorsi, S.P., M.R. Scavizzi, A. Guiyoule, J.H. Amouroux and E. Carniel. 1994. Assessment of a fluoroquinolone, three beta-lactams, two aminoglycosides, and a cycline in treatment of murine *Yersinia pestis* infection. Antimicrob. Agents Chemother. *38*: 481–486.

Bonami, J.R. and R. Pappalardo. 1980. Rickettsial infection in marine crustacea. Experientia (Basel) *36*: 180–181.

Bonar, D.B., R.M. Weiner and R.R. Colwell. 1986. Microbial-invertebrate interactions and potential for biotechnology. Microb. Ecol. *12*: 101–110.

Bonas, U., R. Schulte, S. Fenselau, G.V. Minsavage, B.J. Staskawicz and R.E. Stall. 1991. Isolation of a gene cluster from *Xanthomonas campestris* pv. *vesicatoria* that determines pathogenicity and the hypersensitive response on pepper and tomato. Mol. Plant-Microbe Interact. *4*: 81–88.

Bondar, G. 1915. Molestia bacteriana da mandioca. Biol. Agr. Sao Paulo. *16a*: 513–524.

Bonet, R., B. Magariños, J.L. Romalde, M.D. Simonpujol, A.E. Toranzo and F. Congregado. 1994. Capsular polysaccharide expressed by *Pasteurella piscicida* grown in-vitro. FEMS Microbiol. Lett. *124*: 285–289.

Bonfiglio, G., S. Stefani and G. Nicoletti. 1995. Clinical isolate of a *Xanthomonas maltophilia* strain producing L-1-deficient and L-2-inducible β-lactamases. Chemotherapy. *41*: 121–124.

Bonifassi, E., M.F.L. Saux, N. Boemare, A. Lanois, C. Laumond and G. Smart. 1999. Gnotobiological study of infective juveniles and symbionts of *Steinernema scapterisci*: A model to clarify the concept of the natural occurrence of monoxenic associations in entomopathogenic nematodes. J. Invertebr. Pathol. *74*: 164–172.

Bonin, P., M. Gilewicz and J.C. Bertrand. 1987. Denitrification by a marine bacterium *Pseudomonas nautica* strain 617. Ann. Inst. Pasteur (Paris) *138*: 371–383.

Bonner, D.P., J. O'Sullivan, S.K. Tanaka, J.M. Clark and R.R. Whitney. 1988. Lysobactin, a novel antibacterial agent produced by *Lycobacter* sp.: II. Biological properties. J. Antibiot. (Tokyo). *41*: 1745–1751.

Bonting, C.F.C., B.M.F. Willemsen, W. Akkermans-van Vliet, P.J.M. Bouvet, G.J.J. Kortstee and A.J.B. Zehnder. 1992. Additional characteristics of the polyphosphate-accumulating *Acinetobacter* strain 210A and its identification as *Acinetobacter johnsonii*. FEMS Microbiol. Ecol. *102*: 57–64.

Boocock, G.R. and J.A. Bowley. 1995. Meningitis in infancy caused by *Pasteurella multocida*. J. Infect. *31*: 161–162.

Boot, R. and M. Bisgaard. 1995. Reclassification of 30 *Pasteurellaceae* strains isolated from rodents. Lab. Anim. *29*: 314–319.

Boot, R., H.C. Thuis and F.A. Reubsaet. 1999. Growth medium affects the cellular fatty acid composition of *Pasteurellaceae*. Zentbl. Bakteriol. *289*: 9–17.

Borch, E., M.L. Kant-Muermans and Y. Blixt. 1996. Bacterial spoilage of meat and cured meat products. Int. J. Food Microbiol. *33*: 103–120.

Borman, E.K. 1957. Genus IV. *Paracolobactrum*. *In* Breed, Murray and Smith (Editors), Bergey's Manual of Determinative Bacteriology, 7th Edition, Williams and Wilkins, Baltimore. 346–348.

Bormann, E.J., M. Leissner and B. Beer. 1998. Growth-associated production of poly(hydroxybutyric acid) by *Azotobacter beijerinckii* from organic nitrogen substrates. Appl. Microbiol. Biotechnol. *49*: 84–88.

Borneleit, P. and H.P. Kleber. 1991. The outer membrane of *Acinetobacter*: structure–function relationships. *In* Towner and Bergogne-Bérézin (Editors), The Biology of *Acinetobacter*: Taxonomy, Clinical Importance, Molecular Biology, Physiology, Industrial Relevance, Plenum Press, New York. pp. 259–271.

Bornstein, N., D. Marmet, M. Surgot, M. Nowicki, H. Meugnier, J. Fleurette, E. Ageron, F. Grimont, P.A.D. Grimont, W.L. Thacker, R.F. Benson and D.J. Brenner. 1989a. *Legionella gratiana*, sp. nov., isolated from French spa water. Res. Microbiol. *140*: 541–552.

Bornstein, N., D. Marmet, M. Surgot, M. Nowicki, H. Meugnier, J. Fleurette, E. Ageron, F. Grimont, P.A.D. Grimont, W.L. Thacker, R.F. Benson and D.J. Brenner. 1991. *In* Validation of the publication of new names and new combinations previously effectively published outside the IJSB. List No. 39. Int. J. Syst. Bacteriol. *41*: 580–581.

Bornstein, N., A. Mercatello, D. Marmet, M. Surgot, Y. Deveaux and J. Fleurette. 1989b. Pleural infection caused by *Legionella anisa*. J. Clin. Microbiol. *27*: 2100–2101.

Borowski, J. and M. Zaremba. 1973. Some problems connected with *Yersinia pseudotuberculosis* resistance to antibiotics. Contr. Microbiol. Immunol. *2*: 196–202.

Borr, J.D., D.A.J. Ryan and J.I. McInnes. 1991. Analysis of *Actinobacillus pleuropneumoniae* and related organisms by DNA–DNA hybridization and restriction endonuclease fingerprinting. Int. J. Syst. Bacteriol. *41*: 121–129.

Borrego, J.J., D. Castro, A. Luque, C. Paillard, P. Maes, M.T. Garcia and A. Ventosa. 1996. *Vibrio tapetis* sp. nov., the causative agent of the brown ring disease affecting cultured clams. Int. J. Syst. Bacteriol. *46*: 480–484.

Borrell, N., S.G. Acinas, M.J. Figueras and A.J. Martinez-Murcia. 1997. Identification of *Aeromonas* clinical isolates by restriction fragment length polymorphism of PCR-amplified 16S rRNA genes. J. Clin. Microbiol. *35*: 1671–1674.

Borrelli, S., E.L. Roggen, D. Hendriksen, J. Jonasson, H.J. Ahmed, P. Piot, P.E. Jansson and A.A. Lindberg. 1995. Monoclonal antibodies against *Haemophilus* lipopolysaccharides: clone DP8 specific for *Haemophilus ducreyi* and clone DH24 binding to lacto-N-neotetraose. Infect. Immun. *63*: 2665–2673.

Borriss, M., E. Helmke, R. Hanschke and T. Schweder. 2003. Isolation and characterization of marine psychrophilic phage-host systems from Arctic sea ice. Extremophiles *7*: 377–384.

Boschker, H.T.S., S.C. Nold, P. Wellsbury, D. Bos, W. De Graaf, R. Pel, R.J. Parkees and T.E. Cappenberg. 1998. Direct linking of microbial populations to specific biogeochemical processes by ^{13}C-labelling of biomarkers. Nature *392*: 801–805.

Boswell, T.C.J. and G. Kudesia. 1992. Serological cross-reaction between *Legionella pneumophila* and *Campylobacter* in the indirect fluorescent antibody test. Epidemiol. Infect. *109*: 291–295.

Böttger, E.C., M. Jürs, T. Barrett, K. Wachsmuth, S. Metzger and D. Bitter-Suermann. 1987. Qualitative and quantitative determination of enterobacterial common antigen (ECA) with monoclonal antibodies: expression of ECA by two *Actinobacillus* species. J. Clin. Microbiol. *25*: 377–382.

Bottone, E.J. 1977. *Yersinia enterocolitica*: a panoramic view of a charismatic microorganism. Crit. Rev. Microbiol. *5*: 211–241.

Bottone, E.J. 1992. The genus *Yersinia* (excluding *Y. pestis*). *In* Balows (Editor), The Prokaryotes: a Handbook on the Biology of Bacteria : Ecophysiology, Isolation, Identification, Applications, 2nd ed., Springer-Verlag, New York. 2863–2887.

Bottone, E.J. 1997. *Yersinia enterocolitica*: the charisma continues. Clin. Microbiol. Rev. *10*: 257–276.

Bottone, E.J. 1999. *Yersinia enterocolitica*: overview and epidemiologic correlates. Microbes Infect. *1*: 323–333.

Bottone, E.J., B. Chester, M.S. Malowany and J. Allerhand. 1974. Unusual *Yersinia enterocolitica* isolates not associated with mesenteric lymphadenitis. Appl. Microbiol. *27*: 858–861.

Bottone, E.J., C.R. Gullans and M.F. Sierra. 1987. Disease spectrum of *Yersinia enterocolitica* serogroup 0:3, the predominant cause of human infection in New York City. Contrib. Microbiol. Immunol. *9*: 56–60.

Bottone, E.J. and T. Robin. 1977. *Yersinia enterocolitica*: recovery and characterization of two unusual isolates from a case of acute enteritis. J. Clin. Microbiol. *5*: 341–345.

Bottone, E.J., K.K. Sandh and M.A. Pisano. 1979. *Yersinia intermedia*: Temperature dependent bacteriocin production. J. Clin. Microbiol. *10*: 433–436.

Bouchotroch, S., E. Quesada, A. Del Moral and V. Bejar. 1999. Taxonomic study of exopolysaccharide-producing, moderately halophilic bacteria isolated from hypersaline environments in Morocco. Syst. Appl. Microbiol. *22*: 412–419.

Bouchotroch, S., E. Quesada, A. del Moral, I. Llamas and V. Bejar. 2001. *Halomonas maura* sp nov., a novel moderately halophilic, exopolysaccharide-producing bacterium. Int. J. Syst. Evol. Microbiol. *51*: 1625–1632.

Boudailliez, B., P. Berquin, P. Mariani-Kurkdjian, D.D. Ilef, B. Cuvelier, I. Capek, B. Tribout, E. Bingen and C. Piussan. 1997. Possible person-to-person transmission of *Escherichia coli* O111-associated hemolytic uremic syndrome. Pediatr. Nephrol. *11*: 36–39.

Bouley, G. 1966. Étude d'une souche d'*Actinobacillus suis* (van Dorssen et Jaartsveld) isolée en Normandie. Recl. Med. Vet. Ec. Alfort. *14*: 225–229.

Bousfield, I.J. and P.N. Green. 1985. Reclassification of bacteria of the genus *Protomonas* Urakami and Komagata 1984 in the genus *Methylobacterium* (Patt, Cole, and Hanson) emend. Green and Bousfield 1983. Int. J. Syst. Bacteriol. *35*: 209.

Bouvet, O.M. and P.A. Grimont. 1987a. Diversity of the phosphoenolpyruvate/glucose phosphotransferase system in the *Enterobacteriaceae*. Ann. Inst. Pasteur Microbiol. *138*: 3–13.

Bouvet, O.M.M., P.A.D. Grimont, C. Richard, E. Aldová, O. Hausner and M. Gabrhelová. 1985. *Budvicia aquatica*, gen. nov., sp. nov.: a hydrogen sulfide-producing member of the *Enterobacteriaceae*. Int. J. Syst. Bacteriol. *35*: 60–64.

Bouvet, O.M., P. Lenormand, E. Ageron and P.A.D. Grimont. 1995a. Taxonomic diversity of anaerobic glycerol dissimilation in the *Enterobacteriaceae*. Res. Microbiol. *146*: 279–290.

Bouvet, O.M.M., P. Lenormand and P.A.D. Grimont. 1989. Taxonomic diversity of the D-glucose oxidation pathway in the *Enterobacteriaceae*. Int. J. Syst. Bacteriol. *39*: 61–67.

Bouvet, O.M., P. Lenormand, V. Guibert and P.A. Grimont. 1995b. Differentiation of *Shigella* species from *Escherichia coli* by glycerol dehydrogenase activity. Res. Microbiol. *146*: 787–790.

Bouvet, P.J.M. and P.A.D. Grimont. 1986. Taxonomy of the genus *Acinetobacter*, gen. nov., with the recognition of *Acinetobacter baumannii*, sp. nov., *Acinetobacter haemolyticus*, sp. nov., *Acinetobacter johnsonii*, sp. nov. and *Acinetobacter junii*, sp. nov. and emended descriptions of *Acinetobacter calcoaceticus* and *Acinetobacter lwoffi*. Int. J. Syst. Bacteriol. *36*: 228–240.

Bouvet, P.J.M. and P.A.D. Grimont. 1987b. Identification and biotyping of clinical isolates of *Acinetobacter*. Ann. Inst. Pasteur Microbiol. *138*: 569–578.

Bouvet, P.J.M. and S. Jeanjean. 1989. Delineation of new proteolytic genomic species in the genus *Acinetobacter*. Res. Microbiol. *140*: 291–300.

Bouvet, P.J.M. and S. Jeanjean. 1995. Differentiation of *Acinetobacter calcoaceticus* sensu stricto from related *Acinetobacter* species by electrophoretic polymorphism of malate dehydrogenase, glutamate dehydrogenase and catalase. Res. Microbiol. *146*: 773–785.

Bouvet, P.J.M., S. Jeanjean, J.F. Vieu and L. Dijkshoorn. 1990. Species, biotype and bacteriophage type determinations compared with cell envelope protein profiles for typing *Acinetobacter* strains. J. Clin. Microbiol. *28*: 170–176.

Bouzar, H., J.B. Jones, R.E. Stall, N.C. Hodge, G.V. Minsavage, A.A. Benedict and A.M. Alvarez. 1994. Physiological, chemical, serological, and pathogenic analyses of a worldwide collection of *Xanthomonas campestris* pathovar *vesicatoria* strains. Phytopathology *84*: 663–671.

Bovarnick, M.R., J.C. Miller and J.C. Snyder. 1950. The influence of certain salts, amino acids, sugars and proteins on the stability of rickettsiae. J. Bacteriol. *59*: 509–522.

Bovien, P. 1937. Some types of association between nematodes and insects. Vidensk. Medd. Dan. Naturhist. Foren. Khobenhavn. *101*: 1–114.

Bøvre, K. 1963. Affinities between *Moraxella* spp. and a strain of *Neisseria catarrhalis* as expressed by transformation. Acta Pathol. Microbiol. Scand. *58*: 528.

Bøvre, K. 1964. Studies on transformation in *Moraxella* and organisms assumed to be related to *Moraxella*. 2. Quantitative transformation reactions between *Moraxella nonliquefaciens* strains, with streptomycin resistance marked DNA. Acta Pathol. Microbiol. Scand. *62*: 239–248.

Bøvre, K. 1965a. Studies on transformation in *Moraxella* and organisms assumed to be related to *Moraxella*. 3. Quantitative streptomycin resistance transformation between *Moraxella bovis* and *Moraxella nonliquefaciens* strains. Acta Pathol. Microbiol. Scand. *63*: 42–50.

Bøvre, K. 1965b. Studies on transformation in *Moraxella* and organisms assumed to be related to *Moraxella*. 4. Streptomycin resistance transformation between asaccharolytic *Neisseria* strains. Acta Pathol. Microbiol. Scand. *64*: 229–242.

Bøvre, K. 1965c. Studies on transformation in *Moraxella* and organisms assumed to be related to *Moraxella*. 5. Streptomycin resistance transformation between serum-liquefying, nonhaemolytic moraxellae, *Moraxella bovis* and *Moraxella nonliquefaciens*. Acta Pathol. Microbiol. Scand. *65*: 435–449.

Bøvre, K. 1965d. Studies on transformation in *Moraxella* and organisms assumed to be related to *Moraxella*. 6. A distinct group of *Moraxella nonliquefaciens*-like organisms (the "19116/51" group). Acta Pathol. Microbiol. Scand. *65*: 641–652.

Bøvre, K. 1967. Studies on transformation in *Moraxella* and organisms assumed to be related to *Moraxella*. 8. The relative position of some oxidase negative, immotile diplobacilli (*Achromobacter*) in the transformation system. Acta Pathol. Microbiol. Scand. *69*: 109–122.

Bøvre, K. 1979. Proposal to divide the genus *Moraxella* Lwoff 1939 emend. Henriksen and Bøvre 1968 into two subgenera — subgenus *Moraxella* (Lwoff 1939) Bøvre 1979 and subgenus *Branhamella* (Catlin 1970) Bøvre 1979. Int. J. Syst. Bacteriol. *29*: 403–406.

Bøvre, K. 1980. Progress in classification and identification of *Neisseriaceae* based on genetic affinity. *In* Goodfellow and Board (Editors), Microbial Classification and Identification, Academic Press, London. pp. 55–72.

Bøvre, K., T. Bergan and L.O. Frøholm. 1970. Electron microscopical and serological characteristics associated with colony type in *Moraxella nonliquefaciens*. Acta Pathol. Microbiol. Scand. B Microbiol. Immunol. *78*: 765–779.

Bøvre, K., K. Bryn, O. Closs, N. Hagen and L.O. Frøholm. 1983. Surface polysaccharide of *Moraxella nonliquefaciens* identical to *Neisseria meningitidis* group B capsular polysaccharide: a chemical and immunological investigation. NIPH (Natl. Inst. Public Health) Ann. *6*: 65–74.

Bøvre, K. and L.O. Frøholm. 1971. Competence of genetic transformation correlated with the occurrence of fimbriae in three bacterial species. Nat. New Biol. *234*: 151–152.

Bøvre, K. and L.O. Frøholm. 1972a. Competence in genetic transformation related to colony type and fimbriation in three species of *Moraxella*. Acta Pathol. Microbiol. Scand. B Microbiol. Immunol. *80*: 649–659.

Bøvre, K. and L.O. Frøholm. 1972b. Variation of colony morphology reflecting fimbriation in *Moraxella bovis* and two reference strains of *M. nonliquefaciens*. Acta Pathol. Microbiol. Scand. B Microbiol. Immunol. *80*: 629–640.

Bøvre, K., J.E. Fuglesang, N. Hagen, E. Jantzen and L.O. Frøholm. 1976. *Moraxella atlantae* sp. nov. and its distinction from *Moraxella phenylpyruvica*. Int. J. Syst. Bacteriol. *26*: 511–521.

Bøvre, K., J.E. Fuglesang, S.D. Henriksen, S.P. Lapage, H. Lautrop and J.J.S. Snell. 1974. Studies on a collection of Gram-negative bacterial strains showing resemblance to *Moraxellae*. Examination by conventional bacteriological methods. Int. J. Syst. Bacteriol. *24*: 438–446.

Bøvre, K. and N. Hagen. 1981. The family *Neisseriaceae*: rod-shaped species of the genera *Moraxella*, *Acinetobacter*, *Kingella*, and *Neisseria*, and the *Branhamella* group of cocci. *In* Starr, Stolp, Trüper, Balows and Schlegel (Editors), The Prokaryotes. A Handbook on Habitats, Isolation and Identification of Bacteria, 1st Ed., Vol. 2, Springer-Verlag, New York. pp. 1506–1529.

Bøvre, K. and N. Hagen. 1984. *In* Validation of the publication of new names and new combinations previously effectively published outside the IJSB. List No. 15. Int. J. Syst. Bacteriol. *34*: 355–357.

Bøvre, K., N. Hagen, B.P. Berdal and E. Jantzen. 1977. Oxidase positive rods from cases of suspected gonorrhoea. A comparison of conventional, gas chromatographic and genetic methods of identification. Acta Pathol. Microbiol. Scand. B. *85*: 27–37.

Bøvre, K. and S.D. Henriksen. 1962. An approach to transformation studies in *Moraxella*. Acta Pathol. Microbiol. Scand. *56*: 223–228.

Bøvre, K. and S.D. Henriksen. 1967a. A new *Moraxella* species, *Moraxella osloensis*, and a revised description of *Moraxella nonliquefaciens*. Int. J. Syst. Bacteriol. *17*: 127–135.

Bøvre, K. and S.D. Henriksen. 1967b. A revised description of *Moraxella polymorpha* Flamm 1957, with a proposal of a new name, *Moraxella phenylpyruvica* for this species. Int. J. Syst. Bacteriol. *17*: 343–360.

Bowditch, R.D., L. Baumann and P. Baumann. 1984a. Description of *Oceanospirillum kriegii* sp. nov. and *Oceanospirillum jannaschii* sp. nov. and assignment of 2 species of *Alteromonas* to this genus as *Oceanospirillum commune* comb. nov. and *Oceanospirillum vagum* comb. nov. Curr. Microbiol. *10*: 221–229.

Bowditch, R.D., L. Baumann and P. Baumann. 1984b. Validation of the publication of new names and combinations previously effectively published outside the IJSB. List No.16. Int. J. Syst. Bacteriol. *34*: 503–504.

Bowles, R.E., J.L. Pahoff, B.N. Smith and P.J. Blackall. 2000. Ribotype diversity of porcine *Pasteurella multocida* from Australia. Aust. Vet. J. *78*: 630–635.

Bowman, J.P. 1992. The systematics of methane-utilizing bacteria., Doctoral thesis, University of Queensland, Brisbane, Australia.

Bowman, J.P. 1998. *Pseudoalteromonas prydzensis* sp. nov., a psychotrophic, halotolerant bacterium from Antarctic sea ice. Int. J. Syst. Bacteriol. *48*: 1037–1041.

Bowman, J.P., J. Cavanagh, J.J. Austin and K. Sanderson. 1996. Novel *Psychrobacter* species from Antarctic ornithogenic soils. Int. J. Syst. Bacteriol. *46*: 841–848.

Bowman, J.P., J.J. Gosink, S.A. McCammon, T.E. Lewis, D.S. Nichols, P.D. Nichols, J.H. Skerratt, J.T. Staley and T.A. McMeekin. 1998. *Colwellia demingiae* sp. nov., *Colwellia hornerae* sp. nov., *Colwellia rossensis* sp. nov., and *Colwellia psychrotropica* sp. nov.: psychrophilic Antarctic species with the ability to synthesize docosahexaenoic acid (22:6ω3). Int. J. Syst. Bacteriol. *48*: 1171–1180.

Bowman, J.P., S.A. McCammon, J.L. Brown and T.A. McMeekin. 1998b. *Glaciecola punicea* gen. nov., sp. nov. and *Glaciecola pallidula* gen. nov., sp. nov.: psychrophilic bacteria from Antarctic sea-ice habitats. Int. J. Syst. Bacteriol. *48*: 1213–1222.

Bowman, J.P., S.A. McCammon, M.V. Brown, D.S. Nichols and T.A. McMeekin. 1997a. Diversity and association of psychrophilic bacteria in Antarctic sea ice. Appl. Environ. Microbiol. *63*: 3068–3078.

Bowman, J.P., S.A. McCammon, T. Lewis, J.H. Skerratt, D.S. Nichols and T.A. McMeekin. 1998c. *Psychroflexus torquis* gen. nov. sp. nov., a psychrophilic species from Antarctic Sea ice, and reclassification of *Flavobacterium gondwanense* (Dobson et al. 1993) as *Psychroflexus gondwanense* gen. nov. comb. nov. Microbiology *144*: 1601–1609.

Bowman, J.P., S.A. McCammon, D.S. Nichols, J.H. Skerratt, S.M. Rea, P.D. Nichols and T.A. McMeekin. 1997b. *Shewanella gelidimarina* sp. nov. and *Shewanella frigidimarina* sp. nov., novel Antartic species with the ability to produce eicosapentaenoic acid (20:5ω3) and grow anaerobically by dissimilatory Fe(III) reduction. Int. J. Syst. Bacteriol. *47*: 1040–1047.

Bowman, J.P., S.A. McCammon and J.H. Skerratt. 1997c. *Methylosphaera hansonii* gen. nov., sp. nov., a psychrophilic, group I methanotroph from antarctic marine-salinity, meromictic lakes. Microbiology *143*: 1451–1459.

Bowman, J.P., S.A. McCammon and J.H. Skerratt. 1998d. *In* Validation of the publication of new names and new combinations previously effectively published outside the IJSB. List No. 64. Int. J. Syst. Bacteriol. *48*: 327–328.

Bowman, J.P., D.S. Nichols and T.A. McMeekin. 1997d. *Psychrobacter glacincola* sp. nov., a halotolerant, psychrophilic bacterium isolated from Antarctic Sea ice. Syst. Appl. Microbiol. *20*: 209–215.

Bowman, J.P., D.S. Nichols and T.A. McMeekin. 1997e. *In* Validation of the publication of new names and new combinations previously effectively published outside the IJSB. List No. 63. Int. J. Syst. Bacteriol. *47*: 1274.

Bowman, J.P., J.H. Skerratt, P.D. Nichols and L.I. Sly. 1991a. Phospholipid fatty acid and lipopolysaccharide fatty acid signature lipids in methane-utilising bacteria. FEMS Microbiol. Ecol. *85*: 15–22.

Bowman, J.P., L.I. Sly, J.M. Cox and A.C. Hayward. 1990a. *Methylomonas fodinarum* sp. nov. and *Methylomonas aurantiaca* sp. nov.: two closely related type I obligate methanotrophs. Syst. Appl. Microbiol. *13*: 279–287.

Bowman, J.P., L.I. Sly, J.M. Cox and A.C. Hayward. 1990b. *In* Validation of the publication of new names and new combinations previously effectively published outside the IJSB. List No. 35. Int. J. Syst. Bacteriol. *40*: 470–471.

Bowman, J.P., L.I. Sly and A.C. Hayward. 1988. *Pseudomonas mixta* sp. nov., a bacterium from soil with degradative activity on a variety of complex polysaccharides. Syst. Appl. Microbiol. *11*: 53–59.

Bowman, J.P., L.I. Sly and A.C. Hayward. 1991b. Contribution of genome characteristics to assessment of taxonomy of obligate methanotrophs. Int. J. Syst. Bacteriol. *41*: 301–305.

Bowman, J.P., L.I. Sly, A.C. Hayward, Y. Spiegel and E. Stackebrandt. 1993a. *Telluria mixta* (*Pseudomonas mixta* Bowman, Sly, and Hayward

1988) gen. nov., comb. nov., and *Telluria chitinolytica* sp. nov., soil-dwelling organisms which actively degrade polysaccharides. Int. J. Syst. Bacteriol. *43*: 120–124.

Bowman, J.P., L.I. Sly, P.D. Nichols and A.C. Hayward. 1993b. Revised taxonomy of the methanotrophs: description of *Methylobacter* gen. nov., emendation of *Methylococcus*, validation of *Methylosinus* and *Methylocystis* species, and a proposal that the family *Methylococcaceae* includes only the group I methanotrophs. Int. J. Syst. Bacteriol. *43*: 735–753.

Bowman, J.P., L.I. Sly and E. Stackebrandt. 1995. The phylogenetic position of the family *Methylococcaceae*. Int. J. Syst. Bacteriol. *45*: 182–185.

Boyce, J.D. and B. Adler. 2000. The capsule is a virulence determinant in the pathogenesis of *Pasteurella multocida* M1404 (B:2). Infect. Immun. *68*: 3463–3468.

Boyce, J.D., J.Y. Chung and B. Adler. 2000. Genetic organisation of the capsule biosynthetic locus of *Pasteurella multocida* M1404 (B:2). Vet. Microbiol. *72*: 121–134.

Boyce, J.M.H., J. Frazer and K. Zinnemann. 1969. The growth requirements of *Haemophilus aphrophilus*. J. Med. Microbiol. *2*: 55–62.

Boyce, T.G., D.L. Swerdlow and P.M. Griffin. 1995. *Escherichia coli* O157:H7 and the hemolytic-uremic syndrome. N. Engl. J. Med. *333*: 364–368.

Boyd, C. and N.T. Keen. 1993. Characterization of the *prtA* and *prtB* genes of *Erwinia chrysanthemi* EC16. Gene *133*: 115–118.

Boyd, J.S.K. 1931. Some investigations into so-called "non-agglutinable" dysentery bacilli. J. Roy. Army Med. Corps. *57*: 161–186.

Boyd, J.S.K. 1932a. Further investigations into the characters and classification of the mannite-fermenting dysentery bacilli. J. Roy. Army Med. Corps. *59*: 241–251.

Boyd, J.S.K. 1932b. Further investigations into the characters and classification of the mannite-fermenting dysentery bacilli. J. Roy. Army Med. Corps. *59*: 331–342.

Boyd, J.S.K. 1938. The antigenic structure of the mannitol-fermenting group of dysentery bacilli. J. Hyg. *38*: 477–499.

Boyd, J.S.K. 1946. Laboratory findings in clinical dysentery in Middle East force between August 1940 and June 1943. J. Pathol. *58*: 237–241.

Boyer, G. and F. Lambert. 1893. Sur deux nouvelles maladies du mûrier. C. R. Hebd. Séances Acad. Sci. *117*: 342–343.

Bozal, N., A. Manresa, J. Castellvi and J. Guinea. 1994. A new bacterial strain of Antarctica, *Alteromonas* sp. that produces a heteropolymer slime. Polar Biol. *14*: 561–567.

Bozal, N., E. Tudela, R. Rosselló-Mora, J. Lalucat and J. Guinea. 1997. *Pseudoalteromonas antarctica* sp. nov., isolated from an Antarctic coastal environment. Int. J. Syst. Bacteriol. *47*: 345–351.

Bozeman, F.M., J.W. Humphries and J.M. Campbell. 1968. A new group of rickettsia-like agents recovered from guinea pigs. Acta Virol. (Praha). *12*: 87–93.

Bozsó, J., J. Durst, A. Pásztor, A. Svidró and J. Szita. 1986. Interpretation of the presence of *Plesiomonas shigelloides* in faecal samples from patients with enteric disease. Acta Microbiol. Hung. *33*: 351–354.

Braak, H.R. 1928. Onderzoekingen over Vergisting van Glycerine, Thesis, Delft.

Bracho, A.M., D. Martínez-Torres, A. Moya and A. Latorre. 1995. Discovery and molecular characterization of a plasmid localized in *Buchnera* sp. bacterial endosymbiont of the aphid *Rhopalosiphum padi*. J. Mol. Evol. *41*: 67–73.

Brackee, G., R. Gunther and C.S. Gillett. 1992. Diagnostic exercise: high mortality in red-eared slider turtles (*Pseudemys scripta elegans*). Lab. Anim. Sci. *42*: 607–609.

Bradbury, J.F. 1986. Guide to Plant Pathogenic Bacteria, CAB International Mycological Institute, Kew.

Brade, H. and H. Brunner. 1979. Serological cross-reactions between *Acinetobacter calcoaceticus* and chlamydiae. J. Clin. Microbiol. *10*: 819–822.

Bradley, D.E. 1972a. Evidence for the retraction of *Pseudomonas aeruginosa* RNA phage pili. Biochem. Biophys. Res. Comm. *47*: 142–149.

Bradley, D.E. 1972b. Stimulation of pilus formation in *Pseudomonas ae-*

ruginosa by RNA bacteriophage adsorption. Biochem. Biophys. Res. Comm. *47*: 1080–1087.

Bradley, D.E. 1974. The adsorption of *Pseudomonas aeruginosa* pilus-dependent bacteriophages to a host mutant with non-retractile pili. Virology *58*: 149–163.

Bradley, D.E. 1980a. Function of *Pseudomonas aeruginosa* PAO polar pili - twitching motility. Can. J. Microbiol. *26*: 146–154.

Bradley, D.E. 1980b. Mobilization of chromosomal determinants for the polar pili of *Pseudomonas aeruginosa* PAO by FP plasmids. Can. J. Microbiol. *26*: 155–160.

Bradley, D.E., S.P. Howard and H. Lior. 1991. Colicinogeny of O157:H7 enterohemorrhagic *Escherichia coli* and the shielding of colicin and phage receptors by their O-antigenic side chains. Can. J. Microbiol. *37*: 97–104.

Bradley, D.E. and T.L. Pitt. 1975. An immunological study of the pili of *Pseudomonas aeruginosa*. J. Hyg. (London). *74*: 419–430.

Bradshaw, M.W., R.P. Schneerson, J.C. and J.B. Robbins. 1971. Bacterial antigens cross-reactive with the capsular polysaccharide of *Haemophilus influenzae* type b. Lancet *1*: 1095–1096.

Bragg, R.R., J.M. Greyling and J.A. Verschoor. 1997. Isolation and identification of NAD-independent bacteria from chickens with symptoms of infectious coryza. Avian Pathol. *26*: 595–606.

Brahmbhatt, H.N., A.A. Lindberg and K.N. Timmis. 1992. *Shigella* lipopolysaccharide: structure, genetics, and vaccine development. Curr. Top. Microbiol. Immunol. *180*: 45–64.

Branca, G., C. Galasso, G. Cornaglia, R. Fontana and G. Satta. 1996. A simple method to detect bacteriolytic enzymes produced by *Enterobacteriaceae*. Microbios. *86*: 205–212.

Brandreth, S.R. and I.M. Smith. 1986. Lack of pathogenicity of haemophili of the "minor group" taxon for the gnotobiotic piglet. Res. Vet. Sci. *40*: 273–275.

Branefors-Helander, P. 1972. Serological studies of *Haemophilus influenzae*. II. Variation in the capsular antigen of type e strains. Int. Arch. Allergy Appl. Immunol. *43*: 908–920.

Branefors-Helander, P. 1977. The structure of the capsular antigen from *Haemophilus influenzae* type A. Carbohydr. Res. *56*: 117–121.

Branefors-Helander, P. 1979. Cross reactivity of the O-antigens among *Haemophilus influenzae* type b strains. Int. Arch. Allergy Appl. Immunol. *52*: 150–154.

Branefors-Helander, P., B. Clausen, L. Kenne and B. Lindberg. 1979. Structural studies of the capsular antigen of *Haemophilus influenzae* type c. Carbohydr. Res. *76*: 197–202.

Branefors-Helander, P., L. Keene and B. Lindqvist. 1980. Structural studies of the capsular antigen from *Haemophilus influenzae* type f. Carbohydr. Res. *79*: 308–312.

Branefors-Helander, P., L. Kenne, B. Lindberg, K. Petersson and P. Unger. 1981. Structural studies of the capsular polysaccharaide elaborated by *Haemophilus influenzae* type D. Carbohydr. Res. *97*: 285–291.

Branson, E.J. and D.A. Diaz-Munoz. 1991. Description of a new disease condition occurring in farmed coho salmon, *Oncorhynchus kisutch* (Walbaum), in South America. J. Fish Dis. *14*: 147–156.

Bratina, B.J., G.A. Brusseau and R.S. Hanson. 1992. Use of 16S rRNA analysis to investigate phylogeny of methylotrophic bacteria. Int. J. Syst. Bacteriol. *42*: 645–648.

Braun, C. and W.G. Zumft. 1992. The structural genes of the nitric-oxide reductase complex from *Pseudomonas stutzeri* are part of a 30-kilobase gene-cluster for denitrification. J. Bacteriol. *174*: 2394–2397.

Braun, E.J. 1982. Ultrastructural investigation of resistant and susceptible maize inbreds infected with *Erwinia stewartii*. Phytopathology *72*: 159–166.

Braun-Howland, E.B., P.A. Vescio and S.A. Nierzwicki-Bauer. 1993. Use of a simplified cell blot technique and 16S ribosomal-RNA- directed probes for identification of common environmental isolates. Appl. Environ. Microbiol. *59*: 3219–3224.

Braunstein, H., M. Tomasulo, S. Scott and M.P. Chadwick. 1980. A biotype of *Enterobacteriaceae* intermediate between *Citrobacter* and *Enterobacter*. Amer. J. Clin. Pathol. *73*: 114–116.

Bravo, S. 1994. Piscirickettsiosis in freshwater. Bull. Eur. Assoc. Fish Pathol. *14*: 137–138.

Bravo, S. and M. Campos. 1989. Coho salmon in freshwater. FHS Newsletter *17*: 3.

Brayer, G.D., L.T.J. Delbaere and M.N.G. James. 1979. Molecular structure of the α-lytic protease from *Myxobacter* 495 at 2.8 Å resolution. J. Mol. Biol. *131*: 743–775.

Brayton, P.R., R.B. Bode, R.R. Colwell, M.T. MacDonell, H.L. Hall, D.J. Grimes, P.A. West and T.N. Bryant. 1986a. *In* Validation of new names and new combinations previously effectively published outside the IJSB. List No. 20. Int. J. Syst. Bacteriol. *36*: 354–356.

Brayton, P.R., R.B. Bode, R.R. Colwell, M.T. MacDonell, H.L. Hall, D.J. Grimes, P.A. West and T.N. Bryant. 1986b. *Vibrio cincinnatiensis* sp. nov., a new human pathogen. J. Clin. Microbiol. *23*: 104–108.

Breed, R.S. 1948. Genus I. *Pseudomonas*. *In* Breed, Murray and Hitchens (Editors), Bergey's Manual of Determinative Bacteriology, 6th Ed., The Williams & Wilkins Co., Baltimore. 82–150.

Breed, R.S. , E.G.D. Murray and A.P. Hitchens (Editors). 1948. Bergey's Manual of Determinative Bacteriology, 6th Ed., The Williams & Wilkins Co., Baltimore.

Breed, R.S., E.G.D. Murray and N.R. Smith (Editors). 1957. Bergey's Manual of Determinative Bacteriology, 7th Ed., The Williams & Wilkins Co., Baltimore.

Brehélin, M., A. Cherqui, L. Drif, J. Luciani, R. Akhurst and N. Boemare. 1993. Ultrastructural study of surface components of *Xenorhabdus* sp. in different cell phases and culture conditions. J. Invertebr. Pathol. *61*: 188–191.

Brenden, R. and J.M. Janda. 1987. Detection, quantitation and stability of the beta haemolysin of *Aeromonas* spp. J. Med. Microbiol. *24*: 247–251.

Brenden, R.A., M.A. Miller and J.M. Janda. 1988. Clinical disease spectrum and pathogenic factors associated with *Plesiomonas shigelloides* infections in humans. Rev. Infect. Dis. *10*: 303–316.

Brenner, D.J. 1978. Characterization and clinical identification of *Enterobacteriaceae* by DNA hybridization. Prog. Clin. Pathol. *7*: 71–117.

Brenner, D.J. 1981. Introduction to the family *Enterobacteriaceae*. *In* Starr, Stolp, Trüper, Balows and Schlegel (Editors), The Prokaryotes. A handbook on Habitats, Isolation, and Identification of Bacteria, 1st Ed., Vol. 2, Springer-Verlag, New York. pp. 1105–1127.

Brenner, D.J. 1984a. Family I. *Enterobacteriaceae*. *In* Krieg and Holt (Editors), Bergey's Manual of Systematic Bacteriology, 1st Ed., Vol. 1, The Williams & Wilkins Co., Baltimore. pp. 408–420.

Brenner, D.J. 1984b. Recommendations on recent proposals for the classification of shigellae. Int. J. Syst. Bacteriol. *34*: 87–88.

Brenner, D.J. 1986. Classification of *Legionellaceae*. Current status and remaining questions. Isr. J. Med. Sci. *22*: 620–632.

Brenner, D.J. 1992a. Additional genera of *Enterobacteriaceae*. *In* Balows, Trüper, Dworkin, Harder and Schleifer (Editors), The Prokaryotes, 2nd Ed., Vol. 3, Springer-Verlag, New York. pp. 2922–2937.

Brenner, D.J. 1992b. Introduction to the family *Enterobacteriaceae*. *In* Balows, A., H.G. Trüper, M. Dworkin, W. Harder and K.-H. Schleifer (Editors), The Prokaryotes. A Handbook on the Biology of Bacteria: Ecophysiology, Isolation, Identification, Applications, 2nd Ed., Vol. III, Springer-Verlag, New York. pp. 2673–2695.

Brenner, D.J., H. Bercovier, J. Ursing, J.M. Alonso, A.G. Steigerwalt, G.R. Fanning, G.P. Carter and H.H. Mollaret. 1980a. *Yersinia intermedia* a new species of *Enterobacteriaceae* composed of rhamnose-positive strains (formerly called atypical *Yersinia enterocolitica* or *Yersinia enterocolitica*-like). Curr. Microbiol. *4*: 207–212.

Brenner, D.J., H. Bercovier, J. Ursing, J.M. Alonso, A.G. Steigerwalt, G.R. Fanning, G.P. Carter and H.H. Mollaret. 1981. *In* Validation of the publication of new names and new combinations previously effectively published outside the IJSB. List No. 6. Int. J. Syst. Bacteriol. *31*: 215–218.

Brenner, D.J., B.R. Davis, A.G. Steigerwalt, C.F. Riddle, A.C. McWhorter, S.D. Allen, J.J. Farmer, III, Y. Saitoh and G.R. Fanning. 1982a. Atypical biogroups of *Escherichia coli* found in clinical specimens and description of *Escherichia hermannii* sp. nov. J. Clin. Microbiol. *15*: 703–713.

Brenner, D.J., B.R. Davis, A.G. Steigerwalt, C.F. Riddle, A.C. McWhorter, S.D. Allen, J.J. Farmer, Y. Saitoh and G.R. Fanning. 1983a. *In* Validation of the publication of new names and new combinations previously effectively published outside the IJSB. List No. 10. Int. J. Syst. Bacteriol. *33*: 438–440.

Brenner, D.J., G.R. Fanning, F.W. Hickmann-Brenner, J.V. Lee, A.G. Steigerwalt, B.R. Davis and J.J. Farmer. 1983b. DNA relatedness among *Vibrionaceae*, with emphasis on the *Vibrio* species associated with human infection. INSERM Colloq. *114*: 175–184.

Brenner, D.J., G.R. Fanning, J.K. Leete Knutson, A.G. Steigerwalt and M.I. Krichevsky. 1984a. Attempts to classify *Herbicola* group-*Enterobacter agglomerans* strains by deoxyribonucleic hybridization and phenotypic tests. Int. J. Syst. Bacteriol. *34*: 45–55.

Brenner, D.J., G.R. Fanning, G.V. Miklos and A.G. Steigerwalt. 1973a. Polynucleotide sequence relatedness among *Shigella* species. Int. J. Syst. Bacteriol. *23*: 1–7.

Brenner, D.J., G.R. Fanning, F.J. Skerman and S. Falkow. 1972a. Polynucleotide sequence divergence among strains of *Escherichia coli* and closely related organisms. J. Bacteriol. *109*: 933–965.

Brenner, D.J., G.R. Fanning and A.G. Steigerwalt. 1974a. Deoxyribonucleic acid relatedness among erwiniae and other *Enterobacteriaceae*: the gall, wilt and dry necrosis organisms (genus *Erwinia* Winslow et al., sensu stricto). Int. J. Syst. Bacteriol. *24*: 197–204.

Brenner, D.J., G.R. Fanning and A.G. Steigerwalt. 1974b. Polynucleotide sequence relatedness in *Edwardsiella tarda*. Int. J. Syst. Bacteriol. *24*: 186–190.

Brenner, D.J., G.R. Fanning, A.G. Steigerwalt, I. Ørskov and F. Ørskov. 1972b. Polynucleotide sequence relatedness among three groups of pathogenic *Escherichia coli* strains. Infect. Immun. *6*: 308–315.

Brenner, D.J. and J.J. Farmer, III. 1981. The genus *Citrobacter*. *In* Starr, Stolp, Trüper, Balows and Schlegel (Editors), The Prokaryotes: A Handbook on Habitats, Isolation and Identification of Bacteria, Springer Verlag, New York. pp. 1140–1147.

Brenner, D.J. and J.J. Farmer, III. 1982. *In* Validation of the publication of new names and new combinations previously published outside of the IJSB. List No. 8. Int. J. Syst. Bacteriol. *32*: 266–268.

Brenner, D.J., J.J. Farmer, III, G.R. Fanning, A.G. Steigerwalt, P. Klykken, H.G. Wathen, F.W. Hickman and W.H. Ewing. 1978. Deoxyribonucleic acid relatedness in species of *Proteus* and *Providencia*. Int. J. Syst. Bacteriol. *28*: 269–282.

Brenner, D.J., J.J. Farmer, III, F.W. Hickman, M.A. Asbury and A.G. Steigerwalt. 1977. Taxonomic and nomenclature changes in *Enterobacteriaceae*. *In* HEW Publication No. (CDC) 79-8356, Center for Disease Control, Atlanta.

Brenner, D.J., P.A.D. Grimont, A.G. Steigerwalt, G.R. Fanning, E. Ageron and C.F. Riddle. 1993. Classification of citrobacteria by DNA hybridization: designation of *Citrobacter farmeri* sp. nov., *Citrobacter youngae* sp. nov., *Citrobacter braakii* sp. nov., *Citrobacter werkmanii* sp. nov., *Citrobacter sedlakii* sp. nov., and three unnamed *Citrobacter* genomospecies. Int. J. Syst. Bacteriol. *43*: 645–658.

Brenner, D.J., F.W. Hickman-Brenner, B. Holmes, P.M. Hawkey, J.L. Penner, P.A. Grimont and C.M. O'Hara. 1995. Replacement of NCTC 4175, the current type strain of *Proteus vulgaris*, with ATCC 29905. Request for an opinion. Int. J. Syst. Bacteriol. *45*: 870–871.

Brenner, D.J., F.W. Hickman-Brenner, J.V. Lee, A.G. Steigerwalt, G.R. Fanning, D.G. Hollis, J.J. Farmer, III, R.E. Weaver, S.W. Joseph and R.J. Seidler. 1983c. *Vibrio furnissii* (formerly aerogenic biogroup of *Vibrio fluvialis*), a new species isolated from human feces and the environment. J. Clin. Microbiol. *18*: 816–824.

Brenner, D.J., F.W. Hickman-Brenner, J.V. Lee, A.G. Steigerwalt, G.R. Fanning, D.G. Hollis, J.J. Farmer, III, R.E. Weaver, S.W. Joseph and R.J. Seidler. 1984b. *In* Validation of new names and new combinations previously effectively published outside the IJSB. List No. 13. Int. J. Syst. Bacteriol. *34*: 91–92.

Brenner, D.J., L.W. Mayer, G.M. Carlone, L.H. Harrison, W.F. Bibb, M.C. Brandileone, F.O. Sottnek, K. Irino, M.W. Reeves and J.M. Swenson. 1988a. Biochemical, genetic, and epidemiologic characterization of *Haemophilus influenzae* biogroup aegyptius (*Haemophilus aegyptius*) strains associated with Brazilian purpuric fever. J. Clin. Microbiol. *26*: 1524–1534.

Brenner, D.J., A.C. McWhorter, A. Kai, A.G. Steigerwalt and J.J. Farmer, III. 1986. *Enterobacter asburiae* sp. nov., a new species found in clinical specimens, and reassignment of *Erwinia dissolvens* and *Erwinia nimipressuralis* to the genus *Enterobacter* as *Enterobacter dissolvens* comb. nov. and *Enterobacter nimipressuralis* comb. nov. J. Clin. Microbiol. *23*: 1114–1120.

Brenner, D.J., A.C. McWhorter, A. Kai, A.G. Steigerwalt and J.J. Farmer. 1988b. *In* Validation of the publication of new names and new combinations previously effectively published outside the IJSB, List No. 25. Int. J. Syst. Bacteriol. *38*: 220.

Brenner, D.J., A.C. McWhorter, J.K.L. Knutson and A.G. Steigerwalt. 1982b. *Escherichia vulneris*: a new species of *Enterobacteriaceae* associated with human wounds. J. Clin. Microbiol. *15*: 1133–1140.

Brenner, D.J., A.C. McWhorter, J.K.L. Knutson and A.G. Steigerwalt. 1983d. *In* Validation of the publication of new names and new combinations previously effectively published outside the IJSB. List No. 10. Int. J. Syst. Bacteriol. *33*: 438–440.

Brenner, D.J., H.E. Muller, A.G. Steigerwalt, A.M. Whitney, C.M. O'Hara and P. Kämpfer. 1998. Two new *Rahnella* genomospecies that cannot be phenotypically differentiated from *Rahnella aquatilis*. Int. J. Syst. Bacteriol. *48*: 141–149.

Brenner, D.J., C.M. O'Hara, P.A.D. Grimont, J.M. Janda, E. Falsen, E. Aldová, E. Ageron, J. Schindler, S.L. Abbott and A.G. Steigerwalt. 1999. Biochemical identification of *Citrobacter* species defined by DNA hybridization and description of *Citrobacter gillenii* sp. nov. (formerly *Citrobacter* genomospecies 10) and *Citrobacter murliniae* sp. nov. (formerly *Citrobacter* genomospecies 11). J. Clin. Microbiol. *37*: 2619–2624.

Brenner, D.J., C.M. O'Hara, P.A.D. Grimont, J.M. Janda, E. Falsen, E. Aldová, E. Ageron, J. Schindler, S.L. Abbott and A.G. Steigerwalt. 2000. *In* Validation of the publication of new names and new combinations previously effectively published outside the IJSEM. List No. 73. Int. J. Syst. Evol. Microbiol. *50*: 423–424.

Brenner, D.J., C. Richard, A.G. Steigerwalt, M.A. Asbury and M. Mandel. 1980b. *Enterobacter gergoviae* sp. nov. a new species of *Enterobacteriaceae* found in clinical specimens and the environment. Int. J. Syst. Bacteriol. *30*: 1–6.

Brenner, D.J., A.G. Steigerwalt, P. Epple, W.F. Bibb, R.M. McKinney, R.W. Starnes, J.M. Colville, R.K. Selander, P.H. Edelstein and C.W. Moss. 1988c. *Legionella pneumophila* serogroup Lansing 3 isolated from a patient with fatal pneumonia, and descriptions of *Legionella pneumophila* subsp. *pneumophila*, subsp. nov., *Legionella pneumophila* subsp. *fraseri*, subsp. nov. and *Legionella pneumophila* subsp. *pascullei*, subsp. nov. J. Clin. Microbiol. *26*: 1695–1703.

Brenner, D.J., A.G. Steigerwalt, P. Epple, W.F. Bibb, R.M. McKinney, R.W. Starnes, J.M. Colville, R.K. Selander, P.H. Edelstein and C.W. Moss. 1989. *In* Validation of the publication of new names and new combinations previously effectively published outside the IJSB. List No. 29. Int. J. Syst. Bacteriol. *39*: 205–206.

Brenner, D.J., A.G. Steigerwalt, D.P. Falcão, R.E. Weaver and G.R. Fanning. 1976. Characterization of *Yersinia enterocolitica* and *Yersinia pseudotuberculosis* by deoxyribonucleic acid hybridization and by biochemical reactions. Int. J. Syst. Bacteriol. *26*: 180–194.

Brenner, D.J., A.G. Steigerwalt and G.R. Fanning. 1972c. Differentiation of *Enterobacter aerogenes* from klebsiellae by deoxyribonucleic acid reassociation. Int. J. Syst. Bacteriol. *22*: 193–200.

Brenner, D.J., A.G. Steigerwalt, G.W. Gorman, R.E. Weaver, J.C. Feeley, L.G. Cordes, H.W. Wilkinson, C. Patton, B.M. Thomason and K.R. Lewallen-Sasseville. 1980c. *Legionella bozemanii* sp. nov. and *Legionella dumoffii* sp nov.: classification of two additional species of *Legionella* associated with human pneumonia. Curr. Microbiol. *4*: 111–116.

Brenner, D.J., A.G. Steigerwalt, G.W. Gorman, R.E. Weaver, J.C. Feeley, L.G. Cordes, H.W. Wilkinson, C. Patton, B.M. Thomason and K.R. Lewallen-Sasseville. 1980d. *In* Validation of the publication of new names and new combinations previously effectively published outside the IJSB. List No. 5. Int. J. Syst. Bacteriol. *30*: 676–677.

Brenner, D.J., A.G. Steigerwalt, G.W. Gorman, H.W. Wilkinson, W.F. Bibb,

M. Hackel, R.L. Tyndall, J. Campbell, J.C. Feeley, W.L. Thacker, P. Skaliy, W.T. Martin, B.J. Brake, B.S. Fields, H.V. McEachern and L.K. Corcoran. 1985. Ten new species of *Legionella*. Int. J. Syst. Bacteriol. *35*: 50–59.

Brenner, D.J., A.G. Steigerwalt and J.E. McDade. 1979. Classification of the Legionnaires' disease bacterium: *Legionella pneumophila*, genus novum, species nova of the family *Legionellceae*, familia nova. Ann. Intern. Med. *90*: 656–658.

Brenner, D.J., A.G. Steigerwalt, G.V. Miklos and G.R. Fanning. 1973b. Deoxyribonucleic acid relatedness among erwiniae and other *Enterobacteriaceae*: the soft-rot organisms (genus *Pectobacterium* Waldee). Int. J. Syst. Bacteriol. *23*: 205–216.

Brenner, D.J., A.G. Steigerwalt, H.G. Wathen, R.J. Gross and B. Rowe. 1982c. Confirmation of aerogenic strains of *Shigella boydii* 13 and further study of *Shigella* serotypes by DNA relatedness. J. Clin. Microbiol. *16*: 432–436.

Brenner, D.J., J. Ursing, H. Bercovier, A.G. Steigerwalt, G.R. Fanning, J.M. Alonso and H.H. Mollaret. 1980e. Deoxyribonucleic acid relatedness in *Yersinia enterocolitica* and *Yersinia enterocolitica*-like organisms. Curr. Microbiol. *4*: 195–200.

Brenner, F.W. and A.C. McWhorter-Murlin. 1998. Identification and serotyping of *Salmonella* and an update of the Kauffmann–White scheme, Atlanta, GA. Centers for Disease Control and Prevention.

Brentjens, R.J., M. Ketterer, M.A. Apicella and S.M. Spinola. 1996. Fine tangled pili expressed by *Haemophilus ducreyi* are a novel class of pili. J. Bacteriol. *178*: 808–816.

Brettar, I. and M.G. Höfle. 1993. Nitrous-oxide producing heterotrophic bacteria from the water column of the central Baltic: abundance and molecular identification. Mar. Ecol. Prog. Ser. *94*: 253–265.

Breuil, C. and D.J. Kushner. 1975. Lipase and esterase formation by psychrophilic and mesophilic *Acinetobacter* species. Can. J. Microbiol. *21*: 423–433.

Breuil, C., T.J. Novitsky and D.J. Kushner. 1975. Characteristics of a facultatively psychrophilic *Acinetobacter* species isolated from river sediment. Can. J. Microbiol. *21*: 2103–2108.

Brewer, J.W., M.D. Harrison and J.A. Winston. 1981. Survival of two varieties of *Erwinia carotovora* on *Drosophila melanogaster* and *Drosophila busckii* (Diptera: Drosophilidae), vectors of potato blackleg in Colorado, USA. Am. Potato J. *58*: 439–449.

Brewer, R.J., R.B. Galland and H.C. Polk, Jr.. 1982. Amelioration by muramyl dipeptide of the effect of induced hyperferremia upon *Klebsiella* infection in mice. Infect. Immun. *38*: 175–178.

Brezina, R. 1978. Phase variation phenomenom in *Coxiella burnetti*. *In* Kazar, Ormsbee and Tarasevich (Editors), Rickettsiae and Rickettsial Diseases, VEDA, Bratislava. pp. 221–235.

Briedis, D.J. and H.G. Robson. 1976. Comparative activity of netilmicin, gentamicin, amikacin and tobramycin against *Pseudomonas aeruginosa* and *Enterobacteriaceae*. Antimicrob. Agents Chemother. *10*: 592–597.

Brigmon, R.L., G. Bitton, S.G. Zam and B. O'Brien. 1995. Development and application of a monoclonal-antibody against *Thiothrix* ssp. Appl. Environ. Microbiol. *61*: 13–20.

Brigmon, R.L. and C. De Ridder. 1998. Symbiotic relationship of *Thiothrix* spp. with an echinoderm. Appl. Environ. Microbiol. *64*: 3491–3495.

Brigmon, R.L., H.W. Martin, T.L. Morris, G. Bitton and S.G. Zam. 1994. Biogeochemical ecology of *Thiothrix* spp. in underwater limestone caves. Geomicrobiol. J. *12*: 141–159.

Brindle, R.J., P.J. Stannett and R.N. Cunliffe. 1987. *Legionella pneumophila*: comparison of isolation from water specimens by centrifugation and filtration. Epidemiol. Infect. *99*: 241–248.

Brink, A.J., A. Vanstraten and A.J. Vanrensburg. 1995. *Shewanella* (*Pseudomonas*) *putrefaciens* bacteremia. Clin. Infect. Dis. *20*: 1327–1332.

Brinkhoff, T. and G. Muyzer. 1997. Increased species diversity and extended habitat range of sulfur-oxidizing *Thiomicrospira* spp. Appl. Environ. Microbiol. *63*: 3789–3796.

Brinkhoff, T., G. Muyzer, C.O. Wirsen and J. Kuever. 1999a. *Thiomicrospira chilensis* sp. nov., a mesophilic obligately chemolithoautotrophic sulfur-oxidizing bacterium isolated from a *Thioploca* mat. Int. J. Syst. Bacteriol. *49*: 875–879.

Brinkhoff, T., G. Muyzer, C.O. Wirsen and J. Kuever. 1999b. *Thiomicrospira kuenenii* sp. nov. and *Thiomicrospira frisia* sp. nov., two mesophilic obligately chemolithoautotrophic sulfur-oxidizing bacteria isolated from an intertidal mud flat. Int. J. Syst. Bacteriol. *49*: 385–392.

Brinkhoff, T., C.M. Santegoeds, K. Sahm, J. Kuever and G. Muyzer. 1998. A polyphasic approach to study the diversity and vertical distribution of sulfur-oxidizing *Thiomicrospira* species in coastal sediments of the German Wadden Sea. Appl. Environ. Microbiol. *64*: 4650–4657.

Brinkhoff, T., S. Sievert, J. Kuever and G. Muyzer. 1999c. Distribution and diversity of sulfur-oxidizing *Thiomicrospira* spp. at a shallow-water hydrothermal vent in the Aegena Sea (Milos, Greece). Appl. Environ. Microbiol. *65*: 3843–3849.

Brinkmann, U., J.L. Ramos and W. Reineke. 1994. Loss of the TOL meta-cleavage pathway functions of *Pseudomonas putida* strain PaW1 (pWWO) during growth on toluene. J. Basic Microbiol. *34*: 303–309.

Brinton, C.C., Jr., M.J. Carter, D.B. Derber, S. Kar, J.A. Kramarik, A.C. To, S.C. To and S.W. Wood. 1989. Design and development of pilus vaccines for *Haemophilus influenzae* diseases. Pediatr. Infect. Dis. J. *8* (*Suppl*): S54–61.

Brion, A. and H. Kayser. 1902. Ueber eine Erkrankung mit dem Befund eines typhus-ähn lichen Bakteriums (Paratyphus). Muenchen Med. Wochenschr. *49*: 611–615.

Brisou, B., C. Richard and A. Lenriot. 1972. Intérêt taxonomique de la recherche de la β-xylosidase chez les *Enterobacteriaceae*. Ann. Inst. Pasteur (Paris) *123*: 341–347.

Brisou, J. and A.R. Prévot. 1954. Étude de systématique bactérienne. X. Révision des espèces réunies dans le genre *Achromobacter*. Ann. Inst. Pasteur (Paris) *86*: 722–728.

Brisse, S. and J. Verhoef. 2001. Phylogenetic diversity of *Klebsiella pneumoniae* and *Klebsiella oxytoca* clinical isolates revealed by randomly amplified polymorphic DNA, *gyrA* and *parC* genes sequencing and automated ribotyping. Int. J. Syst. Evol. Microbiol. *51*: 915–924.

Britigan, B.E., G.T. Rasmussen and C.D. Cox. 1997. Augmentation of oxidant injury to human pulmonary epithelial cells by the *Pseudomonas aeruginosa* siderophore pyochelin. Infect. Immun. *65*: 1071–1076.

Britt, A.J., N.C. Bruce and C.R. Lowe. 1992a. Identification of a cocaine esterase in a strain of *Pseudomonas maltophilia*. J. Bacteriol. *174*: 2087–2094.

Britt, A.J., N.C. Bruce and C.R. Lowe. 1992b. Purification and characterisation of an NAD$^+$-dependent secondary alcohol dehydrogenase from *Pseudomonas maltophilia* MB11L. FEMS Microbiol. Lett. *72*: 49–55.

Brock, T.D. 1964. Knots in *Leucothrix mucor*, a widespread marine microorganism. Science *144*: 870–872.

Brock, T.D. 1966. The habitat of *Leucothrix mucor*, a widespread marine microorganism. Limnol. Oceanogr. *11*: 303–307.

Brock, T.D. 1967. Mode of filamentous growth of *Leucothrix mucor* in pure culture and in nature, as studied by tritiated thymidine autoradiography. J. Bacteriol. *93*: 985–990.

Brock, T.D. 1969. The neotype of *Leucothrix* Oersted 1844 (emend. mut. char. Harold and Stanier 1955). Int. J. Syst. Bacteriol. *19*: 281–282.

Brock, T.D. 1992. The genus *Leucothrix*. *In* Balows, Trüper, Dworkin, Harder and Schleifer (Editors), The Prokaryotes-A Handbook on the Biology of Bacteria: Ecophysiology, Isolation, Identification, Applications., 2nd Ed., Springer-Verlag, New York. pp. 3247–3255.

Brock, T.D. and S.F. Conti. 1969. Electron microscope studies on *Leucothrix mucor*. Arch. Mikrobiol. *66*: 79–90.

Brock, T.D. and M. Mandel. 1966. Deoxyribonucleic acid base composition of geographically diverse strains of *Leucothrix mucor*. J. Bacteriol. *91*: 1659–1660.

Brocklebank, J.R., D.J. Speare, R.D. Armstrong and T. Evelyn. 1992. British Columbia-Septicemia suspected to be caused by a rickettsia-like agent in farmed Atlantic salmon. Can. Vet. J. *33*: 407–408.

Brogden, K.A. 1992. Ovine pulmonary surfactant induces killing of *Pasteurella haemolytica*, *Escherichia coli*, and *Klebsiella pneumoniae* by normal serum. Infect. Immun. *60*: 5182–5189.

Brogden, K.A. and R.A. Packer. 1979. Comparison of *Pasteurella multocida* serotyping systems. Am. J. Vet. Res. *40*: 1332–1335.

Brokopp, C.D. and J.J. Farmer. 1979. Typing methods for *Pseudomonas aeruginosa*. *In* Doggett (Editor), *Pseudomonas aeroginosa*. Clinical manifestations of infection and current therapy, Academic Press, New York. 89–133.

Brondz, I., V. Myhrvold and I. Olsen. 1995. Heterogeneity of *Actinobacillus actinomycetemcomitans* demonstrated with multivariate analysis of fatty acid data from outer membrane vesicles. *In* Donachie, Lainson and Hodgson (Editors), *Haemophilus, Actinobacillus, and Pasteurella*, Plenum Press, New York. pp. 211–212.

Brondz, I. and I. Olsen. 1983. Differentiation of *Actinobacillus actinomycetemcomitans* from *Hemophilus aphrophilus* by gas chromatography of hexane extracts from whole cells. J. Chromatogr. *278*: 13–23.

Brondz, I. and I. Olsen. 1984a. Carbohydrates of whole defatted cells as a basis for differentiation between *Actinobacillus actinomycetemcomitans* and *Hemophilus aphrophilus*. J. Chromatogr. *311*: 31–38.

Brondz, I. and I. Olsen. 1984b. Determination of bound cellular fatty acids in *Actinobacillus actinomycetemcomitans* and *Haemophilus aphrophilus* by gas chromatography and gas chromatography mass spectrometry. J. Chromatogr. *308*: 282–288.

Brondz, I. and I. Olsen. 1984c. Differentiation between *Actinobacillus actinomycetemcomitans* and *Haemophilus aphrophilus* based on carbohydrates in lipopolysaccharide. J. Chromatogr. *310*: 261–272.

Brondz, I. and I. Olsen. 1984d. Whole-cell methanolysis as a rapid method for differentiation between *Actinobacillus actinomycetemcomitans* and *Haemophilus aphrophilus*. J. Chromatogr. *311*: 347–353.

Brondz, I. and I. Olsen. 1985a. Differentiation between major species of the *Actinobaciluus-Haemophilus-Pasteurella* group by gas chromatography of trifluoroacetic acid anhydride derivatives from whole-cell methanolysates. J. Chromatogr. *342*: 13–23.

Brondz, I. and I. Olsen. 1985b. Sugar composition of lipopolysaccharide from *Haemophilus paraphrophilus*. J. Chromatogr. *345*: 119–124.

Brondz, I. and I. Olsen. 1989. Chemical differences in lipopolysaccharides from *Actinobacillus* (*Haemophilus*) *actinomycetemcomitans* and *Haemophilus aphrophilus*: clues to differences in periodontopathogenic potential and taxonomic distinction. Infect. Immun. *57*: 3106–3109.

Brondz, I. and I. Olsen. 1990. Multivariate analyses of carbohydrate data from lipopolysaccharides of *Actinobacillus* (*Haemophilus*) *actinomycetemcomitans*, *Haemophilus aphrophilus*, *Haemophilus paraphrophilus*. Int. J. Syst. Bacteriol. *40*: 405–408.

Brondz, I. and I. Olsen. 1993. Multivariate chemosystematics demonstrate two groups of *Actinobacillus actinomycetemcomitans* strains. Oral Microbiol. Immunol. *8*: 129–133.

Brondz, I., I. Olsen and T. Greibrokk. 1983. Direct analysis of free fatty acids in bacteria by gas chromatography. J. Chromatogr. *274*: 299–304.

Brondz, I., I. Olsen and M. Sjöström. 1990. Multivariate analysis of quantitative chemical and enzymatic characterization data in classification of *Actinobacillus, Haemophilus* and *Pasteurella* spp. J. Gen. Microbiol. *136*: 507–513.

Brook, M.G., H.R. Smith, B.A. Bannister, M. McConnell, H. Chart, S.M. Scotland, A. Sawyer, M. Smith and B. Rowe. 1994. Prospective study of verocytotoxin-producing, enteroaggregative and diffusely adherent *Escherichia coli* in different diarrhoeal states. Epidemiol. Infect. *112*: 63–67.

Brooks, J.B., W.B. Cherry, L. Thacker and C.C. Alley. 1972. Analysis by gas chromatography of amines and nitrosamines produced *in vivo* and *in vitro* by *Proteus mirabilis*. J. Infect. Dis. *126*: 145–153.

Brooks, K. and T. Sodeman. 1974. Clinical studies on a transformation test for identification of *Acinetobacter* (*Mima* and *Herellea*). Appl. Microbiol. *27*: 1023–1026.

Broome, C.V., S.A. Goings, S.B. Thacker, R.L. Vogt, H.N. Beaty and D.W. Fraser. 1979. The Vermont epidemic of Legionnaires' disease. Ann. Intern. Med. *90*: 573–577.

Brorson, J.E., E. Falsen, H. Ehle-Nilsson, S. Rodjer and J. Westin. 1983. Septicemia due to *Moraxella nonliquefaciens* in a patient with multiple myeloma. Scand. J. Infect. Dis. *15*: 221–223.

Brosch, R., M. Lefévre, F. Grimont and P.A.D. Grimont. 1996. Taxonomic diversity of pseudomonads revealed by computer interpretation of ribotyping data. Syst. Appl. Microbiol. *19*: 541–555.

Brosius, J., T.J. Dull, D.D. Sleeter and H.F. Noller. 1981. Gene organization and primary structure of a ribosomal RNA operon from *Escherichia coli*. J. Mol. Biol. *148*: 107–128.

Brosius, J., M.L. Palmer, P.J. Kennedy and H.F. Noller. 1978. Complete nucleotide sequence of a 16S ribosomal RNA gene from *Escherichia coli*. Proc. Natl. Acad. Sci. U.S.A. *75*: 4801–4805.

Brotherton, T., T. Lees and R.D. Feigin. 1976. Susceptibility of *Haemophilus influenzae* type b to cefatrizine, ampicillin, and chloramphenicol. Antimicrob. Agents Chemother. *10*: 322–324.

Brough, C.N. and A.F.G. Dixon. 1990. Ultrastructural features of egg development in oviparae of the vetch aphid, *Megoura viciae* buckton. Tissue Cell *22*: 51–64.

Brown, A., G.M. Garrity and R.M. Vickers. 1981. *Fluoribacter dumoffii* comb. nov. and *Fluoribacter gormanii* comb. nov. Int. J. Syst. Bacteriol. *31*: 111–115.

Brown, A.T. and C. Wagner. 1970. Regulation of enzymes involved in the conversion of tryptophan to nicotinamide adenine dinucleotide in a colorless strain of *Xanthomonas pruni*. J. Bacteriol. *101*: 456–463.

Brown, C.L., M.J. Hill and P. Richards. 1971. Bacterial ureases in uremic men. Lancet *2*: 406–408.

Brown, F., C. Hahn, R. Chehey and R. Hudson. 1998. Detection of an outbreak of *Escherichia coli* O26:H11 in Idaho through surveillance of non-O157 verotoxigenic *E. coli*. International Conference on Emerging Infectious Diseases, Atlanta, Georgia. Centers for Disease Control. Vol. 16: 85.

Brown, G.R., I.C. Sutcliffe and S.P. Cummings. 2001. Reclassification of [*Pseudomonas*] *doudoroffii* (Baumann et al. 1983) in the genus *Ocenaomonas* gen. nov. as *Oceanomonas doudoroffii* comb. nov., and description of a phenol-degrading bacterium from estuarine water as *Oceanomonas baumannii* sp. nov. Int. J. Syst. Evol. Microbiol. *51*: 67–72.

Brown, J.E., P.R. Brown and P.H. Clarke. 1969. Butyramide-utilizing mutants of *Pseudomonas aeruginosa* 8602 which produce an amidase with altered substrate specificity. J. Gen. Microbiol. *57*: 273–285.

Brown, J.E. and P.H. Clarke. 1972. Amino acid substitution in an amidase produced by an acetanilide-utilizing mutant of *Pseudomonas aeruginosa*. J. Gen. Microbiol. *70*: 287–298.

Brown, L.N., R.C. Dillman and R.E. Dierks. 1970. The *Haemophilus somnus* complex. U.S. Animal Hlth. Assoc. Proc. *74*: 94–108.

Brown, L.R. and R.J. Strawinski. 1958. Intermediates in the oxidation of methane. Bacteriol. Proc. *58*: 96–132.

Brown, N.A. 1918. Some bacterial diseases of lettuce. J. Agr. Res. *13*: 367–388.

Brown, N.A. 1923. Bacterial leafspot of geranium in the eastern United States. J. Agr. Res. *23*: 361–372.

Brown, N.A. and C.O. Jamieson. 1913. A bacterium causing a disease of sugar-beet and nasturtium leaves. J. Agr. Res. *1*: 189–210.

Brubaker, R.R. 1972. The genus *Yersinia*: Biochemistry and genetics of virulence. Curr. Top. Microbiol. Immunol. *57*: 111–158.

Brubaker, R.R. 1991. Factors promoting acute and chronic diseases caused by yersiniae. Clin. Microbiol. Rev. *4*: 309–324.

Brubaker, R.R. and M.J. Surgalla. 1962. Pesticins II. Production of pesticin I and II. J. Bacteriol. *84*: 539–545.

Bruce, S.K., D.G. Schick, L. Tanaka, E.M. Jimenez and J.Z. Montgomerie. 1981. Selective medium for isolation of *Klebsiella pneumoniae*. J. Clin. Microbiol. *13*: 1114–1116.

Bruggeman, G., D. Smeets, E.J. Vandamme, T. Vervust and G. Takerkart. 1998. Optimization of xanthan production by *Xanthomonas campestris* NRRL B-1459 using peptone PS as sole nitrogen source. Meded. Fac. Landbouwkun. Toegep. Biol. Wet., Gent. *63*: 1369–1372.

Brumfitt, W. 1959. Some growth requirements of *Haemophilus influenzae* and *Haemophilus pertussis*. J. Pathol. Bacteriol. *77*: 95–100.

Brumpt, E. 1910. Précis de Parasitologie, 1st Ed., Masson and Co., Paris.

Brun, J.G.W. and M.O. Texeira. 1992a. Acaricidal activity of *Cedecea lapagei* on engorged females of *Boophilus microplus* exposed to the environment. Arq. Bras. Med. Vet. Zootec. *44*: 543–544.

Brun, J.G.W. and M.O. Texeira. 1992b. Disease of engorged female *Boophilus microplus* (AcariIxodidae) caused by *Cedecea lapagei* and *Escherichia coli*. Arq. Bras. Med. Vet. Zootec. *44*: 441–443.

Brunder, W., H. Schmidt and H. Karch. 1996. KatP, a novel catalase-peroxidase encoded by the large plasmid of enterohaemorrhagic *Escherichia coli* O157:H7. Microbiology (Reading) *142*: 3305–3315.

Brunel, B., A. Givaudan, A. Lanois, R.J. Akhurst and N.E. Boemare. 1997. Fast and accurate identification of *Xenorhabdus* and *Photorhabdus* species by restriction analysis of PCR-amplified 16S rRNA genes. Appl. Environ. Microbiol. *63*: 574–580.

Bruner, D.W., P.R. Edwards and A.S. Hopson. 1949. The Ballerup group of paracolon bacteria. J. Infect. Dis. *85*: 290–294.

Bruner, D.W. and J.H. Gillespie. 1973. Hagan's Infectious Diseases of Domestic Animals, 6th Ed., Cornell University Press, Ithaca. 1385.

Brunker, P., J. Altenbuchner, K.D. Kulbe and R. Mattes. 1997. Cloning, nucleotide sequence and expression of a mannitol dehydrogenase gene from *Pseudomonas fluorescens* DSM 50106 in *Escherichia coli*. Biochim. Biophys. Acta-Gene Struct. Expression. *1351*: 157–167.

Bruns, A. and L. Berthe-Corti. 1999. *Fundibacter jadensis* gen. nov., sp. nov., a new slightly halophilic bacterium, isolated from intertidal sediment. Int. J. Syst. Bacteriol. *49*: 441–448.

Brunton, J.L., I. MacLean, A.R. Ronald and W.L. Albritton. 1979. Plasmid mediated ampicillin resistance in *H. ducreyi*. Antimicrob. Agents Chemother. *15*: 294–299.

Brunton, J.L., M. Meier, N. Ehrman, I. Maclean, L. Slaney and W.L. Albritton. 1982. Molecular epidemiology of beta-lactamase specifying plasmids of *Haemophilus ducreyi*. Antimicrob. Agents Chemother. *21*: 857–863.

Brusseau, G.A., E.S. Bulygina and R.S. Hanson. 1994. Phylogenetic analysis and development of probes for differentiating methylotrophic bacteria. Appl. Environ. Microbiol. *60*: 626–636.

Brusseau, G.A., H.-C. Tsien, R.S. Hanson and L.P. Wackett. 1990. Optimization of trichloroethylene oxidation by methanotrophs and the use of a colorimetric assay to detect soluble methane monooxygenase activity. Biodegradation *1*: 19–29.

Bruun, B., J.J. Christensen and M. Kilian. 1984a. Bacteremia caused by a beta-lactamase producing *Haemophilus parainfluenzae* strain of a new biotype. A case report. Acta Pathol. Microbiol. Immunol. Scand [B]. *92*: 135–138.

Bruun, B., Y. Ying, E. Kirkegaard and W. Frederiksen. 1984b. Phenotypic differentiation of *Cardiobacterium hominis*, *Kingella indologenes* and CDC group EF-4. Eur. J. Clin. Microbiol. *3*: 230–235.

Bryan, L.E. 1978. Transferable chloramphenicol and ampicillin resistance in a strain of *Haemophilus influenzae*. Antimicrob. Agents Chemother. *14*: 154–156.

Bryan, L.E. 1979. Resistance to antimicrobial agents: the general nature of the problem and the basis of resistance. *In* Doggett (Editor), *Pseudomonas aeroginosa*. Clinical manifestations of infection and current therapy, Academic Press, New York. 219–270.

Bryan, L.E., S.D. Semaka, H.M. van Den Wizen, J.E. Kinnear and R.L.S. Whitehouse. 1973. Characteristics of R931 and other *Pseudomonas aeruginosa* R factors. Antimicrob. Agents Chemother. *3*: 625–637.

Bryan, L.E., M.S. Shahrabadi and H.M. van Den Wizen. 1974. Gentamicin resistance in *Pseudomonas aeruginosa*: R-factor mediated resistance. Antimicrob. Agents Chemother. *6*: 191–199.

Bryan, L.E., H.M. van Den Wizen and J.T. Tseng. 1972. Transferable drug resistance in *Pseudomonas aeruginosa*. Antimicrob. Agents Chemother. *1*: 22–29.

Bryan, M.K. 1926.. Bacterial leaf-spot on hubbard squash. Science *63*: 165.

Bryan, M.K. 1932. Color variations in bacterial plant pathogens. Phytopathology *22*: 787–788.

Bryan, M.K. 1933. Bacterial speck of tomatoes. Phytopathology *23*: 897–904.

Bryan, M.K. and F.P. McWhorter. 1930. Bacterial blight of poppy caused by *Bacterium papavericola* sp. nov. J. Agr. Res. *40*: 1–9.

Bryant, M.P. 1972. Commentary on the Hungate technique for culture of anaerobic bacteria. Am. J. Clin. Nutr. *25*: 1324–1328.

Bryant, M.P. and L.A. Burkey. 1953. Cultural methods of some characteristics of some of the numerous groups of bacteria in the bovine rumen. J. Dairy Sci. *36*: 205–217.

Bryant, M.P. and I.M. Robinson. 1961. Some nutritional requirements of the genus *Ruminobacter*. Appl. Microbiol. *9*: 91–95.

Bryant, M.P. and I.M. Robinson. 1962. Some nutritional characteristics of predominant culturable ruminal bacteria. J. Bacteriol. *84*: 605–614.

Bryant, M.P., I.M. Robinson and I.L. Lindahl. 1961. A note on the flora and fauna in the rumen of steers fed a feedlot bloat-provoking ration and the effect of penicillin. Appl. Microbiol. *9*: 511–515.

Bryant, M.P. and N. Small. 1956. Characteristics of two new genera of anaerobic curved rods isolated from the rumen of cattle. J. Bacteriol. *72*: 22–26.

Bryant, M.P., N. Small, C. Bouma and H. Chu. 1958. *Bacteroides ruminicola* sp. nov. and *Succinimonas amylolytica* gen. nov., species of succinic acid-producing anaerobic bacteria of the bovine rumen. J. Bacteriol. *76*: 1–23.

Bryant, R.D., K.M. McGroarty, J.W. Costerton and E.J. Laishley. 1988. *In* Validation of the publication of new names and new combinations previously effectively published outside the IJSB. List No. 25. Int. J. Syst. Bacteriol. *38*: 220–222.

Bryantseva, I.A., V.M. Gorlenko, E.I. Kompantseva and J.F. Imhoff. 2000. *Thioalkalicoccus limnaeus* gen. nov., sp. nov., a new alkaliphilic purple sulfur bacterium with bacteriochlorophyll *b*. Int. J. Syst. Evol. Microbiol. *50*: 2157–2163.

Bryantseva, I., V.M. Gorlenko, E.I. Kompantseva, J.F. Imhoff, J. Süling and L. Mityushina. 1999. *Thiorhodospira sibirica* gen. nov., sp. nov., a new alkaliphilic purple sulfur bacterium from a Siberian soda lake. Int. J. Syst. Bacteriol. *49*: 697–703.

Bryn, K., E. Jantzen and K. Bøvre. 1977. Occurrence and patterns of waxes in *Neisseriaceae*. J. Gen. Microbiol. *102*: 33–43.

Brzin, B. 1968. Tellurite reduction in *Yersinia*. Experientia (Basel) *24*: 405.

Buchan, L. 1983. Possible biological mechanism of phosphorus removal. Water Sci. Technol. *15*: 87–103.

Buchanan, R.E. 1925. General Systematic Bacteriology, The Williams & Wilkins Co, Baltimore.

Buchanan, R.E. 1957. Order VII. *Beggiatoales* Buchanan, Ordo nov. *In* Breed, Murray and Smith (Editors), Bergey's Manual of Determinative Bacteriology, The Williams & Wilkins Co., Baltimore. 837–851.

Buchanan, R.E. and W.E. Gibbons (Editors). 1974. Bergey's Manual of Determinative Bacteriology, 8th Ed., The Williams & Wilkins Co., Baltimore.

Buchanan, T.M. and W.A. Pearce. 1979. Pathogenic aspects of outer membrane components of Gram negative bacteria. *In* Inouye (Editor), Bacterial Outer Membrances. Biogenesis and Function, John Wiley & Sons, New York. 475–514.

Bucher, C. and A. von Graevenitz. 1982. Evaluation of three differential media for detection of *Enterobacter agglomerans* (*Erwinia herbicola*). J. Clin. Microbiol. *15*: 1164–1166.

Buchner, P. 1965. Endosymbiosis of animals with plant microorganisms, Interscience Publishers, New York.

Buchrieser, C., P. Glaser, C. Rusniok, H. Nedjari, H. D'Hauteville, F. Kunst, P. Sansonetti and C. Parsot. 2000. The virulence plasmid pWR100 and the repertoire of proteins secreted by the type III secretion apparatus of *Shigella flexneri*. Mol. Microbiol. *38*: 760–771.

Buck, J.D., S.P. Meyers and E. Leifson. 1963. *Pseudomonas* (*Flavobacterium*) *piscicida* Bein comb. nov. J. Bacteriol. *86*: 1125–1126.

Buckley, J.T. 1983. Mechanism of action of bacterial glycerophospholipid: cholesterol acyltransferase. Biochemistry *22*: 5490–5493.

Buckley, J.T., L.N. Halasa, K.D. Lund and S. Macintyre. 1981. Purification and some properties of the hemolytic toxin acrolysin. Can. J. Biochem. *59*: 430–435.

Budzikiewicz, H. 1993. Secondary metabolites from fluorescent pseudomonads. FEMS Microbiol. Rev. *104*: 209–228.

Buesching, W.J., R.A. Brust and L.W. Ayers. 1983. Enhanced primary isolation of *Legionella pneumophila* from clinical specimens by low pH treatment. J. Clin. Microbiol. *17*: 1153–1155.

Bugert, P. and K. Geider. 1995. Molecular analysis of the ams operon required for exopolysaccharide synthesis of *Erwinia amylovora*. Mol. Microbiol. *15*: 917–933.

Bugert, P. and K. Geider. 1997. Characterization of the *amsI* gene product as a low molecular weight acid phosphatase controlling exopolysaccharide synthesis of *Erwinia amylovora*. FEBS Lett. *400*: 252–256.

Buhariwalla, F., B. Cann and T.J. Marrie. 1996. A dog-related outbreak of Q fever. Clin. Infect. Dis. *23*: 753–755.

Buissière, J., G. Brault and L. Le Minor. 1981. Intérêt taxonornique de la recherche de l'utilisation du 2-cétogluconate par les *Enterobacteriaceae*. Ann. Microbiol. Inst. Pasteur (Paris) *132A*: 191–195.

Burbach, S. 1987. Reklassifizierung der Gattung *Haemophilus* Winslow et al. 1917 auf Grund der DNA Basensequezhomologie, Philipps-Universität Marburg, Marburg, FRG.

Burgdorfer, W., L.P. Brinton and L.E. Hughes. 1973. Isolation and characterization of symbiotes from the Rocky Mountain wood tick, *Dermacentor andersoni*. J. Invert. Pathol. *22*: 424–434.

Burgess, G.W. 1982. Ovine contagious epididymitis: a review. Vet. Microbiol. *7*: 551–575.

Burgess, N.R., S.N. McDermott and J. Whiting. 1973. Aerobic bacteria occurring in the hind-gut of the cockroach, *Blatta orientalis*. J. Hyg. *71*: 1–7.

Burgher, L.W., G.W. Loomis and F. Ware. 1973. Systemic infection due to *Actinobacillus actinomycetemcomitans*. Am. J. Clin. Pathol. *60*: 412–415.

Burke, J.P., D. Ingall, J.O. Klein, H.M. Gezon and M. Finland. 1971. *Proteus mirabilis* infections in a hospital nursery traced to a human carrier. N. Eng. J. Med. *184*: 115–121.

Burke, V., J. Robinson, M. Gracey, D. Peterson, N. Meyer and V. Haley. 1984a. Isolation of *Aeromonas* spp. from an unchlorinated domestic water supply. Appl. Environ. Microbiol. *48*: 367–370.

Burke, V., J. Robinson, M. Gracey, D. Peterson and K. Partridge. 1984b. Isolation of *Aeromonas hydrophila* from a metropolitan water supply: seasonal correlation with clinical isolates. Appl. Environ. Microbiol. *48*: 361–366.

Burkholder, W.H. 1930. The bacterial diseases of the bean. Cornell Agr. Expt. Sta. Mem. *127*: 1–88.

Burkholder, W.H. 1941. The black rot of *Barbarea vulgaris*. Phytopathology *31*: 347–348.

Burkholder, W.H. 1944. *Xanthomonas vignicola* sp. nov. pathogenic on cowpeas and beans. Phytopathology *34*: 430–432.

Burkholder, W.H. 1948a. Genus I. *Erwinia*. *In* Breed, Murray and Hitchens (Editors), Bergey's Manual of Determinative Bacteriology, 6th Ed., The Williams & Wilkins Co., Baltimore. pp. 463–478.

Burkholder, W.H. 1948b. Genus I. *Pseudomonas* Migula. *In* Breed, Murray and Hitchens (Editors), Bergey's Manual of Determinative Bacteriology, 6th ed., The Williams & Wilkins Co., Baltimore. 82–150.

Burkholder, W.H. 1960a. A bacterial brown rot of parsnip roots. Phytopathology *50*: 280–282.

Burkholder, W.H. 1960b. Some observations on *Erwinia tracheiphila*, the causal agent of the cucurbit wilt. Phytopathology *50*: 179–180.

Burkholder, W.H., L.A. McFadden and A.W. Dimock. 1953. A bacterial blight of chrysanthemum. Phytopathology *43*: 522–526.

Burns, G., A. Ramos and A. Muchlinski. 1996. Fever response in North American snakes. J. Herpetol. *30*: 133–139.

Burns, J.L., P.M. Mendelman, J. Levy, T.L. Stull and A.L. Smith. 1985. A permeability barrier as a mechanism of chloramphenicol resistance in *Haemophilus influenzae*. Antimicrob. Agents Chemother. *27*: 46–54.

Burrill, T.J. 1882. The Bacteria: an account of their nature and effects, together with a systematic description of the species, Illinois Indust. Univ.. Report nr 11th Rep.

Burrill, T.J. 1883. New species of *Micrococcus*. Amer. Naturalist *17*: 319–320.

Burrows, L.L. and R.Y.C. Lo. 1992. Molecular characterization of an RTX toxin determinant from *Actinobacillus suis*. Infect. Immun. *60*: 2166–2173.

Burrows, L.L., E. Olah-Winfield and R.Y. Lo. 1993. Molecular analysis of the leukotoxin determinants from *Pasteurella haemolytica* serotypes 1 to 16. Infect. Immun. *61*: 5001–5007.

Burrows, T.W. 1963. Virulence of *Pasteurella pestis* and immunity to plague. Ergebn. Mikrobiol. *37*: 59–113.

Burrows, T.W. and W.A. Gillett. 1966. The nutritional requirement of some *Pasteurella* species. J. Gen. Microbiol. *45*: 333–346.

Burton, P.R., J. Stueckemann, R.M. Welsh and D. Paretsky. 1978. Some ultrastructural effects of persistent infections by the rickettsia *Coxiella burnetii* in mouse L cells and green monkey kidney (Vero) cells. Infect. Immun. *21*: 556–566.

Burton, S.D. and J.D. Lee. 1978. Improved enrichment and isolation procedures for obtaining pure cultures of *Beggiatoa*. Appl. Environ. Microbiol. *35*: 614–617.

Burton, S.D. and R.Y. Morita. 1964. Effect of catalase and cultural conditions on growth of *Beggiatoa*. J. Bacteriol. *88*: 1755–1761.

Burton, S.D., R.Y. Morita and W. Miller. 1966. Utilization of acetate by *Beggiatoa*. J. Bacteriol. *91*: 1192–1200.

Busse, H.J. and G. Auling. 1988. Polyamine pattern as a chemotaxonomic marker within the *Proteobacteria*. Syst. Appl. Microbiol. *11*: 1–8.

Busse, H.J., S. Bunka, A. Hensel and W. Lubitz. 1997. Discrimination of members of the family *Pasterellaceae* based on polyamine patterns. Int. J. Syst. Bacteriol. *47*: 698–708.

Busse, H.J., T. El-Banna and G. Auling. 1989. Evaluation of different approaches for identification of xenobiotic-degrading Pseudomonads. Appl. Environ. Microbiol. *55*: 1578–1583.

Busse, H.J., P. Kämpfer, E.R.B. Moore, J. Nuutinen, I.V. Tsitko, E.B.M. Denner, L. Vauterin, M. Valens, R. Rosselló-Mora and M.S. Salkinoja-Salonen. 2002. *Thermomonas haemolytica* gen. nov., sp. nov., a g -proteobacterium from kaolin slurry. Int. J. Syst. Evol. Microbiol. *52*: 473–483.

Butler, L.O. 1962. A defined medium for *Haemophilus influenzae* and *Haemophilus parainfluenzae*. J. Gen. Microbiol. *27*: 51–60.

Butler, P.D. and E.R. Moxon. 1990. A physical map of the genome of *Haemophilus influenzae* type b. J. Gen. Microbiol. *136*: 2333–2342.

Butler, T. 1983. Plague and Other *Yersinia* Infections, Plenum Medical Book Co., New York. 220.

Butt, A.A., J. Figueroa and D.H. Martin. 1997a. Ocular infection caused by three unusual marine organisms. Clin. Infect. Dis. *24*: 740.

Butt, H.L., A.W. Cripps and R.L. Clancy. 1997b. In vitro susceptibility patterns of nonserotypable *Haemophilus influenzae* from patients with chronic bronchitis. Pathology *29*: 72–75.

Buttenschøn, J. and S. Rosendal. 1990. Phenotypical and genotypic characteristics of paired isolates of *Pasteurella multocida* from the lungs and kidneys of slaughtered pigs. Vet. Microbiol. *25*: 67–75.

Buttiaux, R., R. Osteux, R. Fresnoy and J. Moriamez. 1954. Les propriétés biochemiques caractèristiques du genre *Proteus*. Inclusion souhaitable des *Providencia* dans celui-ci. Ann. Inst.Pasteur (Paris). *87*: 375–386.

Button, D.K., B.R. Robertson, P.W. Lepp and T.M. Schmidt. 1998. A small, dilute-cytoplasm, high-affinity, novel bacterium isolated by extinction culture and having kinetic constants compatible with growth at ambient concentrations of dissolved nutrients in seawater. Appl. Environ. Microbiol. *64*: 4467–4476.

Button, D.K., F. Schut, P. Quang, R. Martin and B.R. Robertson. 1993. Viability and isolation of marine bacteria by dilution culture: theory, procedures, and initial results. Appl. Environ. Microbiol. *59*: 881–891.

Buxton, A.E., R.L. Anderson, D. Werdegar and E. Atlas. 1978. Nosocomial respiratory tract infection and colonization with *Acinetobacter calcoaceticus*. Epidemiologic characteristics. Am. J. Med. *65*: 507–513.

Buxton, P. 1955. The Natural History of Tsetse Flies: An Account of the Biology of the Genus *Glossina* (Diptera), H.K. Lewis and Co. Ltd., London. pp. 816.

Buyer, J.S., V. Delorenzo and J.B. Neilands. 1991. Production of the siderophore aerobactin by a halophilic pseudomonad. Appl. Environ. Microbiol. *57*: 2246–2250.

Byng, G.S., A. Berry and R.A. Jensen. 1985. Evolutionary implications of features of aromatic amino acids biosynthesis in the genus *Acinetobacter*. Arch. Microbiol. *143*: 122–129.

Byng, G.S., A. Berry and R.A. Jensen. 1986. Evolution of aromatic biosynthesis and fine-tuned phylogenetic positioning of *Azomonas, Azotobacter* and ribosomal RNA group I pseudomonads. Arch. Microbiol. *144*: 222–227.

Byng, G.S., J.L. Johnson, R.J. Whitaker, R.L. Gherna and R.A. Jensen. 1983. The evolutionary pattern of aromatic amino acid biosynthesis and the emerging phylogeny of pseudomonad bacteria. J. Mol. Evol. *19*: 272–282.

Byng, G.S., R.J. Whitaker, R.L. Gherna and R.A. Jensen. 1980. Variable enzymological patterning in tyrosine biosynthesis as a means of determining natural relatedness among the *Pseudomonadaceae*. J. Bacteriol. *144*: 247–257.

Byrd, R.A., W. Egan, M.F. Summer and A. Bax. 1987. New NMR-spectroscopic approaches for structural studies of polysaccharides: applications to the *Haemophilus influenzae* type a capsular polysaccharide. Carbobydr. Res. *166*: 47–58.

Cabadajova, D. and L. Kudrna. 1988. *Moellerella wisconsensis*—the first isolation and identification of a new genus and species of the family *Enterobacteriaceae* in Czechoslovakia. Cesk. Epidemiol. Mikrobiol. Imunol. *37*: 45–48.

Caccavo, F., Jr., R.P. Blakemore and D.R. Lovley. 1992. A hydrogen-oxidizing, Fe(III)-reducing microorganism from the Great Bay Estuary, New Hampshire. Appl. Environ. Microbiol. *58*: 3211–3216.

Caccavo, F., Jr., D.J. Lonergan, D.R. Lovley, M. Davis, J.F. Stolz and M.J. McInerney. 1994. *Geobacter sulfurreducens* sp. nov., a hydrogen- and acetate-oxidizing dissimilatory metal-reducing microorganism. Appl. Environ. Microbiol. *60*: 3752–3759.

Caccavo, F., Jr., N.B. Ramsing and J.W. Costerton. 1996. Morphological and metabolic responses to starvation by the dissimilatory metal-reducing bacterium *Shewanella alga* BrY. Appl. Environ. Microbiol. *62*: 4678–4682.

Cadmus, M.C., C.A. Knutson, A.A. Lagoda, J.E. Pittsley and K.A. Burton. 1978. Synthetic media for production and quality of xanthan gum in 20 liter fermentors. Biotechnol. Bioeng. *20*: 1003–1014.

Cafferkey, M.T., K. McClean and M.E. Drumm. 1989. Production of bacteriocin like antagonism by clinical isolates of *Yersinia enterocolitica*. J. Clin. Microbiol. *27*: 677–680.

Cahill, M.M. 1990. Virulence factors in motile *Aeromonas* species. J. Appl. Bacteriol. *69*: 1–16.

Cai, M., X. Dong, J. Wei, F. Yang, D. Xu, H. Zhang, X. Zheng, S. Wang and H. Jin. 1992. Isolation and identification of *Leclercia adecarboxylate* in clinical isolates in China (Translation). Acta Microbiol. Sin. *32*: 119–123.

Calderwood, S.B., D.W.K. Acheson, T.J. Barrett, J.B. Kaper, B.S. Kaplan, H. Karch, A.D. O'Brien, T.G. Obrig, Y. Takeda, P.I. Tarr and I.K. Wachsmuth. 1996. Proposed new nomenclature for SLT (VT) family. ASM News. *62*: 118–119.

Caldwell, D.E. and S.J. Caldwell. 1974. The response of littoral communities of bacteria to variations in sulfide and thiosulfate. Abstr. Annu. Meet. Am. Soc. Microbiol. *74*: 59.

Caldwell, D.R., M. Keeney, J.S. Barton and J.F. Kelley. 1973. Sodium and other inorganic growth requirements of *Bacteroides amylophilus*. J. Bacteriol. *114*: 782–789.

Caldwell, D.R., M. Keeney and P.J. van Sorest. 1969. Effects of carbon dioxide on growth and maltose fermentation by *Bacteroides amylophilus*. J. Bacteriol. *98*: 668–676.

Caldwell, D.R., D.C. White, M.P. Bryant and R.N. Doetsch. 1965. Specificity of the heme requirement for growth of *Bacteroides ruminicola*. J. Bacteriol. *90*: 1645–1654.

Calhoon, D.A., W.R. Mayberry and J. Slots. 1981. Cellular fatty acid and soluble protein composition of *Actinobacillus actinomycetemcomitans* and related organisms. J. Clin. Microbiol. *14*: 376–382.

Calhoun, A. and G.M. King. 1998. Characterization of root-associated methanotrophs from three freshwater macrophytes: *Pontederia cordata, Sparganium eurycarpum*, and *Sagittaria latifolia*. Appl. Environ. Microbiol. *64*: 1099–1105.

Calia, K.E., M.K. Waldor and S.B. Calderwood. 1998. Use of representational difference analysis to identify genomic differences between pathogenic strains of *Vibrio cholerae*. Infect. Immun. *66*: 849–852.

Callahan, W.S. and K. Koroma. 1979. Use of the Juni-Janik genetic homology technique in the identification of *Yersinia enterocolitica*. J. Amer. Med. Technol. *41*: 229–231.

Callister, S.M. and W.A. Agger. 1987. Enumeration and characterization of *Aeromonas hydrophila* and *Aeromonas caviae* isolated from grocery store produce. Appl. Environ. Microbiol. *53*: 249–253.

Cameron, S., C. Walker, M. Beers, N. Rose and E. Anear. 1995. Enterohaemorrhagic *Escherichia coli* outbreak in South Australia associated with the consumption of mettwurst. Comm. Dis. Intell. *19*: 70–71.

Campbell, A.M. 1996. Bacteriophages. *In* Neidhardt, Curtiss, Ingraham, Lin, Low, Magasanik, Reznikoff, Riley, Schaechter and Umbarger (Editors), *Escherichia coli* and *Salmonella*: Cellular and Molecular Biology, 2nd Ed., ASM Press, Washington, D.C. pp. 2325–2338.

Campbell, J., W.F. Bibb, M.A. Lambert, S. Eng, A.G. Steigerwalt, J. Allard, C.W. Moss and D.J. Brenner. 1984a. *Legionella sainthelensi*: a new species of *Legionella* isolated from water near Mount St Helens (Washington, USA). Appl. Environ. Microbiol. *47*: 369–373.

Campbell, J., W.F. Bibb, M.A. Lambert, S. Eng, A.G. Steigerwalt, J. Allard, C.W. Moss and D.J. Brenner. 1984b. *In* Validation of the publication of new names and new combinations previously effectively published outside the IJSB. List No. 15. Int. J. Syst. Bacteriol. *34*: 355–357.

Campbell, L.L. 1957. Genus *Beneckea*. *In* Breed, Murray and Smith (Editors), Bergey's Manual of Determinative Bacteriology, 7th Ed., The Williams &d Wilkins Co., Baltimore. pp. 328–332.

Campbell, L.M. and I.L. Roth. 1975. Methyl violet: a selective agent for differentiation of *Klebsiella pneumoniae* from *Enterobacter aerogenes* and other Gram-negative organisms. Appl. Microbiol. *30*: 258–261.

Campbell, L.M., I.L. Roth and R.D. Klein. 1976. Evaluation of double agar in the isolation of *Klebsiella pneumoniae* from river water. Appl. Environ. Microbiol. *31*: 213–215.

Campos, J., S. Garcia-Tornel, J.M. Gairi and I. Fabregues. 1986. Multiply resistant *Haemophilus influenzae* type b causing meningitis: comparative clinical and laboratory study. J. Pediatr. *108*: 897–902.

Campos, J., S. Garcia-Tornel and I. Sanfeliu. 1984. Susceptibility studies of multiply resistant *Haemophilus influenzae* isolated from pediatric patients and contacts. Antimicrob. Agents Chemother. *25*: 706–709.

Campos, M.E., J.M. Martinez-Salazar, L. Lloret, S. Moreno, C. Nunez, G. Espin and G. Soberon-Chavez. 1996. Characterization of the gene coding for GDP-mannose dehydrogenase (*algD*) from *Azotobacter vinelandii*. J. Bacteriol. *178*: 1793–1799.

Canale-Parola, E., S.L. Rosenthal and D.G. Kupfer. 1966. Morphological and physiological characteristics of *Spirillum gracile* sp. n. Antonie van Leeuwenhoek J. Microbiol. Serol. *32*: 113–124.

Cannon, G.C., W.R. Strohl, J.M. Larkin and J.M. Shively. 1979. Cytochromes in *Beggiatoa alba*. Curr. Microbiol. *2*: 263–266.

Canonica, F.P. and M.A. Pisano. 1988. Gas-liquid chromatographic analysis of fatty acid methyl esters of *Aeromonas hydrophila, Aeromonas sobria*, and *Aeromonas caviae*. J. Clin. Microbiol. *26*: 681–685.

Cánovas, D., N. Borges, C. Vargas, A. Ventosa, J.J. Nieto and H. Santos. 1999. Role of N-γ-acetyldiaminobutyrate as an enzyme stabilizer and an intermediate in the biosynthesis of hydroxyectoine. Appl. Environ. Microbiol. *65*: 3774–3779.

Cánovas, D., C. Vargas, M.I. Calderón, A. Ventosa and J.J. Nieto. 1998a. Characterization of the genes for the biosynthesis of the compatible solute ectoine in the moderately halophilic bacterium *Halomonas elongata* DSM 3043. Syst. Appl. Microbiol. *21*: 487–497.

Cánovas, D., C. Vargas, L.N. Csonka, A. Ventosa and J.J. Nieto. 1996. Osmoprotectants in *Halomonas elongata*: high-affinity betaine transport system and choline-betaine pathway. J. Bacteriol. *178*: 7221–7226.

Cánovas, D., C. Vargas, L.N. Csonka, A. Ventosa and J.J. Nieto. 1998b. Synthesis of glycine betaine from exogenous choline in the moderately halophilic bacterium *Halomonas elongata*. Appl. Environ. Microbiol. *64*: 4095–4097.

Cánovas, D., C. Vargas, F. Iglesias-Guerra, L.N. Csonka, D. Rhodes, A. Ventosa and J.J. Nieto. 1997. Isolation and characterization of salt-sensitive mutants of the moderate halophile *Halomonas elongata* and cloning of the ectoine synthesis genes. J. Biol. Chem. *272*: 25794–25801.

Cánovas, D., C. Vargas, S. Kneip, M.J. Morón, A. Ventosa, E. Bremer and J.J. Nieto. 2000. Genes for the synthesis of the osmoprotectant glycine betaine from choline in the moderately halophilic bacterium *Halomonas elongata* DSM 3043. Microbiology *146*: 455–463.

Canteros, B.I., G.V. Minsavage, J.B. Jones and R.E. Stall. 1995. Diversity of plasmids in *Xanthomonas campestris* pathovar *vesicatoria*. Phytopathology *85*: 1482–1486.

Caprioli, A., I. Luzzi, F. Rosmini, C. Resti, A. Edefonti, F. Perfumo, C. Farina, A. Goglio, A. Gianviti and G. Rizzoni. 1994. Communitywide outbreak of hemolytic-uremic syndrome associated with non-O157 verocytotoxin-producing *Escherichia coli.* J. Infect. Dis. *169*: 208–211.

Caraccio, V., A. Rocchetti and P.L. Garavelli. 1994. *Rahnella aquatilis* bacteremia in a patient with chronic renal failure. G. Mal. Infett. Parassit. *46*: 330–331.

Carbonetti, N.H. and P.H. Williams. 1984. A cluster of 5 genes specifying the aerobactin iron uptake system of plasmid Colv-K30. Infect. Immun. *46*: 7–12.

Carey, V.C. and L.O. Ingram. 1983. Lipid composition of *Zymomonas mobilis*: effects of ethanol and glucose. J. Bacteriol. *154*: 1291–1300.

Carlin, N.I., M. Rahman, D.A. Sack, A. Zaman, B. Kay and A.A. Lindberg. 1989. Use of monoclonal antibodies to type *Shigella flexneri* in Bangladesh. J. Clin. Microbiol. *27*: 1163–1166.

Carlone, G.M., L. Gorelkin, L.L. Gheesling, A.L. Erwin, S.K. Hoiseth, M.H. Mulks, S.P. O'Connor, R.S. Weyant, J. Myrick and L. Rubin. 1989. Potential virulence-associated factors in Brazilian purpuric fever. Brazilian Purpuric Fever Study Group. J. Clin. Microbiol. *27*: 609–614.

Carlone, G.M., F.O. Sottnek and B.D. Plikaytis. 1985. Comparison of outer membrane protein and biochemical profiles of *Haemophilus aegyptius* and *Haemophilus influenzae* biotype III. J. Clin. Microbiol. *22*: 708–713.

Carlsson, H.E., A.A. Lindberg, G. Lindberg, B. Hederstedt, K.A. Karlsson and B.O. Agell. 1975. Enzyme-linked immunosorbent assay for immunological diagnosis of human tularemia. J. Clin. Microbiol. *10*: 615–621.

Carlsson, H., D. Stockelberg, I. Tengborn, I. Braide, J. Carneskog and J. Kutti. 1995. Effects of *Erwinia*-asparaginase on the coagulation system. Eur. J. Haematol. *55*: 289–293.

Carlton, R.G. and L.L. Richardson. 1995. Oxygen and sulfide dynamics in a horizontally migrating cyanobacterial mat: Black band disease of corals. FEMS Microbiol. Ecol. *18*: 155–162.

Carmi, R., S. Carmeli, E. Levy and F.J. Gough. 1994. (+)-(S)-dihydroaeruginoic acid, an inhibitor of *Septoria tritici* and other phytopathogenic fungi and bacteria, produced by *Pseudomonas fluorescens*. J. Nat. Prod. *57*: 1200–1205.

Carnahan, A.M. 1992. *Aeromonas* taxonomy: a sea of change. 4th International *Aeromonas/Plesiomonas* Symposium, Atlanta, Georgia. .

Carnahan, A. 2001. Genetic relatedness of *Aeromonas* species based on the DNA sequences of four distinct genomic loci , University of Maryland, College Park, College Park, Maryland, USA, . 132 p.

Carnahan, A.M. and M. Altwegg. 1996. Taxonomy. *In* Austin, Altwegg, Gosling and Joseph (Editors), The Genus *Aeromonas*, 1st Ed., John Wiley & Sons, Ltd., Chichester. pp. 1–38.

Carnahan, A.M., S. Behram and S.W. Joseph. 1991a. Aerokey II: A flexible key for identifying clinical *Aeromonas* spp. J. Clin. Microbiol. *29*: 2843–2849.

Carnahan, A.M., T. Chakraborty, G.R. Fanning, D. Verma, A. Ali, J.M. Janda and S.W. Joseph. 1991b. *Aeromonas trota*, sp. nov.: an ampicillin-susceptible species isolated from clinical specimens. J. Clin. Microbiol. *29*: 1206–1210.

Carnahan, A., T. Chakraborty, G.R. Fanning, D. Verma, A. Ali, J.M. Janda and S.W. Joseph. 1992a. *In* Validation of the publication of new names and new combinations previously effectively published outside the IJSB. List No. 40. Int. J. Syst. Bacteriol. *42*: 191–192.

Carnahan, A., G.R. Fanning and S.W. Joseph. 1991c. *Aeromonas jandaei* (formerly genospecies DNA group 9 *Aeromonas sobria*), a new sucrose-

negative species isolated from clinical specimens. J. Clin. Microbiol. *29*: 560–564.

Carnahan, A., G.R. Fanning and S.W. Joseph. 1992b. *In* Validation of the publication of new names and new combinations previously effectively published outside the IJSB. List No. 40. Int. J. Syst. Bacteriol. *42*: 191–192.

Carnahan, A., L. Hammontree, L. Bourgeois and S.W. Joseph. 1990. Pyrazinamidase activity as a phenotypic marker for several *Aeromonas* spp. isolated from clinical specimens. J. Clin. Microbiol. *28*: 391–392.

Carnahan, A.M. and S.W. Joseph. 1993. Systematic assessment of geographically and clinically diverse aeromonads. Syst. Appl. Microbiol. *16*: 72–84.

Carnahan, A.M., S.W. Joseph and J.M. Janda. 1989a. Species identification of *Aeromonas* strains based on carbon substrate oxidation profiles. J. Clin. Microbiol. *27*: 2128–2129.

Carnahan, A.M., M.A. Marii, G.R. Fanning, M.A. Pass and S.W. Joseph. 1989b. Characterization of *Aeromonas schubertii* strains recently isolated from traumatic wound infections. J. Clin. Microbiol. *27*: 1826–1830.

Carnahan, A.M., M. O'Brien, S.W. Joseph and R.R. Colwell. 1988. Enzymatic characterization of three *Aeromonas* spp. using API Peptidase, API "Osidase," and API Esterase test kits. Diagn. Microbiol. Infect. Dis. *10*: 195–204.

Carpenter, K.P. 1961. The relationship of the Enterobacterium A12 (Sachs) to *Shigella boydii* 14. J. Gen. Microbiol. *26*: 535–542.

Carpenter, K.P. 1963. Report of the Subcommittee on Taxonomy of the Enterobacteriaceae. Int. Bull. Bacteriol. Nomencl. Taxon. *13*: 69–93.

Carpenter, K.P. 1964. The *Proteus-Providence* group. *In* Dyke (Editor), Recent Advances in Clinical Pathology, Series IV Little, Brown and Co, Boston. pp. 13–24.

Carpenter, K.P. and E.R.N. Cooke. 1965. An attempt to find shigellae in wild primates. J. Comp. Pathol. *75*: 21–204.

Carr, W.H. 1996. Pathogenic organisms of penaeid shrimp in the Hawaiian Islands. Bishop Mus. Occas. Pap. *0*: 15–18.

Carrell, D.T., M.E. Hammond and W.D. Odell. 1993. Evidence for an autocrine/paracrine function of chorionic gonadotropin in *Xanthomonas maltophilia*. Endocrinology. *132*: 1085–1089.

Carrell, D.T. and W.D. Odell. 1992. A bacterial binding site which binds human chorionic gonadotropin but not human luteinizing hormone. Endocr. Res. *18*: 51–58.

Carter, G.R. 1955. Studies on *Pasteurella multocida*. I. A hemagglutination test for the identification of serological types. Am. J. Vet. Res. *16*: 481–484.

Carter, G.R. and M.C.L. De Alwis. 1989. Haemorrhagic septicaemia. *In* Adlam and Rutter (Editors), *Pasteurella* and Pasteurelloses, Academic Press, London. 131–160.

Carter, J.C. 1945. Wetwood of elms. Ill. Nat. Hist. Surv. Bull. *23*: 407–448.

Carter, J.S., F.J. Bowden, I. Bastian, G.M. Myers, K.S. Sriprakash and D.J. Kemp. 1999. Phylogenetic evidence for reclassification of *Calymmatobacterium granulomatis* as *Klebsiella granulomatis* comb. nov. Int. J. Syst. Bacteriol. *49*: 1695–1700.

Carter, J., S. Hutton, K.S. Sriprakash, D.J. Kemp, G. Lum, J. Savage and F.J. Bowden. 1997. Culture of the causative organism of donovanosis (*Calymmatobacterium granulomatis*) in HEp-2 cells. J. Clin. Microbiol. *35*: 2915–2917.

Carter, J.B. and L.F. Luff . 1977. *Rickettsia*-like organisms infecting *Harpalus rufipes* (Coleoptera: Carabidae). J. Invert. Pathol. *30*: 99–101.

Casalta, J.P., Y. Peloux, D. Raoult, P. Brunet and H. Gallais. 1989. Pneumonia and meningitis caused by a new nonfermentative unknown gram-negative bacterium. J. Clin. Microbiol. *27*: 1446–1448.

Cascon, A., J. Anguita, C. Hernanz, M. Sanchez, M. Fernandez and G. Naharro. 1996. Identification of *Aeromonas hydrophila* hybridization group 1 by PCR assays. Appl. Environ. Microbiol. *62*: 1167–1170.

Case, A.C. 1965. Conditions controlling *Flavobacterium proteus* in brewery fermentations. J. Inst. Brew. *71*: 250–256.

Caselitz, F.H. 1955. Eine neue Bacterium der Gattung: *Vibrio* Muller, *Vibrio jamaicensis*. Z. Tropenmed. Parasitol. *6*: 62.

Cash, P., E. Argo, P.R. Langford and J.S. Kroll. 1997. Development of a

Haemophilus two-dimensional protein database. Electrophoresis *18*: 1472–1482.

Casin, I., F. Grimont and P.A. Grimont. 1986. Deoxyribonucleic acid relatedness between *Haemophilus aegyptius* and *Haemophilus influenzae*. Ann. Inst. Pasteur. Microbiol. *137*: 155–163.

Casin, I., F. Grimont, P.A.D. Grimont and M.J. Sansonlepors. 1985. Lack of deoxyribonucleic-acid relatedness between *Haemophilus ducreyi* and other *Haemophilus* species. Int. J. Syst. Bacteriol. *35*: 23–25.

Cassel-Beraud, A.M. and C. Richard. 1988. The aerobic intestinal flora of the microchiropteran bat *Chaerephon pumila* in Madagascar. Bull. Soc. Pathol. Exot. Filiales. *81*: 806–810.

Castaneda, M.R. 1947. A practical method for routine blood cultures in brucellosis. Proc. Soc. Exp. Biol. Med. *64*: 114–115.

Castaño, J.J., H.D. Thurston and L.V. Crowder. 1964. Transmissión de gomosis en los pastos Micay e Imperial. Agr. Trop. *20*: 379–387.

Castellani, A. 1912. Observation on some intestinal bacteria found in man. Centralbl. Bakteriol. Parasitenkd. Infektionskr. I. Abt. Orig. *65*: 262–269.

Castellani, A. and A.J. Chalmers. 1919. Manual of Tropical Medicine, 3rd Ed., William Wood and Company, New York.

Castellani, A. and A.J. Chalmers. 1920. Sur la classification de certains groupes de bacilles aerobies de l' intestin humain. Ann. Inst. Past. *34*: 600–621.

Castenholz, R.W. and B.K. Pierson. 1995. Ecology of thermophilic anoxygenic phototrophs. *In* Blankenship, Madigan and Bauer (Editors), Anoxygenic Photosynthetic Bacteria, Kluwer Academic Publishers, The Netherlands. pp. 87–103.

Cataldi, M.S. 1940. Aislamiento de *Beggiatoa alba* en cultivo puro. Rev. Inst. Bacteriol. Dep. Nacion. Hig. *9*: 393–423.

Cate, J.C. 1972. Isolation of *Serratia marcescens* from stools with an antibiotic plate. Advances in Antimicrobial and Antineoplasic Chemotherapy. Progress in Research and Clinical Application., Proceeding of the 7th International Congress of Chemotherapy, Munich. Urban and Schwarzenberg. *Vol. 1/2*: 763–764.

Catlin, B.W. 1964. Reciprocal genetic transformation between *Neisseria catarrhalis* and *Moraxella nonliquefaciens*. J. Gen. Microbiol. *37*: 369–379.

Catlin, B.W. 1970. Transfer of the organism named *Neisseria catarrhalis* to *Branhamella* gen. nov. Int. J. Syst. Bacteriol. *20*: 155–159.

Catlin, B.W. 1990. *Branhamella catarrhalis*: an organism gaining respect as a pathogen. Clin. Microbiol. Rev. *3*: 293–320.

Catlin, B.W. 1991. *Branhamaceae* fam. nov., a proposed family to accommodate the genera *Branhamella* and *Moraxella*. Int. J. Syst. Bacteriol. *41*: 320–323.

Catlin, B.W. and L.S. Cunningham. 1964. Genetic transformation of *Neisseria catarrhalis* by deoxyribonucleate preparations having different average base compositions. J. Gen. Bacteriol. *37*: 341–352.

Catlin, B.W. and V.R. Tartagni. 1969. Delayed multiplication of newly capsulated transformants of *Haemophilus influenzae* detected by immunofluorescence. J. Gen. Microbiol. *56*: 387–401.

Cato, E.P., L.V. Holdeman and W.E.C. Moore. 1979. Proposal of neotype strains for seven non-saccharolytic *Bacteroides* species. Int. J. Syst. Bacteriol. *29*: 427–434.

Cato, E.P., W.E.C. Moore and M.P. Bryant. 1978. Designation of neotype strains for *Bacteroides amylophilus* Hamlin and Hungate 1956 and *Bacteroides succinogens* Hungate 1950. Int. J.Syst. Bacteriol. *28*: 491–495.

Catrenich, C.E. and W. Johnson. 1988. Virulence conversion of *Legionella pneumophila*: a one-way phenomenon. Infect. Immun. *56*: 3121–3125.

Catsaras, M., J. Antoniewski and R. Buttiaux. 1965. Sur la production de colonies a centre orange par *Proteus rettgeri* et *Providencia* sur la gelose au desoxycholate-citrate-lactose. Ann. Inst. Pasteur (Lille). *16*: 99–101.

Caugant, D.A., R.K. Selander and I. Olsen. 1990. Differentiation between *Actinobacillus* (*Haemophilus*) *actinomycetemcomitans*, *Haemophilus aphrophilus* and *Haemophilus paraphrophilus* by multilocus enzyme electrophoresis. J. Gen. Microbiol. *136*: 2135–2141.

Caumette, P., R. Baulaigue and R. Matheron. 1988. Characterization of *Chromatium salexigens* sp. nov., a halophilic *Chromatiaceae* isolated from Mediterranean salinas. Syst. Appl. Microbiol. *10*: 284–292.

Caumette, P., R. Baulaigue and R. Matheron. 1991. *Thiocapsa halophila*, sp. nov., a new halophilic phototrophic purple sulfur bacterium. Arch. Microbiol. *155*: 170–176.

Caumette, P., J.F. Imhoff, J. Suling and R. Matheron. 1997. *Chromatium glycolicum* sp. nov., a moderately halophilic purple sulfur bacterium that uses glycolate as substrate. Arch. Microbiol. *167*: 11–18.

Cavanagh, P., C.A. Morris and N.J. Mitchell. 1975. Chloramphenicol resistance in *Haemophilus* species. Lancet *1*: 696.

Cavanaugh, C.M. 1993. Methanotroph-invertebrate symbioses in the marine environment: ultrastructural, biochemical and molecular studies. *In* Murrell and Kelley (Editors), Microbial Growth on C_1 Compounds, Intercept Press Ltd., Andover. pp. 315–328.

Cavanaugh, C.M., P.R. Levering, J.S. Maki, R. Mitchell and M.E. Lidstrom. 1987. Symbiosis of methylotrophic bacteria and deep-sea mussels. Nature *325*: 346–348.

Cavara, F. 1905. Bacteriosi del fico, Aatti Accad. Gioenia Sci. Matur. Catania Ser.. .

Centers for Disease Control 1977. Legionnaires' disease. Morb. Mortal. Wkly. Rep. *26*: 93.

Centers for Disease Control. 1995. Outbreak of acute gastroenteritis attributable to *Escherichia coli* serotype O104:H21 – Helena, Montana, 1994. Morb. Mortal. Wkly. Rep. *44*: 501–503.

Centers for Disease Control. 1999. Outbreaks of *Shigella sonnei* infection associated with eating fresh parsley—United States and Canada, July-August 1998. Morb. Mortal. Wkly. Rep. *48*: 285–289.

Centers for Disease Control. 2000. *Escherichia coli* O111:H8 outbreak among teenage campers—Texas, 1999. Morb. Mortal. Wkly. Rep. *49*: 321–324.

Centers for Disease Control and Prevention 1992. Procedures for the recovery of *Legionella* from the environment, Atlanta, GA. ., U.S. Department of Health and Human Services.

Cerdà-Cuéllar, M., R.A. Rosselló-Mora, J. Lalucat, J. Jofre and A. Blanch. 1997. *Vibrio scophthalmi* sp. nov., a new species from Turbot (*Scophthalmus maximus*). Int. J. Syst. Bacteriol. *47*: 58–61.

Chae, C.H., M.J. Gentry, A.W. Confer and G.A. Anderson. 1990. Resistance to host immune defense mechanisms afforded by capsular material of *Pasteurella haemolytica*, serotype 1. Vet. Microbiol. *25*: 241–251.

Chagas, C.M., V. Rossetti and M.J.G. Beretta. 1992. Electron microscopy studies of a xylem-limited bacterium in sweet orange affected with citrus variegated chlorosis disease in Brazil. J. Phytopathol. *134*: 306–312.

Chakrabarti, A.K., A.N. Ghosh and B.L. Sarkar. 1997. Isolation of *Vibrio cholerae* O139 phages to develop a phage typing scheme. Indian J. Med. Res. *105*: 254–257.

Chakrabarty, A.M. 1998. Nucleoside diphosphate kinase: role in bacterial growth, virulence, cell signalling and polysaccharide synthesis. Mol. Microbiol. *28*: 875–882.

Chakrabarty, P.K., A. Mahadevan, S. Raj, M.K. Meshram and D.W. Gabriel. 1995. Plasmid-borne determinants of pigmentation, exopolysaccharide production, and virulence in *Xanthomonas campestris* pathovar *malvacearum*. Can. J. Microbiol. *41*: 740–745.

Chakrabarty, R.N., H.N. Patel and S.B. Desai. 1990. Isolation and partial characterization of catechol-type siderophore from *Pseudomonas stutzeri* Rc-7. Curr. Microbiol. *20*: 283–286.

Chakraborty, S., G.B. Nair and S. Shinoda. 1997. Pathogenic vibrios in the natural aquatic environment. Rev. Environ. Health. *12*: 63–80.

Chakraborty, T., B. Huhle, H. Hof, H. Bergbauer and W. Goebel. 1987. Marker exchange mutagenesis of the aerolysin determinant in *Aeromonas hydrophila* demonstrates the role of aerolysin in *Aeromonas hydrophila* associated systemic infections. Infect. Immun. *55*: 2274–2280.

Chakravarti, B.P., B. Sarma, K.L. Jain and C.K.P. Prasad. 1984. A bacterial leaf spot of bael (*Aegle marmelos* Correa) in Rajasthan and a revived name of the bacterium. Curr. Sci. (Bangalore) *53*: 488–489.

Chamberlain, R.E. 1965. Evaluation of live tularemia vaccine prepared in a chemically defined medium. Appl. Microbiol. *13*: 232–235.

Chamoiseau, G. 1967. Note sur le pouvoir pathogène d'*Edwardsiella tarda*. Un cas de septicémie mortelle du pigeon. Rev. Elev. Med. Vet. Pays. Trop. (Paris). *20*: 493–495.

Chamorey, E., M. Forel and M. Drancourt. 1999. An *in vitro* evaluation of the activity of chlorine against environmental and nosocomial isolates of *Aeromonas hydrophila*. J. Hosp. Infect. *41*: 45–49.

Champion, A.B., E.L. Barrett, N.J. Palleroni, K.L. Soderberg, R. Kunisawa, R. Contopoulou, A.C. Wilson and M. Doudoroff. 1980. Evolution in *Pseudomonas fluorescens*. J. Gen. Microbiol. *120*: 485–511.

Champomier-Vergès, M.C., A. Stintzi and J.M. Meyer. 1996. Acquisition of iron by the non-siderophore-producing *Pseudomonas fragi*. Microbiology *142*: 1191–1199.

Chan, K.Y., L. Baumann, M.M. Garza and P. Baumann. 1978. Two new species of *Alteromonas*: *Alteromonas espejiana* and *Alteromonas undina*. Int. J. Syst. Bacteriol. *28*: 217–222.

Chan, K.N., A.D. Phillips, S. Knutton, H.R. Smith and J.A. Walker-Smith. 1994. Enteroaggregative *Escherichia coli*: Another cause of acute and chronic diarrhoea in England? J. Pediatr. Gastroenterol. Nutr. *18*: 87–91.

Chand, R. and P.N. Singh. 1994. *Xanthomonas campestris* pv. *obscurae* pv. nov. causal agent of leaf blight of *Ipomoea obscura* in India. Z. Pflanzenkr. Pflanzenschutz. *101*: 590–593.

Chand, R., B.D. Singh, D. Singh and P.N. Singh. 1995. *Xanthomonas campestris* pv. *parthenii* pathovar nov. incitant of leaf blight of parthenium. Antonie Leeuwenhoek *68*: 161–164.

Chandler, F.W., R.M. Cole, M.D. Hicklin, J.A. Blackmon and C.S. Callaway. 1979. Ultrastructure of the Legionnaires' disease bacterium. A study using transmission electron microscopy. Ann. Intern. Med. *90*: 642–647.

Chandler, F.W., M.D. Hicklin and J.A. Blackmon. 1977. Demonstration of the agent of Legionnaires' disease in tissue. N. Engl. J. Med. *297*: 1218–1220.

Chandler, F.W., I.L. Roth, C.S. Callaway, J.L. Bump, B.M. Thomason and R.E. Weaver. 1980. Flagella on Legionnaires' disease bacteria. Ultrastructural observations. Ann. Intern. Med. *93*: 711–714.

Chandler, M.S. 1992. The gene encoding cAMP receptor protein is required for competence development in *Haemophilus influenzae* Rd. Proc. Natl. Acad. Sci. U. S. A. *89*: 1626–1630.

Chandra, M. and A.K. Jain. 1991. Fine structure of *Calymmatobacterium granulomatis* with particular reference to the surface structure. Indian J. Med. Res. *93*: 225–231.

Chandra, M., A.K. Jain, D.D. Ganguly, A.K. Sharma and N.C. Bhargava. 1989. An ultrastructural study of donovanosis. Indian J. Med. Res. *89*: 158–164.

Chang, B.J. and S.M. Bolton. 1987. Plasmids and resistance to antimicrobial agents in *Aeromonas sobria* and *Aeromonas hydrophila* clinical isolates. Antimicrob. Agents Chemother. *31*: 1281–1282.

Chang, C.J., R. Donaldson, M. Crowley and D. Pinnow. 1991. A new semiselective medium for the isolation of *Xanthomonas campestris* pathovar *campestris* from crucifer seeds. Phytopathology *81*: 449–453.

Chang, C.J., C.D. Robacker and R.P. Lane. 1990. Further evidence for the isolation of *Xylella fastidiosa* on nutrient agar from grapevines showing Pierce's disease symptoms. Can. J. Pathol. *12*: 405–408.

Chang, H.R., L.H. Loo, K. Jeyaseelan, L. Earnest and E. Stackebrandt. 1997. Phylogenetic relationships of *Salmonella typhi* and *Salmonella typhimurium* based on 16S rRNA sequence analysis. Int. J. Syst. Bacteriol. *47*: 1253–1254.

Chang, P.C. and A.C. Blackwood. 1969. Simultaneous production of three phenazine pigments by *Pseudomonas aeruginosa* Mac 436. Can. J. Microbiol. *15*: 439–444.

Chang, Y.F. and E. Adams. 1971. Induction of separate catabolic pathways for L-lysine and D-lysine in *Pseudomonas putida*. Biochem. Biophys. Res. Comm. *45*: 570–577.

Chang, Y.F., D.P. Ma, J. Shi and M.M. Chengappa. 1993. Molecular characterization of a leukotoxin gene from a *Pasteurella haemolytica*-like organism, encoding a new member of the RTX toxin family. Infect. Immun. *61*: 2089–2095.

Chang, Y.F., H.W. Renshaw and R. Young. 1987. Pneumonic pasteurel-

losis: examination of typable and untypable *Pasteurella haemolytica* strains for leukotoxin production, plasmid content, and antimicrobial susceptibility. Am. J. Vet. Res. *48*: 378–384.

Chanter, N. and J.M. Rutter. 1989. Pasteurellosis in pigs and determinants of virulence of toxigenic *Pasteurella multocida*. *In* Adlam and Rutter (Editors), *Pasteurella* and Pasteurellosis, Academic Press, London. 161–195.

Chapman, G., J. Hillier and F.H. Johnson. 1951. Observations on the bacteriophagy of *Erwinia carotovora*. J. Bacteriol. *61*: 261–268.

Charles, H. and H. Ishikawa. 1999. Physical and genetic map of the genome of *Buchnera*, the primary endosymbiont of the pea aphid *Acyrthosiphon pisum*. J. Mol. Evol. *48*: 142–150.

Charles, I.G., S. Harford, J.F. Brookfield and W.V. Shaw. 1985. Resistance to chloramphenicol in *Proteus mirabilis* by expression of a chromosomal gene for chloramphenicol acetyltransferase. J. Bacteriol. *164*: 114–122.

Chart, H., D.H. Shaw, E.E. Ishiguro and T.J. Trust. 1984. Structural and immunochemical homogeneity of *Aeromonas salmonicida* lipopolysaccharide. J. Bacteriol. *158*: 16–22.

Chart, H. and T.J. Trust. 1983. Acquisition of iron by *Aeromonas salmonicida*. J. Bacteriol. *156*: 758–764.

Charudattan , R., R.E. Stall and D.L. Batchelor. 1973. Serotypes of *Xanthomonas vesicatoria* unrelated to its pathotypes. Phytopathology *63*: 1260–1265.

Chateil, J.F., M. Brun, Y. Perel, J.C. Sananes, J.F. Castell and F. Diard. 1998. Granulomatous hepatitis in *Pasteurella multocida* infection. Eur. Radiol. *8*: 588–591.

Chatelain, R., H. Bercovier, A. Guiyole, A. Richard and H.H. Mollaret. 1979. Intérêt du composé vibriostatique 0/129 pour différencier les genres *Pasteurella* et *Actinobacillus* de la famille *Enterobacteriaceae*. Ann. Microbiol. (Paris) *130A*: 449–454.

Chatterjee, A., J.L. McEvoy, J.P. Chambost, F. Blasco and A.K. Chatterjee. 1991. Nucleotide sequence and molecular characterization of *pnlA*, the structural gene for damage-inducible pectin lyase of *Erwinia carotovora* subsp. *carotovora* 71. J. Bacteriol. *173*: 1765–1769.

Chatterjee, A.K., L.M. Ross, J.L. McEvoy and K.K. Thurn. 1985. pULB113, an RP4::mini-Mu plasmid, mediates chromosomal mobilization and R-prime formation in *Erwinia amylovora*, *Erwinia chrysanthemi* and subspecies of *Erwinia carotovora*. Appl. Environ. Microbiol. *50*: 1–9.

Chatterjee, A.K. and M.P. Starr. 1980. Genetics of *Erwinia* species. Annu. Rev. Microbiol. *34*: 645–676.

Chatterjee, A.K., K.K. Thurn and D.J. Tyrell. 1981. Regulation of pectolytic enzymes in soft rot *Erwinia carotovora*, *Erwinia chrysanthemi*. Proceedings of the Fifth International Conference on Plant Pathogenic Bacteria, pp. 252–262.

Chattopadhyay, D.J., B.L. Sarkar, M.Q. Ansari, B.K. Chakrabarti, M.K. Roy, A.N. Ghosh and S.C. Pal. 1993. New phage typing scheme for *Vibrio cholerae* O1 biotype El Tor strains. J. Clin. Microbiol. *31*: 1579–1585.

Chatty, H.B. and T.L. Gavan. 1968. *Edwardsiella tarda*—identification and clinical significance. Report of two cases. Clevel. Clin. Q. *35*: 223–228.

Chaudhury, A., G. Nath, B.N. Shukla and S.C. Sanyal. 1996. Biochemical characterisation, enteropathogenicity and antimicrobial resistance plasmids of clinical and environmental *Aeromonas* isolates. J. Med. Microbiol. *44*: 434–437.

Cheasty, R., R.J. Gross, L.V. Thomas and B. Rowe. 1989. Serogrouping of the *Aeromonas hydrophia* group. J. Diarrh Dis. Res. *6*: 95–98.

Chedid, L., F. Audibert and A.G. Johnson. 1978. Biological activities of muramyl dipeptide, a synthetic glycopeptide analogous to bacterial immunoregulating agents. Prog. Allergy. *25*: 63–105.

Chen, C.H., R.B. Dienst and R.B. Greenblatt. 1949. Skin reaction of patients to Donovania granulomatis. Am. J. Syph. Gonorrhea Vener. Dis. *33*: 60–64.

Chen, D.-Q., B.C. Campbell and A.H. Purcell. 1996. A new rickettsia from a herbivorous insect, the pea aphid *Acyrthosiphon pisum* (Harris). Curr. Microbiol. *33*: 123–128.

Chen, D.-Q. and A.H. Purcell. 1997. Occurrence and transmission of facultative endosymbionts in aphids. Curr. Microbiol. *34*: 220–225.

Chen, J., C.J. Chang and R.L. Jarret. 1992a. Plasmids from *Xylella fastidiosa* strains. Can. J. Microbiol. *38*: 993–995.

Chen, J., C.J. Chang, R.L. Jarret and N. Gawel. 1992b. Genetic variation among *Xylella fastidiosa* strains. Phytopathology *82*: 973–977.

Chen, J.H., Y.Y. Hsieh, S.L. Hsiau, T.C. Lo and C.C. Shau. 1999a. Characterization of insertions of IS476 and two newly identified insertion sequences, IS1478 and IS1479, in *Xanthomonas campestris* pathovar *campestris*. J. Bacteriol. *181*: 1220–1228.

Chen, J., O. Lamikanra, C.J. Chang and D.L. Hopkins. 1995. Randomly amplified polymorphic DNA analysis of *Xylella fastidiosa* Pierce's disease and oak leaf Scorch pathotypes. Appl. Environ. Microbiol. *61*: 1688–1690.

Chen, J., P.D. Roberts and D.W. Gabriel. 1994. Effects of a virulence locus from *Xanthomonas campestris* 528(T) on pathovar status and ability to elicit blight symptoms on crucifers. Phytopathology *84*: 1458–1465.

Chen, M.F., S. Yun, G.D. Marty, T.S. McDowell, M.L. House, J.A. Appersen, T.A. Geuenther, K.D. Arkush and R.P. Hedrick. 2000a. A *Piscirickettsia salmonis*-like bacterium associated with mortality of white seabass *Atractoscion nobilis*. Dis. Aquat. Org. *43*: 117–126.

Chen, S.Y., T.A. Hoover, H.A. Thompson and J.C. Williams. 1990a. Characterization of the origin of DNA replication of the *Coxiella burnetii* chromosome. Ann. N. Y. Acad. Sci. *590*: 491–503.

Chen, S.C., R.H. Lawrence, D.R. Packham and T.C. Sorrell. 1991. Cellulitis due to *Pseudomonas putrefaciens*: possible production of exotoxins. Rev. Infect. Dis. *13*: 642–643.

Chen, S.Y., M. Vodkin, H.A. Thompson and J.C. Williams. 1990b. Isolated *Coxiella burnetii* synthesizes DNA during acid activation in the absence of host cells. J. Gen. Microbiol. *136*: 89–96.

Chen, S.C., P.C. Wang, M.C. Tung, K.D. Thompson and A. Adams. 2000b. A *Piscirickettsia salmonis*-like organism in grouper, *Epinephelus melanostigma*, in Taiwan. J. Fish Dis. *23*: 415–418.

Chen, X., S. Li and S. Aksoy. 1999b. Concordant evolution of a symbiont with its host insect species: Molecular phylogeny of genus *Glossina* and its bacteriome-associated endosymbiont, *Wigglesworthia glossinidia*. J. Mol. Evol. *48*: 49–58.

Chen, Y.S., Y.C. Lin, M.Y. Yen, J.H. Wang, J.H. Wang, S.R. Wann and D.L. Cheng. 1997. Skin and soft-tissue manifestations of *Shewanella putrefaciens* infection. Clin. Infect. Dis. *25*: 225–229.

Chen, Y.P., G. Lopezdevictoria and C.R. Lovell. 1993. Utilization of aromatic compounds as carbon and energy sources during growth and N_2 fixation by free living nitrogen fixing bacteria. Arch. Microbiol. *159*: 207–212.

Cheng, T.C., S.P. Harvey and A.N. Stroup. 1993. Purification and properties of a highly active organophosphorus acid anhydrolase from *Alteromonas undina*. Appl. Environ. Microbiol. *59*: 3138–3140.

Chern, R.S. and C.B. Chao. 1994. Outbreaks of a disease caused by a rickettsia-like organism in cultured tilapias in Taiwan. Fish Pathol. *29*: 61–71.

Cherni, N.I., Z.V. Solov'eva, V.D. Fedorov and E.N. Kondrat'eva. 1969. Ultrastructure of cells of two species of purple serobacteria. Mikrobiologiia *38*: 479–484.

Cherry, W.B., G.W. Gorman, L.R. Orrison, C.W. Moss, A.G. Steigerwalt, H.W. Wilkinson, S.E. Johnson, R.M. McKinney and D.J. Brenner. 1982a. *Legionella jordanis*: a new species of *Legionella* isolated from water and sewage. J. Clin. Microbiol. *15*: 290–297.

Cherry, W.B., G.W. Gorman, L.R. Orrison, C.W. Moss, A.G. Steigerwalt, H.W. Wilkinson, S.E. Johnson, R.M. McKinney and D.J. Brenner. 1982b. *In* Validation of the publication of new names and new combinations previously effectively published outside the IJSB. List No. 9. Int. J. Syst. Bacteriol. *32*: 384–385.

Chester, B. and L.B. Moskowitz. 1987. Rapid catalase supplemental test for identification of members of the family *Enterobacteriaceae*. J. Clin. Microbiol. *25*: 439–441.

Chester, B. and G. Stotzky. 1976. Temperature dependent cultural and biochemical characteristics of rhamnose positive *Yersinia enterocolitica*. J. Clin. Microbiol. *3*: 119–127.

Chester, F.D. 1897. A preliminary arrangement of the species of the genus *Bacterium*, bacteria associated with diseases of plants. Del. Agric. Exp. Stn. Annu. Rep. *9*: 127.

Chester, F.D. 1901. A manual of Determinative Bacteriology, The Macmillan Co., New York. 1–401.

Chester, F.D. 1939. Genus IV. *Erwinia*. *In* Bergey, Breed, Murray and Hitchens (Editors), Bergey's Manual of Determinative Bacteriology, 5th Ed. ed., The Williams and Wilkins Co., Baltimore. pp. 404–420.

Chevallier, B., D. Dugourd, K. Tarasiuk, J. Harel, M. Gottschalk, M. Kobisch and J. Frey. 1998. Chromosome sizes and phylogenetic relationships between serotypes of *Actinobacillus pleuropneumoniae*. FEMS Microbiol. Lett. *160*: 209–216.

Chida, T., N. Okamura, K. Ohtani, Y. Yoshida, E. Arakawa and H. Watanabe. 2000. The complete DNA sequence of the O antigen gene region of *Plesiomonas shigelloides* serotype O17 which is identical to *Shigella sonnei* form I antigen. Microbiol. Immunol. *44*: 161–172.

Childress, J.J., C.R. Fisher, J.M. Brooks, M.C. Kennicutt II, R. Bidigare and A.E. Anderson. 1986. A methanotrophic marine molluscan (Bivalvia: *Mytilidae*) symbiosis: mussels fueled by gas. Science *233*: 1306–1308.

Chilukuri, L.N. and D.H. Bartlett. 1997. Isolation and characterization of the gene encoding single-stranded-DNA-binding protein (SSB) from four marine *Shewanella* strains that differ in their temperature and pressure optima for growth. Microbiology *143*: 1163–1174.

Chin, J. and Y. Dai. 1990. Selective extraction of outer-membrane proteins from membrane complexes of *Pseudomonas maltophilia* by chloroform-methanol. Vet. Microbiol. *22*: 69–78.

Chiou, C.S. and A.L. Jones. 1991. The analysis of plasmid-mediated streptomycin resistance in *Erwinia amylovora*. Phytopathology *81*: 710–714.

Chiou, C.S. and A.L. Jones. 1995. Molecular analysis of high-level streptomycin resistance in *Erwinia amylovora*. Phytopathology *85*: 324–328.

Chiu, V.S.W. and P.D. Hoeprich. 1961. Susceptibility of *Proteus* and *Providence* bacilli to 10 antibacterial agents. Am. J. Med. Sci. *241*: 309–321.

Chladeck, D.W. 1975. Bovine abortion associated with *Haemophilus somnus*. Am. J. Vet. Res. *36*: 1041.

Chladeck, D.W. and G.R. Ruth. 1976. Isolation of *Actinobacillus lignieresii* from an epidural abscess in a horse with progressive paralysis. J. Am. Vet. Med. Assoc. *168*: 64–66.

Cho, J.J., A.C. Hayward and R. K.G.. 1980. Nutritional requirements and biochemical activities of pineapple pink disease bacterial strains from Hawaii. Antonie van Leeuwenhoek J. Microbiol. Serol. *46*: 191–204.

Choi, J.E., K. Tsuchiya, N. Matsuyama and S. Wakimoto. 1981. Bacteriological properties related to virulence in *Xanthomonas campestris* pathovar *oryzae*. Ann. Phytopathol. Soc. Japan *47*: 668–676.

Choi, K.H., S.K. Maheswaran and L.J. Felice. 1989a. Characterization of outer-membrane protein-enriched extracts from *Pasteurella multocida* isolated from turkeys. Am. J. Vet. Res. *50*: 676–683.

Choi, S.H., E.Y. Ardales, H. Leung and E.J. Lee. 1989b. Characterization of indigenous plasmid of *Xanthomonas campestris* pathovar *oryzae*. Korean J. Plant Pathol. *5*: 223–229.

Choi, S.H., Y.S. Choi, K.S. Jin, S.H. Lee and E.J. Lee. 1988. Evaluation of wild type *Xanthomonas campestris* pathovar *oryzae* for nutritional requirement and antibiotic resistance. Korean J. Plant Pathol. *4*: 3–9.

Choi-Kim, K., S.K. Maheswaran, L.J. Felice and T.W. Molitor. 1991. Relationship between the iron regulated outer-membrane proteins and the outer-membrane proteins of in vivo grown *Pasteurella multocida*. Vet. Microbiol. *28*: 75–92.

Chopra, A.K. and C.W. Houston. 1999. Molecular characterization of *Aeromonas* enterotoxins. 6th International *Aeromonas/Plesiomonas* Symposium, Chicago, Illinois. 15.

Chopra, A.K., J.W. Peterson, X.J. Xu, D.H. Coppenhaver and C.W. Houston. 1996. Molecular and biochemical characterization of a heat-labile cytotonic enterotoxin from *Aeromonas hydrophila*. Microb. Pathog. *21*: 357–377.

Choudhury, S.R., R.K. Bhadra and J. Das. 1994. Genome size and restriction fragment length polymorphism analysis of *Vibrio cholerae* strains belonging to different serovars and biotypes. FEMS Microbiol. Lett. *115*: 329–334.

Chow, A.W., P.R. Taylor, T.T. Yoshikawa and L.B. Guze. 1979. A noso-comial outbreak of infections due to multiply resistant *Proteus mirabilis*: role of intestinal colonization as a major reservoir. J. Infect. Dis. *139*: 621–627.

Christensen, H., O. Angen, R. Mutters, J.E. Olsen and M. Bisgaard. 2000. DNA–DNA hybridization determined in micro-wells using covalent attachment of DNA. Int. J. Syst. Evol. Microbiol. *50*: 1095–1102.

Christensen, H., M. Bisgaard, O. Angen and J.E. Olsen. 2002. Final clas-sification of Bisgaard taxon 9 as *Actinobacillus arthritidis* sp. nov. and recognition of a novel genomospecies for the equine strains of *Ac-tinobacillus lignieresii*. Int. J. Syst. Evol. Microbiol. *52*: 1239–1246.

Christensen, H., S. Nordentoft and J.E. Olsen. 1998. Phylogenetic rela-tionships of *Salmonella* based on rRNA sequences. Int. J. Syst. Bacteriol. *48*: 605–610.

Christensen, J.P. and M. Bisgaard. 1997. Avian pasteurellosis: taxonomy of the organisms involved and aspects of pathogenesis. Avian Pathol. *26*: 461–483.

Christensen, J.P. and M. Bisgaard. 2000. Fowl cholera. Rev. Sci. Tech. Off. Int. Epizoot. *19*: 626–637.

Christensen, P. 1977. Synonomy of *Flavobacterium pectinovorum* Dorey with *Cytophaga johnsonae* Stanier. Int. J. Syst. Bacteriol. *27*: 122–132.

Christensen, P. and F.D. Cook. 1972. The isolation and enumeration of cytophagas. Can. J. Microbiol. *18*: 1933–1940.

Christensen, P. and F.D. Cook. 1978. *Lysobacter*, a new genus of nonfruit-ing, gliding bacteria, with a high base ratio. Int. J. Syst. Bacteriol. *28*: 367–393.

Christiansen, M. 1917. En ejendommelig pyaemisk Lidelse hos Faar. Maa-nedsskr. Dyrlaeg. *29*: 449–458.

Chuang, Y.C., J.W. Liu, W.C. Ko, K.Y. Lin, J.J. Wu and K.Y. Huang. 1997. *In vitro* synergism between cefotaxime and minocycline against Vibrio vulnificus. Antimicrob. Agents. Chemother. *41*: 2214–2217.

Chuba, P.J., R. Bock, G. Graf, T. Adam and U. Gobel. 1988. Comparison of 16S rRNA sequences from the family *Pasteurellaceae*: phylogenetic relatedness by cluster analysis. J. Gen. Microbiol. *134*: 1923–1930.

Chun, W., J. Cui and A. Poplawsky. 1997. Purification, characterization and biological role of a pheromone produced by *Xanthomonas cam-pestris* pv. *campestris*. Physiol. Mol. Plant Pathol. *51*: 1–14.

Chung, C., G. Ping, Y. Chen, P.D. Shaw and S.K. Farrand. 1998a. Pro-duction of acyl-homoserine lactone quorum-sensing signals by gram-negative plant-associated bacteria. Mol. Plant-Microbe Interact. *11*: 1119–1129.

Chung, H.J., T.S. Lee, Y.J. Koh, I.S. Nou and B.K. Hwang. 1996. Restriction fragment length polymorphisms of genomic DNA in strains of Xan-thomonas campestris pathovar *vesicatoria* from different geographic ar-eas. Korean J. Plant Pathol. *12*: 162–168.

Chung, J.C. and P.F. Strom. 1997. Filamentous bacteria and protozoa found in the rotating biological contactor. J. Environ. Sci. Health Part A Environ. Sci. Eng. Toxic Hazard. Subst. Control *32*: 671–686.

Chung, J.Y., Y.M. Zhang and B. Adler. 1998b. The capsule biosynthetic locus of *Pasteurella multocida* A:1. FEMS Microbiol. Lett. *166*: 289–296.

Chung, S.Y., T. Yaguchi, H. Nishihara, Y. Igarashi and T. Kodama. 1993a. Purification of form L2 RubisCO from a marine obligately autotrophic hydrogen-oxidizing bacterium, *Hydrogenovibrio marinus* strain MH-110. FEMS Microbiol. Lett. *109*: 49–53.

Chung, Y.R., D.J. Brenner, A.G. Steigerwalt, B.S. Kim, H.T. Kim and K.Y. Cho. 1993b. *Enterobacter pyrinus* sp. nov., an organism associated with brown leaf-spot disease of pear trees. Int. J. Syst. Bacteriol. *43*: 157–161.

Chung, Y.R. and K.Y. Cho. 1993. *Enterobacter pyrinus*, a new bacterial path-ogen of pear brown leaf spot disease. Korean J. Plant Pathol. *9*: 232–235.

Ciesielski, C.A., M.J. Blaser and W.L.L. Wang. 1986. Serogroup specificity of *Legionella pneumophila* is related to lipopolysaccharide characteris-tics. Infect. Immun. *51*: 397–404.

Cilia, V., B. Lafay and R. Christen. 1996. Sequence heterogeneities among 16S ribosomal RNA sequences, and their effect on phylogenetic anal-yses at the species level. Mol. Biol. Evol. *13*: 451–461.

Ciufecu, C., N. Nacescu, A. Israil and C. Cedru. 1990. Isolation of motile aeromonads from foods of animal origin. Arch. Roum. Pathol. Exp. Microbiol. *49*: 119–129.

Civerolo, E.L. 1985. Indigenous plasmids in *Xanthomonas campestris* path-ovar *citri*. Phytopathology *75*: 524–528.

Civerolo, E.L., M. Sasser, C. Helkie and D. Burbage. 1982. Selective me-dium for *Xanthomonas campestris* pathovar *pruni*. Plant Dis. *66*: 39–43.

Claesson, B.E.B., D.E.W. Holmlund, C.A. Lindhagen and T.W. Mätzsch. 1984. *Plesiomonas shigelloides* in acute cholecystitis: a case report. J. Clin. Microbiol. *20*: 985–987.

Claflin, J.L. and C.L. Larson. 1972. Infection-immunity in tularemia: spec-ificity of cellular immunity. Infect. Immun. *5*: 311–318.

Clapham, L., R.J.C. McLean, J.C. Nickel, J. Downey and J.W. Costerton. 1990. The influence of bacteria on struvite crystal habit and its im-portance in urinary stone formation. J. Crystal Growth. *104*: 475–484.

Clara, F.M. 1930. A new bacterial leaf disease of tobacco in the Philip-pines. Phytopathology *20*: 691–706.

Clara, F.M. 1932. A new bacterial disease of pears. Science (Wash. D. C.) *75*: 111.

Clara, F.M. 1934. A comparative study of the green-fluorescent bacterial plant pathogens. Cornell Agr. Exptl. Sta. Mem. *159*: 1–36.

Clark, C.A., J. Beltrame and P.A. Manning. 1991a. The *oac* gene encoding a lipopolysaccharide O-antigen acetylase maps adjacent to the inte-grase-encoding gene on the genome of *Shigella flexneri* bacteriophage Sf6. Gene *107*: 43–52.

Clark, D.A. and P.R. Norris. 1996. *Acidimicrobium ferrooxidans* gen. nov. sp. nov.: mixed-culture ferrous iron oxidation with *Sulfobacillus* spe-cies. Microbiology (Read.) *142*: 785–790.

Clark, M.A., L. Baumann and P. Baumann. 1998a. *Buchnera aphidicola* (aphid endosymbiont) contains genes encoding enzymes of histidine biosynthesis. Curr. Microbiol. *37*: 356–358.

Clark, M.A., L. Baumann and P. Baumann. 1998b. Sequence analysis of a 34.7-kb DNA segment from the genome of *Buchnera aphidicola* (en-dosymbiont of aphids) containing *groEL*, *dnaA*, the atp operon, *gidA*, and *rho*. Curr. Microbiol. *36*: 158–163.

Clark, M.A., L. Baumann, M.A. Munson, P. Baumann, B.C. Campbell, J.E. Duffus, L.S. Osborne and N.A. Moran. 1992. The eubacterial endosymbionts of whiteflies (Homoptera, Aleyrodoidea) constitute a lineage distinct from the endosymbionts of aphids and mealybugs. Curr. Microbiol. *25*: 119–123.

Clark, M.A., Y.J. Tang and J.L. Ingraham. 1989. A NosA-specific bacte-riophage can be used to select denitrification-defective mutants of *Pseudomonas stutzeri*. J. Gen. Microbiol. *135*: 2569–2575.

Clark, P.H. and S.T. Cowan. 1952. Biochemical methods for bacteriology. J. Gen. Microbiol. *6*: 187–197.

Clark, R.B. 1992. Antibiotic susceptibilities of the *Vibrionaceae* to mero-penem and other antimicrobial agents. Diagn. Microbiol. Infect. Dis. *15*: 453–455.

Clark, R.B. and J.M. Janda. 1991. *Plesiomonas* and human disease. Clin. Microbiol. Newsl. *13*: 49–52.

Clark, R.B., P.D. Lister, L. Arneson-Rotert and J.M. Janda. 1990. In vitro susceptibilities of *Plesiomonas shigelloides* to 24 antibiotics and antibi-otic-β-lactamase-inhibitor combinations. Antimicrob. Agents Che-mother. *34*: 159–160.

Clark, R.B., P.D. Lister and J.M. Janda. 1991b. *In vitro* susceptibilities of *Edwardsiella tarda* to 22 antibiotics and antibiotic-beta-lactamase-in-hibitor agents. Diagn. Microbiol. Infect. Dis. *14*: 173–175.

Clark, W.A., D.G. Hollis, R.E. Weaver and P. Riley. 1984. Identification of unusual pathogenic gram-negative aerobic and facultatively anaer-obic bacteria, U.S. Dept. of Health and Human Services, Public Health Service, Centers for Disease Control, Atlanta.

Clarke, A.J. 1993. Extent of peptidoglycan O acetylation in the tribe *Proteeae*. J. Bacteriol. *175*: 4550-4553.

Clarke, A.J. and C. Dupont. 1992. O-acetylated peptidoglycan: its occur-rence, pathobiological significance, and biosynthesis. Can. J. Micro-biol. *38*: 85–91.

Clarke, P.H. and R. Drew. 1988. An experiment in enzyme evolution - studies with *Pseudomonas aeruginosa* amidase. Biosci. Rep. *8*: 103–120.

Clarridge, J.E., D.M. Musher, V. Fainstein and R.J. Wallace, Jr.. 1980. Extraintestinal human infection caused by *Edwardsiella tarda*. J. Clin.Microbiol. *11*: 511–514.

Clarridge, J.E.I., T.J. Raich, A. Sjösted, G. Sandström, R.O. Darouiche, R.M. Shawar, P.R. Georghiou, C. Osting and L. Vo. 1996. Characterization of two unusual clinically significant *Francisella* strains. J. Clin. Microbiol. *34*: 1995–2000.

Clarridge, J.E. and S. Zighelboim-Daum. 1985. Isolation and characterization of two hemolytic phenotypes of *Vibrio damsela* associated with a fatal wound infection. J. Clin. Microbiol. *21*: 302–306.

Claus, H., H. Rotlich and Z. Filip. 1992. DNA fingerprints of *Pseudomonas* spp. using rotating field electrophoresis. Microb. Releases. *1*: 11–16.

Claxton, P.D. 1989. Antigenic Classification of *Bacteriodes nodosus*. *In* Egerton, J.R., W.K. Yong and G.G. Riffkin (Editors), Footrot and Foot Abscess of Ruminants, CRC Press, Boca Raton. 155–166.

Claxton, P.D. and R.E. Everett. 1966. Recovery of an organism resembling *Histophilus ovis* from a ram. Aust. Vet. J. *42*: 457–458.

Cleton-Jansen, A.M., N. Goosen, K. Vink and P. van de Putte. 1989. Cloning of the genes encoding the two different glucose dehydrogenases from *Acinetobacter calcoaceticus*. Antonie Leeuwenhoek *56*: 73–79.

Clinkenbeard, K.D. and M.L. Upton. 1991. Lysis of bovine platelets by *Pasteurella haemolytica* leukotoxin. Am. J. Vet. Res. *52*: 453–457.

Cobb, N.A. 1893. Plant diseases and their remedy. Agr. Gaz. N.S.W. *4*: 777–798.

Cobet, A.B., C. Wirsen and G.E. Jones. 1970. The effect of nickel on a marine bacterium, *Arthrobacter marinus* sp. nov. J. Gen. Microbiol. *62*: 159–169.

Coetzee, J.N. 1963. Lysogeny in *Proteus rettgeri* and the host-range of *P. rettgeri* and *P. hauseri* bacteriophages. J. Gen. Microbiol. *31*: 219–229.

Coetzee, J.N. 1972. Genetics of the *Proteus* group. Annu. Rev. Microbiol. *26*: 23–54.

Coetzee, J.N., G. Pernet and J.J. Theron. 1962. Fimbriae and haemagglutinating properties in strains of *Proteus*. Nature *196*: 497–498.

Coetzee, J.N. and T.G. Sacks. 1960. Morphological variants of *Proteus hauseri*. J. Gen. Microbiol. *23*: 209–216.

Cohen, G.N., R.Y. Stanier and G. LeBras. 1969. Regulation of the biosynthesis of amino acids of the aspartate family in coliform bacteria and pseudomonads. J. Bacteriol. *99*: 791–801.

Cohen, M.B. and R.A. Giannella. 1995. Enterotoxigenic *Escherichia coli*. *In* Blaser, Smith, Ravdin, Greenberg and Guerrant (Editors), Infections of the Gastrointestinal Tract, Raven Press, Ltd., New York. pp. 691–707.

Cohn, F. 1875. Untersuchungen über Bakterien II. Beitr. Biol. Pflanz. *1875 1(Heft 3)*: 141–207.

Cohn, F. 1878. Letter of J. Penn which describes *Micrococcus phosphoreum*. *In* Versameling van stukken betreffende het geneeskundig staats toerzitch, 126–130.

Coimbra, R.S., F. Grimont and P.A.D. Grimont. 1999. Identification of *Shigella* serotypes by restriction of amplified O-antigen gene cluster. Res. Microbiol. *150*: 543–553.

Coimbra, R., F. Grimont, P. Lenormand, P. Burguiere, L. Beutin and P.A.D. Grimont. 2000. Identification of *Escherichia coli* O-serogroups by restriction of the amplified O-antigen gene cluster (*rfb*-RFLP). Res. Microbiol. *151*: 639–654.

Coimbra, R.S., M. Lefevre, F. Grimont and P.A.D. Grimont. 2001. Clonal relationships among *Shigella* serotypes suggested by cryptic flagellin gene polymorphism. J. Clin. Microbiol. *39*: 670–674.

Colebrook, L. 1920. The mycelial and other microorganisms associated with human actinomycosis. Br. J. Exp. Pathol. *1*: 197–212.

Coles, B.M., R.K. Stroud and S. Sheggeby. 1978. Isolation of *Edwardsiella tarda* from three Oregon sea mammals. J. Wildlife Dis. *14*: 339–341.

Collier, L.S., I. Gaines, G.L. and E.L. Neidle. 1998. Regulation of benzoate degradation in *Acinetobacter* sp. strain ADP1 by BenM, a LysR-type transcriptional activator. J. Bacteriol. *180*: 2493–2501.

Collins, M.T., S.N. Cho and J.S. Reif. 1982. Prevalence of antibodies to *Legionella pneumophila* in animal populations. J. Clin. Microbiol. *15*: 130–136.

Collins, M.D. and P.N. Green. 1985. Isolation and characterization of a

novel coenzyme Q from some methane-oxidizing bacteria. Biochem. Biophys. Res. Commun. *133*: 1125–1131.

Collins, M.D., A.J. Martinez-Murcia and J. Cai. 1993. *Aeromonas enteropelogenes* and *Aeromonas ichthiosmia* are identical to *Aeromonas trota* and *Aeromonas veronii*, respectively, as revealed by small-subunit rRNA sequence analysis. Int. J. Syst. Bacteriol. *43*: 855–856.

Collins, M.T., J. McDonald, N. Hoiby and O. Aalund. 1984. Agglutinating antibody titers to members of the family *Legionellaceae* in cystic fibrosis patients as a result of cross-reacting antibodies to *Pseudomonas aeruginosa*. J. Clin. Microbiol. *19*: 757–762.

Collinson, S.K., M.A. Abdallah and W.J. Page. 1990. Temperature-sensitive production of azoverdin, the pyoverdin-like siderophore of *Azomonas macrocytogenes* ATCC-12334. J. Gen. Microbiol. *136*: 2297–2305.

Collinson, S.K., J.L. Doran and W.J. Page. 1987. Production of 3,4-dihydroxybenzoic acid by *Azomonas macrocytogenes* and *Azotobacter paspali*. Can. J. Microbiol. *33*: 169–175.

Collinson, S.K. and W.J. Page. 1989. Production of outer-membrane proteins and an extracellular fluorescent compound by iron-limited *Azomonas macrocytogenes*. J. Gen. Microbiol. *135*: 1229–1241.

Collmer, A. and N.T. Keen. 1986. The role of pectic enzymes in plant pathogenesis. Annu. Rev. Phytopathol. *24*: 383–409.

Colnaghi, R., P. Rudnick, L. He, A. Green, D. Yan, E. Larsen and C. Kennedy. 2001. Lethality of *glnD* null mutations in *Azotobacter vinelandii* is suppressible by prevention of glutamine synthetase adenylation. Microbiology *147*: 1267–1276.

Colwell, R.R. (Editor). 1984. Vibrios in the Environment, John Wiley and Sons, New York. 634 pp.

Colwell, R.R., M.T. Macdonell and J. De Ley. 1986. Proposal to recognize the family *Aeromonadaceae* fam. nov. Int. J. Syst. Bacteriol. *36*: 473–477.

Colwell, R.R. and R.Y. Morita. 1964. Reisolation and emendation of description of *Vibrio marinus* (Russell) Ford. J. Bacteriol. *88*: 831–837.

Colwell, R.R. and A.K. Sparks. 1967. Properties of *Pseudomonas enalia*, a marine bacterium pathogenic for the invertebrate *Crassostrea gigas* (Thunberg). Appl. Microbiol. *15*: 980–986.

Comai, L. and T. Kosuge. 1980. Involvement of plasmid deoxyribonucleic-acid in indoleacetic-acid synthesis in *Pseudomonas savastanoi*. J. Bacteriol. *143*: 950–957.

Comstock, L.E., J.A. Johnson, J.M. Michalski, J.G. Morris, Jr. and J.B. Kaper. 1996. Cloning and sequence of a region encoding a surface polysaccharide of *Vibrio cholerae* O139 and characterization of the insertion site in the chromosome of *Vibrio cholerae* O1. Mol. Microbiol. *19*: 815–826.

Condemine, G., C. Dorel, N. Hugouvieux-Cotte-Pattat and J. Robert-Baudouy. 1992. Some of the out genes involved in the secretion of pectate lyases in *Erwinia chrysanthemi* are regulated by kdgR. Mol. Microbiol. *6*: 3199–3211.

Condemine, G., N. Hugouvieux-Cotte-Pattat and J. Robert-Baudouy. 1986. Isolation of *Erwinia chrysanthemi* kduD mutants altered in pectin degradation. J. Bacteriol. *165*: 937–941.

Confer, A.W. 1993. Immunogens of *Pasteurella*. Vet. Microbiol. *37*: 353–368.

Confer, A.W., K.D. Clinkenbeard and G.L. Murphy. 1995. Pathogenesis and virulence of *Pasteurella haemolytica* in cattle: an analysis of current knowledge and future approaches. *In* Donachie, Lainson and Hodgson (Editors), *Haemophilus*, *Actinobacillus*, and *Pasteurella*, Plenum Press, New York. pp. 51–62.

Conrad, R.S., C. Galanos and F.R. Champlin. 1996. Biochemical characterization of lipopolysaccharides extracted from a hydrophobic strain of *Pasteurella multocida*. Vet. Res. Commun. *20*: 195–204.

Conti, E., A. Flaibani, M. O'Regan and I.W. Sutherland. 1994. Alginate from *Pseudomonas fluorescens* and *P. putida* - production and properties. Microbiology *140*: 1125–1132.

Contreras, A. and J. Casadesus. 1987. Tn10 mutagenesis in *Azotobacter vinelandii*. Mol. Gen. Genet. *209*: 276–282.

Contreras, A., R. Maldonado and J. Casadesus. 1991. Tn5 mutagenesis and insertion replacement in *Azotobacter vinelandii*. Plasmid *25*: 76–80.

Cook, F.D., O.E. Edwards, D.C. Gillespie and E.R. Peterson. 1971. 1-Hydroxy 6-methoxy phenazines, U.S. Patent 3,609,153. September 28, 1971.

Cook, G.T. 1948. Urease and other biochemical reactions of the *Proteus* group. J. Pathol. Bacteriol. *60*: 171–181.

Cooke, E.M., J.C. Brayson, A.S. Edmondson and D. Hall. 1979. An investigation into the incidence and sources of *Klebsiella* infections in hospital patients. J. Hyg. (Lond). *82*: 473–480.

Cooksey, D.A. and J.H. Graham. 1989. Genomic fingerprinting of two pathovars of phytopathogenic bacteria by rare-cutting restriction enzymes and field inversion gel electrophoresis. Phytopathology *79*: 745–750.

Coon, S.L., S. Kotob, B.B. Jarvis, S. Wang, W.C. Fuqua and R.M. Weiner. 1994. Homogentisic acid is the product of MelA, which mediates melanogenesis in the marine bacterium *Shewanella colwelliana* D. Appl. Environ. Microbiol. *60*: 3006–3010.

Cooper-Smith, M.E. and A. von Graevenitz. 1978. Non-epidemic *Erwinia herbicola* (*Enterobacter agglomerans*) in blood cultures: bacteriological analysis of fifteen cases. Curr. Microbiol. *1*: 29–32.

Cope, L.D., S. Lumbley, J.L. Latimer, J. Klesney-Tait, M.K. Stevens, L.S. Johnson, M. Purven, R.S. Munson, Jr., T. Lagergard, J.D. Radolf and E.J. Hansen. 1997. A diffusible cytotoxin of *Haemophilus ducreyi*. Proc. Natl. Acad. Sci. U. S. A. *94*: 4056–4061.

Corbel, M.J. 1975. The serological relationship between *Brucella* spp. *Yersinia enterocolitica IX* and *Salmonella* serotypes of Kauffmann-White group N. J. Hyg. Camb. *75*: 151–171.

Corbell, N. and J.E. Loper. 1995. A global regulator of secondary metabolite production in *Pseudomonas fluorescens* Pf-5. J. Bacteriol. *177*: 6230–6236.

Corboz, L. and J. Nicolet. 1975. Infektionen mit sogenannten *Haemophilus somnus* beim Rind: Isolierung und Characterisierung von Stämmen aus Respirations und Geschlechtsorganen. Schweiz. Arch. Tierheilk. *117*: 493–502.

Corboz, L. and P. Wild. 1981. Epidemiologie der *Haemophilus somnus* Infektion beim Rind: Vergleich von Stämmen in der Polyacrilamid-Elektrophorese Schweiz. Arch. Tierheilk. *123*: 79–88.

Cordes, L.G. and D.W. Fraser. 1980. Legionellosis: Legionnaires' disease, Pontiac fever. Med. Clin. North America. *64*: 395–416.

Corey, R.R. and M.P. Starr. 1957a. Colony types of *Xanthomonas phaseoli*. J. Bacteriol. *74*: 137–140.

Corey, R.R. and M.P. Starr. 1957b. Genetic transformation of colony type in *Xanthomonas phaseoli*. J. Bacteriol. *74*: 141–145.

Corey, R.R. and M.P. Starr. 1957c. Genetic transformation of streptomycin resistance in *Xanthomonas phaseoli*. J. Bacteriol. *74*: 146–150.

Cornaglia, G., B. Dainelli, F. Berlutti and M.C. Thaller. 1988. Commercial identification systems often fail to identify *Providencia stuartii*. J. Clin. Microbiol. *26*: 323–327.

Cornelis, G. and E.P. Abraham. 1975. β-Lactamases from *Yersinia enterocolitica*. J. Gen. Microbiol. *87*: 273–284.

Cornelis, G., P. Bennett and J. Grinsted. 1976. Properties of pGC1, a lac plasmid originating in *Yersinia enterocolitica* 842. J. Bacteriol. *127*: 1058–1062.

Cornelis, G. and C. Colson. 1975. Restriction of DNA in *Yersinia enterocolitica* detected by recipient ability for a derepressed R Factor from *Escherichia coli*. J. Gen. Microbiol. *87*: 285–291.

Cornelis, G., M. Van Bouchaute and G. Wauters. 1981. Plasmid-encoded lysine decarboxylation in *Proteus morganii*. J. Clin. Microbiol. *14*: 365–369.

Cornelis, G.R. and F. Van Gijsegem. 2000. Assembly and function of type III secretory systems. Annu. Rev. Microbiol. *54*: 735–774.

Cornelis, G., G. Wauters and E.G. Bruynogh. 1973. Résistance transférables chez des souches sauvages de *Yersinia enterocolitica*. Ann. Microbiol. (Inst. Pasteur) *124A*: 299–309.

Cornish, A.S. and W.J. Page. 1998. The catecholate siderophores of *Azotobacter vinelandii*: their affinity for iron and role in oxygen stress management. Microbiology-(UK). *144*: 1747–1754.

Coroler, L., M. Elomari, B. Hoste, M. Gillis, D. Izard and H. Leclerc.

1996. *Pseudomonas rhodesiae* sp. nov., a new species isolated from natural mineral waters. Syst. Appl. Microbiol. *19*: 600–607.

Coronado, M.J., C. Vargas, H.J. Kunte, E.A. Galinski, A. Ventosa and J.J. Nieto. 1995. Influence of salt concentration on the susceptibility of moderately halophilic bacteria to antimicrobials and its potential use for genetic transfer studies. Curr. Microbiol. *31*: 365–371.

Corstvet, R.E., R.J. Panciera, H.B. Rinker, B.L. Starks and C. Howard. 1973. Survey of tracheas of feedlot cattle for *Haemophilus somnus* and other selected bacteria. J. Am. Vet. Med. Assoc. *163*: 870–873.

Cosenza, B.J. and J.D. Buck. 1966. Simple device for enumeration and isolation of luminescent bacteria. Appl. Microbiol. *14*: 692.

Cosenza, B.J. and J.D. Podgwaite. 1966. A new species of *Proteus* isolated from larvae of the gypsy moth *Porthetria dispar* (L). Antonie Leeuwenhoek J. Microbiol. Serol. *32*: 187–191.

Costa, J.M. and J.E. Loper. 1994. Derivation of mutants of *Erwinia carotovora* subsp. *betavasculorum* deficient in export of pectolytic enzymes with potential for biological control of potato soft rot. Appl. Environ. Microbiol. *60*: 2278–2285.

Costas, M., B. Holmes, K.A. Frith, C. Riddle and P.M. Hawkey. 1993. Identification and typing of *Proteus penneri* and *Proteus vulgaris* biogroups 2 and 3, from clinical sources, by computerized analysis of electrophoretic protein patterns. J. Appl. Bacteriol. *75*: 489–498.

Costas, M., B. Holmes and L.L. Sloss. 1987. Numerical analysis of electrophoretic protein patterns of *Providencia rustigianii* strains from human diarrhoea and other sources. J. Appl. Bacteriol. *63*: 319–328.

Costas, M., B. Holmes and A.C. Wood. 1990. Numerical analysis of electrophoretic protein patterns of *Providencia stuartii* strains from urine, wound and other clinical sources. J. Appl. Bacteriol. *68*: 505–518.

Costas, M., B. Holmes, A.C. Wood and S.L.W. On. 1989. Numerical analysis of electrophoretic protein patterns of *Providencia rettgeri* strains from human feces, urine and other specimens. J. Appl. Bacteriol. *67*: 441–452.

Costerton, J.W., J.M. Ingram and K.L. Cheng. 1974. Structure and function of the cell envelope of Gram-negative bacteria. Bacteriol. Rev. *38*: 87–110.

Cote, G.L. and L.H. Krull. 1988. Characterization of the exocellular polysaccharides from *Azotobacter chroococcum*. Carbohydr. Res. *181*: 143–152.

Couch, J.A. 1978. Diseases, parasites, and toxic responses of commercial penaeid shrimps of the Gulf of Mexico and South Atlantic coasts of North America. U.S. Natl. Mar. Fish. Serv. Fish. Bull. *76*: 1–44.

Couche, G.A., P.R. Lehrbach, R.G. Forage, G.C. Cooney, D.R. Smith and R.P. Gregson. 1987. Occurrence of intracellular inclusions and plasmids in *Xenorhabdus* spp. J. Gen. Microbiol. *133*: 967–973.

Coudron, P.E. and S.M. Markowitz. 1987. *Cedecea lapagei* isolated from lung tissue. Clin. Microbiol. Newsl. *9*: 71–172.

Coutlee, F., L.A. Saint Jean and R. Plante. 1992. Infection with *Edwardsiella tarda* related to a vascular prosthesis [letter]. Clin. Infect. Dis. *14*: 621–622.

Cowan, S.T., K.J. Steel, G.I. Barrow and R.K.A. Feltham (Editors). 1995. Cowan and Steel's Manual for the Identification of Medical Bacteria, University of Cambridge, Cambridge.

Cowan, S.T., K.J. Steel, C. Shaw and J.P. Duguid. 1960. A classification of the *Klebsiella* group. J. Gen. Microbiol. *23*: 601–612.

Cowley, S.C., S.V. Myltseva and F.E. Nano. 1996. Phase variation in *Francisella tularensis* affecting intracellular growth, lipopolysaccharide antigenicity and nitric oxide production. Mol. Microbiol. *20*: 867–874.

Cox, C.D., K.L. Rinehart, M.L. Moore and J.C. Cook. 1981. Pyochelin: novel ntructure of an iron chelating growth promoter for *Pseudomonas aeruginosa*. Proc. Nat. Acad. Sci. U.S.A. *78*: 4256–4260.

Cox, C.D. and G.I. Wallace. 1948. A study of *Shigella* isolated in India and Burma, with special reference to two previously undescribed serotypes. J. Immunol. *60*: 465–473.

Cox, D.I., D.J. Morrison and G.H. Rae. 1986. Report on a new *Aeromonas* species infecting skin lesions of Atlantic Salmon (*Salmo salar* L.) in sea water. Bull. Eur. Assoc. Fish Pathol. *6*: 100–102.

Coykendall, A.L., J. Setterfield and J. Slots. 1983. Deoxyribonucleic acid relatedness among *Actinobacillus actinomycetemcomitans*, *Hemophilus aphrophilus*, and other *Actinobacillus* species. Int. J. Syst. Bacteriol. *33*: 422–424.

Coyne, V.E., C.J. Pillidge, D.D. Sledjeski, H. Hori, B.A. Ortiz-Conde, D.G. Muir, R.M. Weiner and R.R. Colwell. 1989. Reclassification of *Alteromonas colwelliana* to the genus *Shewanella* by DNA–DNA hybridization, serology and 5S ribosomal RNA sequence data. Syst. Appl. Microbiol. *12*: 275–279.

Coyne, V.E., C.J. Pillidge, D.D. Sledjeski, H. Hori, B.A. Ortiz-Conde, D.G. Muir, R.M. Weiner and R.R. Colwell. 1990. *In* Validation of the publication of new names and new combinations previously effectively published outside the IJSB. List No. 34. Int. J. Syst. Bacteriol. *40*: 320–321.

Crabtree, K. and E. McCoy. 1967. *Zoogloea ramigera* Itzigsohn, identification and description. Request for an opinion as to the status of the generic name *Zoogloea*. Int. J. Syst. Bacteriol. *17*: 1–10.

Cradock-Watson, J.E. 1965. The production of bacteriocines by *Proteus* species. Zentrabl. Bakteriol. Parasitenk. Infektionskr. Hyg. Abt. 1 Orig. *196*: 385–388.

Craigie, J. and C.H. Yen. 1938. The demonstration of types of *B. typhosus* by means of preparation of type II Vi phage. Can. J. Publ. Health. *29*: 448–463 484–496.

Cravioto, A., R.J. Gross, S.M. Scotland and B. Rowe. 1979. An adhesive factor found in strains of *Escherichia coli* belonging to the traditional infantile enteropathogenic serotypes. Curr. Microbiol. *3*: 95–99.

Cravioto, A., A. Tello, A. Navarro, J. Ruiz, H. Villafan, F. Uribe and C. Eslava. 1991. Association of *Escherichia coli* HEp-2 adherence patterns with type and duration of diarrhoea. Lancet *337*: 262–264.

Crisel, R.M., R.S. Baker and D.E. Dorman. 1975. Capsular polymer of *Haemophilus influenzae* type b. I. Structural characterization of the capsular polymer of strain Egan. J. Biol. Chem. *250*: 4926–4930.

Croizier, G. and G. Meynadier. 1971. Recherche d'antigènes de groupe chez des rickettsies de la tribu des Wolbachiae par la technique d'agglutination des corps élémentaires. Entomophaga. *16*: 11–17.

Croizier, G. and G. Meynadier. 1972. Etude en immunofluorescence de l'infection expérimentale de la souris par *Rickettsiella grylli*. Ann. Rech. Vétér. *3*: 373–380.

Croizier, G., G. Meynadier, G. Morel and M. Capponi. 1975. Immunologic comparison between some Wolbachiae and research of antigentic community with other members of the order of *Rickettsiales*. Bull. Soc. Pathol. Exot. Filiales. *68*: 133–141.

Cromartie, W.J. 1981. Arthropathic properties of peptidoglycan-polysaccharide complexes of microbial origin. *In* Deicher and Schulz (Editors), Arthritis-models and mechanisms, Springer-Verlag, Berlin. pp. 24–38.

Crosa, J.H., D.J. Brenner, W.H. Ewing and S. Falkow. 1973a. Molecular relationships among the *Salmonelleae*. J. Bacteriol. *115*: 307–315.

Crosa, J., D.J. Brenner and S. Falkow. 1973b. Use of a single-strand specific nuclease for the analysis of bacterial and plasmid DNA homo- and heteroduplexes. J. Bacteriol. *115*: 904–911.

Crosa, J.H., A.G. Steigerwalt, G.R. Fanning and D.J. Brenner. 1974. Polynucleotide sequence divergence in the genus *Citrobacter*. J. Gen. Microbiol. *83*: 271–282.

Crosse, J.E. and C.M.E. Garrett. 1963. Studies on the bacteriophagy of *Pseudomonas morsprunorum*, *Pseudomonas syringae* and related organisms. J. Appl. Bacteriol. *26*: 159–177.

Crosse, J.E. and R.N. Goodman. 1973. A selective medium for a definitive colony characteristic of *Erwinia amylovora*. Phytopathology *63*: 1425–1426.

Crowder, E.F. 1974. A semi-solid maintenance medium for Salmonellae. Aust. J. Med. Technol. *5*: 147–148.

Crutzen, P.J. 1991. Methane's sinks and sources. Nature *350*: 380–381.

Cryz, S.J., Jr., E. Furer and R. Germanier. 1984. Prevention of fatal experimental burn wound sepsis due to *Klebsiella pneumoniae* KP1-O by immunization with homologous capsular polysaccharide. J. Infect. Dis. *150*: 817–822.

Cui, Y., A. Chatterjee, Y. Liu, C.K. Dumenyo and A.K. Chatterjee. 1995. Identification of a global repressor gene, *rsmA*, of *Erwinia carotovora* subsp. *carotovora* that controls extracellular enzymes, N-(3-oxohex-

anoyl)-*l*-homoserine lactone, and pathogenicity in soft-rotting *Erwinia* spp. J. Bacteriol. *177*: 5108–5115.

Cullmann, W. 1991. Antibiotic susceptibility and outer membrane proteins of clinical *Xanthomonas maltophilia* isolates. Chemotherapy. *37*: 246–250.

Cunliffe, D.A. and P. Adcock. 1989. Isolation of *Aeromonas* spp. from water by using anaerobic incubation. Appl. Environ. Microbiol. *55*: 2138–2140.

Cunliffe, H.E., T.R. Merriman and I.L. Lamont. 1995. Cloning and characterization of *pvdS*, a gene required for pyoverdine synthesis in Pseudomonas aeruginosa - PvdS is probably an alternative sigma-factor. J. Bacteriol. *177*: 2744–2750.

Cuppels, D. and A. Kelman. 1974. Evaluation of selective media for isolation of soft-rot bacteria from soil and plant tissue. Phytopathology *64*: 468–475.

Curreri, P.W., H.M. Bruck, R.B. Lindberg, A.D. Mason, Jr. and B.A. Pruitt, Jr.. 1973. *Providencia stuartii* sepsis: a new challenge in the treatment of thermal injury. Ann. Surg. *177*: 133–138.

Cusack, R., D. Groman and S. Jones. 1997. The first reported rickettsial infections of Atlantic salmon in eastern North America (Abstract). European Association of Fish Pathologists VIIIth International Conference "Diseases of Fish and Shellfish", Edinburgh.September 14–19, 1997, Heriot-Watt University. pp. 109.

Cutlip, R.C., W.C. Amtower and M.R. Zinober. 1972. Septic embolic actinobacillosis of swine: a case report and laboratory reproduction of the disease. Am. J. Vet. Res. *33*: 1621–1626.

Cutter, D.L. and A.S. Kreger. 1990. Cloning and expression of the damselysin gene from *Vibrio damsela*. Infect. Immun. *58*: 266–268.

Cuypers, H. and W.G. Zumft. 1992. Regulatory components of the denitrification gene cluster of *Pseudomonas stutzeri*. *In* Galli, Silver and Witholt (Editors), *Pseudomonas*: Molecular Biology and Biotechnology, American Society for Microbiology, Washington D.C. 188–197.

Cvitanich, J.D., O.N. Garate and C.E. Smith. 1991. The isolation of a rickettsia-like organism causing disease and mortality in Chilean salmonids and its confirmation by Koch postulate. J. Fish Dis. *14*: 121–145.

Cynamon, M.H., T.B. Sorg and A. Patapow. 1988. Utilization and metabolism of NAD by *Haemophilus parainfluenzae*. J. Gen. Microbiol. *134*: 2789–2799.

Cyrus, Z. and A. Sladka. 1970. Several interesting organisms present in activated sludge. Hydrobiol. *35*: 383–396.

da Silva Romeiro, R. and A.B. Moura. 1998. Bacterial blight (*Xanthomonas campestris*) of sunflower (*Helianthus annuus*), a new disease. Rev. Ceres. *45*: 233–243.

Dabboussi, F. , M. Hamze, M. Elomari, S. Verhille, N. Baida, D. Izard and H. Leclerc. 1999a. *Pseudomonas libanensis* sp. nov., a new species isolated from Lebanese spring waters. Int. J. of Syst. Bacteriol. *49*: 1091–1101 .

Dabboussi, F., M. Hamze, M. Elomari, S. Verhille, N. Baida, D. Izard and H. Leclerc. 1999b. Taxonomic study of bacteria isolated from Lebanese spring waters: proposal for *Pseudomonas cedrella* sp. nov. and *P. orientalis* sp. nov. Res. Microbiol. *150*: 303–316.

Dadd, R.H. 1985. Nutrition: organisms. *In* Kerkut and Gilbert (Editors), Comprehensive Insect Physiology, Biochemistry and Pharmacology, Vol. 4, Pergamon Press, Elmsford, NY. pp. 315–319.

Dai, H., Y.S. Lo, A. Kuo, B.Y. Lin and K.S. Chiang. 1991. Inheritable instability of colony morphology of *Xanthomonas campestris* pathovar *citri*. Bot. Bull. Acad. Sin. *32*: 271–278.

Daily, O.P., J.C. Coolbaugh, R.I. Walker, B.R. Merrell, D.M. Rollins, R.J. Seidler, R.R. Colwell, C.R. Lissner and S.W. Joseph. 1981. Association of *Aeromonas sobria* with human infection. J. Clin. Microbiol. *13*: 769–777.

Dale, C. and I. Maudlin. 1999. *Sodalis* gen. nov. and *Sodalis glossinidius* sp. nov., a microaerophilic secondary endosymbiont of the tsetse fly *Glossina morsitans* subsp. *morsitans*. Int. J. Syst. Bacteriol. *49*: 267–275.

Dale, C., S.A. Young, D.T. Haydon and S.C. Welburn. 2001. The insect endosymbiont *Sodalis glossinidius* utilizes a type III secretion system for cell invasion. Proc. Natl. Acad. Sci. U. S. A. *98*: 1883-1888.

Dale, T. and G. Blom. 1987. Epigrowth organisms on reared larvae of the European lobster *Homarus gammarus*. Fauna (Oslo) *40*: 16–19.

Dalsgaard, A., A. Alarcon, C.F. Lanata, T. Jensen, H.J. Hansen, F. Delgado, A.I. Gil, M.E. Penny and D. Taylor. 1996. Clinical manifestations and molecular epidemiology of five cases of diarrhoea in children associated with *Vibrio metschnikovii* in Arequipa, Peru. J. Med. Microbiol. *45*: 494–500.

Dalsgaard, A., P. Glerup, L.L. Hoybye, A.M. Paarup, R. Meza, M. Bernal, T. Shimada and D.N. Taylor. 1997. *Vibrio furnissii* isolated from humans in Peru: a possible human pathogen? Epidemiol. Infect. *119*: 143–149.

Dalsgaard, A., O. Serichantalergs, C. Pitarangsi and P. Echeverria. 1995. Molecular characterization and antibiotic susceptibility of *Vibrio cholerae* non-O1. Epidemiol. Infect. *114*: 51–63.

Dalsgaard, I., B.K. Gudmundsdottir, S. Helgason, S. Hole, O.F. Thoresen, U.P. Wichardt and T. Wiklund. 1998. Identification of atypical *Aeromonas salmonicida*: inter-laboratory evaluation and harmonization of methods. J. Appl. Microbiol. *84*: 999–1006.

Dalsgaard, I., B. Nielsen and J.L. Larsen. 1994. Characterization of *Aeromonas salmonicida* subsp. *salmonicida*: a comparative study of strains of different geographic origin. J. Appl. Bacteriol. *77*: 21–30.

Dalton, H. 1992. Methane oxidation by methanotrophs: physiological and mechanistic implications. *In* Murrell and Dalton (Editors), Methane and Methanol Utilizers, Plenum Press, New York. pp. 85–114.

Dando, P.R., I. Bussmann, S.J. Niven, S.C.M. O'Hara, R. Schmaljohann and L.J. Taylor. 1994. A methane seep area in the Skagerrak, the habitat of the pogonophore *Siboglinum poseidoni* and the bivalve mollusc *Thyasira sarsi*. Mar. Ecol. Prog. Ser. *107*: 157–167.

Dando, P.R. and L.E. Hooper. 1996. Hydrothermal *Thioploca*-more unusual bacterial mats from Milos. The BRIDGE Newsletter *11*: 45–47.

Dando, P.R., M. Thomm, H. Arab, M. Brehmer, L.E. Hooper, B. Jochimsen, H. Schlesner, R. Stöhr, J.-C. Miguel and S.W. Fowler. 1998. Microbiology of shallow hydrothermal sites off Palaeochori Bay, Milos (Hellenic Volcanic Arc). Cah. Biol. Mar. *39*: 369–372.

Dang Van, A., G. Bieth and D.H. Bouanchaud. 1975a. Résistance plasmidique à la tetracycline chez *Haemophilus influenzae*. C.R. Acad. Sci. Paris, Ser. D *280*: 1321–1323.

Dang Van, A., F. Goldstein, J.F. Acar and D.H. Bouanchaud. 1975b. A transferable kanamycin resistance plasmid isolated from *Haemophilus influenzae*. Ann. Microbiol. (Inst. Pasteur) *126A*: 397–399.

Daniels, M.J., C.E. Barber, P.C. Turner, M.K. Sawczyc, R.J.W. Byrde and A.H. Fielding. 1984. Cloning of genes involved in pathogenicity of *Xanthomonas campestris* pathovar *campestris* using the broad host range cosmid pLAFR1. EMBO. *3*: 3323–3328.

Daniels, M.J. and J.E. Leach. 1993. Genetics of *Xanthomonas*. *In* Swings and Civerolo (Editors), *Xanthomonas*, Chapman & Hall, London. pp. 301–339.

Danner, D.B., R.A. Deich, K.L. Sisco and H.O. Smith. 1980. An eleven-base-pair sequence determines the specificity of DNA uptake in *Haemophilus* transformation. Gene *11*: 311–318.

D'Aoust, J.Y. and D.J. Kushner. 1972. *Vibrio psychroerythrus* sp. n.: classification of the psychrophilic marine bacterium, NRC 1004. J. Bacteriol. *111*: 340–342.

Darrasse, A., A. Kotoujansky and Y. Bertheau. 1994a. Isolation by genomic subtraction of DNA probes specific for *Erwinia carotovora* subsp. *atroseptica*. Appl. Environ. Microbiol. *60*: 298–306.

Darrasse, A., S. Priou, A. Kotoujansky and Y. Bertheau. 1994b. PCR and restriction fragment length polymorphism of a *pel* gene as a tool to identify *Erwinia carotovora* in relation to potato diseases. Appl. Environ. Microbiol. *60*: 1437–1443.

Darzins, A. 1994. Characterization of a *Pseudomonas aeruginosa* gene-cluster involved in pilus biosynthesis and twitching motility - sequence similarity to the chemotaxis proteins of enterics and the gliding bacterium *Myxococcus xanthus*. Mol. Microbiol. *11*: 137–153.

Darzins, A. and M.A. Russell. 1997. Molecular genetic analysis of type-4 pilus biogenesis and twitching motility using *Pseudomonas aeruginosa* as a model system - a review. Gene *192*: 109–115.

Daskaleros, P.A., J.A. Stoebner and S.M. Payne. 1991. Iron uptake in *Plesiomonas shigelloides*: cloning of the genes for the heme-iron uptake system. Infect. Immun. *59*: 2706–2711.

Dath, A.P. and S. Devadath. 1982. Studies on mutant colony type of *Xanthomonas campestris* pv. *oryzae*. Indian Phytopathol. *35*: 456–460.

Datta, R. 1989. Recovery and purification of lactate salts from whole fermentation broth by electrophoresis. US Patent 4,885,247.

Dauga, C., M. Gillis, P. Vandamme, E. Ageron, F. Grimont, K. Kersters, C. de Mahenge, Y. Peloux and P.A.D. Grimont. 1993a. *Balneatrix alpica* gen. nov., sp. nov., a bacterium associated with pneumonia and meningitis in a spa therapy centre. Res. Microbiol. *144*: 35–46.

Dauga, C., M. Gillis, P. Vandamme, E. Ageron, F. Grimont, K. Kersters, C. de Mahenge, Y. Peloux and P.A.D. Grimont. 1993b. *In* Validation of the publication of new names and new combinations previously effectively published outside the IJSB. List No. 46. Int. J.Syst. Bacteriol. *43*: 624–625.

Dauga, C., F. Grimont and P.A.D. Grimont. 1990. Nucleotide sequences of 16S rRNA from ten *Serratia* species. Res. Microbiol. *141*: 1139–1149.

Davey, J.F., R. Whittenbury and J.F. Wilkinson. 1972. The distribution in the methylobacteria of some key enzymes concerned with intermediary metabolism. Arch. Mikrobiol. *87*: 359–366.

Davidson, L.S. and J.D. Oliver. 1986. Plasmid carriage in *Vibrio vulnificus* and other lactose-fermenting marine vibrios. Appl. Environ. Microbiol. *52*: 211–213.

Davies, D.A.L. 1958. The smooth and rough antigens of *Pasteurella pseudotuberculosis*. J. Gen. Microbiol. *18*: 118–128.

Davies, R.L., S. Arkinsaw and R.K. Selander. 1997. Genetic relationships among *Pasteurella trehalosi* isolates based on multilocus enzyme electrophoresis. Microbiology (UK) *143*: 2841–2849.

Davies, R.L., B.J. Paster and F.E. Dewhirst. 1996. Phylogenetic relationships and diversity within the *Pasteurella haemolytica* complex based on 16S rRNA sequence comparison and outer membrane protein and lipopolysaccharide analysis. Int. J. Syst. Bacteriol. *46*: 736–744.

Davies, R.L. and M. Quirie. 1996. Intra-specific diversity within *Pasteurella trehalosi* based on variation of capsular polysaccharide, lipopolysaccharide and outer-membrane proteins. Microbiology (UK) *142*: 551–560.

Davies, R.L., T.S. Whittam and R.K. Selander. 2001. Sequence diversity and molecular evolution of the leukotoxin (*lktA*) gene in bovine and ovine strains of *Mannheimia (Pasteurella) haemolytica*. J. Bacteriol. *183*: 1394–1404.

Davies, S.L. and R. Whittenbury. 1970. Fine structure of methane- and other hydrocarbon-utilizing bacteria. J. Gen. Microbiol. *61*: 227–232.

Davin-Regli, A., C. Bosi, R. Charrel, E. Ageron, L. Papazian, P.A.D. Grimont, A. Cremieux and C. Bollet. 1997. A nosocomial outbreak due to *Enterobacter cloacae* strains with the *E. hormaechei* genotype in patients treated with fluoroquinolones. J. Clin. Microbiol. *35*: 1008–1010.

Davis, B.R., G.R. Fanning, J.M. Madden, A.G. Steigerwalt, H.B. Bradford, Jr., H.L. Smith, Jr. and D.J. Brenner. 1981a. Characterization of biochemically atypical *Vibrio cholerae* strains and designation of a new pathogenic species, *Vibrio mimicus*. J. Clin. Microbiol. *14*: 631–639.

Davis, B.R., G.R. Fanning, J.M. Madden, A.G. Steigerwalt, H.B. Bradford, Jr., H.L. Smith, Jr. and D.J. Brenner. 1982. *In* Validation of new names and new combinations previously effectively published outside the IJSB. List No. 8. Int. J. Syst. Bacteriol. *32*: 266–268.

Davis, C.P., D. Cleven, J. Brown and E. Balish. 1976. *Anaerobiospirillum*, a new genus of spiral-shaped bacteria. Int. J. Syst. Bacteriol. *26*: 498–504.

Davis, C.M. and C. Collins. 1969. Granuloma inguinale: ultrastructural study of *Calymmatobacterium granulomatis*. J. Investig. Dermatol. *53*: 315–321.

Davis, D.J., M. Pitmann and J.J. Griffitts. 1950. Hemagglutination by the Koch-Weeks bacillus (*Hemophilus aegyptius*). J. Bacteriol. *59*: 427–431.

Davis, G.E. and H.R. Cox. 1938. A filter-passing infectious agent isolated from ticks. I. Isolation from *Dermacentor andersoni*, reactions in animals, and filtration experiments. Pub. Health Rep. *53*: 2259–2267.

Davis, J.B., V.F. Coty and J.P. Stanley. 1964. Atmospheric nitrogen fixation by methane-oxidizing bacteria. J. Bacteriol. *88*: 468–472.

Davis, J.A. and M.J. Eyles. 1992. Discolouration of cottage cheese caused by *Rahnella aquatilis* in the presence of glucono delta-lactone. Aust. J. Dairy Technol. *47*: 62–63.

Davis, M. 1994. Update—Multistate outbreak of *Escherichia coli* O157:H7 infections from hamburgers—Western United States, 1992–1993 (reprinted from MMWR, Vol 269, pg. 2194, 1993). JAMA. *271*: 341.

Davis, M.J., W.J. French and N.W. Schaad. 1981b. Axenic culture of the bacteria associated with phony disease of peach and plum leaf scald. Curr. Microbiol. *6*: 309–314.

Davis, M.J., A.H. Purcell and S.V. Thomson. 1980. Isolation media for the Pierce's disease bacterium. Phytopathology *70*: 425–429.

Davis, M.J., P. Rott, C.J. Warmuth, M. Chatenet and P. Baudin. 1997. Intraspecific genomic variation within *Xanthomonas albilineans*, the sugarcane leaf scald pathogen. Phytopathology *87*: 316–324.

Davis, M.J., R.F. Whitcomb and A.G. Gillaspie, Jr.. 1981c. Fastidious bacteria of plant vascular tissue and invertebrates (including so-called *Rickettsia*-like bacteria). *In* Starr, Stolp, Trüper, Balows and Schlegel (Editors), The Prokaryotes. A Handbook on Habitats, Isolation, and Identification of Vacteria, 1st Ed., Vol. 2, Springer-Verlag, Berlin. pp. 2172–2188.

Davis W.A. II, J.H. Chretien, V.F. Garagusi and M.A. Goldstein. 1978. Snake-to-human transmission of *Aeromonas shigelloides* resulting in gastroenteritis. South. Med. J. *71*: 474–476.

Day, N.P., S.M. Scotland and B. Rowe. 1981. Comparison of an HEp-2 tissue culture test with the Sereny test for detection of enteroinvasiveness in *Shigella* spp. and *Escherichia coli*. J. Clin. Microbiol. *13*: 596–597.

Day, W.R. 1924. The watermark disease of the cricket-bat willow. Oxford For. Mem. *3*: 1–30.

Daza, R.M., J. Iborra, N. Alonso, I. Vera, F. Portero and P. Mendaza. 1993. Isolation of *Leclercia adecarboxylata* in a cirrhotic patient (Translation). Enferm. Infecc. Microbiol. Clin. *11*: 53–54.

De, E., R. De Mot, N. Orange, N. Saint and G. Molle. 1995. Channel-forming properties and structural homology of major outer-membrane proteins from *Pseudomonas fluorescens* MFO and OE 28.3. FEMS Microbiol. Lett. *127*: 267–272.

De, E., N. Orange, N. Saint, J. Guerillon, R. De Mot and G. Molle. 1997. Growth temperature dependence of channel size of the major outer-membrane protein (OprF) in psychrotrophic *Pseudomonas fluorescens* strains. Microbiology *143*: 1029–1035.

De Alwis, M.C.L. 1995. Haemorrhagic septicaemia (*Pasteurella multocida* serotype B:2 and E:2 infection) in cattle and buffaloes. *In* Donachie, Lainson and Hodgson (Editors), *Haemophilus*, *Actinobacillus*, and *Pasteurella*, Plenum Press, London. 9–24.

De Baere, T., G. Wauters, A. Huylenbroeck, G. Claeys, R. Peleman, G. Verschraegen, D. Allemeersch and M. Vaneechoutte. 2001. Isolations of *Leclercia adecarboxylata* from a patient with a chronically inflamed gallbladder and from a patient with sepsis without focus. J. Clin Microbiol. *39*: 1674–1675.

De Boer, S.H., R.J. Copeman and H. Vruggink. 1979. Serogroups of *Erwinia carotovora* potato strains determined with diffusible somatic antigens. Phytopathology *69*: 316–319.

De Boer, S.H. and M.E. McNaughton. 1987. Monoclonal antibodies to the lipopolysaccharide of *Erwinia carotovora* ssp. *atroseptica* serogroup I. Phytopathology *77*: 828–832.

De Boer, S.H. and L.J. Ward. 1995. PCR detection of *Erwinia carotovora* subsp. *atroseptica* associated with potato tissue. Phytopathology *85*: 854–858.

De Boer, W.E. and W. Hazeu. 1972. Observations on the fine structure of a methane-oxidizing bacterium. Antonie Leeuwenhoek *38*: 33–47.

De Boer, W.E., J.W. La Riviere and K. Schmidt. 1971. Some properties of *Achromatium oxaliferum*. Antonie Leeuwenhoek *37*: 553–563.

De Champs, C., S. Le Seaux, J.J. Dubost, S. Boisgard, B. Sauvezie and J. Sirot. 2000. Isolation of *Pantoea agglomerans* in two cases of septic monoarthritis after plant thorn and wood sliver injuries. J. Clin. Microbiol. *38*: 460–461.

de Crecy-Lagard, V., O.M.M. Bouvet, P. Lejeune and A. Danchin. 1991a. Fructose catabolism in *Xanthomonas campestris* pv. *campestris*: Sequence of the PTS operon, characterization of the fructose-specific enzymes. Jo. Biol. Chem. *266*: 18154–18161.

de Crecy-Lagard, V., P. Lejeune, O.M.M. Bouvet and A. Danchin. 1991b. Identification of two fructose transport and phosphorylation pathways in *Xanthomonas campestris* pv. *campestris*. Mol. Gen. Genet. *227*: 465–472.

de Feyter, R. and D.W. Gabriel. 1991. At least six avirulence genes are clustered on a 90-kilobase plasmid in *Xanthomonas campestris* pv. *malvacearum*. Mol. Plant-Microbe Interact. *4*: 423–432.

De Gheldre, Y., M.J. Struelens, Y. Glupczynski, P. De Mol, N. Maes, C. Nonhoff, H. Chetoui, C. Sion, O. Ronveaux and M. Vaneechoutte. 2001. National epidemiologic surveys of *Enterobacter aerogenes* in Belgian hospitals from 1996 to 1998. J. Clin. Microbiol. *39*: 889–896.

De Graaff, J., L.P. Elwell and S. Falkow. 1976. Molecular nature of two beta-lactamase-specifying plasmids isolated from *Haemophilus influenzae* type b. J. Bacteriol. *126*: 439–446.

de Inchaustegui, P.J., E.R. de Valera and I.B. de Caventi. 1976. Aislamento microorganismo *Edwardsiella tarda* en visceras de pescado fresco. Ciencia. *3*: 6–9.

De Kam, M. 1976. *Erwinia salicis*: its metabolism and variability *in vitro*, and a method to demonstrate the pathogen in the host. Antonie Leeuwenhoek J. Microbiol. Serol. *42*: 421–428.

De Kam, M. 1982. Detection of soluble antigens of *Erwinia salicis* in leaves of *Salix alba* by enzyme-linked immunosorbent assay. Eur. J. For. Pathol. *12*: 1–6.

De Kam, M. 1984. *Xanthomonas campestris* pv. *populi*, the causal agent of bark necrosis in poplar. Neth. J. Plant Pathol. *90*: 13–22.

De Kam, M. 1986. Development of a method for testing the susceptibility of *Salix* to *Erwinia salicis*. Energy from Biomass. *1*: 77–79.

de Kievit, T.R. and J.S. Lam. 1994. Monoclonal-antibodies that distinguish inner-core, outer-core, and lipid-A regions of *Pseudomonas aeruginosa* lipopolysaccharide. J. Bacteriol. *176*: 7129–7139.

De Kort, G., A. Bolton, G. Martin, J. Stephen and J.A.M. Vandeklundert. 1994. Invasion of rabbit ileal tissue by *Enterobacter cloacae* varies with the concentration of OmpX in the outer membrane. Infect. Immun. *62*: 4722–4726.

de la Peña, L.D., K. Momoyama, T. Nakai and K. Muroga. 1992. Detection of the causative bacterium of vibriosis in kuruma prawn, *Penaeus japonicus*. Fish Pathol. *27*: 223–228.

de la Vega, M.G., F.J. Cejudo and A. Paneque. 1991. Production of exocellular polysaccharide by *Azotobacter chroococcum*. Appl. Biochem. Biotechnol. *30*: 273–284.

De Ley, J. 1968. DNA base composition and classification of some more free-living nitrogen-fixing bacteria. Antonie Leeuwenhoek *34*: 66–70.

De Ley, J. 1978. Modern molecular methods in bacterial taxonomy: evaluation, application, prospects. Proc. 4th Int. Conf. Plant Path. Bact., Tours, France.1978 Gilbert-Clarey. pp. 347–357.

De Ley, J. 1992. The *Proteobacteria* ribosomal RNA cistron similarities and bacterial taxonomy. *In* Balows, Trüper, Dworkin, Harder and Schleifer (Editors), The Prokaryotes-A Handbook on the Biology of Bacteria: Ecophysiology, Isolation, Identification, Applications., 2nd Ed., Vol. 2, Springer-Verlag, New York. pp. 2111–2140.

De Ley, J., H. Cattoir and A. Reynaerts. 1970. The quantitative measurement of DNA hybridization from renaturation rates. Eur. J. Biochem. *12*: 133–142.

De Ley, J., W. Mannheim, R. Mutters, K. Piechulla, R. Tytgat, P. Segers, M. Bisgaard, W. Frederiksen, K.-H. Hinz and M. Vanhoucke. 1990. Inter- and intrafamilial similarities of rRNA cistrons of the *Pasteurellaceae*. Int. J. Syst. Bacteriol. *40*: 126–137.

De Lima, J.E.O., V.S. Miranda, S.R. Roberto, A. Coutinho, R.R. Palma and A.C. Pizzolitto. 1997. Diagnosis of citrus variegated chlorosis through light microscopy. Fitopatol. Bras. *22*: 370–374.

De Louvois, J. 1969. Serotyping and the Dienes reaction on *Proteus mirabilis* from hospital infections. J. Clin. Pathol. *22*: 263–268.

de Mondino, S.S., M.P. Nunes and I.D. Ricciardi. 1995. Occurrence of *Plesiomonas shigelloides* in water environments of Rio de Janeiro city. Mem. Inst. Oswaldo Cruz *90*: 1–4.

De Mot, R., T. Laeremans, G. Schoofs and J. Vanderleyden. 1993. Characterization of the *recA* gene from *Pseudomonas fluorescens* OE 28.3 and construction of a *recA* mutant. J. Gen. Microbiol. *139*: 49–57.

De Mot, R., G. Schoofs, A. Roelandt, P. Declerck, P. Proost, J. Vanderleyden and J. Van Damme. 1994. Molecular characterization of the major outer-membrane protein OprF from plant root-colonizing *Pseudomonas fluorescens*. Microbiology *140*: 1377–1387.

De Petris, S. 1967. Ultrastructure of the cell wall of *Escherichia coli* and the chemical nature of its constituent layers. Ultrastruct. Res. *19*: 45–83.

de Salles Gomes, L. 1944. Sobre una nova especie do genero *Eberthella* Buchanan, isolada de fezes patologicas de crianca. Rev. Inst. Lutz. *4*: 183–195.

De Smedt, J., M. Banwens, R. Tijtgat and J. De Ley. 1980. Intra- and intergeneric similarities of ribosomal ribonucleic acid cistrons of free living, nitrogen-fixing bacteria. Int. J. Syst. Bacteriol. *30*: 106–122.

de Smet, M.J., G. Eggink, B. Witholt, J. Kingma and H. Wynberg. 1983. Characterization of intracellular inclusions formed by *Pseudomonas oleovorans* during growth on octane. J. Bacteriol. *154*: 870–878.

De Soete, G. 1983. A least squares alogorithm for fitting additive trees to proximity data. Psychometrika. *48*: 621–626.

De Vos, D., A. Lim, P. De Vos, A. Sarniguet, K. Kersters and P. Cornelis. 1993. Detection of the outer-membrane lipoprotein-I and its gene in fluorescent and nonfluorescent pseudomonads - implications for taxonomy and diagnosis. J. Gen. Microbiol. *139*: 2215–2223.

De Vos, P. 1980. Intrageneric and intergeneric similarities of ribosomal RNA cistrons of the genus *Pseudomonas* and the implications for taxonomy. Antonie Leeuwenhoek J. Microbiol. Serol. *46*: 96.

De Vos, P. and J. De Ley. 1983. Intra- and intergeneric similarities of *Pseudomonas* and *Xanthomonas* ribosomal ribonucleic acid cistrons. Int. J. Syst. Bacteriol. *33*: 487–509.

De Vos, P., M. Goor, M. Gillis and J. De Ley. 1985. Ribosomal ribonucleic acid cistron similarities of phytopathogenic *Pseudomonas* species. Int. J. Syst. Bacteriol. *35*: 169–184.

De Vos, P., A. Van Landschoot, P. Segers, R. Tytgat, M. Gillis, M. Bauwens, R. Rossau, M. Goor, B. Pot, K. Kersters, P. Lizzaraga and J. De Ley. 1989. Genotypic relationships and taxonomic localization of unclassified *Pseudomonas* and *Pseudomonas*-like strains by deoxyribonucleic acid: ribonucleic acid hybridizations. Int. J. Syst. Bacteriol. *39*: 35–49.

de Vries, J. and W. Wackernagel. 1998. Detection of *nptII* (kanamycin resistance) genes in genomes of transgenic plants by marker-rescue transformation. Mol. Gen. Genet. *257*: 606–613.

de Vuyst, L. and A. Vermeire. 1994. Use of industrial medium components for xanthan production by *Xanthomonas campestris* NRRL-B-1459. Appl. Microbiol. Biotechnol. *42*: 187–191.

De Wet, J.A. and J.A. Erasmus. 1984. Epididymitis of rams in the central and southern districts of the Orange Free State. J. S. Afr. Vet. Assoc. *55*: 173–179.

de Zwart, J.M.M. and J.G. Kuenen. 1997. Aerobic conversion of dimethyl sulfide and hydrogen sulfide by *Methylophaga sulfidovorans*: implications for modeling DMS conversion in a microbial mat. FEMS Microbiol. Ecol. *22*: 155–165.

de Zwart, J.M.M., P.N. Nelisse and J.G. Kuenen. 1996. Isolation and characterization of *Methylophaga sulfidovorans* sp nov: an obligately methylotrophic, aerobic, dimethylsulfide oxidizing bacterium from a microbial mat. FEMS Microbiol. Ecol. *20*: 261–270.

de Zwart, J.M.M., P.N. Nelisse and J.G. Kuenen. 1998. *In* Validation of the publication of new names and new combinations previously effectively published outside the IJSB. List No. 67. Int. J. Syst. Bacteriol. *48*: 1083–1084.

de Zwart, J.M.M., J.M.R. Sluis and J.G. Kuenen. 1997. Competition for dimethyl sulfide and hydrogen sulfide by *Methylophaga sulfidovorans* and *Thiobacillus thioparus* T5 in continuous cultures. Appl. Environ. Microbiol. *63*: 3318–3322.

Deacon, W.E. 1952. Antigenic study of certain slow lactose fermenting *Aerobacter cloacae* cultures. Proc. Soc. Exp. Biol. Med. *81*: 165–170.

Dealler, S.F., P.M. Hawkey and M.R. Millar. 1988. Enzymatic degradation of urinary indoxyl sulfate by *Providencia stuartii* and *Klebsiella pneu-*

moniae causes the purple urine bag syndrome. J. Clin. Microbiol. *26*: 2152–2156.

Dean, C.R., S. Neshat and K. Poole. 1996. PfeR, an enterobactin-responsive activator of ferric enterobactin receptor gene expression in *Pseudomonas aeruginosa*. J. Bacteriol. *178*: 5361–5369.

Dean, H.F. and A.F. Morgan. 1983. Integration of R91-5::Tn501 into the *Pseudomonas putida* Ppn chromosome and genetic circularity of the chromosomal map. J. Bacteriol. *153*: 485–497.

DeAngelis, P.L. 1999. Molecular directionality of polysaccharide polymerization by the *Pasteurella multocida* hyaluronan synthase. J. Biol. Chem. *274*: 26557–26562.

Debette, J. 1991. Isolation and characterization of an extracellular proteinase produced soil strain of *Xanthomonas maltophilia*. Curr. Microbiol. *22*: 85–90.

Debette, J. and R. Blondeau. 1977. Characterisation des bacteries telluriques assimilables a *Pseudomonas maltophila*. Can. J. Microbiol. *23*: 1123–1127.

Debette, J. and R. Blondeau. 1980. Presence de *Pseudomonas maltophilia* dans la rhizosphere de quelques plantes cultivees. Can. J. Microbiol. *26*: 460–463.

Debette, J. and G. Prensier. 1989. Immunoelectron microscopic demonstration of an esterase on the outer membrane of *Xanthomonas maltophilia*. Appl. Environ. Microbiol. *55*: 233–239.

Debois, J., H. Degreef, J. Vandepitte and J. Spaepen. 1975. *Pseudomonas putrefaciens* as a cause of infection in humans. J. Clin. Pathol. *28*: 993–996.

Decostere, A., F. Haesebrouck, L. Devriese and R. Ducatelle. 1996. Identification and pathogenic significance of *Shewanella* sp. from pond fish. Vlaams Diergeneeskundig Tijdschrift. *65*: 82–85.

Dedysh, S.N., N.S. Panikov and J.M. Tiedje. 1998. Acidophilic methanotrophic communities from Sphagnum peat bogs. Appl. Environ. Microbiol. *64*: 922–929.

Dees, S.B., J. Powell, C.W. Moss, D.G. Hollis and R.E. Weaver. 1981. Cellular fatty-acid composition of organisms frequently associated with human infections resulting from dog bites - *Pasteurella multocida* and groups Ef-4, Iij, M-5, and Df-2. J. Clin. Microbiol. *14*: 612–616.

Degrandis, S.A. and R.M.W. Stevenson. 1985. Antimicrobial susceptibility patterns and R plasmid mediated resistance of the fish pathogen *Yersinia ruckeri*. Antimicrob. Agents Chemother. *27*: 938–942.

Degtyarev, S.K., M.A. Abdurashitov, A. Kolyhalov and N.I. Rechkunova. 1992. *Aci*II, a new restriction endonuclease for *Acinetobacter calcoaceticus* recognizing 5′-A↓CGTT-3′. Nucleic Acids Res. *20*: 3787.

Deich, R.A. and H.O. Smith. 1980. Mechanism of homospecific DNA uptake in *Haemophilus influenzae* transformation. Mol. Gen. Genet. *177*: 369–374.

Dekker, R.F.H. and G.P. Candy. 1979. The β-mannanases elaborated by the phytopathogen *Xanthomonas campestris*. Arch. Microbiol. *122*: 297–299.

del Moral, A., B. Prado, E. Quesada, T. Garcia, R. Ferrer and A. Ramos-Cormenzana. 1988. Numerical taxonomy of moderately halophilic Gram-negative rods from an inland saltern. J. Gen. Microbiol. *134*: 733–741.

Delaporte, B., M. Raynaud and P. Daste. 1965. Une bactérie du sol capable d'utiliser, comme source de carbone, la fraction fixe de certain oléorésines, *Pseudomonas resinovorans* n. sp. C. R. Hebd. Séances Acad. Sci. Paris. *252*: 1073–1075.

Delepelaire, P. 1994. PrtD, the integral membrane ATP-binding cassette component of the *Erwinia chrysanthemi* metalloprotease secretion system, exhibits a secretion signal-regulated ATPase activity. J. Biol. Chem. *269*: 27952–27957.

Delforge, M.L., J. Devriendt, Y. Glupczynski, W. Hansen and N. Douat. 1995. *Plesiomonas shigelloides* septicemia in a patient with primary hemochromatosis. Clin. Infect. Dis. *21*: 692–693.

Dellacasagrande, J., E. Ghigo, C. Capo, D. Raoult and J.L. Mege. 2000. *Coxiella burnetii* survives in monocytes from patients with Q fever endocarditis: Involvement of tumor necrosis factor. Infect. Immun. *68*: 160–164.

Delmas, F. and P. Timon-David. 1985. Effect of invertebrate rickettsiae on vertebrates: experimental infection of mice by *Rickettsiella grylli*. C. R. Acad. Sci. III. *300*: 115–117.

DeLong, E.F., D.G. Franks and A.L. Alldredge. 1993. Phylogenetic diversity of aggregate-attached vs. free-living bacterial assemblages. Limnol. Oceanogr. *38*: 924–934.

DeLong, E.F., D.G. Franks and A.A. Yayanos. 1997. Evolutionary relationships of cultivated psychrophilic and barophilic deep-sea bacteria. Appl. Environ. Microbiol. *63*: 2105–2108.

DeLong, E.F. and A.A. Yayanos. 1986. Biochemical function and ecological significance of novel bacterial lipids in deep-sea prokaryotes. Appl. Environ. Microbiol. *51*: 730–737.

DeLucca, A.J., K.A. Brogden and R. Engen. 1988. *Enterobacter agglomerans* lipopolysaccharide-induced changes in pulmonary surfactant as a factor in the pathogenesis of byssinosis. J. Clin. Microbiol. *26*: 778–780.

DeLucca, A.J. and M.S. Palmgren. 1986. Mesophilic microorganisms and endotoxin levels on developing cotton plants. Am. Ind. Hyg. Assoc. J. *47*: 437–442.

Delwiche, C.F. and J.D. Palmer. 1996. Rampant horizontal transfer and duplication of Rubisco genes in eubacteria and plastids. Mol. Biol. Evol. *13*: 873–882.

DeMarta, A. and R. Peduzzi. 1984. Etude epidemiologique des *Aeromonas* par lysotypie. Estratto dalla Rivista Italiana di Piscicoltura e Ittipatologia. *19*: 148–155.

Deming, J.W. 2002. Unusual or extreme high-pressure environments. *In* Hurst, Crawford, Knudsen, McInerney and Stetzenbach (Editors), ASM Manual of Environmental Microbiology, 2nd Ed., ASM Press, Washington, D.C. 478–490.

Deming, J.W. and J.A. Baross. 2000. Survival, dormancy and non-culturable cells in deep-sea environments. *In* Colwell and Grimes (Editors), Non-culturable Microorganisms in the Environment, ASM Press, Washington, D.C. 147–197.

Deming, J.W. and J.A. Baross. 2002. Search for and discovery of microbial enzymes from thermally extreme environments in the ocean. *In* Burns and Dick (Editors), Enzymes in the Environment: Activity, Ecology, and Applications, Marcel Dekker, New York. 327–362.

Deming, J.W., L.K. Somers, W.L. Staube, D.G. Swartz and M.T. MacDonell. 1988a. Isolation of an obligately barophilic bacterium and description of a new genus, *Colwellia* gen. nov. Syst. Appl. Microbiol. *10*: 152–160.

Deming, J.W., L.K. Somers, W.L. Staube, D.G. Swartz and M.T. MacDonell. 1988b. *In* Validation of the publication of new names and new combinations previously effectively published outside the IJSB. List No. 26. Int. J. Syst. Bacteriol. *38*: 328–329.

d'Émpaire, M. 1969. Les facteurs de croissance des *Edwardsiella tarda*. Ann. Inst. Pasteur (Paris) *116*: 63–68.

Den Dooren de Jong, L.E. 1926. Bijdrage tot de kennis van het mineralisatieproces, Nijgh and van Ditmar Uitgevers-Mij, Rotterdam. pp. 1–200.

Deneer, H.G., L. Slaney, I.W. Maclean and W.L. Albritton. 1982. Mobilization of nonconjugative antibiotic resistance plasmids in *Haemophilus ducreyi*. J. Bacteriol. *149*: 726–732.

Denner, E.B.M., B. Mark, H.J. Busse, M. Turkiewicz and W. Lubitz. 2001a. *Psychrobacter proteolyticus* sp nov., a psychrotrophic, halotolerant bacterium isolated from the antarctic krill *Euphausia superba* Dana, excreting a cold-adapted metalloprotease. Syst. Appl. Microbiol. *24*: 44–53.

Denner, E.B.M., B. Mark, H.J. Busse, M. Turkiewicz and W. Lubitz. 2001b. *In* Validation of the publication of new names and new combinations previously effectively published outside the IJSEM. List No. 82. Int. J. Syst. Evol. Microbiol. *51*: 1619–1620.

Dennis, P.J.L., D.J. Brenner, W.L. Thacker, R. Wait, G. Vesey, A.G. Steigerwalt and R.F. Benson. 1993. Five new *Legionella* species isolated from water. Int. J. Syst. Bacteriol. *43*: 329–337.

Denton, M., N.J. Todd and J.M. Littlewood. 1996. Role of anti-pseudomonal antibiotics in the emergence of *Stenotrophomonas maltophilia* in cystic fibrosis patients. Eur. J. Clin. Microbiol. Infect. Dis. *15*: 402–405.

DePaola, A., G.M. Capers and D. Alexander. 1994. Densities of *Vibrio*

vulnificus in the intestines of fish from the U.S. Gulf Coast. Appl. Environ. Microbiol. *60*: 984–988.

Depiazzi, L.J., J. Henderson and W.J. Penhale. 1990. Measurement of protease thermostability, twitching motility and colony size of *Bacteroides nodosus*. Vet. Microbiol. *22*: 353–363.

Dequeker, J., R. Jamar and M. Walravens. 1980. HLA-B27, arthritis and *Yersinia enterocolitica* infection. J. Rheumatol. *7*: 706–710.

Derby, H.A. and B.W. Hammer. 1931. Bacteriology of butter. IV. Bacteriological studies of surface taint butter. Iowa Agr. Exp. Sta. Res. Bull. *145*: 387–416.

Derrick, E.H. 1939. *Rickettsia burneti*: The cause of "Q" fever. Med. J. Aust. *1*: 14.

Dervartanian, D.V., Y.I. Shethna and H. Beinert. 1969. Purification and properties of two iron-sulfur proteins from *Azotobacter vinelandii*. Biochim. Biophys. Acta *194*: 548–563..

Derx, H.G. 1951. L'accumulation spécifique de l'*Azotobacter agile* Beijerinck et de l'*Azotobacter vinelandii* Lipman. Proc. Sect. Sci. K. Ned. Akad. Wet. *C54*: 342–350.

Desai, M.V. and H.M. Shah. 1959. A new bacterial leaf spot of *Crotalaria juncea* L. Curr. Sci. (Bangalore) *28*: 377–378.

Desai, M.V. and H.M. Shah. 1960. Bacterial leaf spot disease of *Desmodium rotundifolium* DC. Curr. Sci. (Bangalore) *29*: 65–66.

Desai, M.V., M.J. Thirumalachar and M.K. Patel. 1965. Bacterial blight disease of *Eleusine coracana* Gaertn. Indian Phytopathol. *18*: 384–386.

Desai, S.G., A.B. Gandi, M.K. Patel and W.V. Kotasthane. 1966. A new bacterial leaf-spot and blight of *Azadirachta indica*. A. Juss. Indian Phytopathol. *19*: 322–323.

Deschenes, G., C. Casenave, F. Grimont, J.C. Desenclos, S. Benoit, M. Collin, S. Baron, P. Mariani, P.A.D. Grimont and H. Nivet. 1996. Cluster of cases of haemolytic uraemic syndrome due to unpasteurised cheese. Pediatr. Nephrol. *10*: 203–205.

Desenclos, J.C., L. Conti, S. Junejo and K.C. Klontz. 1990. A cluster of *Edwardsiella tarda* infection in a day-care center in Florida. J. Infect. Dis. *162*: 782–783.

Desjardins, P., B. Picard, B. Kaltenbock, J. Elion and E. Denamur. 1995. Sex in *Escherichia coli* does not disrupt the clonal structure of the population: Evidence from random amplified polymorphic DNA and restriction-fragment-length polymorphism. J. Mol. Evol. *41*: 440–448.

Desmarchelier, P.M. and J.L. Reichelt. 1981. Phenotypic characterization of clinical and environmental isolates of *Vibrio cholerae* from Australia. Curr. Microbiol. *5*: 123–127.

Desmond, E. and J.M. Janda. 1986. Growth of *Aeromonas* spp. on enteric agars. J. Clin. Microbiol. *23*: 1065–1067.

Devauchelle, G., G. Meynadier and V. C.. 1972. Étude ultrastructurale du cycle de multiplication de *Rickettsiella melolonthae* (Krieg), Philip, dans les hémocytes de son hôte. J. Ultrastruc. Res. *38*: 134–148.

DeVault, J.D., A. Berry, T.K. Misra, A. Darzins and A.M. Chakrabarty. 1989. Environmental sensory signals and microbial pathogenesis - *Pseudomonas aeruginosa* infection in cystic-fibrosis. Bio-Technology. *7*: 352–357.

DeVreese, K., G. Claeys and G. Verschraegen. 1992. Septicemia with *Ewingella americana*. J. Clin. Microbiol. *30*: 2746–2747.

Devriese, L.A., M. Bisgaard, J. Hommez, E. Uyttebroek, R. Ducatelle and F. Haesebrouck. 1991. Taxon 20 (Fam. *Pasteurellaceae*) infections in European brown hares (*Lepus europaeus*). J. Wildlife Dis. *27*: 685–687.

Dewanti, A.R. and J.A. Duine. 1998. Reconstitution of membrane-integrated quinoprotein glucose dehydrogenase apoenzyme with PQQ and the holoenzyme's mechanism of action. Biochemistry *37*: 6810–6818.

DeWeerd, K.A., L. Mandelco, R.S. Tanner, C.R. Woese and J.M. Suflita. 1990. *Desulfomonile tiedjei* gen. nov. and sp. nov., a novel anaerobic, dehalogenating, sulfate-reducing bacterium. Arch. Microbiol. *154*: 23–30.

Dewhirst, F.E., B.J. Paster, G.J. Fraser and G.J. Olsen. 1995. Taxonomy of the *Pasteurellaceae*: a difficult problem even when the phylogeny is known. *In* Donachie, Lainson and Hodgson (Editors), *Haemophilus*, *Actinobacillus*, and *Pasteurella*, Plenum Press, New York. 199.

Dewhirst, F.E., B.J. Paster, S. La Fontaine and J.I. Rood. 1990. Transfer of *Kingella indologenes* (Snell and Lapage 1976) to the genus *Suttonella* gen. nov. as *Suttonella indologenes* comb. nov.; transfer of *Bacteroides nodosus* (Beveridge 1941) to the genus *Dichelobacter* gen. nov. as *D. nodosus* comb. nov.; and assignment of the genera *Cardiobacterium*, *Dichelobacter*, and *Suttonella* to *Cardiobacteriaceae* fam. nov. in the gamma division of *Proteobacteria* on the basis of 16S rRNA sequence comparisons. Int. J. Syst. Bacteriol. *40*: 426–433.

Dewhirst, F.E., B.J. Paster, I. Olsen and G.J. Fraser. 1992. Phylogeny of 54 representative strains of species in the family *Pasteurellaceae* as determined by comparison of 16S rRNA sequences. J. Bacteriol. *174*: 2002–2013.

Dewhirst, F.E., B.J. Paster, I. Olsen and G.J. Fraser. 1993. Phylogeny of the *Pasteurellaceae* as determined by comparison of 16S ribosomal ribonucleic acid sequences. Zentbl. Bakteriol. *279*: 35–44.

Dhanvantari, B.N. 1977. Taxonomic study of *Pseudomonas papulans*-Rose 1917. N. Z. J. Agric. Res. *20*: 557–561.

Di Cello, F., M. Pepi, F. Baldi and R. Fani. 1997. Molecular characterization of an n-alkane-degrading bacterial community and identification of a new species, *Acinetobacter venetianus*. Res. Microbiol. *148*: 237–249.

Di Fabio, J.L., M.B. Perry and D.R. Bundle. 1987. Analysis of the lipopolysaccharide of *Pseudomonas maltophilia* 555. Biochem. Cell Biol. *65*: 968–977.

Dianese, J.C. and N.W. Schaad. 1982. Isolation and characterization of inner and outer membranes of *Xanthomonas campestris* pv. *campestris*. Phytopathology *72*: 1284–1289.

Diaz, E., M. Munthali, V. De Lorenzo and K.N. Timmis. 1994. Universal barrier to lateral spread of specific genes among microorganisms. Mol. Microbiol. *13*: 855–861.

Dibb, W.L., A. Digranes and S. Tønjum. 1981. *Actinobacillus ligniersii* infection after a horse bite. Br. Med. J. *283*: 583–584.

DiChristina, T.J. and E.F. DeLong. 1993. Design and application of rRNA-targeted oligonucleotide probes for the dissimilatory iron- and manganese-reducing bacterium *Shewanella putrefaciens*. Appl. Environ. Microbiol. *59*: 4152–4160.

Dickey, R.S. 1981. *Erwinia chrysanthemi*: reaction of eight plant species to strains from several hosts and to strains of other *Erwinia* species. Phytopathology *71*: 23–29.

Dickey, R.S. and C.H. Zumoff. 1987. Bacterial leaf blight of *Syngonium* caused by a pathovar of *Xanthomonas campestris*. Phytopathology *77*: 1257–1262.

Dickey, R.S. and C.H. Zumoff. 1988. Emended description of *Enterobacter cancerogenus*, new combination (formerly *Erwinia cancerogena*). Int. J. Syst. Bacteriol. *38*: 371–374.

Dienes, L. 1946. Reproductive processes in *Proteus* cultures. Proc. Soc. Exp. Biol. Med. *63*: 265–270.

Dienst, F.T. 1963. Tularemia—a perusal of three hundred thirty-nine cases. J. Louisiana State M. Soc. *115*: 114–127.

Dienst, R.B., R.B. Greenblatt and C.H. Chen. 1948. Laboratory diagnosis of Granuloma Inguinale and studies on the cultivation of the Donovan Body. Am. J. Syph. Gonorrhea Vener. Dis. *32*: 301–306.

Dienst, R.B., R.B. Greenblatt and E.S. Sanderson. 1938. Cultural studies on the "Donovan bodies" of granuloma inguinale. J. Infect. Dis. *62*: 112–114.

Dietz, A.S. and A.A. Yayanos. 1978. Silica gel media for isolating and studying bacteria under hydrostatic pressure. Appl. Environ. Microbiol. *36*: 966–968.

Diez, A., M.J. Alvarez, M.I. Prieto, J.M. Bautista and A. Garridopertierra. 1995. Monochloroacetate dehalogenase activities of bacterial strains isolated from soil. Can. J. Microbiol. *41*: 730–739.

DiGiacomo, R.F., B.J. Deeb, W.E. Giddens, B.L. Bernard and M.M. Chengappa. 1989. Atrophic rhinitis in New Zealand white-rabbits infected with *Pasteurella multocida*. Am. J. Vet. Res. *50*: 1460–1465.

Dijkshoorn, L. 1996. *Acinetobacter* microbiology. *In* Bergogne-Bérézin, Joly-Guillou and Towner (Editors), *Acinetobacter*. Microbiology, Epidemiology, Infections, Management, CRC Press, Boca Raton. pp. 37–69.

Dijkshoorn, L., H. Aucken, P. Gerner-Smidt, P. Janssen, M.E. Kaufmann,

J. Garaizar, J. Ursing and T.L. Pitt. 1996. Comparison of outbreak and nonoutbreak *Acinetobacter baumannii* strains by genotypic and phenotypic methods. J. Clin. Microbiol. *34*: 1519–1525.

Dijkshoorn, L., H.M. Aucken, P. Gerner-Smidt, M.E. Kaufmann, J. Ursing and T.L. Pitt. 1993. Correlation of typing methods for *Acinetobacter* isolates from hospital outbreaks. J. Clin. Microbiol. *31*: 702–705.

Dijkshoorn, L., I. Tjernberg, B. Pot, M.F. Michel, J. Ursing and K. Kersters. 1990. Numerical analysis of cell envelope protein profiles in *Acinetobacter* strains classified by DNA–DNA hybridization. Syst. Appl. Microbiol. *13*: 338–344.

Dijkshoorn, L., B. van Harsselaar, I. Tjernberg, P.J.M. Bouvet and M. Vaneechoutte. 1998. Evaluation of amplified ribosomal DNA restriction analysis for identification of *Acinetobacter* genomic species. Syst. Appl. Microbiol. *21*: 33–39.

Dijkshoorn, L., W. Van Vianen, J.E. Degener and M.F. Michel. 1987. Typing of *Acinetobacter calcoaceticus* strains isolated from hospital patients by cell envelope protein profiles. Epidemiol. Infect. *93*: 659–667.

Dillard, H.R. and W.L. Kline. 1989. An outbreak of Stewart's bacterial wilt of corn in New York state. Plant Dis. *73*: 273.

Dilling, W., W. Liesack and N. Pfennig. 1995. *Rhabdochromatium marinum* gen. nom. rev., sp. nov., a purple sulfur bacterium from a salt marsh microbial mat. Arch. Microbiol. *164*: 125–131.

Dilling, W., W. Liesack and N. Pfennig. 1996. *In* Validation of the publication of new names and new combinations previously effectively published outside the IJSB. List No. 56. Int. J. Syst. Bacteriol. *46*: 362–363.

Dillon, R.J., C.T. Vennard and A.K. Charnley. 2002. A note: gut bacteria produce components of a locust cohesion pheromone. J. Appl. Microbiol. *92*: 759–763.

Dilworth, M.J., R.R. Eady and M.E. Eldridge. 1988. The vanadium nitrogenase of *Azotobacter chroococcum* - reduction of acetylene and ethylene to ethane. Biochem. J. *249*: 745–751.

Dimarco, A.A. and L.N. Ornston. 1994. Regulation of *p*-hydroxybenzoate hydroxylase synthesis by PobR bound to an operator in *Acinetobacter calcoaceticus*. J. Bacteriol. *176*: 4277–4284.

Diorio, C., J. Cai, J. Marmor, R. Shinder and M.S. DuBow. 1995. An *Escherichia coli* chromosomal *ars* operon homolog is functional in arsenic detoxification and is conserved in Gram-negative bacteria. J. Bacteriol. *177*: 2050–2056.

DiRienzo, J.M. and J. Slots. 1990. Genetic approach to the study of epidemiology and pathogenesis of *Actinobacillus actinomycetemcomitans* in localized juvenile periodontitis. Arch. Oral Biol. *35*: 79S–84S.

DiRienzo, J.M. and E.L. Spieler. 1983. Identification and characterization of the major cell-envelope proteins of oral strains of *Actinobacillus actinomycetemcomitans*. Infect. Immun. *39*: 253–261.

Distel, D.L. and C.M. Cavanaugh. 1994. Independent phylogenetic origins of methanotrophic and chemoautotrophic bacterial endosymbioses in marine bivalves. J. Bacteriol. *176*: 1932–1938.

Dixon, L.G., W.L. Albritton and P.J. Willson. 1994. An analysis of the complete nucleotide sequence of the *Haemophilus ducreyi* broad-host-range plasmid pLS88. Plasmid *32*: 228–232.

Dixon, R. 1998. The oxygen-responsive NifL-NifA complex: a novel two-component regulatory system controlling nitrogenase synthesis in *gamma- proteobacteria*. Arch. Microbiol. *169*: 371–380.

Do Nascimento, G.G.F. and F.C.A. Tavares. 1987. Transformation with *nif* genes in *Azotobacter paspali*. Rev. Microbiol. *18*: 41–45.

do Valle, G.R.F., T.A.T. Gomes, K. Irino and L.R. Trabulsi. 1997. The traditional enteropathogenic *Escherichia coli* (EPEC) serogroup O125 comprises serotypes which are mainly associated with the category of enteroaggregative *E. coli*. FEMS Microbiol. Lett. *152*: 95–100.

Döbereiner, J. 1966. *Azotobacter paspali* sp. n. uma bacteria fixadora de nitrogenio na rizosfera de *Paspalum*. Pesqui. Agropecu. Bras. *1*: 357–365.

Döbereiner, J. 1995. Isolation and Identification of aerobic nitrogen-fixing bacteria from soil and plants. *In* Alef and Nannipieri (Editors), Methods in Applied Soil Microbiology and Biochemistry, Academic Press, London. pp. 134–141.

Dobritsa, S.V. 1985. Restriction analysis of the *Frankia* spp. genome. FEMS Microbiol. Lett. *29*: 123–128.

Dobson, S.J. and P.D. Franzmann. 1996. Unification of the genera *Deleya* (Baumann et al. 1983), *Halomonas* (Vreeland et al. 1980), and *Halovibrio* (Fendrich 1988) and the species *Paracoccus halodenitrificans* (Robinson and Gibbons 1952) into a single genus, *Halomonas*, and placement of the genus *Zymobacter* in the family *Halomonadaceae*. Int. J. Syst. Bacteriol. *46*: 550–558.

Dobson, S.J., S.R. James, P.D. Franzmann and T.A. McMeekin. 1990. Emended description of *Halomonas halmophila* (NCMB 1971T). Int. J. Syst. Bacteriol. *40*: 462–463.

Dobson, S.R., J.S. Kroll and E.R. Moxon. 1992. Insertion sequence IS1016 and absence of *Haemophilus* capsulation genes in the Brazilian purpuric fever clone of *Haemophilus influenzae* biogroup aegyptius. Infect. Immun. *60*: 618–622.

Dobson, S.J., T.A. McMeekin and P.D. Franzmann. 1993. Phylogenetic relationships between some members of the genera *Deleya*, *Halomonas*, and *Halovibrio*. Int. J. Syst. Bacteriol. *43*: 665–673.

Dodgson, C., P. Amor and C. Whitfield. 1996. Distribution of the *rol* gene encoding the regulator of lipopolysaccharide O-chain length in *Escherichia coli* and its influence on the expression of group I capsular K antigens. J. Bacteriol. *178*: 1895–1902.

Dodson, R.F., G.S. Fritz, W.R. Hubler, A.H. Rudolph, J.M. Knox and L.W.-F. Chu. 1973. Donovanosis: a morphologic study. J. Investig. Dermatol. *62*: 611–614.

Doern, G.V. 1992. The *Moraxella* and *Branhamella* subgenera of the genus *Moraxella*. *In* Balows, Trüper, Dworkin, Harder and Schleifer (Editors), The Prokaryotes. A Handbook on the Biology of Bacteria: Ecophysiology, Isolation, Identification, Applications, 2nd Ed., Vol. 4, Springer-Verlag, New York. pp. 3276–3280.

Doern, G.V. and F.L.A. Buckmire. 1976. Ultrastructural characterization of capsulated *Haemophilus influenzae* type b and two spontaneous nontypable mutants. J. Bacteriol. *127*: 523–535.

Doern, G.V. and K.C. Chapin. 1987. Determination of biotypes of *Haemophilus influenzae* and *Haemophilus parainfluenzae* a comparison of methods and a description of a new biotype (VIII) of *H. parainfluenzae*. Diagn. Microbiol. Infect. Dis. *7*: 269–272.

Doetsch, R.N., T.M. Cook and Z. Vaituzis. 1967. On the uniqueness of the flagellum of *Thiobacillus thiooxidans*. Antonie Leeuwenhoek J. Microbiol. *33*: 196–202.

Doggett, R.G. 1969. Incidence of mucoid *Pseudomonas aeruginosa* from clinical sources. Appl. Microbiol. *18*: 936–937.

Doi, R.H. and R.T. Igarashi. 1965. Conservation of ribosomal and messenger ribonucleic acid cistrons in *Bacillus* species. J. Bacteriol. *90*: 384–390.

Doidge, E.M. 1920. A tomato canker. J. Dep. Agr. S. Afr. *1*: 718–721.

Doig, P., T. Todd, P.A. Sastry, K.K. Lee, R.S. Hodges, W. Paranchych and R.T. Irvin. 1988. Role of pili in adhesion of *Pseudomonas aeruginosa* to human respiratory epithelial-cells. Infect. Immun. *56*: 1641–1646.

Doll, J.M., P.S. Zeitz, P. Ettestad, A.L. Bucholtz, T. Davis and K. Gage. 1994. Cat-transmitted fatal pneumonic plague in a person who traveled from Colorado to Arizona. Am. J. Trop. Med. Hyg. *51*: 109–114.

Dollberg, S., A. Gandacu and A. Klar. 1990. Acute pyelonephritis due to a *Kluyvera* species in a child. Eur. J. Clin. Microbiol. Infect. Dis. *9*: 281–283.

Dolzani, L., E. Tonin, C. Lagatolla, L. Prandin and C. Monti-Bragadin. 1995. Identification of *Acinetobacter* isolates in the *A. calcoaceticus–A. baumannii* complex by restriction analysis of the 16S–23S rRNA intergenic spacer sequences. J. Clin. Microbiol. *33*: 1108–1113.

Dominguez, H., B.F. Vogel, L. Gram, S. Hoffmann and S. Schaebel. 1996. *Shewanella alga* bacteremia in two patients with lower leg ulcers. Clin. Infect. Dis. *22*: 1036–1039.

Dominguez, M., G. Gonzalez, H. Bello, A. Garcia, S. Mella, M.E. Pinto, M.A. Martinez and R. Zemelman. 1995. Identification and biotyping of *Acinetobacter* spp. isolated in Chilean hospitals. J. Hosp. Infect. *30*: 267–271.

Donnenberg, M.S. 1995. Enteropathogenic *Escherichia coli*. *In* Blaser, Smith, Ravdin, Greenberg and Guerrant (Editors), Infections of the Gastrointestinal Tract, Raven Press, New York. 709–729.

Donnenberg, M.S., R.A. Donohue and G.T. Keusch. 1989. Epithelial cell invasion: an overlooked property of enteropathogenic *Escherichia coli* (EPEC) associated with the EPEC adherence factor. J. Infect. Dis. *160*: 452–459..

Donnenberg, M.S. and T.S. Whittam. 2001. Pathogenesis and evolution of virulence in enteropathogenic and enterohemorrhagic *Escherichia coli*. J. Clin. Investig. *107*: 539–548.

Donnenberg, M.S., H.Z. Zhang and K.D. Stone. 1997. Biogenesis of the bundle-forming pilus of enteropathogenic *Escherichia coli*: Reconstitution of fimbriae in recombinant *E. coli* and role of DsbA in pilin stability: A review. Gene *192*: 33–38.

Donovan, C. 1905. Ulcerating granuloma of the pudenda. Indian Med. Gaz. *40*: 414.

Donowitz, G.R. and K.I. Earnhardt. 1993. Azithromycin inhibition of intracellular *Legionella micdadei*. Antimicrob. Agents Chemother. *37*: 2261–2264.

Dooley, J.S.G., R. Lallier, D.H. Shaw and T.J. Trust. 1985. Electrophoretic and immunochemical analyses of the lipopolysaccharides from various strains of *Aeromonas hydrophila*. J. Bacteriol. *164*: 263–269.

Doran, J.L., W.H. Bingle, K.L. Roy, K. Hiratsuka and W.J. Page. 1987. Plasmid transformation of *Azotobacter vinelandii* OP. J. Gen. Microbiol. *133*: 2059–2072.

Doring, G., S. Jansen, H. Noll, H. Grupp, F. Frank, K. Botzenhart, K. Magdorf and U. Wahn. 1996. Distribution and transmission of *Pseudomonas aeruginosa* and *Burkholderia cepacia* in a hospital ward. Pediatr. Pulmonol. *21*: 90–100.

Dorn, C.R., S.M. Scotland, H.R. Smith, G.A. Willshaw and B. Rowe. 1989. Properties of Vero cytotoxin-producing *Escherichia coli* of human and animal origin belonging to serotypes other than O157:H7. Epidemiol. Infect. *103*: 83–96.

Dorofe'ev, K.A. 1947. Classification of the causative agent of tularemia. Symp. Res. Works Inst. Epidemiol. Mikrobiol. (Chita) (Russ.). *1*: 170–180.

Doronina, N.V., V.I. Krauzova and Y.A. Trotsenko. 1997. *Methylophaga limanica* sp. nov.: a new species of moderately halophilic, aerobic, methylotrophic bacteria. Mikrobiologiya *66*: 434–449.

Doronina, N.V. and Y.A. Trotsenko. 1997. Aerobic methylotrophic bacterial communities of hypersaline ecosystems. Mikrobiologiya *66*: 111–117.

Dorsch, M., N.J. Ashbolt, P.T. Cox and A.E. Goodman. 1994. Rapid identification of *Aeromonas* species using 16S rDNA targeted oligonucleotide primers: a molecular approach based on screening of environmental isolates. J. Appl. Bacteriol. *77*: 722–726.

Dorsch, M., D. Lane and E. Stackebrandt. 1992. Towards a phylogeny of the genus *Vibrio* based on 16S rRNA sequences. Int. J. Syst. Bacteriol. *42*: 58–63.

dos Santos, R.M.D.B. and J.C. Dianese. 1985. Comparative membrane characterization of *Xanthomonas campestris* pv. *cassavae* and *Xanthomonas campestris* pv. *manihotis*. Phytopathology *75*: 581–587.

Doten, R.C., K.L. Ngai, D.J. Mitchell and L.N. Ornston. 1987. Cloning and genetic organization of the *pca* gene cluster from *Acinetobacter calcoaceticus*. J. Bacteriol. *169*: 3168–3174.

Doudoroff, M. and N.J. Palleroni. 1974. Genus I. *Pseudomonas*. *In* Buchanan and Gibbons (Editors), Bergey's Manual of Determinative Bacteriology, 8th Ed., The Williams & Wilkins Co., Baltimore. pp. 217–243.

Dougherty, B.A. and H.O. Smith. 1999. Identification of *Haemophilus influenzae* Rd transformation genes using cassette mutagenesis. Microbiology *145*: 401–409.

Doughty, S.W., C.G. Ruffolo and B. Adler. 2000. The type 4 fimbrial subunit gene of *Pasteurella multocida*. Vet. Microbiol. *72*: 79–90.

Douglas, A.E. 1988. Sulfate utilization in an aphid symbiosis. Insect Biochem. *18*: 599–606.

Douglas, A.E. 1998. Nutritional interactions in insect-microbial symbioses: aphids and their symbiotic bacteria *Buchnera*. Annu. Rev. Entomol. *43*: 17–37.

Douglas, A.E. and A.F.G. Dixon. 1987. The mycetocyte symbiosis of aphids: variation with age and morph in virginoparae of *Megoura viciae* and *Acyrthosiphon pisum.* J. Insect Physiol. *33:* 109–114.

Douglas, A.E. and W.A. Prosser. 1992. Synthesis of the essential amino acid tryptophan in the pea aphid (*Acyrthosiphon pisum*) symbiosis. J. Insect Physiol. *38:* 565–568.

Dournon, E., W.F. Bibb, P. Rajagopalan, N. Desplaces and R.M. McKinney. 1988. Monoclonal antibody reactivity as a virulence marker for *Legionella pneumophila* serogroup 1 strains. J. Infect. Dis. *157:* 496–501.

Dow, J.M., B.R. Clarke, D.E. Milligan, J.L. Tang and M.J. Daniels. 1990. Extracellular proteases from *Xanthomonas campestris* pv. *campestris*, the black rot pathogen. Appl. Environ. Microbiol. *56:* 2994–2998.

Dow, J.M., D.E. Milligan, L. Jamieson, C.E. Barber and M.J. Daniels. 1989. Molecular cloning of a polygalacturonate lyase gene from *Xanthomonas campestris* pv. *campestris* and role of the gene product in pathogenicity. Physiol. Mol. Plant Pathol. *35:* 113–120.

Dow, J.M., A.E. Osbourn, T.J.G. Wilson and M.J. Daniels. 1995. A locus determining pathogenicity *Xanthomonas campestris* is involved in lipopolysaccharide biosynthesis. Mol. Plant-Microbe Interact. *8:* 768–777.

Dowling, J.N., F.J. Kroboth, M. Karpf, R.B. Yee and A.W. Pasculle. 1983. Pneumonia and multiple lung abscesses caused by dual infection with *Legionella micdadei* and *Legionella pneumophila.* Am. Rev. Respir. Dis. *127:* 121–125.

Dowson, W.J. 1939. On the systematic position and generic names of the Gram negative bacterial plant pathogens. Zentbl. Bakteriol. Parasitenkd. Infektkrankh. Hyg. Abt. II *100:* 177–193.

Dowson, W.J. 1943. On the generic names *Pseudomonas, Xanthomonas* and *Bacterium* for certain bacterial plant pathogens. Trans. Brit. Mycol. Soc. *26:* 1–14.

Drancourt, M., C. Bollet, A. Carta and P. Rousselier. 2001. Phylogenetic analyses of *Klebsiella* species delineate *Klebsiella* and *Raoultella* gen. nov., with description of *Raoultella ornithinolytica* comb. nov., *Raoultella terrigena* comb. nov. and *Raoultella planticola* comb. nov. Int. J. Syst. Evol. Microbiol. *51:* 925–932.

Drancourt, M., C. Bollet and D. Raoult. 1997. *Stenotrophomonas africana* sp. nov., an opportunistic human pathogen in Africa. Int. J. Syst. Bacteriol. *47:* 160–163.

Drancourt, M., L. Niel and D. Raoult. 1995. Unknown multiresistant gram-negative bacterium causing meningitis in HIV-positive patient in Africa. Lancet *346:* 1168.

Drasar, B.S. and M.J. Hill. 1974. Human Intestinal Flora, Academic Press, New York.

Drasar, F.A., W. Farrell, J. Maskell and J.D. Williams. 1976. Tobramycin, amikacin, sissomicin, and gentamicin resistant Gram-negative rods. Br. Med. J. *2:* 1284–1287.

Drechsel, H., A. Thieken, R. Reissbrodt, G. Jung and G. Winkelmann. 1993. α-keto acids are novel siderophores in the genera *Proteus, Providencia*, and *Morganella* and are produced by amino acid deaminases. J. Bacteriol. *175:* 2727–2733.

Drew, R. and S.A. Wilson. 1992. Regulation of amidase expression in *Pseudomonas aeruginosa. In* Galli, Silver and Witholt (Editors), *Pseudomonas:* Molecular Biology and Biotechnology, American Society for Microbiology, Washington D.C. 207–213.

Drobne, D., J. Strus, N. Znidarsic and P. Zidar. 1999. Morphological description of bacterial infection of digestive glands in the terrestrial isopod porcellio scaber (Isopoda, crustacea). J. Invertebr. Pathol. *73:* 113–119.

Droniuk, R., P.T.S. Wong, G. Wisse and R.A. Macleod. 1987. Variation in quantitative requirements for sodium for transport of metabolizable compounds by the marine bacteria *Alteromonas haloplanktis* 214 and *Vibrio fischeri.* Appl. Environ. Microbiol. *53:* 1487–1496.

Drouet, E.B., G.A. Denoyel, M.M. Boude, G. Boussant and H.P. de Montclos. 1989. Distribution of *Haemophilus influenzae* and *Haemophilus parainfluenzae* biotypes isolated from the human genitourinary tract. Eur. J. Clin. Microbiol. Infect. Dis. *8:* 951–955.

Drozanski, W.J. 1991. *Sarcobium lyticum*, gen. nov., sp. nov., an obligate intracellular bacterial parasite of small free-living amoebae. Int. J. Syst. Bacteriol. *41:* 82–87.

Du, L.S. and K.H. Tibelius. 1994. The *hupB* gene of the *Azotobacter chroococcum* hydrogenase gene cluster is involved in nickel metabolism. Curr. Microbiol. *28:* 21–24.

Du, L.S., K.H. Tibelius, E.M. Souza, R.P. Garg and M.G. Yates. 1994. Sequences, organization and analysis of the *hupZMNOQRTV* genes from the *Azotobacter chroococcum* hydrogenase gene cluster. J. Mol. Biol. *243:* 549–557.

Dubinina, G.A. and M.Y. Grabovich. 1983. Isolation of pure *Thiospira* cultures and investigation of their physiology and sulfur metabolism. Mikrobiologiya *52:* 5–12.

Dubinina, G.A., M.Y. Grabovich, A.M. Lysenko, N.A. Chernykh and V.V. Churikova. 1993. Revision of taxonomic position of colorless sulfur spirilla of the genus *Thiospira* and description of a new species *Aquaspirillum bipunctata* comb. nov. Microbiology *62:* 368–644.

Dubnau, D., I. Smith, P. Morell and J. Marmur. 1965. Gene conservation in *Bacillus* species. I. Conserved genetic and nucleic acid base sequence homologies. Proc. Natl. Acad. Sci. U.S.A. *54:* 491–498.

DuBow, M.S. and T. Blumenthal. 1975. Host Factor for coliphage QβRNA replication is present in *Pseudomonas putida.* Molec. Gen. Genet. *141:* 113–119.

DuBow, M.S. and T. Ryan. 1977. Host Factor for coliphage QbRNA replication as an aid in elucidating phylogenetic relationships: the genus *Pseudomonas.* J. Gen. Microbiol. *102:* 263–268.

Dubreuil, J.D., A. Letellier, E. Stenbaek and M. Gottschalk. 1996. Serotyping of *Actinobacillus pleuropneumoniae* serotype 5 strains using a monoclonal-based polystyrene agglutination test. Cana. J. Vet. Res. *60:* 69–71.

Duckworth, A.W., W.D. Grant, B.E. Jones, D. Meijer, M.C. Marquez and A. Ventosa. 2000a. *Halomonas magadii* sp. nov., a new member of the genus *Halomonas*, isolated from a soda lake of the East African Rift Valley. Extremophiles *4:* 53–60.

Duckworth, A.W., W.D. Grant, B.E. Jones, D. Meijer, M.C. Marquez and A. Ventosa. 2000b. *In* Validation of the publication of new names and new combinations previously effectively published outside the IJSEM, List No. 75. Int. J. Syst. Evol. Microbiol. *50:* 1415–1417.

Duckworth, A.W., W.D. Grant, B.E. Jones and R. van Steenbergen. 1996. Phylogenetic diversity of soda lake alkaliphiles. FEMS Microbiol. Ecol. *19:* 181–191.

Dudkiewicz, B. and E. Szewczyk. 1993. Etiology of bacterial endocarditis in materials of departments of cardiology and cardiac surgery of the Lodz medical academy. Med. Dosw. MIkrobiol. *45:* 357–359.

Duetz, W.A., S. Marqués, C. De Jong, J.L. Ramos and J.G. Van Andel. 1994. Inducibility of the TOL catabolic pathway in *Pseudomonas putida* (pWW0) growing on succinate in continuous-culture - evidence of carbon catabolite repression control. J. Bacteriol. *176:* 2354–2361.

Duff, D.C.B. 1937. Dissociation in *Bacillus salmonicida*, with special reference to the appearance of a G form of culture. J. Bacteriol. *33:* 49–67.

Dufresne, J., G. Vezina and R.C. Levesque. 1988. Molecular cloning and expression of the imipenem-hydrolyzing beta-lactamase gene from *Pseudomonas maltophilia* in *Escherichia coli.* Rev. Infect. Dis. *10:* 806–817.

Düggeli, M. 1904. Die Bakterienflora gesunder Samen und daraus gezogener Keimpflanzchen. Zentbl. Bakteriol. Abt. II. *12:* 602–614.

Duguid, J.P. 1959. Fimbriae and adhesive properties in *Klebsiella* strains. J. Gen. Microbiol. *21:* 271–286.

Duguid, J.P. and R.R. Gillies. 1957. Fimbriae and adhesive properties in dysentery bacilli. J. Pathol. Bacteriol. *74:* 397–411.

Duim, B., L. Van Alphen, P. Eijk, H.M. Jansen and J. Dankert. 1994. Antigenic drift of non-encapsulated *Haemophilus influenzae* major outer membrane protein P2 in patients with chronic bronchitis is caused by point mutations. Mol. Microbiol. *11:* 1181–1189.

Duine, J.A. 1991. Energy generation and the glucose dehydrogenase pathway in *Acinetobacter. In* Towner and Bergogne-Bérézin (Editors), The Biology of *Acinetobacter*: Taxonomy, Clinical Importance, Molecular

Biology, Physiology, Industrial Relevance, Plenum Press, New York. pp. 295–312.

Duine, J.A., J. Frank and R. van der Meer. 1982. Different forms of quinoprotein aldose-(glucose) dehydrogenase in *Acinetobacter calcoaceticus*. Arch. Microbiol. *131*: 27–31.

Dulaney, A.D., K. Guto and H. Packer. 1948. Donovania granulomatis: cultivation antigenic preparation, and immunological tests. J. Immunol. *59*: 335–340.

Dulaney, A.D. and H. Packer. 1947. Complement-fixation studies with pus antigen in granuloma inguinale. Proc. Soc. Exp. Biol. Med. *65*: 254–256.

Dul'tseva, N.M. and G.A. Dubinina. 1994. *Thiothrix arctophila* sp. nov. - a new species of filamentous colorless sulfur bacteria. Microbiology *63*: 147–153.

Dul'tseva, N.M., G.A. Dubinina and A.M. Lysenko. 1996. Isolation of marine filamentous sulfur bacteria and description of the new species *Leucothrix thiophila* sp. nov. Microbiology *65*: 79–87.

Dumoff, M. 1979. Direct *in vitro* isolation of the Legionnaires' disease bacterium in two fatal cases. Cultural and staining characteristics. Ann. Intern. Med. *90*: 694–696.

Duncan, A., G.E. Vasiliadis, R.C. Bayly, J.W. May and W.G.C. Raper. 1988. Genospecies of *Acinetobacter* isolated from activated sludge showing enhanced removal of phosphate during pilot-scale treatment of sewage. Biotechnol. Lett. *10*: 831–836.

Dunphy, G.B. 1994. Interaction of mutants of *Xenorhabdus nematophilus* (*Enterobacteriaceae*) with antibacterial systems of *Galleria mellonella* larvae (Insecta: Pyralidae). Can. J. Microbiol. *40*: 161–168.

Dupont, C. and A.J. Clarke. 1991. Dependence of lysozyme-catalyzed solubilization of *Proteus mirabilis* peptidoglycan on the extent of O-acetylation. Eur. J. Biochem. *195*: 763–769.

Dupont, H.L. 2000. Shigella species (bacillary dysentery). *In* Mandell, Bennett and Dolin (Editors), Principles and Practice of Infectious Diseases, 5th Ed., Churchill Livingston, Philadelphia. pp. 2363–2369.

DuPont, H.L., M.M. Levine, R.B. Hornick and S.B. Formal. 1989. Inoculum size in shigellosis and implications for expected mode of transmission. J. Infect. Dis. *159*: 1126–1128.

Duport, C., C. Baysse and Y. Michel-Briand. 1995. Molecular characterization of pyocin S3, a novel S-type pyocin from *Pseudomonas aeruginosa*. J. Biol. Chem. *270*: 8920–8927.

Dupuis, G., J. Petite, O. Peter and M. Vouilloz. 1987. An important outbreak of human Q fever in a Swiss Alpine valley. Int. J. Epidemiol. *16*: 282–287.

Durand, P., A.L. Reysenbach, D. Prieur and N.R. Pace. 1993. Isolation and characterization of *Thiobacillus hydrothermalis*, sp. nov., a mesophilic obligately chemolithotrophic bacterium isolated from a deep-sea hydrothermal vent in Fiji Basin. Arch. Microbiol. *159*: 39–44.

Durbin, R.D. 1992. Role of toxins for plant-pathogenic pseudomonads. *In* Galli, Silver and Witholt (Editors), *Pseudomonas*: Molecular Biology and Biotechnology, American Society for Microbiology, Washington D.C. 43–55.

Durbin, R.D., T.F. Uchytil, J.A. Steele and R.d.D. Ribeiro. 1978. Tabtoxinine beta-lactam from *Pseudomonas tabaci*. Phytochemistry. *17*: 147–147.

Durgapal, J.C. 1977. Albinism in *Xanthomonas sesami*. Curr. Sci. (Bangalore) *46*: 274.

Durgapal, J.C. and B.M. Trivedi. 1976. Bacterial blight of "four-o'clock"— a new disease in India. Curr. Sci. (Bangalore) *45*: 111–112.

Dutka, B.J., K. Jones and H. Bailey. 1987. Enumeration of *Klebsiella* spp. in cold water by using MacConkey inositol-potassium tellurite medium. Appl. Environ. Microbiol. *53*: 1716–1717.

Dutky, S.R. 1959. Insect microbiology. Adv. Appl. Microbiol. *1*: 175–200.

Dutky, S.R. and E.L. Gooden. 1952. *Coxiella popilliae*, n. sp., a rickettsia causing blue disease of Japanese beetle larvae. J. Bacteriol. *63*: 743–750.

Dutton, A.A.C. and M. Ralston. 1957. Urinary tract infection in a male urological ward with special reference to the mode of infection. Lancet *1*: 115–119.

Duval, C.W. 1904. A member of the dysentery group. JAMA (J. Am. Med. Assoc.). *43*: 381–383.

Dworkin, M. and S.M. Gibson. 1964. A system for studying microbial morphogenesis: rapid formation of microcysts in *Myxococcus xanthus*. Science (Wash. D. C.) *146*: 243–244.

Dyar, H.G. 1895. On certain bacteria from the air of New York City. Ann. N. Y. Acad. Sci. *8*: 322–380.

Dye, D.W. 1960. Pectolytic activity in *Xanthomonas*. N. Z. J. Sci. *3*: 61–69.

Dye, D.W. 1962. The inadequacy of the usual determinative tests for the identification of *Xanthomonas* spp. N. Z. J. Sci. *5*: 393–416.

Dye, D.W. 1963a. A bacterial disease of pukatea (*Laurelia nouae-zelandiae* A Cunn.) caused by *Xanthomonas laureliae* n. sp. N. Z. J. Sci. *6*: 179–185.

Dye, D.W. 1963b. The taxonomic position of *Xanthomonas stewartii* (Erw. Smith 1914) Dowson 1939. N. Z. J. Sci. *6*: 495–506.

Dye, D.W. 1963c. The taxonomic position of *Xanthomonas uredovorus* Pon et al., 1954. N. Z. J. Sci. *6*: 146–149.

Dye, D.W. 1964. The taxonomic position of *Xanthomonas trifolii* (Huss, 1907) James 1955. N. Z. J. Sci. *7*: 261–269.

Dye, D.W. 1966a. A comparative study of some atypical "xanthomonads". N. Z. J. Sci. *9*: 843–854.

Dye, D.W. 1966b. Cultural and biochemical reactions of additional *Xanthomonas* spp. N. Z. J. Sci. *9*: 913–919.

Dye, D.W. 1968. A taxonomic study of the genus *Erwinia*. I. The "amylovora" group. N. Z. J. Sci. *11*: 590–607.

Dye, D.W. 1969a. A taxonomic study of the genus *Erwinia*. II. The "carotovora" group. N. Z. J. Sci. *12*: 81–97.

Dye, D.W. 1969b. A taxonomic study of the genus *Erwinia*. III. The "herbicola" group. N. Z. J. Sci. *12*: 223–236.

Dye, D.W. 1978. Genus IX *Xanthomonas* Dowson 1939. *In* J.M. Young, D.W. Dye, J.F. Bradbury, C.G. Panagopoulos and C.F. Robbs. A proposed nomenclature and classification for plant pathogenic bacteria. N.Z.J. Agr. Res. *21*: pp. 153–177.

Dye, D.W. 1980. *Xanthomonas*. *In* Schaad (Editor), Laboratory Guide for Identification of Plant Pathogenic Bacteria, American Phytopathological Society, St. Paul, Minn. pp. 45–49.

Dye, D.W., J.F. Bradbury, R.S. Dickey, M.S. Goto, C.N. Hale, A.C. Hayward, A. Kelman, R.A. Lelliott, P.N. Patel, D.C. Sands, M.N. Schroth, W. D.R.W. and J.M. Young. 1975. Proposals for a reappraisal of the status of names of plant-pathogenic *Pseudomonas* species. Int. J. Syst. Bacteriol. *25*: 252–257.

Dye, D.W., J.F. Bradbury, M. Goto, A.C. Hayward, R.A. Lelliott and M.N. Schroth. 1980. International standards for naming pathovars of phytopathogenic bacteria and a list of pathovar names and pathotype strains. Rev. Plant Pathol. *59*: 153–168.

Dyksterhouse, S.E., J.P. Gray, R.P. Herwig, J.C. Lara and J.T. Staley. 1995. *Cycloclasticus pugetii* gen. nov., sp. nov., an aromatic hydrocarbon-degrading bacterium from marine sediments. Int. J. Syst. Bacteriol. *45*: 116–123.

Eady, R.R. and R.L. Robson. 1984. Characteristics of N$_2$ fixation in Mo-limited batch and continuous cultures of *Azotobacter vinelandii*. Biochem. J. *224*: 853–862.

Eagon, R.G. 1962. *Pseudomonas natriegens*, a marine bacterium with a generation time of less than 10 minutes. J. Bacteriol. *83*: 736–737.

East, A.K., D. Allaway and M.D. Collins. 1992. Analysis of DNA encoding 23S rRNA and 16S–23S rRNA intergenic spacer regions from *Plesiomonas shigelloides*. FEMS Microbiol. Lett. *74*: 57–62.

East, P.D., A. Cao and R.J. Akhurst. 1998. Toxin genes from the bacterium *Xenorhabdus nematophilus* and *Photorhabdus luminescens*., International Patent Application PCT/AU98/00562.

Easwaramurthy, R., V. Kaviyarasan and S.S. Gnanamanickam. 1984. A bacterial disease of ornamental cannas caused by *Xanthomonas campestris* pv. *cannae* pv. nov. Curr. Sci. (Bangalore) *53*: 708–709.

Eaton, R.W. 1996. *p*-Cumate catabolic pathway in *Pseudomonas putida* F1: Cloning and characterization of DNA carrying the *cmt* operon. J. Bacteriol. *178*: 1351–1362.

Eaton, R.W. 1997. *p*-Cymene catabolic pathway in *Pseudomonas putida* F1:

Cloning and characterization of DNA encoding conversion of *p*-cymene to *p*-cumate. J. Bacteriol. *179*: 3171–3180.

Eaves, L.E., D.G. Rogers and P.J. Blackall. 1989. Comparison of hemagglutinin and agglutinin schemes for the serological classification of *Haemophilus paragallinarum* and proposal of a new hemagglutinin serovar. J. Clin. Microbiol. *27*: 1510–1513.

Eberhard, C., C.O. Wirsen and H.W. Jannasch. 1995. Oxidation of polymetal sulfides by chemolithoautotrophic bacteria from deep-sea hydrothermal vents. Geomicrobiol. J. *13*: 145–164.

Eberth, C.J. 1880. Die organismen in den organen bei typhus abdominalis. Arch. Pathol. Anat. Physiol. Klin. Med. *81*: 58–74.

Ebright, J.R., J.R. Lentino and E. Juni. 1982. Endophthalmitis caused by *Moraxella nonliquefaciens*. Am. J. Clin. Pathol. *77*: 362–363.

Ebringer, A., T. Ptaszynska, M. Corbett, C. Wilson, Y. MacAfee, H. Avakian, P. Baron and D.C. James. 1985. Antibodies to *Proteus* in rheumatoid arthritis. Lancet *2*: 305–307.

Echandi, E. and J.W. Moyer. 1979. Production, properties, and morphology of bacteriocins from *Erwinia chrysanthemi*. Phytopathology *69*: 1204–1207.

Echenique, J.R., H. Arienti, M.E. Tolmasky, R.R. Read, R.J. Staneloni, J.H. Crosa and L.A. Actis. 1992. Characterization of a high-affinity iron transport system in *Acinetobacter baumannii*. J. Bacteriol. *174*: 7670–7679.

Echeverria, P., F. Ørskov, I. Ørskov, S. Knutton, F. Scheutz, J.E. Brown and U. Lexomboon. 1991. Attaching and effacing enteropathogenic *Escherichia coli* as a cause of infantile diarrhea in Bangkok, Thailand. J. Infect. Dis. *164*: 550–554.

Echeverria, P., D.N. Taylor, K.A. Bettelheim, A. Chatkaeomorakot, S. Changchwawalit, A. Thongcharoen and U. Leksomboon. 1987. HeLa cell-adherent enteropathogenic *Escherichia coli* in children under 1 year of age in Thailand. J. Clin. Microbiol. *25*: 1472–1475.

Eddy, B.P. 1960. Cephalotrichous, fermentative Gram-negative bacteria: the genus *Aeromonas*. J. Appl. Bacteriol. *23*: 216–249.

Eddy, B.P. 1962. Further studies on *Aeromonas*. I. Additional strains and supplementary biochemical tests. J. Appl. Bacteriol. *25*: 137–146.

Eddy, B.P. and K.P. Carpenter. 1964. Further studies on *Aeromonas*. II. Taxonomy of *Aeromonas* and C-27 strains. J. Appl. Bacteriol. *27*: 96–109.

Edelman, R. and M.M. Levine. 1983. From the National Institute of Allergy and Infectious Diseases. Summary of a workshop on enteropathogenic *Escherichia coli*. J. Infect. Dis. *147*: 1108–1118.

Edelstein, P.H. 1981. Improved semiselective medium for isolation of *Legionella pneumophila* from contaminated clinical and environmental specimens. J. Clin. Microbiol. *14*: 298–303.

Edelstein, P.H. 1982. Comparative study of selective media for isolation of *Legionella pneumophila* from potable water. J. Clin. Microbiol. *16*: 697–699.

Edelstein, P.H., K.B. Beer, J.C. Sturge, A.J. Watson and L.C. Goldstein. 1985. Clinical utility of a monoclonal direct fluorescent reagent specific for *Legionella pneumophila*: comparative study with other reagents. J. Clin. Microbiol. *22*: 419–421.

Edelstein, P.H., W.F. Bibb, G.W. Gorman, W.L. Thacker, D.J. Brenner, H.W. Wilkinson, C.W. Moss, R.S. Buddington, C.J. Dunn, P.J. Roos and P.L. Meenhorst. 1984. *Legionella pneumophila* serogroup 9: a cause of human pneumonia. Ann. Intern. Med. *101*: 196–198.

Edelstein, P.H., D.J. Brenner, C.W. Moss, A.G. Steigerwalt, E.M. Francis and W.L. George. 1982a. *Legionella wadsworthii* species nova: a cause of human pneumonia. Ann. Intern. Med. *97*: 809–813.

Edelstein, P.H., D.J. Brenner, C.W. Moss, A.G. Steigerwalt, E.M. Francis and W.L. George. 1983. *In* Validation of the publication of new names and new combinations previously effectively published outside the IJSB. List No. 11. Int. J. Syst. Bacteriol. *33*: 672–674.

Edelstein, P.H. and M.A. Edelstein. 1989. Evaluation of the Merifluor-Legionella immunofluorescent reagent for identifying and detecting 21 *Legionella* species. J. Clin. Microbiol. *27*: 2455–2458.

Edelstein, P.H. and S.M. Finegold. 1979. Use of a semiselective medium to culture *Legionella pneumophila* from contaminated lung specimens. J. Clin. Microbiol. *10*: 141–143.

Edelstein, P.H., R.D. Meyer and S.M. Finegold. 1980. Laboratory diagnosis of Legionnaires' disease. Am. Rev. Respir. Dis. *121*: 317–328.

Edelstein, P.H., C. Nakahama, J.O. Tobin, K. Calarco, K.B. Beer, J.R. Joly and R.K. Selander. 1986. Paleoepidemiologic investigation of Legionnaires' disease at Wadsworth Veterans Administration Hospital by using three typing methods for comparison of legionellae from clinical and environmental sources. J. Clin. Microbiol. *23*: 1121–1126.

Edelstein, P.H. and E.P. Pryor. 1985. A new biotype of *Legionella dumoffii*. J. Clin. Microbiol. *21*: 641–642.

Edelstein, P.H., J.B. Snitzer and J.A. Bridge. 1982b. Enhancement of recovery of *Legionella pneumophila* from contaminated respiratory tract specimens by heat. J. Clin. Microbiol. *16*: 1061–1065.

Ederer, G.M. and M.L. Schurr. 1971. Optimal bacitracin concentration for selective isolation medium for *Haemophilus*. Am. J. Med. Technol. *37*: 304–305.

Editorial Board of the Judicial Commission of the International Committee on Systematic Bacteriology. 1973. Opinion 48: Rejection of the name *Aerobacter liquefaciens* Beijerinck and conservation of the name *Aeromonas* Stanier with *Aeromonas hydrophila* as the type species. Int. J. Syst. Bacteriol. *23*: 473–474.

Edmonds, C., G.E. Griffin and A.P. Johnstone. 1989. Demonstration and partial characterization of ADP-ribosylation in P*seudomonas maltophilia*. Biochem. J. *261*: 113–118.

Edmondson, A.S. and E.M. Cooke. 1979. The development and assessment of a bacteriocin typing method for *Klebsiella*. J. Hyg. London. *82*: 207–223.

Edwards, P.R. and W.H. Ewing. 1972. Identification of *Enterobacteriaceae*, 3rd Ed., Burgess Publishing Co., Minneapolis.

Edwards, P.R. and M.A. Fife. 1955. Studies on the *Klebsiella–Aerobacter* group of bacteria. J. Bacteriol. *70*: 382–390.

Edwards, P.R., M.G. West and D.W. Bruner. 1948. Antigenic studies of a paracolon bacteria (Bethesda group). J. Bacteriol. *55*: 711–719.

Effendi, I. and B. Austin. 1991. Survival of the fish pathogen *Aeromonas salmonicida* in seawater. FEMS Microbiol. Lett. *84*: 103–106.

Efuet, E.T., L. Pulakat and N. Gavini. 1996. Investigations on the cell volumes of *Azotobacter vinelandii* by scanning electron microscopy. J. Basic Microbiol. *36*: 229–234.

Egan, W., F.P. Tsui, P.A. Climenson and R. Scbneerson. 1980a. Structural and immunological studies of the *Haemophilus influenzae* type c capsular polysaccharide. Carbohydr. Res. *80*: 305–316.

Egan, W., F. Tsui and R. Schneerson. 1980b. Structural studies of the *Haemophilus influenzae* type f capsular polysaccharide. Carbobydr. Res. *79*: 271–277.

Egerton, J.R. 1989. Footrot of cattle, goats, and deer. *In* Egerton, J.R., W.K. Yong and G.G. Riffkin (Editors), Footrot and Foot Abscess of Ruminants, CRC Press, Boca Raton. 47–56.

Egerton, J.R., P.T. Cox, B.J. Anderson, C. Kristo, M. Norman and J.S. Mattick. 1987. Protection of sheep against footrot with a recombinant DNA-based fimbrial vaccine. Vet. Microbiol. *14*: 393–409.

Egidius, E., R. Wiik, K. Andersen, K.A. Hoff and B. Hjeltnes. 1986. *Vibrio salmonicida* sp. nov., a new fish pathogen. Int. J. Syst. Bacteriol. *36*: 518–520.

Egli, T., M. Goto and D. Schmidt. 1975. Bacterial wilt, a new forage grass disease. Phytopathol. Z. *82*: 111–121.

Egli, T. and D. Schmidt. 1982. Pathogenic variation among the causal agents of bacterial wilt of forage grasses. Phytopathol. Z. *104*: 138–150.

Ehrenberg, C.G. 1838. Die Infusionthierchen als vollkommene Organismen: ein Blick in das tiefere organische Leben der Natur, L. Voss, Leipzig. pp. i–xvii; 1–547.

Ehrenberg, C.G. 1840. Charakteristik von 274 neuen Arten von Infusorien, Vol. 1840, Ber Bekannt Verhandl. Königl. Preuss Akad. Wiss., Berlin. 197–219.

Ehrenkranz, N.J., B.C. Alfonso, D.G. Eckert and L.B. Moskowitz. 1989. Proteeae species bacteriuria accompanying Proteeae species groin skin carriage in geriatric outpatients. J. Clin. Microbiol. *27*: 1988–1991.

Ehrenstein, B., A.T. Bernards, L. Dijkshoorn, P. Gerner-Smidt, K.J.

Towner, P.J.M. Bouvet, F.D. Daschner and H. Grundmann. 1996. *Acinetobacter* species identification by using tRNA spacer fingerprinting. J. Clin. Microbiol. *34*: 2414–2420.

Ehrich, S., D. Behrens, E. Lebedeva, W. Ludwig and E. Bock. 1995. A new obligately chemolithoautotrophic, nitrite-oxidizing bacterium, *Nitrospira moscoviensis* sp. nov. and its phylogenetic relationship. Arch. Microbiol. *164*: 16–23.

Eichler, B. and N. Pfennig. 1986. Characterization of a new platelet-forming purple sulfur bacterium, *Amoebobacter pedioformis* sp. nov. Arch. Microbiol. *146*: 295–300.

Eichler, B. and N. Pfennig. 1988. A new purple sulfur bacterium from stratified freshwater lakes, *Amoebobacter purpureus*, sp. nov. Arch. Microbiol. *149*: 395–400.

Eichler, B. and N. Pfennig. 1991. Isolation and characteristics of *Thiopedia rosea* (neotype). Arch. Microbiol. *155*: 210–216.

Eicholz, W. 1902. Erdbeerbacillus (*Bacterium fragi*). Zentralbl. Bakteriol. Parasitenkd. Infektionskr. Hyg. Abt. II. *9*: 425–428.

Eigelsbach, H.T., W. Braun and R.D. Herring. 1951. Studies on the variation of *Bacterium tularense*. J. Bacteriol. *61*: 557–569.

Eigelsbach, H.T. and V.G. McGann. 1981. The genus *Francisella*. *In* Starr, Stolp, Trüper, Balows and Schlegel (Editors), The Prokaryotes: A Handbook on Habitats, Isolation and Identification of Bacteria, Springer-Verlag, Berlin. pp. 620–633.

Eigelsbach, H.T. and V.G. McGann. 1984. Genus *Francisella*. *In* Krieg and Holt (Editors), Bergey's Manual of Systematic Bacteriology, 1 Ed., Vol. 1, The Williams & Wilkins Co., Baltimore. pp. 394–399.

Eikelboom, D.H. 1975. Filamentous organisms observed in activated sludge. Water Res. *9*: 365–388.

Eikelboom, D.H. 1977. Identification of filamentous organisms in bulking activated sludge. Prog. Water Technol. *8*: 153–161.

Eikelboom, D.H. and H.J.J. van Buijsen. 1981. Microscopic sludge investigation manual, Delft. TNO Research institute for environmental hygiene. Report nr A 94A. .

Eilers, H., J. Pernthaler, F.O. Glöckner and R. Amann. 2000. Culturability and *in situ* abundance of pelagic bacteria from the North Sea. Appl. Environ. Microbiol. *66*: 3044–3051.

Eimhjellen, K.E. 1970. *Thiocapsa pfennigii* sp. nov., a new species of phototrophic sulfur bacteria. Arch. Mikrobiol. *73*: 193–194.

Eimhjellen, K.E., H. Steensland and J. Traetteberg. 1967. A *Thiococcus* sp. nov. gen., its pigments and internal membrane system. Arch. Mikrobiol. *59*: 82–92.

Eisenberg, R.C., S.J. Butters, S.C. Quay and S.B. Friedman. 1974. Glucose uptake and phosphorylation in *Pseudomonas fluorescens*. J. Bacteriol. *120*: 147–153.

El Banoby, F.E. and K. Rudolph. 1979. Induction of water-soaking in plant leaves by extracellular polysaccharides from phytopathogenic pseudomonads and xanthomonads. Physiol. Plant Pathol. *15*: 341–349.

El Harrif-Heraud, Z., C. Arpin, S. Benliman and C. Quentin. 1997. Molecular epidemiology of a nosocomial outbreak due to SHV-4-producing strains of *Citrobacter diversus*. J. Clin. Microbiol. *35*: 2561–2567.

El Hendawy, H.H., I.M. Zeid and Z.K. Mohamed. 1998. The biological control of soft rot disease in melon caused by *Erwinia carotovora* subsp. *carotovora* using *Pseudomonas fluorescens*. Microbiol. Res. *153*: 55–60.

el Khizzi, N., S.A. Kasab and A.O. Osoba. 1997. HACEK group endocarditis at the Riyadh Armed Forces Hospital. J. Infect. *34*: 69–4.

El-Helaly, A.F., M.K. Abo-El-Dahab, M.A. Goorani and G. M.R.M.. 1979. Production of cellulase (Cx) by different species of *Erwinia*. Zentbl. Bakteriol. Parasitenkd. Infektkrankh. Hyg. Abt. II *134*: 187–192.

Elazari-Volcani, B. 1939. On *Pseudomonas indigofera* (Voges) Migula and its pigment. Arch. Mikrobiol. *10*: 343–358.

Elazari-Volcani, B. 1940. Studies on the microflora of the Dead Sea, Hebrew University, Jerusalem.

Elisha, B.G. and L.M. Steyn. 1991. Identification of an *Acinetobacter baumannii* gene region with sequence and organizational similarity to Tn2670. Plasmid *25*: 96–104.

Elleman, T.C. 1988. Pilins of *Bacteroides nodosus*: molecular basis of serotypic variation and relationships to other bacterial pilins. Microbiol. Rev. *52*: 233–247.

Ellingboe, A.H. 1981. Changing concepts in host-pathogen genetics. Annu. Rev. Phytopathol. *19*: 125–143.

Elliott, C. 1920. Halo-blight of oats. J. Agr. Res. *19*: 139–172.

Elliott, C. 1923. A bacterial stripe disease of proso millet. J. Agr. Res. *26*: 151–160.

Elliott, C. 1927. Bacterial stripe blight of oats. J. Agr. Res. *35*: 811–824.

Elomari, M., L. Coroler, B. Hoste, M. Gillis, D. Izard and H. Leclerc. 1996. DNA relatedness among *Pseudomonas* strains isolated from natural mineral waters and proposal of *Pseudomonas veronii* sp. nov. Int. J. Syst. Bacteriol. *46*: 1138–1144.

Elomari, M., L. Coroler, D. Izard and H. Leclerc. 1995. A numerical taxonomic study of fluorescent *Pseudomonas* strains isolated from natural mineral waters. J. Appl. Bacteriol. *78*: 71–81.

Elomari, M., L. Coroler, S. Verhille, D. Izard and H. Leclerc. 1997. *Pseudomonas monteilii* sp. nov., isolated from clinical specimens. Int. J. Syst. Bacteriol. *47*: 846–852.

Elomari, M., D. Izard, P. Vincent, L. Coroler and H. Leclerc. 1994. Comparison of ribotyping analysis and numerical taxonomy studies of *Pseudomonas putida* biovar A. Syst. Appl. Microbiol. *17*: 361–369.

Elrod, R.P. 1946. The serological relationship between *Erwinia tracheiphila* and species of *Shigella*. J. Bacteriol. *52*: 405–410.

Elrod, R.P. and A.C. Braun. 1947a. Serological studies of the genus *Xanthomonas*. I. Cross agglutination relationships. J. Bacteriol. *53*: 509–518.

Elrod, R.P. and A.C. Braun. 1947b. Serological studies of the genus *Xanthomonas*. II. *Xanthomonas translucens* group. J. Bacteriol. *53*: 519–524.

Elsden, S.R., M.G. Hilton and J.M. Waller. 1976. The end products of the metabolism of aromatic amino acids by *Clostridia*. Arch. Microbiol. *107*: 283–288.

Elsner, H.A., U. Duhrsen, B. Hollwitz, P.M. Kaulfers and D.K. Hossfeld. 1997. Fatal pulmonary hemorrhage in patients with acute leukemia and fulminant pneumonia caused by *Stenotrophomonas maltophilia*. Ann. Hematol. *74*: 155–161.

Elwell, L.P., J.D. DeGraaff, D. Seibert and S. Falkow. 1975. Plasmid-linked ampicillin resistance in *Haemophilus influenzae* type b. Infect. Immun. *12*: 404–410.

Elwell, L.P., J.R. Saunders, M.H. Richmond and S. Falkow. 1977. Relationships among some R plasmids found in *Haemophilus influenzae*. J. Bacteriol. *131*: 356–362.

Ely, B. 1985. Vectors for transposon mutagenesis of nonenteric bacteria. Mol. Gen. Genet. *200*: 302–304.

Emerson, D. and J.A. Breznak. 1997. The response of microbial populations from oil-brine contaminated soil to gradients of NaCl and sodium *p*-toluate is a diffusion gradient chamber. FEMS Microbiol. Ecol. *23*: 285–300.

Emslie-Smith, A.H. 1961. *Hafnia alvei* strains possessing alpha antigen of Stamp and Stone. J. Pathol. Bacteriol. *81*: 534–536.

Endo, Y., K. Tsurugi, T. Yutsudo, Y. Takeda, T. Ogasawara and K. Igarashi. 1988. Site of action of a vero toxin (VT2) from *Escherichia coli* O157:H7 and of Shiga toxin on eukaryotic ribosomes. RNA *N*-glycosidase activity of the toxins. Eur. J. Biochem. *171*: 45–50.

Eneroth, A., A. Christiansson, J. Brendehaug and G. Molin. 1998. Critical contamination sites in the production line of pasteurised milk, with reference to the psychrotrophic spoilage flora. Int. Dairy J. *8*: 829–834.

Engelhard, E., R.M. Kroppenstedt, R. Mutters and W. Mannheim. 1991. Carbohydrate patterns, cellular lipoquinones, fatty acids and phospholipids of the genus *Pasteurella* sensu stricto. Med. Microbiol. Immunol. *180*: 79–92.

Engelhard, E., R.M. Kroppenstedt, R. Mutters and W. Mannheim. 1992. Cytochemische charakterisierung der humanen *Haemophilus*-spezies sowie von (*Actinobacillus*) actinomycetemcomitans. *In* Flores-de-Jacoby and Mannheim (Editors), 2nd Workshop Mikrobiologie und Immunologie der Paradontalen Erkrankungen: Berlin, 22. und 23. Februar 1991, Quintessenz, Berlin. 273–288.

Engelhardt, H., W. Baumeister and W.O. Saxton. 1983. Electron microscopy of photosynthetic membranes containing bacteriochlorophyll *b*. Arch. Microbiol. *135*: 169–175.

Enger, O., H. Nygaard, M. Solberg, G. Schei, J. Nielsen and I. Dundas. 1987. Characterization of *Alteromonas denitrificans* sp. nov. Int. J. Syst. Bacteriol. *37*: 416–421.

England, A.C.I. and D.W. Fraser. 1981. Sporadic and epidemic nosocomial legionellosis in the United States. Epidemiologic features. Am. J. Med. *70*: 707–711.

England, A.C.I., R.M. McKinney, P. Skaliy and G.W. Gorman. 1980. A fifth serogroup of *Legionella pneumophila*. Ann. Intern. Med. *93*: 58–59.

Englund, G.W. 1969. Persistent septicemia due to *Hafnia alvei*. Report of a case. Am. J. Clin. Pathol. *51*: 717–719.

Enright, M.C., P.E. Carter, I.A. MacLean and H. McKenzie. 1994. Phylogenetic relationships between some members of the genera *Neisseria*, *Acinetobacter*, *Moraxella*, and *Kingella* based on partial 16S ribosomal DNA sequence analysis. Int. J. Syst. Bacteriol. *44*: 387–391.

Enright, M.C. and H. McKenzie. 1997. *Moraxella (Branhamella) catarrhalis*: clinical and molecular aspects of a rediscovered pathogen. J. Med. Microbiol. *46*: 360–371.

Entner, N. and M. Doudoroff. 1952. Glucose and gluconic acid oxidation of *Pseudomonas saccharophila*. J. Biol. Chem. *196*: 853–862.

Entwistle, P.F., J.S. Robertson and B.E. Juniper. 1968. The ultrastructure of a rickettsia pathogenic to a saturnid moth. J. Gen. Microbiol. *54*: 97–104.

Erasmus, J.A., J.A. De Wet and L. Prozesky. 1982. *Actinobacillus seminis* infection in a Walrich ram. J. S. Afr. Vet. Assoc. *53*: 129.

Ercolani, G.L. and M. Caldarola. 1972. *Pseudomonas ciccaronei* sp. n. agente de una maculatura fogliare del carrubo in Puglia. Phytopathol. Mediterr. *11*: 71–73.

Ercolani, G.L., D.J. Hagedorn, A. Kelman and R.E. Rand. 1974. Epiphytic survival of *Pseudomonas syringae* on hairy vetch in relation to epidemiology of bacterial brown spot of bean in Wisconsin. Phytopathology *64*: 1330–1339.

Ericsson, M., I. Golovliov, G. Sandström, A. Tärnvik and A. Sjöstedt. 1997. Characterization of the nucleotide sequence of the *groE* operon encoding heat shock proteins chaperone-60 and -10 of *Francisella tularensis* and determination of the T-cell response to the proteins in individuals vaccinated with *F. tularensis*. Infect. Immun. *65*: 1824–1829.

Eriksen, L., B. Aalbaek, P.S. Leifsson, A. Basse, T. Christiansen, E. Eriksen and R.B. Rimler. 1999. Hemorrhagic septicemia in fallow deer (*Dama dama*) caused by *Pasteurella multocida* subsp. *multocida*. J. Zoo Wildl. Med. *30*: 285–292.

Ertesvag, H., B. Doseth, B. Larsen, G. Skjakbraek and S. Valla. 1994. Cloning and expression of an *Azotobacter vinelandii* mannuronan C-5-epimerase gene. J. Bacteriol. *176*: 2846–2853.

Ertesvag, H., H.K. Hoidal, I.K. Hals, A. Rian, B. Doseth and S. Valla. 1995. A family of modular type mannuronan C5-epimerase genes controls alginate structure in *Azotobacter vinelandii*. Mol. Microbiol. *16*: 719–731.

Escande, F., A. Bailly, S. Bone and J. Lemozy. 1996. *Actinobacillus suis* infection after a pig bite. Lancet *378*: 888.

Escande, F., F. Grimont, P.A.D. Grimont and H. Bercovier. 1984. Deoxyribonucleic acid relatedness among strains of *Actinobacillus* spp. and *Pasteurella ureae*. Int. J. Syst. Bacteriol. *34*: 309–315.

Escande, F. and C. Lion. 1993. Epidemiology of human infections by *Pasteurella* and related groups in France. Zentrabl. Bakteriol. *279*: 131–139.

Escande, F., E. Vallee and F. Aubart. 1997. *Pasteurella caballi* infection following a horse bite. Zentrabl. Bakteriol. *285*: 440–444.

Escherich, T. 1885. Die Darmbakterien des Neugeborenen und Säuglingen. Fortschr Med. *3*: 515–528 547–554.

Eskola, J., H. Peltola, A.K. Takala, H. Kayhty, M. Hakulinen, V. Karanko, E. Kela, P. Rekola, P.R. Ronnberg, J.S. Samuelson, I.K. Gordon and P.H. Mäkela. 1987. Efficacy of *Haemophilus influenzae* type b polysaccharide-diphtheria toxoid conjugate vaccine in infancy. N. Engl. J. Med. *317*: 717–722.

Espinosa, M., B. Deodato, B. Cernigoi and A. Seijo. 1996. Cross reactivity between *Chlamydia psittaci* and *Acinetobacter* in indirect fluorescent antibody assays. Rev. Argent. Microbiol. *28*: 142–146.

Esselman, M.T. and P.V. Liu. 1961. Lecithinase production by gram-negative bacteria. J. Bacteriol. *81*: 939–945.

Esteve, C., M.C. Gutierrez and A. Ventosa. 1995a. *Aeromonas encheleia* sp. nov., isolated from European eels. Int. J. Syst. Bacteriol. *45*: 462–466.

Esteve, C., M.C. Gutierrez and A. Ventosa. 1995b. DNA relatedness among *Aeromonas allosaccharophila* strains and DNA hybridization groups of the genus *Aeromonas*. Int. J. Syst. Bacteriol. *45*: 390–391.

Euzéby, J.P. 1998. Taxonomic note: necessary correction of specific and subspecific epithets according to Rules 12c and 13b of the International Code of Nomenclature of Bacteria (1990 revision). Int. J. Syst. Bacteriol. *48*: 1073–1075.

Euzéby, J.P. 1999. Revised *Salmonella* nomenclature: designation of *Salmonella enterica* (ex Kauffmann and Edwards 1952) Le Minor and Popoff 1987 sp. nov., nom. rev. as the neotype species of the genus *Salmonella* Lignieres 1900 (Approved Lists 1980), rejection of the name *Salmonella choleraesuis* (Smith 1894) Weldin 1927 (Approved Lists 1980), and conservation of the name *Salmonella typhi* (Schroeter 1886) Warren and Scott 1930 (Approved Lists 1980). Request for an Opinion. Int. J. Syst. Bacteriol. *49*: 927–930.

Euzéby, J.P. and N.E. Boemare. 2000. The modern latin word rhabdus belongs to the feminine gender, inducing necessary corrections according to rules 65(2), 12c(1) and 13b of the bacteriological code (1990 revision). Int. J. Syst. Evol. Microbiol. *50*: 1691–1692.

Evans, D., R. Jones, P. Woodley and R. Robson. 1988. Further analysis of nitrogen fixation (*nif*) genes in *Azotobacter chroococcum* - identification and expression in *Klebsiella pneumoniae* of *nifS*, *nifV*, *nifM* and *nifB* genes and localization of *nifE/N*-like, *nifU*-like, *nifA*-like and *fixABC*-like genes. J. Gen. Microbiol. *134*: 931–942.

Evans, M.E., D.W. Gregory, W. Schaffner and Z.A. McGee. 1985. Tularemia: a 30-year experience with 88 cases. Medicine. *64*: 251–269.

Evans, N.M., D.D. Smith and A.J. Wicken. 1974. Haemin and nicotinamide adenine dinucleotide requirements of *Haemophilus influenzae* and *H. parainfluenzae*. J. Med. Microbiol. *7*: 359–365.

Evanylo, L.P., S. Kadis and J.R. Maudsley. 1984. Siderophore production by *Proteus mirabilis*. Can. J. Microbiol. *30*: 1046–1051.

Eveland, W.C. and J.E. Faber. 1953. Antigenic studies on a group of paracolon bacteria (32011 group). J. Infect Dis. *93*: 226–236.

Evelyn, T.P.T. and M.L. Kent. 1992. Salmonid rickettsial septicemia. *In* Kent (Editor), Diseases of Seawater Netpen-reared Salmonid Fishes in the Pacific Northwest, Canadian Special Publication of Fisheries and Aquatic Sciences 116, Department of Fisheries and Oceans, Nanaimo. pp. 18–19.

Every, D. 1982. Proteinase isoenzyme patterns of *Bacteroides nodosus*: distinction between ovine virulent isolates, ovine benign isolates and bovine isolates. J. Gen. Microbiol. *128*: 809–812.

Evins, G.M., D.N. Cameron, J.G. Wells, K.D. Greene, T. Popovic, S. Gionocerezo, I.K. Wachsmuth and R.V. Tauxe. 1995. The emerging diversity of the electrophoretic types of *Vibrio cholerae* in the western hemisphere. J. Infect. Dis. *172*: 173–179.

Ewbank, E. and H. Maraite. 1990. Conversion of methionine to phytotoxic 3-methylthiopropionic acid by *Xanthomonas campestris* pv. *manihotis*. J. Gen. Microbiol. *136*: 1185–1190.

Ewing, W.H. 1946. An additional *Shigella paradysenteriae* serotype. J. Bacteriol. *1*: 433–445.

Ewing, W.H. 1949. *Shigella* nomenclature. J. Bacteriol. *57*: 633–638.

Ewing, W.H. 1953. Serological relationships between *Shigella* and coliform cultures. J. Bacteriol. *66*: 333–340.

Ewing, W.H. 1960. Biochemical method for group differentiation, Center for Disease Control, Atlanta.

Ewing, W.H. 1962. The Tribe *Proteae*: Its nomenclature and taxonomy. Int. Bull. Bacteriol. Nomencl. Taxon. *12*: 93–102.

Ewing, W.H. 1963. An outline of nomenclature for the family *Enterobacteriaceae*. Int. Bull. Bacteriol. Nomencl. Taxon. *13*: 95–110.

Ewing, W.H. 1969. Excerpts from: an evaluation of the *Salmonella* problem, Center for Disease Control, Atlanta.

Ewing, W.H. 1971. Summary of four species. *In* Ewing (Editor), Biochemical Reactions of Shigella, U.S. Department of Health, Education and Welfare, Public Health Service, DHEW Publication No. (CDC) 72-8081, Atlanta. pp. V3–V12.

Ewing, W.H. 1986a. Edwards and Ewing's identification of *Enterobacteriaceae*, 4th Ed., Elsevier, New York.

Ewing, W.H. 1986b. The Genus *Escherichia*. *In* Edwards and Ewing (Editors), Edwards and Ewing's Identification of *Enterobacteriaceae*, 4th Ed., Elsevier Science Publishing Co., New York. pp. 93–122.

Ewing, W.H. 1986c. The genus *Shigella*. *In* Ewing (Editor), Edwards and Ewing's Identification of *Enterobacteriaceae*, 4th Ed., Elsevier Science Publishing Co., Inc., New York. pp. 135–172.

Ewing, W.H. 1986d. Media and reagents. *In* Ewing (Editor), Edwards and Ewing's Identification of *Enterobacteriaceae*, 4th Ed., Elsevier Science Publishing Co., Inc., New York. pp. 509–530.

Ewing, W.H. and B.R. Davis. 1972a. Biochemical characterization of *Citrobacter diversus* (Burkey) Werkman and Gillen and designation of the neotype strain. Int. J. Syst. Bacteriol. *22*: 12–18.

Ewing, W.H. and R.B. Davis. 1972b. Biochemical characterization of the species of *Proteus*. Public Health Lab. *30*: 46–57.

Ewing, W.H., B.R. Davis, M.A. Fife and E.F. Lessel. 1973. Biochemical characterization of *Serratia liquefaciens* (Grimes and Hennerty) Bascomb et al. (formerly *Enterobacter liquefaciens*) and *Serratia rubidaea* (Stapp) comb nov. and designation of type and neotype strains. Int. J. Syst. Bacteriol. *23*: 217–225.

Ewing, W.H., B.R. Davis and J.V. Sikes. 1972. Biochemical characterization of *Providencia*. Public Health Lab. *30*: 25–28.

Ewing, W.H., J.J. Farmer and D.J. Brenner. 1980. Proposal of *Enterobacteriaceae* nom. rev. to replace *Enterobacteriaceae* Rahn 1937, Nom. fem. cons Opin 15, Jud. Comm. 1958, which lost standing in nomenclature on January 1, 1980. Int. J. Syst. Bacteriol. *30*: 674–675.

Ewing, W.H. and M.A. Fife. 1968. *Enterobacter hafniae* (the *Hafnia* group). Int. J. Syst. Bacteriol. *18*: 263–271.

Ewing, W.H. and M.A. Fife. 1972. *Enterobacter agglomerans* (Beijerinck) comb nov. (the Herbicola-Lathyri bacteria). Int. J. Syst. Bacteriol. *22*: 4–11.

Ewing, W.H. and M.C. Hucks. 1952. Four new provisional serotypes of *Shigella*. J. Immunol. *69*: 575–580.

Ewing, W.H., M.C. Hucks and M.W. Taylor. 1951. Provisional *Shigella boydii* 9. Public Health Rep. *66*: 1579–1586.

Ewing, W.H., M.C. Hucks and M.W. Taylor. 1952a. Interrelationship of certain *Shigella* and *Escherichia* cultures. J. Bacteriol. *63*: 319–325.

Ewing, W.H., R. Hugh and J.C. Johnson. 1961. Studies on the *Aeromonas* group, U.S. Department of Health, Education and Welfare, Communicable Disease Center, Atlanta.

Ewing, W.H. and J.G. Johnson. 1960. The differentiation of *Aeromonas* and C27 strains from *Enterobacteriaceae*. Int. Bull. Bacteriol. Nomencl. Taxon. *10*: 223–230.

Ewing, W.H., A.C. McWhorter, M.R. Escobar and A.H. Lubin. 1965. *Edwardsiella*, a new genus of *Enterobacteriaceae* based on a new species, *E. tarda*. Int. Bull. Bacteriol. Nomen. Taxon. *15*: 33–38.

Ewing, W.H., R.W. Reavis and B.R. Davis. 1958. Provisional *Shigella* serotypes. Can. J. Microbiol. *4*: 89–107.

Ewing, W.H., A.J. Ross, D.J. Brenner and G.R. Fanning. 1978. *Yersinia ruckeri* sp. nov., the redmouth (RM) bacterium. Int. J. Syst. Bacteriol. *28*: 37–44.

Ewing, W.H., K.E. Tanner and D.A. Dennard. 1954. The Providence group: an intermediate group of enteric bacteria. J. Infect. Dis. *94*: 134–140.

Ewing, W.H. and M.W. Taylor. 1951. Two provisional *Shigella boydii* serotypes. Public Health Rep. *66*: 1327–1331.

Ewing, W.H., J. Vandepitte, A. Fain and M. Schoetter. 1952b. Provisional *Shigella dysenteriae* 8. Ann. Soc. Belge. Med. Trop. *32*: 585–590.

Expert, D., C. Enard and C. Masclaux. 1996. The role of iron uptake in plant host-pathogen interactions. Trends Microbiol. *4*: 232–237.

Expert, D. and A. Toussaint. 1985. Bacteriocin-resistant mutants of *Erwinia chrysanthemi*: possible involvement of iron acquisition in phytopathogenicity. J. Bacteriol. *163*: 221–227.

Eykyn, S. and I. Phillips. 1978. Carbon dioxide-dependent *Escherichia coli*. Br. Med. J. *1*: 576.

Eyre, J.W. 1900. A clinical and bacteriological study of diplobacillary conjunctivitis. J. Pathol. Bacteriol. *6*: 1–13.

Ezavin, M., R. Vazquez and J. Benitez. 1992. Detection of a plasmid of *Xanthomonas campestris* pv. *vasculorum*. Proteccion de Plantas. *2*: 15–20.

Ezura, Y., M. Kawabata, H. Miyashita and T. Kimura. 1988a. Changes of bacterial population in the nursery tanks for the forced cultivation of Makonbu *Laminaria japonica*. Nippon Suisan Gakkaishi. *54*: 655–663.

Ezura, Y., H. Yamamoto and T. Kimura. 1988b. Isolation of a marine bacterium that produces red spots on the culture bed of Makonbu *Laminaria japonica* cultivation. Nippon Suisan Gakkaishi. *54*: 665–672.

Faden, H., L. Duffy, A. Williams, D.A. Krystofik and J. Wolf. 1995. Epidemiology of nasopharyngeal colonization with nontypeable *Haemophilus influenzae* in the first 2 years of life. J. Infect. Dis. *172*: 132–135.

Fader, R.C., K. Gondesen, B. Tolley, D.G. Ritchie and P. Moller. 1988. Evidence that *in vitro* adherence of *Klebsiella pneumoniae* to ciliated hamster tracheal cells is mediated by type 1 fimbriae. Infect. Immun. *56*: 3011–3013.

Fahy, P.C. 1981. The taxonomy of the bacterial plant pathogens of mushroom culture. *In* Mushroom Science XI. Proceedings of the Eleventh International Scientific Congress on the Cultivation of Edible Fungi, Australia 293–311.

Faibich, M.M. 1959. The problem of increasing the antigenicity of live tularemia vaccine. J. Microbiol. Epidemiol. Immunobiol. *30*: 23–27.

Fainstein, V., R.L. Hopfer, K. Mills and G.P. Bodey. 1982a. Colonization by or diarrhea due to *Kluyvera* species. J. Infect. Dis. *145*: 127.

Fainstein, V., S. Weaver and G.P. Bodey. 1982b. *In vitro* susceptibilities of *Aeromonas hydrophila* against new antibiotics. Antimicrob. Agents Chemother. *22*: 513–514.

Falkenstein, H., W. Zeller and K. Geider. 1989. The 29 kb plasmid, common in strains of *Erwinia amylovora*, modulates development of fireblight symptoms. J. Gen. Microbiol. *135*: 2643–2650.

Falkow, S., I.R. Ryman and O. Washington. 1962. Deoxyribonucleic acid base composition of *Proteus* and *Providence* organisms. J. Bacteriol. *83*: 1318–1321.

Falkow, S., H. Schneider, L.S. Baron and S.B. Formal. 1963. Virulence of *Escherichia–Shigella* genetic hybrids for the guinea pig. J. Bacteriol. *86*: 1251–1258.

Fallik, E., Y.K. Chan and R.L. Robson. 1991. Detection of alternative nitrogenases in aerobic Gram-negative nitrogen-fixing bacteria. J. Bacteriol. *173*: 365–371.

Fallik, E., P.G. Hartel and R.L. Robson. 1993. Presence of a vanadium nitrogenase in *Azotobacter paspali*. Appl. Environ. Microbiol. *59*: 1883–1886.

Fallik, E. and R.L. Robson. 1990. Completed sequence of the region encoding the structural gene for the vanadium nitrogenase of *Azotobacter chroococcum*. Nucleic Acids Res. *18*: 4616.

Faltus, I.I. and Y.G. Kishko. 1980. Certain biological and physicochemical properties of the virulent and temperate phages of *Erwinia carotovora* (Russian). Mikrobiol. Zh. *42*: 226–231.

Famintzin, A. 1892. Eine neue Bacterienform: *Nevskia ramosa*. Bull. Acad. Sci. St. Petersb. New Ser. 2. *34*: 481–486.

Fang, C.T., H.C. Ren, T.Y. Chen, Y.K. Chu, H.C. Faan and S.C. Wu. 1957. A comparison of the rice bacterial leaf blight organism with the bacterial leaf streak organism of rice and *Leersia hexandra* Swartz. Acta Pbytopathol. Sinica. *3*: 99–124.

Fanning, G.R., J.J. Farmer, III, J.N. Parker, G.P. Huntley-Carter and D.J. Brenner. 1979. *Kluyvera*: A new genus in *Enterobacteriaceae*. Abstr. Ann Meet. Amer. Soc. Microbiol., p. 100.

Fanning, G.R., F.W. Hickman-Brenner, J.J. Farmer, III and D.J. Brenner. 1985. DNA relatedness and phenotypic analysis of the genus *Aeromonas*. Abstracts of the 85th General Meeting of the American Society for Microbiology, Amcrican Society for Microbiology. p. 319.

Farmer, J.J., III 1974. Lysotypie de *Serratia marcescens*. Arch. Roum. Pathol. Exp. Microbiol. *34*: 189.

Farmer, J.J., III 1980. Revival of the name *Vibrio vulnificus*. Int. J. Syst. Bacteriol. *30*: 656.

Farmer, J.J., III 1984a. Genus *Obesumbacterium* Shimwell 1963, 759[AL]. *In* Kreig and Holt (Editors), Bergey's Manual of Systematic Bacteriology, Vol. 1, Williams & Wilkins, Baltimore. 506–509.

Farmer, J.J., III 1984b. Other genera of the family *Enterobacteriaceae*. *In* Kreig and Holt (Editors), Bergey's Manual of Systematic Bacteriology, 1st Ed., Vol. 1, The Williams & Wilkins Co, Baltimore. pp. 506–516.

Farmer, J.J., III 1992. The family *Vibrionaceae*. *In* Balows, Trüper, Dworkin, Harder and Schleifer (Editors), The Prokaryotes. A Handbook on the Biology of Bacteria: Ecophysiology, Isolation, Identification, Applications, 2nd Ed., Vol. 3, Springer-Verlag, New York. pp. 2938–2951.

Farmer, J.J., III 1995. *Enterobacteriaceae*: introduction and identification. *In* Murray, Baron, Pfaller, Tenover and Yolken (Editors), Manual of Clinical Microbiology, American Society for Microbiology, Washington, D.C. pp. 438–449.

Farmer, J.J., III 1999. *Enterobacteriaceae*: introduction and identification. *In* Murray, Baron, Pfaller, Tenover and Yolken (Editors), Manual of Clinical Microbiology, 7th Ed., American Society for Microbiology, Washington D.C. 442–458.

Farmer, J.J., III 2003. *Enterobacteriaceae*: introduction and identification. *In* Murray, Baron, Pfaller, Jorgensen and Yolken (Editors), Manual of Clinical Microbiology, 8th Ed., American Society for Microbiology, Washington, D.C. pp. 636–653.

Farmer, J.J., III, M.J. Arduino and F.W. Hickman-Brenner. 1992. The genera *Aeromonas* and *Plesiomonas*. *In* Balows, Trüper, Dworkin, Harder and Schleifer (Editors), The Prokaryotes: a handbook on the biology of bacteria: ecophysiology, isolation, identification, applications, 2nd Ed., Vol. 3, Springer-Verlag, New York. pp. 3012–3045.

Farmer, J.J., III, M.A. Asbury, F.W. Hickman, D.J. Brenner and the *Enterobacteriaceae* study group. 1980a. *Enterobacter sakazakii*: A new species of *Enterobacteriaceae* isolated from clinical specimens. Int. J. Syst. Bacteriol. *30*: 569–584.

Farmer, J.J., III and D.J. Brenner. 1977. Concept of bacterial species. *In* Hoadley and Dutka (Editors), Bacterial Indicators/Health Hazards Associated with Water, Amer. Soc. Testing and Materials, Philadelphia. pp. 37–47.

Farmer, J.J., III, D.J. Brenner and W.A. Clark. 1976. Proposal to conserve the specific epithet *tarda* over the specific epithet *anguillimortiferum* in the name of the organism presently known as *Edwardsiella tarda*: request for an opinion. Int. J. Syst. Bacteriol. *26*: 293–294.

Farmer, J.J., III, B.R. Davis, F.W. Hickman-Brenner, A. McWhorter, G.P. Huntley-Carter, M.A. Asbury, C. Riddle, H.G. Wathen-Grady, C. Elias, G.R. Fanning, A.G. Steigerwalt, C.M. O'Hara, G.K. Morris, P.B. Smith and D.J. Brenner. 1985a. Biochemical identification of new species and biogroups of *Enterobacteriaceae* isolated from clinical specimens. J. Clin. Microbiol. *21*: 46–76.

Farmer, J.J., III, G.R. Fanning, B.R. Davis, C.M. O'Hara, C. Riddle, F.W. Hickman-Brenner, M.A. Asbury, V.A. Lowery, III and D.J. Brenner. 1985b. *Escherichia fergusonii* sp. nov. and *Enterobacter taylorae* sp. nov., 2 enterobacteriaceae isolated from clinical specimens. J. Clin. Microbiol. *21*: 77–81.

Farmer, J.J., III, G.R. Fanning, B.R. Davis, C.M. O'Hara, C. Riddle, F.W. Hickman-Brenner, M.A. Asbury, V.A. Lowery, III and D.J. Brenner. 1985c. *In* Validation of the pulication of new names and new combinations previously effectively published outside the IJSB. List No. 17. Int. J. Syst. Bacteriol. *35*: 223–225.

Farmer, J.J., III, G.R. Fanning, G.P. Huntley-Carter, B. Holmes, F.W. Hickman, C. Richard and D.J. Brenner. 1981a. *Kluyvera*: a new (redefined) genus in the family *Enterobacteriaceae*. Identification of *Kluyvera ascorbata* sp. nov. and *Kluyvera cryocrescens* sp. nov. in clinical specimens. J. Clin. Microbiol. *13*: 919–933.

Farmer, J.J., III, G.R. Fanning, G.P. Huntley-Carter, B. Holmes, F.W. Hickman, C. Richard and D.J. Brenner. 1981b. *In* Validation of the publication of new names and new combinations previously effectively published outside the IJSB. List No. 7. Int. J. Syst. Bacteriol. *31*: 382–383.

Farmer, J.J., III, P.A.D. Grimont, F. Grimont and M.A. Asbury. 1980b. *Cedecea*: A new genus in *Enterobacteriaceae*. Abstract C 123. Abstr. Ann. Meet. Am. Soc. Microbiol, 295.

Farmer, J.J., III, F.W. Hickman, D.J. Brenner, M. Schreiber and D.G. Rickenbach. 1977. Unusual *Enterobacteriaceae*: "*Proteus rettgeri*" that "change" into *Providencia stuartii*. J. Clin. Microbiol. *6*: 373–378.

Farmer, J.J., III and F.W. Hickman-Brenner. 1992. The genera *Vibrio* and *Photobacterium*. *In* Balows, Trüper, Dworkin, Harder and Schleifer (Editors), The Prokaryotes. A Handbook on the Biology of Bacteria: Ecophysiology, Isolation, Identification, Applications, 2nd ed., Vol. 3, Springer-Verlag, New York. 2952–3011.

Farmer, J.J., III, F.W. Hickman-Brenner, G.R. Fanning, M.J. Arduino and D.J. Brenner. 1986. Analysis of *Aeromonas* and *Plesiomonas* by DNA–DNA hybridization and phenotype. 1st International *Aeromonas-/Plesiomonas* Symposium, Manchester, England. p. 1.

Farmer, J.J., III, F.W. Hickman-Brenner, G.R. Fanning, C.M. Gordon and D.J. Brenner. 1988. Characterization of *Vibrio metschnikovii* and *Vibrio gazogenes* by DNA–DNA hybridization and phenotype. J. Clin. Microbiol. *26*: 1993–2000.

Farmer, J.J., III, F.W. Hickman-Brenner and M.T. Kelly. 1985d. *Vibrio*. *In* Lemmette, Balows, Hausler and Shadomy (Editors), Manual of Clinical Microbiology, American Society for Microbiology, Washington, D.C. pp. 282–310.

Farmer, J.J., III, J.H. Jorgensen, P.A.D. Grimont, R.J. Akhurst, E. Ageron, G.V. Pierce, J.A. Smith, G.P. Carter, K.L. Wilson and F.W. Hickman-Brenner. 1989. *Xenorhabdus luminescens* (DNA hybridization group 5) from human clinical specimens. J. Clin. Microbiol. *27*: 1594–1600.

Farmer, J.J., III, N.K. Sheth, J.A. Hudzinski, H.D. Rose and M.F. Asbury. 1982. Bacteremia due to *Cedecea neteri* sp. nov. J. Clin. Microbiol. *16*: 775–778.

Farmer, J.J., III, N.K. Sheth, J.A. Hudzinski, H.D. Rose and M.F. Asbury. 1983. *In* Validation of the publication of new names and new combinations previously effectively published outside the IJSB. List No. 10. Int. J. Syst. Bacteriol. *33*: 438–440.

Farmer, J.J., III, F. Silva and D.R. Williams. 1973. Isolation of *Serratia marcescens* on deoxyribonuclease-toluidine blue-cephalothin agar. Appl. Microbiol. *25*: 151–152.

Farquhar, G.J. and W.C. Boyle. 1971. Occurrence of filamentous microorganisms in activated sludge. J. Water Pollut. Control Fed. *43*: 779–798.

Faruque, S.M., A.R. Abdul Alim, S.K. Roy, F. Khan, G.B. Nair, R.B. Sack and M.J. Albert. 1994. Molecular analysis of rRNA and cholera toxin genes carried by the new epidemic strain of toxigenic *Vibrio cholerae* O139 synonym Bengal. J. Clin. Microbiol. *32*: 1050–1053.

Faruque, S.M., K. Haider, M.M. Rahman, A.R. Alim, Q.S. Ahmad, M.J. Albert and R.B. Sack. 1992. Differentiation of *Shigella flexneri* strains by rRNA gene restriction patterns. J. Clin. Microbiol. *30*: 2996–2999.

Fasano, A., F.R. Noriega, D.R.J. Maneval, S. Chanasongcram, R. Russell, S. Guandalini and M.M. Levine. 1995. *Shigella* enterotoxin 1: an enterotoxin of *Shigella flexneri* 2a active in rabbit small intestine *in vivo* and *in vitro*. J. Clin. Invest. *95*: 2853–2861.

Fass, R.J. and J. Barnishan. 1981. *In vitro* susceptibilies of *Aeromonas hydrophila* to 32 antimicrobial agents. Antimicrob. Agents Chemother. *19*: 357–358.

Faust, L. and R.S. Wolfe. 1961. Enrichment and cultivation of *Beggiatoa alba*. J. Bacteriol. *81*: 99–106.

Faust, U., P. Prave and D.A. Sukatsch. 1977. Continuous biomass production from methanol by *Methylomonas clara*. J. Ferment. Technol. *55*: 609–614.

Favero, M.S., L.A. Carson, W.W. Bond and N.J. Petersen. 1971. *Pseudomonas aeruginosa*: growth in distilled water from hospitals. Science (Wash. D. C.) *173*: 836–838.

Fazal, B.A., J.E. Justman, G.S. Turett and E.E. Telzak. 1997. Community-acquired *Hafnia alvei* infection. Clin. Infect. Dis. *24*: 527–528.

Federici, B.A. 1980. Reproduction and morphogenesis of *Rickettsiella chironomi*, an unusual intracellular procaryotic parasite of midge larvae. J. Bacteriol. *143*: 995–1002.

Federici, B.A., E.I. Hazard and D.W. Anthony. 1974. *Rickettsia*-like organism causing disease in a crangonid amphipod from Florida. Appl. Microbiol. *28*: 885–886.

Fedi, S., D. Brazil, D.N. Dowling and F. O'Gara. 1996. Construction of a modified mini-Tn5 *lacZY* non-antibiotic marker cassette: Ecological evaluation of a *lacZY* marked *Pseudomonas* strain in the sugarbeet rhizosphere. FEMS Microbiol. Lett. *135*: 251–257.

Fedorko, D.P. 1993. The medicinal leech as a source of infection. Clin. Microbiol. Newsl. *15*: 164–165.

Feeley, J.C., R.J. Gibson, G.W. Gorman, N.C. Langford, J.K. Rasheed, D.C. Mackel and W.B. Baine. 1979. CYE agar: a primary isolation medium for *Legionella pneumophila*. J. Clin. Microbiol. *10*: 436–441.

Feeley, J.C., G.W. Gorman, R.E. Weaver, D.C. Mackel and H.W. Smith. 1978. Primary isolation media for the Legionnaires' disease bacterium. J. Clin Microbiol. *8*: 320–325.

Fegan, N., P.J. Blackall and J.L. Pahoff. 1995. Phenotypic characterisation of *Pasteurella multocida* isolates from Australian poultry. Vet. Microbiol. *47*: 281–286.

Fein, J.E. and R.A. MacLeod. 1975. Characterization of neutral amino acid transport in a marine pseudomonad. J. Bacteriol. *124*: 1177–1190.

Feist, C.F. and G.D. Hegeman. 1969. Phenol and benzoate metabolism by *Pseudomonas putida*: regulation of tangential pathways. J. Bacteriol. *100*: 869–877.

Feistner, G., H. Korth, H. Budzikiewicz and G. Pulverer. 1984. Rubrifacine from *Erwinia rubrifaciens*. Curr. Microbiol. *10*: 169–172.

Feistner, G.J., A. Mavridis and K. Rudolph. 1997. Metabolites of *Erwinia*.15. Proferrorosamines and phytopathogenicity in *Erwinia* spp. Biometals *10*: 1–10.

Fekete, F.A., R.A. Lanzi, J.B. Beaulieu, D.C. Longcope, A.W. Sulya, R.N. Hayes and G.A. Mabbott. 1989. Isolation and preliminary characterization of hydroxamic acids formed by nitrogen-fixing *Azotobacter chroococcum* B-8. Appl. Environ. Microbiol. *55*: 298–305.

Felici, A. and G. Amicosante. 1995. Kinetic analysis of extension of substrate specificity with *Xanthomonas maltophilia*, *Aeromonas hydrophila*, and *Bacillus cereus* metallo-beta-lactamases. Antimicrob. Agents Chemother. *39*: 192–199.

Felix, A. and B.R. Callow. 1943. Typing of paratyphoid B bacilli by means of Vi bacteriophage. Br. Med. J. *2*: 127.

Feller, G., T. Lonhienne, C. Deroanne, C. Libioulle, J. Van Beeumen and C. Gerday. 1992. Purification, characterization, and nucleotide sequence of the thermolabile alpha-amylase from the antarctic psychrotroph *Alteromonas haloplanctis* A23. J. Biol. Chem. *267*: 5217–5221.

Felsenstein, J. 1993. PHYLIP (Phylogeny Inference Package), Version 3.5c, ed. Department of Genetics, University of Washington, Seattle.

Fendrich, C. 1988. *Halovibrio variabilis* gen. nov. sp. nov., *Pseudomonas halophila* sp. nov. and a new halophilic aerobic coccoid eubacterium from Great Salt Lake, Utah, USA. Syst. Appl. Microbiol. *11*: 36–43.

Fendrich, C. 1989. *In* Validation of the publication of new names and new combinations previously effectively published outside the IJSB. List No. 29. Int. J. Syst. Bacteriol. *39*: 205–206.

Fenical, W. 1993. Chemical studies of marine bacteria: developing a new resource. Chem. Rev. *93*: 1673–1683.

Fenollar, F., P.E. Fournier, M.P. Carrieri, G. Habib, T. Messana and D. Raoult. 2001. Risks factors and prevention of Q fever endocarditis. Clin. Infect. Dis. *33*: 312–316.

Fenstersheib, M.D., M. Miller, C. Diggins, S. Liska, L. Detwiler, S.B. Werner, D. Lindquist, W.L. Thacker and R.R. Benson. 1990. Outbreak of Pontiac fever due to *Legionella anisa*. Lancet *336*: 35–37.

Ferber, D.M. and R.R. Brubaker. 1981. Plasmids in *Yersinia pestis*. Infect. Immun. *31*: 839–841.

Ferdelman, T.G., C. Lee, S. Pantoja, J. Harder, B.M. Bebout and H. Fossing. 1997. Sulfate reduction and methanogenesis in a *Thioploca*-dominated sediment off the coast of Chile. Geochim. Cosmochim. Acta. *61*: 3065–3079.

Ferguson, G., D.R. Pollard, J.M. Robertson, G.O.P. Doherty, N.B. Haynes, D.W. Mathieson and W.B. Whalley. 1980. The bacterial pigment from *Pseudomonas lemonnieri* 1. Structure of a degradation product, 3-n-octanamidopyridine-2,5,6-trione, by X-ray crystallography. J. Chem. Soc.-Perkin Trans. 1. *8*: 1782–1787.

Ferguson, R., C. Feeney and V.A. Chirurgi. 1993. *Enterobacter agglomerans*-associated cotton fever. Arch. Intern. Med. *153*: 2381–2382.

Ferguson, W.W. and N.D. Henderson. 1947. Description of strain C27: a motile organism with the major antigen of *Shigella sonnei* phase I. J. Bacteriol. *54*: 179–181.

Fernandez, J.L., W.J. Simpson and T.M. Dowhanick. 1993. Enumeration of *Obesumbacterium proteus* in brewery yeasts and characterization of isolated strains using biolog Gn microplates and protein fingerprinting. Lett. Appl. Microbiol. *17*: 292–296.

Fernández-Borrero, O. and S. López-Duque. 1970. Pudricion acuosa del suedo tallo del plátana (*Musa paradisiaca*) causada por *Erwinia paradisiaca*, n. sp. CENICAFE. *21*: 3–44.

Fernandez-Castillo, R., C. Vargas, J.J. Nieto, A. Ventosa and F. Ruiz-Berraquero. 1992. Characterization of a plasmid from moderately halophilic eubacteria. J. Gen. Microbiol. *138*: 1133–1137.

Fernandez-Linares, L., M. Acquaviva, J.C. Bertrand and M. Gauthier. 1996a. Effect of sodium chloride concentration on growth and degradation of eicosane by the marine halotolerant bacterium *Marinobacter hydrocarbonoclasticus*. Syst. Appl. Microbiol. *19*: 113–121.

Fernandez-Linares, L., R. Faure, J.C. Bertrand and M. Gauthier. 1996b. Ectoine as the predominant osmolyte in the marine bacterium *Marinobacter hydrocarbonoclasticus* grown on eicosane at high salinities. Lett. Appl. Microbiol. *22*: 169–172.

Fernández-Martínez, J., M.J. Pujalte, J. García-Martínez, M. Mata, E. Garay and F. Rodríguez-Valera. 2003. Description of *Alcanivorax venustensis* sp. nov. and reclassification of *Fundibacter jadensis* DSM 12178(T) (Bruns and Berthe-Corti 1999) as *Alcanivorax jadensis* comb. nov., members of the emended genus *Alcanivorax*. Int. J. Syst. Evol. Microbiol. *53*: 331–338.

Ferragut, C., D. Izard, F. Gavini, K. Kersters, J. Deley and H. Leclerc. 1983. *Klebsiella trevisanii* a new species from water and soil. Int. J. Syst. Bacteriol. *33*: 133–142.

Ferragut, C., D. Izard, F. Gavini, B. Lefebvre and H. Leclerc. 1981. *Buttiauxella*, a new genus of the family *Enterobacteriaceae*. Zentbl. Bakteriol. Mikrobiol. Hyg. 1 Abt. Orig. C *2*: 33–44.

Ferragut, C., D. Izard, F. Gavini, B. Lefebvre and H. Leclerc. 1982. *In* Validation of the publication of new names and new combinations previously effectively published outside the IJSB. List No. 8. Int. J. Syst. Bacteriol. *32*: 266–268.

Ferragut, C., K. Kersters and J. Deley. 1989. Protein electrophoretic and DNA homology analysis of *Klebsiella* strains. Syst. Appl. Microbiol. *11*: 121–127.

Ferraris, T. 1926. Trattato di Patologia e Terapia Vegetale, 3rd ed., Hoepli, Milan.

Ferreira, H., F.J.A. Barrientos, R.L. Baldini and Y.B. Rosato. 1995. Electrotransformation of three pathovars of *Xanthomonas campestris*. Appl. Microbiol. Biotechnol. *43*: 651–655.

Fett, W.F., C. Wijey and E.R. Lifson. 1992. Occurrence of alginate gene-sequences among members of the pseudomonad ribosomal-RNA homology groups I - IV. FEMS Microbiol. Lett. *99*: 151–157.

Fetzner, S. and F. Lingens. 1994. Bacterial dehalogenases - biochemistry, genetics, and biotechnological applications. Microbiol. Rev. *58*: 641–685.

Fewson, C.A. 1991. Metabolism of aromatic compounds by *Acinetobacter*. *In* Towner and Bergogne-Bérézin (Editors), The Biology of *Acinetobacter*. Taxonomy, Clinical Importance, Molecular Biology, Physiology, Industrial Relevance, Plenum Press, New York. pp. 351–390.

Fialho, A.M., N.A. Zielinski, W.F. Fett, A.M. Chakrabarty and A. Berry. 1990. Distribution of alginate gene-sequences in the *Pseudomonas* ribosomal-RNA homology group I - *Azomonas-Azotobacter* lineage of superfamily-B procaryotes. Appl. Environ. Microbiol. *56*: 436–443.

Field, B., M. Uwaydah, L. Kunz and M. Swarts. 1967. The so-called "paracolon" bacteria. A bacteriologic and clinical reappraisal. Am. J. Med. *42*: 89–106.

Fields, B.S. 1993. *Legionella* and protozoa: interaction of a pathogen and its natural host. *In* Barbaree, Breiman and Dufour (Editors), *Legionella*: Current Status and Emerging Perspectives, American Society for Microbiology, Washington, D.C. pp. 1299–1136.

Fields, B.S., J.M. Barbaree, G.N. Sanden and W.E. Morrill. 1990. Virulence of a *Legionella anisa* strain associated with Pontiac fever: an evaluation using protozoan, cell culture, and guinea pig models. Infect. Immun. *58*: 3139–3142.

Fields, P.I., K. Blom, H.J. Hughes, L.O. Helsel, P. Feng and B. Swaminathan. 1997. Molecular characterization of the gene encoding H antigen in *Escherichia coli* and development of a PCR-restriction fragment length polymorphism test for identification of *E. coli* O157:H7 and O157:NM. J. Clin. Microbiol. *35*: 1066–1070.

Figueras, M.J., N. Borrell, A.J. Martinez-Murcia and J. Guarro. 1999. Misclassification of clinical *Aeromonas* with conventional biochemical tests. 6th International *Aeromonas/Plesiomonas* Symposium, Chicago, Illinois. p. 17.

Figueras, M.J., L. Soler, M.R. Chacon, J. Guarro and A.J. Martinez-Murcia. 2000. Extended method for discrimination of *Aeromonas* spp. by 16S rDNA RFLP analysis. Int. J. Syst. Evol. Microbiol. *50*: 2069–2073.

Filichkin, S.A., S. Brumfield, T.P. Filichkin and M.J. Young. 1997. In vitro interactions of the aphid endosymbiotic SymL chaperonin with barley yellow dwarf virus. J. Virol. *71*: 569–577.

Finelli, L., D. Swerdlow, K. Mertz, H. Ragazzoni and K. Spitalny. 1992. Outbreak of cholera associated with crab brought from an area with epidemic disease. J. Infect. Dis. *166*: 1433–1435.

Finke, A., B. Jann and K. Jann. 1990. CMP-KDO-synthetase activity in *Escherichia coli* expressing capsular polysaccharides. FEMS Microbiol. Lett. *57*: 129–133.

Finkelstein, R.A., M. Boesman-Finkelstein, Y. Chang and C.C. Hase. 1992. *Vibrio cholerae* hemagglutinin/protease, colonial variation, virulence, and detachment. Infect. Immun. *60*: 472–478.

Finkelstein, R.A., M. Boesman-Finkelstein, D.K. Sengupta, W.J. Page, C.M. Stanley and T.E. Phillips. 1997. Colonial opacity variations among the choleragenic vibrios. Microbiology *143*: 23–34.

Finkmann, W., K. Altendorf, E. Stackebrandt and A. Lipski. 2000. Characterization of N₂O-producing *Xanthomonas*-like isolates from biofilters as *Stenotrophomonas nitritireducens* sp. nov., *Luteimonas mephitis* gen. nov., sp. nov and *Pseudoxanthomonas broegbernensis* gen. nov., sp. nov. Int. J. Syst. Evol. Microbiol. *50*: 273–282.

Firehammer, B.D. 1987. Inhibition of growth and swarming of *Proteus mirabilis* and *Proteus mirabilis* by triclosan. J. Clin. Microbiol. *25*: 1312–1313.

Firrao, G. and C. Bazzi. 1994. Specific identification of *Xylella fastidiosa* using the polymerase chain reaction. Phytopathol. Mediterr. *33*: 90–92.

Firsov, N.N., I.I. Cherniad'ev, R.N. Ivanovskii, E.N. Kondrat'eva and N.V. Vdovina. 1974. Carbon dioxide assimilation pathways of *Ectothiorhodospira shaposhnikovii*. Mikrobiologiia *43*: 214–219.

Firsov, N.N. and R.N. Ivanovskii. 1974. Metabolism of propionate in *Ectothiorhodospira shaposhnikovii*. Mikrobiologiia *43*: 400–405.

Firsov, N.N. and R.N. Ivanovskii. 1975. Acetate photometabolism in *Ectothiorhodospira shaposhnikovii*. Mikrobiologiia *44*: 197201.

Fischer, B.E. 1986. Localization of hydrolytic enzymes in *Acinetobacter calcoaceticus*. J. Basic Microbiol. *26*: 9–14.

Fischer, B., R. Claus and H.P. Kleber. 1984. Isolation and characterization of the outer membrane of *Acinetobacter calcoaceticus*. J. Biotechnol. *1*: 111–118.

Fischer, K., T. Chakraborty, H. Hof, T. Kirchner and O. Wamsler. 1988. Pseudoappendicitis caused by *Plesiomonas shigelloides*. J. Clin. Microbiol. *26*: 2675–2677.

Fischer, R., J.L. Penner, G. Zurinaga, C. Riddle, W. Sämisch and D.J. Brenner. 1989. Usefulness of trehalose fermentation and L-glutamic acid decarboxylation for identification of biochemically aberrant *Providencia stuartii* strains. J. Clin. Microbiol. *27*: 1969–1972.

Fischer-Le Saux, M., H. Mauleon, P. Constant, B. Brunel and N.E. Boemare. 1998. PCR-ribotyping of *Xenorhabdus* and *Photorhabdus* isolates from the Caribbean region in relation to the taxonomy and geographic distribution of their nematode hosts. Appl. Environ. Microbiol. *64*: 4246–4254.

Fischer-Le Saux, M., V. Viallard, B. Brunel, P. Normand and N.E. Boemare. 1999. Polyphasic classification of the genus *Photorhabdus* and proposal of new taxa: *P. luminescens* subsp. *luminescens* subsp. nov., *P. luminescens* subsp. *akhurstii* subsp. nov., *P. luminescens* subsp. *laumondii* subsp. nov., *P. temperata* sp. nov., *P. temperata* subsp. *temperata* subsp. nov. and *P. asymbiotica* sp. nov. Int. J. Syst. Bacteriol. *49*: 1645–1656.

Fischer-Romero, C., B.J. Tindall and F. Jüttner. 1996. *Tolumonas auensis* gen. nov., sp. nov., a toluene-producing bacterium from anoxic sediments of a freshwater lake. Int. J. Syst. Bacteriol. *46*: 183–188.

Fishbein, D.B. and D. Raoult. 1992. A cluster of *Coxiella burnetii* infections associated with exposure to vaccinated goats and their unpasteurized dairy products. Am. J. Trop. Med. Hyg. *47*: 35–40.

Fisher, C.R., J.M. Brooks, J.S. Vodenichar, J.M. Zande, J.J. Childress and R.A. Burke, Jr.. 1993. The co-occurrence of methanotrophic and chemoautotrophic sulfur-oxidizing bacterial symbionts in a deep-sea mussel. Mar. Ecol. *14*: 277–289.

Fisher, W.S. 1978. Microbial diseases of cultured lobsters: a review. Aquaculture *14*: 115–140.

Fixter, L.M. and M.K. Sherwani. 1991. Energy reserves in *Acinetobacter*. *In* Towner and Bergogne-Bérézin (Editors), The Biology of *Acinetobacter*. Taxonomy, Clinical Importance, Molecular Biology, Physiology, Industrial Relevance, Plenum Press, New York. pp. 273–294.

Flannery, W.L. and D.M. Kennedy. 1962. The nutrition of *Vibrio costicolus*. I. A simplified synthetic medium. Can. J. Microbiol. *8*: 923–928.

Flatauer, F.E. and M.A. Khan. 1978. Septic arthritis caused by *Enterobacter agglomerans*. Arch. Intern. Med. *138*: 788.

Fleck, J., M. Mock, R. Minck and J.M. Ghuysen. 1971. The cell envelope of *Proteus vulgaris* P18. Isolation and characterization of the peptidoglycan component. Biochim. Biophys. Acta *233*: 489–503.

Fleischmann, R.D., M.D. Adams, O. White, R.A. Clayton, E.F. Kirkness, A.R. Kerlavage, C.J. Bult, J.F. Tomb, B.A. Dougherty, J.M. Merrick, K. McKenney, G. Sutton, W. Fitzhugh, C. Fields, J.D. Gocayne, J. Scott, R. Shirley, L.I. Liu, A. Glodek, J.M. Kelley, J.F. Weidman, C.A. Phillips, T. Spriggs, E. Hedblom, M.D. Cotton, T.R. Utterback, M.C. Hanna, D.T. Nguyen, D.M. Saudek, R.C. Brandon, J.D. Fine, J.L. Fritchman, J.L. Fuhrmann, N.S.M. Geoghagen, C.L. Gnehm, L.A. McDonald, K.V. Small, C.M. Fraser, H.O. Smith and J.C. Venter. 1995. Whole-genome random sequencing and assembly of *Haemophilus influenzae*. Science (Wash. D. C.) *269*: 496–512.

Flesher, A.R. and R. Insel. 1978. Characteristics of lipopolysaccharide of *Haemophilus influenzae*. J. Infect. Dis. *138*: 719–730.

Flesher, A.R., I. Susumu, B.J. Mansheim and D.L. Kasper. 1979. The cell envelope of the Legionnaires' disease bacterium. Ann. Intern. Med. *90*: 628–630.

Fletcher, J.N., H.E. Embaye, B. Getty, R.M. Batt, C.A. Hart and J.R. Saunders. 1992. Novel invasion determinant of enteropathogenic *Escherichia coli* plasmid pLV501 encodes the ability to invade intestinal epithelial cells and HEp-2 cells. Infect. Immun. *60*: 2229–2236.

Fletcher, J.N., J.R. Saunders, R.M. Batt, H. Embaye, B. Getty and C.A. Hart. 1990. Attaching effacement of the rabbit enterocyte brush border is encoded on a single 96.5-kilobase-pair plasmid in an enteropathogenic *Escherichia coli* O111 strain. Infect. Immun. *58*: 1316–1322.

Fliermans, C.B., W.B. Cherry, L.H. Orrison, S.J. Smith, D.L. Tison and D.H. Pope. 1981. Ecological distribution of *Legionella pneumophila*. Appl. Environ Microbiol. *41*: 9–16.

Floderus, E., T. Pal, K. Karlsson and A.A. Lindberg. 1995. Identification of *Shigella* and enteroinvasive *Escherichia coli* strains by a virulence-specific, monoclonal antibody-based enzyme immunoassay. Eur. J. Clin. Microbiol. Infect. Dis. *14*: 111–117.

Flournoy, D.J., K.A. Belobraydic, S.L. Silberg, C.H. Lawrence and P.J. Guthrie. 1988. False positive *Legionella pneumophila* direct immunofluorescent monoclonal antibody test caused by *Bacillus cereus* spores. Diagn. Microbiol. Infect. Dis. *9*: 123–126.

Flügge, C. 1886. Die Microorganismen, F. C. W. Vogel, Leipzig.

Flynn, P. and A.K. Vidaver. 1995. *Xanthomonas campestris* pv. *asclepiadis*, pv. nov., causative agent of bacterial blight of milkweed (*Asclepias* spp.). Plant Dis. *79*: 1176–1180.

Fodor, E., E. Szallas, Z. Kiss, A. Fodor, L.I. Horvath, D.J. Chitwood and T. Farkas. 1997. Composition and biophysical properties of lipids in *Xenorhabdus nematophilus* and *Photorhabdus luminescens*, symbiotic bacteria associated with entomopathogenic nematodes. Appl. Environ. Microbiol. *63*: 2826–2831.

Fonnesbech Vogel, B., K. Venkateswaran, H. Christensen, E. Falsen, G. Christiansen and L. Gram. 2000. Polyphasic taxonomic approach in the description of *Alishewanella fetalis* gen. nov., sp. nov., isolated from a human foetus. Int. J. Syst. Evol. Microbiol. *50*: 1133–1142.

Fontaine, A., J. Arondel and P. Sansonetti. 1988. Role of Shiga toxin in the pathogenesis of bacillary dysentery, studied by using a Tox-mutant of *Shigella dysenteriae* 1. Infect. Immun. *56*: 3099–3109.

Formal, S.B., P.J. Gemski, L.S. Baron and E.H. LaBrec. 1970. Genetic transfer of *Shigella flexneri* antigens to *Escherichia coli* K-12. Infect. Immun. *1*: 279–287.

Formal, S.B., P.J. Gemski, L.S. Baron and E.H. LaBrec. 1971. A chromosomal locus which controls the ability of *Shigella flexneri* to evoke keratoconjunctivitis. Infect. Immun. *3*: 73–79.

Forsberg, C.W. 1978. Effects of heavy metals and other trace elements on the fermentative activity of the rumen microflora and growth of functionally important rumen bacteria. Can. J. Microbiol. *24*: 298–306.

Forsberg, C.W., J.W. Costerton and R.A. MacLeod. 1970. Separation and localization of cell wall layers of a gram-negative bacterium. J. Bacteriol. *104*: 1338–1353.

Forsman, M., K. Kuoppa, A. Sjöstedt and A. Tärnvik. 1990. Use of RNA hybridization in the diagnosis of a case of ulceroglandular tularemia. Eur. J. Clin. Microbiol. Infect. Dis. *9*: 784–785.

Forsman, M., G. Sandström and A. Sjöstedt. 1994. Analysis of 16S ribosomal DNA sequences of *Francisella* strains and utilization for determination of the phylogeny of the genus and for identification of strains by PCR. Int. J. Syst. Bacteriol. *44*: 38–46.

Forst, S., B. Dowds, N.E. Boemare and E. Stackebrandt. 1997. *Xenorhabdus* and *Photorhabdus* spp.: bugs that kill bugs. Annu. Rev. Microbiol. *51*: 47–72.

Forst, S. and K. Nealson. 1996. Molecular biology of the symbiotic-pathogenic bacteria *Xenorhabdus* spp. and *Photorhabdus* spp. Microbiol. Rev. *60*: 21–43.

Foshay, L. and W. Hesselbrock. 1945. Some observations on the filtrability of *Bacterium tularense*. J. Bacteriol. *49*: 233–236.

Fossing, H., V.A. Gallardo, B.B. Jørgensen, M. Hüttel, L.P. Nielsen, H. Schulz, D.E. Canfield, S. Forster, R.N. Glud, J.K. Gundersen, J. Küver, N.B. Ramsing, A. Teske, B. Thamdrup and O. Ulloa. 1995. Concentration and transport of nitrate by the mat-forming sulphur bacterium *Thioploca*. Nature *374*: 713–715.

Foster, G., M.D. Collins, P.A. Lawson, D. Buston, F.J. Murray and A. Sime. 1999. *Actinobacillus seminis* as a cause of abortion in a UK sheep flock. Vet. Rec. *144*: 479–480.

Foster, G., H.M. Ross, H. Malnick, A. Willems, P. Garcia, R.J. Reid and M.D. Collins. 1996. *Actinobacillus delphinicola* sp. nov., a new member of the family *Pasteurellaceae* Pohl (1979) 1981 isolated from sea mammals. Int. J. Syst. Bacteriol. *46*: 648–652.

Foster, G., H.M. Ross, H. Malnick, A. Willems, R.A. Hutson, R.J. Reid and M.D. Collins. 2000. *Phocoenobacter uteri* gen. nov., sp. nov., a new member of the family *Pasteurellaceae* Pohl (1979) 1981 isolated from a harbour porpoise (*Phocoena phocoena*). Int. J. Syst. Evol. Microbiol. *50*: 135–139.

Foster, G., H.M. Ross, I.A.P. Patterson, R.A. Hutson and M.D. Collins. 1998. *Actinobacillus scotiae* sp. nov., a new member of the family *Pasteurellaceae* Pohl (1979) 1981 isolated from porpoises (*Phocoena phocoena*). Int. J. Syst. Bacteriol. *48*: 929–933.

Foster, J.W. and R.H. Davis. 1966. A methane-dependent coccus, with notes on classification and nomenclature of obligate, methane-utilizing bacteria. J. Bacteriol. *91*: 1924–1931.

Fothergill, J.C. and J.R. Guest. 1977. Catabolism of L-lysine by *Pseudomonas aeruginosa*. J. Gen. Microbiol. *99*: 139–155.

Fountoulakis, M., J.F. Juranville and P. Berndt. 1997. Large-scale identification of proteins of *Haemophilus influenzae* by amino acid composition analysis. Electrophoresis *18*: 2968–2977.

Fountoulakis, M., J.F. Juranville, D. Röder, S. Evers, P. Berndt and H. Langen. 1998a. Reference map of the low molecular mass proteins of *Haemophilus influenzae*. Electrophoresis *19*: 1819–1827.

Fountoulakis, M., B. Takács and H. Langen. 1998b. Two-dimensional map

of basic proteins of *Haemophilus influenzae*. Electrophoresis *19*: 761–766.

Fourquet, R., P. Coulanges, J.L. Boehret and J. Delavamd. 1975. Les infections a *Edwardsiella tarda* a propos de l'isolement des premieres souches a Madagascar. Arch. Inst. Pasteur Madag. *44*: 31–48.

Fouz, B., J.L. Larsen, B. Nielsen, J.L. Barja and A.E. Toranzo. 1992. Characterization of *Vibrio damsela* strains isolated from *Turbot scophthalmus-maximus* in Spain. Dis. Aquat. Org. *12*: 155–166.

Fouz, B., A.E. Toranzo, E.G. Biosca, R. Mazoy and C. Amaro. 1994. Role of iron in the pathogenicity of *Vibrio damsela* for fish and mammals. FEMS Microbiol. Lett. *121*: 181–188.

Fowler, V.J., N. Pfennig, W. Schubert and E. Stackebrandt. 1984. Towards a phylogeny of phototrophic purple sulfur bacteria: 16S ribosomal RNA oligonucleotide cataloging of 11 species of *Chromatiaceae*. Arch. Microbiol. *139*: 382–387.

Fox, E.N. and K. Higuchi. 1958. Synthesis of the fraction 1 antigenic protein by *Pasteurella pestis*. J. Bacteriol. *75*: 209–216.

Fox, G.E., K.R. Pechman and C.R. Woese. 1977. Comparative cataloging of 16S ribosomal ribonucleic acid - molecular approach to procaryotic systematics. Int. J. Syst. Bacteriol. *27*: 44–57.

Fox, G.E., E. Stackebrandt, R.B. Hespell, J. Gibson, J. Maniloff, T.A. Dyer, R.S. Wolfe, W.E. Balch, R.S. Tanner, L.J. Magrum, L.B. Zablen, R. Blakemore, R. Gupta, L. Bonen, B.J. Lewis, D.A. Stahl, K.R. Luehrsen, K.N. Chen and C.R. Woese. 1980. The phylogeny of prokaryotes. Science *209*: 457–463.

Fox, G.E., J.D. Wisotzkey and P. Jurtshuk Jr.. 1992. How close is close: 16S rRNA sequence identity may not be sufficient to guarantee species identity. Int. J. Syst. Bacteriol. *41*: 166–170.

Foxwell, A.R., J.M. Kyd and A.W. Cripps. 1998. Nontypeable *Haemophilus influenzae*: pathogenesis and prevention. Microbiol. Mol. Biol. Rev. *62*: 294–308.

Fraaije, B.A., M. Appels, S.H. De Boer, J.W.L. Van Vuurde and R.W. Van Den Bulk. 1997. Detection of soft rot *Erwinia* spp. on seed potatoes: conductimetry in comparison with dilution plating, PCR anad serological assays. Eur. J. Plant Pathol. *103*: 183–193.

Fraaije, B.A., Y. Birnbaum, A.A.J.M. Franken and R.W. Van Den Bulk. 1996a. The development of a conductimetric assay for automated detection of metabolically active soft rot *Erwinia* spp. in potato tuber peel extracts. J. Appl. Bacteriol. *81*: 375–382.

Fraaije, B.A., Y. Birnbaum and R.W. Van Den Bulk. 1996b. Comparison of methods for detection of *Erwinia carotovora* ssp. atroseptica in progeny tubers derived from inoculated tubers of *Solanum tuberosum* L. J. Phytopathol. (Berl.) *144*: 551–557.

Frackman, S., M. Anhalt and K.H. Nealson. 1990. Cloning, organization, and expression of the bioluminescence genes of *Xenorhabdus luminescens*. J. Bacteriol. *172*: 5767–5773.

France, D.R. and N.P. Markham. 1968. Epidemiological aspects of *Proteus* infections with particular reference to phage typing. J. Clin. Pathol. *21*: 97–102.

Francis, D.H., J.E. Collins and J.R. Duimstra. 1986. Infection of gnotobiotic pigs with an *Escherichia coli* O157:H7 strain associated with an outbreak of hemorrhagic colitis. Infect. Immun. *51*: 953–956.

Francis, M.S., A.F. Parker, R. Morona and C.J. Thomas. 1993. Bacteriophage lambda as a delivery vector for Tn10-derived transposons in *Xenorhabdus bovienii*. Appl. Environ. Microbiol. *59*: 3050–3055.

Frank, G.H. 1989. Pasteurellosis of cattle. *In* Adlam and Rutter (Editors), *Pasteurella* and Pasteurellosis, Academic Press, London. pp. 197–222.

Frankel, G., L. Riley, J.A. Giron, J. Valmassoi, A. Friedmann, N. Strockbine, S. Falkow and G.K. Schoolnik. 1990. Detection of *Shigella* in feces using DNA amplification. J. Infect. Dis. *161*: 1252–1256.

Franken, A.A.J.M. 1992. Comparison of immunofluorescence microscopy and dilution-plating for the detection of *Xanthomonas campestris* pv. *campestris* in crucifer seeds. Neth. J. Plant Pathol. *98*: 169–178.

Franken, A.A.J.M., C. van Zeijl, J.G.P.M. van Bilscn, A. Neuvel, R. de Vogel, Y. van Wingerden, Y.E. Birnbaum, J. van Hateren and P.S. Van der Zouwen. 1991. Evaluation of a plating assay for *Xanthomonas campestris* pathovar *campestris*. Seed Science And Technology. *19*: 215–226.

Franken, A., J.F. Zilverentant, P.M. Boonekamp and A. Schots. 1992. Specificity of polyclonal and monoclonal antibodies for the identification of *Xanthomonas campestris* pv. *campestris*. Neth. J. Plant Pathol. *98*: 81–94.

Frantzen, E. 1950. Biochemical and serological studies on *alkalescens* and *dispar* strains. Acta Pathol. Microbiol. Scand. *27*: 236–248.

Franze de Fernandez, M.T., W.S. Hayward and J.T. August. 1972. Bacterial proteins required for replication of phage Q ribonucleic acid. Pruification and properties of host factor I, a ribonucleic acid- binding protein. J. Biol. Chem. *247*: 824–831.

Franzmann, P.D. 1996. Examination of Antarctic prokaryotic diversity through molecular comparisons. Biodivers. Conserv. *5*: 1295–1305.

Franzmann, P.D., H.R. Burton and T.A. McMeekin. 1987. *Halomonas subglaciescola*, a new species of halotolerant bacteria isolated from antarctica. Int. J. Syst. Bacteriol. *37*: 27–34.

Franzmann, P.D. and B.J. Tindall. 1990. A chemotaxonomic study of members of the family *Halomonadaceae*. Syst. Appl. Microbiol. *13*: 142–147.

Franzmann, P.D., U. Wehmeyer and E. Stackebrandt. 1988. *Halomonadaceae* fam. nov., a new family of the class *Proteobacteria* to accomodate the genera *Halomonas* and *Deleya*. Syst. Appl. Microbiol. *11*: 16–19.

Franzmann, P.D., U. Wehmeyer and E. Stackebrandt. 1989. Validation of the publication of new names and new combinations previously effectively published outside the IJSB. List No. 29. Int. J. Syst. Bacteriol. *39*: 205–206.

Fraser, C.M., S. Casjens, W.M. Huang, G.G. Sutton, R. Clayton, R. Lathigra, O. White, K.A. Ketchum, R. Dodson, E.K. Hickey, M. Gwinn, B. Dougherty, J.-F. Tomb, R.D. Fleischmann, D. Richardson, J. Peterson, A.R. Kerlavage, J. Quackenbush, S. Salzberg, M. Hanson, R. van Vugt, N. Palmer, M.D. Adams, J. Gocayne, J. Weidman, T. Utterback, L. Watthey, L. McDonald, P. Artiach, C. Bowman, S. Garland, C. Fujii, M.D. Cotton, K. Horst, K. Roberts, B. Hatch, H.O. Smith and J.C. Venter. 1997. Genomic sequence of a Lyme disease spirochaete, *Borrelia burgdorferi*. Nature *390*: 580–586.

Fraser, C.M., J.D. Gocayne, O. White, M.D. Adams, R.A. Clayton, R.D. Fleischmann, C.J. Bult, A.R. Kerlavage, G. Sutton, J.M. Kelley, J.L. Fritchman, J.F. Weidman, K.V. Small, M. Sandusky, J. Fuhrmann, D. Nguyen, T.R. Utterback, D.M. Saudek, C.A. Phillips, J.M. Merrick, J.-F. Tomb, B.A. Dougherty, K.F. Bott, P.-C. Hu, T.S. Lucier, S.N. Peterson, H.O. Smith, C.A.I. Hutchison and J.C. Venter. 1995. The minimal gene complement of Mycoplasma genitalium. Science *270*: 397–403.

Fraser, D.W., T.F. Tsai, W. Orenstein, W.E. Parkin, H.J. Beecham, R.G. Sharrer, J. Harris, G.F. Mallinson, S.M. Martin, J.E. McDade, C.C. Shepard, P.S. Brachman and t.F.I. Team. 1977. Legionnaires' disease: description of an epidemic. N. Engl. J. Med. *297*: 1189–1197.

Frazer, J., K. Zinnemann and J.M.H. Boyce. 1975. The agglutination reactions of *Haemophilus paraphrophilus* and *H. paraphrohaemolyticus*, and some observations on the agglutination of *H. aphrophilus* and *H. haemoglobinophilus* (*H. canis*). J. Med. Microbiol. *8*: 89–96.

Frazier, W.C. 1926. A method for the detection of changes in gelatin due to bacteria. J. Infect. Dis. *39*: 302–309.

Fredericq, P. 1948. Actions antibiotiques reciproques chez les *Enterobacteriaceae*. Rev. Belge Pathol. Med. Exp. *19 (suppl. 4)*: 1–107.

Frederiksen, W. 1964. A study of some *Yersinia pseudotuberculosis*-like bacteria (*Bacterium enterocoliticum* and *Pasteurella* X). Proc. XIV Scand. Cong. Pathol. Microbiol., Oslo. Universitetsforlaget. 103–104.

Frederiksen, W. 1970. *Citrobacter koseri* (n. sp.), a new species within the genus *Citrobacter*, with a comment on the taxonomic position of *Citrobacter intermedium* (Werkman and Gillen). Publ. Fac. Sci. Univ. J. E. Purkyne. *47*: 89–94.

Frederiksen, W. 1973. *Pasteurella* taxonomy and nomenclature. *In* Winblad (Editor), Contributions to microbiology and immunology. Vol. 2. *Yersinia, Pasteurella* and *Francisella*, Karger, Basel. pp. 170–176.

Frederiksen, W. 1989. Pasteurellosis of man. *In* Adlam and Rutter (Editors), *Pasteurella* and Pasteurellosis, Academic Press, London. 303–320.

Frederiksen, W. 1990. Correct names of the species *Citrobacter koseri, Levinea malonatica*, and *Citrobacter diversus*. Request for an opinion. Int. J. Syst. Bacteriol. *40*: 107–108.

Frederiksen, W. 1993. Ecology and significance of *Pasteurellaceae* in man - an update. Zentbl. Bakteriol. *279*: 27–34.

Frederiksen, W. and M. Kilian. 1981. *Haemophilus, Pasteurella* and *Actinobacillus*: Their significance in human medicine. *In* Kilian, Frederiksen and Biberstein (Editors), *Haemophilus, Pasteurella* and *Actinobacillus*, Academic Press, London. 39–55.

Frederiksen, W. and P. Søgaard. 1992. The genus *Citrobacter*. *In* Balows, Trüper, Dworkin, Harder and Schleifer (Editors), The Prokaryotes. A Handbook on the Biology of Bacteria: Ecophysiology, isolation, identification, applications, 2nd Ed., Springer-Verlag, pp. 2744–2753.

Frederiksen, W. and B. Tønning. 2001. Possible misidentification of *Haemophilus aphrophilus* as *Pasteurella gallinarum*. Clin. Infect. Dis. *32*: 987–989.

Frei, E. and R.D. Preston. 1964. Non-cellulosic structural polysaccharide in algal cell walls. II. Association of xylan and mannan in *Porphyra umbilicalis*. Proc. R. Soc. Lond. Ser. B Biol. Sci. *160*: 314–327.

Freinkel, A.L., Y. Dangor, H.J. Koornhof and R.C. Ballard. 1992. A serological test for granuloma inguinale. Genitourin. Med. *68*: 269–272.

Freitag, V. and O. Friedrich. 1981. Bacteriocin typing of *Enterobacter cloacae* strains. Zntbl. Fur Bakteriol. Mikrobiol. Hyg. 1 Abt. Orig. A Med. Mikrobiol. Infktrankh. Virol. Parasitol. *249*: 63–75.

French, G.L., M.L. Woo, Y.W. Hui and K.Y. Chan. 1989. Antimicrobial susceptibilities of halophilic vibrios. J. Antimicrob. Chemother. *24*: 183–194.

Freney, J., M.O. Husson, F. Gavini, S. Madier, A. Martra, D. Izard, H. Leclerc and J. Fleurette. 1988. Susceptibilities to antibiotics and antiseptics of new species of the family *Enterobacteriaceae*. Antimicrob. Agents Chemother. *32*: 873–876.

Freney, J., P. Laban, M. Desmonceaux and J. Fleurette. 1983. Micromethod for carbon substrate assimilation by *Pseudomonas maltophilia*. Zentralbl. Bakteriol. Parastienkd. Infektionskrak. Hyg. Abt 1 Orig. *255*: 479–488.

Frey, J., M. Beck and J. Nicolet. 1994. RTX toxins of *Actinobacillus pleuropneumoniae*. Zentbl. Bakteriol. *Suppl. 24*: 322–332.

Frey, J., M. Beck and J. Nicolet. 1996. Typisierung der Apx-Toxin-Gene von *Actinobacillus pleuropneumoniaemittels* PCR. Schweiz. Arch. Tierheilk. *138*: 121–124.

Frey, J., M. Beck, J.F. van den Bosch, R.P. Seegers and J. Nicolet. 1995. Development of an efficient PCR method for toxin typing of *Actinobacillus pleuropneumoniae* strains. Mol. Cell. Probes *9*: 277–282.

Frey, J., J.T. Bosse, Y.-F. Chang, J.M. Cullen, B. Fenwick, G.F. Gerlach, D. Gygi, F. Haesebrouck, T.J. Inzana, R. Jansen, E.M. Kamp, J. Macdonald, J.I. MacInnes, K.R. Mittal, J. Nicolet, A. Rycroft, R.P. Seegers, M.A. Smith, E. Stenbaek, D.K. Struck, J.F. van den Bosch, P.J. Wilson and R. Young. 1993. *Actinobacillus pleuropneumoniae* RTX-toxins: uniform designation of haemolysins, cytolysins, pleurotoxin and their genes. J. Gen. Microbiol. *139*: 1723–1728.

Fricker, C.R. 1987. Serotyping of mesophilic *Aeromonas* spp. on the basis of lipopolysaccharide antigens. Lett. Appl. Microbiol. *4*: 113–116.

Fricker, C.R. and S. Tompsett. 1989. *Aeromonas* spp. in foods: a significant cause of food poisoning? Int. J. Food Microbiol. *9*: 17–23.

Friedländer, C. 1882. Über die Schizomyceten bei der acuten fibrosen pneumonie. Arch. Pathol. Anat. Physiol. *87*: 319–324.

Friedman, B.A. and P.R. Dugan. 1968. Identification of *Zoogloea* species and the relationship to zoogloeal matrix and floc formation. J. Bacteriol. *95*: 1903–1909.

Friis-Møller, A. 1981. A new *Actinobacillus* species from the human respiratory tract: *Actinobacillus hominis* nov. sp. *In* Kilian, Frederiksen and Biberstein (Editors), *Haemophilus, Pasteurella* and *Actinobacillus*, Academic Press, London. pp. 151–157.

Friis-Møller, A. 1985. *In* Validation of the publication of new names and new combinations previously effectively published outside the IJSB. List No. 18. Int. J. Syst. Bacteriol. *35*: 375–376.

Friis-Møller, A. and A. Kharazmi. 1988. Neutrophil response, serum opsonic activity, and precipitating antibodies in human infection with *Actinobacillus hominis*. APMIS *96*: 1023–1028.

Frosch, P. and W. Kolle. 1896. Die Mikrokokken. *In* Flügge (Editor), Die

Mikroorganismen, 3 Aufl., 2 Teil, Verlag von Vogel, Leipzig. pp. 154–155.

Frost, J.A., T. Cheasty, A. Thomas and B. Rowe. 1993. Phage typing of Vero cytotoxin-producing *Escherichia coli* O157 isolated in the United Kingdom: 1989–1991. Epidemiol. Infect. *110*: 469–475.

Frutos, R., B.A. Federici, B. Revet and M. Bergoin. 1994. Taxonomic studies of *Rickettsiella*, *Rickettsia*, and *Chlamydia* using genomic DNA. J. Invertebr. Pathol. *63*: 294–300.

Frutos, R., M. Pages, M. Bellis, G. Roizes and M. Bergoin. 1989. Pulsed-field gel electrophoresis determination of the genome size of obligate intracellular bacteria belonging to the genera *Chlamydia*, *Rickettsiella*, and *Porochlamydia*. J. Bacteriol. *171*: 4511–4513.

Fry, N.K., T.J. Rowbotham, N.A. Saunders and T.M. Embley. 1991a. Direct amplification and sequencing of the 16S ribosomal DNA of an intracellular *Legionella* sp. recovered by amoebal enrichment from sputum of a patient with pneumonia. FEMS Microbiol. Lett. *83*: 165–168.

Fry, N.K., S. Warwick, N.A. Saunders and T.M. Embley. 1991b. The use of 16S ribosomal RNA analyses to investigate the phylogeny of the family *Legionellaceae*. J. Gen. Microbiol. *137*: 1215–1222.

Fry, S.M., J.S. Huang and R.D. Milholland. 1994. Isolation and preliminary characterization of extracellular proteases produced by strains of *Xylella fastidiosa* from grapevines. Phytopathology *84*: 357–363.

Fry, S.M. and R.D. Milholland. 1990. Response of resistant, tolerant, and susceptible grapevine tissues to invasion by the Pierce's disease bacterium, *Xylella fastidiosa*. Phytopathology *80*: 66–69.

Fry, S.M., R.D. Milholland and P.Y. Huang. 1990. Isolation and growth of strains of *Xylella fastidiosa* from infected grapevines on nutrient agar media. Plant Dis. *74*: 522–524.

Fryer, J.L., C.N. Lannan, L.H. Garcés, J.J. Larenas and P.A. Smith. 1990. Isolation of a rickettsiales-like organism from diseased coho salmon (*Oncorhynchus kisutch*) in Chile. Fish Pathology *25*: 107–114.

Fryer, J.L., C.N. Lannan, J. Giovannoni and N.D. Wood. 1992. *Piscirickettsia salmonis* gen. nov., sp. nov., the causative agent of an epizootic disease in salmonid fishes. Int. J. Syst. Bacteriol. *42*: 120–126.

Fryer, J.L. and M.J. Mauel. 1997. The rickettsia: an emerging group of pathogens in fish. Emerg. Infect. Dis. *3*: 137–144.

Fuerst, J.A. and A.C. Hayward. 1969. Surface appendages similar to fimbriae (pili) on *Pseudomonas* species. J. Gen. Microbiol. *58*: 227–237.

Fuhs, G.W. and M. Chen. 1975. Microbiological basis of phosphate removal in the activated sludge process for the treatment of wastewater. Microb. Ecol. *2*: 119–138.

Fujii, T., M. Hayashi and M. Okuzumi. 1990. New device for isolation of spiral bacteria from kusaya gravy. Nippon Suisan Gakk. *56*: 161–161.

Fujii, T., M. Takeo and Y. Maeda. 1997. Plasmid-encoded genes specifying aniline oxidation from *Acinetobacter* sp. strain YAA. Microbiology (Reading) *143*: 93–99.

Fujino, T., Y. Okuno, D. Nakada, A. Aoyama, K. Fukai, T. Mukai and T. Ueho. 1951. On the bacteriological examination of shirasu food poisoning (in Japanese). J. Jpn. Assoc. Infect. Dis. *25*: 11–12.

Fukatsu, T., S. Aoki, U. Kurosu and H. Ishikawa. 1994. Phylogeny of *Ceratiphidini* aphids revealed by their symbiotic microorganisms and basic structure of their galls: implications for host-symbiont coevolution and evolution of sterile soldier castes. Zool. Sci. (Tokyo). *11*: 613–623.

Fukatsu, T. and H. Ishikawa. 1992a. A novel eukaryotic extracellular symbiont in an aphid, *Astegopteryx styraci* (Homoptera, *Aphididae*, Hormaphidinae). J. Insect Physiol. *38*: 765–773.

Fukatsu, T. and H. Ishikawa. 1992b. Synthesis and localization of symbionin, an aphid endosymbiont protein. Insect Biochem. Mol. Biol. *22*: 167–174.

Fukatsu, T. and H. Ishikawa. 1993. Occurrence of chaperonin 60 and chaperonin 10 in primary and secondary bacterial symbionts of aphids: implications for the evolution of an endosymbiotic system in aphids. J. Mol. Evol. *36*: 568–577.

Fukatsu, T. and H. Ishikawa. 1996. Phylogenetic position of yeast-like symbiont of *Hamiltonaphis styraci* (Homoptera, *Aphididae*) based on 18S rDNA sequence. Insect Biochem. Mol. Biol. *26*: 383–388.

Fuksa, M., S. Krajden and A. Lee. 1984. Susceptibility of 45 clinical isolates of *Proteus penneri*. Antimicrob. Agents Chemother. *26*: 419–420.

Fukui, M., A. Teske, B. Assmus, G. Muyzer and F. Widdel. 1999. Physiology, phylogenetic relationships, and ecology of filamentous sulfate-reducing bacteria (genus *Desulfonema*). Arch. Microbiol. *172*: 193–203.

Fukushima, H. 1985. Direct isolation of *Yersinia enterocolitica* and *Yersinia pseudotuberculosis* from meat. Appl. Environ. Microbiol. *50*: 710–712.

Fukushima, H. 1987. New selective agar medium for isolation of virulent *Yersinia enterocolitica*. J. Clin. Microbiol. *25*: 1068–1073.

Fukushima, H. and M. Gomyoda. 1986. Growth of *Yersinia pseudotuberculosis* and *Yersinia enterocolitica* biotype 3B *O3* inhibited on cefsulodin-Irgasan-novobiocin agar. J. Clin. Microbiol. *24*: 116–120.

Fukushima, H., M. Gomyoda, S. Ishikura, T. Nishio, S. Moriki, J. Endo, S. Kaneko and M. Tsubokura. 1989. Cat-contaminated environmental substances lead to *Yersinia pseudotuberculosis* infection in children. J. Clin. Microbiol. *27*: 2706–2709.

Fuller, T.E., M.J. Kennedy and D.E. Lowery. 2000. Identification of *Pasteurella multocida* virulence genes in a septicemic mouse model using signature-tagged mutagenesis. Microb. Pathog. *29*: 25–38.

Fulton, M. 1943. The identity of *Bacterium columbensis* Castellani. J. Bacteriol. *46*: 79–82.

Funke, G. and H. Rosner. 1995. *Rahnella aquatilis* bacteremia in an HIV-infected intravenous drug abuser. Diagn. Microbiol. Infect. Dis. *22*: 293–296.

Fuqua, W.C. and R.M. Weiner. 1993. The *melA* gene is essential for melanin biosynthesis in the marine bacterium *Shewanella colwelliana*. J. Gen. Microbiol. *139*: 1105–1114.

Furet, Y.X. and J.C. Pechere. 1991. Newly documented antimicrobial activity of quinolones. Eur. J. Clin. Microbiol. Infect. Dis. *10*: 249–254.

Furniss, A.L., J.V. Lee and T.J. Donovan. 1977. Group F, a new *Vibrio*? Lancet *2*: 565–566.

Furniss, A.L., J.V. Lee and T.J. Donovan. 1978. The Vibrios. Public Health Laboratory Service Monograph Series, Her Majesty's Stationery Office, London.

Furth, A. 1975. Purification and properties of a constitutive beta-lactamase from *Pseudomonas aeruginosa* strain Dalgleish. Biochim. Biophys. Acta *377*: 431–443.

Furuwatari, C., Y. Kawakami, T. Akahane, E. Hidaka, Y. Okimura, J. Nakayama, F. K. and T. Katsuyama. 1994. Proposal for an Aeroscheme (modified Aerokey II) for the identification of clinical *Aeromonas* species. Med. Sci. Res. *22*: 617–619.

Fussing, V. 1998. Genomic relationships of *Actinobacillus pleuropneumoniae* serotype 2 strains evaluated by ribotyping, sequence analysis of ribosomal intergenic regions, and pulsed-field gel electrophoresis. Lett. Appl. Microbiol. *27*: 211–215.

Fussing, V., K. Barford, R. Nielsen, K. Møller, J.P. Nielsen, H.C. Wegener and M. Bisgaard. 1998a. Evaluation and application of ribotyping for epidemiological studies of *Actinobacillus pleuropneumoniae* in Denmark. Vet. Microbiol. *62*: 145–162.

Fussing, V., J.P. Nielsen, M. Bisgaard and A. Meyling. 1999. Development of a typing system for epidemiological studies of porcine toxin-producing *Pasteurella multocida* subsp. *multocida* in Denmark. Vet. Microbiol. *65*: 61–74.

Fussing, V., B.J. Paster, F.E. Dewhirst and L.K. Poulsen. 1998b. Differentiation of *Actinobacillus pleuropneumoniae* strains by sequence analysis of 16S rDNA and ribosomal intergenic regions, and development of a species specific oligonucleotide for in situ detection. Syst. Appl. Microbiol. *21*: 408–418.

Fussing, V. and H.C. Wegener. 1993. Characterization of bovine *Haemophilus somnus* by biotyping, plasmid profiling, REA-patterns and ribotyping. Zentbl. Bakteriol. *279*: 60–74.

Fyfe, J.A.M. and J.R.W. Govan. 1980. Alginate synthesis in mucoid *Pseudomonas aeruginosa* - a chromosomal locus involved in control. J. Gen. Microbiol. *119*: 443–450.

Gaastra, W. and A.M. Svennerholm. 1996. Colonization factors of human enterotoxigenic *Escherichia coli* (ETEC). Trends Microbiol. *4*: 444–452.

Gabriel, D.W., M.T. Kingsley, J.E. Hunter and T. Gottwald. 1989. Reinstatement of *Xanthomonas citri* new species (ex Hasse) and *Xantho-*

monas phaseoli new species (ex Smith) to species and reclassification of all *Xanthomonas campestris* pv. *citri* strains. Int. J. Syst. Bacteriol. *39*: 14–22.

Gacesa, P. 1998. Bacterial alginate biosynthesis - recent progress and future prospects. Microbiology *144*: 1133–1143.

Gadaleta, P., M.E. Pavan and J. Zorzopulos. 1996. Proposal of a new *Kluyvera* species and the relationship of this genus with other members of the *Enterobacteriaceae* family. Abst. Annu. Mtg. Amer. Soc. Microbiol, p. 474.

Gaertner, A. 1888. Ueber die Fleischvergiftung un Frankenhausen a. Kyffh. und den Erreger derselben Correspondenz-Blätter des allgem. ärztl. Vereins von Thuringen. *17*: 573–600.

Gaffney, T.D., S.T. Lam, J. Ligon, K. Gates, A. Frazelle, J. Di Maio, S. Hill, S. Goodwin, N. Torkewitz, A.M. Allshouse, H.J. Kempf and J.O. Becker. 1994. Global regulation of expression of antifungal factors by a *Pseudomonas fluorescens* biological-control strain. Mol. Plant-Microbe Interact. *7*: 455–463.

Gagnevin, L., J.E. Leach and O. Pruvost. 1997. Genomic variability of the *Xanthomonas* pathovar *mangiferaeindicae*, agent of mango bacterial black spot. Appl. Environ. Microbiol. *63*: 246–253.

Gaillot, O., L. Guilbert, C. Maruejouls, F. Escande and M. Simonet. 1995. In-vitro susceptibility to 13 antibiotics of *Pasteurella* spp. and related bacteria isolated from humans. J. Antimicrob. Chemother. *36*: 878–880.

Galan, J.E. 1996a. Molecular and cellular bases of *Salmonella* entry into host cells. Curr. Top. Microbiol. Immunol. *209*: 43–60.

Galan, J.E. 1996b. Molecular genetic bases of *Salmonella* entry into host cells. Mol. Microbiol. *20*: 263–271.

Gal'chenko, V.F. 1994. Sulfate reduction, methane generation and methane oxidation in different Bunger Hills reservoirs (Antarctica). Mikrobiologiya *63*: 388–396.

Gal'chenko, V.F., F.N. Abramochkina, L.V. Bezrukova, E.N. Sokolova and M.V. Ivanov. 1988. The species composition of aerobic methanotrophic microflora in the Black Sea. Mikrobiologiya *57*: 248–253.

Gal'chenko, V.F. and A.I. Nesterov. 1981. Numerical analysis of protein electrophoretograms for obligate methanotrophic bacteria. Mikrobiologiya *50*: 725–730.

Galdbart, J.O., F. Lemann, D. Ainouz, P. Feron, N. Lambert-Zechovsky and C. Branger. 2000. TEM-24 extended spectrum β-lactamase producing *Enterobacter aerogenes*: long term clonal dissemination in French hospitals. Clin. Microbiol. Infect. *6*: 316–323.

Galdiero, M., E. Palomba, L. De, M. Vitiello and P. Pagnini. 1998. Effects of the major *Pasteurella multocida* porin on bovine neutrophils. Am. J. Vet. Res. *59*: 1270–1274.

Gale, E.F. and H.M.R. Epps. 1943. L-Lysine decarboxylase: preparation of specific enzyme and coenzyme. Nature (Lond.) *152*: 327–328.

Galinski, E.A., H.P. Pfeiffer and H.G. Trüper. 1985. 1,4,5,6-Tetrahydro-2-methyl-4-pyrimidinecarboxylic acid: a novel cyclic amino acid from halophilic phototrophic bacteria of the genus *Ectothiorhodospira*. Eur. J. Biochem. *149*: 135–139.

Galinski, E.A. and H.G. Trüper. 1982. Betaine, a compatible solute in the extremely halophilic phototropic bacterium *Ectothiorhodospira halochloris*. FEMS Microbiol. Lett. *13*: 357–360.

Gallacher, S. and T.H. Birkbeck. 1993. Effect of phosphate concentration on production of tetraodonis by *Alteromonas tetraodonis*. Appl. Environ. Microbiol. *59*: 3981–3983.

Gallagher, I.H. 1971. Occurrence of waxes in *Acinetobacter*. J. Gen. Microbiol. *68*: 245–247.

Gallardo, V.A. 1977. Large benthic microbial communities in sulphide biota under Peru-Chile subsurface countercurrent. Nature *268*: 331–332.

Gallardo, V.A., E. Klingelhoeffer, W. Arntz and M. Graco. 1998. First report of the bacterium *Thioploca* in the Benguela ecosystem off Namibia. J. Mar. Biol. Assoc. U.K. *78*: 1007–1010.

Gallois, A. and P.A.D. Grimont. 1985. Pyrazines responsible for the potato-like odor produced by some *Serratia* and *Cedecea* strains. Appl. Environ. Microbiol. *50*: 1048–1051.

Gallois, A., R. Samson, E. Ageron and P.A.D. Grimont. 1992. *Erwinia*

carotovora odorifera, subsp. nov. associated with odorous soft rot of chicory (*Cichorium intybus* L.). Int. J. Syst. Bacteriol. *42*: 582–588.

Gamaléia, M.N. 1888. *Vibrio metschnikovi* (n. sp.) et ses rapports avec le microbe du choléra asiatique. Ann. Inst. Pasteur (Paris) *2*: 482–488.

Gaman, W., C. Cates, C.F.T. Snelling, B. Lank and A.R. Ronald. 1976. Emergence of gentamicin- and carbenicillin-resistant *Pseudomonas aeruginosa* in a hospital environment. Antimicrob. Agents Chemother. *9*: 474–480.

Gander, R.M. and M.T. LaRocco. 1989. Detection of pilus-like structures on clinical and environmental isolates of *Vibrio vulnificus*. J. Clin. Microbiol. *27*: 1015–1021.

Ganesh, R., K.G. Robinson, G.D. Reed and G.S. Sayler. 1997. Reduction of hexavalent uranium from organic complexes by sulfate- and iron-reducing bacteria. Appl. Environ. Microbiol. *63*: 4385–4391.

Gangeswaran, R., D.J. Lowe and R.R. Eady. 1993. Purification and characterization of the assimilatory nitrate reductase of *Azotobacter vinelandii*. Biochem. J. *289*: 335–342.

Ganiere, J.P., F. Escande, G. Andre and M. Larrat. 1993. Characterization of *Pasteurella* from gingival scrapings of dogs and cats. Comp. Immunol. Microbiol. Infect. Dis. *16*: 77–85.

Gannon, V.P.J., M. Rashed, R.K. King and E.J.G. Thomas. 1993. Detection and characterization of the *eae* gene of Shiga-like toxin-producing *Escherichia coli* using polymerase chain reaction. J. Clin. Microbiol. *31*: 1268–1274.

Gannon, V.P.J., C. Teerling, S.A. Masri and C.L. Gyles. 1990. Molecular cloning and nucleotide sequence of another variant of the *Escherichia coli* Shiga-like toxin II family. J. Gen. Microbiol. *136*: 1125–1135.

Garcés, L.H., J.J. Larenas, P.A. Smith, S. Sandino, C.N. Lannan and J.L. Fryer. 1991. Infectivity of a rickettsia isolated from coho salmon *Oncorhynchus kisutch*. Dis. Aquat. Org. *11*: 93–97.

Garcia, A., Jr., J.R.S. Lopes and M.J.G. Beretta. 1997. Population survey of leafhopper vectors of *Xylella fastidiosa* in citrus nurseries, in Brazil. Fruits (Paris) *52*: 371–374.

Garcia, D.C., M.M. Nociari, D.O. Sordelli, A. Di Martino and M. Catalano. 1996. The use of plasmid profile analysis and ribotyping for typing *Acinetobacter baumannii* isolates. J. Hosp. Infect. *34*: 139–144.

Garcia, D., P. Parot, A. Vermeglio and M.T. Madigan. 1986. The light-harvesting complexes of a thermophilic purple sulfur photosynthetic bacterium *Chromatium tepidum*. Biochim. Biophys. Acta *850*: 390–395.

Garcia, M.T., J.J. Nieto, A. Ventosa and F. Ruiz-Berraquero. 1987a. The susceptibility of the moderate halophile *Vibrio costicola* to heavy metals. J. Appl. Bacteriol. *63*: 63–66.

Garcia, M.T., A. Ventosa, F. Ruiz-Berraquero and M. Kocur. 1987b. Taxonomic study and amended description of *Vibrio costicola*. Int. J. Syst. Bacteriol. *37*: 251–256.

Garcia-Delgado, G.A., P.B. Little and D.A. Barnum. 1976. A comparison of various *Haemophilus somnus* strains. Canad. J. Comp. Med. *41*: 380–388.

García-Rodríguez, J.A., J.E. Garcia Sanchez, M.I. Garcia Garcia, E. Garcia Sanchez and J.L. Munoz Bellido. 1991. Antibiotic susceptibility profile of *Xanthomonas maltophilia*. In vitro activity of β-lactam/β-lactamase inhibitor combinations. Diagn. Microbiol. Infect. Dis. *14*: 239–243.

Gardan, L., C. Bollet, M. Abu-Ghorrah, F. Grimont and P.A.D. Grimont. 1992. DNA relatedness among the pathovar strains of *Pseudomonas syringae* subsp. *savastanoi* Janse (1982) and proposal of *Pseudomonas savastanoi* sp. nov. Int. J. Syst. Bacteriol. *42*: 606–612.

Gardan, L., H. Shafik, S. Belouin, R. Broch, F. Grimont and P.A.D. Grimont. 1999. DNA relatedness among the pathovars of *Pseudomonas syringae* and description of *Pseudomonas tremae* sp. nov. and *Pseudomonas cannabina* sp. nov. (*ex* Sutic and Dowson 1959). Int. J. Syst. Bacteriol. *49*: 469–478.

Gardner, G.A. 1980. Identification and ecology of salt requiring *Vibrio* associated with cured meats. Meat Sci. *5*: 71–81.

Gardner, I.A., R. Kasten, G.J. Eamns, K.P. Snipes and R.J. Anderson. 1994. Molecular fingerprinting of *Pasteurella multicoida* associated with progressive atrophic rhinitis in swine herds. J. Vet. Diagn. Invest. *6*: 442–447.

Gardner, J.M. and C.I. Kado. 1972. Comparative base sequence homologies of the deoxyribonucleic acids of *Erwinia* species and other *Enterobacteriaceae*. Int. J. Syst. Bacteriol. *22*: 201–209.

Gardner, J.M. and C.I. Kado. 1976. Polygalacturonic acid trans-eliminase in the osmotic shock fluid of *Erwinia rubrifaciens*: characterization of the purified enzyme and its effect on plant cells. J. Bacteriol. *127*: 451–460.

Gardner, M.W. and J.B. Kendrick. 1923. Bacterial spot of cowpea. Science (Wash. D. C.) *57*: 275.

Gardner, S.E., S.E. Fowlston and W.L. George. 1987. In vitro production of cholera toxin-like activity by *Plesiomonas shigelloides*. J. Infect. Dis. *156*: 720–722.

Garg, R.P., A.L. Menon, K. Jacobs, R.M. Robson and R.L. Robson. 1994. The *hypE* gene completes the gene cluster for H$_2$ oxidation in *Azotobacter vinelandii*. J. Mol. Biol. *236*: 390–396.

Garibaldi, A. and D.F. Bateman. 1973. Pectolytic, cellulolytic and proteolytic enzymes produced by isolated of *Erwinia chrysanthemi* Burkh., McFad. et Dim. Phytopathol. Mediterr. *12*: 30–35.

Garibaldi, J.A. 1967. Media for the enhancement of fluorescent pigment production by *Pseudomonas* species. J. Bacteriol. *94*: 1296–1299.

Garland, S.M. and M.G. Prichard. 1983. *Actinobacillus actinomycetemcomitans* causing a mediastinal abscess. Thorax *38*: 472–473.

Garner, J.G. 1979. Isolation of *Actinobacillus actinomycetemcomitans* and *Hemophilus aphrophilus* at Auckland hospital. N. Z. Med. J. *89*: 384–386.

Garrett, C.M.E., C.G. Panagopoulos and J.E. Crosse. 1966. Comparison of plant pathogenic pseudomonads from fruit trees. J. Appl. Bacteriol. *29*: 342–356.

Garriga, M., M.A. Ehrmann, J. Arnau, M. Hugas and R.F. Vogel. 1998. *Carnimonas nigrificans* gen. nov., sp. nov., a bacterial causative agent for black spot formation on cured meat products. Int. J. Syst. Bacteriol. *48*: 677–686.

Garrity, G.M., A. Brown and R.M. Vickers. 1980. *Tatlockia micdadei*, gen. nov., comb. nov., and *Fluoribacter bozemanae*, gen. nov., sp. nov.: two new organisms resembling *Legionella pneumophila*. Int. J. Syst. Bacteriol. *30*: 609–614.

Garrote, A., R. Bonet, S. Merino, M.D. Simon-Pujol and F. Congregado. 1992. Occurrence of a capsule in *Aeromonas salmonicida*. FEMS Microbiol. Lett. *95*: 127–132.

Gaston, M.A. 1987a. Evaluation of a bacteriophage-typing scheme for *Enterobacter cloacae*. J. Med. Microbiol. *24*: 291–295.

Gaston, M.A. 1987b. Isolation and selection of a bacteriophage-typing set for *Enterobacter cloacae*. J. Med. Microbiol. *24*: 285–290.

Gaston, M.A., C. Bucher and T.L. Pitt. 1983. O-Serotyping scheme for *Enterobacter cloacae*. J. Clin. Microbiol. *18*: 1079–1083.

Gaston, M.A. and T.L. Pitt. 1989a. Improved O-serotyping method for *Serratia marcescens*. J. Clin. Microbiol. *27*: 2702–2705.

Gaston, M.A. and T.L. Pitt. 1989b. O-antigen specificities of the serotype strains of *Serratia marcescens*. J. Clin. Microbiol. *27*: 2697–2701.

Gaston, M.A., T.A. Vale, B. Wright, P. Cox and T.L. Pitt. 1986. Monoclonal-antibodies to the surface-antigens of *Pseudomonas aeruginosa*. FEMS Microbiol. Lett. *37*: 357–361.

Gatz, C., J. Altschmied and W. Hillen. 1986. Cloning and expression of the *Acinetobacter calcoaceticus* mutarotase gene in *Escherichia coli*. J. Bacteriol. *168*: 31–39.

Gatz, C. and W. Hillen. 1986. *Acinetobacter calcoaceticus* encoded mutarotase: nucleotide sequence analysis of the gene and characterization of its secretion in *Escherichia coli*. Nucleic Acids Res. *14*: 4309–4323.

Gäumann, E. 1923. Über zwei Bananenkrankheiten in Niederländisch Indien. Z. Pflanzenkr. Pflanzenpathol. Pflanzenschutz. *33*: 1–17.

Gauthier, G., M. Gauthier and R. Christen. 1995a. Phylgenetic analysis of the genera *Alteromonas*, *Shewanella*, and *Moritella* using genes coding for small-subunit rRNA sequences and division of the genus *Alteromonas* into two genera, *Alteromonas* (emend.) and *Pseudoalteromonas* gen. nov., and proposal of twelve new species combinations. Int. J. Syst. Bacteriol. *45*: 755–761.

Gauthier, G., B. Lafay, R. Ruimy, V. Breittmayer, J.L. Nicolas, M. Gauthier and R. Christen. 1995b. Small-subunit rRNA sequences and whole DNA relatedness concur for the reassignment of *Pasteurella piscicida* (Snieszko et al.) Janssen and Surgalla to the genus *Photobacterium* as

Photobacterium damsela subsp. *piscicida* comb. nov. Int. J. Syst. Bacteriol. *45*: 139–144.

Gauthier, M.J. 1976a. *Alteromonas rubra* sp. nov., a new marine antibiotic-producing bacterium. Int. J. Syst. Bacteriol. *26*: 459–466.

Gauthier, M.J. 1976b. Modification of bacterial respiration by a macro-molecular polyanionic antibiotic produced by a marine *Alteromonas*. Antimicrob. Agents Chemother. *9*: 361–366.

Gauthier, M.J. 1976c. Morphological, physiological, and biochemical characteristics of some violet-pigmented bacteria isolated from sea-water. Can. J. Microbiol. *22*: 138–149.

Gauthier, M.J. 1977. *Alteromonas citrea*, a new Gram-negative, yellow-pigmented species from seawater. Int. J. Syst. Bacteriol. *27*: 349–354.

Gauthier, M.J. 1982. Validation of the name *Alteromonas luteoviolacea*. Int. J Syst. Bacteriol. *32*: 82–86.

Gauthier, M.J. and V.A. Breittmayer. 1979. A new antibiotic-producing bacterium from seawater: *Alteromonas aurantia* sp. nov. Int. J. Syst. Bacteriol. *29*: 366–372.

Gauthier, M.J. and V.A. Breittmayer. 1992. The genera *Alteromonas* and *Marinomonas*. *In* Balows, Trüper, Dworkin, Harder and Schleifer (Editors), The Prokaryotes. A Handbook on the Biology of Bacteria: Ecophysiology, Isolation, Identification, Applications, 2nd Ed., Vol. 3, Springer-Verlag, New York. 3046–3070.

Gauthier, M., L. Fernandez-Linares, M. Acquaviva and J.C. Bertrand. 1997. Influence of glycine betaine on degradation of eicosane by *Marinobacter hydrocarbonoclasticus* at high salinity. Syst. Appl. Microbiol. *20*: 150–153.

Gauthier, M.J. and G.N. Flatau. 1976. Antibacterial activity of marine violet-pigmented *Alteromonas* with special reference to the production of brominated compounds. Can. J. Microbiol. *22*: 1612–1619.

Gauthier, M.J., B. Lafay, R. Christen, L. Fernandez, M. Acquaviva, P. Bonin and J.C. Bertrand. 1992. *Marinobacter hydrocarbonoclasticus* gen. nov., sp. nov., a new, extremely halotolerant, hydrocarbon-degrading marine bacterium. Int. J. Syst. Bacteriol. *42*: 568–576.

Gavini, F., C. Ferragut, D. Izard, P.A. Trinel, H. Leclerc, B. Lefebvre and D.A.A. Mossel. 1979. *Serratia fonticola*, a new species from water. Int. J. Syst. Bacteriol. *29*: 92–101.

Gavini, F., C. Ferragut, B. Lefebre and H. Leclerc. 1976a. Étude taxonomique d'enterobactéries appartenant ou apparentées au genre *Enterobacter*. Ann. Inst. Pasteur Microbiol. *127B*: 317–335.

Gavini, F., B. Holmes, D. Izard, A. Beji, A. Bernigaud and E. Jakubczak. 1989a. Numerical taxonomy of *Pseudomonas alcaligenes*, *Pseudomonas pseudoalcaligenes*, *Pseudomonas mendocina*, *Pseudomonas stutzeri*, and related bacteria. Int. J. Syst. Bacteriol. *39*: 135–144.

Gavini, F., D. Izard, C. Ferragut, J.J. Farmer, III and H. Leclerc. 1983a. Separation of *Kluyvera* and *Buttiauxella* by biochemical and nucleic acid methods. Int. J. Syst. Bacteriol. *33*: 880–882.

Gavini, F., D. Izard, P.A.D. Grimont, A. Beji, E. Ageron and H. Leclerc. 1986. Priority of *Klebsiella planticola* Bagley, Seidler, and Brenner 1982 over *Klebsiella trevisanii* Ferragut, Izard, Gavini, Kersters, Deley, and Leclerc 1983. Int. J. Syst. Bacteriol. *36*: 486–488.

Gavini, F., H. Leclerc, B. Lefèbvre, C. Ferragut and D. Izard. 1977. Étude taxonomique d'entérobactéries appartenant ou apparentées au genre *Klebsiella*. Ann. Microbiol. (Inst. Pasteur) *128B*: 45–49.

Gavini, F., B. Lefebvre and H. Leclerc. 1976b. Positions taxonomiques d'entérobactéries H$_2$S-par rapport au genre *Citrobacter*. Ann. Microbiol. (Paris) *127a*: 275–295.

Gavini, F., B. Lefebvre and H. Leclerc. 1983b. Taxonomic study of strains belonging or related to the genus *Erwinia*, Herbicola group, and the species *Enterobacter agglomerans*. Syst. Appl. Microbiol. *4*: 218–235.

Gavini, F., J. Mergaert, A. Beji, C. Mielcarek, D. Izard, K. Kersters and J. De Ley. 1989b. Transfer of *Enterobacter agglomerans* (Beijerinck 1888) Ewing and Fife 1972 to *Pantoea* gen. nov. as *Pantoea agglomerans* comb. nov. and description of *Pantoea dispersa* sp. nov. Int. J. Syst. Bacteriol. *39*: 337–345.

Gavrilovic, L., R.W. O'Brien and R.L. Sanders. 1982. Secretion of amylase by the marine bacterium *Alteromonas rubra*. Aust. J. Biol. Sci. *35*: 111–124.

Gebhard, F. and K. Smalla. 1998. Transformation of *Acinetobacter* sp. strain

BD413 by transgenic sugar beet DNA. Appl. Environ. Microbiol. *64*: 1550–1554.

Geier, G. and K. Geider. 1993. Characterization and influence on virulence of the levansucrase gene from the fireblight pathogen *Erwinia amylovora*. Physiol. Mol. Plant Pathol. *42*: 387–404.

Geiges, O., B. Staehlin and B. Baumann. 1990. The microbiological evaluation of prepared salad vegetables and sprouts. Mitt. Geb. Lebensmittelunters. Hyg. *81*: 684–271.

Geiselbrecht, A.D. 1998. The distribution and PAH-degradative potential of *Cycloclasticus* spp. in the marine environment, Thesis, University of Washington, Seattle.

Geiselbrecht, A.D., B.P. Hedlund, M.A. Tichi and J.T. Staley. 1998. Isolation of marine polycyclic aromatic hydrocarbon (PAH)-degrading *Cycloclasticus* strains from the Gulf of Mexico and comparison of their PAH degradation with that of Puget Sound *Cycloclasticus* strains. Appl. Environ. Microbiol. *64*: 4703–4710.

Geiselbrecht, A.D., R.P. Herwig, J.W. Deming and J.T. Staley. 1996. Enumeration and phylogenetic analysis of polycyclic aromatic hydrocarbon-degrading marine bacteria from Puget Sound sediments. Appl. Environ. Microbiol. *62*: 3344–3349.

Geissdörfer, W., S.C. Frosch, G. Haspel, S. Ehrt and W. Hillen. 1995. Two genes encoding protein with similarities to rubredoxin and rubredoxin reductase are required for conversion of dodecane to lauric acid in *Acinetobacter calcoaceticus* ADP1. Microbiology (Reading) *141*: 1425–1432.

Geissdörfer, W., A. Ratajczak and W. Hillen. 1998. Transcription of *ppk* from *Acinetobacter* sp. strain ADP1, encoding a putative polyphosphate kinase, is induced by phosphate starvation. Appl. Environ. Microbiol. *64*: 896–901.

Gelev, I., E. Gelev, A.G. Steigerwalt, G.P. Carter and D.J. Brenner. 1990. Identification of the bacterium associated with haemorrhagic septicaemia in rainbow trout as *Hafnia alvei*. Res. Microbiol. *141*: 573–576.

Gelinas, R.E., P.A. Myers and R.J. Roberts. 1977. Two sequence-specific endonucleases from *Moraxella bovis*. J. Mol. Biol. *114*: 169–179.

Gemski, P.J., D.E. Koeltzow and S.B. Formal. 1975. Phage conversion of *Shigella flexneri* group antigens. Infect. Immun. *11*: 685–691.

Gemski, P., J.R. Lazere and T. Casey. 1980a. Plasmid associated with pathogenicity and calcium dependency of *Yersinia enterocolitica*. Infect. Immun. *27*: 682–685.

Gemski, P., J.R. Lazere, T. Casey and I.A. Wohlhieter. 1980b. Presence of a virulence-associated plasmid in *Yersinia pseudotuberculosis*. Infect. Immun. *28*: 1044–1047.

Gennari, M. and P. Lombardi. 1993. Comparative characterization of *Acinetobacter* strains isolated from different foods and clinical sources. Zentbl. Bakteriol. *279*: 553–564.

Gennari, M., M. Parini, D. Volpon and M. Serio. 1992. Isolation and characterization by conventional methods and genetic transformation of *Psychrobacter* and *Acinetobacter* from fresh and spoiled meat, milk and cheese. Int. J. Food Microbiol. *15*: 61–75.

Gennari, M. and F. Stegagno. 1985. Isolation and characterization of *Acinetobacter calcoaceticus* from fresh, frozen and stored fish products. Microbiol. Alim. Nutrit. *3*: 247–259.

Gennari, M. and F. Stegagno. 1986. Isolation and characterization of *Acinetobacter calcoaceticus* from raw, washed and frozen vegetables. Arch. Vet. Ital. *37*: 131–137.

Genne, D., H.H. Siegrist, P. Monnier, M. Nobel, L. Humair and A. deTorrente. 1996. *Pasteurella multocida* endocarditis: Report of a case and review of the literature. Scand. J. Infect. Dis. *28*: 95–97.

Genthner, F.J., L.A. Hook and W.R. Strohl. 1985. Determination of the molecular mass of bacterial genomic DNA and plasmid copy number by high-pressure liquid chromatography. Appl. Environ. Microbiol. *50*: 1007–1013.

Georgala, D.S.L. 1958. The bacteria of the skin of North Sea cod. J. Gen. Microbiol. *18*: 84–91.

George, J.R., L. Pine, M.W. Reeves and W.K. Harrell. 1980. Amino acid requirements of *Legionella pneumophila*. J. Clin. Microbiol. *11*: 286–291.

George, S.J., A.J.M. Richards, A.J. Thomson and M.G. Yates. 1984. *Azo-tobacter chroococcum* 7Fe ferredoxin. Two pH-dependent forms of the reduced 3Fe clusters and its conversion to a 4Fe cluster. Biochem. J. *224*: 247–251.

Georges, C. and J.M. Meyer. 1995. High-molecular-mass, iron-repressed cytoplasmic proteins in fluorescent *Pseudomonas* - potential peptide-synthetases for pyoverdine biosynthesis. FEMS Microbiol. Lett. *132*: 9–15.

Gerin, C. and M. Goutx. 1993. Separation and quantification of phospholipids from marine-bacteria with the Iatroscan Mark IV TLC-FID. JPC-J. Planar Chromatogr.-Mod. TLC. *6*: 307–312.

Gerischer, U., D.A. D'Argenio and L.N. Ornston. 1996. IS1236, a newly discovered member of the IS3 family, exhibits varied patterns of insertion into the *Acinetobacter calcoaceticus* chromosome. Microbiology (Reading) *142*: 1825–1831.

Gerischer, U., A. Saguraa and L.N. Ornston. 1998. *PcaU*, a transcriptional activator gene for protocatechuate utilization in *Acinetobacter*. J. Bacteriol. *180*: 1512–1524.

Gerner-Smidt, P. 1989. Frequency of plasmids in strains of *Acinetobacter calcoaceticus*. J. Hosp. Infect. *14*: 23–28.

Gerner-Smidt, P. 1992. Ribotyping of the *Acinetobacter calcoaceticus-Acinetobacter baumannii* complex. J. Clin. Microbiol. *30*: 2680–2685.

Gerner-Smidt, P. and W. Frederiksen. 1993. *Acinetobacter* in Denmark: I. Taxonomy, antibiotic susceptibility and pathogenicity of 112 clinical strains. APMIS *101*: 815–825.

Gerner-Smidt, P. and I. Tjernberg. 1993. *Acinetobacter* in Denmark: II. Molecular studies of the *Acinetobacter calcoaceticus-Acinetobacter baumannii* complex. APMIS *101*: 826–832.

Gerner-Smidt, P., I. Tjernberg and J. Ursing. 1991. Reliability of phenotypic tests for identification of *Acinetobacter* spp. J. Clin. Microbiol. *29*: 277–282.

Gerrard, J.G., S. McNevin, D. Alfredson, R. Forgan-Smith and N. Fraser. 2003. *Photorhabdus* species: Bioluminescent bacteria as emerging human pathogens? Emerg. Infect. Dis. *9*: 251–254.

Gerritsen, L.J., G. de Raay and P.H. Smits. 1992. Characterization of form variants of *Xenorhabdus luminescens*. Appl. Environ. Microbiol. *58*: 1975–1979.

Gcyid, A., J. Fletcher, B.A. Gashe and A. Ljungh. 1996. Invasion of tissue culture cells by diarrhoeagenic strains of *Escherichia coli* which lack the enteroinvasive *inv* gene. FEMS Immunol, Med. Microbiol. *14*: 15–24.

Gherna, R.L., J.H. Werren, W. Weisburg, R. Cote, C.R. Woese, L. Mandelco and D.J. Brenner. 1991. *Arsenophonus nasoniae*, gen. nov., sp. nov., the causative agent of the son-killer trait in the parasitic wasp *Nasonia vitripennis*. Int. J. Syst. Bacteriol. *41*: 563–565.

Ghigo, E., C. Capo, C.H. Tung, D. Raoult, J.P. Gorvel and J.L. Mege. 2002. *Coxiella burnetii* survival in THP-1 monocytes involves the impairment of phagosome maturation: IFN-gamma mediates its restoration and bacterial killing. J. Immunol. *169*: 4488–4495.

Ghigo, J.M. and C. Wandersman. 1994. A carboxyl-terminal four-amino acid motif is required for secretion of the metalloprotease PrtG through the *Erwinia chrysanthemi* protease secretion pathway. J. Biol. Chem. *269*: 8979–8985.

Giammanco, G., J. Buissière, M. Toucas, G. Brault and L. Le Minor. 1980. Intrérêt taxonomique de la recherche de la γ-glutamyltransférase chez les *Enterobacteriaceae*. Ann. Microbiol. Inst. Pasteur (Paris) *131A*: 181–187.

Gibbons, N.E. 1969. Isolation, growth, and requirements of halophilic bacteria. *In* Norris and Gibbons (Editors), Methods in Microbiology, Vol. 3B, Academic Press, London. 169–183.

Gibert, I., J. Barbe and J. Casadesus. 1990. Distribution of insertion sequence IS200 in *Salmonella* and *Shigella*. J. Gen. Microbiol. *136*: 2555–2560.

Gibson, F. and D.I. Magrath. 1969. The isolation and characterization of a hydroxamic acid (aerobactin) formed by *Aerobacter aerogenes* 62-I. Biochim. Biophys. Acta *192*: 175–184.

Gibson, J., E. Stackebrandt, L.B. Zablen, R. Gupta and C.R. Woese. 1979. A phylogenetic analysis of the purple photosynthetic bacteria. Curr. Microbiol. *3*: 59–64.

Gilardi, G.L. 1971. Characterization of *Pseudomonas* species isolated from clinical specimens. Appl. Microbiol. *21*: 414–419.

Gilardi, G.L. 1972. Infrequently encountered *Pseudomonas* species causing infection in humans. Ann. Intern. Med. *77*: 211–215.

Gilardi, G.L. and E. Bottone. 1971. *Erwinia* and yellow-pigmented *Enterobacter* isolates from human sources. Antonie Leeuwenhoek *37*: 529–535.

Gilchrist, M.J.R. 1995. *Enterobacteriaceae*: opportunistic pathogens and other genera. *In* Murray, Pfaller, Tenover, Yoklken and Baron (Editors), Manual of Clinical Microbiology, 6th Ed., ASM Press, Washington, D.C. 457–464.

Gilder, H. and S. Granick. 1947. Studies on the *Haemophilus* group of organisms. Quantitative aspects of growth on various porphin compounds. J. Gen. Physiol. *31*: 103–117.

Giles, J.S., H. Hariharan and S.B. Heaney. 1995. The plasmid profiles of fish pathogenic isolates of *Aeromonas salmonicida*, *Vibrio anguillarum*, and *Vibrio ordalii* from the Atlantic and Pacific coasts of Canada. Can. J. Microbiol. *41*: 209–216.

Gill, S., J. Belles-Isles, G. Brown, S. Gagne, C. Lemieux, J.P. Mercier and P. Dion. 1994. Identification of variability of ribosomal DNA spacer from *Pseudomonas* soil isolates. Can. J. Microbiol. *40*: 541–547.

Gilleland, H.E. and R.G.E. Murray. 1976. Ultrastructure study of polymyxin-resistant isolates of *Pseudomonas aeruginosa*. J. Bacteriol. *125*: 267–281.

Gilleland, H.E., J.D. Stinnett, I.L. Roth and R.G. Eagon. 1973. Freeze-etch study of *Pseudomonas aeruginosa*: localization within the cell wall of an ethylene-diaminetetraacetate-extractable component. J. Bacteriol. *113*: 417–432.

Gillis, M. and J. De Ley. 1980. Intra- and intergeneric similarities of the ribosomal ribonucleic acid cistrons of *Acetobacter* and *Gluconobacter*. Int. J. Syst Bacteriol. *30*: 7–27.

Gilman, R.H., M. Madasamy, E. Gan, M. Mariappan, C.E. Davis and K.A. Kyser. 1971. *Edwardsiella tarda* in jungle diarrhoea and a possible association with *Entamoeba histolytica*. Southeast Asian J. Trop. Med. Public Health *2*: 186–189.

Gilmour, N.J.L. and J.S. Gilmour. 1989. Pasteurellosis of sheep. *In* Adlam and Rutter (Editors), *Pasteurella* and Pasteurellosis, Academic Press, London. pp. 223–262.

Gilsdorf, J.R. 1998. Antigenic diversity and gene polymorphisms in *Haemophilus influenzae*. Infect. Immun. *66*: 5053–5059.

Gilsdorf, J.R., K.W. McCrea and C.F. Marrs. 1997. Role of pili in *Haemophilus influenzae* adherence and colonization. Infect. Immun. *65*: 2997–3002.

Gimenez, D.F. 1964. Staining rickettsiae in yolk-sac cultures. Stain Technol. *39*: 135–140.

Ginard, M., J. Lalucat, B. Tummler and U. Romling. 1997. Genome organization of *Pseudomonas stutzeri* and resulting taxonomic and evolutionary considerations. Int. J. Syst. Bacteriol. *47*: 132–143.

Ginoza, H.S. and T.A. Matney. 1963. Transmission of a resistance transfer factor from *Escherichia coli* to two species of *Pasteurella*. J. Bacteriol. *85*: 1177–1178.

Ginther, C.L. 1978. Genetic analysis of *Acinetobacter calcoaceticus* proline auxotrophs. J. Bacteriol. *133*: 439–441.

Giordano-Dias, C.M., A.R. Figueiredo, S.M.P. Oliveira, J.M. D'Almeida, E. Hofer and R.P. Brazil. 1997. Isolation and identification of bacteria from sandflies of *Lutzomyia longipalpis* (Diptera: Psychodidae) maintained in laboratory. Mem. Inst. Oswalo Cruz *92*: 320.

Giovannoni, S.J. and L. Margulis. 1981. A red *Benekea* from Laguna Figueroa, Baja California. Microbios. *30*: 47–63.

Girard, G. 1953. Méthodes permettant de différencier *P. pestis* de *P. pseudotuberculosis*. Bull. Wld. Hlth. Org. *9*: 645–653.

Girod, J.C., R.C. Reichman, W.C. Winn, Jr., D.N. Klaucke, R.L. Uogt and R. Dolin. 1982. Pneumonic and nonpneumonic forms of legionellosis: result of a common-source exposure to *Legionella pneumophila*. Arch. Intern. Med. *142*: 545–547.

Girón, J.A. 1995. Expression of flagella and motility by *Shigella*. Mol. Microbiol. *18*: 63–75.

Giroud, P., N. Dumas and B. Hurpin. 1958. Essais d'adaptation a la souris

blanche de la rickettsie agent de la maladie bleue de *Melolontha melolontha* L.: voie pulmonaire et voie buccale. C.R. Acad. Science, Series D. *247*: 2499–2501.

Gitaitis, R.D., C.J. Chang, K. Sijam and C.C. Dowler. 1991. A differential medium for semiselective isolation of *Xanthomonas campestris* pv. *vesicatoria* and other cellulolytic xanthomonads from various natural sources. Plant Dis. *75*: 1274–1278.

Gitaitis, R.D., J.D. Hamm and P.F. Bertrand. 1988. Differentiation of *Xanthomonas campestris* pv. *pruni* from other yellow-pigmented bacteria by the refractive quality of bacterial colonies on an agar medium. Plant Dis. *72*: 416–417.

Gitaitis, R.D., M.J. Sasser, R.W. Beaver, T.B. McInnes and R.E. Stall. 1987. Pectolytic xanthomonads in mixed infections with *Pseudomonas syringae* pv. *syringae*, *Pseudomonas syringae* pv. *tomato*, and *Xanthomonas campestris* pv. *vesicatoria* in tomato and pepper transplants. Phytopathology *77*: 611–615.

Givaudan, A., S. Baghdiguian, A. Lanois and N. Boemare. 1995. Swarming and swimming changes concomitant with phase variation in *Xenorhabdus nematophilus*. Appl. Environ. Microbiol. *61*: 1408–1413.

Givaudan, A., A. Lanois and N.E. Boemare. 1996. Cloning and nucleotide sequence of a flagellin encoding genetic locus from *Xenorhabdus nematophilus*: phase variation leads to differential transcription of two flagellar genes (*fliCD*). Gene *183*: 243–253.

Gladman, G., P.J. Connor, R.F. Williams and T.J. David. 1992. Controlled study of *Pseudomonas cepacia* and *Pseudomonas maltophilia* in cystic fibrosis. Arch. Dis. Child. *67*: 192–195.

Glare, T.R. and M.R.H. Hurst. 2002. A new family of bacterial insecticidal toxins: the Tc family from *Photorhabdus*, *Xenorhabdus*, *Serratia* and other bacteria. *In* Akhurst, Beard and Hughes (Editors), Biotechnology of *Bacillus thuringiensis* and its Environmental Impact, CSIRO, Canberra. 129–134.

Glennon, J., P.J. Ryan, C.T. Keane and J.P. Rees. 1988. Circumcision and periurethral carriage of *Proteus mirabilis* in boys. Arch. Dis. Child. *63*: 556–557.

Glew, R.H., R.C. Moellering, Jr. and L.J. Kunz. 1977. Infections with *Acinetobacter calcoaceticus* (*Herellea vaginicola*): clinical and laboratory studies. Medicine. *56*: 79–97.

Glick, B.R., H.E. Brooks and J.J. Pasternak. 1986. Physiological effects of plasmid DNA transformation on *Azotobacter vinelandii*. Can. J. Microbiol. *32*: 145–148.

Glick, B.R., B.J. Butler, C.I. Mayfield and J.J. Pasternak. 1989. Effect of transformation of *Azotobacter vinelandii* with the low copy number plasmid pRK290. Curr. Microbiol. *19*: 143–146.

Glick, T.H., M.B. Gregg, B. Berman, G. Mallison, W.W.J. Rhodes and I. Kassanoff. 1978. Pontiac fever. An epidemic of unknown etiology in a health department. I. Clinical and epidemiologic aspects. Am. J. Epidemiol. *107*: 149–160.

Glockner, A.B. and W.G. Zumft. 1996. Sequence analysis of an internal 9.72 kb segment from the 30 kb denitrification gene cluster of *Pseudomonas stutzeri*. Biochim. Biophys. Acta-Bioenerg. *1277*: 6–12.

Glöckner, F.O., H.D. Babenzien and R. Amann. 1998. Phylogeny and identification *in situ* of *Nevskia ramosa*. Appl. Environ. Microbiol. *64*: 1895–1901.

Glöckner, F.O., H.-D. Babenzien, J. Wulf and R. Amann. 1999. Phylogeny and diversity of *Achromatium oxaliferum*. Syst. Appl. Microbiol. *22*: 28–38.

Glynn, M.K., C. Bopp, W. Dewitt, P. Dabney, M. Mokhtar and F.J. Angulo. 1998. Emergence of multidrug-resistant *Salmonella enterica typhimurium* DT104 infections in the United States. New Engl. J. Med. *338*: 1333–1338.

Gmeiner, J. and H.P. Kroll. 1981. Murein biosynthesis and O-acetylation of N-acetylmuramic acid during the cell division cycle of *Proteus mirabilis*. Eur. J. Biochem. *117*: 171–177.

Gmür, R., H. McNabb, T.J. van Steenbergen, P. Baehni, A. Mombelli, A.J. van Winkelhoff and B. Guggenheim. 1993. Seroclassification of hitherto nontypeable *Actinobacillus actinomycetemcomitans* strains: evidence for a new serotype e. Oral Microbiol. Immunol. *8*: 116–120.

Gobat, P.F. and T. Jemmi. 1995. Comparison of seven selective media for the isolation of mesophilic *Aeromonas* species in fish and meat. Int. J. Food Microbiol. *24*: 375–384.

Goebel, B.M., P.R. Norris and N.P. Burton. 2000. Acidophiles in biomining. *In* Priest and Goodfellow (Editors), Applied Microbial Systematics, Kluwer, Dordrecht. pp. 293–314.

Goel, U., T. Kauri, H.W. Ackermann and D.J. Kushner. 1996. A moderately halophilic *Vibrio* from a Spanish saltern and its lytic bacteriophage. Can. J. Microbiol. *42*: 1015–1023.

Goldberg, D.J., P.W. Collier, R.J. Fallon, T.M. McKay, T.A. Markwick, J.G. Wrench, J.A. Emslie, G.I. Forbes, A.C. Macpherson and D. Reid. 1989. Lochgoilhead fever: outbreak of non-pneumonic legionellosis due to *Legionella micdadei*. Lancet *1*: 316–318.

Goldberg, J. 1959. Studies on granuloma inguinale. IV. Growth requirements of Donovania granulomatis and its relationship to the natural habitat of the organism. Br. J. Vener. Dis. *35*: 266–268.

Goldberg, J. 1962. Studies on granuloma inguinale. V. Isolation of a bacterium resembling Donovania granulomatis from the feces of a patient with Granuloma inguinale. Br. J. Vener. Dis. *38*: 99–102.

Goldberg, J.B. 1992. Regulation of alginate volume in *Pseudomonas aeruginosa*. *In* Galli, Silver and Witholt (Editors), *Pseudomonas*: Molecular Biology and Biotechnology, American Society for Microbiology, Washington D.C. 75–82.

Goldberg, J.B., W.L. Gorman, J.L. Flynn and D.E. Ohman. 1993. A mutation in *algN* permits trans activation of alginate production by *algT* in *Pseudomonas* species. J. Bacteriol. *175*: 1303–1308.

Goldberg, J., R.H. Weaver, H. Packer and W.G. Simpson. 1953. Studies on granuloma inguinale. II. The complement fixation test in the diagnosis of granuloma inguinale. Am. J. Syphilis, Gonorrhea and Venereal Dis. *37*: 71–76.

Goldberg, M.B., V.J. DiRita and S.B. Calderwood. 1990. Identification of an iron-regulated virulence determinant in *Vibrio cholerae*, using Tn*phoA* mutagenesis. Infect. Immun. *58*: 55–60.

Goldberg, M. and H. Gilboa. 1978. Sodium exchange between two sites. The binding of sodium to halotolerant bacteria. Biochim. Biophys. Acta *538*: 268–283.

Goldstein, E.J.C., D.M. Citron, C.V. Merriam, K. Tyrrell and Y. Warren. 1999. Activities of gemifloxacin (SB 265805, LB 20304) compared to those of other oral antimicrobial agents against unusual anaerobes. Antimicrob. Agents Chemother. *43*: 2726–2730.

Goldstein, J.D., J.L. Keller, W.C. Winn, Jr. and R.L. Myerowitz. 1982. Sporadic *Legionellaceae* pneumonia in renal transplant recipients: a survey of 70 autopsies, 1964–1979. Arch. Pathol. Lab. Med. *106*: 108–111.

Golland, L.C., D.R. Hodgson, J.L. Hodgson, M.A. Brownlow, D.R. Hutchins, R.J. Rawlinson, M.B. Collins, S.A. McClintock and A.L. Raisis. 1994. Peritonitis associated with *Actinobacillus equuli* in horses: 15 cases (1982–1992). J. Am. Vet. Med. Assoc. *205*: 340–343.

Golovacheva, R.S. and G.I. Karavaiko. 1979. *Sulfobacillus* — a new genus of thermophilic spore-forming bacteria. Mikrobiologiya *47*: 815–822.

Golovliov, I., G. Ericsson, G. Sandström, A. Tärnvik and A. Sjöstedt. 1997. Identification of proteins of *Francisella tularensis* induced during growth in macrophages and cloning of the gene encoding a prominently induced 23-kDa protein. Infect. Immun. *65*: 2183–2189.

Goluszko, P., V. Popov, R. Selvarangan, S. Nowicki, T. Pham and B.J. Nowicki. 1997. Dr fimbriae operon of uropathogenic *Escherichia coli* mediate microtubule-dependent invasion to the HeLa epithelial cell line. J. Infect. Dis. *176*: 158–167.

Gomes, T.A.T., P.A. Blake and L.R. Trabulsi. 1989a. Prevalence of *Escherichia coli* strains with localized diffuse and aggregative adherence to Hela cells in infants with diarrhea and matched controls. J. Clin. Microbiol. *27*: 266–269.

Gomes, T.A., M.A. Vieira, I.K. Wachsmuth, P.A. Blake and L.R. Trabulsi. 1989b. Serotype-specific prevalence of *Escherichia coli* strains with EPEC adherence factor genes in infants with and without diarrhea in São Paulo, Brazil. J. Infect. Dis. *160*: 131–135.

Gomez-Alarcon, R.A., C. O'Dowd, A.Z. Leedle and M.P. Bryant. 1982. 1,4-Napthoquinone and other nutrient requirements of *Succinivibrio dextrinosolvens*. Appl. Environ. Microbiol. *44*: 346–350.

Gómez-Duarte, O.G. and J.B. Kaper. 1995. A plasmid-encoded regulatory region activates chromosomal *eaeA* expression in enteropathogenic *Escherichia coli*. Infect. Immun. *63*: 1767–1776.

Gomez-Lus, P., B.S. Fields, R.F. Benson, W.T. Martin, S.P. O'Connor and C.M. Black. 1993. Comparison of arbitrarily primed polymerase chain reaction, ribotyping, and monoclonal antibody analysis for subtyping *Legionella pneumophila* serogroup 1. J. Clin. Microbiol. *31*: 1940–1942.

Gonzales, H.F. and D.P. Bingham. 1983. Genetic relatedness of *Haemophilus somnus* to select genera of bacteria. Am. J. Vet. Res. *44*: 1793–1795.

Gonzalez, A.B. and E.H. Ruffolo. 1966. *Edwardsiella tarda*: etiologic agent in a post-traumatic subgaleal abscess. South. Med. J. *59*: 340, 346.

Gonzalez, C., C. Gutierrez and T. Grande. 1987. Bacterial flora in bottled uncarbonated mineral drinking water. Can. J. Microbiol. *33*: 1120–1125.

González, C.F. and A.K. Vidaver. 1979. Syringomycin production and holcus spot disease of maize - plasmid-associated properties in *Pseudomonas syringae*. Curr. Microbiol. *2*: 75–80.

Gonzalez, G., K. Sossa, H. Bello, M. Dominguez, S. Mella and R. Zemelman. 1998. Presence of integrons in isolates of different biotypes of *Acinetobacter baumannii* from Chilean hospitals. FEMS Microbiol. Lett. *161*: 125–128.

González, J.M., F. Mayer, M.A. Moran, R.E. Hodson and W.B. Whitman. 1997. *Microbulbifer hydrolyticus* gen. nov., sp. nov., and *Marinobacterium georgiense* gen. nov., sp. nov., two marine bacteria from a lignin-rich pulp mill waste enrichment community. Int. J. Syst. Bacteriol. *47*: 369–376.

González , J.M. and W.B. Whitman. 2002. *Oceanospirillum* and related genera. *In* Dworkin (Editor), The Prokaryotes: An Evolving Electronic Resource for the Microbiological Community, 3rd Ed., Release 3.9, Springer-Verlag, New York. http://link.springer-ny.com/link/service/books/10125/.

Gonzalez, L., G.A. McKinley, D.H. Pincus and S.M. Iovino. 1986. Addition of ten newly described gram-negative organisms to the API-20E and API UNISCEPT 20E data-base. 86th General Meeting of the American Society for Microbiology, American Society for Microbiology.

Goodwin, P.H. 1989. Cloning and expression of *Xylella fastidiosa* antigens in *Escherichia coli* and *Erwinia stewartii*. Can. J. Microbiol. *35*: 487–491.

Goodwin, P.H., J.E. De Vay and C.P. Meredith. 1988a. Roles of water stress and phytotoxins in the development of Pierce's disease of the grapevine. Physiol. Mol. Plant Pathol. *32*: 1–16.

Goodwin, P.H., J.E. DeVay and C.P. Meredith. 1988b. Physiological responses of *Vitis vinifera* cultivar "Chardonnay" to infection by the Pierce's disease bacterium. Physiol. Mol. Plant Pathol. *32*: 17–32.

Goodwin, P.H. and C.P. Meredith. 1988. New clues in understanding Pierce's disease. Calif. Agric. *42*: 6–7.

Goosen, N., H.P. Horsman, R.G. Huinen, A. de Groot and P. van de Putte. 1989a. Genes involved in the biosynthesis of PQQ from *Acinetobacter calcoaceticus*. Antonie Leeuwenhoek *56*: 85–91.

Goosen, N., H.P.A. Horsman, R.G.M. Huinen and P. Van De Putte. 1989b. *Acinetobacter calcoaceticus* genes involved in biosynthesis of the coenzyme pyrrolo-quinoline-quinone: nucleotide sequence and expression in *Escherichia coli* K-12. J. Bacteriol. *171*: 447–455.

Goosen, N., D.A.M. Vermaas and P. van de Putte. 1987. Cloning of the genes involved in synthesis of coenzyme pyrrolo-quinoline-quinone from *Acinetobacter calcoaceticus*. J. Bacteriol. *169*: 303–307.

Gorby, Y.A., F. Caccavo and H. Bolton. 1998. Microbial reduction of cobalt(III)EDTA(−) in the presence and absence of manganese(IV) oxide. Environ. Sci. Technol. *32*: 244–250.

Gordon, D.L., C.R. Philpot and C. McGuire. 1983. *Plesiomonas shigelloides* septic arthritis complicating rheumatoid arthritis. Aust. N. Z. J. Med. *13*: 275–276.

Gordon, J. and P.L.C. Small. 1993. Acid resistance in enteric bacteria. Infect. Immun. *61*: 364–367.

Goris, J., K. Kersters and P. De Vos. 1998. Polyamine distribution among authentic pseudomonads and *Azotobacteraceae*. Syst. Appl. Microbiol. *21*: 285–290.

Gorlenko, V.M. 1974. Oxidation of thiosulfate by *Amoebobacter roseus* in darkness in microaerobic conditions. Mikrobiologiya *43*: 729–731.

Gorlenko, V.M., E.N. Krasil'nikova, O.G. Kikina and N.Y. Tatarinova. 1979. The new motile purple sulfur bacterium *Lamprobacter modestohalophilus* nov. gen., nov. sp. with gas vacuoles. Izv. Akad. Nauk S.S.S.R. S. Biol. *5*: 755–767.

Gorlenko, V.M., E.N. Krasil'nikova, O.G. Kikina and N.Y. Tatarinova. 1988. *In* Validation of the publication of new names and new combinations previously effectively published outside the IJSB. List No. 25. Int. J.Syst. Bacteriol. *38*: 220–222.

Gorman, G.W., J.C. Feeley, A. Steigerwalt, P.H. Edelstein, C.W. Moss and D.J. Brenner. 1985a. *Legionella anisa*, a new species isolated from potable waters and a cooling tower. Appl. Environ. Microbiol. *49*: 305–309.

Gorman, G.W., J.C. Feeley, A. Steigerwalt, P.H. Edelstein, C.W. Moss and D.J. Brenner. 1985b. *In* Validation of the publication of new names and new combinations previously effectively published outside the IJSB. List No. 18. Int. J. Syst. Bacteriol. *35*: 375–376.

Gorris, M.T., B. Alarcon, M.M. Lopez and M. Cambra. 1994. Characterization of monoclonal antibodies specific for *Erwinia carotovora* atroseptica and comparison of serological methods for its sensitive detection on potato tubers. Appl. Environ. Microbiol. *60*: 2076–2085.

Gorshkova, R.P., E.L. Nazarenko, V.A. Zubkov, E.P. Ivanova, Y.S. Ovodov, V.S. Shashkov and Y.A. Knirel. 1993. Structure of the repeating unit of the acidic polysaccharide from *Alteromonas haloplanktis* KMM 156. Bioorg. Khim. *19*: 327–336.

Gorshkova, R.P., E.L. Nazarenko, V.A. Zubkov, V.S. Shashkov, Y.A. Knirel, N.A. Paramonov, S.V. Meshkov and E.P. Ivanova. 1997. Structure of the capsular polysaccharide from *Alteromonas nigrifaciens* IAM 13010T containing 2-acetamido-2,6-dideoxy-l-talose and 3-deoxy-D-manno-octulosonic acid. Carbobydr. Res. *299*: 69–76.

Gosink, J.J. and J.T. Staley. 1995. Biodiversity of gas vacuolate bacteria from Antarctic sea ice and water. Appl. Environ. Microbiol. *61*: 3486–3489.

Gosling, P.J. 1996a. *Aeromonas* species in diseases of animals. *In* Austin, Altwegg, Gosling and Joseph (Editors), The Genus *Aeromonas*, John Wiley and Sons, Ltd., Chichester. pp. 175–121.

Gosling, P.J. 1996b. Pathogenic mechanisms. *In* Austin, Altwegg, Gosling and Joseph (Editors), The Genus *Aeromonas*, John Wiley and Sons, Ltd., Chichester. pp. 245–265.

Gossling, J. 1966. The bacteria isolated from lesions of embolic meningoencephalitis of cattle. Illinois Vet. *9*: 14–18.

Goto, M. 1976. *Erwinia mallotivora* sp. nov., the causal organism of bacterial leaf spot of *Mallotus japonicus* Muell. Int. J. Syst. Bacteriol. *26*: 467–473.

Goto, M. 1983a. *Pseudomonas ficuserectae* sp. nov., the causal agent of bacterial leaf-spot of *Ficus erecta* Thunb. Int. J. Syst. Bacteriol. *33*: 546–550.

Goto, M. 1983b. *Pseudomonas pseudoalcaligenes* subsp. *konjaci* subsp. nov., the causal agent of bacterial leaf blight of konjac (*Amorphopallus konjac* Koch). Int. J. Syst. Bacteriol. *33*: 539–545.

Goto, M. 1988. Sensitivity of plant pathogenic bacteria to 2,4-diamino-6,7-diisopropyl-pteridine and its taxonomic significance. Ann. Phytopathol. Soc. Jpn. *54*: 64–67.

Goto, M. 1993. Bacterial brown spot of *Mallotus japonicus* Muell. Arg. caused by *Xanthomonas campestris* pv. *malloti* pv. nov. Ann. Phytopathol. Soc. Japan *59*: 678–680.

Goto, M. and H. Kuwata. 1988. *Rhizobacter daucus* gen. nov., sp. nov., the causal agent of carrot bacterial dall. Int. J. Syst. Bacteriol . *38*: 233–239.

Goto, M. and T. Makino. 1977. Emendation of *Pseudomonas cissicola*, the causal organism of bacterial leaf spot of *Cayratia japonica* (Thunb.) Gagn. and designation of the neotype strain. Ann. Phytopathol. Soc. Japan *43*: 40–45.

Goto, M. and K. Matsumoto. 1987. *Erwinia carotovora* subsp. *wasabiae*, subsp. nov. isolated from diseased rhizomes and fibrous roots of Japanese horseradish (*Eutrema wasabi* maxim.). Int. J. Syst. Bacteriol. *37*: 130–135.

Goto, M. and N. Okabe. 1958. Bacterial plant diseases in Japan IX. 1. Bacterial stem rot of pea. 2. Halo blight of bean. 3. Bacterial spot of physalis plant, Rept. Fac. Agr. Shizuoka Univ. . pp. 33–49.

Goto, M., A. Toyoshima and S. Tanaka. 1978. Studies on saprophytic survival of *Xanthomonas citri* (Hasse) Dowson. 3. Inoculum density of the bacterium surviving in the saprophytic form. Ann. Phytopathol. Soc. Japan *44*: 197–201.

Gottschalk, M., E. Altman, S. Lacouture, F. De Lasalle and J.D. Dubreuil. 1997. Serodiagnosis of swine pleuropneumonia due to *Actinobacillus pleuropneumoniae* serotypes 7 and 4 using long-chain lipopolysaccharides. Can. J. Vet. Res. *61*: 62–65.

Gottwald, T.R., A.M. Alvarez, J.S. Hartung and A.A. Benedict. 1991. Diversity of *Xanthomonas campestris* pv. *citrumelo* strains associated with epidemics of citrus bacterial spot in Florida (USA) citrus nurseries: correlation of detached leaf, monoclonal antibody, and restriction fragment length polymorphism assays. Phytopathology *81*: 749–753.

Götz, P. 1972. "*Rickettsiella chironomi*": an unusual bacterial pathogen which reproduces by multiple cell division. J. Invertebr. Pathol. *20*: 22–30.

Götz, P. and H.G. Boman. 1985. Insect Immunity. *In* Kerkut and Gilbert (Editors), Comprehensive Insect Physiology, Biochemistry and Pharmacology, Vol. 3-Integument, Respiration and Circulation, Pergamon Press, Oxford. pp. 453–485.

Götz, P., A. Boman and H.G. Boman. 1981. Interactions between insect immunity and an insect-pathogenic nematode with symbiotic bacteria. Proc. R. Soc. Lond. Ser. B Biol. Sci. *212*: 333–350.

Goubau, P., F. Van Aelst, J. Verhaegen and M. Boogaerts. 1988. Septicaemia caused by *Rahnella aquatilis* in an immunocompromised patient. Eur. J. Clin. Microbiol. Infect. Dis. *7*: 697–699.

Gough, C.L., J.M. Dow, C.E. Barber and M.J. Daniels. 1988. Cloning of two endoglucanase genes of *Xanthomonas campestris* pv. *campestris*: analysis of the role of the major endoglucanase in pathogenesis. Mol. Plant-Microbe Interact. *1*: 275–281.

Gould, W.D., C. Hagedorn, T.R. Bardinelli and R.M. Zablotowicz. 1985. New selective media for enumeration and recovery of fluorescent pseudomonads from various habitats. Appl. Environ. Microbiol. *49*: 28–32.

Goullet, P. 1980. Esterase electrophoretic pattern relatedness between *Shigella* species and *Escherichia coli*. J. Gen. Microbiol. *117*: 493–500.

Goullet, P. and B. Picard. 1987. Differentiation of *Shigella* by esterase electrophoretic polymorphism. J. Gen. Microbiol. *133*: 1005–1017.

Goullet, P. and B. Picard. 1989. Comparative electrophoretic polymorphism of esterases and other enzymes in *Escherichia coli*. J. Gen. Microbiol. *135*: 135–143.

Govan, J.R. 1974a. Studies on the pyocins of *Pseudomonas aeruginosa*: morphology and mode of action of contractile pyocins. J. Gen. Microbiol. *80*: 1–15.

Govan, J.R. 1974b. Studies on the pyocins of *Pseudomonas aeruginosa*: production of contractile and flexuous pyocins in *Pseudomonas aeruginosa*. J. Gen. Microbiol. *80*: 17–30.

Govan, J.R.W. 1978. Pyocin typing in *Pseudomonas aeruginosa*. *In* Bergan and Norris (Editors), Methods Microbiol., Vol. 10, Academic Press, London. 61–91.

Govan, J.R.W. and J.A.M. Fyfe. 1978. Mucoid *Pseudomonas aeruginosa* and cystic-fibrosis - resistance of mucoid form to carbenicillin, flucloxacillin and tobramycin and isolation of mucoid variants *in vitro*. J. Antimicrob. Chemother. *4*: 233–240.

Govan, J.R.W., J.A.M. Fyfe and T.R. Jarman. 1981. Isolation of alginate-producing mutants of *Pseudomonas fluorescens*, *Pseudomonas putida* and *Pseudomonas mendocina*. J. Gen. Microbiol. *125*: 217–220.

Govorukhina, N.I. and Y.A. Trotsenko. 1991. *Methylovorus*, a new genus of restricted facultatively methylotrophic bacteria. Int. J. Syst. Bacteriol. *41*: 158–162.

Goward, C.R., G.B. Stevens, R. Tattersall and T. Atkinson. 1992. Rapid large-scale preparation of recombinant *Erwinia chrysanthemi* L-asparaginase. Bioseparation. *2*: 335–341.

Grabovich, M.Y., M.S. Muntyan, V.Y. Lebedeva, V.S. Ustiyan and G.A. Dubinina. 1999. Lithoheterotrophic growth and electron transfer chain components of the filamentous gliding bacterium *Leucothrix mucor* DSM 2157 during oxidation of sulfur compounds. FEMS Microbiol. Lett. *178*: 155–161.

Gradin, J.L. and J.A. Schmitz. 1977. Selective medium for isolation of *Bacteroides nodosus*. J. Clin. Microbiol. *6*: 298–302.

Graf, J. 1999a. Diverse restriction fragment length polymorphism patterns of the PCR-amplified 16S rRNA genes in *Aeromonas veronii* strains and possible misidentification of *Aeromonas* species. J. Clin. Microbiol. *37*: 3194–3197.

Graf, J. 1999b. Symbiosis of *Aeromonas veronii* biovar sobria and *Hirudo medicinalis*, the medicinal leech: a novel model for digestive tract associations. Infect. Immun. *67*: 1–7.

Graham, D.C. 1958. The status of *Erwinia lathyri* and related species. Commonwealth Phytopath. News. *4*: 49–51.

Graham, D.C. 1964. Taxonomy of the soft rot coliform bacteria. Annu. Rev. Phytopathol. *2*: 13–42.

Graham, D.R., J.D. Band, C. Thornsberry, D.G. Hollis and R.E. Weaver. 1990. Infections caused by *Moraxella*, *Moraxella urethralis*, *Moraxella*-like groups M-5 and M-6, and *Kingella kingae* in the USA, 1953–1980. Rev. Infect. Dis. *12*: 423–431.

Graham, D.C. and W. Hodgkiss. 1967. Identity of gram negative, yellow pigmented, fermentative bacteria isolated from plants and animals. J. Appl. Bacteriol. *30*: 175–189.

Gralton, E.M., A.L. Campbell and E.N. Neidle. 1997. Directed introduction of DNA cleavage sites to produce a high-resolution genetic and physical map of the *Acinetobacter* sp. strain ADP1 (BD413UE) chromosome. Microbiology (Reading) *143*: 1345–1357.

Gram, L. and H.H. Huss. 1996. Microbiological spoilage of fish and fish products. Int. J. Food Microbiol. *33*: 121–137.

Gram, T. and P. Ahrens. 1998. Improved diagnostic PCR assay for *Actinobacillus pleuropneumoniae* based on the nucleotide sequence of an outer membrane lipoprotein. J. Clin. Microbiol. *36*: 443–448.

Granato, P.A., E.A. Jurek and L.B. Weiner. 1983. Biotypes of *Haemophilus influenzae*: relationship to clinical source of isolation, serotype, and antibiotic susceptibility. Am J Clin Pathol. *79*: 73–77.

Grangeasse, C., P. Doublet, C. Vincent, E. Vaganay, M. Riberty, B. Duclos and A.J. Cozzone. 1998. Functional characterization of the low-molecular-mass phosphotyrosine-protein phosphatase of *Acinetobacter johnsonii*. J. Mol. Biol. *278*: 339–347.

Granoff, D.M. and R.S. Munson, Jr.. 1986. Prospects for prevention of *Haemophilus influenzae* type b disease by immunization. J. Infect. Dis. *153*: 448–461.

Grant, A.N. and L.A. Laidler. 1993. Assessment of the antimicrobial sensitivity of *Aeromonas salmonicida* isolates from farmed Atlantic salmon in Scotland. Vet. Rec. *133*: 389–391.

Grant, F.A., J.I. Prosser, K. Killham and L.A. Glover. 1992. Luminescence based detection of *Erwinia carotovora* ssp. carotovora in soil. Soil Biol. Biochem. *24*: 961–967.

Grant, M.A., S.D. Weagant and P. Feng. 2001. Glutamate decarboxylase genes as a prescreening marker for detection of pathogenic *Escherichia coli* groups. Appl. Environ. Microbiol. *67*: 3110–3114.

Grant, R.B., J.L. Penner, J.N. Hennessy and B.J. Jackowski. 1981. Transferable urease activity in *Providencia stuartii*. J. Clin. Microbiol. *13*: 561–565.

Grassberger, R. 1897. Beiträge zür Bakteriologie der Influenza. Z. Hyg. Infektionskr. *25*: 453–475.

Gratten, M. 1983. *Haemophilus influenzae* biotype VII. J. Clin. Microbiol. *18*: 1015–1016.

Gratten, M., J. Barker, F. Shann, G. Gerega, J. Montgomery, M. Kajoi and T. Lupiwa. 1985. Non-type b *Haemophilus influenzae* meningitis. Lancet *1*: 1343–1344.

Gray, N.D., R. Howarth, A. Rowan, R.W. Pickup, J.G. Jones and I.M. Head. 1999. Natural communities of *Achromatium oxaliferum* comprise genetically, morphologically, and ecologically distinct subpopulations. Appl. Environ. Microbiol. *65*: 5089–5099.

Gray, N.D., R.W. Pickup, J.G. Jones and I.M. Head. 1997. Ecophysiological evidence that *Achromatium oxaliferum* is responsible for the oxidation of reduced sulfur species in a freshwater sediment. Appl. Environ. Microbiol. *63*: 1905–1910.

Gray, S.J. 1984. *Aeromonas hydrophila* in livestock: incidence, biochemical characteristics and antibiotic susceptibility. J. Hyg. *92*: 365–376.

Grebe, H.H. and K.H. Hinz. 1975. Vorkommen von Bakterien der Gattung *Haemophilus* bei verschiedenen Vogelarten. Zentralbl. Veterinaermed. Riehe B. *22*: 749–757.

Green, M.J., D.M. Anderson, D.M. Norris and S.L. Gubbins. 1979. Antimicrobial resistance in *Haemophilus* species. NZ. Med. J. *90*: 29.

Green, P.N. and I.J. Bousfield. 1983. Emendation of *Methylobacterium* Patt, Cole, and Hanson 1976; *Methylobacterium rhodinum* (Heumann 1962) comb. nov. corrig.; *Methylobacterium radiotolerans* (Ito and Iizuka 1971) comb. nov., corrig.; and *Methylobacterium mesophilicum* (Austin and Goodfellow 1979) comb. nov. Int. J. Syst. Bacteriol. *33*: 875–877.

Greenberg, E.P. and E. Canale-Parola. 1977. Motility of flagellated bacteria in viscous environments. J. Bacteriol. *132*: 356–358.

Greer, P.W., F.W. Chandler and M.D. Hicklin. 1980. Rapid demonstration of *Legionella pneumophila* in unembedded tissue: an adaptation of the Giménez stain. Am. J. Clin. Pathol. *73*: 788–790.

Gregg-Jolly, L.A. and L.N. Ornston. 1994. Properties of *Acinetobacter calcoaceticus recA* and its contribution to intracellular gene conversion. Mol. Microbiol. *12*: 985–992.

Greif, Z., M. Moscona, D. Loeb and H. Spira. 1986. Puerperal *Pasteurella multocida* septicemia. Eur. J. Clin. Microbiol. Infect. Dis. *5*: 657–658.

Greiner, M. and G. Winkelmann. 1991. Fermentation and isolation of herbicolin A, a peptide antibiotic produced by *Erwinia herbicola* strain A 111. Appl. Microbiol. Biotechnol. *34*: 565–569.

Grenier, E., E. Bonifassi, P. Abad and C. Laumond. 1996. Use of species-specific satellite DNAs as diagnostic probes in the identification of *Steinernematidae* and *Heterorhabditidae* entomopathogenic nematodes. Parasitology. *113*: 483–489.

Gress, F.M., R.L. Myerowitz, A.W. Pasculle, C.R.J. Rinaldo and J.N. Dowling. 1980. The ultrastructural morphologic features of Pittsburgh pneumonia agent. Am. J. Pathol. *101*: 63–69.

Griffin, P.M. 1995. *Escherichia coli* O157:H7 and other enterohemorrhagic *Escherichia coli*. *In* Blaser, Smith, Ravdin, Greenberg and Guerrant (Editors), Infections of the Gastrointestinal Tract, Raven Press, Ltd., New York. pp. 739–761.

Griffin, P.J., S.F. Snieszko and S.B. Friddle. 1953. A more comprehensive description of *Bacterium salmonicida*. Trans. Amer. Fish. Soc. *82*: 129–138.

Griffith, M.E., D.S. Lindquist, R.F. Benson, W.L. Thacker, D.J. Brenner and H.W. Wilkinson. 1988. First isolation of *Legionella gormanii* from human disease. J. Clin. Microbiol. *26*: 380–381.

Griffiths, G.W. and S.D. Beck. 1973. Intracellular symbiotes of the pea aphid *Acyrthosiphon pisum*. J. Insect Physiol. *19*: 75–84.

Griffiths, G.W. and S.D. Beck. 1974. Effects of antibiotics on intracellular symbiotes in the pea aphid, *Acyrthosiphon pisum*. Cell Tissue Res. *148*: 287–300.

Grimbert, L. and G. Legros. 1900. De l'identite du bacille lactique aerogene. Ann. Inst. Past. *14*: 479–486.

Grimes, D.J., P.R. Brayton, P.A. West, F.L. Singleton and R.R. Colwell. 1986. The probabilistic identification of *Vibrio* spp. isolated from surface seawater with special reference to *Vibrio campbellii*. Lett. Appl. Microbiol. *2*: 93–95.

Grimes, D.J., J. Stemmler, H. Hada, E.B. May, D. Maneval, F. Hetrick, R.T. Jones, M. Stoskopf and R.R. Colwell. 1984. *Vibrio* species associated with mortality of sharks held in captivity. Microb. Ecol. *10*: 271–282.

Grimes, D.J., J. Stemmler, H. Hada, E.B. May, D. Maneval, F. Hetrick, R.T. Jones, M. Stoskopf and R.R. Colwell. 1985. *In* Validation of new names and new combinations previously effectively published outside the IJSB. List No. 17. Int. J. Syst. Bacteriol. *35*: 223–225.

Grimes, M. 1961. Classification of the *Klebsiella–Aerobacter* group with special reference to the cold tolerant mesophilic *Aerobacter* types. Bull. Bact. Nomen. Taxon. Intern. *11*: 111–129.

Grimes, M. and A.J. Hennerty. 1931. A study of bacteria belonging to the sub-genus *Aerobacter*. Sci. Proc. R. Dublin Soc. *20*: 89–97.

Grimont, F. 1977a. Les bactériophages des *Serratia* et bactéries voisines. Taxonomie et Lysotypie, Doctorate in Pharmacy thesis. Bordeaux II.

Grimont, F. and P.A.D. Grimont. 1986. Ribosomal ribonucleic-acid gene restriction patterns as potential taxonomic tools. Ann. Inst. Pasteur Microbiol. *137B*: 165–175.

Grimont, F. and P.A.D. Grimont. 1992. The Genus *Enterobacter. In* Balows, Trüper, Dworkin, Harder and Schleifer (Editors), The Prokaryotes. A Handbook on the Biology of Bacteria: Ecophysiology, Isolation, Identification, Applications., 2nd Ed., Vol. 3, Springer-Verlag, New York. pp. 2797–2815.

Grimont, F. and P.A.D. Grimont. 1995. The genus *Serratia. In* Balows, Trüper, Dworkin, Harder and Schleifer (Editors), The Prokaryotes. A Handbook on the Biology of Bacteria: Ecophysiology, Isolation, Identification, Applications, Springer-Verlag, New York. pp. 2822–2848.

Grimont, F., P.A.D. Grimont and C. Richard. 1991. The genus *Klebsiella. In* Balows, Trüper, Dworkin, Harder and Schleifer (Editors), The Prokaryotes. A Handbook on Biology of Bacteria: Ecophysiology, Isolation, Identification, Applications, 2nd Ed., Springer Verlag, New York. 2775–2796.

Grimont, P.A.D. 1977b. Le genre *Serratia.* Taxonomie et approche ecologique, Thesis, University of Bordeaux, 377 pp.

Grimont, P.A.D. and E. Ageron. 1989. *Enterobacter cancerogenus* (Urosevic, 1966) Dickey and Zumoff 1988, a senior subjective synonym of *Enterobacter taylorae* Farmer et al. (1985). Res. Microbiol. *140*: 459–465.

Grimont, P.A.D. and O.M. Bouvet. 1989. Diversity of glucose entry routes in the *Enterobacteriaceae.* FEMS Microbiol. Rev. *5*: 109–114.

Grimont, P.A.D. and C. Deval. 1982. Somatic and flagellar antigens of *Serratia ficaria* from the United States and the Mediterranean region. Curr. Microbiol. *7*: 363–366.

Grimont, P.A.D. and H.L.C. Dulong de Rosnay. 1972. Numerical study of 60 strains of *Serratia.* J. Gen. Microbiol. *72*: 259–268.

Grimont, P.A., J.J. Farmer, III, F. Grimont, M.A. Asbury, D.J. Brenner and C. Deval. 1983a. *Ewingella americana* gen.nov., sp.nov., a new *Enterobacteriaceae* isolated from clinical specimens. Ann. Microbiol. Inst. Pasteur (Paris) *134A*: 39–52.

Grimont, P.A., J.J. Farmer, III, F. Grimont, M.A. Asbury, D.J. Brenner and C. Deval. 1984a. *In* Validation of the publication of new names and new combinations previously effectively published outside the IJSB. List No. 13. Int. J. Syst. Bacteriol. *34*: 91–92.

Grimont, P.A.D. and F. Grimont. 1978a. Biotyping of *Serratia marcescens* and its use in epidemiological studies. J. Clin. Microbiol. *8*: 73–83.

Grimont, P.A.D. and F. Grimont. 1978b. The genus *Serratia.* Annu. Rev. Microbiol. *32*: 221–248.

Grimont, P.A.D. and F. Grimont. 1978c. Proteinase zymograms of *Serratia marcescens* as an epidemiological tool. Curr. Microbiol. *1*: 15–18.

Grimont, P.A.D. and F. Grimont. 1984. Genus *Serratia. In* Krieg and Holt (Editors), Bergey's Manual of Systematic Bacteriology, The Williams & Wilkins Co., Baltimore. pp. 477–484.

Grimont, P.A.D., F. Grimont and H.L.C. Dulong de Rosnay. 1977a. Characterization of *Serratia marcescens, S. liquefaciens, S. plymuthica,* and *S. marinorubra* by electrophoresis of their proteinases. J. Gen. Microbiol. *99*: 301–310.

Grimont, P.A.D., F. Grimont, H.L.C. Dulong de Rosnay and P.H.A. Sneath. 1977b. Taxonomy of the genus *Serratia.* J. Gen. Microbiol. *98*: 39–66.

Grimont, P.A.D., F. Grimont, J.J. Farmer, III and M.A. Asbury. 1981a. *Cedecea davisae,* gen. nov., sp. nov. and *Cedecea lapagei,* sp. nov., new *Enterobacteriaceae* from clinical specimens. Int. J. Syst. Bacteriol. *31*: 317–326.

Grimont, P.A.D., F. Grimont and K. Irino. 1982a. Biochemical characterization of *Serratia liquefaciens sensu stricto, Serratia proteamaculans,* and *Serratia grimesii* sp. nov. Curr. Microbiol. *7*: 69–74.

Grimont, P.A.D., F. Grimont and K. Irino. 1983b. *In* Validation of the publication of new names and new combinations previously effectively published outside the IJSB. List No. 10. Int. J. Syst. Bacteriol. *33*: 438–440.

Grimont, P.A.D., F. Grimont, S. Le Minor, B. Davis and F. Pigache. 1979a. Compatible results obtained from biotyping and serotyping in *Serratia marcescens.* J. Clin. Microbiol. *10*: 425–432.

Grimont, P.A.D., F. Grimont, C. Richard, B.R. Davis, A.G. Steigerwalt and D.J. Brenner. 1978a. Deoxyribonucleic acid relatedness between *Serratia plymuthica* and other *Serratia* species with a description of *Serratia odorifera* sp. nov. (holotype: ICPB 3995). Int. J. Syst. Bacteriol. *28*: 453–463.

Grimont, P.A.D., F. Grimont, C. Richard and R. Sakazaki. 1980. *Edwardsiella hoshinae,* a new species of *Enterobacteriaceae.* Curr. Microbiol. *4*: 347–351.

Grimont, P.A.D., F. Grimont, C. Richard and R. Sakazaki. 1981b. *In* Validation of the publication of nNew names and new combinations previously effectively published outside the IJSB. List No.6. Int. J. Syst. Bacteriol. *31*: 215–218.

Grimont, P.A.D., F. Grimont and M.P. Starr. 1978b. *Serratia proteamaculans* (Paine and Stanfield) comb. nov., a senior subjective synonym of *Serratia liquefaciens* (Grimes and Hennerty) Bascomb et al. Int. J. Syst. Bacteriol. *28*: 503–510.

Grimont, P.A.D., F. Grimont and M.P. Starr. 1979b. *Serratia ficaria* sp. nov., a bacterial species associated with Smyrna figs and the fig wasp *Blastophaga psenes.* Curr. Microbiol. *2*: 277–282.

Grimont, P.A.D., F. Grimont and M.P. Starr. 1981c. *Serratia* species isolated from plants. Curr. Microbiol. *5*: 317–322.

Grimont, P.A.D., F. Grimont and M.P. Starr. 1981d. *In* Validation of the publication of new names and new combinations previously effectively published outside the IJSB. List No. 6. Int. J. Syst. Bacteriol. *31*: 215–218.

Grimont, P.A.D., K. Irino and F. Grimont. 1982b. The *Serratia liquejaciens–S. proteamaculans–S. grimesii* complex: DNA relatedness. Curr. Microbiol. *7*: 63–68.

Grimont, P.A.D., T.A. Jackson, E. Ageron and M.J. Noonan. 1988. *Serratia entomophila* sp. nov. associated with amber disease in the New Zealand grass grub *Costelytra zealandica.* Int. J. Syst. Bacteriol. *38*: 1–6.

Grimont, P.A.D., A.G. Steigerwalt, N. Boemare, F.W. Hickman-Brenner, C. Deval, F. Grimont and D.J. Brenner. 1984b. Deoxyribonucleic acid relatedness and phenotypic study of the genus *Xenorhabdus.* Int. J. Syst. Bacteriol. *34*: 378–388.

Grimont, P.A.D., M. Vancanneyt, M. Lefèvre, K. Vandemeulebroecke, L. Vauterin, R. Brosch, K. Kersters and F. Grimont. 1996. Ability of Biolog and Biotype-100 systems to reveal the taxonomic diversity of the pseudomonads. Syst. Appl. Microbiol. *19*: 510–527.

Grisez, L. and F. Ollevier. 1995. Comparative serology of the marine fish pathogen *Vibrio anguillarum.* Appl. Environ. Microbiol. *61*: 4367–4373.

Groisman, E.A. and H. Ochman. 1996. Pathogenicity islands: Bacterial evolution in quantum leaps. Cell *87*: 791–794.

Gromkova, R. and H. Koornhof. 1990. Naturally occurring NAD-independent *Haemophilus parainfluenzae.* J. Gen. Microbiol. *136*: 1031–1035.

Gross, C.A. 1996. Function and regulation of the heat shock proteins. *In* Neidhardt, Curtis, III, Ingraham, Lin, Low, Magasanik, Reznikoff, Riley, Schaechter and Umberger (Editors), *Escherichia coli* and *Salmonella*: Cellular and Molecular Biology, 2nd Ed., ASM Press, Washington, DC. pp. 1382–1399.

Gross, D.C., M.L. Powelson, K.M. Regner and G.K. Radamaker. 1991. A bacteriophage-typing system for surveying the diversity and distribution of strains of *Erwinia carotovora* in potato fields. Phytopathology *81*: 220–226.

Gross, M., G. Geier, K. Rudolph and K. Geider. 1992. Levan and levansucrase synthesized by the fireblight pathogen *Erwinia amylovora.* Physiol. Mol. Plant Pathol. *40*: 371–381.

Gross, R.J., T. Cheasty and B. Rowe. 1977. Isolation of bacteriophages specific for the K1 polysaccharide antigen of *Escherichia coli.* J. Clin. Microbiol. *6*: 548–550.

Gross, R.J. and B. Rowc. 1974. Thc scrology of *Citrobacter koseri, Levinea malonatica,* and *Levinea amalonatica.* J. Med. Microbiol. *7*: 155–162.

Gross, R.J. and B. Rowe. 1983. *Citrobacter koseri* (syn. *C. diversus*): biotype, serogroup and drug resistance patterns of 517 strains. J. Hyg. *90*: 233–239.

Gross, R.J. and B. Rowe. 1985. Serotyping of *Escherichia coli*. *In* Sussman (Editor), The Virulence of *Escherichia coli*: Reviews and Methods, Academic Press, London. pp. 345–363.

Gross, R.J., L.V. Thomas, T. Cheasty, B. Rowe and A.A. Lindberg. 1989. Four new provisional serovars of *Shigella*. J. Clin. Microbiol. *27*: 829–831.

Gross, R.J., L.V. Thomas, N.P. Day, T. Cheasty and B. Rowe. 1982. New provisional serovar of *Shigella boydii*. J. Clin. Microbiol. *16*: 1000–1002.

Gross, R.J., L.V. Thomas and B. Rowe. 1980. New provisional serovar (E10163) of *Shigella boydii*. J. Clin. Microbiol. *12*: 167–169.

Grossberg, S.E., R.G. Petersdorf, J.A. Curtin and I.L. Bennett. 1962. Factors influencing the species and antimicrobial resistance of urinary pathogens. Am. J. Med. *32*: 44–55.

Grothues, D. and K. Rudolph. 1991. Macrorestriction analysis of plant pathogenic *Pseudomonas* species and pathovars. FEMS Microbiol. Lett. *79*: 83–88.

Grothues, D. and B. Tümmler. 1991. New approaches in genome analysis by pulsed-field gel- electrophoresis - application to the analysis of *Pseudomonas* species. Mol. Microbiol. *5*: 2763–2776.

Groudieva, T., R. Grote and G. Antranikian. 2003. *Psychromonas arctica* sp nov., a novel psychrotolerant, biofilm-forming bacterium isolated from Spitzbergen. Int. J. Syst. Evol. Microbiol. *53*: 539–545.

Grover, S., Z.A. McGee and W.D. Odell. 1991a. Isolation of a 30 kDa immunoglobulin binding protein from *Pseudomonas maltophilia*. J. Immunol. Methods. *141*: 187–197.

Grover, S., Z.A. McGee and W.D. Odell. 1991b. Isolation of a 48.5-kDa membrane protein from *Pseudomonas maltophilia* which exhibits immunologic cross-reaction to the β-subunit of human chorionic gonadotropin. Endocrinology. *128*: 3096–3104.

Grover, S. and W.D. Odell. 1992. Partial characterization of the 30 kD Ig-binding protein from *Pseudomonas maltophilia*. Biochem. Biophys. Res. Commun. *182*: 1075–1081.

Grover, S., S.R. Woodward, O. Caticha, D.T. Carrell and W.D. Odell. 1993a. Partial nucleotide sequence of the *Xanthomonas maltophilia* chorionic gonadotropin-like receptor. Biochem. Biophys. Res. Commun. *190*: 371–376.

Grover, S., S.R. Woodward and W.D. Odell. 1993b. A bacterial protein has homology with human chorionic gonadotropin (hCG). Biochem. Biophys. Res. Commun. *193*: 841–847.

Grover, S., S.R. Woodward and W.D. Odell. 1995. Complete sequence of the gene encoding a chorionic gonadotropin-like protein from *Xanthomonas maltophilia*. Gene *156*: 75–78.

Gruber, T. 1905. Ein weiterer Beiträg zür Aromabildung speziell zür Bildung des Erdbeergeruches in der Gruppe "*Pseudomonas*". *Pseudomonas fragariae* II. Zentralbl. Bakteriol. Parasitenk. Infektionskr. Hyg. Abt. II. *14*: 122–123.

Guan, S., D.A. Bastin and N.K. Verma. 1999. Functional analysis of the O antigen glucosylation gene cluster of *Shigella flexneri* bacteriophage SfX. Microbiology *145*: 1263–1273.

Guarino, A., G. Capano, B. Malamisura, M. Alessio, S. Guandalini and A. Rubino. 1987. Production of *Escherichia coli* STa-like heat-stable enterotoxin by *Citrobacter freundii* isolated from humans. J. Clin. Microbiol. *25*: 110–114.

Guarino, A., R. Giannella and M.R. Thompson. 1989. *Citrobacter freundii* produces an 18-amino-acid heat-stable enterotoxin identical to the 18-amino-acid *Escherichia coli* heat-stable enterotoxin (ST Ia). Infect. Immun. *57*: 649–652.

Guckert, J.B., D.B. Ringelberg, D.C. White, R.S. Hanson and B.J. Bratina. 1991. Membrane fatty acids as phenotypic markers in the polyphasic taxonomy of methylotrophs within the *Proteobacteria*. J. Gen. Microbiol. *137*: 2631–2641.

Güde, H., W.R. Strohl and J.M. Larkin. 1981. Mixotrophic and heterotrophic growth of *Beggiatoa alba* in continuous culture. Arch. Microbiol. *129*: 357–360.

Guerinot, M.L. 1994. Microbial iron transport. Annu. Rev. Microbiol. *48*: 743–772.

Guerinot, M.L., P.A. West, J.V. Lee and R.R. Colwell. 1982. *Vibrio diaza-trophicus* sp. nov., a marine nitrogen-fixing bacterium. Int. J. Syst. Bacteriol. *32*: 350– 357.

Guerrant, G.O., M.S. Lambert and C.W. Moss. 1979. Identification of diaminopimelic acid in the Legionnaires' bacterium. J. Clin. Microbiol. *10*: 815–818.

Guerrero, M.G., J.M. Vega, E. Leadbetter and M. Losada. 1973. Preparation and characterization of a soluble nitrate reductase from *Azotobacter chroococcum*. Arch. Mikrobiol. *91*: 287–304..

Guettler, M.V., D. Rumler and M.K. Jain. 1999. *Actinobacillus succinogenes* sp. nov., a novel succinic-acid-producing strain from the bovine rumen. Int. J. Syst. Bacteriol. *49*: 207–216.

Guezennec, J. and A. Fiala-Medioni. 1996. Bacterial abundance and diversity in the Barbados Trench determined by phospholipid analysis. FEMS Microbiol. Ecol. *19*: 83–93.

Guidolin, A. and P.A. Manning. 1987. Genetics of *Vibrio cholerae* and its bacteriophages. Microbiol. Rev. *51*: 285–298.

Guignard, L. and C. Sauvageau. 1894. Sur un nouveau microbe chromogène, le *Bacillus chlororaphis*. C. R. Soc. Biol. Paris. Sér. 10. *1*: 841–843.

Guinee, P.A.M. and W.H. Jansen. 1987. Serotyping of *Aeromonas* species using passive hemagglutination. Zentbl. Bakteriol. Mikrobiol. Hyg. Ser. A *265*: 305–313.

Guinée, P.A.M., W.H. Jansen, H.M.E. Maas, L. Le Minor and R. Beaud. 1981. An unusual H antigen (z_{66}) in strains of *S. typhi*. Ann. Microbiol. (Inst Pasteur). *132A*: 331–334.

Guinée, P.A.M. and J.J. Valkenburg. 1968. Diagnostic value of a *Hafnia* specific bacteriophage. J. Bacteriol. *96*: 564.

Gulig, P.A., H. Danbara, D.G. Guiney, A.J. Lax, F. Norel and M. Rhen. 1993. Molecular analysis of *spv* virulence genes of the *Salmonella* virulence plasmids. Mol. Microbiol. *7*: 825–830.

Gunawardana, G.A., K.M. Townsend and A.J. Frost. 2000. Molecular characterisation of avian *Pasteurella multocida* isolates from Australia and Vietnam by REP-PCR and PFGE. Vet. Microbiol. *72*: 97–109.

Gundersen, J.K., B.B. Jørgensen, E. Larsen and H.W. Jannasch. 1992. Mats of giant sulphur bacteria on deep-sea sediments due to fluctuating hydrothermal flow. Nature *360*: 454–456.

Gunnison, J.B., A. Larson and A.S. Lazarus. 1951. Rapid differentiation between *Pasteurella pestis* and *Pasteurella pseudotuberculosis* by action of bacteriophage. J. Infect. Dis. *88*: 251–255.

Gunsalus, I.C., C.A. Tyson, R.L. Tsai and J.D. Lipscomb. 1971. P-450$_{cam}$ hydroxylase: substrate-effector and electron-transport reactions. Chem. Biol. Interact. *4*: 75–78.

Gunzer, F., H. Böhm, H. Rüssmann, M. Bitzan, S. Aleksic and H. Karch. 1992. Molecular detection of sorbitol-fermenting *Escherichia coli* O157 in patients with hemolytic-uremic syndrome. J. Clin. Microbiol. *30*: 1807–1810.

Gürtler, V. and V.A. Stanisich. 1996. New approaches to typing and identification of bacteria using the 16S-23S rDNA spacer region. Microbiology *142*: 3–16.

Gussin, G.N., C.W. Ronson and F.M. Ausubel. 1986. Regulation of nitrogen fixation genes. Annu. Rev. Genet. *20*: 567–591.

Gutell, R.R. 1993. Collection of small subunit (16S- and 16S-like) ribosomal RNA structures. Nucleic Acids Res. *21*: 3051–3054.

Guth, B.E. and E. Perrella. 1996. Prevalance of invasive ability and other virulence-associated characteristics in *Providencia alcalifaciens* strains isolated in São Paulo, Brazil. J. Med. Microbiol. *45*: 459–462.

Guthrie, J.W. 1968. The serological relationship of races of *Pseudomonas phaseolicola*. Phytopathology *58*: 716–717.

Gutiérrez, M.C., M.T. Garcia, A. Ventosa and F. Ruiz-Berraquero. 1989. Relationships among *Vibrio costicola* strains assessed by DNA–DNA hybridization. FEMS. Microbiol. Lett. *61*: 37–40.

Gutnick, D.L., R. Allon, C. Levy, R. Petter and W. Minas. 1991. Applications of *Acinetobacter* as an industrial microorganism. *In* Towner and Bergogne-Bérézin (Editors), The Biology of *Acinetobacter*: Taxonomy, Clinical Importance, Molecular Biology, Physiology, Industrial Relevance, Plenum Press, New York. pp. 411–441.

Gutnick, D., J.M. Calvo, T. Klopotowski and B.N. Ames. 1969. Compounds which serve as the sole source of carbon or nitrogen for *Salmonella typhimurium* LT-2. J. Bacteriol. *100*: 215–219.

Guyoneaud, R., R. Matheron, W. Liesack, J.F. Imhoff and P. Caumette. 1997. *Thiorhodococcus minus*, gen. nov., sp. nov., a new purple sulfur bacterium isolated from coastal lagoon sediments. Arch. Microbiol. *168*: 16–23.

Guyoneaud, R., R. Matheron, W. Liesack, J.F. Imhoff and P. Caumette. 1998a. *In* Validation of the publication of new names and new combinations previously effectively published outside the IJSB. List No. 64. Int. J. Syst. Bacteriol. *48*: 327–328.

Guyoneaud, R., J. Süling, R. Petri, R. Matheron, P. Caumette, N. Pfennig and J.F. Imhoff. 1998b. Taxonomic rearrangements of the genera *Thiocapsa* and *Amoebobacter* on the basis of 16S rDNA sequence analyses, and description of *Thiolamprovum* gen. nov. Int. J. Syst. Bacteriol. *48*: 957–964.

Gwinn, M.L., R. Ramanathan, H.O. Smith and J.F. Tomb. 1998. A new transformation-deficient mutant of *Haemophilus influenzae* Rd with normal DNA uptake. J. Bacteriol. *180*: 746–748.

Gwinn, M.L., A.E. Stellwagen, N.L. Craig, J.F. Tomb and H.O. Smith. 1997. In vitro Tn7 mutagenesis of *Haemophilus influenzae* Rd and characterization of the role of *atpA* in transformation. J. Bacteriol. *179*: 7315–7320.

Gwynn, C.M. and R.H. George. 1973. Neonatal citrobacter meningitis. Arch. Dis. Child. *48*: 455–458.

Gygi, D., M.M. Rahman, H.C. Lai, R. Carlson, J. Guard-Petter and C. Hughes. 1995. A cell-surface polysaccharide that facilitates rapid population migration by differentiated swarm cells of *Proteus mirabilis*. Mol. Microbiol. *17*: 1167–1175.

Gyles, C.L. 1994. *Escherichia coli* Verotoxins and other cytotoxins. *In* Gyles (Editor), *Escherichia coli* in Domestic Animals and Humans, 1st Ed., CAB International, Wallingford. pp. 365–398.

Gyles, C.L., S.A. De Grandis, C. MacKenzie and J.L. Brunton. 1988. Cloning and nucleotide sequence analysis of the genes determining verocytotoxin production in a porcine edema disease isolate of *Escherichia coli*. Microb. Pathog. *5*: 419–426.

Ha, D.M. and K. Komagata. 1984. Electrophoretic comparison of enzymes in the strains in biovars of *Pseudomonas maltophilia*. J. Gen. Appl. Microbiol. *30*: 277–288.

Haahtela, K. and T.K. Korhonen. 1985. *In vitro* adhesion of N_2-fixing enteric bacteria to roots of grasses and cereals. Appl. Environ. Microbiol. *49*: 1186–1190.

Haber, C.L., L.N. Allen and R.S. Hanson. 1984. Methylotrophic bacteria: biochemical diversity and genetics. Science *221*: 1147–1153.

Habs, H. and R.H.W. Schubert. 1962. Über die biochemischen Merkmale und die taxonomische Stellung von Pseudomonas shigelloides (Bader). Zentbl. Bakteriol. Parasitenkd. Infektkrankh. Hyg. Abt. I Orig. *186*: 316–327.

Habs, I. 1957. Untersuchungen über die O-antigene von *Pseudomonas aeruginosa*. Z. Hyg. Infektionskr. *144*: 218–228.

Hacker, J., G. Blum-Oehler, I. Mühldorfer and H. Tschäpe. 1997. Pathogenicity islands of virulent bacteria: structure, function and impact on microbial evolution. Mol. Microbiol. *23*: 1089–1097.

Hacking, A.J., I.W.F. Taylor, T.R. Jarman and J.R.W. Govan. 1983. Alginate biosynthesis by *Pseudomonas mendocina*. J. Gen. Microbiol. *129*: 3473–3480.

Hackstadt, T. 1986. Antigenic variation in the phase I lipopolysaccharide of *Coxiella burnetii* isolates. Infect. Immun. *52*: 337–340.

Hackstadt, T. 1988. Steric hindrance of antibody binding to surface proteins of *Coxiella burnetii* by phase I lipolysaccharide. Infect. Immun. *56*: 802–807.

Hackstadt, T. 1990. The role of lipopolysaccharides in the virulence of *Coxiella burnetii*. Ann. N. Y. Acad. Sci. *590*: 27–32.

Hackstadt, T., M.G. Peacock, P.J. Hitchcock and R.L. Cole. 1985. Lipopolysaccharide variation in *Coxiella burnetti*: Intrastrain heterogeneity in structure and antigenicity. Infect. Immun. *48*: 359–365.

Hackstadt, T. and J.C. Williams. 1981. Biochemical stratagem for obligate parasitism of eukaryotic cells by *Coxiella burnetii*. Proc. Natl. Acad. Sci. U.S.A. *78*: 3240–3244.

Hada, H.S., P.A. West, J.V. Lee, J. Stemmler and R.R. Colwell. 1984. *Vibrio tubiashii* sp. nov., a pathogen of bivalve mollusks. Int. J. Syst. Bacteriol. *34*: 1–4.

Hadley, P., M.W. Elkins and D.N. Caldwell. 1918. The colon-typhoid intermediates as causative agents of disease in birds. 1. The paratyphoid bacteria. Bull. R. I. Agr. Exp. Sta. *174*: 1–216.

Haefeli, C., C. Franklin and K. Hardy. 1984. Plasmid-determined silver resistance in *Pseudomonas stutzeri* isolated from a silver mine. J. Bacteriol. *158*: 389–392.

Haesebrouck, F., K. Chiers, I. Van Overbeke and R. Ducatelle. 1997. *Actinobacillus pleuropneumoniae* infections in pigs: the role of virulence factors in pathogenesis and protection. Vet. Microbiol. *58*: 239–249.

Haffajee, A.D. and S.S. Socransky. 1994. Microbial etiological agents of destructive periodontal disease. Periodontal 2000 *5*: 78–111.

Hagborg, W.A.F. 1942. Classification revision in *Xanthomonas translucens*. Can. J. Res. *20*: 312–326.

Hagen, K.D. and D.C. Nelson. 1996. Organic carbon utilization by obligately and facultatively autotrophic *Beggiatoa* strains in homogeneous and gradient cultures. Appl. Environ. Microbiol. *62*: 947–953.

Hagen, K.D. and D.C. Nelson. 1997. Use of reduced sulfur compounds by *Beggiatoa* spp.: enzymology and physiology of marine and freshwater strains in homogeneous and gradient cultures. Appl. Environ. Microbiol. *63*: 3957–3964.

Häggblom, M.M. 1992. Microbial breakdown of halogenated aromatic pesticides and related-compounds. FEMS Microbiol. Rev. *103*: 29–72.

Haglind, P., M. Lundholm and R. Rylander. 1981. Prevalence of byssinosis in Swedish cotton mills. Br. J. Ind. Med. *38*: 138–143.

Hahn, H.P. 1997. The type-4 pilus is the major virulence-associated adhesin of *Pseudomonas aeruginosa* - A review. Gene *192*: 99–108.

Haigh, R., T. Baldwin, S. Knutton and P.H. Williams. 1995. Carbon dioxide regulated secretion of the EaeB protein of enteropathogenic *Escherichia coli*. FEMS Microbiol. Lett. *129*: 63–67.

Hajna, A.A. 1955. A new enrichment broth medium for Gram-negative organisms of the intestinal group. Public Health Lab. *13*: 83–89.

Hales, B.A., C.A. Hart, R.M. Batt and J.R. Saunders. 1992. The large plasmids found in enterohemorrhagic and enteropathogenic *Escherichia coli* constitute a related series of transfer-defective Inc F-IIA replicons. Plasmid *28*: 183–193.

Hall, G.A., C.R. Dorn, N. Chanter, S.M. Scotland, H.R. Smith and B. Rowe. 1990. Attaching and effacing lesions in vivo and adhesion to tissue culture cells of Vero-cytotoxin-producing *Escherichia coli* belonging to serogroups O5 and O103. J. Gen. Microbiol. *136*: 779–786.

Hall, I.M. and M.E. Badgley. 1957. A rickettsial disease of larvae of species of *Stethorus* caused by *Rickettsiella stethorae*, n. sp. J. Bacteriol. *74*: 452–455.

Hall, M.M., C.A. Mueske, D.M. Ilstrup and J.A.I. Washington. 1979. Evaluation of a biphasic medium for blood cultures. J. Clin. Microbiol. *10*: 673–676.

Hall, R.H., F.M. Khambaty, M.H. Kothary, S.P. Keasler and B.D. Tall. 1994. *Vibrio cholerae* non-O1 serogroup associated with cholera gravis genetically and physiologically resembles O1 E1 Tor cholera strains. Infect. Immun. *62*: 3859–3863.

Hall, R.H., P.A. Vial, J.B. Kaper, J.J. Mekalanos and M.M. Levine. 1988. Morphological studies on fimbriae expressed by *Vibrio cholerae* 01. Microb. Pathog. *4*: 257–265.

Hall, W.J. 1965. Fowl typhoid. *In* Biester and Schwarte (Editors), Diseases of Poultry, 5th Ed., Iowa State University Press, Ames. pp. 329–358.

Hall, W.J., K.L. Heddleston, D.H. Legenhausen and R.W. Hughes. 1955. Studies on pasteurellosis. I. A new species of *Pasteurella* encountered in chronic fowl cholera. Am. J. Vet. Res. *16*: 598–604.

Hallberg, K.B. and E.B. Lindström. 1995. *In* Validation of the publication of new names and new combinations previously effectively published outside the IJSB. List No. 54. Int. J. Syst. Bacteriol. *45*: 619–620.

Halle, F. and J.M. Meyer. 1992. Ferrisiderophore reductases of *Pseudomonas* - purification, properties and cellular location of the *Pseudomonas aeruginosa* ferripyoverdine reductase. Eur. J. Biochem. *209*: 613–620.

Haltalin, K.C., J.D. Nelson, R. Ring, 3rd., M. Sladoje and L.V. Hinton. 1967. Double-blind treatment study of shigellosis comparing ampicillin, sulfadiazine, and placebo. J. Pediatr. *70*: 970–981.

Hamamoto, Y., Y. Soejima, M. Ogasawara, H. Okimura, K. Nagai and C. Asagami. 1995. Necrotizing fasciitis due to *Pasteurella multocida* infection. Dermatology *190*: 145–147.

Hamana, K. 1996. Distribution of diaminopropane and acetylspermidine in *Enterobacteriaceae*. Can. J. Microbiol. *42*: 107–114.

Hamana, K. 1997. Polyamine distribution patterns within the families *Aeromonadaceae*, *Vibrionaceae*, *Pasteurellaceae*, and *Halomonadaceae*, and related genera of the gamma subclass of the *Proteobacteria*. J. Gen. Appl. Microbiol. *43*: 49–59.

Hamana, K. and S. Matsuzaki. 1993. Polyamine distribution patterns serve as a phenotypic marker in the chemotaxonomy of the *Proteobacteria*. Can. J. Microbiol. *39*: 304–310.

Hamana, K., T. Sakane and A. Yokota. 1994. Polyamine analysis of the genera, *Aquaspirillum*, *Magnetospirillum*, *Oceanospirillum* and *Spirillum*. J. Gen. Appl. Microbiol. *40*: 75–82.

Hamedani, P., J. Ali, S. Hafeez, R.J. Bachand, G. Dawood, S. Quereshi, R. Raza and Z. Yab. 1991. The safety and efficacy of clarithromycin in patients with *Legionella pneumonia*. Chest *100*: 1503–1506.

Hamilton, R.L. and W.J. Brown. 1972. Bacteriophage typing of clinically isolated *Serratia marcescens*. Appl. Microbiol. *24*: 899–906.

Hamlin, L.J. and R.E. Hungate. 1956. Culture and physiology of a starch-digesting bacterium (*Bacteroides amylophilus* n. sp.) from the bovine rumen. J. Bacteriol. *72*: 548–554.

Hammarström, E. 1947. Bacteriophage classification of *Shigella sonnei*. Lancet *1*: 102.

Hammarström, E. 1949. Phage typing of *Shigella sonnei*. Acta. Med. Scand. (Suppl. 233) *133*: 1–132.

Hammer, P.E., D.S. Hill, S.T. Lam, K.H. Van Pée and J.M. Ligon. 1997. Four genes from *Pseudomonas fluorescens* that encode the biosynthesis of pyrrolnitrin. Appl. Environ. Microbiol. *63*: 2147–2154.

Hammond, G.W., C.J. Lian, J.C. Wilt, W. Albritton and R. A.R.. 1978a. Determination of the hemin requirement of *Haemophilus ducreyi*: evaluation of the porphyrin test and media used in the satellite growth test. J. Clin. Microbiol. *7*: 243–246.

Hammond, G.W., C.J. Lian, J.C. Wilt and A.R. Ronald. 1978b. Antimicrobial susceptibility of *Haemophilus ducreyi*. Antimicrob. Agents Chemother. *13*: 608–612.

Hamon, Y., A. Kayser, L. Le Minor and J. Maresz. 1969. Les bactériocines d'*Edwardsiella tarda*. Intérêt taxonomique de l'étude de ces antibiotiques. C. R. Acad. Sci. Paris. *268*: 2517–2520.

Hamon, Y., L. Le Minor and Y. Péron. 1970. Les bactériocines d'Enterobacter liquefaciens. Intérêt taxonomique de leur étude. C. R. Acad. Sci. Serie D. *270*: 886–889.

Hamon, Y. and Y. Péron. 1961. Étude de la propriété bactériocinogène dans le genre *Serratia*. Ann. Inst. Pasteur (Paris) *100*: 818.

Hamon, Y. and Y. Péron. 1966. La propriete bacteriocinogene dans la tribu des *Salmonelleae*. Ann. Inst. Pasteur (Paris) *110*: 389–402.

Hamon, Y. and Y. Péron. 1979. Bacteriocines et (ou) phages létaux de *Serratia marcescens*, *S. Iiquefaciens*, et *S. marinorubra*. Ann. Microbiol. *130A*: 403.

Hamon, Y., C. Richard and Y. Péron. 1974. Study of bacteriocinogeny and lysogeny among *Levinea* strains. Ann. Microbiol. *125A*: 35–44.

Hampson, D.J., P.J. Blackall, J.M. Woodward and A.J. Lymbery. 1993. Genetic analysis of *Actinobacillus pleuropneumoniae*, and comparison with *Haemophilus* spp. Taxon "minor group" and Taxon C. Zentbl. Bakteriol. *279*: 83–91.

Hamze, M., J. Mergaert, H.J. van Vuuren, F. Gavini, A. Beji, D. Izard and K. Kersters. 1991. *Rahnella aquatilis*, a potential contaminant in lager beer breweries. Int. J. Food Microbiol. *13*: 63–68.

Han, N., C.I. Hoover, J.R. Winkler, C.Y. Ng and G.C. Armitage. 1991. Identification of genomic clonal types of *Actinobacillus actinomycetemcomitans* by restriction endonuclease analysis. J. Clin. Microbiol. *29*: 1574–1578.

Han, R., W.M. Wouts and L. Li. 1990. Development of *Heterorhabditis* spp.

strains as characteristics of possible *Xenorhabdus luminescens* subspecies. Rev. Nematol. *13*: 411–415.

Hanan, M.S., M. Sobhy, E.M. Riad and N.A. El-Khouly. 2000. Antibacterial efficacy and pharmacokinetic studies of ciprofloxacin on *Pasteurella multocida* infected rabbits. Dtsch. Tierarztl. Wochenschr. *107*: 151–155.

Hancock, R.E.W. and A.M. Carey. 1980. Protein D1 - a glucose-inducible, pore-forming protein from the outer-membrane of *Pseudomonas aeruginosa*. FEMS Microbiol. Lett. *8*: 105–109.

Hancock, R.E.W. and L. Chan. 1988. Outer membranes of environmental isolates of *Pseudomonas aeruginosa*. J. Clin. Microbiol. *26*: 2423–2424.

Hancock, R.E.W., A.A. Wieczorek, L.M. Mutharia and K. Poole. 1982. Monoclonal-antibodies against *Pseudomonas aeruginosa* outer- membrane antigens - isolation and characterization. Infect. Immun. *37*: 166–171.

Hanefeld, U., H.G. Floss and H. Laatsch. 1994. Biosynthesis of the marine antibiotic pentabromopseudilin. 1. The benzene ring. J. Org. Chem. *59*: 3604–3608.

Haneline, S., C.J. Connelly and T. Melton. 1991. Chemotactic behavior of *Azotobacter vinelandii*. Appl. Environ. Microbiol. *57*: 825–829.

Hänninen, M.L. 1994. Phenotypic characteristics of the three hybridization groups of *Aeromonas hydrophila* complex isolated from different sources. J. Appl. Bacteriol. *76*: 455–462.

Hänninen, M.L. and V. Hirvelä-Koski. 1997. Molecular and phenotypic methods for the characterization of atypical *Aeromonas salmonicida*. Vet. Microbiol. *56*: 147–158.

Hänninen, M.L., P. Oivanen and V. Hirvelä-Koski. 1997. *Aeromonas* species in fish, fish-eggs, shrimp and freshwater. Int. J. Food Microbiol. *34*: 17–26.

Hänninen, M.L., J. Ridell and V. Hirvelä-Koski. 1995. Phenotypic and molecular characteristics of *Aeromonas salmonicida* subsp. *salmonicida* isolated in southern and northern Finland. J. Appl. Bacteriol. *79*: 12–21.

Hänninen, M.L. and A. Siitonen. 1995. Distribution of *Aeromonas* phenospecies and genospecies among strains isolated from water, foods or from human clinical samples. Epidemiol. Infect. *115*: 39–50.

Hansen, A.J., A. Ingebritsen and O.B. Weeks. 1963. Flagellation of *Flavabacterium piscicida*. J. Bacteriol. *86*: 602–603.

Hansen, A.J., O.B. Weeks and R.R. Colwell. 1965. Taxonomy of *Pseudomonas piscicida* (Bein) Buck, Meyers and Leifson. J. Bacteriol. *89*: 752–761.

Hansen, G.H. and J.A. Olafsen. 1989. Bacterial colonization of cod (*Gadus morhua* L.) and halibut (*Hippoglossus hippoglossus*) eggs in marine aquaculture. Appl. Environ. Microbiol. *55*: 1435–1446.

Hansen, M.W. and G.Y. Glupczynski. 1984. Isolation of an unusual *Cedecea* species from a cutaneous ulcer. Eur. J. Clin. Microbiol. *3*: 152–153.

Hansen, W., J. Freney, M. Labbe, F. Renaud, E. Yourassowsky and J. Fleurette. 1991. Gas-liquid chromatographic analysis of cellular fatty acid methyl esters in *Aeromonas* spp. Zentbl. Bakteriol. *275*: 1–10.

Hanson, R.S. and T.E. Hanson. 1996. Methanotrophic bacteria. Microbiol. Rev. *60*: 439–471.

Hanson, R.S., A.I. Netrusov and K. Tsuji. 1992. The obligate methanotrophic bacteria *Methylococcus*, *Methylomonas*, and *Methylosinus*. *In* Balows, Trüper, Dworkin, Harder and Schleifer (Editors), The Prokaryotes. A Handbook of Bacteria: Ecophysiology, Isolation, Identification, Applications, 2nd Ed., Vol. 3, Springer-Verlag, New York. pp. 2350–2364.

Hao, M.V., D.J. Brenner, A.G. Steigerwalt, T. Kosako and K. Komagata. 1990. *Erwinia persicinus*, a new species isolated from plants. Int. J. Syst. Bacteriol. *40*: 379–383.

Hara, E. and H. Ishikawa. 1990. Purification and partial characterization of symbionin, an aphid endosymbiont-specific protein. Insect Biochem. *20*: 421–427.

Harada, H., H. Oyaizu, Y. Kosako and H. Ishikawa. 1997. *Erwinia aphidicola*, a new species isolated from pea aphid, *Acyrthosiphon pisum*. J. Gen. Appl. Microbiol. *43*: 349–354.

Harada, H., H. Oyaizu, Y. Kosako and H. Ishikawa. 1998. *In* Validation of the publication of new names and new combinations previously effectively published outside the IJSB. List No. 67. Int. J. Syst. Bacteriol. *48*: 1083–1084.

Harada, K., K. Shimizu and T. Matsuyama. 1957. *Hafnia* isolated from men. Gunma J. Med. Sci. *6*: 109–112.

Harada, S., S. Tsubotani, T. Hida, K. Koyama, M. Kondo and H. Ono. 1988. Chemistry of a new antibiotic: Lactivicin. Tetrahedron. *44*: 6589–6606.

Harayama, S. 1994. Codon usage patterns suggest independent evolution of 2 catabolic operons on toluene-degradative plasmid TOL pWW0 of *Pseudomonas putida*. J. Mol. Evol. *38*: 328–335.

Harayama, S., H. Kishira, Y. Kasai and K. Shutsubo. 1999. Petroleum biodegradation in marine environments. J. Molec. Microbiol. Biotechnol. *1*: 63–70.

Hargreaves, J.E. and D.R. Lucey. 1990. Life-threatening *Edwardsiella tarda* soft-tissue infection associated with catfish puncture wound. J. Infect. Dis. *162*: 1416–1417.

Harhoff, N. 1949. Studies on bacteria of the Ballerup group. Acta Pathol. Microbiol. Scand. *26*: 167–174.

Harold, R. and R.Y. Stanier. 1955. The genera *Leucothrix* and *Thiothrix*. Bacteriol. Rev. *19*: 49–58.

Harper, J.J. and M.H. Tilse. 1991. Biotypes of *Haemophilus influenzae* that are associated with noninvasive infections. J. Clin. Microbiol. *29*: 2539–2542.

Harper, R.E. and P. Talbot. 1984. Analysis of the epibiotic bacteria of lobster (*Homarus*) eggs and their influence of the loss of eggs from the pleopods. Aquaculture *36*: 9–26.

Harrell, L.J., M.L. Cameron and C.M. O'Hara. 1989. *Rahnella aquatilis*, an unusual gram-negative rod isolated from the bronchial washing of a patient with acquired immunodeficiency syndrome. J. Clin. Microbiol. *27*: 1671–1672.

Harrison, A.P. 1982. Genomic and physiological diversity amongst strains of thiobacillus ferrooxidans and genomic comparison with *Thiobacillus thiooxidans*. Arch. Microbiol. *131*: 68–76.

Harrison, A.P. and P.R. Norris. 1985. *Leptospirillum ferrooxidans* and similar bacteria: some characteristics and genomic diversity. FEMS Microbiol. Lett. *30*: 99–102.

Harrison, C.P., A.E. Douglas and A.F.G. Dixon. 1989. A rapid method to isolate symbiotic bacteria from aphids. J. Invertebr. Pathol. *53*: 427–428.

Harrison, T.G., N.A. Saunders, N. Doshi, R. Wait and A.G. Taylor. 1988. Serological diversity within the species *Legionella spiritensis*. J. Appl. Bacteriol. *65*: 425–431.

Harrison, T.G., N.A. Saunders, A. Haththotuwa, G. Hallas, R.J. Birtles and A.G. Taylor. 1990. Phenotypic variation among genotypically homogeneous *Legionella pneumophila* serogroup 1 isolates: implications for the investigation of outbreaks of Legionnaires' disease. Epidemiol. Infect. *104*: 171–180.

Hartingsveldt, J., M.G. Marinus and A.H. Stouthamer. 1971. Mutants of *Pseudomonas aeruginosa* blocked in nitrate or nitrite dissimilation. Genet. *67*: 469–472.

Hartl, D.L. and D.E. Dykhuizen. 1984. The population genetics of *Escherichia coli*. Annu. Rev. Genet. *18*: 31–68.

Hartman, J.R., B.C. Eshenaur and U.E. Jarlfors. 1995. Bacterial leaf scorch caused by *Xylella fastidiosa*: a Kentucky survey: a unique pathogen; and bur oak, a new host. Journal of Arboric. *21*: 77–82.

Hartmann, L., W. Schroder and A. LubkeBecker. 1996. A comparative study of the major outer membrane proteins of the avian haemophili and *Pasteurella gallinarum*. Zentrabl. Bakteriol. *284*: 47–51.

Hartmann, R.K., H.Y. Toschka, N. Ulbrich and V.A. Erdmann. 1986. Genomic organization of rDNA in *Pseudomonas aeruginosa*. FEBS Lett. *195*: 187–193.

Hartnett, G.B. and L.N. Ornston. 1994. Acquisition of apparent DNA slippage structures during extensive evolutionary divergence of *pcaD* and *catD* genes encoding identical catalytic activities in *Acinetobacter calcoaceticus*. Gene *142*: 23–29.

Hartstein, A.I., A.L. Rashad, J.M. Liebler, L.A. Actis, J. Freeman, J.W.J. Rourke, T.B. Stibolt, M.E. Tolmasky, G.R. Ellis and J.H. Crosa. 1988. Multiple intensive care unit outbreak of *Acinetobacter calcoaceticus*

subsp. *anitratus* respiratory infection and colonization associated with contaminated, reusable ventilator circuits and resuscitation bags. Am. J. Med. *85*: 624–631.

Hartung, J.S., J. Beretta, R.H. Brlansky, J. Spisso and R.F. Lee. 1994. Citrus variegated chlorosis bacterium: Axenic culture, pathogenicity, and serological relationships with other strains of *Xylella fastidiosa*. Phytopathology *84*: 591–597.

Hartung, J.S. and E.L. Civerolo. 1987. Genomic fingerprints of *Xanthomonas campestris* pathovar *citri* strains from Asia, South America, and Florida (USA). Phytopathology *77*: 282–285.

Hartung, J.S. and E.L. Civerolo. 1989. Restriction fragment length polymorphisms distinguish *Xanthomonas campestris* strains isolated from Florida (USA) citrus nurseries from *Xanthomonas campestris* pv. *citri*. Phytopathology *79*: 793–799.

Hartung, J.S. and E.L. Civerolo. 1991. Variation among strains of *Xanthomonas campestris* causing citrus bacterial spot. Plant Dis. *75*: 622–626.

Hartung, J.S., M.R. Pooler and H.A.T. van der Scheer. 1997. Immunocapture and multiplexed-PCR assay for *Xanthomonas fragariae*, causal agent of angular leafspot disease. Acta Hortic. *439*: 821–828.

Hartung, J.S., O.P. Pruvost, I. Villemot and A. Alvarez. 1996. Rapid and sensitive colorimetric detection of *Xanthomonas axonopodis* pv. *citri* by immunocapture and a nested-polymerase chain reaction assay. Phytopathology *86*: 95–101.

Harwood, C.S. 1978. *Beneckea gazagenes* sp. nov., a red, facultatively anaerobic, marine bacterium. Curr. Microbiol. *1*: 233–238.

Harwood, C.R. 1993. Plasmids, transposons, and gene flux. *In* Goodfellow and O'Donnell (Editors), Handbook of New Bacterial Systematics, Academic Press, New York. pp. 115–150.

Harwood, C.S., S.S. Bang, S. Baumann and K.H. Nealson. 1980. *Photobacterium logei* sp. nov., nom. rev., *Beneckea nereida* sp. nov., nom. rev. and *Beneckea gazogenes* sp. nov., nom. rev. Int. J. Syst. Bacteriol. *30*: 655.

Hase, S. and E.T. Rietschel. 1976. Isolation and analysis of the lipid A backbone. Lipid A structure of lipopolysaccharides from various bacterial groups. Eur. J. Biochem. *63*: 101–107.

Haseley, S.R., O. Holst and H. Brade. 1998. Structural studies of the O-antigen isolated from the phenol-soluble lipopolysaccharide of *Acinetobacter baumannii* (DNA group 2) strain 9. Eur. J. Biochem. *251*: 189–194.

Haseley, S.R., R. Pantophlet, L. Brade, O. Holst and H. Brade. 1997. Structural and serological characterisation of the O-antigenic polysaccharide of the lipopolysaccharide from *Acinetobacter junii* strain 65. Eur. J. Biochem. *245*: 477–481.

Haseley, S.R. and S.G. Wilkinson. 1997. Structural studies of the putative O-specific polysaccharide of *Acinetobacter baumannii* O24 containing 5,7-diamino-3,5,7,9-tetradeoxy-L-glycero-D-galacto-nonulosonic acid. Eur. J. Biochem. *250*: 617–623.

Hassan, A.K., S. Moriya, P. Baumann, H. Yoshikawa and N. Ogasawara. 1996. Structure of the *dnaA* region of the endosymbiont, *Buchnera aphidicola*, of aphid *Schizaphis graminum*. DNA Res. *3*: 415–419.

Hasse, C.H. 1915. *Pseudomonas citri*, the cause of citrus canker. J. Agr. Res. *4*: 97–100.

Hastings, J.W. and K.H. Nealson. 1977. Bacterial bioluminescence. Annu. Rev. Microbiol. *31*: 549–595.

Hastings, J.W. and K.H. Nealson. 1979. Bacterial bioluminescence - its control and ecological significance. Microbiol. Rev. *43*: 496–518.

Hastings, J.W. and K.H. Nealson. 1981. The symbiotic luminous bacteria. *In* Starr, Stolp, Trüper, Balows and Schlegel (Editors), The Prokaryotes, a Handbook on Habitats, Isolation and Identification of Bacteria, Springer-Verlag, New York. pp. 1332–1345.

Hatchette, T.F., R.C. Hudson, W.F. Schlech, N.A. Campbell, J.E. Hatchette, S. Ratnam, D. Raoult, C. Donovan and T.J. Marrie. 2001. Goat-associated Q fever: A new disease in Newfoundland. Emerg. Infect. Dis. *7*: 413–419.

Hattingh, M.J. and D.F. Walters. 1981. Stalk and leaf necrosis of onion caused by *Erwinia herbicola*. Plant Dis. *65*: 615–618.

Hauben, L., E.R.B. Moore, L. Vauterin, M. Steenackers, J. Mergaert, L.

Verdonck and J. Swings. 1998a. Phylogenetic position of phytopathogens within the *Enterobacteriaceae*. Syst. Appl. Microbiol. *21*: 384–397.

Hauben, L., E.R.B. Moore, L. Vauterin, M. Steenackers, J. Mergaert, L. Verdonck and J. Swings. 1999a. *In* Validation of the publication of new names and new combinations previously effectively published outside the IJSB. List No. 68. Int. J. Syst. Bacteriol. *49*: 1–3.

Hauben, L., M. Steenackers and J. Swings. 1998b. PCR-Based detection of the causal agent of watermark disease in willows (*Salix* spp.). Appl. Environ. Microbiol. *64*: 3966–3971.

Hauben, L., L. Vauterin, E.R. Moore, B. Hoste and J. Swings. 1999b. Genomic diversity of the genus *Stenotrophomonas*. Int. J. Syst. Bacteriol. *49*: 1749–1760.

Hauben, L., L. Vauterin, J. Swings and E.R.B. Moore. 1997. Comparison of 16S ribosomal DNA sequences of all *Xanthomonas* species. Int. J. Syst. Bacteriol. *47*: 328–335.

Hauduroy, P., G. Ehringer, A. Urbain, G. Guillot and J. Magrou. 1937. Dictionnaire des bactéries pathogènes, Masson and Co., Paris.

Hauge, J.G. 1960. Kinetics and specificity of glucose dehydrogenase from *Bacterium anitratum*. Biachim. Biophys. Acta. *45*: 263–269.

Haupt, H. 1934. Zür Frage der Verwandtschaft des *Actinobacillus* Lignièresii Brumpt 1910, des *Bacillus equuli* van Straaten 1918 und des *Bacillus mallei* Flügge 1886. Arch. Wiss. Prakt. Tierheilk. *67*: 513–524.

Hauser, C. 1885. Über Faulnisbakterien und deren Beziehungen zür Septicamie. Ein Beiträg zür Morphologie der Spaltpilze, Vogel, Leipzig.

Hausner, O., Z. Kocmoud and M. Gabrhelová. 1986. The incidence and diagnosis of a new type of bacteria, *Budvicia aquatica*, in specimens from the laboratory for water study. Cesk. Epidemiol. Mikrobiol. Imunol. *35*: 336–340.

Havelaar, A.H., F.M. Schets, A. van Silfhout, W.H. Jansen, G. Wieten and D. van der Kooij. 1992. Typing of *Aeromonas* strains from patients with diarrhoea and from drinking water. J. Appl. Bacteriol. *72*: 435–444.

Havelaar, A.H., J.F. Versteegh and M. During. 1990. The presence of *Aeromonas* in drinking water supplies in The Netherlands. Zentralbl. Hyg. Umweltmed. *190*: 236–256.

Hawke, J.P., A.C. McWhorter, A.G. Steigerwalt and D.J. Brenner. 1981. *Edwardsiella ictaluri* sp. nov., the causative agent of enteric septicemia of catfish. Int. J. Syst. Bacteriol. *31*: 396–400.

Hawkey, P.M. 1984. *Providencia stuartii*: a review of a multiply antibiotic-resistant bacterium. J. Antimicrob. Chemother. *13*: 209–226.

Hawkey, P.M. 1997. *Proteus, Providencia* and *Morganella*. *In* Emmerson, Hawkey and Gillespie (Editors), Principles and Practice of Clinical Bacteriology, John Wiley and Sons Ltd., Chichester. pp. 429–438.

Hawkey, P.M., P.M. Bennett and C.A. Hawkey. 1985. Evolution of an R plasmid from a cryptic plasmid by transposition of two copies of Tn1 in *Providencia stuartii*. J. Gen. Microbiol. *131*: 927–933.

Hawkey, P.M. and C.A. Hawkey. 1984. Comparative in vitro activity of quinolone carboxylic acids against *Proteeae*. J. Antimicrob. Chemother. *14*: 485–489.

Hawkey, P.M., A. McCormick and R.A. Simpson. 1986a. Selective and differential medium for the primary isolation of memebers of the Proteeae. J. Clin. Microbiol. *23*: 600–603.

Hawkey, P.M., S.J. Pedler and A. Turner. 1983. Comparative *in vitro* activity of semisynthetic penicillins against Proteeae. Antimicrob. Agents Chemother. *23*: 619–621.

Hawkey, P.M., J.L. Penner, A.H. Linton, C.A. Hawkey, L.J. Crisp and M. Hinton. 1986b. Speciation, serotyping, antimicrobial sensitivity and plasmid content of *Proteeae* from the environment of calf-rearing units in South West England. J. Hyg. (Lond.). *97*: 405–417.

Hawkey, P.M., J.L. Penner, M.R. Potten, M. Stephens, L.J. Barton and D.C. Speller. 1982a. Prospective study of fecal, urinary tract, and environmental colonization by *Providencia stuartii* in two geriatric wards. J. Clin. Microbiol. *16*: 422–426.

Hawkey, P.M., M.R. Potten and M. Stephens. 1982b. The use of pre-enrichment for the isolation of small numbers of gentamicin-resistant *Providencia stuartii* from faeces. J. Hosp. Infect. *3*: 369–374.

Hayashi, N.R., A. Oguni, T. Yaguchi, S.Y. Chung, H. Nishihara, T. Kodama and Y. Igarashi. 1998. Different properties of gene products of three sets ribulose 1,5-bisphosphate carboxylase/oxygenase from a marine obligately autotrophic hydrogen-oxidizing bacterium, *Hydrogenovibrio marinus* strain MH-110. J. Ferment. Bioeng. *85*: 150–155.

Hayashi, T., K. Makino, M. Ohnishi, K. Kurokawa, K. Ishii, K. Yokoyama, C.G. Han, E. Ohtsubo, K. Nakayama, T. Murata, M. Tanaka, T. Tobe, T. Iida, H. Takami, T. Honda, C. Sasakawa, N. Ogasawara, T. Yasunaga, S. Kuhara, T. Shiba, M. Hattori and H. Shinagawa. 2001. Complete genome sequence of enterohemorrhagic *Escherichia coli* O157:H7 and genomic comparison with a laboratory strain K-12. DNA Res. *8*: 11–22.

Hayat, U., G.P. Reddy, C.A. Bush, J.A. Johnson, A.C. Wright and J.G. Morris, Jr.. 1993. Capsular types of *Vibrio vulnificus*: An analysis of strains from clinical and environmental sources. J. Infect. Dis. *168*: 758–762.

Hayes, M.V., C.J. Thomson and S.G.B. Amyes. 1994. Three beta-lactamases isolated from *Aeromonas salmonicida*, including a carbapenemase not detectable by conventional methods. Eur. J. Clin. Microbiol. Infect. Dis. *13*: 805–811.

Haynes, J. and P.M. Hawkey. 1989. *Providencia alcalifaciens* and travellers' diarrhoea. Br. Med. J. *299*: 94–95.

Haynes, W.C. and W.H. Burkholder. 1957. Genus I *Pseudomonas. In* Breed, Murray and Smith (Editors), Bergey's Manual of Determinative Bacteriology, 7th Ed., The Williams & Wilkins Co., Baltimore. pp. 89–152.

Hayward, A.C. 1964. Characteristics of *Pseudomonas solanacearum*. J. Appl. Bacteriol. *27*: 265–277.

Hayward, A.C. 1974. Latent infection by bacteria. Ann. Rev. Phytopathol. *12*: 87–97.

Hayward, A.C. 1977. Occurrence of glycoside hydrolases in plant pathogenic and related bacteria. J. Appl. Bacteriol. *43*: 407–411.

Hayward, A.C. 1979. Isolation and characterization of *Xanthomonas. In* Skinner and Lovelock (Editors), Identification Methods for Microbiologists, Technical Series No. 14, Society for Applied Bacteriology, pp. 15–32.

Hazen, T.C., C.B. Flierman, R.P. Hirsch and G.W. Esch. 1978. Prevalence and distribution of *Aeromonas hydrophila* in the United States. Appl. Environ. Microbiol. *36*: 731–738.

Hazeu, W., W.H. Batenburg-van der Vegte and J.C. de Bruyn. 1980a. Some characteristics of *Methylococcus mobilis*, sp. nov. Arch. Microbiol. *124*: 211–220.

Hazeu, W., W.H. Batenburg-van der Vegte and J.C. de Bruyn. 1980b. *In* Validation of the publication of new names and new combinations previously effectively published outside the IJSB. List No. 5. Int. J. Syst. Bacteriol. *30*: 676–677.

Head, I.M., N.D. Gray, H.D. Babenzien and F.O. Glöckner. 2000a. Uncultured giant sulfur bacteria of the genus *Achromatium*. FEMS Microbiol. Ecol. *33*: 171–180.

Head, I.M., N.D. Gray , K.J. Clarke, R.W. Pickup and J.G. Jones. 1996. The phylogenetic position and ultrastructure of the uncultured bacterium *Achromatium oxaliferum*. Microbiology *142*: 2341–2354.

Head, I.M., N.D. Gray, R. Howarth, R.W. Pickup, K.J. Clarke and J.G. Jones. 2000b. *Achromatium oxaliferum*. Understanding the unmistakable. Adv. Microb. Ecol. *16*: 1–40.

Head, I.M., W.D. Hiorns, T.M. Embley, A.J. McCarthy and J.R. Saunders. 1993. The phylogeny of autotrophic ammonia-oxidizing bacteria as determined by analysis of 16S ribosomal RNA gene sequences. J. Gen. Microbiol. *139*: 1147–1153.

Heald, S.C. and R.O. Jenkins. 1996. Expression and substrate specificity of the toluene dioxygenase of *Pseudomonas putida* NCIMB 11767. Appl. Microbiol. Biotechnol. *45*: 56–62.

Heath, C.H., D.I. Grove and D.F.M. Looke. 1996. Delay in appropriate therapy of *Legionella* pneumonia associated with increased mortality. Eur. J. Clin. Microbiol. Infect. Dis. *15*: 286–290.

Hebert, A.M. and R.H. Vreeland. 1987. Phenotypic comparison of halotolerant bacteria *Halomonas halodurans* sp. nov., nom., rev., comb. nov. Int. J. Syst. Bacteriol. *37*: 347–350.

Hébert, G.A. 1980. Room temperature storage of *Legionella* cultures. J. Clin. Microbiol. *12*: 807–809.

Hébert, G.A. 1981. Hippurate hydrolysis by *Legionella pneumophila*. J. Clin. Microbiol. *13*: 240–242.

Hébert, G.A., C.W. Moss, L.K. McDougal, F.M. Bozeman, R.M. McKinney and D.J. Brenner. 1980a. The rickettsia-like organisms TATLOCK (1943) and HEBA (1959): bacteria phenotypically similar to but generally distinct from *Legionella pneumophila* and the WIGA bacterium. Ann. Intern. Med. *92*: 45–52.

Hébert, G.A., A.G. Steigerwalt and D.J. Brenner. 1980b. *Legionella micdadei* species nova: classification of a third species of *Legionella* associated with human pneumonia. Curr. Microbiol. *3*: 255–258.

Hébert, G.A., A.G. Steigerwalt and D.J. Brenner. 1980c. *In* Validation of the publication of new names and new combinations previously effectively published outside the IJSB. List No. 5. Int. J. Syst. Bacteriol. *30*: 676–677.

Hébert, G.A., B.M. Thomason, P.P. Harris, M.D. Hicklin and R.M. McKinney. 1980d. "Pittsburgh pneumonia agent": a bacterium phenotypically similar to *Legionella pneumophila* and identical to the TATLOCK bacterium. Ann. Intern. Med. *92*: 53–54.

Heddleston, K.L. 1976. Physiologic characteristics of 1,268 cultures of *Pasteurella multocida*. Am. J. Vet. Res. *37*: 745–747.

Heddleston, K.L., J.E. Gallagher and P.A. Rebers. 1972. Fowl cholera: gel diffusion precipitin test for serotyping *Pasteurella multocida* from avian species. Avian Dis. *16*: 925–936.

Hedegaard, J., H. Okkels, B. Bruun, M. Kilian, K.K. Mortensen and N. Nørskov-Lauritsen. 2001. Phylogeny of the genus *Haemophilus* as determined by comparison of partial *infB* sequences. Microbiology *147*: 2599–2609.

Hedges, A. and R.S. Wolfe. 1974. Extracellular enzyme from *Myxobacter* AL-1 that exhibits both β-1,4 glucanase and chitosanase activities. J. Bacteriol. *120*: 844–853.

Hedges, R.W. 1980. R factors of *Serratia*. *In* Graevenitz, V. and Rubin (Editors), The Genus *Serratia*, CRC Press, Boca Raton. 139–153.

Hedges, R.W., N. Datta, J.N. Coetzee and S. Dennison. 1973. R factors from *Proteus morganii*. J. Gen. Microbiol. *77*: 249–259.

Hedlund, B.P., A.D. Geiselbrecht, T.J. Bair and J.T. Staley. 1999a. Polycyclic aromatic hydrocarbon degradation by a new marine bacterium, *Neptunomonas naphthovorans* gen. nov., sp. nov. Appl. Environ. Microbiol. *65*: 251–259.

Hedlund, B.P., A.D. Geiselbrecht, T.J. Bair and J.T. Staley. 1999b. *In* Validation of the publication of new names and new combinations previously effectively published outside the IJSB. List No. 71. Int. J. Syst. Bacteriol. *49*: 1325-1326.

Hedlund, B.P., A.D. Geiselbrecht and J.T. Staley. 1996. Dioxygenase and phylogenetic diversity among marine PAH-degrading bacteria, Q339. Abstr. Annu. Meet. Am. Soc. Microbiol, 206.

Hedlund, B.P. and J.T. Staley. 2001. *Vibrio cyclotrophicus* sp. nov., a polycyclic aromatic hydrocarbon (PAH)-degrading marine bacterium. Int. J. Syst. Evol. Microbiol. *51*: 61–66.

Hedoin, H., A. Coute, P. Kaiser and R. Laugier. 1996. Nature and occurrence of sulfoxidizing bacteria in Baregine developing in sulfurated thermal waters at Bareges (France). Hydrobiologia *323*: 75–81.

Hegazi, N.A. and V. Jensen. 1973. Study of *Azotobacter* bacteriophage in Egyptian soil. Soil Biol. Biochem. *5*: 231–243.

Heidelberger, M., A. Das and E. Juni. 1969. Immunochemistry of the capsular polysaccharide of an acinetobacter. Proc. Natl. Acad. Sci. U.S.A. *63*: 47–50.

Heimbrook, M.E., W.L.L. Wang and G. Campbell. 1989. Staining bacterial flagella easily. J. Clin. Microbiol. *27*: 2612–2615.

Heinrich, D.W. and A.C. Glasgow. 1997. Transcriptional regulation of type 4 pilin genes and the site-specific recombinase gene, *piv*, in *Moraxella lacunata* and *Moraxella bovis*. J. Bacteriol. *179*: 7298–7305.

Heinrich, S. and G. Pulverer. 1959a. Zür ätiologie und mkrobiologie der aktinomykose. II. Definition und praktische dagnostik des *Actinobacillus actinomycetemcomitans*. Zentralbl. Bakteriol. Parasitenkd. Infektionskr. Hyg. Abt. 1 Orig. *174*: 123–135.

Heinrich, S. and G. Pulverer. 1959b. Zür ätiologie und mkrobiologie der aktinomykose. III. Die pathogène bdeutung des *Actinobacillus actinomycetemcomitans* unter den "Begleitbakterien" des *Actinomyces israeli*.

Zentralbl. Bakteriol. Parasitenkd. Infektionskr. Hyg. Abt. 1 Orig. *176*: 91–101.

Heinzen, R.A., M.E. Frazier and L.P. Mallavia. 1992. *Coxiella burnetii* superoxide dismutase gene: Cloning, sequencing, and expression in *Escherichia coli*. Infect. Immun. *60*: 3814–3823.

Heinzen, R.A. and T. Hackstadt. 1996. A developmental stage-specific histone H1 homolog of *Coxiella burnetii*. J. Bacteriol. *178*: 5049–5052.

Heinzen, R.A. and L.P. Mallavia. 1987. Cloning and functional expression of the *Coxiella burnetii* citrate synthase gene in *Escherichia coli*. Infect. Immun. *55*: 848–855.

Heipieper, H.J., G. Meulenbeld, Q. Van Oirschot and J.A.M. De Bont. 1996. Effect of environmental factors on the *trans/cis* ratio of unsaturated fatty acids in *Pseudomonas putida* S12. Appl. Environ. Microbiol. *62*: 2773–2777.

Heipieper, H.J., F.J. Weber, J. Sikkema, H. Keweloh and J.A.M. Debont. 1994. Mechanisms of resistance of whole cells to toxic organic-solvents. Trends Biotechnol. *12*: 409–415.

Heitzer, A., K. Malachowsky, J.E. Thonnard, P.R. Bienkowski, D.C. White and G.S. Sayler. 1994. Optical biosensor for environmental online monitoring of naphthalene and salicylate bioavailability with an immobilized bioluminescent catabolic reporter bacterium. Appl. Environ. Microbiol. *60*: 1487–1494.

Heizmann, W.R. and R. Michel. 1991. Isolation of *Ewingella americana* from a patient with conjunctivitis. Eur. J. Clin. Microbiol. Infect. Dis. *10*: 957–959.

Helander, K.H. and P. Persson. 1996. Identification of *Erwinia carotovora* subsp. *atroseptica* with a non-radioactive DNA probe. Acta Agric. Scand. Sect. B Soil Plant Sci. *46*: 224–229.

Helbig, J.H., P.C. Lueck, Y.A. Knirel, W. Witzleb and U. Zaehringer. 1995. Molecular characterization of a virulence-associated epitope on the lipopolysaccharide of *Legionella pneumophila* serogroup 1. Epidemiol. Infect. *115*: 71–78.

Heldal, M. and O. Tumyr. 1986. Morphology and content dry matter and some elements in cells and stalks of *Nevskia* from an eutrophic lake. Can. J. Microbiol. *32*: 89–92.

Helinski, D.R., A.E. Tiukdarian and R.P. Novick. 1996. Replication control and other stable maintenance mechanisms of plasmids. *In* Neidhardt, Curtiss, Ingraham, Lin, Low, Magasanik, Reznikoff, Riley, Schaechter and Umbarger (Editors), *Escherichia coli* and *Salmonella*: Cellular and Molecular Biology, 2nd Ed., ASM Press, Washington, D.C. pp. 2295–2324.

Heller, H.M., G. Tortora and H. Burger. 1990. *Pseudomonas putrefaciens* bacteremia associated with shellfish contact. Am. J. Med. *88*: 85–86.

Hendrie, M.S., W. Hodgkiss and J.M. Shewan. 1970. The identification, taxonomy and classification of luminous bacteria. J. Gen. Microbiol. *64*: 151–169.

Hendrie, M.S., A.J. Holding and J.M. Shewan. 1974. Emended descriptions of the genus *Alcaligenes* and of *Alcaligenes faecalis* and proposal that the generic name *Achromobacter* be rejected; status of the named species of *Alcaligenes* and *Achromobacter*. Int. J. Syst. Bacteriol. *24*: 534–550.

Hendrix, L. and L.P. Mallavia. 1984. Active transport of proline by *Coxiella burnetii*. J. Gen. Microbiol. *130*: 2857–2863.

Hendrix, L.R., J.E. Samuel and L.P. Mallavia. 1990. Identification and cloning of a 27-kDa *Coxiella burnetii* immunoreactive protein. Ann. N. Y. Acad. Sci. *590*: 534–540.

Hendrix, L.R., J.E. Samuel and L.P. Mallavia. 1991. Differentiation of *Coxiella burnetii* isolates by analysis of restriction-endonuclease-digested DNA separated by SDS-PAGE. J. Gen. Microbiol. *137*: 269–276.

Hennessy, K.J., J.J. Iandolo and B.W. Fenwick. 1993. Serotype identification of *Actinobacillus pleuropneumoniae* by arbitrarily primed polymerase chain reaction. J. Clin. Microbiol. *31*: 1155–1159.

Henrichs, S.M. and J.W. Farrington. 1984. Peru upwelling region sediments near 15°S. 1. Remineralization and accumulation of organic matter. Limnol. Oceanogr. *29*: 1–19.

Henrichsen, J. 1975. The occurrence of twitching motility among Gram-negative bacteria. Acta Pathol. Microbiol. Scand. Sect. B Microbiol. *83*: 171–178.

Henrichsen, J., L.O. Frøholm and K. Bøvre. 1972. Studies on bacterial surface translocation. 2. Correlation of twitching motility and fimbriation in colony variants of *Moraxella nonliquefaciens, M. bovis,* and *M. kingii.* Acta Pathol. Microbiol. Scand. B Microbiol. Immunol. *80*: 445–452.

Henrici, A.T. and D.E. Johnson. 1935. Studies on freshwater bacteria. II. Stalked bacteria, a new order of schizomycetes. J. Bacteriol. *30*: 61–93.

Henricks, P.A., G.J. Binkhorst, A.A. Drijver and F.P. Nijkamp. 1992. *Pasteurella haemolytica* leukotoxin enhances production of leukotriene B4 and 5-hydroxyeicosatetraenoic acid by bovine polymorphonuclear leukocytes. Infect. Immun. *60*: 3238–3243.

Henriksen, S.D. 1950. A comparison of the phenylpyruvic acid reaction and the urease test in the differentiation of *Proteus* from other enteric organisms. J. Bacteriol. *60*: 225–231.

Henriksen, S.D. 1952. *Moraxella*: classification and taxonomy. J. Gen. Microbiol. *6*: 318–328.

Henriksen, S.D. 1969. Proposal of a neotype strain for *Moraxella lacunata.* Int. J. Syst. Bacteriol. *19*: 263–265.

Henriksen, S.D. 1973. *Moraxella, Acinetobacter,* and the *Mimeae.* Bacteriol. Rev. *37*: 522–561.

Henriksen, S.D. and K. Bøvre. 1968. The taxonomy of the genera *Moraxella* and *Neisseria.* J. Gen. Microbiol. *51*: 387–392.

Henriksen, S.D. and K. Jyssum. 1961. A study of some *Pasteurella* strains from the human respiratory tract. Acta Pathol. Microbiol. Scand. *51*: 354–368.

Herbert, M.A., D. Crook and E.R. Moxon. 1998. Molecular methods for *Haemophilus influenza. In* Woodford and Johnson (Editors), Molecular Bacteriology: Protocols and Clinical Applications, Humana Press, Totowa, N.J. 243–263.

Herbert, R.B. and R.G. Holliman. 1964. Aeruginosin B. A naturally occurring phenazinesulfonic acid. Proc. Chem. Soc. *1964*: 19.

Herbst, E.J. and E.E. Snell. 1949. The nutritional requirements of *Haemophilus parainfluenzae.* J. Bacteriol. *58*: 379–386.

Herendeen, S.L., R.A. VanBogelen and F.C. Neidhardt. 1979. Levels of major proteins of *Escherichia coli* during growth at different temperatures. J. Bacteriol. *139*: 185–194.

Herman, N.J. and E. Juni. 1974. Isolation and characterization of a generalized transducing bacteriophage for *Acinetobacter.* J. Virol. *13*: 46–52.

Hernández, J., M.A. Ferrus, M. Hernandez and R.J. Owen. 1997. Arbitrary primed PCR fingerprinting and serotyping of clinical *Pseudomonas aeruginosa* strains. FEMS Immunol. Med. Microbiol. *17*: 37–47.

Heron, S.J.E., J.F. Wilkinson and C.M. Duffus. 1993. Enterobacteria associated with grass and silages. J. Appl. Bacteriol. *75*: 13–17.

Herrington, D.A., S. Tzipori, R.M. Robins-Browne, B.D. Tall and M.M. Levine. 1987. In vitro and in vivo pathogenicity of *Plesiomonas shigelloides.* Infect. Immun. *55*: 979–985.

Herriott, R.M., E.Y. Meyer, M. Vogt and M. Modan. 1970. Defined medium for growth of *Haemophilus influenzae.* J. Bacteriol. *101*: 513–516.

Hertman, I. 1964. Bacteriophage common to *Pasteurella pestis* and *Escherichia coli.* J. Bacteriol. *88*: 1002–1005.

Herwaldt, L.A., G.W. Gorman, T. McGrath, S. Toma, B. Brake, A.W. Hightower, J. Jones, A.L. Reingold, P.A. Boxer, P.W. Tang, C.W. Moss, H. Wilkinson, D.J. Brenner, A.G. Steigerwalt and C.V. Broome. 1984a. A new *Legionella* species, *Legionella feeleii* species nova, causes Pontiac fever in an automobile plant. Ann. Intern. Med. *100*: 333–338.

Herwaldt, L.A., G.W. Gorman, T. McGrath, S. Toma, B. Brake, A.W. Hightower, J. Jones, A.L. Reingold, P.A. Boxer, P.W. Tang, C.W. Moss, H. Wilkinson, D.J. Brenner, A.G. Steigerwalt and C.V. Broome. 1984b. *In* Validation of the publication of new names and new combinations previously effectively published outside the IJSB. List No. 15. Int. J. Syst. Bacteriol. *34*: 355–357.

Herzer, P.J., S. Inouye, M. Inouye and T.S. Whittam. 1990. Phylogenetic distribution of branched RNA-linked multicopy single-stranded DNA among natural isolates of *Escherichia coli.* J. Bacteriol. *172*: 6175–6181.

Hespell, R.B. 1977. *Serpens flexibilis* gen. nov., sp. nov., an unusually flexible lactate-oxidizing bacterium. Int. J. Syst. Bacteriol. *27*: 371–381.

Heulin, T., O. Berge, P. Mavingui, L. Gouzou, K.P. Hebbar and J. Balandreau. 1994. *Bacillus polymyxa* and *Rahnella aquatilis,* the dominant N_2-fixing bacteria associated with wheat rhizosphere in French soils. Eur. J. Soil Biol. *30*: 35–42.

Hewetson, L., H.M. Dunn and N.W. Dunn. 1978. Evidence for a transmissible catabolic plasmid in *Pseudomonas putida* encoding the degradation of *p*-cresol *via* the protocatechuate *ortho*-cleavage pathway. Genet. Res. *32*: 249–255.

Hickman, F.W. and J.J. Farmer, III. 1976. Differentiation of *Proteus mirabilis* by bacteriophage typing and the Dienes reaction. J. Clin. Microbiol. *3*: 350–353.

Hickman, F.W. and J.J. Farmer, III. 1978. *Salmonella typhi*:Identification, antibiograms, serology, and bacteriophage typing. Amer. J. Med. Technol. *44*: 1149–1159.

Hickman, F.W., J.J. Farmer, III, D.G. Hollis, G.R. Fanning, A.G. Steigerwalt, R.E. Weaver and D.J. Brenner. 1982a. Identification of *Vibrio hollisae* sp. nov. from patients with diarrhea. J. Clin. Microbiol. *15*: 395–401.

Hickman, F.W., J.J. Farmer, III, D.G. Hollis, G.R. Fanning, A.G. Steigerwalt, R.E. Weaver and D.J. Brenner. 1982b. *In* Validation of new names and new combinations previously effectively published outside the IJSB. List No. 9. Int. J. Syst. Bacteriol. *32*: 384–385.

Hickman, F.W., J.J. Farmer, III, A.G. Steigerwalt and D.J. Brenner. 1980. Unusual groups of *Morganella* ("*Proteus*") *morganii* isolated from clinical specimens: lysine-positive and ornithine-negative biogroups. J. Clin. Microbiol. *12*: 88–94.

Hickman, F.W., A.G. Steigerwalt, J.J. Farmer, III and D.J. Brenner. 1982c. Identification of *Proteus penneri* sp. nov., formerly known as *Proteus vulgaris* indole negative or as *Proteus vulgaris* biogroup 1. J. Clin. Microbiol. *15*: 1097–1102.

Hickman, F.W., A.G. Steigerwalt, J.J. Farmer, III and D.J. Brenner. 1983. *In* Validation of the publication of new names and new combinations previously effectively published outside the IJSB. List No. 10. Int. J. Syst. Bacteriol. *33*: 438–440.

Hickman, J. and G. Ashwell. 1966. Isolation of bacterial lipopolysaccharide from *Xanthomonas campestris* containing 3-acetamido-3,6-dideoxy-D-alactose and D-hamnose. J. Biol. Chem. *241*: 1424–1428.

Hickman-Brenner, F.W., G.R. Fanning, M.J. Arduino, D.J. Brenner and J.J. Farmer, III. 1988a. *Aeromonas schubertii,* a new mannitol-negative species found in human clinical specimens. J. Clin. Microbiol. *26*: 1561–1564.

Hickman-Brenner, F.W., G.R. Fanning, M.J. Arduino, D.J. Brenner and J.J. Farmer, III. 1989. *In* Validation of the publication of new names and new combinations previously effectively published outside the IJSB. List No. 29. Int. J. Syst. Bacteriol. *39*: 205–206.

Hickman-Brenner, F.W., G.R. Fanning, H.E. Müller and D.J. Brenner. 1986. Priority of *Providencia rustigianii* Hickman-Brenner, Farmer, Steigerwalt, and Brenner 1985 over *Providencia friedericiana* Muller 1983. Int. J. Syst. Bacteriol. *36*: 565.

Hickman-Brenner, F.W., J.J. Farmer, III, A.G. Steigerwalt and D.J. Brenner. 1983a. *In* Validation of the publication of new names and new combinations previously effectively published outside the IJSB. List No. 11. Int. J. Syst. Bacteriol. *33*: 672–674.

Hickman-Brenner, F.W., J.J. Farmer, III, A.G. Steigerwalt and D.J. Brenner. 1983b. *Providencia rustigianii:* a new species in the family *Enterobacteriaceae* formerly known as *Providencia alcalifaciens* biogroup 3. J. Clin. Microbiol. *17*: 1057–1060.

Hickman-Brenner, F.W., G.P. Huntley-Carter, G.R. Fanning, D.J. Brenner and J.J. Farmer. 1985a. *Koserella trabulsii,* a new genus and species of *Enterobacteriaceae* formerly known as enteric group-45. J. Clin. Microbiol. *21*: 39–42.

Hickman-Brenner, F.W., G.P. Huntley-Carter, Y. Saitoh, A.G. Steigerwalt, J.J. Farmer, III and D.J. Brenner. 1984a. *Moellerella wisconsensis,* a new genus and species of *Enterobacteriaceae* found in human stool specimens. J. Clin. Microbiol. *19*: 460–463.

Hickman-Brenner, F.W., G.P. Huntley-Carter, Y. Saitoh, A.G. Steigerwalt,

J.J. Farmer, III and D.J. Brenner. 1984b. *Moellerella wisconsensis*, gen. nov., spec. nov., a new member of *Enterobacteriaceae* found in human stools. 84th Annual Meeting of the American Society for Microbiology, American Society for Microbiology. C246.

Hickman-Brenner, F.W., G.P. Huntley-Carter, Y. Saitoh, A.G. Steigerwalt, J.J. Farmer, III and D.J. Brenner. 1984c. *In* Validation of the publication of new names and new combinations previously effectively published outside the IJSB. List No. 15. Int. J. Syst. Bacteriol. *34*: 355–357.

Hickman-Brenner, F.W., K.L. Macdonald, A.G. Steigerwalt, G.R. Fanning, D.J. Brenner and J.J. Farmer, III. 1987. *Aeromonas veronii*, a new ornithine decarboxylase-positive species that may cause diarrhea. J. Clin. Microbiol. *25*: 900–906.

Hickman-Brenner, F.W., K.L. Macdonald, A.G. Steigerwalt, G.R. Fanning, D.J. Brenner and J.J. Farmer, III. 1988b. *In* Validation of the publication of new names and new combinations previously effectively published outside the IJSB. List No. 25. Int. J. Syst. Bacteriol. *38*: 220–222.

Hickman-Brenner, F.W., M.P. Vohra, G.P. Huntley-Carter, G.R. Fanning, V.A.I. Lowery, D.J. Brenner and J.J. Farmer, III. 1985b. *Leminorella*, a new genus of *Enterobacteriaceae*: identification of *Leminorella grimontii* sp. nov. and *Leminorella richardii* sp. nov. found in clinical specimens. J. Clin. Microbiol. *21*: 234–239.

Hickman-Brenner, F.W., M.P. Vohra, G.P. Huntley-Carter, G.R. Fanning, V.A.I. Lowery, D.J. Brenner and J.J. Farmer, III. 1985c. *In* Validation of the publication of new names and new combinations previously effectively published outside the IJSB. List No. 18. Int. J. Syst. Bacteriol. *35*: 375–376.

High, N.J., M.P. Jennings and E.R. Moxon. 1996. Tandem repeats of the tetramer 5′-CAAT-3′ present in *lic2A* are required for phase variation but not lipopolysaccharide biosynthesis in *Haemophilus influenzae*. Mol. Microbiol. *20*: 165–174.

Higuchi, K. and J.L. Smith. 1961. Studies on the nutrition and physiology of *Pasteurella pestis*: VI. A differential plating medium for the estimation of the mutation rate to avirulence. J. Bacteriol. *81*: 605–608.

Hildebrand, D.C. 1971. Pectate and pectin gels for differentiation of *Pseudomonas* sp. and other bacterial plant pathogens. Phytopathology *61*: 1430–1436.

Hildebrand, D.C., O.C. Huisman and M.N. Schroth. 1984. Use of DNA hybridization values to construct 3-dimensional models of fluorescent pseudomonad relationships. Can. J. Microbiol. *30*: 306–315.

Hildebrand, D.C., N.J. Palleroni, M. Hendson, J. Toth and J.L. Johnson. 1994. *Pseudomonas flavescens* sp. nov., isolated from walnut blight cankers. Int. J. Syst. Bacteriol. *44*: 410–415.

Hildebrand, D.C., N.J. Palleroni and M.N. Schroth. 1990. DNA relatedness of 24 xanthomonad strains representing 23 *Xanthomonas campestris* pathovars and *Xanthomonas fragariae*. J. Appl. Bacteriol. *68*: 263–270.

Hildebrand, D.C. and M.N. Schroth. 1967. A new species of *Erwinia* causing the drippy nut disease of live oaks. Phytopathology *57*: 250–253.

Hill, B.L. and A.H. Purcell. 1995a. Acquisition and retention of *Xylella fastidiosa* by an efficient vector, *Graphocephala atropunctata*. Phytopathology *85*: 209–212.

Hill, B.L. and A.H. Purcell. 1995b. Multiplication and movement of *Xylella fastidiosa* within grapevine and four other plants. Phytopathology *85*: 1368–1372.

Hill, B.L. and A.H. Purcell. 1997. Populations of *Xylella fastidiosa* in plants required for transmission by an efficient vector. Phytopathology *87*: 1197–1201.

Hill, K.R., F.H. Caselitz and L.M. Moody. 1954. A case of acute, metastatic myosistis caused by a new organism of the family *Pseudomonodaceae*: a preliminary report. W. Ind. Med. J. *3*: 9–11.

Hill, P., D.S. Saunders and J.A. Campbell. 1973. The production of "symbiont-free" *Glossina morsitans* and an associated loss of female fertility. Trans. R. Soc. Trop. Med. Hyg. *67*: 727–728.

Hill, S., G.L. Turner and F.J. Bergersen. 1984. Synthesis and activity of nitrogenase in *Klebsiella pneumoniae* exposed to low concentrations of oxygen. J. Gen. Microbiol. *130*: 1061–1067.

Himmelreich, C.A., S.J. Barenkamp and G.A. Storch. 1985. Comparison of methods for serotyping isolates of *Haemophilus influenzae*. J. Clin. Microbiol. *21*: 158–160.

Hinchliffe, E. and A. Vivian. 1980. Naturally occurring plasmids in *Acinetobacter calcoaceticus*: a P class R factor of restricted host range. J. Gen. Microbiol. *116*: 75–80.

Hinde, R. 1971a. The control of the mycetome symbiotes of the aphids *Brevicoryne brassicae*, *Myzus persicae* and *Macrosiphum rosae*. J. Insect Physiol. *17*: 1791–1800.

Hinde, R. 1971b. The fine structure of the mycetome symbiotes of the aphids *Brevicoryne brassicae*, *Myzus persicae*, and *Macrosiphum rosae*. J. Insect Physiol. *17*: 2035–2050.

Hingorani, M.K. and N.J. Singh. 1959. *Xanthomonas punicae* sp. nov. on *Punica granatum* L. Indian J. Agr. Sci. *29*: 45–48.

Hinojosa-Ahumada, M., B. Swaminathan, S.B. Hunter, D.N. Cameron, J.A. Kiehlbauch, I.K. Wachsmuth and N.A. Strockbine. 1991. Restriction fragment length polymorphisms in rRNA operons for subtyping *Shigella sonnei*. J. Clin. Microbiol. *29*: 2380–2384.

Hinton, D.M. and C.W. Bacon. 1995. *Enterobacter cloacae* is an endophytic symbiont of corn. Mycopathologia. *129*: 117–125.

Hinz, K.H. 1973. Beiträg zür Differenzierung von *Haemophilus* Stämmen aus Huhnern. 1. Mitteilung: Kulturelle und biochemische Untersuchungen. Avian Pathol. *2*: 211–229.

Hinz, K.H. 1980. Heat-stable antigenic determinants of *Haemophilus paragallinarum*. Zentralbl. Veterinamed. *27*: 668–676.

Hinz, K.H. and C. Kunjara. 1977. *Haemophilus avium*, a new species from chickens. Int. J. Syst. Bacteriol. *27*: 324–329.

Hinz, K.H. and H.E. Müller. 1977. Neuraminidase und N-Acylneuraminat-Pyruvat-Lyase bei *Haemophilus paragallinarum* und *Haemophilus paravium* n. sp. Zentralbl. Bakteriol. Parasitenkd. Infektionskr. Hyg. I Abt. Orig. A *237*: 72–79.

Hinze, G. 1903. *Thiophysa volutans*, ein neues Schwefelbakterium. Ber. Dtsch. Bot. Ges. *31*: 309–316.

Hinze, H. 1901. Über den Bau der Zellen von *Beggiatoa mirabilis* Cohn. Ber. Dtsch. Bot. Ges. *19*: 369–374.

Hippe, H., A. Hagelstein, I. Kramer, J. Swiderski and E. Stackebrandt. 1999. Phylogenetic analysis of *Formivibrio citricus*, *Propionivibrio dicarboxylicus*, *Anaerobiospirillum thomasii*, *Succinimonas amylolytica* and *Succinivibrio dextrinosolvens* and proposal for *Succinivibrionaceae* fam. nov. Int. J. Syst. Bacteriol. *49*: 779-782.

Hiraishi, A. and Y. Ueda. 1994. Intrageneric structure of the genus *Rhodobacter*: transfer of *Rhodobacter sulfidophilus* and related marine species to the genus *Rhodovulum* gen. nov. Int. J. Syst. Bacteriol. *44*: 15–23.

Hirsch, P. 1973. Fine structure of *Thiopedia* spp. Abstracts of Symposium on Prokaryotic Photosynthetic Organisms, Freiburg. pp. 184–185.

Hirsch, P. 1981. the genus *Nevskia*. *In* Starr, Stolp, Trüper, Balows and Schlegel (Editors), The Prokaryotes: A Handbook on Habitats, Isolation, and Identification of Bacteria, Springer-Verlag, Berlin. 521–523.

Hirsch, P. 1992. The genus *Nevskia*. *In* Balows, Trüper, Dworkin, Harder and Schleifer (Editors), The Prokaryotes. A Handbook on the Biology of Bacteria: Ecophysiology, Isolation, Identification, Applications., Vol. 4, Springer-Verlag, New York. 4089–4092.

Hirschfeld, L. 1919. A new germ of paratyphoid. Lancet *196*: 296–297.

Hirvelä-Koski, V., P. Koski and H. Niiranen. 1994. Biochemical properties and drug resistance of *Aeromonas salmonicida* in Finland. Dis. Aquat. Org. *20*: 191–196.

Hisatsune, K., S. Kondo, Y. Isshiki, T. Iguchi, Y. Kawamata and T. Shimada. 1993. O-antigenic lipopolysaccharide of *Vibrio cholerae* O139 Bengal, a new epidemic strain for recent cholera in the Indian subcontinent. Biochem. Biophys. Res. Commun. *196*: 1309–1315.

Hitchins, V.M. and H.L. Sadoff. 1970. Morphogenesis of cysts in *Azotobacter vinelandii*. J Bacteriol. *104*: 492–498..

Hlady, W.G. and K.C. Klontz. 1996. The epidemiology of *Vibrio* infections in Florida, 1981–1993. J. Infect. Dis. *173*: 1176–1183.

Ho, A.S., I. Sohel and G.K. Schoolnik. 1992. Cloning and characterization of *fxp*, the flexible pilin gene of *Aeromonas hydrophila*. Mol. Microbiol. *6*: 2725–2732.

Ho, T., K.K. Htwe, N. Yamasaki, G.Q. Zhang, M. Ogawa, T. Yamaguchi, H. Fukushi and K. Hirai. 1995. Isolation of *Coxiella burnetii* from dairy cattle and ticks, and some characteristics of the isolates in Japan. Microbiol. Immunol. *39*: 663–671.

Hobbs, M., B.P. Dalrymple, P.T. Cox, S.P. Livingstone, S.F. Delaney and J.S. Mattick. 1991. Organization of the fimbrial gene region of *Bacteroides nodosus*: class I and class II strains. Mol. Microbiol. *5*: 543–560.

Hobbs, M. and P.R. Reeves. 1994. The JUMPstart sequence: a 39 bp element common to several polysaccharide gene clusters. Mol. Microbiol. *12*: 855–856.

Hochster, R.M. and N.B. Madsen. 1959. The breakdown of adenosine phosphates in extracts of *Xanthomonas phaseoli*. Can. J. Biochem. Physiol. *37*: 639–649.

Hochster, R.M. and C.G. Nozzolillo. 1960. Respiratory carriers and the nature of the reduced diphosphopyridine nucleotide oxidase system in *Xanthomonas phaseoli*. Can. J. Biochem. Physiol. *38*: 79–93.

Hodinka, N.E., L.A. Eriquez and A.P. Jones. 1991. The enzymatic characterization of *Leminorella* species. 91st General Meeting of the American Society for Microbiology, Dallas, Texas. p. 376.

Hoffman, P.S., L. Pine and S. Bell. 1983. Production of superoxide and hydrogen peroxide in medium used to culture *Legionella pneumophila*: catalytic decomposition by charcoal. Appl. Environ. Microbiol. *45*: 784–791.

Höfle, M.G. 1988. Identification of bacteria by low-molecular weight RNA profiles - a new chemotaxonomic approach. J. Microbiol. Methods *8*: 235–248.

Höfle, M.G. 1990. Transfer RNA as genotypic fingerprints of eubacteria. Arch. Microbiol. *153*: 299–304.

Höfle, M.G. 1991. Rapid genotyping of pseudomonads by using low-molecular-weight RNA profiles. *In* Galli, Silver and Witholt (Editors), *Pseudomonas* Molecular Biology and Biotechnology, American Society for Microbiology, Washington D.C. 116–126.

Höfle, M.G. and I. Brettar. 1996. Genotyping of heterotrophic bacteria from the central Baltic sea by use of low-molecular-weight RNA profiles. Appl. Environ. Microbiol. *62*: 1383–1390.

Hogenhout, S.A., F. van der Wilk, M. Verbeek, R.W. Goldbach and J.F.J.M. van den Heuvel. 1998. Potato leafroll virus binds to the equatorial domain of the aphid endosymbiotic groEL homolog. J. Virol. *72*: 358–365.

Høiby, N. 1975. Cross-reactions between *Pseudomonas aeruginosa* and thirty-six other bacterial species. Scand. J. Immunol. *4*: 187–196.

Høiby, N. 1979. Immunity-humoral response. *In* Doggett (Editor), *Pseudomonas aeruginosa*. Clinical manifestations of infection and current therapy, Academic Press, New York. 157–189.

Hoie, S., M. Heum and O.F. Thoresen. 1997. Evaluation of a polymerase chain reaction-based assay for the detection of *Aeromonas salmonicida* subsp. *salmonicida* in Atlantic salmon Salmo salar. Dis. Aquat. Org. *30*: 27–35.

Hoiseth, S.K., P.G. Corn and J. Anders. 1992. Amplification status of capsule genes in *Haemophilus influenzae* type b clinical isolates. J. Infect. Dis. *165 (Suppl 1)*: S114.

Hokawat, S. and K. Rudolph. 1991. Variation in pathogenicity and virulence of strains of *Xanthomonas campestris* pv. *glycines*, the incitant of bacterial pustule of soybean. J. Phytopathol. (Berl.) *131*: 73–83.

Holdeman, L.V., E.P. Cato and W.E.C. Moore. 1977. Anaerobe Laboratory Manual, Virginia Polytechnic Institute and State University, Blacksburg, VA.

Holdeman, L., R.W. Kelly and W.E.C. Moore. 1984a. Family I. *Bacteroidaceae*. *In* Krieg and Holt (Editors), Bergey's Manual of Systematic Bacteriology, 1st Ed., Vol. 1, The Williams & Wilkins Co., Baltimore. pp. 602–631.

Holdeman, L.V., R.W. Kelly and W.E.C. Moore. 1984b. Genus I. *Bacteroides*. *In* Krieg and Holt (Editors), Bergey's Manual of Systematic Bacteriology, 1st Ed., Vol. 1, The Williams & Wilkins Co., Baltimore. pp. 604–631.

Holdeman, L.V. and W.E.C. Moore. 1974. *Bacteroides* Castellani and Chal-

mers 1919. *In* Buchanan and Gibbons (Editors), Bergey's Manual of Determinative Bacteriology, 8th Ed., The Williams & Wilkins Co., Baltimore. pp. 385–404.

Holding, A.J. and J.G. Collee. 1971. Routine biochemical tests. *In* Norris and Ribbons (Editors), Methods Microbiol., Vol. 6A, Academic Press, New York. 1–33.

Holland, D.F. 1920. V. Generic index of the commoner forms of bacteria. *In* Winslow, C.E.A., J. Broadhurst, R.E. Buchanan, L.A. Rogers and G.H. Smith (Editors), The families and genera of the bacteria, Vol. 5, J. Bacteriol. 191–229.

Holländer, R. 1976. Energy metabolism of some representatives of the *Haemophilus* group. Antonie van Leeuwenhoek J. Microbiol. Serol. *42*: 429–444.

Hollick, G.E., F.S. Nolte, B.J. Calnan, J.L. Penner, L.J. Barton and A. Spellacy. 1984. Characterization of endemic *Providencia stuartii* isolates from patients with urinary devices. Eur. J. Clin. Microbiol. *3*: 521–525.

Holliman, F.G. 1957. Pigments of a red strain of *Pseudomonas aeruginosa*. Chem. Ind. *28*: 1668.

Hollis, D.G., F.W. Hickman and G.R. Fanning. 1982. *In* Validation of the publication of new names and new combinations previously effectively published outside the IJSB. List No. 8. Int. J. Syst. Bacteriol. *32*: 266–268.

Hollis, D.G., F.W. Hickman, G.R. Fanning, D.J. Brenner and R.E. Weaver. 1980. EF-9: A newly described group of *Enterobacteriaceae*. Abstr. Ann. Meet. Am. Soc. Microbiol., Abstract C 122, p. 295.

Hollis, D.G., F.W. Hickman, G.R. Fanning, J.J. Farmer, III, R.E. Weaver and D.J. Brenner. 1981. *Tatumella ptyseos* gen. nov., sp. nov., a member of the family *Enterobacteriaceae* found in clinical specimens. J. Clin. Microbiol. *14*: 79–88.

Hollis, D.G., R.E. Weaver, C.N. Baker and C. Thornsberry. 1976. Halophilic *Vibrio* species isolated from blood cultures. J. Clin. Microbiol. *3*: 426–431.

Hollis, D.G., R.E. Weaver, A.G. Steigerwalt, J.D. Wenger, C.W. Moss and D.J. Brenner. 1989. *Francisella philomiragia*, comb. nov. (formerly *Yersinia philomiragia*) and *Francisella tularensis* biovar novicida (formerly *Franciscella novicida*) associated with human disease. J. Clin. Microbiol. *27*: 1601–1608.

Hollis, D.G., R.E. Weaver, A.G. Steigerwalt, J.D. Wenger, C.W. Moss and D.J. Brenner. 1990. *In* Validation of the publication of new names and new combinations previously effectively published outside the IJSB. List No. 32. Int. J. Syst. Bacteriol. *40*: 105–106.

Holloway, B.W., U. Römling and B. Tümmler. 1994. Genomic mapping of *Pseudomonas aeruginosa* PAO. Microbiology *140*: 2907–2929.

Holm, P. 1954. The influence of carbon dioxide on the growth of *Actinobacillus actinomycetemcomitans* (*Bacterium actinomycetemcomitans* (Klinger 1912)). Acta Pathol. Microbiol. Scand. *34*: 235–248.

Holmberg, S.D. and J.J. Farmer, III. 1984. *Aeromonas hydrophila* and *Plesiomonas shigelloides* as causes of intestinal infections. Rev. Infect. Dis. *6*: 633–639.

Holmberg, S.D., W.L. Schell, G.R. Fanning, I.K. Wachsmuth, F.W. Hickman-Brenner, P.A. Blake, D.J. Brenner and J.J. Farmer, III. 1986a. *Aeromonas* intestinal infections in the USA. Ann. Intern. Med. *105*: 683–689.

Holmberg, S.D., I.K. Wachsmuth, F.W. Hickman-Brenner, P.A. Blake and J.J. Farmer, III. 1986b. *Plesiomonas* enteric infections in the United States. Ann. Intern. Med. *105*: 690–694.

Holmes, A.J., A. Costello, M.E. Lidstrom and J.C. Murrell. 1995a. Evidence that particulate methane monooxygenase and ammonia monooxygenase may be evolutionarily related. FEMS Microbiol. Lett. *132*: 203–208.

Holmes, A.J., N.J. Owens and J.C. Murrell. 1995b. Detection of novel marine methanotrophs using phylogenetic and functional gene probes after methane enrichment. Microbiology *141*: 1947–1955.

Holmes, B. 1998. *Actinobacillus*, *Pasteurella* and *Eikenella*. *In* Collier, Balows and Sussman (Editors), Topley & Wilson's Microbiology and Microbial Infections, 9th Ed., Vol. 2, Arnold, London. pp. 1191–1215.

Holmes, B., M. Costas and L.L. Sloss. 1988. Numerical analysis of electrophoretic protein patterns of *Providencia alcalifaciens* strains from human faeces and veterinary specimens. J. Appl. Bacteriol. *64*: 27–35.

Holmes, B., S.P. Lapage and H. Maluick. 1975. Strains of *Pseudomonas putrefaciens* from clinical material. J. Clin. Pathol. *28*: 149–155.

Holmes, B., R.J. Owen and T.A. McMeekin. 1984. Genus *Flavobacterium*. *In* Krieg and Holt (Editors), Bergey's Manual of Systematic Bacteriology, 1st Ed., Vol. 1, The Williams & Wilkins Co., Baltimore. pp. 353–360.

Holmes, B., A.G. Steigerwalt, R.E. Weaver and D.J. Brenner. 1987. *Chryseomonas luteola*, comb. nov. and *Flavimonas oryzihabitans*, gen.nov., comb. nov., *Pseudomonas*-like species from human clinical specimens and formerly known, respectively, as groups Ve-1 and Ve-2. Int. J. Syst. Bacteriol. *37*: 245–250.

Holmes, P., L.M. Niccols and D.P. Sartory. 1996. The ecology of mesophilic *Aeromonas* in the aquatic environment. *In* Austin, Altwegg, Gosling and Joseph (Editors), The Genus *Aeromonas*, John Wiley and Sons, Ltd., Chichester. pp. 127–150.

Holmes, P. and L.N. Nicolls. 1995. Aeromonads in drinking water supplies — their occurrence and significance. J. Chartered Inst. Water Environ. Manag. *9*: 464–469.

Holmes, P. and D.P. Sartory. 1993. An evaluation of media for the membrane filtration enumeration of *Aeromonas* from drinking water. Lett. Appl. Microbiol. *17*: 58–60.

Holmquist, L. and S. Kjelleberg. 1993. Changes in viability, respiratory activity and morphology of the marine *Vibrio* sp. strain S14 during starvation of individual nutrients and subsequent recovery. FEMS Microbiol. Ecol. *12*: 215–224.

Holst, E., J. Rollof, L. Larsson and J.P. Nielsen. 1992. Characterization and distribution of *Pasteurella* species recovered from infected humans. J. Clin. Microbiol. *30*: 2984–2987.

Holström, C., S. James, B.A. Neilan, D.C. White and S. Kjelleberg. 1998. *Pseudoalteromonas tunicata* sp. nov., a bacterium that produces antifouling agents. Int. J. Syst. Bacteriol. *48*: 1205–1212.

Holt, J.G., N.R. Krieg, P.H.A. Sneath, J.T. Staley and S.T. Williams (Editors). 1994. Bergey's Manual of Determinative Bacteriology, 9th Ed., The Williams & Wilkins Co., Baltimore.

Holt, S.C., A.C.R. Tanner and S.S. Socransky. 1980. Morphology and ultrastructure of oral strains of *Actinobacillus actinomycetemcomitans* and *Haemophilus aphrophilus*. Infect. Immun. *30*: 588–600.

Holtel, A., S. Marqués, I. Möhler, U. Jakubzik and K.N. Timmis. 1994. Carbon source-dependent inhibition of *xyl* operon expression of the *Pseudomonas putida* TOL plasmid. J. Bacteriol. *176*: 1773–1776.

Holtwick, R., F. Meinhardt and H. Keweloh. 1997. *Cis-trans* isomerization of unsaturated fatty acids: Cloning and sequencing of the *cti* gene from *Pseudomonas putida* P8. Appl. Environ. Microbiol. *63*: 4292–4297.

Home, M.T., M.F. Tather and R.J. Roberts. 1984. Enteric redmouth: a threat or a promise? Fish Farming Int. *2*: 12.

Honda, S.L., I. Goto, I. Minematsu, N. Ikeda, N. Asano, M. Ishibashi, Y. Kinoshita, M. Nishibuchi, T. Honda and T. Miwatani. 1987. Gastroenteritis due to Kanagawa negative *Vibrio parahaemolyticus*. Lancet *1*: 331–332.

Honda, T., K. Kasemsuksakul, T. Oguchi, M. Kohda and T. Miwatani. 1988. Production and partial characterization of pili on non-O1 *Vibrio cholerae*. J. Infect. Dis. *157*: 217–218.

Honma, Y., M. Ikema, C. Toma, M. Ehara and M. Iwanaga. 1997. Molecular analysis of a filamentous phage (fsl) of *Vibrio cholerae* O139. Biochim. Biophys. Acta *1362*: 109–115.

Honma, Y. and N. Nakasone. 1990. Pili of *Aeromonas hydrophila*: purification, characterization, and biological role. Microbiol. Immunol. *34*: 83–98.

Hood, A.M. 1977. Virulence factors of *Francisella tularensis*. J. Hyg. *79*: 47–60.

Hood, D.W., M.E. Deadman, T. Allen, H. Masoud, A. Martin, J.R. Brisson, R. Fleischmann, J.C. Venter, J.C. Richards and E.R. Moxon. 1996. Use of the complete genome sequence information of *Haemophilus influenzae* strain Rd to investigate lipopolysaccharide biosynthesis. Mol. Microbiol. *22*: 951–965.

Hood, M.A., J.B. Guckert, D.C. White and F. Deck. 1986. Effect of nutrient deprivation on lipid, carbohydrate, DNA, RNA, and protein levels in *Vibrio cholerae*. Appl. Environ. Microbiol. *52*: 788–793.

Hood, M.A., G.E. Ness, G.E. Rodrick and N.J. Blake. 1984. The ecology of *Vibrio cholerae* in two Florida estuaries. *In* Colwell (Editor), Vibrios in the Environment, John Wiley and Sons, New York. pp. 399–409.

Hoogkamp-Korstanje, J.A. 1987. Antibiotics in *Yersinia enterocolitica* infections. J. Antimicrob. Chemother. *20*: 123–131.

Hookey, J.V., R.J. Birtles and N.A. Saunders. 1995. Intergenic 16S rRNA gene (rDNA)–23S rDNA sequence length polymorphisms in members of the family *Legionellaceae*. J. Clin. Microbiol. *33*: 2377–2381.

Hookey, J.V., N.A. Saunders, N.K. Fry, R.J. Birtles and T.G. Harrison. 1996. Phylogeny of Legionellaceae based on small-subunit ribosomal DNA sequences and proposal of Legionella lytica comb. nov. for Legionella-like amoebal pathogens. Int. J. Syst. Bacteriol. *46*: 526–531.

Hoover, T.A. and J.C. Williams. 1990. Characterization of *Coxiella burnetii pyrB*. Ann. N. Y. Acad. Sci. *590*: 485–490.

Hopkins, D.L. 1984. Variability of virulence in grapevine among isolates of the Pierce's disease bacterium. Phytopathology *74*: 1395–1398.

Hopkins, D.L. 1988a. Natural hosts of *Xylella fastidiosa* in Florida. Plant Dis. *72*: 429–431.

Hopkins, D.L. 1988b. *Xylella fastidiosa* and other fastidious bacteria of uncertain affiliation. *In* Schaad (Editor), Laboratory guide for identification of plant pathogenic bacteria, 2nd Ed., APS Press, St. Paul. pp. 95–103.

Hopkins, J.C.F. and W.J. Dowson. 1949. A bacterial leaf and flower disease of Zinnia in Southern Rhodesia. Trans. Brit. Mycol. Soc. *32*: 252–254.

Hopkins, K.L. and A.C. Hilton. 2001. Optimization of random amplification of polymorphic DNA analysis for molecular subtyping of *Escherichia coli* O157. Lett. Appl. Microbiol. *32*: 126–130.

Hoppe, J.E., M. Herter, S. Aleksic, T. Klingebiel and D. Niethammer. 1993. Catheter-related *Rahnella aquatilis* bacteremia in a pediatric bone marrow transplant recipient. J. Clin. Microbiol. *31*: 1911–1912.

Hori, S. 1911. A bacterial leaf-disease of tropical orchids. Zentbl. Bakteriol. Parasitenkd. Infektkrankh. Hyg. Abt. II *31*: 85–92.

Hori, S. 1915. An important disease of tea plants caused by a bacterium. J. Plant Protect. *Tokyo 2*: 1–7.

Horie, S., Y. Yanagida, K. Saheki, A. Hiraishi and K. Cho. 1985. Occurrence of *Rahnella aquatilis*, psychrotrophic coliforms, in mountains soils. J. Food Hyg. Soc. Jpn. *26*: 573–578.

Horino, O. 1973. Ultrastructure of *Xanthomonas oryzae* and its morphological changes by chemical treatment. Ann. Phytopathol. Soc. Jpn. *39*: 14–26.

Horisberger, M. 1977. Structure of the peptidoglycans of *Moraxella glucidolytica* and *Moraxella lwoffi* grown on hydrocarbons. Arch. Microbiol. *112*: 297–302.

Hormaeche, E. and P.R. Edwards. 1958. Observations on the genus *Aerobacter* with a description of two species. Int. Bull. Bacteriol. Nomencl. Taxon. *8*: 111–115.

Hormaeche, E. and P.R. Edwards. 1960a. Proposal for the rejection of the generic name *Cloaca* Castellani and Chambers, and proposal of *Enterobacter* as a generic name with designation of type species and its type culture. Int. Bull. Bacteriol. Nomencl. Taxon. *10*: 75–76.

Hormaeche, E. and P.R. Edwards. 1960b. A proposed genus *Enterobacter*. Int. Bull. Bacteriol. Nomen. Taxon. *10*: 71–74.

Hormaeche, E. and M. Munilla. 1957. Biochemical tests for the differentiation of *Klebsiella* and *Cloaca*. Int. Bull. Bacteriol. Nomencl. Taxon. *7*: 1–20.

Hornstein, M.J., A.M. Jupeau, M.R. Scavizzi, A.M. Philippon and P.A. Grimont. 1985. In vitro susceptibilities of 126 clinical isolates of *Yersinia enterocolitica* to 21 beta-lactam antibiotics. Antimicrob. Agents Chemother. *27*: 806–811.

Hoshina, T. 1962. On a new bacterium, *Paracolobactrum anguillimortiferum* n. sp. Bull. Jpn. Soc. Sci. Fish. *28*: 162–164.

Hosseini, P.K. and K.H. Nealson. 1995. Symbiotic luminous soil bacteria: unusual regulation for an unusual niche. Photochem. Photobiol. *62*: 633–640.

Hou, C.T. (Editor). 1984. Methylotrophs: microbiology, biochemistry, and genetics, CRC Press, Boca Raton, Florida.

Houk, E.J. and G.W. Griffiths. 1980. Intracellular symbiotes of the Homoptera. Annu. Rev. Entomol. 25: 161–187.

Houk, E.J., G.W. Griffiths, N.E. Hadjokas and S.D. Beck. 1977. Peptidoglycan in the cell wall of the primary intracellular symbiote of the pea aphid. Science 198: 401–403.

Houng, H.S., O. Sethabutr and P. Echeverria. 1997. A simple polymerase chain reaction technique to detect and differentiate Shigella and enteroinvasive Escherichia coli in human feces. Diagn. Microbiol. Infect. Dis. 28: 19–25.

Houng, H.S. and M.M. Venkatesan. 1998. Genetic analysis of Shigella sonnei form I antigen: identification of a novel IS630 as an essential element for the form I antigen expression. Microb. Pathog. 25: 165–173.

House, M.L., J.L. Bartholomew, J.R. Winton and J.L. Fryer. 1999. Relative virulence of three isolates of Piscirickettsia salmonis for coho salmon Oncorhynchus kisutch. Dis. Aquat. Org. 35: 107–113.

Hovig, B. and E.H. Aandahl. 1969. A selective method for the isolation of Haemophilus in material from the respiratory tract. Acta Pathol. Microbiol. Scand. 77: 676–684.

Hovland, M. and A.G. Judd. 1988. Seabed Pockmarks and Seepages. Impact on Geology, Biology, and the Marine Environment, Grahm and Trotman, London.

Howard, S.P. 1999. Secretion of aerolysin and other extracellular proteins by the aeromonad. 6th International Aeromonas/Plesiomonas Symposium, Chicago, Illinois. p. 14.

Howard, S.P., J. Critch and A. Bedi. 1993. Isolation and analysis of eight exe genes and their involvement in extracellular protein secretion and outer membrane assembly in Aeromonas hydrophila. J. Bacteriol. 175: 6695–6703.

Howard, S.P., S. MacIntyre and J.T. Buckley. 1996. Toxin. In Austin, Altwegg, Gosling and Joseph (Editors), The Genus Aeromonas, John Wiley and Sons, Ltd., Chichester. pp. 267–286.

Howarth, R., R.F. Unz, E.M. Seviour, R.J. Seviour, L.L. Blackall, R.W. Pickup, J.G. Jones, J. Yaguchi and I.M. Head. 1999. Phylogenetic relationships of filamentous sulfur bacteria (Thiothrix spp. and Eikelboom type 021N bacteria) isolated from wastewater-treatment plants and description of Thiothrix eikelboomii sp. nov., Thiothrix unzii sp. nov., Thiothrix fructosivorans sp. nov., and Thiothrix defluvii sp. nov. Int. J. Syst. Bacteriol. 49: 1817–1827.

Hsu, T.C., W.D.I. Waltman and E.B. Shotts. 1981. Correlation of extracellular enzymatic activity and biochemical characterisitics with regard to virulence of Aeromonas hydrophila. Int. Symp. Fish. Biol. 49: 101–111.

Hu, F.P., J.M. Young, D.E. Stead and M. Goto. 1997. Transfer of Pseudomonas cissicola (Takimoto 1939) Burkholder 1948 to the genus Xanthomonas. Int. J. Syst. Bacteriol. 47: 228–230.

Hu, F.P., J.M. Young and C.M. Triggs. 1991. Numerical analysis and determinative tests for nonfluorescent plant-pathogenic Pseudomonas spp. and genomic analysis and reclassification of species related to Pseudmonas avenae Manns 1909. Int. J. Syst. Bacteriol. 41: 516–525.

Hu, N.T., M.N. Hung, S.J. Chiou, F. Tang, D.C. Chiang, H.Y. Huang and C.Y. Wu. 1992. Cloning and characterization of a gene required for the secretion of extracellular enzymes across the outer membrane by Xanthomonas campestris pv. campestris. J. Bacteriol. 174: 2679–2687.

Huan, P.T., D.A. Bastin, B.L. Whittle, A.A. Lindberg and N.K. Verma. 1997a. Molecular characterization of the genes involved in O-antigen modification, attachment, integration and excision in Shigella flexneri bacteriophage SfV. Gene 195: 217–227.

Huan, P.T., B.L. Whittle, D.A. Bastin, A.A. Lindberg and N.K. Verma. 1997b. Shigella flexneri type-specific antigen V: cloning, sequencing and characterization of the glucosyl transferase gene of temperate bacteriophage SfV. Gene 195: 207–216.

Huang, C.Y., J.L. Garcia, B.K.C. Patel, J.L. Cayol, L. Baresi and R.A. Mah. 2000. Salinivibrio costicola subsp. vallismortis subsp. nov., a halotolerant facultative anaerobe from Death Valley, and emended description of Salinivibrio costicola. Int. J. Syst. Evol. Microbiol. 50: 615–622.

Huang, M., F.B. Oppermann and A. Steinbüchel. 1994. Molecular characterization of the Pseudomonas putida 2,3-butanediol catabolic pathway. FEMS Microbiol. Lett. 124: 141–150.

Huang, T.C. and M.C. Chang. 1975. Studies on xanthobacidin, a new antibiotic from Bacillus subtilis active against Xanthomonas. Botan. Bull. Acad. Sin. 16: 137–148.

Huang, T.-C., F.-H. Lin and T.-T. Kuo. 1975. Properties of membrane-bound adenosine triphosphatase from Xanthomonas oryzae Bot. Bull. Acad. Sinica. 16: 36–44.

Huang, W.M. 1996. Bacterial diversity based on type II DNA topoisomerase genes. Annu. Rev. Genet. 30: 79–107.

Hubert, B., A. de Mahenge, F. Grimont, C. Richard, Y. Peloux, C. de Mahenge, J. Fleurette and P.A.D. Grimont. 1991. An outbreak of pneumonia and meningitis caused by a previously undescribed gram-negative bacterium in a hot spring spa. Epidemiol. Infect. 107: 373–381.

Hudson, H.P., A.A. Lindberg and B.A.D. Stocker. 1978. Lipopolysaccharide core defects in Salmonella typhimurium mutants which are resistant to Felix 0 phage but retain smooth character. J. Gen. Microbiol. 109: 97–112.

Hudson, M.J., D.G. Hollis, R.E. Weaver and C.G. Galvis. 1987. Relationship of CDC group EO-2 and Psychrobacter immobilis. J. Clin. Microbiol. 25: 1907–1910.

Huebner, R.J. and J.A. Bell. 1951. Q fever studies is southern California : Summary of current results and a discussion of possible control measures. JAMA. 145: 301–305.

Huettel, M., S. Forster, S. Klöser and H. Fossing. 1996. Vertical migration in the sediment-dwelling sulfur bacteria Thioploca spp. in overcoming diffusion limitations. Appl. Environ. Microbiol. 62: 1863–1872.

Hugas, M. and J. Arnau. 1987. Aparicíon de manchas de color en la corteza y grasa del jamón durante el post-salado. In Arnau, Hugas and Monfort (Editors), Jamón Curado, Aspectos Técnicos, Institut de Recerca i Tecnología Agroalimentáries, Monells, Spain. pp. 179–182.

Huger, A. 1959. Histological observations on the development of crystalline inclusions of the rickettsial disease of Tipula paludosa Meigen. J. Insect Pathol. 1: 60–66.

Huger, A. 1962. Zür Genese der Begleitkristalle bei Rickettsiella-Infektionen von Insekten. Naturwissenschaften 49: 358–360.

Huger, A. 1964. Eine Rickettsiose der Orientalischen Schabe, Blatta orientalis L. verursacht durch Rickettsiella blattae nov. spec. Naturwissenschaften 51: 22.

Huger, A.M., S.W. Skinner and J.H. Werren. 1985. Bacterial infections associated with the son-killer trait in the parasitoid wasp Nasonia vitripennis (synonym Mormoniella vitripennis) (Hymenoptera: Pteromalidae). J. Invertebr. Pathol. 46: 272–280.

Hugh, R. 1981. Pseudomonas maltophilia sp. nov., nom. rev. Int. J. Syst. Bacteriol. 31: 195.

Hugh, R. and G.L. Gilardi. 1980. Pseudomonas. In Lenette, Balows, Hausler and Truant (Editors), Manual of Clinical Microbiology, 3rd Ed., American Society for Microbiology, Washington, D.C. pp. 289–317.

Hugh, R. and P. Ikari. 1964. The proposed neotype strain of Pseudomonas alcaligenes Monias (1928). Int. Bull. Bacteriol. Nomencl. Taxon. 14: 103–107.

Hugh, R. and E. Leifson. 1953. The taxonomic significance of fermentative versus oxidative metabolism of carbohydrates by various Gram-negative bacteria. J. Bacteriol. 66: 24–26.

Hugh, R. and E. Ryschenkow. 1960. An Alcaligenes-like Pseudomonas species. In Bacteriol. Proc., pp. 78.

Hugh, R. and E. Ryschenkow. 1961. Pseudomonas maltophila, an Alcaligenes-like species. J. Gen. Microbiol. 26: 123–132.

Hugh, R. and R. Sakazaki. 1972. Minimal number of characters for the identification of Vibrio species, Vibrio cholerae, and Vibrio parahaemolyticus. J. Conf. Public Health Lab. Dir. 30: 133–137.

Hughes, D.E. and G.W. Pugh, Jr.. 1970. Isolation and description of a Moraxella from horses with conjunctivitis. Am. J. Vet. Res. 31: 457–462.

Hughes, H.P., M. Campos, L. McDougall, T.K. Beskorwayne, A.A. Potter and L.A. Babiuk. 1994. Regulation of major histocompatibility complex class II expression by *Pasteurella haemolytica* leukotoxin. Infect. Immun. *62*: 1609–1615.

Hugouvieux-Cotte-Pattat, N., G. Condemine, W. Nasser and S. Reverchon. 1996. Regulation of pectinolysis in *Erwinia chrysanthemi*. Annu. Rev. Microbiol. *50*: 213–257.

Hugouvieux-Cotte-Pattat, N., W. Nasser and J. Robert-Baudouy. 1994. Molecular characterization of the *Erwinia chrysanthemi kdgK* gene involved in pectin degradation. J. Bacteriol. *176*: 2386–2392.

Hugouvieux-Cotte-Pattat, N. and J. Robert-Baudouy. 1985. Isolation of *kdgK-lac* and *kdgA-lac* gene fusions in the phytopathogenic bacterium *Erwinia chrysanthemi*. J. Gen. Microbiol. *131*: 1205–1212.

Huisman, G.W., O. Deleeuw, G. Eggink and B. Witholt. 1989. Synthesis of poly-3-hydroxyalkanoates is a common feature of fluorescent pseudomonads. Appl. Environ. Microbiol. *55*: 1949–1954.

Hull, S. 1997. *Escherichia coli* lipopolysaccharide in pathogenesis and virulence. *In* Sussman (Editor), *Escherichia coli*: Mechanisms of Virulence, Cambridge University Press, Cambridge. pp. 145–167.

Hummerjohann, J., E. Küttel, M. Quadroni, J. Ragaller, T. Leisinger and M.A. Kertesz. 1998. Regulation of the sulfate starvation response in *Pseudomonas aeruginosa*: role of cysteine biosynthetic intermediates. Microbiology *144*: 1375–1386.

Humphreys, N.J. and A.E. Douglas. 1997. Partitioning of symbiotic bacteria between generations of an insect: a quantitative study of a *Buchnera* sp. in the pea aphid (*Acyrthosiphon pisum*) reared at different temperatures. Appl. Environ. Microbiol. *63*: 3294–3296.

Hunger, M., R. Schmucker, V. Kishan and W. Hillen. 1990. Analysis and nucleotide sequence of an origin of DNA replication in *Acinetobacter calcoaceticus* and its use for *Escherichia coli* shuttle plasmids. Gene *87*: 45–51.

Hunt, J.C. and P.V. Phibbs. 1981. Failure of *Pseudomonas aeruginosa* to form membrane-associated glucose-dehydrogenase activity during anaerobic growth with nitrate. Biochem. Biophys. Res. Commun. *102*: 1393–1399.

Hunt, M.L., B. Adler and K.M. Townsend. 2000. The molecular biology of *Pasteurella multocida*. Vet. Microbiol. *72*: 3–25.

Hunt, M.L., C.G. Ruffolo, K. Rajakumar and B. Adler. 1998. Physical and genetic map of the *Pasteurella multocida* A:1 chromosome. J. Bacteriol. *180*: 6054–6058.

Hunter, P.R. 1993. The microbiology of bottled natural mineral waters. J. Appl. Bacteriol. *74*: 345–352.

Huppertz, H.I., S. Rutkowski, S. Aleksic and H. Karch. 1997. Acute and chronic diarrhoea and abdominal colic associated with enteroaggregative *Escherichia coli* in young children living in western Europe. Lancet *349*: 1660–1662.

Huq, A., A. Akhtar, M.A.R. Chowdhury and D.A. Sack. 1991. Optimal growth temperature for the isolation of *Plesiomonas shigelloides*, using various selective and differential agars. Can. J. Microbiol. *37*: 800–802.

Huq, A., M. Alam, S. Parveen and R.R. Colwell. 1992. Occurrence of resistance to vibriostatic compound 0/129 in *Vibrio cholerae* 01 isolated from clinical and environmental samples in Bangladesh. J. Clin. Microbiol. *30*: 219–221.

Huq, A., R.R. Colwell, M.A. Chowdhury, B. Xu, S.M. Moniruzzaman, M.S. Islam, M. Yunus and M.J. Albert. 1995. Coexistence of *Vibrio cholerae* O1 and O139 Bengal in plankton in Bangladesh. Lancet *345*: 1249.

Huq, M.I., A.K.M.J. Alam, D.J. Brenner and G.K. Morris. 1980. Isolation of *Vibrio*-like group, EF-6, from patients with diarrhea. J. Clin. Microbiol. *11*: 621–624.

Hurpin, B. and P.H. Robert. 1972. Comparison of the activity of certain pathogens of the cockchafer *Melolontha melolontha* in plots of natural meadowland. J. Invert. Pathol. *19*: 291–298.

Hurpin, B. and P.H. Robert. 1976. Conservation dans le sol de trois germes pathogènes pour les larvaes de *Melolontha melolontha* (Col: Scarabaeidae). Entomophaga. *21*: 73–80.

Hurpin, B. and P.H. Robert. 1977. Effets en population naturelle de *Melolontha melolontha* (Col: Scarabaeidae) d' une introduction de *Rickettsiella melolonthae* et de *Entomopoxvirus melolonthae*. Entomophaga. *22*: 85–92.

Hurtado, A. and F. Rodriguez-Valera. 1999. Accessory DNA in the genomes of representatives of the *Escherichia coli* reference collection. J. Bacteriol. *181*: 2548–2554.

Hurvell, B. and A.A. Lindberg. 1973. Serological cross-reactions between different *Brucella* species and *Yersinia enterocolitica*. Immunochemical studies on phenol-water-extracted lipopolysaccharides from *Brucella abortus* and *Yersinia enterocolitica* type IX. Acta Pathol. Microbiol. Scand. *B81*: 113–119.

Husain, D.R., M. Goutx, M. Acquaviva, M. Gilewicz and J.C. Bertrand. 1997a. The effect of temperature on eicosane substrate uptake modes by a marine bacterium *Pseudomonas nautica* strain 617: relationship with the biochemical content of cells and supernatants. World J. Microbiol. Biotechnol. *13*: 587–590.

Husain, D.R., M. Goutx, C. Bezac, M. Gilewicz and J.C. Bertrand. 1997b. Morphological adaptation of *Pseudomonas nautica* strain 617 to growth on eicosane and modes of eicosane uptake. Lett. Appl. Microbiol. *24*: 55–58.

Hüss, H. 1907. Morphologisch-physiologische Studien über zwei aromabildende Bakterien. Zentbl. Bakteriol. Parasitenkd. Infekrankh. Hyg. 2 Abt. Natwiss. Allg. Landwirtsch. Tech. Bakteriol. Garungsphysiol. Pflan. *19*: 50–70.

Huss, V.A.R., H. Festl and K.H. Schleifer. 1983. Studies on spectrophotometric determination of DNA hybridization from renaturation rates. Syst. Appl. Microbiol. *4*: 184–192.

Husslein, V., T. Chakraborty, A. Carnahan and S.W. Joseph. 1992. Molecular studies on the aerolysin gene of *Aeromonas* species and discovery of a species-specific probe for *Aeromonas trota* sp. nov. Clin. Infect. Dis. *14*: 1061–1068.

Huston, A.L. 2003. Bacterial adaptation to the cold: *in situ* activities of extracellular enzymes in the North Water polynya and characterization of a cold-active aminopeptidase from *Colwellia psychrerythraea* strain 34H. PhD Dissertation, University of Washington, 168 p.

Huston, A.L., B.B. Krieger-Brockett and J.W. Deming. 2000. Remarkably low temperature optima for extracellular enzyme activity from Arctic bacteria and sea ice. Environ. Microbiol. *2*: 383–388.

Hutcheson, K.A. and M. Magbalon. 1999. Periocular abscess and cellulitis from *Pasteurella multocida* in a healthy child. Am. J. Ophthalmol. *128*: 514–515.

Hutchinson, P.B. 1949. A bacterial disease of *Dysoxylum spectabile* caused by the pathogen *Pseudomonas dysoxyli* n. sp. N. Z. J. Sci. Technol. *B30*: 274–286.

Huval, J.H., R. Latta, R. Wallace, D.J. Kushner and R.H. Vreeland. 1995. Description of two new species of *Halomonas*: *Halomonas israelensis* sp. nov. and *Halomonas canadensis* sp. nov. Can. J. Microbiol. *41*: 1124–1131.

Huval, J.H., R. Latta, R. Wallace, D.J. Kushner and R.H. Vreeland. 1996. Validation of the publication of new names and new combinations previously effectively published outside the IJSB. List No. 59. Int. J. Syst. Bacteriol. *46*: 1189–1190.

Huys, G., M. Altwegg, M.L. Hänninen, M. Vancanneyt, L. Vauterin, R. Coopman, U. Torck, J. Lüthy-Hottenstein, P. Janssen and K. Kersters. 1996a. Genotypic and chemotaxonomic description of two subgroups in the species *Aeromonas eucrenophila* and their affiliation to *A. encheleia* and *Aeromonas* DNA hybridization group 11. Syst. Appl. Microbiol. *19*: 616–623.

Huys, G., R. Coopman, P. Janssen and K. Kersters. 1996b. High-resolution genotypic analysis of the genus *Aeromonas* by AFLP fingerprinting. Int. J. Syst. Bacteriol. *46*: 572–580.

Huys, G., R. Denys and J. Swings. 2002a. DNA-DNA reassociation and phenotypic data indicate synonymy between *Aeromonas enteropelogenes* Schubert et al. 1990 and *Aeromonas trota* Carnahan et al. 1991. Int. J. Syst. Evol. Microbiol. *52*: 1969–1972.

Huys, G., P. Kämpfer, M.J. Albert, I. Kuhn, R. Denys and J. Swings. 2002b. *Aeromonas hydrophila* subsp. *dhakensis* subsp. nov., isolated from children with diarrhoea in Bangladesh, and extended description of *Aeromonas hydrophila* subsp. *hydrophila* (Chester 1901) Stanier 1943 (Approved Lists 1980). Int. J. Syst. Evol. Microbiol. *52*: 705–712.

Huys, G., P. Kämpfer, M. Altwegg, R. Coopman, P. Janssen, M. Gillis and

K. Kersters. 1997a. Inclusion of *Aeromonas* DNA hybridization group 11 in *Aeromonas encheleia* and extended descriptions of the species *Aeromonas eucrenophila* and *A. encheleia*. Int. J. Syst. Bacteriol. *47*: 1157–1164.

Huys, G., P. Kämpfer, M. Altwegg, I. Kersters, A. Lamb, R. Coopman, J. Lüthy-Hottenstein, M. Vancanneyt, P. Janssen and K. Kersters. 1997b. *Aeromonas popoffii* sp. nov., a mesophilic bacterium isolated from drinking water production plants and reservoirs. Int. J. Syst. Bacteriol. *47*: 1165–1171.

Huys, G., P. Kämpfer and J. Swings. 2001. New DNA-DNA hybridization and phenotypic data on the species *Aeromonas ichthiosmia* and *Aeromonas allosaccharophila*: *A. ichthiosmia* Schubert et al. 1990 is a later synonym of *A. veronii* Hickman-Brenner et al. 1987. Syst. Appl. Microbiol. *24*: 177–182.

Huys, G., P. Kämpfer, M. Vancanneyt, R. Coopman, P. Janssen and K. Kersters. 1997c. Effect of the growth medium on the cellular fatty acid composition of aeromonads: consequences for the chemotaxonomic differentiation of DNA hybridization groups in the genus *Aeromonas*. J. Microbiol. Methods *28*: 89–97.

Huys, G., I. Kersters, M. Vancanneyt, R. Coopman, P. Janssen and K. Kersters. 1995. Diversity of *Aeromonas* sp. in Flemish drinking water production plants as determined by gas-liquid chromatographic analysis of cellular fatty acid methyl esters (FAMEs). J. Appl. Bacteriol. *78*: 445–455.

Huys, G., M. Pearson, P. Kämpfer, R. Denys, M. Cnockaert, V. Inglis and J. Swings. 2003. *Aeromonas hydrophila* subsp. *ranae* subsp. nov., isolated from septicaemic farmed frogs in Thailand. Int. J. Syst. Evol. Microbiol. *53*: 885–891.

Huys, G., M. Vancanneyt, R. Coopman, P. Janssen, E. Falsen, M. Altwegg and K. Kersters. 1994. Cellular fatty acid composition as a chemotaxonomic marker for the differentiation of phenospecies and hybridization groups in the genus *Aeromonas*. Int. J. Syst. Bacteriol. *44*: 651–658.

Hwang, B.K., J.T. Lee, B.G. Hwang and Y.J. Koh. 1995. Restriction fragment length polymorphism analyses of the plasmid DNAs in strains of *Xanthomonas campestris* pv. *vesicatoria* from different geographic areas. J. Phytopathol. (Berl.) *143*: 185–191.

Hwang, I.G. and Y.S. Cho. 1986. Preservation methods of *Xanthomonas campestris* pv. *oryzae* in relation to virulence and colony-type variation. Korean J. Plant-Pathol. *2*: 150–157.

Hyder, S.L., A. Mayers and M.L. Cayer. 1979. Membrane modulation in a methylotrophic bacterium *Methylococcus capsulatus* (Texas) as a function of growth substrate. Tissue Cell *111*: 597–603.

Hylemon, P.B., J.S. Wells, Jr., N.R. Krieg and H.W. Jannasch. 1973. The genus *Spirillum*: a taxonomic study. Int. J. Syst. Bacteriol. *23*: 340–380.

Hyman, L.J., A. Wallace, M.M. Lopez, M. Cambra, M.T. Gorris and M.C. Perombelon. 1995. Characterization of monoclonal antibodies against *Erwinia carotovora* subsp. *atroseptica* serogroup I: specificity and epitope analysis. J. Appl. Bacteriol. *78*: 437–444.

Hynes, M. 1942. The isolation of intestinal pathogens by selective media. J. Pathol. Bacteriol. *54*: 193–207.

Hypsa, V. and C. Dale. 1997. *In vitro* culture and phylogenetic analysis of "*Candidatus* Arsenophonus triatominarum," an intracellular bacterium from the triatomine bug, *Triatoma infestans*. Int. J. Syst. Bacteriol. *47*: 1140–1144.

Iaconis, J.P. and C.C. Sanders. 1990. Purification and characterization of inducible beta-lactamases in *Aeromonas* spp. Antimicrob. Agents Chemother. *34*: 44–51.

Ibrahim, A., P. Gerner-Smidt and W. Liesack. 1997. Phylogenetic relationship of the twenty-one DNA groups of the genus *Acinetobacter* as revealed by 16S ribosomal DNA sequence analysis. Int. J. Syst. Bacteriol. *47*: 837–841.

Ibrahim, A., P. Gerner-Smidt and A. Sjöstedt. 1996. Amplification and restriction endonuclease digestion of a large fragment of genes coding for rRNA as a rapid method for discrimination of closely related pathogenic bacteria. J. Clin. Microbiol. *34*: 2894–2896.

Ibrahim, A., B.M. Goebel, W. Liesack, M. Griffiths and E. Stackebrandt.

1993. The phylogeny of the genus *Yersinia* based on 16S rDNA sequences. FEMS Microbiol. Lett. *114*: 173–177.

Igarashi, A. 1978. Isolation of a Singhs *Aedes albopictus* cell clone sensitive to dengue and chikungunya viruses. J. Gen. Virol. *40*: 531-544.

Iguchi, T., S. Kondo and K. Hisatsune. 1995. *Vibrio parahaemolyticus* O serotypes from O1 to O13 all produce R-type lipopolysaccharide: SDS-PAGE and compositional sugar analysis. FEMS Microbiol. Lett. *130*: 287–292.

Iida, T., G.Q. Tang, S. Suttikulpitug, K. Yamamoto, T. Miwatani and T. Honda. 1995. Isolation of mutant toxins of *Vibrio parahaemolyticus* hemolysin by *in vitro* mutagenesis. Toxicon. *33*: 209–216.

Iimura, K. and A. Hosono. 1996. Biochemical characteristics of *Enterobacter agglomerans* and related strains found in buckwheat seeds. Int. J. Food Microbiol. *30*: 243–253.

Iino, T. and J. Lederberg. 1964. Genetics of *Salmonella*. *In* Van Oye (Editor), The World Problem of Salmoneilosis, Junk, The Hague. pp. 110–142.

Iizuka, H. and K. Komagata. 1963a. An attempt at grouping of the genus *Pseudomonas*. J. Gen. Appl. Microbiol. *9*: 73–82.

Iizuka, H. and K. Komagata. 1963b. New species of *Pseudomonas* belonging to fluorescent group. (Studies on the microorganisms of cereal grains. Part V). J. Agr. Chem. Soc. Jpn. *37*: 137–141.

Iizuka, H. and K. Komagata. 1964a. Microbiological studies on petroleum and natural gas. I. Determination of hydrocarbon-utilizing bacteria. J. Gen. Appl. Microbiol. *10*: 207–221.

Iizuka, H. and K. Komagata. 1964b. Microbiological studies on petroleum and natural gas. II. Determination of pseudomonads isolated from oil-brines and related materials. J. Gen. Appl. Microbiol. *10*: 223–231.

Iizuka, T., S. Yamanaka and A. Hiraishi. 1998. Isolation and phylogenetic anaysis of aerobic copiotrophic ultramicrobacteria from urban soil. J. Gen. Appl. Microbiol. *44*: 75–84.

Ikeda, J.S. and D.C. Hirsh. 1990. Possession of identical nonconjugative plasmids by different isolates of *Pasteurella multocida* does not imply clonality. Vet. Microbiol. *22*: 79–87.

Ikemoto, S., K. Katoh and K. Komagata. 1978. Cellular fatty-acid composition in methanol-utilizing bacteria. J. Gen. Appl. Microbiol. *24*: 41–49.

Ikemoto, S., K. Suzuki, T. Kaneko and K. Komagata. 1980. Characterization of strains of *Pseudomonas maltophilia* which do not require methionine. Int. J. Syst. Bacteriol. *30*: 437–447.

Imada, C., S. Hara, M. Maeda and U. Simidu. 1986. Amino acid sequences of marinostatins C-1 and C-2 from marine *Alteromonas* sp. Bull. Jpn. Soc. Sci. Fish. *52*: 1455–1459.

Imada, C., M. Maeda and N. Taga. 1985a. Purification and characterization of the protease inhibitor monastatin from a marine *Alteromonas* sp. with reference to inhibition of the protease produced by a bacterium pathogenic to fish. Can. J. Microbiol. *31*: 1089–1094.

Imada, C., U. Simidu and N. Taga. 1985b. Isolation and characterization of marine bacteria producing alkaline protease inhibitor. Bull. Jpn. Soc. Sci. Fish. *51*: 799–803.

Imai, I., Y. Ishida, K. Sakaguchi and Y. Hata. 1995. Algicidal marine bacteria isolated from northern Hiroshima Bay, Japan. Fish. Sci. *61*: 628–636.

Imhoff, J.F. 1984a. Quinones of phototrophic purple bacteria. FEMS Microbiol. Lett. *256*: 85–89.

Imhoff, J.F. 1984b. Reassignment of the genus *Ectothiorhodospira* Pelsh 1936 to a new family, *Ectothiorhodospiraceae* fam. nov., and emended description of the *Chromatiaceae* Bavendamm 1924. Int. J. Syst. Bacteriol. *34*: 338–339.

Imhoff, J.F. 1986. Osmoregulation and compatible solutes in eubacteria. FEMS Microbiol. Rev. *39*: 57–66.

Imhoff, J.F. 1988a. Anoxygenic phototrophic bacteria. *In* Austin (Editor), Methods in Aquatic Bacteriology, John Wiley & Sons Ltd., Chichester. pp. 207–240.

Imhoff, J.F. 1988b. Halophilic phototrophic bacteria. *In* Rodriguez-Valera (Editor), Halophilic Bacteria, CRC Press, Boca Raton. pp. 85–108.

Imhoff, J.F. 1989. Family *Ectothiorhodospiraceae*. *In* Staley, Bryant, Pfennig and Holt (Editors), Bergeys Manual of Systematic Bacteriology, 1st Ed., Vol. 3, The Williams & Wilkins Co., Baltimore. pp. 1654–1658.

Imhoff, J.F. 1991. Polar lipids and fatty acids in the genus *Rhodobacter*. Syst. Appl. Microbiol. *14*: 228–234.

Imhoff, J.F. 1992. Taxonomy, phylogeny, and general ecology of anoxygenic phototrophic bacteria. *In* Mann and Carr (Editors), Biotechnology Handbooks: Photosynthetic Prokaryotes, Vol. 6, Plenum Press, New York. pp. 53–92.

Imhoff, J.F. 1993. Osmotic adaptation in halophilic and halotolerant microorganisms. *In* Vreeland and Hochstein (Editors), The Biology of Halophilic Bacteria, The CRC Press, Boca Raton. pp. 211–253.

Imhoff, J.F. 2001a. The anoxygenic phototrophic purple bacteria. *In* Boone, Castenholz and Garrity (Editors), Bergey's Manual of Systematic Bacteriology, 2nd Ed., Vol. 1, Springer-Verlag, New York. 631–637.

Imhoff, J.F. 2001b. Transfer of *Pfennigia purpurea* Tindall 1999 (*Amoebobacter purpureus* Eichler and Pfennig 1988) to the genus *Lamprocystis* as *Lamprocystis purpurea* comb. nov. Int. J. Syst. Evol. Microbiol. *51*: 1699–1701.

Imhoff, J. F. 2001c. Transfer of *Rhodopseudomonas acidophila* to the new genus *Rhodoblastus* as *Rhodoblastus acidophilus* gen. nov., comb. nov. Int. J. Syst. Evol. Microbiol. *51*: 1863–1866.

Imhoff, J.F. 2001d. True marine and halophilic anoxygenic phototrophic bacteria. Arch. Microbiol. *176*: 243–254.

Imhoff, J.F. and U. Bias-Imhoff. 1995. Lipids, quinones and fatty acids of anoxygenic phototrophic bacteria. *In* Blankenship, Madigan and Bauer (Editors), Anoxygenic Photosynthetic Bacteria, Kluwer Academic Publishing, The Netherlands. pp. 179–205.

Imhoff, J.F., T. Ditandy and B. Thiemann. 1991. Salt adaption of *Ectothiorhodospira*. *In* Rodriguez-Valera (Editor), General and Applied Aspects of Halophilic Microorganisms, Plenum Press, New York. pp. 113–119.

Imhoff, J.F., F. Hashwa and H.G. Trüper. 1978. Isolation of extremely halophilic phototropic bacteria from the alkaline Wadi Natrun, Egypt. Arch. Hydrobiol. *84*: 381–388.

Imhoff, J.F., D.J. Kushner, S.C. Kushwaha and M. Kates. 1982a. Polar lipids in phototrophic bacteria of the R*hodospirillaceae* and *Chromatiaceae* families. J. Bacteriol. *150*: 1192–1201.

Imhoff, J.F., R. Petri and J. Süling. 1998a. Reclassification of species of the spiral-shaped phototrophic purple non-sulfur bacteria of the α-Proteobacteria: description of the new genera *Phaeospirillum* gen. nov., *Rhodovibrio* gen. nov., *Rhodothalassium* gen. nov. and *Roseospira* gen. nov. as well as transfer of *Rhodospirillum fulvum* to *Phaeospirillum fulvum* comb. nov., of *Rhodospirillum molischianum* to *Phaeospirillum molischianum* comb. nov., of *Rhodospirillum salinarum* to *Rhodovibrio salinarum* comb. nov., of *Rhodospirillum sodomense* to *Rhodovibrio sodomensis* comb. nov., of *Rhodospirillum salexigens* to *Rhodothalassium salexigens* comb. nov. and of *Rhodospirillum mediosalinum* to *Roseospira mediosalina* comb. nov. Int. J. Syst. Bacteriol. *48*: 793–798.

Imhoff, J.F. and N. Pfennig. 2001. *Thioflavicoccus mobilis* gen. nov., sp. nov., a novel purple sulfur bacterium with bacteriochlorophyll b. Int. J. Syst. Evol. Microbiol. *51*: 105–110.

Imhoff, J.F. and T. Riedel. 1989. Requirements for, and cytoplasmic concentrations of, sulfate and chloride, and cytoplasmic volume spaces in the halophilic bacterium *Ectothiorhodospira mobilis*. J. Gen. Microbiol. *135*: 237–244.

Imhoff, J.F. and J. Süling. 1996. The phylogenetic relationship among *Ectothiorhodospiraceae*: a reevaluation of their taxonomy on the basis of 16S rDNA analyses. Arch. Microbiol. *165*: 106–113.

Imhoff, J.F. and J. Süling. 1997. *In* Validation of the publication of new names and new combinations previously effectively published outside the IJSB. List No. 62. Int. J.Syst. Bacteriol. *47*: 915–916.

Imhoff, J.F., H.G. Sahl, G.S.H. Soliman and H.G. Trüper. 1979. The Wadi Natrun: chemical composition and microbial mass developments in alkaline brines of eutrophic desert lakes. Geomicrobiol. J. *1*: 219–234.

Imhoff, J.F., J. Süling and R. Petri. 1998b. Phylogenetic relationships among the *Chromatiaceae*, their taxonomic reclassification and description of the new genera *Allochromatium, Halochromatium, Isochromatium, Marichromatium, Thiococcus, Thiohalocapsa* and *Thermochromatium*. Int. J. Syst. Bacteriol. *48*: 1129–1143.

Imhoff, J.F. and B. Thiemann. 1991. Influence of salt concentration and temperature on the fatty acid compositions of *Ectothiorhodospira* and other halophilic phototrophic purple bacteria. Arch. Microbiol. *156*: 370–375.

Imhoff, J.F., B.J. Tindall, W.D. Grant and H.G. Trüper. 1981. *Ectothiorhodospira vacuolata* sp. nov., a new phototropic bacterium from soda lakes. Arch. Microbiol. *130*: 238–242.

Imhoff, J.F., B.J. Tindall, W.D. Grant and H.G. Trüper. 1982b. *In* Validation of the publication of new names and new combinations previously effectively published outside the IJSB. List No. 8. Int. J.Syst. Bacteriol. *32*: 266–268.

Imhoff, J.F. and H.G. Trüper. 1977. *Ectothiorhodospira halochloris* sp. nov. new extremely halophilic phototropic bacterium containing bacteriochlorophyll b. Arch. Microbiol. *114*: 115–121.

Imhoff, J.F. and H.G. Trüper. 1979. *In* Validation of the publication of new names and new combinations previously effectively published outside the IJSB. List No. 2. Int. J.Syst. Bacteriol. *39*: 79.

Imhoff, J.F. and H.G. Trüper. 1980. *Chromatium purpuratum* sp. nov., a new species of the *Chromatiaceae*. Zentbl. Bakteriol. Parasitenkd. Infektkrankh. Hyg. Abt. I Orig. Reihe C. *1*: 61–69.

Imhoff, J.F. and H.G. Trüper. 1982. *In* Validation of the publication of new names and new combinations previously effectively published outside the IJSB. List No. 8. Int. J.Syst. Bacteriol. *32*: 266–268.

Infectious Disease Surveillance Center, National Institute of Infectious Diseases and Infectious Diseases Control Division, Ministry of Health and Welfare, Japan 1997. Verocytotoxin-producing *Escherichia coli* (enterohemorrhagic *E. coli*) infections, Japan, 1996-June, 1997. Infectious Agents Surveillance Report. *18*: 153–154.

Inglis, P.W., J.L. Burden and J.F. Peberdy. 1996. Evidence for the association of the enteric bacterium *Ewingella americana* with internal stipe necrosis of *Agaricus bisporus*. Microbiology (Reading) *142*: 3253–3260.

Inglis, P.W. and J.F. Peberdy. 1996. Isolation of *Ewingella americana* from the cultivated mushroom, *Agaricus bisporus*. Curr. Microbiol. *33*: 334–337.

Ingraham, J.L. and A.G. Marr. 1996. Effect of temperature, pressure, pH, and osmotic stress on growth. *In* Neidhardt, Curtiss, Ingraham, Lin, Low, Magasanik, Reznikoff, Riley, Schaecter and Umbarger (Editors), *Escherichia coli* and *Salmonella*: Cellular and Molecular Biology, 2nd Ed., Vol. 2, ASM Press, Washington D.C. pp. 1570–1578.

Ingram, C.W., A.J. Morrison, Jr. and R.E. Levitz. 1987. Gastroenteritis, sepsis, and osteomyelitis caused by *Plesiomonas shigelloides* in an immunocompetent host: case report and review of the literature. J. Clin. Microbiol. *25*: 1791–1793.

Ingram, L.O., P.F. Gomez, X. Lai, M. Moniruzzaman, B.E. Wood, L.P. Yomano and S.W. York. 1998. Metabolic engineering of bacteria for ethanol production. Biotechnol. Bioeng. *58*: 204–214.

Ingram, M. and J.M. Shewan. 1960. Introductory reflections on the *Pseudomonas–Achromobacter* group. J. Appl. Bacteriol. *23*: 373–378.

Ino, T. and Y. Nishimura. 1989. Taxonomic studies of *Acinetobacter* species based on outer membrane protein patterns. J. Gen. Appl. Microbiol. *35*: 213–224.

Inoue, K., Y. Kosako, K. Suzuki and T. Shimada. 1991. Peritrichous flagellation in *Plesiomonas shigelloides* strains. Jpn. J. Med. Sci. Biol. *44*: 141–146.

Inoue, K. and T. Shimada. 1990. Evaluation of a simple staining method for flagella using "flagella Staining Solution-Shionogi". Jpn. J. Med. Sci. Biol. *43*: 23–27.

Inoue, K., K. Sugiyama, Y. Kosako, R. Sakazaki and S. Yamai. 2000. *Enterobacter cowanii* sp nov., a new species of the family Enterobacteriaceae. Curr. Microbiol. *41*: 417–420.

Inoue, K., K. Sugiyama, Y. Kosako, R. Sakazaki and S. Yamai. 2001. *In* Validation of the publication of new names and new combinations previously effectively published outside the IJSB, List No. 82. Int. J. Syst. Evol. Microbiol. *51*: 1619.

International *Salmonella* Subcommittee. 1934. The genus *Salmonella* Lignières 1900. J. Hyg. *34:* 333–350.

Inward, C.D., J. Williams, I. Chant, J. Crocker, D.V. Milford, P.E. Rose and C.M. Taylor. 1995. Verocytotoxin-1 induces apoptosis in vero cells. J. Infect. *30:* 213–218.

Inzana, T.J. 1995. Simplified procedure for preparation of sensitized latex particles to detect capsular polysaccharides: application to typing and diagnosis of *Actinobacillus pleuropneumoniae*. J. Clin. Microbiol. *33:* 2297–2303.

Inzana, T.J., J.L. Johnson, L. Shell, K. Moller and M. Kilian. 1992. Isolation and characterization of a newly identified *Haemophilus* species from cats: *Candidatus* Haemophilus felis. J. Clin. Microbiol. *30:* 2108–2112.

Inzana, T.J., J.L. Johnson, L. Shell, K. Moller and M. Kilian. 1999. *In* Validation of publication of new names and new combinations previously published outside the IJSB, List No. 69. Int. J. Syst. Bacteriol. *49:* 341–342.

Irazabal, N., I. Marín and R. Amils. 1997. Genomic organization of the acidophilic chemolithoautotrophic bacterium *Thiobacillus ferrooxidans* ATCC 21834. J. Bacteriol. *179:* 1946–1950.

Irey, M.S. and R.E. Stall. 1982. Value of xanthomonadins for identification of pigmented *Xanthomonas campestris* pathovars. Proceedings of the Fifth International Conference on Plant Pathogenic Bacteria, August 16–23, CIAT, Cali, Colombia. pp. 85–95.

Irino, K., F. Grimont, I. Casin and P.A. Grimont. 1988. rRNA gene restriction patterns of *Haemophilus influenzae* biogroup aegyptius strains associated with Brazilian purpuric fever. J. Clin. Microbiol. *26:* 1535–1538.

Isaacson, R.E. and E. Trigo. 1995. Pili of *Pasteurella multocida* of porcine origin. FEMS Microbiol. Lett. *132:* 247–251.

Isberg, R.R. and S. Falkow. 1985. A single genetic locus encoded by *Yersinia pseudotuberculosis* permits invasion of cultured animal cells by *Escherichia coli* K-12. Nature (Lond.) *317:* 262–264.

Iseki, S.K. and K. Kashiwagi. 1955. Induction of somatic 1 antigen by bacteriophage in *Salmonella* B group. Proc. Jpn. Acad. *31:* 558–563.

Iseki, S.K. and K. Kashiwagi. 1957. Lysogenic conversions and transduction of genetic characters by temperate phage iota in *Salmonella*. Proc. Jpn. Acad. *33:* 481–485.

Ishikawa, H. 1982. Isolation of the intracellular symbionts and partial characterizations of their RNA species of the elder aphid, *Acyrthosiphon magnoliae*. Comp. Biochem. Physiol. *72:* 239–248.

Ishikawa, H. 1987. Nucleotide composition and kinetic complexity of the genomic DNA of an intracellular symbiont in the pea aphid *Acyrthosiphon pisum*. J. Mol. Evol. *24:* 205–211.

Ishimaru, C. and E.J. Klos. 1984. New medium for detecting *Erwinia amylovora* and its use in epidemiological studies. Phytopathology *74:* 1342–1345.

Ishimaru, K., M. Akagawa-Matsushita and K. Muroga. 1995. *Vibrio penaeicida* sp. nov., a pathogen of kuruma prawns (Penaeus japonicus). Int. J. Syst. Bacteriol. *45:* 134–138.

Ishimaru, K. , M. Akagawa-Matsushita and K. Muroga. 1996. *Vibrio ichthyoenteri* sp. nov, a pathogen of Japanese flounder (*Paralichthys olivaceus*) larvae. Int. J. of Syst. Bacteriol. *46:* 155–159.

Ishiyama, S. 1922. Studies of bacterial leaf blight of rice. Rept. Imperial Agr. Sta. Konosu. *45:* 233–261.

Isken, S. and J.A.M. De Bont. 1996. Active efflux of toluene in a solvent-resistant bacterium. J. Bacteriol. *178:* 6056–6058.

Islam, D., S. Tzipori, M. Islam and A.A. Lindberg. 1993. Rapid detection of *Shigella dysenteriae* and *Shigella flexneri* in faeces by an immunomagnetic assay with monoclonal antibodies. Eur. J. Clin. Microbiol. Infect. Dis. *12:* 25–32.

Ismaili, A., B. Bourke, J.C. de Azavedo, S. Ratnam, M.A. Karmali and P.M. Sherman. 1996. Heterogeneity in phenotypic and genotypic characteristics among strains of *Hafnia alvei*. J. Clin. Microbiol. *34:* 2973–2979.

Isshiki, Y., S. Kondo, T. Iguchi, Y. Sano, T. Shimada and K. Hisatsune. 1996. An immunochemical study of serological cross-reaction between lipopolysaccharides from *Vibrio cholerae* O22 and O139. Microbiology *142:* 1499–1504.

Ito, H. and H. Iizuka. 1983. Genetic transformation of *Moraxella*-like psychrotrophic bacteria and their radiation sensitivity. Agric. Biol. Chem. *47:* 603–605.

Ito, H., T. Sato and H. Iizuka. 1976. Study of the intermediate type of *Moraxella* and *Acinetobacter* occurring in radurized vienna sausages. Agric. Biol. Chem. *40:* 867–873.

Ito, H., A. Terai, H. Kurazono, Y. Takeda and M. Nishibuchi. 1990. Cloning and nucleotide sequencing of Vero toxin 2 variant genes from *Escherichia coli* O91:H21 isolated from a patient with the hemolytic uremic syndrome. Microb. Pathog. *8:* 47–60.

Ito, H., I. Uchida, T. Sekizaki and N. Terakado. 1995. A specific oligonucleotide probe based on 5S ribosomal-RNA sequences for identification of *Vibrio anguillarum* and *Vibrio ordalii*. Vet. Microbiol. *43:* 167–171.

Itoh, T., H. Funabashi, Y. Katayama-Fujimura, S. Iwasaki and H. Kuraishi. 1985. Structure of methylmenaquinone-7 isolated from *Alteromonas putrefaciens* IAM 12079. Biochim. Biophys. Acta *840:* 51–55.

Itoh, T., T. Higuchi, M. Hirobe, K. Hiramatsu and T. Yokota. 1994. Identification of a novel sugar, 4-amino-4,6-dideoxy-2-*O*-methylmannose, in the lipopolysaccharide of *Vibrio cholerae* O1 serotype Ogawa. Carbohydr. Res. *256:* 113–128.

Itoh, Y., K. Izaki and H. Takahashi. 1978. Purification and characterization of a bacteriocin from *Erwinia carotovora*. J. Gen. Appl. Microbiol. *24:* 27–39.

Itoh, Y., K. Izaki and H. Takahashi. 1982. Mode of action of a bacteriocin from *Erwinia carotovora*. IV. Effects on macromolecule synthesis, ATP level and nutrient transport. J. Gen. Appl. Microbiol. *28:* 95–99.

Ivanoff, B. and M.M. Levine. 1997. Typhoid fever: continuing challenges from a resilient foe. Bull. Inst. Pasteur. *95:* 129–142.

Ivanova, E.P., J. Chun, L.A. Romanenko, M.E. Matte, V.V. Mikhailov, G.M. Frolova, A. Huq and R.R. Colwell. 2000a. Reclassification of *Alteromonas distincta* Romanenko et al. 1995 as *Pseudoalteromonas distincta* comb. nov. Int. J. Syst. Evol. Microbiol. *50:* 141-144.

Ivanova, E.P., R.P. Gorshkova, V.V. Mikhailov, E.L. Nazarenko, V.A. Zubkov, E.A. Kiprianova, G.F. Levanova, A.D. Garagulya, E.A. Kolesova and N.M. Gorshkova. 1994. Capsular polysaccharides of marine bacteria of the genus *Alteromonas*. Microbiology *63:* 120–123.

Ivanova, E.P., E.A. Kiprianova, V.V. Mikhailov, G.F. Levanova, A.D. Garagulya, N.M. Gorshkova, M.V. Vysotskii, D.V. Nicolau, N. Yumoto, T. Taguchi and S. Yoshikawa. 1998. Phenotypic diversity of *Pseudoalteromonas citrea* from different marine habitats and emendation of the description. Int. J. Syst. Bacteriol. *48:* 247–256.

Ivanova, E.P., E.A. Kiprianova, V.V. Mikhailov, G.F. Levanova, A.D. Garagulya, N.M. Gorshkova, N. Yumoto and S. Yoshikawa. 1996a. Characterization and identification of marine *Alteromonas nigrifaciens* strains and emendation of the description. Int. J. Syst. Bacteriol. *46:* 223–228.

Ivanova, E.P. and V.V. Mikhailov. 2001a. A new family, *Alteromonadaceae* fam. nov., including marine proteobacteria of the genera *Alteromonas*, *Pseudoalteromonas*, *Idiomarina*, and *Colwellia*. Microbiology *70:* 10–17.

Ivanova, E.P. and V.V. Mikhailov. 2001b. *In* Validation of the publication of new names and new combinations previously effectively published outside the IJSEM. List No. 81. Int. J. Syst. Evol. Microbiol. *51:* 1229.

Ivanova, E.P., V.V. Mikhailov, E.A. Kiprianosu, G.F. Levanova, A.D. Garagulya, G.M. Frolova and V.I. Svetashev. 1996b. *Alteromonas elyakovii* sp. nov., a new bacterium isolated from marine mollusks. Biol. Morya (Vladivost.). *22:* 213–237.

Ivanova, E.P., V.V. Mikhailov, E.A. Kiprianova, G.F. Levanova, A.D. Garagulya, G.M. Frolova and V.I. Svetashev. 1997. *In* Validation of the publication of new names and new combinations previously effectively published outside the IJSB. List No. 61. Int. J. Syst. Bacteriol. *47:* 601–602.

Ivanova, E.P., L.A. Romanenko, J. Chun, M.H. Matte, G.R. Matte, V.V. Mikhailov, V.I. Svetashev, A. Huq, T. Maugel and R.R. Colwell. 2000b. *Idiomarina* gen. nov., comprising novel indigenous deep-sea bacteria from the Pacific Ocean, including descriptions of two species, *Idiomarina abyssalis* sp. nov. and *Idiomarina zobellii* sp. nov. Int. J. Syst. Evol. Microbiol. *50:* 901–907.

Ivanova, E.P., L.A. Romanenko, M.H. Matté, G.R. Matté, A.M. Lysenko, U. Simidu, K. Kita-Tsukamoto, T. Sawabe, M.V. Vysotskii, G.M. Frolova, V. Mikhailov, R. Christen and R.R. Colwell. 2001. Retrieval of the species *Alteromonas tetraodonis* Simidu et al. 1990 as *Pseudoalteromonas tetraodonis* comb. nov. and emendation of description. Int. J. Syst. Evol. Microbiol. *51*: 1071–1078.

Ivanova, E.P., N.V. Zhukova, V.I. Svetashev, N.M. Gorshkova, V.V. Kurilenko, G.M. Frolova and V.V. Mikhailov. 2000c. Evaluation of phospholipid and fatty acid compositions as chemotaxonomic markers of *Alteromonas*-like *proteobacteria*. Curr. Microbiol. *41*: 341–345.

Ivanova, T.L., T.P. Turova and A.S. Antonov. 1985. DNA–DNA and rRNA–DNA hybridization studies in the genus *Ectothiorhodospira* and other purple sulfur bacteria. Arch. Microbiol. *143*: 154–156.

Iveson, J.B. 1971. Strontium chloride B and E.E. enrichment broth media for the isolation of *Edwardsiella, Salmonella* and *Arizona* species from tiger snakes. J. Hyg. *69*: 323–330.

Iveson, J.B. 1973. Enrichment procedures for the isolation of *Salmonella, Arizona, Edwardsiella* and *Shigella* from feces. J. Hyg. *71*: 349–361.

Ivins, B.E. and R.K. Holmes. 1980. Isolation and characterization of melanin-producing (mel) mutants of *Vibrio cholerae*. Infect. Immun. *27*: 721–729.

Iwamoto, Y., Y. Suzuki, A. Kurita, Y. Watanabe, T. Shimizu, H. Ohgami and Y. Yanagihara. 1995. *Vibrio trachuri* sp. nov., a new species isolated from diseased Japanese horse mackerel. Microbiol. Immunol. *39*: 831–837.

Iwamoto, Y., Y. Suzuki, A. Kurita, Y. Watanabe, T. Shimizu, H. Ohgami and Y. Yanagihara. 1996. *In* Validation of new names and new combinations previously effectively published outside the IJSB. List No. 57. Int. J. Syst. Bacteriol. *46*: 625–626.

Iwanaga, M. and A. Hokama. 1992. Characterization of *Aeromonas sobria* TAP13 pili: a possible new colonization factor. J. Gen. Microbiol. *138*: 1913–1919.

Iwanami, H., T. Yamaguchi and M. Takeuchi. 1995. Fatty acid metabolism in bacteria that produce eicosapentaenoic acid isolated from sea urchin *Strongylocentrotus nudus*. Nippon Suisan Gakkaishi. *61*: 205–210.

Iyoda, S., A. Wada, J. Weller, S.J. Flood, E. Schreiber, B. Tucker and H. Watanabe. 1999. Evaluation of AFLP, a high-resolution DNA fingerprinting method, as a tool for molecular subtyping of enterohemorrhagic *Escherichia coli* O157:H7 isolates. Microbiol. Immunol. *43*: 803–806.

Izard, D., C. Ferragut, F. Gavini, K. Kersters, J. De Ley and H. Leclerc. 1981a. *Klebsiella terrigena*, a new species from soil and water. Int. J. Syst. Bacteriol. *31*: 116–127.

Izard, D., F. Gavini, P.A. Trinel, F. Krubwa and H. Leclerc. 1980. Contribution of DNA–DNA hybridization to the transfer of *Enterobacter aerogenes* to the genus *Klebsiella* as *K. mobilis*. Zentbl. Bakteriol. Mikrobiol. Parasitenkd. Infektkrankh. Hyg. Abt. I. Orig. C. *1*: 257–263.

Izard, D., F. Gavini, P.A. Trinel and H. Leclerc. 1979. *Rahnella aquatilis*, nouveau membre de la famille des *Enterobacteriaceae*. Ann. Inst. Pasteur Microbiol. *130A*: 163–177.

Izard, D., F. Gavini, P.A. Trinel and H. Leclerc. 1981b. Deoxyribonucleic acid relatedness between *Enterobacter cloacae* and *Enterobacter amnigenus* sp. nov. Int. J. Syst. Bacteriol. *31*: 35–42.

Izard, D., F. Gavini, P.A. Trinel and H. Leclerc. 1981c. *In* Validation of the publication of new names and new combinations previously effectively published outside the IJSB. List No. 7. Int. J. Syst. Bacteriol. *31*: 382–383.

Izard, D., J. Mergaert, F. Gavini, A. Beji, K. Kersters, J. De Ley and H. Leclerc. 1985. Separation of *Escherichia adecarboxylata* from the *Erwinia herbicola* and *Enterobacter agglomerans* complex and from the other enterobacteriaceae by nucleic acid and protein electrophoretic techniques. Ann. Inst. Pasteur Microbiol. *136B*: 151–168.

Izdebska-Szymona, K., E. Monczak and B. Lemczak. 1971. Preliminary scheme of phage typing of *Proteus mirabilis* strains. Exp. Med. Microbiol. (Engl. Transl. Med. Dosw. Mikrobiol.). *23*: 18–22. (Originally published in Polish.).

Jackman, P.J.H. 1985. Bacterial taxonomy based on electrophoretic whole-cell protein patterns. *In* Goodfellow and Minnikin (Editors),

Chemical Methods in Bacterial Systematics, Academic Press, London. 115–129.

Jackman, P.J.H. 1987. Microbial systematics based on electrophoretic whole-cell protein-patterns. Methods Microbiol. *19*: 209–225.

Jackson, E.B., T.T. Crocker and J.E. Smadel. 1952. Studies on two rickettsia-like agents probably isolated from guinea pigs. Bacteriol. Proc: 119.

Jackson, J.K., R.L. Murphree and M.L. Tamplin. 1997. Evidence that mortality from *Vibrio vulnificus* infection results from single strains among heterogeneous populations in shellfish. J. Clin. Microbiol. *35*: 2098–2101.

Jackson, M.M., C.G. Jackson, Jr. and M. Fulton. 1969. Investigation of the enteric bacteria of the testudinata. I: occurrence of the genera *Arizona, Citrobacter, Edwardsiella* and *Salmonella*. Bull. Wildl. Dis. Assoc. *5*: 328–329.

Jackson, R.L. and G.R. Matsueda. 1970. *Myxobacter* AL-1 protease. Methods Enzymol. *19*: 591–599.

Jackson, R.L. and R.S. Wolfe. 1968. Composition, properties and substrate specificities of *Myxobacter* AL-1 protease. J. Biol. Chem. *243*: 879–888.

Jacobitz, S. and P.E. Bishop. 1992. Regulation of nitrogenase-2 in *Azotobacter vinelandii* by ammonium, molybdenum, and vanadium. J. Bacteriol. *174*: 3884–3888.

Jacobs, F., C. Liesnard, J.P. Goldstein, M.J. Struelens, G. Primo, J.L. Leclerc and J.P. Thys. 1990. Asymptomatic *Legionella pneumophila* infections in heart transplant recipients. Transplantation. *50*: 174–175.

Jacobsen, M.J. and J.P. Nielsen. 1995. Development and evaluation of a selective and indicative medium for isolation of *Actinobacillus pleuropneumoniae* from tonsils. Vet. Microbiol. *47*: 191–197.

Jacobsen, M.J., J.P. Nielsen and R. Nielsen. 1996. Comparison of virulence of different *Actinobacillus pleuropneumoniae* serotypes and biotypes using an aerosol infection model. Vet. Microbiol. *49*: 159–168.

Jacobson, M.R., K.E. Brigle, L.T. Bennett, R.A. Setterquist, M.S. Wilson, V.L. Cash, J. Beynon, W.E. Newton and D.R. Dean. 1989. Physical and genetic map of the major *nif* gene cluster from *Azotobacter vinelandii*. J. Bacteriol. *171*: 1017–1027.

Jacobson, M.R., R. Premakumar and P.E. Bishop. 1986. Transcriptional regulation of nitrogen fixation by molybdenum in *Azotobacter vinelandii*. J. Bacteriol. *167*: 480–486.

Jacoby, G.A. 1974a. Properties of an R plasmid in *Pseudomonas aeruginosa* producing amikicin (BB-K8), butirosin, kanamycin, tobramycin, and sisomicin resistance. Antimicrob. Agents Chemother. *6*: 807–810.

Jacoby, G.A. 1974b. Properties of R plasmids determining gentamicin resistance by acetylation in *Pseudomonas aeruginosa*. Antimicrob. Agents Chemother. *6*: 239–252.

Jacoby, G.A. 1977. Classification of plasmids in *Pseudomonas aeruginosa*. *In* Schlessinger (Editor), Microbiology 1977, American Society for Microbiology, Washington, D.C. 271–309.

Jacoby, G.A. 1979. Plasmids of *Pseudomonas aeruginosa. In* Doggett (Editor), *Pseudomonas aeruginosa*. Clinical Manifestations of Infection and Current Therapy, Academic Press, New York. 271–309.

Jacoby, G.A. and M. Matthew. 1979. Distribution of beta-lactamase genes on plasmids found in *Pseudomonas*. Plasmid *2*: 41–47.

Jacoby, G.A. and J.A. Shapiro. 1977. Plasmids studied in *Pseudomonas aeruginosa* and other pseudomonads. *In* Bukhari, Shapiro and Adhya (Editors), DNA insertion elements, plasmids, and episomes, Cold Spring Harbor Laboratory, Cold Spring Harbor, New York. 639–656.

Jacques, M. and S.E. Paradis. 1998. Adhesin-receptor interactions in *Pasteurellaceae*. FEMS Microbiol. Rev. *22*: 45–59.

Jaeger, K.E., A. Steinbuchel and D. Jendrossek. 1995. Substrate specificities of bacterial polyhydroxyalkanoate depolymerases and lipases - bacterial lipases hydrolyze poly(omega-hydroxyalkanoates). Appl. Environ. Microbiol. *61*: 3113–3118.

Jaenecke, S., V. De Lorenzo, K.N. Timmis and E. Díaz. 1996. A stringently controlled expression system for analysing lateral gene transfer between bacteria. Mol. Microbiol. *21*: 293–300.

Jäger, C., H. Willems, D. Thiele and G. Baljer. 1998. Molecular characterization of *Coxiella burnetii* isolates. Epidemiol. Infect. *120*: 157–164.

Jagger, I.C. 1921. Bacterial leaf spot disease of celery. J. Agr. Res. *21*: 185–188.

Jahnke, L.L. and P.D. Nichols. 1986. Methyl sterol and cyclopropane fatty acid composition of *Methylococcus capsulatus* grown at low oxygen tensions. J. Bacteriol. *167*: 238–242.

Jahnke, L.L., H. Stan-Lotter, K. Kato and L.I. Hochstein. 1992. Presence of methyl sterol and bacteriohopanepolyol in an outer-membrane preparation from *Methylococcus capsulatus* (Bath). J. Gen. Microbiol. *138*: 1759–1766.

Jahnke, L.L., R.E. Summons, L.M. Dowling and K.D. Zahiralis. 1995. Identification of methanotrophic lipid biomarkers in cold-seep mussel gills: chemical and isotopic analysis. Appl. Environ. Microbiol. *61*: 576–582.

Jain, K.L., S.R.S. Dange and B.S. Siradhana. 1975. Bacterial leaf spot of *Datura metel* caused by *Xanthomonas campestris* f. sp. daturi f. spec. nov. Curr. Sci. (Bangalore) *44*: 447.

Jain, K., K. Radsak and W. Mannheim. 1974. Differentiation of the Oxytocum group from *Klebsiella* by deoxyribonucleic acid-deoxyribonucleic acid hybridization. Int. J. Syst. Bacteriol. *24*: 402–407.

Jain, K.C. and W.B. Whalley. 1980. The bacterial pigment from *Pseudomonas lemonnieri* .2. The synthesis of 3-n-octanamidopyridine-2,5,6-trione - the structure and synthesis of lemonnierin. J. Chem. Soc.-Perkin Trans. 1. *8*: 1788–1794.

Jakobi, M., G. Winkelmann, D. Kaiser, C. Kempler, G. Jung, G. Berg and H. Bahl. 1996. Maltophilin: a new antifungal compound produced by *Stenotrophomonas maltophilia* R3089. J. Antibiot. (Tokyo). *49*: 1101–1104.

James, S.R., S.J. Dobson, P.D. Franzmann and T.A. McMeekin. 1990a. *Halomonas meridiana*, a new species of extremely halotolerant bacteria isolated from antarctic saline lakes. Syst. Appl. Microbiol. *13*: 270–278.

James, S.R., S.J. Dobson, P.D. Franzmann and T.A. McMeekin. 1990b. Validation of the publication of new names and new combinations previously effectively published outside the IJSB. List No. 35. Int. J. Syst. Bacteriol. *40*: 470–471.

Jamieson, A.F., R.L. Bieleski and R.E. Mitchell. 1981. Plasmids and phaseolotoxin production in *Pseudomonas syringae* pv. *phaseolicola*. J. Gen. Microbiol. *122*: 161–165.

Jana, S.C., P.K. Chakrabartty and A.K. Mishra. 1992. Taxonomic relationship of some members of *Azotobacteraceae* based on their protein profiles. J. Basic Microbiol. *32*: 29–33.

Janda, I. and M. Opekarova. 1989. Long-term preservation of active luminous bacteria by lyophilization. J. Biolumin. Chemilumin. *3*: 27–29.

Janda, J.M. 1985. Biochemical and exoenzymatic properties of *Aeromonas* species. Diagn. Microbiol. Infect. Dis. *3*: 223–232.

Janda, J.M. 1991. Recent advances in the study of the taxonomy, pathogenicity, and infectious syndromes associated with the genus *Aeromonas*. Clin. Microbiol. Rev. *4*: 397–410.

Janda, J.M. 1998. *Vibrio*, *Aeromonas* and *Plesiomonas*. *In* Balows and Duerden (Editors), Topley & Wilson's Microbiology and Microbial Infections, 9th Ed., Vol. 2, Arnold, London. pp. 1065–1089.

Janda, J.M. and S.L. Abbott. 1993a. Expression of an iron-regulated hemolysin by *Edwardsiella tarda*. FEMS Microbiol. Lett. *111*: 275–280.

Janda, J.M. and S.L. Abbott. 1993b. Expression of hemolytic activity by *Plesiomonas shigelloides*. J. Clin. Microbiol. *31*: 1206–1208.

Janda, J.M. and S.L. Abbott. 1996. Human pathogens. *In* Austin, Altwegg, Gosling and Joseph (Editors), The Genus *Aeromonas*, John Wiley and Sons, Ltd., Chichester. pp. 397–410.

Janda, J.M. and S.L. Abbott. 1998a. The Enterobacteria, Lippincott-Raven, Philadelphia. pp. 387.

Janda, J.M. and S.L. Abbott. 1998b. Evolving concepts regarding the genus *Aeromonas*: an expanding panorama of species, disease presentations, and unanswered questions. Clin. Infect. Dis. *27*: 332–344.

Janda, J.M., S.L. Abbott and M.J. Albert. 1999. Prototypal diarrheagenic strains of *Hafnia alvei* are actually members of the genus *Escherichia*. J. Clin. Microbiol. *37*: 2399–2401.

Janda, J.M., S.L. Abbott, W.K. Cheung and D.F. Hanson. 1994. Biochem-

ical identification of citrobacteria in the clinical laboratory. J. Clin. Microbiol. *32*: 1850–1854.

Janda, J.M., S.L. Abbott, S. Khashe, G.H. Kellogg and T. Shimada. 1996a. Further studies on biochemical characteristics and serologic properties of the genus *Aeromonas*. J. Clin. Microbiol. *34*: 1930–1933.

Janda, J.M., S.L. Abbott, S. Khashe and T. Robin. 1996b. Biochemical investigations of biogroups and subspecies of *Morganella morganii*. J. Clin. Microbiol. *34*: 108–113.

Janda, J.M., S.L. Abbott, S. Kroske-Bystrom, W.K.W. Cheung, C. Powers, R.P. Kokka and K. Tamura. 1991a. Pathogenic properties of *Edwardsiella* species. J. Clin. Microbiol. *29*: 1997–2001.

Janda, J.M., S.L. Abbott and L.S. Oshiro. 1991b. Penetration and replication of *Edwardsiella* spp. in HEp-2 cells. Infect. Immun. *59*: 154–161.

Janda, J.M., A. Dixon, B. Raucher, R.B. Clark and E.J. Bottone. 1984. Value of blood agar for primary plating and clinical implication of simultaneous isolation of *Aeromonas hydrophila* and *Aeromonas caviae* from a patient with gastroenteritis. J. Clin. Microbiol. *20*: 1221–1222.

Janda, J.M. and P.S. Duffey. 1988. Mesophilic aeromonads in human disease: current taxonomy, laboratory identification, and infectious disease spectrum. Rev. Infect. Dis. *10*: 980–997.

Janda, J.M. and R.P. Kokka. 1991. The pathogenicity of *Aeromonas* strains relative to genospecies and phenospecies identification. FEMS Microbiol. Lett. *90*: 29–34.

Janda, J.M. and M.R. Motyl. 1985. Cephalothin susceptibility as a potential marker for the *Aeromonas sobria* group. J. Clin. Microbiol. *22*: 854–855.

Janda, J.M., L.S. Oshiro, S.L. Abbott and P.S. Duffey. 1987. Virulence markers of mesophilic aeromonads: association of the autoagglutination phenomenon with mouse pathogenicity and the presence of a peripheral cell-associated layer. Infect. Immun. *55*: 3070–3077.

Janda, J.M., C. Powers, R.G. Bryant and S.L. Abbott. 1988. Current perspectives on the epidemiology and pathogenesis of clinically significant *Vibrio* spp. Clin. Microbiol. Rev. *1*: 245–267.

Jang, E.B. and K.A. Nishijimi. 1990. Identification and attractancy of bacteria associated with *Dactus dorsalis* Diptera tephritidae. Environ. Entomol. *19*: 1726–1731.

Janke, A. 1924. Allgemeine Technische Mikrobiologie. I. Teil: Die Mikroorganismen, T. Steinkopf, Dresden.

Jann, A., H. Matsumoto and D. Haas. 1988. The 4th arginine catabolic pathway of *Pseudomonas aeruginosa*. J. Gen. Microbiol. *134*: 1043–1053.

Jannasch, H.W. 1957. Die bakterielle rotfärbung der salzseen des Wadi Natrun. Arch. Hydrobiol. *53*: 425–433.

Jannasch, H.W. 1963. Studies on the ecology of a marine spirillum in the chemostat. 1st International Symposium on Marine Microbiology, C.C. Thomas, Springfield. 558–566.

Jannasch, H.W. 1967. Enrichments of aquatic spirilla in continuous culture. Arch. Mikrobiol. *59*: 165–173.

Jannasch, H.W. 1984. Microbial processes at deep sea hydrothermal vents. *In* Rona, Bostrom, Laubier and Smith (Editors), Hydrothermal Processes at Seafloor Spreading Centers, Plenum Publishing, New York. pp. 667–709.

Jannasch, H.W., D.C. Nelson and C.O. Wirsen. 1989. Massive natural occurrence of unusually large bacteria (*Beggiatoa* sp.) at a hydrothermal deep-sea vent site. Nature *342*: 834–836.

Jannasch, H.W., C.O. Wirsen, D.C. Nelson and L.A. Robertson. 1985. *Thiomicrospira crunogena*, sp. nov., a colorless, sulfur-oxidizing bacterium from a deep-sea hydrothermal vent. Int. J. Syst. Bacteriol. *35*: 422–424.

Jannes, G., M. Vaneechoutte, M. Lannoo, M. Gillis, M. Vancanneyt, P. Vandamme, G. Verschraegen, H. Van Heuverswyn and R. Rossau. 1993. Polyphasic taxonomy leading to the proposal of *Moraxella canis*, sp. nov. for *Moraxella catarrhalis* like strains. Int. J. Syst. Bacteriol. *43*: 438–449.

Janse, J.D. 1982. *Pseudomonas syringae* subsp. *savastanoi* (ex Smith) subsp. nov., nom. rev., the bacterium causing excrescences on *Oleaceae* and *Nerium oleander* L. Int. J. Syst. Bacteriol. *32*: 166–169.

Janse, J.D., P. Rossi, L. Angelucci, M. Scortichini, J.H.J. Derks, A.D.L.

Akkermans, R. De Vrijer and P.G. Psallidas. 1996. Reclassification of *Pseudomonas syringae* pv. *avellanae* as *Pseudomonas avellanae* (spec. nov.), the bacterium causing canker of hazelnut (*Corylus avellana* L.). Syst. Appl. Microbiol. *19*: 589–595.

Janse, J.D., P. Rossi, L. Angelucci, M. Scortichini, J.H.J. Derks, A.D.L. Akkermans, R. DeVrijer and P.G. Psallidas. 1997. *In* Validation of the publication of new names and new combinations previously effectively published outside the IJSB. List. No. 61. Int. J. Syst. Bacteriol. *47*: 601–602.

Janse, J.D. and P.H. Smits. 1990. Whole cell fatty acid patterns of *Xenorhabdus* species. Lett. Appl. Microbiol. *10*: 131–135.

Jansen, R. 1994. The RTX toxins of *Actinobacillus pleuropneumoniae*, Thesis, Utrecht University, Utrecht, . pp. 172.

Janssen, P., R. Coopman, G. Huys, J. Swings, M. Bleeker, P. Vos, M. Zabeau and K. Kersters. 1996. Evaluation of the DNA fingerprinting method AFLP as a new tool in bacterial taxonomy. Microbiology *142*: 1881–1893.

Janssen, P. and L. Dijkshoorn. 1996. High resolution DNA fingerprinting of *Acinetobacter* outbreak strains. FEMS Microbiol. Lett. *142*: 191–194.

Janssen, P., K. Maquelin, R. Coopman, I. Tjernberg, P. Bouvet, K. Kersters and L. Dijkshoorn. 1997. Discrimination of *Acinetobacter* genomic species by AFLP fingerprinting. Int. J. Syst. Bacteriol. *47*: 1179–1187.

Janssen, W.A. and H.J. Surgalla. 1968. Morphology, physiology and serology of a *Pasteurella* species pathogenic for white perch. J. Bacteriol. *96*: 1606–1610.

Jantzen, E., B.P. Berdal and T. Omland. 1979. Cellular fatty acid composition of *Francisella tularensis*. J. Clin. Microbiol. *10*: 928–930.

Jantzen, E., B.P. Berdal and T. Omland. 1981. Cellular fatty acid taxonomy of *Haemophilus*, *Pasteurella* and *Actinobacillus*. *In* Kilian, Frederiksen and Biberstein (Editors), *Haemophilus*, *Pasteurella* and *Actinobacillus*, Academic Press, London. pp. 197–203.

Jantzen, E., K. Bryn, T. Bergan and K. Bøvre. 1974. Gas chromatography of bacterial whole cell methanolysates; V. Fatty acid composition of *Neisseriae* and *Moraxellae*. Acta Pathol. Microbiol. Scand. B Microbiol. Immunol. *82*: 767–779.

Jantzen, E. and J. Lassen. 1980. Characterization of *Yersinia* species by analysis of whole cell fatty acids. Int. J. Syst. Bacteriol. *30*: 421–428.

Janvier, M., J. Frank, M. Luttik and F. Gasser. 1992. Isolation and phenotypic characterization of methanol oxidation mutants of the restricted facultative methylotroph *Methylophaga marina*. J. Gen. Microbiol. *138*: 2113–2123.

Janvier, M., C. Frehel, F. Grimont and F. Gasser. 1985. *Methylophaga marina* gen. nov., sp. nov. and *Methylophaga thalassica* sp. nov., marine methylotrophs. Int. J. Syst. Bacteriol. *35*: 131–139.

Janvier, M. and F. Gasser. 1987. Purification and properties of methanol dehydrogenase from *Methylophaga marina*. Biochimie *69*: 1169–1174.

Janvier, M. and P.A.D. Grimont. 1995. The genus *Methylophaga*, a new line of descent within phylogenetic branch gamma of *Proteobacteria*. Res. Microbiol. *146*: 543–550.

Jarlier, V. 1985. Enterobactéries et bétalactamases. *In* Courvalin, P., F. Goldstein, A. Philippon and J. Sirot (Editors), L'antibiogramme, MPC Videocom, Paris. 87–101.

Jarvis, K.G., J.A. Giron, A.E. Jerse, T.K. Mcdaniel, M.S. Donnenberg and J.B. Kaper. 1995. Enteropathogenic *Escherichia coli* contains a putative type III secretion system necessary for the export of proteins involved in attaching and effacing lesion formation. Proc. Natl. Acad. Sci. U.S.A. *92*: 7996–8000.

Jarvis, W.R., J.W. White, V.P. Munn, J.L. Mosser, T.G. Emori, D.H. Culver, C. Thornsberry and J.M. Hughes. 1984. Nosocomial infection surveillance, 1983. MMWR CDC Surveill. Summ. *33*: 9–21.

Jaulhac, B., M. Nowicki, N. Bornstein, O. Meunier, G. Prevost, Y. Piemont, J. Fleurette and H. Monteil. 1992. Detection of *Legionella* spp. in bronchoalveolar lavage fluids by DNA amplification. J. Clin. Microbiol. *30*: 920–924.

Javor, B.J., D.B. Wilmont and R.D. Vetter. 1990. pH-dependent metabolism of thiosulfate and sulfur globules in the chemolithotrophic marine bacterium *Thiomicrospira crunogena*. Arch. Microbiol. *154*: 231–238.

Jawad, A., P.M. Hawkey, J. Heritage and A.M. Snelling. 1994. Description of Leeds *Acinetobacter* Medium, a new selective and differential medium for isolation of clinically important *Acinetobacter* spp., and comparison with herellea agar and Holton's agar. J. Clin. Microbiol. *32*: 2353–2358.

Jawad, A., A.M. Snelling, J. Heritage and P.M. Hawkey. 1998. Comparison of ARDRA and recA-RFLP analysis for genomic species identification of *Acinetobacter* spp. FEMS Microbiol. Lett. *165*: 357–362.

Jawetz, E. 1950. A pneumotropic *Pasteurella* of laboratory animals. I. Bacteriological and serological characteristics of the organism. J. Infect. Dis. *86*: 172–183.

Jean-Jacques, W., K.R. Rajashekaraiah, J.J. Farmer, III, F.W. Hickman, J.G. Morris and C.A. Kallick. 1981. *Vibrio metschnikovii* bacteremia in a patient with cholecystitis. J. Clin. Microbiol. *14*: 711–712.

Jeannes, A. 1974. Applications of extracellular microbial polysaccharide-polyelectrolytes: review of literature, including patents. J. Polym. Sci. Polym. Symp. No. *45*: 209–227.

Jellison, W.L. 1974. Tularemia in North America 1930–1974, University of Montana Foundation, Missoula.

Jellum, E., V. Tingelstad and I. Olsen. 1984. Differentiation between *Actinobacillus actinomycetemcomitans* and *Hemophilus aphrophilus* by high-resolution, two-dimensional protein electroophoresis. Int. J. Syst. Bacteriol. *34*: 478–483.

Jenkins, C.L. and M.P. Starr. 1982a. The brominated aryl-polyene (xanthomonadin) pigments of *Xanthomonas juglandis* protect against photobiological damage. Curr. Microbiol. *7*: 323–326.

Jenkins, C.L. and M.P. Starr. 1982b. The pigment of *Xanthomonas populi* is a nonbrominated aryl-heptaene belonging to xanthomonadin pigment group II. Curr. Microbiol. *7*: 195–198.

Jenkins, D., M.G. Richard and G.T. Daigger. 1993. Manual on the causes and control of activated sludge bulking and foaming, 2nd ed., Lewis Publishers, Boca Raton.

Jenkins, J.A. and P.W. Taylor. 1995. An alternative bacteriological medium for the isolation of *Aeromonas* spp. J. Wildl. Dis. *31*: 272–275.

Jenkins, O., D. Byrom and D. Jones. 1987. *Methylophilus*: a new genus of methanol-utilizing bacteria. Int. J. Syst. Bacteriol. *37*: 446–448.

Jenny, D.B., P.W. Letendre and G. Iverson. 1987. Endocarditis caused by *Kingella indologenes*. Rev. Infect. Dis. *9*: 787–789.

Jensen, H.L. 1955a. *Azotobacter macrocytogenes* n. sp. a nitrogen-fixing bacterium resistant to acid reactions. Acta Agric. Scand. *5*: 280–294.

Jensen, H.L. 1965. Non-symbiotic nitrogen fixation. *In* Bartholomew and Clark (Editors), Soil Nitrogen, American Society of Agronomy, Inc., Madison. pp. 436–480.

Jensen, J. and E. Thofern. 1953. Chlorhämin als Bakterienwuchsstoff I. Z. Naturforsch. *8b*: 599–603.

Jensen, K.T., W. Frederiksen, F.W. Hickman Brenner, A.G. Steigerwalt, C.F. Riddle and D.J. Brenner. 1992. Recognition of *Morganella* subspecies, with proposal of *Morganella morganii* subsp. *morganii* subsp. nov. and *Morganella morganii* subsp. *sibonii* subsp. nov. Int. J. Syst. Bacteriol. *42*: 613–620.

Jensen, L.B., J.L. Ramos, Z. Kaneva and S. Molin. 1993. A substrate-dependent biological containment system for *Pseudomonas putida* based on the *Escherichia coli gef* gene. Appl. Environ. Microbiol. *59*: 3713–3717.

Jensen, M.J., B.M. Tebo, P. Baumann, M. Mandel and K.H. Nealson. 1980. Characterization of *Alteromonas hanedai* (sp. nov.), a nonfermentative luminous species of marine origin. Curr. Microbiol. *3*: 311–315.

Jensen, M.J., B.M. Tebo, P. Baumann, M. Mandel and K.H. Nealson. 1981. *In* Validation of the publication of new names and new combinations previously effectively published outside the IJSB. List No. 7. Int. J. Syst. Bacteriol. *31*: 382–383.

Jensen, P.R., C.D. Harvell, K. Wirtz and W. Fenical. 1996. Anitmicrobial activity of extracts of Caribbean gorgonian corals. Mar. Biol. *125*: 411–419.

Jensen, R. 1974. Diseases of Sheep, Lea and Febiger, Philadelphia. 389 pp.

Jensen, R.A., D.S. Nasser and E.W. Nester. 1967. Comparative control of a branchpoint enzyme in microorganisms. J. Bacteriol. *94*: 1582–1593.

Jensen, V. 1955b. The *Azotobacter* flora of some Danish watercourses. Bot. Tidsskr. *52*: 143–157.

Jensen, W.I., C.R. Owen and W.J. Jellison. 1969. *Yersinia philomiragia* sp. n., a new member of the *Pasteurella* group of bacteria, naturally pathogenic for the muskrat (*Ondatra zibethica*). J. Bacteriol. *100*: 1237–1241.

Jernigan, D.B., L.I. Sanders, K.B. Waites, E.S. Brookings, R.F. Benson and P.G. Pappas. 1994. Pulmonary infection due to *Legionella cincinnatiensis* in renal transplant recipients: two cases and implications for laboratory diagnosis. Clin. Infect. Dis. *18*: 385–389.

Jerse, A.E. and J.B. Kaper. 1991. The *eae* gene of enteropathogenic *Escherichia coli* encodes a 94-kilodalton membrane protein, the expression of which is influenced by the EAF plasmid. Infect. Immun. *59*: 4302–4309.

Jerse, A.E., J. Yu, B.D. Tall and J.B. Kaper. 1990. A genetic locus of enteropathogenic *Escherichia coli* necessary for the production of attaching and effacing lesions on tissue culture cells. Proc. Natl. Acad. Sci. U.S.A. *87*: 7839–7843.

Jertborn, M. and A.M. Svennerholm. 1991. Enterotoxin-producing bacteria isolated from Swedish travellers with diarrhoea. Scand. J. Infect. Dis. *23*: 473–479.

Jessen, O. 1965. *Pseudomonas aeruginosa* and other green fluorescent pseudomonads. A taxonomic study, Munksgaard, Copenhagen. 1–244.

Ji, J., N. Hugouvieux-Cotte-Pattat and J. Robert-Baudouy. 1989. Molecular cloning of the *outJ* gene involved in pectate lyase secretion by *Erwinia chrysanthemi*. Mol. Microbiol. *3*: 285–293.

Jin, Q., Z.H. Yuan, J.G. Xu, Y. Wang, Y. Shen, W.C. Lu, J.H. Wang, H. Liu, J. Yang, F. Yang, X.B. Zhang, J.Y. Zhang, G.W. Yang, H.T. Wu, D. Qu, J. Dong, L.L. Sun, Y. Xue, A.L. Zhao, Y.S. Gao, J.P. Zhu, B. Kan, K.Y. Ding, S.X. Chen, H.S. Cheng, Z.J. Yao, B.K. He, R.S. Chen, D.L. Ma, B.Q. Qiang, Y.M Wen, Y.D. Hou and J. Yu. 2002. Genome sequence of *Shigella flexneri* 2a: insights into pathogenicity through comparison with genomes of *Escherichia coli* K12 and O157. Nucleic Acids Res. *30*: 4432–4441.

Jin, T. and R.G. Murray. 1987. Urease activity related to the growth and differentiation of swarmer cells of *Proteus mirabilis*. Can. J. Microbiol. *33*: 300–303.

Jindal, J.K. and P.N. Patel. 1984. Variability in xanthomonads of grain legumes. IV. Variations in bacteriological properties of 83 isolates and pathogenic behavior of cultural variants. Phytopathol. Z. *110*: 63–68.

Jindal, J.K., P.N. Patel and R. Singh. 1972. Bacterial leaf spot disease on *Amorphophallus campanulatus*. Indian Phytopathol. *25*: 374–377.

Joerger, R.D. and P.E. Bishop. 1988. Nucleotide sequence and genetic analysis of the *nifB-nifQ* region from *Azotobacter vinelandii*. J. Bacteriol. *170*: 1475–1487.

Joerger, R.D., M.R. Jacobson, R. Premakumar, E.D. Wolfinger and P.E. Bishop. 1989. Nucleotide sequence and mutational analysis of the structural genes (*anfHDGK*) for the second alternative nitrogenase from *Azotobacter vinelandii*. J. Bacteriol. *171*: 1075–1086.

Joerger, R.D., T.M. Loveless, R.N. Pau, L.A. Mitchenall, B.H. Simon and P.E. Bishop. 1990. Nucleotide sequences and mutational analysis of the structural genes for nitrogenase-2 of *Azotobacter vinelandii*. J. Bacteriol. *172*: 3400–3408.

Johns, R.B. and G.J. Perry. 1977. Lipids of marine bacterium *Flexibacter polymorphus*. Arch. Microbiol. *114*: 267–271.

Johnsen, J. 1977. Utilization of benzylpenicillin as carbon, nitrogen and energy-source by a *Pseudomonas fluorescens* strain. Arch. Microbiol. *115*: 271–275.

Johnson, F.H. and I.V. Shunk. 1936. An interesting new species of luminous bacteria. J. Bacteriol. *31*: 585–592.

Johnson, J. 1923. A bacterial leafspot of tobacco. J. Agr. Res. *23*: 481–494.

Johnson, J.C. 1956. Pod twist: a previously unrecorded bacterial disease of French bean *Phaseolus vulgaris* L. Quart. J. Agr. Sci. *13*: 127–158.

Johnson, J.R. 1991. Virulence factors in *Escherichia coli* urinary tract infection. Clin. Microbiol. Rev. *4*: 80–128.

Johnson, J.L. 1994a. Similarity analysis of DNAs. *In* Gerhardt, Murray, Wood and Krieg (Editors), Methods for General and Molecular Bacteriology, American Society for Microbiology, Washington, DC. pp. 655–682.

Johnson, J.L. 1994b. Similarity analysis of rRNAs. *In* Gerhardt, Murray, Wood and Krieg (Editors), Methods for General and Molecular Bacteriology, American Society for Microbiology, Washington, DC. pp. 683–700.

Johnson, J.L., R.S. Anderson and E.J. Ordal. 1970. Nucleic acid homologies among oxidase-negative *Moraxella* species. J. Bacteriol. *101*: 568–573.

Johnson, J.L. and B.S. Francis. 1975. Taxonomy of the clostridia: ribosomal ribonucleic acid homologies among the species. J. Gen. Microbiol. *88*: 229–244.

Johnson, J.R. and T.T. O'Bryan. 2000. Improved repetitive-element PCR fingerprinting for resolving pathogenic and nonpathogenic phylogenetic groups within *Escherichia coli*. Clin. Diagn. Lab. Immunol. *7*: 265–273.

Johnson, J.L. and E.J. Ordal. 1968. Deoxyribonucleic acid homology in bacterial taxonomy: effect of incubation temperature on reaction specificity. J. Bacteriol. *95*: 893–900.

Johnson, J.L. and N.J. Palleroni. 1989. Deoxyribonucleic acid similarities among *Pseudomonas* species. Int. J. Syst. Bacteriol. *39*: 230–235.

Johnson, J.A., C.A. Salles, P. Panigrahi, M.J. Albert, A.C. Wright, R.J. Johnson and J.G. Morris, Jr. 1994. *Vibrio cholerae* O139 synonym bengal is closely related to *Vibrio cholerae* El Tor but has important differences. Infect. Immun. *62*: 2108–2110.

Johnson, K.G., I.J. McDonald and M.B. Perry. 1976. Studies on the cellular and free lipopolysaccharides from *Branhamella catarrhalis*. Can. J. Microbiol. *22*: 460–467.

Johnson, P.W., Sieburth, J.M., Sastry, A., Arnold, C.R. and Doty, M.S.. 1971. *Leucothrix mucor* infestation of benthic crustacea, fish eggs, and tropical algae. Limnol. Oceanogr. *16*: 962–969.

Johnson, R., R.R. Colwell, R. Sakazaki and K. Tamura. 1975. Numerical taxonomy study of the family *Enterobacteriaceae*. Int. J. Syst. Bacteriol. *25*: 12–37.

Johnson, W.M., H. Lior and G.S. Bezanson. 1983. Cytotoxic *Escherichia coli* O157:H7 associated with haemorrhagic colitis in Canada. Lancet *1*: 76.

Johnson, W.M., S.D. Tyler, G. Wang and H. Lior. 1991. Amplification by the polymerase chain reaction of a specific target sequence in the gene coding for *Escherichia coli* verotoxin (VTe variant). FEMS Microbiol. Lett. *68*: 227–230.

Johnston, J.L., S.J. Billington, V. Haring and J.I. Rood. 1998. Complementation analysis of the *Dichelobacter nodosus fimN*, *fimO*, and *fimP* genes in *Pseudomonas aeruginosa* and transcriptional analysis of the *fimNOP* gene region. Infect. Immun. *66*: 297–304.

Johnston, J.L., S.J. Billington, V. Haring and J.I. Rood. 1995. Identification of fimbrial assembly genes from *Dichelobacter nodosus*: evidence that *fimP* encodes the type-IV prepilin peptidase. Gene *161*: 21–26.

Jolie, R.A., M.H. Mulks and B.J. Thacker. 1994. Antigenic differences within *Actinobacillus pleuropneumoniae* serotype 1. Vet. Microbiol. *38*: 329–349.

Joly, J.R. and W.C. Winn. 1984. Correlation of subtypes of *Legionella pneumophila* defined by monoclonal antibodies with epidemiological classification of cases and environment sources. J. Infect. Dis. *150*: 667–671.

Jones, B.D. and H.L. Mobley. 1989. *Proteus mirabilis* urease: nucleotide sequence determination and comparison with jack bean urease. J. Bacteriol. *171*: 6414–6422.

Jones, C.W. and R.K. Poole. 1985. The analysis of cytochromes. Methods Microbiol. *18*: 285–328.

Jones, D.M. 1962. A pasteurella-like organism from the human respiratory tract. J. Pathol. Bacteriol. *83*: 143–151.

Jones, D.A.C., A. McLeod, L.J. Hyman and M.C.M. Perombelon. 1993. Specificity of an antiserum against *Erwinia carotovora* ssp. *atroseptica* in indirect ELISA. J. Appl. Bacteriol. *74*: 620–624.

Jones, L.R. 1901. *Bacillus carotovorus* n. sp., die Ursache einer weichen Faulnis der Mohre. Zentbl. Bakteriol. Parasitenkd. Infektkrankh. Hyg. Abt. II *7*: 12–21.

Jones, L.R., A.G. Johnson and C.S. Reddy. 1917. Bacterial blight of barley. J. Agr. Res. *11*: 625–644.

Jones, L.R., M.M. Williamson, F.A. Wolf and L. McCulloch. 1923. Bacterial leafspot of clovers. J. Agr. Res. *25*: 471–490.

Jones, R.N. 1998. Important and emerging beta-lactamase-mediated resistances in hospital-based pathogens: the Amp C enzymes. Diagn. Microbiol. Infect. Dis. *31*: 461–466.

Jones, R.N., J. Slepack and J. Bigelow. 1976. Ampicillin resistant *Haemophilus paraphrophilus* epiglottitis. J. Clin. Microbiol. *4*: 405–407.

Jones, P., P. Woodley and R. Robson. 1984. Cloning and organization of some genes for nitrogen fixation from *Azotobacter chroococcum* and their expression in *Klebsiella pneumoniae*. Mol. Gen. Genet. *197*: 318–327.

Jordan, E.O. 1890. A report on certain species of bacteria observed in sewage, Massachusetts Board of Public Health. pp. 821–844.

Jordan, G.W. and W.K. Hadley. 1969. Human infections with *Edwardsiella tarda*. Ann. Intern. Med. *70*: 283–288.

Jordens, J.Z. and M.P. Slack. 1995. *Haemophilus influenzae*: then and now. Eur. J. Clin. Microbiol. Infect. Dis. *14*: 935–948.

Jørgensen, B.B. 1978. Distribution of colorless sulfur bacteria (*Beggiatoa* spp.) n a coastal marine sediment. Mar. Biol. *41*: 19–28.

Jørgensen, B.B. and V.A. Gallardo. 1999. *Thioploca* spp: filamentous sulfur bacteria with nitrate vacuoles. FEMS Microbiol. Ecol. *28*: 301–313.

Jørgensen, B.B. and J.R. Postgate. 1982. Ecology of the bacteria of the sulphur cycle with special reference to anoxic-oxic interface environments. Philos. Trans. R. Soc. Lond. Philos. Trans. R. Soc. Lond. B. Biol. Sci. *298*: 543–561.

Jørgensen, B.B. and N.P. Revsbech. 1983. Colorless sulfur bacteria, *Beggiatoa* spp. and *Thiovulum* spp. in oxygen and hydrogen sulfide microgradients. Appl. Environ. Microbiol. *45*: 1261–1270.

Jorgensen, J.H., L.A. Maher and A.W. Howell. 1991. Activity of meropenem against antibiotic-resistant or infrequently encountered Gramnegative bacilli. Antimicrob. Agents Chemother. *35*: 2410–2414.

Joseph, S.W. 1996. *Aeromonas* gastrointestinal disease: a case study in causation? *In* Austin, Altwegg, Gosling and Joseph (Editors), The Genus *Aeromonas*, John Wiley and Sons, Ltd., Chichester. pp. 311–336.

Joseph, S.W. and A.M. Carnahan. 1994. The isolation, identification, and systematics of the motile *Aeromonas* species. Ann. Rev. Fish Dis. *4*: 315–343.

Joseph, S.W., A.M. Carnahan, P.R. Brayton, G.R. Fanning, R. Almazan, C. Drabick, E.W.J. Trudo and R.R. Colwell. 1991. *Aeromonas jandaei* and *Aeromonas veronii* dual infection of a human wound following aquatic exposure. J. Clin. Microbiol. *29*: 565–569.

Joseph, S.W., R.R. Colwell and J.B. Kaper. 1982. Vibrio parahaemolyticus and related halophilic Vibrios. Crit. Rev. Microbiol. *10*: 77–124.

Joseph, S.W., M. Janda and A. Carnahan. 1988. Isolation, enumeration and identification of *Aeromonas* spp. J. Food Saf. *9*: 23–36.

Joshi, M.M. and J.P. Hollis. 1976. Rapid enrichment of *Beggiatoa* from soil. J. Appl. Bacteriol. *40*: 223–224.

Joshi, M.M. and J.P. Hollis. 1977. Interaction of *Beggiatoa* and rice plant: detoxification of hydrogen sulfide in the rice rhizosphere. Science *195*: 179–180.

Joshi-Tope, G. and A.J. Francis. 1995. Mechanisms of biodegradation of metal-citrate complexes by *Pseudomonas fluorescens*. J. Bacteriol. *177*: 1989–1993.

Jostensen, J.P. and B. Landfald. 1997. High prevalence of polyunsaturated-fatty-acid producing bacteria in arctic invertebrates. FEMS Microbiol. Lett. *151*: 95–101.

Joubert, J.J. and S.J. Truter. 1972. A variety of *Xanthomonas campestris* patbogenic to *Zantedeschia aethiopica*. Netherlands J. Plant Pathol. *78*: 212–217.

Jouravleva, E.A., G.A. McDonald, C.F. Garon, M. Boesman-Finkelstein and R.A. Finkelstein. 1998. Characterization and possible functions of a new filamentous bacteriophage from *Vibrio cholerae* O139. Microbiology *144*: 315–324.

Judd, A.K., M. Schneider, M.J. Sadowsky and F.J. De Bruijn. 1993. Use of repetitive sequences and the polymerase chain-reaction technique to classify genetically related *Bradyrhizobium japonicum* serocluster 123 strains. Appl. Environ. Microbiol. *59*: 1702–1708.

Judicial Commission. 1952. Opinion 5. Conservation of the generic name *Pseudomonas* Migula 1894 and designation of *Pseudomonas aeruginosa* (Schroeter) Migula 1900 as type species. Int. Bull. Bacteriol. Nomen. Taxon. *2*: 121–122.

Judicial Commission 1954a. Opinion 11. Nomenclature of species in the bacterial genus *Shigella*. Int. Bull. Bacteriol. Nomencl. Taxon. *4*: 148–149.

Judicial Commission 1954b. Opinion 13. Conservation and rejection of names of genera of bacteria proposed by Trevisan 1842–1890. Int. Bull. Bacteriol. Nomencl. Taxon. *4*: 151–154.

Judicial Commission. 1958a. Opinion 15. Conservation of the family name *Enterobacteriaceae*, of the name of the type genus, and designation of the type species. Int. Bull. Bacteriol. Nomen. Taxon. *8*: 73–74.

Judicial Commission. 1958b. Rejection of the generic names *Nitromonas* Winogradsky 1890 and *Nitromonas* Orla-Jensen 1909, conservation of the generic names *Nitrosomonas* Winogradsky 1892, *Nitrosococcus* Winogradsky 1892, and *Nitrobacter* Winogradsky 1892, and the designation of the type species of these gener. Int. Bull. Bacteriol. Nomencl. Taxon. *8*: 169–170.

Judicial Commission. 1963. Opinion 26. Designation of neotype strains (cultures) of type species of the bacterial genera *Salmonella*, *Shigella*, *Arizona*, *Escherichia*, *Citrobacter*, and *Proteus* of the family Enterobacteriaceae. Int. Bull. Bacteriol. Nomencl. Taxon. *13*: 35–36.

Judicial Commission 1963. Opinion 28. Rejection of the bacterial generic name *Cloaca* Castellani and Chalmers and acceptance of *Enterobacter* Hormaeche and Edwards as a bacterial generic name with type species *Enterobacter cloacae* (Jordan) Hormaeche and Edwards. Int. Bull. Bacteriol. Nomencl. Taxon. *13*: 38.

Judicial Commission 1971. Conservation of the specific epithet "*phenylpyruvica*" in the name *Moraxella phenylpyruvica* Bøvre and Henriksen. Int. J. Syst. Bacteriol. *21*: 107.

Judicial Commission. 1981. Present standing of the family name *Enterobacteriaceae* Rahn 1937. Int. J. Syst. Bacteriol. *31*: 104.

Judicial Commission 1991. Minutes of the Meetings, 14 September 1990, Osaka, Japan. (Minutes 17 (i) and 18 (iii)). Int. J. Syst. Bacteriol. *41*: 185–187.

Judicial Commission. 1993. Opinion 67. Rejection of the name *Citrobacter diversus* Werkman and Gillen 1932. Int. J. Syst. Bacteriol. *43*: 392.

Jukes, T.H. and R.R. Cantor. 1969. Evolution of protein molecules. *In* Munzo (Editor), Mammalian Protein Metabolism, Academic Press, New York. pp. 21–132.

Julianelle, L.A. 1926. A biological classification of *Encapsulatus pneumoniae* (Friedländer's bacillus). J. Exp. Med. *44*: 113–128.

Jung, Y.S., H.S. Gao-Sheridan, J. Christiansen, D.R. Dean and B.K. Burgess. 1999. Purification and biophysical characterization of a new [2Fe-2S] ferredoxin from *Azotobacter vinelandii*, a putative [Fe-S] cluster assembly/repair protein. J. Biol. Chem. *274*: 32402–32410.

Junge, K., H. Eicken and J.W. Deming. 2003. Motility of *Colwellia psychrerythraea* strain 34H at subzero temperatures. Appl. Environ. Microbiol. *69*: 4282–4284.

Junge, K., J.F. Imhoff, J.T. Staley and J.W. Deming. 2002. Phylogenetic diversity of numerically important arctic sea-ice bacteria cultured at subzero temperature. Microb. Ecol. *43*: 315–328.

Juni, E. 1972. Interspecies transformation of *Acinetobacter*: genetic evidence for a ubiquitous genus. J. Bacteriol. *112*: 917–931.

Juni, E. 1974. Simple genetic transformation assay for rapid diagnosis of *Moraxella osloensis*. Appl. Microbiol. *27*: 16–24.

Juni, E. 1977. Genetic transformation assays for identification of strains of *Moraxella urethralis*. J. Clin. Microbiol. *5*: 227–235.

Juni, E. 1978. Genetics and physiology of *Acinetobacter*. Annu. Rev. Microbiol. *32*: 349–371.

Juni, E. 1982. *Acinetobacter*: a tale of two genera. *In* Hollaender, DeMoss, Kaplan, Konisky, Savage and Wolfe (Editors), Genetic Engineering of Microorganisms for Chemicals, Plenum Press, New York. pp. 259–269.

Juni, E. 1990. Application of genetic transformation in identification of Gram-negative bacteria. *In* Olsvik and Bukholm (Editors), Application of Molecular Biology in Diagnosis of Infectious Diseases, Norwegian College of Veterinary Medicine, Oslo. pp. 61–68.

Juni, E. 1992. The genus *Psychrobacter*. *In* Balows, Trüper, Dworkin, Harder and Schleifer (Editors), The Prokaryotes. A Handbook on the Biology of Bacteria: Ecophysiology, Isolation, Identification, Applications, 2nd Ed., Vol. 4, Springer-Verlag, New York. pp. 3241–3246.

Juni, E. and G.A. Heym. 1980. Transformation assay for identification of psychrotrophic achromobacters. Appl. Environ. Microbiol. *40*: 1106–1114.

Juni, E. and G.A. Heym. 1986a. Defined medium for *Moraxella bovis*. Appl. Environ. Microbiol. *52*: 966–968.

Juni, E. and G.A. Heym. 1986b. *Psychrobacter immobilis*, new genus new species: Genospecies composed of gram-negative, aerobic, oxidase-positive coccobacilli. Int. J. Syst. Bacteriol. *36*: 388–391.

Juni, E., G.A. Heym and M. Avery. 1986. Defined medium for *Moraxella catarrhalis*. Appl. Environ. Microbiol. *52*: 546–551.

Juni, E., G.A. Heym and R.A. Bradley. 1984. General approach to bacterial nutrition: growth factor requirements of *Moraxella nonliquefaciens*. J. Bacteriol. *160*: 958–965.

Juni, E., G.A. Heym, M.J. Maurer and M.L. Miller. 1987. Combined genetic transformation and nutritional assay for identification of *Moraxella nonliquefaciens*. J. Clin. Microbiol. *25*: 1691–1694.

Juni, E., G.A. Heym and R.D. Newcomb. 1988. Identification of Moraxella bovis by qualitative genetic transformation and nutritional assays. Appl. Environ. Microbiol. *54*: 1304–1306.

Juni, E. and A. Janik. 1969. Transformation of *Acinetobacter calco-aceticus* (*Bacterium anitratum*). J. Bacteriol. *98*: 281–288.

Jüttner, F. 1988. Quantitative trace analysis of volatile organic compounds. Methods Enzymol. *167*: 609–616.

Jyssum, K. and P.E. Joner. 1965. Growth of *Bacterium anitratum* (B5W) with nitrate or nitrite as nitrogen source. Acta Pathol. Microbiol. Scand. *64*: 381–386.

Kaars Sijpesteijn, A. 1949. Cellulose-decomposing bacteria from the rumen of cattle. J. Microbiol. Serol. *15*: 49–52.

Kaaya, G.P. and M.A. Okech. 1990. Microorganisms associated with tsetse in nature: preliminary results on isolation, identification, and pathogenicity. International Study Workshop on Tsetse Population and Behavior, Duduville, Nairobi. Insect Sci. Appl. *11*: 443–448.

Kabashima, T., K. Ito and T. Yoshimoto. 1996. Dipeptidyl peptidase IV from *Xanthomonas maltophilia*: sequencing and expression of the enzyme gene and characterization of the expressed enzyme. J. Biochem. (Tokyo) *120*: 1111–1117.

Kado, C.I. and M.G. Heskett. 1970. Selective media for isolation of *Agrobacterium*, *Corynebacterium*, *Erwinia*, *Pseudomonas*, and Xanthomonasu. Phytopathology *60*: 969–976.

Kado, C.I. and S.-T., Liu. 1981. Rapid procedure for detection and isolation of large and small plasmids. J. Bacteriol. *145*: 1365–1373.

Kadota, H. 1951. Studies on the biochemical activities of marine bacteria. I. On the agar-decomposing bacteria in the sea. Memoirs Coll. Agric. Kyoto University. *59*: 54–67.

Kageyama, B., M. Nakae, S. Yagi and T. Sonoyama. 1992. *Pantoea punctata* sp. nov., *Pantoea citrea* sp. nov., and *Pantoea terrea* sp. nov. isolated from fruit and soil samples. Int. J. Syst. Bacteriol. *42*: 203–210.

Kahn, M.E. and R. Gromkova. 1981. Occurrence of pili on and adhesive properties of *Haemophilus parainfluenzae*. J. Bacteriol. *145*: 1075–1078.

Kain, K.C., R.L. Barteluk, M.T. Kelly, X. He, G. de-Hua, Y.A. Ge, E.M. Proctor, S. Byrne and H.G. Stiver. 1991. Etiology of childhood diarrhea in Beijing, China. J. Clin. Microbiol. *29*: 90–95.

Kain, K.C. and M.T. Kelly. 1989. Antimicrobial susceptibility of *Plesiomonas shigelloides* from patients with diarrhea. Antimicrob. Agents Chemother. *33*: 1609–1610.

Kainz, A., W. Lubitz and H.J. Busse. 2000. Genomic fingerprints, ARDRA profiles and quinone systems for classification of *Pasteurella* sensu stricto. Syst. Appl. Microbiol. *23*: 191–503.

Kaiser, G.M., P.C. Tso, R. Morris and D. McCurdy. 1997. *Xanthomonas maltophilia* endophthalmitis after cataract extraction. Am. J. Opthalmol. *123*: 410–411.

Kakeda, K. and H. Ishikawa. 1991. Molecular chaperon produced by an intracellular symbiont. J. Biochem. (Tokyo) *110*: 583–587.

Kakimoto, D., H. Maeda, T. Sakata, W. Sharp and R.M. Johnson. 1980. Marine pigmented bacteria. I. Distribution and characteristics of pigmented bacteria. Mem. Fac. Fish., Kagoshima Univ. *29*: 339–347.

Kaldorf, M., K.H. Linne von Berg, U. Meier, U. Servos and H. Bothe. 1993. The reduction of nitrous oxide to dinitrogen by *Escherichia coli*. Arch. Microbiol. *160*: 432–439.

Kalina, G.P., A.S. Antonov, T.P. Turova and T.I. Grafova. 1984. *Allomonas enterica* gen. nov., sp. nov.: deoxyribonucleic acid homology between *Allomonas* and some other members of the *Vibrionaceae*. Int. J. Syst. Bacteriol. *34*: 150–154.

Kallings, L.O. 1967. Sensitivity of various Salmonella strains to Felix 01 phage. Acta Pathol. Microbiol. Scand. *70*: 446–454.

Kamekura, M., R. Wallace, A.R. Hipkiss and D.J. Kushner. 1985. Growth of *Vibrio costicola* and other moderate halophiles in a chemically defined minimal medium. Can. J. Microbiol. *31*: 870–872.

Kameyama, T., A. Takahashi, S. Kurasawa, M. Ishizuka, Y. Okami, T. Takeuchi and H. Umezawa. 1987. Bisucaberin, a new siderophore, sensitizing tumor cells to macrophage-mediated cytolysis. I. Taxonomy of the producing organism, isolation and biological properties. J. Antibiot. (Tokyo). *40*: 1664–1670.

Kamme, C. 1969. Susceptibility in vitro of *Haemophilus influenzae* to penicillin G., penicillin V. and ampicillin. Incubation of strains from acute otitis media in air and in CO_2. atmosphere. Acta Pathol. Microbiol. Scand. *75*: 611–621.

Kamp, E.M., T.M. Vermeulen, M.A. Smits and J. Haagsma. 1994. Production of Apx toxins by field strains of *Actinobacillus pleuropneumoniae* and *Actinobacillus suis*. Infect. Immun. *62*: 4063–4065.

Kämpf, C. and N. Pfennig. 1980. Capacity of *Chromatiaceae* for chemotrophic growth. Specific respiration rates of *Thiocystis violacea* and *Chromatium vinosum*. Arch. Microbiol. *127*: 125–135.

Kämpf, C. and N. Pfennig. 1986. Isolation and characterization of some chemoautotrophic *Chromatiaceae*. J. Basic Microbiol. *26*: 507–515.

Kämpfer, P. 1993. Grouping of *Acinetobacter* genomic species by cellular fatty acid composition. Med. Microbiol. Lett. *2*: 394–400.

Kämpfer, P., A. Albrecht, S. Buczolits and H.J. Busse. 2002a. *Psychrobacter faecalis* sp nov., a new species from a bioaerosol originating from pigeon faeces. Syst. Appl. Microbiol. *25*: 31–36.

Kämpfer, P., A. Albrecht, S. Buczolits and H.J. Busse. 2002b. *In* Validation of the publication of new names and new combinations previously effectively published outside the IJSEM. List No. 87. Int. J. Syst. Evol. Microbiol. *52*: 1437–1438.

Kämpfer, P. and M. Altwegg. 1992. Numerical classification and identification of Aeromonad genospecies. J. Appl. Bacteriol. *72*: 341–351.

Kämpfer, P., K. Blasczyk and G. Auling. 1994. Characterization of *Aeromonas* genomic species by using quinone, polyamine, and fatty acid patterns. Can. J. Microbiol. *40*: 844–850.

Kämpfer, P., S. Meyer and H.E. Müller. 1997. Characterization of *Buttiauxella* and *Kluyvera* species by analysis of whole cell fatty acid patterns. Syst. Appl. Microbiol. *20*: 566–571.

Kämpfer, P., I. Tjernberg and J. Ursing. 1993. Numerical classification and identification of *Acinetobacter* genomic species. J. Appl. Bacteriol. *75*: 259–268.

Kamps, A.M.I.E., E.M. Kamp and M.A. Smits. 1990. Cloning and expression of the dermonecrotic toxin gene of *Pasteurella multocida* subsp. *multocida* in *Escherichia coli*. FEMS Microbiol. Lett. *67*: 187–190.

Kanagawa, T., Y. Kamagata, S. Aruga, T. Kohno, M. Horn and M. Wagner. 2000. Phylogenetic analysis of and oligonucleotide probe development for Eikelboom type 021N filamentous bacteria isolated from bulking activated sludge. Appl. Environ. Microbiol. *66*: 5043–5052.

Kanagawa, T. and E. Mikami. 1989. Removal of methanethiol, dimethyl sulfide, dimethyl disulfide, and hydrogen sulfide from contaminated air by *Thiobacillus thioparus* TK-m. Appl. Environ. Microbiol. *55*: 555–558.

Kanazawa, Y. and K. Ikemura. 1979. Isolation of *Yersinia enterocolitica* and *Yersinia pseudotuberculosis* from human specimens and their drug resistance in the Niigata district in Japan. Contr. Microbiol. Immunol. 5: 106–114.

Kaneko, S. and T. Maruyama. 1987. Pathogenicity of *Yersinia enterocolitica* O:3, biotype 3 strains. J. Clin. Microbiol. 25: 454–455.

Kangatharalingam, N. and J.C. Priscu. 1993. Nitrapyrin-ammonium combination induces rapid multiplication of mixed cultures of the stalked bacterium *Nevskia ramosa* Famintzin and other heterotrophic bacteria. Arch. Microbiol. 159: 48–50.

Kaper, J.B. 1996. Defining EPEC. Rev. Microbiol. São Paulo. 27: 130–133.

Kaper, J.B., H. Lockman, R.R. Colwell and S.W. Joseph. 1981. *Aeromonas hydrophila*: ecology and toxigenicity of isolates from an estuary. J. Appl. Bacteriol. 50: 359–378.

Kaper, J.B., J.G. Morris, Jr. and M.M. Levine. 1995. Cholera. Clin. Microbiol. Rev. 8: 48–86.

Kaper, J.B. and A.D. O'Brien (Editors). 1998. *Escherichia coli* O157:H7 and other Shiga Toxin-Producing Strains, ASM Press, Washington, D.C. 465 pp.

Kaper, J., R.J. Seidler, H. Lockman and R.R. Colwell. 1979. Medium for the presumptive identification of *Aeromonas hydrophila* and *Enterobacteriaceae*. Appl. Environ. Microbiol. 38: 1023–1026.

Kaplan, R.W. and M. Brendel. 1969. Formation of prototrophs in mixtures of two auxotropic mutants of *Serratia marcescens* HY by a transducing bacteriophage produced by some auxotrophs. Mol. Gen. Genetics. 104: 27–39.

Kapperud, G. 1977. *Yersinia enterocolitica* and *Yersinia* like microbes isolated from mammals and water in Norway and Denmark. Acta Pathol. Microbiol. Scand. Sect. B Microbiol. 85: 129–135.

Kapperud, G. 1980. Studies on the pathogenicity of *Yersinia enterocolitica* and *Y. enterocolitica* like bacteria. I. Enterotoxin production at 22°C and 37°C by environmental and human isolates from Scandinavia. Acta. Pathol. Microbiol. Scand. Sect. B. 88: 287–291.

Kappos, T., M.A. John, Z. Hussain and M.A. Valvano. 1992. Outer membrane profiles and multilocus enzyme electrophoresis analysis for differentiation of clinical isolates of *Proteus mirabilis* and *Proteus vulgaris*. J. Clin. Microbiol. 30: 2632–2637.

Karaolis, D.K., J.A. Johnson, C.C. Bailey, E.C. Boedeker, J.B. Kaper and P.R. Reeves. 1998. A *Vibrio cholerae* pathogenicity island associated with epidemic and pandemic strains. Proc. Natl. Acad. Sci. U.S.A. 95: 3134–3139.

Karaolis, D.K., R. Lan and P.R. Reeves. 1994. Sequence variation in *Shigella sonnei* (Sonnei), a pathogenic clone of *Escherichia coli*, over four continents and 41 years. J. Clin. Microbiol. 32: 796–802.

Karaolis, D.K., S. Somara, D.R. Maneval, Jr., J.A. Johnson and J.B. Kaper. 1999. A bacteriophage encoding a pathogenicity island, a type-IV pilus and a phage receptor in cholera bacteria. Nature 399: 375–379.

Karch, H., H. Böhm, H. Schmidt, F. Gunzer, S. Aleksic and J. Heesemann. 1993. Clonal structure and pathogenicity of Shiga-like toxin-producing, sorbitol-fermenting *Escherichia coli* O157:H –. J. Clin. Microbiol. 31: 1200–1205.

Karch, H., J. Heesemann, R. Laufs, A.D. O'Brien, C.O. Tacket and M.M. Levine. 1987. A plasmid of enterohemorrhagic *Escherichia coli* O157:H7 is required for expression of a new fimbrial antigen and for adhesion to epithelial cells. Infect. Immun. 55: 455–461.

Karlsson, K.A. and O. Söderlind. 1973. Studies of the diagnosis of tularemia. Contributions to Microbiology and Immunology. Vol. 2. *Yersinia, Pasteurella* and *Francisella*, Malmo, Sweden. 224–230.

Karmali, M.A. 1989. Infection by verocytotoxin-producing *Escherichia coli*. Clin. Microbiol. Rev. 2: 15–38.

Karmali, M.A., C.A. Lingwood, M. Petric, J. Brunton and C. Gyles. 1996. Maintaining the existing phenotype nomenclatures for *E. coli* cytotoxins. ASM News. 62: 167–169.

Karpati, F. and J. Jonasson. 1996. Polymerase chain reaction for the detection of *Pseudomonas aeruginosa*, *Stenotrophomonas maltophilia* and *Burkholderia cepacia* in sputum of patients with cystic fibrosis. Mol. Cell. Probes 10: 397–403.

Karr, D.E., W.F. Bibb and C.W. Moss. 1982. Isoprenoid quinones of the genus *Legionella*. J. Clin. Microbiol. 15: 1044–1048.

Kashbur, I.M., R.H. George and G.A. Ayliffe. 1974. Resistotyping of *Proteus mirabilis* and a comparison with other methods of typing. J. Clin. Pathol. 27: 572–577.

Kaska, M. 1976. The toxicity of extracellular protease of the bacterium *Serratia marcescens* for larvae of greater wax moth *Galleria mellonella*. J. Invertebr. Pathol. 27: 271.

Kater, J.C., S.C. Marshall and W.J. Hartley. 1962. A specific suppurative synovitis and pyaemia in lambs. NZ. Vet. J. 10: 143–144.

Kato, C., L. Li, Y. Nogi, Y. Nakamura, J. Tamaoka and K. Horikoshi. 1998. Extremely barophilic bacteria isolated from the Mariana Trench, Challenger Deep, at a depth of 11,000 meters. Appl. Environ. Microbiol. 64: 1510–1513.

Kato, C., N. Masui and K. Horikoshi. 1996. Properties of obligately barophilic bacteria isolated from a sample of deep-sea sediment from the Izu-Bonin trench. J. Mar. Biotechnol. 4: 96–99.

Kato, C., T. Sato and K. Horikoshi. 1995. Isolation and properties of barophilic and barotolerant bacteria from deep-sea mud samples. Biodivers. Conserv. 4: 1–9.

Kato, C., M. Smorawinska, L. Li and K. Horikoshi. 1997. Comparison of the gene expression of aspartate β-D-semialdehyde dehydrogenase at elevated hydrostatic pressure in deep-sea bacteria. J. Biochem. (Tokyo) 121: 717–723.

Kato, K. and H. Tsubahara. 1962. Infectious coryza of chickens. II. Identification of isolates. Bull. Nat. Inst. An. Health 45: 21–26.

Katoh, K. and T. Suzuki. 1979. Microflora of manured soils. Bull. Natl. Inst. Agric. Sci. Ser. B. (Soils Fert.) 30: 73–135.

Katz, M.E., P.M. Howarth, W.K. Yong, G.G. Riffkin, L.J. Depiazzi and J.I. Rood. 1991. Identification of three gene regions associated with virulence in *Dichelobacter nodosus*, the causative agent of ovine footrot. J. Gen. Microbiol. 137: 2117–2124.

Katz, W. and J.L. Strominger. 1967. Structure of the cell wall of *Micrococcus lysodeikticus*. II. Study of the structure of the peptides produced after lysis with the myxobacterium enzyme. Biochemistry 6: 930–937.

Katzen, F., A. Becker, A. Zorreguieta, A. Puhler and L. Lelpi. 1996. Promoter analysis of the *Xanthomonas campestris* pv. *campestris* gum operon directing biosynthesis of the xanthan polysaccharide. J. Bacteriol. 178: 4313–4318.

Katznelson, H. 1955. The metabolism of phytopathogenic bacteria. I. Comparative studies on the metabolism of representative species. J. Bacteriol. 70: 469–475.

Katznelson, H. 1958. Metabolism of phytopathogenic bacteria. II. Metabolism of carbohydrates by cell-free extracts. J. Bacteriol. 75: 540–543.

Kauc, L. and S.H. Goodgal. 1989. The size and a physical map of the chromosome of *Haemophilus parainfluenzae*. Gene 83: 377–380.

Kauffman, C.A., A.G. Bergman and C.S. Hertz. 1979. Antimicrobial resistance of *Haemophilus* species in patients with chronic bronchitis. Am. Rev. Resp. Dis. 120: 1382–1385.

Kauffmann, F. 1935. Weitere erfahrungen mit den Kombinierten Anreichurungsverfahren für Salmonellabazillus. Z. Hyg. Infektionskr. 117: 26–32.

Kauffmann, F. 1941. Über mehrere neue Salmonella types. Acta Pathol. Microbiol. Scand. 18: 351–366.

Kauffmann, F. 1954. *Enterobacteriaceae*, 2nd Ed., E. Munksgaard, Copenhagen.

Kauffmann, F. 1956. Zür biochemischen und serologischen Gruppen- und Typen-Einteilung der *Enterobacteriaceae*. Zentbl. Bakteriol. Parasitenkd. Infektkrankh. Hyg. Abt. I Orig. 165: 344–354.

Kauffmann, F. 1960. Two biochemical subdivisions of the genus *Salmonella*. Acta Pathol. Microbiol. Scand. 49: 393–396.

Kauffmann, F. 1961. The species-definition in the *Enterobacteriaceae*. Int. Bull. Bacteriol. Nomencl. Taxon. 11: 5–6.

Kauffmann, F. 1962. Supplement to the Kauffmann-White scheme (V). Acta Pathol. Microbiol. Scand. 55: 349–354.

Kauffmann, F. 1963a. On the species definition. Int. Bull. Bacteriol. Nomencl. Taxon. 13: 181–186.

Kauffmann, F. 1963b. Zür differential diagnose der *Salmonella* sub-genera I, II und III. Acta Pathol. Microbiol. Scand. 58: 109–113.

Kauffmann, F. 1964. Vereinfachtes Antigen-Schema der *Salmonella* subgenera II, III. Acta Pathol. Microbiol. Scand. *62*: 68–72.

Kauffmann, F. 1966. The bacteriology of *Enterobacteriaceae*, The Williams & Wilkins Co., Baltimore.

Kauffmann, F. and A. Dupont. 1950. *Escherichia* strains from infantile epidemic gastroenteritis. Acta Pathol. Microbiol. Immunol. Scand. *27*: 552–564.

Kauffmann, F. and P.R. Edwards. 1952. Classification and nomenclature of *Enterobacteriaceae*. Int. Bull. Bacteriol. Nomencl. Taxon. *2*: 2–8.

Kauffmann, F. and E. Møller. 1940. A new type of *Salmonella* (*S. ballerup*) with Vi antigen. J. Hyg. *40*: 246–251.

Kaulfers, P.M., R. Laufs and G. Lahn. 1978. Molecular properties of transmissible R factors of *Haemophilus influenzae* determining tetracycline resistance. J. Gen. Microbiol. *105*: 243–252.

Kawahara, E., Y. Fukuda and R. Kusuda. 1998. Serological differences among *Photobacterium damsela* subsp. *piscicida* isolates. Fish Pathol. *33*: 281–285.

Kawai, Y. and E. Yabuuchi. 1975. *Pseudomonas pertucinogena* sp. n., an organism previously misidentified as *Bordetella pertussis*. Int. J. Syst. Bacteriol. *25*: 317–323.

Kawakama, W. and S. Yoshida. 1920. Bacterial gall on Milletia plant (*Bacillus milletiae* n. sp.). Bot. Mag. Tokyo. *34*: 110–115.

Kawakami, B., H. Christophe, M. Nagatomo and M. Oka. 1991. Cloning and nucleotide sequences of the *AccI* restriction-1 modification genes in *Acinetobacter calcoaceticus*. Agric. Biol. Chem. *55*: 1553–1560.

Kawamoto, E., T. Sawada and T. Maruyama. 1997. Evaluation of transport media for *Pasteurella multocida* isolates from rabbit nasal specimens. J. Clin. Microbiol. *35*: 1948–1951.

Kawamura, E. 1934. Bacterial leaf spot of sunflower. Ann. Phytopathol. Soc. Japan *4*: 25–28.

Kawasaki, H., K. Yamasato and J. Sugiyama. 1997. Phylogenetic relationships of the helical-shaped bacteria in the *α Proteobacteria* inferred from 16S rDNA sequences. J. Gen. Appl. Microbiol. *43*: 89–95.

Kawasaki, K., Y. Nogi, M. Hishinuma, Y. Nodasaka, H. Matsuyama and I. Yumoto. 2002. *Psychromonas marina* sp nov., a novel halophilic, facultatively psychrophilic bacterium isolated from the coast of the Okhotsk Sea. Int. J. Syst. Evol. Microbiol. *52*: 1455–1459.

Kawauchi, K., K. Shibutani, H. Yagisawa, H. Kamata, S. Nakatsuji, H. Anzai, Y. Yokoyama, Y. Ikegami, Y. Moriyama and H. Hirata. 1997. A possible immunosuppressant, cycloprodigiosin hydrochloride, obtained from *Pseudoalteromonas denitrificans*. Biochem. Biophys. Res. Commun. *237*: 543–547.

Kay, B.A., K.L. Kotloff, L.D. Guers and B.D. Tall. 1990. *Kluyvera* species: Lack of association with pediatric diarrheal disease. 90th Annual Meeting of the American Society for Microbiology, American Society for Microbiology, . p. 366.

Kayser, H. 1902. Das Wachstum der zurischen Bacterium typhi und coli stehenden Spaltpilze auf dem v. Drigalski-Conradi'schen Agarboden. Zentrabl. Bakteriol. Parasitenk. Infektionskr. Hyg. Abt. I Orig. *31*: 426–429.

Kaysner, C.A., C. Abeyta, Jr., M.M. Wekell, A. DePaola, Jr., R.F. Stott and J.M. Leitch. 1987. Incidence of *Vibrio cholerae* from estuaries of the United States West Coast. Appl. Environ. Microbiol. *53*: 1344–1348.

Kaysner, C.A., M.L Tamplin, M.M. Wekell, R.F. Stott and K.G Colburn. 1989. Survival of *Vibrio vulnificus* in shellstock and shucked oysters (*Crassostrea gigas* and *Crassostrea virginica*) and effects of isolation medium on recovery. Appl Environ Microbiol. *55*: 3072-3079.

Kazandjian, D., R. Chiew and G.L. Gilbert. 1997. Rapid diagnosis of *Legionella pneumophila* serogroup 1 infection with the binax enzyme immunoassay urinary antigen test. J. Clin. Microbiol. *35*: 954–956.

Kearns, L.P. and C.N. Hale. 1995. Incidence of bacteria inhibitory to *Erwinia amylovora* from blossoms in New Zealand apple orchards. Plant Pathol. (Oxf.) *44*: 918–924.

Kearns, L.P. and C.N. Hale. 1996. Partial characterization of an inhibitory strain of *Erwinia herbicola* with potential as a biocontrol agent for *Erwinia amylovora*, the fire blight pathogen. J. Appl. Bacteriol. *81*: 369–374.

Keathley, J.D. and W.C. Winn, Jr. 1985. Comparison of media for recovery of clinical isolates of *Legionella pneumophila*. Am. J. Clin. Pathol. *83*: 498–499.

Keeble, J.R. and T. Cross. 1977. An improved medium for the enumeration of *Chromobacterium* in soil and water. J. Appl. Bacteriol. *43*: 325–327.

Keel, J.A., W.R. Finnerty and J.C. Feeley. 1979. Fine structure of the Legionnaires' disease bacterium. Ann. Int. Med. *90*: 652–655.

Keeling, P.J. and W.F. Doolittle. 1997. Evidence that eukaryotic triose-phosphate isomerase is of alpha-proteobacterial origin. Proc. Natl. Acad. Sci. U.S.A. *94*: 1270–1275.

Keil, F. 1912. Beiträge zur Physiologie der farblosen Schwefelbakterien. Beitr. biol. Pflanz. *11*: 335–372.

Kellen, W.R., J.E. Lindegren and D.F. Hoffmann. 1972. Developmental stages and structure of a *Rickettsiella* in the navel orangeworm, *Poramyelois transitella* (Lepidoptera: Phycitidae). J. Invert. Pathol. *20*: 193–199.

Keller, R., M.Z. Pedroso, R. Ritchmann and R.M. Silva. 1998. Occurrence of virulence-associated properties in *Enterobacter cloacae*. Infect. Immun. *66*: 645–649.

Kelley, S.K., V.E. Coyne, D.D. Sledjeski, W.C. Fuqua and R.M. Weiner. 1990. Identification of a tyrosinase from a periphytic marine bacterium. FEMS Microbiol. Lett. *67*: 275–280.

Kelln, R.A. and R.A. Warren. 1971. Isolation and properties of a bacteriophage lytic for a wide range of pseudomonads. Can. J. Microbiol. *17*: 677–682.

Kellogg, D.S., Jr., W.L. Peacock, Jr., W.E. Deacon, L. Brown and C.I. Pirkle. 1963. *Neisseria gonorrhoeae*. I. Virulence genetically linked to clonal variation. J. Bacteriol. *85*: 1274–1279.

Kelly, B.L. 1959. PhD Dissertation, University of Southern California

Kelly, D.P. and A.P. Harrison. 1989. Genus *Thiobacillus*. *In* Staley, Bryant, Pfennig and Holt (Editors), Bergey's Manual of Systematic Bacteriology, 1st Ed., Vol. 3, The Williams & Williams Co., Baltimore. pp. 1842–1858.

Kelly, D.P., E. Stackebrandt, J. Burghardt and A.P. Wood. 1998. Confirmation that *Thiobacillus halophilus* and *Thiobacillus hydrothermalis* are distinct species within the gamma-sublass of the *Proteobacteria*. Arch. Microbiol. *170*: 138–140.

Kelly, D.P. and A.P. Wood. 2000. Reclassification of some species of *Thiobacillus* to the newly designated genera *Acidithiobacillus* gen. nov., *Halothiobacillus* gen. nov. and *Thermithiobacillus* gen. nov. Int. J. Syst. Evol. Microbiol. *50*: 511–516.

Kelly, K.A., J.M. Koehler and L.R. Ashdown. 1993. Spectrum of extraintestinal disease due to *Aeromonas* species in tropical Queensland, Australia. Clin. Infect. Dis. *16*: 574–579.

Kelly, M.T. and T.D. Brock. 1969a. Molecular heterogeneity of isolates of the marine bacterium *Leucothrix mucor*. J. Bacteriol. *100*: 14–21.

Kelly, M.T. and T.D. Brock. 1969b. Warm-water strain of *Leucothrix mucor*. J. Bacteriol. *98*: 1402–1403.

Kelly, M.T. and A. Dinuzzo. 1985. Uptake and clearance of *Vibrio vulnificus* from Gulf coast oysters (*Crassostrea virginica*). Appl. Environ. Microbiol. *50*: 1548–1549.

Kelly, M.D. and J.E. Mortensen. 1995. A low-copy number plasmid mediating β-lactamase production by *Xanthomonas maltophilia*. Adv. Exp. Med. Biol. *390*: 71–80.

Kelly, M.J.S., R.K. Poole, M.G. Yates and C. Kennedy. 1990. Cloning and mutagenesis of genes encoding the cytochrome *bd* terminal oxidase complex in *Azotobacter vinelandii* - mutants deficient in the cytochrome *d* complex are unable to fix nitrogen in air. J. Bacteriol. *172*: 6010–6019.

Kelly, M.T., E.M.D. Stroh and J. Jessop. 1988. Comparison of blood agar, ampicillin blood agar, MacConkey-ampicillin-Tween agar, and modified cefsulodin-Irgasan-novobiocin agar for isolation of *Aeromonas* spp. from stool specimens. J. Clin. Microbiol. *26*: 1738–1740.

Kelm, O., C. Kiecker, K. Geider and F. Bernhard. 1997. Interaction of the regulator proteins RcsA and RcsB with the promoter of the operon for amylovoran biosynthesis in *Erwinia amylovora*. Mol. Gen. Genet. *256*: 72–83.

Kelman, A. and R.S. Dickey. 1980. Soft rot of 'carotovora' group. *In*

Schaad (Editor), Laboratory guide for identification of plant pathogenic bacteria, American Phytopathological Society, St. Paul. pp. 31–35.

Kendrick, J.B. 1934. Bacterial blight of carrot. J. Agr. Res. *49*: 493–510.

Kendrick, J.B. and K.F. Baker. 1942. Bacterial blight of garden stocks and its control by hot water seed treatment. Cali Agr. Exp. Sta. Bull. *665*: 1–23.

Kennan, R.M., S.J. Billington and J.I. Rood. 1998. Electroporation-mediated transformation of the ovine footrot pathogen *Dichelobacter nodosus*. FEMS Microbiol. Lett. *169*: 383–389.

Kennan, R.M., O.P. Dhungyel, R.J. Whittington, J.R. Egerton and J.I. Rood. 2001. The type IV fimbrial subunit gene (*fimA*) of *Dichelobacter nodosus* is essential for virulence, protease secretion and natural competence. J. Bacteriol. *183*: 4451–4458.

Kenne, L., B. Lindberg, K. Petersson, E. Katzenellenbogen and E. Romanowska. 1980. Structural studies of the O-specific side-chains of *Shigella sonnei* phase I lipopolysaccharide. Carbohydr. Res. *78*: 119–126.

Kennedy, B.W. and T.H. King. 1962. Angular leaf spot of strawberry caused by *Xanthomonas fragariae* sp. nov. Phytopathology *52*: 873–875.

Kennedy, C. and D. Dean. 1992. The *nifU*, *nifS* and *nifV* gene products are required for activity of all three nitrogenases of *Azotobacter vinelandii*. Mol. Gen. Genet. *231*: 494–498.

Kennedy, C., R. Gamal, R. Humphrey, J. Ramos, K. Brigle and D. Dean. 1986. The *nifH*, *nifM* and *nifN* genes of *Azotobacter vinelandii* - characterization by Tn5 mutagenesis and isolation from pLAFR1 gene banks. Mol. Gen. Genet. *205*: 318–325.

Kennedy, C.A., M.B. Goetz and G.E. Mathisen. 1990. Postoperative pancreatic abscess due to *Plesiomonas shigelloides*. Rev. Infect. Dis. *12*: 813–816.

Kennedy, C. and R.L. Robson. 1983. Activation of *nif* gene expression in *Azotobacter* by the *nifA* gene product of *Klebsiella pneumoniae*. Nature (Lond.) *301*: 626–628.

Kennedy, C. and A. Toukdarian. 1987. Genetics of azotobacters: applications to nitrogen fixation and related aspects of metabolism. Annu. Rev. Microbiol. *41*: 227–248.

Kennedy, P.C., E.L. Biberstein, J.A. Howarth, L.M. Frazier and D.L. Dungworth. 1960. Infectious meningoencephalitis in cattle, caused by a *Haemophilus* like organism. Am. J. Vet. Res. *21*: 403–409.

Kennedy, P.C., L.M. Frazier, G.H. Theilen and E.L. Biberstein. 1958. A septicemic disease of lambs caused by *Haemophilus agni* (new species). Am. J. Vet. Res. *19*: 645–654.

Kenny, B. and B.B. Finlay. 1995. Protein secretion by enteropathogenic *Escherichia coli* is essential for transducing signals to epithelial cells. Proc. Natl. Acad. Sci. U.S.A. *92*: 7991–7995.

Kerr, G.R., K.J. Forbes, A. Williams and T.H. Pennington. 1993. An analysis of the diversity of *Haemophilus parainfluenzae* in the adult human respiratory tract by genomic DNA fingerprinting. Epidemiol. Infect. *111*: 89–98.

Kersters, K. and J. De Ley. 1968. The occurrence of the Entner–Doudoroff pathway in bacteria. Antonie Van Leeuwenhoek J. Serol. Microbiol. *34*: 393–408.

Kersters, K., W. Ludwig, M. Vancanneyt, P. De Vos, M. Gillis and K.-H. Schleifer. 1996. Recent changes in the classification of the pseudomonads: an overview. Syst. Appl. Microbiol. *19*: 465–477.

Kersters, K., B. Pot, D. Dewettinck, U. Torck, M. Vancanneyt, L. Vauterin and P. Vandamme. 1994. Identification and typing of bacteria by protein electrophoresis. *In* Priest, Ramos-Cormenzana and Tindall (Editors), Bacterial Diversity and Systematics, Plenum Press, New York. 51–66.

Kessler, B. and N.J. Palleroni. 2000. Taxonomic implications of synthesis of poly-β-hydroxybutyrate and other poly-β-hydroxyalkanoates by aerobic pseudomonads. Int. J. Syst. Evol. Bacteriol. *50*: 711–713.

Kessler, H.H., F.F. Reinthaler, A. Pschaid, K. Pierer, B. Kleinhappl, E. Eber and E. Marth. 1993. Rapid detection of *Legionella* species in bronchoalveolar lavage fluids with the enviroAmp *Legionella* PCR amplification and detection kit. J. Clin. Microbiol. *31*: 3325–3328.

Ketchum, P.A. and W.J. Payne. 1992. Purification of two nitrate reductases

from *Xanthomonas maltophilia* grown in aerobic cultures. Appl. Environ. Microbiol. *58*: 3586–3592.

Keusch, G.T. and M.L. Bennish. 1998. Shigellosis. *In* Evans and Brachman (Editors), Bacterial Infections of Humans. Epidemiology and Control, 3rd Ed., Plenum Medical Book Co., New York. pp. 631–656.

Khairat, O. 1940. Endocarditis due to a new species of *Haemophilus*. J. Pathol. Bacteriol. *50*: 497–505.

Khakhria, R., D. Duck and H. Lior. 1990. Extended phage-typing scheme for *Escherichia coli* O157:H7. Epidemiol. Infect. *105*: 511–520.

Khalil-Rizvi, S., S.I. Toth, D. Van der Helm, I. Vidavsky and M.L. Gross. 1997. Structures and characteristics of novel siderophores from plant deleterious *Pseudomonas fluorescens* A225 and *Pseudomonas putida* ATCC 39167. Biochemistry *36*: 4163–4171.

Khalyuzhnaya, M., V.N. Khmelenina, S. Kotelnikova, L. Holmquist, K. Pedersen and Y.A. Trotsenko. 1999. *Methylomonas scandinavica* sp. nov., a new methanotrophic psychrotrophic bacterium isolated from deep igneous rock ground water of Sweden. Syst. Appl. Microbiol. *22*: 565–572.

Khalyuzhnaya, M., V.N. Khmelenina, S. Kotelnikova, L. Holmquist, K. Pedersen and Y.A. Trotsenko. 2000. *In* Validation of the publication of new names and new combinations previously effectively published outside the IJSEM. List No. 50. Int. J. Syst. Evol. Microbiol. *50*: 949–950.

Khan, A.A. and C.E. Cerniglia. 1997. Rapid and sensitive method for the detection of *Aeromonas caviae* and *Aeromonas trota* by polymerase chain reaction. Lett. Appl. Microbiol. *24*: 233–239.

Khan, F.G., A. Rattan, I.A. Khan and A. Kalia. 1996. A preliminary study of fingerprinting of *Pseudomonas aeruginosa* by whole cell protein analysis by SDS-PAGE. Indian J. Med. Res. *104*: 342–348.

Kharat, A.S. and S. Mahadevan. 2000. Analysis of the beta-glucoside utilization (*bgl*) genes of *Shigella sonnei*: evolutionary implications for their maintenance in a cryptic state. Microbiology *146*: 2039–2049.

Khardori, N., L. Elting, E. Wong, B. Schable and G.P. Bodey. 1990. Nosocomial infectious due to *Xanthomonas maltophilia* (*Pseudomonas maltophilia*) in patients with cancer. Rev. Infect. Dis. *12*: 997–1003.

Khardori, N. and V. Fainstein. 1988. *Aeromonas* and *Plesiomonas* as etiological agents. Annu. Rev. Microbiol. *42*: 395–419.

Kharsany, A.B., A.A. Hoosen, P. Kiepiela, R. Kirby and A.W. Sturm. 1999. Phylogenetic analysis of *Calymmatobacterium granulomatis* based on 16S rRNA gene sequences. J. Med. Microbiol. *48*: 841–847.

Kharsany, A.B.M., A.A. Hoosen, P. Kiepiela, T. Naicker and A.W. Sturm. 1996. Culture of *Calymmatobacterium granulomatis*. Clin. Infect. Dis. *22*: 391.

Kharsany, A.B.M., A.A. Hoosen, P. Kiepiela, T. Naicker and A.W. Sturm. 1997. Growth and cultural characteristics of *Calymmatobacterium granulomatis*—the aetiological agent of granuloma inguinale (Donovanosis). J. Med. Microbiol. *46*: 579–585.

Khashe, S. and J.M. Janda. 1996. Iron utilization studies in *Citrobacter* species. FEMS Microbiol. Lett. *137*: 141–146.

Khashe, S. and J.M. Janda. 1998. Biochemical and pathogenic properties of *Shewanella alga* and *Shewanella putrefaciens*. J. Clin. Microbiol. *36*: 783–787.

Khetawat, G., R.K. Bhadra, S. Nandi and J. Das. 1999. Resurgent *Vibrio cholerae* O139: rearrangement of cholera toxin genetic elements and amplification of rrn operon. Infect. Immun. *67*: 148–154.

Khetmalas, M.B., A.K. Bal, L.D. Noble and J.A. Gow. 1996. *Pantoea agglomerans* is the etiological agent for black spot necrosis on beach peas. Can. J. Microbiol. *42*: 1252–1257.

Khmelenina, V.N., M.G. Kalyuzhnaya, N.G. Starostina, N.E. Suzina and Y.A. Trotsenko. 1997. Isolation and characterization of halotolerant alkaliphilic methanotrophic bacteria from Tuva soda lakes. Curr. Microbiol. *35*: 257–261.

Kidambi, S.P., G.W. Sundin, D.A. Palmer, A.M. Chakrabarty and C.L. Bender. 1995. Copper as a signal for alginate synthesis in *Pseudomonas syringae* pathovar *syringae*. Appl. Environ. Microbiol. *61*: 2172–2179.

Kidby, D., P. Sandford, A. Herman and M. Cadmus. 1977. Maintenance procedures for the curtailment of genetic instability: *Xanthomonas campestris* NRRL B-1459. Appl. Environ. Microbiol. *33*: 840–845.

Kielstein, P. and V.J. Rapp-Gabrielson. 1992. Designation of 15 serovars of *Haemophilus parasuis* on the basis of immunodiffusion using heat-stable antigen extracts. J. Clin. Microbiol. *30*: 862–865.

Kiernicka, J., C. Seignez and P. Peringer. 1999. *Escherichia hermanii*—a new bacterial strain for chlorobenzene degradation. Lett. Appl. Microbiol. *28*: 27–30.

Kilian, M. 1974. A rapid method for the differentiation of *Haemophilus* strains. The porphyrin test. Acta Pathol. Microbiol. Scand. Sect. B Microbiol. *82*: 835–842.

Kilian, M. 1976. A taxonomic study of the genus Haemophilus with the proposal of a new species. J. Gen. Microbiol. *93*: 9–62.

Kilian, M. 1978. Rapid identification of *Actinomycetaceae* and related bacteria. J. Clin. Microbiol. *8*: 127–133.

Kilian, M. and P. Bülow. 1976. Rapid diagnosis of *Enterobacteriaceae*. I. Detection of bacterial glycosidases. Acta Pathol. Microbiol. Scand. [B]. *84B*: 245–251.

Kilian, M. and E.L. Biberstein. 1984. Genus II. *Haemophilus. In* Krieg and Holt (Editors), Bergey's Manual of Systematic Bacteriology, 1st Ed., Vol. 1, The Williams & Wilkins Co., Baltimore. pp. 558–569.

Kilian, M. and W. Frederiksen. 1981a. Ecology of *Haemophilus; Pasteurella* and *Actinobacillus. In* Kilian, Frederiksen and Biberstein (Editors), *Haemophilus, Pasteurella* and *Actinobacillus*, Academic Press, London. 11–38.

Kilian, M. and W. Frederiksen. 1981b. Identification tables for the *Haemophilus-Pasteurella-Actinobacillus* group. *In* Kilian, Frederiksen and Biberstein (Editors), *Haemophilus, Pasteurella* and *Actinobacillus*, Academic Press, London. pp. 281–290.

Kilian, M., C.H. Mordhorst, C.R. Dawon and H. Lautrop. 1976. The taxonomy of haemophili isolated from conjunctivae. Acta Pathol. Microbiol. Scand. Sect. B Microbiol. *84*: 132–138.

Kilian, M., J. Nicolet and E.L. Biberstein. 1978. Biochemical and serological characterization of *Haemophilus pleuropneumoniae* (Matthews and Pattison 1961) Shope 1964 and proposal of a neotype strain. Int. J. Syst. Bacteriol. *28*: 20–26.

Kilian, M., J. Reinholdt, H. Lomholt, K. Poulsen and E.V. Frandsen. 1996. Biological significance of IgA1 proteases in bacterial colonization and pathogenesis: critical evaluation of experimental evidence. APMIS *104*: 321–338.

Kilian, M. and C.R. Schiøtt. 1975. Haemophili and related bacteria in the human oral cavity. Arch. Oral Biol. *20*: 791–796.

Kilian, M., I. Sorensen and W. Frederiksen. 1979. Biochemical characteristics of 130 recent isolates from *Haemophilus influenzae* meningitis. J. Clin. Microbiol. *9*: 409–412.

Kilian, M. and J. Theilade. 1975. Cell wall ultrastructure of strains of *Haemophilus ducreyi* and *Haemophilus piscium*. Int. J. Syst. Bacteriol. *25*: 351–356.

Kilian, M. and J. Theilade. 1978. Amended description of *Haemophilus segnis* Kilian 1977. Int. J. Syst. Bacteriol. *28*: 411–415.

Kilpatrick, M.E., J. Escamilla, A.L. Bourgeois, H.J. Adkins and R.C. Rockhill. 1987. Overview of four United States Navy overseas research studies on *Aeromonas*. Experientia (Basel) *43*: 365–366.

Kilpper-Bälz, R. 1991. DNA-rRNA hybridization. *In* Stackebrandt and Goodfellow (Editors), Nucleic Acid Techniques in Bacterial Systematics, John Wiley & Sons, Chichester. 45–68.

Kim, B.H. 1976. Studies on *Actinobacillus equuli*, Thesis, University of Edinburgh.

Kim, B.H., J.E. Phillips and J.G. Atherton. 1976. *Actinobacillus suis* in the horse. Vet. Rec. *98*: 239.

Kim, E. and T. Aoki. 1993. Drug-resistance and broad geographical-distribution of identical R-plasmids of *Pasteurella piscicida* isolated from cultured yellowtail in Japan. Microbiol. Immunol. *37*: 103–109.

Kim, H.J., Y. Tamanoue, G.H. Jeohn, A. Iwamatsu, A. Yokota, Y.T. Kim, T. Takahashi and K. Takahashi. 1997a. Purification and characterization of an extracellular metalloprotease from *Pseudomonas fluorescens*. J. Biochem. *121*: 82–88.

Kim, J.H., R.A. Cooper, K.E. Welty-Wolf, L.J. Harrell, P. Zwadyk and M.E. Klotman. 1989. *Pseudomonas putrefaciens* bacteremia. Rev. Infect. Dis. *11*: 97–104.

Kim, J.F., Z.M. Wei and S.V. Beer. 1997b. The *hrpA* and *hrpC* operons of *Erwinia amylovora* encode components of a type III pathway that secretes harpin. J. Bacteriol. *179*: 1690–1697.

Kim, K.Y., D. Jordan and H.B. Krishnan. 1998. Expression of genes from *Rahnella aquaticus* that are necessary for mineral phosphate solubilization in *Escherichia coli*. FEMS Microbiol. Lett. *159*: 121–127.

Kim, W.S., L. Gardan, S.L. Rhim and K. Geider. 1999. *Erwinia pyrifoliae* sp. nov., a novel pathogen that affects Asian pear trees (*Pyrus pyrifolia* Nakai). Int. J. Syst. Bacteriol. *49*: 899–906.

Kim, Y.S. and E.J. Kim. 1994. A plasmid responsible for malonate assimilation in *Pseudomonas fluorescens*. Plasmid *32*: 219–221.

Kimura, B., S. Hokimoto, H. Takahashi and T. Fujii. 2000. *Photobacterium histaminum* Okuzumi et al. 1994 is a later subjective synonym of *Photobacterium damselae* subsp. *damselae* (Love et al. 1981) Smith et al. 1991. Int. J. Syst. Evol. Microbiol. *50*: 1339–1342.

Kimura, T. 1969. A new subspecies of *Aeromonas salmonicida* as an etiological agent of furnculosis on "Sakuramasu" (*Oncorhynchus masou*) and Pink Salmon (*O. gorbuscha*) rearing for maturity. Part 1. On the morphological and physiological properties. Part 2. On the serological properties. Fish Pathol. (Tokyo). *3*: 45–52.

Kimura, T., I. Sugahara, K. Hanai and T. Asahi. 1995. Purification and characterization of a new γ-glutamylmethylamide-dissimilating enzyme system from *Methylophaga* sp. AA-30. Biosci. Biotechnol. Biochem. *59*: 648–655.

Kimura, T., I. Sugahara and K. Hayashi. 1990a. Use of short chain amines and amino acids as sole sources of nitrogen in a marine methylotrophic bacterium, *Methylophaga* sp. AA-30. Agric. Biol. Chem. *54*: 1873–1874.

Kimura, T., I. Sugahara, K. Hayashi, M. Kobayashi and M. Ozeki. 1990b. Primary metabolic pathway of methylamine in *Methylophaga* sp. AA-30. Agric. Biol. Chem. *54*: 2819–2826.

King, B.M. and D.L. Adler. 1964. A previously undescribed group of *Enterobacteriaceae*. Am. J. Clin. Pathol. *41*: 230–232.

King, E.O. and H.W. Tatum. 1962. *Actinobacillus actinomycetemcomitans* and *Haemophilus aphrophilus*. J. Infect. Dis. *111*: 85–94.

King, E.O., W.K. Ward and D.E. Raney. 1954. Two simple media for the demonstration of pyocyanin and fluorescein. J. Lab. Clin. Med. *11*: 301–307.

King, G.M. 1994. Associations of methanotrophs with the roots and rhizomes of aquatic vegetation. Appl. Environ. Microbiol. *60*: 3220–3227.

King, S. and W.I. Metzger. 1968. A new plating medium for the isolation of enteric pathogens. Appl. Microbiol. *16*: 577–578.

Kingsbury, D.T. and E. Weiss. 1968. Deoxyribonucleic acid homology between species of the genus *Chlamydia*. J. Bacteriol. *96*: 1421–1423.

Kingsland, R.C. and D.A. Guss. 1995. *Actinobacillus ureae* meningitis: case report and review of the literature. J. Emerg. Med. *13*: 623–627.

Kingsley, M.T., D.W. Gabriel, G.C. Marlow and P.D. Roberts. 1993. The *opsX* locus of *Xanthomonas campestris* affects host range and biosynthesis of lipopolysaccharide and extracellular polysaccharide. J. Bacteriol. *175*: 5839–5850.

Kippax, P.W. 1957. A study of *Proteus* infections in a male urological ward. J. Clin. Pathol. *10*: 211–214.

Kiprianova, E.A., G.F. Levanova, E.V. Novova, V.V. Smirnov and A.D. Garagulia. 1985. Taxonomic study of *Pseudomonas aurantiaca* Nakhimovskaya, 1948 and the proposal of a neotype strain of this species. Mikrobiologiia *54*: 434–440.

Kirakosyan, A.V. and Z.S. Melkonyan. 1964. New *Azotobacter* agile varieties from the soils of ARMSSR (R). Dokl. Akad. Nauk. Armyan. S.S.R. *17*: 33–42.

Kirby, B.D., K.M. Snyder, R.D. Meyer and S.M. Finegold. 1978. Legionnaires' disease: clinical features of 24 cases. Ann. Intern. Med. *89*: 297–309.

Kirov, S., M. 1993. The public health significance of *Aeromonas* spp. in foods. Int. J. Food Microbiol. *20*: 179–198.

Kirov, S.M., M.J. Anderson and T.A. McMeekin. 1990. A note on *Aeromonas* spp. from chickens as possible food-borne pathogens. J. Appl. Bacteriol. *68*: 327–334.

Kirov, S.M., J.A. Hudson, L.J. Hayward and S.J. Mott. 1994. Distribution of *Aeromonas hydrophila* hybridization groups and their virulence properties in Australasian clinical and environmental strains. Lett. Appl. Microbiol. *18*: 71–73.

Kirov, S.M., L.A. O'Donovan and K. Sanderson. 1999. Functional characterization of type IV pili expressed on diarrhea-associated isolates of *Aeromonas* species. Infect. Immun. *67*: 5447–5454.

Kirov, S.M. and K. Sanderson. 1996. Characterization of a type IV bundle-forming pilus (SFP) from a gastroenteritis-associated strain of *Aeromonas veronii* biovar sobria. Microb. Pathog. *21*: 23–34.

Kita, K., N. Hiraoka, A. Oshima, S. Kadonishi and A. Obayashi. 1985. *Acd*III, a new restriction endonuclease from *Acinetobacter calcoaceticus*. Nucleic Acids Res. *13*: 8685–8694.

Kita-Tsukamoto, K., H. Oyaizu, K. Nanba and U. Simidu. 1993. Phylogenetic relationships of marine bacteria, mainly members of the family *Vibrionaceae*, determined on the basis of 16S rRNA sequences. Int. J. Syst. Bacteriol. *43*: 8–19.

Kitch, T.T., M.R. Jacobs and P.C. Appelbaum. 1994. Evaluation of RapID onE system for identification of 379 strains in the family *Enterobacteriaceae* and oxidase-negative, gram-negative nonfermenters. J. Clin. Microbiol. *32*: 931–934.

Klapholz, A., K.D. Lessnau, B. Huang, W. Talavera and J.F. Boyle. 1994. *Hafnia alvei*. Respiratory tract isolates in a community hospital over a three-year period and a literature review. Chest *105*: 1098–1100.

Klas, Z. 1936. Über den Formenkreis von *Beggiatoa mirabilis*. Arch. Mikrobiol. *8*: 312–320.

Kleeberger, A., H. Castorph and W. Klingmüller. 1983. The rhizosphere microflora of wheat and barley with special reference to gram-negative bacteria. Arch. Microbiol. *136*: 306–311.

Klein, D. and G.H. Luginbuhl. 1979. Simplified media for growth of *Haemophilus influenzae* from clinical and normal flora sources. J. Gen. Microbiol. *113*: 409–411.

Klein, E. 1889. Ueber epidemische Krankheit der Hühner; verursacht durch einen Bacillus - *Bacillus gallinarum*. Zentrabl. Bakteriol. Parasitenk. Infektionskr. Hyg. Abt. I Orig. *5*: 689–693.

Klein, G.C. 1980. Cross-reaction to *Legionella pneumophila* antigen in sera with elevated titers to *Pseudomonas pseudomallei*. J. Clin. Microbiol. *11*: 27–29.

Klein, J.R. 1940. The oxidation of 1(–) aspartic and 1(+) glutamic acids by *Haemophilus parainfluenzae*. Note on the preparation of pyridine nucleotides from baker's yeast by the method of Warburg and Christian. J. Biol. Chem. *134*: 43–57.

Klein, M.G. and T.A. Jackson. 1992. Bacterial diseases of scarabs. *In* Jackson and Glare (Editors), Use of Pathogens in Scarab Pest Management, Intercept Ltd., Andover. pp. 43–61.

Klein, M.G. and H.K. Kaya. 1995. *Bacillus* and *Serratia* species for Scarab control. Mem. Inst. Oswaldo Cruz *90*: 87–95.

Kline, M.W. 1988. *Citrobacter meningitis* and brain abscess in infancy: epidemiology, pathogenesis, and treatment. J. Pediatr. *113*: 430–434.

Kline, M.W., E.O. Mason, Jr. and S.L. Kaplan. 1988. Characterization of *Citrobacter diversus* strains causing neonatal meningitis. J. Infect. Dis. *157*: 101–105.

Klinger, R. 1912. Untersuchungen über menschliche Aktinomykose. Zentralbl. Bakteriol. Parasitenkd. Infektionskr. Hyg. Abt. 1, Orig. *62*: 191–200.

Kloepper, J.W., J. Leong, M. Teintze and M.N. Schroth. 1980. Enhanced plant-growth by siderophores produced by plant growth-promoting rhizobacteria. Nature (Lond.) *286*: 885–886.

Klontz, K.C., S. Lieb, M. Schreiber, H.T. Janowski, L.M. Baldy and R.A. Gunn. 1988. Syndromes of *Vibrio vulnificus* infections. Clinical and epidemiologic features in Florida cases, 1981–1987. Ann. Intern. Med. *109*: 318–323.

Kloos, D.U., A.A. Dimarco, D.A. Elsemore, K.N. Timmis and L.N. Ornston. 1995. Distance between allcles as a determinant of linkage in natural transformation of *Acinetobacter calcoaceticus*. J. Bacteriol. *177*: 6015–6017.

Kluyver, A.J. 1956. *Pseudomonas aureofaciens* nov. spec. and its pigments. J. Bacteriol. *72*: 406–411.

Kluyver, A.J. and M.T. van den Bout. 1936. Notiz über *Azotobacter agilis* Beijerinck. Arch. Microbiol. *7*: 261–263.

Kluyver, A.J. and C.B. van Niel. 1936. Prospects for a natural system of classification of bacteria. Zentbl. Bakteriol. Parasitenkd. Infektkrankh. Hyg. Abt. II *94*: 369–403.

Knapp, W. 1968. Seroiogische krenzreaktionen zwischen *Pasteurella pseudotuberculosis* (syn. *Yersinia pseudotuberculosis*), *Escherichia coli* and *Enterobacter cloacae*. Prog. Immunobiol. Standard. *2*: 179–186.

Knapp, W. and G. Lebek. 1967. Übertragung der infektiösen resistenz auf Pasteurellen. Pathol. Microbiol. *30*: 103–121.

Knight, V. and R. Blakemore. 1998. Reduction of diverse electron acceptors by *Aeromonas hydrophila*. Arch. Microbiol. *169*: 239–248.

Knirel, Y.A., G. Widmalm, S.N. Senchenkova, P.E. Jansson and A. Weintraub. 1997. Structural studies on the short-chain lipopolysaccharide of *Vibrio cholerae* O139 Bengal. Eur. J. Biochem. *247*: 402–410.

Knösel, D. 1961. Eine an Kohl blattfleckenerzeugende Varietas von *Xanthomonas carnpestris* (Pammel) Dowson. Z. Pflanzenkr. Pflanzenpathol. Pflanzenschutz. *68*: 1–6.

Knosp, O., M. Vontigerstrom and W.J. Page. 1984. Siderophore mediated uptake of iron in *Azotobacter vinelandii*. J. Bacteriol. *159*: 341–347.

Knudson, G.B. and P. Mikesell. 1980. A plasmid in *Legionella pneumophila*. Infect. Immunol. *29*: 1092–1095.

Knutton, S., T. Baldwin, P.H. Williams and A.S. McNeish. 1989. Actin accumulation at sites of bacterial adhesion to tissue culture cells: basis of a new diagnostic test for enteropathogenic and enterohemorrhagic *Escherichia coli*. Infect. Immun. *57*: 1290–1298.

Knutton, S., D.R. Lloyd and A.S. McNeish. 1987. Adhesion of enteropathogenic *Escherichia coli* to human intestinal enterocytes and cultured human intestinal mucosa. Infect. Immun. *55*: 69–77.

Knutton, S., A.D. Phillips, H.R. Smith, R.J. Gross, R. Shaw, P. Watson and E. Price. 1991. Screening for enteropathogenic *Escherichia coli* in infants with diarrhea by the fluorescent-actin staining test. Infect. Immun. *59*: 365–371.

Ko, W.C., K.W. Yu, C.Y. Liu, C.T. Huang, H.S. Leu and Y.C. Chuang. 1996. Increasing antibiotic resistance in clinical isolates of *Aeromonas* strains in Taiwan. Antimicrob. Agents Chemother. *40*: 1260–1262.

Kobayashi, D.Y., M. Guglielmoni and B.B. Clarke. 1995. Isolation of the chitinolytic bacteria *Xanthomonas maltophilia* and *Serratia marcescens* as biological control agents for summer patch disease of turfgrass. Soil Biol. Biochem. *27*: 1479–1487.

Koblavi, S., F. Grimont and P.A. Grimont. 1990. Clonal diversity of *Vibrio cholerae* O1 evidenced by rRNA gene restriction patterns. Res. Microbiol. *141*: 645–657.

Kocka, F.E., S. Srinivasan, M. Mowjood and H.S. Kantor. 1980. Nosocomial multiply resistant *Providencia stuartii*: a long-term outbreak with multiple biotypes and serotypes at one hospital. J. Clin. Microbiol. *11*: 167–169.

Kocur, M. 1984. Genus *Paracoccus* (Davis 1969). *In* Kreig and Holt (Editors), Bergey's Manual of Systematic Bacteriology, Vol. 1, The Williams & Wilkins Co., Baltimore. 399–402.

Kodaka, H., A.Y. Armfield, G.L. Lombard and V.R. Dowell, Jr.. 1982. Practical procedure for demonstrating bacterial flagella. J. Clin. Microbiol. *16*: 948–952.

Kodama, K., N. Kimura and K. Komagata. 1985. Two new species of *Pseudomonas - Pseudomonas oryzihabitans* isolated from rice paddy and clinical specimens and *Pseudomonas luteola* isolated from clinical specimens. Int. J. Syst. Bacteriol. *35*: 467–474.

Kodjo, A., Y. Richard and T. Tønjum. 1997. *Moraxella boevrei* sp. nov., a new *Moraxella* species found in goats. Int. J. Syst. Bacteriol. *47*: 115–121.

Kodjo, A., T. Tønjum, Y. Richard and K. Bøvre. 1995. *Moraxella caprae* sp. nov., a new member of the classical moraxellae with very close affinity to *Moraxella bovis*. Int. J. Syst. Bacteriol. *45*: 467–471.

Kodjo, A., L. Villard, C. Bizet, J.L. Martel, R. Sanchis, E. Borges, D. Gauthier, F. Maurin and Y. Richard. 1999. Pulsed-field gel electrophoresis is more efficient than ribotyping and random amplified polymorphic DNA analysis in discrimination of *Pasteurella haemolytica* strains. J. Clin. Microbiol. *37*: 380–385.

Koehler, J.M. and L.R. Ashdown. 1993. *In vitro* susceptibilities of tropical strains of *Aeromonas* species from Queensland, Australia, to 22 antimicrobial agents. Antimicrob. Agents Chemother. *37*: 905–907.

Koehm, B. and G. Eggers-Schumacher. 1995. Rapid and simple detection of plant pathogens by reverse passive haemagglutination (RPH): detection of the bacteria *Clavibacter michiganensis* (ssp. *sepedonicus*) and *Erwinia carotovora* (ssp. *carotovora*) with RPH in potato tubers. Z. Pflanzenkr. Pflanzenschutz. *102*: 58–62.

Koh, S.-C., J.P. Bowman and G.S. Sayler. 1993. Soluble methane monooxygenase production and trichloroethylene degradation by a type I methanotroph, *Methylomonas methanica* 68-1. Appl. Environ. Microbiol. *59*: 960–967.

Kohler, R.B., W.C.J. Winn and L.J. Wheat. 1984. Onset and duration of urinary antigen excretion in Legionnaires' disease. J. Clin. Microbiol. *20*: 605–607.

Kohlert, R. 1968. Untersuchungen zür Ätiologie der Eileiterentzündung beim Huhn. Monatsh Veterinaermed. *23*: 392–395.

Kok, R.G., D.A. D'Argenio and L.N. Ornston. 1997. Combining localized PCR mutagenesis and natural transformation in direct genetic analysis of a transcriptional regulator gene, *pobR*. J. Bacteriol. *179*: 4270–4276.

Kok, R.G., C.B. Nudel, R.H. Gonzalez, I.M. Nugteren-Roodzant and K.J. Hellingwerf. 1996. Physiological factors affecting production of extracellular lipase (LipA) in *Acinetobacter calcoaceticus* BD413: fatty acid repression of *lipA* expression and degradation of LipA. J. Bacteriol. *178*: 6025–6035.

Kok, R.G., J.J. van Thor, I.M. Nugteren-Roodzant, M.B. Bouwer, M.R. Egmond, C.B. Nudel, B. Vosman and K.J. Hellingwerf. 1995a. Characterization of the extracellular lipase, *lipA*, of *Acinetobacter calcoaceticus* BD413 and sequence analysis of the cloned structural gene. Mol. Microbiol. *15*: 803–818.

Kok, R.G., J.J. van Thor, I.M. Nugteren-Roodzant, B. Vosman and K.J. Hellingwerf. 1995b. Characterization of lipase-deficient mutants of *Acinetobacter calcoaceticus* BD413: identification of a periplasmic lipase chaperone essential for the production of extracellular lipase. J. Bacteriol. *177*: 3295–3307.

Kokai-Kun, J.F., A.R. Melton-Celsa and A.D. O'Brien. 2000. Elastase in intestinal mucus enhances the cytotoxicity of Shiga toxin type 2d. J. Biol. Chem. *275*: 3713–3721.

Kokka, R.P., J.M. Janda, L.S. Oshiro, M. Altwegg, T. Shimada, R. Sakazaki and D.J. Brenner. 1991a. Biochemical and genetic characterization of autoagglutinating phenotypes of *Aeromonas* spp. associated with invasive and noninvasive disease. J. Infect. Dis. *163*: 890–894.

Kokka, R.P., D. Lindquist, S.L. Abbott and J.M. Janda. 1992. Structural and pathogenic properties of *Aeromonas schubertii*. Infect. Immun. *60*: 2075–2082.

Kokka, R.P., N.A. Vedros and J.M. Janda. 1990. Electrophoretic analysis of the surface components of autoagglutinating surface array protein-positive and surface array protein-negative *Aeromonas hydrophila* and *Aeromonas sobria*. J. Clin. Microbiol. *28*: 2240–2247.

Kokka, R.P., N.A. Vedros and J.M. Janda. 1991b. Characterization of classic and atypical serogroup O:11 *Aeromonas*: evidence that the surface array protein is not directly involved in mouse pathogenicity. Microb. Pathog. *10*: 71–80.

Kokoskova, B. 1992. The appearance of *Erwinia rhaponticia* on hyacinth in Czechoslovakia. Ochr. Rostl. *28*: 146–148.

Kolberg, J., E.A. Hoiby and E. Jantzen. 1997. Detection of the phosphorylcholine epitope in Streptococci, *Haemophilus* and pathogenic *Neisseria* by immunoblotting. Microb. Pathog. *22*: 321–329.

Kolibachuk, D. and E.P. Greenberg. 1993. The *Vibrio fischeri* luminescence gene activator LuxR is a membrane-associated protein. J. Bacteriol. *175*: 7307–7312.

Kolkwitz, R. 1955. Über die schwefelbakterie *Thioploca ingrica* Wislouch. Ber. Dtsch. Bot. Ges. *68*: 374–380.

Kolyvas, E., S. Sorger, M.I. Marks and C.H. Pai. 1978. *Pasteurella ureae* meningoencephalitis. J. Pediatr. *92*: 81–82.

Komagata, K., E. Yabuuchi, Y. Tamagawa and A. Ohyama. 1974. *Pseudomonas melanogena* Iizuka and Komagata 1963, a later subjective synonym of *Pseudomonas maltophilia* Hugh and Ryschenkow 1960. Int. J. Syst. Bacteriol. *24*: 242–247.

Kondo, K., A. Takade and K. Amako. 1994. Morphology of the viable but nonculturable *Vibrio cholerae* as determined by the freeze fixation technique. FEMS Microbiol. Lett. *123*: 179–184.

Kondo, S., Y. Haishima and K. Hisatsune. 1990. Analysis of the 2-keto-3-deoxyoctonate (KDO) region of lipopolysaccharides isolated from non-01 *Vibrio cholerae* 05R. FEMS Microbiol. Lett. *56*: 155–158.

Kondrat'eva, E.N., R.N. Ivanovsky and E.N. Krasil'nikova. 1981. Light and dark metabolism in purple sulfur bacteria. Sov. Sci. Rev. Amsterdam: 325–364.

Kondrat'eva, E.N., V.G. Zhukov, R.N. Ivanovsky, U.P. Petushkova and E.Z. Monosov. 1976. The capacity of phototrophic sulfur bacterium *Thiocapsa roseopersicina* for chemosynthesis. Arch. Microbiol. *108*: 287–292.

Kondrat'eva, T.F. and G.I. Karavaiko. 1997. Genomic variability in *Thiobacillus ferrooxidans* and its role in biohydrometallurgical processes. Mikrobiologiya *66*: 735–743.

Koning, H.C. 1938. Bacterial canker of the poplar. Chron. Bot. *4*: 11–12.

Konowalchuk, J., N. Dickie, S. Stavric and J.I. Speirs. 1978. Comparative studies of five heat-labile toxic products of *Escherichia coli*. Infect. Immun. *22*: 644–648.

Konowalchuk, J., J.I. Speirs and S. Stavric. 1977. Vero response to a cytotoxin of *Escherichia coli*. Infect. Immun. *18*: 775–779.

Kontrohr, T. 1977. The identification of 2-amino-2-deoxy-L-altruronic acid as a constituent of *Shigella sonnei* phase I lipopolysaccharide. Carbohydr. Res. *58*: 498–500.

Koops, H.P., B. Böttcher, U.C. Möller, A. Pommerening-Röser and G. Stehr. 1990. Description of a new species of *Nitrosococcus*. Arch. Microbiol. *154*: 244–248.

Koops, H.P., H. Harms and H. Wehrmann. 1976. Isolation of a moderate halophilic ammonia-oxidizing bacterium, *Nitrosococcus mobilis* nov. sp. Arch. Microbiol. *107*: 277–282.

Kopecko, D.J., O. Washington and S.B. Formal. 1980. Genetic and physical evidence for plasmid control of *Shigella sonnei* form I cell surface antigen. Infect. Immun. *29*: 207–214.

Kopf, P.-O. and V. Freitag. 1979. Krankenhausinfektionen mit *Providencia stuartii*. Dtsch. Med. Wochenschr. *104 (Pt. 3)*: 1129–1132.

Köplin, R., W. Arnold, B. Hoette, R. Simon, G. Wang and A. Pühler. 1992. Genetics of xanthan production in *Xanthomonas campestris*: the *xanA* and *xanB* genes are involved in UDP-glucose and GDP-mannose biosynthesis. J. Bacteriol. *174*: 191–199.

Kopp, R., J. Müller and R. Lemme. 1966. Inhibition of *Proteus* by sodium tetradecyl sulphate, β-phenethyl alcohol and *p*-nitrophenylglycerol. Appl. Microbiol. *14*: 872–878.

Koppe, F. 1924. Die Schlammflora der ostholsteinischen Seen und des Bodensees. Arch. Hydrobiol. *14*: 619–672.

Korfhagen, T.R., L. Sutton and G.A. Jacoby. 1978. Classification and physical properties of *Pseudomonas* plasmids. *In* Schlessinger (Editor), Microbiology 1978, American Society for Microbiology, Washington, D.C. 221–224.

Kornberg, H.L. and N.B. Madsen. 1958. The metabolism of C_2 compounds in microorganisms. 3. Synthesis of malate from acetate via the glyoxylate cycle. Biochem. J. *68*: 549–557.

Körner, R.J., A.P. MacGowan and B. Warner. 1992. The isolation of *Plesiomonas shigelloides* in polymicrobial septicaemia originating from the biliary tree. Zentbl. Bakteriol. *277*: 334–339.

Koronakis, V., M. Cross, B. Senior, E. Koronakis and C. Hughes. 1987. The secreted hemolysins of *Proteus mirabilis*, *Proteus vulgaris*, and *Morganella morganii* are genetically related to each other and to the alpha-hemolysin of *Escherichia coli*. J. Bacteriol. *169*: 1509–1515.

Kortstee, G.J., K.J. Appeldoorn, C.F. Bonting, E.W. van Niel and H.W. van Veen. 1994. Biology of polyphosphate-accumulating bacteria involved in enhanced biological phosphorus removal. FEMS Microbiol. Rev. *15*: 137–153.

Kortt, A.A., J.E. Burns and D.J. Stewart. 1983. Detection of the extracellular proteases of *Bacteroides nodosus* in polyacrylamide gels: a rapid method of distinguishing virulent and benign ovine isolates. Res. Vet.Sci. *35*: 171–174.

Kosako, Y. and R. Sakazaki. 1991. Priority of *Yokenella regensburgei* Kosako, Sakazaki, and Yoshizaki 1985 over *Koserella trabulsii* Hickman-Brenner, Huntley-Carter, Brenner, and Farmer 1985. Int. J. Syst. Bacteriol. *41*: 185–187.

Kosako, Y., R. Sakazaki, G.P. Huntleycarter and J.J. Farmer. 1987. *Yokenella regensburgei* and *Koserella trabulsii* are subjective synonyms. Int. J. Syst. Bacteriol. *37*: 127–129.

Kosako, Y., R. Sakazaki and E. Yoshizaki. 1984. *Yokenella regensburgei* gen. nov., sp. nov.: a new genus and species in the family *Enterobacteriaceae*. Japan J. Med. Sci. Biol. *37*: 117–124.

Kosako, Y., R. Sakazaki and E. Yoshizaki. 1985. *In* Validation of the publication of new names and new combinations previously effectively published outside the IJSB. List No. 17. Int. J. Syst. Bacteriol. *35*: 223–225.

Kosako, Y., K. Tamura and K. Miki. 1996. *Enterobacter kobei* sp. nov., a new species of the family *Enterobacteriaceae* resembling *Enterobacter cloacae*. Curr. Microbiol. *33*: 261–265.

Kosako, Y., K. Tamura and K. Miki. 1997. *In* Validation of the publication of new names and new combinations previously effectively published outside the IJSB, List No. 62. Int. J. Syst. Bacteriol. *47*: 915.

Koshi, G. and M.K. Lalitha. 1976. *Edwardsiella tarda* in a variety of human infections. Indian J. Med. Res. *64*: 1753–1759.

Koskela, P. and A. Salminen. 1984. Humoral immunity against *Francisella tularensis* after natural infection. J. Clin. Microbiol. *22*: 973–979.

Koster, M., W. Ovaa, W. Bitter and P. Weisbeek. 1995. Multiple outer-membrane receptors for uptake of ferric pseudobactins in *Pseudomonas putida* Wcs358. Mol. Gen. Genet. *248*: 735–743.

Kotasthane, W.V., A.C. Padhya and M.K. Patel. 1965. Utilization of amino acids as sole source of carbon and nitrogen by some xanthomonads. Indian Phytopathol. *18*: 154–159.

Kotelko, K. 1986. *Proteus mirabilis*: taxonomic position, peculiarities of growth, components of the cell envelope. Curr. Top. Microbiol. Immunol. *129*: 181–215.

Kotloff, K.L., J.P. Winickoff, B. Ivanoff, J.D. Clemens, D.L. Swerdlow, P.J. Sansonetti, G.K. Adak and M.M. Levine. 1999. Global burden of *Shigella infections*: implications for vaccine development and implementation of control strategies. Bull. World Health Organ. *77*: 651–666.

Kotob, S.I., S.L. Coon, E.J. Quintero and R.M. Weiner. 1995. Homogentisic acid is the primary precursor of melanin synthesis in *Vibrio cholerae*, a *Hyphomonas* strain, and *Shewanella colwelliana*. Appl. Environ. Microbiol. *61*: 1620–1622.

Kourany, M., M.A. Vasquez and R. Saenz. 1977. Edwardsiellosis in man and animals in Panama: clinical and epidemiological characteristics. Am. J.Trop. Med. Hyg. *26*: 1183–1190.

Koushik, S.V., B. Sundararaju, R.A. McGraw and R.S. Phillips. 1997. Cloning, sequence, and expression of kynureninase from *Pseudomonas fluorescens*. Arch. Biochem. Biophys. *344*: 301–308.

Kousik, C.S. and D.F. Ritchie. 1996. Race shift in *Xanthomonas campestris* pv. *vesicatoria* within a season in field-grown pepper. Phytopathology *86*: 952–958.

Kovács, N. 1956. Identification of *Pseudomonas pyocyanea* by the oxidase reaction. Nature *178*: 703.

Kowalchuk, G.A., L.A. Gregg-Jolly and L.N. Ornston. 1995. Nucleotide sequences transferred by gene conversion in the bacterium *Acinetobacter calcoaceticus*. Gene *153*: 111–115.

Kowallik, U. and E.G. Pringsheim. 1966. The oxidation of hydrogen sulfide by *Beggiatoa*. Am. J. Bot. *53*: 801–806.

Kozloff, L.M., M.A. Schofield and M. Lute. 1983. Ice nucleating activity of *Pseudomonas syringae* and *Erwinia herbicola*. J. Bacteriol. *153*: 222–231.

Kozubek, A., S. Pietr and A. Czerwonka. 1996. Alkylresorcinols arc abundant lipid components in different strains of *Azotobacter chroococcum* and *Pseudomonas* spp. J. Bacteriol. *178*: 4027–4030.

Krasil'nikov, N.A. 1949. Guide to the bacteria and actinomycetes, Akad. Nauk SSSR, Moscow. 1–830.

Krasil'nikova, E.N. 1975. Enzymes of carbohydrate metabolism in phototrophic bacteria. Mikrobiologiia *44*: 5–10.

Krasil'nikova, E.N. 1981. Assimilation of sulfates by purple sulfur bacteria. Mikrobiologiya *50*: 338–344.

Krasil'nikova, E.N. 1985. Enzymes involved in carbon metabolism in a purple sulfur bacterium *Lamprobacter modestohalophilus*. Mikrobiologiya *53*: 592–594.

Krasil'nikova, E.N. and E.N. Kondrat'eva. 1979. Possible pathways of acetyl-CoA formation by purple bacteria. Mikrobiologiya *48*: 779–884.

Krasil'nikova, E.N., V.G. Zhukov and E.N. Kondrat'eva. 1979. Glycerol metabolism in purple sulfur bacteria. Mikrobiologiya *48*: 586–591.

Krause, B., T.J. Beveridge, C.C. Remsen and K.H. Nealson. 1996. Structure and properties of novel inclusions in *Shewanella putrefaciens*. FEMS Microbiol. Lett. *139*: 63–69.

Krause, T., H.U. Bertschinger, L. Corboz and R. Mutters. 1987. V-factor dependent strains of *Pasteurella multocida* subsp. *multocida*. Zentbl. Bakteriol. Mikrobiol. Hyg. Abt. Orig. A Med. Mikrobiol. Infektkrankh. Parasitol. *266*: 255–260.

Krause, T., I. Kunstyr and R. Mutters. 1989. Characterization of some previously unclassified *Pasteurellaceae* isolated from hamsters. J. Appl. Bacteriol. *67*: 171–175.

Krauss, H. 1989. Clinical aspects and prevention of Q fever in animals. Eur. J. Epidemiol. *5*: 454–455.

Kraut, M.S., J.R. Attenberry, S.M. Finegold and V.L. Sutter. 1972. Detection of *Haemophilus aphrophilus* in the human oral flora with a selective medium. J. Infect. Dis. *126*: 189–192.

Krcmery, V., L. Langsadl, M. Antal and A. Seckarova. 1985. Transferable amikacin and cefamandole resistance: *Pseudomonas maltophilia* and *Acinetobacter* strains as possible reservoirs of R plasmids. J. Hyg. Epidemiol. Microbiol. Immunol. (Prague) *29*: 141–146.

Kreger, A.S., A.W. Bernheimer, L.A. Etkin and L.W. Daniel. 1987. Phospholipase D activity of *Vibrio damsela* cytolysin and its interaction with sheep erythrocytes. Infect. Immun. *55*: 3209–3212.

Krieg, A. 1955a. Licht- und elektronenmikroskopische Untersuchungen zür Pathologie der "Lorscher Erkrankung" von Engerlingen und zür Zytologie der *Rickettsia melolonthae* nov. spec. Z. Naturforsch. *10b*: 34–37.

Krieg, A. 1955b. Untersuchungen zür Wirbeltier-Pathogenität und zum serologischen Nachweis der *Rickettsia melolonthae* im Arthropod-Wirt. Naturwissenschaften *42*: 609–610.

Krieg, A. 1958. Vergleichende taxonomische, morphologische und serologische Untersuchungen an insektenpathogènen Rickettsien. Z. Naturforsch. *13b*: 555–557.

Krieg, A. 1959. On the problem of crystals associated with *Rickettsiella* infections. J. Insect Pathol. *1*: 95.

Krieg, A. 1965. Über eine neue Rickettsie aus Coleopteren, *Rickettsiella tenebrionis* nov. spec. Naturwissenschaften *52*: 144–145.

Krieg, N.R. 1974. The genus *Spirillum*. *In* Buchanan and Gibbons (Editors), Bergey's Manual of Determinative Bacteriology, 8th ed., The Williams & Wilkins Co., Baltimore. 196–207.

Krieg, N.R. 1984a. Aerobic/microaerophilic, motile, helical/vibroid gram-negative bacteria. *In* Krieg and Holt (Editors), Bergey's Manual of Systematic Bacteriology, Vol. 1, The Williams & Wilkins Co., Baltimore. 71–93.

Krieg, N.R. 1984b. Genus *Oceanospirillum*. *In* Krieg and Holt (Editors), Bergey's Manual of Systematic Bacteriology, 1st Ed., Vol. 1, The Williams & Wilkins Co., Baltimore. pp. 104–110.

Krieg, N.R. and P. Gerhardt. 1981. Solid Culture. *In* Gerhardt, Murray, Costilow, Nester, Wood, Krieg and Phillips (Editors), Manual of Methods for General Bacteriology, American Society for Microbiology, Washington, DC. pp. 143–150.

Krieg, N.R. and J.G. Holt (Editors). 1984. Bergey's Manual of Systematic Bacteriology, 1st Ed., Vol. 1, The Williams & Wilkins Co., Baltimore.

Krikler, M.S. 1953. The serology of *Proteus vulgaris*, Thesis, University of London, London.

Krishnapillai, V. and L.S. Baron. 1964. Alterations in the mouse virulence of *Salmonella typhimurium* by genetic recombination. J. Bacteriol. *87*: 593–605.

Kroll, J.S., S. Ely and E.R. Moxon. 1991. Capsular typing of *Haemophilus influenzae* with a DNA probe. Mol. Cell. Probes *5*: 375–379.

Kroll, J.S., E.R. Moxon and B.M. Loynds. 1993. An ancestral mutation enhancing the fitness and increasing the virulence of *Haemophilus influenzae* type b. J. Infect. Dis. *168*: 172–176.

Kroll, J.S., E.R. Moxon and B.M. Loynds. 1994. Natural genetic transfer of a putative virulence-enhancing mutation to *Haemophilus influenzae* type a. J. Infect. Dis. *169*: 676–679.

Krone, W.J.A., G. Koningstein, F.K. Degraaf and B. Oudega. 1985. Plasmid-determined cloacin Df13-susceptibility in *Enterobacter cloacae* and *Klebsiella edwardsii*: Identification of the cloacin-Df13 aerobactin outer-membrane receptor proteins. Antonie Leeuwenhoek J. Microbiol. *51*: 203–218.

Kroppenstedt, R.M. and W. Mannheim. 1989. Lipoquinones in members of the family *Pasteurellaceae*. Int. J. Syst. Bacteriol. *39*: 304–308.

Krotzky, A. and D. Werner. 1987. Nitrogen-fixation in *Pseudomonas stutzeri*. Arch. Microbiol. *147*: 48–57.

Krumwiede, E. and A.G. Kuttner. 1938. A growth-inhibiting substance for the influenza group of organisms in the blood of various animal species. J. Exp. Med. *67*: 429–441.

Kruse, W. 1896. Systematik der streptothricheen and bakterien, Vol. 2, pp. 48–66, 67–96, 185–526.

Kuberski, T., J.M. Papadimitriou and P. Phillips. 1980. Ultrastructure of *Calymmatobacterium granulomatis* in lesions of granuloma inguinale. J. Infect. Dis. *142*: 744–749.

Kudoh, Y.A., A. Kai, H. Obata, J. Kusunoki, C. Monma, M. Shingaki, Y. Yanagawa, S. Yamada, S. Matsushita, T. Ito and K. Ohta. 1994. Epidemiologic surveys on verocytotoxin-producing *Escherichia coli* infections in Japan. *In* Karmali and Goglio (Editors), Recent Advances in Verocytotoxin-Producing *Escherichia coli* infections, Elsevier Science, The Netherlands. pp. 53–56.

Kuenen, J.G. and R.F. Beudeker. 1982. Microbiology of thiobacilli and other sulphur-oxidizing autotrophs, microtrophs and heterotrophs. Philos. Trans. R. Soc. Lond. Philos. Trans. R. Soc. Lond. B. Biol. Sci. *298*: 473–497.

Kuenen, J.G. and L.A. Robertson. 1989. Genus *Thiomicrospira*. *In* Stalcy, Bryant, Pfennig and Holt (Editors), Bergey's Manual of Systematic Bacteriology, 1st Ed., Vol. 3, The Williams & Wilkins Co., Baltimore. pp. 1858–1861.

Kuenen, J.G., L.A. Robertson and O.H. Tuovinen. 1992. The genera *Thiobacillus*, *Thiomicrospira*, and *Thiosphaera*. *In* Balows, Trüper, Dworkin, Harder and Schleifer (Editors), The Prokaryotes. A Handbook on the Biology of Bacteria: Ecophysiology, Isolation, Identification, Applications, 2nd Ed., Vol. III, Springer Verlag, New York. pp. 2638–2657.

Kuenen, J.G. and H. Veldkamp. 1972. *Thiomicrospira pelophila*, gen. n., sp. n., a new obligately chemolithotrophic colourless sulfur bacterium. Antonie Leeuwenhoek *38*: 241–256.

Kuenen, J.G. and H. Veldkamp. 1973. Effects of organic compounds on growth of chemostat cultures of *Thiomicrospira pelophila*, *Thiobacillus thioparus* and *Thiobacillus neapolitanus*. Arch. Microbiol. *94*: 173–190.

Kuhn, I., M.J. Albert, M. Ansaruzzaman, N.A. Bhuiyan, S.A. Alabi, M.S. Islam, P.K.B. Neogi, G. Huys, P. Janssen, K. Kersters and R. Mollby. 1997a. Characterization of *Aeromonas* spp. isolated from humans with diarrhea, from healthy controls, and from surface water in Bangladesh. J. Clin. Microbiol. *35*: 369–373.

Kuhn, I., G. Allestam, G. Huys, P. Janssen, K. Kersters, K. Krovacek and T.A. Stenstrom. 1997b. Diversity, persistence, and virulence of Aeromonas strains isolated from drinking water distribution systems in Sweden. Appl. Environ. Microbiol. *63*: 2708–2715.

Kuhn, R., M.P. Starr, D.A. Kuhn, H. Bauer and H.J. Knackmuss. 1965. Indigoiodine and other pigments related to 3,3'-bipyridyl. Arch. Mikrobiol. *51*: 71–84.

Kühn, W. 1979. Untersuchungen zur Vergärung von Maltose durch *Bacteroides amylophilus* und zu der daran beteiligten Fumarat Reduktase, Thesis, Georg-August University, Göttingen, Germany.

Kuhnert, P., B. HeybergerMeyer, A.P. Burnens, J. Nicolet and J. Frey.

1997. Detection of RTX toxin genes in gram-negative bacteria with a set of specific probes. Appl. Environ. Microbiol. *63*: 2258–2265.

Kuijper, E.J., A.G. Steigerwalt, B.S.C.I.M. Schoenmakers, M.F. Peeters, H.C. Zanen and D.J. Brenner. 1989a. Phenotypic characterization and DNA relatedness in human fecal isolates of *Aeromonas* spp. J. Clin. Microbiol. *27*: 132–138.

Kuijper, E.J., L. van Alphen, E. Leenders and H.C. Zanen. 1989b. Typing of *Aeromonas* strains by DNA restriction endonuclease analysis and polyacrylamide gel electrophoresis of cell envelopes. J. Clin. Microbiol. *27*: 1280–1285.

Kuijper, E.J., A. van Eeden, B. de Wever, R. van Ketel and J. Dankert. 1997. Nonserotypeable *Shigella dysenteriae* isolated from a Dutch patient returning from India. Eur. J. Clin. Microbiol. Infect. Dis. *16*: 553–554.

Kukkonen, M., T. Raunio, R. Virkola, K. Lahteenmaki, P.H. Mäkelä, P. Klemm, S. Clegg and T.K. Korhonen. 1993. Basement membrane carbohydrate as a target for bacterial adhesion: Binding of type 1 fimbriae of *Salmonella enterica* and *Escherichia coli* to laminin. Mol. Microbiol. *7*: 229–237.

Kuklinska, D. and M. Kilian. 1984. Relative proportions of *Haemophilus* species in the throat of healthy children and adults. Eur. J. Clin. Microbiol. *3*: 249–252.

Kulkarni, Y.S., M.K. Patel and S.G. Abhyankar. 1950. A new bacterial leaf spot and stem canker of pigeon pea. Curr. Sci. (Bangalore) *19*: 384.

Kulkarni, Y.S., M.K. Patel and G.W. Dhande. 1951. *Xanthomonas cassiae* a new bacterial disease of *Cassia tora* L. Curr. Sci. (Bangalore) *20*: 47.

Kume, K., A. Sawata, T. Nakai and M. Matsumoto. 1983. Serological classification of *Haemophilus paragallinarum* with a hemagglutinin system. J. Clin. Microbiol. *17*: 958–964.

Kume, K., A. Sawata and Y. Nakase. 1980. Immunologic relationship between Page's and Sawata's serotype strains of *Haemophilus paragallinarum*. Am. J. Vet. Res. *41*: 757–760.

Kume, T. 1962. A case of abortion possibly due to *Hafnia* organism. J. Hokkaido Vet. Assoc. *6*: 1–4.

Kunkle, R.A. and R.B. Rimler. 1998. Early pulmonary lesions in turkeys produced by nonviable *Aspergillus fumigatus* and/or *Pasteurella multocida* lipopolysaccharide. Avian Dis.˙*42*: 770–780.

Kunsman, J.E. and D.R. Caldwell. 1974. Comparison of the sphingolipid content of rumen *Bacteroides* species. Appl. Microbiol. *28*: 1088–1089.

Kuono, K., T. Oki, H. Nomura and A. Ozaki. 1973. Isolation of new methanol-utilizing bacteria and its thiamine requirement for growth. J. Gen. Appl. Microbiol. *19*: 11–21.

Kurazono, H., S. Yamasaki, O. Ratchtrachenchai, G.B. Nair and Y. Takeda. 1996. Analysis of *Vibrio cholerae* O139 Bengal isolated from different geographical areas using macrorestriction DNA analysis. Microbiol. Immunol. *40*: 303–305.

Kurtti, T.J., A.T. Palmer and J.H.J. Oliver. 2002. *Rickettsiella*-like bacteria in *Ixodes woodi* (Acari: Ixodidae). J. Med. Entomol. *39*: 534–540.

Kurz, R.W., W. Graninger, T.P. Egger, H. Pichler and K.H. Tragl. 1988. Failure of treatment of legionella pneumonia with ciprofloxacin. J. Antimicrob. Chemother. *22*: 389–391.

Kusek, J.W. and L.G. Herman. 1980. Typing of *Proteus mirabilis* by bacteriocin production and sensitivity as a possible epidemiological marker. J. Clin. Microbiol. *12*: 112–120.

Kushner, D.J. 1978. Life in high salt and solute concentrations: halophilic bacteria. *In* Kushner (Editor), Microbial Life in Extreme Environments, Academic Press, London. pp. 317–368.

Kusuda, R., T. Itami, M. Munekiyo and H. Nakajima. 1977. Characteristics of *Edwardsiella* sp. from an epidemic in cultured crimson sea beams. Bull. Japan. Soc. Sci. Fish. *43*: 129–134.

Kusuda, R., J. Yokoyama and K. Kawai. 1986. Bacteriological study on cause of mass mortalities in cultured black sea bream fry. Bull. Jpn. Soc. of Sci. Fish. *52*: 1745–1751.

Kützing, F.T. 1849. Species Algarum, I–IV, Leipzig.

Kuzyk, M.A., J.C. Thorton and W.W. Kay. 1996. Antigenic characterization of the salmonid pathogen *Piscirickettsia salmonis*. Infect. Immun. *64*: 5205–5210.

Kvasnikov, E.I., V.V. Stepanyuk, T.M. Klyushnikova, N.S. Serpokrylov, G.A.

Simonova, T.P. Kasatkina and L.P. Panchenko. 1985. A new gram-variable bacterium reducing chromium and having a mixed type of flagellation. Mikrobiologiya *54*: 83–88.

Kwon, S.W., S.-J. Go, H.-W. Kang, J.-C. Ryu and J.-K. Jo. 1997. Phylogenetic analysis of *Erwinia* species based on 16S rRNA gene sequences. Int. J. Syst. Bacteriol. *47*: 1061–1067.

La Fontaine, S. and J.I. Rood. 1996. Physical and genetic map of the chromosome of *Dichelobacter nodosus* strain A198. Gene *184*: 291–298.

la Rivière, J.W.M. and J.G. Kuenen. 1989a. Genus *Thiobacterium* (ex Janke 1924) Nom Rev. *In* Staley, Bryant, Pfennig and Holt (Editors), Bergey's Manual of Systematic Bacteriology, 1st Ed., Vol. 3, The Williams & Wilkins Co., Baltimore. pp. 1838.

la Rivière, J.W.M. and J.G. Kuenen. 1989b. *In* Validation of the publication of new names and new combinations previously effectively published outside the IJSB. List No. 31. Int. J. Syst. Bacteriol. *39*: 495–497.

la Rivière, J.W.M. and K. Schmidt. 1981. Morphologically conspicuous sulfur-oxidizing bacteria. *In* Starr, Stolp, Trüper, Balows and Schlegel (Editors), The Prokaryotes: a Handbook on Habits, Isolation and Identification of Bacteria, Springer-Verlag, Berlin. pp. 1037–1048.

La Rivière, J.W.M. and K. Schmidt. 1989. The Genus *Achromatium*. *In* Staley, Bryant, Pfennig and Holt (Editors), Bergey's Manual of Systematic Bacteriology, Vol. 3, The Williams & Wilkins Co., Baltimore. 2131-2133.

la Riviere, J.W.M. and K. Schmidt. 1992. Morphologically conspicuous sulfur-oxidizing Eubacteria. *In* Balows, Trüper, Dworkin, Harder and Schleifer (Editors), The Prokaryotes. A Handbook on the Biology of Bacteria: Ecophysiology, Isolation, Identification, Applications, Springer-Verlag, New York. pp. 3934–3947.

La Scola, B., H. Lepidi, M. Maurin and D. Raoult. 1998. A guinea pig model for Q fever endocarditis. J. Infect. Dis. *178*: 278–281.

La Scola, B., H. Lepidi and D. Raoult. 1997. Pathologic changes during acute Q fever: Influence of the route of infection and inoculum size in infected guinea pigs. Infect. Immun. *65*: 2443–2447.

Laakso, T., T. Ojanen, I.M. Helander, R. Karjalainen, T.K. Korhonen and K. Haahtela. 1990. Comparison of outer membrane proteins and lipopolysaccharides of *Xanthomonas campestris* pathovars. Proceedings of the VIIth International Conference on Plant Pathogenic Bacteria, Budapest, Hungary. Akadémiai Kiadó. pp. 149–154.

LaBrec, E.H., H. Schneider, T.J. Magnani and S.B. Formal. 1964. Epithelial cell penetration as an essential step in the pathogenesis of bacillary dysentery. J. Bacteriol. *88*: 1503–1518.

Lacey, S.L., S. Mehmet and G.W. Taylor. 1995. Inhibition of *Helicobacter pylori* growth by 4-hydroxy-2-alkyl- quinolines produced by *Pseudomonas aeruginosa*. J. Antimicrob. Chemother. *36*: 827–831.

Lackey, J.B. and E.W. Lackey. 1961. The habitat and description of a new genus of sulphur bacterium. J. Gen. Microbiol. *26*: 29–39.

Lackey, J.B., E.W. Lackey and G.B. Morgan. 1965. Taxonomy and ecology of the sulfur bacteria. Engineering Progress at the University of Florida *19*: 2–23.

Lacy, G.H. and J.V. Leary. 1979. Genetic systems in phytopathogenic bacteria. Annu. Rev. Phytopathol. *17*: 181–202.

Ladha, J.K., W.L. Barraquio and I. Watanabe. 1983. Isolation and identification of nitrogen-fixing *Enterobacter cloacae* and *Klebsiella planticola* associated with rice plants. Can. J. Microbiol. *29*: 1301–1308.

Laemmli, U.K. 1970. Cleavage of structural proteins during the assembly of the head of bacteriophage T4. Nature (Lond.) *227*: 680–685.

Lageveen, R.G., G.W. Huisman, H. Preusting, P. Ketelaar, G. Eggink and B. Witholt. 1988. Formation of polyesters by *Pseudomonas oleovorans* - effect of substrates on formation and composition of poly-(*R*)-3-hydroxyalkanoates and poly-(*R*)-3-hydroxyalkenoates. Appl. Environ. Microbiol. *54*: 2924–2932.

Lahellec, C., C. Meurier, G. Bennejean and M. Catsaras. 1975. A study of 5920 strains of psychrotrophic bacteria isolated from chickens. J. Appl. Bacteriol. *38*: 89–97.

Lai, C.-Y. and P. Baumann. 1992. Genetic analysis of an aphid endosymbiont DNA fragment homologous to the *rnpA-rpmH-dnaA-dnaN-gyrB* region of eubacteria. Gene *113*: 175–181.

Lai, C.-Y., L. Baumann and P. Baumann. 1994. Amplification of *trpEG*: adaptation of *Buchnera aphidicola* to an endosymbiotic association with aphids. Proc. Natl. Acad. Sci. USA. *91*: 3819–3823.

Lai, C.-Y., P. Baumann and N. Moran. 1995. Genetics of the tryptophan biosynthetic pathway of the prokaryotic endosymbiont (*Buchnera*) of the aphid *Schlechtendalia chinensis*. Insect Mol. Biol. *4*: 47–59.

Lai, C.-Y., P. Baumann and N. Moran. 1996. The endosymbiont (*Buchnera* sp.) of the aphid *Diuraphis noxia* contains plasmids consisting of *trpEG* and tandem repeats of *trpEG* pseudogenes. Appl. Environ. Microbiol. *62*: 332–339.

Lai, M., N.J. Panopoulos and S. Shaffer. 1977a. Transmission of R Plasmids among *Xanthomonas* spp. and other plant pathogenic bacteria. Phytopathology *67*: 1044–1050.

Lai, M., S. Shaffer and N.J. Panopoulos. 1977b. Stability of plasmid-borne antibiotic resistance in *Xanthomonas vesicatoria* in infected tomato leaves. Phytopathology *67*: 1527–1530.

Lai, V., L. Wang and P.R. Reeves. 1998. *Escherichia coli* clone Sonnei (*Shigella sonnei*) had a chromosomal O-antigen gene cluster prior to gaining its current plasmid-borne O-antigen genes. J. Bacteriol. *180*: 2983–2986.

Laird, W.J. and D.C. Cavanaugh. 1980. Correlation of autoagglutination and virulence of Yersiniae. J. Clin. Microbiol. *11*: 430–432.

Laivenieks, M., C. Vieille and J.G. Zeikus. 1997. Cloning, sequencing, and overexpression of the *Anaerobiospirillum succiniciproducens* phosphoenolpyruvate carboxykinase (*pckA*) gene. Appl. Environ. Microbiol. *63*: 2273–2280.

Lakso, J.U. and M.P. Starr. 1970. Comparative injuriousness to plants of *Erwinia* spp. and other enterobacteria from plants and animals. J. Appl. Bacteriol. *33*: 692–707.

Lambert, C., J.L. Nicolas, V. Cilia and S. Corre. 1998. *Vibrio pectenicida* sp. nov., a pathogen of scallop (*Pecten maximus*) larvae. Int. J. Syst. Bacteriol. *48*: 481–487.

Lambert, M.A., F.W. Hickman-Brenner, J.J. Farmer and C.W. Moss. 1983. Differentiation of *Vibrionaceae* species by their cellular fatty acid composition. Int. J. Syst. Bacteriol. *33*: 777–792.

Lambert, M.A. and C.W. Moss. 1989. Cellular fatty acid compositions and isoprenoid quinone contents of 23 *Legionella* spp. J. Clin. Microbiol. *27*: 465–473.

Lampis, G., D. Deidda, C. Maullu, S. Petruzzelli, R. Pompei, F. Delle Monache and G. Satta. 1996a. Karalicin, a new biologically active compound from *Pseudomonas fluorescens/putida* I. Production, isolation, physico-chemical properties and structure elucidation. J. Antibiot. *49*: 260–262.

Lampis, G., D. Deidda, C. Maullu, S. Petruzzelli, R. Pompei, F. Delle Monache and G. Satta. 1996b. Karalicin, a new biologically active compound from *Pseudomonas fluorescens/putida* II. Biological properties. J. Antibiot. *49*: 263–266.

Landre, J.P.B., A.A. Gavriel and A.J. Lamb. 1998. False-positive coliform reaction mediated by *Aeromonas* in the Colilert defined substrate technology system. Lett. Appl. Microbiol. *26*: 352–354.

Lane, D.J., A.P. Harrison, Jr., D. Stahl, B. Pace, S.J. Giovannoni, G.J. Olsen and N.R. Pace. 1992. Evolutionary relationships among sulfur- and iron-oxidizing eubacteria. J. Bacteriol. *174*: 269–278.

Lang, G.H. 1990. Coxiellosis (Q fever) in animals. *In* Marrie (Editor), Q fever. The Disease, Vol. 1, CRC Press, Inc., Boca Raton. .

Lang, M.L. 1992. Fatal *Vibrio damsela* bacteremia. Clin. Microbiol. Newsletter. *14*: 166–167.

Lange, E. and D. Knösel. 1970. Zür Bedeutung pektolytischer, cellulolytischer und proteolytischer Enzyme für die Virulenz phytopathogèner Bakterien. Phytopathol. Z. *69*: 315–329.

Langley, J.M., T.J. Marrie, A. Covert, D.M. Waag and J.C. Williams. 1988. Poker players' pneumonia. An urban outbreak of Q fever following exposure to a parturient cat. N. Engl. J. Med. *319*: 354–356.

Lannan, C.N., S.A. Ewing and J.L. Fryer. 1991. A fluorescent antibody test for detection of the rickettsia causing disease in Chilean salmonids. J. Aquat. Anim. Health. *3*: 229–234.

Lannan, C.N. and J.L. Fryer. 1994. Extracellular survival of *Piscirickettsia salmonis*. J. Fish Dis. *17*: 545–548.

Lanoil, B.D., L.M. Ciuffetti and S.J. Giovannoni. 1996. The marine bacterium *Pseudoalteromonas haloplanktis* has a complex genome structure composed of two separate genetic units. Genome Res. *6*: 1160–1169.

Lányi, B. 1956. Serological typing of *Proteus* strains from infantile enteritis and other sources. Acta Microbiol. Acad. Sci. Hung. *3*: 417–428.

Lányi, B. 1970. Serological properties of *Pseudomonas aeruginosa*. II. Type-specific thermolabile (flagellar) antigens. Acta Microbiol. Acad. Sci. Hung. *17*: 35–48.

Lapage, S.P., P.H.A. Sneath, E.F. Lessel Jr., V.B.D. Skerman, H.P.R. See-liger and W.A. Clark (Editors). 1992. International Code of Nomenclature of Bacteria (1990) Revision. Bacteriological Code, American Society for Microbiology, Washington, DC.

Larenas, J.J., C. Astorga, J. Contreras and P. Smith. 1996. *Piscirickettsia salmonis* in ova obtained from rainbow trout (*Oncorhynchus mykiss*) experimentally inoculated. Arch. Med. Vet. *28*: 161–166.

Large, D.T.M. and O.K. Sankaran. 1934. Dysentery among troups in Quetta. Part IID. The non-mannite fermenting group of organisms. J. Roy. Army Corp. *63*: 231–237.

Larkin, J.M. 1980. Isolation of *Thiothrix* in pure culture and observation of a filamentous epiphyte on *Thiothrix*. Curr. Microbiol. *4*: 155–158.

Larkin, J.M. 1989. Genus II. *Thiothrix* Winogradsky 1888. *In* Staley, Bryant, Pfennig and Holt (Editors), Bergey's Manual of Systematic Bacteriology, 1st ed., Vol. 3, The Williams & Wilkins Co., Baltimore. pp. 2098–2101.

Larkin, J., P. Aharon and M.C. Henk. 1994. *Beggiatoa* in microbial mats at hydrocarbon vents in the Gulf of Mexico and Warm Mineral Springs, Florida. Geo-Mar. Lett. *14*: 97–103.

Larkin, J. and M.C. Henk. 1989. Is "hollowness" an adaptation of large prokaryotes to their largeness? Microbios Lett. *42*: 69–72.

Larkin, J.M. and M.C. Henk. 1996. Filamentous sulfide-oxidizing bacteria at hydrocarbon seeps of the Gulf of Mexico. Microsc. Res. Tech. *33*: 23–31.

Larkin, J.M., M.C. Henk and S.D. Burton. 1990. Occurrence of a *Thiothrix* sp. attached to mayfly larvae and presence of a parasitic bacteria in the *Thiothrix* sp. Appl. Environ. Microbiol. *56*: 357–361.

Larkin, J.M. and D.L. Shinabarger. 1983. Characterization of *Thiothrix nivea*. Int. J. Syst. Bacteriol. *33*: 841–846.

Larkin, J.M. and W.R. Strohl. 1983. *Beggiatoa*, *Thiothrix*, and *Thioploca*. Ann. Rev. Microbiol. *37*: 341–367.

Laroche, M. and M. Verhoyen. 1984. Adaptation and application of the ELISA test, indirect method, to the detection of *Erwinia amylovora* (Burrill) Winslow et al. Parasitica (Gembloux). *40*: 197–210.

Larson, C.L., C.B. Philip, W.C. Wicht and L.E. Hughes. 1951. Precipitin reactions with soluble antigens from suspensions of *Pasteurella pestis* or from tissues of animals dead of plague. J. Immunol. *67*: 289–298.

Larson, C.L., W. Wicht and W.L. Jellison. 1955. A new organism resembling *P. tularensis* isolated from water. Public Health Rep. *70*: 253–258.

Larson, E. 1984. A decade of nosocomial *Acinetobacter*. Am. J. Infect. Control. *12*: 14–18.

Larson, T.G. and S.H. Goodgal. 1991. Sequence and transcriptional regulation of *com101A*, a locus required for genetic transformation in *Haemophilus influenzae*. J. Bacteriol. *173*: 4683–4691.

Larsson, P. 1980. O-antigens of *Proteus mirabilis* and *Proteus vulgaris* strains isolated from patients with bacteremia. J. Clin. Microbiol. *12*: 490–492.

Larsson, P. 1984. Serology of *Proteus mirabilis* and *Proteus vulgaris*. *In* Bergan (Editor), Methods in Microbiology, Vol. 14, Academic Press, London. pp. 187–214.

Larsson, P. and S. Olling. 1977. O antigen distribution and sensitivity to the bactericidal effect of normal human serum of *Proteus* strains from clinical specimens. Med. Microbiol. Immunol. *163*: 77–82.

Laskin, A.I. and H.A. Lechevalier. 1981. Guanine-plus-cytosine (GC) composition of the DNA of bacteria, fungi, algae and protozoa. *In* Laskin and Lechevalier (Editors), CRC Handbood of Microbiology: Microbial Composition: Amino Acids, Proteins, and Nucleic Acids, 2nd Ed., Vol. 3, Chemical Rubber Co. Press, Inc., Boca Raton. pp. 559–729.

Lasseur, P. 1913. Contribution à l'étude de *Bacillus lemonnieri*, nov. spec. C. R. Soc. Biol. Paris. *74*: 47–48.

Laufs, R. and P.M. Kaulfers. 1977. Molecular characterization of a plasmid specifying ampicillin resistance and its relationship to other R factors from *Haemophilus influenzae*. J. Gen. Microbiol. *103*: 277–286.

Laufs, R., P.M. Kaulfers and G. Jahn. 1978. Infektiöse Antibiotikaresistenz bei *Haemophilus influenzae*. Dtsch. Med. Wochenschr. *103*: 658–662.

Laufs, R., P.M. Kaulfers, G. Jahn and U. Teschner. 1979. Molecular characterization of a small *Haemophilus influenzae* plasmid specifying β-lactamase and its relationship to R factors from *Neisseria gonorrhoeae*. J. Gen. Microbiol. *111*: 223–231.

Laughlin, T., D. Waag, J. Williams and T.J. Marrie. 1991. Q fever: from deer to dog to man. Lancet *337*: 676–677.

Laurent, F., A. Kotoujansky, G. Labesse and Y. Bertheau. 1993. Characterization and overexpression of the pem gene encoding pectin methylesterase of *Erwinia chrysanthemi* strain 3937. Gene *131*: 17–25.

Lautenschlager, S., H. Willems, C. Jager and G. Baljer. 2000. Sequencing and characterization of the cryptic plasmid QpRS from *Coxiella burnetii*. Plasmid *44*: 85–88.

Lauterborn, R. 1907. Eine neue Gattung der Schwefelbakterien (*Thioploca schmidlei* nov. gen. nov. spec.). Ber. Dtsch. Bot. Ges. *25*: 238–242.

Lautrop, H. 1956. Gelatin liquefying *Klebsiella* strains (*Bacterium oxytoca*). Acta Pathol. Microbiol. Scand. *39*: 375–384.

Lautrop, H. and O. Jessen. 1964. On the distinction between polar monotrichous and lophotrichous flagellation in green fluorescent pseudomonads. Acta Pathol. Microbiol. Scand. *60*: 588–598.

Lautrop, H. and B.W. Lacy. 1960. Laboratory diagnosis of whooping-cough or *Bordetella* infections. Bull. W.H.O. *23*: 15–35.

Lautrop, H., I. Ørskov and K. Gaarslev. 1971. Hydrogensulphide producing variants of *Escherichia coli*. Acta Pathol. Microbiol. Scand. [B] Microbiol. Immunol. *79*: 641–650.

Law, D. and H. Chart. 1998. Enteroaggregative *Escherichia coli*. J. Appl. Microbiol. *84*: 685–697.

Lawani, L.O., L.E. Aririatu, A.J. Goven and A.S. Kester. 1990. Pigment-associated proteins of *Xanthomonas campestris* pv. *juglandis*. Plant Pathol. (Oxf.) *39*: 294–300.

Lawn, A.M., I. Ørskov and F. Ørskov. 1977. Morphological distinction between different H serotypes of *Escherichia coli*. J. Gen. Microbiol. *101*: 111–119.

Lawrence, J.G. and H. Ochman. 1998. Molecular archaeology of the *Escherichia coli* genome. Proc. Natl. Acad. Sci. U.S.A. *95*: 9413–9417.

Lawry, N.H., V. Jani and T.E. Jensen. 1981. Identification of the sulfur inclusion body in *Beggiatoa alba* B18LD by energy -dispersive x-ray microanalysis. Curr. Microbiol. *6*: 71–74.

Lawton, W.D., G.M. Fukiji and M.J. Surgalla. 1960. Studies on the antigens of *Pasteurella pestis* and *Pasteurella pseudotuberculosis*. J. Immunol. *84*: 475–479.

Lawton, W.D., B.C. Morris and T.W. Burrows. 1968a. Gene transfer by conjugation in *P. pseudotuberculosis* and *P. pestis*. Progr. Immunobiol. Standard. *9*: 285–292.

Lawton, W.D., B.C. Morris and T.W. Burrows. 1968b. Gene transfer in strains of *Pasteurella pseudotuberculosis*. J. Gen. Microbiol. *52*: 25–34.

Lawton, W.D. and H.B. Stull. 1971. Chromosome mapping of *Pasteurella pseudotuberculosis* by interrupted mating. J. Bacteriol. *105*: 855–863.

Lax, A.J. and N. Chanter. 1990. Cloning of the toxin gene from *Pasteurella multocida* and its role in atrophic rhinitis. J. Gen. Microbiol. *136*: 81–87.

Layne, P., A.S. Hu, A. Balows and B.R. Davis. 1971. Extrachromosomal nature of hydrogen sulfide production in *Escherichia coli*. J. Bacteriol. *106*: 1029–1030.

Lazlo, V.G., H. Milch and A. Hayna. 1973. Phage typing of *Shigella flexneri* 1. Classification of type phages on the basis of their serological properties, lysogenic ability and lytic activity. Acta Microbiol. Acad. Sci. Hung. *20*: 135–146.

Lazo, G.R. and D.W. Gabriel. 1987. Conservation of plasmid DNA sequences and pathovar identification of strains of *Xanthomonas campestris*. Phytopathology *77*: 448–453.

Le Frock, J.L., A.S. Klainer and K. Zuckerman. 1976. *Edwardsiella tarda* bacteremia. South. Med. J. *69*: 188–190.

Le Goffic, F.L., N. Moreau, S. Siegrist, F.W. Goldstein and J. Acar. 1977. La résistance plasmidique de *Haemophilus* sp. aux antibiotiques aminoglycosidiques: isolément et étude d'une nouvelle phosphotransférase. Ann. Microbiol. (Paris) *128A*: 383–391.

Le Minor, L. 1968. Conversions antigéniques chez les *Salmonella*. Ann. Inst. Pasteur (Paris) *109*: 505–515.

Le Minor, L., J. Buissière and G. Brault. 1979. Intérêt de la recherche de la fermentation du galacturonate pour différencier les *Salmonella* des sousgenres I et III monophasiques des autres *Salmonella* des soulgenres II, III diphasiques, IV, et de *Citrobacter* et *Hafnia alvei*. Ann. Microbiol. Inst. Pasteur (Paris) *130B*: 305–312.

Le Minor, L., A.M. Chalon and M. Véron. 1972. Recherches sur la présence de l'antigéne commun des *Enterobacteriaceae* (antigéne Kunin) chez les *Yersinia*, *Levinea*, *Aeromonas* et *Vibrio*. Ann. Inst. Pasteur. *123*: 761–774.

Le Minor, L., G. Chamoiseau, E. Barbe, C. Charié-Marsaines and L. Egrou. 1969. Dix nouveaux sérotypes de Salmonella isolés au Tchad. Ann. Inst. Pasteur (Paris) *116*: 775–780.

Le Minor, L., M. Chippaux, F. Pichinoty, C. Coynault and M. Piéchaud. 1970a. Méthodes simples permettant de rechercher la tétrathionate-reductase en cultures liquides ou sur colonies isolées. Ann. Inst. Pasteur (Paris) *119*: 733–737.

Le Minor, L. and C. Coynault. 1976. Déterminisme plasmidique du caractère atypique lactose positif de *Enterobacter hafniae* et de *Proteus morganii*. Ann. Inst. Pasteur Microbiol. *127A*: 213–221.

Le Minor, L., C. Coynault and G. Pessoa. 1974. Déterminisme plasmidique du caractère atypique "lactose positif" de souches de *S. typhimurium* et *S. oranienburg* isolées au Brésil lors d'épidémies de 1971à 1973. Ann. Inst. Pasteur Microbiol. *125A*: 261–285.

Le Minor, L., C. Coynault, R. Rohde, B. Rowe and S. Aleksic. 1973. Localisation plasmidique du déterminant génétique du caractère atypique "saccharose +" des *Salmonella*. Ann. Inst. Pasteur Microbiol. *124B*: 295–306.

Le Minor, L. and M.Y. Popoff. 1987. Designation of *Salmonella enterica* sp. nov. nom. rev. as the type and only species of the genus *Salmonella*. Int. J. Syst. Bacteriol. *37*: 465–468.

Le Minor, L., M.Y. Popoff, B. Laurent and D. Hermant. 1986. Individualisation d'une septieme sous-espece de *Salmonella*: *S. choleraesuis* subsp. *indica* subsp. nov. Ann. Inst. Pasteur Microbiol. *137B*: 211–217.

Le Minor, L., M. Popoff, B. Laurent and D. Hermant. 1987. *In* Validation of the publication of new names and new combinations previously effectively published outside the IJSB. List No. 23. Int. J. Syst. Bacteriol. *37*: 179–180.

Le Minor, L. and C. Richard. 1993. Méthodes de laboratoire de laboratoire pour l'identification des entérobacteries, 4, 6, 88, 175–177.

Le Minor, L., R. Rohde and J. Taylor. 1970b. Nomenclature des *Salmonella*. Ann. Inst. Pasteur (Paris) *119*: 206–210.

Le Minor, L., M. Véron and M. Popoff. 1982. Nomenclature des *Salmonella*. Ann. Microbiol. (Paris) *133B*: 245–254.

Le Minor, L., M. Véron and M. Popoff. 1985. *In* Validation of the publication of new names and new combinations previously effectively published outside the IJSB. List No. 18. Int. J. Syst. Bacteriol. *35*: 375–376.

Le Minor, S. and F. Pigache. 1977. Étude antigénique de souches de *Serratia marcescens* isolées en France. I. Antigènes H: individualisation de six nouveaux facteurs H. Ann. Microbiol. *128B*: 207–214.

Le Minor, S. and F. Pigache. 1978. Étude antigénique de souches de *Serratia marcescens* isolées en France. II. Caractérisation des antigènes O et individualisation de 5 nouveaux facteurs, frequence des sérotypes et désignation des nouveaux facteurs H. Ann. Microbiol. *129B*: 407–423.

Le Minor, S. and F. Sauvageot-Pigache. 1981. Nouveaux facteurs antigéniques H (H21–H25) et O (O21) de *Serratia marcescens*. Subdivision des facteurs O5, O10, O16. Ann. Microbiol. *132A*: 239–252.

Leach, J.G. 1964. Observations on cucumber beetles as vectors of cucurbit wilt. Phytopathology *54*: 606–607.

Leach, J.E., M.L. Rhoads, C.M. Vera Cruz, F.F. White, T.W. Mew and H. Leung. 1992. Assessment of genetic diversity and population structure of *Xanthomonas oryzae* pv. *oryzae* with a repetitive DNA element. Appl. Environ. Microbiol. *58*: 2188–2195.

Leadbetter, E.R. 1974a. Family II. *Beggiatoaceae*. *In* Buchanan and Gibbons (Editors), Bergey's Manual of Determinative Bacteriology, 8th Ed., The Williams & Wilkins Co., Baltimore. 112–116.

Leadbetter, E.R. 1974b. Family *Methylomonadaceae*. *In* Buchanan and Gibbons (Editors), Bergey's Manual of Determinative Bacteriology, 8th Ed., The Williams & Wilkins Co., Baltimore. pp. 267-269.

Leahy, J.G., J.M. Jones-Meehan, E.L. Pullias and R.R. Colwell. 1993. Transposon mutagenesis in *Acinetobacter calcoaceticus* RAG-1. J. Bacteriol. *175*: 1838–1840.

Leaves, N.I., T.J. Falla and D.W. Crook. 1995. The elucidation of novel capsular genotypes of *Haemophilus influenzae* type b with the polymerase chain reaction. J. Med. Microbiol. *43*: 120–124.

Lechevallier, M.W., T.M. Babcock and R.G. Lee. 1987. Examination and characterization of distribution system biofilms. Appl. Environ. Microbiol. *53*: 2714–2724.

Lechevallier, M.W., T.M. Evans, R.J. Seidler, O.P. Daily, B.R. Merrell, D.M. Rollins and S.W. Joseph. 1982. *Aeromonas sobria* in chlorinated drinking water supplies. Microb. Ecol. *8*: 325–334.

Leclerc, H. 1962. Étude biochemique d'*Enterobacteriaceae* pigmentées. Ann. Inst. Pasteur (Paris) *102*: 726–741.

Lecso-Bornet, M., J. Pierre, D. Sarkis-Karam, S. Lubera and E. Bergogne-Berezin. 1992. Susceptibility of *Xanthomonas maltophilia* to six quinolones and study of outer membrane proteins in resistant mutants selected in vitro. Antimicrob. Agents Chemother. *36*: 669–671.

Lederberg, J. 2000. 2000. Biological warfare and bioterrorism. *In* Mandell, Bennet and Dolin (Editors), Principles and Practice of Infectious Diseases, Churchill Livingstone, Philadelphia. pp. 3235–3238.

Ledesma, E., M.L. Camaro, E. Carbonell, T. Sacristan, A. Marti, S. Pellicer, J. Llorca, P. Herrero and M.A. Dasi. 1995. Subtyping of *Legionella pneumophila* isolates by arbitrarily primed polymerase chain reaction. Can. J. Microbiol. *41*: 846–848.

Ledyard, K.M. and A. Butler. 1997. Structure of putrebactin, a new dihydroxamate siderophore produced by *Shewanella putrefaciens*. J. Biol. Inorg. Chem. *2*: 93–97.

Ledyard, K.M., E.F. Delong and J.W.H. Dacey. 1993. Characterization of a DMSP-degrading bacterial isolate from the Sargasso Sea. Arch. Microbiol. *160*: 312–318.

Lee, C.W., P.E. Shewen, W.M. Cladman, J.A. Conlon, A. Mellors and R.Y. Lo. 1994. Sialoglycoprotease of *Pasteurella haemolytica* A1: detection of antisialoglycoprotease antibodies in sera of calves. Can. J. Vet. Res. *58*: 93–98.

Lee, C.W., I.W. Wilkie, K.M. Townsend and A.J. Frost. 2000. The demonstration of *Pasteurella multocida* in the alimentary tract of chickens after experimental oral infection. Vet. Microbiol. *72*: 47–55.

Lee, H.A. 1917. A new bacterial citrus disease. J. Agr. Res. *9*: 1–8.

Lee, H.K., J. Chun, E.Y. Moon, S.H. Ko, D.S. Lee, H.S. Lee and K.S. Bae. 2001. *Hahella chejuensis* gen. nov., sp nov., an extracellular-polysaccharide-producing marine bacterium. Int. J. Syst. Evol. Microbiol. *51*: 661–666.

Lee, J.V. 1987. Identification of *Aeromonas* in the routine laboratory. Experientia (Basel) *43*: 355–357.

Lee, J.V., T.J. Donovan and A.L. Furniss. 1978. Characterization, taxonomy, and emended description of *Vibrio metschnikovii*. Int. J. Syst. Bacteriol. *28*: 99–111.

Lee, J.V., D.M. Gibson and J.M. Shewan. 1977. A numerical taxonomic study of some *Pseudomonas*-like marine bacteria. J. Gen. Microbiol. *98*: 439–451.

Lee, J.V., D.M. Gibson and J.M. Shewan. 1981a. *In* Validation of the publication of new names and new combinations previously effectively published outside the IJSB. List No. 6. Int. J. Syst. Bacteriol. *31*: 215–218.

Lee, J.V., P. Shread, A.L. Furniss and T. Bryant. 1981b. Taxonomy and description of *Vibrio fluvialis* sp. nov. (synonym group F vibrios, group EF6). J. Appl. Bacteriol. *50*: 73–94.

Lee, J.V., P. Shread, A.L. Furniss and T. Bryant. 1981c. *In* Validation of the publication of new names and combinations previously effectively published outside the IJSB. List No. 6. Int. J. Syst. Bacteriol. *31*: 215–218.

Lee, K.F., J.M. Ling, K.M. Kam, D.R. Clark and P.C. Shaw. 1997. Restriction endonucleases in clinical isolates of *Shigella* spp. J. Med. Microbiol. *46*: 949–952.

Lee, M. and A.C. Chandler. 1941. A study of the nature, growth and control of bacteria in cutting compounds. J. Bacteriol. *41*: 373–386.

Lee, M.D. and A.D. Henk. 1996. Tn10 insertional mutagenesis in *Pasteurella multocida*. Vet. Microbiol. *50*: 143–148.

Lee, N.A. and D.P. Clark. 1993. A natural isolate of *Pseudomonas maltophila* which degrades aromatic sulfonic acids. FEMS Microbiol. Lett. *107*: 151–155.

Lee, T.C., R.M. Vickers, V.L. Yu and M.M. Wagener. 1993. Growth of 28 *Legionella* species on selective culture media: a comparative study. J. Clin. Microbiol. *31*: 2764–2768.

Lee, W.H., M.E. Harris, D. McClain, R.E. Smith and R.W. Johnston. 1980. Two modified selenite media for the recovery of *Yersinia enterocolitica* from meats. Appl. Environ. Microbiol. *39*: 205–209.

Lee, Y.A., D.C. Hildebrand and M.N. Schroth. 1992. Use of quinate metabolism as a phenotypic property to identify members of *Xanthomonas campestris* DNA homology group 6. Phytopathology *82*: 971–973.

Lees, V.N., J.P. Owens and J.C. Murrell. 1991. Nitrogen metabolism in marine methanotrophs. Arch. Microbiol. *157*: 60–65.

Legakis, N.J., J.T. Papavassiliou and M.E. Xilinas. 1976. Inositol as a selective substrate for the growth of *Klebsiellae* and *Serratiae*. Zentralbl. Bakteriol. Infektionskr. Parasitenkd. Hyg. Abt. I, Orig. A. *235*: 453–458.

Lehmann, K.B. and R. Neumann. 1896. Atlas and Grundress der Bakteriologie und Lehrbuch der speciellen bakteriologischen Diagnostik, 1st Ed., J.F. Lehmann, Munich.

Lehmann, K.B. and R. Neumann. 1899. Lehmann's Medizin, Handatlanten. X. Atlas und Grundriss der Bakteriologie und Lehrbuch der speziellen Bakteriologischen Diagnostik. 3. Aufl,

Lehmann, K.B. and R. Neumann. 1907. Atlas und Grundriss der Bakteriologie und Lehrbuch der speciellen Bakteriologischen Diagnostik. 4. Aufl. Teil 2, Lehrmann, München.

Lehmann, V. 1971. Phospholipase activity of *Acinetobacter calcoaceticus*. Acta Path Microbiol. Scand. Sec. B *79*: 372–376.

Lehmann, V. 1973. Haemolytic activity of various strains of *Acinetobacter*. Acta Pathol. Microbiol. Scand. B Microbiol. Immunol. *81*: 427–432.

Lei, S.P., H.C. Lin, L. Heffernan and G. Wilcox. 1985. Evidence that polygalacturonase is a virulence determinant in *Erwinia carotovora*. J. Bacteriol. *164*: 831–835.

Lei, S.P., H.C. Lin, S.S. Wang, P. Higaki and G. Wilcox. 1992. Characterization of the *Erwinia carotovora peh* gene and its product polygalacturonase. Gene *117*: 119–124.

Leidy, G., E. Hahn and H.E. Alexander. 1956. On the specificity of the desoxyribonucleic acid which induces streptomycin resistance in *Haemophilus*. J. Exp. Med. *104*: 305–320.

Leidy, G., E. Hahn and H.E. Alexander. 1959. Interspecific transformation in *Hemophilus*: A possible index of relationship between *H. influenzae* and *H. aegyptius*. Proc. Soc. Exp. Biol. Med. *102*: 86–88.

Leidy, G., I. Jaffee and H.E. Alexander. 1965. Further evidence of a high degree of genetic homology between *H. influenzae* and *H. aegyptius*. Proc. Soc. Exp. Biol. Med. *118*: 671–679.

Leifson, E. 1936. New Selenite typhoid and paratyphoid (*Salmonella*) bacilli. Am. J. Hyg. *24*: 423–432.

Leifson, E. 1960. Atlas of Bacterial Flagellation, Academic Press, New York London.

Leifson, E. 1962a. The bacterial flora of distilled and stored water. III. New species of the genera *Corynebacterium*, *Flavobacterium*, *Spirillum* and *Pseudomonas*. Int. Bull. Bacteriol. Nomencl. Taxon. *12*: 161–170.

Leifson, E. 1962b. *Pseudomonas spinosa* n. sp. Int. Bull. Bacteriol. Nomencl. Taxon. *12*: 88–92.

Leifson, E. 1963. Determination of carbohydrate metabolism of marine bacteria. J. Bacteriol. *85*: 1183–1184.

Leifson, E., S.R. Carhart and M. Fulton. 1955. Morphological characteristics of flagella of *Proteus* and related bacteria. J. Bacteriol. *69*: 73–82.

Leigh, J.A. and D.L. Coplin. 1992. Exopolysaccharides in plant-bacterial interactions. Annu. Rev. Microbiol. *46*: 307–346.

Leisinger, T. and R. Margraff. 1979. Secondary metabolites of the fluorescent pseudomonads. Microbiol. Rev. *43*: 422–442.

Leite, R.P., Jr., D.S. Egel and R.E. Stall. 1994a. Genetic analysis of *hrp*-related DNA sequences of *Xanthomonas campestris* strains causing diseases of citrus. Appl. Environ. Microbiol. *60*: 1078–1086.

Leite, R.M.V.B.C., R.P. Leite Jr and P.C. Ceresini. 1997. Alternative hosts of *Xylella fastidiosa* in plum orchards with the leaf scald disease. Fitopatol. Bras. *22*: 54–57.

Leite, R.P., Jr., G.V. Minsavage, U. Bonas and R.E. Stall. 1994b. Detection and identification of phytopathogenic *Xanthomonas* strains by amplification of DNA sequences related to the *hrp* genes of *Xanthomonas campestris* pv. *vesicatoria*. Appl. Environ. Microbiol. *60*: 1068–1077.

Lelliott, R.A. 1965. The preservation of plant pathogenic bacteria. J. Appl. Bacteriol. *28*: 181–193.

Lelliott, R.A. 1968. The diagnosis of fireblight (*Erwinia amylovora*) and some diseases caused by *Pseudomonas syringae*, pp. 27–34.

Lelliott, R.A. 1974. Genus XII. *Erwinia* Winslow, Broadhurst, Buchanan, Krumweide, Rogers and Smith 1920, 209. *In* Buchanan and Gibbons (Editors), Bergey's Manual of Determinative Bacteriology, 8th Ed., The Williams and Wilkins Co., Baltimore. pp. 332–339.

Lelliott, R.A., E. Billing and A.C. Hayward. 1966. A determinative scheme for the fluorescent plant pathogenic pseudomonads. J. Appl. Bacteriol. *29*: 470–489.

Lelliott, R.A. and R.S. Dickey. 1984. Genus VII. *Erwinia*. *In* Kreig and Holt (Editors), Bergey's Manual of Systematic Bacteriology, Vol. 1, The Williams & Wilkins Co., Baltimore. pp. 469–476.

Lemos, M.L., A.E. Toranzo and J.L. Barja. 1985. Antibiotic activity of epiphytic bacteria of intertidal seaweeds. Microb. Ecol. *11*: 149–163.

Lemozy, J., R. Bismuth and P. Courvalin. 1985. Entérobactéries et aminosides. *In* Courvalin, Goldstein, Philippon and Sirot (Editors), L'antibiogramme, MPC Videocom, Paris. pp. 111–126.

Leonardo, M.R., D.P. Moser, E. Barbieri, C.A. Brantner, B.J. Macgregor, B.J. Paster, E. Stackebrandt and K.H. Nealson. 1999. *Shewanella pealeana* sp. nov., a member of the microbial community associated with the accessory nidamental gland of the squid *Loligo pealei*. Int. J. Syst. Bacteriol. *49*: 1341–1351.

Leong, J., W. Bitter, M. Koster, V. Venturi and P.J. Weisbeek. 1992. Molecular analysis of iron assimilation in plant growth-promoting *Pseudomonas putida* WCS358. *In* Galli, Silver and Witholt (Editors), *Pseudomonas*: Molecular Biology and Biotechnology, American Society for Microbiology, Washington D.C. 30–36.

Leoni, L., A. Ciervo, N. Orsi and P. Visca. 1996. Iron-regulated transcription of the *pvdA* gene in *Pseudomonas aeruginosa*: Effect of *fur* and PvdS on promoter activity. J. Bacteriol. *178*: 2299–2313.

Leopold, K., S. Jacobsen and O. Nybroe. 1997. A phosphate-starvation-inducible outer-membrane protein of *Pseudomonas fluorescens* Ag1 as an immunological phosphate-starvation marker. Microbiology *143*: 1019–1027.

Leroi, F., J.J. Joffraud, F. Chevalier and M. Cardinal. 1998. Study of the microbial ecology of cold-smoked salmon during storage at 8°C. Int. J. Food Microbiol. *39*: 111–121.

Lessel, E.F. 1971. Minutes of the meeting. International committee on nomenclature of bacteria. Subcommittee on the taxonomy of *Moraxella* and allied bacteria. Int. J. Syst. Bacteriol. *21*: 213–214.

Lessie, T.G. and P.V. Phibbs. 1984. Alternative pathways of carbohydrate utilization in Pseudomonads. Annu. Rev. Microbiol. *38*: 359–387.

Leu, H.H., L.S. Leu and C.P. Lin. 1998. Development and application of monoclonal antibodies against *Xylella fastidiosa*, the causal bacterium of pear leaf scorch. J. Phytopathol. *146*: 31–37.

Leu, L.S. and C.C. Su. 1993. Isolation, cultivation, and pathogenicity of *Xylella fastidiosa*, the causal bacterium of pear leaf scorch disease in Taiwan. Plant Dis. *77*: 642–646.

Levin, L., J. Gage, P. Lamont, L. Cammidge, C. Martin, A. Patience and J. Crooks. 1997. Infaunal community structure in a low-oxygen, organic-rich habitat on the Oman continental slope, NW Arabian Sea. *In* Hawkins and Hutchinson (Editors), The Responses of Marine Organisms to their Environments. Proceedings of the 30th European Marine Biology Symposium, Southampton Oceanography Centre, University of Southamptom, Southampton. pp. 223–230.

Levin, R.E. 1972. Correlation of DNA base composition and metabolism of *Pseudomonas putrefaciens* isolates from food, human clinical specimens and other sources. Antonie Leeuwenhoek J. Microbiol. Serol. *38*: 121–127.

Levine, M. 1920. Dysentery and allied bacilli. J. Infect. Dis. *27*: 31–39.

Levine, M.M. 1987. *Escherichia coli* that cause diarrhea: enterotoxigenic, enteropathogenic, enteroinvasive, enterohemorrhagic, and enteroadherent. J. Infect. Dis. *155*: 377–389.

Levine, M. and D.Q. Anderson. 1932. Two new species of bacteria causing mustiness in eggs. J. Bacteriol. *23*: 337–347.

Levine, M.M., E.J. Bergquist, D.R. Nalin, D.H. Waterman, R.B. Hornick, C.R. Young and S. Sotman. 1978. *Escherichia coli* strains that cause diarrhoea but do not produce heat-labile or heat-stable enterotoxins and are non-invasive. Lancet *1*: 1119–1122.

Levine, M.M., H.L. DuPont, M. Khodabandelou and R.B. Hornick. 1973. Long-term *Shigella*-carrier state. N. Engl J. Med. *288*: 1169–1171.

Levine, M.M. and R. Edelman. 1984. Enteropathogenic *Escherichia coli* of classic serotypes associated with infant diarrhea: epidemiology and pathogenesis. Epidemiol. Rev. *6*: 31–51.

Levine, M.M., J.P. Nataro, H. Karch, M.M. Baldini, J.B. Kaper, R.E. Black, M.L. Clements and A.D. O'Brien. 1985. The diarrheal response of humans to some classic serotypes of enteropathogenic *Escherichia coli* is dependent on a plasmid encoding an enteroadhesiveness factor. J. Infect. Dis. *152*: 550–559.

Levine, M.M., J.G. Xu, J.B. Kaper, H. Lior, V. Prado, B. Tall, J. Nataro, H. Karch and K. Wachsmuth. 1987. A DNA probe to identify enterohemorrhagic *Escherichia coli* of O157:H7 and other serotypes that cause hemorrhagic colitis and hemolytic uremic syndrome. J. Infect. Dis. *156*: 175–182.

Levine, O.S. and M.M. Levine. 1991. Houseflies (*Musca domestica*) as mechanical vectors of shigellosis. Rev. Infect. Dis. *13*: 688–696.

Levison, M. 1977. Factors influencing colonization of the gastro intestinal tract with *Pseudomonas auruginosa*. *In* Young (Editor), *Pseudomonas aeruginosa*: Ecological Aspects and Patient Colonization, Raven Press, New York. 97–109.

Levy, H.L. and D. Ingall. 1967. Meningitis in neonates due to *Proteus mirabilis*. Am. J. Dis. Child. *114*: 320–324.

Levy, J., G. Verhaegen, P. De Mol, M. Couturier, D. Dekegel and J.P. Butzler. 1993. Molecular characterization of resistance plasmids in epidemiologically unrelated strains of multiresistant *Haemophilus influenzae*. J. Infect. Dis. *168*: 177–187.

Levy, P.Y. and J.L. Tessier. 1998. Arthritis due to *Shewanella putrefaciens*. Clin. Infect. Dis. *26*: 536.

Lewallen, K.R., R.M. McKinney, D.J. Brenner, C.W. Moss, D.H. Dail, B.M. Thomason and R.A. Bright. 1979. A newly identified bacterium phenotypically resembling, but genetically distinct from, *Legionella pneumophila*: an isolate in a case of pneumonia. Ann. Intern. Med. *91*: 831–834.

Lewis, I.M. 1930. Growth of plant pathogenic bacteria in synthetic culture media with special reference to *Phytomonas malvaceara*. Phytopathology *20*: 723–731.

Leyns, F., M. de Cleene, J.G. Swings and J. de Ley. 1984. The host range of the genus *Xanthomonas*. Bot. Rev. *50*: 308–356.

Li, J., G. Chen and J.M. Webster. 1996. *N*-(Indol-3-ylethyl)-2′-hydroxy-3′-methylpentamide, a novel indol derivative from *Xenorhabdus nematophilus*. J. Nat. Prod. *59*: 1157–1158.

Li, J., J.M. Musser, P. Beltran, M.W. Kline and R.K. Selander. 1990. Genotypic heterogeneity of strains of *Citrobacter diversus* expressing a 32-kilodalton outer membrane protein associated with neonatal meningitis. J. Clin. Microbiol. *28*: 1760–1765.

Li, K. and C. Miller. 1970. Pathogenic bacteria and their sensitivity patterns in a hospital population of geriatric patients with chronic disease. J. Am. Geriatrics Soc. *18*: 286–294.

Li, L., C. Kato, Y. Nogi and K. Horikoshi. 1998. Distribution of the pressure-regulated operons in deep-sea bacteria. FEMS Microbiol. Lett. *159*: 159–166.

Li, X.Z., D.M. Livermore, D. Ma and H. Nikaido. 1994. Role of efflux pump in intrinsic resistance of *Pseudomonas aeruginosa*: Active efflux as a contributing factor to beta-lactam resistance. Antimicrob. Agents Chemother. *38*: 1742–1752.

Li, X.Z., H. Nikaido and K. Poole. 1995. Role of *mexA-mexB*-OprM in antibiotic efflux in *Pseudomonas aeruginosa*. Antimicrob. Agents Chemother. *39*: 1948–1953.

Liadouze, I., G. Febvay, J. Guillaud and G. Bonnot. 1996. Metabolic fate of energetic amino acids in the aposymbiotic pea aphid *Acyrthosiphon pisum* (Harris) (Homoptera: Aphididae). Symbiosis *21*: 115–127.

Liao, X.W., I. Charlebois, C. Ouellet, M.J. Morency, K. Dewar, J. Lightfoot, J. Foster, R. Siehnel, H. Schweizer, J.S. Lam, R.E.W. Hancock and R.C. Levesque. 1996. Physical mapping of 32 genetic markers on the *Pseudomonas aeruginosa* PAO1 chromosome. Microbiology *142*: 79–86.

Librach, I.M. 1968. *Proteus* meningitis. Develop. Med. Child. Neurol. *10*: 392–394.

Lickfield, K.G., H. Achterrath, F. Hentrich, L. Kolehmainen-Seveus and A. Persson. 1972. Die Feinstrukturen von *Pseudomonas aeruginosa* in ihrer Deutung durch die Gefrierätztechnik, Ultramikrotomie und Kryo-Ultramikrotomie. J. Ultrastruct. Res. *38*: 27–45.

Lidstrom, M.E. 1988. Isolation and characterization of marine methanotrophs. Antonie Leeuwenhoek *54*: 189–200.

Lidstrom, M.E. and D.I. Stirling. 1990. Methylotrophs: genetics and commercial applications. Annu. Rev. Microbiol. *44*: 27–58.

Lightner, D.V., C.T. Fontaine and K. Hanks. 1975. Some forms of gill disease in penaeid shrimp. 6th Annual Workshop, Seattle. World Mariculture society. pp. 347–365.

Lignières, J. 1900. Maladies du porc. Bull. Soc. Cent. Med. Vet. *18*: 389–431.

Lignières, J. and G. Spitz. 1902. L'actinobacillose. Bull. Mém. Soc. Centr. Méd. Vét. *20*: 487–535; 546–565.

Likhosherstov, L.M., S.N. Senchenkova, A.S. Shashkov, V.A. Derevitskaya, I.V. Danilova and I.V. Botvinko. 1991. Structure of the major exopolysaccharide produced by *Azotobacter beijerinckii* B-1615. Carbohydr. Res. *222*: 233–238.

Liljemark, W.F., C.G. Bloomquist, L.A. Uhl, E.M. Schaeffer, L.F. Wolff, B.L. Pihlstrom and C.L. Bandt. 1984. Distribution of oral *Haemophilus* in dental plaque from a large adult population. Infect. Immun. *46*: 778–786.

Lim, H.S., Y.S. Kim and S.D. Kim. 1991. *Pseudomonas stutzeri* YPL-1 genetic transformation and antifungal mechanism against *Fusarium solani*, an agent of plant-root rot. Appl. Environ. Microbiol. *57*: 510–516.

Lim, Y.S., L.J. Young and S. Balakrishnan. 1987. *Plesiomonas shigelloides* associated with diarrhoeal disease in Malaysian children. Singap. Med. J. *28*: 534–536.

Lin, B.C., H.J. Day, S.J. Chen and M.C. Chien. 1979. Isolation and characterization of plasmids in *Xanthomonas manihotis*. Bot. Bull. Acad. Sinica. *20*: 157–171.

Lin, C.P., T.A. Chen, J.M. Wells and T. van der Zwet. 1987. Identification and detection of *Erwinia amylovora* with monoclonal antibodies. Phytopathology *77*: 376–380.

Lin, C.J., C.T. Chiu, D.Y. Lin, I.S. Sheen and J.M. Lien. 1996. Non-O1 *Vibrio cholerae* bacteremia in patients with cirrhosis: 5-year experience from a single medical center. Am. J. Gastroenterol. *91*: 336–340.

Lin, L.P. and H.L. Sadoff. 1968. Encystment and polymer production by *Azotobacter vinelandii* in the presence of beta-hydroxybutyrate. J. Bacteriol. *95*: 2336–2343..

Lin, L.P. and H.L. Sadoff. 1969. Chemical composition of *Azotobacter vinelandii* cysts. J. Bacteriol. *100*: 480–486..

Lin, W., K.J. Fullner, R. Clayton, J.A. Sexton, M.B. Rogers, K.E. Calia, S.B. Calderwood, C. Fraser and J.J. Mekalanos. 1999. Identification of a *Vibrio cholerae* RTX toxin gene cluster that is tightly linked to the cholera toxin prophage. Proc. Natl. Acad. Sci. U.S.A. *96*: 1071–1076.

Lin, Z., H. Kurazono, S. Yamasaki and Y. Takeda. 1993a. Detection of various variant verotoxin genes in *Escherichia coli* by polymerase chain reaction. Microbiol. Immunol. *37*: 543–548.

Lin, Z., S. Yamasaki, H. Kurazono, M. Ohmura, T. Karasawa, T. Inoue, S. Sakamoto, T. Suganami, T. Takeoka, Y. Taniguchi and Y. Takeda. 1993b. Cloning and sequencing of two new Verotoxin 2 variant genes of *Escherichia coli* isolated from cases of human and bovine diarrhea. Microbiol. Immunol. *37*: 451–459.

Lincoln, S.P., T.R. Fermor and B.J. Tindall. 1999. *Janthinobacterium agaricidamnosum* sp. nov., a soft rot pathogen of *Agaricus bisporus*. Int. J. Syst. Bacteriol. *49*: 1577–1589.

Lind, E. and J. Ursing. 1986. Clinical strains of *Enterobacter agglomerans* (synonyms: *Erwinia herbicola*, *Erwinia milletiae*) identified by DNA–DNA hybridization. Acta Pathol. Microbiol. Immunol. Scand. [B]. *94*: 205–213.

Lindberg, A.A. and T. Holme. 1969. Influence of O side chain on the attachment of the Felix 01 bacteriophage of *Salmonella* bacteria. J. Bacteriol. *99*: 513–519.

Lindberg, A.M., A. Ljungh, S. Ahrne, S. Lofdahl and G. Molin. 1998. *Enterobacteriaceae* found in high numbers in fish, minced meat and pasteurised milk or cream and the presence of toxin encoding genes. Int. J. Food Microbiol. *39*: 11–17.

Lindberg, A.A., R. Wollin, P. Gemski and J.A. Wohlhieter. 1978. Interaction between bacteriophage Sf6 and *Shigella flexneri*. J. Virol. *27*: 38–44.

Lindh, E. and W. Frederiksen. 1990. Ornithine decarboxylating strains of *Klebsiella pneumoniae* demonstrated by DNA–DNA hybridization. Apmis *98*: 358–362.

Lindler, L.E. and B.D. Tall. 1993. *Yersinia pestis* Ph-6 antigen forms fimbriae and is induced by intracellular association with macrophages. Mol. Microbiol. *8*: 311–324.

Lindow, S.E., D.C. Arny and C.D. Upper. 1978. *Erwinia herbicola*: A bacterial ice nucleus active increasing frost injury to corn. Phytopathology *68*: 523–527.

Lindquist, D.S., G. Nygaard, W.L. Thacker, R.F. Benson, D.J. Brenner and H.W. Wilkinson. 1988. Thirteenth serogroup of *Legionella pneumophila* isolated from patients with pneumonia. J. Clin. Microbiol. *26*: 586–587.

Lindqvist, K. 1960. A *Neisseria* species associated with infectious keratoconjunctivitis of sheep — *Neisseria ovis* nov. spec. J. Infect. Dis. *106*: 162–165.

Lindsey, J.O., W.T. Martin, A.C. Sonnenwirth and J.V. Bennett. 1976. An outbreak of nosocomial *Proteus rettgeri* urinary tract infection. Am. J. Epidemiol. *103*: 2461–2469.

Lingens, F., P. Vollprecht and V. Gildemeister. 1966. Zür biosynthese der nicotinsaure in *Xanthomonas* und *Pseudomonas*-Arten, Mycobacterium phlei und Rotaigen. Biochem. Z. *344*: 462–477.

Lingwood, C.A., H. Law, S. Richardson, M. Petric, J.L. Brunton, S. De Grandis and M. Karmali. 1987. Glycolipid binding of purified and recombinant *Escherichia coli* produced verotoxin in vitro. J. Biol. Chem. *262*: 8834–8839.

Link, C., S. Eickernjager, D. Porstendørfer and B. Averhoff. 1998. Identification and characterization of a novel competence gene, *comC*, required for DNA binding and uptake in *Acinetobacter* sp. strain BD413. J. Bacteriol. *180*: 1592–1595.

Linker, A. and R.S. Jones. 1966. A new polysaccharide resembling alginic acid isolated from pseudomonads. J. Biol. Chem. *241*: 3845–3851.

Linkerhagner, K. and J. Oelze. 1995. Hydrogenase does not confer significant benefits to *Azotobacter vinelandii* growing diazotrophically under conditions of glucose limitation. J. Bacteriol. *177*: 6018–6020.

Linn, D.M. and N.R. Krieg. 1978. Occurrence of two organisms in cultures of the type strain of *Spirillum lunatum*: proposal for rejection of the name *Spirillum lunatum* and characterization of *Oceanospirillum maris* subsp. *williamsae* and an unclassified vibrioid bacterium. Int. J. Syst. Bacteriol. *28*: 132–138.

Linn, D.M. and N.R. Krieg. 1984. Validation of the publucation of new names and new combinations previously effectively published outside the IJSB. List No.15. Int. J. Syst. Bacteriol. *34*: 355.

Lipman, J.G. 1903. Experiments on the transformation and fixation of nitrogen by bacteria. Rep. N.J. St. Agric. Exp. Stat. *24*: 217–285.

Lipman, J.G. 1904. Soil bacteriological studies. Further contributions to the physiology and morphology of the members of *Azotobacter* group. Rep. N.J. St. Agric. Exp. Stat. *25*: 237–289.

Lipp, E.K. and J.B. Rose. 1997. The role of seafood in foodborne diseases in the United States of America. Rev. Sci. Tech. O. I. E. (Off. Int. Epizoot.). *16*: 620–640.

Lipp, R.L., A.M. Alvarez, A.A. Benedict and J. Berestecky. 1992. Use of monoclonal antibodies and pathogenicity tests to characterize strains of *Xanthomonas campestris* pv. *dieffenbachiae* from aroids. Phytopathology *82*: 677–682.

Lipscomb, J.D. 1994. Biochemistry of the soluble methane monooxygenase. Annu. Rev. Microbiol. *48*: 371–399.

Lipski, A. and K. Altendorf. 1997. Identification of heterotropic bacteria isolated from ammonia-supplied experimental biofilters. Syst. Appl. Microbiol. *20*: 448–457.

Lipski, A., S. Klatte, B. Bendinger and K. Altendorf. 1992. Differentiation of gram-negative, nonfermentative bacteria isolated from biofilters on the basis of fatty acid composition, quinone system, and physiological reaction profiles. Appl. Environ. Microbiol. *58*: 2053–2065.

Lipski, A., E. Spieck, A. Makolla and K. Altendorf. 2001. Fatty acid profiles of nitrite-oxidizing bacteria reflect their phylogenetic heterogeneity. Syst. Appl. Microbiol. *24*: 377–384.

Lipsky, B.A., E.W. Hook, III, A.A. Smith and J.J. Plorde. 1980. *Citrobacter* infections in humans: experience at the Seattle Veterans Administration Medical Center and a review of the literature. Rev. Infect. Dis. *2*: 746–760.

Liptak, A., L. Szabo, J. Kerekgyarto, J. Harangi, P. Nanasi and H. Duddeck. 1986. Synthesis of the tetrasaccharide repeating-unit of the lipopolysaccharide isolated from *Pseudomonas maltophilia*. Carbohydr. Res. *150*: 187–197.

Litwin, C.M. and S.B. Calderwood. 1993. Role of iron in regulation of virulence genes. Clin. Microbiol. Rev. *6*: 137–149.

Litwin, C.M., R.B. Leonard, K.C. Carroll, W.K. Drummond and A.T. Pavia. 1997. Characterization of endemic strains of *Shigella sonnei* by use of plasmid DNA analysis and pulsed-field gel electrophoresis to detect patterns of transmission. J. Infect. Dis. *175*: 864–870.

Liu, D. 1994. Development of gene probes of *Dichelobacter nodosus* for differentiating strains causing virulent, intermediate or benign ovine footrot. Brit. Vet. J. *150*: 451–462.

Liu, J., R. Berry, G. Poinar and A. Moldenke. 1997a. Phylogeny of *Photorhabdus* and *Xenorhabdus* species and strains as determined by comparison of partial 16S rRNA gene sequences. Int. J. Syst. Bacteriol. *47*: 948–951.

Liu, P.Y., Y.J. Lau, B.S. Hu, J.M. Shyr, Z.Y. Shi, W.S. Tsai, Y.H. Lin and C.Y. Tseng. 1995a. Analysis of clonal relationships among isolates of *Shigella sonnei* by different molecular typing methods. J. Clin. Microbiol. *33*: 1779–1783.

Liu, P.C., K.K. Lee and S.N. Chen. 1997b. Susceptibility of different isolates of *Vibrio harveyi* to antibiotics. Microbios. *91*: 175–180.

Liu, P.C., K.K. Lee, C.C. Tu and S.N. Chen. 1997c. Purification and characterization of a cysteine protease produced by pathogenic luminous *Vibrio harveyi*. Curr. Microbiol. *35*: 32–39.

Liu, P.C., K.K. Lee, K.C. Yii, G.H. Kou and S.N. Chen. 1996a. Isolation of *Vibrio harveyi* from diseased kuruma prawns *Penaeus japonicus*. Curr. Microbiol. *33*: 129–132.

Liu, P.Y.F., Z.Y. Shi, Y.J. Lau, B.S. Hu, J.M. Shyr, W.S. Tsai, Y.H. Lin and C.Y. Tseng. 1995b. Comparison of different PCR approaches for characterization of *Burkholderia (Pseudomonas) cepacia* isolates. J. Clin. Microbiol. *33*: 3304–3307.

Liu, P.V. and F. Shokrani. 1978. Biological-activities of pyochelins - iron-chelating agents of *Pseudomonas aeruginosa*. Infect. Immun. *22*: 878–890.

Liu, S.L., A. Hessel, H.Y.M. Cheng and K.E. Sanderson. 1994a. The *XbaI-BlnI-CeuI* genomic cleavage map of *Salmonella paratyphi* B. J. Bacteriol. *176*: 1014–1024.

Liu, S.L., A. Hessel and K.E. Sanderson. 1993a. The *XbaI-BlnI-CeuI* genomic cleavage map of *Salmonella enteritidis* shows an inversion relative to *Salmonella typhimurium* LT2. Mol. Microbiol. *10*: 655–664.

Liu, S.L., A. Hessel and K.E. Sanderson. 1993b. The *XbaI-BlnI-CeuI* genomic cleavage map of *Salmonella typhimurium* LT2 determined by double digestion, end labelling, and pulsed-field gel electrophoresis. J. Bacteriol. *175*: 4104–4120.

Liu, S.L. and K.E. Sanderson. 1992. A physical map of the *Salmonella typhimurium* LT2 genome made by using *XbaI* analysis. J. Bacteriol. *174*: 1662–1672.

Liu, S.L. and K.E. Sanderson. 1995a. The chromosome of *Salmonella paratyphi* A is inverted by recombination between *rrnH* and *rrnG*. J. Bacteriol. *177*: 6585–6592.

Liu, S.L. and K.E. Sanderson. 1995b. Genomic cleavage map of *Salmonella typhi* Ty2. J. Bacteriol. *177*: 5099–5107.

Liu, S.L., A.B. Schryvers, K.E. Sanderson and R.N. Johnston. 1999. Bacterial phylogenetic clusters revealed by genome structure. J. Bacteriol. *181*: 6747–6755.

Liu, S.C., D.A. Webster, M.L. Wei and B.C. Stark. 1996b. Genetic engineering to contain the *Vitreoscilla* hemoglobin gene enhances degradation of benzoic acid by *Xanthomonas maltophilia*. Biotechnol. Bioeng. *49*: 101–105.

Liu, W.-T., T.L. Marsh, H. Cheng and L.J. Forney. 1997d. Characterization of microbial diversity by determining terminal restriction fragment length polymorphisms of genes encoding 16S rRNA. Appl. Environ. Microbiol. *63*: 4516–4522.

Liu, Y., A. Chatterjee and A.K. Chatterjee. 1994b. Nucleotide sequence, organization and expression of *rdgA* and *rdgB* genes that regulate pectin lyase production in the plant pathogenic bacterium *Erwinia carotovora* subsp. *carotovora* in response to DNA-damaging agents. Mol. Microbiol. *14*: 999–1010.

Liu, Y., Y. Cui, A. Mukherjee and A.K. Chatterjee. 1997e. Activation of the *Erwinia carotovora* carotovora pectin lyase structural gene *pnlA*: a role for *RdgB*. Microbiology *143*: 705–712.

Liu, Y., H. Murata, A. Chatterjee and A.K. Chatterjee. 1993c. Characterization of a novel regulatory gene *aepA* that controls extracellular enzyme production in the phytopathogenic bacterium *Erwinia carotovora* subsp. *carotovora*. Mol. Plant-Microbe Interact. *6*: 299–308.

Livermore, D.M. 1995. β-lactamases in laboratory and clinical resistance. Clin. Microbiol. Rev. *8*: 557–584.

Livrelli, V., C. DeChamps, P. DiMartino, A. Darefeuille-Michaud, C. Forestier and B. Joly. 1996. Adhesive properties and antibiotic resistance of *Klebsiella*, *Enterobacter*, and *Serratia* clinical isolates involved in nosocomial infections. J. Clin. Microbiol. *34*: 1963–1969.

Llanes, C., M. Couturier, L. Asfeld, F. Grimont and Y. Michel-Briand. 1994. Replicon typing of 71 multiresistant *Serratia marcescens* strains. Res. Microbiol. *145*: 17–25.

Lloyd, J.R. and L.E. McCaskie. 1996. A novel PhosphorImager-based technique for monitoring the microbial reduction of technetium. Appl. Environ. Microbiol. *62*: 578–582.

Lo, R.Y.C. 1995. Molecular studies of antigens in HAP organisms. *In* Donachie, Lainson and Hodgson (Editors), *Haemophilus, Actinobacillus and Pasteurella*, Plenum Press, New York. pp. 129–142.

Lo, T.M., C.K. Ward and T.J. Inzana. 1998. Detection and identification of *Actinobacillus pleuropneumoniae* serotype 5 by mulitplex PCR. J. Clin. Microbiol. *36*: 1704–1710.

Lo Presti, F., S. Riffard, H. Meugnier, M. Reyrolle, Y. Lasne, P.A.D. Grimont, F. Grimont, F. Vandenesch, J. Etienne, J. Fleurette and J. Freney. 1999. *Legionella taurinensis* sp. nov., a new species antigenically similar to *Legionella spiritensis*. Int. J. Syst. Bacteriol. *49*: 397–403.

Lo Presti, F., S. Riffard, F. Vandenesch, M. Reyrolle, E. Ronco, P. Ichai and J. Etienne. 1997. The first clinical isolate of *Legionella parisiensis*, from a liver transplant patient with pneumonia. J. Clin. Microbiol. *35*: 1706–1709.

Lo Verde, P.T., C. Amento and G.I. Higashi. 1980. Parasite interaction of *Salmonella typhimurium* and *Shistosoma*. J. Infect. Dis. *141*: 177–185.

Lobb, C.J., S.H. Ghaffari, J.R. Hayman and D.T. Thompson. 1993. Plasmid and serological differences between *Edwardsiella ictaluri* strains. Appl. Environ. Microbiol. *59*: 2830–2836.

Lobb, C.J. and M. Rhoades. 1987. Rapid plasmid analysis for identification of *Edwardsiella ictaluri* from infected channel catfish (*Ictalurus punctatus*). Appl. Environ. Microbiol. *53*: 1267–1272.

Lobue, T.D., T.A. Deutsch and R.M. Stein. 1985. *Moraxella nonliquefaciens* endophthalmitis after trabeculectomy. American JAm. J. Opthamol. *99*: 343–345.

Lockwood, D.E., A.S. Kreger and S.H. Richardson. 1982. Detection of toxins produced by *Vibrio fluvialis*. Infect. Immun. *35*: 702–708.

Lockwood, L.B., B. Tabenkin and G.E. Ward. 1941. The production of gluconic acid and 2-keto gluconic acid from glucose by species of *Pseudomonas* and *Phytomonas*. J. Bacteriol. *42*: 51–61.

Lode, E.T. and M.J. Coon. 1971. Enzymatic ω-oxidation. V. Forms of *Pseudomonas oleovorans* rubredoxin containing one or two iron atoms: structure and function in ω-oxidation. J. Biol. Chem. *246*: 791–802.

Loeb, M.R. and D.H. Smith. 1982. Properties and immunogenicity of *Haemophilus influenzae* outer membrane proteins. *In* Sell and Wright (Editors), *Haemophilus influenzae* : Epidemiology, Immunology, and Prevention of Disease, Elsevier Biomedical, New York. 207–217.

Loeb, M.R., A.L. Zachary and D.H. Smith. 1981. Isolation and partial characterization of outer and inner membranes from encapsulated *Haemophilus influenzae* type b. J. Bacteriol. *145*: 596–604.

Loeffler, F. 1892. Über Epidemin unter den im hygienischen Institute zu Greifswald gehaltenen Maüsen und über die Bekämpfung der Feldmausplage. Zentrab. Bakteriol. Parasitenk. Infektionskr. Hyg. Abt. I Orig. *11*: 129–141.

Löhnis, F. 1911. Landwirtschaftlich-bakteriologisches Praktikum, Gebrüder Borntrneger, Berlin.

Lomholt, H. and M. Kilian. 1995. Distinct antigenic and genetic properties of the immunoglobulin A1 protease produced by *Haemophilus influenzae* biogroup aegyptius associated with Brazilian purpuric fever in Brazil. Infect. Immun. *63*: 4389–4394.

Long, G.W., J.J. Oprandy, R.B. Narayanan, A.H. Fortier, K.R. Porter and C.A. Nacy. 1993. Detection of *Francisella tularensis* in blood by polymerase chain reaction. J. Clin. Microbiol. *31*: 152–154.

Long, H.F. and B.W. Hammer. 1941. Distribution of *Pseudomonas putrefaciens*. J. Bacteriol. *41*: 100–101.

Loomes, L.M., B.W. Senior and M.A. Kerr. 1990. A proteolytic enzyme secreted by *Proteus mirabilis* degrades immunoglobulins of the immunoglobulin A1 (IgA1), IgA2, and IgG isotypes. Infect. Immun. *58*: 1979–1985.

Lopes, S.A. and K.E. Damann. 1996. Immunocapture and PCR detection of *Xanthomonas albilineans* from vascular sap of sugarcane leaves. Summa Phytopathol. *22*: 244–247.

Lorenz, H., K. Reipschlager and W. Wackernagel. 1992. Plasmid transformation of naturally competent *Acinetobacter calcoaceticus* in nonsterile soil extract and groundwater. Arch. Microbiol. *157*: 355–360.

Losonsky, G.A., M. Santosham, V.M. Sehgal, A. Zwahlen and E.R. Moxon. 1984. *Haemophilus influenzae* disease in the White Mountain Apaches: molecular epidemiology of a high risk population. Pediatr. Infect. Dis. *3*: 539–547.

Loubinoux, J., A. Lozniewski, C. Lion, D. Garin, M. Weber and A.E. Le Faou. 1999. Value of enterobacterial repetitive intergenic consensus PCR for study of *Pasteurella multocida* strains isolated from mouths of dogs. J. Clin. Microbiol. *37*: 2488–2492.

Louie, M., J. de Azavedo, R. Clarke, A. Borczyk, H. Lior, M. Richter and J. Brunton. 1994. Sequence heterogeneity of the *eae* gene and detection of verotoxin-producing *Escherichia coli* using serotype-specific primers. Epidemiol. Infect. *112*: 449–461.

Louis, C., G. Croizier and G. Meynadier. 1977a. Trame cristalline des inclusions proteiques chez une *Rickettsiella*. Biol. Cell *29*: 77–80.

Louis, C., M. Jourdan and M. Cabanac. 1986. Behavioral fever and therapy in a rickettsia-infected Orthoptera. Am. J. Physiol. *250*: R991–R995.

Louis, C., G. Morel, G. Nicholas and G. Kuhl. 1979. Étude comparée des caractères ultrastructuraux de rickettsies d'arthropodes, révélés par cryodécapage et cytochimie. J. Untrastruc. Res. *66*: 243–253.

Louis, C., A. Yousfi, C. Vago and G. Nicolas. 1977b. Étude par cytochimie et cryodécapage de l'ultrastructure d'une *Rickettsiella* de crustacé. Ann. Microbiol. *128B*: 177–205.

Louws, F.J., D.W. Fulbright, C.T. Stephens and F.J. de Bruijn. 1994. Specific genomic fingerprints of phytopathogenic *Xanthomonas* and *Pseudomonas* pathovars and strains generated with repetitive sequences and PCR. Appl. Environ. Microbiol. *60*: 2286–2295.

Louws, F.J., D.W. Fulbright, C.T. Stephens and F.J. de Bruijn. 1995. Differentiation of genomic structure by rep-PCR fingerprinting to rapidly classify *Xanthomonas campestris* pv. *vesicatoria*. Phytopathology *85*: 528–536.

Love, M., D. Teebken-Fisher, J.E. Hose, J.J. Farmer, F.W. Hickman and G.R. Fanning. 1981. *Vibrio damsela*, a marine bacterium, causes skin ulcers on the damselfish *Chromis punctipinnis*. Science (Wash. D. C.) *214*: 1139–1140.

Love, M., D. Teebken-Fisher, J.E. Hose, J.J. Farmer, F.W. Hickman and G.R. Fanning. 1982. *In* Validation of new names and new combinations previously effectively published outside the IJSB. List No. 8. Int. J. Syst. Bacteriol. *32*: 266–268.

Lovejoy, C., J.P. Bowman and G.M. Hallegraeff. 1998. Algicidal effects of a novel marine *Pseudoalteromonas* isolate (class *Proteobacteria*, gamma subdivision) on harmful algal bloom species of the genera *Chattonella*, *Gymnodinium*, and *Heterosigma*. Appl. Environ. Microbiol. *64*: 2806–2813.

Loveless, T.M. and P.E. Bishop. 1999. Identification of genes unique to Mo-independent nitrogenase systems in diverse diazotrophs. Can. J. Microbiol. *45*: 312–317.

Loveless, T.M., J.R. Saah and P.E. Bishop. 1999. Isolation of nitrogen-fixing bacteria containing molybdenum-independent nitrogenases from natural environments. Appl. Environ. Microbiol. *65*: 4223–4226.

Lovley, D.R. 1993. Dissimilatory metal reduction. Annu. Rev. Microbiol. *47*: 263–290.

Lovley, D.R. 1997. Microbial FE(III) reduction in subsurface environments. FEMS Microbiol. Rev. *20*: 305–313.

Lovley, D.R. and E.J.P. Phillips. 1986. Organic matter mineralization with reduction of ferric iron in anaerobic sediments. Appl. Environ. Microbiol. *51*: 683–689.

Lovley, D.R. and E.J.P. Phillips. 1988. Novel mode of microbial energy metabolism: organic carbon oxidation coupled to dissimilatory reduction of iron or manganese. Appl. Environ. Microbiol. *54*: 1472–1480.

Lovrekovich, L. and Z. Klement. 1965. Serological and bacteriophage sensitivity studies on *Xanthomonas vesicatoria* strains isolated from tomato and pepper. Phytopathol. Z. *52*: 222–228.

Low, D., B. Braaten and M. van der Woude. 1996. Fimbriae. *In* Neidhardt, Curtiss, Ingraham, Lin, Low, Magasanik, Reznikoff, Riley, Schaechter and Umbarger (Editors), *Escherichia coli* and *Salmonella* : Cellular and Molecular Biology, 2nd Ed., ASM Press, Washington, D.C. pp. 146–157.

Lowbury, E.J.L., H.A. Lilly, A. Kidson, G.A.J. Ayliffe and R.J. Jones. 1969. Sensitivity of *Pseudomonas aeruginosa* to antibiotics: emergence of strains highly resistant to carbenicillin. Lancet *1969*: 448–452.

Lowry, P.W., L.M. McFarland and H.K. Threefoot. 1986. *Vibro hollisae* septicemia after consumption of catfish. J. Infect. Dis. *154*: 730–731.

Lu, H.M. and S. Lory. 1996. A specific targeting domain in mature exotoxin A is required for its extracellular secretion from *Pseudomonas aeruginosa*. Embo J. *15*: 429–436.

Lubke, A., L. Hartmann, W. Schroder and E. Hellmann. 1994. Isolation and partial characterization of the major protein of the outer-membrane of *Pasteurella haemolytica* and *Pasteurella multocida*. Zentbl. Bakteriol. Mikrobiol. Hyg. Abt. Orig. A Med. Mikrobiol. Infektkrankh. Parasitol. *281*: 45–54.

Lucas, A.H. 1988. Expression of crossreactive idiotypes by human antibodies specific for the capsular polysaccharide of *Haemophilus influenzae* B. J. Clin. Invest. *81*: 480–486.

Lucas, L.T. and R.G. Grogan. 1969a. Serological variation and identification of *Pseudomonas lachrymans* and other phytopathogenic *Pseudomonas* nomenspecies. Phytopathology *59*: 1908–1912.

Lucas, L.T. and R.G. Grogan. 1969b. Some properties of specific antigens of *Pseudomonas lachrymans* and other *Pseudomonas* nomenspecies. Phytopathology *59*: 1913–1917.

Luck, P.C., J.H. Helbig, W. Ehret and M. Ott. 1995. Isolation of a *Legionella pneumophila* strain serologically distinguishable from all known serogroups. Zentbl. Bakteriol. *282*: 35–39.

Lüderitz, O., A.M. Staub and O. Westphal. 1966. Immunochemistry of O and R antigens of *Salmonella* and related *Enterobacteriaceae*. Bacteriol. Rev. *30*: 192–255.

Lüderitz, O., O. Westphal, A.M. Staub and H. Nikaido. 1971. Isolation and chemical and immunological characterization of bacterial lipopolysaccharides. *In* Weinbaum, Kadis and Ajl (Editors), Microbial Toxins, Vol. IV, Academic Press, New York. pp. 145–233.

Ludwig, W., R. Rossello-Mora, R. Aznar, S. Klugbauer, S. Spring, K. Reetz, C. Beimfohr, E. Brockmann, G. Kirchhof, S. Dorn, M. Bachleitner, N. Klugbauer, N. Springer, D. Lane, R. Nietupsky, M. Weiznegger and K.H. Schleifer. 1995. Comparative sequence analysis of 23S rRNA from *Proteobacteria*. Syst. Appl. Microbiol. *18*: 164–188.

Ludwig, W. and E. Stackebrandt. 1983. A phylogenetic analysis of *Legionella*. Arch. Microbiol. *135*: 45–50.

Ludwig, W. and O. Strunk. 1996. ARB: a software environment for sequence data: ⟨http://www.mikro.biologie.tu-muenchen.de⟩.

Lugtenberg, J.J. and L.A. De Weger. 1992. Plant root colonization by *Pseudomonas* spp. *In* Galli, Silver and Witholt (Editors), *Pseudomonas*: Molecular Biology and Biotechnology, American Society for Microbiology, Washington D.C. 13–19.

Luisetti, J., J.P. Prunier and L. Gardan. 1972. Un milieu pour la mise en évidence de la production d'un pigment fluorescent par *Pseudomonas mors-prunorum* f. sp. persicae. Ann. Phytopathol. *4*: 295–296.

Lukacova, M., D. Valkova, M. Quevedo Diaz, D. Perecko and I. Barak. 1999. Green fluorescent protein as a detection marker for *Coxiella burnetii* transformation. FEMS Microbiol. Lett. *175*: 255–260.

Lukomski, S., B. Muller and G. Schmidt. 1990. Extracellular hemolysin of *Proteus penneri* coded by chromosomal hly genes is similar to the α-hemolysin of *Escherichia coli*. Zentbl. Bakteriol. *273*: 150–155.

Lunder, T., O. Evensen, G. Holstad and T. Håstein. 1995. Winter ulcer in the Atlantic salmon *Salmo salar* pathological and bacteriological investigations and transmission experiments. Dis. Aquat. Org. *23*: 39–49.

Lunder, T., H. Sørum, G. Holstad, A.G. Steigerwalt, P. Mowinckel and D.J. Brenner. 2000. Phenotypic and genotypic characterization of *Vibrio viscosus* sp. nov. and *Vibrio wodanis* sp. nov. isolated from Atlantic salmon (*Salmo salar*) with "winter ulcer". Int. J. Syst. Evol. Microbiol. *50*: 427–450.

Luo, K., X.L. Liao and Z. Chen. 1988. On brown sheath disease of rice. II. Determination of physiological, biochemical and molecular biological characteristics of the causal bacterium. Acta Phytopathol. Sinica. *18*: 29–33.

Luque, F., L.A. Mitchenall, M. Chapman, R. Christine and R.N. Pau. 1993. Characterization of genes involved in molybdenum transport in *Azotobacter vinelandii*. Mol. Microbiol. *7*: 447–459.

Luria, S.E. and J.W. Burrous. 1957. Hybridization between *Escherichia coli* and *Shigella*. J. Bacteriol. *74*: 461–476.

Luttrell, R.E., G.A. Rannick, J.L. Soto-Hernandez and A. Verghese. 1988. *Kluyvera* species soft tissue infection: case report and review. J. Clin. Microbiol. *26*: 2650–3651.

Lwoff, A. 1939. Revision et démembrement des Hemophilae le genre *Moraxella* nov. gen. Ann. Inst. Pasteur. *62*: 168–176.

Lwoff, A. and M. Lwoff. 1937. Rôle physiologique de l'hémine pour *Haemophilus influenzae* Pfeiffer. Ann. Inst. Pasteur. *59*: 129–136.

Lynch, M.J., S. Swift, L. fish, D.F. Kirke, C.E.R. Dodd, G.S.A.B. Stewart, C.W. Keevil and P. Williams. 1999. Puorum sensing in *Aeromonas hydrophila* biofilms. 6th International *Aeromonas/Plesiomonas* Symposium, Chicago, Illinois. p. 24.

Lysenko, A.M., V.F. Gal'chenko and N.A. Chernykh. 1988. Taxonomic study of obligate methanotrophic bacteria using the DNA–DNA hybridization technique. Mikrobiologiya *57*: 653–658.

Lysenko, O. 1976. Chitinase of *Serratia marcescens* and its toxicity for insects. J. Invertebr. Pathol. *27*: 385–386.

Lysenko, O. and J. Weiser. 1974. Bacteria associated with the nematode

Neoplectana carpocapsae and the pathogenicity of this complex for *Galleria mellonella* larvae. J. Invertebr. Pathol. *24*: 332–336.

Lytikäinen, O., S. Kjöljag, M. Härmä and J. Vuopio-Varkila. 1995. Outbreak caused by two multiresistant *Acinetobacter baumannii* clones in a burn unit: emergence of resistance to imipenem. J. Hosp. Infect. *31*: 41–54.

Ma, W.-C. and D.L. Denlinger. 1974. Secretory discharge and microflora of milk gland in tsetse flies. Nature (Lond.) *247*: 301–303.

Macchiavello, A. 1937. Estudios sobre tifus exantematico. III. Un nuevo metodo pare tenir *Rickettsia*. Rev. Chilena Hig. Med. Prev. *1*: 101–106.

MacDonell, M.T. and R.R. Colwell. 1985. Phylogeny of the *Vibrionaceae*, and recommendation for two new genera, *Listonella* and *Shewanella*. Syst. Appl. Microbiol. *6*: 171–182.

MacDonell, M.T. and R.R. Colwell. 1986. *In* Validation of the publication of new names and new combinations previously effectively published outside the IJSB. List No. 20. Int. J. Syst. Bacteriol. *36*: 354–356.

MacDonell, M.T., D.G. Swartz, B.A. Ortiz-Conde, G.A. Last and R.R. Colwell. 1986. Ribosomal RNA phylogenies for the vibrio-enteric group of eubacteria. Microbiol. Sci. *3*: 172–178.

Machado, J., F. Grimont and P.A.D. Grimont. 1998. Computer identification of *Escherichia coli* rRNA gene restriction patterns. Res. Microbiol. *149*: 119–135.

Machado, J., F. Grimont and P.A.D. Grimont. 2000. Identification of *Escherichia coli* flagellar types by restriction of the amplified *fliC* gene. Res. Microbiol. *151*: 535–546.

Machado, W.C. and J. Döbereiner. 1969. Estudos complementares sobre a fisiologia de *Azotobacter paspali* e sue dependencia da planta (*Paspalum notatum*). Pesqm. Agropecu. Brasil. *4*: 53–58.

Machiels, P., J.P. Haxhe, J.P. Trigaux, M. Delos, J.C. Shoevaerdts and O. Vandenplas. 1995. Chronic lung abscess due to *Pasteurella multocida*. Thorax *50*: 1017–1018.

Macierevicz, M. 1966. A proposal of a new group (genus) of *Enterobacteriaceae*. Exp. Med. Microbiol. (Engl. Transl. Med. Dosw. Mikrobiol.) *18*: 333–339.

Maclagan, R.M. and D.C. Old. 1980. Haemagglutins and fimbriae in different serotypes and biotypes of *Yersinia enterocolitica*. J. Appl. Bacteriol. *49*: 353–360.

Maclean, I.W., L. Slaney, J.M. Juteau, R.C. Levesque, W.L. Albritton and A.R. Ronald. 1992. Identification of a ROB-1 beta-lactamase in *Haemophilus ducreyi*. Antimicrob. Agents Chemother. *36*: 467–469.

MacLeod, D.L., C.L. Gyles and B.P. Wilcock. 1991. Reproduction of edema disease of swine with purified Shiga-like toxin-II variant. Vet. Pathol. *28*: 66–73.

MacLeod, P.R. and R.A. MacLeod. 1992. Identification and sequence of a Na(+)-linked gene from the marine bacterium *Alteromonas haloplanktis* which functionally complements the *dagA* gene of *Escherichia coli*. Mol. Microbiol. *6*: 2673–2681.

MacLeod, R.A. 1968. On the role of inorganic ions in the physiology of marine bacteria. Adv. Microbiol. Sea. *1*: 95–126.

Macnab, R.M. 1996. Flagella. *In* Neidhardt, Curtiss, Ingraham, Lin, Low, Magasanik, Reznikoff, Riley, Schaechter and Umbarger (Editors), *Escherichia coli* and *Salmonella* : Cellular and Molecular Biology, 2nd Ed., ASM Press, Washington, D.C. pp. 123–145.

MacPhee, D.G., V. Krishnapillai, R.J. Roantree and B.A.D. Stocker. 1975. Mutations in *Salmonella typhimurium* conferring resistance to Felix O phage without loss of smooth character. J. Gen. Microbiol. *87*: 1–10.

Macy, J.M. and I. Probst. 1979. The biology of gastrointestinal bacteroides. Annu. Rev. Microbiol. *33*: 561–594.

Madigan, M.T. 1984. A novel photosynthetic purple bacterium isolated from a Yellowstone hot spring. Science *225*: 313–315.

Madigan, M.T. 1986. *Chromatium tepidum* sp. nov., a thermophilic photosynthetic bacterium of the family *Chromatiaceae*. Int. J. Syst. Bacteriol. *36*: 222–227.

Madsen, N.B. and R.M. Hochster. 1959. The tricarboxylic acid and glyoxylate cycles in *Xanthomonas phaseoli* (XP8). Can. J. Microbiol. *5*: 1–8.

Maeland, J.A. and A. Digranes. 1975. Common enterobacterial antigen in *Yersinia enterocolitica*. Acta Path. Microbiol. Scand. Sect. B. *83*: 382–386.

Maeng, J.H., Y. Sakai, Y. Tani and N. Kato. 1996. Isolation and characterization of a novel oxygenase that catalyzes the first step of n-alkane oxidation in *Acinetobacter* sp. strain M-1. J. Bacteriol. *178*: 3695–3700.

Maes, M. 1993. Fast classification of plant-associated bacteria in the *Xanthomonas* genus. FEMS Microbiol. Lett. *113*: 161–165.

Maes, M., P. Garbeva and C. Crepel. 1996. Identification and sensitive endophytic detection of the fire blight pathogen *Erwinia amylovora* with 23S ribosomal DNA sequences and the polymerase chain reaction. Plant Pathol. (Oxf.) *45*: 1139–1149.

Magariños, B., R. Bonet, J.L. Romalde, M.J. Martínez, F. Congregado and A.E. Toranzo. 1996. Influence of the capsular layer on the virulence of *Pasteurella piscicida* for fish. Microb. Pathog. *21*: 289–297.

Magariños, B., C.R. Osorio, A.E. Toranzo and J.L. Romalde. 1997. Applicability of ribotyping for intraspecific classification and epidemiological studies of *Photobacterium damsela* subsp. *piscicida*. Syst. Appl. Microbiol. *20*: 634–639.

Magariños, B., J.L. Romalde, I. Bandín, B. Fouz and A.E. Toranzo. 1992. Phenotypic, antigenic, and molecular characterization of *Pasteurella piscicida* strains Isolated from fish. Appl. Environ. Microbiol. *58*: 3316–3322.

Magnum, M.E. and D. Radisch. 1982. *Cedecea* species: unusual clinical isolate. Clin. Microbiol. Newsl. *4*: 117–119.

Magnusson, H. 1928. The commonest forms of actinomycosis in domestic animals and their etiology. Acta Pathol. Microbiol. Scand. *5*: 170–245.

Magrou, J. 1937. Genre Erwinia. *In* Hauduroy, Ehringer, Urbain, Guillot and Magrou (Editors), Dictionnaire des bactéries pathogènes, Masson, Paris. 195-220.

Mague, T.H. and R.A. Lewin. 1974. *Leucothrix*: absence of demonstrable fixation of N_2. J. Gen. Microbiol. *85*: 365–367.

Mahaffee, W.F. and J.W. Kloepper. 1997. Bacterial communities of the rhizosphere and endorhiza associated with field-grown cucumber plants inoculated with a plant growth-promoting rhizobacterium or its genetically modified derivative. Can. J. Microbiol. *43*: 344–353.

Mahalingam, S., Y.M. Cheong, S.K. Kan and T. Pang. 1993. Occurrence of *Vibrio cholerae* 01 strains in Southeast Asia resistant to vibriostatic compound 0/129. Southeast Asian J. Trop. Med. Public Health *24*: 779–780.

Mahalingam, S., Y.M. Cheong, S. Kan, R.M. Yassin, J. Vadivelu and T. Pang. 1994. Molecular epidemiologic analysis of *Vibrio cholerae* O1 isolates by pulsed-field gel electrophoresis. J. Clin. Microbiol. *32*: 2975–2979.

Mahe, B., C. Masclaux, L. Rauscher, C. Enard and D. Expert. 1995. Differential expression of two siderophore-dependent iron-acquisition pathways in *Erwinia chrysanthemi* 3937: characterization of a novel ferrisiderophore permease of the ABC transporter family. Mol. Microbiol. *18*: 33–43.

Maidak, B.L., J.R. Cole, C.T. Parker, Jr., G.M. Garrity, N. Larsen, B. Li, T.G. Lilburn, M.J. McCaughey, G.J. Olsen, R. Overbeek, S. Pramanik, T.M. Schmidt, J.M. Tiedje and C.R. Woese. 1999. A new version of the RDP (Ribosomal Database Project). Nucleic Acids Res. *27*: 171–173.

Maidak, B.L., G.J. Olsen, N. Larsen, R. Overbeek, M.J. McCaughey and C.R. Woese. 1996. The ribosomal database project (RDP). Nucleic Acids Res. *24*: 82–85.

Maidak, B.L., G.J. Olsen, N. Larsen, R. Overbeek, M.J. McCaughey and C.R. Woese. 1997. The RDP (Ribosomal Database Project). Nucleic Acids Res. *25*: 109–111.

Maiden, M.C.J., J.A. Bygraves, E. Feil, G. Morelli, J.E. Russell, R. Urwin, Q. Zhang, J.J. Zhou, K. Zurth, D.A. Caugant, I.M. Feavers, M. Achtman and B.G. Spratt. 1998. Multilocus sequence typing: A portable approach to the identification of clones within populations of pathogenic microorganisms. Proc. Natl. Acad. Sci. U. S. A. *95*: 3140–3145.

Maier, R.J. and F. Moshiri. 2000. Role of the *Azotobacter vinelandii* nitrogenase-protective Shethna protein in preventing oxygen-mediated cell death. J. Bacteriol. *182*: 3854–3857.

Maier, S. 1984. Description of *Thioploca* ingrica sp. nov., nom. rev. Int. J. Syst. Bacteriol. *34*: 344–345.

Maier, S. and V.A. Gallardo. 1984a. Nutritional characteristics of two marine thioplocas determined by autoradiography. Arch. Microbiol. *139*: 218–220.

Maier, S. and V.A. Gallardo. 1984b. *Thioploca araucae* sp. nov. and *Thioploca chileae* sp. nov. Int. J. Syst. Bacteriol. *34*: 414–418.

Maier, S. and R.G.E. Murray. 1965. The fine structure of *Thioploca ingrica* and a comparison with *Beggiatoa*. Can. J. Microbiol. *11*: 645–655.

Maier, S. and W.C. Preissner. 1979. Occurrence of *Thioploca* in Lake Constance and Lower Saxony, Germany. Microb. Ecol. *5*: 117–119.

Maier, S., H. Völker, M. Beese and V.A. Gallardo. 1990. The fine structure of *Thioploca araucae* and *Thioploca chileae*. Can. J. Microbiol. *36*: 438–448.

Maino, A.L., N.N. Schroth and N.J. Palleroni. 1974. Degradation of xylan by bacterial plant pathogens. Phytopathology *64*: 881–885.

Mair, N.S. and E. Fox. 1973. An antigenic relationship between *Yersinia pseudotuberculosis* type 6 and *Escherichia coli* O-group 55. Contr. Microbiol. Immunol. *2*: 180–183.

Mair, N.S., C.J. Randall, G.W. Thomas, J.F. Harbourne, C.T. McCrea and K.P. Cowl. 1974. *Actinobacillus suis* infection in pigs: a report of four outbreaks and two sporadic cases. J. Comp. Pathol. *84*: 113–119.

Mäkela, P.H. and H. Mayer. 1976. Enterobacterial common antigen. Bacteriol. Rev. *40*: 591–632.

Mäkela, P.H., V.V. Valtonen and M. Valtonen. 1973. Role of O-antigens (lipopolysaccharide) factors in the virulence of *Salmonella*. J. Infect. Dis. *128S*: 81–85.

Makemson, J.C., N.R. Fulayfil, W. Landry, L.M. Van Ert, C.F. Wimpee, E.A. Widder and J.F. Case. 1997. *Shewanella woodyi* sp. nov., an exclusively respiratory luminous bacterium isolated from the Alboran Sea. Int. J. Syst. Bacteriol. *47*: 1034–1039.

Maki, D.G. and W.T. Martin. 1975. Nationwide epidemic of septicemia caused by contaminated infusion products. IV. Growth of microbial pathogens in fluids for intravenous infusions. J. Infect. Dis. *131*: 267–272.

Maki, D.G., F.S. Rhame, D.C. Mackel and J.V. Bennett. 1976. Nationwide epidemic of septicemia caused by contaminated intravenous products. I. Epidemiologic and clinical features. Am. J. Med. *60*: 471–485.

Maki-Valkama, T. and R. Karjalainen. 1994. Differentiation of *Erwinia carotovora* subsp. *atroseptica* and subsp. *carotovora* by RAPD-PCR. Ann. Appl. Biol. *125*: 301–309.

Makin, S.A. and T.J. Beveridge. 1996. The influence of A-band and B-band lipopolysaccharide on the surface characteristics and adhesion of *Pseudomonas aeruginosa* to surfaces. Microbiology *142*: 299–307.

Makino, S., C. Sasakawa, K. Kamata, T. Kurata and M. Yoshikawa. 1986. A genetic determinant required for continuous reinfection of adjacent cells on large plasmid in *S. flexneri* 2a. Cell *46*: 551–555.

Makula, R.A. 1978. Phospholipid composition of methane-utilizing bacteria. J. Bacteriol. *134*: 771–777.

Makulu, A., F. Gatti and J. Vandepitte. 1973. *Edwardsiella tarda* infections in Zaire. Ann. Soc. Belg. Med. Trop. *53*: 165–172.

Malashenko, Y.R., Y.U. Khaier, E.N. Budkova, Y.U. Isagulova, U. Berger, T.P. Krishtab, D.V. Chernyshenko and V.A. Romanovskaya. 1987. Methane-oxidizing microflora in fresh and saline water reservoirs. Mikrobiologiya *56*: 115–120.

Malashenko, Y.R., V.A. Romanovskaya and V.N. Bogachenko. 1975a. Method of isolating pure cultures of mesophilic, thermotolerant and thermophilic methane-utilizing bacteria. Mikrobiologiya *44*: 707–713.

Malashenko, Y.R., V.A. Romanovskaya, V.N. Bogachenko and A.D. Shved. 1975b. Thermophilic and thermotolerant methane-assimilating bacteria. Mikrobiologiya *44*: 773–779.

Malashenko, Y.R., V.A. Romanovskaya and E.I. Kvasnikov. 1972. Taxonomy of bacteria utilizing gaseous hydrocarbons. Mikrobiologiya *41*: 871–879.

Maldonado, R., J. Jimenez and J. Casadesus. 1994. Changes of ploidy during the *Azotobacter vinelandii* growth cycle. J. Bacteriol. *176*: 3911–3919.

Malinowski, F. 1966. A primary isolation medium for the differentiation of genus *Proteus* from other non-lactose and lactose fermenters. Can. J. Med. Technol. *28*: 118–121.

Malkoff, K. 1906. Weitere Untersuchungen über die Bakterienkrankheit auf Sesamum orientale. Zentralbl. Bakteriol. Parasitenk. Infektionskr. Hyg. Abt. II. *16*: 664–666.

Mallavia, L.P. 1991. Genetics of rickettsiae. Eur. J. Epidemiol. *7*: 213–221.

Malnick, H. 1997. *Anaerobiospirillum thomasii* sp. nov., an anaerobic spiral bacterium isolated from the feces of cats and dogs and from diarrheal feces of humans, and emendation of the genus *Anaerobiospirillum*. Int. J. Syst. Bacteriol. *47*: 381–384.

Malnick, H., K. Williams, J. Philebosie and A.S. Levy. 1990. Description of a medium for isolating *Anaerobiospirillum* spp., a possible cause of zoonotic disease, from diarrheal feces and blood of humans and use of the medium in a survey of human, canine, and feline feces. J. Clin. Microbiol. *28*: 1380–1384.

Malofeeva, I.V., E.N. Kondratieva and A.B. Rubin. 1975. Ferredoxin linked nitrate reductase from the phototrophic bacterium *Ectothiorhodospira shaposhnikovii*. FEBS Lett. *53*: 188–189.

Maltezou, H.C. and D. Raoult. 2002. Q fever in children. Lancet Infect. Dis. *2*: 686–691.

Mammeri, H., G. Laurans, M. Eveillard, S. Castelain and F. Eb. 2001. Coexistence of SHV-4 and TEM-24 producing *Enterobacter aerogenes* strains before a large outbreak of TEM-24 producing strains in a French hospital. J. Clin. Microbiol. *39*: 2184–2190.

Mamolen, M., R.F. Breiman, J.M. Barbaree, R.A. Gunn, K.M. Stone, J.S. Spika, D.T. Dennis, S.H. Mao and R.L. Vogt. 1993. Use of multiple molecular subtyping techniques to investigate a Legionnaires' disease outbreak due to identical strains at two tourist lodges. J. Clin. Microbiol. *31*: 2584–2588.

Manafi, M. and M.L. Rotter. 1992. Enzymatic profile of *Plesiomonas shigelloides*. J. Microbiol. Meth. *16*: 175–180.

Mandel, A.D., K. Wright and J.M. McKinnon. 1964. Selective medium for isolation of *Mima* and *Herellea* organisms. J. Bacteriol. *88*: 1524–1525.

Mandel, M. 1966. Deoxyribonucleic acid base composition in the genus *Pseudomonas*. J. Gen. Microbiol. *43*: 273–292.

Mandel, M., E.R. Leadbetter, N. Pfennig and H.G. Trüper. 1971. Deoxyribonucleic acid base composition of phototrophic bacteria. Int. J. Syst. Bacteriol. *21*: 222–230.

Mandic-Mulec, I., J. Weiss and A. Zychlinsky. 1997. *Shigella flexneri* is trapped in polymorphonuclear leukocyte vacuoles and efficiently killed. Infect. Immun. *65*: 110–115.

Manjarrez-Hernandez, H.A., T.J. Baldwin, P.H. Williams, R. Haigh, S. Knutton and A. Aitken. 1996. Phosphorylation of myosin light chain at distinct sites and its association with the cytoskeleton during enteropathogenic *Escherichia coli* infection. Infect. Immun. *64*: 2368–2370.

Mann, J.M., H.F. Hull, G.P. Schmid and W.E. Droke. 1984. Plague and the peripheral smear. J. Amer. Med. Assoc. *251*: 953.

Mann, S. 1969. Melanin-forming strains of *Pseudomonas aeruginosa*. Arch. Mikrobiol. *65*: 359–379.

Manna, A.C. and H.K. Das. 1993. Determination of the size of the *Azotobacter vinelandii* chromosome. Mol. Gen. Genet. *241*: 719–722.

Manna, A.C. and H.K. Das. 1994. The size of the chromosome of *Azotobacter chroococcum*. Microbiology-(UK). *140*: 1237–1239.

Manna, A.C. and H.K. Das. 1997. Characterization and mutagenesis of the leucine biosynthetic genes of *Azotobacter vinelandii*: An analysis of the rarity of amino acid auxotrophs. Mol. Gen. Genet. *254*: 207–217.

Mannheim, W. 1981. Taxonomic implications of DNA relatedness and quinone patterns in *Actinobacillus*, *Haemophilus*, and *Pasteurella*. *In* Kilian, Frederiksen and Biberstein (Editors), *Haemophilus*, *Pasteurella* and *Actinobacillus*, Academic Press, London. 265–280.

Mannheim, W. 1983. Taxonomy of the family *Pasteurellaceae* Pohl 1981 as revealed by DNA/DNA hybridization. Colloques de l' INSERM. *114*: 211–226.

Mannheim, W. 1984. Family III. *Pasteurellaceae*. *In* Krieg and Holt (Editors), Bergey's Manual of Systematic Bacteriology, 1st Ed., Vol. 1, The Williams & Wilkins Co., Baltimore. pp. 550–575.

Mannheim, W. and S. Pohl. 1980. Deoxyribonucleic acid relatedness among members of the genera *Actinobacillus, Haemophilus,* and *Pasteurella* (*Pasteurellaceae* fam. nov.). *In* Conference on Taxonomy, Computer Identification of Bacteria and Diagnostic Methods. Abstracts of papers, Czechoslovakian Society for Microbiology, Liblice. pp. 35.

Mannheim, W., S. Pohl and R. Holländer. 1980. Zür systematik von *Actinobacillus, Haemophilus* und *Pasteurella*: basenzusammensetzung der DNS, atmungschinone und kulturell-biochemische eigenschaften repräsentativer sammlungsstämme. Zentralbl. Bakteriol. Parasitenkd. Infektionskr. Hyg. Abt. I Orig. A. *246*: 512–540.

Manning, P.J., R.F. DiGiacomo and D. DeLong. 1989. Pasteurellosis is laboratory animals. *In* Adlam and Rutter (Editors), *Pasteurella* and Pasteurellosis, Academic Press, London. 263–302.

Manning, P.A., U.H. Stroeher and R. Morona. 1994. Molecular basis for O-antigen switching in *Vibrio cholerae* O1: Ogawa-Inaba switching. *In* Wachsmuth, Blake and Olsvik (Editors), *Vibrio cholerae* and Cholera, American Society for Microbiology, Washington, D.C. pp. 77–94.

Manns, T.F. 1909. The blade blight of oats. A bacterial disease. Bull. Ohio Agr. Expt. Sta. *210*: 91–167.

Manns, T.F. and J.J. Taubenhaus. 1913. Streak: a bacterial disease of the sweet pea and clovers. Gardener's Chronicle. *53*: 215–216.

Manorama, V. Taneja, R.K. Agarwal and S.C. Sanyal. 1983. Enterotoxins of *Plesiomonas shigelloides*: partial purification and characterization. Toxicon (Suppl). *3*: 269-272.

Manten, A., B. van Klingeren and M. Dessens-Kroon. 1976. Chloramphenicol resistance in *Haemophilus influenzae*. Lancet *1*: 702.

Manz, W., R. Amann, W. Ludwig, M. Wagner and K.H. Schleifer. 1992. Phylogenetic oligodeoxynucleotide probes for the major subclasses of *Proteobacteria*: Problems and solutions. System. Appl. Microbiol. *15*: 593–600.

Maraki, S., G. Samonis, E. Marnelakis and Y. Tselentis. 1994. Surgical wound infection caused by *Rahnella aquatilis*. J. Clin. Microbiol. *32*: 2706–2708.

Marandi, M.V. and K.R. Mittal. 1995. Identification and characterization of outer-membrane proteins of *Pasteurella multocida* serotype-*D* by using monoclonal antibodies. J. Clin. Microbiol. *33*: 952–957.

Marandi, M.V. and K.R. Mittal. 1996. Characterization of an outer membrane protein of *Pasteurella multocida* belonging to the OmpA family. Vet. Microbiol. *53*: 303–314.

Marchette, N.J. and P.S. Nicholes. 1961. Virulence and citrulline ureidase activity of *Pasteurella tularensis*. J. Bacteriol. *82*: 26–32.

Marconi, A.M., F. Beltrametti, G. Bestetti, F. Solinas, M. Ruzzi, E. Galli and E. Zennaro. 1996. Cloning and characterization of styrene catabolism genes from *Pseudomonas fluorescens* ST. Appl. Environ. Microbiol. *62*: 121–127.

Marcos, M.A., S. Abdalla, F. Pedraza, A. Andreu, F. Fernandez, R. Gomez-Lus, M.T. Jimenez De Anta and J. Vila. 1994. Epidemiological markers of *Acinetobacter baumannii* clinical isolates from a spinal cord injury unit. J. Hosp. Infect. *28*: 39–48.

Marcus, B.B., S.B. Samuels, B. Pittman and W.B. Cherry. 1969. A serologic study of *Herellea vaginicola* and its identification by immunofluorescent staining. Am. J. Clin. Pathol. *52*: 309–319.

Marcus, H., J.M. Ketley, J.B. Kaper and R.K. Holmes. 1990. Effects of DNase production, plasmid size, and restriction barriers on transformation of *Vibrio cholerae* by electroporation and osmotic shock. FEMS Microbiol. Lett. *56*: 149–154.

Marcus, L.C. 1971. Infectious diseases of reptiles. J. Am. Vet. Med. Assoc. *159*: 1626–1631.

Marcus, L.C. 1981. Veterinary Biology and Medicine of Captive Amphibians and Reptiles, Lea & Febiger, Philadelphia.

Marengo, G.W. 1983. Encystment using different carbon substrates in *Azotobacter chroococcum*. Rev Argent Microbiol. *15*: 143–146.

Margesin, R. and F. Schinner. 1991. Characterization of a metalloprotease from psychrophilic *Xanthomonas maltophilia*. FEMS Microbiol. Lett. *79*: 257–262.

Mariani-Kurkdjian, P., H. Cavé, J. Elion, C. Loirat and E. Bingen. 1997. Direct detection of verotoxin genes in stool samples by polymerase chain reaction in hemolytic uremic syndrome patients in France. Clin. Microbiol. Infect. *3*: 117–119.

Mariani-Kurkdjian, P., E. Denamur, A. Milon, B. Picard, H. Cavé, N. Lambert Zechovsky, C. Loirat, P. Goullet, P.J. Sansonetti and J. Elion. 1993. Identification of a clone of *Escherichia coli* O103:H2 as a potential agent of hemolytic-uremic syndrome in France. J. Clin. Microbiol. *31*: 296–301.

Marin, M.E., A.V. Carrascosa and I. Cornejo. 1996. Characterization of *Enterobacteriaceae* strains isolated during industrial processing of dry-cured hams. Food Microbiol. *13*: 375–381.

Markel, D.E., M.J. Fowler and C. Eklund. 1975. Phage-host specificity tests using *Levinea* phages and isolates of *Levinea* spp. and *Citrobacter freundii*. Int. J. Syst. Bacteriol. *25*: 215–218.

Markowitz, S.M. 1980. Isolation of an ampicillin-resistant, non-beta-lactamase-producing strain of *Haemophilus influenzae*. Antimicrob. Agents Chemother. *17*: 80–83.

Marmur, J. 1961. A procedure for the isolation of DNA from microorganisms. J. Mol. Biol. *3*: 208–218.

Marmur, J. and P. Doty. 1962. Determination of the base composition of deoxyriboniucleic acid from its thermal denaturation temperature. J. Mol. Biol. *5*: 109–118.

Marques, L.R.M., J.S.M. Peiris, S.J. Cryz and A.D. O'Brien. 1987. *Escherichia coli* strains isolated from pigs with edema disease produce a variant of Shiga-like Toxin II. FEMS Microbiol. Lett. *44*: 33–38.

Marques, L.R.M., M.R.F. Toledo, N.P. Silva, M. Magalhaes and L.R. Trabulsi. 1984. Invasion of HeLa cells by *Edwardsiella tarda*. Curr. Microbiol. *10*: 129–132.

Márquez, M.C., A. Ventosa and F. Ruiz-Berraquero. 1987. A taxonomic study of heterotrophic halophilic and non-halophilic bacteria from a solar saltern. J. Gen. Microbiol. *133*: 45–56.

Marrs, C.F., F.W. Rozsa, M. Hackel, S.P. Stevens and A.C. Glasgow. 1990. Identification, cloning and sequencing of *piv*, a new gene involved in inverting the pilin genes of *Moraxella lacunata*. J. Bacteriol. *172*: 4370–4377.

Marrs, C.F., G. Schoolnik, J.M. Koomey, J. Hardy, J. Rothbard and S. Falkow. 1985. Cloning and sequencing of a *Moraxella bovis* pilin gene. J. Bacteriol. *163*: 132–139.

Mars, A.E., T. Kasberg, S.R. Kaschabek, M.H. van Agteren, D.B. Janssen and W. Reineke. 1997. Microbial degradation of chloroaromatics: Use of the meta-cleavage pathway for mineralization of chlorobenzene. J. Bacteriol. *179*: 4530–4537.

Marshall, A.R., I.J. Al Jumaili and A.J. Bint. 1986. The isolation of *Moellerella wisconsensis* from stool samples in the U.K. J. Infect. *12*: 31–33.

Marshall, D.L., J.J. Kim and S.P. Donnelly. 1996. Antimicrobial susceptibility and plasmid-mediated streptomycin resistance of *Plesiomonas shigelloides* isolated from blue crab. J. Appl. Bacteriol. *81*: 195–200.

Marshall, W.F., M.R. Keating, J.P. Anhalt and J.M. Steckelberg. 1989. *Xanthomonas maltophilia*: an emerging nosocomial pathogen. Mayo Clin. Proc. *64*: 1097–1104.

Marston, B.J., H.B. Lipman and R.F. Breiman. 1994. Surveillance for Legionnaires' disease: risk factors for morbidity and mortality. Arch. Intern. Med. *154*: 2417–2422.

Martel, A.Y., G. St. Laurent, L.A. Dansereau and M.G. Bergeron. 1989. Isolation and biochemical characterization of *Haemophilus* species isolated simultaneously from the oropharyngeal and anogenital areas. J. Clin. Microbiol. *27*: 1486–1489.

Martens, B., H. Spiegl and E. Stackebrandt. 1987. Sequence of a 16S ribosomal RNA gene of: the relation between homology values and similarity coefficients. Syst. Appl. Microbiol. *9*: 224–230.

Martin, B.F., B.M. Derby, G.N. Budzilovich and J. Ransohoff. 1967. Brain abscesses due to *Actinobacillus actinomycetemcomitans*. Neurology *17*: 833–837.

Martin, G. and F. Jacob. 1962. Transfert de l'épisome sexuel d'*Escherichia coli* à *Pasteurella pestis*. C.R. Acad. Sci. *254*: 3589–3590.

Martin, J.P., J. Fleck, M. Mock and J.M. Ghuysen. 1973. The wall peptidoglycans of *Neisseria perflava, Moraxella glucidolytica, Pseudomonas alcaligenes* and *Proteus vulgaris* strain P18. Eur. J. Biochem. *38*: 301–306.

Martin, P.R., R.J. Shea and M.H. Mulks. 2001. Identification of a plasmid-encoded gene from *Haemophilus ducreyi* which confers NAD independence. J. Bacteriol. *183*: 1168–1174.

Martinez, L.M. 1987. *Edwardsiella tarda* bacteremia. Eur. J. Clin. Microbiol. *6*: 599–600.

Martinez, M.M., G. Sanchez, J. Gomez, P. Mendaza and R.M. Daza. 1998. Isolation of *Leclercia adecarboxylata* in ulcer exudate (Translation). Enferm. Infecc. Microbiol. Clin. *16*: 345.

Martinez, M.J., D. Simon-Pujol, F. Congregado, S. Merino, X. Rubires and J.M. Tomas. 1995. The presence of capsular polysaccharide in mesophilic *Aeromonas hydrophila* serotypes 0:11 and 0:34. FEMS Microbiol. Lett. *128*: 69–74.

Martinez-Murcia, A.J. 1999. Phylogenetic positions of *Aeromonas encheleia*, *Aeromonas popoffii Aeromonas* DNA hybridization Group 11 and *Aeromonas* Group 501. Int. J. Syst. Bacteriol. *49*: 1403–1408.

Martinez-Murcia, A.J., A.I. Antón and F. Rodriguez-Valera. 1999. Patterns of sequence variation in two regions of the 16S rRNA multigene family of *Escherichia coli*. Int. J. Syst. Bacteriol. *49*: 601–610.

Martinez-Murcia, A.J., S. Benlloch and M.D. Collins. 1992a. Phylogenetic interrelationships of members of the genera *Aeromonas* and *Plesiomonas* as determined by 16S ribosomal DNA sequencing: lack of congruence with results of DNA-DNA hybridizations. Int. J. Syst. Bacteriol. *42*: 412–421.

Martinez-Murcia, A.J., C. Esteve, E. Garay and M.D. Collins. 1992b. *Aeromonas allosaccharophila* sp. nov., a new mesophilic member of the genus *Aeromonas*. FEMS Microbiol. Lett. *91*: 199–206.

Martinez-Murcia, A.J., C. Esteve, E. Garay and M.D. Collins. 1992c. *In* Validation of the publication of new names and new combinations previously effectively published outside the IJSB. List No. 42. Int. J. Syst. Bacteriol. *42*: 511.

Martínez-Picado, J., A.R. Blanch and J. Jofre. 1994. Rapid detection and identification of *Vibrio anguillarum* by using a specific oligonucleotide probe complementary to 16S rRNA. Appl. Environ. Microbiol. *60*: 732–737.

Martinez-Salazar, J.M., S. Moreno, R. Najera, J.C. Boucher, G. Espin, G. Soberon-Chavez and V. Deretic. 1996. Characterization of the genes coding for the putative sigma factor AlgU and its regulators MucA, MucB, MucC, and MucD in *Azotobacter vinelandii* and evaluation of their roles in alginate biosynthesis. J. Bacteriol. *178*: 1800–1808.

Martinez-Salazar, J.M., A.N. Palacios, R. Sanchez, A.D. Caro and G. Soberon-Chavez. 1993. Genetic stability and xanthan gum production in *Xanthomonas campestris* pv. *campestris* NRRL B1459. Mol. Microbiol. *8*: 1053–1061.

Martinez-Toledo, M.V., T. Delarubia, J. Moreno and J. Gonzalezlopez. 1988. Root exudates of *Zea mays* and production of auxins, gibberellins and cytokinins by *Azotobacter chroococcum*. Plant Soil *110*: 149–152.

Martinez-Toledo, M.V., J. Gonzalezlopez, B. Rodelas, C. Pozo and V. Salmeron. 1995. Production of poly-beta-hydroxybutyrate by *Azotobacter chroococcum* H23 in chemically-defined medium and alpechin medium. J. Appl. Bacteriol. *78*: 413–418.

Maruyama, A., D Honda, H. Yamamoto, K. Kitamura and T. Higashihara. 2000. Phylogenetic analysis of psychrophilic bacteria isolated from the Japan Trench, including a description of the deep-sea species *Psychrobacter pacificensis* sp nov. Int. J. Syst. Evol. Microbiol. *50*: 835–846.

Maserati, R., C. Farina, L. Valenti and C. Filice. 1985. Liver abscess caused by *Edwardsiella tarda*. Boll. Ist. Sieroter. Milan. *64*: 419–421.

Mashima, T.Y., T.E. Cornish and G.A. Lewbart. 1997. Amyloidosis in a Jack Dempsey cichlid, *Cichlasoma biocellatum* Regan. J. Fish Dis. *20*: 73–75.

Maskell, D.J., M.J. Szabo, P.D. Butler, A.E. Williams and E.R. Moxon. 1992. Molecular biology of phase-variable lipopolysaccharide biosynthesis by *Haemophilus influenzae*. J. Infect. Dis. *165 (Suppl 1)*: S90–92.

Maslog, F.S., M. Motobu, N. Hayashida, K. Yoshihara, T. Morozumi, M. Matsumura and Y. Hirota. 1999. Effects of the lipopolysaccharide-protein complex and crude capsular antigens of *Pasteurella multocida* A on antibody responses and delayed type hypersensitivity responses in the chicken. J. Vet. Med. Sci. *61*: 565–567.

Massa, S., R. Armuzzi, M. Tosques, F. Canganella and L.D. Trovatelli.

1999. Note: Susceptibility to chlorine of *Aeromonas hydrophila* strains. J. Appl. Microbiol. *86*: 169–173.

Massad, G., F.K. Bahrani and H.L. Mobley. 1994a. *Proteus mirabilis* fimbriae: identification, isolation, and characterization of a new ambient-temperature fimbria. Infect. Immun. *62*: 1989–1994.

Massad, G., C.V. Lockatell, D.E. Johnson and H.L. Mobley. 1994b. *Proteus mirabilis* fimbriae: construction of an isogenic *pmfA* mutant and analysis of virulence in a CBA mouse model of ascending urinary tract infection. Infect. Immun. *62*: 536–542.

Massad, G., H. Zhao and H.L. Mobley. 1995. *Proteus mirabilis* amino acid deaminase: cloning, nucleotide sequence, and characterization of *aad*. J. Bacteriol. *177*: 5878–5883.

Masumura, K., H. Yasunobu, N. Okada and K. Muroga. 1989. Isolation of a *Vibrio* sp. the causative bacterium of intestinal necrosis of Japanese flounder larvae. Fish Pathol. *24*: 135–141.

Mates, A., D. Eyny and S. Philo. 2000. Antimicrobial resistance trends in *Shigella* serogroups isolated in Israel, 1990–1995. Eur. J. Clin. Microbiol. Infect. Dis. *19*: 108–111.

Matheron, R. and R. Baulaigue. 1972. Marine phototrophic sulfur bacteria, assimilation of organic and mineral substances, and influnce of the NaCl content of the medium upon growth. Arch. Mikrobiol. *86*: 291–304.

Matheson, A.T., G.D. Sprott, I.J. McDonald and H. Tessier. 1976. Some properties of an unidentified halophile: growth characteristics, internal salt concentration, and morphology. Can. J. Microbiol. *22*: 780–786.

Mathias, R.G., A.R. Ronald, M.J. Garwith, D.W. McCullough, H.G. Stiver, J. Berger, C.Y. Cates, L.M. Fox and B.A. Lank. 1976. Clinical evaluation of amikacin in treatment of infections due to Gram-negative aerobic bacilli. J. Infect. Dis. *134: Suppl.*: S394–S401.

Mathur, R.S., J. Swarup and S.K. Sinha. 1964. *Xanthomonas boerhaaviae* sp. nov. on *Boerhaavia repens* L. J. Sci. Technol. *2*: 257.

Matsui, T., K. Kayashima, M. Kito and T. Ono. 1996. Three cases of *Pasteurella multocida* skin infection from pet cats. J. Dermatol. *23*: 502–504.

Matsui, Y., S. Suzuki, T. Suzuki and K. Takama. 1991. Phospholipid and fatty acid compositions of *Alteromonas putrefaciens* and *A. haloplanktis*. Lett. Appl. Microbiol. *12*: 51–53.

Matsukura, H., K. Katayama, N. Kitano, K. Kobayashi, C. Kanegane, A. Higuchi and S. Kyotani. 1996. Infective endocarditis caused by an unusual gram-negative rod, *Rahnella aquatilis*. Pediatr. Cardiol. *17*: 108–111.

Matsumoto, H. 1963. Studies on the *Hafnia* isolated from normal human. Japan J. Microbiol. *7*: 105–114.

Matsumoto, H. 1964. Additional new antigens of *Hafnia* group. Jpn. J. Microbiol. *8*: 139–141.

Matsumoto, H. and T. Tazaki. 1970. Genetic recombination in *Klebsiella pneumoniae*. An approach to genetic linkage mapping. Jpn. J. Microbiol. *14*: 129–141.

Matsumoto, H., T. Tazaki and S. Hosogaya. 1973. A generalized transducing phage of *Serratia marcescens*. Jpn. J. Microbiol. *17*: 473–479.

Matsushita, K., E. Shinagawa, O. Adachi and M. Ameyama. 1989. Quinoprotein D-glucose dehydrogenase of the *Acinetobacter calcoaceticus* respiratory chain: membrane-bound and soluble forms are different molecular species. Biochemistry *28*: 6276–6280.

Matsushita, S., Y. Kudoh and M. Ohashi. 1984. Transferable resistance to the vibriostatic agent 2,4-diamino-6,7-diisopropyl-pteridine (O/129) in *Vibrio cholerae*. Microbiol. Immunol. *28*: 1159–1162.

Matsushita, S., Y. Noguchi, Y. Yanagawa, H. Igarashi, Y. Ueda, S. Hashimoto, S. Yano, K. Morita, M. Kanamori and Y.A. Kudoh. 1998. *Shigella dysenteriae* strains possessing a new serovar (204/96) isolated from imported diarrheal cases in Japan. Kansenshogaku Zasshi. *72*: 499–503.

Matsushita, S., Y. Noguchi, Y. Yanagawa, K. Kobayashi, H. Nakaya, H. Igarashi and Y. Kudoh. 1997. *Shigella dysenteriae* strains possessing a new serovar isolated from imported diarrheal cases in Japan. Kansenshogaku Zasshi. *71*: 412–416.

Matsushita, S., S. Yamada and Y. Kudoh. 1992a. *Shigella flexneri* strains having a new type antigen. Kansenshogaku Zasshi. *66*: 503–507.

Matsushita, S., S. Yamada and Y. Kudoh. 1992b. *Shigella flexneri* strains having a new type antigen 89-141. Kansenshogaku Zasshi. *66*: 1628–1633.

Matsutani, S. and E. Ohtsubo. 1993. Distribution of the *Shigella sonnei* insertion elements in *Enterobacteriaceae*. Gene *127*: 111–115.

Matte, M.H., G.R. Matte, A.I. Gil, F. Lanata, A. Huq and R.R. Colwell. 1999. Molecular epidemiology of *"Aeromonas arequipensis"* isolated from diarrhea in Arequipa, Peru. 6th International Aeromonas-/Plesiomonas Symposium, Chicago, Illinois. 19.

Matthews, B.G., H. Douglas and D.G. Guiney. 1988. Production of a heat stable enterotoxin by *Plesiomonas shigelloides*. Microb. Pathog. *5*: 207–213.

Matthews, P.R.J. and I.H. Pattison. 1961. The identification of a *Hemophilus*-like organism associated with pneumonia and pleurisy in the pig. J. Comp. Pathol. *71*: 44–52.

Mattick, J.S. 1989. The molecular biology of the fimbriae (pili) of *Bacteriodes nodosus* and the development of a recombinant-DNA-based vaccine. *In* Egerton, Yong and Riffkin (Editors), Footrot and Foot Abscess in Ruminants, CRC Press, Boca Raton. 195–218.

Mattick, J.S., B.J. Anderson, P.T. Cox, B.P. Dalrymple, M.M. Bills, M. Hobbs and J.R. Egerton. 1991. Gene sequences and comparison of the fimbrial subunits representative of *Bacteroides nodosus* serotypes A to I: class I and class II strains. Mol. Microbiol. *5*: 561–573.

Mattick, J.S., C.B. Whitchurch and R.A. Alm. 1996. The molecular genetics of type-4 fimbriae in *Pseudomonas aeruginosa* - A review. Gene *179*: 147–155.

Mauel, M.J., S.J. Giovannoni and J.L. Fryer. 1999. Phylogenetic analysis of *Piscirickettsia salmonis* by 16S, internal transcribed spacer (ITS) and 23S ribosomal DNA sequencing. Dis. Aquat. Org. *35*: 115–123.

Mauff, A.C., S. Miller, V. Kuhnle and M. Carmichael. 1983. Infections due to *Actinobacillus actinomycetemcomitans*. S. Afr. Med. J. *63*: 580–581.

Maurelli, A.T., B. Baudry, H. d'Hauteville, T.L. Hale and P.J. Sansonetti. 1985. Cloning of plasmid DNA sequences involved in invasion of HeLa cells by *Shigella flexneri*. Infect. Immun. *49*: 164–171.

Maurelli, A.T., B. Blackmon and R. Curtiss. 1984a. Loss of pigmentation in *Shigella flexneri* 2a is correlated with loss of virulence and virulence-associated plasmid. Infect. Immun. *43*: 397–401.

Maurelli, A.T., B. Blackmon and R. Curtiss. 1984b. Temperature-dependent expression of virulence genes in *Shigella* species. Infect. Immun. *43*: 195–201.

Maurelli, A.T., R.E. Fernandez, C.A. Bloch, C.K. Rode and A. Fasano. 1998. "Black holes" and bacterial pathogenicity: a large genomic deletion that enhances the virulence of *Shigella* spp. and enteroinvasive *Escherichia coli*. Proc. Natl. Acad. Sci. U.S.A. *95*: 3943–3948.

Maurin, M., A.M. Benoliel, P. Bongrand and D. Raoult. 1992. Phagolysosomal alkalinization and the bactericidal effect of antibiotics: The *Coxiella burnetii* paradigm. J. Infect. Dis. *166*: 1097–1102.

Maurin, M. and D. Raoult. 1999. Q fever. Clin. Microbiol. Rev. *12*: 518–553.

Mavris, M., P.A. Manning and R. Morona. 1997. Mechanism of bacteriophage SfII-mediated serotype conversion in *Shigella flexneri*. Mol. Microbiol. *26*: 939–950.

May, B.J., Q. Zhang, L.L. Li, M.L. Paustian, T.S. Whittam and V. Kapur. 2001. Complete genomic sequence of *Pasteurella multocida*, Pm70. Proc. Natl. Acad. Sci. U.S.A. *98*: 3460–3465.

May, T.B. and A.M. Chakrabarty. 1994. *Pseudomonas aeruginosa*: genes and enzymes of alginate synthesis. Trends Microbiol. *2*: 151–157..

Mayo, J.B. and L.R. McCarthy. 1977. Antimicrobial susceptibility of *Haemophilus parainfluenzae*. Antimicrob. Agents Chemother. *11*: 844–847.

Mazloum, H.A., M. Kilian, Z.M. Mohamed and M.D. Said. 1982. Differentiation of *Haemophilus aegyptius* and *Haemophilus influenzae*. Acta Pathol. Microbiol. Immunol. Scand. [B]. *90*: 109–112.

Mazoy, R., L.M. Botana and M.L. Lemos. 1997. Iron uptake from ferric citrate by *Vibrio anguillarum*. FEMS Microbiol. Lett. *154*: 145–150.

McAllen, R. and F. Hannah. 1999. Biofouling of the high-shore rockpool harpacticoid copepod *Tigriopus brevicornis*. J. Nat. Hist. *33*: 1781–1787.

McAllen, R. and G.W. Scott. 2000. Behavioural effects of biofouling in a marine copepod. J. Mar. Biol. Assoc. U. K. *80*: 369–370.

McAllister, H.A. and G.R. Carter. 1974. An aerogenic *Pasteurella*-like organism recovered from swine. Am. J. Vet. Res. *35*: 917–922.

McCallum, K.L., G. Schoenhals, D. Laakso, B. Clarke and C. Whitfield. 1989. A high molecular weight fraction of smooth lipopolysaccharide in *Klebsiella* serotype O1:K20 contains a unique O-antigen epitope and determines resistance to nonspecific serum killing. Infect. Immun. *57*: 3816–3822.

McCarter, L.L. 1995. Genetic and molecular characterization of the polar flagellum of *Vibrio parahaemolyticus*. J. Bacteriol. *177*: 1595–1609.

McCarter, L. and M. Silverman. 1990. Surface-induced swarmer cell differentiation of *Vibrio parahaemolyticus*. Mol. Microbiol. *4*: 1057–1062.

McCarthy, D.H. 1975. The bacteriology and taxonomy of *Aeromonas liquefaciens*, Weymouth, Dorset, . Fish Diseases Laboratory, Ministry of Agriculture, . 2 pp.

McCaul, T.F. 1991. The development cycle of *Coxiella burnetii*. *In* Williams and Thompson (Editors), Q fever: The Biology of *Coxiella burnetii*, CRC Press, Inc., Boca Raton. pp. 223–258.

McCaul, T.F. and J.C. Williams. 1991. Developmental cycle of *Coxiella burnetii*: Structure and morphogenesis of vegetative and sporogenic differentiations. J. Bacteriol. *147*: 1063–1076.

McClelland, M., R. Jones, Y. Patel and M. Nelson. 1987. Restriction endonucleases for pulsed field-mapping of bacterial genomes. Nucleic Acids Res. *15*: 5985–6005.

McClelland, M., M. Nelson and C.R. Cantor. 1985. Purification of *Mbo*II methylase (GAAGmA) from *Moraxella bovis*: site specific cleavage of DNA at nine and ten base pair sequences. Nucleic Acids Res. *13*: 7171–7182.

McClure, H.E., W.C. Eveland and A. Kase. 1957. The occurrence of certain *Enterobacteriaceae* in birds. Am. J. Vet. Res. *18*: 207–209.

McCoy, G.W. and C.W. Chapin. 1912. Further observations on a plague-like disease of rodents with a preliminary note on the causative agent: *Bacterium tularense*. J. Infect. Dis. *10*: 61–72.

McCrumb, F. 1961. Aerosol infection of man with *Pasteurella tularensis*. Bacteriol. Rev. *25*: 262–267.

McCulloch, L. 1911. A spot disease of cauliflower. Bull. U.S. Depart. Agr. Bur. Plant Ind. No. *225*: 1–15.

McCulloch, L. 1920. Basal glumerot of wheat. J. Agr. Res. *18*: 543–552.

McCulloch, L. 1924. A bacterial blight of gladioli. J. Agr. Res. *27*: 225–230.

McCulloch, L. 1929. A bacterial leaf spot of horse-radish caused by *Bacterium campestre* biovar armoraciae n. var. J. Agr. Res. *38*: 269–287.

McCulloch, L. 1937. An iris leaf disease caused by *Bacterium tardicrescens* n. sp. Phytopathology *27*: 135.

McCulloch, L. and P.P. Pirone. 1939. Bacterial leaf spot of dieffenbachia. Phytopathology *29*: 956–962.

McDade, J.E., D.J. Brenner and F.M. Bozeman. 1979. Legionnaires' disease bacterium isolated in 1947. Ann. Intern. Med. *90*: 659–661.

McDade, J.E. and C.C. Shepard. 1979. Virulent to avirulent conversion of Legionnaires' disease bacterium (*Legionella pneumophila*): its effect on isolation techniques. J. Infect. Dis. *139*: 707–711.

McDade, J.E., C.C. Shepard, D.W. Fraser, T.R. Tsai, M.A. Redus, W.R. Dowdle and L.I. Team. 1977. Legionnaires' disease: isolation of a bacterium and demonstration of its role in other respiratory disease. N. Engl. J. Med. *297*: 1197–1203.

McDaniel, T.K., K.G. Jarvis, M.S. Donnenberg and J.B. Kaper. 1995. A genetic locus of enterocyte effacement conserved among diverse enterobacterial pathogens. Proc. Natl. Acad. Sci. U.S.A. *92*: 1664–1668.

McDaniels, A.E., E.W. Rice, A.L. Reyes, C.H. Johnson, R.A. Haugland and G.N.J. Stelma. 1996. Confirmational identification of *Escherichia coli*, a comparison of genotypic and phenotypic assays for glutamate decarboxylase and β-D-glucuronidase. Appl. Environ. Microbiol. *62*: 3350–3354.

McDermott, C. and J.M. Mylotte. 1984. *Morganella morganii*: epidemiology of bacteremic disease. Infect. Control. *5*: 131–137.

McDonald, I.R., D.P. Kelly, J.C. Murrell and A.P. Wood. 1997. Taxonomic relationships of *Thiobacillus halophilus*, *Thiobacillus aquaesulis*, and other species of *Thiobacillus*, as determined using 16S rRNA sequencing. Arch. Microbiol. *166*: 394–398.

McDonald, I.R. and J.C. Murrell. 1997. The methanol dehydrogenase structural gene *mxaF* and its use as a functional gene probe for methanotrophs and methylotrophs. Appl. Environ. Microbiol. *63*: 3218–3224.

McDonough, M.A. and J.R. Butterton. 1999. Spontaneous tandem amplification and deletion of the shiga toxin operon in *Shigella dysenteriae* 1. Mol. Microbiol. *34*: 1058–1069.

McElroy, L.J. and N.R. Krieg. 1972. A serological method for the identification of spirilla. Can. J. Microbiol. *18*: 57–64.

McGlannan, M.F. and J.C. Makemson. 1990. HCO₃⁻ Fixation by naturally-occurring tufts and pure cultures of *Thiothrix nivea*. Appl. Environ. Microbiol. *56*: 730–738.

McGowan, S., M. Sebaihia, S. Jones, B. Yu, N. Bainton, P.F. Chan, B. Bycroft, G.S. Stewart, P. Williams and G.P. Salmond. 1995. Carbapenem antibiotic production in *Erwinia carotovora* is regulated by *CarR*, a homologue of the *LuxR* transcriptional activator. Microbiology *141*: 541–550.

McHale, P.J., F. Walker, B. Scully, L. English and C.T. Keane. 1981. *Providencia stuartii* infections: a review of 117 cases over an eight year period. J. Hosp. Infect. *2*: 155–165.

McHatton, S.C., J.P. Barry, H.W. Jannasch and D.C. Nelson. 1996. High nitrate concentrations in vacuolate, autotrophic marine *Beggiatoa* spp. Appl. Environ. Microbiol. *62*: 954–958.

McInerney, B.V., R.P. Gregson, M.J. Lacey, R.J. Akhurst, G.R. Lyons, S.H. Rhodes, D.R. Smith, L.M. Engelhardt and A.H. White. 1991a. Biologically active metabolites from *Xenorhabdus* spp., Part 1. Dithiolopyrrolone derivatives with antibiotic activity. J. Nat. Prod. (Lloydia) *54*: 774–784.

McInerney, B.V., W.C. Taylor, M.J. Lacey, R.J. Akhurst and R.P. Gregson. 1991b. Biologically active metabolites from *Xenorhabdus* spp., Part 2. Benzopyran-1-one derivatives with gastroprotective activity. J. Nat. Prod. (Lloydia) *54*: 785–795.

McIntosh, D. and B. Austin. 1990. Recovery of cell wall deficient forms (L forms) of the fish pathogens *Aeromonas salmonicida* and *Yersinia ruckeri*. Syst. Appl. Microbiol. *13*: 378–381.

McKay, S.G., D.W. Morck, J.K. Merrill, M.E. Olson, S.C. Chan and K.M. Pap. 1996. Use of tilmicosin for treatment of pasteurellosis in rabbits. Am. J. Vet. Res. *57*: 1180–1184.

McKell, J. and D. Jones. 1976. A numerical taxonomic study of *Proteus-Providence* bacteria. J. Appl. Bacteriol. *41*: 143–161.

McKinney, R.M., R.K. Porschen, P.H. Edelstein, M.L. Bissett, P.P. Harris, S.P. Bondell, A.G. Steigerwalt, R.E. Weaver, M.E. Ein, D.S. Lindquist, R.S. Kops and D.J. Brenner. 1981. *Legionella longbeachae* sp. nov., another etiologic agent of human pneumonia. Ann. Intern. Med. *94*: 739–743.

McKinney, R.M., R.K. Porschen, P.H. Edelstein, M.L. Bissett, P.P. Harris, S.P. Bondell, A.G. Steigerwalt, R.E. Weaver, M.E. Ein, D.S. Lindquist, R.S. Kops and D.J. Brenner. 1982. *In* Validation of the publication of new names and new combinations previously effectively published outside the IJSB. List No. 8. Int. J. Syst. Bacteriol. *32*: 266–268.

McKinney, R.M., L. Thacker, P.P. Harris, K.R. Lewallen, G.A. Hebert, P.H. Edelstein and B.M. Thomason. 1979a. Four serogroups of Legionnaires' disease bacteria defined by direct immunofluorescence. Ann. Intern. Med. *90*: 621–624.

McKinney, R.M., B.M. Thomason, P.P. Harris, L. Thacker, K.R. Lewallen, H.W. Wilkinson, G.A. Hebert and C.W. Moss. 1979b. Recognition of a new serogroup of Legionnaires' disease bacterium. J. Clin. Microbiol. *9*: 103–107.

McLaughlin, B.G., S.C. Greer, M.M. Chengappa, S. Singh, R.L. Maddux, W.L. Kadel and P.S. McLaughlin. 1991. Association of a *Pasteurella haemolytica*-like organism with enteritis in swine. J. Vet. Diagn. Investig. *3*: 324–327.

McLean, D.L. and E.J. Houk. 1973. Phase contrast and electron microscopy of the mycetocytes and symbiotes of the pea aphid, *Acyrthosiphon pisum*. J. Cell Physiol. *19*: 625–633.

McLean, R.J.C., J.C. Nickel, K.-J. Cheng and J.W. Costerton. 1988. The ecology and pathogenicity of urease-producing bacteria in the urinary tract. Crit. Rev. Microbiol. *16*: 37–79.

McLean, R.A., W.L. Sulzbacher and S. Mudd. 1951. *Micrococcus cryophilus* spec. nov., a large coccus especially suitable for cytologic study. J. Bacteriol. *62*: 723–728.

McLennan, K., L.A. Glover, K. Killham and J.I. Prosser. 1992. Luminescence-based detection of *Erwinia carotovora* associated with rotting potato tubers. Lett. Appl. Microbiol. *15*: 121–124.

McLeod, A. and M.C.M. Pérombélon. 1992. Rapid detection and identification of *Erwinia carotovora* subsp. *atroseptica* by a conjugated *Staphylococcus aureus* slide agglutination test. J. Appl. Bacteriol. *72*: 274–280.

McMahon, P.C. 1973. Mapping the chromosome of *Yersinia pseudotuberculosis* by interrupted mating. J. Gen. Microbiol. *77*: 61–69.

McManus, P.S. and A.L. Jones. 1994. Epidemiology and genetic analysis of streptomycin-resistant *Erwinia amylovora* from Michigan and evaluation of oxytetracycline for control. Phytopathology *84*: 627–633.

McManus, P.S. and A.L. Jones. 1995. Detection of *Erwinia amylovora* by nested PCR and PCR–dot-blot and reverse-blot hybridizations. Phytopathol. *85*: 618–623.

McMaster, C., E.A. Roch, G.A. Willshaw, A. Doherty, W. Kinnear and T. Cheasty. 2001. Verocytotoxin-producing *Escherichia coli* serotype O26:H11 outbreak in an Irish creche. Eur. J. Clin. Microbiol. Infect. Dis. *20*: 430–432.

McMorran, B.J., M.E. Merriman, I.T. Rombel and I.L. Lamont. 1996. Characterisation of the *pvdE* gene which is required for pyoverdine synthesis in *Pseudomonas aeruginosa*. Gene *176*: 55–59.

McNeil, M.M., B.J. Davis, S.L. Solomon, R.L. Anderson, S.T. Shulman, S. Gardner, K. Kabat and W.J. Martone. 1987a. *Ewingella americana*: recurrent pseudobacteremia from a persistent environmental reservoir. J. Clin. Microbiol. *25*: 498–500.

McNeil, M.M., W.J. Martone and V.R. Dowell. 1987b. Bacteremia with *Anaerobiospirillum succiniciproducens*. Rev. Infect. Dis. *9*: 737–742.

McWethy, S.J. and P.A. Hartman. 1977. Purification and some properties of an extracellular alpha-amylase from *Bacteroides amylophilus*. J. Bacteriol. *129*: 1537–1544.

McWhorter, A.C., R.L. Haddock, F.A. Nocon, A.G. Steigerwalt, D.J. Brenner, S. Aleksic, J. Bockemuehl and J.J. Farmer, III. 1991. *Trabulsiella guamensis*, a new genus new species of the family *Enterobacteriaceae* that resembles *Salmonella* subgroups 4 and 5. J. Clin. Microbiol. *29*: 1480–1485.

McWhorter, A.C., R.L. Haddock, F.A. Nocon, A.G. Steigerwalt, D.J. Brenner, S. Aleksic, J. Bockemuehl and J.J. Farmer, III. 1992. *In* Validation of the publication of new names and new combinations previously effectively published outside the IJSB. List No. 41. Int. J.Syst. Bacteriol. *42*: 327–328.

Mdluli, K.E., L.S.D. Anthony, G.S. Baron, M.K. McDonald, S.V. Myltseva and F.E. Nano. 1994. Serum-sensitive mutation of *Francisella novicida*: association with an ABC transporter gene. Microbiology (Reading) *140*: 3309–3318.

Mead, P.S., L. Slutsker, V. Dietz, L.F. McCaig, J.S. Bresee, C. Shapiro, P.M. Griffin and R.V. Tauxe. 1999. Food-related illness and death in the United States. Emerg. Infect. Dis. *5*: 607-625.

Medeiros, A.A., R. Levesque and G.A. Jacoby. 1986. An animal source for the ROB-1 beta-lactamase of *Haemophilus influenzae* type b. Antimicrob. Agents Chemother. *29*: 212–215.

Medema, G. and C. Schets. 1993. Occurrence of *Plesiomonas shigelloides* in surface water: relationship with faecal pollution and trophic state. Zentbl. Hyg. Umweltmed. *194*: 398–404.

Médigue, C., T. Rouxel, P. Vigier, A. Hénaut and A. Danchin. 1991. Evidence for horizontal gene transfer in *Escherichia coli* speciation. J. Mol. Biol. *222*: 851–856.

Meenhorst, P.L., A.L. Reingold, D.G. Groothuis, G.W. Gorman, H.W. Wilkinson, R.M. McKinney, J.C. Feeley, D.J. Brenner and R. van Furth. 1985. Water-related nosocomial pneumonia caused by *Legionella pneumophila* serogroups 1 and 10. J. Infect. Dis. *152*: 356–364.

Mehta, R.J. 1977. Methylamine dehydrogenase from the obligate methylotroph *Methylomonas methylovora*. Can. J. Microbiol. *23*: 402–408.

Mekalanos, J.J. 1992. Environmental signals controlling expression of virulence determinants in bacteria. J. Bacteriol. *174*: 1–7.

Mekalanos, J.J., E.J. Rubin and M.K. Waldor. 1997. Cholera: molecular basis for emergence and pathogenesis. FEMS Immunol. Med. Microbiol. *18*: 241–248.

Meletzus, D., P. Rudnick, N. Doetsch, A. Green and C. Kennedy. 1998. Characterization of the *glnK-amtB* operon of *Azotobacter vinelandii*. J. Bacteriol. *180*: 3260–3264.

Melkerson-Watson, L.J., C.K. Rode, L. Zhang, B. Foxman and C.A. Bloch. 2000. Integrated genomic map from uropathogenic *Escherichia coli* J96. Infect. Immun. *68*: 5933–5942.

Mellado, E., J.A. Asturias, J.J. Nieto, K.N. Timmis and A. Ventosa. 1995a. Characterization of the basic replicon of pCM1, a narrow-host-range plasmid from the moderate halophile *Chromohalobacter marismortui*. J. Bacteriol. *177*: 3443–3450.

Mellado, E., M.T. Garcia, J.J. Nieto, S. Kaplan and A. Ventosa. 1997. Analysis of the genome of *Vibrio costicola*: pulsed field gel electrophoretic analysis of genome size and plasmid content. Syst. Appl. Microbiol. *20*: 20–26.

Mellado, E., M.T. Garcia, E. Roldan, J.J. Nieto and A. Ventosa. 1998. Analysis of the genome of the gram-negative moderate halophiles *Halomonas* and *Chromohalobacter* by using pulsed-field gel electrophoresis. Extremophiles 2: 435–438.

Mellado, E., E.R.B. Moore, J.J. Nieto and A. Ventosa. 1995b. Phylogenetic inferences and taxonomic consequences of 16S ribosomal DNA-sequence comparison of *Chromohalobacter marismortui*, *Volcaniella eurihalina*, and *Deleya salina* and reclassification of *Volcaniella eurihalina* as *Halomonas eurihalina* comb. nov. Int. J. Syst. Bacteriol. *45*: 712–716.

Mellado, E., E.R.B. Moore, J.J. Nieto and A. Ventosa. 1996. Analysis of 16S rRNA gene sequences of *Vibrio costicola* strains: description of *Salinivibrio costicola* gen. nov., comb. nov. Int. J. Syst. Bacteriol. *46*: 817–821.

Mellado, E., J.J. Nieto and A. Ventosa. 1995c. Construction of novel shuttle vectors for use between moderately halophilic bacteria and *Escherichia coli*. Plasmid *34*: 157–164.

Melnikov, A. and P.J. Youngman. 1999. Random mutagenesis by recombinational capture of PCR products in *Bacillus subtilis* and *Acinetobacter calcoaceticus*. Nucleic Acids Res. *27*: 1056–1062.

Melton-Celsa, A.R. and A.D. O'Brien. 1998. Structure, biology, and relative toxicity of Shiga toxin family members for cells and animals. *In* Kaper and O'Brien (Editors), *Escherichia coli* O157:H7 and Other Shiga Toxin-Producing *E. coli* Strains, ASM Press, Washington, D.C. pp. 121–128.

Mendelman, P.M., D.O. Chaffin and G. Kalaitzoglou. 1990. Penicillin-binding proteins and ampicillin resistance in *Haemophilus influenzae*. J. Antimicrob. Chemother. *25*: 525–534.

Menn, F.M., B.M. Applegate and G.S. Sayler. 1993. NAH plasmid-mediated catabolism of anthracene and phenanthrene to naphthoic acids. Appl. Environ. Microbiol. *59*: 1938–1942.

Menon, A.L., L.E. Mortenson and R.L. Robson. 1992. Nucleotide sequences and genetic analysis of hydrogen oxidation (*hox*) genes in *Azotobacter vinelandii*. J. Bacteriol. *174*: 4549–4557.

Mercier, J. and S.E. Lindow. 1996. A method involving ice nucleation for the identification of microorganisms antagonistic to *Erwinia amylovora* on pear flowers. Phytopathology *86*: 940–945.

Mergaert, J., L. Hauben, M. Cnockaert and J. Swings. 1999. Reclassification of nonpigmented *Erwinia herbicola* strains from trees as *Erwinia billingiae* sp. nov. Int. J. Syst. Bacteriol. *49*: 377–383.

Mergaert, J., A. Schirmer, L. Hauben, M. Mau, B. Hoste, K. Kersters, D. Jendrossek and J. Swings. 1996. Isolation and identification of poly(3-hydroxyvalerate) degrading strains of *Pseudomonas lemoignei*. Int. J. Syst. Bacteriol. *46*: 769–773.

Mergaert, J., L. Verdonck and K. Kersters. 1993. Transfer of *Erwinia ananas* (synonym, *Erwinia uredovora*) and *Erwinia stewartii* to the genus *Pantoea* emend. as *Pantoea ananas* comb. nov. (Serrano 1928) and *Pantoea stewartii* comb. nov. (Smith 1898), respectively, and description of *Pantoea stewartii* subsp. *indologenes* subsp. nov. Int. J. Syst. Bacteriol. *43*: 162–173.

Merino, S., A. Aguilar, X. Rubires, N. Abitiu, M. Regue and J.M. Tomas. 1997. The role of the capsular polysaccharide of *Aeromonas hydrophila* serogroup O:34 in the adherence to and invasion of fish cell lines. Res. Microbiol. *148*: 625–631.

Merino, S., X. Rubires, A. Aguilar, S. Alberti, S. Hernandez-Alles, V.J. Benedi and J.M. Tomas. 1996. Mesophilic *Aeromonas* sp. serogroup O:11 resistance to complement-mediated killing. Infect. Immun. *64*: 5302–5309.

Merkel, J.R., E.D. Traganza, B.B. Mukherjee, T.B. Griffin and J.M. Prescott. 1964. Proteolytic activity and general characteristics of a marine bacterium, *Aeromonas proteolytica* sp. n. J. Bacteriol. *87*: 1227–1233.

Merriman, T.R., M.E. Merriman and I.L. Lamont. 1995. Nucleotide-sequence of *pvdD*, a pyoverdine biosynthetic gene from *Pseudomonas aeruginosa* - *pvdD* has similarity to peptide synthetases. J. Bacteriol. *177*: 252–258.

Mesbah, M., U. Premachandran and W.B. Whitman. 1989. Precise measurement of the G + C content of deoxyribonucleic acid by high-performance liquid chromatography. Int. J. Syst. Bacteriol. *39*: 159–167.

Meulenberg, R., M. Pepi and J.A.M. De Bont. 1996. Degradation of 3-nitrophenol by *Pseudomonas putida* B2 occurs via 1,2,4-benzenetriol. Biodegradation 7: 303–311.

Meyer, F.P. and G.L. Bullock. 1973. *Edwardsiella tarda*, a new pathogen of channel catfish (*Ictalurus punctatus*). Appl. Microbiol. *25*: 155–156.

Meyer, J.M. and M.A. Abdallah. 1978. Fluorescent pigment of *Pseudomonas fluorescens* - biosynthesis, purification and physicochemical properties. J. Gen. Microbiol. *107*: 319–328.

Meyer, J.M. and M.A. Abdallah. 1980. The siderochromes of non-fluorescent Pseudomonads - production of nocardamine by *Pseudomonas stutzeri*. J. Gen. Microbiol. *118*: 125–129.

Meyer, J.M., P. Azelvandre and C. Georges. 1992. Iron metabolism in *Pseudomonas*: Salicylic acid, a siderophore of *Pseudomonas fluorescens* CHA0. Biofactors. *4*: 23–27.

Meyer, J., R. Haubold, J. Heyer and W. Bockel. 1986. Contribution to the taxonomy of methanotrophic bacteria: correlation between membrane type and GC-value. Z. Allg. Mikrobiol. *26*: 155–160.

Meyer, J.M. and J.M. Hornsperger. 1978. Role of pyoverdine$_{pf}$, iron-binding fluorescent pigment of *Pseudomonas fluorescens*, in iron transport. J. Gen. Microbiol. *107*: 329–331.

Meyer, J.M., A. Neely, A. Stintzi, C. Georges and I.A. Holder. 1996. Pyoverdin is essential for virulence of *Pseudomonas aeruginosa*. Infect. Immun. *64*: 518–523.

Meyer, J.M., A. Stintzi, D. De Vos, P. Cornelis, R. Tappe, K. Taraz and H. Budzikiewicz. 1997. Use of siderophores to type pseudomonads: The three *Pseudomonas aeruginosa* pyoverdine systems. Microbiology *143*: 35–43.

Meyer, O., E. Stackebrandt and G. Auling. 1993. Reclassification of ubiquinone Q-10 containing carboxidotrophic bacteria: transfer of "[*Pseudomonas*] *carboxydovorans*" OM5 ^{T} to *Oligotropha*, gen. nov., as *Oligotropha carboxidovorans*, comb. nov., transfer of "[*Alcaligenes*] *carboxydus*" DSM 1086^T to *Carbophilus*, gen. nov., as *Carbophilus carboxidus*, comb. nov., transfer of "[*Pseudomonas*] *compansoris*" DSM 1231^T to *Zavarzinia*, gen., nov., as *Zavarzinia compransoris*, comb. nov., and amended descriptions of the new genera. Syst. Appl. Microbiol. *16*: 390–395.

Meyers, B.R., B.L. Berson, M. Gilbert and S.Z. Hirschman. 1973. Clinical patterns of osteomyelitis due to gram-negative bacteria. Arch. Intern. Med. *131*: 228–233.

Meyers, B.R., E. Bottone, S.Z. Hirschman, S.S. Schneierson and K. Gershengorn. 1971. Infection due to *Actinobacillus actinomycetemcomitans*. Am. J. Clin. Pathol. *56*: 204–211.

Meyers, S.P., M.H. Baslow, S.J. Bein and C.E. Marks. 1959. Studies of *Flavabacterium piscicida* Bein. I. Growth, toxicity, and ecological considerations. J. Bacteriol. *78*: 225–230.

Meynadier, G., A. Lopez and J.-L. Duthoit. 1974. Mise en évidence de *Rickettsiales* chez une araignée (*Argyrodes gibbosus* Lucas), Ananeae, Theridiidae. C.R. Acad. Science, Series D. *278*: 2365–2367.

Meynadier, G. and P. Monsarrat. 1969. Une rickettsiose chez une cétoine de Madagascar. Entomophaga. *14*: 401–406.

Meynadier, G., J.M. Quiot and C. Vago. 1967. Infection "*in vitro*" de cellules cardiaques d'invertébrés. Second Int. Coll. Invert. Tissue Culture, Milano. Fondazione Baselli. 218–226.

Meynell, E.W. 1961. A phage, phi chi, which attacks motile bacteria. J. Gen. Microbiol. *25*: 253–290.

Mezzino, M.J., W.R. Strohl and J.M. Larkin. 1984. Characterization of *Beggiatoa alba*. Arch. Microbiol. *137*: 139–144.

Michalke, R., K. Taraz and H. Budzikiewicz. 1996. Azoverdin - An isopyoverdin. Z.Naturforsch.(C). *51*: 772–780.

Michan, C., A. Delgado, A. Haidour, G. Lucchesi and J.L. Ramos. 1997. In vivo construction of a hybrid pathway for metabolism of 4-nitrotoluene in *Pseudomonas fluorescens*. J. Bacteriol. *179*: 3036–3038.

Midgley, J., S.P. LaPage, B.A.G. Jenkins, G.I. Barrow, M.E. Roberts and A.G. Buck. 1970. *Cardiobacterium hominis* endocarditis. J. Med.Microbiol. *3*: 91–98.

Mielenz, J.R., L.E. Jackson, F. O'Gara and K.T. Shanmugan. 1979. Fingerprinting bacterial chromosomal DNA with restriction endonuclease *Eco*RI - comparison of *Rhizobium* spp. and identification of mutants. Can. J. Microbiol. *25*: 803–807.

Miflin, J.K., R.F. Horner, P.J. Blackall, X. Chen, G.C. Bishop, C.J. Morrow, T. Yamaguchi and Y. Iritani. 1995. Phenotypic and molecular characterization of V-factor (NAD)-independent *Haemophilus paragallinarum*. Avian Dis. *39*: 304–308.

Miguez, C.B., D. Bourque, J.A. Sealy, C.W. Greer and D. Groleau. 1997. Detection and isolation of methanotrophic bacteria possessing soluble methane monooxygenase (sMMO) genes using the polymerase chain reaction (PCR). Microb. Ecol. *33*: 21–31.

Migula, W. 1894. Über ein neues System der Bakterien. Arb. Bakteriol. Inst. Karlsruhe. *1*: 235–238.

Migula, W. 1895. Bacteriaceae (Stabchenbactérien). *In* Engler and Prantl (Editors), Pflanzenfamilien, Vol. Teil I, Abt. 1a, W. Engelmann, Leipzig. 20–30.

Migula, W. 1900. System der Bakterien, Vol. 2, Gustav Fischer, Jena.

Mikesell, P., J.W. Ezzell and G.B. Knudson. 1981. Plasmid isolation in *Legionella pneumophila* and *Legionella*-like organisms. Infect. Immun. *31*: 1270–1272.

Miki, K., K. Tamura, R. Sakazaki and Y. Kosako. 1996. Re-speciation of the original reference strains of serovars in the *Citrobacter freundii* (Bethesda-Ballerup group) antigenic scheme of West and Edwards. Microbiol. Immunol. *40*: 915–921.

Miles, D.G., H.D. Anthony and S.M. Dennis. 1972. *Haemophilus somnifer* infection in sheep. Am. J. Vet. Res. *33*: 431–435.

Millard, W.A. 1924. Crown rot of rhubarb. Yorks Council Agr. Ed. Bull. *134*: 1–28.

Miller, D.L. and V.W. Rodwell. 1971. Metabolism of basic amino acids in *Pseudomonas putida*. J. Biol. Chem. *246*: 2758–2764.

Miller, M.L. and J.A. Koburger. 1985. *Plesiomonas shigelloides*: an opportunistic food and waterborne pathogen. J. Food Prot. *48*: 449–457.

Miller, M.L. and J.A. Koburger. 1986a. Evaluation of inositol green bile salts and *Plesiomonas* agars for recovery of *Plesiomonas shigelloides* from aquatic samples in a seasonal survey of the Suwannee River estuary. J. Food Prot. *49*: 274–277.

Miller, M.L. and J.A. Koburger. 1986b. Tolerance of *Plesiomonas shigelloides* to pH, sodium chloride and temperature. J. Food Prot. *49*: 877–879.

Miller, P.W., W.B. Bollen, J.E. Simmons, H.N. Gross and H.P. Barss. 1940. The pathogen of filbert nut bacteriosis compared with *Phytomonas juglandis*, the cause of walnut blight. Phytopathology *30*: 713–733.

Miller, R.J., H.W. Renshaw and J.A. Evans. 1975. *Haemophilus somnus* complex: antigenicity and specificity of fractions of *H. somnus*. Am. J. Vet. Res. *36*: 1123–1128.

Miller, T.D. and M.N. Schroth. 1972. Monitoring the epiphytic population of *Erwinia amylovora* on pear with a selective medium. Phytopathology *62*: 1175–1182.

Millership, S.E. 1996. Identification. *In* Austin, Altwegg, Gosling and Joseph (Editors), The Genus *Aeromonas*, John Wiley and Sons, Ltd., Chichester. pp. 85–108.

Millership, S.E., M.R. Barer and S. Tabaqchali. 1986. Toxin production by *Aeromonas* spp. from different sources. J. Med. Microbiol. *22*: 311–314.

Millership, S.E. and B. Chattopadhyay. 1984. Methods for the isolation of *Aeromonas hydrophila* and *Plesiomonas shigelloides* from faeces. J. Hyg. *92*: 145–152.

Millership, S.E. and B. Chattopadhyay. 1985. *Aeromonas hydrophila* in chlorinated water supplies. J. Hosp. Infect. *6*: 75–80.

Millership, S.E., S.R. Curnow and B. Chattopadhyay. 1983. Faecal carriage rate of *Aeromonas hydrophila*. J. Clin. Pathol. *36*: 920–923.

Mills, S.D., C.A. Jasalavich and D.A. Cooksey. 1993. A two component regulatory system required for copper inducible expression of the copper resistance operon of *Pseudomonas syringae*. J. Bacteriol. *175*: 1656–1664.

Milstoc, M. and P. Steinberg. 1973. Fatal septicemia due to Providence group bacilli. J. Am. Geriatr. Soc. *21*: 159–163.

Milton, D.L., R. O'Toole, P. Horstedt and H. Wolf-Watz. 1996. Flagellin A is essential for the virulence of *Vibrio anguillarum*. J. Bacteriol. *178*: 1310–1319.

Minas, W. and D.L. Gutnick. 1993. Isolation, characterization, and sequence analysis of cryptic plasmids from *Acinetobacter calcoaceticus* and their use in the construction of *Escherichia coli* shuttle plasmids. Appl. Environ. Microbiol. *59*: 2807–2816.

Minges, C.G., J.A. Titus and W.R. Strohl. 1983. Plasmid DNA in colorless filamentous gliding bacteria. Arch. Microbiol. *134*: 38–44.

Miniats, O.P., M.T. Spinato and S.E. Sanford. 1989. *Actinobacillus suis* septicemia in mature swine: two outbreaks resembling erysipelas. Can. Vet. J. *30*: 943–947.

Minnick, M.F., R.A. Heinzen, R. Douthart, L.P. Mallavia and M.E. Frazier. 1990. Analysis of QpRS-specific sequences from *Coxiella burnetii*. Ann. N. Y. Acad. Sci. *590*: 514–522.

Minsavage, G.V. and N.W. Schaad. 1983. Characterization of membrane proteins of *Xanthomonas campestris* subsp. *campestris*. Phytopathology *73*: 747–755.

Minsavage, G.V., C.M. Thompson, D.L. Hopkins, R.M.V.B.C. Leite and R.E. Stall. 1994. Development of a polymerase chain reaction protocol for detection of *Xylella fastidiosa* in plant tissue. Phytopathology *84*: 456–461.

Misaghi, I. and R.G. Grogan. 1969. Nutritional and biochemical comparisons of plant-pathogenic and saprophytic fluorescent pseudomonads. Phytopathology *59*: 1436–1450.

Mise, K., K. Nakajima, N. Terakado and M.J. Ishidate. 1986. Production of restriction endonucleases using multicopy Hsd plasmids occurring naturally in pathogenic *Escherichia coli* and *Shigella boydii*. Gene *44*: 165–169.

Mishra, A.K. and O. Wyss. 1968. Induced mutations in *Azotobacter* and isolation of an adenine requiring mutant. Nucleus. *111*: 96–105.

Mishra, P., B. Roy, R. Prasad and H.K. Das. 1991. Amino acid transport in *Azotobacter vinelandii* - implications of nonavailability of amino acid auxotrophs. FEMS Microbiol. Lett. *79*: 41–44.

Mishra, S., G.B. Nair, R.K. Bhadra, S.N. Sikder and S.C. Pal. 1987. Comparison of selective media for primary isolation of *Aeromonas* species from human and animal feces. J. Clin. Microbiol. *25*: 2040–2043.

Mishu, B., J. Koehler, L.A. Lee, D. Rodrigue, F.H. Brenner, P. Blake and R.V. Tauxe. 1994. Outbreaks of *Salmonella enteritidis* infections in the United States, 1985–1991. J. Infect. Dis. *169*: 547–552.

Mitchell, C.G. 1996. Identification of a multienzyme complex of the tricarboxylic acid cycle enzymes containing citrate synthase isoenzymes from *Pseudomonas aeruginosa*. Biochem. J. *313*: 769–774.

Mitchell, R.G. and W.A. Gillespie. 1964. Bacterial endocarditis due to an actinobacillus. J. Clin. Pathol. *17*: 511–512.

Mitra, R., A. Basu, D. Dutta, G.B. Nair and Y. Takeda. 1996. Resurgence of *Vibrio cholerae* O139 Bengal with altered antibiogram in Calcutta, India. Lancet *348*: 1181..

Mittal, K.R., R. Higgins and S. La Riviere. 1988a. Serologic studies of *Actinobacillus* (*Haemophilus*) *pleuropneumoniae* strains of serotype-3 and their antigenic relationships with other *A. pleuropneumoniae* serotypes in swine. Am. J. Vet. Res. *49*: 152–155.

Mittal, K.R., R. Higgins and S. La Riviere. 1988b. Serological studies of *Actinobacillus* (*Haemophilus*) *pleuropneumoniae* strains of serotype-6 and their antigenic relationship with other serotypes. Vet. Rec. *122*: 199–203.

Mittal, K.R., R. Higgins, S. La Riviere and M. Nadeau. 1992. Serological characterization of *Actinobacillus pleuropneumoniae* strains isolated from pigs in Quebec. Vet. Microbiol. *32*: 135–148.

Mittal, K.R., E.M. Kamp and M. Kobisch. 1993. Serological characterisation of *Actinobacillus pleuropneumomiae* strains of serotypes 1, 9 and 11. Res. Vet. Sci. *55*: 179–184.

Miura, H., M. Horiguchi and T. Matsumoto. 1980. Nutritional interdependence among rumen bacteria, *Bacteroides amylophilus*, *Megasphaera elsdenii* and *Ruminococcus albus*. Appl. Environ. Microbiol. *40*: 294–300.

Miyagawa, E., R. Azuma and T. Suto. 1978. Distribution of sphingolipids in *Bacteroides* species. J. Gen. Appl. Microbiol. *24*: 341–348.

Miyagawa, E., R. Azuma and T. Suto. 1979. Cellular fatty acid composition in Gram-negative obligately anaerobic rods. J. Gen. Microbiol. *25*: 41–51.

Miyagawa, E., R. Azuma and T. Suto. 1981. Peptidoglycan composition of Gram-negative obligately anaerobic rods. J. Gen. Appl. Microbiol. *27*: 199–208.

Miyahara, M., K. Nakajima, T. Shimada and K. Mise. 1990. Restriction endonuclease *Psh*AI from *Plesiomonas shigelloides* with the novel recognition site 5'-GACNN/NNGTC. Gene *87*: 119–122.

Miyajima, K., A. Tanii and T. Akita. 1983. *Pseudomonas fuscovaginae* sp. nov., nom. rev. Int. J. Syst. Bacteriol. *33*: 656–657.

Miyamoto, Y., K. Nakamura and K. Takizawa. 1961. Pathogenic halophiles. Proposals of a new genus "*Oceanomonas*" and of the amended species names. Jpn. J. Microbiol. *5*: 477–486.

Miyoshi-Akiyama, T., M. Hayashi and T. Unemoto. 1993. Purification and properties of cytochrome *bo*-type ubiquinol oxidase from a marine bacterium *Vibrio alginolyticus*. Biochim. Biophys. Acta *1141*: 283–287.

Mo, Y.Y. and L.P. Mallavia. 1994. A *Coxiella burnetii* gene encodes a sensor-like protein. Gene *151*: 185–190.

Mobley, D.F. 1971. *Hafnia* septicemia. South. Med. J. *64*: 505–506.

Mobley, H.L.T. 1996. Virulence of *Proteus mirabilis*. *In* Mobley and Warren (Editors), Urinary Tract Infections. Molecular Pathogenesis and Clinical Management, ASM Press, Washington, D.C. pp. 245–269.

Mobley, H.L. and G.R. Chippendale. 1990. Hemagglutinin, urease, and hemolysin production by *Proteus mirabilis* from clinical sources. J. Infect. Dis. *161*: 525–530.

Mobley, H.L., G.R. Chippendale, M.H. Fraiman, J.H. Tenney and J.W. Warren. 1985. Variable phenotypes of *Providencia stuartii* due to plasmid-encoded traits. J. Clin. Microbiol. *22*: 851–853.

Mobley, H.L., G.R. Chippendale, J.H. Tenney, A.R. Mayrer, L.J. Crisp, J.L. Penner and J.W. Warren. 1988. MR/K hemagglutination of *Providencia stuartii* correlates with adherence to catheters and with persistence in catheter-associated bacteriuria. J. Infect. Dis. *157*: 264–271.

Mobley, H.L., G.R. Chippendale, J.H. Tenney and J.W. Warren. 1986a. Adherence to uroepithelial cells of *Providencia stuartii* isolated from the catheterized urinary tract. J. Gen. Microbiol. *132*: 2863–2872.

Mobley, H.L. and R.P. Hausinger. 1989. Microbial ureases: significance, regulation, and molecular characterization. Microbiol. Rev. *53*: 85–108.

Mobley, H.L., M.D. Island and R.P. Hausinger. 1995. Molecular biology of microbial ureases. Microbiol. Rev. *59*: 451–480.

Mobley, H.L., B.D. Jones and A.E. Jerse. 1986b. Cloning of urease gene sequences from *Providencia stuartii*. Infect. Immun. *54*: 161–169.

Mobley, H.L., B.D. Jones and J.L. Penner. 1987. Urease activity of *Proteus penneri*. J. Clin. Microbiol. *25*: 2302–2305.

Mobley, H.L. and J.W. Warren. 1987. Urease-positive bacteriuria and obstruction of long-term urinary catheters. J. Clin. Microbiol. *25*: 2216–2217.

Mohan, K., P.J. Kelly, F.W.G. Hill, P. Muvavarirwa and A. Pawandiwa. 1997. Phenotype and serotype of *Pasteurella multocida* isolates from diseases of dogs and cats in Zimbabwe. Comp. Immunol. Microbiol. Infect. Dis. *20*: 29–34.

Mohn, W.W., A.E. Wilson, P. Bicho and E.R.B. Moore. 1999a. Physiological and phylogenetic diversity of bacteria growing on resin acids. Syst. Appl. Microbiol. *22*: 68–78.

Mohn, W.W., A.E. Wilson, P. Bicho and E.R.B. Moore. 1999b. *In* Validation of publication of new names and new combinations previously effectively published outside the IJSB. List No. 70. Int. J. Syst. Bacteriol. *49*: 935–936.

Moillo, A.M. 1973. Isolation of a transducing phage forming plaques on *Pseudomonas maltophilia* and *Pseudomonas aeruginosa*. Genet. Res. *21*: 287–289.

Mojtabaee, A. and A. Siadati. 1978. *Enterobacter hafniae* meningitis. J. Pediatr. *93*: 1062–1063.

Molin, G. and A. Ternström. 1982. Numerical taxonomy of psychrotropic pseudomonads. J. Gen. Microbiol. *128*: 1249–1264.

Molin, G., A. Ternström and J. Ursing. 1986. *Pseudomonas lundensis*, a new bacterial species isolated from meat. Int. J. Syst. Bacteriol. *36*: 339–342.

Molisch, H. 1906. Zwei neue Purpurbakterien mit Schwebe körperchen. Bot. Ztg. Abt. 1. *64*: 223–232.

Molisch, H. 1907. Die Purpurbakterien Nach Neuen Untersuchungen, G. Fischer, Jena.

Molisch, H. 1912. Neue farblose Schwefelbakterien. Zentbl. Bakteriol. Parasitenkd. Infektkrankh. Hyg. Abt. II *33*: 55–62.

Mollaret, H.H. 1991. Tularaemia. *In* Standards Commission. Manual of Recommended Diagnostic techniques and Requirements for Biological Products for List A and B Diseases, Vol. 3, Office International Des Epizooties, Paris. pp. 1–7.

Mollaret, H.H., H. Bercovier and J.M. Alonso. 1979. Summary of tbe data received at the W.H.O. reference center for *Yersinia enterocolitica*. Contr. Microbiol. Immunol. *5*: 174–184.

Mollaret, W.H., Y. Karimi, M. Eftekhari and M. Baltazard. 1963. La peste de Fouissement. Bull. Soc. Path. Exot. *56*: 1186–1193.

Mollenhauer, H.H. and D.L. Hopkins. 1974. Ultrastructural study of Pierce's disease bacterium in grape xylem tissue. J. Bacteriol. *119*: 612–618.

Møller, K., L.V. Andersen, G. Christensen and M. Kilian. 1993. Optimalization of the detection of NAD dependent *Pasteurellaceae* from the respiratory tract of slaughterhouse pigs. Vet. Microbiol. *36*: 261–271.

Møller, K., V. Fussing, P.A.D. Grimont, B.J. Paster, F.E. Dewhirst and M. Kilian. 1996a. *Actinobacillus minor* sp. nov., *Actinobacillus porcinus* sp. nov., and *Actinobacillus indolicus* sp. nov., three new V factor-dependent species from the respiratory tract of pigs. Int. J. Syst. Bacteriol. *46*: 951–956.

Møller, K. and M. Kilian. 1990. V factor-dependent members of the family *Pasteurellaceae* in the porcine upper respiratory tract. J. Clin. Microbiol. *28*: 2711–2716.

Møller, K., R. Nielsen, L.V. Andersen and M. Kilian. 1992. Clonal analysis of the *Actinobacillus pleuropneumoniae* population in a geographically restricted area by multilocus enzyme electrophoresis. J. Clin. Microbiol. *30*: 623–627.

Möller, M.M., L.P. Nielsen and B.B. Jørgensen. 1985. Oxygen responses and mat formation by *Beggiatoa* spp. Appl. Environ. Microbiol. *50*: 373–382.

Moller, S., A.R. Pedersen, L.K. Poulsen, E. Arvin and S. Molin. 1996b. Activity and three-dimensional distribution of toluene- degrading *Pseudomonas putida* in a multispecies biofilm assessed by quantitative in situ hybridization and scanning confocal laser microscopy. Appl. Environ. Microbiol. *62*: 4632–4640.

Møller, V. 1954. Distribution of amino acid decarboxylase in *Enterobacteriaceae*. Acta Pathol. Microbiol. Scand. *35*: 259–277.

Møller, V. 1955. Simplified test for some amino acid decarboxylases and arginine dihydrolase system. Acta Pathol. Microbiol. Scand. *36*: 158–172.

Momol, M.T., E.A. Momol, W.F. Lamboy, J.L. Norelli, S.V. Beer and H.S. Aldwinckle. 1997. Characterization of *Erwinia amylovora* strains using random amplified polymorphic DNA fragments (RAPDs). J. Appl. Microbiol. *82*: 389–398.

Monias, B.L. 1928. Classification of *Bacterium alcaligenes, pyocyaneum* and *fluorescens*. J. Infect. Dis. *43*: 330–334.

Moniz, L. and M.K. Patel. 1958. Three new bacterial diseases of plants from Bombay State. Curr. Sci. (Bangalore) *27*: 494–495.

Moniz, L., J.E. Sabley and W.D. More. 1964. A new bacterial canker of *Carissa congesta* in Maharashtra. Indian Phytopathol. *17*: 256.

Monpert, G. 1996. Relation between denitrification and biodegradation of *n*-heptadecane in a marine bacterium. C. R. Acad. Sci. Ser. III Sci. Vie. *319*: 805–809.

Monteiro-Neto, V., L.C. Campos, A.J. Ferreira, T.A. Gomes and L.R. Trabulsi. 1997. Virulence properties of *Escherichia coli* O111:H12 strains. FEMS Microbiol. Lett. *146*: 123–128.

Monteoliva-Sanchez, M. and A. Ramos-Cormenzana. 1987. Cellular fatty acid composition in moderately halophilic gram-negative rods. J. Appl. Bacteriol. *62*: 361–366.

Montfort, P. and B. Baleux. 1991. Distribution and survival of motile *Aeromonas* spp. in brackish water receiving sewage treatment effluent. Appl. Environ. Microbiol. *57*: 2459–2467.

Montgomery, L., B. Flesher and D. Stahl. 1988. Transfer of *Bacteroides succinogenes* (Hungate) to *Fibrobacter*, gen. nov. as *Fibrobacter succinogenes*, comb. nov., and description of *Fibrobacter intestinalis*, sp. nov. Int. J. Syst. Bacteriol. *38*: 430–435.

Montie, T.C. and G.B. Stover. 1983. Isolation and characterization of flagellar preparations from *Pseudomonas* species. J. Clin. Microbiol. *18*: 452–456.

Moore, E.R.B., A.S. Krüger, L. Hauben, S.E. Seal, R. De Baere, R. De Wachter, K.N. Timmis and J. Swings. 1997. 16S rRNA gene sequence analyses and inter- and intrageneric relationships of *Xanthomonas* species and *Stenotrophomonas maltophilia*. FEMS Microbiol. Lett. *151*: 145–153.

Moore, E.R.B., M. Mau, A. Arnscheidt, E.C. Böttger, R.A. Hutson, M.D. Collins, Y. van de Peer, R. De Wachter and K.N. Timmis. 1996. The determination and comparison of the 16S rRNA gene sequences of species of the genus *Pseudomonas* (sensu stricto) and estimation of the natural intrageneric relationships. Syst. Appl. Microbiol. *19*: 478–492.

Moore, L.V., D.M. Bourne and W.E. Moore. 1994a. Comparative distribution and taxonomic value of cellular fatty acids in thirty-three genera of anaerobic gram-negative bacilli. Int. J. Syst. Bacteriol. *44*: 338–347.

Moore, M.K., L. Cicnjakchubbs and R.J. Gates. 1994b. A new selective enrichment procedure for isolating *Pasteurella multocida* from avian and environmental-samples. Avian Dis. *38*: 317–324.

Morabito, S., H. Karch, P. Mariani-Kurkdjian, H. Schmidt, F. Minelli, E. Bingen and A. Caprioli. 1998. Enteroaggregative, Shiga toxin-producing *Escherichia coli* O111:H2 associated with an outbreak of hemolytic-uremic syndrome. J. Clin. Microbiol. *36*: 840–842.

Moran, A.B. and D.W. Bruner. 1949. Further studies on the Bethesda group of paracolon bacteria. J. Bacteriol. *58*: 695–700.

Moran, N.A. 1996. Accelerated evolution and Muller's rachet in endosymbiotic bacteria. Proc. Natl. Acad. Sci. USA. *93*: 2873–2878.

Moran, N.A. and P. Baumann. 1994. Phylogenetics of cytoplasmically inherited microorganisms of arthropods. Trends Ecol. Evol. *9*: 15–20.

Moran, N.A., M.E. Kaplan, M.J. Gelsey, T.G. Murphy and E.A. Scholes. 1999. Phylogenetics and evolution of the aphid genus *Uroleucon* based on mitochondrial and nuclear DNA sequences. Syst. Entomol. *24*: 85–93.

Moran, N.A., M.A. Munson, P. Baumann and H. Ishikawa. 1993. A molecular clock in endosymbiotic bacteria is calibrated using the insect hosts. Proc. R. Soc. Lond. B Biol. Sci. *253*: 167–171.

Moran, N.A., C.D. Von Dohlen and P. Baumann. 1995. Faster evolutionary rates in endosymbiotic bacteria than in cospeciating insect hosts. J. Mol. Evol. *41*: 727–731.

Morax, V. 1896. Note sur un diplobacille pathogène pour la conjonctive humains. Ann. Inst. Pasteur. *10*: 337–345.

Morel, G. 1976. Studies on *Porochlamydia buthi* g. n., sp. n., an intracellular pathogen of the scorpion *Buthus occitanus*. J. Invert. Pathol. *28*: 167–175.

Morel, G. 1977. Étude d' une *Rickettsiella* (Rickettsie) se développant chez un arachnide, l' araignée *Pisaura mirabilis*. Ann. Microbiol. *128A*: 49–59.

Morel, G. 1980. Surface projections of a chlamydia-like parasite of midge larvae. J. Bacteriol. *144*: 1174–1175.

Moreno, S., R. Najera, J. Guzman, G. Soberon-Chavez and G. Espin. 1998. Role of alternative a factor AlgU in encystment of *Azotobacter vinelandii*. J. Bacteriol. *180*: 2766–2769.

Morgan, D.R., P.C. Johnson, H.L. Dupont, T.K. Satterwhite and L.V. Wood. 1985. Lack of correlation between known virulence properties of *Aeromonas hydrophila* and enteropathogenicity for humans. Infect. Immun. *50*: 62–65.

Morgan, G.M., C. Newman, S.R. Palmer, J.B. Allen, W. Shepherd, A.M. Rampling, R.E. Warren, R.J. Gross, S.M. Scotland and H.R. Smith. 1988. First recognized community outbreak of haemorrhagic colitis due to verotoxin-producing *Escherichia coli* O157:H7 in the UK. Epidemiol. Infect. *101*: 83–91.

Morgan, J.A.W., V. Kuntzelmann, S. Tavernor, M.A. Ousley and C. Winstanley. 1997. Survival of *Xenorhabdus nematophilus* and *Photorhabdus luminescens* in water and soil. J. Appl. Microbiol. *83*: 665–670.

Morgan, M.G. and J.M. Hamilton-Miller. 1990. *Haemophilus influenzae* and *H. parainfluenzae* as urinary pathogens. J. Infect. *20*: 143–145.

Morgan, R.D., M. Dalton and R. Stote. 1987. A unique type II restriction endonuclease from *Acinetobacter lwoffi* N. Nucleic Acids Res. *15*: 7201.

Morihara, K. 1964. Production of elastase and proteinase by *Pseudomonas aeruginosa*. J. Bacteriol. *88*: 745–757.

Morihara, K. and J.Y. Homma. 1985. *Pseudomonas* proteases. *In* Holder (Editor), Bacterial Enzymes and Virulence, CRC Press, Boca Raton, FL. 41–75.

Morihara, K., H. Tsuzuki, T. Oka, H. Inoue and M. Ebata. 1965. *Pseudomonas aeruginosa* elastase. Isolation, crystallization, and preliminary characterization. J. Biol. Chem. *240*: 3295–3304.

Morinaga, Y., S. Yamanaha, S. Otsuka and Y. Hirose. 1976. Characteristics of a newly isolated methane-oxidizing bacterium, *Methylomonas flagellata* nov. sp. Agric. Biol. Chem. *40*: 1539–1545.

Morishita, Y., K. Hasegawa, Y. Matsuura, Y. Katsube, M. Kubota and S. Sakai. 1997. Crystal structure of a maltotetraose-forming exo-amylase from *Pseudomonas stutzeri*. J. Mol. Biol. *267*: 661–672.

Morita, K., N. Watanabe, S. Kurata and M. Kanamori. 1994. β-lactam resistance of motile *Aeromonas* isolates from clinical and environmental sources. Antimicrob. Agents Chemother. *38*: 353–355.

Morita, R.Y. 1975. Psychrophilic bacteria. Bacteriol. Rev. *39*: 144–167.

Morita, R.Y., R. Iturriaga and V.A. Gallardo. 1981. *Thioploca*: methylotroph and significance in the food chain. Kiel. Meeresforsch. Sonderh. *5*: 384–389.

Morita, R.Y. and P.W. Stave. 1963. Electron micrograph of an ultrathin section of *Beggiatoa*. J. Bacteriol. *85*: 940–942.

Mormile, M.R., M.F. Romine, M.T. Garcia, A. Ventosa, T.J. Bailey and B.M. Peyton. 1999. *Halomonas campisalis* sp. nov., a denitrifying, moderately haloalkaliphilic bacterium. Syst. Appl. Microbiol. *22*: 551–558.

Mormile, M.R., M.F. Romine, M.T. Garcia, A. Ventosa, T.J. Bailey and B.M. Peyton. 2000. *In* Validation of the publication of new names and new combinations previously effectively published outside the IJSEM, List No. 74. Int. J. Syst. Evol. Microbiol. *50*: 949–950.

Mörner, T. 1981. The use of FA-technique for detecting *Francisella tularensis* in formalin fixed material. A method useful in routine post mortem work. Acta Vet. Scand. *22*: 296–306.

Mörner, T., G. Sandström, R. Mattsson and P.O. Nilsson. 1988. Infections with *Francisella tularensis* biovar palaearctica in hares (*Lepus timidus*, *Lepus europaeus*) from Sweden. J. Wildl. Dis. *24*: 422–433.

Morozumi, T. and J. Nicolet. 1986. Morphological variations of *Haemophilus parasuis* strains. J. Clin. Microbiol. *23*: 138–142.

Morrill, W.E., J.M. Barbaree, B.S. Fields, G.N. Sanden and W.T. Martin. 1990. Increased recovery of *Legionella micdadei* and *Legionella bozemanii* on buffered charcoal yeast extract agar supplemented with albumin. J. Clin. Microbiol. *28*: 616–618.

Morris, E.J. 1958. Selective media for some *Pasteurella* species. J. Gen. Microbiol. *19*: 305–311.

Morris, E.R., D.A. Rees, G. Young, M.D. Walkinshaw and A. Darke. 1977. Order-disorder transition for a bacterial polysaccharide in solution. a role for polysaccharide conformation in recognition between *Xanthomonas* pathogen and its plant host. J. Mol. Biol. *110*: 1–16.

Morris, G.K., M.H. Merson, I. Huq, A. Kibrya and R. Black. 1979. Comparison of four plating media for isolating *Vibrio cholerae*. J. Clin. Microbiol. *9*: 79–83.

Morris, G.K., A. Steigerwalt, J.C. Feeley, E.S. Wong, W.T. Martin, C.M. Patton and D.J. Brenner. 1980a. *Legionella gormanii* sp. nov. J. Clin. Microbiol. *12*: 718–721.

Morris, G.K., A. Steigerwalt, J.C. Feeley, E.S. Wong, W.T. Martin, C.M. Patton and D.J. Brenner. 1980b. *In* Validation of the publication of new names and new combinations previously effectively published outside the IJSB. List No. 5. Int. J. Syst. Bacteriol. *30*: 676–677.

Morris, J.G., Jr. 1995. *Vibrio cholerae* O139 Bengal: emergence of a new epidemic strain of cholera. Infect. Agents Dis. *4*: 41–46.

Morris, J.G., Jr. and R. Black. 1985. Cholera and other vibrioses in the United States. N. Engl. J. Med. *312*: 343–350.

Morris, J.G., Jr., H. Miller, R. Wilson, C.O. Tacket, D.G. Hollis, F.W. Hickman, R.E. Weaver and P.A. Blake. 1982. Illness caused by *Vibrio damsela* and *Vibrio hollisae*. Lancet *1*: 1294–1297.

Morris, J.G., Jr., M.B. Sztein, E.W. Rice, J.P. Nataro, G.A. Losonsky, P. Panigrahi, C.O. Tacket and J.A. Johnson. 1996. *Vibrio cholerae* O1 can assume a chlorine-resistant rugose survival form that is virulent for humans. J. Infect. Dis. *174*: 1364–1368.

Morris, J.G., Jr., J.H. Tenney and G.L. Drusano. 1985. *In vitro* susceptibility of pathogenic *Vibrio* species to norfloxacin and six other antimicrobial agents. Antimicrob. Agents Chemother. *28*: 442–445.

Morris, J.G., Jr., A.C. Wright, D.M. Roberts, P.K. Wood, L.M. Simpson and J.D. Oliver. 1987. Identification of environmental *Vibrio vulnificus* isolates with a DNA probe for the cytotoxin-hemolysin gene. Appl. Environ. Microbiol. *53*: 193–195.

Morse, S.A. 1980. Sexually transmitted diseases, 3rd Ed., American Society for Microbiology, Washington, D.C.

Morton, R.J., K.R. Simons and A.W. Confer. 1996. Major outer membrane proteins of *Pasteurella haemolytica* serovars 1-15: comparison of separation techniques and surface-exposed proteins on selected serovars. Vet. Microbiol. *51*: 319–330.

Moser, D.P. and K.H. Nealson. 1996. Growth of the facultative anaerobe *Shewanella putrefaciens* by elemental sulfur reduction. Appl. Environ. Microbiol. *62*: 2100–2105.

Moshiri, F., J.W. Kim, C.L. Fu and R.J. Maier. 1994. The FeSII Protein of *Azotobacter vinelandii* is not essential for aerobic nitrogen fixation, but confers significant protection to oxygen-mediated inactivation of nitrogenase *in vitro* and *in vivo*. Mol. Microbiol. *14*: 101–114.

Moss, C.W. and M.I. Daneshvar. 1992. Identification of some uncommon monounsaturated fatty acids of bacteria. J. Clin. Microbiol. *30*: 2511–2512.

Moss, C.W. and S.B. Dees. 1979. Further studies of the cellular fatty acid composition of Legionnaires' disease bacteria. J. Clin. Microbiol. *9*: 648–649.

Moss, C.W., S.B. Samuels, J. Liddle and R.M. McKinney. 1973. Occurrence of branched-chain hydroxy fatty acids in *Pseudomonas maltophilia*. J. Bacteriol. *114*: 1018–1024.

Moss, C.W., S.B. Samuels and R.E. Weaver. 1972. Cellular fatty acid composition of selected *Pseudomonas* species. Appl. Microbiol. *24*: 596–598.

Moss, C.W., P.L. Wallace, D.G. Hollis and R.E. Weaver. 1988. Cultural and chemical characterization of CDC groups EO-2, M-5, and M-6, *Moraxella* spp. *Oligella urethralis*, *Acinetobacter* sp. and *Psychrobacter immobilis*. J. Clin. Microbiol. *26*: 484–492.

Moss, M.O. 1983. A note on a prodigiosin producing pseudomonad isolated from a lowland river. J. Appl. Bacteriol. *55*: 373–375.

Motes, M.L., A. DePaola, D.W. Cook, J.E. Veazey, J.C. Hunsucker, W.E. Garthright, R.J. Blodgett and S.J. Chirtel. 1998. Influence of water temperature and salinity on *Vibrio vulnificus* in Northern Gulf and Atlantic Coast oysters (*Crassostrea virginica*). Appl. Environ. Microbiol. *64*: 1459–1465.

Motyl, M.R., G. McKinley and J.M. Janda. 1985. *In vitro* susceptibilities of *Aeromonas hydrophila*, *Aeromonas sobria*, and *Aeromonas caviae* to 22 antimicrobial agents. Antimicrob. Agents. Chemother. *28*: 151–153.

Mouahid, M., M. Bisgaard, A.J. Morley, R. Mutters and W. Mannheim. 1992. Occurrence of V-factor (NAD) independent strains of *Haemophilus paragallinarum*. Vet. Microbiol. *31*: 363–368.

Moule, A.L. and S.G. Wilkinson. 1987. Polar lipids, fatty acids, and isoprenoid quinones of Alteromonas putrefaciens. Syst. Appl. Microbiol. *9*: 192–198.

Moule, A.L. and S.G. Wilkinson. 1989. Composition of lipopolysaccharides from *Alteromonas putrefaciens* (*Shewanella putrefaciens*). J. Gen. Microbiol. *135*: 163–173.

Moulsdale, M.T. 1983. Isolation of *Aeromonas* from faeces. Lancet *1*: 351.

Mountfort, D.O., F.A. Rainey, J. Burghardt, H.F. Kaspar and E. Stackebrandt. 1998a. *Psychromonas antarcticus* gen. nov., sp. nov., a new aerotolerant anaerobic, halophilic psychrophile isolated from pond sediment of the McMurdo Ice Shelf, Antarctica. Arch. Microbiol. *169*: 231–238.

Mountfort, D.O., F.A. Rainey, J. Burghardt, H.F. Kaspar and E. Stackebrandt. 1998b. *In* Validation of new names and new combinations previously effectively published outside the IJSB. List No. 66. Int. J. Syst. Evol. Microbiol. *48*: 631.

Moureaux, N., T. Karjalainen, A. Givaudan, P. Bourlioux and N. Boemare. 1995. Biochemical characterization and agglutinating properties of *Xenorhabdus nematophilus* F1 fimbriae. Appl. Environ. Microbiol. *61*: 2707–2712.

Moxon, E.R. 1986. The carrier state: *Haemophilus influenzae*. J. Antimicrob. Chemother. *18 (Suppl A)*: 17–24.

Moyenuddin, M., I.K. Wachsmuth, S.L. Moseley, C.A. Bopp and P.A. Blake. 1989. Serotype, antimicrobial resistance, and adherence properties of *Escherichia coli* strains associated with outbreaks of diarrheal illness in children in the United States. J. Clin. Microbiol. *27*: 2234–2239.

Moyer, C.L., F.C. Dobbs and D.M. Karl. 1995. Phylogenetic diversity of the bacterial community from a microbial mat at an active, hydrothermal vent system, Loihi Seamount, Hawaii. Appl. Environ. Microbiol. *61*: 1555–1562.

Moyer, N.P. 1996. Isolation and enumeration of aeromonads. *In* Austin, Altwegg, Gosling and Joseph (Editors), The Genus *Aeromonas*, John Wiley & Sons, Ltd., Chichester. pp. 39–84.

Moyer, N.P., H.K. Geiss, M. Marinescu, A. Rigby, J. Robinson and M. Altwegg. 1991. Media and methods for isolation of aeromonads from fecal specimens. A multilaboratory study. Experientia (Basel) *47*: 409–412.

Moyer, N.P., G.M. Luccini, L.A. Holcomb, N.H. Hall and M. Altwegg. 1992. Application of ribotyping for differentiating aeromonads isolated from clinical and environmental sources. Appl. Environ. Microbiol. *58*: 1940–1944.

Mracek, Z. 1977. Steinernema kraussei, a parasite of the body cavity of the sawfly, *Cephaleia abietis*, in Czechoslovakia. J. Invertebr. Pathol. *30*: 87–94.

Mráz, O. 1963. Schizomycetes. *In* Mráz (Editor), Nomina und Synonyma der pathogenen and saprophytaren mikroben, isoliertaus den Wirtschafttlich odor epidemiologish bedutenden Wirbeltieren und lebensmittein tierischer herkunft, VEB Gustav Fisher Verlag, Jena. 53–334.

Mráz, O. 1968. Reevaluation of original strains *Actinobacillus suis* and haemolytic strains *Actinobacillus lignieresii* ATCC isolated from organs of diseased pigs. Acta Univ. Agric. Fac. Vet. *37*: 277–290.

Mráz, O. 1969a. Vergleichende studie der arten *Actinobacillus lignieresii* und *Pasteurella haemolytica*. I. *Actinobacillus lignieresii* Brumpt, 1910; emend. Zentralbl. Bakteriol. Parasitenkd. Infektionskr. Hyg. Abt. 1, Orig. *209*: 212–232.

Mráz, O. 1969b. Vergleichende studie der arten *Actinobacillus lignieresii* und *Pasteurella haemolytica*. III. *Actinobacillus haemolyticus* (Newsom und

Cross 1932) comb. nov. Zentralbl. Bakteriol. Parasitenkd. Infektionskr. Hyg. Abt. 1, Orig. *209*: 349–364.

Mráz, O. 1977. Antigenni vztahy mezi pasteurelami a aktinobacily. Vet. Med. (Prague) *22*: 121–132.

Mráz, O., P. Vladík and J. Bohácek. 1976. Actinobacilli in domestic fowl. Zentralbl. Bakteriol. Parasitenkd. Infektionskr. Hyg. Abt. I, Orig. A. *236*: 294–307.

Muder, R.R., A.P. Harris, S. Muller, M. Edmond, J.W. Chow, K.A. Papadakis, M.W. Wagener, G.P. Bodey and J.M. Steckelberg. 1996. Bacteremia due to *Stenotrophomonas* (*Xanthomonas*) *maltophilia*: a prospective, multicenter study of 91 episodes. Clin. Infect. Dis. *22*: 508–512.

Muhairwa, A.P., J.P. Christensen and M. Bisgaard. 2000. Investigations on the carrier rate of *Pasteurella multocida* in healthy commercial poultry flocks and flocks affected by fowl cholera. Avian Pathol. *29*: 133–142.

Muhairwa, A.P., M.M.A. Mtambo, J.P. Christensen and M. Bisgaard. 2001. Occurrence of *Pasteurella multocida* and related species in village free ranging chickens and their animal contacts in Tanzania. Vet. Microbiol. *78*: 139–153.

Muhle, I., J. Rau and J. Ruskin. 1979. Vertebral osteomyelitis due to *Actinobacillus actinomycetemcomitans*. J. Am. Med. Assoc. *241*: 1824–1825.

Mujica, O.J., R.E. Quick, A.M. Palacios, L. Beingolea, R. Vargas, D. Moreno, T.J. Barrett, N.H. Bean, L. Seminario and R.V. Tauxe. 1994. Epidemic cholera in the Amazon: The role of produce in disease risk and prevention. J. Infect. Dis. *169*: 1381–1384.

Mukherjee, A., Y. Cui, Y. Liu, C.K. Dumenyo and A.K. Chatterjee. 1996. Global regulation in *Erwinia* species by *Erwinia carotovora rsmA*, a homologue of *Escherichia coli csrA*: repression of secondary metabolites, pathogenicity and hypersensitive reaction. Microbiology *142*: 427–434.

Mukhopadhyay, A.K., S. Garg, R. Mitra, A. Basu, K. Rajendran, D. Dutta, S.K. Bhattacharya, T. Shimada, T. Takeda, Y. Takeda and G.B. Nair. 1996. Temporal shifts in traits of *Vibrio cholerae* strains isolated from hospitalized patients in Calcutta: a 3-year (1993 to 1995) analysis. J. Clin. Microbiol. *34*: 2537–2543.

Mukhopadhyay, A.K., S. Garg, G.B. Nair, S. Kar, R.K. Ghosh, S. Pajni, A. Ghosh, T. Shimada, T. Takeda and Y. Takeda. 1995. Biotype traits and antibiotic susceptibility of *Vibrio cholerae* serogroup O1 before, during and after the emergence of the O139 serogroup. Epidemiol. Infect. *115*: 427–434.

Mukoo, H. 1955. On the bacterial blacknode of barley and wheat and its causal bacteria. *In* Jubilee Publication in Commemoration of the Sixtieth Birthdays of Prof. Yoshihiko Tochinai and Prof. Teikichi Fukushi, Sapporo, Japan. 153–157.

Mulholland, V., J.C. Hinton, J. Sidebotham, I.K. Toth, L.J. Hyman, M.C. Perombelon, P.J. Reeves and G.P. Salmond. 1993. A pleiotropic reduced virulence (Rvi⁻) mutant of *Erwinia carotovora* subsp. *atroseptica* is defective in flagella assembly proteins that are conserved in plant and animal bacterial pathogens. Mol. Microbiol. *9*: 343–356.

Mulla, R. and S.E. Millership. 1993. Typing of *Aeromonas* spp. by numerical analysis of immunoblotted SDS-PAGE gels. J. Med. Microbiol. *39*: 325–333.

Müller, H.E. 1972. The aerobic fecal flora of reptiles with special reference to the enterobacteria of snakes. Zentbl. Bakteriol. Parasitenkd. Infektkrankh. Hyg. Abt. I Orig. *222*: 487–495.

Müller, H.E. 1983. *Providencia friedericiana*, a new species isolated from penguins. Int. J. Syst. Bacteriol. *33*: 709–715.

Müller, H.E. 1986a. Occurrence and pathogenic role of *Morganella-Proteus-Providencia* group bacteria in human feces. J. Clin. Microbiol. *23*: 404–405.

Müller, H.E. 1986b. Production and degradation of indole by Gram-negative bacteria. Zentralbl. Bakteriol. Parasitenk. Infektionskr. Hyg. Abt. I Orig. A. *261*: 1–11.

Müller, H.E., D.J. Brenner, G.R. Fanning, P.A.D. Grimont and P. Kämpfer. 1996. Emended description of *Buttiauxella agrestis* with recognition of six new species of *Buttiauxella* and two new species of *Kluyvera*: *Buttiauxella ferragutiae* sp. nov., *Buttiauxella gaviniae* sp. nov., *Buttiauxella*

brennerae sp. nov., *Buttiauxelle izardii* sp. nov., *Buttiauxella noackiae* sp. nov., *Buttiauxella warmboldiae* sp. nov., *Kluyvera cochleae* sp. nov., and *Kluyvera georgiana* sp. nov. Int. J. Syst. Bacteriol. *46*: 50–63.

Müller, H.E., G.R. Fanning and D.J. Brenner. 1995a. Isolation of *Ewingella americana* from mollusks. Curr. Microbiol. *31*: 287–290.

Müller, H.E., G.R. Fanning and D.J. Brenner. 1995b. Isolation of *Serratia fonticola* from mollusks. Syst. Appl. Microbiol. *18*: 279–284.

Müller, H.E. and K.H. Hinz. 1977. The occurrence of neuraminidase and N-acetylneuraminate-pyruvate lyase in pathogenic haemophili of man. Zentralbl. Bakteriol. [Orig A]. *239*: 231–239.

Müller, H.E., C.M. O'Hara, G.R. Fanning, F.W. Hickman-Brenner, J.M. Swenson and D.J. Brenner. 1986. *Providencia heimbachae*, a new species of *Enterobacteriaceae* isolated from animals. Int. J. Syst. Bacteriol. *36*: 252–256.

Muller, L. 1923. Un nouveau milieu d'enrichissement pour la recherche des bacilles typhiques et paratyphiques. C. R. Soc. Biol. *89*: 434–437.

Müller-Kogler, E. 1958. Eine Rickettsiose von Tipula paludosa Meig. durch *Rickettsiella tipulae* nov. spec. Naturwissenschaften *45*: 248–250.

Mulyukin, A.L., K.A. Lusta, M.N. Gryaznova, A.N. Kozlova, M.V. Duzha, V.I. Duda and G.I. El'Registan. 1997. Formation of resting cells by *Bacillus cereus* and *Micrococcus luteus*. Mikrobiologiya *66*: 32–38.

Munch, S., S. Grund and M. Kruger. 1992. Fimbriae and membranes on *Haemophilus parasuis*. Zentralbl. Veterinarmed. [B]. *39*: 59–64.

Munoz-Centeno, M.C., M.T. Ruiz, A. Paneque and F.J. Cejudo. 1996. Posttranslational regulation of nitrogenase activity by fixed nitrogen in *Azotobacter chroococcum*. Biochim. Biophys. Acta *1291*: 67–74.

Munro, P.M., G. Brahic and R.L. Clement. 1994. Seawater effects on various *Vibrio* species. Microbios. *77*: 191–198.

Munson, M.A. and P. Baumann. 1993. Molecular cloning and nucleotide sequence of a putative *trpDC(F)BA* operon in *Buchnera aphidicola* (endosymbiont of the aphid *Schizaphis graminum*). J. Bacteriol. *175*: 6426–6432.

Munson, M.A., P. Baumann, M.A. Clark, L. Baumann, N.A. Moran, D.J. Voegtlin and B.C. Campbell. 1991a. Evidence for the establishment of aphid-eubacterium endosymbiosis in an ancestor of four aphid families. J. Bacteriol. *173*: 6321–6324.

Munson, M.A., P. Baumann and M.G. Kinsey. 1991b. *Buchnera* gen. nov. and *Buchnera aphidicola* sp. nov., a taxon consisting of the mycetocyte-associated, primary endosymbionts of aphids. Int. J. Syst. Bacteriol. *41*: 566–568.

Muraschi, T.F., M. Friend and D. Bolles. 1965. *Erwinia*-like microorganisms isolated from animals and human hosts. Appl. Microbiol. *13*: 128–131.

Murata, H., A. Chatterjee, Y. Liu and A.K. Chatterjee. 1994. Regulation of the production of extracellular pectinase, cellulase, and protease in the soft rot bacterium *Erwinia carotovora* subsp. *carotovora*: evidence that *aepH* of *E. carotovora* subsp. *carotovora* 71 activates gene expression in *E. carotovora* subsp. *carotovora*, *E. carotovora* subsp. *atroseptica*, and *Escherichia coli*. Appl. Environ. Microbiol. *60*: 3150–3159.

Murata, N. and M.P. Starr. 1974. Intrageneric clustering and divergence of *Erwinia* strains from plants and man in the light of desoxyribonucleic and segmental homology. Can. J. Microbiol. *20*: 1545–1565.

Murlin, A.M., D.J. Brenner, A.G. Steigerwalt and J.J. Farmer, III. 1988. Enteric Group 90: a new group of *Enterobacteriaceae* biochemically similar to *Salmonella*. 88th Annual Meeting of the American Society for Microbiology, Washington, D.C. American Society for Microbiology. p. 240.

Muroga, K., Y. Jo and K. Masumura. 1986. Vibrio ordalii isolated from diseased ayu (*Plecoglossus altivelis*) and rockfish (*Sebastes schlegeli*). Fish Pathol. *21*: 239–243.

Murphy, T.F. 1994. Antigenic variation of surface proteins as a survival strategy for bacterial pathogens. Trends Microbiol. *2*: 427–428.

Murphy, T.F. 1996. *Branhamella catarrhalis*: epidemiology, surface antigenic structure, and immune response. Microbiol. Rev. *60*: 267–279.

Murphy, T.V., G.H. McCracken, Jr., T.C. Zweighaft and E.J. Hansen. 1981. Emergence of rifampin-resistant *Haemophilus influenzae* after prophylaxis. J. Pediatr. *99*: 406–409.

Murray, B.E. and R.C. Moellering, Jr.. 1979. Aminoglycoside-modifying enzymes among clinical isolates of *Acinetobacter calcoaceticus* subsp. *anitratus* (*Herellea vaginicola*): explanation for high-level aminoglycoside resistance. Antimicrob. Agents Chemother. *15*: 190–199.

Murray, E.G.D. 1939. Family *Parnobacteriaceae*. *In* Bergey, Breed, Murray and Hitchens (Editors), Bergey's Manual of Determinative Bacteriology, 5th ed., The Williams & Wilkins Co., Baltimore. p. 309.

Murray, E.G.D. 1948. Genus II. *Moraxella* Lwoff. *In* Breed, Murray and Hitchens (Editors), Bergey's Manual of Determinative Bacteriology, 6th Ed., The Williams & Wilkins Co., Baltimore. pp. 590–592.

Murray, J.E., R.C. Davies, F.A. Lainson, C.F. Wilson and W. Donachie. 1992. Antigenic analysis of iron-regulated proteins in *Pasteurella haemolytica* A and T biotypes by immunoblotting reveals biotype-specific epitopes. J. Gen. Microbiol. *138*: 283–288.

Murray, J., L.M. Fixter, I.D. Hamilton, M.C.M. Perombelon, C.E. Quinn and D.C. Graham. 1990a. Serogroups of potato pathogenic *Erwinia carotovora* strains: identification by lipopolysaccharide electrophoretic patterns. J. Appl. Bacteriol. *68*: 231–240.

Murray, R.G.E. and A. Birch-Andersen. 1963. Specialized structure in the region of the flagella tuft in *Spirillum serpens*. Can. J. Microbiol. *9*: 393–401.

Murray, R.G.E., D.J. Brenner, R.R. Colwell, P. DeVos, M. Goodfellow, P.A.D. Grimont, N. Pfennig, E. Stackebrandt and G.A. Zavarzin. 1990b. Report of the ad hoc committee on approaches to taxonomy within the proteobacteria. Int. J. Syst. Bacteriol. *40*: 213–215.

Murrell, J.C. 1992. Genetics and molecular biology of methanotrophs. FEMS Microbiol. Rev. *8*: 233-248.

Murrell, J.C. and H. Dalton. 1983. Nitrogen fixation in obligate methanotrophs. J. Gen. Microbiol. *129*: 3481–3486.

Musser, J.M., S.J. Barenkamp, D.M. Granoff and R.K. Selander. 1986. Genetic relationships of serologically nontypable and serotype b strains of *Haemophilus influenzae*. Infect. Immun. *52*: 183–191.

Musser, J.M., J.S. Kroll, D.M. Granoff, E.R. Moxon, B.R. Brodeur, J. Campos, H. Dabernat, W. Frederiksen, J. Hamel and G. Hammond. 1990. Global genetic structure and molecular epidemiology of encapsulated *Haemophilus influenzae*. Rev. Infect. Dis. *12*: 75–111.

Musser, J.M., J.S. Kroll, E.R. Moxon and R.K. Selander. 1988. Evolutionary genetics of the encapsulated strains of *Haemophilus influenzae*. Proc. Natl. Acad. Sci. U. S. A. *85*: 7758–7762.

Musser, J.M. and R.K. Selander. 1990. Brazilian purpuric fever: evolutionary genetic relationships of the case clone of *Haemophilus influenzae* biogroup aegyptius to encapsulated strains of *Haemophilus influenzae*. J. Infect. Dis. *161*: 130–133.

Mutharia, L.M. and R.E.W. Hancock. 1985. Monoclonal-antibody for an outer-membrane lipoprotein of the *Pseudomonas fluorescens* group of the family *Pseudomonadaceae*. Int. J. Syst. Bacteriol. *35*: 530–532.

Mutters, R., W. Frederiksen and W. Mannheim. 1984. Lack of evidence for the occurence of *Pasteurella ureae* in rodents. Vet. Microbiol. *9*: 83–89.

Mutters, R., P. Ihm, S. Pohl, W. Frederiksen and W. Mannheim. 1985a. Reclassification of the genus *Pasteurella* Trevisan 1887 on the basis of deoxyribonucleic acid homology, with proposals for the new species *Pasteurella dagmatis*, *Pasteurella canis*, *Pasteurella stomatis*, *Pasteurella anatis* and *Pasteurella langaa*. Int. J. Syst. Bacteriol. *35*: 309–322.

Mutters, R., W. Mannheim and M. Bisgaard. 1989. Taxonomy of the group. *In* Adlam and Rutter (Editors), *Pasteurella* and Pasteurellosis, Academic Press, London. pp. 3–34.

Mutters, R., M. Mouahid, E. Engelhard and W. Mannheim. 1993. Characterization of the family *Pasteurellaceae* on the basis of cellular lipids and carbohydrates. Zentbl. Bakteriol. *279*: 104–113.

Mutters, R., K. Piechulla, K.H. Hinz and W. Mannheim. 1985b. *Pasteurella avium* (Hinz and Kunjara 1977) comb. nov. and *Pasteurella volantium* sp. nov. Int. J. Syst. Bacteriol. *35*: 5–9.

Mutters, R., S. Pohl and W. Mannheim. 1986. Transfer of *Pasteurella ureae* Jones 1962 to the genus *Actinobacillus* Brumpt 1910: *Actinobacillus ureae* comb. nov. Int. J. Syst. Bacteriol. *36*: 343–344.

Muyembe, T., J. Vandepitte and J. Desmyter. 1973. Natural colistin resistance in *Edwardsiella tarda*. Antimicrob. Agents Chemother. *4*: 521–524.

Muyzer, G., A. Teske, C.O. Wirsen and H.W. Jannasch. 1995. Phylogenetic relationship of *Thiomicrospira* species and their identification in deep-sea hydrothermal vent samples by denaturing gradient gel electrophoresis of 16S rDNA fragments. Arch. Microbiol. *164*: 165–172.

Myers, C.R. and J.M. Myers. 1993. Role of menaquinone in the reduction of fumarate, nitrate, iron(III) and manganese(IV) by *Shewanella putrefaciens* MR-1. FEMS Microbiol. Lett. *114*: 215–222.

Myers, C.R. and J.M. Myers. 1997. Isolation and characterization of a transposon mutant of *Shewanella putrefaciens* MR-1 deficient in fumarate reductase. Lett. Appl. Microbiol. *25*: 162–168.

Myers, C.R. and K.H. Nealson. 1988. Bacterial manganese reduction and growth with manganese oxide as the sole electron acceptor. Science *240*: 1319–1321.

Myhrvold, V., I. Brondz and I. Olsen. 1992. Application of multivariate analyses of enzymic data to classification of members of the *Actinobacillus-Haemophilus-Pasteurella* group. Int. J. Syst. Bacteriol. *42*: 12–18.

Mylona, P.V., R. Premakumar, R.N. Pau and P.E. Bishop. 1996. Characteristics of orf 1 and orf 2 in the *nifHDGK* genomic region encoding nitrogenase 3 of *Azotobacter vinelandii*. J. Bacteriol. *178*: 204–208.

Mylroie, J.R., D.A. Friello, T.V. Siemens and A.M. Chakrabarty. 1977. Mapping of *Pseudomonas putida* chromosomal genes with a recombinant sex-factor plasmid. Mol. Gen. Genet. *157*: 231–237.

Naemura, L.G., S.T. Bagley, R.J. Seidler, J.B. Kaper and R.R. Colwell. 1979. Numerical taxonomy of *Klebsiella pneumoniae* strains isolated from clinical and nonclinical sources. Curr. Microbiol. *2*: 175–180.

Naganuma, T. and H. Seki. 1991. Sulfur bacteria isolated from the hydrothermal fluid and plume in the North Fiji Basin, southwest Pacific Ocean. Program and Abstracts. Second International Marine Biotechnology Conference IMBC '91. *1991*: 86.

Nagarkoti, M.S., A.K. Banerjee and J. Swarup. 1973. *Xanthomonas convolouli* spec. nov. causing leaf spot of *Conuoloulus arvensis* in India. Indian J. Mycol. Plant Pathol. *3*: 105.

Naidu, A.J. and M. Yadav. 1997. Influence of iron, growth temperature and plasmids on siderophore production in *Aeromonas hydrophila*. J. Med. Microbiol. *46*: 833–838.

Nair, G.B., P.K. Bag, T. Shimada, T. Ramamurthy, T. Takeda, S. Yamamoto, H. Kurazono and Y. Takeda. 1995. Evaluation of DNA probes for specific detection of *Vibrio cholerae* O139 Bengal. J. Clin. Microbiol. *33*: 2186–2187.

Nair, S., K. Kita-Tsukamoto and U. Simidu. 1988. Bacterial flora of healthy and abnormal chaetognaths. Nippon Suisan Gakkaishi. *54*: 491–496.

Nair, S. and U. Simidu. 1987. Distribution and significance of heterotrophic marine bacteria with antibacterial activity. Appl. Environ. Microbiol. *53*: 2957–2962.

Nakada, N. and K. Takimoto. 1923. Bacterial blight of hibiscus. Ann. Phytopathol. Soc. Japan *1*: 13–19.

Nakagawa, Y., T. Sakane and A. Yokota. 1996. Transfer of "*Pseudomonas riboflavina*" (Foster 1944), a gram-negative, motile rod with long-chain 3-hydroxy fatty acids, to *Devosia ribiflavina* gen. nov., sp. nov., nom. rev. Int. J. Syst. Bacteriol. *46*: 16–22.

Nakai, C., H. Uyeyama, H. Kagamiyama, T. Nakazawa, S. Inouye, F. Kishi, A. Nakazawa and M. Nozaki. 1995. Cloning, DNA-sequencing, and amino-acid sequencing of catechol 1,2-dioxygenases (pyrocatechase) from *Pseudomonas putida* Mt-2 and *Pseudomonas arvilla* C-1. Arch. Biochem. Biophys. *321*: 353–362.

Nakamiya, K. and S. Kinoshita. 1995. Isolation of polyacrylamide-degrading bacteria. J. Ferment. Bioeng. *80*: 418–420.

Nakano, K. 1919. Soybean leaf spot. J. Plant Prot. *6*: 217–221.

Nakasone, N. and M. Iwanaga. 1990. Pili of a *Vibrio parahaemolyticus* strain as a possible colonization factor. Infect. Immun. *58*: 61–69.

Nakata, N., T. Tobe, I. Fukuda, T. Suzuki, K. Komatsu, M. Yoshikawa and C. Sasakawa. 1993. The absence of a surface protease, OmpT, determines the intercellular spreading ability of *Shigella*: the relationship between the *ompT* and *kcpA* loci. Mol. Microbiol. *9*: 459–468.

Nakazawa, M., Y. Kagemori and R. Azuma. 1979. Serological variants of *Actinobacillus lignieresii* in slaughtered cattle. Jpn. J. Vet. Sci. *41*: 89–90.

Nakhimovskaya, M.I. 1948. *Pseudomonas aurantiaca* nov. sp. Mikrobiologiya *17*: 58–65.

Nalin, R., P. Simonet, T.M. Vogel and P. Normand. 1999. *Rhodanobacter lindaniclasticus* gen. nov., sp. nov., a lindane-degrading bacterium. Int. J. Syst. Bacteriol. *49*: 19–23.

Namdari, H. and E.J. Bottone. 1990. Microbiologic and clinical evidence supporting the role of *Aeromonas caviae* as a pediatric enteric pathogen. J. Clin. Microbiol. *28*: 837–840.

Namioka, S. 1978. *Pasteurella multocida* - biochemical characteristics and serotypes. *In* Bergan and Norris (Editors), Methods in Microbiology, Vol. 10, Academic Press, London. 271–292.

Namioka, S. and R. Sakazaki. 1958. Étude sur les *Rettgerella*. Ann. Inst. Pasteur (Paris) *94*: 485–499.

Namioka, S. and R. Sakazaki. 1959. New K antigen (C antigen) possessed by *Proteus* and *Rettgerella* cultures. J. Bacteriol. *78*: 301–306.

Namsaraev, B.B., L.E. Dulov, G.A. Dubinina, T.I. Zemskaya, L.Z. Granina and E.V. Karabanov. 1994. Bacterial synthesis and destruction of organic matter in microbial mats of Lake Baikal. Microbiology *63*: 193–197.

Nandy, R.K., T.K. Sengupta, S. Mukhopadhyay and A.C. Ghose. 1995. A comparative study of the properties of *Vibrio cholerae* O139, O1 and other non-O1 strains. J. Med. Microbiol. *42*: 251–257.

Nanninga, N. 1985. Molecular cytology of *Escherichia coli*, Academic Press, Inc., Orlando.

Nano, F.E. 1988. Identification of a heat-modifiable protein of *Francisella tularensis* and molecular cloning of the encoding gene. Microb. Pathog. *5*: 109–120.

Nassar, A., Y. Bertheau, C. Dervin, J.P. Narcy and M. Lemattre. 1994. Ribotyping of *Erwinia chrysanthemi* strains in relation to their pathogenic and geographic distribution. Appl. Environ. Microbiol. *60*: 3781–3789.

Nassar, A., A. Darrasse, M. Lemattre, A. Kotoujansky, C. Dervin, R. Vedel and Y. Bertheau. 1996a. Characterization of *Erwinia chrysanthemi* by pectinolytic isozyme polymorphism and restriction fragment length polymorphism analysis of PCR-amplified fragments of *pel* genes. Appl. Environ. Microbiol. *62*: 2228–2235.

Nassar, A., J.P. Narcy and M. Lemattre. 1996b. Detection of *Erwinia chrysanthemi* pv. *dianthicola* (*Ech*) by the DAS-ELISA method in symptomless carnation cuttings. Agronomie (Paris). *16*: 143–151.

Nasser, W., S. Reverchon, G. Condemine and J. Robert-Baudouy. 1994. Specific interactions of *Erwinia chrysanthemi* KdgR repressor with different operators of genes involved in pectinolysis. J. Mol. Biol. *236*: 427–440.

Nassif, X. and P.J. Sansonetti. 1986. Correlation of the virulence of *Klebsiella pneumoniae* K1 and K2 with the presence of a plasmid encoding aerobactin. Infect. Immun. *54*: 603–608.

Nastasi, A., S. Pignato, C. Mammina and G. Giammanco. 1993. rRNA gene restriction patterns and biotypes of *Shigella sonnei*. Epidemiol. Infect. *110*: 23–30.

Nataro, J.P. 1995. Enteroaggregative and diffusely adherent *Escherichia coli*. *In* Blaser, Smith, Ravdin, Greenberg and Guerrant (Editors), Infections of the Gastrointestinal Tract, Raven Press, Ltd., New York. pp. 727–737.

Nataro, J.P., M.M. Baldini, J.B. Kaper, R.E. Black, N. Bravo and M.M. Levine. 1985a. Detection of an adherence factor of enteropathogenic *Escherichia coli* with a DNA probe. J. Infect. Dis. *152*: 560–565.

Nataro, J.P., Y. Deng, D.R. Maneval, A.L. German, W.C. Martin and M.M. Levine. 1992. Aggregative adherence fimbriae I of enteroaggregative *Escherichia coli* mediate adherence to HEp-2 cells and hemagglutination of human erythrocytes. Infect. Immun. *60*: 2297–2304.

Nataro, J.P. and J.B. Kaper. 1998. Diarrheagenic *Escherichia coli*. Clin. Microbiol. Rev. *11*: 142–201.

Nataro, J.P., J.B. Kaper, R. Robins-Browne, V. Prado, P. Vial and M.M. Levine. 1987. Patterns of adherence of diarrheagenic *Escherichia coli* to HEp-2 cells. Pediatr. Infect. Dis. J. *6*: 829–831.

Nataro, J.P., I.C. Scaletsky, J.B. Kaper, M.M. Levine and L.R. Trabulsi. 1985b. Plasmid-mediated factors conferring diffuse and localized adherence of enteropathogenic *Escherichia coli*. Infect. Immun. *48*: 378–383.

Nataro, J.P., J. Seriwatana, A. Fasano, D.R. Maneval, L.D. Guers, F. Noriega, F. Dubovsky, M.M. Levine and J.G. Morris, Jr.. 1995. Identification and cloning of a novel plasmid-encoded enterotoxin of enteroinvasive *Escherichia coli* and *Shigella* strains. Infect. Immun. *63*: 4721–4728.

Nataro, J.P., D. Yikang, J.A. Girón, S.J. Savarino, M.H. Kothary and R. Hall. 1993. Aggregative adherence fimbria I expression in enteroaggregative *Escherichia coli* requires two unlinked plasmid regions. Infect. Immun. *61*: 1126–1131.

Navon-Venezia, S., Z. Zosim, A. Gottlieb, R. Legmann, S. Carmeli, E.Z. Ron and E. Rosenberg. 1995. Alasan, a new bioemulsifier from *Acinetobacter radioresistens*. Appl. Environ. Microbiol. *61*: 3240–3244.

Nayudu, M.V. 1972. *Pseudomonas viticola* sp. nov. incitant of a new bacterial disease of grape vine. Phytopathol. Z. *73*: 183–186.

Nazarenko, E.L., V.A. Zubkov, V.S. Shashkov, Y.A. Knirel, R.P. Gorshkova, E.P. Ivanova and Y.S. Ovodov. 1993. Structure of the repeating unit of the acidic polysaccharide from *Alteromonas macleodii* 2MM6. Bioorg. Khim. *19*: 740–751.

Nazarowec-White, M. and J.M. Farber. 1997. *Enterobacter sakazakii*: A review. Int. J. Food Microbiol. *34*: 103–113.

Neal, D.J. and S.G. Wilkinson. 1979. Lipopolysaccharides from *Pseudomonas maltophilia*: structural studies of the side-chain polysaccharide from strain NCTC 10257. Carbohydr. Res. *69*: 191–201.

Neal, D.J. and S.G. Wilkinson. 1982. Lipopolysaccharides from *Pseudomonas maltophilia*. Structural studies of the side-chain, core, and lipid-A regions of the lipopolysaccharide from strain NCTC 10257. Eur. J. Biochem. *128*: 143–149.

Nealson, K.H., C.R. Myers and B. Wimpee. 1991. Isolation and identification of manganese-reducing bacteria and estimates of microbial Mn-reducing potential in the Black Sea. Deep-Sea Res. *38*: S907–S920.

Nealson, K.H. and D. Saffarini. 1994. Iron and manganese in anaerobic respiration: environmental significance, physiology, and regulation. Annu. Rev. Microbiol. *48*: 311–343.

Neidhardt, F.C. and H.E. Umbarger. 1996. Chemical composition of *Escherichia coli*. *In* Neidhardt, Curtiss, Ingraham, Lin, Low, Magasanik, Reznikoff, Riley, Schaechter and Umbarger (Editors), *Escherichia coli* and *Salmonella* : Cellular and Molecular Biology, 2nd Ed., ASM Press, Washington, D.C. pp. 13–16.

Neish, A.C., A.C. Blackwood, F.M. Robertson and G.A. Ledingham. 1948. Production and properties of 2,3-butanediol. XXV. Dissimilation of glucose by bacteria of the genus *Serratia*. Can. J. Res. *26*: 335–342.

Nelson, D.C. 1989. Physiology and biochemistry of filamentous bacteria. *In* Schlegel and Bowien (Editors), Autotrophic Bacteria, Science Tech Publishers, Madison. 219–238.

Nelson, D.C. 1992. The Genus *Beggiatoa*. *In* Balows, Trüper, Dworkin, Harder and Schleifer (Editors), The Prokaryotes. A Handbook on the Biology of Bacteria: Ecophysiology, Isolation, Identification, Applications, 2nd Ed., Vol. 4, Springer-Verlag, New York. 3171–3180.

Nelson, D.C. and R.W. Castenholz. 1981a. Organic nutrition of *Beggiatoa* sp. J. Bacteriol. *147*: 236–247.

Nelson, D.C. and R.W. Castenholz. 1981b. Use of reduced sulfur compounds by *Beggiatoa* sp. J. Bacteriol. *147*: 140–154.

Nelson, D.C. and R.W. Castenholz. 1982. Light responses of *Beggiatoa*. Arch. Microbiol. *131*: 146–155.

Nelson, D.C., B.B. Jørgensen and N.P. Revsbech. 1986a. Growth pattern and yield of a chemoautotrophic *Beggiatoa* sp. in oxygen-sulfide microgradients. Appl. Environ. Microbiol. *52*: 225–233.

Nelson, D.C. and H.W. Jannasch. 1983. Chemoautotrophic growth of a marine *Beggiatoa* in sulfide-gradient cultures. Arch. Microbiol. *136*: 262–269.

Nelson, D.C., N.P. Revsbach and B.B. Jørgensen. 1986b. The microoxic/anoxic niche of *Beggiatoa*: a microelectrode survey of marine and freshwater strains. Appl. Environ. Microbiol. *52*: 161–168.

Nelson, D.R., Y. Sadlowski, M. Eguchi and S. Kjelleberg. 1997. The starvation-stress response of *Vibrio* (*Listonella*) *anguillarum*. Microbiology *143*: 2305–2312.

Nelson, D.C., J.B. Waterbury and H.W. Jannasch. 1982. Nitrogen fixation and nitrate utilization by marine and freshwater *Beggiatoa*. Arch. Microbiol. *133*: 172–177.

Nelson, D.C., C.A. Williams, B.A. Farah and J.M. Shively. 1989a. Occurrence and regulation of Calvin cycle enzymes in non-autotrophic *Beggiatoa* strains. Arch. Microbiol. *151*: 15–19.

Nelson, D.C., C.O. Wirsen and H.W. Jannasch. 1989b. Characterization of large, autotrophic *Beggiatoa* spp. abundant at hydrothermal vents of the Guaymas Basin, (Gulf of California, USA). Appl. Environ. Microbiol. *55*: 2909–2917.

Nelson, M.B., T.F. Murphy, H. van Keulen, D. Rekosh and M.A. Apicella. 1988. Studies on P6, an important outer-membrane protein antigen of *Haemophilus influenzae*. Rev. Infect. Dis. *10 (Suppl 2)*: S331–336.

Nemec, A., L. Janda, O. Melter and L. Dijkshoorn. 1999. Genotypic and phenotypic similarity of multiresistant *Acinetobacter baumannii* isolates in the Czech Republic. J. Med. Microbiol. *48*: 287–296.

Nesme, X., M. Vaneechoutte, S. Orso, B. Hoste and J. Swings. 1995. Diversity and genetic relatedness within genera *Xanthomonas* and *Stenotrophomonas* using restriction endonuclease site differences of PCR-amplified 16S rRNA gene. Syst. Appl. Microbiol. *18*: 127–135.

Nesterov, A.I., A.V. Koshelev, V.F. Gal'chenko and M.V. Ivanov. 1986. The survival of obligate methanotrophous bacteria upon lyophilization and subsequent storage. Mikrobiologiya *55*: 215–221.

Neter, E., O. Westphal, O. Lüderitz, R.M. Gino and E.A. Gorzynski. 1955. Demonstration of antibodies against enteropathogenic *Escherichia coli* in sera of children of various ages. Pediatrics. *16*: 801–807.

Neunlist, S. and M. Rohmer. 1985. Novel hopanoids from the methylotrophic bacteria *Methylococcus capsulatus* and *Methylomonas methanica* (22S)-35-aminobacteriohopane- 30,31,32,33,34-pentol and (22S)-35-amino-3 β-methylbacteriohopane- 30,31,32,33,34-pentol. Biochem. J. *231*: 635–639.

Neuwirth, C., E. Siebor, J.M. Duez, A. Pechinot and A. Kazmierczak. 1995. Imipenem resistance in clinical isolates of *Proteus mirabilis* associated with alterations in penicillin-binding proteins. Antimicrob. Agents Chemother. *36*: 335–342.

Neveu-Lemaire, M. 1921. Précis de la Parasitologie Humaine, 5th ed., J. Lemaire, Paris.

New, P.B. and Y.T. Tchan. 1982. *Azomonas macrocytogenes* (Ex Baillie, Hodgkiss, and Norris 1962, 118) nom rev. Int. J. Syst. Bacteriol. *32*: 381–382.

Newland, J.W., M.J. Voll and L.A. McNicol. 1984. Serology and plasmid carriage in *Vibrio cholerae*. Can. J. Microbiol. *30*: 1149–1156.

Newman, M.A., M.J. Daniels and J.M. Dow. 1995. Lipopolysaccharide from *Xanthomonas campestris* induces defense-related gene expression in *Brassica campestris*. Mol. Plant-Microbe Interact. *8*: 778–780.

Newman, M.A., M.J. Daniels and J.M. Dow. 1997. The activity of lipid A and core components of bacterial lipopolysaccharides in the prevention of the hypersensitive response in pepper. Mol. Plant-Microbe Interact. *10*: 926–928.

Newsom, I.E. and F. Cross. 1932. Some bipolar organisms found in pneumonia of sheep. J. Am. Vet. Med. Assoc. *80*: 711–719.

Nguyen, B.H., E.B.M. Denner, T.C.H. Dang, G. Wanner and H. Stan-Lotter. 1999. *Marinobacter* sp. nov., a halophilic bacterium isolated from a Vietnamese oil-producing well. Int. J. Syst. Bacteriol. *49*: 367–375.

Nguyen, M.H. and R.R. Muder. 1994. Meningitis due to *Xanthomonas maltophilia*: case report and review. Clin. Infect. Dis. *19*: 325–326.

Nguyen, S.V. and K. Hirai. 1999. Differentiation of *Coxiella burnetii* isolates by sequence determination and PCR-restriction fragment length polymorphism analysis of isocitrate dehydrogenase gene. FEMS Microbiol. Lett. *180*: 249–254.

Nguyen-Van-Ai, D.H, Nguyen, L.T. Van, V.L. Nguyen and T.L.H. Nguyen. 1975. Contribution a l' étude des *Edwardsiella tarda* isolés au Viet-Nam. Bull. Soc. Pathol. Exot. *68*: 355–359.

Nichols, D.S., J.L. Brown, P.D. Nichols and T.A. McMeekin. 1997a. Production of eicosapentaenoic and arachidonic acids by an Antarctic bacterium: response to growth temperatures. FEMS Microbiol. Lett. *152*: 349–354.

Nichols, D.S., A.R. Greenhill, C.T. Shadbolt, T. Ross and T.A. McMeekin. 1999. Physicochemical parameters for growth of the sea ice bacteria *Glaciecola punicea* ACAM 611[T] and *Gelidibacter* sp. strain IC158. Appl. Environ. Microbiol. *65*: 3757–3760.

Nichols, D.S., P.D. Nichols and T.A. McMeekin. 1995. Ecology and physiology of psychrophilic bacteria from Antarctic saline lakes and sea-ice. Sci. Prog. *78*: 311–347.

Nichols, D.S., P.D. Nichols, N.J. Russell, N.W. Davies and T.A. McMeekin. 1997b. Polyunsaturated fatty acids in the psychrophilic bacterium *Shewanella gelidimarina* ACAM 456[T]: molecular species analysis of major phospholipids and biosynthesis of eicosapentaenoic acid. Biochim. Biophys. Acta *1347*: 164–176.

Nichols, N.N. and C.S. Harwood. 1997. PcaK, a high-affinity permease for the aromatic compounds 4-hydroxybenzoate and protocatechuate from *Pseudomonas putida*. J. Bacteriol. *179*: 5056–5061.

Nichols, P.D., J.M. Henson, C.P. Antworth, J. Parsons, J.T. Wilson and D.C. White. 1987. Detection of a microbial consortium, including type II methanotrophs, by use of phospholipid fatty acids in an aerobic halogenated hydrocarbon-degrading soil column enriched with natural gas. Environ. Toxicol. Chem. *6*: 89–98.

Nichols, P.D., G.A. Smith, C.P. Antworth, R.S. Hanson and D.C. White. 1985. Phospholipid and lipopolysaccharide normal and hydroxy fatty acids as potential signature for methane-utilizing bacteria. FEMS Microbiol. Ecol. *32*: 327–335.

Nickel, J.C., J. Emtage and J.W. Costerton. 1985. Ultrastructural microbial ecology of infection-induced urinary stones. J. Urol. *133*: 622–627.

Nicklas, W. 1989. *Haemophilus* infection in a colony of laboratory rats. J. Clin. Microbiol. *27*: 1636–1639.

Nicolet, J. 1971. Sur l' hémophilose du pore. III. Differentiation sérologique de *Haemophilus parahaemolyticus*. Zentralbl. Bakteriol. Parasitenkd. Infektionskr. Hyg. Abt. I, Orig. *216*: 487–495.

Nicolet, J. and M. Krawinkler. 1981. Polyacrylamide gel electrophoresis, a possible taxonomical tool for *Haemophilus*. In Kilian, Frederiksen and Biberstein (Editors), *Haemophilus, Pasteurella* and *Actinobacillus*, Academic Press, London. 205–212.

Nicolle, P., H.H. Mollaret and J. Brault. 1973. Recherches sur la lysogénie, la lysosensibilité, la lysotypie et la sérologic de *Yersinia enterocolitica*. Contr. Microbiol. Immunol. *2*: 54–58.

Niebel, H., M. Dorsch and E. Stackebrandt. 1987. Cloning and expression in *Escherichia coli* of *Proteus vulgaris* genes for 16S ribosomal RNA. J. Gen. Microbiol. *133*: 2401–2409.

Niebylski, M.L., M.G. Peacock, E.R. Fischer, S.F. Porcella and T.G. Schwan. 1997. Characterization of an endosymbiont infecting wood ticks, *Dermacentor andersoni*, as a member of the genus *Francisella*. Appl. Environ. Microbiol. *63*: 3933–3940.

Niederhauser, J.S. 1943. A bacterial leaf spot and blight of the Russian dandelion. Phytopathology *33*: 959–961.

Nielsen, J.P. and V.T. Rosdahl. 1990. Development and epidemiological applications of a bacteriophage-typing system for typing *Pasteurella multocida*. J. Clin. Microbiol. *28*: 103–107.

Nielsen, P.H., K. Andreasen, M. Wagner, L.L. Blackall, H. Lemmer and R.J. Seviour. 1998. Variability of type 021N in activated sludge as determined by in situ substrate uptake pattern and in situ hybridization with fluorescent rRNA targeted probes. Water Sci. Technol. *37*: 423–440.

Nielsen, P.H., M.A. de Muro and J.L. Nielsen. 2000. Studies on the in situ physiology of *Thiothrix* spp. present in activated sludge. Environ. Microbiol. *2*: 389–398.

Nielsen, R. 1990. New diagnostic techniques: a review of the HAP group of bacteria. Can. J. Vet. Res. *54*: S68–S72.

Nielsen, R., L.O. Andresen, T. Plambeck, J.P. Nielsen, L.T. Krarup and S.E. Jorsal. 1997. Serological characterization of *Actinobacillus pleuropneumoniae* biotype 2 strains isolated from pigs in two Danish herds. Vet. Microbiol. *54*: 35–46.

Nieto, J.J., R. Fernandez-Castillo, M.T. Garcia, E. Mellado and A. Ventosa. 1993. Survey of antimicrobial susceptibility of moderately halophilic eubacteria and extremely halophilic aerobic archaeobacteria—utilization of antimicrobial resistance as a genetic marker. Syst. Appl. Microbiol. *16*: 352–360.

Nieto, J.J., C. Vargas and A. Ventosa. 2000. Osmoprotection mechanisms in the moderately halophilic bacterium *Halomonas elongata*. Recent Res. Devel. Microbiol. *4*: 43–54.

Nikaido, H. 1992. Nonspecific and specific permeation channels of the *Pseudomonas aeruginosa* outer membrane. *In* Galli, Silver and Witholt (Editors), *Pseudomonas*: Molecular Biology and Biotechnology, American Society for Microbiology, Washington D.C. 146–153.

Nikaido, H. 1996. Outer membrane. *In* Neidhardt, Curtiss, Ingraham, Lin, Low, Magasanik, Reznikoff, Riley, Schaechter and Umbarger (Editors), *Escherichia coli* and *Salmonella*: Cellular and Molecular Biology, 2nd Ed., ASM Press, Washington, D.C. pp. 29–47.

Nikaido, N., K. Ito, K. Izaki and H. Takahashi. 1985. DNA sequence of the promoter and Nh2-terminal regions of the pectate lyase I gene from *Erwinia carotovora*. J. Gen. Appl. Microbiol. *31*: 573–576.

Nilehn, B. 1969. Studies on *Yersinia enterocolitica* with special reference to bacterial diagnosis and occurrence in human enteric disease. Acta Pathol. Microbiol. Scand. Suppl. *206*: 1–48.

Nimmich, W. 1994. Detection of *Escherichia coli* K95 strains by bacteriophages. J. Clin. Microbiol. *32*: 2843–2845.

Nimmich, W., U. Krallmann-Wenzel, B. Müller and G. Schmidt. 1992. Isolation and characterization of bacteriophages specific for capsular antigens K3, K7, K12, and K13 of *Escherichia coli*. Zentbl. Bakteriol. *276*: 213–220.

Nishibuchi, M., J.M. Janda and T. Ezaki. 1996. The thermostable direct hemolysin gene (*tdh*) of *Vibrio hollisae* is dissimilar in prevalence to and phylogenetically distant from the *tdh* genes of other vibrios: Implications in the horizontal transfer of the *tdh* gene. Microbiol. Immunol. *40*: 59–65.

Nishibuchi, M. and J.B. Kaper. 1995. Thermostable direct hemolysin gene of *Vibrio parahaemolyticus*: A virulence gene acquired by a marine bacterium. Infect. Immun. *63*: 2093–2099.

Nishibuchi, M., N.C. Roberts, H.B. Bradford, Jr. and R.J. Seidler. 1983. Broth medium for enrichment of *Vibrio fluvialis* from the environment. Appl. Environ. Microbiol. *46*: 425–429.

Nishihara, H., Y. Igarashi and T. Kodama. 1989. Isolation of an obligately chemolithoautotrophic, halophilic and aerobic hydrogen-oxidizing bacterium from marine environment. Arch. Microbiol. *152*: 39–43.

Nishihara, H., Y. Igarashi and T. Kodama. 1991a. Growth characteristics and high cell-density cultivation of a marine obligately chemolithoautotrophic hydrogen-oxidizing bacterium *Hydrogenovibrio marinus* strain MH-110 under a continuous gas-flow system. J. Ferment. Bioeng. *72*: 358–361.

Nishihara, H., Y. Igarashi and T. Kodama. 1991b. *Hydrogenovibrio marinus*, gen. nov., sp. nov., a marine obligately chemolithoautotrophic hydrogen-oxidizing bacterium. Int. J. Syst. Bacteriol. *41*: 130–133.

Nishihara, H., Y. Igarashi, T. Kodama and T. Nakajima. 1993. Production and properties of glycogen in the marine obligate chemolithoautotroph, *Hydrogenovibrio marinus*. J. Ferment. Bioeng. *75*: 414–416.

Nishihara, H., Y. Miyashita, K. Aoyama, T. Kodama, Y. Igarashi and Y. Takamura. 1997. Characterization of an extremely thermophilic and oxygen-stable membrane-bound hydrogenase from a marine hydrogen-oxidizing bacterium *Hydrogenovibrio marinus*. Biochem. Biophys. Res. Commun. *232*: 766–770.

Nishihara, H., T. Yaguchi, S.Y. Chung, K.-I. Suzuki, M. Yanagi, K. Yamasato, T. Kodama and Y. Igarashi. 1998. Phylogenetic position of an obligately chemoautotrophic, marine hydrogen-oxidizing bacterium, *Hydrogenovibrio marinus*, on the basis of 16S rRNA gene sequences and two form I RuBisCO gene sequences. Arch. Microbiol. *169*: 364–368.

Nishikawa, Y., J. Ogasawara, A. Helander and K. Haruki. 1999. An outbreak of gastroenteritis in Japan due to *Escherichia coli* O166. Emerg. Infect. Dis. *5*: 300.

Nishimura, Y., A. Hagiwara, T. Suzuki and S. Yamanaka. 1994. *Xenorhabdus japonicus* sp. nov. associated with the nematode *Steinerema kushidai*. World J. Microbiol. Biotechnol. *10*: 207–210.

Nishimura, Y., A. Hagiwara, T. Suzuki and S. Yamanaka. 1995. *In* Validation of the publication of new names and new combinations previously effectively published outside the IJSB. List no. 54. Int. J. Syst. Bacteriol. *45*: 619–620.

Nishimura, Y., T. Ino and H. Iizuka. 1986a. Isolation and characterization of the outer membrane of radiation-resistant *Acinetobacter* sp. FO-1. J. Gen. Appl. Microbiol. *32*: 177–184.

Nishimura, Y., T. Ino and H. Iizuka. 1988. *Acinetobacter radioresistens*, new species isolated from cotton and soil. Int. J. Syst. Bacteriol. *38*: 209–211.

Nishimura, Y., K. Kanbe and H. Iizuka. 1986b. Taxonomic studies of aerobic coccobacilli from seawater. J. Gen. Appl. Microbiol. *32*: 1–11.

Nishimura, Y., M. Kinpara and H. Iizuka. 1989. *Mesophilobacter marinus* gen. nov., sp. nov.: An aerobic coccobacillus isolated from seawater. Int. J. Syst. Bacteriol. *39*: 378–381.

Nishimura, Y., H. Yamamoto and H. Iizuka. 1979. Taxonomical studies of *Acinetobacter* species—cellular fatty acid composition. Z. Allg. Mikrobiol. *19*: 307–308.

Nishino, M., M. Fukui and T. Nakajima. 1998. Dense mats of *Thioploca*, gliding filamentous sulfur-oxidizing bacteria in Lake Biwa, central Japan. Water Res. *32*: 953–957.

Nissen, H. 1987. Long term starvation of a marine bacterium, *Alteromonas denitrificans*, isolated from a Norwegian fjord. FEMS Microbiol. Ecol. *45*: 173–183.

Niven, D.F. and T. O'Reilly. 1990. Significance of V-factor dependency in the taxonomy of *Haemophilus* species and related organisms. Int. J. Syst. Bacteriol. *40*: 1–4.

Noble, P.A., P.E. Dabinett and J.A. Gow. 1990. A numerical taxonomic study of pelagic and benthic surface-layer bacteria in seasonally-cold coastal waters. Syst. Appl. Microbiol. *13*: 77–85.

Noble, R.C. and S.B. Overman. 1994. *Pseudomonas stutzeri* infection - a review of hospital isolates and a review of the literature. Diagn. Microbiol. Infect. Dis. *19*: 51–56.

Noda, H., U.G. Munderloh and T.J. Kurtti. 1997. Endosymbionts of ticks and their relationship to *Wolbachia* spp. and tick-borne pathogens of humans and animals. Appl. Environ. Microbiol. *63*: 3926–3932.

Nogge, G. 1976. Sterility in tsetse flies (*Glossina morsitans* Westwood) caused by loss of symbionts. Experientia (Basel) *32*: 995.

Nogge, G. 1978. Aposymbiotic tsetse flies, *Glossina morsitans morsitans* obtained by feeding on rabbits immunized specifically with symbionts. J. Insect Physiol. *24*: 299–304.

Nogge, G. 1980. Elimination of symbionts of tsetse flies (*Glossina m. morsitans* Westwood) by help of specific antibodies. *In* Schwemmler and Schenk (Editors), Endocytobiology. Endosymbiosis and Cell Biology. A Synthesis of Recent Research, W. deGruyter, Berlin New York. pp. 445–452.

Nogge, G. 1981. Significance of symbionts for the maintenance of an optimal nutritional state for successful reproduction in hematophagous arthropods. Parasitology. *82*: 101–104.

Nogi, Y. and C. Kato. 1999a. Taxonomic studies of extremely barophilic bacteria isolated from the Mariana Trench and description of *Moritella yayanosii* sp. nov., a new barophilic bacterial isolate. Extremophiles *3*: 71–77.

Nogi, Y. and C. Kato. 1999b. *In* Validation of the publication of new names and new combinations previously effectively published outside the IJSB. List No. 71. Int. J. Syst. Bacteriol. *49*: 1325–1326.

Nogi, Y., C. Kato and K. Horikoshi. 1998a. *Moritella japonica* sp. nov., a novel barophilic bacterium isolated from a Japan Trench sediment. J. Gen. Appl. Microbiol. *44*: 289–295.

Nogi, Y., C. Kato and K. Horikoshi. 1998b. Taxonomic studies of deepsea barophilic *Shewanella* strains and description of *Shewanella violacea* sp. nov. Arch. Microbiol. *170*: 331–338.

Nogi, Y., C. Kato and K. Horikoshi. 1999. *In* Validation of the publication of new names and new combinations previously effectively published outside the IJSB. List No. 69. Int. J. Syst. Bacteriol. *49*: 341–342.

Nogi, Y., C. Kato and K. Horikoshi. 2002. *Psychromonas kaikoae* sp nov., a novel piezophilic bacterium from the deepest cold-seep sediments in the Japan Trench. Int. J. Syst. Evol. Microbiol. *52*: 1527–1532.

Nogi, Y., N. Masui and C. Kato. 1998c. *Photobacterium profundum* sp. nov., a new, moderately barophilic bacterial species isolated from a deep-sea sediment. Extremophiles 2: 1–7.

Nogi, Y. , N. Masui and C. Kato. 1998d. In Validation of publication of new names and new combinations previously effectively published outside the IJSB, List No. 66. Int. J. Syst. Bacteriol. 48: 631–632.

Nomura, J. and T. Aoki. 1985. Morphological analysis of lipopolysaccharide from gram-negative fish pathogenic bacteria. Fish Pathol. 20: 193–197.

Noonan, B. and T.J. Trust. 1997. The synthesis, secretion and role in virulence of the paracrystalline surface protein layers of Aeromonas salmonicida and A. hydrophila. FEMS Microbiol. Lett. 154: 1–7.

Nordeen, R.O. and B.W. Holloway. 1990. Chromosome mapping in Pseudomonas syringae pathovar syringae strain Ps224. J. Gen. Microbiol. 136: 1231–1239.

Norris, J.R. and H.L. Jensen. 1958. Calcium requirement of Azotobacter. Arch. Mikrobiol. 31: 198–205.

Norris, J.R. and W.H. Kingham. 1968. The classification of Azotobacter. In Festskrift til Hans Laurits Jensen, Gadgaard Nielsens Bogtrykkeri, Demvig, Denmark. 95–105.

Norris, P.R., D.A. Clark, J.P. Owen and S. Waterhouse. 1996. Characteristics of Sulfobacillus acidophilus sp. nov. and other moderately thermophilic mineral-sulphide-oxidizing bacteria. Microbiology (Reading) 142: 775–783.

Noterdaeme, L., S. Bigawa, A.G. Steigerwalt, D.J. Brenner and F. Ollevier. 1996. Numerical taxonomy and biochemical identification of fish associated motile Aeromonas spp. Syst. Appl. Microbiol. 19: 624–633.

Nourrisseau, J.G., M. Lansac and M. Garnier. 1993. Marginal chlorosis, a new disease of strawberries associated with a bacterium-like organism. Plant Dis. 77: 1055–1059.

Novick, N.J. and M.E. Tyler. 1985. Isolation and characterization of Alteromonas luteoviolacea strains with sheathed flagella. Int. J. Syst. Bacteriol. 35: 111–113.

Novikova, L.M. 1971. Formation and consumption of stored products by Ectothiorhodospira shaposhnikovii. Mikrobiologiia 40: 28–33.

Novo, M.T.M., A.P. De Souza, O.J. Garcia and L.M.M. Ottoboni. 1996. RAPD genomic fingerprinting differentiates Thiobacillus ferrooxidans strains. Syst. Appl. Microbiol. 19: 91–95.

Nowak, A., A. Burkiewicz and J. Kur. 1995. PCR differentiation of seventeen genospecies of Acinetobacter. FEMS Microbiol. Lett. 126: 181–187.

Nowak, A., J. Kur, E. Gospodarek and K. Bielawski. 1994. Characterization of restriction endonuclease AjoI from Acinetobacter johnsonii. FEMS Microbiol. Lett. 117: 97–102.

Nowak-Thompson, B. and S.J. Gould. 1994. A simple assay for fluorescent siderophores produced by Pseudomonas species and an efficient isolation of pseudobactin. Biometals 7: 20–24.

Nowicki, B., A. Labigne, S. Moseley, R. Hull, S. Hull and J. Moulds. 1990. The Dr hemagglutinin, afimbrial adhesins AFA-I and AFA-III, and F1845 fimbriae of uropathogenic and diarrhea-associated Escherichia coli belong to a family of hemagglutinins with Dr receptor recognition. Infect. Immun. 58: 279–281.

Nozue, H., T. Hayashi, Y. Hashimoto, T. Ezaki, K. Hamasaki, K. Ohwada and Y. Terawaki. 1992. Isolation and characterization of Shewanella alga from human clinical specimens and emendation of the description of S. alga Simidu et al., 1990, 335. Int. J. Syst. Bacteriol. 42: 628–634.

Nunes, C., J. Usall, N. Teixido and I. Vinas. 2001. Biological control of postharvest pear diseases using a bacterium, Pantoea agglomerans CFA-2. Int. J. Food Microbiol. 70: 53–61.

Nuñez, C., S. Moreno, G. Soberon-Chavez and G. Espin. 1999. The Azotobacter vinelandii response regulator AlgR is essential for cyst formation. J. Bacteriol. 181: 141–148.

Nurminen, M., E. Wahlstrom, M. Kleemola, M. Leinonen, P. Saikku and P.H. Makela. 1984. Immunologically related ketodeoxyoctonate-containing structures in Chlamydia trachomatis, Re mutants of Salmonella spp. and Acinetobacter calcoaceticus var. anitratus. Infect. Immun. 44: 609–613.

Nwigwe, C. 1973. Variation of colony morphology and its relation to the virulence of Xanthomonas oryzae. Plant Dis. Rep. 57: 955–956.

Nyland, G., A.C. Goheen, S.K. Lowe and H.C. Kirkpatrick. 1973. The ultrastructure of a rickettsialike organism from a peach tree affected with phony disease. Phytopathology 63: 1275–1278.

Nyman, K., K. Nakamura, H. Ohtsubo and E. Ohtsubo. 1981. Distribution of the insertion sequence IS1 in gram-negative bacteria. Nature 289: 609–612.

Oakey, H.J., J.T. Ellis and L.F. Gibson. 1996a. A biochemical protocol for the differentiation of current genomospecies of Aeromonas. Zentbl. Bakteriol. 284: 32–46.

Oakey, H.J., J.T. Ellis and L.F. Gibson. 1996b. Differentiation of Aeromonas genomospecies using random amplified polymorphic DNA polymerase chain reaction (RAPD-PCR). J. Appl. Bacteriol. 80: 402–410.

Oakey, H.J., J.T. Ellis and L.F. Gibson. 1998. The development of random DNA probe specific for Aeromonas salmonicida. J. Appl. Microbiol. 84: 37–46.

Oakley, C.J. and J.C. Murrell. 1993. nifH genes in the obligate methane oxidizing bacteria. FEMS Microbiol. Lett. 49: 53–57.

Oaks, E.V., T.L. Hale and S.B. Formal. 1986. Serum immune response to Shigella protein antigens in rhesus monkeys and humans infected with Shigella spp. Infect. Immun. 53: 57–63.

Oaks, E.V., M.E. Wingfield and S.B. Formal. 1985. Plaque formation by virulent Shigella flexneri. Infect. Immun. 48: 124–129.

Obendorf, D.L., B. Peel, R.J. Akhurst and L.A. Miller. 1983. Non-susceptibility of mammals to the entomopathogenic bacterium Xenorhabdus nematophilus. Environ. Entomol. 12: 368–370.

Oberhofer, T.R. and A.E. Back. 1979. Biotypes of Haemophilus encountered in clinical laboratories. J. Clin. Microbiol. 10: 168–174.

Obradors, N. and J. Aguilar. 1991. Efficient biodegradation of high-molecular-weight polyethylene glycols by pure cultures of Pseudomonas stutzeri. Appl. Environ. Microbiol. 57: 2383–2388.

O'Brien, A.D., M.A. Karmali and S.M. Scotland. 1994. A proposal for rationalization of the Escherichia coli cytotoxins. In Karmali and Goglio (Editors), Recent Advances in Verocytotoxin-producing Escherichia coli Infections, Elsevier Science, Amsterdam. pp. 147–149.

O'Brien, A.D., J.W. Newland, S.F. Miller, R.K. Holmes, H.W. Smith and S.B. Formal. 1984. Shiga-like toxin-converting phages from Escherichia coli strains that cause hemorrhagic colitis or infantile diarrhea. Science 226: 694–696.

O'Brien, A.D., V.L. Tesh, A. Donohue-Rolfe, M.P. Jackson, S. Olsnes, K. Sandvig, A.A. Lindberg and G.T. Keusch. 1992. Shiga toxin: biochemistry, genetics, mode of action, and role in pathogenesis. Curr. Top. Microbiol. Immunol. 180: 65–94.

O'Brien, M. and R.R. Colwell. 1987. A rapid test for chitinase activity that uses 4-methylumbelliferyl-N-acetyl-β-D-glucosaminide. Appl. Environ. Microbiol. 53: 1718–1720.

Obrig, T.G., C.B. Louise, C.A. Lingwood, B. Boyd, L. Barley-Maloney and T.O. Daniel. 1993. Endothelial heterogeneity in Shiga toxin receptors and responses. J. Biol. Chem. 268: 15484–15488.

Ochi, K. 1995. Comparative ribosomal-protein sequence analyses of a phylogenetically defined genus, Pseudomonas, and its relatives. Int. J. Syst. Bacteriol. 45: 268–273.

Ochiai, K., T. Yamanaka, K. Kimura and O. Sawada. 1959. Studies on inheritance of drug resistance between Shigella strains and E. coli strains. Nihon Ija Shimpo. 1861: 34–46.

Ochman, H. and R.K. Selander. 1984. Standard reference strains of Escherichia coli from natural populations. J. Bacteriol. 157: 690–693.

Ochman, H., T.S. Whittam, D.A. Caugant and R.K. Selander. 1983. Enzyme polymorphism and genetic population structure in Escherichia coli and Shigella. J. Gen. Microbiol. 129: 2715–2726.

Ochman, H. and A.C. Wilson. 1987. Evolution in bacteria: evidence for a universal substitution rate in cellular genomes. J. Mol. Evol. 26: 74–86.

Ochsner, U.A., Z. Johnson, I.L. Lamont, H.E. Cunliffe and M.L. Vasil. 1996. Exotoxin A production in Pseudomonas aeruginosa requires the iron-regulated pvdS gene encoding an alternative sigma factor. Mol. Microbiol. 21: 1019–1028.

Ochsner, U.A. and J. Reiser. 1995. Autoinducer-mediated regulation of rhamnolipid biosurfactant synthesis in *Pseudomonas aeruginosa*. Proc. Natl. Acad. Sci. U. S. A. *92*: 6424–6428.

O'Connor, K., C.M. Buckley, S. Hartmans and A.D.W. Dobson. 1995. Possible regulatory role for nonaromatic carbon-sources in styrene degradation by *Pseudomonas putida* Ca-3. Appl. Environ. Microbiol. *61*: 544–548.

Odintsova, E.V. and G.A. Dubinina. 1990. A new colorless filamentous sulfur bacterium, *Thiothrix ramosa* nov. sp. Microbiology *59*: 437–445.

Odintsova, E.V., A.P. Wood and D.P. Kelly. 1993. Chemolithoautotrophic growth of *Thiothrix ramosa*. Arch. Microbiol. *160*: 152–157.

Odumeru, J.A., A.R. Ronald and W.L. Albritton. 1983. Characterization of cell proteins of *Haemophilus ducreyi* by polyacrylamide gel electrophoresis. J. Infect. Dis. *148*: 710–714.

Oelze, J. 2000. Respiratory protection of nitrogenase in *Azotobacter* species: is a widely held hypothesis unequivocally supported by experimental evidence? FEMS Microbiol. Rev. *24*: 321–333.

Oersted, A.S. 1844. De regionibus marines, elementa topographiae historiconaturalis freti oeresund, J.C. Carling, Copenhagen.

Ogawa, H., A. Nakamura, R. Nakaya, K. Mise, S. Honjo, M. Takasaka, T. Fujiwara and K. Imaizumi. 1967. Virulence and epithelial cell invasiveness of dysentery bacilli. Jpn. J. Med. Sci. Biol. *20*: 315–328.

Ogawa, J. and Y. Amano. 1987. Electron microprobe X-ray analysis of polyphosphate granules in *Plesiomonas shigelloides*. Microbiol. Immunol. *31*: 1121 = 1125.

Ogawa, T., A. Shinohara, H. Ogawa and J. Tomizawa. 1992. Functional structures of the RecA protein found by chimera analysis. J. Mol. Biol. *226*: 651–660.

Ogimi, C. 1977. Studies on bacterial gall chinaberry *Melia azedarach* Lin. caused by *Pseudomonas meliae* n. sp. Bull. Coll. Agric. Univ. Ryukyus. *24*: 497–556.

Ogimi, C. 1981. *In* Validation of the publication of new names and new combinations previously effectively published outside the IJSB. List No. 7. Int. J. Syst. Bacteriol. *31*: 382–383.

Ogunnariwo, J.A. and A.B. Schryvers. 1990. Iron acquisition in *Pasteurella haemolytica*: expression and identification of a bovine-specific transferrin receptor. Infect. Immun. *58*: 2091–2097.

Oh, H.M. and L. Tay. 1995. Bacteraemia caused by *Rahnella aquatilis*: report of two cases and review. Scand. J. Infect. Dis. *27*: 79–80.

Ohanessian, J.H., N. Fourcade, B. Priolet, C.I. Richard, G. Bashour and M. Dugelay. 1987. A propos d' une infection vesiculaire par *Moellerella wisconsensis*. Med. Mal. Infect. *6/7*: 414–416.

O'Hara, C.M., A.G. Steigerwalt, B.C. Hill, J.J. Farmer, G.R. Fanning and D.J. Brenner. 1989. *Enterobacter hormaechei*, a new species of the Family Enterobacteriaceae formerly known as enteric group-75. J. Clin. Microbiol. *27*: 2046–2049.

O'Hara, C.M., A.G. Steigerwalt, B.C. Hill, J.J. Farmer, G.R. Fanning and D.J. Brenner. 1990. *In* Validation of the publication of new names and new combinations previously effectively published outside the IJSB, List No. 32. Int. J. Syst. Bacteriol. *40*: 105.

O'Hara, C.M., A.G. Steigerwalt, B.C. Hill, J.M. Miller and D.J. Brenner. 1998. First report of a human isolate of *Erwinia persicinus*. J. Clin. Microbiol. *36*: 248–250.

Ohnishi, H., T. Nishida, A. Yoshida, Y. Kamio and K. Izaki. 1991. Nucleotide sequence of *pnl* gene from *Erwinia carotovora* Er. Biochem. Biophys. Res. Commun. *176*: 321–327.

Ohtaka, C., H. Nakamura and H. Ishikawa. 1992. Structures of chaperonins from an intracellular symbiont and their functional expression in *Escherichia coli groE* mutants. J. Bacteriol. *174*: 1869–1874.

Ohtsubo, H., K. Nyman, W. Doroszkiewicz and E. Ohtsubo. 1981. Multiple copies of *iso*-insertion sequences of IS1 in *Shigella dysenteriae* chromosome. Nature *292*: 640–643.

Ohuchi, A. and T. Tominaga. 1973. Pectolytic enzymes secreted by soft rot and saprophytic pseudomonads. Ann. Phytopathol. Soc. Japan *39*: 417–424.

Ohuchi, A. and T. Tominaga. 1975. Histochemical changes of cell walls during the macerating action by pectolytic enzyme, endo-PTE, of a soft rot pseudomonad. Bull. Natl. Inst. Agric. Sci. Ser. C, No. *29*: 45–63.

Okabe, N. 1933. Bacterial diseases of plants occurring in Formosa. II. Bacterial leaf spot of tomato. J. Soc. Trop. Agric., Taiwan. *5*: 26–36.

Okada, N., C. Sasakawa, T. Tobe, K.A. Talukder, K. Komatsu and M. Yoshikawa. 1991. Construction of a physical map of the chromosome of *Shigella flexneri* 2a and the direct assignment of nine virulence-associated loci identified by Tn5 insertions. Mol. Microbiol. *5*: 2171–2180.

Okamoto, K., T. Inoue, H. Ichikawa, Y. Kawamoto and A. Miyama. 1981. Partial purification and characterization of heat-stable enterotoxin produced by *Yersinia enterocolitica*. Infect. Immunol. *31*: 554–559.

Okamoto, T., H. Taguchi, K. Nakamura, H. Ikenaga, H. Kuraishi and K. Yamasato. 1993. *Zymobacter palmae* gen. nov., sp. nov., a new ethanol-fermenting peritrichous bacterium isolated from palm sap. Arch. Microbiol. *160*: 333–337.

Okamoto, T., H. Taguchi, K. Nakamura, H. Ikenaga, H. Kuraishi and K. Yamasato. 1995. Validation of the publication of new names and new combinations previously effectively published outside the IJSB. List No. 53. Int. J. Syst. Bacteriol. *45*: 418–419.

Okonya, J.F., T. Kolasa and M.J. Miller. 1995. Synthesis of the peptide fragment of pseudobactin. J. Org. Chem. *60*: 1932–1935.

Okpokwasili, G.C., C.C. Somerville, D.J. Grimes and R.R. Colwell. 1984. Plasmid-associated phenanthrene degradation by Chesapeake Bay sediment bacteria. Colloq. Inst. Fr. Rech. Exploit. Mer. *3*: 601–610.

Okrend, A.J.G., B.E. Rose and B. Bennett. 1987. Incidence and toxigenicity of *Aeromonas* spp. in retail poultry, beef and pork. J. Food Prot. *50*: 509–513.

Oku, T., Y. Wakasaki, N. Adachi, C.I. Kado, K. Tsuchiya and T. Hibi. 1998. Pathogenicity, non-host hypersensitivity and host defence non-permissibility regulatory gene *hrpX* is highly conserved in *Xanthomonas* pathovars. J. Phytopathol. (Berl.) *146*: 197–200.

Okubadejo, O.A. and K.O. Alausa. 1968. Neonatal meningitis caused by *Edwardsiella tarda*. Br. Med. J. *3*: 357–358.

Okuda, J., M. Ishibashi, E. Hayakawa, T. Nishino, Y. Takeda, A.K. Mukhopadhyay, S. Garg, S.K. Bhattacharya, G.B. Nair and M. Nishibuchi. 1997. Emergence of a unique O3:K6 clone of *Vibrio parahaemolyticus* in Calcutta, India, and isolation of strains from the same clonal group from Southeast Asian travelers arriving in Japan. J. Clin. Microbiol. *35*: 3150–3155.

Okujo, N., Y. Sakakibara, T. Yoshida and S. Yamamoto. 1994. Structure of acinetoferrin, a new citrate-based dihydroxamate siderophore from *Acinetobacter haemolyticus*. Biometals *7*: 170–176.

Okujo, N. and S. Yamamoto. 1994. Identification of the siderophores from *Vibrio hollisae* and *Vibrio mimicus* as aerobactin. FEMS Microbiol. Lett. *118*: 187–192.

Okuzumi, M., A. Hiraishi, T. Kobayashi and T. Fujii. 1994. *Photobacterium histaminum* sp. nov., a histamine-producing marine bacterium. Int. J. Syst. Bacteriol. *44*: 631–636.

Old, D.C. and R.A. Adegbola. 1982. Haemagglutinins and fimbriae of *Morganella*, *Proteus* and *Providencia*. J. Med. Microbiol. *15*: 551–564.

Old, D.C. and S.S. Scott. 1981. Hemagglutinins and fimbriae of *Providencia* spp. J. Bacteriol. *146*: 404–408.

Olenginski, T.P., D.C. Bush and T.M. Harrington. 1991. Plant thorn synovitis: an uncommon cause of monoarthritis. Semin. Arthritis Rheum. *21*: 40–46.

Oliver, J.D. 1987. Heterotrophic bacterial populations of the Black Sea. Biol. Oceanogr. *4*: 83–97.

Oliver, J.D. 1995. The viable but non-culturable state in the human pathogen *Vibrio vulnificus*. FEMS Microbiol. Lett. *133*: 203–208.

Oliver, J.D., D.M. Roberts, V.K. White, M.A. Dry and L.M. Simpson. 1986. Bioluminescence in a strain of the human pathogenic bacterium *Vibrio vulnificus*. Appl. Environ. Microbiol. *52*: 1209–1211.

Olsen, A.B., H.P. Melby, L. Speilberg, Ø. Evensen and T. Håstein. 1997. *Piscirickettsia salmonis* infection in Atlantic salmon Salmo salar in Norway-epidemiological, pathological and microbiological findings. Dis. Aquat. Org. *31*: 35–48.

Olsen, G.J., N. Larsen and C.R. Woese. 1991. The ribosomal RNA database project. Nuc. Acids Res. *19 Suppl.*: 2017–2021.

Olsen, G.J. and C.R. Woese. 1993. Ribosomal RNA: a key to phylogeny. FASEB J. *7*: 113–123.

Olsen, I. 1993. Recent approaches to the chemotaxonomy of the *Actinobacillus-Haemophilus-Pasteurella* group (family *Pasteurellaceae*). Oral Microbiol. Immunol. *8*: 327–336.

Olsen, I. and I. Brondz. 1985. Differentiation among closely related organisms of the *Actinobacillus-Haemophilus-Pasteurella* group by means of lysozyme and EDTA. J. Clin. Microbiol. *22*: 629–636.

Olsen, I., V. Myhrvold and I. Brondz. 1995. Multivariate analysis of fatty acid contents of outer membrane vesicles from *Actinobacillus*, *Haemophilus* and *Pasteurella* spp. *In* Donachie, Lainson and Hodgson (Editors), *Haemophilus*, *Actinobacillus* and *Pasteurella*, Plenum Press, New York. pp. 211 (Abstract T236).

Olsen, I., S.K. Rosseland, A.K. Thorsrud and E. Jellum. 1987. Differentiation between *Haemophilus paraphrophilus*, *H. aphrophilus*, *H. influnzae*, *Actinobacillus actinomycetemcomitans*, *Pasteurella multocida*, *P. haemolytica*, and *P. ureae* by high resolution two-dimensional protein electrophoresis. Electrophoresis *8*: 532–535.

Olsen, I., H.N. Shah and S.E. Gharbia. 1999. Taxonomy and biochemical characteristics of *Actinobacillus actinomycetemcomitans* and *Porphyromonas gingivalis*. Periodontal 2000 *20*: 14–52.

Olson, M.O.J., N. Nagabhushan, M. Dzwiniel, L.B. Smillie and D.R. Whitaker. 1970. Primary structure of α-lytic protease: a bacterial homologue of the pancreatic serine proteases. Nature (Lond.) *228*: 438–442.

Olsthoorn, R.C.L., G. Garde, T. Dayhuff, J.F. Atkins and J. Van Duin. 1995. Nucleotide-sequence of a single-stranded RNA phage from *Pseudomonas aeruginosa* - kinship to coliphages and conservation of regulatory RNA structures. Virology *206*: 611–625.

Olsufiev, N.G., O.S. Emelyanova and T.N. Dunayeva. 1959. Comparative study of strains of *B. tularense*. J. Hyg. Epidemiol. Microbiol. Immunol. (Prague) *3*: 138–149.

Olsufjev, N.G. 1970. Taxonomy and characteristic of the genus *Francisella* Dorofeev, 1947. J. Hyg. Epidemiol. Microbiol. Immunol. (Prague) *14*: 67–74.

Olsufjev, N.G. 1974. Tularemia, Moscow.

Olsufjev, N.G. and I.S. Meshcheryakova. 1983. Subspecific taxonomy of *Francisella tularensis* McCoy and Chapin 1912. Int. J. Syst. Bacteriol. *33*: 872–874.

Ølsvik, Ø., K. Wachsmuth, B. Kay, K.A. Birkness, A. Yi and B. Sack. 1990. Laboratory observations on *Plesiomonas shigelloides* strains isolated from children with diarrhea in Peru. J. Clin. Microbiol. *28*: 886–889.

Omel'chenko, M.V., L.V. Vasily'eva, G.A. Zavarzin, N.D. Savel'eva, A.M. Lysenko, L.L. Mityushina, V.N. Khmelenina and Y.A. Trotsenko. 1996. A novel psychrophilic methanotroph of the genus *Methylobacter*. Microbiology *65*: 339–343.

Omel'chenko, M.V., L.V. Vasily'eva, G.A. Zavarzin, N.D. Savel'eva, A.M. Lysenko, L.L. Mityushina, V.N. Khmelenina and Y.A. Trotsenko. 2000. *In* Validation of the publication of new names and new combinations previously effectively published outside the IJSEM. List No. 50. Int. J. Syst. Evol. Microbiol. *50*: 423–424.

Omelianski, W. 1905. Über eine neue art farbloser Thiospirillen. Zentbl. Bakteriol. Parasitenkd. Infektkrankh. Hyg. Abt. II *14*: 769–772.

Omland, T. 1964. Serological Studies of *Haemophilus influenzae* and related species. Vlll. Examination of ultrasonically prepared *Haemophilus* antigens by means of immunoelectrophoresis. Acta Pathol. Microbiol. Scand. *62*: 83–106.

Onarheim, A.M., R. Wiik, J. Burghardt and E. Stackebrandt. 1994. Characterization and identification of two *Vibrio* species indigenous to the intestine of fish in cold sea water; description of *Vibrio iliopiscarius* sp. nov. System. Appl. Microbiol. *17*: 370–379.

O'Neill, E.A., G.M. Kiely and R.A. Bender. 1984. Transposon Tn5 encodes streptomycin resistance in nonenteric bacteria. J. Bacteriol. *159*: 388–389.

O'Neill, K.R., S.H. Jones and D.J. Grimes. 1990. Incidence of *Vibrio vulnificus* in northern New England water and shellfish. FEMS Microbiol. Lett. *60*: 163–167.

O'Neill, S.L., R.H. Gooding and S. Aksoy. 1993. Phylogenetically distant symbiotic microorganisms reside in *Glossina* midgut and ovary tissues. Med. Vet. Entomol. *7*: 377–383.

Onogawa, T., T. Terayama, H. Zen-yoji, Y. Amano and K. Suzuki. 1976. Distribution of *Edwardsiella tarda* and hydrogen sulfide-producing *Escherichia coli* in healthy persons. J. Jpn. Assoc. Infect. Dis. *50*: 10–17.

Opgenorth, D.C., C.D. Smart, F.J. Louws, F.J. de Bruijn and B.C. Kirkpatrick. 1996. Identification of *Xanthomonas fragariae* field isolates by rep-PCR genomic fingerprinting. Plant Dis. *80*: 868–873.

Oppenheim, J. and L. Marcus. 1970. Correlation of ultrastructure in *Azotobacter vinelandii* with nitrogen source for growth. J. Bacteriol. *101*: 286–291..

Oppenheimer, C.H. and H.W. Jannasch. 1962. Some bacterial populations in turbid and clear sea water near Port Aransas, Texas. Publ. Inst. Mar. Sci. Univ. Tex. *8*: 56–60.

O'Reilly, T. and D.F. Niven. 1986. Defining the metabolic and growth responses of porcine haemophili to exogenous pyridine nucleotides and precursors. J. Gen. Microbiol. *132*: 807–818.

Oren, A., M. Kessel and E. Stackebrandt. 1989. *Ectothiorhodospira marismortui* sp. nov. an obligately anaerobic, moderately halophilic purple sulfur bacterium from a hypersaline sulfur spring on the shore of the dead sea. Arch. Microbiol. *151*: 524–529.

Oren, A., M. Kessel and E. Stackebrandt. 1990. *In* Validation of the publication of new names and new combinations previously effectively published outside the IJSB. List No. 32. Int. J. Syst. Bacteriol. *40*: 105–106.

Oren, A., G. Simon and E.A. Galinski. 1991. Intracellular salt and solute concentrations in *Ectothiorhodospira marismortui* glycine betaine and N-αα-carbamoyl glutamineamide as osmotic solutes. Arch. Microbiol. *156*: 350–355.

Orla-Jensen, S. 1921. The main lines of the bacterial system. J. Bacteriol. *6*: 263–273.

Ormsbee, R.A. 1952. The growth of *Coxiella burnetii* in embryonnated eggs. J. Bacteriol. *63*: 73.

Ormsbee, R. and B. Marmion. 1990. Prevention of *Coxiella burnetii* infection: vaccines and guidelines for those at risk. *In* Marrie (Editor), Q fever. The Disease, CRC Press, Boca Raton. p. 226.

Ornstein, M. 1921. Zür Bakteriologie des Schmitzbacillus. Z. Hyg. Infektionskr. *91*: 152–178.

Ornston, L.N. and E.L. Neidle. 1991. Evolution of genes for the β-ketoadipate pathway in *Acinetobacter calcoaceticus*. *In* Towner and Bergogne-Bérézin (Editors), The Biology of *Acinetobacter*. Taxonomy, Clinical Importance, Molecular Biology, Physiology, Industrial Relevance, Plenum Press, New York. pp. 201–237.

Orrison, L.H., W.B. Cherry and D. Milan. 1981. Isolation of *Legionella* from cooling tower water by filtration. Appl. Environ. Microbiol. *41*: 1202–1205.

Orrison, L.H., W.B. Cherry, R.L. Tyndall, C.B. Fliermans, S.B. Gough, M.A. Lambert, L.K. McDougal, W.F. Bibb and D.J. Brenner. 1983a. *Legionella oakridgensis*, sp. nov.: unusual new species isolated from cooling tower water. Appl. Environ. Microbiol. *45*: 536–545.

Orrison, L.H., W.B. Cherry, R.L. Tyndall, C.B. Fliermans, S.B. Gough, M.A. Lambert, L.K. McDougal, W.F. Bibb and D.J. Brenner. 1983b. *In* Validation of the publication of new names and new combinations previously effectively published outside the IJSB. List No. 11. Int. J. Syst. Bacteriol. *33*: 672–674.

Ørskov, F. and I. Ørskov. 1975. *Escherichia coli* O:H serotypes isolated from human blood. Acta Pathol. Microbiol. Scand. B. *83*: 595–600.

Ørskov, F. and I. Ørskov. 1984a. Serotyping of *Escherichia coli*. Meth. Microbiol. *14*: 43–112.

Ørskov, F. and I. Ørskov. 1992. *Escherichia coli* scrotyping and disease in man and animals. Can. J. Microbiol. *38*: 699–704.

Ørskov, I. 1974. The genus *Klebsiella*. *In* Buchanan and Gibbons (Editors), Bergey's Manual of Determinative Bacteriology, 8th Ed., The Williams & Wilkins Co., Baltimore. pp. 321–324.

Ørskov, I. 1981. The genus *Klebsiella* (Medical aspects). *In* Starr, Stolp, Trüper, Balows and Schlegel (Editors), The Prokaryotes: A Handbook on Habitats, Isolation, and Identification of Bacteria, Springer-Verlag, Berlin. pp. 1160–1165.

Ørskov, I. 1984a. Genus V. *Klebsiella. In* Krieg and Holt (Editors), Bergey's Manual of Systematic Bacteriology, 1st Ed., l. Vol. 1, The Williams & Wilkins Co., Baltimore. pp. 481–485.

Ørskov, I. 1984b. *Klebsiella* Trevisan 1885. *In* Krieg and Holt (Editors), Bergey's Manual of Systematic Bacteriology, Vol. 1, The Williams & Wilkins Co., Baltimore. pp. 461–465.

Ørskov, I. 1984c. *In* Validation of the publication of new names and new combinations previously effectively published outside the IJSB. List No. 15. Int. J. Syst. Bacteriol. *34*: 355–357.

Ørskov, I., A. Ferencz and F. Ørskov. 1980a. Tamm-Horsfall protein or uromucoid is the normal urinary slime that traps type 1 fimbriated *Escherichia coli.* Lancet *1*: 887.

Ørskov, I. and F. Ørskov. 1977. Special O:K:H serotypes among enterotoxigenic *E. coli* strains from diarrhea in adults and children. Occurrence of the CF (colonization factor) antigen and of hemagglutinating abilities. Med. Microbiol. Immunol. *163*: 99–110.

Ørskov, I. and F. Ørskov. 1984b. Serotyping of *Klebsiella. In* Bergan (Editor), Methods in Microbiology, Vol. 14, Academic Press, London. pp. 143–164.

Ørskov, I. and F. Ørskov. 1985. *Escherichia coli* in extra-intestinal infections. J. Hyg. (Lond). *95*: 551–575.

Ørskov, I. and F. Ørskov. 1990. Serologic classification of fimbriae. Curr. Top. Microbiol. Immunol. *151*: 71–90.

Ørskov, I., F. Ørskov and A. Birch-Andersen. 1980b. Comparison of *Escherichia coli* fimbrial antigen F7 with type 1 fimbriae. Infect. Immun. *27*: 657–666.

Ørskov, I., F. Ørskov, A. Birch-Andersen, M. Kanamori and C. Svanborg-Eden. 1982. O, K, H and fimbrial antigens in *Escherichia coli* serotypes associated with pyelonephritis and cystitis. Scand. J. Infect. Dis. Suppl. *33*: 18–25.

Ørskov, I., F. Ørskov, B. Jann and K. Jann. 1977. Serology, chemistry, and genetics of O and K antigens of *Escherichia coli.* Bacteriol. Rev. *41*: 667–710.

Ørskov, I., J. Ørskov, W.J. Sojka and J.M. Leach. 1961. Simultaneous occurrence of *E. coli* and L antigens in strains from diseased swine. Acta Pathol. Microbiol. Immunol. Scand. *53*: 404–422.

Ortigosa, M., E. Garay and M.J. Pujalte. 1994. Numerical taxonomy of aerobic, Gram-negative bacteria associated with oysters and surrounding seawater of the Mediterranean coast. Syst. Appl. Microbiol. *17*: 589–600.

Orvos, D.R., G.H. Lacy and J. Cairns, Jr.. 1990. Genetically engineered *Erwinia carotovora*: survival, intraspecific competition, and effects upon selected bacterial genera. Appl. Environ. Microbiol. *56*: 1689–1694.

Ory, J.M., C. Chuard and C. Regamey. 1998. *Pasteurella multocida* pneumonia with empyema. Scand. J. Infect. Dis. *30*: 313–314.

Osawa, R. 1990. Formation of a clear zone on tannin-treated brain heart infusion agar by a *Streptococcus* sp. isolated from feces of koalas. Appl. Environ. Microbiol. *56*: 829–831.

Osawa, R. 1992. Tannin-protein complex-degrading enterobacteria isolated from the alimentary tracts of koalas and a selective medium for their enumeration. Appl. Environ. Microbiol. *58*: 1754–1759.

Osawa, R., P.S. Bird, D.J. Harbrow, K. Ogimoto and G.J. Seymour. 1993a. Microbiological studies of the intestinal microflora of the koala, *Phascolarctos cinereus.* I. Colonization of the cecal wall by tannin-protein-complex-degrading enterobacteria. Aust. J. Zool. *41*: 599–609.

Osawa, R., W.H. Blanshard and P.G. O'Callaghan. 1993b. Microbiological studies of the intestinal microflora of the koala, *Phascolarctos cinereus.* II. Pap, a special maternal faeces consumed by juvenile koalas. Aust. J. Zool. *41*: 611–620.

Osawa, R., T. Fujisawa and L.I. Sly. 1995a. *Streptococcus gallolyticus* sp. nov.; gallate degrading organisms formerly assigned to *Streptococcus bovis.* Syst. Appl. Microbiol. *18*: 74–78.

Osawa, R., F. Rainey, T. Fujisawa, E. Lang, H.J. Busse, T.P. Walsh and E. Stackebrandt. 1995b. *Lonepinella koalarum* gen. nov., sp. nov., a new tannin-protein complex degrading bacterium. Syst. Appl. Microbiol. *18*: 368–373.

Osawa, R., F. Rainey, T. Fujisawa, E. Lang, H.J. Busse, T.P. Walsh and E. Stackebrandt. 1996. *In* Validation of the publication of new names and new combinations previously effectively published outside the IJSB. List No. 56. Int. J. Syst. Bacteriol. *46*: 362–363.

Osawa, R. and T.P. Walsh. 1993. Visual reading method for detection of bacterial tannase. Appl. Environ. Microbiol. *59*: 1251–1252.

Osawa, R. and T.P. Walsh. 1995. Detection of bacterial gallate decarboxylation by visual color discrimination. J. Gen. Appl. Microbiol. *41*: 165–170.

Osawa, R., T.P. Walsh and S.J. Cork. 1993c. Metabolism of tannin-protein complex by facultatively anaerobic bacteria isolated from koala feces. Biodegradation *4*: 91–99.

Oshiro, R.K., T. Picone and B.H. Olson. 1994. Modification of reagents in the EnviroAmp kit to increase recovery of *Legionella* organisms in water. Can. J. Microbiol. *40*: 495–499.

Ostroff, S. 1995. *Yersinia* as an emerging infection: epidemiologic aspects of yersiniosis. Contrib. Microbiol. Immunol. *13*: 5–10.

O'Sullivan, D.J. and F. O'Gara. 1992. Traits of fluorescent *Pseudomonas* spp. involved in suppression of plant-root pathogens. Microbiol. Rev. *56*: 662–676.

O'Sullivan, J., J.E. McCullough, A.A. Tymiak, D.R. Kirsh, W.H. Trejo and P.A. Principe. 1988. Lysobactin, a novel antibacterial agent produced by *lysobacter* sp.: I. Taxonomy, isolation, and partial characterization. J. Antibiot. (Tokyo). *41*: 1740–1744.

Otani, E. and D.A. Bruckner. 1991. *Leclercia adecarboxylata* isolated from a blood culture. Clin. Microbiol. Newslett. *13*: 157–158.

Otis, V.S. and J.L. Behler. 1973. The occurrence of Salmonellae and *Edwardsiella* in the turtles of the New York Zoological park. J. Wildl. Dis. *9*: 4–6.

O'Toole, R., D.L. Milton and H. Wolf-Watz. 1996. Chemotactic motility is required for invasion of the host by the fish pathogen *Vibrio anguillarum.* Mol. Microbiol. *19*: 625–637.

Otta, J.D. 1977. Occurrence and characteristics of isolates of *Pseudomonas syringae* on winter-wheat. Phytopathology *67*: 22–26.

Otta, J.D. and H. English. 1971. Serology and pathology of *Pseudomonas syringae.* Phytopathology *61*: 443–452.

Otte, S., J.G. Kuenen, L.P. Nielsen, H.W. Paerl, J. Zopfi, H.N. Schulz, A. Teske, B. Strotmann, V.A. Gallardo and B.B. Jørgensen. 1999. Nitrogen, carbon, and sulfur metabolism in natural *Thioploca* samples. Appl. Environ. Microbiol. *65*: 3148–3157.

Otten, S., S. Iyer, W. Johnson and R. Montgomery. 1986. Serospecific antigens of *Legionella pneumophila.* J. Bacteriol. *167*: 893–904.

Ouchterlony, O. 1968. Handbook of Immunodiffusion and Immunoelectrophoresis, Ann Arbor Science Publishers, Ann Arbor. 135 pp.

Ovartlarnporn, B., P. Chayakul and S. Suma. 1986. *Edwardsiella tarda* infection in Hat Yai Hospital. J. Med. Assoc. Thai. *69*: 599–603.

Overman, T.L. 1980. Antimicrobial susceptibility of *Aeromonas hydrophila.* Antimicrob. Agents Chemother. *17*: 612–614.

Overman, T.L. and J.M. Janda. 1999. Antimicrobial susceptibility patterns of *Aeromonas jandaei, A. schubertii, A. trota,* and *A. veronii* biotype *veronii.* J. Clin. Microbiol. *37*: 706–708.

Overmann, J., U. Fischer and N. Pfennig. 1992. A new purple sulfur bacterium from saline littoral sediments, *Thiorhodovibrio winogradskyi* gen. nov. and sp. nov. Arch. Microbiol. *157*: 329–335.

Overmann, J., U. Fischer and N. Pfennig. 1993. *In* Validation of the publication of new names and new combinations previously effectively published outside the IJSB. List No. 44. Int. J. Syst. Bacteriol. *43*: 188–189.

Owen, C.R., E.O. Buker, W.L. Jellison, D.B. Lackman and J.F. Bell. 1964. Comparative studies of *Francisella tularensis* and *Francisella novicida.* J. Bacteriol. *87*: 676–683.

Owen, D.J. and A.C. Ward. 1985. Transfer of transposable drug resistance elements Tn5, Tn7, and Tn76 to *Azotobacter beijerinckii* - use of plasmid RP4-Tn76 as a suicide vector. Plasmid *14*: 162–166.

Owen, R.J., A.U. Ahmed and C.A. Dawson. 1987. Guanine-plus-cytosine contents of type strains of the genus *Providencia*. Int. J. Syst. Bacteriol. *37*: 449–450.

Owen, R.J., A. Beck, P.A. Dayal and C. Dawson. 1988. Detection of genomic variation in *Providencia stuartii* clinical isolates by analysis of DNA restriction fragment length polymorphisms containing rRNA cistrons. J. Clin. Microbiol. *26*: 2161–2166.

Owen, R.J., R.M. Legros and S.P. Lapage. 1978. Base composition, size and sequence similarities of genome deoxyribonucleic acids from clinical isolates of *Pseudomonas putrefaciens*. J. Gen. Microbiol. *104*: 127–138.

Owens, D.R., S.L. Nelson and J.B. Addison. 1974. Isolation of *Edwardsiella tarda* from swine. Appl. Microbiol. *27*: 703–705.

Oyaizu, H. and K. Komagata. 1983. Grouping of *Pseudomonas* species on the basis of cellular fatty acid composition and the quinone system with special reference to the existence of 3-hydroxy fatty acids. J. Gen. Appl. Microbiol. *29*: 17–40.

Ozaki, M., S. Mizushima and M. Nomura. 1969. Identification and functional characterization of the protein controlled by the streptomycin-resistant locus in *E. coli*. Nature *222*: 333–339.

Pacini, F. 1854. Osservazione microscopiche e deduzioni patologiche sul Cholera Asiatico. Gaz. Med. Ital. Toscana Firenze. *6*: 405–412.

Packer, H. and J. Goldberg. 1950. Studies of the antigenic relationship of *D. granulomatis* to members of the tribe *Eschericheae*. Am. J. Syphilis, Gonorrhea and Venereal Dis. *34*: 342–350.

Padhya, A.C. and M.K. Patel. 1962. A new bacterial leaf-spot on *Alangium lamarckii* Thw. Curr. Sci. (Bangalore) *31*: 196–197.

Padhya, A.C. and M.K. Patel. 1963. A new bacterial leaf-spot on *Ionidum heterophyllum* Vent. Indian Phytopathol. *16*: 98–99.

Padhya, A.C. and M.K. Patel. 1964. Bacterial leaf spot on *Triumfetta pilosa* Roth. Curr. Sci. (Bangalore) *33*: 342.

Padhya, A.C., M.K. Patel and W.V. Kotasthane. 1965a. A new bacterial leaf spot disease of *Bauhinia racemosa* Lamk. Curr. Sci. (Bangalore) *34*: 224–225.

Padhya, A.C., M.K. Patel and W.V. Kotasthane. 1965b. A new bacterial leaf-spot on *Vitis trifolia*. Curr. Sci. (Bangalore) *34*: 462–463.

Padhya, H.C. and M.K. Patel. 1963. A new bacterial leaf spot on *Corchorus acutangulus* Lam. Curr. Sci. (Bangalore) *32*: 326.

Padilla, E., P. Tudela, M. Gimenez and J.M. Gimeno. 1997. *Kluyvera ascorbata* bacteremia. Midicina Clinica. *108*: 479.

Page, L.A. 1962. *Haemophilus* in chickens. I. Characteristics of 12 *Haemophilus* isolates recovered from diseased chickens. Am. J. Vet. Res. *23*: 85–95.

Page, M.I. and E.O. King. 1966. Infection due to *Actinobacillus actinomycetemcomitans* and *Haemophilus aphrophilus*. N. Engl. J. Med. *275*: 181–188.

Page, W.J. 1985. Genetic transformation of molybdenum starved *Azotobacter vinelandii* - increased transformation frequency and recipient range. Can. J. Microbiol. *31*: 659–662.

Page, W.J. and S.K. Collinson. 1987. Characterization of *Azomonas macrocytogenes* strains isolated from Alberta soils. Can. J. Microbiol. *33*: 830–833.

Page, W.J. and G.A. Grant. 1987. Effect of mineral iron on the development of transformation competence in *Azotobacter vinelandii*. FEMS Microbiol. Lett. *41*: 257–261.

Page, W.J. and S. Shivprasad. 1991a. *Azotobacter salinestris* sp nov, a sodium-dependent, microaerophilic, and aeroadaptive nitrogen fixing bacterium. Int. J. Syst. Bacteriol. *41*: 369–376.

Page, W.J. and S. Shivprasad. 1991b. *In* Validation of the publication of new names and new combinations previously effectively published outside the IJSB. List No. 38. Int. J. Syst. Bacteriol. *41*: 374.

Page, W.J. and S. Shivprasad. 1995. Iron binding to *Azotobacter salinestris* melanin, iron mobilization and uptake mediated by siderophores. Biometals 8: 59–64.

Pai, C.H. and V. Mors. 1978. Production of enterotoxin by *Yersinia enterocolitica*. Infect. Immunol. *19*: 909–911.

Paine, S.G. 1919. Studies on bacteriosis. II. A brown blotch disease of cultivated mushrooms. Ann. Appl. Biol. *5*: 206–219.

Paine, S.G. and H. Stanfield. 1919. Studies in bacteriosis. III. A bacterial

leafspot disease of *Protea cynaroides* exhibiting a host reaction of possibly bacteriolytic nature. Ann. Appl. Biol. *6*: 27–39.

Paju, S., M. Saarela, S. Alaluusua, P. Fives-Taylor and S. Asikainen. 1998. Characterization of serologically nontypeable *Actinobacillus actinomycetemcomitans* isolates. J. Clin. Microbiol. *36*: 2019–2022.

Pal, T., N.A. Al-Sweih, M. Herpay and T.D. Chugh. 1997. Identification of enteroinvasive *Escherichia coli* and *Shigella* strains in pediatric patients by an IpaC-specific enzyme-linked immunosorbent assay. J. Clin. Microbiol. *35*: 1757–1760.

Palleroni, N.J. 1975. General properties and taxonomy of the genus *Pseudomonas*. *In* Clarke and Richmond (Editors), Genetics and Biochemistry of *Pseudomonas*, John Wiley & Sons, London. 1–36.

Palleroni, N.J. 1977. *Pseudomonas*. *In* Laskin and Lechevalier (Editors), CRC Handbook of Microbiology, 2nd Ed., Vol. 1, CRC Press, Inc., Cleveland, Ohio. pp. 247–258.

Palleroni, N.J. 1984. Genus I *Pseudomonas*. *In* Krieg and Holt (Editors), Bergey's Manual of Systematic Bacteriology, 1st Ed., Vol. 1, The Williams & Wilkins Co., Baltimore. pp. 141–199.

Palleroni, N.J. 1986. Taxonomy of the pseudomonads. *In* Sokatch (Editor), The Biology of *Pseudomonas*, Academic Press, Orlando. 3–25.

Palleroni, N.J. 1992a. Introduction to the family *Pseudomonadaceae*. *In* Balows, Truper, Dworkin, Harder and Schleifer (Editors), The Prokaryotes: A Handbook on the Biology of Bacteria: Ecophysiology, Isolation, Identification, Applications, Vol. 2, Springer-Verlag, New York. 3071–3085.

Palleroni, N.J. 1992b. Present situation of the taxonomy of the aerobic pseudomonads. *In* Galli, Silver and Witholt (Editors), *Pseudomonas*: Molecular Biology and Biotechnology, American Society for Microbiology, Washington D.C. 105–115.

Palleroni, N.J. 1993. *Pseudomonas* classification - a new case-history in the taxonomy of Gram-negative bacteria. Antonie Leeuwenhoek Int. J. Gen. Mol. Microbiol. *64*: 231–251.

Palleroni, N.J., R.W. Ballard, E. Ralston and M. Doudoroff. 1972. Deoxyribonucleic acid homologies among some *Pseudomonas* species. J. Bacteriol. *110*: 1–11.

Palleroni, N.J. and J.F. Bradbury. 1993. *Stenotrophomonas*, a new bacterial genus for *Xanthomonas maltophilia* (Hugh 1980) Swings et al. 1983. Int. J. Syst. Bacteriol. *43*: 606–609.

Palleroni, N.J. and M. Doudoroff. 1972. Some properties and taxonomic subdivisions of the genus *Pseudomonas*. Annu. Rev. Phytopathol. *10*: 73–100.

Palleroni, N.J., M. Doudoroff, R.Y. Stanier, R.E. Solanes and M. Mandel. 1970. Taxonomy of the aerobic pseudomonads: the properties of the *Pseudomonas stutzeri* group. J. Gen. Microbiol. *60*: 215–231.

Palleroni, N.J., D.C. Hildebrand, M.N. Schroth and M. Hendson. 1993. Deoxyribonucleic acid relatedness of 21 strains of *Xanthomonas* species and pathovars. J. Appl. Bacteriol. *75*: 441–446.

Palleroni, N.J., R. Kunisawa, R. Contopoulou and M. Doudoroff. 1973. Nucleic acid homologies in the genus *Pseudomonas*. Int. J. Syst. Bacteriol. *23*: 333–339.

Palmen, R. and K.J. Hellingwerf. 1997. Uptake and processing of DNA by *Acinetobacter calcoaceticus*: A review. Gene *192*: 179–190.

Palmen, R., B. Vosman, P. Buijsman, C.K.D. Breek and K.J. Hellingwerf. 1993. Physiological characterization of natural transformation in *Acinetobacter calcoaceticus*. J. Gen. Microbiol. *139*: 295–305.

Palmer, E.L., B.L. Teviotdale and A.L. Jones. 1997. A relative of the broad-host-range plasmid RSF1010 detected in *Erwinia amylovora*. Appl. Environ. Microbiol. *63*: 4604–4607.

Palmer, G.G. 1981. Haemophili in faeces. J. Med. Microbiol. *14*: 147–150.

Palmer, M.A. 1993. A gelatin test to detect activity and stability of proteases produced by *Dichelobacter* (*Bacteroides*) *nodosus*. Vet. Microbiol. *36*: 113–122.

Palumbo, S.A. 1993. The occurrence and significance of organisms of the *Aeromonas hydrophila* group in food and water. Med. Microbiol. Lett. *2*: 339–346.

Palumbo, S.A. 1996. The *Aeromonas hydrophila* group in food. *In* Austin,

Altwegg, Gosling and Joseph (Editors), The Genus *Aeromonas*, John Wiley & Sons, Ltd., Chichester. pp. 287–310.

Palumbo, S.A., C. Abeyta and G. Stelma. 1992. *Aeromonas hydrophila* group. *In* Vanderzant and Splittstoesser (Editors), Compendium of Methods for the Microbiological Examination of Foods, 3rd Ed., American Public Health Association, Washington, D.C. pp. 497–515.

Palumbo, S., M. Golden, L. Yu and C. Briggs. 1999. Identification of motile *Aeromonas* spp. isolated from a swine slaughter plant. 6th International *Aeromonas/Plesiomonas* Symposium, Chicago, Illinois. p. 13.

Palumbo, S.A., F. Maxino, A.C. Williams, R.L. Buchanan and D.W. Thayer. 1985a. Starch-ampicillin agar for the quantitative detection of *Aeromonas hydrophila*. Appl. Environ. Microbiol. *50*: 1027–1030.

Palumbo, S.A., D.R. Morgan and R.L. Buchanan. 1985b. Influence of temperature, sodium chloride and pH on the growth of *Aeromonas hydrophila*. J. Food Sci. *50*: 1417–1421.

Pammel, L.H. 1895. Bacteriosis of rutabaga (*Bacillus campestris* n. sp.). Iowa State Coll. Agr. Exp. Sta. Bull. *27*: 130–134.

Panagopoulos, C.G. 1969. The disease "Tsilik Marasi" of grapevine, its description and identification of the causal agent (*Xanthomonas ampelina* sp. nov.). Ann. Inst. Phytopathol. Benaki. *9*: 59–81.

Panciera, R.J., R.R. Dahlgren and H.B. Rinker. 1968. Observations on septicemia of cattle caused by a *Hemophilus*-like organism. Pathol. Vet. *5*: 212–226.

Pandit, V.M. and Y.S. Kulkarni. 1979. Bacterial leaf-spot of *Clitoria biflora* Dalz. Biovigyanam. *5*: 9–20.

Pankey, G.A. and M.B. Seshul. 1969. Septicemia caused by *Edwardsiella tarda*. J. La. State. Med. Soc. *121*: 41–43.

Pant, N.M. and Y.S. Kulkarni. 1976a. Bacterial leaf-spot of *Desmodium laxiflorum* DC. Biovigyanam. *2*: 97–98.

Pant, N.M. and Y.S. Kulkarni. 1976b. Bacterial leaf-spot of *Merremia gangetica* (L.) Cufod. Biovigyanam. *2*: 207–208.

Pantophlet, R., L. Brade, L. Dijkshoorn and H. Brade. 1998. Specificity of rabbit antisera against lipopolysaccharide of *Acinetobacter*. J. Dermatol. (Tokyo) *25*: 1245–1250.

Papadakis, K.A., S.E. Vartivarian, M.E. Vassilaki and E.J. Anaissie. 1996. Septic prepatellar bursitis caused by *Stenotrophomonas* (*Xanthomonas*) *maltophilia*. Clin. Infect. Dis. *22*: 388–389.

Papadakis, K.A., S.E. Vartivarian, M.E. Vassilaki and E.J. Anaissie. 1997. *Stenotrophomonas maltophilia* meningitis. Report of two cases and review of the literature. J. Neurosurg. *87*: 106–108.

Papasian, C.J., J. Kinney, S. Coffman, R.J. Hollis and M.A. Pfaller. 1996. Transmission of *Citrobacter koseri* from mother to infant documented by ribotyping and pulsed-field gel electrophoresis. Diagn. Microbiol. Infect. Dis. *26*: 63–67.

Parche, S., W. Geissdørfer and W. Hillen. 1997. Identification and characterization of *xcpR* encoding a subunit of the general secretory pathway necessary for dodecane degradation in *Acinetobacter calcoaceticus* ADP1. J. Bacteriol. *179*: 4631–4634.

Parent, J.G., M. Lacroix, D. Page, L. Vezina and S. Vegiard. 1996. Identification of *Erwinia carotovora* from soft rot diseased plants by random amplified polymorphic DNA (RAPD) analysis. Plant Dis. *80*: 494–499.

Park, E.H. and Y.S. Cho. 1996. Isolation of plasmid from Korean copper-resistant *Xanthomonas campestris* pathovar *vesicatoria*. Korean Plant Pathol. *12*: 156–161.

Park, J.T. 1996. The murein sacculus. *In* Neidhardt, Curtiss, Ingraham, Lin, Low, Magasanik, Reznikoff, Riley, Schaechter and Umbarger (Editors), *Escherichia coli* and *Salmonella*: Cellular and Molecular Biology, 2nd Ed., ASM Press, Washington, D.C. pp. 48–57.

Parker, C.D. 1945. The corrosion of concrete. I. The isolation of a species of bacterium associated with the corrosion of concrete exposed to atmosphere containing hydrogen sulphide. Aust. J. Exp. Biol. Med. Sci. *23*: 81–90.

Parker, C.D. 1957. Genus V. *Thiobacillus* Beijerinck 1904. *In* Breed, Murray and Smith (Editors), Bergey's Manual of Determinative Bacteriology, 7th Ed., The Williams & Wilkins Co., Baltimore. pp. 83–88.

Parsot, C. and P.J. Sansonetti. 1996. Invasion and the pathogenesis of *Shigella* infections. Curr. Top. Microbiol. Immunol. *209*: 25–42.

Pasculle, A.W., J.C. Feeley, R.J. Gibson, L.G. Cordes, R.L. Myerowitz, C.M. Patton, G.W. Gorman, C.L. Carmack, J.W. Ezzell and J.N. Dowling. 1980. Pittsburgh pneumonia agent: direct isolation from human lung tissue. J. Infect. Dis. *141*: 727–732.

Pasculle, A.W., R.L. Myerowitz and C.R.J. Rinaldo. 1979. New bacterial agent of pneumonia isolated from renal-transplant recipients. Lancet *2*: 58–61.

Paster, B.J., F.E. Dewhirst, I. Olsen, G.J. Fraser and S.S. Socransky. 1995. Gram-negative anaerobes: 16S rRNA sequences, phylogeny, and DNA probes. *In* Duerden, Wade, Brazier, Eley, Wren and Hudson (Editors), Medical and Dental Aspects of Anaerobes, Science Reviews, Northwood, UK. 373–386.

Paster, B.J., W. Ludwig, W.G. Weisburg, E. Stackebrandt, R.B. Hespell, C.M. Hahn, H. Reichenbach, K.O. Stetter and C.R. Woese. 1985. A phylogenetic grouping of the *Bacteroides*, cytophagas and certain flavobacteria. Syst. Appl. Microbiol. *6*: 34–42.

Pastian, M.R. and M.C. Bromel. 1984. Inclusion bodies in *Plesiomonas shigelloides*. Appl. Environ. Microbiol. *47*: 216–218.

Patel, A.M., J.M. Chanhan, W.V. Kotasthane and M.V. Desai. 1969. A new bacterial disease of *Biophytum sensitivum*. Curr. Sci. (Bangalore) *38*: 274–275.

Patel, A.M. and W.V. Kotasthane. 1969a. Bacterial blight of Leea edgeworthii incited by *Xanthomonas leeanum* nov. sp. Curr. Sci. (Bangalore) *38*: 519–520.

Patel, A.M. and W.V. Kotasthane. 1969b. Bacterial leaf-spot disease of *Corchorus fascicularis* caused by *Xanthomonas nakatae* var. *fascicularis*. Curr. Sci. (Bangalore) *38*: 596–597.

Patel, M.K. 1948. *Xanthomonas uppalli* sp. nov. pathogenic on *Ipomoea muricata*. Indian Phytopathol. *1*: 67–69.

Patel, M.K. 1949. *Xanthomonas desmodii*, a new bacterial leaf-spot of *Desmodium diffusum* DC. Curr. Sci. (Bangalore) *18*: 213.

Patel, M.K., V.V. Bhatt and K. Y.S.. 1951a. Three new bacterial diseases of plants from Bombay. Curr. Sci. (Bangalore) *20*: 326–327.

Patel, M.K., S.G. Desai and A.J. Patel. 1968. A new bacterial leaf-spot on *Veronia cinerea* Less. Sci. Cult. *34*: 220–221.

Patel, M.K., G.W. Dhande and Y.S. Kulkarni. 1953. Bacterial leaf-spot of *Cyamopsis tetragonoloba* (L.) Taub. Curr. Sci. (Bangalore) *22*: 183.

Patel, M.K. and Y.S. Kulkarni. 1949. Nitrogen utilization by *Xanthomonas malvacearum* (Sm.) Dowson. Indian Phytopathol. *2*: 62–64.

Patel, M.K. and Y.S. Kulkarni. 1951a. A new bacterial leaf spot on *Vitis woodrowii* Stapf. Curr. Sci. (Bangalore) *20*: 132.

Patel, M.K. and Y.S. Kulkarni. 1951b. Nomenclature of bacterial plant pathogens. Indian Phytopathol. *4*: 74–84.

Patel, M.K., Y.S. Kulkarni and G.W. Dhande. 1950. *Xanthomonas badrii* sp. nov., on *Xanthium strumarium* L in India. Indian Phytopathol. *3*: 103–104.

Patel, M.K., Y.S. Kulkarni and G.W. Dhande. 1951b. Three bacterial diseases of plants. Curr. Sci. (Bangalore) *20*: 106.

Patel, M.K., Y.S. Kulkarni and G.W. Dhande. 1952a. Some new bacterial diseases of plants. Curr. Sci. (Bangalore) *21*: 345–346.

Patel, M.K., Y.S. Kulkarni and G.W. Dhande. 1952b. Two new bacterial diseases of plants. Curr. Sci. *21*: 74–75.

Patel, M.K. and L. Moniz. 1948. *Xanthomonas desmodii* pathovar *gangeticii*, sp. nov., Uppal, Patel and Moniz; a new bacterial leaf-spot of *Desmodium gangeticum* DC. Curr. Sci. (Bangalore) *17*: 268.

Patel, M.K., L. Moniz and Y.S. Kulkarni. 1948. A new bacterial disease of *Mangifera indica* L. Curr. Sci. (Bangalore) *17*: 189–190.

Patel, M.K., B.N. Wankar and Y.S. Kulkarni. 1952c. Bacterial leaf-spot of *Amaranthus viridis* L. Curr. Sci. (Bangalore) *21*: 346–347.

Patel, P.N. and J.K. Jindal. 1972. Bacterial leaf spot on *Pedalium murex* L caused by a new albino species of *Xanthomonas*. Indian Phytopathol. *25*: 318–320.

Patel, P., C.F. Marrs, J.S. Mattick, W.W. Ruehl, R.K. Taylor and M. Koomey. 1991. Shared antigenicity and immunogenicity of type-4 pilins expressed by *Pseudomonas aeruginosa*, *Moraxella bovis*, *Neisseria gonorrhoeae*, *Dichelobacter nodosus*, and *Vibrio cholerae*. Infect. Immun. *59*: 4674–4676.

Paterson, W.D., D. Douey and D. Desautels. 1980. Relationships between selected strains of typical and atypical *Aeromonas salmonicida*, *Aeromonas hydrophila* and *Haemophilus piscium*. Can. J. Microbiol. *26*: 588–598.

Pati, B.R., S. Sengupta and A.K. Chandra. 1995. Role of nitrogen fixing bacteria on the phyllosphere of wheat seedlings. Acta Microbiol Immunol Hung. *42*: 427–433.

Patil, A.S. and Y.S. Kulkarni. 1981. A new bacterial leaf-spot disease of *Thespesia populnea* Sol. Curr. Sci. (Bangalore) *50*: 1040–1041.

Patil, S.S. 1974. Toxins produced by phytopathogenic bacteria. Annu. Rev. Phytopathol. *12*: 259–279.

Paton, A.M. 1959. An improved method for preparing pectate gels. Nature *183*: 1812–1813.

Paton, A.W., L. Beutin and J.C. Paton. 1995a. Heterogeneity of the amino-acid sequences of *Escherichia coli* Shiga-like toxin type-I operons. Gene *153*: 71–74.

Paton, A.W., A.J. Bourne, P.A. Manning and J.C. Paton. 1995b. Comparative toxicity and virulence of *Escherichia coli* clones expressing variant and chimeric Shiga-like toxin type II operons. Infect. Immun. *63*: 2450–2458.

Paton, A.W., J.C. Paton, P.N. Goldwater, M.W. Heuzenroeder and P.A. Manning. 1993a. Sequence of a variant Shiga-like toxin type-I operon of *Escherichia coli* O111:H−. Gene *129*: 87–92.

Paton, A.W., J.C. Paton, M.W. Heuzenroeder, P.N. Goldwater and P.A. Manning. 1992. Cloning and nucleotide sequence of a variant Shiga-like toxin II gene from *Escherichia coli* OX3:H21 isolated from a case of sudden infant death syndrome. Microb. Pathog. *13*: 225–236.

Paton, A.W., J.C. Paton and P.A. Manning. 1993b. Polymerase chain reaction amplification, cloning and sequencing of variant *Escherichia coli* Shiga-like toxin type II operons. Microb. Pathog. *15*: 77–82.

Paton, A.W., M.C. Woodrow, R. Doyle, J.A. Lanser and J.C. Paton. 1999. Molecular characterization of a Shiga toxigenic *Escherichia coli* O113:H21 strain lacking *eae* responsible for a cluster of cases of hemolytic-uremic syndrome. J. Clin. Microbiol. *37*: 3357–3361.

Paton, R., R.S. Miles and S.G. Amyes. 1994. Biochemical properties of inducible β-lactamases produced from *Xanthomonas maltophilia*. Antimicrob. Agents Chemother. *38*: 2143–2149.

Paul, V.J., S. Frautschy, W. Fenical and K.H. Nealson. 1981. Antibiotics in microbial ecology: Isolation and structure assignment of several new antibacterial compounds from the insect-symbiotic bacteria *Xenorhabdus* spp. J. Chem. Ecol. *7*: 589–598.

Paula, S.J., P.S. Duffey, S.L. Abbott, R.P. Kokka, L.S. Oshiro, J.M. Janda, T. Shimada and R. Sakazaki. 1988. Surface properties of autoagglutinating mesophilic aeromonads. Infect. Immun. *56*: 2658–2665.

Paulin, J.P. and N.A. Nassan. 1978. Lysogenic strains and phage-typing in *Erwinia chrysanthemi*s. Proc. IVth Conf. on Plant Pathogenic Bactera, Gilbert-Clarey, Tours, France. pp. 539–545.

Paulsen, I.T., M.H. Brown and R.A. Skurray. 1996. Proton-dependent multidrug efflux systems. Microbiol. Rev. *60*: 575–608.

Pavan, M.E., S.L. Abbott, J. Zorzópulos and J.M. Janda. 2000. *Aeromonas salmonicida* subsp. *pectinolytica* subsp. nov., a new pectinase-positive subspecies isolated from a heavily polluted river. Int. J. Syst. Evol. Microbiol. *50*: 1119–1124.

Pavia, A.T., J.A. Bryan, K.L. Maher, T.R. Hester, Jr. and J.J. Farmer, III. 1989. *Vibrio carchariae* infection after a shark bite. Ann. Intern. Med. *111*: 85–86.

Pavlov, V.M., I.V. Rodionova, A.N. Mokrievich and I.S. Meshcheryakova. 1994. Isolation and molecular genetic characterization of a cryptic plasmid from the strain *Francisella novicida*-like F6168. Mol. Gen. Mikrobiol. Virusol. *3*: 39–40.

Pavlovich, N.V. and B.N. Mishankin. 1987. Transparent nutrient medium for the cultivation of *Francisella tularensis*. Antibiot. Med. Biotekhnol. *32*: 133–137.

Pavlovich, N.V. and B.N. Mishankin. 1992. Phosphatase and penicillinase activities as stable traits for the differentiation of the racial classification of *Francisella tularensis*. Zh. Mikrobiol. Epidemiol. Immunobiol. *11–12*: 5–7.

Payne, M.P. and R.J. Morton. 1992. Effect of culture media and incubation temperature on growth of selected strains of *Francisella tularensis*. J. Vet. Diagn. Invest. *4*: 264–269.

Payne, S.M. 1988. Iron and virulence in the family Enterobacteriaceae. Crit. Rev. Microbiol. *16*: 81–111.

Payne, W.J., R.G. Eagon and A.K. Williams. 1961. Some observations on the physiology of *Pseudomonas natriegens* nov. spec. Antonie van Leeuwenhoek J. Microbiol. Serol. *27*: 121–128.

Pearce, R. and I.S. Roberts. 1995. Cloning and analysis of gene clusters for production of the *Escherichia coli* K10 and K54 antigens: identification of a new group of serA-linked capsule gene clusters. J. Bacteriol. *177*: 3992–3997.

Pease, P. 1979. Observations on L-forms of *Yersinia enterocolitica*. J. Med. Microbiol. *12*: 337–346.

Pease, P.E., J.W. Lawson, R.L. Bartlett, M. Lane, J.E. Tallack and R. Allan. 1989. Observations on cell-wall deficient forms of *Pseudomonas maltophilia*. Microbios. *57*: 21–26.

Peciña, A., A. Pascual and A. Paneque. 1999. Cloning and expression of the *algL* gene, encoding the *Azotobacter chroococcum* alginate lyase: Purification and characterization of the enzyme. J. Bacteriol. *181*: 1409–1414.

Pecknold, P.C. and R.G. Grogan. 1973. Deoxyribonucleic acid homology groups among phytopathogenic *Pseudomonas* species. Int. J. Syst. Bacteriol. *23*: 111–121.

Pedersen, K.B. 1977. *Actinobacillus infections* in swine. Nordisk Veterinaer Medicin. *29*: 137–140.

Pedersen, K. 1997. Microbial life in deep granitic rock. FEMS Microbiol. Rev. *20*: 399–414.

Pedersen, K., J. Arlinger, S. Ekendahl and L. Hallbeck. 1996a. 16S rRNA gene diversity of attached and unattached bacteria in boreholes along the access tunnel to the Aspo Hard Rock Laboratory, Sweden. FEMS Microbiol. Ecol. *19*: 249–262.

Pedersen, K., I. Dalsgaard and J.L. Larsen. 1997. *Vibrio damsela* associated with diseased fish in Denmark. Appl. Environ. Microbiol. *63*: 3711–3715.

Pedersen, K.B., L.O. Frøholm and K. Bøvre. 1972. Fimbriation and colony type of *Moraxella bovis* in relation to conjunctival colonization and development of keratoconjunctivitis in cattle. Acta Pathol. Microbiol. Scand. B Microbiol. Immunol. *80*: 911–918.

Pedersen, K., S. Koblavi, T. Tiainen and P.A. Grimont. 1996b. Restriction fragment length polymorphism of the pMJ101-like plasmid and ribotyping in the fish pathogen *Vibrio ordalii*. Epidemiol. Infect. *117*: 385–391.

Pedersen, K., T. Tiainen and J.L. Larsen. 1996c. Plasmid profiles, restriction fragment length polymorphisms and O- serotypes among *Vibrio anguillarum* isolates. Epidemiol. Infect. *117*: 471–478.

Pedersen, K., L. Verdonck, B. Austin, D.A. Austin, A.R. Blanch, P.A.D. Grimont, J. Jofre, S. Koblavi, J.L. Larsen, T. Tiainen, M. Vigneulle and J. Swings. 1998. Taxonomic evidence that *Vibrio carchariae* Grimes et al. 1985 is a junior synonym of *Vibrio harveyi* (Johnson and Shunk 1936) Baumann et al, 1981. Int. J. Syst. Bacteriol. *48*: 749–758.

Pedroso, D.M.M., S.T. Iaria, M.L. Cerqueira Campos, S. Heidtmann, V.L.M. Rall, F. Pimenta and S.M.I. Saad. 1997. Virulence factors in motile *Aeromonas* spp isolated from vegetables. Rev. Microbiol. *28*: 49–54.

Peel, M.M., D.A. Alfredson, J.G. Gerrard, J.M. Davis, J.M. Robson, R.J. McDougall, B.L. Scullie and R.J. Akhurst. 1999. Isolation, identification, and molecular characterization of strains of *Photorhabdus luminescens* from infected humans in Australia. J. Clin. Microbiol. *37*: 3647–3653.

Peel, M.M., K.A. Hornidge, M. Luppino, A.M. Stacpoole and R.E. Weaver. 1991. *Actinobacillus* spp. and related bacteria in infected wounds of humans bitten by horses and sheep. J. Clin. Microbiol. *29*: 2535–2538.

Peerbooms, P.G., A.M. Verweij and D.M. MacLaren. 1983. Investigation of the haemolytic activity of *Proteus mirabilis* strains. Antonie Leeuwenhoek *49*: 1–11.

Peerbooms, P.G., A.M. Verweij and D.M. MacLaren. 1984. Vero cell invasiveness of *Proteus mirabilis*. Infect. Immun. *43*: 1068–1071.

Peerbooms, P.G., A.M. Verweij and D.M. MacLaren. 1985. Uropathogenic properties of *Proteus mirabilis* and *Proteus vulagaris*. J. Med. Microbiol. *19*: 55–60.

Pelayo, J.S., I.C. Scaletsky, M.Z. Pedroso, V. Sperandio, J.A. Giron, G.

Frankel and L.R. Trabulsi. 1999. Virulence properties of atypical EPEC strains. J. Med. Microbiol. *48*: 41–49.

Pelaz, C., L.G. Albert and C.M. Bourgon. 1987. Cross-reactivity among *Legionella* spp. and serogroups. Epidemiol. Infect. *99*: 641–646.

Pelczar, M.J. 1953. *Neisseria caviae* nov. spec. J. Bacteriol. *65*: 744.

Pellegrini, G., E. Levre, P. Valentini and M. Cadoni. 1992. Cockroaches: infestation and possible contribution in the spreading of some enterobacteria. Ig. Mod. *97*: 19–30.

Pelsh, A.D. 1936. Hydrobiology of Karabugaz Bay of the Caspian Sea. Tr. Vses. Nauchno-Issled. Inst. Galurgii Leningrad *5*: 49–126.

Pelsh, A.D. 1937. Photosynthetic sulfur bacteria of the eastern reservoir of Lake Sakskoe. Mikrobiologiya *6*: 1090–1100.

Peltola, H. 2000. Worldwide *Haemophilus influenzae* type b disease at the beginning of the 21st century: global analysis of the disease burden 25 years after the use of the polysaccharide vaccine and a decade after the advent of conjugates. Clin Microbiol. Rev. *13*: 302–317.

Pena, C., M.A. Trujillo-Roldan and E. Galindo. 2000. Influence of dissolved oxygen tension and agitation speed on alginate production and its molecular weight in cultures of *Azotobacter vinelandii*. Enzyme Microb. Technol. *27*: 390–398.

Penn, R.G., D.K. Giger, F.C. Knoop and L.C. Preheim. 1982. *Plesiomonas shigelloides* overgrowth in the small intestine. J. Clin. Microbiol. *15*: 869–872.

Penner, J.L. 1984. Genus XIII. *Morganella*. *In* Krieg and Holt (Editors), Bergey's Manual of Systemic Bacteriology, 1st Ed., Vol. 1, The Williams & Wilkins Co., Baltimore. pp. 497–498.

Penner, J.L., P.C. Fleming, G.R. Whiteley and J.N. Hennessy. 1979a. O-serotyping *Providencia alcalifaciens*. J. Clin. Microbiol. *10*: 761–765.

Penner, J.L. and J.N. Hennessy. 1977. Reassignment of the intermediate strains of *Proteus rettgeri* biovar 5 to *Providencia stuartii* on basis of somatic (O) antigens. Int. J. Syst. Bacteriol. *27*: 71–74.

Penner, J.L. and J.N. Hennessy. 1979a. Application of O-serotyping in a study of *Providencia rettgeri* (*Proteus rettgeri*) isolated from human and non-human sources. J. Clin. Microbiol. *10*: 834–840.

Penner, J.L. and J.N. Hennessy. 1979b. O antigen grouping of *Morganella morganii* (*Proteus morganii*) by slide agglutination. J. Clin. Microbiol. *10*: 8–13.

Penner, J.L. and J.N. Hennessy. 1980. Separate O-grouping schemes for serotyping clinical isolates of *Proteus vulgaris* and *Proteus mirabilis*. J. Clin. Microbiol. *12*: 304–309.

Penner, J.L. and N.A. Hinton. 1973. A study of the serotyping of *Proteus rettgeri*. Can. J. Microbiol. *19*: 271–279.

Penner, J.L., N.A. Hinton, I.B.R. Duncan, J.N. Hennessy and G.R. Whiteley. 1979b. O-serotyping of *Providencia stuartii* isolates collected from twelve hospitals. J. Clin. Microbiol. *9*: 11–14.

Penner, J.L., N.A. Hinton, L.J. Hamilton and J.N. Hennessy. 1981. Three episodes of nosocomial urinary tract infections caused by one O-serotype of *Providencia stuartii*. J. Urol. *125*: 668–671.

Penner, J.L., N.A. Hinton and J. Hennessy. 1974. Serotyping of *Proteus rettgeri* on the bais of O antigens. Can. J. Microbiol. *20*: 777–789.

Penner, J.L., N.A. Hinton and J. Hennessy. 1975. Biotypes of *Proteus rettgeri*. J. Clin. Microbiol. *1*: 136–142.

Penner, J.L., N.A. Hinton and J.N. Hennessy. 1976a. Evaluation of a *Proteus rettgeri* O-serotyping system for epidemiological investigation. J. Clin. Microbiol. *3*: 385–389.

Penner, J.L., N.A. Hinton, J.N. Hennessy and G.R. Whiteley. 1976b. Reconstitution of the somatic (O-) antigenic scheme for *Providencia* and preparation of O-typing antisera. J. Infect. Dis. *133*: 283–292.

Penner, J.L., N.A. Hinton, G.R. Whiteley and J.N. Hennessy. 1976c. Variation in urease activity of endemic hospital strains of *Proteus rettgeri* and *Providencia stuartii*. J. Infect. Dis. *134*: 370–376.

Penner, J.L. and M.A. Preston. 1980. Differences among *Providencia* species in their in vitro susceptibilities to five antibiotics. Antimicrob. Agents Chemother. *18*: 868–871.

Penner, J.L., M.A. Preston, J.N. Hennessy, L.J. Barton and M.M. Goodbody. 1982. Species differences in susceptibilities of *Proteeae* spp. to six cephalosporins and three aminoglycosides. Antimicrob. Agents Chemother. *22*: 218–221.

Pepe, C.M., M.W. Eklund and M.S. Strom. 1996. Cloning of an *Aeromonas hydrophila* type IV pilus biogenesis gene cluster: complementation of pilus assembly functions and characterization of a type IV leader peptidase/N-methyltransferase required for extracellular protein secretion. Mol. Microbiol. *19*: 857–869.

Perch, B. 1948. On the serology of the *Proteus* group. Acta Pathol. Microbiol. Scand. *25*: 703–714.

Pereira, A.L.G. 1969. Uma nova doença bacteriana do maracujá (*Passiflora edulis* Sims) causada por *Xanthomonas passiflorae* n. sp. Arq. Inst. Biol. São Paulo. *36*: 163–174.

Pereira, A.L.G., F.O. Paradella and A.G. Zagetto. 1971. Uma nova doença bacteriana da mandioquinha salsa (Arracacia Xanthorrhiza) causada por *Xanthomonas arracaciae* n. sp. Arq. Inst. Biol. São Paulo. *38*: 99–108.

Perkins, S.R., T.A. Beckett and C.M. Bump. 1986. *Cedecea davisae* bacteremia. J. Clin. Microbiol. *24*: 675–676.

Perna, N.T., G.I. Plunkett, V. Burland, B. Mau, J.D. Glasner, D.J. Rose, G.F. Mayhew, P.S. Evans, J. Gregor, H.A. Kirkpatrick, G. Posfai, J. Hackett, S. Klink, A. Boutin, Y. Shao, L. Miller, E.J. Grotbeck, N.W. Davis, A. Lim, E.T. Dimalanta, K.D. Potamousis, J. Apodaca, T.S. Anantharaman, J. Lin, G. Yen, D.C. Schwartz, R.A. Welch and F.R. Blattner. 2001. Genome sequence of enterohaemorrhagic *Escherichia coli* O157:H7. Nature *409*: 529–533.

Pernelle, J.J., E. Cotteux and P. Duchène. 1998. Effectiveness of oligonucleotide probes targeted against *Thiothrix nivea* and type 021N 16S rRNA for *in situ* identification and population monitoring of activated sludges. Water Sci. Technol. *37*: 431–440.

Pernezny, K., R.N. Raid, R.E. Stall, N.C. Hodge and J. Collins. 1995. An outbreak of bacterial spot of lettuce in Florida caused by *Xanthomonas campestris* pv. *vitians*. Plant Dis. *79*: 359–360.

Pérombélon, M.C.M. and L.J. Hyman. 1995. Serological methods to quantify potato seed contamination by *Erwinia carotovora* subsp. *atroseptica*. Bull. OEPP *25*: 195–202.

Pérombelon, M.C.M. and A. Kelman. 1980. Ecology of the soft rot erwinias. Ann. Rev. Phytopathol. *18*: 361–387.

Peros, J.P. 1988. Variability in colony type and pathogenicity of the causal agent of sugarcane gumming *Xanthomonas campestris* pv. *vasculorum* (Cobb) Dye. Z. Pflanzenkr. Pflanzenschutz. *95*: 591–598.

Perrin, T.K. and I.A. Bengtson. 1942. The histopathology of experimental Q fever in mice. Public Health Rep. *57*: 790–794.

Perry, K.A., J.E. Kostka, G.W. Luther, III and K.H. Nealson. 1993. Mediation of sulfur speciation by a Black Sea facultative anaerobe. Science *259*: 801–803.

Perry, R.D. and J.D. Fetherston. 1997. *Yersinia pestis* etiologic agent of plague. Clin. Microbiol. Rev. *10*: 35–66.

Persley, G.J. 1978. Epiphytic survival of *Xanthomonas manihotis* in relation to the disease cycle of cassava bacterial blight. Proc. 4th Internat. Conf. on Plant Pathol. Bact, *II*: pp. 401–429.

Persmark, M., D. Expert and J.B. Neilands. 1989. Isolation, characterization, and synthesis of chrysobactin, a compound with siderophore activity from *Erwinia chrysanthemi*. J. Biol. Chem. *264*: 3187–3193.

Perty, M. 1852. Zur Kenntnis kleinster lebensformen, Vol. I–VIII, Jent and Reinert, Bern.

Pessagno, E.A. 1969. The *Neosciadiocapsidae*: a new family of upper Cretaceous radiolaria. Bull. Amer. Paleontol. *56*: 377–437.

Petersen, K.D., H. Christensen, M. Bisgaard and J.E. Olsen. 2001. Genetic diversity of *Pasteurella multocida* fowl cholera isolates as demonstrated by ribotyping and 16S rRNA and partial *atpD* sequence comparisons. Microbiology (UK) *147*: 2739–2748.

Petersen, S.K. and N.T. Foged. 1989. Cloning and expression of the *Pasteurella multocida* toxin gene, *toxA*, in *Escherichia coli*. Infect. Immun. *57*: 3907–3913.

Peterson, J.A. 1970. Cytochrome content of two pseudomonads containing mixed-function oxidase systems. J. Bacteriol. *103*: 714–721.

Petri, R. and J.F. Imhoff. 2000. The relationship of nitrate reducing bacteria on the basis of *narH* gene sequences and the congruent phylogeny of *narH* and 16S rDNA baed on pure culture studies and analyses of environmental DNA. Syst. Appl. Microbiol. *23*: 47–57.

Petrovskaya, V.G. and V.M. Bondarenko. 1977. Recommended corrections to the classification of *Shigella flexneri* on a genetic basis. Int. J. Syst. Bacteriol. *27*: 171–175.

Petrovskaya, V.C. and N.A. Khomenko. 1979. Proposals for improving the classification of members of the genus *Shigella*. Int. J. Syst. Bacteriol. *29*: 400–402.

Petrovskis, E.A., T.M. Vogel and P. Adriaens. 1994. Effects of electron acceptors and donors on transformation of tetrachloromethane by *Shewanella putrefaciens* MR-1. FEMS Microbiol. Lett. *121*: 357–363.

Pettersson, B., A. Kodjo, M. Ronaghi, M. Uhlen and T. Tønjum. 1998. Phylogeny of the family *Moraxellaceae* by 16S rDNA sequence analysis, with special emphasis on differentiation of *Moraxella* species. Int. J. Syst. Bacteriol. *48*: 75–89.

Petushkov, V.N. and J. Lee. 1997. Purification and characterization of flavoproteins and cytochromes from the yellow bioluminescence marine bacterium *Vibrio fischeri* strain Y1. Eur. J. Biochem. *245*: 790–796.

Pfeiffer, A. 1889. Ueber die bäcillare Pseudotuberkulose bei Nagethieren, Thieme, Leipzig.

Pfennig, N. 1962. Beobachtungen über das Schwärmen von *Chromatium okenii*. Arch. Mikrobiol. *42*: 90–95.

Pfennig, N. 1965. Anreicherungskulturen fur rote und grune Schwefelbakterien. Zentbl. Bakteriol. Parasitenkd. Infektkrankh. Hyg. Abt. I Orig. *Suppl. I.*: 179–189, 503–505.

Pfennig, N. 1977. Phototropic green and purple bacteria comparative, systematic survey. Annu. Rev. Microbiol. *31*: 275–290.

Pfennig, N. 1989a. Genus *Amoebobacter*. *In* Staley, Bryant, Pfennig and Holt (Editors), Bergey's Manual of Systematic Bacteriology, 1st Ed., Vol. 3, The Williams & Wilkins Co., Baltimore. pp. 1651–1652.

Pfennig, N. 1989b. Genus *Chromatium*. *In* Staley, Bryant, Pfennig and Holt (Editors), Bergey's Manual of Systematic Bacteriology, 1st Ed., Vol. 3, The Williams & Wilkins Co., Baltimore. pp. 1639–1643.

Pfennig, N. 1989c. Genus *Thiopedia*. *In* Staley, Bryant, Pfennig and Holt (Editors), Bergey's Manual of Systematic Bacteriology, 1st Ed., Vol. 3, The Williams & Wilkins Co., Baltimore. pp. 1652–1653.

Pfennig, N. 1989d. Genus *Thiospirillum*. *In* Staley, Bryant, Pfennig and Holt (Editors), Bergey's Manual of Systematic Bacteriology, 1st Ed., Vol. 3, The Williams & Wilkins Co., Baltimore. pp. 1644–1645.

Pfennig, N. and K.D. Lippert. 1966. Über das Vitamin B$_{12}$-Bedürfnis phototropher Schwefelbakterien. Arch. Microbiol. *55*: 245–256.

Pfennig, N., M.C. Markham and S. Liaaen-Jensen. 1968. Carotenoids of thiorhodaceae. 8. Isolation and characterization of a *Thiothece, Lamprocystis* and *Thiodictyon* strain and their carotenoid pigments. Arch. Mikrobiol. *62*: 178–191.

Pfennig, N. and H.G. Trüper. 1971a. Higher taxa of the phototrophic bacteria. Int. J. Syst. Bacteriol. *21*: 17–18.

Pfennig, N. and H.G. Trüper. 1971b. New nomenclatural combinations in the phototrophic sulfur bacteria. Int. J. Syst. Bacteriol. *21*: 11–14.

Pfennig, N. and H.G. Trüper. 1974. The phototrophic bacteria. *In* Buchanan and Gibbons (Editors), Bergey's Manual of Determinative Bacteriology, 8th Ed., The Williams & Wilkins Co., Baltimore. pp. 24–60.

Pfennig, N. and H.G. Trüper. 1981. Isolation of the members of the families Chromatiaceae and Chlorobiaceae. *In* Starr, Stolp, Truper, Balows and Schlegel (Editors), The Prokaryotes: A Handbook on Habitats, Isolation and Identification of Bacteria, 1st Ed., Vol. 1, Springer Verlag, New York. pp. 279–289.

Pfennig, N. and H.G. Trüper. 1989. Family I. *Chromatiaceae*. *In* Staley, Bryant, Pfennig and Holt (Editors), Bergey's Manual of Systematic Bacteriology, 1st Ed., Vol. 3, The Williams & Wilkins Co., Baltimore. pp. 1637–1653.

Pfennig, N. and H.G. Trüper. 1992. The family *Chromatiaceae*. *In* Balows, Trüper, Dworkin, Harder and Schleifer (Editors), The Prokaryotes. A Handbook on the Biology of Bacteria: Ecophysiology, Isolation, Identification, Applications, 2nd Ed., Vol. 4, Springer-Verlag, New York. pp. 3200–3221.

Philip, C.B. 1943. Nomenclature of the pathogenic rickettsiae. Am. J. Hyg. *37*: 301–309.

Philip, C.B. 1948. Comments on the name of the Q fever organism. Pub. Health Rep. *63*: 58.

Philip, C.B. 1956. Comments on the classification of the order *Rickettsiales*. Can. J. Microbiol. *2*: 261–270.

Phillips, J.E. 1960. The characterisation of *Actinobacillus lignieresii*. J. Pathol. Bacteriol. *79*: 331–336.

Phillips, J.E. 1961. The commensal role of *Actinobacillus lignieresii*. J. Pathol. Bacteriol. *82*: 205–208.

Phillips, J.E. 1964. Commensal actinobacilli from the bovine tongue. J. Pathol. Bacteriol. *87*: 442–444.

Phillips, J.E. 1966. *Actinobacillus lignieresii*: a study of the organism and its association with its hosts, Thesis, University of Edinburgh, Edinburgh.

Phillips, J.E. 1967. Antigenic structure and serological typing of *Actinobacillus lignieresii*. J. Pathol. Bacteriol. *93*: 463–475.

Phillips, J.E. 1984. Genus III. *Actinobacillus*. *In* Krieg and Holt (Editors), Bergey's Manual of Systematic Bacteriology, 1st Ed., , Vol. 1, The Williams & Wilkins Co., Baltimore. pp. 570–575.

Phillips, N.J., M.A. Apicella, J.M. Griffiss and B.W. Gibson. 1993. Structural studies of the lipooligosaccharides from *Haemophilus influenzae* type b strain A2. Biochemistry *32*: 2003–2012.

Pianetti, A., W. Baffone, F. Bruscolini, E. Barbieri, A. Giudice and L. Salvaggio. 1997. Recovery of *Aeromonas* spp. from mussels collected directly from the marine environment and purchased at the market. Industrie Alimentari. *36*: 175–177.

Picard, B., J.S. Garcia, S. Gouriou, P. Duriez, N. Brahimi, E. Bingen, J. Elion and E. Denamur. 1999. The link between phylogeny and virulence in *Escherichia coli* extraintestinal infection. Infect. Immun. *67*: 546–553.

Picard, B. and P. Goullet. 1985. Comparative electrophoretic profiles of esterases, and of glutamate, lactate and malate dehydrogenases, from *Aeromonas hydrophila, Aeromonas caviae* and *Aeromonas sobria*. J. Gen. Microbiol. *131*: 3385–3392.

Picard, B., P. Goullet, P.J.M. Bouvet, G. Decoux and J.B. Denis. 1989. Characterization of bacterial genospecies by computer-assisted statistical analysis of enzyme electrophoretic data. Electrophoresis *10*: 680–685.

Picardal, F., R.G. Arnold and B.B. Huey. 1995. Effects of electron donor and acceptor conditions on reductive dehalogenation of tetrachloromethane by *Shewanella putrefaciens* 200. Appl. Environ. Microbiol. *61*: 8–12.

Piccolomini, R., L. Cellini, N. Allocati, A. Di Girolamo and G. Ravagnan. 1987. Comparative *in vitro* activities of 13 antimicrobial agents against *Morganella-Proteus-Providencia* group bacteria from urinary tract infections. Antimicrob. Agents Chemother. *31*: 1644–1647.

Pichinoty, F., E. Azoulay, P. Couchoud-Beaumont, L. Le Minor, C. Rigano, J. Bigliardi-Rouvier and M. Piéchaud. 1969. Recherche des nitrate-reductases bactériennes A et B: resultats. Ann. Inst. Pasteur. *116*: 27–42.

Pichinoty, F. and M. Piéchaud. 1968. Recherche des nitrate-réductases bactériennes A et B: methodes. Ann. Inst. Pasteur. *114*: 77–98.

Pickard, D., J. Li, M. Roberts, D. Maskell, D. Hone, M. Levine, G. Dougan and S. Chatfield. 1994. Characterization of defined *ompR* mutants of *Salmonella typhi*: ompR is involved in the regulation of Vi polysaccharide expression. Infect. Immun. *62*: 3984–3993.

Pidcock, K.A., J.A. Wooten, B.A. Daley and T.L. Stull. 1988. Iron acquisition by *Haemophilus influenzae*. Infect. Immun. *56*: 721–725.

Pidiyar, V., A. Kaznowski, N.B. Narayan, M. Patole and Y.S. Shouche. 2002. *Aeromonas culicicola* sp. nov., from the midgut of *Culex quinquefasciatus*. Int. J. Syst. Evol. Microbiol. *52*: 1723–1728.

Piéchaud, M. 1961. Le groupe *Moraxella*. A propos des B5W-*Bacterium anitratum*. Ann. Inst. Pasteur. *100*: 74–85.

Piechulla, K., M. Bisgaard, H. Gerlach and W. Mannheim. 1985a. Taxonomy of some recently described avian *Pasteurella/Actinobacillus*-like organisms as indicated by deoxyribonucleic acid relatedness. Avian Pathology *14*: 281–311.

Piechulla, K., K.H. Hinz and W. Mannheim. 1985b. Genetic and phenotypic comparison of three new avian *Haemophilus*-like taxa and of *Haemophilus paragallinarum* Biberstein and White 1969 with other members of the family *Pasteurellaceae* Pohl 1981. Avian Dis. *29*: 601–612.

Piechulla, K., R. Mutters, S. Burbach, R. Klussmeier, S. Pohl and W. Mannheim. 1986. Deoxyribonucleic-acid relationships of *Histophilus ovis, Haemophilus somnus, Haemophilus haemoglobinophilus,* and *Actinobacillus seminis.* Int. J. Syst. Bacteriol. *36*: 1–7.

Pien, F.D. and A.E. Bruce. 1986. Nosocomial *Ewingella americana* bacteremia in an intensive care unit. Arch. Intern. Med. *146*: 111–112.

Pien, F.D., J.J. Farmer, III and R.E. Weaver. 1983. Polymicrobial bacteremia caused by *Ewingella americana* (family *Enterobacteriaceae*) and an unusual *Pseudomonas* species. J. Clin. Microbiol. *18*: 727–729.

Pien, F., K. Lee and H. Higa. 1977. *Vibrio alginolyticus* infections in Hawaii. J. Clin. Microbiol. *5*: 670–672.

Pien, F.D., W.J. Martin, P.E. Hermans and J.A. Washington. 1972. Clinical and bacteriologic observations on the proposed species, *Enterobacter agglomerans* (the *Herbicola lathyri* bacteria). Mayo. Clin. Proc. *47*: 739–745.

Piérard, D., G. Muyldermans, L. Moriau, D. Stevens and S. Lauwers. 1998. Identification of new verocytotoxin type 2 variant B-subunit genes in human and animal *Escherichia coli* isolates. J. Clin. Microbiol. *36*: 3317–3322.

Piérard, D., D. Stevens, L. Moriau, H. Lior and S. Lauwers. 1997. Isolation and virulence factors of verocytotoxin-producing *Escherichia coli* in human stool samples. Clin. Microbiol. Infect. *3*: 531–540.

Pierce, L. and A.H. McCain. 1992. Selective medium for isolation of pectolytic *Erwinia* sp. Plant Dis. *76*: 382–384.

Pierce, N.B. 1901. Walnut bacteriosis. Bot. Gaz. *31*: 272–273.

Pillay, D., B. Pillay, E.A. Wachters and L. Korsten. 1995. Electrophoretic and immunological analysis of lipopolysaccharides of *Xanthomonas albilineans* from three geographical regions. Lett. Appl. Microbiol. *21*: 210–214.

Pillich, J., Z. Hradecna and M. Kocur. 1964. An attempt at phage typing in the genus *Serratia.* J. Appl. Bacteriol. *27*: 65–68.

Pin, C., P. Morales, M.L. Marin, M.D. Selgas, M.L. Garcia and C. Casas. 1997. Virulence factors-pathogenicity relationships for *Aeromonas* species from clinical and food isolates. Folia Microbiol. *42*: 385–389.

Pindar, D.F. and C. Bucke. 1975. The biosynthesis of alginic acid by *Azotobacter vinelandii.* Biochem. J. *152*: 617–622..

Pine, L., I.R. George, M.W. Reeves and W.K. Harrell. 1979. Development of a chemically defined liquid medium for the growth of Legionnaires' disease bacterium. J. Clin. Microbiol. *9*: 615–626.

Pinkart, H.C. and D.C. White. 1997. Phospholipid biosynthesis and solvent tolerance in *Pseudomonas putida* strains. J. Bacteriol. *179*: 4219–4226.

Pinkart, H.C., J.W. Wolfram, R. Rogers and D.C. White. 1996. Cell envelope changes in solvent-tolerant and solvent-sensitive *Pseudomonas putida* strains following exposure to o-xylene. Appl. Environ. Microbiol. *62*: 1129–1132.

Pinnock, D.E. and R.T. Hess. 1974. The occurrence of intracellular rickettsia-like organisms in the tsetse flies, *Glossina morsitans, G. fuscipes, G. brevipalpis* and *G. pallidipes.* Acta Trop. *31*: 70–79.

Piot, P., E. van Dyck and S.R. Patty. 1977. Sensibilité *d'Haemophilus influenza* à 5 antibiotiques et détection rapide de sa résistance à 1'ampicilline. Pathol. Biol. (Paris) *25*: 83–87.

Pirhonen, M. and E.T. Palva. 1988. Occurrence of bacteriophage T4 receptor in *Erwinia carotovora.* Mol. Gen. Genet. *214*: 170–172.

Pitarangsi, C., P. Echeverria, R. Whitmire, C. Tirapat, S. Formal, G.J. Dammin and M. Tingtalapong. 1982. Enteropathogenicity of *Aeromonas hydrophila* and *Plesiomonas shigelloides*: prevalence among individuals with and without diarrhea in Thailand. Infect. Immun. *35*: 666–673.

Pitt, T.L. and D.E. Bradley. 1975. The antibody response to the flagella of *Pseudomonas aeruginosa.* J. Med. Microbiol. *8*: 97–106.

Pittman, M. 1931. Variation and type specificity in the bacterial species *Haemophilus influenzae.* J. Exp. Med. *53*: 471–492.

Pittman, M. 1953. A classification of the hemolytic bacteria of the genus *Haemophilus*: *Haemophilus haemolyticus* Bergey et al. . and *Haemophilus parahaemolyticus* nov. spec. J. Bacteriol. *65*: 750–751.

Pittman, M. and D.J. Davis. 1950. Identification of the Koch-Weeks bacillus (*Hemophilus aegyptius*). J. Bacteriol. *59*: 413–426.

Pitts, G., A.I. Allam and J.P. Hollis. 1972. *Beggiatoa*: occurrence in the rice rhizosphere. Science *178*: 990–992.

Pivnick, H. 1955. *Pseudomonas rubescens,* a new species from soluble oil emulsions. J. Bacteriol. *70*: 1–6.

Podschun, R. and U. Ullmann. 1998. *Klebsiella* spp. as nosocomial pathogens: epidemiology, taxonomy, typing methods, and pathogenicity factors. Clin. Microbiol. Rev. *11*: 589–603.

Poffe, R. and E. Op de Beeck. 1991. Enumeration of *Aeromonas hydrophila* from domestic wastewater treatment plants and surface waters. J. Appl. Bacteriol. *71*: 366–370.

Poffe, R.J., J. Vanderleyden and H. Verachtert. 1979. Characterization of a *Leucothrix*-type bacterium causing sludge bulking during petrochemical waste-water treatment. Eur. J. Appl. Microbiol. Biotechnol. *8*: 229–237.

Pohl, S. 1979. Reklassifizierung der Gattung *Actinobacillus* Brumpt 1910, *Haemophilus* Winslow et al. 1917 und *Pasteurella* Trevisan 1887 anhand phänotypischer und molekular Daten, insbesondere der DNS-Verwandtschaften bei DNS:DNS-Hybridisierung *in vitro* und Vorschlag einer neuen Familie *Pasteurellaceae,* Thesis, Phillips-Universität, Marburg/Lahn. Mauersberger, Marburg.

Pohl, S. 1981a. DNA relatedness among members of *Actinobacillus, Haemophilus* and *Pasteurella. In* Kilian, Frederiksen and Biberstein (Editors), *Haemophilus, Pasteurella* and *Actinobacillus,* Academic Press, 246–253.

Pohl, S. 1981b. *In* Validation of publication of new names and new combinations previously effectively published outside the IJSB, List No. 7. Int. J. Syst. Bacteriol . *31*: 382–383.

Pohl, S., H.U. Bertschinger, W. Frederiksen and W. Mannheim. 1983. Transfer of *Hemophilus pleuropneumoniae* and the *Pasteurella haemolytica*-like organism causing porcine necrotic pleuropneumoniae to the genus *Actinobacillus* (*Actinobacillus pleuropneumoniae* comb. nov.) on the basis of phenotypic and deoxyribonucleic acid relatedness. Int. J. Syst. Bacteriol. *33*: 510–514.

Poinar, G.O., Jr. 1975. Description and biology of a new insect parasitic rhabditoid, *Heterorhabditis bacteriophora* n. gen. n. sp. (Rhabditida; Heterorhabditidae n. Fam.). Nematologica. *21*: 463–470.

Poinar, G.O. and P.T. Himsworth. 1967. *Neoaplectana* parasitism of larvae of the greater wax moth. J. Invertebr. Pathol. *9*: 241–246.

Poinar, G.O., Jr. and R. Leutenegger. 1968. Anatomy of the infective and normal third-stage juveniles of *Neoaplectana carpocapsae* Weiser (Steinernematidae: Nematoda). J. Parasitol. *54*: 340–350.

Poinar, G.O. and G.M. Thomas. 1965. A new bacterium, *Achromobacter nematophilus* sp. nov. (*Achromobacteraceae*: *Eubacteriales*), associated with a nematode. Int. Bull. Bact. Nomen. Taxon. *15*: 249–252.

Poinar, G.O., Jr. and G.M. Thomas. 1966. Significance of *Achromobacter nematophilus* Poinar and Thomas (*Achromobacteraceae*: *Eubacteriales*) in the development of the nematode, DD-136 (*Neoaplectana* sp. Steinernematidae). Parasitology. *56*: 385–390.

Poinar, G.O., Jr. and G.M. Thomas. 1967. The nature of *Achromobacter nematophilus* as an insect pathogen. J. Invertebr. Pathol. *9*: 510–514.

Poinar, G.O., Jr., G.M. Thomas and R. Hess. 1977. Characteristics of the specific bacterium associated with *Heterorhabditis bacteriophora* (Heterorhabditidae; Rhabditida). Nematologica. *23*: 97–102.

Poinar, G.O., G.M. Thomas, S.B. Presser and J.L. Hardy. 1982. Inoculation of entomogenous nematodes, *Neoaplectana* and *Heterorhabditis,* and their associated bacteria, *Xenorhabdus* spp., into chicks and mice. Environ. Entomol. *11*: 137–138.

Pokhil, S.I. 1996. Species of enterobacteria new to medicine. Mikrobiol. Zh. *58*: 94–103.

Poland, J.D. 1989. Plague. *In* Hoeprich and Jordan (Editors), Infectious Diseases: a Modern Treatise of Infectious Processes, 4th ed., Lippincott, Philadelphia. 1296–1306.

Pollitzer, R. 1954. Plague, W.H.O. Monograph Series 22, World Health Organization, Geneva.

Polman, J.K. and J.M. Larkin. 1988. Properties of *in vivo* nitrogenase activity in *Beggiatoa alba.* Arch. Microbiol. *150*: 126–130.

Polster, M. and M. Svobodová. 1964. Production of reddish-brown pigment from DL-tryptophan by enterobacteria of the *Proteus-Providencia* group. Experientia (Basel) *20*: 637–638.

Polz, M.F., E.V. Odintsova and C.M. Cavanaugh. 1996. Phylogenetic relationships of the filamentous sulfur bacterium *Thiothrix ramosa* based 16s rRNA sequence analysis. Int. J. Syst. Bacteriol. *46*: 94–97.

Pommerening-Röser, A. 1993. Untersuchen zur phylogenie ammoniak oxidierender bakterien, University of Hamburg

Pon, D.S., C.E. Townsend, G.E. Wessman, C.G. Schmitt and C.H. Kingsolver. 1954. A *Xanthomonas* parasite on uredia of cereal rusts. Phytopathology *44*: 707–710.

Pons, J.L., A. Rimbault, J.C. Darbord and G. Leluan. 1984. Biosynthesis of toluene by *Clostridium aerofoetidum* strain WS. Ann. Microbiol. (Paris) *135B*: 219–222.

Poole, K., N. Gotoh, H. Tsujimoto, Q.X. Zhao, A. Wada, T. Yamasaki, S. Neshat, J.I. Yamagishi, X.Z. Li and T. Nishino. 1996. Overexpression of the *mexC-mexD-oprJ* efflux operon in *nfxB*-type multidrug-resistant strains of *Pseudomonas aeruginosa*. Mol. Microbiol. *21*: 713–724.

Poole, K., D.E. Heinrichs and S. Neshat. 1993a. Cloning and sequence-analysis of an EnvCD homolog in *Pseudomonas aeruginosa* - regulation by iron and possible involvement in the secretion of the siderophore pyoverdine. Mol. Microbiol. *10*: 529–544.

Poole, K., K. Krebes, C. McNally and S. Neshat. 1993b. Multiple antibiotic-resistance in *Pseudomonas aeruginosa* - evidence for involvement of an efflux operon. J. Bacteriol. *175*: 7363–7372.

Pooler, M.R. and J.S. Hartung. 1995a. Genetic relationships among strain of *Xylella fastidiosa* from RAPD-PCR data. Curr. Microbiol. *31*: 134–137.

Pooler, M.R. and J.S. Hartung. 1995b. Specific PCR detection and identification of *Xylella fastidiosa* strains causing citrus variegated chlorosis. Curr. Microbiol. *31*: 377–381.

Pooler, M.R., J.S. Hartung and R.G. Fenton. 1997a. Sequence analysis of a 1296-nucleotide plasmid from *Xylella fastidiosa*. FEMS Microbiol. Lett. *155*: 217–222.

Pooler, M.R., I.S. Myung, J. Bentz, J. Sherald and J.S. Hartung. 1997b. Detection of *Xylella fastidiosa* in potential insect vectors by immunomagnetic separation and nested polymerase chain reaction. Lett. Appl. Microbiol. *25*: 123–126.

Popham, P.L., S.M. Pike and A. Novacky. 1995. The effect of harpin from *Erwinia amylovora* on the plasmalemma of suspension-cultured tobacco cells. Physiol. Mol. Plant Pathol. *47*: 39–50.

Poplawsky, A.R. and W. Chun. 1997. *pigB* determines a diffusible factor needed for extracellular polysaccharide slime and xanthomonadin production in *Xanthomonas campestris* pv. *campestris*. J. Bacteriol. *179*: 439–444.

Poplawsky, A.R. and W. Chun. 1998. *Xanthomonas campestris* pv. *campestris* requires a functional *pigB* for epiphytic survival and host infection. Mol. Plant-Microbe Interact. *11*: 466–475.

Poplawsky, A.R., M.D. Kawalek and N.W. Schaad. 1993. A xanthomonadin-encoding gene cluster for the identification of pathovars of *Xanthomonas campestris*. Mol. Plant-Microbe Interact. *6*: 545–552.

Popoff, M. 1969. Étude sur le *Aeromonsd dalmonicida*. I. Caractéres biochimique et antigeniques. Rech. Vet. *3*: 49–57.

Popoff, M. 1984a. Genus III. *Aeromonas*. *In* Krieg and Holt (Editors), Bergey's Manual of Systematic Bacteriology, 1 Ed., Vol. 1, The Williams & Wilkins Co., Baltimore. 545–548.

Popoff, M. 1984b. *In* Validation of the publication of new names and new combinations previously effectively published outside the IJSB. List No. 15. Int. J. Syst. Bacteriol. *34*: 355–357.

Popoff, M.Y., J. Bockemühl and F.W. Brenner. 1998. Supplement 1997 (no. 41) to the Kauffmann-White scheme. Res. Microbiol. *149*: 601–604.

Popoff, M., C. Coynault, M. Kiredjian and M. Lemelin. 1981. Polynucleotide sequence relateness among motile *Aeromonas* species. Curr. Microbiol. *5*: 109–114.

Popoff, M. and C. Richard. 1975. O and H antigens of *Levinea malonatica*. Ann. Microbiol. *126B*: 17–23.

Popoff, M. and M. Veron. 1976. A taxonomic study of the *Aeromonas hydrophila–Aeromonas punctata* group. J. Gen. Microbiol. *94*: 11–22.

Popoff, M. and M. Veron. 1981. *In* Validation of the publication of new names and new combinations previously effectively published outside the IJSB. List No. 6. Int. J. Syst. Bacteriol. *31*: 215–218.

Popov, V., G. Sutakova, J. Rehacek, N. Smirnova and A. Daiter. 1991. *Coxiella* and *Rickettsiella*: comparison of ultrastructure with special reference to their envelope. Acta Virol. *35*: 573–579.

Popovic, T., P.I. Fields, O. Olsvik, J.G. Wells, G.M. Evins, D.N. Cameron, J.J. Farmer, III, C.A. Bopp, K. Wachsmuth, R.B. Sack, M.J. Albert, G.B. Nair, T. Shimada and J.C. Feeley. 1995. Molecular subtyping of toxigenic *Vibrio cholerae* O139 causing epidemic cholera in India and Bangladesh, 1992–1993. J. Infect. Dis. *171*: 122–127.

Poppe, C., N. Smart, R. Khakhria, W. Johnson, J. Spika and J. Prescott. 1998. *Salmonella typhimurium* DT104: A virulent and drug-resistant pathogen. Can. Vet. J.-Rev. Vet. Can. *39*: 559–565.

Poquet, Y., M. Kroca, F. Halary, S. Stenmark, M.A. Peyrat, M. Bonneville, J.J. Fournié and A. Sjöstedt. 1998. Expansion of Vgamma9Vdelta2 T cells is triggered by *Francisella tularensis*-derived phosphoantigens in tularemia but not after tularemia vaccination. Infect. Immun. *66*: 2107–2114.

Porras, O., D.A. Caugant, B. Gray, T. Lagergard, B.R. Levin and C. Svanborg-Eden. 1986. Difference in structure between type b and nontypable *Haemophilus influenzae* populations. Infect. Immun. *53*: 79–89.

Porschen, R.K. and P. Chan. 1977. Anaerobic vibrio-like organisms cultured from blood: *Desulfovibrio desulfuricans* and *Succinivibrio* species. J. Clin. Microbiol. *5*: 444–447.

Portnoy, D.A., S.L. Moseley and S. Falkow. 1981. Characterization of plasmids and plasmid-associated determinants of *Yersinia enterocolitica* pathogenesis. Infect. Immunol. *31*: 775–782.

Porto, M.H., G.J. Noel, P.J. Edelson and Brazilian Purpuric Fever Study Group. 1989. Resistance to serum bactericidal activity distinguishes Brazilian purpuric fever (BPF) case strains of *Haemophilus influenzae* biogroup aegyptius (*H. aegyptius*) from non-BPF strains. J. Clin. Microbiol. *27*: 792–794.

Postma, P.W., J.W. Lengeler and G.R. Jacobson. 1993. Phosphoenolpyruvate:carbohydrate phosphotransferase systems of bacteria. Microbiol. Rev. *57*: 543–594.

Pot, B. 1996. De fylogenie van chemo-organotrofe spirillen. (The phylogeny of chemoorganotrophic spirilla) Proefschrift ingediend tot het behalen van de graad van Doctor in de wetenschappen, University of Gent, Gent, Belgium.

Pot, B., M. Gillis, B. Hoste, A. Van De Velde, F. Bekaert, K. Kersters and J. De Ley. 1989. Intra- and intergeneric relationships of the genus *Oceanospirillum*. Int. J. Syst. Bacteriol. *39*: 23–34.

Pot, B., A. Willems, M. Gillis and J. De Ley. 1992. Intra- and intergeneric relationships of the genus *Aquaspirillum*: *Prolinoborus*, a new genus for *Aquaspirillum fasciculus*, with the species *Prolinoborus fasciculus* comb. nov. Int. J. Syst. Bacteriol. *42*: 44–57.

Potee, K.G., S.S. Wright and M. Finland. 1954. *In vitro* susceptibility of recently isolated strains of *Proteus* to 10 antibiotics. J. Lab. Clin. Med. *44*: 463–477.

Potrikus, C.J. and J.A. Breznak. 1977. Nitrogen-fixing *Enterobacter agglomerans* isolated from guts of wood-eating termites. Appl. Environ. Microbiol. *33*: 392–399.

Potts, T.V., J.J. Zambon and R.J. Genco. 1985. Reassignment of *Actinobacillus actinomycetemcomitans* to the genus *Haemophilus* as *Haemophilus actinomycetemcomitans* comb. nov. Int. J. Syst. Bacteriol. *35*: 337–341.

Poulos, C.D., S.O. Matsumura, B.M. Willey, D.E. Low and A. McGeer. 1995. In vitro activities of antimicrobial combinations against *Stenotrophomonas* (*Xanthomonas*) *maltophilia*. Antimicrob. Agents Chemother. *39*: 2220–2223.

Pourquier, M., J. Mandin and C. Vago. 1963. Développement d'une rickettsie de coléoptère en culture de tissus de vertébrés. Ann. Epiphyties. *14*: 193–197.

Pourrat, H., F. Regerat, P. Morvan and A. Pourrat. 1987. Microbiological production of gallic acid from *Rhus coriaria* L. Biotechnol. Lett. *9*: 731–734.

Powell, M. 1988. Antimicrobial resistance in *Haemophilus influenzae*. J. Med. Microbiol. *27*: 81–87.

Powell, M., S.F. Yeo, A. Seymour, M. Yuan, J.D. Williams and Y.S. Fah. 1992. Antimicrobial resistance in *Haemophilus influenzae* from England and Scotland in 1991. J. Antimicrob. Chemother. *29*: 547–554.

Powell, P.E., G.R. Cline, C.P.P. Reid and P.J. Szaniszlo. 1980. Occurrence of hydroxamate siderophore iron chelators in soils. Nature (Lond.) *287*: 833–834.

Praillet, T., W. Nasser, J. Robert-Baudouy and S. Reverchon. 1996. Purification and functional characterization of *PecS*, a regulator of virulence-factor synthesis in *Erwinia chrysanthemi*. Mol. Microbiol. *20*: 391–402.

Praillet, T., S. Reverchon and W. Nasser. 1997. Mutual control of the PecS/PecM couple, two proteins regulating virulence-factor synthesis in *Erwinia chrysanthemi*. Mol. Microbiol. *24*: 803–814.

Prakash, O., S. Dayal and S.L. Kalra. 1966. Bacterial aetiology of infantile diarrhoea in a village population with observations on some Providence strains isolated from diarrhoea and non-diarrhoea cases. Ind. J. Med. Res. *54*: 705–713.

Premakumar, R., M.R. Jacobson, T.M. Loveless and P.E. Bishop. 1992. Characterization of transcripts expressed from nitrogenase-3 structural genes of *Azotobacter vinelandii*. Can. J. Microbiol. *38*: 929–936.

Prere, M.F., M. Chandler and O. Fayet. 1990. Transposition in *Shigella dysenteriae*: isolation and analysis of IS911, a new member of the IS3 group of insertion sequences. J. Bacteriol. *172*: 4090–4099.

Prest, A.G., J.R.M. Hammond and G.S.A.B. Stewart. 1994. Biochemical and molecular characterization of *Obesumbacterium proteus*, a common contaminant of brewing yeasts. Appl. Environ. Microbiol. *60*: 1635–1640.

Preston, M.A., W. Johnson, R. Khakhria and A. Borczyk. 2000. Epidemiologic subtyping of *Escherichia coli* serogroup O157 strains isolated in Ontario by phage typing and pulsed-field gel electrophoresis. J. Clin. Microbiol. *38*: 2366–2368.

Prévot, A.R. 1966. Manual for the Classification and Determination of the Anaerobic Bacteria, 1st Amer. Ed. ed., Lea and Febiger, Philadelphia.

Price, E.H. and G.H. Hunt. 1986. *Aeromonas* in hospital — methods of isolation. J. Hosp. Infect. *8*: 309–311.

Priest, F.G., H.J. Somerville, J.A. Cole and J.S. Hough. 1973. The taxonomic position of *Obesumbacterium proteus*, a common brewery contaminant. J. Gen. Microbiol. *75*: 295–307.

Prieto, M., M.R. García-Armesto, M.L. García-López, C. Alonso and A. Otero. 1992. Species of Pseudomonas obtained at 7°C and 30°C during aerobic storage of lamb carcasses. J. Appl. Bacteriol. *73*: 317–323.

Prince, R.C., K.E. Stokley, C.E. Haith and H.W. Jannasch. 1988. The cytochromes of a marine *Beggiatoa*. Arch. Microbiol. *150*: 193–196.

Pringsheim, E.G. 1949. The relationship between bacteria and the *Myxophyceae*. Bacteriol. Rev. *13*: 47–91.

Pringsheim, E.G. 1951. The *Vitreoscillaceae*: a family of colourless, gliding, filamentous organisms. J. Gen. Microbiol. *5*: 124–149.

Pringsheim, E.G. 1957. Observations on *Leucothrix mucor* and *Leucothrix cohaerens* nov. sp. Bacteriol. Rev. *21*: 69–76.

Pringsheim, E.G. 1964. Heterotrophism and species concepts in *Beggiatoa*. Am. J. Bot. *51*: 898–913.

Pringsheim, E.G. 1967. Die Mixotrophie von *Beggiatoa*. Arch. Mikrobiol. *59*: 247–254.

Pringsheim, E.G. 1970. Prefatory chapter: contributions toward the development of general microbiology. Annu. Rev. Microbiol. *24*: 1–16.

Pringshcim, E.G. and W. Wiessner. 1963. Minimum requirements for heterotrophic growth and reserve substance in *Beggiatoa*. Nature *197*: 102.

Prinsloo, H.E. 1966. Bacteriocins and phages produced by *Serratia marcescens*. J. Gen. Microbiol. *45*: 205–212.

Prior, S.D. and H. Dalton. 1985. The effect of copper ions on membrane content and methane monooxygenase activity in methanol-grown cells of *Methylococcus capsulatus* (Bath). J. Gen. Microbiol. *131*: 155–164.

Prochazka, O. 1966. Preparation of conjugates of 19-S and 7-S globulins of antitularemic sera for the determination of specific fluorescence of *Pasteurella tularensis*. Folia Microbiol. (Praha). *11*: 337–346.

Proctor, H.M., J.R. Norris and D.W. Ribbons. 1969. Fine structure of methane-utilizing bacteria. J. Appl. Bacteriol. *32*: 118–121.

Proom, H. and A.J. Woiwod. 1951. Amine production in the genus *Proteus*. J. Gen. Microbiol. *5*: 930–938.

Provasoli, L. 1964. Growing marine seaweeds. Proceedings of the 4th International Seaweed Symposium, Biarritz, France. Pergamon Press, New York,. Vol. *4*: 9–17.

Provenza, J.M., S.A. Klotz and R.L. Penn. 1986. Isolation of *Francisella tularensis* from blood. J. Clin. Microbiol. *24*: 453–455.

Pruneda, R.C. and J.J. Farmer, III. 1977. Bacteriophage typing of *Shigella sonnei*. J. Clin. Microbiol. *5*: 66–74.

Prunier, J.P., J. Luisetti and L. Gardan. 1970. Études sur les bactérioses des arbres fruitiers. II. Caractérisation d'un *Pseudomonas* non-fluorescent agent d'une bactériose nouvelle du pêcher. Ann. Phytopathol. *2*: 181–197.

Pruvost, O., A. Couteau, X. Perrier and J. Luisetti. 1998. Phenotypic diversity of *Xanthomonas* sp. *mangiferaeindicae*. J. Appl. Microbiol. *84*: 115–124.

Pruvost, O., J.S. Hartung, E.L. Civerolo, C. Dubois and X. Perrier. 1992. Plasmid DNA fingerprints distinguish pathotypes of *Xanthomonas campestris* pv. *citri*, the causal agent of citrus bacterial canker disease. Phytopathology *82*: 485–490.

Pryamukhina, N.S. and N.A. Khomenko. 1988. Suggestion to supplement *Shigella flexneri* classification scheme with the subserovar *Shigella flexneri* 4c: phenotypic characteristics of strains. J. Clin. Microbiol. *26*: 1147–1149.

Psallidas, P.G. 1993. *Pseudomonas syringae* pv. *avellanae* pathovar nov. the bacterium causing canker disease on *Corylus avellana*. Plant Pathol. (Oxf.) *42*: 358–363.

Psallidas, P.G. and C.G. Panagopoulos. 1975. A new bacteriosis of almond caused by *Pseudomonas amygdali* sp. nov. Ann. Inst. Phytopathol. Benaki (N.S.). *11*: 94–108.

Pshenin, L.N. 1964. *Azotobacter miscellum* nov. sp. an inhabitant of the Black Sea. Microbiology (Engl. Transl.). *33*: 615–620.

Puchkova, N.N., J.F. Imhoff and V.M. Gorlenko. 2000. *Thiocapsa litoralis* sp. nov., a new purple sulfur bacterium from microbial mats from the White Sea. Int. J. Syst. Evol. Microbiol. *50*: 1441–1447.

Puente, M.E. and Y. Bashan. 1994. The desert epiphyte *Tillandsia recurvata* harbors the nitrogen fixing bacterium *Pseudomonas stutzeri*. Can. J. Bot.-Rev. Can. Bot. *72*: 406–408.

Pujalte, M.J. and E. Garay. 1986. Proposal of *Vibrio mediterranei* sp. nov., a new marine member of the genus *Vibrio*. Int. J. Syst. Bacteriol. *36*: 278–281.

Pujalte, M.J., M. Ortigosa, M.C. Urdaci, E. Garay and P.A.D. Grimont. 1993. *Vibrio mytili* sp. nov., from mussels. Int. J. Syst. Bacteriol. *43*: 358–362.

Pujol, C.J. and C.I. Kado. 2000. Genetic and biochemical characterization of the pathway in *Pantoea citrea* leading to pink disease of pineapple. J. Bacteriol. *182*: 2230–2237.

Pulverer, G. and H.L. Ko. 1970. *Actinobacillus actinomycetemcomitans*: fermentative capabilities of 140 strains. Appl. Microbiol. *20*: 693–695.

Pulverer, G. and H.L. Ko. 1972. Serological studies on *Actinobacillus actinomycetem-comitans*. Appl. Microbiol. *23*: 207–210.

Punsalang, A., Jr., R. Edinger and F.S. Nolte. 1987. Identification and characterization of *Yersinia intermedia* isolated from human feces. J. Clin. Microbiol. *25*: 859–862.

Pupo, G.M., D.K. Karaolis, R. Lan and P.R. Reeves. 1997. Evolutionary relationships among pathogenic and nonpathogenic *Escherichia coli* strains inferred from multilocus enzyme electrophoresis and *mdh* sequence studies. Infect. Immun. *65*: 2685–2692.

Pupo, G.M., R. Lan and P.R. Reeves. 2000. Multiple independent origins of *Shigella* clones of *Escherichia coli* and convergent evolution of many of their characteristics. Proc. Natl. Acad. Sci. U.S.A. *97*: 10567–10572.

Purcell, A.H. and S. Saunders. 1995. Harvested grape clusters as inoculum for Pierce's disease. Plant Dis. *79*: 190–192.

Py, B., M. Chippaux and F. Barras. 1993. Mutagenesis of cellulase EGZ for studying the general protein secretory pathway in *Erwinia chrysanthemi*. Mol. Microbiol. *7*: 785–793.

Qadri, F., S. Haq and I. Ciznár. 1989. Hemagglutinating properties of *Shigella dysenteriae* type 1 and other *Shigella* species. Infect. Immun. *57*: 2909–2911.

Qhobela, M. and L.E. Claflin. 1988. Characterization of *Xanthomonas campestris* pv. *pennamericanum* new pathovar, causal agent of bacterial leaf streak of pearl millet. Int. J. Syst. Bacteriol. *38*: 362–366.

Qhobela, M., L.E. Claflin and D.C. Nowell. 1990. Evidence that *Xanthomonas campestris* pv. *zeae* can be distinguished from other pathovars capable of infecting maize by restriction fragment length polymorphism of genomic DNA. Can. J. Plant Pathol. *12*: 183–186.

Quadt-Hallmann, A. and J.W. Kloepper. 1996. Immunological detection and localization of the cotton endophyte *Enterobacter asburiae* JM22 in different plant species. Can. J. Microbiol. *42*: 1144–1154.

Quan, S.F., W. Knapp, M.I. Goldenberg, B.W. Hudson, W.D. Lawton, T.H. Chen and L. Kartman. 1965. Isolation of a strain of *Pasteurella pseudotuberculosis* from Alaska identified as *Pasteurella pestis*; an immunofluorescent false positive. Am. J. Trop. Med. Hyg. *14*: 424–432.

Quentin, R., I. Dubarry, C. Martin, B. Cattier and A. Goudeau. 1992. Evaluation of four commercial methods for identification and biotyping of genital and neonatal strains of *Haemophilus* species. Eur. J. Clin. Microbiol. Infect. Dis. *11*: 546–549.

Quentin, R., A. Goudeau, R.J. Wallace, Jr., A.L. Smith, R.K. Selander and J.M. Musser. 1990. Urogenital, maternal and neonatal isolates of *Haemophilus influenzae*: identification of unusually virulent serologically non-typable clone families and evidence for a new *Haemophilus* species. J. Gen. Microbiol. *136*: 1203–1209.

Quentin, R., C. Martin, J.M. Musser, N. Pasquier-Picard and A. Goudeau. 1993. Genetic characterization of a cryptic genospecies of *Haemophilus* causing urogenital and neonatal infections. J. Clin. Microbiol. *31*: 1111–1116.

Quentin, R., J.M. Musser, M. Mellouet, P.Y. Sizaret, R.K. Selander and A. Goudeau. 1989. Typing of urogenital, maternal, and neonatal isolates of *Haemophilus influenzae* and *Haemophilus parainfluenzae* in correlation with clinical source of isolation and evidence for a genital specificity of *H. influenzae* biotype IV. J. Clin. Microbiol. *27*: 2286–2294.

Quentin, R., R. Ruimy, A. Rosenau, J.M. Musser and R. Christen. 1996. Genetic identification of cryptic genospecies of *Haemophilus* causing urogenital and neonatal infections by PCR using specific primers targeting genes coding for 16S rRNA. J. Clin. Microbiol. *34*: 1380–1385.

Quesada, E., M.J. Valderrama, V. Bejar, A. Ventosa, M.C. Gutierrez, F. Ruiz-Berraquero and A. Ramos-Cormenzana. 1990. *Volcaniella eurihalina* gen. nov., sp. nov., a moderately halophilic nonmotile Gram-negative rod. Int. J. Syst. Bacteriol. *40*: 261–267.

Quesada, E., A. Ventosa, F. Rodriguez-Valera and A. Ramos-Cormenzana. 1982. Types and properties of some bacteria isolated from hypersaline soils. J. Appl. Bacteriol. *53*: 155–162.

Quesada, E., A. Ventosa, F. Ruizberraquero and A. Ramoscormenzana. 1984. *Deleya halophila*, a new species of moderately halophilic bacteria. Int. J. Syst. Bacteriol. *34*: 287–292.

Quintela, J.C., M. Caparrós and M.A. de Pedro. 1995. Variability of peptidoglycan structural parameters in gram-negative bacteria. FEMS Microbiol. Lett. *125*: 95–100.

Quintiliani, R.J. and R. Courvalin. 1995. Mechanisms of resistance to antimicrobial agents. *In* Murray, Baron, Pfaller, Tenover and Yolken (Editors), Manual of Clinical Microbiology, 6th Ed., ASM Press, Washington, D.C. pp. 1308–1326.

Quirie, M., W. Donachie and N.J. Gilmour. 1986. Serotypes of *Pasteurella haemolytica* from cattle. Vet. Rec. *119*: 93–94.

Rabenhorst, L. 1865. Flora Europaea Algarum aguae dulcis et submarinae. Sectio II, Algas physochromaceas complectens, Leipzig. 1–319.

Rabsch, W. and G. Winkelmann. 1991. The specificity of bacterial siderophore receptors probed by bioassays. Biol. Met. *4*: 244–250.

Radek, R. 2000. Light and electron microscopic study of a *Rickettsiella* species from the cockroach *Blatta orientalis*. J. Invertebr. Pathol. *76*: 249–256.

Rademaker, J.L., B. Hoste, F.J. Louws, K. Kersters, J. Swings, L. Vauterin, P. Vauterin and F.J. de Bruijn. 2000. Comparison of AFLP and rep-PCR genomic fingerprinting with DNA–DNA homology studies: *Xanthomonas* as a model system. Int. J. Syst. Evol. Microbiol. *50*: 665–677.

Raetz, C.R. 1996. Bacterial lipopolysaccharides: remarkable family of bioreactive macroamphiphiles. *In* Neidhardt, Curtiss, Ingraham, Lin, Low, Magasanik, Reznikoff, Riley, Schaechter and Umbarger (Editors), *Escherichia coli* and *Salmonella*: Cellular and Molecular Biology, 2nd Ed., ASM Press, Washington, D.C. pp. 1035–1063.

Rafaeli-Eshkol, D. 1968. Studies on halotolerance in a moderately halophilic bacterium: Effect of growth conditions on salt resistance of the respiratory system. Biochem. J. *109*: 679–685.

Rafaeli-Eshkol, D. and Y. Avi-Dor. 1968. Studies on halotolerance in a moderately halophilic bacterium. Effect of betaine on salt resistance of the respiratory system. Biochem. J. *109*: 687–691.

Raguénès, G., R. Christen, J. Guezennec, P. Pignet and G. Barbier. 1997a. *Vibrio diabolicus* sp. nov., a new polysaccharide-secreting organism isolated from a deep-sea hydrothermal vent polychaete annelid, *Alvinella pompejana*. Int. J. Syst. Bacteriol. *47*: 989–995.

Raguénès, G.H., A. Peres, R. Ruimy, P. Pignet, R. Christen, M. Loaec, H. Rougeaux, G. Barbier and J.G. Guezennec. 1997b. *Alteromonas infernus* sp. nov., a new polysaccharide-producing bacterium isolated from a deep-sea hydrothermal vent. J. Appl. Microbiol. *82*: 422–430.

Raguénès, G., P. Pignet, G. Gauthier, A. Peres, R. Christen, H. Rougeaux, G. Barbier and J. Guezennec. 1996. Description of a new polymer-secreting bacterium from a deep-sea hydrothermal vent, *Alteromonas macleodii* subsp. *fijiensis*, and preliminary characterization of the polymer. Appl. Environ. Microbiol. *62*: 67–73.

Rahaley, R.S. and W.E. White. 1977. *Histophilus ovis* infection in sheep in western Victoria. Aust. Vet. J. *53*: 124–127.

Rahmati-Bahram, A., J.T. Magee and S.K. Jackson. 1995. Growth temperature-dependent variation of cell envelope lipids and antibiotic susceptibility in *Stenotrophomonas* (*Xanthomonas*) *maltophilia*. J. Antimicrob. Chemother. *36*: 317–326.

Rahmati-Bahram, A., J.T. Magee and S.K. Jackson. 1996. Temperature-dependent aminoglycoside resistance in *Stenotrophomonas* (*Xanthomonas*) *maltophilia*; alterations in protein and lipopolysaccharide with growth temperature. J. Antimicrob. Chemother. *37*: 665–676.

Rahn, O. 1937. New principles for the classification of bacteria. Zentralbl. Bakteriol. Parasitenkd. Infektionskr. Hyg. Abt. II. *96*: 273–286.

Raina, R., U.K. Bageshwar and H.K. Das. 1993. The ORF encoding a putative ferredoxin-like protein downstream of the *vnfH* gene in *Azotobacter vinelandii* is involved in the vanadium-dependent alternative pathway of nitrogen fixation. Mol. Gen. Genet. *236*: 459–462.

Rainey, F.A., E. Lang and E. Stackebrandt. 1994a. The phylogenetic structure of the genus *Acinetobacter*. FEMS Microbiol. Lett. *124*: 349–353.

Rainey, F.A., R.-U. Ehlers and E. Stackebrandt. 1995. Inability of the polyphasic approach to systematics to determine the relatedness of the genera *Xenorhabdus* and *Photorhabdus*. Int. J. Syst. Bacteriol. *45*: 379–381.

Rainey, P.B. and M.J. Bailey. 1996. Physical and genetic map of the *Pseudomonas fluorescens* SBW25 chromosome. Mol. Microbiol. *19*: 521–533.

Rainey, P.B., C.L. Brodey and K. Johnstone. 1993. Identification of a gene-cluster encoding 3 high-molecular weight proteins, which is required for synthesis of tolaasin by the mushroom pathogen *Pseudomonas tolaasii*. Mol. Microbiol. *8*: 643–652.

Rainey, P.B., I.P. Thompson and N.J. Palleroni. 1994b. Genome and fatty-acid analysis of *Pseudomonas stutzeri*. Int. J. Syst. Bacteriol. *44*: 54–61.

Raj, H.D. 1977. *Leucothrix*. Crit. Rev. Microbiol. *5*: 270–304.

Rake, G. 1948. The antigenic relationships of *Donovania granulomatis* (Anderson) and the significance of this organism in granuloma inguinale. Am. J. Syph. Gonorrhea Vener. Dis. *32*: 150–158.

Rakovsky, J. and E. Aldová. 1965. Isolation of strains of the new Enterobacteriaceae group "Bartholomew" in Cuba. J. Hyg. Epidemiol. Microbiol. Immunol. *9*: 112.

Ralston-Barrett, E., N.J. Palleroni and M. Doudorof. 1976. Phenotypic characterization and deoxyribonucleic acid homologies of the "*Pseudomonas alcaligenes*" group. Int. J. Syst. Bacteriol. *26*: 421–426.

Ramamurthy, T., A. Pal, S.C. Pal and G.B. Nair. 1992. Taxonomical implications of the emergence of high frequency of occurrence of 2,4-diamino-6,7-diisoprophylpteridine-resistant strains of *Vibrio cholerae* from clinical cases of cholera in Calcutta, India. J. Clin. Microbiol. *30*: 742–743.

Ramia, S., E. Neter and D.J. Brenner. 1982. Production of enterobacterial common antigen as an aid to classification of newly identified species of the families *Enterobacteriaceae* and *Vibrionaceae*. Int. J. Syst. Bacteriol. *32*: 395–398.

Ramirez, M.E., L. Fucikovsky, F. Garcia-Jimenez, R. Quintero and E. Galindo. 1988. Xanthan gum production by altered pathogenicity variants of *Xanthomonas campestris*. Appl. Microbiol. Biotechnol. *29*: 5–10.

Ramos, F., G. Blanco, J.C. Gutierrez, F. Luque and M. Tortolero. 1993. Identification of an operon involved in the assimilatory nitrate reducing system of *Azotobacter vinelandii*. Mol. Microbiol. *8*: 1145–1153.

Ramos, J.L., E. Duque, M.J. Huertas and A. Haïdour. 1995. Isolation and expansion of the catabolic potential of a *Pseudomonas putida* strain able to grow in the presence of high concentrations of aromatic hydrocarbons. J. Bacteriol. *177*: 3911–3916.

Ramos, J.L., E. Duque, J.J. Rodriguez-Herva, P. Godoy, A. Haïdour, F. Reyes and A. Fernandez-Barrero. 1997. Mechanisms for solvent tolerance in bacteria. J. Biol. Chem. *272*: 3887–3890.

Ramos, J. and R.L. Robson. 1987. Cloning of the gene for phosphoenolpyruvate carboxylase from *Azotobacter chroococcum*, an enzyme important in aerobic nitrogen fixation. Mol. Gen. Genet. *208*: 481–484.

Randhawa, P.S. and N.W. Schaad. 1984. Selective isolation of *Xanthomonas campestris* pv. *campestris* from crucifer seeds. Phytopathology *74*: 268–272.

Rangaraj, P., C. Ruttiman-Johnson, V.K. Shah and P. Ludden. 2000. Biosynthesis of the iron–molybdenum and iron–vanadium cofactors of the *nif* and *vnf*-encoded nitrogenases. *In* Triplett (Editor), Prokaryotic Nitrogen Fixation: A Model System fof the Analysis of a Biological Process, Horizon Scientific Press, Wymondham. pp. 55–80.

Rangaswami, G. and K.S.S. Easwaran. 1962. A bacterial leafspot disease of bhendi or okra. Andhra Agr. J. *9*: 1–2.

Rani, N.L. and D. Lalithakumari. 1994. Degradation of methyl parathion by *Pseudomonas putida*. Can. J. Microbiol. *40*: 1000–1006.

Ranson, S.E. and R.J. Huebner. 1951. Studies on the resistance of *Coxiella burnetii* to physical and chemical agents. Am. J. Hyg. *53*: 110–119.

Rao, C.S. and S. Devadath. 1977. Variation in the aggressiveness of colony types of *Xanthomonas translucens* f. sp. oryzicola. Indian Phytopathol. *30*: 233–236.

Rao, K.R., J. Shah, K.R. Rajashekaraiah, A.R. Patel, D.B. Miskew and P.S. Fennewald. 1981. *Edwardsiella tarda* osteomyelitis in a patient with SC hemoglobinopathy. South. Med. J. *74*: 288–292.

Rao, Y.P. and S.K. Mohan. 1970. A new bacterial leaf stripe disease of arecanut (*Areca catechu*) in Mysore State. Indian Phytopathol. *23*: 702–704.

Raoult, D., M. Drancourt and G. Vestris. 1990. Bactericidal effect of doxycycline associated with lysosomotropic agents on *Coxiella burnetii* in P388D1 cells. Antimicrob. Agents Chemother. *34*: 1512–1514.

Raoult, D., H. Tissot-Dupont, C. Foucault, J. Gouvernet, P.E. Fournier, E. Bernit, A. Stein, M. Nesri, J.R. Harle and P.J. Weiller. 2000. Q fever 1985–1998. Clinical and epidemiologic features of 1,383 infections. Medicine (Baltimore). *79*: 109–123.

Ratajczak, A., W. Geissdoerfer and W. Hillen. 1998. Alkane hydroxylase from *Acinetobacter* sp. strain ADP1 is encoded by alkM and belongs to a new family of bacterial integral-membrane hydrocarbon hydroxylases. Appl. Environ. Microbiol. *64*: 1175–1179.

Ratiner, I.A. 1967. Mutation of *E. coli* with regard to the H-antigen. Isolation of H-antigen mutants from test H-strains of *Escherichia* cultures. Zh. Mikrobiol. Epidemiol. Immunobiol. *44*: 23–29.

Ratiner, Y.A. 1982. Phase variation of the H antigen in *Escherichia coli* strain Bi7327-41, the standard strain for *Escherichia coli* flagellar antigen 3. FEMS Microbiol. Lett. *15*: 33–36.

Ratiner, Y.A. 1999. Temperature-dependent flagellar antigen phase variation in *Escherichia coli*. Res. Microbiol. *150*: 457–463.

Ratnam, S., R.W. Butler, S. March, S. Parsons, P. Clarke, A. Bell and K.

Hogan. 1979. *Enterobacter hafniae*-associated gastroenteritis-Newfoundland. Can. Dis. Wkly. Rep. *5*: 231–232.

Ratto, P., D.O. Sordelli, E. Abeleira, M. Torrero and M. Catalano. 1995. Molecular typing of *Acinetobacter baumannii–Acinetobacter calcoaceticus* complex isolates from endemic and epidemic nosocomial infections. Epidemiol. Infect. *114*: 123–132.

Rauss, K., T. Kontrohr, A. Vertenyi and L. Szendrei. 1970. Serological and chemical studies of *Sh. sonnei*, *Pseudomonas shigelloides* and C27 strains. Acta Microbiol. Acad. Sci. Hung. *17*: 157–166.

Rautelin, H., M.L. Hänninen, A. Sivonen, U. Turunen and V. Valtonen. 1995a. Chronic diarrhea due to a single strain of *Aeromonas caviae*. Eur. J. Clin. Microbiol. Infect. Dis. *14*: 51–53.

Rautelin, H., A. Sivonen, A. Kuikka, O.-V. Renkonen, V. Valtonen and T.U. Kosunen. 1995b. Enteric *Plesiomonas shigelloides* infections in Finnish patients. Scand. J. Infect. Dis. *27*: 495–498.

Ravel, J., I.T. Knight, C.E. Monahan, R.T. Hill and R.R. Colwell. 1995. Temperature-induced recovery of *Vibrio cholerae* from the viable but nonculturable state: Growth or resuscitation? Microbiology *141*: 377–383.

Ravin, A.W. 1963. Experimental approaches to the study of bacterial phylogeny. Am. Natur. *97*: 307–318.

Rawling, E.G., N.L. Martin and R.E.W. Hancock. 1995. Epitope mapping of the *Pseudomonas aeruginosa* major outer membrane porin protein OprF. Infect. Immun. *63*: 38–42.

Rayan, G.M., D.J. Flournoy and S.L. Cahill. 1987. Aerobic mouth flora of the rhesus monkey. J. Hand. Surg. [Am]. *12*: 299–301.

Rayman, M.K. and R.A. MacLeod. 1975. Interaction of Mg^{2+} with peptidoglycan and its relation to the prevention of lysis of a marine pseudomonad. J. Bacteriol. *122*: 650–659.

Raymond, J.C. and W.R. Sistrom. 1969. *Ectothiorhodospira halophila*: a new species ofthe genus *Ectothiorhodospira*. Arch. Mikrobiol. *69*: 121–126.

Read, T.D., S.W. Satola, J.A. Opdyke and M.M. Farley. 1998. Copy number of pilus gene clusters in *Haemophilus influenzae* and variation in the *hifE* pilin gene. Infect. Immun. *66*: 1622–1631.

Reams, A.B. and E.L. Neidle. 2003. Genome plasticity in *Acinetobacter*: new degradative capabilities acquired by the spontaneous amplification of large chromosomal scgments. Molec. Microbiol. *47*: 1291–1304.

Reary, B.W. and S.A. Klotz. 1988. Enhancing recovery of *Francisella tularensis* from blood. Diagn. Microbiol. Infect. Dis. *11*: 117–119.

Reasoner, D.J. and E.E. Geldreich. 1985. A new medium for the enumeration and subculture of bacteria from potable water. Appl. Environ. Microbiol. *49*: 1–7.

Rebers, P.A., A.E. Jensen and G.A. Laird. 1988. Expression of pili and capsule by the avian strain P-1059 of *Pasteurella multocida*. Avian Dis. *32*: 313–318.

Reboli, A.C., E.D. Houston, J.S. Monteforte, C.A. Wood and R.J. Hamill. 1994. Discrimination of epidemic and sporadic isolates of *Acinetobacter baumannii* by repetitive element PCR-mediated DNA fingerprinting. J. Clin. Microbiol. *32*: 2635–2640.

Rechtman, D.J. and J.P. Nadler. 1991. Abdominal abscess due to *Cardiobacterium hominis* and *Clostridium bifermentans*. Rev. Infect. Dis. *13*: 418–419.

Reddy, C.A. and M.P. Bryant. 1977. Deoxyribonucleic acid base composition of certain species of the genus *Bacteroides*. Can. J. Microbiol. *23*: 1252–1256.

Reddy, C.S. and J. Godkin. 1923. A bacterial disease of brome-grass. Phytopathology *13*: 75–86.

Reddy, C.S., J. Godkin and A.G. Johnson. 1924. Bacterial blight of rye. J. Agr. Res. *28*: 1039–1040.

Reddy, P.G., R. Allon, M. Mevarech, S. Mendelovitz, Y. Sato and D.L. Gutnick. 1989. Cloning and expression in *Escherichia coli* of an esterase-coding gene from the oil-degrading bacterium *Acinetobacter calcoaceticus* RAG-1. Gene *76*: 145–152.

Redfield, R.J. 1991. *sxy-1*, a *Haemophilus influenzae* mutation causing greatly enhanced spontaneous competence. J. Bacteriol. *173*: 5612–5618.

Reeburgh, W.S., S.C. Whalen and M.L. Alpern. 1993. The role of methylotrophy in the global methane budget. *In* Murrell and Kelly (Editors), Microbial growth on C_1 compounds, Intercept Press Ltd., Andover. pp. 1–14.

Reed, W.M. and P.R. Dugan. 1978. Distribution of *Methylomonas methanica* and *Methylosinus trichosporium* in Cleveland Harbor as determined by an indirect fluorescent antibody-membrane filter technique. Appl. Environ. Microbiol. *35*: 422–430.

Reeve, E.C.R. and J.A. Braithwaite. 1973. Lac$^+$ plasmids are responsible for the strong lactose-positive phenotype found in many strains of *Klebsiella* species. Genet. Res. *22*: 329–333.

Reeves, M.W., G.M. Evins, A.A. Heiba, B.D. Plikaytis and J.J. Farmer, III. 1989a. Clonal nature of *Salmonella typhi* and its genetic relatedness to other salmonellae as shown by multilocus enzyme electrophoresis, and proposal of *Salmonella bongori* comb. nov. J. Clin. Microbiol. *27*: 313–320.

Reeves, M.W., G.M. Evins, A.A. Heiba, B.D. Plikaytis and J.J. Farmer, III. 1989b. *In* Validation of the publication of new names and new combinations previously effectively published outside the IJSB. List no. 30. Int. J. Syst. Bacteriol. *39*: 371.

Reeves, P.J., D. Whitcombe, S. Wharam, M. Gibson, G. Allison, N. Bunce, R. Barallon, P. Douglas, V. Mulholland, S. Stevens, D. Walker and G.P.C. Salmond. 1993. Molecular cloning and characterization of 13 *out* genes from *Erwinia carotovora* subspecies *carotovora*: genes encoding members of a general secretion pathway (GSP) widespread in gram-negative bacteria. Mol. Microbiol. *8*: 443–456.

Rehm, B.H., H. Ertesvåg and S. Valla. 1996. New *Azotobacter vinelandii* mannuronan C5-epimerase gene (*algG*) is part of an *alg* gene cluster physically organized in a manner similar to that in *Pseudomonas aeruginosa*. J. Bacteriol. *178*: 5884–5889.

Rehm, B.H. and R.E.W. Hancock. 1996. Membrane topology of the outer membrane protein OprH from *Pseudomonas aeruginosa*: PCR-mediated site-directed insertion and deletion mutagenesis. J. Bacteriol. *178*: 3346–3349.

Reich, K.A., T. Biegel and G.K. Schoolnik. 1997. The light organ symbiont *Vibrio fischeri* possesses two distinct secreted ADP-ribosyltransferases. J. Bacteriol. *179*: 1591–1597.

Reichelt, J.L. and P. Baumann. 1973a. Change of the name *Alteromonas marinopraesens* (ZoBell and Upham) Baumann et al. to *Alteromonas haloplanktis* (ZoBell and Upham) comb. nov. and assignment of strain ATCC 23821 (*Pseudomonas enalia*) and strain c-A1 of De Voe and Oginsky to this species. Int. J. Syst. Bacteriol. *23*: 438–441.

Reichelt, J.L. and P. Baumann. 1973b. Taxonomy of the marine, luminous bacteria. Arch. Mikrobiol. *94*: 283–330.

Reichelt, J.L., P. Baumann and L. Baumann. 1976. Study of genetic relationships among marine species of the genera *Beneckea* and *Photobacterium* by means of in vitro DNA/DNA hybridization. Arch. Microbiol. *110*: 101–120.

Reichelt, J.L., P. Baumann and L. Baumann. 1979. Validation of the publication of new names and combinations previously effectively published outside the IJSB, List No. 2. Int. J. Syst. Bacteriol. *29*: 79–80.

Reichenbach, H., W. Kohl and H. Achenbach. 1981. The flexirubin-type pigments, chemosystematically useful compounds. *In* Reichenbach and Weeks (Editors), The *Flavobacterium-Cytophaga* Group, Verlag Chemie, Weinheim, F.R.G. 101–108.

Reichenbach, H., W. Ludwig and E. Stackebrandt. 1986. Lack of relationship between gliding cyanobacteria and filamentous gliding heterotrophic eubacteria: comparison of 16S ribosomal RNA catalogues of *Spirulina*, *Saprospira*, *Vitreoscilla*, *Leucothrix*, and *Herpetosiphon*. Arch. Microbiol. *145*: 391–395.

Reid, D.H. 1938. Grease-spot of passion-fruit. N.Z. J. Sci. Technol. *A20*: 260–265.

Reid, G.A. and E.H. Gordon. 1999. Phylogeny of marine and freshwater *Shewanella*: reclassification of *Shewanella putrefaciens* NCIMB 400 as *Shewanella frigidimarina*. Int. J. Syst. Bacteriol. *49*: 189–191.

Reid, J.D., S.D. Stoufer and D.M. Ogrydziak. 1982. Efficient transformation of *Serratia marcescens* with pBR322 plasmid DNA. Gene *17*: 107–112.

Reid, R.T. and A. Butler. 1991. Investigation of the mechanism of iron acquisition by the marine bacterium *Alteromonas luteoviolceus*: characterization of siderophore production. Limnol. Oceanogr. *36*: 1783–1792.

Reid, R.T., D.H. Live, D.J. Faulkner and A. Butler. 1993. A siderophore from a marine bacterium with an exceptional ferric ion affinity constant. Nature *366*: 455–458.

Reid, S.D., R.K. Selander and T.S. Whittam. 1999. Sequence diversity of flagellin (*fliC*) alleles in pathogenic *Escherichia coli*. J. Bacteriol. *181*: 153–160.

Reilly, T.J., G.S. Baron, F.E. Nano and K. M.S.. 1996. Characterization and sequencing of a respiratory burst-inhibiting acid phosphatase from *Francisella tularensis*. J. Biol. Chem. *271*: 10973–10983.

Reilly, W.R. and F.T. Carter. 1997. Surveillance of *E. coli* O157 in Scotland (No. V128/I). VTEC '97: 3rd International Symposium and Workshop on Shiga Toxin (Verocytotoxin)-Producing *Escherichia coli* Infections, p. 16.

Reimmann, C., M. Beyeler, A. Latifi, H. Winteler, M. Foglino, A. Lazdunski and D. Haas. 1997. The global activator GacA of *Pseudomonas aeruginosa* PAO positively controls the production of the autoinducer N-butyryl-homoserine lactone and the formation of the virulence factors pyocyanin, cyanide, and lipase. Mol. Microbiol. *24*: 309–319.

Reina, J. and A. Lopez. 1996a. Clinical and microbiological characteristics of *Rahnella aquatilis* strains isolated from children. J. Infect. *33*: 135–137.

Reina, J. and A. Lopez. 1996b. Gastroenteritis caused by *Aeromonas trota* in a child. J. Clin. Pathol. (Lond.). *49*: 173–175.

Reinhardt, C., R. Steiger and H. Hecker. 1972. Ultrastructural study of the midgut mycetome-bacteroids of the tsetse flies *Glossina morsitans*, *G. fuscipes* and *G. brevipalpis* (Diptera, Brachycera). Acta Trop. *29*: 280–288.

Reinhardt, J.F., S. Fowlston, J. Jones and W.L. George. 1985. Comparative *in vitro* activities of selected antimicrobial agents against *Edwardsiella tarda*. Antimicrob. Agents Chemother. *27*: 966–967.

Reinhardt, J.F. and W.L. George. 1985. Comparative in vitro activities of selected antimicrobial agents against *Aeromonas* species and *Plesiomonas shigelloides*. Antimicrob. Agents Chemother. *27*: 643–645.

Reissbrodt, R. and W. Rabsch. 1988. Further differentiation of *Enterobacteriaceae* by means of siderophore pattern analysis. Zntbl. Fur Bakteriol. Mikrobiol. Hyg. 1 Abt. Orig. A Med. Mikrobiol. Infktrankh. Virol. Parasitol. *268*: 306–317.

Remsen, C.C. 1978. Comparative subcellular architecture of photosynthetic bacteria. *In* Clayton and Sistrom (Editors), The Photosynthetic Bacteria, Plenum Press, New York and London. pp. 31–60.

Ren, X.Z. and Z.D. Fang. 1981. *Xanthomonas zingibericola* n. sp., the causal organism of bacterial leaf blight of ginger. Acta Phylopathol. Sinica. *11*: 37–40.

Renaud, F., J. Freney, J.M. Boeufgras, D. Monget, A. Sedaillan and J. Fleurette. 1988. Carbon substrate assimilation patterns of clinical and environmental strains of *Aeromonas hydrophila*, *Aeromonas sobria* and *Aeromonas caviae* observed with a micromethod. Zentbl. Bakteriol. Mikrobiol. Hyg. Ser. A *269*: 323–330.

Renders, N., U. Römling, H. Verbrugh and A. Van Belkum. 1996. Comparative typing of *Pseudomonas aeruginosa* by random amplification of polymorphic DNA or pulsed-field gel electrophoresis of DNA macrorestriction fragments. J. Clin. Microbiol. *34*: 3190–3195.

Reniero, D., E. Galli and P. Barbieri. 1995. Cloning and comparison of mercury-resistance and organomercurial-resistance determinants from a *Pseudomonas stutzeri* plasmid. Gene *166*: 77–82.

Replogle, M.L., D.W. Fleming and P.R. Cieslak. 2000. Emergence of antimicrobial-resistant shigellosis in Oregon. Clin. Infect. Dis. *30*: 515–519.

Resibois, A., M. Colet, M. Faelen, E. Schoonejans and A. Toussaint. 1984. Phi-EC2, a new generalized transducing phage of *Erwinia chrysanthemi*. Virology *137*: 102–112.

Restrepo, S., M. Duque, J. Tohme and V. Verdier. 1999. AFLP fingerprinting: an efficient technique for detecting genetic variation of

Xanthomonas axonopodis pv. *manihotis*. Microbiology (Read.) *145*: 107–114.

Restrepo, S. and V. Verdier. 1997. Geographical differentiation of the population of *Xanthomonas axonopodis* pv. *manihotis* in Colombia. Appl. Environ. Microbiol. *63*: 4427–4434.

Retailliau, H.F., A.W. Hightower, R.E. Dixon and J.R. Allen. 1979. *Acinetobacter calcoaceticus*: a nosocomial pathogen with an unusual seasonal pattern. J. Infect. Dis. *139*: 371–375.

Rettger, L.F. 1909. Further studies on fatal septicaemia in young chickens or "white diarrhoea". J. Med. Res. *21*: 115–123.

Reusch, R.N. and H.L. Sadoff. 1983a. D-(-)-poly-beta-hydroxybutyrate in membranes of genetically competent bacteria. J. Bacteriol. *156*: 778–788.

Reusch, R.N. and H.L. Sadoff. 1983b. Novel lipid components of the *Azotobacter vinelandii* cyst membrane. Nature (Lond.) *302*: 268–270.

Reverchon, S., W. Nasser and J. Robert-Baudouy. 1994. *pecS*: a locus controlling pectinase, cellulase and blue pigment production in *Erwinia chrysanthemi*. Mol. Microbiol. *11*: 1127–1139.

Reverchon, S. and J. Robert-Baudouy. 1987. Molecular cloning of an *Erwinia chrysanthemi* oligogalacturonate lyase gene involved in pectin degradation. Gene *55*: 125–133.

Rey, L. and R.J. Maier. 1997. Cytochrome *c* terminal oxidase pathways of *Azobacter vinelandii*: Analysis of cytochrome c_4 and c_5 mutants and up-regulation of cytochrome *c*-dependent pathways with N_2 fixation. J. Bacteriol. *179*: 7191–7196.

Reyn, A., A. Birch-Andersen and R.G.E. Murray. 1971. The fine structure of *Cardiobacterium hominis*. Acta Pathol. Scand. Sect. B Microbiol. Immunol. *79*: 51–60.

Rhim, S.L., B. Völksch, L. Gardan, J.P. Paulin, C. Langlotz, W.S. Kim and K. Geider. 1999. *Erwinia pyrifoliae*, an *Erwinia* species different from *Erwinia amylovora*, causes a necrotic disease of Asian pear trees. Plant Pathol. (Oxf.) *48*: 514–520.

Rhodes, A.N., J.W. Urbance, H. Youga, H. Corlew-Newman, C.A. Reddy, M.J. Klug, J.M. Tiedje and D.C. Fisher. 1998. Identification of bacterial isolates obtained from intestinal contents associated with 12,000-year-old mastodon remains. Appl. Environ. Microbiol. *64*: 651–658.

Rhodes, J.B., D. Schweitzer and J.E. Ogg. 1985. Isolation of non-O1 *Vibrio cholerae* associated with enteric disease of herbivores in western Colorado. J. Clin. Microbiol. *22*: 572–575.

Rhodes, K.R. and R.B. Rimler. 1989. Fowl cholera. *In* Adlam and Rutter (Editors), *Pasteurella* and Pasteurellosis, Academic Press, London. 95–113.

Ribeiro, G.A., G.R. Carter, W. Frederiksen and F. Riet-Correa. 1989. *Pasteurella haemolytica*-like bacterium from a progressive granuloma of cattle in Brazil. J. Clin. Microbiol. *27*: 1401–1402.

Ribeiro, G.A., G.R. Carter, W. Frederiksen and F. Riet-Correa. 1990. *In* Validation of the publication of new names and new combinations previously effectively published outside the IJSB. List No. 32. Int. J. Syst. Bacteriol. *40*: 105–106.

Ribi, E. and C. Shepard. 1955. Morphology of *Bacterium tularense* during its growth cycle in liquid medium as revealed by the electron microscope. Exp. Cell Res. *8*: 474–487.

Rice, E.W., M.J. Allen, D.J. Brenner and S.C. Edberg. 1991. Assay for β-glucuronidase in species of the genus *Escherichia* and its applications for drinking-water analysis. Appl. Environ. Microbiol. *57*: 592–593.

Rice, E.W., M.J. Allen and S.C. Edberg. 1990. Efficacy of β-glucuronidase assay for identification of *Escherichia coli* by the defined-substrate technology. Appl. Environ. Microbiol. *56*: 1203–1205.

Richard, C. 1972. Méthode rapide pour l'étude des réactions de rouge de méthyle et Voges–Proskauer. Ann. Inst. Pasteur (Paris) *122*: 979–986.

Richard, C. 1977. La tétrathionate-réductase (TTR) chez les bacilles à gram négatif: intérêt diagnostique et épidémiologique. Bull. Inst. Pasteur (Paris). *75*: 369–382.

Richard, C. 1979. Enterobactéries inhabituelles. Bull. Inst. Pasteur. *77*: 83–98.

Richard, C. 1982. Bactériologie et epidémiologie des espèces du genre *Klebsiella*. Bull. Inst. Pasteur. *80*: 127–145.

Richard, C. 1984. Genus VI. *Enterobacter*. *In* Kreig and Holt (Editors), Bergey's Manual of Systematic Bacteriology, Vol. 1, The Williams & Wilkins Co., Baltimore. pp. 465–469.

Richard, C. 1989. New *Enterobacteriaceae* found in medical bacteriology: *Moellerella wisconsensis, Koserella trabulsii, Leclercia adecarboxylata, Escherichia fergusonii, Enterobacter asburiae, Rahnella aquatilis*. Ann. Biol. Clin. *47*: 231–236.

Richard, C., B. Joly, J. Sirot, G.H. Stoleru and M. Popoff . 1976. Étude de souches de *Enterobacter* appartenant à un groupe particulier proche de *E. aerogenes*. Ann. Inst. Pasteur (Paris) *127A*: 545–548.

Richard, C., M. Kiredjian and I. Guilvout. 1985. Characteristics of phenotypes of *Alteromonas putrefaciens*. Study of 123 strains. Ann. Biol. Clin. *43*: 732–738.

Richard, C., M. Popoff and G. Prats Pastor. 1974. Étude bactériologique d'infections urinaires intrahospitalières à *Proteus rettgeri* fermentant le lactose. Ann. Biol. *32*: 149–154.

Richardson, L.L. 1996. Horizontal and vertical migration patterns of *Phormidium corallyticum* and *Beggiatoa* spp. associated with black-band disease of corals. Microb. Ecol. *32*: 323–335.

Richardson, M.J. 1990. An annotated list of seed-borne diseases. 4th Ed, Zurich, Switzerland, . International Seed Testing Association..

Ridé, M. 1958. Sur l'étiologie du chancre suintant du peuplier. C. R. Hebd. Séances Acad. Sci. Paris. *246*: 2795–2798.

Ridé, M. and S. Ridé. 1978. *Xanthomonas populi* (Ridé) comb. nov. (syn. *Aplanobacter populi* Ridé), specificité, variabilité et absence de relations avec *Erwinia cancerogena* Ur. Eur. J. For. Path. *8*: 310–333.

Ridé, M. and S. Ridé. 1992. *Xanthomonas populi*, new species (ex Ridé 1958) nom. rev. Int. J. Syst. Bacteriol. *42*: 652–653.

Ridell, J. and H. Korkeala. 1997. Minimum growth temperatures of *Hafnia alvei* and other *Enterobacteriaceae* isolated from refrigerated meat determined with a temperature gradient incubator. Int. J. Food Microbiol. *35*: 287–292.

Ridell, J., A. Siitonen, L. Paulin, O. Lindroos, H. Korkeala and M.J. Albert. 1995. Characterization of *Hafnia alvei* by biochemical tests, random amplified polymorphic DNA PCR, and partial sequencing of 16S rRNA gene. J. Clin. Microbiol. *33*: 2372–2376.

Ridell, J., A. Siitonen, L. Paulin, L. Mattila, H. Korkeala and M.J. Albert. 1994. *Hafnia alvei* in stool specimens from patients with diarrhea and healthy controls. J. Clin. Microbiol. *32*: 2335–2337.

Ried, J.L. and A. Collmer. 1986. Comparison of pectic enzymes produced by *Erwinia chrysanthemi, Erwinia carotovora* subsp. *carotovora*, and *Erwinia carotovora* subsp. *atroseptica*. Appl. Environ. Microbiol. *52*: 305–310.

Riet-Correa, F., M.C. Méndez, A.L. Schild, G.A. Ribeiro and S.M. Almeida. 1992. Bovine focal proliferative fibrogranulomatous panniculitis (Lechiguana) associated with *Pasteurella granulomatis*. Vet. Pathol. *29*: 93–103.

Rietschel, E.T., O. Luderitz and W.A. Volk. 1975. Nature, type of linkage, and absolute configuration of (hydroxy) fatty acids in lipopolysaccharides from *Xanthomonas sinensis* and related strains. J. Bacteriol. *122*: 1180–1188.

Riffard, S., F. Lo Presti, P. Normand, F. Forey, M. Reyrolle, J. Etienne and F. Vandenesch. 1998. Species identification of *Legionella* via intergenic 16S–23S ribosomal spacer PCR analysis. Int. J. Syst. Bacteriol. *48*: 723–730.

Riffkin, M.C., L.F. Wang, A.A. Kortt and D.J. Stewart. 1995. A single amino-acid change between the antigenically different extracellular serine proteases V2 and B2 from *Dichelobacter nodosus*. Gene *167*: 279–283.

Riggio, M.P. and A. Lennon. 1997. Rapid identification of *Actinobacillus actinomycetemcomitans, Haemophilus aphrophilus*, and *Haemophilus paraphrophilus* by restriction enzyme analysis of PCR-amplified 16S rRNA genes. J. Clin. Microbiol. *35*: 1630–1632.

Rihs, J.D., V.L. Yu, J.J. Zuravleff, A. Goetz and R.R. Muder. 1985. Isolation of *Legionella pneumophila* from blood with the BACTEC system: a prospective study yielding positive results. J. Clin. Microbiol. *22*: 422–424.

Riker, A.J., F.R. Jones and M.C. Davis. 1935. Bacterial leaf spot of alfalfa. J. Agr. Res. *51*: 177–182.

Riley, P.A., N. Parasakthi and W.A. Abdullah. 1996. *Plesiomonas shigelloides* bacteremia in a child with leukemia. Clin. Infect. Dis. *23*: 206–207.

Riley, P.S., H.W. Tatum and R.E. Weaver. 1972. *Pseudomonas putrefaciens* isolates from clinical specimens. Appl. Microbiol. *24*: 798–800.

Riley, P.S. and R.E. Weaver. 1974. Observation of nitrate reduction in some non-saccharolytic strains of *Acinetobacter*. Appl. Microbiol. *28*: 1071–1072.

Rimler, R.B. 1990. Comparisons of *Pasteurella multocida* lipopolysaccharides by sodium dodecyl sulfate-polyacrylamide gel-electrophoresis to determine relationship between group-B and group-E hemorrhagic septicemia strains and serologically related group-A strains. J. Clin. Microbiol. *28*: 654–659.

Rimler, R.B. 1994. Presumptive identification of *Pasteurella multocida* serogroup-A, serogroup-D and serogroup-F by capsule depolymerization with mucopolysaccharidases. Vet. Rec. *134*: 191–192.

Rimler, R.B., P.A. Rebers and M. Phillips. 1984. Lipopolysaccharides of the Heddleston serotypes of *Pasteurella multocida*. Am. J. Vet. Res. *45*: 759–763.

Rimler, R.B. and K.R. Rhoades. 1987. Serogroup-F, a new capsule serogroup of *Pasteurella multocida*. J. Clin. Microbiol. *25*: 615–618.

Rimler, R.B. and K.R. Rhoades. 1989. *Pasteurella multocida*. *In* Adlam and Rutter (Editors), *Pasteurella* and Pasteurellosis, Academic Press, London. 37–73.

Rimler, R.B., E.B. Shotts, J. Brown and R.B. Davis. 1976. The effect of atmospheric conditions on the growth of *Haemophilus gallinarum* in a defined medium. J. Gen. Microbiol. *92*: 405–409.

Rimler, R.B., E.B. Shotts, J. Brown and R.B. Davis. 1977. The effect of sodium chloride and NADH on the growth of 6 strains of *Haemophilus* species pathogenic to chickens. J. Gen. Microbiol. *98*: 349–354.

Rina, M., F. Caufrier, M. Markaki, K. Mavromatis, M. Kokkinidis and V. Bouriotis. 1997. Cloning and characterization of the gene encoding PspPI methyltransferase from the Antarctic psychrotroph *Psychrobacter* sp. strain TA137. Predicted interactions with DNA and organization of the variable region. Gene *197*: 353–360.

Rippey, S.R. and V.J. Cabelli. 1979. Membrane filter procedure for enumeration of *Aeromonas hydrophila* in fresh waters. Appl. Environ. Microbiol. *38*: 106–113.

Rippey, S.R. and V.J. Cabelli. 1980. Occurrence of *Aeromonas hydrophila* in limnetic environments: relationship of the organism to trophic state. Microb. Ecol. *6*: 45–54.

Rippey, S.R. and V.J. Cabelli. 1985. Growth characteristics of *Aeromonas hydrophila* in limnetic waters of varying trophic state. Arch. Hydrobiol. *104*: 311–320.

Riquelme, C., G. Hayashida, R. Araya, A. Uchida, M. Satomi and Y. Ishida. 1996. Isolation of a native bacterial strain from the scallop *Argopecten purpuratus* with inhibitory effects against pathogenic vibrios. J. Shellfish Res. *15*: 369–374.

Risberg, A., E.K. Schweda and P.E. Jansson. 1997. Structural studies of the cell-envelope oligosaccharide from the lipopolysaccharide of *Haemophilus influenzae* strain RM.118-28. Eur. J. Biochem. *243*: 701–707.

Rische, H., W. Beer, G. Seltmann, E. Thal and G. Horn. 1973. Die zusammensetzung der lipopolysaccharide von *Yersinia enterocolitica* und *Yersinia pseudotuberculosis* und die empfindlichkeit gegenüber bakteriophagen. Contr. Microbiol. Immunol. *2*: 23–26.

Ristroph, J.D., K.W. Hedlund and R.G. Allen. 1980. Liquid medium for growth of *Legionella pneumophila*. J. Clin. Microbiol. *11*: 19–21.

Ritter, D.B. and R.K. Gerloff. 1966. Deoxyribonucleic acid hybridization among some species of the genus *Pasteurella*. J. Bacteriol. *92*: 1838–1839.

Rivers, T.M. 1922. Influenza-like bacilli. Growth of influenza-like bacilli on media containing only an autoclave-labile substance as an accessory food factor. Johns Hopkins Hosp. Bull. *33*: 429–431.

Robbs, C.F. 1956. Uma nova doenca bacteriana do mamoeiro. Rev. Soc. Brasil. Agron. *12*: 73–76.

Robbs, C.F., O. Kimura and R.L.D. de Ribeiro. 1981. Descrição de um novo patovar de *Xanthomonas campestris* em beterraba hortícola e estudo comparativo com *Xanthomonas beticola*. Fitopatol. Bras. *6*: 387–394.

Robbs, C.F., A.G. Medeiros and O. Kimura. 1982. Mancha bacteriana das folhas do guaranazeiro causada por um novo patovar de *Xanthomonas campestris*. Arq. Univ. Fed. Rural, do Rio de Janeiro. *5*: 195–201.

Robbs, C.F., J.R. Neto, V.A.J. Malavolta and O. Kimura. 1989. Bacterial spot and blight of yellow shrimp (*Pacystachys lutea*) caused by a new pathovar of *Xanthomonas campestris*. Summa Phytopathol. *15*: 174–179.

Robbs, C.F., R.d.L.D. Ribiero and O. Kimura. 1974. Sobre a posição taxonômica de *Pseudomonas mangiferaeindicae* Patel et al. 1948, agente causal da "mancha bacteriana" das folhas da mangueira (*Mangifera indica* L.). Arq. Univ. Fed. Rural, Rio de Janeiro. *4*: 11–14.

Roberts, D.S. 1956. A new pathogen from a ewe with mastitis. Aust. Vet. J. *32*: 330–332.

Roberts, D.E., E. Higgs and P.J. Cole. 1987. Selective medium that distinguishes *Haemophilus influenzae* from *Haemophilus parainfluenzae* in clinical specimens: its value in investigating respiratory sepsis. J. Clin. Pathol. *40*: 75–76.

Roberts, D.E., A. Ingold, S.V. Want and J.R. May. 1974. Osmotically stable L forms of *Haemophilus influenzae* and their signiflcance in testing sensitivity to penicillins. J. Clin. Pathol. *27*: 560–564.

Roberts, I.S. and M.J. Coleman. 1991. The virulence of *Erwinia amylovora*: molecular genetic perspectives. J. Gen. Microbiol. *137*: 1453–1457.

Roberts, J.A., E.N. Fussell and M.B. Kaack. 1990. Bacterial adherence to urethral catheters. J. Urol. *144*: 264–269.

Roberts, M.C., C.D. Swenson, L.M. Owens and A.L. Smith. 1980. Characterization of chloramphenicol-resistant *Haemophilus influenzae*. Antimicrob. Agents Chemother. *18*: 610–615.

Roberts, P. 1974. *Erwinia rhapontici* (Millard) Burkholder associated with pink grain of wheat. J. Appl. Bacteriol. *37*: 353–358.

Roberts, P.D., N.C. Hodge, H. Bouzar, J.B. Jones, R.E. Stall, R.D. Berger and A.R. Chase. 1998. Relatedness of strains of *Xanthomonas fragariae* by restriction fragment length polymorphism, DNA–DNA reassociation, and fatty acid analyses. Appl. Environ. Microbiol. *64*: 3961–3965.

Roberts, P. and C.M. Scarlett. 1981. *Pseudomonas corrugata* sp. nov. In Validation of the publication of new names and new combinations previously effectively published outside the IJSB. List No. 6. Int. J. Syst. Bacteriol. *31*: 216.

Roberts, R.S. 1947. An immunologic study of *Pasteurella septica*. J. Comp. Pathol. *57*: 261–278.

Roberts, R.J. 1985. Restriction and modification enzymes and their recognition sequences. Nucleic Acids Res. *13 Suppl*: r165–200.

Robins-Browne, R.M. 1987. Traditional enteropathogenic *Escherichia coli* of infantile diarrhea. Rev. Infect. Dis. *9*: 28–53.

Robins-Browne, R.M., S. Cianciosi, A.M. Bordun and G. Wauters. 1991. Pathogenicity of *Yersinia kristensenii* for mice. Infect. Immun. *59*: 162–167.

Robins-Browne, R.M. and J.K. Pipic. 1985. Effects of iron and desferrioxamine on infections with *Yersinia enterocolitica*. Infect. Immun. *47*: 774–779.

Robinson, J., J. Beaman, L. Wagener and V. Burke. 1986. Comparison of direct plating with the use of enrichment culture for isolation of *Aeromonas* spp. from feces. J. Med. Microbiol. *22*: 315–318.

Robinson, J., V. Burke, P.J. Worthy, J. Beaman and L. Wagener. 1984. Media for isolation of *Aeromonas* spp. from feces. J. Med. Microbiol. *18*: 405–412.

Robinson, J. and N.E. Gibbons. 1952. The effect of salts on the growth of *Micrococcus halodenitrificans* (n. sp.). Can. J. Bot. *30*: 147–154.

Robson, R.L. 1979. Characterization of an oxygen stable nitrogenase complex isolated from *Azotobacter chroococcum*. Biochem. J. *181*: 569–575.

Robson, R.L., J.A. Chesshyre, C. Wheeler, R. Jones, P.R. Woodley and J.R. Postgate. 1984. Genome size and complexity in *Azotobacter chroococcum*. J. Gen. Microbiol. *130*: 1603–1612.

Robson, R.L. and J.R. Postgate. 1980. Oxygen and nitrogen in biological nitrogen fixation. Ann. Rev. Microbiol. *34*: 183–207.

Robson, R.L., P. Woodley and R. Jones. 1986. Second gene (*nifH**) coding for a nitrogenase iron protein in *Azotobacter chroococcum* is adjacent to a gene coding for a ferredoxin-like protein. EMBO (Eur. Mol. Biol. Organ.) J. *5*: 1159–1163.

Robson, R.L., P.R. Woodley, R.N. Pau and R.R. Eady. 1989. Structural genes for the vanadium nitrogenase from *Azotobacter chroococcum*. Embo J. *8*: 1217–1224.

Rocchetta, H.L. and J.S. Lam. 1997. Identification and functional characterization of an ABC transport system involved in polysaccharide export of A-band lipopolysaccharide in *Pseudomonas aeruginosa*. J. Bacteriol. *179*: 4713–4724.

Roche, R.J. and E.R. Moxon. 1995. Phenotypic variation of carbohydrate surface antigens and the pathogenesis of *Haemophilus influenzae* infections. Trends Microbiol. *3*: 304–309.

Rodger, H.D. and E.M. Drinan. 1993. Observation of a rickettsia-like organism in Atlantic salmon, *Salmo sala* L., in Ireland. J. Fish Dis. *16*: 361–369.

Rodicheva, E.K., I.N. Trubachev, S.E. Medvedeva, O.I. Egorova and L.Y. Shitova. 1993. Growth and luminescence of luminous bacteria promoted by agents of microbial origin. J. Biolumin. Chemilumin. *8*: 293–299.

Rodrigues-Neto, J., C.F. Robbs and T. Yamashiro. 1987. A bacterial disease of guava (*Psidium guajava*) caused by *Erwinia psidii*, sp. nov. Fitopatol. Bras. *12*: 345–350.

Rodrigues-Neto, J., C.F. Robbs and T. Yamashiro. 1988. *In* Validation of the publication of new names and new combinations previously effectively published outside the IJSB. List No. 26. Int. J.Syst. Bacteriol. *38*: 328–329.

Rodriguez, L.A., A.E. Ellis and T.P. Nieto. 1992. Purification and characterisation of an extracellular metalloprotease, serine protease and haemolysin of *Aeromonas hydrophila* strain B-32: all are lethal for fish. Microb. Pathog. *13*: 17–24.

Rodriguez-Valera, F., A. Ventosa, G. Juez and J.F. Imhoff. 1985. Variation of environmental features and microbial populations with salt concentrations in a multi-pond saltern. Microb. Ecol. *11*: 107–116.

Roggendorf, M. and H.E. Müller. 1976. Enterobacteria of reptiles (translation). Zentralbl. Bakteriol. Parasitenkd. Hyg. Infektionskr., Abt. I Orig A. *236*: 22–35.

Rohde, R. 1965. The identification, epidemiology and pathogenicity of the salmonellae of subgenus II. J. Appl. Bacteriol. *28*: 368–372.

Rohde, R. 1966. Neue serologische Befunde hinsichtlich der Subgenus Einteilung der Salmonellen. Zentralbl. Bakteriol. Parasitenk. Infektionskr. Hyg. Abt. I Orig. *202*: 484–503.

Rohde, R. 1967. Zür serologischen Differential-diagnose der *Salmonella* Subgenera I-IV. Zentrab. Bakteriol. Parasitenk. Infektionskr. Hyg. Abt. I Orig. *205*: 404–424.

Rohmer, M., P. Bouvier and G. Ourisson. 1979. Molecular evolution of biomembranes: structural equivalents and phylogenetic precursors of sterols. Proc. Natl. Acad. Sci. U.S.A. *76*: 847–851.

Roland, F.P. 1970. Leg gangrene and endotoxin shock due to *Vibrio parahaemolyticus*—an infection acquired in New England coastal waters. N. Engl. J. Med. *282*: 1306.

Roland, F.P. 1971. *Vibrio parahaemolyticus*: A case report. Clin. Med. *78*: 26–33.

Rolland, K., N. Lambert-Zechovsky, B. Picard and E. Denamur. 1998. *Shigella* and enteroinvasive *Escherichia coli* strains are derived from distinct ancestral strains of *E. coli*. Microbiology *144*: 2667–2672.

Romalde, J.L. and B. Magariños. 1997. Immunization with bacterial antigens: Pasteurellosis. *In* Gudding, Lillehaug, Midtlyng and Brown (Editors), Fish Vaccinology, Dev. Biol. Stand., Vol. 90, Basel, Karger. 167–177.

Romalde, J.L., B. Magarinos, K.D. Turnbull, A.M. Baya, J.M. Barja and A.E. Toranzo. 1995. Fatty-acid profiles of *Pasteurella piscicida* comparison with other fish pathogenic Gram-negative bacteria. Arch. Microbiol. *163*: 211–216.

Roman, M.J., P.D. Coriz and O.G. Baca. 1986. A proposed model to explain persistent infection of host cells with *Coxiella burnetii*. J. Gen. Microbiol. *132*: 1415–1422.

Romanenko, L.A., A.M. Lysenko, V.V. Mikhailov and A.V. Kurika. 1994. A new species of brown agar-digesting bacteria of the genus *Alteromonas*. Microbiology. *63*: 1081–1087.

Romanenko, L.A., V.V. Mikhailov, A.M. Lysenko and V.I. Stapanenko.

1995a. *In* Validation of the publication of new names and new combinations previously effectively published outside the IJSB. List No. 55. Int. J. Syst. Bacteriol. *45*: 879–880.

Romanenko, L.A., V.V. Mikhailov, A.M. Lysenko and V.I. Stepanenko. 1995b. A new species of melanin-producing bacteria of the genus *Alteromonas*. Microbiology *64*: 60–62.

Romanenko, L.A., P. Schumann, M. Rohde, A.M. Lysenko, V.V. Mikhailov and E. Stackebrandt. 2002. *Psychrobacter submarinus* sp nov and *Psychrobacter marincola* sp nov., psychrophilic halophiles from marine environments. Int. J. Syst. Evol. Microbiol. *52*: 1291–1297.

Romano, I., B. Nicolaus, L. Lama, M.C. Manca and A. Gambacorta. 1996. Characterization of a haloalkalophilic strictly aerobic bacterium, isolated from Pantelleria Island. System. Appl. Microbiol. *19*: 326–333.

Romano, I., B. Nicolaus, L. Lama, M.C. Manca and A. Gambacorta. 1997. Validation of the publication of new names and new combinations previously effectively published outside the IJSB. List No. 61. Int. J. Syst. Bacteriol. *47*: 601–602.

Romanovskaya, V.A., Y.R. Malashenko and V.N. Bogachenko. 1978. Corrected diagnoses of the genera and species of methane-utilizing bacteria. Mikrobiologiia *47*: 120–130.

Romanovskaya, V.A., A.G. Titov, I.V. Kharchenko and N.A. Ugryumova. 1992. Interrelation of monocultures in the microbe associations *Methylococcus capsulatus* and *Frateuria aurantia*. Mikrobiol. Zh. *54*: 3–8.

Romeiro, R.S. and O. Kimura. 1997. Induced resistance in pepper leaves infiltrated with purified bacterial elicitors from *Xanthomonas campestris* pv. *vesicatoria*. J. Phytopathol. (Berl.) *14*: 495–498.

Romero, X., J.F. Turnbull and R. Jimenez. 2000. Ultrastructure and cytopathology of a rickettsia-like organism causing systemic infection in the redclaw crayfish, *Cherax quadricarinatus* (Crustacea: decapoda), in Ecuador. J. Invertebr. Pathol. *76*: 95–104.

Römling, U., K.D. Schmidt and B. Tümmler. 1997. Large genome rearrangements discovered by the detailed analysis of 21 *Pseudomonas aeruginosa* clone C isolates found in environment and disease habitats. J. Mol. Biol. *271*: 386–404.

Rontani, J.F., M.J. Gilewicz, V.D. Michotey, T.L. Zheng, P.C. Bonin and J.C. Bertrand. 1997. Aerobic and anaerobic metabolism of 6,10,14-trimethylpentadecan-2-one by a denitrifying bacterium isolated from marine sediments. Appl. Environ. Microbiol. *63*: 636–643.

Rood, J.I., P.A. Howarth, V. Haring, S.J. Billington, W.K. Yong, D. Liu, M.A. Palmer, D.R. Pitman, I. Links, D.J. Stewart and J.A. Vaughan. 1996. Comparison of gene probe and conventional methods for the differentiation of ovine footrot isolates of *Dichelobacter nodosus*. Vet. Microbiol. *52*: 127–141.

Rosato, Y.B., S.A.L. Destefano and M.J. Daniels. 1994. Cloning of a locus involved in pathogenicity and production of extracellular polysaccharide and protease in *Xanthomonas campestris* pv. *campestris*. FEMS Microbiol. Lett. *117*: 41–45.

Rose, D.H. 1917. Blister spot of apples and its relation to a disease of apple bark. Phytopathology *7*: 198–208.

Rosen, H.R. 1922. The bacterial pathogen of corn stalk rot . Phytopathology *12*: 497–499.

Rosenau, A., P.Y. Sizaret, J.M. Musser, A. Goudeau and R. Quentin. 1993. Adherence to human cells of a cryptic *Haemophilus* genospecies responsible for genital and neonatal infections. Infect. Immun. *61*: 4112–4118.

Rosenberg, H., A.H. Ennor and V.F. Morrison. 1956. The estimation of arginine. Biochem. J. *63*: 153–159.

Rosenbusch, C.T. and L.A. Merchant. 1939. A study of the hemmorrhagic septicemia Pasteurellae. J. Bacteriol. *37*: 69–89.

Rosendal, S., D.A. Boyd and K.A. Gilbride. 1985. Comparative virulence of porcine *Haemophilus* bacteria. Can. J. Comp. Med. *49*: 68–74.

Rosenshine, I., M.S. Donnenberg, J.B. Kaper and B.B. Finlay. 1992. Signal transduction between enteropathogenic *Escherichia coli* (EPEC) and epithelial cells: EPEC induces tyrosine phosphorylation of host cell proteins to initiate cytoskeletal rearrangement and bacterial uptake. Embo J. *11*: 3551–3560.

Ross, A.J., R.R. Rucker and W.H. Ewing. 1966. Description of a bacterium associated with redmouth disease of rainbow trout (*Salmo gairdneri*). Can. J. Microbiol. *12*: 763–770.

Ross, J.L., P.I. Boon, P. Ford and B.T. Hart. 1997. Detection and quantification with 16S rRNA probes of planktonic methylotrophic bacteria in a floodplain lake. Microb. Ecol. *34*: 97–108.

Ross, R.F., J.E. Hall, A.P. Orning and S.E. Dale. 1972. Characterization of an *Actinobacillus* isolated from the sow vagina. Int. J. Syst. Bacteriol. *22*: 39–46.

Rossau, R., A. Van Landschoot, M. Gillis and J. De Ley. 1991. Taxonomy of *Moraxellaceae* fam. nov., a new bacterial family to accommodate the genera *Moraxella*, *Acinetobacter*, and *Psychrobacter* and related organisms. Int. J. Syst. Bacteriol. *41*: 310–319.

Rossau, R., A. Van Landschoot, W. Mannheim and J. De Ley. 1986. Intergeneric and intrageneric similarities of ribosomal RNA cistrons of the *Neisseriaceae*. Int. J. Syst. Bacteriol. *36*: 323–332.

Rossau, R., G. Vandenbusche, S. Thielemans, P. Segers, H. Grosch, E. Goethe, W. Mannheim and J. De Ley. 1989. Ribosomal RNA cistron similarities and DNA homologies of *Neisseria*, *Kingella*, *Eikenella*, *Simonsiella*, *Alysiella*, and Centers for Disease Control groups EF-4 and M-5 in the emended family *Neisseriaceae*. Int. J. Syst. Bacteriol. *39*: 185–198.

Rosselló-Mora, R.A., F. Caccavo, Jr., K. Osterlehner, N. Springer, S. Spring, D. Schuler, W. Ludwig, R. Amann, M. Vancanneyt and K.H. Schleifer. 1994a. Isolation and taxonomic characterization of a halotolerant, facultatively iron-reducing bacterium. Syst. Appl. Microbiol. *17*: 569–573.

Rosselló-Mora, R.A., E. García-Valdés, J. Lalucat and J. Ursing. 1991. Genotypic and phenotypic diversity of *Pseudomonas stutzeri*. Syst. Appl. Microbiol. *14*: 150–157.

Rosselló-Mora, R.A., J. Lalucat, W. Dott and P. Kämpfer. 1994b. Biochemical and chemotaxonomic characterization of *Pseudomonas stutzeri* genomovars. J. Appl. Bacteriol. *76*: 226–233.

Rosselló-Mora, R.A., J. Lalucat and E. Garcia-Valdés. 1994c. Comparative biochemical and genetic-analysis of naphthalene degradation among *Pseudomonas stutzeri* strains. Appl. Environ. Microbiol. *60*: 966–972.

Rosselló-Mora, R.A., W. Ludwig, P. Kämpfer, R. Amann and K.-H. Schleifer. 1996. *In* Validation of the publication of new names and new combinations previously published outside the IJSB. List No. 56. Int. J. Syst. Bacteriol. *46*: 362–363.

Rosselló-Mora, R.A., W. Ludwig, P. Kämpfer, R. Amann and K.-H. Schleifer. 1995. *Ferrimonas balearica* gen. nov., sp. nov., a new marine facultative Fe(III)-reducing bacterium. Syst. Appl. Microbiol. *18*: 196–202.

Rothenpieler, U., R. Mutters, W. Frederiksen, R. Rossau, P. Segers, J. De Ley and W. Mannheim. 1986. DNA relationships of *Cardiobacterium hominis*. *In* Microbe 86 : XIV International Congress of Microbiology, 7–13 September, 1986, Manchester, England, Vol. 2, International Congress for Microbiology, Manchester. pp. B4–B8.

Rouboud, E. 1919. Les particularités de la nutrition et la vie symbiotique chez les mouches tsétsés. Ann. Inst. Pasteur (Paris) *33*: 489–537.

Rouhbakhsh, D. and P. Baumann. 1995. Characterization of a putative 23S–5S rRNA operon of *Buchnera aphidicola* (endosymbiont of aphids) unlinked to the 16S rRNA-encoding gene. Gene *155*: 107–112.

Rouhbakhsh, D., C.-Y. Lai, C.D. von Dohlen, M.A. Clark, L. Baumann, P. Baumann, N.A. Moran and D.J. Voegtlin. 1996. The tryptophan biosynthetic pathway of aphid endosymbionts (*Buchnera*): genetics and evolution of plasmid-associated anthranilate synthase (*trpEG*) within the aphididae. J. Mol. Evol. *42*: 414–421.

Rouhbakhsh, D., M.A. Clark, L. Baumann, N.A. Moran and P. Baumann. 1997. Evolution of the tryptophan biosynthetic pathway in *Buchnera* (aphid endosymbionts): studies of plasmid-associated *trpEG* within the genus *Uroleucon*. Mol. Phylogenet. Evol. *8*: 167–176.

Rouhbakhsh, D., N.A. Moran, L. Baumann, D.J. Voegtlin and P. Baumann. 1994. Detection of *Buchnera*, the primary prokaryotic endosymbiont of aphids, using the polymerase chain reaction. Insect Mol. Biol. *3*: 213–217.

Roux, V., M. Bergoin, N. Lamaze and D. Raoult. 1997. Reassessment of the taxonomic position of *Rickettsiella grylli*. Int. J. Syst. Bacteriol. *47*: 1255–1257.

Rowbotham, T.J. 1983. Isolation of *Legionella pneumophila* from clinical specimens via amoebae, and the interaction of those and other isolates with amoebae. J. Clin. Pathol. *36*: 978–986.

Rowbotham, T.J. 1986. Current views on the relationships between amoebae, legionellae and man. Isr. J. Med. Sci. *22*: 678–689.

Rowe, B., R.J. Gross and E. van Oye. 1975. An organism differing from *Shigella boydii* 13 only in its ability to produce gas from glucose. Int. J. Syst. Bacteriol. *25*: 301–303.

Roy, T.M., D. Fleming and W.H. Anderson. 1989. Tularemic pneumonia mimicking Legionnaires' disease with false-positive direct fluorescent antibody stains for *Legionella*. South. Med. J. *82*: 1429–1431.

Royle, P.L., H. Matsumoto and B.W. Holloway. 1981. Genetic circularity of the *Pseudomonas aeruginosa* PAO chromosome. J. Bacteriol. *145*: 145–155.

Rózalski, A. and K. Kotelko. 1987. Hemolytic activity and invasiveness in strains of *Proteus penneri*. J. Clin. Microbiol. *25*: 1094–1096.

Rózalski, A., Z. Sidorczyk and K. Kotelko. 1997. Potential virulence factors of *Proteus bacilli*. Microbiol. Mol. Biol. Rev. *61*: 65–89.

Rozsa, F.W. and C.F. Marrs. 1991. Interesting sequence differences between the pilin gene inversion regions of *Moraxella lacunata* ATCC 17956 and *Moraxella bovis* Epp63. J. Bacteriol. *173*: 4000–4006.

Rubin, L.G., A.A. Medeiros, R.H. Yolken and E.R. Moxon. 1981. Ampicillin treatment failure of apparently beta-lactamase-negative *Haemophilus influenzae* type b meningitis due to novel beta-lactamase. Lancet *2*: 1008–1010.

Rubin, S.J. and R.C. Tilton. 1975. Isolation of *Vibrio alginolyticus* from wound infections. J. Clin. Microbiol. *2*: 556–558.

Ruby, E.G., E.P. Greenberg and J.W. Hastings. 1980. Planktonic marine luminous bacteria: species distribution in the water column. Appl. Environ. Microbiol. *39*: 302–306.

Ruby, E.G. and H.W. Jannasch. 1982. Physiological characteristics of *Thiomicrospira* sp. strain L-12 isolated from deep-sea hydrothermal vents. J. Bacteriol. *149*: 161–165.

Ruby, E.G. and K.H. Lee. 1998. The *Vibrio fischeri Euprymna scolopes* light organ association: Current ecological paradigms. Appl. Environ. Microbiol. *64*: 805–812.

Ruby, E.G., C.O. Wirsen and H.W. Jannasch. 1981. Chemolithotrophic sulfur-oxidizing bacteria from the Galapagos Rift (Pacific Ocean) hydrothermal vents. Appl. Environ. Microbiol. *42*: 317–324.

Ruckelshausen, R. and R. Holländer. 1978. On the phenotypical characteristics of *Haemophilus* isolates from human respiratory tracts. Zentralbl. Bakteriol. Parasitenkd. Infektionskr. Hyg. Abt. I Orig. A. *242*: 500–511.

Rucker, R.R. 1966. Redmouth disease in rainbow trout (*Salmo gairdneri*). Bull. Off. Int. Epiz. *65*: 825–830.

Rudolph, K. 1993. Infection of the plant by *Xanthomonas*. *In* Swings and Civerolo (Editors), *Xanthomonas*, Chapman & Hall, London. pp. 193–264.

Ruehl, W.W., C.F. Marrs, R. Fernandez, S. Falkow and G.K. Schoolnik. 1988. Purification, characterization, and pathogenicity of *Moraxella bovis* pili. J. Exp. Med. *168*: 983–1002.

Ruffolo, C.G., J.M. Tennent, W.P. Michalski and B. Adler. 1997. Identification, purification, and characterization of the type 4 fimbriae of *Pasteurella multocida*. Infect. Immun. *65*: 339–343.

Ruimy, R., V. Breittmayer, P. Elbaze, B. Lafay, O. Boussemart, M. Gauthier and R. Christen. 1994. Phylogenetic analysis and assessment of the genera *Vibrio*, *Photobacterium*, *Aeromonas*, and *Plesiomonas* deduced from small-subunit rRNA sequences. Int. J. Syst. Bacteriol. *44*: 416–426.

Rules Revision Committee, J.C., International Committee on Systematic Bacteriology 1985. Proposal to emend the international code of nomenclature of bacteria. Int. J. Syst. Bacteriol. *35*: 123.

Russell, N.J. and J.K. Volkman. 1980. The effect of growth temperature on wax ester composition in the psychrophilic bacterium *Micrococcus cryophilus* ATCC 15174. J. Gen. Microbiol. *118*: 131–142.

Russo, A., Y. Moenne-Loccoz, S. Fedi, P. Higgins, A. Fenton, D.N. Dowling, M. O'Regan and F. O'Gara. 1996. Improved delivery of biocontrol

Pseudomonas and their antifungal metabolites using alginate polymers. Appl. Microbiol. Biotechnol. *44*: 740–745.

Russo, T.A. and J.R. Johnson. 2000. Proposal for a new inclusive designation for extraintestinal pathogenic isolates of *Escherichia coli*: ExPEC. J. Infect. Dis. *181*: 1753–1754.

Rust, L., E.C. Pesci and B.H. Iglewski. 1996. Analysis of the *Pseudomonas aeruginosa* elastase (*lasB*) regulatory region. J. Bacteriol. *178*: 1134–1140.

Rustigian, R. and C.A. Stuart. 1943. Taxonomic relationships in the genus *Proteus*. Proc. Soc. Exp. Biol. Med. *53*: 241–243.

Rustigian, R. and C.A. Stuart. 1945. The biochemical and serological relationships of the organisms of the genus *Proteus*. J. Bacteriol. *49*: 419–436.

Rutala, W.A., F.A. Sarubi, Jr., C.S. Finch, J.N. McCormack and G.E. Steinkraus. 1982. Oyster-associated outbreak of diarrhoeal disease possibly caused by *Plesiomonas shigelloides*. Lancet *1*: 739.

Ruttimann-Johnson, C., C.R. Staples, P. Rangaraj, V.K. Shah and P.W. Ludden. 1999. A vanadium and iron cluster accumulates on VnfX during iron- vanadium-cofactor synthesis for the vanadium nitrogenase in *Azotobacter vinelandii*. J. Biol. Chem. *274*: 18087–18092.

Ruzafa, C., F. Solano and A. Sanchez-Amat. 1994. The protein encoded by the *Shewanella colwelliana melA* gene is *p*-hydroxyphenylpyruvate dioxygenase. FEMS Microbiol. Lett. *124*: 179–184.

Ryan, W.J. 1964. *Moraxella* commonly present on the conjunctiva of guinea pigs. J. Gen. Microbiol. *35*: 361–372.

Rychlik, I., M. Bartos and K. Sestak. 1994. Use of DNA fingerprinting for accurate typing of *Actinobacillus pleuropneumoniae*. Vet. Med. (Praha) *39*: 167–174.

Rye, A.J., J.W. Drozd, C.W. Jones and J.D. Linton. 1988. Growth efficiency of *Xanthomonas campestris* in continuous culture. J. Gen. Microbiol. *134*: 1055–1062.

Ryley, H.C., L. Millar-Jones, A. Paull and J. Weeks. 1995. Characterization of *Burkholderia cepacia* from cystic-fibrosis patients living in Wales by PCR ribotyping. J. Med. Microbiol. *43*: 436–441.

Ryll, M., R. Mutters and W. Mannheim. 1991. Study of the genetic classification of the *Pasteurella pneumotropica* complex. Berliner Munchener Tierarztl. Wochenschr. *104*: 243–245.

Saarela, M., S. Asikainen, S. Alaluusua, L. Pyhala, C.H. Lai and H. Jousimies-Somer. 1992. Frequency and stability of mono- or poly-infection by *Actinobacillus actinomycetemcomitans* serotypes a, b, c, d or e. Oral Microbiol. Immunol. *7*: 277–279.

Saarela, M., S. Asikainen, H. Jousimies-Somer, T. Asikainen, B. von Troil-Linden and S. Alaluusua. 1993. Hybridization patterns of *Actinobacillus actinomycetemcomitans* serotype a-e detected with an rRNA gene probe. Oral. Microbiol. Immunol. *8*: 111–115.

Saari, M., T. Cheasty, K. Leino and A. Siitonen. 2001. Phage types and genotypes of Shiga toxin-producing *Escherichia coli* O157 in Finland. J. Clin. Microbiol. *39*: 1140–1143.

Sabath, L.D., M. Jago and E.P. Abraham. 1965. Cephalosporinase and penicillinase activities of a beta-lactamase from *Pseudomonas pyocyanea*. Biochem. J. *96*: 739–752.

Sabbaj, J., V.L. Sutter and S.M. Finegold. 1970. Urease and deaminase activities of fecal bacteria in hepatic coma. Antimicrob. Agents Chemother. *10*: 181–185.

Sabet, K.A. 1957. Studies in the bacterial diseases of Sudan crops I. Bacterial leaf spot of jute (*Corchorus olitorius* L.). Ann. Appl. Biol. *45*: 516–520.

Sabet, K.A. 1959. Studies in the bacterial diseases of Sudan crops IV. Bacterial leaf spot and canker disease of mahogany (*Khaya senegalensis* (Desr.) A. Juss and *K. grandifoliola* C. DC). Ann. Appl. Biol. *47*: 658–665.

Sabet, K.A. and W.J. Dowson. 1960. Bacterial leaf spot of sesame (*Sesamum orientale* L.). Phytopathol. Z. *37*: 252–258.

Sabet, K.A., F. Ishag and O. Khalil. 1969. Studies on the bacterial diseases of Sudan crops VII. New records. Ann. Appl. Biol. *63*: 357–369.

Sabra, W., A.P. Zeng, H. Lunsdorf and W.D. Deckwer. 2000. Effect of oxygen on formation and structure of *Azotobacter vinelandii* alginate

and its role in protecting nitrogenase. Appl. Environ. Microbiol. *66*: 4037–4044.

Sabry, S.A., K.M. Ghanem and W.A. Sabra. 1996. Effect of nutrients on alginate synthesis in *Azotobacter vinelandii* and characterization of the produced alginate. Microbiologia *12*: 593–598..

Sachs, J.M., M. Pacin and G.W. Counts. 1974. Sickle hemoglobinopathy and *Edwardsiella tarda* meningitis. Am. J.Dis. Child. *128*: 387–388.

Sack, R.B., M. Rahman, M. Yunus and E.H. Khan. 1997. Antimicrobial resistance in organisms causing diarrheal disease. Clin. Infect. Dis. *24 Suppl 1*: S102–105.

Sackett, W.G. 1916. A bacterial stem blight of field and garden peas. Bull. Colo. Agr. Sta. No. *218*: 1–43.

Sader, H.S., A.C. Pignatari, R. Frei, R.J. Hollis and R.N. Jones. 1994. Pulsed-field gel electrophoresis of restriction-digested genomic DNA and antimicrobial susceptibility of *Xanthomonas maltophilia* strains from Brazil, Switzerland and the USA. J. Antimicrob. Chemother. *33*: 615–618.

Sadoff, H.L., B. Shimei and S. Ellis. 1979. Characterization of *Azotobacter vinelandii* deoxyribonucleic acid and folded chromosomes. J. Bacteriol. *138*: 871–877.

Sadusky, T.J. and R.A. Bullis. 1994. Experimental disinfection of lobster eggs infected with *Leucothrix mucor*. Biol. Bull. (Woods Hole) *187*: 254–255.

Saffarini, D.A., T.J. DiChristina, D. Bermudes and K.H. Nealson. 1994. Anaerobic respiration of *Shewanella putrefaciens* requires both chromosomal and plasmid-borne genes. FEMS Microbiol. Lett. *119*: 271–277.

Safrin, S., J.G. Morris, Jr., M. Adams, V. Pons, R. Jacobs and J.E. Conte, Jr.. 1988. Non-O:1 *Vibrio cholerae* bacteremia: Case report and review. Rev. Infect. Dis. *10*: 1012–1017.

Sahin, F. and S.A. Miller. 1996. Characterization of Ohio strains of *Xanthomonas campestris* pv. *vesicatoria*, causal agent of bacterial spot of pepper. Plant Dis. *80*: 773–778.

Sahl, H.G. and H.G. Trüper. 1977. Enzymes of CO_2 fixation in *Chromatiaceae*. FEMS Microbiol. Lett. *2*: 129–132.

Saint-Onge, A., F. Romeyer, P. Lebel, L. Masson and R. Brousseau. 1992. Specificity of the *Pseudomonas aeruginosa* PAO1 lipoprotein-I gene as a DNA probe and PCR target region within the *Pseudomonadaceae*. J. Gen. Microbiol. *138*: 733–741.

Saito, A., H. Koga, H. Shigeno, K. Watanabe, K. Mori, S. Kohno, Y. Shigeno, Y. Suzuyama, K. Yamaguchi, M. Hirota and K. Hara. 1986. The antimicrobial activity of ciprofloxacin against *Legionella* species and the treatment of experimental *Legionella* pneumonia in guinea pigs. J. Antimicrob. Chemother. *18*: 251–260.

Saito, A., R.D. Rolfe, P.H. Edelstein and S.M. Finegold. 1981. Comparison of liquid growth media for *Legionella pneumophila*. J. Clin. Microbiol. *14*: 623–627.

Saitou, N. and M. Nei. 1987. The neighbor-joining method: a new method for reconstructing phylogenetic trees. Mol. Biol. Evol. *4*: 406–425.

Sakane, T. and A. Yokota. 1994. Chemotaxonomic investigation of heterotrophic, aerobic and microaerophilic spirilla, the genera *Aquaspirillum*, *Magnetopirillum* and *Oceanospirillum*. Syst. Appl. Microbiol. *17*: 128–134.

Sakata, T., K. Sakaguchi and D. Kakimoto. 1982. Antibiotic production by marine pigmented bacteria. I. Antibacterial effect of *Alteromonas luteoviolaceus*. Mem. Fac. Fish., Kagoshima Univ. *31*: 243–250.

Sakata, T., K. Sakaguchi and D. Kakimoto. 1986. Antibiotic production by marine pigmented bacteria. II. Purification and characterization of antibiotic substances of *Alteromonas luteoviolacea*. Mem. Fac. Fish., Kagoshima Univ. *35*: 29–37.

Sakazaki, R. 1961. Studies on the *Hafnia* group of *Enterobacteriaceae*. Japan J. Med. Sci. Biol. *14*: 223–241.

Sakazaki, R. 1967. Studies on the Asakusa group of *Enterobacteriaceae* (*Edwardsiella tarda*). Jpn. J. Med. Sci. Biol. *20*: 205–212.

Sakazaki, R. 1968. Proposal of *Vibrio alginolyticus* for the biotype 2 of *Vibrio parahaemolyticus*. Jpn. J. Med. Sci. Biol. *21*: 359–362.

Sakazaki, R. 1984. Serology and epidemiology of *Plesiomonas shigelloides*. Methods Microbiol. *16*: 259–269.

Sakazaki, R. 1987. Serology of mesophilic *Aeromonas* spp. and *Plesiomonas shigelloides*. Experientia (Basel) *43*: 357–358.

Sakazaki, R. and A. Balows. 1981. The genera *Vibrio, Pleisomonas*, and *Aeromonas. In* Starr, Stolp, Trüper and Schlegel (Editors), The Prokaryotes. A Handbook on Habitats, Isolation, and Identification of Bacteria, Vol. 2, Springer-Verlag, New York. pp. 1272–1301.

Sakazaki, R., S. Iwanami and H. Fukumi. 1963. Studies on the enteropathogenic facultatively halophilic bacteria *Vibrio parahaemolyticus*. I. Morphological, cultural and biochemical properties and its taxonomical position. Jpn. J. Med. Sci. Biol. *16*: 161–188.

Sakazaki, R., S. Kuramochi, Y. Kosako and K. Tamura. 1983. Independency of *Eschericia adecarboxylata* from *Enterobacter agglomerans. In* Leclerc (Editor), Les bacilles a Gram-negatif medical et en sante publique: taxonomie-identification-applications, INSERM, Paris. 157–166.

Sakazaki, R. and Y. Murata. 1962. The new group of the *Enterobacteriaceae*, the Asakusa group. Jpn. J. Bacteriol. *17*: 617–618.

Sakazaki, R. and S. Namioka. 1957. Biochemical studies on Voges-Proskauer positive enteric bacteria. Japan. J. Exp. Med. *27*: 273–282.

Sakazaki, R. and S. Namioka. 1960. Serological studies on the *Cloaca* (*Aerobacter*) group of enteric bacteria. Jpn. J. Med. Sci. Biol. *13*: 1–12.

Sakazaki, R., S. Namioka, A. Osada and C. Yamada. 1960. A problem on the pathogenic role of *Citrobacter* of enteric bacteria. Jpn. J. Exp. Med. *30*: 13–21.

Sakazaki, R. and T. Shimada. 1984. O-serogrouping for mesophilic *Aeromonas* strains. Jpn. J. Med. Sci. Biol. *37*: 247–255.

Sakazaki, R. and K. Tamura. 1975. Priority of the specific epithet *anguillimortiferum* over the specific epithet *tarda* in the name of the organism presently known as *Edwardsiella tarda*. Int. J. Syst. Bacteriol. *25*: 219–220.

Sakazaki, R., K. Tamura, R. Johnson and R.R. Colwell. 1976. Taxonomy of some recently described species in the family *Enterobacteriaceae*. Int. J. Syst. Bacteriol. *26*: 158–179.

Sakazaki, R., K. Tamura, Y. Kosako and E. Yoshizaki. 1989a. *Klebsiella ornithinolytica* sp. nov., formerly known as ornithine-positive *Klebsiella oxytoca*. Curr. Microbiol. *18*: 201–206.

Sakazaki, R., K. Tamura, Y. Kosako and E. Yoshizaki. 1989b. *In* Validation of the publication of new names and new combinations previously effectively published outside the IJSB. Validation List No. 31. Int. J. Syst. Bacteriol. *39*: 495–497.

Sakazaki, R., K. Tamura, L.M. Prescott, Z. Bencic, S.C. Sanyal and R. Sinha. 1971. Bacteriological examination of diarrheal stools in Calcutta. Indian J. Med. Res. *59*: 1025–1034.

Sakellaris, H., N.K. Hannink, K. Rajakumar, D. Bulach, M. Hunt, C. Sasakawa and B. Adler. 2000. Curli loci of *Shigella* spp. Infect. Immun. *68*: 3780–3783.

Sakiyama, T. and K. Ohwada. 1997. Isolation and growth characteristics of deep-sea barophilic bacteria from the Japan Trench. Fish. Sci. *63*: 228–232.

Salam, M.A. and M.L. Bennish. 1991. Antimicrobial therapy for shigellosis. Rev. Infect. Dis. *13 Suppl 4*: S332–341.

Salati, F., S. Kawai and R. Kusuda. 1989. Characteristics of the lipopolysaccharide from *Pasteurella piscicida*. Fish Pathol. *24*: 143–147.

Salmon, R.L., I.D. Farrell, J.G. Hutchison, D.J. Coleman, R.J. Gross, N.K. Fry, B. Rowe and S.R. Palmer. 1989. A christening party outbreak of haemorrhagic colitis and haemolytic uraemic syndrome associated with *Escherichia coli* O157.H7. Epidemiol. Infect. *103*: 249–254.

Samain, E., M. Milas, L. Bozzi, G. Dubreucq and M. Rinaudo. 1997. Simultaneous production of two different gel-forming exopolysaccharides by an *Alteromonas* strain originating from deep sea hydrothermal vents. Carbohydr. Polym. *34*: 235–241.

Sambrook, J., E.F. Fritsch and T. Maniatis. 1989. Molecular cloning: a laboratory manual, Cold Spring Harbor Press, Cold Spring Harbor.

Samitz, E.M. and E.L. Biberstein. 1991. *Actinobacillus suis*-like organisms and evidence of hemolytic strains of *Actinonbacillus lignieresii* in horses. Am. J. Vet. Res. *52*: 1245–1251.

Samson, R., A. Arfi and N. Carvil. 1989. Criteria for identification of *Xanthomonas campestris* causing wilt of *Gramineae*. EPPO Bull. *19*: 43–50.

Samson, R. and N. Nassan-Agha. 1978. Biovar and serovars among 129 strains of *Erwinia chrysanthemi*. Station de Pathologie Végétale et Phytobactériologie, Proc. IVth Int. Conf. on Plant Pathogenic Bacteria, Tours, France. Gilbert-Clarey. pp. 547–553.

Samuel, J.E., M.E. Frazier, M.L. Kahn, L.S. Thomashow and L.P. Mallavia. 1983. Isolation and characterization of a plasmid from phase I *Coxiella burnetii*. Infect. Immun. *41*: 488–493.

Samuel, J.E., M.E. Frazier and L.P. Mallavia. 1985. Correlation of plasmid type and disease caused by *Coxiella burnetii*. Infect. Immun. *49*: 775–779.

Sanarelli, G. 1891. Über enine neuen Mikroorganismus des Wassers, welcher für Thiere mit veraenderlichen und konstanter temperature pathogen ist. Zentbl. Bakteriol. Parasitenkd. *9*: 222–228.

Sanchez-Amat, A. and F. Solano. 1997. A pluripotent polyphenol oxidase from the melanogenic marine *Alteromonas* sp. shares catalytic capabilities of tyrosinases and laccases. Biochem. Biophys. Res. Commun. *240*: 787–792.

Sanden, G.N., W.E. Morrill, B.S. Fields, R.F. Breiman and J.M. Barbaree. 1992. Incubation of water samples containing amoebae improves detection of legionellae by the culture method. Appl. Environ. Microbiol. *58*: 2001–2004.

Sanders, W.E. and C.C. Sanders. 1997. *Enterobacter* spp.: Pathogens poised to flourish at the turn of the century. Clin. Microbiol. Rev. *10*: 220–241.

Sanderson, K.E., A. Hessel, S.L. Liu and K.E. Rudd. 1996. The genetic map of *Salmonella typhimurium*, edition VIII. *In* Curtiss., Ingraham, Lin, Low, Magasanik, Reznikoff, Riley, Schaechter and Umbarger (Editors), *Escherichia coli* and *Salmonella cellular* and Molecular Biology, 2 Ed., Vol. 2, ASM Press, Washington, DC. pp. 1903–1999.

Sands, D.C., F.H. Gleason and D.C. Hildebrand. 1967. Cytochromes of *Pseudomonas syringae*. J. Bacteriol. *94*: 1785–1786.

Sands, D.C., L. Hankin and M. Zucker. 1972. A selective medium for pectolytic fluorescent pseudomonads. Phytopathology *62*: 998–1000.

Sands, D.C. and A.D. Rovira. 1970. Isolation of fluorescent pseudomonads with a selective medium. Appl. Microbiol. *20*: 513–514.

Sands, D.C. and A.D. Rovira. 1971. *Pseudomonas fluorescens* biotype G, the dominant fluorescent pseudomonad in south Australian soils and wheat rhizospheres. J. Appl. Bacteriol. *34*: 261–275.

Sands, D.C., M.N. Schroth and D.C. Hildebrand. 1970. Taxonomy of phytopathogenic pseudomonads. J. Bacteriol. *101*: 9–23.

Sands, D.C., M.N. Schroth and D.C. Hildebrand. 1980. *Pseudomonas. In* Schaad (Editor), Laboratory Guide for Identification of Plant Pathogenic Bacteria, American Phytopathological Society, St. Paul, Minn. 36–44.

Sandström, G., S. Löfgren and A. Tärnvik. 1988. A capsule-deficient mutant of *Francisella tularensis* LVS exhibits enhanced sensitivity to killing by serum but diminished sensitivity to killing by polymorphonuclear leukocytes. Infect. Immun. *56*: 1194–1202.

Sandström, G., A. Sjöstedt, T. Johansson, K. Kuoppa and J.C. Williams. 1992. Immunogenicity and toxicity of lipopolysaccharide from *Francisella tularensis* LVS. FEMS (Fed. Eur. Microbiol. Soc.) Microbiol. Immunol. *105*: 201–210.

Sandström, G.E., H. Wolf-Watz and A. Tärnvik. 1986. Duct ELISA for detection of bacteria in fluid samples. J. Microbiol. Methods *5*: 41–48.

Sandström, J. and J. Pettersson. 1994. Amino acid composition of phloem sap and the relation to intraspecific variation in pea aphid (*Acyrthosiphon pisum*) performance. J. Insect Physiol. *40*: 947–955.

Sansonetti, P.J., J. Arondel, A. Fontaine, H. d'Hauteville and M.L. Bernardini. 1991. *ompB* (osmo-regulation) and *icsA* (cell-to-cell spread) mutants of *Shigella flexneri*: vaccine candidates and probes to study the pathogenesis of shigellosis. Vaccine *9*: 416–422.

Sansonetti, P.J., T.L. Hale, G.J. Dammin, C. Kapfer, H.H. Collins, Jr. and S.B. Formal. 1983. Alterations in the pathogenicity of *Escherichia coli* K-12 after transfer of plasmid and chromosomal genes from *Shigella flexneri*. Infect. Immun. *39*: 1392–1402.

Sansonetti, P.J., T.L. Hale and E.V. Oaks. 1985. Genetics of virulence in enteroinvasive *Escherichia coli*. *In* Schlessinger (Editor), Microbiology, American Society for Microbiology, Washington, D.C. pp. 74–77.

Sansonetti, P.J., D.J. Kopecko and S.B. Formal. 1981. *Shigella sonnei* plasmids: evidence that a large plasmid is necessary for virulence. Infect. Immun. *34*: 75–83.

Sansonetti, P.J., D.J. Kopecko and S.B. Formal. 1982. Involvement of a plasmid in the invasive ability of *Shigella flexneri*. Infect. Immun. *35*: 852–860.

Santos, J.A., T.M. Lopez-Diaz, M.C. Garcia-Fernandez, M.L. Garcia-Lopez and A. Otero. 1996. Villalon, a fresh ewe's milk Spanish cheese, as a source of potentially pathogenic *Aeromonas* strains. J. Food Prot. *59*: 1288–1291.

Santos, J., T.M. Lopez-Diaz, M.L. Garcia-Lopez, M.C. Garcia-Fernandez and A. Otero. 1994. Minimum water activity for the growth of *Aeromonas hydrophila* as affected by strain, temperature and humectant. Lett. Appl. Microbiol. *19*: 76–78.

Sar, N. and E. Rosenberg. 1989. Colonial differentiation and hydrophobicity of a *Vibrio* sp. Curr. Microbiol. *18*: 331–334.

Sarafian, S.K., S.R. Johnson, M.L. Thomas and J.S. Knapp. 1991. Novel plasmid combinations in *Haemophilus ducreyi* isolates from Thailand. J. Clin. Microbiol. *29*: 2333–2334.

Saralov, A.I., I.N. Krylova, E.E. Saralova and S.I. Kuznetsov. 1984. The distribution and species composition of methane-oxidizing bacteria in lake water. Mikrobiologiya *53*: 695–700.

Sareneva, T., H. Holthöfer and T.K. Korhonen. 1990. Tissue-binding affinity of *Proteus mirabilis* fimbriae in the human urinary tract. Infect. Immun. *58*: 3330–3336.

Sarff, L.D., G.H. McCracken, M.S. Schiffer, M.P. Glode, J.B. Robbins, I. Ørskov and F. Ørskov. 1975. Epidemiology of *Escherichia coli* K1 in healthy and diseased newborns. Lancet *1*: 1099–1104.

Sarker, R.I., W. Ogawa, M. Tsuda, S. Tanaka and T. Tsuchiya. 1994. Characterization of a glucose transport system in *Vibrio parahaemolyticus*. J. Bacteriol. *176*: 7378–7382.

Sarniguet, A., J. Kraus, M.D. Henkels, A.M. Muehlchen and J.E. Loper. 1995. The sigma factor sigmas affects antibiotic production and biological control activity of *Pseudomonas fluorescens* Pf-5. Proc. Natl. Acad. Sci. U. S. A. *92*: 12255–12259.

Sasaki, T., T. Aoki, H. Hayashi and H. Ishikawa. 1990. Amino acid composition of the honeydew of symbiotic and aposymbiotic pea aphids *Acyrthosiphon pisum*. J. Insect Physiol. *36*: 35–40.

Sasaki, T., H. Hayashi and H. Ishikawa. 1991. Growth and reproduction of the symbiotic and aposymbiotic pea aphids, *Acyrthosiphon pisum* maintained on artificial diets. J. Insect Physiol. *37*: 749–756.

Sasaki, T. and H. Ishikawa. 1995. Production of essential amino acids from glutamate by mycetocyte symbionts of the pea aphid, *Acyrthosiphon pisum*. J. Insect Physiol. *41*: 41–46.

Saslaw, S. and H.N. Carlisle. 1961. Studies with tularemia vaccine in volunteers. IV. *Brucella agglutinins* in vaccinated and nonvaccinated volunteers challenged with *Pasteurella tularensis*. Am. J. Med. Sci. *242*: 166–172.

Saslaw, S., H.T. Eigelsbach, J.A. Prior, H.E. Wilson and S. Carhart. 1961. Tularemia vaccine study. II. Respiratory challenge. Arch. Intern. Med. *107*: 702–714.

Sato, G., M. Asagi, C. Oka, N. Ishiguro and N. Terakado. 1978. Transmissible citrate-utilizing ability in *Escherichia coli* isolated from pigeons, pigs and cattle. Microbiol. Immunol. *22*: 357–360.

Sato, S. and H. Ishikawa. 1997a. Expression and control of an operon from an intracellular symbiont which is homologous to the *groE* operon. J. Bacteriol. *179*: 2300–2304.

Sato, S. and H. Ishikawa. 1997b. Structure and expression of the *dnaKJ* operon of *Buchnera*, an intracellular symbiotic bacteria of aphid. J. Biochem. (Tokyo) *122*: 41–48.

Satomi, M. , B. Kimura, T. Hamada, S. Harayama and T. Fujii. 2002. Phylogenetic study of the genus *Oceanospirillum* based on 16S rRNA and *gyrB* genes: emended description of the genus *Oceanospirillum*, description of *Pseudospirillum* gen. nov., *Oceanobacter* gen. nov and *Terasakiella* gen. nov and transfer of *Oceanospirillum jannaschii* and

Pseudomonas stanieri to *Marinobacterium* as *Marinobacterium jannaschii* comb. nov and *Marinobacterium stanieri* comb. nov. Int. J. Syst. Evol. Microbiol. *52*: 739–747.

Satomi, M., B. Kimura, M. Hayashi, Y. Shouzen, M. Okuzumi and T. Fujii. 1998. *Marinospirillum* gen. nov., with descriptions of *Marinospirillum megaterium* sp. nov., isolated from kusaya gravy, and transfer of *Oceanospirillum minutulum* to *Marinospirillum minutulum* comb. nov. Int. J. Syst. Bacteriol. *48*: 1341–1348.

Satomi, M., B. Kimura, G. Takahashi and T. Fujii. 1997. Microbial diversity in kusaya gravy. Fish. Sci. *63*: 1019–1023.

Satta, G., R. Pompei, G. Grazi and G. Cornaglia. 1988. Phosphatase activity is a constant feature of all isolates of all major species of the family Enterobacteriaceae. J. Clin. Microbiol. *26*: 2637–2641.

Saunders, N.A., N. Doshi and T.G. Harrison. 1992. A second serogroup of *Legionella erythra* serologically indistinguishable from *Legionella rubrilucens*. J. Appl. Bacteriol. *72*: 262–265.

Saunders, N.A., T.G. Harrison, A. Haththotuwa and A.G. Taylor. 1991. A comparison of probes for restriction fragment length polymorphism (RFLP) typing of *Legionella pneumophila* serogroup 1 strains. J. Med. Microbiol. *35*: 152–158.

Savage, D.D., R.L. Kagan, N.A. Young and A.E. Horvath. 1977. *Cardiobacterium hominis endocarditis*: description of two patients and characterization of the organism. J. Clin. Microbiol. *5*: 75–80.

Savard, L., J.N.G. Hutchinson and T.M. Dowhanick. 1994. Characterization of different isolates of *Obesumbacterium proteus* using random amplified polymorphic DNA. J. Am. Soc. Brew. Chem. *52*: 62–65.

Savarino, S.J., A. Fasano, D.C. Robertson and M.M. Levine. 1991. Enteroaggregative *Escherichia coli* elaborate a heat-stable enterotoxin demonstrable in an *in vitro* rabbit intestinal model. J. Clin. Invest. *87*: 1450–1455.

Savarino, S.J., A. Fasano, J. Watson, B.M. Martin, M.M. Levine, S. Guandalini and P. Guerry. 1993. Enteroaggregative *Escherichia coli* heat-stable enterotoxin 1 represents another subfamily of *E. coli* heat-stable toxin. Proc. Natl. Acad. Sci. U.S.A. *90*: 3093–3097.

Savarino, S.J., A. McVeigh, J. Watson, A. Cravioto, J. Molina, P. Echeverria, M.K. Bhan, M.M. Levine and A. Fasano. 1996. Enteroaggregative *Escherichia coli* heat-stable enterotoxin is not restricted to enteroaggregative *E. coli*. J. Infect. Dis. *173*: 1019–1022.

Savinelli, E.A. and L.P. Mallavia. 1990. Comparison of *Coxiella burnetii* plasmids to homologous chromosomal sequences present in a plasmidless endocarditis-causing isolate. Ann. N. Y. Acad. Sci. *590*: 523–533.

Săvulescu, T. 1947. Contribution à la classification des bactériacées phytopathogénes. Anal. Acad. Romane Ser. III *22*: 1–26.

Sawabe, T., H. Makino, M. Tatsumi, K. Nakano, K. Tajima, M.M. Iqbal, I. Yumoto, Y. Ezura and R. Christen. 1998a. *Pseudoalteromonas bacteriolytica* sp. nov., a marine bacterium that is the causative agent of red spot disease of *Laminaria japonica*. Int. J. Syst. Bacteriol. *48*: 769–774.

Sawabe, T., I. Sugimura, M. Ohtsuka, K. Nakano, K. Tajima, Y. Ezura and R. Christen. 1998b. *Vibrio halioticoli* sp. nov., a non-motile alginolytic marine bacterium isolated from the gut of the abolone *Haliotis discus hannai*. Int. J. Syst. Bacteriol. *48*: 573–580.

Sawabe, T., R. Tanaka, M.M. Iqbal, K. Tajima, Y. Ezura, E.P. Ivanova and R. Christen. 2000. Assignment of *Alteromonas elyakovii* KMM162[T] and five strains isolated from spot-wounded fronds of *Laminaria japonica* to *Pseudoalteromonas elyakovii* comb. nov. and the extended description of the species. Int. J. Syst. Evol. Bacteriol. *50*: 265–271.

Sawle, G.V., B.C. Das, P.R. Acland and D.A. Heath. 1986. Fatal infection with *Aeromonas sobria* and *Plesiomonas shigelloides*. Br. Med. J. Clin. Res. *292*: 525–526.

Sawula, R.V. and I.P. Crawford. 1972. Mapping of the tryptophan genes of *Acinetobacter calcoaceticus* by transformation. J. Bacteriol. *112*: 797–805.

Sawyer, M.H., P. Baumann and L. Baumann. 1977a. Pathways of D-fructose and D-glucose catabolism in marine species of *Alcaligenes*, *Pseudomonas marina* and *Alteromonas communis*. Arch. Microbiol. *112*: 169–172.

Sawyer, M.H., P. Baumann, L. Baumann, S.M. Berman, J.L. Canovas and

R.H. Berman. 1977b. Pathways of D-fructose metabolism in species of *Pseudomonas*. Arch. Microbiol. *112*: 49–55.

Scaletsky, I.C.A., M.L.M. Silva, M.R.F. Toledo, B.R. Davis,, P.A. Blake and L.R. Trabulsi. 1985. Correlation between adherence to HeLa cells and serogroups, serotypes, and bioserotypes of *Escherichia coli*. Infect. Immun. *49*: 528-532.

Scaletsky, I.C., M.L. Silva and L.R. Trabulsi. 1984. Distinctive patterns of adherence of enteropathogenic *Escherichia coli* to HeLa cells. Infect. Immun. *45*: 534–536.

Scally, C.M. and F.G. Winder. 1991. Deoxyribonuclease from *Pseudomonas maltophilia*. Biochem. Soc. Trans. *19*: 40S.

Scanferlato, V.S., D.R. Orvos, J.J. Cairns and G.H. Lacy. 1989. Genetically engineered *Erwinia carotovora* in aquatic microcosms: survival and effects on functional groups of indigenous bacteria. Appl. Environ. Microbiol. *55*: 1477–1482.

Scarlett, C.M., J.T. Fletcher, P. Roberts and R.A. Lelliott. 1978. Tomato pith necrosis caused by *Pseudomonas corrugata* n. sp. Ann. Appl. Biol. *88*: 105–114.

Scarlett, M. 1916. Infections cornéennes a diplobacilles. Notes sur deux diplobacilles non encore décrit (*Bacillus duplex nonliquefaciens* et *Bacillus duplex josefi*). Ann. Ocul. *153*: 100–111.

Schaad, N.W. 1976. Immunological comparison and characterization of ribosomes of *Xanthomonas vesicatoria*. Phytopathology *66*: 770–776.

Schaad, N.W. 1978. Use of direct and indirect immunofluorescence tests for identification of *Xanthomonas campestris*. Phytopathology *68*: 249–252.

Schaad, N.W. 1979. Serological identification of plant pathogenic bacteria. Annu. Rev. Phytopathol. *17*: 123–147.

Schaad, N.W., G. Sowell, Jr., R.W. Goth, R.R. Colwell and R.E. Webb. 1978. *Pseudomonas pseudoalcaligenes* subsp. *citrulli* subsp. nov. Int. J. Syst. Bacteriol. *28*: 117–125.

Schaad, N.W. and R.E. Stall. 1988. *Xanthomonas*. *In* Schaad (Editor), Laboratory Guide for Identification of Plant Pathogenic Bacteria, 2nd Ed., APS Press, Minneapolis. pp. 81–94..

Schable, B., D.L. Rhoden, R. Hugh, R.E. Weaver, N. Khardori, P.B. Smith, G.P. Bodey and R.L. Anderson. 1989. Serological classification of *Xanthomonas maltophilia* (*Pseudomonas maltophilia*) based on heat-stable O antigens. J. Clin. Microbiol. *27*: 1011–1014.

Schadow, K.H., D.K. Giger and C.C. Sanders. 1993. Failure of the Vitek AutoMicrobic system to detect beta-lactam resistance in *Aeromonas* species. Am. J. Clin. Pathol. *100*: 308–310.

Schaeffer, P. 1958. Interspecific reactions in bacterial transformation. Gene *11*: 311–318.

Schauer, D.B. and S. Falkow. 1993. Attaching and effacing locus of a *Citrobacter freundii* biotype that causes transmissible murine colonic hyperplasia. Infect. Immun. *61*: 2486–2492.

Schauer, D.B., B.A. Zabel, I.F. Pedraza, C.M. O'Hara, A.G. Steigerwalt and D.J. Brenner. 1995. Genetic and biochemical characterization of *Citrobacter rodentium* sp. nov. J. Clin. Microbiol. *33*: 2064–2068.

Schauer, D.B., B.A. Zabel, I.F. Pedraza, C.M. O'Hara, A.G. Steigerwalt and D.J. Brenner. 1996. *In* Validation of the publication of new names and new combinations previously effectively published outside the IJSB. List No. 56. Int. J. Syst. Bacteriol. *46*: 362–363.

Scheel, O., T. Hoel, T. Sandvik and B.P. Berdal. 1993. Susceptibility pattern of Scandinavian *Francisella tularensis* isolates with regard to oral and parenteral antimicrobial agents. APMIS *101*: 33–36.

Scheel, O., R. Reiersen and T. Hoel. 1992. Treatment of tularemia with ciprofloxacin. Eur. J. Clin. Microbiol. Infect. Dis. *11*: 447–448.

Scheminzky, F., Z. Klas and C. Job. 1972. Über das Vorkommen von Thiobacterium bovista in Thermalwässern. Int. Rev. Gestamen Hydrobiol. *57*: 801–813.

Scheutz, F., B. Olesen, J. Engberg, A.M. Petersen, K. Mølbak, P. Schiellerup and P. Gerner-Smidt. 2001. Clinical features and epidemiology of infections by vero cytotoxigenic *E. coli* (VTEC) from Danish patients 1997–2000, and characterisation of VTEC isolates by serotypes and virulence factors. Proceedings from the EU Concerted Action Conference on Epidemiology of VTEC and Workshop on Typing methods for VTEC Strains, Malahide, Ireland. *Vol. 5*: 58–66.

Schewiakoff, W. 1893. Über einen neuen bacterienähnlichen Organismus des Susswassers., Habilitationsschrift, C. Winter, Universität Heidelberg. 1–36.

Schiemann, D. 1979. Synthesis of a selective agar medium for *Yersinia enterocolitica*. Can. J. Microbiol. *25*: 1298–1304.

Schiewe, M.H., J.H. Crosa and E.J. Ordal. 1977. Deoxyribonucleic acid relationships among marine vibrios pathogenic to fish. Can. J. Microbiol. *23*: 954–958.

Schiewe, M.H., T.J. Trust and J.H. Crosa. 1981. *Vibrio ordalii* sp. nov.: A causative agent of vibriosis in fish. Curr. Microbiol. *6*: 343–348.

Schiewe, M.H., T.J. Trust and J.H. Crosa. 1982. *In* Validation of new names and new combinations previously effectively published outside the IJSB. List No. 9. Int. J. Syst. Bacteriol. *32*: 384–385.

Schilf, W. and V. Krishnapillai. 1986. Genetic analysis of insertion mutations of the promiscuous IncP-1 plasmid R18 mapping near *oriT* which affect its host range. Plasmid *15*: 48–56.

Schillinger, I. 1997. Ulcerative infectious stomatitis in ophidians. Point Vet. *28*: 35–39.

Schindler, J., B. Potužníková and E. Aldová. 1992. Classification of strains of *Pragia fontium*, *Budvicia aquatica* and of *Leminorella* by whole-cell protein pattern. J. Hyg. Epidemiol. Microbiol. Immunol. (Prague) *36*: 207–216.

Schink, B. and N. Pfennig. 1982a. Fermentation of trihydroxybenzenes by *Pelobacter acidigallici*, new genus new species: a new strictly anaerobic, non-spore-forming bacterium. Arch. Microbiol. *133*: 195–201.

Schink, B. and N. Pfennig. 1982b. *Propionigenium modestum*, gen. nov., sp. nov.: a new strictly anaerobic, non-spore-forming bacterium growing on succinate. Arch. Microbiol. *133*: 209–216.

Schink, B., T.E. Thompson and J.G. Zeikus. 1982. Characterization of *Propionispira arboris*, gen. nov., sp. nov., a nitrogen-fixing anaerobe common to wetwoods of living trees. J. Gen. Microbiol. *128*: 2771–2780.

Schiøtz, P.O., N. Høiby and J.B. Hertz. 1979. Cross-reaction between *Haemophilus influenzae* and nineteen other bacterial species. Acta Pathol. Microbiol. Scand. Sect. B Microbiol. *87*: 337–344.

Schirmer, A. and D. Jendrossek. 1994. Molecular characterization of the extracellular poly(3-Hydroxyoctanoic Acid) [P(3HO)] depolymerase gene of *Pseudomonas fluorescens* Gk13 and of its gene product. J. Bacteriol. *176*: 7065–7073.

Schirmer, A., D. Jendrossek and H.G. Schlegel. 1993. Degradation of poly(3-hydroxyoctanoic acid) [P(3HO)] by bacteria - purification and properties of a P(3HO) depolymerase from *Pseudomonas fluorescens* Gk13. Appl. Environ. Microbiol. *59*: 1220–1227.

Schirmer, A., C. Matz and D. Jendrossek. 1995. Substrate specificities of poly(hydroxyalkanoate)-degrading bacteria and active-site studies on the extracellular poly(3-hydroxyoctanoic acid) depolymerase of *Pseudomonas fluorescens* Gk13. Can. J. Microbiol. *41*: 170–179.

Schirmer, F., S. Ehrt and W. Hillen. 1997. Expression, inducer spectrum, domain structure, and function of MopR, the regulator of phenol degradation in *Acinetobacter calcoaceticus* NCIB8250. J. Bacteriol. *179*: 1329–1336.

Schlater, L.K., D.J. Brenner, A.G. Steigerwalt, C.W. Moss, M.A. Lambert and R.A. Packer. 1989. *Pasteurella caballi*, a new species from equine clinical specimens. J. Clin. Microbiol. *27*: 2169–2174.

Schlater, L.K., D.J. Brenner, A.G. Steigerwalt, C.W. Moss, M.A. Lambert and R.A. Packer. 1990. *In* Validation of publication of new names and new combinations previously effectively published outside the IJSEM. List No. 34. Int. J. Syst. Bacteriol. *40*: 320–321.

Schlegel, H.G. and R.M. Lafferty. 1971. Novel energy and carbon sources. A. The production of biomass from hydrogen and carbon dioxide. Adv. Biochem. Eng. *1*: 143–168.

Schleifer, K.H., R. Amann, W. Ludwig, C. Rothemund, N. Springer and S. Dorn. 1992. Nucleic acid probes for the identification and in situ detection of pseudomonads. *In* Galli, Silver and Witholt (Editors), *Pseudomonas*: Molecular Biology and Biotechnology, American Society for Microbiology, Washington D.C. 127–134.

Schleifer, K.H., W. Ludwig and R. Amann. 1993. Nucleic acid probes. *In*

Goodfellow and O'Donnell (Editors), Handbook of new bacterial systematics, Academic Press Limited, London. 464–512.

Schleifstein, J. and M.B. Coleman. 1943. *Bacterium enterocoliticum*, N.Y. State Dep. Health Div. Lab. Res. Annu. Report. . 56

Schleissner, C., A. Reglero and J.M. Luengo. 1997. Catabolism of D-glucose by *Pseudomonas putida* U occurs via extracellular transformation into D-gluconic acid and induction of a specific gluconate transport system. Microbiology *143*: 1595–1603.

Schmid, H., M. Hartung and E. Hellmann. 1991. Crossed Immunoelectrophoresis applied to representative strains from 11 different *Pasteurella* species under taxonomic aspects. Zentbl. Bakteriol. Mikrobiol. Hyg. Abt. Orig. A Med. Mikrobiol. Infektkrankh. Parasitol. *275*: 16–27.

Schmidt, G.L., G. Nicolson and M.D. Kamem. 1971. Composition of the sulfur particle of *Chromatium vinosum* strain D. J. Bacteriol. *105*: 1137–1141.

Schmidt, H., L. Beutin and H. Karch. 1995a. Molecular analysis of the plasmid-encoded hemolysin of *Escherichia coli* O157:H7 strain EDL 933. Infect. Immun. *63*: 1055–1061.

Schmidt, H., H. Karch and L. Beutin. 1994. The large-sized plasmids of enterohemorrhagic *Escherichia coli* O157 strains encode hemolysins which are presumably members of the *E. coli* α-hemolysin family. FEMS Microbiol. Lett. *117*: 189–196.

Schmidt, H., C. Kernbach and H. Karch. 1996a. Analysis of the EHEC *hly* operon and its location in the physical map of the large plasmid of enterohaemorrhagic *Escherichia coli* O157:H7. Microbiology *142*: 907–914.

Schmidt, H., C. Knop, S. Franke, S. Aleksic, J. Heesemann and H. Karch. 1995b. Development of PCR for screening of enteroaggregative *Escherichia coli*. J. Clin. Microbiol. *33*: 701–705.

Schmidt, H., E. Maier, H. Karch and R. Benz. 1996b. Pore-forming properties of the plasmid-encoded hemolysin of enterohemorrhagic *Escherichia coli* O157:H7. Eur. J. Biochem. *241*: 594–601.

Schmidt, H., J. Scheef, S. Morabito, A. Caprioli, L.H. Wieler and H. Karch. 2000. A new Shiga toxin 2 variant (Stx2f) from *Escherichia coli* isolated from pigeons. Appl. Environ. Microbiol. *66*: 1205–1208.

Schmidt, K. 1978. Biosynthesis of carotenoids. *In* Clayton and Sistrom (Editors), The Photosynthetic Bacteria, Plenum Press, New York. pp. 729–750.

Schmidt, K., N. Pfennig and S. Liaaen-Jensen. 1965. Carotenoids of *Thiorodaceae*. IV. The carotenoid composition of 25 pure isolates. Arch. Mikrobiol. *52*: 132–146.

Schmidt, K.D., B. Tümmler and U. Römling. 1996c. Comparative genome mapping of *Pseudomonas aeruginosa* PAO with *P. aeruginosa* C, which belongs to a major clone in cystic fibrosis patients and aquatic habitats. J. Bacteriol. *178*: 85–93.

Schmidt, K. and H.G. Trüper. 1971. Carotenoid composition in the genus *Ectothiorhodospira* Pelsh. Arch. Mikrobiol. *80*: 38–42.

Schmidt, T.M., B. Arieli, Y. Cohen, E. Padan and W.R. Strohl. 1987. Sulfur metabolism in *Beggiatoa alba*. J. Bacteriol. *169*: 5466–5472.

Schmidt, T.M. and A.A. Dispirito. 1990. Spectral characterization of c-type cytochromes purified from *Beggiatoa alba*. Arch. Microbiol. *154*: 453–458.

Schmidt, T.M., V.A. Vinci and W.R. Strohl. 1986. Protein synthesis by *Beggiatoa alba* B18LD in the presence and absence of sulfide. Arch. Microbiol. *144*: 158–162.

Schmidt, W.C. and C.D. Jeffries. 1974. Bacteriophage typing of *Proteus mirabilis*, *Proteus vulgaris* and *Proteus morganii*. Appl. Microbiol. *27*: 47–53.

Schmitt, C.K., M.L. McKee and A.D. O'Brien. 1991. Two copies of Shiga-like toxin II-related genes common in enterohemorrhagic *Escherichia coli* strains are responsible for the antigenic heterogeneity of the O157:H− strain E32511. Infect. Immun. *59*: 1065–1073.

Schmitz, K.E.F. 1917. Ein neuer Typus ans der Gruppe der Ruhrbazillen als Erreger einer groesser Epidemie. Ztschr. Hyg. Infektionskr. *84*: 449–516.

Schnaitman, C.A. 1970. Comparison of the envelope protein compositions of several Gram-negative bacteria. J. Bacteriol. *104*: 1404–1405.

Schnaitman, C.A. and J.D. Klena. 1993. Genetics of lipopolysaccharide biosynthesis in enteric bacteria. Microbiol. Rev. *57*: 655–682.

Schnathorst, W.C. 1966. Unaltered specificity in several xanthomonads after repeated passage through *Phaseolus vulgaris*. Phytopathology *56*: 58–60.

Schneider, K., A. Muller, U. Schramm and W. Klipp. 1991. Demonstration of a molybdenum-independent and vanadium-independent nitrogenase in a *nifHDK* deletion mutant of *Rhodobacter capsulatus*. Eur. J. Biochem. *195*: 653–661.

Schober, B.M. and J.W.L. Van Vuurde. 1997. Detection and enumeration of *Erwinia carotovora* subsp. atroseptica using spiral plating and immunofluorescence colony staining. Can. J. Microbiol. *43*: 847–853.

Schoner, B. and R.G. Schoner. 1981. Distribution of IS5 in bacteria. Gene *16*: 347–352.

Schoonejans, E., D. Expert and A. Toussaint. 1987. Characterization and virulence properties of *Erwinia chrysanthemi* lipopolysaccharide-defective, vphi-EC2-resistant mutants. J. Bacteriol. *169*: 4011–4017.

Schoonmaker, D., T. Heimberger and G. Birkhead. 1992. Comparison of ribotyping and restriction enzyme analysis using pulsed-field gel electrophoresis for distinguishing *Legionella pneumophila* isolates obtained during a nosocomial outbreak. J. Clin. Microbiol. *30*: 1491–1498.

Schramek, S. and H. Mayer. 1982. Different sugar compositions of lipopolysaccharides isolated from phase I and pure phase II cells of *Coxiella burnetii*. Infect. Immun. *38*: 53–57.

Schramek, S., J. Radziejewska-Lebrecht and H. Mayer. 1985. 3-C-branched aldoses in lipopolysaccharide of phase I *Coxiella burnetii* and their role as immunodominant factors. Eur. J. Biochem. *148*: 455–461.

Schröder, D., H. Deppisch, M. Obermayer, G. Krohne, E. Stackebrandt, B. Hölldobler, W. Goebel and R. Gross. 1996. Intracellular endosymbiotic bacteria of *Camponotus* species (carpenter ants): systematics, evolution and ultrastructural characterization. Mol. Microbiol. *21*: 479–489.

Schroeter, J. 1872. Ueber einige durch Bacterien gebildete Pigmente. *In* Cohn (Editor), Beiträge zür Biologie der Pflanzen, J.U. Kern's Verlag, Breslau. 109–126.

Schroeter, J. 1885–1889. Kryptogamenflora von Schlesien. Bd. 3 Heft 3, Pilze. *In* Cohn (Editor), Breslau, J.U. Kern's Verlag. .

Schroth, M.N. and D.C. Hildebrand. 1980. *E. amylovora* or true erwiniae group. *In* Schaad (Editor), Laboratory Guide for Identification of Plant Pathogenic Bacteria, American Phytopathological Society, St. Paul. pp. 26–30.

Schroth, M.N., D.C. Hildebrand and M.P. Starr. 1981. Phytopathogenic members of the genus *Pseudomonas*. *In* Starr, Stolp, Trüper and Balows (Editors), The Prokaryotes, A handbook on habitats, isolation and identification of the bacteria, Springer-Verlag, Berlin. 701–718.

Schubert, R.H.W. 1967a. The occurrence of *Aeromonas hydrophila* in surface waters. Arch. Hyg. Bakteriol. *150*: 688–708.

Schubert, R.H.W. 1967b. The taxonomy and nomenclature of the genus *Aeromonas* Kluyver and Van Niel 1936. Part I. Suggestions on the taxonomy and nomenclature of aerogenic *Aeromonas* species. Int. J. Syst. Bacteriol. *17*: 23–37.

Schubert, R.H.W. 1967c. The taxonomy and nomenclature of the genus *Aeromonas* Kluyver and Van Niel 1936. Part II. Suggestions on the taxonomy and nomenclature of the anaerogenic *Aeromonas* species. Int. J. Syst. Bacteriol. *17*: 273–279.

Schubert, R.H. 1969. *Aeromonas hydrophila* subsp. *proteolytica* (Merkel et al. 1964) comb. nov. Zentbl. Bakteriol. *211*: 409–412.

Schubert, R.H.W. 1971. Status of the names *Aeromonas* and *Aerobacter liquefaciens* Beijerinck and designation of a neotype strain for *Aeromonas hydrophila*. Request for an opinion. Int. J. Syst. Bacteriol. *21*: 87–90.

Schubert, R.H.W. 1974. Genus II. *Aeromonas*. *In* Buchanan and Gibbons (Editors), Bergey's Manual of Determinative Bacteriology, 8th Ed., The Williams & Wilkins Co., Baltimore. 345–348.

Schubert, R.H.W. 1975. The relationship of aerogenic to anaerogenic aeromonads of the "hydrophila-punctata-group" in river water depending on the load of waste. Zentbl. Bakteriol. Mikrobiol. Hyg. Ser. B. *160*: 237–245.

Schubert, R.H.W. 1984. Genus IV. *Plesiomonas*. *In* Krieg and Holt (Editors), Bergey's Manual of Systematic Bacteriology, 1st Ed., Vol. 1, The Williams & Wilkins Co., Baltimore. pp. 548–550.

Schubert, R.H.W. 1987. Ecology of aeromonads and isolation from environmental samples. Experientia (Basel) *43*: 351–354.

Schubert, R.H.W. and S. Groeger-Söhn. 1998. Detection of *Budvicia aquatica* and *Pragia fontium* and occurrence in surface waters. Zentbl.Hyg. Umweltmed. *201*: 371–376.

Schubert, R.H.W. and M. Hegazi. 1988a. *Aeromonas eucrenophila*, new species *Aeromonas caviae* a later and illegitimate synonym of *Aeromonas punctata*. Zentbl. Bakteriol. Mikrobiol. Hyg. Ser. A *268*: 34–39.

Schubert, R.H.W. and M. Hegazi. 1988b. *In* Validation of the publication of new names and new combinations previously effectively published outside the IJSB. List No. 27. Int. J. Syst. Bacteriol. *38*: 449.

Schubert, R.H.W., M. Hegazi and W. Wahlig. 1990a. *Aeromonas enteropelogenes* species nova. Hyg. Med. *15*: 471–472.

Schubert, R.H.W., M. Hegazi and W. Wahlig. 1990b. *Aeromonas ichthiosmia* species nova. Hyg. Med. *15*: 477–479.

Schuch, R. and A.T. Maurelli. 1997. Virulence plasmid instability in *Shigella flexneri* 2a is induced by virulence gene expression. Infect. Immun. *65*: 3686–3692.

Schulte, R. and U. Bonas. 1992a. Expression of the *Xanthomonas campestris* pv. *vesicatoria hrp* gene cluster, which determines pathogenicity and hypersensitivity on pepper and tomato, is plant inducible. J. Bacteriol. *174*: 815–823.

Schulte, R. and U. Bonas. 1992b. A *Xanthomonas* pathogenicity locus is induced by sucrose and sulfur-containing amino acids. Plant Cell *4*: 79–86.

Schulz, H.N., T. Brinkhoff, T.G. Ferdelman, M. Hernández Mariné, A. Teske and B.B. Jørgensen. 1999a. Dense populations of a giant sulfur bacterium in Namibian shelf sediments. Science *284*: 493–495.

Schulz, H.N., T. Brinkhoff, T.G. Ferdelman, M. Hernández Mariné, A. Teske and B.B. Jørgensen. 1999b. *In* Validation of the publication of new names and new combinations previously effectively published outside the IJSB. List No. 71. Int. J. Syst. Bacteriol. *49*: 1325–1326.

Schulz, H.N. and D. De Beer. 2002. Uptake rates of oxygen and sulfide measured with individual *Thiomargarita namibiensis* cells by using microelectrodes. Appl. Environ. Microbiol. *68*: 5746–5749.

Schulz, H.N. and B.B. Jørgensen. 2001. Big bacteria. Annu. Rev. Microbiol. *55*: 105–137.

Schulz, H.N., B.B. Jørgensen, H.A. Fossing and N.B. Ramsing. 1996. Community structure of filamentous, sheath-building sulfur bacteria, *Thioploca* spp., off the coast of Chile. Appl. Environ. Microbiol. *62*: 1855–1862.

Schulz, H.N., B. Strotmann, V.A. Gallardo and B.B. Jørgensen. 2000. Population study of the filamentous sulfur bacteria *Thioploca* spp. off the Bay of Concepción, Chile. Mar. Ecol. Prog. Ser. *200*: 117–126.

Schuster, M.I. and D.P. Coyne. 1974. Survival mechanisms of phytopathogenic bacteria. Ann. Rev. Phytopathol. *12*: 199–221.

Schwach, T. 1979. Case report. Clin. Microbiol. Newsletter. *1*: 4–5.

Scotland, S.M., B. Rowe, H.R. Smith, G.A. Willshaw and R.J. Gross. 1988. Vero cytotoxin-producing strains of *Escherichia coli* from children with haemolytic uraemic syndrome and their detection by specific DNA probes. J. Med. Microbiol. *25*: 237–243.

Scotland, S.M., H.R. Smith, T. Cheasty, B. Said, G.A. Willshaw, N. Stokes and B. Rowe. 1996. Use of gene probes and adhesion tests to characterise *Escherichia coli* belonging to enteropathogenic serogroups isolated in the United Kingdom. J. Med. Microbiol. *44*: 438–443.

Scotland, S.M., H.R. Smith, B. Said, G.A. Willshaw, T. Cheasty and B. Rowe. 1991. Identification of enteropathogenic *Escherichia coli* isolated in Britain as enteroaggregative or as members of a subclass of attaching-and-effacing *E. coli* not hybridising with the EPEC adherence-factor probe. J. Med. Microbiol. *35*: 278–283.

Scotland, S.M., G.A. Willshaw, T. Cheasty and B. Rowe. 1992. Strains of *Escherichia coli* O157:H8 from human diarrhoea belong to attaching and effacing class of *E coli*. J. Clin. Pathol. *45*: 1075–1078.

Scotland, S.M., G.A. Willshaw, T. Cheasty, B. Rowe and J.E. Hassall. 1994. Association of enteroaggregative *Escherichia coli* with travellers' diarrhoea. J. Infect. *29*: 115–116.

Scotland, S.M., G.A. Willshaw, H.R. Smith, R.J. Gross and B. Rowe. 1989. Adhesion to cells in culture and plasmid profiles of enteropathogenic *Escherichia coli* isolated from outbreaks and sporadic cases of infant diarrhoea. J. Infect. *19*: 237–249.

Scotland, S.M., G.A. Willshaw, H.R. Smith and B. Rowe. 1990. Properties of strains of *Escherichia coli* O26:H11 in relation to their enteropathogenic or enterohemorrhagic classification. J. Infect. Dis. *162*: 1069–1074.

Scott, C.C., S.R. Makula and W.R. Finnerty. 1976. Isolation and characterization of membranes from a hydrocarbon-oxidizing *Acinetobacter* sp. J. Bacteriol. *127*: 469–480.

Scott, D., J. Brannan and I.J. Higgins. 1981. The effect of growth conditions on intracytoplasmic membranes and methane monooxygenase activities in *Methylosinus trichosporium* OB3b. J. Gen. Microbiol. *125*: 63–72.

Scott, D.L., R-G. Zhang and E.M. Westbrook. 1996. The cholera family of enterotoxins. *In* Parker (Editor), Protein Toxin Structure, R.G. Landes and Chapman and Hall, Austin. pp. 123–146.

Scott, J.H. and K.H. Nealson. 1994. A biochemical study of the intermediary carbon metabolism of *Shewanella putrefaciens*. J. Bacteriol. *176*: 3408–3411.

Scotten, H.L. and J.L. Stokes. 1962. Isolation and properties of *Beggiatoa*. Arch. Mikrobiol. *42*: 353–368.

Scriver, S.R., S.L. Walmsley, C.L. Kau, D.J. Hoban, J. Brunton, A. McGeer, T.C. Moore and E. Witwicki. 1994. Determination of antimicrobial susceptibilities of Canadian isolates of *Haemophilus influenzae* and characterization of their beta-lactamases. Canadian *Haemophilus* Study Group. Antimicrob. Agents Chemother. *38*: 1678–1680.

Sears, C.L. and J.B. Kaper. 1996. Enteric bacterial toxins: Mechanisms of action and linkage to intestinal secretion. Microbiol. Rev. *60*: 167–215.

Sebald, M. and M. Véron. 1963. Teneur en bases de l'ADN et classification des vibrions. Ann. Inst. Pasteur (Paris) *105*: 897–910.

Sechter, I., M. Shmilovitz, G. Altmann, R. Seligmann, B. Kretzer, I. Braunstein and C.B. Gerichter. 1983. *Edwardsiella tarda* isolated in Israel between 1961 and 1980. J. Clin. Microbiol. *17*: 669–671.

Sedláak, J. 1973. Present knowledge and aspects of *Citrobacter*. Curr. Top. Microbiol. Immunol. *62*: 41–59.

Sedlácek, I., P. Gerner-Smidt, J. Schmidt and W. Frederiksen. 1993. Genetic relationship of strains of *Haemophilus aphrophilus*, *H. paraphrophilus*, and *Actinobacillus actinomycetemcomitans* studied by ribotyping. Zentbl. Bakteriol. *279*: 51–59.

Sedlák, J. and M. Slajsova. 1966. On the antigenic relationships of certain *Citrobacter* and *Hafnia* cultures. J. Gen. Microbiol. *43*: 151–158.

Seidler, R., D.A. Allen, H. Lockman, R.R. Colwell, S.W. Joseph and O.P. Daily. 1980. Isolation, enumeration and characterization of *Aeromonas* from polluted waters encountered in diving operations. Appl. Environ. Microbiol. *39*: 1010–1018.

Seidler, R.J., M.D. Knittel and C. Brown. 1975. Potential pathogens in the environment: Cultural reactions and nucleic acid studies of *Klebsiella pneumoniae* from clinical and environmental sources. Appl. Microbiol. *29*: 819–825.

Seifert, H., R. Baginski, A. Schulze and G. Pulverer. 1993a. Antimicrobial susceptibility of *Acinetobacter* species. Antimicrob. Agents Chemother. *37*: 750–753.

Seifert, H., R. Baginski, A. Schulze and G. Pulverer. 1993b. The distribution of *Acinetobacter* species in clinical culture materials. Zentbl. Bakteriol. *279*: 544–552.

Seifert, H., L. Dijkshoorn, P. Gerner-Smidt, N. Pelzer, I. Tjernberg and M. Veneechoutte. 1997. Distribution of *Acinetobacter* species on human skin: comparison of phenotypic and genotypic identification methods. J. Clin. Microbiol. *35*: 2819–2825.

Seifert, H., A. Schulze, R. Baginski and G. Pulverer. 1994a. Comparison of four different methods for epidemiologic typing of *Acinetobacter baumannii*. J. Clin. Microbiol. *32*: 1816–1819.

Seifert, H., A. Schulze, R. Baginski and G. Pulverer. 1994b. Plasmid DNA fingerprinting of *Acinetobacter* species other than *Acinetobacter bauman-nii*. J. Clin. Microbiol. *32*: 82–86.

Sekeyova, Z., V. Roux and D. Raoult. 1999. Intraspecies diversity of *Coxiella burnetii* as revealed by *com1* and *mucZ* sequence comparison. FEMS Microbiol. Lett. *180*: 61–67.

Sektas, M., T. Kaczorowski and A.J. Podhajska. 1995. Interaction of the *Mbo*II restriction endonuclease with DNA. Gene *157*: 181–185.

Selander, R.K., D.A. Caugant, H. Ochman, J.M. Musser, M.N. Gilmour and T.S. Whittam. 1986. Methods of multilocus enzyme electrophoresis for bacterial population genetics and systematics. Appl. Environ. Microbiol. *51*: 873–884.

Selander, R.K., D.A. Caugant and T.S. Whittam. 1987. Genetic structure and variation in natural populations of *Escherichia coli*. *In* Neidhardt, Ingraham, Low, Magasanik, Schaechter and Umbarger (Editors), *Escherichia coli* and *Salmonella*: Cellular and Molecular Biology, 1st Ed., American Society for Microbiology, Washington, D.C. pp. 1625–1648.

Selenska-Pobell, S., A. Otto and S. Kutschke. 1998. Identification and discrimination of thiobacilli using ARDREA, RAPD and rep-APD. J. Appl. Microbiol. *84*: 1085–1091.

Sellwood, J. and R.A. Lelliott. 1978. Internal browning of hyacinth caused by *Erwinia rhapontici*. Plant Pathol. (Oxf.) *27*: 120–124.

Semancik, J.S., A.K. Vidaver and J.L. van Etten. 1973. Characterization of a segmented double-helical RNA from bacteriophage ø6. J. Molec. Biol. *78*: 617–625.

Semple, K.M., J.L. Doran and D.W.S. Westlake. 1989. DNA relatedness of oil-field isolates of *Shewanella putrefaciens*. Can. J. Microbiol. *35*: 925–931.

Semple, K.M. and D.W.S. Westlake. 1987. Characterization of iron-reducing *Alteromonas putrefaciens* strains from oil-field fluids. Can. J. Microbiol. *33*: 366–371.

Semrau, J.D., A. Chistoserdov, J. Lebron, A. Costello, J. Davagnino, E. Kenna, A.J. Holmes, R. Finch, J.C. Murrell and M.E. Lidstrom. 1995. Particulate methane monooxygenase genes in methanotrophs. J. Bacteriol. *177*: 3071–3079.

Senda, K., Y. Arakawa, S. Ichiyama, K. Nakashima, H. Ito, S. Ohsuka, K. Shimokata, N. Kato and M. Ohta. 1996. PCR detection of metallo-beta-lactamase gene *blaIMP* in gram-negative rods resistant to broad-spectrum beta-lactams. J. Clin. Microbiol. *34*: 2909–2913.

Senerwa, D., Ø. Olsvik, L.N. Mutanda, K.J. Lindqvist, J.M. Gathuma, K. Fossum and K. Wachsmuth. 1989. Enteropathogenic *Escherichia coli* serotype O111:HNT isolated from preterm neonates in Nairobi, Kenya. J. Clin. Microbiol. *27*: 1307–1311.

Sengha, S.S., A.J. Anderson, A.J. Hacking and E.A. Dawes. 1989. The production of alginate by *Pseudomonas mendocina* in batch and continuous culture. J. Gen. Microbiol. *135*: 795–804.

Senior, B.W. 1977a. The Dienes phenomenon: identification of the determinants of compatibility. J. Gen. Microbiol. *102*: 236–244.

Senior, B.W. 1977b. Typing of *Proteus* strains by proticine production and sensitivity. J. Med. Microbiol. *10*: 7–17.

Senior, B.W. 1983. *Proteus morganii* is less frequently associated with urinary tract infections than *Proteus mirabilis*—an explanation. J. Med. Microbiol. *16*: 317–322.

Senior, B.W. 1993. The production of HlyA toxin by *Proteus penneri* strains. J. Med. Microbiol. *39*: 282–289.

Senior, B.W. 1997. Media for the detection and recognition of the enteropathogen *Providencia alcalifaciens* in faeces. J. Med. Microbiol. *46*: 524–527.

Senior, B.W., M. Albrechtsen and M.A. Kerr. 1988. A survey of IgA protease production among clinical isolates of Proteeae. J. Med. Microbiol. *25*: 27–31.

Senior, B.W. and P. Larsson. 1983. A highly discriminatory multi-typing scheme for *Proteus mirabilis* and *Proteus vulgaris*. J. Med. Microbiol. *16*: 193–202.

Senior, B.W. and S. Vörös. 1989. Discovery of new morganocin types of *Morganella morganii* in strains of diverse serotype and the apparent independence of bacteriocin type from serotype of strains. J. Med. Microbiol. *29*: 89–93.

Serbinov, E.L. 1915. Bacterial necrosis of the cortex of fruit trees incited by *Bacterium amylovorum* (Burrill). Serb. Belonzi Rast. *9*: 131–145.

Sérèny, B. 1957. Experimental kerato-conjunctivitis shigellosa. Acta Microbiol. Acad. Sci. Hung. *4*: 367–376.

Serino, L., C. Reimmann, H. Baur, M. Beyeler, P. Visca and D. Haas. 1995. Structural genes for salicylate biosynthesis from chorismate in *Pseudomonas aeruginosa*. Mol. Gen. Genet. *249*: 217–228.

Serrano, F.B. 1928. Bacterial fruitlet brown-rot of pineapple in the Philippines. Philipp. J. Sci. *36*: 271–305.

Sertic, V. and N.A. Boulgakov. 1936. Bactériophages spécifiques pour des variétés bactériennes flagellées. C. R. Soc. Biol. Paris. *123*: 887–888.

Seshadri, R. and J.E. Samuel. 2001. Characterization of a stress-induced alternate sigma factor, RpoS, of *Coxiella burnetii* and its expression during the development cycle. Infect. Immun. *69*: 4874–4883.

Sethabutr, O., M. Venkatesan, G.S. Murphy, B. Eampokalap, C.W. Hoge and P. Echeverria. 1993. Detection of Shigellae and enteroinvasive *Escherichia coli* by amplification of the invasion plasmid antigen H DNA sequence in patients with dysentery. J. Infect. Dis. *167*: 458–461.

Sethabutr, O., M. Venkatesan, S. Yam, L.W. Pang, B.L. Smoak, W.K. Sang, P. Echeverria, D.N. Taylor and D.W. Isenbarger. 2000. Detection of PCR products of the *ipaH* gene from *Shigella* and enteroinvasive *Escherichia coli* by enzyme linked immunosorbent assay. Diagn. Microbiol. Infect. Dis. *37*: 11–16.

Seubert, W. 1960. Degradation of isoprenoid compounds by microorganisms. I. Isolation and characterization of an isoprenoid-degrading bacterium, *Pseudomonas citronellolis* n. sp. J. Bacteriol. *79*: 426–434.

Severin, J., A. Wohlfarth and E.A. Galinski. 1992. The predominant role of recently discovered tetrahydropyrimidines for the osmoadaptation of halophilic eubacteria. J. Gen. Microbiol. *138*: 1629–1638.

Severin, V. 1978. Ein neues pathogènes Bakterium an Hanf—*Xanthomonas campestris* pathovar *cannabis*. Arch. Phytopathol. Pflanzenschutz. *14*: 7–15.

Sezer, M.T., M. Gultekin, F. Gunseren, M. Erkilic and F. Ersoy. 1996. A case of *Kluyvera cryocrescens* peritonitis in a CAPD patient. Peritoneal Dialysis International. *16*: 326–327.

Shafi, M.S. and N. Datta. 1975. Infection caused by *Proteus mirabilis* strains with transferable gentamicin-resistance factors. Lancet 2: 1355–1357.

Shah, H.N. and M.D. Collins. 1983. Genus *Bacteroides*. A chemotaxonomical perspective. J. Appl. Bacteriol. *55*: 403–416.

Shah, H.N. and D.M. Collins. 1990. *Prevotella*, a new genus to include *Bacteroides melaninogenicus* and related species formerly classified in the genus *Bacteroides*. Int. J. Syst. Bacteriol. *40*: 205–208.

Shanahan, P.M.A., C.J. Thomson and S.G.B. Amyes. 1996. Antibiotic susceptibilities of *Haemophilus influenzae* in central Scotland. Clin. Microbiol. Infect. *1*: 168–174.

Shane, S.M. and D.H. Gifford. 1985. Prevalence and pathogenicity of *Aeromonas hydrophila*. Avian Dis. *29*: 681–689.

Shane, S.M., K.S. Harrington, M.S. Montrose and R.G. Roebuck. 1984. The occurrence of *Aeromonas hydrophila* in avian diagnostic submissions. Avian. Dis. *28*: 804–807.

Sharma, V.K., Y.K. Kaura and I.P. Singh. 1974. Frogs as carriers of *Salmonella* and *Edwardsiella*. Antonie Leeuwenhoek *40*: 171–175.

Shashkov, A.S., S.N. Senchenkova, E.L. Nazarenko, V.A. Zubkov, N.M. Gorshkova, Y.A. Knirel and R.P. Gorshkova. 1997. Structure of a phosphorylated polysaccharide from *Shewanella putrefaciens* strain S29. Carbohydr. Res. *303*: 333–338.

Shaw, B.G. and J.B. Latty. 1982. A numerical taxonomic study of *Pseudomonas* strains from spoiled meat. J. Appl. Bacteriol. *52*: 219–228.

Shaw, B.G. and J.B. Latty. 1988. A numerical taxonomic study of non-motile non-fermentative gram-negative bacteria from foods. J. Appl. Bacteriol. *65*: 7–22.

Shaw, C. and P.H. Clarke. 1955. Biochemical classification of *Proteus* and Providence cultures. J. Gen. Microbiol. *13*: 155–161.

Shaw, D.H. and H.J. Hodder. 1978. Lipopolysaccharides of the motile aeromonads; core oligosaccharide analysis as an aid to taxonomic classification. Can. J. Microbiol. *24*: 864–886.

Shaw, K.N. and S.M. Kirov. 1999. Diarrhea-associated *Aeromonas* species produce lateral flagella: accessory colonisation factors? 6th International *Aeromonas/Plesiomonas* Symposium, Chicago, Illinois. p. 26.

Shaw, M.K., A.G. Marr and J.L. Ingraham. 1971. Determination of the minimal temperature for growth of *Escherichia coli*. J. Bacteriol. *105*: 683–684.

Shaw, W.V., D.H. Bouanchaud and F.W. Goldstein. 1978. Mechanism of transferable resistance to chloramphenicol in *Haemophilus parainfluenzae*. Antimicrob. Agents Chemother. *13*: 326–330.

Sheela, P., G. Amuthan and A. Mahadevan. 1994. Plasmids in *Xanthomonas campestris* pv. *sesami*. Z. Pflanzenkr. Pflanzenschutz. *101*: 482–486.

Shelly, D.C., J.M. Quarles and I.M. Warner. 1980. Identification of fluorescent *Pseudomonas* species. Clin. Chem. *26*: 1127–1132.

Shelton, R.G.J., P.M.J. Shelton and A.S. Edwards. 1975. Observations with the scanning electron microscope on a filamentous bacterium present on the aesthetic setae of the brown shrimp *Crangon crangon* (L.). J. Mar. Biol. Assoc. U.K. *55*: 795–800.

Shen, P., M. Huang and Z. Peng. 1992. Studies of plasmids of *Pseudomonas maltophilia*. I Ch'uan Hsueh Pao. *19*: 355–361.

Shen, R.N., C.L. Yu, Q.Q. Ma and S.B. Li. 1997. Direct evidence for a soluble methane monooxygenase from type I methanotrophic bacteria: purification and properties of a soluble methane monooxygenase from *Methylomonas* sp. GYJ3. Arch. Biochem. Biophys. *345*: 223–229.

Shepherd, J.G., L. Wang and P.R. Reeves. 2000. Comparison of O-antigen gene clusters of *Escherichia coli*, (*Shigella*) *sonnei* and *Plesiomonas shigelloides* O17: sonnei gained its current plasmid-borne O-antigen genes from *P. shigelloides* in a recent event. Infect. Immun. *68*: 6056–6061.

Sherald, J.L. and J.D. Lei. 1991. Evaluation of a rapid ELISA test kit for detection of *Xylella fastidiosa* in landscape trees. Plant Dis. *75*: 200–203.

Sherman, P., R. Soni and M. Karmali. 1988. Attaching and effacing adherence of Vero cytotoxin-producing *Escherichia coli* to rabbit intestinal epithelium *in vivo*. Infect. Immun. *56*: 756–761.

Sherris, J.C., N.W. Preston and J.G. Shoesmith. 1957. The influence of oxygen on the motility of a strain of *Pseudomonas* sp. J. Gen. Microbiol. *16*: 86–96.

Sherwin, R.P. and J. Wilkins. 1973. The ultrastructure of *Hemophilus influenzae*. *In* Sell (Editor), *Hemophilus influenzae*, Vanderbilt University Press, Nashville, Tennessee. 143–151.

Sheth, N.K. and V.P. Kurup. 1975. Evaluation of tyrosine medium for the identification of *Enterobacteriaceae*. J. Clin. Microbiol. *1*: 483–485.

Shevchik, V.E., J. Robert-Baudouy and N. Hugouvieux-Cotte-Pattat. 1997. Pectate lyase PelI of *Erwinia chrysanthemi* 3937 belongs to a new family. J. Bacteriol. *179*: 7321–7330.

Shewan, J.M. 1971. The microbiology of fish and fishery products—a progress report. J. Appl. Bacteriol. *34*: 299–315.

Shewan, J.M. and M. Véron. 1974. Genus *Vibrio*. *In* Buchanan and Gibbons (Editors), Bergey's Manual of Determinative Bacteriology, 8th Ed., The Williams and Wilkins Co., Baltimore. pp. 340–345.

Shewen, P.E. and B.N. Wilkie. 1982. Cyto-toxin of *Pasteurella haemolytica* acting on bovine leukocytes. Infect. Immun. *35*: 91–94.

Shiaris, M.P. and J.J. Cooney. 1983. Replica plating method for estimating phenanthrene-utilizing and phenanthrene-co-metabolizing microorganisms. Appl. Environ. Microbiol. *45*: 706–710.

Shieh, W.Y., A.L. Chen and H.H. Chiu. 2000. *Vibrio aerogenes* sp. nov., a facultatively anaerobic marine bacterium that ferments glucose with gas production. Int. J. Syst. Evol. Microbiol. *50*: 321–329.

Shieh, W.Y. and W.D. Jean. 1998. *Alterococcus agarolyticus*, gen. nov., sp. nov., a halophilic thermophilic bacterium capable of agar degradation. Can. J. Microbiol. *44*: 637–645.

Shieh, W.Y. and W.D. Jean. 1999. Validation of publication of new names and new combinations previously effectively published outside the IJSB. List No. 69. Int. J. Syst. Bacteriol. *49*: 341–342.

Shiga, K. 1898. Über den Dysenteriebacillus (*Bacillus dysenteriae*). Zentralbl. Bakteriol. Parasitenkd. Infektionskr. Hyg. Abt. I Orig. *24*: 817–828.

Shigidi, M.A. and A.B. Hoerlein. 1970. Characterization of the *Haemo-philus*-like organism of infectious thromboembolic meningoencephalitis of cattle. Am. J. Vet. Res. *31*: 1017–1022.

Shimada, T., E. Arakawa, K. Itoh, Y. Kosako, K. Inoue, Y. Zhengshi and E. Aldová. 1994a. New O and H antigens of *Plesiomonas shigelloides* and their O antigenic relationships to *Shigella boydii*. Curr. Microbiol. *28*: 351–354.

Shimada, T., E. Arakawa, K. Itoh, T. Okitsu, A. Matsushima, Y. Asai, S. Yamai, T. Nakazato, G.B. Nair, M.J. Albert and Y. Takeda. 1994b. Extended serotyping scheme for *Vibrio cholerae*. Curr. Microbiol. *28*: 175–178.

Shimada, T. and Y. Kosako. 1991. Comparison of two O-serogrouping systems for mesophilic *Aeromonas* spp. J. Clin. Microbiol. *29*: 197–199.

Shimada, T., Y. Kosako, K. Inoue, M. Ohtomo, S. Matsushita, S. Yamada and Y. Kudoh. 1991. *Vibrio fluvialis* and *V. furnissii* serotyping scheme for international use. Curr. Microbiol. *22*: 335–337.

Shimada, T. and R. Sakazaki. 1984. On the serology of *Vibrio vulnificus*. Jpn. J. Med. Sci. Biol. *37*: 241–246.

Shimada, T., R. Sakazaki and K. Tobita. 1987. *Vibrio fluvialis*: A new serogroup (19) possessing the Inaba factor antigen of *Vibrio cholerae* O1. Jpn. J. Med. Sci. Biol. *40*: 153–157.

Shimizu, S., S. Iyobe and S. Mitsuhashi. 1977. Inducible high resistance to colistin in *Proteus* strains. Antimicrob. Agents Chemother. *12*: 1–3.

Shimodori, S., K. Iida, F. Kojima, A. Takade, M. Ehara and K. Amako. 1997. Morphological features of a filamentous phage of *Vibrio cholerae* O139 Bengal. Microbiol. Immunol. *41*: 757–763.

Shimwell, J.L. 1936. A study of the common rod bacteria of brewers' yeast. J. Inst. Brew. *42*: 119–127.

Shimwell, J.L. 1948. Brewing bacteriology V. Gram-negative wort, yeast and beer bacteria. Wallerstein Lab. Commun. *11*: 135–145.

Shimwell, J.L. 1963. *Obesumbacterium* gen. nov. Brewers' J. *99*: 759–760.

Shimwell, J.L. 1964. *Obseumbacterium*, a new genus for the inclusion of "*Flavobacterium Proteus*". J. Inst. Brewing. *70*: 247–248.

Shimwell, J.L. and M. Grimes. 1936. The distinguishing characters of *Flavobacterium proteum* (sp. nov.), the common rod bacterium of brewers' yeast. J. Inst. Brew. *42*: 348–350.

Shin, J.H., M.G. Shin, S.P. Suh, D.W. Ryang, J.S. Rew and F.S. Nolte. 1996. Primary *Vibrio damsela* septicemia. Clin. Infect. Dis. *22*: 856–857.

Shinde, P.A. and F.L. Lukezic. 1974a. Characterization and serological comparisons of bacteria of the genus *Erwinia* associated with discolored alfalfa roots. Phytopathology *64*: 871–876.

Shinde, P.A. and F.L. Lukezic. 1974b. Isolation, pathogenicity and characterization of fluorescent pseudomonads associated with discolored alfalfa roots. Phytopathology *64*: 865–871.

Shinoda, S. and K. Okamoto. 1977. Formation and function of *Vibrio parahaemolyticus* lateral flagella. J. Bacteriol. *129*: 1266–1271.

Shirahata, K., T. Deguchi, T. Hayashi, I. Matsubara and T. Suzuki. 1970. The structures of fluopsins C and F. J. Antibiot. *23*: 546–550.

Shively, J.M., W. Devore, L. Stratford, L. Porter, L. Medlin and S.E. Stevens. 1986. Molecular evolution of the large subunit of ribulose-1,5-bisphosphate carboxylase oxygenase (RuBisCo). FEMS Microbiol. Lett. *37*: 251–257.

Shkedy-Vinkler, C. and Y. Avi-Dor. 1975. Betaine-induced stimulation of respiration at high osmolarities in a halotolerant bacterium. Biochem. J. *150*: 219–226.

Shmilovitz, M., B. Kretzer and S. Levi. 1985. A new provisional serovar of *Shigella dysenteriae*. J. Clin. Microbiol. *21*: 240–242.

Shoji, J., T. Kato, H. Hinoo, T. Hattori, K. Hirooka, K. Matsumoto, T. Tanimoto and E. Kondo. 1986. Production of fosfomycin (phosphonomycin) by *Pseudomonas syringae*. J. Antibiot. *39*: 1011–1012.

Shoji, J., R. Sakazaki, T. Hattori, K. Matsumoto, N. Uotani and T. Yoshida. 1989. Isolation and characterization of agglomerins A, B, C and D. J. Antibiot. (Tokyo). *42*: 1729–1733.

Shope, R.E. 1964. Porcine contagious pleuropneumonia. I. Experimental transmission, etiology and pathology. J. Exp. Med. *119*: 357–368.

Shortland-Webb, W.R. 1968. *Proteus* and coliform meningo-encephalitis in neonates. J. Clin. Pathol. *21*: 422–431.

Shotts, E.B., V.S. Blazer and W.D. Waltman. 1986. Pathogenesis of experimental *Edwardsiella ictaluri* infections in channel catfish (*Ictalurus punctatus*). Can. J. Fish. Aquat. Sci. *43*: 36–42.

Shotts, E.B., Jr., J.L. Gaines, Jr., L. Martin and A.K. Prestwood. 1972. *Aeromonas*-induced deaths among fish and reptiles in an eutrophic inland lake. J. Am. Vet. Med. Assoc. *161*: 603–607.

Shotts, E.B., Jr. and R. Rimler. 1973. Medium for the isolation of *Aeromonas hydrophila*. Appl. Microbiol. *26*: 550–553.

Shotts, E.B. and S.F. Snieskzo. 1976. Selected bacterial fish diseases. *In* Page (Editor), Wildlife Diseases, Plenum Publishing Co., New York. pp. 143–151.

Shread, P., T.J. Donovan and J.A. Lee. 1981. A survey of the incidence of *Aeromonas* in human faeces. Soc. Gen. Microbiol. Q. *8*: 184.

Shu, S., E. Setianingrum, L. Zhao, Z. Li, H. Xu, Y. Kawamura and T. Ezaki. 2000. I-CeuI fragment analysis of the *Shigella* species: evidence for large-scale chromosome rearrangement in *S. dysenteriae* and *S. flexneri*. FEMS Microbiol. Lett. *182*: 93–98.

Shuttleworth, K.L. and R.F. Unz. 1991. Influence of metals and metal speciation on the growth of filamentous bacteria. Water Res. *25*: 1177–1186.

Shuttleworth, K.L. and R.F. Unz. 1993. Sorption of heavy-metals to the filamentous bacterium *Thiothrix* strain A1. Appl. Environ. Microbiol. *59*: 1274–1282.

Sias, S.R., A.H. Stouthamer and J.L. Ingraham. 1980. The assimilatory and dissimilatory nitrate reductases of *Pseudomonas aeruginosa* are encoded by different genes. J. Gen. Microbiol. *118*: 229–234.

Siboni, K. 1976. Correlation of the characters fermentation of trehalose, non-transmissible resistance to tetracycline, and relatively long flagellar wavelength in *Proteus morganii*. Acta Pathol. Microbiol. Scand. B Microbiol. *84*: 421–427.

Sidibé, M., S. Messier, S. Larivière, M. Gottschalk and K.R. Mittal. 1993. Detection of *Actinobacillus pleuropneumoniae* in the porcine upper respiratory tract as a complement to serological tests. Can. J. Vet. Res. *57*: 204–208.

Sidorczyk, Z., W. Kaca, H. Brade, E.T. Rietschel, V. Sinnwell and U. Zähringer. 1987. Isolation and structural characterization of an 8-O-(4-amino-4-deoxy-β-L-arabinopyranosyl)-3-deoxy-D-manno-octulosonic acid disaccharide in the lipopolysaccharide of *Proteus mirabilis* deep rough mutant. Eur. J. Biochem. *168*: 269–273.

Sidorczyk, Z., W. Kaca and K. Kotelko. 1975. Studies on lipopolysaccharides of *Proteus vulgaris* serogroups. Chemotypes of genus *Proteus* lipopolysaccharides. Bull. Acad. Pol. Sci. Ser. Sci. Biol. *23*: 603–309.

Sidorczyk, Z., A. Rózalski, M. Deka and K. Kotelko. 1978. Immunochemical studies on free lipid A from *Proteus mirabilis* 1959. Arch. Immunol. Ther. Exp. (Warsz). *26*: 239–243.

Sidorczyk, Z., A. Swierzko, Y.A. Knirel, E.V. Vinogradov, A.Y. Chernyak, L.O. Kononov, M. Cedzynski, A. Rózalski, W. Kaca, A.S. Shashkov and N.K. Kochetkov. 1995. Structure and epitope specificity of the O-specific polysaccharide of *Proteus penneri* 12 (ATCC 33519) containing the amide of D-galacturonic acid with L-threonine. Eur. J. Biochem. *230*: 713–721.

Sidorczyk, Z., U. Zähringer and E.T. Rietschel. 1983. Chemical structure of the lipid A component of the lipopolysaccharide from a *Proteus mirabilis* Re-mutant. Eur. J. Biochem. *137*: 15–22.

Sidorczyk, Z. and K. Zych. 1986. Lipopolysaccharides of flagellated and non-flagellated *Proteus vulgaris* strains. Arch. Immunol. Ther. Exp. *34*: 461–469.

Sidorczyk, Z., K. Zych, A. Swierzko, E.V. Vinogradov and Y.A. Knirel. 1996. The structure of the O-specific polysaccharide of *Proteus penneri* 52. Eur. J. Biochem. *240*: 245–251.

Sidorova, T.N., Z.K. Makhneva, N.N. Puchkova, V.M. Gorlenko and A.A. Moskalenko. 1998. Pigment-protein complexes of purple photosynthetic bacterium *Lamprobacter* sp. containing okenon as main carotenoid. Doklady Akademii Nauk. *361*: 415–418.

Sieburth, J.M., P.W. Johnson, V.M. Church and D.C. Laux. 1993. C₁ bacteria in the water column of Chesapeake Bay, USA. III. Immunologic relationships of the type species of marine monomethylamine- and methane-oxidizing bacteria to wild estuarine and oceanic cultures. Mar. Ecol. Prog. Ser. *95*: 91–102.

Sieburth, J.M., P.W. Johnson, M.A. Eberhardt, M.E. Sieracki, M. Lidstrom and D. Laux. 1987. The first methane-oxidizing bacterium from the upper mixing layer of the deep ocean: *Methylomonas pelagica* sp. nov. Curr. Microbiol. *14*: 285–294.

Siefert, E. and N. Pfennig. 1984. Convenient method to prepare neutral sulfide solution for cultivation of phototrophic sulfur bacteria. Arch. Microbiol. *139*: 100–101.

Siehnel, R., N.L. Martin and R.E.W. Hancock. 1990. Sequence and relatedness in other bacteria of the *Pseudomonas aeruginosa oprP* gene coding for the phosphate-specific porin P. Mol. Microbiol. *4*: 831–838.

Sierra, G. 1957. A simple method for the detection of lipolytic activity of microorganisms and some observations on the influence of the contact between cells and fatty substrates. Antonie Leeuwenhoek J. Microbiol. Serol. *23*: 15–22.

Sierra-Madero, J., K. Pratt, G.S. Hall, R.W. Stewart, J.J. Scerbo and D.L. Longworth. 1990. *Kluyvera mediastinitis* following open-heart surgery: A case report. J. Clin. Microbiol. *28*: 2848–2849.

Sievert, S.M., T. Heidorn and J. Kuever. 2000. *Halothiobacillus kellyi* sp. nov., a mesophilic, obligately chemolithoautotrophic, sulfur-oxidizing bacterium isolated from a shallow-water hydrothermal vent in the Aegean Sea, and emended description of the genus *Halothiobacillus*. Int. J. Syst. Evol. Microbiol. *50*: 1229–1237.

Siitonen, A. and H. Mattila. 1990. Effect of transport medium on recovery of *Aeromonas* species in intestinal infections. Eur. J. Clin. Microbiol. Infect. Dis. *9*: 368–370.

Sijam, K., C.J. Chang and R.D. Gitaitis. 1991. An agar medium for the isolation and identification of *Xanthomonas campestris* pv. *vesicatoria* from seed. Phytopathology *81*: 831–834.

Sijam, K., C.J. Chang and R.D. Gitaitis. 1992. A medium for differentiating tomato and pepper strains of *Xanthomonas campestris* pv. *vesicatoria*. Can. J. Plant Pathol. *14*: 182–184.

Sijderius, R. 1946. Heterotrophe bacteriën, die thiosulfaat oxydeeren, Univ. Amsterdam p. 1–146.

Silva, R.M., S. Saadi and W.K. Maas. 1988. A basic replicon of virulence-associated plasmids of *Shigella* spp. and enteroinvasive *Escherichia coli* is homologous with a basic replicon in plasmids of IncF groups. Infect. Immun. *56*: 836–842.

Simberkoff, M.S. 1980. Experimental *Serratia marcescens* infection and defense mechanisms. *In* Graevenitz, v. and Rubin (Editors), The genus *Serratia*, CRC Press, Boca Raton, Florida. pp. 157–164.

Simidu, U., K. Kita-Tsukamoto, T. Yasumoto and M. Yotsu. 1990. Taxonomy of four marine bacterial strains that produce tetrodotoxin. Int. J. Syst. Bacteriol. *40*: 331–336.

Simidu, U., T. Noguchi, D.F. Hwang, Y. Shida and K. Hashimoto. 1987. Marine bacteria which produce tetrodotoxin. Appl. Environ. Microbiol. *53*: 1714–1715.

Simidu, U. and K. Tsukamoto. 1985. Habitat segregation and biochemical activities of marine members of the Family *Vibrionaceae*. Appl. Environ. Microbiol. *50*: 781–790.

Simon, J., R. Gross, O. Klimmek, M. Ringel and A. Kroeger. 1998. A periplasmic flavoprotein in *Wolinella succinogenes* that resembles the fumarate reductase of *Shewanella putrefaciens*. Arch. Microbiol. *169*: 424–433.

Simonetta, A.C., L.G. Moragues de Velasco and L.N. Frison. 1997. Antibacterial activity of enterococci strains against *Vibrio cholerae*. Lett. Appl. Microbiol. *24*: 139–143.

Simons, M., A.J. Van der Bij, I. Brand, L.A. De Weger, C.A. Wijffelman and B.J.J. Lugtenberg. 1996. Gnotobiotic system for studying rhizosphere colonization by plant growth-promoting *Pseudomonas* bacteria. Mol. Plant Microbe Interact. *9*: 600–607.

Simonson, J.G. and R.J. Siebeling. 1988. Coagglutination of *Vibrio cholerae*, *Vibrio mimicus*, and *Vibrio vulnificus* with anti-flagellar monoclonal antibody. J. Clin. Microbiol. *26*: 1962–1966.

Simoons-Smit, A.M., A.M. Verweij-van Vught and D.M. MacLaren. 1986. The role of K antigens as virulence factors in *Klebsiella*. J. Med. Microbiol. *21*: 133–137.

Simoons-Smit, A.M., A.M. Verwey-van Vught, I.Y. Kanis and D.M. Mac-

Laren. 1984. Virulence of *Klebsiella* strains in experimentally induced skin lesions in the mouse. J. Med. Microbiol. *17*: 67–77.

Simpson, D.A., R. Ramphal and S. Lory. 1995. Characterization of *Pseudomonas aeruginosa fliO*, a gene involved in flagellar biosynthesis and adherence. Infect. Immun. *63*: 2950–2957.

Simpson, L.M. and J.D. Oliver. 1987. Ability of *Vibrio vulnificus* to obtain iron from transferrin and other iron-binding proteins. Curr. Microbiol. *15*: 155–157.

Simpson, L.M., V.K. White, S.F. Zane and J.D. Oliver. 1987. Correlation between virulence and colony morphology in *Vibrio vulnificus*. Infect. Immun. *55*: 269–272.

Sims, W. 1970. Oral haemophili. J. Med. Microbiol. *3*: 615–625.

Sinclair, N.A. 1972. Cell division frequency: microorganisms, Part II. Viruses and bacteria. *In* Altman and Dittmer (Editors), Biology Data Book, 2nd Ed., Vol. 1, Federation of American Societies for Experimental Biology, Bethesda. pp. 117–118.

Singer, E. and J. Debette. 1993. Nutritional factors controlling exocellular proteinase production in a soil-isolated *Xanthomonas maltophilia* strain. J. Basic Microbiol. *33*: 113–121.

Singer, E., J. Debette, A. Lepretre and J. Swings. 1994. Comparative esterase electrophoretic polymorphism and phenotypic analysis of *Xanthomonas maltophilia* and related strains. Syst. Appl. Microbiol. *17*: 387–394.

Singer, J.T. and W.R. Finnerty. 1984. Insertional specificity of transposon Tn5 *Acinetobacter* sp. J. Bacteriol. *157*: 607–611.

Singer, J.T., J.J. van Tuijl and W.R. Finnerty. 1986. Transformation and mobilization of cloning vectors in *Acinetobacter* spp. J. Bacteriol. *165*: 301–303.

Singer, J. and B.E. Volcani. 1955. An improved ferric chloride test for differentiating *Proteus-Providence* group from other *Enterobacteriaceae*. J. Bacteriol. *69*: 303–306.

Singh, D.V. and S.C. Sanyal. 1994. Antibiotic resistance in clinical and environmental isolates of *Aeromonas* spp. J. Antimicrob. Chemother. *33*: 368–369.

Singh, R.A., N.R. Choudhury and H.K. Das. 2000. The replication origin of *Azotobacter vinelandii*. Mol. Gen. Genet. *262*: 1070–1080.

Singh, S., B. Koehler and W.F. Fett. 1992. Effect of osmolarity and dehydration on alginate production by fluorescent pseudomonads. Curr. Microbiol. *25*: 335–339.

Singh, U. 1997. Isolation and identification of *Aeromonas* spp. from ground meats in eastern Canada. J. Food Prot. *60*: 125–130.

Sirois, M. and R. Higgins. 1991. Biochemical typing of *Actinobacillus pleuropneumoniae*. Vet. Microbiol. *27*: 397–401.

Sisco, K.L. and H.O. Smith. 1979. Sequence-specific DNA uptake in *Haemophilus* transformation. Proc. Natl. Acad. Sci. U. S. A. *76*: 972–976.

Sismeiro, O., P. Trotot, F. Biville, C. Vivares and A. Danchin. 1998. *Aeromonas hydrophila* adenylyl cyclase 2: a new class of adenylyl cyclases with thermophilic properties and sequence similarities to proteins from hyperthermophilic archaebacteria. J. Bacteriol. *180*: 3339–3344.

Sisti, M., A. Albano and G. Brandi. 1998. Bactericidal effect of chlorine on motile *Aeromonas* spp. in drinking water supplies and influence of temperature on disinfection efficacy. Lett. Appl. Microbiol. *26*: 347–351.

Sjästedt, A., K. Kuoppa, T. Johansson and G. Sandström. 1992. The 17 kDa lipoprotein and encoding gene of *Francisella tularensis* LVS are conserved in strains of *Francisella tularensis*. Microb. Pathog. *13*: 243–249.

Sjöstedt, A., U. Eriksson, L. Berglund and A. Tärnvik. 1997. Detection of *Francisella tularensis* in ulcers of patients with tularemia by PCR. J. Clin. Microbiol. *35*: 1045–1048.

Sjöstedt, A., G. Sandström, A. Tärnvik and B. Jaurin. 1990. Nucleotide sequence and T cell epitopes of a membrane protein of *Francisella tularensis*. J. Immunol. *145*: 311–317.

Skerman, T.M. 1975. Determination of some *in vitro* growth requirements of *Bacteroides nodosus*. J. Gen. Microbiol. *87*: 107–119.

Skerman, T.M. 1989. Isolation and identification of *Bacteriodes nodusus*. *In* Egerton, Yong and Riffkin (Editors), Footrot and Foot Abscess in Ruminants, CRC Press, Boca Raton. 85–104.

Skerman, V.B.D. 1967. A Guide for the Identification of the Genera of Bacteria, 2nd Ed., The Williams & Wilkins Co., Baltimore.

Skerman, V.B.D., G. Dementjeva and B.J. Carey. 1957. Intracellular deposition of sulfur by *Sphaerotilus natans*. J. Bacteriol. *73*: 504–521.

Skerman, V.B.D., V. McGowan and P.H.A. Sneath. 1980. Approved lists of bacterial names. Int. J. Syst. Bacteriol. *30*: 225–420.

Skerman, V.B.D., V. McGowan and P.H.A. Sneath. 1989. Approved lists of bacterial names, amended edition, American Society for Microbiology, Washington, D.C.

Skirrow, M.B. 1969. The Dienes (mutual inhibition) test in the investigation of *Proteus infections*. J. Med. Microbiol. *2*: 471–477.

Slack, M.P.E. and J.Z. Jordens. 1998. *Haemophilus*. *In* Topley, Wilson, Collier, Balows and Sussman (Editors), Topley & Wilson's Microbiology and Microbial Infections, Vol. 2, Systematic Bacteriology, Arnold; Oxford University Press, London New York. 1167–1190.

Slade, H.D., C.C. Doughty and W.C. Slamp. 1954. The synthesis of high-energy phosphate in the citrulline ureidase reaction by soluble enzymes of *Pseudomonas*. Arch. Biochem. Biophys. *48*: 338–346.

Slade, M.B. and A.I. Tiffin. 1984. Biochemical and serological characterization of *Erwinia*. *In* Bergan (Editor), Methods in Microbiology, Vol. 15, Academic Press, London. pp. 228–293.

Slawson, R.M., J.T. Trevors and H. Lee. 1992. Silver accumulation and resistance in *Pseudomonas stutzeri*. Arch. Microbiol. *158*: 398–404.

Sledjeski, D.D. and R.M. Weiner. 1991. *Hyphomonas* spp., *Shewanella* spp., and other marine bacteria lack heterogeneous (ladderlike) lipopolysaccharides. Appl. Environ. Microbiol. *57*: 2094–2096.

Sleesman, J.P. and C. Leben. 1978. Preserving phytopathogenic bacteria at −70°C or with silica gel. Plant Dis. Rep. *62*: 910–913.

Slopek, S. 1978. Phage typing of *Klebsiella*. Methods Microbiol. *11*: 193–222.

Slopek, S., I. Durlakowa, A. Kucharewicz-Krubkowska, T. Krzywy, A. Slopek and B. Weber. 1973. Phage typing of *Shigella sonnei*. Arch. Immunol. Ther. Exp. *21*: 1–161.

Slopek, S. and J. Maresz-Babbczyszyn. 1967. A working scheme for typing *Klebsiella bacilli* by means of pneumocins. Arch. Immunol. Ther. Exp. *15*: 525–529.

Slopek, S., M. Mulczyk and A. Krukowska. 1968. Phage typing of *Shigella flexneri*. Arch. Immunol. Ther. Exp. *16*: 512–518.

Slopek, S., A. Przonko-Hessek, A. Milch and S. Deak. 1967. A working scheme for bacteriophage typing of *Klebsiella* bacilli. Arch. Immunol. Ther. Exp. *15*: 589–599.

Slotnick, I.J. 1968. *Cardiobacterium hominis* in genitourinary specimens. J. Bacteriol. *95*: 1175.

Slotnick, I.J. and M. Dougherty. 1964. Further characterization of an unclassified group of bacteria causing endocarditis in man: *Cardiobacterium hominis* gen. et sp. n. Antonie Leeuwenhoek J. Microbiol. Serol. *30*: 261–272.

Slotnick, I.J. and M. Dougherty. 1965. Unusual toxicity of riboflavin and flavin mononucleotide for *Cardiobacterium hominis*. Antonie Leeuwenhoek J. Microbiol. Serol. *31*: 355–360.

Slotnick, I.J., J.A. Mertz and M. Dougherty. 1964. Fluorescent antibody detection of human occurrence of an unclassified bacterial group causing endocarditis. J. Infect. Dis. *114*: 503–505.

Slots, J. 1981. Enzymatic characterization of some oral and nonoral Gram-negative bacteria with the API ZYM System. J. Clin. Microbiol. *14*: 288–294.

Slots, J. 1982a. Salient biochemical characters of *Actinobacillus actinomycetemcomitans*. Arch. Microbiol. *131*: 60–67.

Slots, J. 1982b. Selective medium for isolation of *Actinobacillus actinomycetemcomitans*. J. Clin. Microbiol. *15*: 606–609.

Slots, J., H.S. Reynolds and R.J. Genco. 1980. *Actinobacillus actinomycetemcomitans* in human periodontal disease: a cross-sectional microbiological investigation. Infect. Immun. *29*: 1013–1020.

Small, P., D. Blankenhorn, D. Welty, E. Zinser and J.L. Slonczewski. 1994. Acid and base resistance in *Escherichia coli* and *Shigella flexneri*: role of *rpoS* and growth pH. J. Bacteriol. *176*: 1729–1737.

Šmarda, J. 1987. Production of bacteriocin-like agents of *Budvicia aquatica* and *Pragia fontium.* Zentbl. Bakteriol. Mikrobiol. Hyg. Ser. A *265*: 74–81.

Smart, N.L. and O.P. Miniats. 1989. Preliminary assessment of a *Haemophilus parasuis* bacterium for use in specific pathogen free swine. Can. J. Vet. Res. *53*: 390–393.

Smets, B.F., B.E. Rittmann and D.A. Stahl. 1993. The specific growth-rate of *Pseudomonas putida* PAW1 influences the conjugal transfer rate of the TOL Plasmid. Appl. Environ. Microbiol. *59*: 3430–3437.

Smibert, R.M. and N.R. Krieg. 1994. Phenotypic characterization. *In* Gerhardt, Murray, Wood and Krieg (Editors), Methods for General and Molecular Bacteriology, American Society for Microbiology, Washington, D.C. pp. 607–654.

Smid, E.J., A.H.J. Hansen and L.G.M. Gorris. 1995. Detection of *Erwinia carotovora* subsp. *atroseptica* and *Erwinia chrysanthemi* in potato tubers using polymerase chain reaction. Plant Pathol. (Oxf.) *44*: 1058–1069.

Smigielski, A.J. and R.J. Akhurst. 1992. Is phase variation in *Xenorhabdus nematophilus* mediated by genomic rearrangement? Proc. XIX Int. Cong. Entomol., Beijing. pp. 270.

Smigielski, A.J. and R.J. Akhurst. 1994. Megaplasmids in *Xenorhabdus* and *Photorhabdus* spp., bacterial symbionts of entomopathogenic nematodes (families *Steinernematidae* and *Heterorhabditidae*). J. Invertebr. Pathol. *64*: 214–220.

Smigielski, A.J., R.J. Akhurst and N.E. Boemare. 1994. Phase variation in *Xenorhabdus nematophilus* and *Photorhabdus luminescens*: differences in respiratory activity and membrane energization. Appl. Environ. Microbiol. *60*: 120–125.

Smirnov, V.V., E.A. Kiprianova, A.D. Garagulya, S.E. Esipov and S.A. Dovjenko. 1997. Fluviols, bicyclic nitrogen-rich antibiotics produced by *Pseudomonas fluorescens.* FEMS Microbiol. Lett. *153*: 357–361.

Smirnov, V.V., E.A. Kiprianova, A.D. Garagulya, V.I. Ruban and T.A. Dodatko. 1984. Bacteriocins of some *Pseudomonas* species. Antibiotiki. *29*: 730–735.

Smit, E., A.C. Wolters, H. Lee, J.T. Trevors and J.D. Van Elsas. 1996. Interactions between a genetically marked *Pseudomonas fluorescens* strain and bacteriophage Phi R2f in soil: Effects of nutrients, alginate encapsulation, and the wheat rhizosphere. Microb. Ecol. *31*: 125–140.

Smit, M. and A.G. Clark. 1971. The observation of myxobacterial fruiting bodies. J. Appl. Bacteriol. *34*: 399–401.

Smith, A.W., S. Freeman, W.G. Minett and P.A. Lambert. 1990. Characterisation of a siderophore from *Acinetobacter calcoaceticus.* FEMS Microbiol. Lett. *58*: 29–32.

Smith, C.O. 1913. Black pit of lemon. Phytopathology *3*: 277–281.

Smith, E.F. 1895. *Bacillus tracheiphilus* sp. nov. die Ursache des Verwelkens verschiedener Curcurbitaceen. Zentbl. Bakteriol. Parasitenkd. Infektkrankh. Hyg. Abt. II *1*: 364–373.

Smith, E.F. 1897. Description of *Bacillus phaseoli* n. sp. Bot. Gaz. *24*: 192.

Smith, E.F. 1898. Notes on Stewart's sweet corn germ, *Pseudomonas stewarti* n sp. Proc. Amer. Assoc. Adv. Sci. *47*: 422–426.

Smith, E.F. 1901. The cultural characters of *Pseudomonas hyacinth*, *Ps. campestris*, *Ps. phaseoli* and *Ps. stewarti*—four one-flagellate yellow bacteria parasitic on plants. U.S. Dept. Agr. Div. Veg. Phys. Pathol. Bull. *28*: 1–153.

Smith, E.F. 1903. Observations on a hitherto unreported bacterial disease, the cause of which enters the plant through ordinary stomata. Science (Wash. D.C.). *17*: 456–457.

Smith, E.F. 1904. Bacterial leaf spot diseases. Science (Wash. D.C.). *19*: 417–418.

Smith, E.F. and M.K. Bryan. 1915. Angular leaf-spot of cucumbers. J. Agr. Res. *5*: 465–476.

Smith, E.F., L.R. Jones and C.S. Reddy. 1919. The black chaff of wheat. Science (Wash. D.C.). *50*: 48.

Smith, E., P. Leeflang and K. Wernars. 1997a. Detection of shifts in microbial community structure and diversity in soil caused by copper contamination using amplified ribosomal DNA restriction analysis. FEMS Microbiol. Ecol. *23*: 249–261.

Smith, F.B. 1938. An investigation of a taint in rib bones of bacon. The determination of halophilic vibrios. Proc. Roy. Soc, Queensland. *49*: 29–52.

Smith, G.R. 1962. An unusual haemolytic effect produced by *Pasteurella haemolytica.* J. Pathol. Bacteriol. *83*: 501–508.

Smith, H.R., T. Cheasty and B. Rowe. 1997b. Enteroaggregative *Escherichia coli* and outbreaks of gastroenteritis in UK. Lancet *350*: 814–815.

Smith, H.R., B. Rowe, R.J. Gross, N.K. Fry and S.M. Scotland. 1987. Haemorrhagic colitis and Vero-cytotoxin-producing *Escherichia coli* in England and Wales. Lancet *1*: 1062–1065.

Smith, H.R., S.M. Scotland, G.A. Willshaw, B. Rowe, A. Cravioto and C. Eslava. 1994. Isolates of *Escherichia coli* O44:H18 of diverse origin are enteroaggregative. J. Infect. Dis. *170*: 1610–1613.

Smith, H.R., G.A. Willshaw, S.M. Scotland, A. Thomas and B. Rowe. 1993. Properties of Vero cytotoxin-producing *Escherichia coli* isolated from human and non-human sources. Zentbl. Bakteriol. *278*: 436–444.

Smith, I.W. 1963. The classification of *Bacterium salmonicida.* J. Gen. Microbiol. *33*: 263–274.

Smith, J.E. and E. Thal. 1965. A taxonomic study of the genus *Pasteurella* using a numerical technique. Acta Pathol. Microbiol. Scand. *64*: 213–223.

Smith, K.S., A.M. Costello and M.E. Lidstrom. 1997c. Methane and trichloroethylene oxidation by an estuarine methanotroph, *Methylobacter* sp. strain BB5.1. Appl. Environ. Microbiol. *63*: 4617–4620.

Smith, S.K., D.C. Sutton, J.A. Fuerst and J.L. Reichelt. 1991. Evaluation of the genus *Listonella* and reassignment of *Listonella damsela* (Love et al.) MacDonell and Colwell to the genus *Photobacterium* as *Photobacterium damsela* comb. nov. with an emended description. Int. J. Syst. Bacteriol. *41*: 529–534.

Smith, T. 1894. The hog-cholera group of bacteria. U.S. Bur. Anim. Ind. Bull. *6*: 6–40.

Smith, T. and M.S. Taylor. 1919. Some morphological and biochemical characteristics of the spirilla (*Vibrio fetus*, n. sp.) associated with disease of the fetal membranes in cattle. J. Exp. Med. *30*: 299–311.

Smith-Vaughan, H.C., K.S. Sriprakash, J.D. Mathews and D.J. Kemp. 1997. Nonencapsulated *Haemophilus influenzae* in Aboriginal infants with otitis media: prolonged carriage of P2 porin variants and evidence for horizontal P2 gene transfer. Infect. Immun. *65*: 1468–1474.

Smyth, C.J., M.B. Marron, J.M.G.I. Twohig and S.G.J. Smith. 1996. Fimbrial adhesins: Similarities and variations in structure and biogenesis. FEMS Immunol. Med. Microbiol. *16*: 127–139.

Sneath, P.H. 1993. Evidence from *Aeromonas* for genetic crossing-over in ribosomal sequences [letter]. Int. J. Syst. Bacteriol. *43*: 626–629.

Sneath, P.H.A. and R. Johnson. 1973. Numerical taxonomy of *Haemophilus* and related bacteria. Int. J. Syst. Bacteriol. *23*: 405–418.

Sneath, P.H.A. and V.B.D. Skerman. 1966. A list of type and reference strains of bacteria. Int. J. Syst. Bacteriol. *16*: 1–133.

Sneath, P.H. and M. Stevens. 1985. A numerical taxonomic study of *Actinobacillus, Pasteurella* and *Yersinia.* J. Gen. Microbiol. *131*: 2711–2738.

Sneath, P.H. and M. Stevens. 1990. *Actinobacillus rossii* sp. nov., *Actinobacillus seminis* sp. nov., nom. rev., *Pasteurella bettii* sp. nov., *Pasteurella lymphangitidis* sp. nov., *Pasteurella mairi* sp. nov., and *Pasteurella trehalosi* sp. nov. Int. J. Syst. Bacteriol. *40*: 148–153.

Sneath, P.H.A., M. Stevens and M.J. Sackin. 1981. Numerical taxonomy of *Pseudomonas* based on published records of substrate utilization. Anton. Leeuwenhoek J. Microbiol. *47*: 423–448.

Snell, J.J., L.R. Hill and S.P. Lapage. 1972. Identification and characterization of *Moraxella phenylpyruvica.* J. Clin. Pathol. *25*: 959–965.

Snell, J.J.S. and S.P. Lapage. 1976. Transfer of some saccharolytic *Moraxella* species to *Kingella* Henriksen and Bøvre 1976, with descriptions of *Kingella indologenes* sp. nov. and *Kingella denitrificans* sp. nov. Int. J. Syst. Bacteriol. *26*: 451–458.

Snellen, J.E. and H.D. Raj. 1970. Morphogenesis and fine structure of *Leucothrix mucor* and effects of calcium deficiency. J. Bacteriol. *101*: 240–249.

Snellings, N.J., B.D. Tall and M.M. Venkatesan. 1997. Characterization of *Shigella* type 1 fimbriae: expression, FimA sequence, and phase variation. Infect. Immun. *65*: 2462–2467.

Snieszko, S.F., P.J. Griffin and S.B. Fridle. 1950. A new bacterium (*Hemophilus piscium* n. sp.) from ulcer disease of trout. J. Bacteriol. *59*: 699–710.

Snipes, K.P. and E.L. Biberstein. 1982. *Pasteurella testudinis* sp. nov. a parasite of desert tortoises (*Gopherus agassizi*). Int. J. Syst. Bacteriol. *32*: 201–210.

Snipes, K.P., R.W. Kasten, J.M. Calagoan and J.T. Boothby. 1995. Molecular characterization of *Pasteurella testudinis* isolated from desert tortoises (*Gopherus agassizii*) with and without upper respiratory-tract disease. J. Wildl. Dis. *31*: 22–29.

Sobel, J., D.N. Cameron, J. Ismail, N. Strockbine, M. Williams, P.S. Diaz, B. Westley, M. Rittmann, J. DiCristina, H. Ragazzoni, R.V. Tauxe and E.D. Mintz. 1998. A prolonged outbreak of *Shigella sonnei* infections in traditionally observant Jewish communities in North America caused by a molecularly distinct bacterial subtype. J. Infect. Dis. *177*: 1405–1409.

Socolofsky, M.D. and O. Wyss. 1962. Resistance of the *Azotobacter* cyst. J. Bacteriol. *84*: 119–124.

Soddell, J.A., A.M. Beacham and R.J. Seviour. 1993. Phenotypic identification of non-clinical isolates of *Acinetobacter* species. J. Appl. Bacteriol. *74*: 210–214.

Solangi, M.A., R.M. Overstreet and A.L. Gannam. 1979. A filamentous bacterium on the brine shrimp (*Artemia salina*) and its control. Gulf Res. Rep. *6*: 275–282.

Solano, F., E. García, E.P. de Egea and A. Sanchez-Amat. 1997. Isolation and characterization of strain MMB-1 (CECT 4803), a novel melanogenic marine bacterium. Appl. Environ. Microbiol. *63*: 3499–3506.

Solano, F., P. Lucas-Elío, E. Fernández and A. Sanchez-Amat. 2000. *Marinomonas mediterranea* MMB-1 transposon mutagenesis: Isolation of a multipotent polyphenol oxidase mutant. J. Bacteriol. *182*: 3754–3760.

Solano, F. and A. Sanchez-Amat. 1999. Studies on the phylogenetic relationships of melanogenic marine bacteria: proposal of *Marinomonas mediterranea* sp. nov. Int. J. Syst. Bacteriol. *49*: 1241–1246.

Soldati, L. and J.C. Piffaretti. 1991. Molecular typing of *Shigella* strains using pulsed field gel electrophoresis and genome hybridization with insertion sequences. Res. Microbiol. *142*: 489–498.

Sompolinsky, D., I. Boldur, R.A. Goldwasser, H. Kahana, R. Kazak, A. Keysary and A. Pik. 1986. Serological cross-reactions between *Rickettsia typhi*, *Proteus vulgaris* OX19 and *Legionella bozmanii* in a series of febrile patients. Isr. J. Med. Sci. *22*: 745–752.

Sompolinsky, D., J.B. Hertz, N. Høiby, K. Jensen, B. Mansa, V.B. Pedersen and Z. Samra. 1980a. An antigen common to a wide-range of bacteria 2. A biochemical-study of a common antigen from *Pseudomonas aeruginosa*. Acta Pathol. Microbiol. Scand. B. *88*: 253–260.

Sompolinsky, D., J.B. Hertz, N. Høiby, K. Jensen, B. Mansa and Z. Samra. 1980b. An antigen common to a wide-range of bacteria 1. The isolation of a common antigen from *Pseudomonas aeruginosa*. Acta Pathol. Microbiol. Scand. B. *88*: 143–149.

Sonne, C. 1915. Über die Bakteriologie der giftarmen Dysenteriebazillen (Paradysenteriebazillen). Zentbl. Bakteriol. Orig. *75*: 408–456.

Sonnenshein, C. 1927. Die Mucosus-Form des Pyocyaneus-Bakteriums, Bacterium pyocyaneum mucosum. Zentraibl. Bakteriol. Parasitenk. Infektionskr. Hyg. Abt. I, Orig. *104*: 365–373.

Sonnenwirth, A.C. and B.A. Kallus. 1968. Meningitis due to *Edwardsiella tarda*. First report of meningitis caused by E. tarda. Am. J. Clin. Pathol. *49*: 92–95.

Sorensen, B., E.S. Falk, E. Wisloffnilsen, B. Bjorvatn and B.E. Kristiansen. 1985. Multivariate-analysis of *Neisseria* DNA restriction endonuclease patterns. J. Gen. Microbiol. *131*: 3099–3104.

Sorokin, D.Y., A.M. Lysenko, L.L. Mityushina, T.P. Tourova, B.E. Jones, F.A. Rainey, L.A. Robertson and G.J. Kuenen. 2001a. *Thioalkalimicrobium aerophilum* gen. nov., sp. nov. and *Thioalkalimicrobium sibericum* sp nov., and *Thioalkalivibrio versutus* gen. nov., sp. nov., *Thioalkalivibrio nitratis* sp. nov. and *Thioalkalivibrio denitrificans* sp. nov., novel obligately alkaliphilic and obligately chemolithoautotrophic sulfur-oxidizing bacteria from soda lakes. Int. J. Syst. Evol. Microbiol. *51*: 565–580.

Sorokin, D.Y., G. Muyzer, T. Brinkhoff, J.G. Kuenen and M.S.M. Jetten. 1998. Isolation and characterization of a novel facultatively alkaliphilic *Nitrobacter* species, N. *alkalicus* sp. nov. Arch. Microbiol. *170*: 345–352.

Sorokin, D.Y., G. Muyzer, T. Brinkhoff, J.G. Kuenen and M.S.M. Jetten.

2001b. *In* Validation of publication of new names and new combinations previously effectively published outside the IJSEM. List No. 78. Int. J. Syst. Evol. Microbiol. *51*: 1–2.

Sorokin, D.Y., T.P. Tourova, A.M. Lysenko and J.G. Kuenen. 2001c. Microbial thiocyanate utilization under highly alkaline conditions. Appl. Environ. Microbiol. *67*: 528–538.

Sottnek, F.O. and W.L. Albritton. 1984. *Haemophilus influenzae* biotype VIII. J. Clin. Microbiol. *20*: 815–816.

Sourek, J. 1968. On some findings concerning Dienes' phenomenon in swarming *Proteus* strains. Zentbl. Bakteriol. Parasitenkd. Infektkrankh. Hyg. Abt. I Orig. *208*: 419–427.

Sourek, J. and E. Aldová. 1976. Serotyping of strains belonging to the *Citrobacter-Levinea* group isolated from diagnostic material. Zentbl. Bakteriol. Parasitenkd. Infektkrankh. Hyg. Erste Abt. Orig. Reihe A Med. Mikrobiol. Parasitol. *234*: 480–490.

Sourek, J. and E. Aldová. 1988. Importance of serological tests in the diagnosis of *Citrobacter diversus* and *Citrobacter amalonaticus*. Syst. Appl. Microbiol. *11*: 60–66.

Southern, P.M., Jr. 1975. Bacteremia due to *Succinivibrio dextrinosolvens*. Am. J. Clin. Pathol. *64*: 540–543.

Southward, A.J., E.C. Southward, P.R. Dando, G.H. Rau, H. Felbeck and H. Flugel. 1981. Bacterial symbionts and low $^{13}C/^{12}C$ ratios in tissues of *Pogonophora* indicate unusual nutrition and metabolism. Nature *293*: 616–620.

Spangenberg, C., R. Fislage, W. Sierralta, B. Tümmler and U. Römling. 1995. Comparison of type IV-pilin genes of *Pseudomonas aeruginosa* of various habitats has uncovered a novel unusual sequence. FEMS Microbiol. Lett. *125*: 265–273.

Spangenberg, C., T. Heuer, C. Bürger and B. Tümmler. 1996. Genetic diversity of flagellins of *Pseudomonas aeruginosa*. FEBS Lett. *396*: 213–217.

Spangler, B.D. 1992. Structure and function of cholera toxin and the related *Escherichia coli* heat-labile enterotoxin. Microbiol. Rev. *56*: 622–647.

Spaulding, A.W. and C.D. von Dohlen. 2001. Psyllid endosymbionts exhibit patterns of co-speciation with hosts and destabilizing substitutions in ribosomal RNA. Insect Mol. Biol. *10*: 57–67.

Spaulding, W.B. and J.N. Hennessy. 1960. Cat scratch fever. A study of eighty-three cases. Am. J. Med. *28*: 504–509.

Spencer, R. 1963. Bacterial viruses in the sea. *In* Oppenheimer (Editor), Symposium on Marine Microbiology, Thomas, Springfield, Illinois. 350–365.

Spiegl, H., W. Ludwig, K.H. Schleifer and E. Stackebrandt. 1988. Complete nucleotide sequence of a 23S ribosomal RNA gene from *Ruminobacter amylophilus*. Nucleic Acids Res. *16*: 2345.

Spierings, G., C. Ockhuijsen, H. Hofstra and J. Tommassen. 1993. Polymerase chain reaction for the specific detection of *Escherichia coli/Shigella*. Res Microbiol. *144*: 557–564.

Spinola, S.M., J. Peacock, F.W. Denny, D.L. Smith and J.G. Cannon. 1986. Epidemiology of colonization by nontypable *Haemophilus influenzae* in children: a longitudinal study. J. Infect. Dis. *154*: 100–109.

Sponenberg, D.P., M.E. Carter, G.R. Carter, D.O. Cordes, S.E. Stevens and H.P. Veit. 1983. Supportive epididymitis in a ram infected with *Actinobacillus seminis*. J. Am. Vet. Med. Assoc. *182*: 990–991.

Spreng, M., J. Deleforge, V. Thomas, B. Boisramé and H. Drugeon. 1995. Antibacterial activity of marbofloxacin - a new fluoroquinolone for veterinary use against canine and feline isolates. J. Vet. Pharmacol. Therapeutics *18*: 284–289.

Springer, G.F., J.C. Adye, A. Bezkorovainy and J.R. Murthy. 1973. Functional aspects and nature of the lipopolysaccharide-receptor of human erythrocytes. *In* Kass and Wolff (Editors), Bacterial Lipopolysaccharides. The Chemistry, Biology and Clinical Significance of Endotoxins, The University of Chicago Press, Chicago. pp. 194–204.

Spröer, C., E. Lang, P. Hobeck, J. Burghardt, E. Stackebrandt and B.J. Tindall. 1998. Transfer of *Pseudomonas nautica* to *Marinobacter hydrocarbonoclasticus*. Int. J. Syst. Bacteriol. *48*: 1445–1448.

Spröer, C., U. Mendrock, J. Swiderski, E. Lang and E. Stackebrandt . 1999. The phylogenetic position of *Serratia*, *Buttiauxella* and some other genera of the family *Enterobacteriaceae*. Int. J. Syst. Bacteriol. *49*: 1433–1438..

Srinivasan, M.C. and M.K. Patel. 1956. Three undescribed species of *Xanthomonas*. Curr. Sci. (Bangalore) *25*: 366–367.

Srinivasan, M.C. and M.K. Patel. 1957. Two new phytopathogenic bacteria on verbenaceous hosts. Curr. Sci. (Bangalore) *26*: 90–91.

Srinivasan, M.C., M.K. Patel and M.J. Thirumalachar. 1961a. A bacterial blight disease of coriander. Proc. Indian Acad. Sci. Sect. B. *53*: 298–301.

Srinivasan, M.C., M.K. Patel and M.J. Thirumalachar. 1961b. A new bacterial blight disease of *Argemone mexicana*. Proc. Indian Acad. Sci. Sect. B. *27*: 104–107.

Srinivasan, M.C., M.K. Patel and M.J. Thirumalachar. 1962. Two bacterial leaf-spot diseases on *Physalis minima* and studies on their relationship to *Xanthomonas vesicatoria* (Doidge) Dowson. Proc. Indian Acad. Sci. Sect. B. *56*: 93–96.

St. Geme, J.W., 3rd, A. Takala, E. Esko and S. Falkow. 1994. Evidence for capsule gene sequences among pharyngeal isolates of nontypeable *Haemophilus influenzae*. J. Infect. Dis. *169*: 337–342.

Stackebrandt, E., V.J. Fowler, W. Schubert and J.F. Imhoff. 1984. Towards a phylogeny of phototrophic purple sulfur bacteria the genus *Ectothiorhodospira*. Arch. Microbiol. *137*: 366–370.

Stackebrandt, E., W. Frederiksen, G.M. Garrity, P.A.D. Grimont, P. Kämpfer, M.C.J. Maiden, X. Nesme, R. Rossello-Mora, J. Swings, H.G. Trüper, L. Vauterin, A.C. Ward and W.B. Whitman. 2002. Report of the ad hoc committee for the re-evaluation of the species definition in bacteriology. Int. J. Syst. Evol. Microbiol. *52*: 1043–1047.

Stackebrandt, E. and B.M. Goebel. 1994. Taxonomic note: A place for DNA–DNA reassociation and 16S rRNA sequence analysis in the present species definition in bacteriology. Int. J. Syst. Bacteriol. *44*: 846–849.

Stackebrandt, E. and H. Hippe. 1986. Transfer of *Bacteroides amylophilus* to a new genus *Ruminobacter* gen. nov., nom. rev. as *Ruminobacter amylophilus* comb. nov. Syst. Appl. Microbiol. *8*: 204–207.

Stackebrandt, E. and H. Hippe. 1987. *In* Validation of the publication of new names and new combinations previously effectively published outside the IJSB. List No. 23. Int. J.Syst. Bacteriol. *37*: 179–180.

Stackebrandt, E., R.G.E. Murray and H.G. Trüper. 1988. *Proteobacteria* classis nov., a name for the phylogenetic taxon that includes the "purple bacteria and their relatives". Int. J. Syst. Bacteriol. *38*: 321–325.

Stahl, D.A. and R.I. Amann. 1991. Development and application of nucleic acid probes in bacterial systematics. *In* Stackebrandt and Goodfellow (Editors), Nucleic Acid Techniques in Bacterial Systematics, John Wiley & Sons, Chichester. pp. 205–248.

Stahl, D.A., D.J. Lane, G.J. Olsen, D.J. Heller, T.M. Schmidt and N.R. Pace. 1987. Phylogenetic analysis of certian sulfide-oxidizing and related morphologically conspicuous bacteria by 5S ribosomal ribonucleic acid sequences. Int. J. Syst. Bacteriol. *37*: 116–122.

Stahl, S.J., K.R. Stewart and F.D. Williams. 1983. Extracellular slime associated with *Proteus mirabilis* during swarming. J. Bacteriol. *154*: 930–937.

Stainer, D.W. and M.J. Scholte. 1970. A simple chemically defined medium for the production of Phase I *Bordetella pertussis*. J. Gen. Microbiol. *63*: 211–220.

Stainthorpe, A.C., V. Lees, G.P. Salmond, H. Dalton and J.C. Murrell. 1991. Screening of obligate methanotrophs for soluble methane monooxygenase genes. FEMS Microbiol. Lett. *70*: 211–216.

Staley, J.T. 1968. *Prosthecomicrobium* and *Ancalomicrobium*: new prosthecate freshwater bacteria. J. Bacteriol. *95*: 1921–1942.

Staley, J.T. and J.J. Gosink. 1999. Poles apart: biodiversity and biogeography of sea ice bacteria. Annu. Rev. Microbiol. *53*: 189–215.

Staley, J.T., R.L. Irgens and D.J. Brenner. 1987. *Enhydrobacter aerosaccus* gen. nov., sp. nov., a gas-vacuolated, facultatively anaerobic, heterotrophic rod. Int. J. Syst. Bacteriol. *37*: 289–291.

Staley, T.E., E.W. Jones and L.D. Corley. 1969. Attachment and penetration of *Escherichia coli* into intestinal epithelium of the ileum in newborn pigs. Am. J. Pathol. *56*: 371–392.

Stall, R.E., C. Beaulieu, D. Egel, N.C. Hodge, R.P. Leite, G.V. Minsavage, H. Bouzar, J.B. Jones, A.M. Alvarez and A.A. Benedict. 1994. Two genetically diverse groups of strains are included in *Xanthomonas campestris* pv. *vesicatoria*. Int. J. Syst. Bacteriol. *44*: 47–53.

Stall, R.E., T.R. Gottwald, M. Koizumi and N.C. Schaad. 1993. Ecology of plant pathogenic xanthomonads. *In* Swings and Civerolo (Editors), *Xanthomonas*, Chapman & Hall, London. pp. 265–299.

Stall, R.E., D.C. Loschke and J.B. Jones. 1986. Linkage of copper resistance and avirulence loci on a self-transmissible plasmid in *Xanthomonas campestris* pv. *vesicatoria*. Phytopathology *76*: 240–243.

Stalon, V. and A. Mercenier. 1984. *L*-Arginine utilization by *Pseudomonas* species. J. Gen. Microbiol. *130*: 69–76.

Stalon, V., C. Vander Wauven, P. Momin and C. Legrain. 1987. Catabolism of arginine, citrulline and ornithine by *Pseudomonas* and related bacteria. J. Gen. Microbiol. *133*: 2487–2495.

Stamler, D.A., M.A.C. Edelstein and P.H. Edelstein. 1994. Azithromycin pharmacokinetics and intracellular concentrations in *Legionella pneumophila* infected and uninfected guinea pigs and their alveolar macrophages. Antimicrob. Agents Chemother. *38*: 217–222.

Stamp, L. and D.M. Stone. 1944. An agglutinogen common to certain strains of lactose and non-lactose fermenting coliform bacilli. J. Hyg. *43*: 266–272.

Stams, A.J.M. and T.A. Hansen. 1984. Fermentation of glutamate and other compounds by *Acidaminobacter hydrogenoformans*, gen. nov., sp. nov., an obligate anaerobe isolated from black mud: studies with pure cultures and mixed cultures with sulfate-reducing and methanogenic bacteria. Arch. Microbiol. *137*: 329–337.

Stanghellini, M.E., D.C. Sands, W.C. Kronland and M.M. Mendonca. 1977. Serological and physiological differentiation among isolates of *Erwinia carotovora* from potato and sugarbeet. Phytopathology *67*: 1178–1182.

Stanier, R.Y. 1943. A note on the taxonomy of *Proteus hydrophilus*. J. Bacteriol. *46*: 213–214.

Stanier, R.Y. and L.N. Ornston. 1973. The β-ketoadipate pathway. *In* Rose and Tempest (Editors), Advances in Microbial Physiology, Vol. 9, Academic Press, London. pp. 89–151.

Stanier, R.Y., N.J. Palleroni and M. Doudorof . 1966. The aerobic pseudomonads: a taxonomic study. J. Gen. Microbiol. *43*: 159–271.

Stanley, S.H., S.D. Prior, D.J. Leak and H. Dalton. 1983. Copper stress underlies the fundamental change in intracellular location of methane monooxygenase in methane-oxidizing organisms: studies in batch and continuous cultures. Biotechnol. Lett. *5*: 487–492.

Stapp, C. 1928. *Schizomycetes* (Spaltpilze oder Bakterien). *In* Sorauer (Editor), Handbuch der Pflanzenkrankheiten, 5th Ed., Vol. 2, Paul Parey, Berlin. pp. 1–295.

Stapp, C. 1940. *Bacterium rubidaeum* nov. spec. Zentralbl. Bakteriol. Parasitenkd. Infektionskr. Hyg. Abt. II. *102*: 252–260.

Starr, M.P. 1946. The nutrition of phytopathogenic bacteria I. Minimal nutritive requirements of the genus *Xanthomonas*. J. Bacteriol. *51*: 131–143.

Starr, M.P. 1958. The blue pigment of *Corynebacterium insidiosum*. Arch. Mikrobiol. *30*: 325–334.

Starr, M.P. 1981. The genus *Xanthomonas*. *In* Starr, Stolp, Trüper, Balows and Schlegel (Editors), The Prokaryotes: A Handbook on Habitats, Isolation and Identification of Bacteria, 1st Ed., Vol. 1, Springer Verlag, New York. pp. 742–763.

Starr, M.P., W. Blau and G. Cosens. 1960. The blue pigment of *Pseudomonas lemonnieri*. Biochem. Z. *333*: 328–334.

Starr, M.P. and A.K. Chatterjee. 1972. The genus *Erwinia*: Enterobacteria pathogenic to plants and animals. Annu. Rev. Microbiol. *26*: 389–426.

Starr, M.P. and D. Folsom. 1951. Bacterial fireblight of raspberry. Phytopathology *41*: 915–919.

Starr, M.P. and C. Garcés. 1950. El agente causante de la gomosis bacterial del pasto imperial en Colombia. Rev. Fac. Nac. Agron. (Medellin) *11*: 73–83.

Starr, M.P., P.A.D. Grimont, F. Grimont and P.B. Starr. 1976. Caprylatethallous agar medium for selectively isolating *Serratia* and its utility in the clinical laboratory. J. Clin. Microbiol. *4*: 270–276.

Starr, M.P., C.L. Jenkins, L.B. Bussey and A.G. Andrewes. 1977. Chemotaxonomic significance of the xanthomonadins, novel brominated aryl polyene pigments produced by bacteria of the genus *Xanthomonas*. Arch. Microbiol. *113*: 1–9.

Starr, M.P., H.G. Knackmuss and G. Cosens. 1967. The intracellular blue pigment of *Pseudomonas lemonnieri*. Arch. Mikrobiol. *59*: 287–294.

Starr, M.P. and M. Mandel. 1969. DNA base composition and taxonomy of phytopathogenic and other enterobacteria. J. Gen. Microbiol. *56*: 113–123.

Starr, M.P. and W.L. Stephens. 1964. Pigmentation and taxonomy of the genus *Xanthomonas*. J. Bacteriol. *87*: 293–302.

Starr, M.P. and J.E. Weiss. 1943. Growth of phytopathogenic bacteria in a synthetic asparagin medium. Phytopathology *33*: 314–318.

Statner, B., M.J. Jones and W.L. George. 1988. Effect of incubation temperature on growth and soluble protein profiles of motile *Aeromonas* strains. J. Clin. Microbiol. *26*: 393–393.

Stead, D.E. 1989. Grouping of *Xanthomonas campestris* pathovars of cereals and grasses by fatty acid profiling. EPPO Bull. *19*: 57–68.

Stead, D.E. 1992. Grouping of plant-pathogenic and some other *Pseudomonas* spp. by using cellular fatty acid profiles. Int. J. Syst. Bacteriol. *42*: 281–295.

Stead, P., B.A.M. Rudd, H. Bradshaw, D. Noble and M.J. Dawson. 1996. Induction of phenazine biosynthesis in cultures of *Pseudomonas aeruginosa* by L-N-(3-oxohexanoyl)homoserine lactone. FEMS Microbiol. Lett. *140*: 15–22.

Steadman, J.R., C.R. Maier, H.F. Schwartz and E.D. Kerr. 1975. Pollution of surface irrigation waters by plant pathogenic organisms. Water Resour. Bull. *11*: 796–804.

Stecchini, M.L., I. Sarais and P. Giavedoni. 1993. Effect of essential oils on *Aeromonas hydrophila* in a culture medium and in cooked pork. J. Food Prot. *56*: 406–409.

Steele, T.W., J. Lanser and N. Sangster. 1990. Isolation of *Legionella longbeachae* serogroup 1 from potting mixes. Appl. Environ. Microbiol. *56*: 49–53.

Steele, W., P. Larock and J.M. Larkin. 1995. Radioisotopic labeling and microautoradiography of *Beggiatoa* bacterial mats from Warm Mineral Springs. 95th Annual Meeting of the American Society for Microbiology, American Society for Microbiology. p. 340.

Steigerwalt, A.G., G.R. Fanning, M.A. Fife-Asbury and D.J. Brenner. 1976. DNA relatedness among species of *Enterobacter* and *Serratia*. Can. J. Microbiol. *22*: 121–137.

Stein, A. and D. Raoult. 1993. Lack of pathotype specific gene in human *Coxiella burnetii* isolates. Microb. Pathog. *15*: 177–185.

Stein, A., N.A. Saunders, A.G. Taylor and D. Raoult. 1993. Phylogenic homogeneity of *Coxiella burnetii* strains as determined by 16S ribosomal RNA sequencing. FEMS Microbiol. Lett. *113*: 339–344.

Steinberger, E.M., G.Y. Cheng and S.V. Beer. 1990. Characterization of a 56-kb plasmid of *Erwinia amylovora* Ea322: its noninvolvement in pathogenicity. Plasmid *24*: 12–24.

Steinbüchel, A. and H.E. Valentin. 1995. Diversity of bacterial polyhydroxyalkanoic acids. FEMS Microbiol. Lett. *128*: 219–228.

Steiner, R., W. Schafer, I. Blos, H. Wieschhoff and H. Scheer. 1981. Delta-2,10-phytadienol as esterifying alcohol of bacteriochlorophyll *b* from *Ectothiorhodospira halochloris*. Z. Naturforsch. (C) *36*: 417–420.

Steinfeld, S., C. Rossi, N. Bourgeois, I. Mansoor, J.P. Thys and T. Appelboom. 1998. Septic arthritis due to *Aeromonas veronii* biotype *sobria*. Clin. Infect. Dis. *27*: 402–403.

Steinhart, W.L. and R.M. Herriott. 1968. Genetic integration in the heterospecific transformation of *Haemophilus influenzae* cells by *Haemophilus parainfluenzae* DNA. J. Bacteriol. *96*: 1725–1731.

Stenström, I.M. and G. Molin. 1990. Classification of the spoilage flora of fish, with special reference to *Shewanella putrefaciens*. J. Appl. Bacteriol. *68*: 601–618.

Stenström, I.M., A. Zakaria, A. Ternström and G. Molin. 1990. Numerical taxonomy of fluorescent *Pseudomonas* associated with tomato roots. Antonie Van Leeuwenhoek Int. J. Gen. Mol. Microbiol. *57*: 223–236.

Stenzel, W. 1961. *Proteus inconstans* O 13 H 30 als enteritiserreger bei kleinkindern. Zentralbl. Bakteriol. Parastienkd. Infektionskr. Hyg. 1 Abt. Orig. *182*: 178–183.

Stephens, W.L. and M.P. Starr. 1963. Localization of carotenoid pigment in the cytoplasmic membrane of *X. juglandis*. J. Bacteriol. *86*: 1070–1074.

Stephenson, J.R., S.E. Millership and S. Tabaqchali. 1987. Typing of *Aeromonas* species by polyacrylamide-gel electrophoresis of radiolabelled cell proteins. J. Med. Microbiol. *24*: 113–118.

Sternberg, S. 1998. Isolation of *Actinobacillus equuli* from the oral cavity of healthy horses and comparison of isolates by restriction enzyme digestion and Pulsed-Field Gel Electrophoresis. Vet. Microbiol. *59*: 147–156.

Sternberg, S. and B. Brandstrom. 1999. Biochemical fingerprinting and ribotyping of isolates of *Actinobacillus equuli* from healthy and diseased horses. Vet. Microbiol. *66*: 53–65.

Stevens, A.M. and E.P. Greenberg. 1997. Quorum sensing in *Vibrio fischeri*: Essential elements for activation of the luminescence genes. J. Bacteriol. *179*: 557–562.

Stevens, F.L. 1925. Plant Disease Fungi, Macmillan Co., New York. 1–469.

Stevens, J., B.L. Fanburg and J.J. Lanzillo. 1990. Determination of peptidyl dipeptidase activity in 24 bacterial species. Can. J. Microbiol. *36*: 56–59.

Stevenson, G., B. Neal, D. Liu, M. Hobbs, N.H. Packer, M. Batley, J.W. Redmond, L. Lindquist and P. Reeves. 1994. Structure of the O antigen of *Escherichia coli* K-12 and the sequence of its *rfb* gene cluster. J. Bacteriol. *176*: 4144–4156.

Stevenson, R.M. 1997. Immunization with bacterial antigens: yersiniosis. Dev. Biol. Stand. *90*: 117–124.

Stevenson, R.M.W. and D.W. Airdrie. 1984. Serological variation among *Yersinia ruckeri* strains. J. Fish Dis. *7*: 247–254.

Stevenson, R.M.W. and J.G. Daly. 1982. Biochemical and serological characteristics of Ontario isolates of *Yersinia ruckeri*. Can. J. Fish. Aquat. Sci. *39*: 870–876.

Stewart, D.J. 1973. An electron microscopic study of *Fusiformis nodosus*. Res. Vet. Sci. *14*: 132–134.

Stewart, D.J. 1977. Biochemical and biological studies on the lipopolysaccharide of *Bacteroides nodosus*. Res. Vet. Sci. *23*: 319–325.

Stewart, D.J. 1978. The role of various antigenic factors of *Bacteroides nodosus* in eliciting protection against foot rot in vaccinated sheep. Res. Vet. Sci. *24*: 14–19.

Stewart, D.J. 1979. The role of elastase in the differentiation of *Bacteroides nodosus* infections in sheep and cattle. Res. Vet. Sci. *27*: 99–105.

Stewart, D.J. 1989. Footrot of sheep. *In* Egerton, Yong and Riffkin (Editors), Footrot and Foot Abscess in Ruminants, CRC Press, Boca Raton. 5–45.

Stewart, D.J. and J.R. Egerton. 1979. Studies on the ultrastructural morphology of *Bacteroides nodosus*. Res. Vet. Sci. *26*: 227–235.

Stewart, D.J. and T.C. Elleman. 1987. A *Bacteroides nodosus* pili vaccine produced by recombinant DNA for the prevention and treatment of foot-rot in sheep. Aust. Vet. J. *64*: 79–81.

Stewart, D.J., J.E. Peterson, J.A. Vaughan, B.L. Clark, D.L. Emery, J.B. Caldwell and A.A. Kortt. 1986. The pathogenicity and cultural characteristics of virulent, intermediate and benign strains of *Bacteroides nodosus* causing ovine foot-rot. Aust. Vet. J. *63*: 317–326.

Stewart, W.W. 1971. Isolation and proof of structure of wildfire toxin. Nature (Lond.) *229*: 174–178.

Stickler, D.J., C. Fawcett and J.C. Chawla. 1985. *Providencia stuartii*: a search for its natural habitat. J. Hosp. Infect. *6*: 221–223.

Stieb, M. and B. Schink. 1985. A new 3-hydroxybutyrate fermenting anaerobe, *Ilyobacter polytropus*, new genus new species, possessing various fermentation pathways. Arch. Microbiol. *140*: 139–146.

Stintzi, A., P. Cornelis, D. Hohnadel, J.M. Meyer, C. Dean, K. Poole, S. Kourambas and V. Krishnapillai. 1996. Novel pyoverdine biosynthesis gene(s) of *Pseudomonas aeruginosa* PAO. Microbiology *142*: 1181–1190.

Stocker, B.A.D. 1958. Lysogenic conversion by the A phages of *Salmonella typhimurium*. J. Gen. Microbiol. *18*: IX.

Stocker, B.A.D. and P.A. Mäkelä. 1971. Genetic aspect of biosynthesis and structure of *Salmonella lipopolysaccharide*. *In* Weinbaum, Kadis and Ajl (Editors), Microbial Toxins, Vol. IV, Academic Press, New York. pp. 369–438.

Stocker, B.A.D. and P.H. Makela. 1978. Genetics of the (Gram negative) bacterial surface. Proc. R. Soc. Lond. Ser. B Biol. Sci. *202*: 5–30.

Stocks, P.K. and C.S. McCleskey. 1964. Identity of the pink-pigmented methanol-oxidizing bacteria as *Vibrio extorquens.* J. Bacteriol. *88*: 1065–1070.

Stoker, M.G.P. and P. Fiset. 1956. Phase variation of the Nine Mile and other strains of *Rickettsia burnetii.* Can. J. Microbiol. *2*: 310–321.

Stoleru, G.H., L. Le Minor and A.M. Lhéritier. 1976. Polynucleotide sequence divergence among strains of *Salmonella* subgenus IV and closely related organisms. Ann. Inst. Pasteur Microbiol. *127A*: 477–486.

Stolp, H. and D. Gadkari. 1981. Non-pathogenic members of the genus *Pseudomonas. In* Starr, Stolp, Trüper, Balows and Schlegel (Editors), The Prokaryotes, a handbook on the habitats, isolation and identification of the bacteria, Vol. 1, Springer-Verlag, Berlin. 719–741.

Storch, G., W.B. Baine, D.W. Fraser, C.V. Broome, H.W. Clegg, M.L. Cohen, S.A. Goings, B.D. Politi, W.A. Terranova, T.F. Tsai, B.D. Plikaytis, C.C. Shepard and J.V. Bennett. 1979. Sporadic community-acquired Legionnaires' disease in the United States. A case-control study. Ann. Intern. Med. *90*: 596–600.

Story, P. 1954. *Proteus* infections in hospital. J. Pathol. Bacteriol. *68*: 55–62..

Stout, J.D. 1960. Biological studies of some tussock-grassland soils. XV. Bacteria of two cultivated soils. N. Z. J. Agr. Res. *3*: 214–223.

Stouthamer, A.H., W. de Vries and H.G.D. Niekus. 1979. Microaerophily. Antonie Leeuwenhoek J. Microbiol. Serol. *45*: 5–12.

Straley, S.C. 1993. Adhesins in *Yersinia pestis.* Trends Microbiol. *1*: 285–286.

Strandskov, F.B., H.W. Baker and J.B. Bockelmann. 1953. A study of the Gram-negative bacterial rod infection of brewery yeast and brewery fermentations. Wallerstein Lab. Commun. *16*: 261–270.

Strandskov, F.B. and J.B. Bockelmann. 1955. Nutritional requirements of brewing microorganisms. I. The nutritional requirements of *Flavobacterium proteus.* Wallerstein Lab. Commun. *18*: 275–282.

Strätz, M., M. Mau and K.N. Timmis. 1996. System to study horizontal gene exchange among microorganisms without cultivation of recipients. Mol. Microbiol. *22*: 207–215.

Straus, D.C., P.J. Unbehagen and C.W. Purdy. 1993. Neuraminidase production by a *Pasteurella haemolytica* A1 strain associated with bovine pneumonia. Infect. Immun. *61*: 253–259.

Strockbine, N.A., S.M. Faruque, B.A. Kay, K. Haider, K. Alam, A.N. Alam, S. Tzipori and I.K. Wachsmuth. 1992. DNA probe analysis of diarrhoeagenic *Escherichia coli*: detection of EAF-positive isolates of traditional enteropathogenic *E. coli* serotypes among Bangladeshi paediatric diarrhoea patients. Mol. Cell. Probes *6*: 93–99.

Stroeher, U.H., L. Bode, L. Beutin and P.A. Manning. 1993. Characterization and sequence of a 33-kDa enterohemolysin (Ehly 1)-associated protein in *Escherichia coli.* Gene *132*: 89–94.

Strohl, W.R. 1989. Genus I. *Beggiatoa. In* Staley, Bryant, Pfennig and Holt (Editors), Bergey's Manual of Systematic Bacteriology, 1st Ed., Vol. 3, The Williams & Wilkins Co., Baltimore. 2091–2097.

Strohl, W.R., G.C. Cannon, J.M. Shively, H. Güde, L.a. Hook, C.M. Lane and J.M. Larkin. 1981a. Heterotrophic carbon metabolism by *Beggiatoa alba.* J. Bacteriol. *148*: 572–583.

Strohl, W.R., I. Geffers and J.M. Larkin. 1981b. Structure of the sulfur inclusion envelopes from four beggiatoas. Curr. Microbiol. *6*: 75–79.

Strohl, W.R., K.S. Howard and J.M. Larkin. 1982. Ultrastructure of *Beggiatoa alba* strain B15LD. J. Gen. Microbiol. *128*: 73–84.

Strohl, W.R. and J.M. Larkin. 1978a. Cell division and trichome breakage in *Beggiatoa.* Curr. Microbiol. *1*: 151–155.

Strohl, W.R. and J.M. Larkin. 1978b. Enumeration, isolation and characterization of *Beggiatoa* from freshwater sediments. Appl. Environ. Microbiol. *36*: 755–770.

Strohl, W.R. and T.M. Schmidt. 1984. Mixotrophy of the colorless, sulfide-oxidizing, gliding bacteria *Beggiatoa* and *Thiothrix. In* Strohl and Touvinen (Editors), Microbial Chemoautotrophy, Ohio State University Press, Columbus. pp. 79–95.

Strohl, W.R., T.M. Schmidt, N.H. Lawry, M.J. Mezzino and J.M. Larkin.

1986a. Characterization of *Vitreoscilla beggiatoides* and *Vitreoscilla filiforms* sp. nov., nom. rev., and comparison with *Vitreoscilla stercoraria* and *Beggiatoa alba.* Int. J. Syst. Bacteriol. *36*: 302–313.

Strohl, W.R., T.M. Schmidt, V.A. Vinci and J.M. Larkin. 1986b. Electron transport and respiration in *Beggiatoa* and *Vitreoscilla.* Arch. Microbiol. *145*: 71–75.

Strom, D. and A.F. Morgan. 1990. *Pseudomonas putida* PPN. *In* O'Brian (Editor), Genetic Maps. Locus Maps of Complex Genomes, Vol. 5, Cold Spring Harbor Laboratory, Cold Spring Harbor, NY. 2.79–2.87.

Strom, M.S. and S. Lory. 1993. Structure-function and biogenesis of the type IV pili. Annu. Rev. Microbiol. *47*: 565–596.

Strom, M.S. and C.M. Pepe. 1999. Characterization of the *tap* Type IV pilus gene cluster in *Aeromonas hydrophila* and *A. salmonicida.* 6th International *Aeromonas/ Plesiomonas* Symposium, Chicago, Illinois. p. 14.

Strom, T., T. Ferenci and J.R. Quayle. 1974. The carbon assimilation pathways of *Methylococcus capsulatus, Pseudomonas methanica* and *Methylosinus trichosporium* (OB3b) during growth on methane. Biochem. J. *144*: 465–476.

Struelens, M.J., E. Carlier, N. Maes, E. Serruys, W.G.V. Quint and A. van Belkum. 1993. Nosocomial colonization and infection with multiresistant *Acinetobacter baumannii*: outbreak delineation using DNA macrorestriction analysis and PCR-fingerprinting. J. Hosp. Infect. *25*: 15–32.

Strzeszewski, B. 1913. Beiträge zur Kenntnis der Schwefelflora in der Umgebung von Krakau. Bull. Int. Acad. Sci. Ser. V. Cracovie Ser. B Sci. Nat. I : 309–334.

Stuart, C.A. and R. Rustigian. 1943. Further studies on one type of paracolon organisms. Am. J. Public Health. *33*: 1323–1325.

Stuart, C.A., K.M. Wheeler, R. Rustigian and A. Zimmerman. 1943. Biochemical and antigenic relationships of the paracolon bacteria. J. Bacteriol. *45*: 101–109.

Stuhlmann, F. 1907. Beitrage zur Kenntnis der Tsetsefliege (*Glossina fusca* und *Gl. tachinoides*). Arbeit. K. Gesundhamt. *26*: 301–383.

Stull, T.L., J.J. LiPuma and T.D. Edlind. 1988. A broad-spectrum probe for molecular epidemiology of bacteria - ribosomal RNA. J. Infect. Dis. *157*: 280–286.

Stumpf, P.K. and D.E. Green. 1944. L-amino acid oxidase of *Proteus valgaris.* J. Biol. Chem. *153*: 387–399.

Sturm, A.W. 1986. Isolation of *Haemophilus influenzae* and *Haemophilus parainfluenzae* from genital-tract specimens with a selective culture medium. J. Med. Microbiol. *21*: 349–352.

Stürmeyer, H., J. Overmann, H.D. Babenzien and H. Cypionka. 1998. Ecophysiological and phylogenetic studies of *Nevskia ramosa* in pure culture. Appl. Environ. Microbiol. *64*: 1890–1894.

Stuy, J.H. 1978. On the nature of nontypable *Haemophilus influenzae.* Antonie Leeuwenhoek *44*: 367–376.

Stuy, J.H. 1979. Plasmid transfer in *Haemophilus influenzae.* J. Bacteriol. *139*: 520–529.

Stuy, J.H. 1980. Chromosomally integrated conjugative plasmids are common in antibiotic- resistant *Haemophilus influenzae.* J. Bacteriol. *142*: 925–930.

Su, C.J., R. Reusch and H.L. Sadoff. 1979. Fatty acids in phospholipids of cells, cysts, and germinating cysts of *Azotobacter vinelandii.* J. Bacteriol. *137*: 1434–1436.

Su, C.J. and H.L. Sadoff. 1981. Unique lipids in *Azotobacter vinelandii* cysts - synthesis, distribution, and fate during germination. J. Bacteriol. *147*: 91–96.

Su, J.J. and D. Kafkewitz. 1994. Utilization of toluene and xylenes by a nitrate-reducing strain of *Pseudomonas maltophilia* under low oxygen and anoxic conditions. FEMS Microbiol. Ecol. *15*: 249–257.

Subandiyah, S., N. Nikoh, T. S., S. Somowiyarjo and T. Fukatsu. 2000. Complex endosymbiotic microbiota of the citrus psyllid *Diaphorina citri* (Homoptera: Psylloidea). Zool. Sci. *17*: 983–989.

Subasinghe, R.P. and M. Shariff. 1992. Multiple bacteriosis, with special reference to spoilage bacterium *Shewanella putrefaciens*, in cage-cultured barramundi perch in Malaysia. J. Aquat. Anim. Health. *4*: 309–311.

Subcommittee on the Taxonomy of Vibrios 1975. Minutes of the closed meeting, 3 September 1974. Int. J. Syst. Bacteriol. *25*: 389–391.

Sudo, T., K. Shinohara, N. Dohmae, K. Takio, R. Usami, K. Horikoshi and H. Osada. 1996. Isolation and characterization of the gene encoding an aminopeptidase involved in the selective toxicity of ascamycin toward *Xanthomonas campestris* pv. *citri*. Biochem. J. *319*: 99–102.

Suga, K., K. Ito, D. Tsuru and T. Yoshimoto. 1995. Prolylcarboxypeptidase (angiotensinase C): purification and characterization of the enzyme from *Xanthomanas maltophilia*. Biosci. Biotechnol. Biochem. *59*: 298–301.

Sugimoto, C., Y. Isayama, R. Sakazaki and S. Kuramochi. 1983. Transfer of *Haemophilus equigenitalis* Taylor et al. 1978 to the genus *Taylorella*, gen. nov. as *Taylorella equigenitalis*, comb. nov. Curr. Microbiol. *9*: 155–162.

Suhan, M.L., S.Y. Chen and H.A. Thompson. 1996. Transformation of *Coxiella burnetii* to ampicillin resistance. J. Bacteriol. *178*: 2701–2708.

Suhan, M., S.Y. Chen, H.A. Thompson, T.A. Hoover, A. Hill and J.C. Williams. 1994. Cloning and characterization of an autonomous replication sequence from *Coxiella burnetii*. J. Bacteriol. *176*: 5233–5243.

Suitor, E.C., Jr. 1964. Propagation of *Rickettsiella popilliae* (Dutky and Gooden) Philip and *Rickettsiella melolonthae* (Krieg) Philip in cell cultures. J. Insect Pathol. *6*: 31–40.

Suitor, E.C., Jr. and E. Weiss. 1961. Isolation of a rickettsialike microorganism (*Wolbachia persica* n. sp.) from *Argas persicus* (Oken). J. Infect. Dis. *108*: 95–106.

Sukhan, A. and R.E.W. Hancock. 1996. The role of specific lysine residues in the passage of anions through the *Pseudomonas aeruginosa* porin OprP. J. Biol. Chem. *271*: 21239–21242.

Sulzinski, M.A., G.W. Moorman, B. Schlagnhaufer and C.P. Romaine. 1995. Fingerprinting of *Xanthomonas campestris* pv. *pelargonii* and related pathovars using random-primed PCR. J. Phytopathol. (Berl.) *143*: 429–433.

Summers, A.O., G.A. Jacoby, M.N. Swartz, G. McHugh and L. Sutton. 1978. Metal cations and oxyanion resistances in plasmids of Gram-negative bacteria. *In* Schlessinger (Editor), Microbiology 1978, American Society for Microbiology, Washington, D.C. 128–131.

Sun, W.Q., J.G. Cao, K. Teng and E.A. Meighen. 1994. Biosynthesis of poly-3-hydroxybutyrate in the luminescent bacterium *Vibrio harveyi* and regulation by the Lux autoinducer, *N*-(3-Hydroxybutanoyl)homoserine lactone. J. Biol. Chem. *269*: 20785–20790.

Sun, X.Y., M. Griffith, J.J. Pasternak and B.R. Glick. 1995. Low-temperature growth, freezing survival, and production of antifreeze protein by the plant-growth promoting rhizobacterium *Pseudomonas putida* Gr12-2. Can. J. Microbiol. *41*: 776–784.

Sundaram, S. and K.V. Murthy. 1983. Occurrence of 2,4-diamino-6,7-diisopropyl-pteridine (0/129) resistance in human isolates of *Vibrio cholerae*. FEMS Microbiol. Lett. *19*: 115–117.

Sundh, I., M. Nilsson and P. Borga. 1997. Variation in microbial community structure in two boreal peatlands as determined by analysis of phospholipid fatty acid profiles. Appl. Environ. Microbiol. *63*: 1476–1482.

Surgalla, M.J., A.W. Andrews and D.M. Cavanaugh. 1968. Studies on virulence factors of *Pasteurella pestis* and *Pasteurella pseudotuberculosis*. Symp. Ser. Immunol. Stand. *9*: 293–302.

Surgey, N., J. Robert-Baudouy and G. Condemine. 1996. The *Erwinia chrysanthemi pecT* gene regulates pectinase gene expression. J. Bacteriol. *178*: 1593–1599.

Surico, G., L. Mugnai, R. Pastorelli, L. Giovannetti and D.E. Stead. 1996. *Erwinia alni*, a new species causing bark cankers of alder (*Alnus* Miller) species. Int. J. Syst. Bacteriol. *46*: 720–726.

Sussman, M. (Editor). 1985. The Virulence of *Escherichia coli*: Reviews and Methods, Academic Press, London. 473 pp.

Sutakova, G. and F. Ruttgen. 1978. *Rickettsiella phytoseiuli* and virus-like particles in *Phytosfiulus persimilis* (Gamasoidea: Phytoseiidae) mites. Acta Virol. *22*: 333–336.

Sutakova, G., Z. Sekejova, J. Rechacek, I.G. Skripal and L.P. Malinovskaia. 1991. A trial of cultivating *Rickettsiella phytoseiuli* on the SM IMV-72

medium used for growing phytopathogenic mycoplasmas. Mikrobiol. Zh. *53*: 57–61.

Sutherland, I.W. 1993. Xanthan. *In* Swings and Civerolo (Editors), *Xanthomonas*, Chapman & Hall, London. pp. 363–388.

Sutherland, I.W. and D.C. Ellwood. 1979. Microbial exopolysaccharides - industrial polymers of current and future potential. Symp. Soc. Gen. Microbiol. *29*: 107–150.

Sutherland, I.W. and A.F.D. Kennedy. 1986. Comparison of bacterial lipopolysaccharides by high-performance liquid chromatography. Appl. Environ. Microbiol. *52*: 948–950.

Sutherland, I.W. and C.L. Mackenzie. 1977. Glucan common to the microcyst walls of cyst-forming bacteria. J. Bacteriol. *129*: 599–605.

Suthienkul, O., M. Ishibashi, T. Iida, N. Nettip, S. Supavej, B. Eampokalap, M. Makino and T. Honda. 1995. Urease production correlates with possession of the *trh* gene in *Vibrio parahaemolyticus* strains isolated in Thailand. J. Infect. Dis. *172*: 1405–1408.

Šutić, D. and W.J. Dowson. 1959. An investigation of a serious disease of hemp (*Cannabina sativa* L.) in Jugoslavia. Phytopathology *34*: 307–314.

Šutić, D. and Z. Tesic. 1958. Judna nova bakterioza bresta izazivac *Pseudomonas ulmi* n. sp. Zast. Bilja. *45*: 13–25.

Sutinen, S. and H. Syrjälä. 1986. Histopathology of human lymph node tularemia caused by *Francisella tularensis* var. *palaearctica*. Arch. Pathol. Lab. Med. *110*: 42–46.

Sutra, L., F. Siverio, M.M. López, G. Hunault, C. Bollet and L. Gardan. 1997. Taxonomy of *Pseudomonas* strains isolated from tomato pith necrosis: Emended description of *Pseudomonas corrugata* and proposal of three unnamed fluorescent *Pseudomonas* genomospecies. Int. J. Syst. Bacteriol. *47*: 1020–1033.

Sutter, G.R. and V.M. Kirk. 1968. *Rickettsia*-like particles in fat-body cells of carabid beetles. J. Invert. Pathol. *10*: 445–449.

Sutter, V.L. 1968. Identification of *Pseudomonas* species isolated from hospital environment and human sources. Appl. Microbiol. *16*: 1532–1538.

Sutter, V.L. and S.M. Finegold. 1970. *Haemophilus aphrophilus* infections: clinical and bacteriologic studies. Ann. NY Acad. Sci. *174*: 468–487.

Sutton, R.G.A., M.F. O'Keeffe, M.A. Bundock, J. Jeboult and M.P. Tester. 1972. Isolation of a new *Moraxella* from a corneal abscess. J. Med. Bacteriol. *5*: 148–150.

Suzina, N.E., V.I. Duda, L.A. Anisimova, V.V. Dmitriev and A.M. Boronin. 1995. Cytological aspects of resistance to potassium tellurite conferred on *Pseudomonas* cells by plasmids. Arch. Microbiol. *163*: 282–285.

Suzuki, A. and M. Goto. 1971. Isolation and characterization of pteridines from *Pseudomonas ovalis*. Bull. Chem. Soc. (Japan). *44*: 1869–1872.

Suzuki, A. and M. Goto. 1972. The structure of a new pteridine compound produced by *Pseudomonas ovalis*. Bull. Chem. Soc. Japan. *45*: 2198–2199.

Suzuki, K., M. Goodfellow and A.G. O'Donnell. 1993a. Cell envelopes in classification. *In* Goodfellow and O'Donnell (Editors), Handbook of New Bacterial Systematics, Academic Press, London. pp. 195–250.

Suzuki, K., H. Matsunaga, C. Itami and Y. Kimura. 1993b. Polymyxin B enhances formation of a pigment, a peptide-ferropyrimine complex, in *Serratia marcescens*. Biosci. Biotechnol. Biochem. *57*: 1763–1765.

Suzuki, T., H. Yabusaki and Y. Nishimura. 1996. Phylogenetic relationships of entomopathogenic nematophilic bacteria: *Xenorhabdus* spp. and *Photorhabdus* sp. J. Basic Microbiol. *36*: 351–354.

Suzuki, T., S. Yamanaka and Y. Nishimura. 1990. Chemotaxonomic study of *Xenorhabdus* spp.: cellular fatty acids, ubiquinone and DNA–DNA hybridization. J. Gen. Appl. Microbiol. *36*: 393–401.

Svanem, B.I.G., G. Skjak-Braek, H. Ertesvag and S. Valla. 1999. Cloning and expression of three new *Azotobacter vinelandii* genes closely related to a previously described gene family encoding mannuronan C-5-epimerases. J. Bacteriol. *181*: 68–77.

Svetashev, V.I., M.V. Vysotskii, E.P. Ivanova and V.V. Mikhailov. 1995. Cellular fatty acids of *Alteromonas* species. Syst. Appl. Microbiol. *18*: 37–43.

Swaminathan, B., T.J. Barrett, S.B. Hunter and R.V. Tauxe. 2001.

PulseNet: the molecular subtyping network for foodborne bacterial disease surveillance, United States. Emerg. Infect. Dis. *7*: 382–389.

Swanson, J., B. Kearney, D. Dahlbeck and B. Staskawicz. 1988. Cloned avirulence gene of *Xanthomonas campestris* pv. *vesicatoria* complements spontaneous race-change mutants. Mol. Plant-Microbe Interact. *1*: 5–9.

Sweerts, J.-P.R.A., D. De Beer, L.P. Nielsen, H. Verdouw, J.C. Van den Heuvel, Y. Cohen and T.E. Cappenberg. 1990. Denitrification by sulphur oxidzing *Beggiatoa* spp. mats on freshwater sediments. Nature *344*: 762–763.

Swerdlow, D.L. and A.A. Ries. 1993. *Vibrio cholerae* non-O1—the eighth pandemic? Lancet *342*: 382–383.

Swihart, K.G. and R.A. Welch. 1990a. Cytotoxic activity of the *Proteus* hemolysin HpmA. Infect. Immun. *58*: 1861–1869.

Swihart, K.G. and R.A. Welch. 1990b. The HpmA hemolysin is more common than HlyA among *Proteus* isolates. Infect. Immun. *58*: 1853–1860.

Swingle, D.B. 1925. Center rot of "French endive" or wilt of chicory (*Chichorium intybus* L.). Phytopathology *15*: 730.

Swings, J., M. Gillis, K. Kersters, P. De Vos, F. Gosselé and J. De Ley. 1980. *Frateuria*, a new genus for "*Acetobacter aurantius*". Int. J. Syst. Bacteriol. *30*: 547–556.

Swings, J., M. Van den Mooter, L. Vauterin, B. Hoste, M. Gillis, T.W. Mew and K. Kersters. 1990. Reclassification of the causal agents of bacterial blight (*Xanthomonas campestris* pv. *oryzae*) and bacterial leaf streak (*Xanthomonas campestris* pv. *oryzicola*) of rice as pathovars of *Xanthomonas oryzae*, new species (ex Ishiyama 1922) revived name. Int. J. Syst. Bacteriol. *40*: 309–311.

Swings, P., P. De Vos, M. Van den Mooter and J. De Ley. 1983. Transfer of *Pseudomonas maltophilia* Hugh 1981 to the genus *Xanthomonas* as *Xanthomonas maltophilia* (Hugh 1981) comb. nov. Int. J. Syst. Bacteriol. *33*: 409–413.

Swofford, D.L. 1993. PAUP: Phylogenetic Analysis Using Parsimony v3.1.1, Illinois Natural History Survey, Champaign, IL.

Sykes, R.B. and M. Matthew. 1976. The beta-lactamases of of Gram-negative bacteria and their role in resistance to beta-lactam antibiotics. J. Antimicrob. Chemother. *2*: 115–157.

Symbas, P.N., R.C. Schlant, C.R.J. Hatcher and J. Lindsay. 1967. Congenital fistula of right coronary artery to right ventricle complicated by *Actinobacillus actinomycetemcomitans* endarteritis. J. Thorac. Cardiovasc. Surg. *53*: 379–384.

Syrjälä, H., E. Herva, J. Ilonen, K. Saukkonen and A. Salminen. 1984. A whole-blood lymphocyte stimulation test for the diagnosis of human tularemia. J. Infect. Dis. *150*: 912–915.

Syrjälä, H., R. Schildt and S. Raisainen. 1991. *In vitro* susceptibility of *Francisella tularensis* to fluoroquinolones and treatment of tularemia with norfloxacin and ciprofloxacin. Eur. J. Clin. Microbiol. Infect. Dis. *10*: 68–70.

Szállás, E., C. Koch, A. Fodor, J. Burghardt, O. Buss, A. Szentirmai, K.H. Nealson and E. Stackebrandt. 1997. Phylogenetic evidence for the taxonomic heterogeneity of *Photorhabdus luminescens*. Int. J. Syst. Bacteriol. *47*: 402–407.

Szturm, S., M. Piéchard and R. Neél. 1950. Un nouveau type antigénique de *Shigella boydii*. Ann. Inst. Pasteur. *78*: 146–147.

Tabe, Y. and J. Igari. 1994. Susceptibilities of glucose non-fermentative gram-negative bacilli to antibiotics. Jpn. J. Antibiot. *47*: 1030–1040.

Tacket, C.O., F. Brenner and P.A. Blake. 1984. Clinical features and an epidemiological study of *Vibrio vulnificus* infections. J. Infect. Dis. *149*: 558–561.

Tagger, S., N. Truffaunt and J. Le Petit. 1990. Preliminary study of relationships among strains forming a bacterial community selected on naphthalene from a marine sediment. Can. J. Microbiol. *36*: 676–681.

Tai, P.C. and H. Jackson. 1969. Mesophilic mutants of an obligate psychrophile, *Micrococcus cryophilus*. Can. J. Microbiol. *15*: 1145–1150.

Takada, Y., T. Ochiai, H. Okuyama, K. Nishi and S. Sasaki. 1979. An obligately psychrophilic bacterium isolated on the Hokkaido coast. J. Gen. Appl. Microbiol. *25*: 11–19.

Takahashi, A., H. Nakamura, T. Kameyama, S. Kurasawa, H. Naganawa, Y. Okami, T. Takeuchi, H. Umezawa and Y. Iitaka. 1987. Bisucaberin, a new siderophore, sensitizing tumor cells to macrophage-mediated cytolysis. II. Physico-chemical properties and structure determination. J. Antibiot. (Tokyo). *40*: 1671–1676.

Takaichi, S., T. Maoka, S. Hanada and J.F. Imhoff. 2001. Unusual carotenoid glycoside ester, di-hydroxylycopene digulcoside diester, in two species of *Ectothiorhodospiraceae*: *Halorhodospira abdelmalekii* and *Halorhodospira halochloris*. Arch. Microbiol. *175*: 161–167.

Takala, A.K., J. Eskola, M. Leinonen, H. Kayhty, A. Nissinen, E. Pekkanen and P.H. Makela. 1991. Reduction of oropharyngeal carriage of *Haemophilus influenzae* type b (Hib) in children immunized with an Hib conjugate vaccine. J. Infect. Dis. *164*: 982–986.

Takasaka, M., S. Honjo, T. Fujiwara, T. Hagiwara, H. Ogawa and K. Imaizumi. 1964. Shigellosis in cynomolgus monkeys (*Maca irus*). I. Epidemiological surveys on *Shigella* infection rate. Jpn. J. Med. Sci. Biol. *17*: 259–265.

Takeda, K. 1988. Characteristics of a nitrogen-fixing methanotroph, *Methylocystis* T-1. Antonie Leeuwenhoek *54*: 521–534.

Takeda, K., S. Motomatsu, Y. Hachiya, S. Fukuoka and Y. Takahara. 1974. Characterization and culture conditions for a methane-oxidizing bacterium. J. Ferment. Technol. *52*: 793–798.

Takeda, Y., H. Kurazono and S. Yamasaki. 1993. Vero toxins (Shiga-like toxins) produced by enterohemorrhagic *Escherichia coli* (verocytotoxin-producing *E. coli*). Microbiol. Immunol. *37*: 591–599.

Takeyama, H., D. Takeda, K. Yazawa, A. Yamada and T. Matsunaga. 1997. Expression of the eicosapentaenoic acid synthesis gene cluster from *Shewanella* sp. in a transgenic marine cyanobacterium, *Synechococcus* sp. Microbiology *143*: 2725–2731.

Takimoto, S. 1920. On the bacterial leaf-spot of *Antirrhinum majus* L. Bot. Magaz., Tokyo. *34*: 253–257.

Takimoto, S. 1927. Bacterial black spot of burdock. J. Plant Prot. *14*: 519–523.

Takimoto, S. 1931. Bacterial bud rot of loquat. J. Plant. Prot., Tokyo. *18*: 349–355.

Takimoto, S. 1933. The bacterial disease of New Zealand flax. J. Plant Prot. *20*: 774–778.

Takimoto, S. 1934. Leaf spot of begonia. J. Plant Prot. *21*: 258–262.

Talmadge, M.B. and R.M. Herriott. 1960. A chemically defined medium for growth, transformation, and isolation of nutritional mutants of *Haemophilus influenzae*. Biophys. Biochem. Res. Comm. *2*: 203–206.

Tamura, K., R. Sakazaki, Y. Kosako and E. Yoshizaki. 1986. *Leclercia adecarboxylata*, gen. nov., comb. nov., formerly known as *Escherichia adecarboxylata*. Curr. Microbiol. *13*: 179–184.

Tamura, K., R. Sakazaki, Y. Kosako and E. Yoshizaki. 1987. *In* Validation of the publication of new names and new combinations previously effectively published outside the IJSB. List No. 23. Int. J. Syst. Bacteriol. *37*: 179–180.

Tamura, K., R. Sakazaki, A.C. McWhorter and Y. Kosako. 1988. *Edwardsiella tarda* serotyping scheme for international use. J. Clin. Microbiol. *26*: 2343–2346.

Tan, C.K. and L. Owens. 2000. Infectivity, transmission and 16S rRNA sequencing of a rickettsia, *Coxiella cheraxi* sp. nov., from the freshwater crayfish *Cherax quadricarinatus*. Dis. Aquat. Organ. *41*: 115–122.

Tan, R.J., E.W. Lim and B. Ishak. 1978. Intestinal bacterial flora of the household lizard, *Gecko gecko*. Res. Vet. Sci. *24*: 262–263.

Tan, R.J., E.W. Lim and M. Teo. 1977. Occurence and medical importance of *Edwardsiella tarda* in clinical specimens. Ann. Acad. Med. Singapore. *6*: 337–341.

Tan, S., R.K.O. Apenten and J. Knapp. 1996. Low temperature organic phase biocatalysis using cold-adapted lipase from psychrotrophic *Pseudomonas* P38. Food Chem. *57*: 415–418.

Tan, S.C., Y.H. Wong, M. Jegathesan and S.M. Chang. 1989. The first isolate of *Tatumella ptyseos* in Malaysia. Malays J. Pathol. *11*: 25–27.

Tandoi, V., N. Caravaglio, D.D. Balsamo, M. Majone and M.C. Tomei. 1994. Isolation and physiological characterization of *Thiothrix* sp. Water Sci. Technol. *29*: 261–269.

Tang, P.W., S. Toma and L.G. MacMillan. 1985. *Legionella oakridgensis*:

laboratory diagnosis of a human infection. J. Clin. Microbiol. *21*: 462–463.

Tang, P.W., S. Toma, C.W. Moss, A.G. Steigerwalt, D.J. Brenner and T.G. Cooligan. 1984. *Legionella bozemanii* serogroup 2: a new etiological agent. J. Clin. Microbiol. *19*: 30–33.

Tang, Y.W., N.M. Ellis, M.K. Hopkins, D.H. Smith, D.E. Dodge and D.H. Persing. 1998. Comparison of phenotypic and genotypic techniques for identification of unusual aerobic pathogenic gram-negative bacilli. J. Clin. Microbiol. *36*: 3674–3679.

Tanii, A., K. Miyajima and T. Akita. 1976. The sheath brown rot disease of rice plant and its causal bacterium, *Pseudomonas fuscovaginae* A. Tanii, K. Miyajima, et T. Akita sp. nov. Ann. Phytopathol. Soc. Japan *42*: 540–548.

Tankovic, J., P. Legrand, G. De Gatines, V. Chemineau, C. Brun-Buisson and J. Duval. 1994. Characterization of a hospital outbreak of imipenem-resistant *Acinetobacter baumannii* by phenotypic and genotypic typing methods. J. Clin. Microbiol. *32*: 2677–2681.

Tanner, A.C., R.A. Visconti, S.S. Socransky and S.C. Holt. 1982. Classification and identification of *Actinobacillus actinomycetemcomitans* and *Haemophilus aphrophilus* by cluster analysis and deoxyribonucleic acid hybridizations. J. Periodontal Res. *17*: 585–596.

Targowski, S. and H. Targowski. 1979. Characterization of a *Haemophilus paracuniculus* isolated from gastrointestinal tracts of rabbits with mucoid enteritis. J. Clin. Microbiol. *9*: 33–37.

Targowski, S. and H. Targowski. 1984. *In* Validation of the publication of new names and new combinations previously effectively published outside the IJSB. List no.15. Int. J. Syst. Bacteriol. *34*: 355–357.

Tärnvik, A. and S. Löfgren. 1975. Stimulation of human lymphocytes by a vaccine strain of *Francisella tularensis*. Infect. Immun. *12*: 951–957.

Tärnvik, A., G. Sandström and A. Sjöstedt. 1997. Infrequent manifestations of tularaemia in Sweden. Scand. J. Infect. Dis. *29*: 443–446.

Tatlock, H. 1944. A rickettsia-like organism recovered from guinea pigs. Proc. Soc. Exp. Biol. Med. *57*: 95–99.

Tatum, F.M., R.E. Briggs, S.S. Sreevatsan, E.S. Zehr, S. Ling Hsuan, L.O. Whiteley, T.R. Ames and S.K. Maheswaran. 1998. Construction of an isogenic leukotoxin deletion mutant of *Pasteurella haemolytica* serotype 1: characterization and virulence. Microb. Pathog. *24*: 37–46.

Tatusov, R.L., A.R. Mushegian, P. Bork, N.P. Brown, W.S. Hayes, M. Borodovsky, K.E. Rudd and E.V. Koonin. 1996. Metabolism and evolution of *Haemophilus influenzae* deduced from a whole genome comparison with *Escherichia coli*. Curr. Biol. *6*: 279–291.

Taylor, C.M., R.H. White, M.H. Winterborn and B. Rowe. 1986a. Haemolytic-uraemic syndrome: clinical experience of an outbreak in the West Midlands. Br. Med. J. *292*: 1513–1516.

Taylor, D.C., A.W. Cripps, R.L. Clancy, K. Murree-Allen, M.J. Hensley, N.A. Saunders and D.C. Sutherland. 1990. Evaluation of a selective medium for the isolation and differentiation of *Haemophilus influenzae* and *Haemophilus parainfluenzae* from the respiratory tract of chronic bronchitics. Pathology *22*: 162–164.

Taylor, D.C., A.W. Cripps, R.L. Clancy, K. Murree-Allen, M.J. Hensley, N.A. Saunders and D.C. Sutherland. 1992. Biotypes of *Haemophilus parainfluenzae* from the respiratory secretions in chronic bronchitis. J. Med. Microbiol. *36*: 279–282.

Taylor, D.N., P. Echeverria, T. Pál, O. Sethabutr, S. Saiborisuth, S. Sricharmorn, B. Rowe and J. Cross. 1986b. The role of *Shigella* spp., enteroinvasive *Escherichia coli*, and other enteropathogens as causes of childhood dysentery in Thailand. J. Infect. Dis. *153*: 1132–1138.

Taylor, E.W., C. S.T., R. Lovell, J. Taylor, S.B. Thomas, W.A. Cuthbert, S.E. Jacobs and L. Clegg. 1956. The nomenclature of the coliaerogenes bacteria. J. Appl. Bacteriol. *19*: 108–111.

Taylor, J., H.J. Bensted, J.S.K. Boyd, K.P. Carpenter, W.J. Dowson, R. Lovell, E.W. Taylor, H.G. Thornton, G.S. Wilson and C. Shawl. 1952. Classification of the *Bacteriaceae*. Int. Bull. Bacteriol. Nomencl. Taxon. *2*: 137–140.

Taylor, R.K., V.L. Miller, D.B. Furlong and J.J. Mekalanos. 1987. Use of *phoA* gene fusions to identify a pilus colonization factor coordinately regulated with cholera toxin. Proc. Natl Acad. Sci. U.S.A. *84*: 2833–2837.

Taylor, V.I., P. Baumann, J.L. Reichelt and R.D. Allen. 1974. Isolation, enumeration, and host range of marine bdellovibrios. Arch. Microbiol. *98*: 101–114.

Taylor, W.I. and D. Achanzar. 1972. Catalase test as an aid to the identification of *Enterobacteriaceae*. Appl. Microbiol. *24*: 58–61.

Taylor, W.H. and E. Juni. 1961. Pathways for biosynthesis of a bacterial capsular polysaccharide II. Carbohydrate metabolism and terminal oxidation mechanisms of a capsule-producing coccus. J. Bacteriol. *81*: 694–703.

Tchan, Y.T. 1968. Importance of systematics of *Azotobacteriaceae* in the study of its ecology. Trans. 9th Int. Congr. Soil Sci. Adelaide. *2*: 115–124.

Tchan, Y.T. 1984. Family II. *Azotobacteraceae*. *In* Kreig, N.R. and J.G. Holt (Editors), Bergey's Manual of Systematic Bacteriology, Vol. 1, Willliams and Wilkins, Baltimore/London. 219–220.

Tchan, Y.T. and P.B. New. 1984. Genus I. *Azotobacter*. *In* Kreig, N.R. and J.G. Holt (Editors), Bergey's Manual of Systematic Bacteriology, Vol. 1, Willliams and Wilkins, Baltimore/London. 220–229.

Tchan, Y.T., Z. Wyszomirskadreher, P.B. New and J.C. Zhou. 1983. Taxonomy of the *Azotobacteraceae* determined by using immunoelectrophoresis. Int. J. Syst. Bacteriol. *33*: 147–156.

Tchan, Y.T., Z. Wyszomirskadreher and J.M. Vincent. 1980. Preliminary study of taxonomy of *Azotobacter* and *Azomonas* by using rocket line immunoelectrophoresis. Curr. Microbiol. *4*: 265–270.

Telford, J.R. and K.N. Raymond. 1997. Amonobactin: a family of novel siderophores from a pathogenic bacterium. J. Biolog. Inorg. Chem. *2*: 750–761.

Temesgen, Z., D.R. Toal and F.R. Cockerill, III. 1997. *Leclercia adecarboxylata* infections: case report and review. Clin. Infect. Dis. *25*: 79–81.

Temple, K.L. and A.R. Colmer. 1951. The autotrophic oxidation of iron by a new bacterium, *Thiobacillus ferrooxidans*. J. Bacteriol. *62*: 605–611.

Tenover, F.C., L. Carlson, L. Goldstein, J. Sturge and J.J. Plorde. 1985. Confirmation of *Legionella pneumophila* cultures with a fluorescein-labeled monoclonal antibody. J. Clin. Microbiol. *21*: 983–984.

Tenover, F.C., P.H. Edelstein, L.C. Goldstein, J.C. Sturge and J.J. Plorde. 1986. Comparison of cross-staining reactions by *Pseudomonas* spp. and fluorescein-labeled polyclonal and monoclonal antibodies directed against *Legionella pneumophila*. J. Clin. Microbiol. *23*: 647–649.

Tenover, F.C., T. Popovic and Ø. Olsvik. 1995. Genetic methods for detecting antibacterial resistance genes. *In* Murray, Baron, Pfaller, Tenover and Yolken (Editors), Manual of Clinical Microbiology, 6th Ed., ASM Press, Washington, D.C. pp. 1368–1378.

Terasaki, Y. 1963. On the isolation of *Spirillum*. Bull. Suzugamine Women's Coll. Nat. Sci. *10*: 1–10.

Terasaki, Y. 1970. Über die Anhäufung von in Süsswasser und Meerwasser vorkommenden *Spirillum*. Bull. Suzagamine Women's Coll. Nat. Sci. *15*: 1–7.

Terasaki, Y. 1972. Studies on the genus *Spirillum* Ehrenberg. I. Morphological, physiological, and biochemical characteristics of water spirilla. Bull. Suzugamine Women's Coll. Nat. Sci. *16*: 1–146.

Terasaki, Y. 1973. Studies on the genus *Spirillum* Ehrenberg. II. Comments on type and reference strains of *Spirillum* and descriptions of new species and new subspecies. Bull. Suzugamine Women's Coll. Nat. Sci. *17*: 1–71.

Terasaki, Y. 1975. Freeze-dried cultures of water spirilla made on experimental basis. Bull. Suzugamine Women's Coll., Nat. Sci. *19*: 1–10.

Terasaki, Y. 1979. Transfer of five species and two subspecies of *Spirillum* to other genera (*Aquaspirillum* and *Oceanospirillum*), with emended descriptions of the species and subspecies. Int. J. Syst. Bacteriol. *29*: 130–144.

Terasaki, Y. 1980. Enrichment and isolation of aerobic chemoheterotrophic spirilla from mud and sand samples. J. Gen. Appl. Microbiol. *26*: 395–402.

Terry, J.M., S.E. Pina and S.J. Mattingly. 1991. Environmental conditions which influence mucoid conversion *Pseudomonas aeruginosa* PAO1. Infect. Immun. *59*: 471–477.

Teske, A., E. Alm, J.M. Regan, S. Toze, B.E. Rittmann and D.A. Stahl.

1994. Evolutionary relationships among ammonia- and nitrite-oxidizing bacteria. J. Bacteriol. *176*: 6623–6630.

Teske, A., N.B. Ramsing, J. Küver and H. Fossing. 1995. Phylogeny of *Thioploca* and related filamentous sulfide-oxidizing bacteria. Syst. Appl. Microbiol. *18*: 517–526.

Teske, A., M.L. Sogin, L.P. Nielsen and H.W. Jannasch. 1999. Phylogenetic relationships of a large marine *Beggiatoa*. Syst. Appl. Microbiol. *22*: 39–44.

Thacker, L., R.M. McKinney, C.W. Moss, H.M. Sommers, M.L. Spivack and T.F. O'Brien. 1981. Thermophilic sporeforming bacilli that mimic fastidious growth characteristics and colonial morphology of legionellae. J. Clin. Microbiol. *13*: 794–797.

Thacker, W.L., R.F. Benson, L. Hawes, H. Gidding, B. Dwyer, W.R. Mayberry and D.J. Brenner. 1991a. *Legionella fairfieldensis*, sp. nov. isolated from cooling tower waters in Australia. J. Clin. Microbiol. *29*: 475–478.

Thacker, W.L., R.F. Benson, L. Hawes, H. Gidding, B. Dwyer, W.R. Mayberry and D.J. Brenner. 1991b. *In* Validation of the publication of new names and new combinations previously effectively published outside the IJSB. List No. 39. Int. J. Syst. Bacteriol. *41*: 580–581.

Thacker, W.L., R.F. Benson, L. Hawes, W.R. Mayberry and D.J. Brenner. 1990a. Characterization of a *Legionella anisa* strain isolated from a patient with pneumonia. J. Clin. Microbiol. *28*: 122–123.

Thacker, W.L., R.F. Benson, R.B. Schifman, E. Pugh, A.G. Steigerwalt, W.R. Mayberry, D.J. Brenner and H.W. Wilkinson. 1989a. *Legionella tucsonensis*, sp. nov. isolated from a renal transplant recipient. J. Clin. Microbiol. *27*: 1831–1834.

Thacker, W.L., R.F. Benson, R.B. Schifman, E. Pugh, A.G. Steigerwalt, W.R. Mayberry, D.J. Brenner and H.W. Wilkinson. 1990b. *In* Validation of the publication of new names and new combinations previously effectively published outside the IJSB. List No. 32. Int. J. Syst. Bacteriol. *40*: 105–106.

Thacker, W.L., R.F. Benson, J.L. Staneck, S.R. Vincent, W.R. Mayberry, D.J. Brenner and H.W. Wilkinson. 1988a. *Legionella cincinnatiensis*, sp. nov. isolated from a patient with pneumonia. J. Clin. Microbiol. *26*: 418–420.

Thacker, W.L., R.F. Benson, J.L. Staneck, S.R. Vincent, W.R. Mayberry, D.J. Brenner and H.W. Wilkinson. 1989b. *In* Validation of the publication of new names and new combinations previously effectively published outside the IJSB. List No. 29. Int. J. Syst. Bacteriol. *39*: 205–206.

Thacker, W.L., R.F. Benson, H.W. Wilkinson, N.M. Ampel, E.J. Wing, A.G. Steigerwalt and D.J. Brenner. 1986. 11th serogroup of *Legionella pneumophila* isolated from a patient with fatal pneumonia. J. Clin. Microbiol. *23*: 1146–1147.

Thacker, W.L., J.W. Dyke, R.F. Benson, D.H.J. Havlichenk, D.B. Robinson, H. Stiefel, W. Schneider, C.W. Moss, W.R. Mayberry and D.J. Brenner. 1992. *Legionella lansingensis* sp.nov. isolated from a patient with pneumonia and underlying chronic lymphocytic leukemia. J. Clin. Microbiol. *30*: 2398–2401.

Thacker, W.L., J.W. Dyke, R.F. Benson, D.H.J. Havlichenk, D.B. Robinson, H. Stiefel, W. Schneider, C.W. Moss, W.R. Mayberry and D.J. Brenner. 1994. *In* Validation of the publication of new names and new combinations previously effectively published outside the IJSB. List No. 50. Int. J. Syst. Bacteriol. *44*: 595.

Thacker, W.L., H.W. Wilkinson, R.F. Benson and D.J. Brenner. 1987. *Legionella pneumophila* serogroup 12 isolated from human and environmental sources. J. Clin. Microbiol. *25*: 569–570.

Thacker, W.L., H.W. Wilkinson, R.F. Benson, S.C. Edberg and D.J. Brenner. 1988b. *Legionella jordanis* isolated from a patient with fatal pneumonia. J. Clin. Microbiol. *26*: 1400–1401.

Thacker, W.L., H.W. Wilkinson, B.B. Plikaytis, A.G. Steigerwalt, W.R. Mayberry, C.W. Moss and D.J. Brenner. 1985. Second serogroup of *Legionella feeleii* strains isolated from humans. J. Clin. Microbiol. *22*: 1–4.

Thal, E. 1973. Observation on immunity in *Yersinia pseudotuberculosis*. Contr. Microbiol. Immunol. *2*: 190–195.

Thal, E. and W. Knapp. 1971. A revised antigenic scheme of *Yersinia pseudotuberculosis*. Progr. Immunobiol. Standard. *15*: 219–222.

Thaler, J.O., S. Baghdiguian and N. Boemare. 1995. Purification and characterization of xenorhabdicin, a phage tail-like bacteriocin, from the lysogenic strain F1 of *Xenorhabdus nematophilus*. Appl. Environ. Microbiol. *61*: 2049–2052.

Thaller, M.C., F. Berlutti, S. Schippa, P. Iori, C. Passariello and G.M. Rossolini. 1995. Heterogeneous patterns of acid phosphatases containing low-molecular-mass polypeptides in members of the family *Enterobacteriaceae*. Int. J. Syst. Bacteriol. *45*: 255–261.

Thaller, R., F. Berlutti and M.C. Thaller. 1988. A *Kluyvera cryocrescens* strain from a gall-bladder infection. Eur. J. Epidemiol. *4*: 124–126.

Thamdrup, B. and D.E. Canfield. 1996. Pathways of carbon oxidation in continental margin sediments off central Chile. Limnol. Oceanogr. *41*: 1629–1650.

Thao, M.L. and P. Baumann. 1998. Sequence analysis of a DNA fragment from *Buchnera aphidicola* (aphid endosymbiont) containing the genes *dapD-htrA-ilvI-ilvH-ftsL-ftsI-murE*. Curr. Microbiol. *37*: 214–216.

Thao, M.L., L. Baumann, P. Baumann and N.A. Moran. 1998. Endosymbionts (*Buchnera*) from the aphids *Schizaphis graminum* and *Diuraphis noxia* have different copy numbers of the plasmid containing the leucine biosynthetic genes. Curr. Microbiol. *36*: 238–240.

Thao, M.L., M.A. Clark, L. Baumann, E.B. Brennan, N.A. Moran and P. Baumann. 2000a. Secondary endosymbionts of psyllids have been acquired multiple times. Curr. Microbiol. *41*: 300–304.

Thao, M.L., N.A. Moran, P. Abbot, E.B. Brennan, D.H. Burckhardt and P. Baumann. 2000b. Cospeciation of psyllids and their primary prokaryotic endosymbionts. Appl. Environ. Microbiol. *66*: 2898–2905.

Thaveechai, N. and N.W. Schaad. 1986. Immunochemical characterization of a subspecies-specific antigenic determinant of a membrane protein extract of *Xanthomonas campestris* pv. *campestris*. Phytopathology *76*: 148–153.

Thayer, D.W. 1978. Carboxymethylcellulase produced by facultative bacteria from the hind-gut of the termite *Reticulitermes hesperus*. J. Gen. Microbiol. *106*: 13–18.

Thibault, P. and L. Le Minor. 1957. Méthodes simples de recherche de la lysine-décarboxylase et de la tryptophane-desaminase a l'aide des milieux pour différenciation rapide des Entérobactériacees. Ann. Inst. Pasteur (Paris) *92*: 551–554.

Thiele, D. and H. Willems. 1994. Is plasmid based differentiation of *Coxiella burnetii* in "acute" and "chronic' isolates still valid. Eur. J. Epidemiol. *10*: 427–434.

Thiele, D., H. Willems, M. Haas and H. Krauss. 1994. Analysis of the entire nucleotide sequence of the cryptic plasmid QpH1 from *Coxiella burnetti*. Eur. J. Epidemiol. *10*: 413–420.

Thiele, D., H. Willems, G. Köpf and H. Krauss. 1993. Polymorphism in DNA restriction patterns of *Coxiella burnetii* isolates investigated by pulsed field gel electrophoresis and image analysis. Eur. J. Epidemiol. *9*: 419–425.

Thiele, H.H. 1968. Die Verwertung einfacher organischer Substrate durch *Thiorhodaceae*. Arch. Mikrobiol. *60*: 124–138.

Thiemann, B. and J.F. Imhoff. 1996. Differentiation of *Ectothiorhodospiraceae* based on their fatty acid composition. Syst. Appl. Microbiol. *19*: 223–230.

Thjøtta, T. and O.T. Avery. 1921. Studies on bacterial nutrition. II. Growth accessory substances in the cultivation of hemophilic bacilli. III. Plant tissue, as a source of growth accessory substances in the cultivation of *Bacillus influenzae*. J. Exp. Med. *34*: 97–114.

Thjøtta, T. and E. Kåss. 1945. A study of alginic acid destroying bacteria. Avh. Utgitt Nor. Vidensk.-Akad. Oslo Mat.-Naturvidensk. Kl. *5*: 1–20..

Thom, B.T. 1970. *Klebsiella* in faeces. Lancet *2*: 1033.

Thomas, A., T. Cheasty, J.A. Frost, H. Chart, H.R. Smith and B. Rowe. 1996a. Vero cytotoxin-producing *Escherichia coli*, particularly serogroup O157, associated with human infections in England and Wales: 1992–4. Epidemiol. Infect. *117*: 1–10.

Thomas, G.M. and G.O. Poinar. 1979. *Xenorhabdus* gen. nov., a genus of entomopathogenic nematophilic bacteria of the family *Enterobacteriaceae*. Int. J. Syst. Bacteriol. *29*: 352–360.

Thomas, G.M. and G.O. Poinar. 1983. Amended description of the genus *Xenorhabdus* Thomas and Poinar. Int. J. Syst. Bacteriol. *33*: 878–879.

Thomas, J.H. 1958. A simple medium for the isolation and cultivation of *Fusiformis nodosus*. Aust. Vet. J. *34*: 411.

Thomas, J.C., F. Berger, M. Jacquier, D. Bernillon, F. Baud-Grasset, N. Truffaut, P. Normand, T.M. Vogel and P. Simonet. 1996b. Isolation and characterization of a novel γ-hexachlorocyclohexane-degrading bacterium. J. Bacteriol. *178*: 6049–6055.

Thomas, J.D., P.J. Reeves and G.P. Salmond. 1997. The general secretion pathway of *Erwinia carotovora* subsp. *carotovora*: analysis of the membrane topology of OutC and OutF. Microbiology *143*: 713–720.

Thomas, L.V., R.J. Gross, T. Cheasty and B. Rowe. 1990. Extended serogrouping scheme for motile, mesophilic *Aeromonas* species. J. Clin. Microbiol. *28*: 980–984.

Thomen, L.F. and M.J. Frobisher. 1945. A study of *Shigella* by means of bacteriophage. Am. J. Hyg. *42*: 225–253.

Thompson, B.J., M.S. Wagner, E. Domingo and R.C. Warner. 1980. Pseudo-lysogenic conversion of *Azotobacter vinelandii* by phage A21 and the formation of a stably converted form. Virology *102*: 278–285.

Thompson, C.J., C. Daly, T.J. Barrett, J.P. Getchell, M.J. Gilchrist and M.J. Loeffelholz. 1998. Insertion element IS3-based PCR method for subtyping *Escherichia coli* O157:H7. J. Clin. Microbiol. *36*: 1180–1184.

Thompson, F.L., B. Hoste, K. Vandemeulebroecke and J. Swings. 2003. Reclassification of *Vibrio hollisae* as *Grimontia hollisae* gen. nov., comb. nov. Int. J. Syst. Evol. Microbiol. *53*: 1615–1617.

Thompson, J.P. 1989. Counting viable *Azotobacter chroococcum* in vertisols. II. Comparison of media. Plant Soil *117*: 17–29.

Thompson, J., J.W. Costerton and R.A. MacLeod. 1970. K$^+$-Dependent deplasmolysis of a marine pseudomonad plasmolyzed in a hypotonic solution. J. Bacteriol. *102*: 843–854.

Thompson, J. and R.A. MacLeod. 1973. Na$^+$ and K$^+$ gradients and an α-aminoisobutyric acid transport in a marine pseudomonad. J. Biol. Chem. *248*: 7106–7111.

Thompson, J.P. and V.B.D. Skerman. 1979. *Azotobacteraceae*: The Taxonomy and Ecology of the Aerobic Nitrogen-Fixing Bacteria, Academic Press, London.

Thompson, J.P. and V.B.D. Skerman. 1981. *In* Validation of the publication of new names and new combinations previously effectively published outside the IJSB. List No. 6. Int. J.Syst. Bacteriol. *31*: 215–218.

Thomson, A.J. 1991. Does ferredoxin I (*Azotobacter*) represent a novel class of DNA-binding proteins that regulate gene expression in response to cellular iron(II). FEBS Lett. *285*: 230–236.

Thomson, S.V., D.C. Hildebrand and M.N. Schroth. 1981. Identification and nutritional differentiation of the *Erwinia* sugar beet pathogen from members of *Erwinia carotovora* and *Erwinia chrysanthemi*. Phytopathology *71*: 1037–1042.

Thorley, C.M. 1976. A simplified method for the isolation of *Bacteroides nodosus* from ovine foot-rot and studies on its colony morphology and serology. J. Appl. Bacteriol. *40*: 301–309.

Thornberry, H.H. and H.W. Anderson. 1931a. A bacterial disease of barberry caused by *Phytomonas berberidis* n. sp. J. Agr. Res. *43*: 29–36.

Thornberry, H.H. and H.W. Anderson. 1931b. Bacterial leaf spot of viburnum. Phytopathology *21*: 907–912.

Thornberry, H.H. and H.W. Anderson. 1937. Some bacterial diseases of plants in Illinois. Phytopathology *27*: 946–949.

Thorne, G.M. and W.E. Farrar, Jr.. 1975. Transfer of ampicillin resistance between strains of *Haemophilus influenzae* type b. J. Infect. Dis. *132*: 276–281.

Thorne, K.J., M.J. Thornley and A.M. Glauert. 1973. Chemical analysis of the outer membrane and other layers of the cell envelope of *Acinetobacter* sp. J. Bacteriol. *116*: 410–417.

Thorne, L., L. Tansey and T.J. Pollock. 1987. Clustering of mutations blocking synthesis of xanthan gum by *Xanthomonas campestris*. J. Bacteriol. *169*: 3593–3600.

Thornley, M.J. 1960. The differentiation of *Pseudomonas* from other Gram-negative bacteria on the basis of arginine metabolism. J. Appl. Bacteriol. *23*: 37–52.

Thornley, M.J. 1967. A taxonomic study of *Acinetobacter* and related genera. J. Gen. Microbiol. *49*: 211–257.

Thornsberry, C., C.N. Baker and L.A. Kirven. 1978. *In vitro* activity of antimicrobial agents on the Legionnaires' disease bacterium. Antimicrob. Agents Chemother. *13*: 78–80.

Thornsberry, C. and L.A. Kirven. 1978. β-1actamase of the Legionnaires' bacterium. Curr. Microbiol. *1*: 51–54.

Thornton, D.J., R.C. Tustin, B.J. Pienaar, W.N. Pienaar and H.D. Bubb. 1975. Cat bite transmission of *Yersinia pestis* infection to man. J. S. Afr. Vet. Assoc. *46*: 165–169.

Thornton, J.M., D.A. Austin, B. Austin and R. Powell. 1999. Small subunit rRNA gene sequences of *Aeromonas salmonicida* subsp. *smithia* and *Haemophilus piscium* reveal pronounced similarities with *A. salmonicida* subsp. *salmonicida*. Dis. Aquat. Org. *35*: 155–158.

Thurm, V. and E. Ritter. 1993. Genetic diversity and clonal relationships of *Acinetobacter baumannii* strains isolated in a neonatal ward: epidemiological investigations by allozyme, whole-cell protein and antibiotic resistance analysis. Epidemiol. Infect. *111*: 491–498.

Thyssen, A., L. Grisez, R. van Houdt and F. Ollevier. 1998. Phenotypic characterization of the marine pathogen *Photobacterium damselae* subsp. *piscicida*. Int. J. Syst. Bacteriol. *48*: 1145–1151.

Thyssen, A., S. Van Eygen, L. Hauben, J. Goris, J. Swings and F. Ollevier. 2000. Application of AFLP for taxonomic and epidemiological studies of *Photobacterium damselae* subsp. *piscicida*. Int. J. Syst. Evol. Microbiol. *50*: 1013–1019.

Tibelius, K.H., L. Du, D. Tito and F. Stejskal. 1993. The *Azotobacter chroococcum* hydrogenase gene cluster - sequences and genetic analysis of four accessory genes, hupA, hupB, hupY and hupC. Gene *127*: 53–61.

Till, D.H. and F.P. Palmer. 1960. A review of actinobacillosis with a study of the causal organism. Vet. Rec. *72*: 527–534.

Timm, A. and A. Steinbüchel. 1990. Formation of polyesters consisting of medium chain-length 3-hydroxyalkanoic acids from gluconate by *Pseudomonas aeruginosa* and other fluorescent pseudomonads. Appl. Environ. Microbiol. *56*: 3360–3367.

Timmer-ten Hoor, A. 1975. A new type of thiosulfate oxidizing, nitrate reducing microorgansim: *Thiomicrospira denitrificans* sp. nov. Neth. J. Sea Res. *9*: 344–350.

Timmis, K.N., C.L. Clayton and T. Sekizaki. 1985. Localization of Shiga toxin gene in the region of *Shigella dysenteriae* 1 chromosome specifying virulence functions. FEMS Microbiol. Lett. *30*: 301–305.

Tindale, A.E., M. Mehrotra, D. Ottem and W.J. Page. 2000. Dual regulation of catecholate siderophore biosynthesis in *Azotobacter vinelandii* by iron and oxidative stress. Microbiology-(UK). *146*: 1617–1626.

Tindall, B.J. 1980. Phototrophic bacteria from Kenyan soda lakes, Doctoral thessis, University of Leicester, Leicester, England.

Tindall, B.J. 1999. Taxonomic note: transfer of *Amoebobacter purpureus* to the genus *Pfennigia* gen. nov. as *Pfennigia purpurea* comb. nov., on the basis of the illegitimate proposal to make *Amoebobacter purpureus* the type species of the genus *Amoebobacter*. Int. J.Syst. Bacteriol. *49*: 1307–1308.

Tippen, P.S., A. Meyer, E.C. Blank and H.D. Donnell. 1989. Aquarium-associated *Plesiomonas shigelloides* infection-Missouri. Morb. Mortal. Wkly. Rep. *38*: 617–619.

Tipper, D.J., J.L. Strominger and J.C. Ensign. 1967. Structure of the cell walls of *Staphylococcus aureus*, strain Copenhagen. VII. Mode of action of the bacteriolytic peptidase from *Myxobacter* and isolation of intact cell wall polysaccharides. Biochemistry *6*: 906–920.

Tisdale, W.B. and M.M. Williamson. 1923. Bacterial leaf spot of lima bean. J. Agr. Res. *25*: 141–154.

Tison, D.L., M. Nishibuchi, J.D. Greenwood and R.J. Seidler. 1982. *Vibrio vulnificus* biogroup 2: New biogroup pathogenic for eels. Appl. Environ. Microbiol. *44*: 640–646.

Tison, D.L. and R.J. Seidler. 1983. *Vibrio aestuarianus*: A new species from estuarine waters and shellfish. Int. J. Syst. Bacteriol. *33*: 699–702.

Tissot-Dupont, H., S. Torres, M. Nezri and D. Raoult. 1999. Hyperendemic focus of Q fever related to sheep and wind. Am. J. Epidemiol. *150*: 67–74.

Tjernberg, I. and J. Ursing. 1989. Clinical strains of *Acinetobacter* classified by DNA–DNA hybridization. APMIS *97*: 595–605.

Tobe, T., G.K. Schoolnik, I. Sohel, V.H. Bustamante and J.L. Puente. 1996. Cloning and characterization of *bfpTVW*, genes required for the transcriptional activation of *bfpA* in enteropathogenic *Escherichia coli*. Mol. Microbiol. *21*: 963–975.

Tocal, J.V. and C.F. Mezez. 1968. The isolation of *Edwardsiella tarda* from a dog. Philippines J. Vet. Med. *7*: 143–145.

Toenniessen, E. 1914. Über Vererbung und Variabilität bei Bakterien mit besonderer Berücksichtigung der Virulenz. Zentbl. Bakteriol. I. Abt. Orig. *73*: 241–277.

Tolson, D.L., D.L. Barrigar, R.J. McLean and E. Altman. 1995. Expression of a nonagglutinating fimbriae by *Proteus mirabilis*. Infect. Immun. *63*: 1127–1129.

Tomas, J.M., B. Ciurana and J.T. Jofre. 1986. New, simple medium for selective, differential recovery of *Klebsiella* spp. Appl. Environ. Microbiol. *51*: 1361–1363.

Tomb, J.F., H. el-Hajj and H.O. Smith. 1991. Nucleotide sequence of a cluster of genes involved in the transformation of *Haemophilus influenzae* Rd. Gene *104*: 1–10.

Tominaga, A., M.A. Mahmoud, T. Mukaihara and M. Enomoto. 1994. Molecular characterization of intact, but cryptic, flagellin genes in the genus *Shigella*. Mol. Microbiol. *12*: 277–285.

Tompkins, L.S., B.J. Roessler, S.C. Redd, L.E. Markowitz and M.L. Cohen. 1988. *Legionella* prosthetic valve endocarditis. New Engl. J. Med. *318*: 530–535.

Tompkins, L.S., N.J. Troup, T. Woods, W. Bibb and R.M. McKinney. 1987. Molecular epidemiology of *Legionella* sp. by restriction endonuclease and alloenzyme analysis. J. Clin. Microbiol. *25*: 1875–1880.

Tondella, M.L., F.D. Quinn and B.A. Perkins. 1995. Brazilian purpuric fever caused by *Haemophilus influenzae* biogroup aegyptius strains lacking the 3031 plasmid. J. Infect. Dis. *171*: 209–212.

Tønjum, T., G. Bukholm and K. Bøvre. 1989. Differentiation of some species of *Neisseriaceae* and other bacterial groups by DNA—DNA hybridization. APMIS *97*: 395–405.

Tønjum, T., G. Bukholm and K. Bøvre. 1990. Identification of *Haemophilus aphrophilus* and *Actinobacillus actinomycetemcomitans* by DNA–DNA hybridization and genetic transformation. J. Clin. Microbiol. *28*: 1994–1998.

Tønjum, T., D.A. Caugant and K. Bøvre. 1992. Differentiation of *Moraxella nonliquefaciens*, *Moraxella lacunata*, and *Moraxella bovis* by using multilocus enzyme electrophoresis and hybridization with pilin-specific DNA probes. J. Clin. Microbiol. *30*: 3099–3107.

Tønjum, T., C.F. Marrs, F. Rozsa and K. Bøvre. 1991. The type 4 pilin of *Moraxella nonliquefaciens* exhibits unique similarities with the pilins of *Neisseria gonorrhoeae* and *Dichelobacter nodosus*. J. Gen. Microbiol. *137*: 2483–2490.

Topley, W.W.C. and G.S. Wilson. 1936. The Principles of Bacteriology and Immunity, 2nd Ed., W. Wood & Company, Baltimore.

Toranzo, A.E. and J.L. Barja. 1993. Virulence factors of bacteria pathogenic for coldwater fish. *In* Annu. Rev. of Fish Dis, 5–36.

Toranzo, A.E., J.L. Barja, R.R. Colwell and F.M. Hetrick. 1983. Characterization of plasmids in bacterial fish pathogens. Infect. Immun. *39*: 184–192.

Toranzo, A.E., J.M. Cutrín, B.S. Roberson, S. Núnez, J.M. Abell, F.M. Hetrick and A.M. Baya. 1994. Comparison of the taxonomy, serology, drug resistance transfer, and virulence of *Citrobacter freundii* strains from mammals and poikilothermic hosts. Appl. Environ. Microbiol. *60*: 1789–1797.

Toranzo, A.E., Y. Santos, T.P. Nieto and J.L. Barja. 1986. Evaluation of different assay systems for identification of environmental *Aeromonas* strains. Appl. Environ. Microbiol. *51*: 652–656.

Tornabene, T.G. 1973. Lipid composition of selected strains of *Yersinia pestis* and *Yersinia pseudotuberculosis*. Biochim Biophys. Acta. *306*: 173–185.

Torres, J.L. 1990. Studies on motile *Aeromonas* spp. associated with healthy an depizootic ulcerative syndrome-positive fish, Doctoral thesis, Univ. Petanian Malaysia, Serdang Selangor, Malaysia. ..

Torres, J.L., M. Shariff and L. Tajimi. 1991. Serological relationships among motile *Aeromonas* spp. associated with healthy and epizootic ulcerative syndrome (EUS) positive fish. Proceedings from the Symposium on Diseases in Asian Aquaculture, Bali. Asian Fisheries Society. p. 25.

Tóth, E., G. Kovács, P. Schumann, A.L. Kovacs, U. Steiner, A. Halbritter and K. Márialigeti. 2001. *Schineria larvae* gen. nov., sp nov., isolated from the 1st and 2nd larval stages of *Wohlfahrtia magnifica* (Diptera: Sarcophagidae). Int. J. Syst. Evol. Microbiol. *51*: 401–407.

Toth, I., M. Perombelon and G. Salmond. 1993. Bacteriophage phi-KP mediated generalized transduction in *Erwinia carotovora* subspecies *carotovora*. J. Gen. Microbiol. *139*: 2705–2709.

Toukdarian, A., G. Saunders, G. Selmansosa, E. Santero, P. Woodley and C. Kennedy. 1990. Molecular analysis of the *Azotobacter vinelandii glnA* gene encoding glutamine synthetase. J. Bacteriol. *172*: 6529–6539.

Towner, K.J. 1978. Chromosome mapping in *Acinetobacter calcoaceticus*. J. Gen. Microbiol. *104*: 175–180.

Towner, K.J. 1983. Transposon-directed mutagenesis and chromosome mobilization in *Acinetobacter calcoaceticus* EBF65/65. Genet. Res. *41*: 97–102.

Towner, K.J. 1991. Plasmid and transposon behaviour in *Acinetobacter*. *In* Towner and Bergogne-Bérézin (Editors), The Biology of *Acinetobacter*. Taxonomy, Clinical Importance, Molecular Biology, Physiology, Industrial Relevance, Plenum Press, New York. pp. 149–167.

Towner, K.J. and A. Vivian. 1976a. RP4 fertility variants in *Acinetobacter calcoaceticus*. Genet. Res. *28*: 301–306.

Towner, K.J. and A. Vivian. 1976b. RP4-mediated conjugation in *Acinetobacter calcoaceticus*. J. Gen. Microbiol. *93*: 355–360.

Townsend, K.M., J.D. Boyce, J.Y. Chung, A.J. Frost and B. Adler. 2001. Genetic organization of *Pasteurella multocida* cap loci and development of a multiplex capsular PCR typing system. J. Clin. Microbiol. *39*: 924–929.

Townsend, T.R. and J.Y. Gillenwater. 1969. Urinary tract infection due to *Actinobacillus actinomycetemcomitans*. J. Am. Med. Assoc. *210*: 558.

Tozzi, A.E., S. Gorietti and A. Caprioli. 2001. Epidemiology of human infections by *Escherichia coli* O157 and other vero cytotoxin-producing *E. coli*. *In* Duffy, Garvey and McDowell (Editors), Verocytotoxigenic *E. coli*, Food & Nutrition Press, Inc., Conneticut. pp. 161–180.

Tran Van Nhieu, G., R. Bourdet-Sicard, G. Dumenil, A. Blocker and P.J. Sansonetti. 2000. Bacterial signals and cell responses during *Shigella* entry into epithelial cells. Cell.Microbiol. *2*: 187–193.

Traub, W.H. 1972. Studies on group A bacteriocins of *Serratia marcescens*: preliminary characterization of two subgroups of bacteriocins. Zentralbl. Bakteriol. Parasitenkd. Infaktionskr. Hyg. Abt. 1: Orig. Reihe A. *222*: 232–244.

Traub, W.H. 1980. Bacteriocin and phage typing of *Serratia*. *In* von Graevenitz and Rubin (Editors), The Genus *Serratia*, CRC Press, Boca Raton. pp. 79–100.

Traub, W.H. 1991. Serotyping of *Serratia marcescens*: detection of two new O-antigens (O25 and O26). Zentbl. Bakteriol. *275*: 495–499.

Traub, W.H., M.E. Craddock, E.A. Raymond, M. Fox and C.E. McCall. 1971. Charaterization of an unusual strain of *Proteus rettgeri* associated with an outbreak of nosocomial urinary-tract infection. Appl. Microbiol. *22*: 278–283.

Traub, W.H. and P.I. Fukushima. 1979a. Serotyping of *Serratia marcescens*: current status of seven recently described flagellar (H) antigens. J. Clin. Microbiol. *10*: 56–63.

Traub, W.H. and P.I. Fukushima. 1979b. Serotyping of *Serratia marcescens*: simplified tube O-agglutination test and comparison with other serological procedures. Zentralbl. Bakteriol. Orig. A. *244*: 474–493.

Traub, W.H. and B. Leonhard. 1994. Serotyping of *Acinetobacter baumannii* and genospecies 3: an update. Med. Microbiol. Lett. *3*: 120–127.

Traub, W.H., M. Spohr and R. Blech. 1982. Bacteriocin typing of clinical isolates of *Enterobacter cloacae*. J. Clin. Microbiol. *16*: 885–889.

Trees, D.L. and S.A. Morse. 1995. Chancroid and *Haemophilus ducreyi*: an update. Clin. Microbiol. Rev. *8*: 357–375.

Tresselt, H.B. and M.K. Ward. 1964. Blood-free medium for rapid growth of *Pasteurella tularensis*. Appl. Microbiol. *12*: 504–507.

Trevisan, V. 1842. Prospetto della flora Euganea. *In* Coi tipi Del Seminario, Padova. 1–68.

Trevisan, V. 1845. Nomenclator algarum. *In* Impr. du seminaire, Padova. 58–59.

Trevisan, V. 1885. Caratteri di alcuni nuovi generi di Batteriacee. Atti Accad. Fis-Med-Stat. Milano (Ser.4). *3*: 92–107.

Trevisan, V. 1887. Sul micrococco della rabbia e sulla possibilità di riconoscere durante il periode d'incubazione, dell'esame del sangue della persona moricata, de ha contratta l'infezione rabbica. Rend. Ist. Lombardo (Ser.2) *20*: 88–105.

Trevisan, V. 1889a. Bacilli endofitobil destruent. *In* His I Generi e le specie delle Batteriacea, Milano. pp. 19.

Trevisan, V. 1889b. I generi e le specie delle batteriacee, Zanaboni and Gabuzzi, Milan. 1–35.

Trias, J., M. Viñas, J. Guinea and J.G. Loren. 1988. Induction of yellow pigmentation in *Serratia marcescens.* Appl. Environ. Microbiol. *54*: 3138–3141.

Tricot, C., A. Piérard and V. Stalon. 1990. Comparative-studies on the degradation of guanidino and ureido compounds by *Pseudomonas.* J. Gen. Microbiol. *136*: 2307–2317.

Tristram, D.A. and B.A. Forbes. 1988. *Kluyvera*: A case report of urinary tract infection and sepsis. Pediatr. Infect. Dis. J. *7*: 297–298.

Trottier, S., K. Stenberg and C. Svanborg-Eden. 1989. Turnover of nontypable *Haemophilus influenzae* in the nasopharynges of healthy children. J. Clin. Microbiol. *27*: 2175–2179.

Trucksis, M., J. Michalski, Y.K. Deng and J.B. Kaper. 1998. The *Vibrio cholerae* genome contains two unique circular chromosomes. Proc. Natl Acad. Sci. U.S.A. *95*: 14464–14469.

Truman, R. 1974. Die-back of *Eucalyptus citriodora* caused by *Xanthomonas eucalypti* sp. n. Phytopathology *64*: 143–144.

Trüper, H.G. 1968. *Ectothiorhodospira mobilis* Pelsh, a photosynthetic sulfur bacterium depositing sulfur outside the cells. J. Bacteriol. *95*: 1910–1920.

Trüper, H.G. 1970. Culture and isolation of phototrophic sulfer bacteria from the marine environment. Helgol. Wiss. Meersunters *20*: 6–16.

Trüper, H.G. 1978. Sulfur metabolism. *In* Clayton and Sistrom (Editors), The Photosynthetic Bacteria, Plenum Press, New York. pp. 677–690.

Trüper, H.G. and L. De'Clari. 1997. Taxonomic note: necessary correction of specific epithets formed as substantives (nouns) "in apposition". Int. J. Syst. Bacteriol. *47*: 908–909.

Trüper, H.G., U. Fischer and D.P. Kelly. 1982. Anaerobic oxidation of sulfur compounds as electron donors for bacterial photosynthesis. Philos. Trans. R. Soc. Lond. Ser. B-Biol. Sci. *298*: 529–542.

Trüper, H.G. and H. Jannasch. 1968. *Chromatium buderi* nov. spec., eine neue Art der groben *Thiorhodaceae.* Arch. Mikrobiol. *61*: 363–372.

Trust, T.J. and K.H. Bartlett. 1974. Occurrence of potential pathogens in water containing ornamental fishes. Appl. Microbiol. *28*: 35–40.

Trzesicka-Mlynarz, D. and O.P. Ward. 1995. Degradation of polycyclic aromatic-hydrocarbons (Pahs) by a mixed culture and its component pure cultures, obtained from Pah-contaminated soil. Can. J. Microbiol. *41*: 470–476.

Tsai, G.J. and T.H. Chen. 1996. Incidence and toxigenicity of *Aeromonas hydrophila* in seafood. Int. J. Food Microbiol. *31*: 121–131.

Tseng, Y.H., K.T. Choy, C.H. Hung, N.T. Lin, J.Y. Liu, C.H. Lou, B.Y. Yang, F.S. Wen, S.F. Weng and J.R. Wu. 1999. Chromosome map of *Xanthomonas campestris* pv. *campestris* 17 with locations of genes involved in xanthan gum synthesis and yellow pigmentation. J. Bacteriol. *181*: 117–125.

Tsuchiya, K., T.W. Mew and S. Wakimoto. 1982. Bacteriological and pathological characteristics of wild types and induced mutants of *Xanthomonas campestris* pv. *oryzae.* Phytopathology *72*: 43–46.

Tsuda, M., H. Miyazaki and T. Nakazawa. 1995. Genetic and physical mapping of genes involved in pyoverdin production in *Pseudomonas aeruginosa* PAO. J. Bacteriol. *177*: 423–431.

Tsui, F.-P., R. Schneerson, R.A. Boykins, A.B. Karpas and W. Egan. 1981a. Structural and immunological studies of tbe *Haemophilus influenzae* type d capsular polysaccharide. Carbohydr. Res. *97*: 293–306.

Tsui, F.P., R. Schneerson and W. Egan. 1981b. Structural studies of the *Haemophilus influenzae* type e capsular polysaccharide. Carbohydr. Res. *88*: 85–92.

Tsukamoto, T., T. Kimoto, M. Magalhaes and Y. Takeda. 1992. Enteroadherent *Escherichia coli* exhibiting localized pattern of adherence among infants with diarrhoea in Brazil—incidence and prevalence of serotypes. Kansenshogaku Zasshi. *66*: 1538–1542.

Tsukamoto, T., Y. Kinoshita, T. Shimada and R. Sakazaki. 1978. Two epidemics of diarrhoeal disease possibly caused by *Plesiomonas shigelloides.* J. Hyg. *80*: 275–280.

Tsukamoto, T. and Y. Takeda. 1993. Incidence and prevalence of serotypes of enteroaggregative *Escherichia coli* from diarrheal patients in Brazil, Myanmar and Japan. Kansenshogaku Zasshi. *67*: 289–294.

Tucker, D.N., I.J. Slotnick, E.O. King, B. Tynes, J. Nicholson and L. Crevasse. 1962. Endocarditis caused by a *Pasteurella*-like organism. N. Engl. J. Med. *267*: 913–916.

Tunevall, G. 1953. Studies on *Haemophilus influenzae* antigen studied by gel precipitation method. Acta Pathol. Microbiol. Scand. *32*: 193–197.

Tunnicliff, E.A. 1941. A study of *Actinobacillus lignieresii* from sheep affected with actinobacillosis. J. Infect. Dis. *69*: 52–58.

Turk, D.C. and J.R. May. 1967. *Haemophilus influenzae.* Its clinical importance, The English Universities Press Ltd., London.

Tuyau, J.E. and W. Sims. 1974. Neuraminidase activity in human oral strains of haemophili. Arch. Oral Biol. *19*: 817–819.

Tuyau, J.E. and W. Sims. 1975. Occurrence of haemophili in dental plaque and their association with neuraminidase activity. J. Dent. Res. *54*: 737–739.

Twarog, R. and L.E. Blouse. 1968. Isolation and characterization of transducing bacteriophage BP1 for *Bacterium anitratum* (*Achromobacter* sp.). J. Virol. *2*: 716–722.

Twedt, R.M., D.F. Brown and D.L. Zink. 1981. Comparison of plasmid deoxyribonucleic acid contents, culture characteristics, and indices of pathogenicity among selected strains of *Vibrio parahaemolyticus.* Infect. Immun. *33*: 322–325.

Tyler, S.D., W.M. Johnson, H. Lior, G. Wang and K.R. Rozee. 1991. Identification of verotoxin type 2 variant B subunit genes in *Escherichia coli* by the polymerase chain reaction and restriction fragment length polymorphism analysis. J. Clin. Microbiol. *29*: 1339–1343.

Tymiak, A.A., T.J. McCormick and S.E. Unger. 1989. Structure determination of lysobactin, a macrocyclic peptide lactone antibiotic. J. Org. Chem. *54*: 1149–1157.

Tzipori, S., I.K. Wachsmuth, C. Chapman, R. Birden, J. Brittingham, C. Jackson and J. Hogg. 1986. The pathogenesis of hemorrhagic colitis caused by *Escherichia coli* O157:H7 in gnotobiotic piglets. J. Infect. Dis. *154*: 712–716.

Uchida, M., K. Nakata and M. Maeda. 1997. Conversion of *Ulva* fronds to a hatchery diet for *Artemia nauplii* utilizing the degrading and attaching abilities of *Pseudoalteromonas espejiana.* J. Appl. Phycol. *9*: 541–549.

Uchida, T., L. Bonen, H.W. Schaup, B.J. Lewis, L. Zablen and C.R. Woese. 1974. Tbe use of ribonuclease U2 in RNA sequence determination: Some corrections in the catalog of oligomers produced by ribonuclease T1 digestion of *Escherichia coli* 16s ribosomal RNA. J. Mol. Evol. *3*: 63–77.

Uchino, M., Y. Kosako, T. Uchimura and K. Komagata. 2000. Emendation of *Pseudomonas straminea* Iizuka and Komagata 1963. Int. J. Syst. Evol. Microbiol. *50*: 1513–1519.

Udey, L.R. and J.L. Fryer. 1978. Immunization of fish with bacterins of *Aeromonas salmonicida.* Mar. Fish. Rev. *40*: 12–17.

Ueno, T., H. Ito, F. Kimizuka, H. Kotani and K. Nakajima. 1993. Gene structure and expression of the *Mbo*I restriction-modification system. Nucleic Acids Res. *21*: 2309–2313.

Ulaganathan, K. and A. Mahadevan. 1988. Isolation of a plasmid from *Xanthomonas campestris* pv. *vignicola.* J. Phytopathol. (Berl.) *123*: 92–96.

Ulaganathan, K. and A. Mahadevan. 1991. Indigenous plasmids of *Xanthomonas campestris* and characters encoded by a plasmid of *Xanthomonas campestris* pv. *vignicola.* Indian J. Exp. Biol. *29*: 1022–1026.

Ullah, M.A. and T. Arai. 1983a. Exotoxic substances produced by *Edwardsiella tarda*. Fish Pathol. *18*: 71–75.

Ullah, M.A. and T. Arai. 1983b. Pathological activities of the naturally occurring strains of *Edwardsiella tarda*. Fish Pathol. *18*: 65–70.

Umelo, E. and T.J. Trust. 1997. Identification and molecular characterization of two tandemly located flagellin genes from *Aeromonas salmonicida* A449. J. Bacteriol. *179*: 5292–5299.

Unertl, K.E., F.P. Lenhart, H. Forst, G. Vogler, V. Wilm, W. Ehret and G. Ruckdeschel. 1989. Ciprofloxacin in the treatment of legionellosis in critically ill patients including those cases unresponsive to erythromycin. Am. J. Med. *87*: 128s–131s.

Unterman, B.M. and P. Baumann. 1990. Partial characterization of the ribosomal RNA operons of the pea-aphid endosymbionts: evolutionary and physiological implications. *In* Cambell and Eikenbary (Editors), Aphid-Plant Genotype Interactions, Elsevier Biomedical Press, Amsterdam. pp. 329–350.

Unterman, B.M., P. Baumann and D.L. McLean. 1989. Pea aphid symbiont relationships established by analysis of 16S rRNAs. J. Bacteriol. *171*: 2970–2974.

Unz, R.F. 1984. Genus IV. *Zoogloea*. *In* Krieg and Holt (Editors), Bergey's Manual of Systematic Bacteriology, 1st Ed., Vol. 1, The Williams & Wilkins Co., Baltimore. pp. 214–219.

Unz, R.F. and T.M. Williams. 1988. The effect of controlled pH on the development of rosette-forming bacteria in axenic culture and bulking activated sludge. Water Sci. Technol. *20*: 249–255.

Unz, R.F. and T.M. Williams. 1989. Substrate utilization by filamentous sulfer bacteria of activated sludge. *In* Hattori, Ishida, Maruyama, Morita and Uchida (Editors), Recent Advances in Microbial Ecology, Proceedings of the 5th International Symposium on Microbial Ecology (ISME 5), Japan Scientific Societies Press, Tokyo. pp. 412–416.

Uphof, J.C.T. 1927. Zur Oekologie der Schwefelbakterien in den Schwefelquellen Mittelfloridas. Arch. Mikrobiol. *18*: 71–84.

Uphoff, T.S. and R.A. Welch. 1990. Nucleotide sequencing of the *Proteus mirabilis* calcium-independent hemolysin genes (*hpmA* and *hpmB*) reveals sequence similarity with *Serratia marcescens* hemolysin genes (*shlA* and *shlB*). J. Bacteriol. *172*: 1206–1216.

Urakami, T., H. Araki, H. Oyanagi, K.-I. Suzuki and K. Komagata. 1992. Transfer of *Pseudomonas aminovorans* (den Dooren de Jong 1926) to *Aminobacter* gen. nov. as *Aminobacter aminovorans* comb. nov. and description of *Aminobacter aganoensis* sp. nov. and *Aminobacter niigataensis* sp. nov. Int. J. Syst. Bacteriol. *42*: 84–92.

Urakami, T. and K. Komagata. 1986a. Cellular fatty acid composition and coenzyme Q system in Gram-negative methanol-utilizing bacteria. J. Gen. Appl. Microbiol. *25*: 343–360.

Urakami, T. and K. Komagata. 1986b. Emendation of *Methylobacillus* Yordy and Weaver 1977, a genus for methanol-utilizing bacteria. Int. J. Syst. Bacteriol. *36*: 502–511.

Urakami, T. and K. Komagata. 1986c. Occurrence of isoprenoid compounds in gram-negative methanol-utilizing, methane-utilizing, and methylamine-utilizing bacteria. J. Gen. Appl. Microbiol. *32*: 317–341.

Urakami, T. and K. Komagata. 1987. Characterization of species of marine methylotrophs of the genus *Methylophaga*. Int. J. Syst. Bacteriol. *37*: 402–406.

Urakawa, H., K. Kita-Tsukamoto and K. Ohwada. 1999a. Reassessment of the taxonomic position of *Vibrio iliopiscarius* (Onarheim et al. 1994) and proposal for *Photobacterium iliopiscarium* comb. nov. Int. J. Syst. Bacteriol. *49*: 257–260.

Urakawa, H., K. Kita-Tsukamoto, S.E. Steven, K. Ohwada and R.R. Colwell. 1998. A proposal to transfer *Vibrio marinus* (Russell 1891) to a new genus *Moritella* gen. nov. as *Moritella marina* comb. nov. FEMS Microbiol. Lett. *165*: 373–378.

Urakawa, H., K. Kita-Tsukamoto, S.E. Steven, K. Ohwada and R.R. Colwell. 1999b. *In* Validation of the publication of new names and new combinations previously effectively published outside the IJSB. List No. 69. Int. J. Syst. Bacteriol. *49*: 341–342.

Urdaci, M.C., M. Marchand, E. Ageron, J.M. Arcos, B. Sesma and P.A. Grimont. 1991. *Vibrio navarrensis* sp. nov., a species from sewage. Int. J. Syst. Bacteriol. *41*: 290–294.

Urdaci, M.C., M. Marchand and P.A. Grimont. 1990. Characterization of 22 *Vibrio* species by gas chromatography analysis of their cellular fatty acids. Res. Microbiol. *141*: 437–452.

Urosevic, B. 1966. Canker of poplar caused by *Erwinia cancerogena* n. sp. (Czech.). Lesn. Cas. *12*: 493–505.

Ursing, J., D.J. Brenner, H. Bercovier, G.R. Fanning, A.G. Steigerwalt, J. Brault and H.H. Mollaret. 1980a. *Yersinia frederiksenii*: a new species of *Enterobacteriaceae* composed of rhamnose positive strains (formerly called atypical *Yersinia enterocolitica* or *Yersinia enterocolitica*-like). Curr. Microbiol. *4*: 213–217.

Ursing, J., D.J. Brenner, H. Bercovier, G.R. Fanning, A.G. Steigerwalt, J. Brault and H.H. Mollaret. 1981. *In* Validation of the publication of new names and new combinations previously effectively published outside the IJSB. List No. 6. Int. J. Syst. Bacteriol. *31*: 215–218.

Ursing, J., A.G. Steigerwalt and D.J. Brenner. 1980b. Lack of genetic relatedness between *Yersinia philomiragia* (the "Philomiragia" bacterium) and *Yersinia* species. Curr. Microbiol. *4*: 231–233.

Utermöhl, H. 1925. Limnologische Phytoplanktonstudien. Arch. Hydrobiol. Suppl. *5*: 251–277.

Utsunomiya, A., T. Naito, M. Ehara, Y. Ichinose and A. Hamamoto. 1992. Studies on novel pili from *Shigella flexneri*. I. Detection of pili and hemagglutination activity. Microbiol. Immunol. *36*: 803–813.

Vaara, M., T. Vaara, M. Jensen, I. Helander, M. Nurminen, E.T. Rietschel and P.H. Makela. 1981. Characterization of the lipopolysaccharide from the polymyxin-resistant *pmrA* mutants of *Salmonella typhimurium*. FEBS Lett. *129*: 145–149.

Vago, C. and R. Martoja. 1963. Une rickettsiose chez les Gryllidae (Orthoptera). C.R. Acad. Science Ser. D. *256*: 1945–1947.

Vago, C. and G. Meynadier. 1965. Une rickettsiose chez le criquet pelerin (*Schistocerca gregaria* Forsk.). Entomophaga. *10*: 307–310.

Vago, C., G. Meynadier, P. Juchault, J.-J. Legrand, A. Amargier and J.-J. Duthoit. 1970. Une maladie rickettsienne chez les crustaces isopodes. C.R. Acad. Science Ser. D. *271*: 2061–2063.

Väisänen, O.M., A. Weber, A. Bennasar, F.A. Rainey, H.J. Busse and M.S. Salkinoja-Salonen. 1998. Microbial communities of printing paper machines. J. Appl. Microbiol. *84*: 1069–1084.

Valcarcel, J., A. Allardet-Servent, G. Bourg, D. O'Callaghan, P. Michailesco and M. Ramuz. 1997. Investigation of the *Actinobacillus actinomycetemcomitans* genome by pulsed field gel electrophoresis. Oral Microbiol. Immunol. *12*: 33–39.

Valderrama, M.J., E. Quesada, V. Bejar, A. Ventosa, M.C. Gutierrez, F. Ruizberraquero and A. Ramoscormenzana. 1991. *Deleya salina* sp. nov., a moderately halophilic Gram-negative bacterium. Int. J. Syst. Bacteriol. *41*: 377–384.

Vali, L., C.J. Thomson and S.G.B. Amyes. 1994. *Haemophilus influenzae* identification of a novel beta-lactamase. J. Pharm. Pharmacol. *46*: 1041.

Valkova, D. and J. Kazar. 1995. A new plasmid (QpDV) common to *Coxiella burnetii* isolates associated with acute and chronic Q fever. FEMS Microbiol. Lett. *125*: 275–280.

Valleé, A., P. Thibault and L. Second. 1963. Contribution à l'étude d'*A. lignieresii* et d'*A. equuli*. Ann. Inst. Pasteur (Paris) *104*: 108–114.

Valleé, A., R. Tinelli, J.-C. Guillon, a. le Priol and T. Cuong. 1974. Étude d'un actinobacillus isolé chez un cheval. Recl. Med. Vet. Ec. Alfort. *150*: 695–700.

Van Alphen, L. 1993. The molecular epidemiology of *Haemophilus influenzae*. Rev. Med. Microbiol. *4*: 159–166.

Van Alphen, L., D.A. Caugant, B. Duim, M. O'Rourke and L.D. Bowler. 1997. Differences in genetic diversity of nonencapsulated *Haemophilus influenzae* from various diseases. Microbiology *143*: 1423–1431.

Van Alphen, L., M. Klein, L. Geelen van den Broek, T. Riemens, P. Eijk and J.P. Kamerling. 1990. Biochemical characterization and worldwide distribution of serologically distinct lipopolysaccharides of *Haemophilus influenzae* type b. J. Infect. Dis. *162*: 659–663.

van Bijsterveld, O.P. 1970. New *Moraxella* strain isolated from angular conjunctivitis. Appl. Microbiol. *20*: 405–408.

van Bruggen, A.H.C., K.N. Jochimsen and P.R. Brown. 1990. *Rhizomonas suberifaciens* gen. nov., sp. nov., the causal agent of corky root of lettuce. Int. J. Syst. Bacteriol . *40*: 175–188.

Van Damme, L.R. and J. Vandepitte. 1980. Frequent isolation of *Edwardsiella tarda* and *Pleisiomonas shigelloides* from healthy Zairese freshwater fish: a possible source of sporadic diarrhea in the tropics. Appl. Environ. Microbiol. *39*: 475–479.

Van de Peer, Y., M. Vancanneyt and R. DeWachter. 1996. Compilation of pseudomonad sequences present in a database on the structure of ribosomal RNA. Syst. Appl. Microbiol. *19*: 493–500.

Van De Wolf, J.M. 1993. Time-resolved fluoroimmunoassay as a method for detection of *Erwinia chrysanthemi* in potato peel extracts. J. Appl. Bacteriol. *74*: 367–371.

van den Heuvel, J.F.J.M., A. Bruyere, S.A. Hogenhout, V. Ziegler-Graff, V. Brault, M. Verbeek, F. van der Wilk and K. Richards. 1997. The N-terminal region of the luteovirus readthrough domain determines virus binding to *Buchnera* GroEL and is essential for virus persistence in the aphid. J. Virol. *71*: 7258–7265.

van den Heuvel, J.F.J.M., M. Verbeek and F. van der Wilk. 1994. Endosymbiotic bacteria associated with circulative transmission of potato leafroll virus by *Myzus persicae*. J. Gen. Virol. *75*: 2559–2565.

Van den Mooter, M. and J. Swings. 1990. Numerical analysis of 295 phenotypic features of 266 *Xanthomonas* strains and related strains and an improved taxonomy of the genus. Int. J. Syst. Bacteriol. *40*: 348–369.

Van Der Goot, F., J. Ausio, K.R. Wong, F. Pattus and J.T. Buckley. 1993. Dimerization stabilizes the pore-forming toxin aerolysin in solution. J. Biol. Chem. *268*: 18272–18279.

Van der Hofstad, G.A.J.M., J.D. Murugg, G.M.G.M. Verjans and P.J. Weisbeek. 1986. Characterization and structural analysis of the siderophore produced by the PGPR *Pseudomonas putida* strain WCS358. *In* Swinburne (Editor), Iron, Siderophores, and Plant Diseases, Plenum Press, New York. 71–75.

Van der Kooj, D. 1988. Properties of aeromonads and their occurrence and hygienic significance in drinking water. Zentbl. Bakteriol. Mikrobiol. Hyg. Ser. B. *187*: 1–17.

Van der Waaij, D., B.J. Cohen and G.W. Nace. 1974. Colonization patterns of aerobic gram-negative bacteria in the cloaca of *Rana pipiens*. Lab. Anim. Sci. *24*: 307–317.

Van der Werf, M.J., M.V. Guettler, M.K. Jain and J.G. Zeikus. 1997. Environmental and physiological factors affecting the succinate product ratio during carbohydrate fermentation by *Actinobacillus* sp. 130Z. Arch. Microbiol. *167*: 332–342.

van Dijk, K. and E.B. Nelson. 2000. Fatty acid competition as a mechanism by which *Enterobacter cloacae* suppresses *Pythium ultimum* sporangium germination and damping-off. Appl. Environ. Microbiol. *66*: 5340–5347.

van Doorn, J., P.M. Boonekamp and B. Oudega. 1994. Partial characterization of fimbriae of *Xanthomonas campestris* pathovar *hyacinthi*. Mol. Plant-Microbe Interact. *7*: 334–344.

van Dorssen, C.A. and F.H.J. Jaartsveld. 1962. *Actinobacillus suis* (novo species), een bij het varken voorkomende bacterie. Tijdschr. Diergeneeskd. *87*: 450–458.

Van Ert, M. and J.T. Staley. 1971. A new gas vacuolated heterotrophic rod from freshwaters. Arch. Mikrobiol. *80*: 70–77.

Van Eys, J. 1960. Pyridine ribosidase from *Xanthomonas pruni*. J. Bacteriol. *80*: 386–393.

Van Gijsegem, F. and A. Toussaint. 1982. Chromosome transfer and R-prime formation by an RP4::mini-Mu derivative in *Escherichia coli, Salmonella typhimurium, Klebsiella pneumoniae*, and *Proteus mirabilis*. Plasmid *7*: 30–44.

Van Groenestijn, J.W., G.J.F.M. Vlekke, D.M.E. Anink, M.H. Deinema and A.J.B. Zehnder. 1988. Role of cations in accumulation and release of phosphate by *Acinetobacter* strain 210A. Appl. Environ. Microbiol. *54*: 2894–2901.

van Hall, C.J.J. 1902. Bijdragen tot de kennis der Bakterieele Plantenziekten, Amsterdam.

van Ham, R.C., A. Moya and A. Latorre. 1997. Putative evolutionary origin of plasmids carrying the genes involved in leucine biosynthesis in *Buchnera aphidicola* (endosymbiont of aphids). J. Bacteriol. *179*: 4768–4777.

Van Ham, S.M., L. Van Alphen, F.R. Mooi and J.P. van Putten. 1993. Phase variation of *H. influenzae* fimbriae: transcriptional control of two divergent genes through a variable combined promoter region. Cell *73*: 1187–1196.

Van Klingeren, B., J.D.A. Van Embden and M. Dessons-Kroon. 1977. Plasmid-mediated chloramphenicol resistance in *Haemophilus influenzae*. Antimicrob. Agents Chemother. *11*: 383–387.

Van Kregten, E., N.A.C. Westerdaal and J.M.N. Willers. 1984. New, simple medium for selective recovery of *Klebsiella pneumoniae* and *Klebsiella oxytoca* from human feces. J. Clin. Microbiol. *20*: 936–941.

van Landschoot, A. and J. De Ley. 1983. Intra- and intergeneric similarities of the rRNA cistrons of *Alteromonas, Marinomonas* (gen. nov.) and some other Gram-negative bacteria. J. Gen. Microbiol. *129*: 3057–3074.

Van Landschoot, A. and J. De Ley. 1984. Validation of the publication of new names and combinations previously effectively published outside the IJSB. List No.13. Int. J. Syst. Bacteriol. *34*: 91–92.

Van Landschoot, A., R. Rossau and J. De Ley. 1986. Intrageneric and intergeneric similarities of the ribosomal RNA cistrons of *Acinetobacter*. Int. J. Syst. Bacteriol. *36*: 150–160.

Van Loghem, J.J. 1944. The classification of plague bacillus. Antonie van Leeuwenhoek J. Serol. Microbiol. *10*: 15–16.

van Loon, F.P., Z. Rahim, K.A. Chowdhury, B.A. Kay and S.A. Rahman. 1989. Case report of *Plesiomonas shigelloides*-associated persistent dysentery and pseudomembranous colitis. J. Clin. Microbiol. *27*: 1913–1915.

van Niel, C.B. and M. Allen. 1952. A note on *Pseudomonas stutzeri*. J. Bacteriol. *64*: 413–422.

van Orden, A.E. and P.W. Greer. 1977. Modification of the Dieterle spirochete stain. Histotechnology. *1*: 51–53.

Van Ostaaijen, J., J. Frey, S. Rosendal and J.I. MacInnes. 1997. *Actinobacillus suis* strains isolated from healthy and diseased swine are clonal and carry *apxICABD* subsp. *suis* and *apxIICA* subsp. *suis* toxin genes. J. Clin. Microbiol. *35*: 1131–1137.

Van Overbeek, L.S., L. Eberl, M. Givskov, S. Molin and J.D. Vanelsas. 1995. Survival of, and induced stress resistance in, carbon-starved *Pseudomonas fluorescens* cells residing in soil. Appl. Environ. Microbiol. *61*: 4202–4208.

van Oye, E. 1964. The World Problem of Salmonellosis, Junk, The Hague.

van Oye, E., M. Thevelin and C. Richard. 1975. Antigenic relationships between *Levinea amalonatica* and *Shigella dysenteriae* and *boydii*. Ann. Microbiol. *126A*: 187–192.

Van Pée, K.H., O. Salcher, P. Fischer, M. Bokel and F. Lingens. 1983. The biosynthesis of brominated pyrrolnitrin derivatives by *Pseudomonas aureofaciens*. J. Antibiot. *36*: 1735–1742.

Van Pee, W. and J. Stragier. 1979. Evaluation of some cold enrichment and isolation media for the recovery of *Yersinia enterocolitica*. Antonie van Leenwenhoek J. Serol. Microbiol. *45*: 465–477.

van Steenbergen, T.J., C.J. Bosch-Tijhof, A.J. van Winkelhoff, R. Gmür and J. de Graaff. 1994. Comparison of six typing methods for *Actinobacillus actinomycetemcomitans*. J. Clin. Microbiol. *32*: 2769–2774.

van Steenbergen, T.J., S.D. Colloms, P.W. Hermans, J. de Graaff and R.H. Plasterk. 1995. Genomic DNA fingerprinting by restriction fragment end labeling. Proc. Natl. Acad. Sci. U.S.A. *92*: 5572–5576.

van Straaten, H. 1918. Bacteriologische bevindingen bij eenige gevallen van pyosepticaemie (Lähme) der venlens. Verslag van den Werksaambeden der Rijksseruminrichting voor 1916–1917. *In* Verslag van den Werksaambeden der Rijksseruminrichting voor 1916-1917, Rotterdam. pp. 71–76.

Van Tiel-Menkveld, G.J., J.M. Mentjoxvervuurt, B. Oudega and F.K. Degraaf. 1982. Siderophore production by *Enterobacter cloacae* and a common receptor protein for the uptake of aerobactin and cloacin Df13. J. Bacteriol. *150*: 490–497.

van Tonder, E.M. 1973. Infection of rams with *Actinobacillus seminis*. J. S. Afr. Vet. Assoc. *44*: 235–240.

van Tonder, E.M. 1979. *Actinobacillus seminis* infection in sheep in the

Republic of South Africa. II. Incidence and geographical distribution. Onderstepoort J. Vet. Res. *46*: 135–140.

Van Veen, H.W., T. Abee, A.W.F. Kleefsman, B. Melgers, G.J.J. Kortstee, W.N. Konings and A.J.B. Zehnder. 1994. Energetics of alanine, lysine and proline transport in cytoplasmic membranes of the polyphosphate-accumulating *Acinetobacter johnsonii* strain 210A. J. Bacteriol. *176*: 2670–2676.

Van Veen, H.W., T. Abee, G.J.J. Kortstee, W.N. Konings and A.J.B. Zehnder. 1993. Characterization of two phosphate transport systems in *Acinetobacter johnsonii* 210A. J. Bacteriol. *175*: 200–206.

van Vuuren, H.J.J. 1978. Identification and physiology of *Enterobacteriaceae* isolated from South African lager beer breweries, Rijksuniversiteit Gent, Belgium

van Vuuren, H.J.J., K. Kersters, J. De Ley and D.F. Toerien. 1981. The identification of *Enterobacteriaceae* from breweries: Combined use and comparison of API 20E system, gel electrophoresis of proteins and gas chromotography of volatile metabolites. J. Appl. Bacteriol. *51*: 51–65.

Van Zyl, E. and P.L. Steyn. 1990. Differentiation of phytopathogenic *Pseudomonas* spp. and *Xanthomonas* spp. and pathovars by numerical taxonomy and protein gel electrophoregrams. Syst. Appl. Microbiol. *13*: 60–71.

Van Zyl, E. and P.L. Steyn. 1992. Reinterpretation of the taxonomic position of *Xanthomonas maltophilia* and taxonomic criteria in this genus. Request for an opinion. Int. J. Syst. Bacteriol. *42*: 193–198.

Vancanneyt, M., U. Torck, D. Dewettinck, M. Vaerewijck and K. Kersters. 1996a. Grouping of pseudomonads by SDS-PAGE of whole-cell proteins. Syst. Appl. Microbiol. *19*: 556–568.

Vancanneyt, M., S. Witt, W.R. Abraham, K. Kersters and H.L. Fredrickson. 1996b. Fatty acid content in whole cell hydrolysates and phospholipid fractions of pseudomonads: a taxonomic evaluation. Syst. Appl. Microbiol. *19*: 528–540.

Vandamme, P., M. Gillis, M. Vancanneyt, B. Hoste, K. Kersters and E. Falsen. 1993. *Moraxella lincolnii*, new species, isolated from the human respiratory tract, and reevaluation of the taxonomic position of *Moraxella osloensis*. Int. J. Syst. Bacteriol. *43*: 474–481.

Vanden Abeele, P., C. Van Keer, J. Swings, F. Gosselé and J. De Ley. 1980. Browning and rotting of apples caused by acetic acid bacteria. Meded. Fac. Landbouwwet. Rijksuniv. Gent. *45*: 391–397.

Vandenburgh, P.A., A.M. Wright and A.K. Vidaver. 1985. Partial purification and characterization of a polysaccharide depolymerase associated with phage-infected *Erwinia amylovora*. Appl. Environ. Microbiol. *49*: 993–994.

Vandepitte, J., J. Colaert, J. Lamotte-Legrand, C. Lamotte-Legrand and F. Perrin. 1953. Les ostéites à *Salmonella* chez les sicklanémiques: à propos de 5 observations. Ann. Soc. Belge Med. Trop. *33*: 511–522.

Vandepitte, J., P. Lemmens and L. de Swert. 1983. Human Edwardsiellosis traced to ornamental fish. J. Clin. Microbiol. *17*: 165–167.

Vandepitte, J., A. Makulu and F. Gatti. 1974. *Plesiomonas shigelloides*: survey and possible association with diarrhoea in Zaire. Ann. Soc. Belge Med. Trop. *54*: 503–513.

Vander Wauven, C. and V. Stalon. 1985. Occurrence of succinyl derivatives in the catabolism of arginine in *Pseudomonas cepacia*. J. Bacteriol. *164*: 882–886.

VanderMolen, G.E. and F.D. Williams. 1977. Observation of swarming of *Proteus mirabilis* with scanning electron microscopy. Can. J. Microbiol. *23*: 107–112.

Vaneechoutte, M., A. Elaichouni, K. Maquelin, G. Claeys, A. Van Liedekerke, H. Louagie, G. Verschraegen and L. Dijkshoorn. 1995. Comparison of arbitrarily primed polymerase chain reaction and cell envelope protein electrophoresis for analysis of *Acinetobacter baumannii* and *A. junii* outbreaks. Res. Microbiol. *146*: 457–465.

Vaneechoutte, M., I. Tjernberg, F. Baldi, M. Pepi, R. Fani, E.R. Sullivan, J. van der Toorn and L. Dijkshoorn. 1999. Oil-degrading *Acinetobacter* strain RAG-1 and strains described as '*Acinetobacter venetianus*' sp. nov. belong to the same genomic species. Res. Microbiol. *150*: 69–73.

Vaneechoutte, M., G. Verschraegen, G. Claeys and A.M. Van Den Abeele.

1990. Serological typing of *Branhamella catarrhalis* strains on the basis of lipopolysaccharide antigens. J. Clin. Microbiol. *28*: 182–187.

Vanhoof, R., P. Sonck and E. Hannecart-Pokorni. 1995. The role of lipopolysaccharide anionic binding sites in aminoglycoside uptake in *Stenotrophomonas* (*Xanthomonas*) *maltophilia*. J. Antimicrob. Chemother. *35*: 167–171.

Vanneste, J.L., J.P. Paulin and D. Expert. 1990. Bacteriophage Mu as a genetic tool to study *Erwinia amylovora* pathogenicity and hypersensitive reaction on tobacco. J. Bacteriol. *172*: 932–941.

Vargas, A. and W.R. Strohl. 1985a. Ammonium assimilation and metabolism by Beggiatoa alba. Arch. Microbiol. *142*: 272–278.

Vargas, A. and W.R. Strohl. 1985b. Utilization of nitrate by *Beggiatoa alba*. Arch. Microbiol. *142*: 279–284.

Vargas, C., M.J. Coronado, A. Ventosa and J.J. Nieto. 1997. Host range, stability and compatibility of broad host-range-plasmids and a shuttle vector in moderately halophilic bacteria. Evidence of intrageneric and intergeneric conjugation in moderate halophiles. Syst. Appl. Microbiol. *20*: 173–181.

Vargas, C., R. Fernández-Castillo, D. Cánovas, A. Ventosa and J.J. Nieto. 1995a. Isolation of cryptic plasmids from moderately halophilic eubacteria of the genus *Halomonas*. Characterization of a small plasmid from H. elongata and its use for shuttle vector construction. Mol. Gen. Genet. *246*: 411–418.

Vargas, E., S. Gutiérrez, M.E. Ambriz and C. Cervantes. 1995b. Chromosome-encoded inducible copper resistance in *Pseudomonas* strains. Antonie Van Leeuwenhoek Int. J. Gen. Mol. Microbiol. *68*: 225–229.

Vartian, C.V. and E.J. Septimus. 1990. Soft-tissue infection caused by *Edwardsiella tarda* and *Aeromonas hydrophila*. J. Infect. Dis. *161*: 816.

Vartivarian, S.E., K.A. Papadakis and E.J. Anaissie. 1996. *Stenotrophomonas* (*Xanthomonas*) *maltophilia* urinary tract infection. A disease that is usually severe and complicated. Arch. Intern. Med. *156*: 433–435.

Vartivarian, S.E., K.A. Papadakis, J.A. Palacios, J.T. Manning, Jr. and E.J. Anaissie. 1994. Mucocutaneous and soft tissue infections caused by *Xanthomonas maltophilia*. A new spectrum. Ann. Intern. Med. *121*: 969–973.

Vasil, M.L., D. Kabat and B.H. Iglewski. 1977. Structure-activity-relationships of an exotoxin of *Pseudomonas aeruginosa*. Infect. Immun. *16*: 353–361.

Vastine, D.W., C.R. Dawson, I. Hoshiwara, C. Yoneda, T. Daghfous and M. Messadi. 1974. Comparison of media for the isolation of *Haemophilus* species from cases of seasonal conjunctivitis associated with severe endemic trachoma. Appl. Microbiol. *28*: 688–691.

Vasyurenko, Z.P. and Y.N. Chernyavskaya. 1990. Confirmation of *Morganella* distinction from *Proteus* and *Providencia* among *Enterobacteriaceae* on the basis of cellular and lipopolysaccharide fatty acid composition. J. Hyg. Epidemiol. Microbiol. Immunol. (Prague) *34*: 81–90.

Vaucher, J.P. 1803. Histoire des conferves d'eau douce, contenant leurs different modes de reproduction, et la description de leurs principales especes, J. Paschoud, Geneva. 1–285.

Vauterin, L., B. Hoste, K. Kersters and J. Swings. 1995. Reclassification of *Xanthomonas*. Int. J. Syst. Bacteriol. *45*: 472–489.

Vauterin, L. and J. Swings. 1997. Are classification and phytopathological diversity compatible in *Xanthomonas*? J. Ind. Microbiol. Biotechnol. *19*: 77–82.

Vauterin, L., J. Swings and K. Kersters. 1991. Grouping of *Xanthomonas campestris* pathovars by SDS-PAGE of proteins. J. Gen. Microbiol. *137*: 1677–1687.

Vauterin, L., J. Swings, K. Kersters, M. Gillis, T.W. Mew, M.N. Schroth, N.J. Palleroni, D.C. Hildebrand, D.E. Stead, E.L. Civerolo, A.C. Hayward, H. Maraîte, R.E. Stall, A.K. Vidaver and J.F. Bradbury. 1990. Towards an improved taxonomy of *Xanthomonas*. Int. J. Syst. Bacteriol. *40*: 312–316.

Vauterin, L., P. Yang, A. Alvarez, Y. Takikawa, D.A. Roth, A.K. Vidaver, R.E. Stall, K. Kersters and J. Swings. 1996a. Identification of nonpathogenic *Xanthomonas* strains associated with plants. Syst. Appl. Microbiol. *19*: 96–105.

Vauterin, L., P. Yang, B. Hoste, B. Pot, J. Swings and K. Kersters. 1992.

Taxonomy of xanthomonads from cereals and grasses based on SDS-PAGE of proteins, fatty acid analysis and DNA hybridization. J. Gen. Microbiol. *138*: 1467–1477.

Vauterin, L., P. Yang and J. Swings. 1996b. Utilization of fatty acid methyl esters for the differenatiation of new *Xanthomonas* species. Int. J. Syst. Bacteriol. *46*: 298–304.

Veenstra, J., P. Rietra, J.M. Coster, E. Slaats and S. Dirks-Go. 1994. Seasonal variations in the occurrence of *Vibrio vulnificus* along the Dutch coast. Epidemiol. Infect. *112*: 285–290.

Vela, G.R., G.D. Cagle and P.R. Holmgren. 1970. Ultrastructure of *Azotobacter vinelandii*. J. Bacteriol. *104*: 933–939..

Vela, G.R. and O. Wyss. 1964. Improved stain for visualization of *Azotobacter* encystment. J. Bacteriol. *87*: 467–477.

Venezia, R.A. and R.G. Robertson. 1975. Bactericidal substance produced by *Haemophilus influenzae* b. Can. J. Microbiol. *21*: 1587–1594.

Venkatesan, M.M., M.B. Goldberg, D.J. Rose, E.J. Grotbeck, V. Burland and F.R. Blattner. 2001. Complete DNA sequence and analysis of the large virulence plasmid of *Shigella flexneri*. Infect. Immun. *69*: 3271–3285.

Venkatesh, T.V. and H.K. Das. 1992. The *Azotobacter vinelandii recA* gene - sequence analysis and regulation of expression. Gene *113*: 47–53.

Venkateswaran, K. and N. Dohmoto. 2000. *Pseudoalteromonas peptidolytica* sp. nov., a novel marine mussel-thread-degrading bacterium isolated from the Sea of Japan. Int. J. Syst. Evol. Microbiol. *50*: 565–574.

Venkateswaran, K., M.E. Dollhopf, R. Aller, E. Stackebrandt and K.H. Nealson. 1998. *Shewanella amazonensis* sp. nov., a novel metal-reducing facultative anaerobe from Amazonian shelf muds. Int. J. Syst. Bacteriol. *48*: 965–972.

Venkateswaran, K., D.P. Moser, M.E. Dollhopf, P. Lies Douglas, D.A. Saffarini, B.J. Macgregor, D.B. Ringelberg, D.C. White, M. Nishijima, H. Sano, J. Burghardt, E. Stackebrandt and K.H. Nealson. 1999. Polyphasic taxonomy of the genus *Shewanella* and description of *Shewanella oneidensis* sp. nov. Int. J. Syst. Bacteriol. *49*: 705–724.

Ventosa, A., R. Fernández-Castillo, C. Vargas, E. Mellado, M.T. Garcia and J.J. Nieto. 1994. Isolation and characterization of new plasmids from moderately halophilic eubacteria: developing of cloning vectors. Proceedings of the 6th European Congress on Biotechnology, Amsterdam. Elsevier Science. 271–274.

Ventosa, A., M.C. Gutierrez, M.T. Garcia and F. Ruiz-Berraquero. 1989. Classification of "*Chromobacterium marismortui*" in a new genus, *Chromohalobacter* gen. nov. as *Chromohalobacter marismortui*, comb. nov., nov. rev. Int. J. Syst. Bacteriol. *39*: 382–386.

Ventosa, A., J.J. Nieto and A. Oren. 1998. Biology of moderately halophilic aerobic bacteria. Microbiol. Mol. Biol. Rev. *62*: 504–544.

Ventosa, A., E. Quesada, F. Rodriguez-Valera, F. Ruiz-Berraquero and A. Ramos-Cormenzana. 1982. Numerical taxonomy of moderately halophilic Gram-negative rods. J. Gen. Microbiol. *128*: 1959–1968.

Ventura, S., L. Giovannetti, A. Gori, C. Viti and R. Materassi. 1993. Total DNA restriction pattern and quinone composition of members of the family *Ectothiorhodospiraceae*. Syst. Appl. Microbiol. *16*: 405–410.

Ventura, S., C. Viti, R. Pastorelli and L. Giovannetti. 2000. Revision of species delineation in the genus *Ectothiorhodospira*. Int. J. Syst. Evol. Microbiol. *50*: 583–591.

Vera Cruz, C.M., F. Gosselé, K. Kersters, P. Segers, M. Van den Mooter, J. Swings and J. De Ley. 1984. Differentiation between *Xanthomonas campestris* pv. *oryzae*, *Xanthomonas campestris* pv. *oryzicola* and the bacterial "brown blotch" pathogen on rice by numerical analysis of phenotypic features and protein gel electrophoregrams. J. Gen. Microbiol. *130*: 2983–3000.

Verder, E. and J. Evans. 1961. A proposed antigenic schema for the identification of strains of *Pseudomonas aeruginosa*. J. Infect. Dis. *109*: 183–193.

Verdier, V., K. Assigbetse, C.K. Gopal, K. Wydra, K. Rudolph and J.P. Geiger. 1998a. Molecular characterization of the incitant of cowpea bacterial blight and pustule, *Xanthomonas campestris* pv. *vignicola*. Eur. J. Plant Pathol. *104*: 595–602.

Verdier, V., S. Restrepo, G. Mosquera, M.C. Duque, A. Gerstl and R. Laberry. 1998b. Genetic and pathogenic variation of *Xanthomonas*

axonopodis pv. *manihotis* in Venezuela. Plant Pathol. (Oxf.) *47*: 601–608.

Verdonck, L., J. Mergaert, C. Rijckaert, J. Swings, K. Kersters and J. De Ley. 1987. Genus *Erwinia*: numerical analysis of phenotypic features. Int. J. Syst. Bacteriol. *37*: 4–18.

Verhaegen, J., H. Verbraeken, A. Cabuy, J. Vandeven and J. Vandepitte. 1988. *Actinobacillus* (formerly *Pasteurella*) *ureae* meningitis and bacteraemia: report of a case and review of the literature. J. Infect. *17*: 249–253.

Verhille, S., N. Baida, F. Dabboussi, M. Hamze, D. Izard and H. Leclerc. 1999a. *Pseudomonas gessardii* sp. nov. and *Pseudomonas migulae* sp. nov., two new species isolated from natural mineral waters. Int. J. of Syst. Bacteriol. *49*: 1559–1572 .

Verhille, S., N. Baida, F. Dabboussi, D. Izard and H. Leclerc. 1999b. Taxonomic study of bacteria isolated from natural mineral waters: Proposal of *Pseudomonas jessenii* sp. nov. and *Pseudomonas mandelii* sp. nov. Syst. Appl. Microbiol. *22*: 45–58.

Verma, U.K., D.J. Brenner, W.L. Thacker, R.F. Benson, G. Vesey, J.B. Kurtz, P.J.L. Dennis, A.G. Steigerwalt, J.S. Robinson and C.W. Moss. 1992. *Legionella shakespearei*, sp.nov., isolated from cooling tower water. Int. J. Syst. Bacteriol. *42*: 404–407.

Vernon-Shirley, M. and R. Burns. 1992. The development and use of monoclonal antibodies for detection of *Erwinia*. J. Appl. Bacteriol. *72*: 97–102.

Véron, M.M. 1965. La position taxonomique des *Vibrio* et de certaines bactéries comparables. C.R. Acad. Sci. Paris *261*: 5243–5246.

Véron, M. 1975. Nutrition et taxonomie des *Enterobacteriaceae* et bactéries voisines. I. Méthode d' étude des auxanogrammes. Ann. Microbiol. Inst. Pasteur (Paris) *136A*: 267–274.

Véron, M. and P. Berche. 1976. Virulence et antigenes de *Pseudomonas aeruginosa*. Bull. Inst. Pasteur. *74*: 295–337.

Véron, M. and F. Gasser. 1963. Sur la detection de l' hydrogene sulfure produit par certaines enterobacteriacees dans les milieux de diagnostic rapid. Ann. Inst. Pasteur. *105*: 524–534.

Véron, M. and L. Le Minor. 1975a. Nutrition et taxonomie des *Enterobacteriaceae* et bactéries voisines. II. Résultats d' ensemble et classification. Ann. Microbiol. Inst. Pasteur (Paris) *126B*: 111–124.

Véron, M. and L. Le Minor. 1975b. Nutrition et taxonomie des *Enterobacteriaceae*. III. Caracteres nutritionnels et différenciation des groupes taxonomiques. Ann. Microbiol. Inst. Pasteur (Paris) *126B*: 125–147.

Véron, M., A. Lenvoisé-Furet, C. Coustère, C. Ged and F. Grimont. 1993. Relatedness of three species of "false neisseriae," *Neisseria caviae*, *Neisseria cuniculi*, and *Neisseria ovis*, by DNA–DNA hybridization and fatty acid analysis. Int. J. Syst. Bacteriol. *43*: 210–220.

Versalovic, J., T. Koeuth and J.R. Lupski. 1991. Distribution of repetitive DNA-sequences in eubacteria and application to fingerprinting of bacterial genomes. Nucleic Acids Res. *19*: 6823–6831.

Verweyen, H.M., H. Karch, F. Allerberger and L.B. Zimmerhackl. 1999. Enterohemorrhagic *Escherichia coli* (EHEC) in pediatric hemolytic-uremic syndrome: a prospective study in Germany and Austria. Infection *27*: 341–347.

Vial, P.A., R. Robins-Browne, H. Lior, V. Prado, J.B. Kaper, J.P. Nataro, D. Maneval, A. Elsayed and M.M. Levine. 1988. Characterization of enteroadherent-aggregative *Escherichia coli*, a putative agent of diarrheal disease. J. Infect. Dis. *158*: 70–79.

Vicente, M. and J.L. Cánovas. 1973. Glucolysis in *Pseudomonas putida*: physiological role of alternative routes from the analysis of defective mutants. J. Bacteriol. *116*: 908–914.

Vickers, R.M., A. Brown and G.M. Garrity. 1981. Dye-containing buffered charcoal-yeast extract medium for differentiation of members of the family *Legionellaceae*. J. Clin. Microbiol. *13*: 380–382.

Vidaver, A.K. and S. Buckner. 1978. Typing of fluorescent phytopathogenic pseudomonads by bacteriocin production. Can. J. Microbiol. *24*: 14–18.

Vidaver, A.K., R.K. Koski and J.L. van Etten. 1973. Bacteriophage ø6: a lipid-containing virus of *Pseudomonas phaseolicola*. J. Virol. *11*: 799–805.

Vidaver, A.K., M.L. Mathys, M.E. Thomas and M.L. Schuster. 1972. Bacteriocins of the phytopathogens *Pseudomonas syringae*, *P. glycinea* and *P. phaseolicola*. Can. J. Microbiol. *18*: 705–713.

Vieu, J.-F. and M. Capponi. 1965. Lysotypie des *Proteus* OX19, OXK, OX2 et OXL. Ann. Inst. Pasteur (Paris) *108*: 103–106.

Vieu, J.-F., O. Croissant and C. Dauguet. 1965. Structure des bactériophages responsables des phénomènes de conversion chez les *Salmonella*. Ann. Inst. Pasteur (Paris) *109*: 160–166.

Vijayalakshmi, N., R.S. Rao and S. Badrinath. 1997. Minimum inhibitory concentration (MIC) of some antibiotics against *Vibrio cholerae* O139 isolates from Pondicherry. Epidemiol. Infect. *119*: 25–28.

Vila, J., A. Gene, M. Vargas, J. Gascon, C. Latorre and M.T.J. de Anta. 1998. A case-control study of diarrhoea in children caused by *Escherichia coli* producing heat-stable enterotoxin (EAST-1). J. Med. Microbiol. *47*: 889–891.

Viljanen, M.K., T. Nurmi and A. Salminen. 1983. Enzyme-linked immunosorbent assay (ELISA) with bacterial sonicate antigen for immunoglobulin M, immunoglobulin A and immunoglobulin G antibodies to *Francisella tularensis*: comparison with bacterial agglutination test and ELISA with lipopolysaccharide antigen. J. Infect. Dis. *148*: 715–720.

Villalobo, E. and A. Torres. 1998. PCR for detection of *Shigella* spp. in mayonnaise. Appl. Environ. Microbiol. *64*: 1242–1245.

Vincent, P., P. Pignet, F. Talmont, L. Bozzi, B. Fournet, J. Guezennec, C. Jeanthon and D. Prieur. 1994. Production and characterization of an exopolysaccharide excreted by a deep-sea hydrothermal vent bacterium isolated from the polychaete annelid *Alvinella pompejana*. Appl. Environ. Microbiol. *60*: 4134–4141.

Vinogradov, E.V., R. Pantophlet, S.R. Haseley, L. Brade, O. Holst and H. Brade. 1997. Structural and serological characterisation of the O-specific polysaccharide from lipopolysaccharide of *Acinetobacter calcoaceticus* strain 7 (DNA group 1). Eur. J. Biochem. *243*: 167–173.

Vinogradov, E.V., A.S. Shashkov, Y.A. Knirel, N.K. Kochetkov, N.V. Tochtamysheva, S.F. Averin, O.V. Goncharova and V.S. Khlebnikov. 1991. Structure of the O-antigen of *Francisella tularensis* strain 15. Carbohydr. Res. *214*: 289–298.

Virlogeux-Payant, I. and M.Y. Popoff. 1996. The Vi antigen of *Salmonella typhi*. Bull. Inst. Pasteur. *94*: 237–250.

Visalli, M.A., S. Bajaksouzian, M.R. Jacobs and P.C. Appelbaum. 1997. Comparative activity of trovafloxacin, alone and in combination with other agents, against gram-negative nonfermentative rods. Antimicrob. Agents Chemother. *41*: 1475–1481.

Visca, P., A. Ciervo, V. Sanfilippo and N. Orsi. 1993. Iron regulated salicylate synthesis by *Pseudomonas* spp. J. Gen. Microbiol. *139*: 1995–2001.

Viscontini, M. and M. Frater-Schröder. 1968. Isolierung von 6-hydroxymethylpterin aus kulturen von *Pseudomonas roseus-fluorescens* J.C. Marchal 1937. Helv. Chim. Acta. *51*: 1554–1557.

Vishniac, W. 1974. Genus *Thiobacillus*. *In* Buchanan and Gibbons (Editors), Bergey's Manual of Determinative Bacteriology, 8th Ed., The Williams & Wilkins Co., Baltimore. pp. 456–461.

Visloukh, S.M. 1914. *Spirillum kolkwitzii* nov. sp. Zh. Mikrobiol. Epidemiol. Immunol. *1*: 42–51.

Visscher, P.T. and B.F. Taylor. 1994. Demethylation of dimethylsulfoniopropionate to 3-mercaptopropionate by an aerobic marine bacterium. Appl. Environ. Microbiol. *60*: 4617–4619.

Vivian, A. 1991. Genetic organisation of *Acinetobacter*. *In* Towner and Bergogne-Bérézin (Editors), The Biology of *Acinetobacter*: Taxonomy, Clinical Importance, Molecular Biology, Physiology, Industrial Relevance, Plenum Press, New York. pp. 191–200.

Vivian, A. 1992. Avirulence genes in *Pseudomonas syringae* pathovars. *In* Galli, Silver and Witholt (Editors), *Pseudomonas*: Molecular Biology and Biotechnology, American Society for Microbiology, Washington, D.C. 37–42.

Vodkin, M.H. and J.C. Williams. 1986. Overlapping deletion in two spontaneous phase variants of *Coxiella burnetii*. J. Gen. Microbiol. *132*: 2587–2594.

Vodkin, M.H. and J.C. Williams. 1988. A heat shock operon in *Coxiella*

burnetti produces a major antigen homologous to a protein in both mycobacteria and *Escherichia coli*. J. Bacteriol. *170*: 1227–1234.

Vogel, B.F., K. Jorgensen, H. Christensen, J.E. Olsen and L. Gram. 1997. Differentiation of *Shewanella putrefaciens* and *Shewanella alga* on the basis of whole-cell protein profiles, ribotyping, phenotypic characterization, and 16S rRNA gene sequence analysis. Appl. Environ. Microbiol. *63*: 2189–2199.

Vohora, K. and E. Torrijos. 1988. Neonatal sepsis and meningitis due to *Edwardsiella tarda* (E.t.): a rare pathogen. Program and abstract of the 88th annual meeting of the American Society for Microbiology, American Society for Microbiology,. p. 358.

Volgyi, A., A. Fodor, A. Szentirmai and S. Forst. 1998. Phase variation in *Xenorhabdus nematophilus*. Appl. Environ. Microbiol. *64*: 1188–1193.

Volk, W.A. 1966. Cell wall lipopolysaccharides from *Xanthomonas* spp. J. Bacteriol. *91*: 39–42.

Volk, W.A. 1968a. Isolation of D-galacturonic acid 1-phosphate from hydrolysates of cell wall lipopolysaccharide extracted from *Xanthomonas campestris*. J. Bacteriol. *95*: 782–786.

Volk, W.A. 1968b. Quantitative assay of polysaccharide components obtained from cell wall lipopolysaccharides of *Xanthomonas* species. J. Bacteriol. *95*: 980–982.

Volk, W.A., N.L. Salmonsky and D. Hunt. 1972. *Xanthomonas sinensis* cell wall lipopolysaccharides I. Isolation of 4,7-anhydro- and 4,8-anhydro-3-deoxy octulosonic acid following hydrolysis. J. Biol. Chem. *247*: 3881–3887.

Vollack, K.U., J. Xie, E. Härtig, U. Römling and W.G. Zumft. 1998. Localization of denitrification genes on the chromosomal map of *Pseudomonas aeruginosa*. Microbiology *144*: 441–448.

von der Hofstad, G.A., J.D. Marugg, G.M. Verjans and P.J. Weisbeek. 1986. Characterization and structural analysis of the siderophore produced by the PGPR *Pseudomonas putida* strain WC 358. *In* Swinburne (Editor), Iron, Siderophores, and Plant Diseases, Plenum, New York. pp. 36–40.

von Frisch, A. 1882. Zur aetiologie des rhinoskleroms. Wien. Med. Wochenschr. *32*: 969–972.

von Graevenitz, A. 1970. *Erwinia* species isolates. Ann. NY Acad. Sci. *174*: 436–443.

von Graevenitz, A. 1980. Infection and colonization with *Serratia*. *In* von Graevenitz and Rubin (Editors), The Genus *Serratia*, CRC Press, Boca Raton. pp. 167–186.

von Graevenitz, A. 1990. Revised nomenclature of *Campylobacter laridis*, *Enterobacter intermedium*, and "*Flavobacterium branchiophila*". Int. J. Syst. Bacteriol. *40*: 211.

Von Graevenitz, A. and C. Bucher. 1983. Evaluation of differential and selective media for isolation of *Aeromonas* and *Plesiomonas* spp. from human feces. J. Clin. Microbiol. *17*: 16–21.

von Graevenitz, A. and A.H. Mensch. 1968. The genus *Aeromonas* in human bacteriology. New Engl. J. Med. *278*: 245–249.

Von Graevenitz, A. and M. Nourbakhsh. 1972. Antimicrobial resistance of the genera *Proteus*, *Providencia* and *Serratia* with special reference to multiple resistance patterns. Med. Microbiol. Immunol. *157*: 142–148.

von Graevenitz, A. and G. Palermo. 1980. *Erwinia herbicola* (*Enterobacter agglomerans*) in urinary tract infections. Microbios Lett. *14*: 123–124.

Von Graevenitz, A. and G. Simon. 1970. Potentially pathogenic nonfermentative, H_2S-producing gram-negative rod (1b). Appl. Microbiol. *19*: 176.

Von Roekel, H.V. 1965. Pullorum disease. *In* Biester and Schwarte (Editors), Diseases of Poultry, 5th Ed., Iowa State University Press, Ames. pp. 220–259.

von Tigerstrom, R.G. 1980. Extracellular nucleases of *Lysobacter enzymogenes*: production of the enzymes and purification and characterization of an endonuclease. Can. J. Biochem. *26*: 1029–1037.

von Tigerstrom, R.G. 1981. Extracellular nucleases of *Lysobacter enzymogenes*: purification and characterization of a ribonuclease. Can. J. Biochem. *27*: 1080–1086.

von Tigerstrom, R.G. 1983. The effect of magnesium and manganese ion concentration and medium composition on the production of extracellular enzymes by *Lysobacter enzymogenes*. J. Gen. Microbiol. *129*: 2293–2299.

von Tigerstrom, R.G. 1984. Production of two phosphatases by *Lysobacter enzymogenes* and purification and characterization of the extracellular enzymes. Appl. Environ. Microbiol. *47*: 693–698.

von Tigerstrom, R.G. and S. Stelmaschuk. 1985. Localization of the cell-associated phosphatases in *Lysobacter enzymogenes*. J. Gen. Microbiol. *131*: 1611–1618.

von Tigerstrom, R.G. and S. Stelmaschuk. 1986. Purification and characterization of the outer membrane associated alkaline phosphatase of *Lysobacter enzymogenes*. J. Gen. Microbiol. *132*: 1379–1388.

von Tigerstrom, R.G. and S. Stelmaschuk. 1987. Comparison of the phosphatases of *Lysobacter enzymogenes* with those of related bacteria. J. Gen. Microbiol. *133*: 3121–3128.

Vörös, S. and B.W. Senior. 1990. New O antigens of *Morganella morganii* and the relationships between haemolysin production. O antigens and morganocin types of strains. Acta Microbiol. Hung. *37*: 341–349.

Vos, P., R. Hogers, M. Bleeker, M. Reijans, T. van de Lee, M. Hornes, A. Frijters, J. Pot, J. Peleman, M. Kuiper and M. Zabeau. 1995. AFLP: A new technique for DNA fingerprinting. Nucleic Acids Res. *23*: 4407–4414.

Vosman, B. and K.J. Hellingwerf. 1991. Molecular-cloning and functional characterization of a RecA analog from *Pseudomonas stutzeri* and construction of a *P. stutzeri* RecA mutant. Antonie Leeuwenhoek *59*: 115–123.

Vreeland, R.H. 1984. Genus *Halomonas*. *In* Krieg and Holt (Editors), Bergey's Manual of Systematic Bacteriology, 1st Ed., Vol. 1, The Williams & Wilkins Co., Baltimore. pp. 340–343.

Vreeland, R.H. 1993. Taxonomy of halophilic bacteria. *In* Vreeland and Hochstein (Editors), The Biology of Halophilic Bacteria, CRC Press, Boca Raton. 105–135.

Vreeland, R.H., R. Andersen and R.G.E. Murray. 1984. Cell wall and lipid composition and its relationship to the salt tolerance of *Halomonas elongata*. J. Bacteriol. *160*: 879–883.

Vreeland, R.H., C.D. Litchfield, E.L. Martin and E. Elliot. 1980. *Halomonas elongata*, a new genus and species of extremely salt-tolerant bacteria. Int. J. Syst. Bacteriol. *30*: 485–495.

Vreeland, R.H. and E.L. Martin. 1980. Growth-characteristics, effects of temperature, and ion specificity of the halotolerant bacterium *Halomonas elongata*. Can. J. Microbiol. *26*: 746–752.

Vu-Thien, H., D. Moissenet, M. Valcin, C. Dulot, G. Tournier and A. Garbarg-Chenon. 1996. Molecular epidemiology of *Burkholderia cepacia*, *Stenotrophomonas maltophilia*, and *Alcaligenes xylosoxicans* in a cystic fibrosis center. Eur. J. Clin. Microbiol. Infect. Dis. *15*: 876–879.

Wachsmuth, I.K. 1980. *Vibrio*. *In* Lennette, Balows, Hausler and Truant (Editors), Manual of Clinical Microbiology, American Society for Microbiology, Washington, D.C. pp. 226–234.

Wachsmuth, I.K. 1984. Laboratory detection of enterotoxins. *In* Ellner (Editor), Infectious Diarrheal Diseases, Marcel Dekker, New York. pp. 93–115.

Wachsmuth, I.K., B.R. Davis and S.D. Allen. 1979. Ureolytic *Escherichia coli* of human origin: serological, epidemiological, and genetic analysis. J. Clin. Microbiol. *10*: 897–902.

Wackett, L.P., M.J. Sadowsky, L.M. Newman, H.G. Hur and S.Y. Li. 1994. Metabolism of polyhalogenated compounds by a genetically-engineered bacterium. Nature (Lond.) *368*: 627–629.

Waddell, T., S. Head, M. Petric, A. Cohen and C. Lingwood. 1988. Globotriosyl ceramide is specifically recognized by the *Escherichia coli* verocytotoxin 2. Biochem. Biophys. Res. Commun. *152*: 674–679.

Wadowsky, R.W. and R.B. Yee. 1981. Glycine-containing selective medium for isolation of *Legionellaceae* from environmental specimens. Appl. Environ. Microbiol. *42*: 768–772.

Wagner, C. and A.T. Brown. 1970. Regulation of tryptophan pyrrholase activity in *Xanthomonas pruni*. J. Bacteriol. *104*: 90–97.

Wagner, M., R. Amann, P. Kämpfer, B. Assmus, A. Hartmann, P. Hutzler, N. Springer and K.-H. Schleifer. 1994a. Identification and in situ detection of Gram-negative filamentous bacteria in activated sludge. Syst. Appl. Microbiol. *17*: 405–417.

Wagner, M., R. Erhart, W. Manz, R. Amann, H. Lemmer, D. Wedi and K.H. Schleifer. 1994b. Development of an rRNA-targeted oligonucleotide probe specific for the genus *Acinetobacter* and its application for in situ monitoring in activated sludge. Appl. Environ. Microbiol. *60*: 792–800.

Wahlund, T.M., C.R. Woese, R.W. Castenholz and M.T. Madigan. 1991. A thermophilic green sulfur bacterium from New Zealand hot springs, *Chlorobium tepidum* sp. nov. Arch. Microbiol. *156*: 81–90.

Wakabayashi, H. and S. Egusa. 1972. Characteristics of a *Pseudomonas* sp. from an epizootic of pond-cultured eels (*Anguillula japonica*). Bull. Jpn. Soc. Scient. Fisheries. *38*: 577–586.

Wakabayashi, H. and S. Egusa. 1973. *Edwardsiella tarda* (*Paracolobactrum anguillimortiferum*) associated with pond-cultured eel disease. Bull. Japan. Soc. Sci. Fish. *39*: 931–936.

Wakker, J.H. 1883. Vorläufige Mittheilungen über Hyacinthenkrankheiten. Bot. Centrabl. *14*: 315–317.

Waksman, S.A. and J.S. Joffe. 1922. Microorganisms concerned in the oxidation of sulfur in the soil. II. The *Thiobacillus thiooxidans*, a new sulfur oxidizing organism isolated from the soil. J. Bacteriol. *7*: 239–256.

Waldee, E.L. 1945. Comparative studies of some peritrichous phytopathogenic bacteria. Iowa State Coll. J. Sci. *19*: 435–484.

Waldhalm, D.G., R.F. Hall, W.A. Meinershagen, C.S. Card and F.W. Frank. 1974a. *Haemophilus somnus* infection in the cow as a possible contributing factor to weak calf syndrome: isolation and animal inoculation studies. Am. J. Vet. Res. *35*: 1401–1403.

Waldhalm, D.G., R.F. Hall, W.A. Meinershagen, E. Stauber and F.W. Frank. 1974b. Combined effect of neonatal calf diarrhea virus and *Providencia stuartii* on suckling beef calves. Am. J. Vet. Res. *35*: 515–516.

Waldhalm, D.G., W.A. Meinershagen and F.W. Frank. 1969. *Providencia stuartii* as an etiological agent in neonatal diarrhea in calves. Am. J. Vet. Res. *30*: 1573–1575.

Waldhalm, D.G., H.G. Stoenner, R.E. Simmons and L.A. Thomas. 1978. Abortion associated with *Coxiella burnetii* infection in dairy goats. J. Am. Vet. Med. Assoc. *137*: 1580–1581..

Waldor, M.K., R. Colwell and J.J. Mekalanos. 1994. The *Vibrio cholerae* O139 serogroup antigen includes an O-antigen capsule and lipopolysaccharide virulence determinants. Proc. Natl Acad. Sci. U.S.A. *91*: 11388–11392.

Waldor, M.K. and J.J. Mekalanos. 1996. Lysogenic conversion by a filamentous phage encoding cholera toxin. Science *272*: 1910–1914.

Walker, C.N. and P.W. Smith. 1980. Ampicillin resistance in *Haemophilus parainfluenzae*. Am. J. Clin. Pathol. *74*: 229–232.

Walker, J.D., H.F. Austin and R.R. Colwell. 1975. Utilization of mixed hydrocarbon substrate by petroleum degrading microorganisms. J. Gen. Appl. Microbiol. *21*: 27–39.

Walker, P.D., J. Short, R.O. Thomson and D.S. Roberts. 1973. The fine structure of *Fusiformis nodosus* with special reference to the location of antigens associated with immunogenicity. J. Gen. Microbiol. *77*: 351–361.

Walkes, C.M. and L.W. O'Garro. 1996. Role of extracellular polysaccharides from *Xanthomonas campestris* pv. *vesicatoria* in bacterial spot of pepper. Physiol. Mol. Plant Pathol. *48*: 91–104.

Wall, P.G., D. Morgan, K. Lamden, M. Ryan, M. Griffin, E.J. Threlfall, L.R. Ward and B. Rowe. 1994. A case control study of infection with an epidemic strain of multiresistant *Salmonella typhimurium DT104* in England and Wales. Commun. Dis. Rep. CDR Rev. *4*: R130–135.

Wallace, J.J. and R.G. Petersdorf. 1971. Urinary tract infections. Postgrad. Med. *50*: 138–144.

Wallace, L.J., F.H. White and H.L. Gore. 1966. Isolation of *Edwardsiella tarda* from a sea lion and two alligators. J. Am. Vet. Med. Assoc. *149*: 881–883.

Wallace, P.L., D.G. Hollis, R.E. Weaver and C.W. Moss. 1988. Cellular fatty acid composition of *Kingella* species, *Cardiobacterium hominis*, and *Eikenella corrodens*. J. Clin. Microbiol. *26*: 1592–1594.

Wallace, R.J., Jr., C.J. Baker, F.J. Quinones, D.G. Hollis, R.E. Weaver and K. Wiss. 1983. Nontypable *Haemophilus influenzae* (biotype 4) as a neonatal, maternal, and genital pathogen. Rev. Infect. Dis. *5*: 123–136.

Wallet, F., A. Fruchart, P.J. Bouvet and R.J. Courcol. 1994. Isolation of *Moellerella wisconsensis* from bronchial aspirate. Eur. J. Clin. Microbiol. Infect. Dis. *13*: 182–183.

Wallet, F., F. Touré, A. Devalckenaere, D. Pagniez and R.J. Courcol. 2000. Molecular identification of *Pasteurella dagmatis* peritonitis in a patient undergoing peritoneal dialysis. J. Clin. Microbiol. *38*: 4681–4682.

Wallin, J.R. and C.S. Reddy. 1945. A bacterial streak disease of *Phleum pratense* L. Phytopathology *35*: 937–939.

Walmsley, J. and C. Kennedy. 1991. Temperature dependent regulation by molybdenum and vanadium of expression of the structural genes encoding three nitrogenases in *Azotobacter vinelandii*. Appl. Environ. Microbiol. *57*: 622–624.

Walsh, T.R., L. Hall, S.J. Assinder, W.W. Nichols, S.J. Cartwright, A.P. MacGowan and P.M. Bennett. 1994. Sequence analysis of the L1 metallo-β-lactamase from *Xanthomonas maltophilia*. Biochim. Biophys. Acta *1218*: 199–201.

Walsh, T.R., A.P. MacGowan and P.M. Bennett. 1997a. Sequence analysis and enzyme kinetics of the L2 serine β-lactamase from *Stenotrophomonas maltophilia*. Antimicrob. Agents Chemother. *41*: 1460–1464.

Walsh, T.R., R.A. Stunt, J.A. Nabi, A.P. Macgowan and P.M. Bennett. 1997b. Distribution and expression of beta-lactamase genes among *Aeromonas* spp. J. Antimicrob. Chemother. *40*: 171–178.

Waltman, W.D., E.B. Shotts and V.S. Blazer. 1985. Recovery of *Edwarsiella ictaluri* from danio (*Danio devario*). Aquaculture *46*: 63–66.

Waltman, W.D., II, E.B. Shotts and T.C. Hsu. 1982. Enzymatic characterization of *Aeromonas hydrophila* complex by the API-ZYM system. J. Clin. Microbiol. *16*: 692–696.

Waltman, W.D., E.B. Shotts and T.C. Hsu. 1986. Biochemical characteristics of *Edwardsiella ictaluri*. Appl. Environ. Microbiol. *51*: 101–104.

Walton, D.T., S.L. Abbott and J.M. Janda. 1993. Sucrose-positive *Edwardsiella tarda* mimicking a biogroup 1 strain isolated from a patient with cholelithiasis. J. Clin. Microbiol. *31*: 155–156.

Wandersman, C., T. Andro and Y. Bertheau. 1986. Extracellular proteases in *Erwinia chrysanthemi*. J. Gen. Microbiol. *132*: 899–906.

Wang, H. and B.C. Dowds. 1993. Phase variation in *Xenorhabdus luminescens*: cloning and sequencing of the lipase gene and analysis of its expression in primary and secondary phases of the bacterium. J. Bacteriol. *175*: 1665–1673.

Wang, L., D. Rothemund, H. Curd and P.R. Reeves. 2000. Sequence diversity of the *Escherichia coli* H7 *fliC* genes: implication for a DNA-based typing scheme for *E. coli* O157:H7. J. Clin. Microbiol. *38*: 1786–1790.

Wang, T.H., Y.T. Tchan, A.M.M. Zeman and I.R. Kennedy. 1993. Presence of sodium dependent *Azotobacter* in Australia (New South Wales). Soil Biol. Biochem. *25*: 637–639.

Wang, T.W. and Y.H. Tseng. 1992. Electrotransformation of *Xanthomonas campestris* by RF DNA of filamentous phage fLf. Lett. Appl. Microbiol. *14*: 65–68.

Wang, X.Z., B. Li, P.L. Herman and D.P. Weeks. 1997. A three-component enzyme system catalyzes the O demethylation of the herbicide dicamba in *Pseudomonas maltophilia* DI-6. Appl. Environ. Microbiol. *63*: 1623–1626.

Wang, Y., P.C. Lau and D.K. Button. 1996. A marine oligobacterium harboring genes known to be part of aromatic hydrocarbon degradation pathways of soil pseudomonads. Appl. Environ. Microbiol. *62*: 2169–2173.

Wanke, C.A. and R.L. Guerrant. 1987. Small-bowel colonization alone is a cause of diarrhea. Infect. Immun. *55*: 1924–1926.

Warburton, D.W., K.L. Dodds, R. Burke, M.A. Johnston and P.J. Laffey. 1992. A review of the microbiological quality of bottled water sold in Canada between 1981 and 1989. Can. J. Microbiol. *38*: 12–19.

Ward, C.K. and T.J. Inzana. 1997. Identification and characterization of a DNA region involved in the export of capsular polysaccharide by *Actinobacillus pleuropneumoniae* serotype 5a. Infect. Immun. *65*: 2491–2496.

Ward, J.I., T.F. Tsai, G.A. Filice and D.W. Fraser. 1978. Prevalence of ampicillin- and chloramphenicol-resistant strains of *Haemophilus influenzae* causing meningitis and bacteremia: national survey of hospital laboratories. J. Infect. Dis. *138*: 421–424.

Ward, L.J. and S.H. De Boer. 1989. Characterization of a monoclonal antibody against active pectate lyase from *Erwinia carotovora*. Can. J. Microbiol. *35*: 651–655.

Ward, L.J. and S.H. De Boer. 1990. A DNA probe specific for serologically diverse strains of *Erwinia carotovora*. Phytopathology *80*: 665–669.

Ward, L.J. and S.H. De Boer. 1994. Specific detection of *Erwinia carotovora* subsp. *atroseptica* with a digoxigenin-labeled DNA probe. Phytopathology *84*: 180–186.

Wards, B.J., M.A. Joyce, M. Carman, F. Hilbink and G.W. deLisle. 1998. Restriction endonuclease analysis and plasmid profiling of *Actinobacillus pleuropneumoniae* serotype 7 strains. Vet. Microbiol. *59*: 175–181.

Warren, J.W. 1986. *Providencia stuartii*: a common cause of antibiotic-resistant bacteriuria in patients with long-term indwelling catheters. Rev. Infect. Dis. *8*: 61–67.

Warren, J.W. 1996. Clinical presentations and epidemiology of urinary tract infections. *In* Mobley and Warren (Editors), Urinary Tract Infections. Molecular Pathogenesis and Clinical Management, ASM Press, Washington, D.C. pp. 3–27.

Warren, J.W., J.H. Tenney, J.M. Hoopes, H.L. Muncie and W.C. Anthony. 1982. A prospective microbiologic study of bacteriuria in patients with chronic indwelling urethral catheters. J. Infect. Dis. *146*: 719–723.

Warren, S.H. and W.M. Scott. 1930. A new serological type of *Salmonella*. J. Hyg. *29*: 415–417.

Warren, W.J. and R.D. Miller. 1979. Growth of Legionnaires' disease bacterium (*Legionella pneumophila*) in a chemically defined medium. J. Clin. Microbiol. *10*: 50–55.

Warskow, A.L. and E. Juni. 1972. Nutritional requirements of *Acinetobacter* strains isolated from soil, water, and sewage. J. Bacteriol. *112*: 1014–1016.

Washington, J.A., II, R.J. Birk and R.E. Ritts, Jr.. 1971. Bacteriologic and epidemiologic characteristics of *Enterobacter hafniae* and *Enterobacter liquefaciens*. J. Infect. Dis. *124*: 379–386.

Washington, J.A. and J.A. Timm. 1976. Unclassified, citrate-positive member of the family *Enterobacteriaceae* resembling *Escherichia coli*. J. Clin. Microbiol. *4*: 165–167.

Wassif, C., D. Cheek and R. Belas. 1995. Molecular analysis of a metalloprotease from *Proteus mirabilis*. J. Bacteriol. *177*: 5790–5798.

Watanabe, H. and K.N. Timmis. 1984. A small plasmid in *Shigella dysenteriae* 1 specifies one or more functions essential for O antigen production and bacterial virulence. Infect. Immun. *43*: 391–396.

Watanabe, K., C. Ishikawa, K. Yazawa, K. Kondo and A. Kawaguchi. 1996. Fatty acid and lipid composition of an eicosapentaenoic acid-producing marine bacterium. J. Mar. Biotechnol. *4*: 104–112.

Watanabe, K., K. Yazawa, K. Kondo and A. Kawaguchi. 1997. Fatty acid synthesis of an eicosapentaenoic acid-producing bacterium: *de novo* synthesis, chain elongation, and desaturation systems. J. Biochem. (Tokyo) *122*: 467–473.

Watanabe, N. 1959. On four new halophilic species of *Spirillum*. Bot. Mag. (Tokyo). *72*: 77–86.

Watanabe, T., J. Shima, K. Izaki and T. Sugiyama. 1992a. New polyenic antibiotics active against gram-positive and gram-negative bacteria. VII. Isolation and structure of enacyloxin IVa, a possible biosynthetic intermediate of enacyloxin IIa. J. Antibiot. (Tokyo). *45*: 575–576.

Watanabe, T., T. Sugiyama, K. Chino, T. Suzuki, S. Wakabayashi, H. Hayashi, R. Itami, J. Shima and K. Izaki. 1992b. New polyenic antibiotics active against gram-positive and gram-negative bacteria. VIII. Construction of synthetic medium for production of mono-chloro-congeners of enacyloxins. J. Antibiot. (Tokyo). *45*: 476–484.

Watanabe, T., T. Sugiyama, M. Takahashi, J. Shima, K. Yamashita, K. Izaki, K. Furihata and H. Seto. 1992c. New polyenic antibiotics active against gram-positive and gram-negative bacteria. IV. Structural elucidation of enacyloxin IIa. J. Antibiot. (Tokyo). *45*: 470–475.

Watanakunakorn, C. and J. Weber. 1989. *Enterobacter* bacteremia: A review of 58 episodes. Scand. J. Infect. Dis. *21*: 1–8.

Waterman, S.R. and P.L. Small. 1996. Identification of sigma S-dependent genes associated with the stationary-phase acid-resistance phenotype of *Shigella flexneri*. Mol. Microbiol. *21*: 925–940.

Watson, A.A., R.A. Alm and J.S. Mattick. 1996a. Identification of a gene, *pilF*, required for type 4 fimbrial biogenesis and twitching motility in *Pseudomonas aeruginosa*. Gene *180*: 49–56.

Watson, A.A., J.S. Mattick and R.A. Alm. 1996b. Functional expression of heterologous type 4 fimbriae in *Pseudomonas aeruginosa*. Gene *175*: 143–150.

Watson, J.J. and F.H. White. 1979. Hemolysins of *Edwardsiella tarda*. Can. J. Comp. Med. *43*: 78–83.

Watson, S.W. 1965. Characteristics of a marine, nitrifying bacterium, *Nitrocystis oceanus* sp. nov. Limnol. Oceanogr. (Suppl.) *10*: R274–289.

Watson, S.W. 1971. Taxonomic considerations of the family *Nitrobacteraceae* Buchanan. Request for opinions. Int. J. Syst. Bacteriol. *21*: 254–270.

Watson, S.W., E. Bock, H. Harms, H.P. Koops and A.B. Hooper. 1989. Nitrifying bacteria. *In* Staley, Bryant, Pfennig and Holt (Editors), Bergey's Manual of Systematic Bacteriology, 1st Ed., Vol. 3, The Williams & Wilkins Co., Baltimore. pp. 1808–1833.

Watson, S.W. and C.C. Remsen. 1969. Macromolecular subunits in the walls of marine nitrifying bacteria. Science (Wash. D. C.) *163*: 685–686.

Watson, S.W. and C.C. Remsen. 1970. Cell envelope of *Nitrosocystis oceanus*. J. Ultrastruct. Res. *33*: 148–160.

Watson, S.W. and J.B. Waterbury. 1971. Characteristics of two marine nitrite oxidizing bacteria, *Nitrospina gracilis* nov. gen. nov. sp. and *Nitrococcus mobilis* nov. gen. nov. sp. Arch. Microbiol. *77*: 203–230.

Watt, D.A., V. Bamford and M.E. Nairn. 1970. *Actinobacillus seminis* as a cause of polyarthritis and posthitis in sheep. Aust. Vet. J. *46*: 515.

Wauters, G., S. Aleksic, J. Charlier and G. Schulze. 1991. Somatic and flagellar antigens of *Yersinia enterocolitica* and related species. Contr. Microbiol. Immunol. *12*: 239–243.

Wauters, G., M. Janssens, A.G. Steigerwalt and D.J. Brenner. 1988. *Yersinia mollaretii* sp. nov. and *Yersinia bercovieri* sp. nov., formerly called *Yersinia enterocolitica* biogroups 3a and 3b. Int. J. Syst. Bacteriol. *38*: 424–429.

Wauters, G., K. Kandolo and M. Janssens. 1987. Revised biogrouping scheme of *Yersinia enterocolitica*. Contrib. Microbiol. Immunol. *9*: 14–21.

Wauters, G., L. Le Minor, A. Chalon and J. Lassen. 1972. Supplement au schema antigénique de *Yersinia enterocolitica*. Ann. Inst. Pasteur. *122*: 951–956.

Wayne, L.G. 1994. Actions of the Judicial Commission of the International Committee on Systematic Bacteriology on requests for Opinions published between January 1985 and July 1993. Int. J. Syst. Bacteriol. *44*: 177–178.

Wayne, L.G., D.J. Brenner, R.R. Colwell, P.A.D. Grimont, O. Kandler, M.I. Krichevsky, L.H. Moore, W.E.C. Moore, R.G.E. Murray, E. Stackebrandt, M.P. Starr and H.G. Trüper. 1987. Report of the ad hoc committee on reconciliation of approaches to bacterial systematics. Int. J. Syst. Bacteriol. *37*: 463–464.

Weaver, R.E. and N.J. Ehrenkranz. 1975. Letter: *Vibrio parahaemolyticus* septicemia. Arch. Intern. Med. *135*: 197.

Weaver, R.E. and J.C. Feeley. 1979. Cultural and biochemical characterization of the Legionnaires' disease bacterium. *In* Jones and Hébert (Editors), "Legionnaires' " the Disease, the Bacterium and Methodology, Center for Disease Control, Atlanta. pp. 20–25.

Webb, C.D. and W.J. Payne. 1971. Influence of Na$^+$ on synthesis of macromolecules by a marine bacterium. Appl. Microbiol. *21*: 1080–1088.

Weber, J.T., E.D. Mintz, R. Canizares, A. Semiglia, I. Gomez, R. Sempertegui, A. Davila, K.D. Greene, N.D. Puhr, D.N. Cameron, F.C. Tenover, T.J. Barrett, N.H. Bean, C. Ivey, R.V. Tauxe and P.A. Blake. 1994. Epidemic cholera in Ecuador: Multidrug-resistance and transmission by water and seafood. Epidemiol. Infect. *112*: 1–11.

Webster, C.A., K.J. Towner, H. Humphreys, B. Ehrenstein, D. Hartung and H. Grundmann. 1996. Comparison of rapid automated laser fluorescence analysis of DNA fingerprints with four other computer-assisted approaches for studying relationships between *Acinetobacter baumannii* isolates. J. Med. Microbiol. *44*: 185–194.

Wecke, J. and I. Horbach. 1999. Ultrastructural characterization of *Anaerobiospirillum succiniciproducens* and its differentiation from *Campylobacter* species. FEMS Microbiol. Lett. *170*: 83–88.

Weckesser, J., J.G. Drews and H. Mayer. 1979. Lipopolysaccharides of photosyntetic procaryotes. Annu. Rev. Microbiol. *33*: 215–239.

Weckesser, J., G. Drews, J. Roppel, H. Mayer and I. Fromme. 1974. The lipopolysaccharides (O- antigens) of *Rhodopseudomonas viridis*. Arch. Microbiol. *101*: 233–245.

Weckesser, J., H. Mayer and G. Shulz. 1995. Anoxygenic phototrophic bacteria: model organisms for studies on cell wall macromolecules. *In* Blankenship, Madigan and Bauer (Editors), Anoxygenic Photosynthetic Bacteria, Kluwer Academic Publishing, The Netherlands. pp. 207–230.

Weeks, O.B. 1974. Genus *Flavobacterium* Bergey et al., 1923. *In* Buchanan and Gibbons (Editors), Bergey's Manual of Determinative Bacteriology, 8th ed., The Williams & Wilkins Co., Baltimore. 357–364.

Weeks, O.B., S.M. Beck, M.D. Thomas and H.D. Isenberg. 1962. Pigment of *Flavobacterium piscicida*. J. Bacteriol. *84*: 1118.

Wehler, T. and N.I. Carlin. 1988. Structural and immunochemical studies of the lipopolysaccharide from a new provisional serotype of *Shigella flexneri*. Eur. J. Biochem. *176*: 471–476.

Wei, J., M.B. Goldberg, V. Burland, M.M. Venkatesan, W. Deng, G. Fournier, G.F. Mayhew, G. Plunkett, D.J. Rose, A. Darling, B. Mau, N.T. Perna, S.M. Payne, L.J. Runyen-Janecky, S. Zhou, D.C. Schwartz and F.R. Blattner. 2003. Complete genome sequence and comparative genomics of *Shigella flexneri* serotype 2a strain 2457T. Infect. Immun. *71*: 2775–2786.

Wei, Z.M. and S.V. Beer. 1995. *hrpL* activates *Erwinia amylovora hrp* gene transcription and is a member of the *Ecf* subfamily of sigma factors. J. Bacteriol. *177*: 6201–6210.

Wei, Z.M., B.J. Sneath and S.V. Beer. 1992. Expression of *Erwinia amylovora hrp* genes in response to environmental stimuli. J. Bacteriol. *174*: 1875–1882.

Weidner, S., W. Arnold, E. Stackebrandt and A. Pühler. 2000. Phylogenetic analysis of bacterial communities associated with leaves of the seagrass *Halophila stipulacea* by a culture-independent small-subunit rRNA gene approach. Microb. Ecol. *39*: 22–31.

Weimberg, R. 1962. Studies with a constitutive dehydrogenase in *Pseudomonas fragi*. Biochim. Biophys. Acta *67*: 349–358.

Weiner, R.M., V.E. Coyne, P. Brayton, P. West and S.F. Raiken. 1988. *Alteromonas colwelliana* sp. nov., an isolate from oyster habitats. Int. J. Syst. Bacteriol. *38*: 240–244.

Weinstein, D.L., M.P. Jackson, J.E. Samuel, R.K. Holmes and A.D. O'Brien. 1988. Cloning and sequencing of a Shiga-like toxin type II variant from *Escherichia coli* strain responsible for edema disease of swine. J. Bacteriol. *170*: 4223–4230.

Weisburg, W.G., M.E. Dobson, J.E. Samuel, G.A. Dasch, L.P. Mallavia, O. Baca, L. Mandelco, J.E. Sechrest, E. Weiss and C.R. Woese. 1989. Phylogenetic diversity of the *Rickettsiae*. J. Bacteriol. *171*: 4202–4206.

Weisburg, W.G., C.R. Woese, M.E. Dobson and E. Weiss. 1985. A common origin of rickettsiae and certain plant pathogens. Science *230*: 556–558.

Weiser, J. 1963. Diseases of insects of medical importance in Europe. Bull. WHO *28*: 121–127.

Weiser, J.N., S.T. Chong, D. Greenberg and W. Fong. 1995. Identification and characterization of a cell envelope protein of *Haemophilus influenzae* contributing to phase variation in colony opacity and nasopharyngeal colonization. Mol. Microbiol. *17*: 555–564.

Weiser, J.N., J.M. Love and E.R. Moxon. 1989. The molecular mechanism of phase variation of *H. influenzae* lipopolysaccharide. Cell *59*: 657–665.

Weiser, J.N., D.J. Maskell, P.D. Butler, A.A. Lindberg and E.R. Moxon. 1990. Characterization of repetitive sequences controlling phase variation of *Haemophilus influenzae* lipopolysaccharide. J. Bacteriol. *172*: 3304–3309.

Weiser, J.N., M. Shchepetov and S.T. Chong. 1997. Decoration of lipopolysaccharide with phosphorylcholine: a phase variable characteristic of *Haemophilus influenzae*. Infect. Immun. *65*: 943–950.

Weiser, J. and Z. Zizka. 1968. Electron-microscope studies of *Rickettsiella chironomi* in the midge *Camptochironomus tentans*. J. Invert. Pathol. *12*: 222–230.

Weiss, B.D., M.A. Capage, M. Kessel and S.A. Benson. 1994. Isolation and characterization of a generalized transducing phage for *Xanthomonas campestris* pv. *campestris*. J. Bacteriol. *176*: 3354–3359.

Weiss, E., G.A. Dasch and K.P. Chang. 1984a. Genus *Rickettsiella*. *In* Krieg and Holt (Editors), Bergey's Manual of Systematic Bacteriology, Vol. 1, The Williams & Wilkins Co., Baltimore. pp. 713–717.

Weiss, E., G.A. Dasch and K.P. Chang. 1984b. *In* Validation of the publication of new names and new combinations previously effectively published outside the IJSB. List No. 15. Int. J. Syst. Bacteriol. *34*: 355–357.

Weiss, E., W.F.M. Myers, J. Suitor, E.C. and J.E.M. Neptune. 1962. Respiration of a rickettsialike microorganism, *Wolbachia persica*. J. Infect. Dis. *110*: 155–164.

Weissman, J.P., E.J. Gangarosa and H.L. Dupont. 1973. Changing needs in the antimicrobial therapy of shigellosis. J. Infect. Dis. *127*: 611–613.

Weitzman, P.D.J. 1980. Citrate synthase and succinate thiokinase in classification and identification. *In* Goodfellow and Board (Editors), Microbiological Classification and Identification, Academic Press, London New York. pp. 107–125.

Weitzman, P.D.J. 1991. *Acinetobacter*—citric acid cyclist with a difference. *In* Towner and Bergogne-Bérézin (Editors), The Biology of *Acinetobacter*: Taxonomy, Clinical Importance, Molecular Biology, Physiology, Industrial Relevance, Plenum Press, New York. pp. 313–322.

Weitzman, P.D.J. and D. Jones. 1968. Regulation of citrate synthase and microbial taxonomy. Nature (Lond.) *219*: 270–272.

Welburn, S.C., I. Maudlin and D.S. Ellis. 1987. *In vitro* cultivation of *Rickettsia* like-organisms from *Glossina* spp. Ann. Trop. Med. Parasitol. *81*: 331–335.

Welch, R.A. 1987. Identification of two different hemolysin determinants in uropathogenic *Proteus* isolates. Infect. Immun. *55*: 2183–2190.

Welch, R.A. 1991. Pore-forming cytolysins of Gram-negative bacteria. Mol. Microbiol. *5*: 521–528.

Weldin, J.C. 1927. The colon-typhoid group of bacteria and related forms. Relationships and classification. Iowa State J. Sci. *1*: 121–197.

Wells, J.M. and B.C. Raju. 1984. Cellular fatty acid composition of 6 fastidious, gram-negative, xylem-limited bacteria from plants. Curr. Microbiol. *10*: 231–236.

Wells, J.M., B.C. Raju, H.-Y. Hung, W.G. Weisburg, L. Mandelco-Paul and D.J. Brenner. 1987. *Xylella fastidiosa* gen. nov., sp. nov: Gram-negative, xylem-limited, fastidious plant bacteria related to *Xanthomonas* spp. Int. J. Syst. Bacteriol. *37*: 136–143.

Wells, J.M., B.C. Raju, G. Nyland and S.K. Lowe. 1981. Medium for isolation and growth of bacteria associated with plum (*Prunus salicina*) leaf scald and phony peach (*Prunus persica*) diseases. Appl. Environ. Microbiol. *42*: 357–363.

Wells, J.S., W.H. Trejo, P.A. Principe and R.B. Sykes. 1984. Obafluorin, a novel beta-lactone produced by *Pseudomonas fluorescens* - taxonomy, fermentation and biological properties. J. Antibiot. *37*: 802–803.

Welsh, D.T. and R.A. Herbert. 1993. Identification of organic solutes accumulated by purple and green sulphur bacteria during osmotics stress using natural abundance ^{13}C nuclear magnetic resonance spectroscopy. FEMS Microbiol. Ecol. *13*: 145–150.

Wendt-Potthoff, K., F. Niepold and H. Backhaus. 1992. High-efficiency electrotransformation of the plant pathogen *Pseudomonas syringae* pv. *syringae* R32. J. Microbiol. Methods *16*: 33–37.

Wenger, J.D., D.G. Hollis, R.E. Weaver, C.N. Baker, G.R. Brown, D.J. Brenner and C.V. Broome. 1989. Infection caused by *Francisella philomiragia* (formerly *Yersinia philomiragia*): a newly recognized human pathogen. Ann. Intern. Med. *110*: 888–892.

Wenger, P.N., J.I. Tokars, P. Brennan, C. Samel, L. Bland, M. Miller, L. Carson, M. Arduino, P. Edelstein, S. Aguero, C. Riddle, C. Ohara and W. Jarvis. 1997. An outbreak of *Enterobacter hormaechei* infection and colonization in an intensive care nursery. Clin. Infect. Dis. *24*: 1243–1244.

Weniger, B.G., A.J. Warren, V. Forseth, G.W. Shipps, T. Creelman, J. Gorton and A.M. Barnes. 1984. Human bubonic plague transmitted by a domestic cat scratch. J. Amer. Med. Assoc. *251*: 927–928.

Wenner, J.J. and L.F. Rettger. 1919. A systematic study of the *Proteus* group of bacteria. J. Bacteriol. *4*: 331–353.

Wenzel, R.P., K.J. Hunting, C.A. Osterman and M.A. Sande. 1976. *Providencia stuartii*, a hospital pathogen: potential factors for its emergence and transmission. Am. J. Epidemiol. *104*: 170–180.

Werkman, C.H. and G.F. Gillen. 1932. Bacteria producing trimethylene glycol. J. Bacteriol. *23*: 167–182.

Werner, S.B., C.E. Weidmer, B.C. Nelson, G.S. Nygaard, R.M. Goethals and J.D. Poland. 1984. Primary plague pneumonia contracted from a domestic cat at South Lake Tahoe, Calif. J. Amer. Med. Assoc. *251*: 929–931.

Werren, J.H., S.W. Skinner and A.M. Huger. 1986. Male-killing bacteria in a parasitic wasp. Science *231*: 990–992.

West, M.G. and P.R. Edwards. 1954. The Bethesda-Ballerup group of paracolon bacteria. Public Health Monograph No. 22, U.S.D.H.E.W, Atlanta.

West, P.A. and R.R. Colwell. 1984. Identification and classification of *Vibrionaceae*: an overview. *In* Colwell (Editor), Vibrios in the environment, John Wiley and Sons, Ltd., New York. 285–263.

West, P.A., G.C. Okpokwasili, P.R. Brayton, D.J. Grimes and R.R. Colwell. 1984. Numerical taxonomy of phenanthrene-degrading bacteria isolated from the Chesapeake Bay. Appl. Environ. Microbiol. *48*: 988–993.

Westerman, E.L. and J. McDonald. 1983. Tularemia pneumonia mimicking Legionnaires' disease: isolation of organism on CYE agar and successful treatment with erythromycin. South. Med. J. *76*: 1169–1170.

Westfall, H.N., R.A. Goldwasser, E. Weiss and D. Hussong. 1986. Prevalence of antibodies to *Legionella* species in a series of patients in Israel. Isr. J. Med. Sci. *22*: 131–138.

Wetmore, P.W., J.F. Thiel, Y.F. Herman and J.R. Harr. 1963. Comparison of selected *Actinobacillus* species with a hemolytic variety of *Actinobacillus* from irradiated swine. J. Infect. Dis. *113*: 186–194.

Wetzler, T.F. 1970. Animal diseases transmissible to man, pseudotuberculosis. *In* Bodily, Updyke and Mason (Editors), Diagnostic Procedures for Bacterial, Mycotic, and Parasitic Infections, 5th ed., Amer. Pub. Health Assoc. Inc., New York. 449.

Weyant, R.S., C.W. Moss, R.E. Weaver, D.G. Hollis, J.G. Jordan, E.C. Cook and M.I. Daneshvar (Editors). 1996. Identification of Unusual Pathogenic Gram-negative Aerobic and Facultatively Anaerobic Bacteria, 2nd Ed,, The Williams & Wilkins Co., Baltimore.

Whang, H.Y., M.E. Heller and E. Neter. 1972. Production by *Aeromonas* of common enterobacterial antigen and its possible taxonomic significance. J. Bacteriol. *110*: 161–164.

Wheat, R.W. and M. Pittman. 1960. Degradation of certain nucleotides by *Haemphilus influenzae* and *Haemophilus aegyptius*. J. Bacteriol. *79*: 137–141.

Whitaker, D.R. 1965. Lytic enzymes of *Sorangium* sp.: isolation anf enzymatic properties of the α and β-lytic proteases. Can. J. Biochem. *45*: 991–993.

Whitaker, D.R. 1967. Simplified procedures for production and isolation of the bacteriolytic proteases of *Sorangium* sp. Can. J. Biochem. *45*: 991–993.

Whitaker, D.R. 1970. The α-lytic protease of a myxobacterium. Methods Enzymol. *19*: 599–613.

Whitaker, D.R., F.D. Cook and D.C. Gillespie. 1965a. Lytic enzymes of *Sorangium* sp. Some aspects of enzyme production in submerged culture. Can. J. Biochem. *43*: 1927–1933.

Whitaker, D.R., L. Jurasek and K.L. Roy. 1966. The nature of the bacteriolytic proteases of *Sorangium* sp. Biochem. Biophys. Res. Commun. *24*: 173–178.

Whitaker, D.R. and K.L. Roy. 1967. Concerning the nature of the α- and β-lytic proteases of *Sorangium* sp. Can. J. Biochem. *45*: 911–916.

Whitaker, D.R., K.L. Roy, C.S. Tsai and L. Jurasek. 1965b. Lytic enzymes of *Sorangium* sp. A comparison of the proteolytic properties of the α and β-lytic proteases. Can. J. Biochem. *43*: 1961–1970.

Whitaker, R.J., G.S. Byng, R.L. Gherna and R.A. Jensen. 1981a. Comparative allostery of 3-deoxy-D-arabino-heptulosonate 7-phosphate synthetase as an indicator of taxonomic relatedness in pseudomonad genera. J. Bacteriol. *145*: 752–759.

Whitaker, R.J., G.S. Byng, R.L. Gherna and R.A. Jensen. 1981b. Diverse enzymological patterns of phenylalanine biosynthesis in pseudomonad bacteria are conserved in parallel with DNA/DNA homology groupings. J. Bacteriol. *147*: 526–534.

Whitby, J.L. and G.G. Muir. 1961. Bacteriological studies of urinary tract infection. Br. J. Urol. *33*: 130–134.

Whitchurch, C.B., R.A. Alm and J.S. Mattick. 1996. The alginate regulator AlgR and an associated sensor FimS are required for twitching motility in *Pseudomonas aeruginosa*. Proc. Natl. Acad. Sci. U. S. A. *93*: 9839–9843.

White, A.H. 1940. A bacterial discoloration of print butter. Sci. Agr. *20*: 638–645.

White, B. 1926. Further studies of the *Salmonella* Group. Med. Res. Comm. Spec. Rep. *103*: 3–160.

White, D.C. 1963. Respiratory systems in the hemin-requiring *Haemophilus* species. J. Bacteriol. *85*: 84–96.

White, D.C. 1966. The obligatory involvement of the electron transport system in the catabolic metabolism of *Haemophilus parainfluenzae*. Antonie van Leeowenhoek J. Microbiol. Serol. *32*: 139–158.

White, D.C. and S. Granick. 1963. Hemin biosynthesis in *Haemophilus*. J. Bacteriol. *85*: 842–850.

White, D.C. and P.R. Sinclair. 1971. Branched electron transport systems in bacteria. Adv. Microb. Physiol. *5*: 173–211.

White, F.H., F.C. Neal, C.F. Simpson and A.F. Walsh. 1969. Isolation of *Edwardsiella tarda* from an ostrich and an Australian skink. J. Am. Vet. Med. Assoc. *155*: 1057–1058.

White, F.H., C.F. Simpson and L.E. Williams, Jr.. 1973. Isolation of *Edwardsiella tarda* from aquatic animal species and surface waters in Florida. J. Wildlife Dis. *9*: 204–208.

White, H.E. 1930. Bacterial spot of radish and turnip. Phytopathology *20*: 653–662.

White, J.N. and M.P. Starr. 1971. Glucose fermentation end products of *Erwinia* spp. and other enterobacteria. J. Appl. Bacteriol. *34*: 459–475.

White, P.J., I.S. Hunter and C.A. Fewson. 1991. Codon usage in *Acinetobacter* structural genes. *In* Towner and Bergogne-Bérézin (Editors), The Biology of *Acinetobacter*: Taxonomy, Clinical Importance, Molecular Biology Physiology, Industrial Relevance, Plenum Press, New York. pp. 251–257.

Whitehead, L.F. and A. Douglas. 1993. A metabolic study of *Buchnera*, the intracellular bacterial symbionts of the pea aphid *Acyrthosiphon pisum*. J. Gen. Microbiol. *139*: 821–826.

Whiteley, G.R., J.L. Penner, I.O. Stewart, P.C. Stokan and N.A. Hinton. 1977. Nosocomial urinary tract infections caused by two O-serotypes of *Providencia stuartii* in one hospital. J. Clin. Microbiol. *6*: 551–554.

Whitfield, C. 1995. Biosynthesis of lipopolysaccharide O antigens. Trends Microbiol. *3*: 178–185.

Whitfield, C., I.W. Sutherland and R.E. Cripps. 1981. Surface polysaccharides in mutants of *Xanthomonas campestris*. J. Gen. Microbiol. *124*: 385–392.

Whitfield, C., I.W. Sutherland and R.E. Cripps. 1982. Glucose metabolism in *Xanthomonas campestris*. J. Gen. Microbiol. *128*: 981–986.

Whittam, T.S. 1996. Genetic variation and evolutionary processes in natural populations of *Escherichia coli*. *In* Neidhardt, Curtiss, Ingraham, Lin, Low, Magasanik, Reznikoff, Riley, Schaecter and Umbarger (Editors), *Escherichia coli* and *Salmonella*: Cellular and Molecular Biology, 2nd Ed., ASM Press, Washington, D.C. pp. 2708–2720.

Whittam, T.S., H. Ochman and R.K. Selander. 1983. Multilocus genetic structure in natural populations of *Escherichia coli*. Proc. Natl. Acad. Sci. U.S.A. *80*: 1751–1955.

Whittam, T.S., M.L. Wolfe, I.K. Wachsmuth, F. Ørskov, I. Ørskov and R.A. Wilson. 1993. Clonal relationships among *Escherichia coli* strains that cause hemorrhagic colitis and infantile diarrhea. Infect. Immun. *61*: 1619–1629.

Whittenbury, R.A., S.L. Davies and J.F. Davey. 1970a. Exospores and cysts formed by methane-utilizing bacteria. J. Gen. Microbiol. *61*: 219–226.

Whittenbury, R.A. and N.R. Krieg. 1984a. Family IV. *Methylococcaceae*. *In* Krieg and Holt (Editors), Bergey's Manual of Systematic Bacteriology, 1st Ed., Vol. 1, The Williams & Wilkins Co., Baltimore. pp. 256–261.

Whittenbury, R.A. and N.R. Krieg. 1984b. *In* Validation of the publication of new names and new combinations previously effectively published outside the IJSB. List No. 15. Int. J. Syst. Bacteriol. *34*: 355–356.

Whittenbury, R.A., K.C. Phillips and J.F. Wilkinson. 1970b. Enrichment, isolation and some properties of methane-utilizing bacteria. J. Gen. Microbiol. *61*: 205–218.

WHO. 1997. Vaccine research and development. New strategies for accelerating *Shigella* vaccine development. Wkly. Epidemiol. Rec. *72*: 73–79.

WHO. 1999. Zoonotic non-O157 Shiga toxin-producing *Escherichia coli* (STEC). Report of a WHO scientific working group meeting. Berlin, Germany, 1–28.

Widdel, F. and F. Bak. 1992. Gram negative mesophilic sulfate-reducing bacteria. *In* Balows, Trüper, Dworkin, Harder and Schleifer (Editors), The Prokaryotes. A Handbook on the Biology of Bacteria: Ecophysiology, Isolation, Identification, Applications, 2nd ed., Vol. 4, Springer Verlag, New York. pp. 3352-3378.

Widdel, F. and N. Pfennig. 1981. Studies on dissimilatory sulfate-reducing bacteria that decompose fatty acids. I. Isolation of new sulfate-reducing bacteria with acetate from saline environments. Description of *Desulfobacter postgatei* gen. nov., sp. nov. Arch. Microbiol. *129*: 395–400.

Wiehe, P.O. and W.J. Dowson. 1953. A bacterial disease of cassava (*Manihot utilissima*) in Nyasaland. Emp. J. Exp. Agr. *21*: 141–143.

Wiessner, W. 1981. The family *Beggiatoaceae*. *In* Starr, Stolp, Trüper, Balows and Schlegel (Editors), The Prokaryotes. A Handbook on Habitats, Isolation, and Identification of Bacteria, Springer-Verlag, Berlin. 380–389.

Wigglesworth, V.B. 1929. Digestion in the tsetse fly: a study of structure and function. Parasitology. *21*: 288–321.

Wiklund, T. and I. Dalsgaard. 1998. Occurrence and significance of atypical *Aeromonas salmonicida* in non-salmonid and salmonid fish species: a review. Dis. Aquat. Org. *32*: 49–69.

Wilcox, M.H., T.G. Winstanley and R.C. Spencer. 1994. Outer membrane protein profiles of *Xanthomonas maltophilia* isolates displaying temperature-dependent susceptibility to gentamicin. J. Antimicrob. Chemother. *33*: 663–666.

Wilkie, P.J., D.W. Dye and D.R.W. Watson. 1973. Further hosts of *Pseudomonas viridiflava*. N.Z. J. Agr. Res. *16*: 315–323.

Wilkinson, H.W. and B.J. Brake. 1982. Formalin-killed versus heat-killed *Legionella pneumophila* serogroup 1 antigen in the indirect immunofluorescence assay for legionellosis. J. Clin. Microbiol. *16*: 979–981.

Wilkinson, H.W., D.D. Bruce and C.V. Broome. 1981. Validation of *Legionella pneumophila* indirect immunofluorescence assay with epidemic sera. J. Clin. Microbiol. *13*: 139–146.

Wilkinson, H.W., V. Drasar, W.L. Thacker, R.F. Benson, J. Schindler, B. Potuznikova, W.R. Mayberry and D.J. Brenner. 1988a. *Legionella moravica*, sp. nov., and *Legionella brunensis*, sp. nov., isolated from cooling-tower water. Ann. Inst. Pasteur Microbiol. *139*: 393–402.

Wilkinson, H.W., V. Drasar, W.L. Thacker, R.F. Benson, J. Schindler, B. Potuznikova, W.R. Mayberry and D.J. Brenner. 1989. *In* Validation of the publication of new names and new combinations previously effectively published outside the IJSB. List No. 29. Int. J. Syst. Bacteriol. *39*: 205–206.

Wilkinson, H.W., C.E. Farshy, B.J. Fikes, D.D. Cruce and L.P. Yealy. 1979a. Measure of immunoglobulin G-, M-, and A-specific titers against *Legionella pneumophila* and inhibition of titers against nonspecific, gram-negative bacterial antigens in the indirect immunofluorescence test for legionellosis. J. Clin. Microbiol. *10*: 685–689.

Wilkinson, H.W. and B.J. Fikes. 1980. Slide agglutination test for serogrouping Legionella pneumophila and atypical *Legionella*-like organisms. J. Clin. Microbiol. *11*: 99–101.

Wilkinson, H.W., B.J. Fikes and D.D. Cruce. 1979b. Indirect immunofluorescence test for serodiagnosis of Legionnaires' disease: evidence for serogroup diversity of Legionnaires' disease bacterial antigens and for multiple specificity of human antibodies. J. Clin. Microbiol. *9*: 379–383.

Wilkinson, H.W., A.L. Reingold, B.J. Brake, D.L. McGiboney, G.W. Gorman and C.V. Broome. 1983a. Reactivity of serum from patients with suspected legionellosis against 29 antigens of *Legionellaceae* and *Legionella*-like organisms by indirect immunofluorescence assay. J. Infect. Dis. *147*: 23–31.

Wilkinson, H.W., W.L. Thacker, R.F. Benson, S.S. Polt, E. Brookings, W.R. Mayberry, D.J. Brenner, R.G. Gilley and J.K. Kirklin. 1987. *Legionella birminghamensis*, sp. nov., isolated from a cardiac transplant recipient. J. Clin. Microbiol. *25*: 2120–2122.

Wilkinson, H.W., W.L. Thacker, R.F. Benson, S.S. Polt, E. Brookings, W.R. Mayberry, D.J. Brenner, R.G. Gilley and J.K. Kirklin. 1988b. *In* Validation of the publication of new names and new combinations previously effectively published outside the IJSB. List No. 25. Int. J. Syst. Bacteriol. *38*: 220–222.

Wilkinson, H.W., W.L. Thacker, D.J. Brenner and K.J. Ryan. 1985a. Fatal *Legionella maceachernii* pneumonia. J. Clin. Microbiol. *22*: 1055.

Wilkinson, H.W., W.L. Thacker, A.G. Steigerwalt, D.J. Brenner, N.M. Ampel and E.J. Wing. 1985b. Second serogroup of *Legionella hackeliae* isolated from a patient with pneumonia. J. Clin. Microbiol. *22*: 488–489.

Wilkinson, I.J., N. Sangster, R.M. Ratcliff, P.A. Mugg, D.E. Davos and J.A. Lanser. 1990. Problems associated with identification of *Legionella* species from the environment and isolation of six possible new species. Appl. Environ. Microbiol. *56*: 796–802.

Wilkinson, S.G. 1968a. Glycosyl diglycerides from *Pseudomonas rubescens*. Biochim. Biophys. Acta *164*: 148–156.

Wilkinson, S.G. 1968b. Studies on the cell walls of *Pseudomonas* spp. resistant to ethylene diaminetetra-acetic acid. J. Gen. Microbiol. *54*: 195–213.

Wilkinson, S.G. 1970. Cell walls of *Pseudomonas* species sensitive to ethylenediaminetetra-acetic acid. J. Bacteriol. *104*: 1035–1044.

Wilkinson, S.G. 1972. Composition and structure of the ornithine-containing lipid from *Pseudomonas rubescens*. Biochim. Biophys. Acta *270*: 1–17.

Wilkinson, S.G. 1983. Composition and structure of lipopolysaccharides from *Pseudomonas aeruginosa*. Rev. Infect. Dis. *5*: S941–S949.

Wilkinson, S.G. 1988. Gram-negative bacteria. *In* Wilkinson (Editor), Microbial Lipids, Vol. 1, Academic Press, London. pp. 299–408.

Wilkinson, S.G. and P.F. Caudwell. 1980. Lipid compostion and chemotaxonomy of *Pseudomonas putrefaciens* (*Alteromonas putrefaciens*). J. Gen. Microbiol. *118*: 329–341.

Wilkinson, S.G., L. Galbraith and W.J. Anderton. 1983b. Lipopolysaccharides from *Pseudomonas maltophilia*: composition of the lipopolysaccharide and structure of the side-chain polysaccharide from strain N.C.I.B. 9204. Carbohydr. Res. *112*: 241–252.

Wilkinson, S.G., L. Galbraith and G.A. Lightfoot. 1973. Cell walls, lipids, and lipopolysaccharides of *Pseudomonas* species. Eur. J. Biochem. *33*: 158–174.

Willems, A., M. Gillis, K. Kersters, L. Van den Broecke and J. De Ley. 1987. Transfer of *Xanthomonas ampelina* Panagopoulos 1969 to a new genus, *Xylophilus* gen. nov., as *Xylophilus ampelinus* (Panagopoulos 1969) comb. nov. Int. J. Syst. Bacteriol. *37*: 422–430.

Willems, A., M. Goor, S. Thielemans, M. Gillis, K. Kersters and J. De Ley. 1992. Transfer of several phytopathogenic *Pseudomonas* species to *Acidovorax* as *Acidovorax avenae* subsp. *avenae* subsp. nov., comb. nov., *Acidovorax avenae* subsp. *citrulli*, *Acidovorax avenae* subsp. *cattleyae*, and *Acidovorax konjaci*. Int. J. Syst. Bacteriol. *42*: 107–119.

Willems, A., B. Pot, E. Falsen, P. Vandamme, M. Gillis, K. Kersters and J. De Ley. 1991. Polyphasic taxonomic study of the emended genus *Comamonas*: Relationship to *Aquaspirillum aquaticum*, E. Falsen group 10, and other clinical isolates. Int. J. Syst. Bacteriol. *41*: 427–444.

Willems, H., C. Jager and G. Baljer. 1998. Physical and genetic map of the obligate intracellular bacterium *Coxiella burnetii*. J. Bacteriol. *180*: 3816–3822.

Willems, H., D. Thiele, C. Burger, M. Ritter, W. Oswald and H. Krauss. 1996. Molecular biology of *Coxiella burnetii*. *In* Kazar and Toman (Editors), Rickettsiae and Rickettsial Diseases, Slovak Academy of Sciences, Bratislava, Slovakia. pp. 363–378.

William, F. and A. Mahadevan. 1980. Degradation of aromatic compounds by *Xanthomonas* species. Z. Pflanzenkr. Pflanzenschutz. *87*: 738–744.

Williams, F.D. 1973. Abolition of swarming of *Proteus* by *p*-nitrophenyl glycerin: application to blood agar media. Appl. Microbiol. *25*: 751–754.

Williams, F.D. and R.H. Schwarzhoff. 1978. Nature of the swarming phenomenon in *Proteus*. Annu. Rev. Microbiol. *32*: 101–122.

Williams, J.E. 1965. Paratyphoid and Arizona infections. *In* Beister and Schwarte (Editors), Diseases of Poultry, 5th Ed., Iowa State University Press, Ames. pp. 260–328.

Williams, J.E. 1983. Warning on a new potential for laboratory-acquired infections as a result of the new nomenclature for the plague bacillus. Bull. World Health Organ. *61*: 545–548.

Williams, J.D., S. Kattan and P. Cavanagh. 1974. Letter: Penicillinase production by *Haemophilus influenzae*. Lancet 2: 103.

Williams, J.C., L.A. Thomas and M.G. Peacock. 1986a. Humoral immune response to Q fever : Enzyme-linked immunosorbent assay antibody response to *Coxiella burnetii* in experimentally infected guinea pigs. J. Clin. Microbiol. *24*: 935–939.

Williams, M.A. and S.C. Rittenberg. 1956. Microcyst formation and germination in *Spirillum lunatum*. J. Gen. Microbiol. *15*: 205–209.

Williams, M.A. and S.C. Rittenberg. 1957. A taxonomic study of the genus *Spirillum* Ehrenberg. Int. Bull. Bacteriol. Nomencl. Taxon. *7*: 49–111.

Williams, P.H., T.J. Baldwin and S. Knutton. 1997. Enteropathogenic *Escherichia coli*. *In* Sussman (Editor), *Escherichia coli*: Mechanisms of Virulence, Cambridge University Press, Cambridge. pp. 403–420.

Williams, P., H. Chart, E. Griffiths and P. Stevenson. 1987. Expression of high affinity iron uptake systems by clinical isolates of *Klebsiella*. FEMS Microbiol. Lett. *44*: 407–412.

Williams, P., P.A. Lambert, M.R. Brown and R.J. Jones. 1983. The role of the O and K antigens in determining the resistance of *Klebsiella aerogenes* to serum killing and phagocytosis. J. Gen. Microbiol. *129*: 2181–2191.

Williams, P., P.A. Lambert, C.G. Haigh and M.R. Brown. 1986b. The influence of the O and K antigens of *Klebsiella aerogenes* on surface hydrophobicity and susceptibility to phagocytosis and antimicrobial agents. J. Med. Microbiol. *21*: 125–132.

Williams, P. and J.M. Tomas. 1990. The pathogenicity of *Klebsiella pneumoniae*. Rev. Med. Microbiol. *1*: 196–204.

Williams, R.P. and S.M.H. Qadri. 1980. The pigment of *Serratia*. *In* von Graevenitz and Rubin (Editors), The genus *Serratia*, CRC Press, Boca Raton. pp. 31–75.

Williams, T.M. 1985. The Biology of *Thiothrix* spp. and Eikelboom Type 021N Bacteria, Pennsylvania State University, University Park.

Williams, T.M. and R.F. Unz. 1985a. Filamentous sulfur bacteria of activated sludge: characterization of *Thiothrix*, *Beggiatoa* and Eikelboom type 021N strains. Appl. Environ. Microbiol. *49*: 887–898.

Williams, T.M. and R.F. Unz. 1985b. Isolation and characterization of filamentous bacteria present in bulking activated-sludge. Appl. Microbiol. Biotechnol. *22*: 273–282.

Williams, T.M. and R.F. Unz. 1989. The nutrition of *Thiothrix*, type 021N, *Beggiatoa* and *Leucothrix* strains. Water Res. *23*: 15–22.

Williamson, G.M. and K. Zinnemann. 1951. The occurrence of two distinct capsular antigens in *Haemophilus influenzae* type e strains. J. Pathol. Bacteriol. *63*: 695–698.

Williamson, G.M. and K. Zinnemann. 1954. The degradation of *Haemophilus influenzae* type e capsular antigens. J. Pathol. Bacteriol. *68*: 453–457.

Willis, J.W., J.K. Engwall and A.K. Chatterjee. 1987. Cloning of genes for *Erwinia carotovora* carotovora pectolytic enzymes and further characterization of the polygalacturonases. Phytopathology 77: 1199–1205.

Willshaw, G.A., T. Cheasty, H.R. Smith, S.J. O'Brien and G.K. Adak. 2001. Verocytotoxin-producing *Escherichia coli* (VTEC) O157 and other

VTEC from human infections in England and Wales: 1995–1998. J. Med. Microbiol. *50*: 135–142.

Wilson, E.E. and A.R. Magie. 1964. Systemic invasion of the host plant by the tumor inducing bacterium *Pseudomonas savastanoi*. Phytopathology *54*: 576–579.

Wilson, E.E., M.P. Starr and J.A. Berger. 1957. Bark canker, a bacterial disease of the Persian walnut tree. Phytopathology *47*: 669–673.

Wilson, E.E., F.M. Zeitoun and D.L. Fredrickson. 1967. Bacterial phloem canker, a new disease of Persian walnut trees. Phytopathology *57*: 618–621.

Wilson, G.S. and A.A. Miles. 1964. Topley and Wilson's Principles of Bacteriology and Immunity, 5th Ed., The Williams & Wilkins Co., Baltimore.

Wilson, G.S. and A.A. Miles. 1975. Topley and Wilson's Principles of Bacteriology, Virology and Immunity, , E. Arnold Ltd., London.

Wilson, J.P., R.R. Waterer, J.D. Wofford, Jr. and S.W. Chapman. 1989a. Serious infections with *Edwardsiella tarda*. A case report and review of the literature. Arch. Intern. Med. *149*: 208–210.

Wilson, K.H., R. Blitchington, P. Shah, G. McDonald, R.D. Gilmore and L.P. Mallavia. 1989b. Probe directed at a segment of *Rickettsia rickettsii* rRNA amplified with polymerase chain reaction. J. Clin. Microbiol. *27*: 2692–2696.

Wilson, M.E. 1991. The heat-modifiable outer membrane protein of *Actinobacillus actionmycetemcomitans*: relationship to OmpA proteins. Infect. Immun. *59*: 2505–2507.

Wilson, M., H.A.S. Epton and D.C. Sigee. 1992. Interactions between *Erwinia herbicola* and *Erwinia amylovora* on the stigma of Hawthorn blossoms. Phytopathology *82*: 914–918.

Wilson, M. and S.E. Lindow. 1993. Interactions between the biological control agent *Pseudomonas fluorescens* A506 and *Erwinia amylovora* in pear blossoms. Phytopathology *83*: 117–123.

Wilson, M.A., M.J. Morgan and G.E. Barger. 1993. Comparison of DNA fingerprinting and serotyping for identification of avian *Pasteurella multocida* isolates. J. Clin. Microbiol. *31*: 255–259.

Wilson, R.G. and L.M. Henderson. 1963. Tryptophan-niacin relationship in *Xanthomonas pruni*. J. Bacteriol. *85*: 221–229.

Wilson, S.A., S.J.M. Wachira, R.A. Norman, L.H. Pearl and R.E. Drew. 1996. Transcription antitermination regulation of the *Pseudomonas aeruginosa* amidase operon. Embo J. *15*: 5907–5916.

Wilson, W.J., M. Wiedmann, H.R. Dillard and C.A. Batt. 1994. Identification of *Erwinia stewartii* by a ligase chain reaction assay. Appl. Environ. Microbiol. *60*: 278–284.

Windsor, H.M., R.C. Gromkova and H.J. Koornhof. 1991. Plasmid-mediated NAD independence in *Haemophilus parainfluenzae*. J. Gen. Microbiol. *137*: 2415–2421.

Wingard, M., G.R. Matsueda and R.S. Wolfe. 1972. *Myxobacter* AL-1 protease II: Specific peptide bond cleavage on the amino side of lysine. J. Bacteriol. *112*: 940–949.

Wink, M. 1979. The endosymbionts of *Glossina morsitans* and *Glossina palpalis*: cultivation and experiments and some physiological properties. Acta Trop. *36*: 215–222.

Winn, A.M., C.T. Miles and S.G. Wilkinson. 1996. Structure of the O3 antigen of *Stenotrophomonas* (*Xanthomonas* or *Pseudomonas*) *maltophilia*. Carbohydr. Res. *282*: 149–156.

Winn, A.M., A.W. Miller and S.G. Wilkinson. 1995. Structure of the O10 antigen of *Stenotrophomonas* (*Xanthomonas*) *maltophilia*. Carbohydr. Res. *267*: 127–133.

Winn, A.M. and S.G. Wilkinson. 1995. Structure of the O6 antigen of *Stenotrophomonas* (*Xanthomonas* or *Pseudomonas*) *maltophilia*. Carbohydr. Res. *272*: 225–230.

Winn, A.M. and S.G. Wilkinson. 1996. Structure of the O20 antigen of *Stenotrophomonas* (*Xanthomonas* or *Pseudomonas*) *maltophilia*. Carbohydr. Res. *294*: 109–115.

Winn, A.M. and S.G. Wilkinson. 1997. Structure of the O2 antigen of *Stenotrophomonas* (*Xanthomonas* or *Pseudomonas*) *maltophilia*. Carbohydr. Res. *298*: 213–217.

Winn, W.C., Jr., G.S. Davis, D.W. Gump, J.E. Graighead and H.N. Beaty.

1982. Legionnaires' pneumonia after intratracheal inoculation of guinea pigs and rats. Lab. Investig. *47*: 568–578.

Winn, W.C., Jr. and R.L. Myerowitz. 1981. The pathology of the *Legionella* pneumonias. A review of 74 cases and the literature. Hum. Pathol. *12*: 401–422.

Winogradsky, S. 1887. Über Schwefelbakterien. Bot. Ztg. *45*: 489–610.

Winogradsky, S. 1888. Beiträge zur Morphologie und Physiologie der Bakterium. Heft 1. Zur Morphologie und Physiologie der Schwefelbacterien, Arthur Felix, Leipzig. pp. 1–120.

Winogradsky, S. 1892. Contributions a la morphologie des organismes de la nitrification. Arch. Sci. Biol. (St. Petersb.) *1*: 86–137.

Winogradsky, S. 1904. Die Nitrifikation. *In* Handbuch de Technischen Mykologie, Lafar, Jena. 132–181.

Winogradsky, S. 1929. Études sur la microbiologie du sol. Sur la degradation da la cellulose dans la sol. Ann. Inst. Pasteur. *43*: 529–633.

Winogradsky, S. 1938. Sur la morphologie et l'écologie des *Azotobacter*. Ann. Inst. Pasteur (Paris) *60*: 351–400.

Winslow, C.-E.A., J. Broadhurst, R.E. Buchanan, C.J. Krumwiede, L.A. Rogers and G.H. Smith. 1917. The families and genera of the bacteria. Preliminary report of the Committee of the Society of American Bacteriologists on characterization and classification of bacterial types. J. Bacteriol. *2*: 506–566.

Winslow, C.-E.A., J. Broadhurst, R.E. Buchanan, C.J. Krumwiede, L.A. Rogers and G.H. Smith. 1920. The families and genera of the Bacteria. Final report of the Committee of the Society of American Bacteriologists on characterization and classification of bacterial types. J. Bacteriol. *5*: 191–229.

Winslow, C.-E.A., I.J. Kliger and W. Rothberg. 1919. Studies on the classification of the colon-typhoid group of bacteria with special reference to their fermentative reactions. J. Bacteriol. *4*: 429–503.

Winstanley, C., M.A. Coulson, B. Wepner, J.A.W. Morgan and C.A. Hart. 1996. Flagellin gene and protein variation amongst clinical isolates of *Pseudomonas aeruginosa*. Microbiology *142*: 2145–2151.

Winstanley, C., J.A.W. Morgan, R.W. Pickup and J.R. Saunders. 1994. Molecular-cloning of 2 *Pseudomonas* flagellin genes and basal body structural genes. Microbiology *140*: 2019–2031.

Winstanley, C., J.A.W. Morgan, J.R. Saunders and R.W. Pickup. 1991. Plasmid stability and the expression and regulation of a marker gene in *Acinetobacter* and other gram-negative hosts. *In* Towner and Bergogne-Bérézin (Editors), The Biology of *Acinetobacter*: Taxonomy, Clinical Importance, Molecular Biology, Physiology, Industrial Relevance, Plenum Press, New York. pp. 169–181.

Winstanley, C., S.C. Taylor and P.A. Williams. 1987. pWW174: a large plasmid from *Acinetobacter calcoaceticus* encoding benzene catabolism by the β-ketoadipate pathway. Mol. Microbiol. *1*: 219–227.

Wirsen, C.O., T. Brinkhoff, J. Kuever, G. Muyzer, S. Molyneaux and H.W. Jannasch. 1998. Comparison of a new *Thiomicrospira* strain from the Mid-Atlantic Ridge with known hydrothermal vent isolates. Appl. Environ. Microbiol. *64*: 4057–4059.

Wirsen, C.O., H.W. Jannasch and S.J. Molyneaux. 1993. Chemosynthetic microbial activity at Mid-Atlantic Ridge hydrothermal vent sites. J. Geophys. Res. *98*: 9693–9703.

Wise, E.M., Jr., S.P. Alexander and M. Powers. 1973. Adenosine 3':5'-cyclic monophosphate as a regulator of bacterial transformation. Proc. Natl. Acad. Sci. U. S. A. *70*: 471–474.

Wise, M.G., J.V. McArthur and L.J. Shimkets. 1999. Methanotroph diversity in landfill soil: isolation of novel type I and type II methanotrophs whose presence was suggested by culture-independent 16S ribosomal DNA analysis. Appl. Environ. Microbiol. *65*: 4887–4897.

Wise, M.G., J.V. McArthur and L.J. Shimkets. 2001. *Methylosarcina fibrata* gen. nov., sp. nov. and *Methylosarcina quisquiliarum* sp. nov., novel type I methanotrophs. Int. J. Syst. Evol. Microbiol. *51*: 611–621.

Wislouch, S.M. 1912. *Thioploca ingrica* nov. sp. Ber. Dtsch. Bot. Ges. *30*: 470–474.

Wittke, J.W., S. Aleksic and H.H. Wuthe. 1985. Isolation of *Moellerella wisconsensis* from an infected human gallbladder. Eur. J. Clin. Microbiol. *4*: 351–352.

Wiwat, C., M. Lertcanawanichakul, P. Siwayapram, S. Pantuwatana and

A. Bhumiratana. 1996. Expression of chitinase-encoding genes from *Aeromonas hydrophila* and *Pseudomonas maltophilia* in *Bacillus thuringiensis* subsp. *israelensis*. Gene *179*: 119–126.

Wodzinski, R.S., T.E. Umholtz, J.R. Rundle and S.V. Beer. 1994. Mechanisms of inhibition of *Erwinia amylovora* by *Erw. herbicola* in vitro and in vivo. J. Appl. Bacteriol. *76*: 22–29.

Woese, C.R. 1987. Bacterial evolution. Microbiol. Rev. *51*: 221–271.

Woese, C.R., P. Blanz, R.B. Hespell and C.M. Hahn. 1982. Phylogenetic relationships among various helical bacteria. Curr. Microbiol. *7*: 119–124.

Woese, C.R., W.G. Weisburg, C.M. Hahn, B.J. Paster, L.B. Zablen, B.J. Lewis, T.J. Macke, W. Ludwig and E. Stackebrandt. 1985. The phylogeny of purple bacteria: the gamma subdivision. Syst. Appl. Microbiol. *6*: 25–33.

Woese, C.R., W.G. Weisburg, B.J. Paster, C.M. Hahn, R.S. Tanner, N.R. Krieg, H.P. Koops, H. Harms and E. Stackebrandt. 1984. The phylogeny of purple bacteria: the beta subdivision. Syst. Appl. Microbiol. *5*: 327–336.

Wolf, F.A. and A.C. Foster. 1917. Bacterial leaf spot of tobacco. Science (Wash. D.C.). *46*: 361–362.

Wolin, H.L. 1963. Defined medium for *Haemophilus influenzae* type b. J. Bacteriol. *85*: 253–254.

Wong, J.D., M.A. Miller and J.M. Janda. 1989. Surface properties and ultrastructure of *Edwardsiella* species. J. Clin. Microbiol. *27*: 1797–1801.

Wong, P.K. and C.M. So. 1993. Copper accumulation by a strain of *Pseudomonas putida*. Microbios. *73*: 113–121.

Wong, S.H., D.R. Cullimore and D.L. Bruce. 1985. Selective medium for the isolation and enumeration of *Klebsiella* spp. Appl. Environ. Microbiol. *49*: 1022–1024.

Wong, T.Y., L. Graham, E. O'Hara and R.J. Maier. 1986. Enrichment for hydrogen-oxidizing *Acinetobacter* spp. in the rhizosphere of hydrogen-evolving soybean root nodules. Appl. Environ. Microbiol. *52*: 1008–1013.

Wong, V.K. 1987. Broviac catheter infection with *Kluyvera cryocrescens*: A case report. J. Clin. Microbiol. *25*: 1115–1116.

Wong, W.C. and T.F. Preece. 1979. Identification of *Pseudomonas tolaasi* - white line in agar and mushroom tissue block rapid pitting tests. J. Appl. Bacteriol. *47*: 401–407.

Wood, A.P. and D.P. Kelly. 1985. Physiological characteristics of a new thermophilic, obligately chemolithotrophic *Thiobacillus* species, *Thiobacillus tepidarius*. Int. J. Syst. Bacteriol. *35*: 434–437.

Wood, A.P. and D.P. Kelly. 1986. Chemolithotrophic metabolism of the newly isolated moderately thermophilic, obligately autotrophic *Thiobacillus tepidarius*. Arch. Microbiol. *144*: 71–77.

Wood, A.P. and D.P. Kelly. 1989. Isolation and physiological characterisation of *Thiobacillus thyasiris* sp. nov., a novel marine facultative autotroph and the putative symbiont of *Thyasira flexuosa*. Arch. Microbiol. *152*: 160–166.

Wood, A.P. and D.P. Kelly. 1991. Isolation and characterization of *Thiobacillus halophilus* sp. nov., a sulfur-oxidizing autotrophic eubacterium from a western Australian hypersaline lake. Arch. Microbiol. *156*: 277–280.

Wood, A.P. and D.P. Kelly. 1993. Reclassification of *Thiobacillus thyasiris* as *Thiomicrospira thyasirae*, new combination, an organism exhibiting pleomorphism in response to environmental conditions. Arch. Microbiol. *159*: 45–47.

Wood, A.P. and D.P. Kelly. 1995. *In* Validation of the publication of new names and new combinations previously effectively published outside the IJSB. List No. 53. Int. J. Syst. Bacteriol. *15*: 418–419.

Woods, C.R., Jr., J. Versalovic, T. Koeuth and J.R. Lupski. 1992. Analysis of relationships among isolates of *Citrobacter diversus* by using DNA fingerprints generated by repetitive sequence-based primers in the polymerase chain reaction. J. Clin. Microbiol. *30*: 2921–2929.

Woods, D.E., D.C. Straus, W.G. Johanson, V.K. Berry and J.A. Bass. 1980. Role of pili in adherence of *Pseudomonas aeruginosa* to mammalian buccal epithelial-cells. Infect. Immun. *29*: 1146–1151.

Woodward, B.W., M. Carter and R.J. Seidler. 1979. Most nonclinical *Klebsiella* strains are not *K. pneumoniae sensu stricto*. Curr. Microbiol. *2*: 181–185.

Woodward, E.J. and K. Robinson. 1990. An improved formulation and method of preparation of crystal violet pectate medium for detection of pectolytic erwinia. Lett. Appl. Microbiol. *10*: 171–173.

Woodward, L.M., A.R. Bielke, J.F. Eisses and P.A. Ketchum. 1990. Occurrence of nitrate reductase and molybdopterin in *Xanthomonas maltophilia*. Appl. Environ. Microbiol. *56*: 3766–3771.

World Health Organization, Programme for Control of Diarrhoeal Diseases. 1983. Manual for Laboratory Investigations of Acute Enteric Diseases, World Health Organization, Geneva. 113 pp.

Wormald, H. 1930. Bacterial diseases of stone fruit trees in Britain. II. Bacterial shoot wilt of plum trees. Ann. Appl. Biol. *17*: 725–744.

Wormald, H. 1931. Bacterial diseases of stone fruit trees in Britain. III. The symptoms of bacterial canker in plum trees. J. Pomol. Hortic. Sci. *9*: 239– 256.

Wozny, M.A., M.P. Bryant, L.V. Holdeman and W.E.C. Moore. 1977. Urease assay and urease-producing species of anaerobes in the bovine rumen and human feces. Appl. Environ. Microbiol. *33*: 1097–1104.

Wray, C., I.M. McLaren and P.J. Carroll. 1993. *Escherichia coli* isolated from farm animals in England and Wales between 1986 and 1991. Vet. Rec. *133*: 439–442.

Wray, C. and M.J. Woodward. 1997. *Escherichia coli* infections in farm animals. *In* Sussman (Editor), *Escherichia coli*: Mechanisms of virulence, Cambridge University Press, Cambridge. pp. 49–84.

Wright, A.C. and J.G. Morris, Jr.. 1991. The extracellular cytolysin of *Vibrio vulnificus*: Inactivation and relationship to virulence in mice. Infect. Immun. *59*: 192–197.

Wright, J.H. 1895. Report on the results of an examination of the water supply of Philadelphia. Mem. Nat. Acad. Sci. 7: 422–484.

Wright, S.A., C.H. Zumoff, L. Schneider and S.V. Beer. 2001. *Pantoea agglomerans* strain EH318 produces two antibiotics that inhibit *Erwinia amylovora in vitro*. Appl. Environ. Microbiol. *67*: 284–292.

Wu, G.H., S. Hill, M.J.S. Kelly, G. Sawers and R.K. Poole. 1997a. The *cydR* gene product, required for regulation of cytochrome *bd* expression in the obligate aerobe *Azotobacter vinelandii*, is an Fnr-like protein. Microbiology-(UK). *143*: 2197–2207.

Wu, S.S., J. Wu and D. Kaiser. 1997b. The *Myxococcus xanthus pilT* locus is required for social gliding motility although pili are still produced. Mol. Microbiol. *23*: 109–121.

Wu, W.C., H.F. Kuo, Y.K. Hsueh and L.Y. Wang. 1990. A simple method for maintaining *Xanthomonas campestris* pv. *citri*. Chinese Journal Of Microbiology And Immunology. *23*: 162–165.

Wüst, J., J. Gubler, W. Mannheim and A. von Graevenitz. 1991. *Actinobacillus hominis* as a causative agent of septicemia in hepatic failure. Eur. J. Clin. Microbiol. Infect. Dis. *10*: 693–694.

Wyatt, L.E., R. Nickelson and C. Vanderzant. 1979. *Edwardsiella tarda* in freshwater catfish and their environment. Appl. Environ. Microbiol. *38*: 710–714.

Xilinas, M.E., J.T. Papavassiliou and N.J. Legakis. 1975. Selective medium for growth of *Proteus*. J. Clin. Microbiol. *2*: 459–460.

Xu, J., S. Lohrke, I.M. Hurlbert and R.E. Hurlbert. 1989. Transformation of *Xenorhabdus nematophilus*. Appl. Environ. Microbiol. *55*: 806–812.

Xu, J., M.E. Olson, M.L. Kahn and R.E. Hurlbert. 1991. Characterization of Tn5-induced mutants of *Xenorhabdus nematophilus* ATCC 19061. Appl. Environ. Microbiol. *57*: 1173–1180.

Xu, X.J., M.R. Ferguson, V.L. Popov, C.W. Houston, J.W. Peterson and A.K. Chopra. 1998. Role of a cytotoxic enterotoxin in *Aeromonas*-mediated infections: development of transposon and isogenic mutants. Infect. Immun. *66*: 3501–3509.

Xu, Y., Y. Nogi, C. Kato, Z.Y. Liang, H.J. Ruger, D. De Kegel and N. Glansdorff. 2003. *Psychromonas profunda* sp nov., a psychropiezophilic bacterium from deep Atlantic sediments. Int. J. Syst. Evol. Microbiol. *53*: 527–532.

Yadav, A.S. and S.S. Verma. 1998. Occurrence of enterotoxigenic *Aeromonas* in poultry eggs and meat. J. Food Sci. Technol. *35*: 169–170.

Yaeger, M.J. 1996. An outbreak of *Actinobacillus suis* septicemia in grow-/finish pigs. J. Vet. Diagn. Investig. *8*: 381–383.

Yague, M.A., C.H. Castro and A.C. Arranz. 1989. Isolation of Koserella trabulsii from clinical samples (Letter in Spanish). Enfermedades Infecciosas y Microbiologia Clinica. 7: 393–394.

Yakimov, M.M. , L. Giuliano, G. Gentile, E. Crisafi, T.N. Chernikova, W.R. Abraham, H. Lunsdorf, K.N. Timmis and P.N. Golyshin. 2003. Oleispira antarctica gen. nov., sp nov., a novel hydrocarbonoclastic marine bacterium isolated from Antarctic coastal sea water. Int.J. Syst. Evol. Microbiol. 53: 779–785.

Yakimov, M.M., P.N. Golyshin, S. Lang, E.R.B. Moore, W.R. Abraham, H. Lunsdorf and K.N. Timmis. 1998. Alcanivorax borkumensis gen. nov., sp. nov., a new, hydrocarbon-degrading and surfactant-producing marine bacterium. Int. J. Syst. Bacteriol. 48: 339–348.

Yakrus, M. and N.W. Schaad. 1979. Serological relationships among strains of Erwinia chrysanthemi. Phytopathology 69: 517–522.

Yakubu, D.E., D.C. Old and B.W. Senior. 1989. The haemagglutinins and fimbriae of Proteus penneri. J. Med. Microbiol. 30: 279–284.

Yale, M.W. 1939a. Genus Escherichia Castellani and Chalmers. In Breed, Murray and Hitchens (Editors), Bergey's Manual of Determinative Bacteriology, 5th Ed., The Williams & Wilkins Co., Baltimore. pp. 389–396.

Yale, M.W. 1939b. The genus Proteus. In Bergey, Breed, Murray and Hitchens (Editors), Bergey's Manual of Determinative Bacteriology, 5th Ed.., The Williams & Wilkins Co., Baltimore. pp. 430–436.

Yamada, Y., Y. Okada and K. Kondo. 1976. Isolation and characterization of "polarly flagellated intermediate strains" in acetic acid bacteria. J. Gen. Appl. Microbiol. 22: 237–245.

Yamada, Y., H. Takinaminakamura, Y. Tahara, H. Oyaizu and K. Komagata. 1982. Significance of the ubiquinone and menaquinone systems in the classification of Gram-negative and Gram-positive bacteria VIII. The ubiquinone systems in the strains of Pseudomonas species. J. Gen. Appl. Microbiol. 28: 7–12.

Yamai, S., T. Okitsu, T. Shimada and Y. Katsube. 1997. Distribution of serogroups of Vibrio cholerae non-O1 non-O139 with specific reference to their ability to produce cholera toxin, and addition of novel serogroups. J. Jpn Assoc. Infect. Dis. 71: 1037–1045.

Yamamoto, M., H. Iwaki, K. Kouno and T. Inui. 1980. Identification of marine methanol-utilizing bacteria. J. Ferment. Technol. 58: 99–106.

Yamamoto, S., P.J.M. Bouvet and S. Harayama. 1999. Phylogenetic structures of the genus Acinetobacter based on gyrB sequences: comparison with the grouping by DNA–DNA hybridization. Int. J. Syst. Bacteriol. 49: 87–95.

Yamamoto, S., M.A.R. Chowdhury, M. Kuroda, T. Nakano, Y. Koumoto and S. Shinoda. 1991a. Further study on polyamine compositions in Vibrionaceae. Can. J. Microbiol. 37: 148–153.

Yamamoto, S. and S. Harayama. 1996. Phylogenetic analysis of Acinetobacter strains based on the nucleotide sequences of gyrB genes and on the amino acid sequences of their products. Int. J. Syst. Bacteriol. 46: 506–511.

Yamamoto, S. and S. Harayama. 1998. Phylogenetic relationships of Pseudomonas putida strains deduced from the nucleotide sequences of gyrB, ropD and 16S rRNA genes. Int. J. Syst. Bacteriol. 48: 813–819.

Yamamoto, S., N. Okujo and Y. Sakakibara. 1994. Isolation and structure elucidation of acinetobactin, a novel siderophore from Acinetobacter baumannii. Arch. Microbiol. 162: 249–254.

Yamamoto, T., G.B. Nair and Y. Takeda. 1995. Emergence of tetracycline resistance due to a multiple drug resistance plasmid in Vibrio cholerae O139. FEMS Immunol. Med. Microbiol. 11: 131–136.

Yamamoto, Y., T.W. Klein and H. Friedman. 1991b. Legionella pneumophila growth in macrophages from susceptible mice is genetically controlled. Proc. Soc. Exp. Biol. Med. 196: 405–409.

Yamanaka, S., A. Hagiwara, Y. Nishimura, H. Tanabe and N. Ishibashi. 1992. Biochemical and physiological characteristics of Xenorhabdus species, symbiotically associated with entomopathogenic nematodes including Steinernema kushidai and their pathogenicity against Spodoptera litura (Lepidoptera: Noctuidae). Arch. Microbiol. 158: 387–393.

Yamanaka, T. 1964. Identity of Pseudomonas cytochrome oxidase with Pseudomonas nitrite reductase. Nature (Lond.) 204: 253–255.

Yamasaki, S., Z. Lin, H. Shirai, A. Terai, Y. Oku, H. Ito, M. Ohmura, T. Karasawa, T. Tsukamoto, H. Kurazono and Y. Takeda. 1996. Typing of verotoxins by DNA colony hybridization with poly- and oligonucleotide probes, a bead-enzyme-linked immunosorbent assay, and polymerase chain reaction. Microbiol. Immunol. 40: 345–352.

Yamashita, M., Y. Hida, S. Suzuki and F. Gejyo. 1997. Genetic analysis of Pseudomonas aeruginosa by pulsed field gel electrophoresis. Kansenshogaku Zasshi. 71: 607–613..

Yamazaki, E., J. Ishii, K. Sato and T. Nakae. 1989. The barrier function of the outer membrane of Pseudomonas maltophilia in the diffusion of saccharides and β-lactam antibiotics. FEMS Microbiol. Lett. 51: 85–88.

Yang, C.H., H.R. Azad and D.A. Cooksey. 1996a. A chromosomal locus required for copper resistance, competitive fitness, and cytochrome c biogenesis in Pseudomonas fluorescens. Proc. Natl. Acad. Sci. U. S. A. 93: 7315–7320.

Yang, C.H., J.A. Menge and D.A. Cooksey. 1993a. Role of copper resistance in competitive survival of Pseudomonas fluorescens in soil. Appl. Environ. Microbiol. 59: 580–584.

Yang, C.H., T.G. Young, M.Y. Peng and M.C. Weng. 1996b. Clinical spectrum of Pseudomonas putida infection. J. Formos. Med. Assoc. 95: 754–761.

Yang, J., X.Z. Wang, D.S. Hage, P.L. Herman and D.P. Weeks. 1994. Analysis of dicamba degradation by Pseudomonas maltophilia using high-performance capillary electrophoresis. Anal. Biochem. 219: 37–42.

Yang, P., P. De Vos, K. Kersters and J. Swings. 1993b. Polyamine patterns as chemotaxonomic markers for the genus Xanthomonas. Int. J. Syst. Bacteriol. 43: 709–714.

Yang, P., P. Rott, L. Vauterin, B. Hoste, P. Baudin, K. Kersters and J. Swings. 1993c. Intraspecific variability of Xanthomonas albilineans. Syst. Appl. Microbiol. 16: 420–426.

Yang, P., L. Vauterin, M. Vancanneyt, J. Swings and K. Kersters. 1993d. Application of fatty acid methyl esters for the taxonomic analysis of the genus Xanthomonas. Syst. Appl. Microbiol. 16: 47–71.

Yang, S.-E., F.H. Lin and T.-T. Kuo. 1975. The utilization of exogenously supplied nucleotide by Xanthomonas oryzae. Bot. Bull. Acad. Sinica. 16: 61–65.

Yang, Y., L.P. Yeh, Y. Cao, L. Baumann, P. Baumann, J.S.E. Tang and B. Beaman. 1983a. Characterization of marine luminous bacteria isolated off the coast of China and description of Vibrio orientalis sp nov. Curr. Microbiol. 8: 95–100.

Yang, Y., L.P. Yeh, Y. Cao, L. Baumann, P. Baumann, J.S.E. Tang and B. Beaman. 1983b. In Validation of new names and new combinations previously effectively published outside the IJSB. List No. 11. Int. J. Syst. Bacteriol. 33: 672–674.

Yano, T., A.F. Pestana de Castro, J.A. Lauritis and T. Namekata. 1979. Serological differentiation of bacteria belonging to the Xanthomonas campestris group by indirect haemagglutination test. Ann. Phytopathol. Soc. Japan 45: 1–8.

Yano, Y., A. Nakayama and K. Yoshida. 1997. Distribution of polyunsaturated fatty acids in bacteria present in intestines of deep-sea fish and shallow-sea poikilothermic animals. Appl. Environ. Microbiol. 63: 2572–2577.

Yanofsky, C. 1954. The absence of a tryptophan-niacin relationship in Escherichia coli and Bacillus subtilis. J. Bacteriol. 68: 577–584.

Yao, J.D., M. Louie, L. Louie, J. Goodfellow and A.E. Simor. 1995. Comparison of E test and agar dilution for antimicrobial susceptibility testing of Stenotrophomonas (Xanthomonas) maltophilia. J. Clin. Microbiol. 33: 1428–1430.

Yaphe, W. and B. Baxter. 1955. The enzymic hydrolysis of carrageenin. Appl. Microbiol. 3: 380–383.

Yaping, J., L. Xiaoyang and Y. Jiaqi. 1990. Saccharobacter fermentatus gen. nov., sp. nov., a new ethanol-producing bacterium. Int. J. Syst. Bacteriol. 40: 412–414.

Yasumoto, T., D. Yasumura, M. Yotsu, T. Michishita, A. Endo and Y. Kotaki. 1986. Bacterial production of tetrodotoxin and anhydrotetrodotoxin. Agric. Biol. Chem. 50: 793–795.

Yates, M.G. and F.O. Campbell. 1989. The effect of nutrient limitation on the competition between an H_2-uptake hydrogenase positive (Hup$^+$) recombinant strain of *Azotobacter chroococcum* and the Hup$^-$ mutant parent in mixed populations. J. Gen. Microbiol. *135*: 221–226.

Yavzori, M., D. Cohen, R. Wasserlauf, R. Ambar, G. Rechavi and S. Ashkenazi. 1994. Identification of *Shigella* species in stool specimens by DNA amplification of different loci of the *Shigella* virulence plasmid. Eur. J. Clin. Microbiol. Infect. Dis. *13*: 232–237.

Yayanos, A.A., A.S. Dietz and R. Van Boxtel. 1979. Isolation of a deep-sea barophilic bacterium and some of its growth characteristics. Science *205*: 808–810.

Yayanos, A.A., A.S. Dietz and R. Van Boxtel. 1982. Dependence of reproduction rate on pressure as a hallmark of deep-sea bacteria. Appl. Environ. Microbiol. *44*: 1356–1361.

Yazawa, K. 1996. Production of eicosapentaenoic acid from marine bacteria. Lipids. *31*: S297–S300.

Yi, H., Y.H. Chang, H.W. Oh, K.S. Bae and J. Chun. 2003. *Zooshikella ganghwensis* gen. nov., sp nov., isolated from tidal flat sediments. Int. J. Syst. Evol. Microbiol. *53*: 1013–1018.

Yii, K., T. Yang and K.K. Lee. 1997. Isolation and characterization of *Vibrio carchariae*, a causative agent of gastroenteritis in the groupers, *Epinephelus coioides*. Curr. Microbiol. *35*: 109–115.

Yirgou, D. 1964. *Xanthomonas guizotiae* sp. nov. on *Guizatia abyssinica*. Phytopathology *54*: 1490–1491.

Yirgou, D. and J.F. Bradbury. 1968. Bacterial wilt of enset (*Ensete ventricosum*) incited by *Xanthomonas musacearum*. Phytopathology *58*: 111–112.

Yogev, R. and S. Koszlowski. 1990. Peritonitis due to *Kluyvera ascorbata*: Case report and review. Rev. Infect. Dis. *12*: 399–402.

Yohe, S., J.T. Fishbain and M. Andrews. 1997. *Shewanella putrefaciens* abscess of the lower extremity. J. Clin. Microbiol. *35*: 3363.

Yokoyama, M.T. and J.R. Carlson. 1981. Production of skatole and *p*-cresol by a rumen *Lactobacillus* sp. Appl. Environ. Microbiol. *41*: 71–76.

Yoon, J.H., S.H. Choi, K.C. Lee, Y.H. Kho, K.H. Kang and Y.H. Park. 2001. *Halomonas marisflavae* sp. nov., a halophilic bacterium isolated from the Yellow Sea in Korea. Int. J. Syst. Evol. Microbiol. *51*: 1171–1177.

Yorifuji, T., T. Kobayashi, A. Tabuchi, Y. Shiritani and K. Yonaha. 1983. Distribution of amidinohydrolases among *Pseudomonas* and comparative studies of some purified enzymes by one-dimensional peptide-mapping. Agr. Biol. Chem. *47*: 2825–2830.

Yoshida, A., M. Izuta, K. Ito, Y. Kamio and K. Izaki. 1991. Cloning and characterization of the pectate lyase III gene of *Erwinia carotovora* Er. Agric. Biol. Chem. *55*: 933–940.

Yoshii, H. and S. Takimoto. 1928. Bacterial leafspot of castor bean and its pathogen. J. Plant Prot. *15*: 12–18.

Yoshikawa, K., T. Takadera, K. Adachi, M. Nishijima and H. Sano. 1997. Korormicin, a novel antibiotic specifically active against marine gram-negative bacteria, produced by a marine bacterium. J. Antibiot. (Tokyo). *50*: 949–953.

You, K.M. and Y.K. Park. 1998. Cd^{2+} removal by *Azomonas agilis* PY101, a cadmium accumulating strain in continuous aerobic culture. Biotechnol. Lett. *20*: 1157–1159.

Young, G.M., D. Amid and V.L. Miller. 1996a. A bifunctional urease enhances survival of pathogenic *Yersinia enterocolitica* and *Morganella morganii* at low pH. J. Bacteriol. *178*: 6487–6495.

Young, J.M. 1970. Drippy gill: a bacterial disease of cultivated mushrooms caused by *Pseudomonas agarici* n. sp. N. Z. J. Agr. Res. *13*: 977–990.

Young, J.M. 1978. Survival of bacteria on Prunus leaves. Proc. 4th Internat. Conf Plant Path. Bact., *Vol. II*: pp. 779–786.

Young, J.M., J.F. Bradbury, R.E. Davis, R.S. Dickey, G.L. Ercolani, A.C. Hayward and A.K. Vidaver. 1991a. Nomenclatural revisions of plant pathogenic bacteria and list of names 1980–1988. Rev. Plant Path. *70*: 211–221.

Young, J.M., J.F. Bradbury, L. Gardan, R.I. Gvozdyak, D.E. Stead, Y. Takikawa and A.K. Vidaver. 1991b. Comment on the reinstatement of *Xanthomonas citri* (ex Hasse 1915) Gabriel et al. 1989 and *Xanthomonas phaseoli* (ex Smith 1897) Gabriel et al. 1989: indication of the need

for minimal standards for the genus *Xanthomonas*. Int. J. Syst. Bacteriol. *41*: 172–177.

Young, J.M., D.W. Dye, J.F. Bradbury, C.G. Panagopoulos and C.F. Robbs. 1978. A proposed nomenclature and classification for plant pathogenic bacteria. N. Z. J. Agric. Res. *21*: 153–177.

Young, J.M., G.S. Saddler, Y. Takikawa, S.H. De Boer, L. Vauterin, L. Gardan, R.I. Gvozdyak and D.E. Stead. 1996b. Names of plant pathogenic bacteria 1864–1995. Rev. Plant Path. *75*: 721–763.

Young, J.M. and C.M. Triggs. 1994. Evaluation of determinative tests for pathovars of *Pseudomonas syringae* Van Hall 1902. J. Appl. Bacteriol. *77*: 195–207.

Young, V.M., D.M. Kenton, B.J. Hobbs and M.R. Moody. 1971. *Levinea*, a new genus of the family *Enterobacteriaceae*. Int. J. Syst. Bacteriol. *21*: 58–63.

Yousfi, A., C. Louis and G. Meynadier. 1979. An ultrastructural study of the transformation of elementary bodies of *Rickettsiella* into initial bodies. Experientia (Basel) *35*: 1175–1176.

Yu, J. and J.B. Kaper. 1992. Cloning and characterization of the *eae* gene of enterohaemorrhagic *Escherichia coli* O157:H7. Mol. Microbiol. *6*: 411–417.

Yu, X. and D. Raoult. 1994. Serotyping *Coxiella burnetii* isolates from acute and chronic Q fever patients by using monoclonal antibodies. FEMS Microbiol. Lett. *117*: 15–19.

Yuen, G.Y., A.M. Alvarez, A.A. Benedict and K.J. Trotter. 1987. Use of monoclonal antibodies to monitor the dissemination of *Xanthomonas campestris* pv. *campestris*. Phytopathology *77*: 366–370.

Yum, D.Y., Y.P. Lee and J.G. Pan. 1997. Cloning and expression of a gene cluster encoding three subunits of membrane-bound gluconate dehydrogenase from *Erwinia cypripedii* ATCC 29267 in *Escherichia coli*. J. Bacteriol. *179*: 6566–6572.

Yumoto, I.I., H. Iwata, T. Sawabe, K. Ueno, N. Ichise, H. Matsuyama, H. Okuyama and K. Kawasaki. 1999a. Characterization of a facultatively psychrophilic bacterium, *Vibrio rumoiensis* sp. nov., that exhibits high catalase activity. Appl. Environ. Microbiol. *65*: 67–72.

Yumoto, I.I., H. Iwata, T. Sawabe, K. Ueno, N. Ichise, H. Matsuyama, H. Okuyama and K. Kawasaki. 1999b. *In* Validation of new names and new combinations previously effectively published outside the IJSB. List No. 70. Int. J. Syst. Evol. Microbiol. *49*: 935–936.

Yumoto, I., K. Kawasaki, H. Iwata, H. Matsuyama and H. Okuyama. 1998. Assignment of *Vibrio* sp. strain ABE-1 to *Colwellia maris* sp. nov., a new psychrophilic bacterium. Int. J. Syst. Bacteriol. *4*: 1357–1362.

Yun, Y.C., Y.S. Kim and Y.S. Cho. 1995. Isolation and characterization of transposon WKm-mediated nonpathogenic mutants of *Xanthomonas campestris* pv. *vesicatoria*. Korean J. Plant Pathol. *11*: 265–270.

Zablotowicz, R.M., R.E. Hoagland, M.A. Locke and W.J. Hickey. 1995. Glutathione-S-transferase activity and metabolism of glutathione conjugates by rhizosphere bacteria. Appl. Environ. Microbiol. *61*: 1054–1060.

Zaccardelli, M., C. Bazzi, J.F. Chauveau and S. Paillard. 1993. Sero-diagnosis of *Xylella fastidiosa* in grapevine xylem extracts. Phytopathol. Mediterr. *32*: 174–181.

Zagallo, A.C. and C.H. Wang. 1967. Comparative glucose catabolism of *Xanthomonas* species. J. Bacteriol. *93*: 970–975.

Zaharija, I., Z. Modrić, T. Naglić and F. Sanković. 1979. Spontaneous infection with *Actinbacillus lignieresii* in horses: study of the first case in Croatia. Vet. Arh. *49*: 105–112.

Zahorchak, R.J., W.T. Charnetzky, R.V. Little and R.R. Brubaker. 1979. Consequences of Ca^{2+} deficiency on macromolecular synthesis and adenylate energy charge in *Yersinia pestis*. J. Bacteriol. *139*: 792–799.

Zahr, M., B. Fobel, H. Mayer, J.F. Imhoff, P.V. Campos and J. Weckesser. 1992. Chemical composition of the lipopolysaccharides of *Ectothiorhodospira shaposhnikovii*, *Ectothiorhodospira mobilis*, and *Ectothiorhodospira halophila*. Arch. Microbiol. *157*: 499–504.

Zakharenko, V.I., L.V. Komissarova, B.P. Ulanov and A.G. Skavronskaia. 1985. Properties of the plasmid pFT15/10-1 isolated from the vaccine strain of *Francisella tularensis*. Mol. Gen. Mikrobiol. Virusol. *5*: 19–22.

Zakhariev, Z.A. 1971. *Plesiomonas shigelloides* isolated from sea water. J. Hyg. Epidemiol. Microbiol. Immunol. (Prague) *15*: 402–404.

Zambon, J.J. 1985. *Actinobacillus actinomycetemcomitans* in human periodontal disease. J. Clin. Periodontol. *12*: 1–20.

Zambon, J.J., J. Slots and R.J. Genco. 1983. Serology of oral *Actinobacillus actinomycetemcomitans* and serotype distribution in human periodontal disease. Infect. Immun. *41*: 19–27.

Zambon, J.J., T. Umemoto, E. De Nardin, F. Nakazawa, L.A. Christersson and R.J. Genco. 1988. *Actinobacillus actinomycetemcomitans* in the pathogenesis of human periodontal disease. Adv. Dent. Res. *2*: 269–274.

Zamze, S.E. and E.R. Moxon. 1987. Composition of the lipopolysaccharide from different capsular serotype strains of *Haemophilus influenzae*. J. Gen. Microbiol. *133*: 1443–1451.

Zaremba, M. and E. Aldová. 1979. Sensitivity to chemotherapeutics of *Yersinia enterocolitica* strains. Arch. Immunol. Therap. Exp. *27*: 847–852.

Zarett, A.J. and R.N. Doetsch. 1949. A new selective medium for quantitative determination of members of the genus *Proteus* in milk. J. Bacteriol. *57*: 266.

Zarnke, R.L. and L. Schlater. 1988. *Actinobacillosis* in free-ranging snowshoe hares (*Lepus americanus*) from Alaska. J. Wildlife Dis. *24*: 176–177.

Zavarzin, G.A. and A.N. Nozhevnikova. 1977. Aerobic carboxydobacteria. Microb. Ecol. *3*: 305–326.

Zeiter, J.H., D.D. Koch, D.W. Parke and R.L. Font. 1989. Endogenous endophthalmitis with lenticular abscess caused by *Enterobacter agglomerans* (*Erwinia* species). Ophthalmic Surg. *20*: 9–12.

Zeitoun, F.M. and E.E. Wilson. 1969. The relation of bacteriophage to the walnut-tree pathogens, *Erwinia nigrifluens* and *Erwinia rubrijaciens*. Phytopathology *59*: 756–761.

Zelinskaia, N.V., N.P. Kovalevskaia, N.N. Matvienko, L.A. Zheleznaia and N.I. Matvienko. 1996. Novel site-specific endonuclease from *Acinetobacter* species M strain. Biokhimiya *61*: 1471–1182.

Zhang, G.F., C.T. Lu, X.C. Shen, K.R. Wang and H.Q. Wang. 1996. Effects of antibiotic Zhongshenmycin on the incidence of *Xanthomonas oryzae* in rice seedlings. Plant Prot. *22*: 6–8.

Zhang, W.L., M. Bielaszewská, A. Liesegang, H. Tschäpe, H. Schmidt, M. Bitzan and H. Karch. 2000. Molecular characteristics and epidemiological significance of Shiga toxin-producing *Escherichia coli* O26 strains. J. Clin. Microbiol. *38*: 2134–2140.

Zhang, Y. and K. Geider. 1997. Differentiation of *Erwinia amylovora* strains by pulsed-field gel electrophoresis. Appl. Environ. Microbiol. *63*: 4421–4426.

Zhao, G.S., C. Pijoan, K.W. Choi, S.K. Maheswaran and E. Trigo. 1995. Expression of iron-regulated outer-membrane proteins by porcine strains of *Pasteurella multocida*. Can. J. Vet. Res.-Rev. Can. Rech. Vet. *59*: 46–50.

Zhao, H., X. Li, D.E. Johnson and H.L. Mobley. 1999. Identification of protease and rpoN-associated genes of uropathogenic *Proteus mirabilis* by negative selection in a mouse model of ascending urinary tract infection. Microbiology *145*: 185–195.

Zhao, X.J. and K. McEntee. 1990. DNA sequence analysis of the *recA* genes from *Proteus vulgaris*, *Erwinia carotovora*, *Shigella flexneri* and *Escherichia coli* B/r. Mol. Gen. Genet. *222*: 369–376.

Zheng, L.M., V.L. Cash, D.H. Flint and D.R. Dean. 1998. Assembly of iron-sulfur clusters — Identification of an *iscSUA–hscBA–fdx* gene cluster from *Azotobacter vinelandii*. J. Biol. Chem. *273*: 13264–13272.

Zherebilo, O.K. and R.I. Gvozdyak. 1976. Decarboxylation of amino acids by bacteria of the genus *Erwinia* under aerobic conditions. Mikrobiol. Zh. *38*: 3–8.

Zhivotchenko, A.G., E.S. Nikonova and M.H. Jorgensen. 1995a. Copper effect on the growth kinetics of Methylococcus capsulatus (Bath). Biotechnol. Tech. *9*: 163–168.

Zhivotchenko, A.G., E.S. Nikonova and M.H. Jorgensen. 1995b. Effect of fermentation conditions on N_2 fixation by *Methylococcus capsulatus*. Bioprocess Eng. *14*: 9–15.

Zhu, C., J. Harel, F. Dumas and J.M. Fairbrother. 1995. Identification of EaeA protein in the outer membrane of attaching and effacing *Escherichia coli* O45 from pigs. FEMS Microbiol. Lett. *129*: 237–242.

Zide, N., J. Davis and N.J. Ehrenkranz. 1974. Fulminating *Vibrio parahem-olyticus* septicemia. A syndrome of erythemia multiforme, hemolytic anemia, and hypotension. Arch. Intern. Med. *133*: 479–481.

Ziegler, M., M. Lange and W. Dott. 1990. Isolation and morphological and cytological characterization of filamentous bacteria from bulking-sludge. Water Res. *24*: 1437–1451.

Zielke, R., A. Schmidt and K. Naumann. 1993. Comparison of different serological methods for the detection of the fire blight pathogen, *Erwinia amylovora* (Burrill) Winslow et. al. Zentbl. Mikrobiol. *148*: 379–391.

Ziemke, F., I. Brettar and M.G. Höfle. 1997. Stability and diversity of the genetic structure of a *Shewanella putrefaciens* population in the water column of the central baltic. Aquat Microb. Ecol. *13*: 63–74.

Ziemke, F., M.G. Höfle, J. Lalucat and R. Rosselló-Mora. 1998. Reclassification of *Shewanella putrefaciens* Owen's genomic group II as *Shewanella baltica* sp. nov. Int. J. Syst. Bacteriol. *48*: 179–186.

Zighelboim, J., T.W. Williams, M.W. Bradshaw and R.L. Harris. 1992. Successful medical management of a patient with multiple hepatic abscesses due to *Edwardsiella tarda*. Clin. Infect. Dis. *14*: 117–120.

Zimmerman, S.E., M.L.V. French, S.D. Allen, E. Wilson and R.B. Kohler. 1982. Immunoglobulin M antibody titers in the diagnosis of Legionnaires' disease. J. Clin. Microbiol. *16*: 1007–1011.

Zimmermann, O.E.R.. 1890. Die Bakterien unserer Trink- und Nutzwässer insbesondere des Wassers der Chemnitzer Wasserleitung. Elfter Bericht. Naturwiss. Ges. Chemnitz, pp. 53–154.

Zimmermann, T. 1964. Untersuchungen über die Actinobazillose des Schweines. I Mitteilung: Isolierung und Characterisierung der Erreger. Dtsch. Tieraerztl. Wochenschr. *71*: 457–461.

Zinder, N.D. 1957. Lysogenic conversion in *S. typhimurium*. Science *126*: 1237.

Zink, D.L., J.C. Feeley, J.G. Wells, C. Vanderzant, J.C. Vickery, W.D. Roof and G.A. O'Donovan. 1980. Plasmid-mediated tissue invasiveness in *Yersinia enterocolitica*. Nature (Lond.) *283*: 224–226.

Zink, R.T., J.K. Engwall, J.L. McEvoy and A.K. Chatterjee. 1985. *recA* is required in the induction of pectin lyase and carotovoricin in *Erwinia carotovora* subsp. *carotovora*. J. Bacteriol. *164*: 390–396.

Zinnemann, K., K.B. Rogers, J. Frazer and J.M.H. Boyce. 1968. A new V-dependent *Haemophilus* species preferring increased CO_2 tension for growth and named *Haemophilus paraphrophilus*, nov. sp. J. Pathol. Bacteriol. *96*: 413–419.

Zinnemann, K., K.B. Rogers, J. Frazer and S.K. Devaraj. 1971. A haemolytic V-dependent CO_2-preferring *Haemophilus* species *Haemophilus paraphrohae-molyticus* nov. spec. J. Med. Microbiol. *4*: 139–143.

ZoBell, C.E. 1941. Studies on marine bacteria. I. The cultural requirements of heterotrophic aerobes. J. Mar. Res. *4*: 99–106.

ZoBell, C.E. and H.C. Upham. 1944. A list of marine bacteria including descriptions of sixty new species. Bull. Scripps Inst. Oceanogr. Univ. Calif. *5*: 239–292.

Zon, G. and J.D. Robbins. 1983. 31P- and 13C-n.m.r.-spectral and chemical characterization of the end-group and repeating-unit components of oligosaccharides derived by acid hydrolysis of *Haemophilus influenzae* type b capsular polysaccharide. Carbohydr. Res. *114*: 103–21.

Zoon, K.C. and J.J. Scocca. 1975. Constitution of the cell envelope of *Haemophilus influenzae* in relation to competence for genetic transformation. J. Bacteriol. *123*: 666–677.

Zopfi, J., T. Kjaer, L.P. Nielsen and B.B. Jørgensen. 2001. Ecology of *Thioploca* spp.: Nitrate and sulfur storage in relation to chemical microgradients and influence of *Thioploca* spp. on the sedimentary nitrogen cycle. Appl. Environ. Microbiol. *67*: 5530–5537.

Zreik, L., J.M. Bove and M. Garnier. 1998. Phylogenetic characterization of the bacterium-like organism associated with marginal chlorosis of strawberry and proposition of a *Candidatus* taxon for the organism, "*Candidatus* Phlomobacter fragariae". Int. J. Syst. Bacteriol. *48*: 257–261.

Zuber, M., T.A. Hoover and D.L. Court. 1995a. Cloning, sequencing and expression of the *dnaJ* gene of *Coxiella burnetii*. Gene *152*: 99–102.

Zuber, M., T.A. Hoover, M.T. Dertzbaugh and D.L. Court. 1995b. Analysis of the DnaK molecular chaperone system of *Francisella tularensis*. Gene *164*: 149–152.

Zuerner, R.L. and H.A. Thompson. 1983. Protein synthesis by intact *Coxiella burnetii* cells. J. Bacteriol. *156*: 186–191.

Zulty, J.J. and G.J. Barcak. 1995. Identification of a DNA transformation gene required for *com101A+* expression and supertransformer phenotype in *Haemophilus influenzae*. Proc. Natl. Acad. Sci. U. S. A. *92*: 3616–3620.

Zumft, W.G. 1997. Cell biology and molecular basis of denitrification. Microbiol. Mol. Biol. Rev. *61*: 533–616.

Zumft, W.G. and H. Körner. 1997. Enzyme diversity and mosaic gene organization in denitrification. Antonie Leeuwenhoek *71*: 43–58.

Zuravleff, J.J., V.L. Yu, J.W. Shonnard, B.K. Davis and J.R. Rihs. 1983. Diagnosis of Legionnaires' disease: an update of laboratory methods with new emphasis on isolation by culture. JAMA. *250*: 1981–1985.

Zych, K., M. Kowalczyk, F.V. Toukach, N.A. Paramonov, A.S. Shashkov, Y.A. Knirel and Z. Sidorczyk. 1997. Structural and immunochemical studies on the O-specific polysaccharide of *Proteus penneri* strain 15. Arch. Immunol. Ther. Exp. (Warsz). *45*: 435–441.

Zychlinsky, A., M.C. Prevost and P.J. Sansonetti. 1992. *Shigella flexneri* induces apoptosis in infected macrophages. Nature (Lond.) *358*: 167–169.

Zylstra, G.J. 1994. Molecular analysis of aromatic hydrocarbon degradation. *In* Garte (Editor), Molecular Environmental Biology, CRC Press, Boca Raton. 83–115.

Zylstra, G.J. and D.T. Gibson. 1991. Aromatic hydrocarbon degradation: a molecular approach. *In* Setlow (Editor), Genetic Engineering, Vol. 13, Plenum Press, New York. 183–203.

Zylstra, G.J., E. Kim and A.K. Goyal. 1997. Comparative molecular analysis of genes for polycyclic aromatic hydrocarbon degradation. Genetic Engineering (New York) *19*: 257–269.

Zywno, S.R., J.E.L. Arceneaux, M. Altwegg and B.R. Byers. 1992. Siderophore production and DNA hybridization groups of *Aeromonas* spp. J. Clin. Microbiol. *30*: 619–622.

Index of Scientific Names of *Archaea* and *Bacteria*

Key to the fonts and symbols used in this index:

Nomenclature
  Lower case, Roman

Genera, species, and subspecies of bacteria. Every bacterial name mentioned in the *Manual* is listed in the index. Specific epithets are listed individually and also under the genus.*

CAPITALS, ROMAN:

Names of taxa higher than genus (tribes, families, orders, classes, divisions, kingdoms).

Pagination
  Roman:

Pages on which taxa are mentioned.

Boldface:

Indicates page on which the description of a taxon is given.†

---

* Infrasubspecific names, such as serovars, biovars, and pathovars, are not listed in the index.

† A description may not necessarily be given in the *Manual* for a taxon that is considered as *incertae sedis* or that is listed in an addendum or note added in proof; however, the page on which the complete citation of such a taxon is given is indicated in boldface type.

ISBN 0-387-24144-2

EAN

9 780387 241449 >